NEW ATLAS
OF THE
BRITISH &
IRISH FLORA

D0320928

BSBI

Centre for Ecology & Hydrology
NATURAL ENVIRONMENT RESEARCH COUNCIL

DEFRA
Department for Environment, Food & Rural Affairs

An Roinn Ealaíon, Oidhreachta, Gaeltachta agus Oileán
Department of Arts, Heritage, Gaeltacht and the Islands

Dúchas
The Heritage Service

ENVIRONMENT AND HERITAGE SERVICE

NEW ATLAS OF THE BRITISH & IRISH FLORA

An Atlas of the Vascular Plants of Britain, Ireland, the Isle of Man and the Channel Islands

Edited by C. D. PRESTON
Centre for Ecology and Hydrology

D. A. PEARMAN
Botanical Society of the British Isles

T. D. DINES
Botanical Society of the British Isles

Assisted by H. R. Arnold

Jane M. Croft

The Vice-county Recorders of
the Botanical Society of the British Isles

Financial support by The Department for Environment, Food and Rural Affairs

OXFORD
UNIVERSITY PRESS

OXFORD
UNIVERSITY PRESS

Great Clarendon Street, Oxford OX2 6DP

Oxford University Press is a department of the University of Oxford.
It furthers the University's objective of excellence in research, scholarship,
and education by publishing worldwide in

Oxford New York

Auckland Bangkok Buenos Aires Cape Town Chennai
Dar es Salaam Delhi Hong Kong Istanbul Karachi Kolkata
Kuala Lumpur Madrid Melbourne Mexico City Mumbai
Nairobi São Paulo Shanghai Taipei Tokyo Toronto

Oxford is a registered trade mark of Oxford University Press
in the UK and in certain other countries

Published in the United States
by Oxford University Press Inc., New York

© Queen's Printer and Controller of HMSO, 2002

Copyright in the typographical arrangement and design rests with the Crown

Published for the Department for Environment, Food and Rural Affairs

This publication is not subject to terms of the Class Licence for the Reproduction of Crown copyright material

Applications for reproduction should be made in writing to HMSO, The Copyright Unit,
St Clements House, 2–16 Colegate, Norwich NR3 1BQ Fax: 01603 723000 or e-mail:
copyright©hmso.gov.uk

First published 2002
Reprinted 2003

You must not circulate this book in any other binding or cover
and you must impose this same condition on any acquirer

A catalogue record for this title is available from the British Library

Library of Congress Cataloguing in Publication Data
(Data available)

ISBN 0 19 851067 5

10 9 8 7 6 5 4 3 2 1

Printed in Hong Kong on acid-free paper

Contents

Foreword *vii*
F. H. Perring and S. M. Walters

Acknowledgements *ix*
T. D. Dines, D. A. Pearman and C. D. Preston

1 Introduction *1*
C. D. Preston, D. A. Pearman and Andrew Stott

2 The 1962 *Atlas* and later recording projects *3*
C. D. Preston, Jane M. Croft and D. A. Pearman
- The *Atlas of the British Flora* (1962) *3*
- Recording after the 1962 *Atlas*: the publication of county Floras *3*
- Recording after the 1962 *Atlas*: national recording schemes *6*
- Recording after the 1962 *Atlas*: other publications *7*
- The BSBI Monitoring Scheme (1987–88) *7*
- Preparatory work for the *New Atlas* *8*

3 Scope of the *New Atlas* project *9*
T. D. Dines, D. A. Pearman and C. D. Preston
- Aims *9*
- Geographical scope *9*
- Taxonomic scope *9*
- Separation of records into date-classes *9*
- Identification of native and introduced plants in Britain and Ireland *10*
- Archaeophytes, neophytes and casuals *10*
- Collection of data on the native or introduced status of species in 10-km squares *10*
- Collection of data on the persistence of introduced species *11*
- Summary of taxa covered by the project *11*

4 Organisation of the *New Atlas* project *13*
T. D. Dines, D. A. Pearman and C. D. Preston
- Launch of the project *13*
- Recording methodology *13*
- Field survey, 1996–99 *13*
- Collection of data from Vice-county Recorders *14*
- Collection of data from other sources *15*
- Incorporation of data into the Vascular Plant Database *17*
- Selection of taxa to be mapped in the *New Atlas* *17*
- Preparation and editing of the distribution maps *17*
- Preparation and editing of the text *17*
- Notes on the sources of data cited in the text *18*
  - Altitudinal ranges *18*
  - Dates when plants were introduced into cultivation in Britain and Ireland *18*
  - First records of introduced plants in the wild in Britain and Ireland *18*

5 Coverage obtained by the *New Atlas* project *21*
H. R. Arnold, T. D. Dines, D. A. Pearman and C. D. Preston
- Number of 10-km squares recorded *21*
- Assessing the quality of the coverage *21*
- Comparison of species totals in the 1962 *Atlas* and the *New Atlas* *25*
- Summary *26*

6 Vascular plant biodiversity in Britain and Ireland *27*
H. R. Arnold, T. D. Dines, D. A. Pearman and C. D. Preston
- Total species diversity *27*
- Diversity of native and introduced species *27*
- Distribution of species in different floristic elements *27*

7 The changing flora of Britain, 1930–99 *35*
C. D. Preston, M. G. Telfer, H. R. Arnold and P. Rothery
- Historical background *35*
- Methods of data analysis *35*
- Results of the analysis *37*
  - Summary of results *37*
  - The successful species *37*
  - The unsuccessful species *39*
  - Native species, archaeophytes and neophytes *39*
  - Trends in the distribution of native species *41*
- Preliminary comparison with other studies *42*
- The 20th-century changes in an historical perspective *44*

8 Species accounts *47*
Edited by T. D. Dines, D. A. Pearman and C. D. Preston with the assistance of H. R. Arnold, Jane M. Croft, S. J. Leach and M. J. Wigginton
- Introduction to the species accounts *47*
- Selection of species for inclusion in the book and CD-ROM *47*
- Taxonomy and nomenclature *47*
- Notes on the maps *47*
- Notes on the species texts *48*
- Paragraph 1: habitat and altitudinal range *48*
- Paragraph 2: trends in the distribution of the species *48*
- Paragraph 3: the wider distribution of the species *49*
- Paragraph 4: key references *49*
- Species accounts *53*

9 List of contributors *857*

10 Glossary *863*

11 Bibliography *865*

12 Index to the species accounts *885*
R. G. Ellis

Foreword

F. H. Perring and S. M. Walters

When we set out in 1954 to gather data for the first *Atlas of the British Flora*, our primary objective was to produce a comprehensive series of maps to illustrate in detail the distribution of vascular plants in Great Britain and Ireland. The aim was to provide ecologists, taxonomists and phytogeographers with a more accurate source of information than the inadequate vice-county maps which were then available and still being used in early volumes in the *New Naturalist* series such as *Wild Orchids of Britain* (1951). Now, nearly half a century later, the focus is quite different – those opening the pages of this new *Atlas* for the first time will be looking excitedly to see what *changes* have taken place. Understandably, conservationists will dwell upon the sad, continued decline of many of our rarer species of fragile and limited habitats, but the major surprises will undoubtedly be the huge increase in range of some of the aliens in our flora, including species like New Zealand pigmyweed, *Crassula helmsii*, not known until after work on the first *Atlas* had begun, and least duckweed, *Lemna minuta*, only discovered in Britain in 1977.

Discovery of such changes is only possible because of the large number of members of the Botanical Society of the British Isles who have given weeks and months of their time collecting the field data, and the network of dedicated Vice-county Recorders who have checked, collated and, in many cases, put the data into computers for automatic transmission to the central database at the Centre of Ecology and Hydrology, Monks Wood. This network of Recorders is, in part, a legacy from the first *Atlas*. When recording began in 1954 large areas of the British Isles were covered by very few – most of Wales by A. E. Wade and much of Ireland by D. A. Webb! But by the 1960s, after publication, there were Recorders for most vice-counties who, often inspired by the incompleteness of the data from their areas, went on to produce county and other local Floras on a more detailed basis; these have already contributed substantially to a whole series of publications such as *Scarce Plants in Britain* (1994) and the third edition of *British Red Data Books 1: Vascular Plants* (1999).

All this body of work has been incorporated into the new *Atlas*, which is thereby far more comprehensive and complete than the one we helped to compile in the 1950s. The accompanying CD-ROM will also allow the data to be used much more flexibly. The *New Atlas* will surely be of far greater value than the first *Atlas* and enable comparisons to be made with future maps without suspicion that any apparent changes are not real but due to under-recording in the past.

What will this new *Atlas* inspire? It is clearly a conservation tool, the more so because of the succinct commentaries provided for each species included. But this *Atlas* also breaks new ground in recording the distribution of so many alien plants for the first time. This will perhaps persuade more field botanists to take a greater interest in non-native species and follow their future spread, helped particularly by the even-handed treatment of species in the countryside, whatever their origin, adopted by C. A. Stace in his *New Flora of the British Isles* (1991, 1997). The total number of species in our flora has probably never been greater, and field botanists have the means to identify and record them all.

Another area of 'biodiversity investigation' which might be inspired by this *Atlas* is to correct the continued lack of information about the distribution of infraspecific taxa. Here, except for critical families and genera covered by the BSBI's Handbook Series, there are still enormous gaps. However, most of these taxa have been clearly described in the *Plant Crib 1998* (Rich & Jermy, 1998) and the one volume so far published of the new *Flora of Great Britain and Ireland* by Sell & Murrell (1996). What is needed now is a BSBI initiative to train a body of field botanists to identify and record them and to prepare a definitive *Critical Supplement* to stand beside the *New Atlas*.

All those involved in the preparation of this remarkable book deserve the gratitude of the biological community, not only in Britain and Ireland but throughout Europe and in places further afield where our native species are established aliens. The project has been sponsored by the Department for Environment, Food and Rural Affairs, and supported by conservation agencies in both the UK and Ireland. Special thanks are due to the editorial team of Chris Preston, David Pearman and Trevor Dines, and their assistants Jane Croft and Henry Arnold at Monks Wood, for collating all the data and bringing this gigantic cooperative task to fruition so soon after recording ceased in December 1999. But from these we must single out one person – David Pearman. He has been the energy source which has driven the whole scheme from the start – from cajoling people round tables at endless meetings and recording up mountains to the final hours and hours of writing and editing the species accounts, and then bringing it all together in this magnificent volume. And there should surely be a medal for his wife Anita who has stood by him through these whirlwind years. May they both enjoy a period of peace – until the next challenge at least.

What of the future? We need to attract people to the pleasures and rewards of field botany in the good BSBI tradition. An education programme, in which BSBI joins with other organisations to promote such botanical studies, is of supreme importance. Without it our successors will not have a third Atlas project in the mid-21st century.

October 2001

Acknowledgements

T. D. Dines, D. A. Pearman and C. D. Preston

Sponsoring organisations

The support of the three major sponsors of the *New Atlas* project, the Botanical Society of the British Isles (BSBI), the Centre for Ecology and Hydrology (CEH) and the Department for Environment, Food and Rural Affairs (DEFRA), has been essential to its successful completion. We are also very grateful to the other sponsoring organisations: Countryside Council for Wales; Dúchas, the Heritage Service; English Nature; Environment and Heritage Service (Northern Ireland); Joint Nature Conservation Committee; Royal Society for the Protection of Birds; Scottish Natural Heritage; Wild Flower Society. The Environment and Heritage Service funded the preparation of the Vascular Plant Database for Northern Ireland, thus taking responsibility for the compilation of records from this large area. Several of the sponsoring organisations provided funding to help with particular aspects of the project, including grants to support field-work in remote areas (Wild Flower Society), the extraction of records (Scottish Natural Heritage) and editorial work on the text (English Nature, Environment and Heritage Service, the Royal Society for the Protection of Birds and Scottish Natural Heritage).

Field recording

It is a pleasure to thank the recorders who collected the field records on which this *New Atlas* is based. Over 1600 people who made a significant number of recent records are included in the List of Contributors (Chapter 9). In addition to these recorders, we should like to thank those participants in the Botanical Society of the British Isles Monitoring Scheme and in the *New Atlas* recording meetings who are not included in this list. A number of recorders deserve special thanks for undertaking a lot of fieldwork outside their home areas. Mrs D. L. Brookman, C. Dixon, T. W. J. D. Dupree, Mrs S. E. Erskine, Mrs M. Godfrey, D. J. McCosh, J. W. McIntosh, S. A. Maxwell, Miss J. Muscott, G. P. Rothero and R. M. Walls travelled to many of the special meetings held in under-worked areas, and G. M. Kay was an almost constant attender. We are grateful to J. W. McIntosh, G. P. Rothero and K. J. Watson for help in surveying remote areas in Scotland and to L. J. Wolstenholme for single-handedly covering large areas of Co. Mayo.

Collating and checking records for vice-counties

The following BSBI Vice-county Recorders (or, in some cases, deputies who agreed to take on the *New Atlas* work) have not only collated the records for their areas but in many cases have been primarily responsible for carrying out the field survey as well. They have probably made the greatest single contribution to the project, and we are conscious of the extent to which this work has taken over their lives in recent years. Those Recorders indicated by an asterisk also acted as Regional Co-ordinators and helped plan the early stages of the *New Atlas* project, as did R. G. Ellis and D. A. Pearman.

1, West Cornwall	Dr C. N. French
	Mrs R. E. Parslow (Isles of Scilly)
2, East Cornwall	Miss R. J. Murphy
3, South Devon	L. J. Margetts
	L. M. Spalton
4, North Devon	W. H. Tucker
5, South Somerset	P. R. Green
6, North Somerset	I. P. Green
7, North Wiltshire	D. E. Green
8, South Wiltshire	Miss A. M. Hutchison
9, Dorset	Dr H. J. M. Bowen
10, Isle of Wight	Dr C. R. Pope
11, South Hampshire	P. J. Selby
12, North Hampshire	M. F. Wildish
13 & 14, West & East Sussex	Dr A. G. Knapp
14, East Sussex	P. A. Harmes*
15 & 16, East & West Kent	E. G. Philp
17, Surrey	Mrs J. E. Smith
18 & 19, South & North Essex	Dr K. J. Adams
19, North Essex	Mrs T. Tarpey
20, Hertfordshire	T. J. James
21, Middlesex	R. M. Burton*
22, Berkshire	Prof. M. J. Crawley
23, Oxfordshire	H. J. Killick
24, Buckinghamshire	R. Maycock
25 & 26, East & West Suffolk	M. N. Sanford
27, East Norfolk	R. W. Ellis
28, West Norfolk	Mrs G. Beckett
29, Cambridgeshire	Mrs G. Crompton
	D. A. Wells
30, Bedfordshire	C. R. Boon
31, Huntingdonshire	T. C. E. Wells*
32, Northamptonshire	Mrs G. M. Gent
33 & 34, East & West Gloucestershire	M. A. R. Kitchen
	Mrs C. Kitchen
35, Monmouthshire	T. G. Evans
36, Herefordshire	Mrs S. E. Thomson
37, Worcestershire	J. J. Day
38, Warwickshire	Mrs P. Copson
39, Staffordshire	J. E. Hawksford
40, Shropshire	Ms S. J. Whild
	A. J. Lockton
41, Glamorgan	Dr Q. O. N. Kay (West Glamorgan)
	J. P. Woodman (East Glamorgan)
42, Breconshire	M. Porter
43, Radnorshire	Dr D. R. Humphreys
44, Carmarthenshire	R. D. Pryce
45, Pembrokeshire	S. B. Evans
46, Cardiganshire	A. O. Chater
47, Montgomeryshire	Mrs M. Wainwright
48, Merioneth	P. M. Benoit
49, Caernarvonshire	G. H. Battershall
50, Denbighshire	Mrs J. A. Green
51, Flintshire	Dr G. Wynne
52, Anglesey	I. R. Bonner
	N. H. Brown
53 & 54, South & North Lincolnshire	Mrs I. Weston
55, Leicestershire	M. B. Jeeves

56, Nottinghamshire	D. C. Wood
57, Derbyshire	N. J. Moyes
	Dr A. Willmot
58, Cheshire	G. M. Kay*
59, South Lancashire	P. S. Gateley
60, West Lancashire	E. F. Greenwood
61, South-east Yorkshire	P. J. Cook
	Dr F. E. Crackles
62, North-east Yorkshire	T. F. Medd
63, South-west Yorkshire	G. T. D. Wilmore
64, Mid-west Yorkshire	Mrs P. P. Abbott*
65, North-west Yorkshire	Mrs D. J. Millward
66, Co. Durham	A. Coles
67 & 68, South Northumberland & Cheviot	Prof. G. A. Swan
69 & 70, Westmorland & Cumberland	Dr G. Halliday
71, Isle of Man	Dr L. S. Garrad
72, Dumfriesshire	Dr C. J. Miles
73, Kirkcudbrightshire	Mrs O. M. Stewart
74, Wigtownshire	Dr A. J. Silverside
75, Ayrshire	A. McG. Stirling*
76, Renfrewshire	K. J. Watson
77, Lanarkshire	Dr P. Macpherson
78, Peeblesshire	D. J. McCosh
79 & 80, Selkirkshire & Roxburghshire	Dr R. W. M. Corner
81, Berwickshire	M. E. Braithwaite*
82, East Lothian	Miss E. H. Jackson
83, Midlothian	D. R. McKean
84, West Lothian	Miss J. Muscott
85, Fife and Kinross	G. H. Ballantyne
86, Stirlingshire	Mrs E. W. Stewart
87, West Perth	N. W. Taylor
88, Mid Perth	Dr R. E. Thomas
89, East Perth	Dr R. A. H. Smith
90, Angus	Mrs B. G. Hogarth
91, Kincardineshire	D. Welch*
92, South Aberdeenshire	Mrs K. M. Fallowfield
93, North Aberdeenshire	D. Welch*
94, Banffshire	J. R. Edelsten
95, Moray	I. P. Green
96, Easterness	Mrs M. Barron
97, Westerness	Dr I. Strachan
	I. R. Bonner
98, Main Argyll	B. H. Thompson*
99, Dunbartonshire	Miss A. Rutherford
100, Clyde Isles	A. R. Church
101, Kintyre	Mrs P. Batty
	B. D. Batty
102, South Ebudes	Dr R. L. Gulliver
103, Mid Ebudes	Miss L. Farrell
104, North Ebudes	Mrs C. W. Murray
105, West Ross	Prof. D. M. Henderson
106, East Ross	P. C. H. Wortham
	Ms R. Scott
107, East Sutherland	Mrs M. E. Murray
108, West Sutherland	Mrs P. A. Evans
109, Caithness	J. K. Butler*
110, Outer Hebrides	Dr R. J. Pankhurst
	P. A. Smith
111, Orkney	Miss E. R. Bullard
112, Shetland	W. Scott
S, Channel Islands	B. Bonnard (Alderney)
	Mrs B. J. Ozanne (Guernsey)
	Mrs M. L. Long (Jersey)
	Mrs J. Banks (Jersey)
	Dr R. M. Veall (Sark)
H1, South Kerry	Dr E. C. Mhic Daeid
H2, North Kerry	Dr M. B. Wyse Jackson
	Dr P. S. Wyse Jackson
H3, H4 & H5, West, Mid & East Cork	T. O'Mahony
H6, Co. Waterford	P. R. Green
	I. P. Green
H7, South Tipperary	Miss R. Fitzgerald
H8, Co. Limerick	Mrs S. Reynolds
H9, Co. Clare	Miss C. Brady
H10, North Tipperary	Dr D. W. Nash
H11, Co. Kilkenny	R. N. Goodwillie
H12, Co. Wexford	Lady Rosemary FitzGerald
H13, Co. Carlow	Dr S. Parr
H14, Laois	Dr E. Moorkens
H15, South-east Galway	Dr M. Sheehy Skeffington
H16, West Galway	J. Conaghan
H17, North-east Galway	Dr C. M. Roden
H18, Offaly	Mrs A. Austin
H19, Co. Kildare	Dr D. A. Doogue*
H20, Co. Wicklow	Dr D. A. Doogue*
H21, Co. Dublin	Dr D. W. Nash
H22, Meath	Miss M. P. Norton
H23, Westmeath	C. Breen
H24, Co. Longford	S. Howard
H25, Co. Roscommon	J. J. Earley
H26 & H27, East & West Mayo	G. Sharkey
H28 & H29, Co. Sligo & Co. Leitrim	Dr D. C. F. Cotton
H30, Co. Cavan	P. Reilly
H31, Co. Louth	D. M. Synnott
H32, Co. Monaghan	A. G. Hill
H33, Fermanagh	Dr R. S. Forbes
H34, East Donegal	Miss P. Hodson
H35, West Donegal	R. Sheppard
	Dr D. A. Doogue*
H36, Tyrone	I. McNeill
H37, Co. Armagh	Dr J. S. Faulkner
H38, Co. Down	P. Hackney
H39, Co. Antrim	S. Beesley
H40, Co. Londonderry	Dr D. H. Riley

I. P. and P. R. Green made a remarkable contribution: they not only acted as Vice-county Recorders for North and South Somerset respectively, but also took on Moray (Scotland) and Waterford (Ireland) and helped record in other parts of Ireland as well!

Organisation of the project

Dr S. D. Webster played a crucial part in negotiating the contract between DEFRA and CEH which provided the financial basis for the *New Atlas* project. We thank Dr A. Stott and successive DEFRA nominated officers, Dr G. Ovenden, Dr M. Taylor, Dr J. Kirby and D. Jackson, for their advice, direction and support. We also thank the other members of the project Steering Group for their support: Dr C. Ó'Críodáin, Dr T. G. F. Curtis and Mrs J. A. Neff (Dúchas, the Heritage Service), Dr D. W. Gibbons (Royal Society for the Protection of Birds), P. T. Harding (CEH), N. G. Hodgetts (Joint Nature Conservation Committee), Dr E. A. Howe (Countryside Council for Wales), S. J. Leach (English Nature), Dr M. Meharg (Environment and Heritage Service (Northern Ireland)) and Dr C. Sydes (Scottish Natural Heritage).

D. J. McCosh played an important part in setting up the project within BSBI and we are grateful to successive BSBI Honorary Treasurers M. Walpole and M. E. Braithwaite for their support. Michael Braithwaite has been in office for much of the period of the project and we particularly thank him for his unfailing helpfulness. D. E. Green helped with the preparations for the *New Atlas*, including the plans for the first field season, pending the appointment of Dr T. D. Dines as the BSBI Atlas Organiser. We are very grateful to Professor J. E. G. Good for providing facilities for Trevor Dines to work at the CEH station at Bangor, and to A. H. Thomas for help with computing there.

In Northern Ireland P. Hackney has supervised the work on the Vascular Plant Database for the BSBI and answered many of our queries about records from the province. In the Republic of Ireland Dr D. A. Doogue was the key figure in organising the field survey and much of the data input; he deserves particular credit for obtaining coverage of many areas where resident botanists are absent.

Extraction and computerisation of data

We thank the Countryside Council for Wales, English Nature and Scottish Natural Heritage for grants that enabled the BSBI to supply some Vice-County Recorders with computers. The BSBI Co-ordinators C. S. Crook and A. J. Lockton have played a vital part in the *New Atlas* project by advising Recorders on computer matters and helping with the transfer of data to the Biological Records Centre. M. Thurner provided advice to Recorders using the BioBase package.

A. J. Lockton and Ms S. J. Whild computerised recent records from Colonsay and Jura, and R. J. Cooper and J. Mallabar those from Kent and Herefordshire respectively. Alex Lockton also provided details of selected species from the Threatened Plant Database. Mrs S. F. Maitland compiled the database in Northern Ireland at the Centre for Environmental Data and Recording (CEDaR), Ulster Museum, Belfast. D. Mitchel, Dr J. D. Nunn and B. E. Picton provided additional assistance in the computerisation of Northern Ireland records. The recent records for the Republic of Ireland were computerised by Miss R. J. Burton, Miss M. M. Cole, Dr D. A. Doogue, Miss P. Hodson, Dr D. W. Nash and Dr D. R. Humphreys. Mrs J. A. Neff prepared the Dúchas database on scarce species in Ireland which was made available to the *New Atlas* project.

Data on critical genera were provided by G. D. Kitchener (*Epilobium*), R. V. Lansdown (*Callitriche*), R. Maskew and the Rev. A. L. Primavesi (*Rosa*). R. D. Meikle determined many *Salix* specimens for *New Atlas* recorders. E. J. Clement kindly allowed us access to his data on the distribution of introduced species. C. G. Hanson and T. B. Ryves also contributed data on aliens, and Ms W. Atkinson extracted records from the Reading University herbarium.

We thank J. J. Day, D. Fraser and Dr O. Lassière for their help in transferring data from the Scottish Loch Survey and Ms S. Bradley, Mrs S. Jack, Ms E. McLaughlin, L. Ross, G. P. Rothero, Ms R. Scott and R. Youngman for extracting data from the files of Scottish Natural Heritage. Their work was co-ordinated by Dr C. Sydes. A. Austin, W. Bradley and Dr E. C. Mhic Daeid kindly photocopied records held by Dúchas for distribution to Vice-county Recorders.

Compilation of the database at BRC

In preparing the Vascular Plant Database at the Biological Records Centre, Monks Wood, we have relied heavily on Mrs J. M. Croft, who has supervised the computerisation and checking of the data, and H. R. Arnold, who was responsible for management of the Database and for the production of innumerable draft maps and listings. Mrs V. J. Burton and Mrs W. Forrest were primarily responsible for inputting records from record cards. The project could not have been completed without the dedicated work of these four members of the BRC staff. For additional help at BRC we thank Mrs V. J. Appleby, Miss R. J. Burton, Dr P. W. Lambdon, W. R. Meek, Miss C. Pinches, Mrs M. Schofield and Miss S. E. Yates. We are most grateful to the head of BRC, P. T. Harding, for his constant support.

Preparation of the maps and text

The distribution maps for the *New Atlas* were prepared with the DMAP software written by Dr A. J. Morton.

We are very grateful to the following authors of species accounts for their contributions, prepared to tight deadlines:

Dr J. R. Akeroyd, M. E. Braithwaite, R. M. Burton, Dr P. D. Carey, R. J. Cooke, Dr J. H. S. Cox, Mrs J. M. Croft, C. Dixon, Lady Rosemary FitzGerald, Dr M. J. Y. Foley, Dr R. J. Gornall, Dr G. Halliday, Dr A. D. Headley, Miss A. Horsfall, Dr G. Hutchinson, A. C. Jermy, G. M. Kay, H. J. Killick, G. D. Kitchener, R. V. Lansdown, S. J. Leach, P. S. Lusby, R. Maskew, D. J. McCosh, W. R. Meek, Dr F. H. Perring, R. D. Porley, M. S. Porter, Dr M. C. F. Proctor, Dr A. J. Richards, Dr N. K. B. Robson, Dr F. J. Rumsey, M. J. Southam, Prof. C. A. Stace, I. Taylor, K. Walker, Dr M. F. Watson, Dr D. Welch, M. J. Wigginton, G. T. D. Wilmore, Dr P. J. Wilson and R. Wilson.

M. J. Wigginton was responsible for the initial editing of the draft accounts into the *New Atlas* format. C. R. Boon, Mrs M. Briggs, R. W. M. Corner, M. A. Pearman and G. P. Rothero helped with the compilation of data on altitudinal limits or first records of introduced species in cultivation.

Draft maps and species accounts were circulated to the organisations represented on the Steering Group, and we thank the following for their comments: Dr E. A. Howe, Dr B. Jones and R. G. Woods (Countryside Council for Wales), Dr C. Ó'Críodáin, Dr T. G. F. Curtis, Mrs J. A. Neff and Dr M. B. Wyse Jackson (Dúchas), R. J. Cooke, Dr J. H. S. Cox, Dr C. C. Gibson, S. M. Hedley, S. J. Leach, Ms H. J. Lennon, R. D. Porley, Mrs S. J. Sutcliffe and I. Taylor (English Nature), R. S. Weyl (Environment and Heritage Service), N. G. Hodgetts (Joint Nature Conservation Committee) and I. A. MacDonald, A. G. Payne, M. M. Scott, Ms R. Scott, C. Sydes, N. W. Taylor and P. Wortham (Scottish Natural Heritage). We owe particular thanks to Michael Wyse Jackson for commenting on many of the species accounts from an Irish perspective. P. Hackney and B. H. Thompson also commented on the species accounts on behalf of EHS and SNH respectively, and C. R. Boon, Dr D. A. Doogue and Professor C. A. Stace read through them at our request. Many of these also commented on drafts of the introductory chapters, as did D. Jackson, Dr J. Kirby and Dr A. Stott (DEFRA) and A. O. Chater, M. Gurney, P. T. Harding, P. H. Oswald and F. H. Perring. We thank R. G. Ellis for painstakingly checking the bibliography and compiling the index.

The texts for particular genera were also circulated to specialists, and we thank Dr R. M. Bateman for his comments on *Dactylorhiza*, Dr J. P. Bailey and Miss A. P. Conolly (alien Polygonaceae), Dr R. K. Brummitt (*Calystegia*), A. O. Chater (*Populus*), Dr T. A. Cope (*Bromus*), R. Maskew (*Rosa*), P. H. Oswald (*Lactuca*, roadside halophytes), Dr F. H. Perring (*Symphytum*), Rev. A. L. Primavesi (*Rosa*), Dr T. C. G. Rich (*Filago*, *Gnaphalium*), F. J. Roberts (*Alchemilla*), Dr F. Rose (*Salicornia*), Dr F. J. Rumsey (*Trichomanes*) and Dr A. J. Silverside (*Euphrasia*).

We are grateful to Dr M. G. Telfer and Dr P. Rothery for developing the method for comparing the 1962 *Atlas* and the *New Atlas* datasets which provides the basis for Chapter 7.

We should like to pay a particular tribute to the work of Simon Leach (English Nature) in helping with the preparation of both maps and text. He has written many species accounts and commented extensively on many of the other accounts and on the introductory chapters. We greatly appreciate his sustained, critical interest in the project.

Concluding acknowledgements

We are conscious that during this project we have made many demands on the botanists listed above, who have usually been asked to undertake a large amount of work to unreasonably tight deadlines. We are grateful to all who have responded so readily, and often with such good humour, to our requests. We should also like to thank Mrs A. V. Pearman and Miss S. E. Yates for providing hospitality during editorial meetings based at Frome St Quintin and at Monks Wood respectively. Although not called upon to endure editorial visits to Bethel, N. Hughes has also had to suffer the disruption occasioned by Trevor Dines' long absences on recording meetings and other editorial business. We offer our apologies to these and all the other innocent bystanders who found themselves caught up for so long in the *New Atlas* project.

CHAPTER 1
Introduction

C. D. Preston, D. A. Pearman and Andrew Stott

The need for the *New Atlas*

Many organisations and individuals need reliable and up-to-date information on the distribution of plant and animal species in Britain and Ireland. This information is required at a range of scales and levels of detail, from distribution maps of a species across the entire area to lists of the sites where rarer species are found. For forty years the main source of information on the general distribution of flowering plants and ferns in Britain and Ireland has been the *Atlas of the British Flora* (Perring & Walters, 1962). This was the first atlas to map the presence of species in 10 × 10 km grid squares, a mode of presentation which has since become widely used for illustrating the distributions of the better-known plants and animals. However, the 1962 *Atlas* is now outdated. A replacement is required by those charged with responsibility for implementing national and international policies on biological diversity. It is also required by those naturalists and academics whose interest lies in botanical recording and in the scientific study of the changing distribution of British and Irish plants. Their interest in plant recording has stimulated much of the heightened environmental awareness in recent decades, and their work still continues to underpin our scientific understanding and practice.

These compelling reasons for the *New Atlas* are summarised below.

Policy background

In the four decades since the publication of the first *Atlas* in 1962, biodiversity has developed from an interest of naturalists and academics to a major issue of public concern, government policy and international treaties. Legislation passed by the national governments of the United Kingdom and the Republic of Ireland has introduced special measures for the protection of wild plants and the sites on which they occur, with a focus on rare and threatened plants. In England and Wales these measures have recently been further strengthened by the passing of the Countryside and Rights of Way Act 2000. As members of the European Union, both governments are also committed to the 1992 Habitats Directive, which lists species and habitats for which special conservation measures must be taken, including the establishment of a European network of Special Areas of Conservation. Following the 'Earth Summit' at Rio de Janeiro in 1992 both governments signed the UN Convention on Biological Diversity, which in Article 7 commits the parties to identify and monitor components of biological diversity. Since 1994, the Biodiversity Action Plan (UKBAP) has become the focus of a partnership approach to conserving biodiversity in the UK. Assessments have been made of the status of UK biodiversity, and priorities for action, with detailed targets, have been agreed. By May 2001, 63 action plans had been prepared for vascular plants. A Biodiversity Action Plan is currently being prepared in the Republic of Ireland.

Whilst national and international frameworks are important to provide overall direction, most of the actions and decisions which affect biodiversity are taken locally. It is therefore essential that there is information available at this local level so that these decisions can take account of local biodiversity and set it within a national context. Up-to-date and reliable information on plants is a key aspect. Whether in determining national priorities or in deciding on local action, data on species' distribution are crucial.

Botanical background

It has been clear for at least twenty years that the 1962 *Atlas* is no longer an adequate guide to the current distribution of many plant species. The major changes that have taken place in the countryside in recent decades have been well documented by projects such as the Countryside Survey (Barr *et al.*, 1993; Haines-Young *et al.*, 2000), and corresponding changes in the distribution of birds (Gibbons *et al.*, 1993) and, more recently, butterflies (Asher *et al.*, 2001) have been revealed by published atlases. We know that the distribution of some plant species has also changed dramatically since 1962. This has been demonstrated by updated distribution maps of some species that have been published in a wide range of books and in papers in specialist journals, and by numerous local studies. However, there are still many species that have not been mapped since 1962, or have never been mapped at all. There is clearly a need for an updated account of the distribution of all our flowering plants and ferns.

In addition to the changes in the actual distribution of plant species, there are at least five other reasons why the maps in the 1962 *Atlas* are now outdated:

- taxonomic studies have improved our knowledge of the flora of Britain and Ireland, resulting in the discovery of new species and taxonomic revisions of species or groups of related species, and these necessitate a corresponding revision of distribution maps;
- an immense amount of systematic botanical recording in the last forty years means that we now know much more about the detailed distribution of our flora than ever before;
- botanists now adopt a more inclusive approach to recording, and routinely note the presence of many introduced taxa that might formerly have been disregarded as being of little interest;
- many plant hybrids were virtually ignored until interest in them was stimulated by the publication of a major review volume, *Hybridization and the Flora of the British Isles* (Stace, 1975);
- the mapping grid system adopted for Ireland in the 1962 *Atlas* is no longer used for biological recording and does not match the Irish national grid.

Sponsors of the *New Atlas* project

This *New Atlas* is the result of a partnership between several organisations. The main official bodies have been the Department for Environment, Food and Rural Affairs (DEFRA) in Britain, which has acted as the major funding body, the Environment and Heritage Service in Northern Ireland and Dúchas, the Heritage Service, Ireland. The collection and collation of records and the preparation of the *New Atlas* has been the joint responsibility of the Botanical Society of the British Isles (BSBI) and the Centre for Ecology and Hydrology (CEH). The BSBI has co-ordinated the small army of volunteer recorders who have undertaken the necessary fieldwork, and the Society's Vice-county Recorders have collated the records from the vice-counties for which they are responsible.[1] CEH hosts the Biological Records Centre, Monks Wood (BRC), where the database on which the *New Atlas* maps are based has been assembled and managed. Both the BSBI and CEH have contributed to the funding of the project.

Many other organisations have made an important contribution to the *New Atlas* project and these are listed in Table 1.1.

1 'Vice-counties' are areas used for biological recording; they are based on former administrative counties but their boundaries, unlike those of administrative counties, have been stable for over a hundred years. They are mapped in Fig. 8.2.

Table 1.1 Organisations which have made an important contribution to the *New Atlas* project. Some organisations have changed their names in recent years; for simplicity the current names are usually used in the text even when it refers to a period when the earlier name was in use, but the former names are also given in this table. Abbreviations cited in the text are also included in the glossary.

Abbreviation	Name	Notes
BRC	Biological Records Centre	Based at the CEH Monks Wood site; CEH and JNCC have joint responsibility for BRC.
BSBI	Botanical Society of the British Isles	
CCW	Countryside Council for Wales	The statutory nature conservation agency in Wales.
CEH	Centre for Ecology and Hydrology	The Institute of Terrestrial Ecology was subsumed into CEH in April 2000.
Dúchas	Dúchas, the Heritage Service, Department of Arts, Heritage, Gaeltacht and the Islands, Ireland	The statutory body in the Republic of Ireland with responsibility for heritage conservation in general, including nature conservation. This was previously the responsibility of the Office of Public Works.
DEFRA	Department for Environment, Food and Rural Affairs	A UK government department; formerly Department of the Environment, then Department of the Environment, Transport and the Regions (DETR).
EHS	Environment and Heritage Service	The statutory nature conservation agency in Northern Ireland; part of the Department of the Environment (Northern Ireland).
EN	English Nature	The statutory nature conservation agency in England.
JNCC	Joint Nature Conservation Committee	The forum through which CCW, EN and SNH deliver their statutory responsibilities for Great Britain as a whole and internationally.
NERC	Natural Environment Research Council	The parent body of CEH.
RSPB	Royal Society for the Protection of Birds	
SNH	Scottish Natural Heritage	The statutory nature conservation agency in Scotland.
WFS	Wild Flower Society	

Administration of the project

The major funding for the *New Atlas* project was made available by means of a contract from DEFRA in Britain to the Natural Environment Research Council (NERC), the parent body of CEH. This contract made provision for a subcontract to be agreed between NERC and BSBI to cover the BSBI's contribution to the work. Further funding was obtained from BSBI, CEH and the other organisations listed in the Acknowledgements. Although they were not formally involved in the contract, the project was supported by the statutory nature conservation agencies in both Britain and Ireland. The project was officially entitled 'Atlas of Flowering Plants of Great Britain and Ireland', but became widely known informally as 'Atlas 2000'.

For DEFRA Dr A. Stott was in overall charge of the project, with Dr G. Ovenden (1995–98), Dr M. Taylor (1998–99), Dr J. Kirby (2000–01) and Debbie Jackson (2001 onwards) acting as DEFRA's nominated officers for the contract. CEH appointed Dr C. D. Preston as its nominated officer and D. A. Pearman acted as honorary nominated officer for the BSBI. BSBI employed D. E. Green to help with the initial organisation of the *New Atlas* project in the winter of 1995–96 and then appointed Dr T. D. Dines as Atlas Organiser from January 1996 until February 2001.

At the start of the *New Atlas* project the BSBI was already employing a Co-ordinator, whose duties included the development of the Vice-county Recorder network. Although the Co-ordinator post was not formally linked to the *New Atlas* project, successive Co-ordinators played an important role in helping Vice-county Recorders computerise their data and make it available to the project.

The *New Atlas* project was overseen by a Steering Group appointed by DEFRA. The Steering Group met at approximately six-monthly intervals from February 1996. Day-to-day co-ordination of the project was the responsibility of Trevor Dines, David Pearman and Chris Preston, who were in contact almost daily from January 1996 until the *New Atlas* was delivered to Oxford University Press in May 2001.

Arrangement of the book and CD-ROM

The historical background to the current project, including an outline of the work which resulted in the 1962 *Atlas* and details of subsequent recording initiatives, is recounted in Chapter 2. The scope and organisation of the *New Atlas* project are then detailed in Chapters 3 and 4 respectively. The results are summarised in Chapters 5, 6 and 7, including the coverage of the study area, the general patterns of vascular plant diversity and an analysis of changes in distributions since 1962. The detailed results of the *New Atlas* project, consisting of distribution maps of individual species with an accompanying text, are then presented as Chapter 8. The material in the species accounts in Chapter 8 is also available on the accompanying CD-ROM, together with briefer treatments of a further 942 introduced taxa which are too uncommon to merit a printed account.

CHAPTER 2

The 1962 *Atlas* and later recording projects

C. D. Preston, Jane M. Croft and D. A. Pearman

The *Atlas of the British Flora* (1962)

The records mapped in this *New Atlas* incorporate those collected for the 1962 *Atlas of the British Flora* (Perring & Walters, 1962), edited and amended in the light of subsequent knowledge. Records made in later recording projects and those made specifically for the *New Atlas* have been added to form the Vascular Plant Database. Some background information on the 1962 *Atlas* is therefore essential in understanding how the database has been built up. In particular, it helps in interpreting comparisons between the results of the 1962 *Atlas* and those of the current survey. This chapter briefly describes the events leading up to the launch of the BSBI Maps Scheme in 1954 and the sources of the records that were mapped in the 1962 *Atlas*. It also summarises the most important of the subsequent recording schemes. More detailed accounts of the 1962 *Atlas* project are provided by Allen (1986), Perring (1992) and Perring & Walters (1962). The scope and organisation of the *New Atlas* project are described in Chapters 3 and 4.

Origin and organisation

Before the publication of the 1962 *Atlas*, the distribution of vascular plants in Britain and Ireland was usually summarised with reference to the vice-counties devised for biological recording by H. C. Watson and R. L. Praeger. In the 1962 *Atlas* the presence of species was mapped at a finer scale, using the 10 × 10 km squares of the (British) Ordnance Survey National Grid. This grid appeared on Sixth Edition Ordnance Survey maps published immediately after the Second World War, in 1945–47 (Seymour, 1980). Its potential as the basis for national distribution maps was realised almost immediately and was demonstrated by A. R. Clapham at a BSBI conference in March–April 1950 (Clapham, 1951). This conference resolved 'that the BSBI should discuss the possibility of preparing and producing a series of maps of the British flora'. The BSBI established a Maps Committee which obtained a grant from the Nuffield Foundation to cover the costs of the work, with substantial additional support from the Nature Conservancy.

The Maps Scheme project was launched at a further BSBI conference on 9 April 1954 (Walters, 1954), with Dr S. M. Walters as Director. Dr F. H. Perring was employed as Senior Worker from 1 October 1954, and his enthusiasm and ability to enlist recorders was a vital factor in the success of the project. The scheme covered Britain, Ireland, the Isle of Man and the Channel Islands and was run from a central office in Cambridge and regional offices in Scotland and Ireland. Records were collected from the 10-km squares of the British Ordnance Survey grid; as this covered only the eastern half of Ireland, D. A. Webb (1955a) calculated an extension of the grid to cover the whole of Ireland and made arrangements to have it copied onto Irish Ordnance Survey maps. The majority of the fieldwork for the 1962 *Atlas* was completed in six seasons (1955–60), the final maps were plotted between December 1960 and December 1961, and the *Atlas of the British Flora* was published in April 1962.

The taxonomic framework for the 1962 *Atlas* was provided by 'CTW', Clapham, Tutin & Warburg's (1952) *Flora of the British Isles*. The nomenclature used in the final publication, however, followed Dandy (1958).

Sources of records

At the start of the 1962 *Atlas* project the BSBI network of Vice-county Recorders was only rudimentary, particularly in Scotland and Ireland, and appeals for help in surveying the 3500 10-km squares were made in 1954 to BSBI members, natural history societies, university departments and other interested organisations, and the general public. Most of the records in the 1962 *Atlas* were collected by field survey between 1955 and 1960; 1500 people contributed to the scheme although much of the work was done by 250 stalwarts. These field records were supplemented by records from other sources. In some counties botanists had compiled (or were compiling) data for county Floras and were able to make collated historical and recent records available to the scheme. Dorset had been surveyed in the 1930s by Professor Ronald Good, and these records provided the basis of the *Atlas* coverage of that county; in the Outer Hebrides the mapped records were derived from a series of surveys dating from about 1935 onwards. In Ireland there were fewer botanists than in Britain, so much greater use was made of pre-1930 literature and herbarium records to supplement the results of field recording.

The 1962 *Atlas* mapped 'all generally accepted native British species (excluding critical segregates) and most well-established introductions'. The species were divided into three categories. The rarest species, in category 'A', were mainly those listed from 20 or fewer of the 152 vice-counties in Druce's *Comital Flora of the British Isles* (1932). An attempt was made to make maps of these species as complete as possible, by adding records from relevant county Floras and from some herbaria to those collected in the field, and by circulating draft maps to experts. Species recorded from 21–100 vice-counties (and a few more frequent species which had clearly declined) were in the 'B' category. Records of these species were checked to ensure that there was at least one record mapped from each vice-county in which they had been recorded, but no attempt was made to make a complete search for old records. Maps of both 'A' and 'B' species distinguished recent records (made from 1930 onwards) from older records. The 'C' species, which had been recorded from more than 100 vice-counties, were collected and checked for completion at the vice-county level like the 'B' species but they were mapped as 'all records', with no distinction between pre- and post-1930 records. Finally, some critical species were given special treatment. Some maps of species in the genera *Fumaria* and *Potamogeton* were based only on records verified by national experts and others distinguished verified and non-verified records.

Recording after the 1962 *Atlas*: the publication of county Floras

One of the effects of the publication of the 1962 *Atlas* was to stimulate local plant recording and the publication of county Floras. It demonstrated the convenience of the National Grid as a recording framework and showed how volunteers could be recruited to undertake grid square recording. The 1962 *Atlas* project also did much to develop the network of BSBI Vice-county Recorders, so that many counties gained an active, resident Vice-county Recorder capable of leading a local recording team. The growing interest in nature conservation provided an increasing number of potential recruits to such teams, and the greater prosperity in the post-war years, with the concomitant growth in car ownership and in the availability of leisure time, ensured that new recruits were able to play a full part in the recording programme.

Not surprisingly, the first Flora to list records in 10-km squares was Cambridgeshire, the headquarters of the BSBI Mapping Scheme (Perring *et al.*,

Table 2.1 Floras and checklists of vice-counties (or areas of equivalent size) published since 1962. In the case of the Mid Ebudes and the Channel Islands, the vice-county is covered (or almost completely covered) by Floras dealing with separate component islands or island groups. Those Floras that map or list records by Ordnance Survey 1-km, 2-km, 5-km or 10-km grid squares are indicated by the appropriate number under 'Scale'; the inclusion of maps of many or all species is indicated by 'Yes' in the 'Maps' column. The vice-counties listed are the principal vice-counties included in the Floras, but the study areas do not necessarily follow vice-county boundaries. (C) indicates a checklist and (S) a supplement to a previous Flora. Checklists that simply list the species present in an area but provide no distributional information are excluded from the list. For information on Floras published before 1962, see McCosh (1988) and Simpson (1960).

V.c. no.	Vice-county name	Author(s) or editor(s)	Date	Scale (km)	Maps
1 & 2	W. & E. Cornwall	Margetts & David	1981	10	
1 & 2	W. & E. Cornwall	Margetts & Spurgin (S)	1991	10	
1 & 2	W. & E. Cornwall	French *et al.*	1999	2	Yes
3 & 4	S. & N. Devon	Ivimey-Cook	1984	2	Yes
5 & 6	S. & N. Somerset	Roe	1981		
5 & 6	S. & N. Somerset	Green *et al.*	1997	2	Yes
7 & 8	N. & S. Wiltshire	Stearn (S)	1975		
7 & 8	N. & S. Wiltshire	Gillam	1993	2	Yes
9	Dorset	Bowen	2000	2	Yes
10	Isle of Wight	Bevis *et al.*	1978		
11 & 12	S. & N. Hampshire	Brewis *et al.*	1996	2	Yes
13 & 14	W. & E. Sussex	Hall	1980	2	Yes
13 & 14	W. & E. Sussex	Briggs (S)	1990	2	Yes
15 & 16	E. & W. Kent	Philp	1982	2	Yes
17	Surrey	Lousley	1976	2	Yes
17	Surrey	Leslie (C)	1987	10	
18 & 19	S. & N. Essex	Jermyn	1974	10	Yes
20	Hertfordshire	Dony	1967	2	Yes
21	Middlesex	Kent	1975		
21	Middlesex and parts of adjacent counties	Burton	1983	2	Yes
22	Berkshire	Bowen	1968	5	Yes
23	Oxfordshire	Killick *et al.*	1998	2	Yes
25 & 26	E. & W. Suffolk	Simpson	1982	10	
27 & 28	E. & W. Norfolk	Petch & Swann	1968	10	
27 & 28	E. & W. Norfolk	Swann (S)	1975	10	
27 & 28	E. & W. Norfolk	Beckett & Bull	1999	2	Yes
29	Cambridgeshire	Perring *et al.*	1964	10	
29	Cambridgeshire	Crompton & Whitehouse (C)	1983	10	
30	Bedfordshire	Dony	1976	2	Yes
31	Huntingtonshire	Gilbert (C)	1965		
32	Northamptonshire	Gent, Wilson *et al.*	1995	5	Yes
33 & 34	E. & W. Gloucestershire	Holland (S)	1986		
35, 41–52	All Welsh vice-counties	Ellis	1983	10	Yes
35	Monmonthshire	Wade	1970		
36	Herefordshire	Whitehead	1976		
38	Warwickshire	Cadbury *et al.*	1971	2	Yes
39	Staffordshire	Edees	1972	2	Yes
40	Shropshire	Sinker *et al.*	1985	2	Yes
41	Glamorgan	Wade *et al.*	1994	5	Yes
43	Radnorshire	Woods	1993	5	Yes
44	Carmathenshire	May (C)	1967		
45	Pembrokeshire	Davis (C)	1970		
47	Montgomeryshire	Hignett & Lacey (C)	1977		
47	Montgomeryshire	Trueman *et al.*	1995	2	Yes
48	Merioneth	Benoit & Richards	1963		
51	Flintshire	Wynne	1993	2	Yes
52	Anglesey	Roberts	1982		

Table 2.1 *(cont.)*

V.c. no.	Vice-county name	Author(s) or editor(s)	Date	Scale (km)	Maps
53 & 54	S. & N. Lincolnshire	Gibbons	1975		
53 & 54	S. & N. Lincolnshire	Gibbons & Weston (S)	1985		
55	Leicestershire	Primavesi & Evans	1988	2	Yes
56	Nottinghamshire	Howitt & Howitt	1963		
57	Derbyshire	Clapham	1969	10	
57	Derbyshire	Patrick & Hollick (S)	1974	10	
57	Derbyshire	Hollick & Patrick (S)	1980	10	
58	Cheshire	Newton	1971	5	Yes
58	Cheshire	Newton (S)	1991	5	
59	S. Lancashire	Savidge *et al.*	1963		
61	S.E. Yorkshire	Crackles	1990	2	Yes
63	S.W. Yorkshire	Lavin & Wilmore	1994	1	Yes
66	Co. Durham	Graham	1988	2	Yes
67 & 68	S. Northumberland & Cheviot	Swan	1993	5	Yes
69 & 70	Westmorland & Cumberland	Halliday	1997	2	Yes
71	Isle of Man	Allen	1984		
73	Kirkcudbrightshire	Stewart (C)	1990		
81	Berwickshire	Braithwaite & Long (C)	1990		
82	E. Lothian	Silverside & Jackson (C)	1988		
83	Midlothian	McKean (C)	1989		
84	W. Lothian	Muscott (C)	1989		
87–89	W., Mid & E. Perth	Smith *et al.* (C)	1992		
90	Angus	Ingram & Noltie	1981	10	
93	N. Aberdeenshire	Welch	1993	10	
95 & 96	Moray & Easterness	McCallum Webster	1978	10	Yes
96 & 97	Easterness & Westerness	Hadley	1985	5	Yes
98	Main Argyll	Rothero & Thompson (C)	1994		
100	Clyde Isles	Church & Smith (C)	2000	10	
101	Kintyre	Cunningham & Kenneth	1979	10	
103	Mid Ebudes (Mull)	Jermy & Crabbe	1978	10	
103	Mid Ebudes (Coll, Tiree)	Pearman & Preston	2000	10	
104	N. Ebudes	Murray	1980	10	
105	W. Ross	Henderson (C)	1991	10	
106	E. Ross	Duncan	1980	10	
107 & 108	E. & W. Sutherland	Kenworthy	1976		
110	Outer Hebrides	Pankhurst & Mullin	1991		
111	Orkney	Bullard (C)	1995	10	
112	Shetland	Scott & Palmer	1987	10	
S	Channel Islands (Alderney)	Bonnard (C)	1988		
S	Channel Islands (Jersey)	Le Sueur	1984		
S	Channel Islands (Guernsey)	McClintock	1975		
S	Channel Islands (Guernsey)	McClintock (S)	1987		
S	Channel Islands (Herm)	Le Huquet	1993		
S	Channel Islands (Sark)	Marsden	1995		
H9, 15, 16 & 17	Connemara & the Burren	Webb & Scannell	1983		
H9, 15, 16 & 17	Connemara & the Burren	Scannell & Jebb (S)	2000		
H13	Co. Carlow	Booth	1979		
H21	Co. Dublin	Doogue *et al.*	1998	2	
H26 & H27	E. & W. Mayo	Synnott (C)	1986	10	
H38–40	Cos Down, Antrim & Londonderry	Hackney	1992		

1964). However, a more significant development was E. S. Edees' decision to map the flora of Staffordshire in 2 × 2 km squares or 'tetrads'. He started recording along these lines in 1956. J. G. Dony followed suit, and his *Flora of Hertfordshire* (1967) was the first Flora with tetrad maps to reach publication. The county Floras and checklists published since 1962 are listed in Table 2.1, and those that mapped or listed plant records by 10-km or smaller grid squares are indicated. This table demonstrates how popular tetrad recording has proved to be in England and Wales, where the supply of botanists is adequate to ensure the success of such fine-scale recording. Elsewhere 5-km or 10-km squares have been the usual recording units. Table 2.1 includes Floras or checklists of 62 vice-counties or equivalent areas which report results by grid squares, a figure which excludes a synoptic treatment of all the Welsh vice-counties (Ellis, 1983). Not all authors have adopted the National Grid as a framework for recording, and a further 36 vice-counties are covered by Floras or checklists that document plant distribution along more traditional lines. Thus, 98 of the 153 vice-counties have been covered by Floras or checklists since 1962, compared to approximately 50 covered between 1900 and 1962. Other Floras have been published for smaller areas, and examples of these are listed in Table 2.2.

There are at least three significant consequences to the *New Atlas* project of this growth in local recording. First, the field survey has been able to draw upon an experienced and disciplined group of recorders. Secondly, local recording has given rise to a large body of botanical records, although until recently it has not been easy to incorporate these records into a national database either because they were collected on cards and mapped by hand or because of difficulties of computer compatibility. Finally, this intensive local recording has given many Vice-county Recorders a detailed knowledge of their areas, so that they were able to plan the recording in the 1987–1999 period with much greater efficiency than would otherwise have been possible.

Recording after the 1962 *Atlas*: national recording schemes

The Biological Records Centre database

The records on which the *Atlas of the British Flora* were based were transferred to the then Monks Wood Experimental Station of the Nature Conservancy in April 1964, to form the nucleus of the Biological Records Centre, with Dr F. H. Perring as its head. The 10-km square records collected for the *Atlas* were converted to electronic format in 1970–71 (Harding & Sheail, 1992), but the records extracted individually from literature and herbarium sources were not computerised at this stage. These individual records were added to the database in a piecemeal fashion until computerisation of the British records was completed at the start of the *New Atlas* project; some Irish records were also computerised during this period but others have not yet been added to the database.

After the completion of the 1962 *Atlas*, records continued to be sent to BRC as updates to the published maps or for publication as new Vice-county Records in the journal *Watsonia*. However, many additional records received in the period between the end of the recording for the 1962 *Atlas* and the start of the *New Atlas* project were the result of specific national recording projects organised by the BSBI in collaboration with BRC (Croft & Preston, 1999). These projects, and some similar recording or mapping schemes that were organised in other ways, are outlined in the following sections, along with details of some recent projects which ran alongside recording for the *New Atlas*.

Critical Supplement to the Atlas of the British Flora (1968)

The microspecies listed by Dandy (1958) were not mapped in the 1962 *Atlas*. These included species in the critical genera *Alchemilla*, *Euphrasia*, *Hieracium*

Table 2.2 Examples of local Floras and checklists published since 1962, selected to illustrate the range of such studies. The vice-counties listed are the principal ones in which the study areas fall, but the areas covered are smaller than vice-counties. (C) indicates a checklist and (S) a supplement to a previous Flora. The presence of 1-km, 2-km, 5-km or 10-km square records and maps is indicated in the 'Scale' and 'Maps' columns as in Table 2.1.

V.c. no.	Area name	Author(s) or editor(s)	Date	Scale (km)	Maps
1	Isles of Scilly	Lousley	1971		
6 & 34	Bristol	Green *et al.*	2000	1	Yes
11	Christchurch	Woodhead	1994	1	Yes
14	Ashdown Forest	Rich *et al.*	1996	1	Yes
19	N.E. Essex	Tarpey & Heath	1990	1	Yes
26 & 28	Breckland	Trist	1979	2	Yes
55	Rutland	Messenger	1971	2	Yes
57 & 63	Sheffield	Shaw	1988	1	Yes
60	North Lancashire	Livermore & Livermore	1987	2	Yes
62	North York Moors	Sykes	1993	2	Yes
63	Huddersfield	Lucas & Middleton	1985	10	
64	Harrogate	Jowsey	1978	1	Yes
65	Wensleydale	Millward	1988		
76, 77 & 99	Glasgow	Dickson *et al.*	2000	2	Yes
84, 86	Falkirk	Stewart	1988		
85	Kinross	Ballantyne	1985		
102	Colonsay	Clarke & Clarke (C)	1991		
102	Islay	Ogilvie	1995	10	
103	Iona	Millar (C)	1993		
H3	Cape Clear & islands of Roaringwater Bay	Akeroyd	1996		
H9	Burren	Nelson (C)	2000		
H21	Inner Dublin	Wyse Jackson & Sheehy Skeffington	1984		
H21	Phoenix Park	Reilly	1993		
H36–40	Lough Neagh	Harron	1986	2	Yes
H38, 39	Belfast	Beesley & Wilde	1997	1	

(including *Pilosella*), *Limonium*, *Rubus* and *Sorbus*. (In critical genera the distinctions between the species are slight or are obscured by hybridisation, and considerable experience is needed before plants can be named with any degree of reliability.) The critical *Rosa* species were not mapped, and apomictic microspecies of *Taraxacum* were not recognised at that period in Britain or Ireland. Some other segregate species which had recently been described, or which were difficult to identify, were also omitted, as were most hybrids and infraspecific taxa. Treatment of these critical taxa was reserved for the *Critical Supplement to the Atlas of the British Flora* (Perring & Sell, 1968). Data for the *Critical Supplement* were collected from herbarium specimens and from appropriate literature sources; in some cases field records from reliable observers were also included. It proved possible to provide a full treatment of most of the critical genera, although only 16 *Rubus* species were mapped and *Rosa* was not covered. *Taraxacum* was represented by maps of three of the four aggregate species recognised by Dandy (1958). In addition to the treatment of the main critical genera, other recently discovered species, segregates, hybrids and infraspecific taxa were mapped. Each map was accompanied by a brief explanatory text.

Surveys of nationally rare and scarce plants

The publication of the 1962 *Atlas* highlighted the decline of some species and the rarity of others. With the completion of the *Critical Supplement*, recorders in Britain were asked by BRC to help reassess the distribution of the rarest species, represented in 15 or fewer 10-km squares. Data on these species were gathered from 1968 onwards, by collating data already held by BSBI Vice-county Recorders and others and by specially targeted fieldwork. The results were summarised in *British Red Data Books: 1 Vascular Plants* (Perring & Farrell, 1977), with a second edition, based on data collected up to the end of 1980, published in 1983. In Ireland a similar exercise carried out by the then National Parks and Wildlife Service in co-operation with the BSBI Vice-county Recorders in Ireland resulted in *The Irish Red Data Book: 1 Vascular Plants* (Curtis & McGough, 1988). The distribution maps of the rarer species in the 1962 *Atlas* were revised to produce second and third editions of the *Atlas of the British Flora* (Perring & Walters, 1976, 1982).

Although the distribution of nationally rare species in Britain received much attention from the late 1960s onwards, there was for many years no corresponding work on the species which were only slightly more frequent, the 'nationally scarce' species (those that were believed to occur in 16–100 10-km squares). The Nature Conservancy Council (and later the Joint Nature Conservation Committee) therefore supported a joint project with BSBI and BRC to update the data on these species. In 1990–92 BSBI Recorders were asked to provide details of the records they held and to check as many populations as possible in the field. The results were summarised in *Scarce Plants in Britain* (Stewart *et al.*, 1994), which included a detailed text for each species and updated distribution maps. This was followed by a complete revision of the British Red Data Book, co-ordinated by M. J. Wigginton for JNCC, which gave rise to a similar account of rare species (Wigginton, 1999). A reassessment of the distribution of scarce species in the Republic of Ireland has been completed recently by Mrs J. A. Neff on behalf of Dúchas, the Heritage Service.

Data from the British scarce and rare species projects were incorporated in 1999 into a 'Threatened Plants Database', managed for the BSBI by A. J. Lockton. In the course of this project the distribution of some taxa (listed in Chapter 4) has been thoroughly revised, with the incorporation of records from new fieldwork and from herbarium and literature sources.

Atlas of Ferns of the British Isles (1978)

The distribution maps of ferns were the first to be plotted for the 1962 *Atlas* and were therefore less complete than those of the flowering plants. They became more seriously outdated as a result of taxonomic revision, as cytological studies of species complexes in several fern genera (especially *Asplenium*, *Dryopteris* and *Polypodium*) in the 1950s and 1960s revealed the presence of morphologically distinct (but rather cryptic) species, subspecies and hybrids. For these reasons the BSBI and BRC, together with the British Pteridological Society, produced an *Atlas of Ferns of the British Isles* (Jermy *et al.*, 1978) which not only updated the maps of the species included in the 1962 *Atlas*, but also mapped additional segregates, subspecies and hybrids.

Recording after the 1962 *Atlas*: other publications

BSBI handbooks and other monographs of difficult plant groups

The preparation of identification handbooks for difficult plant groups presented opportunities to update the BRC database and to check records of the more critical taxa. Revised maps were published for most species (and some hybrids) in the handbooks on sedges, *Carex* spp. (Jermy *et al.*, 1982), pondweeds, *Potamogeton* spp. (Preston, 1995), and roses, *Rosa* spp. (Graham & Primavesi, 1993), and for selected crucifers, Brassicaceae (Rich, 1991). In the case of *Rosa*, most of the mapped records were based on herbarium specimens newly determined by the authors. The BSBI handbook on dandelions, *Taraxacum* spp. (Dudman & Richards, 1997), and the Ray Society monograph of the British *Rubus* species (Edees & Newton, 1988) included maps derived from the authors' own databases of expertly determined herbarium specimens, as most of these microspecies had never been mapped before.

Additional records were also compiled during the taxonomic revisions of smaller genera, such as the papers published by McAllister & Rutherford (1990) on *Hedera*, Rumsey & Jury (1991) on *Orobanche*, Simpson (1986) on *Elodea*, and Taschereau (1985a, 1989) on *Atriplex*. For a bibliography of the updated distribution maps published between 1962 and 1989, see Preston (1990).

Aquatic Plants in Britain and Ireland (1997)

The preparation of distribution maps for the BSBI *Potamogeton* handbook revealed how much the distribution of these aquatic plants had changed. This led to a joint project by the Environment Agency, JNCC and BRC to update the data on the distribution of some 200 species and subspecies of aquatic plants (Preston & Croft, 1997). Aquatic habitats often tend to be neglected by recorders with predominantly terrestrial interests, but they were the subject of many specialist surveys in the 1970s and 1980s in both Britain and Ireland. The results of most of these studies were incorporated into the BRC database during the Aquatic Plants project. These included major surveys of the lochs of Scotland, undertaken by the Nature Conservancy Council and later by Scottish Natural Heritage, and of the lakes of Northern Ireland, which were surveyed in a joint project organised by the Department of Agriculture and the Department of the Environment (Northern Ireland).

The BSBI Monitoring Scheme (1987–88)

The second and third editions of the 1962 *Atlas* (Perring & Walters, 1976, 1982) contained updated maps of only the rarest species; the maps of the other species were unchanged except for the correction of a few minor errors. As it became increasingly apparent in the early 1980s that the 1962 *Atlas* was outdated, the prospect of a replacement began to be considered. At a BSBI Recorders' Conference in 1983, D. A. Wells formally proposed that the Society should embark on a new *Atlas* project. Although many of those involved in the subsequent discussions felt that the time was right for a new survey, there were also senior members of BSBI who doubted whether the change since 1962 had been sufficiently marked to justify the effort of a new survey or who wondered whether it would be possible to find sufficient volunteers to undertake the recording. It was therefore decided to resurvey a sample of 10-km squares (1 square in 9, or 11% of the total) in Britain and Ireland. This survey had the twofold aim of providing an objective assessment of the changes in frequency of taxa since the 1962 *Atlas* and (by detailed recording in three 2 × 2 km squares in each of the sampled 10-km squares) of establishing a baseline against which further changes could be monitored.

Recording for the Monitoring Scheme took place in 1987–88 under the leadership of Dr T. C. G. Rich. Rich & Woodruff (1990) concluded that 'The Monitoring Scheme has been an unqualified success ... There can be little doubt that recording for an '*Atlas of the British and Irish floras* 2000' would be taken up with equal enthusiasm'. Although there were unexpected problems in comparing Monitoring Scheme data with those collected for the 1962 *Atlas*, there was 'little doubt that the majority of native species have declined during the last 25, 50 or 100 years due to agriculture, forestry, industry, urbanization, etc, and that many introduced plants have spread'. The results of the survey were edited for publication by Palmer & Bratton (1995) and the changes it revealed in the floras of England and Scotland were summarised by Rich & Woodruff (1996). As the Monitoring Scheme was believed to have detected only the more dramatic changes, Rich & Woodruff (1990) recommended a more comprehensive survey to document these changes more clearly. They concluded:

> The primary recommendation of this report, following the documentation of widespread and general change, is that a comprehensive survey should be undertaken to produce a new *Atlas of the British flora* . . . An appropriate timescale for this work would be 1987–99, – an '*Atlas of the British and Irish floras* 2000'.

Preparatory work for the *New Atlas*

In February 1992 the BSBI Council accepted Rich & Woodruff's (1990) recommendation that the Society should undertake a comprehensive survey of the British and Irish floras, in order to produce a replacement for the 1962 *Atlas*. Detailed planning for the *New Atlas* project began with a special meeting of the Society's Records Committee in March 1992. Much preliminary work was done by the Secretary of the Records Committee, D. J. McCosh, in conjunction with 12 Regional Co-ordinators who were asked to report on the work needed in different areas of Britain and Ireland. The Co-ordinators reported back by December 1992, but initial attempts to obtain financial support for what was necessarily a relatively long-term project were unsuccessful. It was not until 1995 that the required funding was secured. This was made available by means of a contract from the then Department of the Environment, Transport and the Regions (DETR) in Britain under the arrangements described in Chapter 1. The scope and organisation of the *New Atlas* project are outlined in the following chapters.

CHAPTER 3

Scope of the *New Atlas* project

T. D. Dines, D. A. Pearman and C. D. Preston

Aims

There were three main aims of the *New Atlas* project:

- To bring together vascular plant records made since the 1962 *Atlas* into a Vascular Plant Database (VPD), based on the existing database at BRC, Monks Wood, and on the wealth of additional records held by BSBI Vice-county Recorders and other botanists.
- To complete the recent (1987–95) coverage of Britain and Ireland by a programme of intensive, targeted fieldwork in the four-year period 1996–99.
- To summarise the distribution of species and selected infraspecific taxa as 10-km square distribution maps in a published atlas, distinguishing on the maps recent from older records and (as far as possible) native from introduced occurrences.

The *New Atlas* project therefore involved the compilation of historical records from many different sources, coupled with an intensive programme of new fieldwork to produce an up-to-date set of recent records for the whole of Britain and Ireland. In this *New Atlas* these records are summarised in the form of 10-km square distribution maps.

Geographical scope

The *New Atlas* project covered the whole of Great Britain, the Isle of Man, Ireland and the Channel Islands. In the interests of brevity this area is referred to as 'Britain and Ireland' in the title and text of this *New Atlas*, although in the text we have often referred to it less formally as 'our area'. The only available name for the entire area, the British Isles, is not used, as the term is considered by some botanists to be ambiguous.

Recorders were asked to survey separately all 10-km squares covered by the project that contained any land or fresh water or any coastal waters supporting the only marine vascular plants in our area, *Zostera* spp. In the 1962 *Atlas*, records for some coastal 10-km squares were amalgamated with records for neighbouring 10-km squares with more land. This practice was not repeated for the *New Atlas*, and our aim was to record separately all 10-km squares, regardless of the amount of land they include.

Recorders were asked to record the taxa covered by the project that were growing *in the wild*. Plants which were deliberately planted in enclosed areas such as parks and gardens were not recorded. Field crops were not recorded where deliberately sown, but plants regenerating in other situations were recorded. Thus, for example, recorders were not expected to record Oil-seed Rape (*Brassica napus* subsp. *oleifera*) planted in a field, but were expected to record plants arising from spilt seed on a roadside or occurring as 'volunteers' in another crop. Trees were treated rather differently, because of their longer life-span. Recorders were asked to record occurrences of deliberate planting in the wild, including crops of forest trees grown on a large scale. They were also asked to record other species which were deliberately planted in a wild situation (e.g. plants sown as 'wild-flower' mixes in the countryside).

Taxonomic scope

The *New Atlas* project covers vascular plants, which comprise pteridophytes (ferns and fern allies) and flowering plants (gymnosperms and angiosperms). It was fortunate that a thoroughly revised Flora covering exactly the same area as the *New Atlas* project, Stace's *New Flora of the British Isles* (1991), was available at the start of the project. The aim of the *New Atlas* project was to collect data on the occurrence in the wild of the following taxa:

- all native vascular plant species, with the exception of the numerous microspecies in the large genera *Hieracium*, *Rubus* and *Taraxacum*;
- all naturalised introductions or frequently recurrent casuals;
- all field crops, forestry crops and ornamental trees planted on a large scale;
- the more distinctive native and introduced subspecies;
- all hybrids.

The list of taxa covered was initially based on the species and subspecies *treated in full* by Stace (1991) and all hybrids listed in this Flora. It excluded subspecies and introduced species that are mentioned in the *New Flora* but not incorporated into its formal numbering system. Stace's *New Flora* was reprinted with minor changes in 1992 and 1995, and a thoroughly revised second edition was published in 1997. The second edition (Stace, 1997) provided the final working list of taxa to be covered by the project (Arnold & Preston, 1997). Since then two species have been added: *Serapias parviflora*, which was discovered in 1989, excluded as a probable introduction from Stace (1997) but regarded as possibly native in this Atlas, and *Callitriche palustris*, which was confirmed as a native Irish species in 1999.

The microspecies in the large genera *Hieracium*, *Rubus* and *Taraxacum* can only be identified by specialists, and the collection of data on them would have required a different approach to that adopted for other species. We did consider collecting expertly determined records and including these taxa in the *New Atlas*, but after wide consultation we rejected the idea as it would have been difficult to include hundreds of maps of these taxa without reducing the space available for species of more general interest. There is no published study of the *Ranunculus auricomus* segregates in our area, and therefore no possibility of mapping them at present.

Separation of records into date-classes

As the BSBI Monitoring Scheme had recorded one 10-km square in nine in 1987 and 1988, the aim of the *New Atlas* project was to extend this coverage to ensure that all 10-km squares were surveyed during the period 1987–99. An exception had to be made for some areas of northern and upland Scotland where the paucity of field recorders and the inaccessibility of the terrain meant that we were unlikely to obtain thorough coverage from 1987 onwards; in these areas, recorders were asked to take into account earlier records to ensure that they obtained thorough coverage from 1970 onwards (see Chapter 5). It followed from these decisions that records on the distribution maps in the *New Atlas* would be shown in three date-classes: pre 1970, 1970–86 and 1987–99. The pre 1970 class includes records dating from the time of the first botanical publications in the 16th century until 1969.

Table 3.1 General characteristics of native and introduced species.

Native species	Introduced species	Exceptions to these generalisations
Distributions in our area are usually a natural extension of the world range.	Main world range is often in distant areas (e.g. N. America, E. Asia, Australia, New Zealand).	1. Some species introduced from mainland Europe. 2. Species which arrived naturally (e.g. from N. America) by long-distance dispersal.
Distributions in Europe are relatively stable.	Have often spread as introductions elsewhere in Europe.	Species which are introduced and persist in a limited area without spreading.
Distributions in our area are more or less continuous, or if clumped or scattered they reflect the clumped or scattered nature of suitable habitats.	Distributions in Britain and Ireland are scattered, reflecting the chances of dispersal and establishment rather than controlling ecological factors.	1. Species which have spread outwards from a single introduction. 2. Introduced species which have been introduced several times but have been present for sufficient time or have sufficient powers of dispersal to occupy a continuous range.
Distributions in our area have reached equilibrium and change only in response to environmental factors.	Often expand gradually or rapidly into previously available habitats.	Introduced species which have been present for sufficient time or have sufficient powers of dispersal to occupy all the available habitat.
Usually grow in semi-natural vegetation, although they may spread from this to artificial habitats.	Usually confined to artificial habitats.	1. Native species which once grew in natural habitats which have now been destroyed. 2. Introduced species which have invaded semi-natural vegetation.
Persist in specific sites or (weedy species) specific areas for long periods.	Fail to persist at many sites.	Some introduced species may be very persistent.
Likely to have been present in our area when botanical recording began in 16th and 17th centuries.	May or may not have been present in our area when botanical recording began in 16th and 17th centuries.	Species which have arrived naturally in recent years.

Identification of native and introduced plants in Britain and Ireland

In considering the plants of Britain and Ireland it is useful to separate *native* species (which reached our area by natural dispersal) from *introduced* (or alien) species (which were brought here by man). These terms have been formally defined by Macpherson *et al.* (1996):

> A *native* species is one which arrived in the study area without intervention by man, whether intentional or unintentional, having come from an area in which it is native *or* one which has arisen *de novo* in the study area.
>
> An *introduced* species is one which was brought to the study area by man, intentionally or unintentionally, even if native to the source area *or* one which has come into the area without man's intervention, but from an area in which it is present as an introduction.

In making the initial classification of species as native or introduced the study area is considered as a whole, and a species that is native anywhere within this area is considered to be native.

There are many species which are very easily classified as either native or introduced. Some can be shown to be native, as they have a continuous fossil record since at least the last glacial period (e.g. *Carex rostrata*, *Corylus avellana*, *Potamogeton natans*). Even in the absence of such evidence, many species can be judged to be native beyond any reasonable doubt on the basis of historical, phytogeographical and ecological evidence. Such species were often recorded in their current localities by the early botanists, they grow in semi-natural habitats, and their presence in our area fits in with their wider distribution in Europe and elsewhere in the northern hemisphere. By contrast, many other species are known to have been brought to Europe from other continents in recent centuries (e.g. *Carpobrotus edulis* from S. Africa, *Epilobium brunnescens* from New Zealand, *Pseudotsuga menziesii* from western N. America). However, there is a sizeable minority of species which cannot be classified so easily. The main criteria we have used in assessing whether such species should be treated as native or introduced are outlined in Table 3.1. Other criteria may be useful in specific cases. As Table 3.1 illustrates, there are exceptions to every generalisation. Decisions often have to be made on the basis of the balance of evidence, and cannot always be proved beyond reasonable doubt.

In the species accounts (Chapter 8) taxa have been classified as native or introduced. We have tried to indicate in the text where other authors have taken a different view from that adopted here. In a few cases where there are some strong reasons for regarding a species as native and others for regarding it as introduced, or where the evidence although more equivocal appears to be evenly balanced, we have classified species as 'native or introduced' or 'native or alien'.

Archaeophytes, neophytes and casuals

Introduced species have been classified as *archaeophytes*, *neophytes* and *casuals*. Both archaeophytes and neophytes are introduced species which are present in the wild as naturalised populations, that is they are spreading vegetatively or reproducing effectively by seed. An *archaeophyte* is a plant which became naturalised before AD 1500. A *neophyte* is one which was first introduced after 1500, or was only present as a casual before 1500 and is naturalised now only because it was re-introduced subsequently. In contrast to archaeophytes and neophytes, a *casual* is a plant which is present only as populations which fail to persist in the wild for periods of more than approximately five years, and such a species is therefore dependent on constant re-introduction (Macpherson *et al.*, 1996).

Because archaeophytes were introduced so long ago, they have often been able to attain the continuous distributions, in equilibrium with environmental factors, normally expected of native plants. Many of the criteria outlined above for separating native and introduced species cannot be expected to apply. The separation of archaeophytes from natives and neophytes relies on a combination of palaeobotanical, archaeological, ecological and historical evidence. Both archaeophytes and neophytes are usually absent from the fossil record in the last glacial period, the late glacial and the early post-glacial. We have identified plants as archaeophytes if they are more or less confined to artificial habitats and are known or suspected to have been naturalised in our area before 1500. They are often known from archaeological evidence to have been present in prehistoric times. There are virtually no botanical records before 1500, but these species were usually recorded before 1700. There is no evidence that plants classified as neophytes were growing in the wild in our area before 1500, and many come from areas such as the Far East or the New World, which in itself virtually precludes the possibility that they arrived before that date. A detailed discussion of archaeophytes in the British and Irish flora is in preparation.

Collection of data on the native or introduced status of plants in 10-km squares

Taxa which occur as natives in our area are not necessarily native throughout their British and Irish range, while those that have been introduced by man often differ in the extent to which they persist in the wild. One subsidiary aim of the project was to gain as much information as possible on the native range of taxa in Britain and Ireland. Recorders were therefore asked to specify whether the taxa in each 10-km square were native or introduced. Some, but not all, recorders complied with this request. The native ranges shown on the maps in Chapter 8 have been derived from their responses combined with other sources of information (notably the maps in the 1962 *Atlas* and subsequent accounts in county Floras and other literature). Many species reach their northern limit in N. England or S. Scotland, and fortunately excellent Floras or checklists covering much of this area have been published in recent years (Graham, 1988; Halliday, 1997; Swan, 1993; Braithwaite & Long, 1990). Further north in Scotland, McCallum Webster (1978) includes helpful comments on the status of many species. For Ireland we have usually followed the assessments of Scannell & Synnott (1987), treating the taxa they describe as certainly or probably introduced as introduced and those that they describe as possibly introduced as native. In N.E. Ireland the assessments of Hackney (1992) have been very helpful. Few authors have approached the problem from anything other than their own regional perspectives. It is therefore reassuring that they often hold a common view, and in attempting to obtain a consistent treatment over our entire area we have only occasionally had to decide between contradictory opinions.

Vice-county Recorders have often reported isolated occurrences of taxa as introductions within the native range. These have usually been accepted without further checking.

It is impossible to separate native and introduced records of some taxa. All records of these have been mapped as if they were native, with an explanatory note in the accompanying text. The maps of some species treated as 'native or alien' show possibly native occurrences as native and undoubted introductions as alien.

Collection of data on the persistence of introduced species

Recorders were asked to allocate records of introduced plants to the following categories, defined by a BSBI Plant Status Working Party (Macpherson *et al.*, 1996; Macpherson, 1997):

- **Established**: present in the wild for at least five years and spreading vegetatively or reproducing effectively by seed;
- **Surviving**: present in the wild for at least five years but neither spreading vegetatively nor reproducing effectively from seed;
- **Casual**: present briefly, i.e. for less than five years, or intermittently;
- **Planted**: planted deliberately in a 'wild' situation but not established.

The response of Vice-county Recorders to this request was highly variable. There were several reasons for this. Many records available to Recorders had been made before these status categories were defined and it was difficult to apply the classification retrospectively. Even for modern records, it was sometimes difficult to assess the correct category on the basis of a single observation. An even greater obstacle to the use of the system was the fact that some of the most widely used computer packages (e.g. Recorder 3.3) did not provide a facility for allocating status to individual records. Only a minority of Recorders were meticulous in submitting status data. We have used these data in helping to delimit the native ranges of taxa, but it has not been possible to display them on the maps in this Atlas.

Summary of the taxa covered by the project.

For a summary of the 4,111 taxa covered by the project, see Table 3.2.

Table 3.2 Summary of the taxa mapped in this book and on the CD-DOM, and covered by the *New Atlas* project. The CD-ROM totals include all the taxa mapped in the book plus additional alien taxa. For the criteria for the selection of taxa to be mapped, see Chapter 8. Aggregates are genera or aggregates of similar species or infraspecific taxa used for recording purposes when identification of the component species or segregates is particularly difficult. 'Generations' covers the two generations of *Trichomanes speciosum*, which are mapped separately. The category 'alien' includes both archaeophytes and neophytes. In calculating the project totals, an aggregate of six species is counted as one aggregate plus six species, and a species with two subspecies as one species plus two subspecies.

		Natives	Natives or aliens	Archaeo-phytes	Archaeo-phytes or neophytes	Neophytes	Casuals	Native × native hybrids	Native × alien hybrids	Alien × alien hybrids	TOTAL
Aggregates	Book	26	2	1	0	7	1	–	–	–	37
	CD-ROM	26	2	1	0	7	1	–	–	–	37
	Project	40	2	1	0	8	1	–	–	–	52
Species	Book	1324	42	146	0	498	39	–	–	–	2049
	CD-ROM	1324	42	146	0	1149	238	–	–	–	2899
	Project	1363	44	149	0	1155	240	–	–	–	2951
Subspecies	Book	134	0	9	0	15	0	–	–	–	158
	CD-ROM	134	0	9	1	34	13	–	–	–	191
	Project	241	3	15	1	48	13	–	–	–	321
Generations	Book	2	0	0	0	0	0	–	–	–	2
	CD-ROM	2	0	0	0	0	0	–	–	–	2
	Project	2	0	0	0	0	0	–	–	–	2
Hybrids	Book	–	–	2	0	55	0	85	22	2	166
	CD-ROM	–	–	2	0	114	0	85	22	2	225
	Project	–	–	2	0	116	0	554	87	26	785
TOTAL	Book	1486	44	158	0	575	40	85	22	2	2412
	CD-ROM	1486	44	158	1	1304	252	85	22	2	3354
	Project	1646	49	167	1	1327	254	554	87	26	4111

CHAPTER 4

Organisation of the *New Atlas* project

T. D. Dines, C. D. Preston and D. A. Pearman

Launch of the project

The detailed preparatory work on the *New Atlas* project in the winter of 1995–6 included the preparation of regional field recording cards and the planning of special field meetings for the 1996 season. The project was launched in April 1996, with the publication of several articles in *BSBI News* (Pearman, 1996; Dines, 1996a) and the project instruction booklet (Dines, 1996b). The latter set out the aims and organisation of the project, provided details of its taxonomic and geographical scope and gave instructions on how to record, how to allocate records to the recognised introduced status categories and how to submit records to the Vice-county Recorders. An article was published in *British Wildlife* (Pearman & Preston, 1996) with the aim of bringing the *New Atlas* to the attention of botanists and other naturalists outside the BSBI, and the project was introduced to European botanists at the VIII Meeting of the Committee for Mapping the Flora of Europe in August 1997 (Croft & Preston, 1999). At a local level many Vice-county Recorders wrote articles publicising the project in appropriate journals and newsletters.

Recording methodology

In developing the methodology for the *New Atlas* survey, the major constraint was the need to achieve coverage of a large and varied area. In Britain most resident botanists live in well-populated lowland areas where the countryside is easily accessible, provided that permission can be obtained to visit private land. The uplands tend not only to have few or no resident botanists, but also to include areas where obtaining complete coverage of a 10-km square presents physical as well as botanical challenges. In Ireland botanists are concentrated in the major cities of Belfast and Dublin.

The Vice-county Recorders were primarily responsible for ensuring thorough coverage of their 10-km squares in the 1987–99 period. Arrangements were made to help those Recorders based in areas where there were too few resident botanists to cover the ground adequately, and these are described later in this chapter. In some of the most remote areas of Scotland it was only possible to obtain thorough coverage in the 1970–99 rather than the 1987–99 period (see Chapter 5). Inevitably, more time will have been spent recording in some areas than in others, and this introduces the possibility that the distribution maps will be affected by recorder bias. This is discussed in Chapter 5.

The results of the *New Atlas* project present a picture of the flora of Britain and Ireland for the period 1987–99. This 13-year period is longer than the snapshot periods of four years over which the birds were mapped by Gibbons *et al.* (1993) and the five years recently taken to map butterflies (Asher *et al.*, 2001). There are many more vascular plants than birds or butterflies, and for this reason alone a longer period is needed to achieve adequate coverage. However, it can also be argued that there is less need for a precise snapshot of the distribution of vascular plants than of the inherently more mobile birds and butterflies.

Our methodology differs greatly from that advocated by Rich *et al.* (1996), who in their highly structured 1-km square survey of the flora of Ashdown Forest, E. Sussex, noted the number of visits to each square and length of time spent recording. In an attempt to increase the 'evenness' of coverage, they also encouraged their recorders to visit as many squares as possible rather than allowing them to concentrate on just one part of the study area. Although these methods ensure a more even coverage of the study area, it would have been impossible to adopt them for the *New Atlas* project. The degree of co-ordination of a small band of recorders within an area such as Ashdown Forest (which covers 71 km^2, or 71% of one 10-km square) could never, in practice, have been achieved for a project that involved more than 1600 recorders surveying over 3800 10-km squares across the whole of Britain and Ireland.

Field survey 1996–99

Planning the fieldwork

Vice-county Recorders were responsible for planning the field survey of their own areas. Many Recorders organised and arranged their own programmes of field meetings, but there were also special recording meetings led by the Atlas Organiser in areas where additional help had been requested (Table 4.1). Most other BSBI field meetings during the period of the project had at least the partial aim of recording for the *New Atlas*. Over 1600 botanists who played a major part in the field survey are included in the List of Contributors (Chapter 9).

Some of the less accessible areas of Scotland required additional targeted fieldwork. G. P. Rothero (in 1997, 1998 and 1999), K. J. Watson (in 1997) and J. McIntosh (in 1999) surveyed upland 10-km squares that would otherwise have been badly under-recorded.

For the final field season, a special effort was made to galvanise support and focus field recording on those areas in most need of help. A strategy for the final year was published as a booklet which listed those 10-km squares and vice-counties requiring additional help, and also included some preliminary distribution maps (Dines, 1999). In order to complete the coverage of Ireland, L. J. Wolstenholme was employed by the BSBI to spend four weeks recording in E. Mayo and W. Mayo, and other groups spent shorter periods recording intensively in Co. Kilkenny, S. E. Galway and Co. Carlow, in Co. Cork and Co. Waterford and in Co. Sligo.

Workshops, conferences and publications

A series of workshops and conferences was arranged to improve recorders' knowledge of difficult taxonomic groups, stimulate recording, report on progress and discuss problems with the project. Workshops were single-day events dealing with difficult plant groups, with short lectures from specialists followed by practical sessions and the identification of specimens. They were held at the Royal Botanic Garden, Edinburgh, and at the University of Reading in 1996 and at Glasgow Museum in 1998. Recorders' conferences were held at St Martin's College, Lancaster, in 1997 and 1998. These 3-day events included taxonomic workshops, lectures, reports on progress, discussion sessions and fieldwork. The BSBI regional Annual General Meetings in Wales, Scotland and Ireland also provided opportunities for the Atlas Organiser to report on the progress of the project.

A series of booklets was produced on various aspects of recording. These included notes and advice on identification books and papers and a summary of some difficult and under-recorded taxa (Preston, 1996), an introduction to collecting and pressing specimens (Chater, 1997) and a beginners' guide to recording plants in the field (Dines, 1997). A much more substantial work,

Table 4.1 Special BSBI meetings held to record for the *New Atlas* project. References to reports refer to *BSBI News* except for those prefixed IBN, which appeared in *Irish Botanical News*.

Vice-county	Date	Report
4, N. Devon	7–8 July 1996	73: 10 (Sept. 1996)
10, Isle of Wight	25–6 May 1996	73: 8–9 (Sept. 1996)
10, Isle of Wight	27–8 May 1998	80: 65–6 (Jan. 1999)
12, N. Hampshire	24–5 May 1997	77: 67–8 (Dec. 1997)
18, S. Essex	12–13 June 1999	
22, Berkshire	14–15 July 1997	77: 75–6 (Dec. 1997)
22, Berkshire	6–7 June 1998	
25, E. Suffolk	22–3 May 1996	73: 8 (Sept. 1996)
25, E. Suffolk	16–17 July 1997	
26, W. Suffolk	22 July 1998	
33, E. Gloucestershire	17–20 July 1997	
39, Staffordshire	17–18 July 1999	83: 60–1 (Jan. 2000)
39, Staffordshire	7–8 August 1999	83: 61–2 (Jan. 2000)
41, Glamorgan	18 July 1999	
42, Breconshire	26 June 1999	
43, Radnorshire	4 July 1999	83: 60 (Jan. 2000)
44, Carmarthen	18–21 June 1999	
46, Cardiganshire	12 June 1999	83: 58–9 (Jan. 2000)
48, Merioneth	21–2 June 1997	77: 70–1 (Dec. 1997)
49, Caernarvonshire	8 August 1999	82: 68–9 (Sept. 1999)
50, Denbighshire	3 July 1999	82: 68 (Sept. 1999)
51, Flintshire	29 May 1999	84: 71–2 (April 2000)
52, Anglesey	13 June 1999	
57, Derbyshire	8–9 August 1998	
61, S.E. Yorkshire	14–15 June 1997	77: 69–70 (Dec. 1997)
62 & 65, N. Yorkshire	29–30 June 1996	73: 10 (Sept. 1996)
62 & 65, N. Yorkshire	12–13 June 1997	77: 69 (Dec. 1997)
65, N.W. Yorkshire	10–11 July 1999	84: 74–5 (April 2000)
66, Co. Durham	13–14 June 1998	
72, Dumfriesshire	1–2 August 1996	74: 71–2 (Jan. 1997)
72, Dumfriesshire	11–12 July 1998	
78, Peebleshire	26–7 June 1999	82: 67–8 (Sept. 1999)
79 & 80, Selkirkshire & Roxburghshire	14–15 July 1998	80: 70–71 (Jan. 1999)
80, Roxburghshire	3–4 July 1997	77: 73 (Dec. 1997)
81, Berwickshire	13 July 1998	84: 71 (April 2000)
81, Berwickshire	5 June 1999	82: 65–6 (Sept. 1999)
86, Stirlingshire	7–9 July 1998	80: 69–70 (Jan. 1999)
90, Angus	26–8 July 1999	
92, S. Aberdeen	22–7 July 1996	74: 69–70 (Jan. 1997)
92, S. Aberdeen	8–10 August 1997	
92, S. Aberdeen	23–5 July 1999	
94, Banffshire	3–4 July 1999	84: 73 (April 2000)
97, Westerness	17–23 July 1998	
100, Clyde Islands	4–5 July 1998	
101, Kintyre	27–31 July 1997	
101, Kintyre	17–19 June 1999	84: 73 (April 2000)
103, Mid Ebudes	22–9 June 1996	73: 9–10 (Sept. 1996)
103, Mid Ebudes	21–4 June 1997	77: 71 (Dec. 1997)
103, Mid Ebudes	20–3 June 1999	83: 59–60 (Jan. 2000)
105, W. Ross	29–30 July 1996	74: 71 (Jan. 1997)
105, W. Ross	2–6 August 1997	77: 79–80 (Dec. 1997)
106, E. Ross	30 July–1 August 1999	84: 75–6 (April 2000)
107, E. Sutherland	3–8 August 1999	
H2, N. Kerry	24–5 July 1999	
H5, E. Cork	26–7 June 1999	IBN 10: 50–3 (March 2000)
H9, Co. Clare	22–5 May 1999	IBN 10: 48 (March 2000)
H11, Kilkenny	19–20 June 1999	
H16, W. Galway	7–8 August 1999	IBN 10: 53–6 (March 2000)
H23, Westmeath	21–2 August 1999	IBN 10: 49 (March 2000)
H27, W. Mayo	17–18 July 1999	
H34, E. Donegal	3–4 July 1999	IBN 10: 49 (March 2000)
H35, W. Donegal	25–31 July 1998	IBN 9: 37 (March 1999)

Plant Crib 1998 (Rich & Jermy, 1998), provided additional information to that available in Stace's *New Flora* (1991, 1997). It drew together previous guides to identification (Wigginton & Graham, 1981; Rich & Rich, 1988; Jermy & Camus, 1991) and revised, updated and augmented this information with much new material.

Progress reports and articles outlining the need for further work appeared in *BSBI News* throughout the course of the project.

Collection of data from Vice-county Recorders

Planning the data flow

The BSBI Vice-county Recorders were primarily responsible for contributing new data to the Vascular Plant Database. It was clear that the project would make severe demands on the Vice-county Recorders, so they were all asked at the start whether they were happy to fulfil the needs of the project. Most recorders were more than enthusiastic; for those that were not, deputies were enrolled alongside the Vice-county Recorders to undertake data collection and submission. In some cases, a deputy was appointed to computerise the records, allowing the Vice-county Recorder to concentrate on field recording. Each Vice-county Recorder or deputy was asked to submit details of the taxa recorded in each of the 10-km squares in their vice-county. All recorders submitted details of recent records; many also contributed a summary of historical records derived from a search of literature sources and, in some cases, herbaria.

At an early stage in the project Vice-county Recorders were asked when they would be able to start submitting data. From their responses, an Atlas strategy was developed for each vice-county. This included annual targets for the number of squares to be recorded each year and the number for which data would be submitted. The aim was to begin data submission in 1998, and thereafter maintain a high level of data flow to BRC. This made it possible to stagger the data-processing and to prevent the delays which would have resulted if all the data had been sent to BRC at the end of the project.

The special arrangements made for data input in Northern Ireland are described later in this chapter. In the Republic of Ireland the co-ordination of Vice-county Recorders and the targeting of recording was undertaken voluntarily by Dr D. A. Doogue.

Submission of data as computer files

In 1995, the BSBI began a policy of computerisation, to encourage as many Vice-county Recorders as possible to computerise their own records and submit them on disk to the project. The BSBI Co-ordinator, C. S. Crook, undertook a survey of Vice-county Recorders in 1996 to determine how many were already using computers to process their records, which recording packages they used, and which of them wanted help with computerisation. At the time, 46 of the 153 Vice-county Recorders were already using computers to store their data, 22 requested help with computerisation and 19 wanted their records computerised on their behalf. Grants were obtained to supply 23 recorders with computers.

In order to facilitate the transfer of records on disk to BRC, a Data Transfer Standard was produced (Crook, 1996). This set out both the minimum and the ideal data that were required, and the format of the ASCII text files to contain these data. The minimum requirements were the BRC code number for the taxon, the grid reference, the vice-county, the date or date-class, the recorder, the status and the locality. The ideal standard included additional data such as altitude, abundance, the name of any determiner and the source of the records.

Many recorders were new to computers and the use of recording software, so it was decided to recommend three commercially available packages that would allow the export of records to files that automatically fulfilled the needs of the Data Transfer Standard; these were Aditsite, BioBase and Recorder. These packages differed in their methods of data entry, handling and export, but a range of packages was recommended to fit the differing needs and abilities of the recorders. Not all recorders used one of the standard packages, however, and a range of other software was employed.

Recorders who submitted data as computer files were able to transfer more details of each record (if those details were on their computer databases) and were spared the detailed checking of manually input records. However, implementation of the Data Transfer Standard was sometimes poor, resulting in time-consuming re-formatting of records between submission and transfer to BRC, or after arrival at BRC.

Submission of data on mastercards

The aim of the mastercard was to allow Recorders to submit non-computerised records in an efficient and concise way, designed to ensure that the minimum amount of data needed for mapping was transferred for manual input. The records for each 10-km square were entered on a 46-page mastercard which listed all the taxa covered by the project, with their names and BRC code numbers and with columns to be completed for the other data (Fig. 4.1). These data were the vice-county, the date-class of the last record (1987 onwards, 1970–86, and pre 1970) and the native or introduced status. Experience with the Monitoring Scheme had shown that, if only the commoner taxa were listed on a card, recorders tended to omit the additional taxa which had to be written in at the end. This might have been even more of a danger in the *New Atlas* project than in the Monitoring Scheme, as many taxa covered by the project had become widely known only after the publication of Stace's *New Flora*. The first version of the mastercard was produced in 1996 listing all the taxa in Stace (1991), and an updated version was produced in 1997 incorporating the taxa added by Stace (1997). A blank page of the mastercard was also issued to allow additional records to be sent in after the rest of the mastercard had been submitted.

Despite their size, expense of production and postage, and some difficulties of handling, the mastercards were extremely successful in transferring data to BRC. They proved straightforward to computerise, and also provided a permanent paper copy of the records submitted for future reference. Their main, and considerable, drawback was the lack of supporting detail for each record. To compensate for this lack of detail, Recorders submitting mastercard data were asked to send more detailed individual record cards for any new records of nationally rare and scarce species. They retain details of the other records in their archives.

Special arrangements for the submission of data from Northern Ireland

In Northern Ireland, the production of the Vascular Plant Database for Northern Ireland (VPD(NI)) began in August 1996. Mrs S. F. Maitland (née McKee) was employed from April 1997 to co-ordinate the Vice-county Recorders, undertake a large amount of data entry and ultimately supply the data to the *New Atlas* project. The VPD(NI) was based at the Centre for Environmental Data and Recording (CEDaR) at the Ulster Museum, Belfast. A full review of the development of the VPD(NI) is given by McKee (1999).

Collection of data from other sources

While most effort was spent on generating new, and therefore up-to-date, field records, other important sources of recent or historical records were also made available to the *New Atlas* project. The major additional sources are listed below.

Rare, threatened and scarce species in Ireland

Dúchas made its database of rare and scarce species in the Republic of Ireland available to the project. Records of rare species were prepared initially for the *Irish Red Data Book 1. Vascular Plants* (Curtis & McGough, 1988) and subsequently updated. Records of scarce species have been compiled in recent years by Mrs J. A. Neff.

Rare and scarce species in Britain

The records that form the basis of *Scarce Plants in Britain* (Stewart *et al.*, 1994) and the third edition of the British vascular plant Red Data Book (Wigginton, 1999) are held in the Vascular Plant Database. However, a detailed revision of the records of some rare and scarce species is currently being undertaken by A. J. Lockton as part of the BSBI Threatened Plants Database project, and updated records of the following taxa were made available to the *New Atlas* project: *Asparagus officinalis* subsp. *prostratus*, *Bromus interruptus*, *Calamagrostis stricta*, *Carex depauperata*, *Fumaria purpurea*, *Galeopsis angustifolia*, *Luronium natans*, *Lycopodiella inundata*, *Melampyrum sylvaticum*, *Pilularia globulifera*, *Saxifraga hirculus*, *Sium latifolium*, *Trichomanes speciosum*.

SNH Scottish Loch Survey

Detailed surveys of a large number of Scottish lochs have been undertaken in a project initially carried out by the former Nature Conservancy Council between 1984 and 1990 and later continued by Scottish Natural Heritage from 1993 to 1997. Records from the earlier surveys had already been computerised at BRC for the published account of *Aquatic Plants in Britain and Ireland* (Preston & Croft, 1997); computerised records from the later period of recording were made available by SNH.

Records from under-recorded areas of Scotland

In consultation with Vice-county Recorders, areas of Scotland with a paucity of recent botanical records were identified. As well as the additional fieldwork undertaken in these areas described above, SNH files were searched and recent records from these areas extracted and forwarded to Vice-county Recorders. In addition, some records of rare, scarce and introduced species were extracted from the herbarium of the Royal Botanic Garden, Edinburgh.

Records of introduced species

The *New Atlas* project, following Stace (1991, 1997), covered many more introduced species than had traditionally been included in British and Irish Floras. However, many records of the less well-known taxa had been accumulated by botanists with a particular interest in aliens and were not necessarily known to Vice-county Recorders. E. J. Clement kindly made available the source material for the definitive checklists of introduced species in Britain and Ireland (Clement & Foster, 1994; Ryves *et al.*, 1996). W. R. Meek computerised records from these files for species which were sufficiently common to merit inclusion in the *New Atlas* but not so familiar that they would already be well-recorded. C. G. Hanson sent details of records of aliens from his own herbarium and field notebooks, T. B. Ryves extracted records of introduced grasses from the herbarium at the Royal Botanic Gardens, Kew, and Ms W. Atkinson extracted details of some relevant specimens from the herbarium of the University of Reading, which includes the herbarium of J. E. Lousley (1907–76), who had a particular interest in aliens.

Records of *Rosa* species and hybrids

The more critical *Rosa* species and the commoner *Rosa* hybrids were first mapped in the BSBI handbook *Roses of Great Britain and Ireland* (Graham &

Master card (version 1.0) for British 10-km square ....NN 22.... VC ....98.... (unless stated)

Name	BRC no.	VC	stat	87+	70+	p70
Abies alba	2395.					
Abies grandis	2396.					
Abies procera	2397.					
Abutilon theophrasti	2418.					
Acacia melanoxylon	5415.					
Acaena anserinifolia	2528.					
A. anserinifolia x inermis	4325.					
Acaena inermis	2530.					
Acaena novae-zelandiae	2527.					
Acaena ovalifolia	2526.					
Acanthus mollis	2.					
Acanthus spinosus	7131.					
Acer campestre	3.					
Acer cappadocicum	7132.					
Acer negundo	7133.					
Acer platanoides	4.					
Acer pseudoplatanus	5.		P	✓		
Acer saccharinum	5387.					
Aceras anthropophorum	6.					
A. anthro. x Orchis simia (x	4350.					
Achillea distans	10.					
Achillea ligustica	7134.					
Achillea millefolium	7.			✓		
Achillea ptarmica	9.			✓		
Aconitum napellus	14.					
A. napellus sens. lat.	3305.					
A. napellus x variegatum (/	2985.					
Aconitum vulparia	7135.					
Acorus calamus	15.					
Acorus gramineus	4517.					
Acroptilon repens	7136.					
Actaea spicata	16.					
Adiantum capillus-veneri	17.					
Adonis annua	18.					
Adoxa moschatellina	19.			✓		
Aegopodium podagraria	20.		E	✓		
Aeonium cuneatum	7137.					
Aesculus carnea	2420.					
Aesculus hippocastanum	2241.					
Aetheorhiza bulbosa	7138.					
Aethusa cynapium	21.					
A. cyn. subsp.agrestis	21.1					
A. cyn. subsp.cynapium	21.2					
Agapanthus praecox	5417.					
Agave americana	7139.					
Ageratum houstonianum	7935.					
Agrimonia eupatoria	22.					

Name	BRC no.	VC	stat	87+	70+	p70
A. eupatoria x procera (A.	2710.					
Agrimonia procera	23.					
Agrostemma githago	34.					
Agrostis avenacea	7140.					
Agrostis canina sens.lat.	35.					
Agrostis canina sens.str.	35.2					
Agrostis capillaris	40.			✓		
A. capillaris x castellana (A	7143.					
A. capillaris x gigantea (A.	2915.					
A. capillaris x stolonifera (A	2916.					
A. capillaris x vinealis	2914.					
Agrostis castellana	7141.					
Agrostis curtisii	38.					
Agrostis gigantea	36.					
A. gigantea x stolonifera	2917.					
Agrostis hyemalis	3022.					
Agrostis lachnantha	3013.					
Agrostis scabra	4371.					
Agrostis stolonifera	39.				✓	
A. stol. x Polypogon monsp	25.					
A. stol. x Polypogon viridis	7673.					
A. stolonifera x vinealis	7142.					
Agrostis vinealis	35.1	88		✓		
Ailanthus altissima	4411.					
Aira caryophyllea	41.					✓
Aira praecox	42.			✓		
Ajuga chamaepitys	43.					
Ajuga pyramidalis	45.					
A. pyramidalis x reptans (A	2819.					
Ajuga reptans	46.			✓		
Alcea rosea	88.					
Alchemilla acutiloba	47.					
Alchemilla alpina	48.			✓		
Alchemilla conjuncta	49.					
Alchemilla filicaulis	4480.					
A. filicaulis subsp.filicaulis	50.			✓		
A. filicaulis subsp.vestita	57.			✓		
Alchemilla glabra	51.			✓		
Alchemilla glaucescens	54.					
Alchemilla glomerulans	52.			✓		
Alchemilla gracilis	2552.					
Alchemilla minima	53.					
Alchemilla mollis	2255.					
Alchemilla monticola	55.					
Alchemilla subcrenata	56.					
Alchemilla tytthantha	2275.					
Alchemilla vulgaris agg.	58.					

Figure 4.1 Example of a completed page of the mastercard designed to facilitate the manual submission of data.

Primavesi, 1993). Since the preparation of these maps, Rev. A. L. Primavesi has continued to revise the identification of *Rosa* specimens in herbaria and the *Rosa* maps included in this *New Atlas* include records resulting from the systematic revision of specimens in most of the major herbaria in our area. R. Maskew has assisted with this work, and both these experts have also sent records they have acquired from specimens they have received as BSBI referees. Once all *Rosa* records had been received, they were scrutinised by Roger Maskew and Tony Primavesi and doubtful or unsubstantiated records of the more critical taxa have been excluded from this atlas.

Incorporation of data into the Vascular Plant Database

The first stage in the development of the Vascular Plant Database (VPD) was the computerisation of a backlog of records held by BRC on record cards. This comprised those British records collected on 'individual record cards' for the 1962 *Atlas* which had not previously been computerised in the course of other projects and some records submitted more recently. As a result of this work, over 450 000 records were added to the database, mostly between March 1996 and August 1998.

The first new records arrived from Vice-county Recorders in April 1997. The bulk of the data was submitted in the autumns of 1998 and 1999, after the end of the recording seasons (Fig. 4.2). The final data were scheduled to arrive in November 1999. In the event, the last records for British vice-counties arrived in January 2000 and for Irish vice-counties (which were computerised before being sent to BRC) in June 2000.

Once the main submission of data for the *New Atlas* had begun, records from Vice-county Recorders were loaded into a separate table of the Oracle database at BRC. A copy of these records was then returned for checking to the Vice-county Recorders. For Recorders submitting their data on mastercard, these verification lists were a direct copy of their data. For records submitted on disk, the verification lists were in the same format but represented a summary of the data. This was because data on disk frequently included more than one record of a taxon for a single 10-km square. The verification list therefore only listed the most recent record for each species in each square. The Vice-county Recorders checked the verification lists for errors and returned them to BRC, where any corrections needed were made to the database.

The records submitted for the *New Atlas* were also compared with the pre-existing records in the BRC database for each 10-km square. A discrepancy list for the 10-km square was then generated, giving details of any records held at BRC for taxa that did not appear in the data from the Vice-county Recorder and of any records at BRC falling into a more recent date-class than those supplied. Vice-county Recorders were thus supplied with records of which they were apparently unaware, allowing them to check the records and either incorporate them into their own records or reject them if they felt it was necessary to do so.

In Ireland, a particular effort was made to check records which had been made for the 1962 *Atlas* in 10-km squares of the extension of the British grid (see Chapter 2). The original record cards held by Dúchas were photocopied by BSBI and circulated to Vice-county Recorders. The locality data on the cards were used to assess which of the records could be assigned to the squares of the current grid and which had to be rejected as insufficiently well-localised. British recorders also had the opportunity to correct anomalies resulting from the amalgamation in the 1962 *Atlas* of coastal squares containing small areas of land with larger neighbouring squares. It is, however, possible that some records have escaped this scrutiny and some pre-1970 records in Ireland or on the coast of Britain may therefore be misplaced by a single 10-km square.

To complete the process of checking, Vice-county Recorders were sent a 'final check list' that summarised the records which would appear on the published maps, with their date-class and status. These were returned with appropriate corrections.

At the end of the *New Atlas* project the Vascular Plant Database included 9 058 358 records, comprising 3 609 314 held in the computer database at BRC at the start of the project, 4 861 421 received from Vice-county Recorders during the course of the project and 587 623 arising from the backlog of uncomputerised records at BRC and from all other sources used for the *New Atlas*.

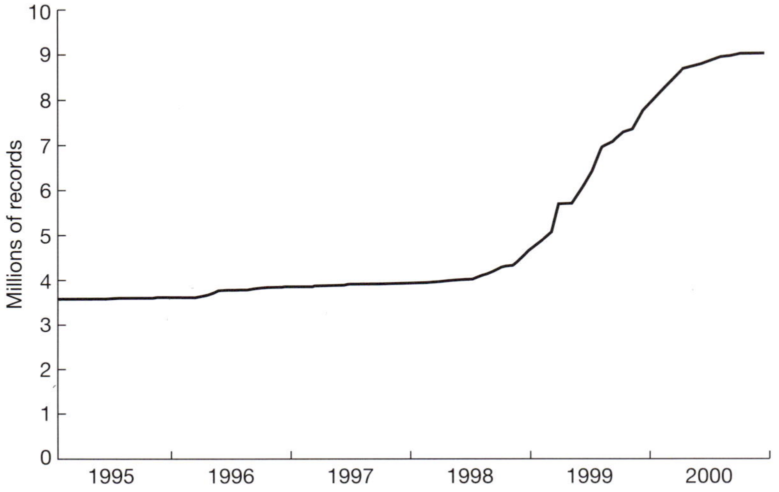

Figure 4.2 Number of records in the Vascular Plant Database 1995–2000.

Selection of taxa to be mapped in the *New Atlas*

From the start of the *New Atlas* project it had been recognised that it would be impossible to include maps of all the taxa covered by the project in a published atlas. After detailed consideration, and the examination of draft maps, it was decided to limit the published maps to native species (except microspecies in the large apomictic genera), selected subspecies, and those introduced species and hybrids which had been recorded in at least 50 10-km squares (see Chapter 8). The preparation and editing of the maps and text for these taxa is outlined below. Records of the alien taxa recorded from fewer than 50 10-km squares are included on the CD-ROM which accompanies the *New Atlas*.

Preparation and editing of the distribution maps

At an early stage of the project 50 taxa were selected to represent the full range of those covered by the project (native and introduced; rare, scarce and common; species, subspecies and hybrids). Maps of these taxa were plotted at regular intervals and circulated to members of the Steering Group to provide a measure of the progress of the project. Once the majority of records had been received, full sets of maps were plotted and scrutinised by the editors. This scrutiny included a comparison with the maps published in the 1962 *Atlas*. Any records that appeared doubtful were tagged on the database with a special identifier which enabled them to be listed and returned to Vice-county Recorders for comment.

Preparation and editing of the text

A preliminary list of 2299 taxa which satisfied the criteria for mapping was prepared in 1998 (Dines, 1998) and a panel of authors was recruited to write the species texts. Each author was asked to take responsibility for an entire family (although some of the larger families were split between authors). Authors were asked to collate the habitat information in the first paragraph, and if possible to provide a draft of the second paragraph as well. They were also asked to specify key references for the bibliography. Towards the end of the project the list of taxa which qualified for mapping was revised in the light of the final results of the project, and authors were asked to provide accounts of any additional taxa in their families. The draft accounts were initially edited by M. J. Wigginton and then loaded into a Microsoft Access database implemented by Trevor Dines for the remaining editing.

The altitudinal data were usually added by the editors, with David Pearman taking primary responsibility for this task.

Paragraph 2, which describes trends in the distribution of the species, was initially drafted from published data by the authors of the species accounts or on the basis of interim distribution maps by David Pearman. It was reviewed by all the editors once the final maps were available. Data on the first records of introduced plants in cultivation and in the wild were also compiled and added centrally, with David Pearman again responsible for collating these data. The world distributions in paragraph 3 were added by the editors, sometimes in collaboration with the authors of the accounts, and Chris Preston took primary responsibility for this aspect of the work.

Once all the species' texts and maps had been edited, they were circulated for comment to the family authors, to the country conservation agencies represented on the Steering Group, to experts on particular genera or families, and to a few botanists with more wide-ranging interests. The final stage was a cross-checking by the editors of the maps against the captions to eliminate any discrepancies between them.

The introductory chapters were drafted by Chris Preston in conjunction with the co-authors of each chapter, and then circulated for comment. The bibliography was checked by R. G. Ellis, who also compiled the index.

Notes on the sources of data cited in the text

Altitudinal ranges

The main sources of information on the altitude ranges of British and Irish plants are Wilson (1956), who compiled records for England, Wales, Scotland and Ireland, and Halliday (1997), who provided altitudinal ranges for all species in Cumbria. In compiling the data for the *New Atlas*, these were supplemented by information in Stewart *et al.* (1994), Preston & Croft (1997) and Wigginton (1999), and from the Vascular Plant Database. Wilson (1956) compiled his altitudinal data from many published sources, dating from 1847 up to his death in 1949, and from his own observations. His work was first published, in parts, from 1931 onwards. Many of the altitudes cited in Wilson (1956) were derived from 19th-century sources and are couched only in general terms, such as '3200 ft in the Scottish Highlands' or '2000 ft in Scotland'. If these remain the highest records we have supplemented them, where possible, by a precise figure from a lower site. From 1949 until 1994 the study of altitudinal limits enjoyed little popularity, although McVean & Ratcliffe (1962) recorded the heights of their quadrats. There is much more work to be done to establish the true altitudinal limits of plants in Britain and Ireland; even a few days spent systematically collecting data in the field can yield substantial results (cf. Preston & Pearman, 1998a).

Dates when plants were introduced into cultivation in Britain and Ireland

There is little information on plants cultivated in gardens before horticultural literature of the 16th century. We have drawn upon Harvey (1981) for information on plants cultivated in medieval times, although the identification of species from medieval manuscripts is always difficult and often impossible to achieve with certainty.

The standard source used for the first records of plants cultivated in the modern period was the second edition of J. C. Loudon's *Encyclopaedia of Plants* (Loudon, 1855). This comprises the first edition, prepared by J. C. Loudon and published in 1829, together with two supplements. In the first edition Loudon provided details of the first record in cultivation and place of origin of 16 712 plants. The first supplement included plants recorded up to 1840, prepared by J. C. Loudon with W. H. Baxter and G. Don. The second supplement, prepared by Mrs J. W. Loudon with G. Don and D. Wooster, covered plants introduced up to March 1855. These two supplements covered 4577 further plants, making a grand total of 21 289.

Loudon did not give his sources, but it is clear that the first edition relied heavily on the work of William Aiton, the first superintendent of Kew Gardens. Aiton's *Hortus Kewensis* (1789) was the first work, as his preface states, 'to trace back, as far as possible, how long each plant has been cultivated in the British Gardens, and to fix, with as much precisions as the nature of the subject would allow, the epoch of its introduction.' Aiton did give his sources, back to the first part of Turner's Herbal, issued in 1551, and, furthermore, included 19 pages listing books cited. He gave, in the main body of his text, the Latin binomial, the pre-Linnaean name and authority, the English name, the country of origin, the date, the grower, and often the source of that information, differentiating between dates when a plant was known to be introduced and dates when plants with no known date of introduction were first recorded in cultivation. A second edition appeared in 1810–13, enlarged by his son W. T. Aiton. We have used Loudon in preference to Aiton because it covers another forty years of introductions and because it incorporated corrections to Aiton's work. A comparison of Loudon's work with later, more accessible sources such as Chittenden (1956), Salisbury (1964) and Kelly (1995) suggests that the later works have largely used Loudon's dates, without acknowledgement. In fact since the second edition of Loudon, in 1855, there has been little systematic attempt to research first dates of introduction, although we have found useful *Johnson's Gardeners' Dictionary* (Fraser & Hemsley, 1917), *Alan Mitchell's Trees of Britain* (1996), Thomas' *Perennial Garden Plants* (1990) and Bean's *Trees and Shrubs Hardy in the British Isles* (1970–80).

There is one problem that we cannot properly resolve. Loudon (and Aiton) not only gave the dates of introduction of species which are clearly garden plants, but also included many of no conceivable horticultural merit. We do not know whether they were trying to cover *all* plants known to have occurred in the British Isles, or whether some of the plants we now consider weeds were originally introduced for ornament or curiosity. A good example of the latter is provided by some species of *Medicago*. Nowadays, species such as *M. scutellata* and *M. intertexta* are recorded as introductions arriving as contaminants of grain or wool shoddy. To Loudon they were seen as 'border flowers [grown] for the curiosity of their pods', and Fraser & Hemsley were still giving cultivation notes for these as border annuals as late as 1917. However, it is difficult to see why species such as *Lepidium virginicum*, *Malva parviflora* and *Matricaria discoidea* were cultivated. Perhaps some were grown only at Kew or in other botanic gardens, for scientific reasons. The dates of cultivation of these species are cited in the text, but this does not necessarily mean that the plants recorded in the wild originated as escapes from cultivation.

Unless otherwise stated, the dates cited for the introduction of species into cultivation refer to Britain or Ireland.

First records of introduced plants in the wild in Britain and Ireland

Clarke (1900) set out 'to extract from printed botanical books published in Great Britain the earliest notice of each distinct species of our native and naturalized flowering plants'. He explained his methods, cited the principal works referred to, and, against each species, referred back to that bibliography, which commenced with Turner's *Libellus de Re Herbaria*, published in 1538. However, Clarke included only a limited number of introduced species.

An important source of first records of introduced plants after Clarke is Dunn's *Alien Flora of Britain* (1905), although this gives general statements about the frequency of species rather than precise dates. Sometimes an entry in Dunn represents the first mention of a plant in the British Isles, since many of the Victorian county Floras omitted all but the most frequently naturalised introduced plants. Druce included more plants in his *List of British plants* (1908) and *British Plant List* (1928) but he gave no details of his sources. Druce's *Comital Flora of the British Isles* (1932) did give first dates and sources for almost all the plants he covered (natives, archaeophytes and about 160 neophytes), but this work of Druce's old age is notoriously unreliable (Allen, 1986) and we have felt more comfortable using his predecessors, Clarke and Dunn, wherever possible.

Since 1932, interest in introduced plants has increased but there has been no systematic compilation of the date of the first records of these species in the wild. Relatively few of the county Floras published since 1945 have specified the first records of native or introduced plants in their area. Clement & Foster (1994) and Ryves *et al.* (1996) cited many references that *might* lead to some idea of a first date, but missed the opportunity of incorporating those data into their other-wise invaluable compilations. We have not been able to make comprehensive searches for first dates in even the most important herbaria, and the dates we give for many plants that were first recorded in the wild in the 20th century must be regarded as indicative rather than definitive. Our reliance on literature rather than herbarium records also introduces the possibility that some of the first records are based on misidentifications. There is still much work to be done in documenting the dates of the introduction and spread of alien plants.

There is also a particular and perhaps insoluble problem in reporting the date at which trees and shrubs were first recorded 'in the wild'. Many recorders currently adopt an inclusive approach and record introduced tree and shrub species which have been deliberately planted in the wild as well as those that have spread naturally, and this practice was adopted in the guidelines for the *New Atlas* project. However, in the past such deliberately planted specimens were usually ignored; recorders either disregarded introduced trees completely or followed other conventions such as ignoring the planted individuals but recording any seedlings arising from them. The dates of first records 'in the wild' therefore often represent the date when the current convention was first applied, rather than reflecting the true appearance of the species outside gardens. The same applies to the common crop plants, which must have occurred as casuals from the time they were first cultivated but which were usually ignored by botanists until recently. In all these cases the date of first cultivation is probably more meaningful than the date of the first record in the wild.

CHAPTER 5
Coverage obtained by the *New Atlas* project

H. R. Arnold, T. D. Dines, D. A. Pearman and C. D. Preston

Number of 10-km squares recorded

The term 'coverage' is used to describe the extent and completeness of the records obtained by an atlas project. The *New Atlas* project has been successful in obtaining records from almost all 10-km squares in the study area (Fig. 5.1). In all, records are mapped in 2823 squares in Britain, 14 in the Isle of Man, 1007 in Ireland and 15 in the Channel Islands. There are records from 99.2% of these squares in the 1987–99 date-class. Almost all the 10-km squares for which there are no records, or no recent records, are on the coast or are off-shore rocks and islands; some may actually be rocks without any vascular plants. There is just one inland square without post-1986 records, NH61, a remote square in Easterness which was recorded for *A Map Fora of Mainland Inverness-shire* (Hadley, 1985) but not subsequently.

The coastal squares which were regarded as too small to be mapped separately in the 1962 *Atlas* but which are included in the *New Atlas* are shown in Fig. 5.2. In the 1962 *Atlas* records were mapped in 2647 squares in Britain and the Isle of Man, compared to 2837 in this *New Atlas*. No direct comparison is possible for the Channel Islands and Ireland because of the change of grid in these areas.

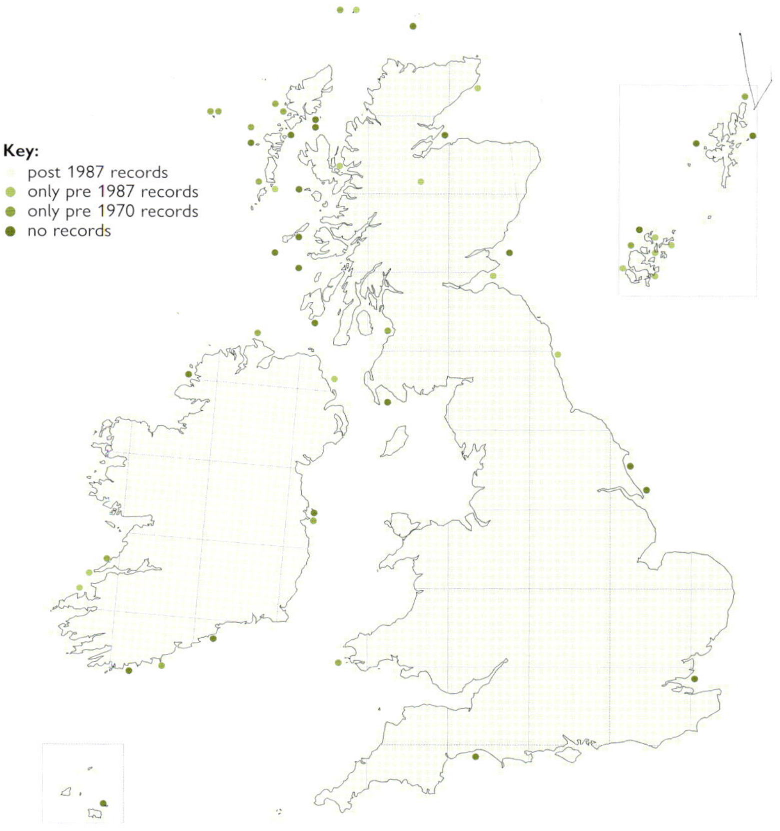

Figure 5.1 Summary of the 10-km square coverage obtained by the *New Atlas* project.

Assessing the quality of the coverage

The *New Atlas* project has achieved a very good geographical coverage. However, it is much more difficult to assess the 'quality' of the recording in each square, i.e. the extent to which the taxa actually present in a square have been recorded in the most recent date-class. There are a number of possible ways in which the recent (1987–99) coverage might be inadequate:

- the total number of species recorded in a square might be too low, if the square has never been recorded thoroughly
- a square might have been well-recorded historically, but not recorded thoroughly in the 1970–86 or 1987–99 date-classes

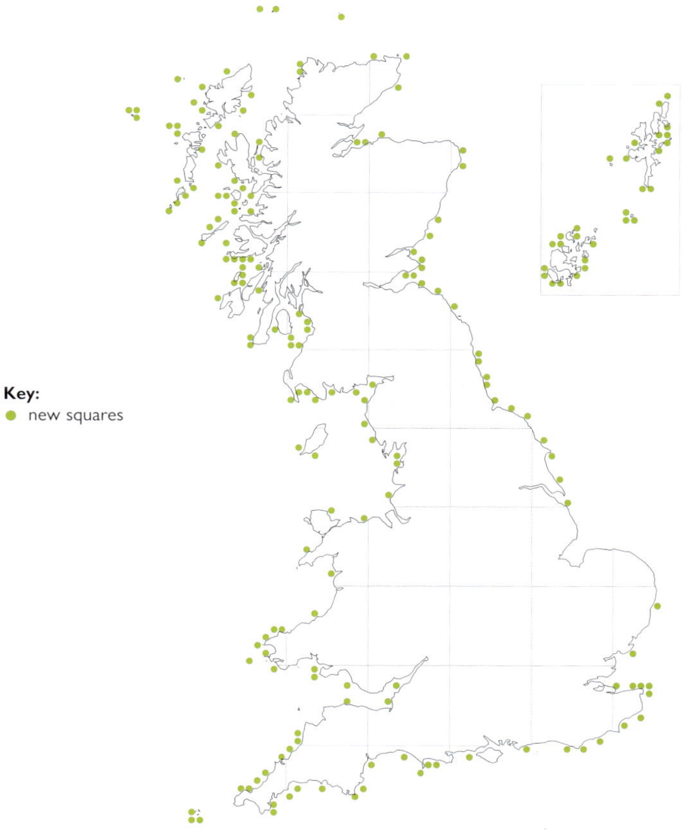

Figure 5.2 The 10-km squares in Britain and the Isle of Man which were not included on the maps in the 1962 *Atlas* but are mapped in the *New Atlas*. (Some of these squares may have been recorded but, if so, the records were amalgamated with those in a neighbouring square.) Information on the squares mapped in the 1962 *Atlas* is derived from Perring & Walters (1962, Fig. 3, p. xiv). Although in the 1962 *Atlas* all records from St Kilda were actually mapped in NF 20 (a landless square) and all those from the Isles of Scilly in SV 90 (a square with very little land), in the above figure they are treated as having been mapped in NF 09 and SV 91 respectively. The records from Ireland and the Channel Islands in the 1962 *Atlas* are mapped on a different grid, so no such comparison is possible.

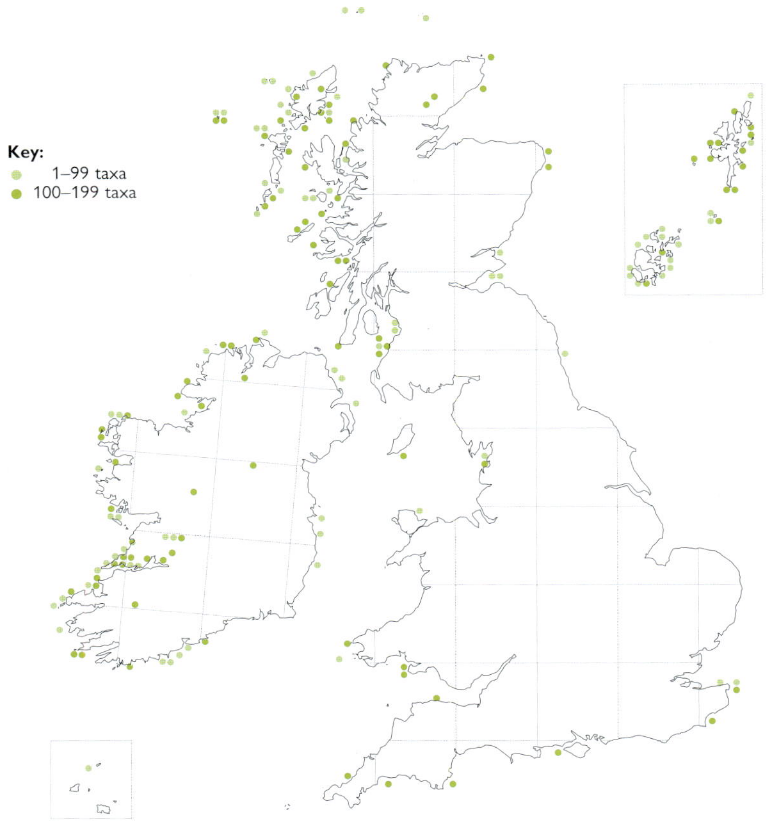

Figure 5.3 The 10-km squares for which fewer than 200 species and hybrids have been recorded. Species that are mapped as aggregates and segregates in this Atlas are counted only once, as the aggregate, even if several segregates are recorded from the square.

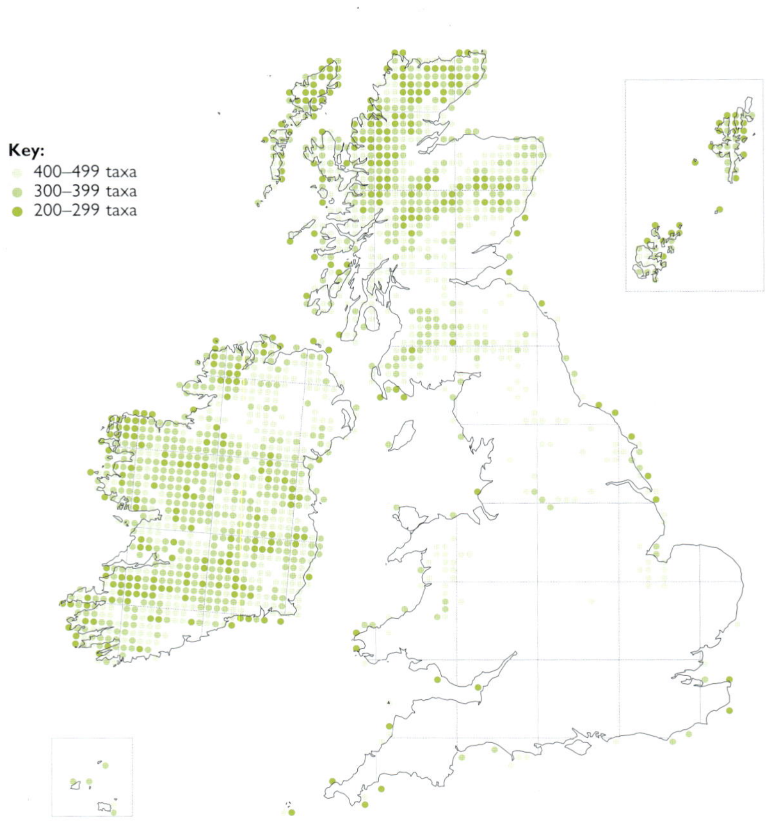

Figure 5.4 The 10-km squares for which 200–500 species and hybrids have been recorded. The totals have been calculated on the same basis as those in Fig. 5.3.

- species which are easy to identify might be well-recorded for the 1987–99 date-class, but there might be a shortage of records of taxa which are difficult to identify (such as subspecies and hybrids)
- native taxa might be well-recorded for the 1987–99 date-class, but introduced taxa might be under-recorded
- individual taxa might be under-recorded in an otherwise well-recorded square, if a recorder has a 'blind-spot' for those plants.

As we have no independent measure of the number of taxa that are present in a square, only indirect assessments of recording quality are possible. The approach used here to assess the quality of coverage is to plot maps to show the number or proportion of species in different groups, and to use these to show variation between neighbouring areas which might indicate recording bias. Unless stated, these maps show aggregate species or hybrids covered by the *New Atlas* project; segregates and subspecies are not counted separately. It is not possible to test for particular 'blind-spots' in this way. However, the individual species accounts in Chapter 8 have also been used to indicate the ways in which recording bias may have affected the distribution maps of particular taxa.

Total number of species and hybrids recorded per square

Squares from which a relatively low total number of taxa have been recorded *might* be under-recorded; alternatively they might just be species-poor. Squares from which fewer than 200 species and hybrids have been recorded are mapped in Fig. 5.3. Almost all of these are coastal squares containing very little land. The number of taxa in many of these squares is probably less than 200, although some squares may have been under-recorded because of their remoteness. Four inland squares in Britain have less than 200 recorded taxa. Two are moorland squares in East Sutherland, NC61 with 190 taxa and NC72 with 185. Both lack any roads or settlements other than isolated homesteads. Each has been visited since 1987 at least twice for *New Atlas* recording, and the database includes additional records from specialist habitat surveys. The other two squares are on Lewis, Outer Hebrides, NB02 with 178 taxa and NB33 with 149. They are moorland areas with many lochs but very little additional habitat. The low totals almost certainly reflect the genuinely species-poor nature of these squares. The seven inland squares with less than 200 taxa in Ireland seem to be no different from many of the surrounding squares and are probably under-recorded.

Squares with 200–499 species and hybrids are mapped in Fig. 5.4. The strong regional (as opposed to vice-comital) variation in the distribution of symbols, with a concentration of species-poor squares in N.W. Scotland and Ireland, suggests that most of these squares are genuinely species-poor. This is supported by the fact that even in a well-recorded and botanically diverse area such as the Isle of Skye there are fewer than 500 taxa recorded in almost all 10-km squares (Murray, 1980). The total flora of Shetland, which is well-recorded but 'very meagre by southern standards', amounts to fewer than 750 taxa, excluding microspecies (Scott & Palmer, 1987). In western Ireland it is unlikely that many 10-km squares contain more than 500 taxa: a checklist of the two vice-counties which comprise Co. Mayo lists fewer than 800 taxa in the entire area, which covers 95 complete or partial 10-km squares (Synnott, 1986).

Most of the 10-km squares in England and Wales with fewer than 500 recorded taxa are coastal squares containing very little land. The inland squares comprise tracts of land which are both species-poor and to a greater or lesser extent under-recorded – these two factors often interact, as botanists record these areas dutifully but are reluctant to spend as much time in them as they do in richer terrain. Most are upland (but not montane) areas of acidic rocks and soils.

Coverage in the 1970–86 and 1987–99 date-classes

A square that has been well-recorded historically but has been poorly covered from 1970 onwards will have a high proportion of taxa which have not been recorded since 1970. However, a square with a high proportion of taxa that have not been recorded since 1970 has not necessarily been under-recorded in recent years – it may simply have lost a high proportion of its species through extinction. Furthermore, the proportion of historical records is influenced by the thoroughness of the historical as well as the recent coverage, and the extent to which historical records have been searched out and incorporated into the Vascular Plant Database.

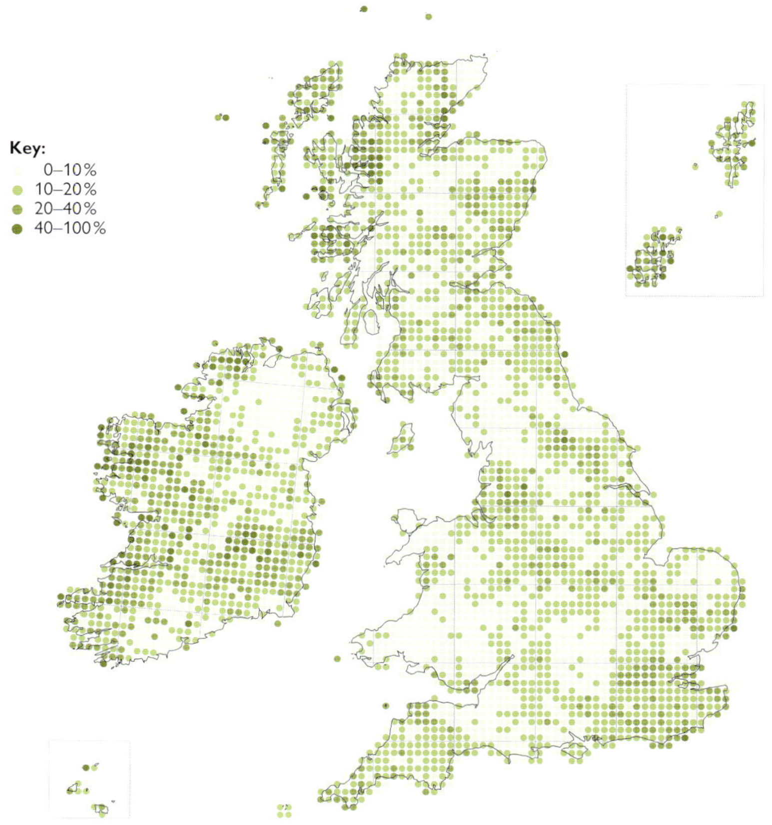

Figure 5.5 The proportion of species and hybrids recorded from each 10-km square which have been recorded in the pre-1970 date-class but not subsequently.

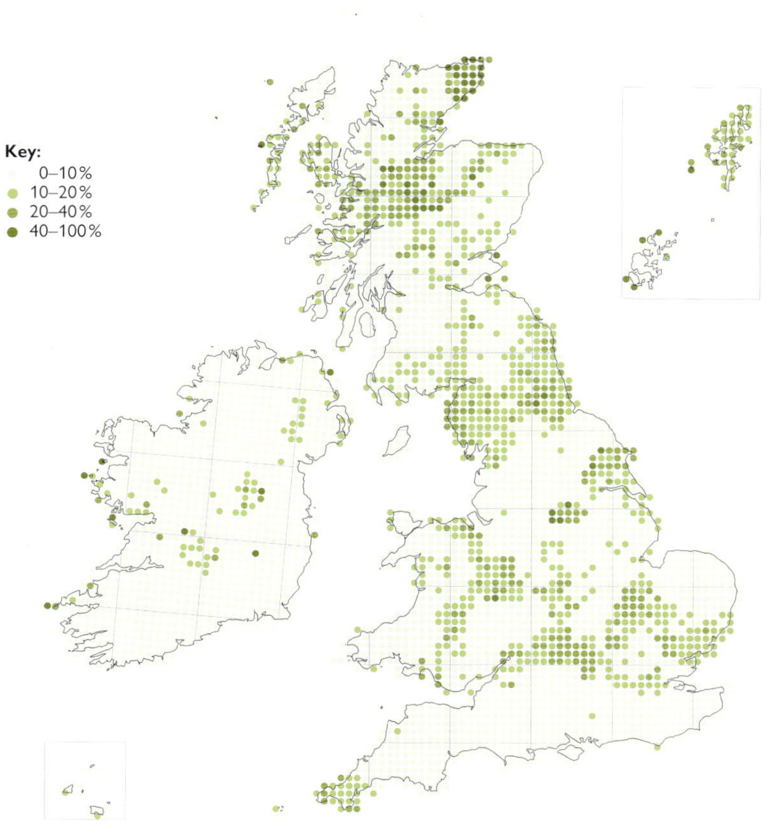

Figure 5.6 The proportion of species and hybrids recorded from each 10-km square between 1970 and 1999 which have been recorded in the 1970–86 but not in the 1987–99 date-class.

The number of species and hybrids that have not been recorded from 1970 onwards in each square, calculated as a proportion of the total number recorded in the square, is plotted in Fig. 5.5. Although the interpretation of this map is complicated, those squares with only 0–20% of taxa that have not been refound have probably been well-recorded from 1970 onwards. Those with 20% or more taxa that have not been refound have *either* lost a relatively high proportion of native taxa by extinction, *or* have lost a large number of transient introductions, *or* have been less well-recorded than other squares from 1970 onwards. It is likely that squares with a high proportion of species that have not been refound in the London area, for example, have genuinely lost many native and introduced species, whereas remote areas of N.W. Scotland have probably been under-recorded in recent years. The island of Mull, for example, was particularly well-surveyed between 1966 and 1970 by the botanical staff of the British Museum (Natural History) for the book *The Island of Mull* (Jermy & Crabbe, 1978).

In Ireland the interpretation of Fig. 5.5 is further complicated by the fact that there are some areas where many pre-1970 records were recorded on the grid used for the 1962 *Atlas* and cannot be assigned to the current grid as further details of the locality are not available. These records have had to be omitted from the *New Atlas* database. Areas where historical records have been rejected will clearly have a lower proportion of pre-1970 records than those where the historical records could be accepted.

For Fig. 5.6 the records dating from 1970 onwards were analysed, and for each square the number of taxa that have been recorded only in the earlier of the two recording periods (1970–86) is shown as a proportion of the total recorded since 1970. Although the same factors apply to the interpretation of this map as to Fig. 5.5, the pattern shown by the map is much clearer. Squares with a high proportion of taxa that have not been recorded in the later date-class are almost certainly under-recorded from 1987 onwards. Many of the concentrations of squares that have a high proportion of taxa in the 1970–86 date-class are in vice-counties that were well-recorded in 1970–86 as part of a Flora project. These include Oxfordshire (intensively recorded between 1968 and 1979), Shropshire (1970–1983), West Yorkshire (1974–1993), the East Riding of Yorkshire (1970–*c.*1988), Cumbria (1974–1996) and Durham (1968–1986). The converse situation is shown by areas for which few 1970–86 records are available, and much of Ireland falls into this category. An exception in Ireland is the group of squares with a high proportion of latest records in the 1970–86 period in Co. Westmeath, where there is a current Flora project which collected many records during this period.

At the start of the project it was agreed that recorders in northern and western Scotland would be unable to achieve full coverage in the 1987–99 period, and they were therefore encouraged to achieve coverage that was as comprehensive as possible for the 1970–99 period. In the event the coverage of this area was much better than expected, and, whereas Westerness, Easterness and Caithness have a high proportion of 1970–86 records, other areas in northern and western Scotland do not.

Records of subspecies and hybrids

The number of subspecies recorded from each square from 1970 onwards is mapped in Fig. 5.7 and the number of hybrids recorded in this period is mapped in Fig. 5.8. Although we have no estimate of the number of subspecies and hybrids that could be expected per square in different regions, the patchy appearance of the maps suggests that these critical taxa have been unevenly recorded. There are some areas that appear to have been well-recorded for both subspecies and hybrids (e.g. N. & S. Somerset, Cardiganshire, Worcestershire, Fife). In other areas hybrids appear to be relatively numerous (and presumably, therefore, well-recorded), but there is no evidence that subspecies have been particularly well-studied (e.g. Dorset, Isle of Man, Co. Dublin), whereas the opposite may be true elsewhere (e.g. in N. Hampshire, Northamptonshire, Wigtownshire).

Records of introduced species

One might expect more variation in the recording of introduced species than of natives. Some botanists are more interested than others in recording aliens, and in particular there are differences in the extent to which individual recorders note the presence of species which are planted or doubtfully naturalised, such as garden plants that spread onto old walls or trees that are planted in the wild. Although recording conventions were laid down for the *New Atlas* project,

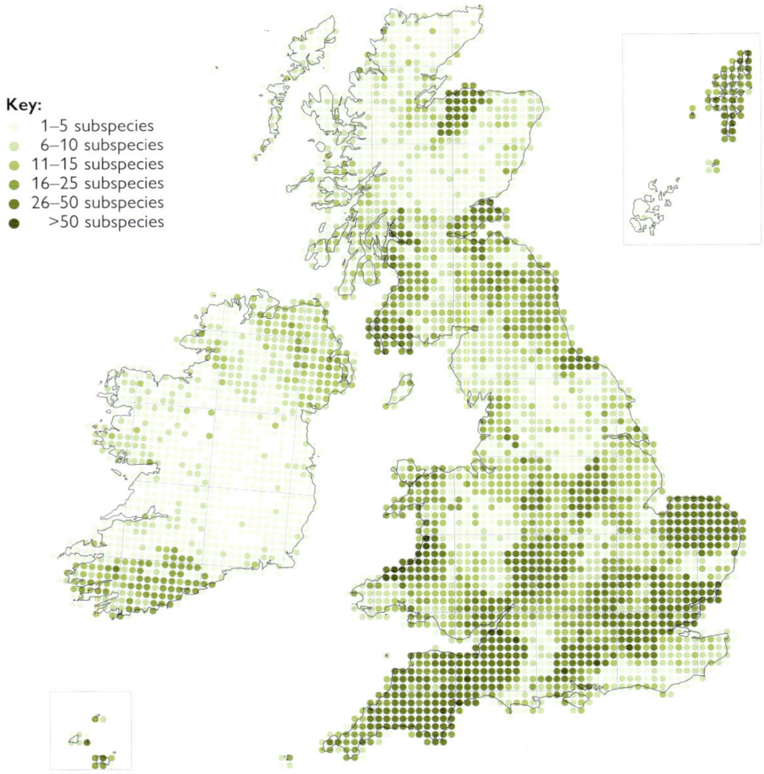

Figure 5.7 The number of subspecies recorded in each 10-km square, 1970–99.

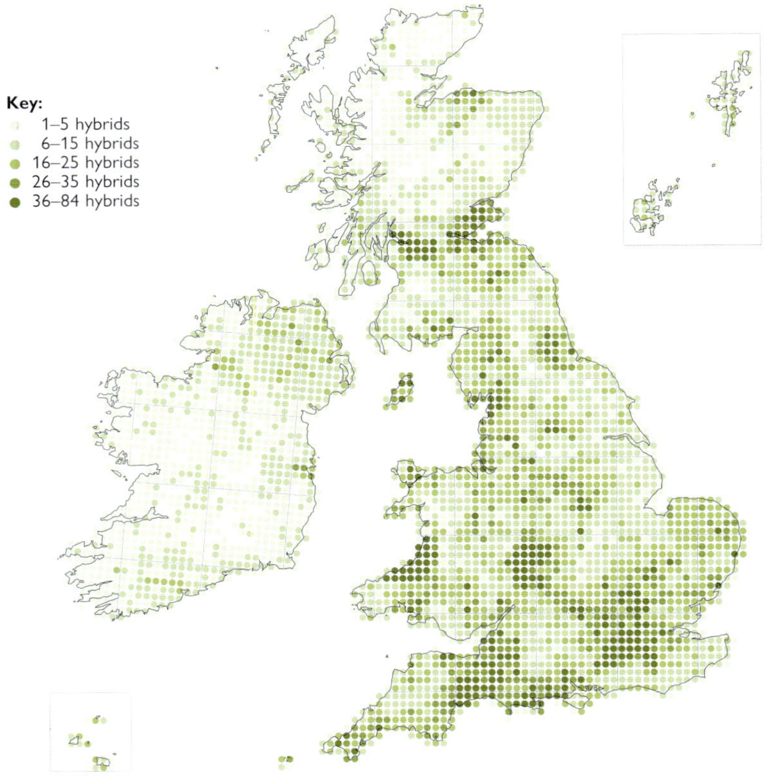

Figure 5.8 The number of hybrids recorded in each 10-km square, 1970–99.

some squares had been recorded before the launch of the project and in other areas recorders were sometimes reluctant to change their established practices.

As a test of the extent to which one group of introduced species have been recorded, a list of 100 introduced trees and shrubs which are often planted (rather than naturalised) in the wild was drawn up (Table 5.1). The number of these species recorded from each square is mapped in Fig. 5.9. This reveals a rather patchy distribution, demonstrating that there has probably been some variation in the recording of these species.

Comparison of species totals in the 1962 *Atlas* and the *New Atlas*

In the 1962 *Atlas* the number of species recorded in each square was summarised in 11 bands from 0 (0–50 species), 1 (50–150), 2 (150–250) and thereafter in successive bands of 100 species up to 10 (950–1050). To compare the data for the *New Atlas* with these totals, we have calculated the number of species now known from each square, and allocated these values to similar bands. As the 1962 *Atlas* data are included in the current database, the number of species (and hence the band number) for any square must remain the same or increase. An increase of 1 band indicates that there are 1–200 additional species in the square, an increase of 2 bands indicates 100–300 additional species, an increase of 3 bands indicates 200–400 additional species, etc. In Fig. 5.10 the increase in band number has been calculated using current totals based solely on those taxa that were mapped in the 1962 *Atlas* (taxa which were not mapped then have not been counted). It is clear that there are substantially more species recorded now in many areas. As one would expect, the differences are even more pronounced if all species and hybrids, rather than just those mapped in 1962, are included in the current totals (Fig. 5.11). The increase in taxa recorded per square is likely to reflect both an increase in recording

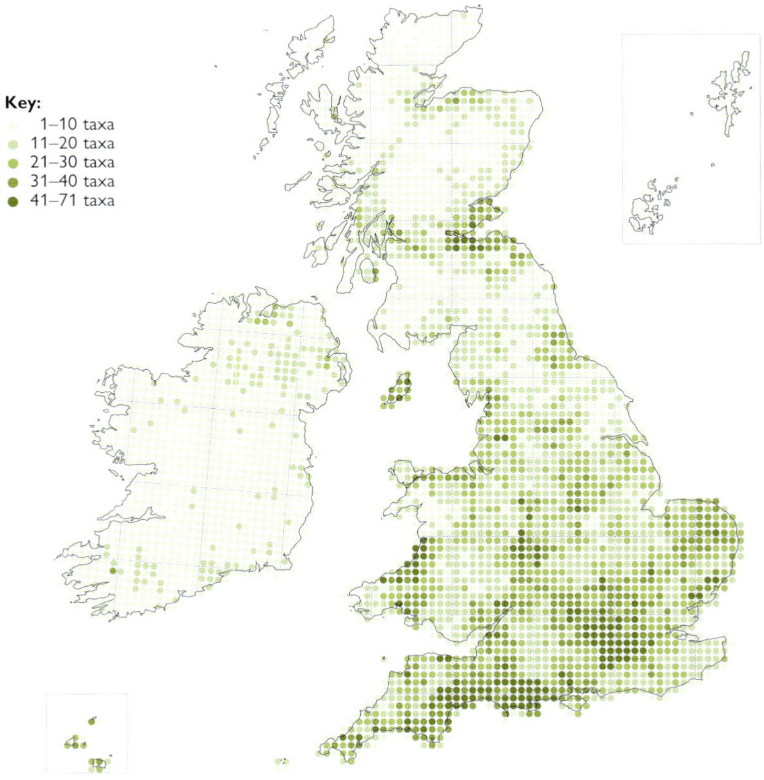

Figure 5.9 The number of commonly planted trees and shrubs recorded in each 10-km square. For a list of the 100 taxa on which this map is based, see Table 5.1.

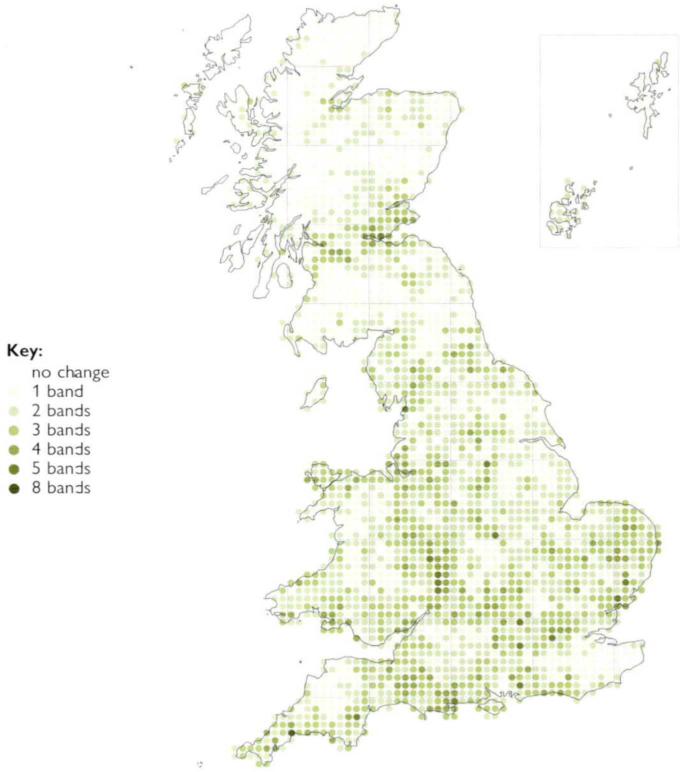

Figure 5.10 The difference between the number of species and hybrids mapped in each square in the 1962 *Atlas* and the number mapped in this Atlas. Only taxa mapped in the 1962 *Atlas* are included in the *New Atlas* totals. An increase of 1 band indicates that there are 1–200 additional species in the square, an increase of 2 bands indicates 100–300 additional species, an increase of 3 bands indicates 200–400 additional species, etc. The comparisons are restricted to 10-km squares in Britain and the Isle of Man which were mapped in both Atlases (see Fig. 5.2).

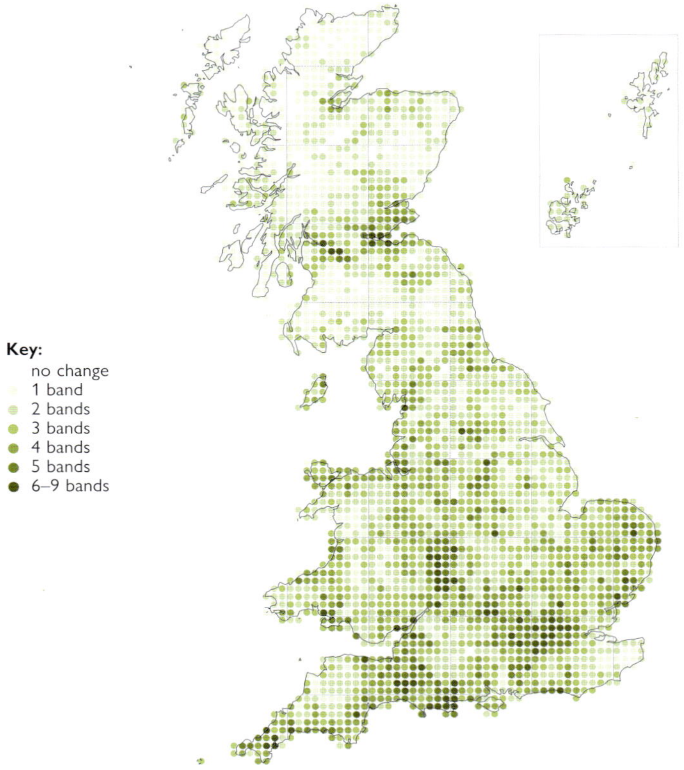

Figure 5.11 The difference between the number of species and hybrids mapped in each square in the 1962 *Atlas* and the number mapped in the *New Atlas*. The totals for taxa mapped in the *New Atlas* include all species and hybrids covered by the project, whether or not they were mapped in the 1962 *Atlas*, although species that are mapped as aggregates and segregates in the *New Atlas* are only counted once, as the aggregate. An increase of 1 band indicates that there are 1–200 additional species in the square, an increase of 2 bands indicates 100–300 additional species, an increase of 3 bands indicates 200–400 additional species, etc. The comparisons are restricted to 10-km squares in Britain and the Isle of Man which were mapped in both Atlases (see Fig. 5.2).

Table 5.1 Introduced trees and shrubs that tend to be planted in the wild. For a coincidence map of the occurrence of these species, see Fig. 5.9.

Abies alba	Cornus alba	Nothofagus nervosa	Quercus rubra
Abies grandis	Cornus mas	Mahonia aquifolium	Rhododendron luteum
Abies procera	Cornus sericea	Nothofagus obliqua	Rhus typhina
Acer cappadocicum	Crataegus persimilis	Olearia macrodonta	Ribes sanguineum
Acer negundo	Cryptomeria japonica	Philadelphus coronarius	Robinia pseudoacacia
Acer platanoides	Cupressus macrocarpa	Picea abies	Rosa ferruginea
Acer pseudoplatanus	Cytisus striatus	Picea sitchensis	Rosa 'Hollandica'
Acer saccharinum	Euonymus japonicus	Pinus contorta	Rosa multiflora
Aesculus carnea	Fuchsia magellanica	Pinus nigra	Rosa rugosa
Aesculus hippocastanum	Genista hispanica	Pinus pinaster	Rosa virginiana
Ailanthus altissima	Griselinia littoralis	Pinus radiata	Rubus spectabilis
Alnus cordata	Hebe salicifolia	Pinus strobus	Salix daphnoides
Alnus incana	Juglans regia	Populus alba	Salix elaeagnos
Amelanchier lamarckii	Kerria japonica	Populus nigra (fastigiate cultivars)	Sequoia sempervirens
Araucaria araucana	Laburnum alpinum	Populus trichocarpa	Sequoiadendron giganteum
Aucuba japonica	Laburnum anagyroides	Prunus cerasifera	Sorbus croceocarpa
Berberis darwinii	Larix decidua	Prunus domestica	Sorbus intermedia
Berberis thunbergii	Larix kaempferi	Prunus laurocerasus	Sorbus latifolia
Buddleja globosa	Leycesteria formosa	Prunus lusitanica	Spartium junceum
Cedrus atlantica	Ligustrum ovalifolium	Prunus serotina	Symphoricarpos albus
Cedrus deodara	Lonicera japonica	Pseudotsuga menziesii	Syringa vulgaris
Cedrus libani	Lonicera nitida	Pyracantha coccinea	Tamarix gallica
Chaenomeles speciosa	Lonicera pileata	Pyrus communis sens. lat.	Thuja plicata
Chamaecyparis lawsoniana	Lycium agg.	Quercus cerris	Tsuga heterophylla
Chamaecyparis pisifera	Mespilus germanica	Quercus ilex	Viburnum tinus

efficiency and the much larger number of introduced taxa that were covered by the *New Atlas* project.

Summary

This review of the quality of recording, coupled with an examination of the maps of individual species in Chapter 8, suggests that the distribution patterns for the more frequent, conspicuous and easily identified taxa are likely to be well displayed on the maps. Almost all 10-km squares have been surveyed for the *New Atlas* project in the 1987–99 recording period, but some areas have been well-recorded for the 1970–86 date-class but less well-recorded from 1987 onwards (Fig. 5.6). Subspecies, hybrids and (to a lesser extent) those introduced trees and shrubs which are deliberately planted in the wild are recorded rather patchily, and in any analysis of the results of the project it might be better to exclude these taxa, or to treat them separately. Such 'noise' is perhaps inevitable in an ambitious survey such as that undertaken for the *New Atlas*, which stretched the recording resources of the British and Irish botanical community to their limits.

CHAPTER 6

Vascular plant diversity in Britain and Ireland

H. R. Arnold, T. D. Dines, D. A. Pearman and C. D. Preston

Total species diversity

The total number of species recorded in each 10-km square from 1970 onwards is shown in Fig. 6.1. This demonstrates the tendency in Britain for species-richness to decrease from south to north, a trend which is found in most (but not all) plant and animal groups. Within Britain some areas are recognisably species-poor, such as the fenland south of the Wash and parts of upland Wales, but some of the variation in the number of species reflects the variation in coverage discussed in Chapter 5. In Ireland there tend to be fewer species recorded than in areas of equivalent latitude in England and Wales; there is a broad similarity between species diversity in Ireland and Scotland.

Some insight into the factors underlying the patterns in Fig. 6.1 can be gained by dividing the species into separate groups and examining the distribution of the groups more closely. The next sections deals with the numbers of native and introduced species, and with the different categories of introductions. Another way of examining the components of biodiversity is to look at the distribution in our area of species with different distributions in the wider world. The native species have been classified into groups based on their distribution in the northern hemisphere (*floristic elements*), and the distribution of these groups is also mapped in this chapter.

Diversity of native and introduced species

The distinction between native species (which reached our area by natural dispersal) and introductions (which were brought here by man) has been discussed in Chapter 3. The number of native species recorded in each 10-km square since 1970 is summarised in Fig. 6.2. This is broadly similar to the map showing total species-richness (Fig. 6.1). An equivalent map showing the diversity of archaeophytes (Fig. 6.3) shows an even greater concentration of species in S.E. England. In part, this reflects the fact that many archaeophytes are arable weeds and are thus concentrated in the areas where arable farming predominates. The number of neophytes is shown in Fig. 6.4. The concentration of neophytes in major urban areas such as London, Birmingham, the Bradford–Leeds conurbation, Edinburgh, Glasgow, Dublin and Belfast is apparent from the maps. However, there are also concentrations in well-covered areas where introduced species have been enthusiastically recorded, such as Dorset, N. & S. Somerset and the Isle of Man. The proportion of neophytes in the total flora (Fig. 6.5) is particularly high in the London area; elsewhere in Britain it decreases from south to north. Casuals (Fig. 6.6) show a similar pattern to neophytes.

Distribution of species in different floristic elements

Native species have been grouped into floristic elements by Preston & Hill (1997). They are classified according to two criteria: their presence as natives in one or more of the major terrestrial biomes and their eastern limits (Tables 6.1, 6.2). Each species is allocated to a major biome category and an eastern limit category and the floristic element represents a combination of the two, *e.g.* the Oceanic Boreal-montane floristic element consists of species in the Oceanic eastern limit category and the Boreal-montane major biome category.

In addition to the elements defined above, there are three special elements: Mediterranean-Atlantic; Submediterranean-Subatlantic and Mediterranean-montane. The Mediterranean-Atlantic species are confined to the Southern (Mediterranean) biome in eastern Europe but extend into the Temperate region in western Europe. The Submediterranean-Subatlantic species show a similar trend, but with more extensive distributions in both the Southern and Temperate biomes. Finally, a small group of Mediterranean-montane species have a montane distribution in the Mediterranean region but are found (usually rather rarely) at lower altitudes in the Temperate zone.

Species-richness maps for the major biome categories defined in Table 6.1 are plotted as Figs 6.7–6.14. The distribution of species in our area closely reflects their wider distribution in the northern hemisphere. Species in the Arctic-montane elements (Fig. 6.7) are more or less restricted to montane areas. The scattered 10-km squares with one Arctic-montane species outside the main upland areas reflect the occurrence of a few native populations of plants such as *Cochlearia pyrenaica* and *Diphasiastrum alpinum* and the presence of species such as *Alchemilla alpina* and *Sedum rosea* as garden escapes. The Boreo-arctic Montane and Boreal-montane species (Figs 6.8, 6.10) have similar, although less restricted, distributions. The relatively few species with very wide world distributions, placed in the Wide-boreal and Wide-temperate elements (Figs 6.9, 6.12), are widespread in both Britain and Ireland and well-represented in most 10-km squares, although the Wide-boreal species are most frequent in Scotland whereas the Wide-temperate species have a slight southerly bias. The three largest major biome categories are the Boreo-temperate (Fig. 6.11), Temperate (Fig. 6.13) and Southern-temperate (Fig. 6.14). Species in all these groups of elements are widespread, but the Boreo-temperate species are slightly better represented in northern areas whereas the Temperate and Southern-temperate species show the opposite trend.

Species in all three Mediterranean elements are mapped in Fig. 6.15, and the Mediterranean-Atlantic and Submediterranean-subatlantic elements in Figs 6.16 and 6.17. The true Mediterranean-Atlantic species have a much more restricted and predominantly coastal distribution than those in the Submediterranean-subatlantic element.

Species in the eastern limit categories (Table 6.2) are mapped in Figs 6.18–6.23. The Oceanic species (Fig. 6.18) have a marked but not extreme western bias in Britain; in Ireland they tend to be scarce in the centre. Suboceanic species (Fig. 6.19) show a similar trend, but European, Eurosiberian and Eurasian species (Figs. 6.20–6.22) do not. The Circumpolar species (Fig. 6.23) have a more northerly distribution, reflecting the predominance of Circumpolar distributions in the Arctic and Boreal major biome categories.

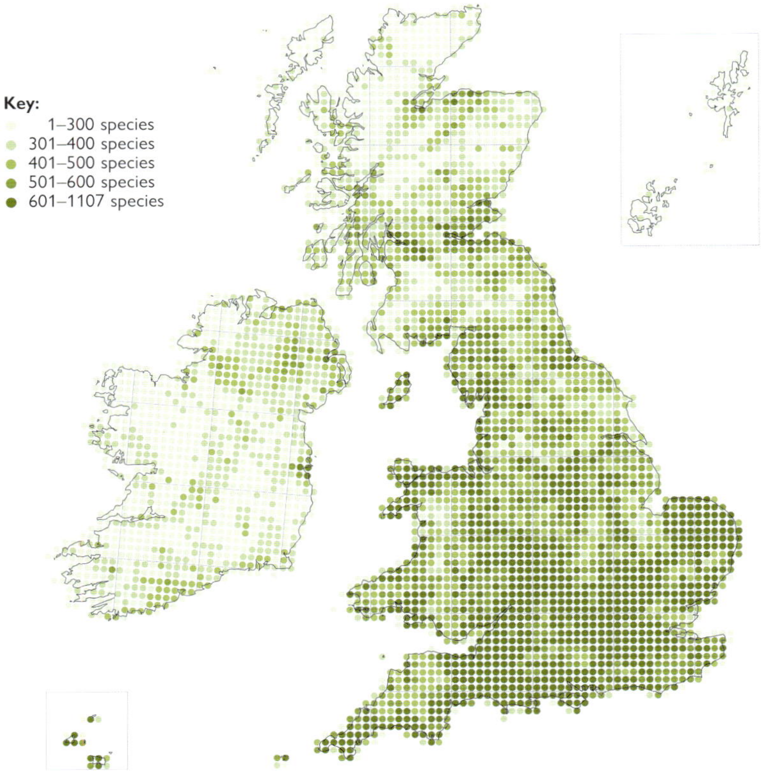

Figure 6.1 The number of species recorded from 1970 onwards in each 10-km square.

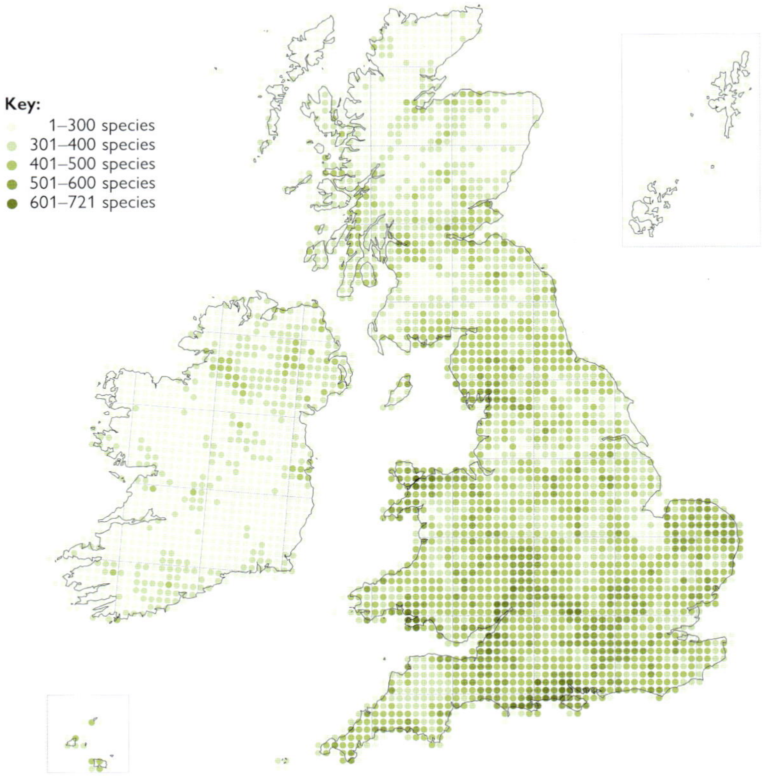

Figure 6.2 The number of native species recorded from 1970 onwards in each 10-km square.

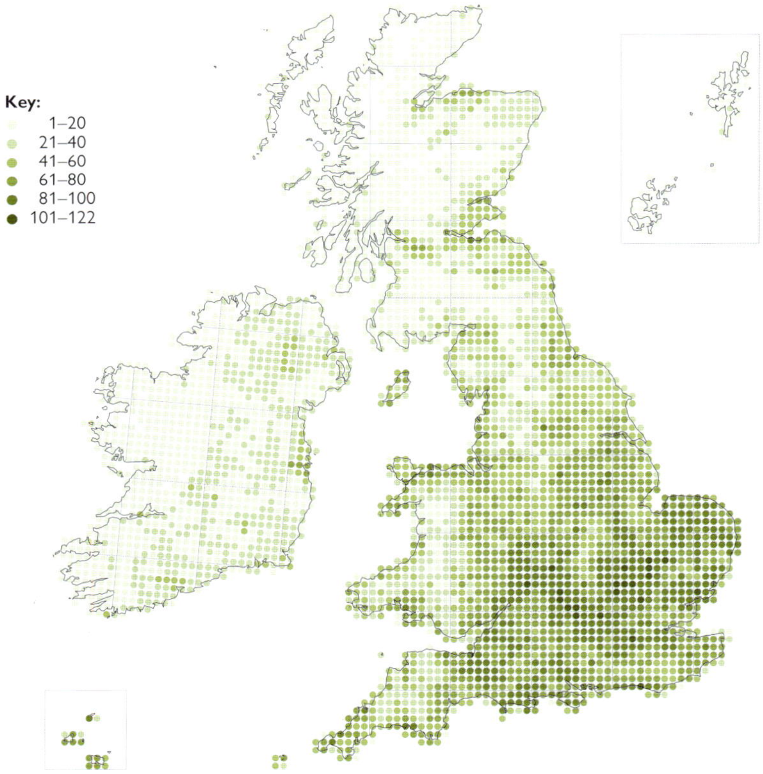

Figure 6.3 The number of archaeophytes recorded from 1970 onwards in each 10-km square.

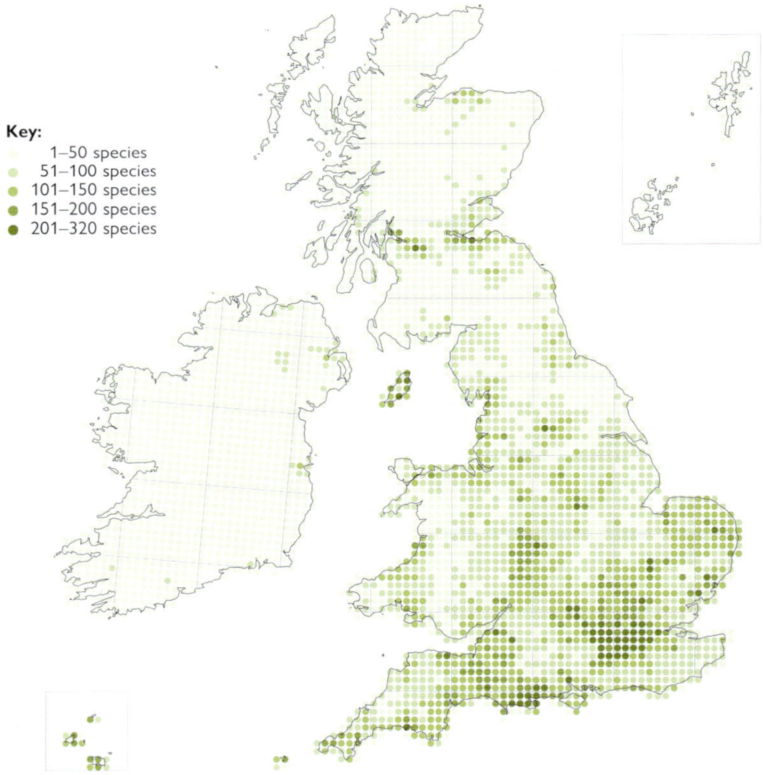

Figure 6.4 The number of neophytes recorded from 1970 onwards in each 10-km square.

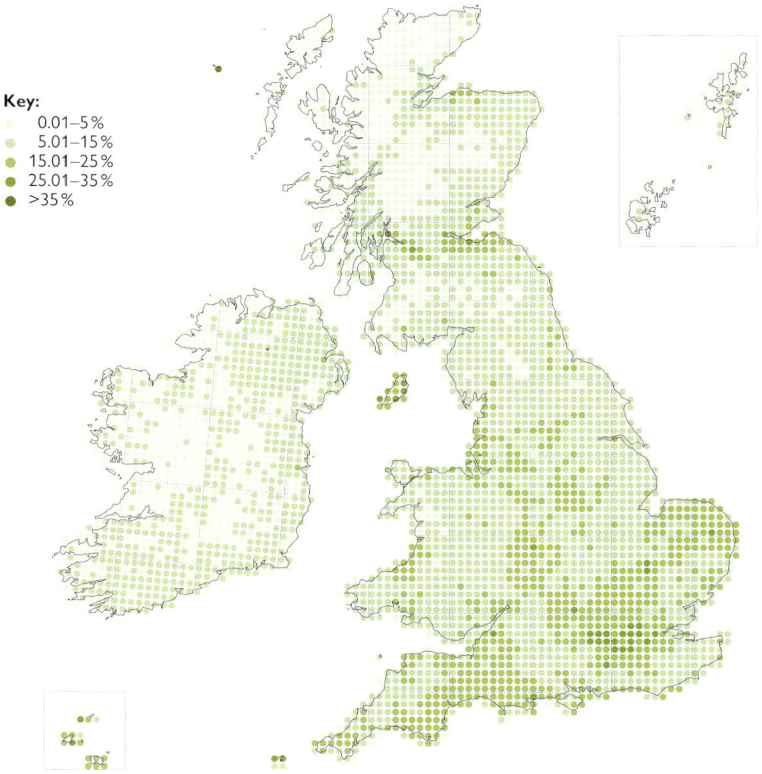

Figure 6.5 The proportion of neophytes in the total flora, based on species recorded from 1970 onwards in each 10-km square.

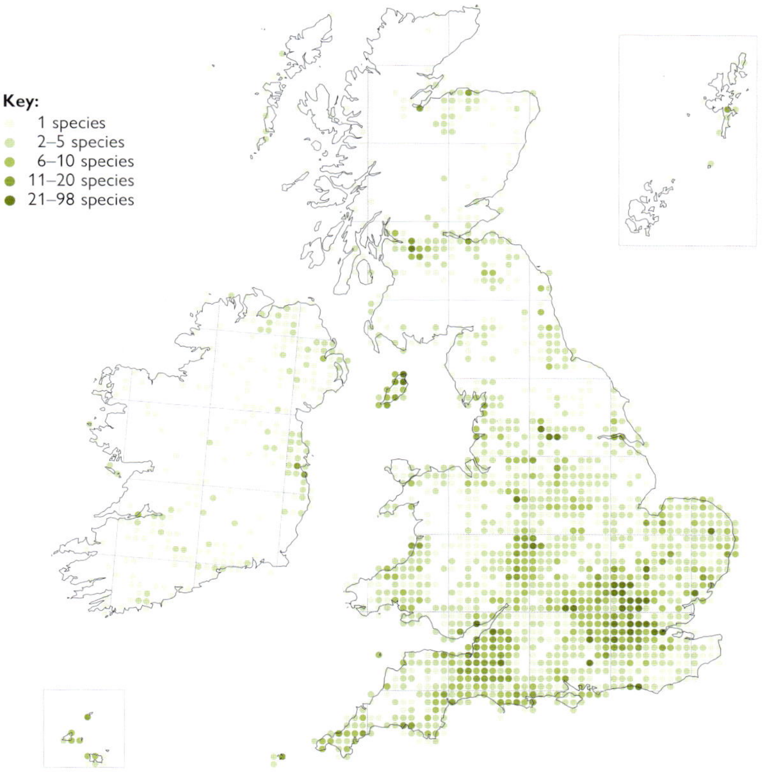

Figure 6.6 The number of casuals recorded from 1970 onwards in each 10-km square.

Table 6.1 Classification of species into major biome categories.

Major biome category	Main distribution of species
Arctic-montane	North of the tree-line or (on mountains) above the tree-line, or both
Boreo-arctic Montane	In both Arctic-montane and Boreal-montane zones
Wide-boreal	In Arctic-montane, Boreal-montane and Temperate zones
Boreal-montane	In the coniferous forest zone, either in the Boreal zonobiome or on mountains further south or both
Boreo-temperate	In both Boreal-montane and Temperate zones
Wide-temperate	In Boreal-montane, Temperate and Southern-temperate zones
Temperate	In the cool-temperate, broad-leaved deciduous forest zone
Southern	In the warm-temperate zone south of the broad-leaved deciduous forest zone, which in Europe is represented by the Mediterranean zone

Table 6.2 Classification of species into eastern limit categories.

Eastern limit category	Main distribution of species
Oceanic	Confined to Western Europe (Norway, W. Denmark, Low Countries, Britain, Ireland, W. France and the Atlantic fringe of Spain and Portugal)
Suboceanic	Confined to Western and Central Europe (occurring west of a line from the Baltic to the Adriatic)
European	Widespread in Europe, but with an Eastern limit west of 60°E
Eurosiberian	Widespread in Europe and western Asia, with an eastern limit between 60°E and 120°E
Eurasian	Widespread in Europe and Asia, with an eastern limit east of 120°E
Circumpolar	Present in Europe, widespread in Asia and also present in North America

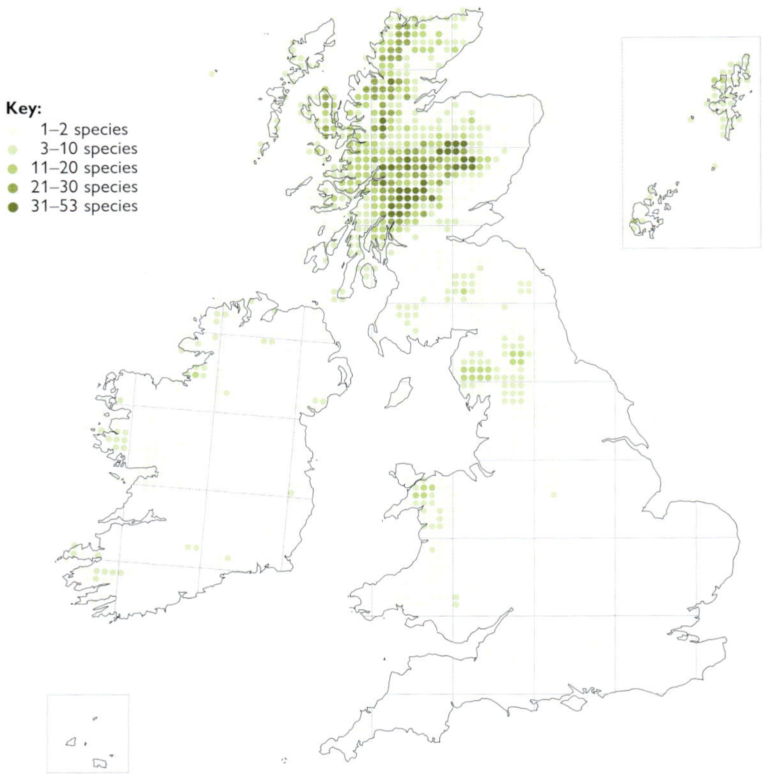

Figure 6.7 The number of Arctic-montane species recorded from 1970 onwards in each 10-km square.

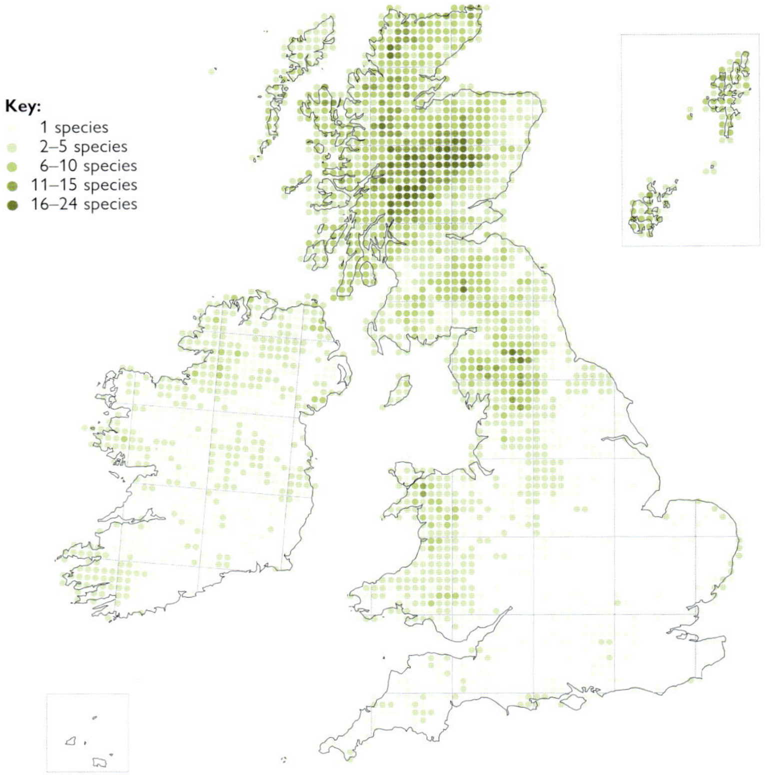

Figure 6.8 The number of Boreo-arctic Montane species recorded from 1970 onwards in each 10-km square.

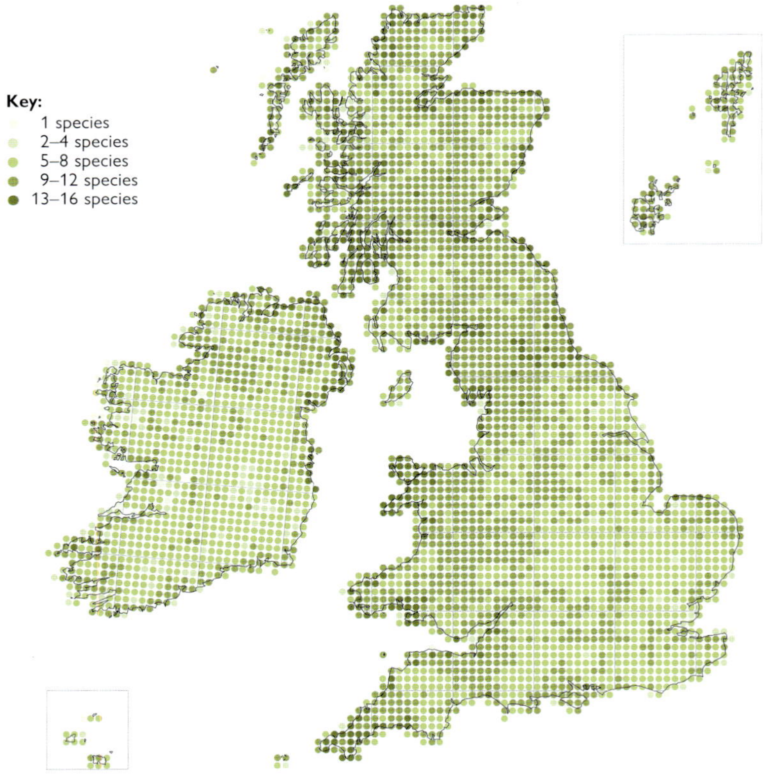

Figure 6.9 The number of Wide-boreal species recorded from 1970 onwards in each 10-km square.

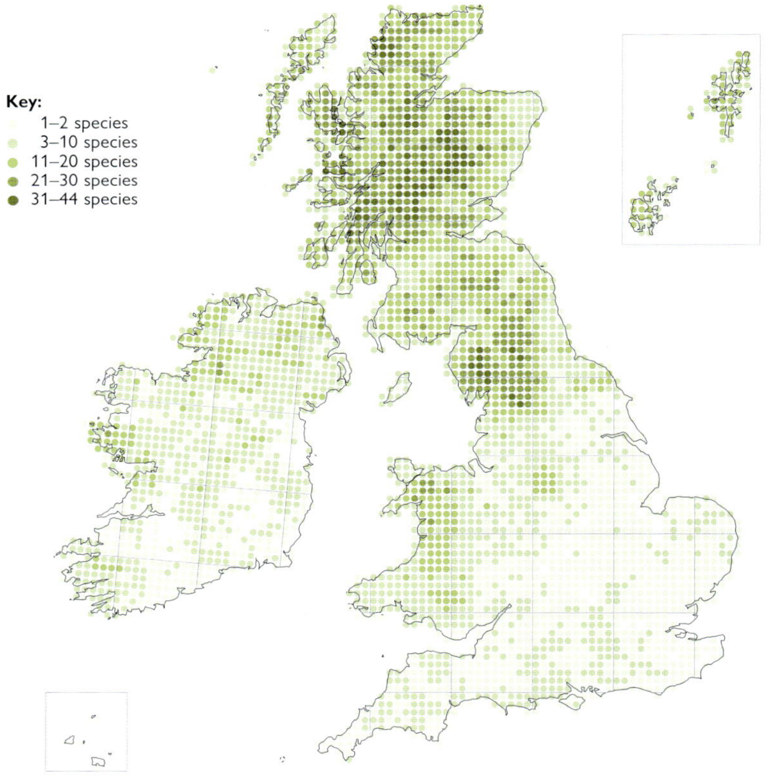

Figure 6.10 The number of Boreal-montane species recorded from 1970 onwards in each 10-km square.

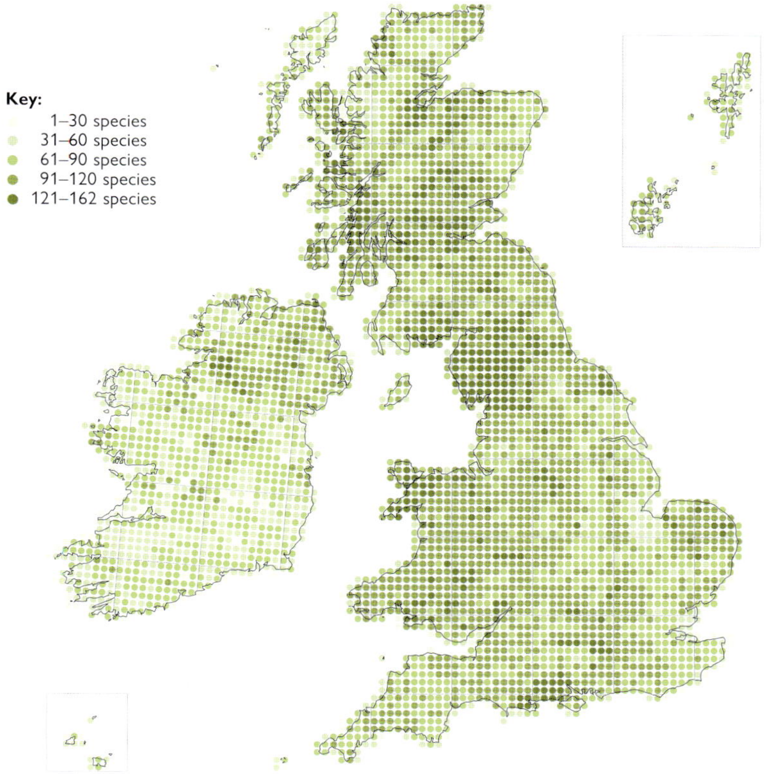

Figure 6.11 The number of Boreo-temperate species recorded from 1970 onwards in each 10-km square.

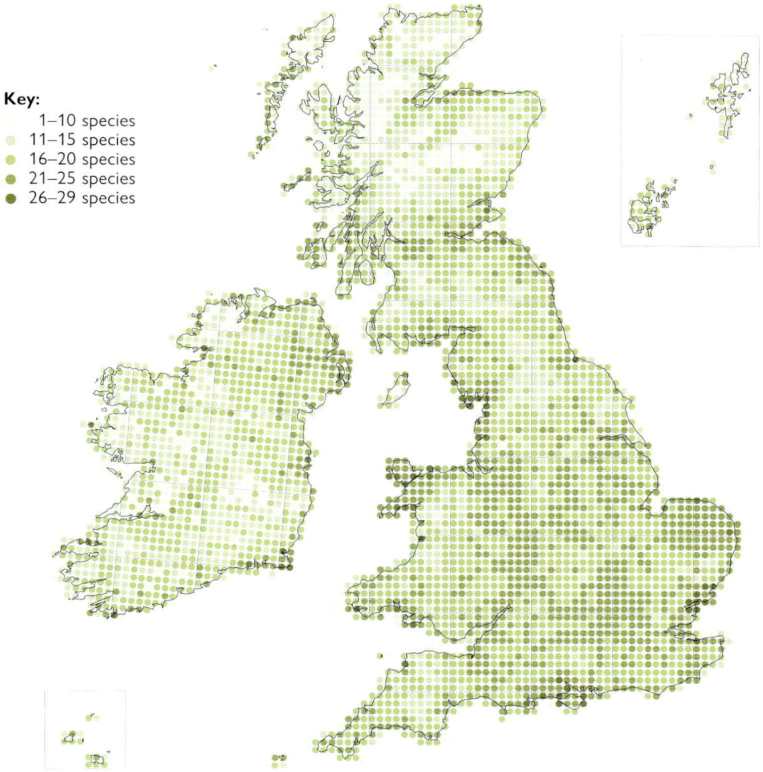

Figure 6.12 The number of Wide-temperate species recorded from 1970 onwards in each 10-km square.

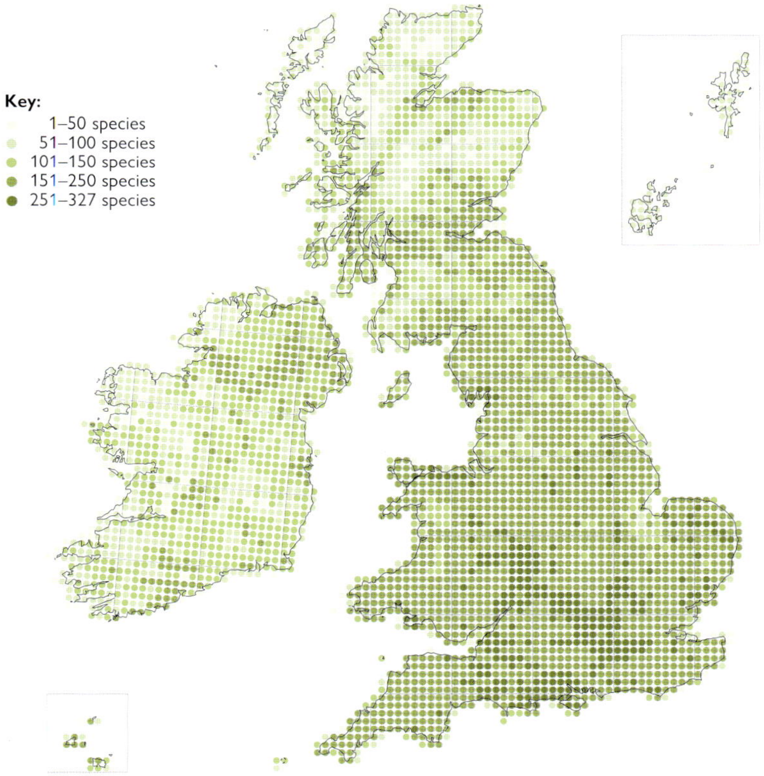

Figure 6.13 The number of Temperate species recorded from 1970 onwards in each 10-km square.

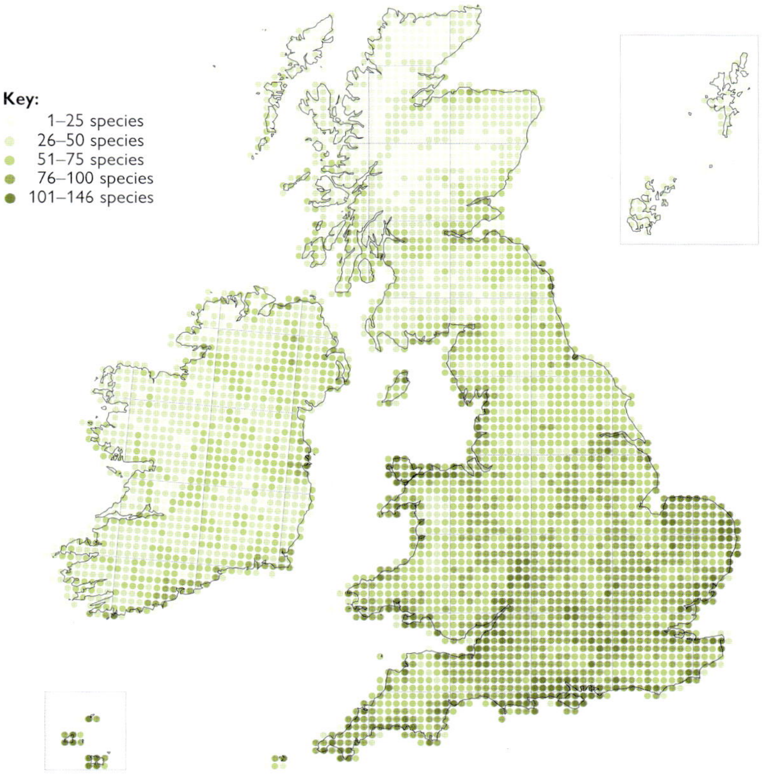

Figure 6.14 The number of Southern-temperate species recorded from 1970 onwards in each 10-km square.

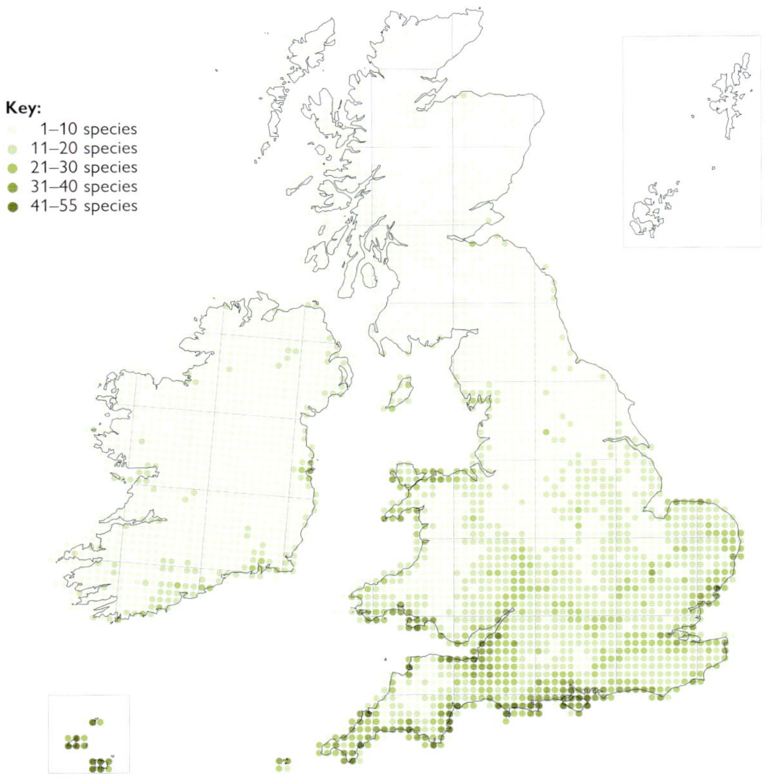

Figure 6.15 The number of Mediterranean species recorded from 1970 onwards in each 10-km square. Mediterranean-Atlantic, Submediterranean-Subatlantic and Mediterranean-montane species are all included in the totals.

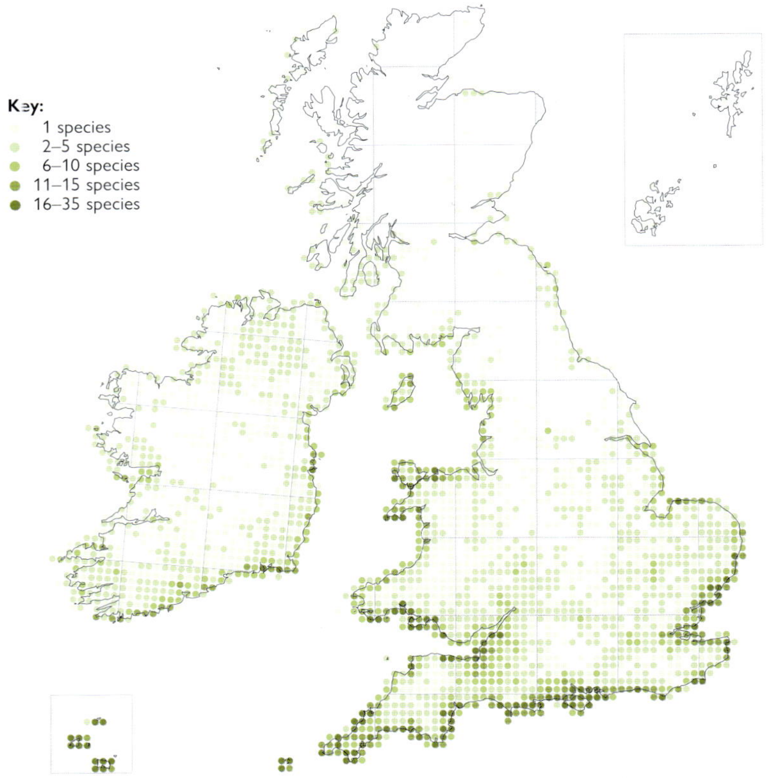

Figure 6.16 The number of Mediterranean-Atlantic species recorded from 1970 onwards in each 10-km square.

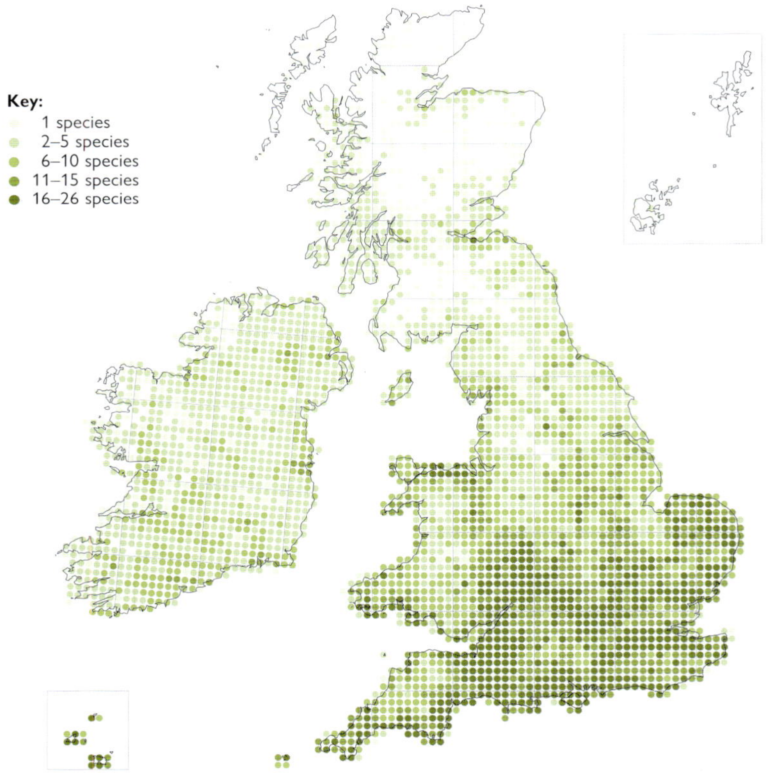

Figure 6.17 The number of Submediterranean-Subatlantic species recorded from 1970 onwards in each 10-km square.

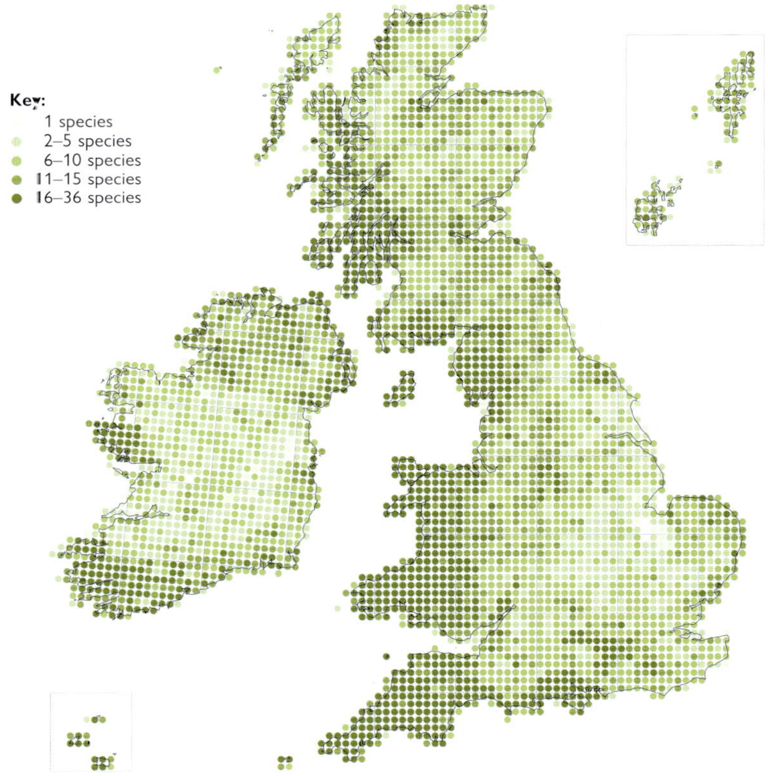

Figure 6.18 The number of Oceanic species recorded from 1970 onwards in each 10-km square.

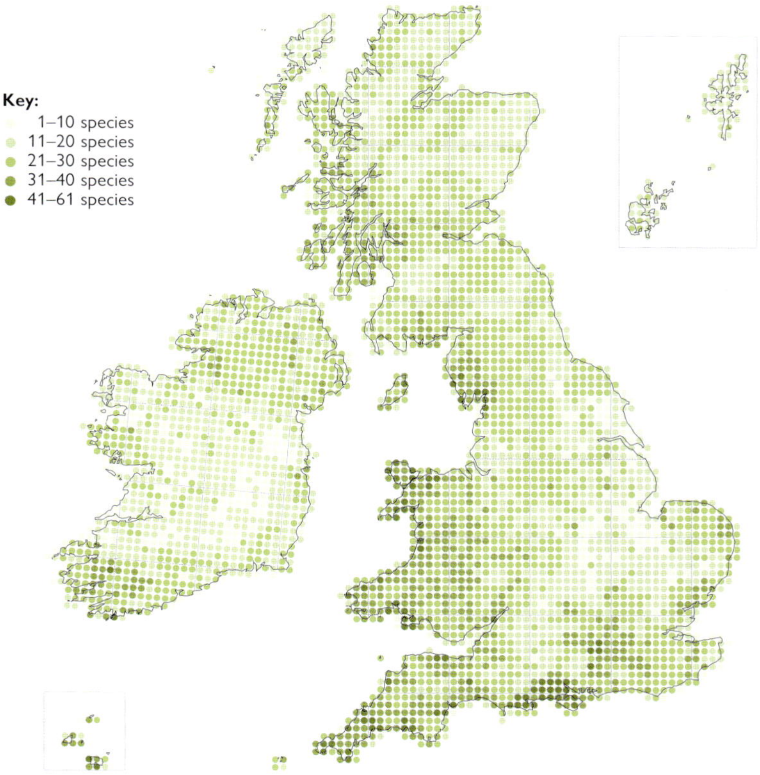

Figure 6.19 The number of Suboceanic species recorded from 1970 onwards in each 10-km square.

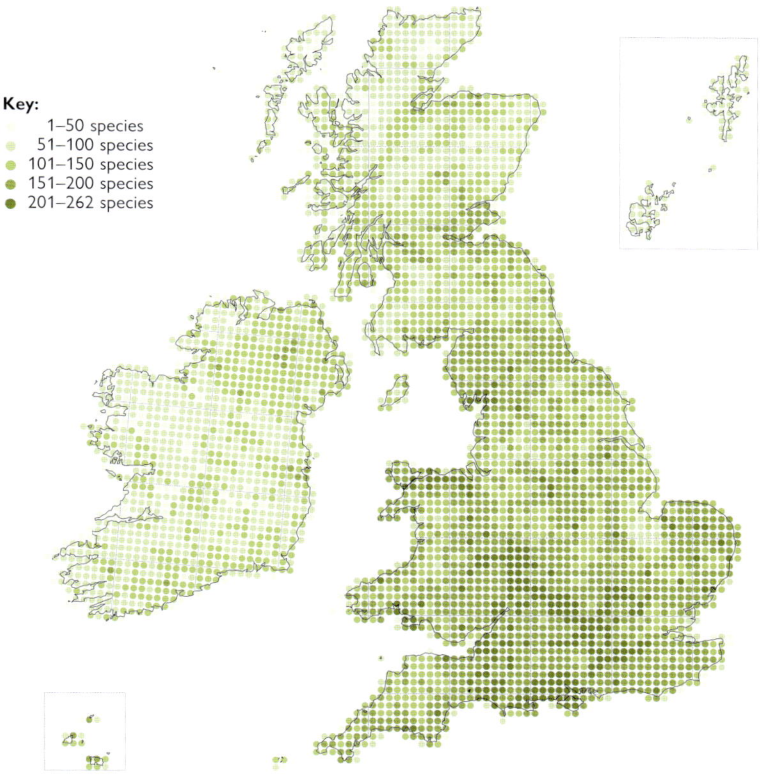

Figure 6.20 The number of European species recorded from 1970 onwards in each 10-km square.

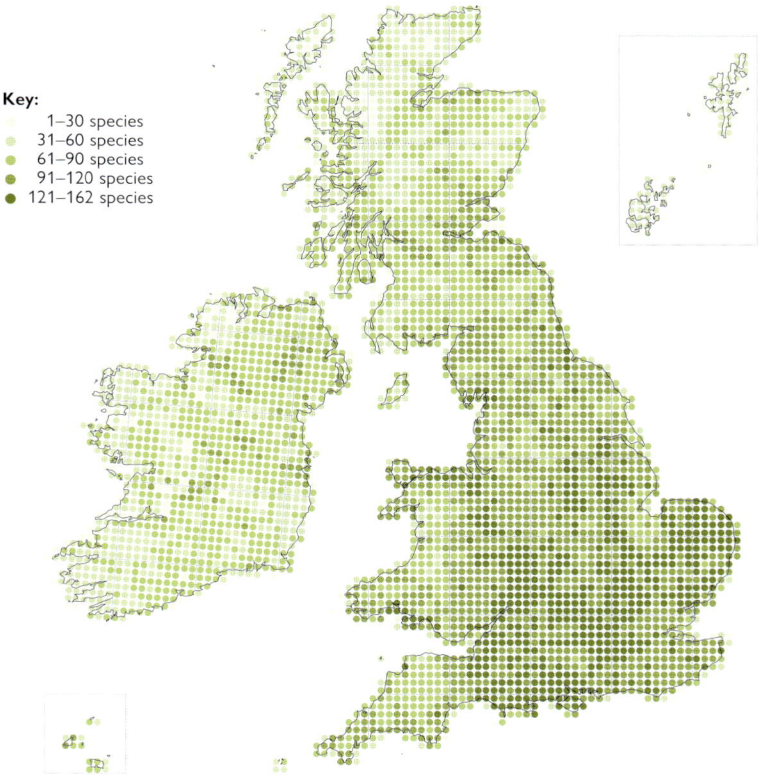

Figure 6.21 The number of Eurosiberian species recorded from 1970 onwards in each 10-km square.

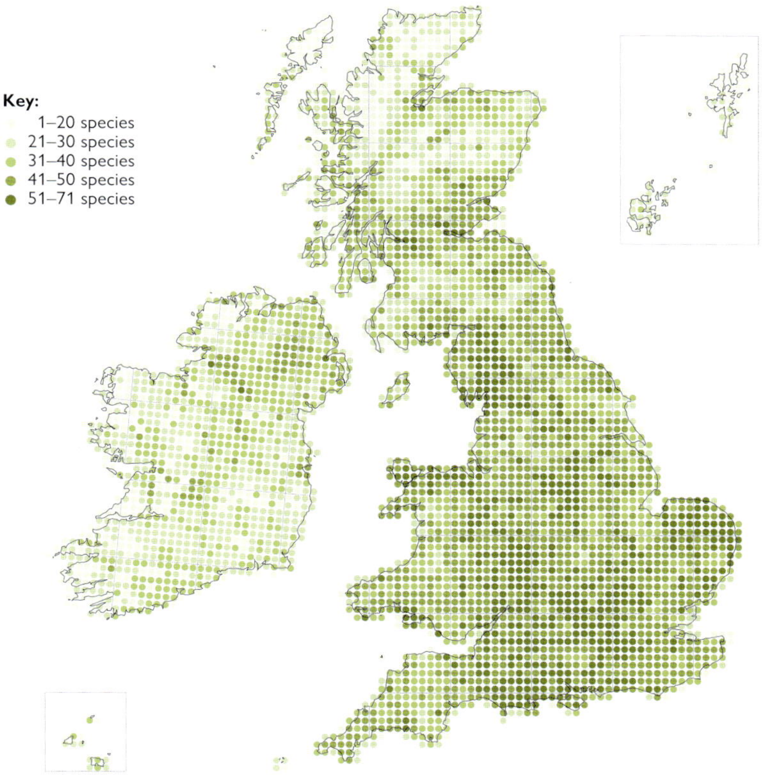

Figure 6.22 The number of Eurasian species recorded from 1970 onwards in each 10-km square.

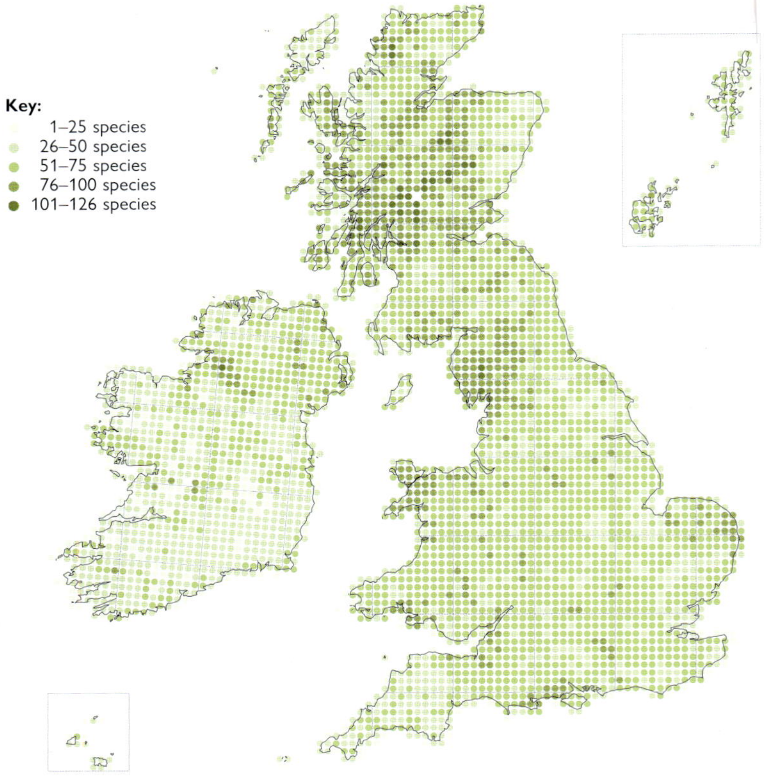

Figure 6.23 The number of Circumpolar species recorded from 1970 onwards in each 10-km square.

CHAPTER 7

The changing flora of Britain, 1930–99

C. D. Preston, M. G. Telfer, H. R. Arnold and P. Rothery

Historical background

The composition of the British and Irish flora and the distributions of its constituent species have been determined by many thousands of years of climatic, vegetational and land-use change. The most dramatic of these changes in geologically recent times have been the successive glacial and inter-glacial periods of the Pleistocene epoch, which began 2.3 million years ago. The most recent glacial period started 120 000 years ago, attained its maximum development between 25 000 and 18 000 years before present (BP) and finally petered out in the climatic fluctuations of the Late-glacial period between 16 000 and 11 500 BP. During this period trees were absent from Britain and Ireland but there was some vegetation south of the ice sheets. Trees recolonised after the Late-glacial period and by 9000 BP temperate vegetation dominated by native tree species that are still familiar to us today had already become established in the lowlands.

Early hominids are known to have been living in Britain 500 000 years ago, and Palaeolithic and Mesolithic 'hunter-gatherer' communities may have had an appreciable influence on our vegetation. However, human influence on the landscape and vegetation became much more pronounced after the advent of animal husbandry and crop cultivation. These developments define the start of the Neolithic period, which began some 6000 years ago in Britain and 5500 years ago in Ireland. Since then the management of the landscape for agriculture and the deliberate and accidental importation of plant species have had a profound effect on our flora and vegetation.

Evidence for the changing flora of Britain and Ireland before AD 1500 comes principally from palaeobotanical and archaeological studies, supplemented from the Roman period onwards by written sources which are few in number and difficult to interpret. It was not until the 16th century that the modern science of botany developed from its medieval precursors. The first 'botanical records' are found in the publications of pioneers such as William Turner (*c.*1510–68) and John Gerarde (1545–1612) who spanned the transition between medieval herbalists and modern botanists. The first 'county Flora', John Ray's catalogue of the plants of Cambridgeshire, was published in 1660. Thereafter, an increasing number of publications offers a rich corpus of historical records. These provide evidence for major changes in the British flora which took place during the 18th and 19th centuries as a result of agricultural improvement, including the drainage of wetlands and the agricultural intensification that followed parliamentary enclosure. Maps in the 1962 *Atlas* show that some species had suffered a marked contraction of range before 1930 (e.g. *Drosera anglica*, *Ophrys sphegodes*, *Orobanche rapum-genistae*, *Pulicaria vulgaris*, *Pulsatilla vulgaris*). The losses sustained during this period are, of course, still apparent on the maps published in this *New Atlas*. The botanical records from the 16th century onwards also document the introduction and spread of non-native species, and many of these were also mapped in the 1962 *Atlas*. The changes in the British flora in the historical period were reviewed by Perring (1970, 1974). There have been more detailed local studies in many county Floras and in papers such as those of Dony (1977), James (1997) and Preston (2000). Studies of particular habitats or species are too numerous to list individually. There are fewer historical county Floras and other plant records for Ireland, especially outside the main urban centres of Belfast and Dublin, and so the changes in the flora of Ireland in the 18th and 19th centuries are less well-documented.

Methods of data analysis

Rather than review the changes in our flora over the last four centuries, this chapter concentrates on an analysis of changes that have occurred in Britain between the periods 1930–69 and 1987–99. The first period is that for which 'recent' records were accepted for the 1962 *Atlas* (1930 to 1959–60) and the years immediately afterwards. The second period is the one in which the majority of records were collected for the *New Atlas*. Although the 1930–69 period is the longer one, there is little doubt that in 1987–99 more intensive recording was achieved in many areas (see Chapter 5). Because of these variations in recording intensity, we have analysed the records for the two periods using a technique which does not attempt to assess the extent to which species have increased or decreased in absolute terms. Instead, we have measured the *relative* performance of the species. In other words, we measure the extent to which species have changed in relation to the 'average' species. It is not possible to assess, using this method, whether the 'average' species has itself increased or decreased. The statistical analysis is restricted to those taxa which were mapped in the 1962 *Atlas*, but we have also commented in more general terms on some that were not mapped then but are included in this book. We have regretfully excluded Ireland from this analysis, for reasons explained below.

Source of records

The records used in this analysis have been taken from the Vascular Plant Database compiled during the *New Atlas* project. Those for the 1930–69 period include records compiled for the 1962 *Atlas*, with subsequent amendments and additions. Some of the records added to the database in the years immediately after the end of recording for the 1962 *Atlas* were dated as '1950 onwards', and they cannot therefore be separated from those which were made in the period of recording. This is the main reason for extending the first period for comparison to 1969. There is still a degree of doubt about some of the records included in the Database for this period, in that some pre-1930 records may have been submitted as recent records to the 1962 *Atlas* and records may have continued to be sent as '1950 onwards' into the 1970s. However, the vast majority of records dated as 1930–69 were actually made during this period and any exceptions can have had only a very insignificant effect on the results of the analysis.

The records for the period 1987–99 have been collected from the sources detailed in Chapter 4. We have chosen to compare the records made in the 1930–69 period with those made between 1987 and 1999, rather than with those made from 1970 to 1999, as we considered that the more recent period gave a more accurate reflection of the current flora. There was a period of particularly rapid agricultural intensification between 1970 and 1988 (Chamberlain *et al.*, 2000) and the date-spans we have analysed, 1930–69 and 1987–99, are well placed to detect any effects this may have had on the British flora. Although some areas were under-recorded in the period from 1987 onwards, it is doubtful whether they were less well-recorded than they were for the 1962 *Atlas*, and in any event the method we have chosen to analyse the data helps correct for such differences. There is a possibility of bias in the results of the analysis for species that are largely restricted to areas which were under-recorded either in the earlier or in the later period. However, the map of squares which were better recorded in 1970–86 than in 1987–99 (Fig. 5.6) shows that although these are concentrated in particular vice-counties, these vice-counties are not concentrated in particular regions.

Geographical coverage

The area covered in this analysis comprises the Watsonian vice-counties 1–112, i.e. Great Britain and the Isle of Man. The Channel Islands are excluded, as records in the 1962 *Atlas* were simply mapped for the five major islands rather than in the 15 grid squares in which they are mapped in this *Atlas*. Ireland is also excluded, for two reasons. The editors of the 1962 *Atlas* reported that 'Because of the scarcity of botanists living in the country and the difficulty of persuading great numbers of volunteers to make a visit, much greater use has had to be made of old (pre-1930) records in Ireland than elsewhere' (Perring & Walters 1962, p. xv). This suggests that the difference in the intensity of recording in Ireland between 1930–69 and 1987–99 might not match that in Britain, and that Ireland might therefore be better treated in a separate analysis. The different grids used for Ireland in the 1962 *Atlas* and this book present further complications.

In the current study the results of the two survey periods are considered for Britain as a whole. However, the distribution maps of individual species in Chapter 8 suggest that there have been strong regional trends in the distribution of species, and these regional trends will require further analysis.

Taxonomic coverage

The analysis includes those native and introduced taxa that were mapped in the 1962 *Atlas* and which are currently recognised as species, aggregates of closely related species or interspecific hybrids. Records of taxa mapped as species in 1962 but treated as only subspecifically distinct by Stace (1997), such as *Empetrum hermaphroditicum* & *E. nigrum*, *Narcissus obvallaris* & *N. pseudonarcissus*, *Raphanus maritimus* & *R. raphanistrum*, are aggregated to provide totals for the species as defined by Stace. *Trifolium incarnatum* is treated as the species rather than simply subsp. *molinerii*, which was mapped as *T. molinerii* in 1962. Where the map in the 1962 *Atlas* is that of an aggregate of closely related species, this aggregate has been used in the analysis. For example, we aggregated records into *Alchemilla vulgaris* agg. (including all *Alchemilla* species except *A. alpina* and *A. conjuncta*), *Dryopteris filix-mas* agg. (*D. affinis*, *D. filix-mas*, *D. oreades*) and *Tripleurospermum maritimum* agg. (*T. inodorum*, *T. maritimum*). Only a few hybrids were mapped in the 1962 *Atlas*, but these are included in the analysis (e.g. *Crocosmia* × *crocosmiiflora*). Taxa previously mapped as species but now interpreted as hybrids (e.g. *Carex stenolepis*, *Fragaria ananassa*, *Hypericum elatum*, *Populus canescens*, *Sagina normaniana*) are also included. Individual decisions have been made on the treatment of taxa with a complex or confused taxonomic history. Thus the records of the three *Mimulus* spp. mapped in 1962 are aggregated to a single *Mimulus* agg. and those of the four *Oenothera* spp. are aggregated to a single *Oenothera* agg. These aggregates also include the hybrids with which these species have sometimes been confused. However, the map of *Circaea alpina* agg. in the 1962 *Atlas* is treated as referring to *Circaea* × *intermedia* and therefore *C.* × *intermedia* (but not the much rarer *C. alpina*) is included in the analysis.

Carex divulsa, *C. muricata*, *C. spicata*, *Potamogeton crispus*, *P. natans*, *P. pectinatus*, *P. perfoliatus* and *P. polygonifolius* have been excluded as the data collected for the 1962 *Atlas* have been replaced by data based on herbarium material which is more reliable taxonomically but does not give a comparable geographical coverage. *Hypericum perforatum* and *Juncus squarrosus* have been excluded as computerised data for the 1962 *Atlas* are not available.

Over a hundred taxa (most of them native species) have been excluded from the analysis as they were recorded in less than five 10-km squares in the 1930–69 period (see below). In addition, species which are restricted to the Channel Islands or Ireland are, of course, excluded as these areas are not included in the analysis.

Treatment of introduced populations of native species

All records are taken into account, regardless of whether they are believed to be of native or introduced populations. Thus all records of species which occur as both natives and introductions (e.g. *Nymphoides peltata*, *Pinus sylvestris* and *Potentilla fruticosa*) are aggregated to provide the data analysed for these species.

Scale of the analysis

The results of any study of the changing range of species are dependent on the scale of the analysis. In general, the coarser the scale, the larger a change will need to be before it is detected. In this study distributions are summarised in 10-km squares, and the only changes which will be detected are those that affect the 10-km square distributions of species. The 10-km square is a particularly convenient unit for studies of Britain as a whole. It is much smaller than the vice-county, but it is still possible to survey all the British squares in a national project. It is perhaps also a suitable scale for an analysis of change, as it is neither a very small nor a very large area – it appears to be a very small area compared to the European distributions of most species, but it is a relatively coarse unit compared to the extent of many plant populations. Changes which are detected at the 10-km square scale are likely to be biologically significant, but important changes in the distribution of species might also take place at finer scales and be undetected by a study of 10-km squares. These could include changes in the number or average size of populations within the existing 10-km square range. Some recent studies of changes in the distribution of British butterflies illustrate the relationship between change at different scales very elegantly (Asher *et al.*, 2001).

Statistical methods

The method used to analyse the data from the two recording periods is outlined here and explained in detail by Telfer *et al.* (in press). Only grid squares for which there are records for both the 1930–69 and the 1987–99 periods are included in the analysis. The proportion of these squares in which each taxon was recorded is calculated for each period. The proportions for each period are independently transformed using a logit transformation. The logit transformation has been shown to be an appropriate method of converting range size values to a series with an approximately normal distribution (Williamson & Gaston, 1999). A linear regression line is fitted to the transformed proportions using the data for the earlier period as the predictor variable. Initial plots showed that the linear relationship broke down for species which were recorded in less than five squares in the initial period, so these taxa were excluded from the analysis. Even after these have been excluded, the rarer species show more variation than the common species. This reflects the fact that a species recorded in (say) five squares in the earlier period and four or six in the later period has changed by only a single square but by a relatively large proportion of the original five squares. A weighted regression technique is therefore employed to allow for this. The regression line obtained represents the behaviour of the 'average' species (Fig. 7.1). Points above the line represent species which are present in a higher proportion of squares than would be expected from the data collected for the earlier period and have therefore increased in relation to the average species; points below the line represent species which have not been refound in as many squares as expected. The vertical distances from the regression line are measured by standardised residuals, which are positive for points above the line and negative for those beneath.

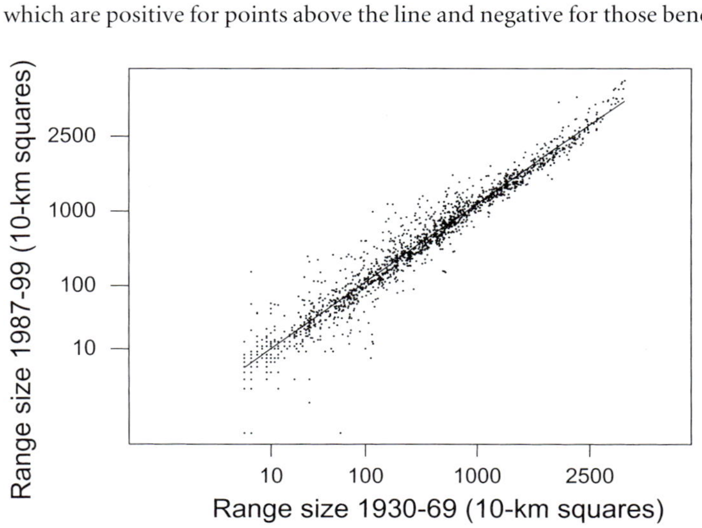

Figure 7.1 Regression of range size in the 1987–99 survey period against range size 1930–69 ($y = 1.05x + 0.29$, $r^2 = 0.94$). Each point represents a single species. The 10-km square range sizes have been expressed as proportions of the total number of squares recorded and logit-transformed (see text).

The standardised residual for each species is described here as the *change index*. The regression line is calculated to produce a mean change index for all species of zero.

The change index therefore measures the relative performance of each species between the two surveys. It does not in itself indicate whether there has been an absolute increase or an absolute decrease in range size. The index allows for overall variation in recording effort between the two surveys. However, factors that affect individual species or groups of species will introduce biases. If, for example, many recorders did not know how to recognise a species in the first survey but all had become familiar with it by the time of the second survey, this might result in a positive change index. Another potential source of bias is variation in the intensity of recording in different regions or in different habitats. If, for example, one area or habitat was relatively well surveyed in the first survey but relatively badly surveyed in the second, this might be reflected in the change indices for species which are concentrated in that area or habitat. Change indices for individual species therefore require interpretation in the light of their taxonomic and recording history. The mean change index for a group of species with no common recording history is likely to be a more robust measure of trends in British biodiversity.

Mean change indices reported below have been analysed by one-way analysis of variance.

Results of the analysis

Summary of results

Records from 2788 British 10-km squares are available from both recording periods. Change indices were obtained for 1524 taxa, including 1142 natives, 32 classified as 'native or introduced', 141 archaeophytes, 208 neophytes and a single recurrent casual (for the distinction between these categories, see Chapter 3). Five of the native taxa and five of the neophytes are hybrids. Change indices have been rounded to two decimal places. The individual indices for the taxa mapped in this *Atlas* are given in the species accounts in Chapter 8.

It is apparent from Fig. 7.1 that none of the commonest species in the 1930–69 period have declined when compared to the average species. The 19 most frequent species in 1930–69 all have a positive change statistic, suggesting that any changes since 1930 have favoured frequent and wide-ranging plants.

The 100 taxa that have shown the greatest relative increases and decreases have been selected rather arbitrarily to illustrate the characteristics of the species which have changed most. They are discussed in the following sections, which are followed by an examination of the trends shown by groups of species with particular morphological, phytogeographical or ecological characteristics. References relating to the ecology or history of particular species are cited in the appropriate accounts in Chapter 8.

The successful species

There are 56 neophytes in the list of 100 taxa that have shown the greatest relative increase (Table 7.1), a much larger proportion than that in the study as a whole (16%). Nine of the top ten species are neophytes, including eight species which were originally introduced as garden plants. There is no doubt that species such as *Buddleja davidii*, *Cotoneaster simonsii*, *Linaria purpurea* and *Lysimachia punctata* have genuinely spread since 1969, although in the case of *Laburnum anagyroides*, *Prunus laurocerasus* and *Syringa vulgaris* the increase in range size is probably at least partly caused by an increased tendency in recent years to record planted shrubs in hedges, plantations and other 'wild' places. *Prunus cerasifera* is a species that was almost certainly overlooked in the 1930–69 period, as only in recent years have recorders appreciated the fact that it can be identified easily by its habit of flowering in early spring, normally well before the native *P. spinosa*. The final introduction in the 'top ten' is *Epilobium ciliatum*, a weed that was accidentally introduced and has a well-documented spread.

Ten archaeophytes are represented in the 100 taxa showing the greatest relative increase. It is surprising that species which were apparently introduced over five hundred years ago have continued to increase at such a rate. Indeed, the absence of a recent rapid spread is one criterion which has been used to define archaeophytes, but for these species it is outweighed by other evidence. At least three of these species are grown as ornamental garden plants and owe their continued spread to introductions from this source (*Euphorbia lathyris*, *Mentha spicata*, *Sedum album*). *Prunus domestica* is grown for its fruit and becomes established when plum stones are discarded into the wild. It is difficult to know whether *Chenopodium ficifolium* and *Lamium hybridum* were under-recorded in the 1962 *Atlas*, have spread subsequently, or appear to have increased because of a combination of both factors. *Lactuca serriola* has spread along new roads, and *Valerianella carinata* and *Vulpia myuros* have certainly spread, but for reasons which are unclear.

The native species that shows the most marked relative increase is *Agrostis stolonifera*, one of three common grasses (the others being *Festuca arundinacea* and *F. rubra* agg.) to show a surprisingly high change index. The reasons for the apparent increase of these species is unclear, but they may have been under-recorded in the 1930–69 period. The other native species in the top 100 fall into a number of groups. Many are plants with a limited native range but which have been widely spread by gardeners, and therefore have ranges made up of native and introduced components. These include *Aquilegia vulgaris*, *Allium schoenoprasum*, *Arum italicum*, *Asparagus officinalis* (where the native plant is at least subspecifically distinct), *Buxus sempervirens*, *Cyperus longus*, *Leucojum aestivum*, *Meconopsis cambrica*, *Myosotis sylvatica*, *Nymphoides peltata*, *Sedum forsterianum*, *Tilia cordata* and *T. platyphyllos*. The increase of these species has been caused by the expansion of the introduced populations, probably coupled in the case of the trees and shrubs (*Tilia* spp. and *Buxus sempervirens*) with the increased tendency for botanists to record woody aliens planted in the wild. Three species in the top 100 which are classified as 'native or alien' also owe their spread to the undoubtedly alien component in their range (*Muscari neglectum*, *Ribes rubrum* and *Stratiotes aloides* – although in the case of *M. neglectum* some records may be errors for the invasive horticultural introduction *M. armeniacum*). Taxonomic uncertainties may have contributed to the under-recording of *Ribes rubrum* in the earlier period. Four species (*Atriplex littoralis*, *Cochlearia danica*, *Puccinellia distans*, *Spergularia marina*) are plants that had predominantly coastal distributions in the 1962 *Atlas* but have since spread dramatically inland along salt-treated roadsides. *Geranium rotundifolium* also appears to have increased, but for reasons which are not so easily understood.

The apparent increase of some native species can be attributed to more effective recording in the 1987–99 period. One of the two ferns in Table 7.1, *Dryopteris affinis*, was not well-understood at the time of the 1962 *Atlas* but was better known by the start of the current recording period. The increase in *Trichomanes speciosum* reflects the recent discovery that the tiny gametophyte is much more widespread than the large but much rarer sporophyte. Other native species that are rather inconspicuous or somewhat critical and that have probably been better recorded in the recent period include *Epilobium tetragonum*, *Fumaria muralis*, *Glyceria declinata*, *Hypericum maculatum*, *Potamogeton berchtoldii*, *Ranunculus fluitans* and *Sparganium angustifolium*. *Ranunculus fluitans* was not well understood in the earlier period and many records made then have been rejected as dubious; there are more records now but some of these may also be doubtful. Recorders for the 1962 *Atlas* were required to divide *Epilobium tetragonum* into two species that were difficult to separate and which are no longer recognised; some recorders must have avoided this challenge and both segregates were almost certainly under-recorded. In the case of *P. berchtoldii* the mapped records in the 1962 *Atlas* were restricted to those determined by experts, and many more records of both this species and *S. angustifolium* have been made in recent years during surveys of aquatic habitats in northern and western Britain. *Hypericum maculatum* has a distribution that is concentrated in areas of Wales which were probably under-recorded in the 1962 *Atlas* but have been well-recorded in the *New Atlas* survey.

Although the change index in itself does not indicate whether a species has undergone an absolute increase or an absolute decline, a review of all the species listed in Table 7.1 suggests that there is independent evidence from county Floras or from studies of individual species that taxa with a change index greater than 1.5 have either undergone a genuine increase in their range or are taxa for which there has been a marked increase in records because of species-specific (rather than overall) increases in recording efficiency.

Table 7.1 The 100 taxa which have shown the greatest relative increase between the 1930–69 and 1987–99 recording periods.

Name	Status	Change index
Prunus laurocerasus	Neo	+4.70
Lysimachia punctata	Neo	+4.62
Syringa vulgaris	Neo	+4.48
Epilobium ciliatum	Neo	+3.88
Buddleja davidii	Neo	+3.73
Laburnum anagyroides	Neo	+3.71
Linaria purpurea	Neo	+3.66
Agrostis stolonifera	Nat	+3.66
Cotoneaster simonsii	Neo	+3.55
Prunus cerasifera	Neo	+3.43
Cochlearia danica	Nat	+3.31
Crocosmia aurea x pottsii (C. x crocosmiiflora)	Neo	+3.11
Puccinellia distans	Nat	+3.02
Galanthus nivalis	Neo	+3.01
Crocus vernus	Neo	+2.99
Cerastium tomentosum	Neo	+2.97
Festuca rubra agg.	Nat	+2.96
Sedum spurium	Neo	+2.94
Larix decidua	Neo	+2.91
Brassica napus	Neo	+2.88
Antirrhinum majus	Neo	+2.84
Nymphoides peltata	Nat	+2.81
Campanula persicifolia	Neo	+2.80
Calystegia pulchra	Neo	+2.78
Azolla filiculoides	Neo	+2.76
Cyclamen hederifolium	Neo	+2.75
Senecio cineraria	Neo	+2.73
Lactuca serriola	Arc	+2.70
Veronica filiformis	Neo	+2.69
Tilia platyphyllos	Nat	+2.57
Papaver somniferum	Arc	+2.54
Buxus sempervirens	Nat	+2.54
Allium triquetrum	Neo	+2.46
Dryopteris affinis	Nat	+2.44
Leucojum aestivum	Nat	+2.42
Sedum album	Arc	+2.41
Quercus ilex	Neo	+2.37
Erigeron karvinskianus	Neo	+2.37
Soleirolia soleirolii	Neo	+2.36
Meconopsis cambrica	Nat	+2.36
Lobularia maritima	Neo	+2.34
Quercus cerris	Neo	+2.32
Trichomanes speciosum	Nat	+2.23
Cyperus longus	Nat	+2.22
Sedum rupestre	Neo	+2.20
Prunus domestica	Arc	+2.19
Myosotis sylvatica	Nat	+2.18
Euphorbia lathyris	Arc	+2.16
Valerianella carinata	Arc	+2.15
Spiraea agg.	Neo	+2.14
Populus deltoides x nigra (P. x canadensis)	Neo	+2.11
Hypericum maculatum	Nat	+2.11
Escallonia macrantha	Neo	+2.09
Ceratochloa carinata	Neo	+2.09
Arum italicum	Nat	+2.09
Heracleum mantegazzianum	Neo	+2.09
Geranium endressii	Neo	+2.07
Ranunculus fluitans	Nat	+1.96
Barbarea intermedia	Neo	+1.92
Chenopodium ficifolium	Arc	+1.90
Impatiens glandulifera	Neo	+1.85
Fuchsia magellanica	Neo	+1.85
Lupinus arboreus	Neo	+1.84
Fallopia japonica	Neo	+1.83
Symphytum orientale	Neo	+1.83
Allium paradoxum	Neo	+1.83
Spergularia marina	Nat	+1.83
Rhododendron ponticum	Neo	+1.83
Pentaglottis sempervirens	Neo	+1.81
Glyceria declinata	Nat	+1.79
Ribes rubrum	Nat/Int	+1.79
Hypericum androsaemum x hircinum (H. x inodorum)	Neo	+1.78
Asparagus officinalis	Nat	+1.78
Coronopus didymus	Neo	+1.77
Pulmonaria officinalis	Neo	+1.77
Ribes nigrum	Neo	+1.76
Fumaria muralis	Nat	+1.75
Symphoricarpos albus	Neo	+1.74
Melissa officinalis	Neo	+1.73
Bromopsis inermis	Neo	+1.71
Festuca arundinacea	Nat	+1.71
Aquilegia vulgaris	Nat	+1.70
Ranunculus lingua	Nat	+1.70
Geranium rotundifolium	Nat	+1.70
Allium schoenoprasum	Nat	+1.69
Mentha spicata	Arc	+1.69
Sparganium angustifolium	Nat	+1.66
Epilobium tetragonum	Nat	+1.66
Potamogeton berchtoldii	Nat	+1.66
Stratiotes aloides	Nat/Int	+1.65
Tilia cordata	Nat	+1.64
Oxalis corniculata	Neo	+1.62
Mahonia aquifolium	Neo	+1.61
Eranthis hyemalis	Neo	+1.59
Atriplex littoralis	Nat	+1.59
Lamium hybridum	Arc	+1.57
Acanthus mollis	Neo	+1.55
Muscari neglectum	Nat/Int	+1.55
Vulpia myuros	Arc	+1.55
Sedum forsterianum	Nat	+1.54

Key to status categories: Arc, archaeophyte; Nat, native; Nat/Int, native or introduced; Neo, neophyte.

The unsuccessful species

The 100 taxa that have shown the greatest decline when compared to the hypothetical 'average species' are listed in Table 7.2. Like the list of successful species, this list is dominated by introduced species, in this case by archaeophytes. Only 9% of the 1524 taxa for which a change index is available are archaeophytes, but there are 39 archaeophytes amongst the 100 species showing the most marked relative declines. The percentage of neophytes in Table 7.2 is the same as that in the sample as a whole (16%) whereas disproportionately fewer native species have undergone a severe relative decline, 44% in Table 7.2 compared to 75% of the total.

Many archaeophytes in Britain are weeds of arable fields. It is because of the decline of these arable species that archaeophytes dominate the list of the least successful species. Eight of the ten species showing the greatest relative decline are largely or almost exclusively weeds of arable fields. The species with the greatest negative change index is *Caucalis platycarpos*, which is perhaps atypical of the group, as at many sites it may have been only a casual. It is no longer found in Britain. *Arnoseris minima*, *Bupleurum rotundifolium*, *Centaurea solstitialis* and *Lolium temulentum* are also extinct as established arable weeds, and *Galium tricornutum* is very rare. *Ranunculus arvensis* and *Scandix pecten-veneris* are very distinctive species which were formerly frequent but declined greatly in the 20th century. Many of the other archaeophytes in Table 7.2 are also characteristic arable weeds, including *Adonis annua*, *Anthemis arvensis*, *A. cotula*, *Chrysanthemum segetum*, *Euphorbia exigua*, *Fallopia convolvulus*, *Galeopsis angustifolia*, *G. speciosa*, *Lithospermum arvense*, *Lythrum hyssopifolia*, *Papaver argemone*, *Silene gallica*, *S. noctiflora*, *Sinapis arvensis*, *Stachys arvensis*, *Torilis arvensis*, *Valerianella dentata* and *V. rimosa*. Other archaeophytes which show a relative decline include species which were formerly deliberately cultivated for food or medicinal use, including *Campanula rapunculus*, *Chenopodium bonus-henricus* and *Cichorium intybus*, or are weeds of disturbed ground, such as *Chenopodium glaucum*, *C. murale*, *C. urbicum*, *C. vulvaria* and *Lamium purpureum*.

The neophytes which show a marked relative decline include some long-established introductions, including *Leonurus cardiaca* which was naturalised by the end of the 16th century. Other neophytes were introduced later, often as contaminants of grain or with wool shoddy or other products, but have declined as improved grain-cleaning or other changes in agricultural practice have reduced the rate at which they are imported. These species include *Alyssum alyssoides*, *Anthoxanthum aristatum*, *Bromus arvensis*, *Centaurea solstitialis*, *Melilotus indicus*, *Poa palustris*, *Prunella laciniata*, *Salvia verticillata* and perhaps the enigmatic, apparently endemic, *Bromus interruptus*.

Some of the native species that appear to have decreased have both native and introduced components to their distribution, and it is the introduced populations which have shown the marked decline. These include *Raphanus raphanistrum* (native in coastal habitats as subsp. *maritimus*, but present as the archaeophyte subsp. *raphanistrum* inland), *Spergula arvensis* (which we take to be native only in the Channel Islands, where it is present as var. *nana*, but widespread as other varieties elsewhere) and *Trifolium incarnatum* (native in the Channel Islands and on the Lizard peninsula, W. Cornwall, as subsp. *molinerii*, but introduced elsewhere as subsp. *incarnatum*). These are the counterpart of the native species mentioned in the last section with ranges which have increased because of the spread of introduced populations.

The native species that have shown the most marked relative decline are drawn from a variety of habitats. They are discussed in relation to the general study of trends in native species in the following sections.

In general, the species listed in Table 7.2 have always been well-recorded by botanists and there are fewer problems of interpretation than for the successful species in Table 7.1. *Bromus lepidus* is perhaps the only species where it is unclear whether the marked relative decline is real or whether recent recorders are less able to distinguish it from the widespread *B. hordeaceus*. Comments on trends in the distribution of the species listed in Table 7.2 in county Floras and other local studies suggest that they have undergone an absolute as well as a relative decline.

Native species, archaeophytes and neophytes

The mean change indices for native taxa, archaeophytes and neophytes are listed in Table 7.3. The analysis of variance indicates significant differences

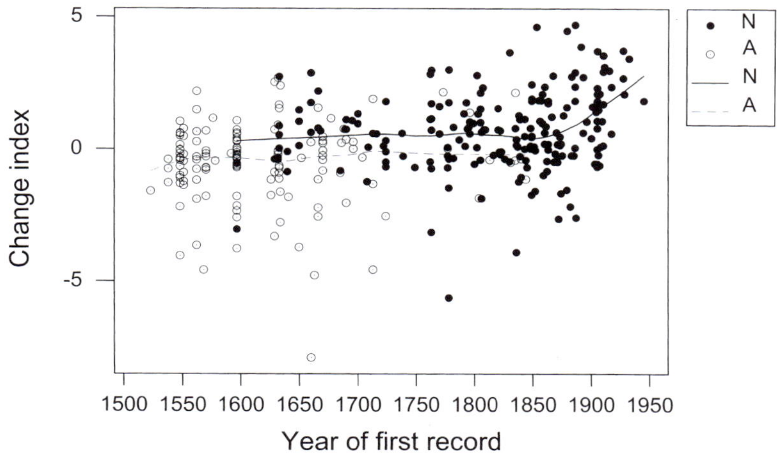

Figure 7.2 The first record of neophytes (N) and archaeophytes (A), plotted against change index. Each point represents a single species; the lines represent the smoothed trend in the change index with year of first record (LOWESS method).

between the groups ($p<0.001$) and the differences between all groups except the native and the 'native or introduced' species are statistically significant. The relative decline of the archaeophytes is striking. The native and 'native or introduced' species have undergone a much less marked relative decline and neophytes a marked increase in relation to the total sample. These figures reinforce the conclusions suggested by the examination of the lists of the most and the least successful species.

The relationship of the change index to the first dates when introduced species were recorded in the wild is shown in Fig. 7.2. In this figure we have used the date of the first botanical record of archaeophytes, although fossil evidence shows that many of these species were present for thousands of years before botanical recording began, and all archaeophytes are (by definition) thought to have been present by 1500. The dates for neophytes are subject to a degree of uncertainty, both because plants may not have been recorded until some time after they became established and because further studies of literature or herbarium sources might reveal earlier dates. Fig. 7.2 shows, perhaps surprisingly, that the average change index for species recorded before 1850 is close to zero, although the average for neophytes is consistently higher than that for archaeophytes. Only the neophytes first recorded after 1850 show a markedly positive average change index.

The spread of recently introduced neophytes since 1962 is probably underestimated in Fig. 7.2, as that diagram excludes those neophytes for which change indices have not been calculated because the species were not mapped in the 1962 *Atlas*. The 1962 *Atlas* mapped 'most well-established introductions' although Perring & Walters (1962, p. xvi) admitted that their choice 'was necessarily arbitrary'. A surprisingly large number of the introductions that were not mapped in 1962 are now frequent in Britain. Some of the taxa now recorded from over 500 10-km squares almost certainly spread markedly in the 20th century, including *Alchemilla mollis* (first recorded in the wild in 1948), *Bidens frondosa* (1918, but not known as a naturalised plant until 1952), *Cicerbita macrophylla* (1915), *Crassula helmsii* (1956), *Elodea nuttallii* (1966), *Hyacinthoides hispanica* (1909) and *H. hispanica* × *H. non-scripta* (1963, but perhaps overlooked earlier as *H. hispanica*), *Lamiastrum galeobdolon* subsp. *argentatum* (1974), *Lemna minuta* (1977), *Leucanthemum lacustre* × *L. maximum* (1913), *Rosa rugosa* (1927), and *Solidago gigantea* (1916). It is difficult to know whether other species that were not mapped in the 1962 *Atlas* were disregarded by many recorders as obvious garden escapes of little interest, or whether they spread subsequently. In most of the following cases it seems likely that both factors apply: *Aster novi-belgii* (first recorded in the wild in 1860), *Borago officinalis* (1777), *Lamium maculatum* (*c.* 1730), *Lathyrus latifolius* (1670), *Lunaria annua* (1597) and *Pilosella aurantiaca* (1793). The absence of other species from the 1962 *Atlas* reflects different recording priorities, as there was much less interest in the 1930–69 period in crop and other food-plant casuals such as *Avena sativa*, *Linum usitatissimum*, *Lycopersicon esculentum* and *Triticum aestivum*, in coniferous trees such as *Picea abies*, *P. sitchensis* and *Pseudotsuga menziesii* (although *Larix decidua* was mapped) and in broadleaved trees and

Table 7.2 The 100 taxa which have shown the greatest relative decrease between the 1930–69 and 1987–99 recording periods.

Name	Status	Change index
Caucalis platycarpos	Arc	–7.86
Centaurea solstitialis	Neo	–5.62
Galium tricornutum	Arc	–4.78
Bupleurum rotundifolium	Arc	–4.58
Chenopodium urbicum	Arc	–4.57
Lolium temulentum	Arc	–4.05
Bromus lepidus	Neo	–3.91
Ranunculus arvensis	Arc	–3.77
Arnoseris minima	Arc	–3.72
Scandix pecten-veneris	Arc	–3.65
Galeopsis angustifolia	Arc	–3.31
Bromus arvensis	Neo	–3.15
Leonurus cardiaca	Neo	–3.05
Avena strigosa	Cas	–3.01
Silene gallica	Arc	–2.78
Scleranthus annuus	Nat	–2.68
Anthoxanthum aristatum	Neo	–2.65
Gnaphalium sylvaticum	Nat	–2.65
Prunella laciniata	Neo	–2.60
Chenopodium vulvaria	Arc	–2.60
Torilis arvensis	Arc	–2.56
Valerianella rimosa	Arc	–2.55
Himantoglossum hircinum	Nat	–2.40
Centaurea calcitrapa	Arc	–2.34
Spergula arvensis	Nat	–2.30
Carum carvi	Arc	–2.22
Adonis annua	Arc	–2.19
Salvia verticillata	Neo	–2.18
Campanula rapunculus	Arc	–2.16
Silene noctiflora	Arc	–2.04
Marrubium vulgare	Nat	–2.02
Medicago minima	Nat	–1.97
Lithospermum arvense	Arc	–1.91
Galium spurium	Neo	–1.87
Valerianella dentata	Arc	–1.86
Sium latifolium	Nat	–1.83
Galeopsis speciosa	Arc	–1.82
Chrysanthemum segetum	Arc	–1.80
Sisyrinchium bermudiana	Neo	–1.80
Papaver argemone	Arc	–1.79
Chenopodium bonus-henricus	Arc	–1.79
Anthemis arvensis	Arc	–1.79
Orchis ustulata	Nat	–1.77
Trifolium incarnatum	Nat	–1.76
Sinapis arvensis	Arc	–1.76
Bromus interruptus	Neo	–1.73
Minuartia hybrida	Nat	–1.70
Potamogeton compressus	Nat	–1.68
Potentilla norvegica	Neo	–1.68
Platanthera bifolia	Nat	–1.67
Chenopodium murale	Arc	–1.63
Euphrasia officinalis agg.	Nat	–1.61
Anthemis cotula	Arc	–1.60
Melilotus indicus	Neo	–1.59
Clinopodium acinos	Nat	–1.59
Poa palustris	Neo	–1.55
Viola tricolor	Nat	–1.52
Lactuca saligna	Nat	–1.51
Euphorbia peplis	Nat	–1.49
Otanthus maritimus	Nat	–1.49
Cynosurus echinatus	Neo	–1.47
Raphanus raphanistrum	Nat	–1.39
Lathyrus aphaca	Nat/Int	–1.38
Hyoscyamus niger	Arc	–1.38
Ranunculus aquatilis sens. lat.	Nat	–1.37
Carex maritima	Nat	–1.34
Medicago polymorpha	Nat	–1.34
Coeloglossum viride	Nat	–1.34
Ophrys insectifera	Nat	–1.34
Chenopodium glaucum	Arc	–1.32
Galium pumilum	Nat	–1.32
Fallopia convolvulus	Arc	–1.31
Dianthus armeria	Nat	–1.31
Mentha arvensis	Nat	–1.30
Blysmus compressus	Nat	–1.28
Pedicularis sylvatica	Nat	–1.28
Cuscuta epithymum	Nat	–1.28
Gentianella campestris	Nat	–1.28
Cichorium intybus	Arc	–1.27
Silene vulgaris	Nat	–1.26
Campanula rapunculoides	Neo	–1.24
Alyssum alyssoides	Neo	–1.24
Nepeta cataria	Arc	–1.23
Groenlandia densa	Nat	–1.23
Iberis amara	Nat	–1.21
Crepis mollis	Nat	–1.20
Filago vulgaris	Nat	–1.20
Euphorbia exigua	Arc	–1.18
Oenanthe fistulosa	Nat	–1.18
Stachys arvensis	Arc	–1.17
Arabis glabra	Nat	–1.16
Anagallis minima	Nat	–1.16
Bromus secalinus	Arc	–1.15
Leucanthemum vulgare	Nat	–1.14
Sagina nodosa	Nat	–1.14
Filago pyramidata	Arc	–1.14
Polygala vulgaris	Nat	–1.14
Lythrum hyssopifolium	Arc	–1.12
Pyrola media	Nat	–1.09
Lamium purpureum	Arc	–1.09

Key to status categories: Arc, archaeophyte; Nat, native; Nat/Int, native or introduced; Neo, neophyte.

shrubs of horticultural origin (*Acer platanoides*, *Ligustrum ovalifolium*, *Populus alba*, *Robinia pseudoacacia* and *Ribes sanguineum* were not mapped in 1962, although some equivalent species such as *Aesculus hippocastanum* and *Prunus laurocerasus* were included).

Although rapidly spreading neophytes understandably receive much publicity, it is worth drawing attention to some species that have been recorded for many years but are still very infrequent. Species that have been known from the wild since the 18th, 19th or early 20th centuries but are still present in fewer than 50 10-km squares include *Calla palustris* (first recorded in the wild in 1861), *Chaerophyllum aureum* (1810), *Dianthus caryophyllus* (1778), *Ledum palustre* (1860), *Linaria supina* (1847), *Persicaria sagittata* (1830) and *Sagittaria rigida* (1908).

Trends in the distribution of native species

To examine the nature of the trends shown by native species, the change indices for groups of species with similar morphological, phytogeographical or ecological characteristics have been examined. Species classified as 'native' and 'native or introduced' have been included in the analysis. The characteristics examined are floristic element, plant height, and requirements for light, moisture, pH and nutrients.

Table 7.3 Mean change indices and standard errors for native and introduced taxa.

Status	No. species	Mean change index ± s.e.
Native	1142	–0.06 ± 0.007
Native or introduced	32	+0.12 ± 0.001
Archaeophyte	141	–0.53 ± 0.001
Neophyte	209	+0.67 ± 0.003

Table 7.4 Mean change indices and standard errors for major biome categories.

Major biome category	No. species	Mean change index ± s.e.
Arctic-montane	60	–0.25 ± 0.004
Boreo-arctic Montane	33	–0.21 ± 0.007
Wide-boreal	18	+0.07 ± 0.004
Boreal-montane	76	+0.07 ± 0.001
Boreo-temperate	203	+0.07 ± 0.000
Wide-temperate	33	+0.45 ± 0.014
Temperate	450	+0.05 ± 0.000
Southern-temperate	204	+0.16 ± 0.001
Mediterranean	90	+0.26 ± 0.003

Table 7.5 Mean change indices and standard errors for Ellenberg light (L) values.

Light value (L)	No. species	Mean change index ± s.e.
1 Plant in deep shade (no British examples)	0	
2 Between 1 and 3	3	+0.41 ± 0.813
3 Shade plant, mostly less than 5% relative illumination, seldom more than 30% illumination when trees are in full leaf	14	–0.10 ± 0.039
4 Between 3 and 5	57	+0.18 ± 0.014
5 Semi-shade plant, rarely in full light, but generally with more than 10% relative illumination when trees are in leaf	82	+0.08 ± 0.009
6 Between 5 and 7	118	–0.03 ± 0.006
7 Plant generally in well lit places, but also in partial shade	414	–0.03 ± 0.002
8 Light-loving plant rarely found where relative illumination in summer is less than 40%	368	–0.14 ± 0.002
9 Plant in full light, found mostly in full sun	113	–0.17 ± 0.006

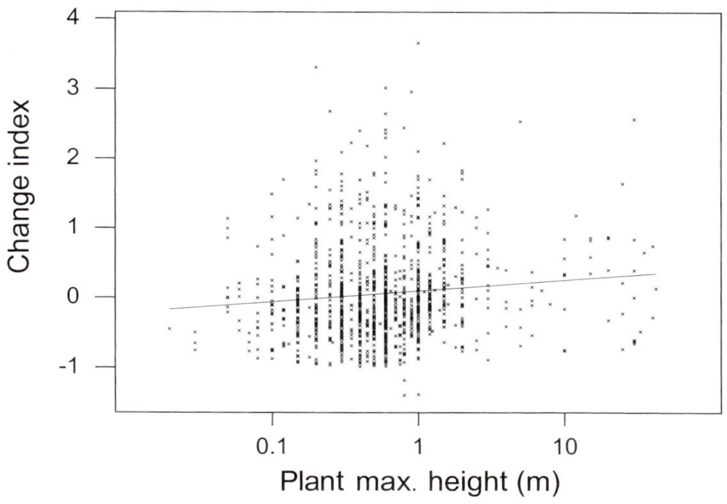

Figure 7.3 Regression of change index against $\log_{10}$ typical maximum height (m) ($y = 0.16x - 0.22$, $r^2 = 0.01$). Each point represents a single native species.

Floristic elements are based on Preston & Hill (1997). Data on height are derived from the Ecological Flora Database (see Fitter & Peat, 1994), with some additions and amendments. The categories for light, moisture, pH and nutrient requirements are based on 'Ellenberg values', a classification of species into groups of similar ecology devised by H. Ellenberg for the central European flora (e.g. Ellenberg, 1988) and modified for Britain by Hill *et al.* (1999). Published Ellenberg values are not available for many aggregate species for which change indices have been calculated, but values for 34 of these aggregates have been derived from the values of their component segregates in consultation with M. O. Hill. In all, Ellenberg values are available for 1168 of the 1174 native or 'native or introduced' taxa for which there are change indices. One-way analysis of variance was used to test for differences between the groups.

Floristic element

The classification of species into floristic elements based on their occurrence in major biome categories and their eastern limits is described in Chapter 6. The differences between the species in major biome categories set out in Table 7.4 are statistically significant ($p<0.001$). The Arctic and Boreo-arctic species show a relative decline. The small group of Wide-temperate species shows the greatest relative increase, and this value is not only significantly greater than that for the two northernmost elements but is also greater than that for the Temperate species. The Wide-temperate species occur from the Boreal to the Mediterranean zones and include some of the most widespread plants in Britain. The two most southerly elements (Southern-temperate and Mediterranean) also have relatively high mean change indices.

A similar analysis revealed no significant variation in change index between the eastern limit categories; the results are not presented in detail here.

Plant height

The typical maximum height of a plant in summer is a useful measure of its competitive ability (Thompson, 1994). There is a very weak but highly significant regression between height and the change index ($p<0.001$), with shorter, less competitive species having declined in comparison to taller and therefore in general more competitive plants (Fig. 7.3). Only terrestrial species are included in this analysis; 61 aquatics are excluded.

Light requirements

The Ellenberg values for light (L) divide the species into nine groups (Hill *et al.*, 1999), from plants of deep shade to those characteristically found in full sunlight. The mean change indices for the eight groups represented in Britain (Table 7.5) show significant differences between the groups ($p<0.01$). With the exception of one small group of plants of heavy shade (L3), the mean change index decreases with successive categories from L2 to L9. These results suggest that plants of shady conditions have done relatively well and those of open conditions relatively badly between 1930–69 and 1987–99.

Moisture requirements

The Ellenberg values for moisture (M) divide the species into 12 groups (Hill *et al.*, 1999). The mean change indices (Table 7.6) again reveal significant differences between the groups ($p<0.001$). There are only two species in M1, but indicators of dry sites (M2–4) have mean change indices which indicate a relative decline. The mean change indices for the species of moist to wet soils are close to zero, but those for the aquatic species (M10–12) are positive, and in particular the value for the emergent and floating species (M11) is significantly higher than those for all the other groups except M1 (where the sample is tiny) and M12. Thus species of dry habitats appear to have declined in comparison with those of aquatic habitats.

pH requirements

The Ellenberg values for pH (R) divide the species into nine groups (Hill *et al.*, 1999). The mean change indices (Table 7.7) show significant differences between the groups ($p<0.001$). There is a clear trend for the species which require the more acidic and the more basic conditions to have declined when compared with those which characteristically occur in the middle of the pH spectrum.

Nutrient requirements

The Ellenberg values for nitrogen (N) are in effect a general indicator of the fertility of the soil or water in which the species grows. Nine groups are recognised (Hill *et al.*, 1999) and there are significant differences between them ($p<0.001$). The relationship between N value and change index is very clear (Table 7.8). If the tiny sample of species in the category N9 is ignored, the species which have been most successful are those with the highest nutrient requirements (N8), and the species with relatively high nutrient requirements (N5–8) all have a positive change index. By contrast, the species which tend to occur in areas with low levels of nutrients (N1–4) all show a relative decline.

A review of the least successful native species

There are 44 native species amongst the 100 species which have shown the greatest relative decline between the two recording periods. These are listed in Table 7.9, with an indication of their height and their Ellenberg L, R and N values. These species include two which became extinct in Britain in the 1930–69 period (*Euphorbia peplis*, *Otanthus maritimus*); the remainder range from plants which are very rare (e.g. *Dianthus armeria*, *Lactuca saligna*) to those which are widespread (e.g. *Pedicularis sylvatica*, *Silene vulgaris*). The sample is too small to reveal many statistically significant trends in their ecological characteristics, although the absence of plants of shaded habitats (Ellenberg L1–6) is significant ($p<0.01$). However, it is interesting to see that some species combine all the attributes of the least successful species (*Orchis ustulata*, *Minuartia hybrida*, *Clinopodium acinos*, *Pedicularis sylvatica*) whereas others have declined even though they have none of the extremes of height, light, pH or nutrient requirements which are correlated with a high negative change index (*Spergula arvensis*, *Sium latifolium*, *Potamogeton compressus*, *Raphanus raphanistrum*, *Ranunculus aquatilis sens. lat.*, *Mentha arvensis*, *Filago vulgaris*, *Oenanthe fistulosa*). Almost all the species in the second group are plants of aquatic or arable habitats.

Preliminary comparison with other studies

The main conclusions of the above analysis can be summarised by contrasting the groups which appear to have shown the most marked relative change (Table 7.10). There are clearly many aspects of recording bias which might influence the results of the current survey. It is therefore useful to compare the results of this study with those of other, more or less independent studies.[1] These comparisons are discussed below and summarised in Table 7.10.

Few investigators have used data that are independent of those analysed here to compare the relative success of native and introduced taxa in a particular time period. However, there are many examples of the spread of introduced species in the 20th century (Manchester & Bullock, 2000). Similarly, the decline of arable weeds has been a clear trend in recent decades, documented by many national and local studies. This decline was noted even before the end of our first recording period in 1969 (e.g. by Salisbury, 1961), but it has continued

Table 7.6 Mean change indices and standard errors for Ellenberg moisture (M) values.

Moisture value (M)	No. species	Mean change index ± s.e.
1 Indicator of extreme dryness, restricted to soils that often dry out for some time	2	+0.07 ± 0.073
2 Between 1 and 3	16	–0.37 ± 0.048
3 Dry-site indicator, more often found on dry ground than in moist places	89	–0.23 ± 0.007
4 Between 3 and 5	175	–0.22 ± 0.004
5 Moist-site indicator, mainly on fresh soils of average dampness	279	–0.05 ± 0.003
6 Between 5 and 7	158	+0.04 ± 0.005
7 Dampness indicator, mainly on constantly moist or damp, but not wet soils	102	–0.06 ± 0.007
8 Between 7 and 9	141	–0.06 ± 0.005
9 Wet-site indicator, often on water-saturated, badly aerated soils	88	–0.06 ± 0.007
10 Indicator of shallow-water sites that may lack standing water for extensive periods	52	+0.13 ± 0.013
11 Plants rooting under water, but at least for a time exposed above, or plant floating on the surface	28	+0.35 ± 0.032
12 Submerged plant, permanently or almost constantly under water	39	+0.17 ± 0.020

Table 7.7 Mean change indices and standard errors for Ellenberg pH (R) values.

pH value (R)	No. species	Mean change index ± s.e.
1 Indicator of extreme acidity, never found on weakly acid or basic soils	9	–0.12 ± 0.015
2 Between 1 and 3	43	–0.42 ± 0.010
3 Acidity indicator, mainly on acid soils, but exceptionally also on nearly neutral ones	42	–0.21 ± 0.005
4 Between 3 and 5	92	–0.10 ± 0.001
5 Indicator of moderately acid soils, only occasionally found on very acid or on neutral to basic soils	142	–0.02 ± 0.000
6 Between 5 and 7	251	+0.02 ± 0.000
7 Indicator of weakly acid to weakly basic conditions; never found on very acid soils	414	+0.02 ± 0.000
8 Between 7 and 9	164	–0.23 ± 0.001
9 Indicator of basic reaction, always found on calcareous or other high-pH soils	12	–0.35 ± 0.032

Table 7.8 Mean change indices and standard errors for Ellenberg nitrogen (N) values.

Nitrogen value (N)	No. species	Mean change index ± s.e.
1 Indicator of extremely infertile sites	65	–0.82 ± 0.008
2 Between 1 and 3	226	–0.29 ± 0.003
3 Indicator of more or less infertile sites	189	–0.23 ± 0.003
4 Between 3 and 5	154	–0.05 ± 0.004
5 Indicator of sites of intermediate fertility	210	+0.06 ± 0.004
6 Between 5 and 7	179	+0.17 ± 0.004
7 Plant often found in richly fertile places	113	+0.03 ± 0.005
8 Between 7 and 9	31	+0.23 ± 0.026
9 Indicator of extremely rich situations, such as cattle resting places or near polluted rivers	2	–0.12 ± 1.098

1 The BSBI Monitoring Scheme has not been included as one of these independent studies as the results are a subset of those analysed in this chapter. However, studies of individual counties have been included even though some of the data on which they are based have been incorporated into the Vascular Plant Database.

Table 7.9 The native species which have shown the greatest relative decrease between the 1930–69 and 1987–99 recording periods. The height class and Ellenberg values for L, R and N are given. Plants are allocated to a height class as follows: 1, 0–10 cm; 2, 11–29 cm; 3, 30–59 cm; 4, 60–99 cm; 5, 100–300 cm; 6, 301–600 cm; 7, 601–1500 cm; 8, over 1500 cm. Bold figures indicate categories which have a high negative mean change index (short plants, indicated here by height classes 1–2, and Ellenberg L=8–9, R=1–3, 8–9 and N=1–3).

Name	Height class	Ellenberg L	Ellenberg R	Ellenberg N	Change index
Scleranthus annuus	**2**	7	4	4	–2.68
Gnaphalium sylvaticum	4	7	4	**3**	–2.65
Himantoglossum hircinum	3	7	**9**	**2**	–2.40
Spergula arvensis	4	7	5	5	–2.30
Marrubium vulgare	4	**9**	7	8	–2.02
Medicago minima	**2**	**9**	7	**2**	–1.97
Sium latifolium	5	7	7	7	–1.83
Orchis ustulata	**2**	**8**	**8**	**2**	–1.77
Trifolium incarnatum	3	**8**	5	**2**	–1.76
Minuartia hybrida	**2**	**9**	**8**	**3**	–1.70
Potamogeton compressus	-	7	7	4	–1.68
Platanthera bifolia	3	6	6	**2**	–1.67
Euphrasia officinalis agg.	3	**8**	5	**3**	–1.61
Clinopodium acinos	**2**	**8**	**8**	**1**	–1.59
Viola tricolor	3	**8**	6	4	–1.52
Lactuca saligna	5	**8**	7	6	–1.51
Euphorbia peplis	**1**	**9**	7	5	–1.49
Otanthus maritimus	3	**9**	5	**2**	–1.49
Raphanus raphanistrum	4	7	6	6	–1.39
Lathyrus aphaca	5	7	**8**	4	–1.38
Ranunculus aquatilis sens. lat.	-	7	7	7	–1.37
Carex maritima	**2**	**9**	7	**2**	–1.34
Medicago polymorpha	4	**9**	5	5	–1.34
Coeloglossum viride	**2**	7	6	**2**	–1.34
Ophrys insectifera	4	**8**	**9**	**2**	–1.34
Galium pumilum	3	7	**8**	**3**	–1.32
Dianthus armeria	4	**8**	5	**3**	–1.31
Mentha arvensis	4	6	7	6	–1.30
Blysmus compressus	3	**8**	**8**	**3**	–1.28
Pedicularis sylvatica	**2**	**8**	**3**	**2**	–1.28
Cuscuta epithymum	-	7	**2**	**2**	–1.28
Gentianella campestris	3	**8**	6	**3**	–1.28
Silene vulgaris	4	7	**8**	5	–1.26
Groenlandia densa	-	**8**	**8**	5	–1.23
Iberis amara	3	7	**8**	**3**	–1.21
Crepis mollis	4	**8**	7	5	–1.20
Filago vulgaris	3	7	6	4	–1.20
Oenanthe fistulosa	4	7	7	6	–1.18
Arabis glabra	5	7	**8**	5	–1.16
Anagallis minima	**1**	**8**	5	**3**	–1.16
Leucanthemum vulgare	5	**8**	7	4	–1.14
Sagina nodosa	**2**	**8**	7	**3**	–1.14
Polygala vulgaris	**1**	**8**	6	**3**	–1.14
Pyrola media	3	5	5	**2**	–1.09

Table 7.10 A summary of the main trends identified by the current study between 1930–69 and 1987–99, with a summary of the extent to which these are supported by the evidence of other studies. Details of the Countryside Survey and vice-county studies are given in the text.

Relative increase	Relative decline	Support from other studies	Conclusion
Neophytes, especially those first recorded in the wild after 1850	Archaeophytes, especially arable weeds	Few comparable studies, but many supporting individual case histories	A very marked change
Species with high nutrient requirements	Species with low nutrient requirements	Strong support from Countryside Survey and vice-county studies	A real change
Taller species	Shorter species	Support from Countryside Survey and vice-county studies	A real change
Shade-tolerant species	Species of open habitats	Support from Countryside Survey and vice-county studies	A real change
Species occurring in middle of pH spectrum	Species of very acidic or very basic habitats	Some support from vice-county studies	Probably represents a real change
Species of aquatic habitats	Species of dry habitats	Little support from other studies	May reflect recording bias (coverage of aquatic habitats) rather than a real change
Native species with a very wide, or southerly, European distribution	Native species with a very northerly European distribution	Supported only by studies in S. England, which may be atypical	Regional trends require investigation

since then, doubtless hastened by the continued agricultural intensification of the 1970s and 1980s (Chamberlain *et al.*, 2000). There has been little comment on the decline of those archaeophytes that were probably introduced deliberately. This may be because archaeophytes have not been recognised as a distinct category by many authors, so that former medicinal and food plants have been included with the mass of neophytes and have attracted little attention.

The relative success of taller, shade-tolerant and nutrient-demanding species has been noted in other studies over a range of spatial scales and different (although overlapping) timescales. Three examples are given below.

- In the Countryside Survey, a national study of sample 1-km squares, Haines-Young *et al.* (2000) reported that between 1978, 1990 and 1998 there was particularly strong evidence for the effects of increased nutrients (eutrophication) and possibly atmospheric pollution, leading to reduced species-richness and increase of already widespread, tall, competitive plants at the expense of slower-growing, more localised plant species.
- In a study of species that have become extinct in Cambridgeshire and Middlesex since the start of botanical recording in the late 16th century, Preston (2000) found that in both vice-counties the species which had suffered disproportionately were those which are short, grow in open habitats or are characteristic of sites where nutrient levels are low. This analysis was restricted to native species, although the concept of native species was rather broader than that adopted in the *New Atlas* and included many archaeophytes.
- The frequency of species in Northamptonshire in the period *c.* 1880–1930, as assessed by G. C. Druce (1930) in *The Flora of Northamptonshire*, has been compared with their frequency in the period 1970–94, as mapped in a more recent Flora (Gent, Wilson *et al.*, 1995). McCollin *et al.* (2000) concluded that 'the major factor correlated with changes in plant status is soil nitrogen. In general, increasing species are those associated with nutrient-rich habitats whereas declining plant species are typically those of habitats with nutrient-poor soils.' The Northamptonshire study also obtained similar results to ours for the relationship of decline to Ellenberg light value.

There is some support for the conclusion that species that have preferences for markedly acidic or basic substrates have done less well than those which occur in intermediate conditions. In Middlesex (as in the current national study) species that are found in very acidic or very basic habitats have declined, whereas in Cambridgeshire (where basic habitats are more frequent) only the species in acidic habitats have suffered disproportionate extinction (Preston, 2000). The Northamptonshire study detected no relationship between pH requirements and changing frequency, but it is unlikely that the method of analysis would have detected the decline of species at both extremes but not in the middle of the pH spectrum, the pattern found nationally and in Middlesex. Both these studies cover restricted areas in S. E. England, but the results of the Countryside Survey also indicate a decrease in species associated with acidic soils in upland areas. There is also evidence from the Countryside Survey of a decrease in habitats such as acidic and calcareous grassland, which would adversely affect species with preferences for markedly acidic or basic soils (Haines-Young *et al.*, 2000).

The relative increase in aquatic species noted by our survey receives no support from the other studies. In Cambridgeshire and Middlesex the proportion of extinct aquatic plants was similar to, or greater than, that in those counties as a whole. In Northamptonshire McCollin *et al.* (2000) drew attention to the decrease of species in a wide range of wetland habitats. The most likely explanation for the relative increase observed in the national dataset is the increased efficiency of recording these species.

The current study suggests that species in the most northerly floristic elements have decreased in relation to those in the most widespread and in the more southerly elements. Britain is so diverse phytogeographically that it is difficult to assess these results without an analysis of regional trends. Although the same patterns detected here are apparent amongst the extinct species of Bedfordshire (Boon, 1999), Cambridgeshire and Middlesex (Preston, 2000), these are all counties in southern England where the more northerly elements are very poorly represented. It seems best to reserve judgement on these conclusions until the results of the survey are analysed regionally.

The above comparisons represent only a preliminary analysis, and there is considerable scope for a more detailed analysis of the *New Atlas* results in relation to other datasets.

The 20th-century changes in an historical perspective

The *New Atlas* project is simply a recording exercise which provides no direct information on the reasons for any observed changes. However, it is not difficult to reconcile the observed changes with our existing knowledge of the history of British vegetation. As open vegetation of the Late-glacial period gave way to a more closed forest, species of open habitats must have become more restricted in their distribution. The subsequent large-scale destruction of native woodland from the Neolithic period onwards, which accompanied agricultural expansion, allowed many such species to recolonise newly reopened habitats. Grasslands and heathlands became more extensive and a new habitat, arable land, was available for colonisation by native species or by plants that were originally introduced deliberately as alternative crops or accidentally (for example, as seed impurities). There followed a long period when conditions were suitable for species of open habitats. Changes that reversed these trends and discriminated against species of open vegetation probably first had a marked effect on the British flora in the late 18th and early 19th centuries, but have certainly had a profound influence since 1930. These have included:

- the increasing availability and use of artificial fertilisers, which has not only resulted in greatly increased nutrient inputs into farmed grassland and arable crops but has also led to higher nutrient levels in hedgerows, woodlands and water courses;
- the loss of grassland and heathland in southern England as arable farming has become more profitable, and the reduction in the grazing of remnants of these habitats in areas where stock are no longer kept and rabbit numbers have been reduced by myxomatosis;
- the intensive management of grassland in pastoral areas, with the frequent application of fertilisers and herbicides to produce dense, species-poor swards which are often harvested for silage;
- the reafforestation of the uplands, and of lowlands in the west, with coniferous trees;
- the loss of small and probably unintensively managed arable plots in northern, western and upland areas, formerly maintained as a source of grain for livestock;
- the increased use of herbicides in arable land and grassland, coupled with the development of more competitive crop varieties, which have directly and indirectly eliminated 'weed' species.

The changes in the British flora identified by the comparison of the results of the 1962 *Atlas* with those of the current survey took place over a long time period, between 1930–69 and 1987–99. Some or all of the trends identified here may be continuing to operate. However, it is possible that some may now have slowed down or ceased, either because the causal factors are no longer in operation or because the change has progressed so far that there is little scope for further change in the same direction. In the most recent report of the Countryside Survey, which examined changes between 1990 and 1998, Haines-Young *et al.* (2000) noted a continuing decline in plant diversity in some habitats (such as the least agriculturally improved grasslands) and continued evidence for increasing levels of nutrient availability, favouring tall, competitive plants. However, they found that negative trends in some key components of countryside quality had slowed or halted during the 1990s – there was no evidence of further loss of hedgerows or ponds, for example, and the survey reported an increase in the diversity of plants growing around the edges of arable fields.

One would not expect the effects of some recent developments in the countryside to be detectable in the current analysis. There has been a recent increase in the planting of broadleaved woodland in lowland areas, for example, but the effects are unlikely to be apparent because of the small scale of the plantings and because many woodland species of restricted distribution are slow to colonise newly available habitats. Similarly, plants are likely to respond slowly to climate change, and any effects of changing climate would be difficult to disentangle from the other, more marked effects discussed above.

The presence of a high proportion of introduced species amongst the plants with both the most positive and the most negative change indices suggests that they are more likely to be subject to marked changes in range size than natives. There is even evidence from the results of the *New Atlas* survey to suggest that for species with both native and introduced populations, the introduced component may be less stable than the native. Changes in the ranges of introduced species may be caused by environmental changes similar to those that have affected the distribution of many native plants. However, they may also reflect changes in the frequency with which the species is introduced into the wild. Plants which were traditionally grown for medicinal or other uses have often been replaced by modern pharmaceutical products, and many seed contaminants are now eliminated by improved seed-cleaning techniques. These species therefore tend to be introduced less frequently than they were. However, many of the most successful introductions originated as garden plants, and as horticultural fashions change new species continue to be introduced to gardens and find their way into the wild. The ecological characteristics of the successful introduced species in Table 7.1 require further study, but Crawley *et al.* (1996) noted that garden plants contain an undue proportion of woody species, and in general alien plants tend to be taller than their native counterparts. The introductions may therefore tend to match the characteristics of the more successful native species.

CHAPTER 8

Species accounts

Edited by T. D. Dines, D. A. Pearman and C. D. Preston

with the assistance of H. R. Arnold, Jane M. Croft, S. J. Leach and M. J. Wigginton

Introduction to the species accounts

This chapter includes distribution maps and accompanying text for 2412 taxa covered by the *New Atlas* project. The criteria for inclusion are laid out in this Introduction, followed by notes on the sources of information which have been used in compiling the species texts. The texts have been written by authors who are acknowledged individually after each account. The format of the species accounts is summarised in Table 8.1.

Selection of species for inclusion in the book and CD-ROM

We have included in this book maps and accounts for taxa in the following categories.

1. All native species treated in Stace (1997), or subsequently added to the British or Irish flora, have been mapped. A few species aggregates are also included where the coverage of the constituent species is inadequate, e.g. a map of species in the genus *Spiraea* in addition to the individual *Spiraea* species and hybrids.
2. Native subspecies have been considered for mapping if they are treated in full by Stace (1997), i.e. if they have a formal entry with a description in Stace's *New Flora*. There is considerable variation in the extent to which these subspecies have been recorded. We have mapped those subspecies for which the available records provide a reasonably informative map, even if the coverage is less good than that of many species. In many cases where a species has one rare and one common subspecies, recorders working in the area where the rare subspecies occurs have reported the distribution of the subspecies separately, but those in areas where only the common subspecies grows have simply reported the species. In these cases we have provided two maps, one for the species as a whole and the other for the rarer subspecies.
3. All introduced species treated in full by Stace (1997) have been mapped if they have been recorded in 50 or more 10-km squares in our area, regardless of date. Introduced subspecies have been considered and mapped in the same way as native subspecies.

   There are some species that are treated here as introductions but which were treated as native in the 1962 *Atlas*, or by other recent authors, and which occur in fewer than 50 10-km squares. These would not qualify for mapping under a strict application of the above criterion. We have, however, mapped these species as it seemed better to present the data on their distribution and the reasons for regarding them as introductions, rather than simply relegate them to the list of alien species which are too uncommon to be mapped. These species are *Bupleurum falcatum*, *Cotoneaster cambricus*, *Equisetum ramosissimum*, *Fumaria reuteri*, *Galeopsis segetum*, *Holosteum umbellatum*, *Juncus subulatus*, *Lavatera cretica*, *Linaria pelisseriana*, *L. supina*, *Petrorhagia prolifera*, *Spergularia bocconei*, *Stachys alpina*, *Teucrium botrys*, *Tordylium maximum*, *Veronica praecox* and *V. triphyllos*.

   Six neophytes were known from fewer than 50 10-km squares when the book was completed but additional records were submitted before work on the CD-ROM was finished and these are now known in 50 or more squares. These taxa, *Acaena ovalifolia*, *Centaurea melitensis*, *Echinochloa frumentacea*, *Phytolacca acinosa*, *Populus balsamifera* × *P. trichocarpa* (*P.* 'Balsam Spire') and *Spiraea douglasii* subsp. *douglasii*, are included only on the CD-ROM. One species which is mapped in the book (*Lysimachia ciliata*) is shown in fewer than 50 10-km squares as a number of doubtful records were deleted at a late stage in data-processing.
4. All hybrids listed by Stace (1997) have been mapped if they have been recorded in 50 or more 10-km squares, regardless of date.

All these taxa are also included on the CD-ROM, as are data for an extra 970 introduced species present in fewer than 50 10-km squares.

Taxonomy and nomenclature

Taxonomy and nomenclature follow Stace's *New Flora of the British Isles* (1997). The order in which the taxa appear also follows Stace (1997), except that hybrids follow the first parent alphabetically rather than the first parent in systematic sequence. Synonyms are not included in the text, but synonyms which appear in Perring & Walters (1962), Perring & Sell (1968), Jermy *et al.* (1978), Clapham *et al.* (1981, 1987), Stace (1991), Kent (1992, 1996, 2000), Sell & Murrell (1996) and Webb *et al.* (1996) are given in the index.

Notes on the maps

Symbols

Records are mapped in the 10 × 10-km squares of the Ordnance Survey National Grid in Great Britain and the Isle of Man and in the Ordnance Survey/Suirbhéireacht Ordanáis National Grid in Ireland. Records from the Channel Islands are mapped in the 10 × 10-km squares of the Universal Transverse Mercator Grid. The Channel Islands are included in one inset on the map and Orkney and Shetland in another. The letter codes for the 100-km squares are given in Fig. 8.1.

Records are mapped as one of the following symbols:

present as a native between 1987 and 1999

present as a native between 1970 and 1986, but not recorded as either a native or an introduction since then

present as a native before 1970, but not recorded as either a native or an introduction since then, *or* records undated

present as an introduction between 1987 and 1999

present as an introduction between 1970 and 1986, but not recorded as either a native or an introduction since then

present as an introduction before 1970, but not recorded as either a native or an introduction since then, *or* records undated

Plants which are classified as 'native or alien' are mapped as if they are native, or the possibly native sites are distinguished from the certainly introduced. Archaeophytes, neophytes and casuals are all mapped as introductions.

Records from the most recent date-class thus take priority over those from earlier date-classes; if the species is known as both a native and an introduction in the most recent date-class in which it has been recorded, it is mapped

Table 8.1 Summary of the format of the species accounts. Explanation of the material included in the text is provided on the cited pages.

Key to symbols on distribution maps:
native, 1987–99
introduced, 1987–99
native, last recorded 1970–86
introduced, last recorded 1970–86
native, last recorded before 1970 or records undated
introduced, last recorded before 1970 or records undated

Symbols are listed in descending order of preference.

Key to text:	
Paragraph 1	Habitat; altitudinal range (p. 18)
Paragraph 2	Native status (p. 10); change statistic (pp. 36–37); summary of known changes in distribution or frequency (pp. 48–49)
Paragraph 3	Wider distribution (p. 27)
Paragraph 4	Key references (p. 49)
Name(s) of author(s)	

as a native. Thus a species that is known in a particular 10-km square only as an introduction in the 1987–99 period but was found there as a native before 1970 will be mapped as a recent introduction. Records of deliberate introductions for conservation purposes have been mapped with other deliberate or accidental introductions and are usually mentioned in the text. However, there is no central register of deliberate introductions for conservation purposes and BSBI Vice-county Recorders are not always notified of them, so there are probably many of which we are unaware.

The order of priority in the *New Atlas* departs from that in the 1962 *Atlas* and in many other national distribution maps which give old native records preference over recent introduced records. We have decided on this course because the principal aim of this Atlas is to map the current flora, but we recognise that one consequence is to obscure the historical native records. This is particularly significant for species that have declined as natives but have spread as introductions (e.g. *Ranunculus lingua*). For a discussion of the replacement of native populations by introductions in the London area, where it is probably most frequent, see Burton (1998).

The number of 10-km squares mapped with each of the above combinations of native or introduced status and date-class is given as an inset to the map for Britain, the Isle of Man and the Channel Islands (as GB) and, separately, for Ireland (IR). The total number of squares in these areas for which records are available is 2852 and 1007 respectively.

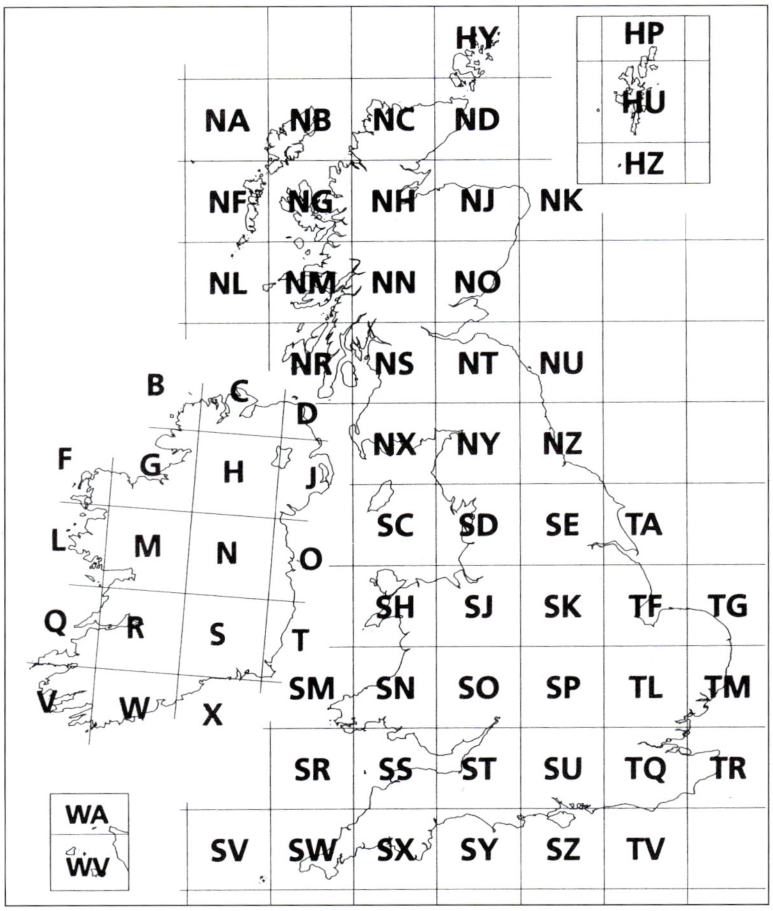

Figure 8.1 Letter codes for the 100-km squares.

Notes on the species texts

Unlike the 1962 *Atlas*, we have included a brief text with each map. We have been encouraged to do this by the examples of recent atlases of other groups such as bryophytes (Hill *et al.*, 1991–4) and molluscs (Kerney, 1999), as even a brief text does much to help the interpretation of a map. However, the large number of species mapped in this Atlas has severely limited the amount of information that we can present about each. More detailed information about a species can often be found in the cited references.

The first paragraph of the species accounts deals with the habitat and altitudinal range of the species, the second with trends in its distribution over time, the third with its distribution outside our area, and the fourth provides key references. Localities cited in the text are usually followed by the vice-counties in which they occur. The vice-counties are mapped in Fig. 8.2

Notes on the sources of information for each paragraph are given below.

Paragraph 1: habitat and altitudinal range

The first paragraph of the species accounts sets out the habitat of the species. This habitat information is based on the personal experience of the author(s) and on many literature sources. There is no single source of ecological data for the entire British and Irish flora, but authors have drawn extensively on the descriptions of rare, scarce and aquatic plants provided by Curtis & McGough (1988), Preston & Croft (1997), Stewart *et al.* (1994) and Wigginton (1999), on the standard vegetation accounts of Burnett (1964), McVean & Ratcliffe (1962) and Rodwell (1991–2000), and on the ecological data for more common species summarised by Grime *et al.* (1988). We have only rarely incorporated unpublished information; this is indicated by '*in litt.*'.

The first paragraph ends with a summary of the altitudinal range of the taxon. This gives the locality and vice-county for the highest point, and sometimes (for montane species) the lowest point as well. 'Lowland' indicates that a species is not found above 300 metres, 'upland' indicates altitudes between 300 and 600 metres and 'montane' is used for altitudes above 600 metres. In an altitudinal range 0 is taken as indicating that the species descends to very low altitudes, but we have not necessarily got a precise record from sea level. All records in feet have been converted to metres, and all altitudes are rounded down to the nearest 5 m.

Paragraph 2: trends in the distribution of the species

The second paragraph begins with the classification of species or subspecies as native, archaeophyte, neophyte or casual (see Chapter 3). Spontaneous hybrids between two native species are described as native. Spontaneous hybrids between a native and an introduced species are described as such in the text and mapped with the symbol used for native taxa. Spontaneous hybrids between two introduced species are also described as such in the text but are mapped with the symbol for introduced occurrences. Hybrids which have been introduced to our area *as hybrids*, or have originated in gardens and subsequently escaped into the wild, are treated as introduced. Three of these four possibilities are found in the genus *Senecio*. The hybrid between the two native species *S. aquaticus* and *S. jacobaea* is described and mapped as native. The spontaneous hybrid between *S. jacobaea* (native) and *S. cinerea* (introduced) is mapped with the native symbols, whereas the hybrid between two introduced species, *S. squalidus* and *S. viscosus*, is mapped with the introduced symbols. In the related genus *Brachyglottis* the hybrid of uncertain parentage known by the cultivar name 'Sunshine' escaped from gardens into the wild and is therefore described as a neophyte and mapped as an introduction.

The description of native status is followed by the change index, indicating the relative change in range size from 1930–69 to 1987–99 (see Chapter 7). There is no change index for species which were not mapped in the 1962 *Atlas*, or which were recorded from less than five 10-km squares in the 1930–69 period. In most cases the change index applies to the taxon mapped both in the 1962 *Atlas* and on the current map. In four cases the taxon is an aggregate, but one species is the most frequent plant throughout the range of the aggregate, so the change index is reported under that segregate species without further comment. These species are *Dryopteris filix-mas* (where the change index also includes *D. affinis* and *D. oreades*), *Ophioglossum vulgatum* (including *O. azoricum* and *O. lusitanicum* in addition to *O. vulgatum*), *Sorbus aria* (including seven apomictic species in Britain in addition to *S. aria*) and *Stellaria media* (including *S. pallida* and *S. neglecta* as well as *S. media*). Change indices which include one relatively common and one much rarer subspecies are reported under the commoner subspecies, but with a note that technically the index applies to the species. In the cases of one aggregate (*Tripleurospermum maritimum* sens. lat.) and species with subspecies which are more equally distributed, the change index is reported under all the relevant taxa, again with a note that it refers to the aggregate or species.

The second paragraph goes on to comment on trends in the distribution of the species. The word 'distribution' is used to cover two separate components, individually described as 'range' (the total geographical spread) and 'frequency' (the proportion of 10-km squares occupied within that range). For native species, we have attempted to summarise, in the very brief space available, the nature and the timing of major changes in distribution. It is important to emphasise that unless otherwise stated these comments refer to the 10-km square scale. The extent to which the distributions of species change depends on the scale at which they are examined: changes are likely to be fewer at a 50-km square scale and more marked at a 1-km square scale. If detailed local studies have demonstrated changes at a finer scale we have cited them, but we have tried to avoid speculating about such changes in the absence of hard evidence. For introduced species we have tried to provide some idea of the timing of their introduction and spread, including the date of introduction of garden plants into cultivation in our area and the date of the first record of neophytes in the wild. The sources of these dates are outlined in Chapter 4. The second paragraph also comments on the extent to which taxa are believed to be under-recorded. We have in general avoided comments on taxonomy, but any taxonomic or other problems which might have resulted in errors on the maps are indicated in this paragraph.

Paragraph 3: the wider distribution of the species

The third paragraph provides an indication of the wider range of the species. For native species and subspecies, the range is given according to the classification proposed by Preston & Hill (1997) and outlined in Chapter 6. The world range of archaeophytes is usually classified by the same system, but the total range of the taxon as an archaeophyte is used rather than the native range, which for archaeophytes is usually indeterminable. Some neophytes are also classified in the same way, but for introductions this classification is often replaced by a simple summary of the countries of origin. Native taxa are described as belonging to the appropriate phytogeographical element, whereas the word 'element' is not used for introductions. Thus the world distribution of the native *Athyrium filix-femina* is given as 'Circumpolar Boreo-temperate element' whereas the introduced *Matthiola struthiopteris* is described as 'A Circumpolar Boreal-montane species.'

There is little summarised information on the world distribution of hybrids, and we have not had time to undertake any research into this topic ourselves. Our summaries of the wider distribution of hybrids are usually taken from the information on their European range provided by Stace (1975). We have not been able to find sufficient information on which to base a summary of the wider distribution of *Rosa* hybrids, partly because there is no European consensus on the taxonomy of the genus.

Paragraph 4: key references

The final paragraph begins with a reference to the distribution map in the 1962 *Atlas* (Perring & Walters, 1962) or its *Critical Supplement* (Perring & Sell, 1968), cited as 'Atlas' or 'Atlas Supp.' respectively. For each book the citation is followed by the page number and a letter to indicate the position on the page. For the 1962 *Atlas* 'a' is used for the map on the top left, 'b' top right, 'c' bottom left and 'd' bottom right. The format of the *Critical Supplement* is less regular, but the maps are numbered in the same sequence. Thus 'Atlas (1d)' indicates that the map (of *Lycopodium clavatum*) is in the bottom right position on p. 1 of Perring & Walters (1962).

After these entries we usually list some key references to provide a lead into the wider literature on the taxonomy, ecology or distribution of the species. These provide only a point of entry into the literature, and we have tried to choose references which themselves cite other sources for the species. Our reference may not be the most important publication on a species – it may be a relatively recent paper which cites some more important but earlier works. Thus for British rare and scarce plants we have usually cited Wigginton (1999) and Stewart *et al.* (1994) respectively, and for aquatic plants we have cited Preston & Croft (1997), and only exceptionally have we also given references listed in these works. We have always cited the accounts in the *Biological Flora of the British Isles* series in the *Journal of Ecology* and the detailed accounts in Grime *et al.* (1988). We have also tried to include references to European or world distribution maps in readily accessible publications. We have cited references to the European maps in *Atlas Florae Europaeae* (Jalas & Suominen, 1972–99), which now covers the taxa in the first volume of *Flora Europaea*, and the maps of the distribution in the northern hemisphere of species which occur in Scandinavia (Hultén & Fries, 1986) and central Europe (Meusel & Jäger, 1992; Meusel *et al.*, 1965, 1978). If species are not mapped in these publications we have given references to the small-scale maps in Hultén (1968) or Bolos & Vigo (1984–95). All these references are cited in the normal form except those to the *Flora of North America*, edited by the Flora of North America Editorial Committee, which to save space is abbreviated to 'FNAEC'. We have been able to add references to works published in 2000 but only to a very few published early in 2001.

We have generally not referred to basic Floras and checklists, such as Stace (1997) and Clement & Foster (1994), as these would need to be cited under most species.

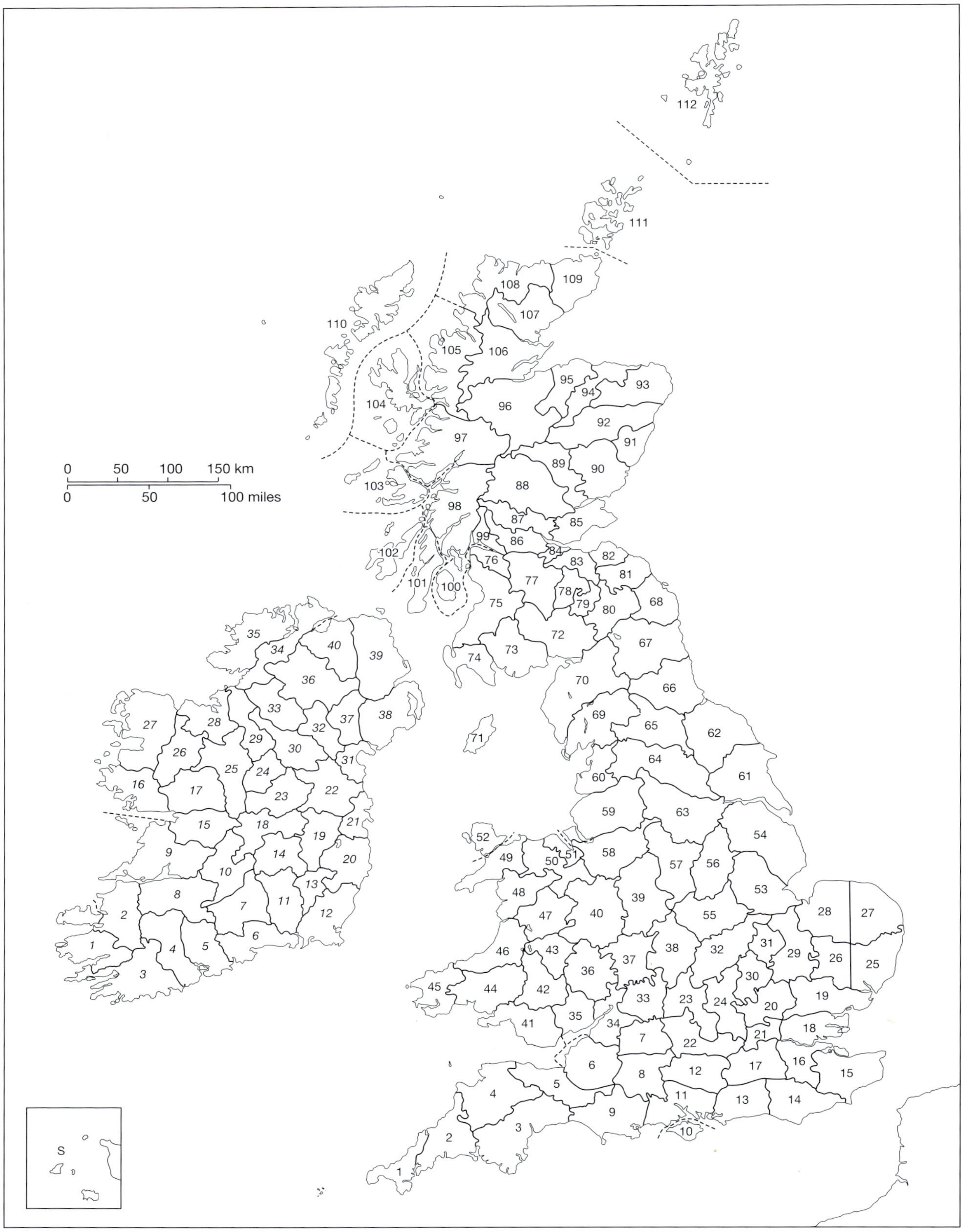

Figure 8.2 Vice-county boundaries. The counties are numbered on the map; for their names, see Table 8.2. For further details of the boundaries, see Dandy (1969) and Webb (1980).

Table 8.2 The numbers and names of the vice-counties mapped in Fig. 8.2.

England

1, W. Cornwall
2, E. Cornwall
3, S. Devon
4, N. Devon
5, S. Somerset
6, N. Somerset
7, N. Wiltshire
8, S. Wiltshire
9, Dorset
10, Isle of Wight
11, S. Hampshire
12, N. Hampshire
13, W. Sussex
14, E. Sussex
15, E. Kent
16, W. Kent
17, Surrey
18, S. Essex
19, N. Essex
20, Hertfordshire
21, Middlesex
22, Berkshire
23, Oxfordshire (Oxon)
24, Buckinghamshire
25, E. Suffolk
26, W. Suffolk
27, E. Norfolk
28, W. Norfolk
29, Cambridgeshire
30, Bedfordshire
31, Huntingtonshire
32, Northamptonshire
33, E. Gloucestershire
34, W. Gloucestershire
36, Herefordshire
37, Worcestershire
38, Warwickshire
39, Staffordshire
40, Shropshire (Salop)
53, S. Lincolnshire
54, N. Lincolnshire
55, Leicestershire
56, Nottinghamshire
57, Derbyshire
58, Cheshire
59, S. Lancashire
60, W. Lancashire
61, S.E. Yorkshire
62, N.E. Yorkshire
63, S.W. Yorkshire
64, Mid-W. Yorkshire
65, N.W. Yorkshire
66, Co. Durham
67, S. Northumberland
68, Cheviot
69, Westmorland
70, Cumberland

Wales

35, Monmouthshire
41, Glamorgan
42, Breconshire
43, Radnorshire
44, Carmarthenshire
45, Pembrokeshire
46, Cardiganshire
47, Montgomeryshire
48, Merioneth
49, Caernarvonshire
50, Denbighshire
51, Flintshire
52, Anglesey

Isle of Man

71, Isle of Man

Scotland

72, Dumfriesshire
73, Kirkcudbrightshire
74, Wigtownshire
75, Ayrshire
76, Renfrewshire
77, Lanarkshire
78, Peeblesshire
79, Selkirkshire
80, Roxburghshire
81, Berwickshire
82, E. Lothian
83, Midlothian
84, W. Lothian
85, Fife
86, Stirlingshire
87, W. Perth
88, Mid Perth
89, E. Perth
90, Angus
91, Kincardineshire
92, S. Aberdeen
93, N. Aberdeen
94, Banffshire
95, Moray
96, Easterness
97, Westerness
98, Main Argyll
99, Dunbarton
100, Clyde Isles
101, Kintyre
102, S. Ebudes
103, Mid Ebudes
104, N. Ebudes
105, W. Ross
106, E. Ross
107, E. Sutherland
108, W. Sutherland
109, Caithness
110, Outer Hebrides
111, Orkney
112, Shetland

Channel Islands

S, Channel Islands

Ireland

H1, S. Kerry
H2, N. Kerry
H3, W. Cork
H4, Mid Cork
H5, E. Cork
H6, Co. Waterford
H7, S. Tipperary
H8, Co. Limerick
H9, Co. Clare
H10, N. Tipperary
H11, Co. Kilkenny
H12, Co. Wexford
H13, Co. Carlow
H14, Laois
H15, S.E. Galway
H16, W. Galway
H17, N.E. Galway
H18, Offaly
H19, Co. Kildare
H20, Co. Wicklow
H21, Co. Dublin
H22, Meath
H23, Westmeath
H24, Co. Longford
H25, Co. Roscommon
H26, E. Mayo
H27, W. Mayo
H28, Co. Sligo
H29, Co. Leitrim
H30, Co. Cavan
H31, Co. Louth
H32, Co. Monaghan
H33, Fermanagh
H34, E. Donegal
H35, W. Donegal
H36, Tyrone
H37, Co. Armagh
H38, Co. Down
H39, Co. Antrim
H40, Co. Londonderry

Huperzia selago Fir Clubmoss

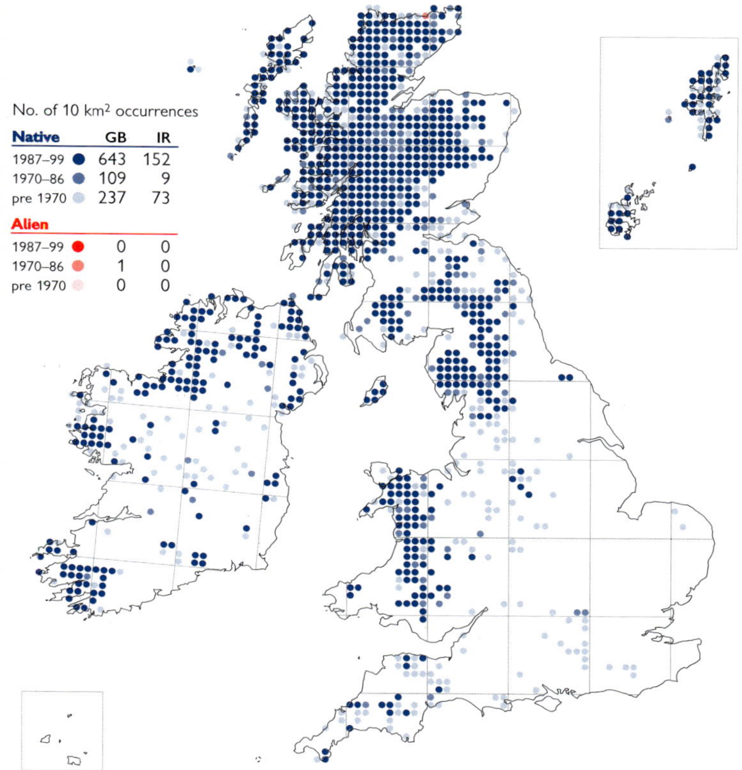

A semi-decumbent evergreen perennial herb found on acidic, nutrient-poor, sandy or peaty soils in grassland, heathland, blanket bog, montane communities and, rarely, sand quarry tips. Generally upland, but from sea level to 1310 m (Ben Nevis, Westerness, and Ben Macdui, S. Aberdeen).

Native (change –0.41). Most of the sites in lowland England were lost before 1930, due to habitat destruction, agricultural improvement and drainage. In the uplands some sites have been lost to heather burning and overgrazing, but its overall distribution is stable.

Circumpolar Boreo-arctic Montane element.

References: Atlas (1a), Headley & Callaghan (1990), Hultén & Fries (1986), Jalas & Suominen (1972), Jermy *et al.* (1978), Meusel *et al.* (1965), Page (1997).

A. D. HEADLEY

Lycopodiella inundata Marsh Clubmoss

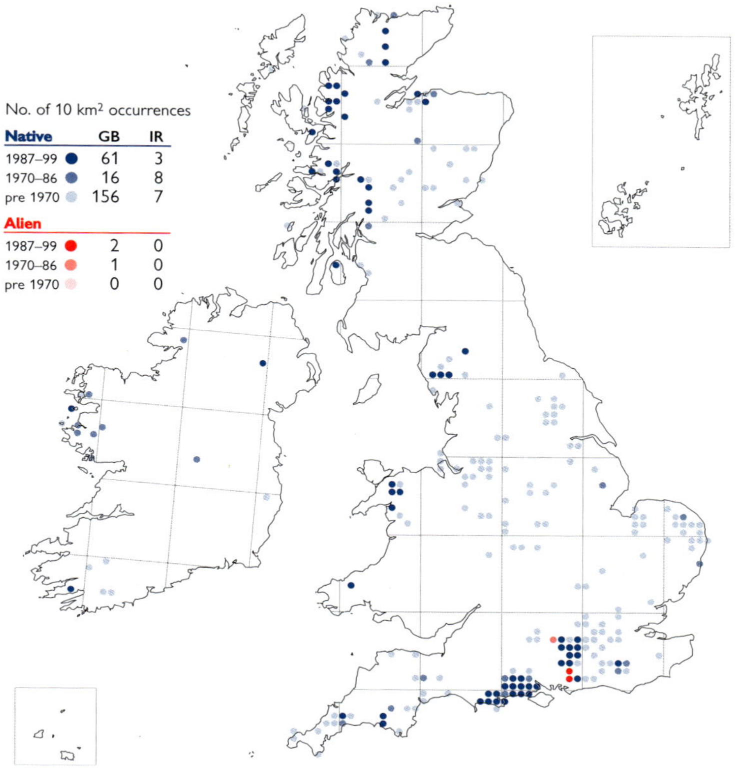

A prostrate perennial herb of wet, bare, peaty or sandy margins of lakes, pools, flushes and trackways. It can rapidly colonise substrates kept open by winter inundation, cattle poaching or peat cutting. 0–390 m (Llyn Cwmffynnon, Caerns.).

Native (change –0.65). Many sites for *L. inundata* were lost before 1930, and losses have continued due to drainage, a lack of grazing and other disturbance, and conversion to scrub, especially in England. However, it is easily overlooked and new sites have been found outside England since the 1962 *Atlas*.

European Boreo-temperate element; also in E. Asia and N. America.

References: Atlas (1b), Byfield & Pearman (1996), Curtis & McGough (1988), Hultén & Fries (1986), Jalas & Suominen (1972), Jermy *et al.* (1978), Meusel *et al.* (1965), Page (1997), Stewart *et al.* (1994).

A. D. HEADLEY

Lycopodium clavatum Stag's-horn Clubmoss

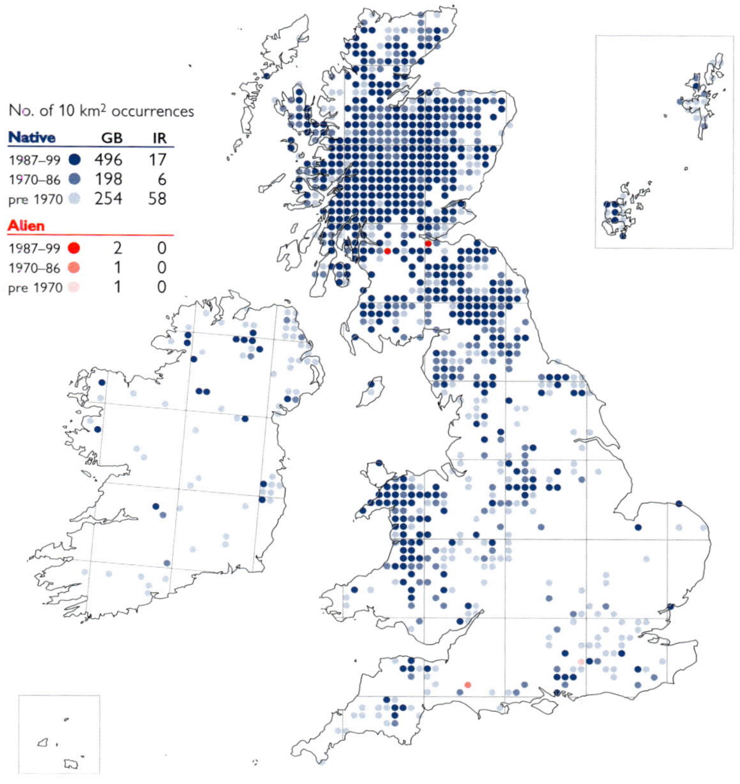

A prostrate, evergreen perennial herb of heaths, moors and mountains. It is often frequent on base-rich micaceous soils, but also occurs on more acidic *Calluna* heath and *Nardus* grassland. Propagation is mostly vegetative, but spores can colonise new sites, particularly the disturbed soil of roadside banks and quarries. Lowland to 840 m in Atholl (E. Perth).

Native (change –0.52). Many lowland sites of *L. clavatum* were lost before 1930. Populations elsewhere are somewhat transient, with losses owing to overgrazing, heather burning, conversion to scrub and agricultural improvement being offset by the establishment of new populations.

Circumpolar Boreo-temperate element.

References: Atlas (1d), Hultén & Fries (1986), Jalas & Suominen (1972), Jermy *et al.* (1978), Page (1997).

A. D. HEADLEY

Lycopodium annotinum Interrupted Clubmoss

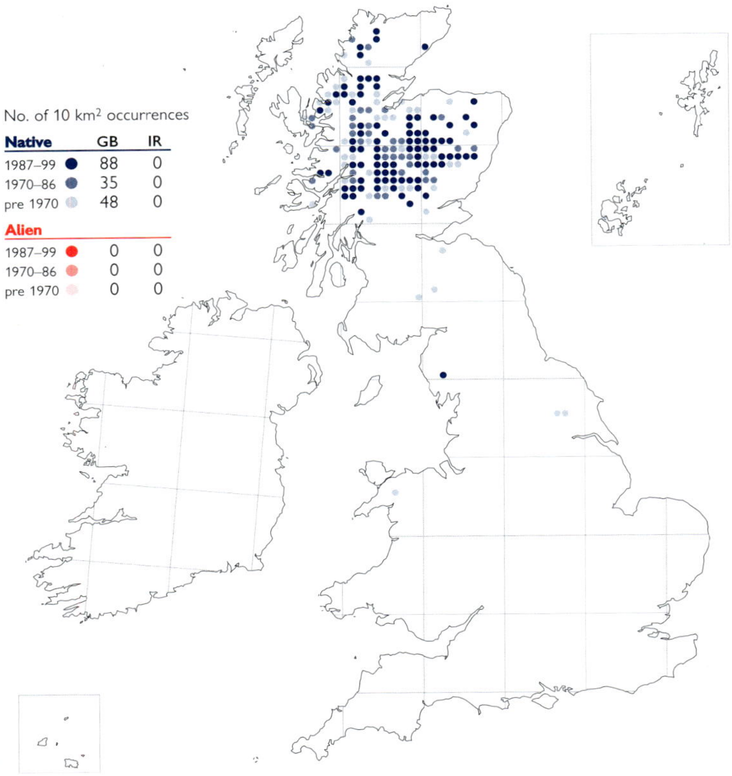

A sprawling, evergreen herb typically found on mountains and moorlands amongst deep *Calluna* on hill slopes, and sometimes in *Pinus sylvestris* woods. It usually grows on acidic peaty soils, often overlying boulders, or in hollows where snow accumulates. From 45 m on Mull (S. Ebudes) to 1000 m (Coire Cheap, Westerness).

Native (change −0.38). There have certainly been losses of this species over the past century due to changes in moorland management, although its distribution now appears to be stable. It may still be present in remote squares for which there are only pre-1987 records.

Circumpolar Boreo-arctic Montane element.

References: Atlas (1c), Callaghan *et al.* (1986), Hultén & Fries (1986), Jalas & Suominen (1972), Jermy *et al.* (1978), Page (1997), Stewart *et al.* (1994). A. D. HEADLEY

Diphasiastrum alpinum Alpine Clubmoss

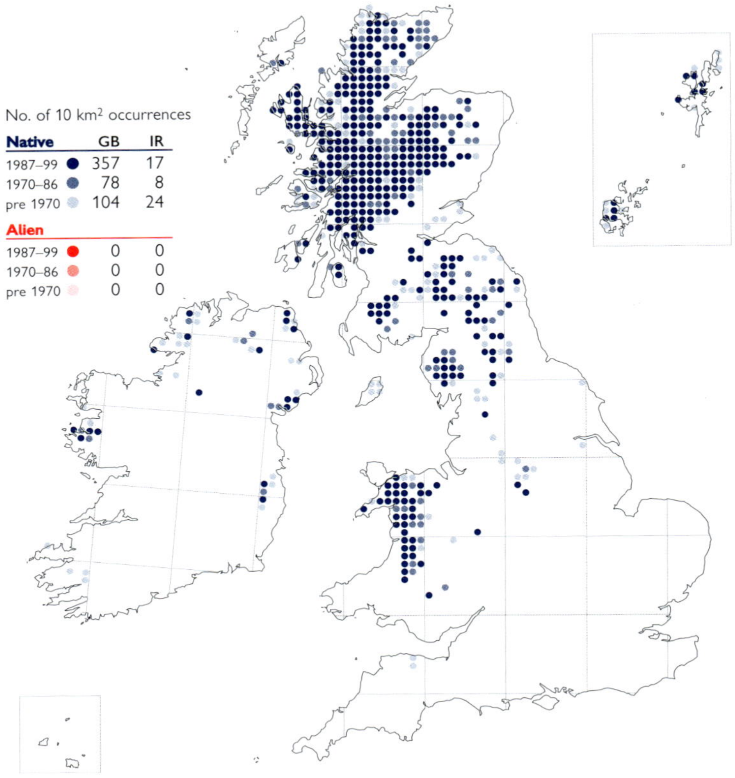

An evergreen herb of short acidic grassland on mountains and moors, where it grows on moist but well-drained, thin peaty soils, especially those directly overlying rocks. More rarely it occurs on acidic sands. Vegetative propagation is much more frequent than sexual reproduction, which usually takes place on bare, disturbed ground. 0–1220 m (Ben Macdui, Banffs.).

Native (change −0.51). There has been little change in the distribution of this species in the uplands since the 1962 *Atlas*. Some of its lowland sites have been lost to agricultural improvement, mainly before 1930.

Circumpolar Arctic-montane element, with a disjunct distribution.

References: Atlas (2a), Hultén & Fries (1986), Jalas & Suominen (1972), Jermy *et al.* (1978), Meusel *et al.* (1965), Page (1997). A. D. HEADLEY

Diphasiastrum complanatum Issler's Clubmoss

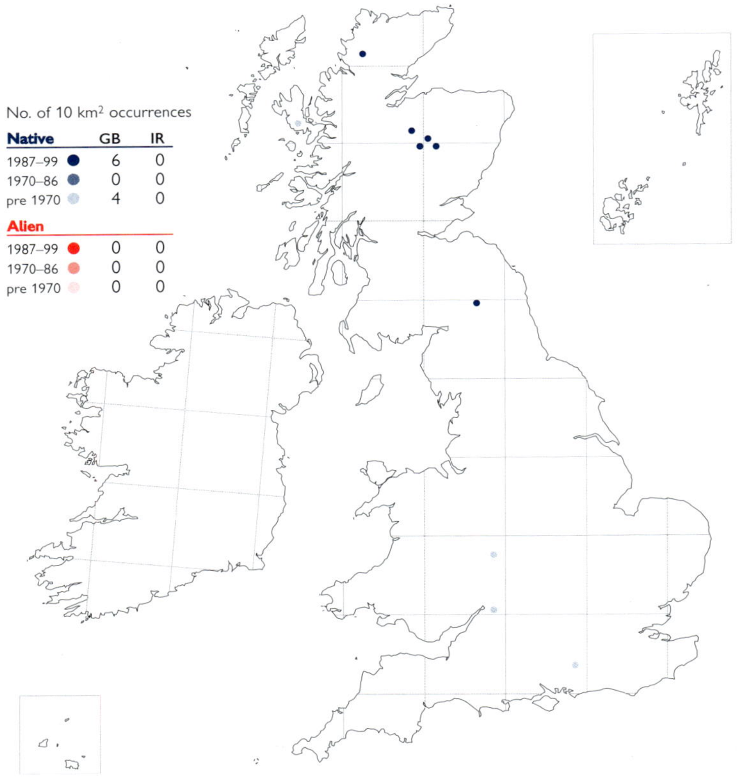

An evergreen, procumbent perennial herb of heaths and moors, where it typically grows on well-drained, sandy, often skeletal soils in dwarf *Calluna* heath. In Scotland its sites are usually *Calluna-Arctostaphylos* heath. Between 270 m (Canisp, W. Sutherland) and 960 m (Lochnagar, S. Aberdeen).

Native. The map shows only confirmed records, as many old records are based on abnormal *D. alpinum*. Burning and grazing of heaths may have caused the extinction of *D. complanatum* in lowland England but it may be under-recorded in montane areas.

Circumpolar Boreal-montane element, with a continental distribution in western Europe.

References: Hultén & Fries (1986), Jalas & Suominen (1972), Jermy (1989), Jermy *et al.* (1978), Meusel *et al.* (1965), Page (1997), Wigginton (1999).

A. D. HEADLEY

Selaginella selaginoides Lesser Clubmoss

Native	**GB**	**IR**
1987–99 ●	754	183
1970–86 ●	103	19
pre 1970 ●	131	86
Alien		
1987–99 ●	0	0
1970–86 ●	0	0
pre 1970 ●	0	0

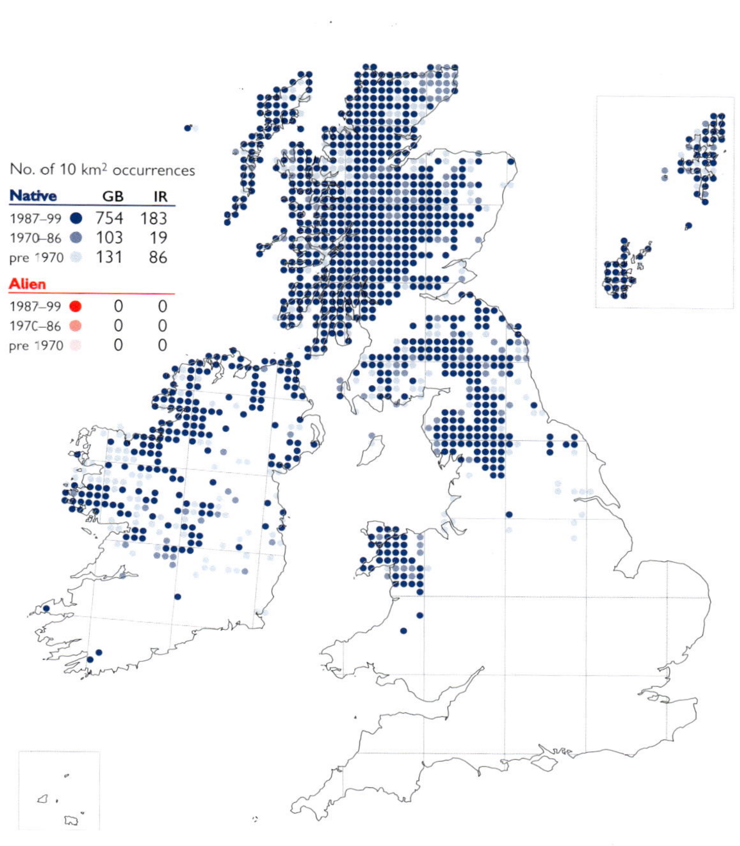

A small, perennial, moss-like herb which is characteristic of damp, base-rich sites where there is little competition. Typical habitats include dune-slacks, fens, flushes, mires, wet cliffs and ledges, and short upland grassland. 0–1065 m (Breadalbanes, Mid Perth).

Native (change −0.47). *S. selaginoides* was eliminated from lowland sites in Britain, mainly before 1930, by drainage and habitat destruction. In the uplands, however, there is little evidence for any significant change in its distribution.

Circumpolar Boreal-montane element.

References: Atlas (2b), Hultén & Fries (1986), Jalas & Suominen (1972), Jermy *et al.* (1978), Meusel *et al.* (1965), Page (1997).

A. C. JERMY & T. D. DINES

Selaginella kraussiana Kraus's Clubmoss

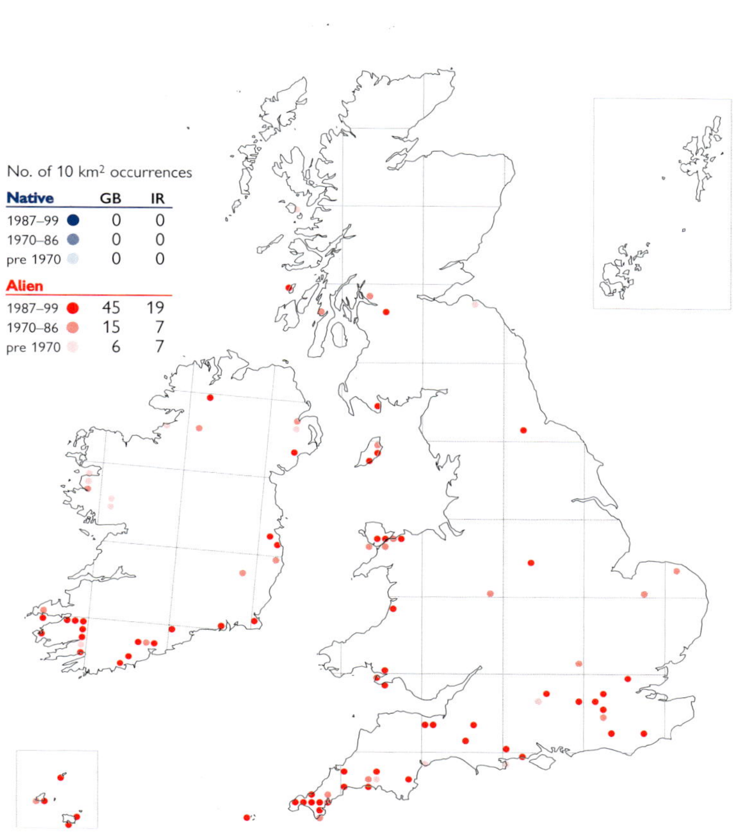

A perennial, moss-like herb which grows in woodland and moist, shaded grassland. It is naturalised in gardens, churchyards and damp valleys near the sea. At Landewednack (W. Cornwall) a population first recorded in 1922 was badly hit by summer drought in 1975 and 1976 but recovered slowly thereafter. Lowland.

Neophyte. *S. kraussiana* was introduced to cultivation in Britain in 1878 and is often an abundant weed in the shady corners of greenhouses. It was first recorded in the wild in Camborne (W. Cornwall) and Tullaghan (Co. Leitrim) in 1917. It has spread slowly since it was first mapped by Jermy *et al.* (1978).

Native of C. & S. Africa; locally naturalised in S.W. Europe and the W. Mediterranean region.

Reference: Jalas & Suominen (1972).

C. D. PRESTON

Isoetes lacustris Quillwort

Native	**GB**	**IR**
1987–99 ●	399	79
1970–86 ●	54	20
pre 1970 ●	69	48
Alien		
1987–99 ●	0	0
1970–86 ●	0	0
pre 1970 ●	0	0

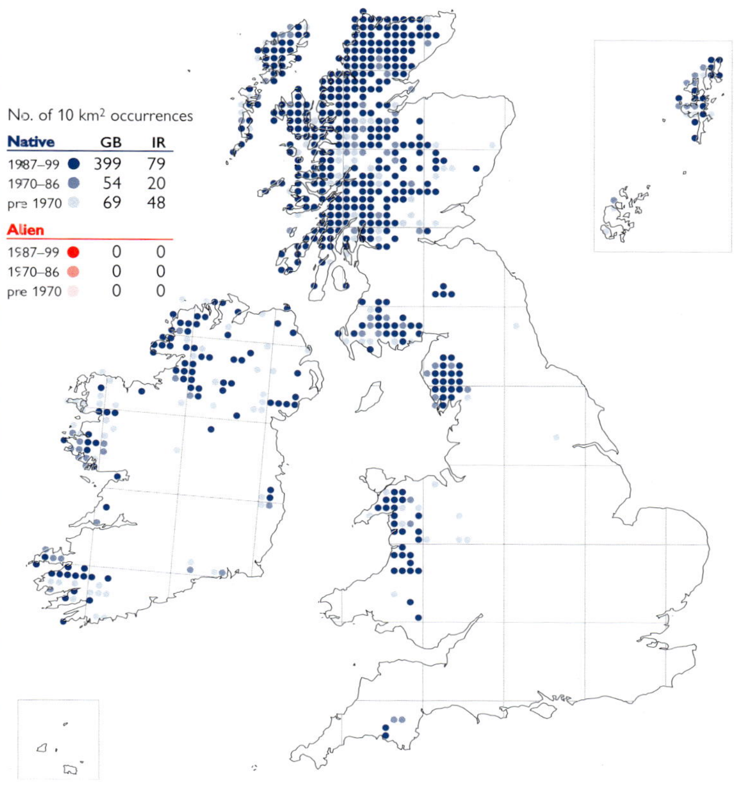

A submerged aquatic perennial found in oligotrophic lakes with a rocky or skeletal substrate. It frequently forms extensive lawns of many square metres at depths up to 2.5 m and rarely deeper (6 m in Loch Lundie, Westerness). *I. lacustris* also occasionally colonises artificial reservoirs. 0–850 m (Carn nam Sac, S. Aberdeen).

Native (change +0.95). *I. lacustris* was under-recorded in the 1962 *Atlas*, and may still be so in some areas. There is little evidence of change in its actual distribution although a few lowland sites have been lost through agricultural eutrophication.

Eurosiberian Boreal-montane element; also in N. America.

References: Atlas (2c), Hultén & Fries (1986), Jalas & Suominen (1972), Jermy *et al.* (1978), Meusel *et al.* (1965), Page (1997), Preston & Croft (1997).

A. C. JERMY & T. D. DINES

Isoetes echinospora Spring Quillwort

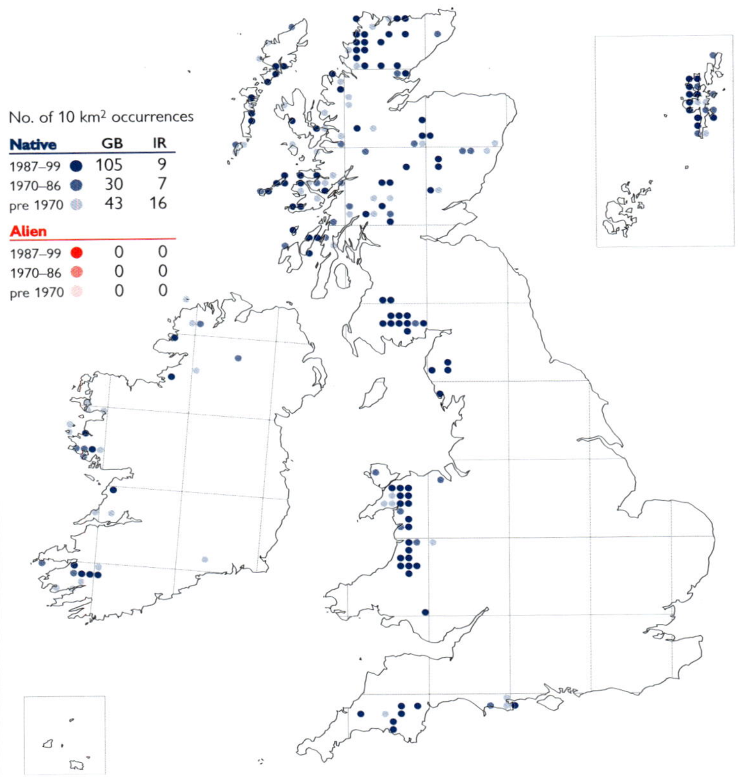

No. of 10 km² occurrences

Native		GB	IR
1987–99	●	105	9
1970–86	●	30	7
pre 1970	●	43	16
Alien			
1987–99	●	0	0
1970–86	●	0	0
pre 1970	●	0	0

A submerged aquatic perennial usually found in nutrient-poor lakes over a wide range of substrates, from rocks and stones to silt and peat. It also grows in more mesotrophic water, such as coastal lakes enriched by wind-borne base salts, lowland reservoirs, slow-flowing rivers and flooded gravel- and clay-pits. It often grows with *I. lacustris*, and not infrequently hybridises with it. 0–500 m (Loch Callater, S. Aberdeen).

Native (change +0.65). *I. echinospora* is now known from many more sites than in the 1962 *Atlas*, and probably remains under-recorded.

Circumpolar Boreal-montane element, with a disjunct distribution.

References: Atlas (2d), Hultén & Fries (1986), Jalas & Suominen (1972), Jermy *et al.* (1978), Meusel *et al.* (1965), Page (1997), Preston & Croft (1997), Stewart *et al.* (1994).

A. C. JERMY & T. D. DINES

Isoetes histrix Land Quillwort

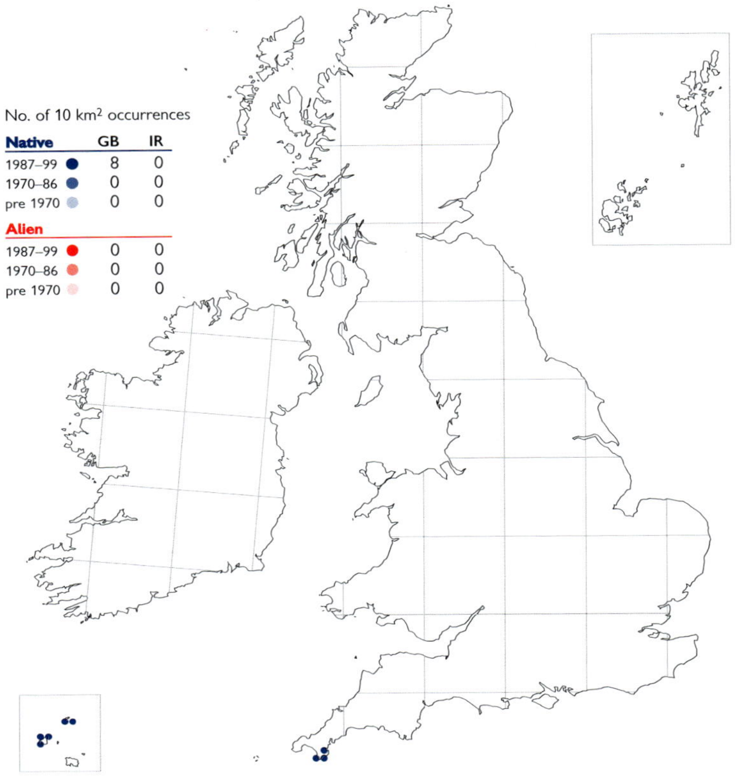

No. of 10 km² occurrences

Native		GB	IR
1987–99	●	8	0
1970–86	●	0	0
pre 1970	●	0	0
Alien			
1987–99	●	0	0
1970–86	●	0	0
pre 1970	●	0	0

A small, summer-deciduous perennial, growing from a 'corm'. It prefers skeletal acidic soils that are moist or flooded in winter but experience summer drought. A poor competitor, it is most frequent on pans of bare soil over S.-facing rock outcrops, but also grows on erosion pans and footpaths. It sometimes occurs in short turf, and in dune-slacks in Guernsey. Lowland.

Native. The distribution of this species is stable. A few sites have been lost due to a lack of grazing, to fire and to conversion to cultivation, but it has colonised areas from which turf has been stripped.

Mediterranean-Atlantic element.

References: Atlas (2d), Bolòs & Vigo (1984), Jalas & Suominen (1972), Jermy *et al.* (1978), Page (1997), Wigginton (1999).

T. D. DINES

Equisetum hyemale Rough Horsetail

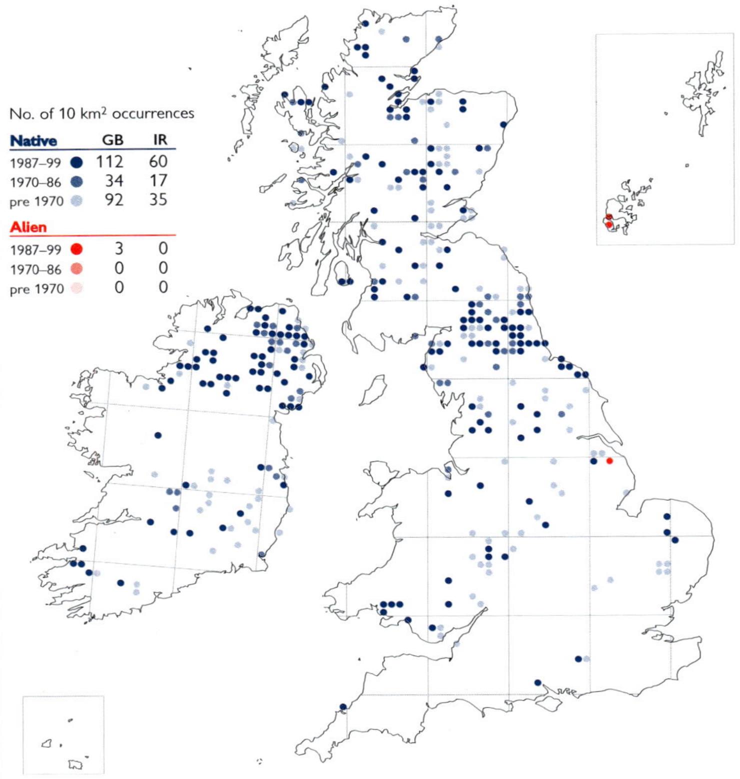

No. of 10 km² occurrences

Native		GB	IR
1987–99	●	112	60
1970–86	●	34	17
pre 1970	●	92	35
Alien			
1987–99	●	3	0
1970–86	●	0	0
pre 1970	●	0	0

A slow-growing, evergreen herb forming colonies of shoots from branching rhizomes. It prefers heavy soils derived from sand or clay which are permanently moist and have a high mineral and silica content. It is usually found in shaded open woodland beside streams and rivers, but also grows in base-rich moorland flushes and sand dunes. 0–610 m (Glen Lyon, Mid Perth).

Native (change +0.30). Most losses of *E. hyemale* occurred before 1930, probably due to drainage and an increase in grazing and trampling of river banks by livestock. However, many new sites have been discovered since the 1962 *Atlas*.

Circumpolar Boreo-temperate element.

References: Atlas (3a), Hultén & Fries (1986), Jalas & Suominen (1972), Jermy *et al.* (1978), Page (1997).

C. DIXON & T. D. DINES

Equisetum hyemale × *E. variegatum* (*E.* × *trachyodon*) Mackay's Horsetail

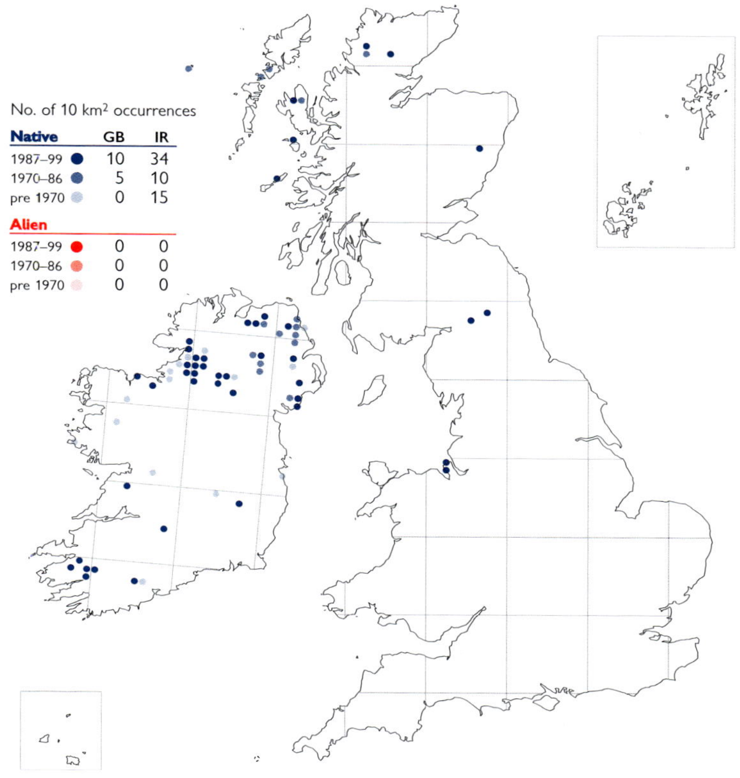

A rhizomatous evergreen herb found in a range of mostly base-rich habitats, often in the absence of both parents. In Ireland it usually occurs in woodland, and on sheltered wooded river banks and lake margins. In Britain it grows in dune-slacks, on flushed sandy river banks, and in peaty turf in coastal machair. It is sterile and spreads by rooting stem and rhizome fragments. Lowland.

Native. The distribution of *E.* × *trachyodon* is apparently stable. Although its distribution is much better known than when mapped by Perring & Sell (1968), it probably remains under-recorded. It is unclear whether losses are genuine or attributable to under-recording.

Widespread in C. & N. Europe and N. America.

References: Atlas Supp. (1a), Jermy *et al.* (1978), Page (1997), Page & Barker (1985), Stace (1975).

T. D. DINES

Equisetum ramosissimum Branched Horsetail

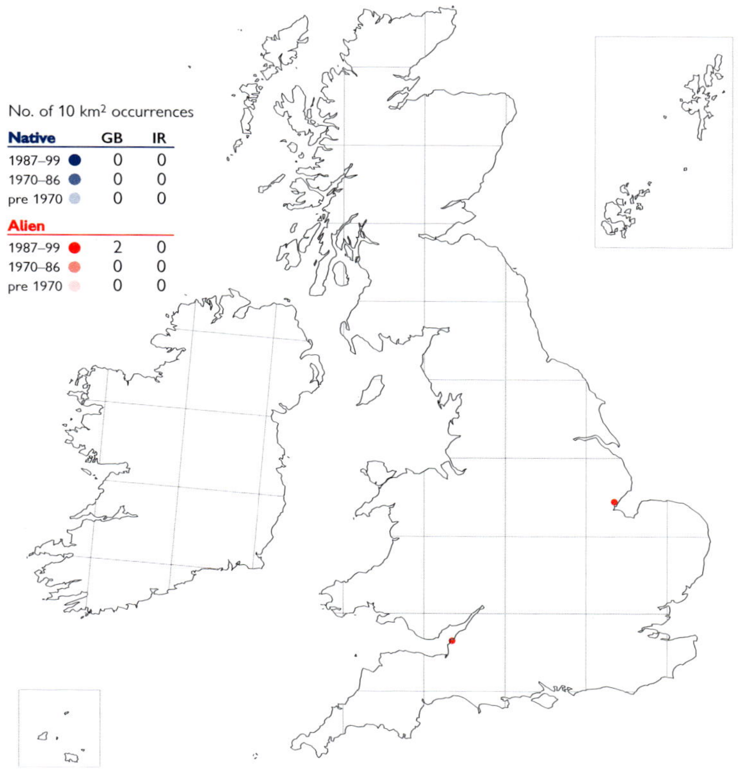

An erect evergreen herb found growing in rough grassland near the sea, on sand or clay soil. Lowland.

Neophyte. *E. ramosissimum* was once regarded as possibly native but is now considered to have been introduced. In N. Lincolnshire it was found in 1947 on a river bank that was straightened between 1880 and 1887 and it is thought to be a ballast alien. In N. Somerset it was not correctly identified until 1986, despite having been known at the site since 1963.

A Eurasian Southern-temperate species.

References: Atlas (3b), Alston (1949), FitzGerald & Jermy (1987), Hultén & Fries (1986), Jalas & Suominen (1972), Jermy *et al.* (1978), Meusel *et al.* (1965).

T. D. DINES

Equisetum variegatum Variegated Horsetail

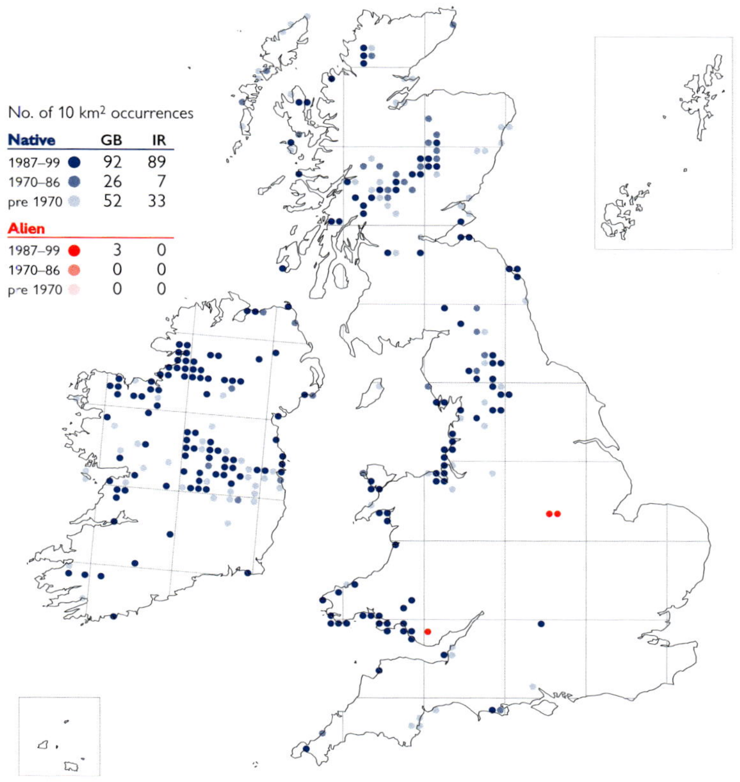

An evergreen, prostrate herb found in a wide variety of habitats, including dune-slacks, river shingle, upland flushes and stony loch-shores. It is a calcicole and a poor competitor; its sites are usually open and often winter-flooded. In Ireland a more vigorous, upright ecotype is found mostly on canal banks. 0–1040 m (Ben Lawers, Mid Perth).

Native (change −0.12). There is some evidence of decline in this species, but the position is obscured as so many sites have been discovered since the 1962 *Atlas*. Upland sites are sensitive to over-stocking, whilst lowland populations have been lost due to drainage and sand dune development.

Circumpolar Boreo-arctic Montane element.

References: Atlas (3c), Hultén & Fries (1986), Jalas & Suominen (1972), Jermy *et al.* (1978), Page (1997), Stewart *et al.* (1994).

C. DIXON & T. D. DINES

Equisetum fluviatile Water Horsetail

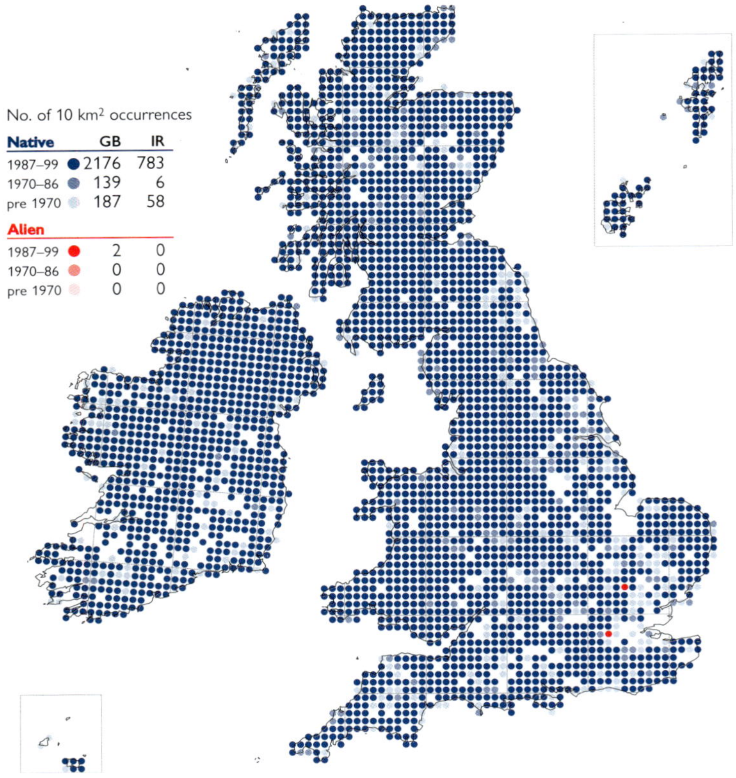
No. of 10 km² occurrences
Native GB IR
1987–99 2176 783
1970–86 139 6
pre 1970 187 58
Alien
1987–99 2 0
1970–86 0 0
pre 1970 0 0

This deciduous herb grows in a wide variety of aquatic and semi-aquatic habitats, from ditches and small ponds to large lakes and sheltered rivers. It tolerates a remarkable range of water and substrate pH, nutrient levels, substrate type and water depth, and is often a pioneer species in freshwater successions. 0–915 m (Breadalbanes, Mid Perth).

Native (change +0.42). Like many aquatic plants, this species is now much better recorded than in the 1962 *Atlas*. Most of the losses in S.E. England have taken place since 1950, reflecting the loss of small wetlands and the unsympathetic management of remaining sites.

Circumpolar Boreo-temperate element.

References: Atlas (3d), Grime *et al.* (1988), Hultén & Fries (1986), Jalas & Suominen (1972), Jermy *et al.* (1978), Page (1997), Preston & Croft (1997).

C. DIXON & T. D. DINES

Equisetum arvense Field Horsetail

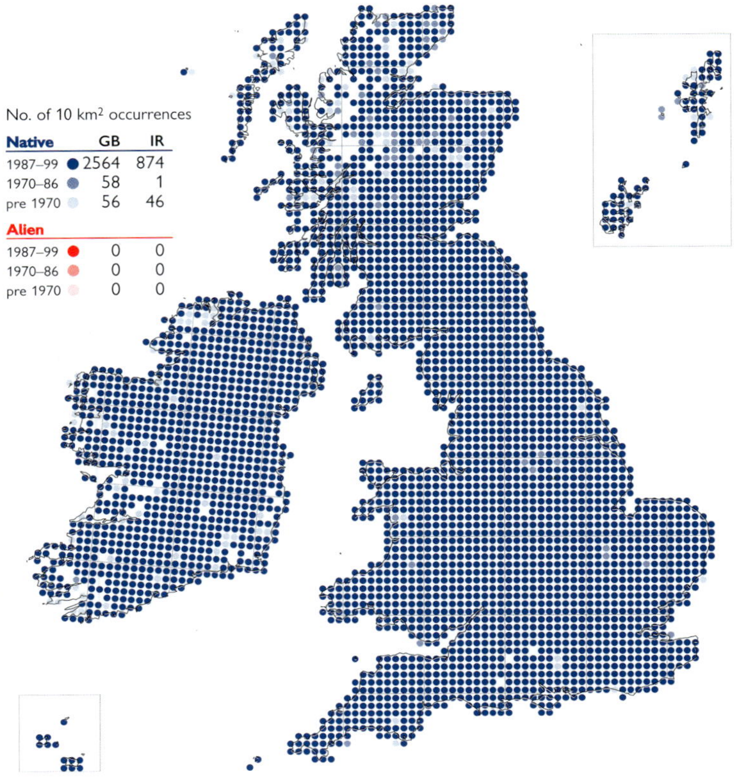
No. of 10 km² occurrences
Native GB IR
1987–99 2564 874
1970–86 58 1
pre 1970 56 46
Alien
1987–99 0 0
1970–86 0 0
pre 1970 0 0

The natural habitats of this deciduous herb include river banks, fixed dune grassland, sea-cliffs and montane flushes, but it has become closely associated with human activity. Being long-lived, vigorous, resistant to herbicides and tolerant of drier conditions than other *Equisetum* species, it is now frequent on roadsides, railways, paths, soil banks and waste ground, and in quarries and gardens, where its spread is assisted by rhizome fragments. 0–1005 m (Beinn Heasgarnich, Mid Perth).

Native (change +0.39). There has been no change in the distribution of this species.

Circumpolar Wide-boreal element.

References: Atlas (4d), Cody & Wagner (1980), Grime *et al.* (1988), Hultén & Fries (1986), Jalas & Suominen (1972), Jermy *et al.* (1978), Meusel *et al.* (1965), Page (1997).

C. DIXON & T. D. DINES

Equisetum arvense × *E. fluviatile* (*E.* × *litorale*) Shore Horsetail

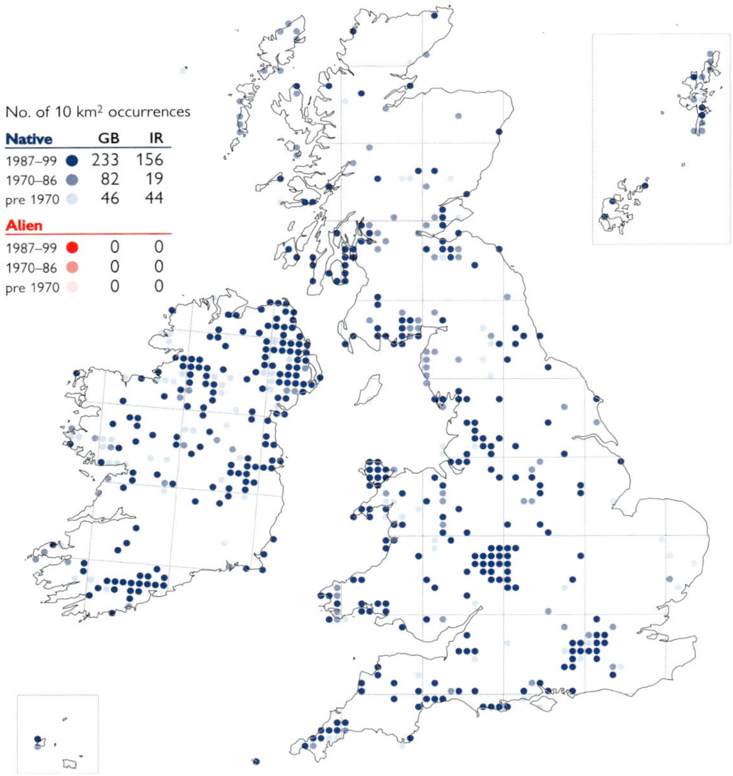
No. of 10 km² occurrences
Native GB IR
1987–99 233 156
1970–86 82 19
pre 1970 46 44
Alien
1987–99 0 0
1970–86 0 0
pre 1970 0 0

A vigorous, deciduous herb found in a wide range of habitats, often in the absence of one or both parents. It has been recorded from open woodland and scrub, pond and lake margins, streams and rivers, canals, ditches, roadsides, open peat on moorland, wet pastures and gravel-pits. Many sites are disturbed or seasonally flooded. Lowland.

Native. This is the most frequent *Equisetum* hybrid, and was mapped in 1968, 1978 and 1997. Despite this, it is still not consistently recorded and the map shows much recorder bias. However, it appears to be genuinely more frequent in Ireland and W. Britain.

Widespread in C. & N. Europe and N. America.

References: Atlas Supp. (1b), Jermy *et al.* (1978), Page (1997), Preston & Croft (1997), Stace (1975).

C. DIXON & T. D. DINES

Equisetum pratense Shady Horsetail

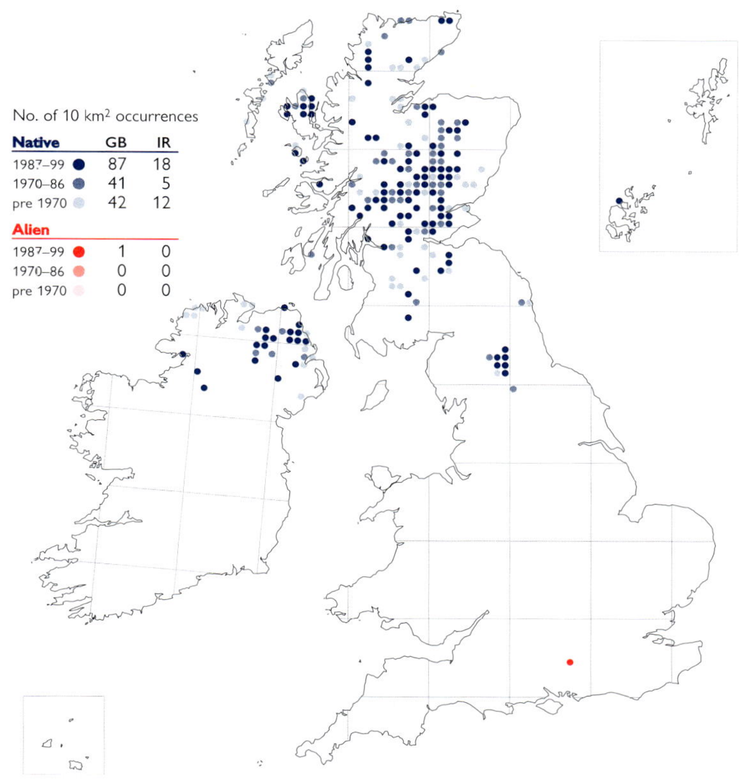

No. of 10 km² occurrences

Native		GB	IR
1987–99	●	87	18
1970–86	●	41	5
pre 1970	●	42	12
Alien			
1987–99	●	1	0
1970–86	●	0	0
pre 1970	●	0	0

An evergreen herb, typically found on sloping sites where the substrate is derived from calcareous alluvial silts or sand, especially lightly wooded stream banks in the lower parts of upland valleys. It can also extend onto open moorland, and is found on grassy slopes beneath base-rich upland cliffs. 0–915 m (Breadalbanes, Mid Perth).

Native (change +0.11). The considerable increase in records since the 1962 *Atlas* is due to better recording. Although most populations are long-lived, cone production is usually very poor, possibly because of climatic conditions, and Page (1997) suggests that the species is in slow decline.

Circumpolar Boreal-montane element.

References: Atlas (4c), Curtis & McGough (1988), Hultén & Fries (1986), Jalas & Suominen (1972), Jermy *et al.* (1978), Stewart *et al.* (1994). C. DIXON & T. D. DINES

Equisetum sylvaticum Wood Horsetail

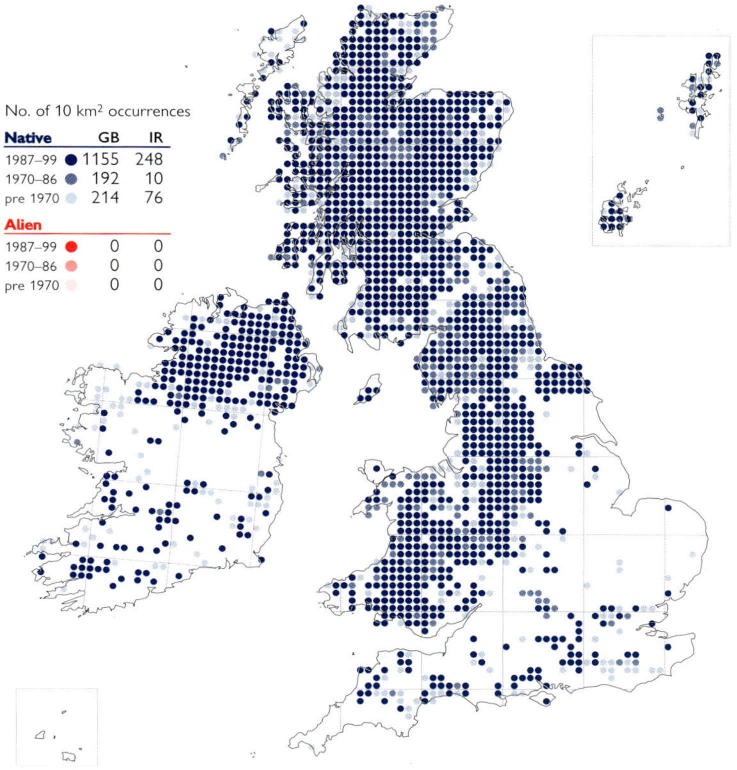

No. of 10 km² occurrences

Native		GB	IR
1987–99	●	1155	248
1970–86	●	192	10
pre 1970	●	214	76
Alien			
1987–99	●	0	0
1970–86	●	0	0
pre 1970	●	0	0

A deciduous, colony-forming herb which generally grows on deep, mildly acidic, often peaty soils that are kept permanently damp by flushing. It occurs on the lower slopes of mountain valleys, steep streamsides, wet ledges and open flushes, beside lakes and on the edges of drainage ditches. It also occurs on wet road verges and railway banks. 0–850 m (Breadalbanes, Mid Perth).

Native (change –0.35). There is little evidence of a decline in upland areas since the 1962 *Atlas*. The decline in the British lowlands was already apparent then, and has continued.

Circumpolar Boreo-temperate element.

References: Atlas (4b), Hultén & Fries (1986), Jalas & Suominen (1972), Jermy *et al.* (1978), Meusel *et al.* (1965), Page (1997).

C. DIXON & T. D. DINES

Equisetum palustre Marsh Horsetail

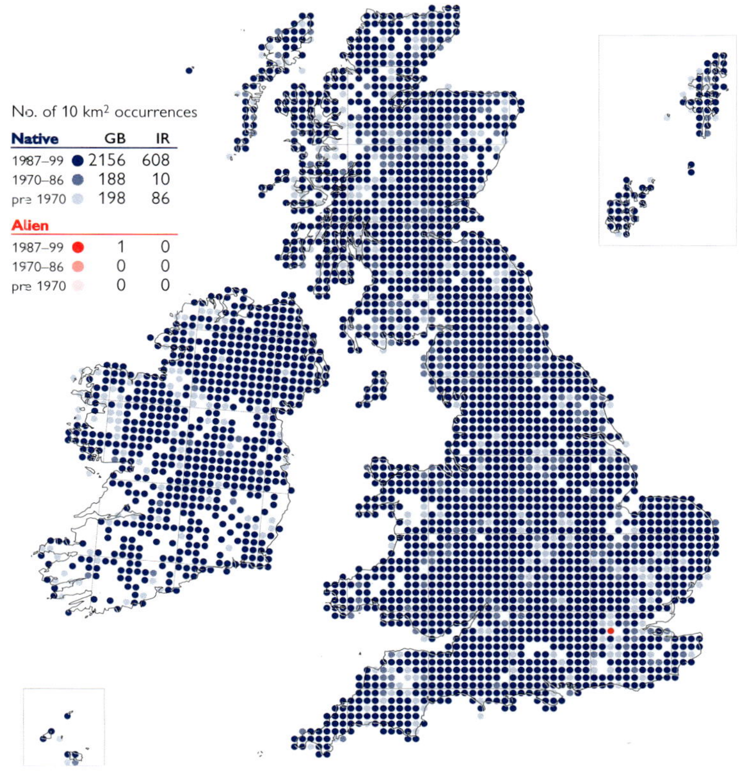

No. of 10 km² occurrences

Native		GB	IR
1987–99	●	2156	608
1970–86	●	188	10
pre 1970	●	198	86
Alien			
1987–99	●	1	0
1970–86	●	0	0
pre 1970	●	0	0

A deciduous herb associated with marshes, damp pastures, ditches, dune-slacks, streams, rivers and mountain flushes. It tolerates a wide range of soil types and substrates, provided that they are permanently damp and adequately base-rich. 0–945 m (Meall nan Tarmachan, Mid Perth).

Native (change +0.18). Mapped as 'all records' in the 1962 *Atlas*, there is little evidence for a change in the distribution of this species, although some lowland sites have been lost to drainage and agricultural improvement.

Circumpolar Boreo-temperate element.

References: Atlas (4a), Grime *et al.* (1988), Hultén & Fries (1986), Jalas & Suominen (1972), Jermy *et al.* (1978), Page (1997).

C. DIXON & T. D. DINES

Equisetum telmateia Great Horsetail

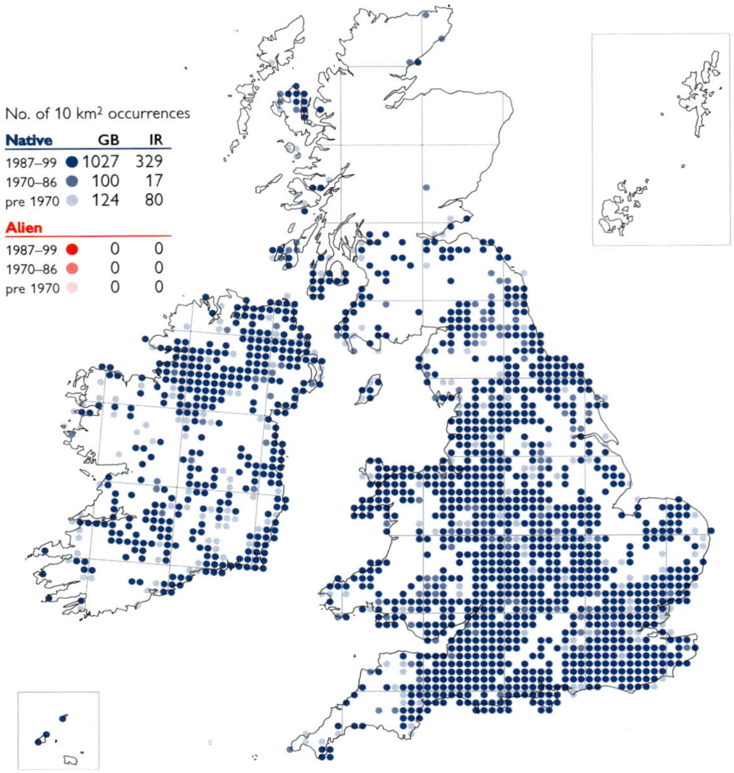

A robust, deciduous, colony-forming herb of base-rich clay soils in sites with spring-lines, permanent seepages and open flushes, especially in areas where porous rocks are interbedded with clays. It prefers open habitats and is particularly frequent on eroding sea- and river-cliffs, but also grows on roadsides and railway banks. Lowland to 365 m in Fossdale (N.W. Yorks.).

Native (change +0.41). This species has apparently become more frequent in inland parts of S.E. England in the last 200 years (Kent, 1975). There is, however, little evidence for any appreciable change since the 1962 *Atlas*.

European Southern-temperate element; also in western N. America (subsp. *braunii*).

References: Atlas (5a), Hultén & Fries (1986), Jalas & Suominen (1972), Jermy *et al.* (1978), Meusel *et al.* (1965), Page (1997).

C. DIXON & T. D. DINES

Ophioglossum vulgatum Adder's-tongue

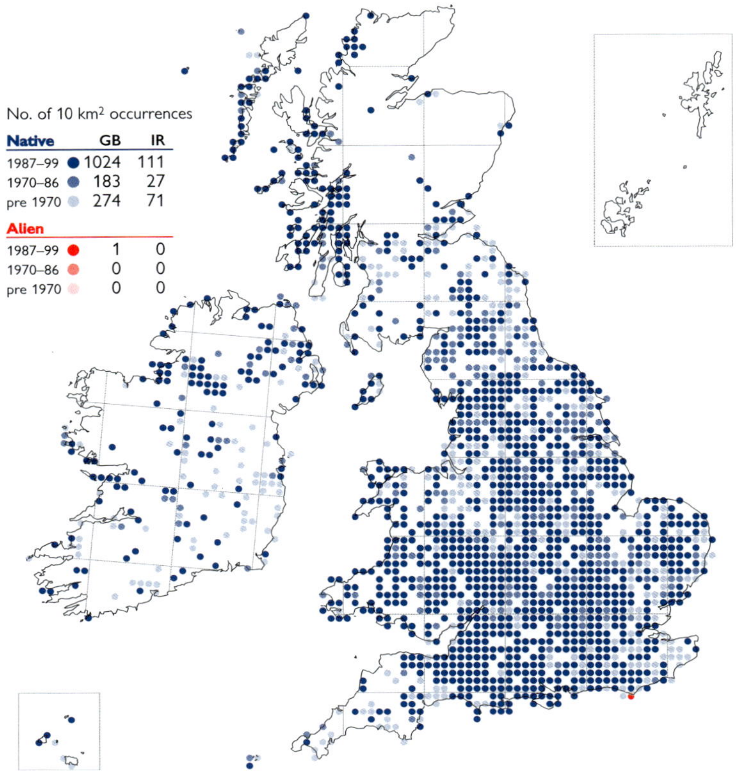

A rhizomatous, deciduous fern found on mildly acidic to base-rich soils in open woodland, meadows and damp pastures, and on sand dunes, under *Pteridium* on heaths, and on peat in regularly mown fen. 0–660 m (Burnhope Seat, Cumberland).

Native (change +0.72). *O. vulgatum* is an inconspicuous species which is very much better recorded than in the 1962 *Atlas*. It has been lost from many lowland sites where the intensification of agriculture, grazing and drainage have contributed to its decline.

Circumpolar Temperate element, with a disjunct distribution.

References: Atlas (16a), Hultén & Fries (1986), Jalas & Suominen (1972), Jermy *et al.* (1978), Meusel *et al.* (1965), Page (1997).

A. C. JERMY

Ophioglossum azoricum Small Adder's-tongue

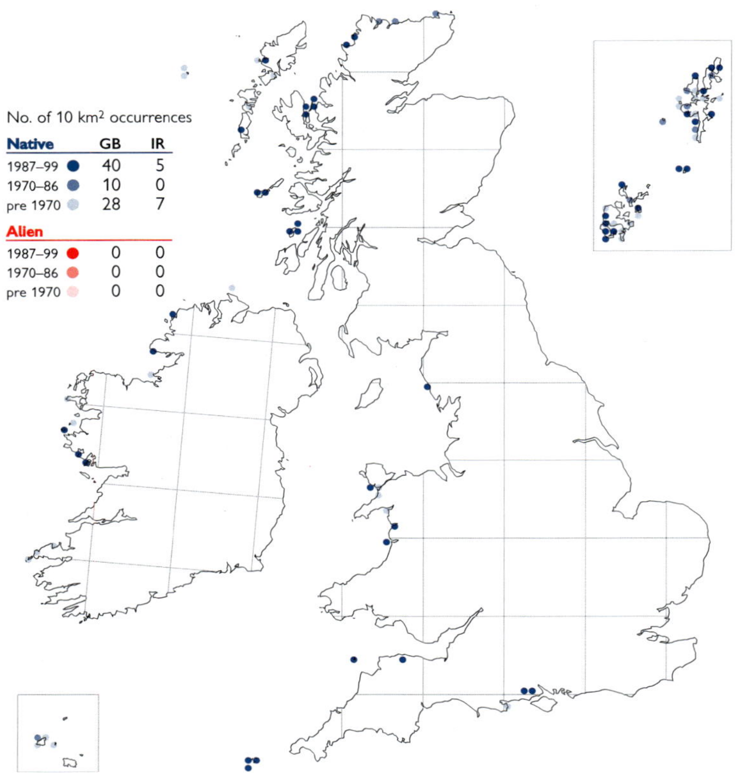

A small rhizomatous, deciduous fern of gently sloping grassland, cliff-tops, damp dune-slacks and sandy maritime heaths on both acidic and alkaline soils. Most sites are frost-free situations near to and facing the sea, with exceptions in the New Forest (S. Hants.) where it grows in highly-grazed damp grassland. Lowland.

Native. In the past small plants of *O. vulgatum* have been mis-identified as *O. azoricum*, and this has led to an unjustified impression that the species is declining. Although the number of sites is low, populations or single clones can spread over many square metres.

Suboceanic Boreo-temperate element.

References: Atlas Supp. (4a), Jalas & Suominen (1972), Jermy *et al.* (1978), Page (1997), Paul (1987), Stewart *et al.* (1994).

A. C. JERMY

Ophioglossum lusitanicum Least Adder's-tongue

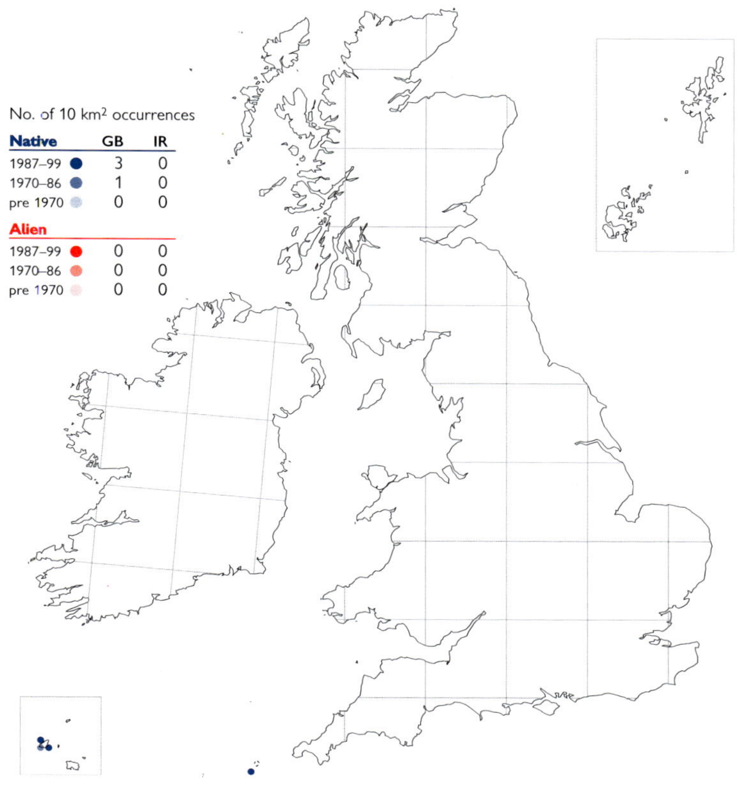

A small, rhizomatous, summer-deciduous fern, growing in open therophyte communities and parched acidic grassland on sea-cliffs and rock promontories. It prefers thin peaty soils, but is also found over shallow blown sand over acidic rocks. All sites are unshaded and exposed, but are warm and S.- or S.W.-facing. Lowland.

Native. *O. lusitanicum* was first discovered in Guernsey in 1853, but not found on the Isles of Scilly until 1950. In Guernsey, many sites have been lost to encroachment by *Ulex europaeus* owing to under-grazing.

Mediterranean-Atlantic element; also in C. Asia.

References: Atlas (16b), Bolòs & Vigo (1984), Jalas & Suominen (1972), Jermy *et al.* (1978), McClintock (1975), Page (1997), Wigginton (1999).

A. C. JERMY

Botrychium lunaria Moonwort

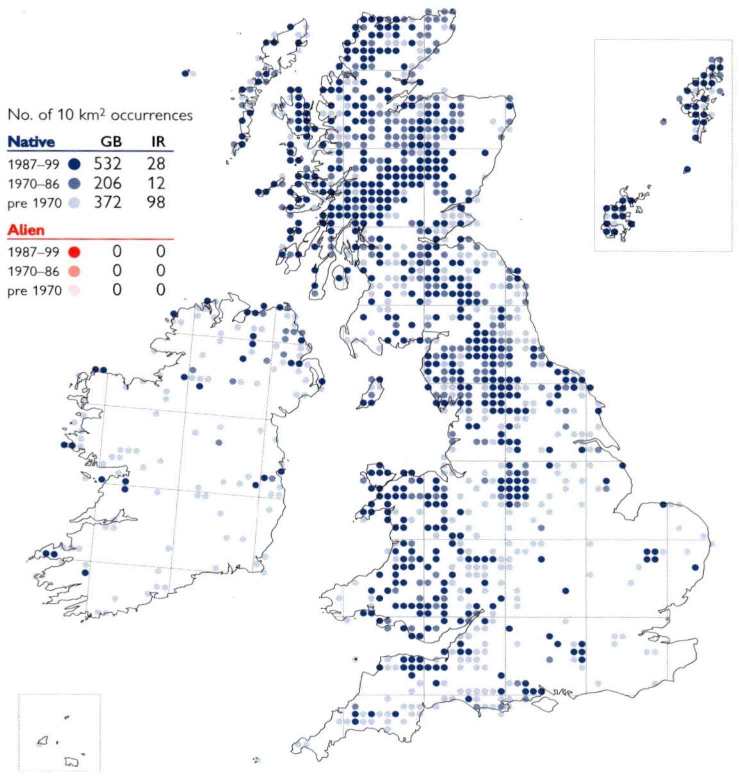

A small fern, often occurring singly or in small populations. It prefers well-drained sites, usually with a high base-content, although it can occur on more acidic substrates. Habitats include meadows, pastures, open woodland, sand dunes and grassy rock ledges. It can also colonise slag heaps and quarry spoil. 0–1065 m (Ben Lawers, Mid Perth).

Native (change –0.43). *B. lunaria* was lost from many lowland sites before 1930, and this loss has continued, particularly in N. England, due to grassland improvement and scrub invasion. There appear to have been some losses in upland areas, but the species can easily be overlooked and may therefore be under-recorded.

Circumpolar Boreo-temperate element.

References: Atlas (15d), Hultén & Fries (1986), Jalas & Suominen (1972), Jermy *et al.* (1978), Meusel *et al.* (1965), Page (1997).

A. C. JERMY

Osmunda regalis Royal Fern

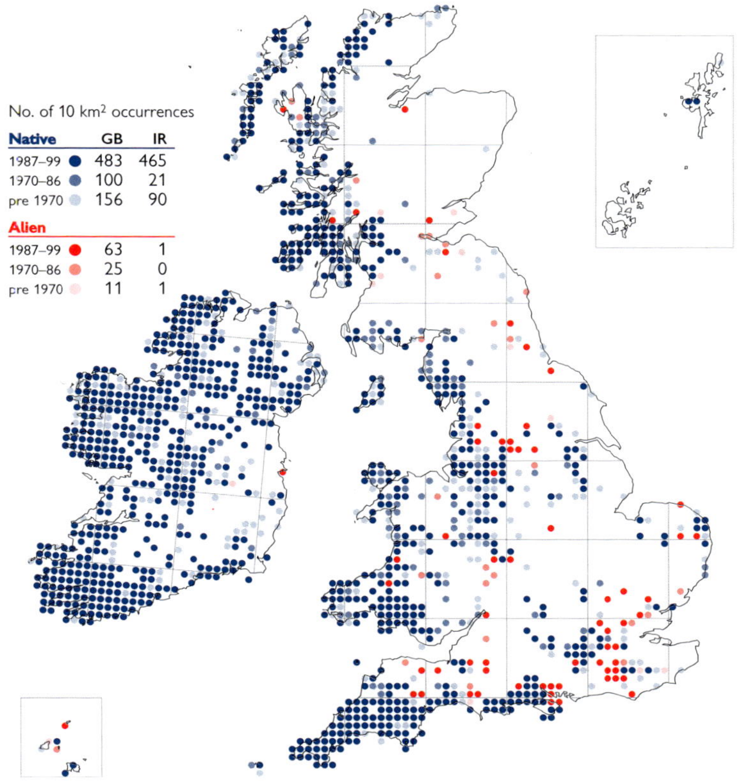

A large fern found on neutral or acidic substrates in fen-carr woodland and ditches, and on river banks and rocky lake shores. In W. Ireland it also grows in wet fields, mires and, more rarely, on limestone sea-cliffs. It is often confined to inaccessible sites in grazed areas. Lowland to 365 m in S. Kerry.

Native (change +0.56). *O. regalis* was heavily collected in Victorian times for cultivation and osmunda fibre. This, and habitat loss, caused its decline, though it is now recovering in some areas and few sites have been lost since the 1962 *Atlas*. It is also planted and occurs as a garden escape.

Suboceanic Southern-temperate element; also in E. Asia and N. America.

References: Atlas (5b), Hultén & Fries (1986), Jalas & Suominen (1972), Jermy *et al.* (1978), Marren (1999), Meusel *et al.* (1965), Page (1997).

R. J. COOKE

Cryptogramma crispa Parsley Fern

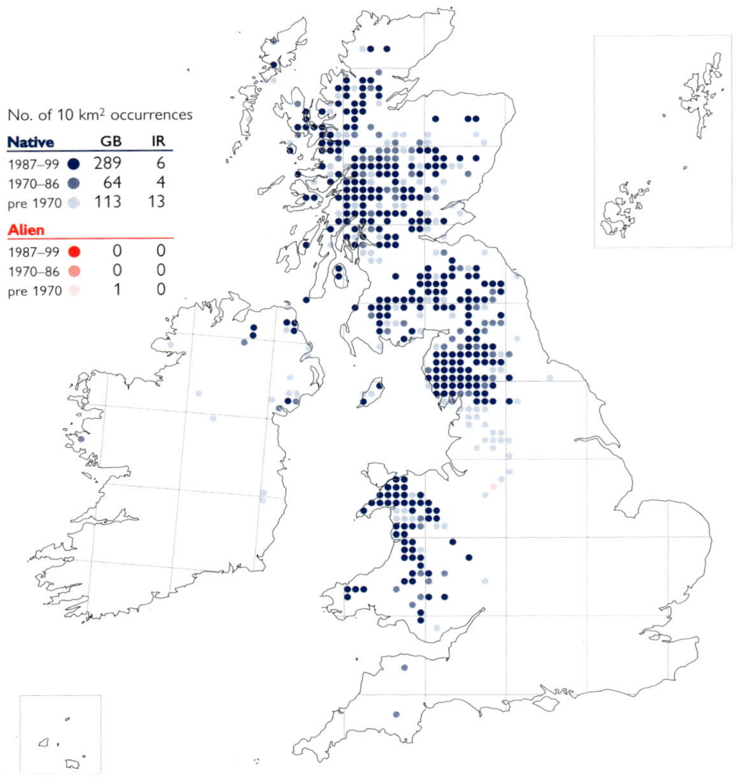

This small, deciduous, long-lived fern is a strong calcifuge and is found in well-drained sites on relatively stable, steep scree slopes, where it is a pioneer species. It also occurs on cliff ledges and mortar-free dry-stone walls. From 80 m (Glen Etive, Main Argyll) to 1280 m (Ben Nevis, Westerness).

Native (change –0.63). Losses in the S. Pennines took place before 1930, and there is little evidence for a change in distribution since the 1962 *Atlas*.

European Boreal-montane element; also in E. Asia.

References: Atlas (6c), Curtis & McGough (1988), Hultén & Fries (1986), Jalas & Suominen (1972), Jermy *et al.* (1978), Meusel *et al.* (1965), Page (1997).

T. D. DINES

Anogramma leptophylla Jersey Fern

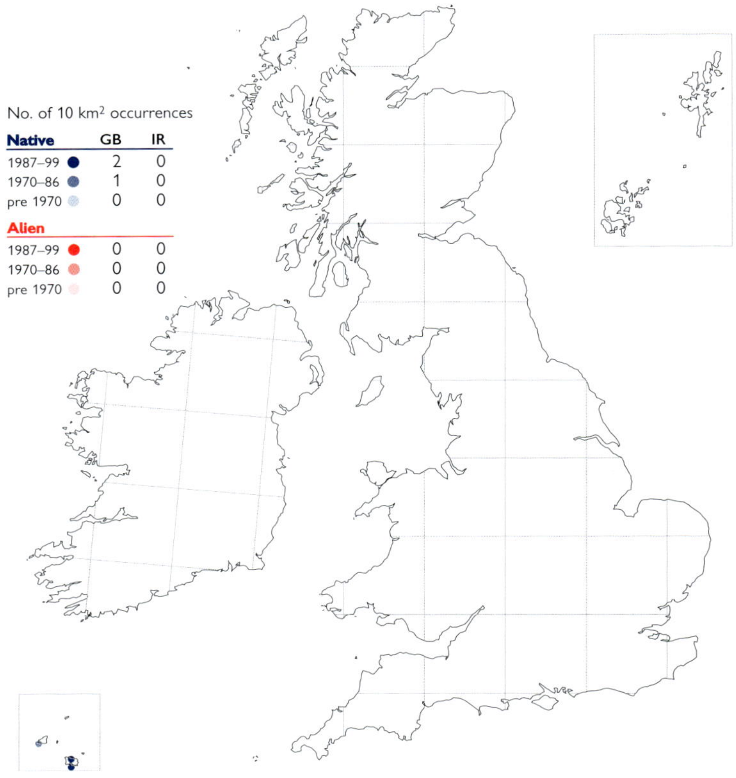

A small fern found on moist but well-drained shady lane banks, especially where granite is used to support the bank. It prefers bare soil where some surface erosion reduces competition. It is the only British fern with an annual sporophyte; its spores mature early (usually April) and plants die soon afterwards. Its prothallus, however, is perennial and may over-winter in warm crevices to produce new sporophytes the following year. Lowland.

Native. Populations of *A. leptophylla* vary in size yearly. They are highly vulnerable to the scraping of lane banks by passing traffic and to over-enthusiastic hedge bank management.

Mediterranean-Atlantic element; widespread in N. America.

References: Atlas (6c), Bolòs & Vigo (1984), Jalas & Suominen (1972), Jermy *et al.* (1978), Page (1997).

A. C. JERMY

Adiantum capillus-veneris Maidenhair Fern

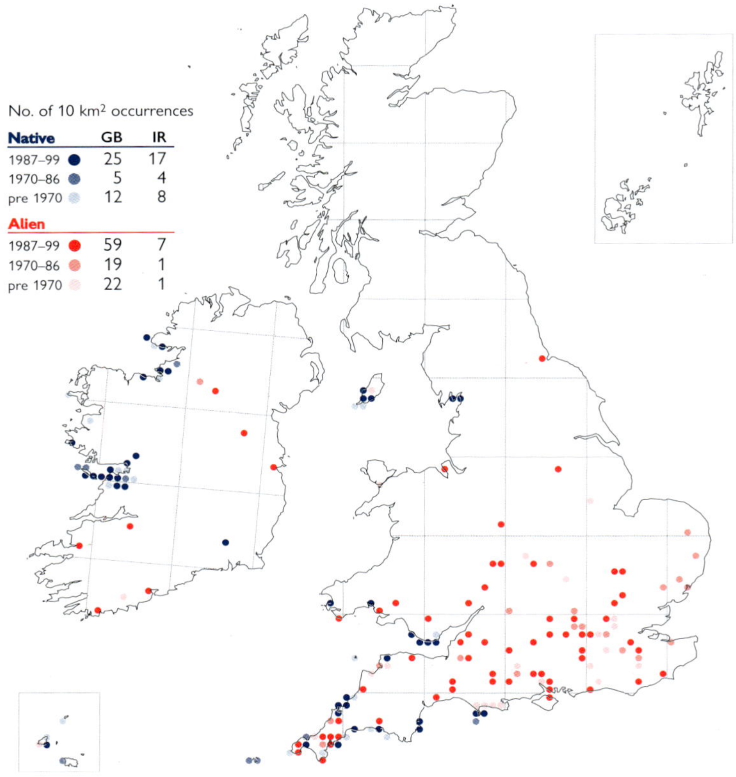

A semi-evergreen fern found in areas with an oceanic climate on wet, calcareous cliffs where its rhizomes are protected in crevices; in the Aran Islands (W. Donegal) and the Burren (Co. Clare) it grows in grikes in limestone pavement. Many inland records in sheltered warm sites, such as damp mortared walls, railway sidings and canal locks, arise from spores derived from cultivated plants. Lowland.

Native (change +0.54). In the 19th century, native populations were jeopardised by collectors; many pre-1970 sites were lost before 1930. There have been few recent losses, however, and many more alien records.

Mediterranean-Atlantic element; also in C. and E. Asia and N. America.

References: Atlas (6d), Hultén & Fries (1986), Jalas & Suominen (1972), Jermy *et al.* (1978), Page (1997), Stewart *et al.* (1994).

A. C. JERMY

Pilularia globulifera Pillwort

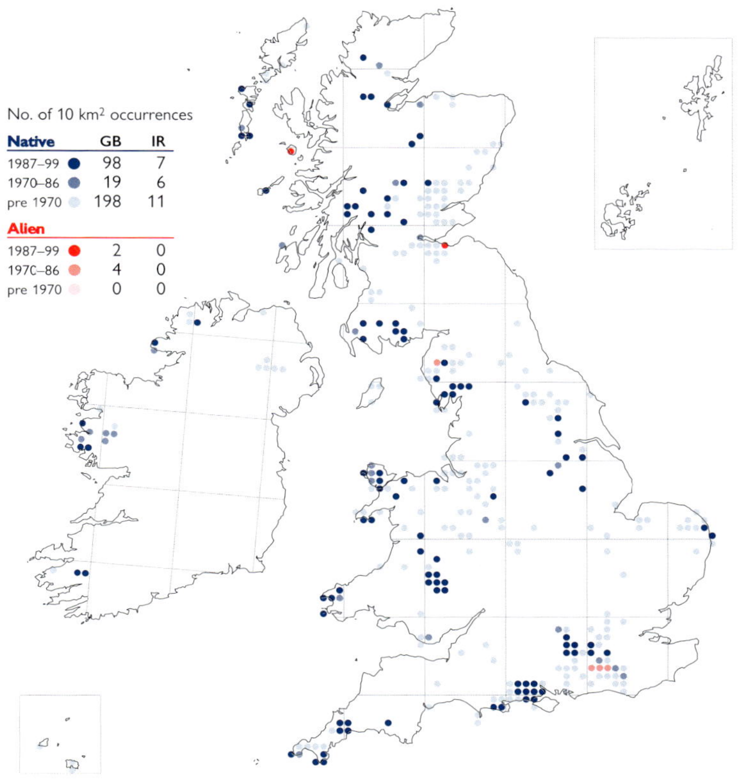

A small, rhizomatous fern growing on the edges of non-calcareous lakes, reservoirs, ponds or slow-flowing rivers, and sometimes on damp mine workings or as a submerged aquatic. It requires areas where competition is reduced by fluctuating water levels or disturbance. 0–450 m (Pant-y-llyn Hill, Brecs.).

Native (change –0.03). *P. globulifera* was lost from many sites before 1930 due to habitat destruction. Eutrophication and reduced disturbance have led to further losses in E. Britain and Ireland. In the west, many new sites have been found since 1980. It has been re-introduced to some former native sites (e.g. Rum).

Suboceanic Temperate element.

References: Atlas (15b), Curtis & McGough (1988), Hultén & Fries (1986), Jalas & Suominen (1972), Jermy *et al.* (1978), Page (1997), Preston & Croft (1997), Scott *et al.* (1999), Stewart *et al.* (1994).

C. D. PRESTON

Hymenophyllum tunbrigense Tunbridge Filmy-Fern

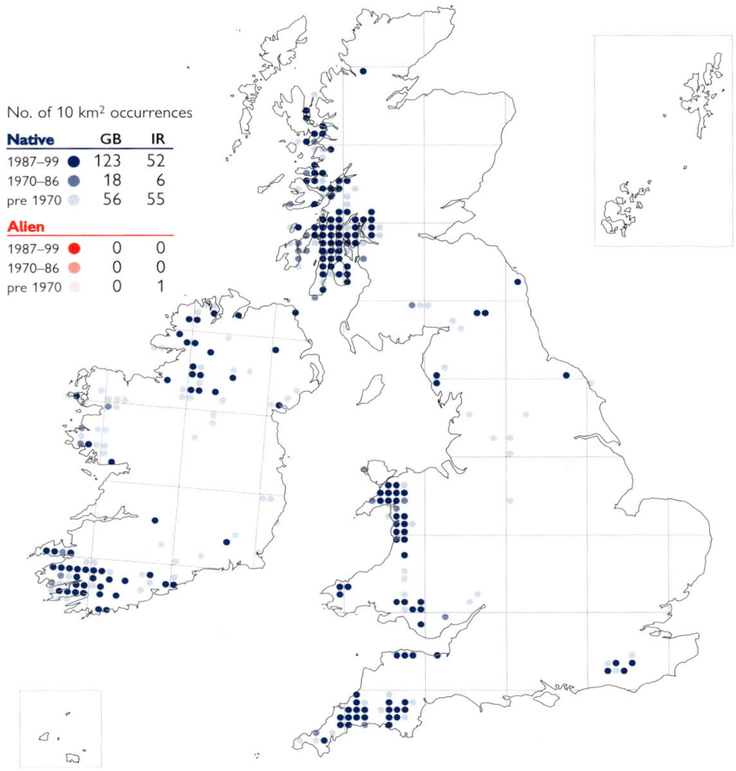

A rhizomatous perennial fern of very sheltered, often deeply shaded, humid habitats; these include acidic rock faces, humic banks and tree trunks, particularly in deep stream valleys, and crevices on upland boulder scree. 0–760 m (Galtee Mountains, Co. Tipperary).

Native (change –0.54). The distribution of *H. tunbrigense* is largely stable, although until recently many small populations have been overlooked. However, 20% of sites in S.E. England have been lost since 1950, largely through woodland loss and shading by *Rhododendron ponticum*.

Oceanic Temperate element; also one site in N. America.

References: Atlas (5d), Hultén & Fries (1986), Jalas & Suominen (1972), Jermy *et al.* (1978), Meusel *et al.* (1965), Page (1997), Rich *et al.* (1995), Richards & Evans (1972).

F. J. RUMSEY

Hymenophyllum wilsonii Wilson's Filmy-Fern

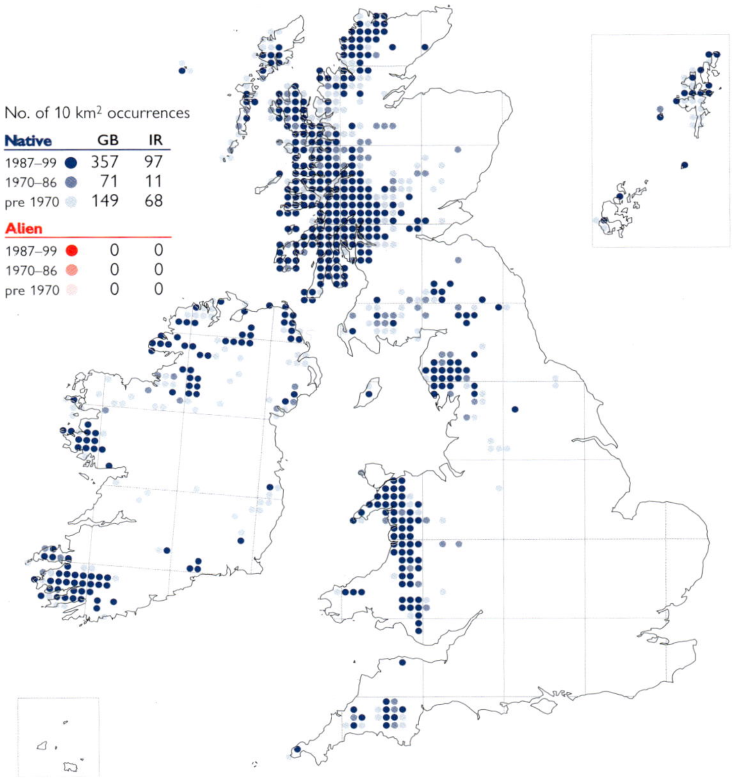

A rhizomatous perennial fern, forming dense colonies on a variety of substrates, including sheltered acidic or, rarely, mildly basic rocks, and trees in humid sites. It also occurs on damp upland cliffs, boulder scree and, rarely, old walls. 0–1005 m (Macgillycuddy's Reeks, S. Kerry).

Native (change –0.87). The distribution of *H. wilsonii* is largely stable, although there have apparently been losses since the 1962 *Atlas*, particularly in upland areas of C. and E. Scotland. Some of this may be due to under-recording.

Oceanic Boreo-temperate element; confined to the hyperoceanic zone of W. Europe and Macaronesia.

References: Atlas (6a), Hultén & Fries (1986), Jalas & Suominen (1972), Jermy *et al.* (1978), Page (1997), Richards & Evans (1972).

F. J. RUMSEY

Trichomanes speciosum (sporophyte)

Killarney Fern

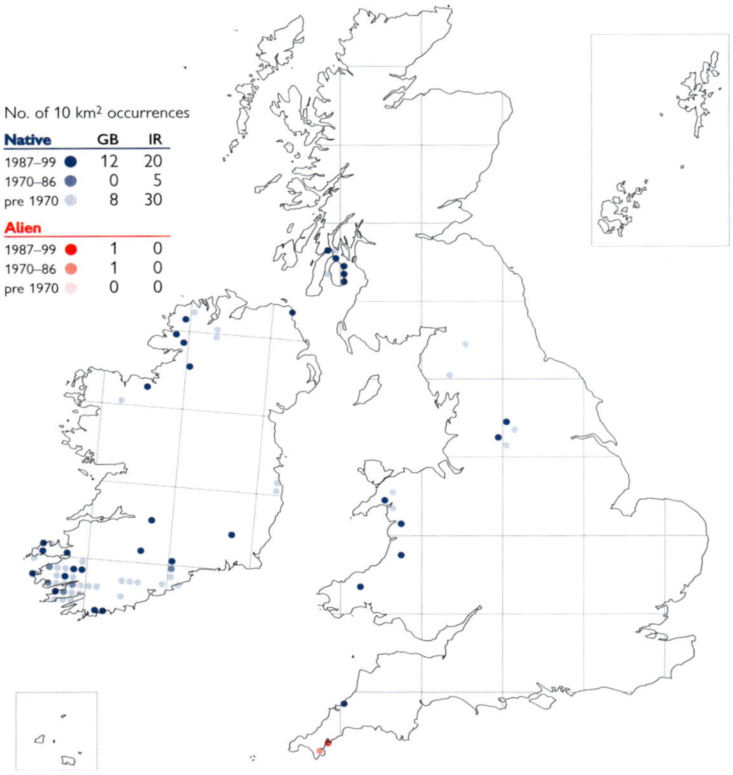

A rhizomatous fern, restricted to humid, winter-warm sites. The sporophyte occurs only in constantly damp, shaded localities, usually on acidic, but often base-flushed rocks, rarely on damp humic banks, and exceptionally as an epiphyte. 0–420 m (Caerns.).

Native. This species has declined due to collecting and habitat disturbance. One site has been lost since the 1962 *Atlas*, but recent fieldwork has revealed several new populations. Other extant sites in S.W. Wales, Cumbria and N.E. Yorkshire are not mapped here, and there are rumours of other sites elsewhere.

Oceanic Temperate element; confined to Macaronesia, W. Europe and N. Italy.

References: Atlas (5c), Curtis & McGough (1988), Jalas & Suominen (1972), Jermy *et al.* (1978), Page (1997), Wigginton (1999).

F. J. RUMSEY

Trichomanes speciosum (gametophyte)

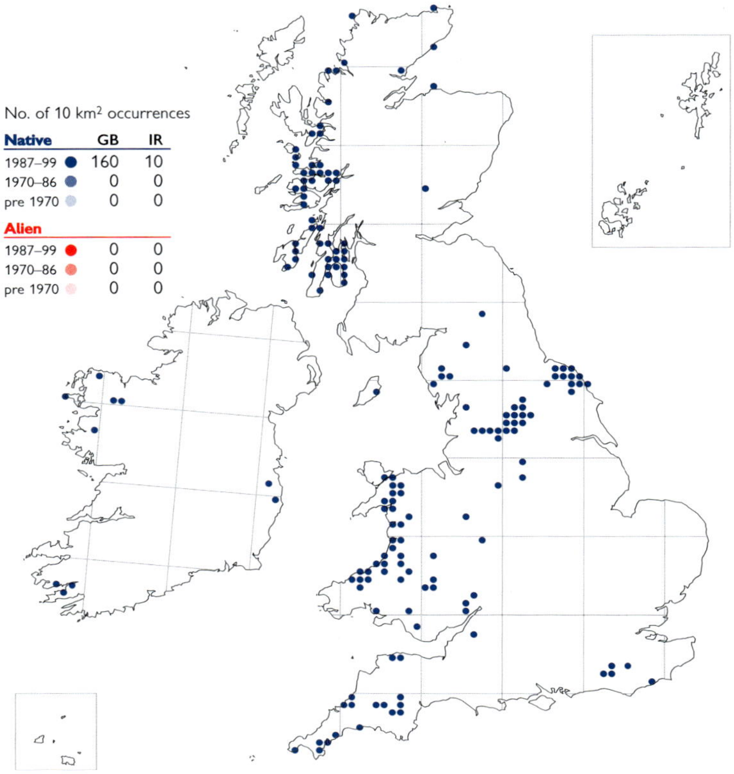

The gametophyte of *T. speciosum* grows deep in clefts, crevices and natural rock hollows on a range of acidic to neutral rocks. Such sites are dark (less than 1% ambient light) and are often humid, being located on sea-cliffs, river-cliffs or streamsides, or are kept damp through soil capillary action. 0–530 m (Moel yr Ogof, Caerns.).

Native (change for species +2.23). Although first described from cultivated material in 1888, the gametophyte was overlooked in the wild until 1989, when it was identified in N. England. Recent fieldwork has revealed an extensive, but still under-recorded, distribution.

Gametophytes occur within the European range of the sporophyte, but also extend eastwards to C. Europe.

References: Page (1997), Rumsey *et al.* (1990, 1998), Wigginton (1999).

F. J. RUMSEY

Polypodium vulgare sens. lat. Polypodies

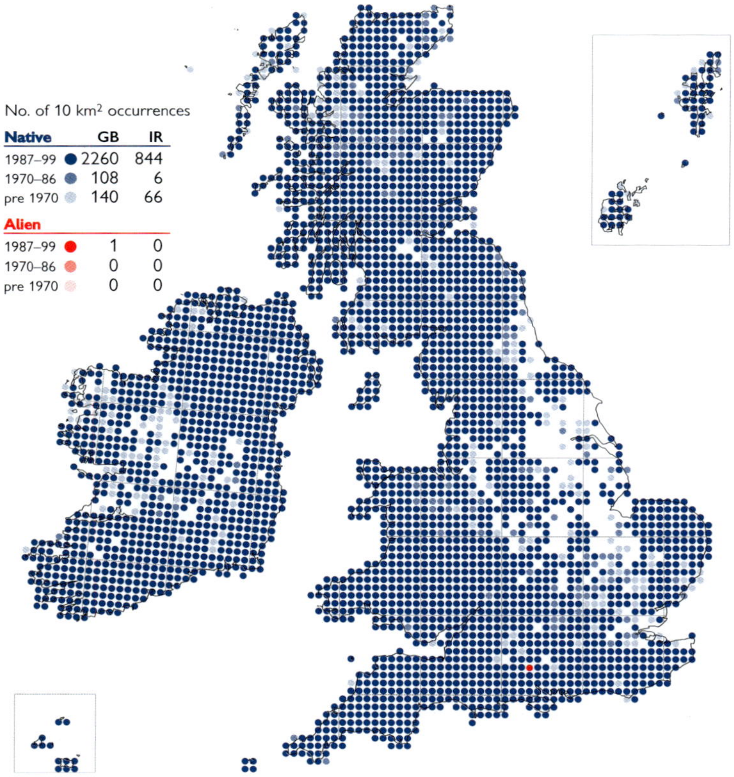

These perennial rhizomatous ferns are found on a wide variety of natural and artificial rocky substrates. They also grow as epiphytes, especially in W. Britain and Ireland. 0–850 m (Breadalbanes, Mid Perth, and Mt Brandon, S. Kerry), possibly higher in W. Ross.

Native (change –0.03). The *P. vulgare* aggregate mapped here consists of diploid *P. cambricum*, tetraploid *P. vulgare* and hexaploid *P. interjectum*, and their hybrids. Although the segregates were described by Shivas (1960) and mapped by Perring & Sell (1968) and Jermy *et al.* (1978), they require microscopic examination for certain identification and are still under-recorded.

The aggregate has a Circumpolar Boreo-temperate distribution.

References: Atlas (15a), Hultén & Fries (1986), Jalas & Suominen (1972), Meusel *et al.* (1965), Page (1997).

R. J. COOKE

Polypodium vulgare Polypody

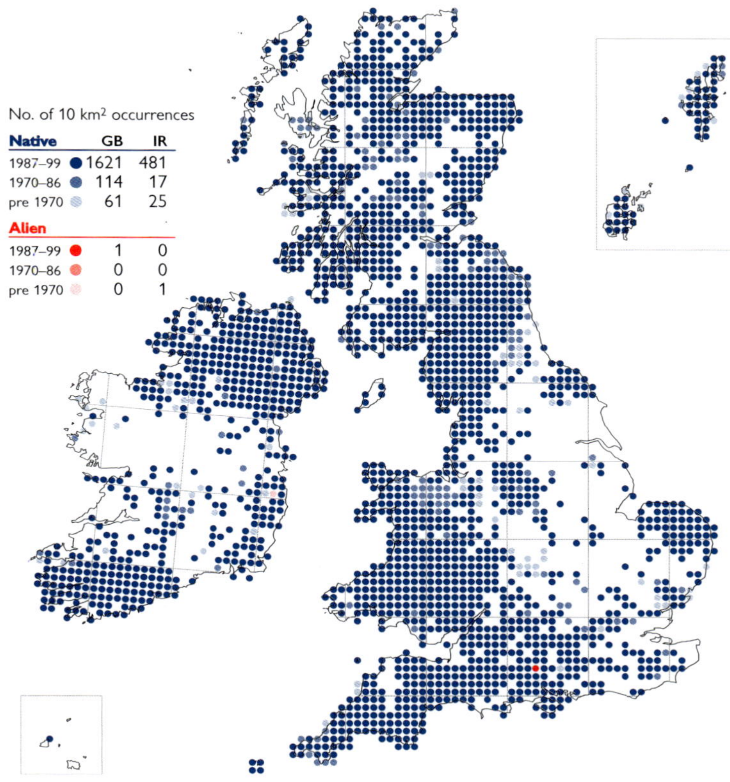

An evergreen, perennial, rhizomatous fern of well-drained, predominantly acidic substrates, including dry-stone walls, roadside banks and rock outcrops. It also occurs as an epiphyte on *Quercus* and other deciduous trees, mainly in W. Britain and Ireland, and is also found in conifer plantations. It is very tolerant of exposure, growing, for example, on montane scree. 0–760 m (Beinn na Socaich, Westerness).

Native. There is little evidence for any change in the distribution of *P. vulgare sens. str.* It is probably under-recorded in Ireland and in some areas of Scotland but is rarer than *P. interjectum* in S.E. England.

European Boreo-temperate element.

References: Atlas Supp. (3a), Jermy *et al.* (1978), Page (1997).

R. J. COOKE

Polypodium interjectum Intermediate Polypody

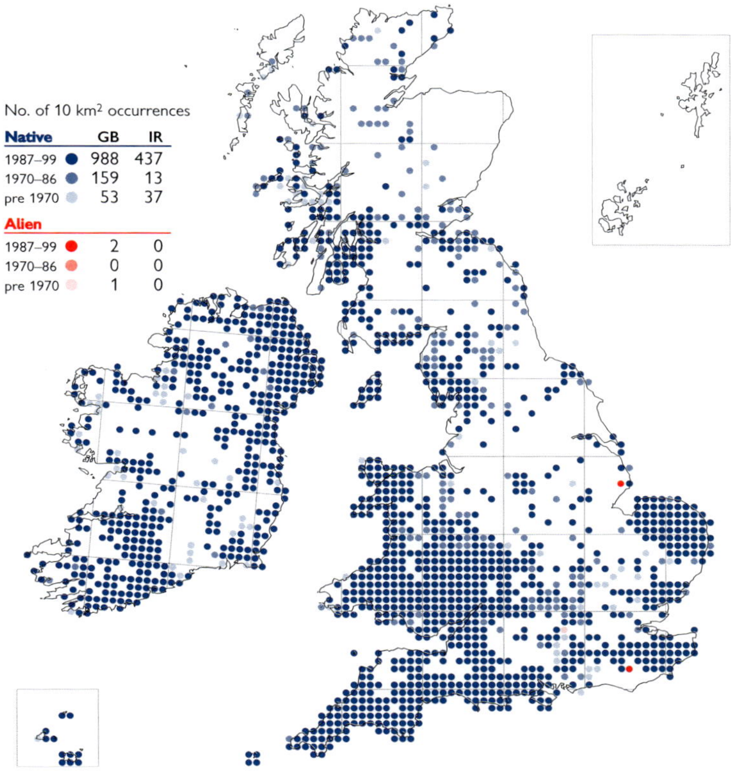

An evergreen, perennial, rhizomatous fern that prefers more basic substrates than *P. vulgare* but can be found in acidic conditions where exposed to salt-laden air. It is found in a wide range of habitats such as mortared stone walls, hedge banks, rock exposures, mature sand dunes and as an epiphyte, especially near the sea. 0–820 m (Creag Meagaidh, Westerness).

Native. The increased number of records of this species since it was mapped by Perring & Sell (1968) and Jermy *et al.* (1978) is due to better recording and not to any real increase in its distribution. It is probably still under-recorded.

Suboceanic Temperate element.

References: Atlas Supp. (3c), Jalas & Suominen (1972), Page (1997).

R. J. COOKE

Polypodium interjectum × *P. vulgare* (*P.* × *mantoniae*)

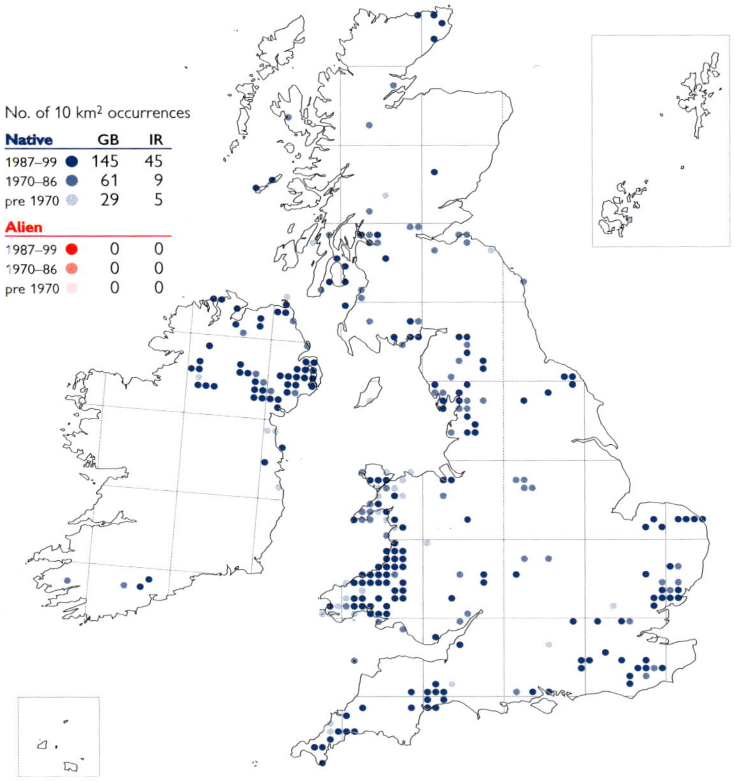

This hybrid is found in the habitats of its parents, on both acidic and basic substrates, including rocks (limestone, granite, sandstone and millstone grit), roadside banks and walls, and on trees in open coastal woodland. It is sterile but can be very vigorous vegetatively, often forming extensive colonies. Generally lowland, but reaching 320 m at Logan Burn (Midlothian).

Native. *P.* × *mantoniae* is a frequent hybrid and is likely to be found throughout the range of the parents; it is probably still greatly under-recorded.

Apparently frequent in those parts of Europe where both parents occur.

References: Jermy *et al.* (1978), Page (1997), Stace (1975).

R. J. COOKE

Polypodium cambricum Southern Polypody

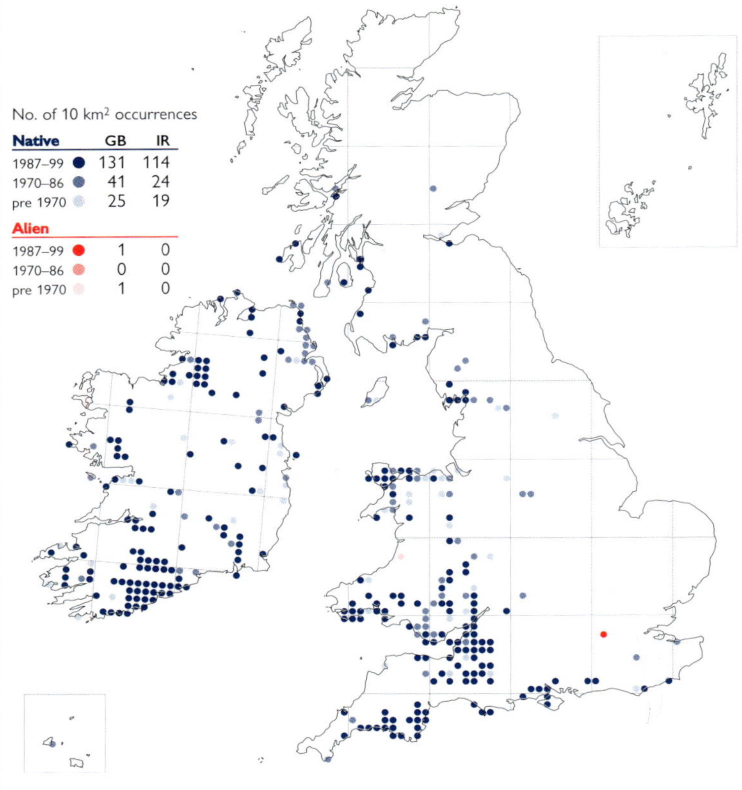

No. of 10 km² occurrences

Native		GB	IR
1987–99	●	131	114
1970–86	●	41	24
pre 1970	●	25	19
Alien			
1987–99	●	1	0
1970–86	●	0	0
pre 1970	●	1	0

A perennial, rhizomatous fern of well-drained base-rich rocky substrates, often found on sheltered limestone cliffs, old quarry faces, castle walls built of limestone, and on old mortared walls. It is also found as an epiphyte, especially in Ireland, and on road banks. 0–460 m (Farreg-y-fran, Merioneth, and Malham Tarn, Mid-W. Yorks.).

Native. Many sites for *P. cambricum* have been found since it was mapped by Perring & Sell (1968). Some populations have been lost to scrub encroachment and masonry cleaning.

Mediterranean-Atlantic element; it reaches its northern limit at Lismore (Main Argyll).

References: Atlas Supp. (3b), Hultén & Fries (1986), Jalas & Suominen (1972), Jermy *et al.* (1978), Meusel *et al.* (1965), Page (1997), Stewart *et al.* (1994).

R. J. COOKE

Pteridium aquilinum Bracken

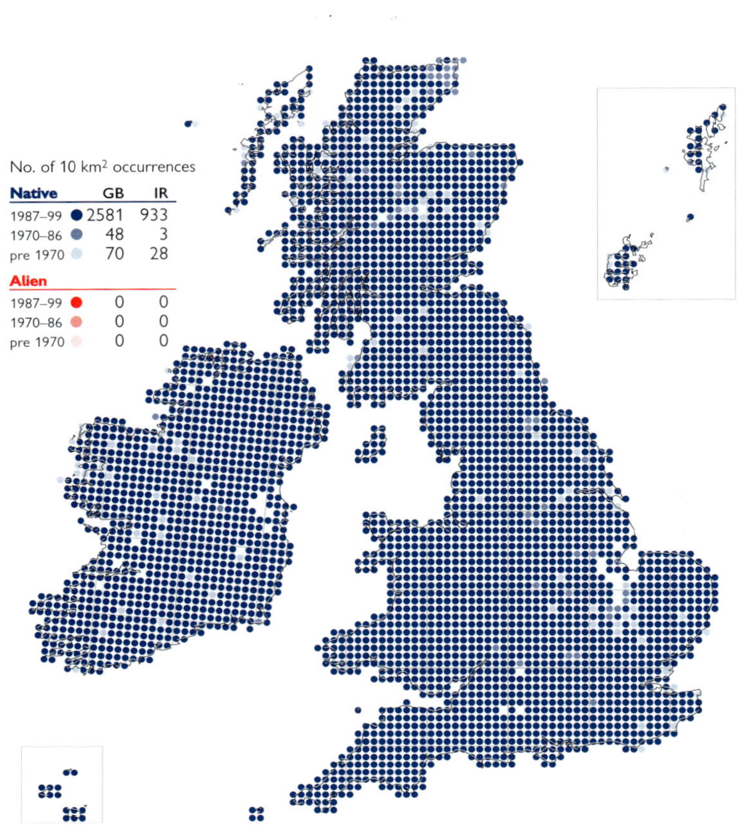

No. of 10 km² occurrences

Native		GB	IR
1987–99	●	2581	933
1970–86	●	48	3
pre 1970	●	70	28
Alien			
1987–99	●	0	0
1970–86	●	0	0
pre 1970	●	0	0

A deciduous fern of moorland, hill pasture and other habitats on acidic soils. It is most vigorous when growing on deep loam, sands or alluvium and is rare on base-rich soils. 0–585 m (Lochnagar, S. Aberdeen), and probably higher elsewhere.

Native (change –0.71). *P. aquilinum* increased markedly in the 20th century, apparently in response to more intensive sheep grazing and more frequent burning of hill vegetation. In the 1970s it was invading more than 10 000 hectares of agricultural land annually. This expansion is not apparent at the 10-km scale as the species was almost ubiquitous in the 1962 *Atlas*.

Circumpolar Temperate element.

References: Atlas (6b), Cody & Crompton (1975), Grime *et al.* (1988), Hultén & Fries (1986), Jalas & Suominen (1972), Jermy *et al.* (1978), Meusel *et al.* (1965), Page (1982, 1997), Smith & Taylor (1995).

A. C. JERMY

Thelypteris palustris Marsh Fern

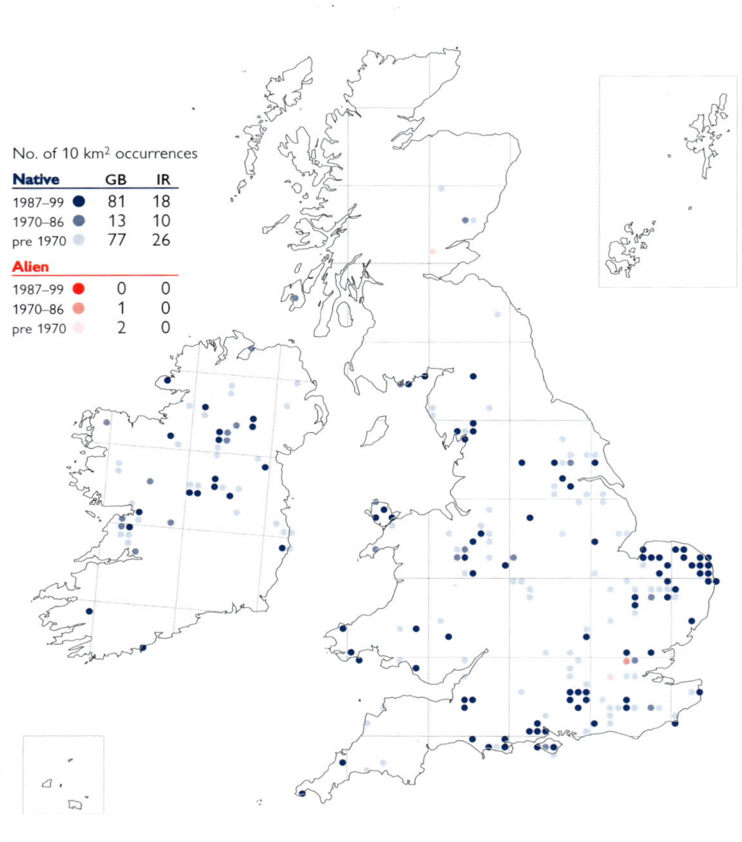

No. of 10 km² occurrences

Native		GB	IR
1987–99	●	81	18
1970–86	●	13	10
pre 1970	●	77	26
Alien			
1987–99	●	0	0
1970–86	●	1	0
pre 1970	●	2	0

A perennial fern of open or recently wooded fen or open carr, where the soil is permanently wet and organic, but not too acidic. It is a characteristic component of *Phragmites-Cladium* fen, but also persists as vigorous colonies in fen *Alnus* woods or *Salix* carr. Generally lowland, but formerly at 335 m (Braemar, S. Aberdeen).

Native (change –0.35). *T. palustris* declined before 1930 due to drainage, but it can be remarkably tenacious where natural succession has occurred, and has been re-found in several of its stations after many decades. There have been few losses since the 1962 *Atlas*.

Circumpolar Temperate element, with a disjunct distribution.

References: Atlas (14a), Hultén & Fries (1986), Jalas & Suominen (1972), Jermy *et al.* (1978), Page (1997), Stewart *et al.* (1994).

R. J. COOKE

Phegopteris connectilis Beech Fern

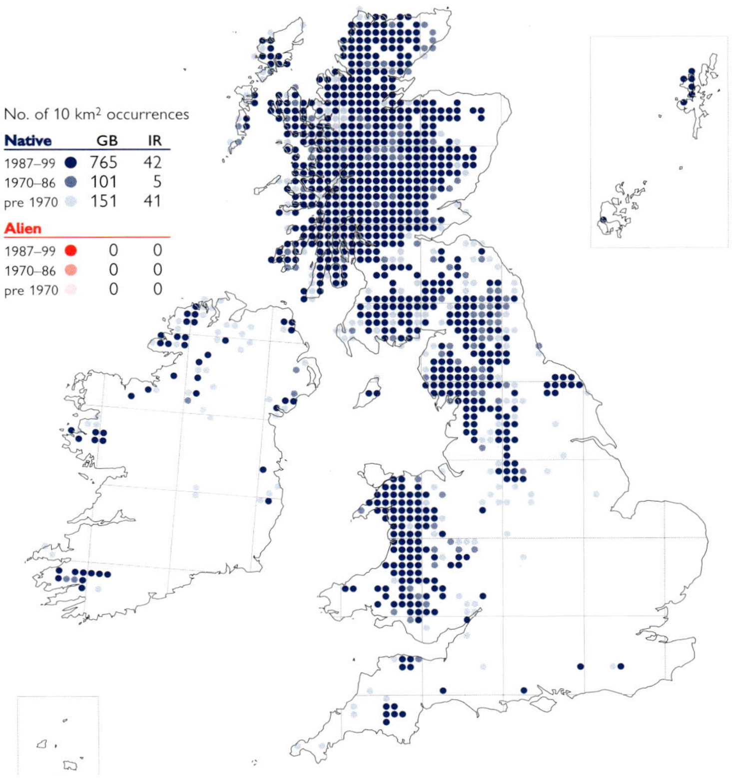

No. of 10 km² occurrences

Native		GB	IR
1987–99	●	765	42
1970–86	●	101	5
pre 1970	●	151	41
Alien			
1987–99	●	0	0
1970–86	●	0	0
pre 1970	●	0	0

A creeping, rhizomatous fern, most common in ancient woodlands dominated by *Quercus petraea* on neutral to acidic soils, where it frequently occurs on deeper soils on gully sides where base-rich water percolates. It can also be found amongst boulders and on wet rock faces in the uplands where it is afforded protection from grazing. 0–1120 m (Breadalbanes, Mid Perth).

Native (change –0.22). The distribution of *P. connectilis* is stable.

Circumpolar Boreo-temperate element.

References: Atlas (14b), Hultén & Fries (1986), Jalas & Suominen (1972), Jermy *et al.* (1978), Meusel *et al.* (1965), Page (1997).

R. J. COOKE

Oreopteris limbosperma Lemon-scented Fern

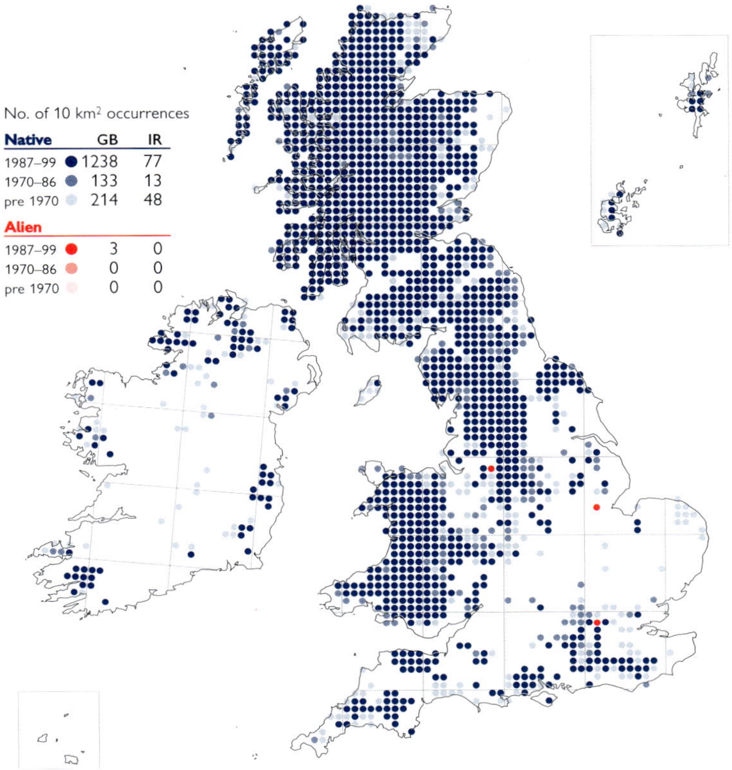

No. of 10 km² occurrences

Native		GB	IR
1987–99	●	1238	77
1970–86	●	133	13
pre 1970	●	214	48
Alien			
1987–99	●	3	0
1970–86	●	0	0
pre 1970	●	0	0

A fern of acidic, peaty or humus-rich soils in open woodland, along drainage ditches and streamsides, and on damp heaths, upland grassland and damp rock ledges. It is especially associated with the edges of watercourses, including man-made ditches, and is therefore more frequent on poorly-drained substrates. 0–1010 m (Ben Ime, Main Argyll).

Native (change –0.18). The distribution of this species in the uplands is stable. Many of the losses in the lowlands occurred before 1930, caused especially by the destruction of heathland.

European Temperate element.

References: Atlas (13d), Hultén & Fries (1986), Jalas & Suominen (1972), Jermy *et al.* (1978), Meusel *et al.* (1965), Page (1997).

T. D. DINES

Phyllitis scolopendrium Hart's-tongue

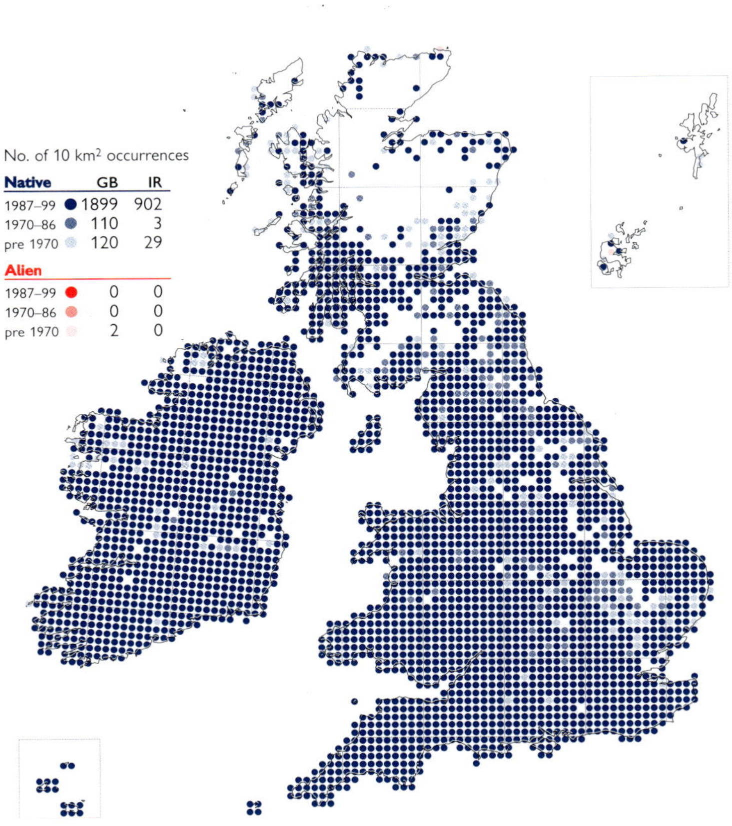

No. of 10 km² occurrences

Native		GB	IR
1987–99	●	1899	902
1970–86	●	110	3
pre 1970	●	120	29
Alien			
1987–99	●	0	0
1970–86	●	0	0
pre 1970	●	2	0

An evergreen perennial fern of sheltered, humid, moist habitats, including rocky woodlands, stream and hedge banks, grikes in limestone pavement, and on brickwork and walls, where it often grows in a stunted form. It avoids the most acidic substrates. 0–700 m (Great Dun Fell, Westmorland).

Native (change +0.45). The range of this species appears to be unchanged since the 1962 *Atlas*, where it was mapped as 'all records', although it is now better recorded. Although treated here as *Phyllitis scolopendrium*, it is clear from molecular evidence that the genus *Phyllitis* should be subsumed into *Asplenium*.

European Temperate element; also in E. Asia and N. America.

References: Atlas (7b), Hultén & Fries (1986), Jalas & Suominen (1972), Jermy *et al.* (1978), Meusel *et al.* (1965), Page (1997).

F. J. RUMSEY

Asplenium adiantum-nigrum Black Spleenwort

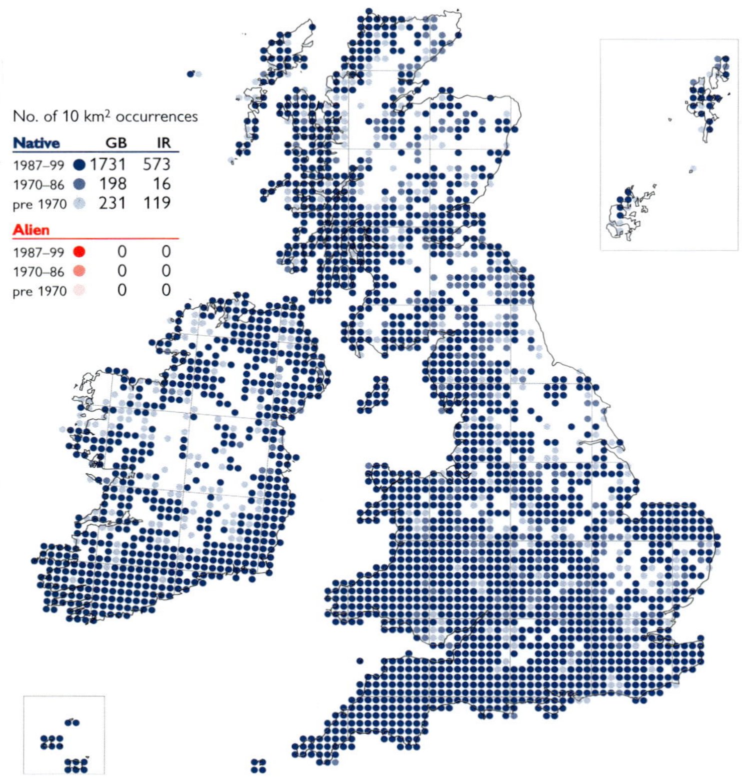

This evergreen perennial fern occurs on a wide range of well-drained, usually basic substrates, in lightly shaded habitats where there is little competition. It is found on cliffs and screes, in quarries, on lane banks and walls. Generally lowland, but reaching 575 m at Moor House (Westmorland) and possibly higher in the Cairngorms.

Native (change +0.35). The distribution of this species appears to be stable. Several variants, one previously confused with the C. European *A. cuneifolium*, are present in the British Isles. They cannot be distinguished on purely edaphic grounds (*pace* Page, 1997) and are the subject of ongoing research.

European Temperate element; also in C. Asia and N. America.

References: Atlas (7c), Hultén & Fries (1986), Jalas & Suominen (1972), Jermy *et al.* (1978).

F. J. RUMSEY

Asplenium onopteris Irish Spleenwort

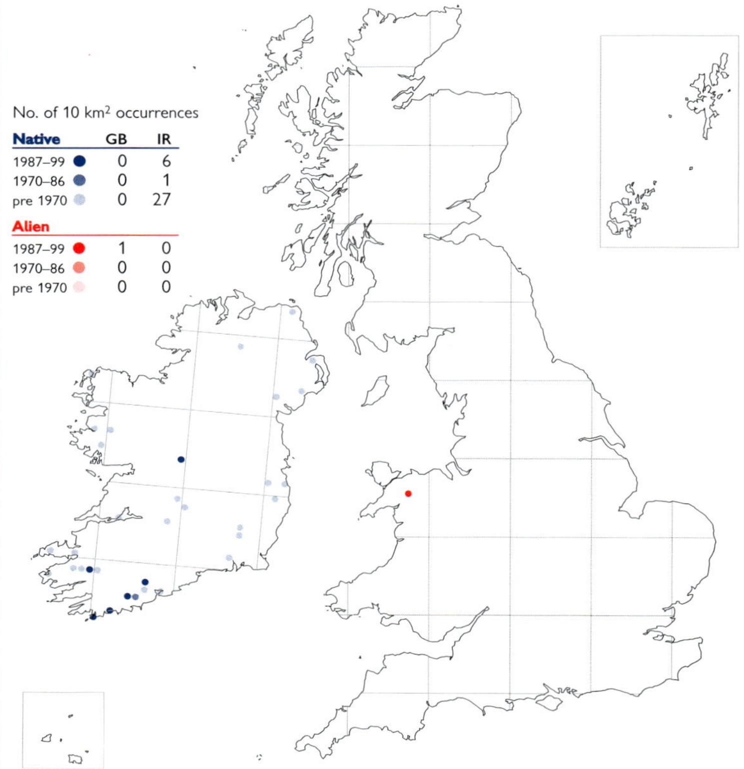

An evergreen perennial fern of dry, warm, lightly shaded, usually basic, earthy banks and rock faces in open deciduous woodland. Lowland.

Native. The distribution of this species needs detailed study. It was treated as part of the *A. adiantum-nigrum* complex in the 1962 *Atlas. A. adiantum-nigrum* (a polyploid derived in part from *A. onopteris*) and *A. onopteris* freely hybridise (*A.* × *ticinense*); hybrid populations, and extreme forms of *A. adiantum-nigrum*, are likely to have produced recording errors.

Mediterranean-Atlantic element; it reaches its northern limit in Ireland where it is morphologically invariable, in contrast to the variation shown in the Mediterranean region.

References: Atlas Supp. (2a), Hultén & Fries (1986), Jalas & Suominen (1972), Jermy *et al.* (1978), Page (1997).

F. J. RUMSEY

Asplenium obovatum Lanceolate Spleenwort

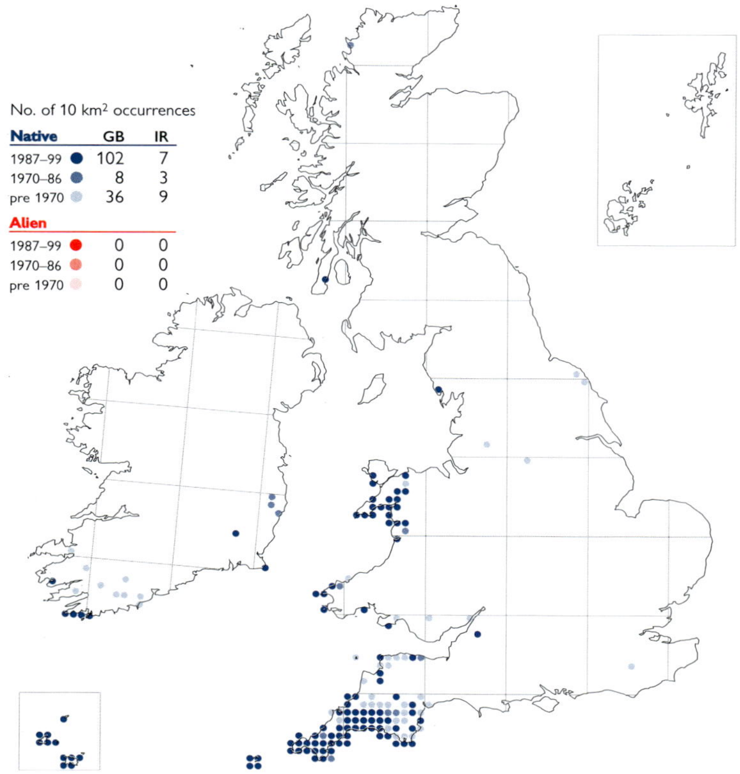

This perennial, evergreen, calcifuge fern is mainly a plant of sheltered, shady crevices and ledges on maritime cliffs and on rock outcrops. It also occurs on well-drained, acidic, loamy lane banks and dry-stone walls. Most of its sites are near the sea, and the plant is not vigorous in its colder inland sites. Lowland.

Native (change −0.18). *A. obovatum* was lost before 1930 from most of the sites which lack a post-1970 record, and its distribution is probably stable.

Mediterranean-Atlantic element.

References: Atlas (7d), Bolòs & Vigo (1984), Curtis & McGough (1988), Jalas & Suominen (1972), Jermy *et al.* (1978), Page (1997), Stewart *et al.* (1994).

F. J. RUMSEY

Asplenium marinum Sea Spleenwort

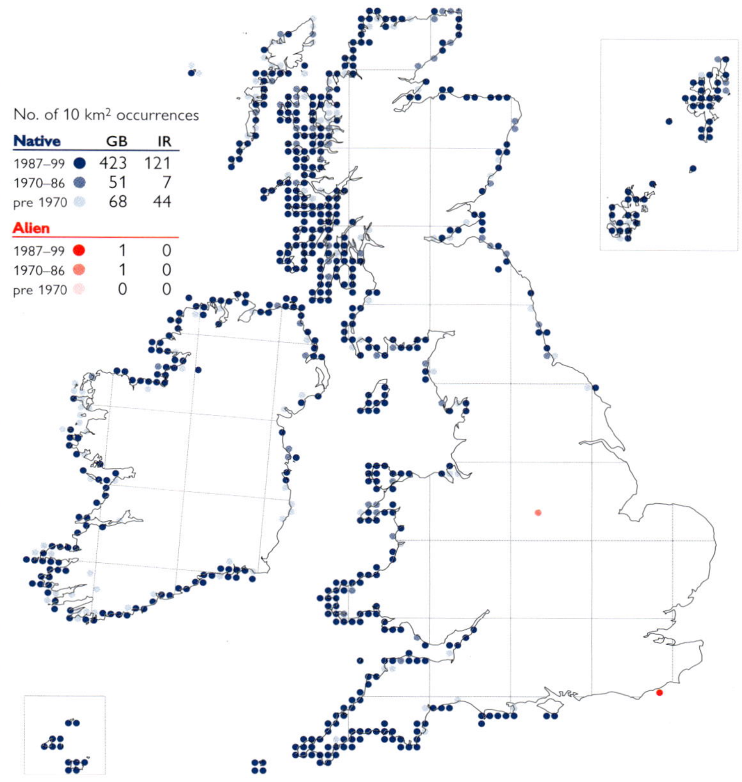

An evergreen perennial fern, predominantly found in cool, moist crevices and fissures in maritime cliffs, and often within range of sea-spray. It occasionally grows on walls in coastal areas, but, because of its requirement for a frost-free environment, it is only exceptionally found on rocks inland. Lowland.

Native (change +0.02). Most of the British sites lacking a post-1970 record were lost before 1930, and the distribution of this species is currently stable.

Suboceanic Southern-temperate element.

References: Atlas (8a), Hultén & Fries (1986), Jalas & Suominen (1972), Jermy *et al.* (1978), Page (1997).

F. J. RUMSEY

Asplenium trichomanes Maidenhair Spleenwort

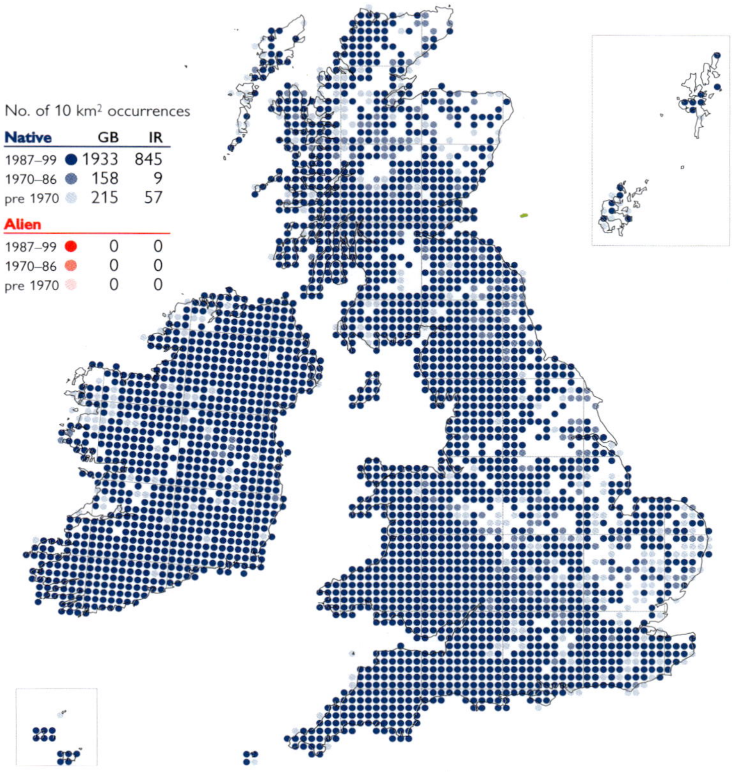

A perennial, evergreen fern which grows in a range of rocky habitats, including cliffs, rock faces, screes, mine waste and, perhaps now most commonly, on walls. 0–870 m (Macgillycuddy's Reeks, S. Kerry).

Native (change +0.07). There is evidence for some decline in East Anglia since 1950. This is a polyploid complex within which three ecologically and morphologically fairly distinct taxa have been recognised within the British Isles, all of which are mapped here separately. However, this taxonomic treatment over-simplifies a far more complex pattern of reticulate evolution.

Circumpolar Southern-temperate element, with a disjunct distribution.

References: Atlas (8b), Hultén & Fries (1986), Jalas & Suominen (1972), Jermy *et al.* (1978), Meusel *et al.* (1965), Page (1997).

F. J. RUMSEY

Asplenium trichomanes subsp. *trichomanes*

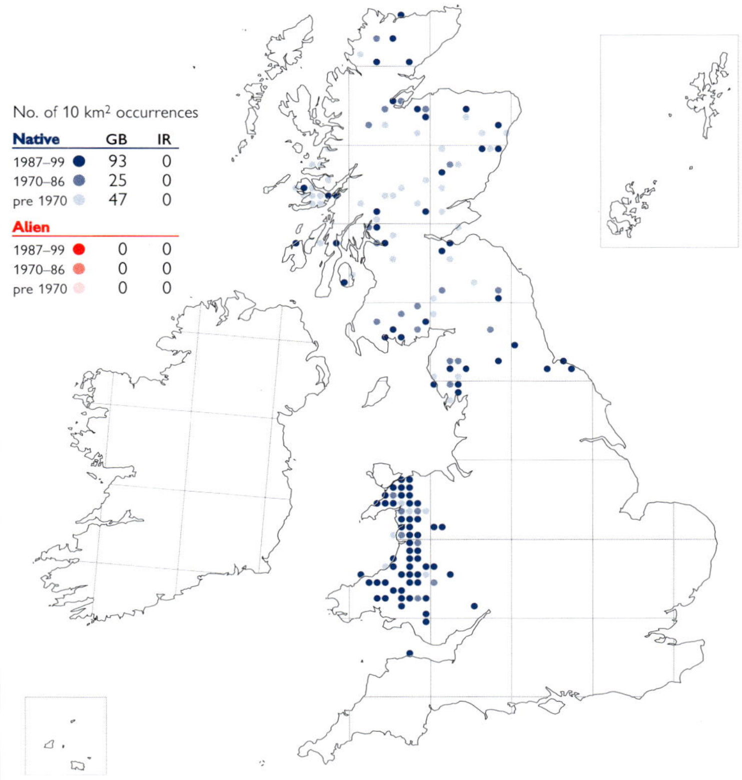

A. trichomanes subsp. *trichomanes* is restricted to acidic, siliceous substrates. It is found rooted into crevices in rock outcrops, and on screes, mine waste and dry-stone walls, particularly in oceanic upland areas. 0–540 m (Hartsop, Westmorland).

Native. This taxon was known to Lovis (1955) and Kent (1965) but did not become widely known until mapped by Jermy *et al.* (1978). It has been much confused with, and is still over-recorded for, forms of the nearly ubiquitous subsp. *quadrivalens*, and it is not possible to assess changes in its distribution.

Circumpolar Temperate element. Subsp. *trichomanes* is more frequent than subsp. *quadrivalens* in Scandinavia, but much rarer in the Mediterranean region.

Reference: Page (1997).

F. J. RUMSEY

Asplenium trichomanes subsp. *quadrivalens*

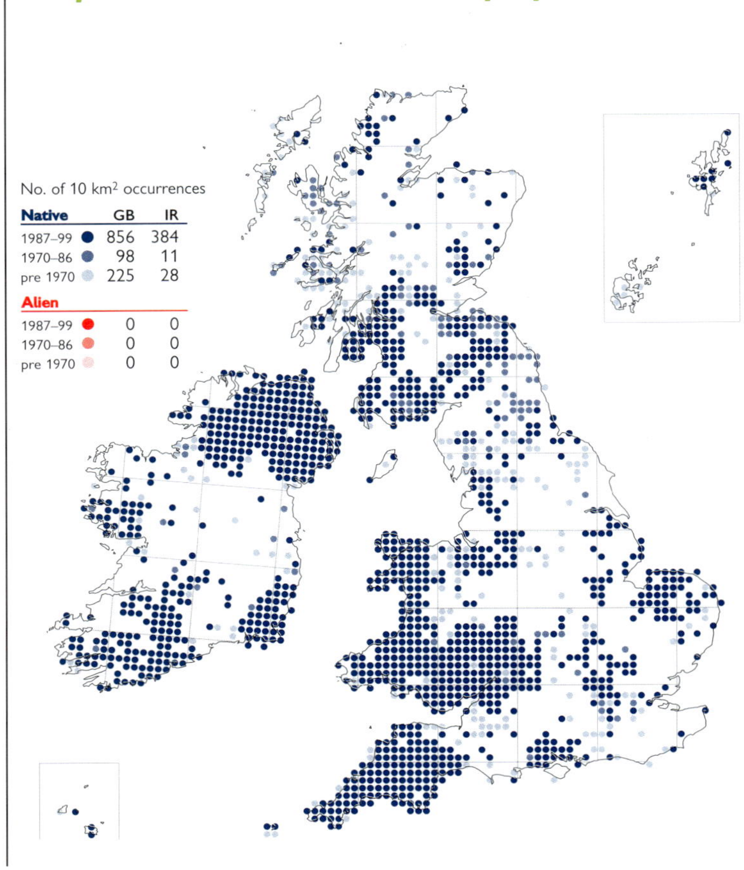

This segregate of *A. trichomanes* is found on mortared walls and calcareous rocks. It is replaced by subsp. *trichomanes* only in the most acidic situations. When growing with the rarer subspecies, noticeably vigorous sterile hybrids may sometimes be found. Generally lowland, but reaching at least 730 m on Cross Fell (Cumberland), and probably 870 m on Macgillycuddy's Reeks (S. Kerry).

Native. Like subsp. *trichomanes*, this taxon was known to Lovis (1955) and Kent (1965) but did not become widely known until mapped by Jermy *et al.* (1978). As currently defined, this is the most variable and by far the most abundant and widespread of the *A. trichomanes* subspecies. It is under-recorded in many areas.

Circumpolar Southern-temperate element.

References: Grime *et al.* (1988), Page (1997).

F. J. RUMSEY

Asplenium trichomanes subsp. *pachyrachis*

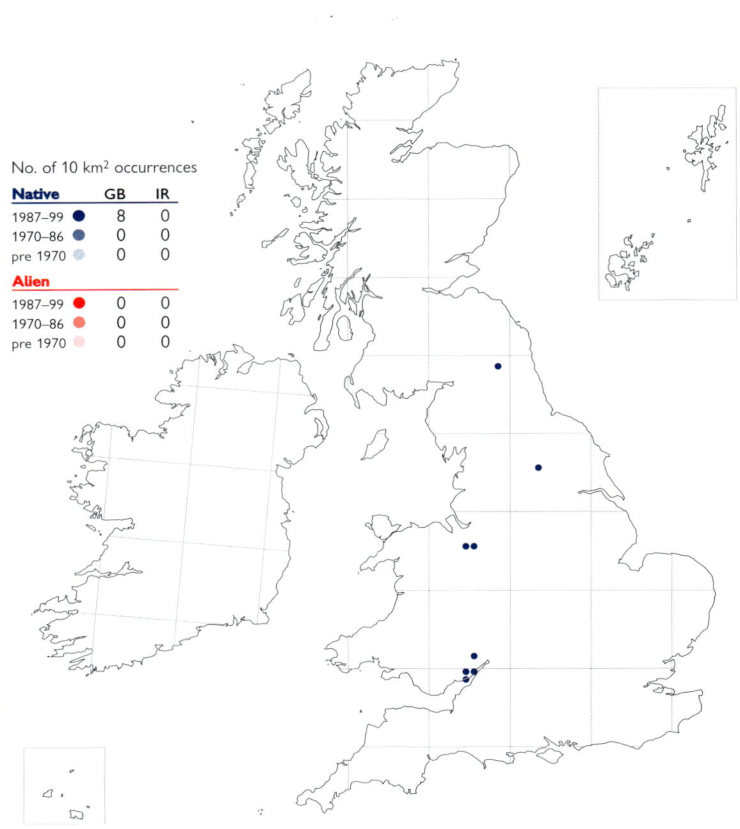

A fern of perpendicular calcareous limestone, usually in shaded, sheltered, humid places, and often under overhangs. It is rarely found on sandstone rock and mortared walls. Lowland.

Native. This plant was known in Victorian times, but it has only recently been identified as the continental subsp. *pachyrachis* (Rickard, 1989). It is treated at specific rank, as *A. csikii*, by Vogel *et al.* (1999). Though it appears to be genuinely rare and strangely absent from many apparently suitable areas, it is undoubtedly still under-recorded. However, it is difficult to separate from similar polyploid lineages which also occur in Britain.

Widespread but scattered in Europe, centred on limestone massifs in S. Europe with rare occurrences further north.

References: Page (1997), Wigginton (1999).

F. J. RUMSEY

Asplenium viride Green Spleenwort

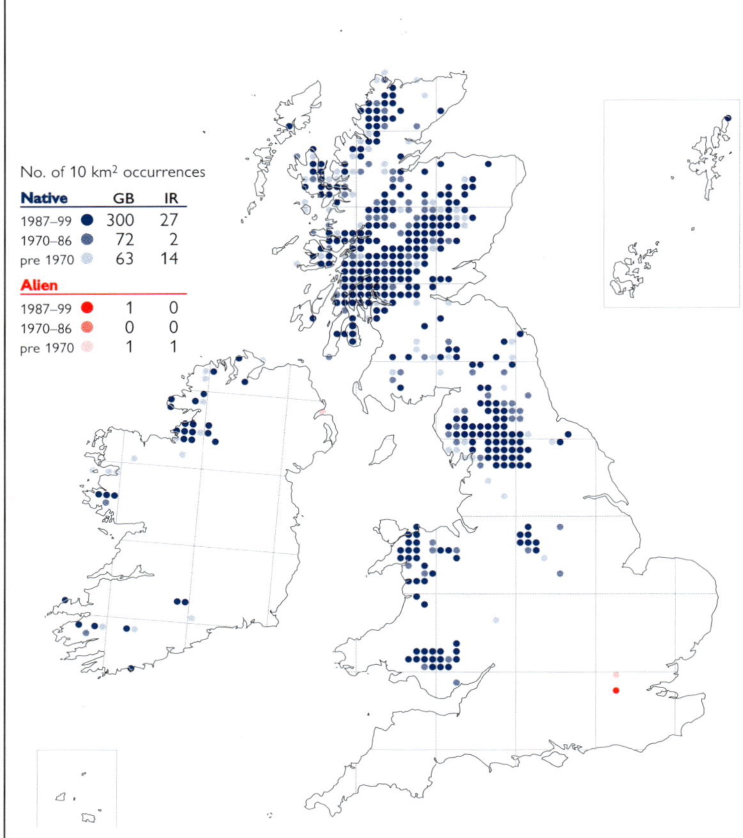

A. viride is an evergreen fern of moist, sheltered crevices in basic rocks, and very rarely also on mortared walls. It is occasionally a colonist of old metal mine workings. From sea level on the coasts of W. Britain and Ireland to 975 m on Ben Lawers (Mid Perth).

Native (change –0.21). This species shows no appreciable change in distribution since the 1962 *Atlas*. Because of its requirement for an environment which is cool and humid in summer (Page, 1997), its occasional occurrences in lowland C. and E. England have all been short-lived.

Circumpolar Boreal-montane element, with a disjunct distribution.

References: Atlas (8c), Hultén & Fries (1986), Jalas & Suominen (1972), Jermy *et al.* (1978), Meusel *et al.* (1965), Rumsey (1997).

F. J. RUMSEY

Asplenium ruta-muraria Wall-rue

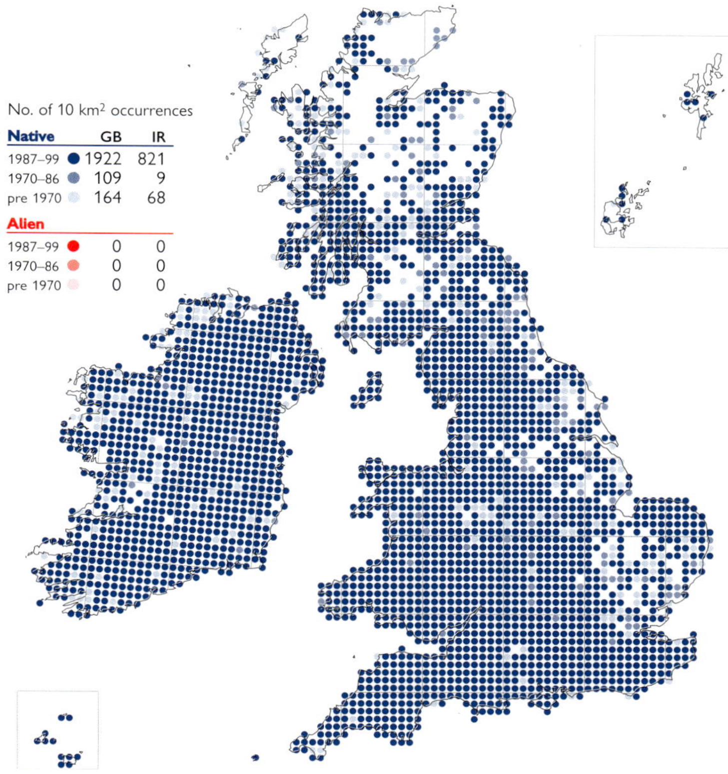

This perennial, evergreen fern occurs naturally on limestone and other basic rocks, where it grows on steep, bare faces and in crevices; it is also found in hollowed clints in limestone pavement. However, in most lowland areas it is now abundant on mortared walls and other man-made structures. 0–625 m (Ingleborough, Mid-W. Yorks.).

Native (change +0.15). Some decline in industrial areas through acidification has undoubtedly occurred (Page, 1997), but in recent years cleaner air may have reversed the decline. Other than this, most losses have occurred since 1950.

Circumpolar Temperate element, with a disjunct distribution.

References: Atlas (8d), Grime *et al.* (1988), Hultén & Fries (1986), Jalas & Suominen (1972), Jermy *et al.* (1978).

F. J. RUMSEY

Asplenium septentrionale Forked Spleenwort

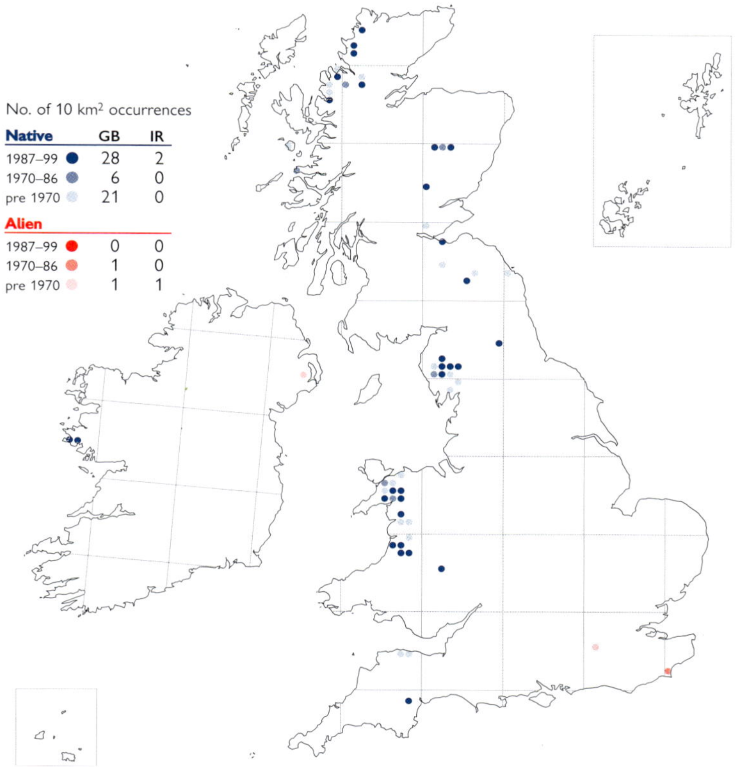

An often long-lived, evergreen fern of well-drained, exposed, sunny, usually acidic rock faces, metalliferous mine spoil and the sides of unmortared stone walls. In Ireland, it grows on ultrabasic rocks. 0–535 m (Moel yr Ogof, Caerns.), formerly to 715 m at Llyn y Cwn (Caerns.).

Native (change –0.08). Losses since 1930 have occurred through tidying and restoring old mining sites and walls, tree planting and scrub growth. It can rapidly colonise walls and mine spoil, but spread on natural rock is rare. New sites have been found in N.W. Scotland and C. Wales since 1987.

European Temperate element; also in C. Asia and western N. America.

References: Atlas (9a), Curtis & McGough (1988), Hultén & Fries (1986), Jalas & Suominen (1972), Jermy *et al.* (1978), Meusel *et al.* (1965), Page (1997), Stewart *et al.* (1994).

F. J. RUMSEY

Ceterach officinarum Rustyback

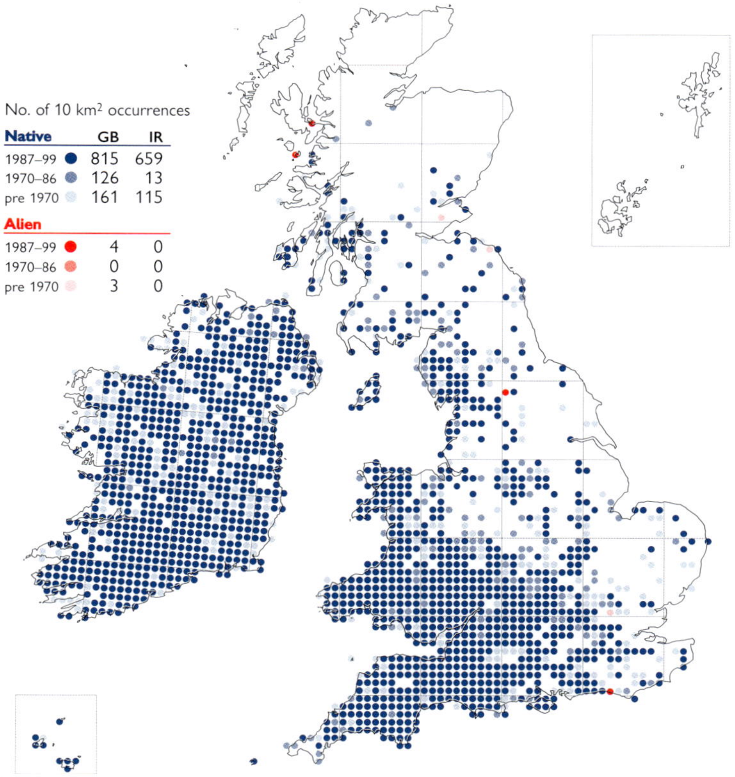

A perennial, calcicole fern found on crags and cliffs of basic rocks, especially limestone, and also on limestone pavements and mortared walls. Generally lowland, reaching *c.* 550 m in Wales.

Native (change –0.30). Historically this species has benefited from the built environment. There are many more records since the 1962 *Atlas*, particularly in E. Britain. Molecular evidence shows that the genus *Ceterach* should be subsumed into *Asplenium*.

Submediterranean-Subatlantic element; also in C. Asia.

References: Atlas (9b), Hultén & Fries (1986), Jalas & Suominen (1972), Jermy *et al.* (1978), Meusel *et al.* (1965), Page (1997).

F. J. RUMSEY

Matteuccia struthiopteris Ostrich Fern

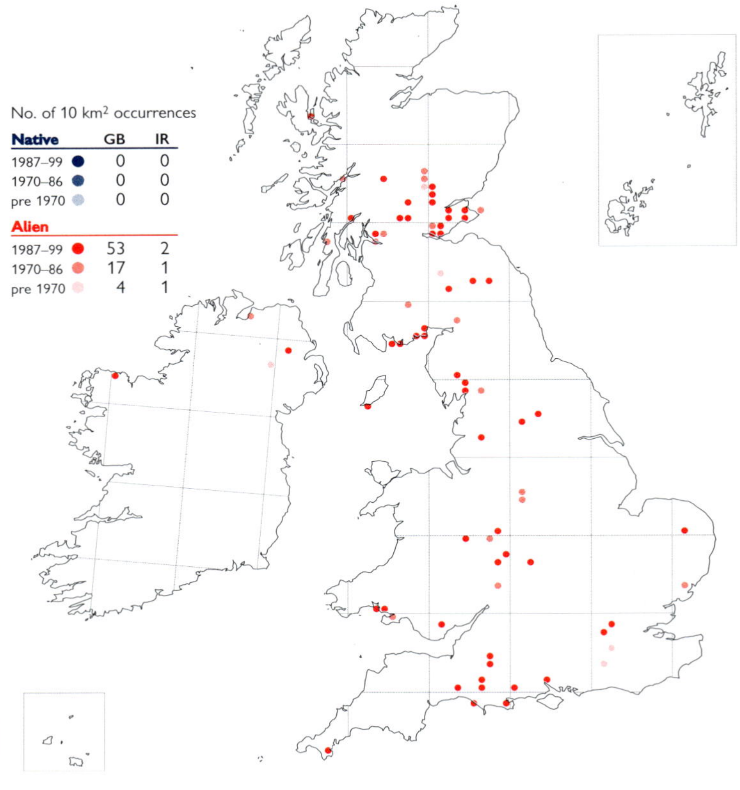

A large deciduous fern found growing in damp woodland, by streams and lakes, and in fen carr under *Betula*. It tolerates a range of soil pH, but prefers waterlogged clay substrates. It spreads through the production of stolons. Lowland.

Neophyte. *M. struthiopteris* was introduced to cultivation in Britain in 1760 and is frequently grown for its distinctive shuttlecock shape. It has been known in the wild since at least 1834, and may be increasing through a rise in its popularity in water garden plantings.

A Circumpolar Boreal-montane species; absent as a native from much of W. Europe.

References: Hultén & Fries (1986), Jalas & Suominen (1972).

T. D. DINES

Athyrium filix-femina Lady-fern

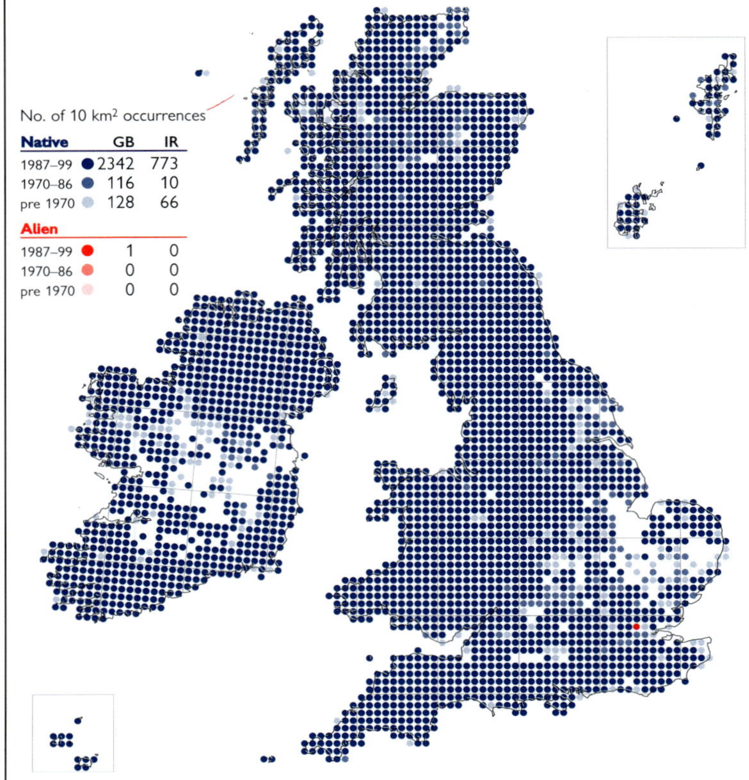

A deciduous fern that prefers moist but well-drained acidic soils, but can tolerate more basic substrates if these are overlain by mildly acidic layers. It is particularly frequent in deciduous woodland, especially on stream banks, and in moist, rocky habitats, but is also found in hedgerows and drainage ditches. It is one of few species able to colonise metalliferous lead and tin mine deposits. 0–1005 m (Carnedd Llewelyn, Caerns.).

Native (change +0.25). There is little evidence for any change in the distribution of this species.

Circumpolar Boreo-temperate element.

References: Atlas (9c), Grime *et al.* (1988), Hultén & Fries (1986), Jalas & Suominen (1972), Jermy *et al.* (1978), Meusel *et al.* (1965), Page (1997).

T. D. DINES

Athyrium distentifolium Alpine Lady-fern

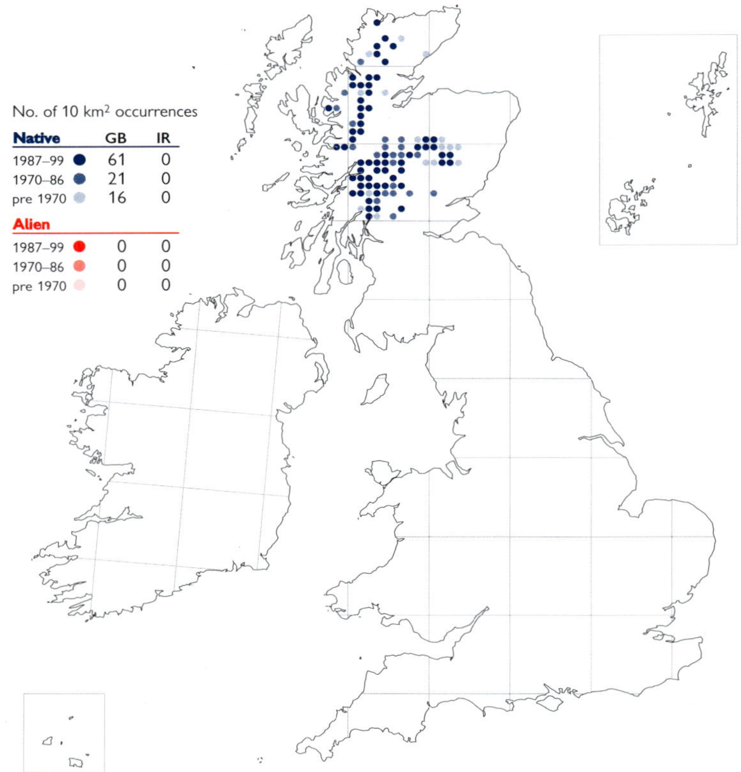

A deciduous fern of the higher mountains, growing on rock ledges, gullies, block screes and in shallow hollows where snow lies late into summer. It prefers more stable, acidic block screes with a N. or N.E. aspect and some degree of soil accumulation. Found from 455 m in the Breadalbanes (Mid Perth) to 1220 m (Ben Macdui, S. Aberdeen).

Native (change +0.38). There is little evidence for a change in the distribution of this species, although it is susceptible to over-grazing. It is probably under-recorded.

Circumpolar Arctic-montane element, with a disjunct distribution.

References: Atlas (9d), Hultén & Fries (1986), Jalas & Suominen (1972), Jermy *et al.* (1978), Meusel *et al.* (1965), Page (1997), Stewart *et al.* (1994).

A. C. JERMY & T. D. DINES

Athyrium flexile Newman's Lady-fern

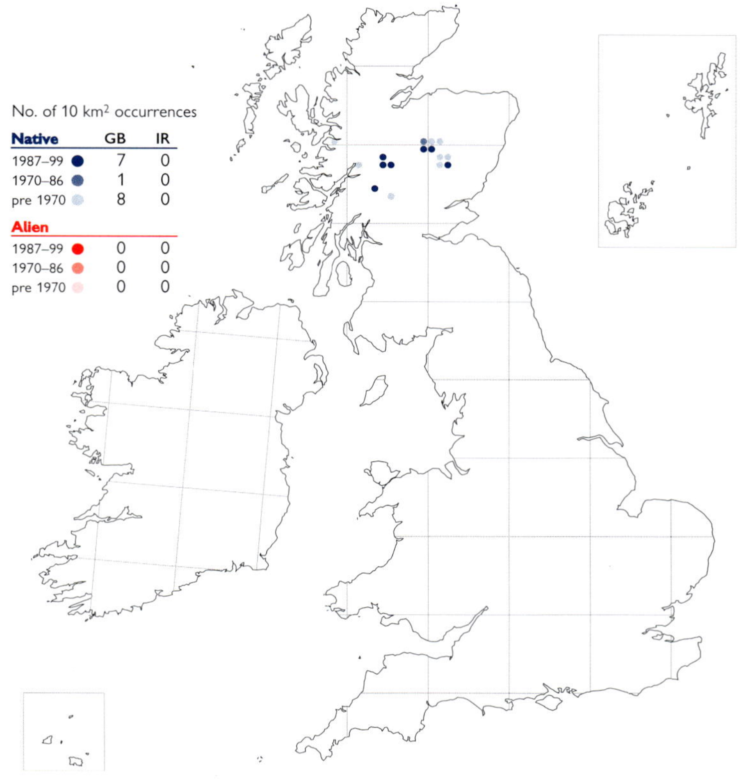

No. of 10 km² occurrences

Native		GB	IR
1987–99	●	7	0
1970–86	●	1	0
pre 1970	●	8	0
Alien			
1987–99	●	0	0
1970–86	●	0	0
pre 1970	●	0	0

A small deciduous fern usually found on cool, shaded, N.E.- to N.W.-facing scree-slopes or amongst block scree of acidic rocks, especially where snow lies late and where melt-water trickles down gullies. From 700 m on Meall Buidhe (Main Argyll) to 900 m (Glen Einich, Easterness), and reportedly to 1140 m.

Native. There is little evidence for any change in the distribution of this species, although it is not only taxonomically critical but difficult to find. Since it was first described in 1853 there has been doubt over its taxonomic status. Recent experimental work has shown that it is probably best regarded as a variety of *A. distentifolium.*

Endemic.

References: Atlas Supp. (2c), Jermy *et al.* (1978), McHaffie (1999), Page (1997), Wigginton (1999).

A. C. JERMY & T. D. DINES

Gymnocarpium dryopteris Oak Fern

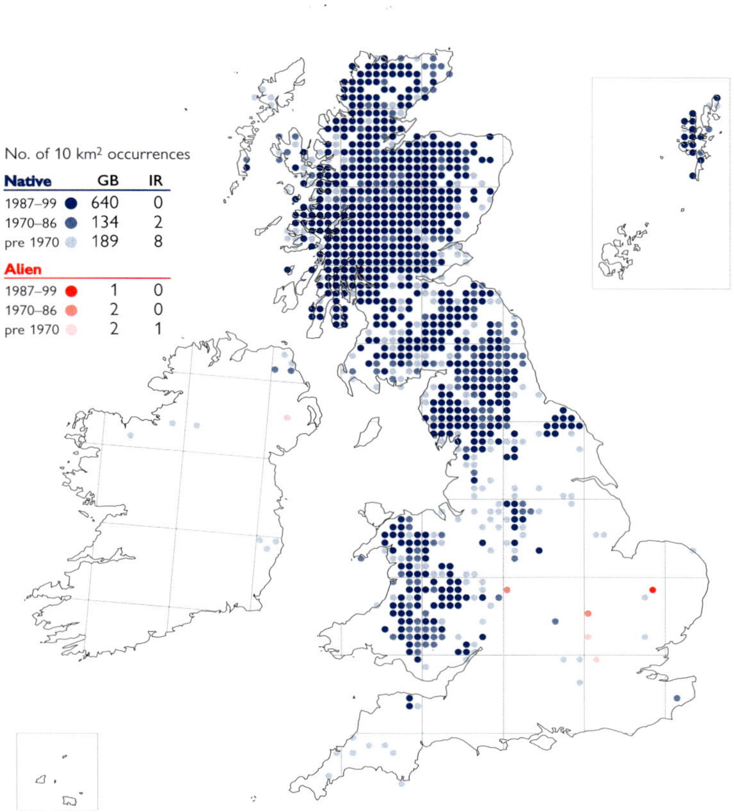

No. of 10 km² occurrences

Native		GB	IR
1987–99	●	640	0
1970–86	●	134	2
pre 1970	●	189	8
Alien			
1987–99	●	1	0
1970–86	●	2	0
pre 1970	●	2	1

A gregarious, deciduous fern growing in rocky deciduous woodland and ravines, along stream banks, and on cliff ledges and stable block screes. It prefers moist but open, light-textured mineral soils with a high humus content, and tolerates a moderate range of pH. 0–915 m (Rannoch, Mid Perth).

Native (change –0.21). This species, which is susceptible to heavy grazing, has declined at the edges of its British range, where it has been lost from lowland woods even though the woods themselves often survive. Most of the losses occurred before 1950. It was last recorded in Ireland in 1986.

Circumpolar Boreo-temperate element.

References: Atlas (14c), Curtis & McGough (1988), Hultén & Fries (1986), Jalas & Suominen (1972), Jermy *et al.* (1978), Meusel *et al.* (1965), Page (1997).

T. D. DINES

Gymnocarpium robertianum Limestone Fern

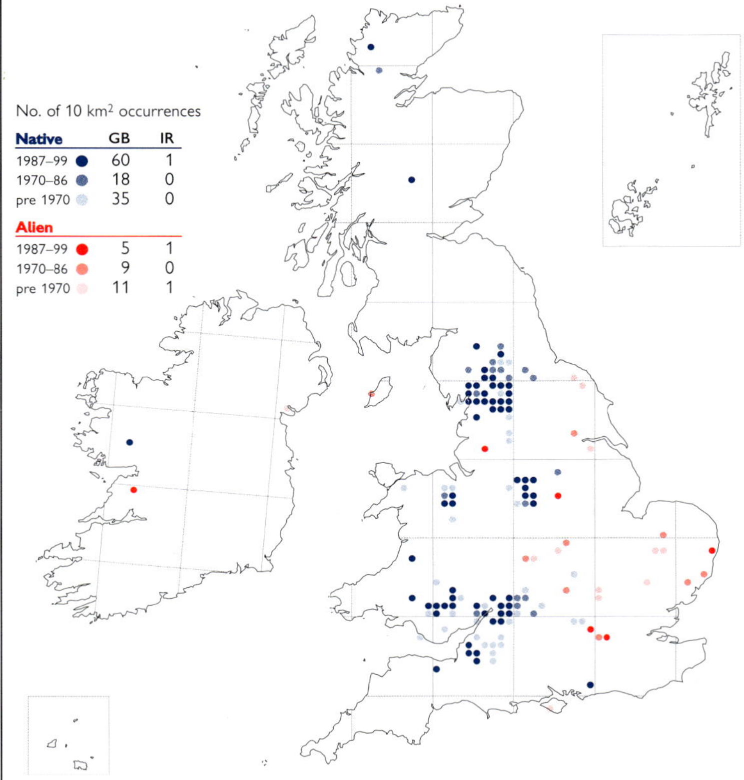

No. of 10 km² occurrences

Native		GB	IR
1987–99	●	60	1
1970–86	●	18	0
pre 1970	●	35	0
Alien			
1987–99	●	5	1
1970–86	●	9	0
pre 1970	●	11	1

A deciduous fern of cracks, fissures and scree in limestone rock, but also found in shallow grikes of limestone pavement, and, rarely, on chalk. It prefers warm, sunny exposures but can tolerate light shading. It has become established as a garden escape on walls and culverts. Lowland to 585 m at Carreg yr Ogof (Carms.).

Native (change –0.37). The distribution of this species is stable. Some sites have, however, been lost through competition with *Mercurialis perennis* and *Crataegus monogyna,* while quarrying and overgrazing have had detrimental effects on others.

Circumpolar Boreo-temperate element, with a disjunct distribution.

References: Atlas (14d), Curtis & McGough (1988), Hultén & Fries (1986), Jalas & Suominen (1972), Jermy *et al.* (1978), Meusel *et al.* (1965), Page (1997), Stewart *et al.* (1994).

T. D. DINES

Cystopteris fragilis Brittle Bladder-fern

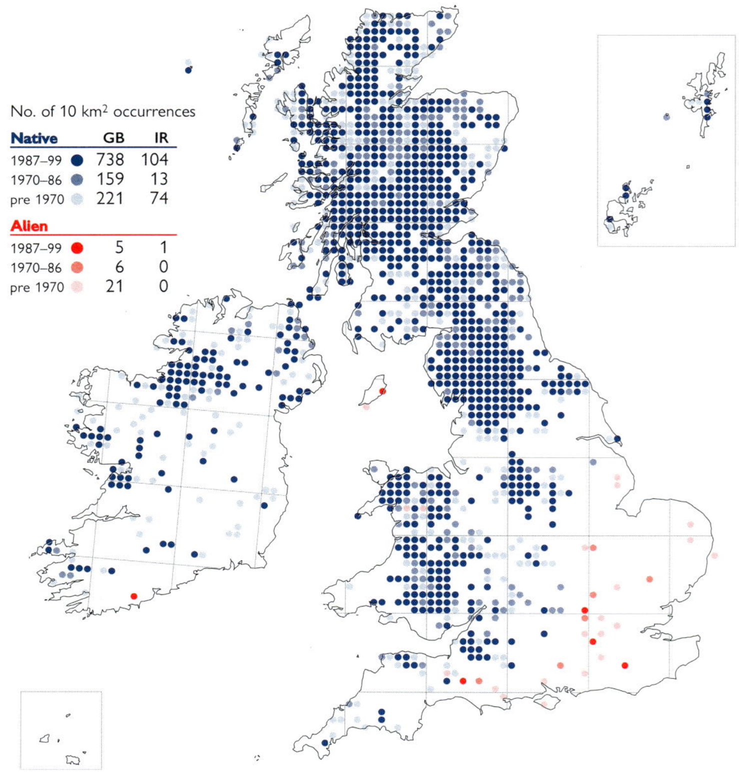

A fern of damp, shaded rock crevices, cliffs, cave entrances, ravines and mortared walls, always growing on a mineral-enriched substrate, and most frequent over limestone. It is also found on field boundary banks where water seeps from improved pasture, and in *Fraxinus* woodland. 0–1220 m (Breadalbanes, Mid Perth).

Native (change –0.69). The distribution of *C. fragilis* in the uplands is stable. Some lowland sites, particularly those on old buildings, bridges, canal locks and railway platforms, have been lost since 1950 through demolition or cleaning.

Circumpolar Wide-boreal element.

References: Atlas (10a), Grime *et al.* (1988), Hultén & Fries (1986), Jalas & Suominen (1972), Jermy & Camus (1991), Jermy *et al.* (1978), Page (1997).

A. C. JERMY & T. D. DINES

Cystopteris dickieana Dickie's Bladder-fern

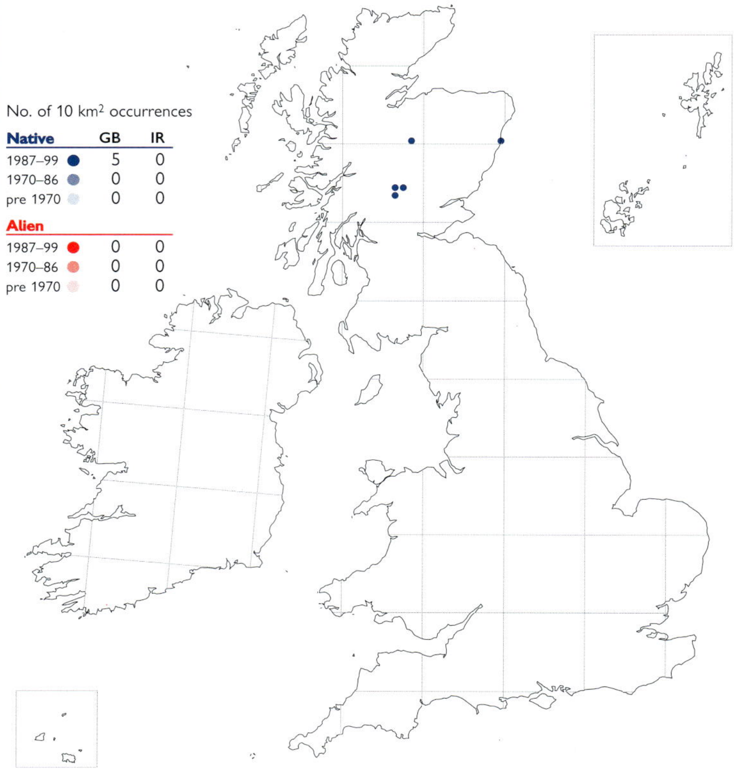

A fern of rock crevices, shady ravines, cave entrances and also under overhangs. It prefers moderately base-rich rocks. 0–380 m (above Loch Tay, Mid Perth).

Native. This species was originally described from the coast south of Aberdeen in 1848. It has recently been found in a few inland sites, and may be overlooked elsewhere. However, its taxonomic status 'remains uncertain and controversial' (Dyer *et al.*, 2000) and recent allozyme studies do not support its recognition as a species distinct from *C. fragilis*.

As the characters which define *C. dickieana* are unresolved, it is impossible to summarise its world distribution.

References: Atlas (10b), Hultén & Fries (1986), Jalas & Suominen (1972), Jermy *et al.* (1978), Marren (1983), Page (1997), Parks *et al.* (2000), Tennant (1996), Wigginton (1999).

A. C. JERMY & T. D. DINES

Cystopteris montana Mountain Bladder-fern

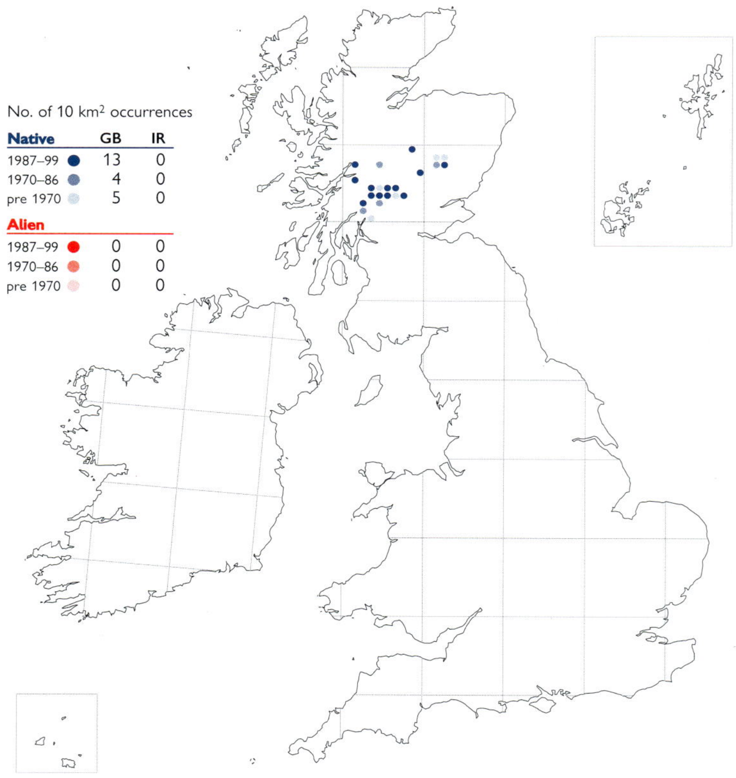

A deciduous fern of sheltered, humid, N.- or E.-facing limestone and mica-schist cliffs where there is periodic irrigation. It prefers dripping rock ledges, cliff bases, gullies and steep, unstable scree slopes. From 490 m on Ben Lui (Main Argyll) to 1125 m on Aonach Beag (Westerness).

Native (change –0.25). Easily overlooked and sometimes growing in inaccessible sites, *C. montana* is probably present in most 10-km squares for which only pre-1987 records exist. It is susceptible to grazing, and had been lost from some sites before 1930 because of collecting.

Circumpolar Boreal-montane element.

References: Atlas (10b), Halliday (1997), Hultén & Fries (1986), Jalas & Suominen (1972), Jermy *et al.* (1978), Meusel *et al.* (1965), Page (1997), Stewart *et al.* (1994), Wigginton (1999).

A. C. JERMY & T. D. DINES

Woodsia ilvensis Oblong Woodsia

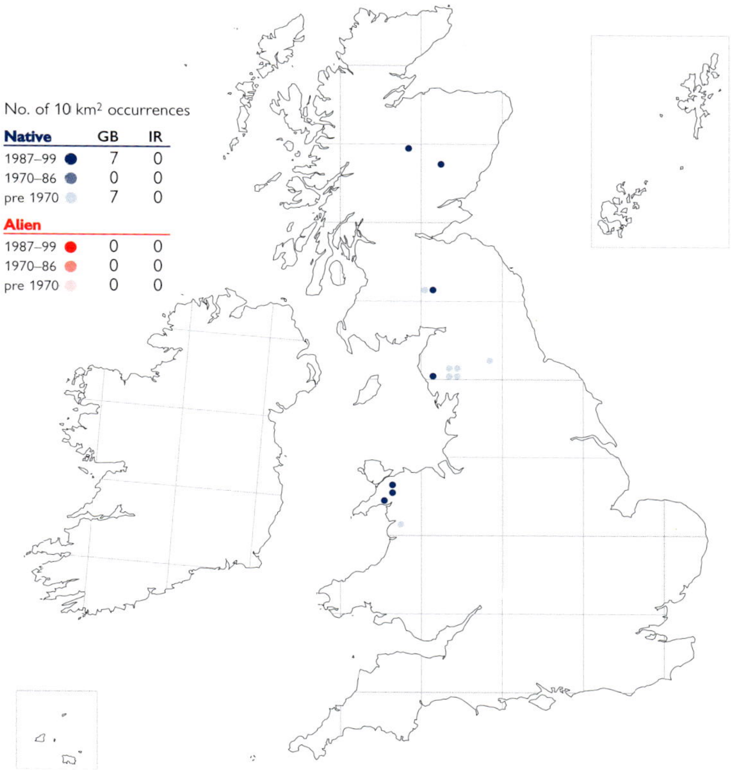

No. of 10 km² occurrences

Native	GB	IR
1987–99	7	0
1970–86	0	0
pre 1970	7	0
Alien		
1987–99	0	0
1970–86	0	0
pre 1970	0	0

An evergreen fern, growing in cracks and fissures in cliffs and crags on rocks ranging from calcareous tuffs and hornblende schists to more acidic tuffs, grits and shales. Sites are very free-draining, with little competition. Reproduction is probably mostly vegetative. From 365 m to 760 m (Cumberland).

Native (change –0.10). Current populations of *W. ilvensis* are probably relics of a more widespread post-glacial distribution. It suffered serious declines due to collecting in the 19th century. Some older records (e.g. in Cumberland) may be erroneous. Re-introductions in Scotland and Teesdale are not mapped.

Circumpolar Boreo-arctic Montane element.

References: Atlas (10c), Hultén & Fries (1986), Jalas & Suominen (1972), Jermy *et al.* (1978), Meusel *et al.* (1965), Page (1997), Wigginton (1999).

T. D. DINES

Woodsia alpina Alpine Woodsia

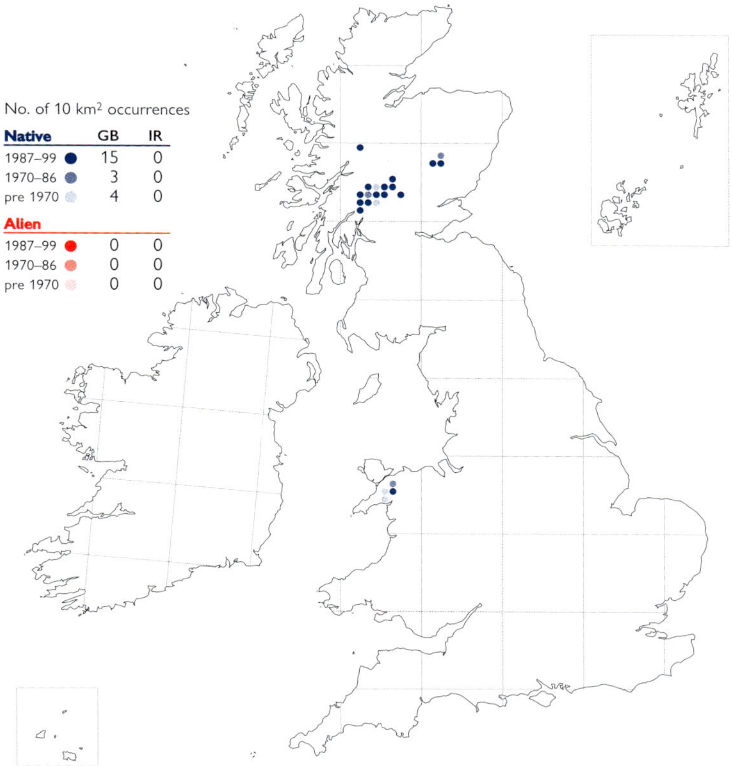

No. of 10 km² occurrences

Native	GB	IR
1987–99	15	0
1970–86	3	0
pre 1970	4	0
Alien		
1987–99	0	0
1970–86	0	0
pre 1970	0	0

W. alpina grows on the steep, free-drained, bare faces of calcareous rocks, including pumice tuffs, basalts, mica- and hornblende schists, slates and limestones. Sites are very free-draining, with little competition. From 525 m to 975 m on Ben Lawers (Mid Perth).

Native (change +0.11). First reported from Snowdonia in 1790, current populations of *W. alpina* are probably relics from more widespread populations in post-glacial times. It suffered a serious decline through collecting in the 19th century. New sites have been discovered since the 1962 *Atlas*, and current populations appear to be relatively stable.

Circumpolar Boreo-arctic Montane element.

References: Atlas (10d), Hultén & Fries (1986), Jalas & Suominen (1972), Jermy *et al.* (1978), Meusel *et al.* (1965), Page (1997), Wigginton (1999).

T. D. DINES

Polystichum setiferum Soft Shield-fern

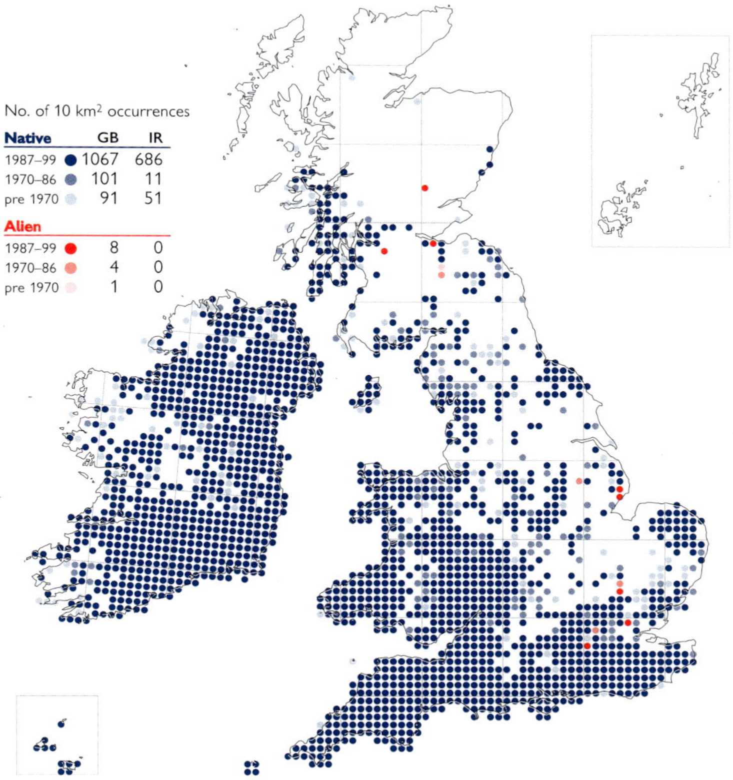

No. of 10 km² occurrences

Native	GB	IR
1987–99	1067	686
1970–86	101	11
pre 1970	91	51
Alien		
1987–99	8	0
1970–86	4	0
pre 1970	1	0

This semi-evergreen fern is a moderate calcicole, occurring in shaded deciduous woodland, hedgerows, lane banks and sheltered streamsides, and also in the peaty bottoms of grikes in limestone pavement. It grows on a wide range of soil types, from those derived from sands to clays, but prefers sloping or well-drained ground. Generally lowland, but reaching 305 m on Walla Crag (Cumberland).

Native (change +1.47). There are many more records for this species than in the 1962 *Atlas*; these can be attributed to more efficient recording and there is little evidence for a change in its distribution.

Submediterranean-Subatlantic element.

References: Atlas (13a), Hultén & Fries (1986), Jalas & Suominen (1972), Jermy *et al.* (1978), Page (1997).

T. D. DINES

Polystichum aculeatum Hard Shield-fern

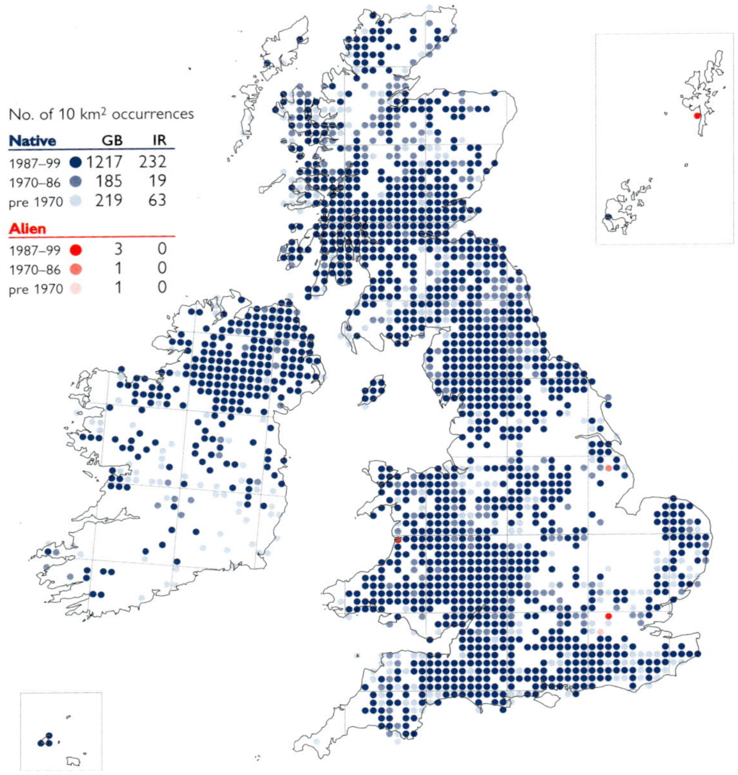

This evergreen species is characteristic of mountain gorges and steep wooded river valleys where it grows in thin but damp, mildly acidic to base-rich soils between rocks and in crevices. It also grows in the grikes of limestone pavement, on shady mortared walls, on hedge banks, and around cave entrances and mine shafts. It is rarely plentiful in S. England, usually occurring as scattered individuals and only becoming common in the north of its range. 0–760 m (Breadalbanes, Mid Perth).

Native (change +0.54). There is little change in the distribution of this species, other than in S.E. England where it appears to have decreased since the 1962 *Atlas*.

Eurasian Temperate element.

References: Atlas (13b), Dupont (1962), Hultén & Fries (1986), Jalas & Suominen (1972), Jermy *et al.* (1978), Page (1997).

T. D. DINES

Polystichum aculeatum × *P. setiferum* (*P.* × *bicknellii*)

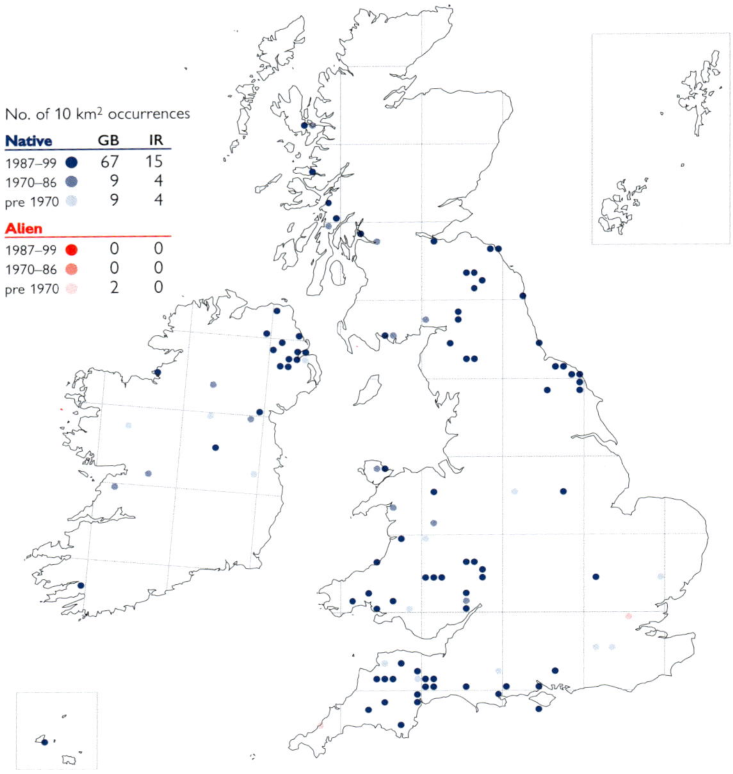

An evergreen fern that can occur as sporadic individuals whenever the parents meet. It is found in rocky woodlands, gorges, on stream banks and in disused quarries. This hybrid is a calcicole and is often associated with disturbed sites. Generally lowland, but reaching 420 m at Esgair Hir (Cards.).

Native. *P.* × *bicknellii* is almost certainly more widespread than indicated by the map; it is often overlooked as one of its parents. The increase in records since it was mapped by Jermy *et al.* (1978) is due to better recording rather than a genuine increase in occurrence.

Apparently frequent in those areas of Europe where both parents occur.

References: Page (1997), Stace (1975).

T. D. DINES

Polystichum lonchitis Holly-fern

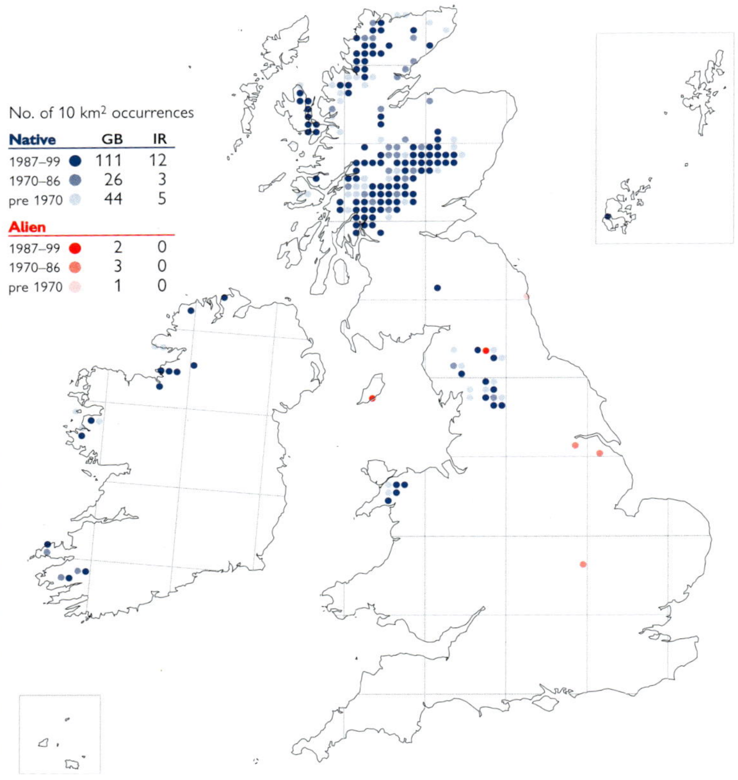

This evergreen species is a calcicole, growing in well-drained, cool and moist positions at the base of cliffs, on rocky ledges, and particularly in stabilised boulder-scree. It also grows in deep grikes of limestone pavements. *P. lonchitis* is a poor competitor, but is long-lived once established. From 180 m at Inchnadamph, W. Sutherland, but generally above 600 m and reaching 1150 m in the Breadalbanes (Mid Perth).

Native (change –0.76). Although some sites were lost to collecting before 1930, there has been little evidence of a change in the distribution since the 1962 *Atlas*.

Circumpolar Boreal-montane element, with a disjunct distribution.

References: Atlas (13c), Curtis & McGough (1988), Hultén & Fries (1986), Jalas & Suominen (1972), Jermy *et al.* (1978), Meusel *et al.* (1965), Page (1997).

T. D. DINES

Dryopteris oreades Mountain Male-fern

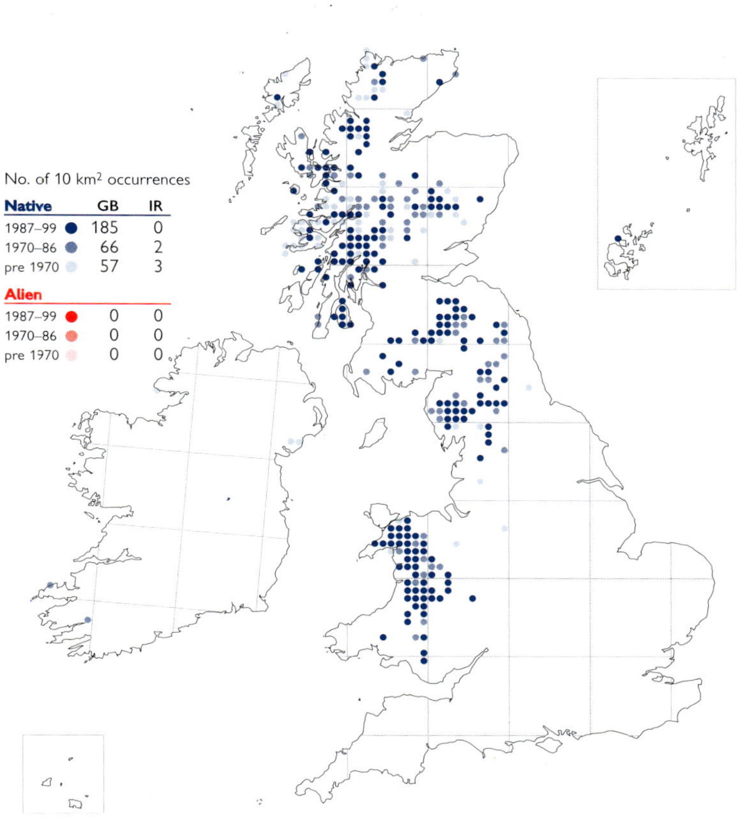

This deciduous fern grows in colonies on well-drained rocky ledges, steep, loose scree slopes and in gullies. Substrates include relatively acidic sandstones, slates and mica-schist. It is very sensitive to grazing, often becoming confined to inaccessible ledges and unstable scree slopes in heavily grazed areas. From 105 m (Llyn Padarn, Caerns.) to 850 m (An Sgurr, Mid Perth and Coire na Creiche, N. Ebudes).

Native (change +0.24). *D. oreades* is now much better recorded than it was in the 1962 *Atlas*, although it is easily overlooked and is probably still under-recorded. There appear to have been some losses in Scotland, possibly through overgrazing.

Suboceanic Boreal-montane element.

References: Atlas (11c), Jalas & Suominen (1972), Jermy *et al.* (1978), Page (1997).

T. D. DINES

Dryopteris filix-mas Male-fern

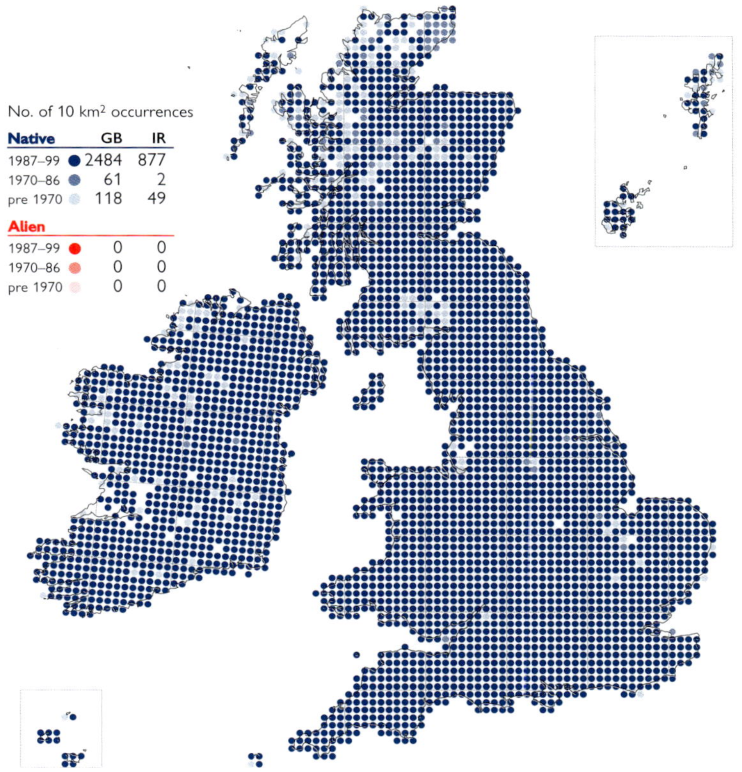

D. filix-mas is a common deciduous fern of woodlands, hedgerows, ditches, roadside verges, stream banks, rocky hillsides, cliff ledges and scree slopes. It prefers light, well-drained but moist soils that are mostly acidic to neutral, though sometimes slightly basic. It is also found in urban habitats, including railway banks, bridges, walls and gardens. 0–960 m (Macgillycuddy's Reeks, S. Kerry).

Native (change +0.03). There is no evidence for an appreciable change in the distribution of this species.

Circumpolar Temperate element.

References: Grime *et al.* (1988), Hultén & Fries (1986), Jalas & Suominen (1972), Jermy *et al.* (1978), Meusel *et al.* (1965), Page (1997).

T. D. DINES

Dryopteris affinis Scaly Male-fern

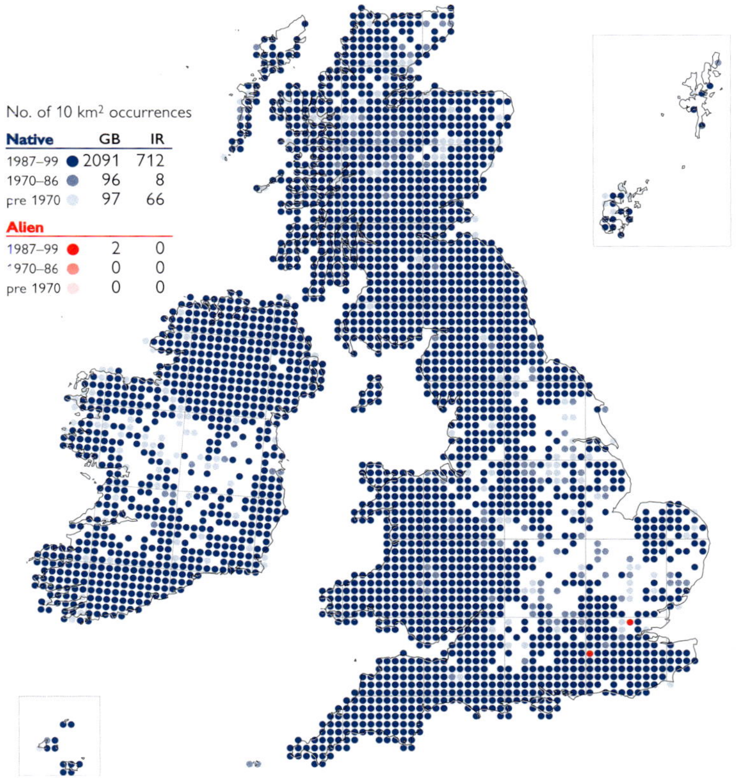

A deciduous or evergreen fern found in deciduous woodland, along rides in coniferous plantations, in ditches, on shady banks and road verges, usually on acidic substrates. In more oceanic areas, it grows in the open on well-drained W.- or S.W.-facing hillsides, scree slopes and mountain ledges. It is also found on brickwork in urban areas. 0–705 m (Little Dun Fell, Westmorland).

Native (change +2.44). *D. affinis* was greatly under-recorded in the 1962 Atlas but its distribution is probably stable. The three subspecies recognised by Stace (1997) are difficult to identify and are poorly recorded; they are not mapped here.

European Temperate element.

References: Atlas (11b), Hultén & Fries (1986), Jalas & Suominen (1972), Jermy *et al.* (1978), Page (1997), Rich & Jermy (1998).

A. C. JERMY & T. D. DINES

Dryopteris affinis × *D. filix-mas* (*D.* × *complexa*)

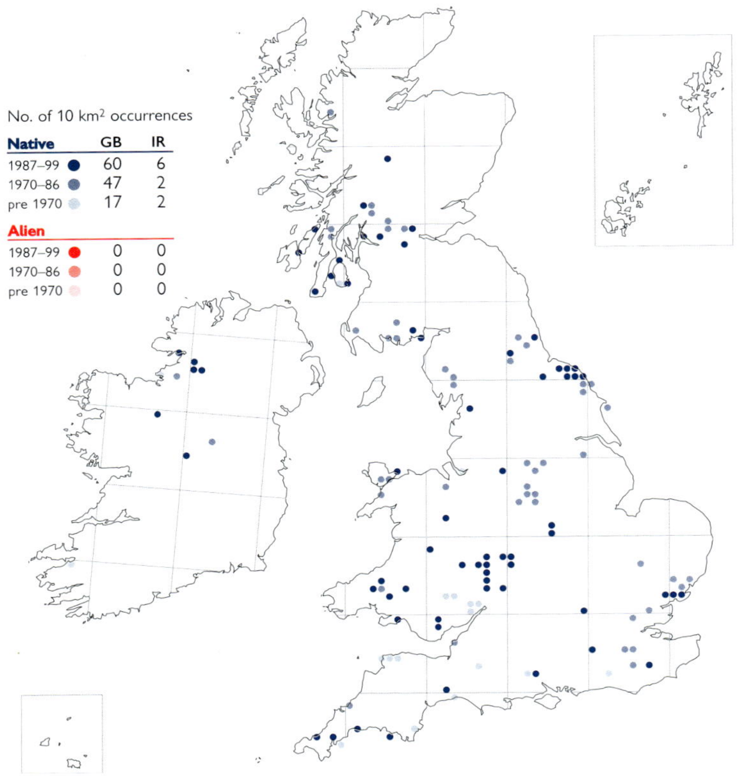

A deciduous fern found on damp, acidic soils in sheltered woodland, river banks, ditches, hedgerows and coniferous forest tracks, particularly where the original habitat has been disturbed. Plants may occur as single individuals, or in small groups, often in the absence of one or both parents. This hybrid is fertile and backcrossing is prevalent. Generally lowland.

Native. Identification of this hybrid is difficult, especially in view of the variability of *D. affinis*, and *D.* × *complexa* is almost certainly under-recorded.

Widespread in W. & C. Europe.

References: Jermy *et al.* (1978), Page (1997), Rich & Jermy (1998), Stace (1975).

T. D. DINES

Dryopteris remota Scaly Buckler-fern

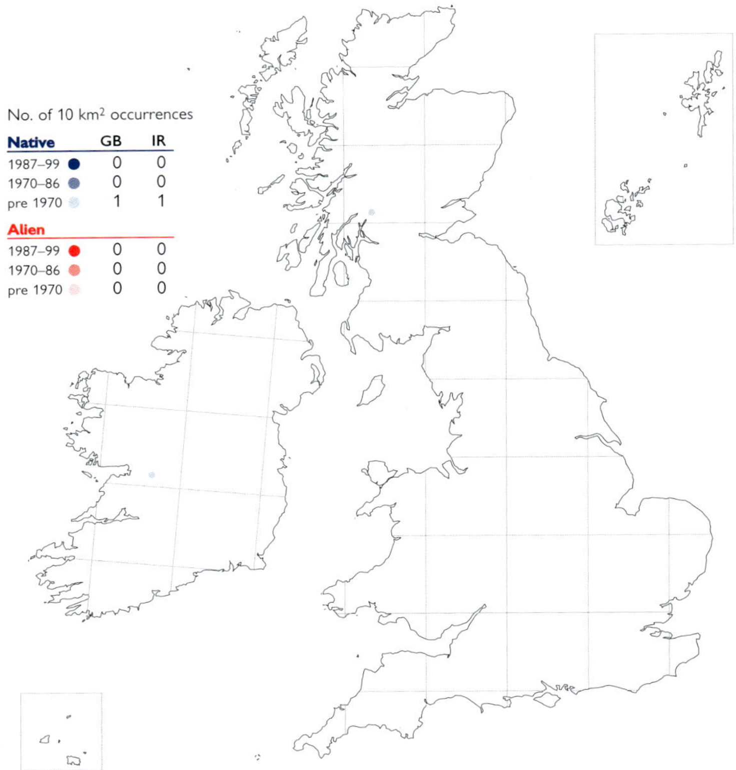

A deciduous fern of damp deciduous and open conifer woodland on more acidic soils. Lowland.

Native. This species, which is a fertile derivative of the hybrid between *D. affinis* and *D. expansa*, is now extinct in Britain and Ireland. Its possible occurrence at Loch Lomond (Dunbarton), where it was collected in 1894, is suspect and whilst the S.E. Galway plant has been well-studied in cultivation, it has never been re-found. A record from Kerry is based on an atypical sterile frond and needs confirmation; it is not mapped.

European Boreal-montane element, but absent from the Boreal zonobiome.

References: Jermy *et al.* (1978), Page (1997), Stace (1975).

A. C. JERMY

Dryopteris aemula Hay-scented Buckler-fern

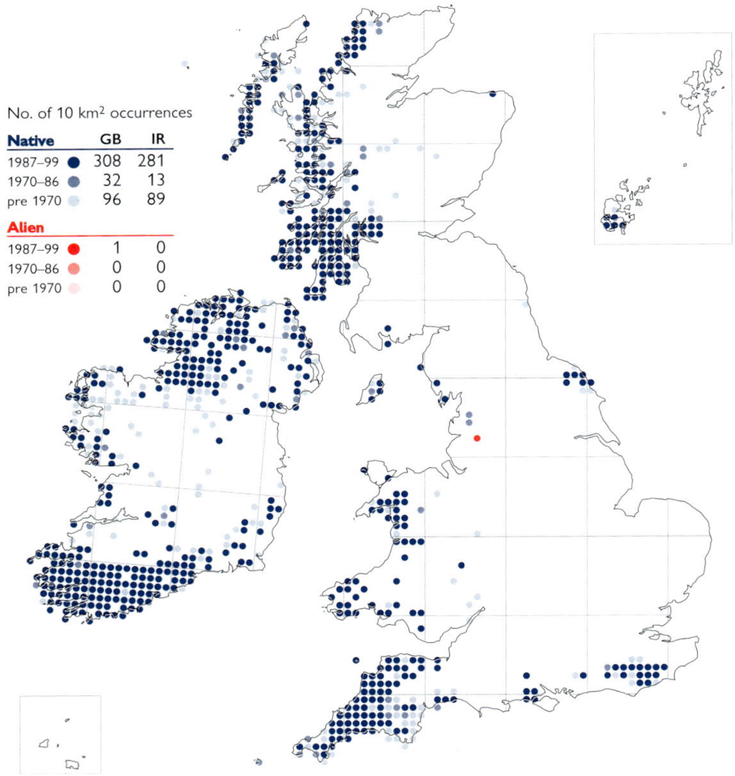

A fern of moist but well-drained acidic to neutral soils of low base content, growing on banks, sea-cliffs and wooded slopes. In the Weald (Sussex) it occurs in deep, steep-sided wooded ravines which emulate its Atlantic habitats. Generally lowland, but reaching 640 m in Macgillycuddy's Reeks (S. Kerry).

Native (change –0.04). *D. aemula* is much better recorded than in the 1962 *Atlas*; it has recently been discovered in the New Forest (S. Hants.) for example. However, it is vulnerable to changes in humidity and clear-felling, and the coppicing of old woodland is responsible for the loss of some populations in the south.

Oceanic Temperate element; confined to the hyperoceanic zone in W. Europe and Macaronesia.

References: Atlas (12d), Jalas & Suominen (1972), Jermy *et al.* (1978), Page (1997).

A. C. JERMY

Dryopteris submontana Rigid Buckler-fern

No. of 10 km² occurrences

Native	GB	IR
1987–99 ●	23	0
1970–86 ●	6	0
pre 1970 ●	4	0
Alien		
1987–99 ●	2	0
1970–86 ●	0	0
pre 1970 ●	0	0

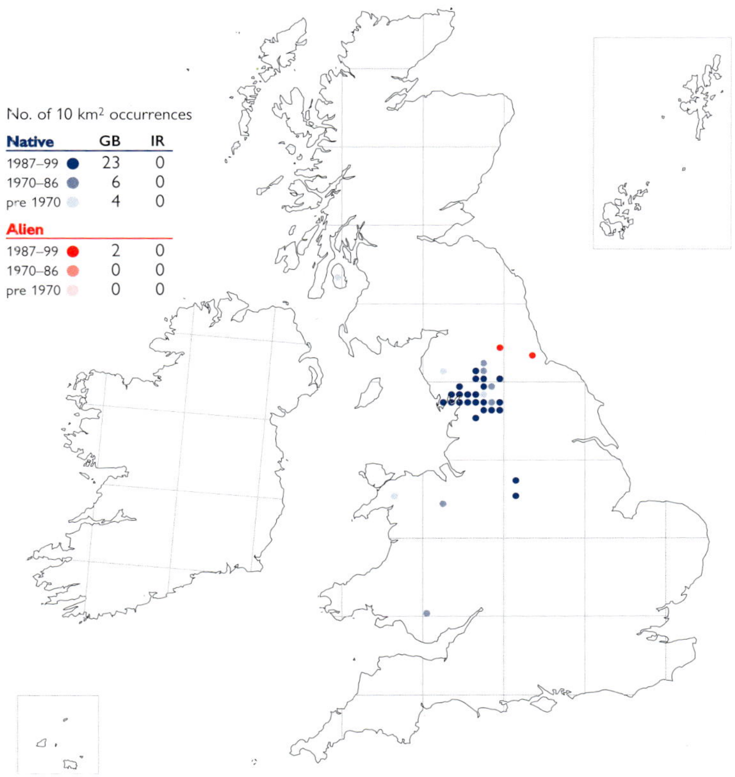

A deciduous fern of limestone pavement, screes and rock crevices, where moist, humus-rich soils develop. It prefers some degree of shelter, often growing in grikes, but is intolerant of shade. It can extend onto more exposed rock, but only where low woody scrub affords some protection. It is also recorded from limestone walls, and from other base-rich rocks. 0–465 m (Highfolds Scar, Mid-W. Yorks.).

Native (change +0.10). *D. submontana* is highly sensitive to grazing, and this may be the reason for its loss from old sites in Snowdonia and Arran. It is also highly vulnerable to quarrying, but otherwise there have been few losses since the 1962 *Atlas*.

Mediterranean-montane element.

References: Atlas (11d), Gilbert (1970), Jalas & Suominen (1972), Jermy *et al.* (1978), Page (1997), Stewart *et al.* (1994).

T. D. DINES

Dryopteris cristata Crested Buckler-fern

No. of 10 km² occurrences

Native	GB	IR
1987–99 ●	6	0
1970–86 ●	4	0
pre 1970 ●	21	0
Alien		
1987–99 ●	1	0
1970–86 ●	0	0
pre 1970 ●	0	0

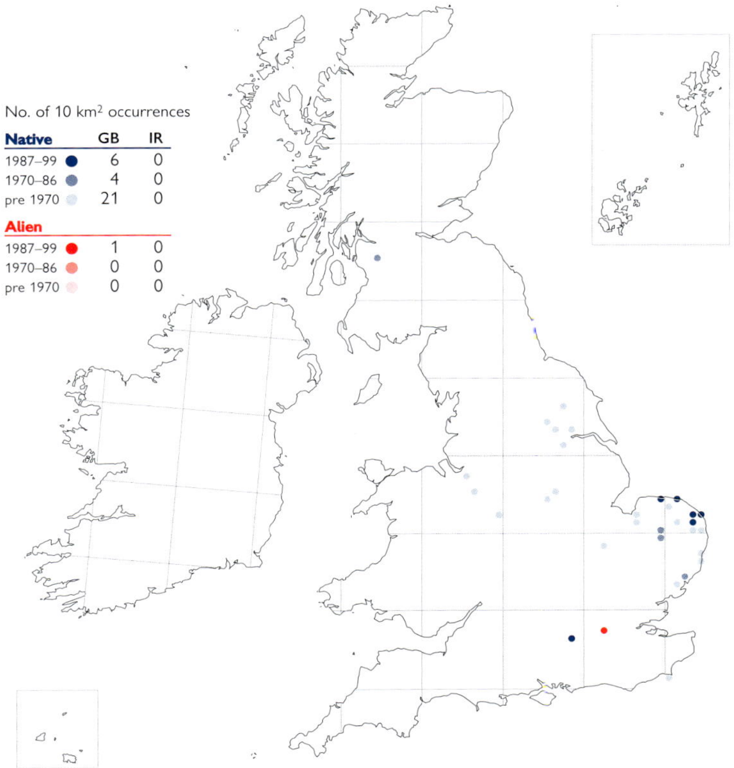

This deciduous fern grows in mildly acidic 'floating' fens that develop within or from more base-rich fens. It is characteristic of *Sphagnum* lawns, where it can tolerate the shade of invading *Phragmites*, and *Salix* and *Betula* scrub. It can persist in fen carr. Lowland.

Native (change –0.68). *D. cristata* was lost from most English sites outside East Anglia by 1970, through drainage, scrub encroachment and the loss of mown fen. Since then, losses have continued in Scotland and Suffolk. An increase in Norfolk records is due to a genuine expansion of its range.

Eurosiberian Temperate element, with a continental distribution in W. Europe; also in N. America.

References: Atlas (12a), Hultén & Fries (1986), Jalas & Suominen (1972), Jermy *et al.* (1978), Meusel *et al.* (1965), Page (1997), Wigginton (1999).

A. C. JERMY & T. D. DINES

Dryopteris carthusiana Narrow Buckler-fern

No. of 10 km² occurrences

Native	GB	IR
1987–99 ●	1171	196
1970–86 ●	209	26
pre 1970 ●	243	91
Alien		
1987–99 ●	0	0
1970–86 ●	0	0
pre 1970 ●	0	0

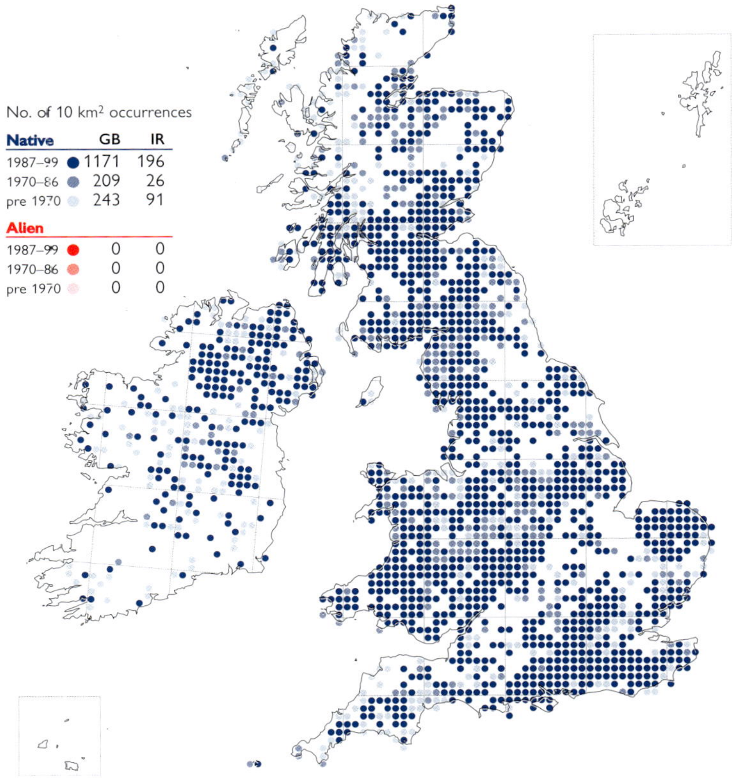

This deciduous fern is found in a range of damp habitats, including wet heaths, fens, mires, raised bogs, carr and wet woodland. It prefers rich alluvial soils with a high water table. More rarely it extends onto open moorland, possibly as a relic of former woodland. Generally lowland, but reaching 730 m in Atholl (E. Perth).

Native (change +1.06). *D. carthusiana* has declined in the last hundred years, wetland drainage, forest removal and agricultural improvement having taken their toll. However, it is much better recorded now than in the 1962 *Atlas*.

Eurosiberian Boreo-temperate element; also in N. America.

References: Atlas (12b), Hultén & Fries (1986), Jalas & Suominen (1972), Jermy *et al.* (1978), Meusel *et al.* (1965), Page (1997).

T. D. DINES

Dryopteris carthusiana × *D. dilatata* (*D.* × *deweveri*)

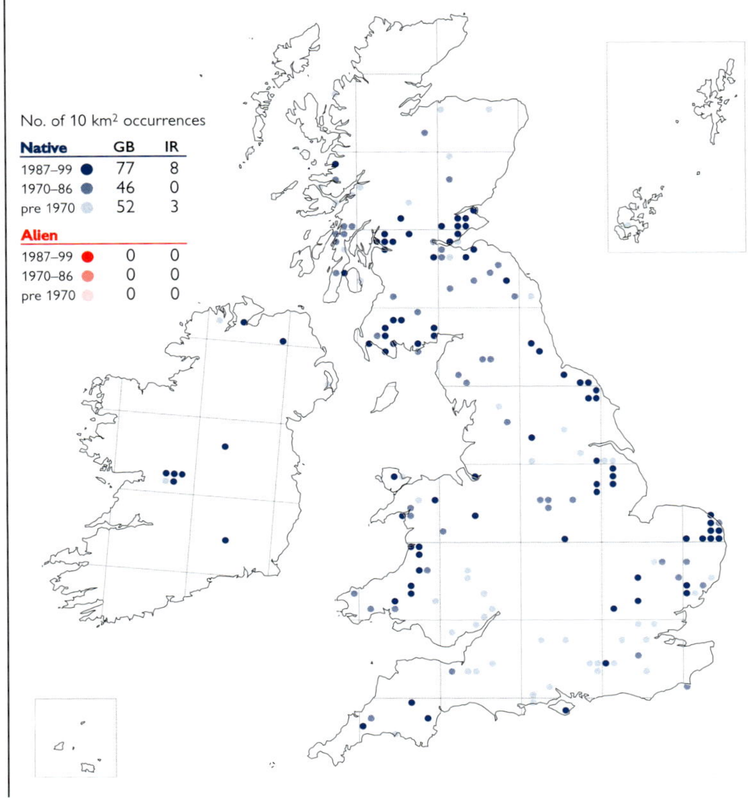

A deciduous fern found on damp, fairly acidic substrates in deciduous woodland, conifer plantations, ditches, *Alnus* and *Salix* carr, and disturbed lowland peat bogs. It is usually found with both parents, but often occurs in the absence of one, usually *D. carthusiana*, as the hybrid is more tolerant of the drier conditions typical of drained woodland. Generally lowland, but reaching 370 m at Llyn Rhos-goch (Cards.).

Native. This hybrid is under-recorded and is probably much more frequent than the map suggests.

Widespread in N., W. & C. Europe.

References: Jermy *et al.* (1978), Page (1997), Stace (1975).

T. D. DINES

Dryopteris dilatata Broad Buckler-fern

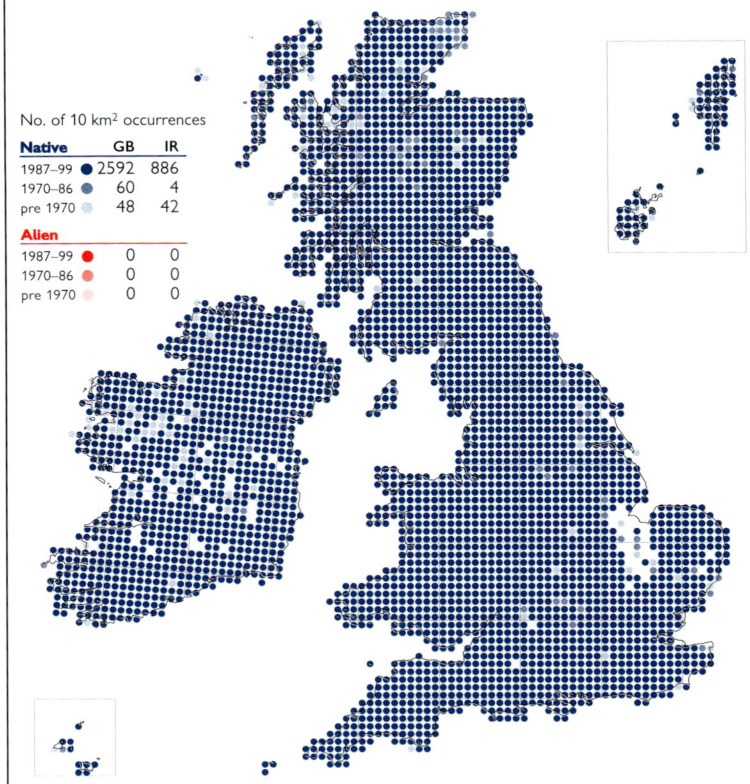

This deciduous fern grows on moderately to very acidic, well- to poorly-drained substrates. Habitats include deciduous and coniferous woodland, hedgerows, ditches, open moorland, rocky slopes, boulder scree and rock fissures. It can also be epiphytic in damp climates. 0–1050 m (N. of Loch Rannoch, Mid Perth), and reportedly to 1125 m in Scotland.

Native (change +1.32). The distribution of this species in the majority of its range is stable. In East Anglia it has become more frequent since 1930, as a decline in woodland management has increased the shade and humidity of woods and as it has invaded conifer plantations (Perring *et al.* 1964; Simpson, 1982).

European Temperate element.

References: Atlas (12c), Grime *et al.* (1988), Hultén & Fries (1986), Jalas & Suominen (1972), Jermy *et al.* (1978), Meusel *et al.* (1965), Page (1997). T. D. DINES

Dryopteris expansa Northern Buckler-fern

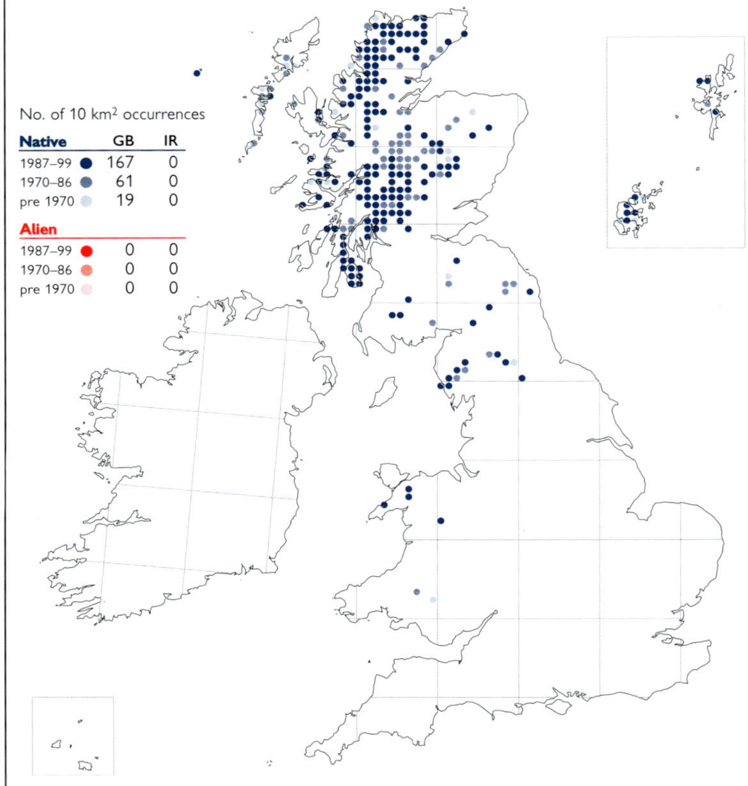

A deciduous fern found growing in open wet woodland and around rock outcrops at low altitudes, and in damp, sheltered hollows of upland boulder scree. Its substrates are usually mildly acidic, but it can grow in scree derived from quite base-rich mica-schists. 0–945 m (Stob Binnein, W. Perth).

Native. *D. expansa* was first mapped by Jermy *et al.* (1978). Some lowland sites have been lost to woodland clearance, but upland sites are relatively stable. The species is, however, difficult to identify and is almost certainly under-recorded.

Circumpolar Boreal-montane element, with a disjunct distribution.

References: Hultén & Fries (1986), Jalas & Suominen (1972), Page (1997).

T. D. DINES

Blechnum spicant Hard-fern

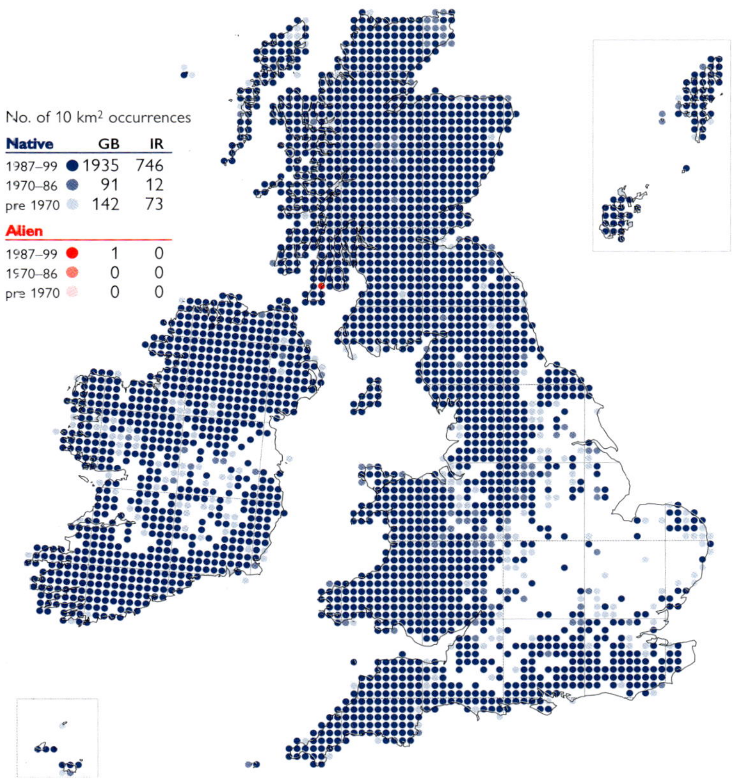

This evergreen calcifuge fern grows on damp peaty or loamy soils in deciduous and coniferous woodland. In suitably wet climates it extends onto open moorland, streamsides and hedgerows. 0–1065 m (Aonach Beag, Westerness), and reportedly to 1185 m elsewhere in Scotland.

Native (change –0.39). Mapped as 'all records' in the 1962 *Atlas*, this species has declined in England and C. Ireland; analysis of the database reveals that most losses have occurred since 1950. Declines are due to woodland clearance and the conversion of lowland heathland to improved pasture.

European Temperate element; also in E. Asia and western N. America.

References: Atlas (7a), Hultén & Fries (1986), Jalas & Suominen (1972), Jermy *et al.* (1978), Meusel *et al.* (1965), Page (1997).

T. D. DINES

Azolla filiculoides Water Fern

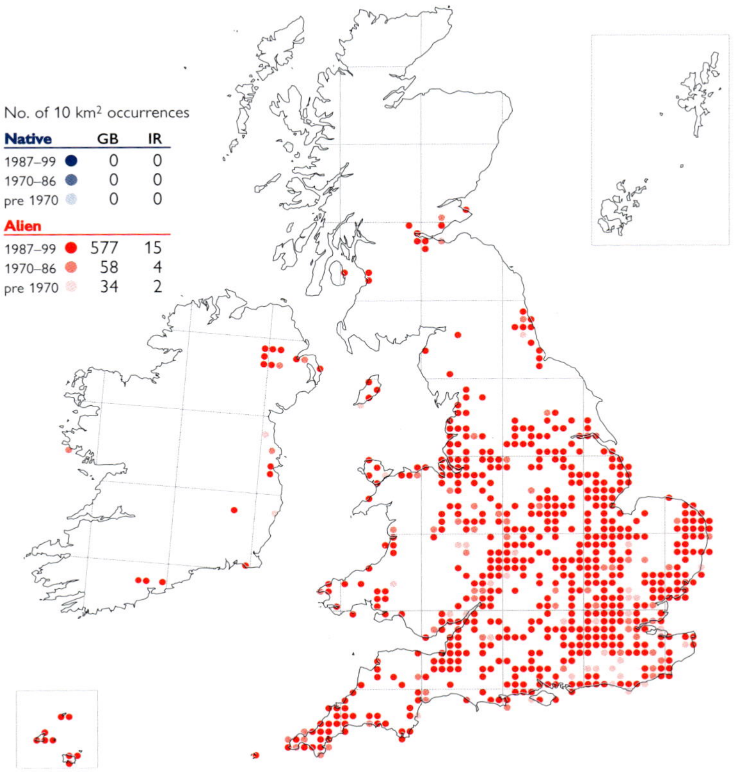

A floating fern of canals, ditches, ponds and sheltered bays in lakes and rivers. It is most frequent in calcareous water, or near the sea. It reproduces vegetatively, often producing dense mats, and sporocarps are not uncommon. Generally lowland, but reaching 450 m above Rydal Water (Westmorland).

Neophyte, though native in previous interglacials (change +2.76). First recorded in Britain in 1883, this species is much more frequent now than it was in the 1962 *Atlas*. In some areas populations fluctuate greatly annually. It is widely cultivated and frequently disposed of into the wild.

Native of western N. & S. America; widely naturalised in temperate and southern Europe and elsewhere.

References: Atlas (15c), FNAEC (1993b), Jalas & Suominen (1972), Jermy *et al.* (1978), Preston & Croft (1997).

C. D. PRESTON

Abies alba European Silver-fir

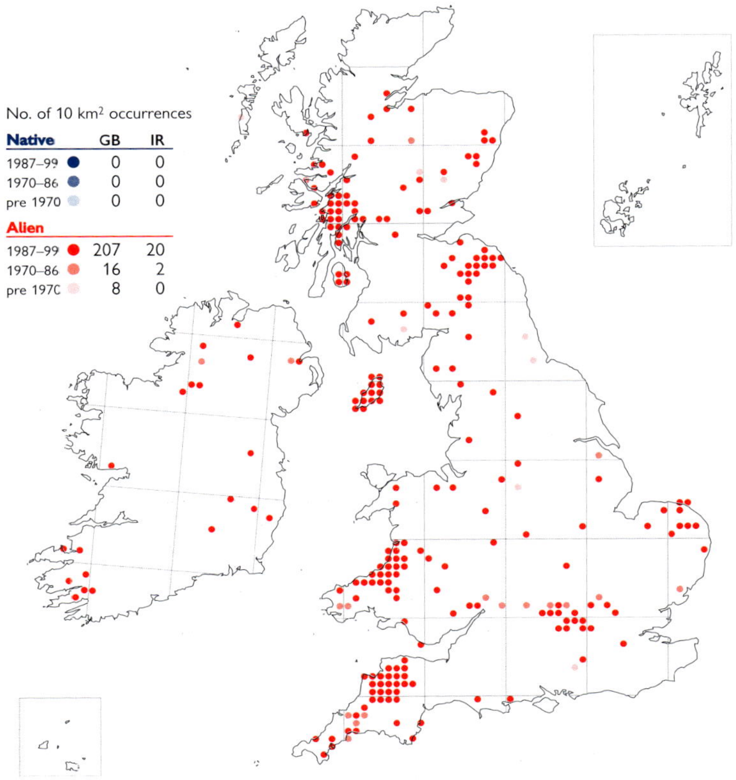

An evergreen tree of estate woodlands and plantations, less commonly found in towns. It regenerates freely from self-sown seed in mixed woodland on fertile soils, and saplings develop even under a woodland canopy. Lowland.

Neophyte. This species was introduced to Britain in 1603, and was formerly widely planted as a specimen tree and in plantations. It was known from the wild by 1914. It is now little planted because it is more susceptible to rust fungus and woolly aphids than are its congeners *A. grandis* and *A. procera*. The gaunt tops of old trees are often noticeable from afar. It is inconsistently recorded.

Native to the mountains of C. & S. Europe.

References: Jalas & Suominen (1973), Meusel *et al.* (1965), Mitchell (1972).

M. E. BRAITHWAITE

Abies grandis Giant Fir

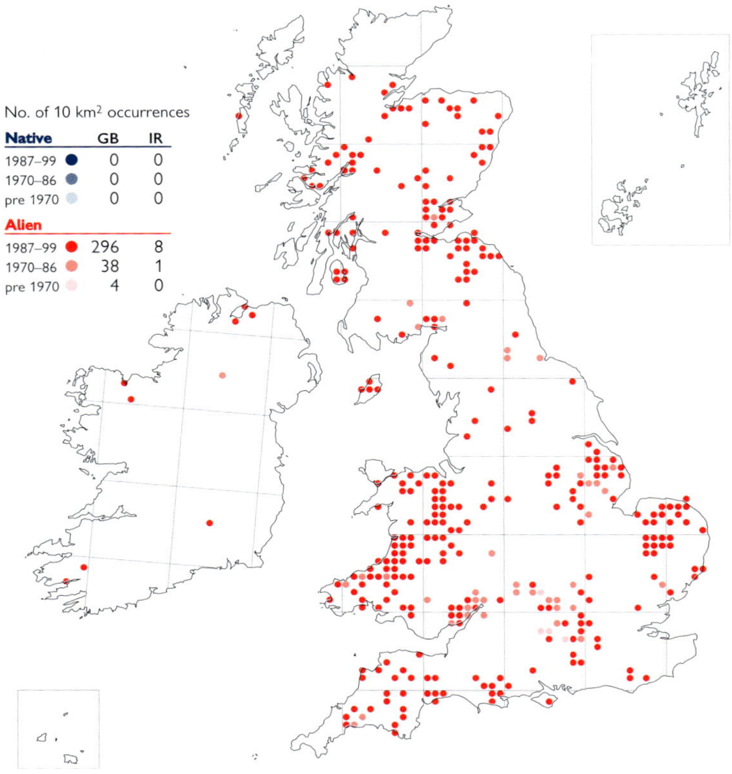

An evergreen tree of plantations and estate woodland, and an ornamental in parks, estates and large gardens. Though mainly grown in fertile soils in sheltered sites, it can thrive on poorer soils and survive in exposed situations. It occasionally regenerates from self-sown seed. Lowland.

Neophyte. This species, introduced to Britain in 1831, is increasingly being grown for forestry, but usually in small blocks. However, it was not recorded from the wild until 1981. The tallest specimen known was a tree at Cairndow (Main Argyll), more than 63 m high, but this tree lost its top in 1990. It is inconsistently recorded.

Native of western N. America; widely planted in N. & C. Europe.

References: FNAEC (1993b), James (1982), Mitchell (1972, 1996).

M. E. BRAITHWAITE

Abies procera Noble Fir

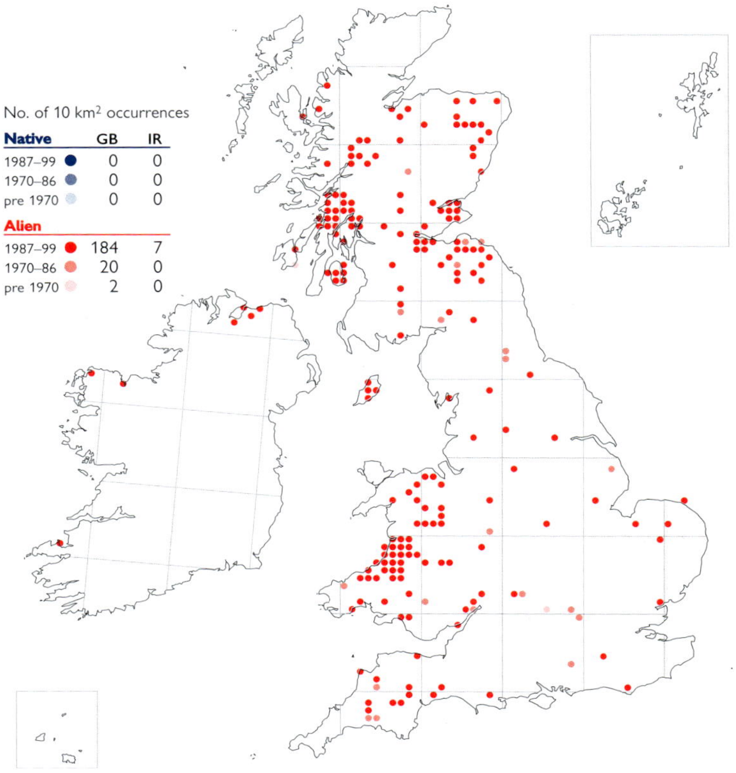

This evergreen tree is grown in parks and estate woodland, and occasionally reproduces from self-sown seed. Generally lowland, but reaching 420 m near Esgair Hir (Cards.).

Neophyte. Introduced to Britain in 1830, this species is particularly favoured in Scotland where magnificent specimens contribute in the backdrop to Scottish Baronial architecture. Although less vigorous than *A. alba* and *A. grandis*, it is a very robust tree which continues to be planted widely. It was not formally recorded from the wild, however, until 1959. Varieties with glaucous-blue foliage are often selected for cultivation. It is inconsistently recorded.

Native in western N. America; a rare tree in its native range.

References: FNAEC (1993b), Mitchell (1972, 1996).

M. E. BRAITHWAITE

Pseudotsuga menziesii Douglas Fir

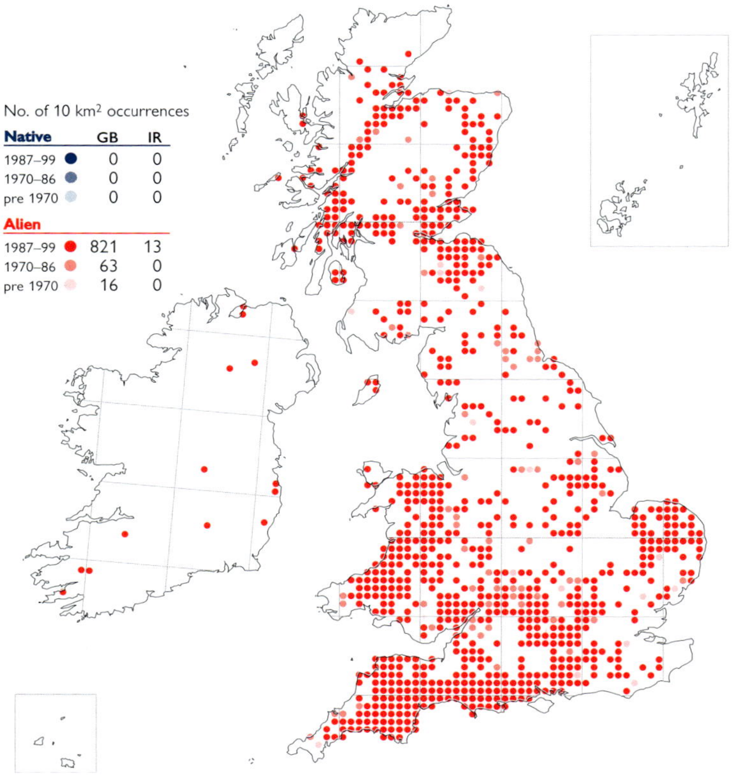

A very tall evergreen tree of plantations and policy woodlands, usually planted in sheltered, well-drained sites; also found in parks and large gardens as a specimen tree. Seed production in Britain is erratic, with long gaps between mast years, but natural regeneration occurs fairly widely. Generally lowland, but upper altitudinal limit unknown.

Neophyte. *P. menziesii* was introduced to Britain in 1826, and was recorded from the wild by 1968. It is locally common in forestry and very widespread in estate woodland. A 63 m specimen at Dunkeld, Perthshire, may currently be the tallest tree of any species in Britain. It is under-recorded.

Native of western N. America; very widely planted in Europe and elsewhere.

References: FNAEC (1993b), James (1982), Mitchell (1972, 1996), Nixon & Worrell (1999).

M. E. BRAITHWAITE

Tsuga heterophylla Western Hemlock-spruce

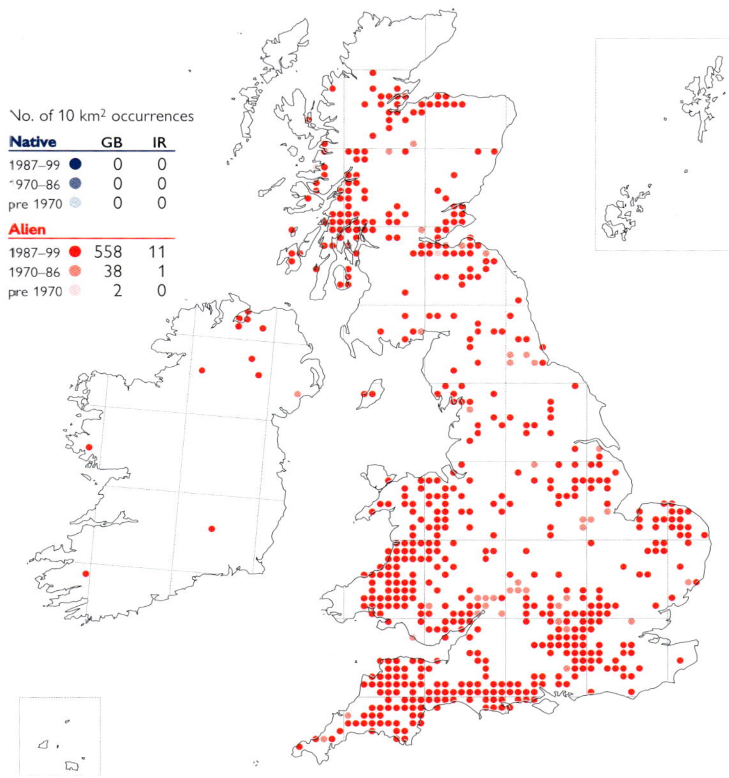

An evergreen tree of plantations and large gardens, grown mainly in moist, sheltered sites. It sometimes regenerates abundantly from seed, but saplings seldom reach maturity. Generally lowland, but upper altitudinal limit unknown.

Neophyte. *T. heterophylla* was introduced to Britain in 1852, but not recorded from the wild until 1959 when it was recorded as regenerating vigorously in S. Somerset. It is very tolerant of shading, and was formerly often planted under hardwoods. However, it is susceptible to butt rot, and its timber is not in demand. It is widespread as an amenity tree, but more frequent in Scotland in forestry plantations. It is under-recorded.

Native of western N. America.

References: FNAEC (1993b), Hultén (1968), James (1982), Mitchell (1972, 1996), Nixon & Worrell (1999).

M. E. BRAITHWAITE

Picea sitchensis Sitka Spruce

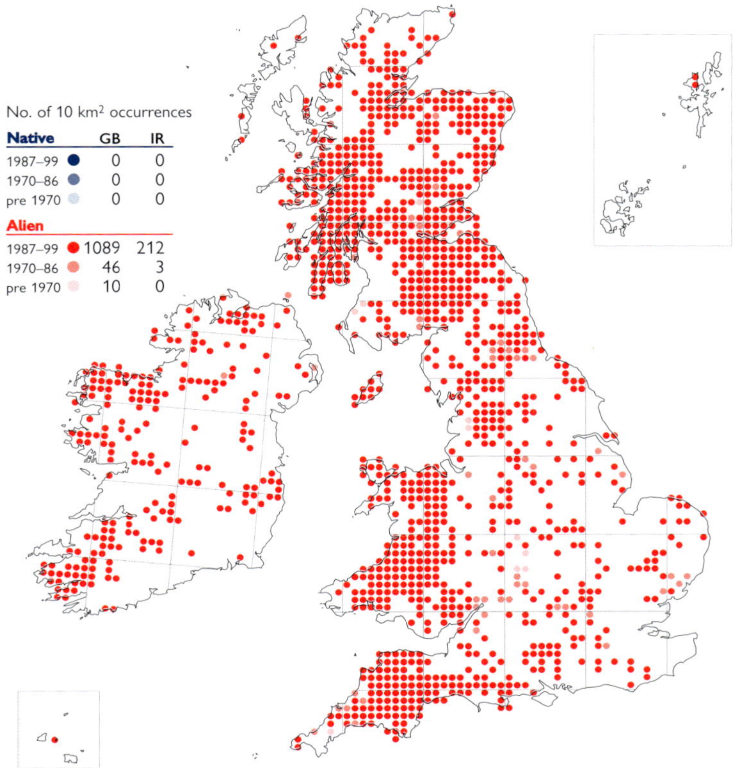

An evergreen tree of plantations, especially on wetter ground, in high rainfall areas and in exposed sites. It regenerates freely in open ground, in areas of clear felling, and on *Calluna* heath. 0–690 m (where it has been reported as regenerating near Stuc a' Chroin, W. Perth), and probably higher.

Neophyte. Introduced in 1832, *P. sitchensis* is now the most frequently planted conifer in our area, dominating huge areas of former moorland and bog. However, it is now being planted more often at lower altitudes. It was known from the wild by 1957.

Native of coastal forests in western N. America; widely planted in N.W. & C. Europe.

References: Cannell (1984), FNAEC (1993b), Hultén (1968), James (1982), Mitchell (1972, 1996), Nixon & Worrell (1999).

M. E. BRAITHWAITE

Picea abies Norway Spruce

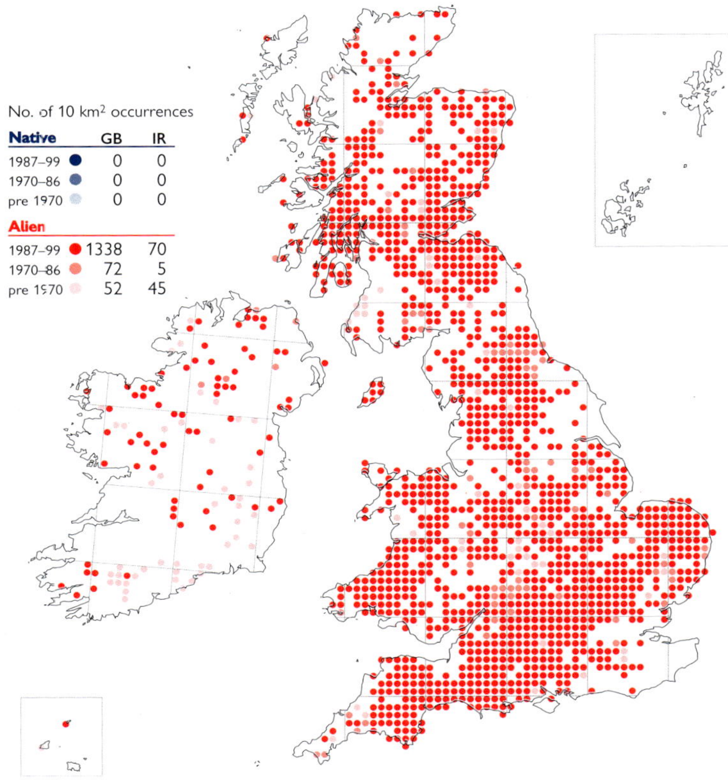

An evergreen tree of plantations and shelter-belts, grown in a wide variety of soils but not suited to deep peat, dry soils and exposed situations. It seeds freely, and natural regeneration occurs in open ground, in clear-felled areas and on heathland. Generally lowland, but upper altitudinal limit unknown.

Neophyte. *P. abies*, native to Britain in previous interglacials, has been cultivated in gardens since 995 (Harvey, 1981) and was recorded in the wild by 1927. Although less popular than *P. sitchensis*, it is still much planted for timber, pulping and Christmas trees.

A Eurasian Boreal-montane species; absent as a native from W. Europe.

References: Godwin (1975), Hultén & Fries (1986), Jalas & Suominen (1973), James (1982), Meusel *et al.* (1965), Mitchell (1972).

M. E. BRAITHWAITE

Larix decidua European Larch

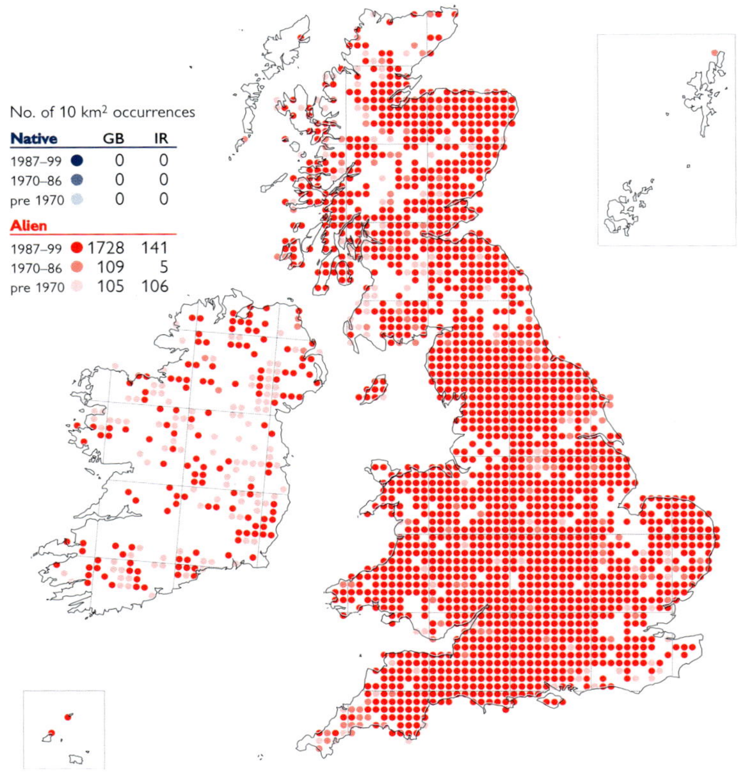

A deciduous tree of plantations, shelter-belts and parkland, grown in open situations, but not in very dry or waterlogged sites. It regenerates from seed freely, especially onto disturbed soil or rocky ground, and may become naturalised on crags. 0–660 m (S. of Garrigill, Westmorland).

Neophyte (change +2.91). *L. decidua* was introduced to Britain by 1629, and was formerly much grown in plantations. It was first recorded from the wild in 1886. It was greatly under-recorded in the 1962 *Atlas*. Though still planted for its timber, it is susceptible to larch canker and other species are now generally preferred.

Native to the Alps and Carpathians; widely planted elsewhere.

References: Atlas (16c), Jalas & Suominen (1973), James (1982), Meusel *et al.* (1965), Mitchell (1972, 1996), Nixon & Worrell (1999).

M. E. BRAITHWAITE

Larix decidua × *L. kaempferi* (*L.* × *marschlinsii*) Hybrid Larch

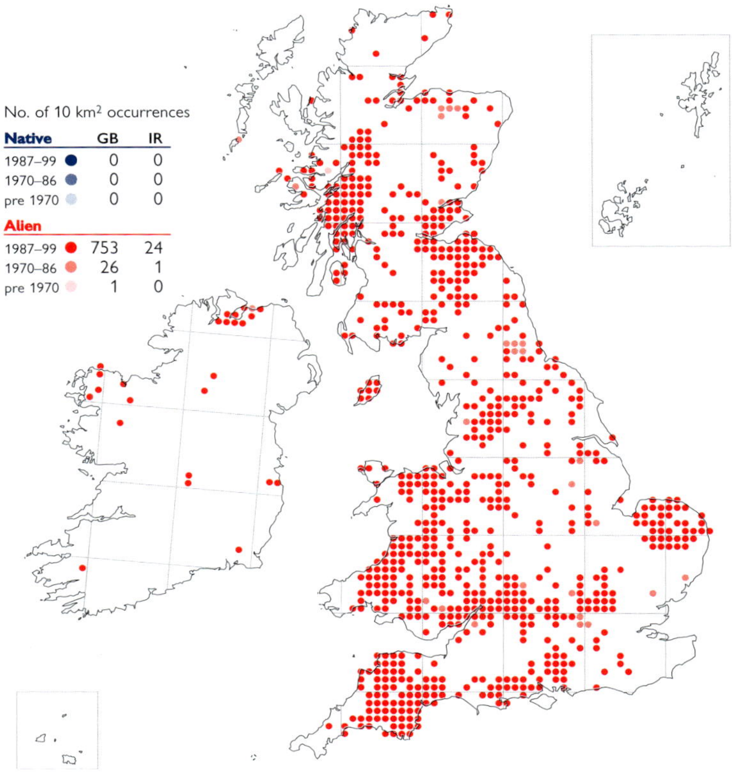

A deciduous tree of plantations and woodland. It is fertile and can back-cross with either parent, and regeneration from seed is frequent. Generally lowland, but upper altitudinal limit unknown.

Neophyte. This hybrid has proved to be a very vigorous tree. F_1 plants are much favoured for forestry and are widely planted. It was not recorded from the wild, however, until 1983. Its distribution is very similar to that of *L. kaempferi*; both taxa are widespread but under-recorded.

A hybrid tree which often originates when the parents are grown together, as at Dunkeld where it arose in about 1897 and was first noticed in 1904.

References: James (1982), Mitchell (1972, 1996), Nixon & Worrell (1999), Stace (1975).

M. E. BRAITHWAITE

Larix kaempferi Japanese Larch

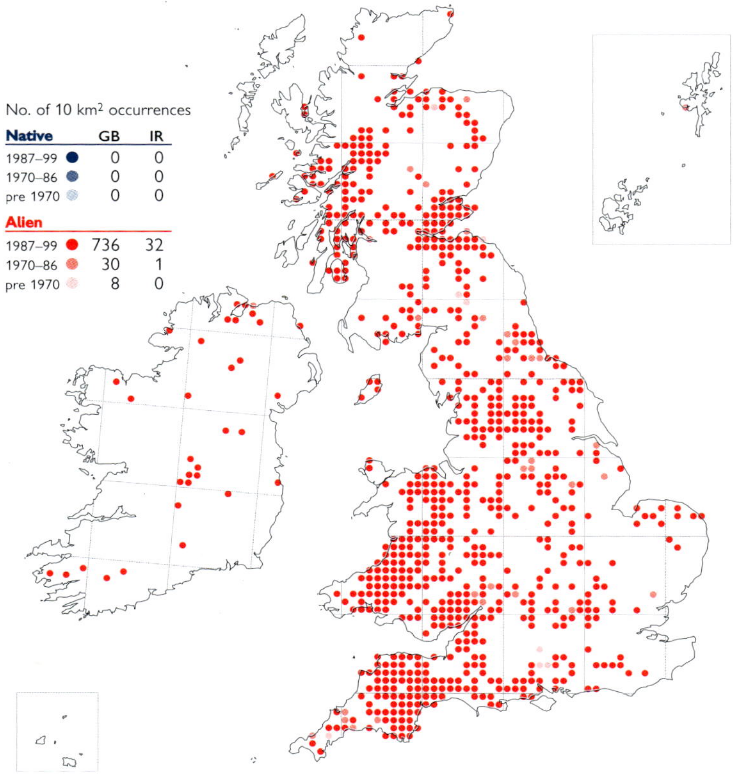

This tree is planted mainly as a forestry crop, but is also found in parks and large gardens. It grows in open situations, but does not thrive in either dry or waterlogged sites. Regeneration from seed can be frequent. Generally lowland, but upper altitudinal limit unknown.

Neophyte. Introduced to Britain in 1861, this species is less susceptible than *L. decidua* to larch canker, grows faster and tolerates poorer ground. It is therefore preferred by foresters, and is widely planted for timber, especially fencing. It was known in the wild by at least 1957. Like *L.* × *marschlinsii*, it is under-recorded.

Native of Japan; widely planted in N.W. Europe.

References: James (1982), Meusel *et al.* (1965), Mitchell (1972, 1996), Nixon & Worrell (1999).

M. E. BRAITHWAITE

Cedrus deodara Deodar

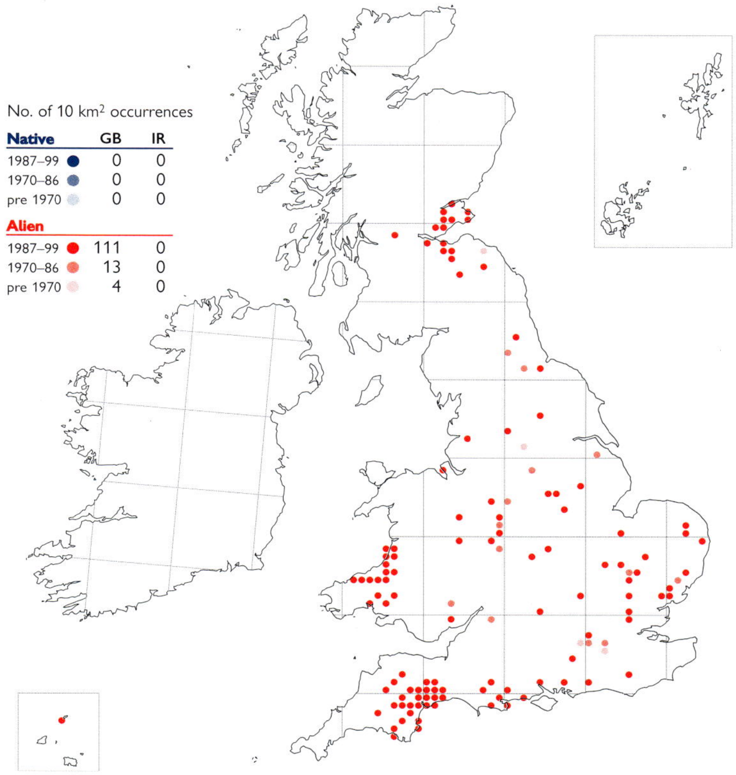

An evergreen tree of parks, gardens and churchyards, but not favoured in areas of high rainfall. It occasionally regenerates from seed. Lowland.

Neophyte. *C. deodara* was introduced to Britain in 1831. It is commonly planted in England and E. Scotland, but much less so in W. Britain. It is currently less popular than *C. atlantica* as a specimen tree, since it is not so visually dramatic until mature. It is under-recorded.

Native of the W. Himalayas.

References: Mitchell (1972, 1996).

M. E. BRAITHWAITE

Cedrus libani Cedar-of-Lebanon

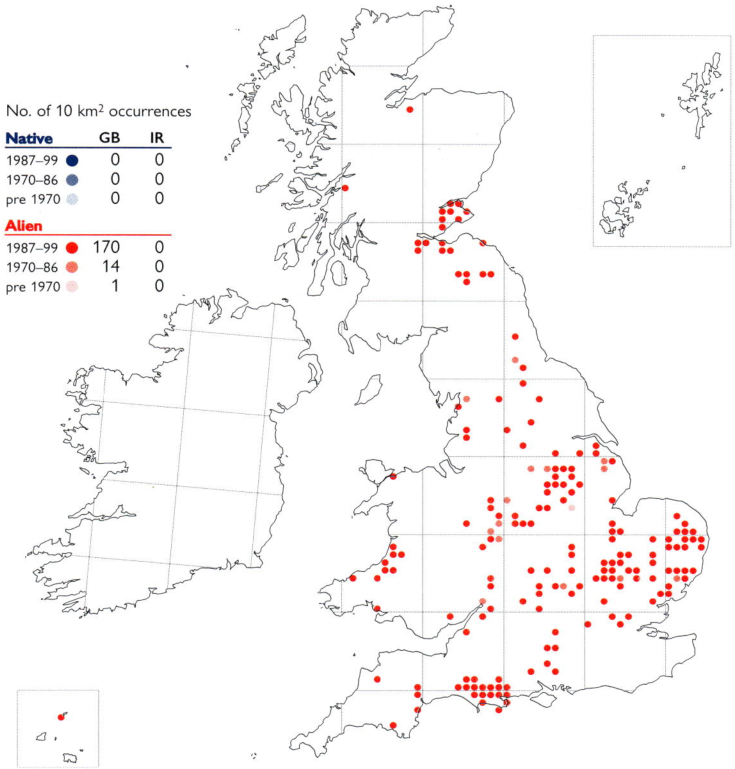

This evergreen tree of parks and gardens thrives best in the warmer, lower rainfall areas of Britain. Abundant seed often ripens, but natural regeneration is unknown. Lowland.

Neophyte. *C. libani* was introduced to Britain in 1638–9. When mature it is a wonderfully spreading tree in open situations, and for a long period was the pre-eminent ornamental tree of large formal gardens, complementing fine English architecture. But it was eclipsed in Victorian times by the variety of newly introduced conifers, and many of the early 19th century trees are now dying. It is under-recorded.

Native of montane forests from Lebanon to S. Turkey.

References: Mitchell (1972, 1996), Walters (1993).

M. E. BRAITHWAITE

Cedrus atlantica Atlas Cedar

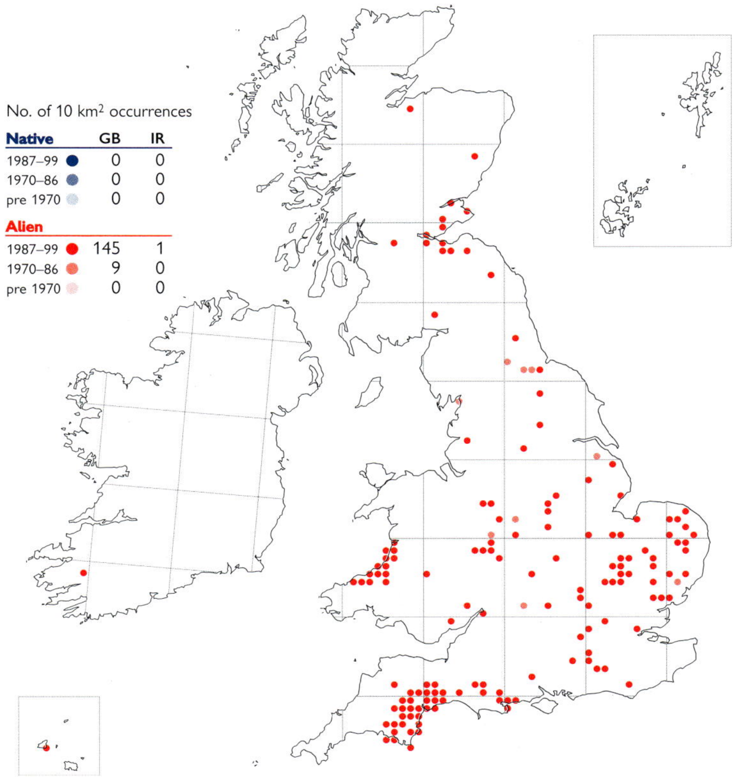

This is the hardiest of the cedars, and is tolerant of the widest range of growing conditions. It is found in parks and large gardens, but rarely regenerates from seed. Lowland.

Neophyte. Introduced into cultivation in Britain in about 1840, the early plantings of the wild type are fairly widespread. A variety with vividly glaucous-blue foliage has been selected from an introduction in 1845 and is now one of the most widely planted of ornamental conifers, although it was not formally recorded from the wild until 1986. It is under-recorded.

Native of the Atlas mountains (Algeria, Morocco); regarded by some authorities as a subspecies of *C. libani*.

References: Mitchell (1972, 1996).

M. E. BRAITHWAITE

Pinus sylvestris Scots Pine

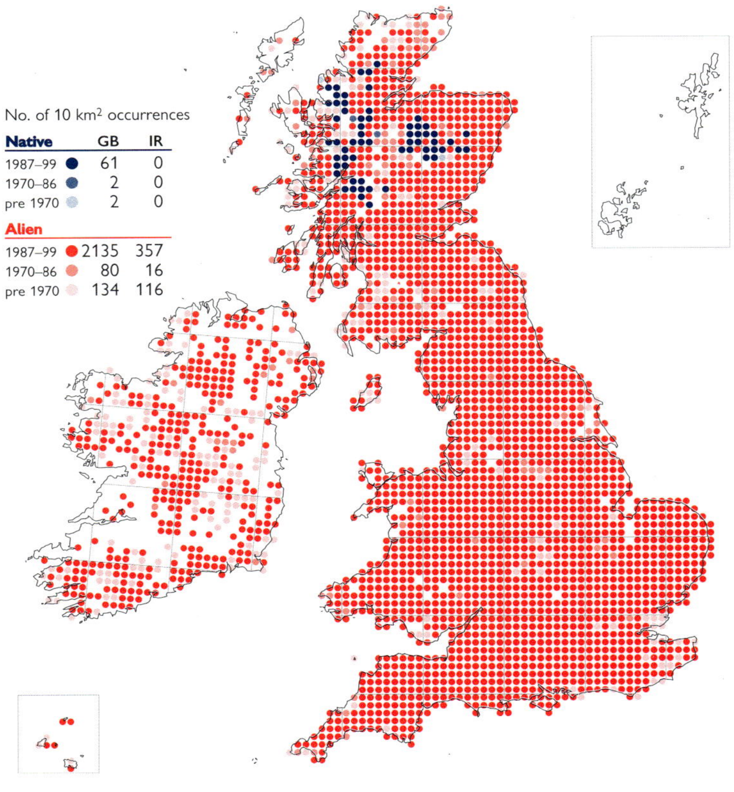

Native	GB	IR
1987–99	61	0
1970–86	2	0
pre 1970	2	0
Alien		
1987–99	2135	357
1970–86	80	16
pre 1970	134	116

P. sylvestris occurs as a native in pure stands or with other trees in mixed woodland. It prefers sandy and stony, acidic soils, though will also grow on waterlogged peats. It is widely planted in woods and shelter-belts, often becoming naturalised on heaths and bogs. 0–675 m as a native (Beinn a' Bhuird, S. Aberdeen).

Native (change +0.40). The native range is defined by the Caledonian Pinewood Inventory (Forestry Authority Scotland, 1998). Much effort has been directed at conserving native populations and controlling introduced populations on heathland. It is out of favour as a forestry tree, except in N.E. Britain.

Eurasian Boreal-montane element.

References: Atlas (16d), Carlisle & Brown (1968), Hultén & Fries (1986), Jalas & Suominen (1973), Meusel *et al.* (1965), Stewart *et al.* (1994).

M. E. BRAITHWAITE

Pinus nigra Austrian Pine, Corsican Pine

Native	GB	IR
1987–99	0	0
1970–86	0	0
pre 1970	0	0
Alien		
1987–99	946	15
1970–86	50	3
pre 1970	19	7

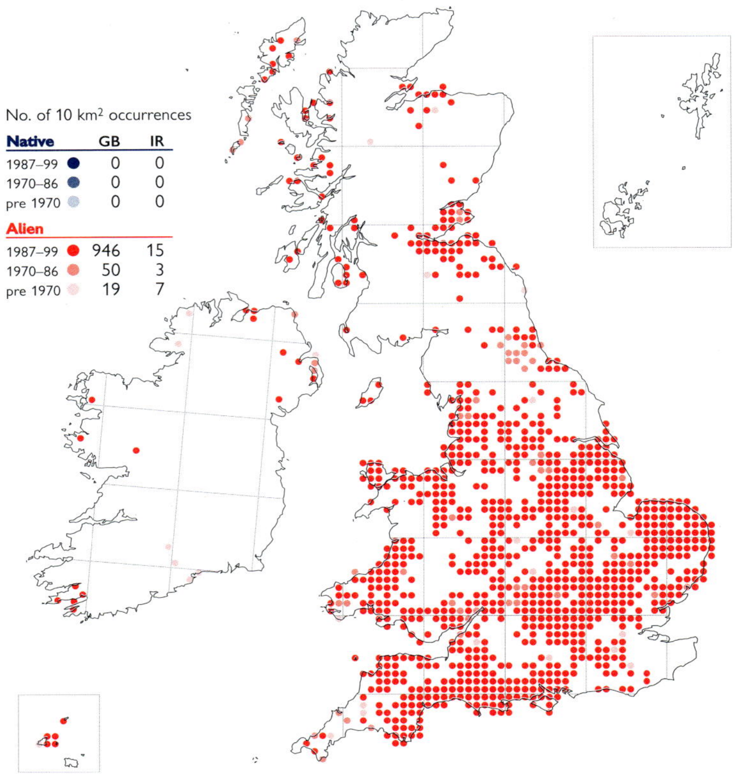

P. nigra is found in plantations, shelter-belts, parks, churchyards and large gardens, locally becoming naturalised on heaths and sand dunes. Subsp. *nigra* prefers light soils; subsp. *laricio* tolerates a wider range of soils. Regeneration from seed is frequent. Lowland.

Neophyte. Subsp. *laricio* was introduced in 1814, subsp. *nigra* in 1835; the former was known in the wild by 1905 and the latter by 1951. *P. nigra* is preferred to *P. sylvestris* for forestry in S. Britain, where it has become widespread on heaths.

The species is a native of S. Europe, N. Africa and S.W. Asia. Subsp. *nigra* occurs in Austria, Italy and the Balkans and subsp. *laricio* in Corsica, S. Italy and Sicily.

References: Jalas & Suominen (1973), James (1982), Meusel *et al.* (1965), Mitchell (1972, 1996).

M. E. BRAITHWAITE

Pinus contorta Lodgepole Pine

Native	GB	IR
1987–99	0	0
1970–86	0	0
pre 1970	0	0
Alien		
1987–99	424	116
1970–86	18	0
pre 1970	4	0

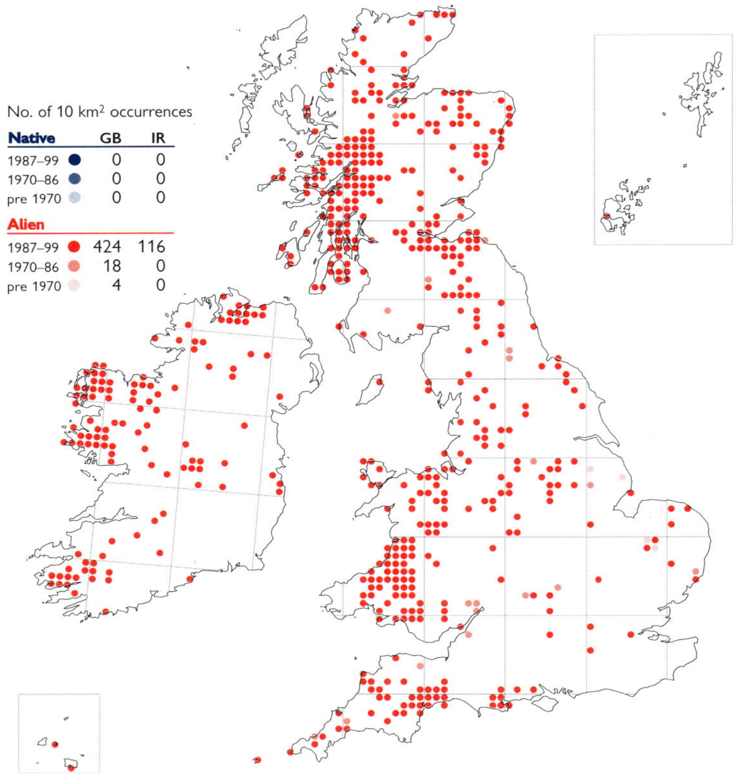

This evergreen tree of plantations is generally grown in more exposed and higher rainfall areas than *P. sylvestris*, where it will cope with poor, wet soils. It regenerates from seed freely and becomes naturalised in suitable habitats, particularly heathland. The altitudinal range is unknown.

Neophyte. *P. contorta* was introduced in 1851 and was recorded from the wild by 1968. Foresters have struggled to develop varieties suited to our climate, but the tree is now extensively and increasingly planted in north and west Britain and Ireland. In many such areas it is currently the only economic alternative to *Picea sitchensis*.

Native of western N. America; widely planted in N. Europe.

References: FNAEC (1993b), Hultén (1968), James (1982), Mitchell (1972, 1996).

M. E. BRAITHWAITE

Pinus pinaster Maritime Pine

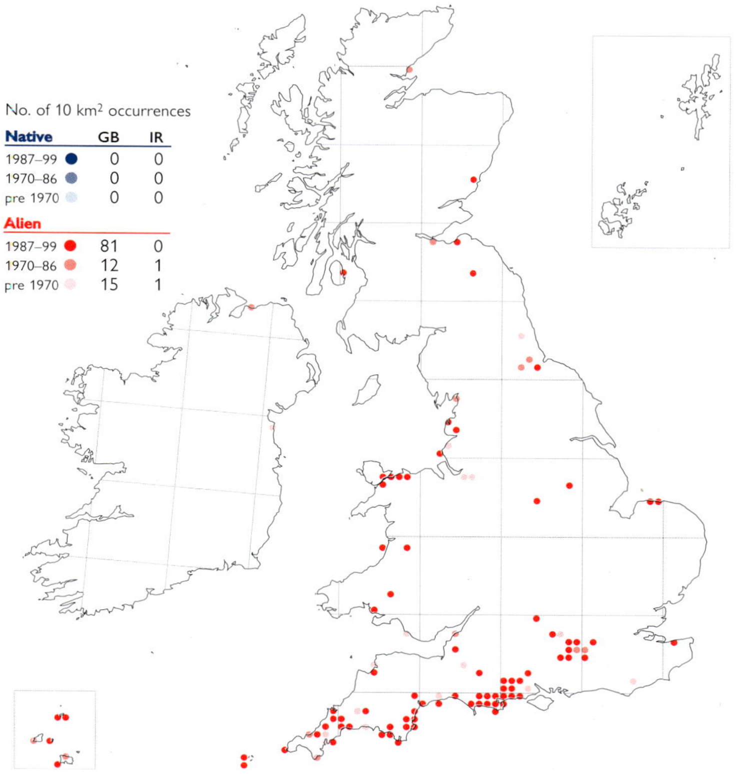

No. of 10 km² occurrences

Native	GB	IR
1987–99	0	0
1970–86	0	0
pre 1970	0	0
Alien		
1987–99	81	0
1970–86	12	1
pre 1970	15	1

An evergreen tree of shelter-belts and small plantations, flourishing only in southern coastal areas where it regenerates freely from seed. Lowland.

Neophyte. *P. pinaster* was introduced before 1596 and has been known from the wild since at least 1850. It has been planted quite widely along the coast of S. England, and is naturalised on heathland in Dorset and Hampshire, where it can be invasive. Away from the south coast, it is uncommon and found only as a specimen tree.

Native of S.W. Europe and the W. Mediterranean area.

References: Bolòs & Vigo (1984), Jalas & Suominen (1973), Mitchell (1972, 1996).

M. E. BRAITHWAITE

Pinus radiata Monterey Pine

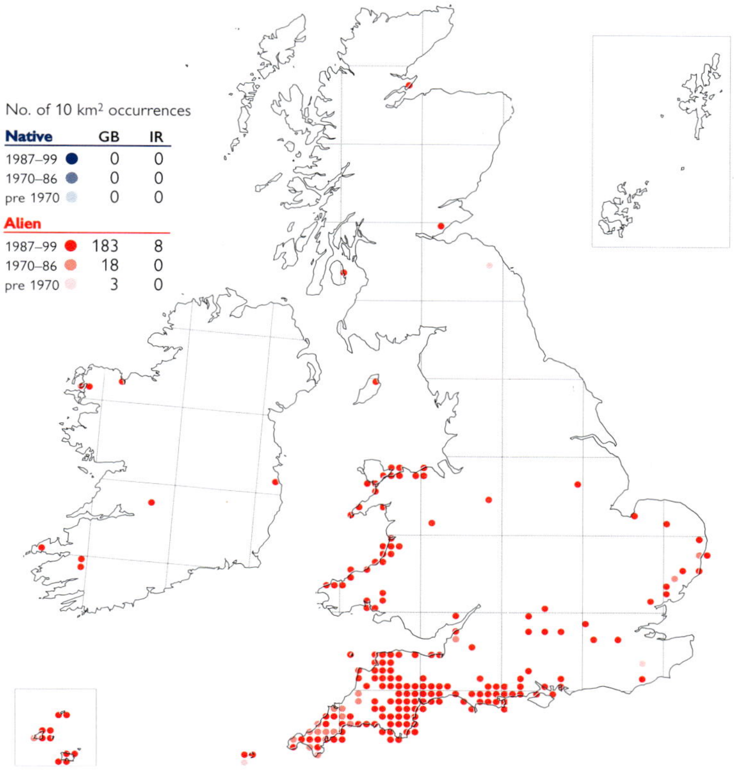

No. of 10 km² occurrences

Native	GB	IR
1987–99	0	0
1970–86	0	0
pre 1970	0	0
Alien		
1987–99	183	8
1970–86	18	0
pre 1970	3	0

An evergreen of shelter-belts, fringing many fields, parklands and estates in S. England. It is occasional elsewhere as a specimen tree in large gardens. It is adapted to regenerate after fire, with the cones retained on the tree and the seeds in the cones for up to twenty years. It rarely regenerates from seed in our area. Lowland.

Neophyte. *P. radiata* was introduced in 1832 and was known from the wild by 1963. As with most conifers, it is under-recorded.

Native of the coastal fog belt of California and Mexico (Baja California), where it grows on sandy substrates. Widely planted in Europe and elsewhere.

References: FNAEC (1993b), Mitchell (1972, 1996).

M. E. BRAITHWAITE

Pinus strobus Weymouth Pine

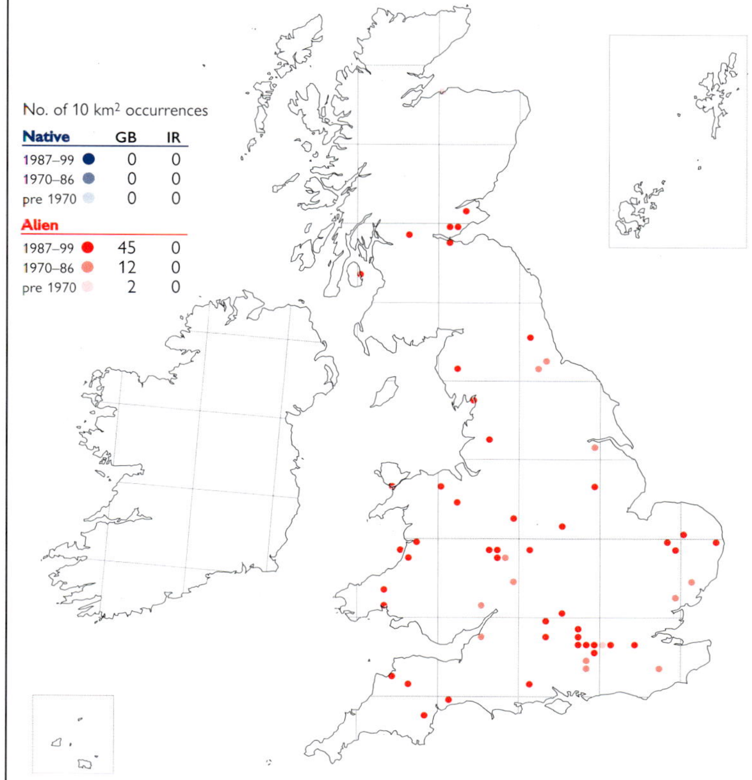

No. of 10 km² occurrences

Native	GB	IR
1987–99	0	0
1970–86	0	0
pre 1970	0	0
Alien		
1987–99	45	0
1970–86	12	0
pre 1970	2	0

An evergreen tree of mixed woodland and estates, occasionally found in parks, churchyards and large gardens. It is most suited to light loams in the milder districts of Britain, and regenerates from seed when mature. Lowland.

Neophyte. *P. strobus* was introduced in 1605, and recorded from the wild by 1957. Old trees are frequent, but younger ones are fewer, since planting is discouraged because of its susceptibility to a blister-rust disease which is killing many of the older trees.

Native of eastern N. & C. America.

References: FNAEC (1993b), Mitchell (1972, 1996).

M. E. BRAITHWAITE

Sequoia sempervirens Coastal Redwood

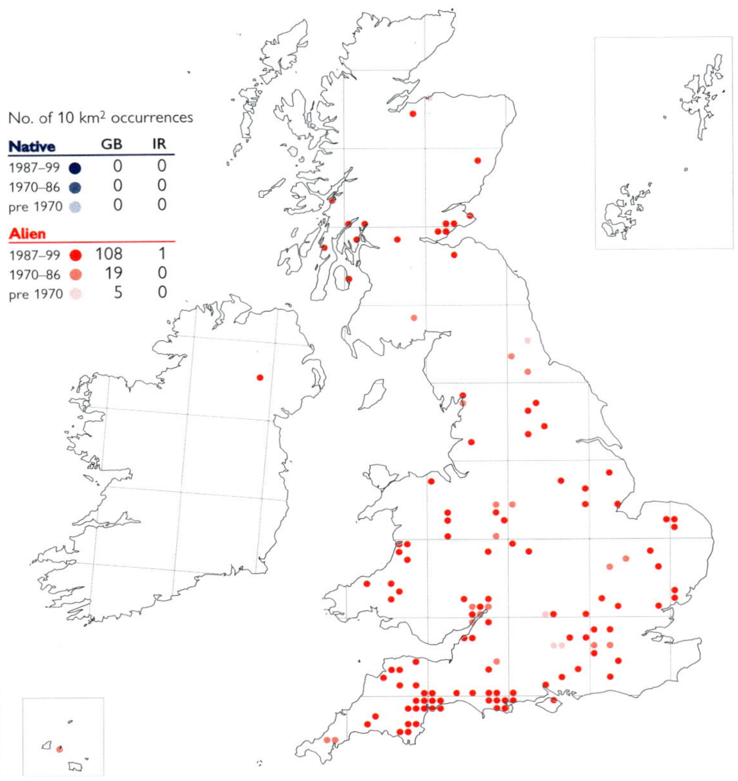

A tall evergreen tree of large gardens and parks. It has not been observed to set seed in Britain but it can produce dense coppice growth when felled. Lowland.

Neophyte. This tree was introduced (via the Crimea) to Britain in 1844 from California, where a native specimen of 114 m is currently the tallest tree in the world. In Britain, especially in the south west, early growth can be exceptionally vigorous but the tallest tree yet known is only 47 m, at Bodnant (Caerns.). It is probably under-recorded.

Native of coastal forests in western N. America (California, Oregon).

References: FNAEC (1993b), Mitchell (1972, 1996).

M. E. BRAITHWAITE

Sequoiadendron giganteum Wellingtonia

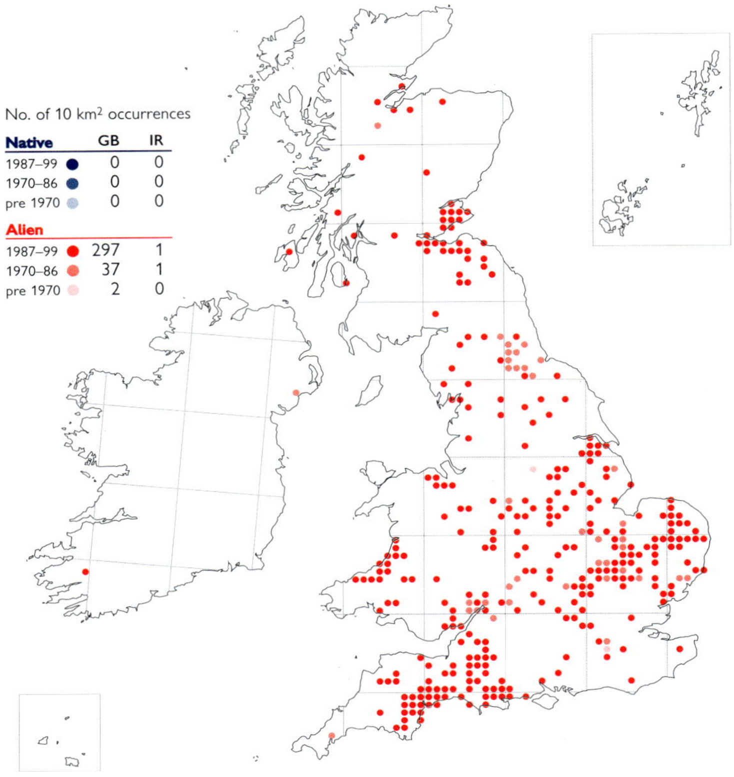

A very tall evergreen tree of parks, estates and large gardens, sometimes planted as avenues, and a specimen tree in lawns; found also in churchyards and occasionally as a roadside tree. It has not been observed to regenerate from seed in Britain. Lowland.

Neophyte. This tree was introduced to Britain in 1853. It arrived with a reputation as the biggest tree in the world and was immediately popular. It features prominently in the landscape but in the north it is consistently overtopped by *Abies alba* and *Pseudotsuga menziesii*. It has been inconsistently recorded.

Native of western N. America, where it occurs in montane coniferous forest in California.

References: FNAEC (1993b), Mitchell (1972, 1996).

M. E. BRAITHWAITE

Cryptomeria japonica Japanese Red-cedar

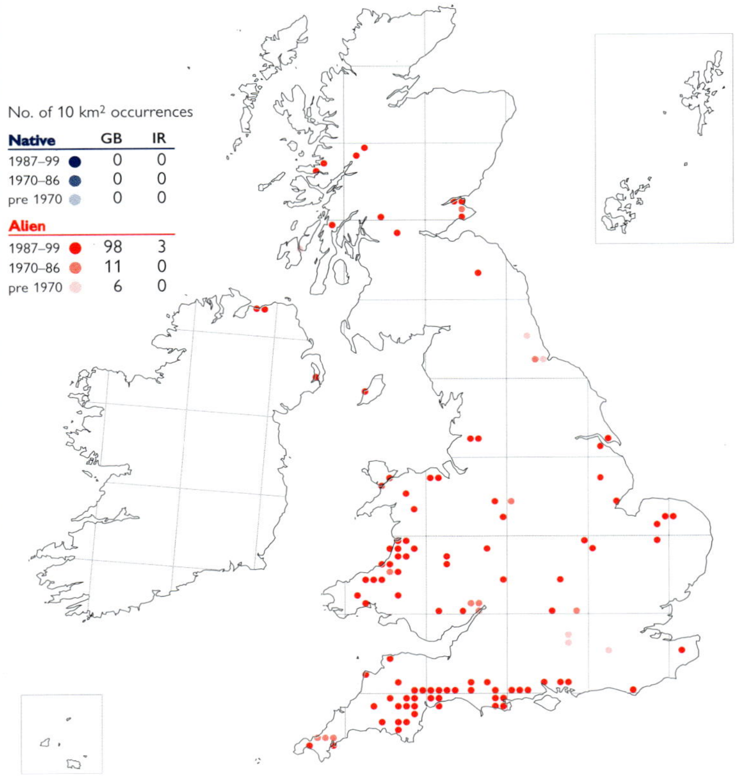

A medium or tall evergreen tree sometimes grown in forestry plots, especially in Wales, but more widely found in parks and large gardens, though seldom in towns. It grows best in W. Britain but rarely regenerates from seed. Lowland.

Neophyte. *C. japonica* was introduced to Britain from China in 1842 and from Japan in 1846. It has been cultivated over a long period in Japan and many cultivars are available, mainly dwarfs with permanently juvenile foliage. It is these cultivars that have found favour in gardens, but the type tree is sometimes included in conifer collections. The species was recorded from the wild in 1965.

Native of E. Asia (China, Japan).

References: Mitchell (1972, 1996).

M. E. BRAITHWAITE

Cupressus macrocarpa Monterey Cypress

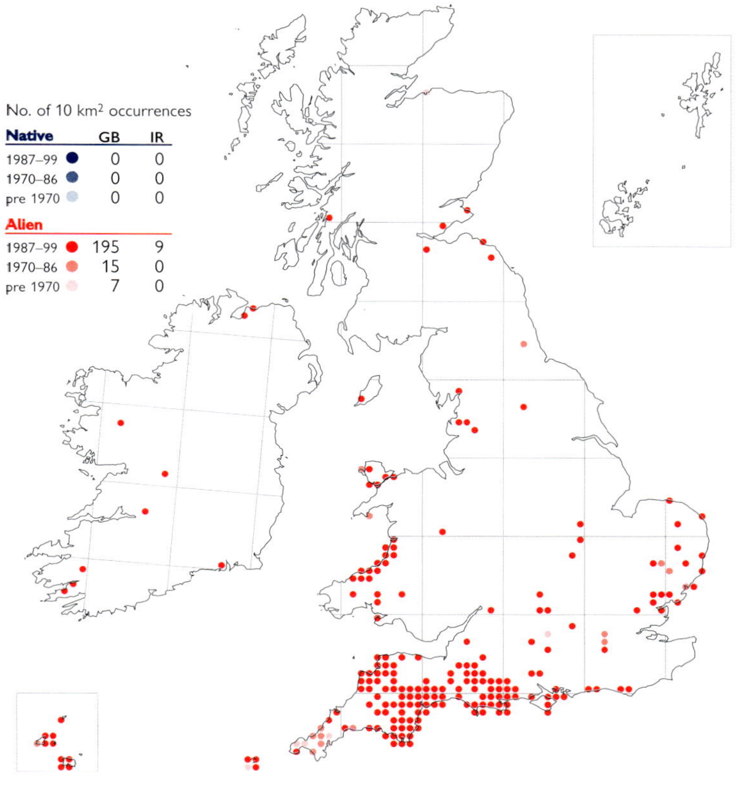

An evergreen tree of parks, gardens and churchyards, most vigorous in S. and W. Britain, where it is especially common. It has also been much used as hedging, though × *Cupressocyparis leylandii* is now preferred. It may regenerate from seed. Lowland.

Neophyte. *C. macrocarpa* was introduced to Britain in 1838, and is a popular ornamental conifer.

Native of California; a palaeoendemic restricted as a native to two coastal groves near Monterey but widely planted in N. America, Europe and elsewhere.

References: FNAEC (1993b), Mitchell (1972, 1996).

M. E. BRAITHWAITE

Chamaecyparis nootkatensis × *Cupressus macrocarpa* (*X Cupressocyparis leylandii*) Leyland Cypress

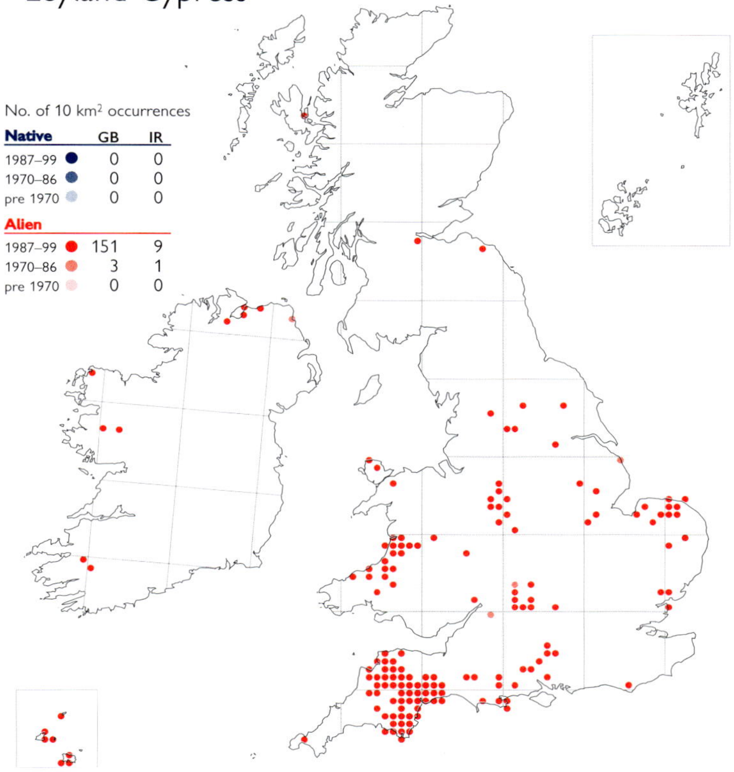

An evergreen tree especially grown as a suburban hedging plant, but also in parks and gardens as a specimen tree where it can exceed 30 m. Flowering is largely confined to the cultivar 'Leighton Green', but viable seed is not formed. Lowland.

Neophyte. Despite its ubiquitous use for hedging, this astonishingly vigorous hybrid is relatively uncommon in the countryside, though it is increasingly used for screening various agricultural and commercial sites in rural areas, as well as in towns. It is very inconsistently recorded.

A hybrid of garden origin which has arisen several times, first in 1888 at Leighton Hall, near Welshpool (Monts.). There are many cultivars, many derived from the trees that arose at Leighton Hall and were sent to Haggerston Castle, Northumberland.

References: Mitchell (1972, 1996).

M. E. BRAITHWAITE

Chamaecyparis lawsoniana Lawson's Cypress

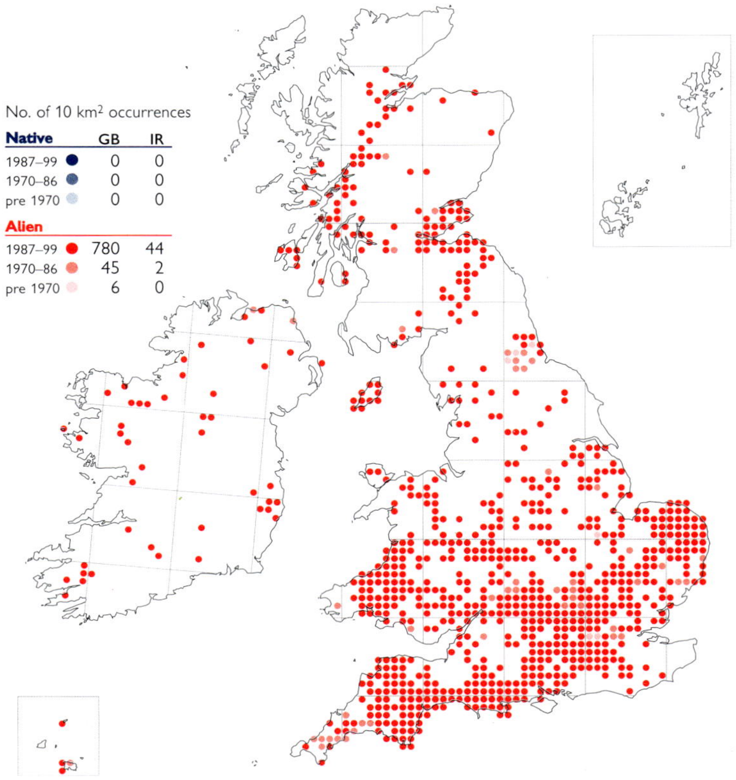

This evergreen tree is widely found in parks, gardens and churchyards, and in shelter-belts. It frequently regenerates from seed, and plants arise opportunistically on banks, walls and woodland margins. Lowland.

Neophyte. *C. lawsoniana* was introduced to Britain in 1854, and is now represented by a wide range of cultivars. It is less common in rural areas than in towns, but is sometimes recommended for underplanting in plantations. It is increasingly widespread, but was not formally recorded from the wild until 1958 (W. Kent).

Native of western N. America (California, Oregon); widely planted in Europe.

References: FNAEC (1993b), Hultén (1968), James (1982), Mitchell (1972, 1996).

M. E. BRAITHWAITE

Chamaecyparis pisifera Sawara Cypress

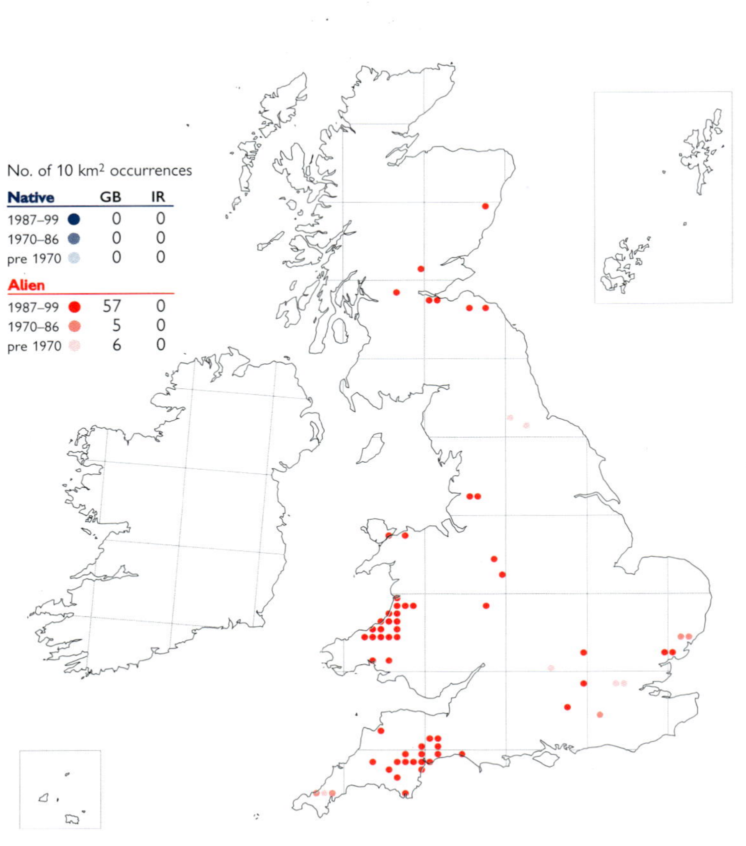

An evergreen shrub or tree of parks and gardens, including suburban amenity areas, occasionally regenerating from seed. Lowland.

Neophyte. *C. pisifera* was first introduced to Britain as the cultivar 'Squarrosa' in 1843, with the type following in 1861. It had long been cultivated in Japan and numerous varieties have been introduced, some with permanently juvenile foliage. It is a very widely used conifer, but is uncommon in the countryside where it is usually found in association with *C. lawsoniana*. It was not formally recorded from the wild until 1982.

Native of Japan.

References: Mitchell (1972, 1996).

M. E. BRAITHWAITE

Thuja plicata Western Red-cedar

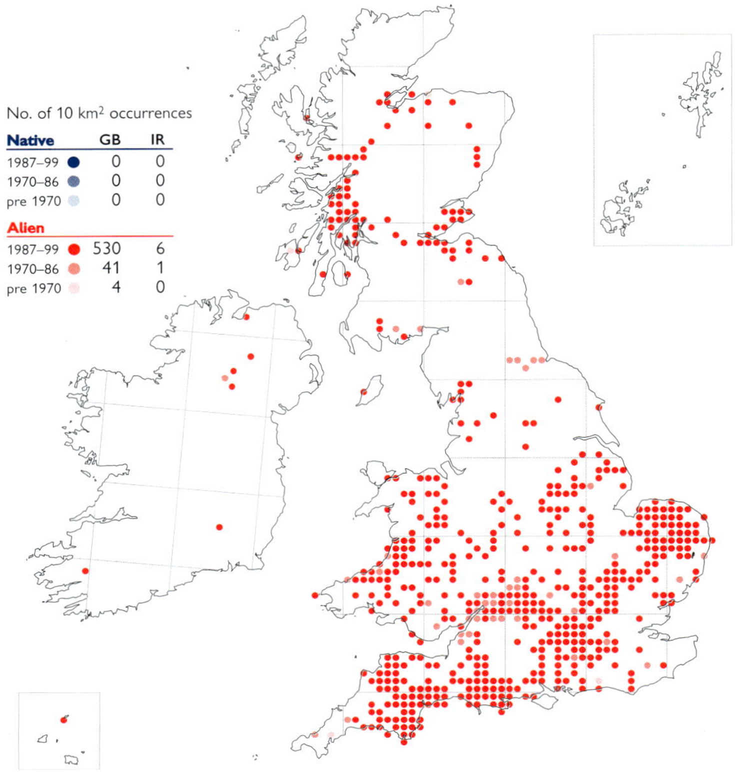

An evergreen tree of parks and gardens, and of plantations, regenerating from seed freely when mature. It prefers fertile conditions, but will grow on chalk and other shallow, infertile soils. Lowland.

Neophyte. *T. plicata* was first introduced to Britain in 1853. It is very widespread as a decorative tree, but also locally in plantations, where it is used for under-planting hardwoods or larch. It was not recorded from the wild until 1954, and is increasing.

Native of western N. America.

References: FNAEC (1993b), Hultén (1968), James (1982), Mitchell (1972, 1996).

M. E. BRAITHWAITE

Juniperus communis Common Juniper

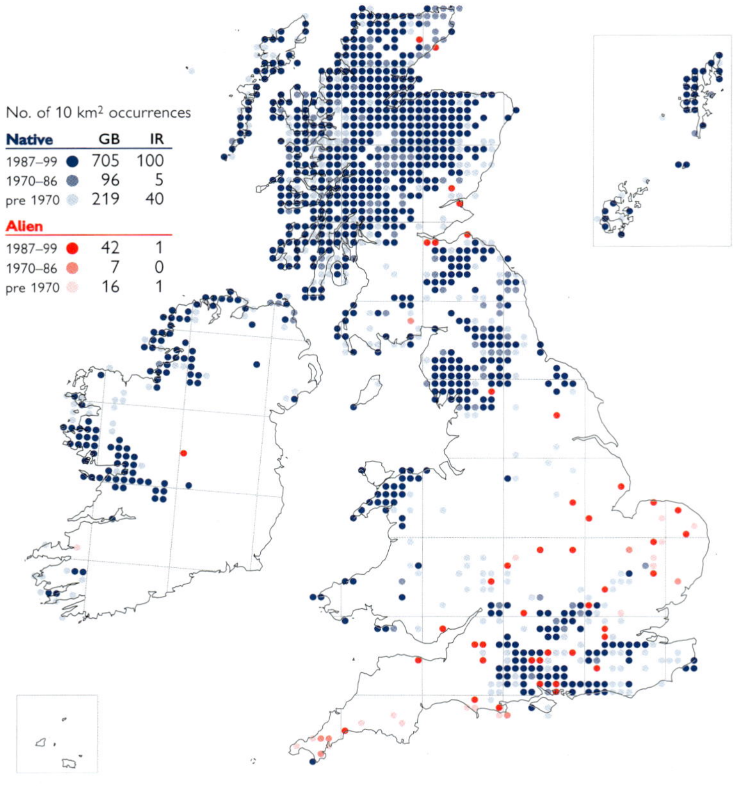

J. communis is a dioecious evergreen conifer found on basic and acidic soils in a wide range of habitats, including chalk downland, heather moorland, oceanic heaths, rocky slopes and in *Betula*, *Quercus* and *Pinus* woods. Regeneration is inhibited by continuous grazing. 0–975 m (Braeriach, S. Aberdeen).

Native (change –0.42). The decline in S. Britain was already apparent in the 1962 *Atlas* and since then losses in other areas have become apparent. Burning has largely eliminated it on grouse moors, and the scrub communities in which it regenerates have declined severely due to grazing, burning, succession to woodland and afforestation.

Circumpolar Boreo-temperate element.

References: Atlas (17a), Dearnley & Duckett (1999), Hultén & Fries (1986), Jalas & Suominen (1973), Meusel *et al.* (1965), Ward (1973, 1981).

M. E. BRAITHWAITE

Juniperus communis subsp. *communis*

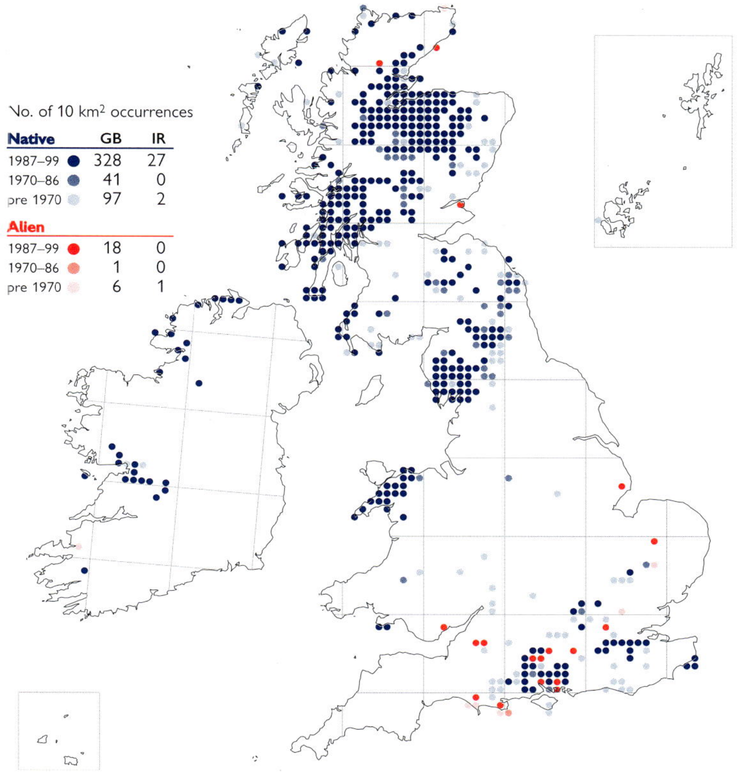

A member of a distinctive shrub community on the chalk in S. England, generally growing on shallow rendzinas. Elsewhere it occurs on both acidic and basic soils on rocky hillsides, moorland and maritime heaths, and in *Betula, Quercus* and *Pinus* woods. 0–975 m (Braeriach, S. Aberdeen).

Native. Many stands have an even-age structure and plants may therefore become moribund together. Regeneration is often poor as reproduction from seed is dependant on disturbance or a sudden cessation of grazing. Stands have been eliminated by overgrazing, burning and afforestation. Some recorders have not identified plants to subspecies.

Eurosiberian Boreo-temperate element.

References: Atlas Supp. (4b), Dearnley & Duckett (1999), Hultén & Fries (1986), Jalas & Suominen (1973), Tansley (1939), Ward (1973). M. E. BRAITHWAITE

Juniperus communis subsp. *hemisphaerica*

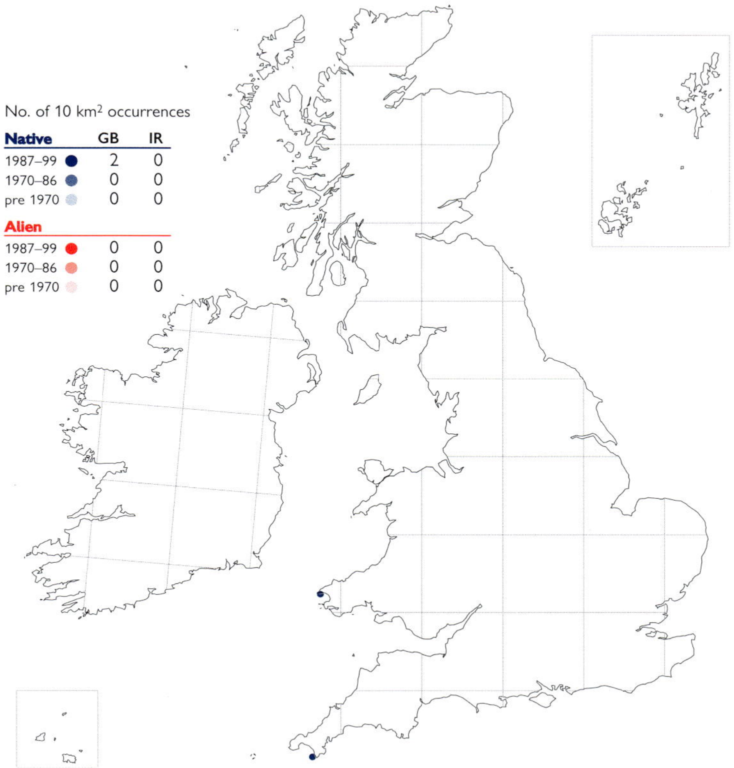

This dwarf evergreen shrub is found in species-rich maritime heath. At the Lizard (W. Cornwall) it grows on broken, rocky slopes over serpentine rocks, where cattle-tracks act as fire-breaks. In Pembrokeshire, it occurs in wind-pruned coastal scrub and dwarf heath. Lowland.

Native. This subspecies is known only as tiny populations on the Lizard peninsula and Pembrokeshire. The Lizard population was first recorded in 1871; it was described as occurring at its single locality 'in abundance' in 1874 but had been reduced to seven plants by 1970.

Mediterranean-montane element.

References: Coombe (1973), Hultén & Fries (1986), Jalas & Suominen (1973), Perring & Sell (1968). M. E. BRAITHWAITE

Juniperus communis subsp. *nana*

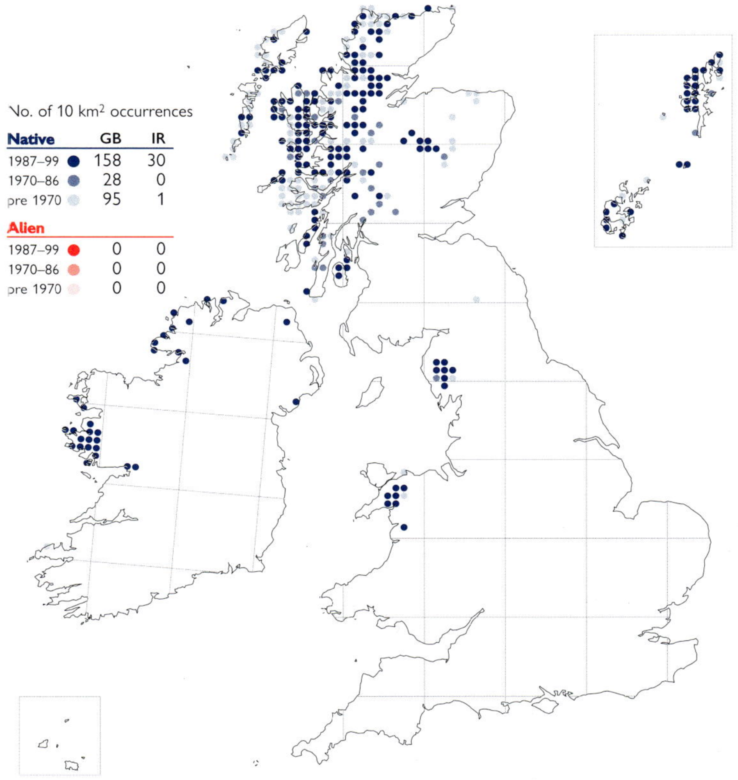

This low shrub is found in a variety of coastal and montane dwarf-shrub heaths, often subject to severe wind-pruning. It also grows on rock ledges and blanket bogs. Although it may be locally plentiful, many populations are very small. 0–800 m (Y Lliwedd, Caerns.), and possibly higher in Scotland.

Native. Subsp. *nana* may once have been more widespread in Scotland, but has contracted because of its sensitivity to burning and grazing (McVean & Ratcliffe, 1962). Some recorders have not identified material to subspecies, and some pre-1970 records may relate to stunted subsp. *communis*, making an assessment of change difficult.

Circumpolar Boreal-montane element.

References: Atlas Supp. (4c), Hultén & Fries (1986), Jalas & Suominen (1973), McVean (1961), Perring & Sell (1968), Rodwell (1991b). M. E. BRAITHWAITE

Araucaria araucana Monkey-puzzle

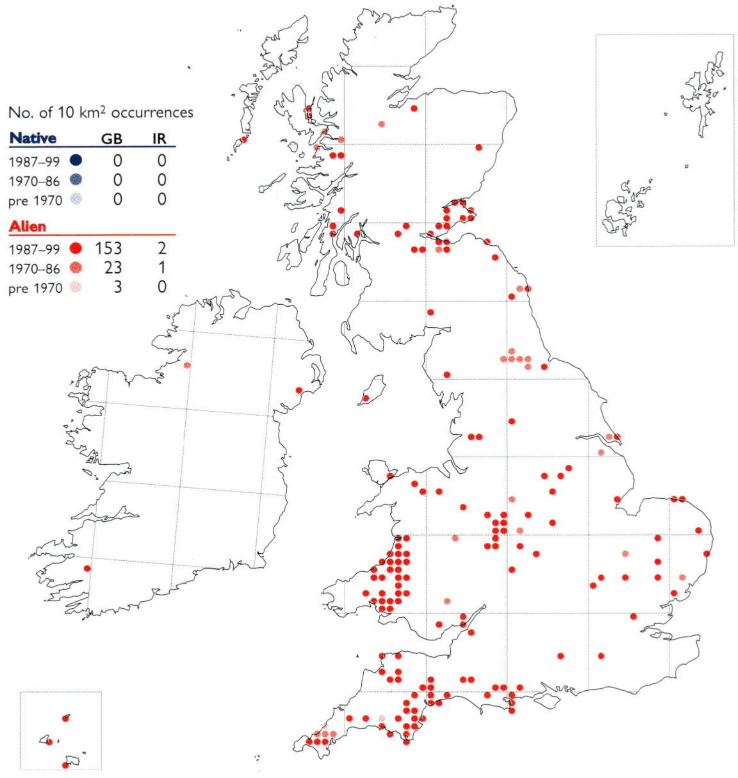

A dioecious evergreen tree of town gardens, parks, policies and plantations. It grows best in fertile soils, especially on the western seaboard, but it is very hardy and will thrive in exposed situations. Seed is abundantly produced, but the species very rarely regenerates in Britain. Lowland.

Neophyte. This tree was introduced to Britain in 1795. It is most commonly grown as a specimen tree, but may be less frequently planted than formerly. It is inconsistently recorded.

Native of S. America (Argentina, Chile).

References: Mitchell (1972, 1996).

M. E. BRAITHWAITE

Taxus baccata Yew

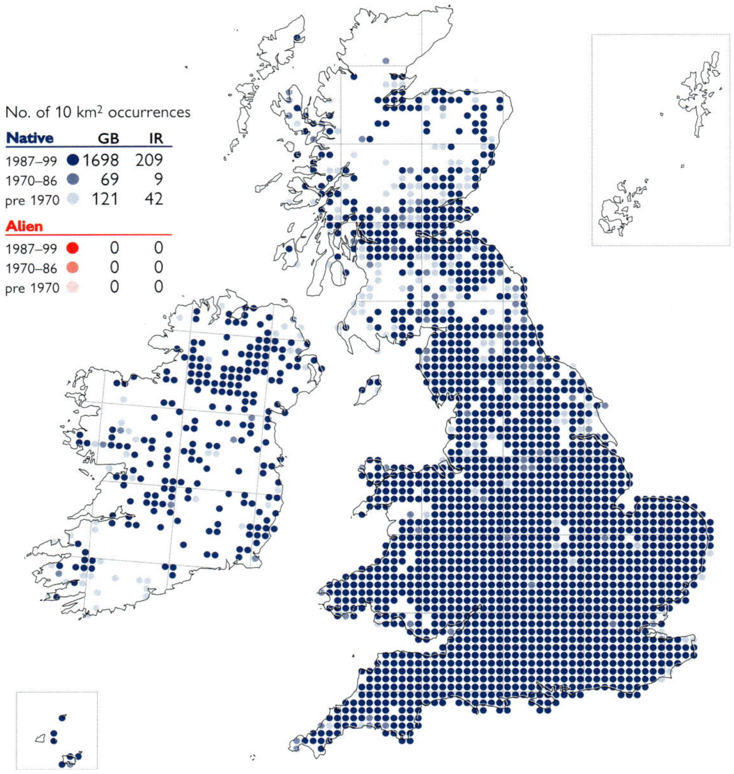

This evergreen tree is mainly a plant of well-drained calcareous soils, but occurs locally over acidic rocks. It grows in mixed deciduous woods on limestone and also forms pure stands; ancient woods of *T. baccata* occur on the chalk in S. England. It is very widely planted in churchyards, parks and large gardens, from where it seeds freely. 0–470 m (Purple Mt., S. Kerry).

Native (change +0.86). For the 1962 *Atlas*, only the presumed native distribution was mapped. However, native and alien records have proved impossible to separate, and all occurrences are mapped here as if they were native.

European Temperate element.

References: Atlas (17b), Halliday (1997), Hultén & Fries (1986), Jalas & Suominen (1973), Meusel *et al.* (1965), Mitchell (1972, 1996), Tansley (1939), Williamson (1978).

M. E. BRAITHWAITE & M. J. WIGGINTON

Laurus nobilis Bay

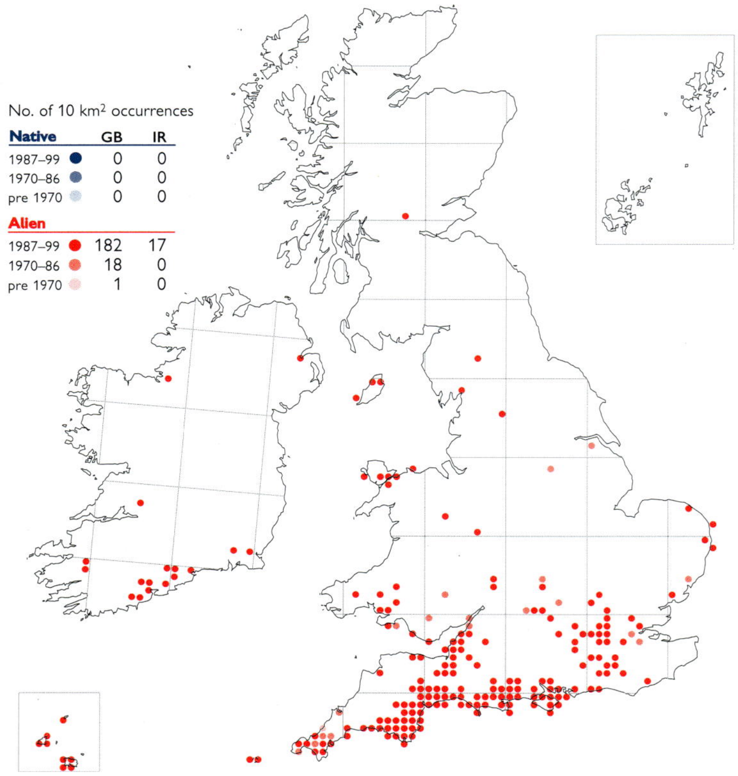

A dioecious evergreen shrub or small tree, arising from bird-sown seed or found as a garden throw-out in woodland and scrub and on sea-cliffs, dunes, roadsides and river banks; it also occurs as a relic of cultivation. Many populations are well-naturalised and reproduce from seed, especially those on coastal cliffs. Lowland.

Neophyte. This species, very popular in gardens for its aromatic leaves, has been cultivated in British gardens since at least 995 (Harvey, 1981). It may fruit prolifically, especially in hot summers, although many plants are highly pruned and may not fruit freely. It was not recorded in the wild until 1924, but appears to be increasing.

Native of the Mediterranean region.

References: Bean (1973), Bolòs & Vigo (1984), Jalas & Suominen (1991), Meusel *et al.* (1965).

T. D. DINES

Asarum europaeum Asarabacca

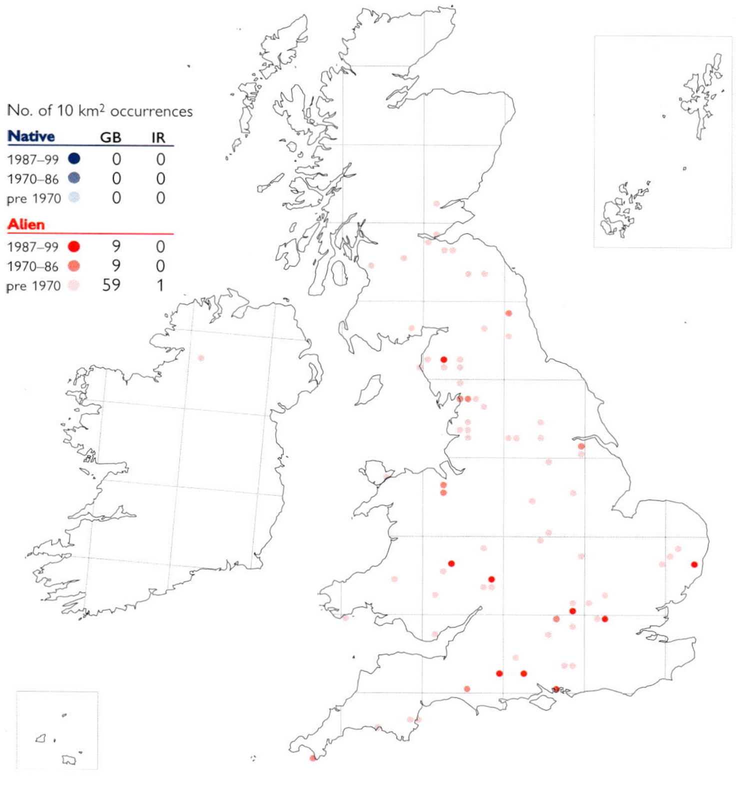

A perennial herb found in shaded places, including woodland, hedges, churchyards and on banks. It sometimes reproduces by seed, at least in S. England, and spreads by means of rhizomes to form a dense mass. Lowland.

Neophyte (change –0.86). Though sometimes claimed to be native, this species is certainly an introduction and has been grown in British gardens as a medicinal herb since at least 1200 (Harvey, 1981). It has been naturalised in the wild since 1640, but had greatly declined by 1930. It has continued to decline gradually since then.

A European Temperate species with a continental distribution; absent as a native from much of W. Europe.

References: Atlas (168d), Coombe (1956b), Hultén & Fries (1986), Jalas & Suominen (1976), Meusel *et al.* (1965).

G. M. KAY

Aristolochia clematitis Birthwort

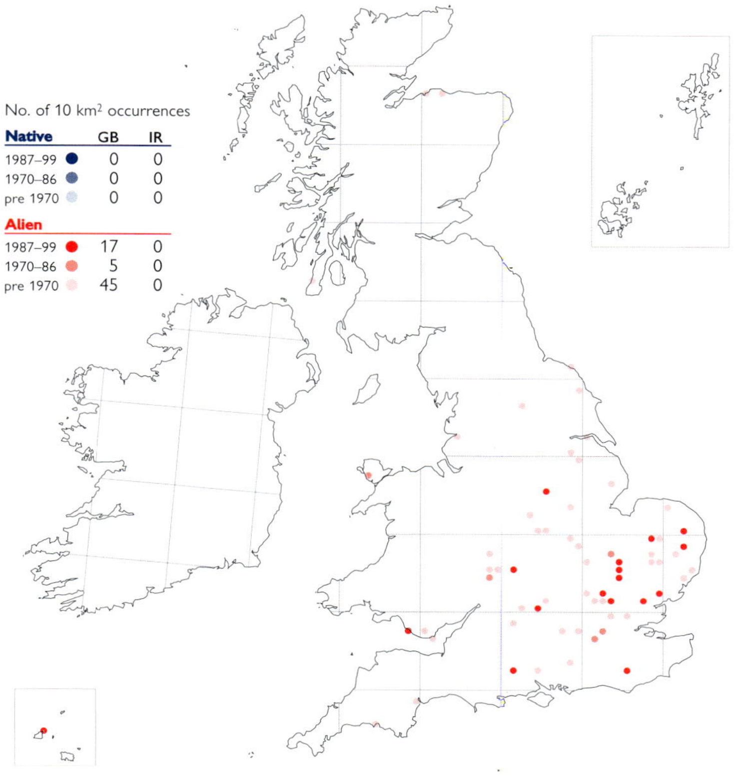

A scrambling or trailing perennial herb found as an escape or a relic of cultivation in waste and rough places, often by old abbeys or nunneries, and in churchyards, woods and on grassy banks. It spreads by rhizomes. Lowland.

Neophyte (change –0.82). This species was grown for its medicinal properties. The date of introduction is unknown, but it was recorded in the wild from Cambridgeshire in 1685. It is now rarely cultivated, and is gradually declining.

Probably native in S.E. Europe, N. Turkey and the Caucasus, but so widely naturalised in temperate Europe that its native range is obscured.

References: Atlas (169a), Bolòs & Vigo (1984), Hultén & Fries (1986), Jalas & Suominen (1976).

G. M. KAY

Nymphaea alba White Water-lily

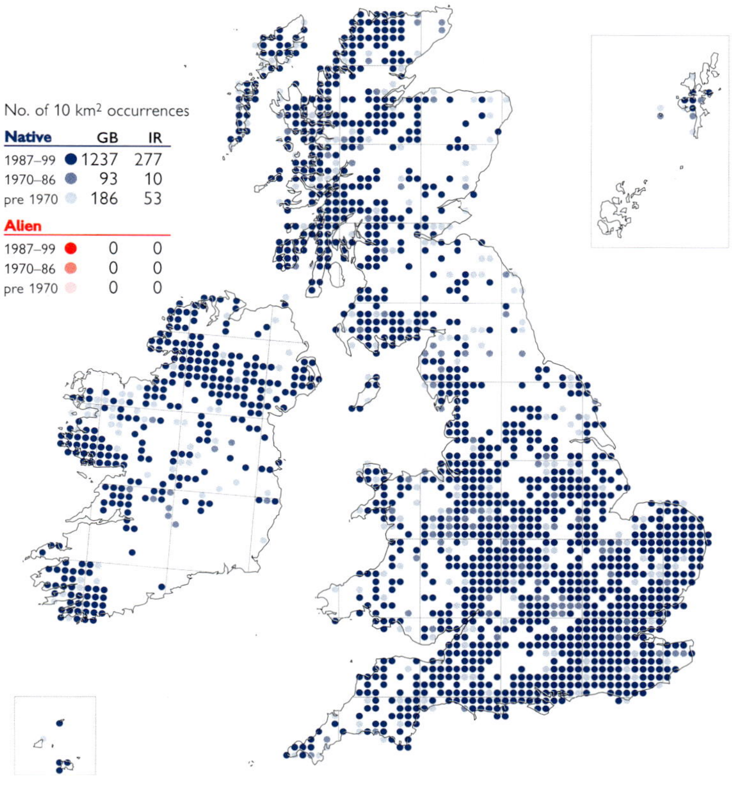

N. alba grows in lakes, ponds, the backwaters of rivers or large ditches, and occasionally in mires. It tolerates a wide range of water chemistry but lacks submerged leaves and is therefore vulnerable to disturbance by boats. Generally lowland, but reaching 405 m at Dock Tarn (Cumberland) and formerly 480 m at Angle Tarn (Westmorland).

Native (change +1.02). The native range of *N. alba* in our area has been obscured by introductions, and all records are mapped as if they are native. The increased frequency in S.E. England since the 1962 *Atlas* perhaps reflects the recording of such alien populations, which also include hybrids and cultivars.

European Temperate element.

References: Atlas (27b), Heslop-Harrison (1955b), Hultén & Fries (1986), Jalas & Suominen (1989), Meusel *et al.* (1965), Preston & Croft (1997).

C. D. PRESTON

Nuphar lutea Yellow Water-lily

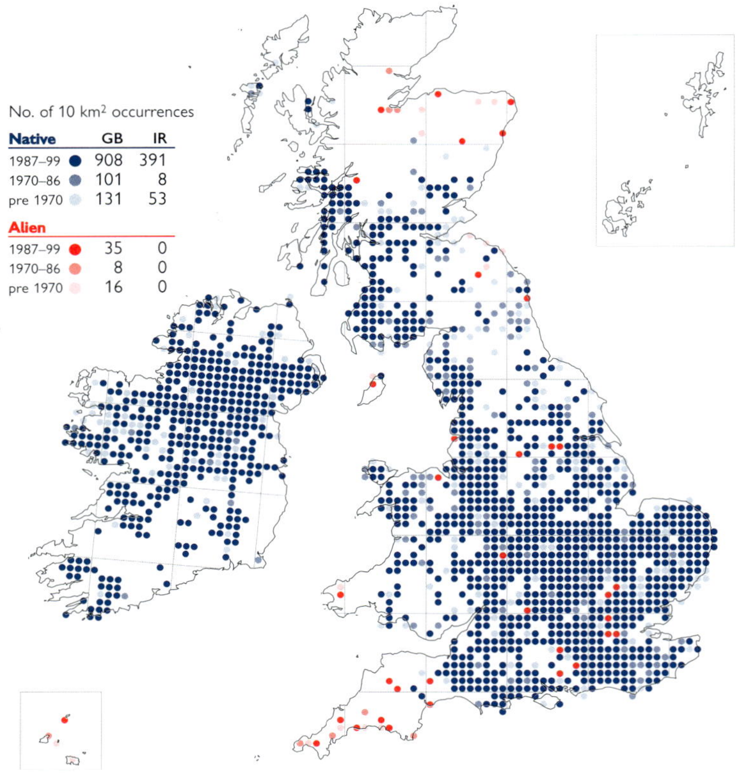

This perennial grows in mildly acidic or basic, mesotrophic or eutrophic water in lakes and slowly flowing rivers, canals and large ditches. Its submerged leaves allow it to persist in disturbed sites where the floating leaves are broken off. Generally lowland, but reaching 510 m at Llyn Crugnant (Cards.).

Native (change –0.13). Although *N. lutea* has been planted in some sites, it is less frequently introduced than *Nymphaea alba* and we have therefore attempted to map its native distribution. There is little evidence of change since the 1962 *Atlas*.

Eurosiberian Boreo-temperate element; closely related taxa occur in N. America.

References: Atlas (27c), Heslop-Harrison (1955a), Hultén & Fries (1986), Jalas & Suominen (1989), Meusel *et al.* (1965), Preston & Croft (1997).

C. D. PRESTON

Nuphar pumila Least Water-lily

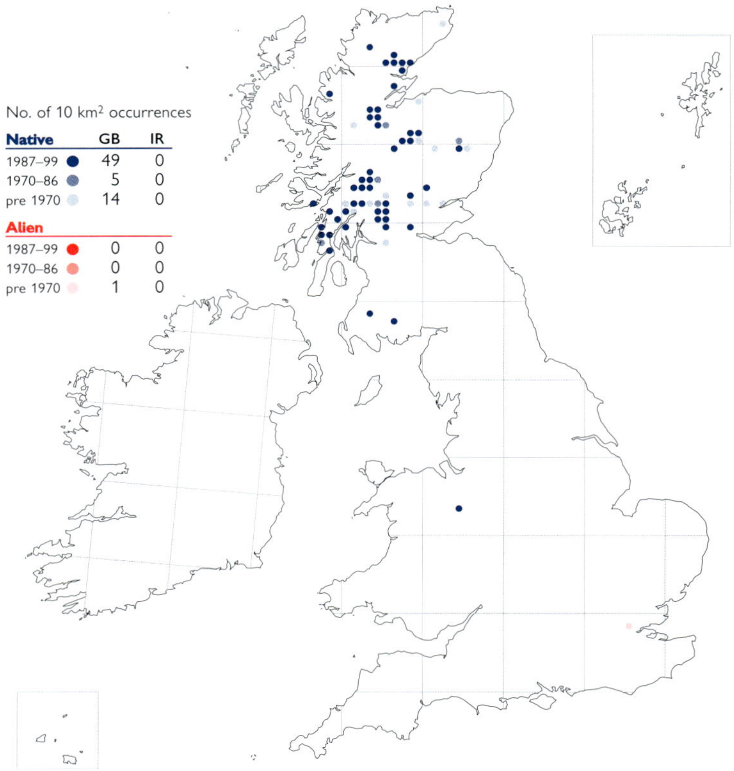

N. pumila grows in oligotrophic or mesotrophic water in lakes, sheltered bays, ditches and pools in marshes and bogs. It persists in one eutrophic lake in Shropshire. Lowland; upland records require confirmation.

Native (change +0.87). This species has long been confused with the hybrid *N. lutea* × *N. pumila* (*N.* × *spenneriana*), and the distribution of both is still poorly known. However, it is much better recorded than in the 1962 *Atlas*.

Circumpolar Boreal element; absent from western N. America and the eastern population is sometimes distinguished as *N. microphylla*.

References: Atlas (27d), Heslop-Harrison (1955a), Hultén & Fries (1986), Jalas & Suominen (1989), Meusel *et al.* (1965), Preston & Croft (1997), Stewart *et al.* (1994).

C. D. PRESTON

Ceratophyllum demersum Rigid Hornwort

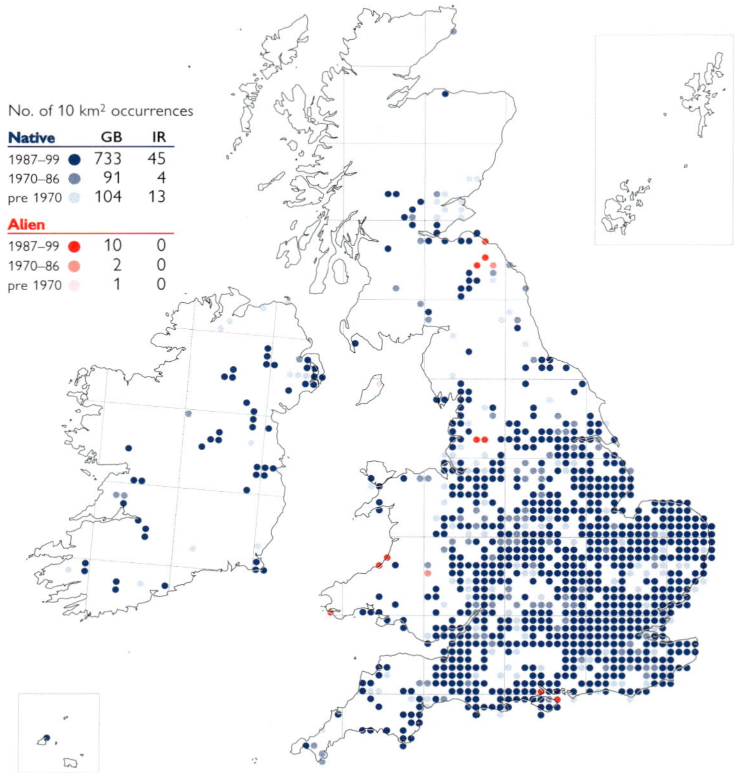

An aquatic which grows submerged in still or slowly flowing, eutrophic water in lakes, ponds, rivers, canals and ditches. It may be so abundant in ponds and ditches that it forms dense masses which rise above the water surface. Reproduction is mostly by vegetative fragmentation, but seeds are produced in still-water habitats in some years. Lowland.

Native (change +0.87). This species appears to be more frequent than shown in the 1962 *Atlas*. This probably reflects better recording of aquatic plants, but there is little doubt that because this species grows in eutrophic water its distribution is at least stable, if not increasing.

Circumpolar Southern-temperate element.

References: Atlas (28a), Hultén & Fries (1986), Jalas & Suominen (1989), Preston & Croft (1997).

C. D. PRESTON

Ceratophyllum submersum Soft Hornwort

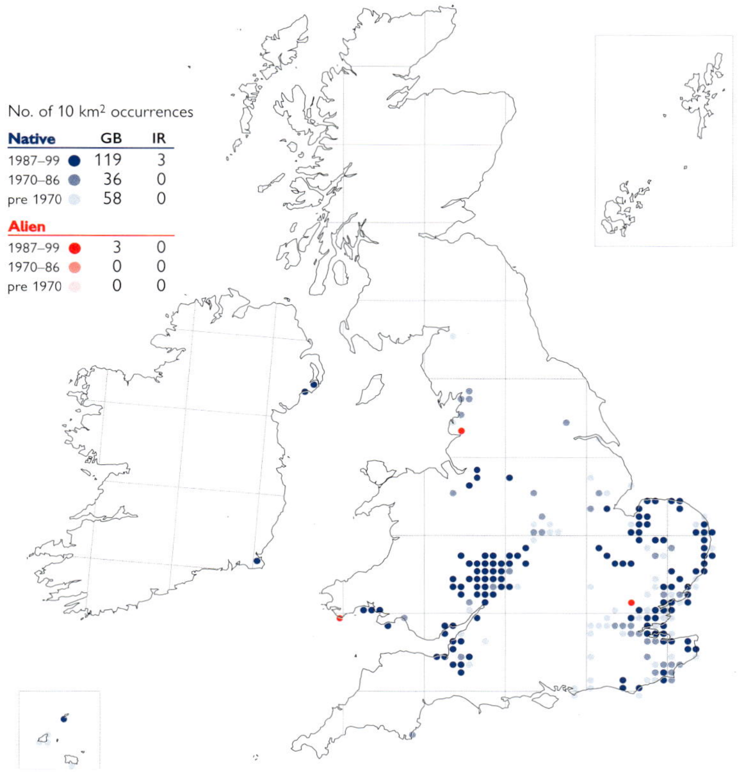

This aquatic grows in eutrophic or slightly brackish water in shallow, sheltered lakes, ponds and ditches. It is particularly frequent in coastal grazing marshes. Like *C. demersum*, reproduction is mostly by vegetative fragmentation and it can occur in dense masses, even in shaded ponds. Lowland.

Native (change +0.39). This species appears to be much more frequent now than at the time of the 1962 *Atlas*. Whereas the increase in records in coastal S.E. England probably reflects more detailed recording of grazing marshes, the increase in inland sites probably results from a genuine expansion. It was discovered in Ireland in 1989.

Eurosiberian Temperate element.

References: Atlas (28b), Hultén & Fries (1986), Jalas & Suominen (1989), Preston & Croft (1997), Stewart *et al.* (1994).

C. D. PRESTON

Caltha palustris Marsh-marigold

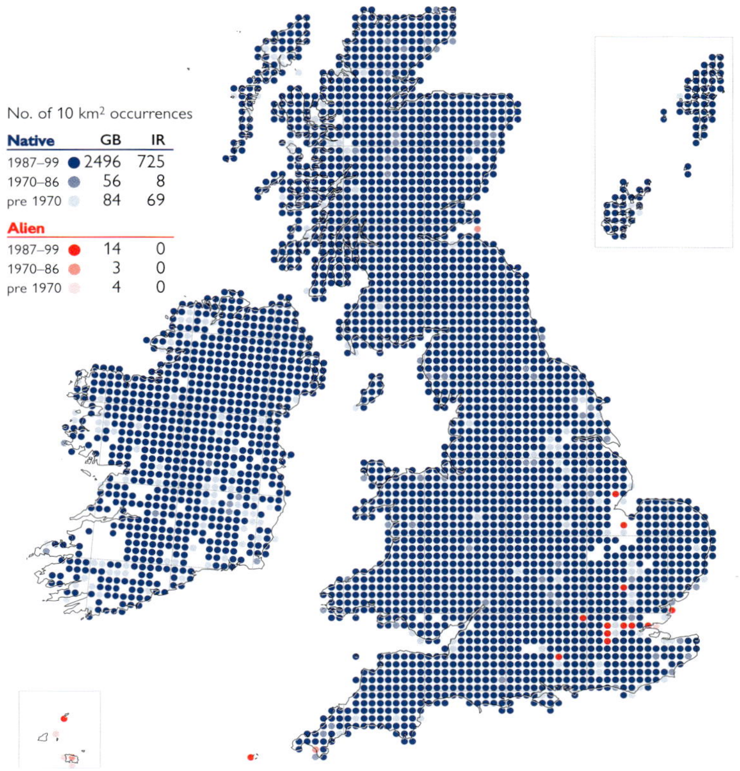

A perennial herb of various wet habitats, usually neutral to base-rich rather than very acidic, including *Alnus* carr, the edges of rivers, streams, canals, lakes and ponds, ditches and winter-wet meadows and pastures. A small form, var. *radicans*, is found in mountain flushes and lake shores in N. & W. Britain and Ireland. 0–1100 m (Braeriach and Lochnagar, S. Aberdeen).

Native (change –0.26). There is little evidence of any change in the 10-km square distribution of this species.

Circumpolar Wide-boreal element.

References: Atlas (17c), Grime *et al.* (1988), Hultén & Fries (1986), Jalas & Suominen (1989), Meusel *et al.* (1965), Smit (1973).

R. A. FITZGERALD

Trollius europaeus Globeflower

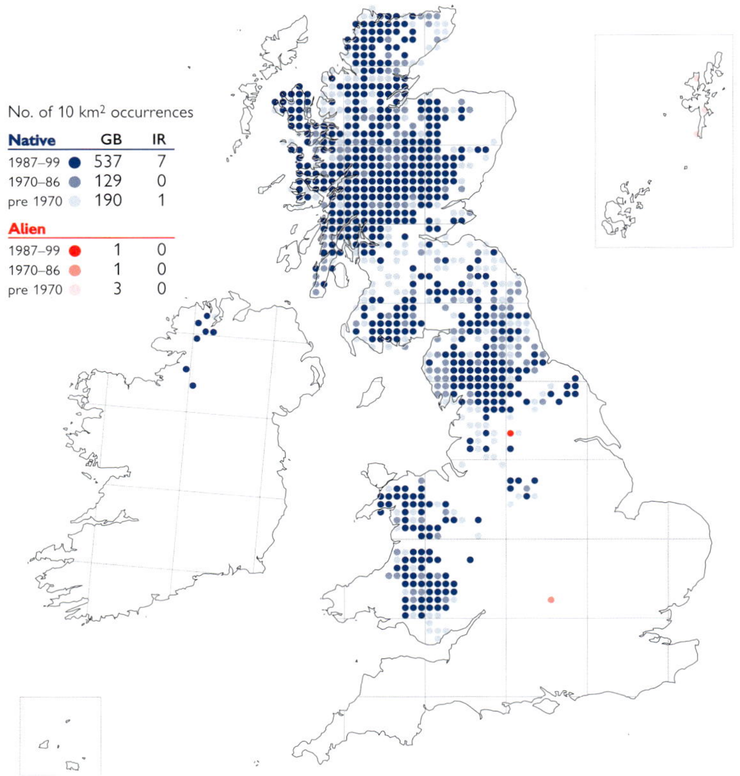

A perennial herb of cool, damp habitats, including hay meadows, stream and river banks, lake margins, open woodland and rock ledges. It prefers basic soils, and is often associated with limestone. It is sensitive to grazing, but can persist as small, non-flowering plants in the uplands. 0–1005 m (Stuic, S. Aberdeen).

Native (change –0.73). The map clearly shows the decline of this species in Britain, especially at the fringes of its range, which began before 1930. The main cause is the agricultural improvement of hill land by drainage and the application of fertiliser.

European Boreal-montane element.

References: Atlas (17d), Curtis & McGough (1988), Halliday (1997), Hultén & Fries (1986), Jalas & Suominen (1989), Meusel *et al.* (1965), Pryce (1999).

R. A. FITZGERALD

Helleborus foetidus Stinking Hellebore

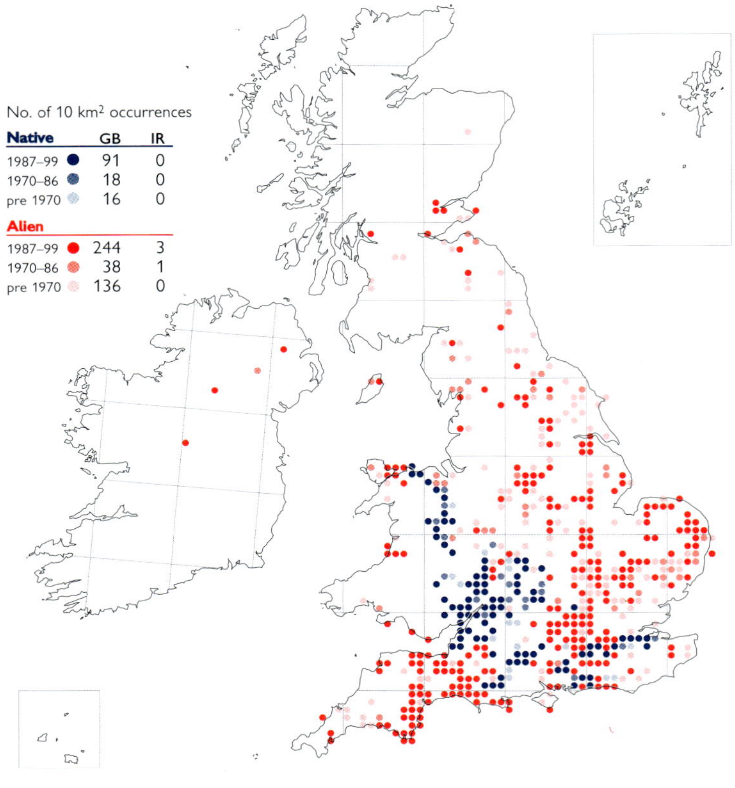

A short-lived perennial herb of shallow calcareous soils. It is a poor competitor, and intolerant of deep shade, so is usually found in small colonies in woodland glades or open scrub, on scree slopes, rock ledges, hedge banks, and as an introduction in churchyards. Adult plants near senescence (4–5 years old) are typically found with a cohort of seedlings. Lowland.

Native (change +0.86). The native range of this species has been obscured by the presence of numerous naturalised populations. It is often grown as a garden plant and escapes can become established in the wild, although sometimes for only a few years.

Suboceanic Southern-temperate element.

References: Atlas (18a), Bolòs & Vigo (1984), Jalas & Suominen (1989), Meusel *et al.* (1965), Stewart *et al.* (1994).

R. A. FITZGERALD

Helleborus viridis Green Hellebore

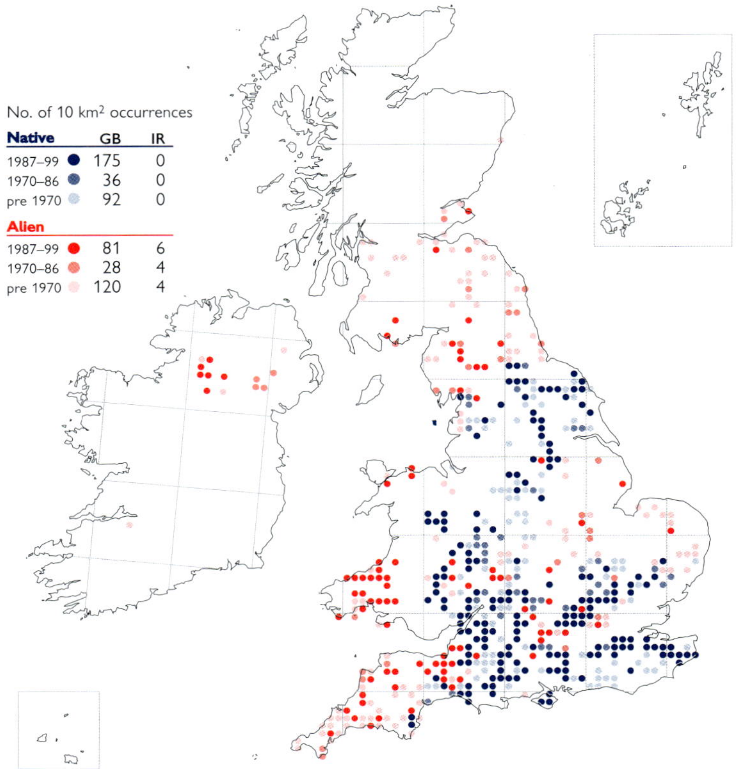

A perennial herb of rather shady habitats, usually on chalk or limestone, found in woodland glades, rocky dingles and old hedge banks. Populations are often small, but persist over many years without obvious changes in numbers. Lowland.

Native or alien (change –0.28). *H. viridis* has been grown in gardens since medieval times, and was recorded in the wild by 1562. It is very difficult to delimit its native range. The decline apparent on the map may be partly attributable to records of introductions which failed to persist, but long-established populations have been lost as a result of the clearance of hedges and copses and the cessation of coppicing.

Suboceanic Temperate element.

References: Atlas (18b), Bolòs & Vigo (1984), Brewis *et al.* (1996), Jalas & Suominen (1989), Swan (1993).

R. A. FITZGERALD

Eranthis hyemalis Winter Aconite

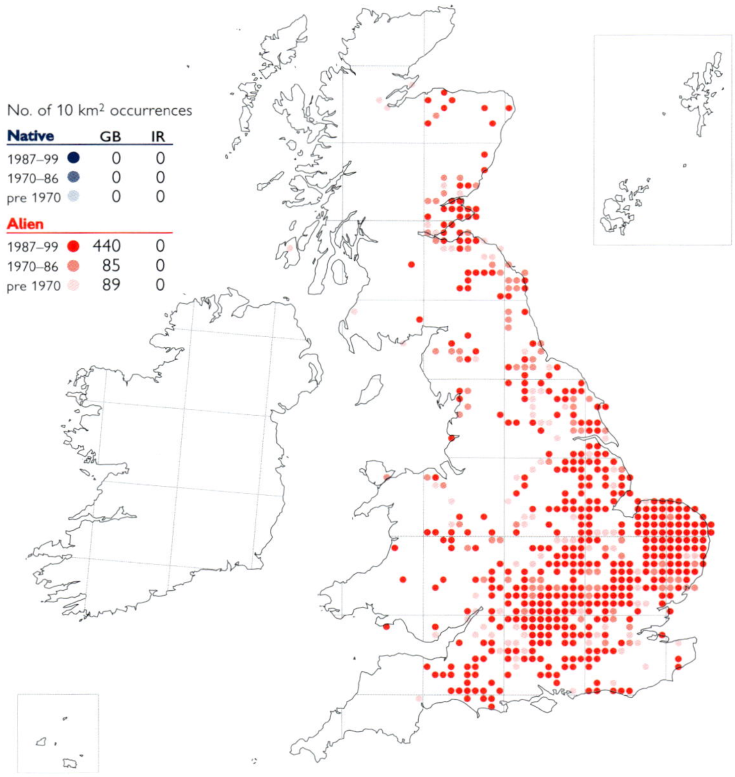

A small, tuberous perennial, dying back in summer. It is naturalised, sometimes in large numbers, in open woodland, grassland and scrub associated with habitation, under park trees, in gardens and on road verges. Lowland.

Neophyte (change +1.59). *E. hyemalis* was introduced as a garden plant by 1596, and has become thoroughly established in some areas; it was first recorded in the wild in 1838. The eastern distribution was already apparent in the 1962 *Atlas*. The great increase since then is probably due to a genuine increase in frequency and the improved recording of aliens.

Native of S. Europe from Italy to Bulgaria, and of Turkey; widely naturalised in Europe outside its native range.

References: Atlas (18c), Jalas & Suominen (1989).

R. A. FITZGERALD

Nigella damascena Love-in-a-mist

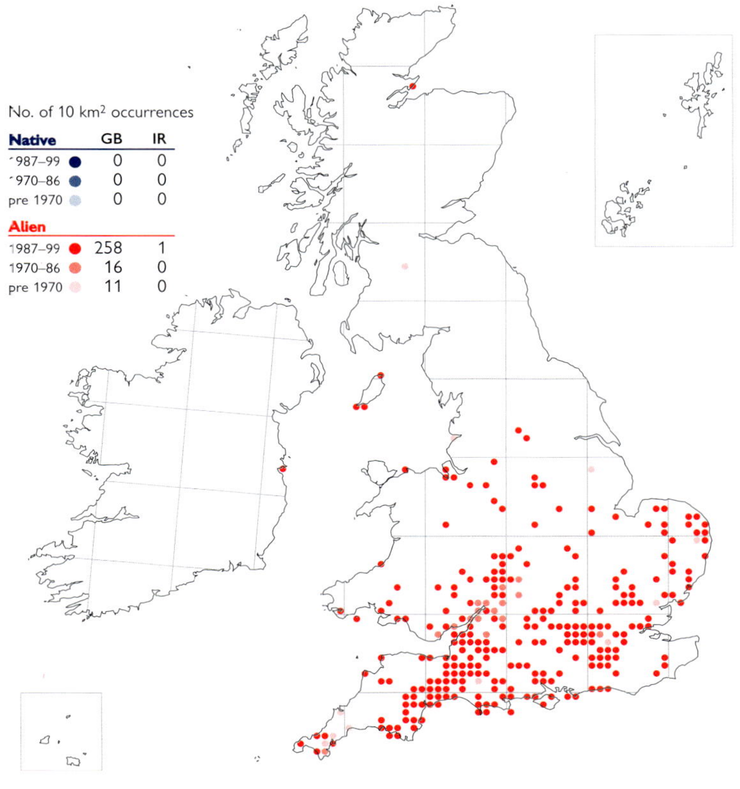

An annual which escapes from cultivation onto rubbish tips, waste ground, old walls and in pavement cracks. It readily regenerates from seed and populations can be persistent, but only in disturbed sites where there is little competition. It was formerly a relatively frequent grain alien (Dunn, 1905). Lowland.

Neophyte. An enduring garden favourite which has been cultivated in Britain since at least 1570. It was known from the wild by 1876 and is likely to be increasing. Four other species of *Nigella* have been recorded in Britain, and may be recorded in error for *N. damascena*.

Native of the Mediterranean region.

References: Bolòs & Vigo (1984), Jalas & Suominen (1989), Wurzell (1990).

R. A. FITZGERALD

Aconitum napellus sens. lat. Monk's-hood

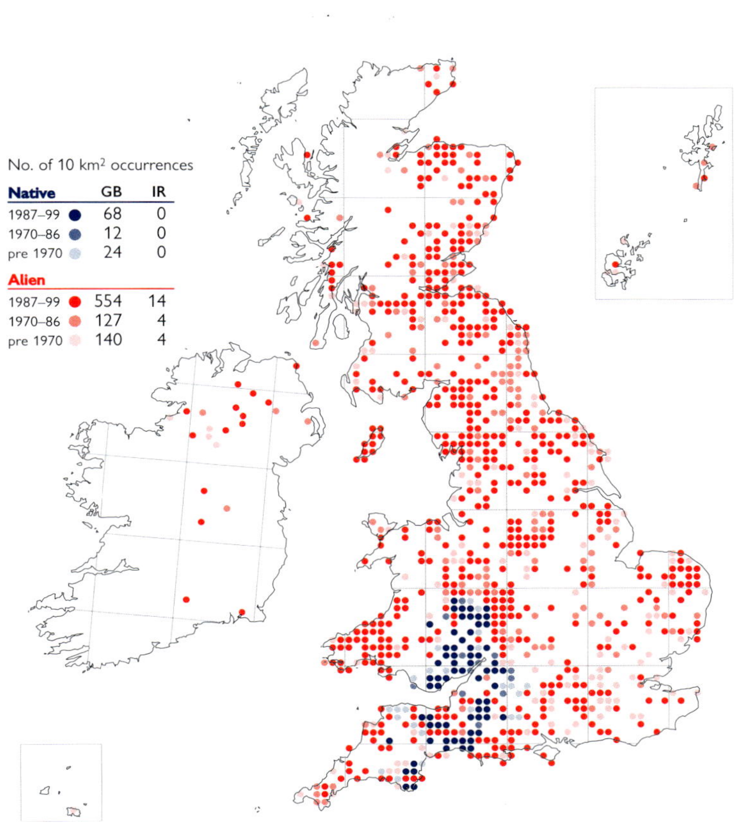

These tuberous perennial herbs grow on calcareous to slightly acidic soil along stream banks, often in shade, in damp, open woodland and sometimes in damp meadows, and as aliens on roadsides, waste ground and rubbish tips. Generally lowland, but reaching 460 m at Quarnford (Staffs.).

Native or alien (change +1.42). All records of *A. napellus sens. str.* and *A. × cammarum* are mapped here. The British *A. napellus* is regarded as the endemic subsp. *napellus*, suggesting that it is native, but the species has long been grown in gardens and was not recorded in the wild until 1821. It may be over-recorded as a supposed native in the Welsh Marches.

European Temperate element.

References: Atlas (18d), Bolòs & Vigo (1984), Jalas & Suominen (1989), Stewart *et al.* (1994), White (1912).

R. A. FITZGERALD

Aconitum napellus × A. variegatum (A. × cammarum) Hybrid Monk's-hood

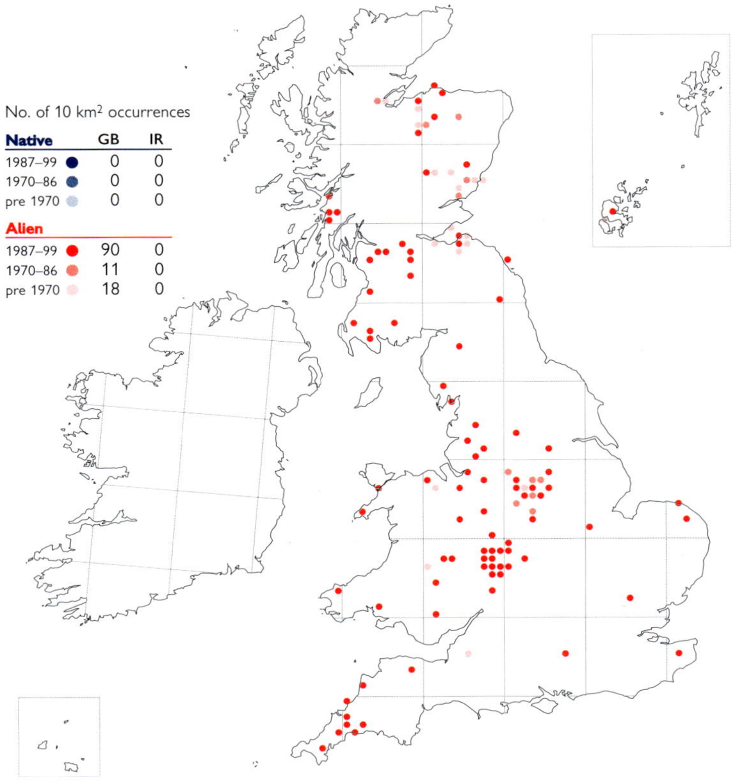

A perennial with annually renewed tuberous rhizomes, found established in damp places on a range of soils, usually in shaded sites or in tall vegetation. Its habitats are more varied than those of other *Aconitum* taxa and include damp roadsides and pastures, waste ground and moist woodland. 0–460 m (Quarnford, Staffs.).

Neophyte. This plant was in cultivation by 1752, and has been known in the wild since at least 1905. Blue, white and bicoloured cultivars exist, the last being the most common. Blue forms can easily be mistaken for *A. napellus*, and the hybrid is probably under-recorded.

A hybrid which occurs naturally in Europe and S.W. Asia, but also in gardens; the cultivar 'Bicolor' is of garden origin.

References: Rich & Jermy (1998), Stewart *et al.* (1994).

R. A. FITZGERALD

Consolida ajacis Larkspur

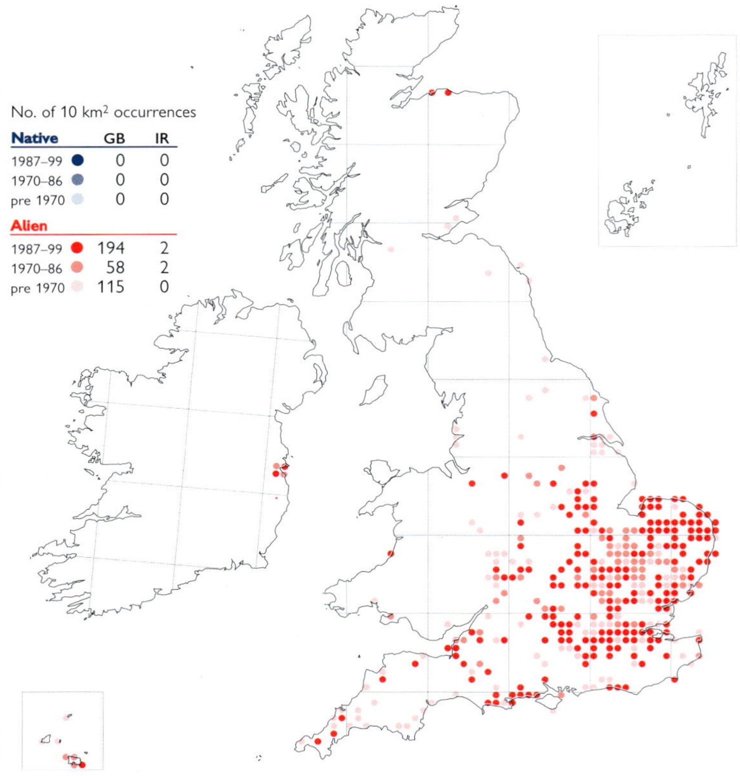

An annual species found on waste ground, rubbish tips and in cultivated fields. As an arable weed it usually occurs on dry soils in chalky or sandy areas. Lowland.

Neophyte. *C. ajacis* was being grown in British gardens by 1573 and was first recorded from the wild in 1650. It is a casual, formerly introduced as a contaminant of imported grain but now much more frequent as a garden escape. Some older records may be misidentifications of *C. orientalis* or *C. regalis*, which were often confused with it but which are not commonly grown in gardens.

Apparently native of S. Europe and S.W. Asia, but its native range is obscured by its occurrence as a naturalised and casual alien.

References: Burton (1983), Jalas & Suominen (1989).

R. A. FITZGERALD

Actaea spicata Baneberry

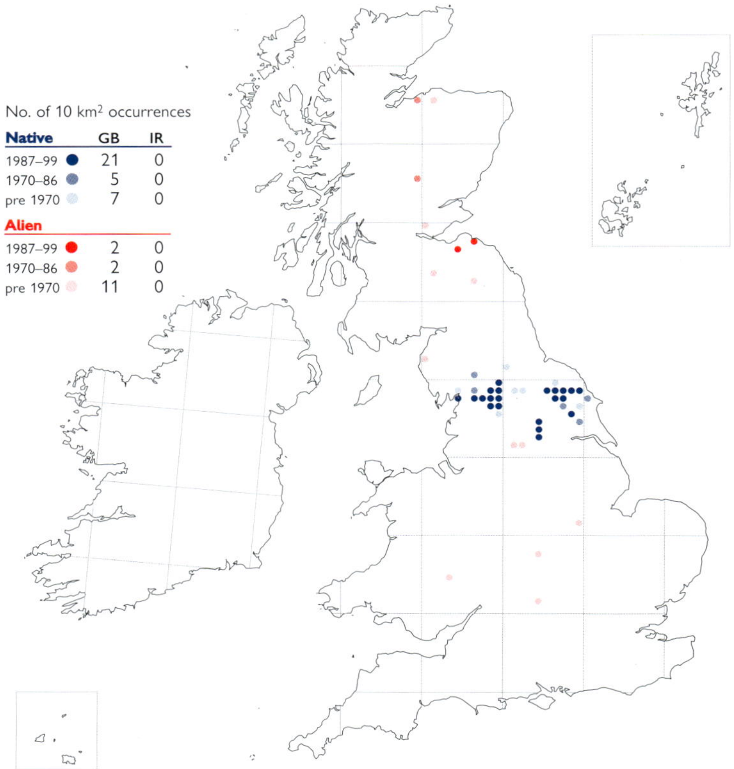

A perennial herb of shaded sites on limestone. Its habitats differ superficially, being found in the grikes of limestone pavement, on rock ledges, and in deciduous woodland, but all have the same characteristics of shade, low competition and a cool, protected root run. 0–450 m (Hawkswick Clowder, Mid-W. Yorks.).

Native (change –0.44). Populations of this species are long-lived and stable. Despite occasional threats such as quarrying or neglect of scrub or woodland causing increased competition, the species maintains its historic range. It is occasionally grown in gardens, and naturalised or relict garden populations can persist for many years.

Circumpolar Boreal-montane element.

References: Atlas (19a), Hultén & Fries (1986), Jalas & Suominen (1989), Meusel *et al.* (1965), Stewart *et al.* (1994).

R. A. FITZGERALD

Anemone nemorosa Wood Anemone

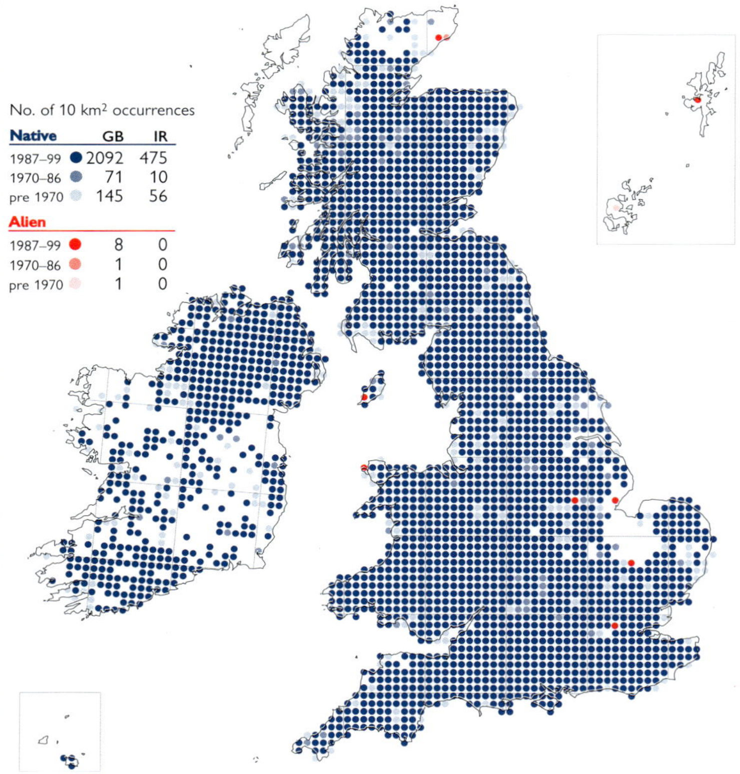

A rhizomatous perennial, found in woodland, on streamsides, under *Pteridium*, on hedge banks, in heathy grassland, on open moorland, in scree and on limestone pavement. It has a wide pH tolerance, but in woodlands it is most abundant where the vigour of more competitive species is reduced by acidity, waterlogging or regular coppicing. 0–1190 m (Ben Lawers, Mid Perth).

Native (change –0.70). The distribution of *A. nemorosa* is stable at the 10-km square scale. Double-flowered cultivars and various colour forms are widely cultivated and occasionally escape into the wild, although not all native populations are white-flowered.

Eurosiberian Temperate element.

References: Atlas (19b), Grime *et al.* (1988), Hultén & Fries (1986), Jalas & Suominen (1989), Meusel *et al.* (1965), Rackham (1980), Shirreffs (1985).

R. A. FITZGERALD

Anemone apennina Blue Anemone

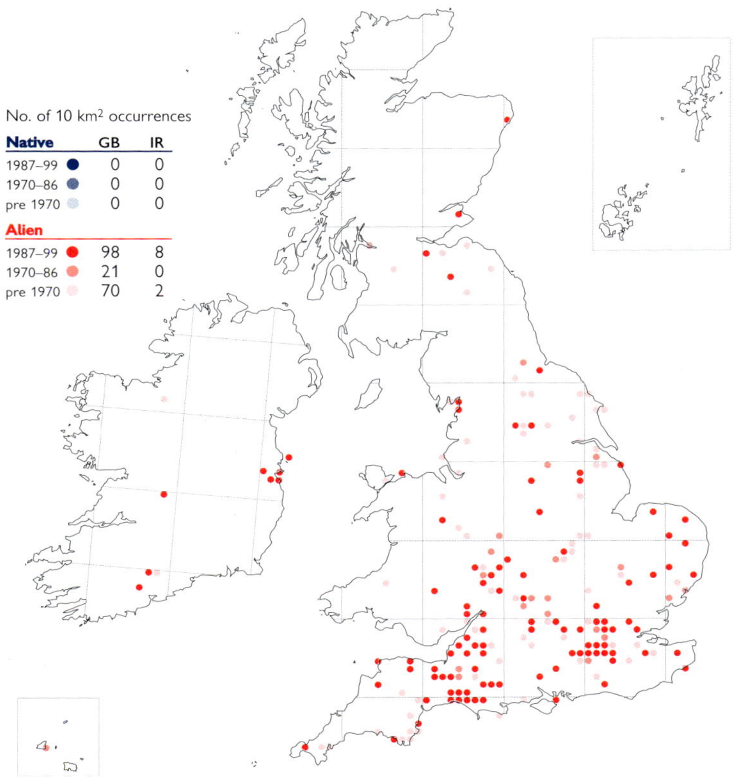

A rhizomatous perennial, found in woodland, open scrub, under park trees, in churchyards and near former habitations. Like the native *A. nemorosa*, it requires light shade. Lowland.

Neophyte. The species has been recorded from the wild in Britain since 1724, and it is sometimes thoroughly naturalised and long-persistent. A range of colour forms (pink, white and blue) occur in cultivation. Native blue-flowered varieties of *A. nemorosa* may be mis-recorded as this species.

Native of central S. Europe from Corsica to the Balkans; widely naturalised elsewhere in Europe.

Reference: Jalas & Suominen (1989).

R. A. FITZGERALD

Anemone blanda Balkan Anemone

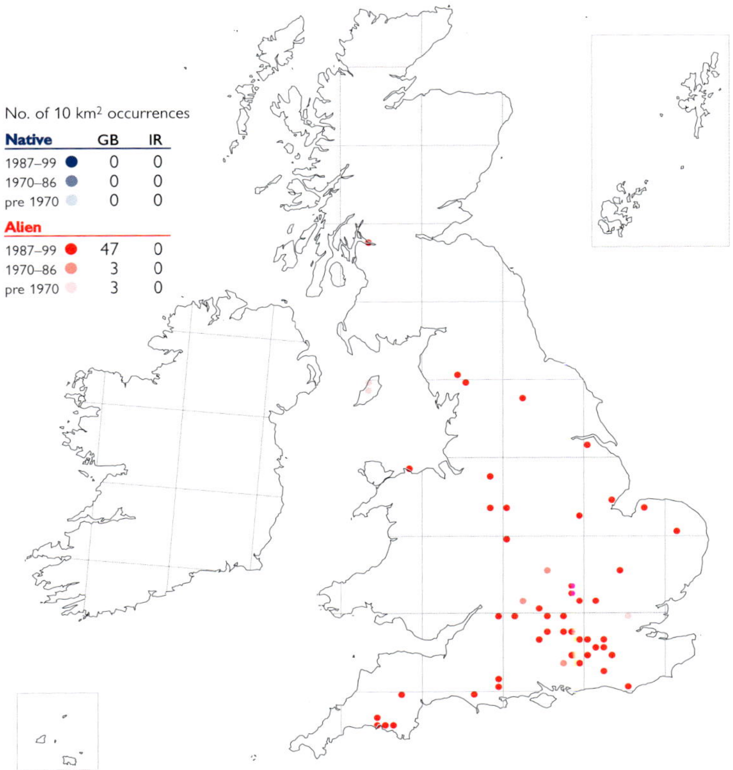

A tuberous perennial herb, found as a garden throw-out in woodland, on roadside verges, under park trees, in churchyards and as a relic of cultivation. Reproduction from seed is rare. Lowland.

Neophyte. *A. blanda* was introduced to cultivation in Britain in 1898. Given its popularity in gardens, where numerous colour variants are grown, it is surprisingly rare in the wild; the first published record appears to have been in 1983.

Native of the Balkans and S.W. Asia; an eastern counterpart of the closely related *A. apennina*.

Reference: Jalas & Suominen (1989).

R. A. FITZGERALD

Anemone ranunculoides Yellow Anemone

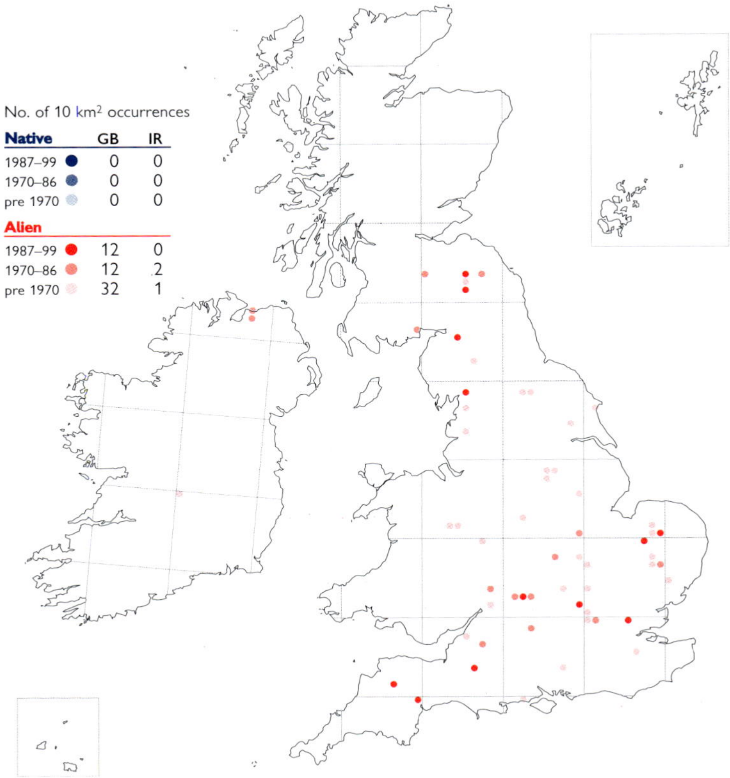

A spring-flowering rhizomatous perennial herb naturalised in shady places, such as in woodland and along paths. Lowland.

Neophyte. *A. ranunculoides* has been grown in British gardens since at least 1596. It was first recorded in the wild in 1778.

A European Temperate species, absent as a native from much of W. Europe.

References: Hultén & Fries (1986), Jalas & Suominen (1989), Meusel *et al.* (1965).

R. A. FITZGERALD

Anemone hupehensis × *A. vitifolia* (*A.* × *hybrida*) Japanese Anemone

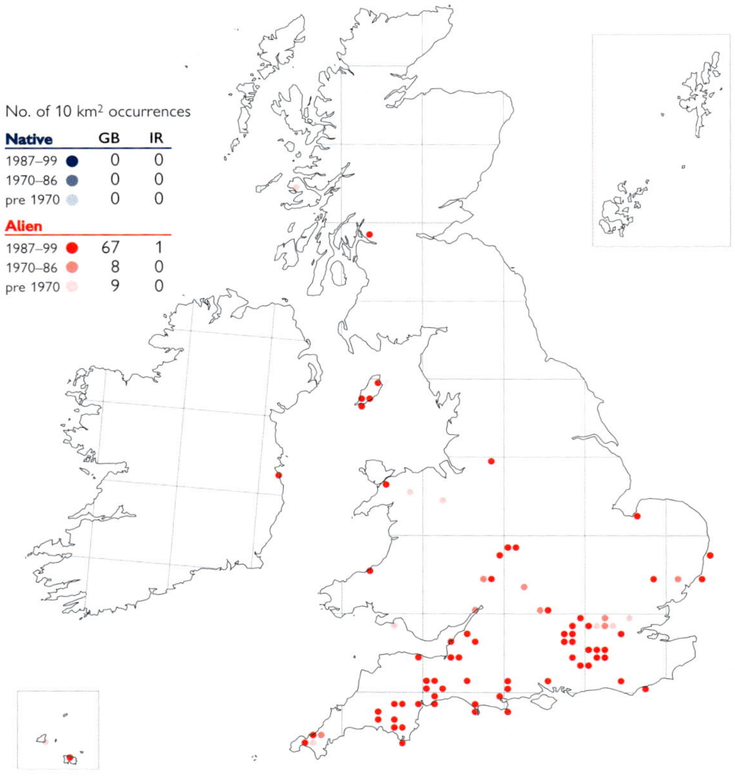

A perennial herb which is occasionally recorded as an escape or persistent horticultural relic, on waste ground or near habitation. Lowland.

Neophyte. This hybrid, which has been in cultivation since 1848, was first recorded in the wild *c.* 1900. It appears to be increasing, at least in parts of S. England.

A hybrid of garden origin.

R. A. FITZGERALD

Pulsatilla vulgaris Pasqueflower

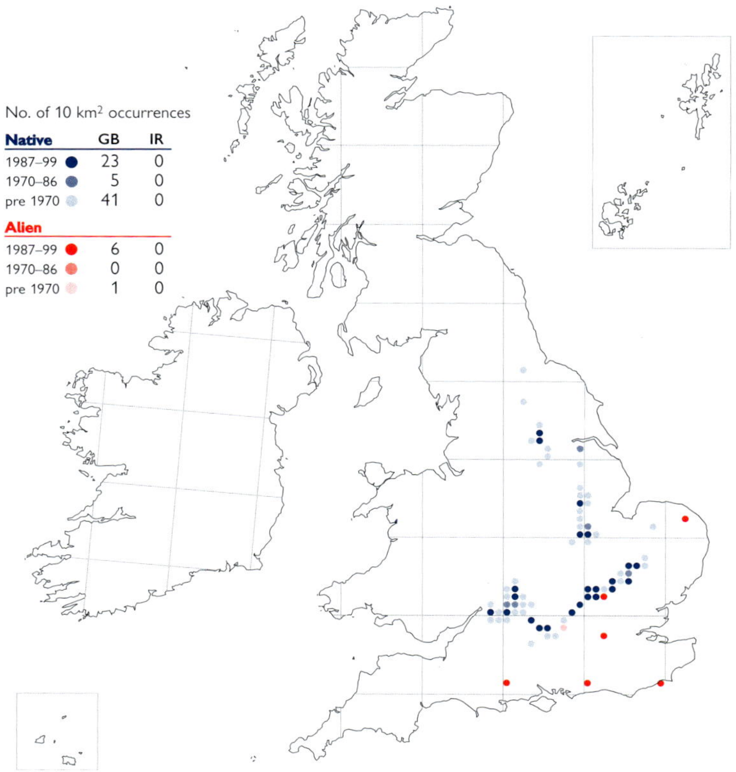

A perennial rhizomatous herb of species-rich turf on the slopes of chalk or oolite escarpments, and the banks of ancient earthworks, usually with a S. or S.W. aspect. Plants produce viable seed, but seedling establishment is rare. Lowland.

Native (change –0.50). This species was lost from many sites as a result of agricultural improvements in the 18th and 19th centuries. The decline continues, even on nature reserves, as relict chalk grassland ceases to be grazed and suffers from fertiliser drift. It also occurs as a garden escape and may have been planted on the Wiltshire/Dorset border.

European Temperate element, with a continental distribution in W. Europe.

References: Atlas (19c), Hultén & Fries (1986), Jalas & Suominen (1989), Meusel *et al.* (1965), Stewart *et al.* (1994), Wells & Barling (1971).

R. A. FITZGERALD

Clematis vitalba Traveller's-joy

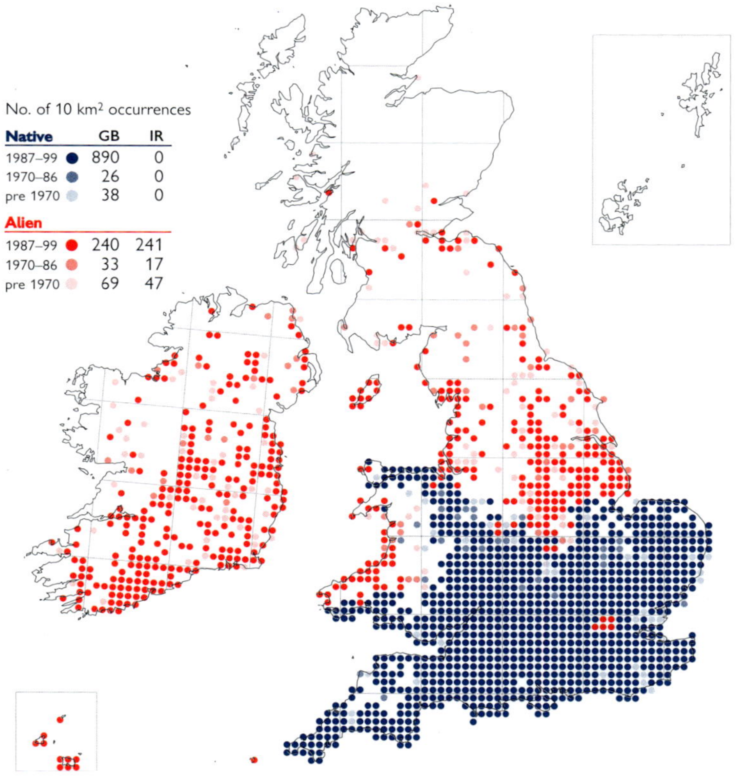

A climbing perennial with liana-like woody stems, often covering large areas on hedge banks, hedges and walls, trees and scrub, sand dunes, disused quarry faces and ruins. It is a classic railway plant. On base-rich soils, or utilising lime mortar, the plant can form virtual monocultures. The familiar feathered propagules, 'Old Man's Beard', disperse readily, and often colonise new cuttings or banks. 0–305 m (N. of Matlock, Derbys.).

Native (change 0.00). Comparison of the current map with the 1962 *Atlas* suggests that the distribution of *C. vitalba* is stable.

European Temperate element; widely naturalised outside its native range.

References: Atlas (19d), Jalas & Suominen (1989), Meusel *et al.* (1965).

R. A. FITZGERALD

Ranunculus acris Meadow Buttercup

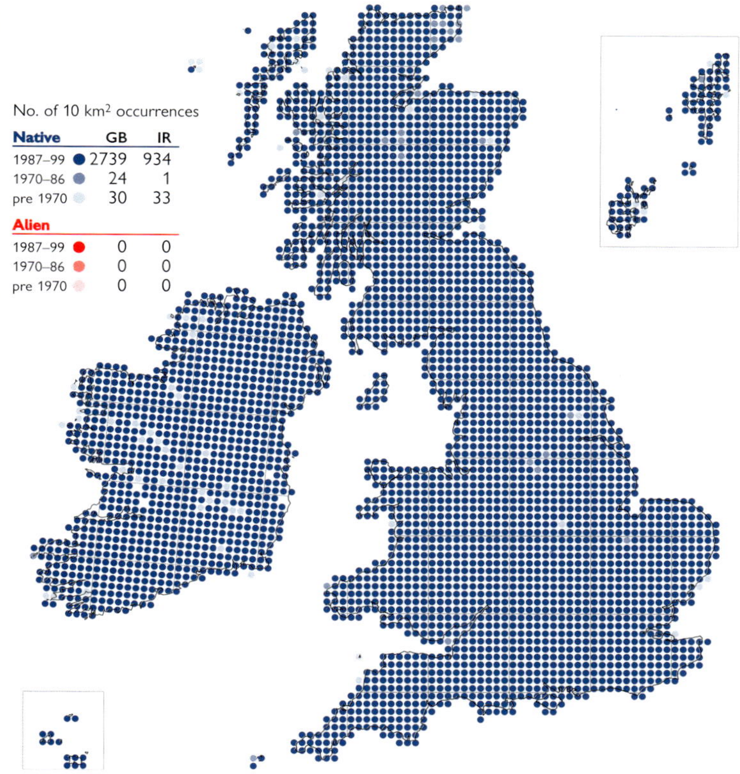

A perennial herb of damp meadows and pastures on a wide variety of soils, only avoiding very dry or acid conditions. It is a characteristic plant of unimproved hay and water-meadow communities, and now of relict herb-rich fragments on damp road verges; it also grows on dune grassland, in montane flushes and in tall-herb communities on rock ledges. It is unpalatable to grazing animals, but easily controlled in intensively managed pastures. 0–1220 m (Cairntoul, S. Aberdeen).

Native (change +0.30). The distribution of *R. acris* is stable.

Eurasian Wide-boreal element, but naturalised in N. America so distribution is now Circumpolar Wide-boreal.

References: Atlas (20a), Grime *et al.* (1988), Harper (1957), Hultén & Fries (1986), Jalas & Suominen (1989), Meusel *et al.* (1965).

R. A. FITZGERALD

Ranunculus repens Creeping Buttercup

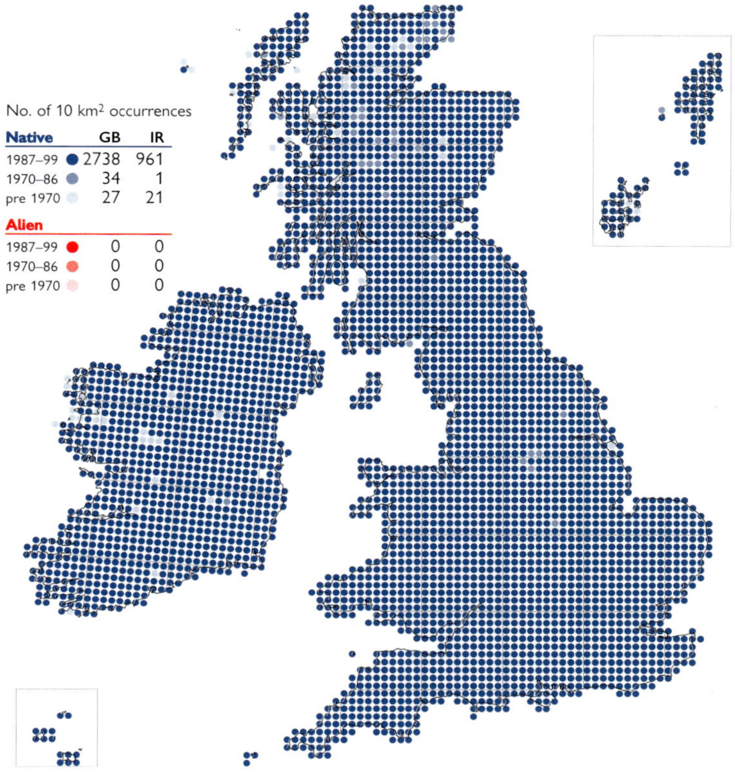

A perennial herb with creeping stems, *R. repens* has a very wide ecological tolerance, but is most typical of disturbed habitats on damp or wet nutrient-rich soils, including woodland rides, ditch sides, farm gateways, gardens and waste ground. It also occurs in damp or periodically flooded grasslands, in dune-slacks and on lake shores. It is absent from very acidic soils. Generally lowland, but reaching 1035 m on Snowdon (Caerns.).

Native (change +0.55). An extremely common species with a stable distribution.

Eurasian Boreo-temperate element, but naturalised in N. America so distribution is now Circumpolar Boreo-temperate.

References: Atlas (20b), Grime *et al.* (1988), Harper (1957), Hultén & Fries (1986), Jalas & Suominen (1989), Lovett-Doust *et al.* (1990), Meusel *et al.* (1965).

R. A. FITZGERALD

Ranunculus bulbosus Bulbous Buttercup

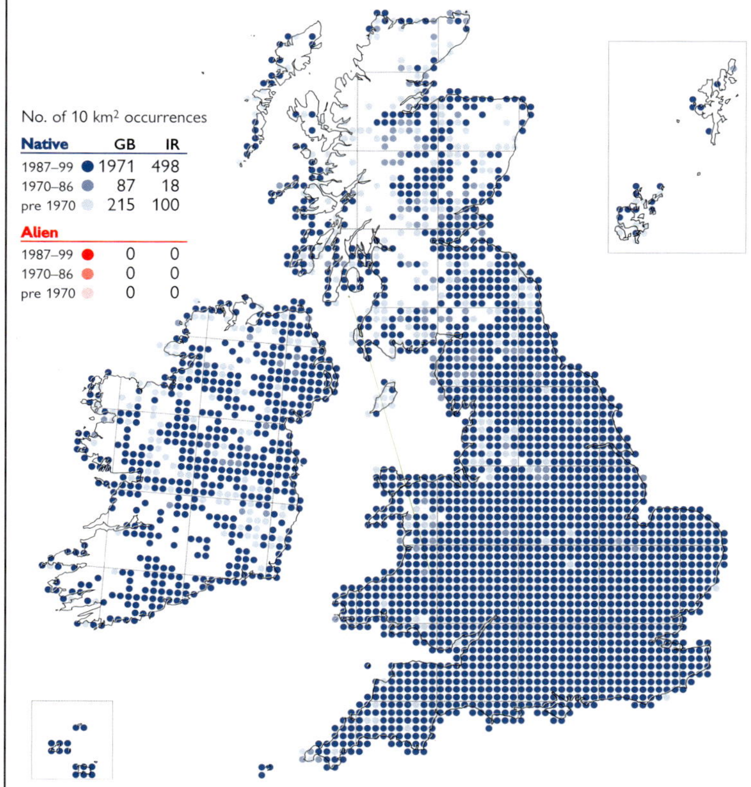

A perennial herb with a corm-like stem-base, found on well-drained, neutral or calcareous soils in meadows, pastures and dunes. It is absent from highly productive, fertile grassland and from strongly acidic soils. Generally lowland, but reaching 580 m on Dartmoor (S. Devon) and at Hartside (Westmorland).

Native (change –0.48). *R. bulbosus* has maintained its range, and can be locally abundant. The map suggests a slight decline towards its northern and western limits, but the species is most readily recorded up to early summer and may have been overlooked in some areas.

European Southern-temperate element; widely naturalised outside its native range.

References: Atlas (20c), Grime *et al.* (1988), Harper (1957), Hultén & Fries (1986), Jalas & Suominen (1989), Meusel *et al.* (1965).

R. A. FITZGERALD

Ranunculus sardous Hairy Buttercup

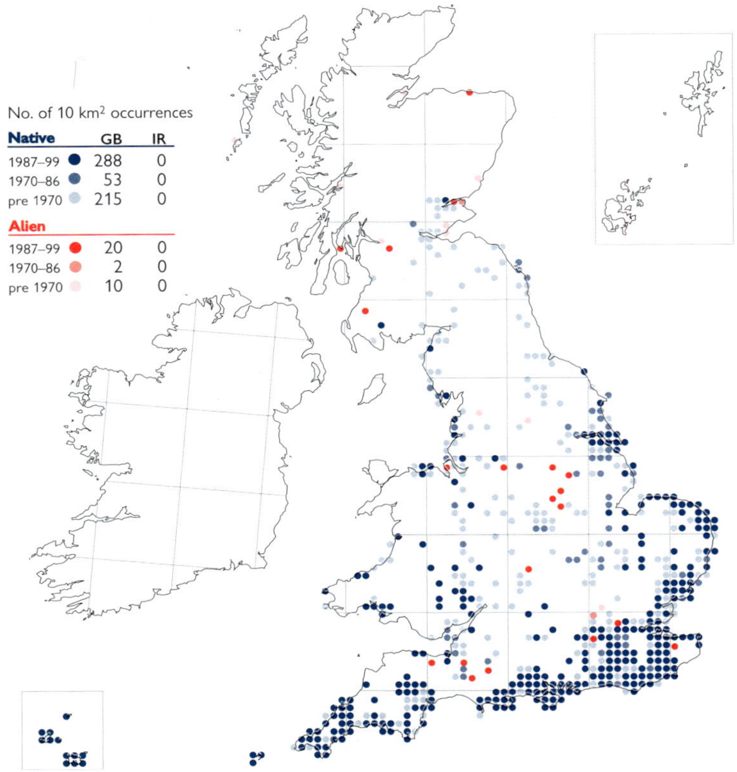

An annual of damp coastal pastures, poached pond edges and wet hollows, road verges, farm tracks and gateways. It is generally restricted to thin turf or disturbed areas on damp, neutral, moderately fertile soils. Lowland.

Native or alien (change +0.24). *R. sardous* was first recorded in Britain in 1663. It was last recorded before 1930 from most of the inland sites for which only pre-1970 records are available. It may have been only a casual introduction at some of these localities, perhaps as an impurity of grass seed in sown leys. It persists in submaritime regions of England and Wales, where it is much better recorded than in the 1962 *Atlas*.

European Temperate element; widely naturalised outside its native range.

References: Atlas (21a), Hultén & Fries (1986), Jalas & Suominen (1989).

R. A. FITZGERALD

Ranunculus parviflorus Small-flowered Buttercup

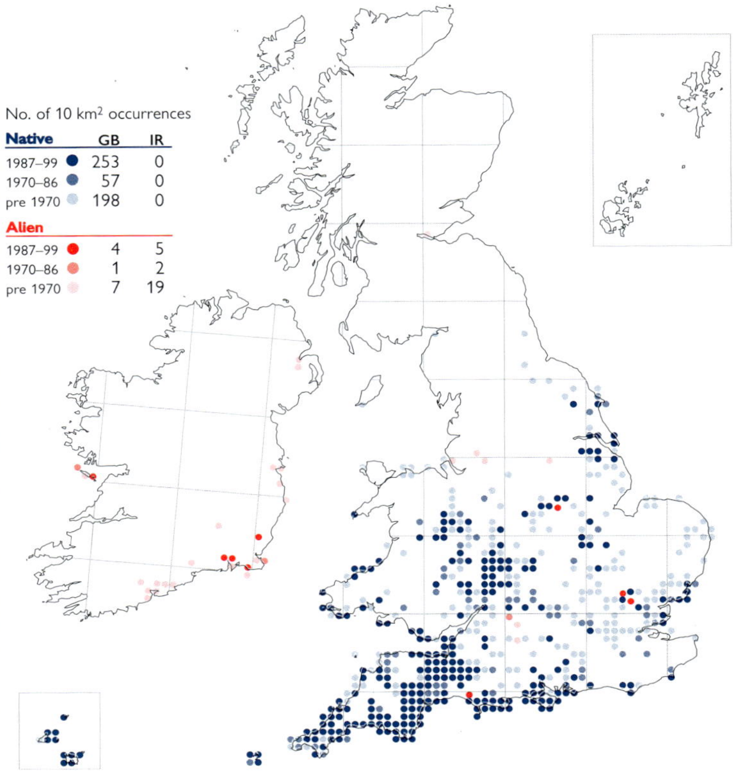

An annual of dry disturbed habitats on a range of neutral and calcareous soils. Typical sites include broken turf on cliff edges, open, droughted slopes and banks, rabbit scrapes, tracks, poached gateways, building sites and gardens. The seeds appear to be long-lived, and populations may reappear after disturbance or persist for many years. Lowland.

Native (change –0.08). The retreat of this species south-westwards in Britain had already taken place by 1930. Its distribution appears to be stable within its core areas. It is rare and decreasing in Ireland.

Suboceanic Southern-temperate element.

References: Atlas (21b), Bolòs & Vigo (1984), Jalas & Suominen (1989), Stewart *et al.* (1994).

R. A. FITZGERALD

Ranunculus arvensis Corn Buttercup

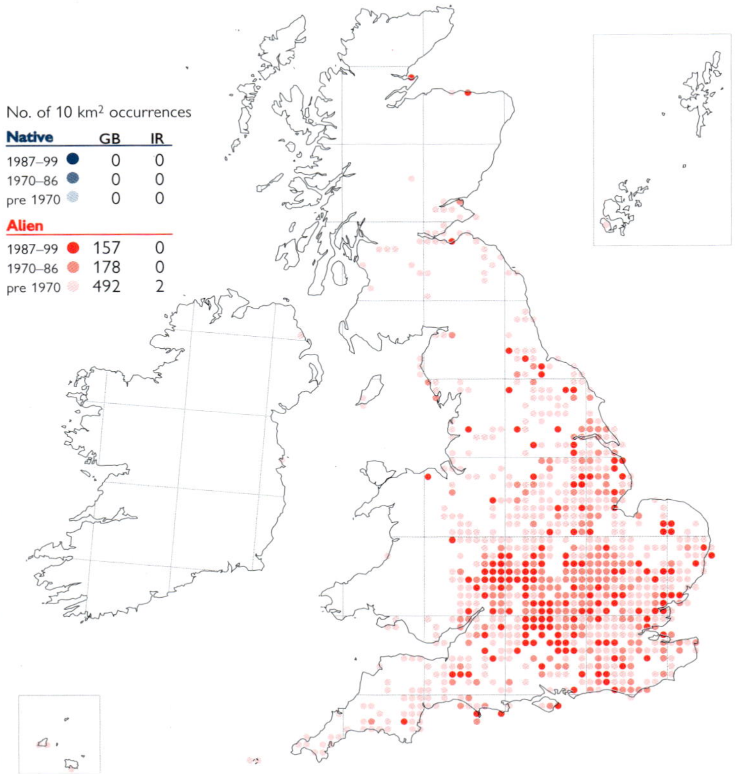

An annual of arable land on loams, sands, clays and chalk. The seeds are long-lived, and plants sometimes reappear on disturbed waste ground, or in gardens or new roadside verges on former arable land. Lowland.

Archaeophyte (change –3.77). *R. arvensis* has been present in Britain since Roman times. In the 1962 *Atlas* a decline in this species was only apparent at the northern fringe of its range but since then it has declined dramatically. The losses reflect the intensification of arable farming, and in particular improved seed screening and herbicide treatments.

As an archaeophyte *R. arvensis* has a Eurosiberian Southern-temperate distribution; it is widely naturalised outside this range.

References: Atlas (20d), Hultén & Fries (1986), Jalas & Suominen (1989), Meusel *et al.* (1965), Stewart *et al.* (1994).

R. A. FITZGERALD

Ranunculus paludosus Jersey Buttercup

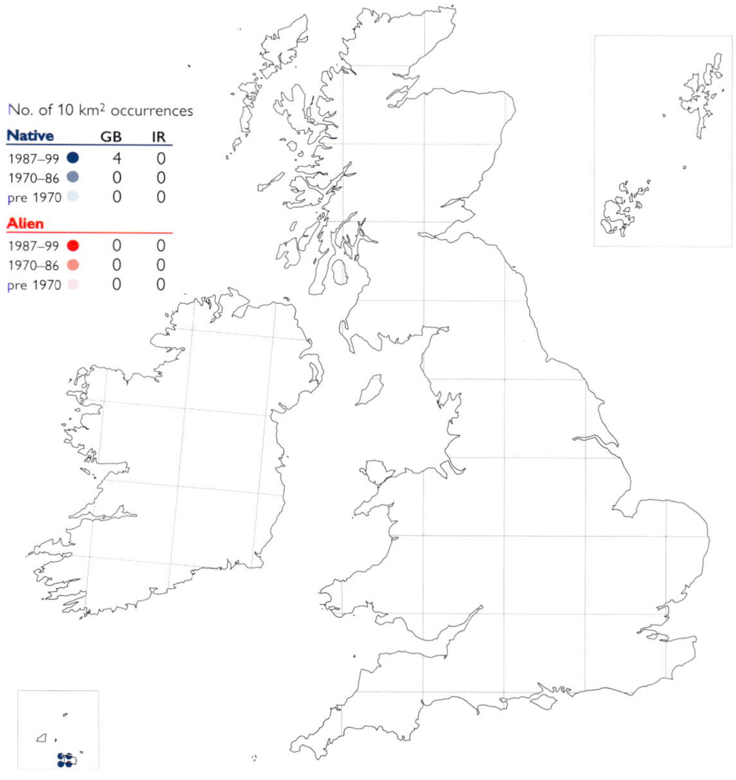

A winter-green perennial herb which dies down to spindle-shaped tubers after flowering in May. It grows in grassland which is wet in winter, but sun-baked in summer. The number of flowering plants in a population may vary considerably from year to year. Lowland.

Native. This species has always been restricted in our area to Jersey, where it was discovered in 1872. It was long thought to be an extreme rarity, but its habitat requirements were not understood until the late 1950s. It was then discovered in a number of sites on the west and south coast of the island.

Mediterranean-Atlantic element, reaching its northern limit in Jersey.

References: Atlas (20d), Bolòs & Vigo (1984), Jalas & Suominen (1989), Le Sueur (1984).

R. A. FITZGERALD

Ranunculus auricomus Goldilocks Buttercup

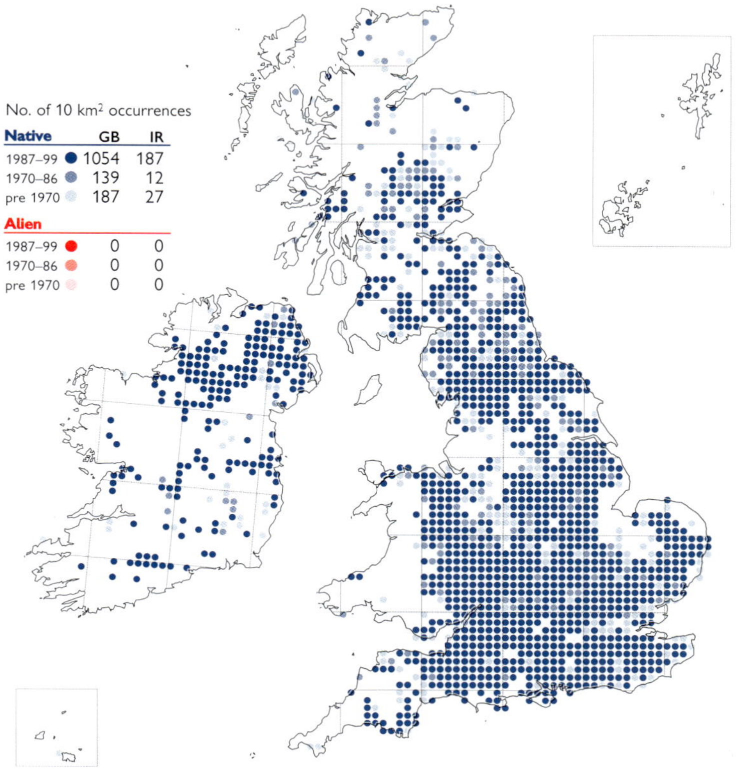

A perennial, characteristic of deciduous woodland on chalk, limestone and other basic soils. It also grows in scrub, on roadsides and in churchyards, and rarely on open moorland sheltered by boulders and on montane ledges. It is apomictic, showing considerable variation throughout Europe, though the agamospecies have not yet been formally described in our area. Generally lowland, but reaching 1090 m on Aonach Beag (Westerness).

Native (change –0.33). The range of *R. auricomus* is remarkably stable, although improved recording has led to the discovery of more sites within the range and a number of outlying populations in N. Scotland.

European Boreo-temperate element.

References: Atlas (21c), Hultén & Fries (1986), Jalas & Suominen (1989), Meusel *et al.* (1965).

R. A. FITZGERALD

Ranunculus sceleratus Celery-leaved Buttercup

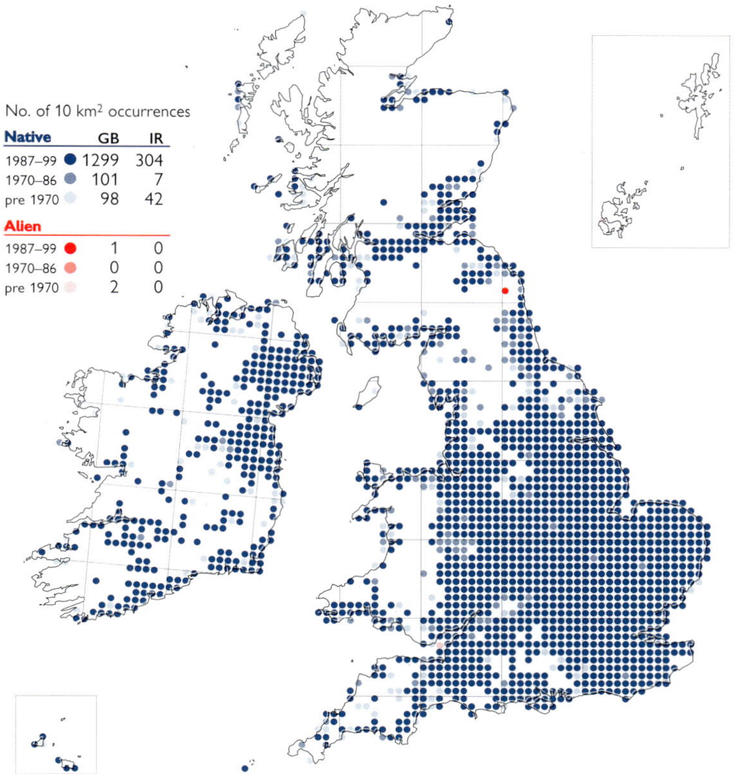

An annual of shallow water or wet, disturbed, nutrient-rich mud, especially at the edges of ponds, ditches, streams or rivers which are poached by drinking livestock. It is salt-tolerant and frequent on grazed estuarine marshes. Its seeds are long-lived and plants can re-appear following disturbance after many years of absence. Lowland.

Native (change –0.05). The distribution of *R. sceleratus* is stable, although the species is now better recorded than it was in the 1962 *Atlas*, especially in Scotland and Ireland.

Circumpolar Boreo-temperate element, with a disjunct distribution; widely naturalised outside its native range.

References: Atlas (22c), Grime *et al.* (1988), Hultén & Fries (1986), Jalas & Suominen (1989).

R. A. FITZGERALD

Ranunculus lingua Greater Spearwort

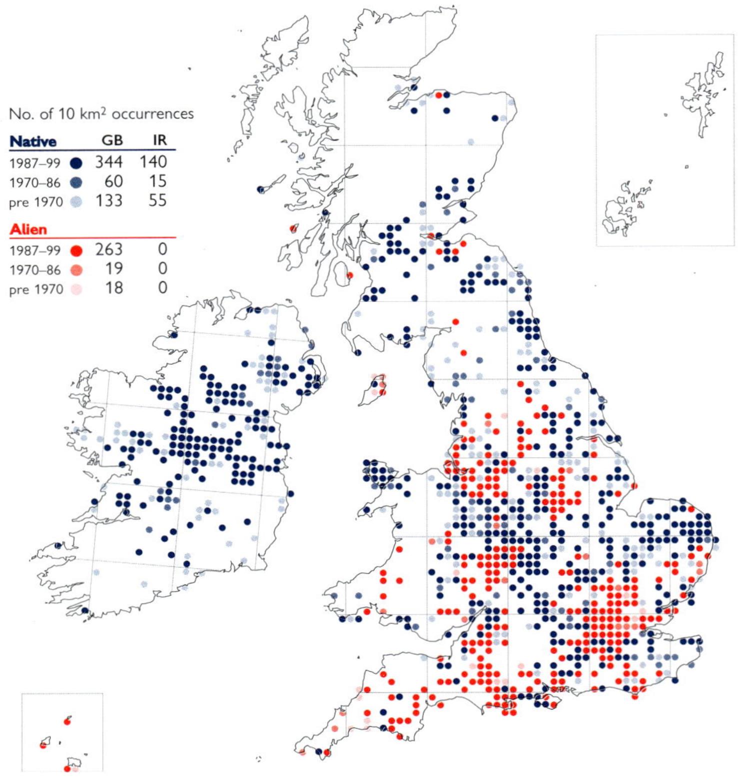

A stoloniferous perennial herb which grows in fens and marshes, on ditch, canal and pond edges, around reservoirs and in flooded gravel-pits and quarries. It is normally found in base-rich, still or slowly flowing water. Lowland.

Native (change +1.70). The 1962 *Atlas* illustrated the decline of native *R. lingua* in Britain, a result of drainage over the last two hundred years. Since 1962, however, the situation has been transformed by its increased popularity as an ornamental plant. It is frequently introduced to ponds and other wetlands in the wild and the distinction between native and alien populations is now hopelessly blurred. In Ireland the native plant is still locally frequent.

Eurosiberian Temperate element.

References: Atlas (21d), Hultén & Fries (1986), Jalas & Suominen (1989), Meusel *et al.* (1965).

R. A. FITZGERALD

Ranunculus flammula Lesser Spearwort

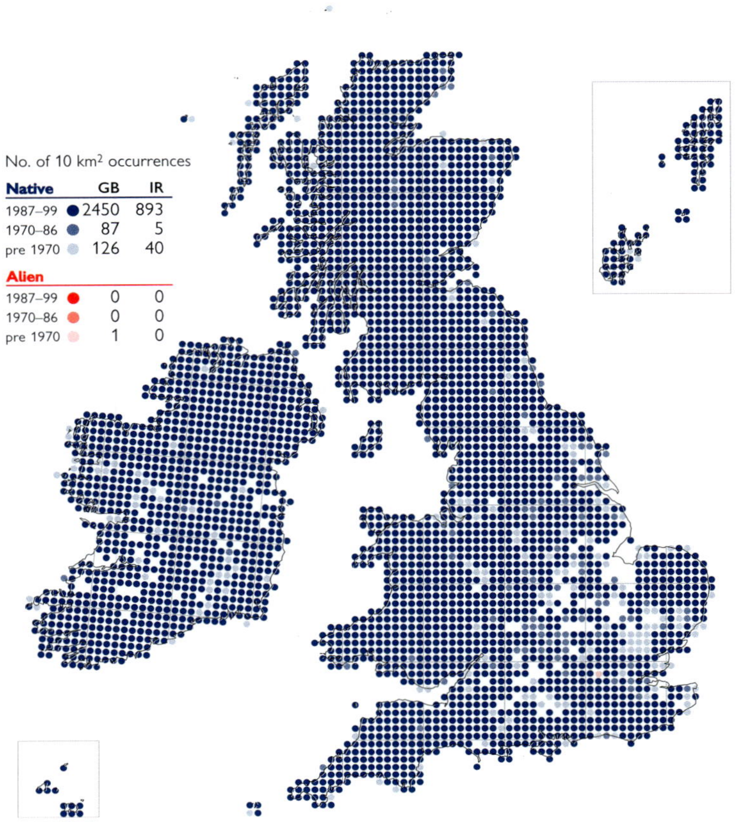

A perennial herb of wet habitats, particularly those with seasonal water level fluctuations. It is found in springs and flushes, around ponds, on lake shores, streamsides, in dune-slacks, marshes, water-meadows, flood pastures, bogs and in ditches and track ruts. It usually grows in oligotrophic or mesotrophic water over neutral to acid substrates. 0–930 m (Carnedd Llewelyn, Caerns.).

Native (change –0.60). There is some evidence for a loss of *R. flammula* in S.E. England, but it remains frequent elsewhere. Subsp. *flammula* occurs throughout the British and Irish range of the species; the two rarer subspecies are mapped separately.

European Temperate element.

References: Atlas (22a), Grime *et al.* (1988), Hultén & Fries (1986), Jalas & Suominen (1989), Meusel *et al.* (1965), Preston & Croft (1997).

R. A. FITZGERALD

Ranunculus flammula subsp. *minimus*

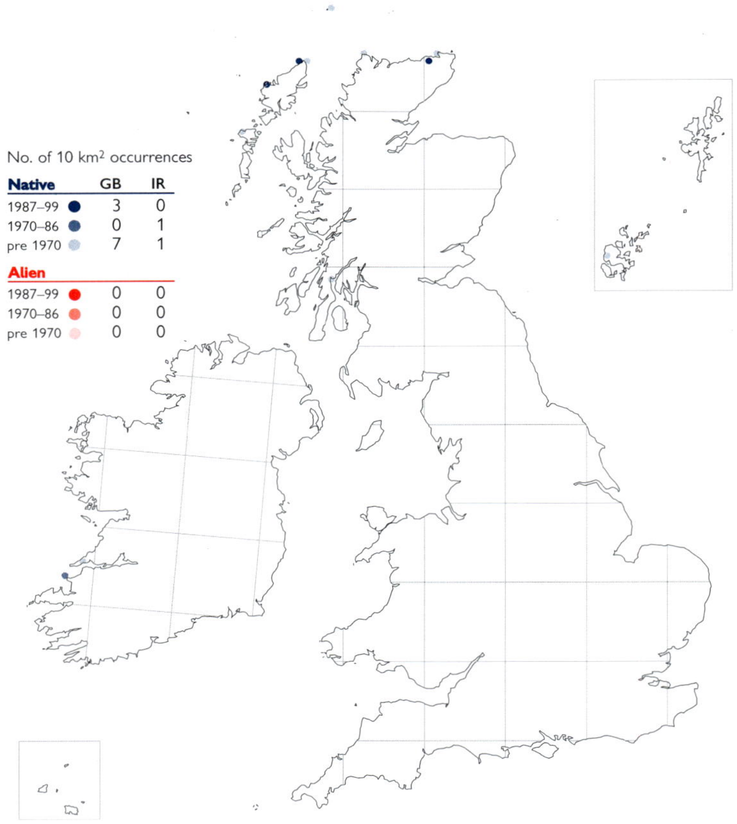

A dwarf and prostrate perennial herb occurring in dense mats on exposed, flushed sea-cliffs, and in short damp turf on moorland. Lowland.

Native. This segregate has been little studied, and is known only from very remote localities. It has been confused with dwarf forms of subsp. *flammula*, but remains distinct in cultivation. It is probably under-recorded.

Endemic.

References: Jalas & Suominen (1989), Preston & Croft (1997).

R. A. FITZGERALD

Ranunculus flammula subsp. *scoticus*

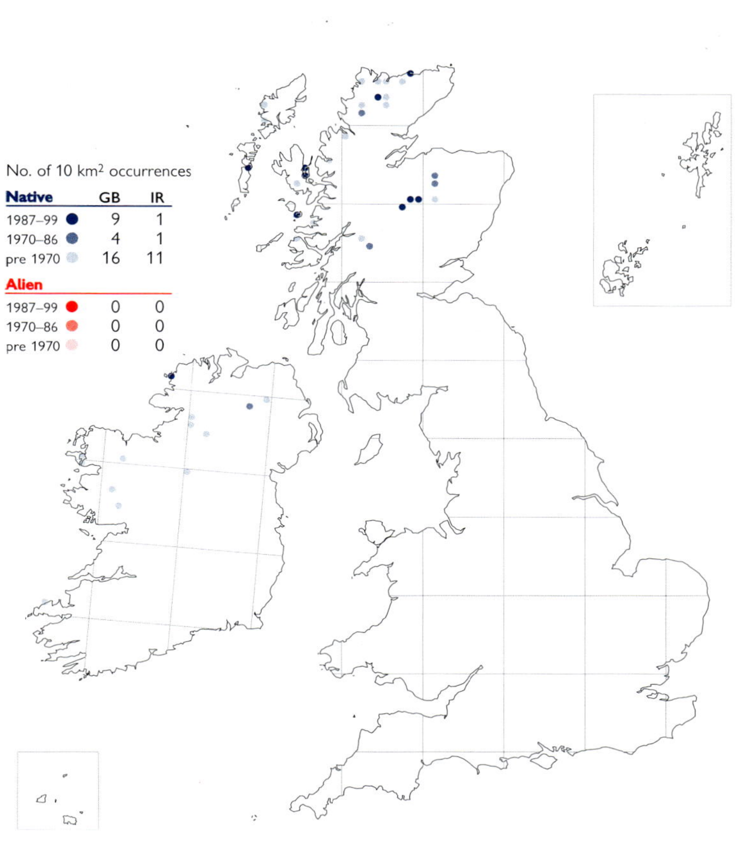

A perennial herb which grows on stony or peaty substrates at the edge of lakes. 0–485 m (Loch Builg, Banffs.).

Native. This segregate is little known, and has been confused in the past with subsp. *flammula*. It is almost certainly under-recorded.

Endemic.

References: Jalas & Suominen (1989), Preston & Croft (1997).

R. A. FITZGERALD

Ranunculus reptans Creeping Spearwort

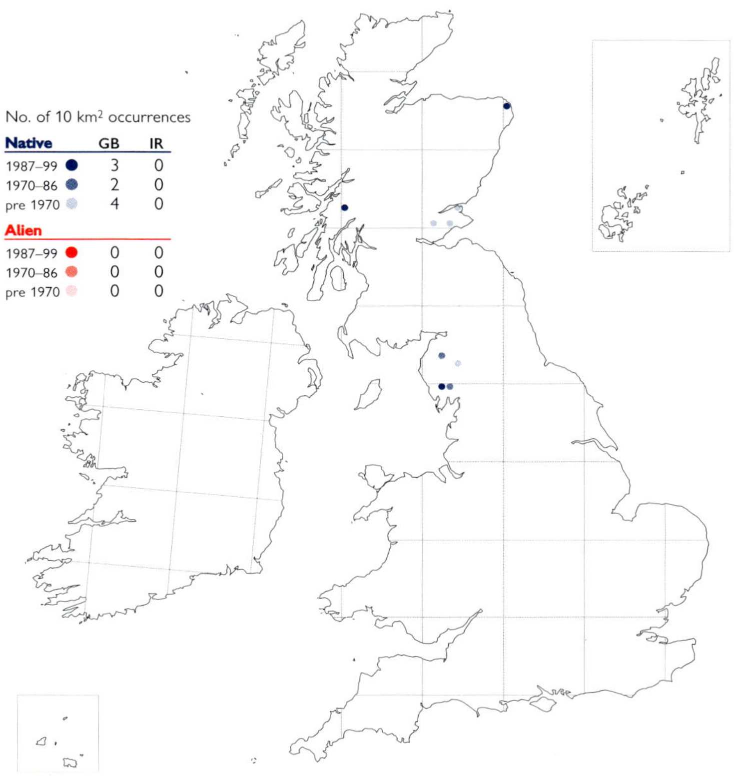

A stoloniferous perennial herb of lake shores, growing on gravel or silty sand. At the Loch of Strathbeg (N. Aberdeen), where it has been known since 1876, it grows in open vegetation in a zone of *Eleocharis palustris* which is intermittently exposed above the water level in summer. Lowland.

Native. At some sites *R. reptans* may be a transient member of the British flora, perhaps introduced as seed by migrating wildfowl but not persisting for long. It can hybridise with the common *R. flammula*, and the hybrid *R.* × *levenensis* forms persistent populations in Britain and is also recorded from Ireland.

Circumpolar Boreal-montane element.

References: Birse (1997), Hultén & Fries (1986), Jalas & Suominen (1989), Meusel *et al.* (1965), Preston & Croft (1997), Wigginton (1999).

R. A. FITZGERALD & C. D. PRESTON

Ranunculus ophioglossifolius
Adder's-tongue Spearwort

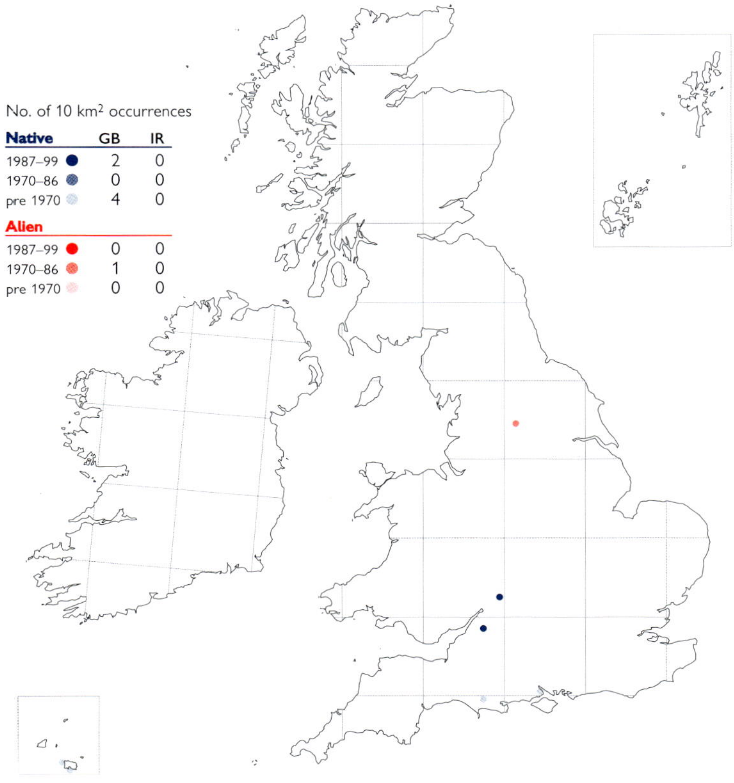

An annual found in a highly specialised marshy habitat. It requires winter inundation, bare, wet mud for seedling establishment, reduced summer water levels and low competition. The substrate at the two extant sites is base-rich Lias clay, with most water input from rain. Lowland.

Native. Sites in S. Hampshire, Dorset and Jersey were lost by the early twenteenth century. The largest surviving population, at Badgeworth (E. Gloucs.), has been dependent on management since 1962, and with appropriate human disturbance a sizeable population of plants flower and fruit every year. A smaller site nearby is still managed by grazing, and that population is more erratic.

European Southern-temperate element.

References: Atlas (22b), Hultén & Fries (1986), Jalas & Suominen (1989), Marren (1999), Wigginton (1999).

R. A. FITZGERALD

Ranunculus ficaria Lesser Celandine

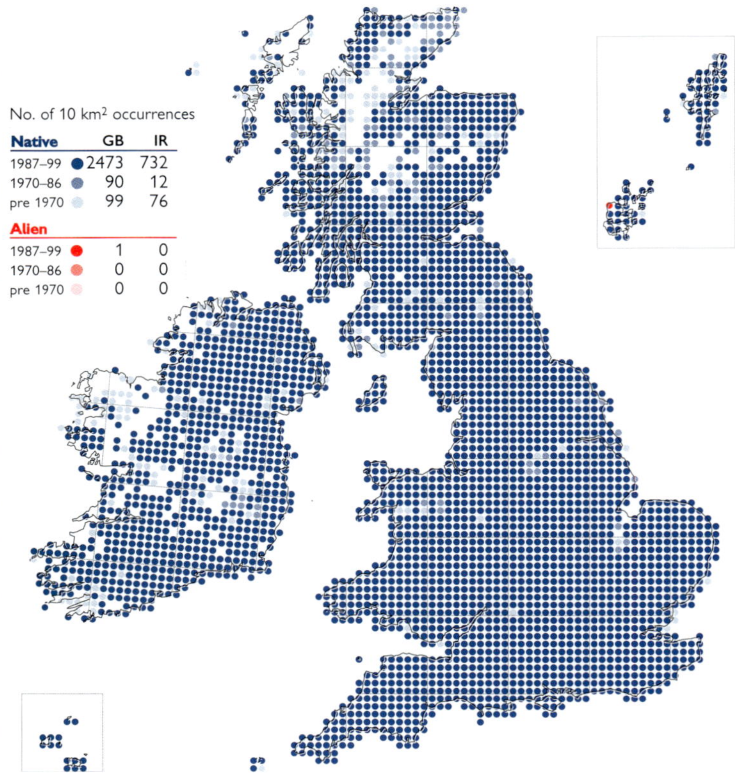

An aestivating perennial herb that grows in woods, hedge banks, meadows, roadsides, maritime grassland, the banks of rivers and streams and shaded waste ground. It prefers damp, loamy or clay soils, and avoids very dry, very acidic or permanently waterlogged sites. 0–750 m (Loch Lochy, Westerness), but probably higher elsewhere in Scotland.

Native (change +0.16). The distribution of *R. ficaria* is stable. Losses at the fringe of its Scottish range probably reflect a lack of recent fieldwork in spring. The species is morphologically and cytologically variable; the two native subspecies are mapped separately.

European Southern-temperate element.

References: Atlas (24d), Grime *et al.* (1988), Hultén & Fries (1986), Jalas & Suominen (1989), Meusel *et al.* (1965), Taylor & Markham (1978).

R. A. FITZGERALD & C. D. PRESTON

Ranunculus ficaria subsp. *ficaria*

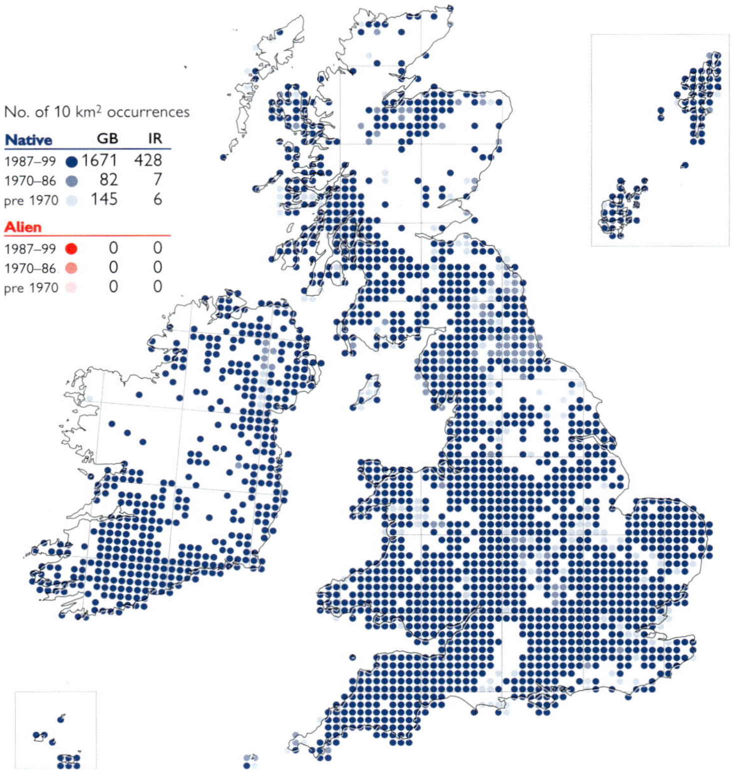

This subspecies occurs in less disturbed habitats than subsp. *bulbilifer*, being the usual variant found in ancient woodland and grassland. The ecological distinction between the two taxa is not absolute, however, and subsp. *ficaria* can be found growing where subsp. *bulbilifer* might be expected. A fertile diploid, it normally sets viable seed. 0–750 m (Loch Lochy, Westerness), but probably higher elsewhere in Scotland.

Native. Subsp. *ficaria* was mapped by Perring & Sell (1968); subsequent recording has improved our knowledge of its distribution but it is still under-recorded, and almost certainly occurs wherever *R. ficaria* is mapped.

Suboceanic Southern-temperate element.

References: Atlas Supp. (5a), Jalas & Suominen (1989), Sell (1994).

R. A. FITZGERALD & C. D. PRESTON

Ranunculus ficaria subsp. *bulbilifer*

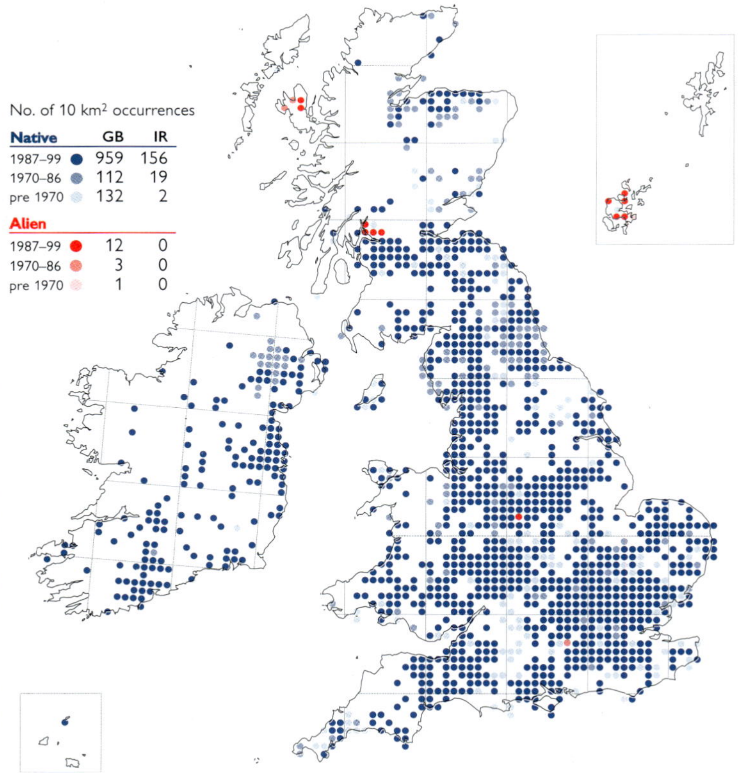

A perennial herb, typically found in more disturbed habitats than subsp. *ficaria*. It is often abundant on the sides of streams and rivers, growing below or just above the level of winter flooding, and is also found as a weed in churchyards and gardens. It is tetraploid and largely sterile, but bears tuberous bulbils in the leaf axils which are its main method of dispersal. Generally lowland, but reaching 390 m (Forest-in-Teesdale, Co. Durham).

Native. Subsp. *bulbilifer* was mapped by Perring & Sell (1968). The current map is much more complete but the subspecies is probably still under-recorded.

European Temperate element.

References: Atlas Supp. (5b), Jalas & Suominen (1989), Sell (1994).

R. A. FITZGERALD & C. D. PRESTON

Ranunculus hederaceus Ivy-leaved Crowfoot

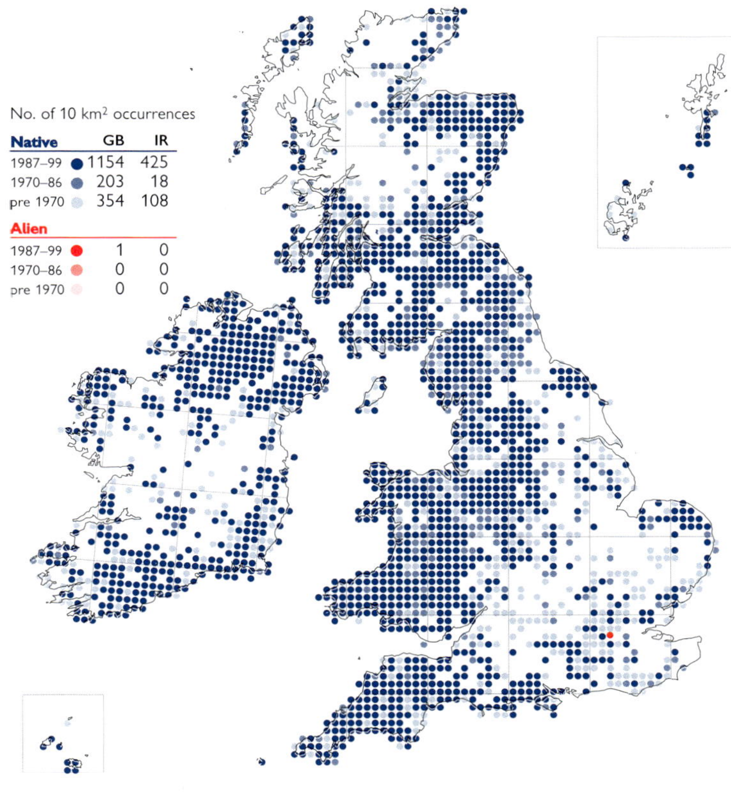

A small annual or short-lived perennial, found at the edge of small water bodies and by the sheltered backwaters of rivers. It often grows on the cattle-poached edges of ponds, ditches and streams, in wet gateways and on paths and tracks. It tolerates a broad range of pH and nutrient levels, including nitrophilous conditions. 0–770 m (Little Dun Fell, Westmorland).

Native (change +0.10). This species was lost from some sites in S.E. England before 1900 as a result of agricultural improvement and urban spread. It has declined since 1950 in areas where the agriculture is now mostly arable, but remains frequent elsewhere.

Suboceanic Southern-temperate element.

References: Atlas (22d), Hultén & Fries (1986), Jalas & Suominen (1989), Meusel *et al.* (1965), Preston & Croft (1997).

C. D. PRESTON

Ranunculus omiophyllus Round-leaved Crowfoot

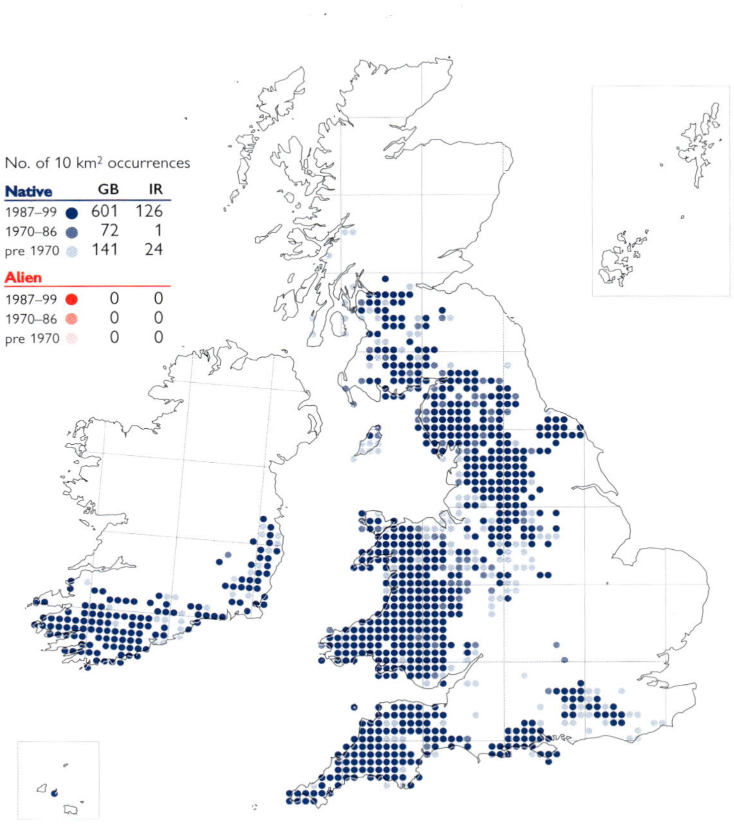

A small annual or short-lived perennial which grows in shallow water or on wet soil. Typical sites include the margins of ponds and ditches, flushes, damp depressions, gateways and tracks in pastures and on heathland, and the sheltered backwaters of rivers. Unlike *R. hederaceus*, it is confined to acidic, mesotrophic or oligotrophic soils. 0–1005 m (Carnedd Llewelyn, Caerns.).

Native (change +0.52). *R. omiophyllus* remains frequent in W. Britain and S. Ireland, where there are many more records than there were in the 1962 *Atlas*. The decline at the fringes of its range was already discernible in the 1962 *Atlas*, and is now much more marked.

Suboceanic Southern-temperate element.

References: Atlas (23a), Jalas & Suominen (1989), Preston & Croft (1997).

C. D. PRESTON

Ranunculus tripartitus Three-lobed Crowfoot

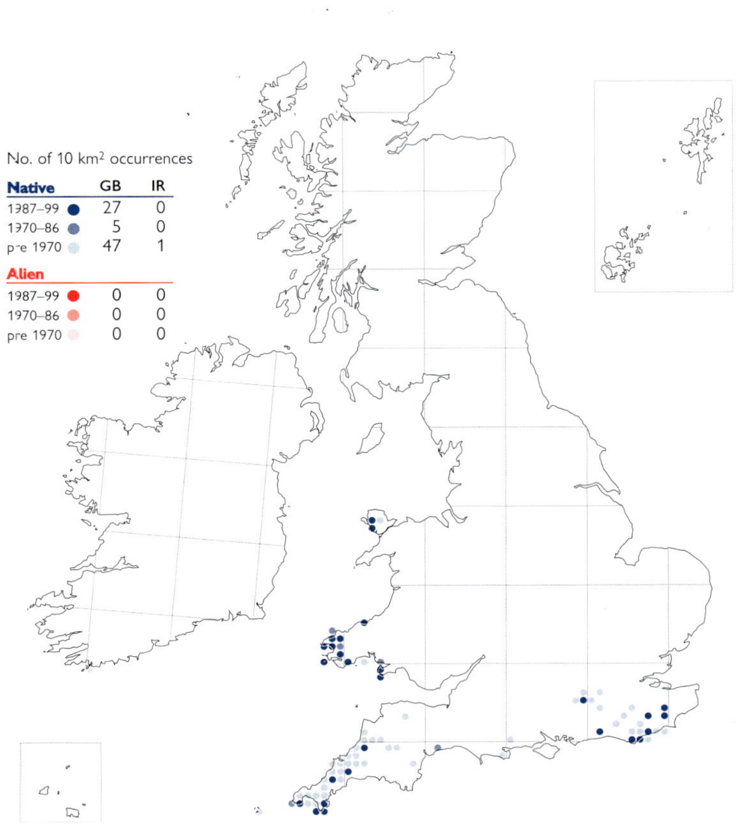

An annual of shallow water bodies over base- and nutrient-poor substrates, in open sites which are flooded in winter but summer-dry. In S.E. England it is also found in pools in coppiced woodland. Lowland to 300 m at Belstone (N. Devon).

Native (change –1.09). Habitat destruction or reduced disturbance have led to the gradual decline of this species from the end of the 19th century onwards. However, it has recently been discovered at some new and historical British sites, and in 2000 it was refound in Ireland. Some mapped records may represent its hybrid with *R. omiophyllus*.

Oceanic Southern-temperate element.

References: Atlas (23b), Bolòs & Vigo (1984), Curtis & McGough (1988), FitzGerald & Stewart (2000), Jalas & Suominen (1989), Kay & John (1995), Preston & Croft (1997), Stewart *et al.* (1994), Wigginton (1999).

C. D. PRESTON

Ranunculus baudotii Brackish Water-crowfoot

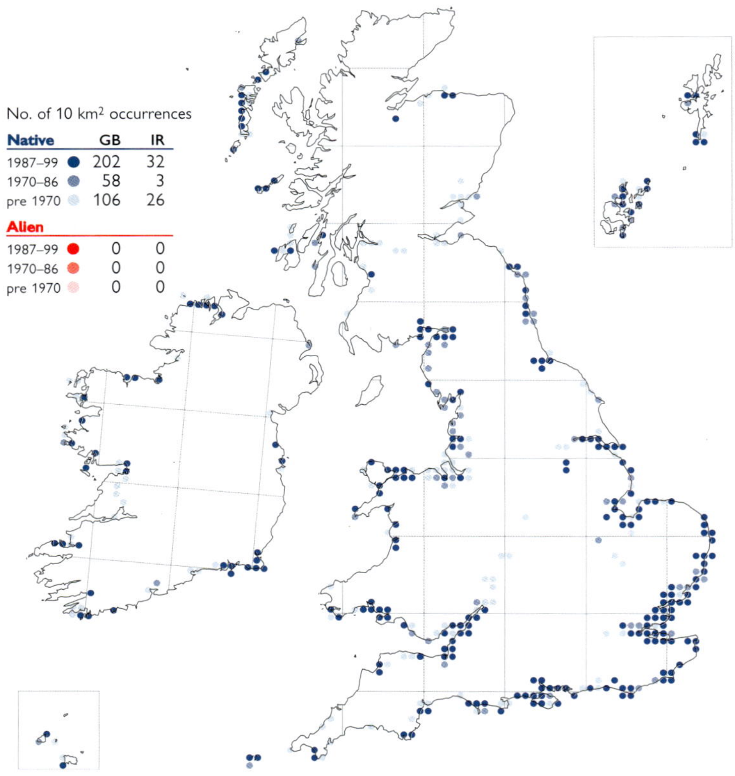

This annual or perennial herb grows in coastal water bodies, including lagoons, machair lochs, ditches, pools, dune-slacks and borrow-pits. It is most frequent in water 0.5–1 m deep, but can grow in shallower water or as a dwarf terrestrial form on wet mud. Its inland sites include flooded mineral workings and canals, some receiving saline drainage water but others lacking any saline influence. Lowland.

Native (change –0.04). *R. baudotii* is now known to be much more frequent than was appreciated at the time of the 1962 *Atlas*. Many of the pre-1970 records date from the 19th century, suggesting a long-term decline in some areas.

European Southern-temperate element.

References: Atlas (24c), Hultén & Fries (1986), Jalas & Suominen (1989), Preston & Croft (1997), Stewart *et al.* (1994).

C. D. PRESTON

Ranunculus trichophyllus Thread-leaved Water-crowfoot

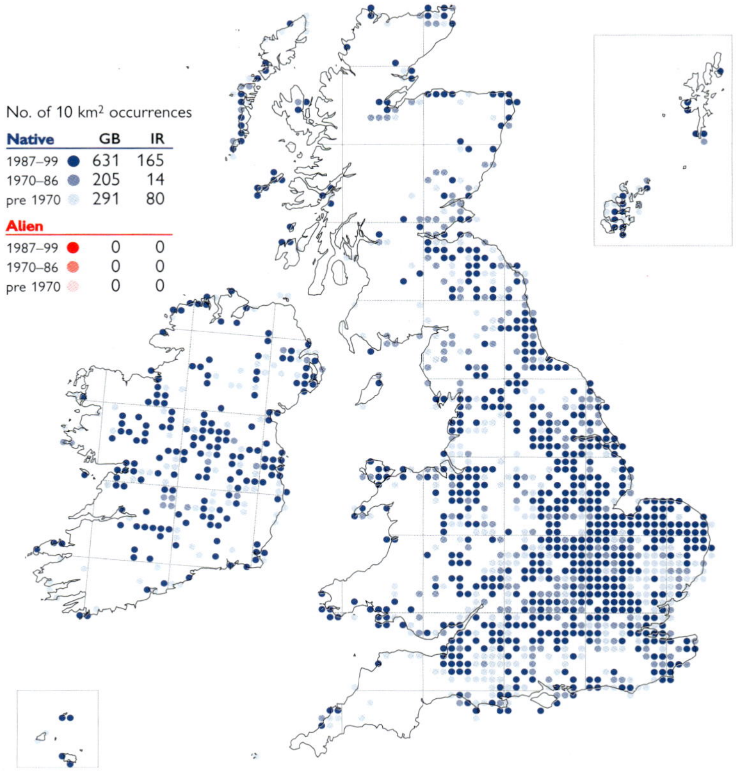

A small annual or perennial which grows in shallow, still or very slowly flowing water. It is most frequent in ponds, dune-slacks and drainage ditches, but it is also found in larger sites if they are sheltered. It tolerates a range of water chemistry but is most frequent in mesotrophic or eutrophic water. 0–310 m (Alston Moor, Cumberland).

Native (change –0.07). The aquatic *Ranunculus* species are difficult to identify, and this early-flowering plant is easily overlooked. Under-recording rather than decline may therefore be the reason for the concentration of pre-1970 records in some areas as there is little evidence of decline in neighbouring counties.

Circumpolar Wide-boreal element.

References: Atlas (24a), Hultén & Fries (1986), Jalas & Suominen (1989), Preston & Croft (1997).

C. D. PRESTON

Ranunculus aquatilis Common Water-crowfoot

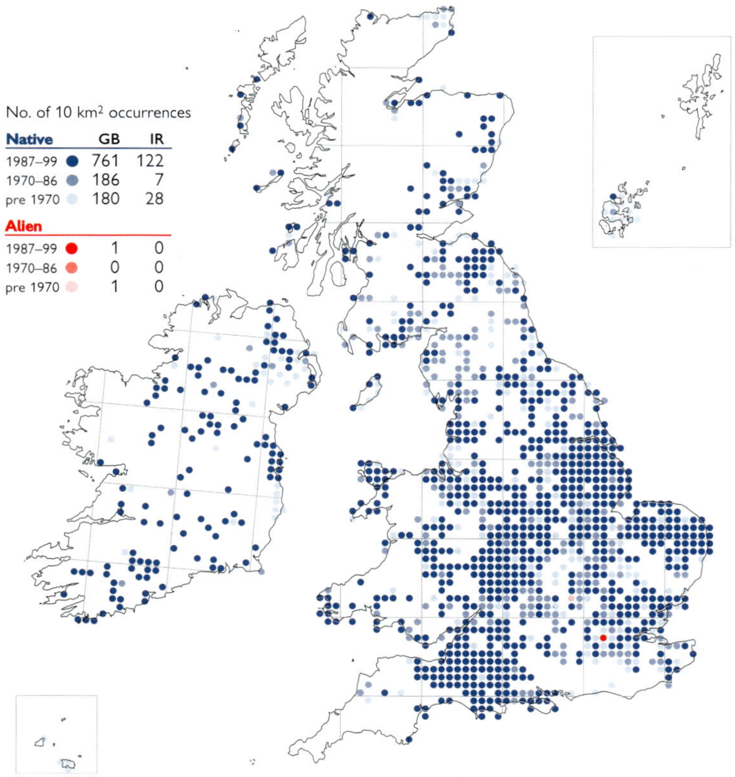

This is an annual or short-lived perennial which grows in shallow water in marshes, ponds and ditches, and at the edge of slow-flowing streams and sheltered lakes. It occurs chiefly in water which is eutrophic and at least mildly base-rich, and is favoured by a degree of disturbance. Generally lowland, but reaching 445 m at Small Water (Cumberland).

Native. The map of *R. aquatilis* in the 1962 *Atlas* included records of *R. peltatus* and *R. penicillatus*, as the taxonomy of these species was then unresolved. It is, therefore, difficult to assess changes in the distribution of the individual species.

European Temperate element; also in C. and E. Asia and western N. America.

References: Hultén & Fries (1986), Jalas & Suominen (1989), Preston & Croft (1997).

C. D. PRESTON

Ranunculus peltatus Pond Water-crowfoot

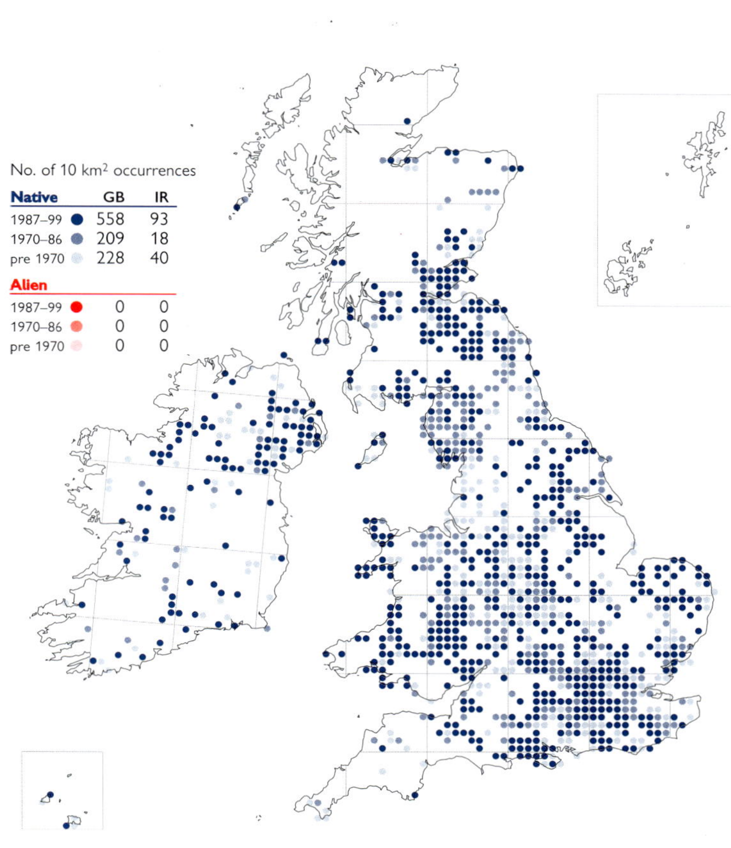

This perennial or sometimes annual species grows in slow-flowing streams and rivers, coastal lagoons, shallow lakes, ditches, ponds and dune-slacks. It is difficult to define its ecological preferences, as it grows in the upper reaches of highly calcareous rivers but in some areas favours base-poor waters; it has a broad trophic range. Generally lowland, but reaching 500 m at Dogber Tarn (Westmorland).

Native. The larger aquatic species of *Ranunculus* are difficult to identify, and the map may contain some errors. In the 1962 *Atlas*, records for this species were included in the map of *R. aquatilis*, so changes in distribution are difficult to assess.

European Wide-temperate element.

References: Grime *et al.* (1988), Hultén & Fries (1986), Jalas & Suominen (1989), Preston & Croft (1997).

C. D. PRESTON

Ranunculus penicillatus Stream Water-crowfoot

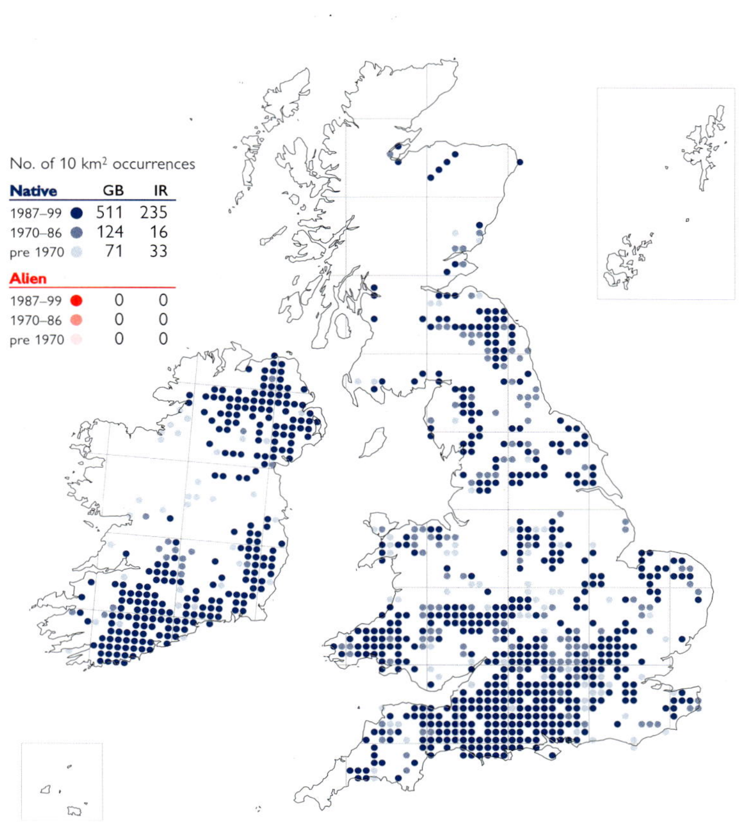

A vigorous perennial species of moderate to rapidly flowing rivers and streams, growing in oligotrophic and mesotrophic waters. Generally lowland, but reaching 310 m in the Carlcroft Burn (Cheviot).

Native. The taxonomy of the *R. penicillatus* aggregate was not elucidated until 1966, and the subspecies were not mapped separately until 1988. The riparian *Ranunculus* taxa still present formidable difficulties of identification, especially as both sterile and fertile hybrids occur in addition to the recognised species. Subsp. *penicillatus* and subsp. *pseudofluitans* are mapped separately.

European Temperate element, but the distribution of the species and its subspecies is clouded by taxonomic uncertainty.

References: Cook (1966), Jalas & Suominen (1989), Webster (1988).

C. D. PRESTON

Ranunculus penicillatus subsp. *penicillatus*

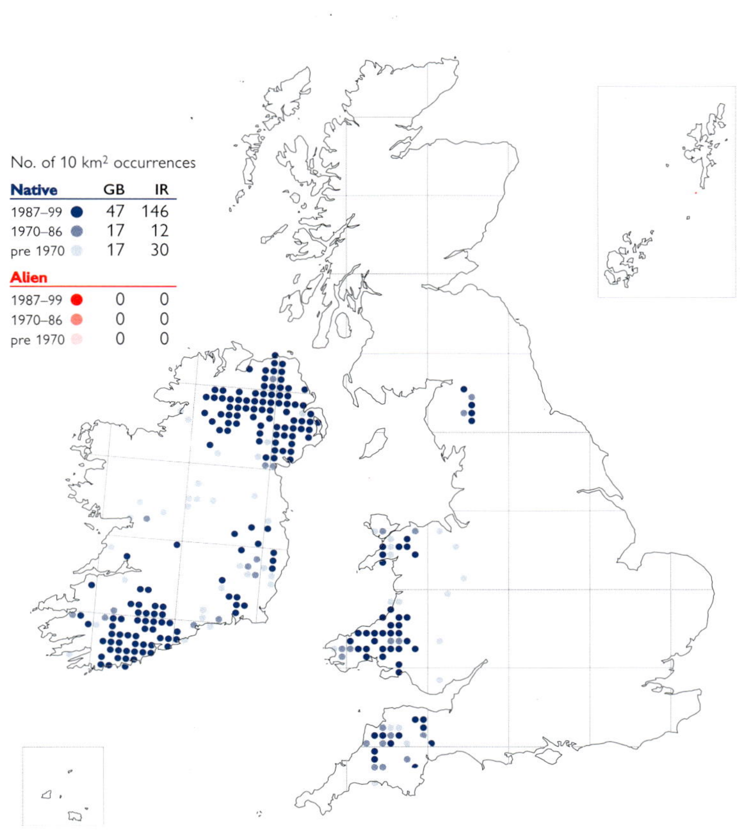

This is a robust perennial species which can grow in dense, dominant stands in rapidly flowing rivers. In Britain it is confined to base-poor and more or less mesotrophic rivers flowing over Palaeozoic or igneous rocks. In Ireland it has a much broader habitat range, being found in both base-rich and base-poor rivers and streams, and sometimes in lakes and canals. Lowland.

Native. *R. penicillatus* subsp. *penicillatus* was not mapped in the 1962 *Atlas*. It is likely to be under-recorded, especially in Ireland.

European Temperate element.

References: Preston & Croft (1997), Webster (1988).

C. D. PRESTON

Ranunculus penicillatus subsp. *pseudofluitans*

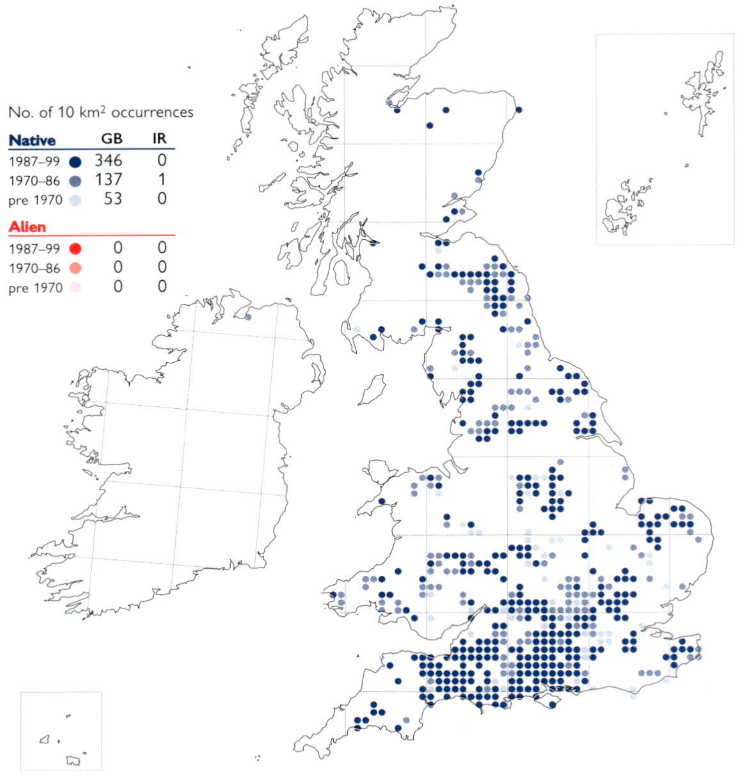

This is the dominant plant in many British base-rich rivers and streams, favouring moderately or rapidly flowing, mesotrophic or meso-eutrophic water, and being most frequent where the water flow is broken by riffles. In sluggish eutrophic rivers it is replaced by other species. In Ireland, it is found in a single, rather base-poor river. 0–310 m (Carlcroft Burn, Cheviot).

Native. The taxonomy of subsp. *pseudofluitans* has only been clarified since the 1962 *Atlas*, and it is probably still under-recorded in some areas of Britain.

Suboceanic Temperate element; known from Britain and Ireland south to the Iberian peninsula but also reported from Greece.

References: Grime *et al.* (1988), Preston & Croft (1997), Webster (1988).

C. D. PRESTON

Ranunculus fluitans River Water-crowfoot

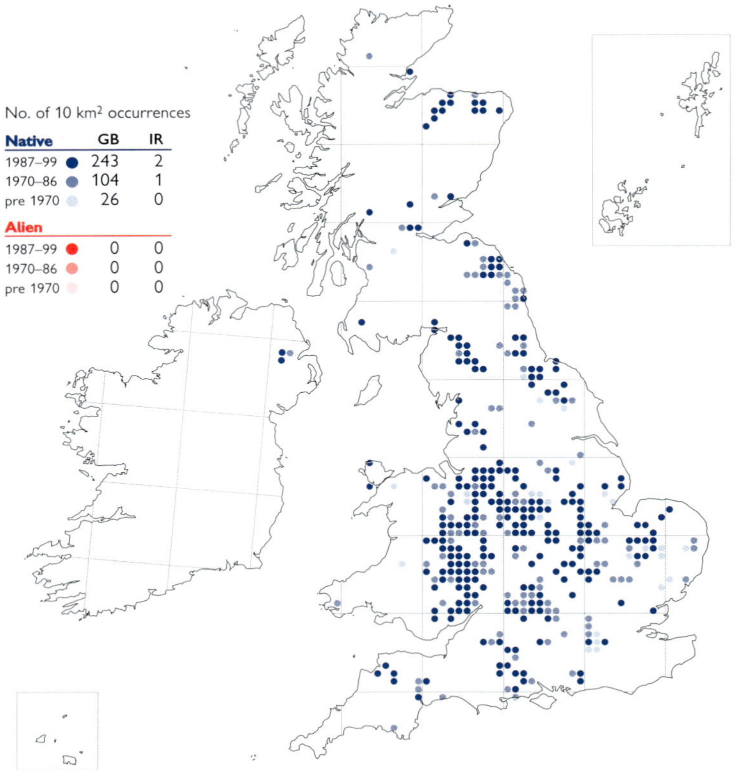

This is a perennial species which grows in large, rapidly flowing rivers with a stable substrate. It is usually found in base-rich and meso-eutrophic water. In Ireland, it is confined to a single, now locally highly polluted, river. Lowland.

Native (change +1.96). This is not a well-understood species; in some areas of Britain it may be under-recorded whereas in others it may be reported in error for other taxa. Historically *R. penicillatus* subsp. *pseudofluitans* may have been recorded as *R. fluitans*, whereas now hybrids with *R. fluitans* as one parent are more likely to be reported as the species.

European Temperate element.

References: Atlas (23c), Curtis & McGough (1988), Hultén & Fries (1986), Jalas & Suominen (1989), Preston & Croft (1997).

C. D. PRESTON

Ranunculus circinatus Fan-leaved Water-crowfoot

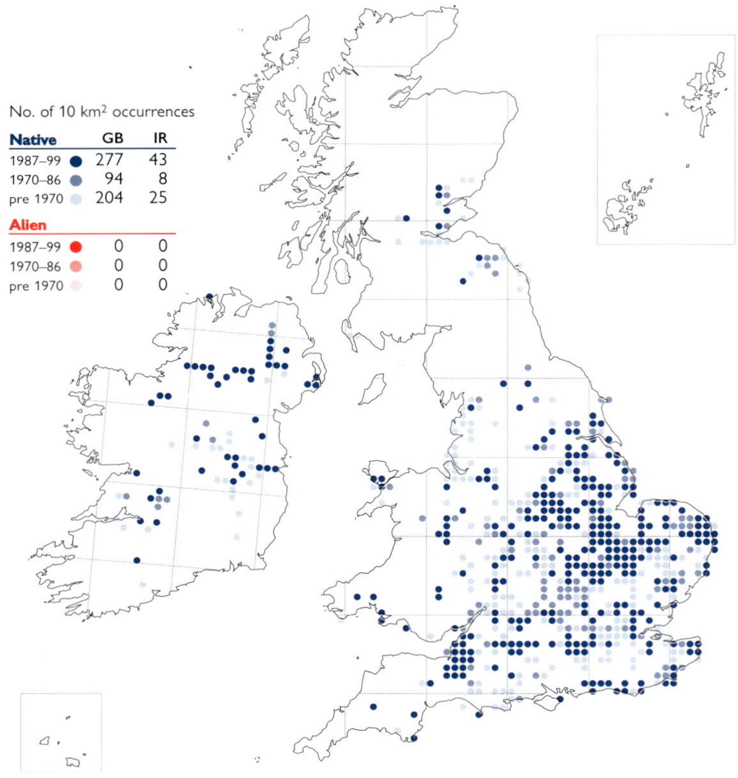

A perennial herb of clear, base-rich, standing or very slowly flowing water, most frequently in lakes, flooded gravel-pits, sluggish streams and rivers, canals and ditches. It usually grows at depths of 1–3 m in meso-eutrophic or eutrophic water; only growing in shallower water if it does not dry up in summer. 0–310 m (Akermoor Loch, Selkirks.).

Native (change –0.34). Unlike most aquatic *Ranunculus* species, this is an easily recognised plant which should be well recorded. It has declined across much of its range because of habitat destruction and eutrophication. It was lost from many sites before 1930, but the decline has continued since the 1962 *Atlas*.

Eurasian Temperate element.

References: Atlas (23d), Hultén & Fries (1986), Jalas & Suominen (1989), Preston & Croft (1997).

C. D. PRESTON

Adonis annua Pheasant's-eye

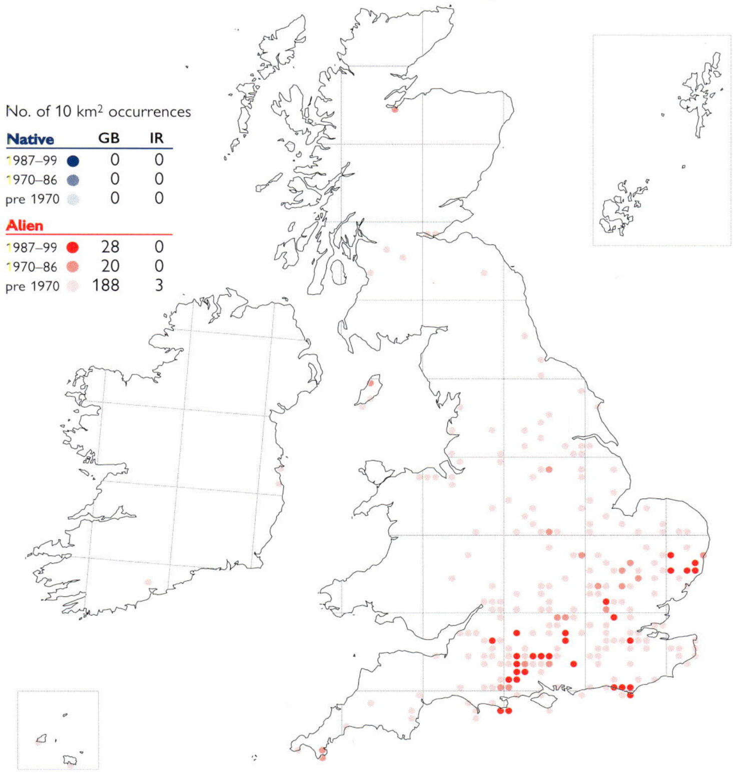

An arable weed of dry soils on chalk and limestone, also recorded from tracks, chalk pits and other disturbed habitats. Seed production is low but there is a long-lived soil seed bank. Most populations are small and restricted to field edges. Lowland.

Archaeophyte (change −2.19). This species is known from Iron Age deposits (Jones, 1984). It underwent a catastrophic decline from *c.* 1880 to 1950, and has been lost from a further 30% of 10-km squares since the 1962 *Atlas*. This decline is due to improved seed cleaning methods, increased use of agrochemicals and the density of modern crops. Some recent records are deliberate introductions.

Native of the Mediterranean region; long naturalised northwards to Britain.

References: Atlas (25a), Jalas & Suominen (1989), Meusel *et al.* (1965), Wigginton (1999).

R. A. FITZGERALD

Myosurus minimus Mousetail

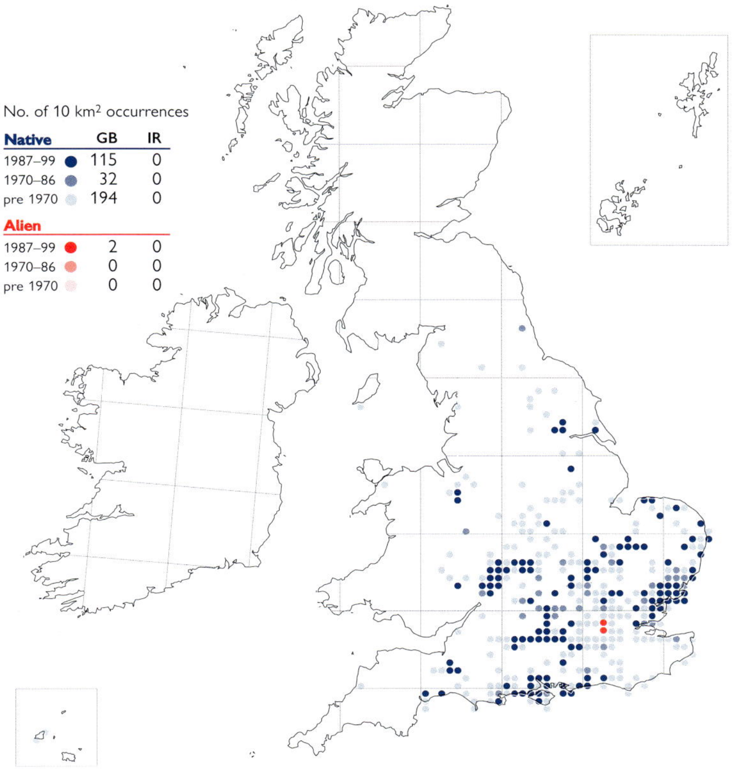

An annual of seasonally flooded, nutrient-rich soils in areas disturbed by machinery or animals, such as hollows on ploughed land, rutted tracks and gateways in pastures. Its seeds appear to be long-lived. Lowland.

Native or alien (change −0.66). It is difficult to assess trends in this inconspicuous and sporadic species, which was known in Britain by 1597. Many sites were lost before 1930, probably through the disuse of commons, re-surfacing of tracks and the drainage and filling of small ponds. It persists in areas such as coastal grazing marshes.

European Temperate element; also in N. America and Australasia and widely naturalised elsewhere.

References: Atlas (25b), Hultén & Fries (1986), Jalas & Suominen (1989), Meusel *et al.* (1965), Stewart *et al.* (1994).

R. A. FITZGERALD

Aquilegia vulgaris Columbine

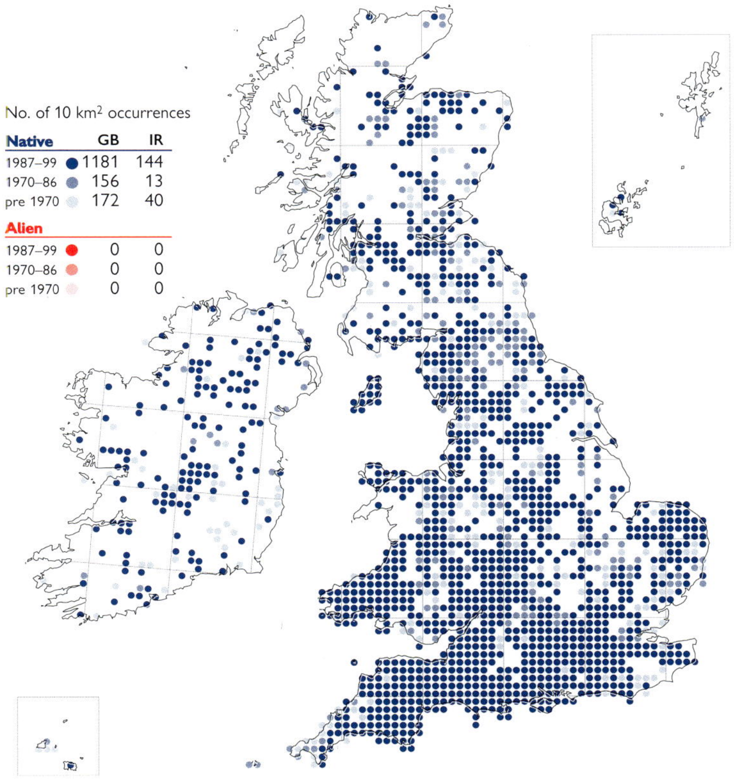

Native populations of this perennial are found on calcareous soil over limestone rocks in England and Wales. It typically grows in woodland glades and open scrub, by woodland rides and streamsides, in damp grassland and fen, and on scree slopes. Garden escapes can be naturalised in quarries, on roadsides, railway banks and old walls. 0–470 m (Sandbed Gill, Cumberland).

Native (change +1.70). *A. vulgaris* has increased since the 1962 *Atlas*, presumably because of the increasing frequency of garden escapes. The native distribution is now totally obscured and all records are mapped as if they were native.

European Temperate element; widely naturalised outside its native range.

References: Atlas (25c), Hultén & Fries (1986), Jalas & Suominen (1989), Meusel *et al.* (1965).

R. A. FITZGERALD

Thalictrum flavum Common Meadow-rue

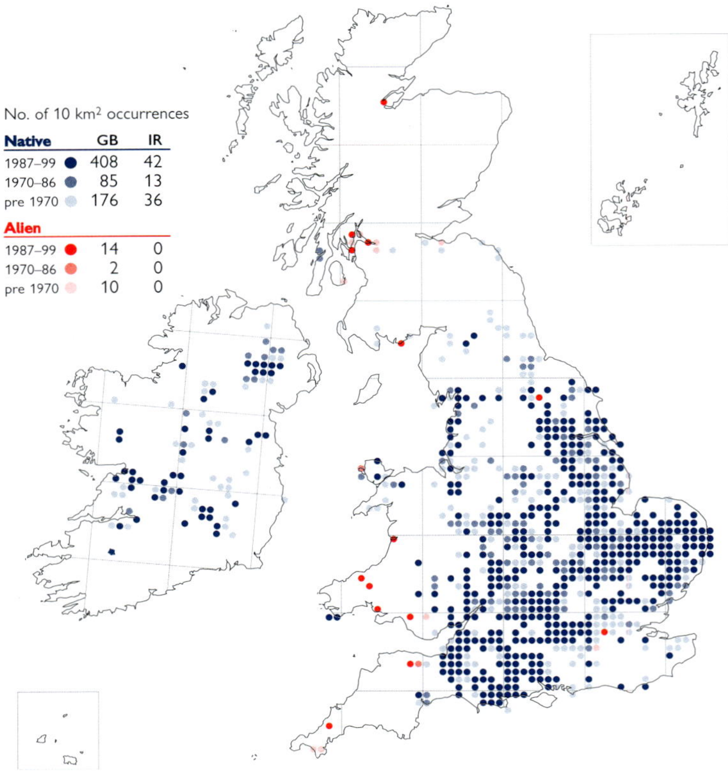

A rhizomatous perennial of fens, ditches and streamsides, and tall vegetation in wet meadows, always found where the substrate or water is base-rich. It is also recorded from open fen carr. Lowland.

Native (change –0.53). This handsome plant has declined since 1930 because of drainage and agricultural intensification on grazing marshes, and it is now often restricted to relict linear habitats such as river banks and roadside ditches. It can, however, still be locally abundant in unimproved wet grasslands with traditionally managed drainage systems.

Eurosiberian Boreo-temperate element.

References: Atlas (25d), Hultén & Fries (1986), Jalas & Suominen (1989), Meusel *et al.* (1965).

R. A. FITZGERALD

Thalictrum minus Lesser Meadow-rue

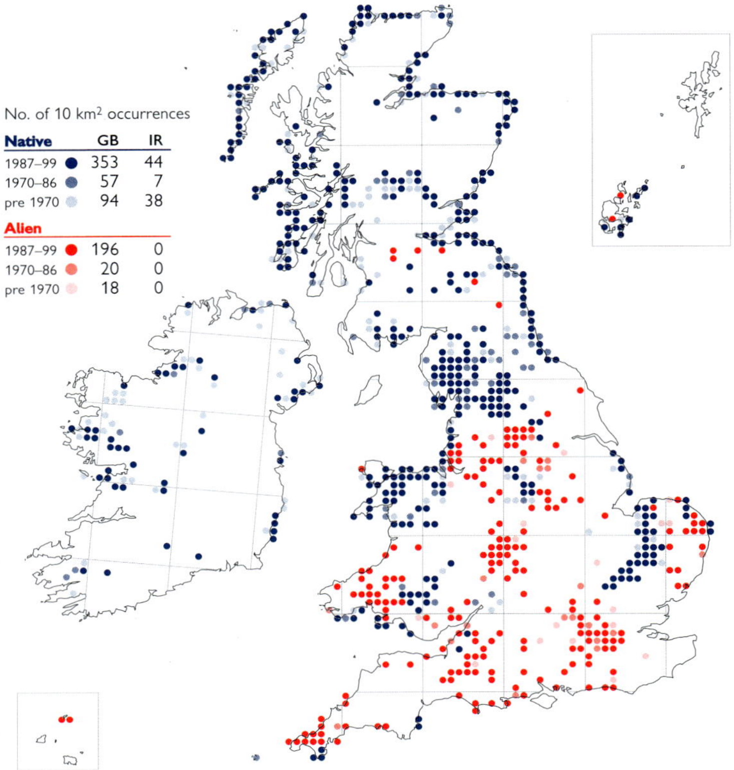

A morphologically variable, perennial herb found in calcareous or other base-rich habitats where competition is low, including fixed dunes, scrubby banks, rocky lake and river edges, limestone and serpentine cliffs, limestone grassland and pavement and montane rock ledges. It also occurs in other habitats, including churchyards, hedge banks and roadsides, as a garden escape. 0–855 m (Snowdon, Caerns.).

Native (change +0.56). The native distribution is stable, and much better recorded now than in the 1962 *Atlas*, but the species is apparently increasing as an alien in S. England.

Eurasian Boreo-temperate element.

References: Atlas (26b), Hultén & Fries (1986), Jalas & Suominen (1989), Meusel *et al.* (1965).

R. A. FITZGERALD

Thalictrum alpinum Alpine Meadow-rue

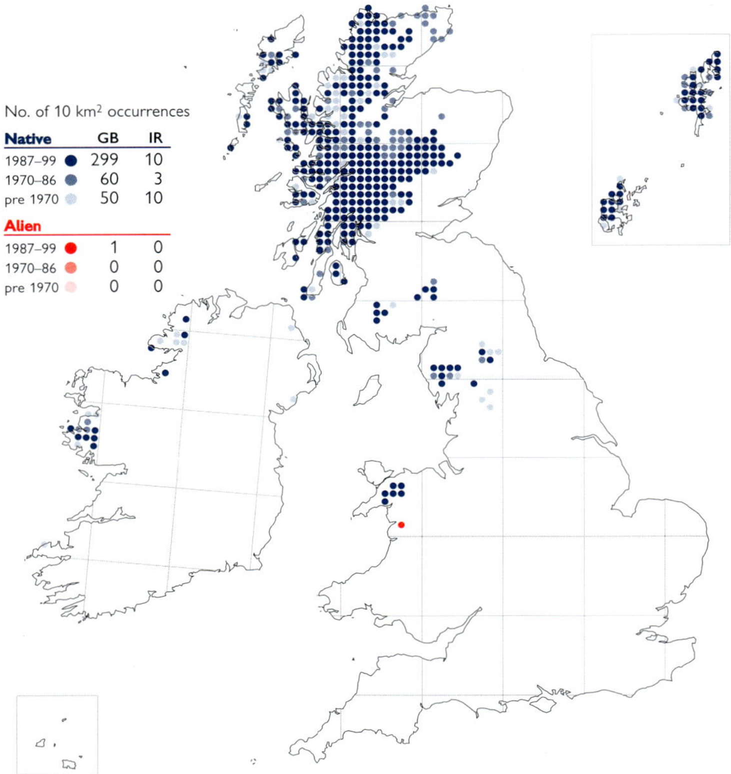

A small, rhizomatous perennial of mountain habitats. It is found on damp rock ledges, at the open edges of stony streams and flushes and in thin grassland. It is a good indicator of substrates which are at least slightly base-rich. From sea level in N. Scotland, but generally above 300 m and reaching 1190 m on Ben Lawers (Mid Perth).

Native (change –0.34). In Britain the distribution of *T. alpinum* is stable; in Ireland it has not been recorded recently at sites in Co. Antrim and Co. Down, despite searching, and further fieldwork is required to establish its current status elsewhere.

Circumpolar Arctic-montane element, with a disjunct distribution.

References: Atlas (26a), Hultén & Fries (1986), Jalas & Suominen (1989), Meusel *et al.* (1965).

R. A. FITZGERALD

Berberis vulgaris Barberry

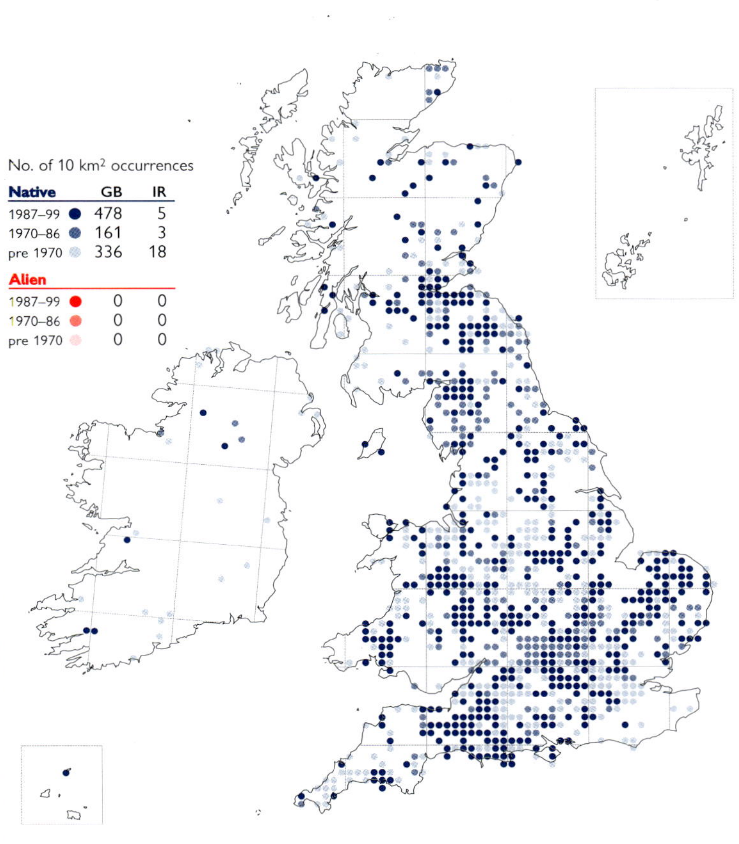

A deciduous shrub found in hedgerows and coppices, and on banks, cliffs and waste ground. Generally lowland, but reaching 395 m on Wanthwaite Crags (Cumberland).

Native or alien (change –0.61). *B. vulgaris* has been recorded from Neolithic deposits at Grimes Graves (W. Norfolk). It might be native in England and Wales, but it was cultivated in medieval times and later widely planted for hedging. All records are mapped as if they are native. Its deleterious effect on wheat crops was appreciated before it was known to be a host of the rust *Puccinia graminis* and consequently eradicated from many hedgerows in the 19th century.

European Temperate element; widely naturalised outside its native range.

References: Atlas (26d), Hultén & Fries (1986), Jalas & Suominen (1991), Mercer (1981), Meusel *et al.* (1965), White (1912).

T. D. DINES

Berberis thunbergii Thunberg's Barberry

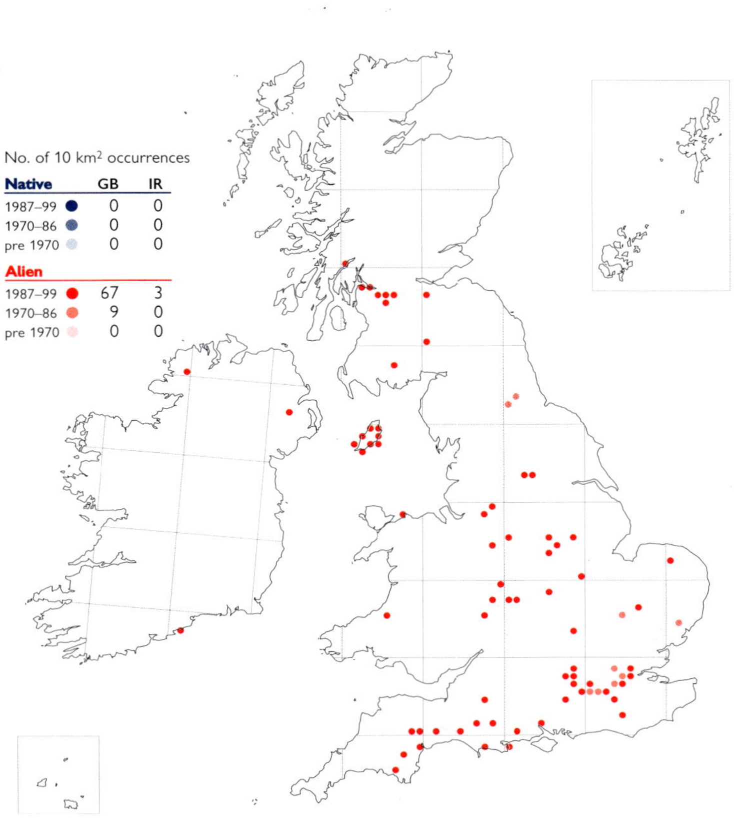

This spiny deciduous shrub occurs in woodland, on roadsides, railway banks and waste ground, and by footpaths. It reproduces by seed and is occasionally bird-sown, and can become abundant in suitable habitats, for example on rocky slopes near Llandudno (Caerns.). Lowland.

Neophyte. *B. thunbergii* was introduced into cultivation in Britain in 1883 and is commonly grown, often as hedging, for its autumn colour and red berries. It was not formally recorded from the wild until 1971. It is likely to become more frequent as it is increasingly planted on roadsides.

Native of Japan.

References: Bean (1970), Coats (1963).

T. D. DINES

Berberis darwinii Darwin's Barberry

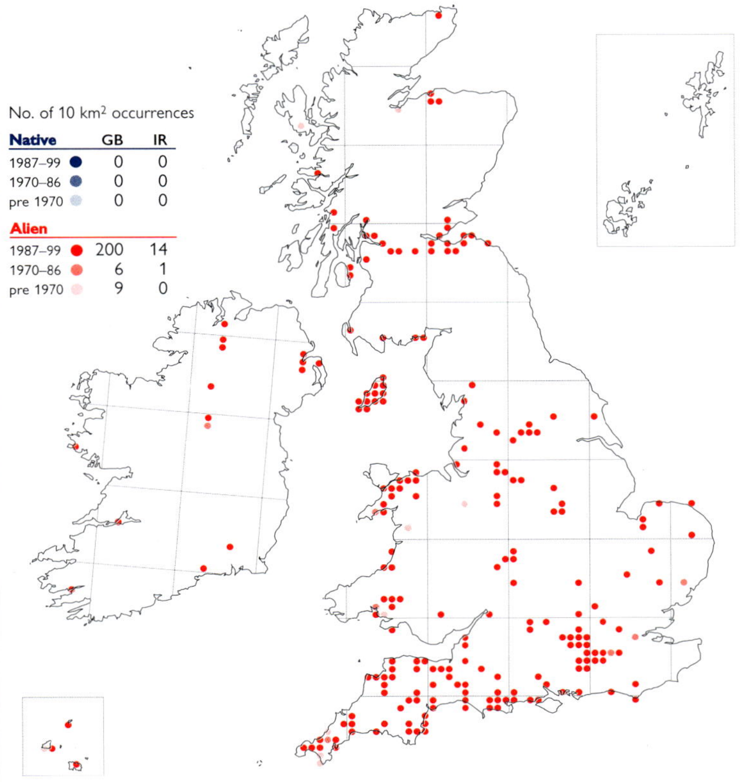

A spiny, evergreen shrub which grows in woodland and scrub, on roadsides, hedge banks and walls, and as an occasional relic of cultivation. Reproduction is by seed, which can be bird-sown. Lowland.

Neophyte. This species was discovered by Charles Darwin on the *Beagle* voyage in 1835, and first cultivated in 1849. It is now widely grown. It was recorded from the wild by 1928, and is likely to be increasing because of its popularity in roadside plantings.

Native of S. America (Argentina, Chile).

References: Bean (1970), Coats (1963).

T. D. DINES

Berberis darwinii × B. empetrifolia (B. × stenophylla) Hedge Barberry

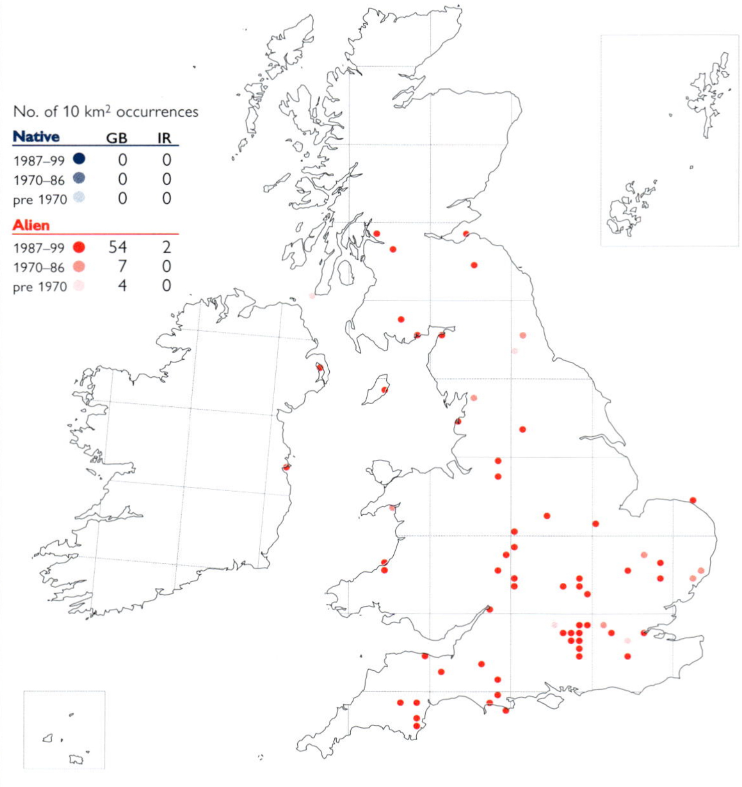

A large, spiny evergreen shrub commonly planted in parks and gardens, and occasionally found naturalised on roadsides and in hedges. Although fertile, fruit is not abundantly produced, and most reproduction is vegetative through suckering. Lowland.

Neophyte. *B. × stenophylla* arose in a garden near Sheffield (S.W. Yorks.) around 1860 and is frequently grown for its leaves and flowers. It was recorded in the wild in 1935 and is probably increasing.

A hybrid of garden origin.

References: Bean (1970), Coats (1963).

T. D. DINES

Mahonia aquifolium Oregon-grape

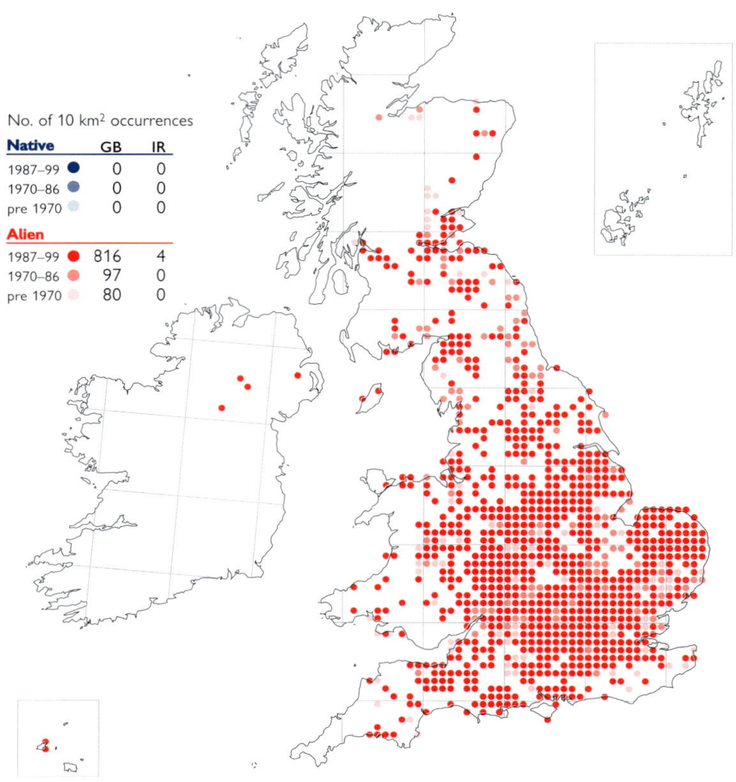

An evergreen shrub which spreads rapidly by stolons and can become well established in hedgerows, road verges and woodland. Lowland.

Neophyte (change +1.61). *M. aquifolium* was introduced in 1823. It is commonly cultivated, and frequently planted in some areas for game-cover. It was known from the wild by 1874. It has increased since the 1962 *Atlas*, probably due to widespread planting and escapes from gardens.

Native of western N. America; widely naturalised in W. & C. Europe.

References: Atlas (27a), Bean (1973), FNAEC (1997).

T. D. DINES

Papaver pseudoorientale Oriental Poppy

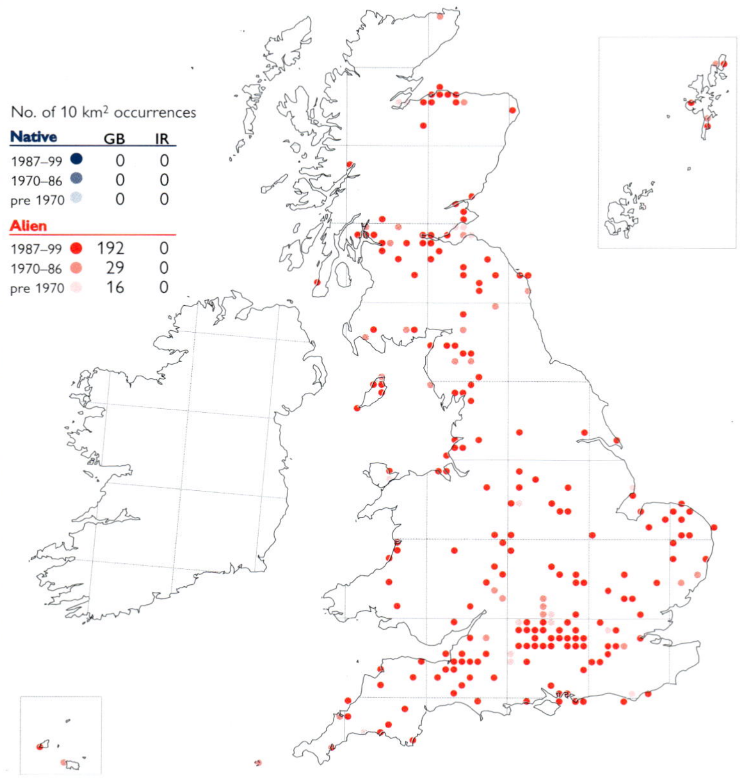

A large, tap-rooted perennial herb found as a garden escape or throw-out on roadsides, railway banks, rough grassland, sand dunes, waste ground and rubbish tips; it also occurs as a relic of cultivation. Populations can become well-naturalised, especially on well-drained soils. Reproduction is mostly by vegetative growth. Lowland.

Neophyte. This imposing species was introduced into cultivation in Britain by 1714 and has long been popular in gardens. It was first recorded from the wild in 1927, and is likely to be increasing. Some records may refer to *P. bracteatum* and *P. orientale*, which are also grown; these may be conspecific as intermediates occur.

Native of S.W. Asia.

T. D. DINES

Papaver atlanticum Atlas Poppy

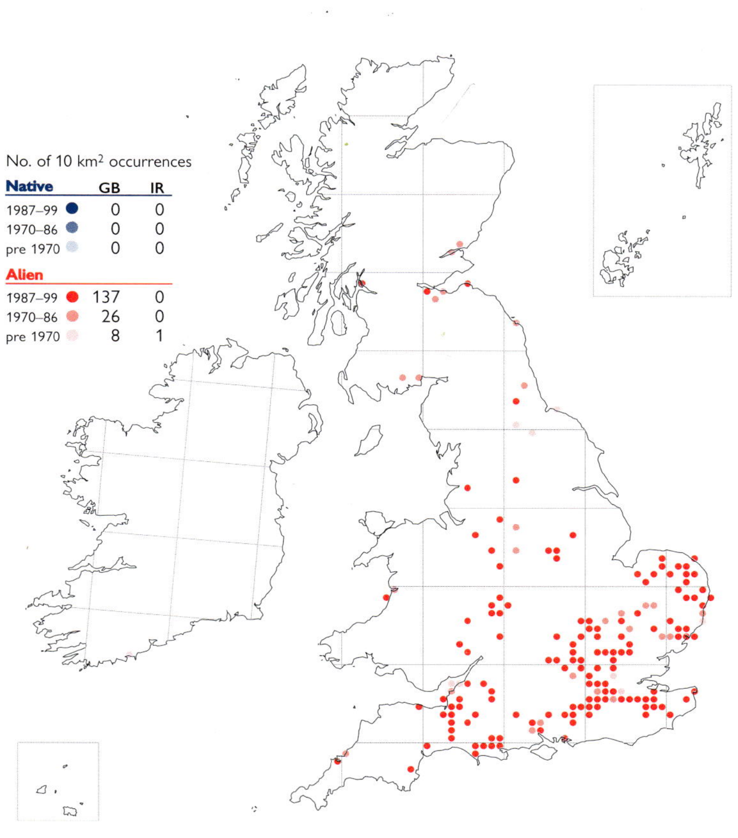

A perennial herb that has become naturalised on walls, roadsides and waste ground. Lowland.

Neophyte. *P. atlanticum*, introduced into cultivation in Britain in 1889, is widely cultivated in gardens. It was recorded in the wild by 1928 and is now a frequent escape.

Native of N.W. Africa (Morocco).

Reference: Jalas & Suominen (1991).

P. J. WILSON

Papaver somniferum Opium Poppy

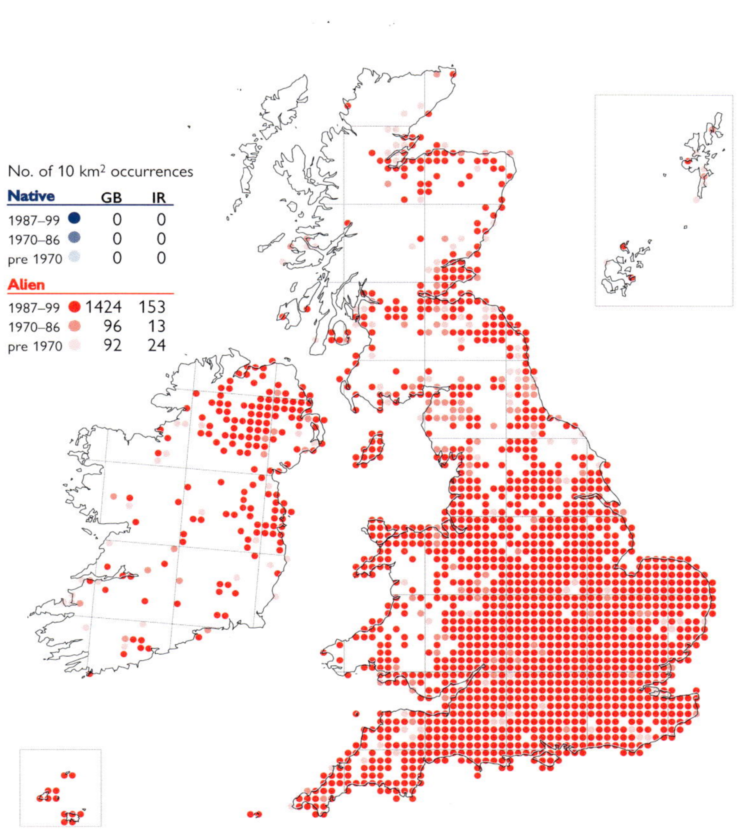

An annual occurring as a casual garden escape on roadsides, waste ground and rubbish tips, and occasionally in arable fields as a relic of cultivation for poppy seed. Lowland, with one record at 410 m (Eisteddfa Gurig, Cards.).

Archaeophyte (change +2.54). This species is a medicinal, culinary and ornamental plant. Its seeds have been found in Bronze Age archaeological deposits and are frequently found from the Iron Age onwards. It is widely cultivated as a garden plant. Subsp. *somniferum* is the common taxon, whilst subsp. *setigerum* is a rare casual which might be a distinct species (Stace, 1997).

Probably native of the eastern Mediterranean region, but native distribution obscured by its spread in cultivation.

References: Atlas (29c), Bolòs & Vigo (1984), Jalas & Suominen (1991), Kadereit (1986a, 1987), Zohary & Hopf (2000).

P. J. WILSON

Papaver rhoeas Common Poppy

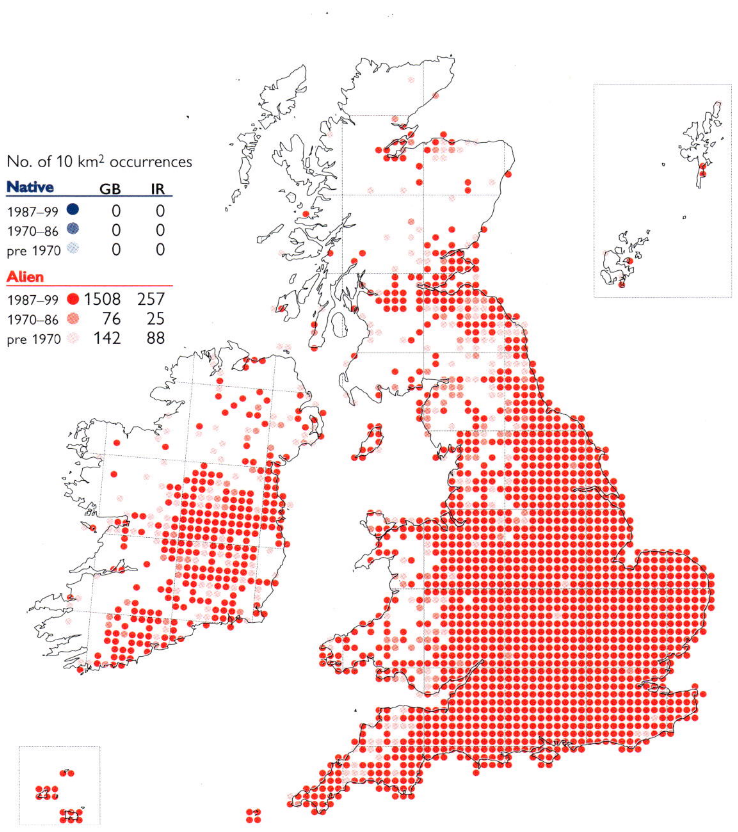

An annual of arable fields and other disturbed and open habitats. It is most frequent on light, calcareous soils. It is sensitive to herbicide, but can be abundant in unsprayed strips in fields. Its seed can be very long-lived. It is also frequent in wild-flower seed mixtures and occurs as a garden escape. Lowland.

Archaeophyte (change –0.41). This species was mapped as 'all records' in the 1962 *Atlas*. Although there have been losses around the edges of its range, the overall distribution is remarkably stable.

As an archaeophyte *P. rhoeas* has a European Southern-temperate distribution; it is widely naturalised outside this range.

References: Atlas (28c), Grime *et al.* (1988), Hultén & Fries (1986), Jalas & Suominen (1991), Kadereit (1989), McNaughton & Harper (1964), Meusel *et al.* (1965), Roberts & Boddrell (1984), Wilson (1990).

P. J. WILSON

Papaver dubium Long-headed Poppy

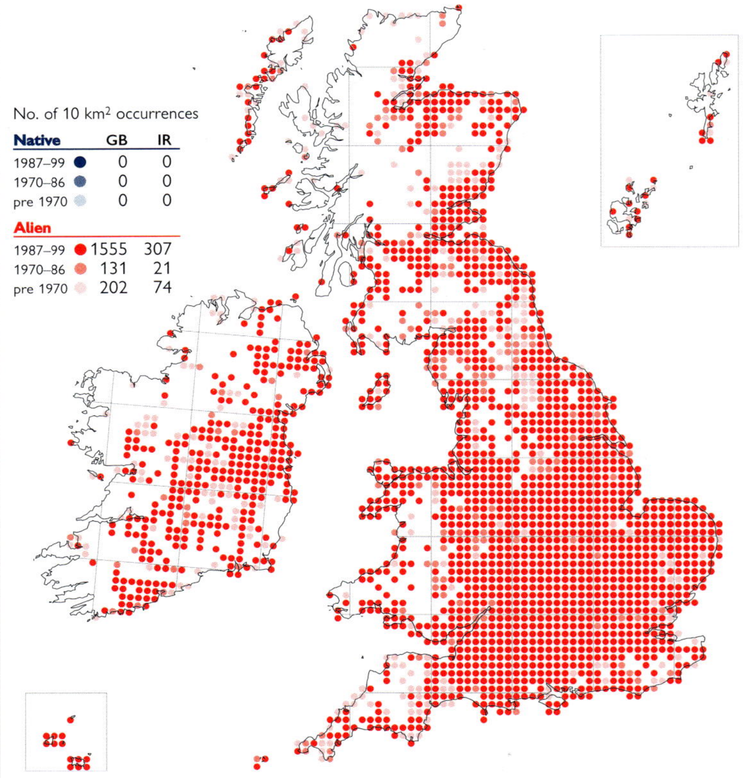

An annual found principally in arable fields, where it can occur on both light and heavy calcareous soils. It is also found on waste ground by roadsides and railways, and in gardens. The seed is very long-lived. Lowland, but with one record at 425 m in Atholl (E. Perth).

Archaeophyte (change +0.23). *P. dubium* was mapped as 'all records' in the 1962 *Atlas*. It has declined locally as a result of agricultural intensification, but its overall distribution is stable. Subsp. *dubium* and subsp. *lecoqii* are mapped separately.

As an archaeophyte *P. dubium* has a Eurosiberian Southern-temperate distribution; it is widely naturalised outside this range.

References: Atlas (28d), Jalas & Suominen (1991), Hultén & Fries (1986), Kadereit (1989), McNaughton & Harper (1964), Roberts & Boddrell (1984), Wilson (1990).

P. J. WILSON

Papaver dubium subsp. *dubium*

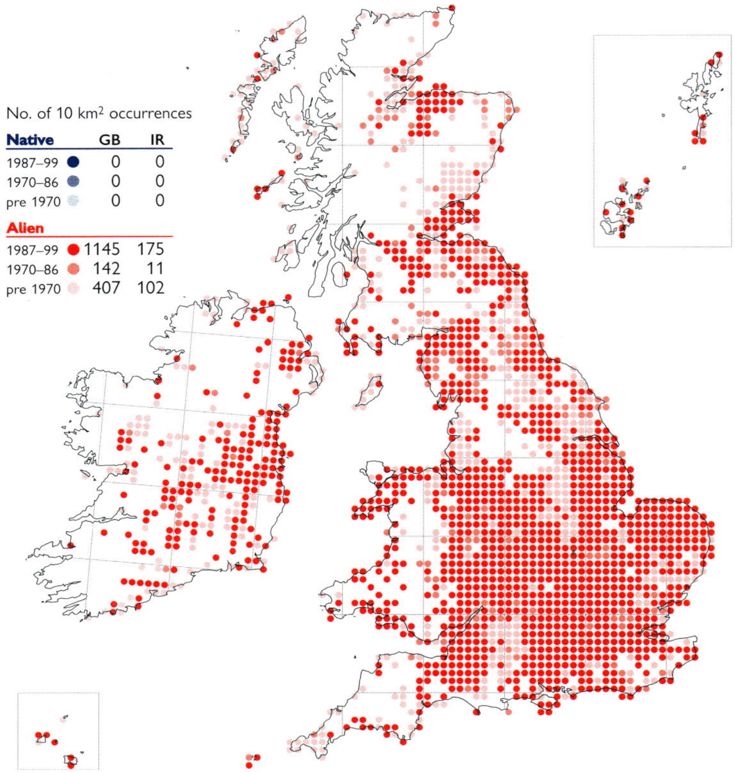

This annual is mainly an associate of arable crops, though also found on other disturbed land, particularly rubbish tips, and is most frequent on light soils. It can be found in both spring- and autumn-sown crops, although most seedlings emerge in the autumn. The seed can be very long-lived. Lowland.

Archaeophyte. This taxon was not mapped separately in the 1962 *Atlas*. It has declined locally as a result of agricultural intensification. Some recorders have failed to distinguish the subspecies of *P. dubium* and subsp. *dubium* is therefore under-recorded.

European Southern-temperate element.

References: Kadereit (1989), McNaughton & Harper (1964), Roberts & Boddrell (1984), Wilson (1990).

P. J. WILSON

Papaver dubium subsp. *lecoqii*

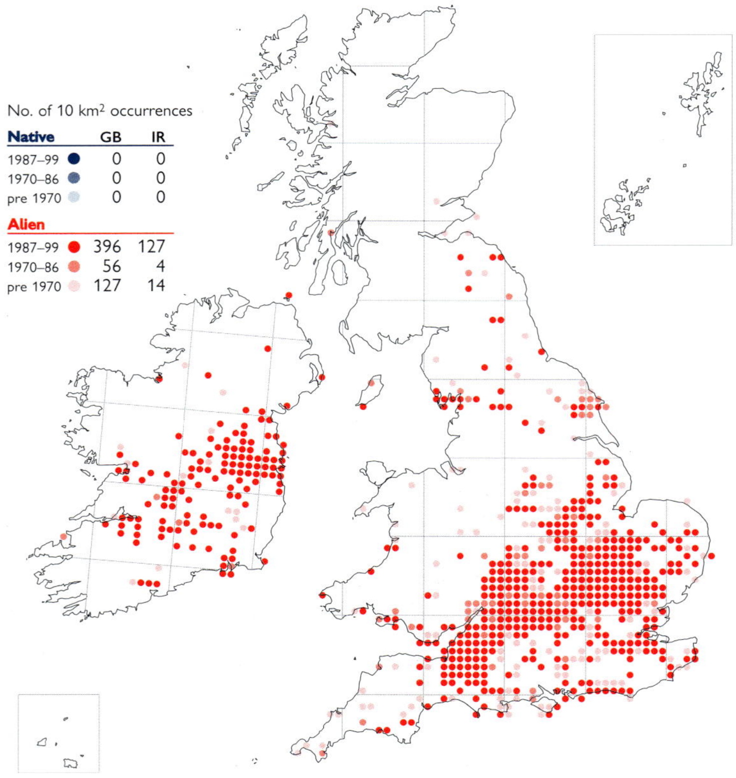

This annual is found in arable fields and in other disturbed open ground such as on roadsides, railway tracks and waste land, in gardens and along hedges. It is most frequent on heavy calcareous soils, and can be found in both spring- and autumn-sown crops. The seed can be very long-lived. Lowland.

Archaeophyte. *P. dubium* subsp. *lecoqii* was mapped by Perring & Sell (1968). There are now many more records, indicating that this subspecies is now more consistently recorded.

European Southern-temperate element.

References: Atlas Supp. (6b), Jalas & Suominen (1991), Kadereit (1989), McNaughton & Harper (1964), Wilson (1990).

P. J. WILSON

Papaver hybridum Rough Poppy

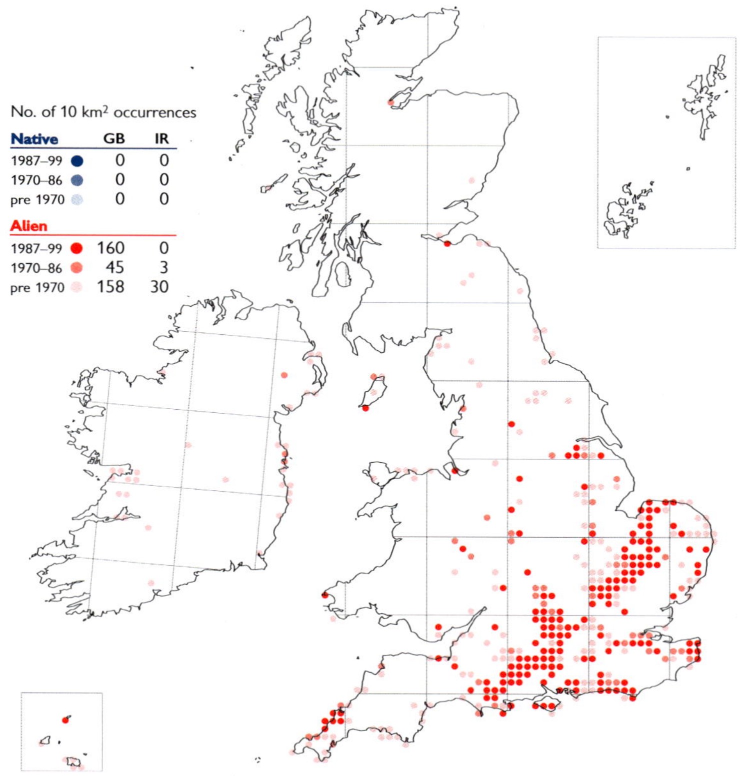

This annual occurs in arable crops, and sometimes in other disturbed habitats. It is most frequent on chalky soils, but also grows on other limestones and on calcareous sands. The seed, which can be long-lived, germinates in both autumn and spring. 0–320 m (Buxton, Derbys.).

Archaeophyte (change –0.35). Many of the losses of this species took place before 1930. Although it is thought to have decreased in abundance since 1950 due to the increased use of herbicides, this is not reflected at the 10-km scale where its distribution is more or less stable.

As an archaeophyte *P. hybridum* has a Submediterranean-Subatlantic distribution.

References: Atlas (29a), Curtis & McGough (1988), Jalas & Suominen (1991), Kadereit (1986b), McNaughton & Harper (1964), Meusel *et al.* (1965), Stewart *et al.* (1994), Wilson (1990).

P. J. WILSON

Papaver argemone Prickly Poppy

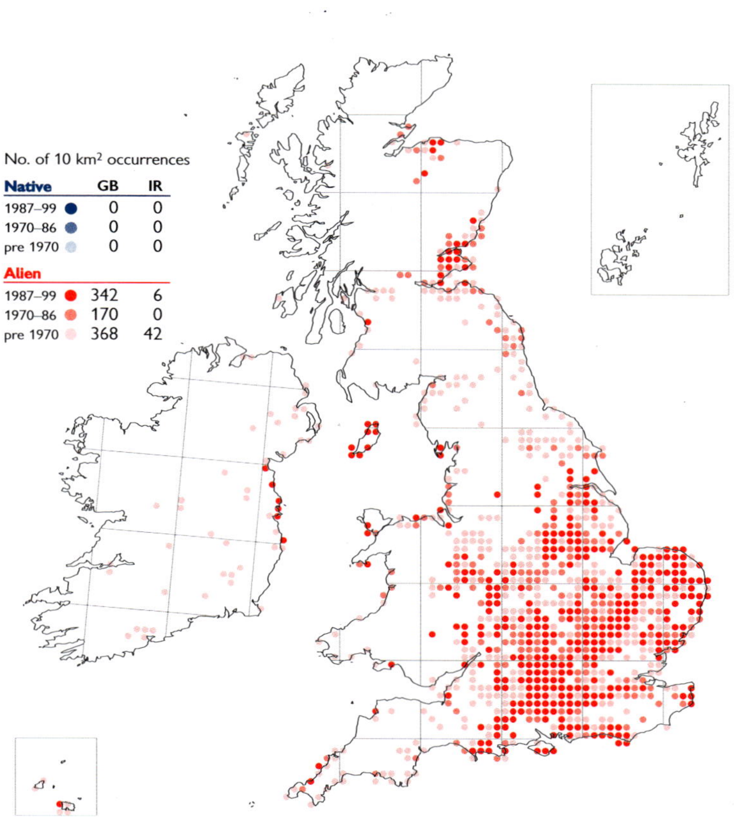

An annual of arable crops, usually found on field edges and in unsprayed corners, often in small numbers. It is most frequent on light sandy, gravelly and chalky soils. Rarely, it occurs on waste ground. Lowland.

Archaeophyte (change –1.79). *P. argemone,* mapped as 'all records' in the 1962 *Atlas,* is susceptible to herbicides and has declined because of this. Some marginal losses may only have been casual populations.

As an archaeophyte *P. argemone* has a European Southern-temperate distribution; it is widely naturalised outside this range.

References: Atlas (29b), Hultén & Fries (1986), Jalas & Suominen (1991), Kadereit (1986b), McNaughton & Harper (1964), Roberts & Boddrell (1984), Stewart *et al.* (1994), Wilson (1990).

P. J. WILSON

Meconopsis cambrica Welsh Poppy

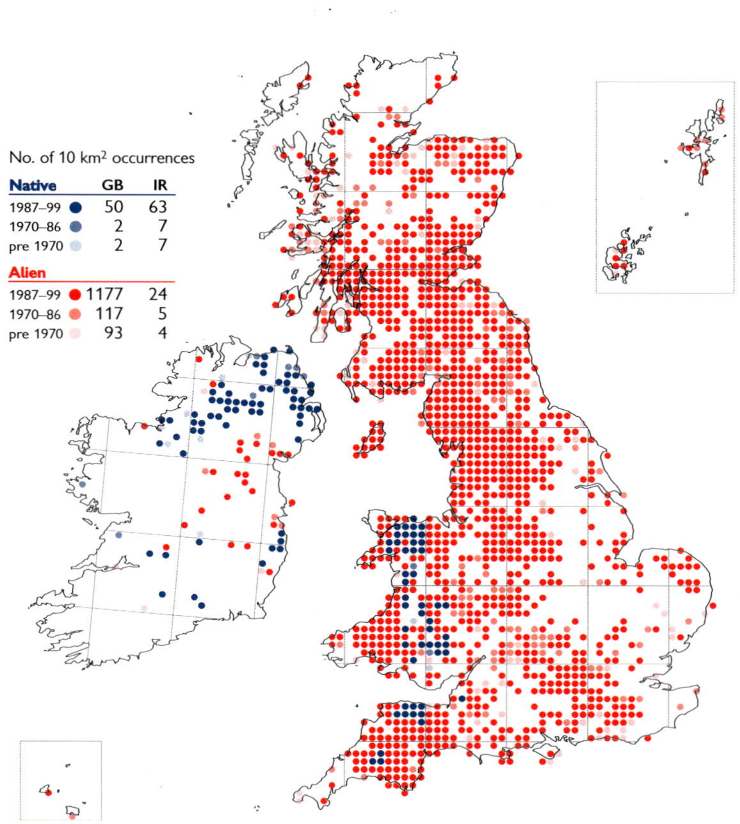

A long-lived perennial herb, native in damp, rocky woodlands and on shaded cliff ledges. It is also grown in gardens and has become naturalised on hedge banks, walls, roadsides and waste ground. Native plants range from the lowlands to 640 m (Cwm Idwal, Caerns.).

Native (change +2.36). Native populations rarely spread into apparently suitable habitat and are probably in slow decline. In contrast, garden plants can spread rapidly and the species is increasing as an established alien. Within its core native areas, it can be difficult to separate native from alien occurrences, particularly in Ireland.

Oceanic Boreo-temperate element; confined as a native to W. Europe but widely naturalised outside its native range.

References: Atlas (29d), Jalas & Suominen (1991), Stewart *et al.* (1994).

P. J. WILSON

Glaucium flavum Yellow Horned-poppy

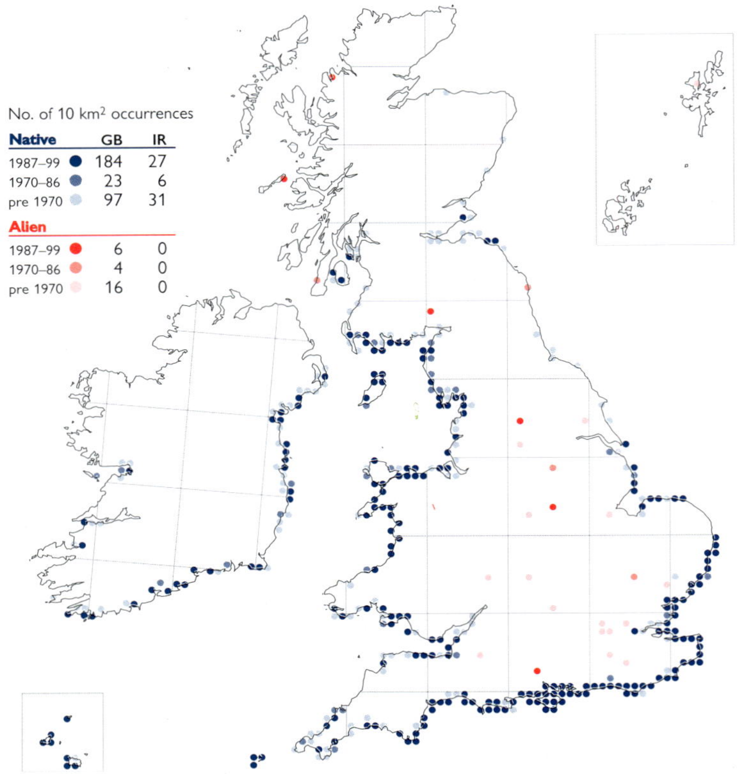

A short-lived perennial herb of shingle banks and stony beaches; also, more rarely, amongst loose rock and on eroding cliffs of sand and clay, and on the bare tops of chalk cliffs. The few inland records are of casual occurrences. Lowland.

Native (change –0.39). *G. flavum* has been lost from some sites in Britain in recent years as a result of coastal defence work and trampling pressures on tourist beaches. However, its overall distribution shows little change since the 1962 *Atlas*.

Mediterranean-Atlantic element; widely naturalised outside its native range.

References: Atlas (30a), Hultén & Fries (1986), Jalas & Suominen (1991), Scott (1963b).

P. J. WILSON

Chelidonium majus Greater Celandine

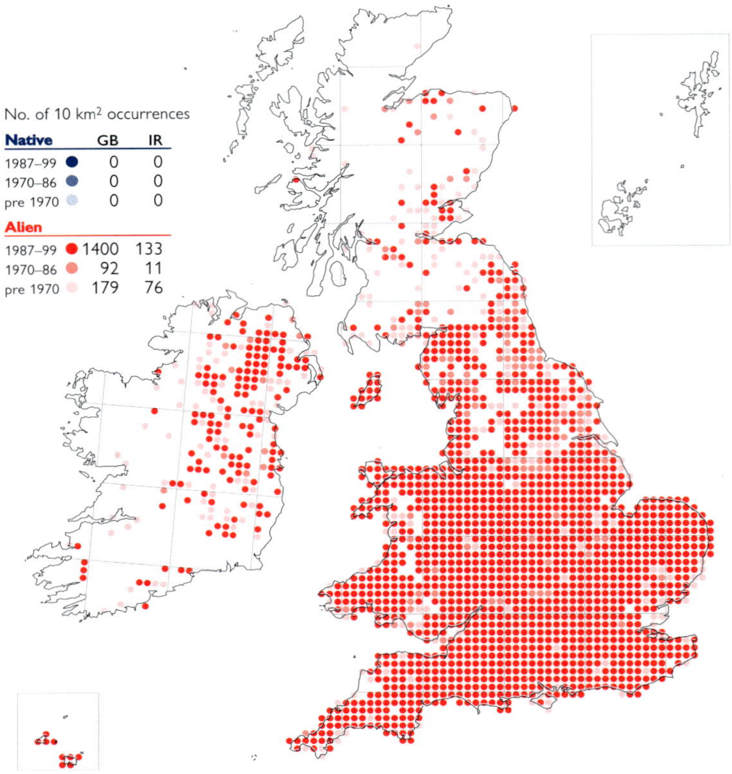

This perennial herb is widely naturalised by roadsides and paths, in the crevices of old walls, on waste ground and in hedge-bottoms. It was at one time cultivated as a medicinal plant, and most localities are near habitation. Lowland.

Archaeophyte (change –0.72). Fossil evidence shows that *C. majus* has been present in Britain since Roman times. Although it is better recorded since the 1962 *Atlas*, the map suggests a decline at the edges of the range.

As an archaeophyte *C. majus* has a Eurasian Temperate distribution; it is widely naturalised outside this range.

References: Atlas (30b), Hultén & Fries (1986), Jalas & Suominen (1991), Meusel *et al.* (1965).

P. J. WILSON

Eschscholzia californica Californian Poppy

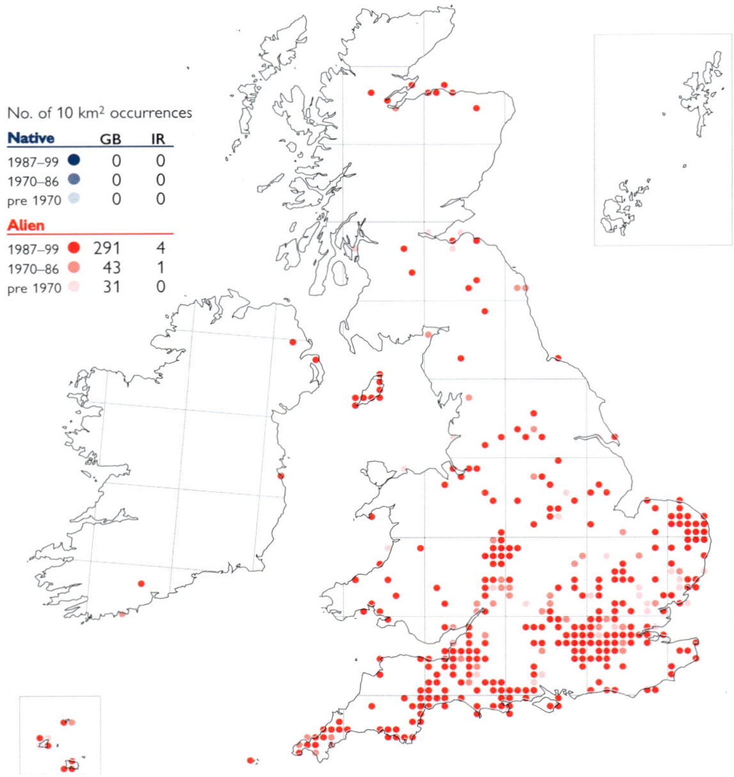

A perennial herb that is frequently encountered as a casual on rubbish tips and roadsides, but sometimes becomes naturalised in quarries, on railway tracks and waste ground. It has long been naturalised on dunes in Guernsey. Lowland.

Neophyte. This widely cultivated species has been grown in Britain since 1826 and is popular as a summer bedding plant, but rarely overwinters in Britain other than in milder areas. It was first recorded from the wild in 1864.

Native of western N. America and Mexico (Baja California).

References: FNAEC (1997), Jalas & Suominen (1991).

P. J. WILSON

Dicentra formosa Bleeding-heart

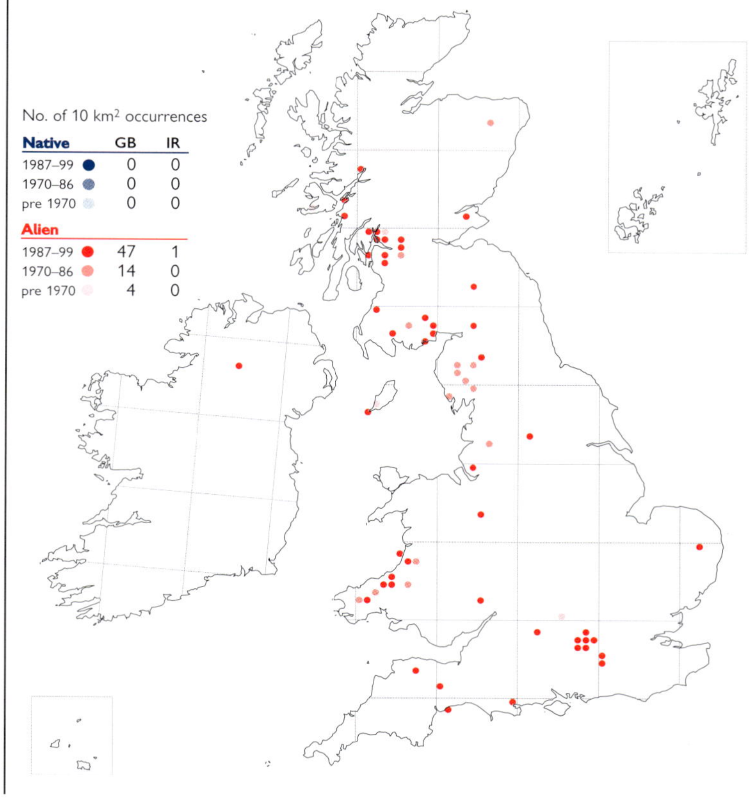

A rhizomatous perennial herb well-naturalised in woodland and on shaded roadsides and stream banks. It originates as a garden escape or throw-out, and reproduces by seed and rhizome extension. Lowland.

Neophyte. *D. formosa* is very popular in gardens and has been cultivated in Britain since 1796. It was described as 'occasionally naturalised near gardens' by Dunn (1905), and may be increasing slowly; it was first recorded in Ireland in 1995 (Co. Tyrone). Some records may refer to *D. eximia*, or its hybrid with *D. formosa*.

Native of western N. America.

Reference: Jalas & Suominen (1991).

T. D. DINES

Corydalis solida Bird-in-a-bush

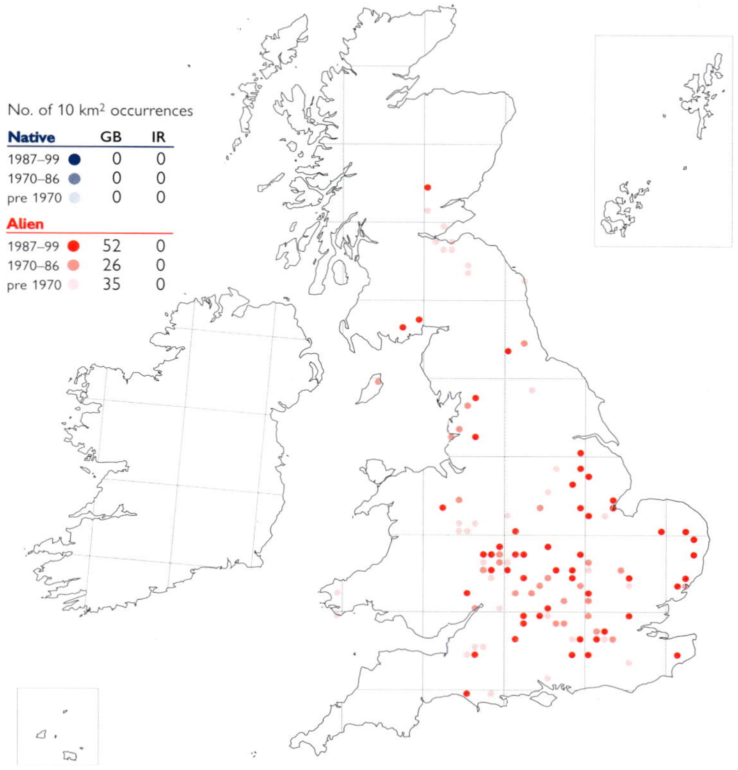

A tuberous perennial herb found in woodland, hedgerows, churchyards and rough grassland, and on roadsides, river banks and walls. It occurs as a garden escape or throw-out, and often becomes naturalised. Reproduction is by seed and tubers. Lowland.

Neophyte. *C. solida* was cultivated in Britain by 1596 and was first recorded as naturalised in the wild in 1796. Its overall distribution is likely to be stable.

A Eurosiberian Boreo-temperate species, absent as a native from much of W. Europe.

References: Hultén & Fries (1986), Jalas & Suominen (1991), Meusel *et al.* (1965).

T. D. DINES

Pseudofumaria lutea Yellow Corydalis

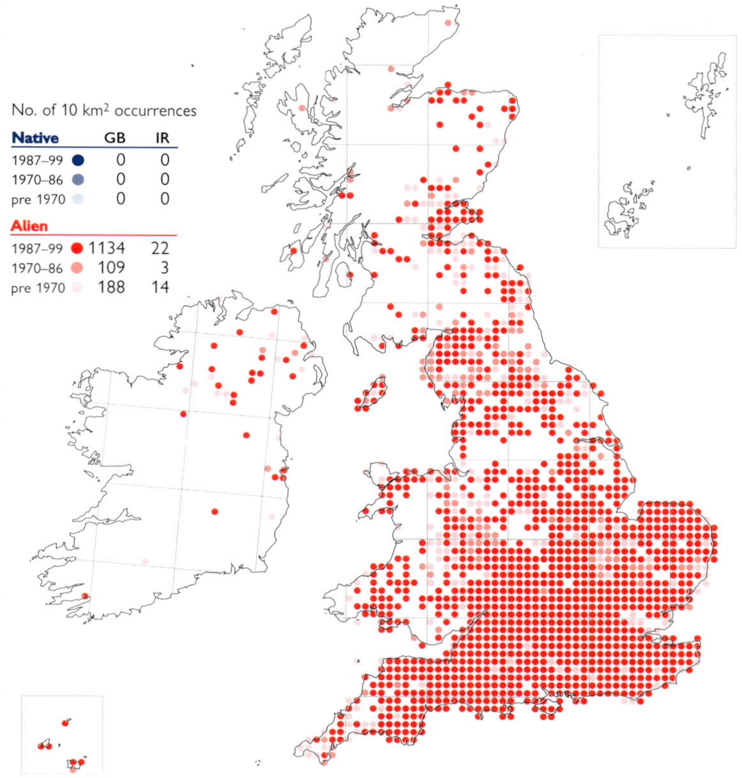

A perennial herb, commonly cultivated and widely naturalised in Britain, less so in Ireland. It is most frequently found rooted into the crevices of old mortared walls, pavements and other masonry, and on brick rubble and stony waste ground. Once established in an area, it can quickly colonise new sites. Generally lowland, but reaching 305 m at Great Hucklow (Derbys.).

Neophyte (change +0.59). *P. lutea* was being grown in Britain by 1596; it was first recorded in the wild in 1796 and became widespread in the early 1800s. Since the 1962 *Atlas* it has clearly increased in some areas, such as Wales and S.W. England.

Native of the southern foothills of the S.W. & C. Alps; widely naturalised elsewhere in Europe.

References: Atlas (30d), Jalas & Suominen (1991).

P. J. WILSON

Ceratocapnos claviculata Climbing Corydalis

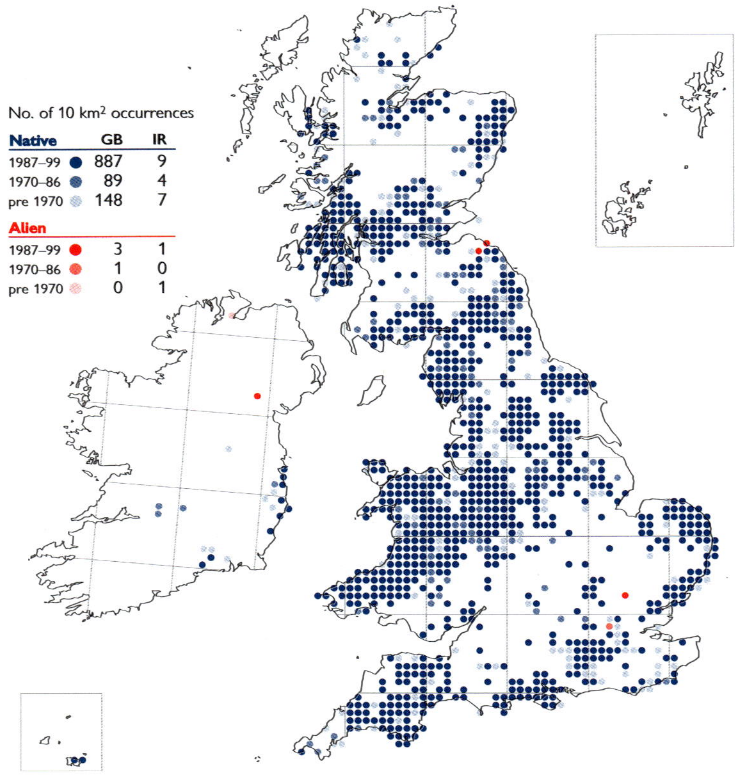

A climbing or scrambling annual of freely-draining acidic, mineral or peaty soils. It occurs in deciduous and coniferous woodland, especially in clearings and in recently felled areas, under *Pteridium* and in scrub, and occasionally over rock outcrops. In Ireland, it occurs on shaded boulder slopes. Generally lowland, but reaching 430 m on Deadwater Fell (S. Northumb.).

Native (change +0.57). The overall distribution of this species has not changed since the 1962 *Atlas*, although it is now much better recorded.

Oceanic Temperate element.

References: Atlas (30c), Hultén & Fries (1986), Jalas & Suominen (1991), Meusel *et al.* (1965).

P. J. WILSON

Fumaria capreolata White Ramping-fumitory

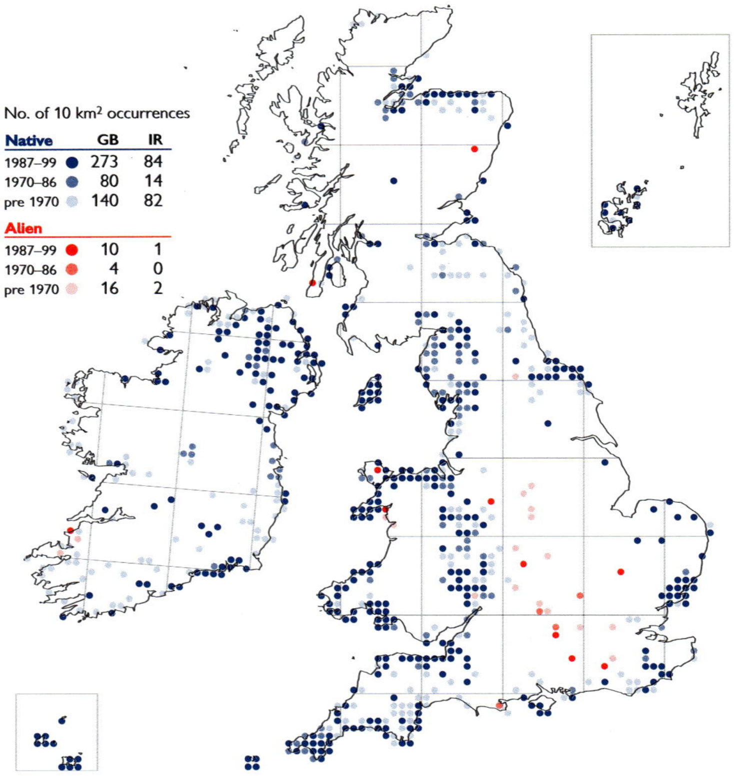

A scrambling annual of open scrub, hedge banks and cliffs, and only occasionally found in arable land and gardens. Unlike most other species of *Fumaria*, it can be a winter-annual. Lowland.

Native (change +0.31). The endemic subsp. *babingtonii* is the widespread segregate, with the continental subsp. *capreolata* apparently confined to the Channel Islands. *F. capreolata* has declined in its inland sites, where it may only have been casual, but the distribution is stable on or near the coast.

Submediterranean-Subatlantic element.

References: Atlas (31b), Bolòs & Vigo (1984), Jalas & Suominen (1991), Stewart *et al.* (1994).

P. J. WILSON

Fumaria occidentalis Western Ramping-fumitory

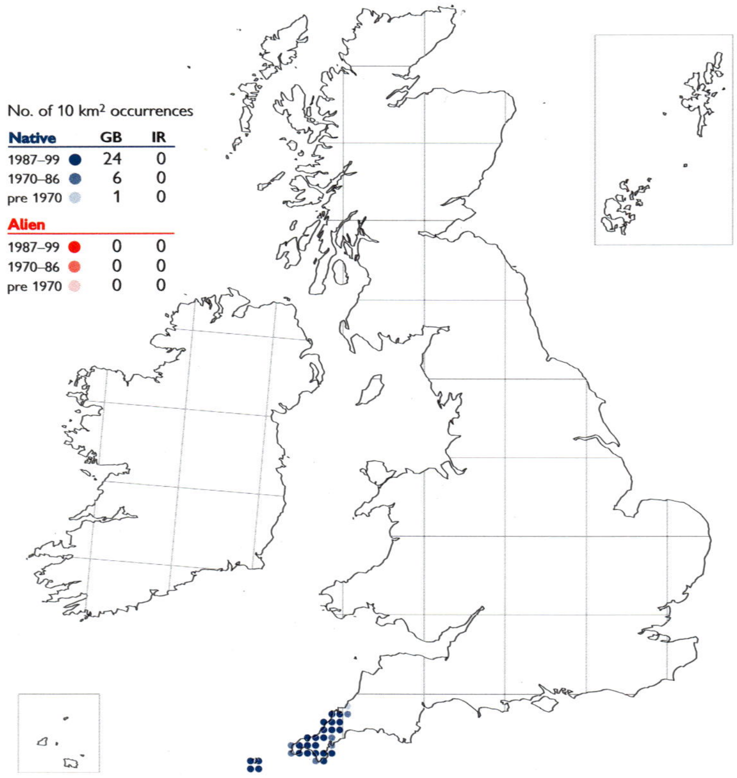

A scrambling annual of freely-draining, base-poor substrates, found in a range of habitats including field margins, hedge banks, road verges, waste ground and cultivated land. Lowland.

Native (change +0.04). *F. occidentalis* can be irregular in appearance, but on the mainland is thought to be spreading as a colonist of waste ground. In contrast, it has apparently suffered a sharp decline since 1970 in bulb-fields in the Isles of Scilly.

Endemic.

References: Atlas (31a), Jalas & Suominen (1991), Wigginton (1999).

P. J. WILSON

Fumaria bastardii Tall Ramping-fumitory

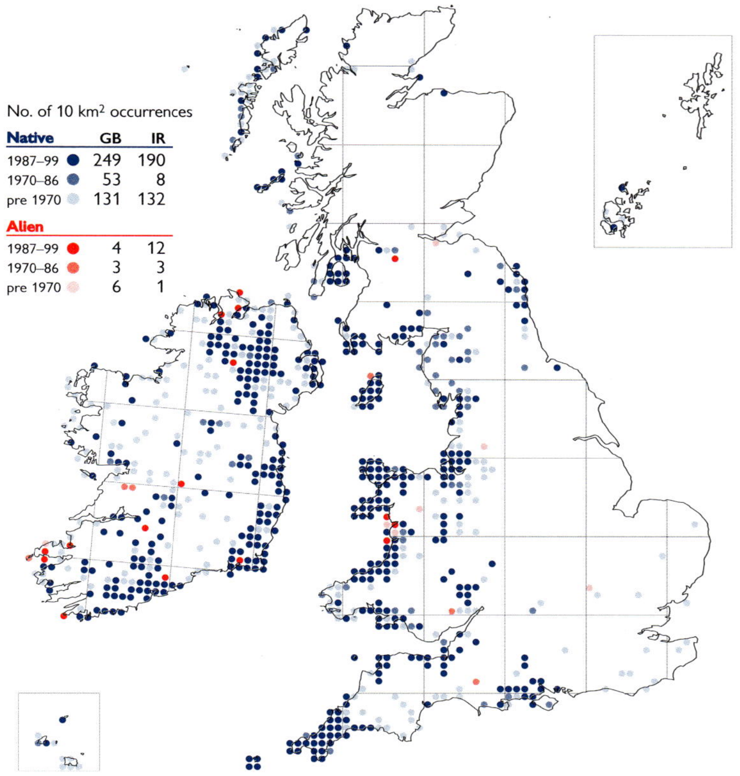

A scrambling annual of arable and horticultural land and, more rarely, hedge banks, usually growing on freely-draining, acidic soils. Probably mainly spring-germinating, it is typically found in spring-sown crops. Lowland.

Native (change +0.39). In contrast to many other less common arable species, its distribution has changed little. It usually occurs mixed with other large-flowered *Fumaria* species, and can be difficult to distinguish from *F. muralis*. It is under-recorded in some areas of Ireland where recorders have been unable to distinguish these two species.

Mediterranean-Atlantic element.

References: Atlas (31d), Bolòs & Vigo (1984), Jalas & Suominen (1991), Stewart *et al.* (1994).

P. J. WILSON

Fumaria reuteri Martin's Ramping-fumitory

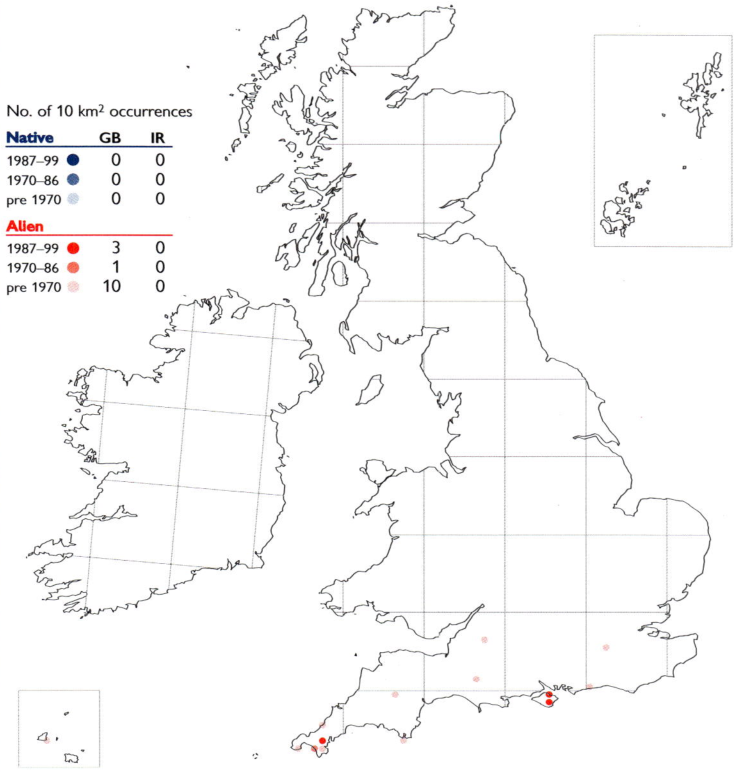

F. reuteri is a scrambling annual of freely-draining acidic soils, which has most recently been recorded in spring- and summer-sown crops on allotments, in gardens and in potato fields; also on the eroded soil of hedge banks. Lowland.

Neophyte (change –0.62). First recorded in 1904, this species has been reported from only two areas since 1980, and appears to be in sharp decline at one of these, in Cornwall, as formerly cultivated land reverts to grassland. It can be very difficult to distinguish from *F. muralis*.

An Oceanic Southern-temperate species.

References: Atlas (32a), Jalas & Suominen (1991), Wilson (1997), Wigginton (1999).

P. J. WILSON

Fumaria muralis Common Ramping-fumitory

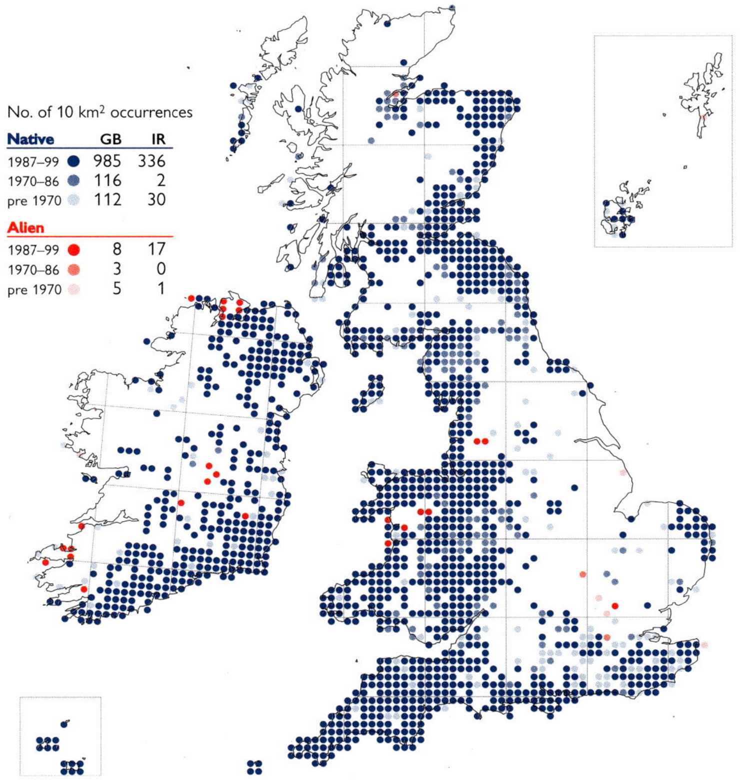

An annual, scrambling herb of arable land, gardens and hedge banks on freely-draining, acidic soils. Lowland.

Native (change +1.75). *F. muralis* is the most common of the large-flowered *Fumaria* species. It was mapped as 'all records' in the 1962 *Atlas*, and greatly under-recorded. It may have become less common in arable habitats in recent years. Three subspecies have been described in our area, but these are taxonomically very close, and indeed the species itself can be difficult to separate from *F. bastardii* and *F. reuteri.*

Oceanic Southern-temperate element; widely naturalised outside its native range.

References: Atlas (32b), Atlas Supp. (7a), Hultén & Fries (1986), Jalas & Suominen (1991).

P. J. WILSON

Fumaria purpurea Purple Ramping-fumitory

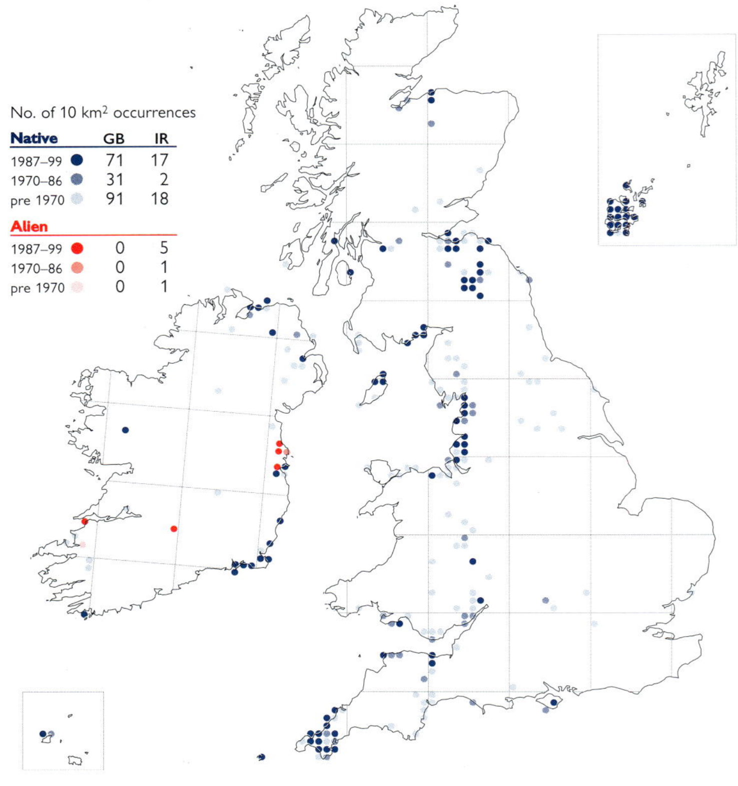

F. purpurea is a scrambling annual of hedge banks, earth-core walls, arable land and gardens on acidic, freely-draining soils, usually most abundant in disturbed places, or in habitats opened up by summer drought. Most occurrences are in spring-sown crops, although in the Isles of Scilly it is found in bulb-fields. Lowland.

Native (change +0.25). This species is erratic in its appearance at some sites, and is probably unfamiliar to many recorders. Trends in its distribution are therefore difficult to ascertain, but it seems to be stable in its core areas. Elsewhere it may be declining or overlooked, and at some sites it may have only been casual.

Endemic.

References: Atlas (31c), Jalas & Suominen (1991), Stewart *et al.* (1994).

P. J. WILSON

Fumaria officinalis Common Fumitory

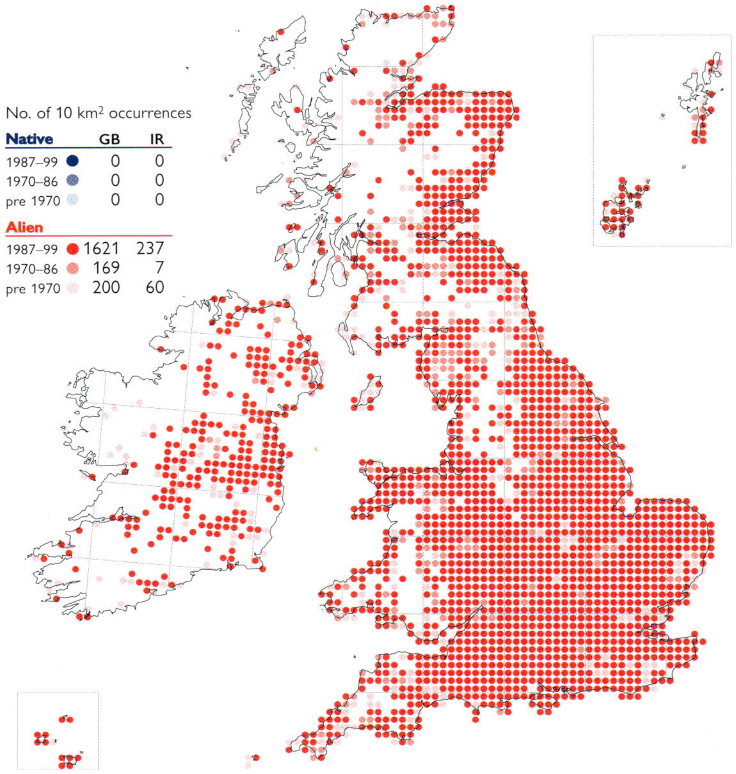

A scrambling annual of arable fields, allotments, gardens and other disturbed land, most commonly found on calcareous soils. Most germination occurs in the spring, and the seed bank is long-lived. Lowland, reaching 305 m near Shap quarries (Westmorland).

Archaeophyte (change –0.34). *F. officinalis* was mapped as 'all records' in the 1962 *Atlas*; its overall distribution at the 10-km scale is stable. Subsp. *officinalis* occurs throughout the range of the species while subsp. *wirtgenii* is most frequent on light soils in the east.

As an archaeophyte *F. officinalis* has a European Southern-temperate distribution; it is widely naturalised outside this range.

References: Atlas (32d), Atlas Supp. (6c), Hultén & Fries (1986), Jalas & Suominen (1991), Meusel *et al.* (1965), Perring & Sell (1968), Roberts & Feast (1973).

P. J. WILSON

Fumaria densiflora Dense-flowered Fumitory

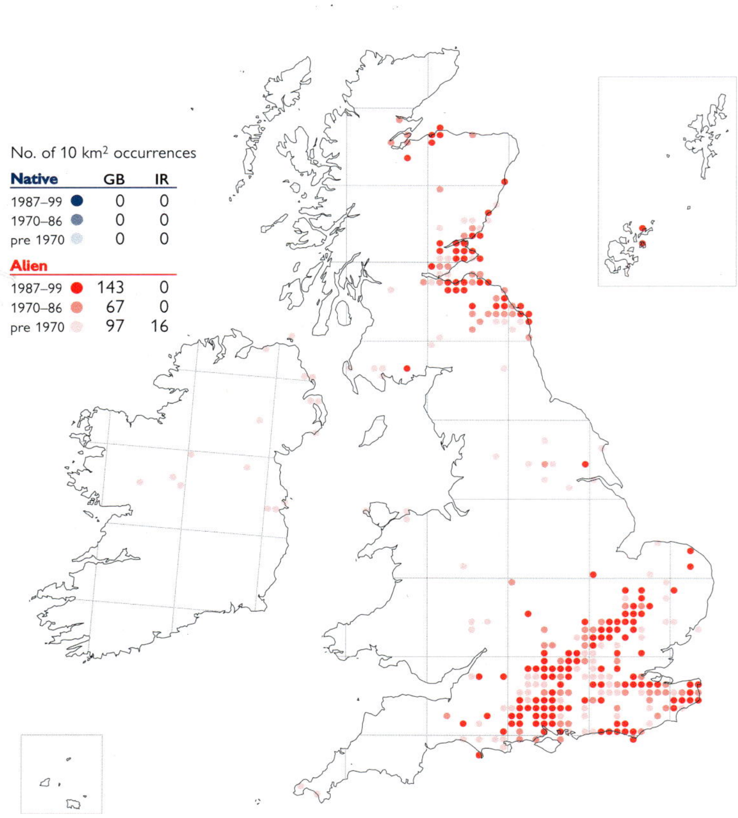

A scrambling annual of arable land. It is most frequently found on chalk but can also occur on other freely-draining soils. It normally grows in species-rich communities, often with other species of *Fumaria*, and is most commonly found in spring-sown cereals and root crops. It has a long-lived seed bank, and germination is mainly in the spring. Lowland.

Archaeophyte (change –0.37). Like many other species of arable land, *F. densiflora* has become less frequent as a result of agricultural intensification, and is increasingly found in vegetable rather than cereal crops. It was last recorded in N. Ireland in 1946.

As an archaeophyte *F. densiflora* has a European Southern-temperate distribution.

References: Atlas (32c), Bolòs & Vigo (1984), Jalas & Suominen (1991), Roberts & Boddrell (1983a), Stewart *et al.* (1994).

P. J. WILSON

Fumaria parviflora Fine-leaved Fumitory

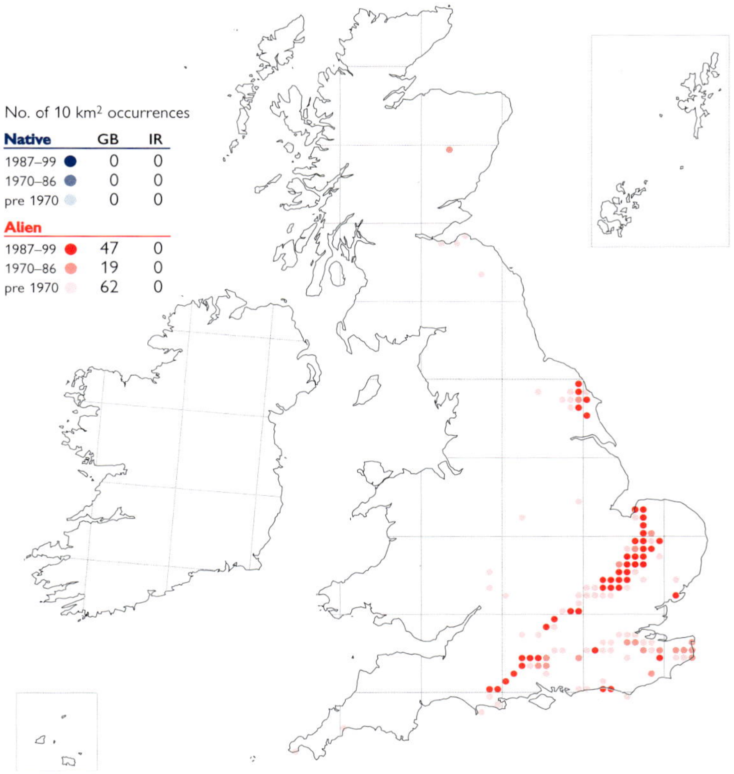

This scrambling annual is almost exclusively found in arable fields on chalky soils, though may occasionally be found on other areas of disturbed ground near arable populations. It is usually associated with other uncommon arable species and, like the other small-flowered *Fumaria* species, is generally found in spring-sown crops. Lowland.

Archaeophyte (change –0.55). *F. parviflora* has never been very frequent, but has declined since the Second World War as a result of agricultural intensification. It is now increasingly restricted to field margins.

European Southern-temperate element.

References: Atlas (31a), Bolòs & Vigo (1984), Jalas & Suominen (1991), Stewart *et al.* (1994).

P. J. WILSON

Fumaria vaillantii Few-flowered Fumitory

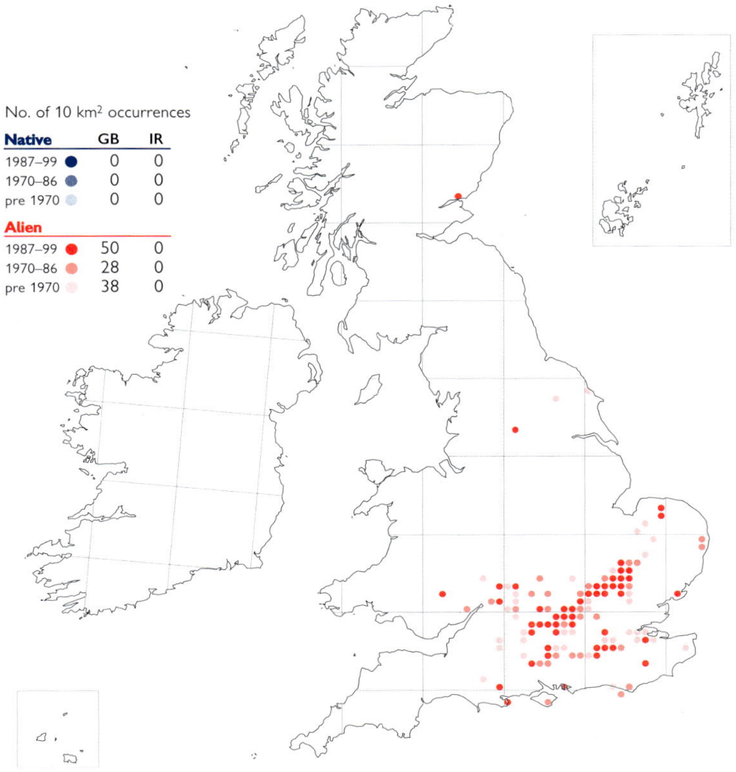

This scrambling annual is almost exclusively found in arable fields on chalky soils, and is usually associated with other uncommon arable species. Like the other small-flowered *Fumaria* species, *F. vaillantii* is most frequently found in spring-sown crops. Lowland.

Archaeophyte (change –0.51). This species has never been very frequent, and has declined since 1950 as a result of agricultural intensification.

As an archaeophyte *F. vaillantii* has a Eurosiberian Temperate distribution.

References: Atlas (33a), Hultén & Fries (1986), Jalas & Suominen (1991), Meusel *et al.* (1965), Stewart *et al.* (1994).

P. J. WILSON

Platanus × *hispanica* (*P. occidentalis* × *P. orientalis*) London Plane

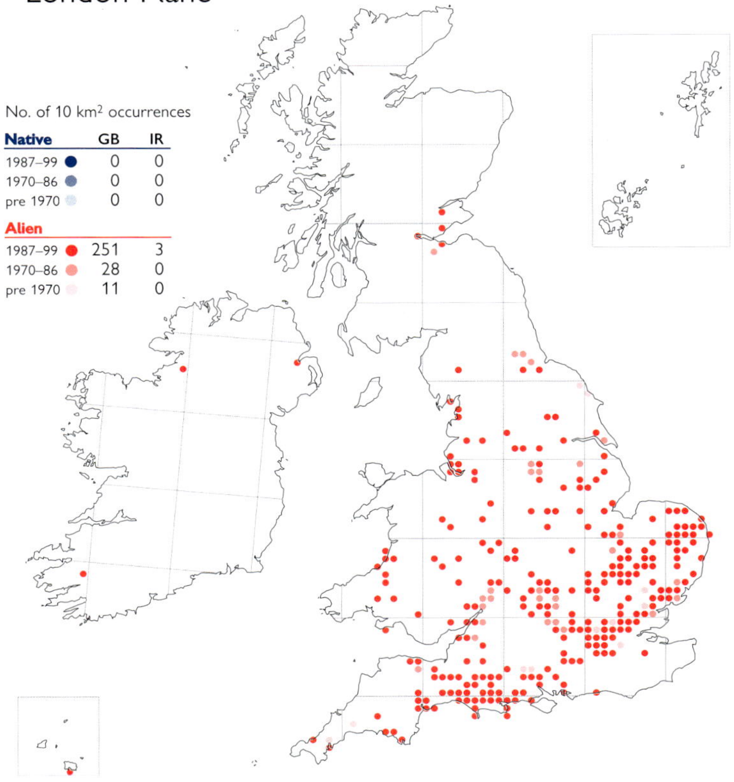

A long-lived tree, extensively planted in streets and parks. Unlike its putative parents, it is extremely hardy in our area. It is vigorous even in polluted air and where root-space is restricted, and can be repeatedly pruned. It is fully fertile and seedlings are frequent in urban areas. Lowland.

Neophyte. *P.* × *hispanica* apparently arose in the 17th century. Some of the trees planted in the 17th century are still growing well. It has been recorded in the wild since at least 1939, and the distribution is probably stable.

There is some doubt whether this is a hybrid of garden origin between Old and New World species or a variant of *P. orientalis*, a plant of the Mediterranean region and S.W. Asia.

References: Bean (1976), Jalas & Suominen (1999), Mitchell (1996).

D. A. PEARMAN

Ulmus glabra Wych Elm

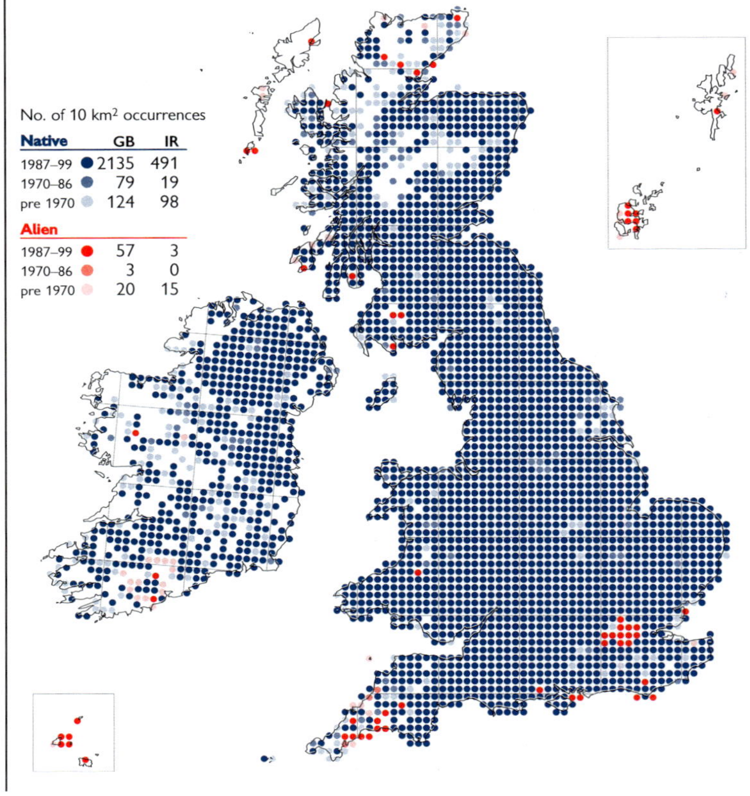

U. glabra is a largely non-suckering tree of hedges, field-borders and streamsides, but also forms mixed or pure woodland, especially on limestone and other base-rich soils. It is also a colonist of ungrazed grassland, rocky ground and waste and spoil heaps, and is also planted. 0–530 m (Atholl, E. Perth).

Native (change –0.28). In C. and S. England, *U. glabra* hybridises freely, especially with *U. minor*, and some intermediates may have been mis-recorded as *U. glabra*. Although more resistant to Dutch Elm disease than *U. procera*, most mature trees outside Scotland have now been killed. The native status of many records is questionable.

European Temperate element.

References: Atlas (183a), Grime *et al.* (1988), Hultén & Fries (1986), Jalas & Suominen (1976), Meusel *et al.* (1965), Richens (1983, 1987).

C. A. STACE

Ulmus glabra × *U. minor* (*U.* × *vegeta*) Huntingdon Elm

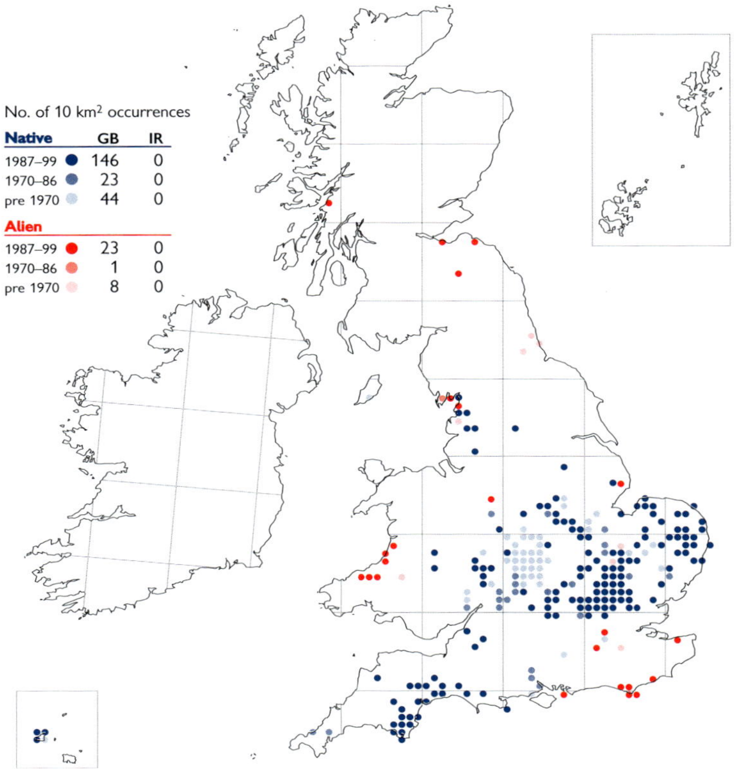

A widely spreading tree often similar in habit to *U. glabra*, found in hedgerows and field-borders but rarely in woodland. It is limited in its natural ecological and geographical ranges to those of its *U. minor* parent. Lowland.

Native. Natural hybrids are common within the range of *U. minor* subsp. *minor* in C. England and S. Wales, but widespread planting has obscured the native distribution. Most planted trees are the cultivar 'Huntingdon Elm', which unlike some of the natural hybrids shows the Dutch Elm disease resistance and lack of suckering of its *U. glabra* parent. This hybrid is probably under-recorded in S. England and S. Wales.

Recorded from W. & C. Europe.

References: Richens (1961, 1983), Stace (1975).

C. A. STACE

Ulmus glabra × *U. plotii* (*U.* × *elegantissima*)

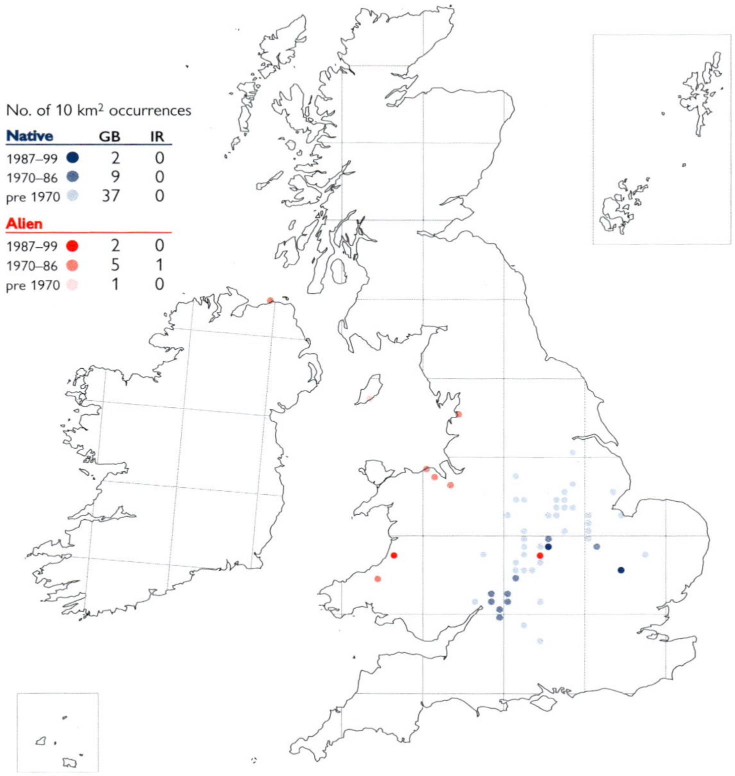

A very variable tree of hedgerows and field-borders, but rarely in woodland, limited in its natural ecological and geographical ranges to those of its *U. plotii* parent. Lowland.

Native. Hybrids of this parentage are frequent within the range of *U. plotii* in the Midlands and East Anglia, extending south-west to Wiltshire. It is now a common constituent of managed hedgerows. The lack of recent records reflects the general under-recording of *Ulmus* taxa, brought about by the death of many mature trees from disease and the absence of any consensus on their taxonomic treatment.

Not known outside Britain.

References: Richens (1983), Stace (1975).

C. A. STACE

Ulmus × *hollandica* (? *U. glabra* × *U. minor* × *U. plotii*) Dutch Elm

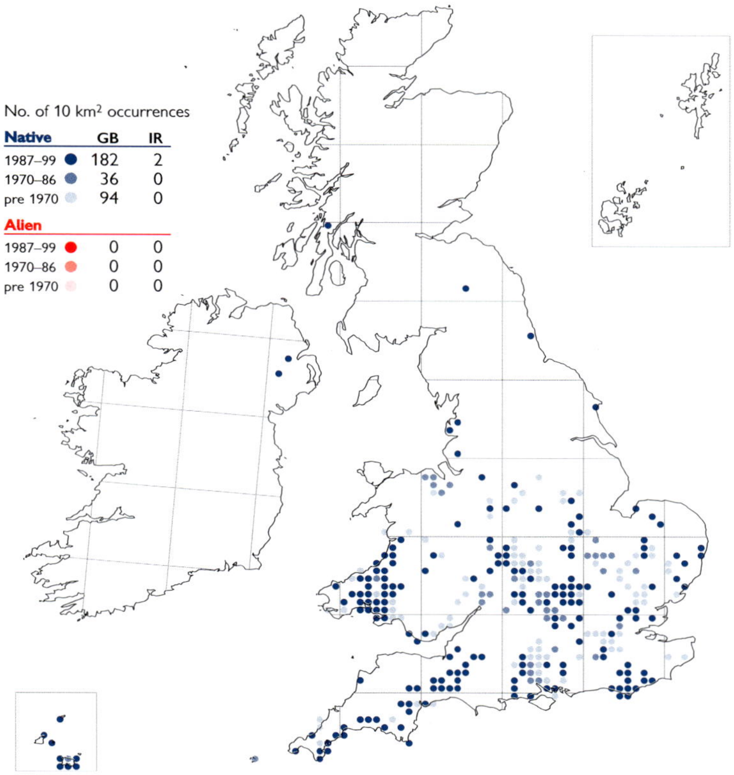

A tall and rather spreading tree of hedgerows and field-borders, but rarely in woodland, limited in its natural ecological and geographical ranges to those of its *U. plotii* parent. This hybrid suckers more freely than *U.* × *vegeta* and is more commonly a constituent of managed hedgerows. Lowland.

Native. Natural hybrids are frequent within the range of *U. plotii* in the Midlands and East Anglia, with a more western distribution than *U.* × *vegeta*. Elsewhere, records are of cultivated trees, mostly the cultivar to which the name 'Dutch Elm' strictly refers. The native distribution has been totally obscured by planting so all records are mapped as if they are native.

Frequent in W. Europe.

References: Richens (1983), Stace (1975).

C. A. STACE

Ulmus procera English Elm

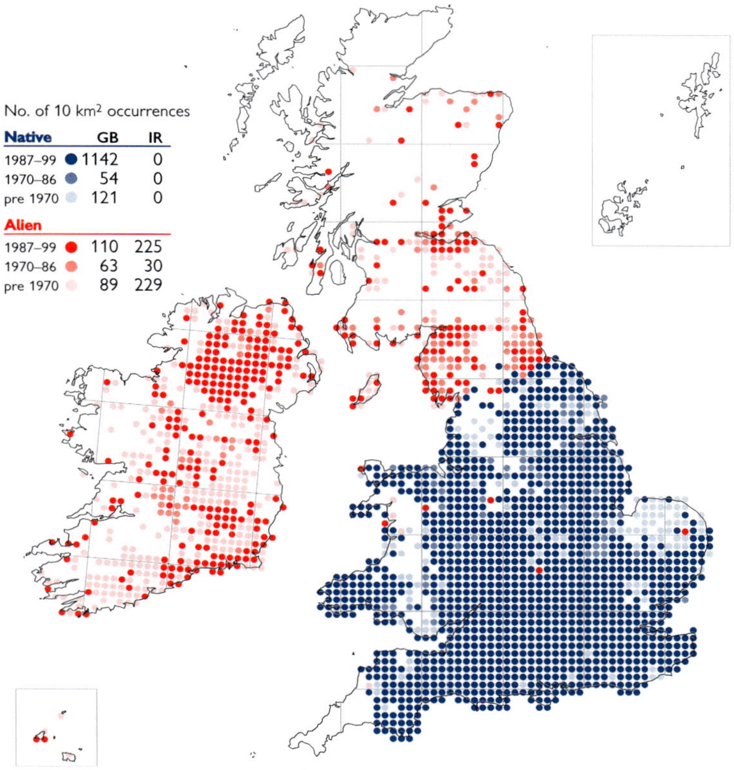

A large wide-topped tree, often considered a definitive component of the English countryside, found in hedgerows and field-borders, rarely in woodland but sometimes forming small copses. It prefers the deep and moist soils of major river systems. Lowland.

Native or alien (change –0.48). In most areas few mature trees remain, being very susceptible to the current outbreak of Dutch Elm disease which began *c.* 1965. New sapling growth still succumbs, but the species remains a major hedgerow constituent, particularly in its core area. The natural distribution is much confused by planting, but probably does not extend much beyond England and Wales. Hybrids with other *Ulmus* taxa are infrequent.

European Temperate element.

References: Atlas (183b), Richens (1983).

C. A. STACE

Ulmus minor

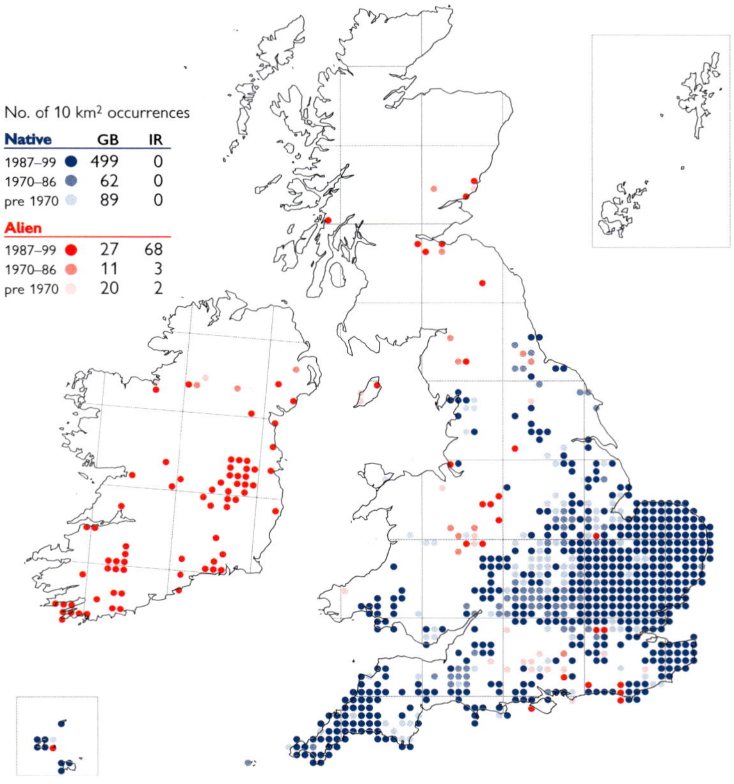

An extremely variable tree, with many named variants, varying from widely spreading to almost fastigiate, occurring in hedgerows, wood margins and field-borders, rarely in woodland but often forming small copses. Lowland.

Native (change +0.75). Few mature trees remain, being very susceptible to the current outbreak of Dutch Elm disease which began *c.* 1965, and new sapling growth still succumbs after a few years. It is, however, a major hedgerow constituent in its native area. Several areas of Britain possess distinctive local variants. The species is widely planted and the limits of its native range are uncertain.

European Temperate element; widely naturalised outside its native range.

References: Atlas (183c), Hultén & Fries (1986), Jalas & Suominen (1976), Meusel *et al.* (1965), Richens (1983).

C. A. STACE

Ulmus plotii Plot's Elm

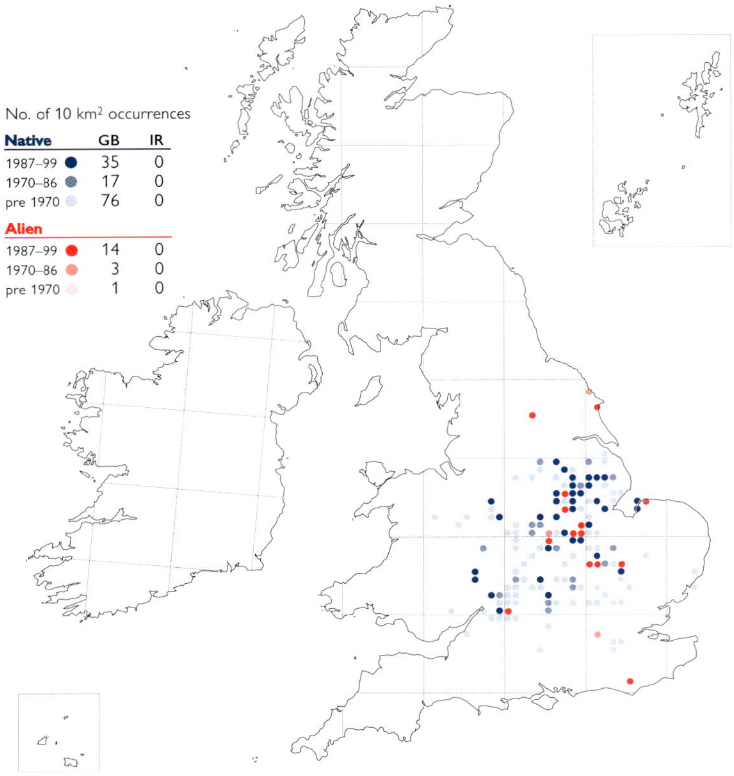

A distinctive narrow tree formerly lending a characteristic appearance to the landscape in its native area, where it occurs in hedgerows and field-borders. It is particularly common on neutral to base-rich soils in the English E. Midlands, mostly in moist, deep-soiled river valleys. Lowland.

Native. Few mature trees remain, being very susceptible to the current outbreak of Dutch Elm disease which began *c.* 1965, and new sapling growth still succumbs after a few years. It was apparently declining even before that disease, but it remains a major hedgerow constituent in its native area. Hybrids with *U. minor* and *U. glabra* are common.

Endemic.

References: Coleman *et al.* (2000), Melville (1940), Richens (1983), Stewart *et al.* (1994).

C. A. STACE

Cannabis sativa Hemp

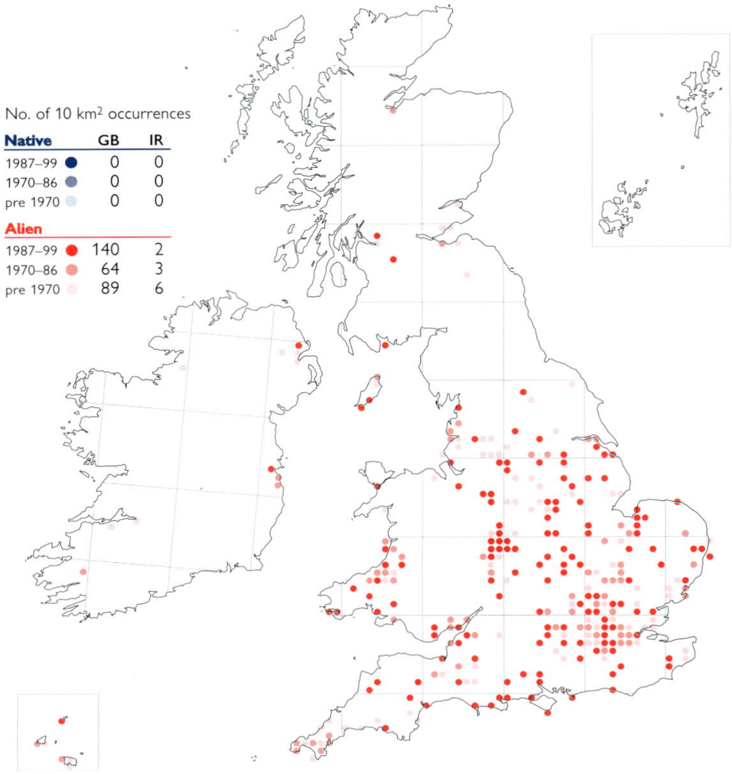

A dioecious annual occurring as a casual from a wide variety of sources, but never persisting. Lowland.

Casual. *C. sativa* has been grown in gardens in Britain since at least 1304 (Thirsk, 1997) and was cultivated up to the 17th and 18th centuries as a fibre crop; there is a small resurgence now in that use. It is also a constituent of bird-seed mixtures, is used as bait by anglers and is cultivated, illegally and in small quantities, as a drug plant. It has been known in the wild since at least 1863.

Probably a native of S.W. & C. Asia, this species has been cultivated from ancient times and is now widely naturalised in warm temperate and tropical regions of the world.

References: Small & Cronquist (1976), Smart & Simmonds (1995), Zohary & Hopf (2000).

D. A. PEARMAN

Humulus lupulus Hop

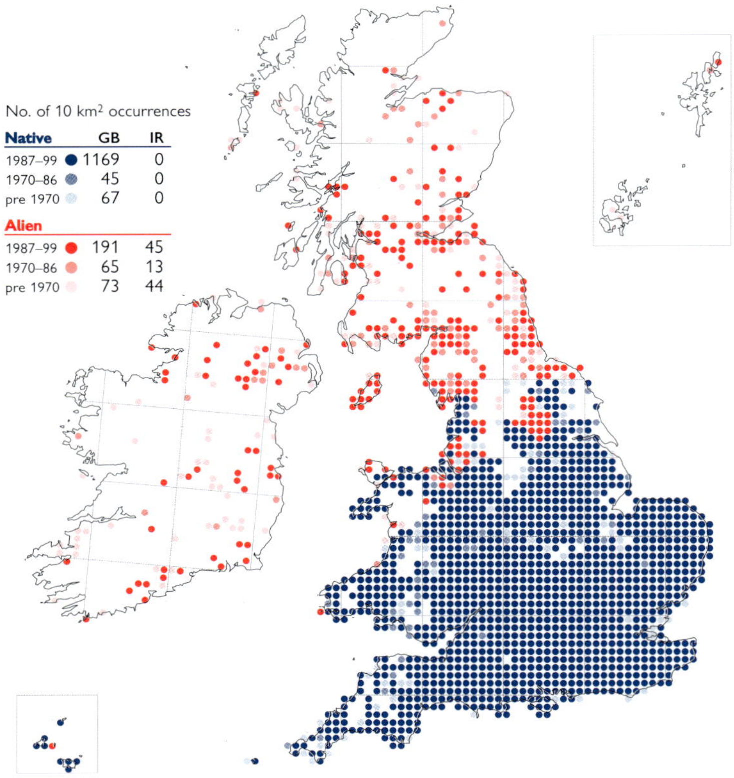

A scrambling, perennial, dioecious climber which is probably native in moist, open woods, fen carr and hedges. It is frequent as an escape from cultivation or as a planted ornamental. Lowland.

Native (change –0.09). *H. lupulus* has long been cultivated for flavouring beer; a widespread cultivar was introduced from Flanders in the 16th century for that purpose. It is very difficult to separate possibly native from alien plants; it is certainly alien in Ireland and north of Yorkshire and Lancashire, and many records in S. Britain are relics of cultivation. Mapped as 'all records' in the 1962 *Atlas*, its overall distribution is stable.

Eurosiberian Temperate element; also in E. Asia and N. America and widely naturalised outside its native range.

References: Atlas (182d), Hultén & Fries (1986), Jalas & Suominen (1976), Vaughan & Geissler (1997).

D. A. PEARMAN

Ficus carica Fig

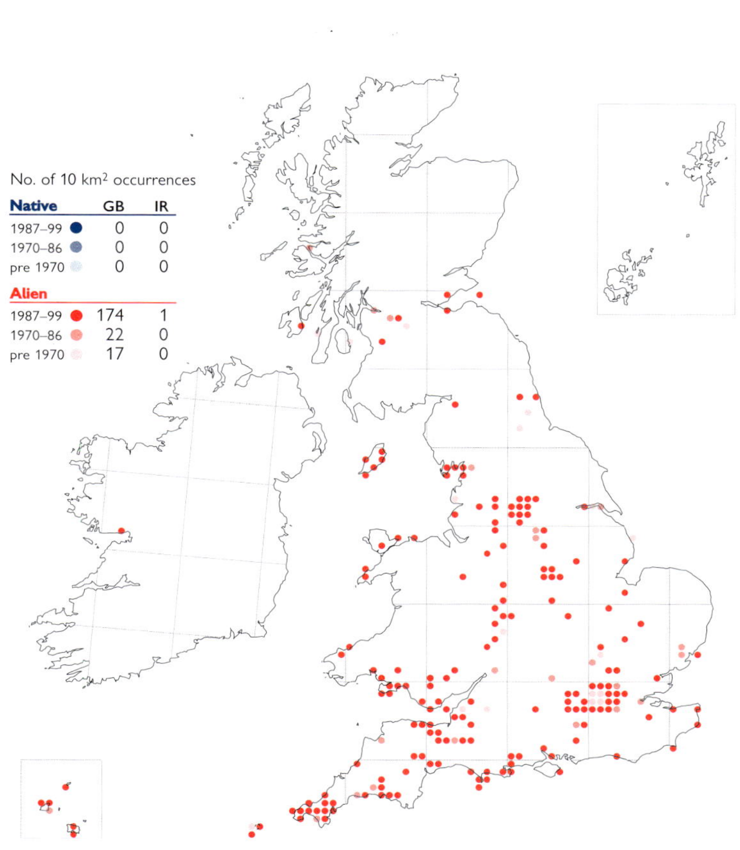

A spreading deciduous shrub or small tree naturalised on waste ground, in churchyards, on railway banks, cliffs and walls, and especially on the banks of urban rivers. Lowland.

Neophyte. Figs have been imported to Britain since Roman times, and the tree has been cultivated here since at least 995 (Harvey, 1981). Plants often arise from discarded fruit or sewage waste, and were first recorded in the wild in 1918. Some colonies are associated with warm water discharge into rivers.

Native of the eastern Mediterranean region and S.W. Asia, but native distribution obscured by 5000 years of cultivation and now naturalised throughout the Mediterranean area.

References: Bolòs & Vigo (1990), Gilbert (1990), Jalas & Suominen (1976), Lousley (1948b), Roach (1985), Vaughan & Geissler (1997), Zohary & Hopf (2000).

T. D. DINES

Urtica dioica Common Nettle

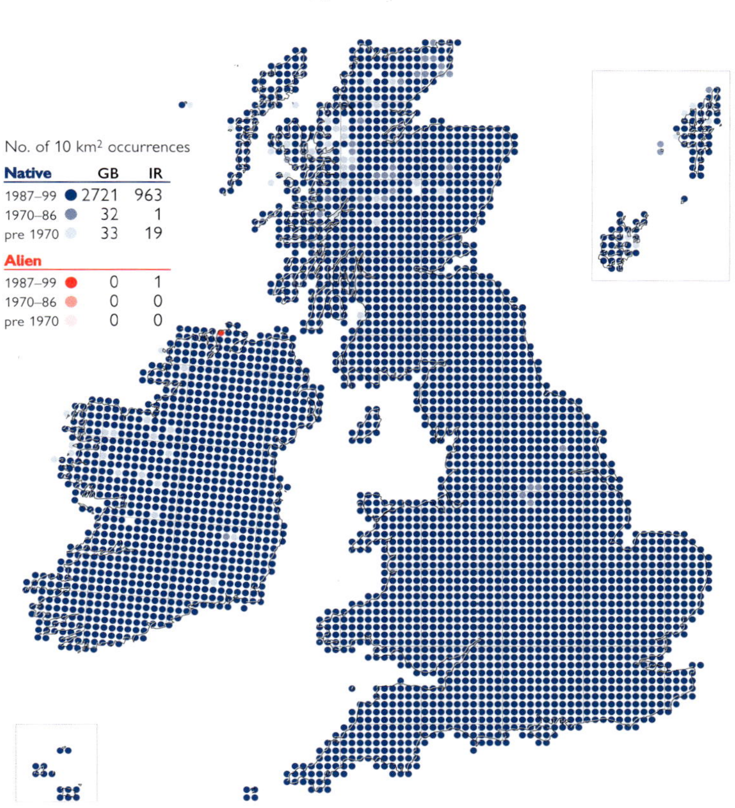

A rhizomatous and stoloniferous perennial herb occurring in a wide range of habitats, including woods, scrub, unmanaged grasslands, fens, river banks, hedgerows, roadsides, manure heaps, cultivated and waste ground. It prefers damp, nutrient-rich soils. 0–850 m (Great Dun Fell, Westmorland).

Native (change +0.28). Since the 1950s, *U. dioica* seems to have increased, perhaps due to the widespread use of artificial fertilisers, but this is not obvious at the 10-km scale. A pubescent, stingless form from fens and carr has been referred to *U. galeopsifolia*.

Eurosiberian Boreo-temperate element; widely naturalised outside its native range.

References: Atlas (182c), Bassett *et al.* (1977), Geltman (1992), Greig-Smith (1948), Grime *et al.* (1988), Hultén & Fries (1986), Jalas & Suominen (1976), Meusel *et al.* (1965), Srutek & Teckelmann (1998).

S. J. LEACH

Urtica urens Small Nettle

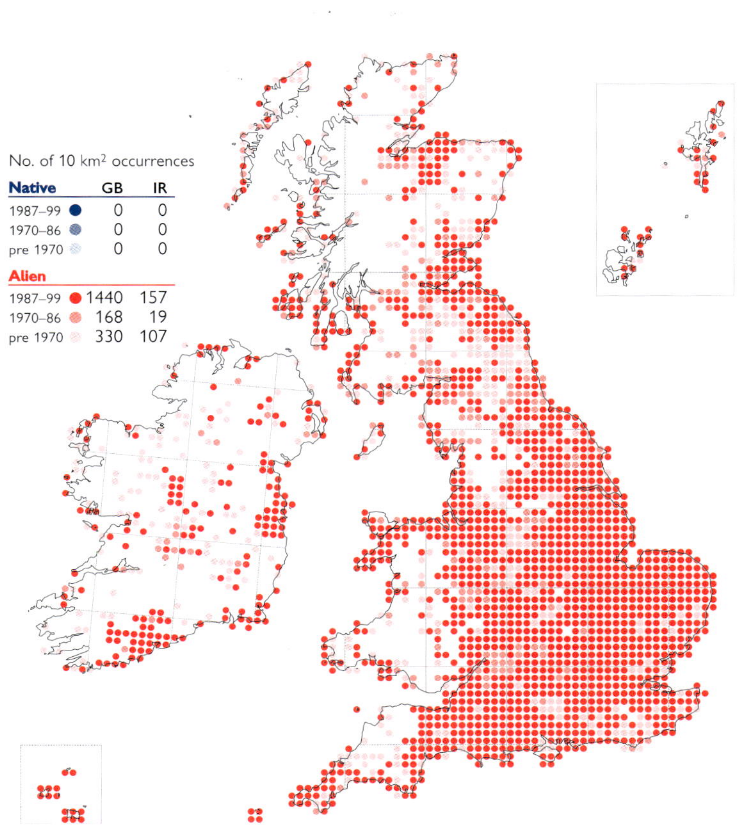

A spring-germinating annual of well-tilled arable land, especially fields of broad-leaved crops, and also allotments, gardens, farmyards and waste ground. It prefers light, often sandy, soils of high fertility. Generally lowland, but reaching 500 m in E. Allendale (S. Northumb.).

Archaeophyte (change −0.70). There has been little change in the distribution of *U. urens* since the 1962 *Atlas*, where it was mapped as 'all records'.

As an archaeophyte *U. urens* has a Eurosiberian Southern-temperate distribution, but it is widely naturalised so that its distribution is now Circumpolar Southern-temperate.

References: Atlas (182b), Greig-Smith (1948), Grime *et al.* (1988), Hultén & Fries (1986) Jalas & Suominen (1976).

S. J. LEACH

Parietaria judaica Pellitory-of-the-wall

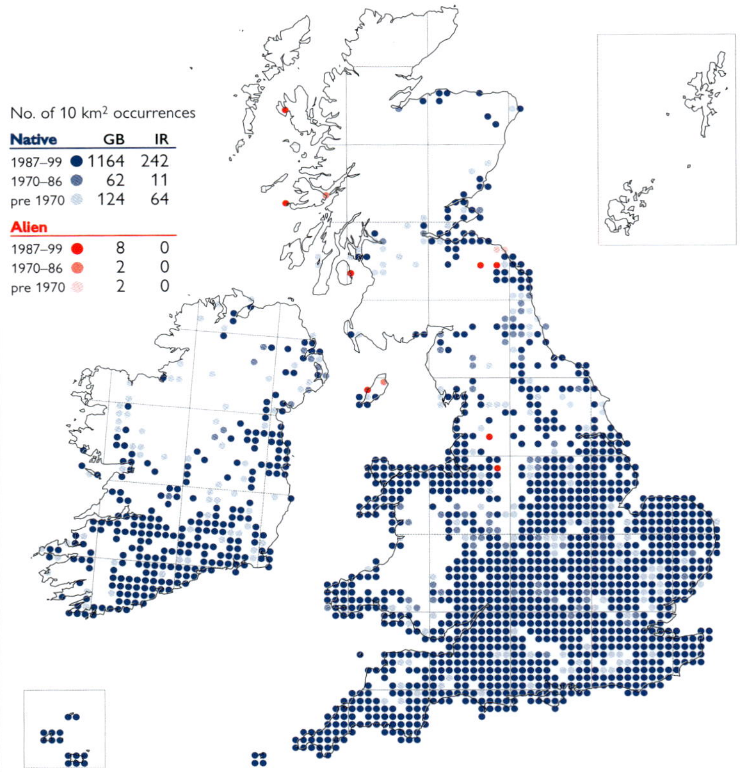

A much-branched perennial herb growing from the cracks and mortar crevices of brick and stone walls, on building rubble, rocks, cliffs and steep-sided hedge banks. It prefers dry, sunny, sheltered spots, and is often found in built-up areas or not far from habitation. Lowland.

Native (change +0.08). There appears to have been little change in distribution since the 1962 *Atlas*, where it was mapped as 'all records'. Its frequent occurrence on the walls of abbeys and priories is possibly connected to its use by medieval herbalists as a remedy for urinary disorders.

Submediterranean-Subatlantic element; also in C. Asia.

References: Atlas (181d), Bolòs & Vigo (1990), Jalas & Suominen (1976).

S. J. LEACH

Soleirolia soleirolii Mind-your-own-business

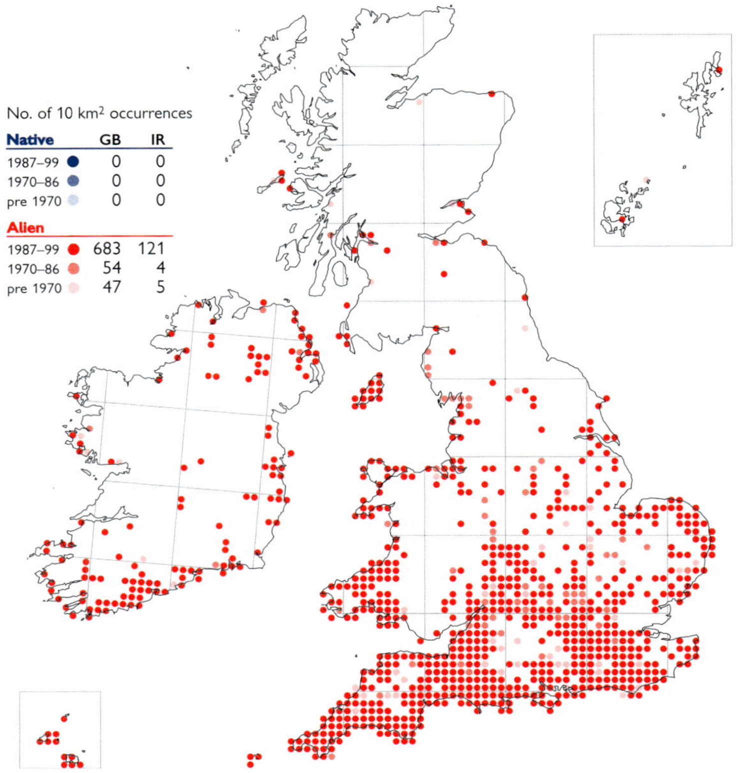

This evergreen, procumbent, carpet-forming perennial herb is found on damp paths, shaded banks and roadside walls, and in sheltered places in churchyards and gardens, usually close to habitation. It is somewhat frost-sensitive. Lowland.

Neophyte (change +2.36). *S. soleirolii* is a frequent escape from gardens and greenhouses. It has been cultivated in Britain since 1905 and was recorded as naturalised by 1917. It has increased markedly since the 1962 *Atlas*, with a considerable extension of range northwards and eastwards from its headquarters in S.W. England.

A native of the W. Mediterranean islands.

References: Atlas (182a), Bolòs & Vigo (1990), Jalas & Suominen (1976).

S. J. LEACH

Juglans regia Walnut

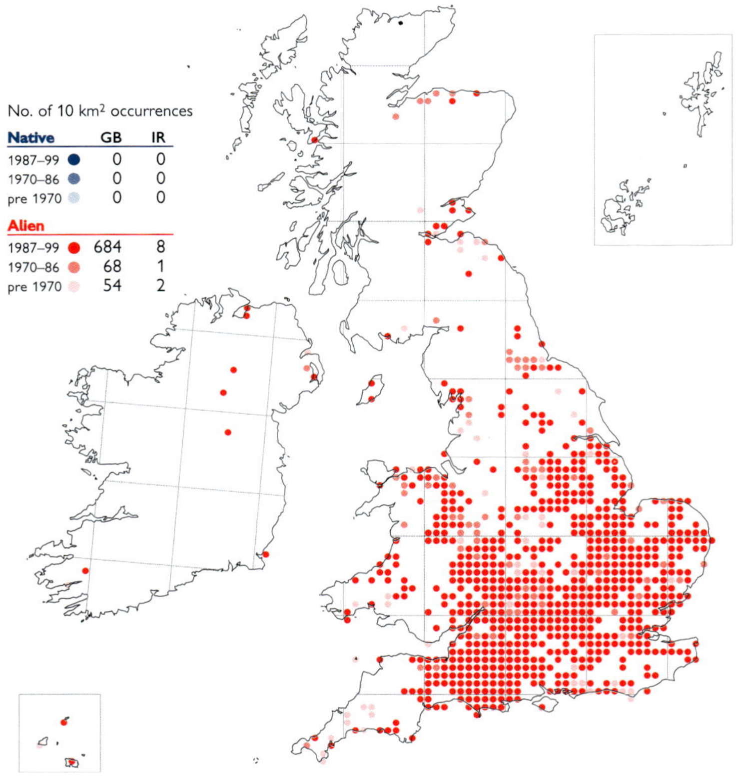

A large deciduous tree, commonly planted in urban areas and sometimes also in the wild in secondary woodland and hedgerows, and on river banks, field-borders and roadsides. It is often self-sown or buried by grey squirrels, especially in the south and west, with plants appearing on waste ground, in copses, rough grassland, gravel-pits and on railway banks. Lowland.

Neophyte. *J. regia* has been grown in British gardens since Roman times and was certainly present by 995 (Godwin, 1975; Harvey, 1981), but it was not recorded formally in the wild until 1836. Its distribution is likely to be stable.

Thought to be native of S.W. & C. Asia, perhaps west to the Balkans; widely planted elsewhere.

References: Jalas & Suominen (1976), Mitchell (1996), Roach (1985), Vaughan & Geissler (1997), Zohary & Hopf (2000).

T. D. DINES

Myrica gale Bog-myrtle

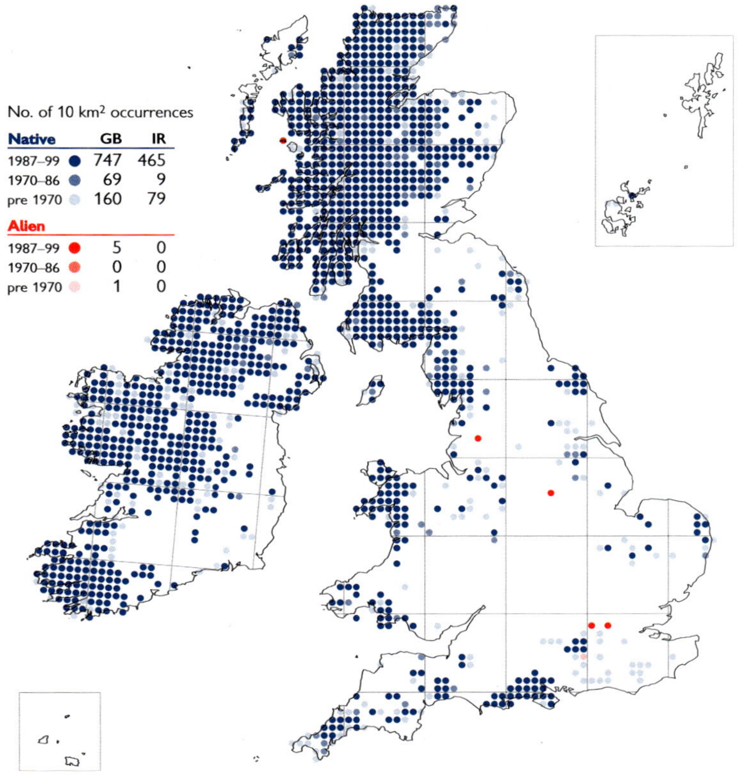

A small shrub which suckers to form dense thickets. It grows in organic soils in base-poor bogs and moorland, lowland raised bogs, wet heaths and acid carr; in all its sites, moving groundwater is a constant feature. It tolerates light shade and is not grazed. Generally lowland, but reaching at least 520 m in the Forest of Drumochter (Easterness).

Native (change –0.75). Some lowland populations have been lost to peat extraction and agricultural reclamation, whilst others have gone as woodland develops on bogs that have been drained or are drying out due to a falling water-table. In the uplands the distribution is stable.

Suboceanic Boreo-temperate element; also in E. Asia and N. America.

References: Atlas (183d), Hultén & Fries (1986), Jalas & Suominen (1976), Meusel *et al.* (1965), Skene *et al.* (2000).

D. A. PEARMAN

Fagus sylvatica Beech

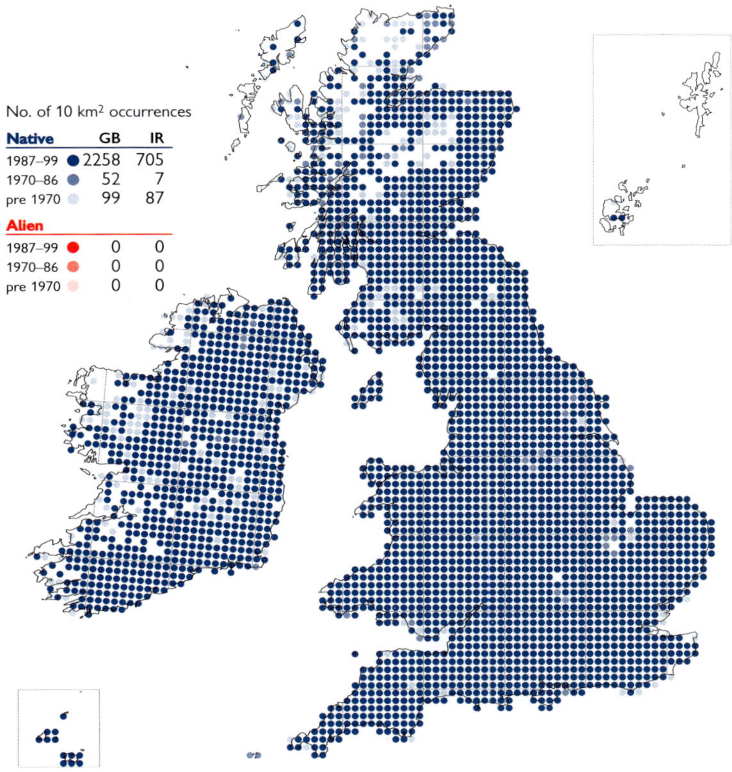

F. sylvatica is a large tree found on a wide variety of base-rich to acidic, free-draining soils. It grows in pure woodlands or as standard trees or pollards in wood pasture. It is widely planted outside its native range as woodland, in avenues and as hedges. 0–650 m S. of Garrigill (Cumberland).

Native (change –0.62). Beech was a late colonist of Britain after the last glacial period; it is restricted as a native to S.E. England and S.E. Wales. The native range has never been precisely defined and all records are mapped as native regardless of status. Its overall range is stable.

European Temperate element.

References: Atlas (185c), Brown (1953), Grime *et al.* (1988), Hultén & Fries (1986), Jalas & Suominen (1976), Meusel *et al.* (1965), Rackham (1980), Watt (1931, 1934).

T. D. DINES

Nothofagus obliqua Roble

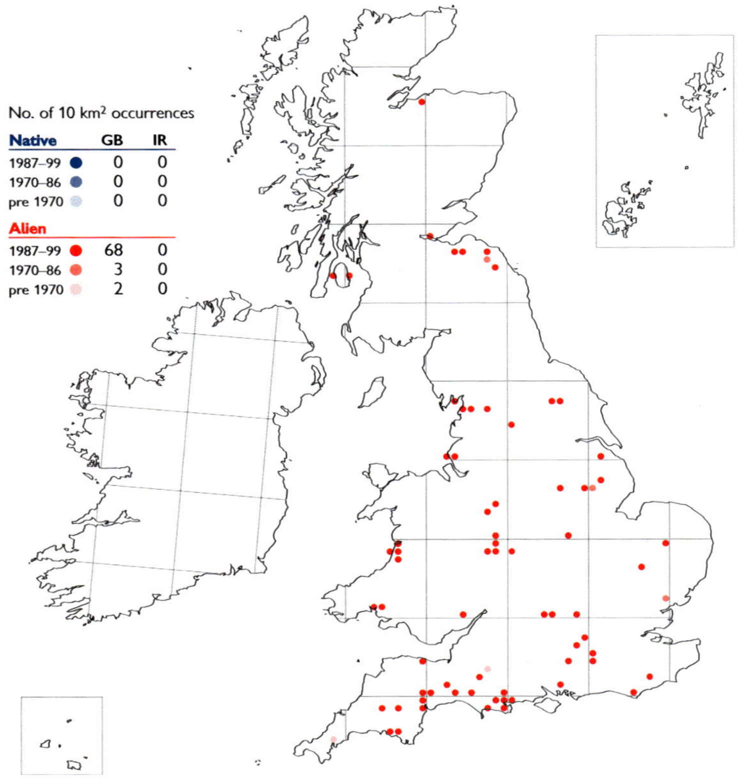

A deciduous hardwood tree grown in pure or mixed plantations, in parks, arboreta, amenity areas and large gardens; sometimes also used as a hedging plant. It grows well on a wide range of soils, except over chalk, setting seed and regenerating freely. Lowland.

Neophyte. *N. obliqua* was possibly introduced to Britain in 1849, but the first confirmed record was in 1902. It is valued for its fast growth and good quality hardwood, and commercial plantations were established from the 1930s onwards. It was recorded from the wild by 1956.

Native of S. America (Argentina, Chile).

References: Bean (1976), Mitchell (1996), Wigston (1979).

T. D. DINES

Nothofagus nervosa Rauli

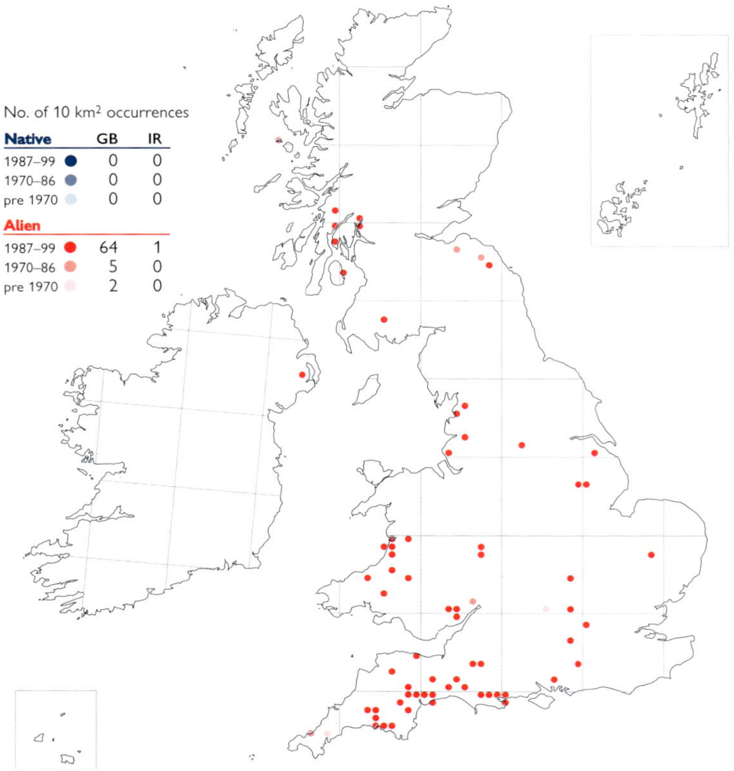

This deciduous hardwood tree grows on a wide range of soils, and seems to do particularly well in areas of high rainfall. It is found in pure and mixed plantations, and is grown as a specimen tree in parks, amenity areas and large gardens. Abundant seed is set, and it regenerates freely at many of its sites in Britain. Lowland.

Neophyte. Introduced in 1910, *N. nervosa* is attractive to commercial foresters for its quick growth and good quality hardwood. Plantations were established from the 1930s, and it continues to be used commercially and for ornament. It was recorded from the wild by 1956.

Native of S. America (Argentina, Chile), where it grows at higher altitudes than *N. obliqua.*

References: Bean (1976), Mitchell (1996), Wigston (1979).

T. D. DINES

Castanea sativa Sweet Chestnut

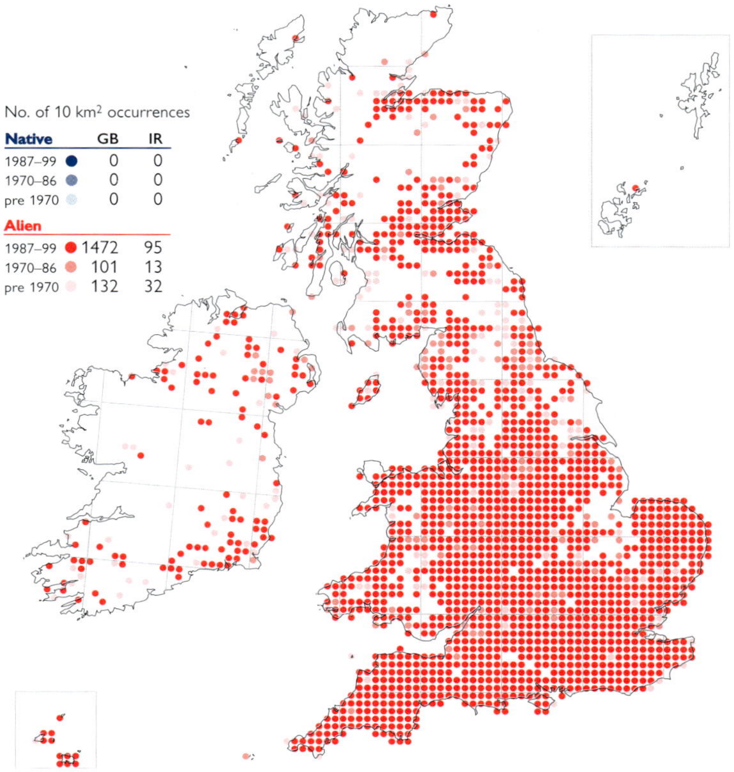

This deciduous tree is a major constituent of coppiced woodland in S.E. England. It is also planted in hedgerows, wood-borders, parkland and amenity areas, and in large gardens. It tolerates a wide range of soils, but thrives on moist, sandy soils. Seed is set freely in the south, but seedlings rarely reach maturity. 0–410 m (Fern Hill, Rads.).

Archaeophyte (change +0.59). *C. sativa* was probably introduced in Roman times, and its later presence is attested by Anglo-Saxon place names and its mention in Medieval documents. The large increase in records since the 1962 *Atlas* is due to better recording and continued planting.

Native of S. Europe, N. Africa & S.W. Asia; widely naturalised in W. Europe.

References: Atlas (185d), Jalas & Suominen (1976), Meusel *et al.* (1965), Rackham (1980), Zohary & Hopf (2000).

T. D. DINES

Quercus cerris Turkey Oak

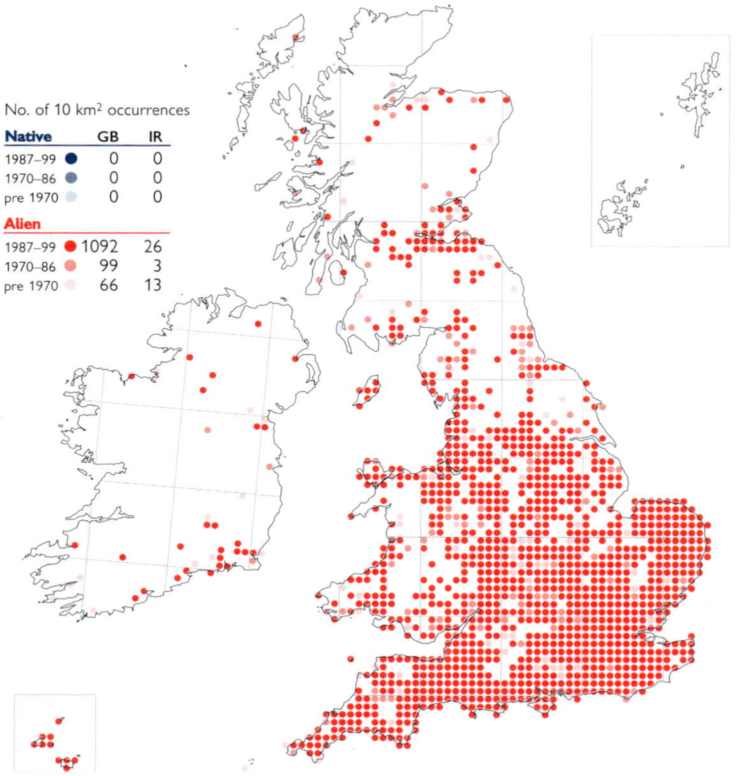

A deciduous tree planted in woodlands, town parks, estates, large gardens and along roads, especially on acidic, sandy soils. It seeds freely, and has become naturalised on free-draining soils in other habitats including railway banks and waste ground, spreading into calcareous grassland and heathland. Lowland.

Neophyte (change +2.32). *Q. cerris* was cultivated in Britain by 1735 and has been known from the wild since at least 1905. There has been a dramatic increase in records since the 1962 *Atlas*, due to both a genuine increase and better recording of aliens. Its free regeneration threatens the native flora at some sites.

A native of S.C. & S.E. Europe and S.W. Asia.

References: Atlas (186a), Jalas & Suominen (1976), E.W. Jones (1959), Meusel *et al.* (1965), Mitchell (1996).

T. D. DINES

Quercus cerris × *Q. suber* (*Q.* × *crenata*)
Lucombe Oak

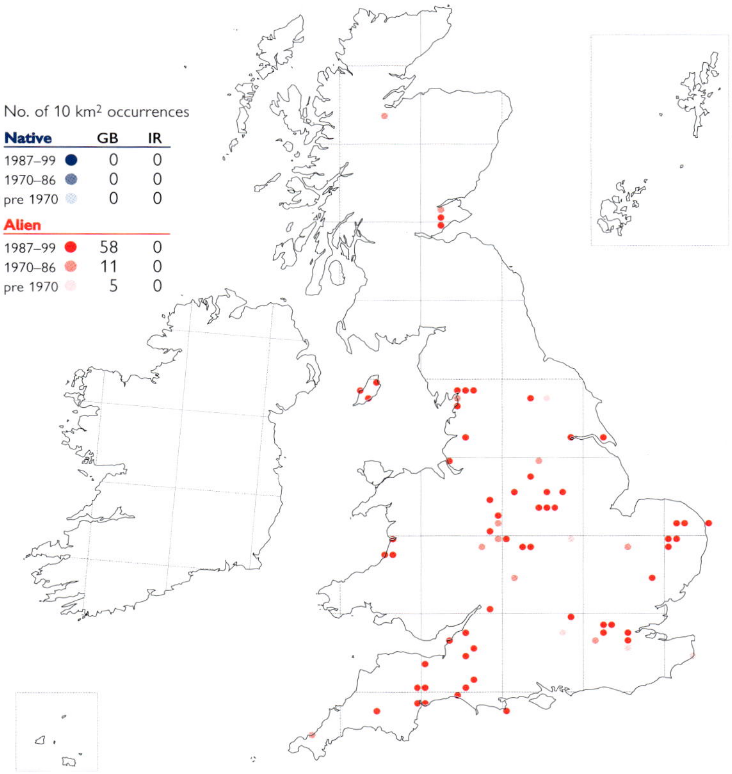

A large, semi-evergreen tree of parks, large gardens, woodland and field boundaries. It is fertile and is occasionally self-sown, the progeny exhibiting a range of variation between the parent species. Lowland.

Neophyte. *Q.* × *crenata* arose in W. Lucombe's nursery in Exeter in 1762, less than thirty years after *Q. cerris* had been introduced there. It has since been widely, if sporadically, planted, but was not recorded formally from the wild until 1964.

The plants normally cultivated are hybrids of garden origin, but the hybrid also occurs naturally with its parents in S. Europe.

References: Bean (1976), Mitchell (1996).

T. D. DINES

Quercus ilex Evergreen Oak

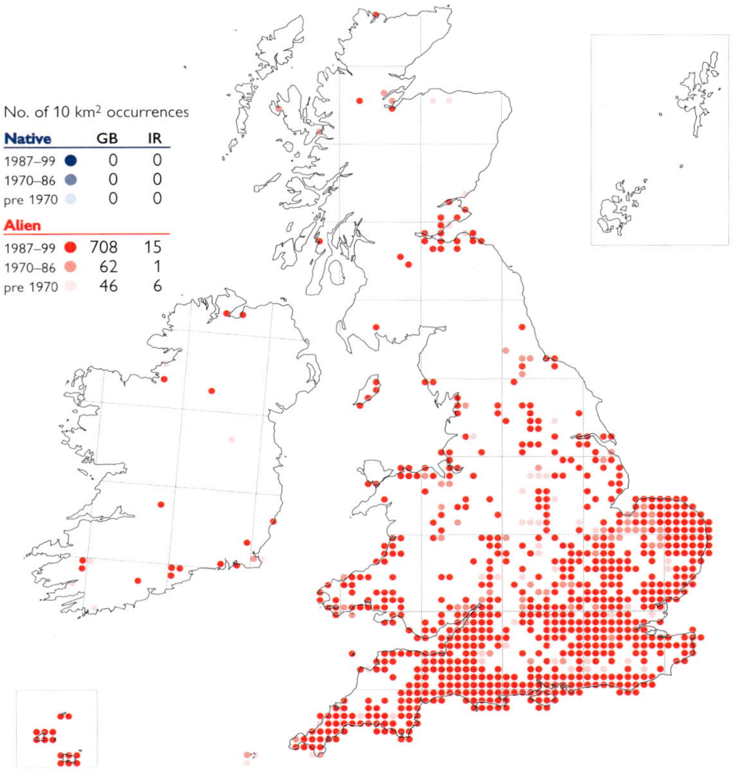

An evergreen tree, planted in parks, large gardens, churchyards and cemeteries, and becoming well-established in copses, woodland and on sand dunes. It prefers light, warm soils, and is frequently planted near coasts. Seed production can be prolific, and it regenerates freely in parts of S. and E. England. Lowland.

Neophyte (change +2.37). *Q. ilex* has been cultivated since the 16th century, and was widely planted in the 18th century. It was recorded in the wild in 1862. It can colonise natural habitats aggressively and replace native vegetation. The significant increase in records since the 1962 *Atlas* is due to better recording and a genuine spread in the south.

Native of the Mediterranean region.

References: Atlas (186b), Jalas & Suominen (1976), James *et al.* (1981), E.W. Jones (1959), Meusel *et al.* (1965).

T. D. DINES

Quercus petraea Sessile Oak

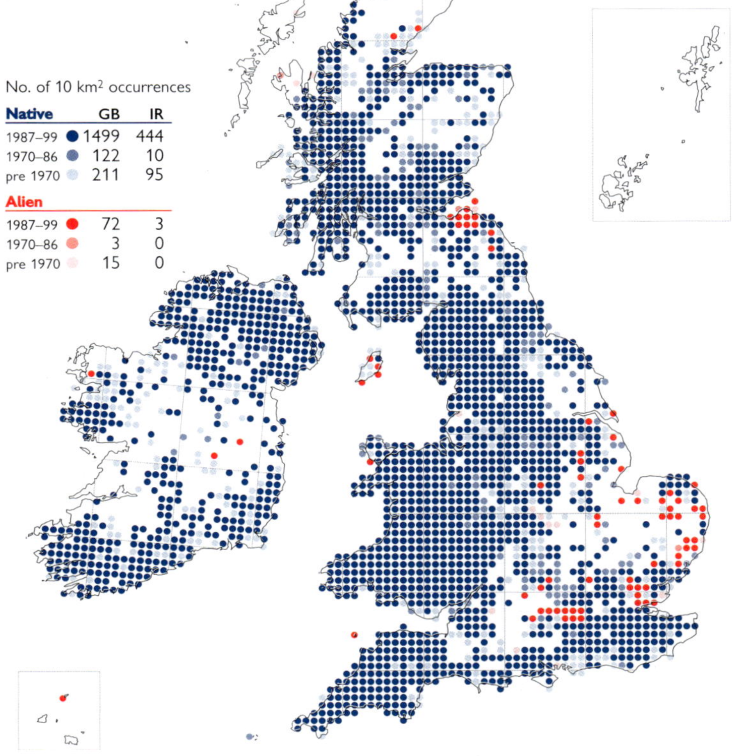

A long-lived, deciduous tree, forming high forest or coppice woodland, especially on well-drained, shallow, moderately to strongly acidic soils. It is the characteristic species of upland oakwoods. Seedlings of *Q. petraea* may be more shade-tolerant than those of *Q. robur* and is therefore more able to regenerate in woodland. 0–450 m (Kesdadale, Cumberland).

Native (change +0.14). *Q. petraea* was not favoured by 18th and 19th century foresters, and it tends to have a relict distribution in Britain and Ireland. The map may underestimate the extent to which it is introduced at the eastern edge of the range.

European Temperate element.

References: Atlas (186d), Gardiner (1974), Grime *et al.* (1988), Hultén & Fries (1986), Jalas & Suominen (1976), E.W. Jones (1959), Meusel *et al.* (1965), Morris & Perring (1974), Rackham (1980).

T. D. DINES

Quercus petraea × *Q. robur* (*Q.* × *rosacea*)

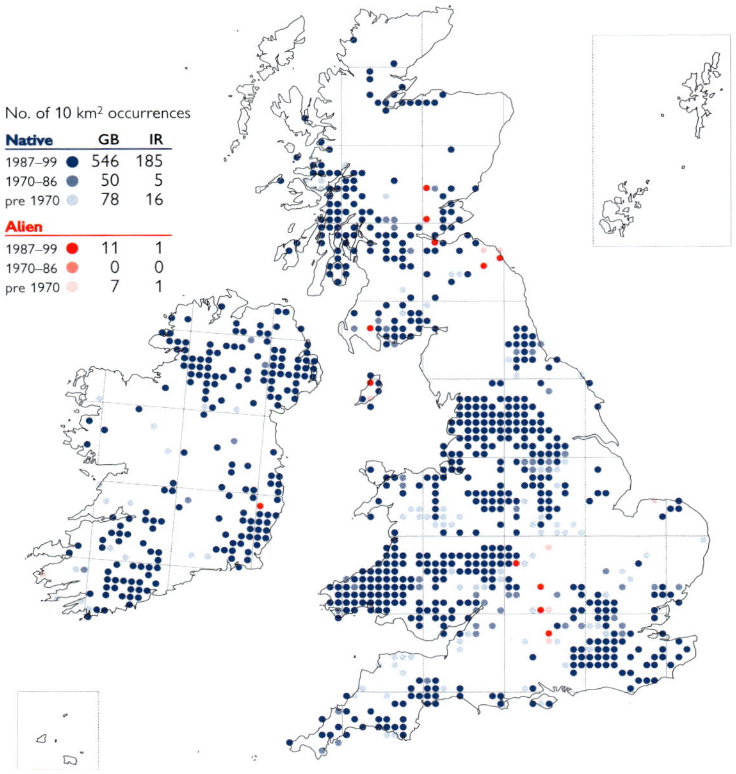

A large deciduous tree, potentially occurring wherever the parents meet and sometimes occurring in the absence of the less frequent *Q. petraea*. It may be present as pure stands, as in some steep coastal river valleys, or as individual trees. Generally lowland, but upper altitudinal limit unknown.

Native. First described in 1909, this hybrid is under-recorded because of the variability of the parents and disagreement over the delimitation of hybrid plants. This taxon is supposedly more common in Britain and Scandinavia than elsewhere in Europe as the flowering time between the parents is closer in the north, but this explanation is disputed in Stace (1975).

Widespread in Europe, and commoner than *Q. petraea* in some areas of S. Scandinavia.

References: Grime *et al.* (1988), Oliver (2000), Rushton (1978a, 1978b).

T. D. DINES

Quercus robur Pedunculate Oak

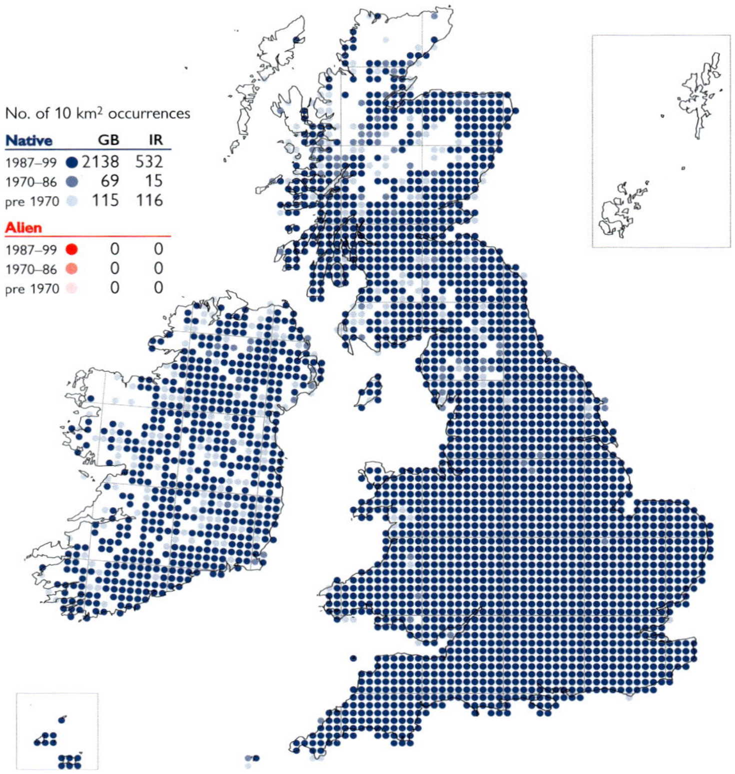

A long-lived, deciduous tree of high forest, coppice woodland and ancient wood-pasture. It grows on a wide range of soils, typically those which are heavy and fertile, but does not thrive on thin soils over limestone or acidic peat. It is fairly tolerant of waterlogging, growing at fen margins and in *Alnus* woodland. It is very widely planted in hedges and woodland. 0–450 m (Talgarth, Brecs.).

Native (change –0.60). The dominance of *Q. robur* in many woods is the result of many centuries of selective woodland management, followed by deliberate planting in recent centuries. All records are mapped as if they are native.

European Temperate element.

References: Atlas (186c), Gardiner (1974), Grime *et al.* (1988), Hultén & Fries (1986), Jalas & Suominen (1976), E.W. Jones (1959), Meusel *et al.* (1965), Morris & Perring (1974), Rackham (1980).

T. D. DINES

Quercus rubra Red Oak

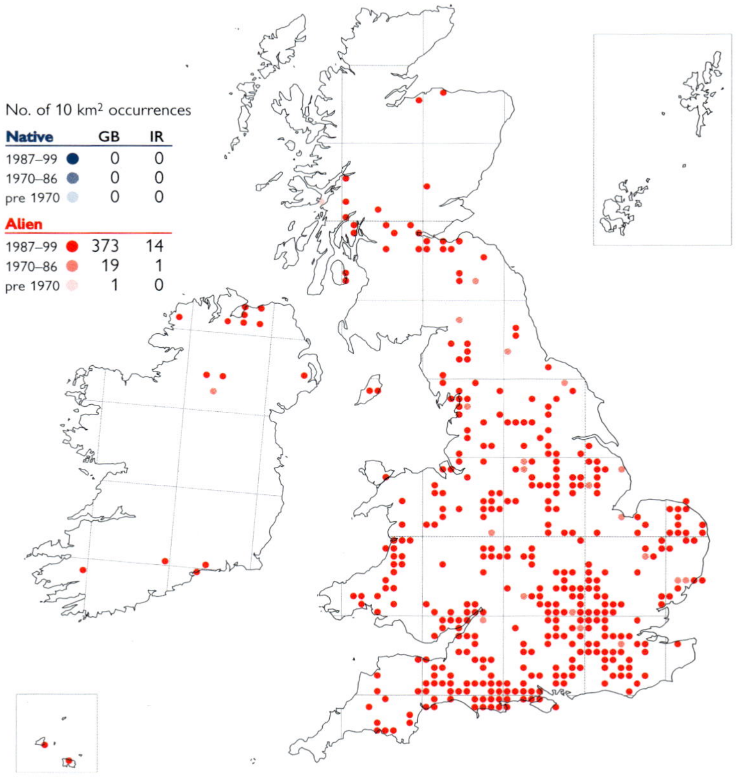

A large deciduous tree, widely planted for ornament in parks, estates, gardens and roadsides, and occasionally for forestry, hedging and screening, especially on light, sandy soils. It is often self-sown and is becoming naturalised in a few places. Lowland.

Neophyte. Cultivated in Britain by 1724, *Q. rubra* is increasingly planted in parks and amenity areas, but to a lesser extent than some other *Quercus* species. It was not recorded from the wild until 1942.

Native of eastern N. America.

References: Bean (1976), FNAEC (1997), E.W. Jones (1959), Mitchell (1996).

T. D. DINES

Betula pendula Silver Birch

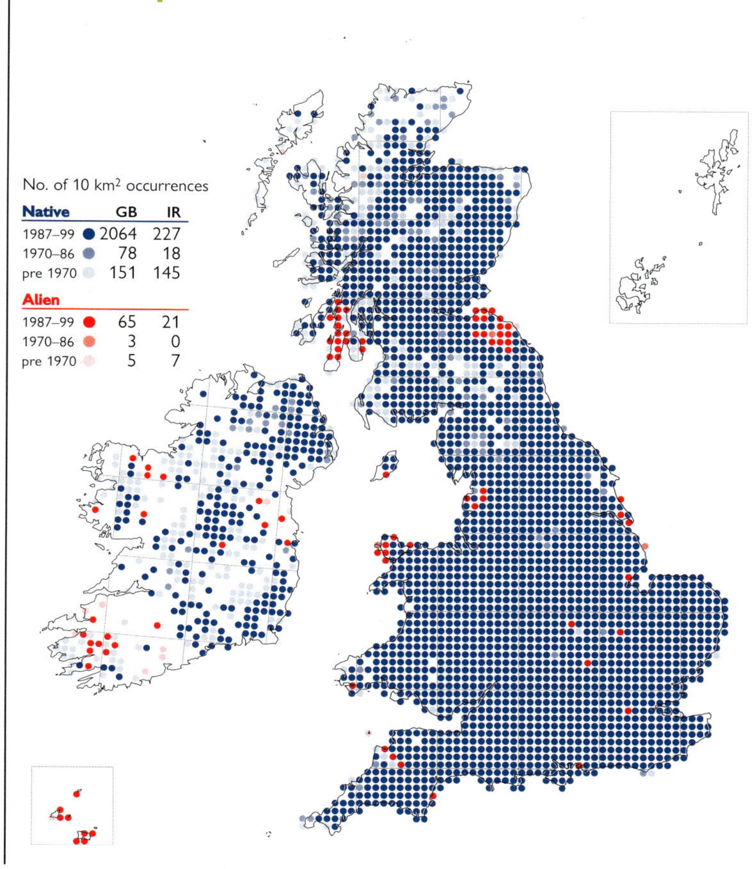

A deciduous tree found as even-aged stands or in mixed woodland on a wide range of light, well-drained, particularly acidic soils. It can rapidly colonise open ground, particularly burned areas, and can become a threat to open heathland. It is also widely planted on roadsides and in parkland. Generally lowland, but upper altitudinal limit unknown.

Native (change −0.23). The distribution of *B. pendula* is apparently stable. Worrell & Malcolm (1998) have suggested that it is over-recorded in much of Scotland, as few recorders can reliably distinguish it from *B. pubescens*. Some introduced populations may also be mapped as native.

Eurosiberian Boreo-temperate element.

References: Atlas (184a), Atkinson (1992), Grime *et al.* (1988), Hultén & Fries (1986), Jalas & Suominen (1976), Meusel *et al.* (1965).

T. D. DINES

Betula pendula × *B. pubescens* (*B.* × *aurata*)

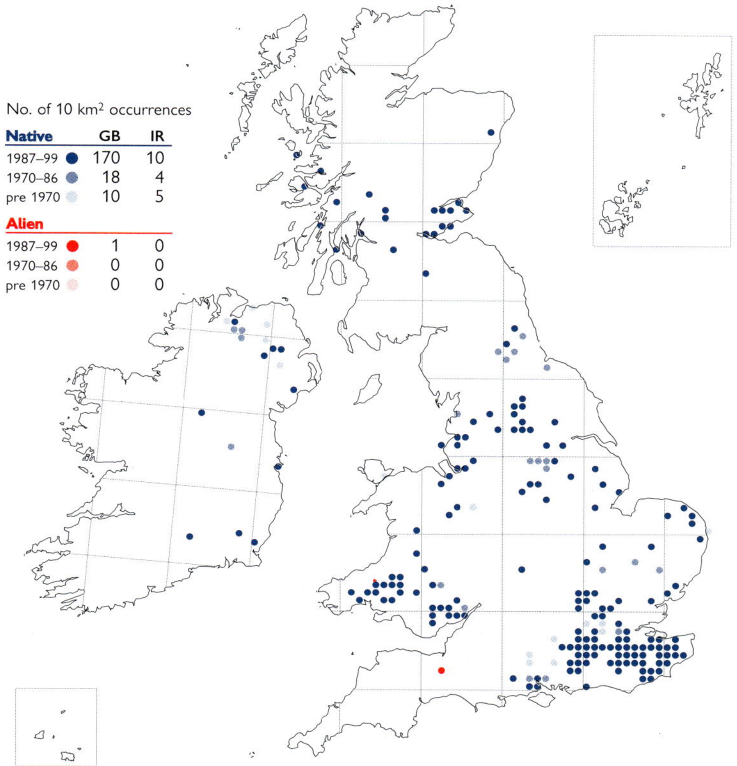

A deciduous tree found in similar habitats to its parents, including woodland, copses, heathland, disused sand- and gravel-pits and waste ground, but showing a preference for disturbed sites. Generally lowland, but upper altitudinal limit unknown.

Native. Often found in the absence of one or both parents, *B.* × *aurata* is very variable in morphology and sterility. The parents themselves are very variable; some forms closely resemble the hybrid and may be recorded in error for it. Because of this confusion, opinions differ on the extent to which hybridisation occurs.

Widespread in N. & C. Europe, but apparently commoner in the British Isles than elsewhere.

References: Brown *et al.* (1982), Kennedy & Brown (1983), Rich & Jermy (1998), Stace (1975).

T. D. DINES

Betula pubescens Downy Birch

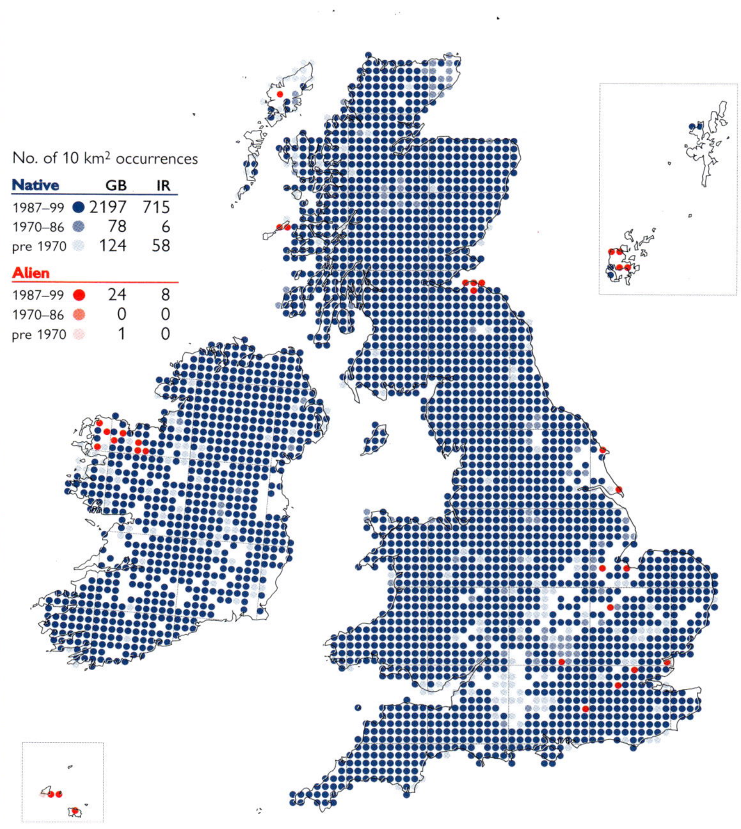

A deciduous tree, usually found in mixed woodland or as isolated trees on roadsides and field boundaries. It grows on a wide range of soils but prefers more acidic, wetter, peatier soils, especially in the uplands. It can rapidly colonise open, unshaded ground, particularly burned areas, open heathland and cut-out peat bogs. It is also widely planted, becoming established in urban areas. 0–685 m (Hilton Fell, Cumberland), and certainly higher in Scotland.

Native (change +0.40). The apparent increase since the 1962 *Atlas* is probably due to more comprehensive recording rather than a genuine spread, and the distribution is unchanged.

Eurosiberian Boreo-temperate element.

References: Atlas (184b), Atkinson (1992), Grime *et al.* (1988), Hultén & Fries (1986), Jalas & Suominen (1976), Meusel *et al.* (1965).

T. D. DINES

Betula nana Dwarf Birch

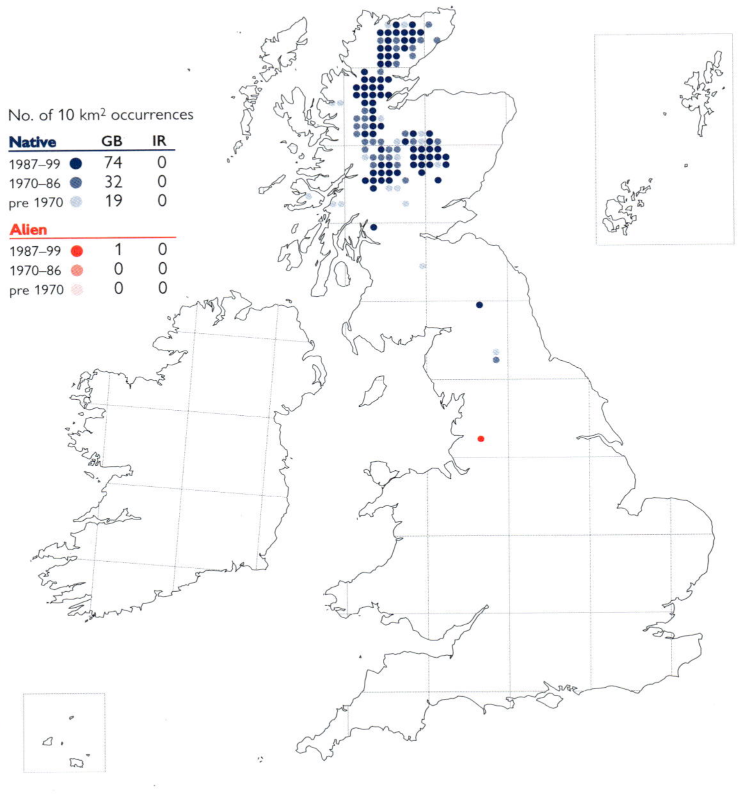

A low-growing, deciduous shrub of upland heaths and blanket bogs, usually found on acidic peat but occasionally rooted into rock crevices. It is found on both moderately dry, sloping sites and on waterlogged, flat ground. Germination may be frequent, but seedlings appear to be removed by grazing. Generally upland, reaching 860 m in Glen Cannich (E. Ross), but descending to 120 m in W. Sutherland.

Native (change –0.09). Although recording since the 1962 *Atlas* has extended its known range, and new populations continue to be found, *B. nana* is declining as a result of over-grazing, afforestation and burning.

Circumpolar Boreo-arctic Montane element.

References: Atlas (184c), Groot *et al.* (1997), Hultén & Fries (1986), Jalas & Suominen (1976), Meusel *et al.* (1965), Stewart *et al.* (1994).

T. D. DINES

Alnus glutinosa Alder

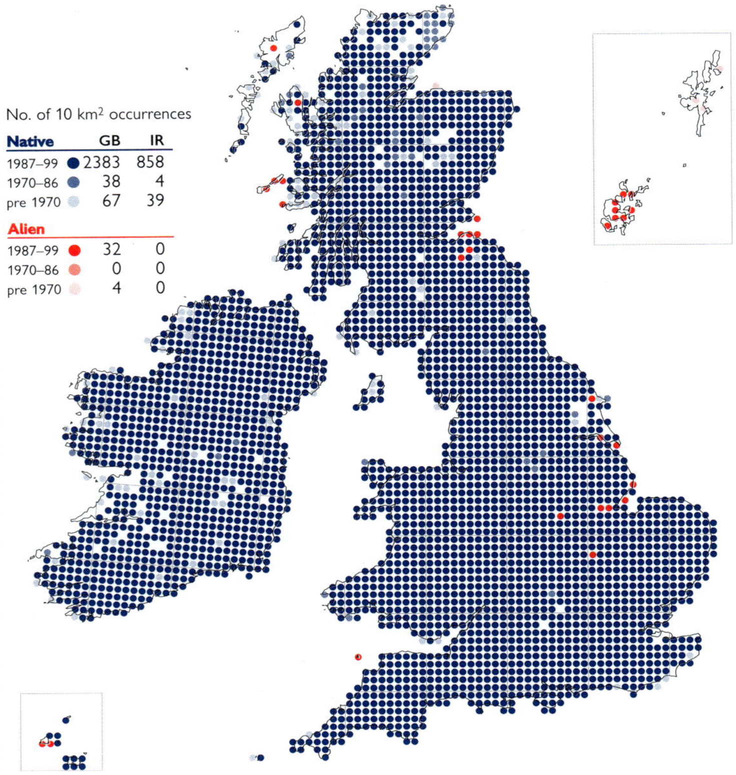

A deciduous tree of damp or wet, basic to moderately acidic soils, found beside rivers, streams, canals, lakes and ditches, and in flood plains, fens and bogs, carr and wooded dune-slacks. It can rapidly seed into open sites, producing even-aged stands of mature trees, but seedlings are very shade- and drought-sensitive, so regeneration in woodland is often poor. It is also widely planted. 0–470 m (Garrigill, Cumberland).

Native (change –0.32). The overall distribution of this species is stable. However, a recently evolved *Phytophthora* fungus has killed 10% of trees in S. England and Wales, and may have a wider impact in the future.

Eurosiberian Temperate element.

References: Atlas (184d), Everett (1999), Grime *et al.* (1988), Hultén & Fries (1986), Jalas & Suominen (1976), McVean (1953, 1955a, 1955b), Meusel *et al.* (1965).

T. D. DINES

Alnus incana Grey Alder

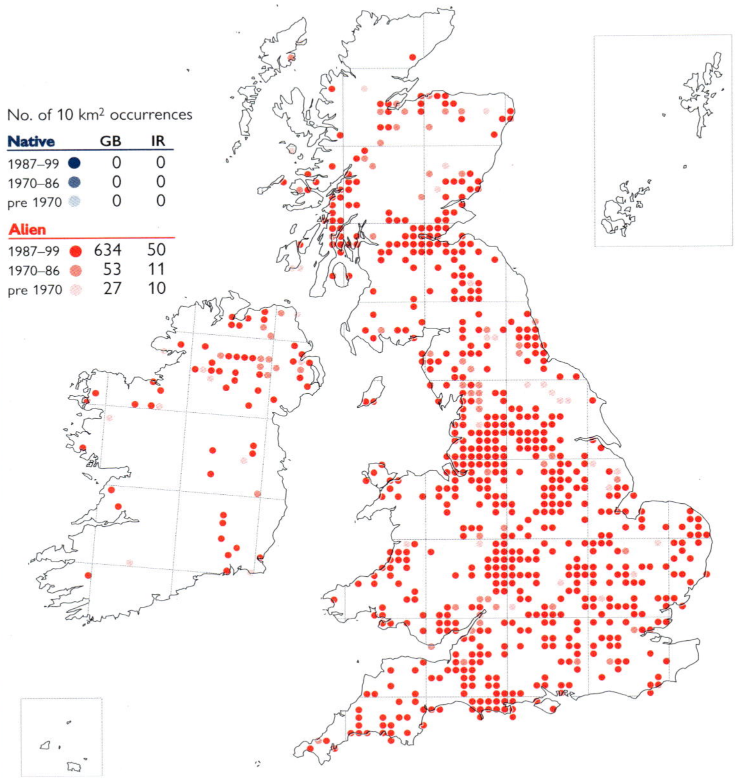

A deciduous tree, widely planted in parks, on roadsides, reclaimed tips and river banks, and sometimes in small plantations. It is very hardy and tolerant of poor, wet soils, making it useful in amenity planting in the north. It sometimes spreads to, and becomes naturalised on, waste ground and railway banks. It reproduces by seed and suckering. Lowland.

Neophyte. Introduced to cultivation in Britain in 1780, *A. incana* was not recorded from the wild until 1922. It is planted with increasing frequency and is probably under-recorded.

A Circumpolar Boreal-montane species, absent as a native from much of W. Europe.

References: Bean (1970), Hultén & Fries (1986), Jalas & Suominen (1976), Meusel *et al.* (1965).

T. D. DINES

Alnus cordata Italian Alder

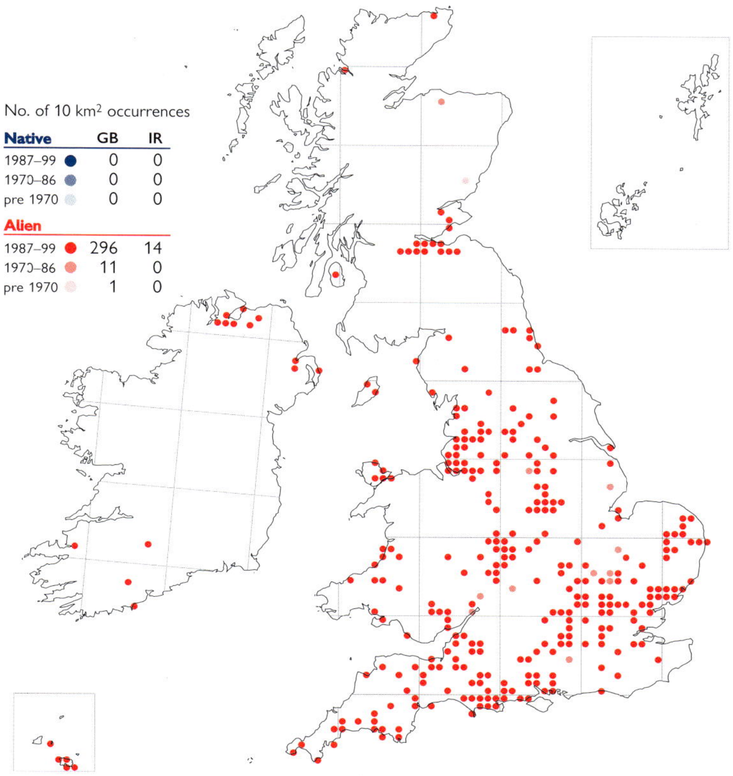

A deciduous tree found on roadsides, in town parks and in amenity areas. Unlike other *Alnus* species, it thrives on poor, dry soils, even those over chalk. Street trees flower and fruit freely; the seed is wind dispersed and seedlings are frequent by pavements and on waste ground. Generally lowland, but reaching 305 m at Shap (Westmorland).

Neophyte. *A. cordata* was introduced to Britain in 1820 and was recorded from the wild in 1935. It is increasingly planted as an ornamental tree.

Native of Corsica, S. Italy and Albania.

References: Bean (1970), Jalas & Suominen (1976), Meusel *et al.* (1965), Mitchell (1996).

T. D. DINES

Carpinus betulus Hornbeam

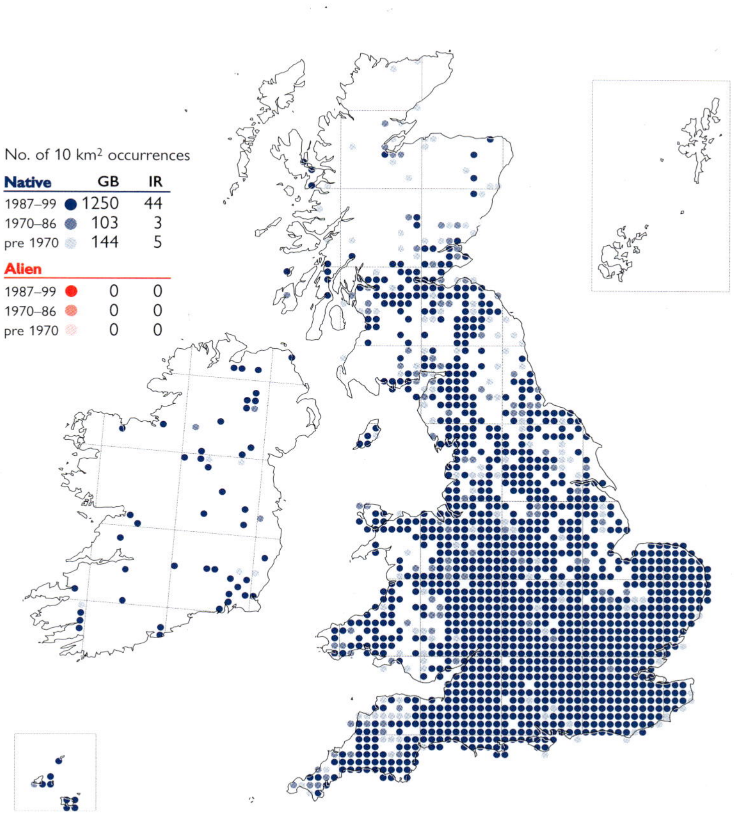

A long-lived deciduous tree, found as native in both pure and mixed woodland on base-poor sandy or loamy clays, or clay-with-flints. Within its native range, coppiced plants are often the dominant member of the shrub layer in *Quercus* woods. It is also extensively planted in woodlands, on roadsides, in amenity areas and for hedging, both within and outside its native range. Lowland as a native, but to 380 m as an alien on Great Mell Fell (Cumberland).

Native (change +0.84). *C. betulus* is confined as a native to S.E. England. Its precise native range has not been defined and all records are mapped as if they are native. There are many more alien records since the 1962 *Atlas*.

European Temperate element.

References: Atlas (185a), Hultén & Fries (1986), Jalas & Suominen (1976), Meusel *et al.* (1965), Rackham (1980).

T. D. DINES

Corylus avellana Hazel

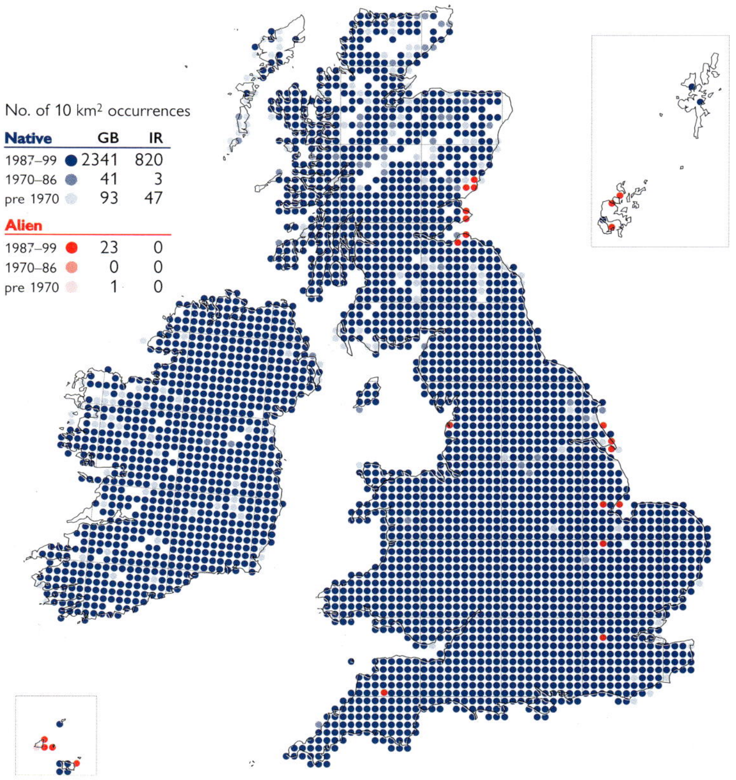

A deciduous suckering shrub of dry or damp, calcareous to mildly acidic soils, but favouring moist, base-rich conditions. It is native in the understorey of many woods, in scrub, hedgerows, on river banks, limestone pavement, cliffs and gullies, but it is also widely planted in copses and hedgerows. 0–640 m (Atholl, E. Perth).

Native (change –0.54). There is little evidence for a change in the distribution of this species; the increase in Scottish records is due to better recording. High numbers of livestock, deer and squirrels can limit regeneration, and conifer planting and the cessation of woodland management may reduce abundance locally.

European Temperate element.

References: Atlas (185b), Hultén & Fries (1986), Jalas & Suominen (1976), Meusel *et al.* (1965), Rackham (1980), Roach (1985), Zohary & Hopf (2000).

T. D. DINES

Carpobrotus edulis Hottentot-fig

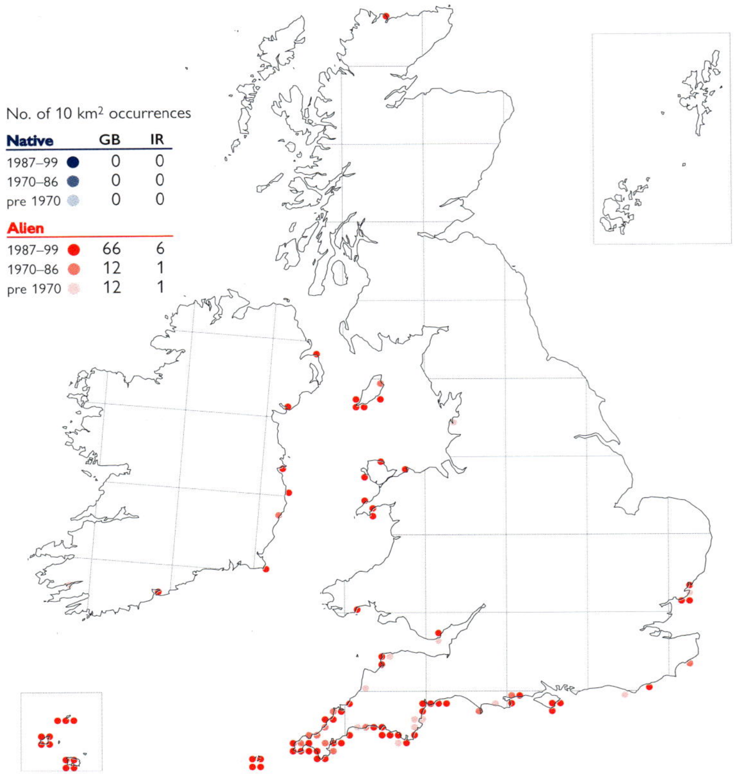

A succulent, mat-forming perennial which may cover sea-cliffs and sand dunes, or hang down rocks and walls. Some colonies were planted to stabilise dunes, while others have arisen from discarded garden material. Reproduction by seed is probably insignificant. Vegetative spread can be assisted by nesting gulls. It is frost-sensitive. Lowland.

Neophyte (change +0.05). *C. edulis* was being cultivated in gardens by *c.* 1690, but was not recorded in the wild until 1886. It has spread since the 1962 *Atlas* but is limited by climate. It is perceived as a serious threat to native species on Cornish cliffs and much effort is spent on its attempted eradication.

Native of S. Africa; widely naturalised in Mediterranean-Atlantic regions of Europe.

References: Atlas (80d), Bolòs & Vigo (1990), Jalas & Suominen (1980), Preston & Sell (1989).

C. D. PRESTON

Chenopodium bonus-henricus Good-King-Henry

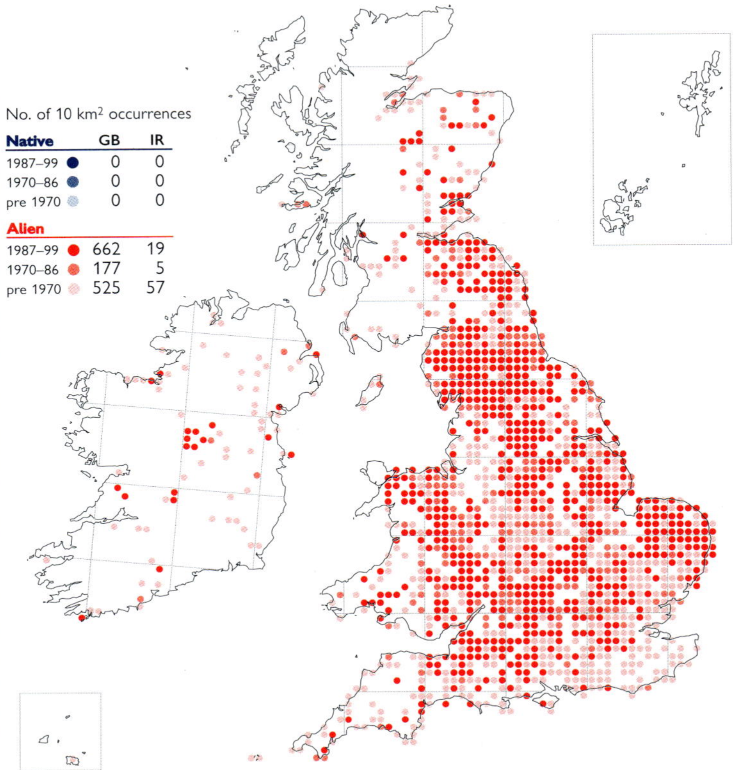

A perennial, forming patches on disturbed, nutrient-rich soil around farm buildings and ruins, and on roadsides and waste ground. It also sometimes occurs in limestone grassland. 0–455 m (Helbeck Fell, Westmorland).

Archaeophyte (change −1.79). This species, present in Roman times, was once grown for its edible leaves. It can persist for many years but has decreased since the 1962 *Atlas*, where it was mapped as 'all records', perhaps because it is no longer being introduced and populations are gradually lost through a general tidying of ruderal habitats.

Native to the mountains of C. & S. Europe, where it grows in late snow-patches; naturalised in much of temperate Europe and in N. America.

References: Atlas (81a), Hultén & Fries (1986), Jalas & Suominen (1980), Meusel *et al.* (1965), Strid (1986).

J. R. AKEROYD

Chenopodium glaucum Oak-leaved Goosefoot

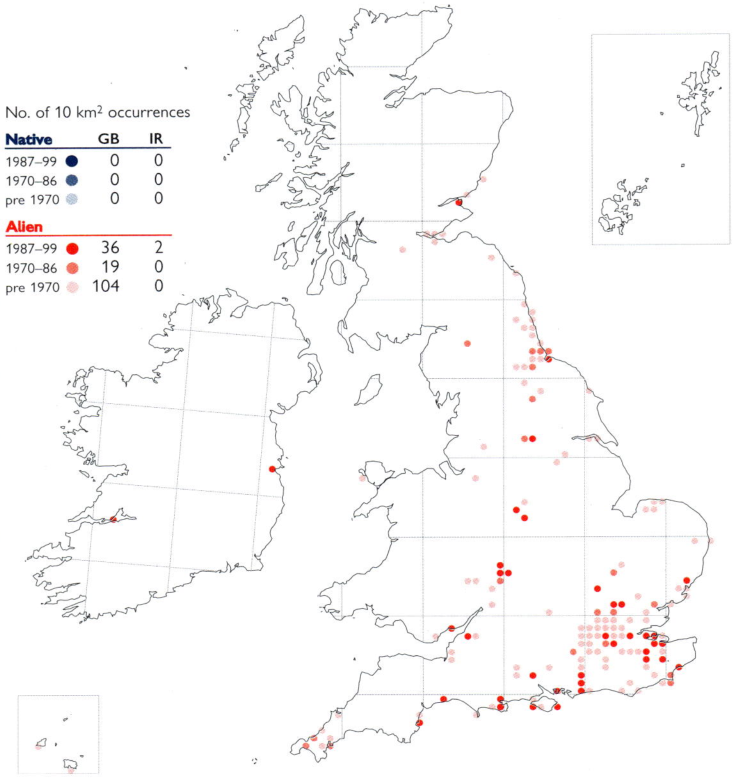

An annual of disturbed, nutrient-rich waste ground and manure heaps, and on damp ground near the sea; it is also recorded on rubbish tips and around docks and wharves. Lowland.

Archaeophyte (change −1.32). This species was first recorded in the wild in 1713. There has been a marked decline since the 1962 *Atlas*. Most records are of small, casual populations or even single plants.

A Circumpolar Temperate species, but absent from eastern N. America; widely naturalised elsewhere.

References: Atlas (83c), Hultén & Fries (1986), Jalas & Suominen (1980), Meusel *et al.* (1965).

J. R. AKEROYD

Chenopodium rubrum Red Goosefoot

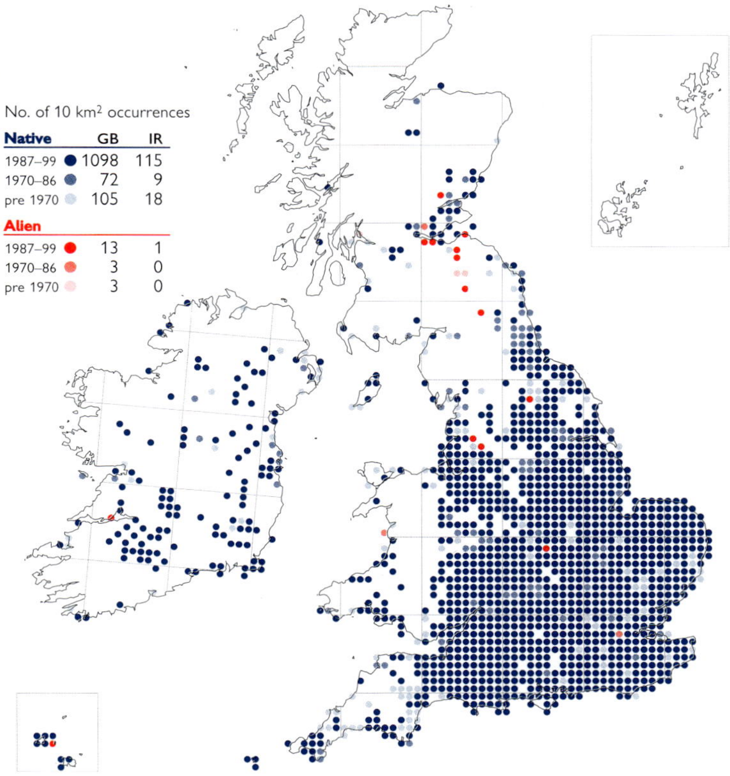

An annual of nutrient-rich mud around the dried-up margins of freshwater or brackish ponds and ditches trampled by livestock; also, more widely, in cultivated and waste ground, on manure heaps and farm tracks and in field gateways. Lowland.

Native (change +1.00). There has been a marked increase in the distribution of this mobile species since the 1962 *Atlas*, particularly in East Anglia, C.S. England, S.W. Scotland and Ireland.

Eurosiberian Temperate element; also present as a native in western N. America and widely naturalised so that the distribution is now Circumpolar Temperate.

References: Atlas (83a), Grime *et al.* (1988), Hultén & Fries (1986), Jalas & Suominen (1980), Williams (1969).

J. R. AKEROYD

Chenopodium chenopodioides Saltmarsh Goosefoot

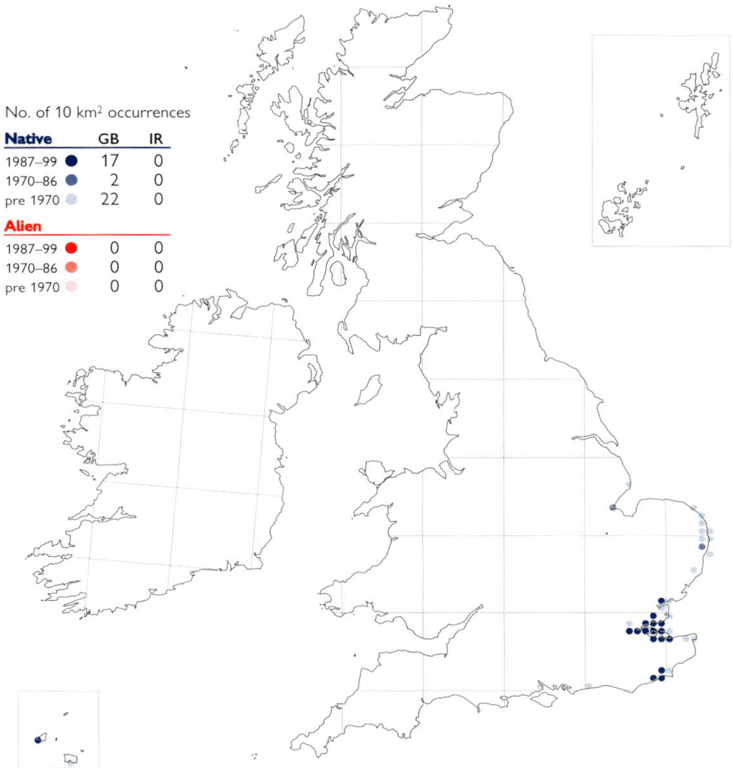

An annual of dry, brackish mud of ditches, salt-pans and hoof-marks in the upper part of saltmarshes and in coastal grazing marshes. Lowland.

Native (change –0.17). Since the 1962 *Atlas*, there has been no change in the distribution of *C. chenopodioides* in its core area of the Thames estuary, but it seems to have become extinct in Suffolk and Norfolk. It requires traditional marsh grazing regimes, and many sites have been converted to arable, or have become unsuitable as water levels in ditches have been stabilised after drainage.

Eurosiberian Temperate element, with a coastal distribution in W. Europe but in inland saline areas further east. It is widely naturalised outside its native range.

References: Atlas (83b), Hultén & Fries (1986), Jalas & Suominen (1980), Stewart *et al.* (1994), Wigginton (1999).

J. R. AKEROYD

Chenopodium polyspermum Many-seeded Goosefoot

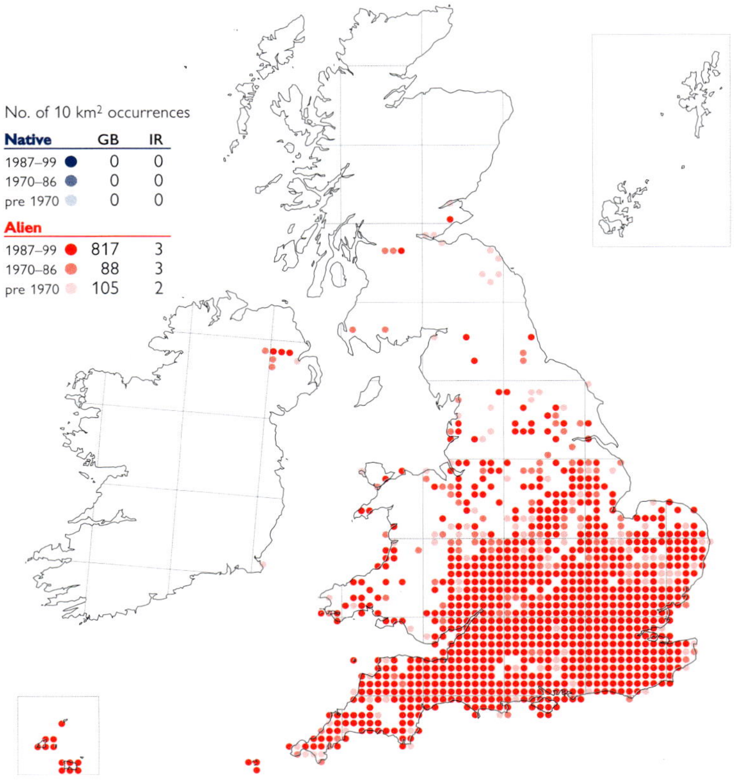

An annual of disturbed cultivated ground, including gardens, especially on light, nutrient-rich soils. It has also been recorded on bare peat, on disturbed soil in newly coppiced woodland and in dried up ponds. Lowland.

Archaeophyte (change +0.62). There has been no change in overall range since the 1962 *Atlas*, but there are many more records.

Eurosiberian Temperate element; widely naturalised outside its native range.

References: Atlas (81b), Hultén & Fries (1986), Jalas & Suominen (1980), Meusel *et al.* (1965).

J. R. AKEROYD

Chenopodium vulvaria Stinking Goosefoot

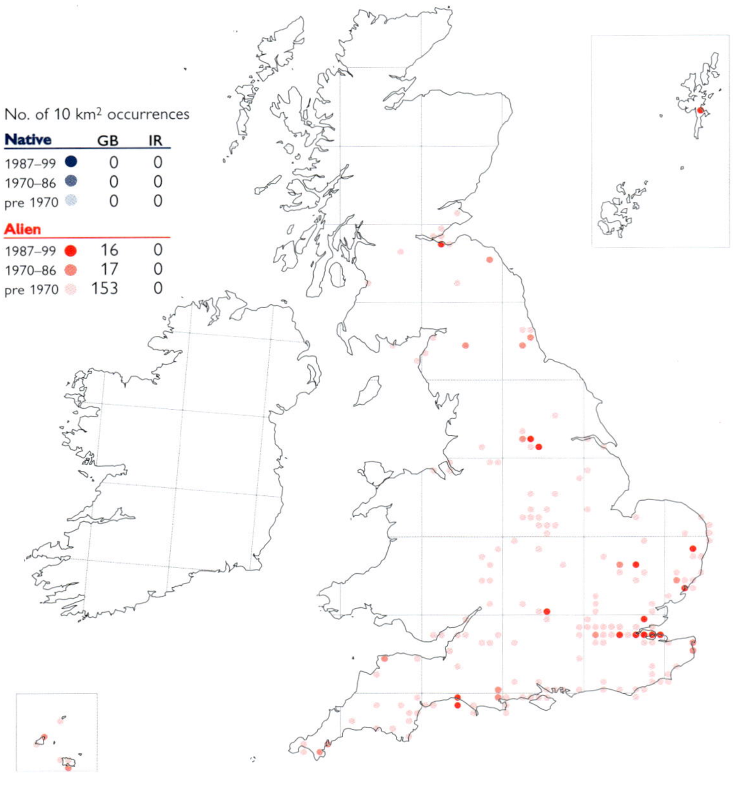

A foetid, often prostrate annual of disturbed, nutrient-rich soil on sandy shingle beaches, sand dunes and coastal cliffs, where the soil is enriched by the droppings of sea-birds. It was formerly a ruderal of places enriched with animal dung. Lowland.

Archaeophyte (change –2.60). This species had declined dramatically before 1930, perhaps because of the change from horse to tractor power, and the declining use of dung as a fertiliser. By 1930 it was virtually confined to coastal habitats, and even here it has continued to decline for reasons which are unclear.

As an archaeophyte *C. vulvaria* has a Eurosiberian Southern-temperate distribution; it is widely naturalised outside its native range.

References: Atlas (81c), Hultén & Fries (1986), Jalas & Suominen (1980), Meusel *et al.* (1965), Wigginton (1999).

J. R. AKEROYD

Chenopodium hybridum Maple-leaved Goosefoot

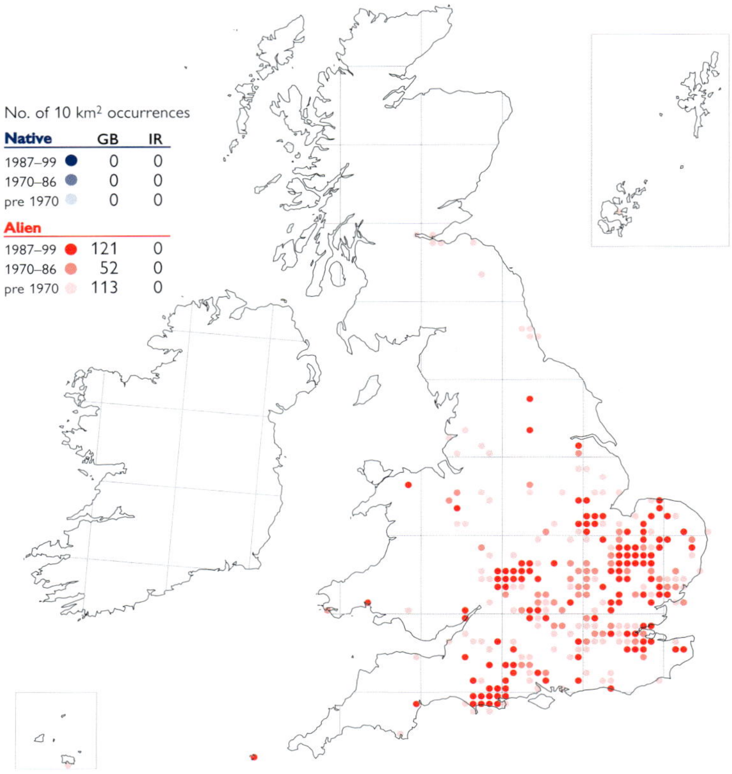

An annual of disturbed, nutrient-rich arable land and waste ground. It is a characteristic weed of humus-rich cultivated soils in the Fens. Lowland.

Archaeophyte (change –0.32). *C. hybridum* is known from archaeological evidence to have been present in Britain since Roman times. There has been no change in distribution since the 1962 *Atlas*.

As an archaeophyte *C. hybridum* has a Circumpolar Temperate distribution.

References: Atlas (82d), Hultén & Fries (1986), Jalas & Suominen (1980), Meusel *et al.* (1965).

J. R. AKEROYD

Chenopodium urbicum Upright Goosefoot

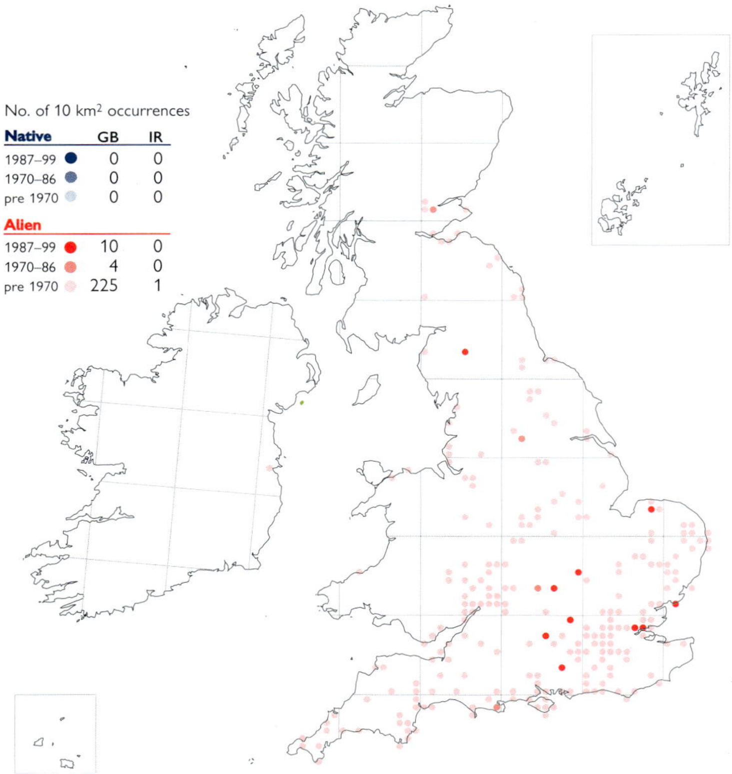

An annual of disturbed, nutrient-rich cultivated and waste ground, often occurring only as a casual. Lowland.

Archaeophyte (change –4.57). Until about 1940 *C. urbicum* occurred frequently as a casual, mainly as a seed impurity, and populations persisted in some areas, such as at Ridge in Dorset where it was present for over fifty years. Since 1970 it has only occurred as a rare casual.

As an archaeophyte *C. urbicum* has a Eurosiberian Temperate distribution.

References: Atlas (82c), Hultén & Fries (1986), Jalas & Suominen (1980).

J. R. AKEROYD

Chenopodium murale Nettle-leaved Goosefoot

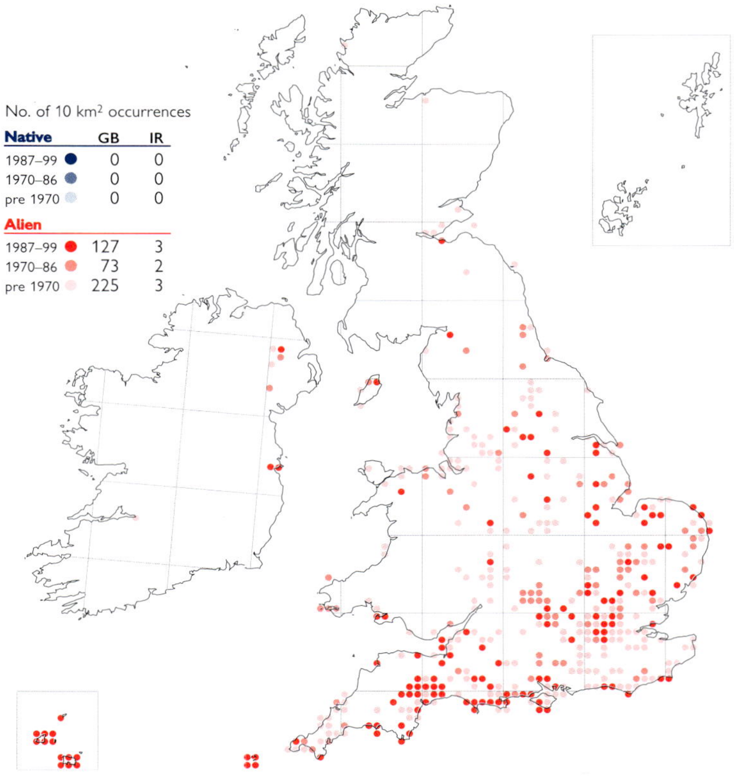

No. of 10 km² occurrences

Native	GB	IR
1987–99	0	0
1970–86	0	0
pre 1970	0	0
Alien		
1987–99	127	3
1970–86	73	2
pre 1970	225	3

An annual of disturbed, nutrient-rich cultivated and waste ground. Usually casual, it is long-established near the sea in S.E. England. Lowland.

Archaeophyte (change –1.63). *C. murale* has been present in Britain since Roman times. Like some other *Chenopodium* species it has declined markedly since the 1962 *Atlas*.

As an archaeophyte *C. murale* has a Eurosiberian Southern-temperate distribution; it is widely naturalised outside this range.

References: Atlas (82b), Hultén & Fries (1986), Jalas & Suominen (1980).

J. R. AKEROYD

Chenopodium ficifolium Fig-leaved Goosefoot

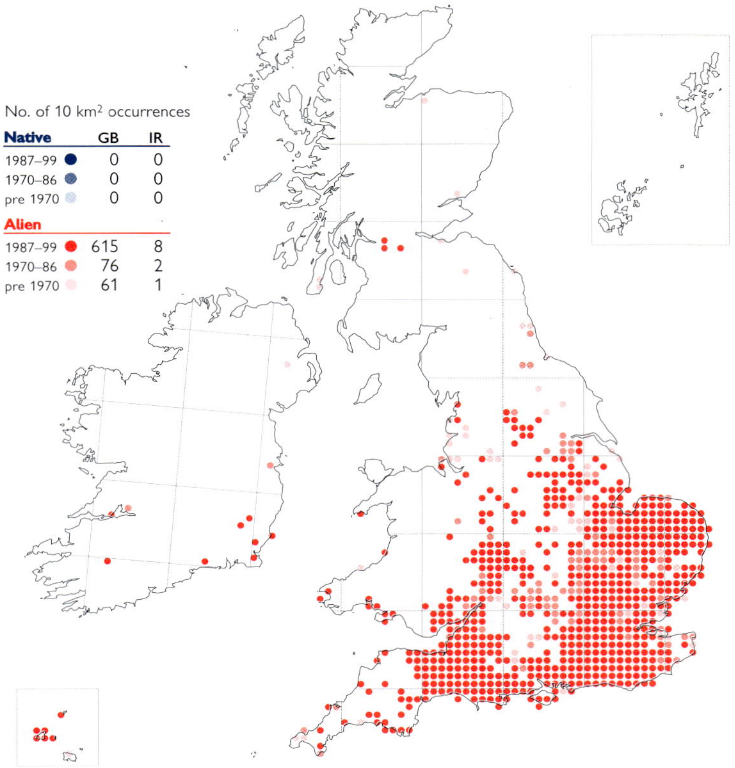

No. of 10 km² occurrences

Native	GB	IR
1987–99	0	0
1970–86	0	0
pre 1970	0	0
Alien		
1987–99	615	8
1970–86	76	2
pre 1970	61	1

An annual of disturbed, nutrient-rich ground, especially cultivated land on deep, fertile soils. It is usually casual but can be persistent in S. England and S. Wales. Lowland.

Archaeophyte (change +1.90). There is a continuous archaeological record of this species in Britain from the Iron Age onwards. Despite this, there were very few records of it in the 1962 *Atlas* away from its strongholds in the Fens and the London area. Even allowing for the possibility that it had been under-recorded elsewhere, it seems likely that this species has undergone a marked expansion of range since the 1950s. The reasons for this are far from clear.

As an archaeophyte *C. ficifolium* has a Eurosiberian Temperate distribution.

References: Atlas (82a), Jalas & Suominen (1980).

J. R. AKEROYD

Chenopodium opulifolium Grey Goosefoot

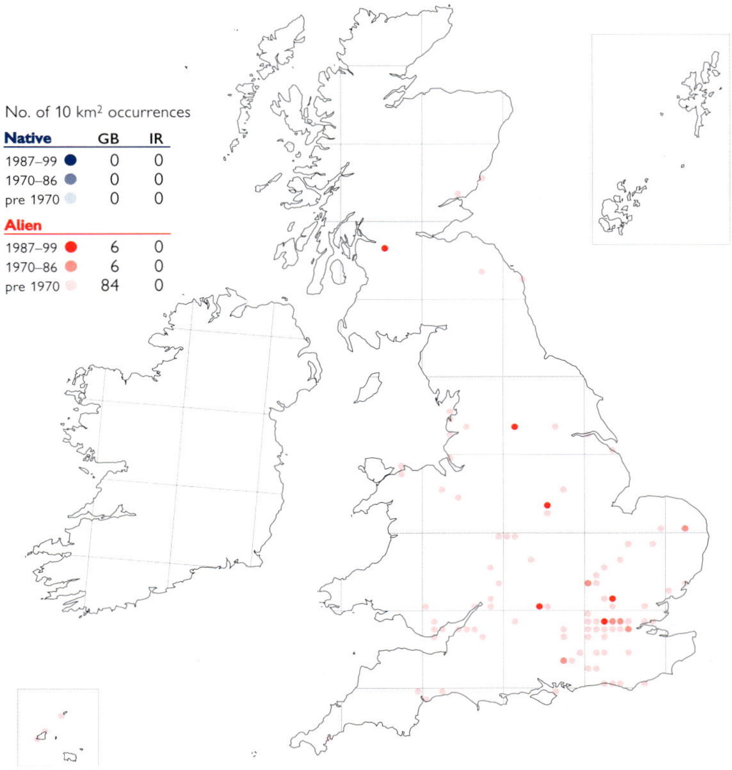

No. of 10 km² occurrences

Native	GB	IR
1987–99	0	0
1970–86	0	0
pre 1970	0	0
Alien		
1987–99	6	0
1970–86	6	0
pre 1970	84	0

An annual of waste ground and rubbish tips, which is introduced with grain, cork, wool and oil- and bird-seed. Lowland.

Neophyte. *C. opulifolium* was first recorded in the wild in Britain in 1853. It appears to be declining, perhaps as a result of improved seed cleaning techniques, but is much confused with variants of *C. album* and is possibly under-recorded.

A European Southern-temperate species, extending south to central Africa. It is difficult in many areas to distinguish native and alien occurrences.

References: Hultén & Fries (1986), Jalas & Suominen (1980).

J. R. AKEROYD

Chenopodium album agg. Fat-hen

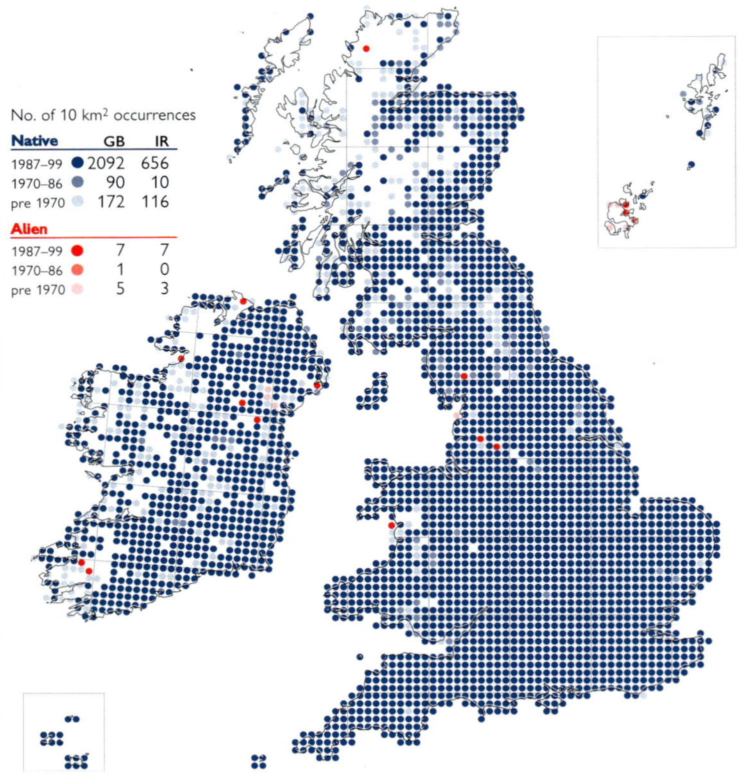

Annuals of disturbed, nutrient-rich habitats, including cultivated fields and gardens, manure heaps, waste ground, rubbish tips and soil heaps. Copious seed is produced, resulting in a long-lived seed bank. Generally lowland, but reaching 435 m in Clun Forest (Salop).

Native (change –0.73). Records of all eight species in the aggregate (species 20–27 in Stace, 1997) are included here, but *C. album* is by far the commonest. The distribution since the 1962 *Atlas* is stable in much of its range, though the decline in arable farming in W. Scotland has made it less frequent there.

Eurasian Wide-temperate element, but naturalised in N. America so distribution is now Circumpolar Wide-temperate.

References: Atlas (81d), Bassett & Crompton (1978), Grime *et al.* (1988), Hultén & Fries (1986), Jalas & Suominen (1980), Williams (1963).

J. R. AKEROYD

Bassia scoparia Summer-cypress

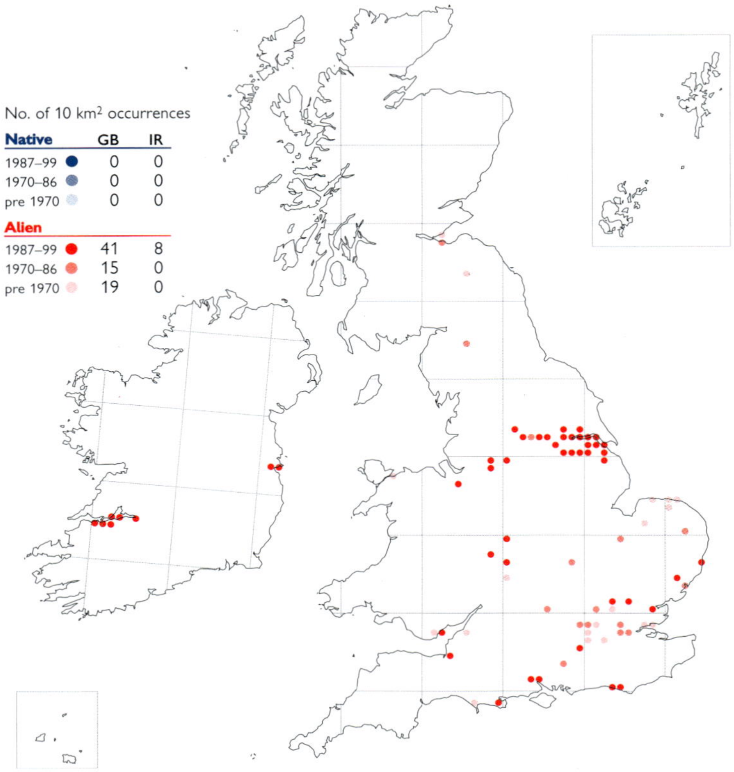

A bushy annual found in hedgerows and on roadsides, rubbish tips and waste ground, especially around ports, and in gravel-pits. It is sometimes naturalised on roadsides but is usually casual, being introduced with grain and carrot seed, oil-seed and wool-shoddy. It also occurs as a garden escape. Lowland.

Neophyte. *B. scoparia* was being cultivated in Britain by 1629, and is now frequent in gardens. Although it was first recorded from the wild in 1866 (Surrey), it remained rare until the early 1900s. It now appears to be increasing, especially on roadsides around the Humber estuary. It was first recorded in Co. Limerick in 1988, but does not appear to be spreading there.

Native of temperate Asia; widely naturalised, especially in S.E. Europe.

References: Bolòs & Vigo (1990), Cook (1997, 1998), Jalas & Suominen (1980).

H. J. KILLICK & J. R. AKEROYD

Atriplex hortensis Garden Orache

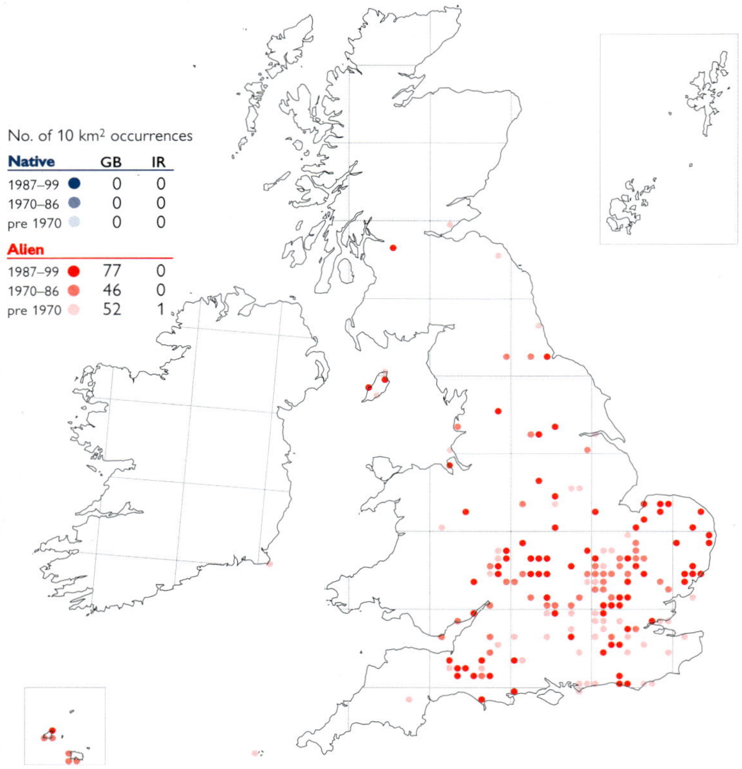

An erect annual occurring as a casual on waste ground, rubbish tips and roadsides. It is an occasional bird-seed alien, sometimes escaping from gardens and allotments. Lowland.

Neophyte. *A. hortensis* was introduced by 1548, and was widely cultivated as a food crop until the 18th century (Vaughan & Geissler, 1997). It is now grown for ornament, and rarely as a leaf-vegetable. It was first recorded from the wild in 1824 (Surrey), but current trends in distribution are difficult to assess.

Supposedly native of C. Asia; cultivated widely and naturalised in C. & S. Europe.

Reference: Hultén & Fries (1986).

S. J. LEACH

Atriplex prostrata Spear-leaved Orache

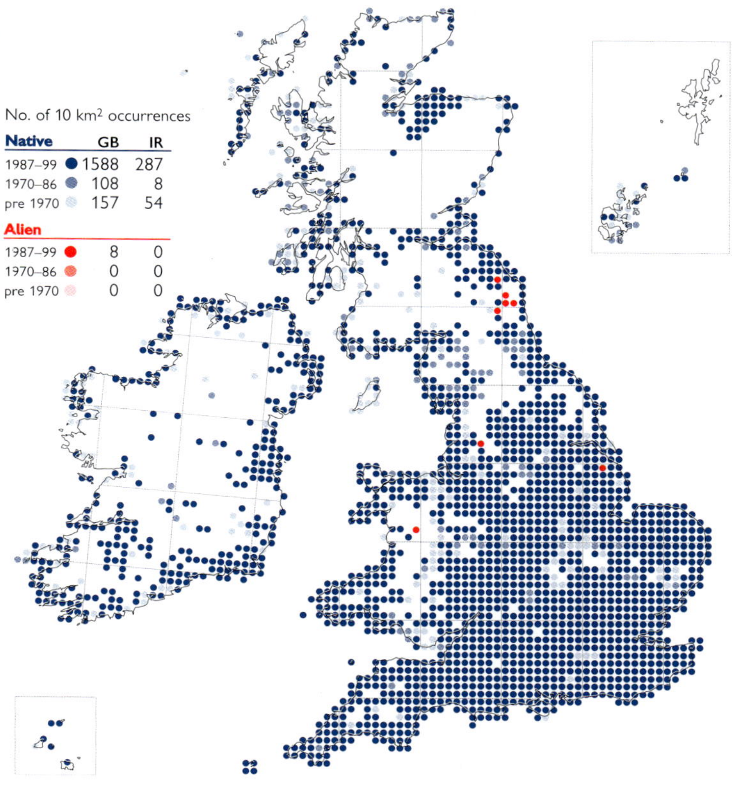

An annual of beaches, saltmarshes and other open, often wet, saline habitats near the sea; also inland in disturbed areas on moist, fertile, neutral soils, such as the trampled margins of ditches and ponds, on cultivated land, tips and waste ground. It also grows in inland saltmarshes and along salt-treated roadsides. Generally lowland, but reaching 415 m at Carter Bar (Roxburghs.).

Native (change +1.10). Since the 1962 *Atlas*, *A. prostrata* has been better recorded on the coasts of N. and W. Britain and Ireland, but its increased frequency inland may partly represent its spread along roads.

Eurosiberian Wide-temperate element; widely naturalised outside its native range.

References: Atlas (84c), Bassett & Munro (1987), Grime *et al.* (1988), Hultén & Fries (1986), Jalas & Suominen (1980), Taschereau (1985a).

S. J. LEACH

Atriplex glabriuscula Babington's Orache

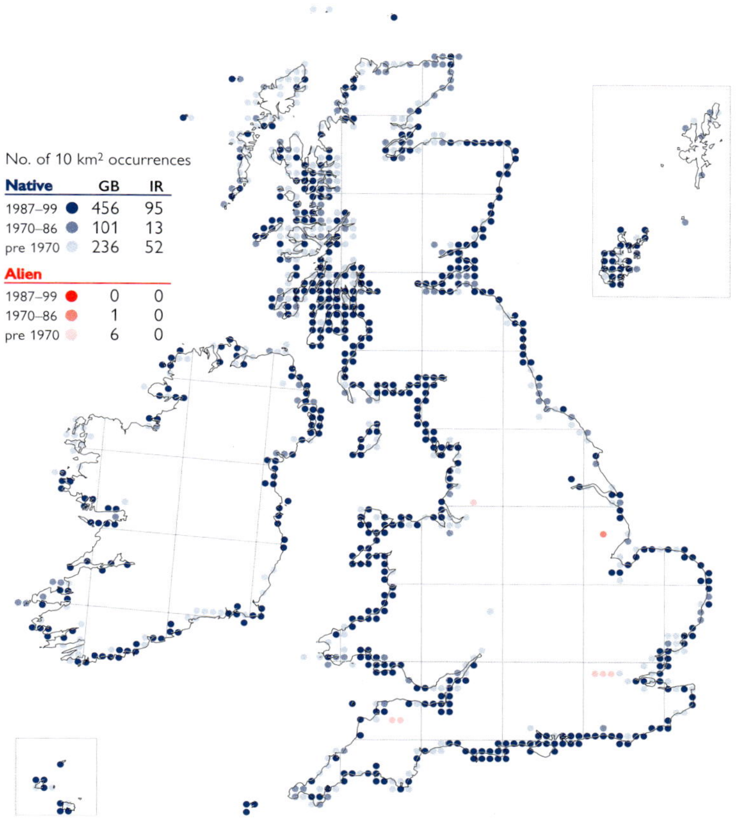

A procumbent annual found close to the strand-line on moderately exposed sand and shingle beaches, and in waste places near the sea. Lowland.

Native (change –0.93). It is difficult to interpret trends in the distribution of this species. *A. glabriuscula* is difficult to identify and is more comprehensively recorded in some areas than others. Some old records are probably unreliable due to confusion with other coastal *Atriplex* taxa, especially *A.* × *taschereaui* (*A. glabriuscula* × *A. longipes*) and *A. prostrata*.

Suboceanic Boreo-temperate element; also in N. America.

References: Atlas (84d), Hultén & Fries (1986), Jalas & Suominen (1980), Meusel *et al.* (1965), Taschereau (1985a).

S. J. LEACH

Atriplex longipes Long-stalked Orache

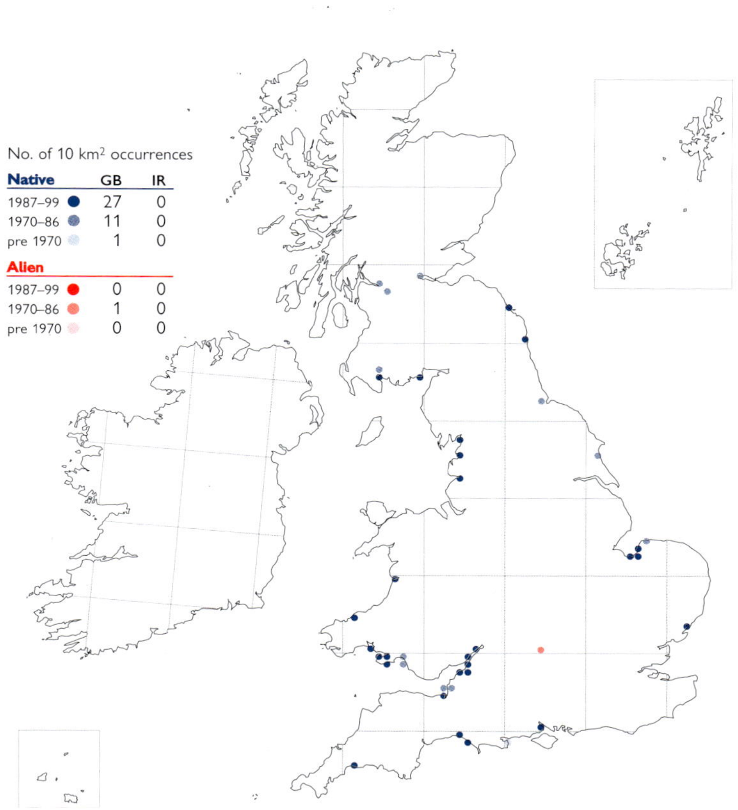

An erect to procumbent annual occurring in the upper parts of silty estuarine salt-marshes, usually in tall, ungrazed vegetation (including stands of *Juncus maritimus* and *Phragmites australis*) inundated by brackish water during high spring tides. Lowland.

Native. *A. longipes* was only confirmed as a British species in 1977, although it was first reported in 1957. It is almost certainly still under-recorded.

European Boreal-montane element.

References: Jalas & Suominen (1980), Jones (1975), Stewart *et al.* (1994), Taschereau (1977, 1985a, 1985b).

S. J. LEACH

Atriplex praecox Early Orache

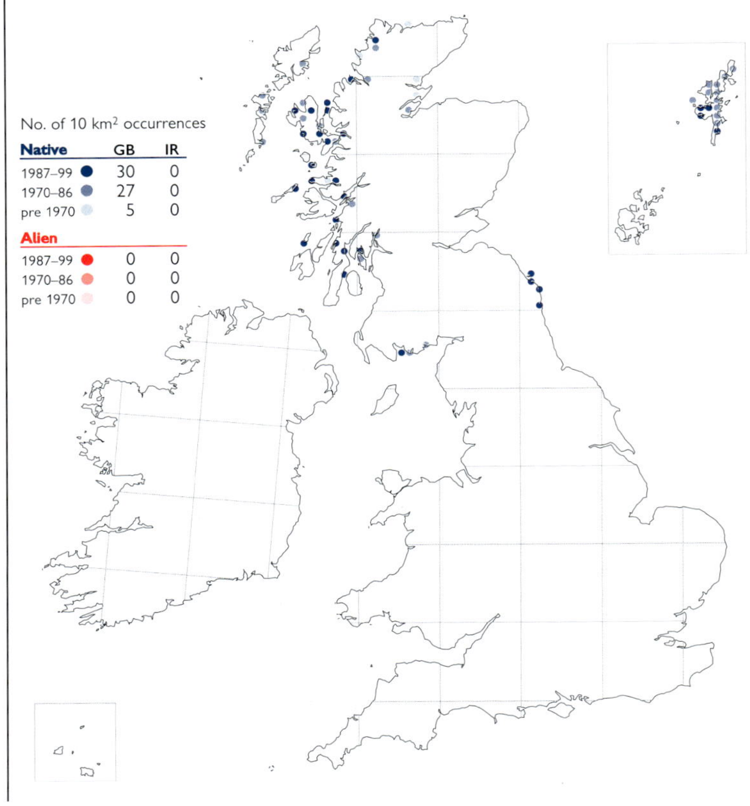

A small annual occurring on sand and shingle beaches around the margins of sea-lochs and other sheltered inlets and bays. It grows in the lowest part of sparsely vegetated strandlines below, for example, the zone normally occupied by *Cakile maritima* and other *Atriplex* species. Lowland.

Native. *A. praecox* was only recognised as a British species in 1975, and is almost certainly under-recorded. In Scandinavia this taxon is usually treated as *A. longipes* subsp. *praecox*.

European Boreal-montane element; also in Greenland.

References: Jalas & Suominen (1980), Stewart *et al.* (1994), Taschereau (1977, 1985a).

S. J. LEACH

Atriplex littoralis Grass-leaved Orache

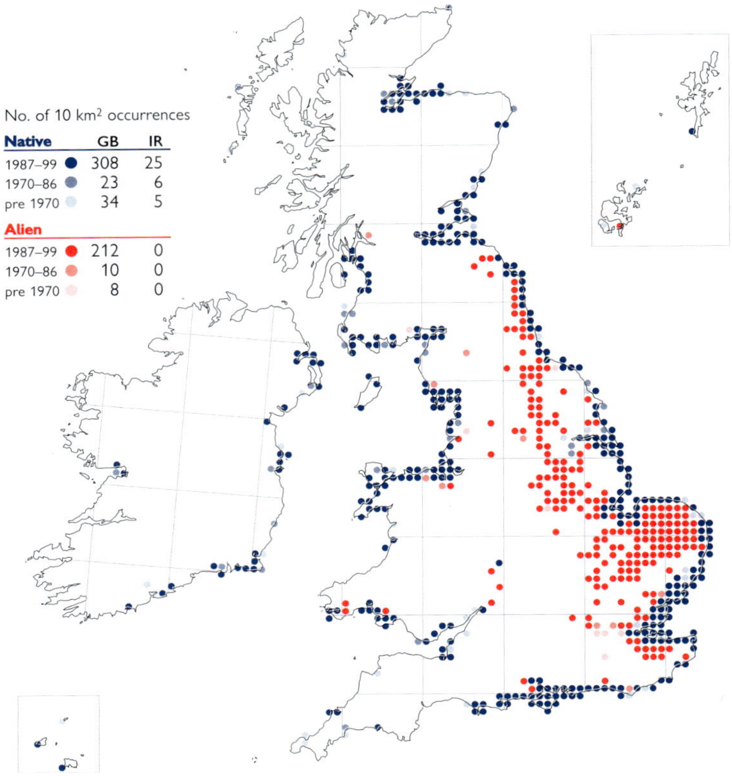

An erect annual of open, usually sandy or silty places near the sea, often forming dense stands along saltmarsh drift-lines, estuarine banks and sea-walls, and on waste ground around docks. It also grows in saline areas inland, and as a colonist by salt-treated roads. Lowland.

Native (change +1.59). The coastal distribution of *A. littoralis* has remained largely unchanged since the 1962 *Atlas*, but from the early 1980s onwards it spread rapidly along inland roadsides, particularly in E. England.

Circumpolar Temperate element, with a disjunct distribution.

References: Atlas (84a), Bungard & Leach (1991), Hultén & Fries (1986), Jalas & Suominen (1980), Leach (1999), Meusel *et al.* (1965), Taschereau (1985a).

S. J. LEACH

Atriplex patula Common Orache

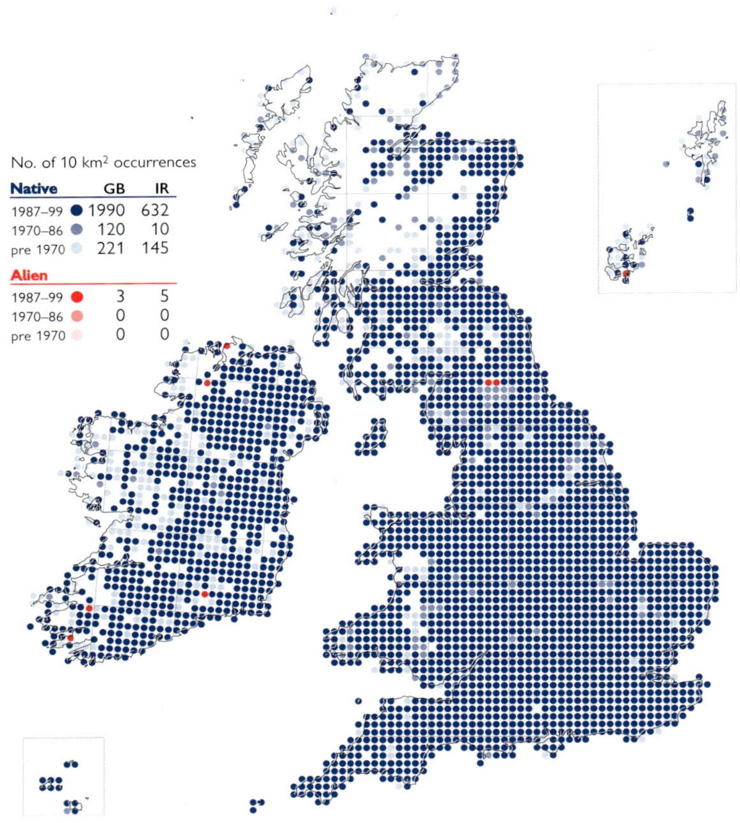

An annual of cultivated ground, manure heaps, roadsides, rubbish tips and waste places in towns and cities; also on fertile soils in a wide range of disturbed semi-natural habitats, such as river banks, pond margins and sea-bird cliffs. *A. patula* is frequent in coastal waste places but rare in littoral zone habitats such as saltmarshes and sand and shingle drift-lines. Mainly lowland, but reaching 435 m in Clun Forest (Salop).

Native (change −0.34). There is little evidence of any change since the 1962 *Atlas*, where *A. patula* was mapped as 'all records'.

Eurosiberian Wide-temperate element; widely naturalised outside its native range.

References: Atlas (84b), Bassett & Munro (1987), Grime *et al.* (1988), Hultén & Fries (1986), Jalas & Suominen (1980), Taschereau (1985a).

S. J. LEACH

Atriplex laciniata Frosted Orache

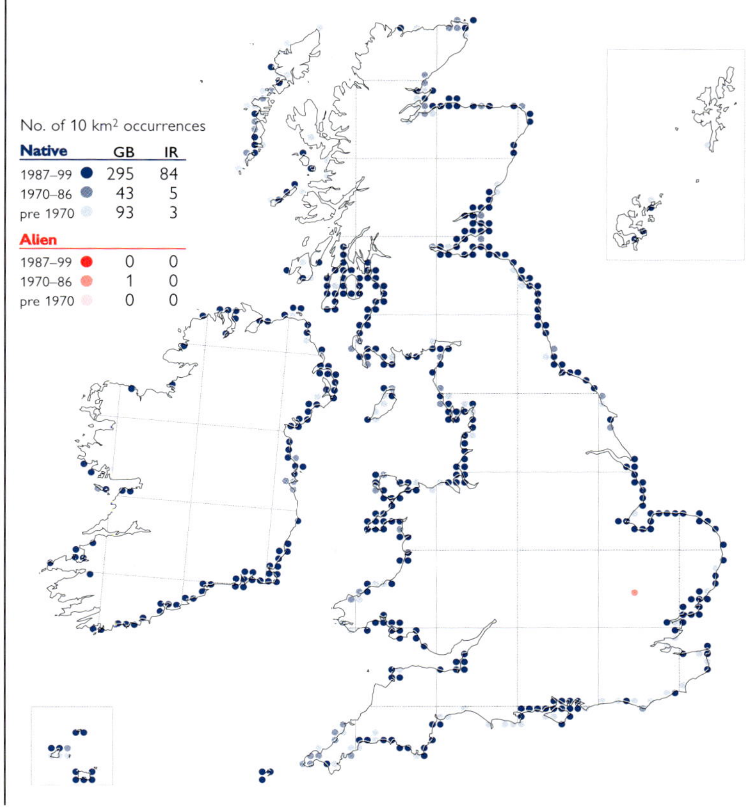

A decumbent, widely spreading annual of sand and shingle beaches, more rarely found on saltmarsh-sand dune transitions and saltmarsh drift-lines. Populations are usually small and often of sporadic appearance. *A. laciniata* typically occurs with *Cakile maritima*, *Salsola kali* and other *Atriplex* species in a mixed strandline community. Lowland.

Native (change +0.38). There has been little change since the 1962 *Atlas*, though many populations, particularly those close to popular coastal resorts in S. England, have dwindled or been lost altogether due to excessive recreational pressure.

Oceanic Temperate element; also in N. America.

References: Atlas (85a), Hultén & Fries (1986), Jalas & Suominen (1980), Taschereau (1985a).

S. J. LEACH

Atriplex portulacoides Sea-purslane

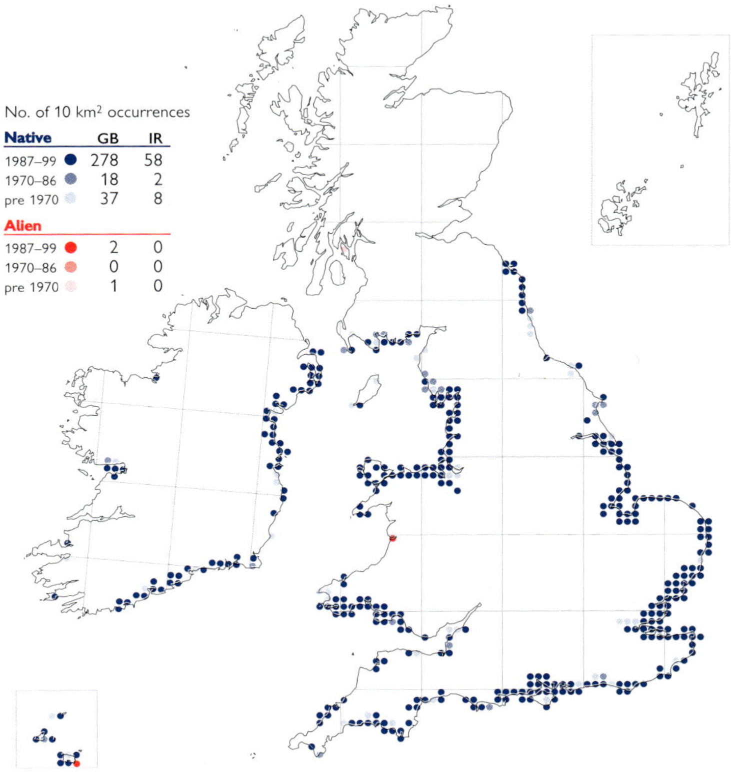

A low shrub of muddy or sandy saltmarshes, commonly fringing intertidal pools and creeks, and often forming extensive stands on ungrazed saltings. In W. Britain and Ireland it also occurs locally on coastal rocks and cliffs. Lowland.

Native (change +0.06). This species markedly extended its range northwards in the British Isles during the 20th century, as shown by its spread in N. Ireland (Leach, 1989) and its recent resurgence on the Isle of Man (Allen, 1984). This has been paralleled by similar changes elsewhere in N. Europe, for example, in Sweden (Blomgren, 1992).

Mediterranean-Atlantic element.

References: Atlas (85b), Akeroyd & Preston (1984), Chapman (1950), Hultén & Fries (1986), Jalas & Suominen (1980), Meusel *et al.* (1965).

S. J. LEACH

Atriplex pedunculata Pedunculate Sea-purslane

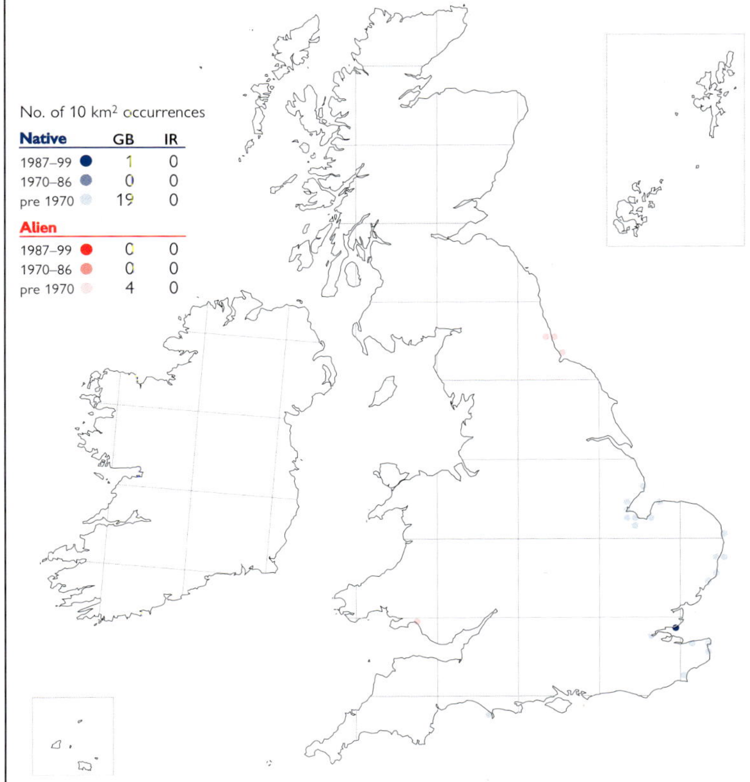

An annual occurring in the drier parts of saltmarshes; elsewhere in N. Europe (and formerly in Britain) in tidally inundated dune-slacks and saltmarsh-shingle transitions, and rarely as a casual in other open, disturbed saline areas near the sea. Lowland.

Native. *A. pedunculata* had been presumed extinct in the British Isles since the late 1930s until it was discovered near Shoeburyness, S. Essex, in 1987 (Leach, 1988). It is threatened there by encroaching stands of *Elytrigia atherica*, but further colonies and sites have been established using pot-grown plants and seeds derived from the original colony.

Eurosiberian Temperate element; at inland saline sites from C. Europe eastwards.

References: Atlas (85c), Gibson (2000), Hultén & Fries (1986), Jalas & Suominen (1980), Meusel *et al.* (1965), Wigginton (1999).

S. J. LEACH

Beta vulgaris subsp. *maritima* Sea Beet

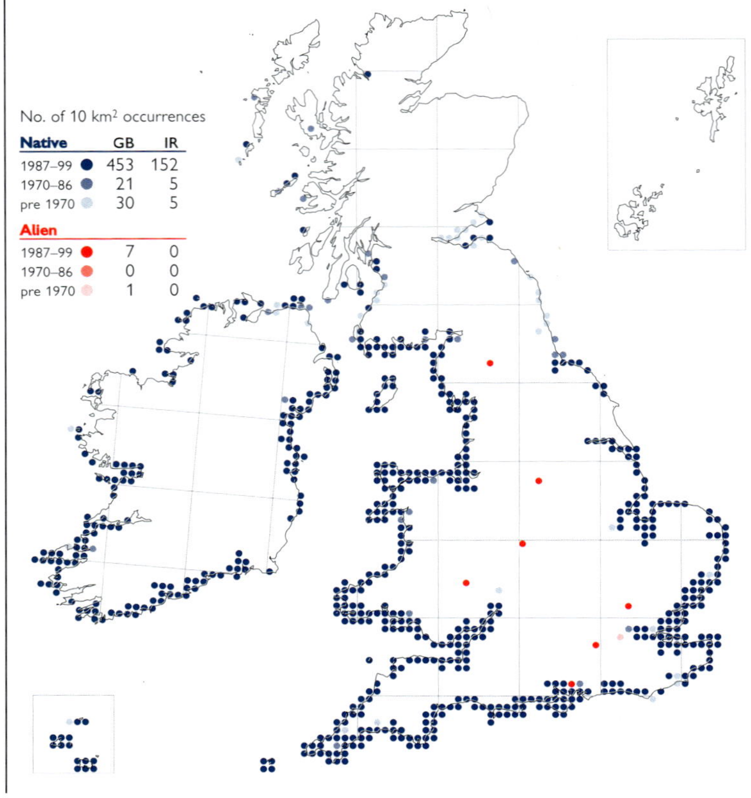

A much-branched perennial herb found on coastal rocks and cliffs, saltmarsh drift-lines, sea-walls, and on sand and shingle beaches, favouring nutrient-enriched sites such as sea-bird cliffs and coastal paths popular with dog-walkers. It also occurs on waste ground near the sea and, rarely, inland as a casual of rubbish tips and roadsides. Lowland.

Native (change for species +1.23). There has been little change in the range of this taxon since the 1962 *Atlas*. It appears to be more frequent throughout much of its range, perhaps because of better recording.

Mediterranean-Atlantic element.

References: Atlas (83d), Hultén & Fries (1986), Jalas & Suominen (1980), Meusel *et al.* (1965).

S. J. LEACH

Beta vulgaris subsp. *vulgaris* Root Beet

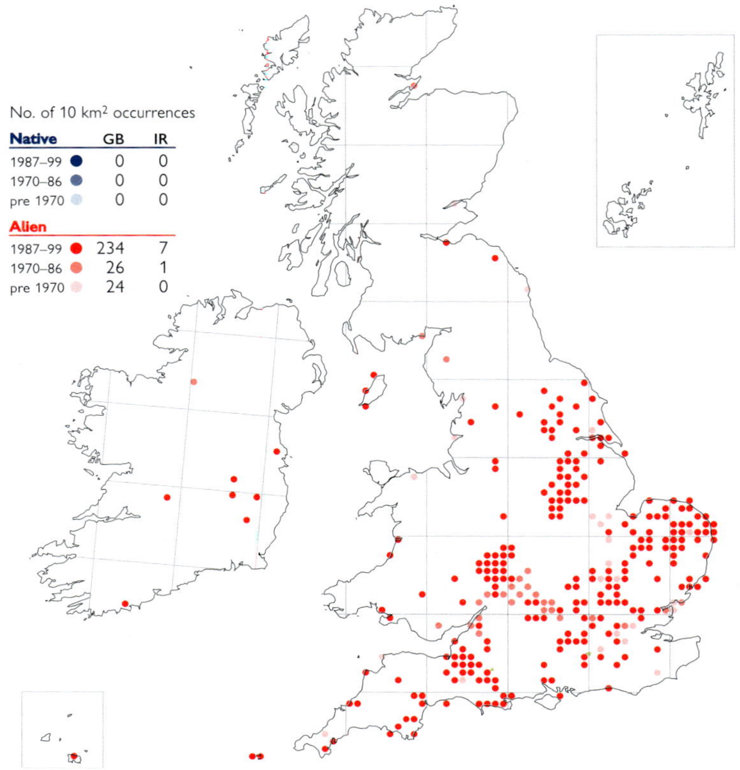

A little-branched annual to biennial herb, found as a casual on rubbish tips and waste ground and as a relic of cultivation. Lowland.

Neophyte (change for species +1.23). This taxon was being cultivated in Britain by 1548 and was known from the wild by at least 1905. Many forms are grown as root vegetables, including Beetroot, Sugar Beet, Fodder Beet and Mangel-wurzel. Recording of the casual occurrences of these crops has been uneven, and the map almost certainly underestimates their distribution.

This subspecies comprises a range of variants which presumably arose by selection in cultivation.

References: Smart & Simmonds (1995), Vaughan & Geissler (1997), Zohary & Hopf (2000).

S. J. LEACH

Sarcocornia perennis Perennial Glasswort

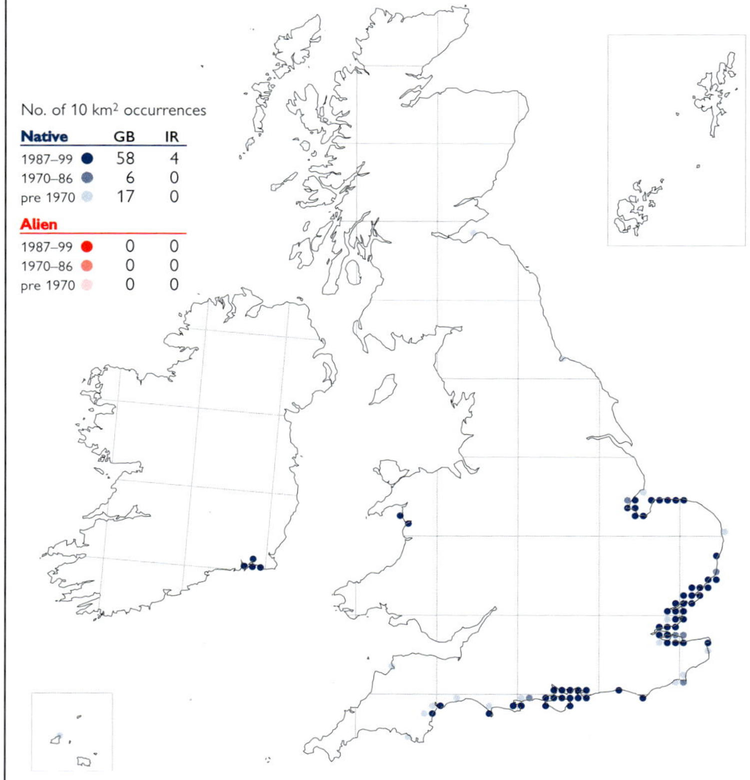

A woody perennial subshrub of saltmarshes, especially in bare or sparsely vegetated areas on firm, muddy sand and gravel. *S. perennis* occurs on eroding lower parts of saltmarshes, at higher elevations on saltmarsh drift-lines and on shell and shingle banks; sometimes also on bare ground behind sea-walls. Lowland.

Native (change –0.22). There has been little change in distribution since the 1962 *Atlas*. At many localities *S. perennis* is found in small and fragmentary stands or as just a few widely scattered bushes, suggesting that it could still be present but undetected in some of the squares for which there are only old records.

Mediterranean-Atlantic element.

References: Atlas (86c), Bolòs & Vigo (1990), Curtis & McGough (1988), Jalas & Suominen (1980), Stewart *et al.* (1994).

S. J. LEACH

Salicornia agg. Glassworts

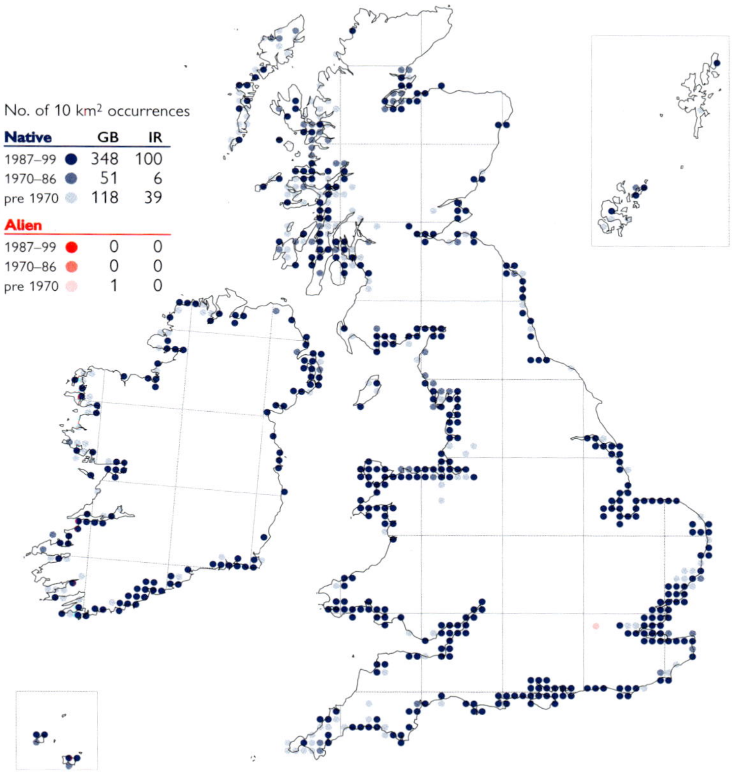

No. of 10 km² occurrences

Native	GB	IR
1987–99 ●	348	100
1970–86 ●	51	6
pre 1970 ●	118	39
Alien		
1987–99 ●	0	0
1970–86 ●	0	0
pre 1970 ●	1	0

These annuals grow in a variety of coastal habitats, including saltmarshes, sand, muddy shingle, creeks and brackish fields behind sea-walls. Lowland.

Native (change –0.82). The genus *Salicornia* is acknowledged to be very difficult taxonomically. Live material is essential for serious study, and it has proved hard to obtain a taxonomical or nomenclatural consensus between different authorities in the British Isles or in N.W. Europe. The following species maps must be regarded as provisional.

The genus has an almost cosmopolitan distribution; the world distributions of the species are given in the accounts below.

References: Atlas (86d), Dalby (1989), Ingrouille (1989), Ingrouille & Pearson (1987), Jalas & Suominen (1980), Rich & Jermy (1998), Rose (1989).

S. J. LEACH

Salicornia pusilla One-flowered Glasswort

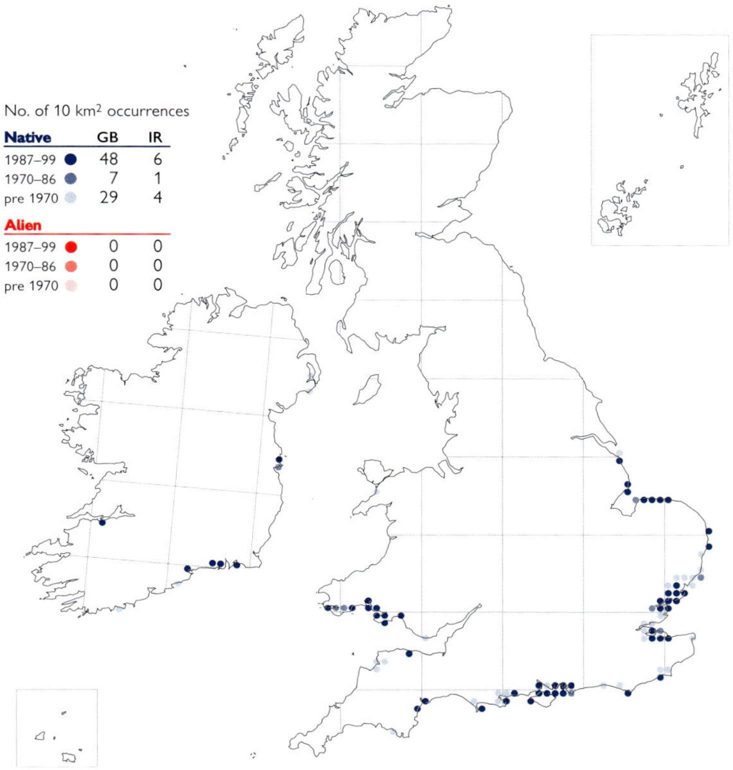

No. of 10 km² occurrences

Native	GB	IR
1987–99 ●	48	6
1970–86 ●	7	1
pre 1970 ●	29	4
Alien		
1987–99 ●	0	0
1970–86 ●	0	0
pre 1970 ●	0	0

An annual restricted to the uppermost parts of saltmarshes, growing on firm mud or sand in salt-pans and on the drift-line; also behind sea-walls in areas subject to saltwater seepage and kept open by vehicles or livestock. Lowland.

Native (change –0.21). The increase in records since the 1962 *Atlas* is due to more comprehensive recording. In parts of S.E. England (e.g. Hampshire), habitat loss, either through successional changes (e.g. spread of *Spartina anglica*) or agricultural land-claim, may have been an important factor in the demise of some populations.

Oceanic Temperate element.

References: Atlas (87a), Ball & Tutin (1959), Dalby (1989), Jalas & Suominen (1980), Ingrouille (1989), Rich & Jermy (1998), Stewart *et al.* (1994).

S. J. LEACH

Salicornia ramosissima Purple Glasswort

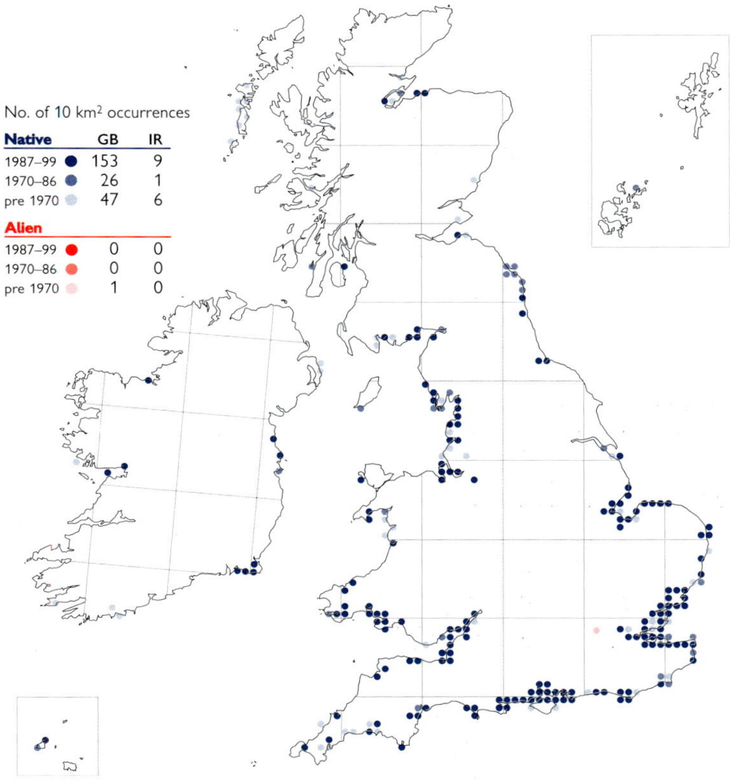

No. of 10 km² occurrences

Native	GB	IR
1987–99 ●	153	9
1970–86 ●	26	1
pre 1970 ●	47	6
Alien		
1987–99 ●	0	0
1970–86 ●	0	0
pre 1970 ●	1	0

This morphologically highly variable annual is usually found in the middle and upper zones of saltmarshes, in closed *Puccinellia maritima* swards, salt-pans, creeks and drift-lines. It also occurs on firm sand and muddy shingle, and behind sea-walls in open areas of brackish grazing marsh. Ball & Tutin (1959) note that this species occurs 'in all parts of saltmarshes except the lower mud-flats'. Lowland.

Native. *S. ramosissima* was not mapped in the 1962 *Atlas* and there is no reliable information on trends in its distribution. There is much confusion between this species and *S. europaea*.

Suboceanic Southern-temperate element.

References: Dalby (1989), Ingrouille (1989), Rich & Jermy (1998).

S. J. LEACH

Salicornia europaea Common Glasswort

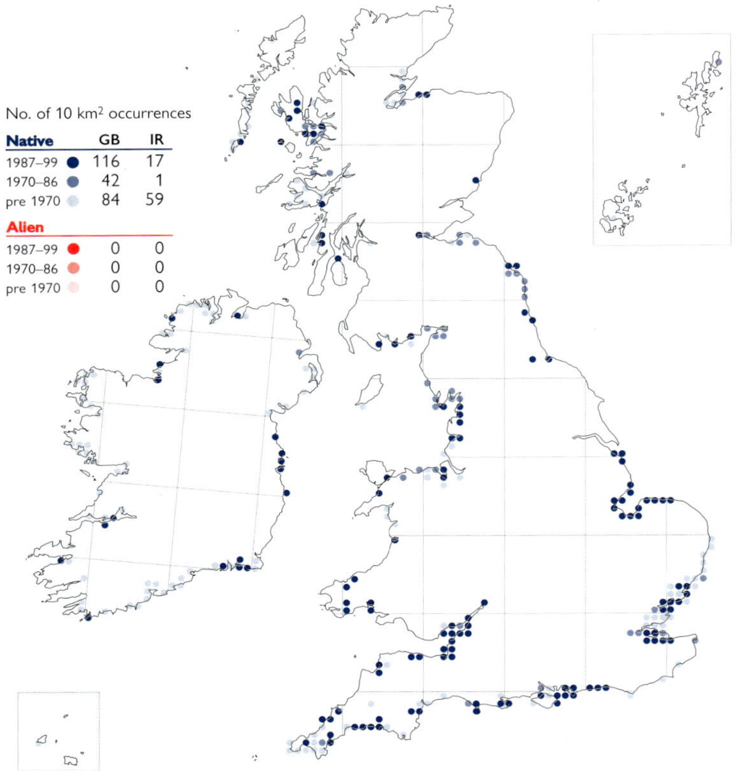

An annual found at all levels of sandy or muddy saltmarshes, in saltmarsh-sand dune transitions and wet, tidally inundated dune-slacks; also, more rarely, in relict saltmarsh and other open saline areas behind sea-walls. Halliday (1997) notes that in Cumbria *S. europaea* is probably the only *Salicornia* species able to persist on grazed marshes. Lowland.

Native. This taxon was not mapped in the 1962 *Atlas* and there is no reliable information on trends in its distribution. It is uncertain whether *S. europaea* has been under- or over-recorded, and in the past it has been confused with *S. ramosissima*.

Circumpolar Wide-temperate element.

References: Ball & Tutin (1959), Dalby (1989), Ingrouille (1989), Rich & Jermy (1998).

S. J. LEACH

Salicornia obscura Glaucous Glasswort

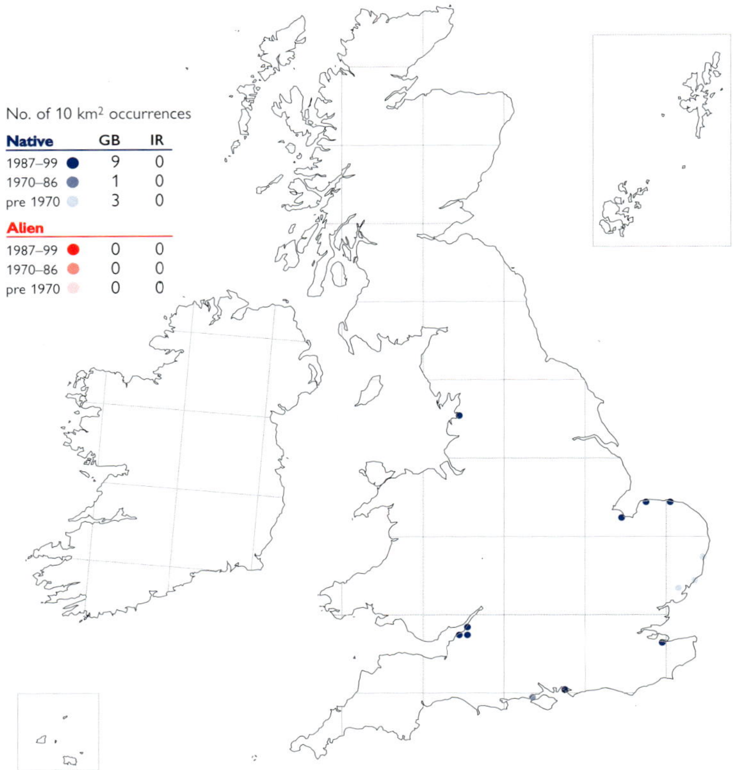

An annual of bare damp mud in saltmarsh pans, creeks and runnels. Brewis *et al.* (1996) report that, in Hampshire, *S. obscura* prefers the sloping sides of upper salt-marsh creeks. Lowland.

Native. This taxon, described as a new species by Ball & Tutin (1959), was not mapped in the 1962 *Atlas*. *S. obscura* has been lost from some sites, including the type locality on Hayling Island (Brewis *et al.*, 1996), due to agricultural land-claim and other coastal developments. It is almost certainly under-recorded, with some populations probably being mis-identified or overlooked as *S. europaea*.

Oceanic Temperate element.

References: Dalby (1989), Ingrouille (1989), Rich & Jermy (1998).

S. J. LEACH

Salicornia nitens Shiny Glasswort

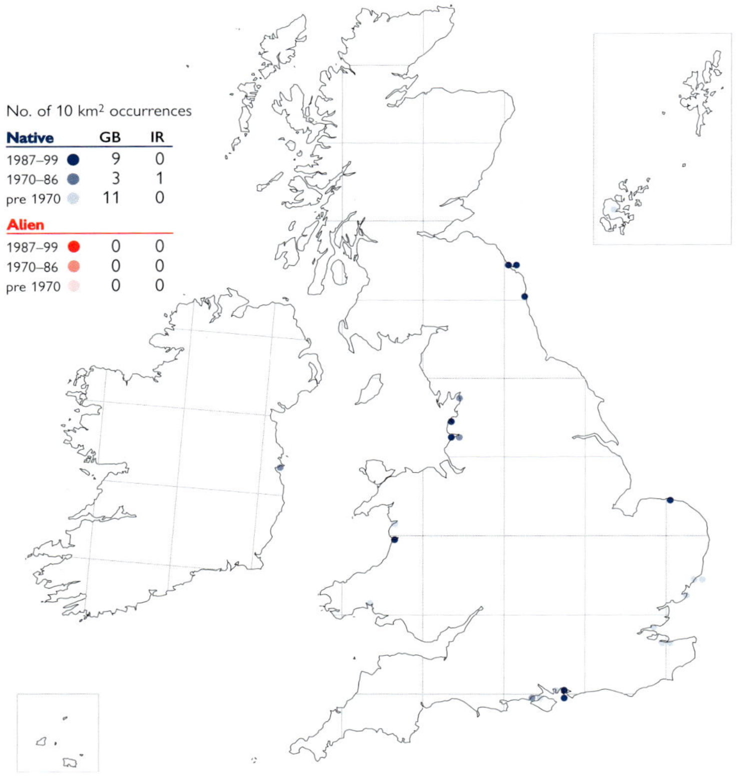

An annual which, in contrast to other members of the *S. procumbens* group (*S. dolichostachya*, *S. fragilis*), occurs mainly in the middle and upper parts of saltmarshes, especially in bare mud associated with salt-pans and high-level runnels. Lowland.

Native. *S. nitens* was described as a new species by Ball & Tutin (1959), but was not mapped in the 1962 *Atlas*. There is no reliable information on trends in its distribution, though it is known to have been lost from several sites in S. England due to saltmarsh reclamation. It is almost certainly under-recorded.

Oceanic Southern-temperate element; its European distribution is incompletely understood because of taxonomic uncertainties.

References: Dalby (1989), Ingrouille (1989), Rich & Jermy (1998).

S. J. LEACH

Salicornia fragilis Yellow Glasswort

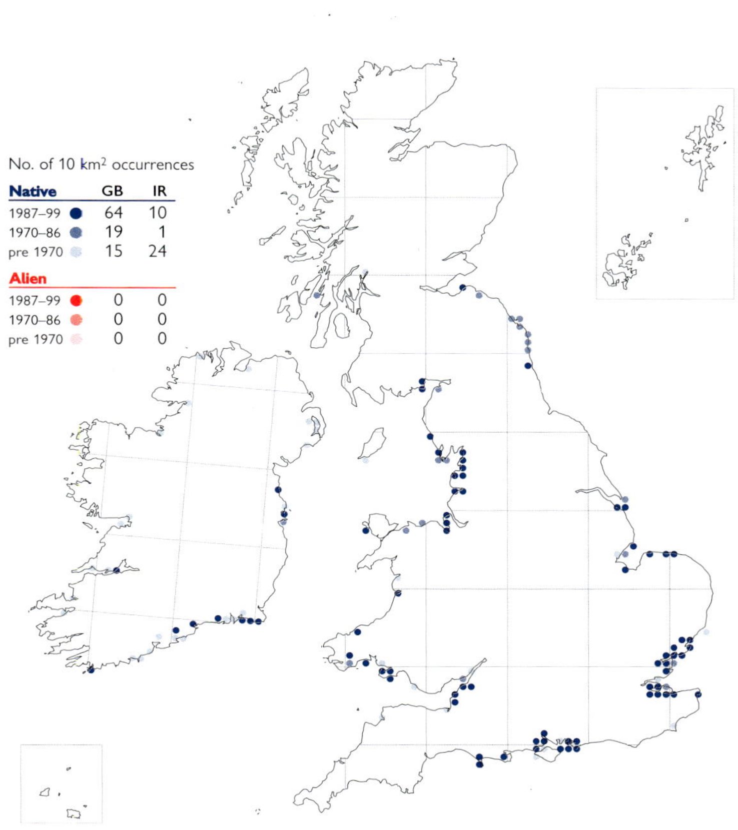

An annual largely restricted to open mud and muddy sand on intertidal flats and in the lowest parts of saltmarshes. Lowland.

Native. *S. fragilis* was described as a new species by Ball & Tutin (1959), but was not mapped in the 1962 *Atlas*. It is almost certainly under-recorded, and perhaps also misidentified as *S. dolichostachya* which occurs in similar habitats. Some authorities consider that it is a variety of the latter species (Rich & Jermy, 1998).

Oceanic Temperate element.

References: Dalby (1989), Ingrouille (1989).

S. J. LEACH

Salicornia dolichostachya Long-spiked Glasswort

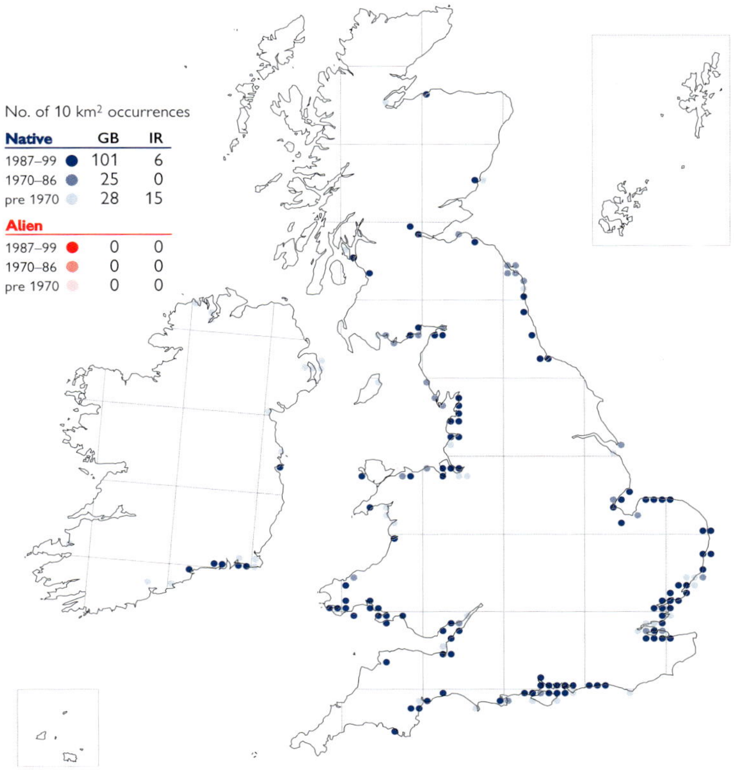

An annual of open mud and muddy sand on intertidal flats and in the lowest parts of saltmarshes, and occasionally in the mid-marsh along the banks of saltmarsh creeks and runnels. Lowland.

Native. *S. dolichostachya* was not mapped in the 1962 *Atlas* and there is no reliable information on trends in its distribution. It is almost certainly under-recorded, being overlooked or else dismissed, along with other glassworts, as '*Salicornia* agg'.

European Boreo-temperate element.

References: Ball & Tutin (1959), Dalby (1989), Ingrouille (1989), Rich & Jermy (1998).

S. J. LEACH

Suaeda vera Shrubby Sea-blite

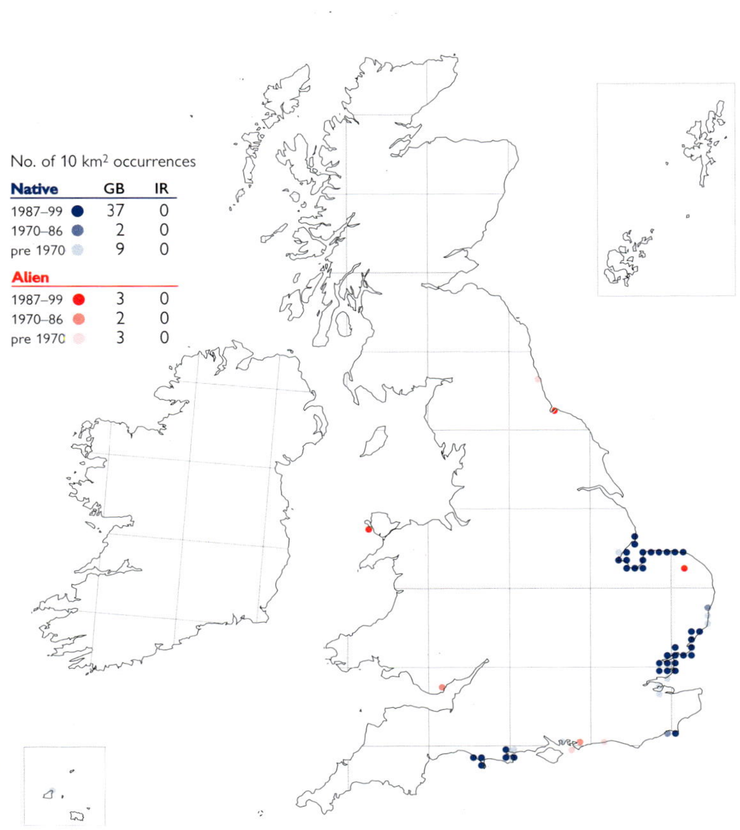

An evergreen shrub of shingle drift-lines and the dry upper zones of saltmarshes, especially where these adjoin shingle banks or sand dunes; also along sea-wall drift-lines and, more rarely, beside brackish creeks and ditches in coastal grazing marshes. Lowland.

Native (change –0.11). There is little evidence of any recent change in distribution. Some populations have probably been lost in the past through agricultural reclamation. It is probably alien in Anglesey; if it were native (Rich & Brown, 2000) it would represent a remarkable extension of its range.

Mediterranean-Atlantic element.

References: Atlas (86a), Bolòs & Vigo (1990), Chapman (1947b), Jalas & Suominen (1980), Stewart *et al.* (1994).

S. J. LEACH

Suaeda maritima Annual Sea-blite

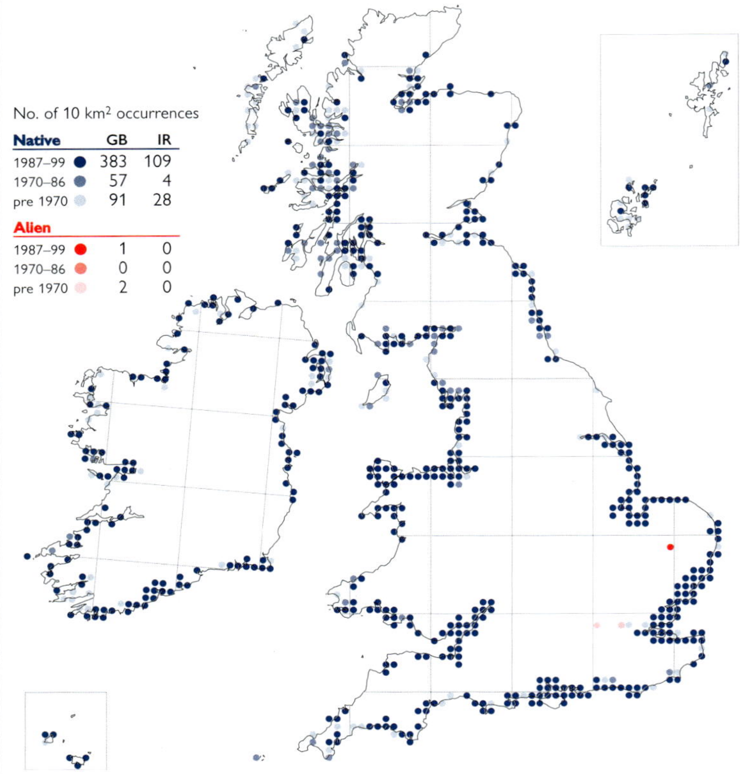

An annual found in the middle and lower parts of saltmarshes, often with *Salicornia* species. It is an early colonist of intertidal mud- and sand-flats, sometimes also occurring higher up in salt-pans and drift-lines, on shell and shingle banks, and in thinly vegetated brackish areas behind sea-walls. Lowland.

Native (change –0.47). There is no evidence of any change in the distribution of this species since the 1962 *Atlas*.

Eurasian Southern-temperate element; widely naturalised outside its native range.

References: Atlas (85d), Adam (1990), Chapman (1947a), Hultén & Fries (1986), Jalas & Suominen (1980).

S. J. LEACH

Salsola kali subsp. kali Prickly Saltwort

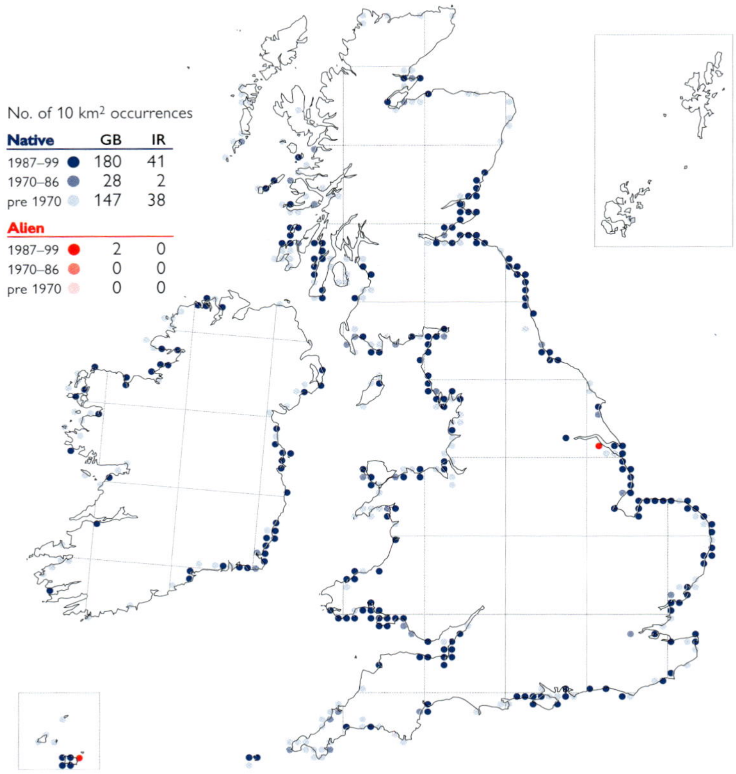

A somewhat woody annual of sand and shingle beaches, usually on the drift-line with *Atriplex laciniata*, *Cakile maritima* and *Honckenya peploides* as typical associates. Lowland.

Native (change for species –0.61). *S. kali* subsp. *kali* has declined since the 1962 *Atlas*. Many populations, especially in S. England, have been lost due to excessive recreational pressures, and in some areas there has been a drastic decline (e.g. Cornwall and Hampshire).

Eurosiberian Southern-temperate element; widely naturalised outside its native range.

References: Atlas (86b), Brewis *et al.* (1996), French *et al.* (1999), Hultén & Fries (1986), Jalas & Suominen (1980), Meusel *et al.* (1965).

S. J. LEACH

Salsola kali subsp. ruthenica Spineless Saltwort

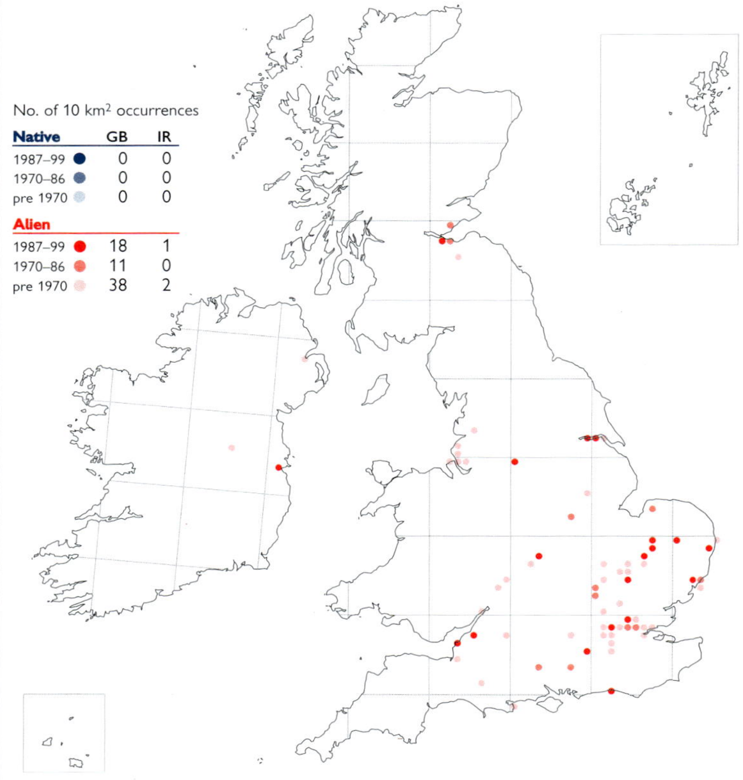

A bushy annual of waste ground, rubbish tips and arable fields, introduced with wool shoddy, bird-seed and as a contaminant of grain and agricultural seed. It is usually casual, but may persist at some sites (e.g. in S. Essex). Lowland.

Neophyte. This subspecies was first recorded in 1875 (Oxon). It remains a rare casual.

This taxon is known to occur in C. & S. Europe and in Asia, but the distributions of the two subspecies of *S. kali* are not well documented.

Reference: Jalas & Suominen (1980).

J. R. AKEROYD

Amaranthus retroflexus Common Amaranth

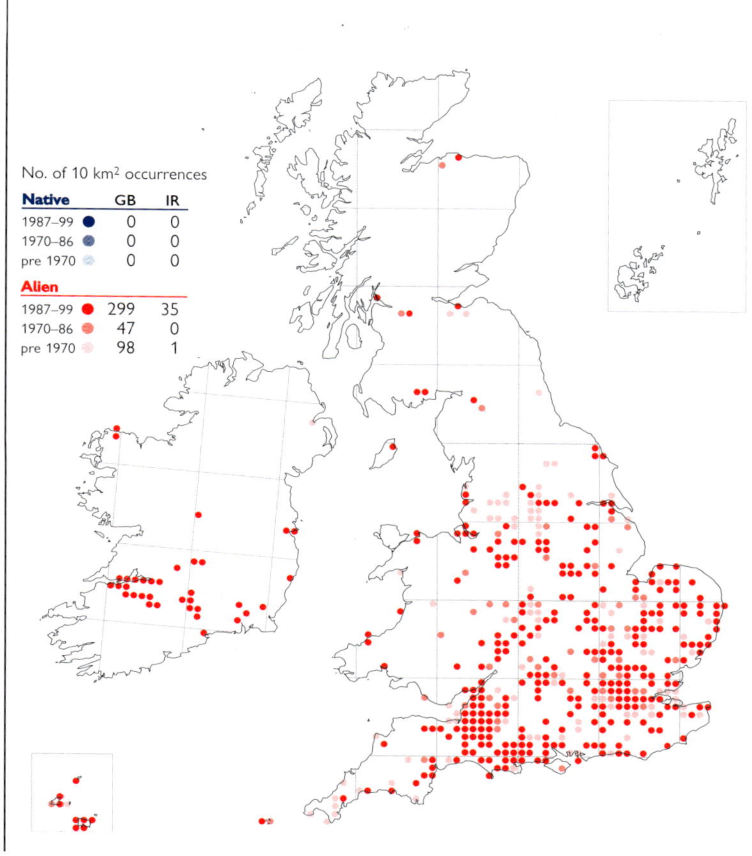

An annual of disturbed, nutrient-rich waste ground, waysides, rubbish tips and cultivated land, usually casual but occasionally persisting in milder areas. Lowland.

Neophyte. *A. retroflexus* was known in gardens by 1759, and escapes, of which the earliest on record was in 1853, have been reinforced by introductions with shoddy, bird-seed and soya-bean waste. Dunn (1905) described it as 'now very common in most parts of Europe, including England', but later changes in distribution are unclear.

Native of temperate and tropical America; naturalised throughout much of temperate and southern Europe and in other continents.

References: Brenan (1961), Hultén & Fries (1986), Jalas & Suominen (1980), Meusel *et al.* (1965), Weaver & McWilliams (1980).

J. R. AKEROYD

Amaranthus hybridus Green Amaranth

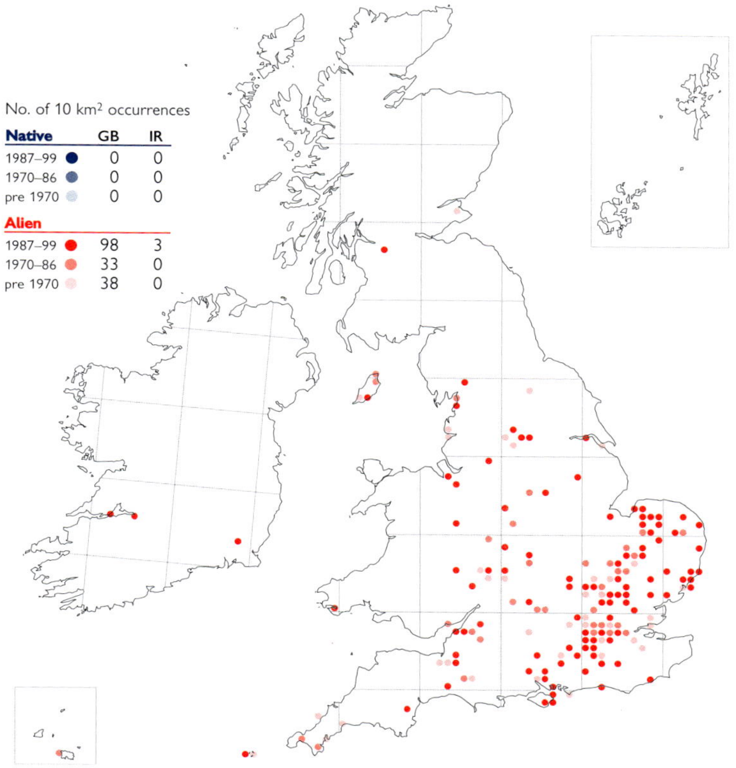

An annual of disturbed, nutrient-rich waste ground, waysides, rubbish tips, market gardens and arable fields. It is usually casual and only very rarely naturalised. Lowland.

Neophyte. *A. hybridus* arises from a wide variety of sources including wool, grain, bird-seed, spices and pet-foods. It was in cultivation in Britain by 1656, and was first recorded in the wild in 1876.

Native to tropical and subtropical America; it is a member of a complex of ill-defined species which are naturalised in temperate and southern Europe and elsewhere.

References: Bolòs & Vigo (1990), Brenan (1961), Jalas & Suominen (1980).

J. R. AKEROYD

Amaranthus albus White Pigweed

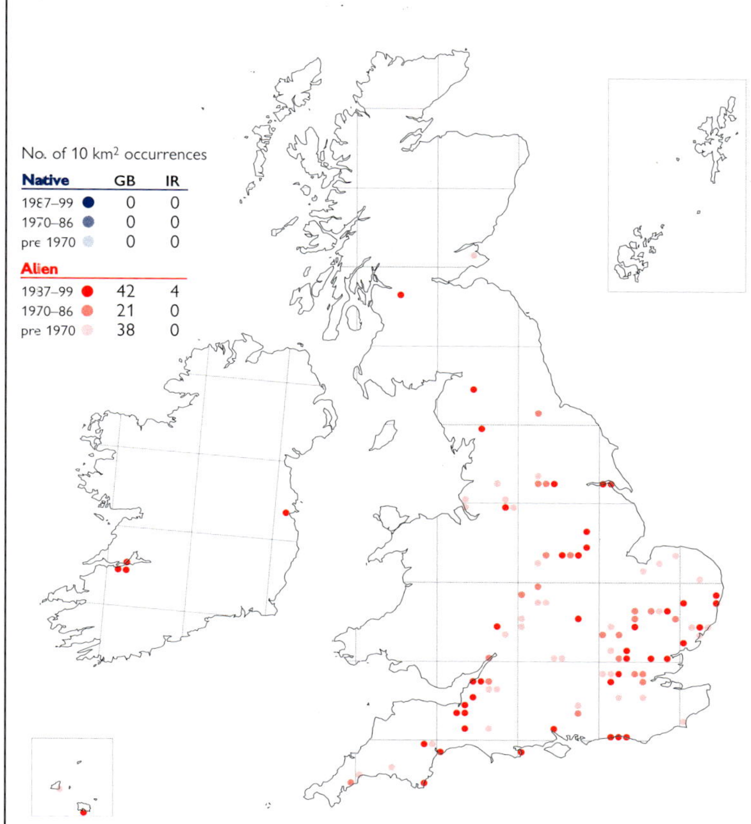

An annual of disturbed, nutrient-rich waste ground and rubbish tips, predominantly casual and very rarely naturalised. It is introduced with fibre, grain, oil- and bird-seed, and with bark for tanning. Lowland.

Neophyte. *A. albus* has been cultivated in Britain since 1778. It was recorded in the wild in 1872, and Dunn (1905) described it as a 'rare casual weed in England'.

Native of N. America; widely naturalised in temperate and southern Europe and elsewhere.

References: Bolòs & Vigo (1990), Brenan (1961), Jalas & Suominen (1980).

J. R. AKEROYD

Portulaca oleracea Common Purslane

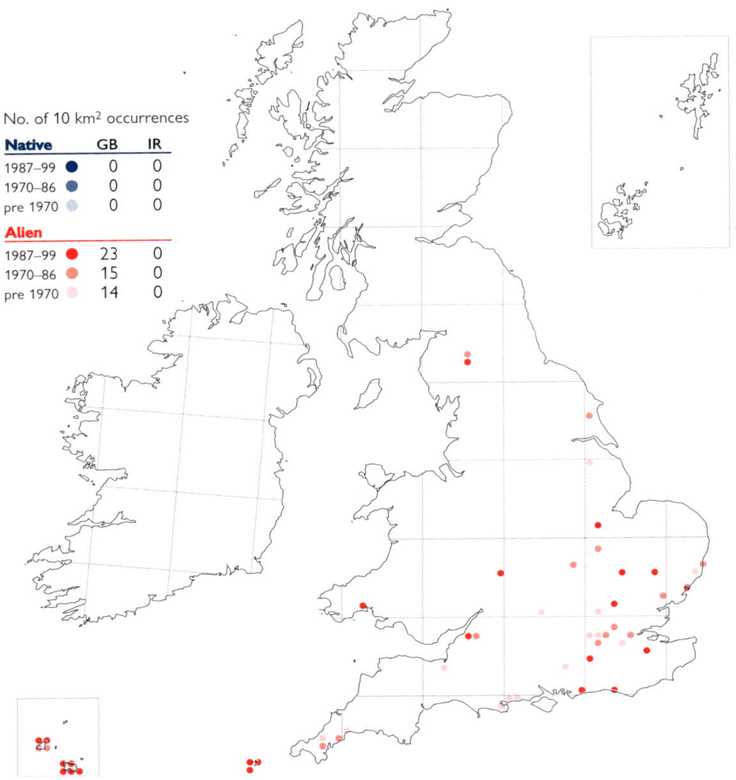

A succulent annual sometimes found as a persistent weed of arable land, especially in the Channel Islands and Isles of Scilly. Elsewhere it is a casual on waste ground, introduced in bird-seed and wool shoddy. Lowland.

Neophyte. This species has been grown in British gardens since at least 1200 (Harvey, 1981), and was recorded in the wild by 1874. Most records of *P. oleracea* in our area are referable to subsp. *oleracea*, although subsp. *nitida* does occur rarely. Subsp. *sativa*, grown on a small scale as a pot-herb, does not appear to have been found in the wild.

The native range of this species is obscure; it is now a weed in tropical and warm-temperate areas throughout the world.

References: Hultén & Fries (1986), Jalas & Suominen (1980).

D. A. PEARMAN

Claytonia perfoliata Springbeauty

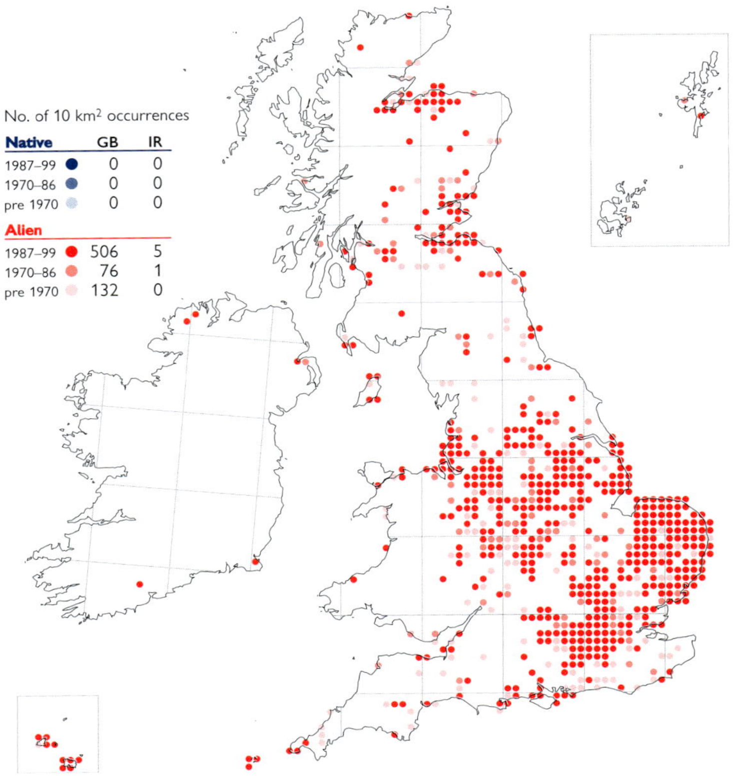

An annual of open, sandy, disturbed ground, occurring as a weed in farmland and in gardens, and also under light shade, such as Breckland pine-belts. It is absent from wet and ill-drained soils and from limestone. Lowland.

Neophyte (change +0.50). *C. perfoliata* was introduced into cultivation in 1794. It was first recorded in the wild in 1849 and by 1853 was the most troublesome weed in the Chelsea Physic Garden (Kent, 1975). However, it was still relatively uncommon up to 1930. Since then it has spread significantly, and still seems to be increasing.

Native of western N. America; widely naturalised in N.W. Europe.

References: Atlas (80b), Hultén (1968), Jalas & Suominen (1980).

D. A. PEARMAN

Claytonia sibirica Pink Purslane

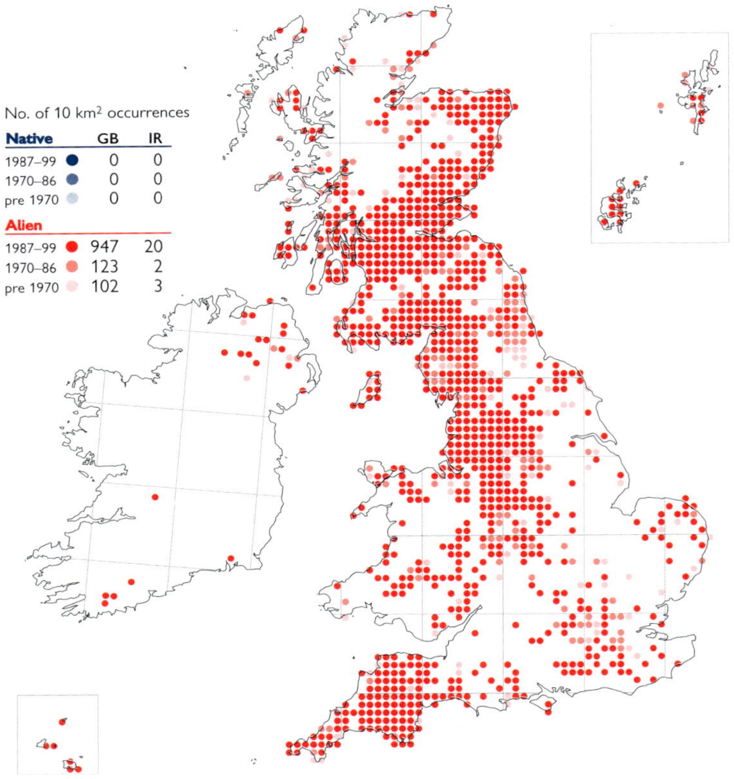

An annual to perennial herb of damp, bare sites, often found in open woodlands or hedgerows, or by shaded streams from where it may be washed downstream to new sites. Predominately lowland, but reaching 425 m at Dockray (Cumberland).

Neophyte (change +1.28). *C. sibirica* was cultivated in Britain by 1768 and was first noted in the wild in 1838. It has spread rapidly since 1930; in Cornwall, for example, it was not recorded until the 1930s but is now known in almost every 10-km square. It can quickly colonise woodland, suppressing other vegetation by its lush mass of spring foliage which then flops over nearby plants.

Native of eastern Asia and western N. America; widely naturalised in N.W. Europe.

References: Atlas (80c), French *et al.* (1999), Hultén (1968), Jalas & Suominen (1980).

D. A. PEARMAN

Montia fontana Blinks

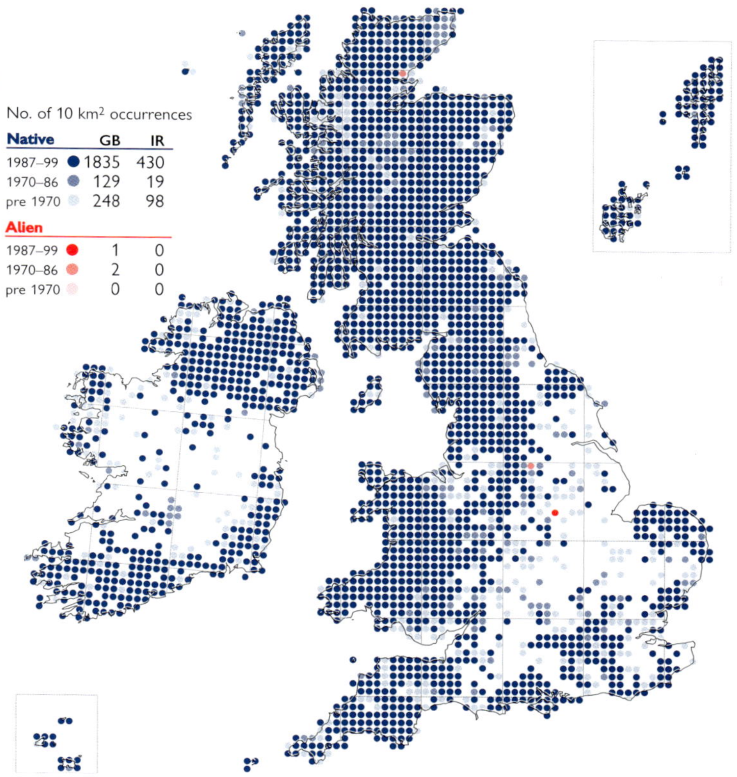

An annual to perennial herb of acidic or neutral, seasonally or permanently wet places, including springs and flushes, where it often grows in bryophyte-rich communities, the sides of lakes, rivers and streams, damp tracks and paths and winter-moist sand or gravel. 0–945 m (Mt Brandon, S. Kerry), with unlocalised records from 970 m in the Scottish Highlands.

Native (change +0.14). The species was mapped as 'all records' in the 1962 *Atlas*. There has probably been some loss in lowland England since then. The four subspecies in our area were mapped by Perring & Sell (1968) and are mapped separately here.

European Boreo-temperate element; also in C. & E. Asia and N. America.

References: Atlas (80a), Hultén & Fries (1986), Jalas & Suominen (1980), Meusel *et al.* (1965), Walters (1953).

D. A. PEARMAN

Montia fontana subsp. *fontana*

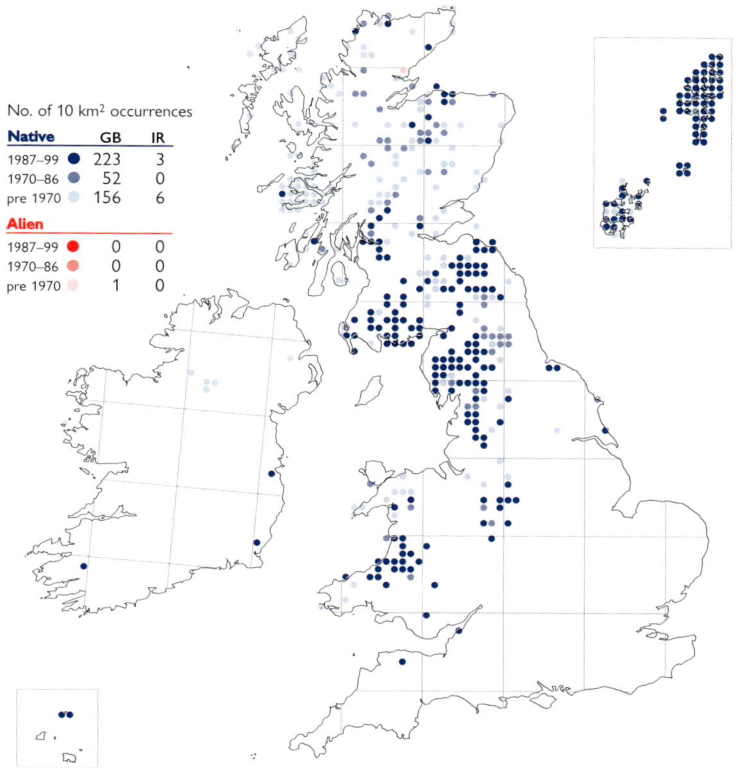

An annual to perennial herb of rills or trickles of water or permanently wet places, usually occurring in bryophyte-rich habitats over acid soils or rocks. 0–915 m (Ben More, Mid Perth), and presumably at the higher altitudes from which *M. fontana* is recorded.

Native. Although there is some geographical variation in recording effort, *M. fontana* subsp. *fontana* is, by and large, better recorded than when it was mapped by Perring & Sell (1968). It shows no change in distribution.

European Boreo-temperate element; also in C. and E. Asia and N. America.

References: Atlas Supp. (15a), Hultén & Fries (1986), Jalas & Suominen (1980), Walters (1953).

D. A. PEARMAN

Montia fontana subsp. *variabilis*

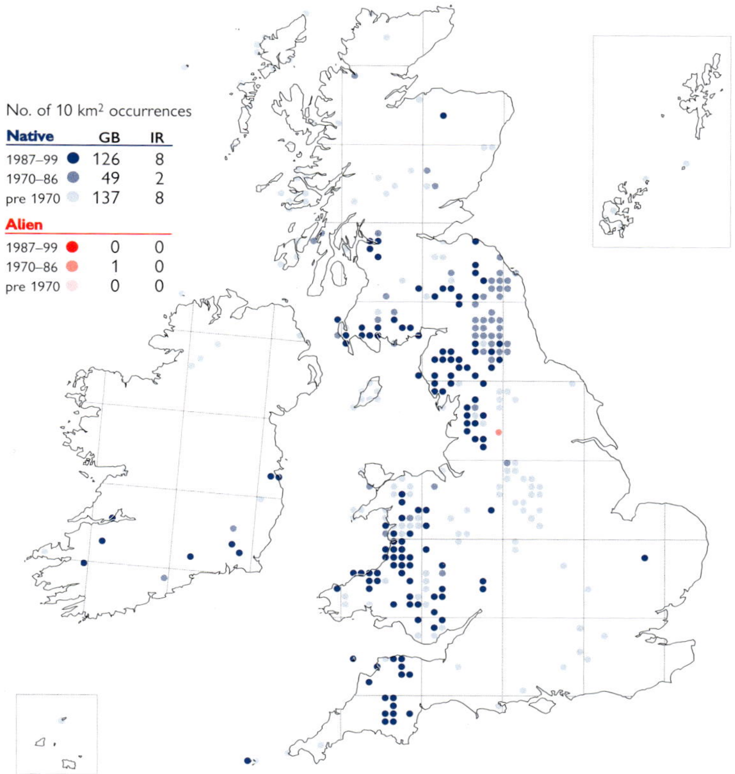

An annual to perennial herb of streams, springs and permanently wet places. 0–450 m (South Stainmore, Westmorland), but certainly higher elsewhere.

Native. The distribution of *M. fontana* subsp. *variabilis* shows no change from that mapped by Perring & Sell (1968), but it is almost certainly still under-recorded.

European Temperate element, but distribution imperfectly known; also in eastern N. America.

References: Atlas Supp. (15b), Hultén & Fries (1986), Jalas & Suominen (1980), Walters (1953).

D. A. PEARMAN

Montia fontana subsp. *amporitana*

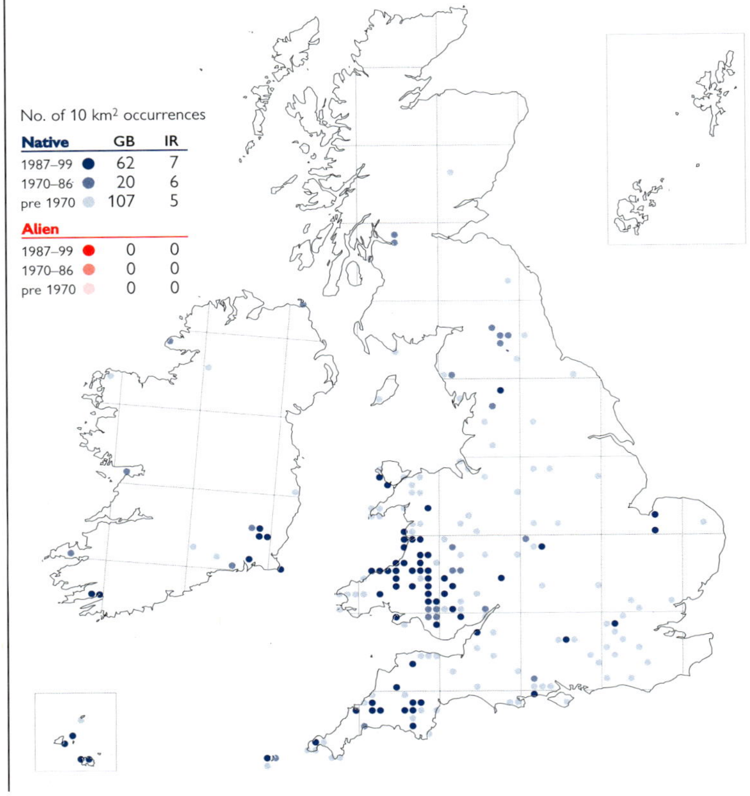

An annual to perennial herb which usually grows in permanently wet places, or as an aquatic. Lowland.

Native. Although *M. fontana* subsp. *amporitana* may not be as well recorded as the other subspecies, it seems to have declined in S. & E. England, presumably because of drainage.

European Temperate element, but distribution imperfectly known.

References: Atlas Supp. (15c), Hultén & Fries (1986), Jalas & Suominen (1980), Meusel *et al.* (1965), Walters (1953).

D. A. PEARMAN

Montia fontana subsp. *chondrosperma*

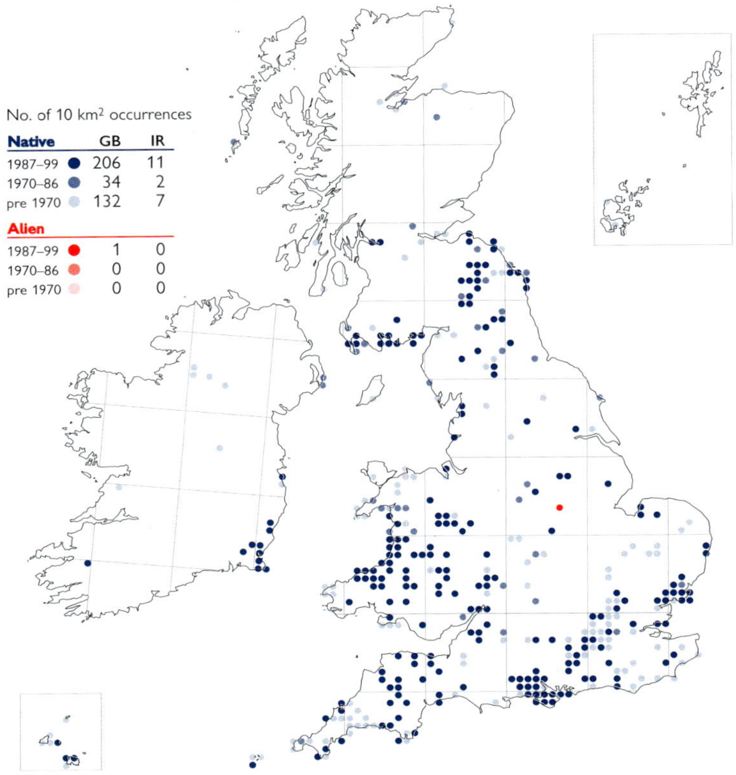

A tufted annual to perennial herb which grows in drier habitats than the other subspecies, usually over sand and gravel in places which are moist or flooded in winter but dry out in summer. It rarely grows submerged in water, and is relatively intolerant of competition. Generally lowland, but reaching 450 m at South Stainmore (Westmorland).

Native. *M. fontana* subsp. *chondrosperma* is a distinctive subspecies which appears to be fairly well recorded. It shows little change in distribution since it was mapped by Perring & Sell (1968).

European Temperate element, but distribution imperfectly known.

References: Atlas Supp. (15d), Hultén & Fries (1986), Jalas & Suominen (1980), Meusel *et al.* (1965), Walters (1953).

D. A. PEARMAN

Arenaria serpyllifolia Thyme-leaved Sandwort

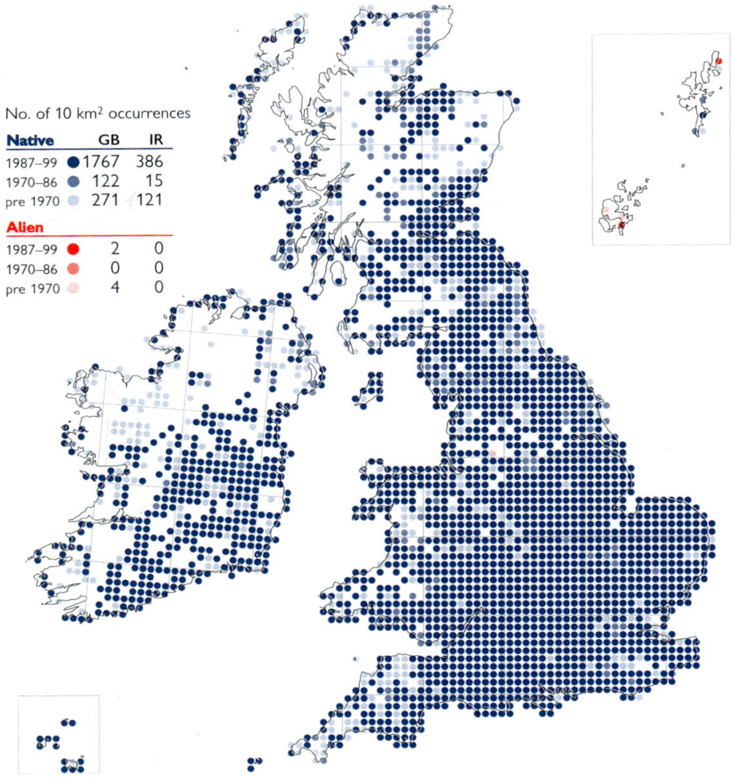

A winter- or rarely summer-annual of dry, usually shallow, neutral to basic soils. It is found in a wide range of open habitats, including rock outcrops, cliffs, screes, walls, spoil heaps from mines and in quarries, railway ballast, waysides and arable field margins. Generally lowland, but reaching at least 610 m at Melmerby Fell (Cumberland).

Native (change –0.76). There appears to be no appreciable change since the 1962 *Atlas*, where it was mapped as 'all records'.

Eurosiberian Southern-temperate element; widely naturalised outside its native range.

References: Atlas (76a), Grime *et al.* (1988), Hultén & Fries (1986), Jalas & Suominen (1983).

P. S. LUSBY

Arenaria serpyllifolia subsp. *serpyllifolia*

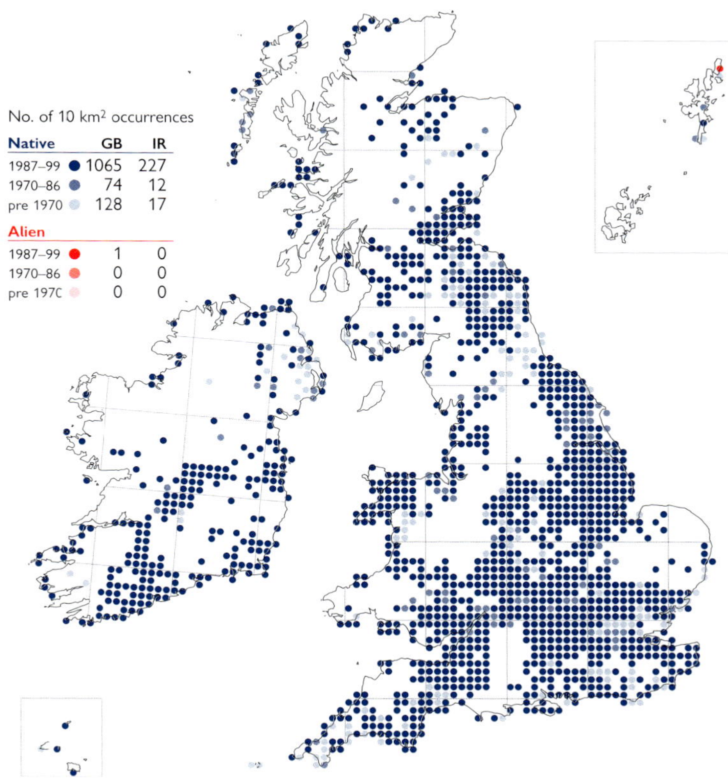

A winter- or rarely summer-annual of dry, open sites, such as disturbed ground and ant-hills on chalk and limestone downs, sand dunes and other sandy and gravelly places, rabbit warrens, arable fields, roadsides, railways, quarries, rock ledges, walls, spoil-tips and waste ground. 0–610 m (Melmerby Fell, Cumberland).

Native. The map shows uneven recording effort, but no change in distribution since 1968. It includes plants referable to subsp. *lloydii*, which were mapped separately by Perring & Sell (1968) but treated as synonymous with subsp. *serpyllifolia* by Stace (1997).

Eurosiberian Southern-temperate element; subsp. *serpyllifolia* extends further north and to higher altitudes than subsp. *leptoclados*.

References: Atlas Supp. (13a), Grime *et al.* (1988), Jalas & Suominen (1983).

P. S. LUSBY

Arenaria serpyllifolia subsp. *leptoclados*

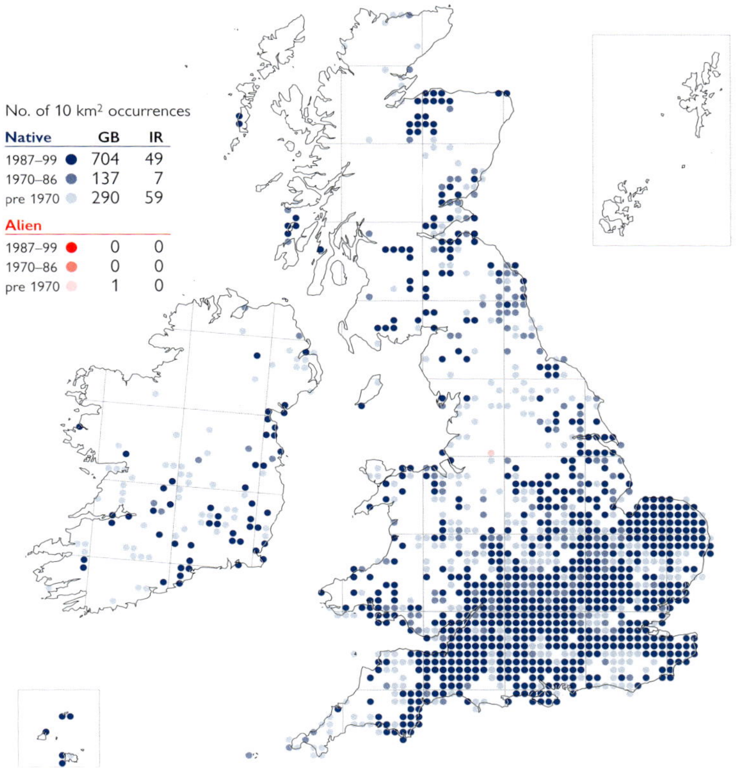

A winter- or rarely summer-annual of dry, open sites in similar habitats to those of subsp. *serpyllifolia* and sometimes growing with it. However, there may be a stronger preference for cultivated and waste ground, old walls and quarries (Grose, 1957). It is also found in bare places in calcareous grassland, on roadsides and railway tracks. 0–500 m (Teesdale, N.W. Yorks.).

Native. The current map shows a much more coherent pattern than that in Perring & Sell (1968), and the distribution seems stable.

Eurosiberian Southern-temperate element, with a more southerly distribution than subsp. *serpyllifolia*.

References: Atlas Supp. (13b), Grime *et al.* (1988), Jalas & Suominen (1983).

P. S. LUSBY

Arenaria norvegica subsp. *norvegica*

Arctic Sandwort

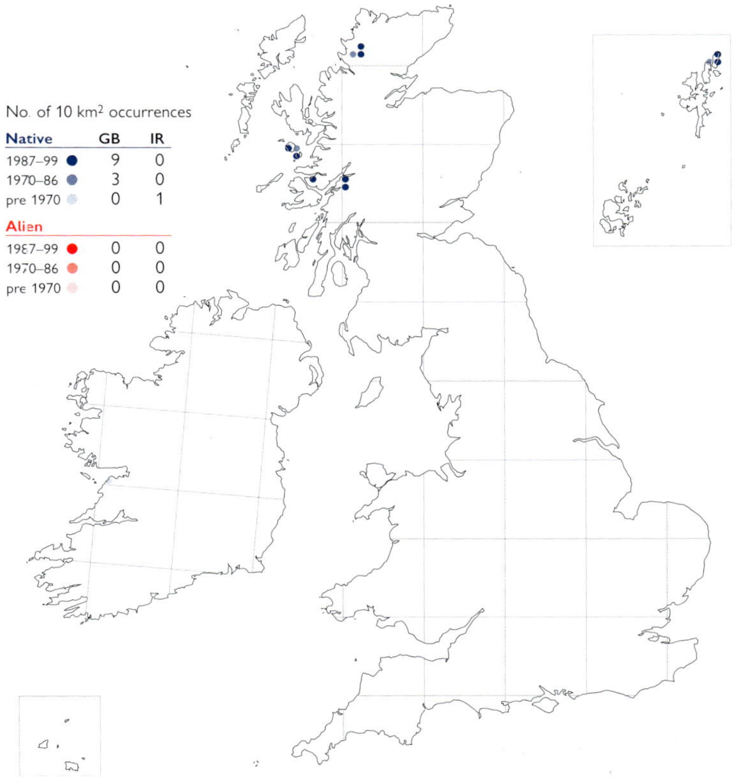

This annual, biennial or perennial herb grows on base-rich substrates over limestone, serpentine and other basic rocks, occurring on rocky knolls, screes, river gravels, fell-field, and occasionally on exposed summit ridges. From *c.* 15 m on Unst (Shetland) to 650 m on Beinn Sgulaird (Main Argyll).

Native (change for species 0.21). Although the number of known sites has almost doubled since the 1962 *Atlas*, this is probably due to increased recording rather than a real spread. The identity of a specimen from the Burren (Co. Clare) in 1961 was confirmed but the site has never been refound.

European Arctic-montane element; it reaches its southern limit in our area.

References: Atlas (76b), Curtis & McGough (1988), Hultén & Fries (1986), Jalas & Suominen (1983), Wigginton (1999).

P. S. LUSBY

Arenaria norvegica subsp. *anglica* English Sandwort

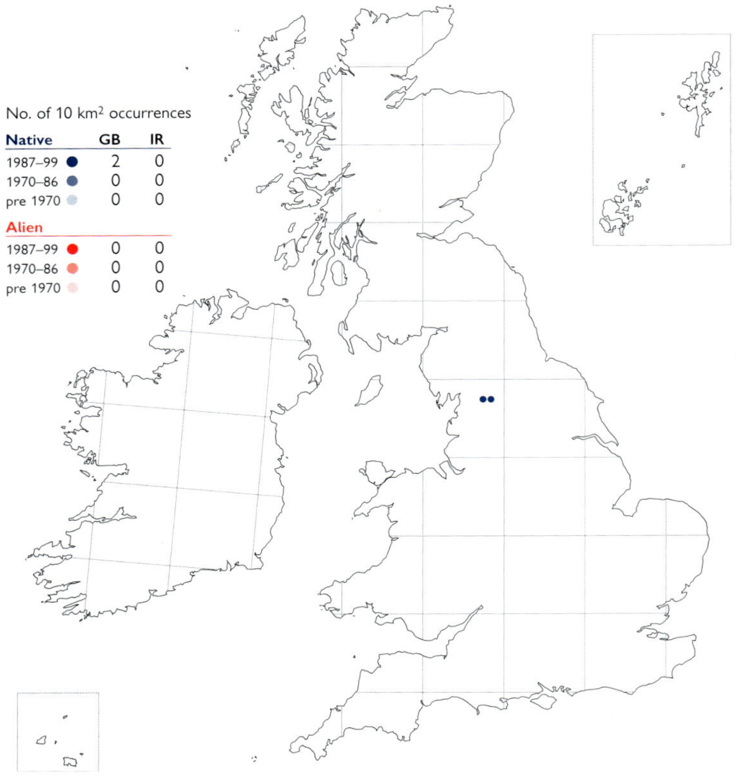

An annual or biennial herb which usually grows in sparsely vegetated, well-drained sites associated with Carboniferous limestone, including peaty depressions on flat slabs, peat in cracks and hollows in limestone pavement, and in open, bryophyte-rich tufaceous flushes. Its seed appears to have considerable longevity. Found only at moderate altitudes between 295 m at Selside and 410 m at Dawson Close (both Mid-W. Yorks.).

Native (change for species +0.21). There is no change in the 10-km distribution since the 1962 *Atlas*, but some sites are threatened by excessive disturbance by walkers and cyclists.

Endemic.

References: Atlas (76b), Jalas & Suominen (1983), Walker (2000), Wigginton (1999).

P. S. LUSBY

Arenaria ciliata Fringed Sandwort

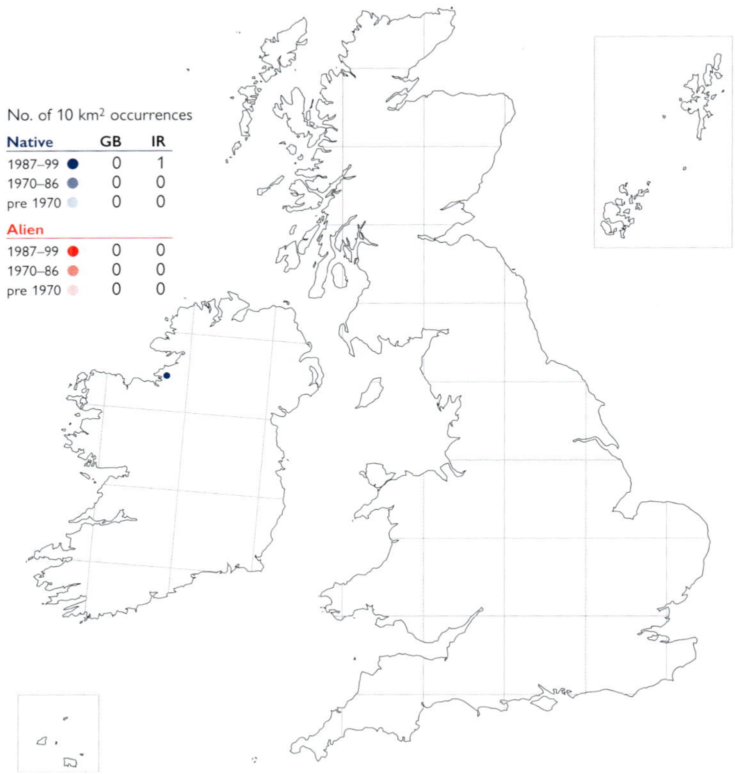

This low-growing perennial calcicole is found in open mountain grassland and on N.-facing Carboniferous limestone cliffs in the Ben Bulben range in Co. Sligo, at altitudes of 400–600 m.

Native. The population appears to have been stable since it was first recorded in 1806.

European Arctic-montane element; also in Greenland.

References: Atlas (76b), Curtis & McGough (1988), Hultén & Fries (1986), Jalas & Suominen (1983), Praeger (1934).

P. S. LUSBY

Arenaria balearica Mossy Sandwort

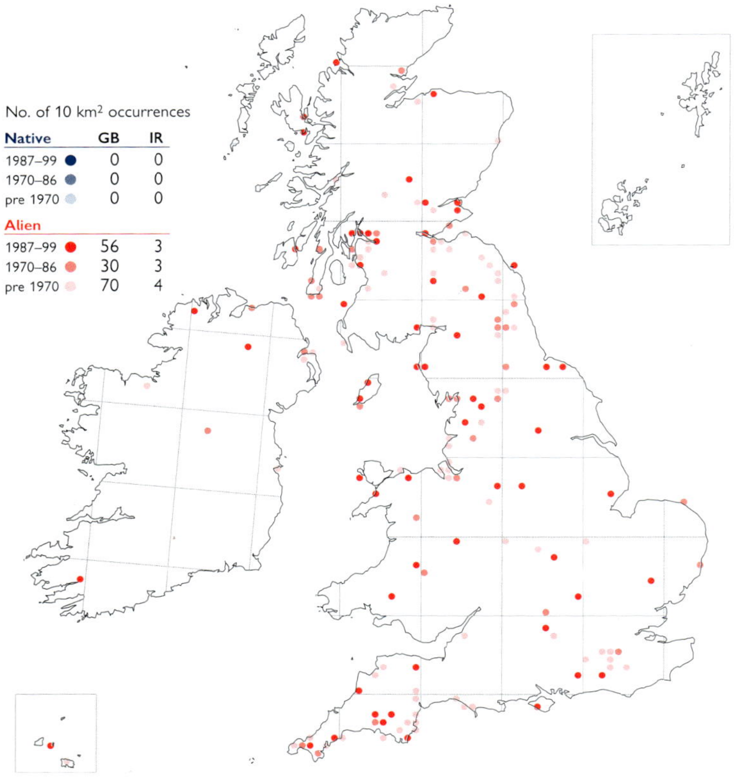

A procumbent perennial herb found as a well-naturalised garden escape on walls, buildings and stony banks, often in damp, shaded sites. Lowland.

Neophyte (change −0.55). *A. balearica* has been cultivated in Britain since 1787 and is widely grown on walls, rockeries and paved areas in gardens. It was first recorded in the wild in 1861 (Dorset), and the distribution is similar to that shown in the 1962 *Atlas*.

Native of the W. Mediterranean islands.

References: Atlas (76c), Bolòs & Vigo (1990), Jalas & Suominen (1983).

P. S. LUSBY

Moehringia trinervia Three-nerved Sandwort

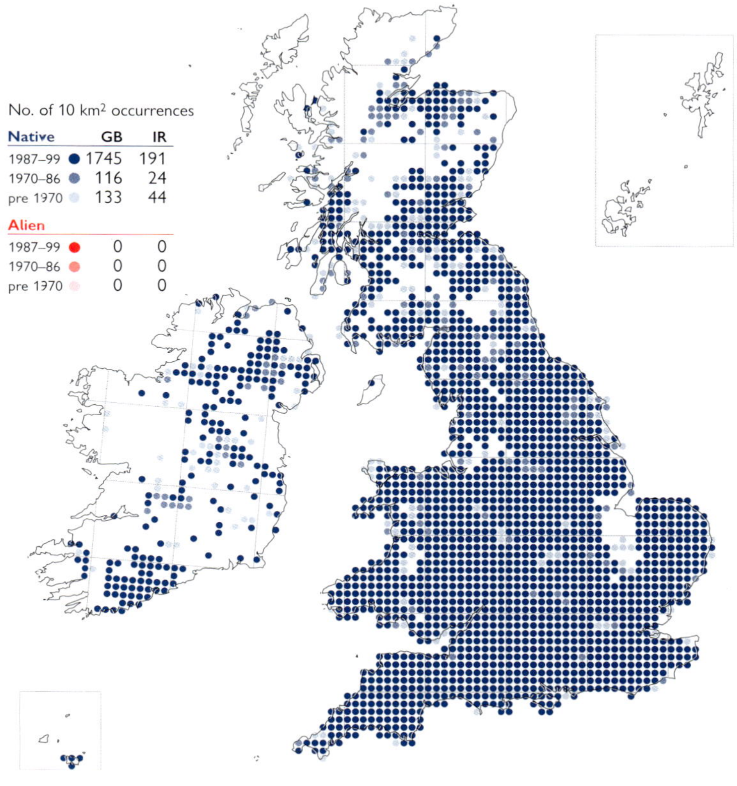

An annual of open, often moist, ground, generally found in woodland but also in shaded hedge banks, and rarely in unshaded places such as on walls and railway banks. It favours slightly acidic substrates, and there is a slight preference for warmer slopes in woodland, which hastens the successful completion of the plant's life-cycle. 0–425 m (Mallowdale Fell, W. Lancs.).

Native (change –0.40). There appears no significant change since the 1962 *Atlas*, where it was mapped as 'all records'.

European Temperate element; also in E. Asia.

References: Atlas (75d), Grime *et al.* (1988), Hultén & Fries (1986), Jalas & Suominen (1983), Meusel *et al.* (1965).

P. S. LUSBY

Honckenya peploides Sea Sandwort

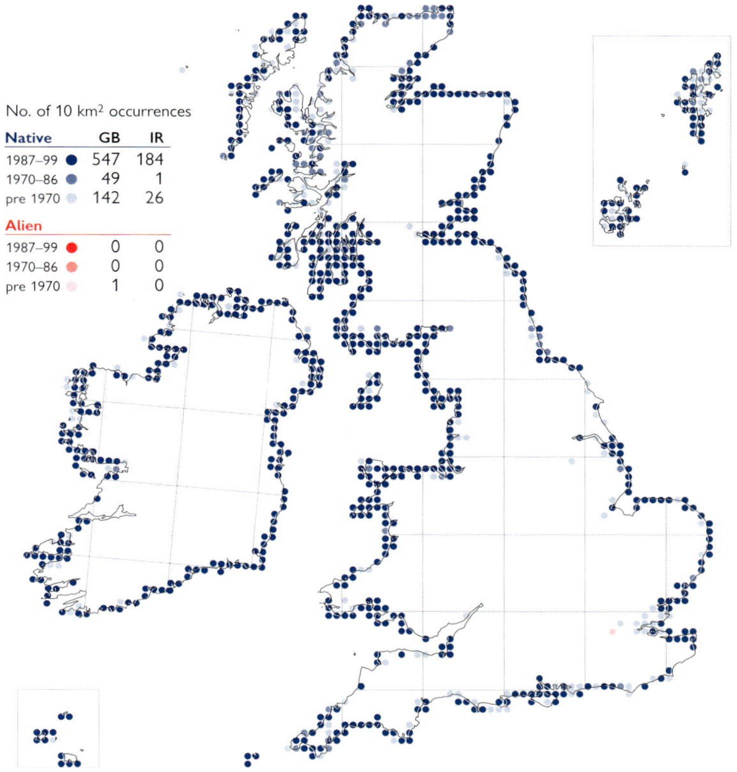

H. peploides is a succulent perennial herb with creeping stolons that make it well adapted to growing in mobile substrates. It often occurs in abundance on shingle beaches and on shifting sand in foreshore communities. It is one of the pioneer colonists of open fore-dunes with plants such as *Elytrigia juncea*. Lowland.

Native (change –0.58). There does not appear to be any significant change since the 1962 *Atlas*, where it was mapped as 'all records'.

Circumpolar Wide-boreal element, with a disjunct distribution.

References: Atlas (75c), Hultén & Fries (1986), Jalas & Suominen (1983), Meusel *et al.* (1965).

P. S. LUSBY

Minuartia recurva Recurved Sandwort

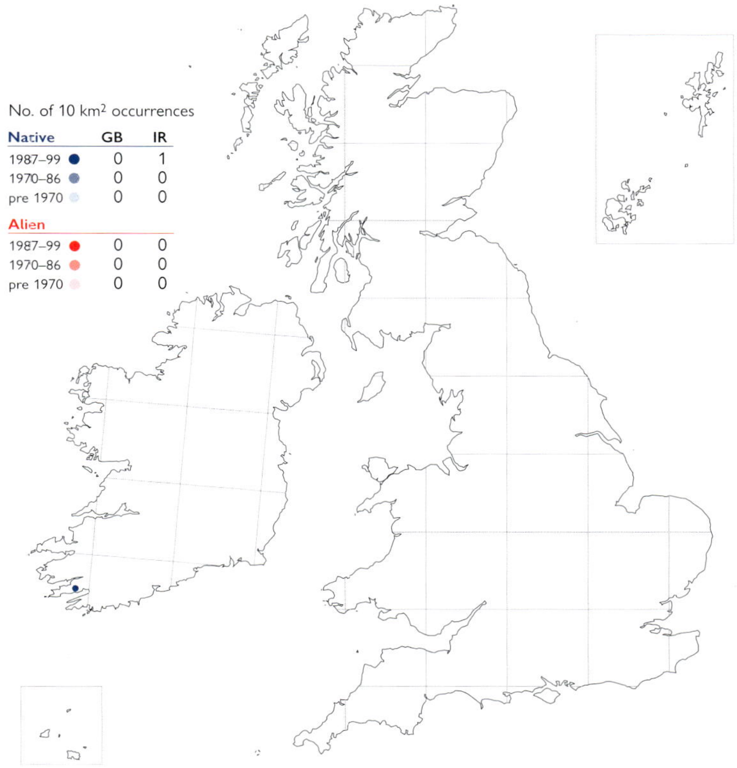

This tufted, calcifugous perennial herb grows in thin humus in dry narrow cracks in slabs of siliceous rocks of Old Red Sandstone age. 510–610 m in the Caha Mountains (S. Kerry and W. Cork).

Native. *M. recurva* was discovered in 1964 in the Caha Mountains, where two populations are found straddling the Cork–Kerry border. These populations appear to be stable.

European Arctic-montane element; confined to the mountains of S.W. Asia, S. & S.C. Europe and Ireland.

References: Atlas Supp. (12b), Bolòs & Vigo (1990), Curtis & McGough (1988), Jalas & Suominen (1983), Moore (1966).

P. S. LUSBY

Minuartia verna Spring Sandwort

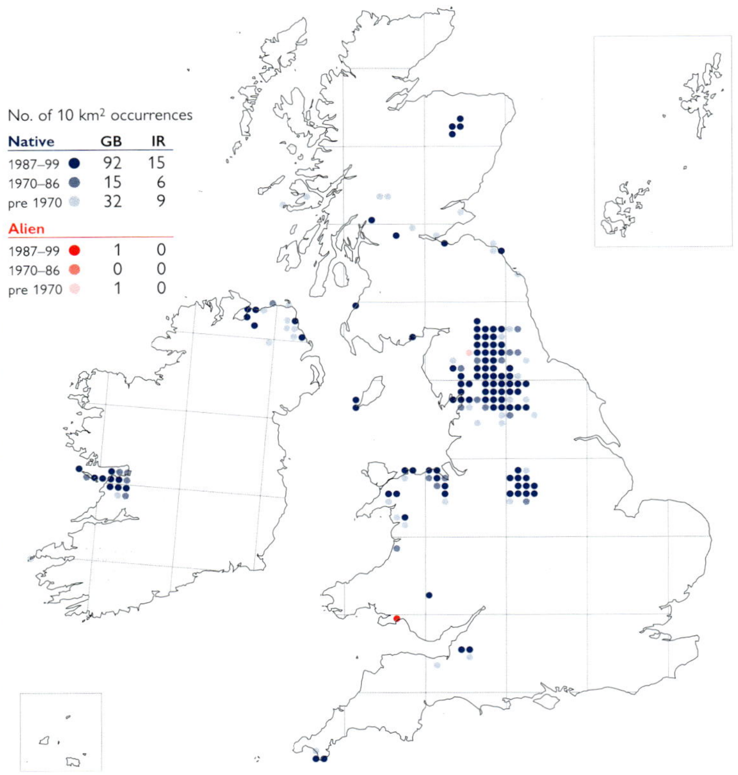

M. verna is a perennial, basicolous, cushion-forming herb, characteristic of Carboniferous limestone districts where it is found in short grassland, on scars, on limestone pavement and scree. It also grows on base-rich volcanic rock in N. Wales and basalt in N. Ireland, on metal-rich soils, including those derived from serpentine, and on mining spoil. It prefers open sites with reduced competition, but may suffer from drought in very exposed conditions. Although seldom above 600 m, it has been recorded at 875 m on Snowdon (Caerns.).

Native (change –0.42). The distribution of *M. verna* is broadly stable.

Eurasian Boreal-montane element, with a disjunct distribution.

References: Atlas (74c), Grime *et al.* (1988), Hultén & Fries (1986), Jalas & Suominen (1983), Meusel *et al.* (1965), Stewart *et al.* (1994).

P. S. LUSBY

Minuartia rubella Mountain Sandwort

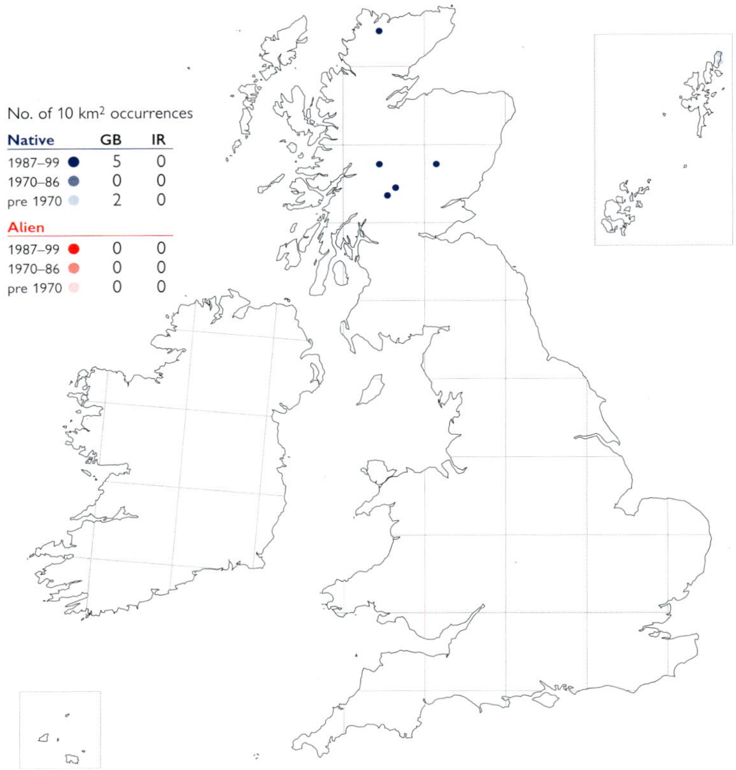

This cushion-forming montane perennial herb is always associated with strongly base-rich rocks, including limestone and soft calcareous schists. The vegetation is usually open, as the ground on which it usually grows is both easily eroded and subjected to frost-heave. It reaches an altitude of 1180 m on Ben Lawers (Mid Perth), and formerly descended to 120 m on Unst (Shetland).

Native (change +0.01). The national population of *M. rubella* is likely to be stable, but its discovery in 1990 in the Ben Alder range (Westerness) extended its known range.

Circumpolar Arctic-montane element; absent from mountains of C. Europe.

References: Atlas (74d), Hultén & Fries (1986), Jalas & Suominen (1983), Wigginton (1999).

P. S. LUSBY

Minuartia stricta Teesdale Sandwort

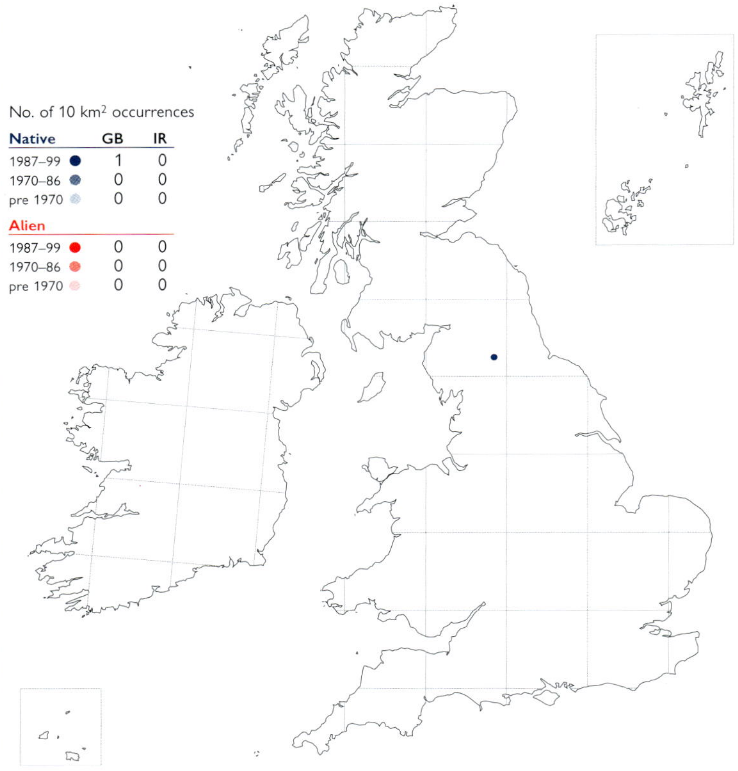

A loosely tufted but slender perennial herb that has only ever been known in Britain from Widdybank Fell, where it grows in open, gravelly flushes and eroding margins of sikes on metamorphic sugar limestone. The plant is not a strong competitor and is mainly associated with hummock-forming mosses and species such as *Carex capillaris*, *Juncus triglumis*, *Minuartia verna* and *Primula farinosa*. Upland, from 490 to 510 m on Widdybank Fell (Co. Durham).

Native. There is no change in distribution from the 1962 *Atlas*.

Circumpolar Arctic-montane element.

References: Atlas (74d), Hultén & Fries (1986), Jalas & Suominen (1983), Wigginton (1999).

P. S. LUSBY

Minuartia hybrida Fine-leaved Sandwort

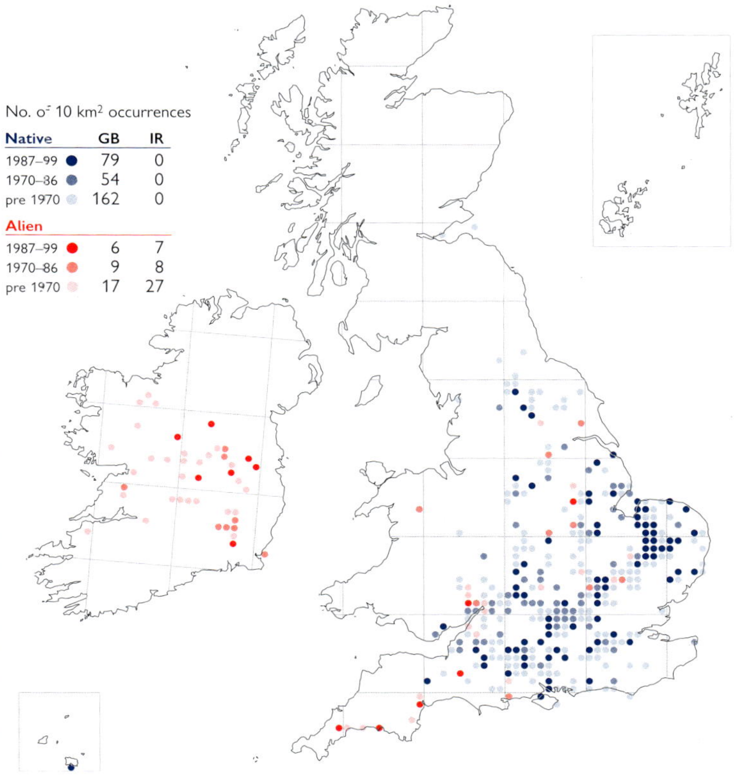

This annual grows on light soils in dry places. Its natural habitat is dry, rocky, calcareous grassland on chalk and limestone. However, it is more frequent in artificial habitats such as abandoned arable fields, quarries, old walls, trackways, railway banks and sidings. Mainly lowland, but reaching 400 m at Langcliffe in the Craven Pennines (Mid-W. Yorks.).

Native (change –1.70). *M. hybrida* has greatly declined in arable and grassland habitats through agricultural intensification, but this has been somewhat offset by records from railways. This loss had mainly occurred before the 1962 *Atlas*, but has continued since.

Submediterranean-Subatlantic element.

References: Atlas (75a), Jalas & Suominen (1983), Meusel *et al.* (1965), Stewart *et al.* (1994).

P. S. LUSBY

Minuartia sedoides Cyphel

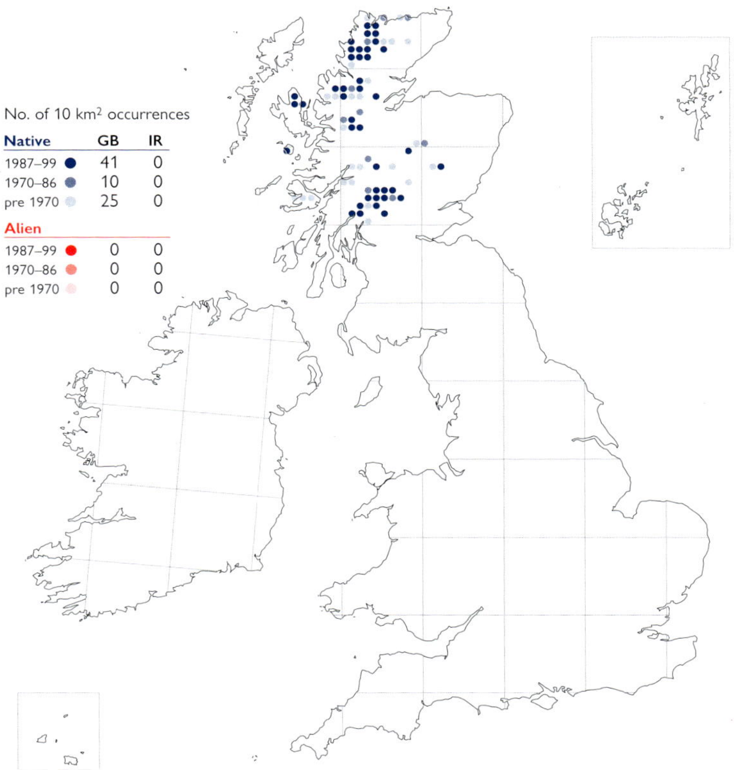

M. sedoides is a mat- or cushion-forming perennial herb of base-rich rocks, flushed grassland, exposed montane heath, and mountain ledges and plateaux. From 335 m on Skye (N. Ebudes) to 1200 m on Ben Lawers (Mid Perth), but has been recorded at 215 m washed down in the stony bed of the R. Fillan at Tyndrum (Mid Perth).

Native (change –0.75). There has probably been little or no change in its 10-km distribution since the 1962 *Atlas*, since it is likely to be extant in most of the squares for which only pre-1970 records are available.

European Arctic-montane element; confined to the mountains of C. Europe and Scotland.

References: Atlas (75b), Hultén & Fries (1986), Jalas & Suominen (1983), Lusby & Wright (1996), Meusel *et al.* (1965), Stewart *et al.* (1994).

P. S. LUSBY

Stellaria nemorum Wood Stitchwort

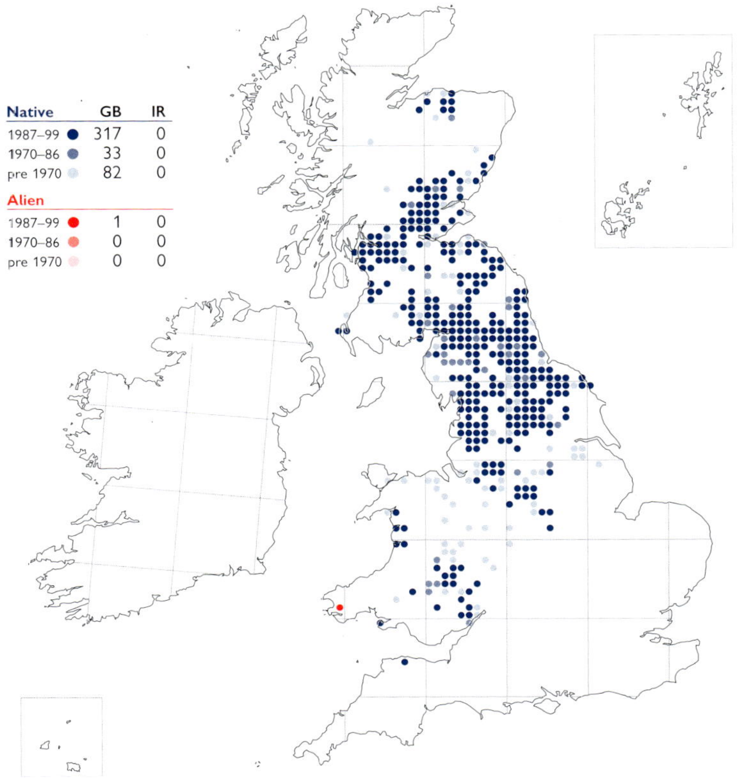

This herbaceous, stoloniferous perennial prefers fertile soils and occurs mostly in damp, shaded habitats, and sometimes on periodically flooded ground. It is usually found by streamsides and ditches and in wet woods and damp hedge banks. Generally lowland, but reaching *c.* 915 m above Coire Kander (S. Aberdeen).

Native (change +0.21). There has been no significant change in the distribution of this species since the 1962 *Atlas*. Subsp. *nemorum* occurs throughout the range of the species; subsp. *montana* is mapped separately.

European Boreo-temperate element.

References: Atlas (69d), Hultén & Fries (1986), Jalas & Suominen (1983), Meusel *et al.* (1965).

P. S. LUSBY

Stellaria nemorum subsp. *montana*

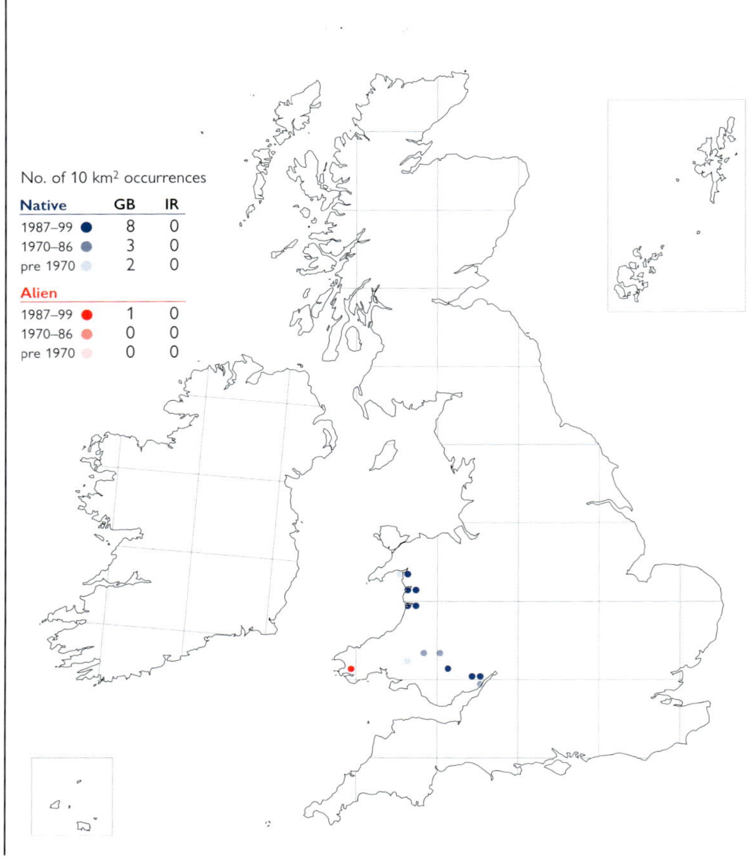

No. of 10 km² occurrences

Native	GB	IR
1987–99 ●	8	0
1970–86 ●	3	0
pre 1970 ●	2	0
Alien		
1987–99 ●	1	0
1970–86 ●	0	0
pre 1970 ●	0	0

This stoloniferous perennial herb is found in similar habitats to subsp. *nemorum*, especially damp woods and hedgerows on basic soils. Generally lowland, but upper altitudinal limit unknown.

Native. The distribution of this subspecies has not changed since it was mapped by Perring & Sell (1968), although it is now better recorded. Putative hybrids between the two subspecies have been recorded, and in Wales populations of subsp. *nemorum* are thought to have been lost through introgression with subsp. *montana*.

European Temperate element.

References: Atlas Supp. (12b), Green (1954), Hultén & Fries (1986), Jalas & Suominen (1983).

P. S. LUSBY

Stellaria media Common Chickweed

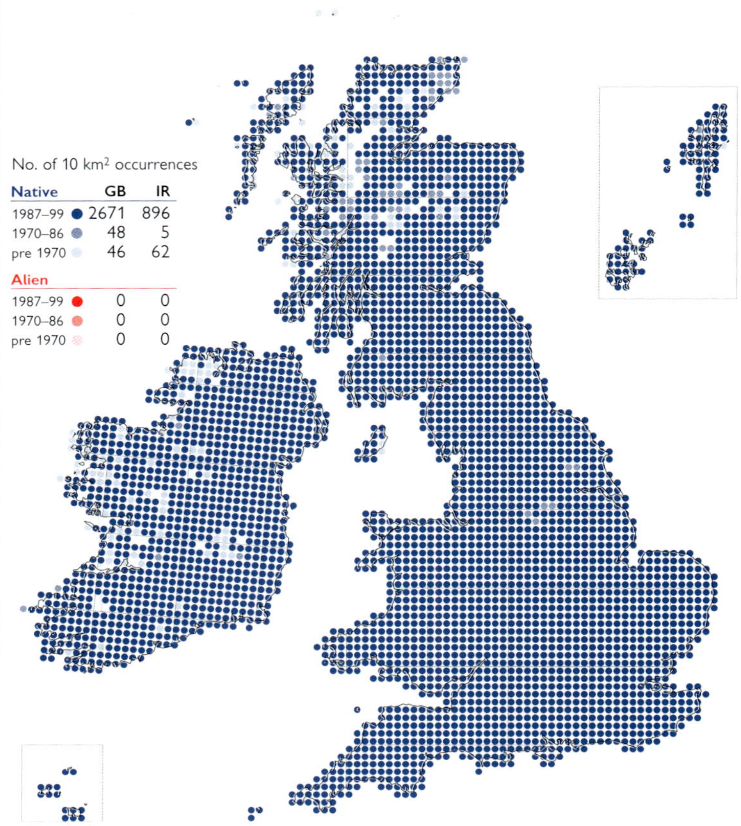

No. of 10 km² occurrences

Native	GB	IR
1987–99 ●	2671	896
1970–86 ●	48	5
pre 1970 ●	46	62
Alien		
1987–99 ●	0	0
1970–86 ●	0	0
pre 1970 ●	0	0

An annual or a short-lived perennial herb, capable of two or three generations in a year, found in a wide range of disturbed habitats, especially in nutrient-enhanced conditions. It is a conspicuous weed of gardens or crops, also found on manure heaps, in sewage works, on walls, and often in recently established plantations. It is also a characteristic plant of coastal strand lines and sea-bird rocks. 0–950 m (Mt Brandon, S. Kerry).

Native (change +0.03). There seems to have been little change in the range of *S. media* since the 1962 *Atlas*.

Eurasian Wide-temperate element, but naturalised in N. America so distribution is now Circumpolar Wide-temperate.

References: Atlas (70a), Grime *et al.* (1988), Hultén & Fries (1986), Jalas & Suominen (1983), Sobey (1981), Turkington *et al.* (1980).

P. S. LUSBY

Stellaria pallida Lesser Chickweed

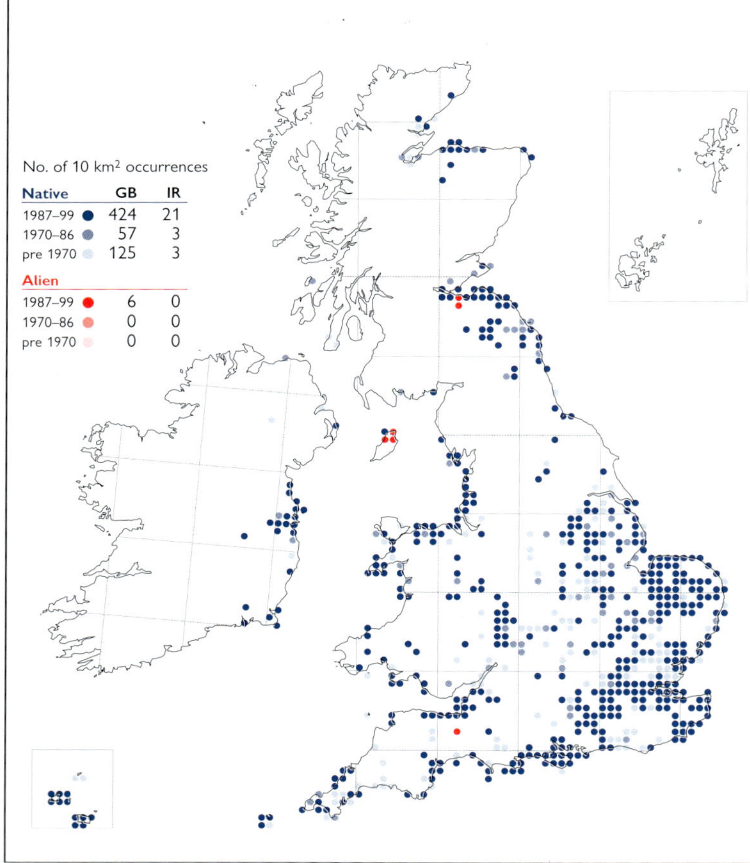

No. of 10 km² occurrences

Native	GB	IR
1987–99 ●	424	21
1970–86 ●	57	3
pre 1970 ●	125	3
Alien		
1987–99 ●	6	0
1970–86 ●	0	0
pre 1970 ●	0	0

An annual growing in open conditions on light, well-drained soils. Near the coast it is found on sand dunes, shingle and in other sandy or stony places; elsewhere it occurs on waste and cultivated ground, in gravel- and sand-pits and on tracks in conifer plantations. It is sometimes also found in lawns and on walls. Lowland.

Native (change +1.17). There are now many more records than in the 1962 *Atlas*, and an extension of its range in Scotland. This is probably a result of better recording of this inconspicuous, spring-flowering species.

Eurosiberian Southern-temperate element; widely naturalised outside its native range.

References: Atlas (70b), Hultén & Fries (1986), Jalas & Suominen (1983).

P. S. LUSBY

Stellaria neglecta Greater Chickweed

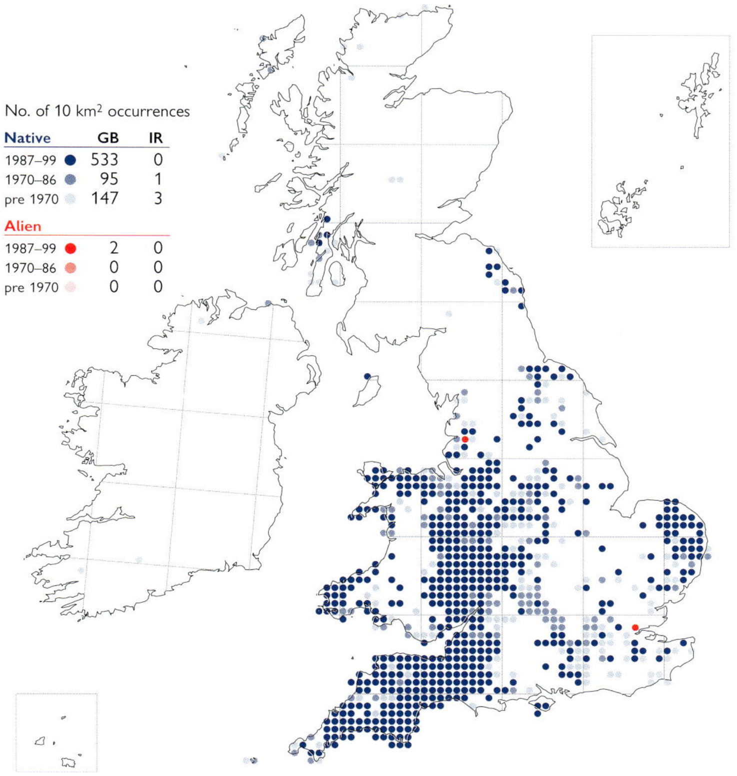

An annual to short-lived perennial herb of damp, shaded places such as hedgerows, wood margins, streamsides and the borders of damp copses, on a range of soils from stiff, poorly-drained clays to damp sand and peaty alluvium. Generally lowland, but reaching 440 m at Kinloch Rannoch (Mid Perth).

Native (change +0.42). *S. neglecta* is now known to be much more frequent than was appreciated in the 1962 *Atlas* in S.W. England, East Anglia and Wales. Conversely, there is some evidence for losses in S.E. England. Records in Ireland have only been accepted if based on critically determined specimens; some British records may be errors for the variable *S. media*.

European Temperate element; also in C. and E. Asia.

References: Atlas (70c), Hultén & Fries (1986), Jalas & Suominen (1983).

P. S. LUSBY

Stellaria holostea Greater Stitchwort

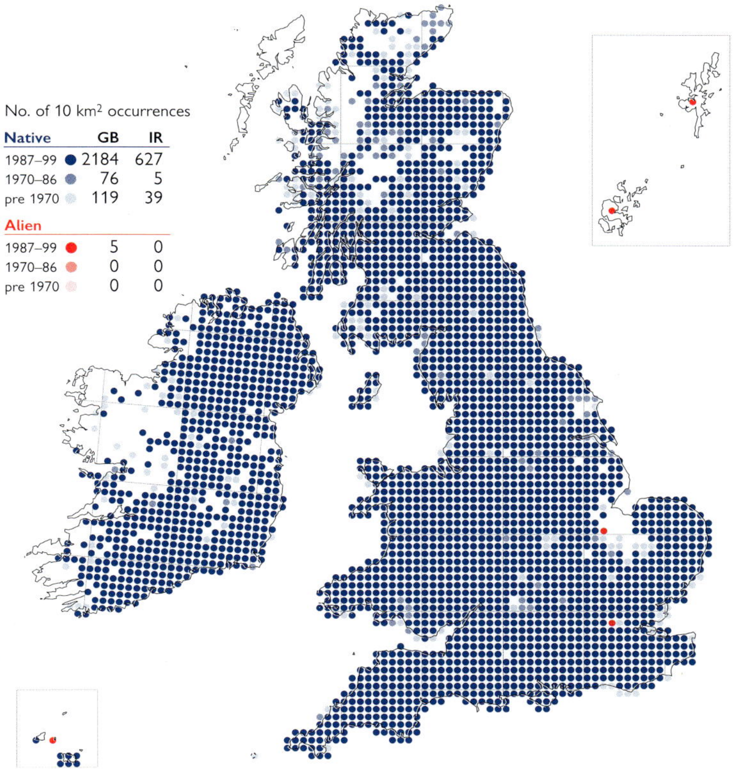

This winter-green perennial herb occurs in hedgerows, copses and woodland margins, and on unmanaged grassy roadsides. It tolerates a wide range of soils, but does best on those that are moist, mildly acid and infertile. It avoids permanently wet conditions and the most freely-drained substrates. 0–915 m (Beinn an Dothaidh, Main Argyll), although this record was for a non-flowering specimen.

Native (change –0.56). There seems to be little change in the national distribution of *S. holostea*, which was mapped as 'all records' in the 1962 *Atlas*.

Eurosiberian Temperate element.

References: Atlas (70d), Grime *et al.* (1988), Hultén & Fries (1986), Jalas & Suominen (1983), Meusel *et al.* (1965).

P. S. LUSBY

Stellaria palustris Marsh Stitchwort

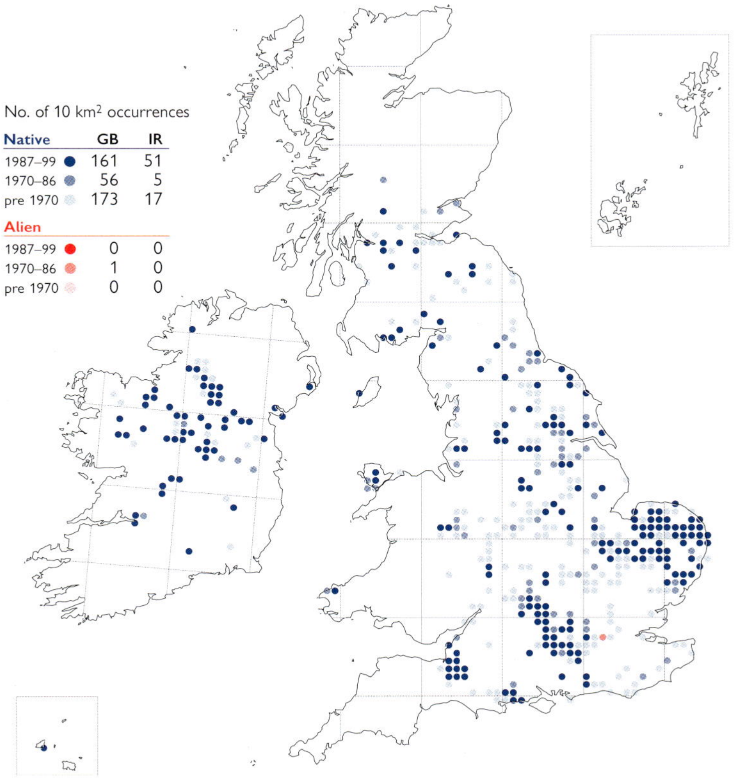

This perennial, rhizomatous herb is a species of damp and wet places, including pastures, grassy fens and marshes, especially in areas with standing water in winter. It is also able to colonise artificial habitats such as old peat diggings. Generally lowland, but reaching 360 m on Cronkley Fell (N.W. Yorks.).

Native (change –0.89). Many sites were lost in C. & E. England before 1930, and losses have continued in most parts of its British range.

Eurasian Boreo-temperate element; widely naturalised outside its native range.

References: Atlas (71a), Hultén & Fries (1986), Jalas & Suominen (1983).

P. S. LUSBY

Stellaria graminea Lesser Stitchwort

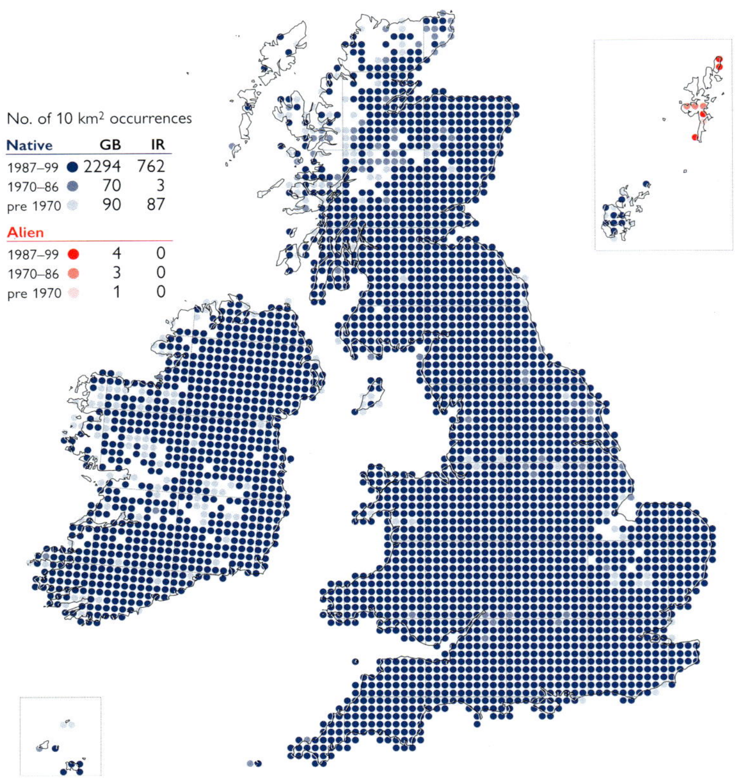

A perennial herb of damp or free-draining, neutral and acidic soils. Habitats include woodland clearings, neglected pastures, hay meadows, grass-heaths, hedge banks and waysides. It is tolerant of some nutrient enrichment, and is often a constituent of neglected pasture. 0–740 m (Knock Fell, Westmorland).

Native (change –0.02). There has been no appreciable change in its distribution since the 1962 *Atlas*.

Eurosiberian Boreo-temperate element; widely naturalised outside its native range.

References: Atlas (71b), Hultén & Fries (1986), Jalas & Suominen (1983).

P. S. LUSBY

Stellaria uliginosa Bog Stitchwort

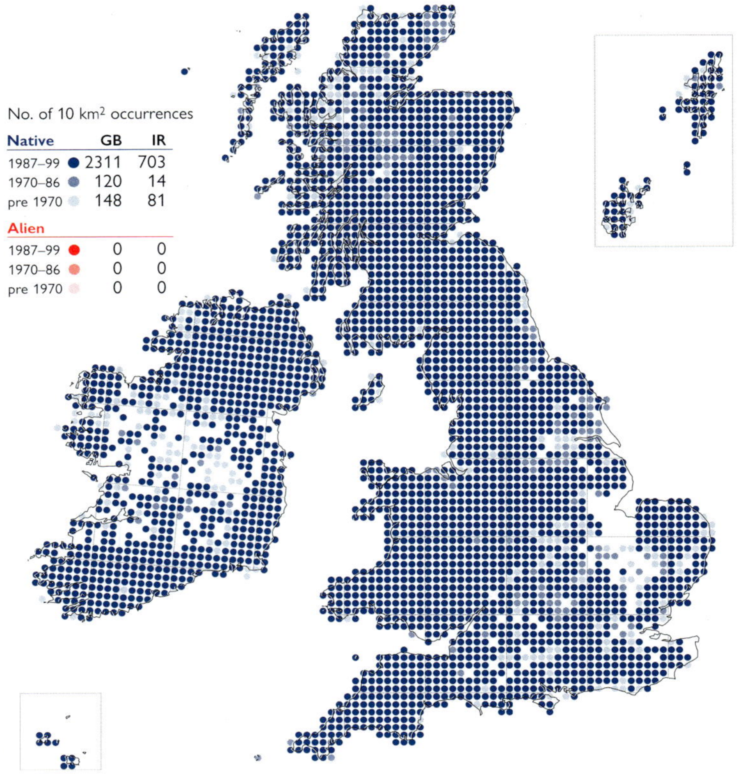

A mat-forming, winter-green perennial, occurring in a range of wet habitats including mires, wet grassland, streamsides, river banks and ditches, and especially characteristic of cattle-poached areas. 0–885 m (Helvellyn, Cumberland) with unlocalised records up to 1005 m in the Scottish Highlands.

Native (change –0.10). This species was mapped as 'all records' in the 1962 *Atlas*. It has declined in S. and E. England, and analysis of the database indicates that this has occurred since 1950, presumably due to drainage, the re-seeding of wet grasslands, or their conversion to arable.

European Temperate element; also in C. and E. Asia and N. America and widely naturalised outside its native range.

References: Atlas (71c), Grime *et al.* (1988), Hultén & Fries (1986), Jalas & Suominen (1983), Meusel *et al.* (1965).

P. S. LUSBY

Holosteum umbellatum Jagged Chickweed

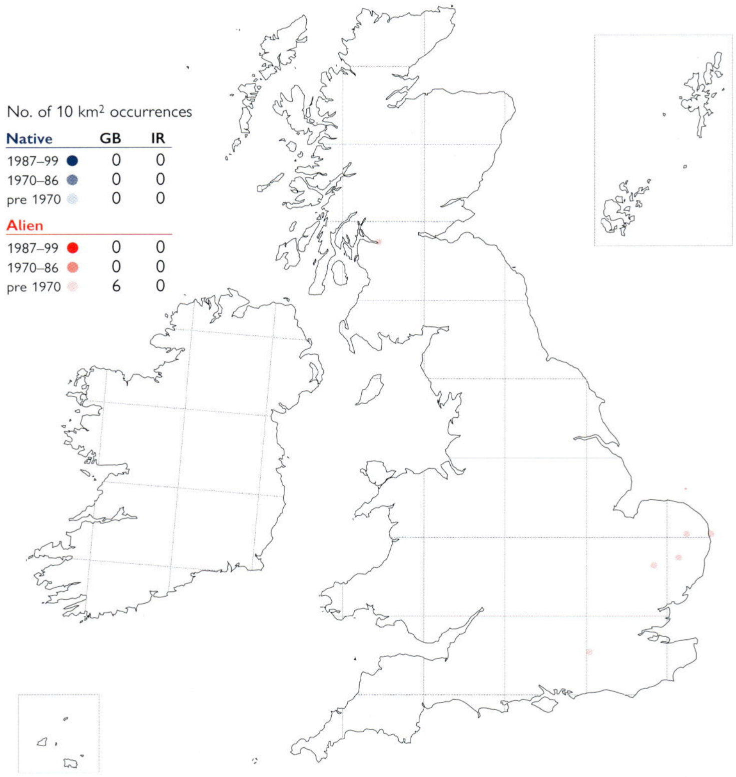

In Britain, recorded habitats of *H. umbellatum* include boundary walls, the walls of old buildings and ruins, thatched roofs, and open ground in a field. Lowland.

Neophyte. This species was recorded in Britain between 1765 and 1887 in Norwich (E. Norfolk), between 1773 and 1889 at several sites in Suffolk, and between 1813 and 1865 on Clydeside (Dunbarton). It was discovered in Surrey on ruins and in a nearby field in 1905, but was last recorded there in 1930. Although often thought to be native in England, it has not been recorded from semi-natural habitats and is treated here as an alien.

A Eurosiberian Southern-temperate species; widely naturalised outside its native range.

References: Atlas (71d), Hultén & Fries (1986), Jalas & Suominen (1983), Lousley (1976a), Meusel *et al.* (1965), Simpson (1982).

P. S. LUSBY

Cerastium cerastoides Starwort Mouse-ear

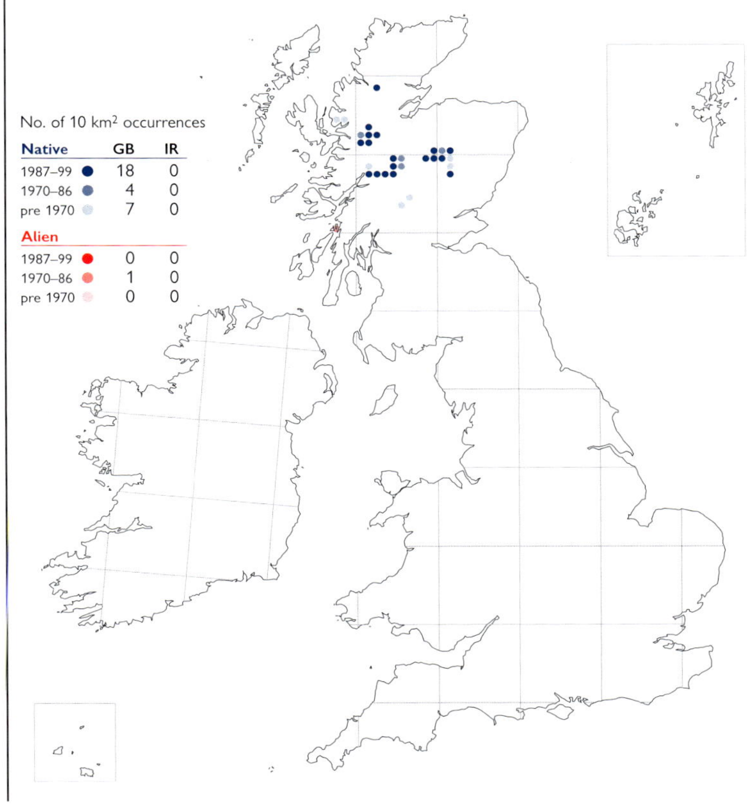

A straggling, mat-forming montane perennial herb that grows on wet acidic rocks, often in areas of late snow-lie. It is usually found above 750 m and reaches an altitude of 1220 m on Ben Macdui (S. Aberdeen), but occurs at 335 m near Mar Lodge (S. Aberdeen).

Native (change –0.05). More sites have been discovered since the 1962 *Atlas*, but the general 10-km distribution seems stable. *C. cerastoides* may still be under-recorded.

European Arctic-montane element; also in C. Asia and N. America.

References: Atlas (67a), Hultén & Fries (1986), Jalas & Suominen (1983), Meusel *et al.* (1965), Stewart *et al.* (1994).

P. S. LUSBY

Cerastium arvense Field Mouse-ear

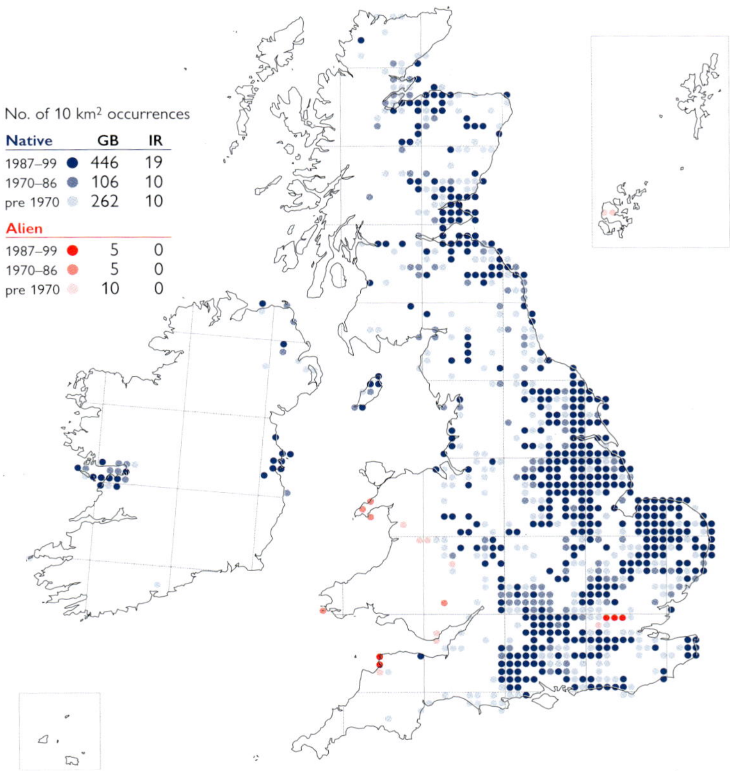

This perennial grows mainly on dry, calcareous to slightly acid, sandy soils, occurring in pastures, on dry roadsides, wayside banks, the margins of arable fields, sandy or gravelly waste ground, sand dunes and sand-pits. 0–300 m (Hounam Law, Roxburghs.).

Native (change –1.05). *C. arvense* has declined, particularly at the edges of its range. The reasons for this are unclear.

Circumpolar Boreo-temperate element; widely naturalised outside its native range.

References: Atlas (67b), Hultén & Fries (1986), Jalas & Suominen (1983), Meusel *et al.* (1965).

P. S. LUSBY

Cerastium tomentosum Snow-in-summer

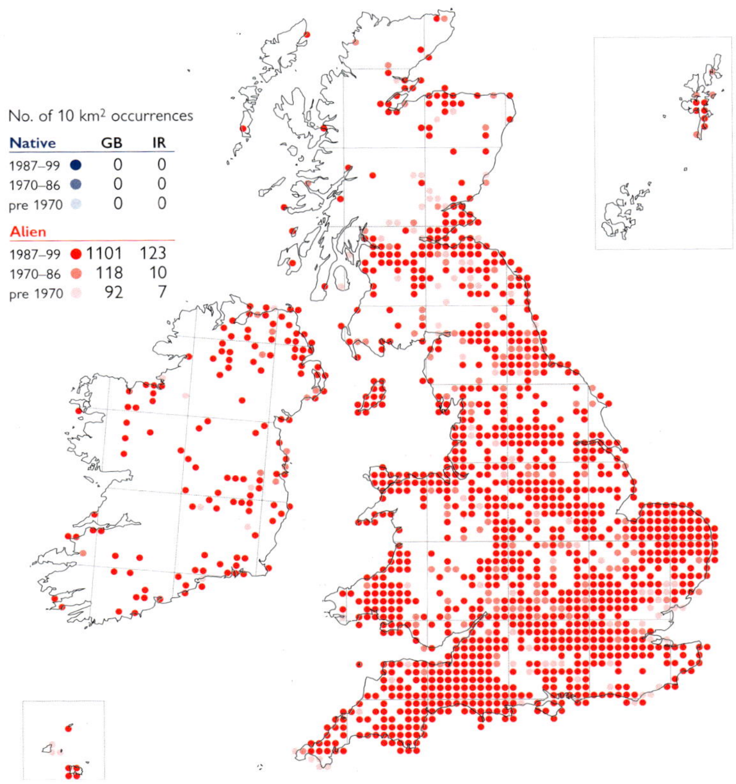

A spreading, mat-forming perennial herb naturalised on roadsides, railway banks, waste ground, tips, dunes and coastal shingle. Lowland.

Neophyte (change +2.97). *C. tomentosum* has been cultivated in Britain since 1648 and is now common in parks and gardens; records from the wild, the first of which was in 1915, are of garden escapes or plants originating from discarded material. There are five times as many 10-km square records of this species as in the 1962 *Atlas* and this probably results from both a genuine increase and the improved recording of aliens.

Native of Italy & Sicily; widely naturalised elsewhere in Europe.

References: Atlas (67c), Jalas & Suominen (1983).

P. S. LUSBY

Cerastium alpinum Alpine Mouse-ear

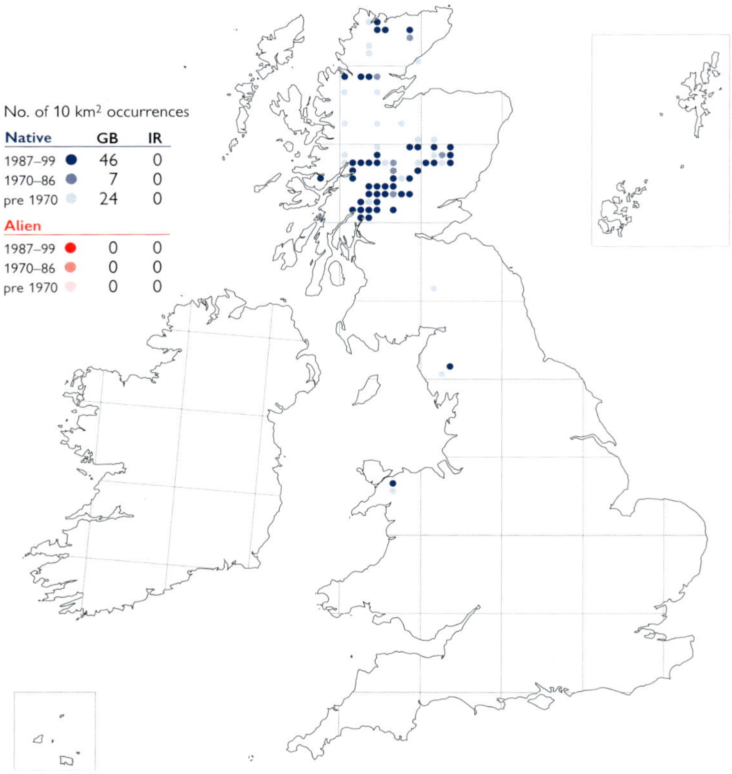

A mat-forming montane perennial herb which grows in similar habitats to *C. arcticum*, but on more strongly basic rocks. It is particularly abundant on soft mica-schists but also occurs on limestone and, rarely, serpentine. It often occurs in species-rich dwarf-herb communities. From 300 m on Seana Bhraigh (E. Ross) to 1210 m on Ben Lawers (Mid Perth).

Native (change –0.84). *C. alpinum* may still be present in many of the 10-km squares for which there are only pre-1987 records.

European Arctic-montane element; also in N. America.

References: Atlas (67d), Hultén & Fries (1986), Jalas & Suominen (1983), Meusel *et al.* (1965), Stewart *et al.* (1994).

P. S. LUSBY

Cerastium arcticum Arctic Mouse-ear

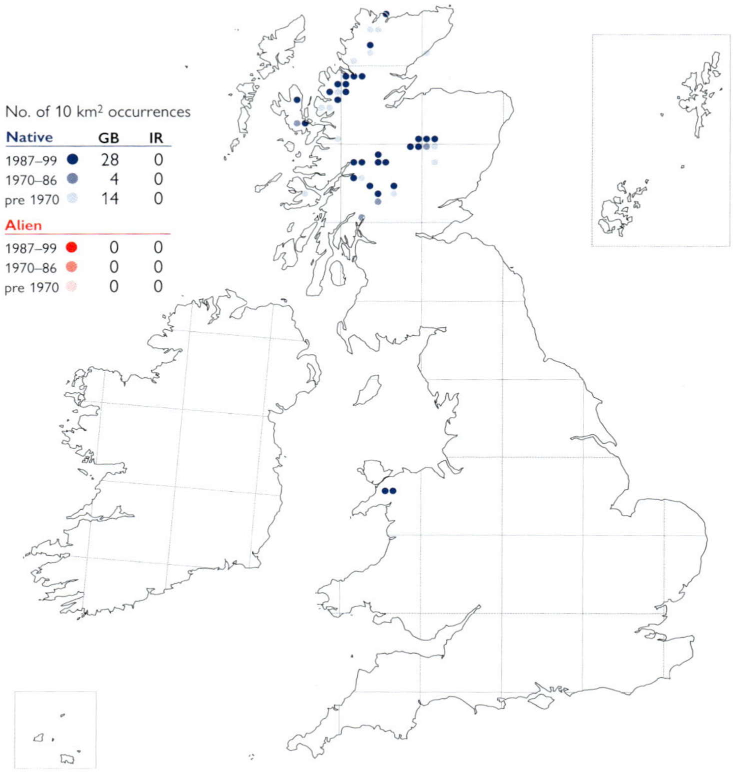

A montane, tufted perennial herb of acidic and hard basic rocks. It normally occurs in wet, thinly-vegetated crevices and on ledges in N.-facing corries, but has been recorded on a montane fell-field on Skye. Plants may sometimes be found rooted at the foot of mountain cliffs. It is rarely found below 700 m and reaches 1200 m on Ben Nevis (Westerness).

Native (change –0.37). The map shows no appreciable change in the distribution of *C. arcticum* since the 1962 *Atlas.*

European Arctic-montane element, but absent from mountains of C. Europe; also in N. America.

References: Atlas (68a), Hultén & Fries (1986), Jalas & Suominen (1983), Stewart *et al.* (1994).

P. S. LUSBY

Cerastium nigrescens Shetland Mouse-ear

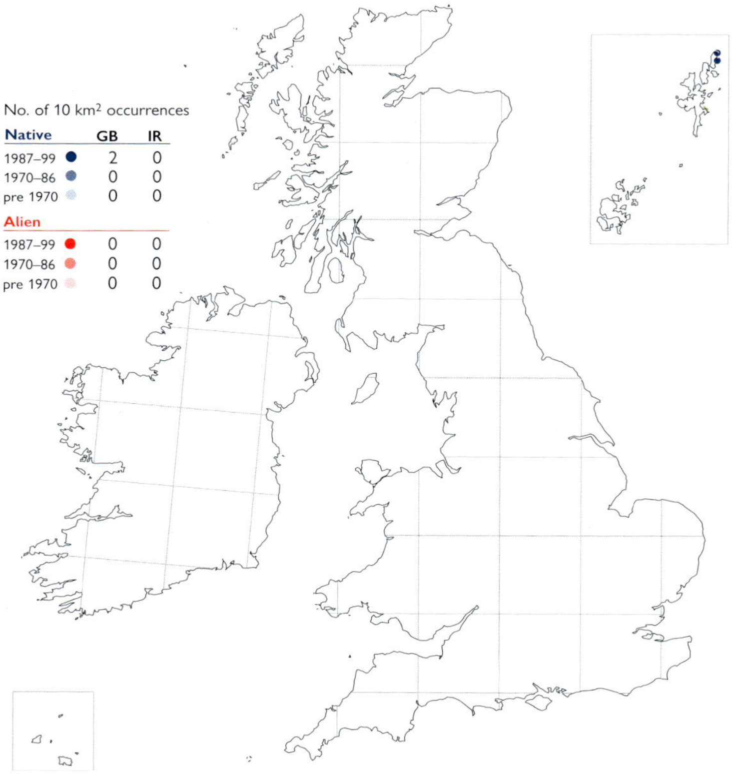

A tufted perennial herb confined to two adjacent hills on very exposed, sparsely-vegetated fell-field of shattered serpentine rock. Although it grows to an altitude of only 80 m, the habitat has many similarities to more montane communities.

Native. The numbers of *C. nigrescens* fluctuate from year to year, for reasons that are unclear, but the underlying trend seems stable, and there has been no change in its distribution.

Endemic.

References: Atlas (68a), Jalas & Suominen (1983), Scott & Palmer (1987), Wigginton (1999).

P. S. LUSBY

Cerastium fontanum Common Mouse-ear

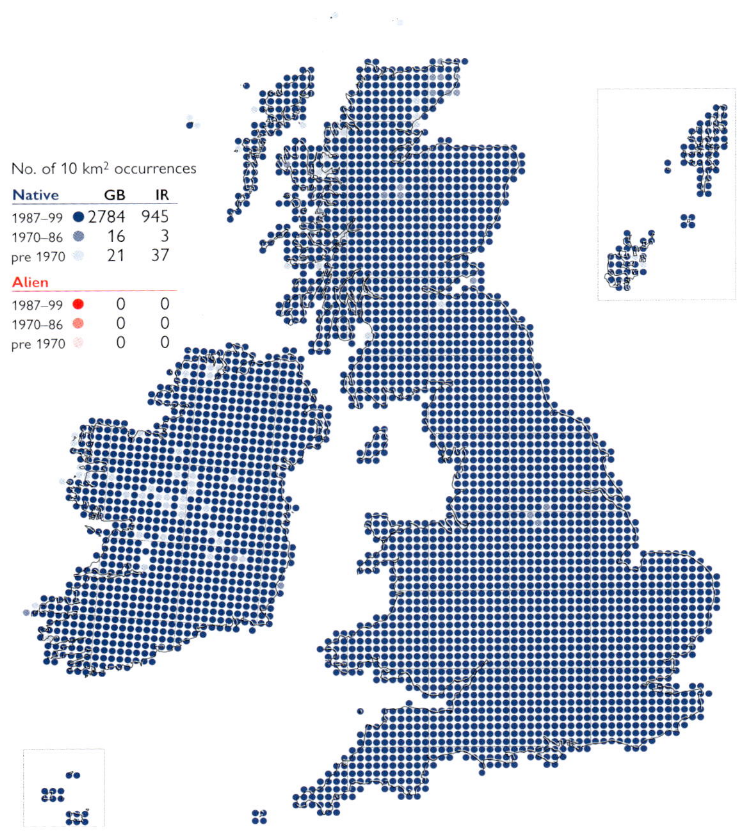

A tufted or mat-forming perennial herb found in a wide range of, usually fertile, habitats including neutral pastures and meadows, calcareous and acidic grassland, fen-meadows, rush-pastures, heaths and mires, springs and flushes, montane grassland, rocks and ledges, sand dunes and shingle, cultivated ground, waste places and walls. 0–1220 m (Aonach Beag, Westerness).

Native (change +1.40). The distribution is similar to that in the 1962 *Atlas*. Subsp. *vulgare* and subsp. *holosteoides* are common throughout the range of the species, although the latter is more frequent in the north; subsp. *scoticum* is mapped separately.

Eurosiberian Boreo-temperate element, but widely naturalised so that distribution is now Circumpolar Boreo-temperate.

References: Atlas (68b), Grime *et al.* (1988), Hultén & Fries (1986), Jalas & Suominen (1983), Peterson (1969).

P. S. LUSBY

Cerastium fontanum subsp. *scoticum*

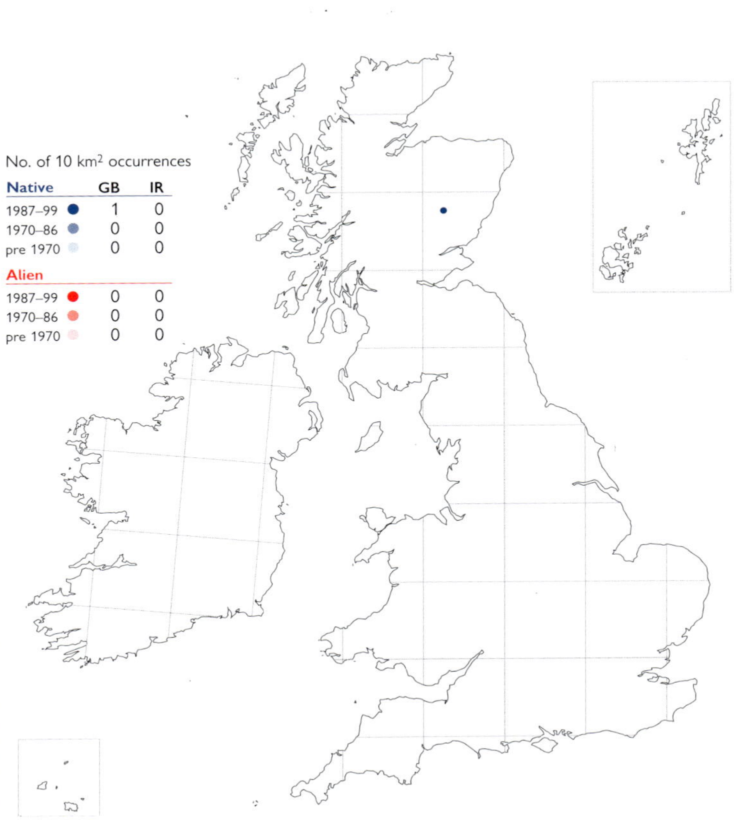

This short-lived perennial herb is known only from Meikle Kilrannoch in Angus, where it occurs principally amongst serpentine debris on exposed montane fell-field; also in montane heath dominated by *Empetrum nigrum* subsp. *hermaphroditum* and *Vaccinium myrtillus*, and rarely in *Nardus stricta* grassland. The altitude is *c.* 860 m.

Native. *C. fontanum* subsp. *scoticum* was not described until 1967. Although the three known populations are small, they are not thought to be under any significant threat.

Endemic.

References: Atlas Supp. (12b), Jalas & Suominen (1983), Wigginton (1999).

P. S. LUSBY

Cerastium glomeratum Sticky Mouse-ear

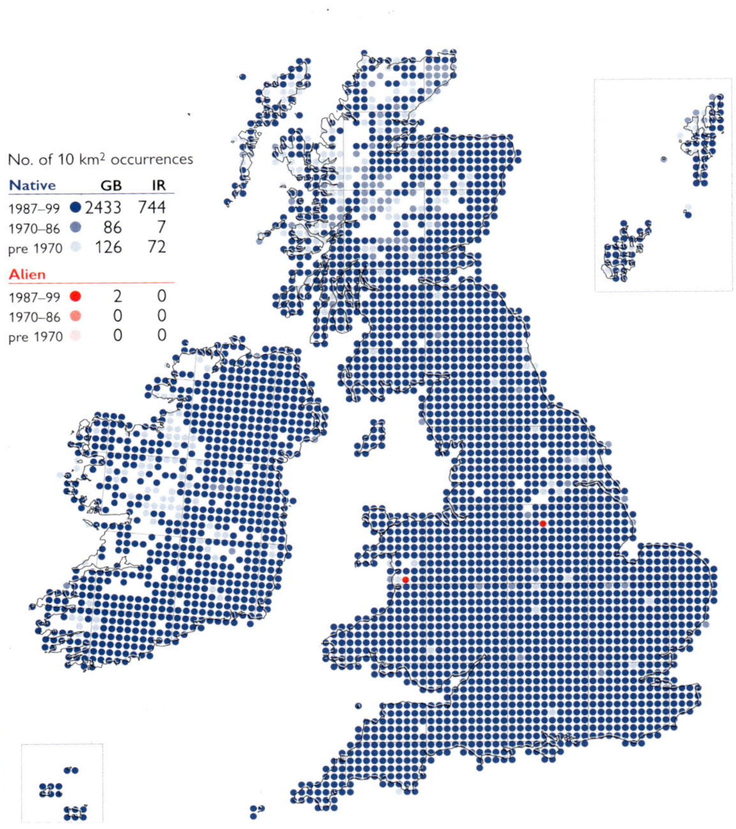

C. glomeratum grows in disturbed areas, often in places where there is some nutrient enrichment. It is fairly tolerant of trampling and is particularly common around farms, in gateways, on field edges, in bare patches in improved grassland, beside tracks and in waste places. It is also frequent on sand dunes and shingle. 0–610 m (Black's Hope, Dumfriess.).

Native (change +1.44). There has been a large increase in records compared to the 1962 *Atlas*, especially in N. & W. Britain and in Ireland. This is probably due both to more comprehensive recording and a genuine spread in disturbed, nutrient-rich habitats.

European Southern-temperate element, but widely naturalised so that distribution is now Circumpolar Southern-temperate.

References: Atlas (68c), Hultén & Fries (1986), Jalas & Suominen (1983).

P. S. LUSBY

Cerastium brachypetalum Grey Mouse-ear

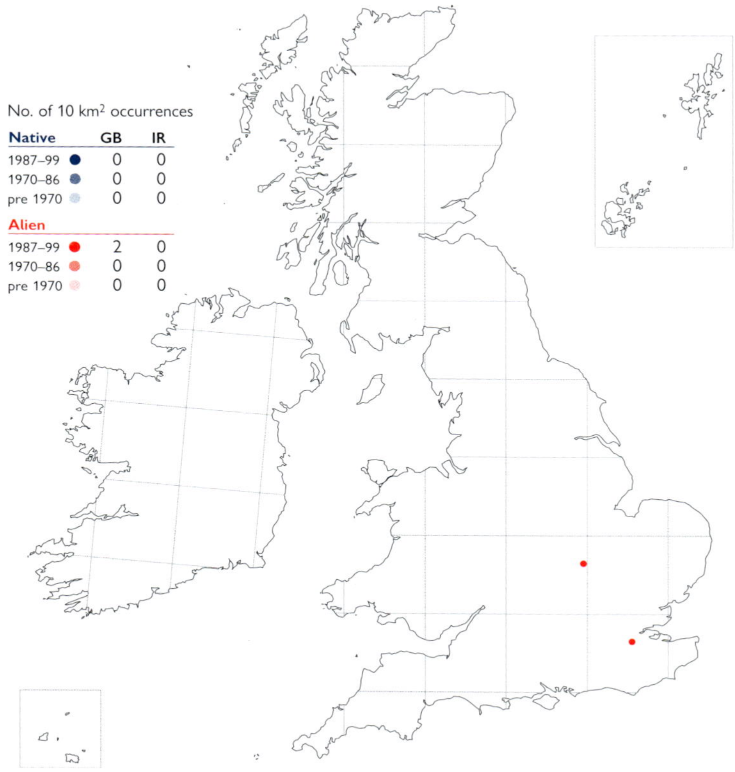

An annual of dry, calcareous, open grassland and railway banks. Seed is freely produced but is poorly dispersed, with colonies appearing in the same place year after year. Lowland.

Neophyte. *C. brachypetalum* was first discovered in Britain in Bedfordshire in 1947; its late discovery and proximity to railways suggest that it is an introduction. However, Palmer (1994) argued that its occurrence in old, *Bromopsis erecta* dominated grassland on chalk in W. Kent indicates that these populations are relics of a more widespread distribution in this area.

A European Southern-temperate species.

References: Atlas (68d), Hultén & Fries (1986), Jalas & Suominen (1983), Wigginton (1999).

P. S. LUSBY

Cerastium diffusum Sea Mouse-ear

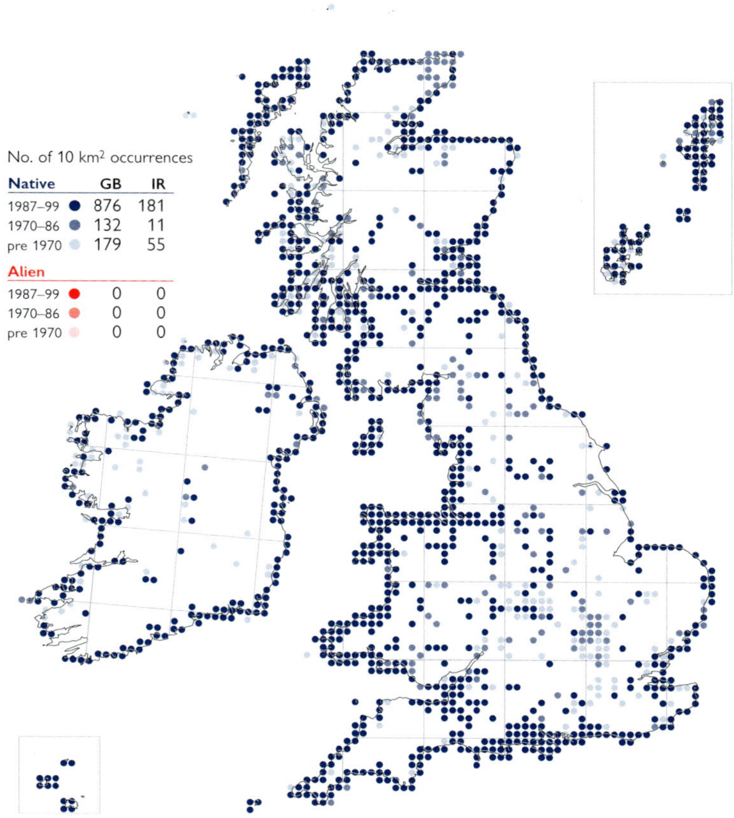

An annual of light, dry, sandy or gravelly soil. Coastal habitats include open grassland, fixed dunes, sandy banks and sheltered rock crevices. Inland, it occurs in a range of open habitats, including dry grassland, by paths, on wall-tops, road verges and railway ballast. 0–455 m (Dalnaspidal, E. Perth).

Native (change +0.38). There has been no change in the coastal distribution since the 1962 *Atlas*. It has declined in many inland railway sites, which it colonised during the Second World War, but this seems to have been offset to some extent by recent records from salt-treated roadsides. It has not been possible to distinguish alien from native records inland and all are mapped as if they were native.

European Temperate element.

References: Atlas (69a), Hultén & Fries (1986), Jalas & Suominen (1983).

P. S. LUSBY

Cerastium pumilum Dwarf Mouse-ear

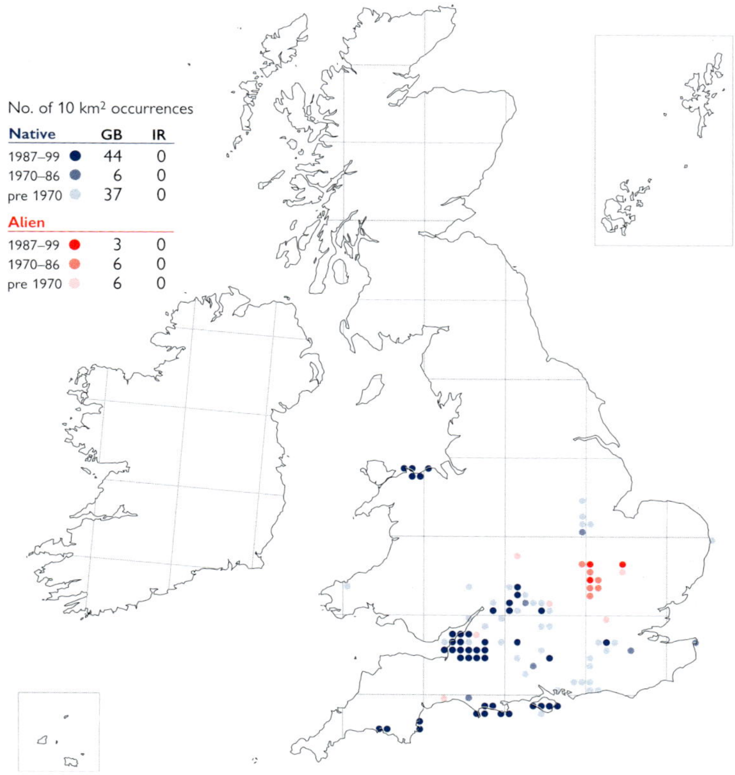

A winter-annual of chalk and limestone substrates, occurring mainly in open, barish patches within short-grazed grassland, especially on sunny banks and cliffs; also found as an alien in quarries, on spoil heaps and on ballast along railway lines (e.g. in Beds.). Lowland.

Native (change –0.17). Many of the sites in the eastern part of its range were lost before 1930, but *C. pumilum* has apparently decreased further here since the 1962 *Atlas*. It seems to be stable elsewhere, and now much better recorded in its core areas. However, because the plant resembles other *Cerastium* species, has a short season and fluctuates in numbers, it may perhaps be overlooked to some extent.

European Temperate element.

References: Atlas (68d), Hultén & Fries (1986), Jalas & Suominen (1983), Stewart *et al.* (1994).

P. S. LUSBY

Cerastium semidecandrum Little Mouse-ear

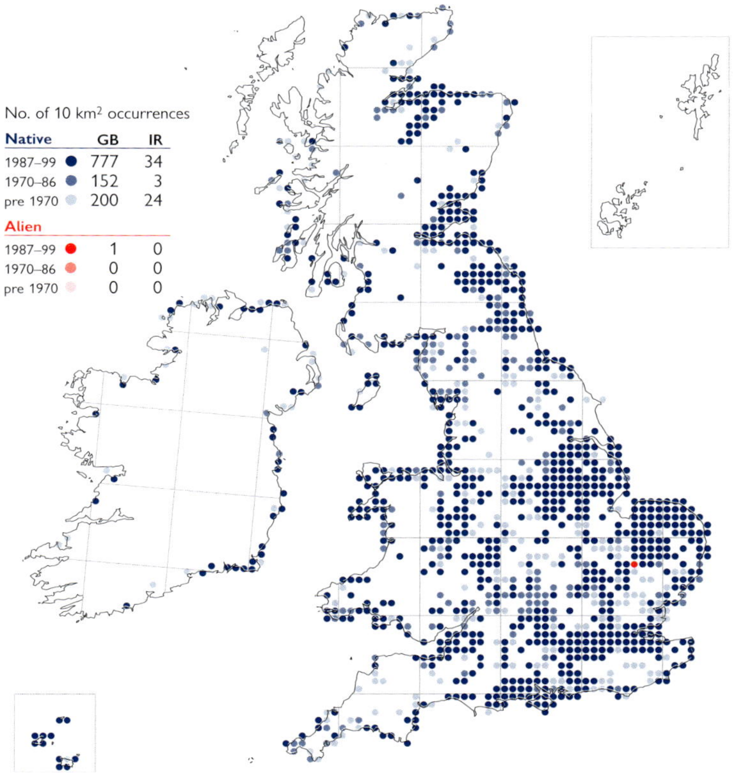

This annual or overwintering herb of well-drained, sandy or calcareous soils is found on dry banks and open, grassy places, on heathy ground, fixed dunes, disturbed sandy areas near the sea and walls. It also grows on rock ledges and bare places on limestone, and occasionally grows in heavier clay soils. 0–485 m (Isla, E. Perth).

Native (change +0.50). This species was under-recorded in the 1962 *Atlas*. The map presents a much improved illustration of its distribution, but it may remain poorly recorded in some areas.

European Temperate element; widely naturalised outside its native range.

References: Atlas (69b), Hultén & Fries (1986), Jalas & Suominen (1983), Meusel *et al.* (1965).

P. S. LUSBY

Myosoton aquaticum Water Chickweed

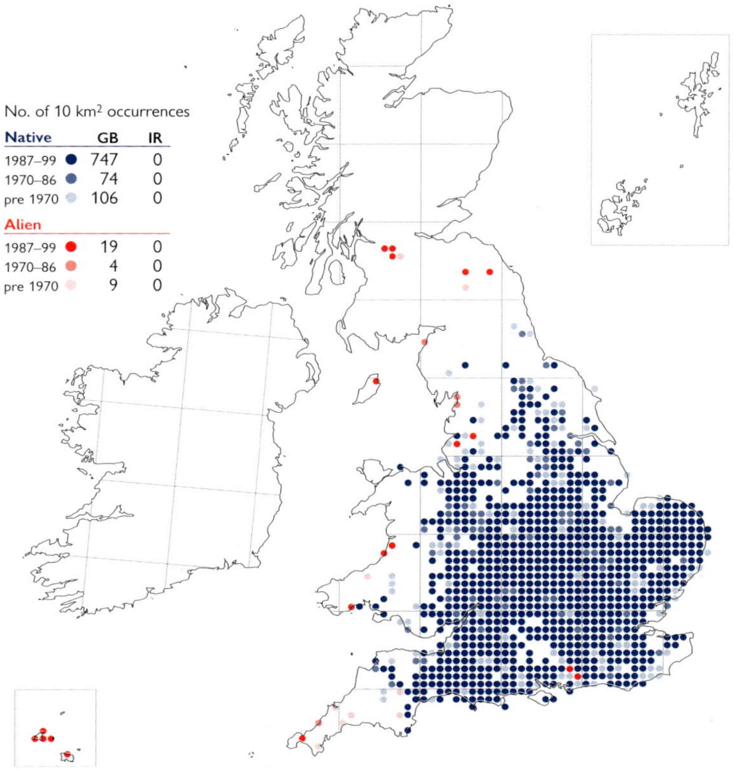

This perennial herb usually grows in damp or wet habitats, including damp woods, *Alnus* and *Salix* carr, the banks of rivers, streams, canals and ditches, by ponds and in marshes and other wet places. It has been recorded on newly surfaced forestry rides, and can also tolerate occasional flooding by brackish water. Lowland.

Native (change 0.00). The distribution has been stable since the 1962 *Atlas*. Some records in the north of its range represent casual occurrences.

Eurosiberian Temperate element; also in E. Asia and widely naturalised outside its native range.

References: Atlas (69c), Hultén & Fries (1986), Jalas & Suominen (1983).

P. S. LUSBY

Moenchia erecta Upright Chickweed

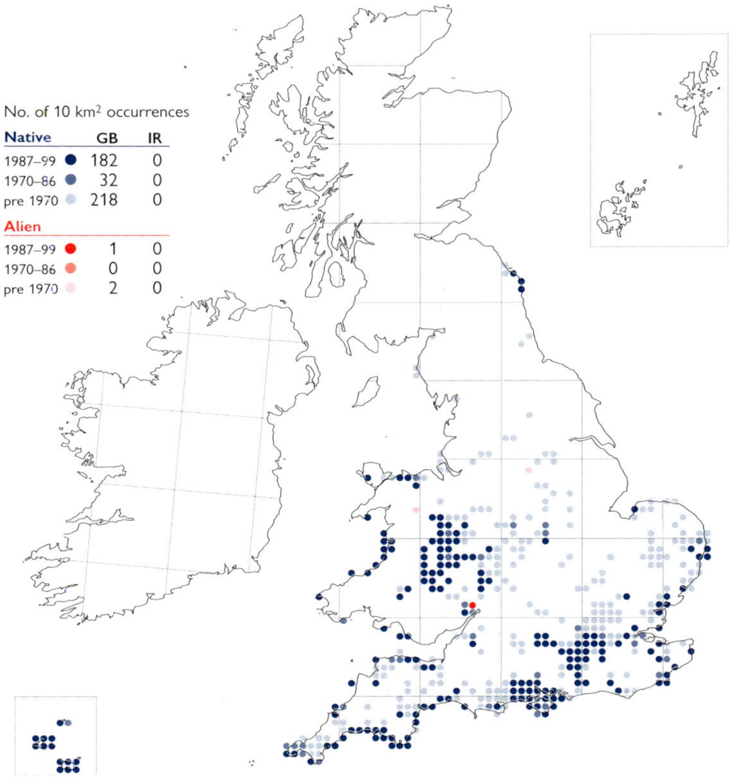

This annual grows on summer-droughted soils in tightly-grazed grasslands and heaths, on cliff-tops, pathsides, coastal dunes and sandy shingle, usually in open patches where competition is limited. It is also found in quarries and sand-pits, and on other disturbed ground. Generally lowland, but reaching 410 m at Widecombe, Dartmoor (S. Devon).

Native (change –0.65). The species suffered a considerable decline before 1930, and this decline may still be continuing. The increased number of 10-km records since the 1962 *Atlas* must be attributed to a greater intensity of recording of this rather inconspicuous species.

Suboceanic Southern-temperate element.

References: Atlas (72a), Jalas & Suominen (1983), Meusel *et al.* (1965), Stewart *et al.* (1994).

P. S. LUSBY

Sagina nodosa Knotted Pearlwort

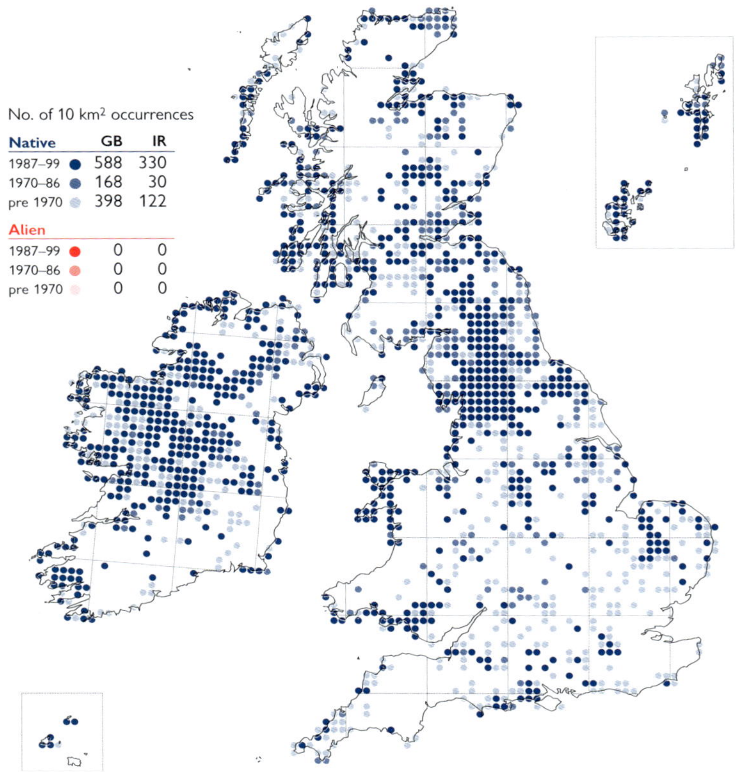

S. nodosa is a plant of damp habitats, principally in mires and springs irrigated with base-rich water, but also in open, calcareous, sandy habitats, especially dunes and dune-slacks and sometimes in drier calcareous grassland. Generally lowland, but reaching 850 m on Glas Moel (Angus).

Native (change –1.14). This species was mapped as 'all records' in the 1962 *Atlas*. There has been a widespread decline in S. England, and analysis of the database reveals that this has occurred since 1950. This reflects the loss of open, calcareous habitats.

Eurosiberian Boreo-temperate element; also in N. America.

References: Atlas (74b), Hultén & Fries (1986), Jalas & Suominen (1983), Meusel *et al.* (1965).

P. S. LUSBY

Sagina nivalis Snow Pearlwort

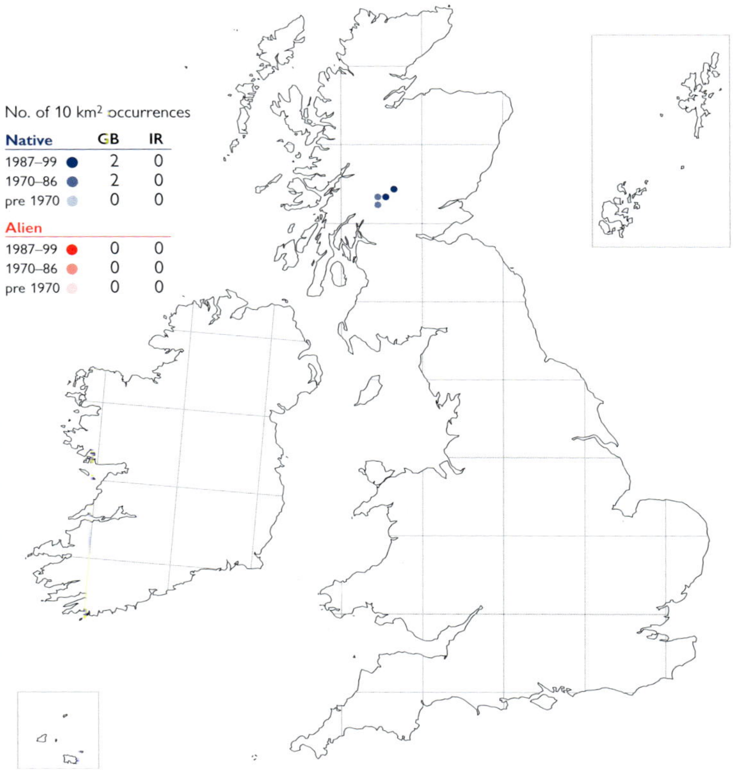

This cushion-forming, perennial herb usually occurs in open, unstable habitats with little vascular plant cover, generally growing on soft calcareous schist. It is found in both dry and damp conditions, often growing through a bryophyte mat on a rock face or on gravelly ground. Its wetter sites are subject to periodic flushing. Between 840 m on Meall nan Tarmachan and 1190 m on Ben Lawers (both Mid Perth).

Native. Though losses occur from rock falls and erosion, new colonies establish on the exposed soil, and its range and distribution is likely to be more or less stable.

Circumpolar Arctic-montane element; absent from mountains of C. Europe.

References: Atlas (73d), Hultén & Fries (1986), Jalas & Suominen (1983), Wigginton (1999).

P. S. LUSBY

Sagina subulata Heath Pearlwort

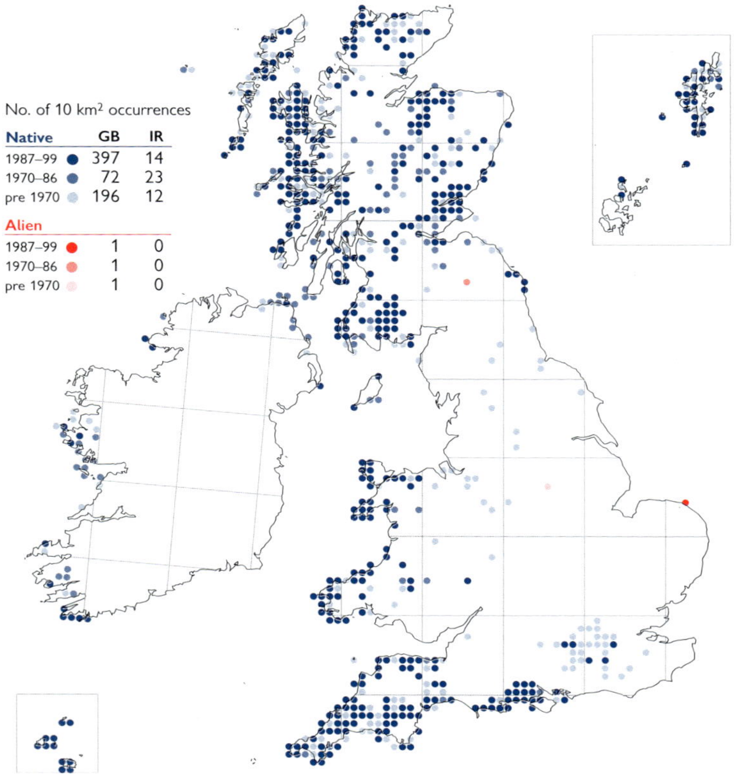

A perennial, mat-forming herb usually of dry, open, sandy or gravelly places. Habitats include heaths, dry pastures, banks and rocks, grassy slopes near the sea, and especially trackways through heaths and moors. It also occurs on basaltic gravel terraces on the Trotternish Mountains (Skye), accompanying *Koenigia islandica*. 0–700 m (Mt Brandon, S. Kerry), and reportedly to 835 m on Ben Lawers (Mid Perth).

Native (change –0.44). Many inland sites had been lost before 1930 and, whilst there have been further losses, especially in W. Ireland, there is little evidence of decline in coastal areas.

European Temperate element.

References: Atlas (74a), Harrold (1978), Hultén & Fries (1986), Jalas & Suominen (1983).

P. S. LUSBY

Sagina saginoides Alpine Pearlwort

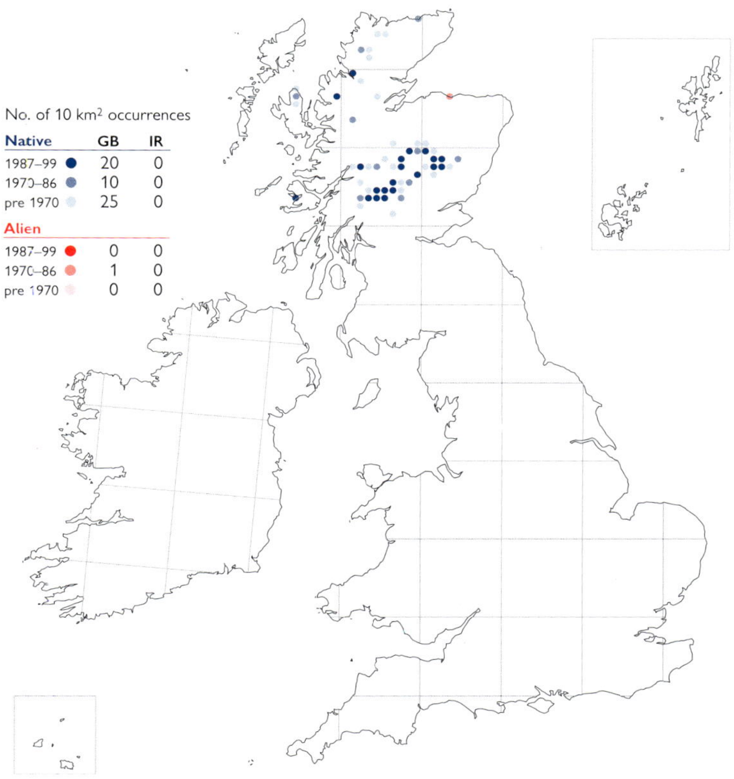

A tufted, perennial herb which usually grows on base-rich, well-drained soils. It is a poor competitor, occurring on steep ground, in areas of late snow-lie and areas of severe wind-scour which provide the open conditions it requires. Generally between 700 and 900 m, but reaching 1190 m on Ben Lawers (Mid Perth), and descending to 460 m on the Old Man of Storr, Skye (N. Ebudes).

Native (change –0.77). There does not appear to be any significant change since the 1962 *Atlas*. The lack of recent records in some squares may only be due to under-recording.

Circumpolar Arctic-montane element.

References: Atlas (73b), Hultén & Fries (1986), Jalas & Suominen (1983), Stewart *et al.* (1994).

P. S. LUSBY

Sagina procumbens Procumbent Pearlwort

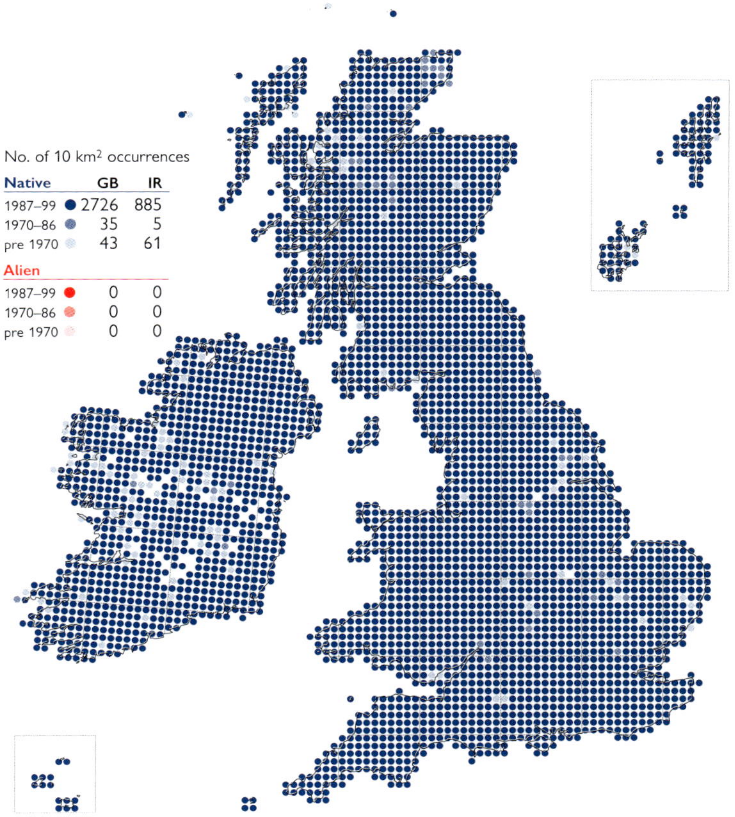

Natural habitats of this mat-forming perennial include rocks, cliffs and river banks, but it also grows in a wide variety of artificial, disturbed and fertile habitats, including spoil heaps, mining waste, paths, roadside verges and urban pavements. It is a common weed of horticulture, especially in lawns, and is a particular nuisance in pots. The plant is tolerant of a wide range of soils and can stand heavy trampling. 0–1150 m (E. Scottish Highlands).

Native (change +1.28). There has been no change in the distribution of this species since the 1962 *Atlas*.

Eurosiberian Boreo-temperate element; widely naturalised outside its native range.

References: Atlas (73a), Grime *et al.* (1988), Hultén & Fries (1986), Jalas & Suominen (1983), Meusel *et al.* (1965).

P. S. LUSBY

Sagina apetala Annual Pearlwort

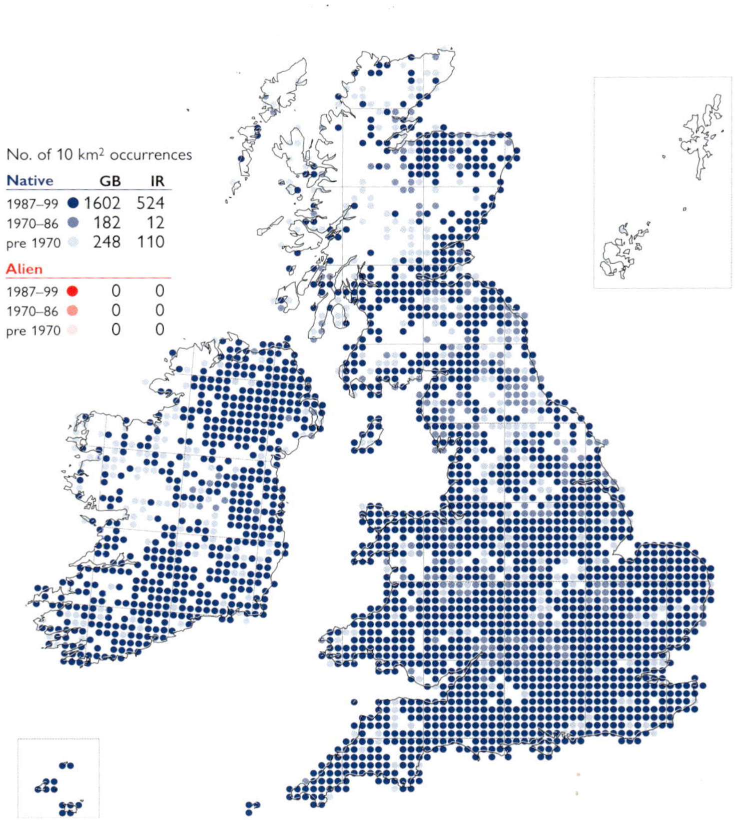

An annual of open situations, predominately in artificial habitats but also in gravelly and sandy places. 0–425 m (Grinton, N.W. Yorks.).

Native (change +0.25). The map shows little change from those of the two component subspecies which were mapped in the 1962 *Atlas*. Comparison of this map with the following maps of the subspecies shows that recorders have only unevenly distinguished between them.

European Southern-temperate element; widely naturalised outside its native range.

References: Atlas (72b, c), Hultén & Fries (1986), Jalas & Suominen (1983).

P. S. LUSBY

Sagina apetala subsp. *apetala*

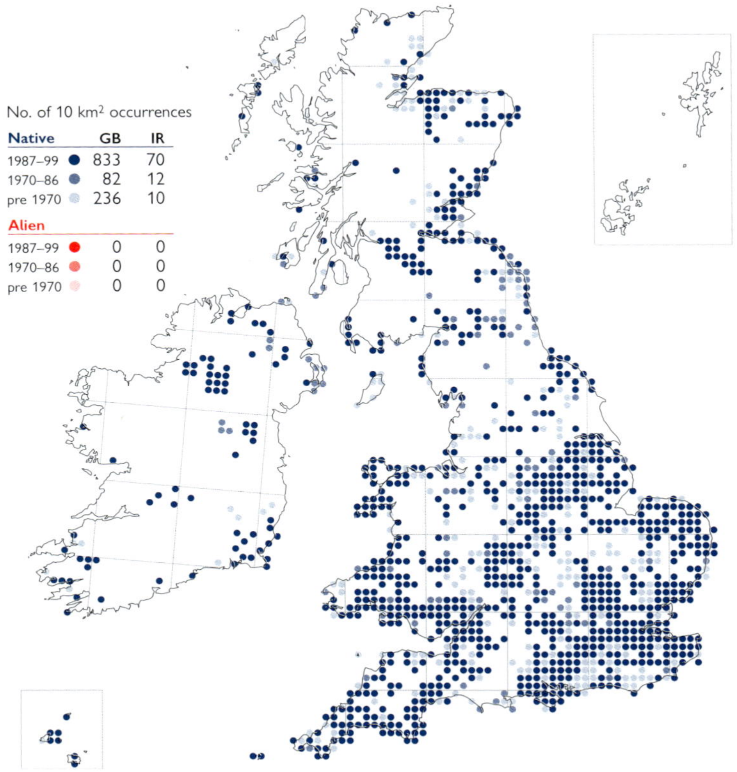

This small overwintering annual is found mainly in very open situations in artificial habitats, including the bases of walls, between paving slabs, on sandy and gravelly paths, forestry tracks and cinders in railway sidings. More natural habitats include gravelly and sandy places on heaths and commons. Generally lowland, but reaching 375 m at Esgair Fraith (Cards.).

Native. Although now better recorded than in the 1962 *Atlas*, subsp. *apetala* may still be under-recorded in some areas, making the assessment of changes in distribution difficult.

European Southern-temperate element. There is little difference between the ranges of the two subspecies, and the distinctions between them tend to break down in the Mediterranean region.

Reference: Atlas (72c).

P. S. LUSBY

Sagina apetala subsp. *erecta*

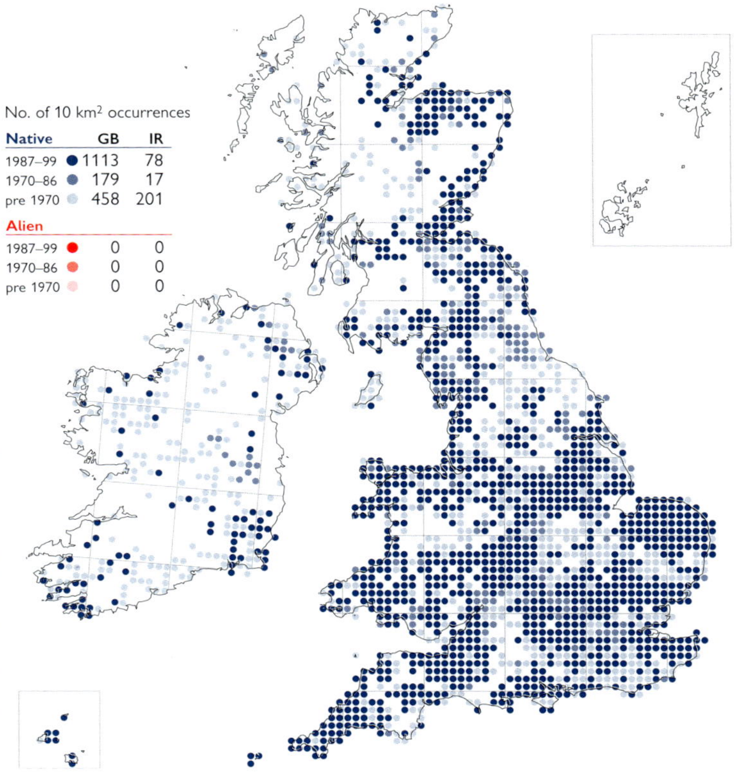

Subsp. *erecta* is found in similar habitats to subsp. *apetala*, including walls and pavements, paths and tracks, forestry rides, railways sidings, and, exceptionally, saltmarsh turf (Halliday, 1997). 0–425 m (Grinton, N.W. Yorks.).

Native. Although now better recorded than in the 1962 *Atlas*, subsp. *erecta* may still be under-recorded in some areas, making the assessment of changes in distribution difficult. There is also the possibility of nomenclatural confusion, as subsp. *erecta* is the plant mapped as *S. apetala* in the 1962 *Atlas*.

European Southern-temperate element.

Reference: Atlas (72b).

P. S. LUSBY

Sagina maritima Sea Pearlwort

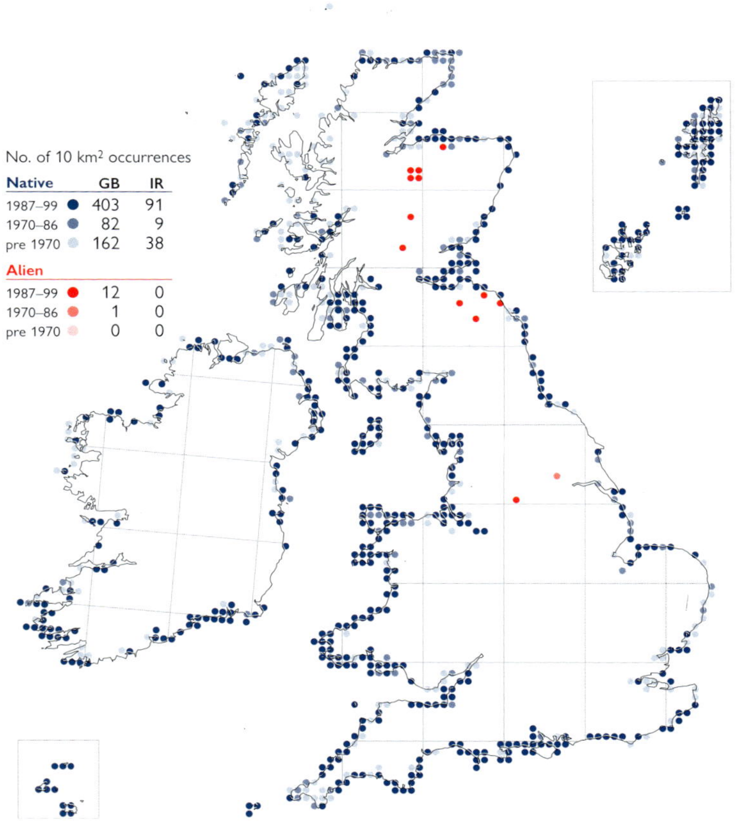

An annual of maritime rock crevices, cliff-tops, stabilised shingle, dune-slacks and disturbed areas in upper saltmarsh on sandy substrates; also on walls and tracks, in pavements and on sandy roadsides near the sea. A form of it, described as var. *alpina*, has been recorded at 1210 m on Ben Nevis (Westerness), though the identity of such plants requires confirmation.

Native (change –0.08). There has been little significant change in the distribution of this species since the 1962 *Atlas*, although it appears to have declined or has possibly been overlooked recently in N.W. Scotland. It has recently been recorded from inland squares beside salt-treated roads.

European Southern-temperate element.

References: Atlas (72d), Hepburn (1952), Hultén & Fries (1986), Jalas & Suominen (1983), McCallum Webster (1978), Meusel *et al.* (1965).

P. S. LUSBY

Scleranthus perennis subsp. *perennis*

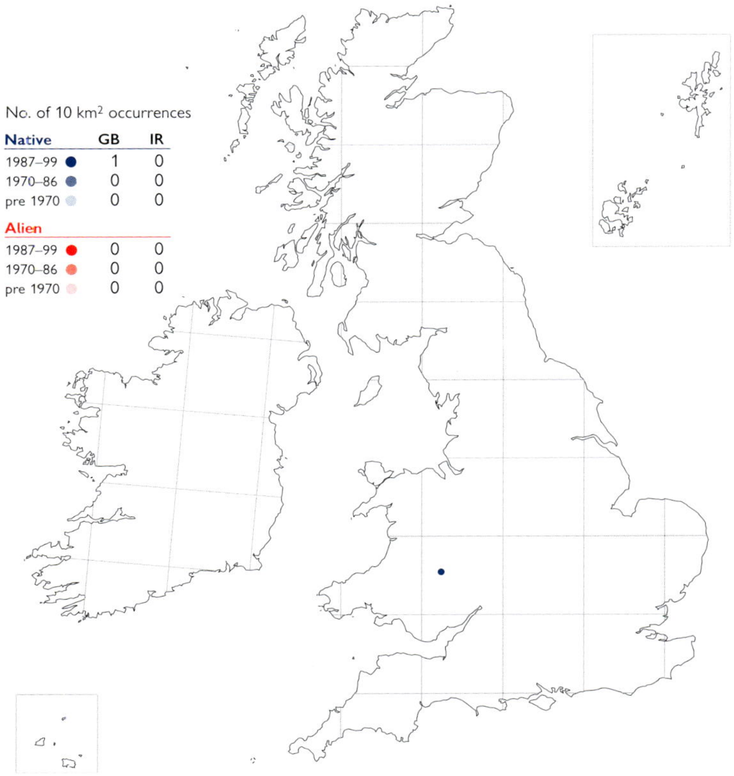

A biennial herb restricted to dry, basic igneous rocks. It grows in pockets of shallow soil at the same site as a distinctive assemblage of other drought-tolerant species, including *Gagea bohemica*. Lowland.

Native (change for species –0.11). Stanner Rocks (Rads.) remains its only known site, where it was first confirmed in 1850. The population size fluctuates greatly from year to year, and flowering is usually sparse and sometimes lacking altogether. However, in the longer term, the population appears to be stable, and plants from this site are being maintained in cultivation.

European Temperate element.

References: Atlas Supp. (14a), Hultén & Fries (1986), Jalas & Suominen (1983), Kay & John (1995), Meusel *et al.* (1965), Wigginton (1999), Woods (1993).

P. S. LUSBY

Scleranthus perennis subsp. *prostratus*

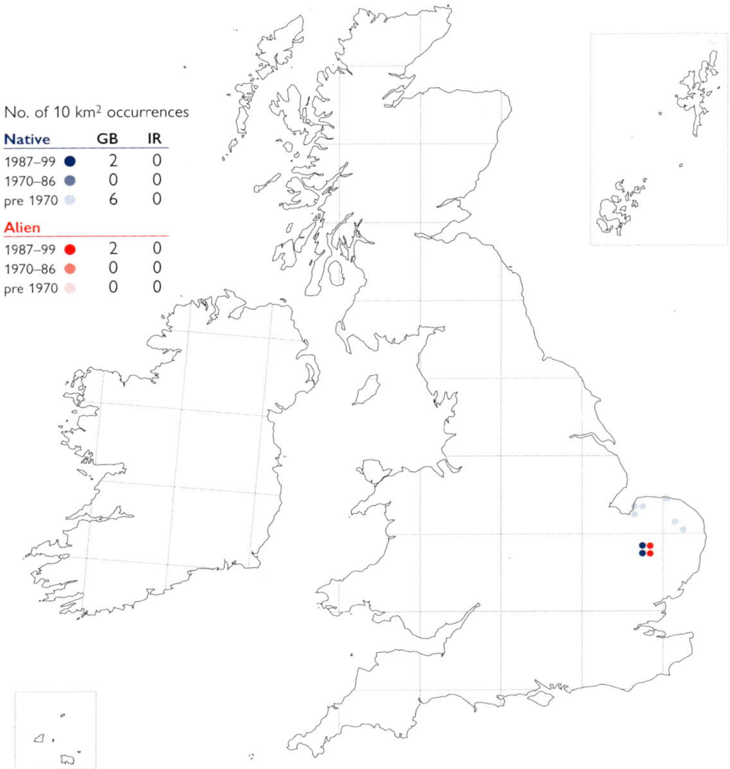

A biennial or short-lived perennial of Breckland, where it grows on acidic sandy soils in open grassland and bare areas beside tracks. It is a poor competitor, but persists in semi-closed grassland by tracks and roadsides. It has also been recorded on abandoned arable land and open fallow. Lowland.

Native (change for species –0.11). *S. perennis* subsp. *prostratus* was lost from its sites outside Breckland by 1961. In recent years the appearance of some Breckland populations has been sporadic, but site management together with deliberate re-introductions to former sites may assist its survival.

Endemic.

References: Atlas Supp. (14a), Jalas & Suominen (1983), Wigginton (1999).

P. S. LUSBY

Scleranthus annuus Annual Knawel

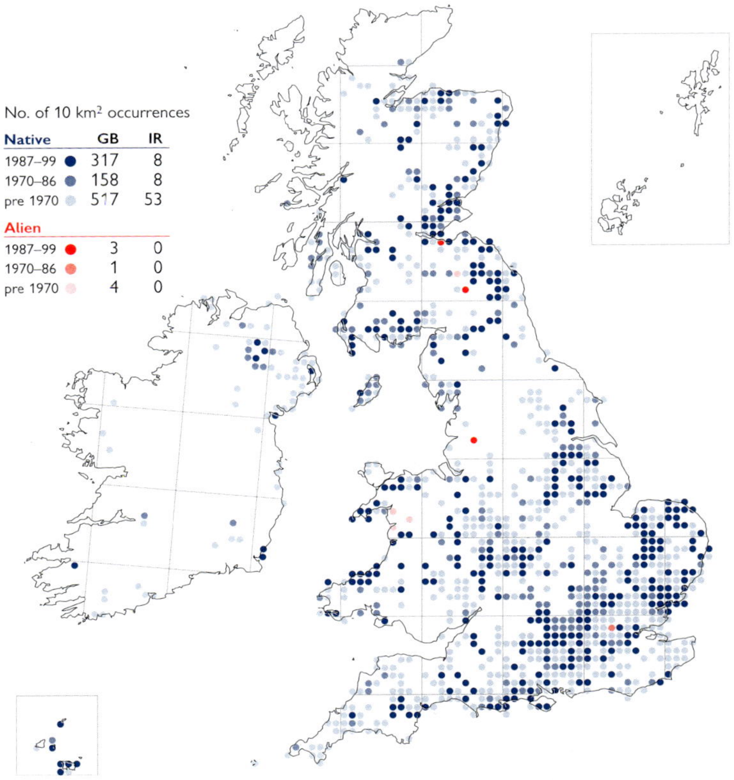

An annual or biennial herb found in soil pockets on summer-droughted rocks, and on disturbed sandy soil on heaths, commons, waste places, arable fields and, rarely, river or maritime shingle. Generally lowland, but reaching 365 m in Aberdeenshire.

Native (change –2.68). *S. annuus* was mapped as 'all records' in the 1962 *Atlas*. It has declined significantly throughout its range and analysis of the database shows that most losses have occurred since 1950. Subsp. *annuus* occurs throughout the range of the species, subsp. *polycarpos* is mapped separately.

European Temperate element; widely naturalised outside its native range.

References: Atlas (79c), Hultén & Fries (1986), Jalas & Suominen (1983).

P. S. LUSBY

Scleranthus annuus subsp. *polycarpos*

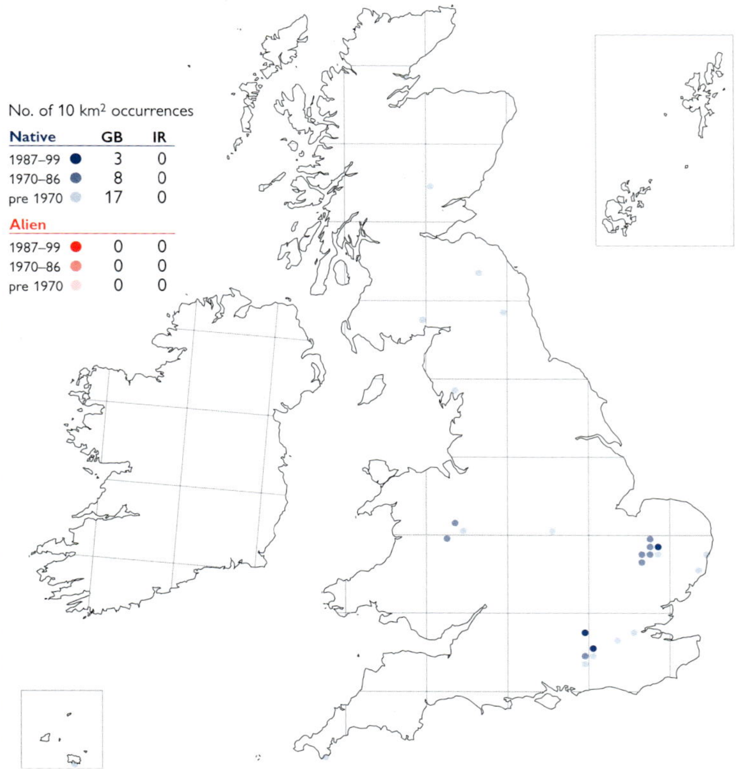

An annual to biennial herb of dry, sandy ground on heaths, commons and forestry tracks in the Breckland. Elsewhere it is recorded from sandy heaths, igneous rock outcrops, wall-tops and, rarely, as a weed of cultivated land. Lowland.

Native. This taxon was tentatively reported (as *S. polycarpos*) by Clapham *et al.* (1962), treated as a subspecies by Tutin *et al.* (1964) and mapped in Britain by Jalas & Suominen (1983), but its presence did not become known to most recorders until it was reported from Suffolk by Stace (1991). Many recorders have probably not realised that it occurs outside Breckland, and it is almost certainly under-recorded.

European Temperate element.

Reference: Trist (1979).

P. S. LUSBY

Corrigiola litoralis Strapwort

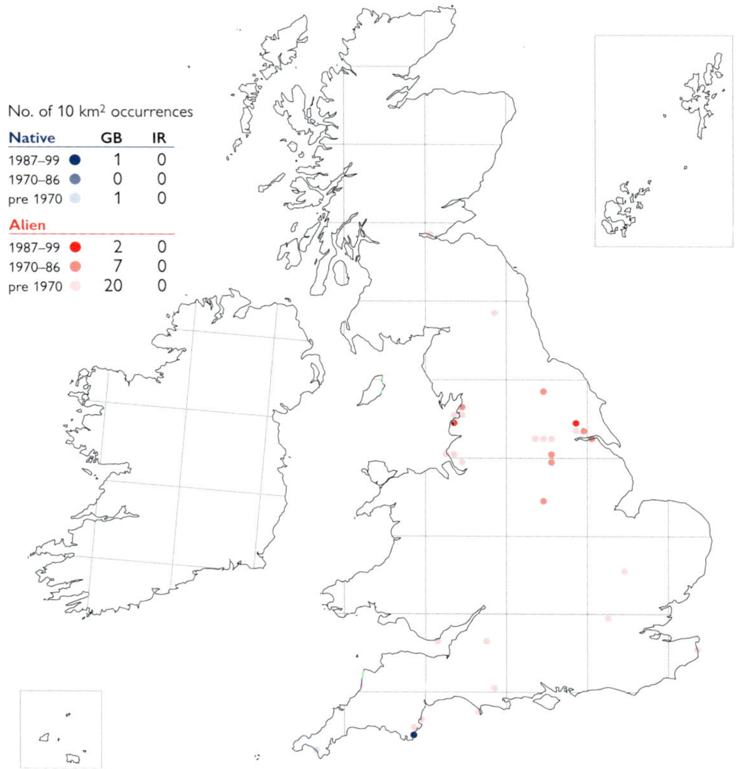

This annual is now confined, as a native plant, to periodically inundated, open, muddy shingle around the margins of Slapton Ley (S. Devon). It formerly grew in a similar site at Loe Pool (W. Cornwall), and casual plants have been recorded elsewhere, from railway ballast and waste ground. Lowland.

Native (change –0.96). At Slapton Ley, populations have declined in recent years, partly due to reduced trampling by cattle, but current conservation management aimed at maintaining open conditions should ensure its survival there. The Loe Pool population was lost because of stabilisation of water levels and invasion by *Phragmites australis.*

Suboceanic Southern-temperate element.

References: Atlas (78c), Coker (1962), Hultén & Fries (1986), Jalas & Suominen (1983), Meusel *et al.* (1965), Wigginton (1999).

P. S. LUSBY

Herniaria glabra Smooth Rupturewort

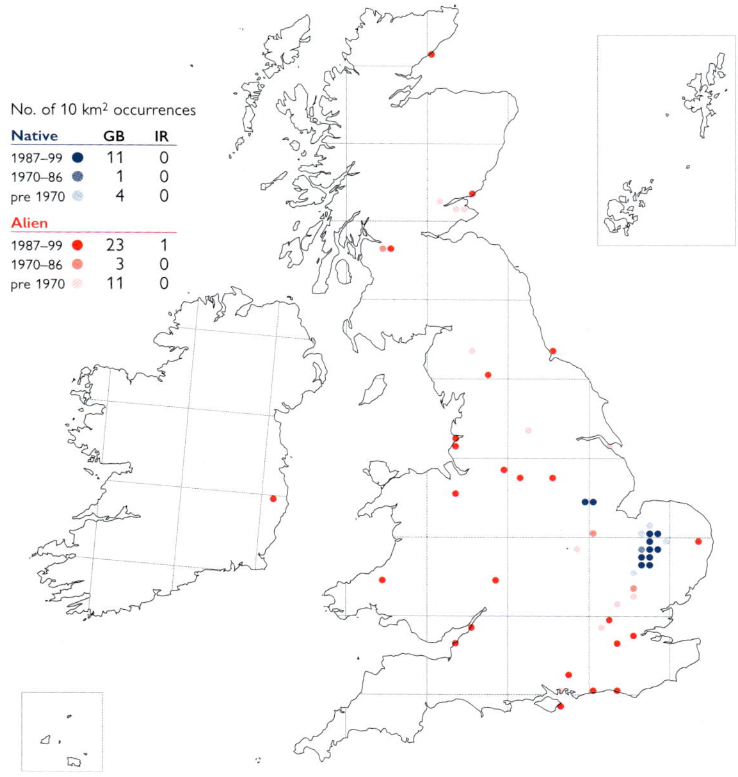

An annual or short-lived perennial of compacted sandy or gravelly soils, often with chalk or limestone fragments. Its habitats are generally kept open by seasonal standing water or other disturbance, and include forestry rides, golf courses, car parks, disused gravel-pits and disturbed areas in short grassland. Lowland.

Native (change +0.83). Intensive surveys since the 1962 *Atlas* have shown *H. glabra* to have a wider distribution in East Anglia, with new sites still being discovered. However, it is presumed extinct in Cambridgeshire, where it was last seen in 1990. It has appeared as a casual in many sites, possibly as a garden escape, and is sometimes naturalised.

Eurosiberian Temperate element.

References: Atlas (78d), Hultén & Fries (1986), Jalas & Suominen (1983), Meusel *et al.* (1965), Wigginton (1999).

P. S. LUSBY

Herniaria ciliolata Fringed Rupturewort

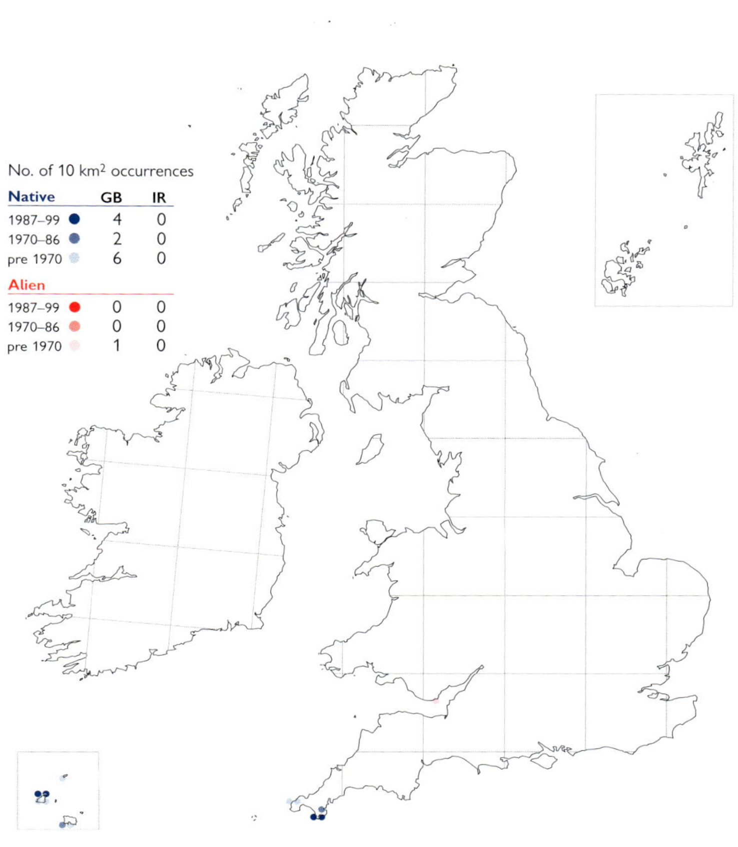

This mat-forming perennial grows in short, open vegetation, usually on dry, base-rich soils or rocks on S.-facing slopes. Its habitats include coastal cliff-slopes, valley sides, dune grassland, rock outcrops and open patches in heathlands. More ruderal habitats include path edges and stone-faced banks. Lowland.

Native. There has been little change in the distribution of this species since the 1962 *Atlas*. In our area there are two endemic subspecies, subsp. *ciliolata* (Cornwall, Alderney and Guernsey) and subsp. *subciliata* (Jersey).

Oceanic Southern-temperate element; confined to the coast of W. Europe.

References: Atlas (79a), Jalas & Suominen (1983), Wigginton (1999).

P. S. LUSBY

Illecebrum verticillatum Coral-necklace

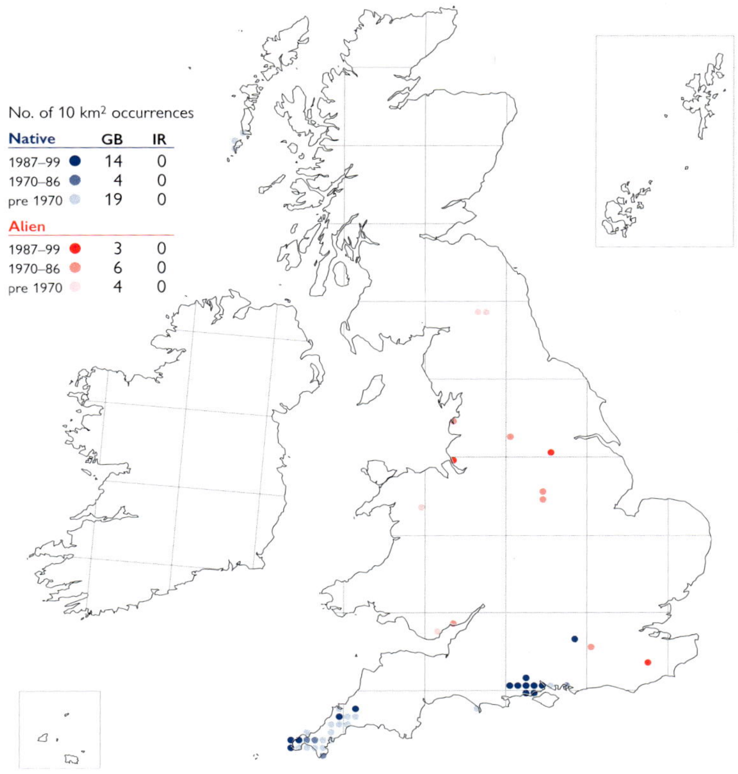

An annual of periodically wet or inundated acidic to neutral soils on gravelly tracks, pool and ditch margins, in very short heathy swards and grassland; also recorded on clinker in railway sidings. Lowland.

Native (change –0.60). This species was not recorded in Hampshire until 1920 and it appears to have spread steadily in the New Forest since 1950. Cornish populations remain fairly stable, with most losses there having occurred before 1930. Old records from the Outer Hebrides have not been confirmed but subfossil pollen has recently been discovered there, suggesting that they may be correct (Whittington & Edwards, 2000).

Suboceanic Southern-temperate element.

References: Atlas (79b), Brewis *et al.* (1996), Hultén & Fries (1986), Jalas & Suominen (1983), Meusel *et al.* (1965), Stewart *et al.* (1994).

P. S. LUSBY

Polycarpon tetraphyllum Four-leaved Allseed

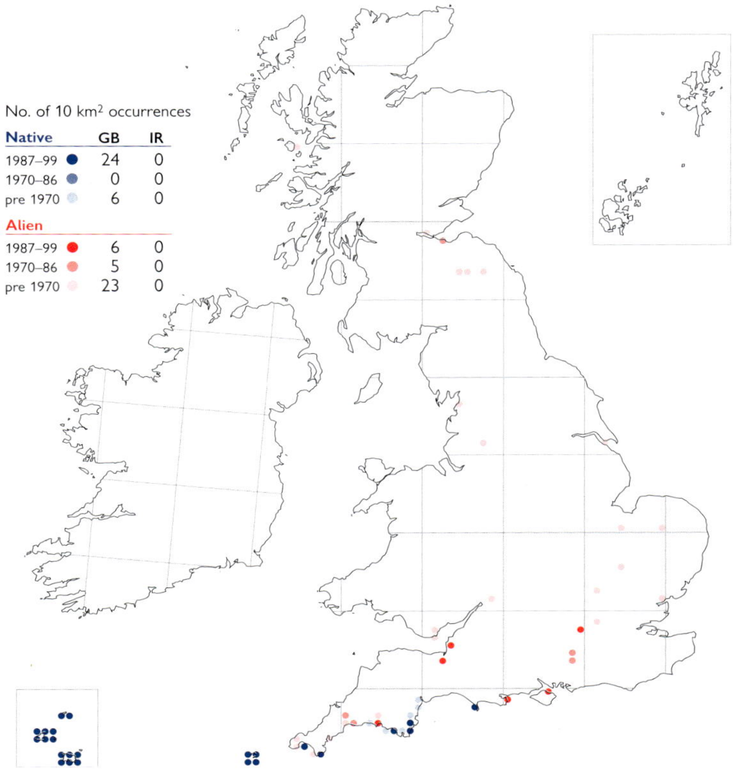

A summer- or occasionally winter-annual found in open, sunny sites that are droughted in summer and relatively frost-free in winter. It grows with other therophytes on steep S.-facing banks, on compacted shingle or sand, in bulb-fields and gardens and at the base of roadside walls. Lowland.

Native or alien (change –0.04). *P. tetraphyllum* has been known in Dorset since 1770 and S. Devon since 1778. Populations can fluctuate dramatically in numbers, and it has been rediscovered at some sites after an apparent absence of many years. Comparison with the 1962 *Atlas* shows little change at the 10-km scale.

Mediterranean-Atlantic element.

References: Atlas (78b), Bolòs & Vigo (1990), Jalas & Suominen (1983), Wigginton (1999).

P. S. LUSBY

Spergula arvensis Corn Spurrey

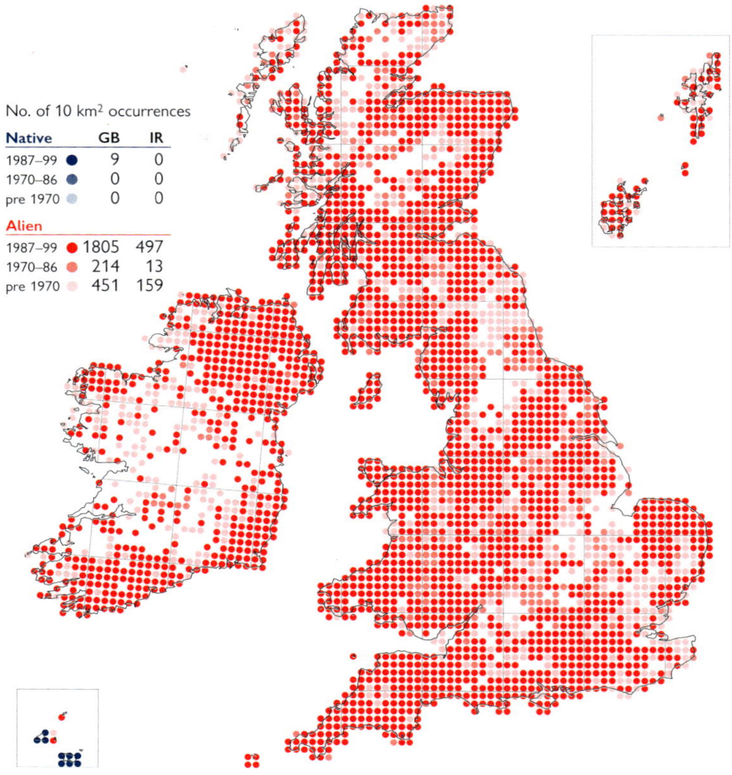

The only native populations of this annual appear to be those of the genetically dwarf var. *nana* on granite cliff ledges in the Channel Islands (McClintock, 1987b). Elsewhere it is an archaeophyte found in open, disturbed habitats on light, often sandy soils, most frequently in arable fields but also on sea-shores, roadsides and waste ground. It tends to avoid calcareous soils. 0–450 m (Clun Forest, Salop).

Native (change –2.30). *S. arvensis* was mapped as 'all records' in the 1962 *Atlas*. Its decline reflects agricultural intensification and the loss of arable land in the uplands.

Eurosiberian Wide-temperate element, but widely naturalised so that distribution is now Circumpolar Wide-temperate.

References: Atlas (76d), Grime *et al.* (1988), Hultén & Fries (1986), Jalas & Suominen (1983), New (1961).

P. S. LUSBY

Spergularia rupicola Rock Sea-spurrey

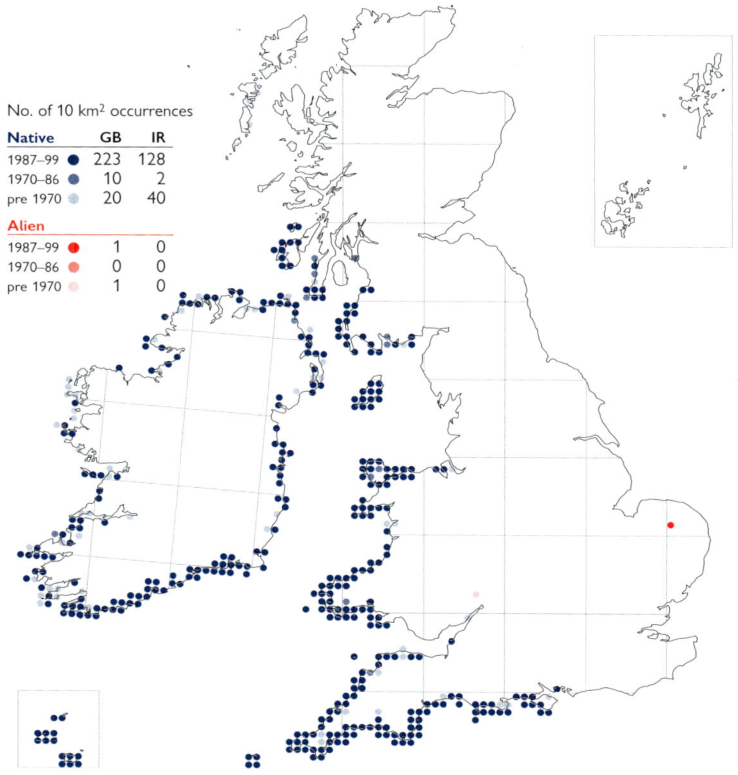

A perennial herb of maritime rocks and cliffs, growing in crevices, on ledges and on friable rock surfaces, sometimes in guano-enriched sites near sea-bird colonies. It also grows in short cliff-top grassland and on the masonry of piers and walls near the sea. It is indifferent to soil reaction. Lowland.

Native (change +0.30). There has been little change in the distribution of *S. rupicola* since the 1962 *Atlas*. It may be under-recorded in W. Ireland.

Oceanic Temperate element; restricted to the coast of W. Europe.

References: Atlas (77c), Jalas & Suominen (1983), Malloch & Okusanya (1979), Okusanya (1979a, b, c).

P. S. LUSBY

Spergularia media Greater Sea-spurrey

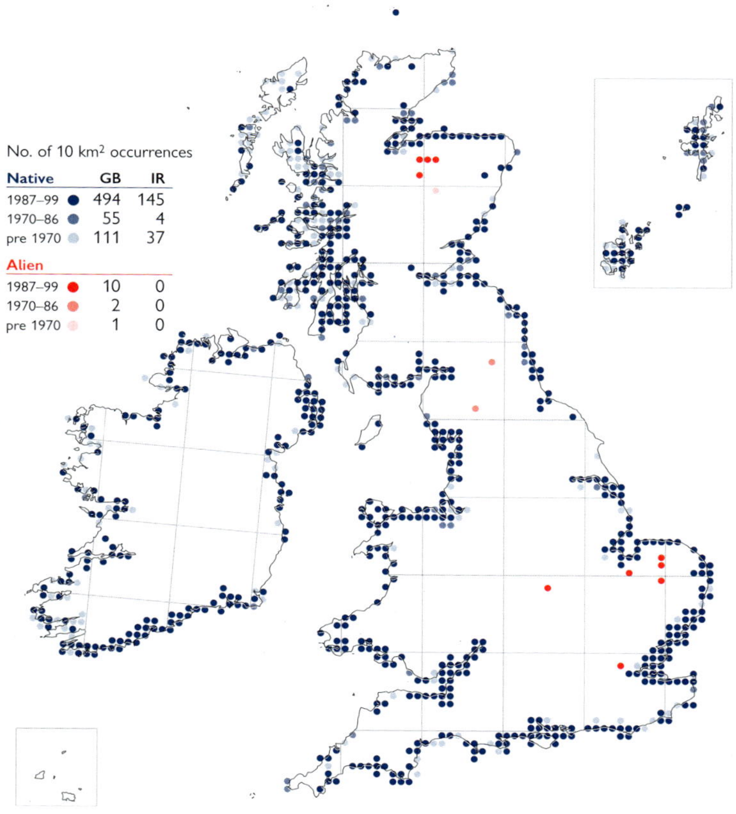

The natural habitats of this perennial herb are strictly maritime. It is found in salt-marshes, on muddy beaches, banks and low cliffs, in tidally inundated dune-slacks, and on the margins of saline ditches in coastal grazing marshes. In saltmarshes, it is generally found at lower elevations than *S. marina*. Inland, it occurs very occasionally as a colonist of salt-treated roadsides. Lowland.

Native (change –0.24). There has been no significant change in the distribution since the 1962 *Atlas* other than its colonisation of road verges, first noted in the late 1970s.

Eurosiberian Southern-temperate element; widely naturalised outside its native range.

References: Atlas (77d), Hultén & Fries (1986), Jalas & Suominen (1983), Meusel *et al.* (1965), Scott & Davison (1982).

P. S. LUSBY

Spergularia marina Lesser Sea-spurrey

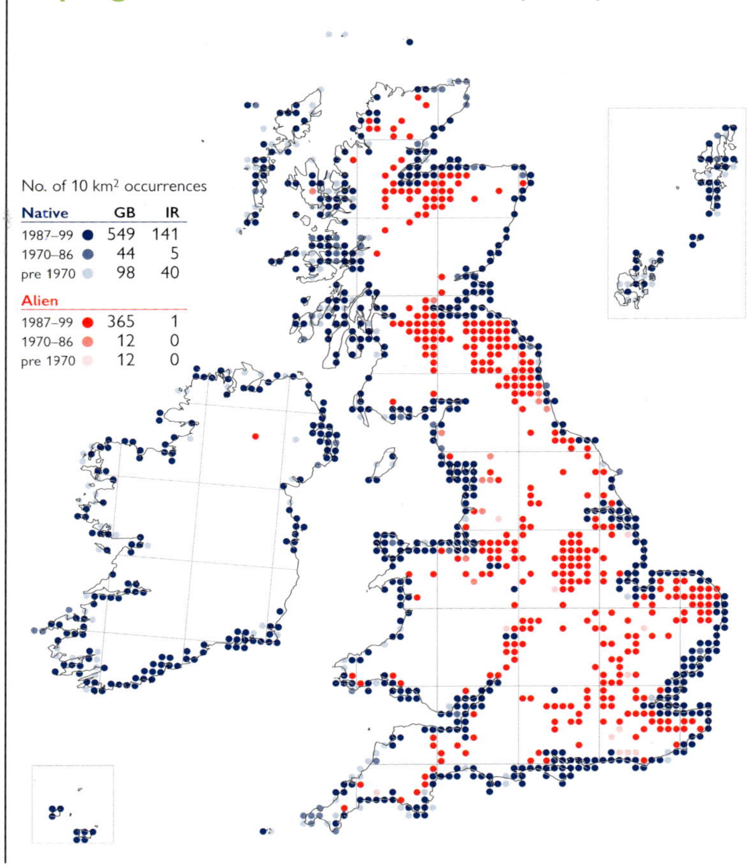

An annual of saltmarshes, sea-walls, muddy shingle, brackish grazing pastures and the base of coastal cliffs. Inland, it is a local colonist of the margins of saline sludge lagoons, and also occurs beside salt-treated roads. Generally lowland, but reaching 520 m on a roadside at Holme Moss (Cheshire).

Native (change +1.83). There are many more coastal records of this species than in the 1962 *Atlas*, probably due to better recording. Since the late 1970s it has undergone a remarkable expansion of range along inland roadsides.

Circumpolar Southern-temperate element; widely naturalised outside its native range.

References: Atlas (78a), Hultén & Fries (1986), Jalas & Suominen (1983), Lee (1977), Meusel *et al.* (1965), Scott (1985), Scott & Davison (1982).

P. S. LUSBY

Spergularia rubra Sand Spurrey

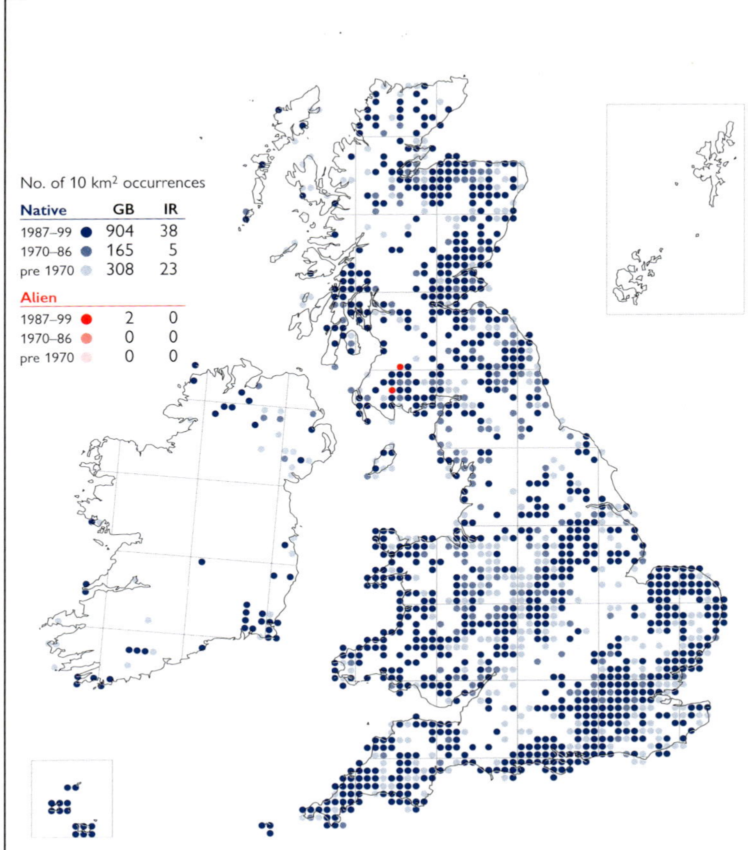

An annual or biennial herb, typically occurring in open habitats on free-draining acidic sands and gravels. Habitats include heaths, commons, tracks (particularly forestry tracks in W. Scotland), quarries, gravel- and sand-pits, railway yards and waste ground. It occasionally grows on stabilised shingle and sand dunes. It is tolerant of trampling. Generally lowland, but recorded at over 560 m on Deadwater Fell (S. Northumb.).

Native (change +0.05). There has been little change in the distribution of this species since the 1962 *Atlas*, although there have been local declines in the English Midlands.

European Southern-temperate element; also in N. America and widely naturalised outside its native range.

References: Atlas (77a), Hultén & Fries (1986), Jalas & Suominen (1983).

P. S. LUSBY

Spergularia bocconei Greek Sea-spurrey

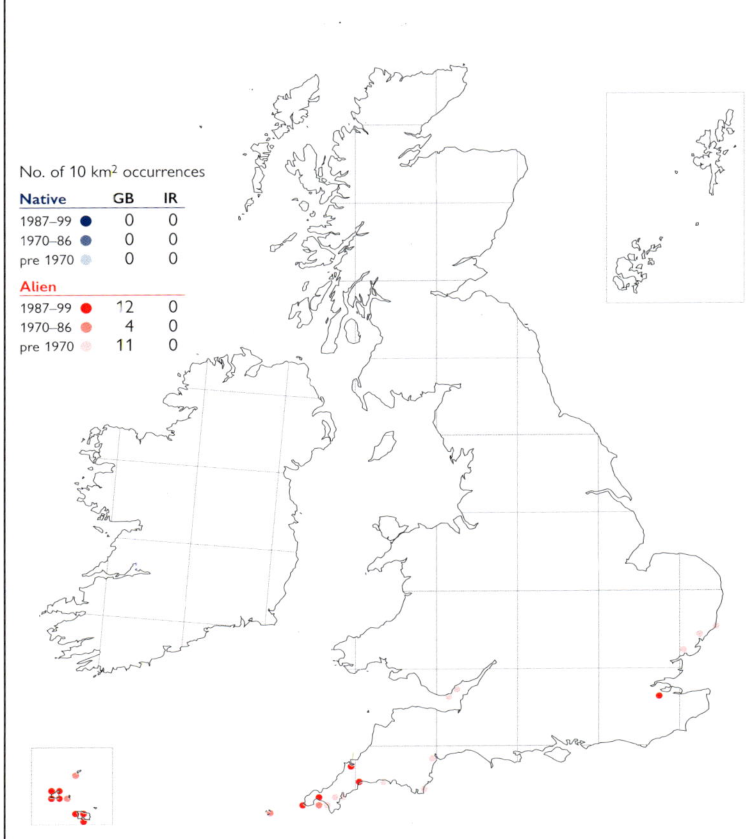

This annual or biennial herb occurs on freely-draining skeletal or sandy soils on waste ground by the sea, including dockyards, roadsides and car parks. Lowland.

Neophyte (change −0.22). *S. bocconei* was first recorded in Britain in 1901. All extant populations are small and vulnerable, although it has been found at new sites in Cornwall and in E. Kent in the last ten years. Although sometimes considered native, its ruderal habitat and the failure of populations to persist suggest that it is more likely to be an introduction.

A Mediterranean-Atlantic species.

References: Atlas (77b), Jalas & Suominen (1983), Wigginton (1999).

P. S. LUSBY

Lychnis coronaria Rose Campion

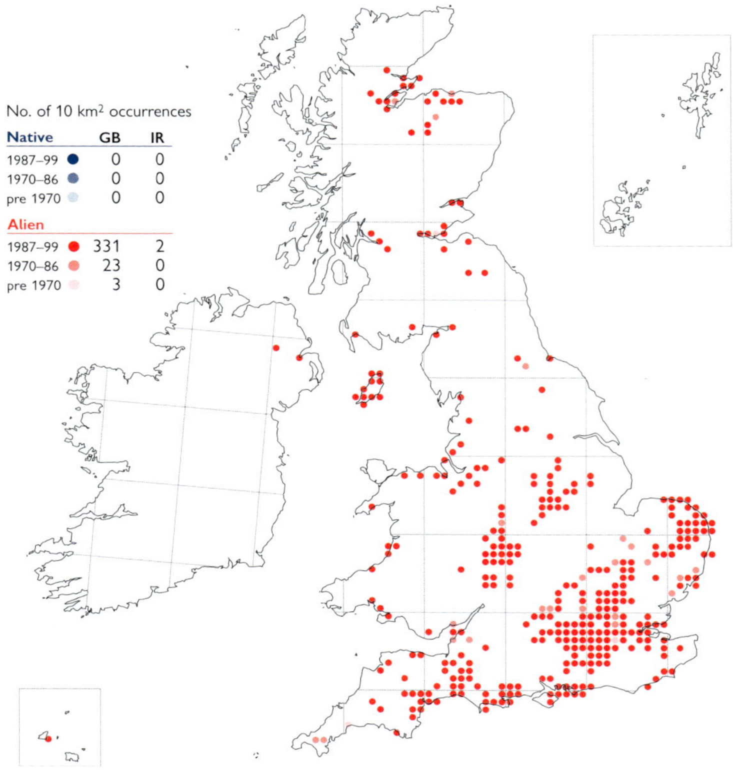

L. coronaria is a persistent garden escape found mainly on light soils in a wide range of habitats, including the edges of heaths, sand dunes, railway banks, roadsides and waste ground. It produces abundant seed. Lowland.

Neophyte. *L. coronaria* was known as a garden plant by the mid 14th century. Dunn (1905) described it as 'recorded as an escape in a few places in Britain'.

Native of S. and especially S.E. Europe, and S.W. Asia.

References: Bolòs & Vigo (1990), Jalas & Suominen (1986).

P. S. LUSBY

Lychnis flos-cuculi Ragged-Robin

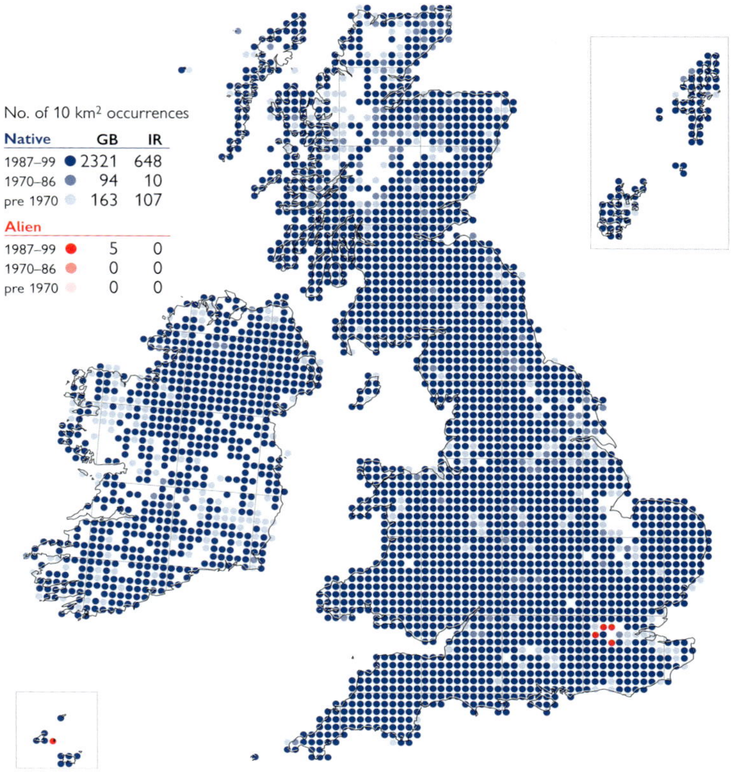

L. flos-cuculi is a herb of damp habitats, found in wet grassland, rush-pasture, fen-meadow, ditches, tall-herb fen and damp woodland margins. A dwarf form (var. *congesta*), known from exposed coastal grassland in E. Kent, Caithness, Sutherland and Shetland, apparently retains this character in cultivation. 0–750 m (Cross Fell, Cumberland).

Native (change –0.79). There has been little change in the distribution of *L. flos-cuculi* since the 1962 *Atlas*, in which it was mapped as 'all records'. However, many county floras record local declines due mainly to drainage and agricultural improvement of wet grassland.

Eurosiberian Temperate element; widely naturalised outside its native range.

References: Atlas (65b), Bezzant (1992), Hultén & Fries (1986), Jalas & Suominen (1986), Meusel *et al.* (1965).

P. S. LUSBY

Lychnis viscaria Sticky Catchfly

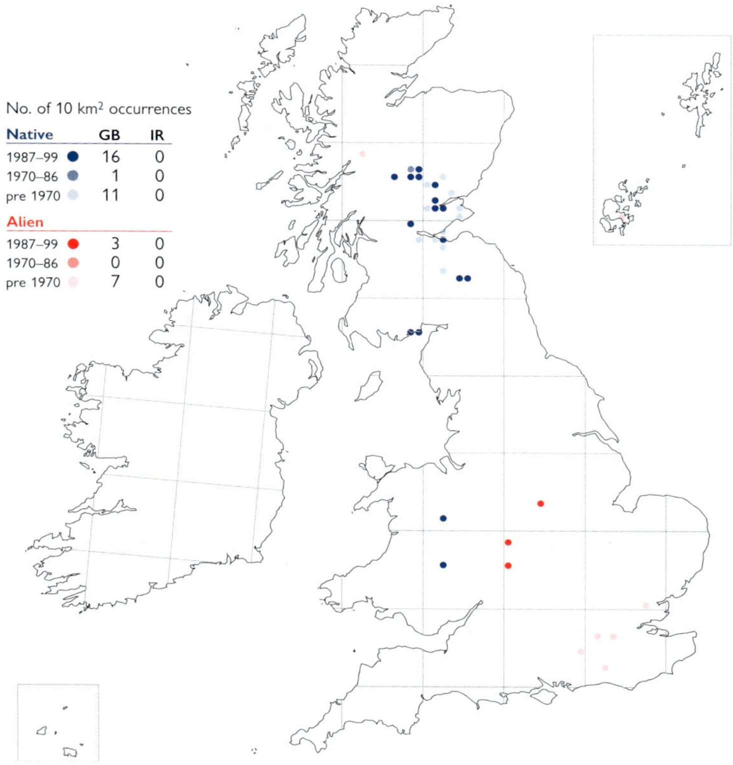

A tufted evergreen perennial herb which grows on rocks and steep dry slopes of predominantly southerly aspect. It prefers mildly acidic to moderately basic soils and many stations are on the more basic volcanic rocks such as basalt and dolerite. From near sea level on undercliffs in Kirkcudbrightshire to 425 m at Kinloch Rannoch (Mid Perth); most populations are above 300 m.

Native (change +0.01). The losses shown on the map took place before 1930 and there has been no significant change in the distribution of this species since the 1962 *Atlas*. It is widely cultivated and sometimes escapes.

Eurosiberian Temperate element, with a continental distribution in W. Europe.

References: Atlas (65a), Hultén & Fries (1986), Jalas & Suominen (1986), Meusel *et al.* (1965), Wigginton (1999), Wilson, Wright *et al.* (1995).

P. S. LUSBY

Lychnis alpina Alpine Catchfly

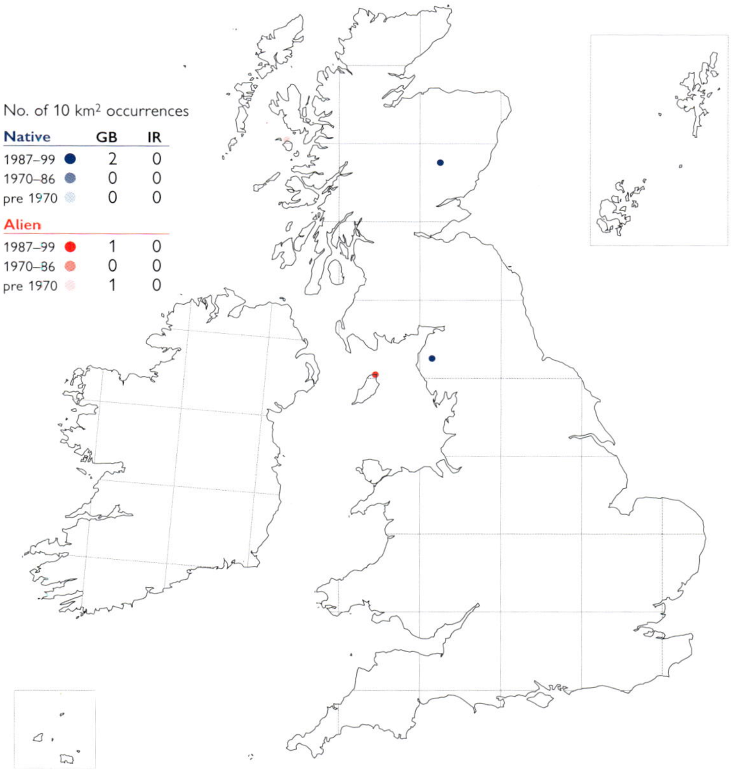

No. of 10 km² occurrences

Native		GB	IR
1987–99	●	2	0
1970–86	●	0	0
pre 1970	●	0	0
Alien			
1987–99	●	1	0
1970–86	●	0	0
pre 1970	●	1	0

L. alpina is a montane rosette-forming perennial herb. In Angus, it grows on serpentine debris rich in magnesium and other metals, and in Cumberland on copper-rich soil over Skiddaw slate. Reproduction is exclusively by seed. At 600–700 m in Cumberland, and 870 m on Meikle Kilrannoch (Angus).

Native. There has been no change since the 1962 *Atlas*. Although regarded as an indicator of metal-rich substrates in Britain, it is not restricted to metalliferous soils in Scandinavia.

European Boreo-arctic Montane element; also in N. America.

References: Atlas (64d), Hultén & Fries (1986), Jalas & Suominen (1986), Meusel *et al.* (1965), Wigginton (1999).

P. S. LUSBY

Agrostemma githago Corncockle

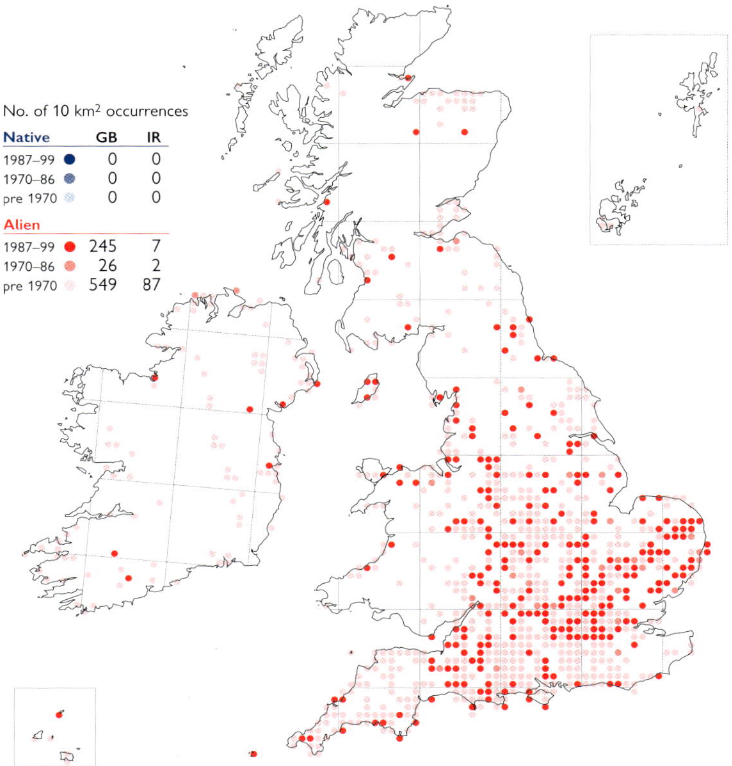

No. of 10 km² occurrences

Native		GB	IR
1987–99	●	0	0
1970–86	●	0	0
pre 1970	●	0	0
Alien			
1987–99	●	245	7
1970–86	●	26	2
pre 1970	●	549	87

An annual weed of cereal and other arable crops, tolerant of various soil types, with its distribution largely mirroring that of the crops in which it grew. Lowland.

Archaeophyte (change –0.75). *A. githago*, introduced as a grain contaminant, has been present in Britain since the Iron Age. Although common until the 20th century, it has dramatically declined with improved seed cleaning. It is now extinct as an arable weed, but is a frequent component of wild-flower seed mixtures.

Native range uncertain, perhaps the E. Mediterranean region; spread with cultivation throughout temperate and S. Europe and to many other areas.

References: Atlas (65c), Curtis & McGough (1988), Firbank (1988), Hultén & Fries (1986), Jalas & Suominen (1986), Meusel *et al.* (1965).

P. S. LUSBY

Silene nutans Nottingham Catchfly

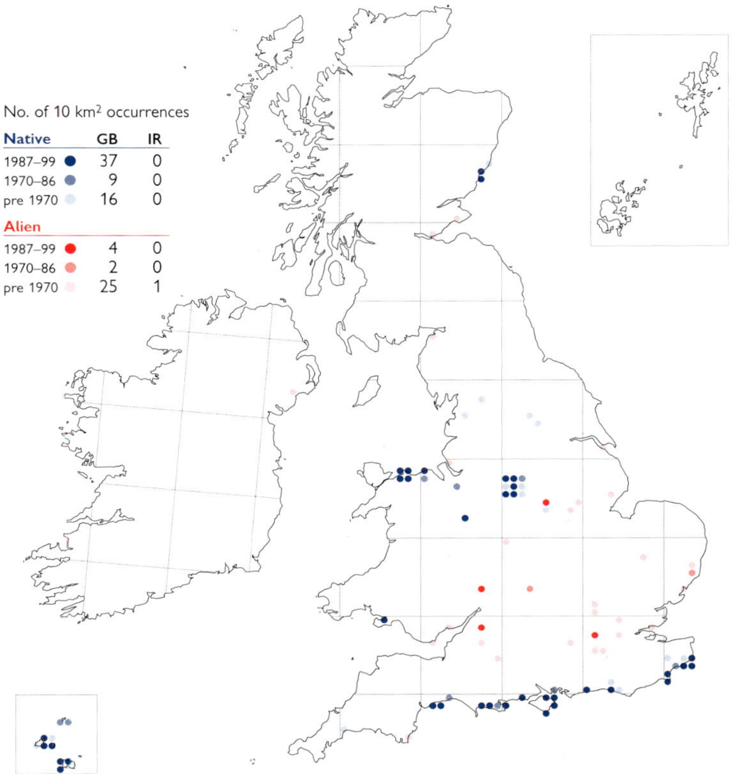

No. of 10 km² occurrences

Native		GB	IR
1987–99	●	37	0
1970–86	●	9	0
pre 1970	●	16	0
Alien			
1987–99	●	4	0
1970–86	●	2	0
pre 1970	●	25	1

The coastal habitats of this long-lived perennial herb are grassy cliffs, sand dunes and shingle. Inland, it grows on limestone rock outcrops and cliff ledges. *S. nutans* is mainly a plant of shallow, drought-prone, calcareous soils on chalk and limestone, but it also occurs on acidic soil overlying shingle. It has occurred as a casual at ports and on railway banks. Reproduction is usually by seed, but can be vegetative by procumbent stems rooting at the nodes. Lowland.

Native (change –0.39). There has been little change in the coastal distribution of this species, but it has declined inland since the 1962 *Atlas*.

Eurosiberian Temperate element.

References: Atlas (63c), Hepper (1956), Hultén & Fries (1986), Jalas & Suominen (1986), Meusel *et al.* (1965), Stewart *et al.* (1994).

P. S. LUSBY

Silene otites Spanish Catchfly

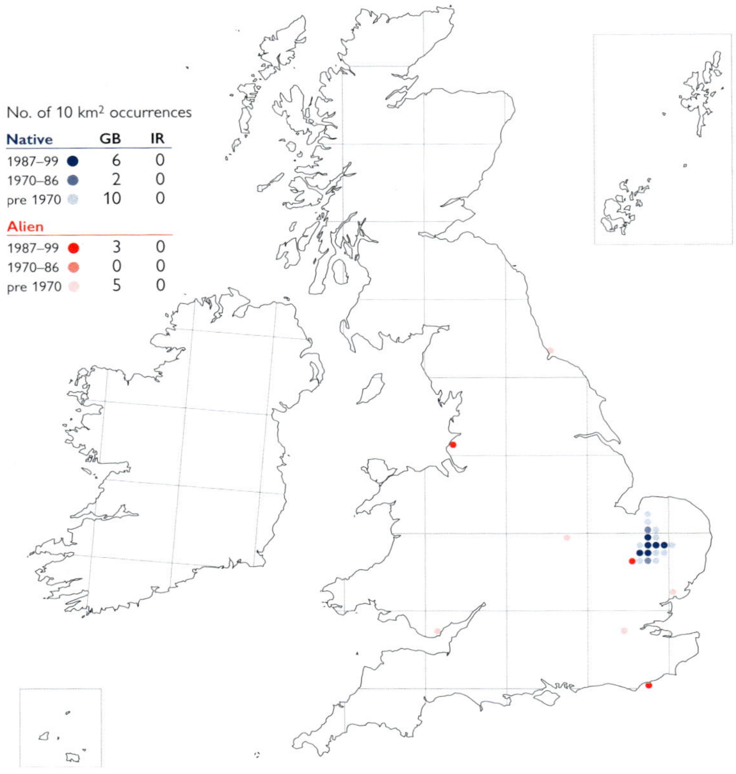

A perennial herb of shallow, well-drained, light calcareous soils. As a native plant, it is confined to Breckland grass-heaths and roadsides, where open, disturbed ground provide sites for seedlings. Mature plants can survive for a while in denser swards, but are eventually excluded. It occurs elsewhere as a casual. Lowland.

Native (change –0.36). *S. otites* has declined significantly since 1930, the chief causes being agricultural intensification and afforestation. The map shows a contraction since the 1962 *Atlas* but its distribution is more or less stable at present.

European Temperate element, with a continental distribution in W. Europe.

References: Atlas (63b), Hultén & Fries (1986), Jalas & Suominen (1986), Meusel *et al.* (1965), Wigginton (1999).

P. S. LUSBY

Silene vulgaris Bladder Campion

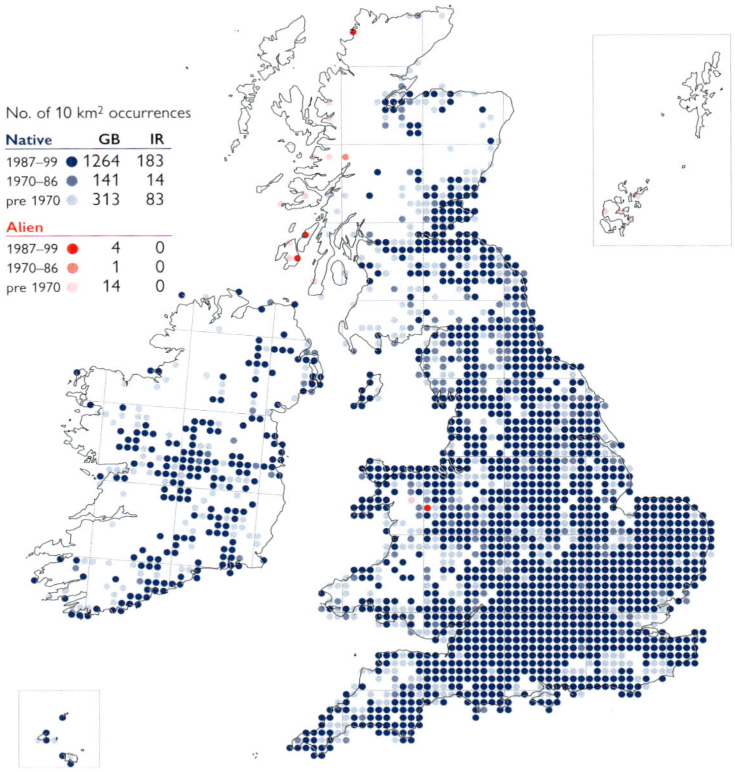

S. vulgaris is a perennial herb found in a wide range of soils in open and grassy habitats, including cultivated and abandoned arable fields, rough pasture, roadside verges, quarries, gravel-pits, railway banks, walls and waste places. It is able to tolerate partial shade and may grow in open woodland and on hedge banks. 0–360 m (Shap Summit, Westmorland).

Native (change –1.26). The map suggests an appreciable decline in the frequency of *S. vulgaris* since the 1962 *Atlas*.

Eurasian Southern-temperate element, but naturalised in N. America so distribution is now Circumpolar Southern-temperate.

References: Atlas (62a), Hultén & Fries (1986), Jalas & Suominen (1986), Marsden-Jones & Turrill (1957).

P. S. LUSBY

Silene uniflora Sea Campion

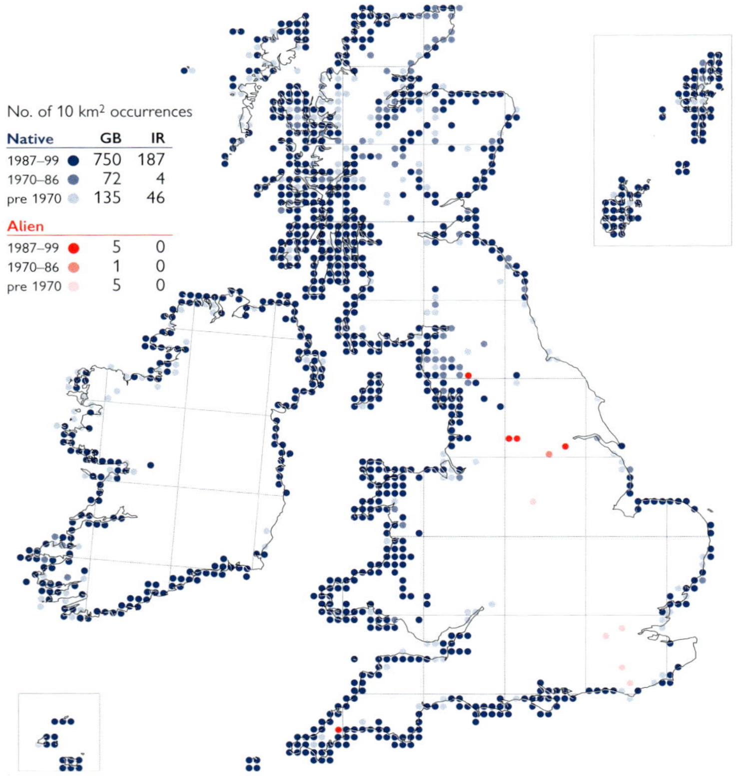

A perennial herb occurring on rocky sea-cliffs from the lowest zone of vascular plants to cliff-top grassland, on seaside walls, shingle banks and on drift-lines. It can tolerate high levels of nutrient enrichment, and can be abundant on cliff-tops adjoining sea-bird colonies. It also occurs rarely on upland lake shores, streamsides, river shingle, cliffs and in gullies. Artificial habitats include metalliferous mine spoil, disused railway lines, ballast and tips. 0–775 m (Fairfield, Westmorland), and to 970 m in the Scottish Highlands.

Native (change –0.39). There has been no significant change in the distribution of *S. uniflora* since the 1962 *Atlas*.

Suboceanic Boreo-temperate element.

References: Atlas (62b), Hultén & Fries (1986), Jalas & Suominen (1986), Marsden-Jones & Turrill (1957).

P. S. LUSBY

Silene acaulis Moss Campion

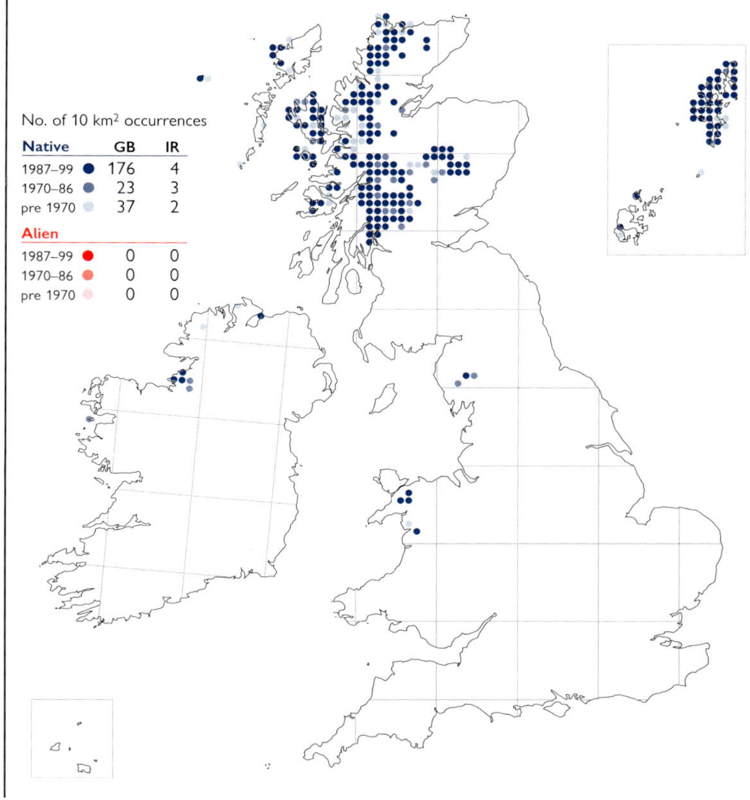

No. of 10 km² occurrences

Native		GB	IR
1987–99	●	176	4
1970–86	●	23	3
pre 1970	●	37	2
Alien			
1987–99	●	0	0
1970–86	●	0	0
pre 1970	●	0	0

This cushion- or mat-forming perennial herb is confined to base-rich substrates. It is characteristic of a species-rich dwarf-herb ledge community on Scottish mountains. However, it also grows in sparse vegetation on exposed mountain plateaux, serpentine fell-fields, cliff-slopes and stabilised sand dunes. From sea level in W. and N. Scotland to 1305 m on Ben Macdui (S. Aberdeen).

Native (change –0.47). There has been no appreciable change in the distribution of *S. acaulis* since the 1962 *Atlas*. It may still be present in many squares for which there are only pre-1987 records.

European Arctic-montane element; also in E. Asia and N. America.

References: Atlas (63a), Curtis & McGough (1988), Hultén & Fries (1986), Jalas & Suominen (1986), Jones & Richards (1962), Lusby & Wright (1996), Meusel *et al.* (1965).

P. S. LUSBY

Silene armeria Sweet-William Catchfly

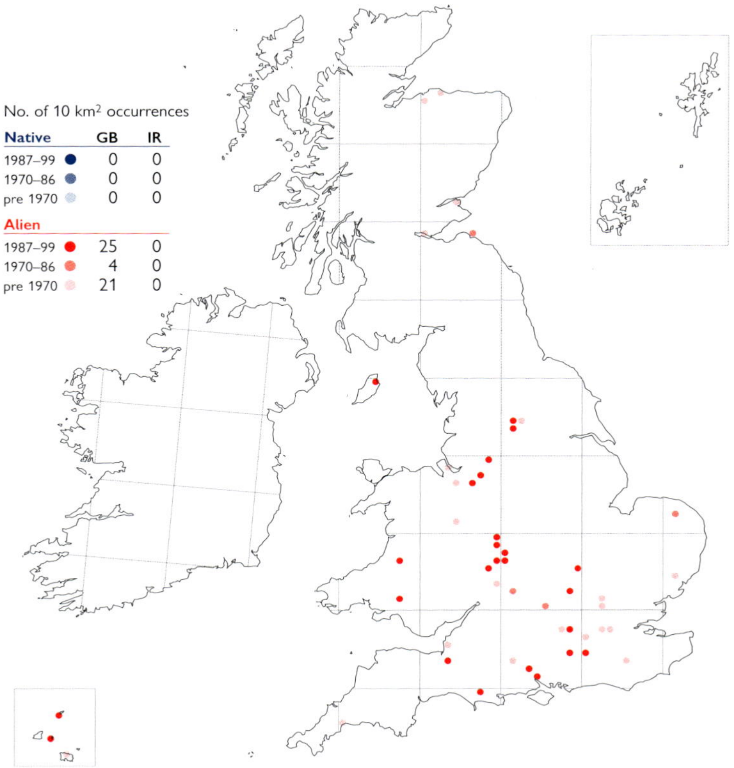

No. of 10 km² occurrences

Native		GB	IR
1987–99	●	0	0
1970–86	●	0	0
pre 1970	●	0	0
Alien			
1987–99	●	25	0
1970–86	●	4	0
pre 1970	●	21	0

An annual occurring as a garden escape or throw-out on rubbish tips and waste places. It is usually casual, but may persist for a few years in disturbed, open sites, reproducing by seed. Lowland.

Neophyte. This attractive species has been cultivated in Britain since before 1800. It was recorded from the wild by 1840, and its overall distribution is stable.

Native of the southern part of the European Temperate zone; it is naturalised further north in Europe and elsewhere.

Reference: Jalas & Suominen (1986).

T. D. DINES

Silene noctiflora Night-flowering Catchfly

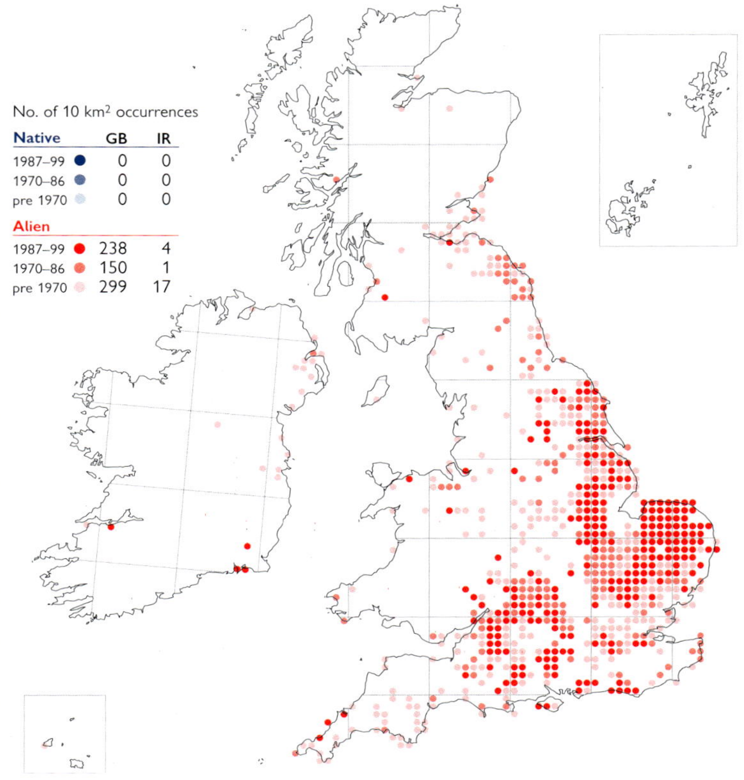

No. of 10 km² occurrences

Native		GB	IR
1987–99	●	0	0
1970–86	●	0	0
pre 1970	●	0	0
Alien			
1987–99	●	238	4
1970–86	●	150	1
pre 1970	●	299	17

This spring-germinating annual occurs mainly on cultivated land, but sometimes also on open waste ground. It is mostly found on dry, sandy and calcareous substrates, but also on heavier soils over oolitic limestone. Lowland.

Archaeophyte (change –2.04). *S. noctiflora* has declined markedly since the 1950s with the increased use of herbicides and fertilisers, and the shift from spring-sown to autumn-sown crops. This loss still continues.

As an archaeophyte *S. noctiflora* has a European Temperate distribution; it is widely naturalised outside this range.

References: Atlas (64a), Hultén & Fries (1986), Jalas & Suominen (1986), McNeill (1980), Meusel *et al.* (1965), Stewart *et al.* (1994), Wilson (1991).

P. S. LUSBY

Silene latifolia White Campion

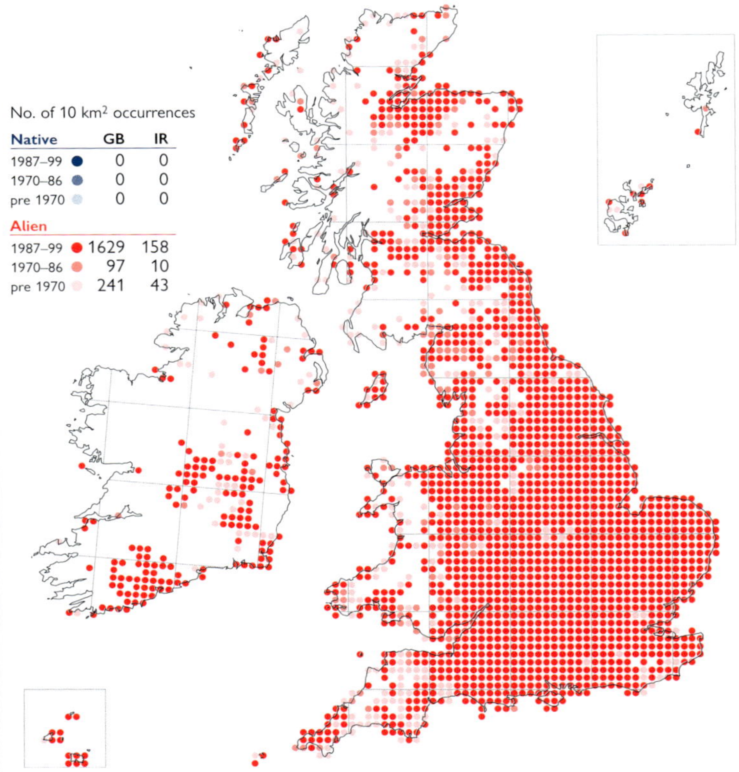

S. latifolia is usually a short-lived perennial, but it may occasionally be annual or biennial. It occurs on arable land, in hedge banks and waste places, being most abundant on deep, well-drained soils. Mainly lowland, but reaching 425 m in Atholl (E. Perth).

Archaeophyte (change −0.88). Fossil records of *S. latifolia* in Britain date from the Bronze Age onwards. It was mapped as 'all records' in the 1962 *Atlas*. The map suggests a decline at the western edge of its British range.

As an archaeophyte *S. latifolia* has a Eurosiberian Southern-temperate distribution, but it is widely naturalised so that its distribution is now Circumpolar Southern-temperate.

References: Atlas (64c), Baker (1947), Hultén & Fries (1986), Jalas & Suominen (1986), McNeill (1977).

P. S. LUSBY

Silene dioica Red Campion

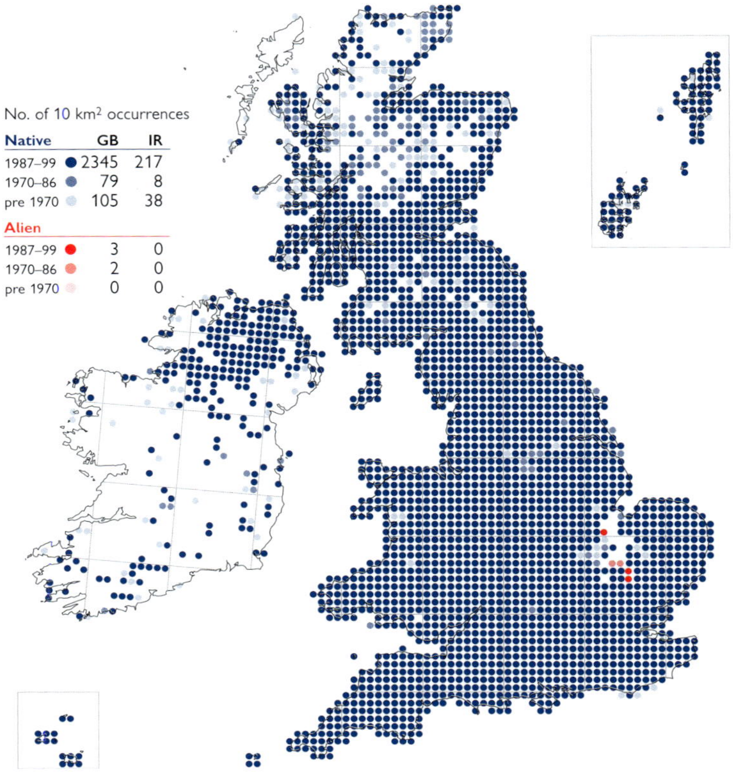

A short-lived perennial herb that is most prominent in lightly shaded habitats such as hedgerows, coppices, and woodland clearings and rides. It can, however, survive in deep shade in a non-flowering form. It also occurs in coastal habitats, including sheltered clifftop grassland and scrub, rock crevices, sea-bird rocks and stabilised shingle, and is also found on montane screes and cliffs. 0–1065 m (Lochnagar, S. Aberdeen).

Native (change −0.44). There is no change in the distribution of *S. dioica* since the 1962 *Atlas*.

European Boreo-temperate element; also in C. Asia.

References: Atlas (64b), Baker (1947), Grime *et al.* (1988), Hultén & Fries (1986), Jalas & Suominen (1986), Meusel *et al.* (1965).

P. S. LUSBY

Silene dioica × *S. latifolia* (*S.* × *hampeana*)

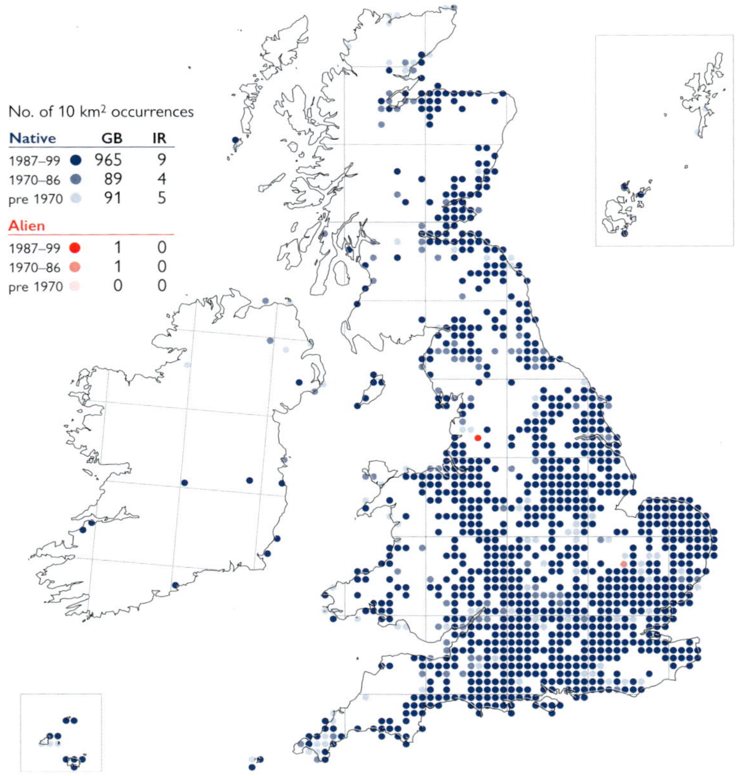

The perennial *S.* × *hampeana* is found predominantly on woodland margins and rides, on scrubby banks and in hedgerows. It shows quite high levels of fertility, and back-crossing to the parents may give rise to hybrid swarms. Lowland.

A spontaneous hybrid between native and alien parents. This hybrid can be found wherever populations of the parent species come into contact. It may also occur in areas where one or other parent is rare or absent. The hybrid is better recorded than when mapped by Perring & Sell (1968) but shows no change in its overall range.

Frequent in mainland Europe wherever the parents come into contact.

References: Atlas Supp. (11b), Baker (1947), Stace (1975).

P. S. LUSBY

Silene gallica Small-flowered Catchfly

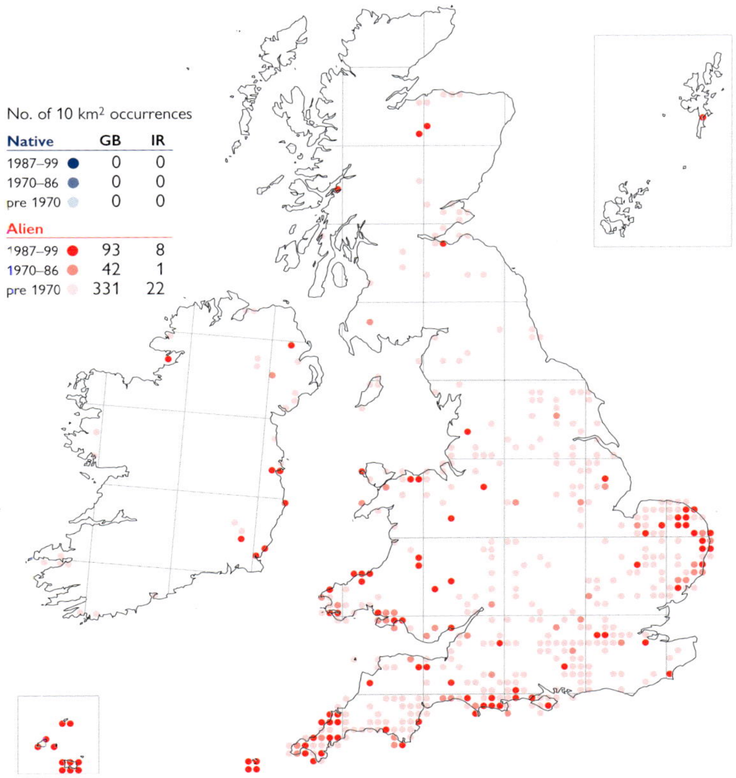

A winter-annual of cultivated and disturbed ground, mainly in arable fields on (often acidic) sandy or gravelly soils, and on old walls and waste ground. It also occurs in open, drought-prone coastal grassland on banks and cliffs, and on sand dunes in the Channel Islands. It is sensitive to low winter temperatures. Lowland.

Archaeophyte (change −2.78). *S. gallica* had been lost by 1930 from many inland sites. Since then it has declined further in response to agricultural intensification, especially from the increased use of herbicides and fertilisers. Many populations are now very small.

As an archaeophyte *S. gallica* has a Submediterranean-Subatlantic distribution; it is widely naturalised outside this range.

References: Atlas (62d), Bolòs & Vigo (1990), Jalas & Suominen (1986), Stewart *et al.* (1994).

P. S. LUSBY

Silene conica Sand Catchfly

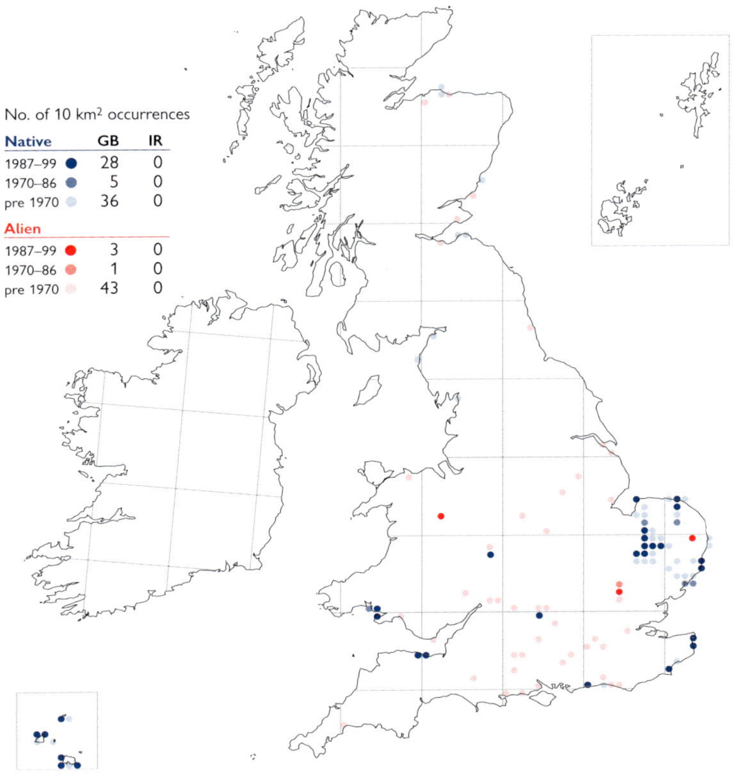

An annual of open habitats on free-draining sandy soils. In coastal regions, it is found on stabilised dunes and sandy shingle, in open pastures and on waste ground; inland it also occurs at the edges of tracks across heathland, in abandoned arable fields and on commons. It flowers freely, but good seed production occurs only in hot summers. Lowland.

Native (change −1.05). Since the 1962 *Atlas*, *S. conica* has declined in England and has become extinct as a native plant in Scotland. Remaining populations are often small and vulnerable.

Eurosiberian Southern-temperate element; widely naturalised outside its native range.

References: Atlas (62c), Hultén & Fries (1986), Jalas & Suominen (1986), Meusel *et al.* (1965), Stewart *et al.* (1994).

P. S. LUSBY

Saponaria officinalis Soapwort

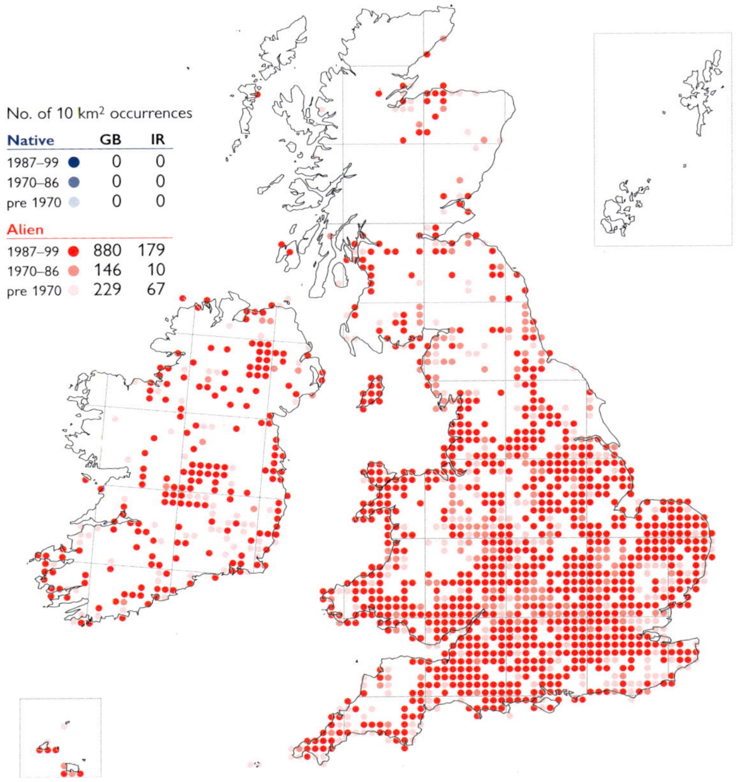

This rhizomatous perennial herb is found in a wide range of man-made and marginal habitats, often near habitation, including hedge banks, quarries, roadsides, railway banks, tips and waste ground. It is thoroughly naturalised by streams and in damp woods, especially in S.W. England and N. Wales, where it has sometimes been considered native. Lowland.

Archaeophyte (change +0.29). This species has long been grown in gardens, and readily establishes from outcast plants. Mapped as 'all records' in the 1962 *Atlas*, there is little evidence for any change in its distribution.

As an archaeophyte *S. officinalis* has a European Temperate distribution; it is widely naturalised outside this range.

References: Atlas (66c), Hultén & Fries (1986), Jalas & Suominen (1986).

P. S. LUSBY

Vaccaria hispanica Cowherb

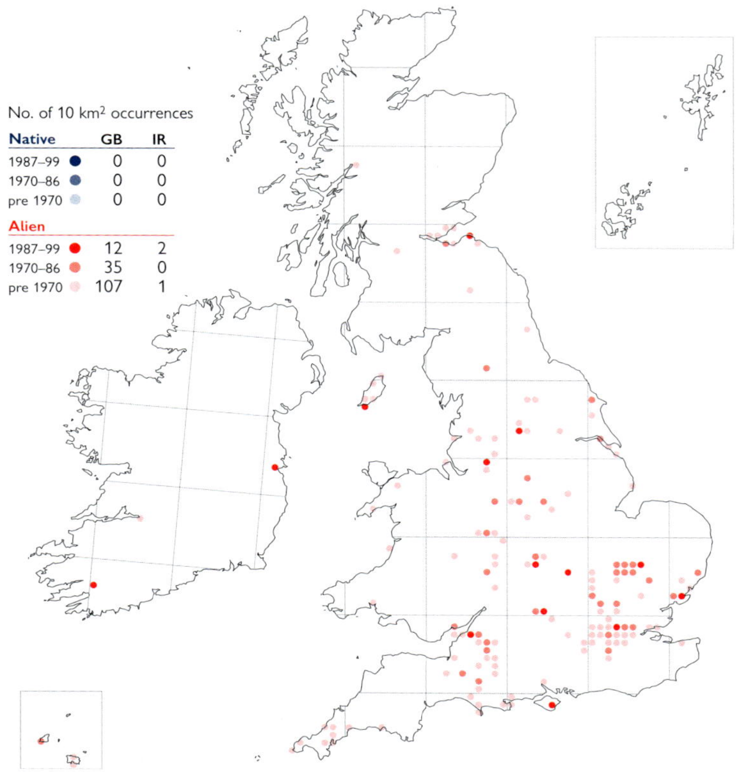

An annual found in hedges, waste places, tips and cultivated ground, usually only as a casual, but occasionally becoming established. Lowland.

Neophyte. *V. hispanica* was known as a garden plant by 1548, and is still grown as an ornamental, escaping into semi-wild habitats. It has been introduced as a grain contaminant, and in wool and bird-seed. It was recorded from the wild by 1832.

Apparently native to S.E. Europe & S.W. Asia, but now a very widespread weed in both N. and S. hemispheres.

References: Hultén & Fries (1986), Jalas & Suominen (1986).

P. S. LUSBY

Petrorhagia nanteuilii Childing Pink

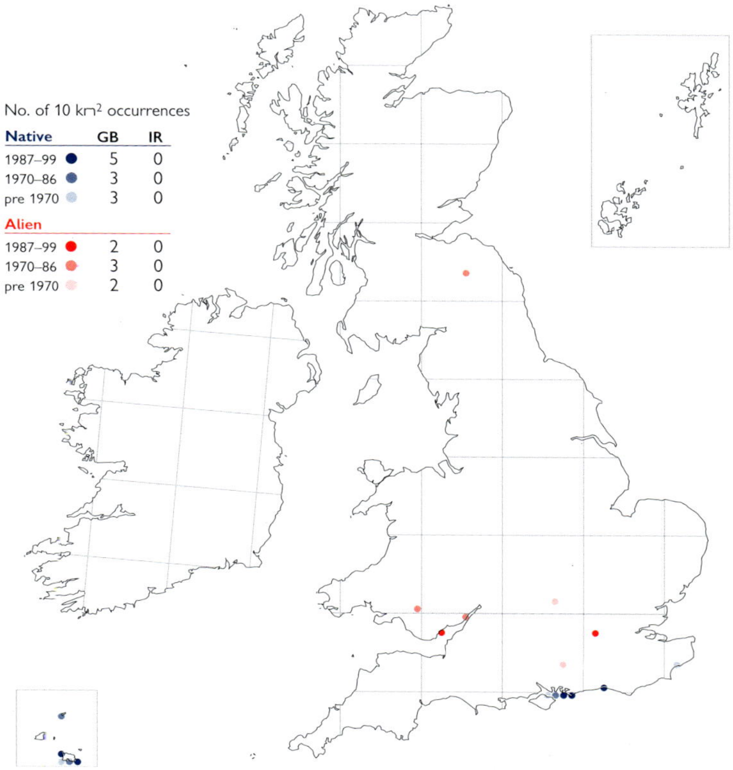

As a native plant, this annual is recorded from thinly vegetated, stabilised shingle in Britain and from stabilised dunes in Jersey. It is also recorded as an introduction around dockyards, in re-seeded grassland and as a garden weed. Lowland.

Native. The map in the 1962 *Atlas* combined records of *P. nanteuilii* and *P. prolifera*; *P. nanteuilii sens. str.* was mapped by Perring & Sell (1968). Some *P. nanteuilii* sites were lost to building development between 1850 and 1930. It was last seen in Kent in 1960 and was thought to have become extinct in Hampshire in 1968, until refound there 30 years later. It is, however, still locally frequent in Jersey.

Suboceanic Southern-temperate element.

References: Atlas Supp. (12a), Hultén & Fries (1986), Jalas & Suominen (1986), Wigginton (1999).

P. S. LUSBY

Petrorhagia prolifera Proliferous Pink

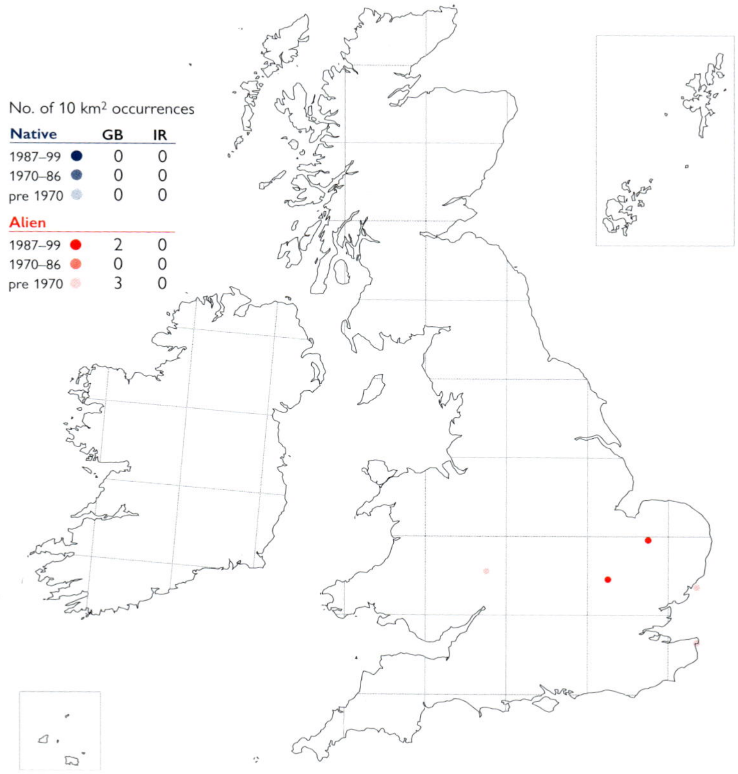

An autumn-germinating annual of freely-draining substrates. In Norfolk, it has most recently been recorded from the open edge of sandy grass-heath where it abuts a roadway, and in Bedfordshire from a sand-pit. Lowland.

Neophyte. Early British records of *Petrorhagia* cannot be identified to the two species currently recognised. Current populations in Norfolk date from 1835; it was known at several sites until the 1950s, but not seen again until 1985. Although it is treated here as an introduction, it is arguably native in Norfolk and possibly in Suffolk (Akeroyd & Beckett, 1995). It is well-established in Bedfordshire but a casual elsewhere.

A European Temperate species.

References: Hultén & Fries (1986), Jalas & Suominen (1986), Wigginton (1999).

P. S. LUSBY

Dianthus gratianopolitanus Cheddar Pink

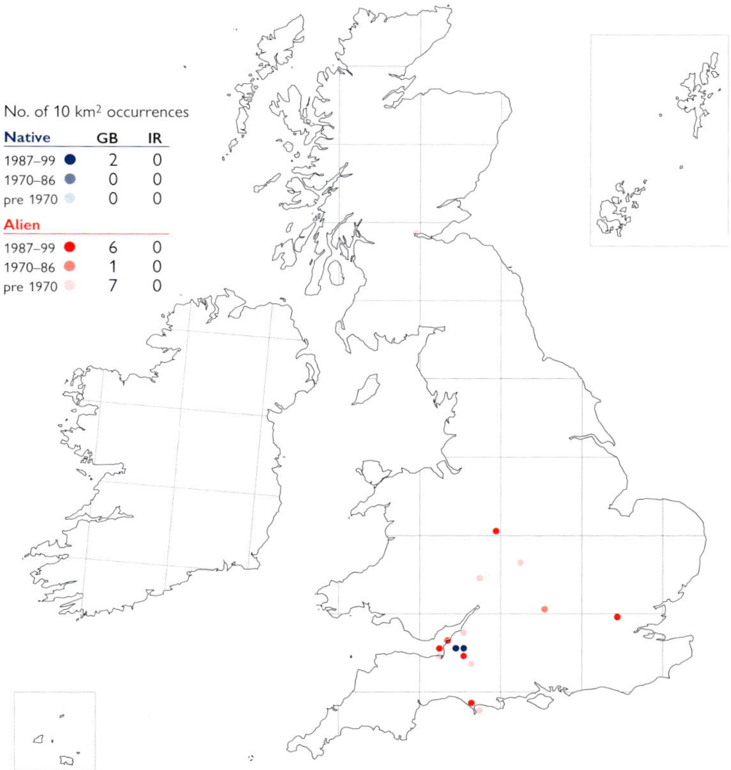

This densely tufted perennial herb is now mainly confined to high, inaccessible crevices and ledges on Carboniferous limestone cliffs, though it is also found in tightly-grazed, species-rich limestone turf. It reproduces only by seed. Lowland.

Native (change +0.19). The species was once abundant in Cheddar Gorge (N. Somerset), but there is a long history of gathering plants, and the lower slopes were stripped of the plant long ago. However, populations have increased recently following statutory protection, scrub clearance and the re-instatement of grazing. It persists at some nearby sites to which it was deliberately introduced, and is a casual elsewhere.

European Temperate element, with a continental distribution in W. Europe.

References: Atlas (66a), Jalas & Suominen (1986), Meusel *et al.* (1965), Wigginton (1999).

P. S. LUSBY

Dianthus plumarius Pink

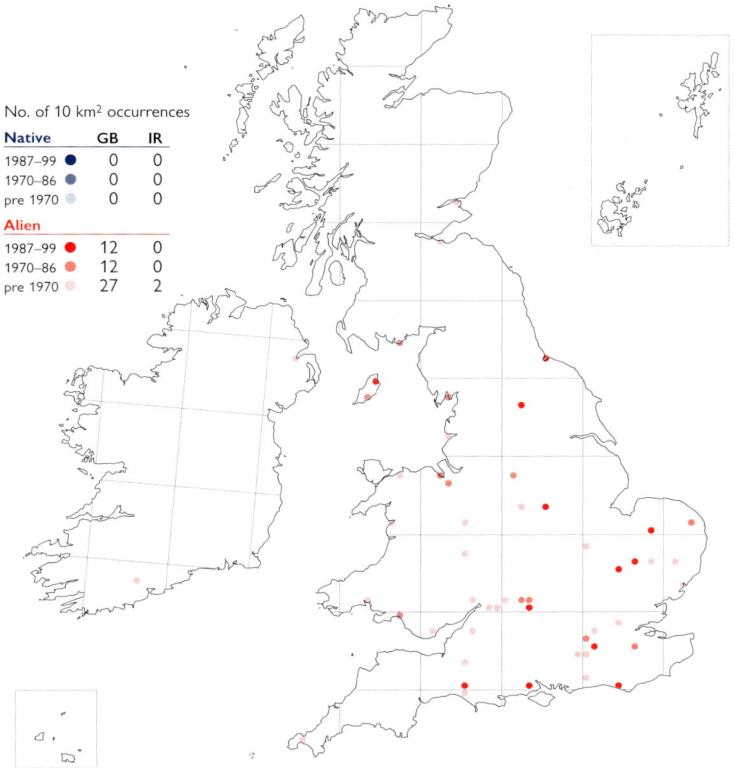

A tufted perennial herb found naturalised on roadside banks and railway cuttings, especially on well-drained soils overlying chalk and limestone, and on old mortared walls. It also occurs rarely as a casual on rubbish tips. Lowland.

Neophyte. This species was being cultivated in Britain by 1629; it is extremely common in gardens and is available in a huge variety of cultivars. It was first recorded in the wild in 1724 (Middlesex), and has been naturalised on walls at Beaulieu Abbey (S. Hants.) since at least 1856. Its overall distribution is probably stable.

Native of the mountains of E.C. Europe.

Reference: Jalas & Suominen (1986).

T. D. DINES

Dianthus deltoides Maiden Pink

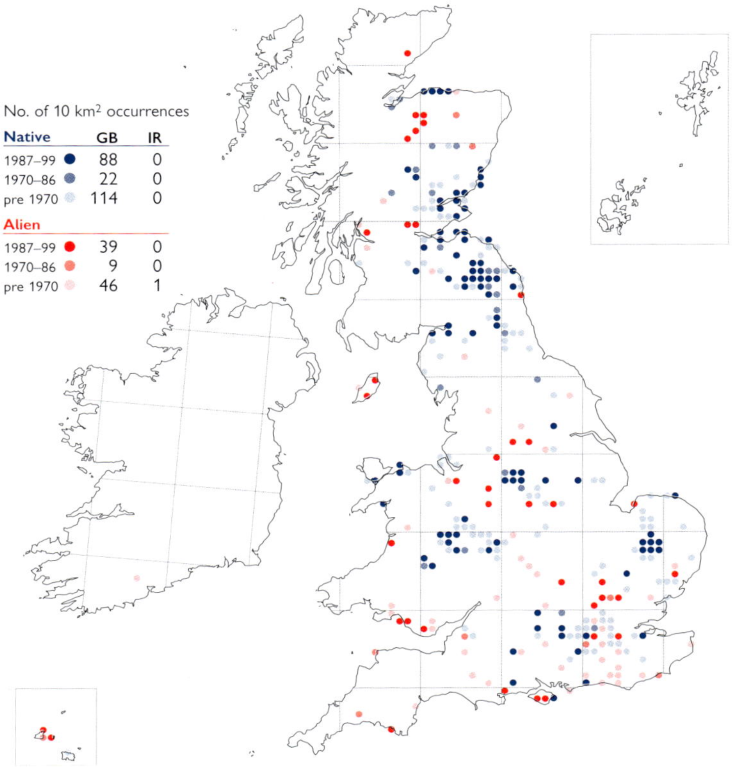

A perennial herb of dry, usually base-rich, soils overlying chalk and limestone, mica-schist or basalt; sometimes on metal-rich mining spoil or sandy soils and dunes. It can occur in short, closed grassland, but prefers an open sward broken by bare rock or soil. It also occurs as a garden escape. 0–355 m (Parsley Hay, Derbys.).

Native (change –0.41). Much of the decline of *D. deltoides* took place before 1930. Further losses have been offset by new discoveries. Nonetheless, many colonies are small and suffer from overgrazing and nutrient enrichment, or undergrazing and scrub encroachment.

Eurosiberian Boreo-temperate element, with a continental distribution in W. Europe; widely naturalised outside its native range.

References: Atlas (66b), Hultén & Fries (1986), Jalas & Suominen (1986), Meusel *et al.* (1965), Stewart *et al.* (1994).

P. S. LUSBY

Dianthus barbatus Sweet-William

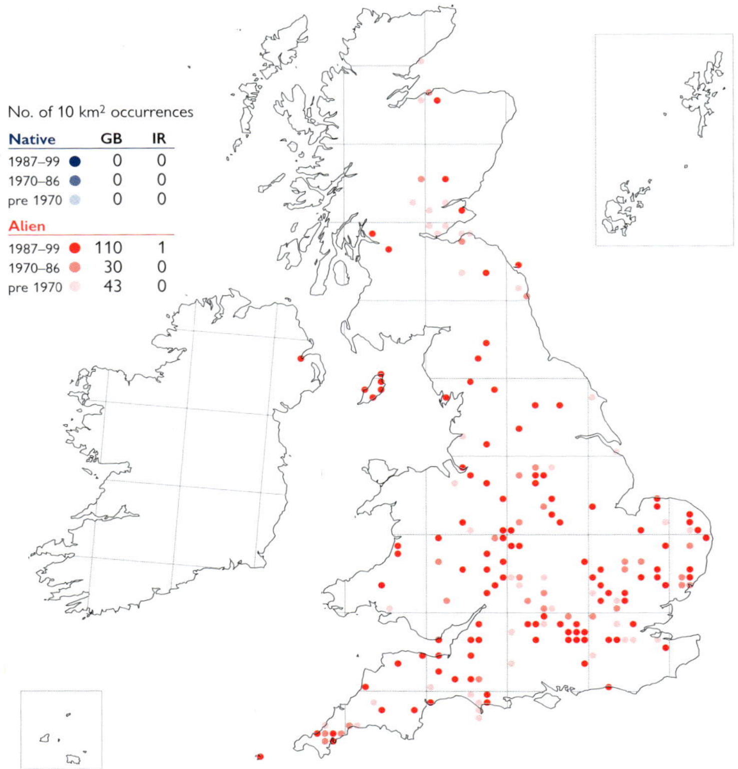

A mat-forming biennial or short-lived perennial herb found as a garden escape or throw-out on roadside banks, railways, sand dunes, waste ground and rubbish tips. Populations are usually casual, but can persist for a few years. Lowland, but reaching 330 m at Garrigill (Cumberland).

Neophyte. This species was in cultivation in Britain by 1573 and has long been popular in gardens. It was known from the wild by 1805, and Dunn (1905) described it as 'occasionally established on old walls'. It may be increasing.

Native of the mountains of S. Europe.

References: Bolòs & Vigo (1990), Jalas & Suominen (1986).

T. D. DINES

Dianthus armeria Deptford Pink

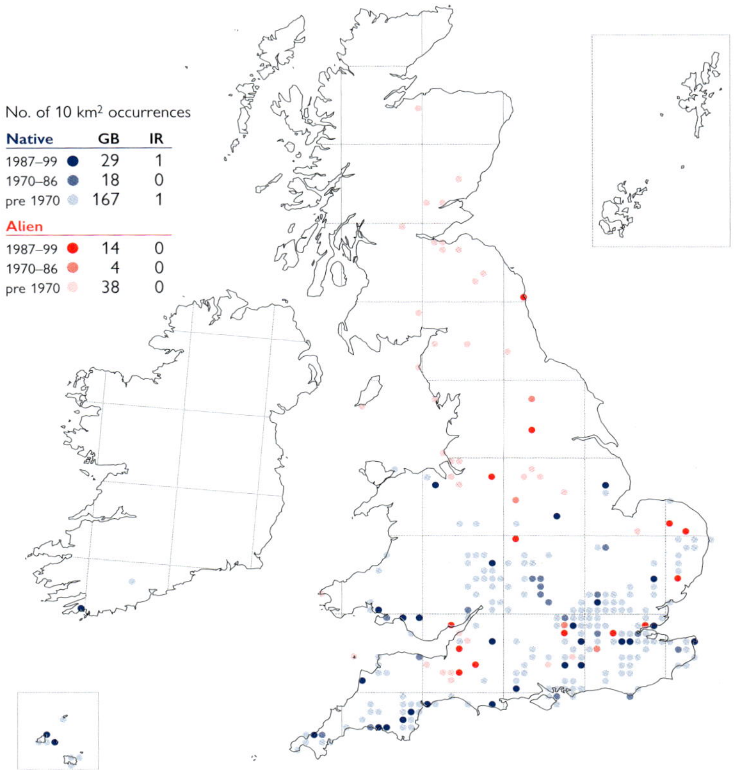

An annual or short-lived perennial herb of open, disturbed sites, occurring in short grassland in pastures, roadsides, waysides and field margins, and as a casual on waste ground. It usually grows on dry, often mildly basic soils, but has been recorded on fen-peat. Lowland.

Native (change –1.31). It is difficult to distinguish native and alien populations in both Britain and Ireland. A marked decline was apparent before 1930 and this has continued, with most losses due to the conversion of pasture to arable or buildings, or from a lack of open areas at its remaining sites. It was first recognised in Ireland in 1993.

European Temperate element; widely naturalised elsewhere.

References: Atlas (65d), Akeroyd & Clark (1993), Hultén & Fries (1986), Jalas & Suominen (1986), Stewart *et al.* (1994), Wigginton (1999), Wilson (1999).

P. S. LUSBY

Persicaria campanulata Lesser Knotweed

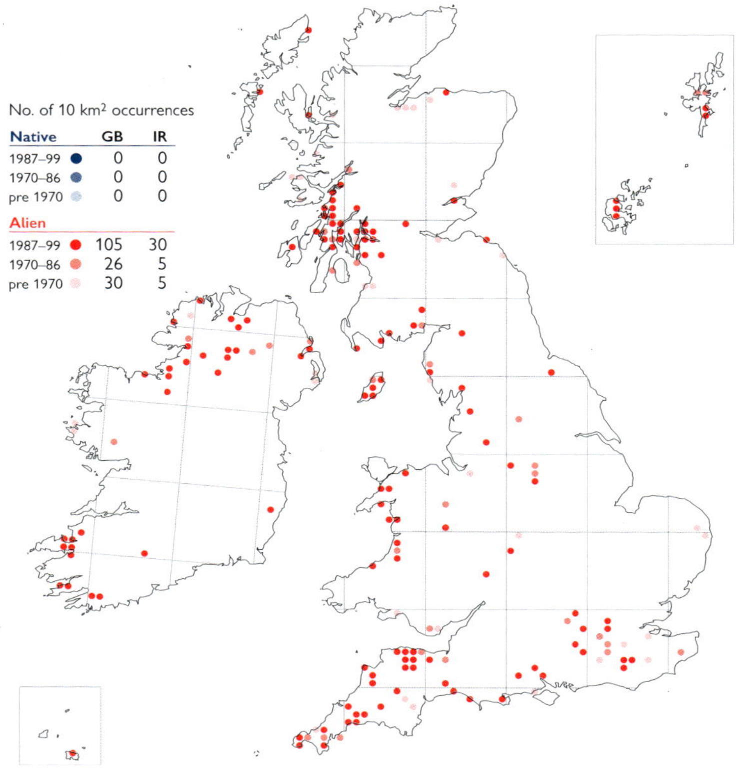

A stoloniferous perennial herb which forms dense patches on damp roadsides, hedge banks and streamsides. Its flowers are heterostylous and seed set has not been reported in Britain, but the species establishes readily from discarded rhizome fragments. Lowland.

Neophyte. *P. campanulata* was introduced to Britain as a garden plant in about 1909. The first record of it in the wild was made in 1933 and its range is continuing to expand.

Native of the Himalayas.

References: Conolly (1977, 1991), Lousley & Kent (1981), Meusel *et al.* (1965).

J. R. AKEROYD

Persicaria wallichii Himalayan Knotweed

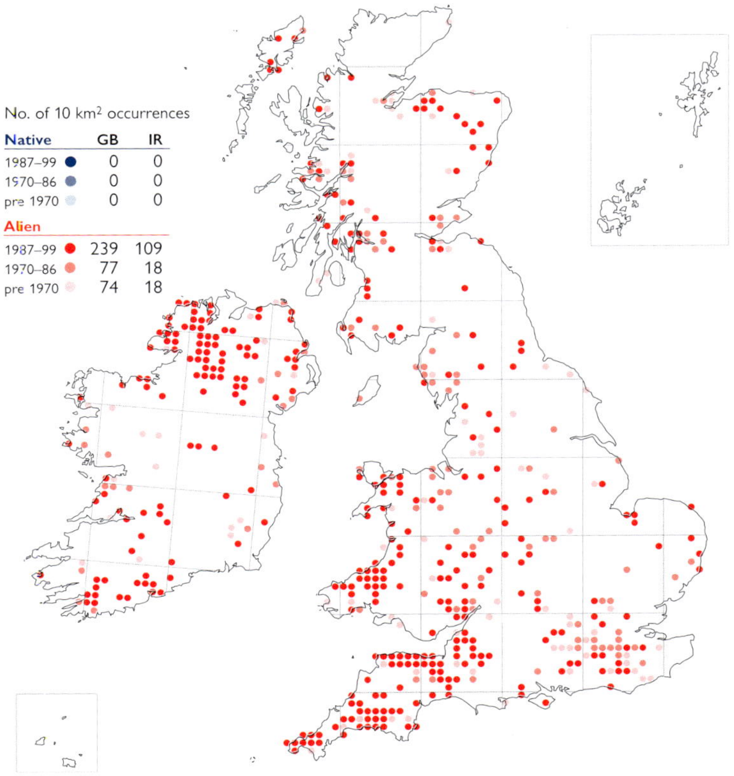

No. of 10 km² occurrences

Native	GB	IR
1987–99 ●	0	0
1970–86 ●	0	0
pre 1970 ●	0	0
Alien		
1987–99 ●	239	109
1970–86 ●	77	18
pre 1970 ●	74	18

A tall, rhizomatous perennial herb of streamsides, hedge banks, roadsides, railway banks and waste ground, growing in dense stands. Its flowers are heterostylous and seed is only occasionally set in Britain, but it establishes readily by vegetative reproduction from outcast rhizomes. Generally lowland, but reaching 330 m at Simonsbath (S. Somerset).

Neophyte (change +0.59). *P. wallichii* was introduced to Britain just before 1900, and was first recorded in the wild in 1917. Its distribution has expanded considerably since the 1962 *Atlas*.

Native of the Himalayas.

References: Atlas (177b), Conolly (1977, 1991), Jalas & Suominen (1979), Lousley & Kent (1981).

J. R. AKEROYD

Persicaria bistorta Common Bistort

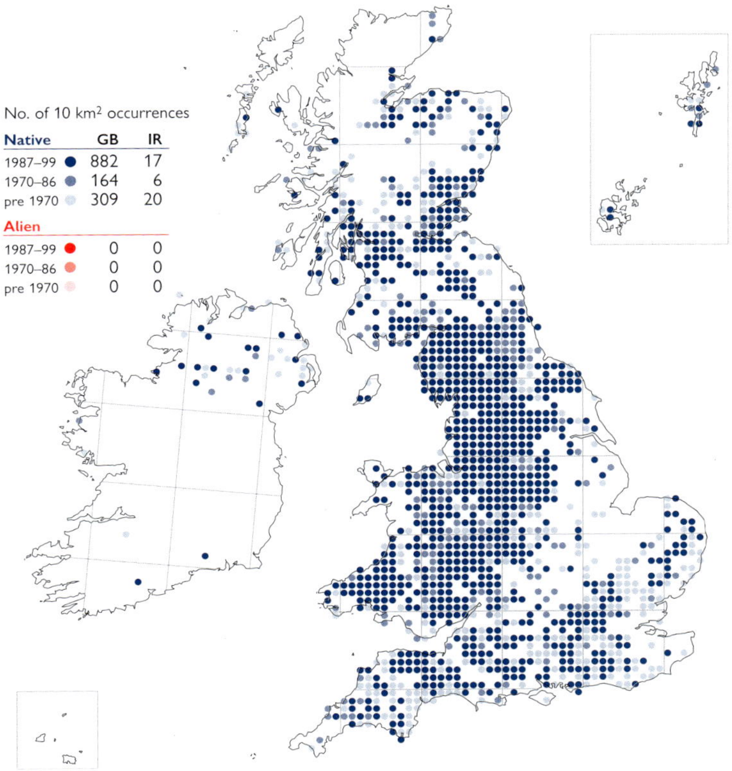

No. of 10 km² occurrences

Native	GB	IR
1987–99 ●	882	17
1970–86 ●	164	6
pre 1970 ●	309	20
Alien		
1987–99 ●	0	0
1970–86 ●	0	0
pre 1970 ●	0	0

A perennial herb of base-poor soils in damp pastures, hay meadows and river banks, in tall-herb communities in river valleys and on mountain ledges and roadsides. Many colonies originate as garden escapes or throw-outs. 0–430 m above Garrigill (Cumberland).

Native (change –0.44). The native range of *P. bistorta* is now largely obscured by alien occurrences, and all records are mapped as if they are native. It was lost from many sites in S. England and Ireland before 1930. Although now better recorded than in the 1962 *Atlas*, its overall range is stable.

Eurasian Boreo-temperate element, in N. America native in west and naturalised in east so now Circumpolar Boreo-temperate.

References: Atlas (174b), Hultén & Fries (1986), Jalas & Suominen (1979), Lousley & Kent (1981), Mabey (1996), Meusel *et al.* (1965).

J. R. AKEROYD

Persicaria amplexicaulis Red Bistort

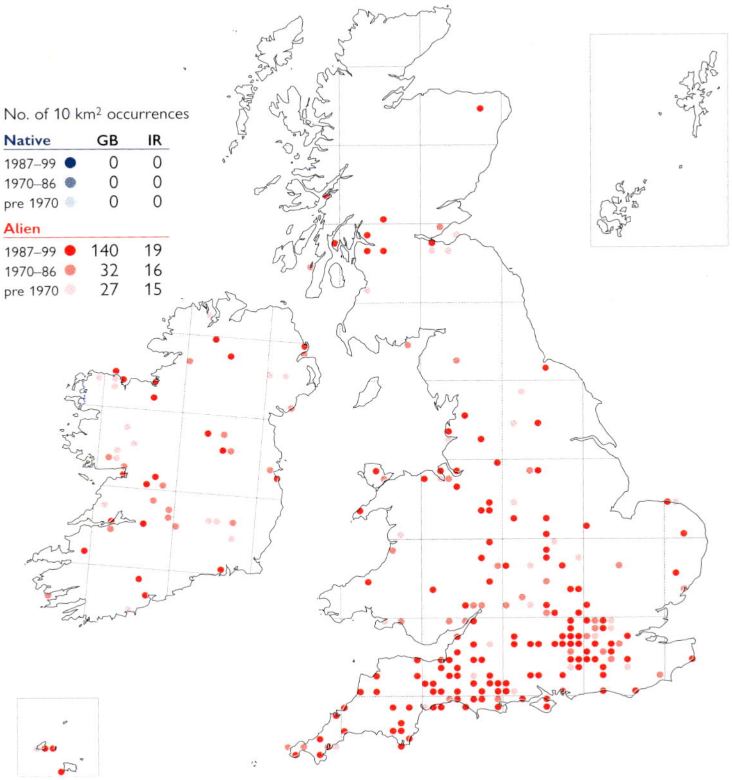

No. of 10 km² occurrences

Native	GB	IR
1987–99 ●	0	0
1970–86 ●	0	0
pre 1970 ●	0	0
Alien		
1987–99 ●	140	19
1970–86 ●	32	16
pre 1970 ●	27	15

A tufted perennial herb, naturalised on roadsides, hedge banks and waste ground. Plants have rarely been observed to set seed in the wild in Britain and Ireland, and presumably spread by vegetative reproduction. Lowland.

Neophyte. The species was introduced into cultivation in 1826; it is widely grown in gardens where it can be invasive and persistent. It has been known in the wild since at least 1908, and its range is continuing to expand.

Native of Asia, from Afghanistan to S.W. China.

References: Jalas & Suominen (1979), Lousley & Kent (1981).

J. R. AKEROYD

Persicaria vivipara Alpine Bistort

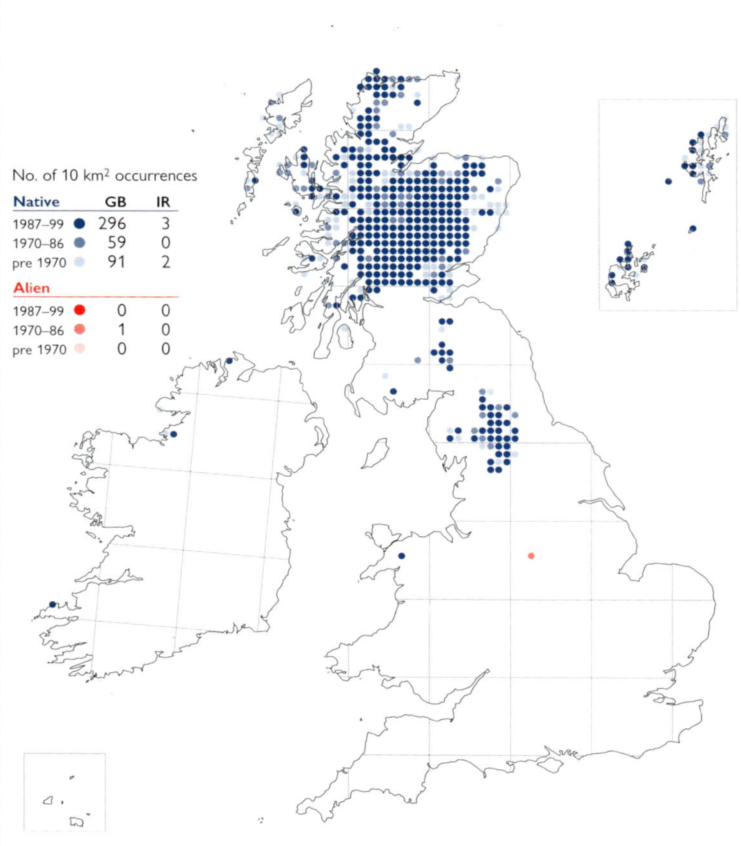
No. of 10 km² occurrences
Native GB IR
1987–99 296 3
1970–86 59 0
pre 1970 91 2
Alien
1987–99 0 0
1970–86 1 0
pre 1970 0 0

A short, tufted perennial herb, usually found on base-rich substrates and less frequently in acidic conditions. It grows on wet rocks, consolidated screes, in grassland and in damp flushes in the mountains, and it is often abundant in montane pastures. Reproduction is mostly by bulbils at the base of the inflorescence, frequently carried down to lower levels by streams. It reaches 1210 m on Ben Lawers (Mid Perth), but descends to near sea level in N. Scotland.

Native (change –0.58). There has been little change in the distribution of this species since the 1962 *Atlas*.

Circumpolar Boreo-arctic Montane element.

References: Atlas (174a), Curtis & McGough (1988), Hultén & Fries (1986), Jalas & Suominen (1979), Lousley & Kent (1981), Meusel *et al.* (1965).

J. R. AKEROYD

Persicaria amphibia Amphibious Bistort

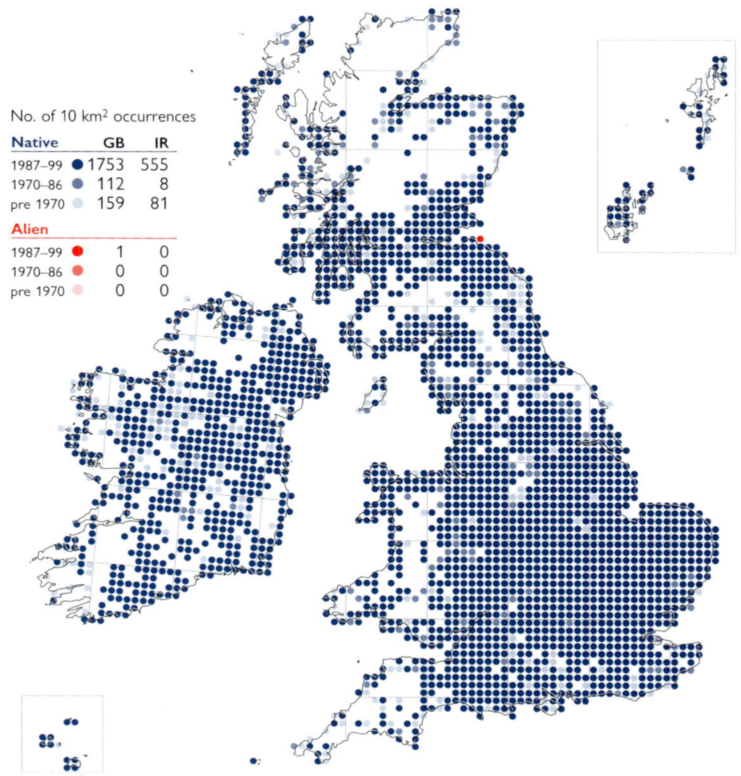
No. of 10 km² occurrences
Native GB IR
1987–99 1753 555
1970–86 112 8
pre 1970 159 81
Alien
1987–99 1 0
1970–86 0 0
pre 1970 0 0

A floating aquatic perennial herb, which sometimes grows in considerable quantity in lakes, ponds, canals, slow-flowing rivers and ditches, or a terrestrial plant found in damp places on watersides, in marshes, wet meadows and dune-slacks, and as a weed of cultivated land. Reproduction is by seed and fragments of rhizome; terrestrial plants are much less floriferous than aquatic ones. 0–570 m (Blind Tarn, Westmorland).

Native (change +0.27). There has been no appreciable change in the distribution of this species since the 1962 *Atlas*.

Circumpolar Boreo-temperate element; widely naturalised outside its native range.

References: Atlas (174c), Grime *et al.* (1988), Hultén & Fries (1986), Jalas & Suominen (1979), Lousley & Kent (1981), Meusel *et al.* (1965), Preston & Croft (1997).

J. R. AKEROYD

Persicaria maculosa Redshank

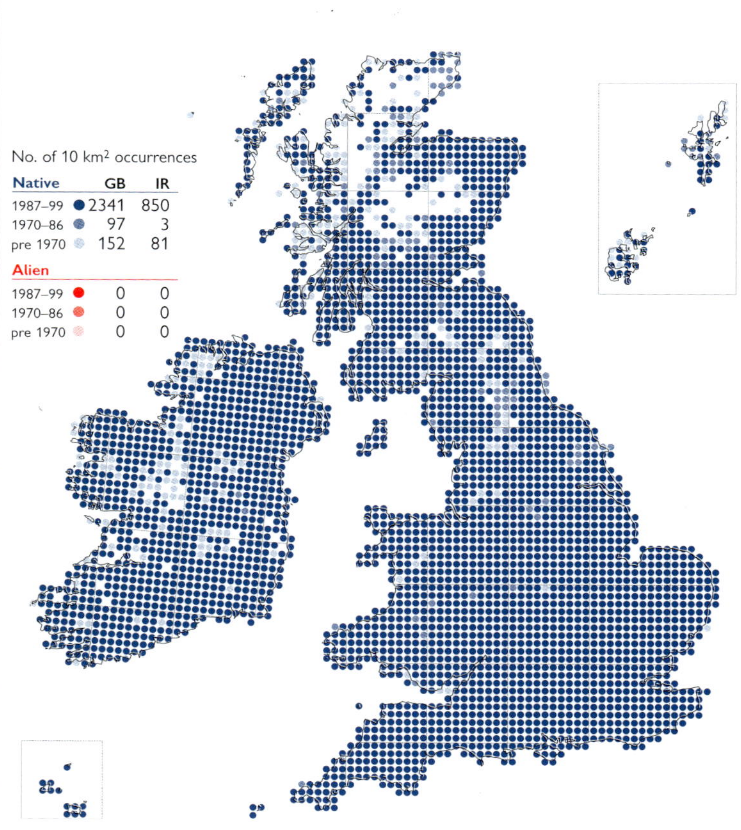
No. of 10 km² occurrences
Native GB IR
1987–99 2341 850
1970–86 97 3
pre 1970 152 81
Alien
1987–99 0 0
1970–86 0 0
pre 1970 0 0

An annual of open ground on a wide range of soils, particularly those which are rich in nutrients. It is found by ponds, lakes, streams and ditches, in waste places, on roadsides and railways, and is sometimes a pestilential weed of cultivated land. Mainly lowland, but ascending to at least 450 m near Garrigill (Cumberland) and in Clun Forest (Salop).

Native (change –0.95). The overall distribution of this species is stable since the 1962 *Atlas*, although there have been some losses at the northern edge of its range.

Eurasian Temperate element, but naturalised in N. America so distribution is now Circumpolar Temperate.

References: Atlas (174d), Grime *et al.* (1988), Hultén & Fries (1986), Jalas & Suominen (1979), Lousley & Kent (1981), Simmonds (1945).

J. R. AKEROYD

Persicaria lapathifolia Pale Persicaria

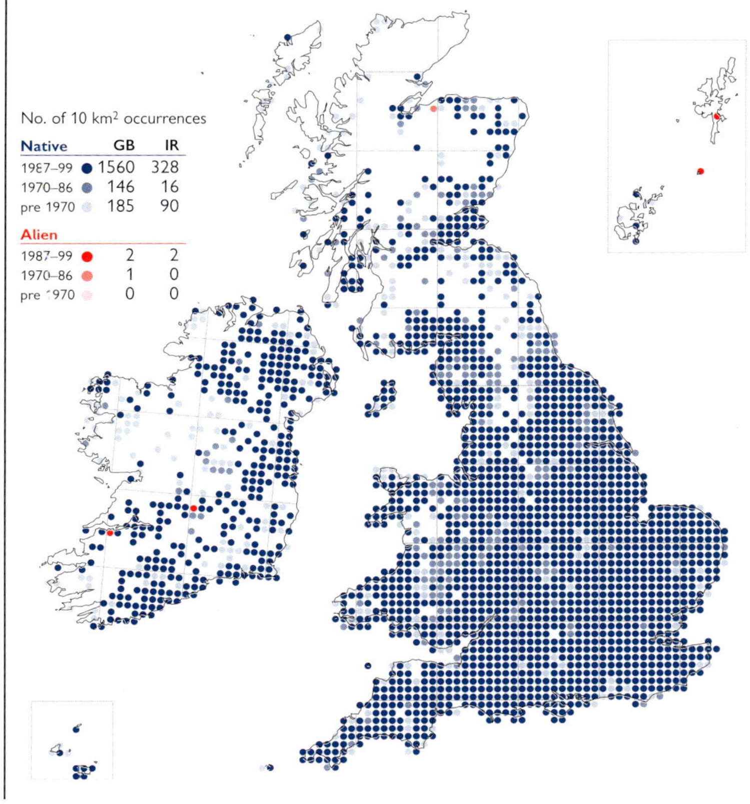

An annual of open and disturbed ground on a wide range of soils ranging from sand to clay and peat. It is a poor competitor, found in cultivated fields, on the open margins of lakes, ponds, streams and rivers, and on waste ground. Robust adventive variants have been recorded in waste places and fields treated with wool shoddy. 0–450 m (Clun Forest, Salop).

Native (change –0.04). The distribution of *P. lapathifolia* has not changed appreciably since the 1962 *Atlas*.

Circumpolar Southern-temperate element; widely naturalised outside its native range.

References: Atlas (175a, b), Hultén & Fries (1986), Jalas & Suominen (1979), Lousley & Kent (1981), Meusel *et al.* (1965), Simmonds (1945), Timson (1963).

J. R. AKEROYD

Persicaria hydropiper Water-pepper

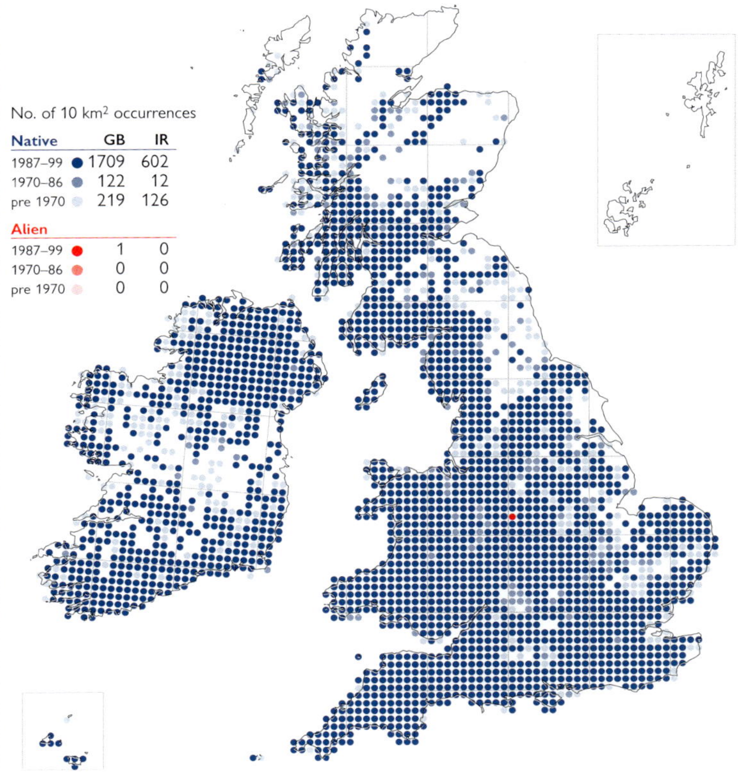

An annual of damp mud on the margins of ponds and lakes, canals, rivers and streams, or shallow depressions such as vehicle tracks and hoof-marks in woodland rides, around field gateways and in wet meadows. It is almost invariably in sites which are waterlogged in winter, often on base-poor soils and sometimes in partial shade. 0–505 m (Llyn Crugnant, Cards.).

Native (change –0.41). The distribution of *P. hydropiper* has not changed markedly since the 1962 *Atlas*, although there have been losses at the eastern edge of its range.

Circumpolar Temperate element; widely naturalised outside its native range.

References: Atlas (175c), Hultén & Fries (1986), Jalas & Suominen (1979), Lousley & Kent (1981), Meusel *et al.* (1965), Timson (1966).

J. R. AKEROYD

Persicaria mitis Tasteless Water-pepper

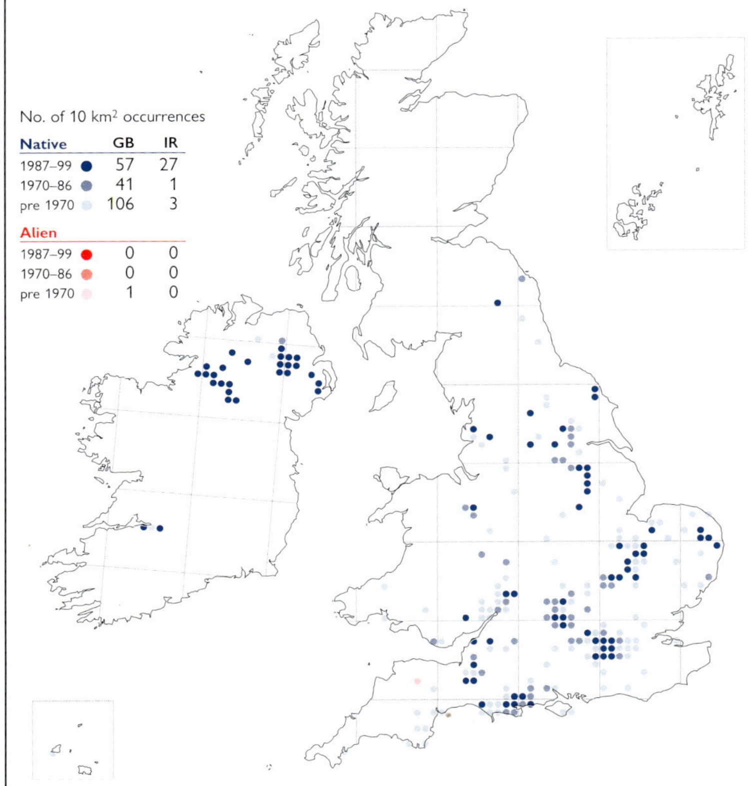

An annual of wet places growing beside ponds, lakes and rivers and in shallow ditches, damp hollows in fields, cattle-trampled places in pasture and abandoned peat cuttings. It grows in nutrient-rich soils, but appears indifferent to soil reaction. Lowland.

Native (change –0.90). *P. mitis* has been confused with *P. hydropiper* and *P. minor*, making interpretation of the map difficult. It is now known from many more 10-km squares since the 1962 *Atlas*. In Britain, however, populations have been lost through the regulation of water levels, the fencing of ditches, and the filling in of ponds. Webb (1984) suggests that this species is a recent introduction to Ireland.

European Temperate element.

References: Atlas (175d), Hultén & Fries (1986), Jalas & Suominen (1979), Lousley & Kent (1981), Stewart *et al.* (1994).

J. R. AKEROYD

Persicaria minor Small Water-pepper

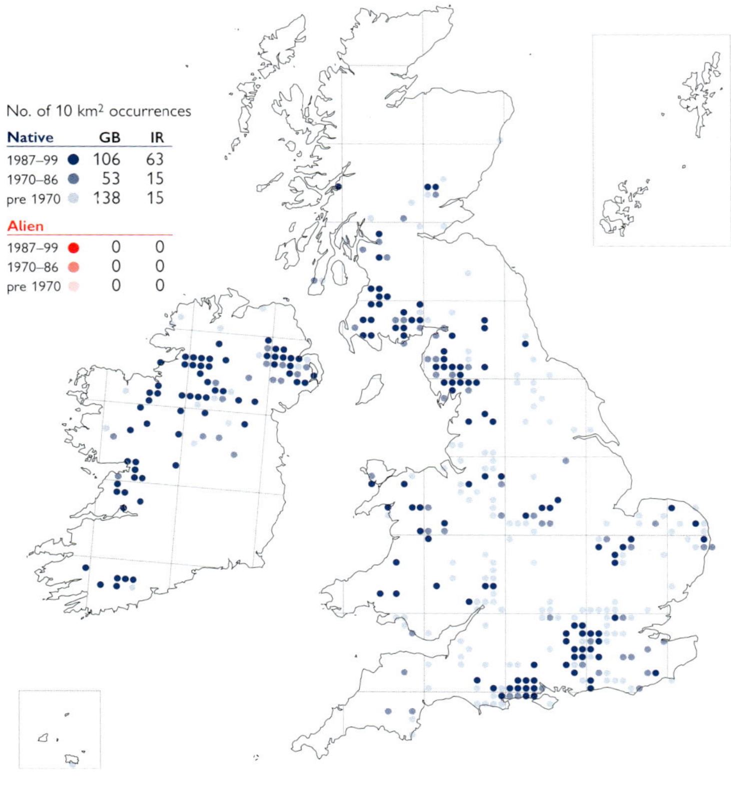

An annual of wet marshy places, winter-flooded ground beside ponds, lakes and ditches, or damp pastures trampled by stock. It is found on a wide range of soils, from nutrient-rich muds in pastures to sandy and gravelly lake shores. 0–315 m (Skeggles Water, Westmorland).

Native (change –0.06). While there has been little change in distribution since the 1962 *Atlas*, *P. minor* is now known from many more 10-km squares. New records from Ireland, S.W. Scotland and Wales increase the known distribution slightly westwards and northwards.

Eurasian Temperate element.

References: Atlas (176a), Hultén & Fries (1986), Jalas & Suominen (1979), Lousley & Kent (1981), Stewart *et al.* (1994).

J. R. AKEROYD

Koenigia islandica Iceland-purslane

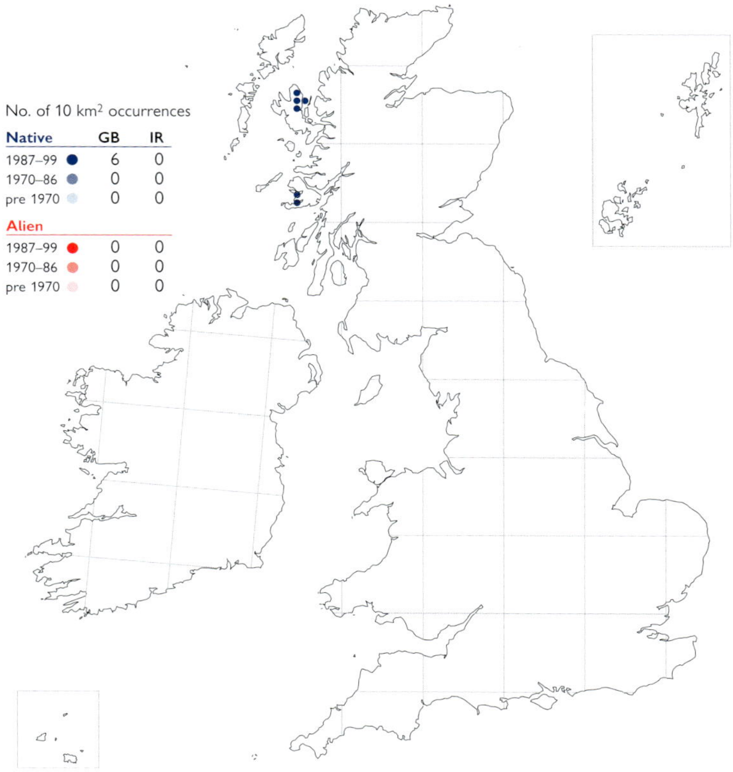

A tiny annual of bare, intermittently flushed or constantly moist, basaltic gravel pans and screes, mainly with a northerly or easterly aspect. Frost action and wind erosion assist in keeping the habitat open. 300–715 m near summit of Storr, Skye (N. Ebudes).

Native (change +0.12). The species was first collected in Britain in 1934 and not correctly identified until 1950. There has been no change in distribution since the 1962 *Atlas*, but severe declines in some populations on Mull (Mid Ebudes) since 1994 are linked to unusually dry weather over several years, especially in the spring.

Circumpolar Arctic-montane element; absent from mountains of C. Europe.

References: Atlas (177d), Hultén & Fries (1986), Jalas & Suominen (1979), Lousley & Kent (1981), Lusby & Wright (1996), D. A. Ratcliffe (1959), Wigginton (1999).

J. R. AKEROYD

Fagopyrum esculentum Buckwheat

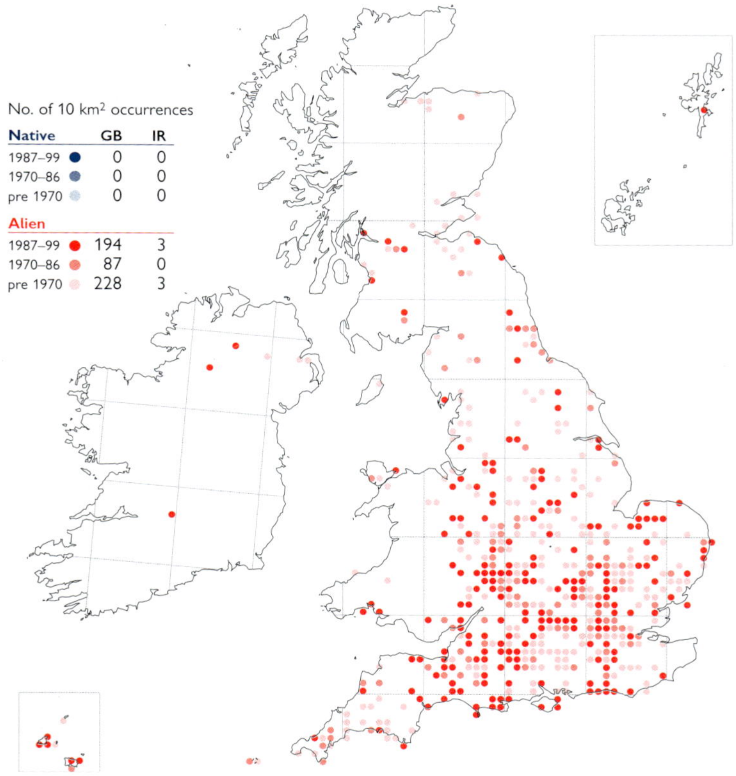

An annual appearing erratically on waste ground, rubbish tips, field margins and in woodland rides. It rarely persists long at any one site. Lowland.

Neophyte (change –0.53). *F. esculentum* was recorded from the wild by 1597. Until the 19th century, it was a significant grain crop in Britain and parts of Ireland, and is still grown on a small scale for green manure, for pheasants and as a bee-plant. It sometimes occurs in sites where it is introduced with pheasant food.

An ancient crop, probably originating in S.W. China but widespread as an introduction in temperate Eurasia and elsewhere.

References: Atlas (177c), Lousley & Kent (1981), Smart & Simmonds (1995), Vaughan & Geissler (1997).

J. R. AKEROYD

Polygonum maritimum Sea Knotgrass

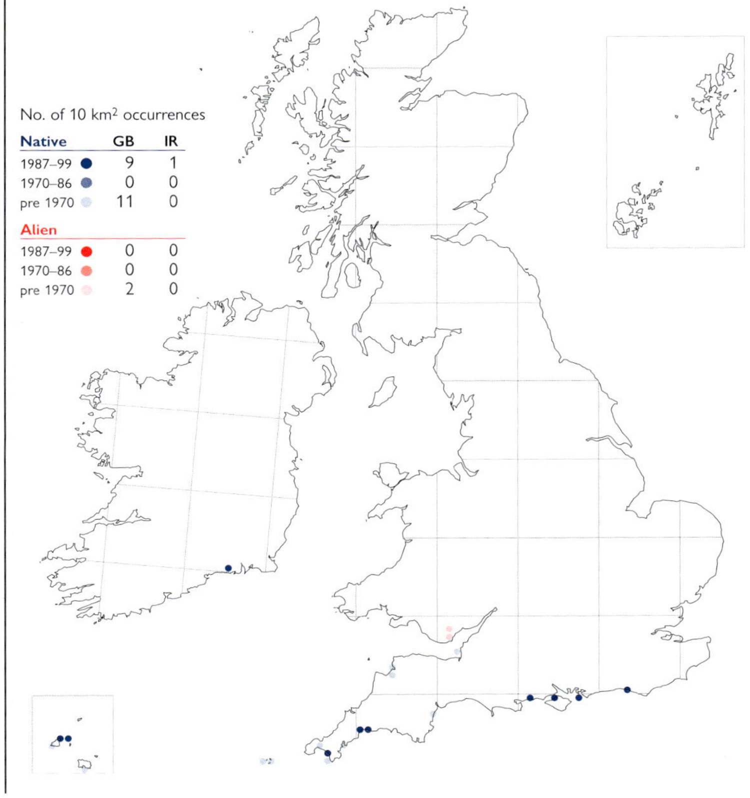

A prostrate perennial herb of sand, shingle or shell beaches, growing above the limit of the highest tides with other strand-line and foredune plants. Lowland.

Native (change +0.21). The 1962 *Atlas* recorded this species from just four post-1930 sites in S.W. England (W. Cornwall, N. Devon and the Channel Islands). Records during the 1990s have considerably extended the known range. Its recent spread in England correlates with a run of mild winters and hot summers. In Ireland it persists at the site where it was discovered in 1973.

Mediterranean-Atlantic element.

References: Atlas (173d), Bolòs & Vigo (1990), Curtis & McGough (1988), Ferguson & Ferguson (1974), Harmes & Spiers (1993), Jalas & Suominen (1979), Lousley & Kent (1981), Wigginton (1999).

J. R. AKEROYD

Polygonum oxyspermum Ray's Knotgrass

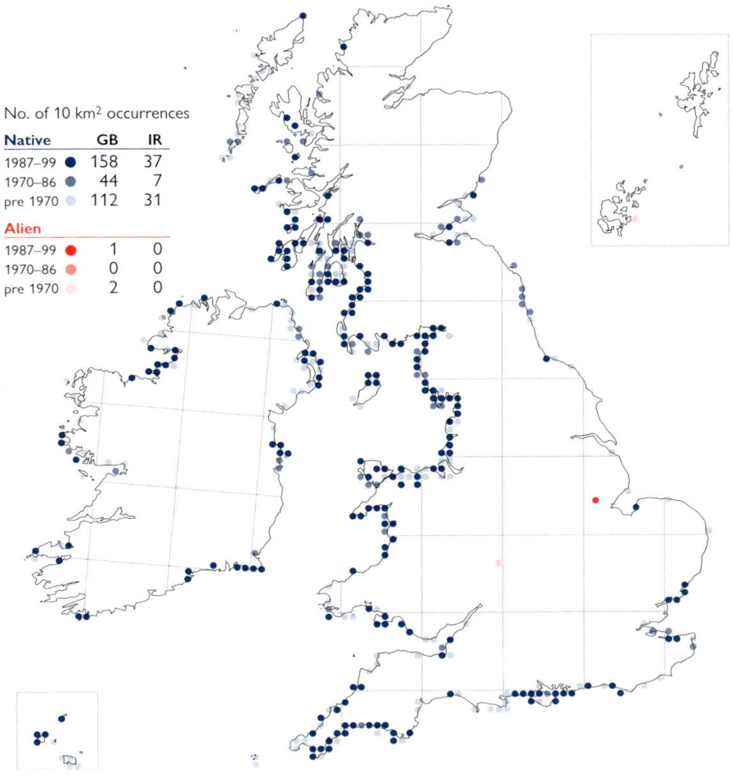

A prostrate annual, biennial or short-lived perennial of sand, shingle or shell beaches, sometimes found on other open sandy ground near the sea, usually just above the limit of the highest tides. Lowland.

Native (change +0.01). There are now many more records for this species than in the 1962 *Atlas*, not only in N. England and Scotland, where it was patently under-recorded, but also Cornwall, Hampshire and Wales. Like other strand-line species, its numbers often fluctuate annually. Most populations are subsp. *raii*. Subsp. *oxyspermum* has been reported as a casual in Scotland, perhaps having arrived naturally from the Baltic region.

European Wide-temperate element; also in N. America.

References: Atlas (173c), Hultén & Fries (1986), Jalas & Suominen (1979), Lousley & Kent (1981), Stewart *et al.* (1994), Styles (1962).

J. R. AKEROYD

Polygonum aviculare agg. Knotgrasses

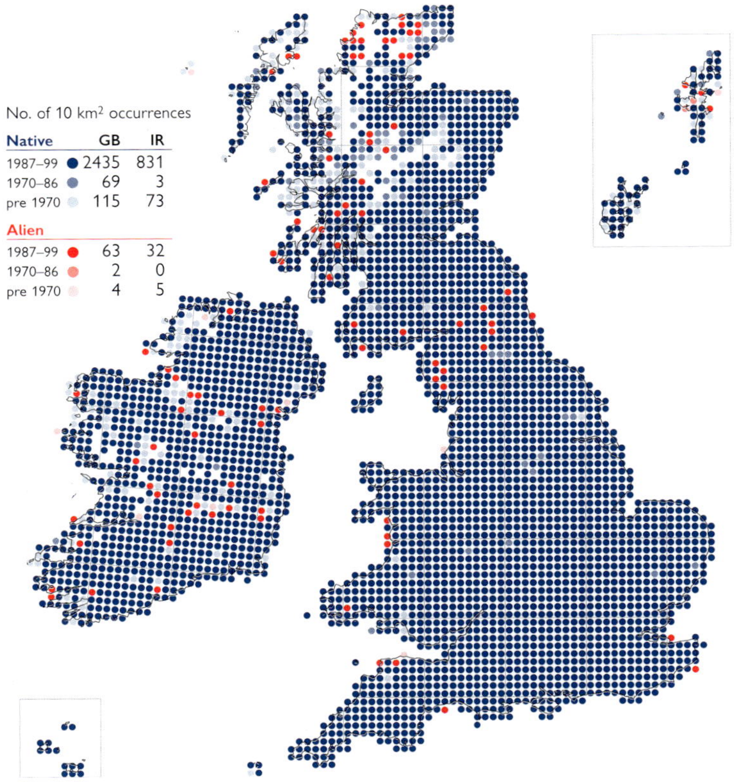

These annuals of open, usually fertile, disturbed soils are commonly found in gardens, on arable land, tracks, paths, pavements, roadsides, waste ground, damp pond margins and on seashores. 0–670 m (Great Dun Fell, Westmorland).

Native (change −0.70). This aggregate includes the native *P. aviculare sens. str.* and *P. boreale* and the archaeophytes *P. arenastrum* and *P. rurivagum*; all four are mapped separately. A fifth species, *P. neglectum*, may be overlooked in Britain. In Norden, all these taxa are treated as subspecies (Jonsell, 2000).

A complex of closely related taxa which together have a Circumpolar Wide-temperate distribution.

References: Atlas (173b), Grime *et al.* (1988), Hultén & Fries (1986), Jalas & Suominen (1979), Lousley & Kent (1981), Styles (1962).

J. R. AKEROYD

Polygonum arenastrum Equal-leaved Knotgrass

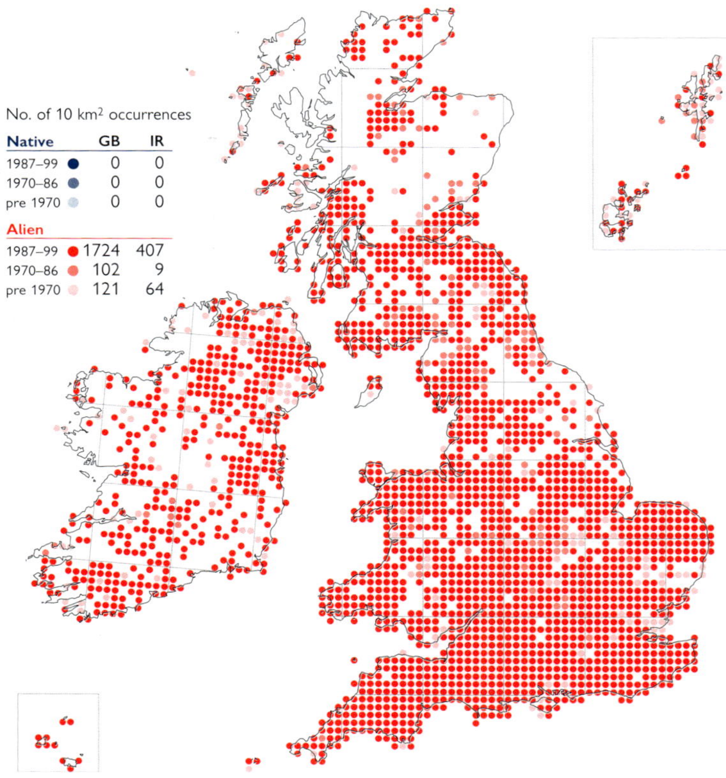

A prostrate annual of open and trampled ground, especially of well-drained soils, on tracks, paths, roadsides and weedy pavements, around ponds and field gateways. 0–450 m (near Birkdale, Westmorland).

Archaeophyte. *P. arenastrum* is widespread throughout Britain and Ireland, although often overlooked as *P. aviculare*. It has not been mapped previously on a national scale, having been included within *P. aviculare* agg. in the 1962 *Atlas*. Its distribution is not likely to have changed appreciably.

As an archaeophyte, *P. arenastrum* has a Eurasian Wide-temperate distribution; in recent centuries it has become widely naturalised outside this range.

References: Lousley & Kent (1981), Styles (1962).

J. R. AKEROYD

Polygonum aviculare Knotgrass

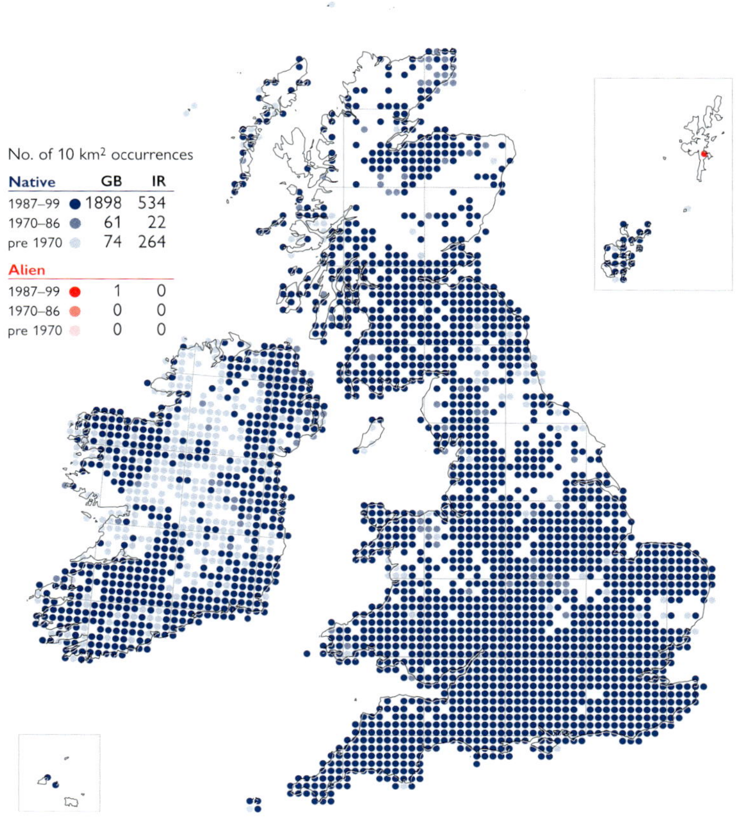

An annual of open and disturbed ground, including arable land, gardens, waste places and seashores. The species remains a significant agricultural weed. 0–550 m (Northumberland), with an exceptional record at 670 m on Great Dun Fell (Westmorland).

Native. *P. aviculare* was mapped with other members of the aggregate in the 1962 *Atlas*, but it is most unlikely that there have been any changes in distribution. It is under-recorded in areas where the segregates of *P. aviculare* have not been distinguished. Almost all records from Shetland are now referred to *P. boreale*.

Circumpolar Wide-temperate element; widely naturalised outside its native range.

References: Lousley & Kent (1981), Styles (1962).

J. R. AKEROYD

Polygonum boreale Northern Knotgrass

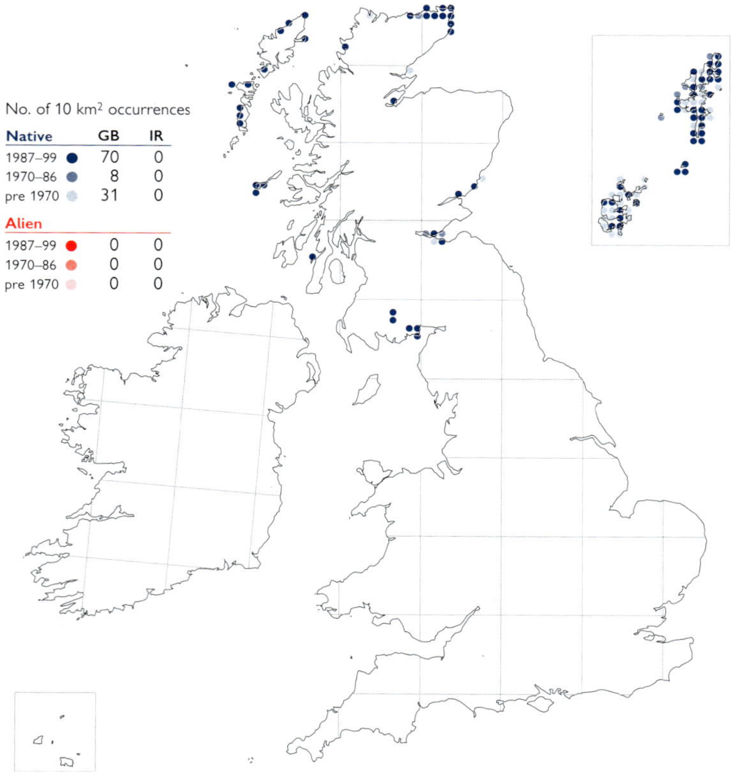

An annual of open and disturbed ground on roadsides and paths, cultivated land, sandy beaches, dunes and coastal shingle. In Orkney and Shetland it replaces *P. aviculare* and *P. arenastrum*, although it was formerly overlooked elsewhere due to confusion with these species. Lowland.

Native. *P. boreale* was not mapped separately in the 1962 *Atlas*, being included within *P. aviculare* agg., but was mapped by Perring & Sell (1968). Subsequent records have extended the known range along the N. coast of Scotland and in the Hebrides, with scattered, perhaps adventive, localities further south. It is likely still to be under-recorded.

Suboceanic Boreal-montane element; also in N. America.

References: Atlas Supp. (43c), Lousley & Kent (1981), Stewart *et al.* (1994), Styles (1962).

J. R. AKEROYD

Polygonum rurivagum Cornfield Knotgrass

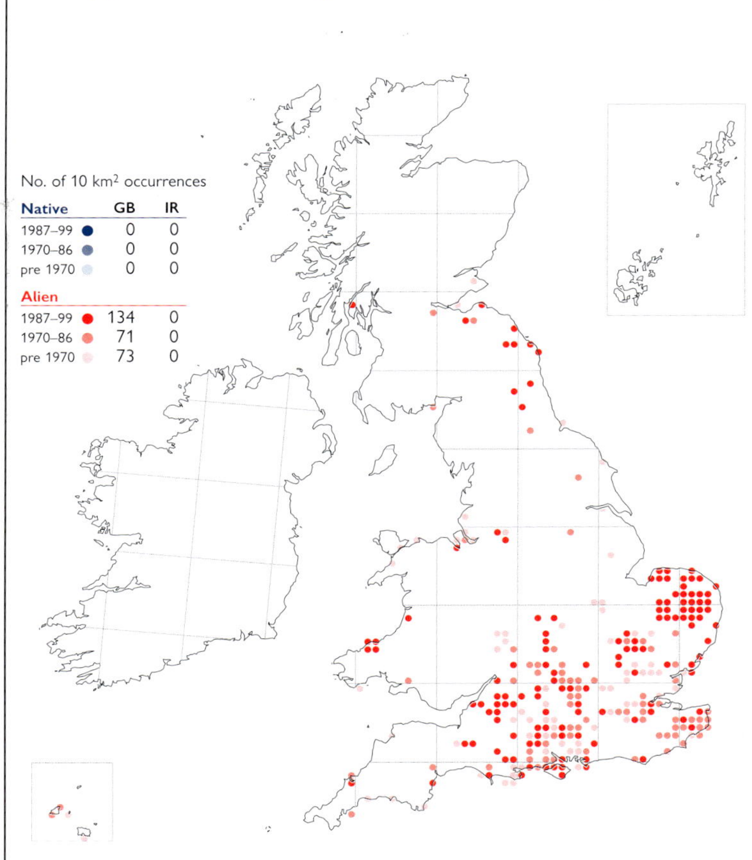

An annual of arable fields and more rarely of ruderal habitats, especially on light chalky soils and calcareous clays. Lowland.

Archaeophyte. The species was first mapped on a national scale by Perring & Sell (1968). Records since then have considerably extended the known range, and it may also be spreading with translocated topsoil. It is probably still under-recorded, as it is easily overlooked as *P. aviculare*.

As an archaeophyte *P. rurivagum* has a European Temperate distribution.

References: Atlas Supp. (43b), Lousley & Kent (1981), Stewart *et al.* (1994), Styles (1962).

J. R. AKEROYD

Fallopia japonica Japanese Knotweed

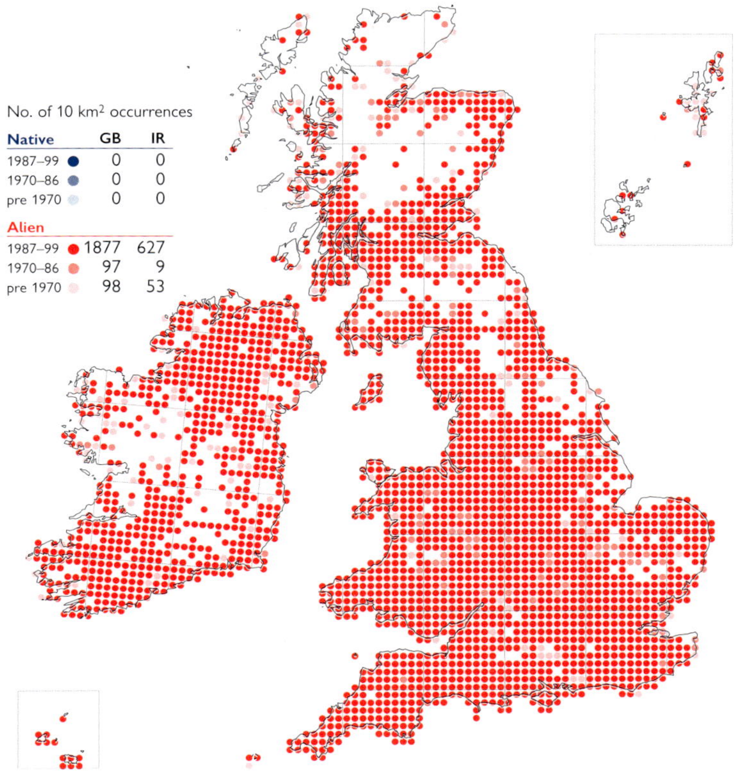

A persistent rhizomatous perennial forming dense thickets on waste ground, rubbish tips, roadsides, railway banks, along canal, stream and river banks, and on sea-loch shores. Rhizome fragments are dispersed in garden and other rubbish, and by river floods. Lowland.

Neophyte (change +1.83). *F. japonica* has been grown in British gardens since 1825. It was first recorded in the wild in 1886 and became well-established between 1920 and 1940. In Ireland it was recorded as planted near Dublin in 1902. It has substantially increased since the 1962 *Atlas*.

Native of E. Asia; widely naturalised as a single, male-sterile clone in temperate Europe.

References: Atlas (176d), Bailey & Conolly (2000), Beerling *et al.* (1994), Conolly (1977), Grime *et al.* (1988), Hollingsworth & Bailey (2000), Jalas & Suominen (1979), Lousley & Kent (1981).

J. R. AKEROYD

Fallopia japonica × *F. sachalinensis* (*F.* × *bohemica*)

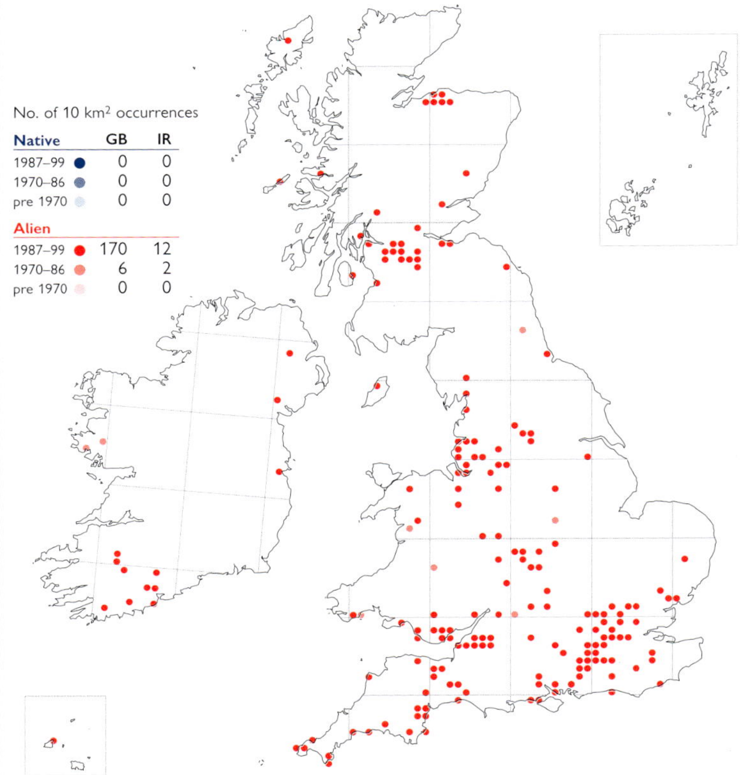

A rampant, rhizomatous perennial herb, locally forming thickets on waste ground, roadsides, railway banks and on riversides. Male plants of this hybrid show some fertility, and more than one clone is present in some sites. Lowland.

Neophyte. This hybrid is known to have been in cultivation in Britain since 1872, but it was not collected from the wild until 1954 and not recognised until the 1980s. Although it has spread from cultivation, it also arises spontaneously in our area. It is almost certainly under-recorded, especially in Ireland.

A hybrid originating in Europe.

References: Bailey *et al.* (1996), Bailey & Conolly (2000), Hollingsworth & Bailey (2000).

J. R. AKEROYD

Fallopia sachalinensis Giant Knotweed

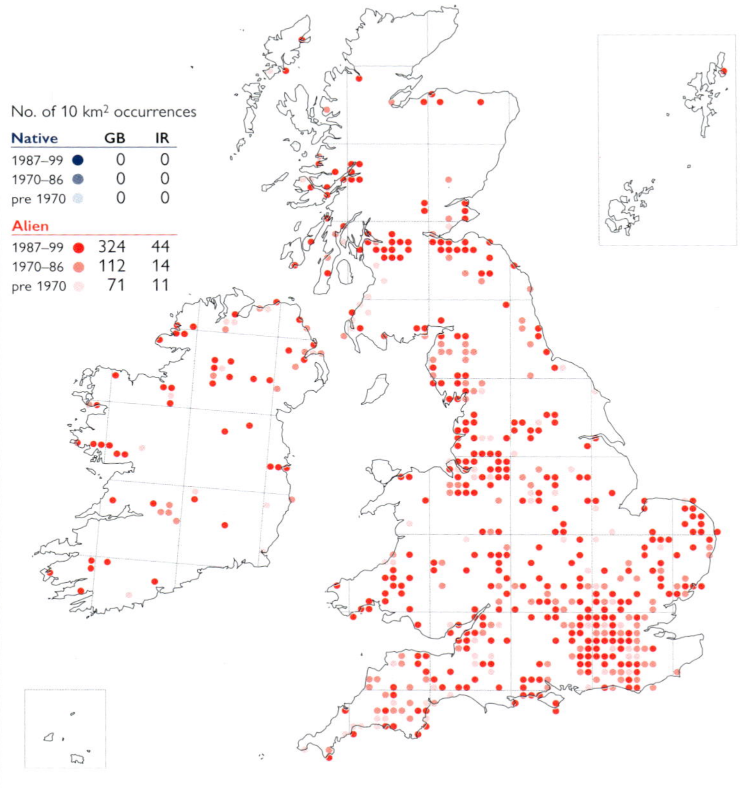

This robust rhizomatous perennial herb forms extensive thickets on waste ground, roadsides, river banks and lake and sea-loch shores. Generally lowland, but upper altitudinal limit unknown.

Neophyte (change +1.05). *F. sachalinensis* became commercially available to British gardeners in 1869, but was much less widely planted than *F. japonica* because of its great size. It was naturalised in Ireland by 1896 and in Britain by 1903. It has substantially increased since the 1962 *Atlas*.

Native of E. Asia (Sakhalin, Japan); widely naturalised in temperate Europe.

References: Atlas (177a), Beerling *et al.* (1994), Conolly (1977), Hollingsworth & Bailey (2000), Jalas & Suominen (1979), Lousley & Kent (1981).

J. R. AKEROYD

Fallopia baldschuanica Russian-vine

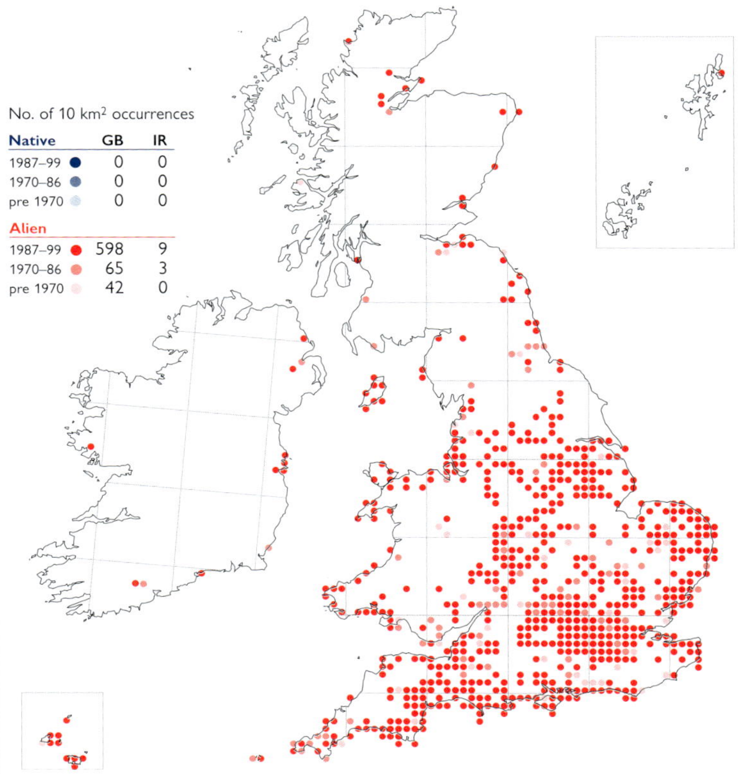

A climbing perennial, with vine-like stems that festoon trees, scrub, hedges and neglected outbuildings. Lowland.

Neophyte. This species was introduced into British gardens in about 1894. It was first recorded in the wild in 1936 and appears to be increasing due to its continued use to screen eyesores and from the discarding of surplus garden material onto roadsides and rubbish tips. It is rarely naturalised away from habitation and some of the increases may be attributable to an increased tendency to record aliens; some records may be of plants rooted in gardens.

Native of C. Asia.

References: Coats (1963), Lousley & Kent (1981).

J. R. AKEROYD

Fallopia convolvulus Black-bindweed

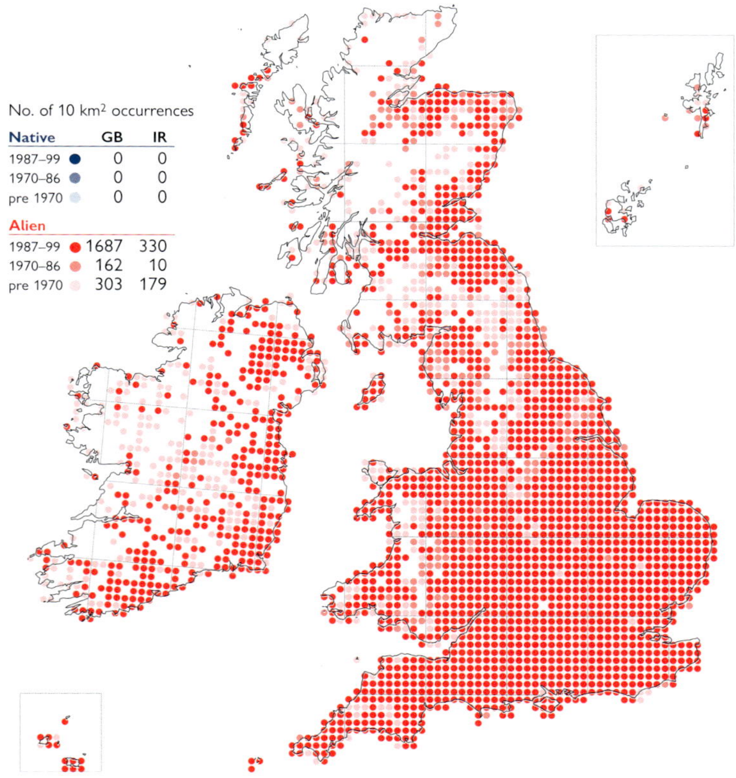

An annual found in arable land, gardens, waste places, rubbish tips and on roadsides. 0–450 m (Clun Forest, Salop).

Archaeophyte (change −1.31). This species has been a weed of cultivation since the Neolithic, and it was formerly the major contaminant of agricultural seed in Britain. It was mapped as 'all records' in the 1962 *Atlas*. Analysis of the database reveals that the widespread decline in N. England and Scotland has occurred since 1950, possibly as marginal cultivations are abandoned.

As an archaeophyte *F. convolvulus* has a Eurosiberian Wide-temperate distribution, but it is widely naturalised so that its distribution is now Circumpolar Wide-temperate.

References: Atlas (176b), Grime *et al.* (1988), Hultén & Fries (1986), Hume *et al.* (1983), Jalas & Suominen (1979), Lousley & Kent (1981).

J. R. AKEROYD

Fallopia dumetorum Copse-bindweed

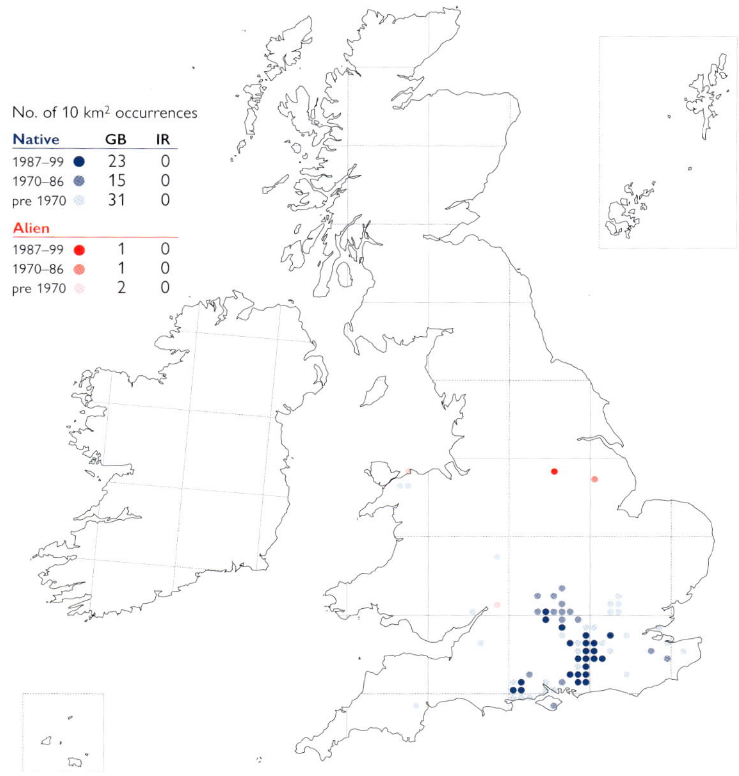

A climbing annual of hedges, thickets and wood-borders on well-drained soils. Erratic in appearance, it sometimes occurs in quantity following the felling, thinning or coppicing of hedgerows and woodland. Lowland.

Native (change –0.33). This species has always been local in distribution. It seems to have declined since the 1962 *Atlas*, probably due to a lack of woodland management. In Oxfordshire, however, where it was first recorded between 1968 and 1973, it has now been found in at least 12 tetrads (Killick *et al.*, 1998).

Eurasian Temperate element.

References: Atlas (176c), Hultén & Fries (1986), Jalas & Suominen (1979), Lousley & Kent (1981), Meusel *et al.* (1965), Stewart *et al.* (1994).

J. R. AKEROYD

Rheum × *hybridum* Rhubarb

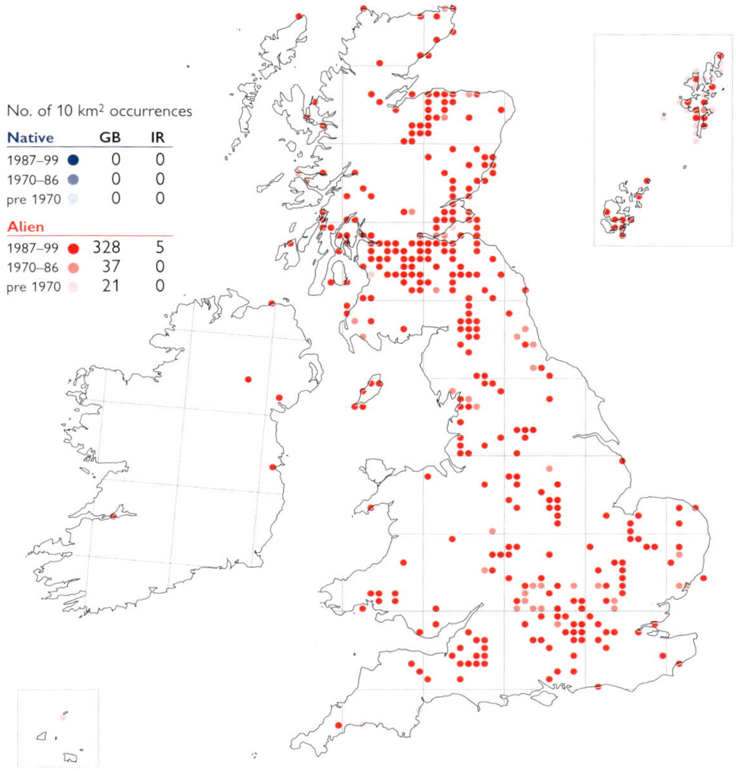

A rhizomatous perennial herb of waste ground, roadsides, railway banks, river and stream banks, and around cottage ruins. Plants are usually outcasts or relics of cultivation; more isolated colonies can originate from rhizome fragments washed down streams and rivers. Generally lowland, but recorded at 550 m (Moor House, Westmorland).

Neophyte. *R* × *hybridum* has been cultivated in our area as a mild laxative since at least 1573, and for its edible petioles since the 18th century. It was recorded from the wild in 1960, but is unlikely to be spreading, as plants do not set seed and colonies are long-lived.

A hybrid of uncertain origin, which probably evolved in N. China or E. Siberia.

References: Foust (1992), Lousley & Kent (1981), Phillips & Rix (1993), Vaughan & Geissler (1997).

J. R. AKEROYD

Rumex acetosella Sheep's Sorrel

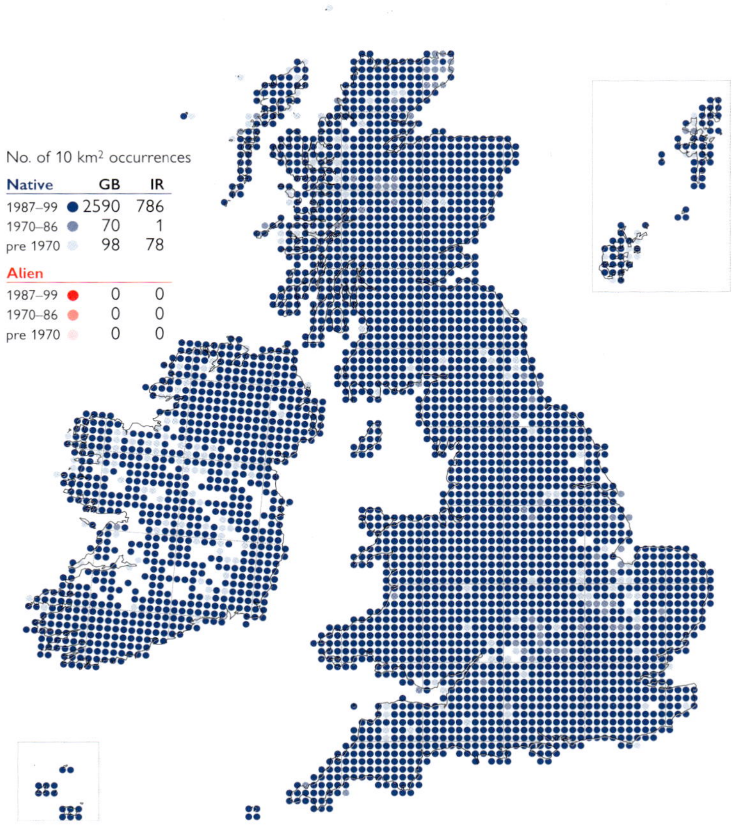

A rhizomatous perennial herb of dry heaths, non-calcareous sand dunes, shingle beaches and other short, open grasslands on acidic, impoverished, sandy or stony soils. It is sometimes found on outcrops of acidic rocks. 0–1050 m (Carnedd Llewelyn, Caerns.).

Native (change –0.62). There has been little change in distribution since the 1962 *Atlas*. Two subspecies occur in Britain: the widespread subsp. *pyrenaicus* and the scarcer subsp. *acetosella*, which replaces it in N. Britain but is not known from Ireland.

Eurosiberian Wide-temperate element, but widely naturalised so that distribution is now Circumpolar Wide-temperate.

References: Atlas (178b), Atlas Supp. (44a), Grime *et al.* (1988), Hultén & Fries (1986), Jalas & Suominen (1979), Lousley & Kent (1981), Meusel *et al.* (1965), Nijs (1984).

J. R. AKEROYD

Rumex acetosa Common Sorrel

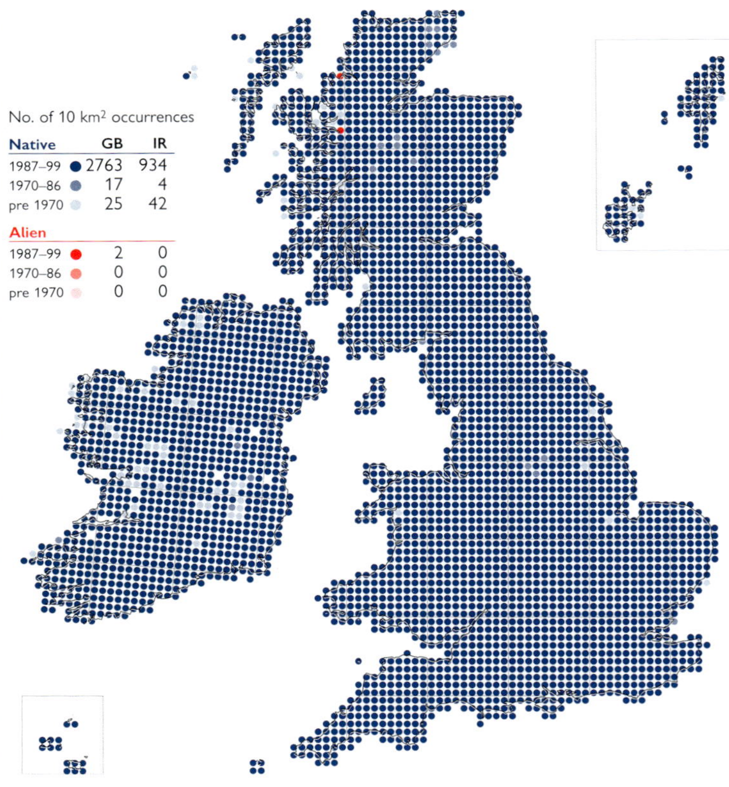

A tufted, short-lived perennial of neutral to slightly acidic soil in meadows, pastures, woodland rides and glades, mountain ledges and shingle beaches; absent from severely improved grasslands and leys. 0–1215 m (Breadalbanes, Mid Perth).

Native (change +1.32). There has been no change in distribution since the 1962 *Atlas*. Four subspecies occur in our area: subsp. *acetosa* is found throughout the range of the species, subsp. *biformis* is restricted to sea-cliffs in W. Britain and W. Ireland, subsp. *ambiguus* is a rare alien in S. England and subsp. *hibernicus* is mapped separately.

Eurosiberian Boreo-temperate element, but widely naturalised so that distribution is now Circumpolar Boreo-temperate.

References: Atlas (178c), Grime *et al.* (1988), Hultén & Fries (1986), Jalas & Suominen (1986), Lousley & Kent (1981), Meusel *et al.* (1965).

J. R. AKEROYD

Rumex acetosa subsp. *hibernicus*

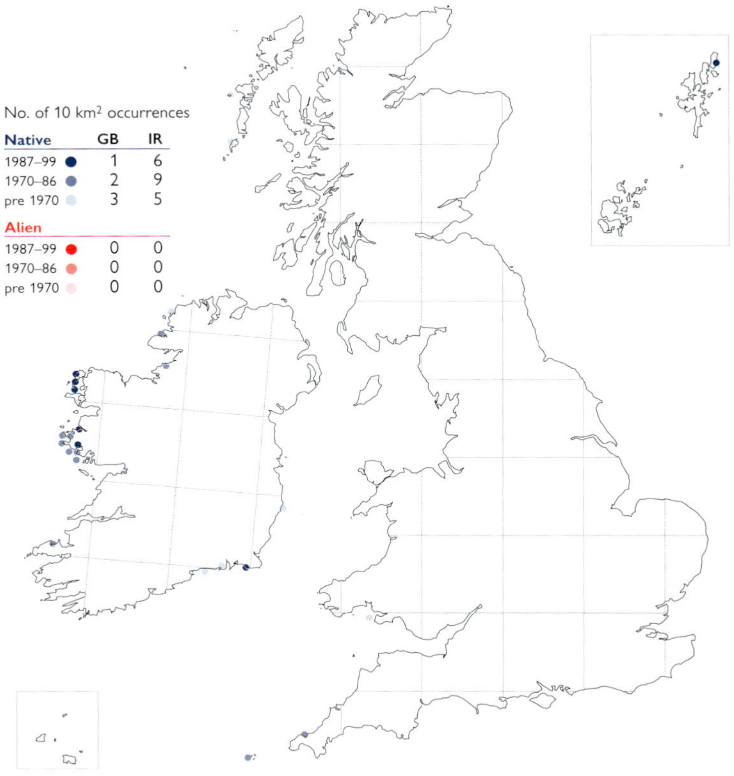

A short-lived perennial herb found on fixed sand dunes, dune grassland and machair. Plants from open, serpentine debris in Shetland cannot be distinguished from plants on sand dunes, and are included within this subspecies. Lowland.

Native. This subspecies was first described as *R. hibernicus* in 1961. Its distribution is likely to be stable, but it is almost certainly under-recorded.

Endemic.

Reference: Rechinger (1961).

J. R. AKEROYD

Rumex pseudoalpinus Monk's-rhubarb

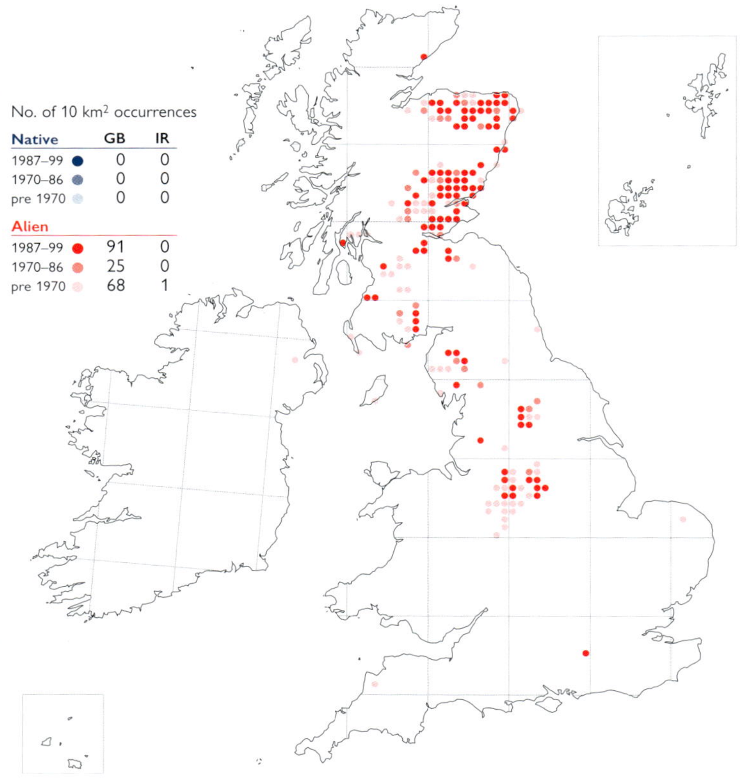

A rhizomatous perennial herb found growing near farm buildings, by streams and on roadsides, especially in places manured by animals. Lowland to 375 m at Dowthwaite Head (Cumberland).

Archaeophyte (change −0.42). There is archaeological evidence that *R. pseudoalpinus* has been present in Scotland since the medieval period. It remains locally common in E. Scotland, but it has declined in the Peak District as farms are converted to non-agricultural residences and tidied up.

Native of the mountains of C. & S. Europe and S.W. Asia.

References: Atlas (179a), Jalas & Suominen (1979), Lousley & Kent (1981), Meusel *et al.* (1965).

J. R. AKEROYD

Rumex aquaticus Scottish Dock

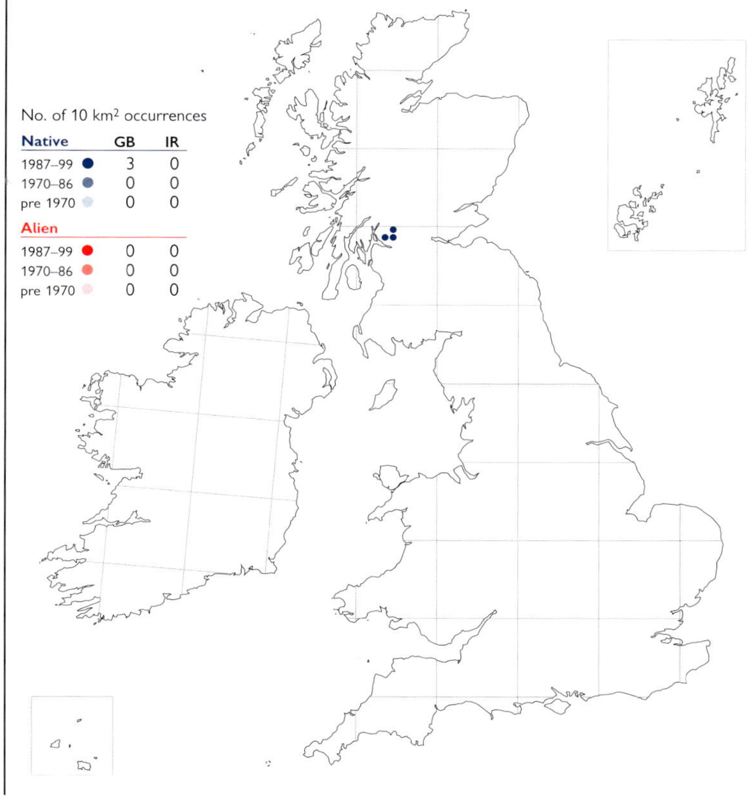

An aquatic perennial herb growing on silty and gravelly lake shores, beside ditches and streams, in marshes, wet fields and woodland clearings. Lowland.

Native. There has been no change in distribution since the 1962 *Atlas*. The species was only discovered in Britain in 1935 and not correctly identified until 1939. It is still holding its own, despite the concern of some conservationists about a high level of introgressive hybridisation with *Rumex obtusifolius*.

Circumpolar Boreo-temperate element, with a continental distribution in W. Europe.

References: Atlas (179b), Hultén & Fries (1986), Jalas & Suominen (1979), Lousley & Kent (1981), Lusby & Wright (1996), Meusel *et al.* (1965), Preston & Croft (1997), Wigginton (1999).

J. R. AKEROYD

Rumex longifolius Northern Dock

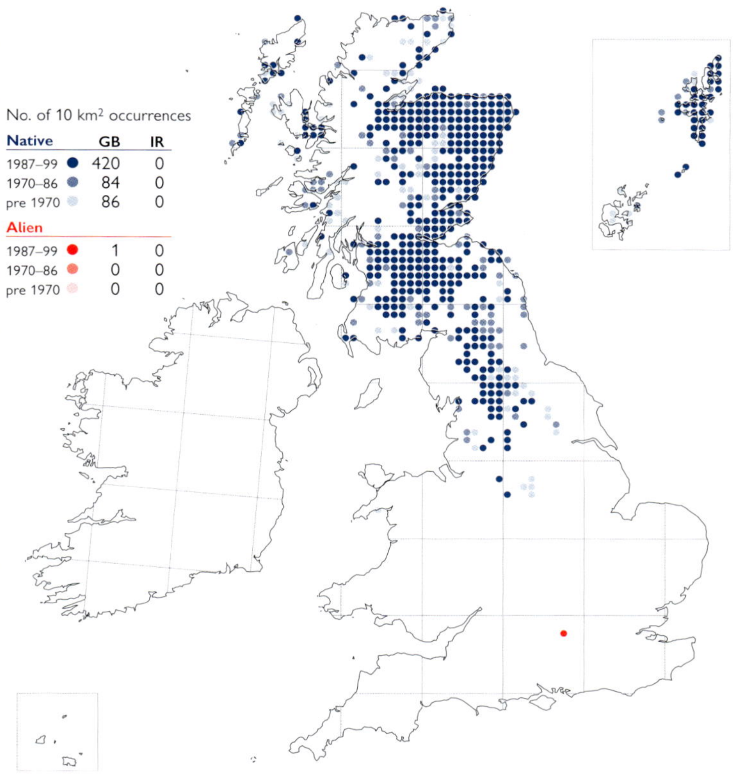

A perennial herb of open, disturbed ground on roadsides, river banks, streamsides and lake shores, in fields and around farms. 0–520 m (Hartside, Cumberland, and in Atholl, E. Perth).

Native (change +0.93). The map demonstrates that *R. longifolius* is much more frequent in the Scottish borders than was shown in the 1962 *Atlas*. It is spreading along roadsides, and new records from England have extended the known distribution south to Cheshire and Staffordshire.

Eurasian Boreal-montane element; widely naturalised outside its native range.

References: Atlas (179c), Hultén & Fries (1986), Jalas & Suominen (1979), Lousley & Kent (1981).

J. R. AKEROYD

Rumex longifolius × *R. obtusifolius* (*R.* × *hybridus*)

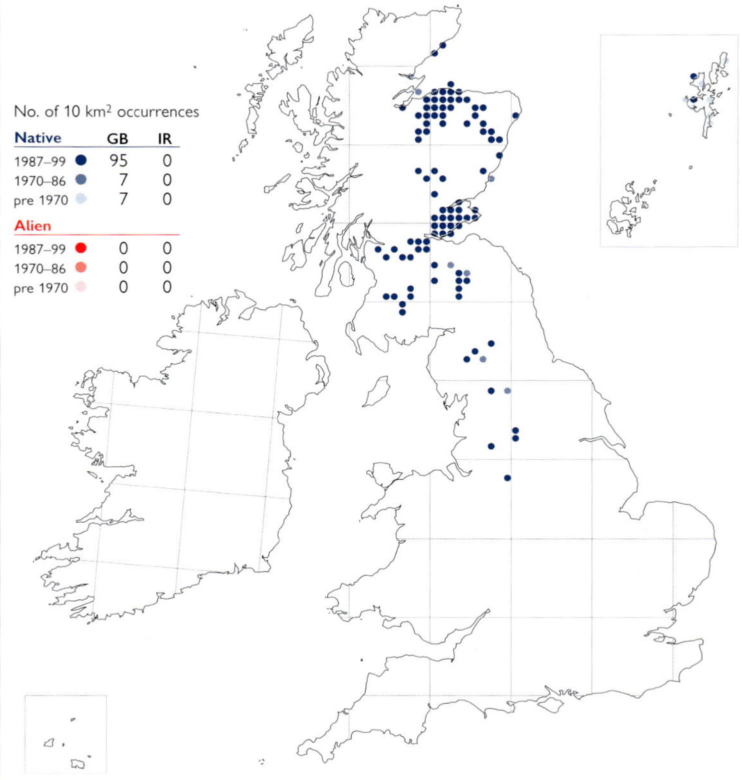

A variable perennial herb of roadsides, waste ground and around buildings. It appears to be fertile as most of the fruiting perianths mature. 0–430 m (Garrigill, Cumberland).

Native. This hybrid has proved to be widespread where both parents occur together. Like other *Rumex* hybrids, it is under-recorded.

Commonly found where both parents occur, and therefore frequent in N. Europe.

References: Lousley & Kent (1981), Stace (1975).

J. R. AKEROYD

Rumex hydrolapathum Water Dock

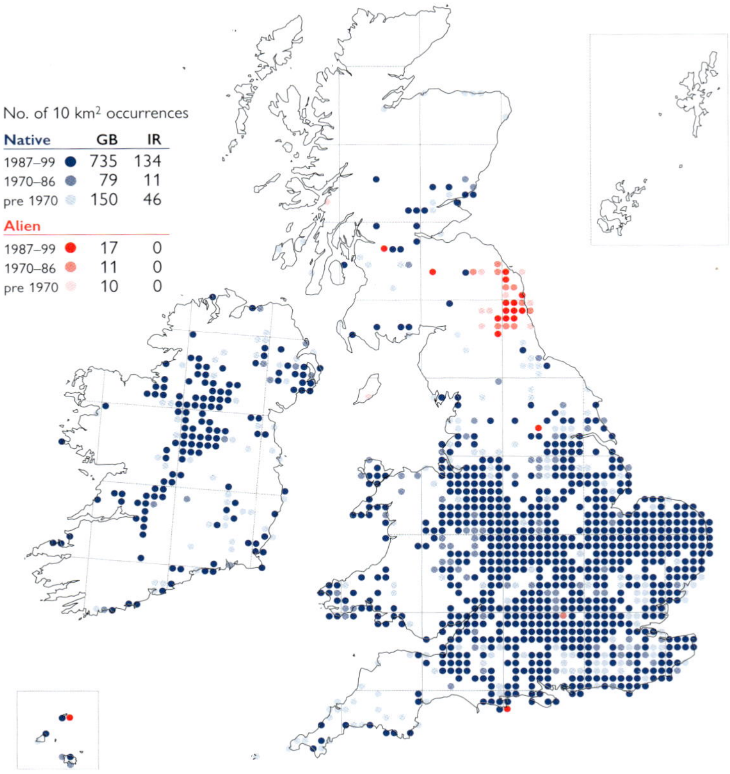

No. of 10 km² occurrences

Native		GB	IR
1987–99	●	735	134
1970–86	●	79	11
pre 1970	●	150	46
Alien			
1987–99	●	17	0
1970–86	●	11	0
pre 1970	●	10	0

A tufted perennial herb found, usually as an emergent, on the margins of slow-flowing rivers and streams, by canals, lakes and ponds, and in ditches. It can also colonise bare ground in marshes and fens, but does not survive in closed vegetation. Lowland.

Native (change –0.13). There has been no appreciable change in the distribution of this species in S. England since the 1962 *Atlas*. However, it has significantly extended its range in N. England, Scotland and Ireland, although some records in these areas may be escapes from sites where it was originally planted for ornament.

European Temperate element.

References: Atlas (178d), Hultén & Fries (1986), Jalas & Suominen (1979), Lousley & Kent (1981), Meusel *et al.* (1965), Preston & Croft (1997).

J. R. AKEROYD

Rumex cristatus Greek Dock

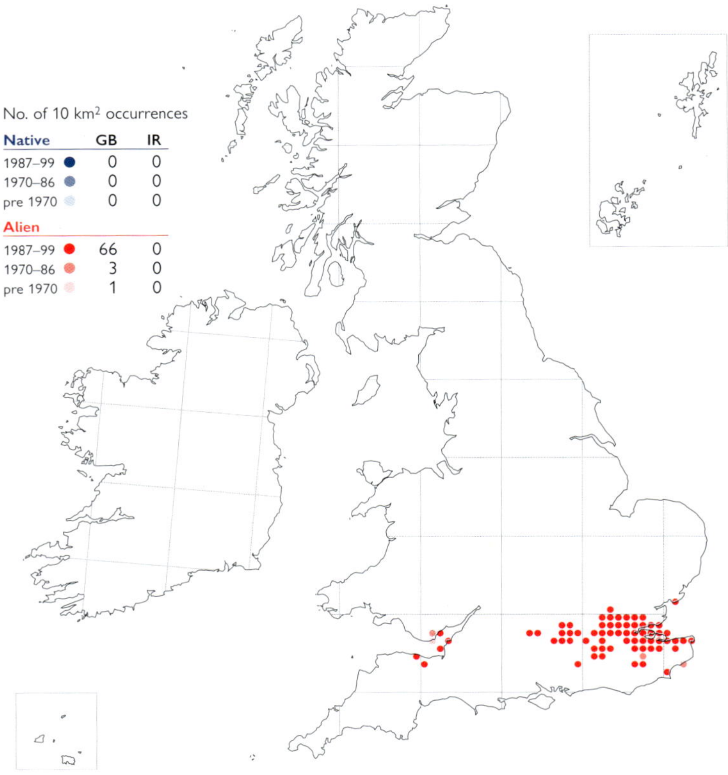

No. of 10 km² occurrences

Native		GB	IR
1987–99	●	0	0
1970–86	●	0	0
pre 1970	●	0	0
Alien			
1987–99	●	66	0
1970–86	●	3	0
pre 1970	●	1	0

A tall perennial herb of river banks, pathsides, sand dunes, waste ground and rubbish tips. It may be introduced with grass and clover seed. Populations can be casual, but are often well-naturalised; reproduction is by seed. Lowland.

Neophyte. This species was reportedly known from the River Rhymney (Mons.) *c.* 1920; it still thrives there today. It was recorded from the Thames at Kew Bridge in 1938 (Surrey) and spread rapidly in S.E. England, being recorded from Hadleigh Marshes (N. Essex) in 1949 and the Rivers Medway and Swale (E. & W. Kent) by 1982. In N. Somerset it was first recorded in 1942 but it does not seem to be spreading there to the same extent.

Native of S.E. Europe and S.W. Asia.

References: Jalas & Suominen (1986), Lousley & Kent (1981).

H. J. KILLICK & J. R. AKEROYD

Rumex crispus Curled Dock

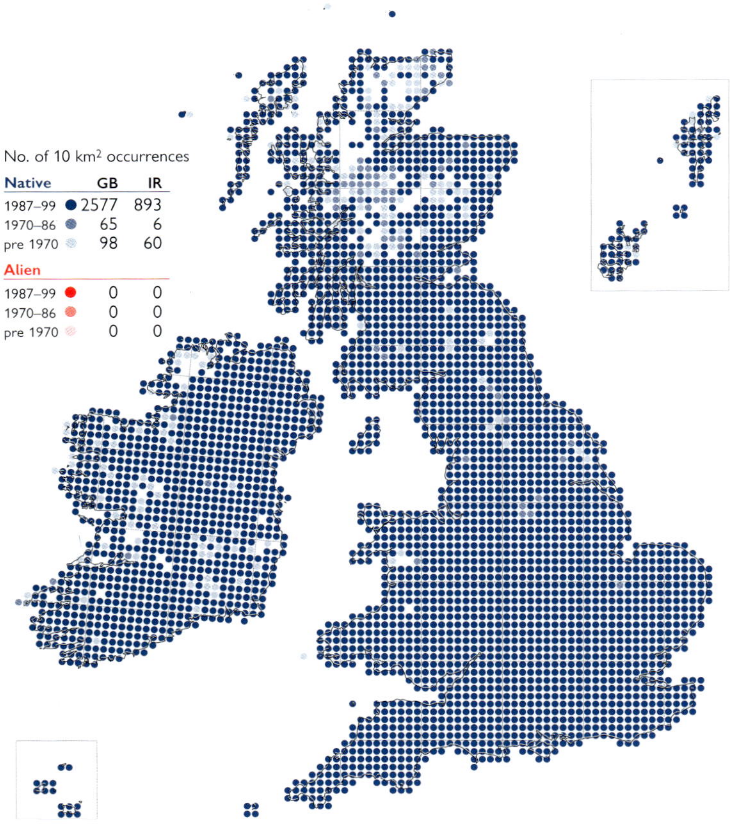

No. of 10 km² occurrences

Native		GB	IR
1987–99	●	2577	893
1970–86	●	65	6
pre 1970	●	98	60
Alien			
1987–99	●	0	0
1970–86	●	0	0
pre 1970	●	0	0

An annual to short-lived perennial herb of waste ground, roadsides, disturbed pastures and arable land; also in a range of coastal habitats including drift-lines, shingle beaches, sand dunes, tidal river banks and the uppermost parts of saltmarshes. 0–845 m (Great Dun Fell, Westmorland).

Native (change +0.11). There has been little change in distribution since the 1962 *Atlas*. Three subspecies occur in Britain: subsp. *crispus* occurs throughout the range of the species and subsp. *littoreus* and subsp. *uliginosus* are mapped separately.

Eurosiberian Southern-temperate element, but widely naturalised so that distribution is now Circumpolar Southern-temperate.

References: Atlas (179d), Akeroyd & Briggs (1983a, b), Cavers & Harper (1964), Grime *et al.* (1988), Hultén & Fries (1986), Jalas & Suominen (1979), Lousley & Kent (1981).

J. R. AKEROYD

Rumex crispus subsp. littoreus

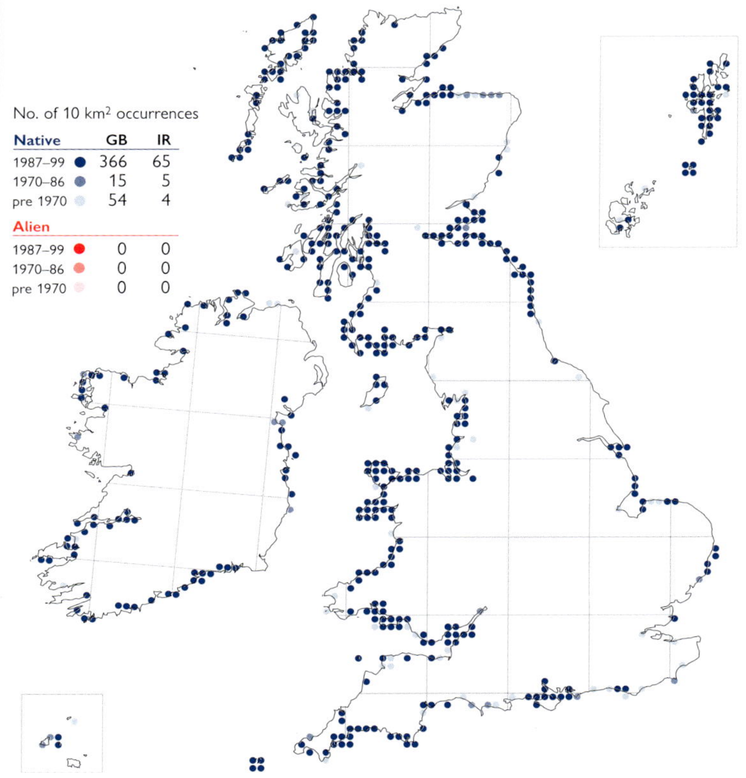

A perennial herb of shingle beaches, sandy and rocky shores, banks by the sea, sand dunes and the upper part of saltmarshes. It is most frequent in strand-line communities on shingle. Lowland.

Native. This taxon has long been known as a variety but it has only been recorded systematically since it was treated as a subspecies in Stace (1991); it is probably under-recorded. Its distribution is almost certainly stable.

Oceanic Temperate element.

References: Akeroyd & Briggs (1983a, b), Cavers & Harper (1967), Lousley & Kent (1981).

J. R. AKEROYD

Rumex crispus subsp. uliginosus

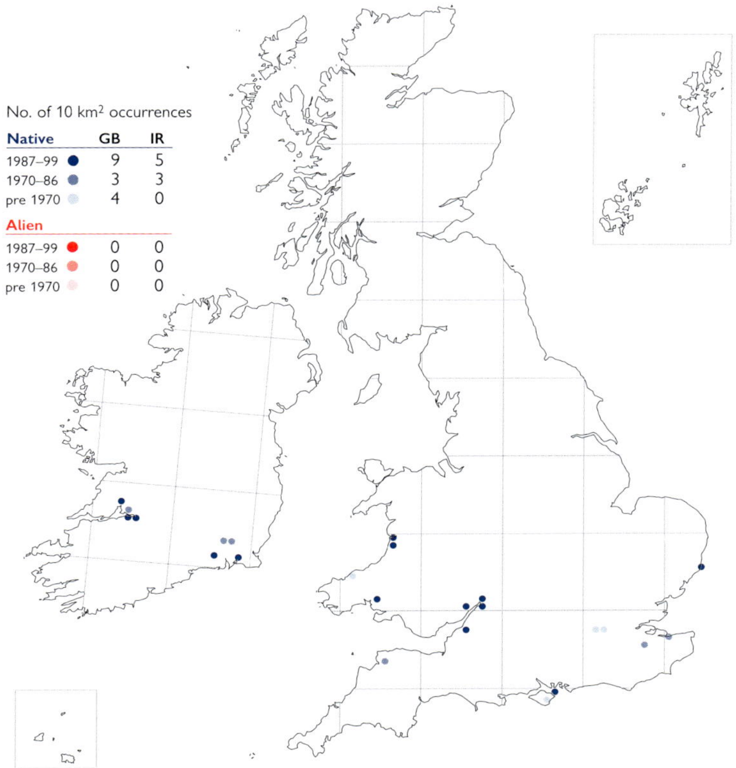

A perennial herb of tidal mud- and river banks near the upper limits of tidal influence, especially where the banks are steep and unstable. Lowland.

Native. The presence of *R. crispus* subsp. *uliginosus* in Britain and Ireland was first recognised by Lousley (1944), but it did not come to the attention of most recorders until it was treated as a subspecies in Stace (1991) after more intensive taxonomic studies. It is almost certainly under-recorded.

Oceanic Temperate element; recorded from Britain, Ireland and N.W. France.

References: Akeroyd & Briggs (1983a, b), Lousley & Kent (1981).

J. R. AKEROYD

Rumex crispus × R. longifolius (R. × propinquus)

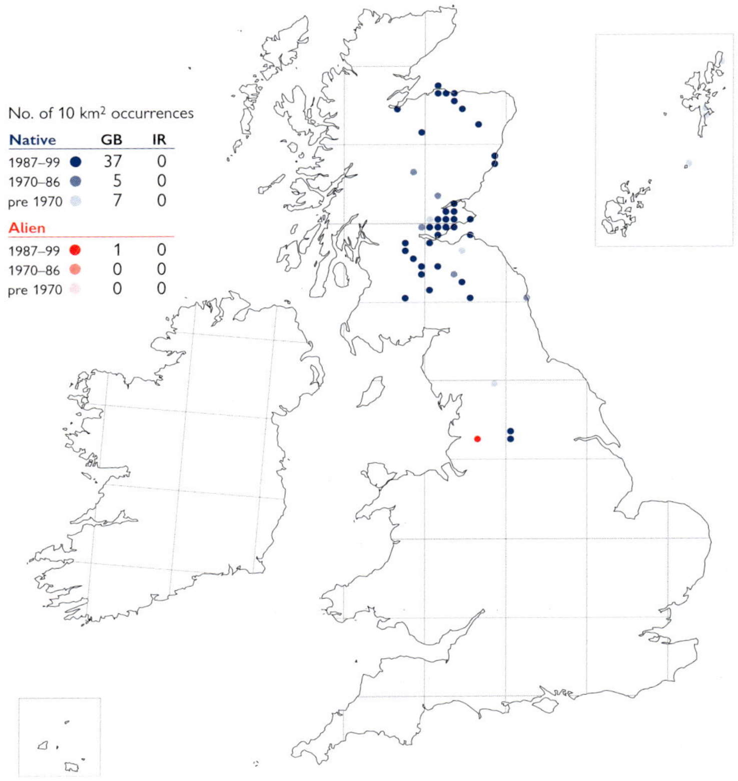

A perennial herb found on roadsides, river banks, in fields and around farms, and on waste ground. It can be locally common where the parents grow together. Generally lowland, but its precise altitudinal range is unknown.

Native. This hybrid was not mapped in the 1962 *Atlas*, and because of its similarity to *R. longifolius* it is almost certainly under-recorded.

Known from C. & N. Europe; common in Fennoscandia.

References: Lousley & Kent (1981), Stace (1975).

J. R. AKEROYD

Rumex crispus × R. obtusifolius (R. × pratensis)

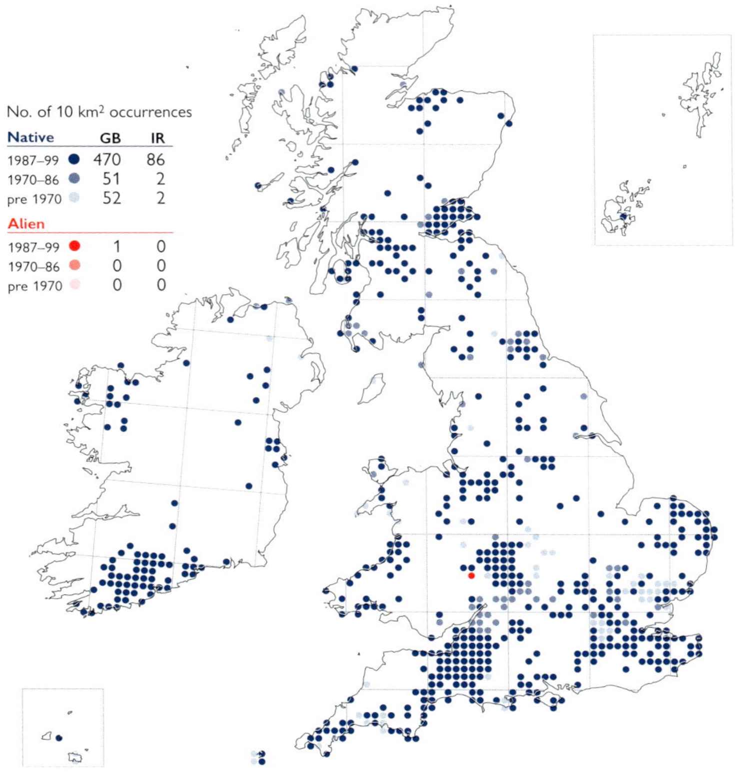

This common hybrid dock is found in fields and waste places, in pastures, by roadsides, on river banks and on other disturbed ground, sometimes in the absence of either parent. It is more vigorous than either parent, and often occurs as a high proportion of a dock population. It is, however, much less fertile than the parents. Generally lowland, but upper altitudinal limit unknown.

Native. This hybrid was not mapped in the 1962 *Atlas*, and is under-recorded, especially in Ireland. Areas of concentrated records indicate the activities of botanists familiar with the hybrid.

The commonest *Rumex* hybrid in Europe.

References: Lousley & Kent (1981), Stace (1975), Ziburski *et al.* (1986).

J. R. AKEROYD

Rumex conglomeratus Clustered Dock

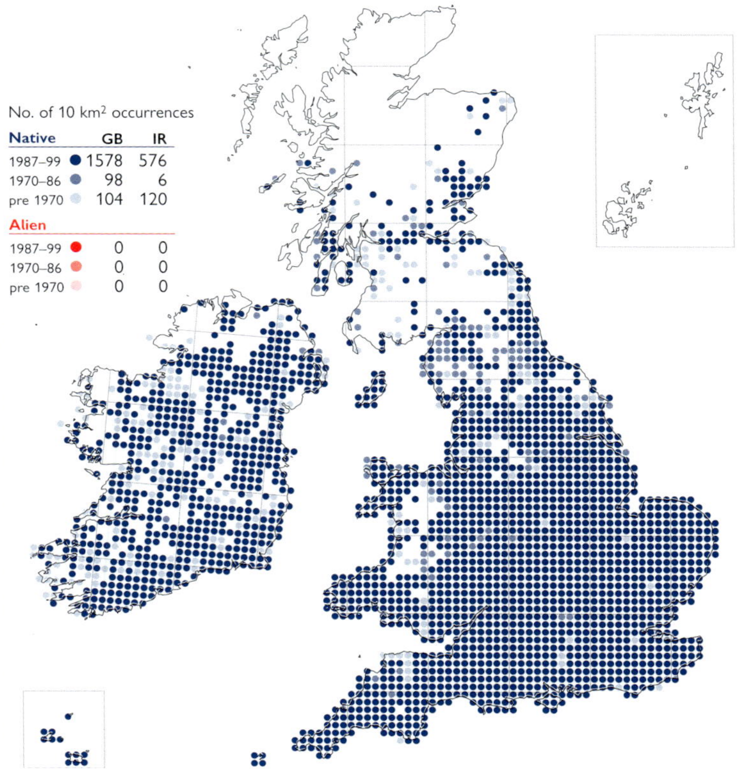

A short-lived perennial herb of wet meadows, stream and river banks, ditches, muddy pathsides and field margins and gateways, often in places flooded or waterlogged in winter. 0–420 m (Swindale, Westmorland).

Native (change +0.20). There has been no change in the distribution of *R. conglomeratus* since the 1962 *Atlas*.

Eurosiberian Southern-temperate element, but widely naturalised so that distribution is now Circumpolar Southern-temperate.

References: Atlas (180d), Hultén & Fries (1986), Jalas & Suominen (1979), Lousley & Kent (1981), Meusel *et al.* (1965).

J. R. AKEROYD

Rumex sanguineus Wood Dock

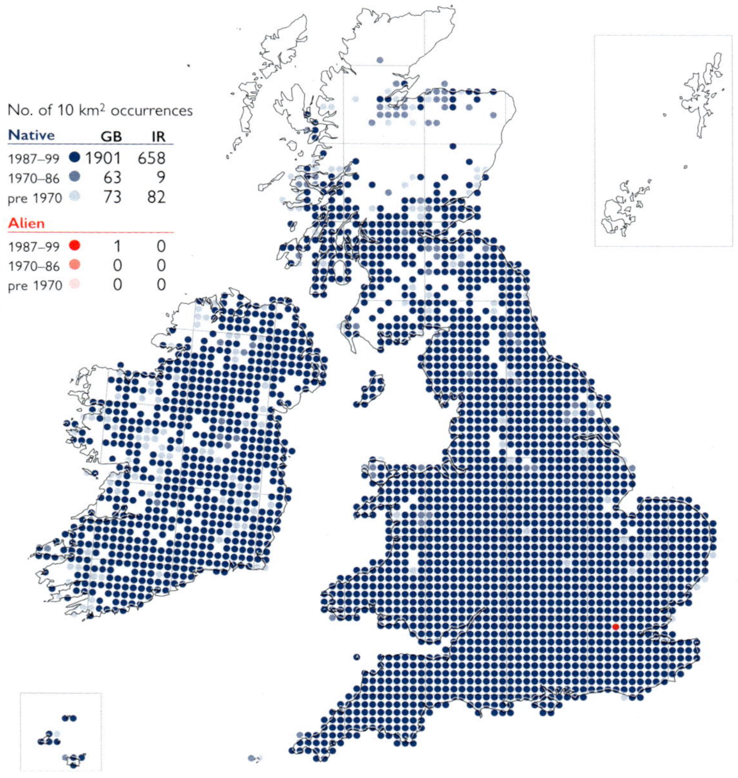

A short-lived perennial herb of woodland margins and rides, hedgerows, roadsides and waste ground. Found on a wide range of soils but favouring damp clay, and usually in more shaded, drier places than *R. conglomeratus*. Mainly lowland, but reaching 350 m in Glen Shee (Angus) and 380 m in W. Yorkshire.

Native (change +0.66). There has been no change in the distribution of this species since the 1962 *Atlas*, but better recording has shown it to be more frequent within its range.

European Temperate element.

References: Atlas (180c), Hultén & Fries (1986), Jalas & Suominen (1979), Lousley & Kent (1981).

J. R. AKEROYD

Rumex rupestris Shore Dock

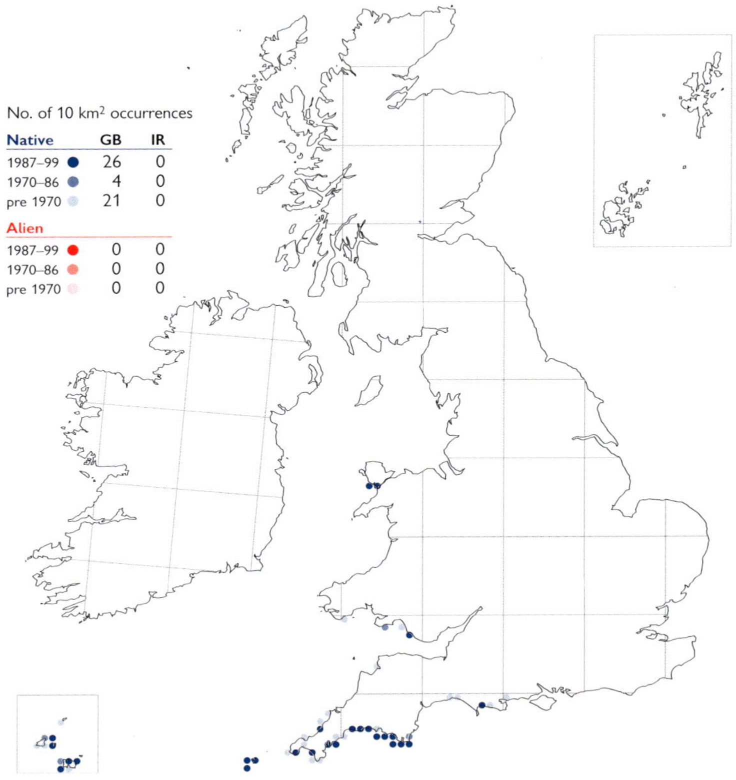

A perennial herb of sand and shingle beaches, the base of often unstable sea-cliffs, amongst coastal rocks and in damp dune-slacks; invariably in places where freshwater trickles, or streams debouch, onto the shore. Lowland.

Native (change –0.28). The range of this species has not expanded since the 1962 *Atlas*, but many new sites have been found and old ones refound in S.W. England. Most populations number only a few individuals, and threats include sea defence works, visitor pressure and winter storms. Three re-introductions have recently been attempted in Devon and Cornwall as part of a recovery plan.

Oceanic Temperate element; restricted to the coast of W. Europe.

References: Atlas (181a), Daniels *et al.* (1998), Jalas & Suominen (1979), Kay & John (1995), Lousley & Kent (1981), Wigginton (1999).

J. R. AKEROYD

Rumex pulcher Fiddle Dock

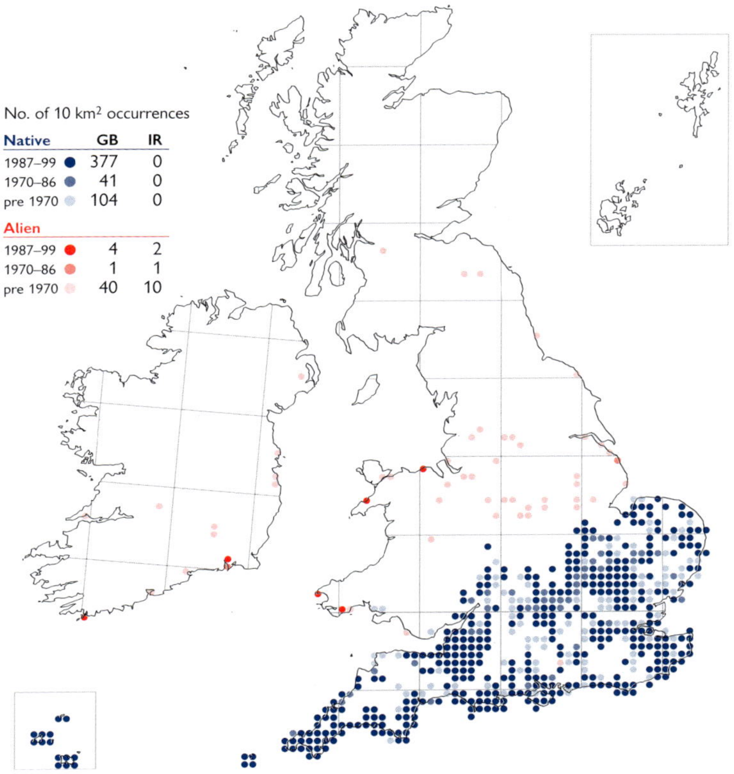

A biennial or short-lived perennial herb of dry coastal pastures and slightly disturbed grassland on commons, village greens, churchyards and roadsides, mainly on lighter soils and often where the habitat is grazed or trampled. Lowland.

Native (change +0.24). There has been little change in the distribution of *R. pulcher* since the 1962 *Atlas*. It is native in S. and E. England and S. Wales but perhaps a long-naturalised introduction in the north. It is alien in Ireland, where it was formerly regarded as a casual but now known to be naturalised at some sites.

European Southern-temperate element.

References: Atlas (180b), Akeroyd (1993), Bolòs & Vigo (1990), Jalas & Suominen (1979), Lousley & Kent (1981).

J. R. AKEROYD

Rumex obtusifolius Broad-leaved Dock

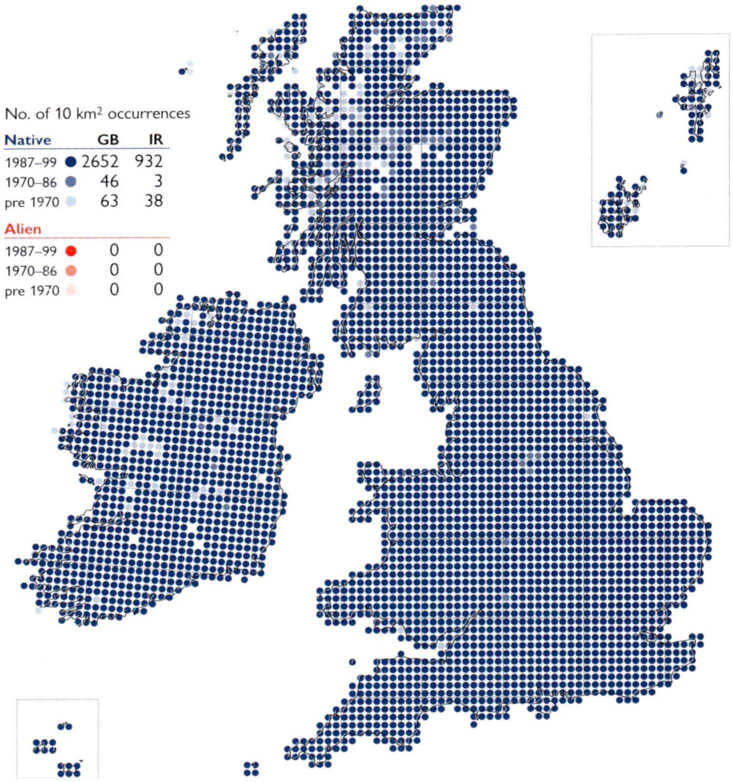

A perennial herb of field margins, hedge banks, roadsides, stream and river banks, ditches and neglected cultivated ground. Mostly lowland, up to 565 m on Dent Crag (Westmorland), but with an exceptional record at 845 m on Great Dun Fell (Westmorland).

Native (change +0.66). There has been no change in the distribution of this species since the 1962 *Atlas*. It is the commonest species of *Rumex* in Britain.

European Temperate element; widely naturalised outside its native range.

References: Atlas (180a), Cavers & Harper (1964), Grime *et al.* (1988), Hultén & Fries (1986), Jalas & Suominen (1979), Lousley & Kent (1981), Meusel *et al.* (1965).

J. R. AKEROYD

Rumex obtusifolius × *R. sanguineus* (*R.* × *dufftii*)

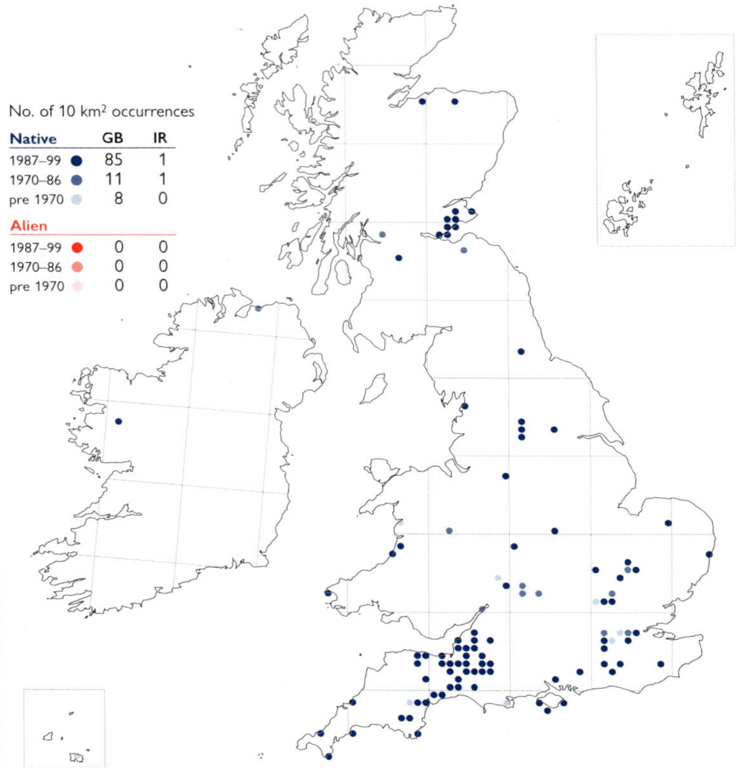

A short-lived perennial herb of woodland edges and rides, roadsides, river banks, rough grassland and waste ground. Lowland.

Native. This hybrid was first noted in 1897 in Berkshire. Most records are recent, but this probably reflects improved recording of hybrids; concentrations of records indicate the activities of botanists familiar with it. It is certainly under-recorded in many areas, especially in Ireland.

Widespread in Europe.

References: Lousley & Kent (1981), Stace (1975).

H. J. KILLICK & J. R. AKEROYD

Rumex palustris Marsh Dock

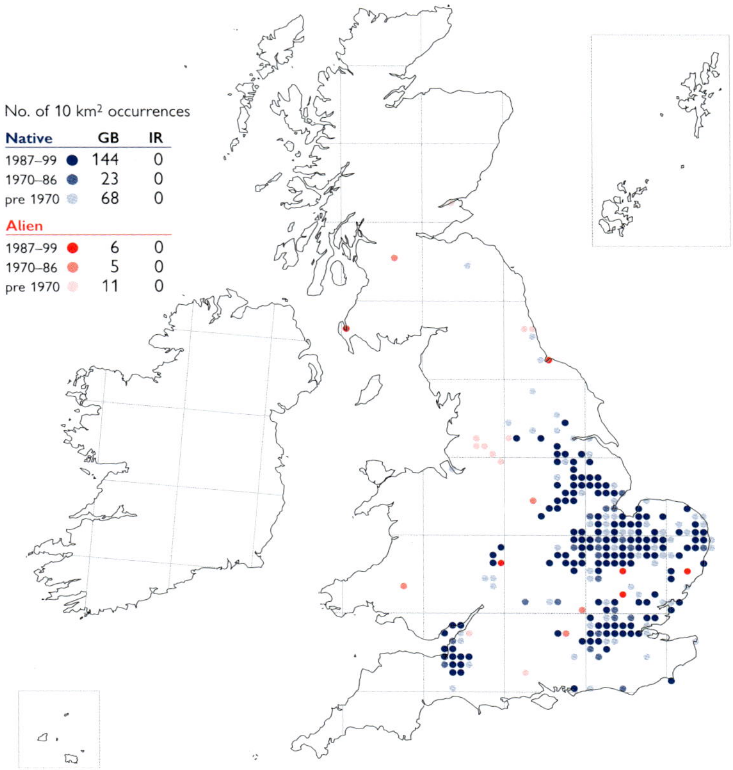

An annual, biennial or short-lived perennial, typical of wet, nutrient-rich mud exposed in summer and autumn, most often in marshes and beside ponds and ditches, but also in clay- and gravel-pits and on damp disturbed ground. It is also an occasional weed of dry open sites, and has been recorded as a ballast alien. Lowland.

Native (change +0.31). The species declined in Britain, largely before 1930. However, the decrease is less than that of *R. maritimus* and its distribution is now stable. The apparent increase in Somerset is attributed to better recording rather than a real increase.

European Temperate element.

References: Atlas (181b), Hultén & Fries (1986), Jalas & Suominen (1979), Lousley & Kent (1981), Stewart *et al.* (1994).

J. R. AKEROYD

Rumex maritimus Golden Dock

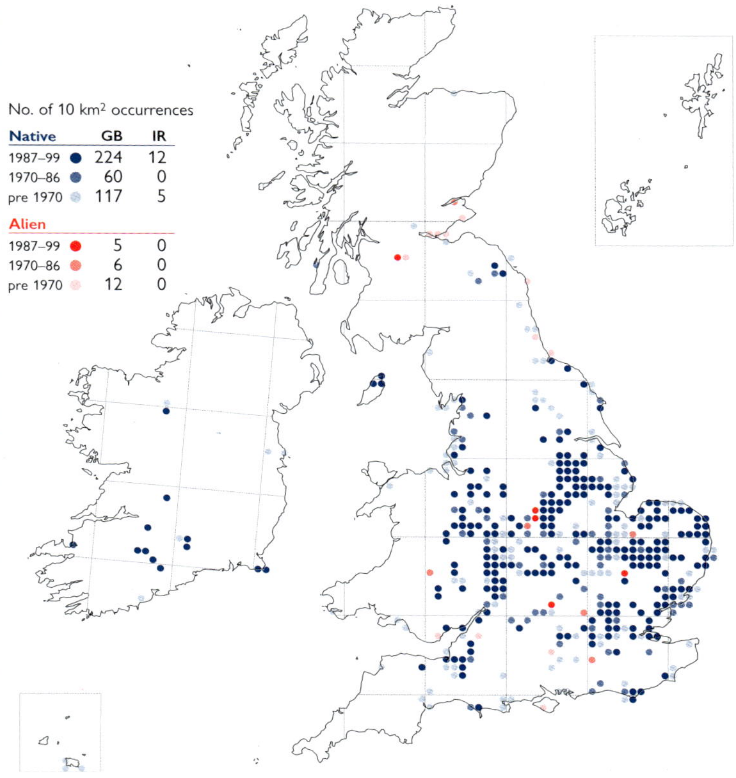

An annual to short-lived perennial herb, growing on the margins of pools, lakes, rivers and ditches, in clay-pits and wet hollows in marshy fields. Its sites are usually waterlogged in winter, but it occasionally occurs on dry ground. It can tolerate mildly saline conditions. Lowland.

Native (change +0.42). *R. maritimus* has been lost from some sites in Britain due to drainage, the loss of ponds and the regulation of water levels. However, there are many more records now than in the 1962 *Atlas*, and the species spreads readily to new sites; its overall distribution is stable. Increasing eutrophication in Ireland has favoured an expansion there.

Circumpolar Temperate element.

References: Atlas (181c), Curtis & McGough (1988), Hultén & Fries (1986), Jalas & Suominen (1979), Lousley & Kent (1981), Stewart *et al.* (1994).

J. R. AKEROYD

Oxyria digyna Mountain Sorrel

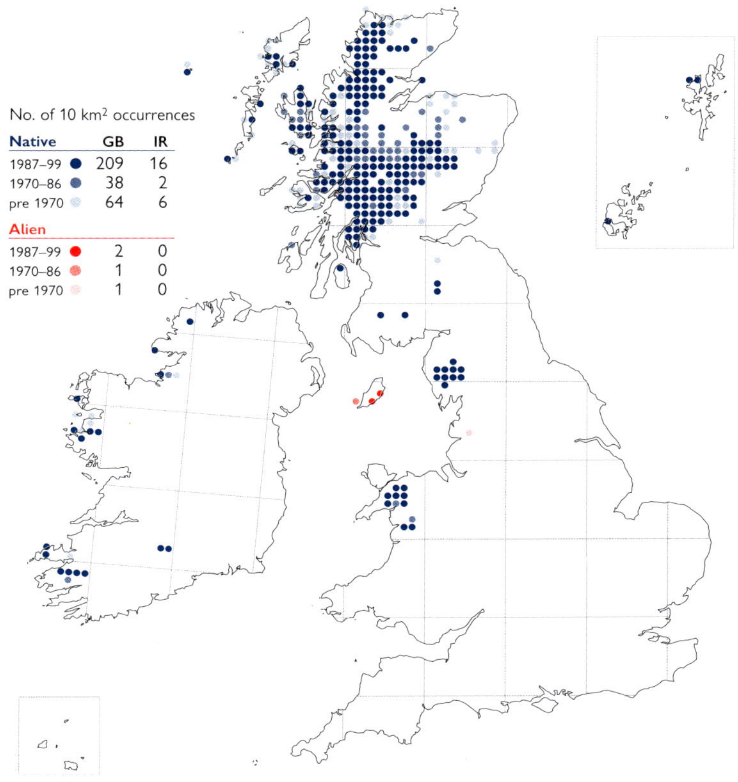

A tufted perennial herb of damp, ungrazed mountain ledges, wet, shaded gullies and the sides of gills and streams. In Scotland it sometimes descends to near sea level along streams, but usually occurs above 150 m, reaching 1190 m on Ben Lawers (Mid Perth).

Native (change –0.71). There has been no change in distribution of *O. digyna* since the 1962 *Atlas*, and it is probably still present in many of the more remote 10-km squares in Scotland for which there are only pre-1970 records.

Circumpolar Arctic-montane element.

References: Atlas (178a), Hultén & Fries (1986), Jalas & Suominen (1979), Lousley & Kent (1981), Meusel *et al.* (1965).

J. R. AKEROYD

Limonium vulgare Common Sea-lavender

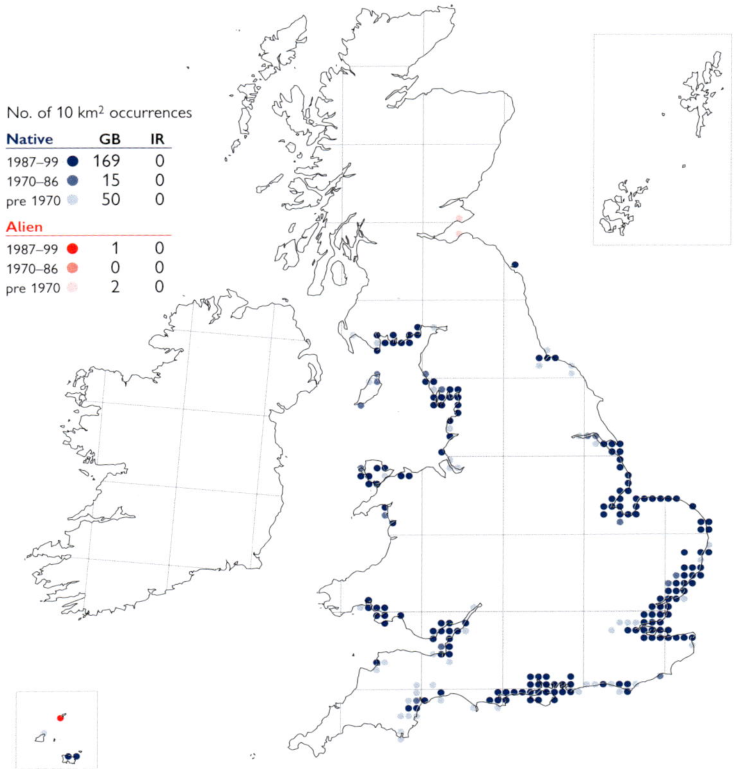

A perennial herb of ungrazed or lightly grazed muddy saltmarshes, occasionally also growing amongst nearby rocks and on the stonework of sea-walls. Its habitats are similar to those of *Limonium humile*, which often grows with it. Lowland.

Native (change –0.31). The map shows little change since the 1962 *Atlas*, but the species has gone from a few 10-km squares, and is declining in many areas as saltings are invaded by *Spartina anglica* or have become more intensively grazed. Intermediates between *L. vulgare* and *L. humile* are common (including the hybrid *L.* × *neumanii*) and some populations are difficult to assign to either species.

Mediterranean-Atlantic element.

References: Atlas (199a), Boorman (1967), Dawson & Ingrouille (1995), Hultén & Fries (1986), Meusel *et al.* (1978).

S. J. LEACH

Limonium humile Lax-flowered Sea-lavender

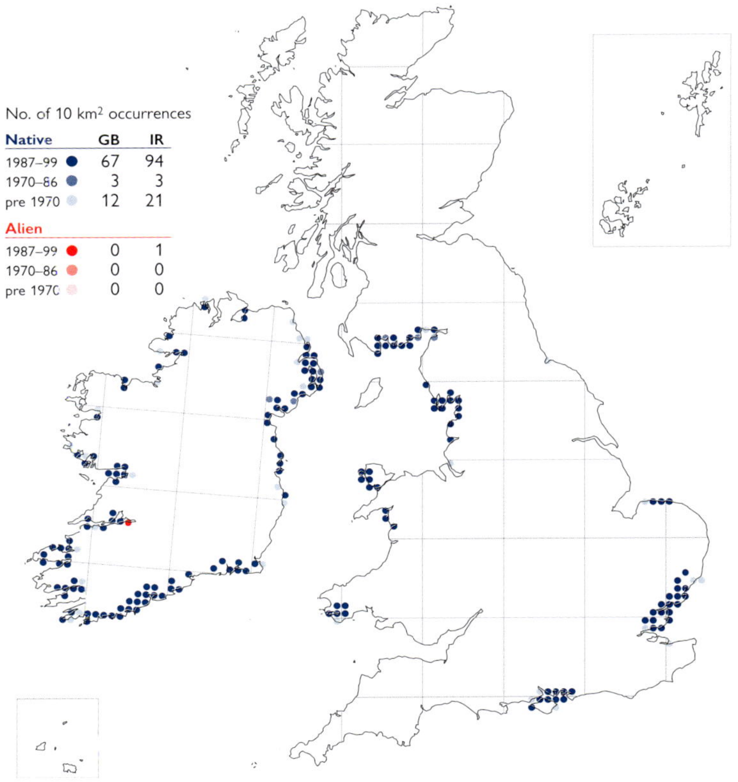

A perennial herb of ungrazed or lightly grazed muddy estuarine saltmarshes, often growing in close proximity to its commoner relative, *L. vulgare*, but replacing it in some areas. It rarely occurs on rocky cliffs. Lowland.

Native (change +0.05). There has been very little change in distribution since the 1962 *Atlas*, though a few local losses have been noted as a result of land-use changes or invasion by *Spartina*. Intermediates between *L. humile* and *L. vulgare* are frequent (including the hybrid *L.* × *neumanii*), and some populations are difficult to assign to either species.

Oceanic Temperate element.

References: Atlas (199b), Boorman (1967), Dawson & Ingrouille (1995), Dupont (1962), Hultén & Fries (1986), Meusel *et al.* (1978), Stewart *et al.* (1994).

S. J. LEACH

Limonium bellidifolium Matted Sea-lavender

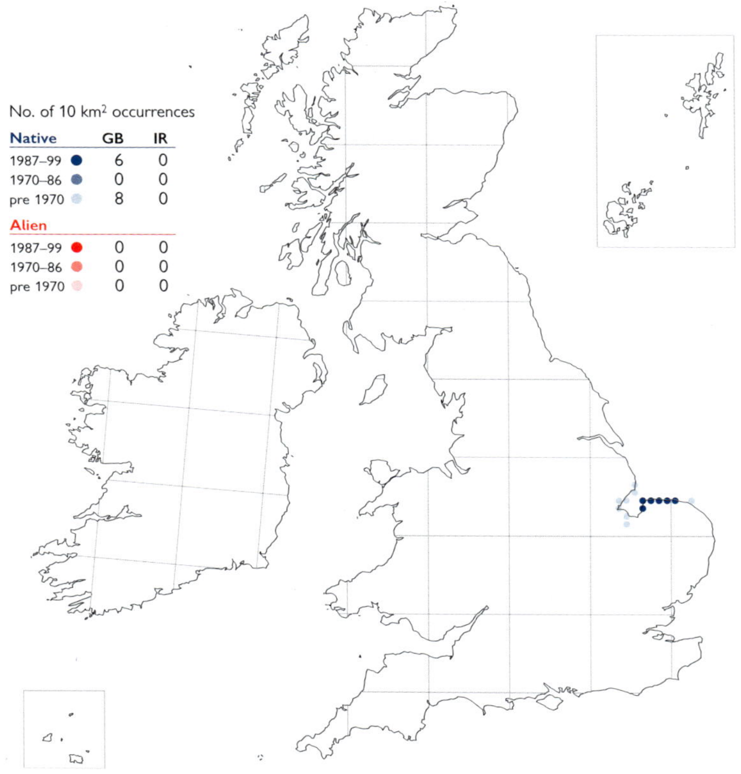

A mat-forming perennial herb of the upper parts of saltmarshes and saltmarsh-sand dune transitions, especially where firm sandy or silty sediments overlie coarser grained material. Lowland.

Native (change +0.01). There has been very little change in distribution since the 1962 *Atlas*. It has not been seen in Lincolnshire since 1967, whilst a few local losses in W. Norfolk have occurred as a result of human trampling or burial under shifting sand dunes.

Eurosiberian Southern-temperate element.

References: Atlas (199c), Bolòs & Vigo (1995), Wigginton (1999).

S. J. LEACH

Limonium auriculae-ursifolium Broad-leaved Sea-lavender

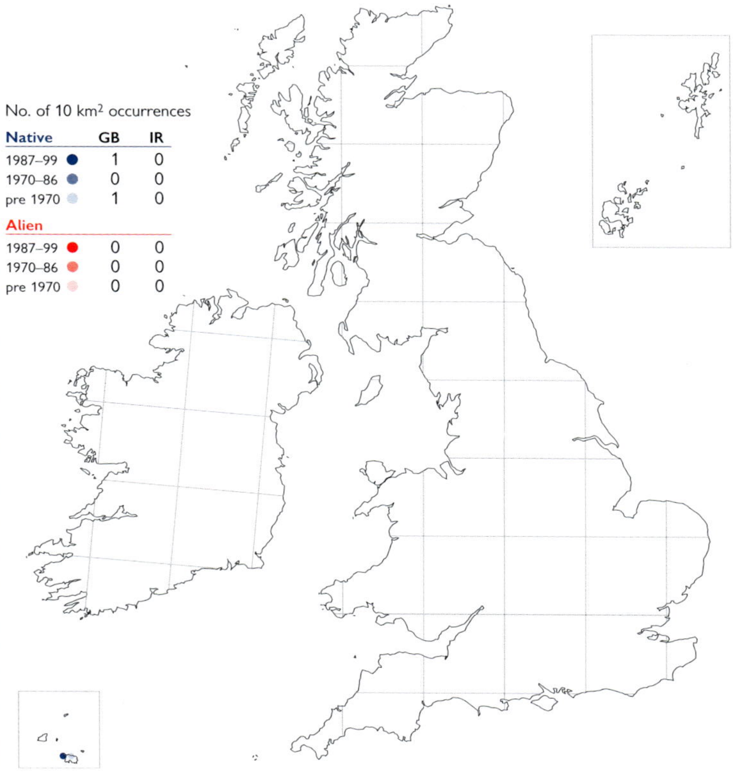

An apomictic perennial herb, now restricted to a single locality (Plémont Point, Jersey) where it grows in rock crevices on the side of a bare granite gully. It used to occur at a second site on Jersey (Rouge Nez), on nearly inaccessible cliffs. Lowland.

Native. The taxonomy of *L. auriculae-ursifolium* was revised in the 1980s, with other populations of this species-complex in the Channel Islands being assigned to *L. normannicum*.

Oceanic Southern-temperate element.

References: Atlas (199c), Bolòs & Vigo (1995), Ingrouille (1985), Le Sueur (1984).

S. J. LEACH

Limonium normannicum Alderney Sea-lavender

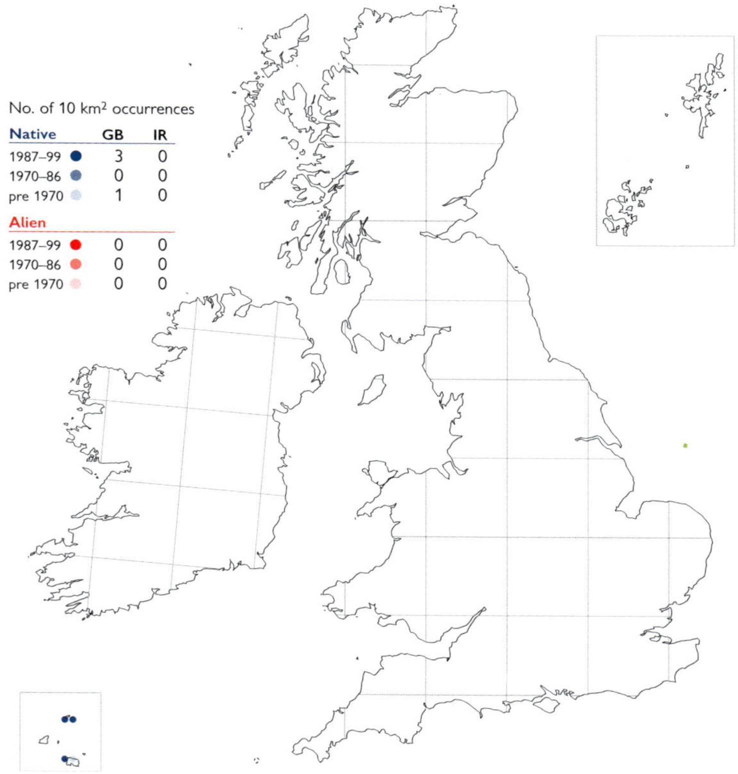

A perennial apomictic herb of stabilised sand dunes (St Ouen's Bay, Jersey) and coastal rocks (St Esquère Bay, Alderney), in the latter site growing with *L. binervosum* subsp. *sarniense*. Lowland.

Native. *L. normannicum* was first distinguished as a separate taxon in 1985, having previously been included within *L. auriculae-ursifolium*. It is known to have been lost from at least one site on Jersey (Rouge Nez), but it has colonised sea defences built during the German occupation of 1940–1945.

Oceanic Temperate element; world range limited to N.W. France and the Channel Islands.

References: Ingrouille (1985), Le Sueur (1984).

S. J. LEACH

Limonium binervosum agg. Rock Sea-lavenders

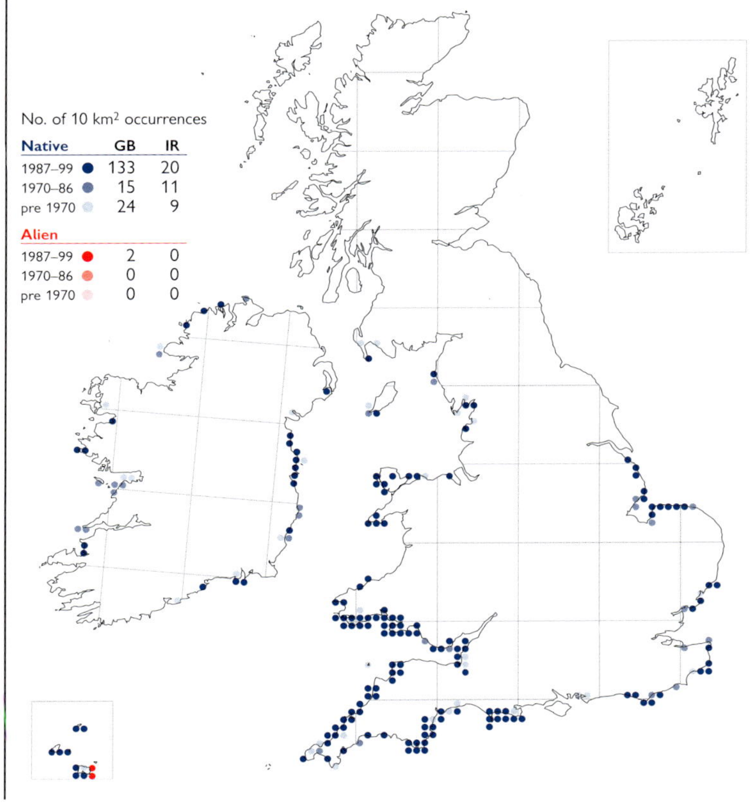

A group of apomictic perennial herbs comprising the following nine species and numerous infraspecific taxa, many of which are British and Irish endemics. They occur in a wide range of coastal habitats including sea-cliffs, dock walls, shingle banks and saltmarshes. Lowland.

Native (change +0.16). Many of the taxa included here were not recognised at the time of the 1962 *Atlas*. However, the map shows that there has been no appreciable change in distribution of the aggregate in the last 30–40 years. Most colonies occur in places unlikely to be damaged by human activities, although losses from some artificial habitats have been noted.

The aggregate has an Oceanic Temperate distribution.

References: Atlas (199d), Curtis & McGough (1988), Dupont (1962), Ingrouille & Stace (1986), Stewart *et al.* (1994), Wigginton (1999).

S. J. LEACH

Limonium binervosum subsp. *binervosum*

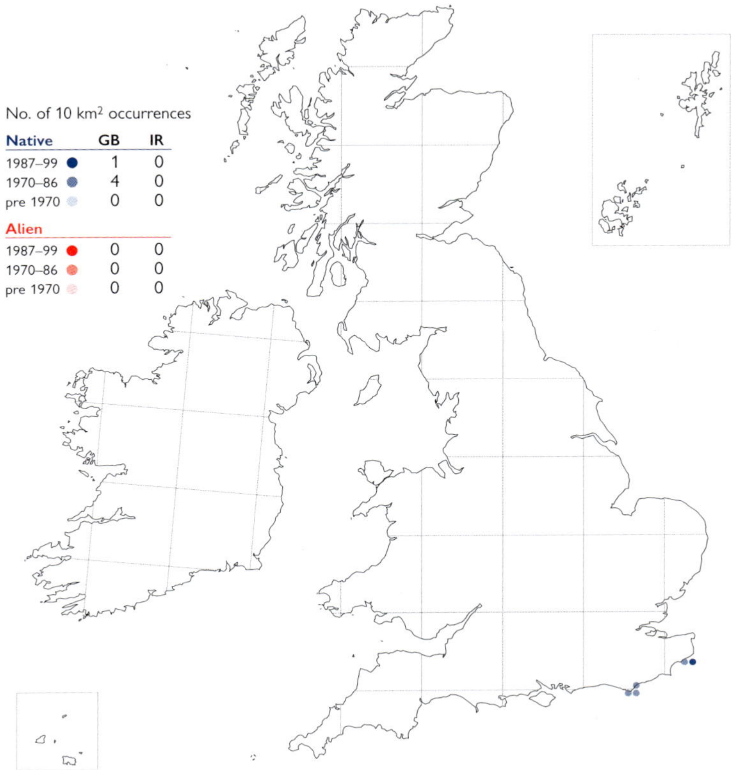

A perennial herb which is almost entirely restricted to chalk sea-cliffs with a few tiny colonies on adjoining shingle and the drier parts of saltmarshes. Lowland.

Native. This taxon was not recognised at the time of the 1962 *Atlas*. However, there are no indications of any recent change in distribution.

Oceanic Temperate element; world range limited to N.W. France and S.E. England.

References: Ingrouille & Stace (1986), Wigginton (1999).

S. J. LEACH

Limonium binervosum subsp. *cantianum*

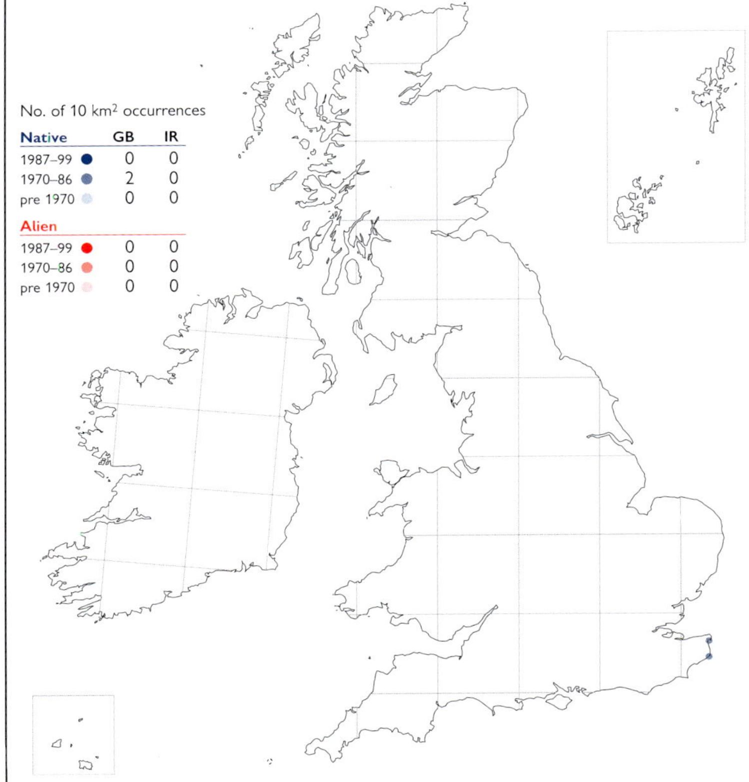

A perennial herb found on chalk sea-cliffs and in the drier parts of saltmarshes. Lowland.

Native. This taxon was not recognised at the time of the 1962 *Atlas*. However, although there are no records since 1987, there is no reason to suspect any recent change in distribution.

Endemic.

References: Ingrouille & Stace (1986), Wigginton (1999).

S. J. LEACH

Limonium binervosum subsp. *anglicum*

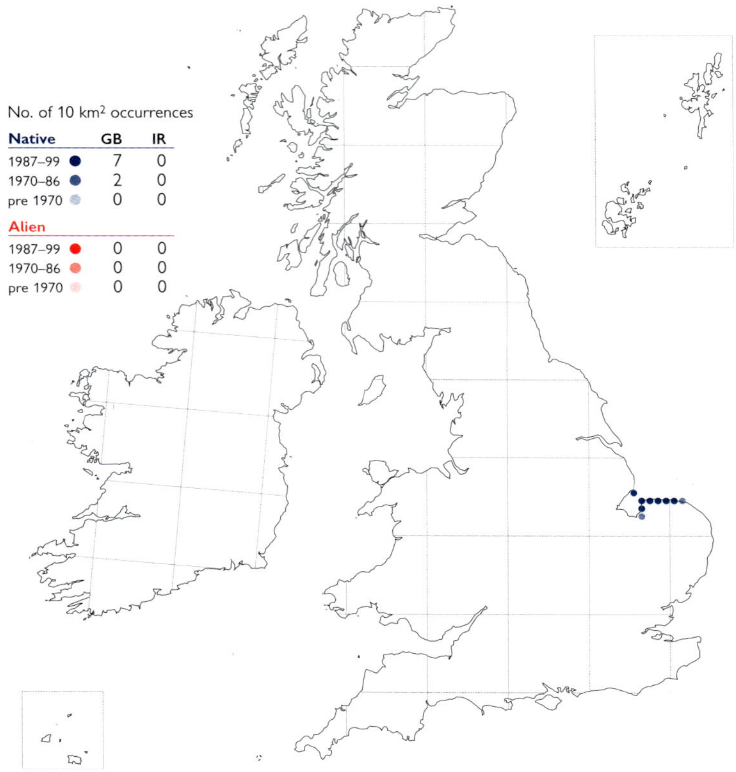

This subspecies occurs in the drier parts of saltmarshes and tidally-inundated dune-slacks, often where blown sand and inwashed silt overlie a firm shingly base. There is also a single record from a stabilised earth cliff. Lowland.

Native. The taxon was not recognised at the time of the 1962 *Atlas*. However, it is frequent within its restricted range and there are no indications of any recent change in distribution.

Endemic.

References: Ingrouille & Stace (1986), Wigginton (1999).

S. J. LEACH

Limonium binervosum subsp. *saxonicum*

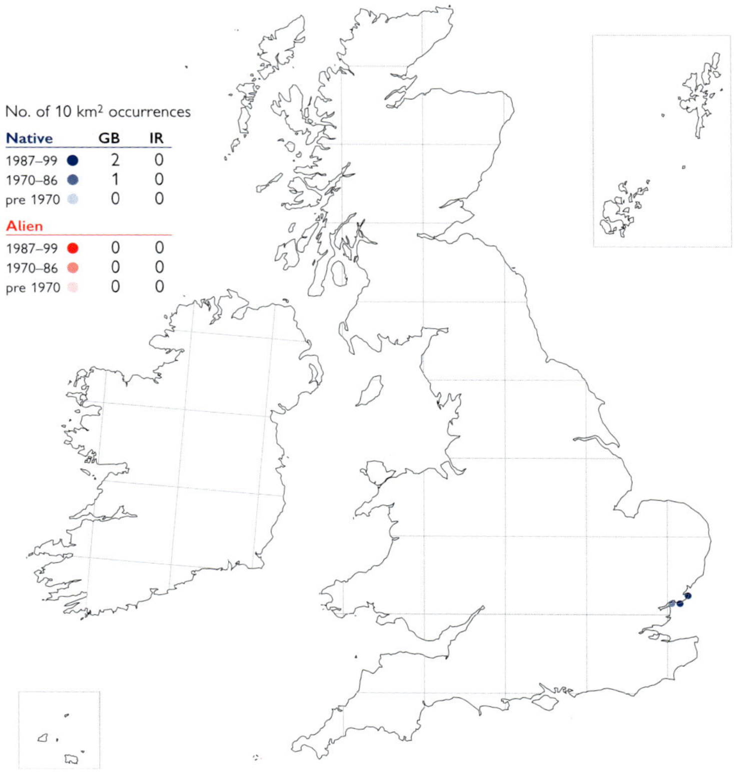

A perennial herb restricted to dry saltmarsh and saltmarsh-sand dune transitions, particularly favouring sites where blown sand and inwashed silt overlie shingle. Lowland.

Native. The separation of this taxon from subsp. *anglicum* is not entirely clear, and it has been suggested that it may be more widely distributed on the East Anglian coast than is shown in the map. This subspecies was not recognised at the time of the 1962 *Atlas* but there are no indications of any recent appreciable changes in distribution.

Endemic.

References: Ingrouille & Stace (1986), Wigginton (1999).

S. J. LEACH

Limonium binervosum subsp. *mutatum*

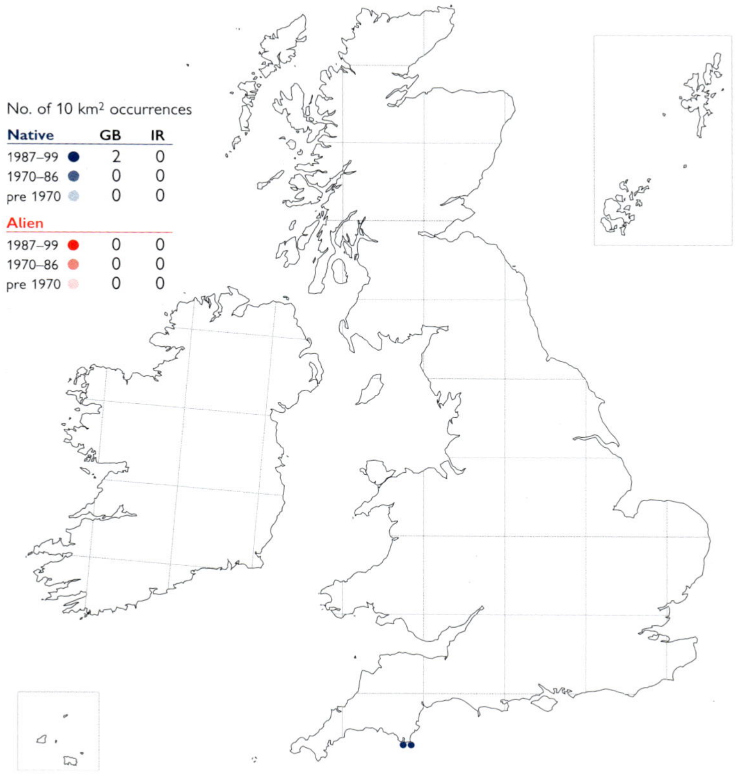

A perennial herb which is restricted to coastal rocks and nearby cliffs of Quaternary head deposits (unconsolidated sand and rock fragments) at Lannacombe (S. Devon). Lowland.

Native. This taxon was not recognised at the time of the 1962 *Atlas*, and it is hard to distinguish in the field.

Endemic.

References: Ingrouille & Stace (1986), Wigginton (1999).

S. J. LEACH

Limonium binervosum subsp. *sarniense*

No. of 10 km² occurrences

Native	GB	IR
1987–99 ●	4	0
1970–86 ●	3	0
pre 1970 ●	0	0
Alien		
1987–99 ●	0	0
1970–86 ●	0	0
pre 1970 ●	0	0

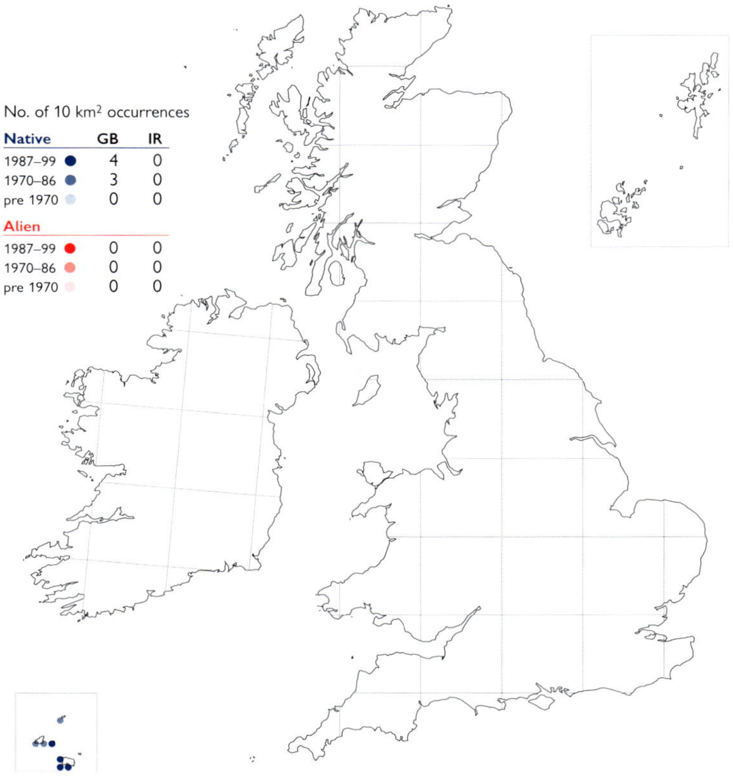

A perennial herb found amongst coastal rocks and screes and on sea-cliffs. Lowland.

Native. The sole representative of the *L. binervosum* aggregate in the Channel Islands, occurring on all the main islands. It was not recognised at the time of the 1962 *Atlas*. Historical records suggest there has been little change in its distribution in Jersey.

Endemic.

References: Ingrouille & Stace (1986), Le Sueur (1984).

S. J. LEACH

Limonium paradoxum

No. of 10 km² occurrences

Native	GB	IR
1987–99 ●	1	0
1970–86 ●	0	0
pre 1970 ●	0	0
Alien		
1987–99 ●	0	0
1970–86 ●	0	0
pre 1970 ●	0	0

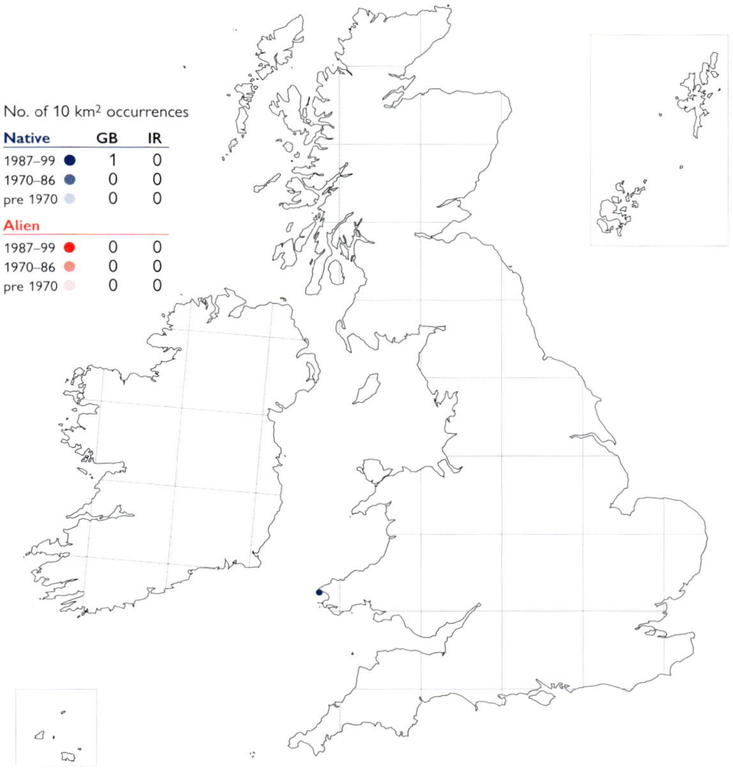

A perennial herb restricted to a single site where it occurs on basic igneous rock outcrops, along the seaward edge of cliff-tops and on inaccessible cliff ledges. Lowland.

Native. The distribution of *L. paradoxum* has remained largely unchanged since the species was first described in 1931, and the population size has been remarkably stable for at least the last twenty years. Perring & Sell (1968) mapped *L. paradoxum* in Ireland (E. Donegal), but this and other N.W. Irish populations are now considered to be a separate taxon, *L. recurvum* subsp. *humile*.

Endemic.

References: Atlas Supp. (46a), Curtis & McGough (1988), Ingrouille & Stace (1986), Wigginton (1999).

S. J. LEACH

Limonium procerum subsp. *procerum*

No. of 10 km² occurrences

Native	GB	IR
1987–99 ●	17	1
1970–86 ●	2	0
pre 1970 ●	0	1
Alien		
1987–99 ●	0	0
1970–86 ●	0	0
pre 1970 ●	0	0

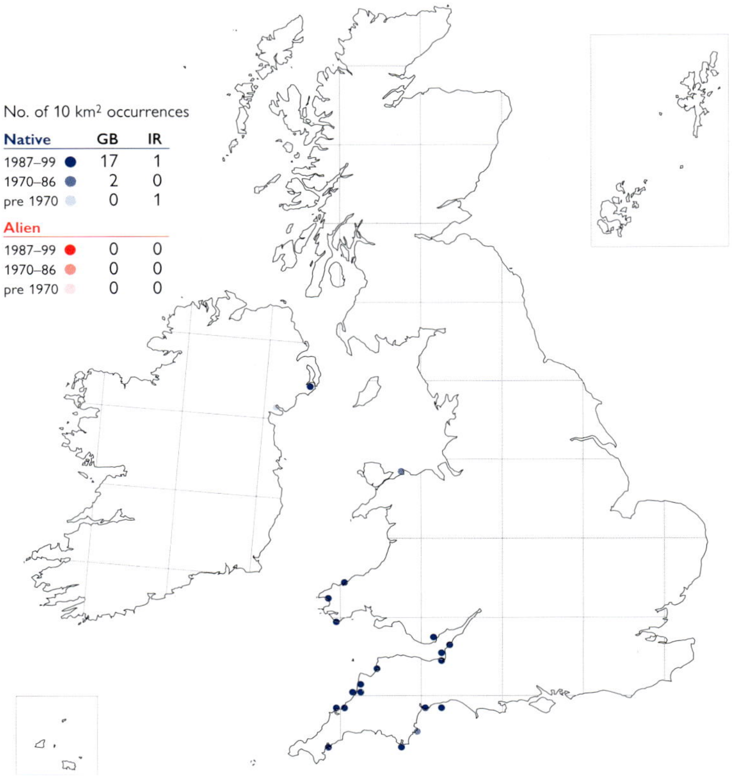

A perennial herb of coastal rocks, sea-cliffs, stabilised shingle, in the upper parts of saltmarshes and saltmarsh-sand dune transitions, and locally on the stonework of sea defences and harbour walls. Lowland.

Native. This subspecies was not recognised at the time of the 1962 *Atlas*, and shows no appreciable change in distribution since it was first described in 1986. There have, however, been a few recorded local losses from dock walls and other artificial habitats.

Endemic.

References: Ingrouille & Stace (1986), Stewart *et al.* (1994).

S. J. LEACH

Limonium procerum subsp. *devoniense*

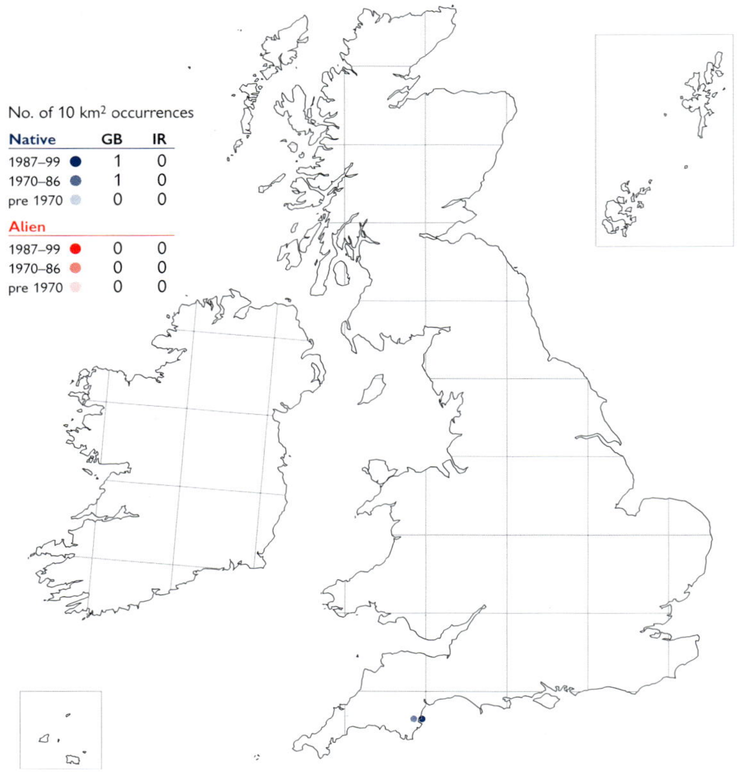

A perennial herb of coastal rocks and sea-cliffs. Lowland.

Native. This taxon was not recognised at the time of the 1962 *Atlas*, and there is no evidence of any populations having been lost since it was first described in 1986.

Endemic.

References: Ingrouille & Stace (1986), Wigginton (1999).

S. J. LEACH

Limonium procerum subsp. *cambrense*

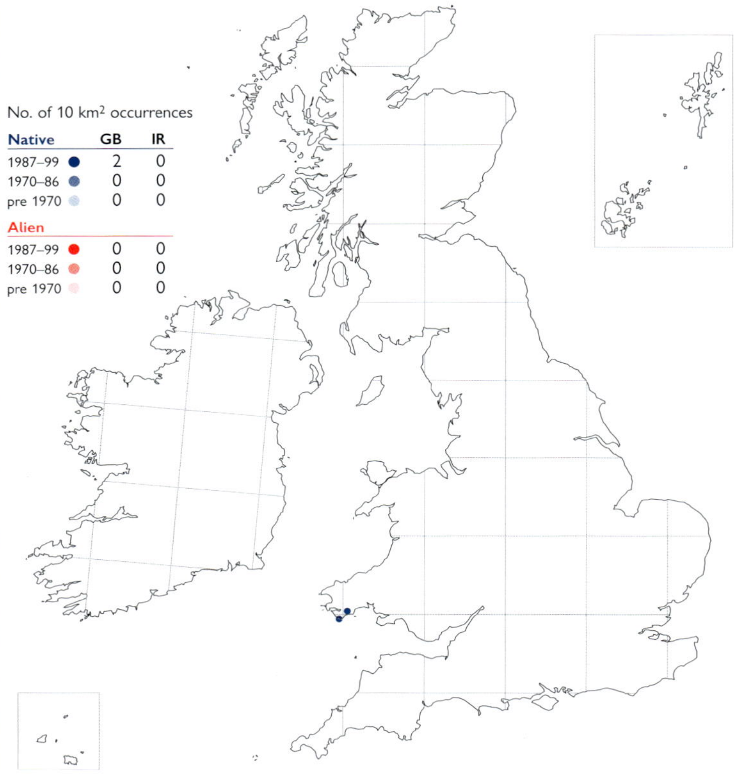

A perennial herb of cliff ledges and steep S.- to S.W.-facing rocky cliff-slopes of Carboniferous limestone, with *Limonium parvum* and *L. britannicum* subsp. *transcanalis* as frequent associates. Lowland.

Native. This taxon was not recognised at the time of the 1962 *Atlas*, but there is no evidence that any populations have been lost since it was first described in 1986. Old records of *L. transwallianum* from the coast between Stackpole and St Govans Head (Pembs.) are probably referable to this taxon.

Endemic.

References: Ingrouille & Stace (1986), Wigginton (1999).

S. J. LEACH

Limonium britannicum subsp. *britannicum*

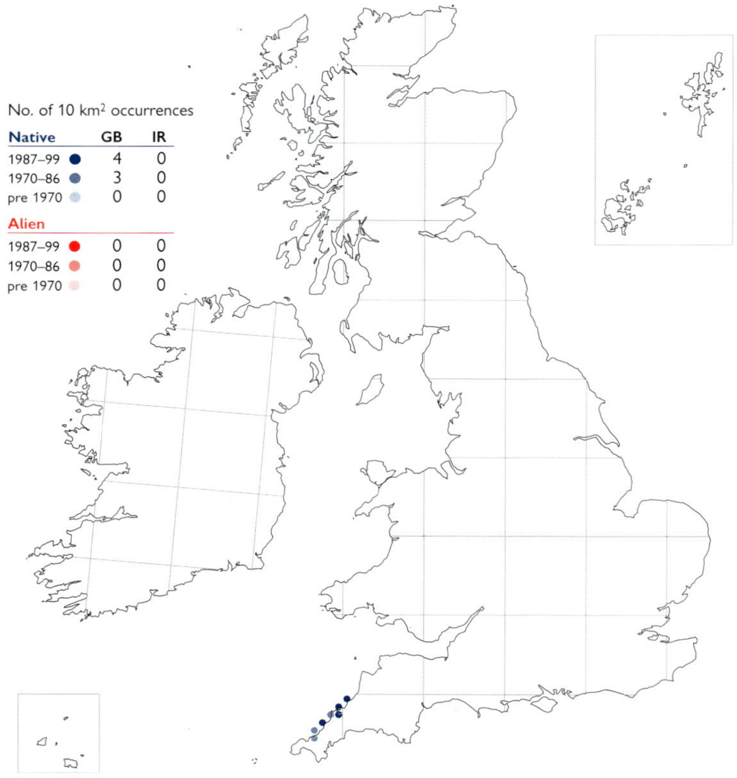

A perennial herb of coastal rocks and sea-cliffs. Lowland.

Native. This taxon was not recognised at the time of the 1962 *Atlas*. It is difficult to distinguish in the field from other *L. britannicum* subspecies in S.W. England (subsp. *coombense* and subsp. *transcanalis*) which, in practice, are separated largely on the basis of their currently known geographical limits. It is probably under-recorded.

Endemic.

References: Ingrouille & Stace (1986), Wigginton (1999).

S. J. LEACH

Limonium britannicum subsp. *coombense*

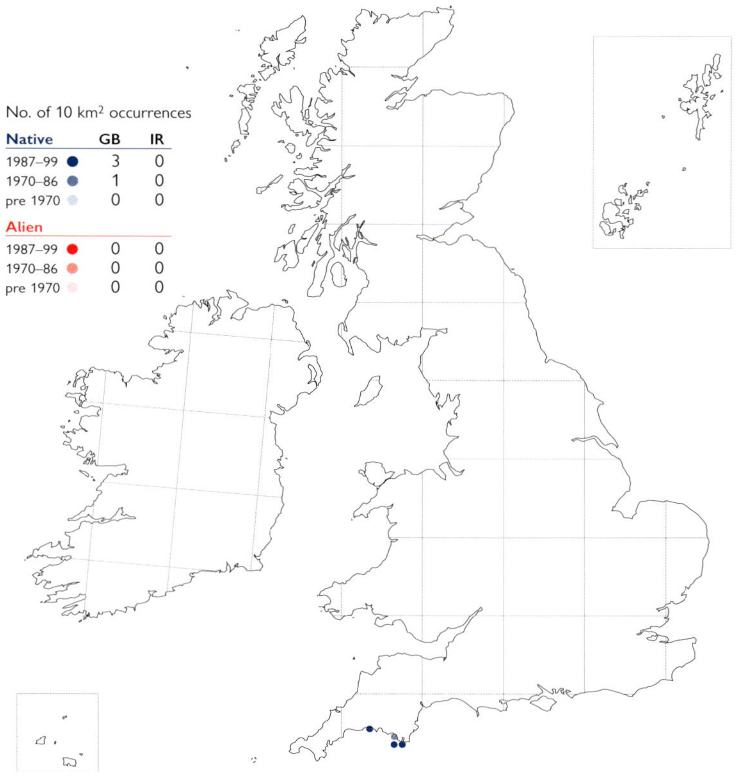

A perennial herb of coastal rocks and sea-cliffs, especially favouring S.-facing cliffs of Quaternary head deposits of unconsolidated sand and rock fragments. Lowland.

Native. Another taxon not recognised at the time of the 1962 *Atlas*, and difficult to distinguish in the field from other *L. britannicum* subspecies in S. W. England (subsp. *britannicum* and subsp. *transcanalis*). There is no evidence of any populations having been lost since it was first described in 1986.

Endemic.

References: Ingrouille & Stace (1986), Wigginton (1999).

S. J. LEACH

Limonium britannicum subsp. *transcanalis*

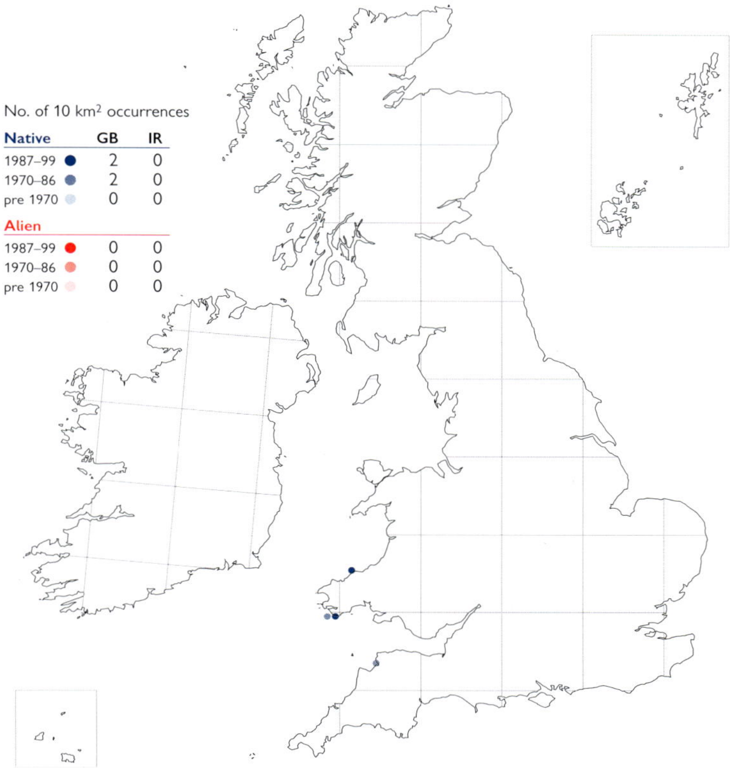

A perennial herb of coastal rocks and sea-cliffs, pebble beaches and the drier parts of saltmarshes. Some of the largest populations are on S.-facing cliffs of Carboniferous limestone in Pembrokeshire. Lowland.

Native. This is another taxon which was not recognised at the time of the 1962 *Atlas*, and hard to separate in the field from other *L. britannicum* subspecies in S. W. England (subsp. *britannicum* and subsp. *coombense*). The distribution is stable, with no evidence of any populations having been lost since it was first described in 1986.

Endemic.

References: Ingrouille & Stace (1986), Wigginton (1999).

S. J. LEACH

Limonium britannicum subsp. *celticum*

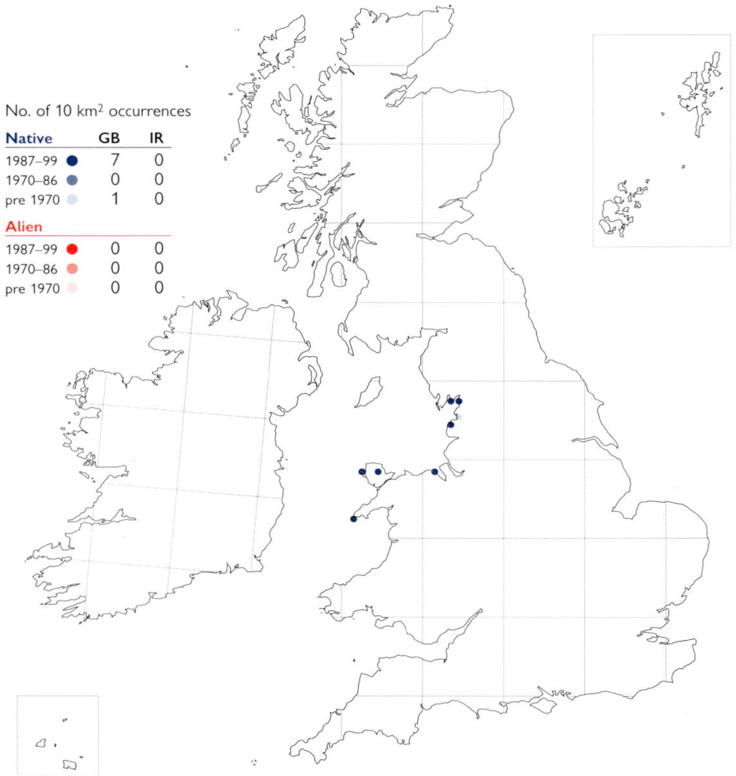

A perennial herb of rocky shores, sea-cliffs and at high elevations in saltmarshes, also very locally in artificial habitats such as sea-walls and the stonework of coastal railway banks. Lowland.

Native. This taxon was not recognised at the time of the 1962 *Atlas* and there are no indications of any recent changes in distribution.

Endemic.

References: Ingrouille & Stace (1986), Wigginton (1999).

S. J. LEACH

Limonium parvum

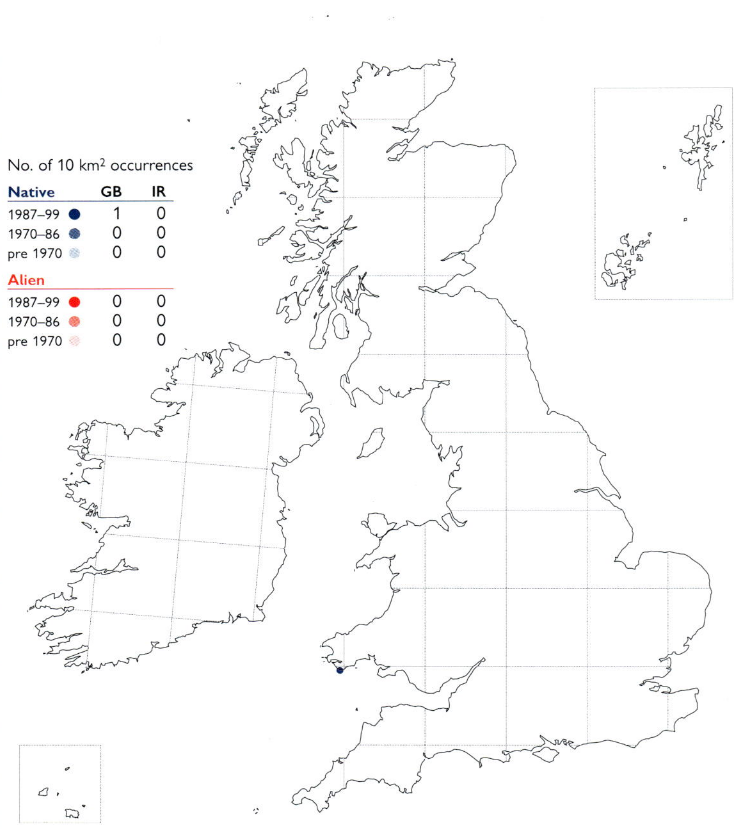

A perennial herb of S.- or S.W.-facing Carboniferous limestone sea-cliffs, in rock crevices, on cliff ledges and in open turf on heavily grazed cliff-slopes, often where there is a thin veneer of blown sand. *L. procerum* subsp. *cambrense* is a frequent associate. Lowland.

Native. *L. parvum* was described as a separate species in 1986, since when several new populations have been located within a few kilometres of the original site.

Endemic.

References: Ingrouille & Stace (1986), Wigginton (1999).

S. J. LEACH

Limonium loganicum

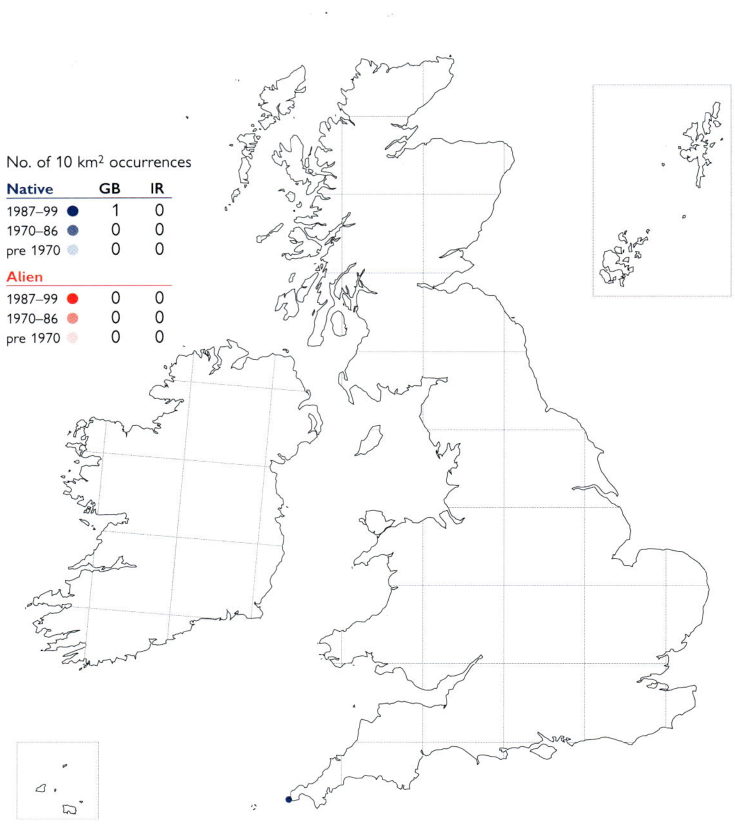

A perennial herb of granitic sea-cliffs, coastal rocks and screes. Lowland.

Native. This taxon was not recognised at the time of the 1962 *Atlas*. It is thought to be one of the most endangered of the members of the *L. binervosum* aggregate. The short stretch of coastline on which it occurs is a popular climbing resort and many colonies are small and vulnerable to trampling.

Endemic.

References: Ingrouille & Stace (1986), Wigginton (1999).

S. J. LEACH

Limonium transwallianum

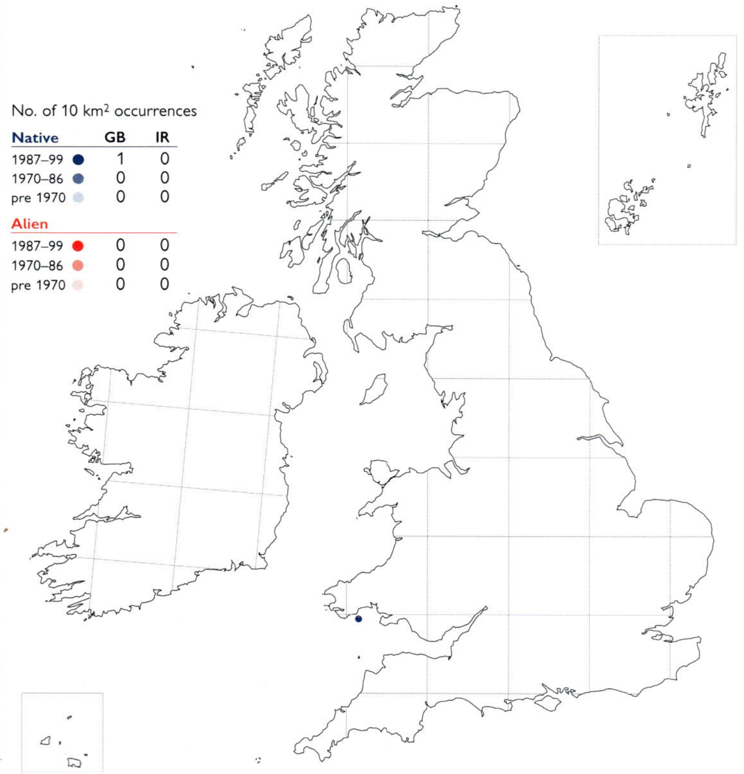

This perennial herb was first described in 1924, and has therefore been recognised longer than many members of the *L. binervosum* aggregate. It is restricted to a single site where it occurs on steep S.-facing Carboniferous limestone cliff-slopes and in rock crevices. Lowland.

Native. There has been no change in the distribution of this taxon since the map in Perring & Sell (1968), although records of *L. transwallianum* in Ireland are now regarded as a separate taxon, *L. recurvum* subsp. *pseudotranswallianum*, while the record from N. Devon is now considered to be a narrow-leaved variant of *L. procerum* subsp. *procerum* (var. *medium*).

Endemic.

References: Atlas Supp. (46a), Ingrouille & Stace (1986), Pugsley (1924), Wigginton (1999).

S. J. LEACH

Limonium dodartiforme

Native	GB	IR
1987–99 ●	8	0
1970–86 ●	0	0
pre 1970 ●	0	0
Alien		
1987–99 ●	0	0
1970–86 ●	0	0
pre 1970 ●	0	0

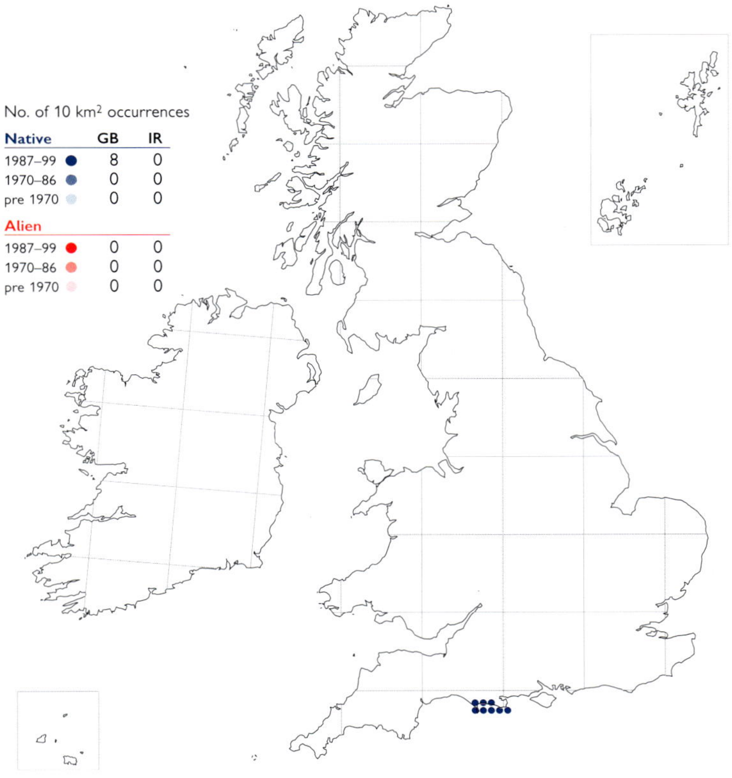

A perennial herb found mainly on chalk sea-cliffs, but with a few small outlying populations on stabilised shingle (Chesil Beach), on limestone at Lulworth and on Purbeck, and on breakwaters in Portland Harbour (all in Dorset). Lowland.

Native. This taxon was not recognised at the time of the 1962 *Atlas* and there are no indications of any recent changes in its distribution.

Endemic.

References: Ingrouille & Stace (1986), Wigginton (1999).

S. J. LEACH

Limonium recurvum subsp. *recurvum*

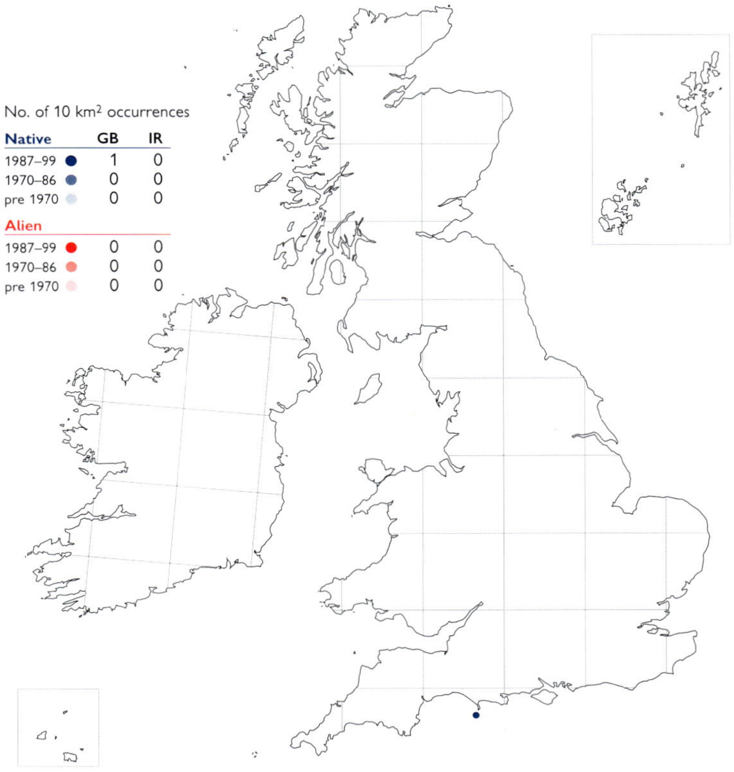

Native	GB	IR
1987–99 ●	1	0
1970–86 ●	0	0
pre 1970 ●	0	0
Alien		
1987–99 ●	0	0
1970–86 ●	0	0
pre 1970 ●	0	0

A perennial herb of limestone cliff-slopes, quarry cliffs and rock ledges, in similar habitats to those of subsp. *portlandicum*. Lowland.

Native. This subspecies was not recognised as taxonomically distinct until 1986. Its distribution is stable.

Endemic.

References: Ingrouille & Stace (1986), Perring & Sell (1968), Wigginton (1999).

S. J. LEACH

Limonium recurvum subsp. *portlandicum*

Native	GB	IR
1987–99 ●	2	2
1970–86 ●	0	0
pre 1970 ●	0	0
Alien		
1987–99 ●	0	0
1970–86 ●	0	0
pre 1970 ●	0	0

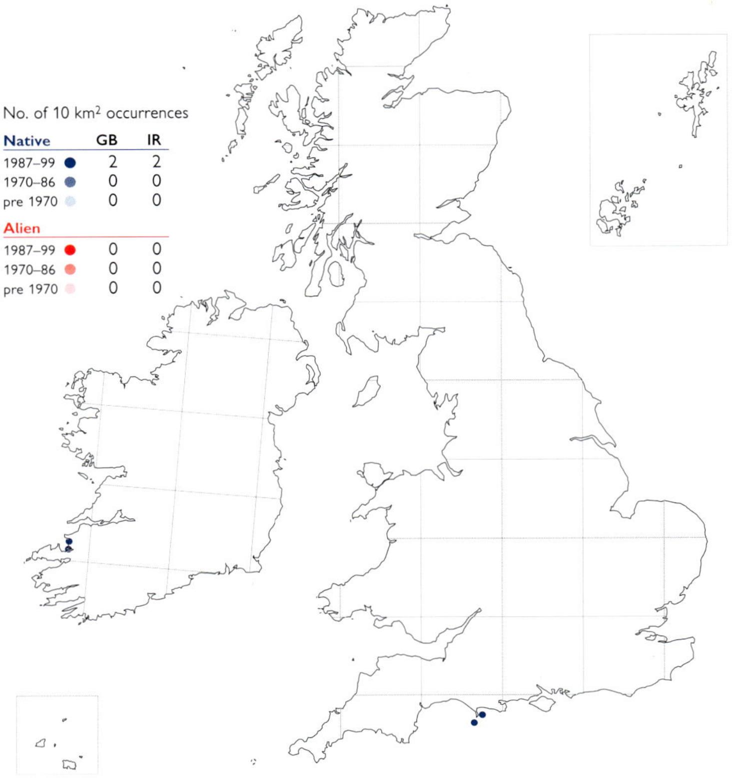

A perennial herb of crumbling limestone cliff-slopes, quarry cliffs and rock ledges (England), and the drier parts of saltmarshes (Ireland). Lowland.

Native. This subspecies was not recognised until 1986. Its distribution is stable.

Endemic.

References: Perring & Sell (1968), Ingrouille & Stace (1986), Wigginton (1999).

S. J. LEACH

Limonium recurvum subsp. *pseudotranswallianum*

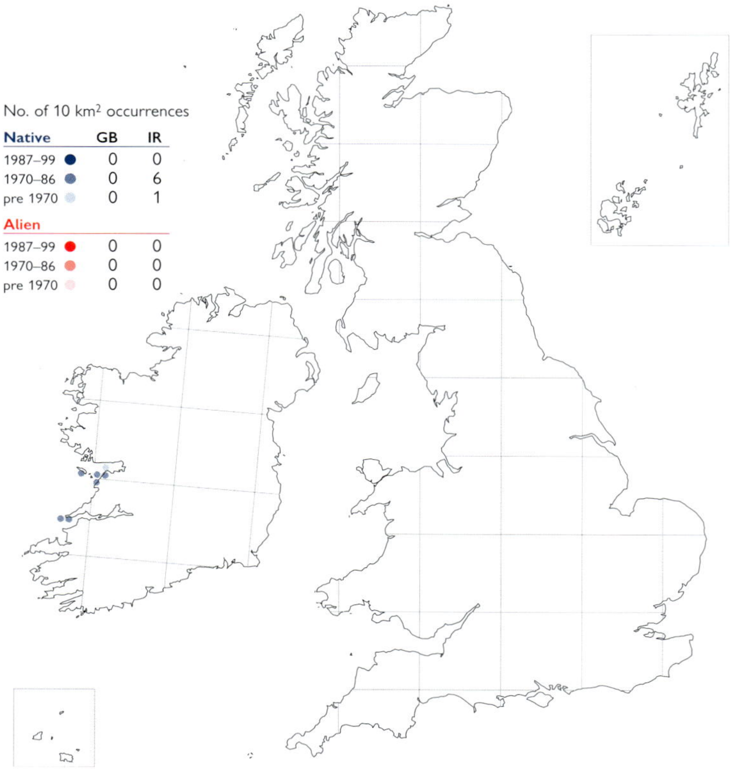

No. of 10 km² occurrences

	GB	IR
Native		
1987–99 ●	0	0
1970–86 ●	0	6
pre 1970 ●	0	1
Alien		
1987–99 ●	0	0
1970–86 ●	0	0
pre 1970 ●	0	0

A perennial herb of carboniferous limestone rocks and sea-cliffs. Lowland.

Native. Until Ingrouille & Stace's (1986) revision of the *L. binervosum* aggregate this taxon was included in a broadly defined *L. transwallianum* and its records were therefore included on the map of that species in Perring & Sell (1968).

Endemic.

Reference: Curtis & McGough (1988).

S. J. LEACH

Limonium recurvum subsp. *humile*

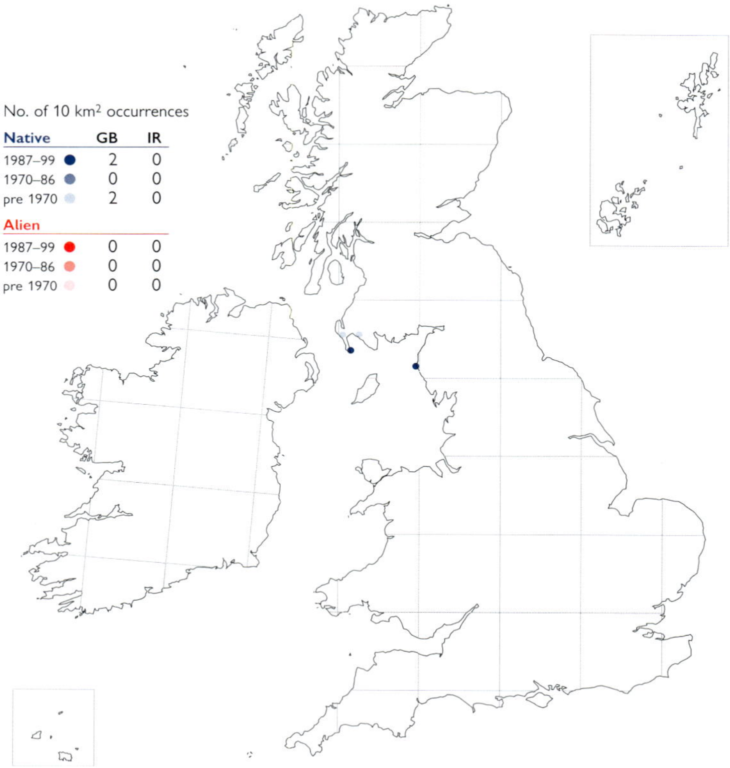

No. of 10 km² occurrences

	GB	IR
Native		
1987–99 ●	2	0
1970–86 ●	0	0
pre 1970 ●	2	0
Alien		
1987–99 ●	0	0
1970–86 ●	0	0
pre 1970 ●	0	0

This is a plant of sea-cliffs and coastal screes, but is also known from a railway bank in Cumberland. Lowland.

Native. There is no indication of any change in distribution since this subspecies was first described in 1986. Irish subsp. *humile* was tentatively mapped as *L. paradoxum* by Perring & Sell (1968).

Endemic.

References: Ingrouille & Stace (1986), Wigginton (1999).

S. J. LEACH

Armeria maritima Thrift

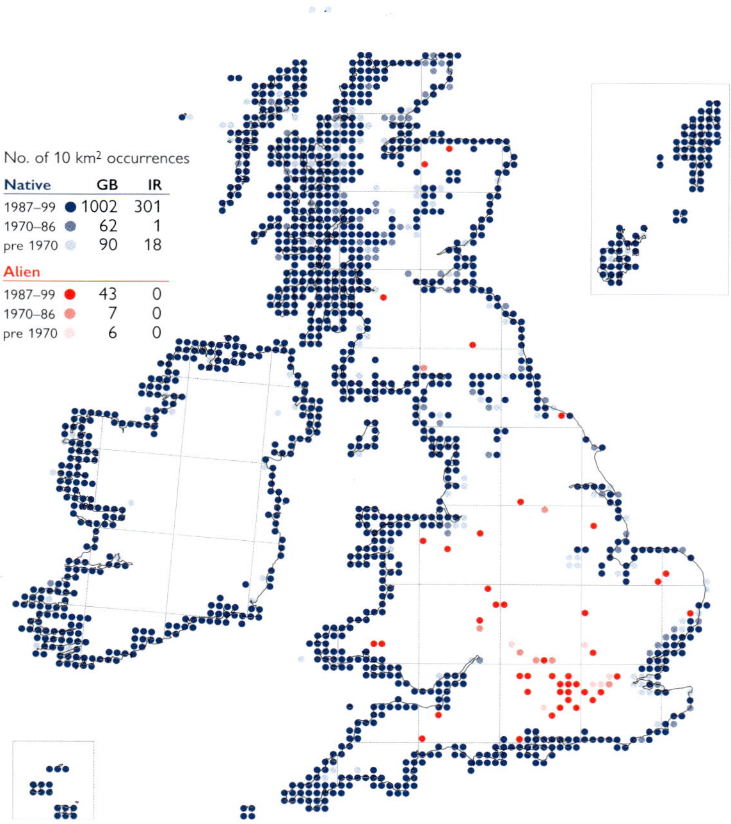

No. of 10 km² occurrences

	GB	IR
Native		
1987–99 ●	1002	301
1970–86 ●	62	1
pre 1970 ●	90	18
Alien		
1987–99 ●	43	0
1970–86 ●	7	0
pre 1970 ●	6	0

A perennial herb of sea-cliffs, stone walls, stabilised shingle and saltmarshes. Inland, it grows on montane rock ledges, stony flushes and windswept moss-heaths, around old lead workings and other metalliferous mine wastes, and on riverside shingle. It also occurs inland beside salt-treated roads and as a garden escape. 0–1270 m (Ben Nevis, Westerness).

Native (change –0.14). There has been little change in distribution since the 1962 *Atlas*, where it was mapped as 'all records', apart from its appearance from the 1980s on roadsides, mainly in S. England. Almost all British and Irish plants are referable to subsp. *maritima*; the rare subsp. *elongata* is mapped separately.

Circumpolar Wide-boreal element.

References: Atlas (200a), Hultén & Fries (1986), Meusel *et al.* (1978), Welch & Welch (1998), Woodell & Dale (1993).

S. J. LEACH

Armeria maritima subsp. *elongata*

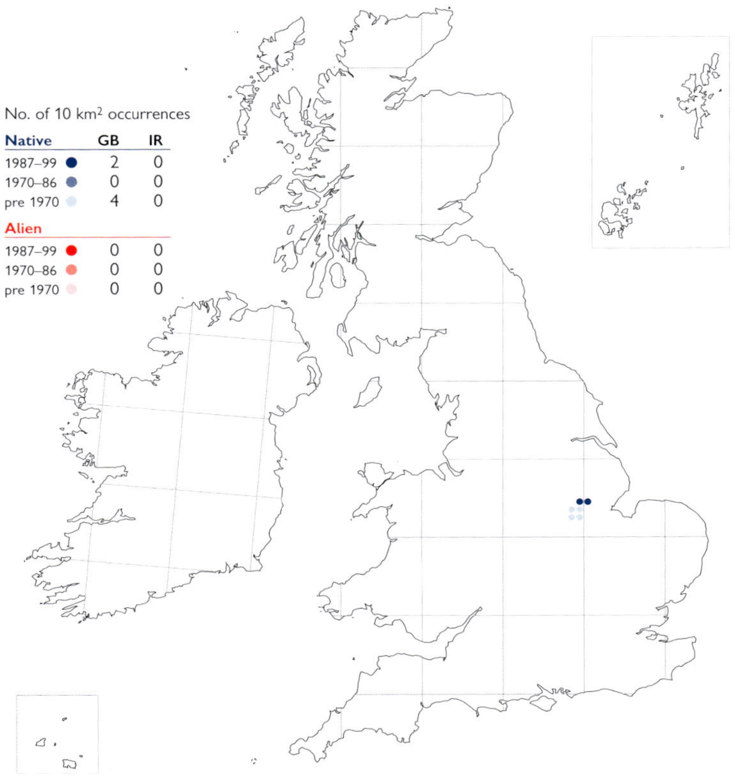

A perennial herb of neutral grasslands overlying river gravels or alluvium, formerly also at one site in limestone grassland. Lowland.

Native. In the 1950s subsp. *elongata* was known from at least twelve localities, all in Lincolnshire, but only two populations are still extant. The other populations were lost to ploughing, re-seeding and other agricultural improvement.

European Temperate element, with a restricted range in N.C. Europe and a continental distribution.

References: Atlas (200b), Gibbons & Lousley (1958), Meusel *et al.* (1978), Wigginton (1999), Woodell & Dale (1993).

S. J. LEACH

Armeria arenaria Jersey Thrift

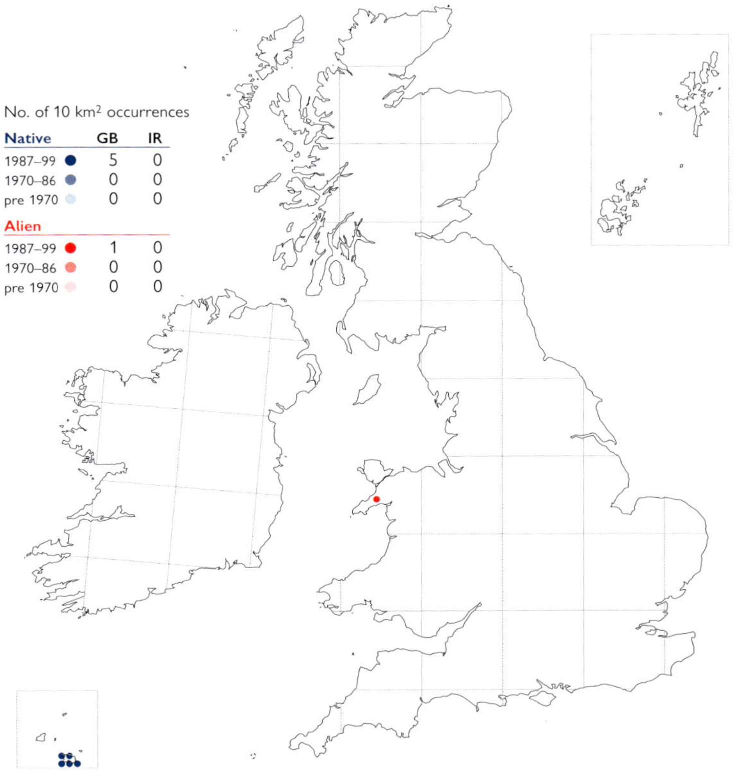

A perennial herb of fixed sand dunes and coastal headlands. It has a more inland distribution than *A. maritima* subsp. *maritima*, and rarely grows with it. It is also recorded as a casual. Lowland.

Native. In our area *A. arenaria* is known as a native only in Jersey. There has been no change in its distribution since the 1962 *Atlas*.

Suboceanic Southern-temperate element.

References: Atlas (200b), Le Sueur (1984), Meusel *et al.* (1978).

C. D. PRESTON

Paeonia officinalis Garden Peony

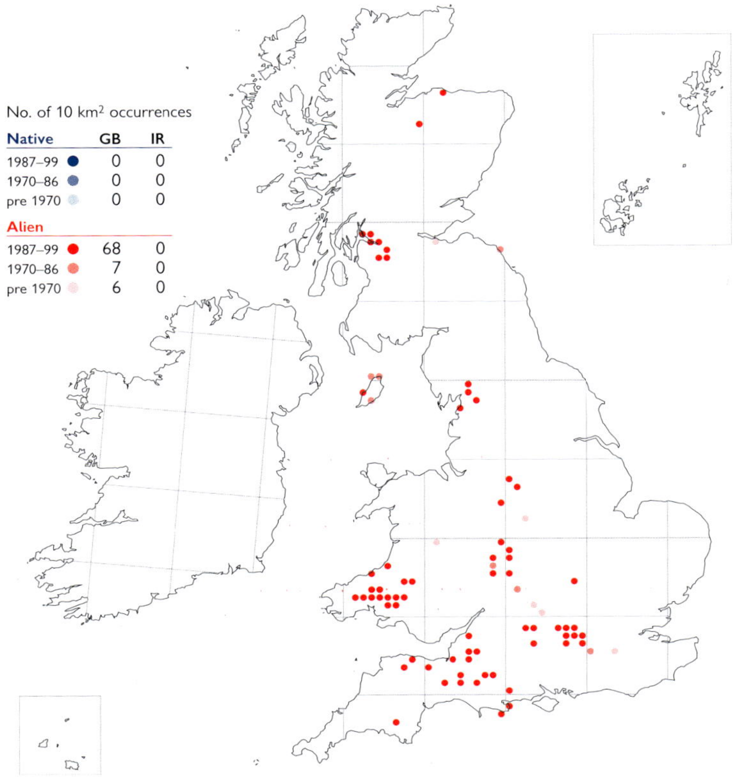

A perennial herb grown in gardens and found as a garden escape or throw-out in woodland and on road verges, rubbish tips and waste ground, and as a relic of cultivation. Lowland.

Neophyte. This species was cultivated in Britain by 1548 and is very popular in gardens. It was first recorded in the wild in 1650. Most records are recent and it is perhaps increasing. It is sometimes identified as *P. mascula*, but this species is apparently found only in the Bristol Channel on Steep Holm (N. Somerset) and Flat Holm (Glamorgan).

Native of S. Europe.

References: Bolòs & Vigo (1984), Jalas & Suominen (1991).

H. J. KILLICK

Elatine hexandra Six-stamened Waterwort

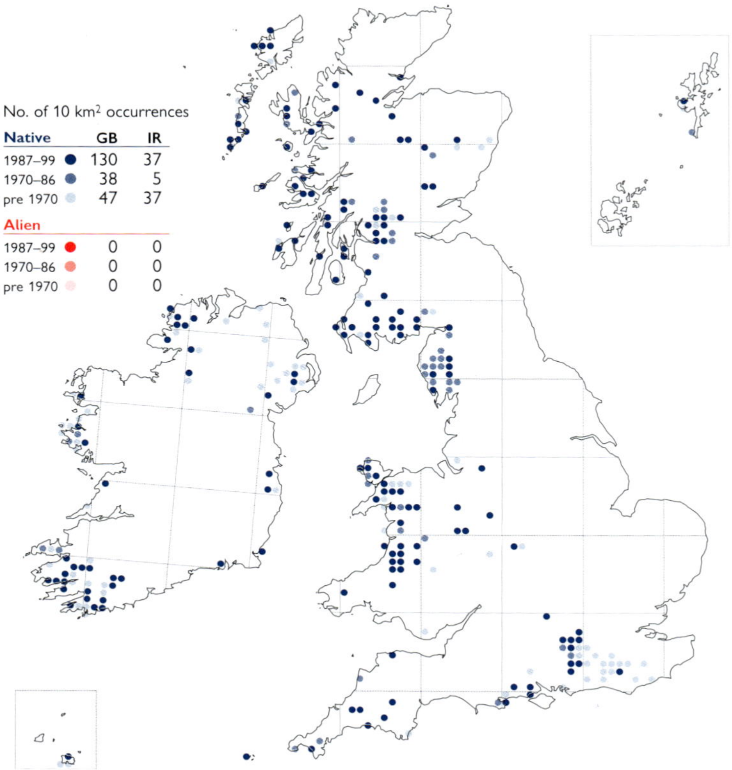

No. of 10 km² occurrences

Native	GB	IR
1987–99 ●	130	37
1970–86 ●	38	5
pre 1970 ●	47	37
Alien		
1987–99 ●	0	0
1970–86 ●	0	0
pre 1970 ●	0	0

E. hexandra grows as an annual on exposed mud at the edge of lakes, reservoirs, ponds and flooded gravel-pits, or submerged on open substrates in shallow, oligotrophic to eutrophic water. When submerged it may sometimes persist as a short-lived perennial. Like many species in this habitat, it is subject to large annual fluctuations in numbers. 0–440 m (Lake Ferta, N. Kerry), and reportedly to 490 m in the Scottish Highlands.

Native (change +1.07). This species is now known to be much more frequent than was apparent in the 1962 *Atlas*. This is almost certainly attributable to more detailed recording of aquatic habitats.

European Temperate element.

References: Atlas (61c), Hultén & Fries (1986), Meusel *et al.* (1978), Preston & Croft (1997), Stewart *et al.* (1994).

C. D. PRESTON

Elatine hydropiper Eight-stamened Waterwort

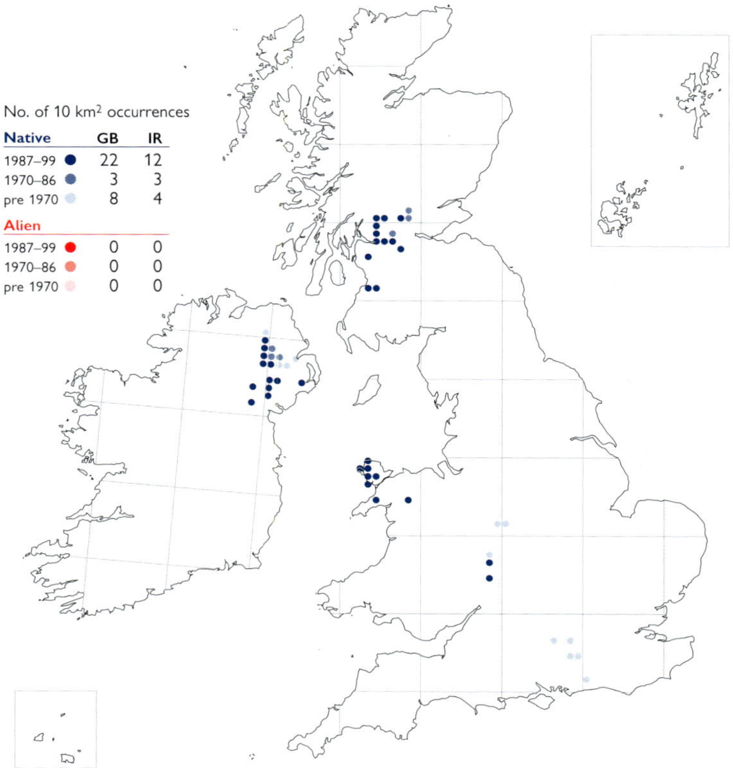

No. of 10 km² occurrences

Native	GB	IR
1987–99 ●	22	12
1970–86 ●	3	3
pre 1970 ●	8	4
Alien		
1987–99 ●	0	0
1970–86 ●	0	0
pre 1970 ●	0	0

Like *E. hexandra*, this is an annual which grows in shallow water or on damp mud or silty sand exposed at the water's edge. Unlike that species it is confined to mesotrophic or eutrophic habitats. Lowland.

Native (change +0.66). Although *E. hydropiper* has disappeared from S.E. England (last recorded 1944), it was first recorded in Scotland in 1968 and has since been found in many more sites there. Mitchell (1981) suggests that it has been spread by birds, but it is possible that, like the equally inconspicuous *E. hexandra*, it may have been under-recorded in the 1962 *Atlas*. In Ireland its distribution has remained stable.

Eurosiberian Boreal-temperate element, with a continental distribution in W. Europe.

References: Atlas (61d), Curtis & McGough (1988), Hultén & Fries (1986), Preston & Croft (1997), Stewart *et al.* (1994).

C. D. PRESTON

Hypericum calycinum Rose-of-Sharon

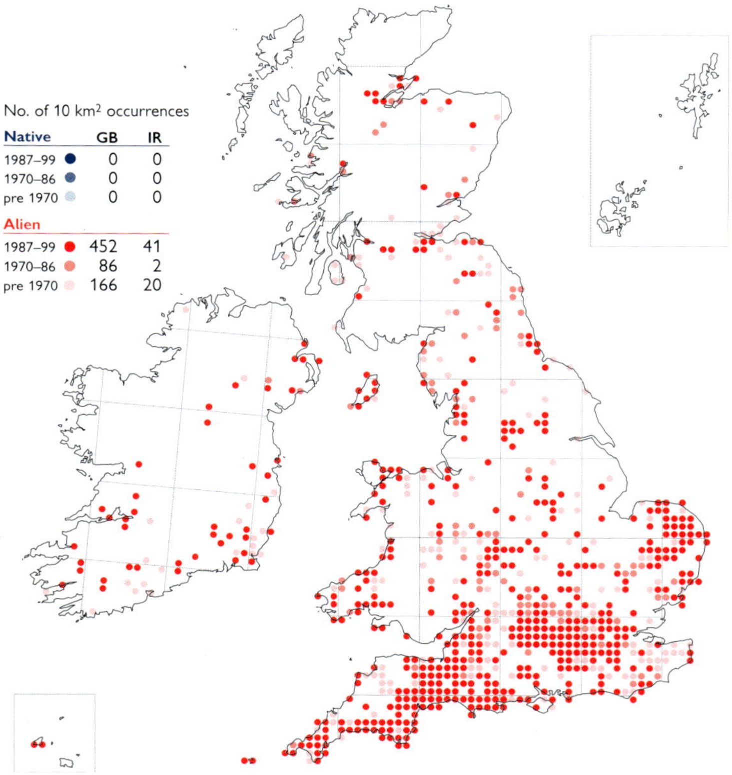

No. of 10 km² occurrences

Native	GB	IR
1987–99 ●	0	0
1970–86 ●	0	0
pre 1970 ●	0	0
Alien		
1987–99 ●	452	41
1970–86 ●	86	2
pre 1970 ●	166	20

A low-growing, rhizomatous shrub, widely cultivated and naturalised in hedgerows and on roadsides and railway banks. Lowland.

Neophyte (change +0.74). *H. calycinum* was introduced into Britain in 1676, but not recorded from the wild until 1809. All naturalised plants appear to have been derived from the original introduction, as fertile seed is sparsely and erratically produced owing to self-incompatibility. Damp autumns also reduce seed-set, and seed and the rare seedlings suffer predation by sparrows. Recent new introductions to cultivation may, however, alter this situation. *H. calycinum* is much better recorded than in the 1962 *Atlas*.

Native of S.E. Bulgaria and Turkey.

References: Atlas (57c), Bean (1973), Robson (1985), Salisbury (1963, 1969a).

N. K. B. ROBSON

Hypericum androsaemum Tutsan

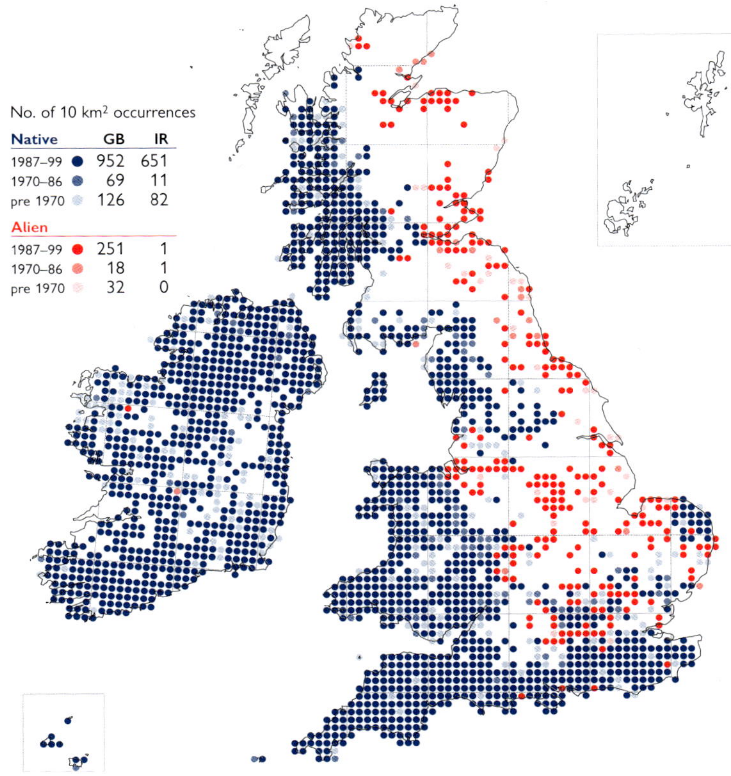

A shrub of damp or shaded habitats including woods and hedgerows. It is also widely cultivated and occurs well outside its native range in semi-natural and artificial, often drier habitats, being apparently spread by birds. 0–630 m (Macgillycuddy's Reeks, S. Kerry).

Native (change +0.78). The native distribution of *H. androsaemum* is now much better documented, and alien occurrences are more frequent, or more frequently reported, than in the 1962 *Atlas*. The boundary of the native distribution cannot be defined with certainty.

Submediterranean-Subatlantic element; widely naturalised outside its native range.

References: Atlas (56d), Bolòs & Vigo (1990), Robson (1985).

N. K. B. ROBSON

Hypericum androsaemum × *H. hircinum* (*H.* × *inodorum*) Tall Tutsan

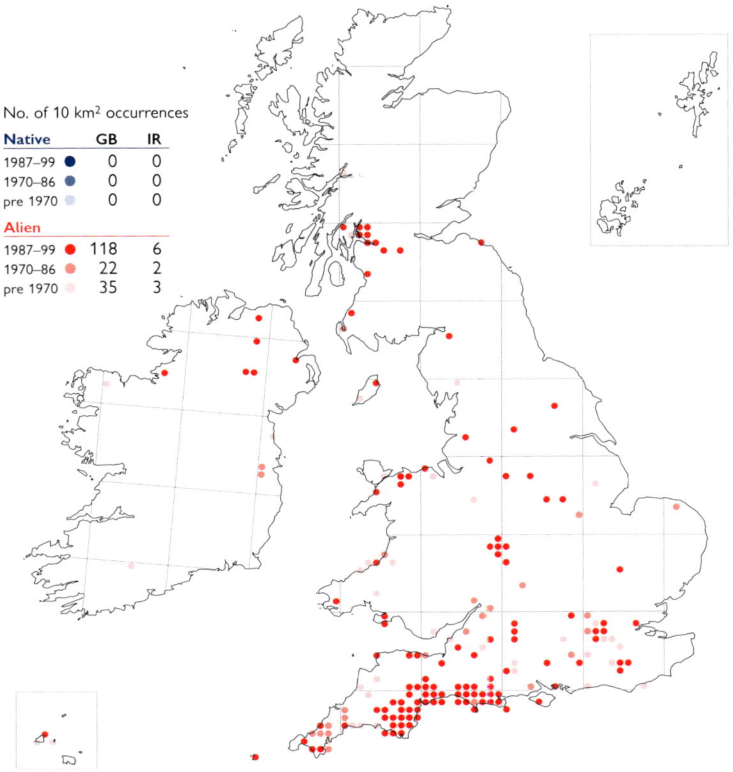

A bushy shrub naturalised in damp or shaded habitats, including hedges, thickets and stream banks. Dispersal by birds may be significant, owing to the succulence of the young fruits, but spread is not rapid, and naturalised populations are usually found near parks and gardens. Lowland.

Neophyte (change +1.78). *H.* × *inodorum* had apparently originated spontaneously in English gardens by about 1760 and was known from the wild by 1882 (Herts.). It is well-established in some areas, and probably still under-recorded.

Native of S.W. Europe.

References: Atlas (57a), Bean (1973), Robson (1985), Stace (1975), Turrill (1962).

N. K. B. ROBSON

Hypericum hircinum Stinking Tutsan

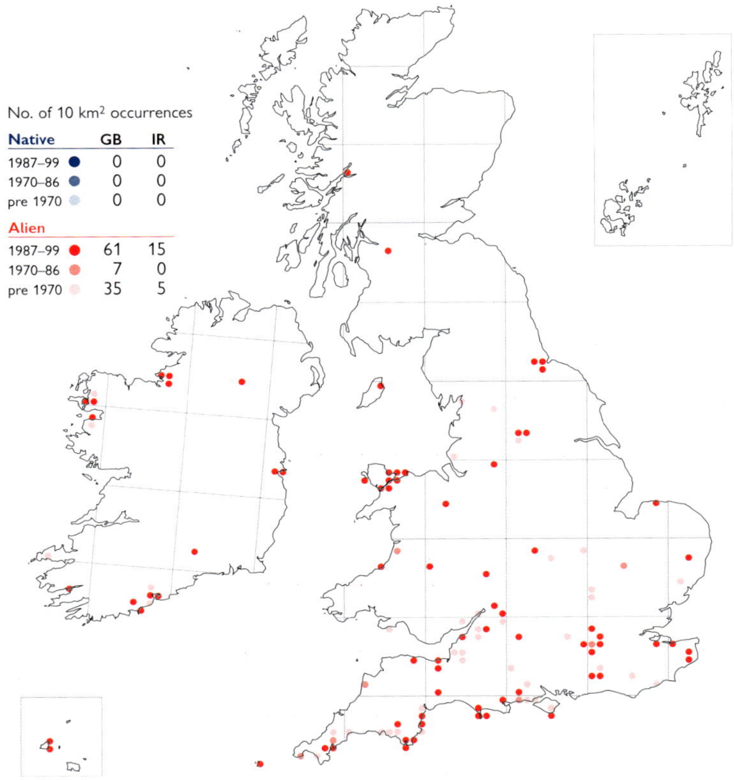

A bushy shrub naturalised in shaded habitats, and occasionally also in more open places where its growth is often stunted. Lowland.

Neophyte (change +1.08). *H. hircinum* was being grown in Britain by 1640, and had been recorded in the wild by 1856. Dunn (1905) described it as 'naturalised in many situations'. It does not appear to spread rapidly into natural vegetation. There are more records than in the 1962 *Atlas*, but there is no real sign of spread.

A native of the Mediterranean region; naturalised in W. Europe.

References: Atlas (57b), Bean (1973), Bolòs & Vigo (1990), Robson (1985).

N. K. B. ROBSON

Hypericum perforatum
Perforate St John's-wort

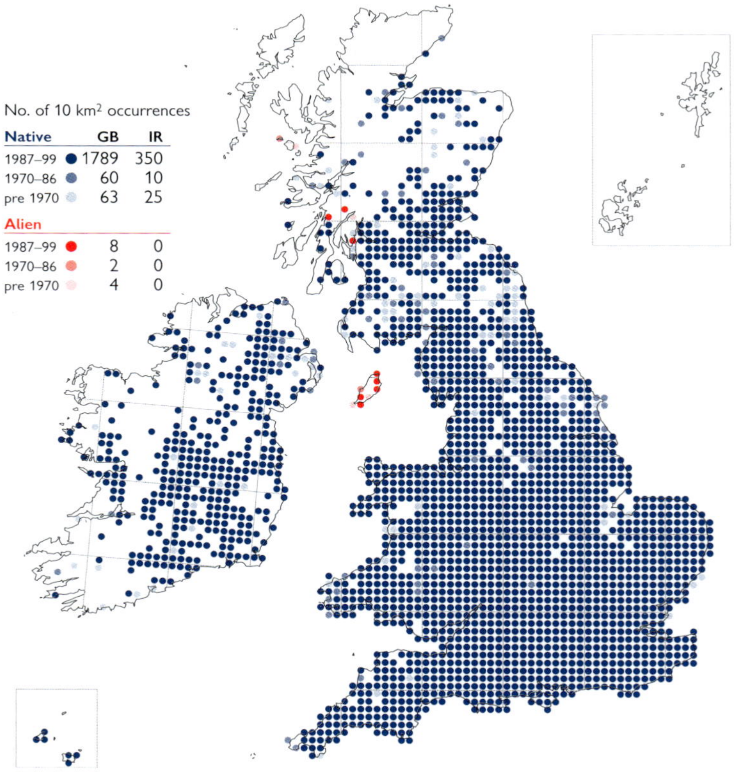

A variable, rhizomatous perennial herb of meadows, hedge banks, open woods, roadsides and along railways. It often grows in drier, more ruderal and more calcareous habitats than *H. maculatum*. It reproduces by seed and also sometimes by root buds. 0–480 m (Great Rhos, Rads.), with an exceptional record at 845 m (Great Dun Fell, Westmorland).

Native. Suitable natural habitat is rare in N. Scotland, but *H. perforatum* extended into this region along railways during the 20th century. Otherwise there has been little change since the 1962 *Atlas*.

Eurosiberian Southern-temperate element; widely naturalised outside its native range.

References: Atlas (57d), Crompton *et al.* (1988), Grime *et al.* (1988), Hultén & Fries (1986), Meusel *et al.* (1978), Salisbury (1952).

N. K. B. ROBSON

Hypericum maculatum Imperforate St John's-wort

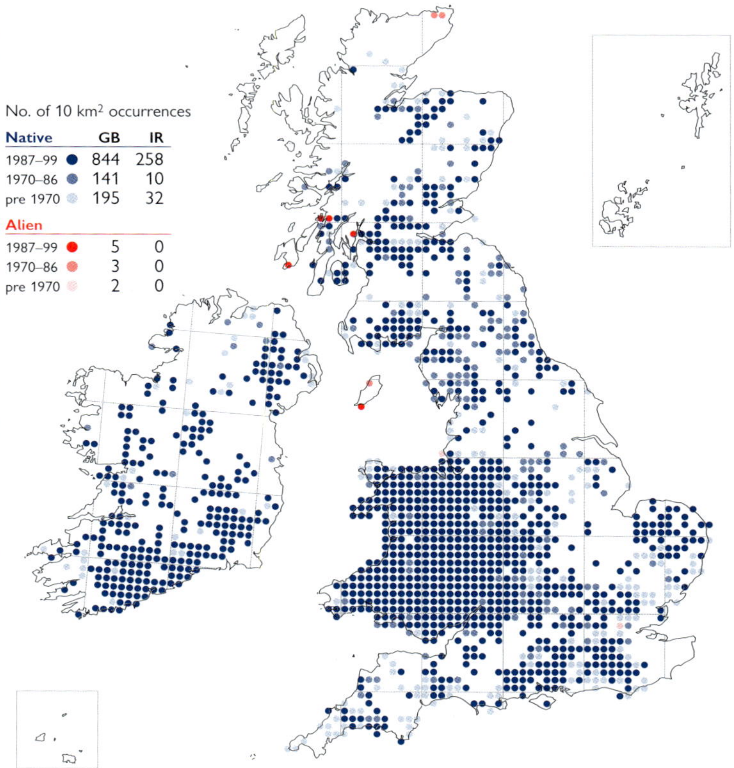

A shortly rhizomatous perennial herb growing in damp shaded places, but also in more ruderal habitats such as quarries, waste ground and rough, grassy places. 0–380 m (Cotterdale, N.W. Yorks.).

Native (change +2.11). Many of the areas in which *H. maculatum* grows were not well recorded in the 1962 *Atlas* and its distribution is now much better known. Two subspecies occur in our area, the diploid subsp. *maculatum* and the tetraploid subsp. *obtusiusculum*, and these are mapped separately.

Eurosiberian Boreo-temperate element.

References: Atlas (58a), Hultén & Fries (1986), Meusel *et al.* (1978).

N. K. B. ROBSON

Hypericum maculatum subsp. *maculatum*

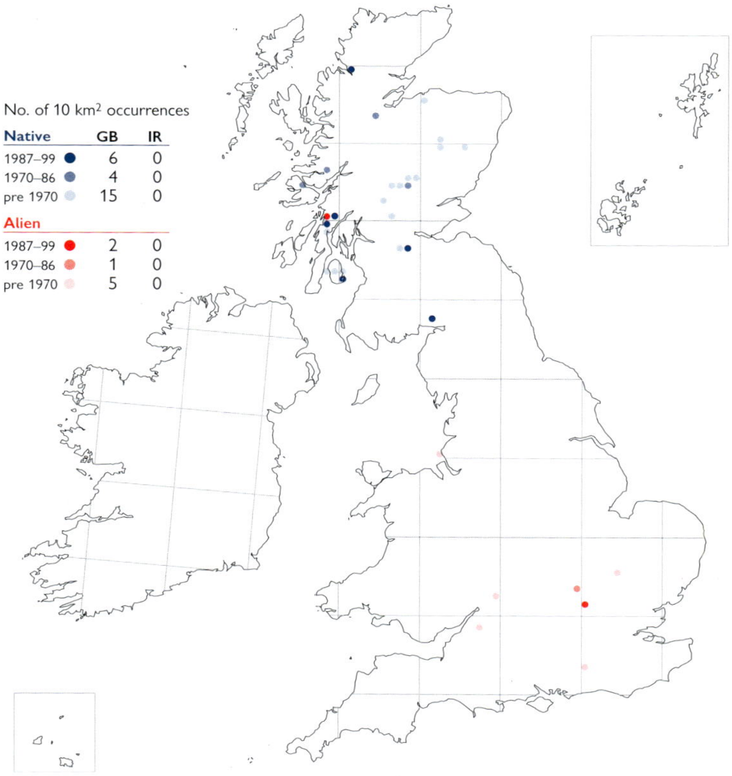

A shortly rhizomatous perennial herb, mainly in damp or shaded habitats such as woodland margins, hedgerows and streamsides. Generally lowland, but its upper altitudinal limit is uncertain.

Native. The map of subsp. *maculatum* in Perring & Sell (1968) was incomplete, being based only on records confirmed by N.K.B. Robson. It remains under-recorded. The status of the scattered English records is doubtful. It is not clear whether they are casual introductions or relics left behind by a northward-retreating population.

Eurosiberian Boreo-temperate element. This diploid subspecies is the only one that occurs in E. Europe and Siberia. In continental W. Europe it is confined to montane or upland regions.

References: Atlas Supp. (10b), Robson (1958a).

N. K. B. ROBSON

Hypericum maculatum subsp. *obtusiusculum*

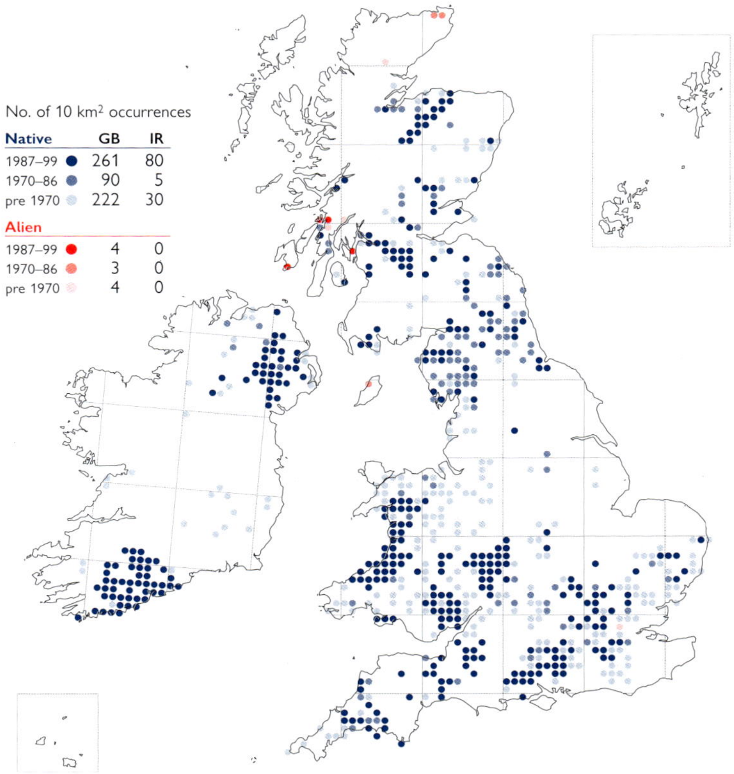

This tetraploid subspecies occurs in similar habitats to those of subsp. *maculatum*, but tends to grow in more ruderal habitats such as rough grassland, scrub, quarries, roadsides, railway banks and waste ground. Generally lowland, but reaching 320 m (Cynwyd, Merioneth).

Native. The map of this taxon in Perring & Sell (1968) was incomplete, being based only on records confirmed by N.K.B. Robson. It remains under-recorded in many areas. Overall, its distribution is probably stable. The relative frequency of the two subspecies in Scotland requires further study.

Suboceanic Temperate element. Subsp. *obtusiusculum* is confined mainly to lowlands of N.W. Europe.

References: Atlas Supp. (10c), Robson (1958a).

N. K. B. ROBSON

Hypericum maculatum × *H. perforatum* (*H.* × *desetangsii*) Des Etangs' St John's-wort

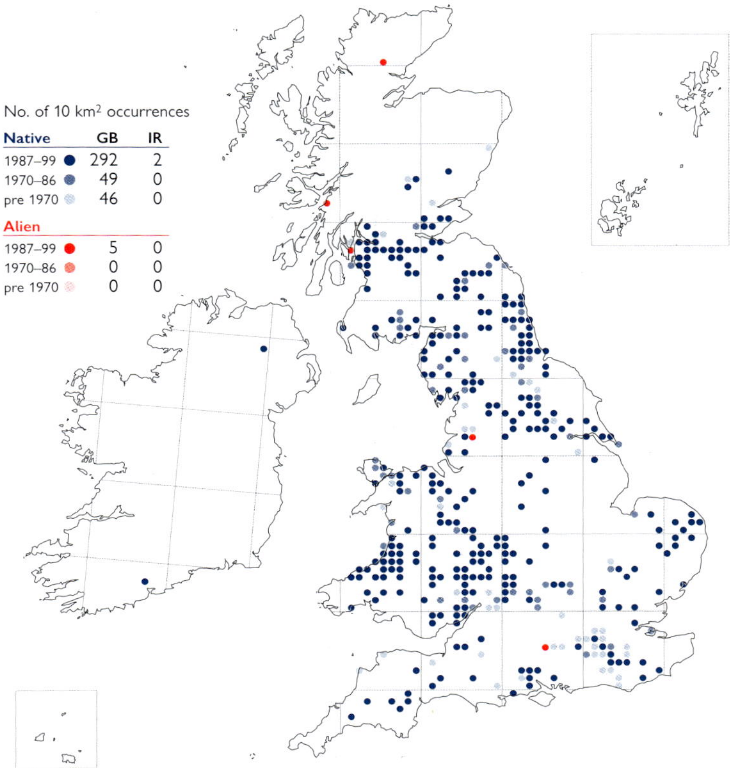

A perennial herb intermediate in form between the parents, occurring in rough grassland, gravel- and colliery-pits, on roadsides and railway banks. Because the habitats of the parents rather widely overlap, it often occurs with one or other of them, but can also be found in their absence. Lowland.

Native. *H. perforatum* forms a continuous series of intermediates with *H. maculatum* subsp. *obtusiusculum*, and these are widespread in Britain. Hybrids with *H. maculatum* subsp. *maculatum* have been recorded only from S.W. Scotland. It has not previously been mapped at a national scale and is almost certainly under-recorded.

Widespread in Europe.

Reference: Stace (1975).

N. K. B. ROBSON

Hypericum undulatum Wavy St John's-wort

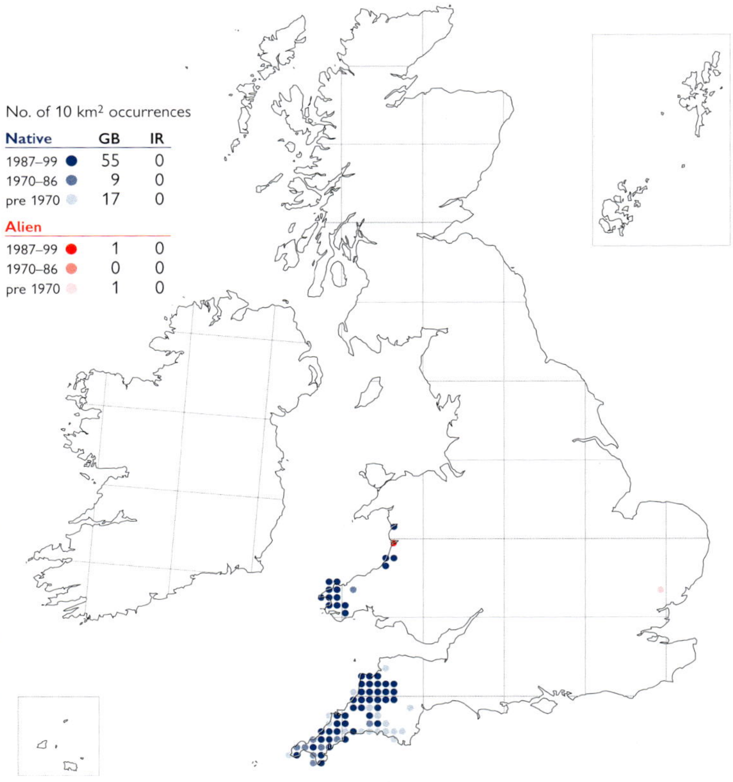

A shortly rhizomatous perennial herb of non-calcareous marshy fields and streamsides, fen-meadows and mildly acidic bogs. It shows a preference for waterlogged areas subject to lateral water movement, giving slight base-enrichment. Seed can be persistent in the soil. Lowland.

Native (change –0.12). There have been many losses due to habitat destruction, especially through agricultural intensification, and this trend has continued in recent years. However, *H. undulatum* is much better recorded than in the 1962 *Atlas*. There is a herbarium specimen from Ireland (Glengarriff, W. Cork) but it may be mis-labelled.

Oceanic Southern-temperate element.

References: Atlas (58b), Kay & John (1994), Robson (1958b), Stewart *et al.* (1994).

N. K. B. ROBSON

Hypericum tetrapterum
Square-stalked St John's-wort

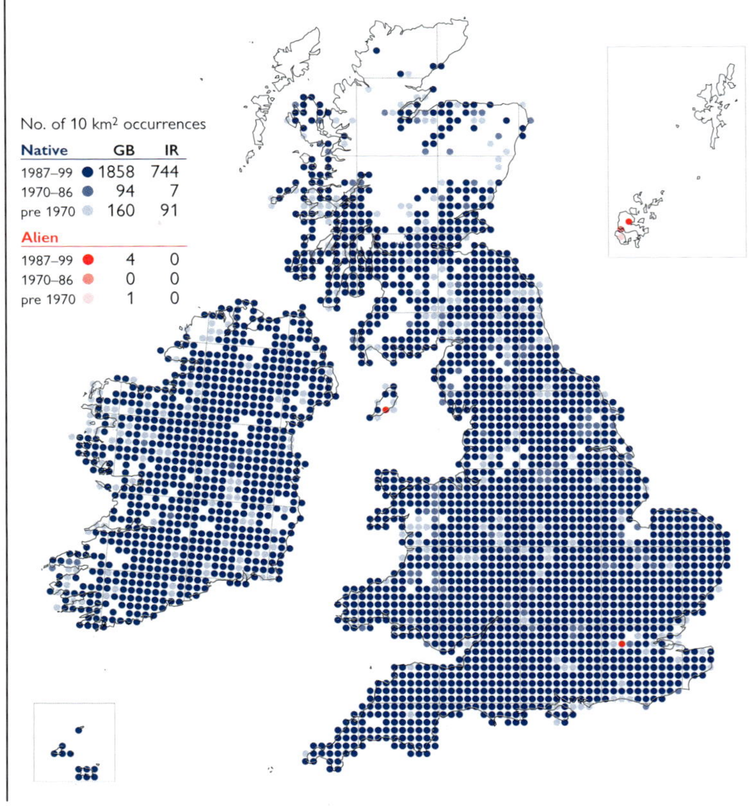

A shortly rhizomatous perennial herb of eutrophic damp or wet habitats including damp meadows, marshes, streamsides and open ditches. Usually lowland, but ascending to 380 m in W. Yorkshire.

Native (change –0.41). There has been little change in the distribution of *H. tetrapterum*, which was mapped as 'all records' in the 1962 *Atlas*.

European Temperate element.

References: Atlas (58c), Hultén & Fries (1986), Meusel *et al.* (1978).

N. K. B. ROBSON

Hypericum humifusum Trailing St John's-wort

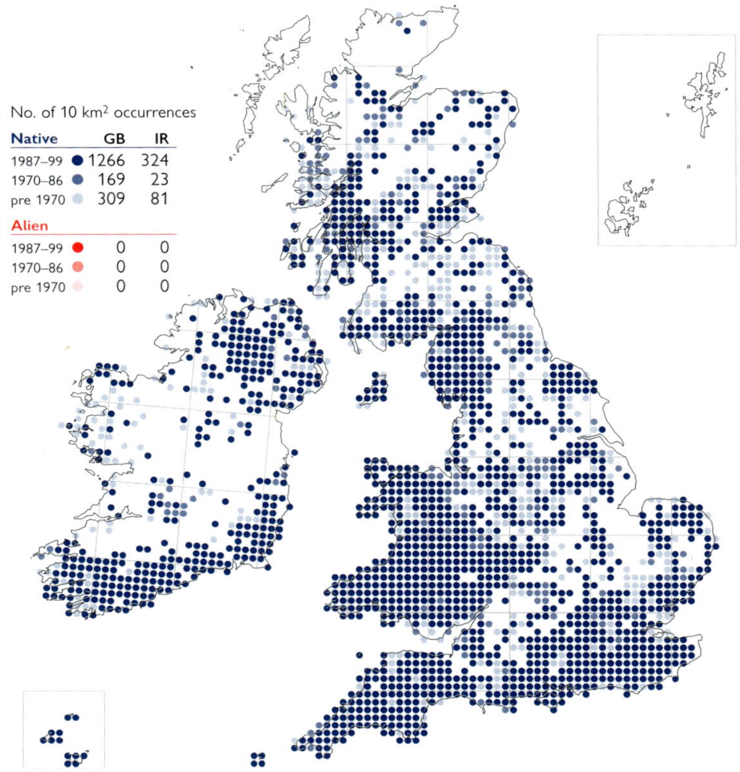

A short-lived perennial herb of open well-drained habitats (heaths, dry moors, open woodlands, tracksides, sometimes roadside banks) on light acidic soils; if apparently on calcareous soils, then rooting in a shallow acidic surface layer. Generally lowland, but reaching 530 m at Cat-and-Fiddle, Cheshire.

Native (change –0.40). Although *H. humifusum* is widespread, it is always local and rarely more than locally abundant. It was mapped as 'all records' in the 1962 *Atlas*. Analysis of the database reveals that the widespread decline in England and Scotland has occurred since 1950. Losses in W. Ireland are probably due to under-recording.

European Temperate element.

References: Atlas (58d), Hultén & Fries (1986), Meusel *et al.* (1978).

N. K. B. ROBSON

Hypericum linariifolium
Toadflax-leaved St John's-wort

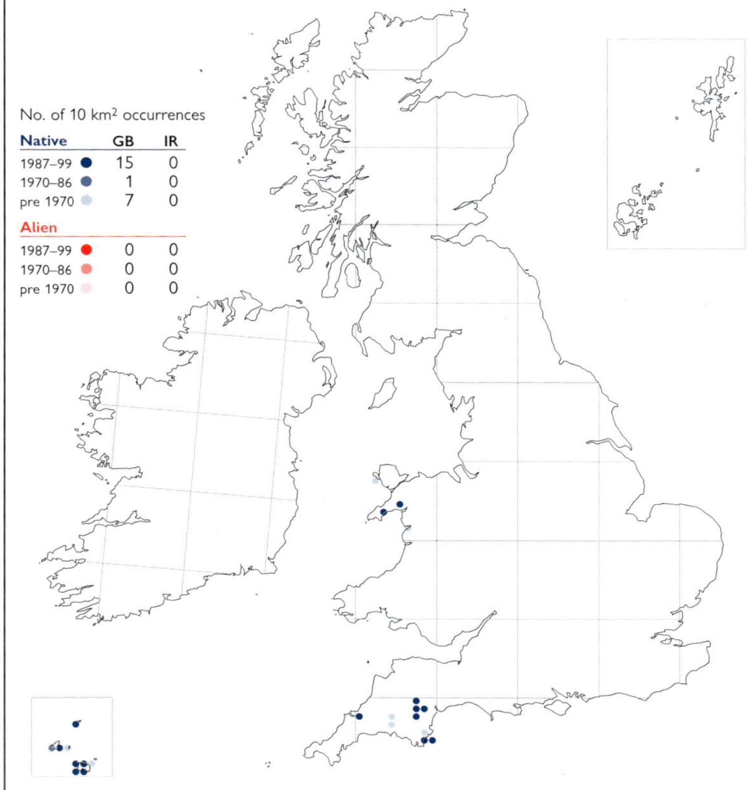

A short-lived perennial, or sometimes biennial, herb of steep rocky slopes and eroded banks with a southerly aspect, on well-drained acidic soils and without shading or competition. It reproduces by seed, which is copiously produced. Lowland.

Native (change +0.09). Recent detailed recording in E. Devon has revealed many new sites. The distribution of populations in such core areas may now be stable, but some outlying populations are small. Intermediates (which might be hybrids) with *H. humifusum* are found in the Channel Islands, and isoenzyme studies suggest there may be introgressed populations in mainland Britain (Kay & John, 1995).

Oceanic Southern-temperate element; it reaches its world northern limit in N. Wales.

References: Atlas (59a), Ivimey-Cook (1963), McDonnell (1995), Meusel *et al.* (1978), Wigginton (1999).

N. K. B. ROBSON

Hypericum pulchrum Slender St John's-wort

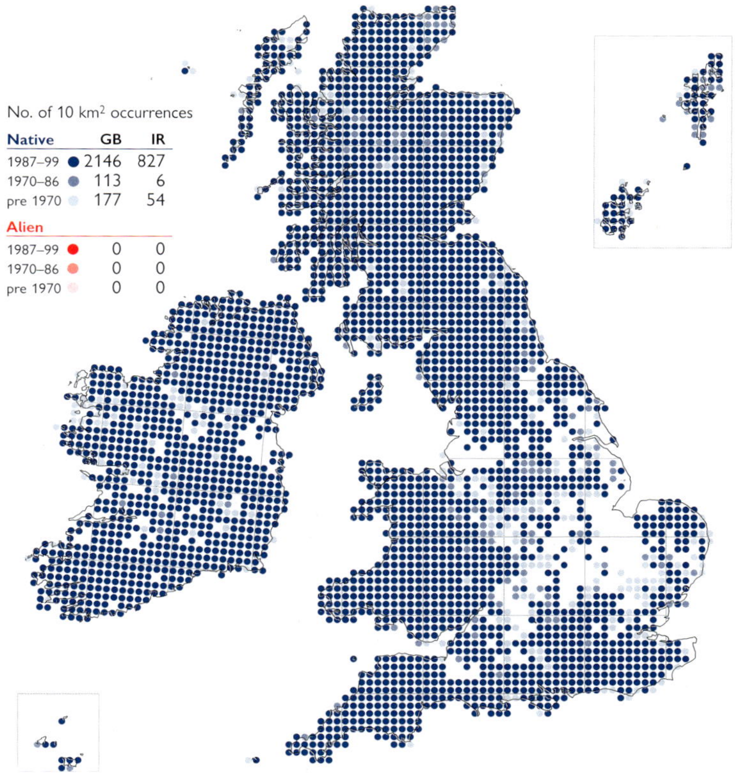

A perennial herb of heaths and open woods on non-calcareous, fairly dry, sandy, peaty, or leached soils. Reproduction is by seed, and budding from the roots sometimes occurs. A dwarf form with prostrate to procumbent stems (var. *procumbens*) occurs in exposed habitats in N. and W. Scotland and in islands off W. Ireland. 0–820 m (Breadalbanes, Mid Perth).

Native (change –0.32). *H. pulchrum* was mapped as 'all records' in the 1962 *Atlas*. Analysis of the database reveals that the decline in C. England has occurred since 1950.

Suboceanic Temperate element.

References: Atlas (59b), Böcher (1940), Dupont (1962), Hultén & Fries (1986), Meusel *et al.* (1978).

N. K. B. ROBSON

Hypericum hirsutum Hairy St John's-wort

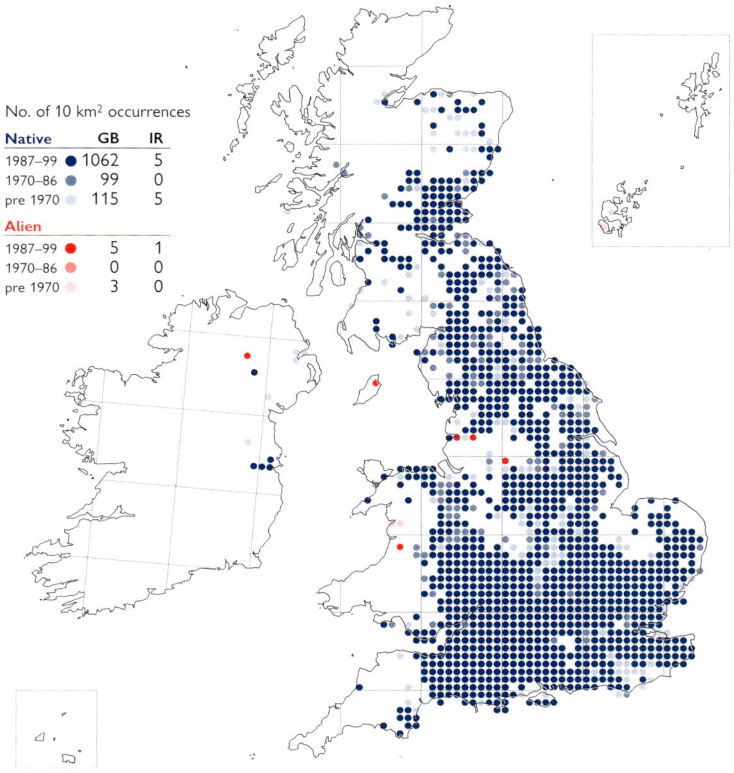

A perennial herb of well-drained, neutral to basic soils in open or partially shaded habitats including rough and ungrazed grassland, woodland rides and clearings, river banks, roadside banks and verges. 0–450 m at Garrigill (Cumberland).

Native (change –0.18). The distribution of *H. hirsutum* in Britain seems stable. In Ireland it appears to be declining; it was last seen in the Belfast area in 1937.

Eurosiberian Temperate element.

References: Atlas (59c), Curtis & McGough (1988), Grime *et al.* (1988), Hultén & Fries (1986), Meusel *et al.* (1978), Salisbury (1952).

N. K. B. ROBSON

Hypericum montanum Pale St John's-wort

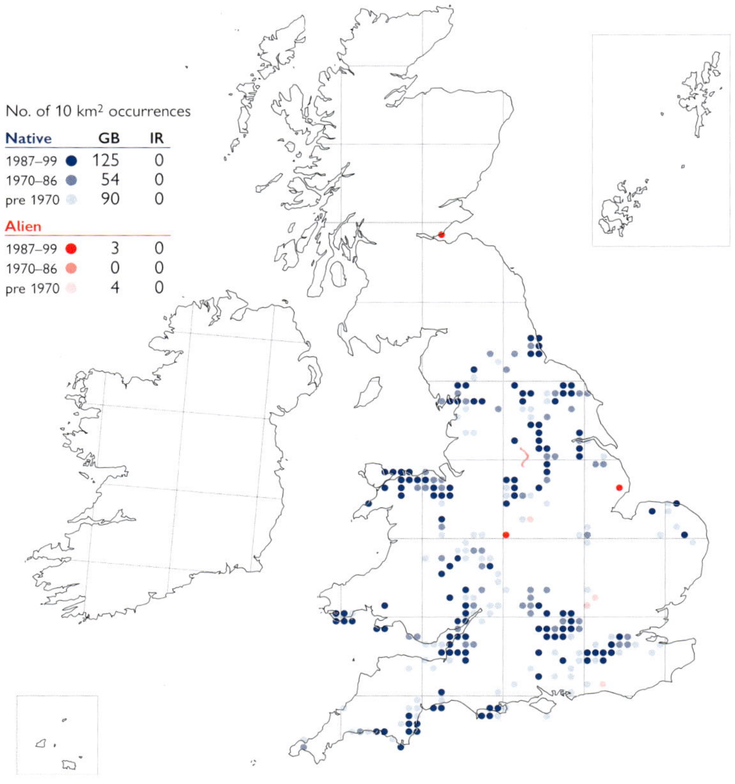

This perennial herb is almost confined to chalk and limestone districts, and only occasionally found over other substrates. It prefers well-drained soils, and grows by hedges and in thickets, amongst scrub, in rough grassland, open woodland and rides, and the grikes of limestone pavement. Reproduction is by seed, and vegetative spread is very slow. 0–330 m (above Great Asby, Westmorland).

Native (change –0.49). Habitat loss or degradation accounts for much of the loss of this species from former sites, most of which occurred before 1930. This remains the main threat to its populations, which tend to be small and therefore vulnerable.

European Temperate element.

References: Atlas (59d), Hultén & Fries (1986), Meusel *et al.* (1978), Robson (1996), Stewart *et al.* (1994).

N. K. B. ROBSON

Hypericum elodes Marsh St John's-wort

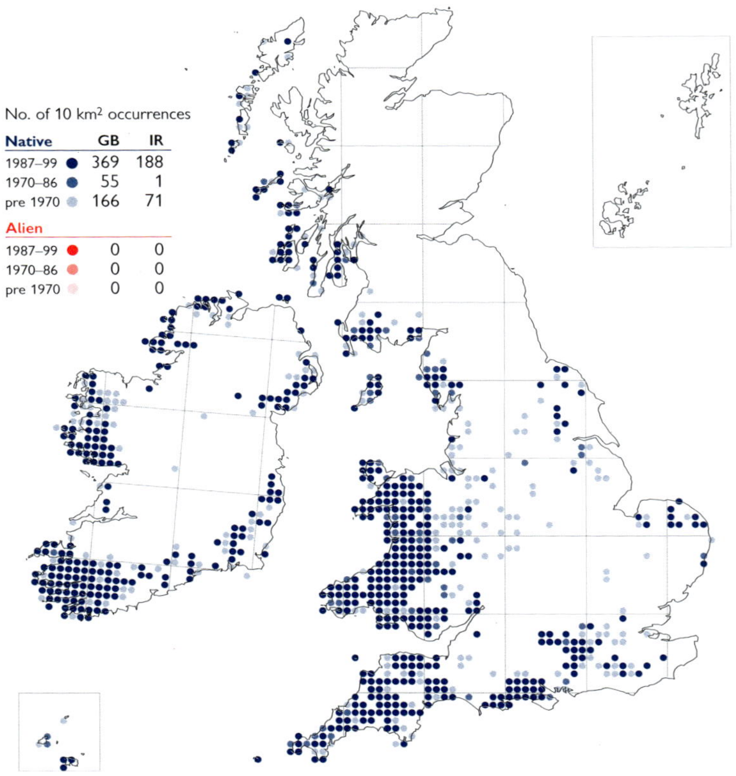

A stoloniferous perennial herb of peat or peaty mineral soils in damp or wet acidic, nutrient-poor habitats, usually found in shallow water, but sometimes terrestrial, or in deeper water where it may form floating mats. It occurs in heathland pools, on the margins of ponds and slow-flowing streams, and along seepages and runnels in mires. 0–425 m (Exmoor, S. Somerset).

Native (change –0.46). The range of *H. elodes* has been considerably reduced in the last hundred and fifty years, and its habitats are increasingly threatened, mostly by drainage. Although many losses in England occurred before 1930, this decline has continued since the 1962 *Atlas*.

Oceanic Temperate element; one remaining site in Italy, extinct in eastern Germany and Austria.

References: Atlas (60a), Meusel *et al.* (1978), Robson (1996). N. K. B. ROBSON

Hypericum canadense Irish St John's-wort

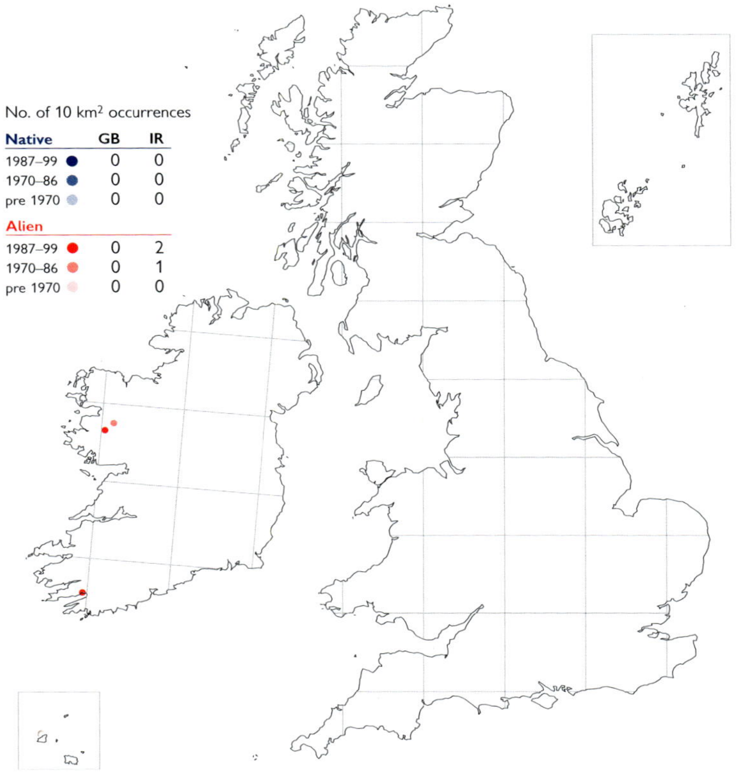

An annual to perennial tap-rooted herb of acidic, sandy or peaty soil in flushes or by small streams on moorland, heathland or in rough grazing. It grows in wet turf which is grazed or trampled by cattle. Lowland.

Neophyte. *H. canadense* was cultivated in 1770, but is not grown as a garden plant. It was first found in 1954 at Lough Mask (W. Galway) and later at Glengarriff (W. Cork). It is almost certainly an introduction from eastern N. America, although no one knows how it arrived.

Native of eastern N. America from Newfoundland to N. Florida and Alabama; in Europe, known only from the Netherlands (first found in 1909) and Ireland.

References: Atlas (59a), Curtis & McGough (1988), Webb (1957, 1958), Webb & Halliday (1973).

N. K. B. ROBSON

Tilia platyphyllos Large-leaved Lime

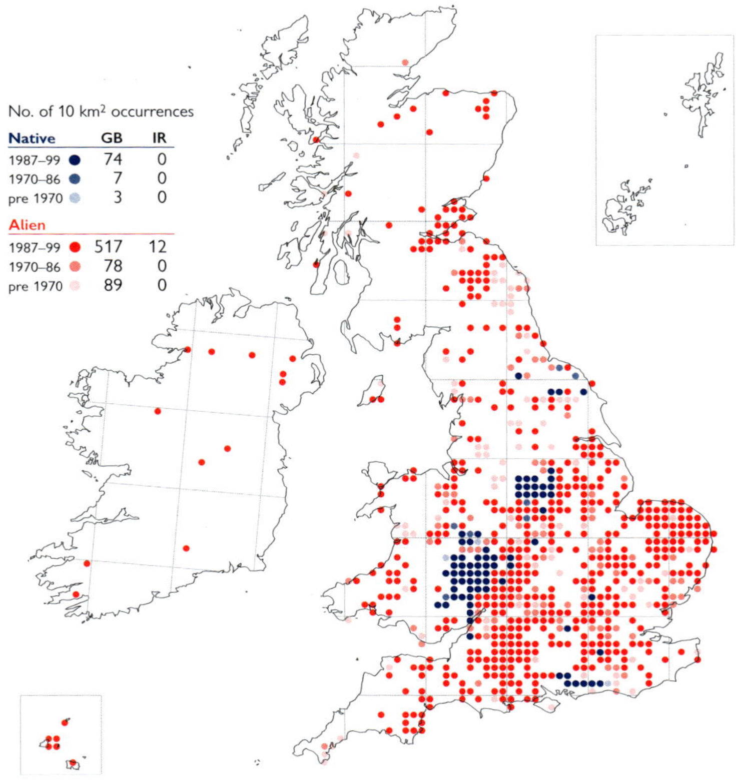

This species occurs as a native in old, mixed deciduous woodland on calcareous or, rarely, acidic soils, typically as a large tree or coppice stool. It also grows on cliff ledges, and as a planted tree on roadsides, in gardens, parkland and plantations. Seedlings are frequent, but saplings rare. Vegetative reproduction is by new shoots from the tree base. 0–400 m (Craig y Cilau, Brecs.).

Native (change +2.57). *T. platyphyllos* has been in cultivation since at least the 16th century, whence it has spread to semi-natural habitats, so that its native status can be doubtful in some areas. Its native distribution is stable. Planted trees were not mapped in the 1962 *Atlas*.

European Temperate element.

References: Atlas (87b), Abraham & Rose (2000), Hultén & Fries (1986), Meusel *et al.* (1978), Pigott (1981), Stewart *et al.* (1994). G. T. D. WILMORE

Tilia cordata Small-leaved Lime

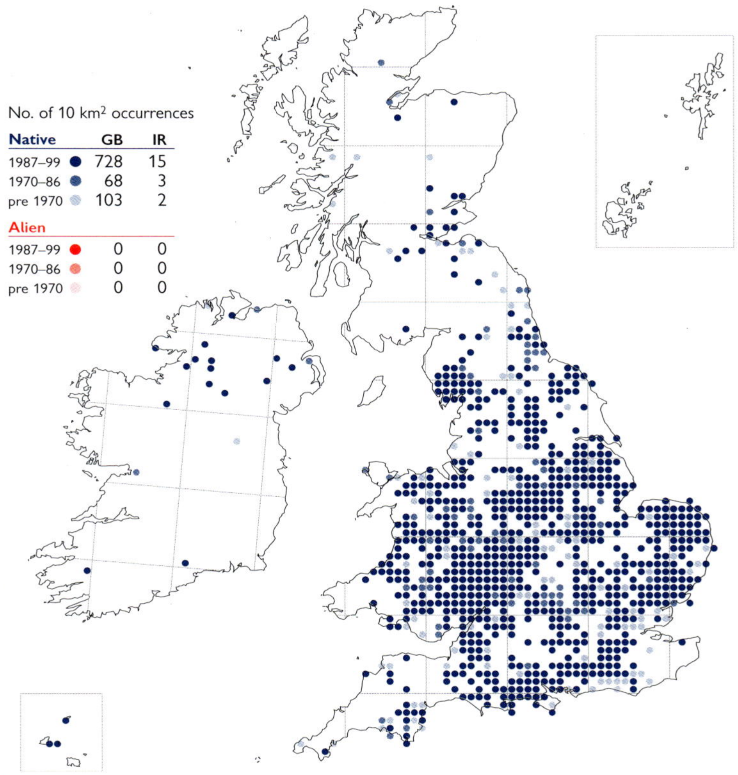

This tree occurs mostly in mixed deciduous *Quercus* or *Fraxinus* woodland on a wide range of soil types, and frequently on steep slopes and cliffs. Regeneration by seed occurs, mainly in S. England, but is rare. Vegetative reproduction is by shoots arising from fallen trees, or by layering, and it often occurs as an ancient coppiced tree. It has also been planted in parks and as a street tree. Generally lowland, but reaching *c.* 600 m in Cumbria.

Native (change +1.64). *T. cordata* is native in England and Wales but it is often planted inside and outside this range. All records are mapped as if they are native as further research is needed to establish the native distribution.

Eurosiberian Temperate element.

References: Atlas (87c), Hultén & Fries (1986), Meusel *et al.* (1978), Pigott (1991).

G. T. D. WILMORE

Tilia cordata × *T. platyphyllos* (*T.* × *europaea*) Lime

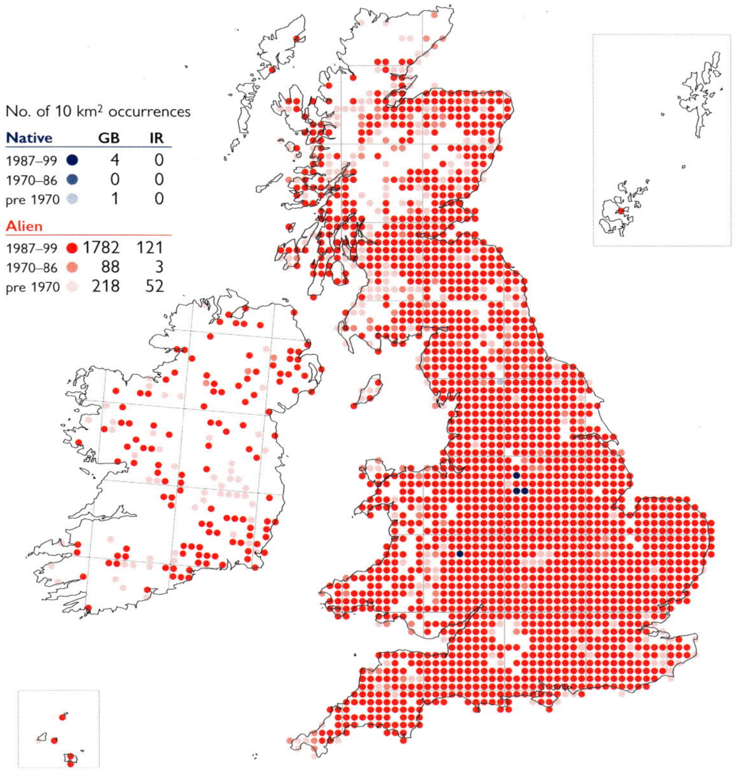

This tree is native in a few woods with both parents, but it is mostly a planted tree. It is easily propagated from suckers, and is therefore common in woods, scrub, shelter-belts, avenues, copses, parkland, roadsides, and as an urban street tree. Generally lowland, but reaching 415 m near Nenthead (Cumberland).

Native (change +0.33). Two clones of this hybrid were widely planted in the late 17th and early 18th centuries, and it remains a popular planted tree. It was mapped as 'all records' in the 1962 *Atlas*, and the current map indicates a stable distribution since then.

Occasional with the parents in Europe, but widely planted outside its native range.

References: Atlas (87d), Stace (1975), Pigott (1969, 1992).

G. T. D. WILMORE

Malva moschata Musk-mallow

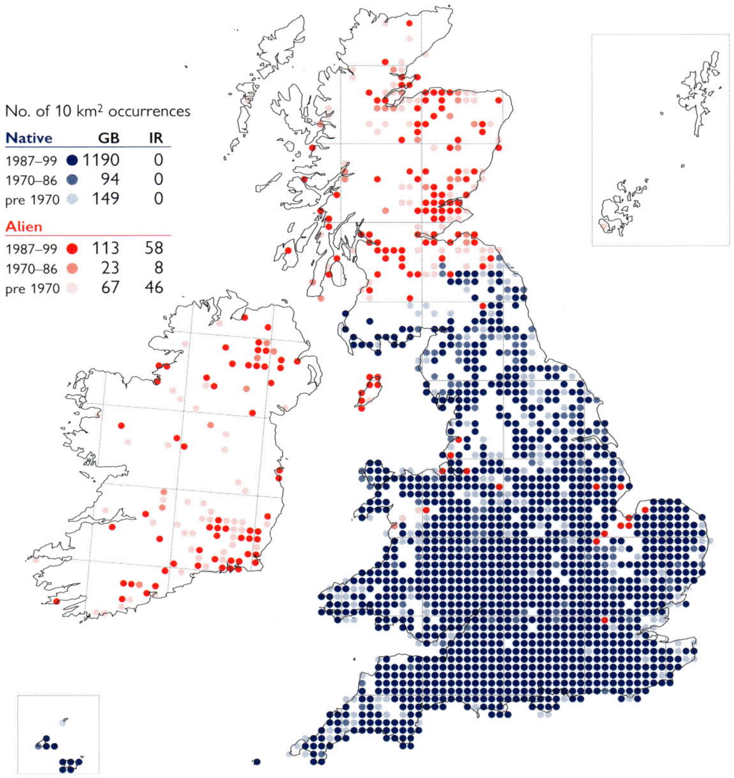

A perennial herb of roadsides, hedge banks, woodland edges, pastures, field-borders, river banks and grassy waste places, preferring well-drained soils in unshaded or lightly shaded situations. It is tolerant of moderate levels of grazing or mowing, and seed is persistent in the soil. Predominantly lowland, but reaching 305 m at Scarth Nick (N.W. Yorks.), with an exceptional record at 845 m (Great Dun Fell, Westmorland).

Native (change –0.04). *M. moschata* is native in S. England, but is probably introduced in N. Britain and Ireland. The distribution is stable, although it may be increasingly introduced with wild-flower seed mixtures and as garden escapes.

European Temperate element; widely naturalised outside its native range.

References: Atlas (88a), Hultén & Fries (1986), Meusel *et al.* (1978).

J. H. S. COX

Malva sylvestris Common Mallow

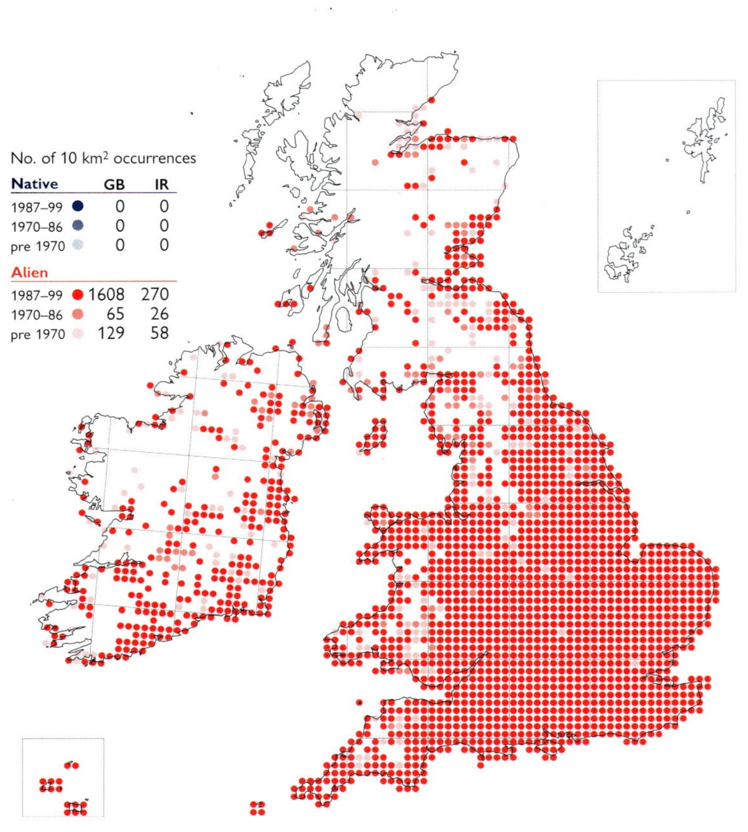

A drought-tolerant perennial herb of well-drained, often nutrient-enriched soils in unshaded situations, found on roadsides, railway banks, waste ground and field-borders, often near settlements, around farms and in the shelter of walls, occasionally on sea-cliffs. It reproduces freely from seed. Lowland.

Archaeophyte (change –0.30). There has been little change in the distribution of this species since the 1962 *Atlas*.

As an archaeophyte *M. sylvestris* has a Eurosiberian Southern-temperate distribution, but it is widely naturalised so that its distribution is now Circumpolar Southern-temperate.

References: Atlas (88b), Hultén & Fries (1986), Meusel *et al.* (1978).

J. H. S. COX

Malva parviflora Least Mallow

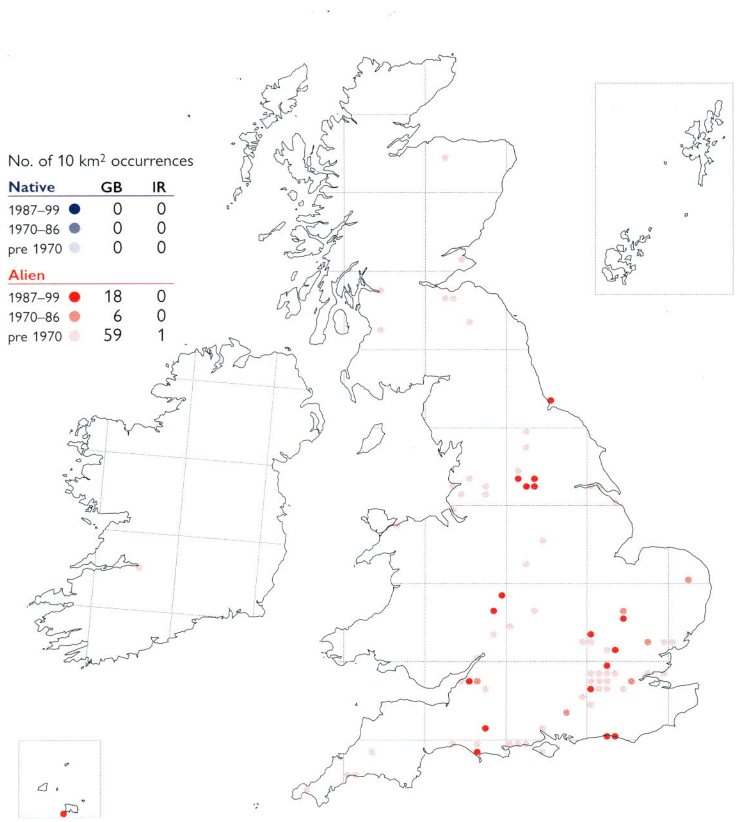

An annual that is casual or sometimes persistent in waste places, including rubbish tips and nursery gardens, and a rare colonist of arable land. Lowland.

Neophyte. *M. parviflora* was first cultivated in Britain in 1779, but is introduced in the wild with wool, grain, bird-seed and esparto. It was first recorded as a casual in 1859. It appears to have declined in recent years.

Native of S.W. Europe and the Mediterranean region, east to C. Asia.

Reference: Bolòs & Vigo (1990).

J. H. S. COX

Malva pusilla Small Mallow

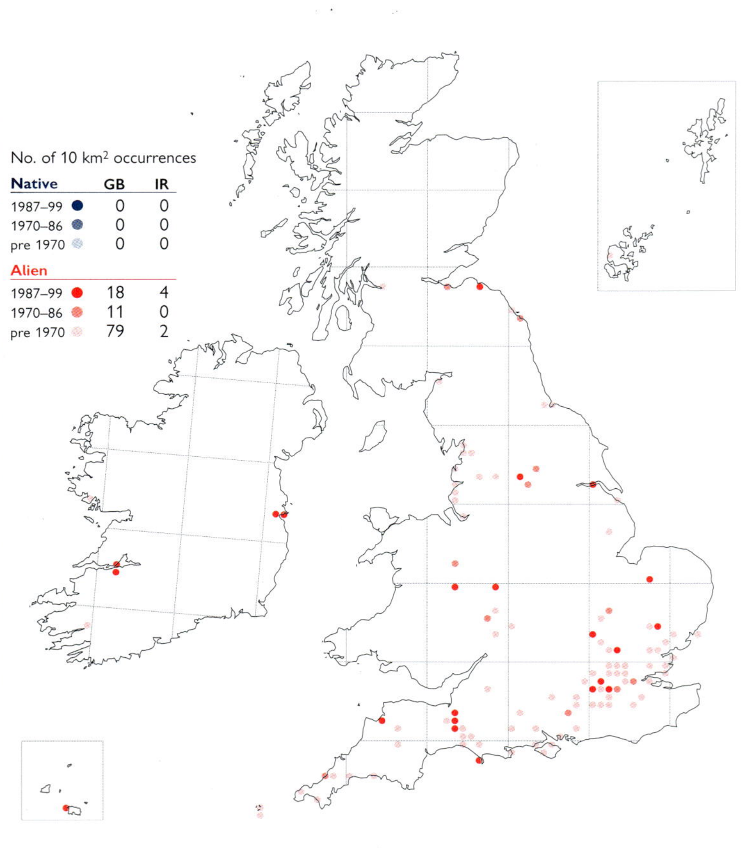

An annual of roadsides, farmyards, waste ground, gravel-pits, rubbish tips and foreshores. Lowland.

Neophyte. This species, first recorded in the wild in 1700, is usually a casual introduction with wool, grain and bird-seed, but populations sometimes persist. It appears to have declined recently and is now of only occasional occurrence.

A Eurosiberian Temperate species, extending west as a native to Scandinavia, Belgium and Italy.

References: Hultén & Fries (1986), Meusel *et al.* (1978).

J. H. S. COX

Malva neglecta Dwarf Mallow

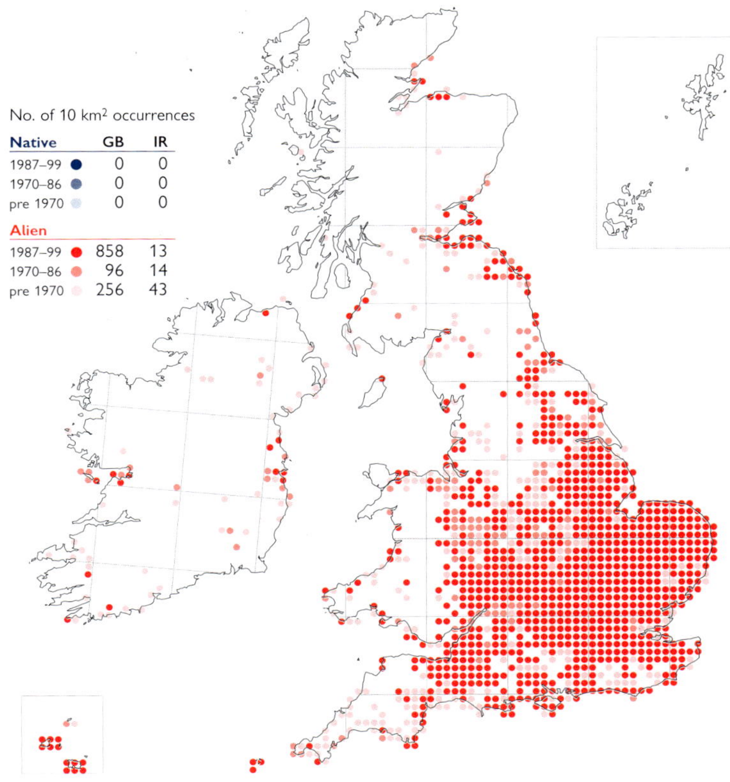

An annual which sometimes overwinters, occurring in waste places, gateways, paths, rough ground and on roadsides (often near habitation), occasionally on coastal drift-lines. It favours shallow, dry soils, and is tolerant of grazing and mowing, but not of competition with more vigorous species. Lowland.

Archaeophyte (change –0.22). *M. neglecta* was present in Britain in Roman times. It is often casual in N. England, Wales, Scotland and Ireland. It was mapped as 'all records' in the 1962 *Atlas*. Analysis of the database reveals that most of the losses have occurred since 1950, and it is probably declining in the west and north of its range.

As an archaeophyte *M. neglecta* has a European Temperate distribution; it is widely naturalised outside this range.

References: Atlas (88c), Hultén & Fries (1986).

J. H. S. COX

Lavatera arborea Tree-mallow

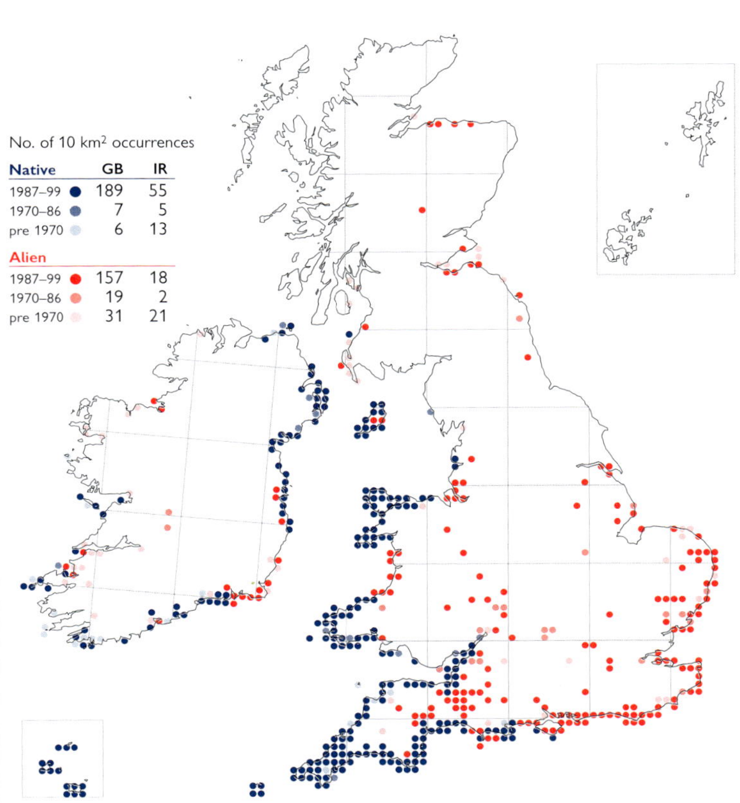

A monocarpic herb, rarely native more than 100 m from the coast. It usually grows in shallow, nutrient-enriched soils, occurring most frequently amongst vegetation in sea-bird roosts, and on ground enriched by garden waste. Plants are killed by severe frost and the species is therefore restricted to mild micro-climates near the sea. Lowland.

Native (change +1.20). It is difficult to define the native range of *L. arborea*, as it is taken from the wild into gardens in some areas and escapes from gardens in others. Although there are local losses, these are more than offset by the spread of plants from coastal gardens into nearby ruderal habitats.

Mediterranean-Atlantic element.

References: Atlas (88d), Bolòs & Vigo (1990), Stewart *et al.* (1994), Malloch & Okusanya (1979), Okusanja (1979a, b, c).

J. H. S. COX

Lavatera cretica Smaller Tree-mallow

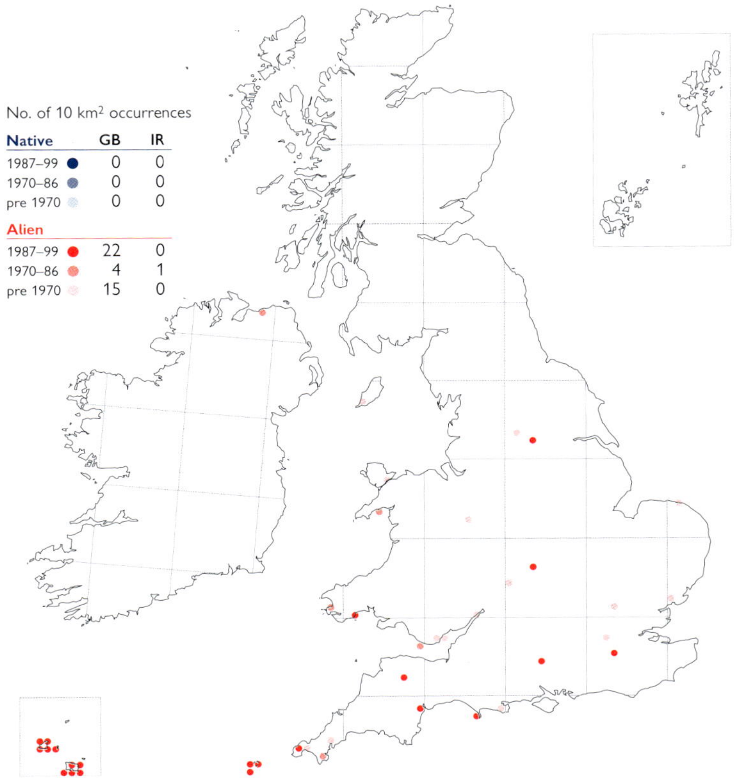

An annual or biennial herb occurring in mostly open habitats near the sea, including bulb fields, old quarries, roadsides and disturbed ground. Lowland.

Neophyte (change +0.15). *L. cretica* was introduced to cultivation in Britain in 1723. It has apparently increased in frequency in Guernsey in the 20th century. In the Isles of Scilly its range has not changed markedly since its discovery in 1873, and its population appears to be relatively stable at present. It occurs only sporadically in W. Cornwall. Some authorities regard *L. cretica* as a native in the Channel Islands and S.W. England, but it is treated here as an introduction.

L. cretica has a Mediterranean-Atlantic distribution.

References: Atlas (89a), Bolòs & Vigo (1990), McClintock (1975), Wigginton (1999).

J. H. S. COX

Lavatera thuringiaca sens. lat.
Garden Tree-mallow

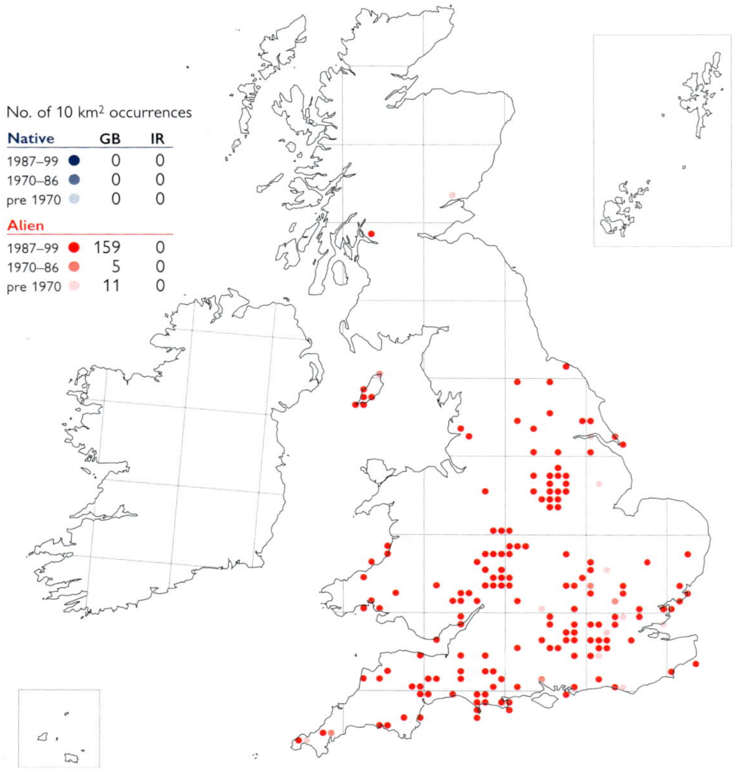

A large, often rather woody perennial herb, occurring as a garden escape or throw-out on roadsides and waste ground, and in other urban habitats. It occasionally also appears in quantity on new road verges. It can be persistent, but is often only casual; reproduction is by seed. Lowland.

Neophyte. This species was cultivated in Britain by 1731, and is now extremely popular in gardens. It was known in the wild by at least 1862. Several taxa and many cultivars are grown, the most frequent derived from *L. olbia* × *L. thuringiaca*, a fertile hybrid. Both parents are also grown, however, and the map probably includes records for all these taxa.

Native of S.E. Europe, from Italy eastwards, W. & C. Asia.

References: Grant & Miller (1998), Hultén & Fries (1986), Meusel *et al.* (1978).

J. H. S. COX

Althaea officinalis Marsh-mallow

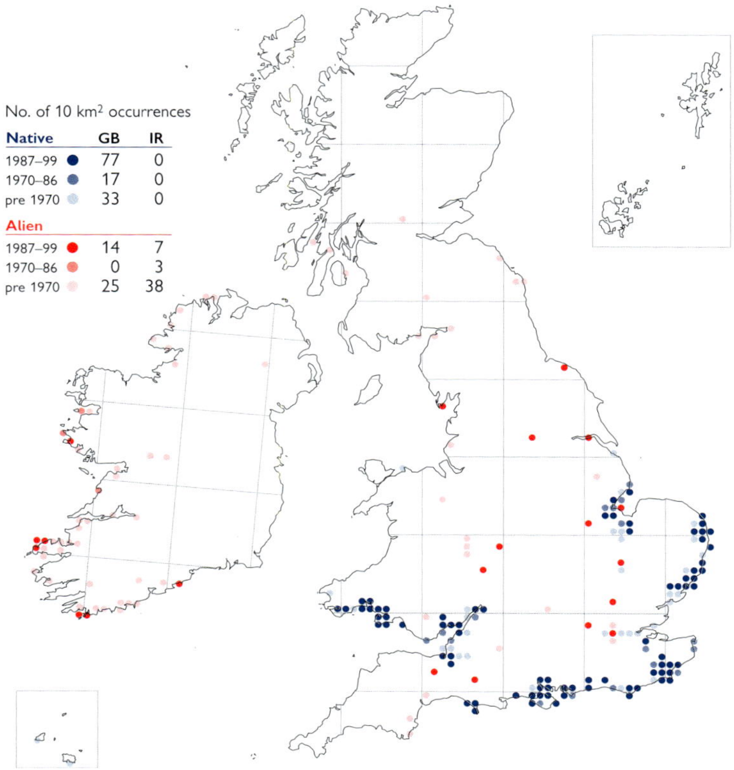

A perennial herb of coastal habitats, growing on the banks of ditches containing brackish water, in brackish pastures, and in the transition zone between the upper saltmarsh and freshwater habitats. It is intolerant of grazing and cutting. It also occurs as a garden escape. Lowland.

Native (change –0.29). *A. officinalis* has declined throughout most of its British range, due to drainage and development in the coastal zone. Much of the loss occurred before 1930, but it has continued since then, particularly in East Anglia.

Eurosiberian Temperate element; widely naturalised outside its native range.

References: Atlas (89b), Hultén & Fries (1986), Meusel *et al.* (1978), Stewart *et al.* (1994).

J. H. S. COX

Althaea hirsuta Rough Marsh-mallow

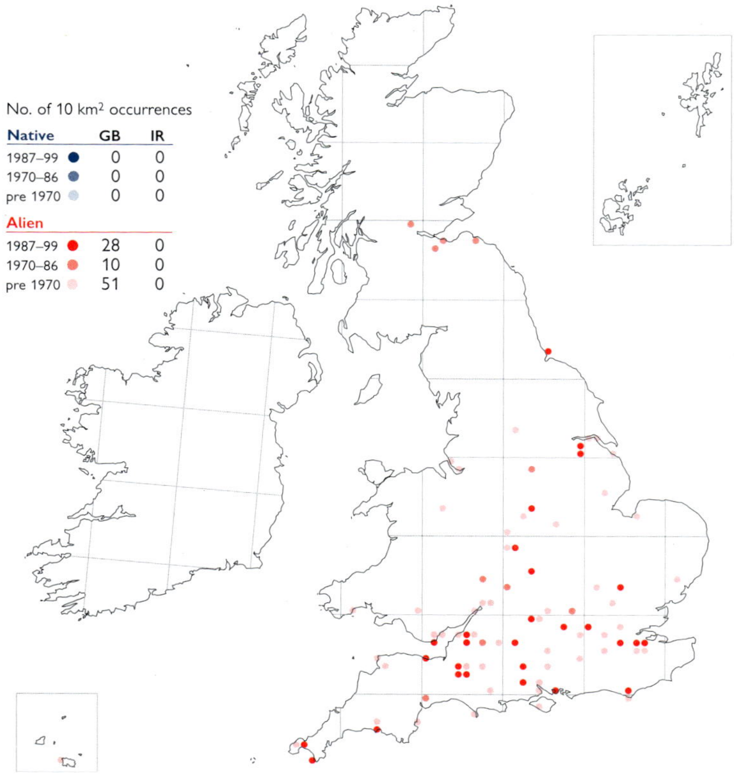

An annual, or rarely biennial, herb naturalised on open, dry calcareous soils, particularly on S.-facing slopes, and in woodland. The long-lived seed requires disturbance for germination, and populations are subject to marked annual fluctuations in numbers. It also occurs as a casual on waste ground. Lowland.

Neophyte (change +0.11). This species was cultivated in Britain by 1683. There are a few long-lived populations; it has been known from Kent since 1792 and is sometimes considered native there and in Oxfordshire and Somerset. Another persistent population in Lincolnshire probably originated from pheasant food.

A European Southern-temperate species, occurring throughout the Mediterranean region and in Europe north to Slovakia.

References: Atlas (89c), Bolòs & Vigo (1990), FitzGerald (1998), Wigginton (1999).

J. H. S. COX

Alcea rosea Hollyhock

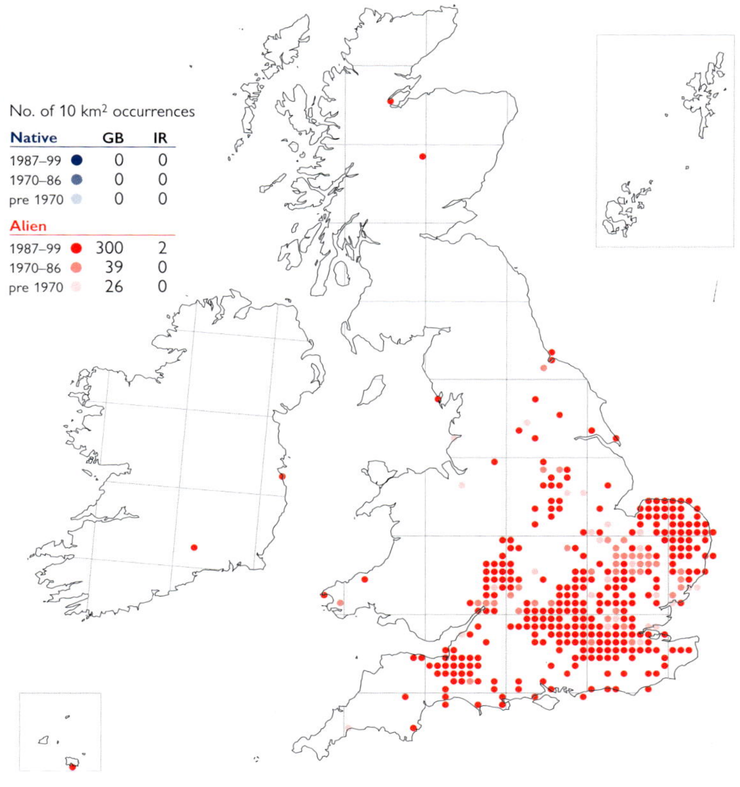

A biennial or perennial herb occurring as a garden escape or throw-out on waste ground, roadsides, railway banks and refuse tips. Lowland.

Neophyte. *A. rosea* is a popular garden plant, known in Britain by 1573 but not recorded from the wild until 1906. Occurrences outside gardens are usually short-lived, but the species sometimes persists at a site for several seasons. It is not known whether there has been any change in its frequency or distribution in the last fifty years.

Unknown in the wild; thought to have originated by hybridisation in cultivation.

J. H. S. COX

Abutilon theophrasti Velvetleaf

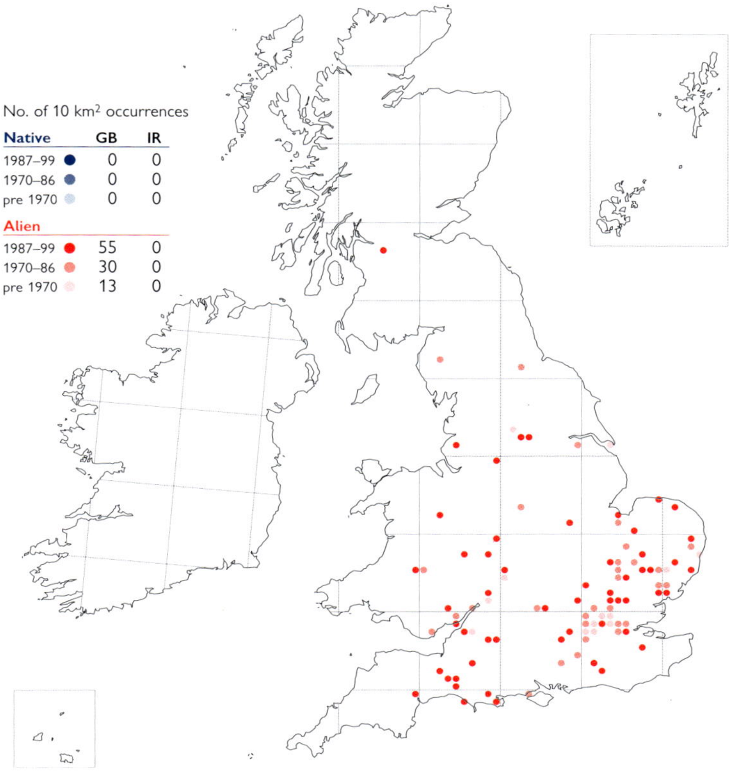

An annual found as a casual in waste places, fields and on rubbish tips. Lowland.

Casual. This species has been known in cultivation in Britain since at least 1596, and was recorded in the wild by 1887. It is introduced with wool shoddy, bird-seed and oil-seed. Trends in its distribution are unknown.

Native of the E. Mediterranean region and S.W. Asia; naturalised in most of the warmer parts of the world, where it is sometimes cultivated for the fibre China Jute.

Reference: Warwick & Black (1988).

J. H. S. COX

Hibiscus trionum Bladder Ketmia

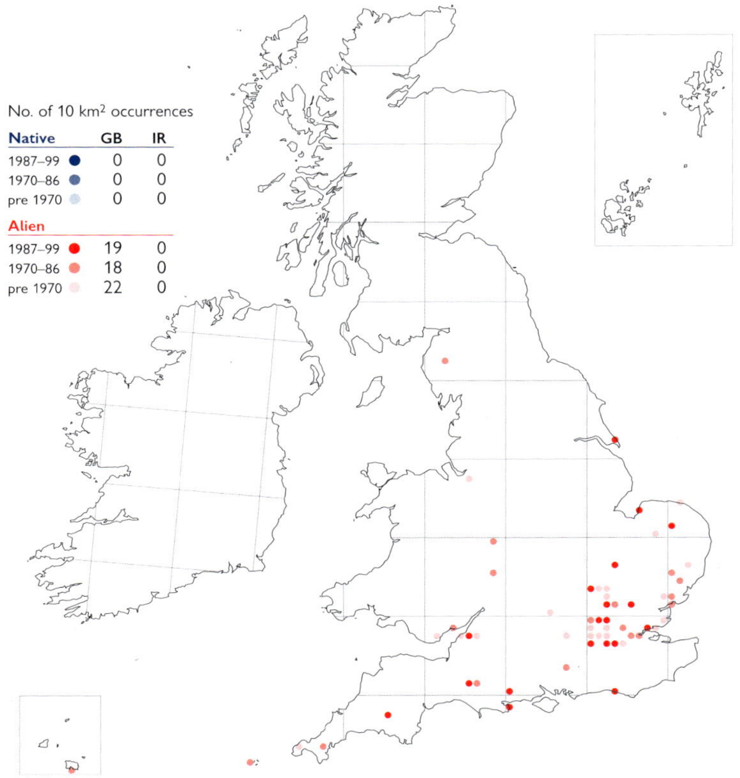

An annual of waste places, pavement cracks and rubbish tips. It is usually casual, being introduced with wool shoddy, bird-seed and oil-seed, but has persisted in Jersey since 1977. It also occurs as a garden escape. Lowland.

Casual. This species was cultivated in Britain before 1596, and is frequently grown in gardens. It was first recorded from the wild in 1859 (Surrey). Its overall distribution is probably stable.

Native of S.E. Europe, N. Africa and S.W. Asia; widely naturalised as a weed outside its native range.

J. H. S. COX

Drosera rotundifolia Round-leaved Sundew

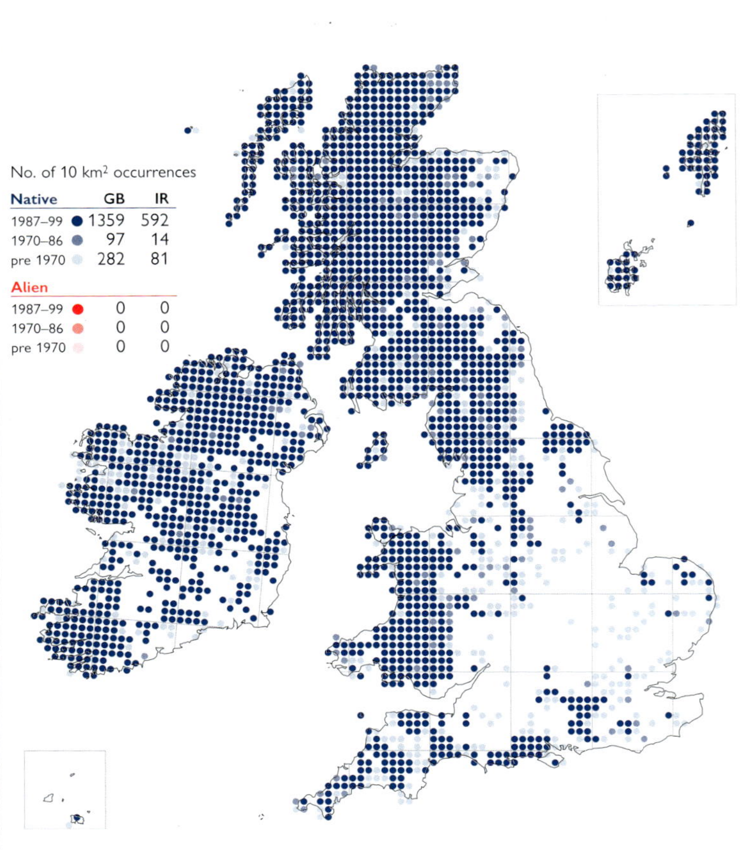

An insectivorous rosette-forming perennial herb of damp acid heath and moorland, bogs and upland flushes, growing among *Sphagnum* or on bare acid peat. It can be an abundant colonist of ditch sides cut through wet, peaty ground. 0–670 m (Snowdonia, Caerns.) and reportedly to 700 m in the Scottish Highlands.

Native (change –0.56). The overall distribution of *D. rotundifolia* has not changed since the 1962 *Atlas*. A decline in S.E. England was already apparent then and has continued in lowland areas because of habitat destruction.

Circumpolar Boreo-temperate element.

References: Atlas (141d), Crowder *et al.* (1990), Hultén & Fries (1986), Jalas & Suominen (1999), Meusel *et al.* (1965).

F. J. RUMSEY

Drosera anglica Great Sundew

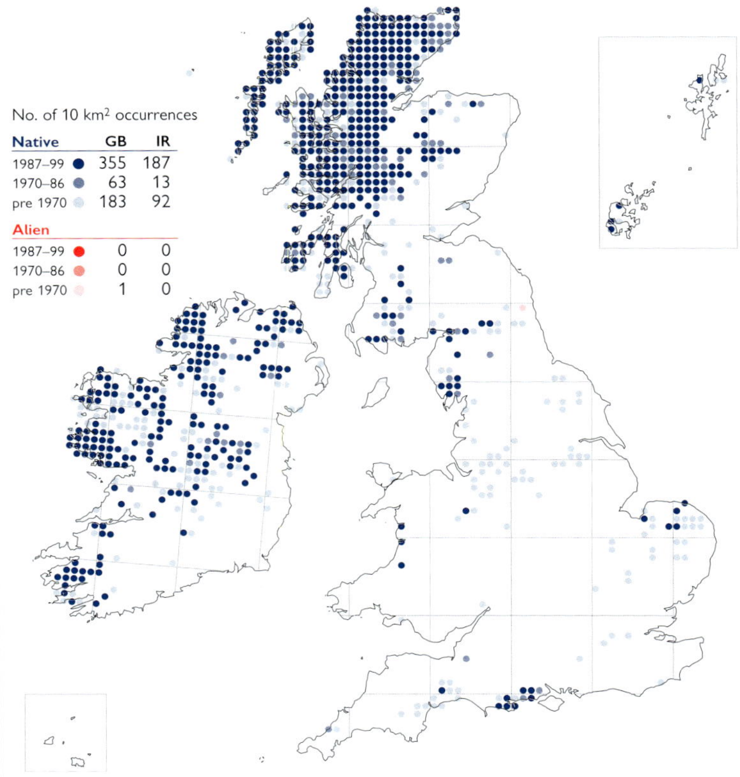

An insectivorous, rosette-forming perennial herb growing in the wetter parts of raised and blanket bogs (often in standing water), in flushed valley bogs, on stony lake shores and, more rarely, in calcareous mires. Generally lowland, but with an exceptional record of 915 m from Glas Maol (Angus).

Native (change –0.85). *D. anglica* is the most striking of our sundews. It has been declining in England and C. Ireland since the 19th century due to drainage, eutrophication and peat extraction. Such losses continue, particularly in the English part of its range.

Circumpolar Boreal-montane element.

References: Atlas (142a), Crowder *et al.* (1990), Hultén & Fries (1986), Jalas & Suominen (1999), Meusel *et al.* (1965).

F. J. RUMSEY

Drosera anglica × *D. rotundifolia* (*D.* × *obovata*)

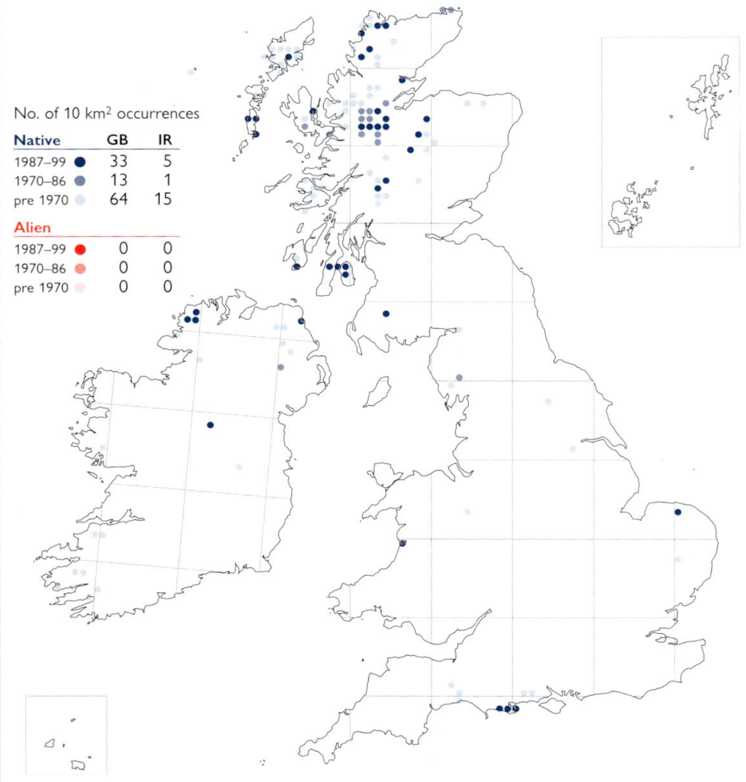

This sterile hybrid is found on damp pool margins and runnel edges in wet acid mires, usually as individual plants, rarely as several together. It occurs infrequently where its parents grow together. Generally lowland, but reaching 365 m N. of Loch Moy (Westerness).

Native. The hybrid is perhaps overlooked due to its similarity to *D. intermedia*, or to small plants of *D. anglica*.

This hybrid has been recorded with the parents in N. & C. Europe, N. Asia and N. America.

References: Atlas Supp. (40b), Crowder *et al.* (1990), Stace (1975).

F. J. RUMSEY

Drosera intermedia Oblong-leaved Sundew

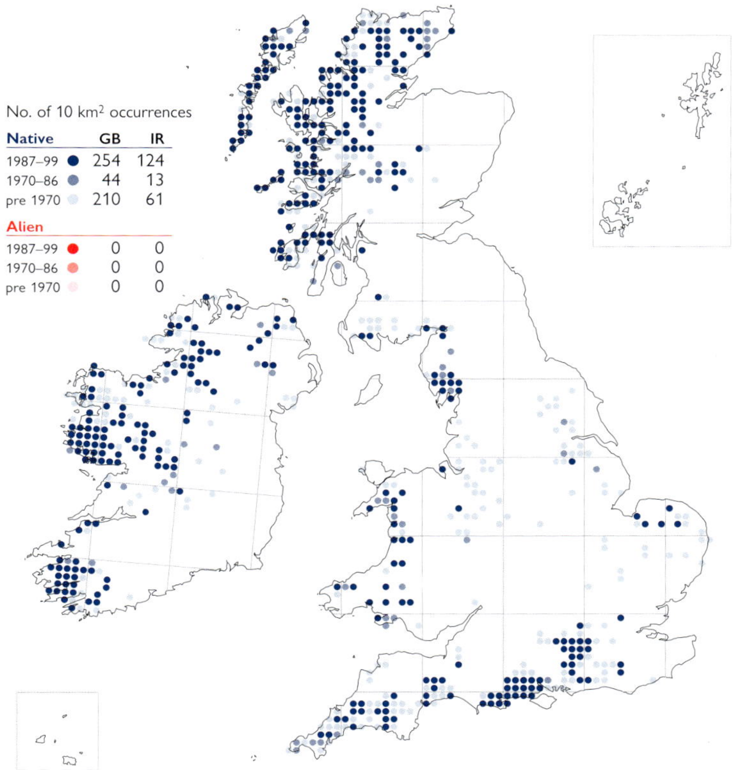

An insectivorous, rosette-forming perennial herb found on wet heaths, valley- and raised bogs, and in a band at the edge of oligotrophic lochs, most often on acidic peat over which water continuously seeps. It is rarely found in *Sphagnum*, and then only when *Sphagnum* forms a fringe around bog pools. 0–335 m (Donegal).

Native (change –0.50). *D. intermedia* was lost from some sites in lowland England and in C. Ireland before 1900. It is decreasing through drainage, afforestation, peat extraction and the loss of lowland heath.

Suboceanic Temperate element; also in N. America.

References: Atlas (142b), Crowder *et al.* (1990), Hultén & Fries (1986), Jalas & Suominen (1999), Meusel *et al.* (1965).

F. J. RUMSEY

Tuberaria guttata Spotted Rock-rose

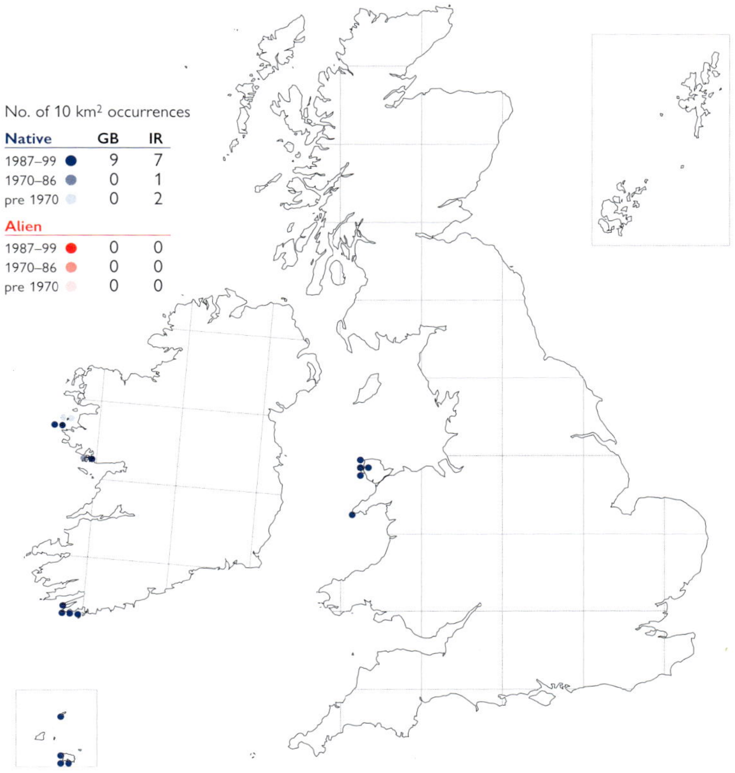

An autumn- and spring-germinating annual found in bare patches of thin, dry soil overlying hard igneous rock in open areas within wind-cut heath near the sea. In Ireland it sometimes grows in areas where burning has taken place the previous year. It may occur with a sparse growth of other small annuals, and is typically found in lichen-rich communities. Lowland.

Native. Although *T. guttata* varies greatly in numbers from year to year, its overall distribution is stable. Most populations are small but show very high levels of genetic variation and diversity.

Mediterranean-Atlantic element.

References: Atlas (60b), Curtis & McGough (1988), Kay & John (1995), Meusel *et al.* (1978), Proctor (1960), Wigginton (1999).

M. C. F. PROCTOR

Helianthemum nummularium Common Rock-rose

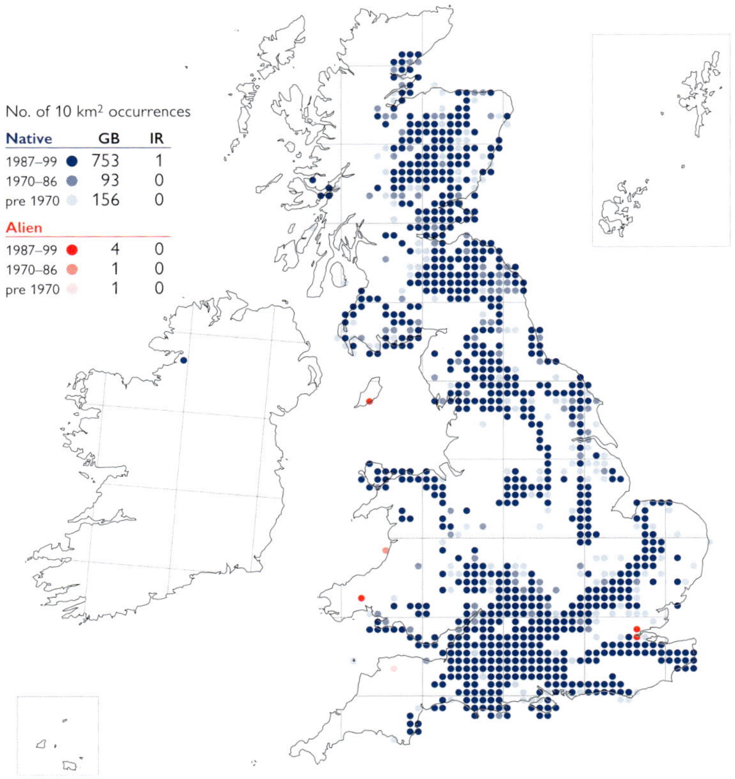

A prostrate sub-shrub occurring in short, dry, calcareous grassland, and rather strictly confined to chalk and limestone in England, but extending into mildly acid pastures and heaths on well-drained soils in E. Scotland, and on base-rich soils over basalt in N.E. England and E. Scotland. 0–640 m (Mickle Fell, N.W. Yorks., and in Atholl, E. Perth).

Native (change –0.70). This species was mapped as 'all records' in the 1962 *Atlas*. Analysis of the database reveals that most losses have occurred since 1950, caused by conversion of chalk grassland to arable and its reversion to scrub. Its rarity in Ireland has long been a phytogeographical conundrum.

European Temperate element.

References: Atlas (60c), Curtis & McGough (1988), Grime *et al.* (1988), Hultén & Fries (1986), Meusel *et al.* (1978), Proctor (1956).

M. C. F. PROCTOR

Helianthemum apenninum White Rock-rose

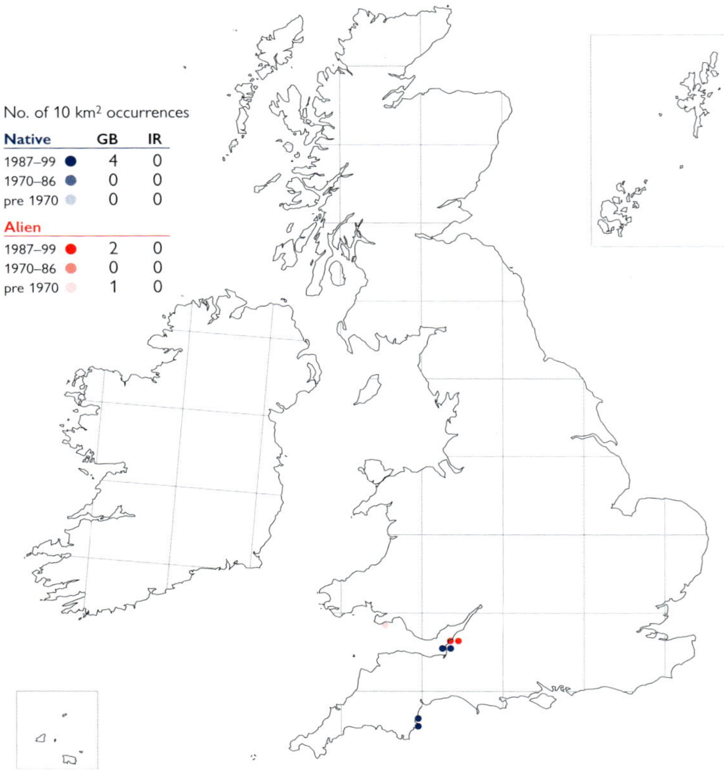

No. of 10 km² occurrences

		GB	IR
Native	1987–99	4	0
	1970–86	0	0
	pre 1970	0	0
Alien	1987–99	2	0
	1970–86	0	0
	pre 1970	1	0

A perennial sub-shrub, found in open, dry rocky limestone grassland (of which it is a conspicuous component where it occurs), generally on S-facing slopes; also limestone crags and cliff edges. It is found on more skeletal soils than *H. nummularium*, and hybrid plants (*H.* × *sulphureum*) may be found in intermediate habitats at sites where both species occur. Lowland.

Native (change +0.12). Although there has been no change in its 10-km distribution, *H. apenninum* is less frequent than formerly in some of its sites, mainly because of scrub colonisation following the decline of rabbits.

Mediterranean-montane element.

References: Atlas (60d), Meusel *et al.* (1978), Proctor (1956), Wigginton (1999).

M. C. F. PROCTOR

Helianthemum oelandicum Hoary Rock-rose

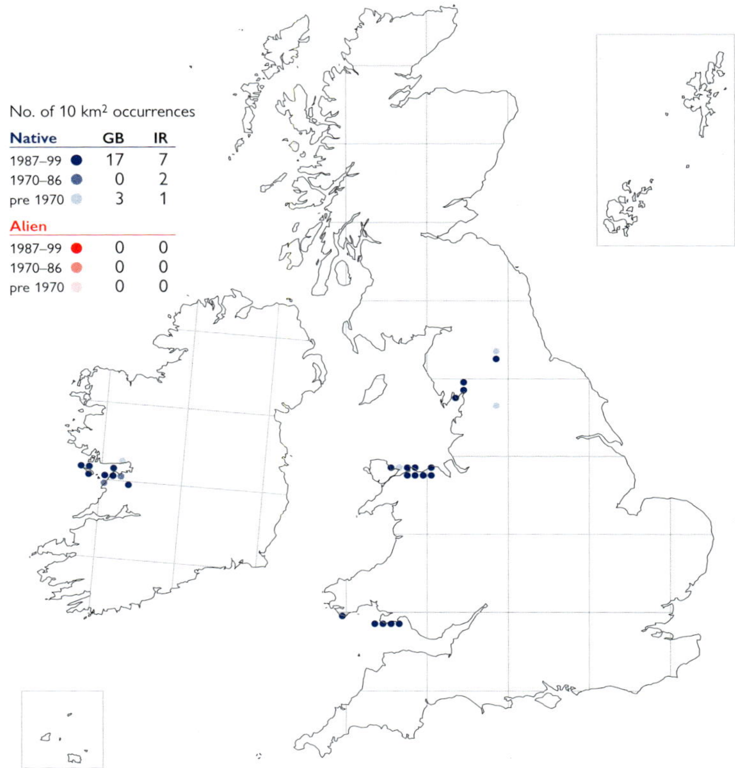

No. of 10 km² occurrences

		GB	IR
Native	1987–99	17	7
	1970–86	0	2
	pre 1970	3	1
Alien	1987–99	0	0
	1970–86	0	0
	pre 1970	0	0

This small, prostrate, perennial sub-shrub is often abundant in short, open, rocky Carboniferous limestone grassland, on rock outcrops and on cliff edges. It typically grows in exposed S.- to W.-facing sites. 0–540 m (Cronkley Fell, N.W. Yorks.).

Native (change +0.03). The distribution of *H. oelandicum* has remained largely unchanged since at least the 1950s. Subsp. *incanum* occurs in Wales and N.W. England, the endemic subsp. *levigatum* occurs on Cronkley Fell and subsp. *piloselloides* grows in W. Ireland.

Mediterranean-montane element. The species is morphologically very variable and has a highly disjunct range in Europe.

References: Atlas (60d), Atlas Supp. (11a), Curtis & McGough (1988), Griffiths & Proctor (1956), Hultén & Fries (1986), Meusel *et al.* (1978), Stewart *et al.* (1994), Wigginton (1999).

M. C. F. PROCTOR

Viola odorata Sweet Violet

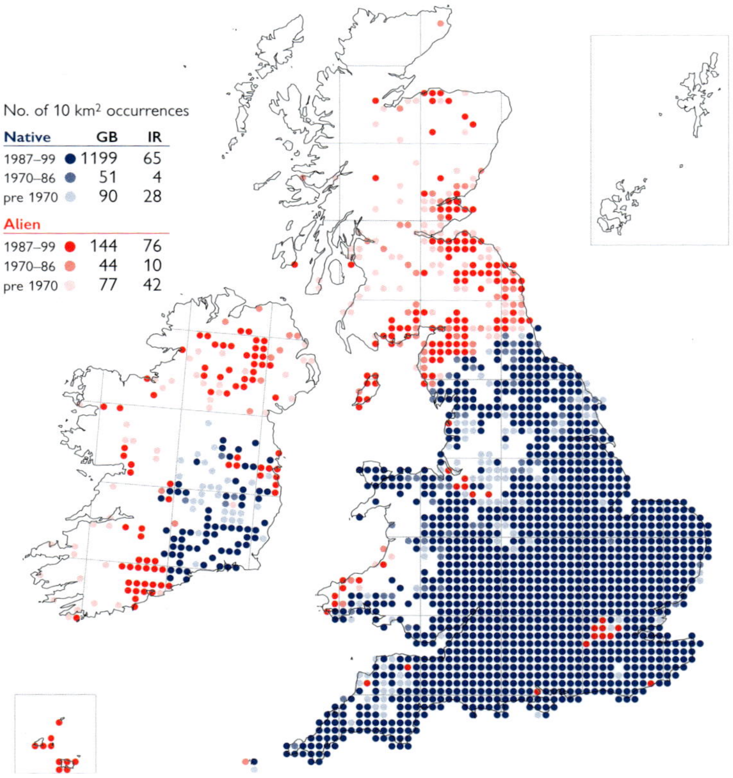

No. of 10 km² occurrences

		GB	IR
Native	1987–99	1199	65
	1970–86	51	4
	pre 1970	90	28
Alien	1987–99	144	76
	1970–86	44	10
	pre 1970	77	42

A perennial herb, usually found on calcareous or other base-rich soils. Its habitats include open woodlands, hedge banks and scrub, and less frequently shady road and railway banks and verges. Alien populations are naturalised in churchyards and elsewhere. Reproduction is by seed and by rooting stolons. Lowland.

Native (change –0.19). The distribution appears to be stable. In the north and west of its British range many plants are alien, and it may not be native north of Westmorland and Durham. In Ireland it is probably alien in the north and west; elsewhere its status is often uncertain and the map probably overstates the native sites.

European Temperate element; widely naturalised outside its native range.

References: Atlas (53a), Hultén & Fries (1986), Meusel *et al.* (1978), Walters (1946).

M. S. PORTER & M. J. Y. FOLEY

Viola hirta Hairy Violet

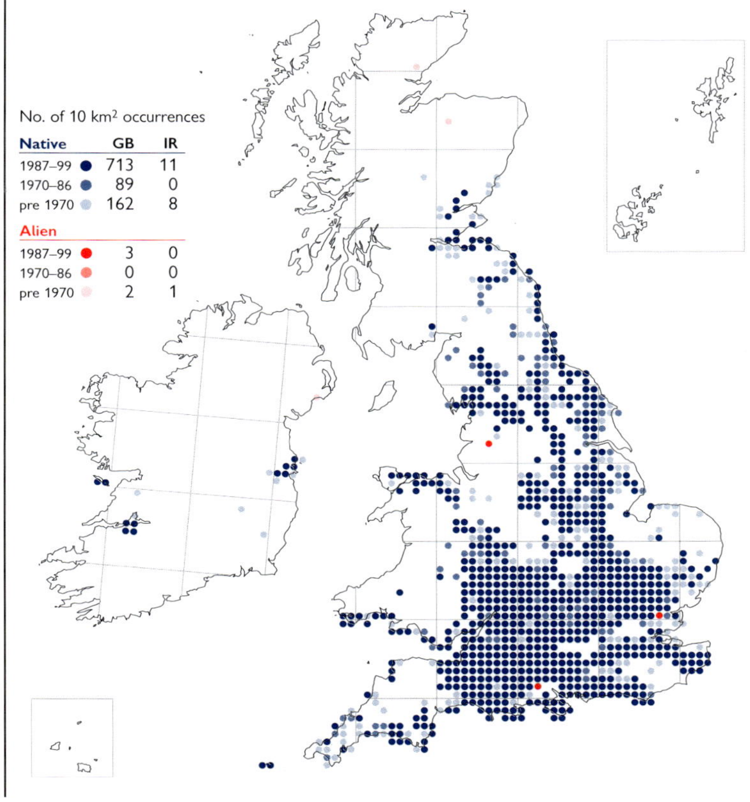

A perennial herb, occurring mainly on calcareous soils, and found in short grassland or open scrub on downland, rocky slopes, limestone pavement, woodland borders and rides, and sometimes on base-flushed but more acidic riverside substrates; also on roadsides and railway banks. Generally lowland, but reaching 610 m on Long Fell (Westmorland).

Native (change –0.46). *V. hirta* shows some decline throughout its range, largely through habitat loss to agricultural or other development.

Eurosiberian Temperate element.

References: Atlas (53b), Beattie (1972), Curtis & McGough (1988), Grime *et al.* (1988), Hultén & Fries (1986), Meusel *et al.* (1978).

M. J. Y. FOLEY & M. S. PORTER

Viola hirta × *V. odorata* (*V.* × *scabra*)

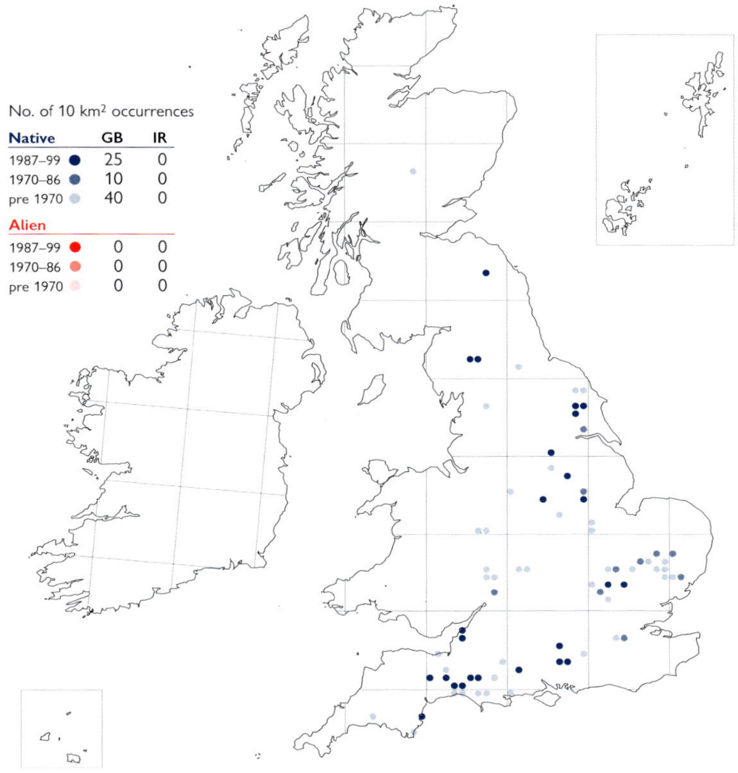

A partially fertile hybrid, usually found in association with both parents in open woodland and open scrub on the slopes of calcareous downs and escarpments, and on roadside banks and tracksides. Lowland.

Native. Trends in the distribution of this hybrid are difficult to ascertain. It is almost certainly under-recorded.

Recorded from W. & C. Europe.

Reference: Stace (1975).

M. J. Y. FOLEY & M. S. PORTER

Viola rupestris Teesdale Violet

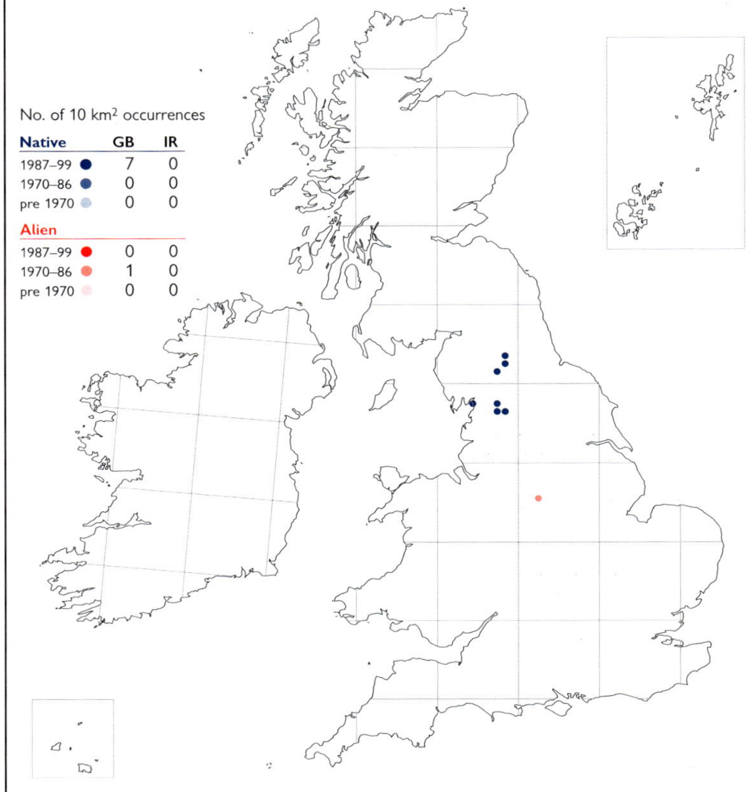

This shy-flowering perennial occurs in exposed, dry, open limestone grassland, invariably on bare or eroded slopes or hummock-tops. From 140 m on Arnside (Westmorland) to 600 m on Long Fell (Westmorland).

Native. Significant populations of *V. rupestris* were discovered in the Craven Pennines by Roberts (1977). However, colonies were lost on Widdybank Fell (Durham) through erosion and when a reservoir was completed in 1970, and have been reduced at Arnside through trampling and under-grazing. Some remaining colonies, however, are very large. Contrary to some claims, it is not being ousted by the hybrid with *V. riviniana* on Long Fell.

Eurasian Temperate element, with a continental distribution in W. Europe.

References: Atlas (53c), Hultén & Fries (1986), Jonsell *et al.* (2000), Meusel *et al.* (1978), Wigginton (1999).

M. S. PORTER & M. J. Y. FOLEY

Viola riviniana Common Dog-violet

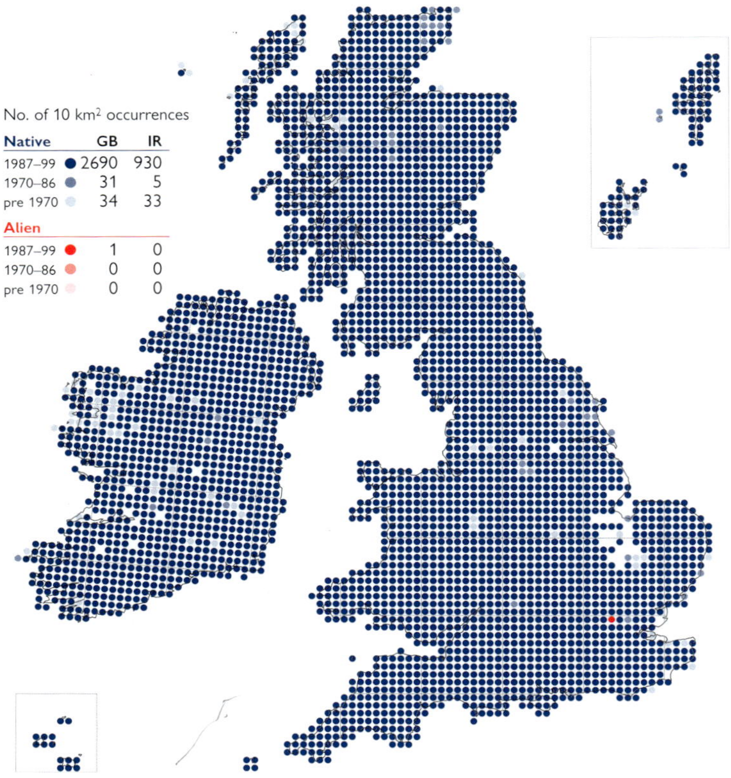

No. of 10 km² occurrences

Native	GB	IR
1987–99 ●	2690	930
1970–86 ●	31	5
pre 1970 ●	34	33
Alien		
1987–99 ●	1	0
1970–86 ●	0	0
pre 1970 ●	0	0

This perennial herb occurs in a wide range of habitats, including open deciduous woodland, hedge banks and road verges, meadows, heaths, moorland, mountain grassland, rocky slopes and cliff ledges; it can become a serious weed in gardens. It avoids wet areas but is generally indifferent to soil type, shunning only the most acidic habitats. 0–1020 m (Stuchd an Lochain, Mid Perth).

Native (change +1.07). There has been no significant change in the distribution of *V. riviniana* since the 1962 *Atlas*.

European Temperate element.

References: Atlas (53d), Beattie (1972), Grime *et al.* (1988), Hultén & Fries (1986), Meusel *et al.* (1978).

M. S. PORTER & M. J. Y. FOLEY

Viola reichenbachiana Early Dog-violet

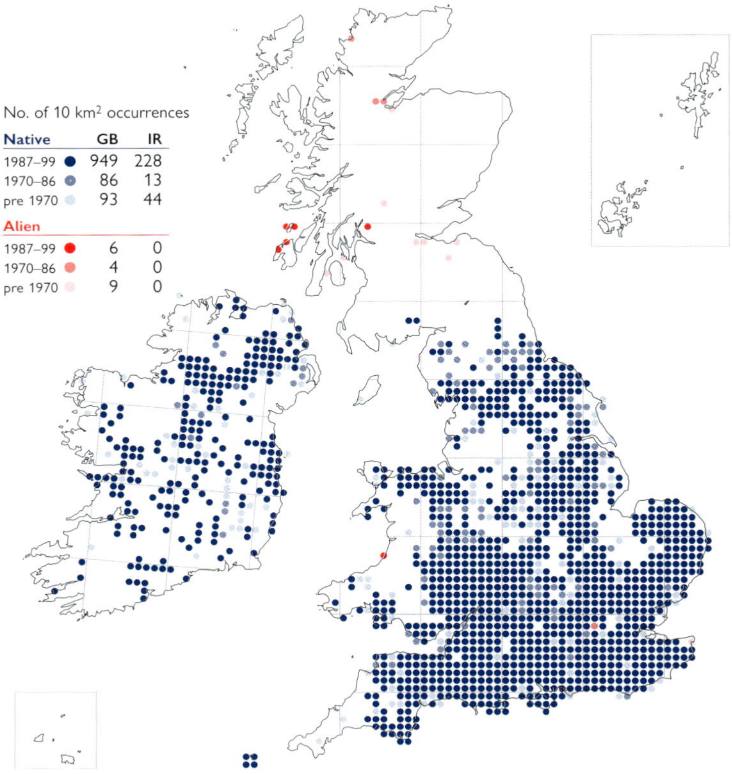

No. of 10 km² occurrences

Native	GB	IR
1987–99 ●	949	228
1970–86 ●	86	13
pre 1970 ●	93	44
Alien		
1987–99 ●	6	0
1970–86 ●	4	0
pre 1970 ●	9	0

This perennial herb is found in deciduous woodland and hedge banks, usually in moderately shaded situations but also in the open where it can sometimes persist following woodland clearance. It is most frequent on calcareous soils, being particularly common in woods over limestone and chalk or base-rich clays. Generally lowland, but formerly reaching 610 m as an alien in Mid Perthshire.

Native (change +0.20). There has been little change in the distribution of this species since the 1962 *Atlas*, although there are some losses at the north of its range.

European Temperate element.

References: Atlas (54a), Beattie (1972), Hultén & Fries (1986), Meusel *et al.* (1978).

M. S. PORTER & M. J. Y. FOLEY

Viola reichenbachiana × *V. riviniana* (*V.* × *bavarica*)

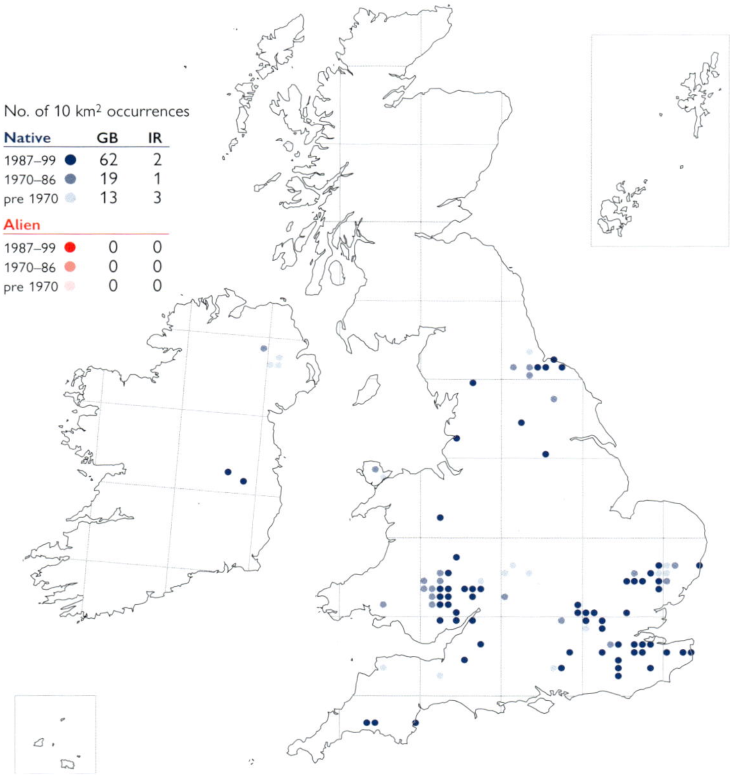

No. of 10 km² occurrences

Native	GB	IR
1987–99 ●	62	2
1970–86 ●	19	1
pre 1970 ●	13	3
Alien		
1987–99 ●	0	0
1970–86 ●	0	0
pre 1970 ●	0	0

A highly sterile hybrid recorded from woods, woodland edges and roadsides, especially those on calcareous soils. Lowland.

Native. This hybrid appears to be rarer than might be expected considering the frequently overlapping habitats of its parents. It may well be under-recorded, often being mistaken for *V. reichenbachiana*. It is impossible to assess changes in its distribution.

Widespread in temperate regions of Europe.

Reference: Stace (1975).

M. S. PORTER & M. J. Y. FOLEY

Viola canina Heath Dog-violet

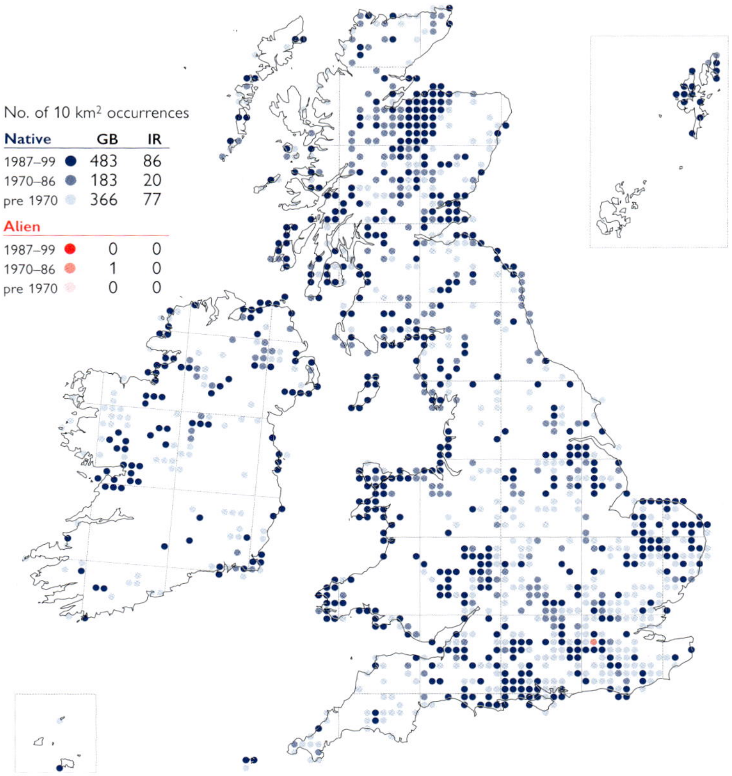

A perennial herb of a variety of acid habitats, including heaths, coastal dunes, stony riversides and lake shores, especially in Scotland. It can also occur on thin, heavily leached substrates overlying chalk and (as subsp. *montana*) in fens. 0–425 m (Isla, E. Perth).

Native (change –0.87). This species has declined severely since 1950, mainly due to habitat loss, drainage and agricultural improvement, but also over- and under-grazing and possibly hybridisation with other *Viola* species. It may also be overlooked for *V. riviniana*. Subsp. *canina* is found throughout range of the species; subsp. *montana* is mapped separately.

Eurosiberian Boreo-temperate element; also in Greenland.

References: Atlas (54b), Corner (1989), Hultén & Fries (1986), Meusel *et al.* (1978).

M. J. Y. FOLEY & M. S. PORTER

Viola canina subsp. *montana* Heath Dog-violet

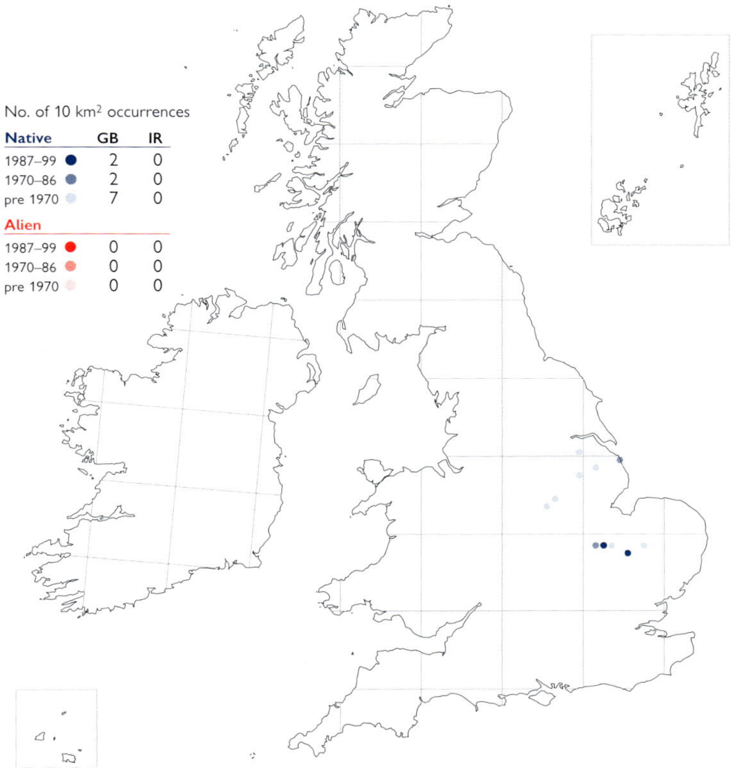

A perennial herb of open, moist, peaty ground in fens and on their margins. Reproduction is by seed, which may lie dormant for many years, germinating in response to disturbance which opens up the habitat and reduces competition. Lowland.

Native. Subsp. *montana* was exterminated before 1930 from some East Anglian sites as a result of habitat destruction. Extant sites at Wicken Fen and Woodwalton Fen (Cambs.) have statutory protection. Morphologically similar plants occur in W. Ireland, but their taxonomic status is uncertain (Webb & Scannell, 1983).

Eurosiberian Boreo-temperate element, with a continental distribution in W. Europe.

References: Hultén & Fries (1986), Wigginton (1999).

M. J. Y. FOLEY & M. S. PORTER

Viola canina × *V. riviniana* (*V.* × *intersita*)

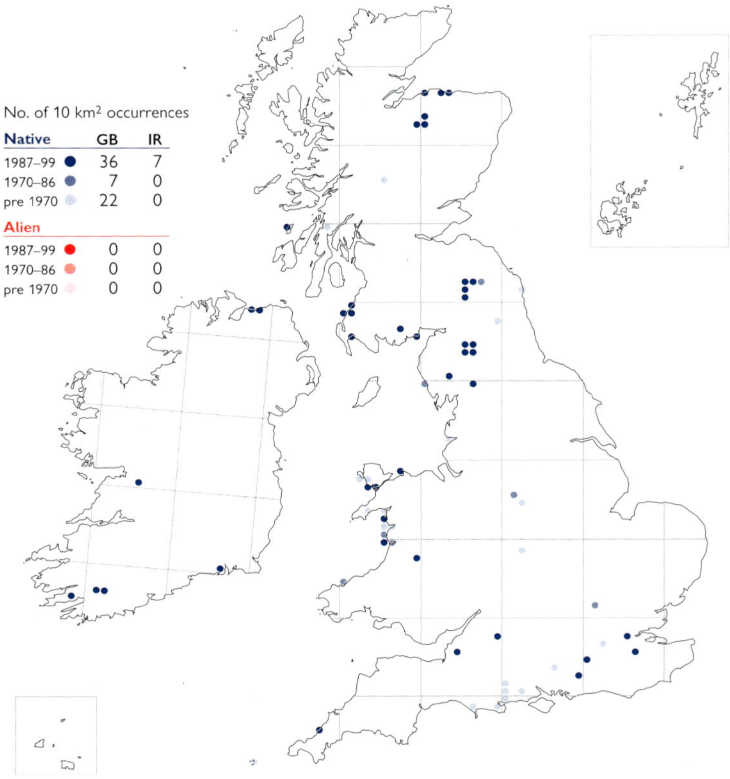

A sterile hybrid found generally in open situations on sandy, acidic soils, sometimes in the absence of one parent. Habitats include dry grassland, heaths, dunes, riverside rocks and roadsides. Lowland.

Native. It is impossible to assess changes in the distribution of this hybrid. However, some sites have been lost since 1960, particularly in C. & S. England. It is likely to be under-recorded.

Widespread in Europe.

Reference: Stace (1975).

M. S. PORTER & M. J. Y. FOLEY

Viola lactea Pale Dog-violet

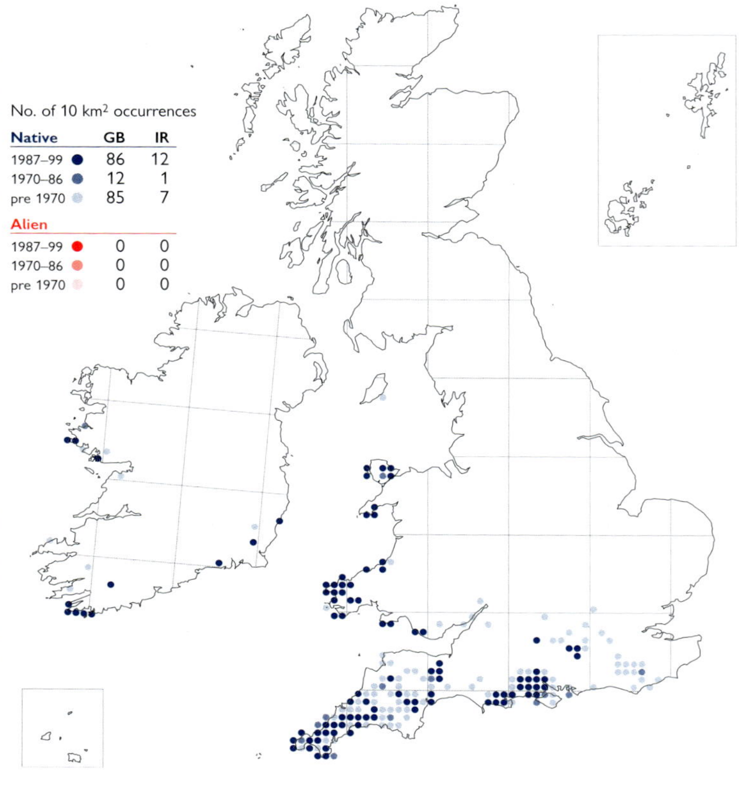

A perennial herb of dry, well-drained soils in heathland habitats. It is a poor competitor, and therefore generally occurs in open habitats, including patchy grassland, tracksides, areas kept open by grazing or rotational burning and other disturbed ground. Lowland.

Native (change –1.08). *V. lactea* suffered a severe decline before 1930, especially in S.E. and S.W. England. There have been further losses since the 1962 *Atlas*, particularly in Dorset and Devon, but this loss has been offset by new discoveries in Wales. Habitat loss and the lack of appropriate management account for its decline.

Oceanic Temperate element.

References: Atlas (53c), Byfield & Pearman (1996), Curtis & McGough (1988), Moore (1958), Stewart *et al.* (1994).

M. J. Y. FOLEY & M. S. PORTER

Viola lactea × *V. riviniana*

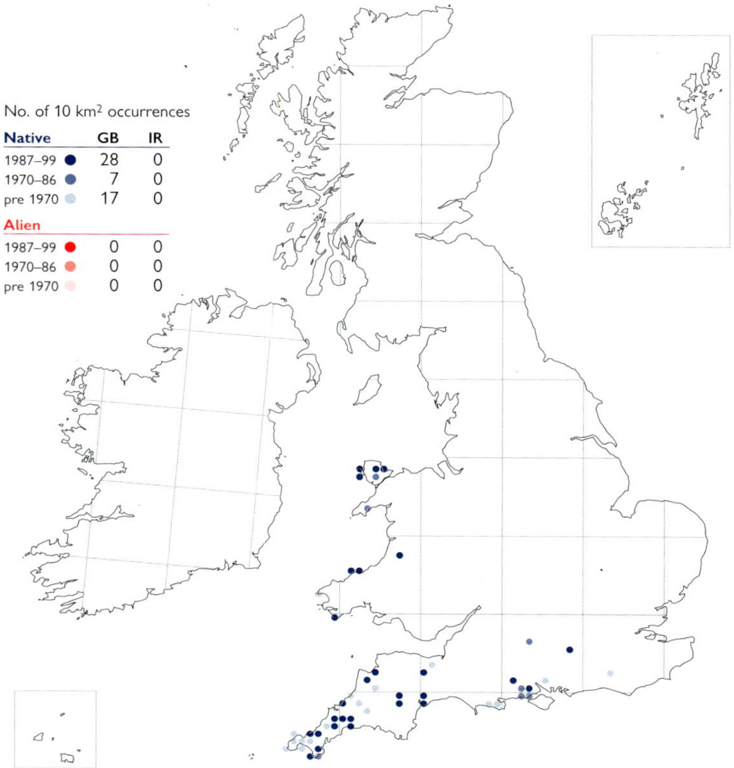

This sterile hybrid occurs with the parents in heathland habitats, often forming large, clonal, free-flowering patches. It is vigorous and often occurs in less open habitats than those preferred by *V. lactea*. Lowland.

Native. Since the hybrid has not been previously mapped at the national level, changes in its distribution cannot be assessed.

This hybrid is also known from the Iberian peninsula and France.

Reference: Stace (1975).

M. J. Y. FOLEY & M. S. PORTER

Viola persicifolia Fen Violet

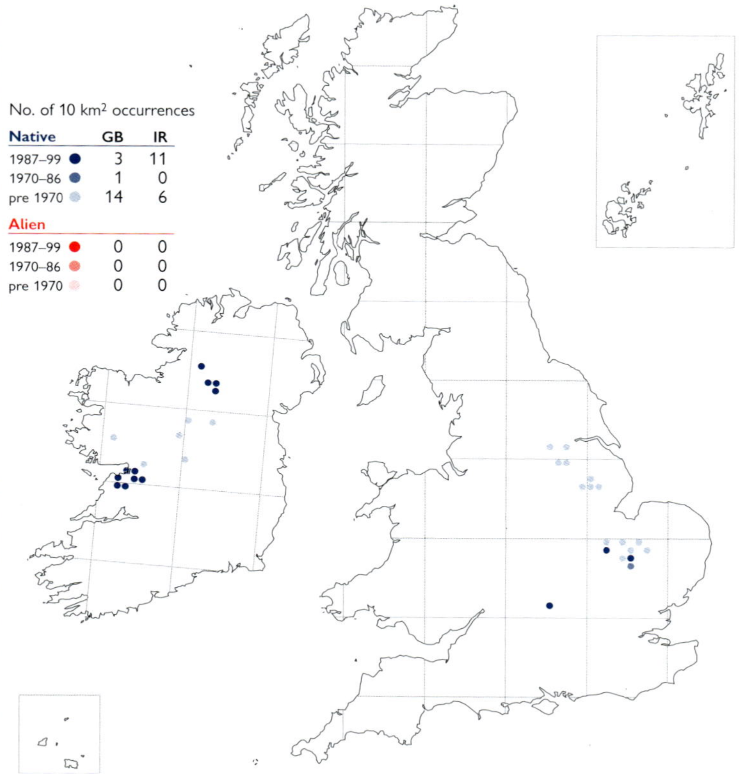

A perennial herb found in Britain on damp peaty or clayey, base-rich soils in seasonally wet fens, and in Ireland on the margins of turloughs. It is a poor competitor, preferring areas subject to fluctuating water levels, cattle trampling or peat-digging. Seed is long-lived. Lowland.

Native (change –0.62). *V. persicifolia* was lost from many sites before 1930. It survives at Wicken Fen and Woodwalton Fen (Cambs.), but its management at Wicken has proved to be difficult. It was thought to be extinct at Otmoor (Oxon), but a small population was found in 1997. Its distribution in Ireland is stable.

Eurosiberian Temperate element, with a continental distribution in W. Europe.

References: Atlas (54c), Croft (2000), Curtis & McGough (1988), Hultén & Fries (1986), Rowell (1984), Rowell *et al.* (1982), Wigginton (1999).

M. J. Y. FOLEY & M. S. PORTER

Viola palustris Marsh Violet

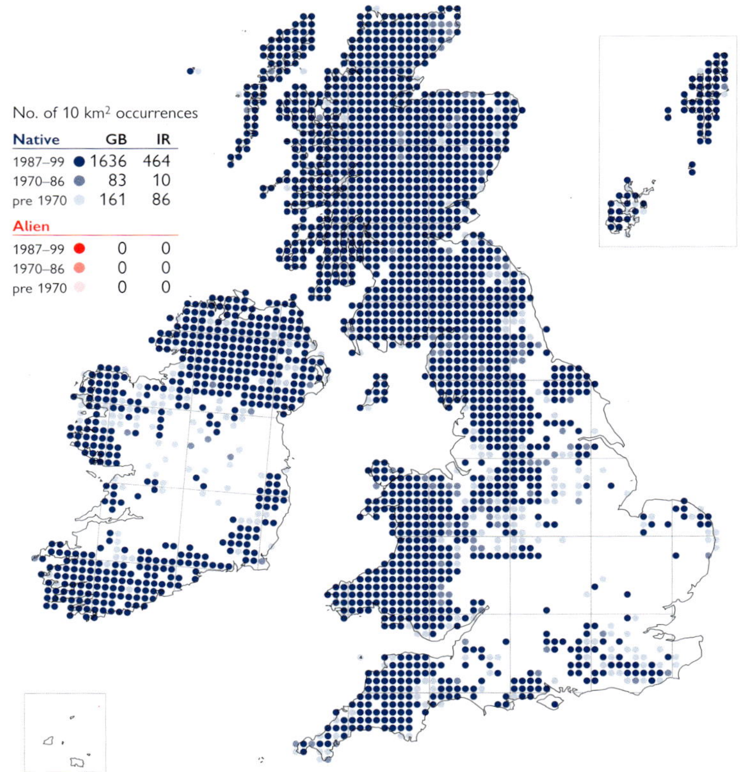

A perennial herb of bogs, wet heaths, marshes, *Alnus* and *Salix* carr and wet woods, especially on acidic soils where there is some flushing. It is frequently associated with *Sphagnum*, and is also found in non-calcareous dune-slacks. 0–1220 m (Ben Macdui, S. Aberdeen).

Native (change –0.30). *V. palustris* was mapped as 'all records' in the 1962 *Atlas*. Analysis of the database reveals that the losses in S. & E. England and C. Ireland have taken place since 1950; they are mostly due to agricultural improvement and peat extraction. Subsp. *palustris* occurs throughout the range of the species; subsp. *juressi* is mapped separately.

European Boreo-temperate element; also in N. America.

References: Atlas (54d), Hultén & Fries (1986), Meusel *et al.* (1978), Perring & Sell (1968).

M. S. PORTER & M. J. Y. FOLEY

Viola palustris subsp. *juressi*

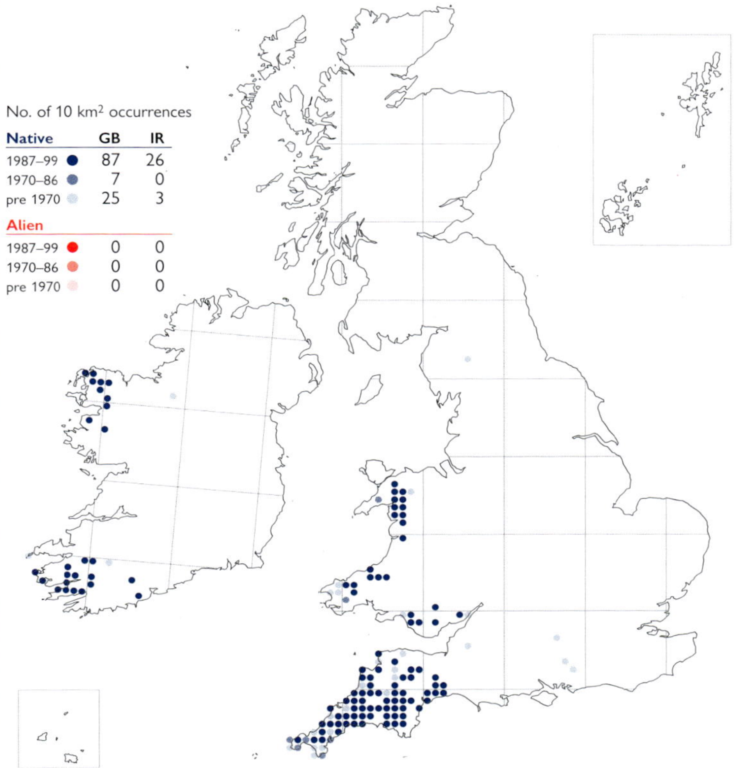

A rhizomatous perennial herb, found in identical habitats to subsp. *palustris*: bogs, marshes, wet heaths, woods and carr, particularly in acidic areas flushed by springs. Lowland.

Native. This subspecies was first mapped by Perring & Sell (1968). It is now much better recorded, especially in Wales, but its overall range has remained stable.

Subsp. *juressi* is restricted to W. Europe, from Britain and Ireland south to the Iberian peninsula.

Reference: Atlas Supp. (10a).

M. S. PORTER & M. J. Y. FOLEY

Viola cornuta Horned Pansy

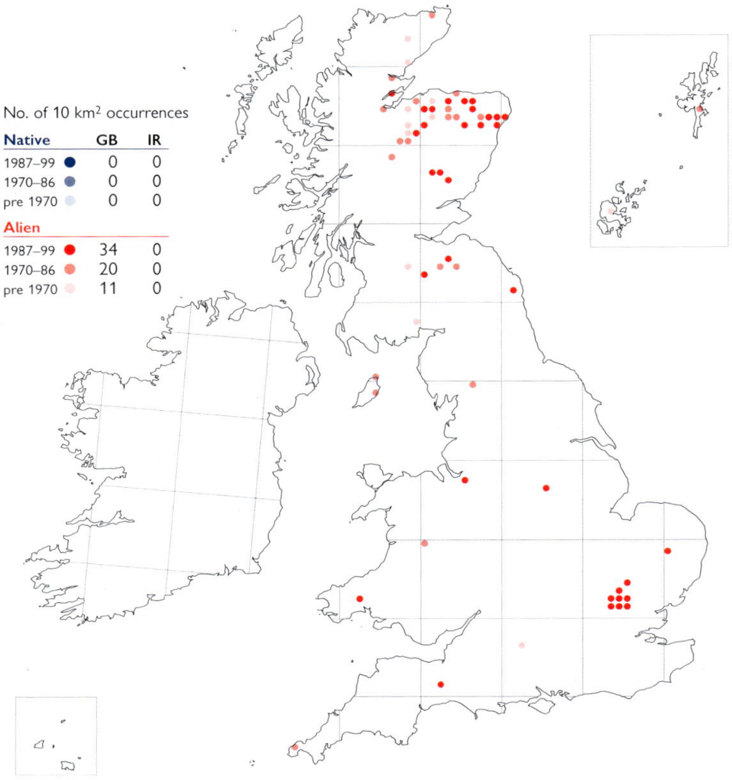

An annual or short-lived perennial herb, widely grown in gardens and sometimes escaping to become established on roadsides, wood-borders and railway banks, and in hedgerows and rough grassland. It also occurs as a casual on waste ground and rubbish tips. Lowland.

Neophyte. This species was first cultivated in Britain in 1776. It was known from the wild in 1878 (Lanarks.), but trends in its distribution are difficult to assess.

Native of montane pastures in the Pyrenees.

Reference: Bolòs & Vigo (1990).

M. S. PORTER & M. J. Y. FOLEY

Viola lutea Mountain Pansy

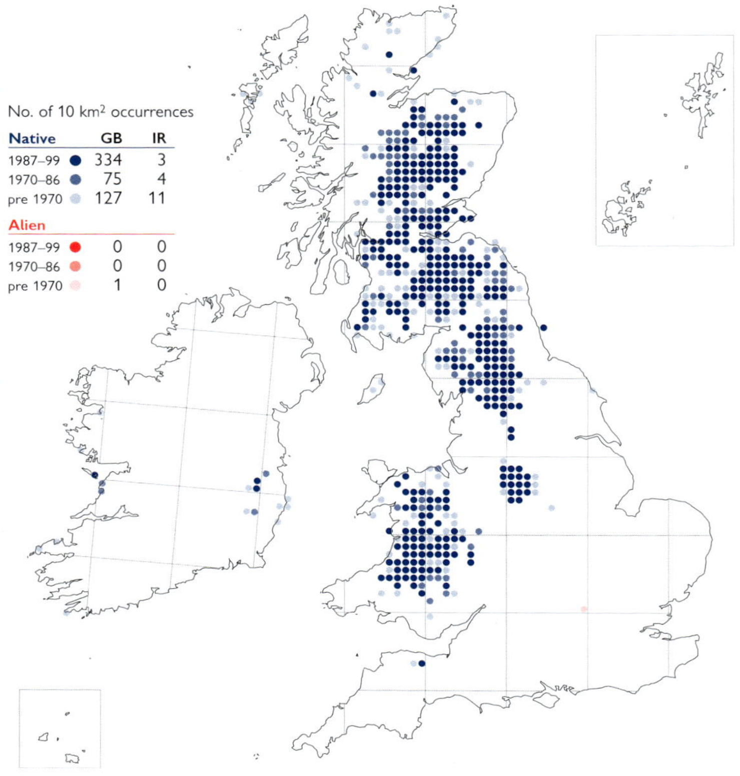

A perennial herb of grazed grassland on hill-slopes and banks, and on rock ledges. Although usually found on calcareous rocks, it is a mild calcifuge, preferring leached soil but avoiding very acidic sites; it also grows on metalliferous soils. In W. Ireland it occurs on coastal dunes. Mainly upland, reaching 1050 m (Breadalbanes, Mid Perth).

Native (change –0.69). *V. lutea* has declined since the 1962 *Atlas* due to agricultural improvement. Its confirmation from Somerset in 1990, where it was first reported in 1901, represents a significant extension of its British range. Irish dune populations have some characteristics of *V. tricolor* subsp. *curtisii.*

European Boreal-montane element, but absent from the Boreal zonobiome.

References: Atlas (55a), Balme (1954), Meusel *et al.* (1978).

M. S. PORTER & M. J. Y. FOLEY

Viola tricolor Wild Pansy

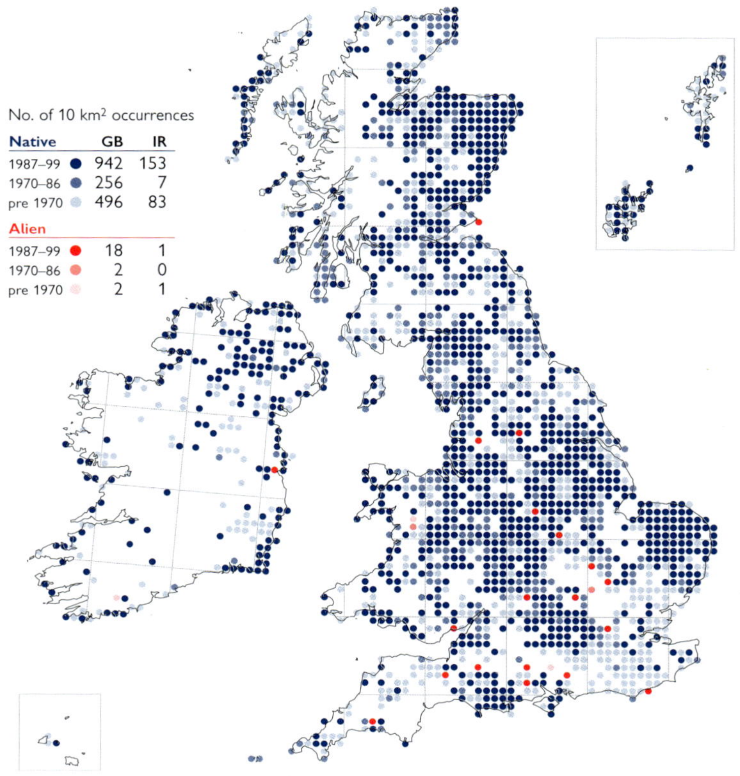

An annual or perennial herb, found on dunes and other sandy areas, on acidic grassland on heaths and hills, and in cultivated ground, gardens and waste places. 0–460 m (Clun Forest, Salop), and reportedly to 575 m in Scotland.

Native (change –1.52). *V. tricolor* was mapped as 'all records' in the 1962 *Atlas*. The current map and local floras suggest that a widespread decline has occurred since then, particularly in S.E. England. Subsp. *tricolor* occurs throughout the range of the species; subsp. *curtisii* is mapped separately.

European Temperate element; widely naturalised outside its native range.

References: Atlas (55b), Hultén & Fries (1986), Meusel *et al.* (1978).

M. S. PORTER & M. J. Y. FOLEY

Viola tricolor subsp. *curtisii*

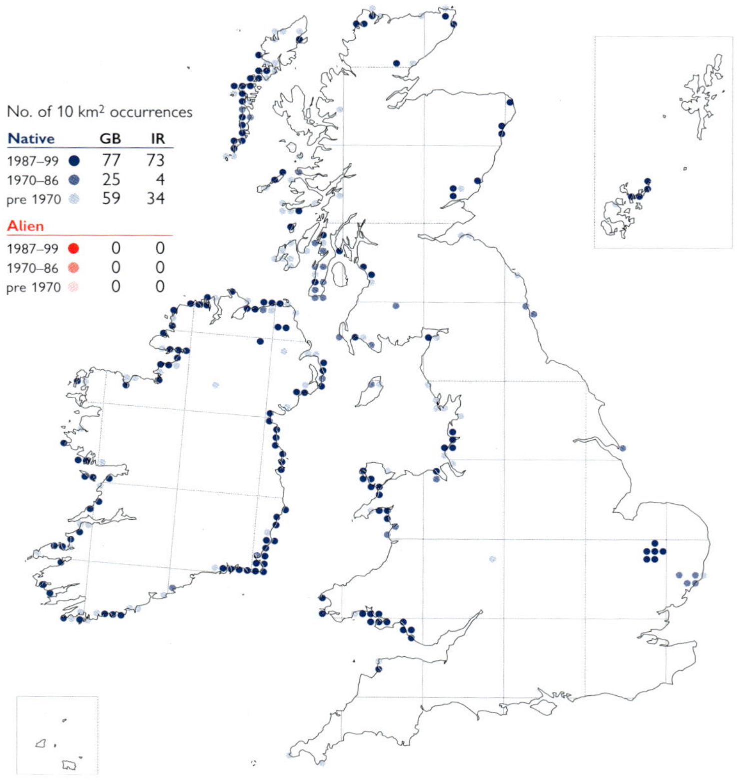

A perennial herb, found mainly in sand dunes and coastal grassland, but also occurring inland by lakes in N. Ireland and on the sandy heaths of the Breckland. Lowland.

Native. Although some losses appear to be genuine, many are probably due to under-recording. Losses may be due to the development and agricultural improvement of coastal grassland.

Suboceanic Temperate element.

M. S. PORTER & M. J. Y. FOLEY

Viola × wittrockiana Garden Pansy

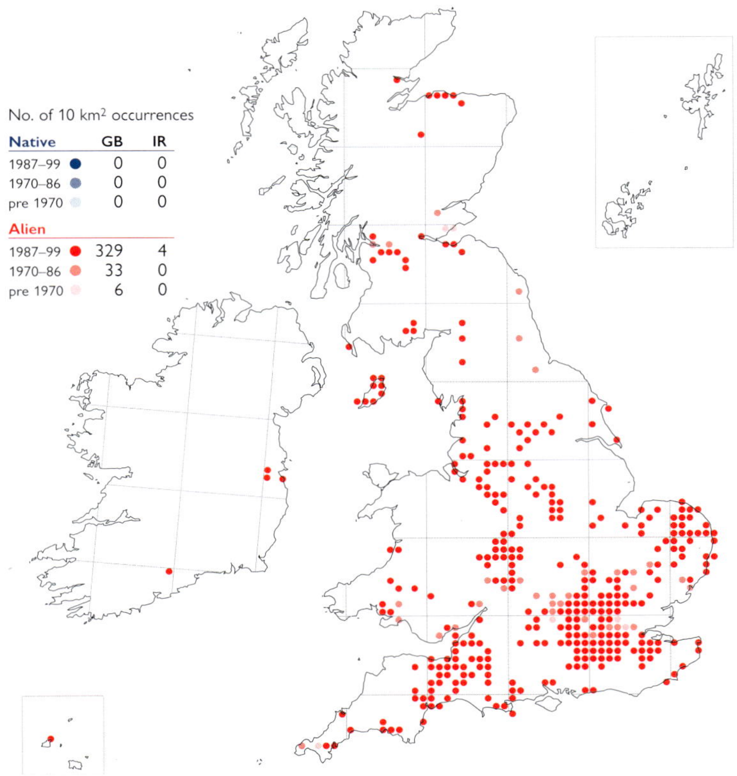

This annual or perennial herb is very commonly grown in gardens and sometimes escapes, typically being found near habitation on rough or cultivated land, waste ground, roadside verges and rubbish tips. It is usually casual, but can persist for a few years. Lowland.

Neophyte. *V.* × *wittrockiana* has been cultivated in Britain since at least 1816, and was known from the wild by 1927. It appears to be increasing, especially in S. England.

A hybrid of garden origin.

Reference: Gorer (1970).

M. S. PORTER & M. J. Y. FOLEY

Viola arvensis Field Pansy

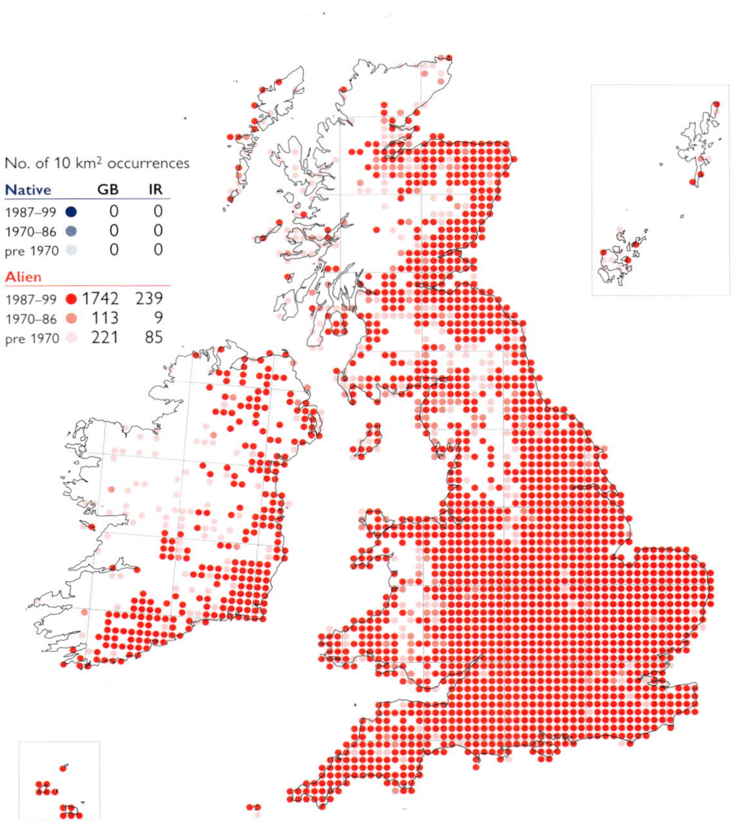

An annual usually found on light, well-drained soils in cultivated fields, in old gravel- and sand-pits, and on waste ground where soil is disturbed. It still occurs in many intensively cultivated arable fields. Lowland.

Archaeophyte (change –0.29). *V. arvensis* was mapped as 'all records' in the 1962 *Atlas*. Its distribution appears to be more or less stable, though the decline in arable farming in W. Scotland has made it less frequent there, and there appear to have been losses in C. and W. Ireland.

As an archaeophyte *V. arvensis* has a Eurosiberian Temperate distribution; it is widely naturalised outside this range.

References: Atlas (55c), Doohan & Monaco (1992), Hultén & Fries (1986), Meusel *et al.* (1978).

M. J. Y. FOLEY & M. S. PORTER

Viola arvensis × *V. tricolor* (*V.* × *contempta*)

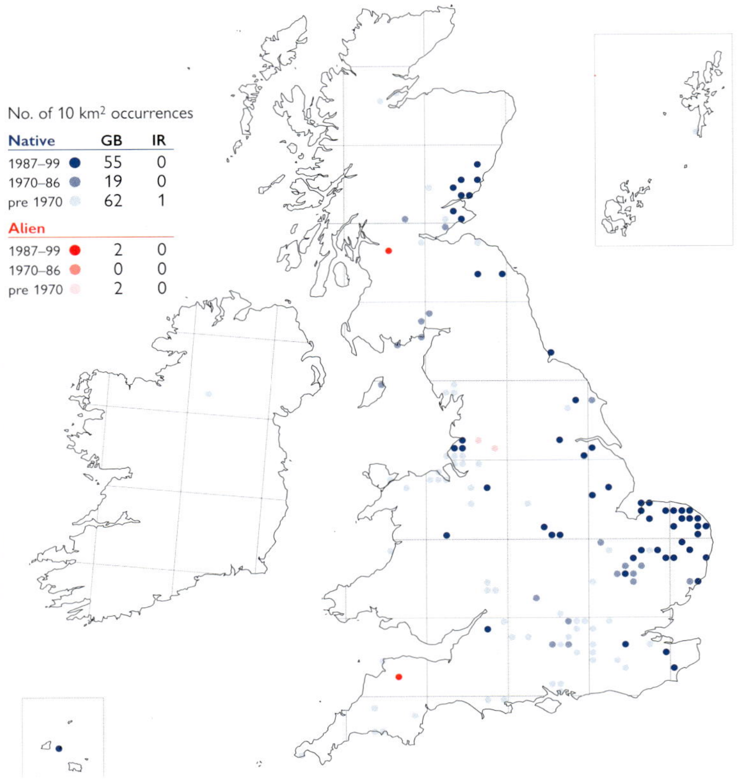

A partially fertile hybrid, generally occurring with the parents, but occasionally in the absence of one or both. It is found on cultivated land and waste ground. Lowland.

Native. Changes in the distribution of this hybrid cannot be easily assessed. Some variants of *V. arvensis* are morphologically very similar to this hybrid, and the map may contain some errors. *V.* × *contempta* is also probably under-recorded.

Intermediates between *V. arvensis* and *V. tricolor* are apparently widespread in Europe.

Reference: Stace (1975).

M. J. Y. FOLEY & M. S. PORTER

Viola kitaibeliana Dwarf Pansy

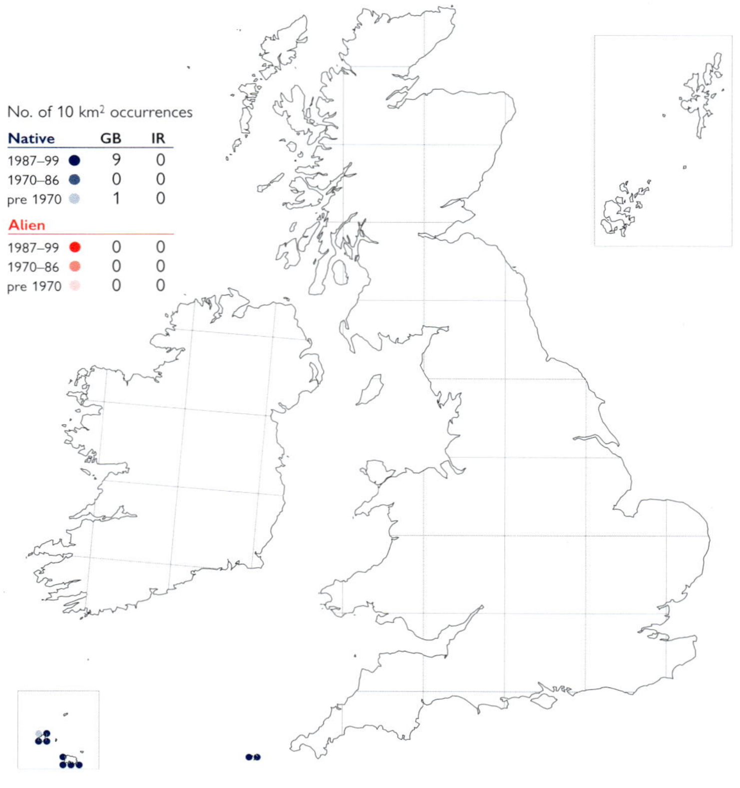

A tiny annual of short coastal turf, open disturbed areas on sandy soils, open sand on eroding coastal dunes, around rabbit burrows, and arable fields. It can also grow under the shelter of coastal *Pteridium* in the Isles of Scilly, and in thin soil on granite sea-cliffs in Guernsey. Lowland.

Native. The 10-km distribution of this species is stable. However, several sites have been lost in the Isles of Scilly since the 1950s, and others are threatened by coastal erosion, building development and by coarse vegetation. Colonies have been known to re-appear on former sites following disturbance, suggesting the presence of a persistent seed bank.

European Southern-temperate element.

References: Atlas (54c), French *et al.* (1999), Lousley (1971), Meusel *et al.* (1978), Wigginton (1999).

M. J. Y. FOLEY & M. S. PORTER

Tamarix gallica Tamarisk

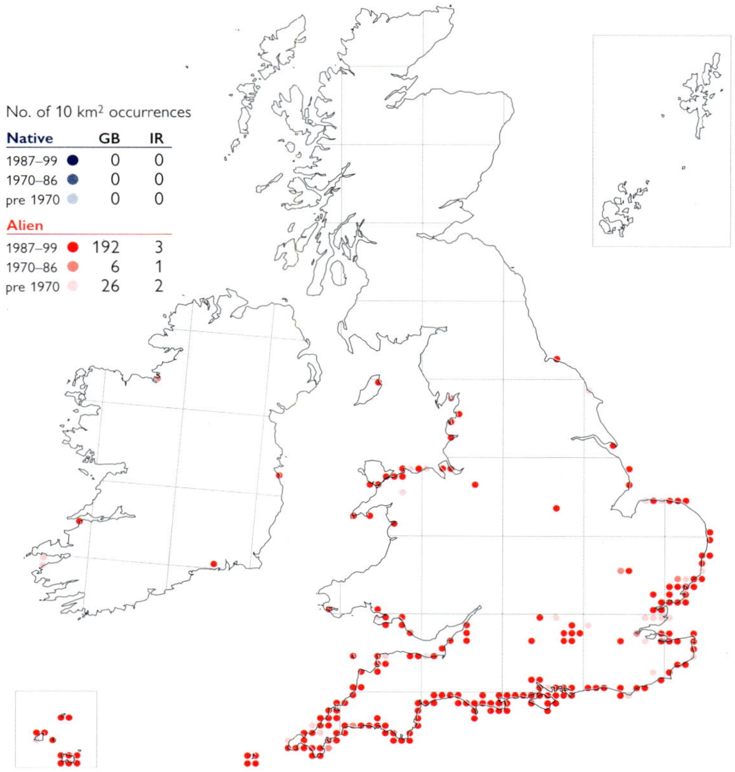

A spreading shrub, or rarely a small tree, extensively planted in coastal habitats as it is very resistant to wind. Most records are of planted specimens, which can be long-lived, but it can spread by suckering. It is also found inland as a garden escape on waste ground and rubbish tips. Lowland.

Neophyte (change +1.04). This species has been grown in British gardens since before 1597, and was known from the wild by 1796. The increase in records since the 1962 *Atlas* is mostly due to better recording of alien trees and shrubs.

Native of the W. Mediterranean region and S.W. Europe.

References: Atlas (61a), Bean (1980), Bolòs & Vigo (1990), Coats (1963).

D. A. PEARMAN

Frankenia laevis Sea-heath

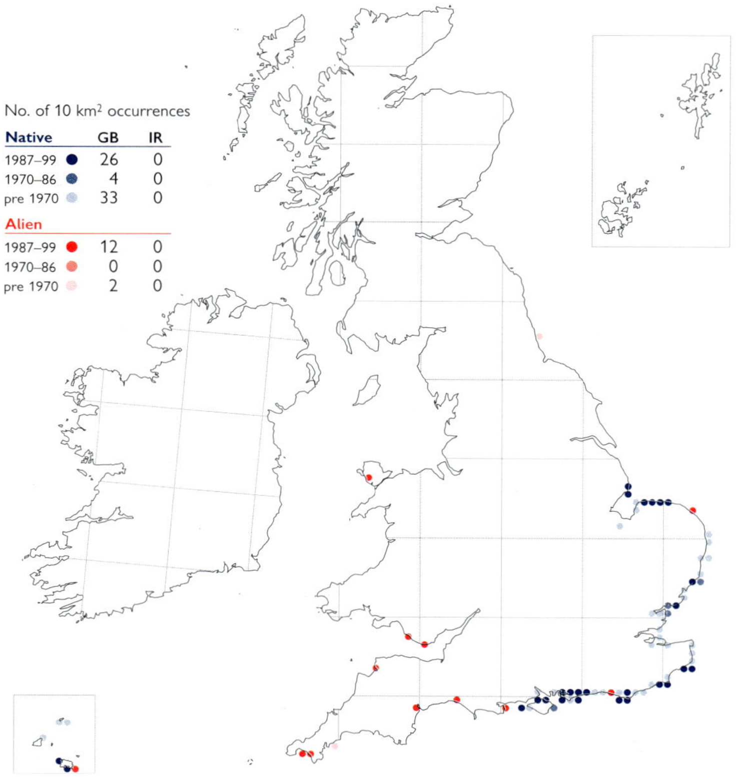

A mat-forming perennial herb of saltmarshes and saltmarsh-sand dune transitions, especially where firm sand or silt overlies coarser grained material; also rarely on shingle beaches and chalk sea-cliffs. Lowland.

Native (change +0.03). *F. laevis* has declined in E. England due to disturbance of saltmarshes and their destruction through improvement of sea defences, urbanisation and industrial development, but these losses, at a 10-km scale, were largely before 1930. It is sometimes grown on rock gardens, and this is probably the origin of the recently discovered sites in S.W. England and Wales.

Suboceanic Southern-temperate element.

References: Atlas (61b), Bolòs & Vigo (1990), Brightmore (1979), Newcombe (1991), Stewart *et al.* (1994).

S. J. LEACH

Bryonia dioica White Bryony

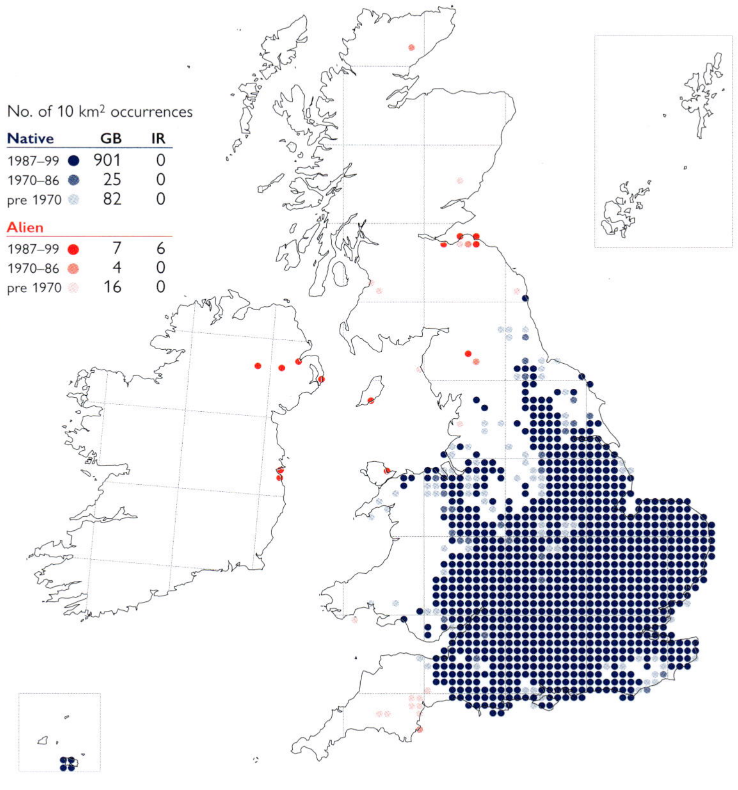

A dioecious scrambling perennial herb with a massive, tuberous root-stock. It grows on well-drained, often base-rich, soils in hedgerows, scrub, woodland borders, and on rough and waste ground. It is unpalatable to rabbits and can be locally frequent around warrens. Lowland.

Native (change –0.50). There has been no substantial change in the distribution since the 1962 *Atlas*. The limits of its native distribution are remarkably stable.

Submediterranean-Subatlantic element.

References: Atlas (168c), Meusel & Jäger (1992).

G. T. D. WILMORE

Cucurbita pepo Marrow

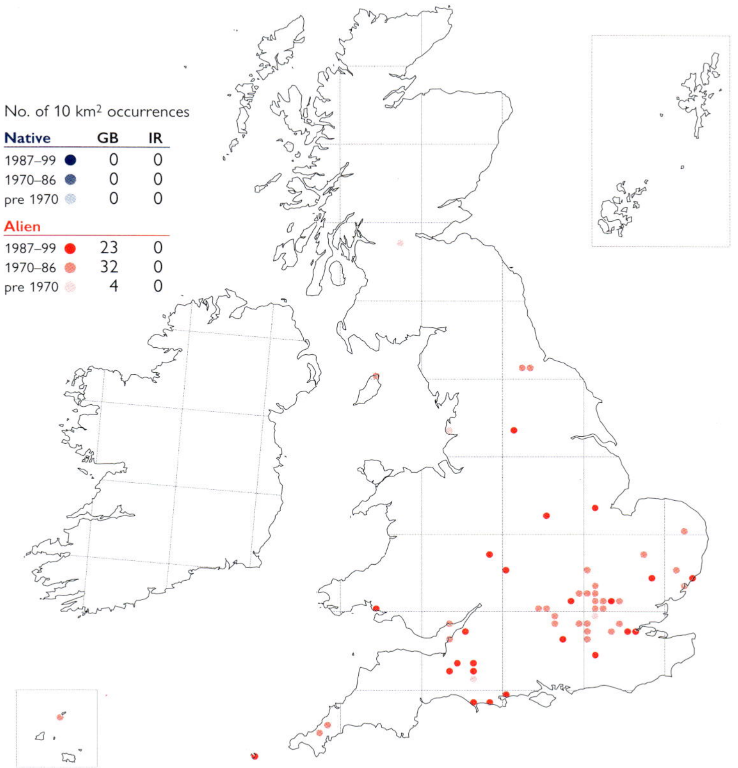

A robust, hispid annual, occurring casually at sewage works, on tips and as a garden throw-out. It is frequently grown as a crop. Lowland.

Casual. *C. pepo* was originally introduced to Britain in 1570, and was known from the wild by 1927. The majority of casually occurring plants originate from kitchen waste or from seeds of ornamental gourds.

Native of C. & N. America.

References: Grenfell (1984), Smart & Simmonds (1995), Vaughan & Geissler (1997).

G. T. D. WILMORE

Populus alba White Poplar

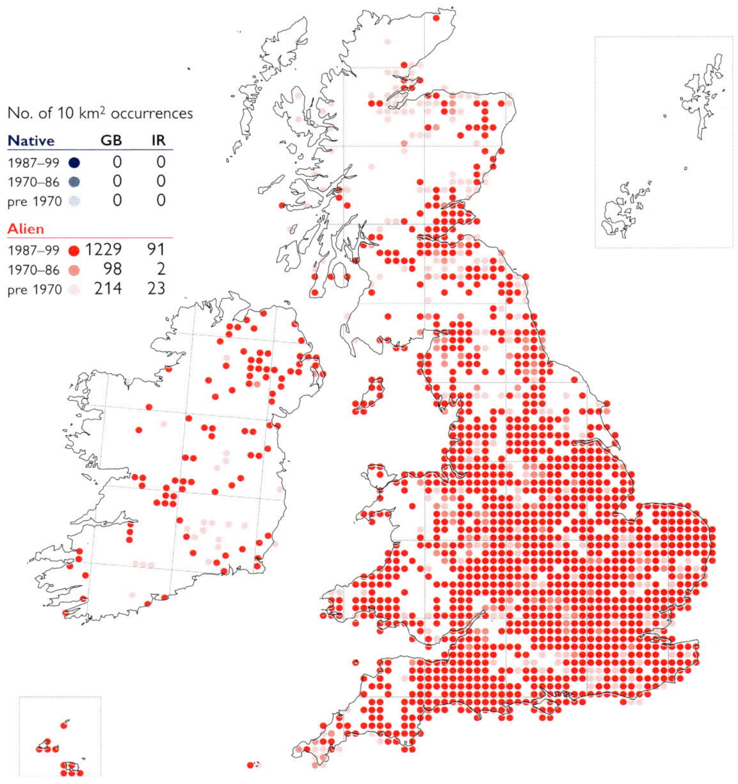

A broad-crowned tree which is most frequent as a female in amenity plantings along roadsides and in parks; also in windbreaks and on coastal dunes, but rarely in plantations. It suckers freely and sometimes becomes well-established, forming dense thickets. It is resistant to salt-laden winds. Lowland.

Neophyte. There is no certain evidence for the presence of *P. alba* in Britain before 1500. It is usually said to have been brought from Holland in the 16th century, and was certainly recorded from the wild by 1597. It is now much planted, sometimes as distinct cultivars.

Native of S., C. & E. Europe, eastwards to C. Asia; widely naturalised north and west of its native range.

References: Bolòs & Vigo (1990), Jalas & Suominen (1976), Jobling (1990), Mabey (1996), Meikle (1984), Rackham (1990).

G. HUTCHINSON

Populus alba × *P. tremula* (*P.* × *canescens*)
Grey Poplar

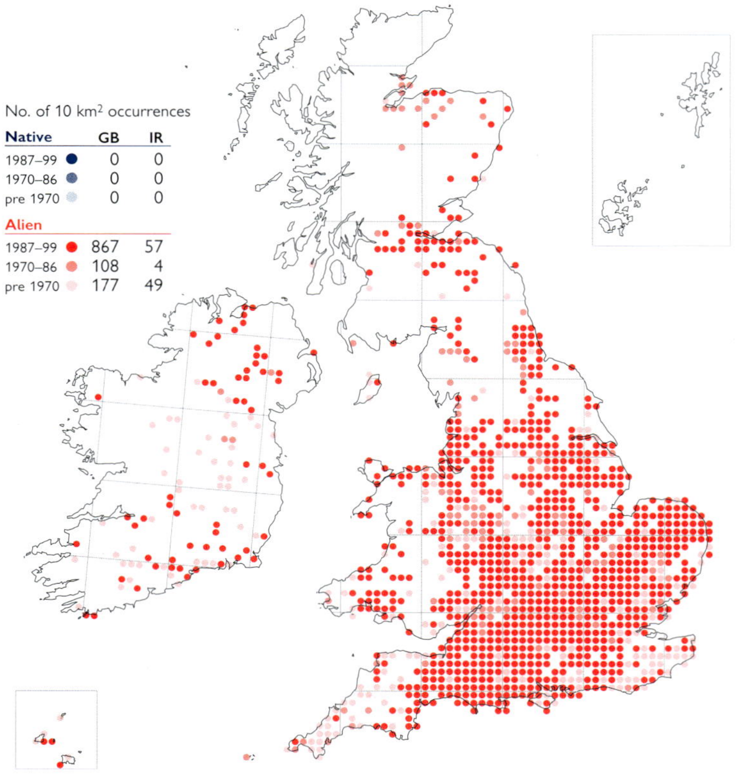

This broad-crowned tree often grows as a solitary, usually male, specimen or among native trees and shrubs. It is planted in windbreaks and as an amenity tree, especially in damp woods and by streams. Lowland.

Neophyte (change +0.97). *P.* × *canescens* was perhaps imported to our area as a hybrid, and was formerly planted for timber. The first certain record was made *c.* 1700. It is now much planted, but rarely becomes naturalised and probably rarely arises anew from the parents. It has increased in Scotland and N. Ireland since the 1962 *Atlas*, but seems stable elsewhere.

A hybrid which occurs naturally within the native range of *P. alba*, and has been spread beyond by planting.

References: Atlas (187a), Jalas & Suominen (1976), Jobling (1990), Meikle (1984), Stace (1975).

G. HUTCHINSON

Populus tremula Aspen

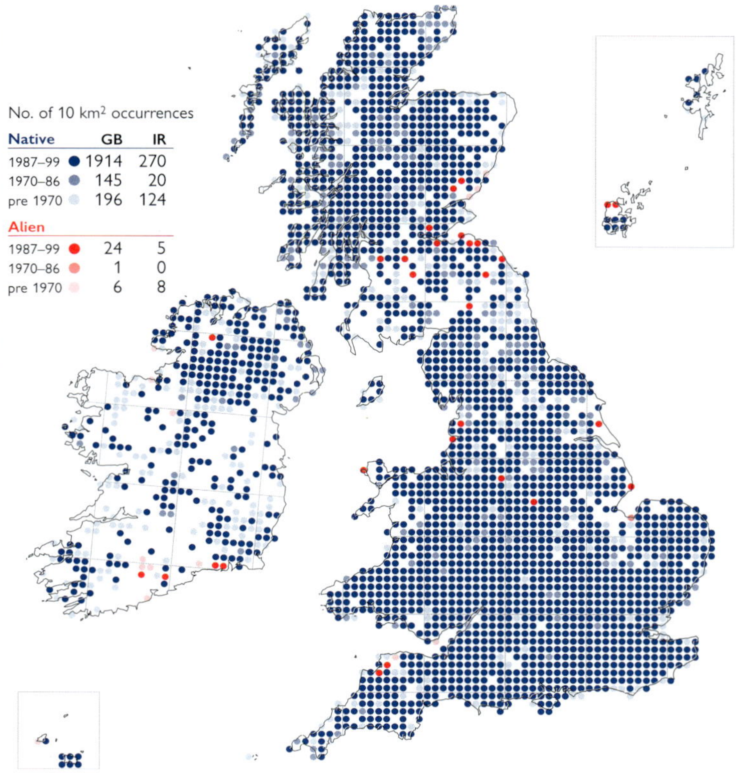

A broad-crowned tree of moist clay or sandy soils in mixed broad-leaved woodlands, hedgerows, on heathland, in disused clay- and sand-pits, and occasionally in pine woods. In the north and west, it grows on cliffs, rocky outcrops and river banks, often as a shrub. It suckers to form thickets, and readily colonises bare ground. 0–640 m (Atholl, E. Perth).

Native (change +0.88). Like many trees, *P. tremula* may have been under-recorded in the 1962 *Atlas*. It has been planted for amenity and as food for browsing deer.

Eurasian Boreo-temperate element.

References: Atlas (187b), Hultén & Fries (1986), Jalas & Suominen (1976), Jobling (1990), Meikle (1984), Meusel *et al.* (1965), Rackham (1980).

G. HUTCHINSON

Populus nigra subsp. *betulifolia* Black-poplar

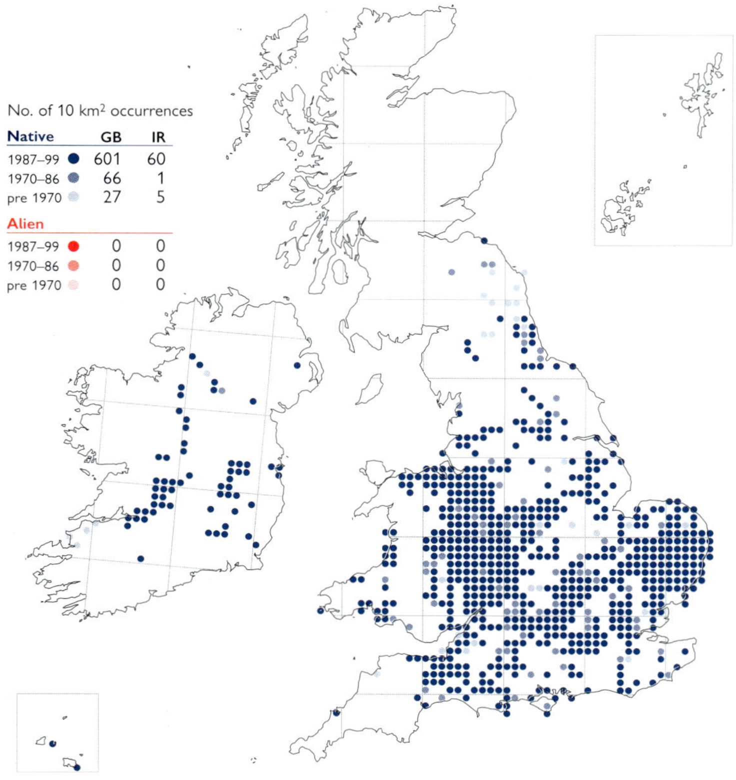

A majestic, broad-crowned tree which grows by watercourses, by ponds and in hedgerows, especially on lowland flood plains, and as an amenity tree in urban areas. Lowland.

Native. Although regarded as native, the original habitat of this taxon has long been modified beyond recognition and it is no longer possible to separate native trees from those planted long ago. All records are mapped as if they are native. Many trees are old and in decline, but it is now being replanted in many areas. Urban trees are probably under-recorded.

P. nigra sens. lat. belongs to the Eurosiberian Temperate element; subsp. *betulifolia* occurs in W. Europe.

References: Cottrell *et al.* (1997), Hobson (1991, 1993), Jalas & Suominen (1976), Mabey (1996), Meikle (1984), Meusel *et al.* (1965), Milne-Redhead (1990), Tabbush & Beaton (1998).

G. HUTCHINSON & C. D. PRESTON

Populus nigra (fastigiate cultivars)
Lombardy-poplars

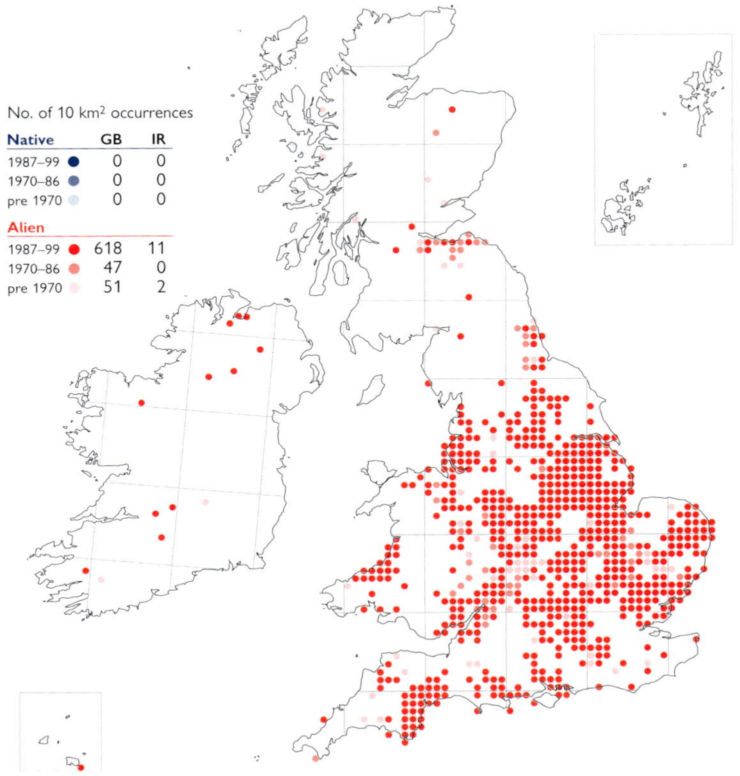

The fastigiate cultivars of *P. nigra* are planted as amenity trees in parks and large gardens, and as screens and windbreaks in the countryside. Although conspicuous as landscape features, the cultivars are single-sex clones and are always planted. Lowland.

Neophyte. A number of fastigiate cultivars are widely planted, including the narrowly fastigiate male 'Italica', the slightly broader male 'Plantierensis' and the much broader female 'Gigantea'. 'Italica', the true Lombardy-poplar, originated in N. Italy and was introduced to Britain in 1758 and was recorded in the wild by 1886. The distribution of these cultivars is probably stable, but they are unevenly recorded.

The fastigiate cultivars of *P. nigra* are of garden origin.

References: Bean (1976), Jobling (1990), Meikle (1984), Mitchell (1996).

G. HUTCHINSON & C. D. PRESTON

Populus deltoides × *P. nigra* (*P.* × *canadensis*)
Hybrid Black-poplar

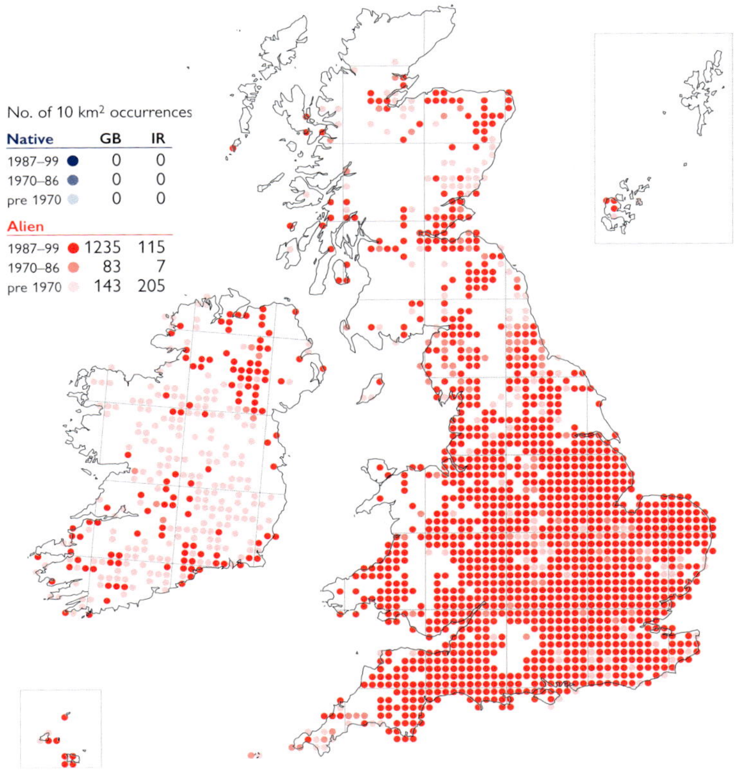

A broad-crowned tree, usually found in plantations, as an amenity tree in parkland and along roadsides and hedgerows, and in screens and windbreaks. Lowland.

Neophyte (change +2.11). This hybrid was introduced to the British Isles in about 1770 and is now represented by several unisexual cultivars. It is almost always planted, but less frequently in recent times due to the preference for faster-growing and more disease-resistant poplars. It was recorded from the wild by 1799. Concentrations of records often reflect recording effort. It was mapped with *P. nigra* in the 1962 *Atlas* as *P. nigra* agg.

A hybrid of garden origin, thought to have originated in France in *c.* 1750; very widely planted in Europe.

References: Heinze (1997), Jobling (1990), Meikle (1984), Mitchell (1996), Stace (1975).

G. HUTCHINSON

Populus balsamifera × *P. deltoides* (*P.* × *jackii*) Balm-of-Gilead

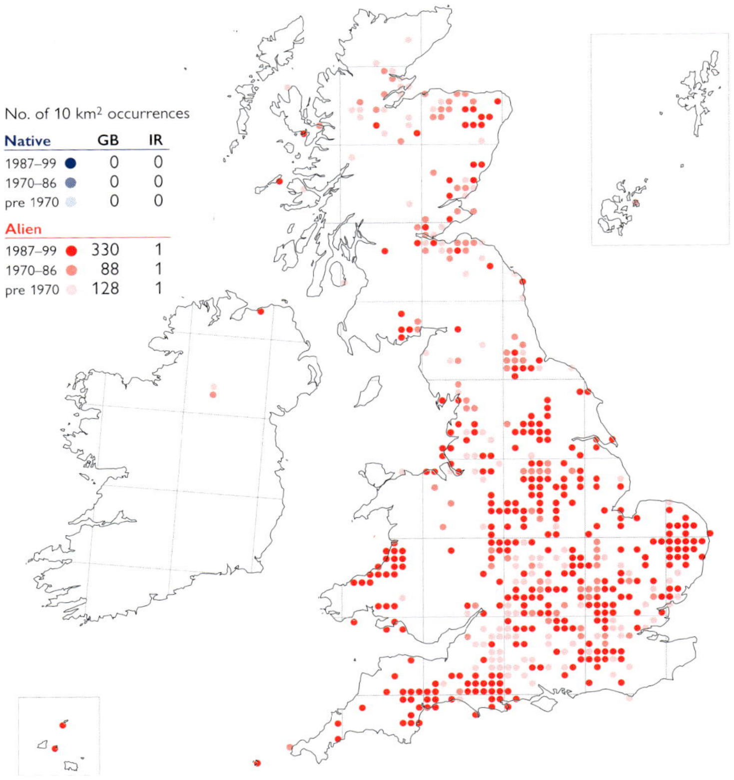

A deciduous tree which is most frequently planted in damp woods and by rivers and ponds, but is increasingly being planted in parks and along roadsides. Only female plants are known in our area but they often become naturalised by suckering. Lowland.

Neophyte. This hybrid was introduced to cultivation in Britain (as *P.* × *candicans*) in 1773 and was recorded from the wild by 1876 (N. Lincs.). It is probably under-recorded.

The hybrid is native of N. America, but the origin of cultivated plants is uncertain. They are treated as a cultivar of *P. balsamifera* by Jonsell (2000).

References: Bean (1976), Jobling (1990), Meikle (1984).

G. HUTCHINSON

Populus trichocarpa Western Balsam-poplar

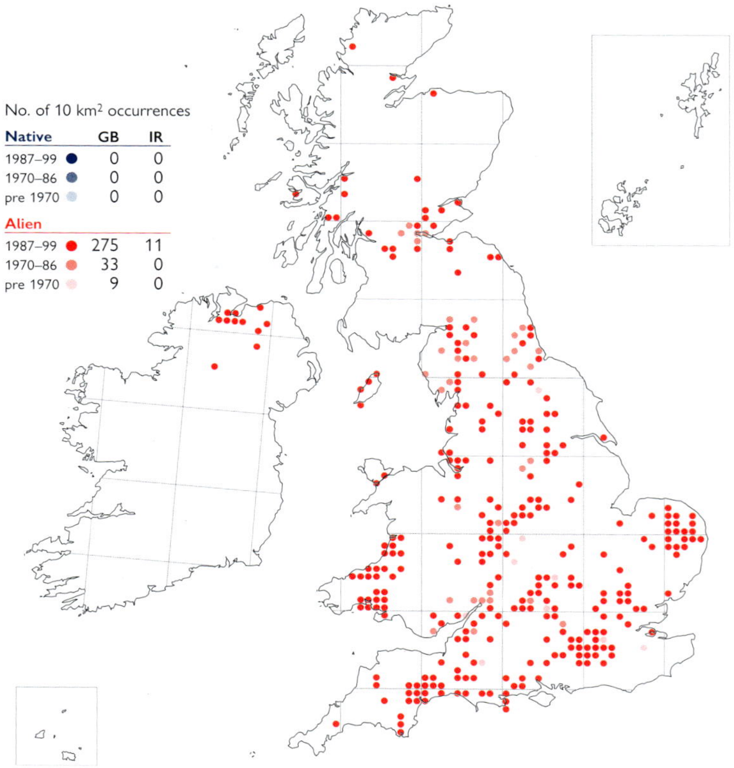

A narrow-outlined tree with spreading lower branches which is often planted for amenity or timber. Most planted trees are male and these cannot reproduce, but they occasionally spread by suckers. The seed will only germinate within a day or two of being shed. Lowland.

Neophyte. This species was introduced to our area in 1892, and was first recorded in the wild in 1935 (E. Kent). Several new cultivars (e.g. 'Trichobel') with very fast growth rates are now being grown. It is probably increasing due to continued planting.

Native of western N. America.

References: Bean (1976), Jobling (1990), Meikle (1984), Mitchell (1996).

G. HUTCHINSON

Salix pentandra Bay Willow

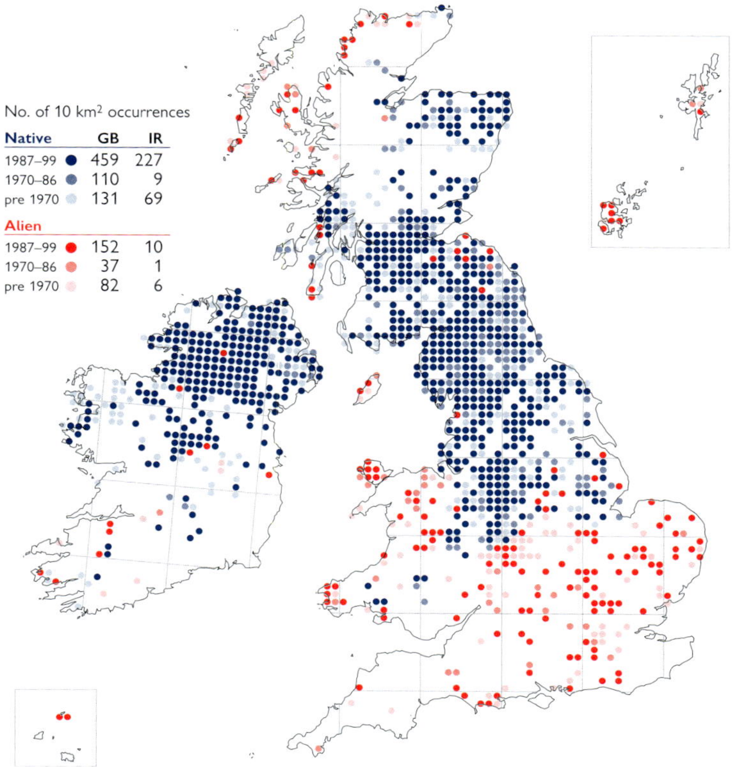

A large shrub or small tree which grows in damp or wet ground, mostly in marshes, fens and wet woods, in winter-flooded dune-slacks and by ponds and streams; sometimes in drier sites such as shaded roadsides. Male plants are widely planted as ornamentals within and outside the native range. 0–410 m (Allendale, S. Northumb.).

Native (change +0.11). The distribution of *S. pentandra* has not significantly changed since the 1962 *Atlas*. Delimitation of the southern limit of its native range is difficult, although the absence of female plants in a population can indicate non-native status.

Eurosiberian Boreo-temperate element; widely naturalised outside its native range.

References: Atlas (187d), Bean (1980), Howitt & Howitt (1990), Hultén & Fries (1986), Jalas & Suominen (1976), Meikle (1984), Meusel *et al.* (1965).

G. HUTCHINSON

Salix fragilis Crack-willow

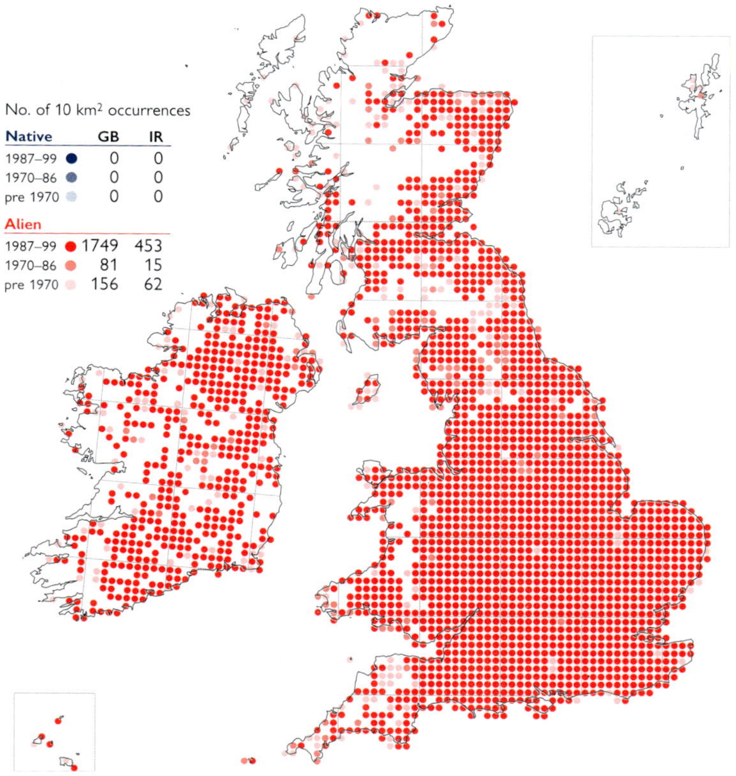

A broad-crowned, often pollarded, tree which grows in hedgerows, marshes, fens, wet woods and hollows, and by ponds, ditches, streams and rivers. It can tolerate a polluted atmosphere and salt-laden winds. 0–410 m (Allendale, S. Northumb.).

Archaeophyte (change +0.26). This species has often been planted, sometimes as distinct unisexual varieties. The distribution is little changed since the 1962 *Atlas*.

As an archaeophyte *S. fragilis* has a Eurosiberian Temperate distribution; it is widely naturalised outside this range. Many of the variants regarded as *S. fragilis* in Britain and Ireland would be regarded as *S.* × *rubens* by botanists in continental Europe.

References: Atlas (188b), Bean (1980), Hultén & Fries (1986), Jalas & Suominen (1976), Jonsell (2000), Meikle (1984, 1992).

G. HUTCHINSON

Salix fragilis × *S. pentandra* (*S.* × *meyeriana*)
Shiny-leaved Willow

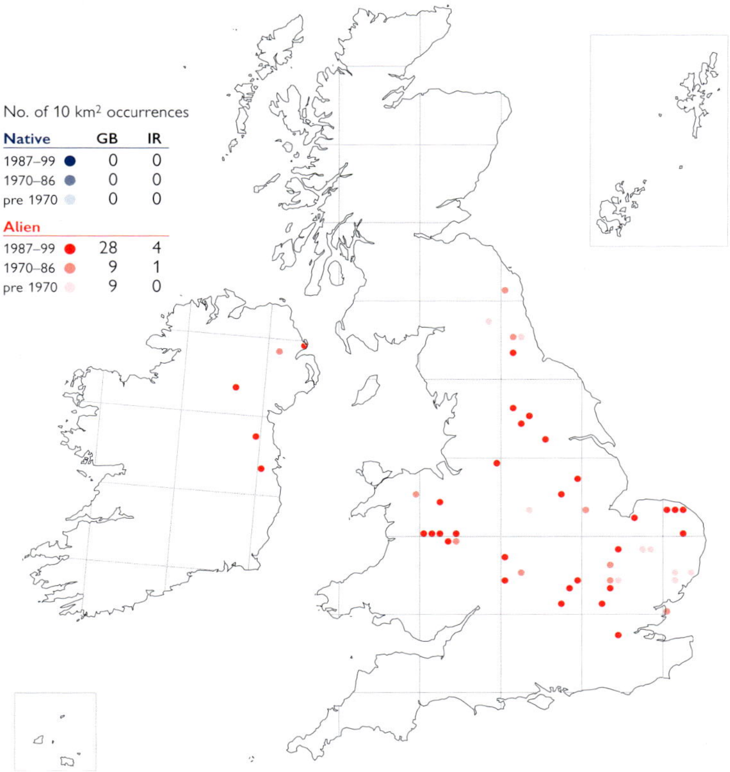

No. of 10 km² occurrences

	GB	IR
Native		
1987–99	0	0
1970–86	0	0
pre 1970	0	0
Alien		
1987–99	28	4
1970–86	9	1
pre 1970	9	0

A tall shrub or small tree of moist, often waterlogged places including river margins and ditches and on wet roadsides and field-borders. Plants of this hybrid are usually of cultivated origin, being planted as an ornamental both inside and outside the range of native *S. pentandra*. The altitudinal range is unknown.

A spontaneous hybrid between native and alien parents. As little evidence is available to assess the status of individual populations all records are mapped as if they are alien. There is no evidence to assess changes in the distribution of this hybrid.

Widespread in C. & N. Europe; in Norden almost all populations consist of male plants (Jonsell, 2000).

References: Maxted & Trueman (1983), Meikle (1984), Stace (1975).

G. HUTCHINSON

Salix alba White Willow

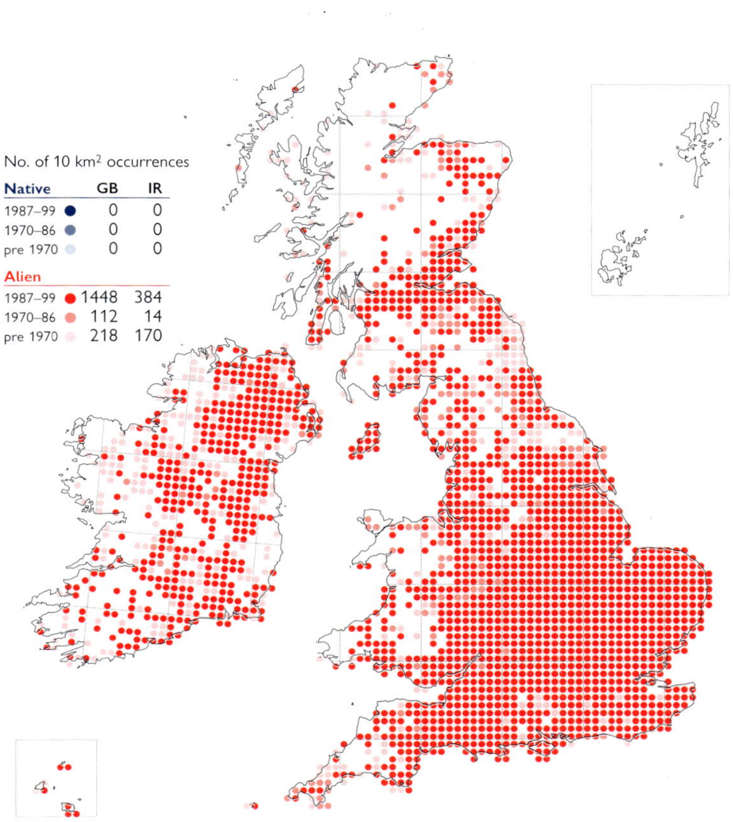

No. of 10 km² occurrences

	GB	IR
Native		
1987–99	0	0
1970–86	0	0
pre 1970	0	0
Alien		
1987–99	1448	384
1970–86	112	14
pre 1970	218	170

A conspicuous tree which grows in marshes and wet hollows and by ponds, ditches, streams and rivers. 0–350 m (S. of Alston, Cumberland).

Archaeophyte (change +0.02). This species has been much planted, sometimes as distinct cultivars. It may have decreased in Ireland, but has not significantly changed in Britain. Var. *caerulea* (Cricket-bat Willow) is particularly vulnerable to Watermark Disease involving the bacterium *Erwinia salicis*.

As an archaeophyte *S. alba* has a Eurosiberian Southern-temperate distribution; it is widely naturalised outside this range.

References: Atlas (188a), Bean (1980), Howitt & Howitt (1990), Jalas & Suominen (1976), Meikle (1984), Meusel *et al.* (1965), White (1992).

G. HUTCHINSON

Salix alba × *S. babylonica* (*S.* × *sepulcralis*)
Weeping Willow

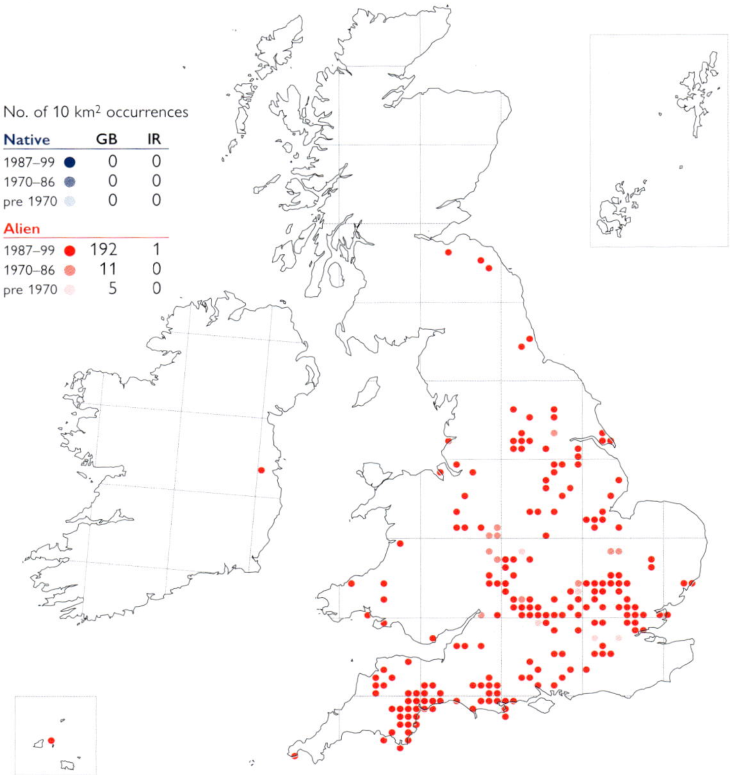

No. of 10 km² occurrences

	GB	IR
Native		
1987–99	0	0
1970–86	0	0
pre 1970	0	0
Alien		
1987–99	192	1
1970–86	11	0
pre 1970	5	0

This tree most frequently occurs in ornamental and amenity plantings in parks and on riversides, and may be long-lived. Lowland.

Neophyte. This hybrid has been grown in Britain since about 1869, and was recorded from the wild soon after, in 1886. It had superseded the less-hardy parent *S. babylonica* by the early part of the 20th century. It is unevenly recorded.

A hybrid of garden origin.

References: Chmelar (1984), Meikle (1984), Mitchell (1996), Stace (1975), White (1992).

G. HUTCHINSON

Salix alba × *S. fragilis* (*S.* × *rubens*)
Hybrid Crack-willow

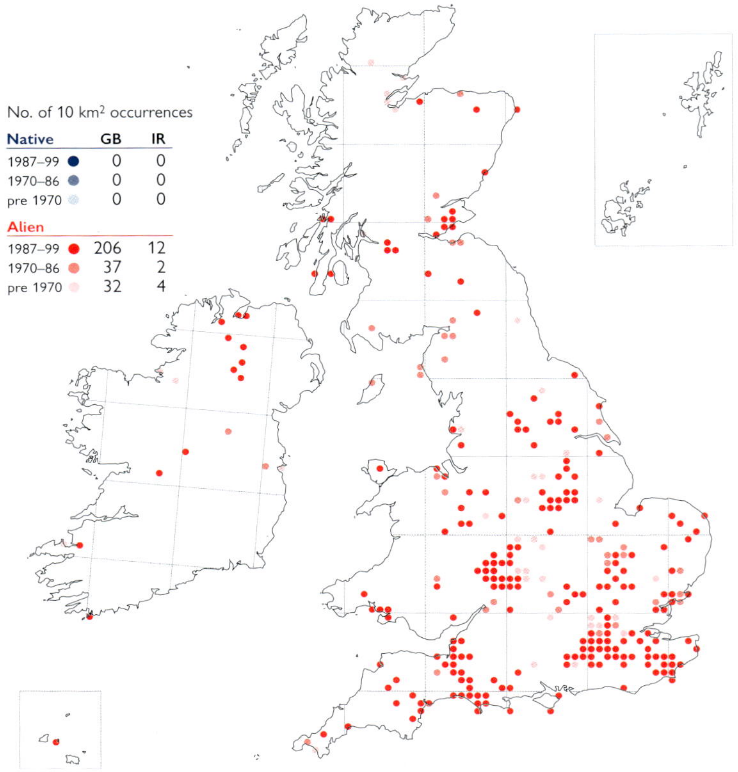

No. of 10 km² occurrences

Native	GB	IR
1987–99	0	0
1970–86	0	0
pre 1970	0	0
Alien		
1987–99	206	12
1970–86	37	2
pre 1970	32	4

A tall broad-crowned, often pollarded, tree which grows in marshes, wet hollows, by ponds, ditches, streams and rivers and as an amenity tree along roadsides and in landscaped areas. It is found mostly in the absence of the parents. Lowland.

Archaeophyte. This hybrid between two alien parents sometimes arises spontaneously, but is usually planted, having originally been introduced for basket making. It is now more frequently planted for amenity and landscaping. A range of ornamental cultivars or varieties are available, some of which have prominent orange-yellow twigs. It is difficult to distinguish from its parents and is under-recorded.

Widespread in Europe.

References: Bean (1980), Howitt & Howitt (1990), Meikle (1984), Stace (1975), Trieste *et al.* (1997).

G. HUTCHINSON

Salix triandra Almond Willow

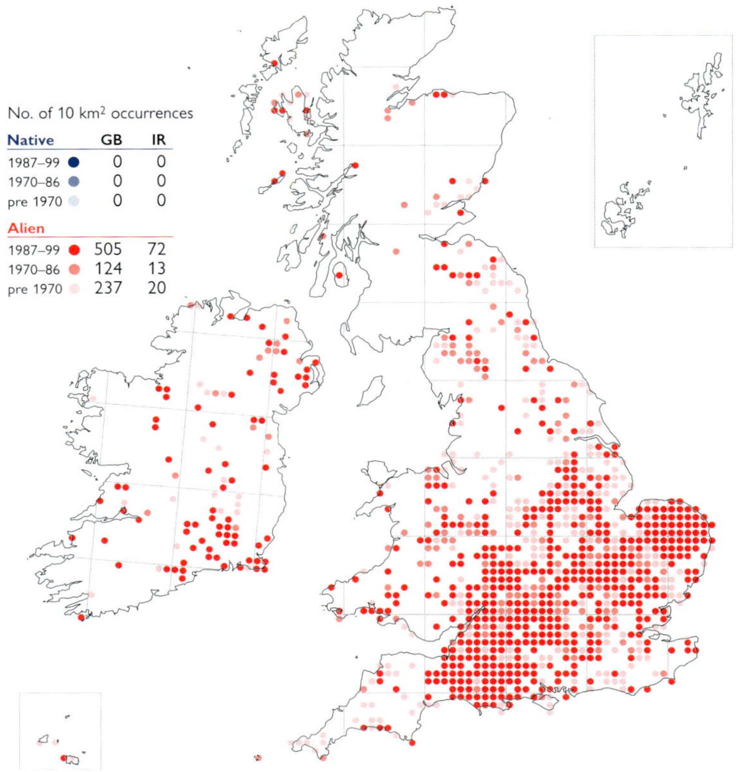

No. of 10 km² occurrences

Native	GB	IR
1987–99	0	0
1970–86	0	0
pre 1970	0	0
Alien		
1987–99	505	72
1970–86	124	13
pre 1970	237	20

A shrub or small tree which grows in damp or wet places, by rivers, streams and ponds and in marshes and osier-beds. Lowland.

Archaeophyte (change –0.06). A species which has been much planted for basketry, often as distinct cultivars. Its distribution has not significantly changed since the 1962 *Atlas.*

As an archaeophyte *S. triandra* has a Eurasian Temperate distribution.

References: Atlas (188c), Bean (1980), Hultén & Fries (1986), Jalas & Suominen (1976), Meikle (1984), Meusel *et al.* (1965).

G. HUTCHINSON

Salix triandra × *S. viminalis* (*S.* × *mollissima*)
Sharp-stipuled Willow

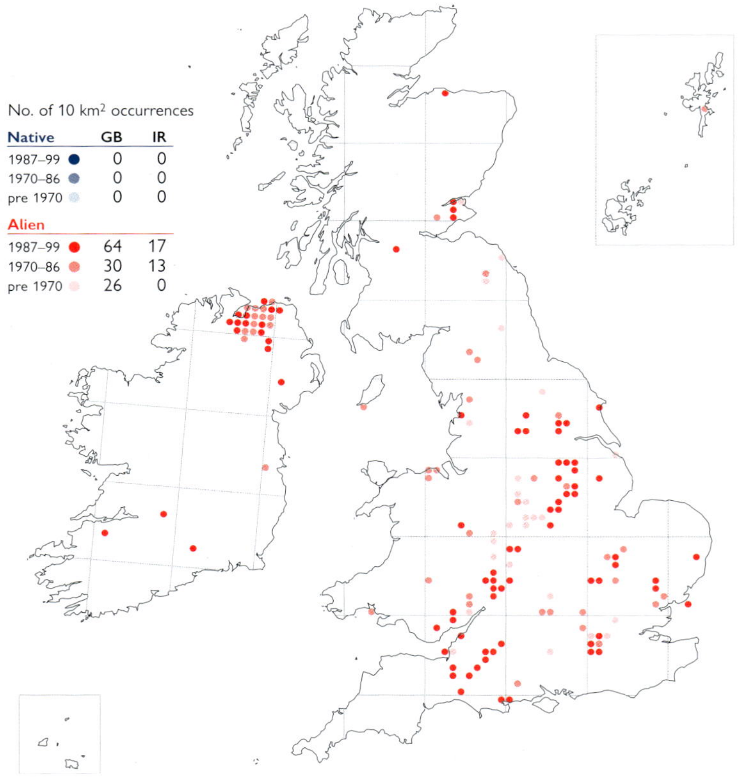

No. of 10 km² occurrences

Native	GB	IR
1987–99	0	0
1970–86	0	0
pre 1970	0	0
Alien		
1987–99	64	17
1970–86	30	13
pre 1970	26	0

A shrub which grows in damp places. It often occurs in the absence of both parents, but in such cases it is introduced, frequently indicating the former presence of osier-beds. Lowland.

Archaeophyte. This hybrid has been much planted for basketry, especially in N. Ireland. More recently it has been planted locally for biomass production. Some records of *S. triandra* may belong here.

Widespread in Europe.

References: Bean (1980), Meikle (1984), Stace (1975).

G. HUTCHINSON

Salix purpurea Purple Willow

Native	**GB**	**IR**
1987–99 ●	758	248
1970–86 ●	148	13
pre 1970 ○	286	118
Alien		
1987–99 ●	0	0
1970–86 ●	0	0
pre 1970 ○	0	0

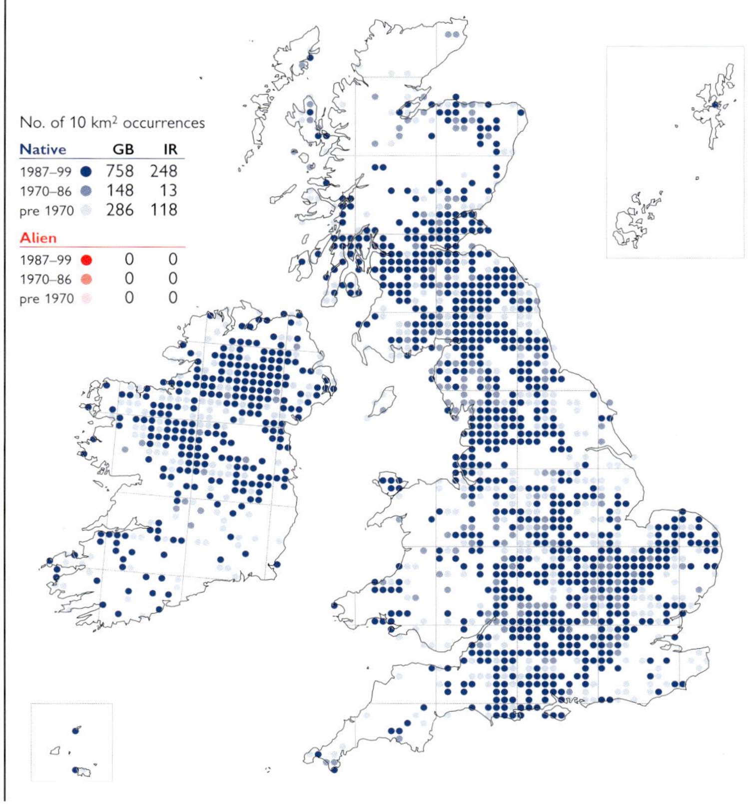

A variable shrub or small tree found on wet ground, at wood margins, on damp hillsides, by streams and rivers, on river shingle, in marshes and fens, and sometimes planted as an osier. 0–440 m (E. Allendale, S. Northumb.).

Native (change −0.01). This species has been much planted for basketry. Native and alien occurrences are now so confused that all records are mapped as if they are native. Its distribution has not significantly changed since the 1962 *Atlas*.

Eurosiberian Temperate element.

References: Atlas (188d), Bean (1980), Jalas & Suominen (1976), Meikle (1984), Meusel *et al.* (1965).

G. HUTCHINSON

Salix purpurea × *S. viminalis* (*S.* × *rubra*) Green-leaved Willow

Native	**GB**	**IR**
1987–99 ●	98	14
1970–86 ●	36	1
pre 1970 ○	41	5
Alien		
1987–99 ●	0	0
1970–86 ●	0	0
pre 1970 ○	0	0

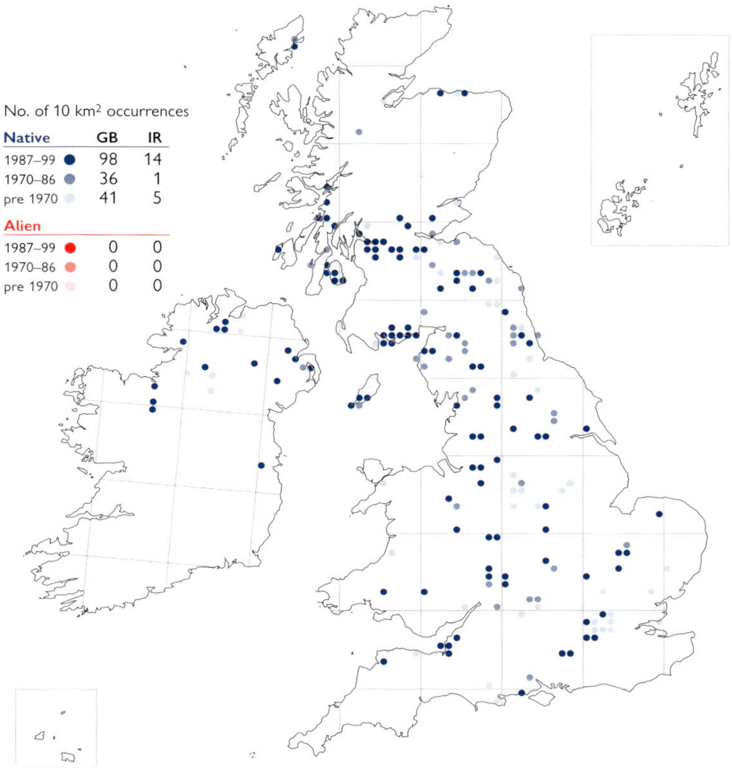

A shrub or small tree which is often found with the parents in osier-beds and thickets, and on roadsides. 0–380 m (Dowthwaite Head, Cumberland).

A spontaneous hybrid between native and alien parents. This hybrid is cultivated in osier-beds, being a popular willow for basketry. It is difficult to separate spontaneous occurrences from planted trees and all records are therefore mapped as if they are native. There is little evidence to assess changes in its distribution.

Widespread in Europe.

References: Bean (1980), Howitt & Howitt (1990), Meikle (1984), Stace (1975).

G. HUTCHINSON

Salix daphnoides European Violet-willow

Native	**GB**	**IR**
1987–99 ●	0	0
1970–86 ●	0	0
pre 1970 ○	0	0
Alien		
1987–99 ●	115	8
1970–86 ●	37	0
pre 1970 ○	32	1

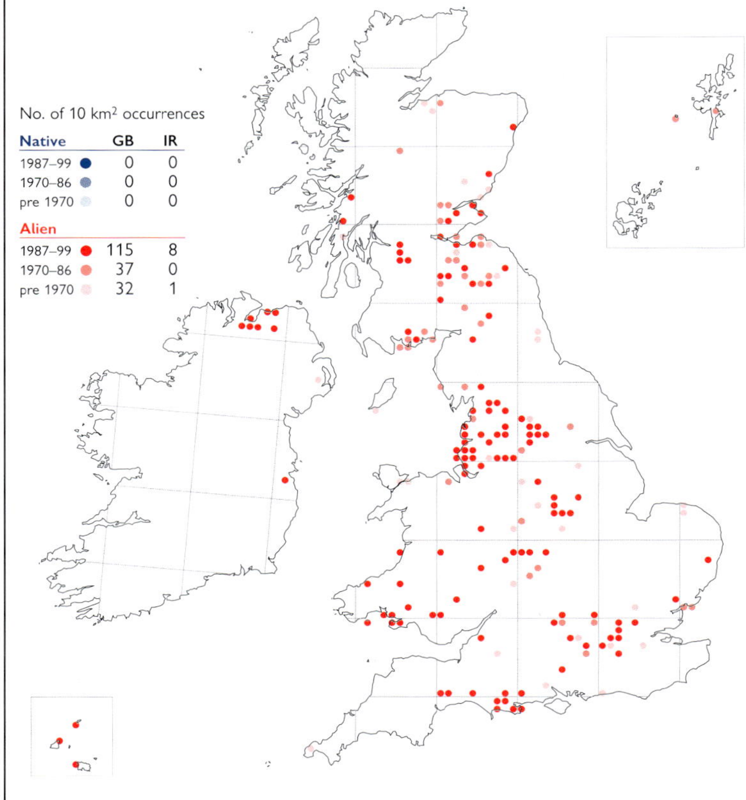

A tall shrub or small tree which is most frequent as an ornamental, being planted for amenity in parks and landscaped areas, but sometimes found naturalised in wild places. Lowland.

Neophyte. This species was introduced into cultivation in Britain in about 1829, and was known from the wild by at least 1905. It is usually planted for its violet-brown twigs which have a dense white bloom. It is certainly increasing as an amenity tree.

Native of Europe (Scandinavia, Baltic States, Alps and Carpathians).

References: Bean (1980), Howitt & Howitt (1990), Hultén & Fries (1986), Jalas & Suominen (1976), Meikle (1984), White (1992).

G. HUTCHINSON

Salix viminalis Osier

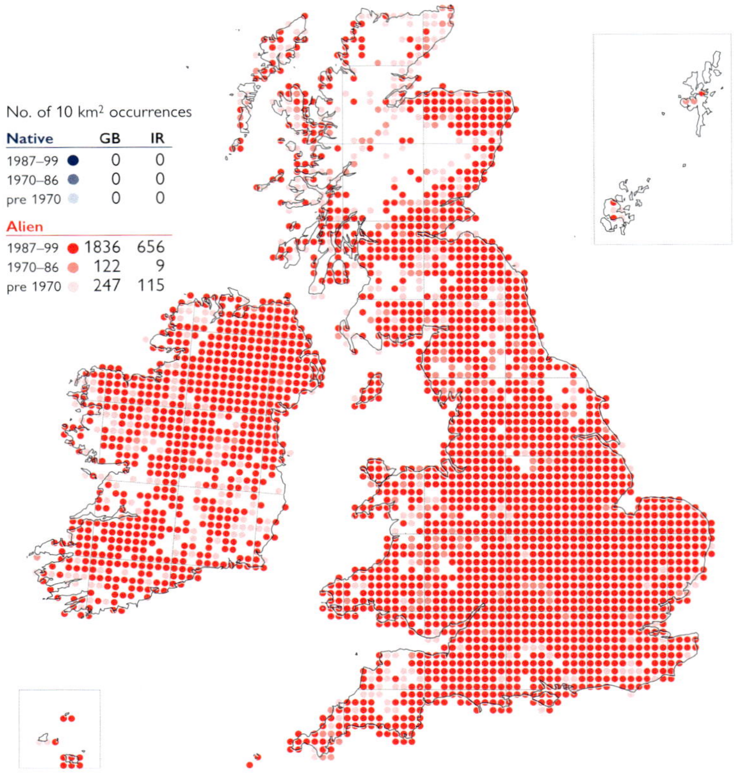

An erect shrub or small tree, frequently coppiced and pollarded, which grows in damp places, by streams and ponds, in marshes, fens, osier-beds and landscaped areas. 0–410 m (E. Allendale, S. Northumb.).

Archaeophyte (change +0.61). A species which has been much planted for basketry and also for amenity, sometimes as distinct cultivars. It is now being planted for biomass production. The range has not significantly changed since the 1962 *Atlas*, although it is clearly more frequent in W. Britain than previously suspected.

As an archaeophyte *S. viminalis* has a Eurasian Temperate distribution; it is widely naturalised outside this range.

References: Atlas (189a), Bean (1980), Howitt & Howitt (1990), Hultén & Fries (1986), Jalas & Suominen (1976), Meikle (1984).

G. HUTCHINSON

Salix elaeagnos Olive Willow

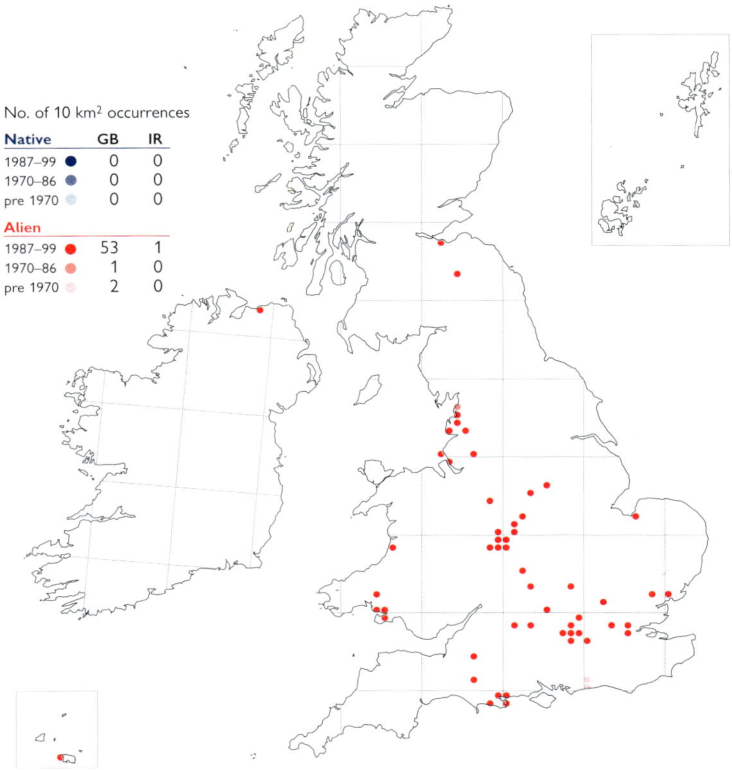

An erect shrub or slender much-branched tree found planted or as a relic of cultivation in damp, open situations such as ditches and riversides; it does not thrive in shade or when competing with other shrubs. It can become more or less naturalised where neglected or thrown out. Lowland.

Neophyte. *S. elaeagnos* was introduced to cultivation in 1820 and is grown in gardens for its densely pubescent twigs and narrow, tomentose leaves. It was recorded from the wild by 1928, and has been increasingly planted for amenity and landscaping.

A European Southern-temperate species; widely planted in temperate Europe north of its native range.

References: Bean (1980), Jalas & Suominen (1976), Meikle (1984) Meusel *et al.* (1965), White (1992).

G. HUTCHINSON

Salix caprea Goat Willow

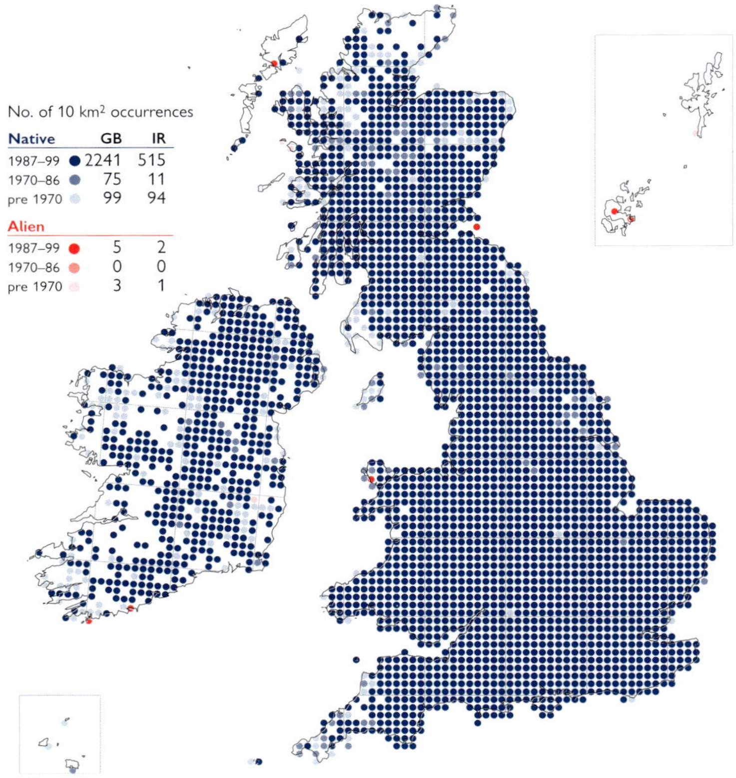

A shrub or tree which grows in open woodland and wood margins, scrub and hedgerows, and around rocky lake and streamsides. It colonises waste ground and can tolerate drier and more base-rich soils than *S. cinerea*. 0–760 m (Breadalbanes, Mid Perth).

Native (change +0.34). The range of this species is similar to that in the 1962 *Atlas*, but local increases in abundance have been noted in England along disused railways. It may have been over-recorded for *S.* × *reichardtii* or large-leaved forms of *S. cinerea*. Subsp. *caprea* occurs throughout the range of the species; subsp. *sphacelata* is mapped separately.

Eurasian Boreo-temperate element.

References: Atlas (189b), Grime *et al.* (1988), Hultén & Fries (1986), Jalas & Suominen (1976), Meikle (1984, 1992), Meusel *et al.* (1965).

G. HUTCHINSON

Salix caprea subsp. *sphacelata*

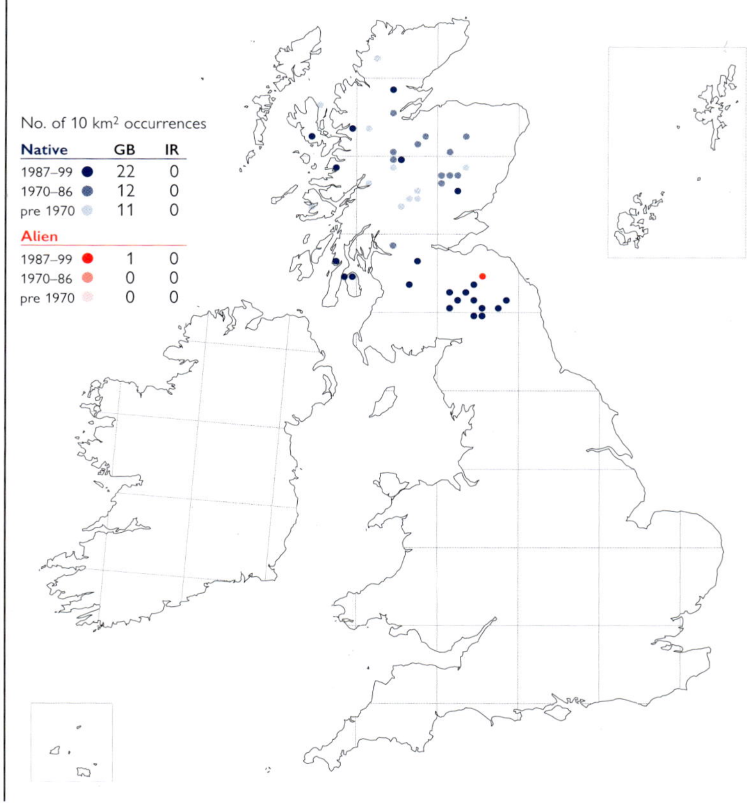
No. of 10 km² occurrences
Native GB IR
1987–99 22 0
1970–86 12 0
pre 1970 11 0
Alien
1987–99 1 0
1970–86 0 0
pre 1970 0 0

A shrub or occasionally a small tree which grows beside lakes, in riverside *Salix* carr, on roadsides and hillsides and most frequently on the sides of rocky streams and in ravines. From 245 m (Carse, Kintyre) to 760 m (Breadalbanes, Mid Perth).

Native. This subspecies is almost certainly under-recorded. It was not mapped in the 1962 *Atlas*, and there is insufficient historical data from which to assess changes.

European Boreal-montane element; widespread in the mountains of C. Europe.

References: Bean (1980), Howitt & Howitt (1990), Meikle (1984, 1992).

G. HUTCHINSON

Salix caprea × *S. cinerea* (*S.* × *reichardtii*)

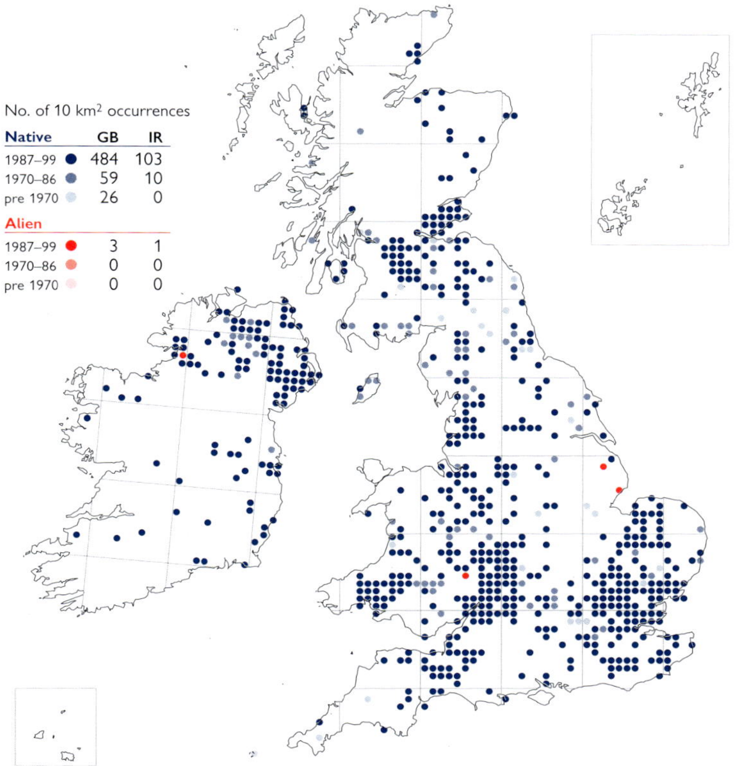
No. of 10 km² occurrences
Native GB IR
1987–99 484 103
1970–86 59 10
pre 1970 26 0
Alien
1987–99 3 1
1970–86 0 0
pre 1970 0 0

A variable shrub or small tree growing in damp places and on disturbed waste ground where the parents coexist, but frequently where one parent is much less abundant than the other. Where woodland has been felled or the habitat otherwise disturbed, this hybrid and *S. cinerea* tend to replace *S. caprea*. An unbroken series of intermediates sometimes links the parents, particularly in highly disturbed areas. 0–450 m (Alston Moor, Cumberland), with an exceptional record at 845 m on Great Dun Fell (Westmorland).

Native. This hybrid was not mapped in the 1962 *Atlas*. It is certainly under-recorded and the map must be regarded as incomplete.

Widespread in Europe.

References: Grime *et al.* (1988), Howitt & Howitt (1990), Meikle (1984), Stace (1975).

G. HUTCHINSON

Salix caprea × *S. cinerea* × *S. viminalis* (*S.* × *calodendron*) Holme Willow

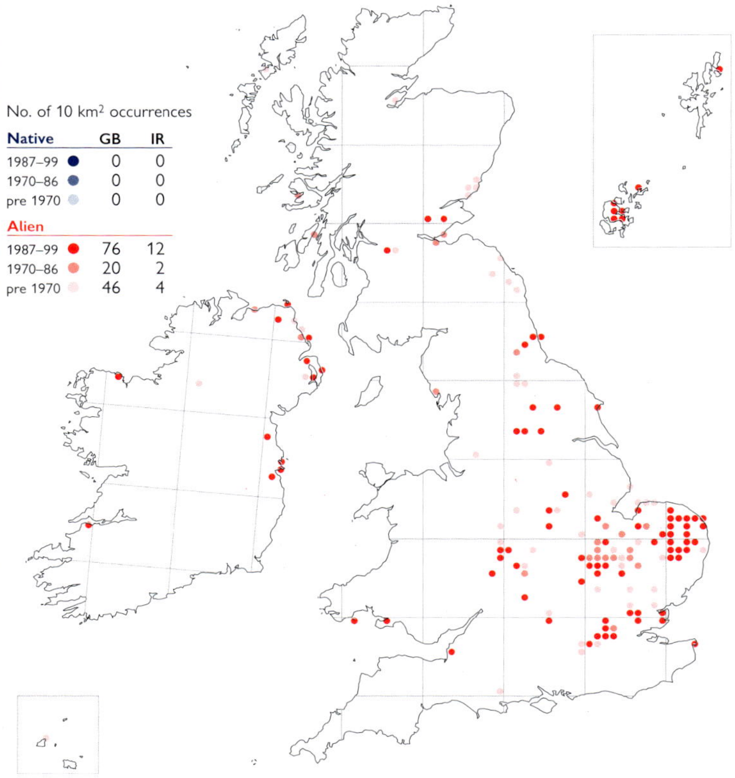
No. of 10 km² occurrences
Native GB IR
1987–99 0 0
1970–86 0 0
pre 1970 0 0
Alien
1987–99 76 12
1970–86 20 2
pre 1970 46 4

An erect shrub or small tree which grows in damp places, but not always near any of the parents. It has recently been grown in plantations for biomass production. Lowland.

Neophyte. This hybrid appears to be represented in our area by a single female clone. Its presence in the wild is presumably the result of introductions, but this is a mysterious tree and the date and reason for its introduction are quite unknown.

Known as a rare relic of cultivation in N. & C. Europe.

References: Bean (1980), Howitt & Howitt (1990), Meikle (1984), Stace (1975).

G. HUTCHINSON

Salix caprea × *S. viminalis* (*S.* × *sericans*)
Broad-leaved Osier

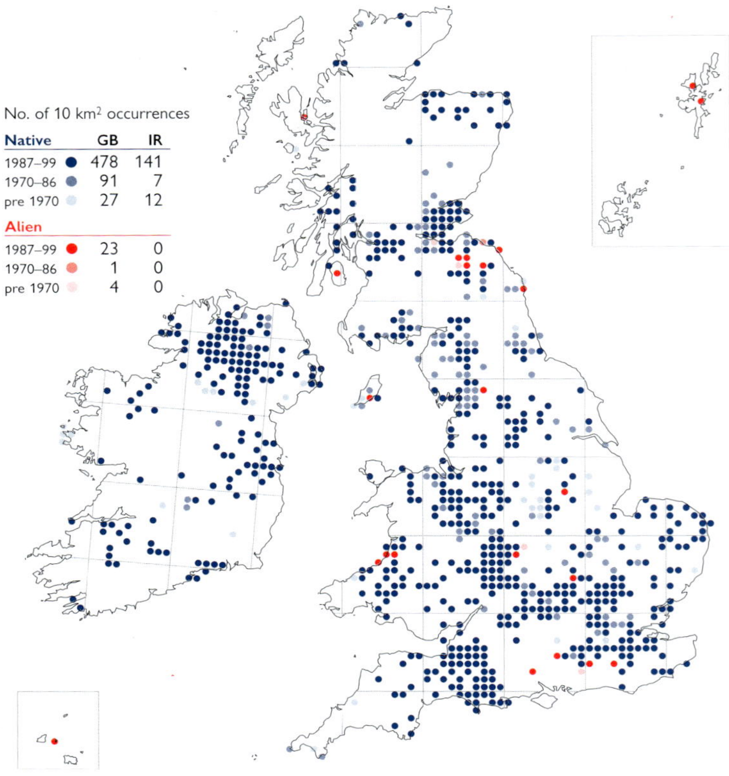

No. of 10 km² occurrences

Native	GB	IR
1987–99 ●	478	141
1970–86 ●	91	7
pre 1970 ●	27	12
Alien		
1987–99 ●	23	0
1970–86 ●	1	0
pre 1970 ●	4	0

A shrub or small tree which grows commonly with the parents in hedgerows, thickets and waste ground, and also as an obvious introduction in osier-beds and plantations. 0–460 m (Gildersdale, Cumberland).

A spontaneous hybrid between native and alien parents. This hybrid is a relic of cultivation: it was formerly considered of some value for coarse basket work and is now being planted for biomass production. It was not mapped in the 1962 *Atlas* and the present map must be regarded as incomplete.

Widespread in Europe.

References: Bean (1980), Meikle (1984, 1992), Stace (1975).

G. HUTCHINSON

Salix cinerea Grey Willow

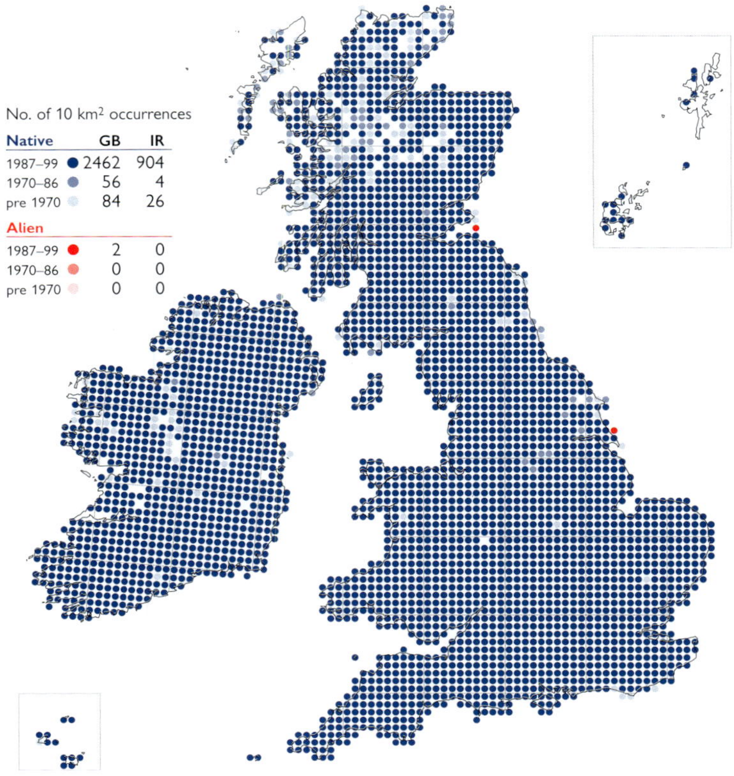

No. of 10 km² occurrences

Native	GB	IR
1987–99 ●	2462	904
1970–86 ●	56	4
pre 1970 ●	84	26
Alien		
1987–99 ●	2	0
1970–86 ●	0	0
pre 1970 ●	0	0

A shrub or small tree which grows in wet places, including woods, marshes and fens, by streams and bogs, and as a colonist of damp places on waste ground and in disused mineral workings. Generally lowland, with an exceptional record at 845 m on Great Dun Fell (Westmorland).

Native (change +0.84). This species may be increasing, especially in England and S. Ireland, but it is so frequent that this is barely obvious at the 10-km scale. There are two subspecies in the British Isles, both of which are mapped separately.

Eurosiberian Boreo-temperate element.

References: Atlas (189c), Alliende & Harper (1989), Grime *et al.* (1988), Hultén & Fries (1986), Jalas & Suominen (1976), Meikle (1984, 1992).

G. HUTCHINSON

Salix cinerea subsp. *cinerea*

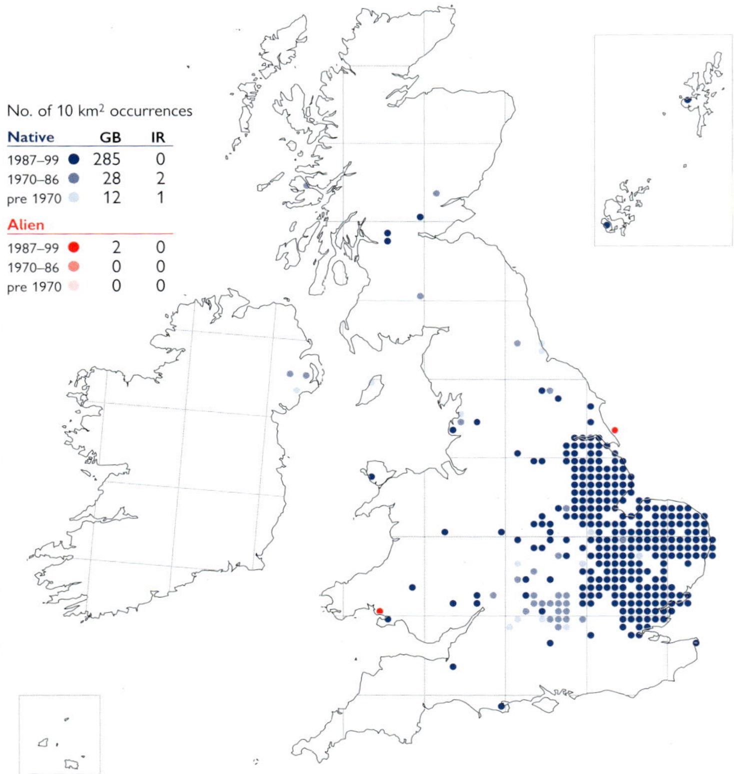

No. of 10 km² occurrences

Native	GB	IR
1987–99 ●	285	0
1970–86 ●	28	2
pre 1970 ●	12	1
Alien		
1987–99 ●	2	0
1970–86 ●	0	0
pre 1970 ●	0	0

A much-branched shrub which grows mostly in base-rich fens and marshes. Lowland.

Native. This subspecies was not mapped in the 1962 *Atlas*. On the western edge of its range it becomes more difficult to distinguish from subsp. *oleifolia*, but it is not clear whether this is because of clinal variation or hybridisation. It is almost certainly under-recorded.

Eurosiberian Boreo-temperate element.

References: Bolòs & Vigo (1990), Grime *et al.* (1988), Howitt & Howitt (1990), Hultén & Fries (1986), Jalas & Suominen (1976), Meikle (1984).

G. HUTCHINSON

Salix cinerea subsp. *oleifolia*

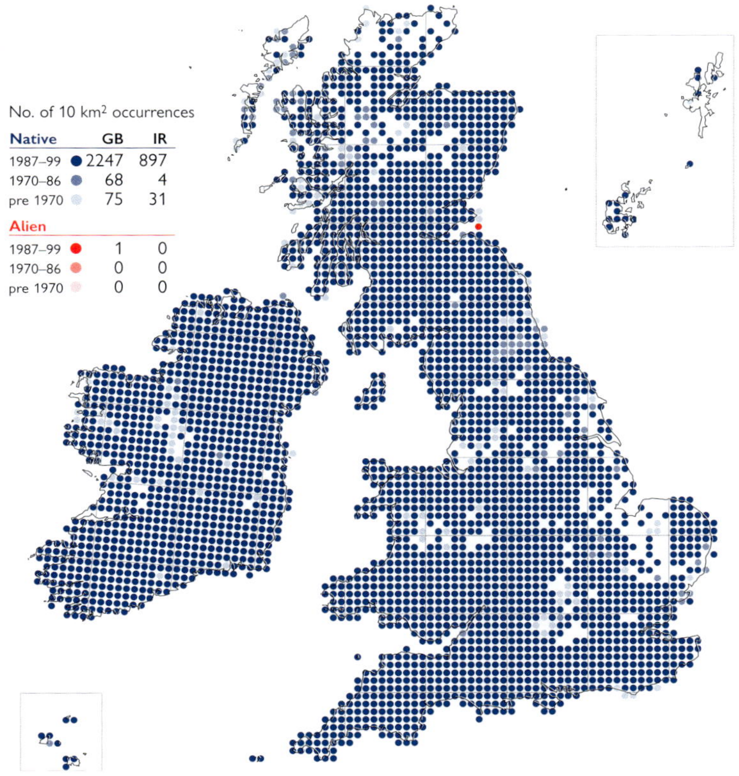

A shrub or small tree which grows on acidic to base-rich soils in wet places by streams, bogs and marshes, and in hedgerows, woods, moist wood margins and other marginal habitats, colonising waste ground especially in damp places. 0–665 m (Devil's Elbow, E. Perth), with an exceptional record at 845 m on Great Dun Fell (Westmorland).

Native. This is the usual subspecies of *S. cinerea* in much of Britain and Ireland. There is no evidence to suggest that there has been any change in its distribution since the 1962 *Atlas*.

Suboceanic Temperate element.

References: Bolòs & Vigo (1990), Grime *et al.* (1988), Howitt & Howitt (1990), Hultén & Fries (1986), Jalas & Suominen (1976), Meikle (1984).

G. HUTCHINSON

Salix cinerea × *S. phylicifolia* (*S.* × *laurina*)

Laurel-leaved Willow

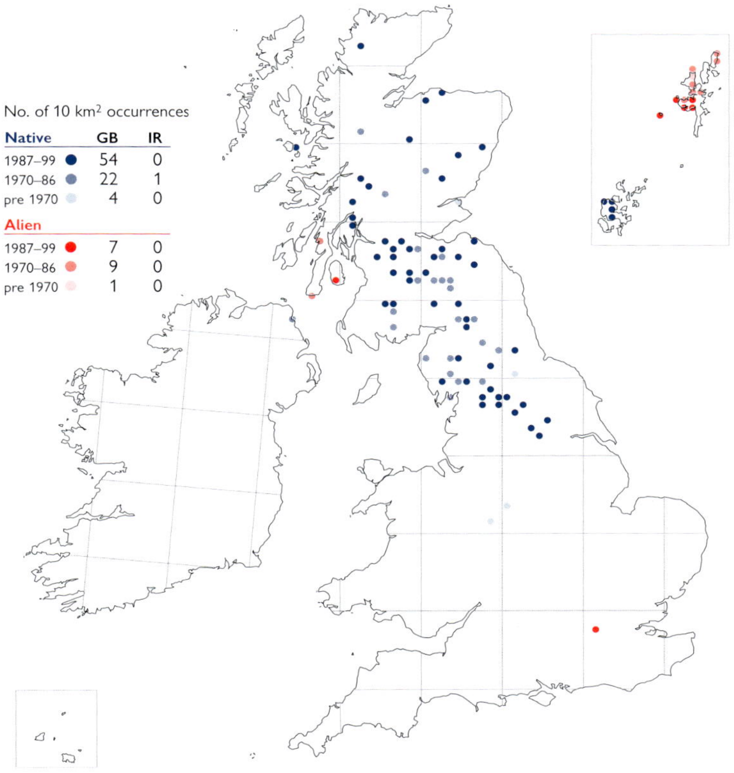

An erect shrub which grows frequently with the parents or as a relic of cultivation. Generally lowland, but reaching 330 m at Garrigill (Cumberland).

Native. This hybrid was not mapped in the 1962 *Atlas*, and there is no evidence to assess changes in its distribution.

Recorded from France and as an escape or a relic of cultivation in Denmark and Scandinavia.

References: Meikle (1984), Stace (1975), Synnott (1983).

G. HUTCHINSON

Salix cinerea × *S. purpurea* × *S. viminalis* (*S.* × *forbyana*) Fine Osier

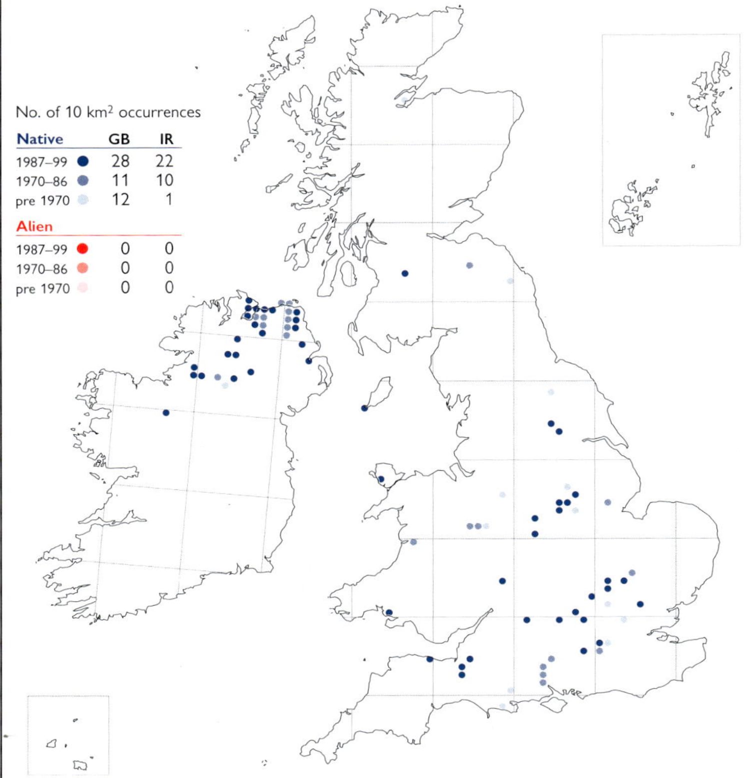

An erect shrub of wet places, especially moist thickets and river banks, and often found as a relic of osier cultivation. Lowland.

A hybrid between native and alien parents. This hybrid was first collected by the Rev. Joseph Forby in Norfolk at the beginning of the 19th century and named after him by Sir James Smith in 1804, but probably occurred before this time as it was used in basketry. The hybrid was not mapped in the 1962 *Atlas*, and there is no evidence to assess changes in its distribution. It is often not possible to separate native and alien populations, and all records are mapped as if they are native.

Outside Britain this is a rare hybrid, recorded from Austria, France, Germany and Switzerland.

References: Bean (1976), Meikle (1984), Stace (1975).

G. HUTCHINSON

Salix cinerea × *S. viminalis* (*S.* × *smithiana*)
Silky-leaved Osier

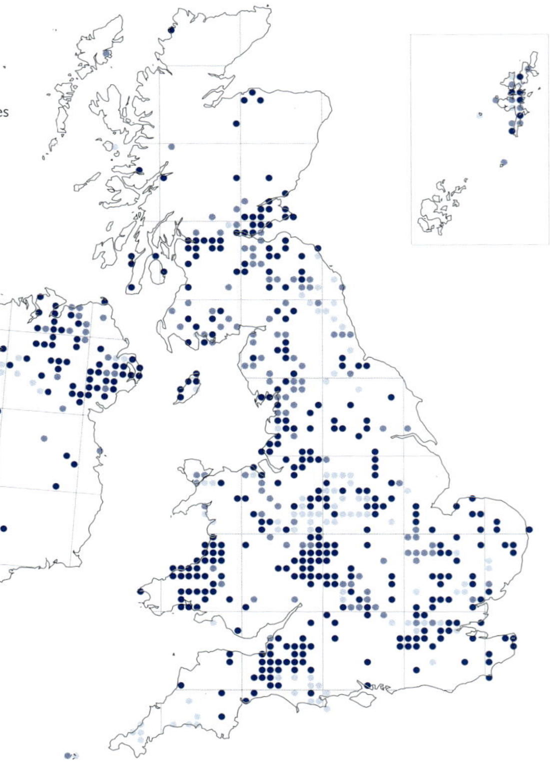

No. of 10 km² occurrences

Native		GB	IR
1987–99	●	336	58
1970–86	●	116	17
pre 1970	●	97	8
Alien			
1987–99	●	0	0
1970–86	●	0	0
pre 1970	●	0	0

An erect shrub or small tree which often grows with the parents, but is sometimes solitary. The uniformity of local male populations suggests that the hybrid can spread vegetatively. Other populations are entirely female. Lowland.

A spontaneous hybrid between native and alien parents. It may sometimes be planted, but it is difficult to distinguish these populations and all are treated as if they are native. This hybrid was not mapped in the 1962 *Atlas*, and there is no evidence to assess changes in its distribution. It is under-recorded.

Widespread in Europe.

References: Howitt & Howitt (1990), Meikle (1984), Stace (1975).

G. HUTCHINSON

Salix aurita Eared Willow

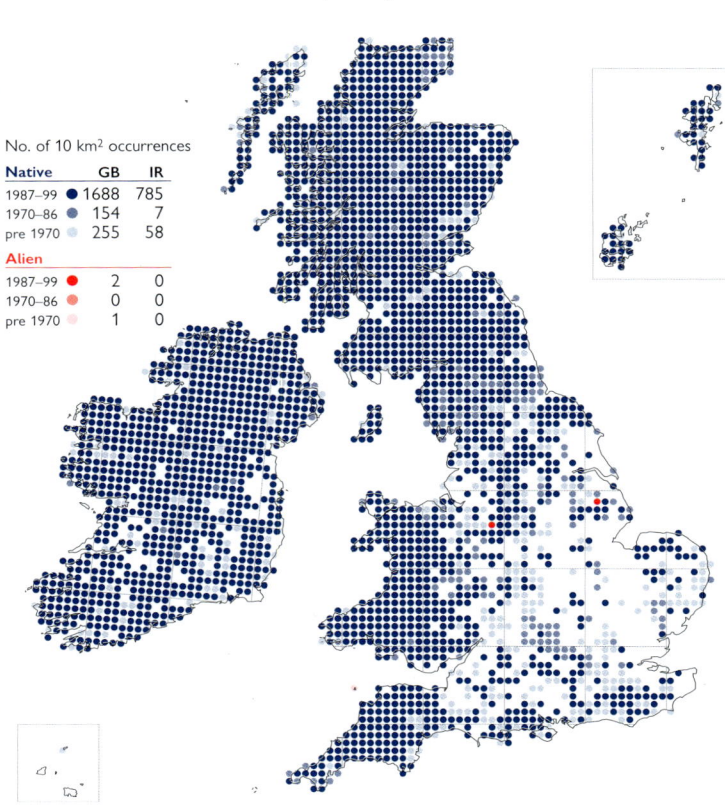

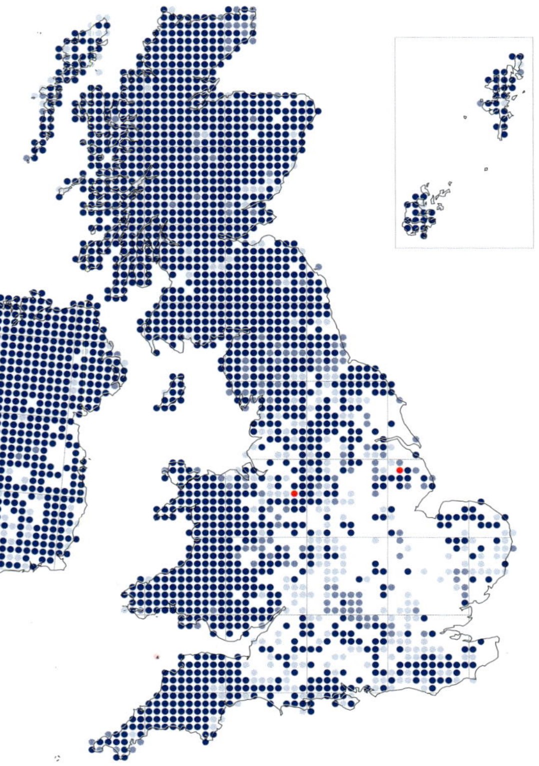

No. of 10 km² occurrences

Native		GB	IR
1987–99	●	1688	785
1970–86	●	154	7
pre 1970	●	255	58
Alien			
1987–99	●	2	0
1970–86	●	0	0
pre 1970	●	1	0

A much-branched shrub which grows on acidic soils on heathland and moorland, in scrub, by watercourses and on rocky streamsides and hills. 0–790 m (Atholl, E. Perth).

Native (change –0.01). This species is frequent or abundant in suitable habitats through much of its range. There have been losses in S.E. England, some of them since 1930. The acidic habitats it favours have been lost to drainage and agricultural improvement. In some areas, the hybrid *S.* × *multinervis* persists in the absence of *S. aurita*.

European Boreo-temperate element.

References: Atlas (189d), Howitt & Howitt (1990), Hultén & Fries (1986), Jalas & Suominen (1976), Mardon (1990), Meikle (1984, 1992).

G. HUTCHINSON

Salix aurita × *S. caprea* (*S.* × *capreola*)

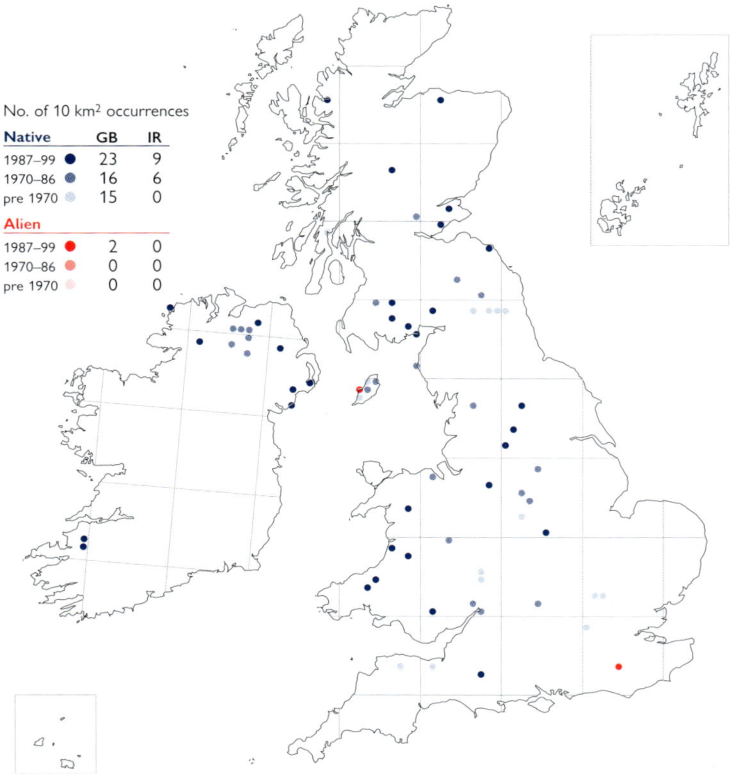

No. of 10 km² occurrences

Native		GB	IR
1987–99	●	23	9
1970–86	●	16	6
pre 1970	●	15	0
Alien			
1987–99	●	2	0
1970–86	●	0	0
pre 1970	●	0	0

An erect shrub or small tree of woodland margins, scrub and hedgerows. Generally lowland, but reaching 335 m at Brassington (Derbys.).

Native. *S.* × *capreola* was not mapped in the 1962 *Atlas* but the distribution has probably not changed significantly since then. It is almost certainly under-recorded.

Widespread in Europe.

Reference: Stace (1975).

G. HUTCHINSON

Salix aurita × *S. caprea* × *S. viminalis* (*S.* × *stipularis*) Eared Osier

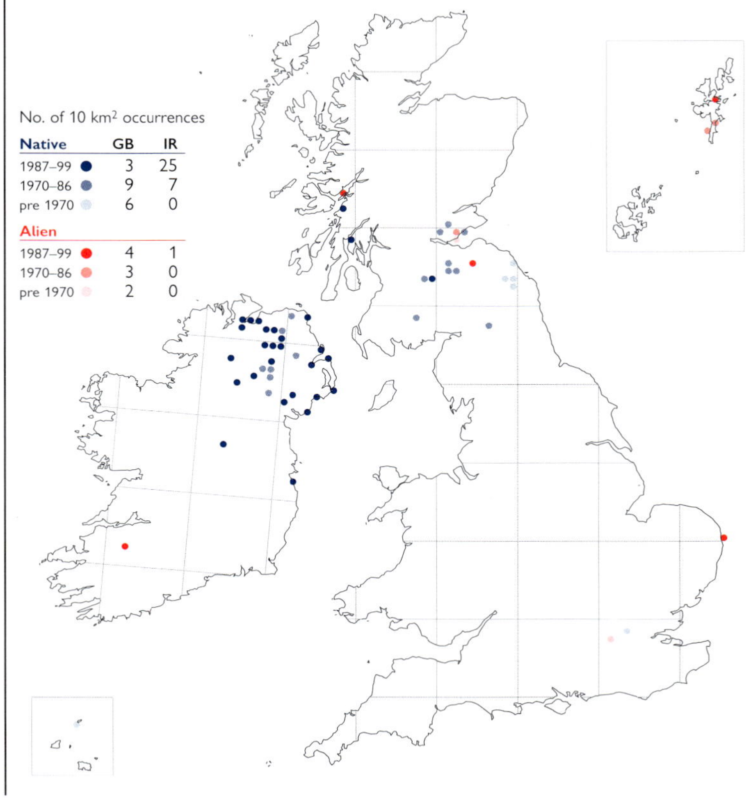

An erect shrub or small tree of damp places, hedgerows and scrub. It is probably planted at many of its sites. Lowland.

A hybrid between native and alien parents. There is no evidence to assess changes in the distribution of this hybrid, which is almost certainly under-recorded.

There are scattered records of this hybrid from mainland Europe; a male plant is known from Norway but all others are female.

Reference: Stace (1975).

G. HUTCHINSON

Salix aurita × *S. cinerea* (*S.* × *multinervis*)

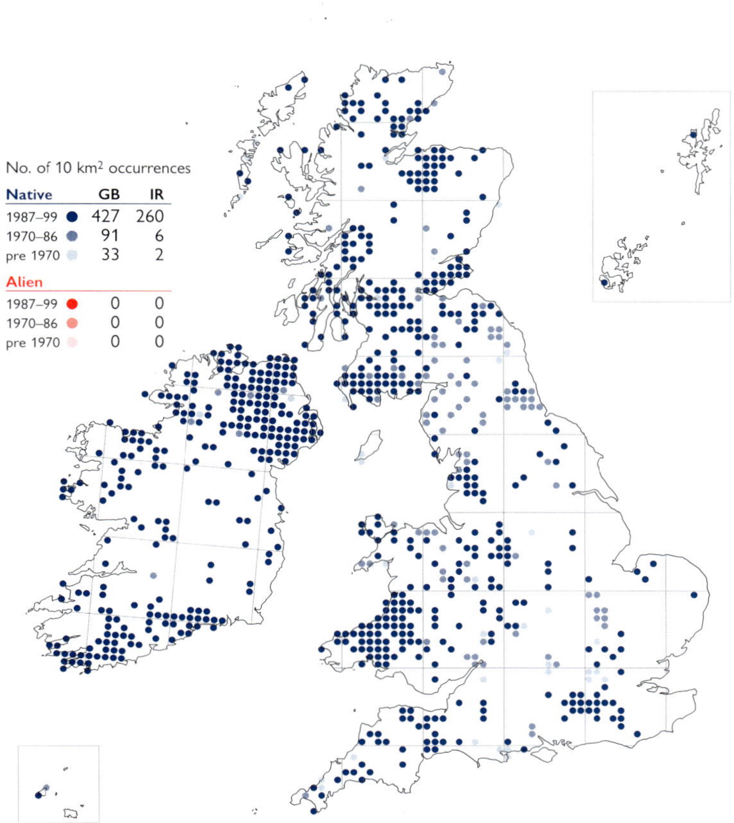

A much-branched shrub or small tree, recorded from hedgerows, heathy scrub and woodland margins on acidic soils, growing where the parent species coexist, and often where *S. aurita* is absent. The hybrid is sometimes linked by an unbroken series of intermediates to its parents, indicative of back-crossing. 0–320 m (north of Howgill Fells, Westmorland).

Native. This hybrid was not mapped in the 1962 *Atlas*, but the distribution has probably not changed significantly since then. The map must be regarded as incomplete.

Widespread in N. & C. Europe, although many records in Norden are referable to one or other of the parents (Jonsell, 2000).

References: Howitt & Howitt (1990), Meikle (1984), Stace (1975).

G. HUTCHINSON

Salix aurita × *S. repens* (*S.* × *ambigua*)

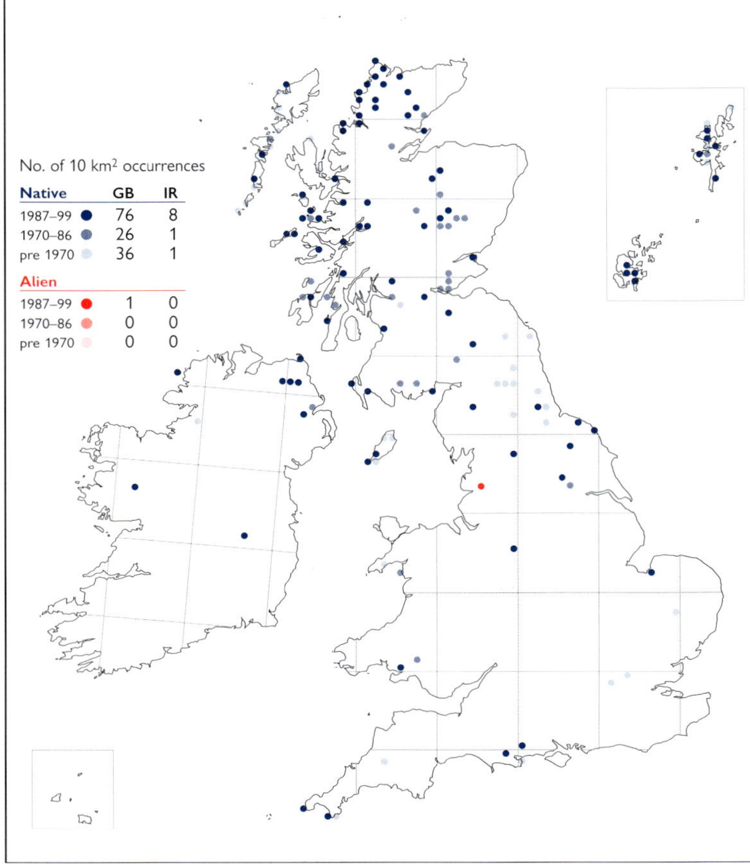

A low, sprawling shrub which grows on acidic or slightly base-enriched heaths and moors and in limestone grassland, occurring where both parents are to be found, though usually where one of them is much less abundant than the other. 0–350 m (Greystoke Forest, Cumberland).

Native. This hybrid was not mapped in the 1962 *Atlas*, and there is no evidence to assess changes in its distribution.

Widespread in Europe.

References: Bean (1980), Howitt & Howitt (1990), Meikle (1984), Stace (1975).

G. HUTCHINSON

Salix aurita × *S. viminalis* (*S.* × *fruticosa*) Shrubby Osier

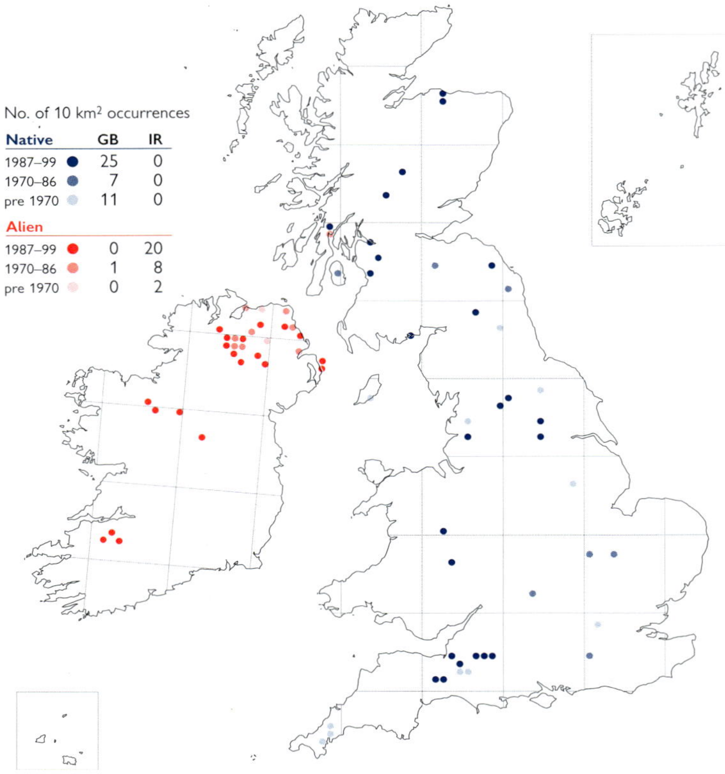

No. of 10 km² occurrences

Native		GB	IR
1987–99	●	25	0
1970–86	●	7	0
pre 1970	●	11	0
Alien			
1987–99	●	0	20
1970–86	●	1	8
pre 1970	●	0	2

An erect shrub or small tree of hedgerows and thickets. Only female plants are known. This hybrid sometimes arises spontaneously, but it has been planted for basketry and more recently for biomass production. Lowland.

A spontaneous hybrid between native and alien parents. This hybrid may have increased in recent years, primarily due to biomass plantings.

Recorded from widely scattered localities in Europe.

References: Meikle (1984), Stace (1975).

G. HUTCHINSON

Salix myrsinifolia Dark-leaved Willow

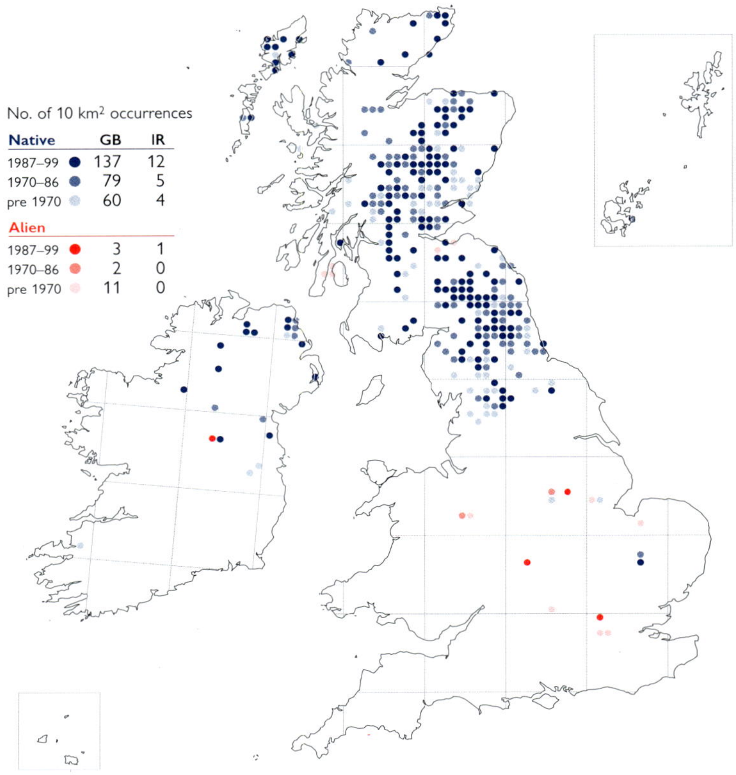

No. of 10 km² occurrences

Native		GB	IR
1987–99	●	137	12
1970–86	●	79	5
pre 1970	●	60	4
Alien			
1987–99	●	3	1
1970–86	●	2	0
pre 1970	●	11	0

A shrub, or less often a small tree, which grows mainly on rocks or on gravelly river banks and lake shores, less frequently in thickets on marshy ground or by wet woodland margins. In Scotland it sometimes grows in wet dune-slacks and as a dwarf, spreading shrub on wet rock ledges. Generally lowland, but reaching 940 m on Stob Binnein (Mid Perth).

Native (change +0.93). There has been no significant change in the distribution of this species since the 1962 *Atlas*, but it is now much better recorded. It does, however, closely resemble *S. phylicifolia* and their hybrid, and some mapped records may be erroneous.

Eurosiberian Boreal-montane element.

References: Atlas (190a), Harron (1992), Hultén & Fries (1986), Jalas & Suominen (1976), Mardon (1990), Meikle (1984, 1992), Stewart *et al.* (1994), Synnott (1983).

G. HUTCHINSON

Salix myrsinifolia × *S. phylicifolia* (*S.* × *tetrapla*)

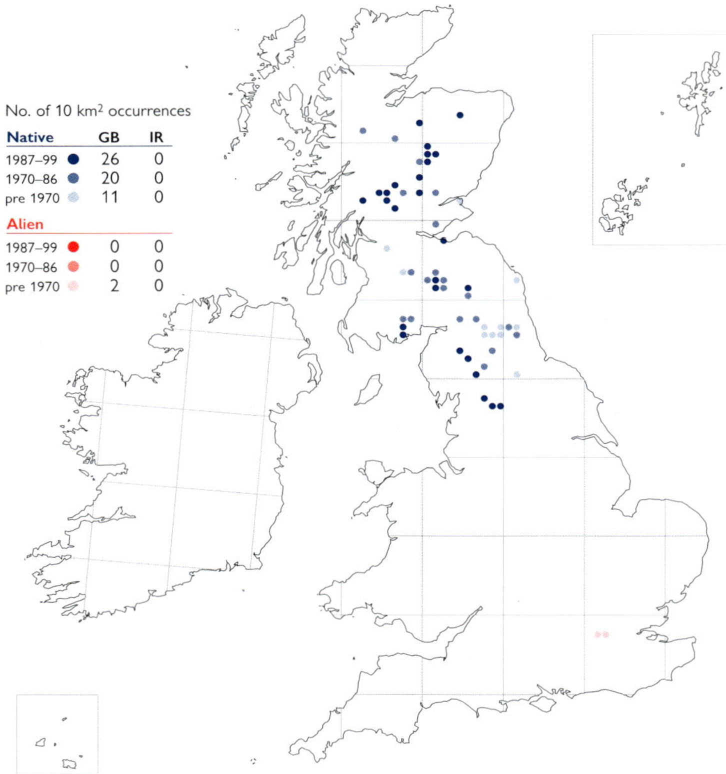

No. of 10 km² occurrences

Native		GB	IR
1987–99	●	26	0
1970–86	●	20	0
pre 1970	●	11	0
Alien			
1987–99	●	0	0
1970–86	●	0	0
pre 1970	●	2	0

An erect shrub or small tree of damp rocky ground, river banks and lake shores. It is found where both parents are present, often in populations showing a complete range of intermediates between them, and also in the absence of *S. phylicifolia* in highland Scotland. Generally lowland, but upper altitudinal limit unknown.

Native. This hybrid was not mapped in the 1962 *Atlas*, and there is no evidence to assess changes in its distribution. It is probably under-recorded.

Frequent in N. Europe. In Fennoscandia, as in Britain, it is fertile and backcrosses to its parents are more frequent in some areas than *S. phylicifolia*, or occur in its absence (Jonsell, 2000).

References: Bean (1976), Meikle (1992), Stace (1975).

G. HUTCHINSON

Salix phylicifolia Tea-leaved Willow

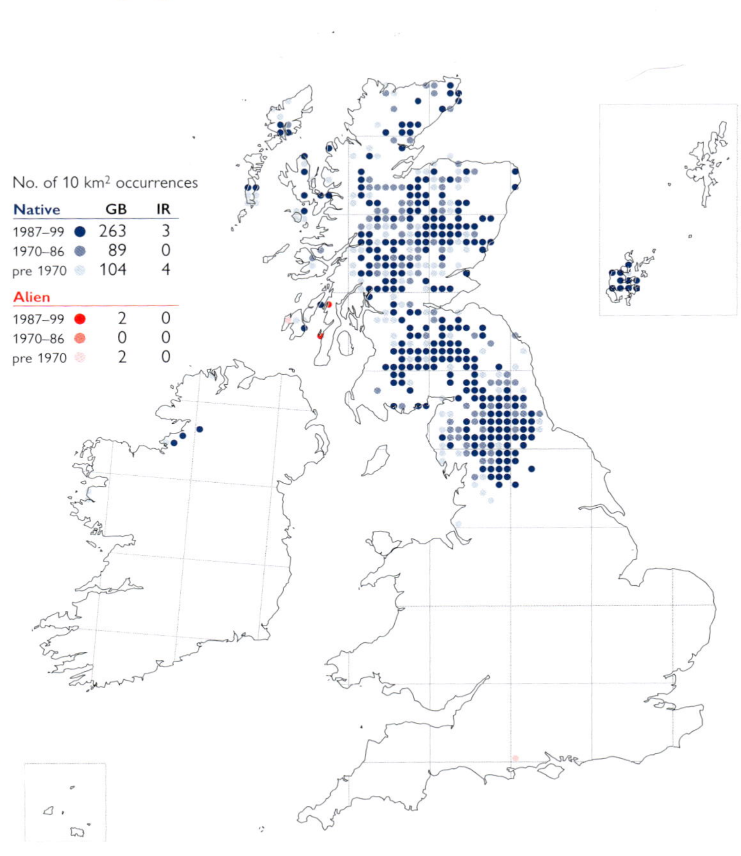

A much-branched shrub or small tree which grows by ponds, streams and rivers, and in damp rocky places, preferring base-rich soils and sometimes associated with Carboniferous limestone. In Ireland it is a montane species. From near sea level to 685 m at Catstycam (Westmorland).

Native (change −0.14). Although now much better recorded, the distribution of this species has not changed significantly since the 1962 *Atlas*. It is very similar to *S. myrsinifolia*, and may be mis-identified for it and the hybrid between them.

Circumpolar Boreo-arctic Montane element.

References: Atlas (190b), Curtis & McGough (1988), Hultén & Fries (1986), Jalas & Suominen (1976), Mardon (1990), Meikle (1984, 1992), Synnott (1983).

G. HUTCHINSON

Salix repens Creeping Willow

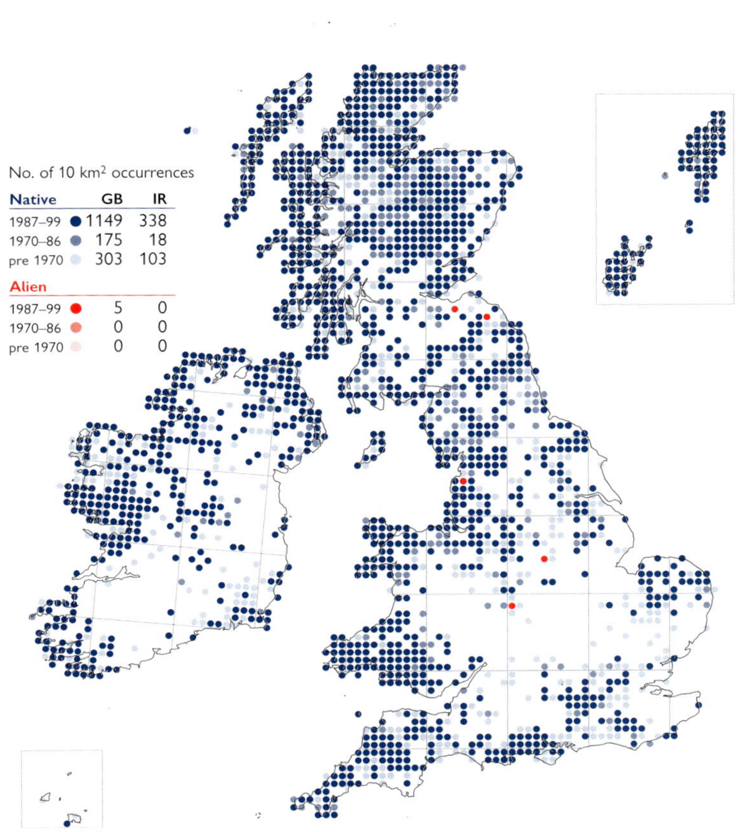

A variable shrub growing in a range of habitats. The prostrate var. *argentea* and var. *repens* are typically found on fixed dunes and especially dune-slacks, in maritime heaths and heathy grassland and on inland heaths and moorland. The erect var. *fusca* is found in fens. The species becomes more confined to moist or wet habitats in the south and east of its range. 0–855 m (Atholl, E. Perth).

Native (change −0.42). The species has declined, especially inland in England, but since it was mapped in the 1962 *Atlas* as 'all records' it is difficult to know when these losses occurred.

Eurosiberian Boreo-temperate element.

References: Atlas (190c), Bean (1980), Fowler *et al.* (1983), Howitt & Howitt (1990), Hultén & Fries (1986), Jalas & Suominen (1976), Meikle (1984).

G. HUTCHINSON

Salix lapponum Downy Willow

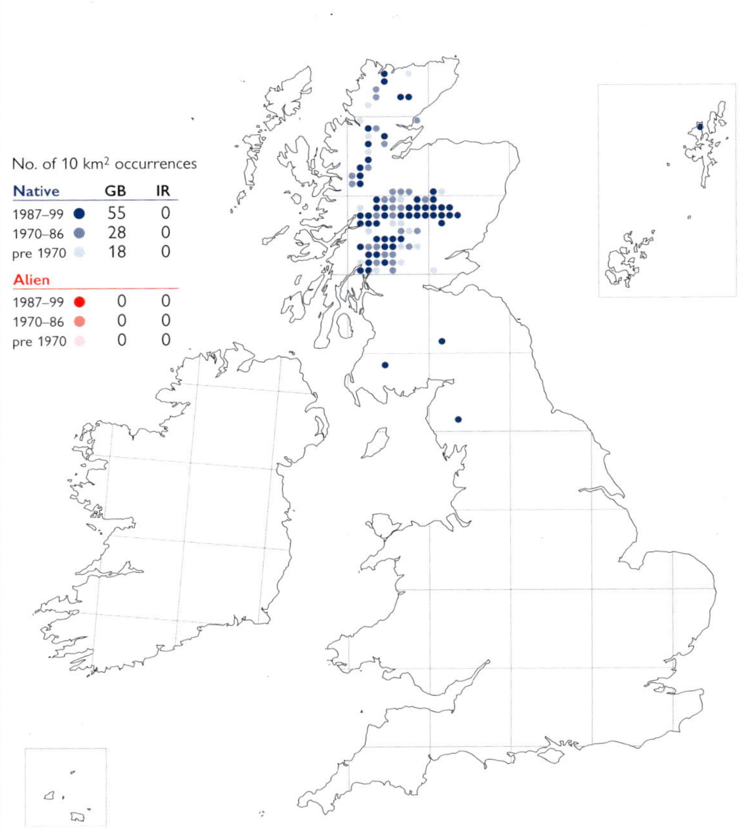

A low shrub of moist or wet, moderately base-enriched sites on rocky mountain slopes and cliffs. This species tolerates a wider range of soil conditions than most montane *Salix*, but is now largely confined to cliffs. From 210 m (Ochil Hills, W. Perth) to 1000 m (Aonach Beag, Westerness).

Native (change −0.73). Trends in the distribution of this species are unclear. Some small, ungrazed colonies appear to be producing no or very few seedlings, and may be in decline. Further work will be needed to establish whether it has been lost from all the 10-km squares for which there are only pre-1987 records.

Eurosiberian Boreo-arctic Montane element.

References: Atlas (190d), Bean (1980), Hultén & Fries (1986), Jalas & Suominen (1976), Mardon (1990), Meikle (1984), Meusel *et al.* (1965), Stewart *et al.* (1994).

G. HUTCHINSON

Salix lanata Woolly Willow

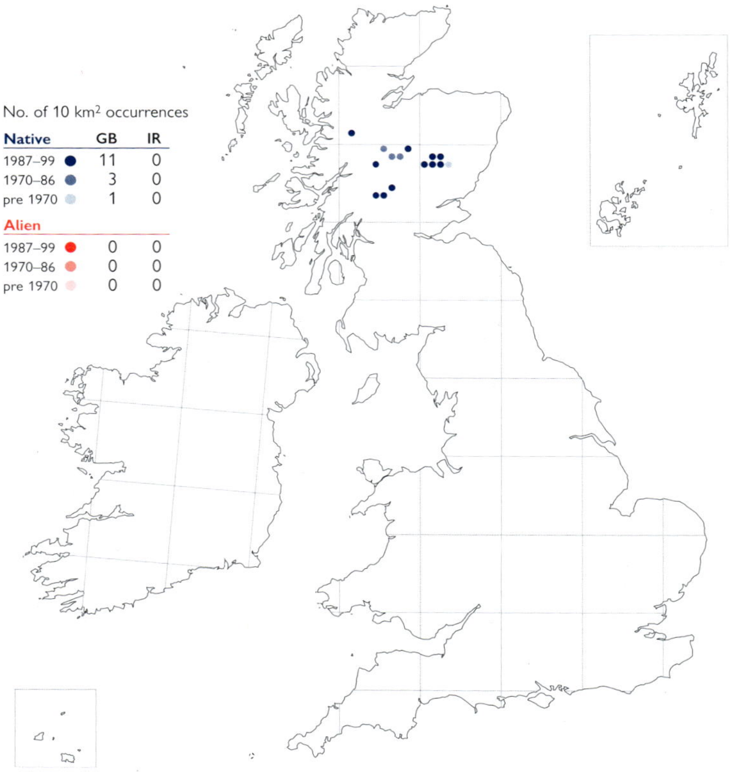

A low shrub of damp base-rich mountain rock ledges and crags in N.-facing corries, often in areas where snow lies late. It is usually on calcareous schist and rarely on limestone. From 620 m to 1035 m (Geal Charn, Westerness).

Native (change +0.07). This species is now known from more 10-km squares than were mapped in the 1962 *Atlas*. However, numbers have declined due to increasing grazing pressure by deer and sheep. Some populations have been reduced to single plants, and others lost altogether. Fencing is being tried to arrest the decline at some sites.

Circumpolar Arctic-montane element; absent from mountains of C. Europe.

References: Atlas (191a), Bean (1980), Hultén & Fries (1986), Jalas & Suominen (1976), Mardon (1990), Meikle (1984), Wigginton (1999).

G. HUTCHINSON

Salix arbuscula Mountain Willow

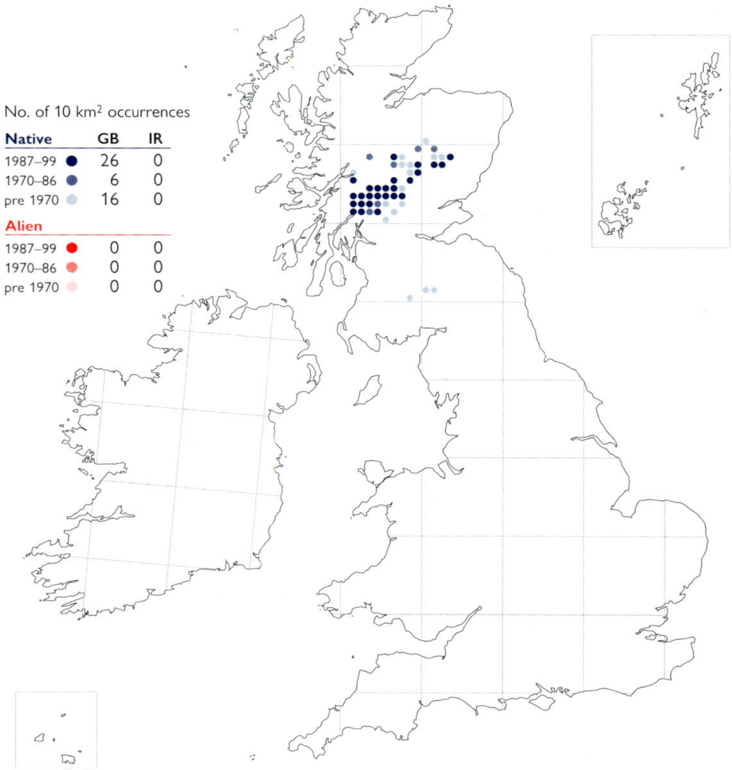

A low shrub of base-rich substrates on mountains, occurring in moist or wet habitats, mostly in flushes, on gravelly soil near burns and on damp ledges of calcareous rock. From 460 m (Ben Lui, Mid Perth) to 870 m (Carn Gorm in Glen Lyon, Mid Perth) but rarely below 600 m.

Native (change –0.12). This species may be more tolerant of grazing than most other montane *Salix*. The distribution has not significantly changed since the 1962 *Atlas*, but some new sites have been discovered. It was lost from two of the squares in S. Scotland before 1930.

European Arctic-montane element; absent from mountains of C. Europe.

References: Atlas (191b), Hultén & Fries (1986), Jalas & Suominen (1976), Mardon (1990), Meikle (1984), Stewart *et al.* (1994).

G. HUTCHINSON

Salix myrsinites Whortle-leaved Willow

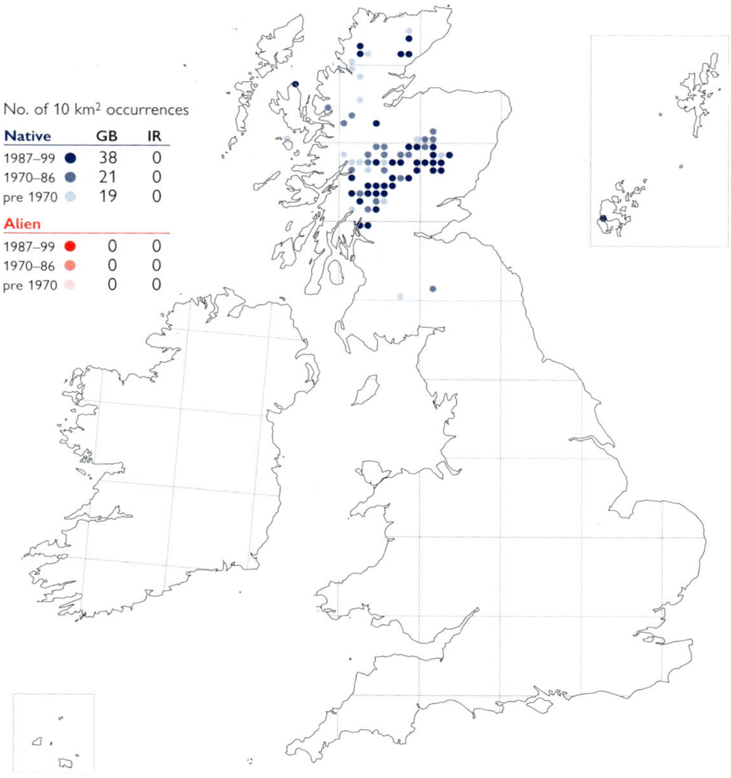

A low, spreading shrub which grows mainly in moist or wet, base-enriched sites on mountains. It is restricted to ungrazed or lightly grazed areas. From 180 m (Inchnadamph, W. Sutherland) to 915 m (Ben Alder and Aonach Beag, Westerness).

Native (change –0.58). Data are insufficient to show current trends, but like *S. lapponum*, isolated colonies of this species are possibly declining. Its procumbent habit may, however, allow it to escape grazing at some sites.

European Arctic-montane element; absent from mountains of C. Europe.

References: Atlas (191c), Bean (1980), Hultén & Fries (1986), Jalas & Suominen (1976), Mardon (1990), Meikle (1984), Stewart *et al.* (1994).

G. HUTCHINSON

Salix herbacea Dwarf Willow

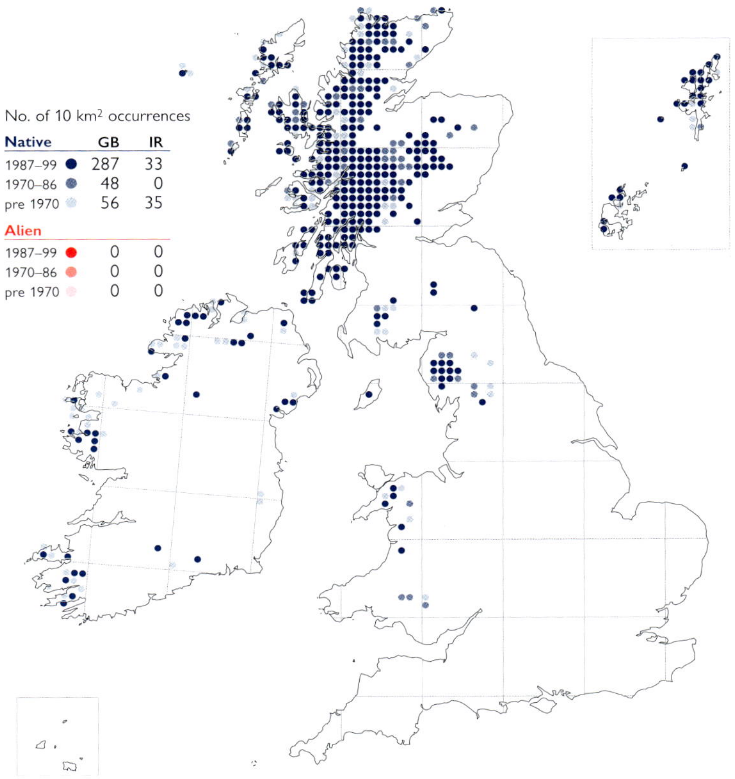

A dwarf shrub of open, often bryophyte-rich communities in areas of late snow-lie, or in conditions of extreme exposure. It grows on erosion surfaces or stony ground on windswept ridges and cols, on screes, in corries and hollows, and locally on ledges and in montane grass-heath. From near sea level (Fethaland, Shetland) to 1310 m (Ben Nevis, Westerness), but mostly not below 600 m.

Native (change –0.33). Populations in England appear to be in decline (Halliday, 1997), and many sites in Ireland were lost before 1930. There has been no significant overall change in Scotland, and populations in Wales are stable.

European Arctic-montane element; also in N. America.

References: Atlas (191d), Beerling (1998), Hultén & Fries (1986), Jalas & Suominen (1976), Meikle (1984), Meusel *et al.* (1965).

G. HUTCHINSON

Salix reticulata Net-leaved Willow

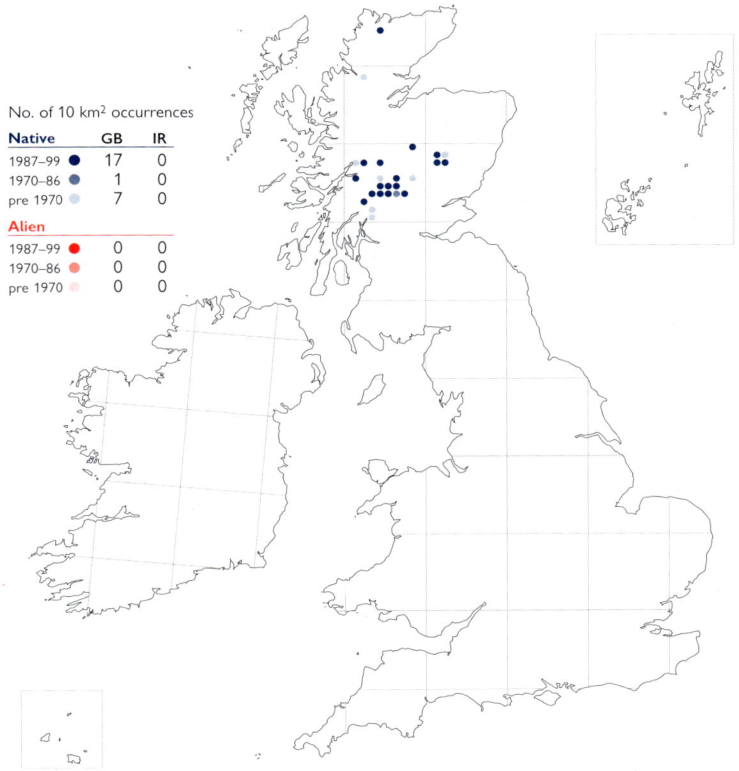

A creeping dwarf shrub which grows on base-rich montane rock ledges of limestone or calcareous schist. From 650 m (Creag Mhor, Mid Perth) to 1125 m (Ben Lawers, Mid Perth), but rarely found at the lower end of this range.

Native (change –0.17). Although lost from several 10-km squares before 1930, the overall distribution of this species now appears to be stable. Local populations, however, may fluctuate in numbers.

Circumpolar Arctic-montane element.

References: Atlas (192a), Hultén & Fries (1986), Jalas & Suominen (1976), Mardon (1990), Meikle (1984), Meusel *et al.* (1965), Stead (1980), Stewart *et al.* (1994).

G. HUTCHINSON

Sisymbrium irio London-rocket

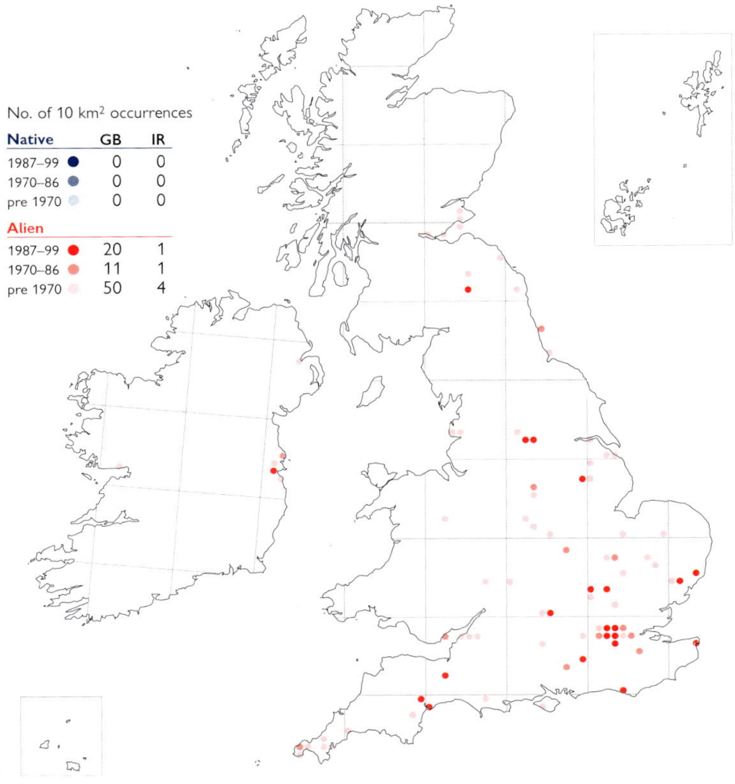

This annual is occasionally naturalised in waste places, in pavement cracks and on roadsides, banks and walls, but is more frequently found as a casual, sometimes with grain imports and formerly as a wool alien. Lowland.

Neophyte (change +0.13). *S. irio* has been known from the wild in Britain since about 1650. It has fluctuated in abundance, being notably frequent in London after the Great Fire of 1666 but absent between the early 19th and mid 20th centuries. There is little change in its distribution since the 1962 *Atlas*.

Native of W. Europe, the Mediterranean region and W. Asia.

References: Atlas (51b), Bolòs & Vigo (1990), Jalas & Suominen (1994), Rich (1991).

D. A. PEARMAN

Sisymbrium loeselii False London-rocket

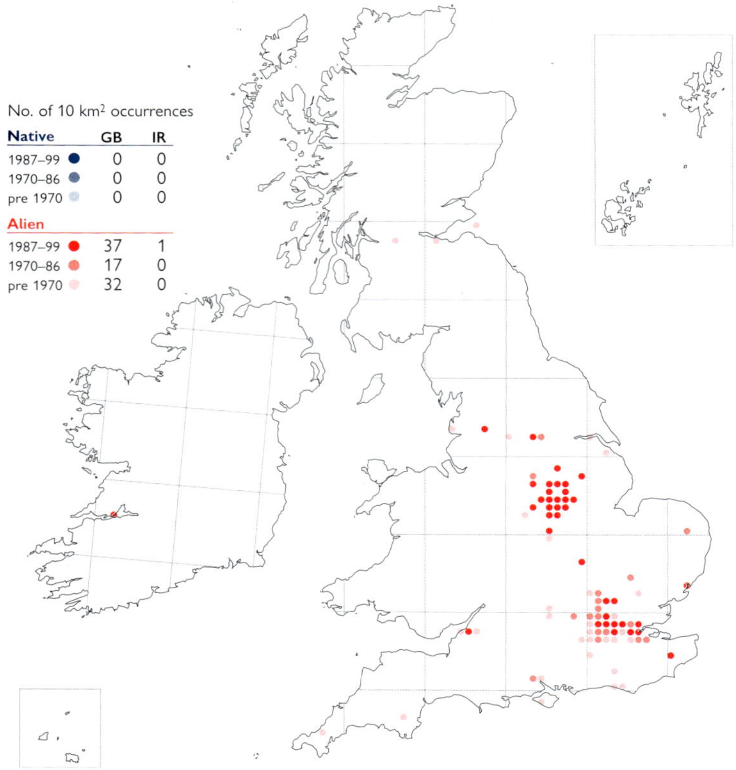

An annual of waste places, railways and roadsides. It is usually casual, but very occasionally naturalised, as in parts of London. It is a common contaminant of bird-seed mixtures. Lowland.

Neophyte. *S. loeselii* was introduced to cultivation in Britain in 1787. It was first recorded in the wild in 1883 (Middlesex), has only become persistent since the 1940s, and is now probably increasing.

Native of C. & E. Europe, W. & C. Asia; widely naturalised in Europe north and west of its native range.

References: Hultén & Fries (1986), Jalas & Suominen (1994), Meusel *et al.* (1965), Rich (1991).

D. A. PEARMAN

Sisymbrium altissimum Tall Rocket

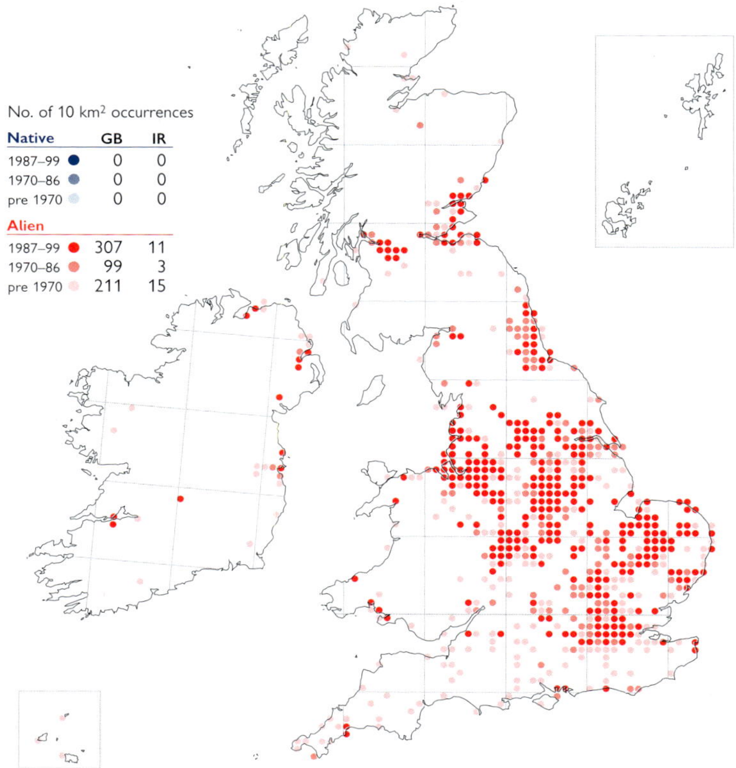

An annual which is frequently naturalised on rubbish tips and waste ground, and by roads and railways. It is a contaminant of bird-seed and grass-seed mixtures. Populations can be short-lived. Lowland.

Neophyte (change –0.84). *S. altissimum* was being cultivated in Britain by 1768. Although it was recorded in the wild by 1862 (Middlesex), it was a rare plant until the population was reinforced by plants introduced by troops returning from the First World War battlefields, where it was apparently abundant (Clarke, 1925).

Native of E. Europe and W. Asia; widely naturalised in temperate W. Europe, N. America and elsewhere.

References: Atlas (51d), Hultén & Fries (1986), Jalas & Suominen (1994), Meusel *et al.* (1965), Rich (1991).

D. A. PEARMAN

Sisymbrium orientale Eastern Rocket

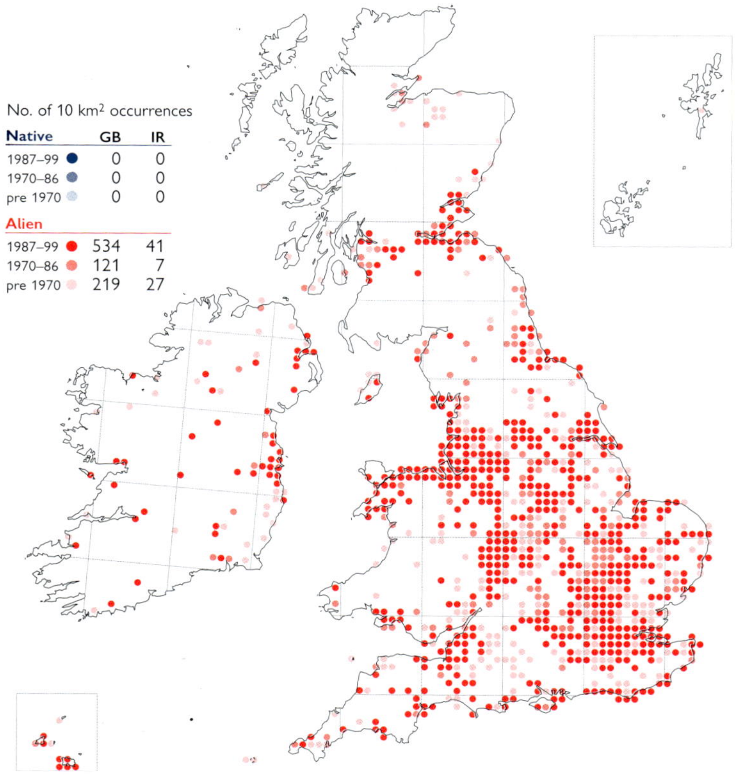

An annual naturalised in rough ground and waste places, including railway land. Lowland.

Neophyte (change –0.24). *S. orientale* was cultivated in Britain by 1739 and is also introduced with grain. It was recorded in the wild by 1859 (Surrey). It has spread since the Second World War, and is more frequent in some areas now than when mapped in the 1962 *Atlas*.

Native of S. Europe, the Mediterranean region and W. Asia; widely naturalised in temperate Europe.

References: Atlas (51c), Bolòs & Vigo (1990), Jalas & Suominen (1994), Rich (1991).

D. A. PEARMAN

Sisymbrium officinale Hedge Mustard

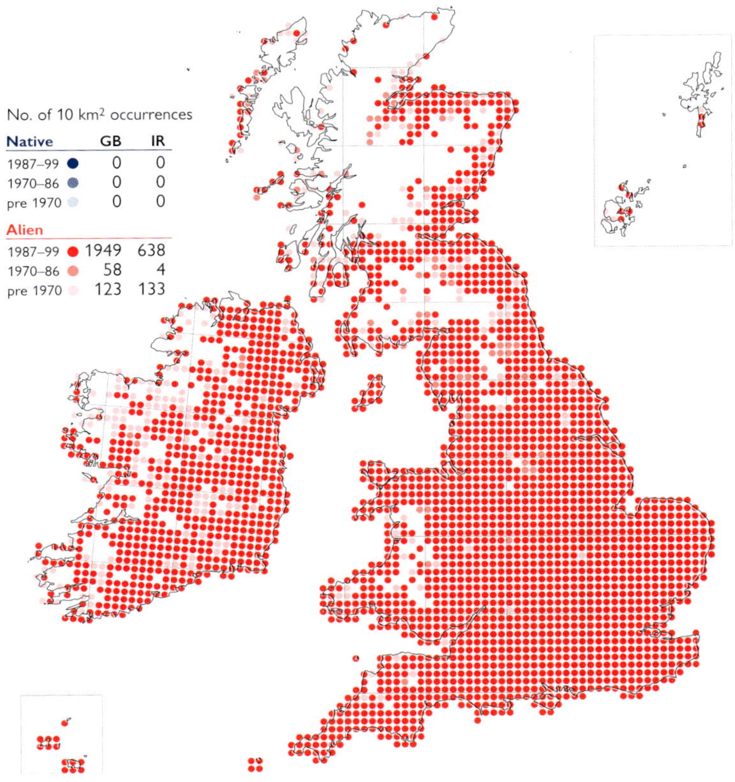

An annual or biennial herb of dry, neutral or base-rich soils, doing best in open situations and frequent in cultivated ground, on roadsides and waste ground. It is almost invariably associated with man, but it also occurs rarely in natural habitats such as river banks. 0–315 m (Garrigill, Cumberland).

Archaeophyte (change –0.21). There has been no significant change in the distribution of *S. officinale* since the 1962 *Atlas*.

As an archaeophyte *S. officinale* has a European Southern-temperate distribution, but it is widely naturalised so that its distribution is now Circumpolar Southern-temperate.

References: Atlas (51a), Hultén & Fries (1986), Jalas & Suominen (1994), Rich (1991).

D. A. PEARMAN

Descurainia sophia Flixweed

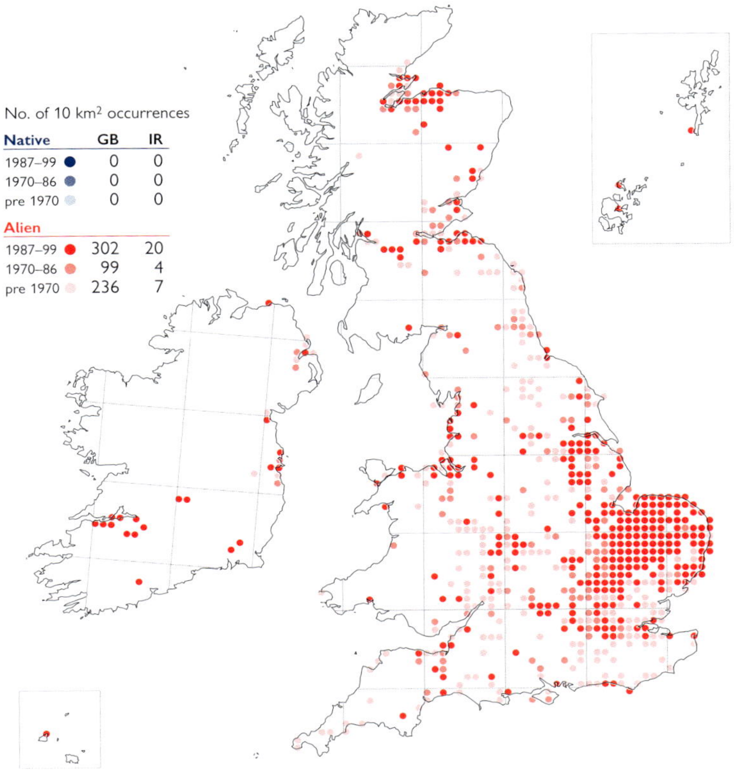

An annual, or rarely biennial, herb which is locally abundant as a weed in arable fields in light soils in E. England. Elsewhere, it can occur in long-established populations, but is usually a casual in waste places. Lowland.

Archaeophyte (change –0.29). *D. sophia* suffered a considerable decline in lowland Britain before 1930, but there has been little change since the 1962 *Atlas*. It appears to be resistant to many herbicides.

The precise native range of *D. sophia* is uncertain; as an archaeophyte it has a Eurosiberian Temperate range and is naturalised in N. America and the S. hemisphere.

References: Atlas (52b), Best (1977), Hultén & Fries (1986), Jalas & Suominen (1994), Meusel *et al.* (1965), Rich (1991), Salisbury (1964).

D. A. PEARMAN

Alliaria petiolata Garlic Mustard

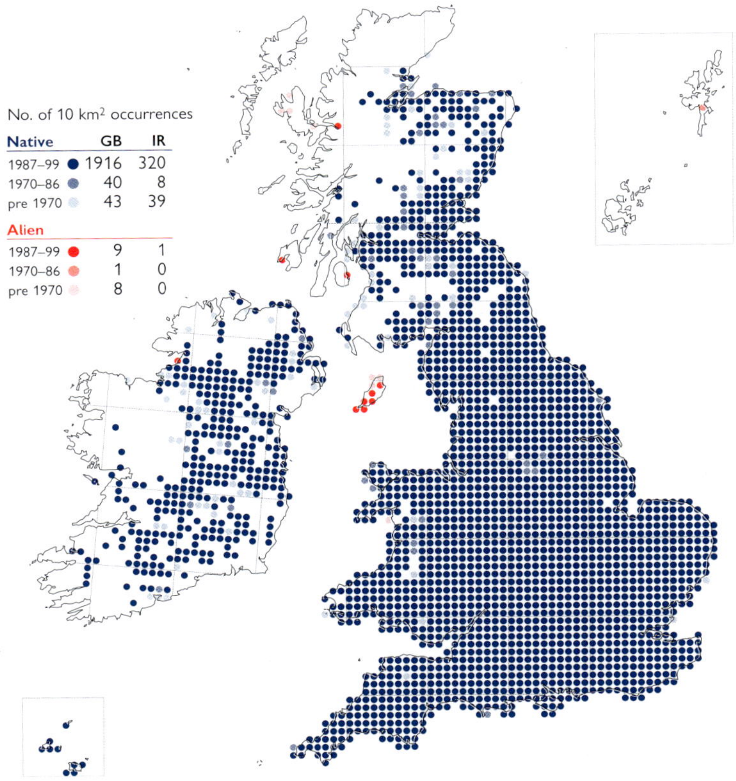

A biennial or monocarpic herb, found in a wide range of habitats including disturbed woodland, woodland edges and clearings, shaded hedge banks, river banks, the base of walls, road verges, waste ground, farmyards and gardens. It grows especially well on relatively fertile, moist soils, but avoids only the most acidic sites. Generally lowland, but reaching 535 m S. of Garsdale Head (N.W. Yorks.).

Native (change +0.03). No change is apparent in the distribution of this species since the 1962 *Atlas*.

European Temperate element; also in C. Asia and widely naturalised outside its native range.

References: Atlas (50d), Anderson *et al.* (1996), Cavers *et al.* (1979), Grime *et al.* (1988), Hultén & Fries (1986), Jalas & Suominen (1994), Meusel *et al.* (1965), Rich (1991).

D. A. PEARMAN

Arabidopsis thaliana Thale Cress

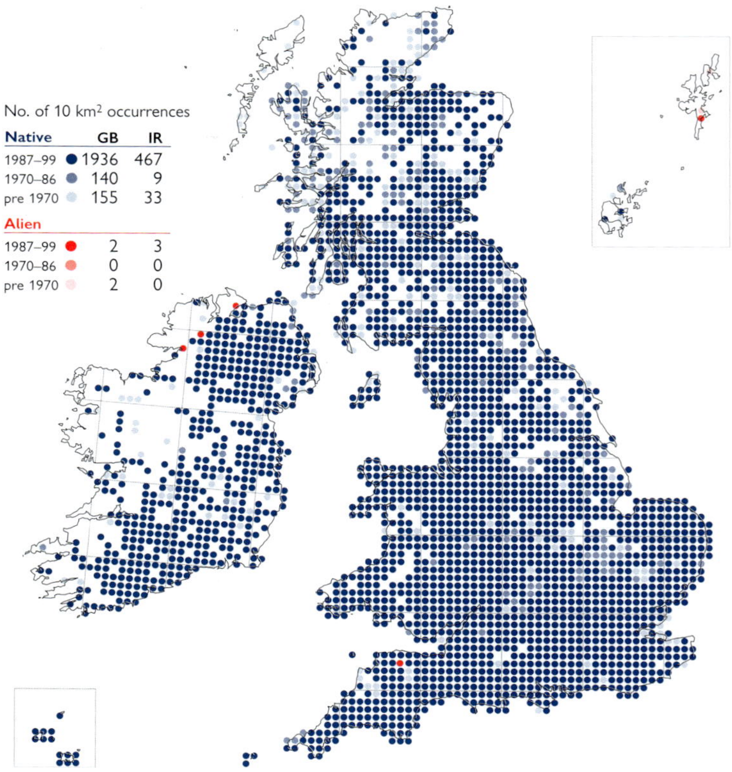

This winter-annual is a pioneer species, intolerant of competition, which is found on rocky ground, dunes and other open sandy or calcareous habitats. It is also very frequent as a weed in gardens and nurseries, and on all sorts of waste ground, especially by railways. 0–850 m (Meall nan Tarmachan, Mid Perth).

Native (change +1.21). The range of *A. thaliana* is stable. Because of its small genome, rapid growth and copious seed production, it is a fundamental research plant and the subject of a multi-national genome project.

Eurosiberian Temperate element, but widely naturalised so that distribution is now Circumpolar Temperate.

References: Atlas (52a), Grime *et al.* (1988), Hultén & Fries (1986), Jalas & Suominen (1994), Meusel *et al.* (1965), Rich (1991), Wilson *et al.* (1991).

D. A. PEARMAN

Isatis tinctoria Woad

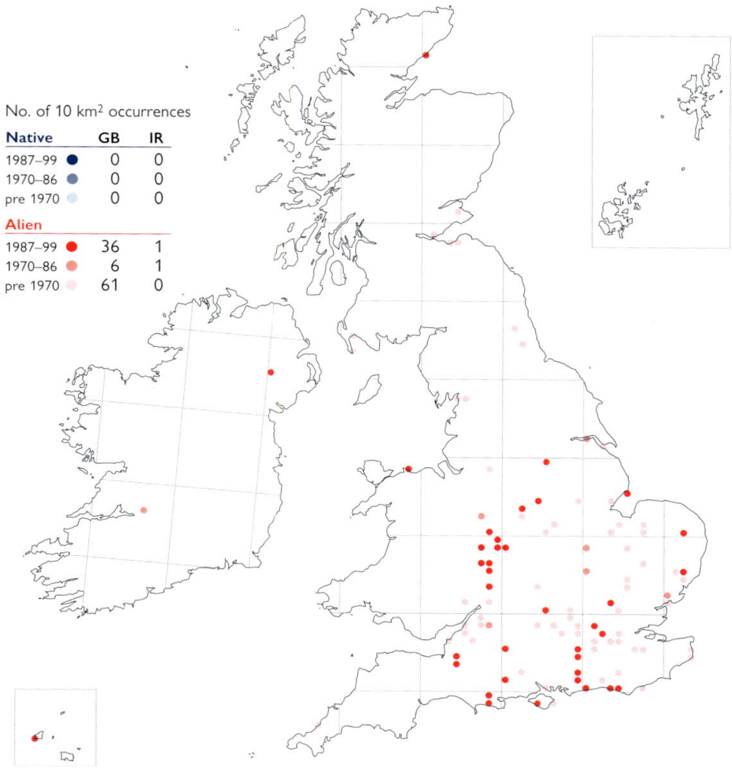

A biennial or perennial herb, found only in ruderal habitats such as quarries, bare cliffs, arable fields, docks and waste places. It is usually casual, but persists at a few sites including Guildford, Surrey (first recorded in 1814) and Tewkesbury, E. Gloucs. (1818). Lowland.

Archaeophyte (change +1.08). *I. tinctoria* has been used as a dye plant in Britain since the Iron Age. In the 1930s, however, the world's last two woad mills, in Lincolnshire, were closed and since then it has been a rare casual.

I. tinctoria sens. lat. is widespread across Eurasia; subsp. *tinctoria* may be native to S.W. Asia but is now widely naturalised in Europe. It is expanding around the Baltic coast.

References: Atlas (39a), Hultén & Fries (1986), Hurry (1930), Jalas & Suominen (1994), Rich (1991).

D. A. PEARMAN

Bunias orientalis Warty-cabbage

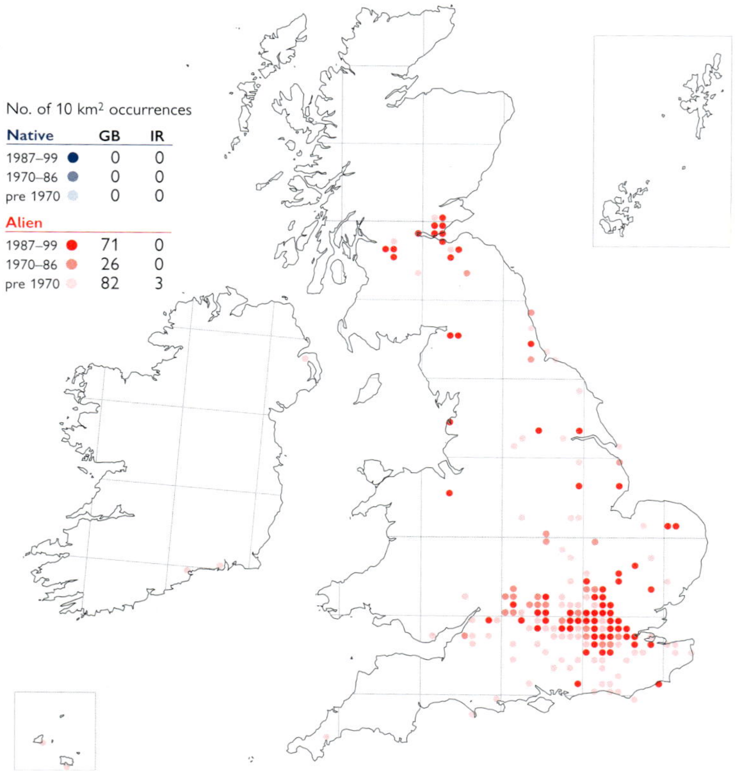

A perennial, or occasionally biennial, herb, very persistent on waste ground, roadsides, docks and railways. Lowland.

Neophyte (change −0.27). This species was cultivated in Britain by 1731, and was recorded from the wild a few years later in 1739. It was apparently frequent in the early 20th century, probably due to introductions from contaminated grain, but it has gradually declined since then. It has recently gained favour as a salad vegetable (Turkish Rocket).

A native of E. Europe and W. Asia, now naturalised throughout most of temperate W. & C. Europe.

References: Atlas (42b), Hultén & Fries (1986), Jalas & Suominen (1994), B.M.G. Jones (1959), Meusel *et al.* (1965), Rich (1991).

D. A. PEARMAN

Erysimum cheiranthoides Treacle-mustard

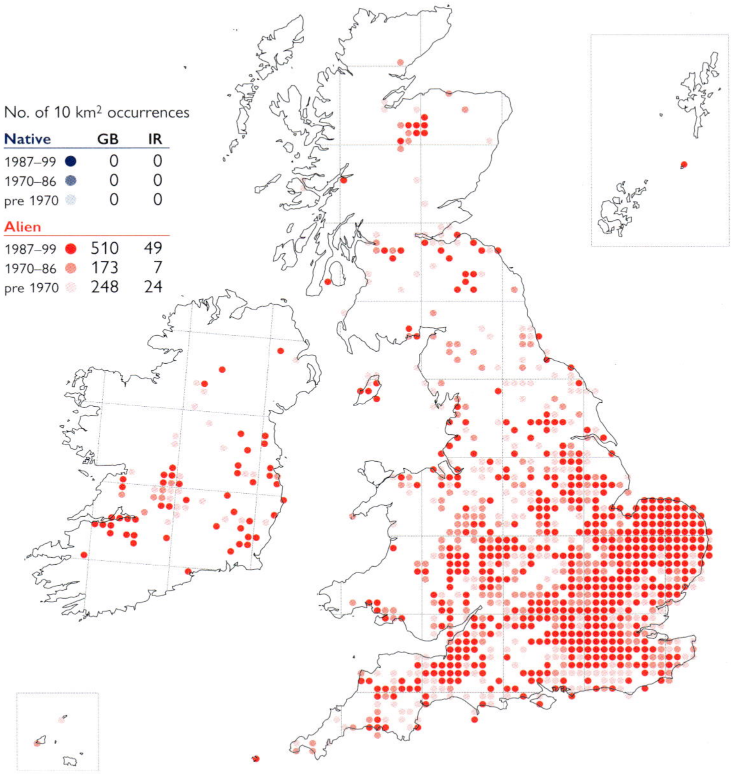

No. of 10 km² occurrences

Native	GB	IR
1987–99	0	0
1970–86	0	0
pre 1970	0	0
Alien		
1987–99	510	49
1970–86	173	7
pre 1970	248	24

An annual, locally frequent in arable fields but common also as a weed of waste ground, roadsides and railways. It prefers sandy ground. Mainly lowland, but recorded at 435 m (Clun Forest, Salop).

Archaeophyte (change –0.65). There is archaeological evidence of this species in Bronze Age and Roman Britain. It is susceptible to herbicides and has declined since 1950 as an arable weed, but in other habitats it seems stable or is even increasing as a contaminant of clover- or grass-seed.

Native range uncertain but thought to include E. Europe, Siberia and western N. America; now a widely naturalised plant with a Circumpolar Boreo-temperate distribution.

References: Atlas (50b), Hultén & Fries (1986), Jalas & Suominen (1994), Meusel *et al.* (1965), Roberts & Boddrell (1983b), Rich (1991), Salisbury (1964). D. A. PEARMAN

Erysimum cheiri Wallflower

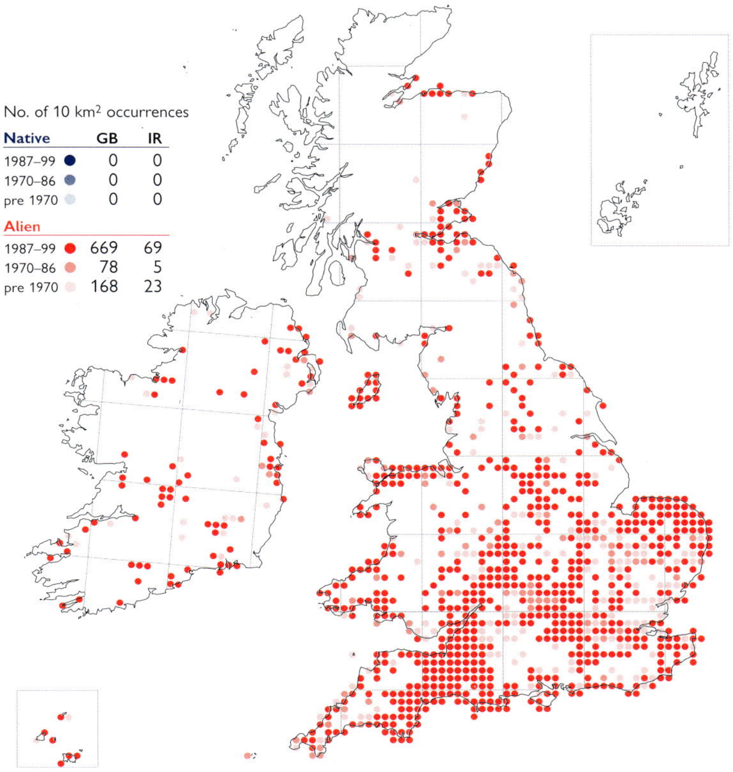

No. of 10 km² occurrences

Native	GB	IR
1987–99	0	0
1970–86	0	0
pre 1970	0	0
Alien		
1987–99	669	69
1970–86	78	5
pre 1970	168	23

A perennial herb widely naturalised on cliffs, old walls and rocks, particularly on calcareous substrates where it is often very persistent. It tolerates poor, thin, dry soils, but a warm site is essential. Lowland.

Archaeophyte (change +1.05). *E. cheiri* has been cultivated since medieval times, and its first record as a wild plant was as long ago as 1548. There has been no change in its range since it was mapped in the 1962 *Atlas*; the increase in frequency may reflect the more enthusiastic recording of alien species in recent years.

A plant of garden origin, widely naturalised in W. & C. Europe.

References: Atlas (50c), Bolòs & Vigo (1990), Jalas & Suominen (1994), Rich (1991).

D. A. PEARMAN

Hesperis matronalis Dame's-violet

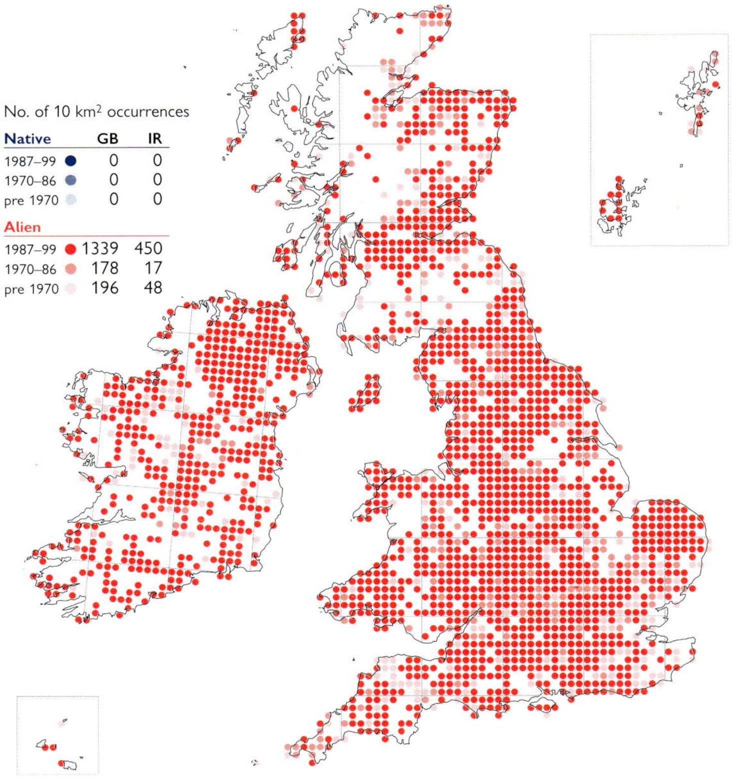

No. of 10 km² occurrences

Native	GB	IR
1987–99	0	0
1970–86	0	0
pre 1970	0	0
Alien		
1987–99	1339	450
1970–86	178	17
pre 1970	196	48

A perennial, or sometimes biennial, herb of shaded moist habitats, found in hedgerows and wood borders, on river banks, roadsides and waste ground, usually near habitation. It is often well-naturalised, but only where there is little competition. Casual plants occur on tips and in waste places. Lowland.

Neophyte (change +1.53). *H. matronalis* has been cultivated in gardens since at least 1375 (Harvey, 1981), and was known in the wild by 1805. The increase in records since the 1962 *Atlas* may indicate a real change, but is more likely to be due to better recording of aliens.

Native of S. Europe and W. Asia, although absent from the Mediterranean area; widely naturalised in C. & N. Europe, N. America and elsewhere.

References: Atlas (50a), Hultén & Fries (1986), Jalas & Suominen (1994), Rich (1991).

D. A. PEARMAN

Malcolmia maritima Virginia Stock

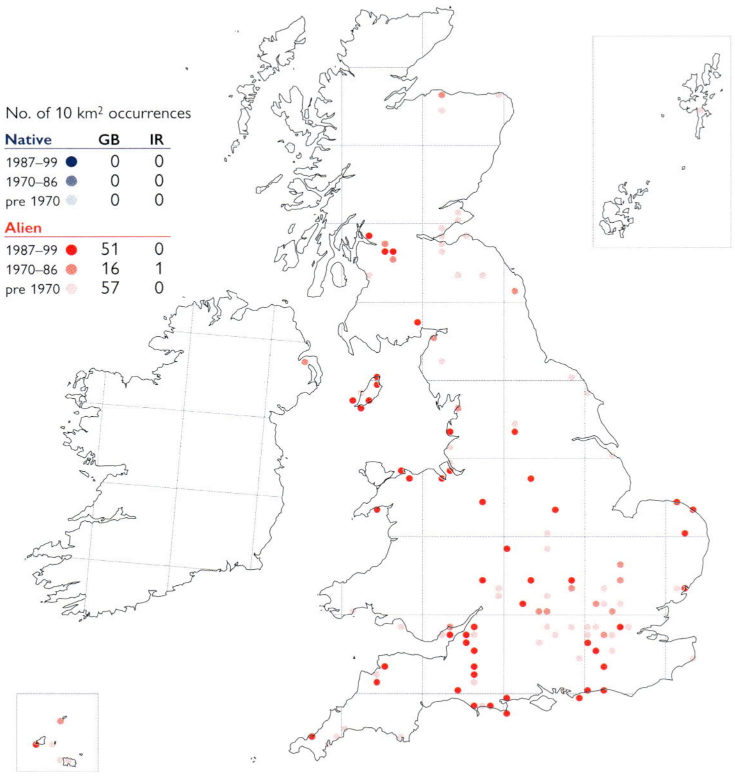

An annual, almost always occurring in our area as a casual of waste ground, rubbish tips and roadsides. Lowland.

Neophyte. This species has been cultivated in Britain since 1713 and was known from the wild by 1866 (Middlesex). It is difficult to assess changes in distribution, but there is no evidence for any change in frequency in recent years.

Native of the coast of Italy and the Balkans; naturalised further west along the Mediterranean coast.

References: Bolòs & Vigo (1990), Jalas & Suominen (1994), Rich (1991).

D. A. PEARMAN

Matthiola incana Hoary Stock

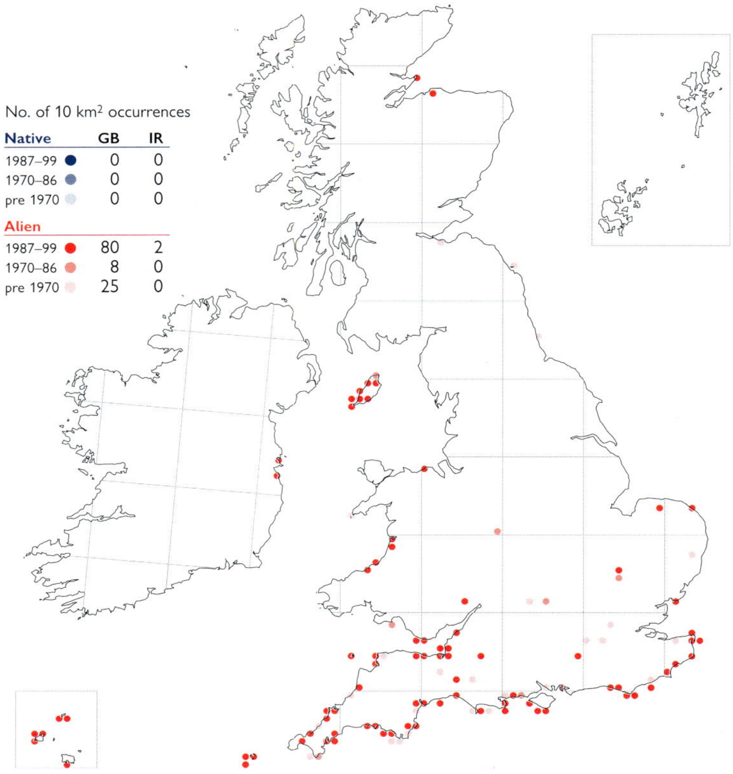

A short-lived perennial, well-naturalised on sea-cliffs, shingle and other habitats by the sea, and occasionally inland where it is more obviously a garden escape. Lowland.

Neophyte (change +0.75). *M. incana* has been widely cultivated in Britain since at least 1596, but may have been introduced long before that date. It has been considered to be native, but it was not recorded in the wild until 1808. It is now more frequent than when mapped in the 1962 *Atlas.*

A Mediterranean-Atlantic species; widely naturalised outside its native range.

References: Atlas (49c), Bolòs & Vigo (1990), Jalas & Suominen (1994), Rich (1991).

D. A. PEARMAN

Matthiola sinuata Sea Stock

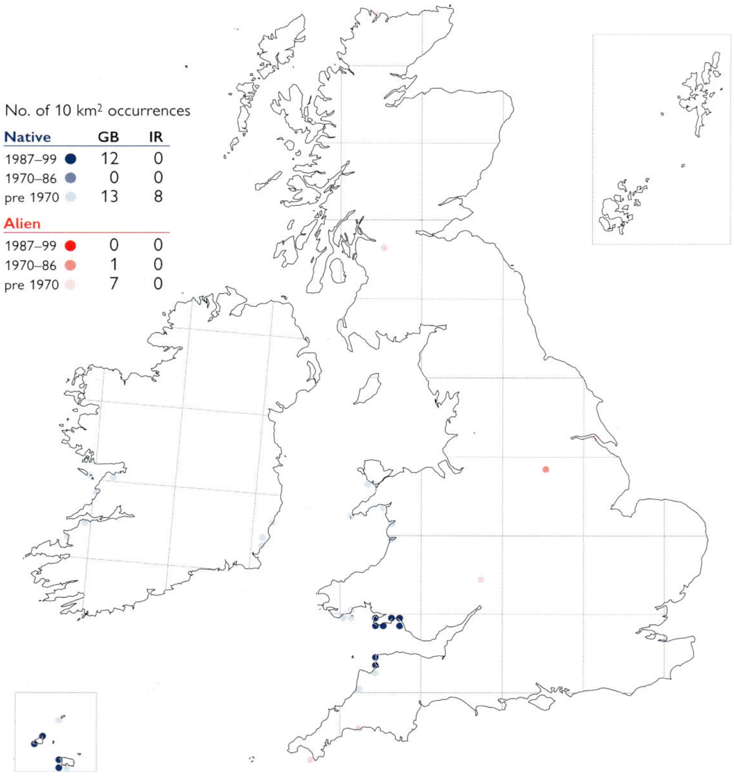

A biennial or short-lived perennial herb of sand dunes and sea-cliffs. Most colonies are on young, fairly mobile dunes, and it has probably spread in the past by seeds floating in sea-water to new sites. Lowland.

Native or alien. *M. sinuata* was first recorded in Britain in 1633. It seems to have declined slowly over the historical period, but its overall distribution is currently stable. However, large populations often disappear for reasons that remain uncertain; the species was thought to have become extinct in Glamorgan in 1848 until it was rediscovered there in 1964.

Mediterranean-Atlantic element.

References: Atlas (49d), Curtis & McGough (1988), Jalas & Suominen (1994), Rich (1991), Wigginton (1999).

D. A. PEARMAN

Matthiola longipetala Night-scented Stock

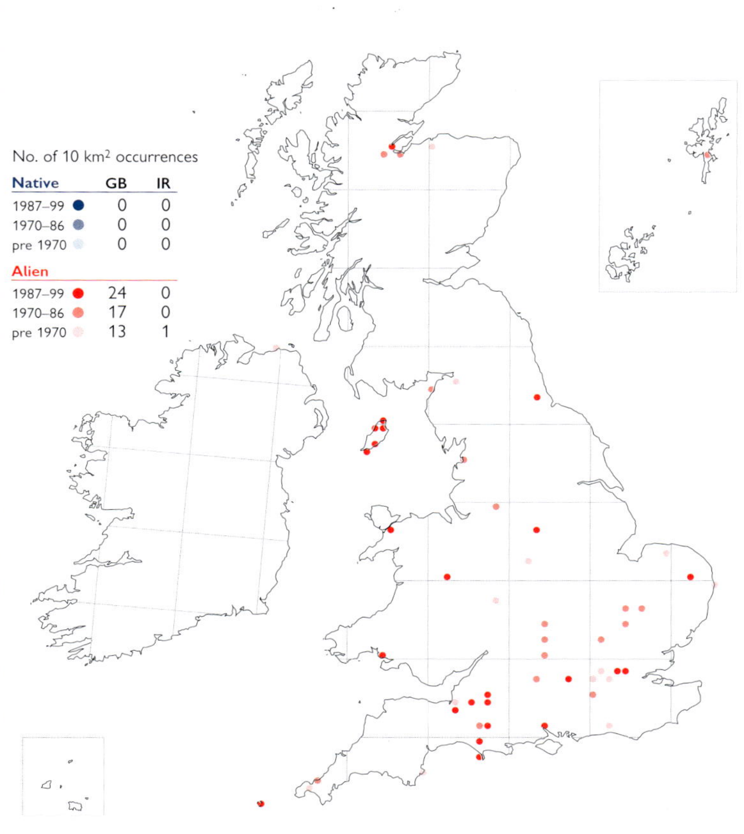

An annual occurring as a casual on rubbish tips and waste ground and in close proximity to gardens. Lowland.

Casual. Introduced by 1818 (and possibly as '*M. odoratissima*' by 1797), *M. longipetala* is commonly grown in gardens for its scented flowers. It was known to occur in the wild by 1905.

Native of Greece, N. Africa & S.W. Asia; it is frequent in the deserts of Arabia.

References: Jalas & Suominen (1994), Rich (1991).

D. A. PEARMAN

Barbarea vulgaris Winter-cress

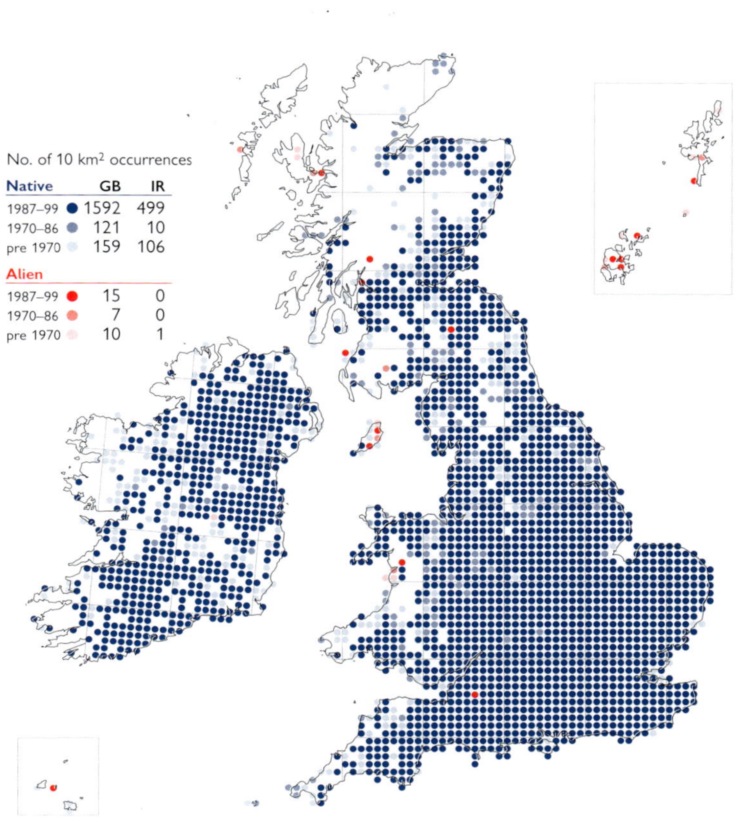

A biennial or perennial herb preferring a damp substrate, and thus widespread by rivers and streams, avoiding only the most acidic sites. It is also found on roadsides, by hedges and in ditches and waste places. It requires a degree of disturbance, which may be provided by seasonal flooding or human activity. Generally lowland, but occurring as a casual at 380 m near Alston (Cumberland).

Native (change –0.02). The distribution of this species shows no appreciable change since the 1962 *Atlas*.

Eurosiberian Temperate element; widely naturalised outside its native range.

References: Atlas (45d), Hultén & Fries (1986), Jalas & Suominen (1994), MacDonald & Cavers (1991), Meusel *et al.* (1965), Rich (1987, 1991).

D. A. PEARMAN

Barbarea stricta Small-flowered Winter-cress

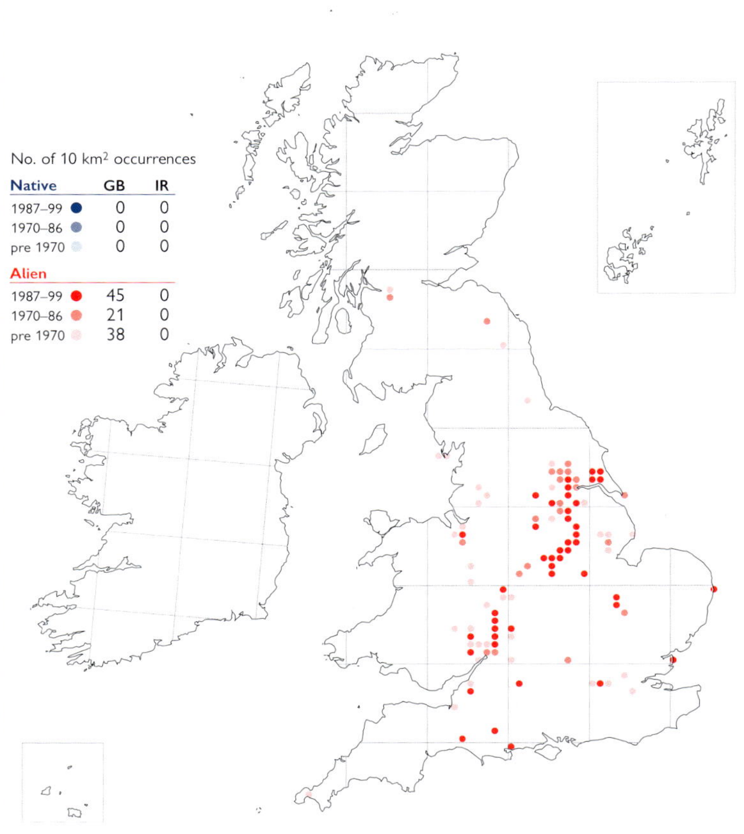

A biennial or perennial herb of moist places by rivers, ditches, canals and marshes, and a rare casual of waste places. Lowland.

Neophyte (change +0.50). *B. stricta* was first recorded in Britain in 1843, although it may have previously been confused with *B. vulgaris*, and it is stable in its distribution.

A Eurosiberian Boreo-temperate species, absent as a native from much of W. Europe.

References: Atlas (46a), Hultén & Fries (1986), Jalas & Suominen (1994), Rich (1987, 1991).

D. A. PEARMAN

Barbarea intermedia
Medium-flowered Winter-cress

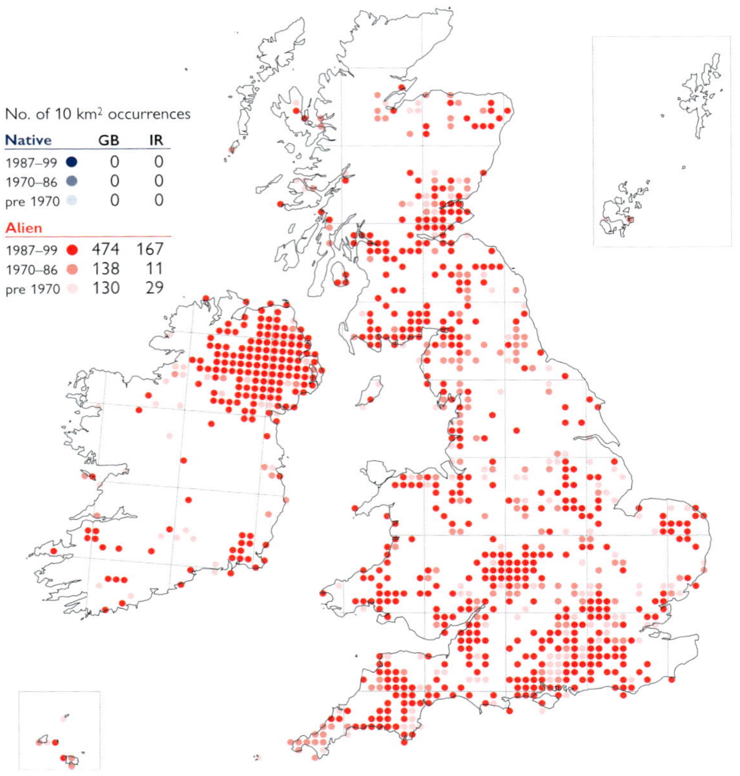

A biennial herb, rarely a perennial or an annual, found in a variety of waste and disturbed places, although it was formerly most frequent as an arable weed. Generally lowland, but occurring as a casual at 340 m N. of Alston (Cumberland).

Neophyte (change +1.92). This species was first recorded in the wild before 1849, although it may have previously been confused with *B. vulgaris.* It has considerably increased since the 1962 *Atlas*, and may be seriously under-recorded in the Irish Republic.

Native of N.W. Africa and W. Europe from Spain to Denmark; naturalised north and east of its native range.

References: Atlas (46b), Bolòs & Vigo (1990), Jalas & Suominen (1994), Rich (1987, 1991).

D. A. PEARMAN

Barbarea verna American Winter-cress

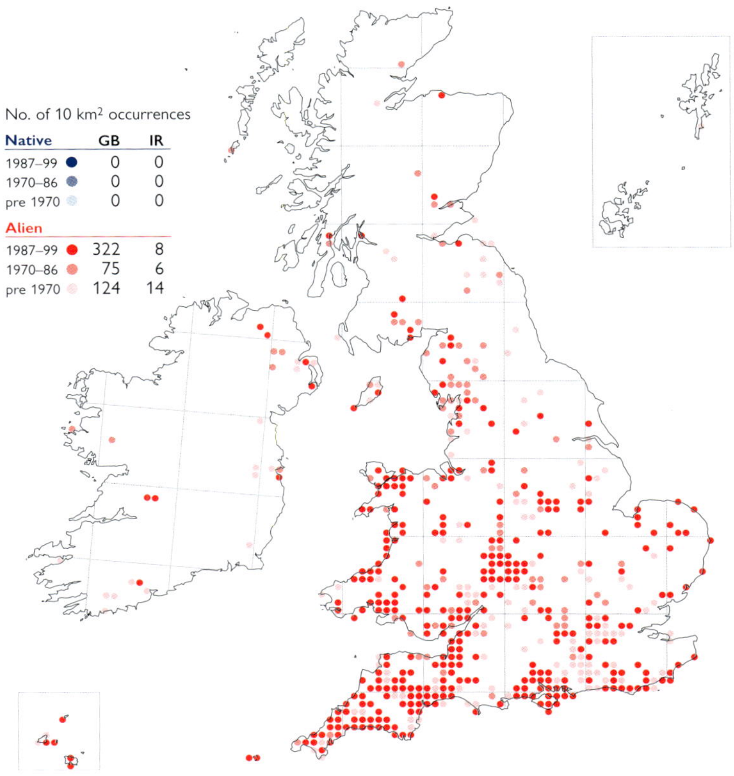

A biennial, or occasionally annual, herb which is most frequent as a garden escape on waste ground, by roads and on railways. It has a long-lived seed bank. Lowland.

Neophyte (change +1.34). *B. verna* was grown for centuries as a substitute for watercress, and is now widely cultivated again after a gap in the 18th and 19th centuries. It was first recorded in the wild in 1803. Although the overall range has not changed since the 1962 *Atlas*, there are many more recent records.

Native of S.W. Europe, east to Italy.

References: Atlas (46c), Bolòs & Vigo (1990), Jalas & Suominen (1994), Rich (1987, 1991).

D. A. PEARMAN

Rorippa nasturtium-aquaticum agg.
Water-cresses

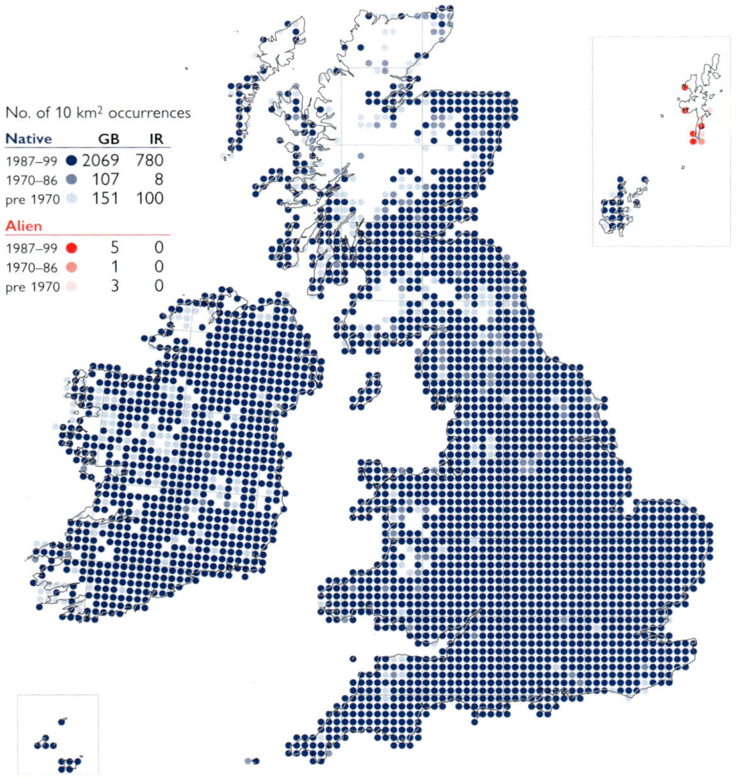

These perennial herbs grow beside streams both in calcareous and acidic areas, and favour waters that are moderately nutrient-rich but not eutrophic. 0–550 m (near Moor House, Westmorland).

Native (change –0.56). The taxonomy of the *R. nasturtium-aquaticum* aggregate, including diploid *R. nasturtium-aquaticum*, tetraploid *R. microphylla* and their triploid hybrid (*R.* × *sterilis*), was not elucidated until 1950 and 1951. Records made since then are probably reliable, but the maps in Perring & Sell (1968) were too provisional to allow comparison with the current map.

The world distribution of the two species in this aggregate and their hybrid is given in the following accounts.

References: Atlas (48a), Grime *et al.* (1988), Howard & Lyon (1952), Hultén & Fries (1986), Preston & Croft (1997), Rich (1991).

D. A. PEARMAN

Rorippa nasturtium-aquaticum Water-cress

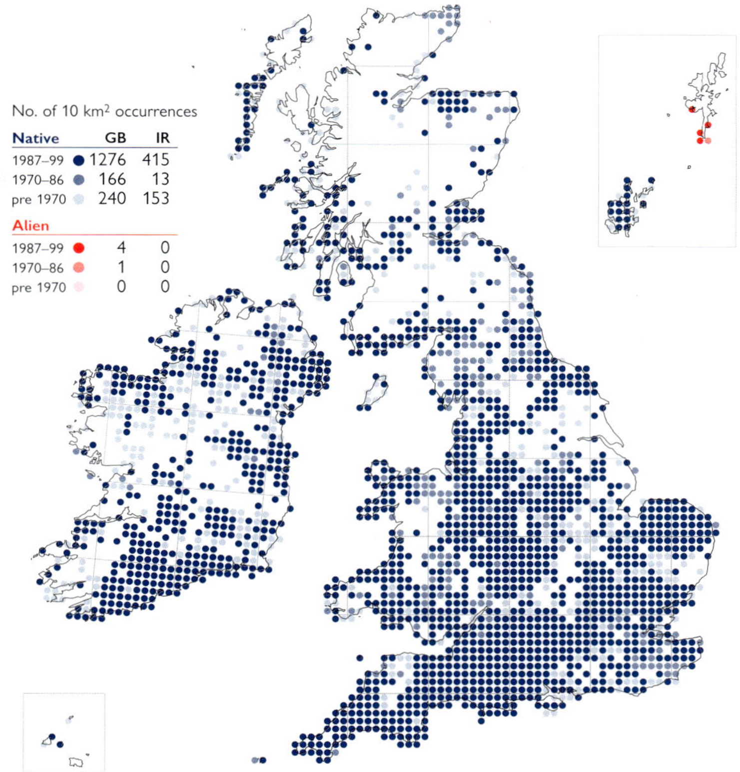

A perennial herb, found both in and beside clear shallow streams and in ditches, ponds, canals and marshes. Although it grows in similar habitats to *R. microphylla*, it tends to be restricted to calcareous substrates. It is also cultivated, but not as often as *R.* × *sterilis*. Lowland.

Native. Although *R. nasturtium-aquaticum* was mapped in Perring & Sell (1968), that map was too provisional for any assessment of subsequent change to be made. Some losses have occurred, but the species is probably still under-recorded in some areas.

Eurosiberian Southern-temperate element, but widely naturalised so that distribution is now Circumpolar Southern-temperate.

References: Atlas Supp. (9a), Grime *et al.* (1988), Howard & Lyon (1952), Jalas & Suominen (1994), Preston & Croft (1997), Rich (1991).

D. A. PEARMAN

Rorippa microphylla Narrow-fruited Water-cress

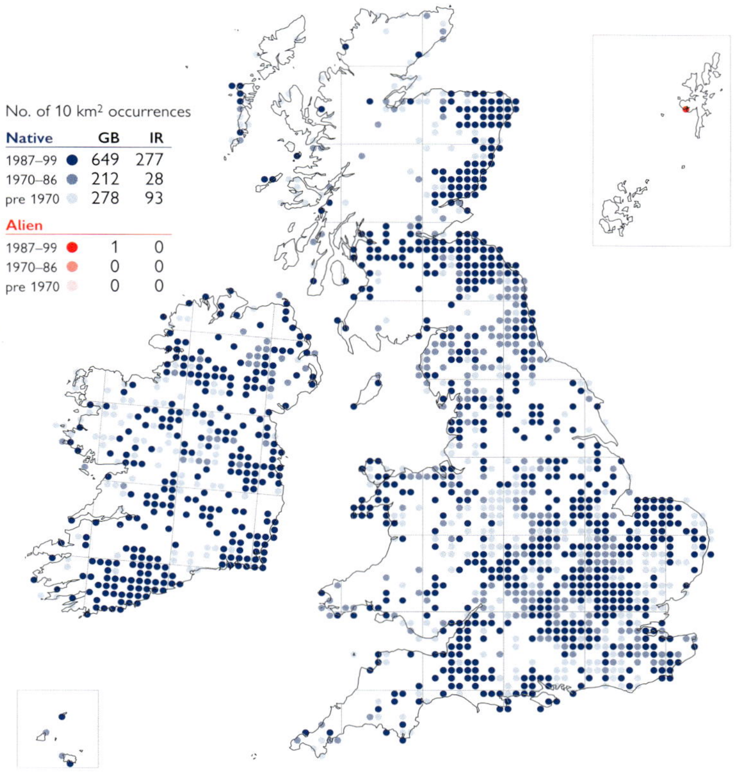

A perennial herb growing in and beside streams, ditches, ponds and canals, and in marshes. This species extends into rather more acidic sites than *R. nasturtium-aquaticum*, and in some areas is the most common segregate on non-calcareous soils. 0–550 m (near Moor House, Westmorland).

Native. This segregate is almost certainly still under-recorded; it may be more frequent in N. England and E. Scotland than our records suggest. It is not commercially cultivated as watercress.

World distribution uncertain; it is known to be widespread in Europe and naturalised in C. & S. Africa, N. America and Australasia.

References: Atlas Supp. (9b), Grime *et al.* (1988), Howard & Lyon (1952), Jalas & Suominen (1994), Preston & Croft (1997), Rich (1991).

D. A. PEARMAN

Rorippa microphylla × *R. nasturtium-aquaticum* (*R.* × *sterilis*) Hybrid Water-cress

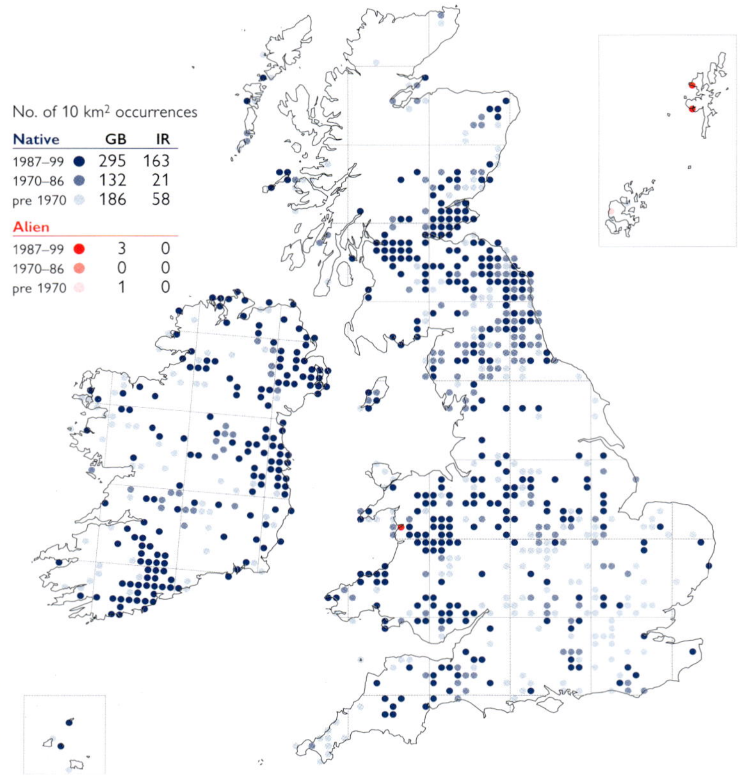

A perennial herb occurring in and beside water, especially in streams, ditches and ponds, either with its parents or independently of them. Few well-formed seeds are produced, and effective reproduction is almost entirely vegetative. 0–395 m (Craig Cerrig-Gleisiad, Brecs.).

Native. *R.* × *sterilis* is the most widely cultivated watercress, since it is usually frost-hardy, and many wild colonies must have arisen from that source. It was first cultivated commercially in about 1808, near Gravesend (W. Kent). It shows no appreciable change since it was mapped by Perring & Sell (1968), but it is probably under-recorded.

Widespread in Europe and naturalised in Japan, N. America, New Zealand and elsewhere.

References: Atlas Supp. (9c), Grime *et al.* (1988), Preston & Croft (1997), Rich (1991), Stace (1975).

D. A. PEARMAN

Rorippa islandica Northern Yellow-cress

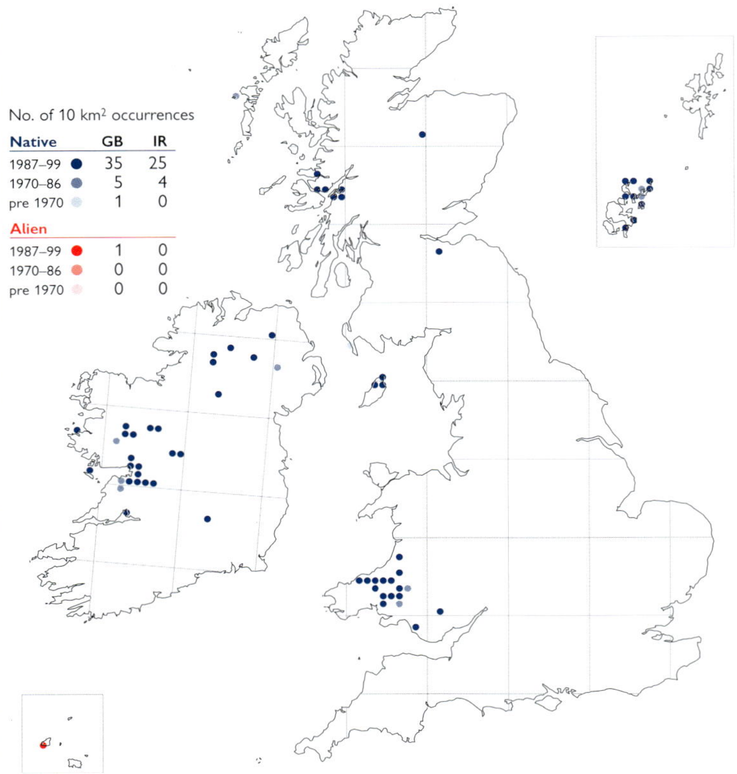

An annual or short-lived perennial herb found in open, muddy habitats such as lake, pond and pool margins, ditch banks, depressions in pasture, in turloughs and rarely on rocks by rivers. There are also records from waste ground and tips. Lowland.

Native. *R. islandica* was first recognised as distinct from *R. palustris* in 1968. Formerly, both species were included under one taxon, '*R. islandica*', old records of which still cause confusion. *R. islandica sens. str.* is the rarer plant, albeit more frequent than initially supposed. Some sites have been lost through the canalisation of rivers and pond drainage.

Eurosiberian Boreal-montane element.

References: Chater & Rich (1995), Curtis & McGough (1988), Hultén & Fries (1986), Jalas & Suominen (1994), Jonsell (1968), Rich (1991), Wigginton (1999).

D. A. PEARMAN

Rorippa palustris Marsh Yellow-cress

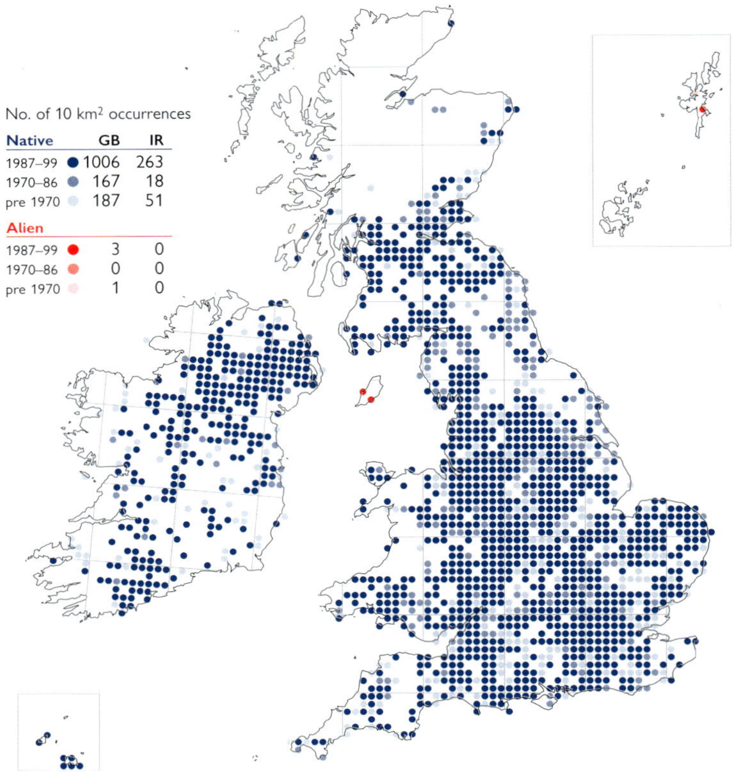

A summer-annual, or rarely perennial, herb growing on river banks, on wet mud exposed above receding lake margins, and in ponds that dry out in summer. It is an early colonist of new wetland sites, and there is evidence for longevity of the seed bank. It is an uncommon weed of arable fields, railway cinders, waste ground and gardens. Generally lowland, but reaching 320 m at Buxton (Derbys.).

Native (change +0.44). *R. palustris* was mapped as 'all records' in the 1962 *Atlas*. Analysis of the database reveals that many of the losses shown on the map have occurred since 1950.

Circumpolar Boreo-temperate element.

References: Atlas (48d), Grime *et al.* (1988), Hultén & Fries (1986), Jalas & Suominen (1994), Jonsell (1968), Rich (1991).

D. A. PEARMAN

Rorippa sylvestris Creeping Yellow-cress

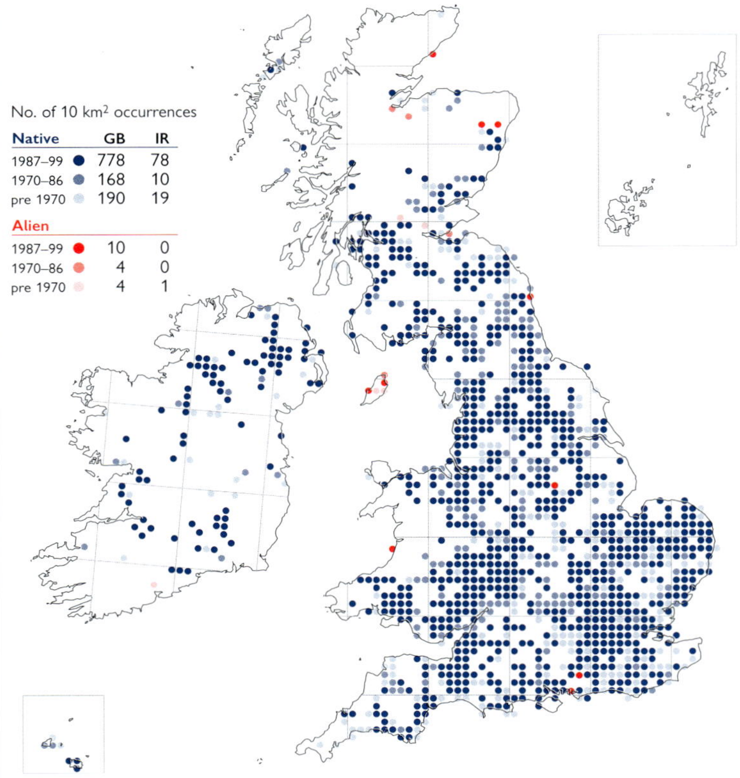

A perennial herb growing on damp bare ground, often in sites flooded in winter, on the margins and banks of rivers, streams, canals and ditches, by lakes and ponds and in depressions in pastures. It is also a weed of cultivated ground, often in drier situations than in its semi-natural habitats. It is a vigorous pioneer species, intolerant of competition, spreading by seed and by broken pieces of rhizome. Lowland.

Native (change +0.73). The distribution of *R. sylvestris* is probably stable in semi-natural habitats, but it has become more widespread as a weed since the 1950s. It is now recorded in many more squares than in the 1962 *Atlas*.

European Temperate element; widely naturalised outside its native range.

References: Atlas (48c), Hultén & Fries (1986), Jalas & Suominen (1994), Jonsell (1968), Rich (1991).

D. A. PEARMAN

Rorippa amphibia Great Yellow-cress

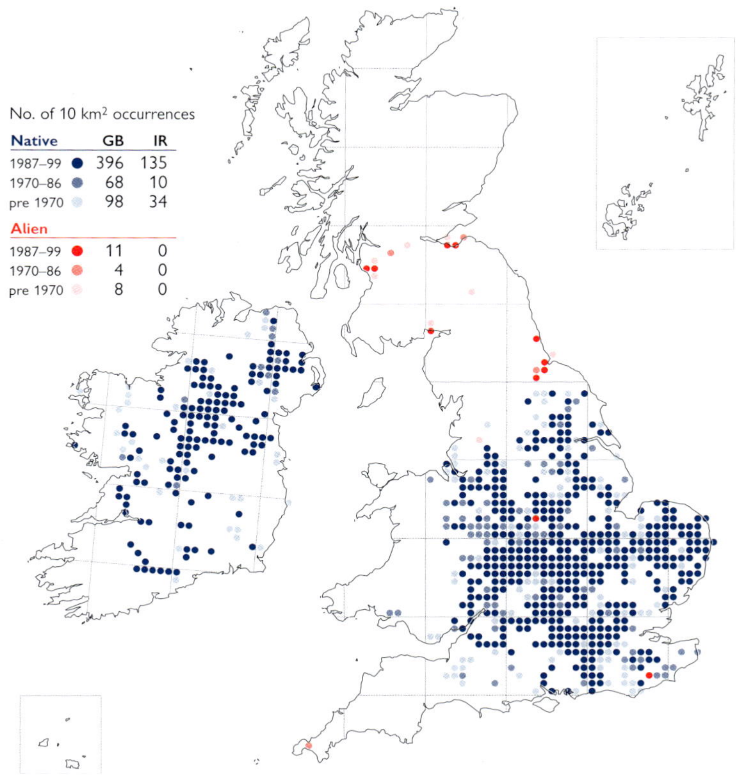

A perennial herb of emergent vegetation along the edges of streams and rivers, by lakes and ponds and in other swampy ground. It often grows in sites which are flooded in winter and where some water remains in the summer, and is usually found where the water is calcareous and eutrophic. Seed set is often poor, possibly because plants are highly self-incompatible, and spread is mainly by fragmentation of mature plants. Lowland.

Native (change +0.03). *R. amphibia* is better recorded both in Britain and especially in Ireland than it was in the 1962 *Atlas*. However, it has declined in some areas and was lost from Somerset after 1972.

Eurasian Temperate element; widely naturalised outside its native range.

References: Atlas (49a), Hultén & Fries (1986), Jalas & Suominen (1994), Jonsell (1968), Meusel *et al.* (1965), Preston & Croft (1997), Rich (1991).

D. A. PEARMAN

Rorippa amphibia × *R. sylvestris* (*R.* × *anceps*) Hybrid Yellow-cress

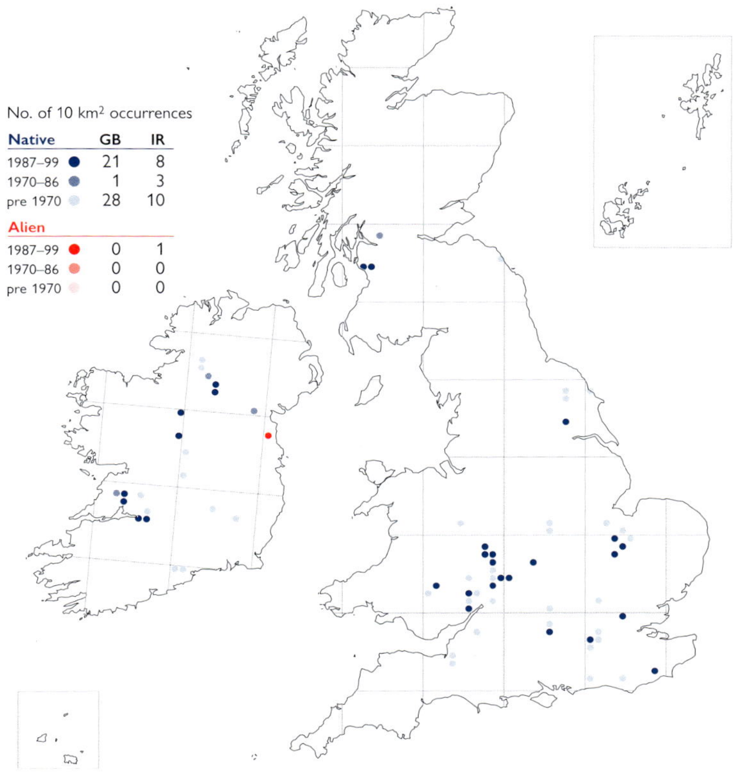

A fertile, perennial hybrid found in damp places by rivers, lakes, ponds and on bare ground, often in the absence of one or other parent. In Britain most records are from river banks, but in Ireland it is just as frequent on lake shores. Lowland.

Native. This hybrid is easily overlooked, especially as some populations show evidence of backcrossing with the parents. Although it was mapped by Rich (1991) and Preston & Croft (1997), it is probably still under-recorded.

Frequent in Europe, and introduced in N. America.

References: Jonsell (1968), Stace (1975).

D. A. PEARMAN

Armoracia rusticana Horse-radish

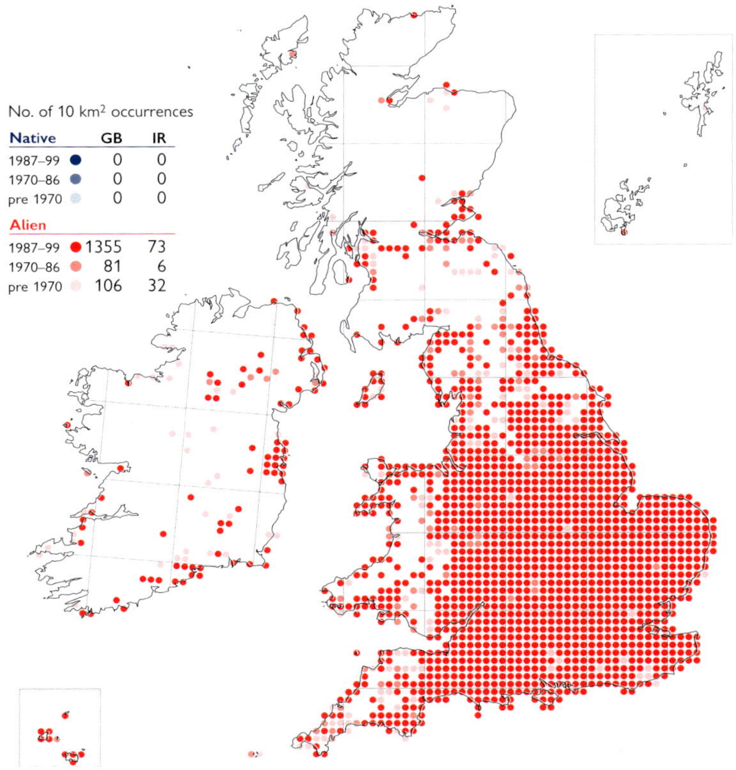

A long-lived perennial herb, persisting in old gardens and allotments and spreading by root fragments to roadsides, waste ground, railways, sandy seashores and river banks. The plant is highly sterile, and seed-set is unknown in our area. Lowland.

Archaeophyte (change +0.05). *A. rusticana* was introduced before 1500, initially as a medicinal herb, but by 1650 it had replaced *Lepidium latifolium* as a vegetable cultivated for hot relishes. It spreads wherever there is cultivation or dumping of soil, but its range now seems to be stable.

Not known in the wild; naturalised throughout temperate Europe, whence it has spread to N. America and elsewhere.

References: Atlas (44a), Grieve (1974), Hultén & Fries (1986), Jalas & Suominen (1994), Rich (1991).

D. A. PEARMAN

Cardamine bulbifera Coralroot

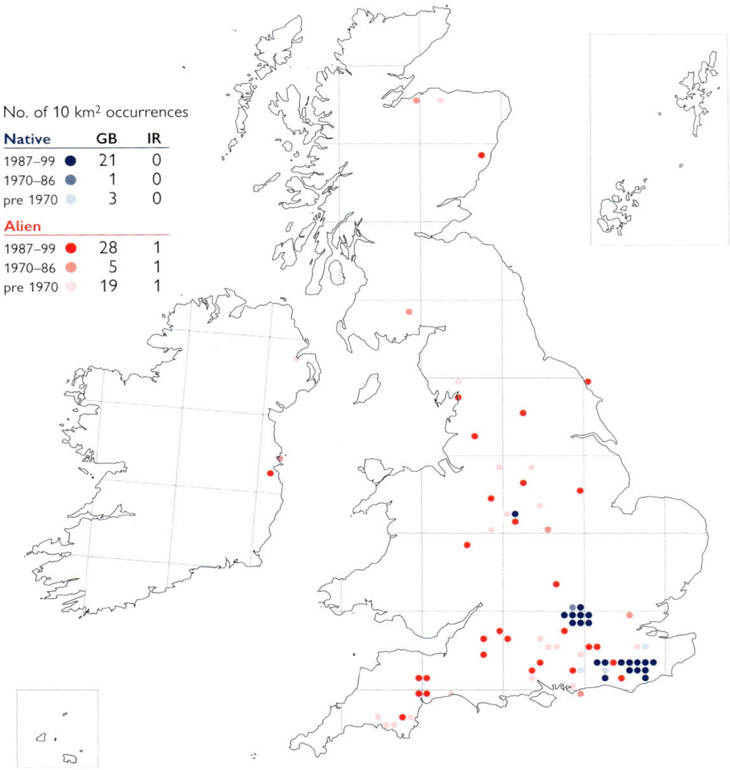

A rhizomatous perennial herb which grows in Britain in two habitats: on dry woodland slopes over chalk in the Chilterns, and in damp woodlands over clay in the Weald. Elsewhere it is an escape from cultivation by roads and in woodland and parkland. Lowland.

Native (change +0.36). The distribution of *C. bulbifera* in its major strongholds is stable. It is not known whether the disjunct, apparently native populations reflect a wider earlier range, ancient long-distance dispersal or more recent colonisation. *C. bulbifera* sometimes escapes from cultivation, but these plants differ in leaf-shape.

European Temperate element, with a continental distribution in W. Europe.

References: Atlas (45c), Hultén & Fries (1986), Jalas & Suominen (1994), Meusel *et al.* (1965), Rich (1991), Showler & Rich (1993), Stewart *et al.* (1994).

D. A. PEARMAN

Cardamine amara Large Bitter-cress

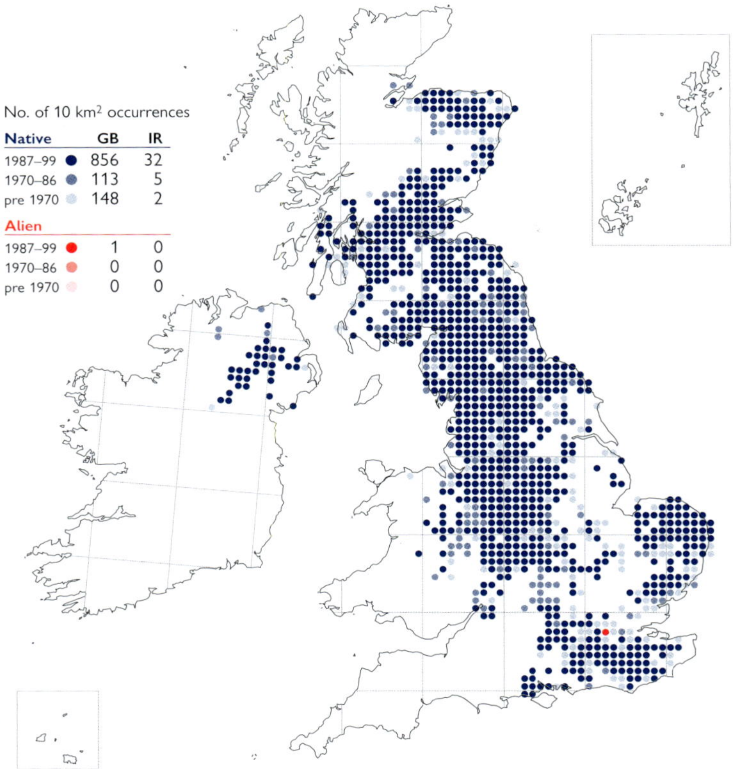

A perennial winter-green herb of streamsides and marshes, wet meadows and wet woodland, often growing in slow-moving or still water, preferring an acidic substrate and tolerant of shade. Generally lowland, but reaching 640 m in the Ochil Hills (W. Perth).

Native (change 0.00). The distribution of *C. amara* is stable in Britain, with some local losses in S.E. England which are presumably attributable to agricultural improvements. In Ireland the dramatic increase in records since the 1962 *Atlas* appears to result from both a genuine spread and better recording.

European Temperate element.

References: Atlas (44c), Curtis & McGough (1988), Grime *et al.* (1988), Hultén & Fries (1986), Jalas & Suominen (1994), Rich (1991).

D. A. PEARMAN

Cardamine raphanifolia Greater Cuckooflower

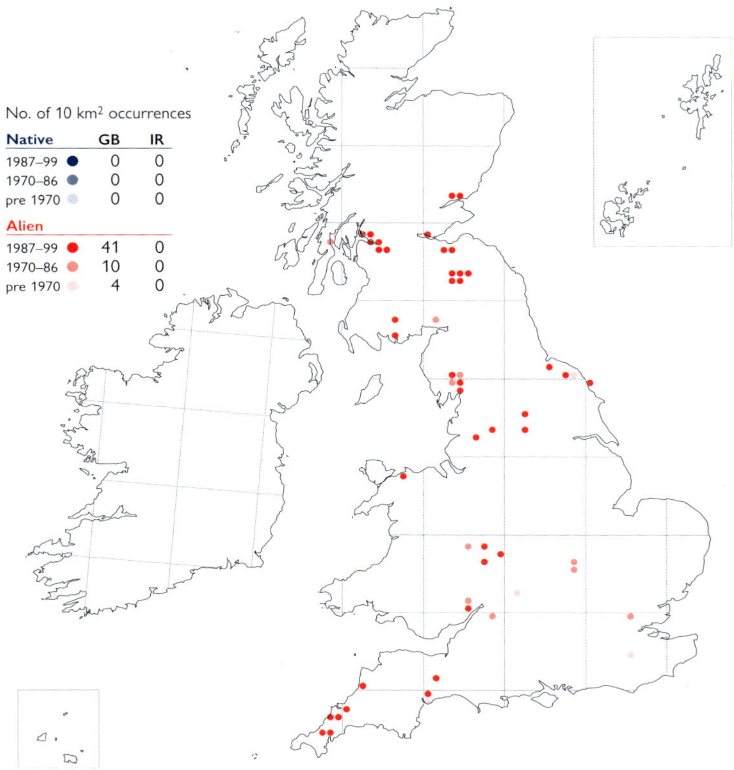

A rhizomatous perennial herb, naturalised in damp, shaded places by rivers and lakes. Lowland.

Neophyte. Introduced into cultivation in 1710, *C. raphanifolia* was not recorded from the wild until 1930 (N.E. Yorks.). It seems to have increased recently: for example, it was not recorded in Cornwall until 1981 but has spread since then along several rivers.

Native of the mountains of S. Europe, N. Turkey and the Caucasus.

References: Bolòs & Vigo (1990), French *et al.* (1999), Jalas & Suominen (1994), Rich (1991).

D. A. PEARMAN

Cardamine pratensis Cuckooflower

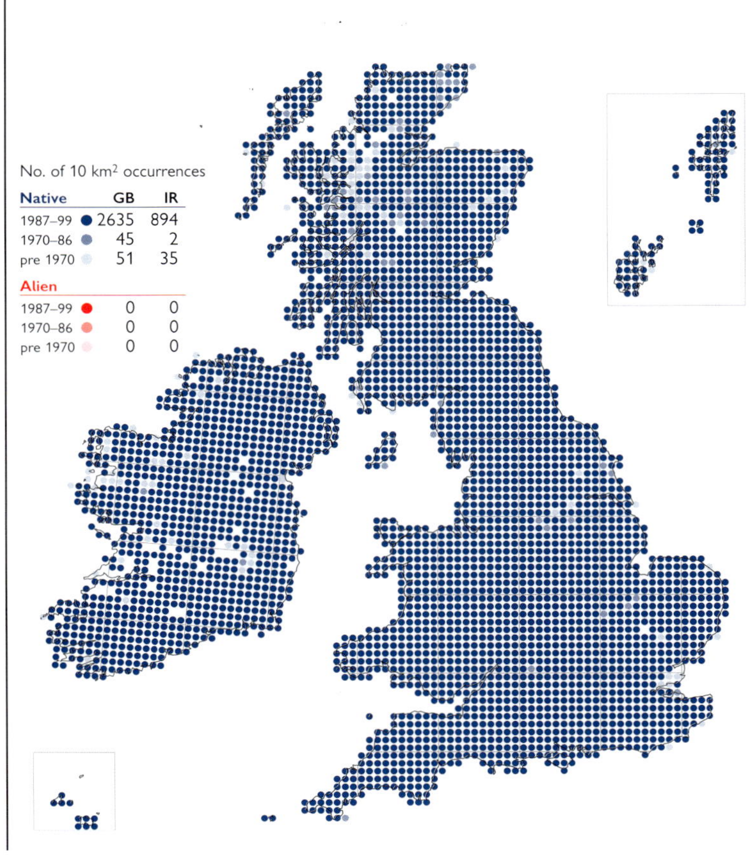

A perennial herb of wet grassy places on moderately fertile, seasonally waterlogged soils in woods, wet meadows, fens and flushes. In upland areas it is also found in rush pasture and bryophyte-rich springs. It occasionally persists in gardens and lawns. It is morphologically and cytologically variable, many clones reproducing by rooting from the leaves. 0–1080 m (Ben Lawers, Mid Perth).

Native (change +0.42). There has been no change in the overall distribution since the 1962 *Atlas*. Although many of its lowland habitats have been improved, *C. pratensis* is resistant to some herbicides and can survive in semi-improved pastures.

Circumpolar Wide-boreal element.

References: Atlas (44b), Grime *et al.* (1988), Hultén & Fries (1986), Jalas & Suominen (1994), Meusel *et al.* (1965), Rich (1991).

D. A. PEARMAN

Cardamine impatiens Narrow-leaved Bitter-cress

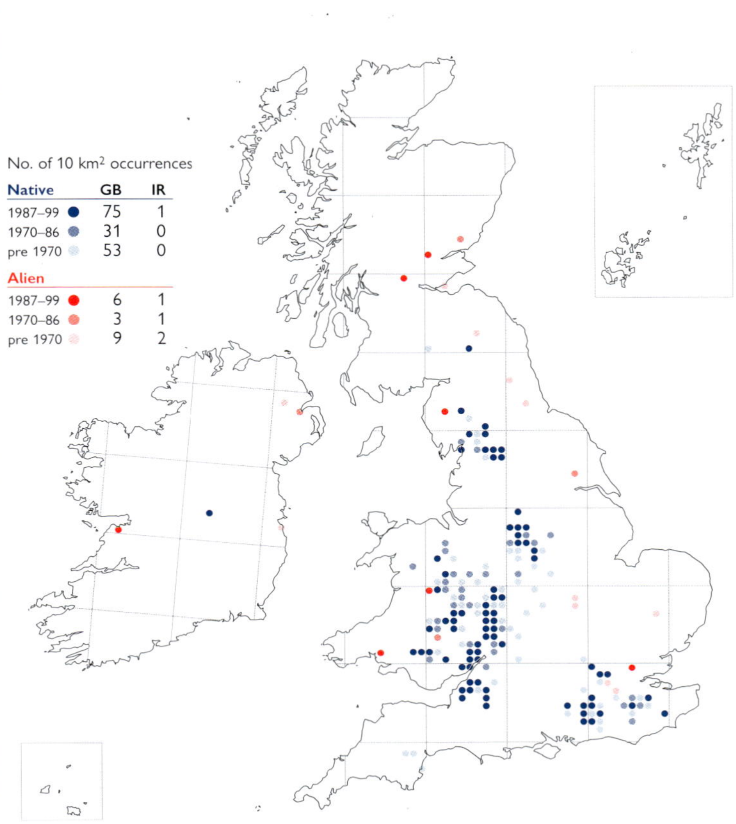

A biennial herb found in woodland (particularly under *Fraxinus*), on moist limestone rocks (including the grikes of limestone pavement) and stable screes, by rivers and on damp roadsides; rarely found as a garden escape. It is intolerant of competition, but can be invasive in recently disturbed habitats. 0–610 m (Ingleborough, Mid-W. Yorks.).

Native (change –0.09). Populations fluctuate markedly from year to year and from site to site, but the overall distribution of this species is stable. In Ireland it is probably only native in Co. Westmeath.

Eurasian Temperate element; widely naturalised outside its native range.

References: Atlas (44d), Curtis & McGough (1988), Hultén & Fries (1986), Jalas & Suominen (1994), Meusel *et al.* (1965), Rich (1991), Stewart *et al.* (1994), Williams (2000).

D. A. PEARMAN

Cardamine flexuosa Wavy Bitter-cress

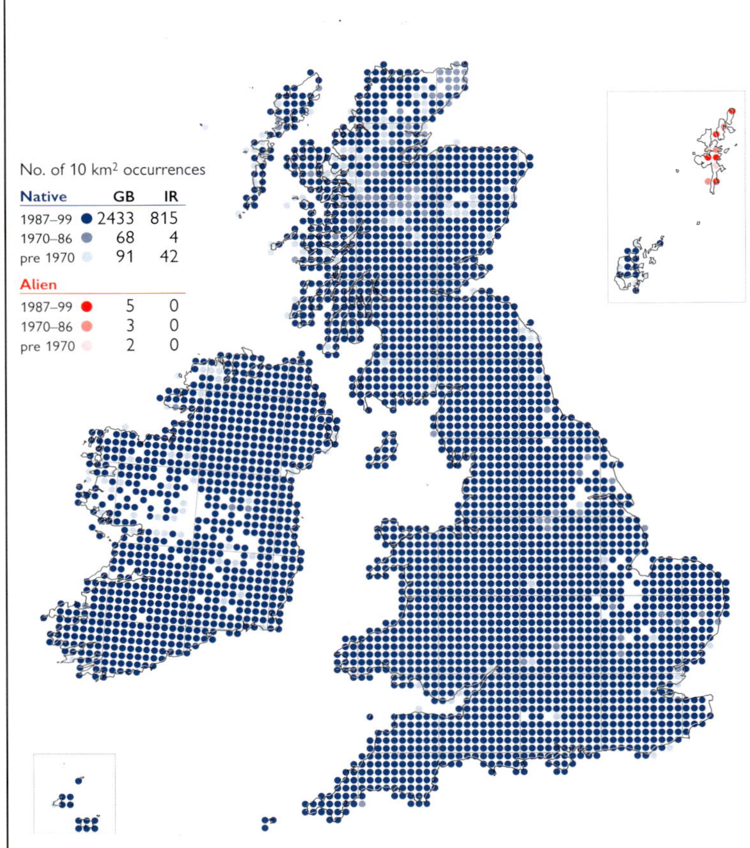

A winter- or summer-annual, or rarely a short-lived perennial, most frequent in open, moist, shaded vegetation in marshland, by rivers and streams, and in gardens. It prefers soils which are at least mildly basic, and is absent from those that are strongly acidic. It is an effective colonist of disturbed, fertile habitats. Generally lowland, but reaching 830 m on Snowdon (Caerns.) and 1190 m in the Breadalbanes (Mid Perth).

Native (change +1.06). There has been no change in the range of *C. flexuosa* since the 1962 *Atlas*.

European Temperate element; also in C. & E. Asia and N. America.

References: Atlas (45a), Grime *et al.* (1988), Hultén & Fries (1986), Jalas & Suominen (1994), Rich (1991).

D. A. PEARMAN

Cardamine hirsuta Hairy Bitter-cress

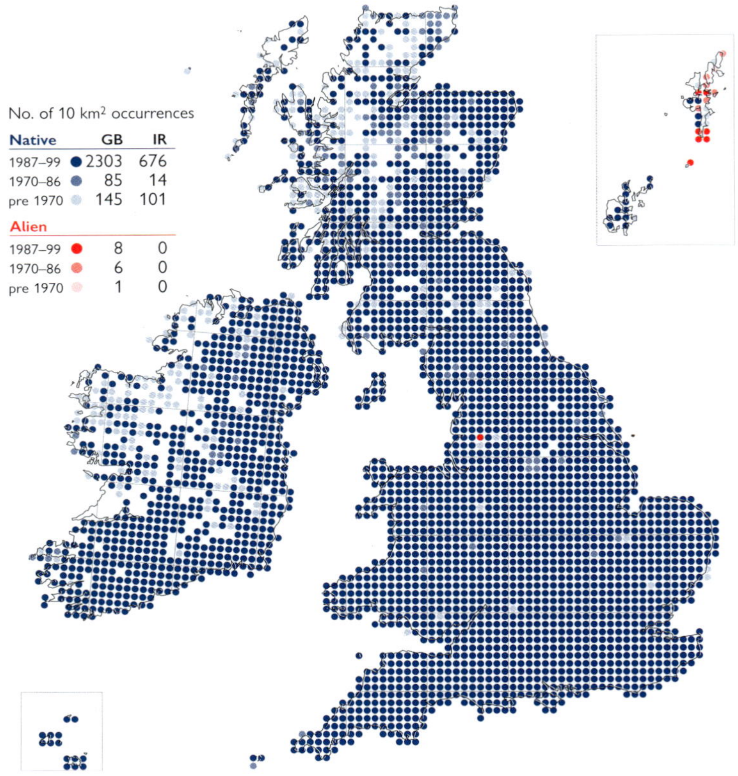

A winter-annual which, under favourable conditions, can flower and seed again in autumn. *C. hirsuta* is particularly common as a weed of cultivation and in other ruderal habitats, but it also grows on rock outcrops, by streams and in woods. 0–1190 m (Breadalbanes, Mid Perth).

Native (change +0.69). The distribution of *C. hirsuta* is stable, other than in N. Scotland and W. Ireland where it is uncertain whether the concentration of pre-1970 records indicates a real decline or a lack of recording in spring. It was mapped as 'all records' in the 1962 *Atlas*.

Eurosiberian Southern-temperate element, but widely naturalised so that distribution is now Circumpolar Southern-temperate.

References: Atlas (45b), Grime *et al.* (1988), Hultén & Fries (1986), Jalas & Suominen (1994), Rich (1991).

D. A. PEARMAN

Arabis petraea Northern Rock-cress

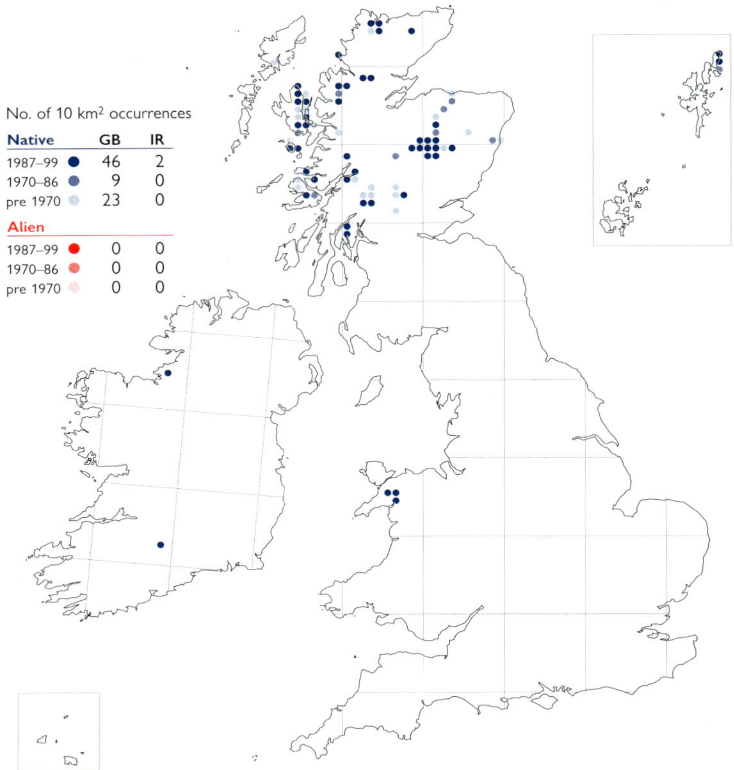

A perennial herb of very open sites on acidic and basic rocks and rock ledges, on montane cliff faces and screes and on sea-cliffs. It is also found on river shingle and on serpentine fellfield in Shetland. Being a colonist of open habitats, populations can be transient at particular sites, and the species has a curiously disjunct distribution. From near sea level (Shetland) to at least 1220 m (Braeriach, S. Aberdeen).

Native (change –0.64). It is difficult to assess trends in the frequency of this species, as it may still be present in some of the squares for which there are only pre-1987 records.

Eurasian Arctic-montane element.

References: Atlas (46d), Curtis & McGough (1988), Hultén & Fries (1986), Jalas & Suominen (1994), Meusel *et al.* (1965), Rich (1991), Stewart *et al.* (1994).

D. A. PEARMAN

Arabis glabra Tower Mustard

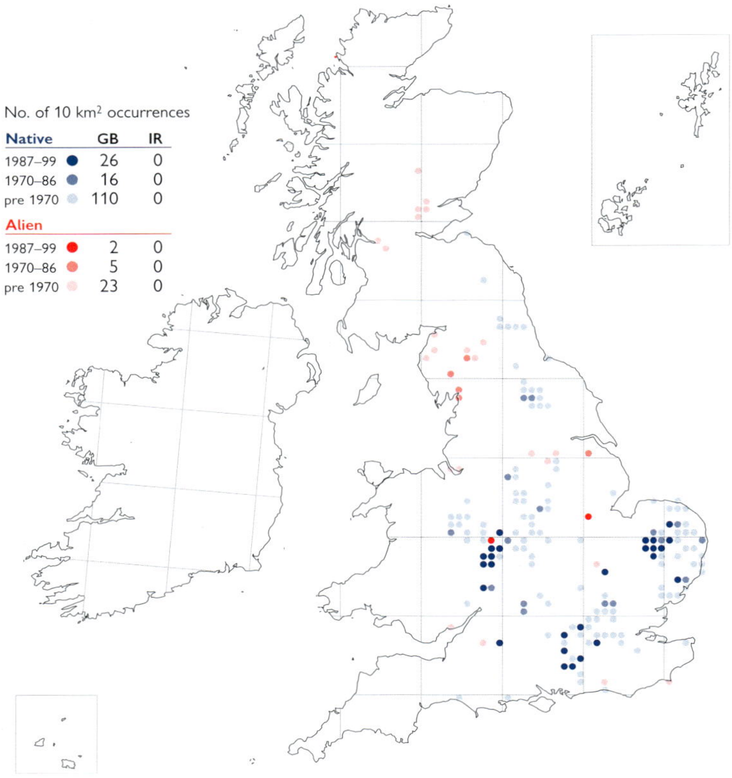

A biennial, rarely perennial, herb of grassy places and on disturbed ground on free-draining sandy soils over chalk and limestone. Lowland.

Native (change –1.16). *A. glabra* has declined in many areas, with most of the losses occurring before 1930. Since open ground is required for germination, it will not survive when the habitat becomes overgrown. However, the seed seems to be long-lasting, so populations have the potential to reappear on newly opened ground, and in Breckland it is thriving in disturbed areas.

Eurosiberian Temperate element, but widely naturalised so that distribution is now Circumpolar Temperate.

References: Atlas (47d), Hultén & Fries (1986), Jalas & Suominen (1994), Meusel *et al.* (1965), Rich (1991), Stewart *et al.* (1994), Wigginton (1999).

D. A. PEARMAN

Arabis alpina Alpine Rock-cress

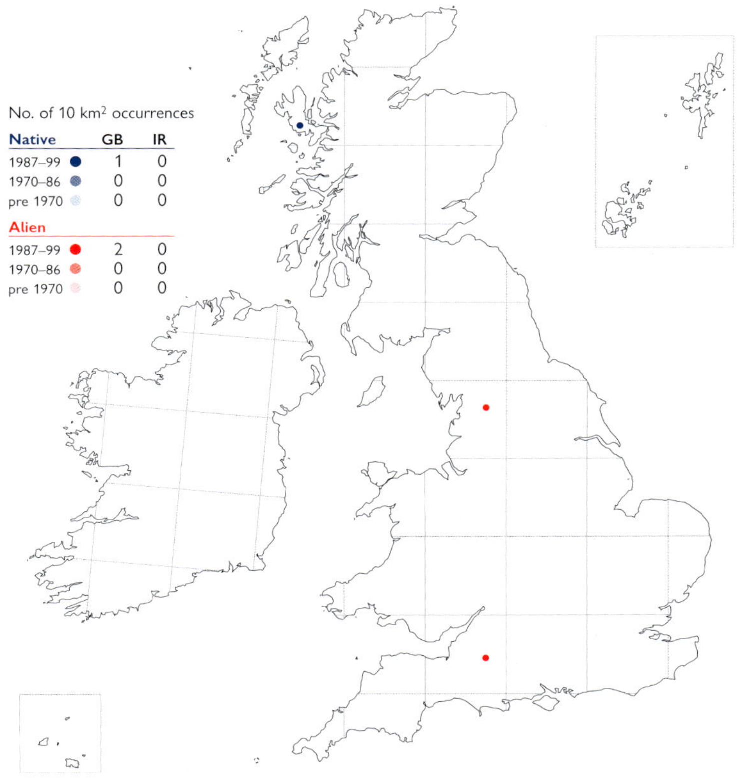

A perennial, mat-forming herb of shaded ledges on basic cliffs. It is also a very rare garden escape. The native populations are montane, growing at 820–850 m (N. Ebudes).

Native. *A. alpina* was discovered in Britain in 1887, and the population at the original site on Skye has generally remained stable. A second population has recently been found in the same area. The Somerset alien site is on an old wall, where it has possibly persisted since *c.* 1900. In Yorkshire it survives in small quantity on a limestone scar on Reginald Farrar's former estate.

Eurosiberian Arctic-montane element; also in N. America.

References: Atlas (47b), Rich (1991), Hultén & Fries (1986), Wigginton (1999).

D. A. PEARMAN

Arabis caucasica Garden Arabis

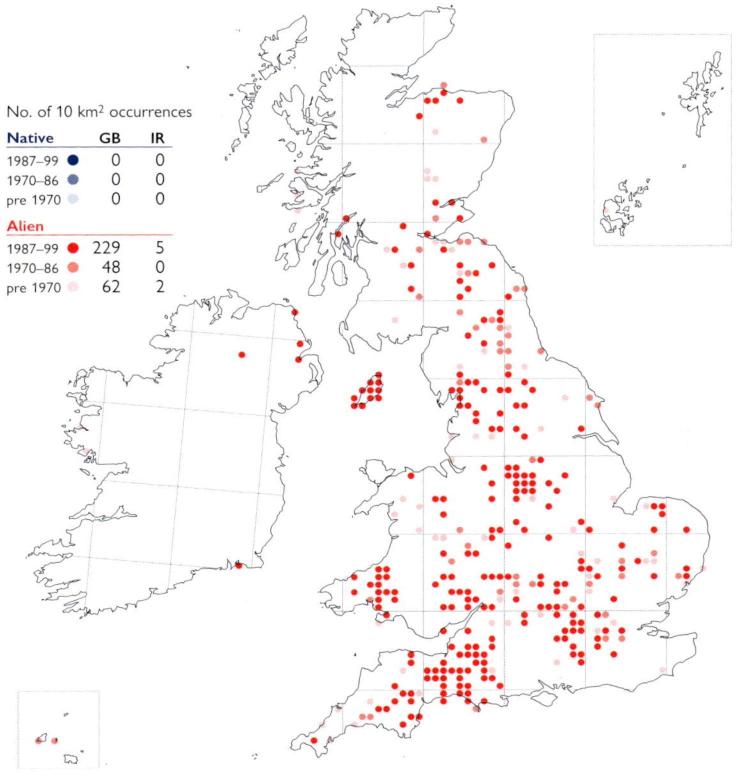

A perennial, mat-forming herb, naturalised on walls, rocks and cliffs. Lowland.

Neophyte. *A. caucasica* was introduced into cultivation by 1798 and is very popular in gardens. It was first recorded in the wild in 1855 (Kew, Surrey) and persists in many places, but usually not far from habitation. It is unevenly recorded.

Native of S. Europe, N. Africa and S.W. Asia; not clearly distinct from *A. alpina* and treated by recent southern European Floras as synonymous with it.

References: Hultén & Fries (1986), Rich (1991).

D. A. PEARMAN

Arabis hirsuta Hairy Rock-cress

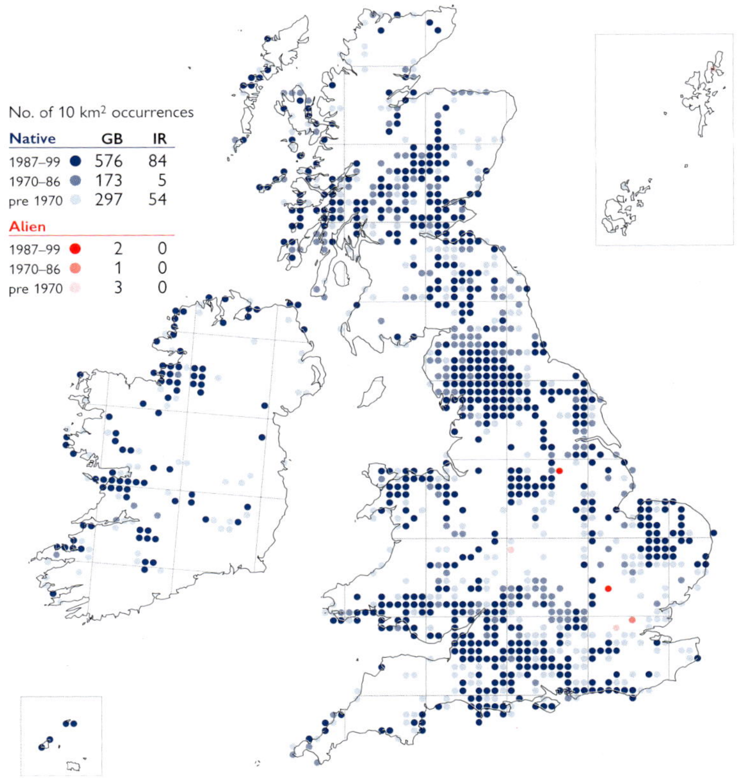

A biennial or perennial herb growing in dry, sunny, exposed situations on rock outcrops and in grassland on base-rich substrates, particularly chalk and limestone; also occurring on sand dunes and on bridges and walls. 0–1005 m (Aonach Beag, Westerness).

Native (change –1.02). *A. hirsuta* is still widespread in many habitats. Many of the losses were before 1950, probably largely as a result of agricultural improvement, and these have continued, particularly in S.E. England.

Circumpolar Boreo-temperate element.

References: Atlas (47b, c), Grime *et al.* (1988), Hultén & Fries (1986), Jalas & Suominen (1994), Meusel *et al.* (1965), Rich (1991).

D. A. PEARMAN

Arabis scabra Bristol Rock-cress

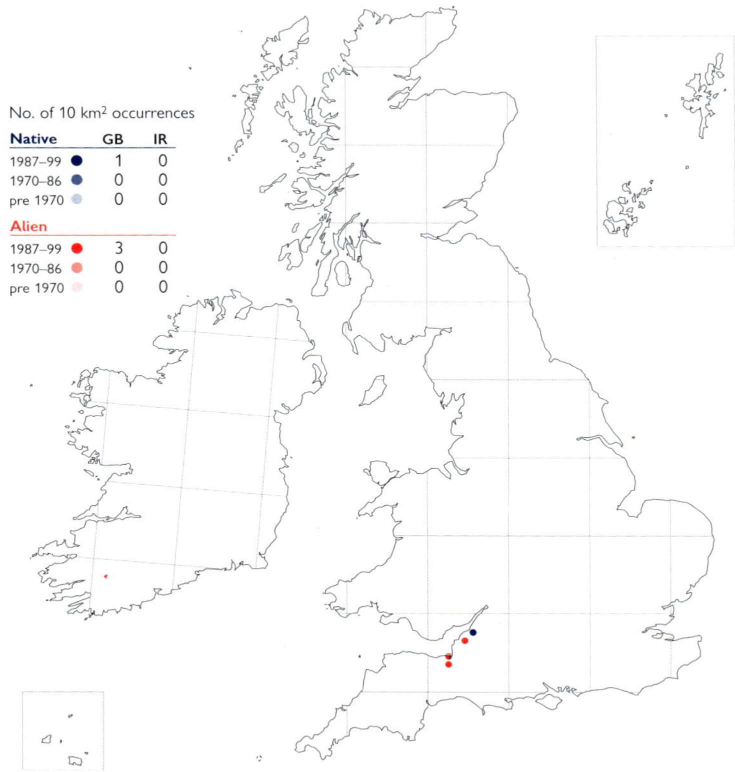

A short-lived perennial herb of Carboniferous limestone, growing in shallow soils, on scree, and on rock ledges. It spreads by seed. Lowland.

Native. *A. scabra* has been known in the Avon Gorge since 1686. Although invasion of scrub and wire-netting to prevent rock falls have eliminated some sites, and reduced the open ground needed for germination, populations are still healthy. Introductions elsewhere have proved to be short-lived failures, except those at Combwich (S. Somerset).

Mediterranean-montane element.

References: Atlas (47a), Jalas & Suominen (1994), Meusel *et al.* (1965), Pring (1961), Rich (1991), Wigginton (1999).

D. A. PEARMAN

Aubrieta deltoidea Aubretia

Native	**GB**	**IR**
1987–99	0	0
1970–86	0	0
pre 1970	0	0
Alien		
1987–99	381	13
1970–86	28	0
pre 1970	7	0

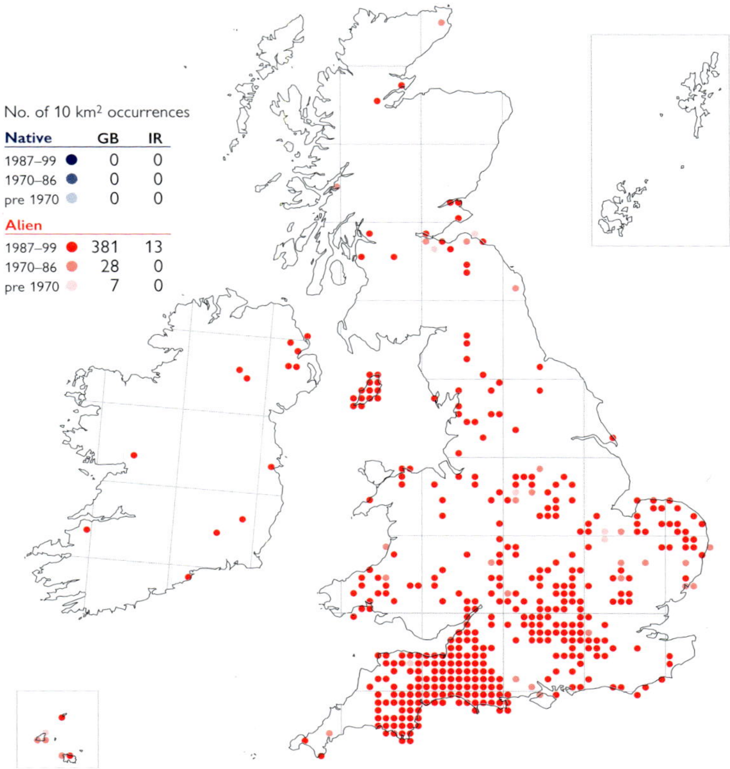

A mat-forming perennial herb, naturalised near habitations on walls, paths and waste ground. Lowland.

Neophyte. *A. deltoidea* has been cultivated in Britain since at least 1710, and is frequent as an escape from gardens. It was recorded from the wild by 1928. It appears to be increasing but it is unevenly recorded.

Native of Sicily, the southern Balkan peninsula and S.W. Asia.

References: Jalas & Suominen (1994), Rich (1991).

D. A. PEARMAN

Lunaria annua Honesty

Native	**GB**	**IR**
1987–99	0	0
1970–86	0	0
pre 1970	0	0
Alien		
1987–99	1396	56
1970–86	78	4
pre 1970	29	1

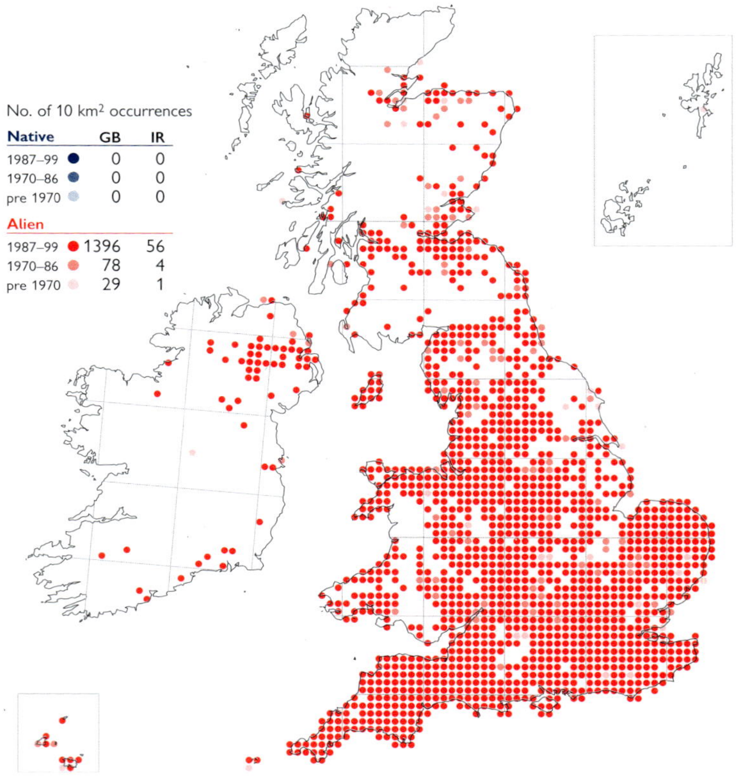

A biennial herb, found naturalised near habitations on scrubby banks, beside paths, in hedgerows and secondary woodland, and on waste ground. It is also a frequent casual in similar habitats. Lowland.

Neophyte. *L. annua* has been cultivated in Britain since at least 1570. It was known from the wild by 1597. Clapham *et al.* (1952) describe it as 'often found as an escape', suggesting that its overall distribution is stable.

Our plant is subsp. *annua*, a cultivated plant of unknown origin, widely naturalised in Europe; subsp. *pachyrhiza* is a perennial native to Italy and the Balkans.

References: Jalas & Suominen (1996), Rich (1991).

D. A. PEARMAN

Alyssum alyssoides Small Alison

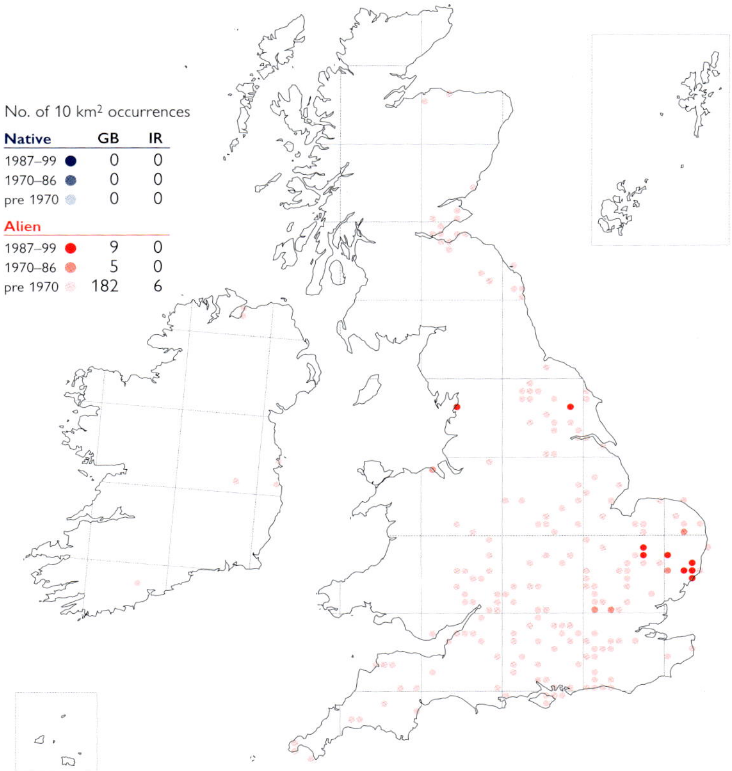

No. of 10 km² occurrences

Native	GB	IR
1987–99 ●	0	0
1970–86 ●	0	0
pre 1970 ●	0	0
Alien		
1987–99 ●	9	0
1970–86 ●	5	0
pre 1970 ●	182	6

A casual annual, appearing erratically in arable fields, sandy tracks, pits, waste ground and docks. Very rarely it persists in the same habitats, but it is a poor competitor and needs regular disturbance to provide open soil for seedlings. Lowland.

Neophyte (change –1.24). *A. alyssoides* has been grown in Britain since 1740 and was first recorded in the wild in 1838. It has declined as a casual since the 19th century. Since 1980, small but persistent populations have been recorded at two sites in Breckland, where it was first recorded in 1912. Recent records elsewhere represent deliberate plantings.

Native in much of temperate Europe, rarer in the Mediterranean region; naturalised in S. Scandinavia.

References: Atlas (42c), Hultén & Fries (1986), Jalas & Suominen (1996), Meusel *et al.* (1965), Rich (1991).

D. A. PEARMAN

Alyssum saxatile Golden Alison

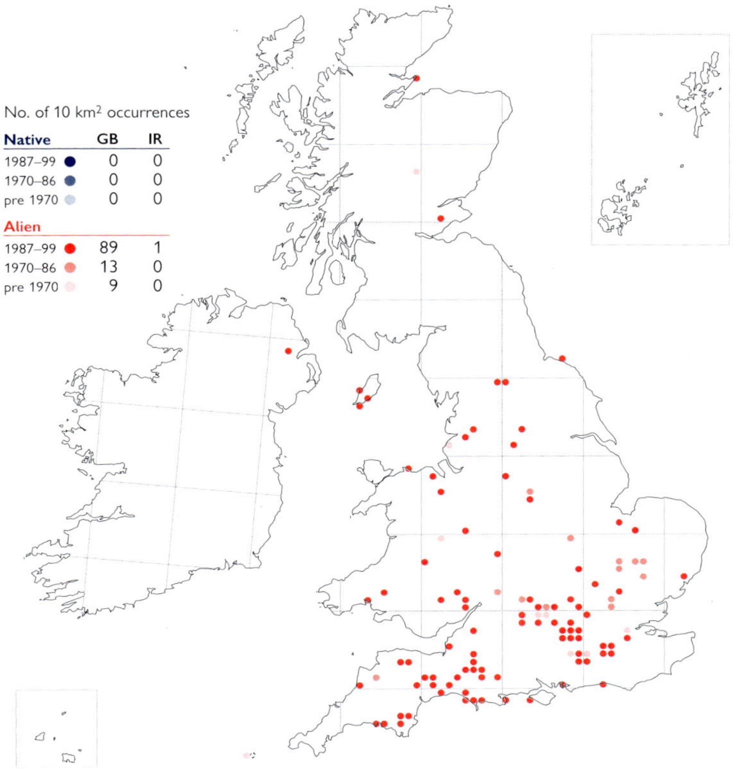

No. of 10 km² occurrences

Native	GB	IR
1987–99 ●	0	0
1970–86 ●	0	0
pre 1970 ●	0	0
Alien		
1987–99 ●	89	1
1970–86 ●	13	0
pre 1970 ●	9	0

A perennial herb which occurs as a casual on walls, rocks and dry waste ground. Despite reproducing from seed, it only rarely persists. Lowland.

Neophyte. Introduced into cultivation in 1710 and first recorded in the wild in 1912, *A. saxatile* is very commonly grown in gardens. It is unevenly recorded, as botanists differ in the tendency to which they record species established on garden walls.

Native of C. & S.E. Europe and S.W. Asia.

References: Jalas & Suominen (1996), Meusel *et al.* (1965), Rich (1991).

D. A. PEARMAN

Berteroa incana Hoary Alison

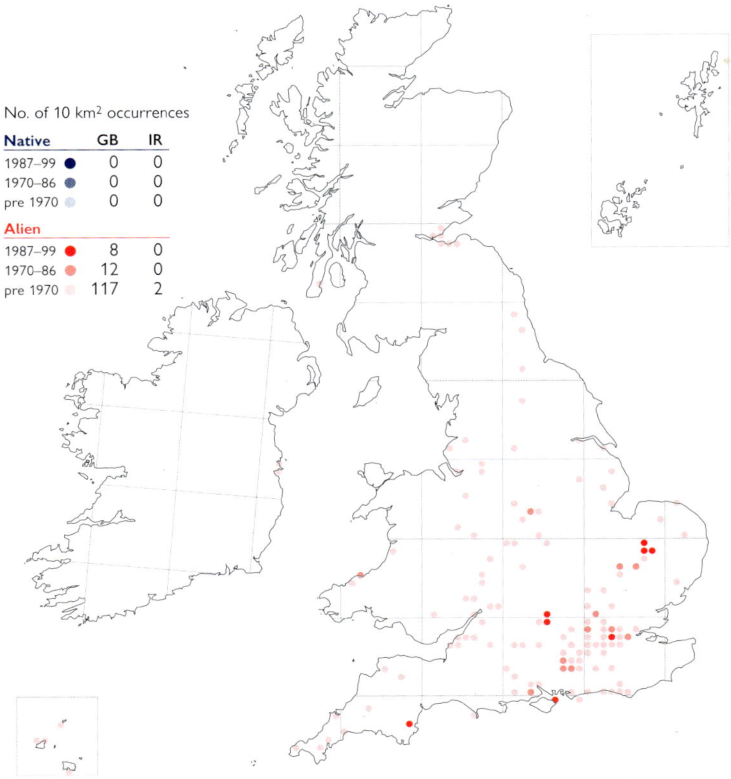

No. of 10 km² occurrences

Native	GB	IR
1987–99 ●	0	0
1970–86 ●	0	0
pre 1970 ●	0	0
Alien		
1987–99 ●	8	0
1970–86 ●	12	0
pre 1970 ●	117	2

A biennial, but occasionally annual or perennial herb, rarely naturalised on waste ground, and predominantly occurring as a casual in arable fields, on waste ground, around docks and in newly sown grass or clover leys. Lowland.

Neophyte. *B. incana*, which was cultivated in Britain by 1640, has arisen mainly as a grain impurity and was recorded as a casual in the wild by 1870. Some naturalised populations are stable, but casual records have declined due to improved seed cleaning techniques.

A Eurosiberian Temperate species, absent as a native from W. Europe but naturalised in Fennoscandia, W. France and N. America.

References: Hultén & Fries (1986), Jalas & Suominen (1996), Meusel *et al.* (1965), Rich (1991).

D. A. PEARMAN

Lobularia maritima Sweet Alison

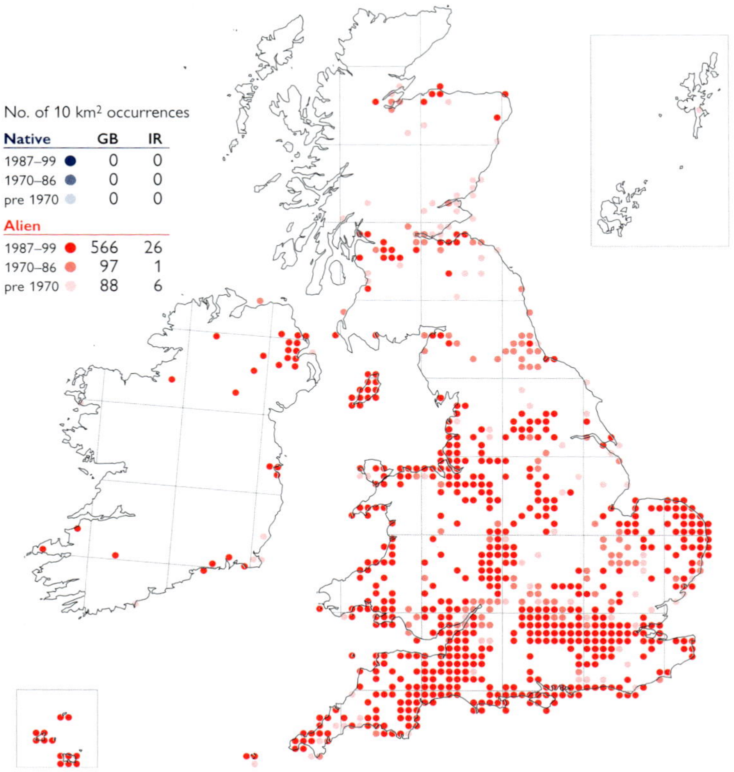

No. of 10 km² occurrences

Native	GB	IR
1987–99	0	0
1970–86	0	0
pre 1970	0	0
Alien		
1987–99	566	26
1970–86	97	1
pre 1970	88	6

An annual, biennial or perennial herb, naturalised on sea-cliffs, sand dunes and open ground near the sea, and occurring as a casual inland in a variety of waste ground habitats. It sets seed readily but persists only in mild regions. Lowland.

Neophyte (change +2.34). *L. maritima* was cultivated in Britain by 1722 and is very widely grown in gardens. It was known from the wild by 1807. It is now much better recorded than in the 1962 *Atlas*, but is probably also increasing.

Native of the coasts of S.W. Europe and the Mediterranean region; naturalised elsewhere in Europe, especially in coastal areas.

References: Atlas (42d), Bolòs & Vigo (1990), Jalas & Suominen (1996), Rich (1991).

D. A. PEARMAN

Draba aizoides Yellow Whitlowgrass

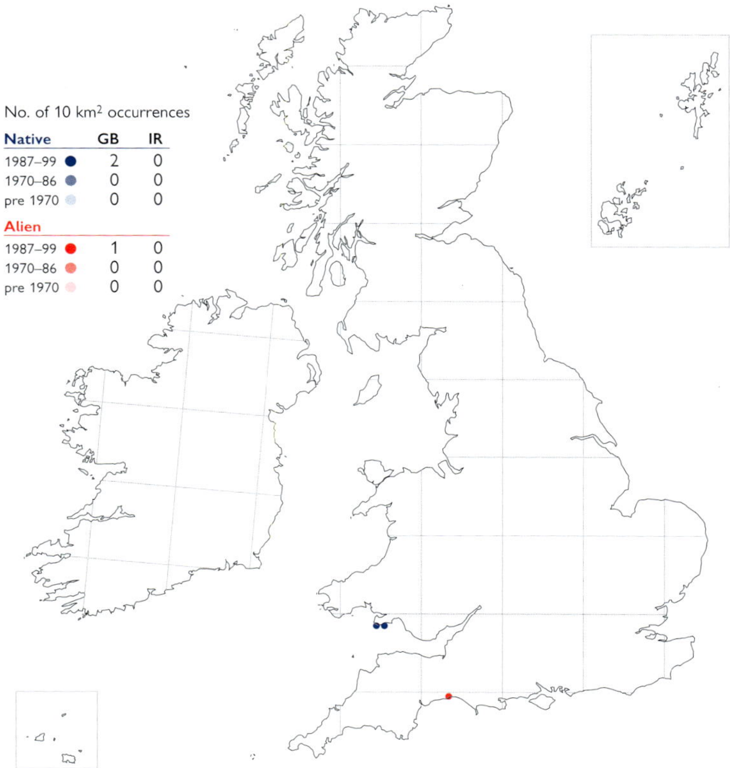

No. of 10 km² occurrences

Native	GB	IR
1987–99	2	0
1970–86	0	0
pre 1970	0	0
Alien		
1987–99	1	0
1970–86	0	0
pre 1970	0	0

This short-lived, cushion-forming, perennial herb is restricted to limestone rocks, where it grows in crevices in humic calcareous soils. Plants that occur in grassland and in bare soil away from rocks seldom survive. Lowland.

Native. Although *D. aizoides* was discovered as recently as 1795, genetic analysis has shown that the Gower (Glamorgan) populations are very different from those in continental Europe, supporting its claim to be native to Britain (John, 1992). Its overall distribution has remained stable, although some accessible colonies have been lost, up to quite recently, to collectors. Others have decreased in size for reasons that are unclear.

Mediterranean-montane element.

References: Atlas (43a), Jalas & Suominen (1996), Kay & Harrison (1970), Meusel *et al.* (1965), Rich (1991), Wigginton (1999).

D. A. PEARMAN

Draba norvegica Rock Whitlowgrass

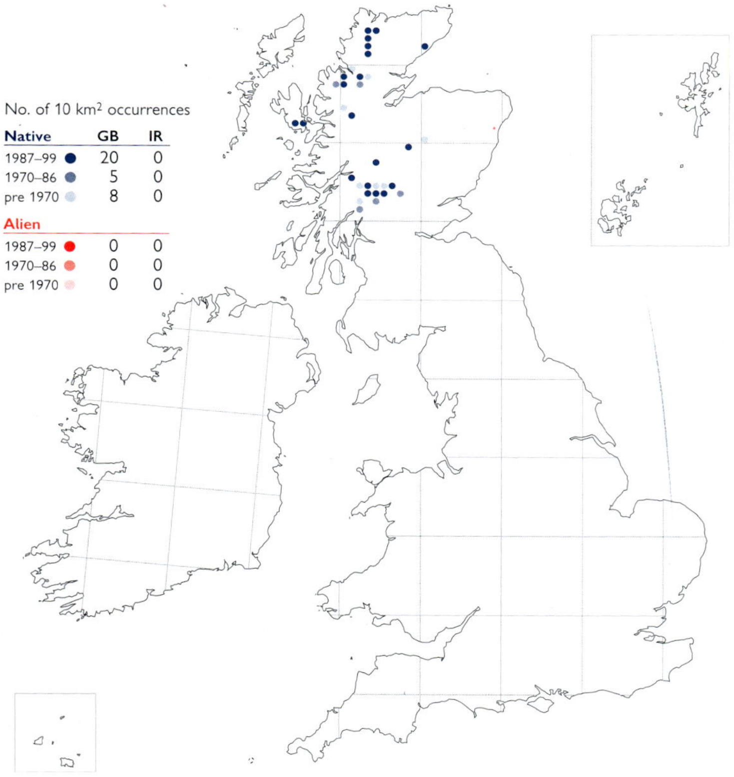

No. of 10 km² occurrences

Native	GB	IR
1987–99	20	0
1970–86	5	0
pre 1970	8	0
Alien		
1987–99	0	0
1970–86	0	0
pre 1970	0	0

A perennial tufted herb of base-rich rocks, occurring on rock ledges, in crevices in cliffs, on consolidated scree and in other bare places. Upland, from 310 m in Glendhu Forest (W. Sutherland) to 1160 m on Ben Lawers (Mid Perth), and more frequent at the higher end of that range.

Native (change 0.00). A few more sites have been discovered in recent years, suggesting that *D. norvegica* has previously been under-recorded, but in general the distribution is stable.

European Arctic-montane element, but absent from mountains of C. Europe; also in N. America.

References: Atlas (43a), Corner (1999), Hultén & Fries (1986), Jalas & Suominen (1996), Rich (1991), Stewart *et al.* (1994).

D. A. PEARMAN

Draba incana Hoary Whitlowgrass

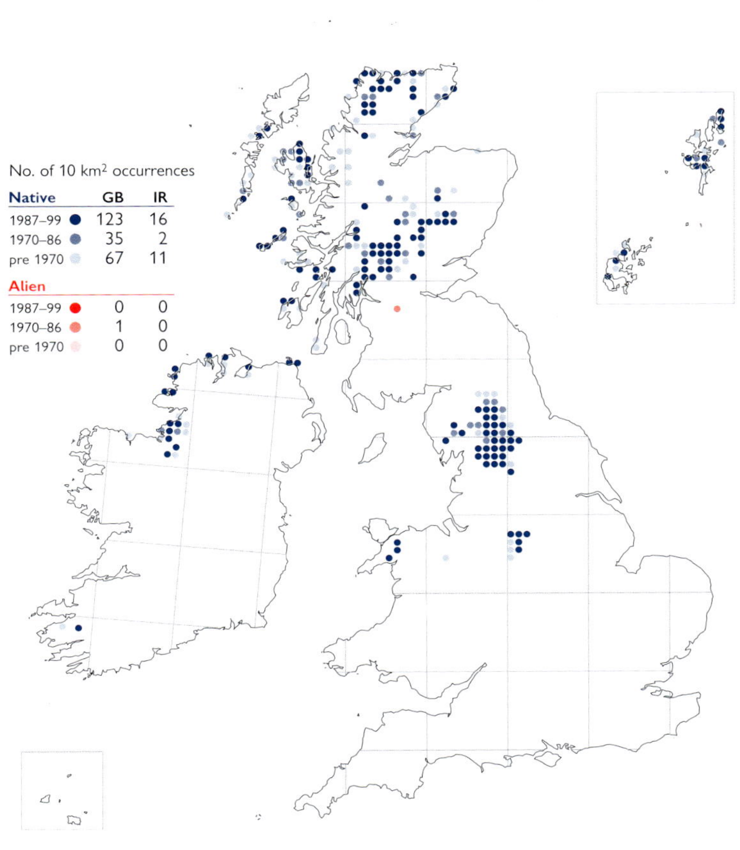

No. of 10 km² occurrences

Native	GB	IR
1987–99 ●	123	16
1970–86 ●	35	2
pre 1970 ●	67	11
Alien		
1987–99 ●	0	0
1970–86 ●	1	0
pre 1970 ●	0	0

A morphologically variable biennial or perennial tufted herb, which is usually found on limestone rock ledges, screes and pavements, and occasionally in open grassland on thin droughted soils. It also occurs on sand dunes and, more rarely, on base-rich mica-schists and igneous rocks, and on sandstone cliffs. From sea level, but more commonly an upland plant, reaching 1080 m in the Breadalbanes (Mid Perth).

Native (change –0.75). *D. incana* is much better recorded in its core areas than in the 1962 *Atlas*. The reason for the apparent loss of outlying sites is unclear.

European Boreo-arctic Montane element; also in N. America.

References: Atlas (43b), Curtis & McGough (1988), Hultén & Fries (1986), Jalas & Suominen (1996), Rich (1991).

D. A. PEARMAN

Draba muralis Wall Whitlowgrass

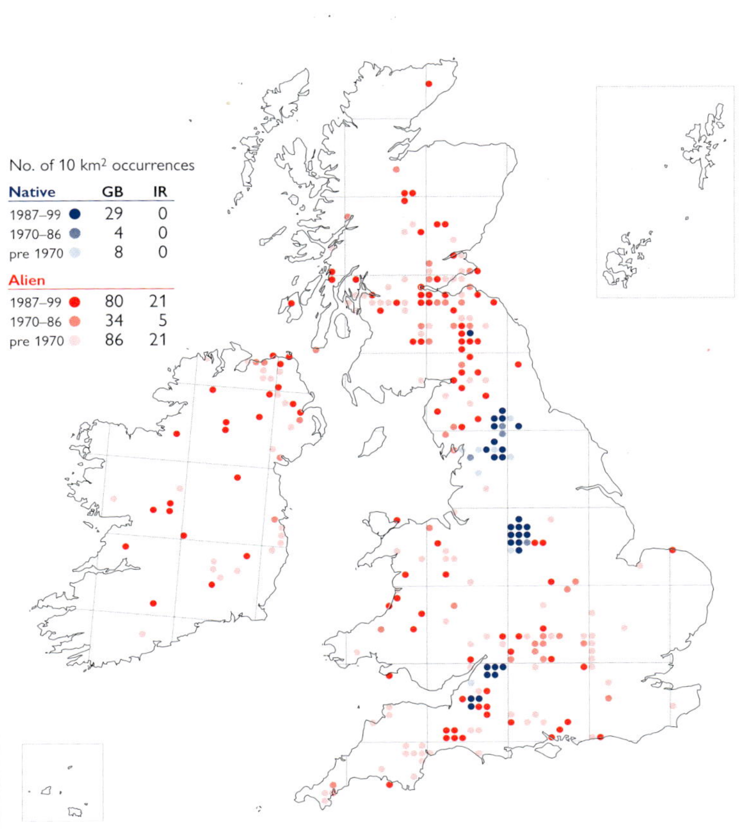

No. of 10 km² occurrences

Native	GB	IR
1987–99 ●	29	0
1970–86 ●	4	0
pre 1970 ●	8	0
Alien		
1987–99 ●	80	21
1970–86 ●	34	5
pre 1970 ●	86	21

A winter-annual, found as a native on limestone rocks on open skeletal soils, and on S.-facing ledges and screes. It is also a colonist on old walls, forest tracks and railways, and has been recorded as a garden weed where the conditions of its summer-dry, winter-moist, native habitat are mimicked. Generally lowland, but reaching 490 m in the Craven Pennines (Mid-W. Yorks.).

Native (change –0.17). Native populations of *D. muralis* are probably mostly stable; many other records are only casual, but it can be long persistent on some old walls.

European Temperate element; also in C. Asia and widely naturalised outside its native range.

References: Atlas (43c), Hultén & Fries (1986), Jalas & Suominen (1996), Ratcliffe (1960), Rich (1991), Stewart *et al.* (1994).

D. A. PEARMAN

Erophila verna agg. Common Whitlowgrasses

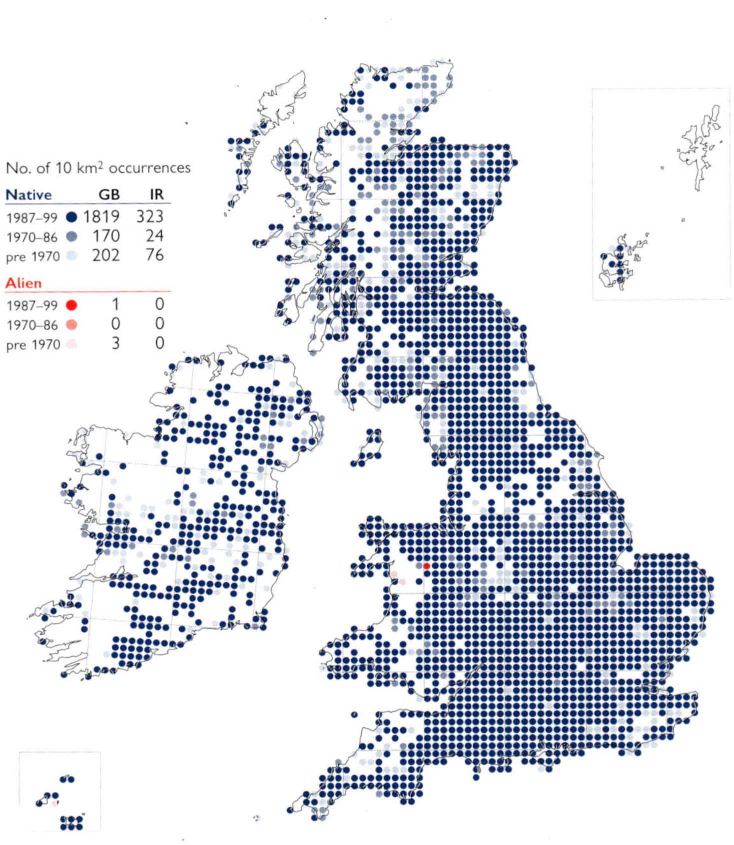

No. of 10 km² occurrences

Native	GB	IR
1987–99 ●	1819	323
1970–86 ●	170	24
pre 1970 ●	202	76
Alien		
1987–99 ●	1	0
1970–86 ●	0	0
pre 1970 ●	3	0

These ephemeral annuals grow in all sorts of dry open ground on shallow soils. They are particularly frequent on calcareous soils, including dunes, walls, rocks and open grassland, but are also found in cracks in pavements and on waste ground. 0–845 m (Great Dun Fell, Westmorland).

Native (change +0.52). The 1962 *Atlas* map of *E. verna* agg. is equivalent to the aggregate mapped here, and subsp. *spathulata*, mapped by Perring & Sell (1968), is no longer recognised. The current map shows no change in the range of the aggregate. The three segregate species that are currently recognised are mapped separately.

Eurosiberian Southern-temperate element.

References: Atlas (43d), Atlas Supp. (8c), Hultén & Fries (1986), Jalas & Suominen (1996), Rich (1991).

D. A. PEARMAN & C. D. PRESTON

Erophila majuscula Hairy Whitlowgrass

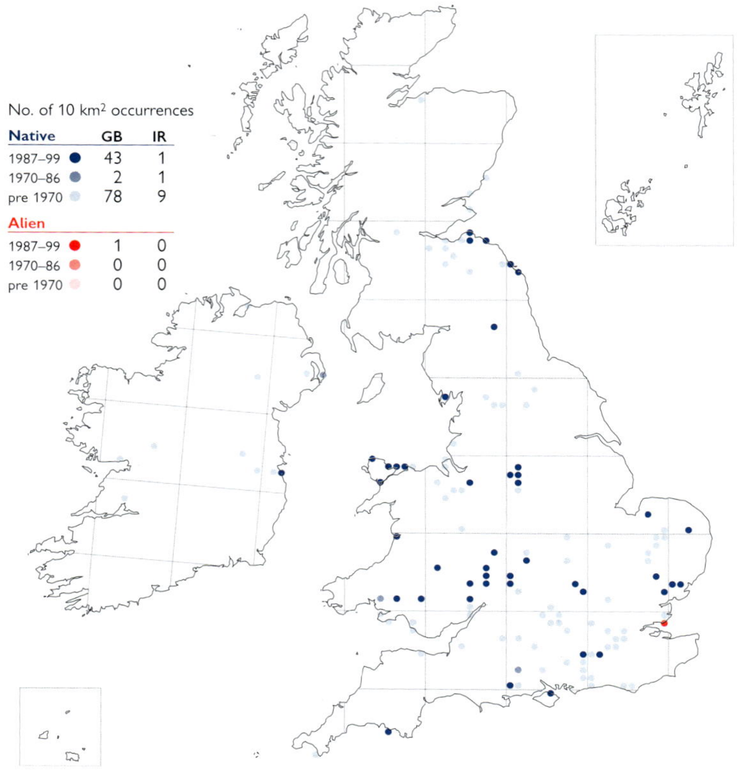

The habitats of *E. majuscula* are typical of those of the aggregate as a whole: limestone rocks and thin limestone turf, chalk downland (where it may grow on anthills), sand dunes and sandy ground inland, walls, railway lines and gravel paths. Lowland.

Native. The three species of *Erophila* currently recognised in Britain were not described in national floras until 1987 (Clapham *et al.*, 1987). The diploid *E. majuscula* is the most distinctive and appears to be the rarest, especially outside S. England. It is certainly under-recorded.

World range uncertain.

References: Rich (1991), Rich & Jermy (1998), Rich & Lewis (1999).

D. A. PEARMAN & C. D. PRESTON

Erophila verna Common Whitlowgrass

Native	GB	IR
1987–99	764	103
1970–86	70	15
pre 1970	246	59
Alien		
1987–99	1	0
1970–86	0	0
pre 1970	1	0

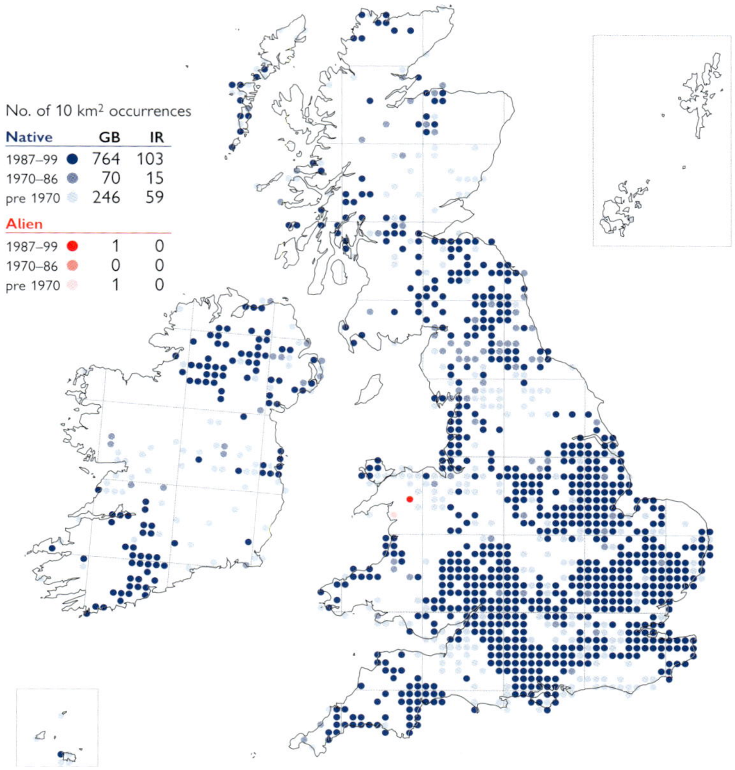

Typical semi-natural habitats of *E. verna* include limestone cliffs and pavements, sand dunes, sandy banks and coastal and riparian shingle. It is also frequent in quarries, sand- and gravel-pits and on lime-mine spoil heaps, walls, railway clinker, rubble and waste ground and cracks in pavements. 0–770 m (Great Dun Fell, Westmorland).

Native. This polyploid is the most frequent of the three segregate *Erophila* species. The high proportion of pre-1970 records reflects the fact that many records are derived from herbarium specimens studied by Rich & Lewis (1999); it does not signify a decline. The species is still under-recorded.

World range uncertain.

References: Rich (1991), Rich & Jermy (1998).

D. A. PEARMAN & C. D. PRESTON

Erophila glabrescens Glabrous Whitlowgrass

Native	GB	IR
1987–99	216	14
1970–86	29	11
pre 1970	117	15
Alien		
1987–99	0	0
1970–86	0	0
pre 1970	0	0

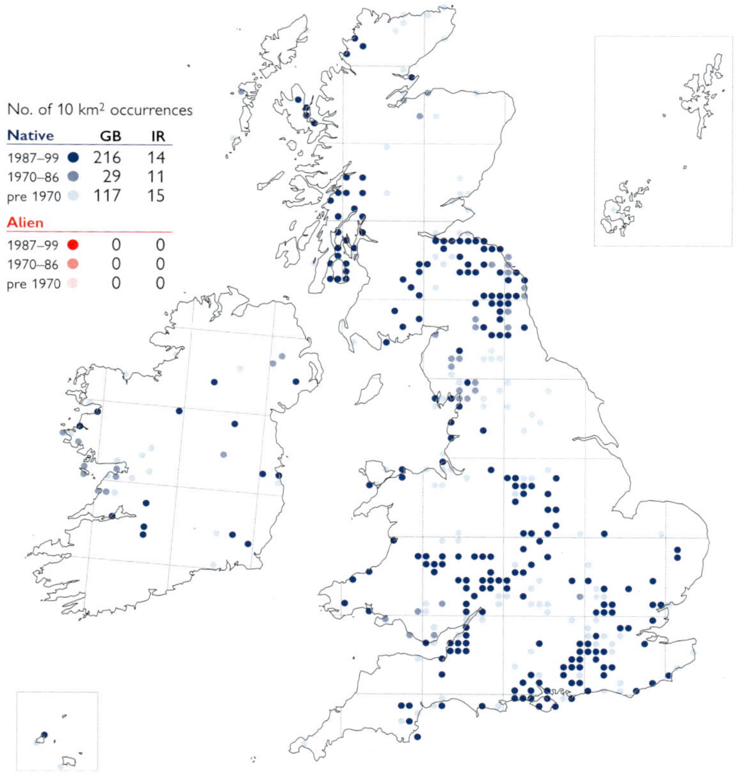

This segregate has been recorded from limestone rocks, pavements and hill-slopes, chalk downland, rocky river- and streamsides, river shingle, sand dunes and sandy grassland, walls, roadside verges, crevices in concrete and in pavements, loose road metal and gravelly paths. Lowland.

Native. *E. glabrescens* was first recognised in national floras in 1987 (Clapham *et al.*, 1987). It is not always easy to separate this polyploid species from *E. verna sens. str.* It may be as frequent as *E. verna* in N. England and Scotland. It is very under-recorded.

World range uncertain.

References: Rich (1991), Rich & Jermy (1998), Rich & Lewis (1999).

D. A. PEARMAN & C. D. PRESTON

Cochlearia anglica English Scurvygrass

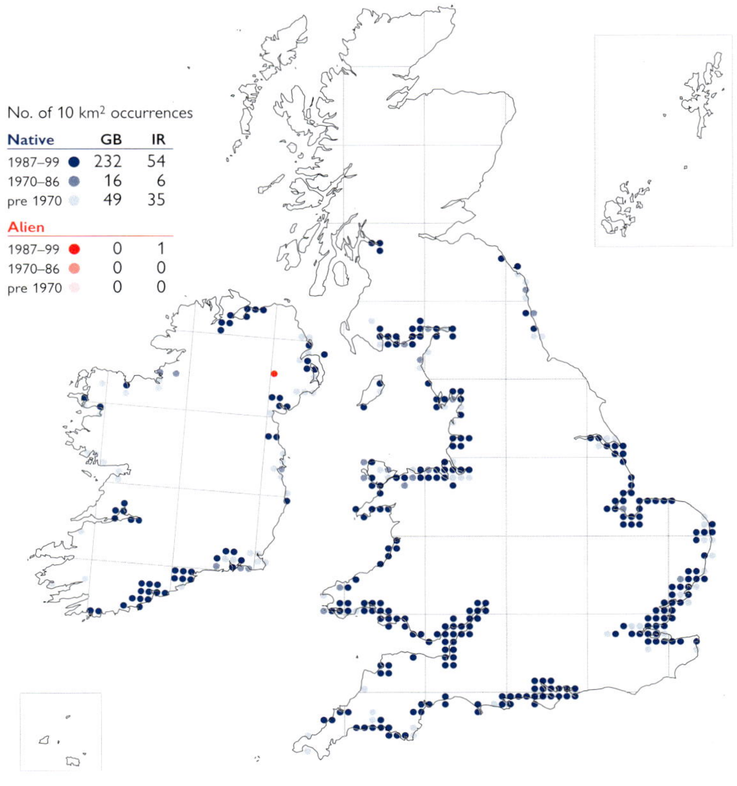

A biennial to perennial herb found in saltmarshes on soft, silty substrates, and in firmer areas of mud (and on sea-walls) near the high water mark of estuaries and tidal rivers. Unlike *C. officinalis* and *C. danica*, it is not found by roads. Lowland.

Native (change +0.02). The distribution of *C. anglica* appears to be stable. Scottish records in the 1962 *Atlas* from north of the Forth-Clyde line have all been referred to the *C. officinalis* aggregate. Some plants from N.W. England appear to be intermediate between *C. anglica* and *C. officinalis* (Halliday, 1997).

Oceanic Temperate element.

References: Atlas (41d), Jalas & Suominen (1996), Rich (1991).

D. A. PEARMAN

Cochlearia officinalis agg. Common Scurvygrass

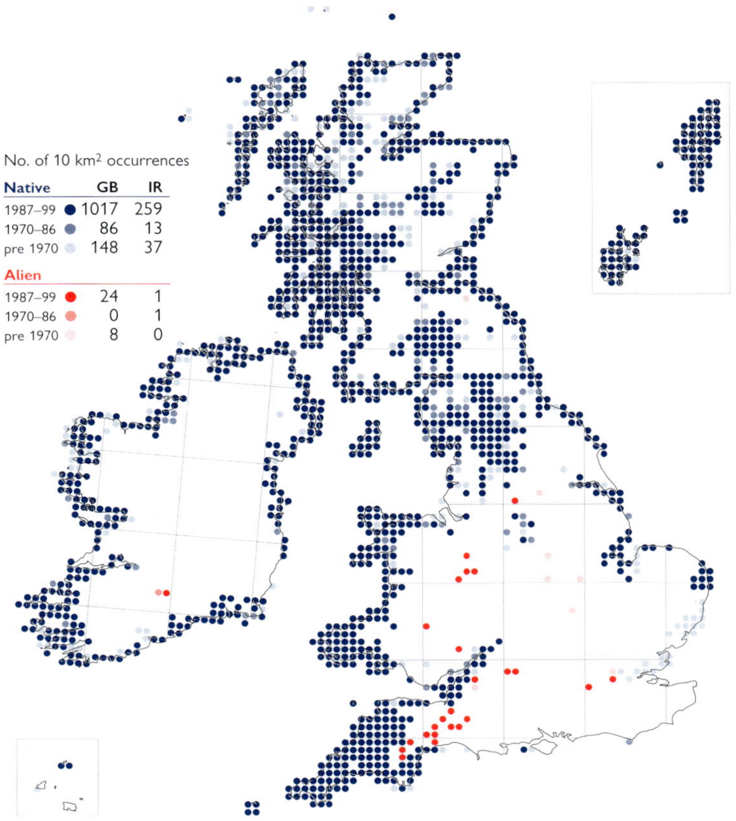

Biennial or perennial herbs found in a range of moist, open, coastal and montane habitats. They also grow on hedge banks, for example in S.W. England and Wales, but are only occasional recent colonists of salted roadsides. 0–1155 m (Ben Lawers, Mid Perth).

Native (change –0.18). This aggregate (which includes species 3–5 in Stace, 1997), comprises cytologically, genetically and phenotypically variable taxa which probably hybridise freely. The taxonomic treatments of Stace (1991, 1997) and Rich (1991) are significantly different, and many recorders have only recorded the aggregate. The distribution is stable, other than in East Anglia where there is evidence of a decline.

Circumpolar Wide-boreal element.

References: Atlas (41a, b), Hultén & Fries (1986), Jalas & Suominen (1996), Meusel *et al.* (1965).

D. A. PEARMAN & C. D. PRESTON

Cochlearia pyrenaica Pyrenean Scurvygrass

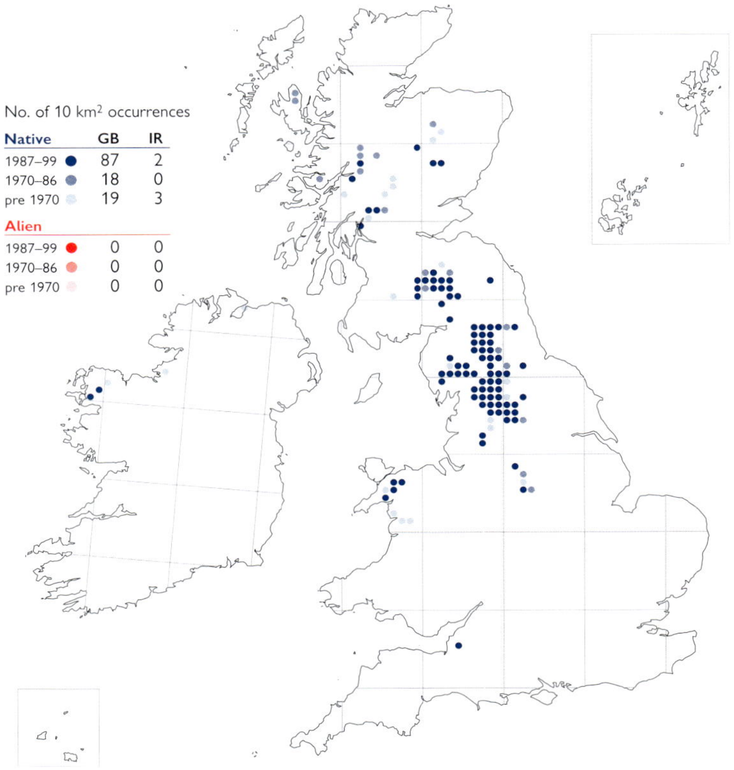

A biennial or perennial herb of damp, open habitats, including montane cliffs, wet gullies, bryophyte-dominated flushes and spoil heaps by old lead and zinc mines. Subsp. *pyrenaica* is usually found on more basic soils than subsp. *alpina*. Generally upland, reaching 960 m on Carnedd Llewelyn (Caerns.) and perhaps higher in Scotland.

Native. The diploid subsp. *pyrenaica* is only known from N. England and Skye; the tetraploid subsp. *alpina* is more widespread. The latter is sometimes considered as a subspecies of *C. officinalis*. Because of the varying taxonomic treatments in recent years, *C. pyrenaica* is under-recorded, especially in Scotland.

European Arctic-montane element, but absent from the Boreal zonobiome.

References: Jalas & Suominen (1996), Rich (1991).

D. A. PEARMAN

Cochlearia officinalis subsp. *scotica*

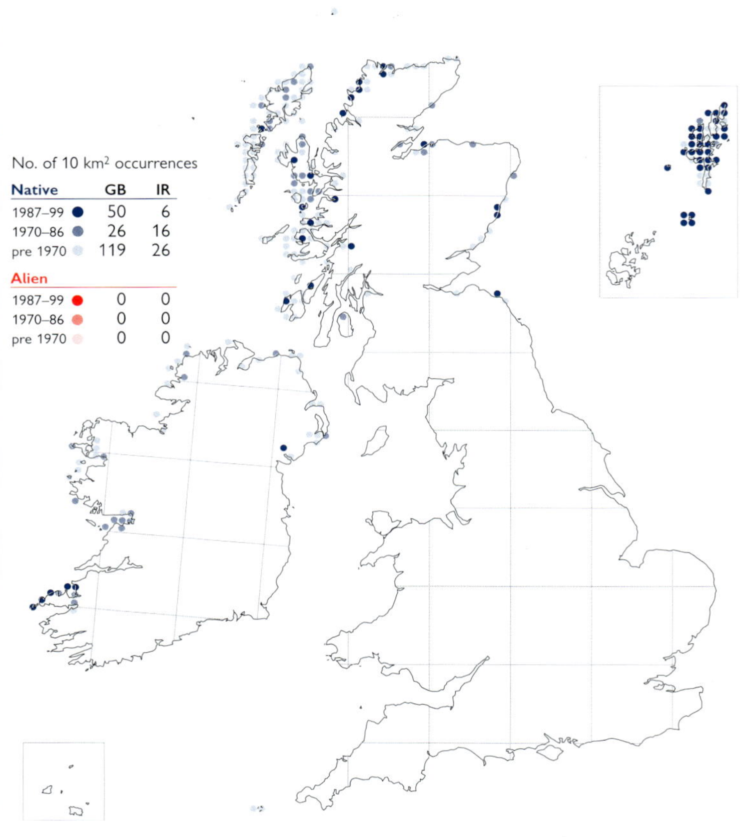

A biennial or perennial herb which grows in a variety of coastal habitats, including open, stony shores, the crevices between rock and boulders near the sea, shingle spits, sand dunes and short, grazed grassland on cliff-tops and saltmarshes. Lowland.

Native. The current disagreement over the taxonomic status of subsp. *scotica* has discouraged recorders from attempting to identify it, and resulted in a paucity of recent records. Wyse Jackson (1991) and Stace (1997) recognise it as a subspecies of *C. officinalis*, whereas D.H. Dalby suggests that the populations referred to subsp. *scotica* might be a mixture of environmentally dwarfed individuals of other taxa (Rich, 1991).

Endemic.

References: Atlas (41b), Jalas & Suominen (1996), Stewart *et al.* (1994).

D. A. PEARMAN & C. D. PRESTON

Cochlearia micacea Mountain Scurvygrass

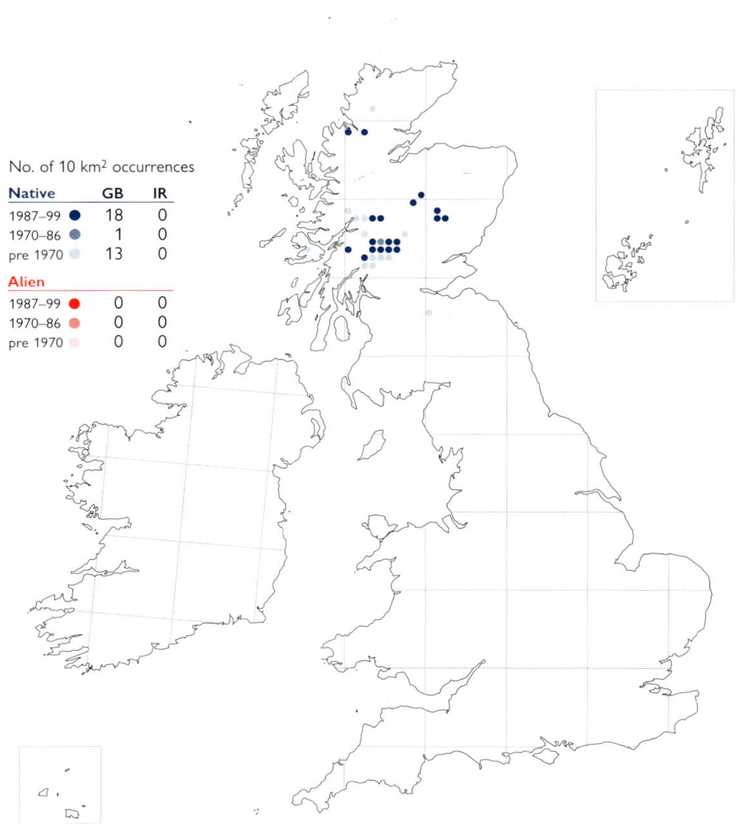

A perennial herb of micaceous soils, growing mainly in flushes, by springs and on streamsides, but also found on ledges, cliffs and in ravines. Upland, from 600 m on Meall nan Tarmachan (Mid Perth) to 1155 m on Ben Lawers (Mid Perth).

Native. This species was described by E.S. Marshall in 1894 and the records were recently revised by Dalby & Rich (1995). It is almost certainly still under-recorded, and there is no reason to believe that its distribution is other than stable.

Endemic.

References: Jalas & Suominen (1996), Wigginton (1999).

D. A. PEARMAN

Cochlearia danica Danish Scurvygrass

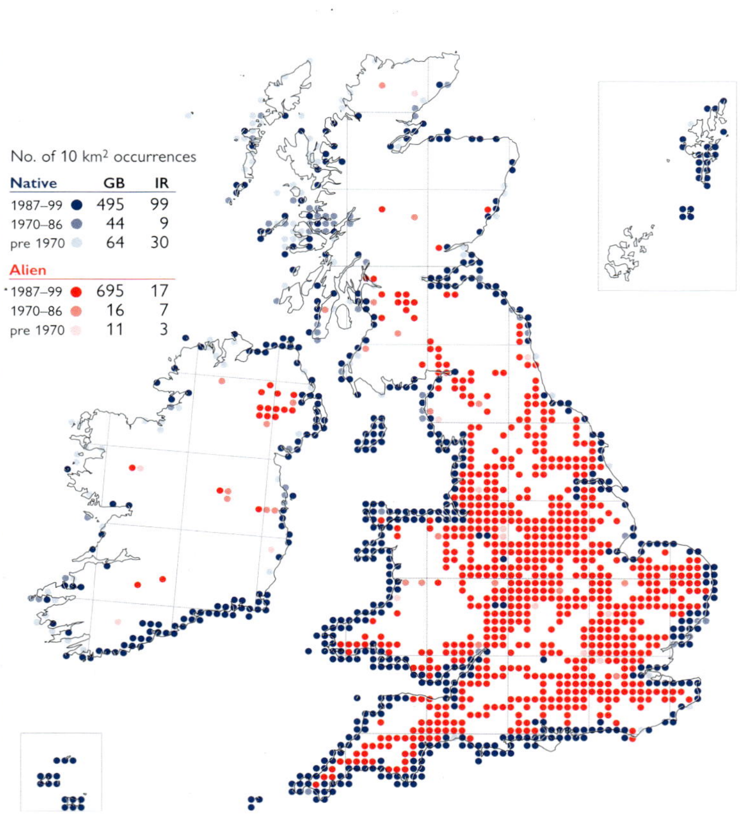

A winter-annual of cliff-tops, sand dunes and sea-walls, and on old walls and pavements in coastal towns; generally preferring open ground on well-drained sandy soils or bare rock. It used to occur on railway ballast, and is now frequent along inland roadsides. Lowland, but above 300 m on roadsides in N. England.

Native (change +3.31). Since the early 1980s, *C. danica* has spread rapidly along salt-treated roads in Britain and N. Ireland, but not in S. Ireland where grit is used. A former preference for the central reservations of motorways and dual carriageways is becoming less obvious, and it now occurs on many single-carriageway A roads in England and Wales.

Oceanic Temperate element.

References: Atlas (41c), Coombe (1994), Dupont (1962), Hultén & Fries (1986), Jalas & Suominen (1996), Leach (1990), Rich (1991), Scott (1985).

S. J. LEACH

Camelina sativa Gold-of-pleasure

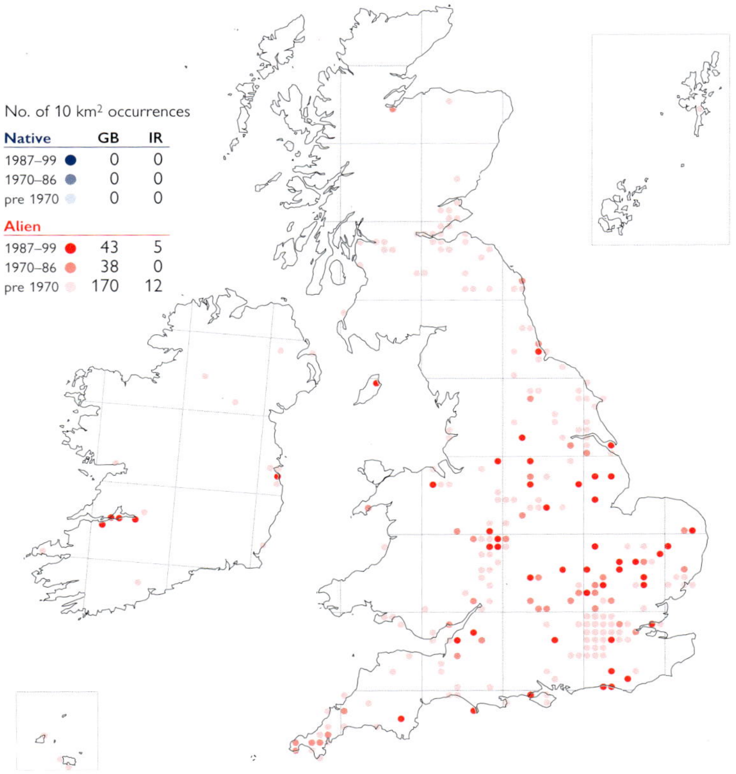

An annual, now occurring mainly as a casual on rubbish tips and waste ground, and in gardens. It was formerly a frequent weed of arable fields and a contaminant of flax seed. Lowland.

Archaeophyte. *C. sativa*, cultivated in Europe since prehistoric times, was formerly a frequent casual. Most 19th century records were of escapes from cultivation for oil and fibre, and of grain aliens, but nearly all recent records arise from bird-seed mixtures.

Native range uncertain, but now widespread in Eurasia and introduced to N. America and elsewhere.

References: Hanson & Mason (1985), Hultén & Fries (1986), Jalas & Suominen (1996), Rich (1991), Zohary & Hopf (2000).

D. A. PEARMAN

Neslia paniculata Ball Mustard

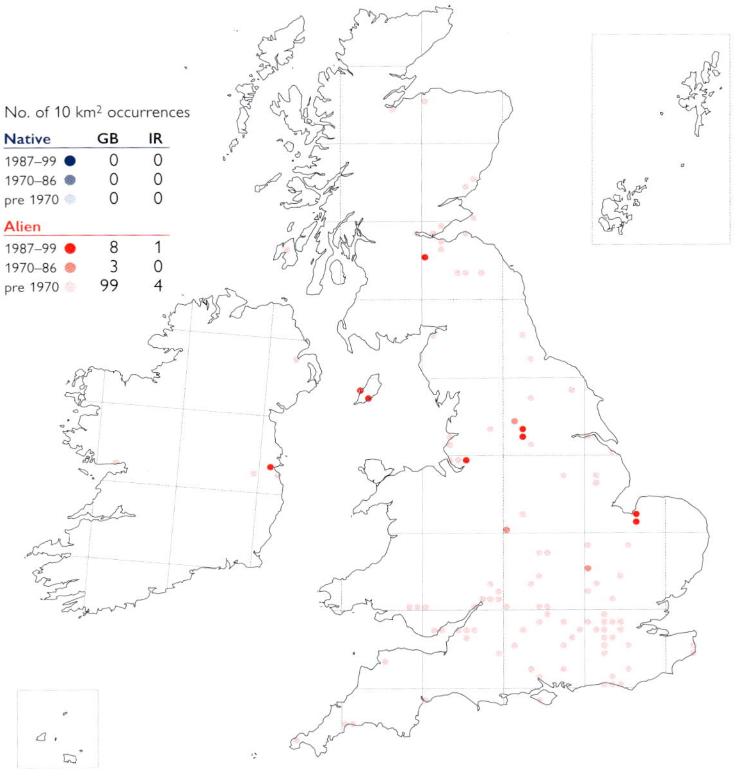

An annual which is introduced to rubbish tips and waste ground as an impurity of grain and bird-seed. Lowland.

Casual. Although cultivated in Britain by 1683, *N. paniculata* was first recorded in the wild as a casual from grain in 1859. It is now of only rare occurrence, presumably because of improved seed cleaning techniques.

N. paniculata sens. lat. (including *N. apiculata*) is a Eurosiberian Southern-temperate species which occurs outside its native range as a naturalised or casual plant in Europe, E. Asia, N. America and elsewhere.

References: Hultén & Fries (1986), Jalas & Suominen (1996), Meusel *et al.* (1965), Rich (1991).

D. A. PEARMAN

Capsella bursa-pastoris Shepherd's-purse

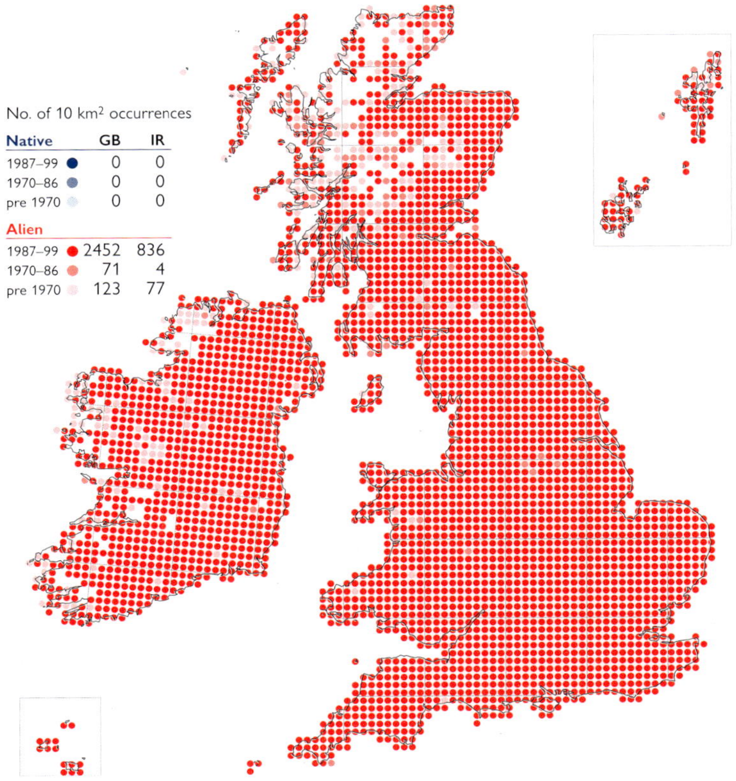

An annual, germinating throughout the year in suitable conditions and ubiquitous in many disturbed and nutrient-rich habitats. It is abundant on waste ground and in gardens, and is frequent in cultivated fields, particularly amongst broad-leaved crops. It avoids the wettest and most acidic soils. It has a very long-lived seed bank. 0–780 m (Knock Fell, Westmorland).

Archaeophyte (change –1.01). The overall range of *C. bursa-pastoris* is stable, and the reason for the apparent losses in Scotland are unclear.

As an archaeophyte this species has a Eurosiberian Wide-temperate distribution, but it is widely naturalised so that its distribution is now Circumpolar Wide-temperate.

References: Atlas (40c), Aksoy *et al.* (1998), Grime *et al.* (1988), Hultén & Fries (1986), Jalas & Suominen (1996), Rich (1991).

D. A. PEARMAN

Hornungia petraea Hutchinsia

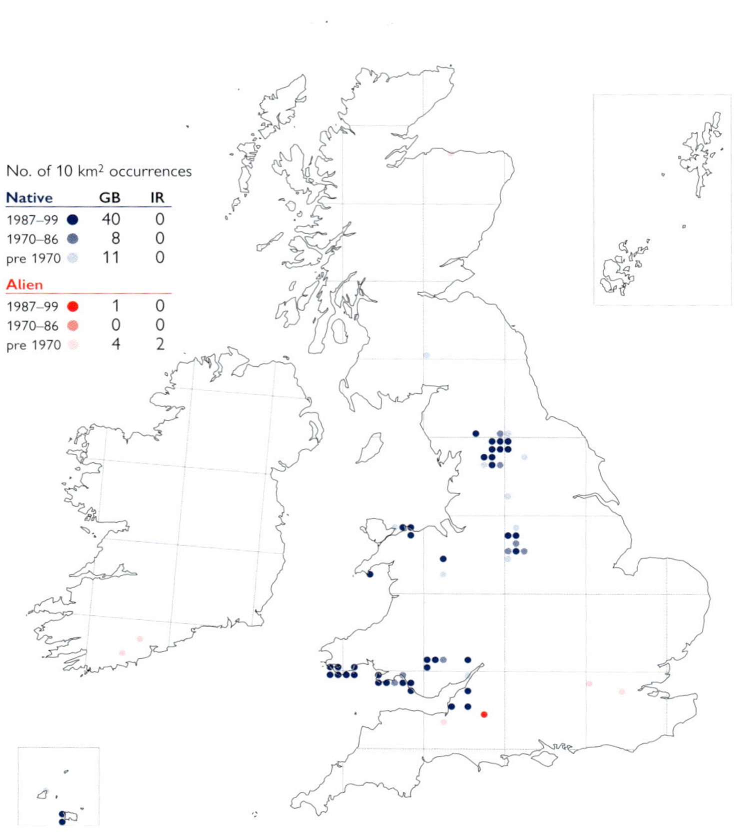

A winter-annual of very open habitats on calcareous soils and rocks which are subject to summer drought, especially on rocky slopes on Carboniferous limestone and on fixed but open sand dunes. It also occurs as an alien on garden walls and in chalk-pits. Generally lowland, but reaching 490 m near Hawes (N.W. Yorks.).

Native (change +0.31). The distribution of *H. petraea* is stable, and scrub invasion of its open habitats is the only reported threat.

European Temperate element.

References: Atlas (40d), Hultén & Fries (1986), Jalas & Suominen (1996), Meusel *et al.* (1965), D. Ratcliffe (1959), Rich (1991), Stewart *et al.* (1994).

D. A. PEARMAN

Teesdalia nudicaulis Shepherd's Cress

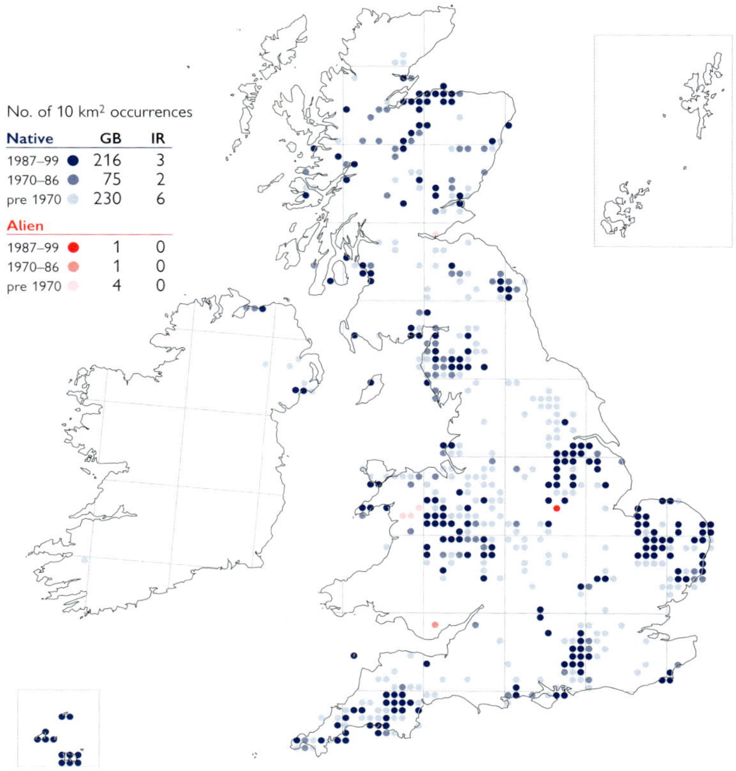

A winter-annual of acidic, well-drained sandy soils on heathlands, sand dunes, shingle and gravels, on sandy lake shores in Ireland, by railways and on coal and cinder tips. It prefers bare or disturbed ground. Generally lowland, but reaching 455 m on Ben More (Mid Ebudes) and Wasdale Screes (Cumberland).

Native (change –0.81). *T. nudicaulis* has a very short-lived seed bank and this may have contributed to losses arising from scrub invasion and afforestation, as well as from urbanisation. Although most of the losses were before 1930, it has continued to decline in E. and N.E. England. It is now much better recorded in Scotland.

European Temperate element.

References: Atlas (40b), Curtis & McGough (1988), Hultén & Fries (1986), Jalas & Suominen (1996), Meusel *et al.* (1965), Rich (1991), Stewart *et al.* (1994).

D. A. PEARMAN

Thlaspi arvense Field Penny-cress

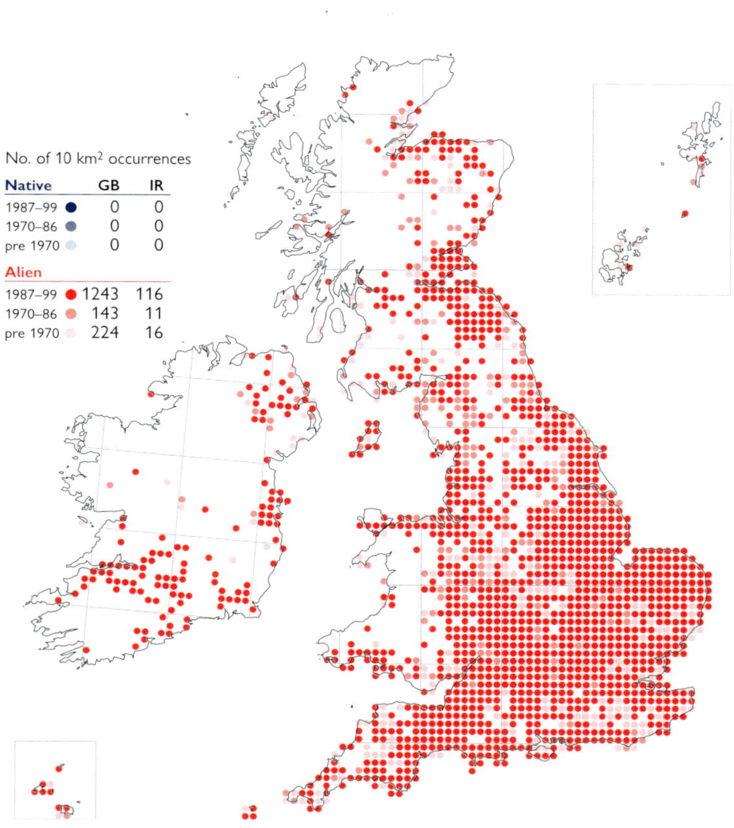

An annual found as an arable weed, particularly with broad-leaved crops and mainly on heavier soils. It is also a frequent weed on disturbed roadsides, and in waste places and gardens. Lowland, but there is a casual record at 330 m at Tomintoul (Banffs.).

Archaeophyte (change +0.16). *T. arvense* was mapped as 'all records' in the 1962 *Atlas*. The map suggests that it has increased in frequency in the north and west of its range since then.

As an archaeophyte *T. arvense* has a Eurasian Temperate distribution, but it is naturalised in N. America so its distribution is now Circumpolar Temperate.

References: Atlas (39c), Best & McIntyre (1975), Hultén & Fries (1986), Jalas & Suominen (1996), Meusel *et al.* (1965), Rich (1991).

D. A. PEARMAN

Thlaspi perfoliatum Perfoliate Penny-cress

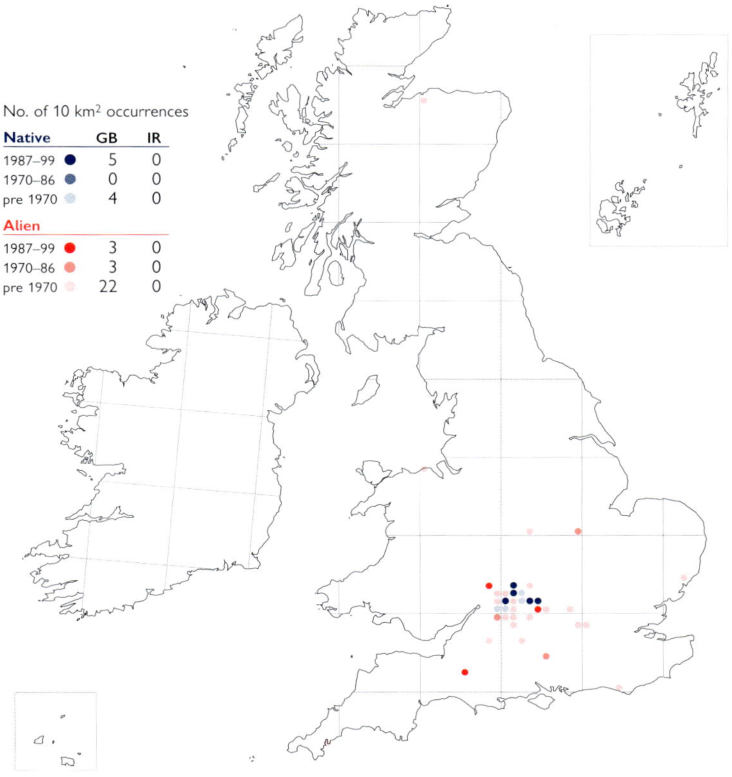

An annual of bare or sparsely vegetated habitats on oolitic limestone, found on screes, stony banks and open pastures; also found in old quarries and on broken rocks. Elsewhere it is usually a casual of waste places, although populations have persisted on railway banks. Lowland.

Native (change –0.94). *T. perfoliatum* has declined because of loss of open habitat due to scrub invasion, the lack of grazing or the cessation of quarrying. It has poor seed dispersal, and it has been suggested that many colonies away from the Cotswolds result from seed spread in the slipstream of trains.

Eurosiberian Southern-temperate element; widely naturalised outside its native range.

References: Atlas (40a), Hultén & Fries (1986), Jalas & Suominen (1996), Meusel *et al.* (1965), Rich, Lambrick *et al.* (1998), Wigginton (1999).

D. A. PEARMAN

Thlaspi caerulescens Alpine Penny-cress

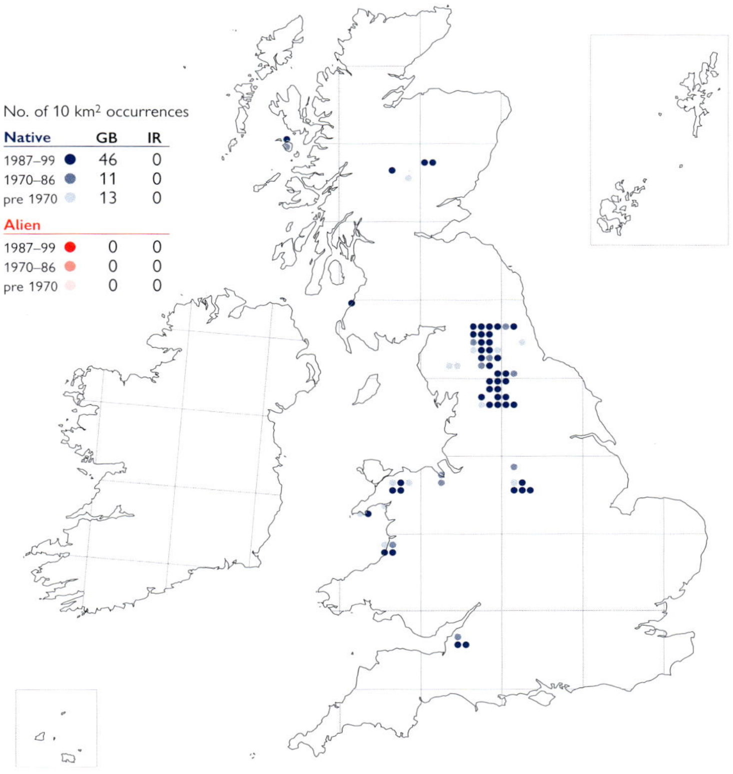

A perennial, or rarely biennial, herb almost confined in Britain to rocks or soils enriched with lead or zinc, being found on spoil heaps and mine waste and on metalliferous river gravels. It is also found, rarely, on outcrops and scree of limestone and other base-rich rocks, particularly in Scotland. Generally upland, reaching 940 m on Caenlochan (Angus), but descending to 100 m in Caernarvonshire.

Native (change +0.01). The distribution of *T. caerulescens* appears to be stable, although reworking of lead mine spoils may have destroyed some populations.

European Boreal-montane element, but absent from the Boreal zonobiome.

References: Atlas (39d), Hultén & Fries (1986), Ingrouille & Smirnoff (1986), Jalas & Suominen (1996), Meusel *et al.* (1965), Rich (1991), Stewart *et al.* (1994).

D. A. PEARMAN

Iberis sempervirens Perennial Candytuft

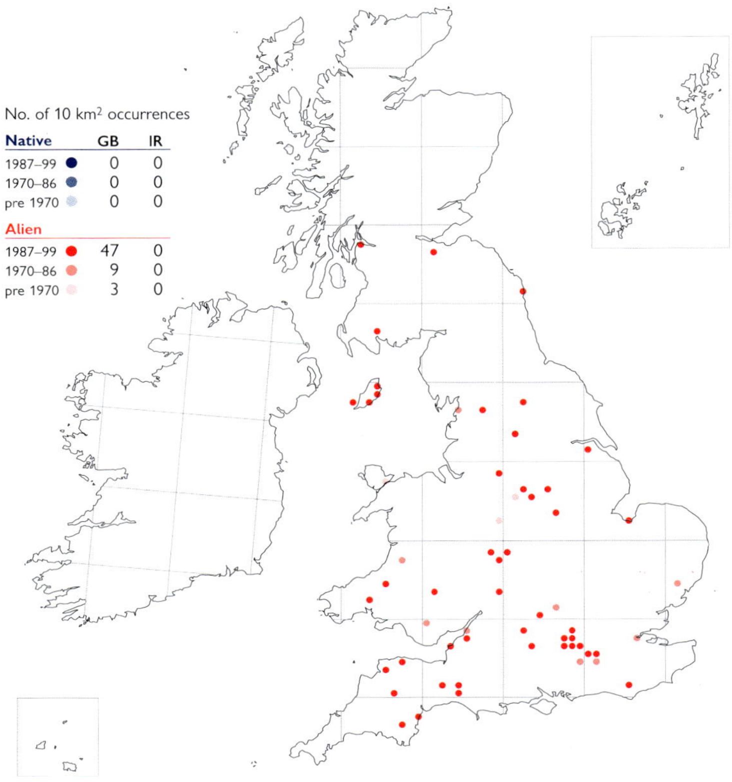

A perennial dwarf shrub which occasionally escapes onto waste ground or walls where it sometimes persists. Lowland.

Neophyte. *I. sempervirens* was cultivated in Britain by 1731. Although it is much grown in gardens, it is still rare as an escape. It was known in the wild by 1928 and its distribution is probably stable.

Native of the mountains of S. Europe, N. Africa and S.W. Asia.

References: Bolòs & Vigo (1990), Jalas & Suominen (1996), Rich (1991).

D. A. PEARMAN

Iberis amara Wild Candytuft

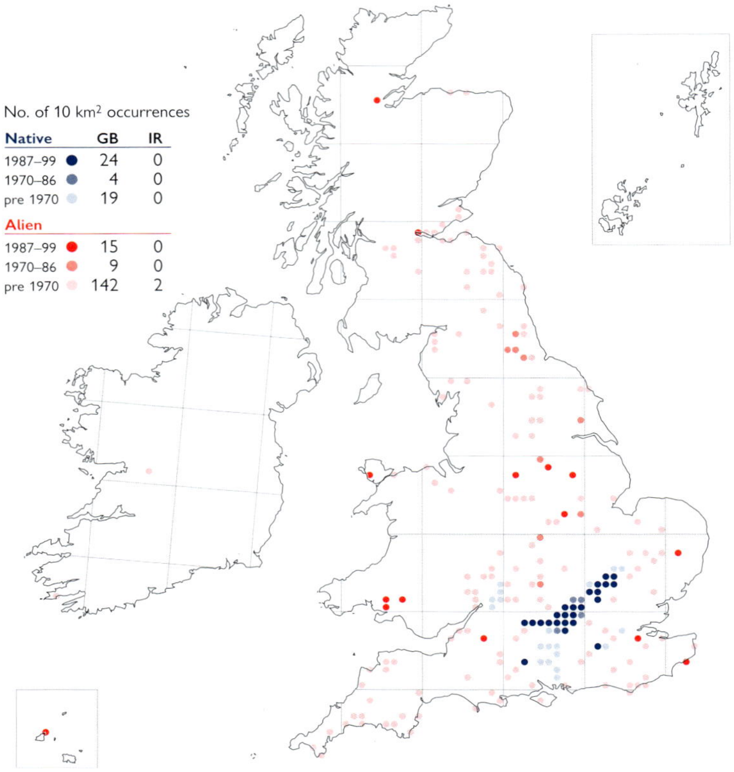

An annual, rarely biennial, herb of bare, open ground on S.-facing slopes on chalk, being found in bare places in grassland, particularly rabbit scrapes, and in quarries. It also occurs as an arable weed, and as a casual in a wide variety of ruderal habitats. Lowland.

Native (change –1.21). Rabbits were important in maintaining an open habitat for this species, and after myxomatosis some sites were lost. However, seed is apparently long-lived, and plants can reappear in native sites following disturbance or clearance. Many alien records may be misidentifications of the similar *I. umbellata*.

Suboceanic Southern-temperate element; widely naturalised outside its native range.

References: Atlas (39b), Bolòs & Vigo (1990), Jalas & Suominen (1996), Rich (1991), Stewart *et al.* (1994).

D. A. PEARMAN

Iberis umbellata Garden Candytuft

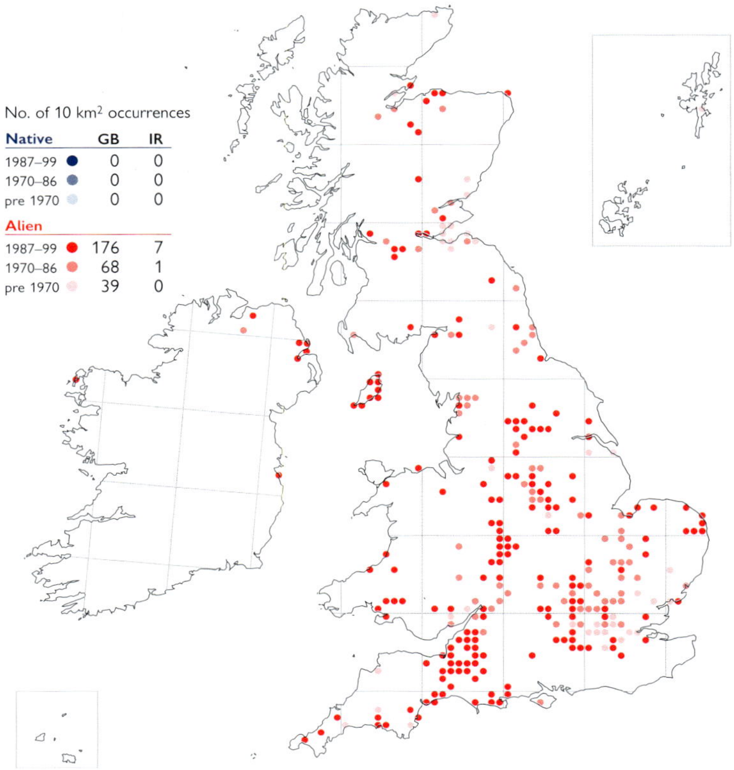

An annual, widely distributed as a casual of tips, waste ground and gardens. It is never persistent in the wild. Lowland.

Casual. *I. umbellata* was known as a garden plant by 1596. It was first recorded in the wild in 1858. There has been a large increase in records since it was mapped by Rich (1991), probably largely as a result of improved recording. All records in our area are of the garden cultivar rather than the wild species which is native to S. Europe.

Native of the European Mediterranean region; widely cultivated and naturalised elsewhere.

Reference: Jalas & Suominen (1996).

D. A. PEARMAN

Lepidium sativum Garden Cress

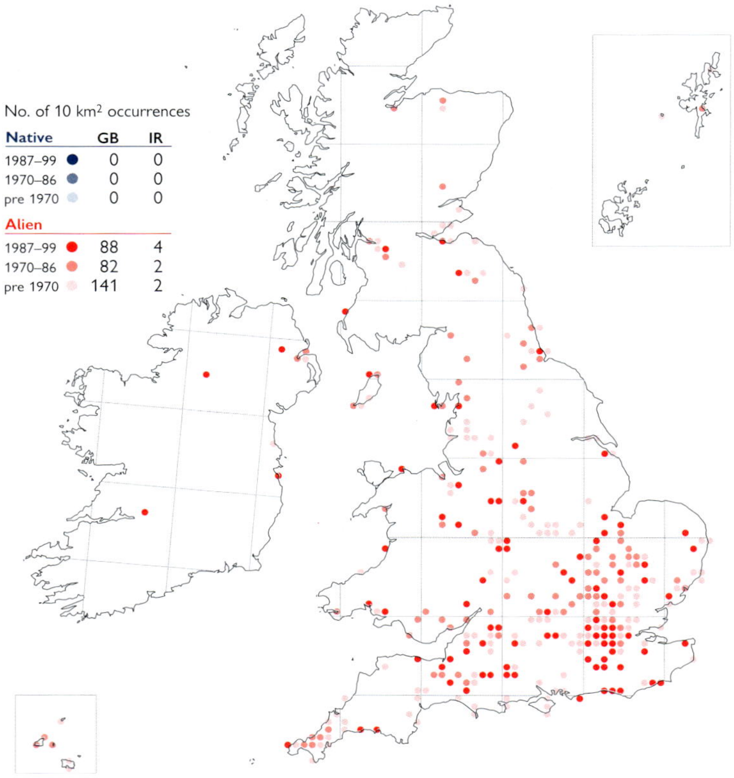

An annual of waste and ruderal habitats, arising principally from bird-seed and culinary sources, and usually occurring as a casual. Lowland.

Casual. *L. sativum* has been grown in gardens since at least 995 (Harvey, 1981) and was known from the wild by about 1860. It is the original 'cress' of 'mustard-and-cress', but is now very largely replaced in salads by *Brassica napus*. There is no indication of a change in frequency in recent years.

Probably native of Egypt and S.W. Asia; as a naturalised or more frequently casual plant it is widespread in Europe and elsewhere.

References: Hultén (1968), Rich (1988b, 1991), Vaughan & Geissler (1997).

D. A. PEARMAN

Lepidium campestre Field Pepperwort

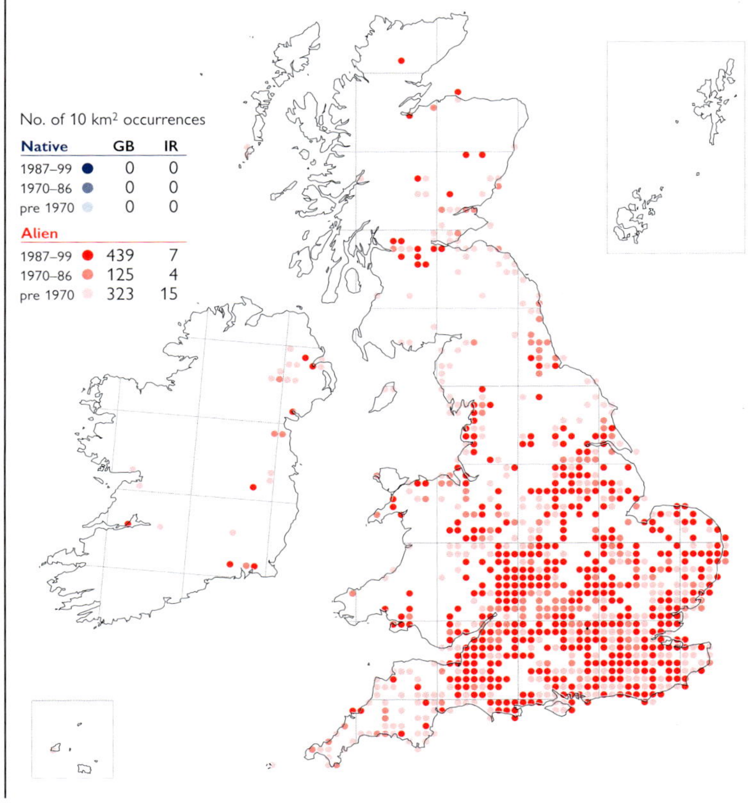

An annual, or occasionally biennial, herb of open grassland and arable fields, particularly on sandy or gravelly soils; also found on roadsides and walls, in gardens and waste places. It is often persistent, but occurs just as frequently as a casual. Lowland.

Archaeophyte (change –0.70). *L. campestre* was mapped as 'all records' in the 1962 *Atlas*. Analysis of the database reveals that most of the losses have occurred since 1950, with a marked decline in some areas.

As an archaeophyte *L. campestre* has a European Temperate distribution; it is widely naturalised outside this range.

References: Atlas (37b), Hultén & Fries (1986), Jalas & Suominen (1996), Rich (1991).

D. A. PEARMAN

Lepidium heterophyllum Smith's Pepperwort

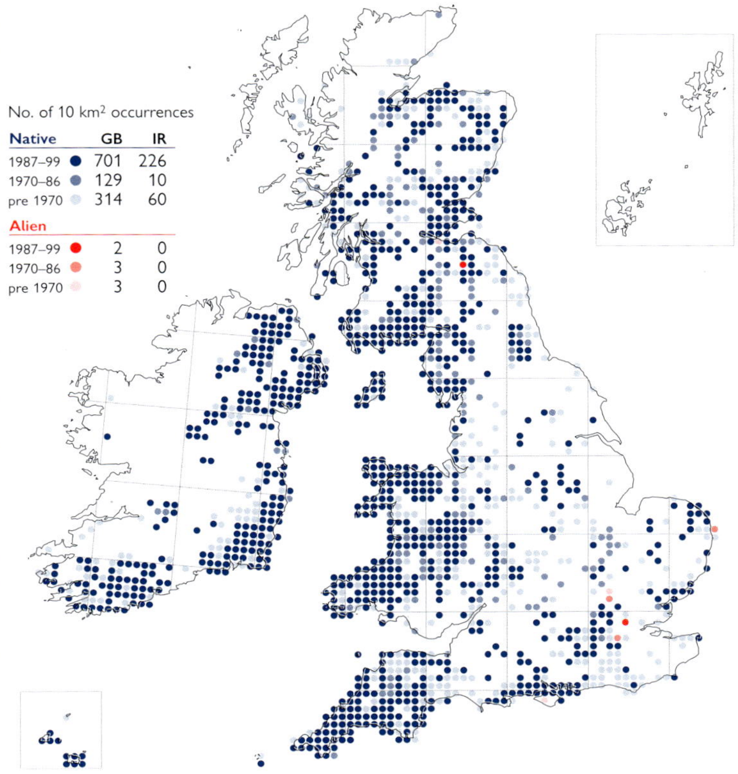

A perennial, or rarely biennial, herb of acidic soils in dry heathy and gravelly places. It is also frequent on shingle, railway ballast and banks, and, less commonly, in arable fields. It is tolerant of grazing. Generally lowland, but reaching 425 m (Sow of Atholl, E. Perth).

Native (change –0.51). *L. heterophyllum* was mapped as 'all records' in the 1962 *Atlas*. There is little evidence of any change in its distribution except in S.E. England where it appears to be decreasing.

Oceanic Southern-temperate element.

References: Atlas (37c), Bolòs & Vigo (1990), Jalas & Suominen (1996), Rich (1991).

D. A. PEARMAN

Lepidium virginicum Least Pepperwort

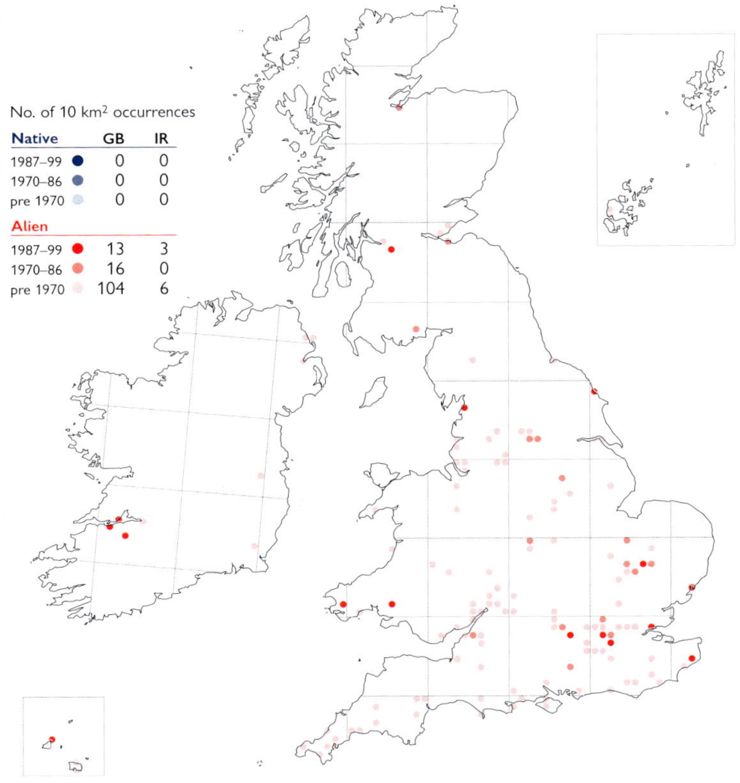

An annual or biennial herb occurring on railways, rubbish tips, waste ground and in arable fields, arising from impurities in grain, and in wool shoddy and bird-seed. It is usually casual, occasionally persisting for a few years. Lowland.

Neophyte. Although cultivated in Britain by 1713, *L. virginicum* has arisen as a grain impurity and was first recorded from the wild in 1881 (Surrey). It has declined and is now only a rare casual.

Native of N. America; now a widespread alien in Europe and the S. Hemisphere.

References: Hultén & Fries (1986), Jalas & Suominen (1996), Meusel *et al.* (1965), Rich (1991).

D. A. PEARMAN

Lepidium ruderale Narrow-leaved Pepperwort

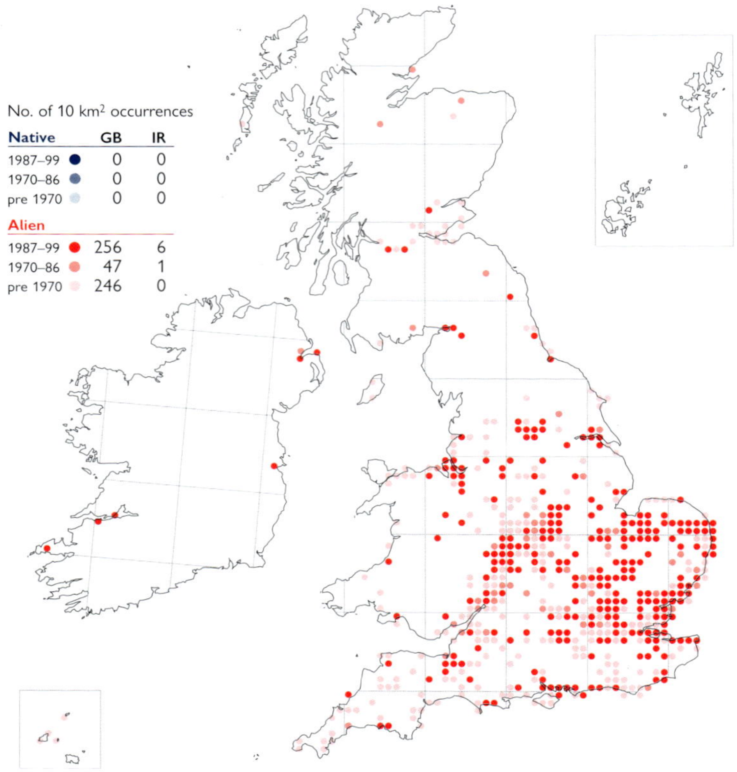

An annual, or rarely biennial, herb of banks and bare waste land near the sea, and of salted road verges. It is also frequent as a casual of roadsides, rubbish tips, gardens and waste places. Lowland.

Archaeophyte (change –0.04). *L. ruderale* is increasingly found as a halophyte along trunk roads, but is apparently less frequent as a casual than it was before the Second World War.

As an archaeophyte *L. ruderale* has a Eurosiberian Temperate distribution; it is widely naturalised outside this range.

References: Atlas (37d), Coombe (1994), Hultén & Fries (1986), Jalas & Suominen (1996), Meusel *et al.* (1965), Rich (1991).

D. A. PEARMAN

Lepidium perfoliatum Perfoliate Pepperwort

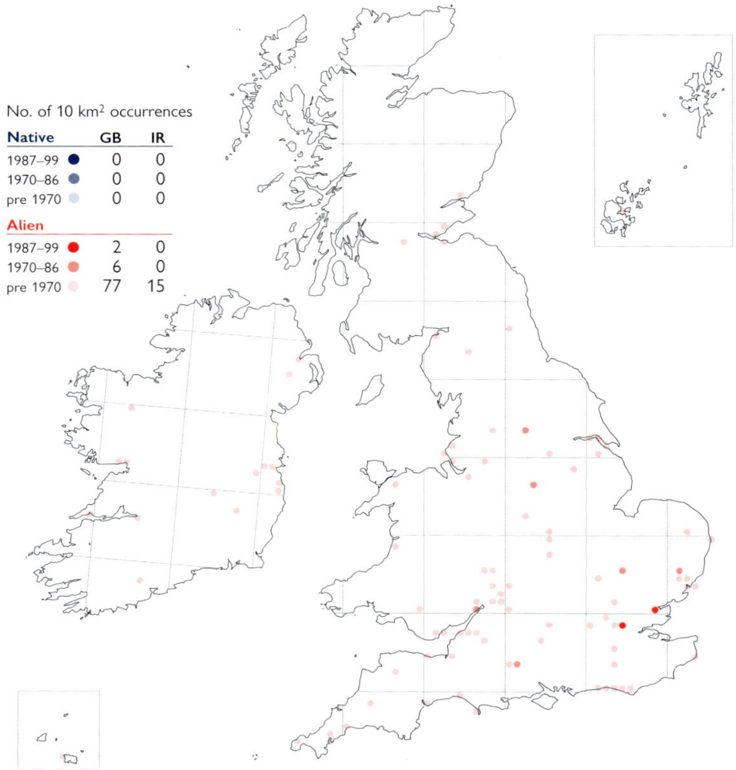

An annual or biennial herb found as a casual on rubbish tips, roadsides and by docks, where it is introduced as an impurity of grain and grass-seed. It occasionally persists for a few years, and is very rarely naturalised. Lowland.

Neophyte. Introduced into cultivation in Britain by 1640, *L. perfoliatum* was first recorded from the wild in 1888. It was formerly more frequent, but is now an extremely rare casual. It has persisted at Northey Island (S. Essex) since 1976.

A Eurosiberian Southern-temperate species; widely naturalised in E. Asia, N. America, Australasia and elsewhere.

References: Bolòs & Vigo (1990), Jalas & Suominen (1996), Rich (1991).

D. A. PEARMAN

Lepidium latifolium Dittander

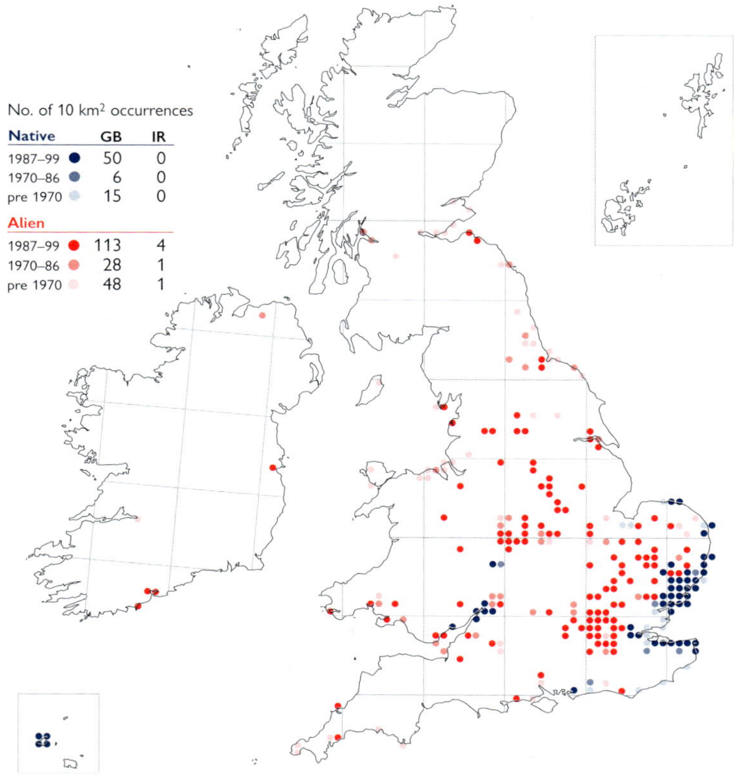

A rhizomatous perennial herb native on creek-sides, ditches, sea-walls, open brackish grassland and the upper fringes of estuarine saltmarshes. It is also naturalised in disturbed areas such as waste ground, dockland, railways and roadsides. Lowland.

Native (change +1.23). *L. latifolium* was cultivated up to the 17th century as a hot flavouring before *Armoracia rusticana* was used for that purpose. Its ruderal habitat and the persistence of relics of cultivation make its native range difficult to delimit. The coastal distribution remains stable, but there are many more inland records than in the 1962 *Atlas*.

Eurosiberian Southern-temperate element; widely naturalised outside its native range.

References: Atlas (38a), Hultén & Fries (1986), Jalas & Suominen (1996), Rich (1991), Stewart *et al.* (1994).

D. A. PEARMAN

Lepidium draba Hoary Cress

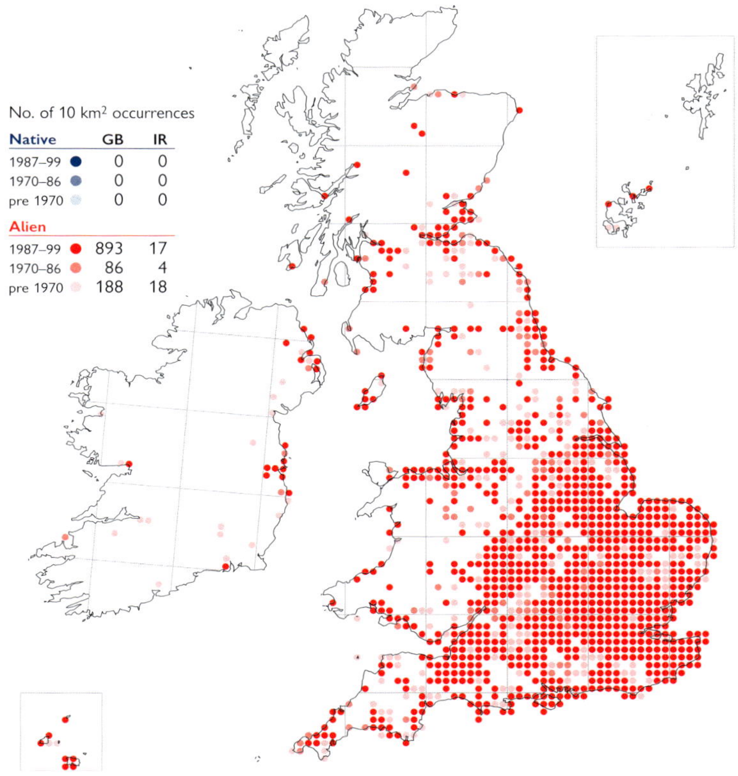

A perennial rhizomatous herb of roadsides, and on dry limestone or clinker ballast of railways. It also grows on waste ground, in arable fields on light soils, on sand dunes and other sandy ground, particularly near the sea, and in the uppermost zone of saltmarshes. Lowland.

Neophyte (change +0.06). Accidentally introduced to Swansea (Glamorgan) in 1802, and subsequently to other ports, *L. draba* spread rapidly in the 19th century, particularly in urban and industrial areas. Its range has expanded in Britain since the 1962 *Atlas*, but not in Ireland.

Native of S. Europe and S.W. Asia; naturalised throughout temperate W. Europe, N. America and the S. hemisphere.

References: Atlas (38d), Hultén & Fries (1986), Jalas & Suominen (1996), Meusel *et al.* (1965), Mulligan & Findlay (1974), Rich (1991), Scurfield (1962). D. A. PEARMAN

Coronopus squamatus Swine-cress

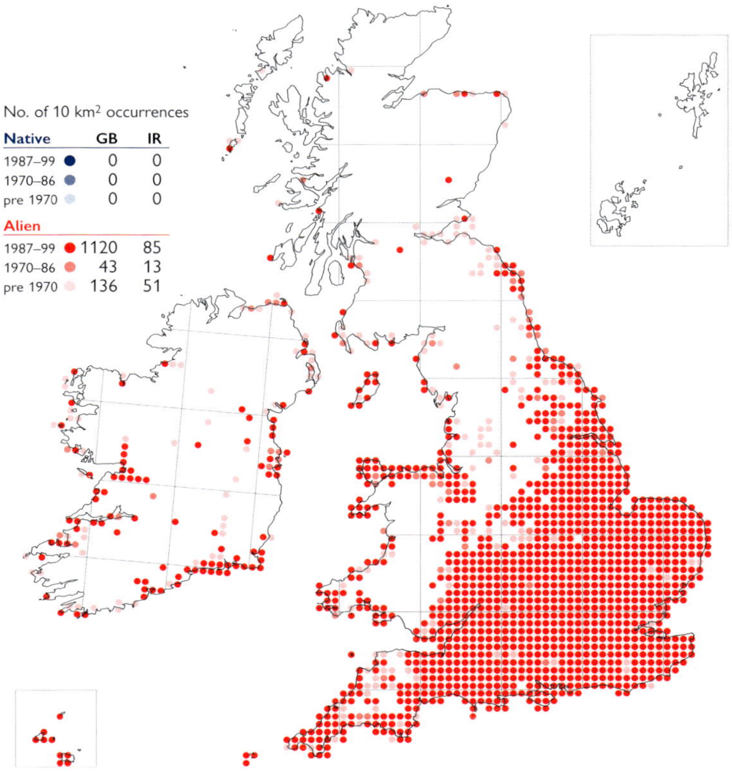

A spring-germinating annual, rarely biennial, herb of nutrient-rich, often compacted soils in open, dry or winter-wet habitats. Typical sites include farmyards, waste ground, paths and particularly gateways. Lowland.

Archaeophyte (change +0.33). The distribution of *C. squamatus* is stable.

As an archaeophyte *C. squamatus* has a European Southern-temperate distribution; it is widely naturalised outside this range.

References: Atlas (38b), Hultén & Fries (1986), Jalas & Suominen (1996), Meusel *et al.* (1965), Rich (1991).

D. A. PEARMAN

Coronopus didymus Lesser Swine-cress

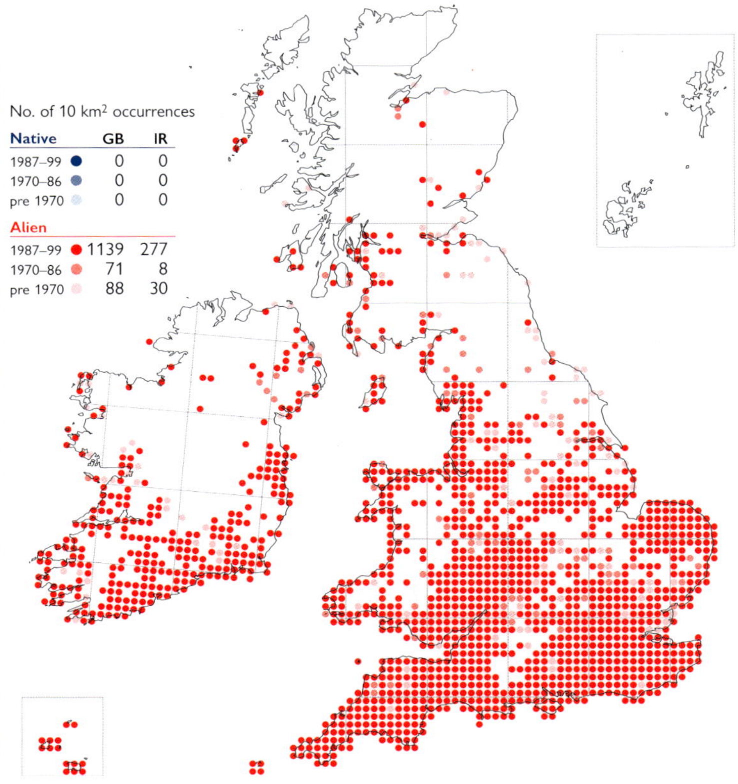

An annual or biennial herb of damp, often winter-wet soils, occurring on cultivated and waste ground, and frequently found in gardens and lawns, by paths and roadsides and on rubbish tips. Lowland.

Neophyte (change +1.77). *C. didymus* reached Britain in the early 18th century, being recorded from the wild by 1778. It is now frequent in urban and industrial areas, and is still spreading into rural areas where it is widespread but scattered.

A widespread alien in W. Europe, N. America and the S. hemisphere; origin uncertain, often cited as S. America.

References: Atlas (38c), Hultén & Fries (1986), Jalas & Suominen (1996), Rich (1991).

D. A. PEARMAN

Subularia aquatica Awlwort

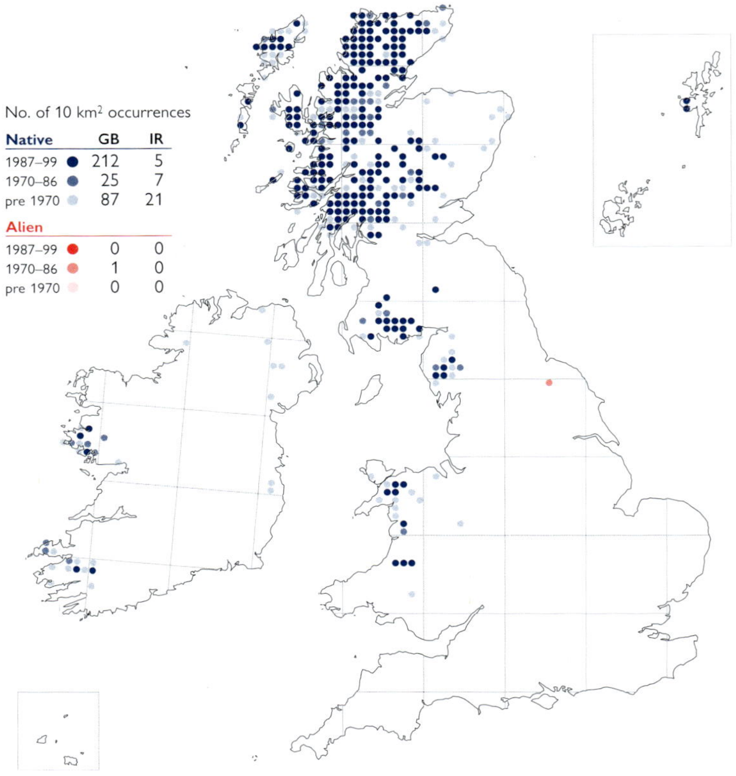

An annual aquatic plant, sometimes overwintering as a rosette, growing on silt, gravel or stony substrates in acidic, oligotrophic lakes. It is normally a plant of water shallower than one metre, and is only rarely found in other water bodies, such as outfall streams. 0–825 m (Ffynnon Llyffaint, Caerns.).

Native (change +0.73). Because of better exploration, *S. aquatica* is now known from many more sites than were mapped in the 1962 *Atlas*. However, eutrophication seems to have caused declines along the eastern fringe of its range, and also in the Lake District and N. Ireland.

Circumpolar Boreal-montane element, with a disjunct distribution.

References: Atlas (42a), Hultén & Fries (1986), Jalas & Suominen (1996), Meusel *et al.* (1965), Preston & Croft (1997), Rich (1991), Stewart *et al.* (1994), Woodhead (1951c).

D. A. PEARMAN

Conringia orientalis Hare's-ear Mustard

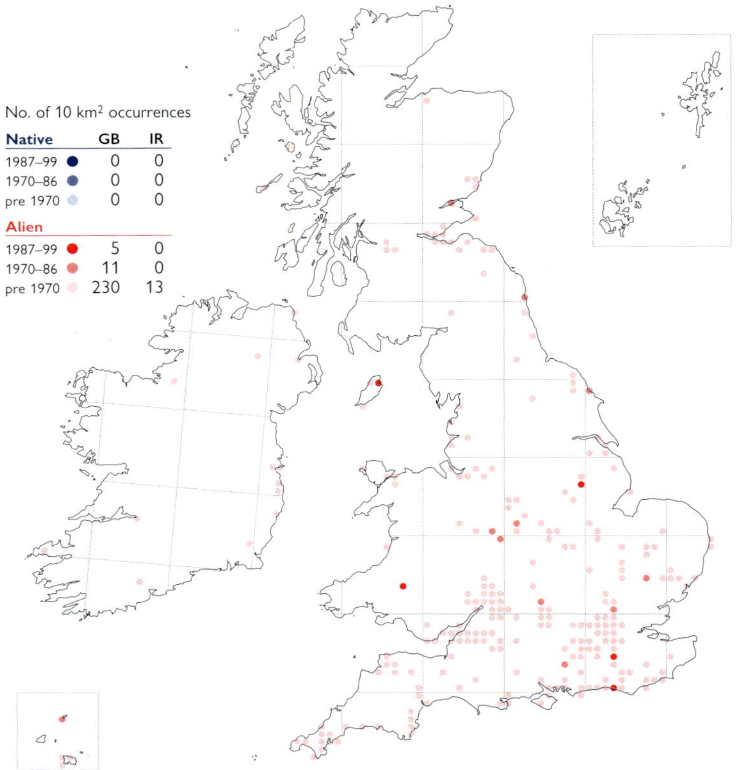

An annual occurring as a casual in arable fields, and on waste ground and tips, usually arising from bird-seed. Lowland.

Casual. *C. orientalis*, which was first recorded in the wild in 1778, is now much less common than it was in the 19th and early 20th centuries, when it was a frequent contaminant of cereal and clover seed from C. Europe.

Native of C. & S. Europe, the Mediterranean region and W. Asia; frequently naturalised outside its native range in Europe.

References: Jalas & Suominen (1996), Meusel *et al.* (1965), Rich (1991).

D. A. PEARMAN

Diplotaxis tenuifolia Perennial Wall-rocket

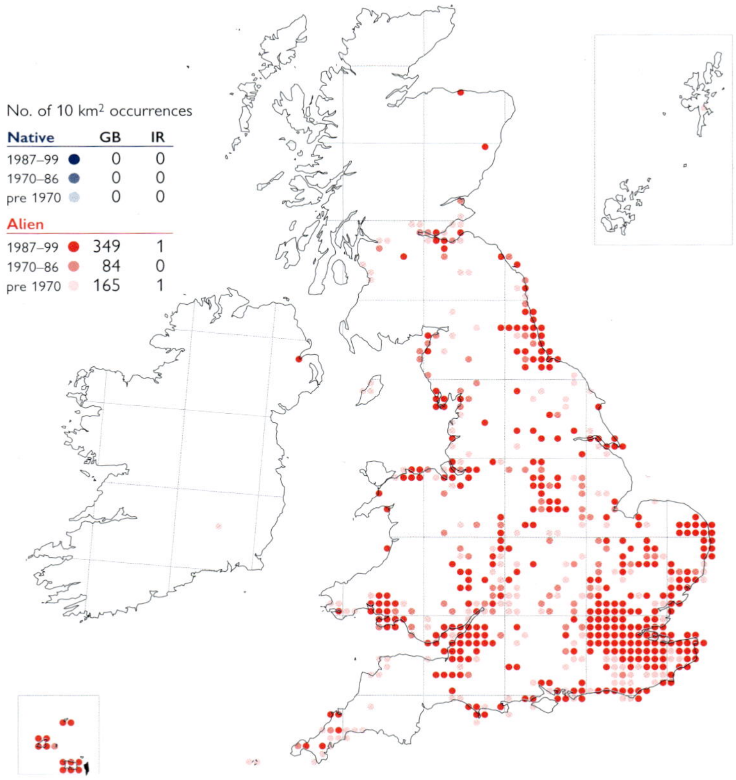

A perennial herb, most common in warm, dry habitats, occurring in waste ground, on walls and banks, and in quarries and railway sidings. Lowland.

Archaeophyte (change −0.13). The range of *D. tenuifolia* underwent a marked spread westwards in the late 19th century, and this is still continuing, albeit at a slower rate. Remarkably, the distribution is still centred on ports and industrial areas.

As an archaeophyte *D. tenuifolia* has a European Temperate distribution; it is widely naturalised outside this range.

References: Atlas (35d), Hultén & Fries (1986), Jalas & Suominen (1996), Meusel *et al.* (1965), Rich (1991).

D. A. PEARMAN

Diplotaxis muralis Annual Wall-rocket

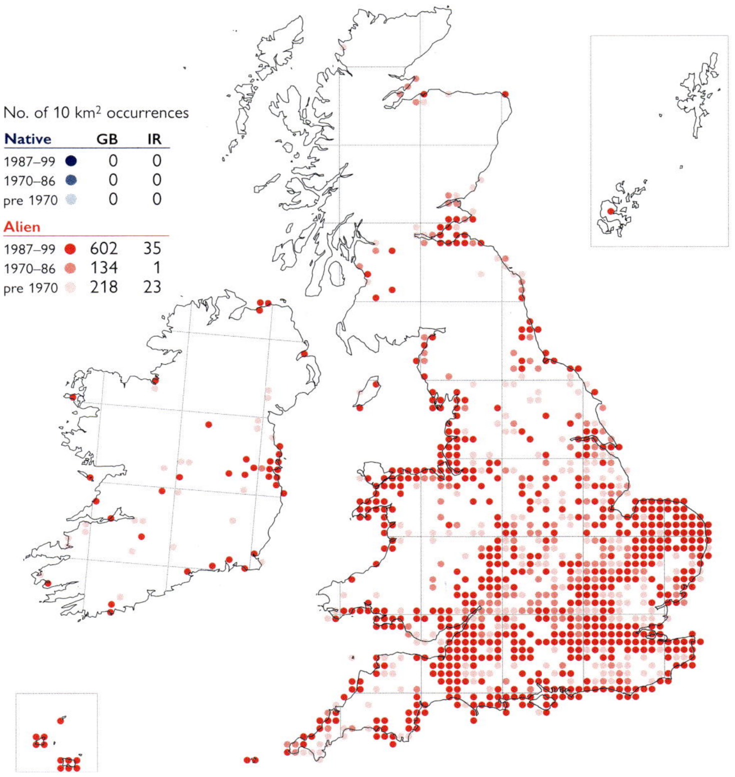

No. of 10 km^2 occurrences

Native		GB	IR
1987–99	●	0	0
1970–86	●	0	0
pre 1970	●	0	0
Alien			
1987–99	●	602	35
1970–86	●	134	1
pre 1970	●	218	23

An annual or short-lived perennial herb found in a variety of dry, open habitats. It is most frequent in waste places such as by railways, roads and on tips, but is also found on rocks, cliffs, walls and in gardens. It is occasionally cultivated and ploughed in as a 'green manure'. Lowland.

Neophyte (change –0.37). First recorded in 1778 in a field of oats raised from imported seeds from a ship wrecked on the Kent coast, *D. muralis* has spread steadily, probably bolstered by other accidental introductions. It is still invading rural areas.

Native of C. & S. Europe and N. Africa; naturalised in Europe north of its native range.

References: Atlas (35c), Hultén & Fries (1986), Jalas & Suominen (1996), Rich (1991).

D. A. PEARMAN

Brassica oleracea Cabbage

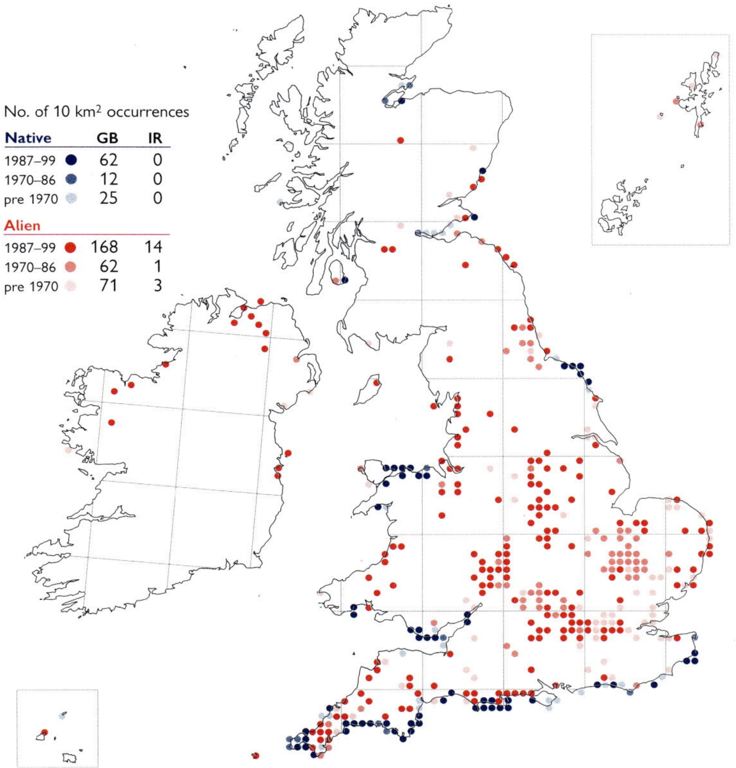

No. of 10 km^2 occurrences

Native		GB	IR
1987–99	●	62	0
1970–86	●	12	0
pre 1970	●	25	0
Alien			
1987–99	●	168	14
1970–86	●	62	1
pre 1970	●	71	3

A perennial herb, found as an apparent native on sea-cliffs, predominantly on chalk and limestone but also on other base-rich substrates. It is most frequent on bare cliff edges, but also grows in maritime grassland and in quarries inland. Elsewhere it is a casual garden escape in waste places and on roadsides. Lowland.

Native or alien (change +0.90). *B. oleracea* was first recorded in Britain in 1548. On sea-cliffs it is impossible to distinguish native from alien populations and all are mapped as if they are native. These coastal plants are var. *oleracea*; other varieties and cultivars occur as casuals.

Suboceanic Southern-temperate element.

References: Atlas (33b), Jalas & Suominen (1996), Meusel *et al.* (1965), Mitchell & Richards (1979), Rich (1991), Stewart *et al.* (1994), Zohary & Hopf (2000).

D. A. PEARMAN

Brassica napus Rape

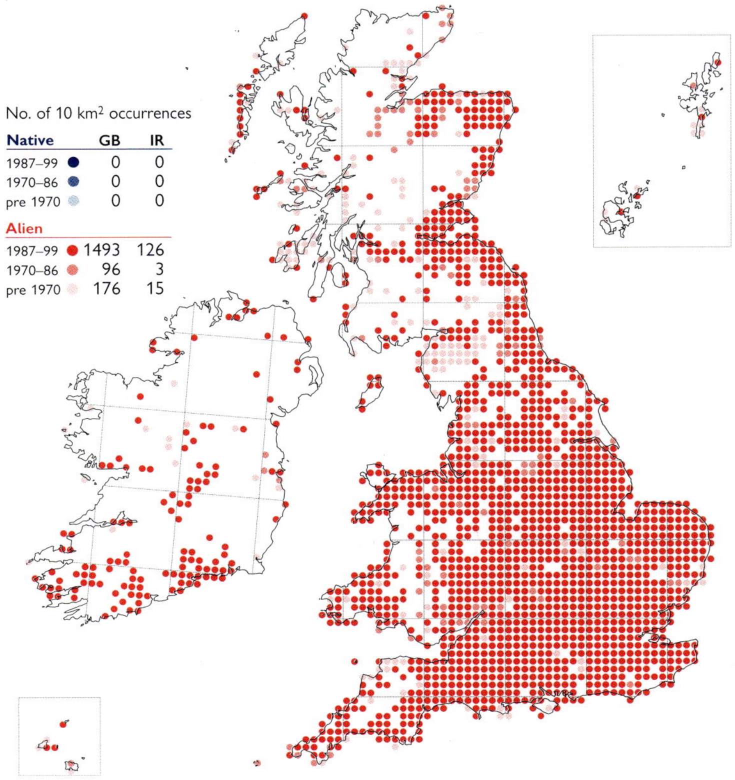

No. of 10 km^2 occurrences

Native		GB	IR
1987–99	●	0	0
1970–86	●	0	0
pre 1970	●	0	0
Alien			
1987–99	●	1493	126
1970–86	●	96	3
pre 1970	●	176	15

An annual or biennial, rarely perennial, herb of disturbed ground on roadsides, waste and cultivated ground, rubbish tips and docks. Lowland, but casual at 420 m at Stainmore (Westmorland).

Neophyte (change +2.88). *B. napus* was first recorded in the wild in 1660. Two subspecies are widely cultivated in our area. The acreage of subsp. *oleifera* (Oil-seed Rape) has dramatically increased since 1980 and it is now widespread as a casual. Subsp. *rapifera* (Swede) is a rare relic of cultivation. The maps in the 1962 *Atlas* are unreliable, since this species was confused with *B. rapa*.

This allotetraploid, derived from *B. oleracea* and *B. rapa*, is not known in the wild but is naturalised in temperate regions throughout the world.

References: Atlas (33c), Bunting (1988), Hultén (1968), Rich (1991).

D. A. PEARMAN

Brassica rapa Turnip

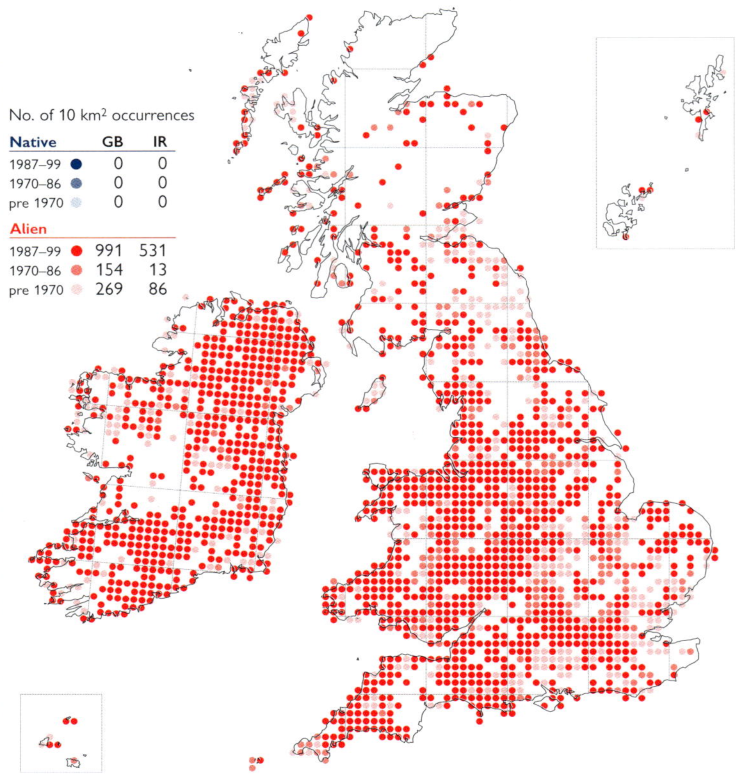

An annual or biennial herb, found as long-established populations on river and canal banks, and as a casual on roadsides, in arable fields and on tips. Lowland.

Archaeophyte (change +0.74). Subsp. *campestris* is the taxon found in semi-natural habitats. Two other subspecies are widely cultivated in Britain: subsp. *oleifera* is a bird-seed or oil-seed species; subsp. *rapa* (Turnip) is a frequent relic of cultivation. The maps in the 1962 *Atlas* are unreliable, since this species was confused with *B. napus.*

A native of Eurasia; its precise native range is completely obscured by its spread in cultivation. It is also naturalised in Africa, N. & S. America and Australasia.

References: Atlas (33d), Hultén & Fries (1986), Jalas & Suominen (1996), Rich (1991), Zohary & Hopf (2000).

D. A. PEARMAN

Brassica juncea Chinese Mustard

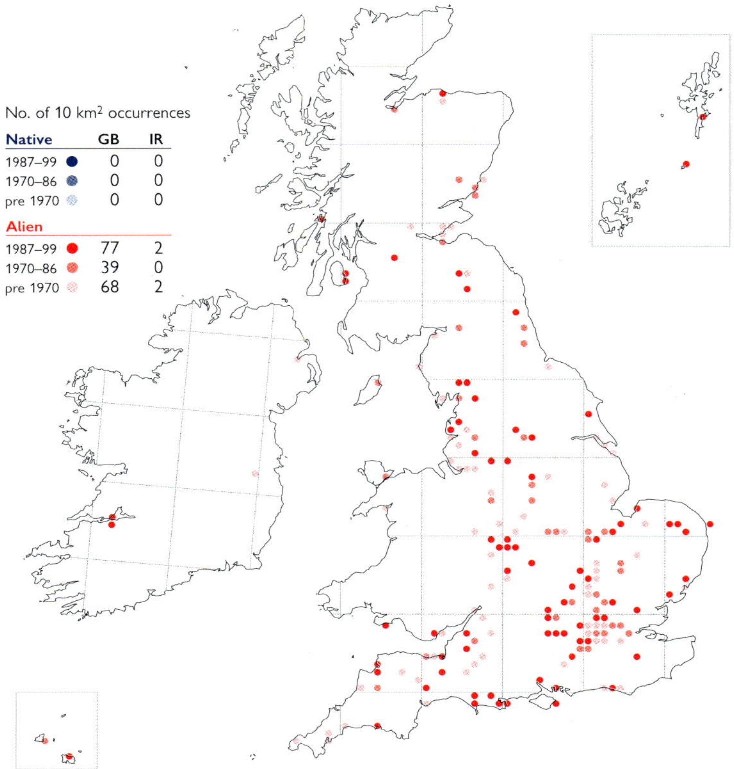

An annual, only ever occurring as a casual from bird-seed or from imports of the seed for mustard. It is found on tips, docks and waste ground. Lowland.

Casual. This species was grown here in 1710 but was introduced much later as a casual, being recorded from the wild by about 1876. It has increased in the 20th century. The seeds of *B. juncea* have largely replaced those of *B. nigra* in commercial mustard (Vaughan & Geissler, 1997).

An allotetraploid derived from *B. nigra* and *B. rapa* which has probably originated more than once and occurs as a naturalised or casual alien scattered throughout the N. hemisphere.

References: Bunting (1988), Hultén & Fries (1986), Jalas & Suominen (1996), Rich (1991).

D. A. PEARMAN

Brassica nigra Black Mustard

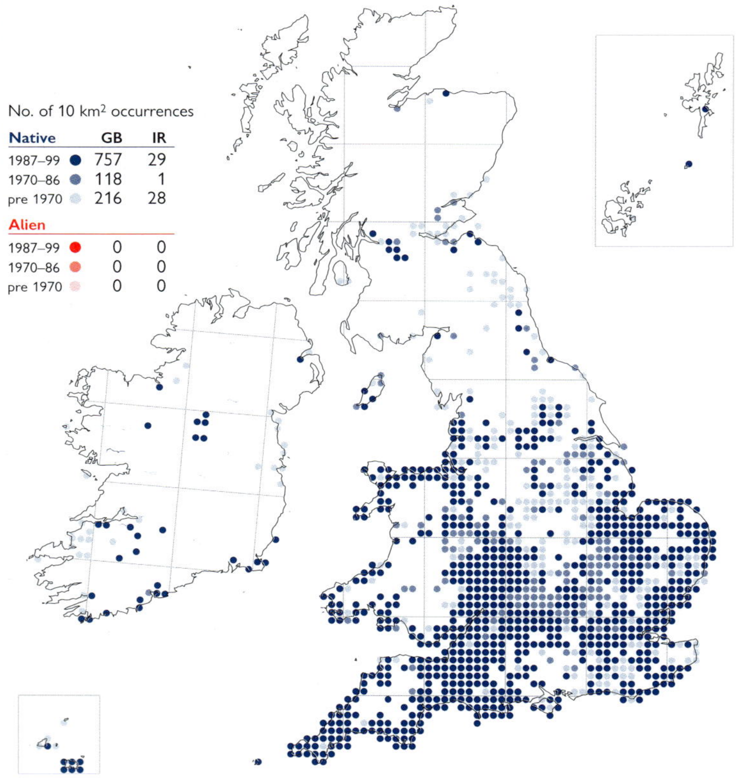

An annual forming persistent populations by rivers, where it is a member of the tall-herb community in the flood zone, and on sea-cliffs and shingle. It is also widespread as a casual on roadsides and waste ground, and can occur at the edge of arable fields. Generally lowland, but with a casual record at 380 m from Withypool (S. Somerset).

Native or alien (change –0.02). *B. nigra* was first recorded in Britain in 1640. Its native range is uncertain and all records are therefore mapped as if they are native. The losses have largely occurred since the 1962 *Atlas* because it is now rarely grown for seed.

European Temperate element; widely naturalised outside its native range.

References: Atlas (34a), Hultén & Fries (1986), Jalas & Suominen (1996), Meusel *et al.* (1965), Rich (1991).

D. A. PEARMAN

Sinapis arvensis Charlock

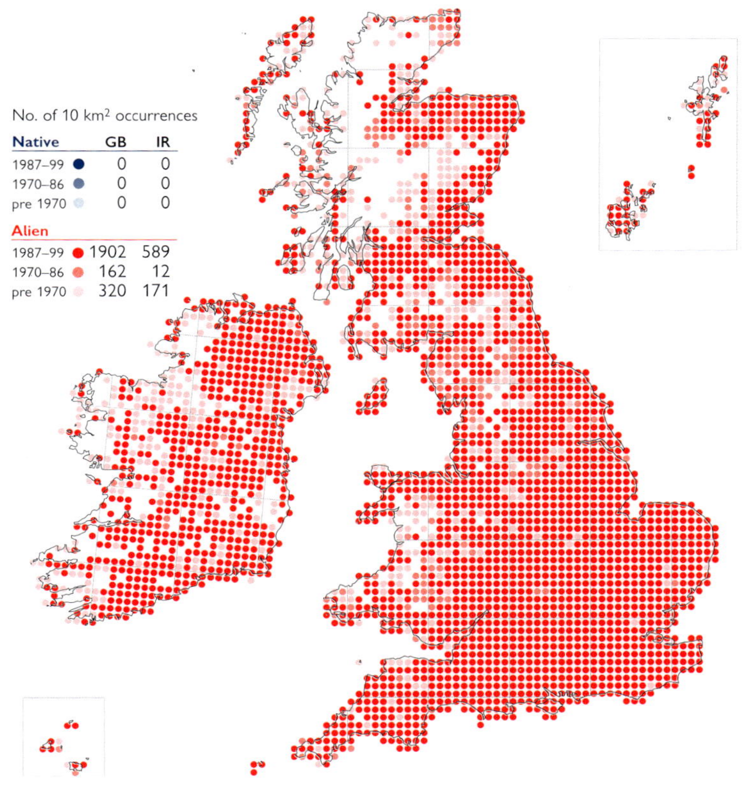

An annual, abundant as a weed of cultivation and also found on roadsides, railways, tips and waste ground. It is now a frequent weed associated with roadworks. Generally lowland, but reaching 450 m in Clun Forest (Salop).

Archaeophyte (change –1.76). There is little change in the range of *S. arvensis*, which was mapped as 'all records' in the 1962 *Atlas*. It is easily controlled by herbicides, and thus is less frequent than formerly in arable fields. The seed is, however, long-lived, and populations can readily re-appear on disturbed ground.

As an archaeophyte *S. arvensis* has a Eurosiberian Temperate distribution; it is widely naturalised outside this range.

References: Atlas (35a), Fogg (1950), Grime *et al.* (1988), Hultén & Fries (1986), Jalas & Suominen (1996), Mulligan & Bailey (1975), Rich (1991).

D. A. PEARMAN

Sinapis alba White Mustard

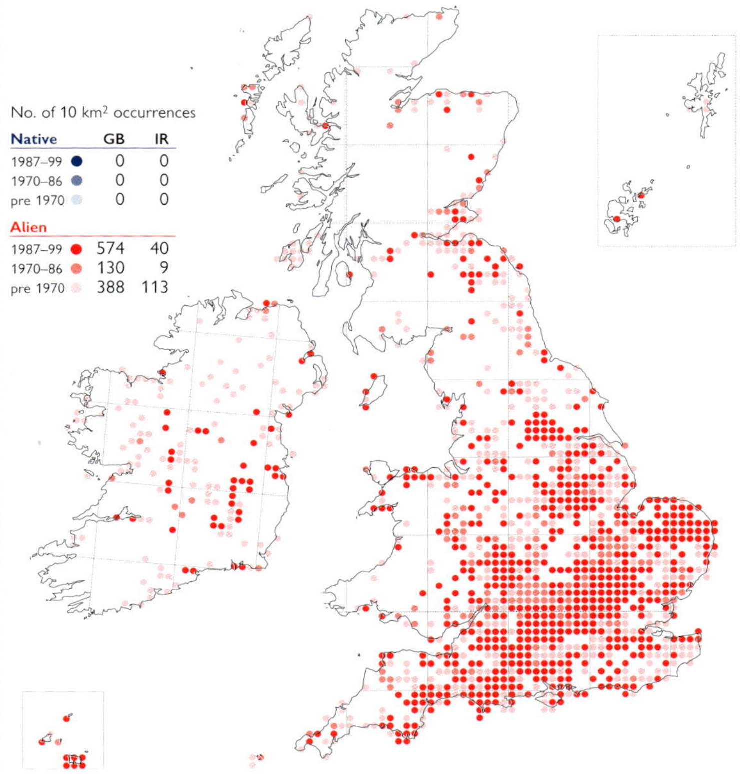

An annual, abundant as a persistent weed or a casual in arable fields, and on waste ground and roadsides, often on calcareous soils. Lowland.

Archaeophyte (change –0.90). *S. alba* was formerly more widely grown for mustard, but is still frequent as an escape from that source. It is also grown as a 'green manure' and a salad plant, and occurs as a bird-seed alien.

The native distribution of this species has been completely obscured by its spread as a cultivated plant; as an archaeophyte it has a European Southern-temperate distribution and is widely naturalised elsewhere.

References: Atlas (35b), Hultén & Fries (1986), Jalas & Suominen (1996), Rich (1991), Vaughan & Geissler (1997).

D. A. PEARMAN

Eruca vesicaria Garden Rocket

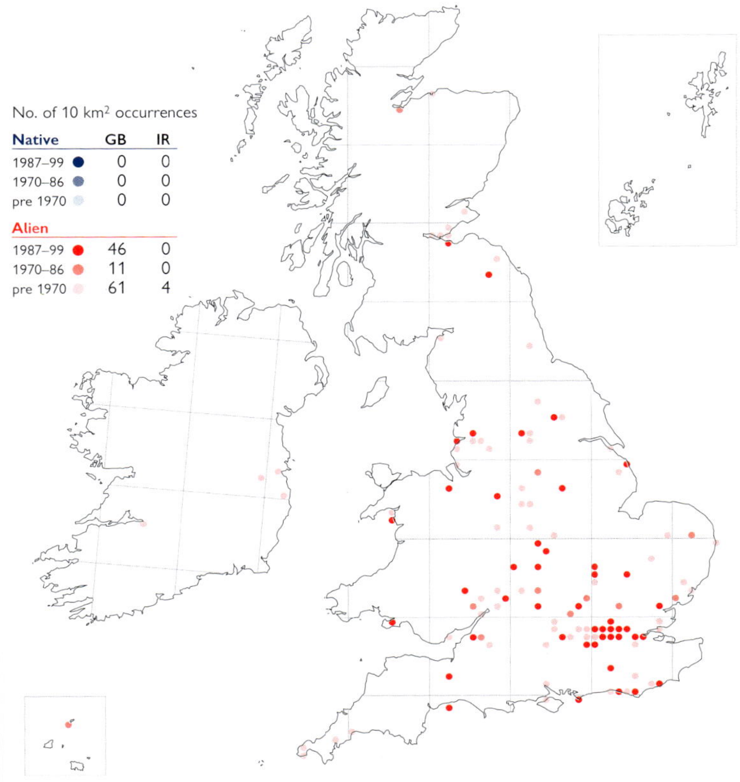

An annual found on waste ground and rubbish tips which occurs as small, short-lived populations. It is introduced from crop, bird-seed and spice impurities, and as a garden escape. Lowland.

Casual. *E. vesicaria* has been grown in gardens since at least 1200 (Harvey, 1981) as a medicinal plant; the oil from the seeds was said to be an aphrodisiac. It was first recorded from the wild in 1859 and is perhaps increasing in occurrence because of its current popularity as a salad vegetable.

Native of the Mediterranean region and S.W. Asia but limits of native range obscured by its spread in cultivation to temperate Europe, C. & E. Asia, C. America, tropical S. Africa and elsewhere.

References: Bolòs & Vigo (1990), Jalas & Suominen (1996), Rich (1991).

D. A. PEARMAN

Erucastrum gallicum Hairy Rocket

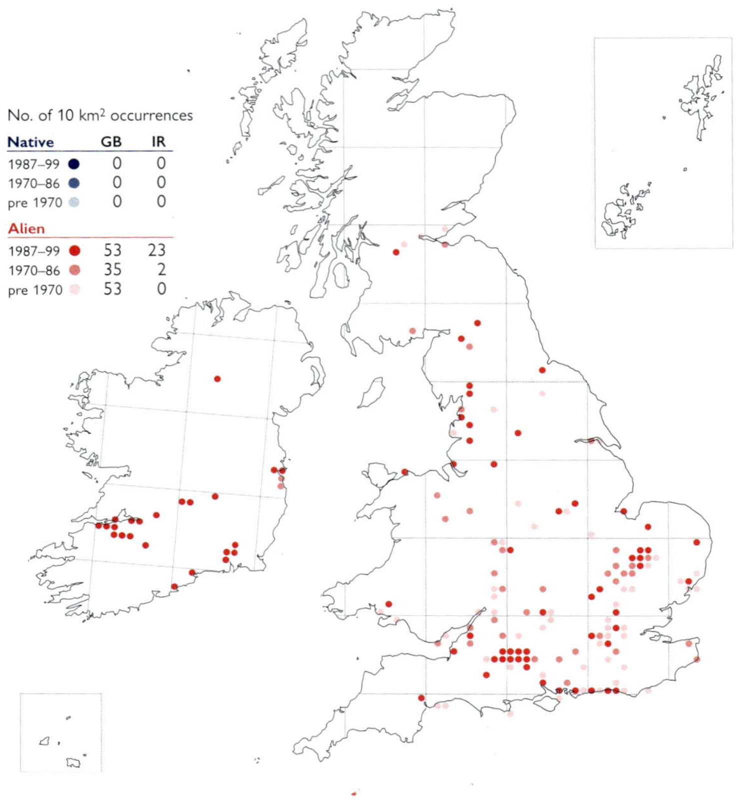

An annual, occasionally persisting in quarries and along tracks on chalk soils, but almost always casual on roadsides and waste ground. Lowland.

Neophyte (change –0.02). *E. gallicum* was first recorded from the wild in 1863. In Ireland it is more frequent now than it was in the 1962 *Atlas*. It has spread widely on Salisbury Plain (S. Wilts.) as a result of disturbance from army activity. It is apparently sown in order to bind and stabilise steep chalk road cuttings (Brewis *et al.*, 1996).

Native of the Pyrenees and C. Europe; widely naturalised elsewhere in Europe and in N. America.

References: Atlas (34b), Hultén & Fries (1986), Jalas & Suominen (1996), Meusel *et al.* (1965), Rich (1991).

D. A. PEARMAN

Coincya monensis subsp. *monensis*
Isle of Man Cabbage

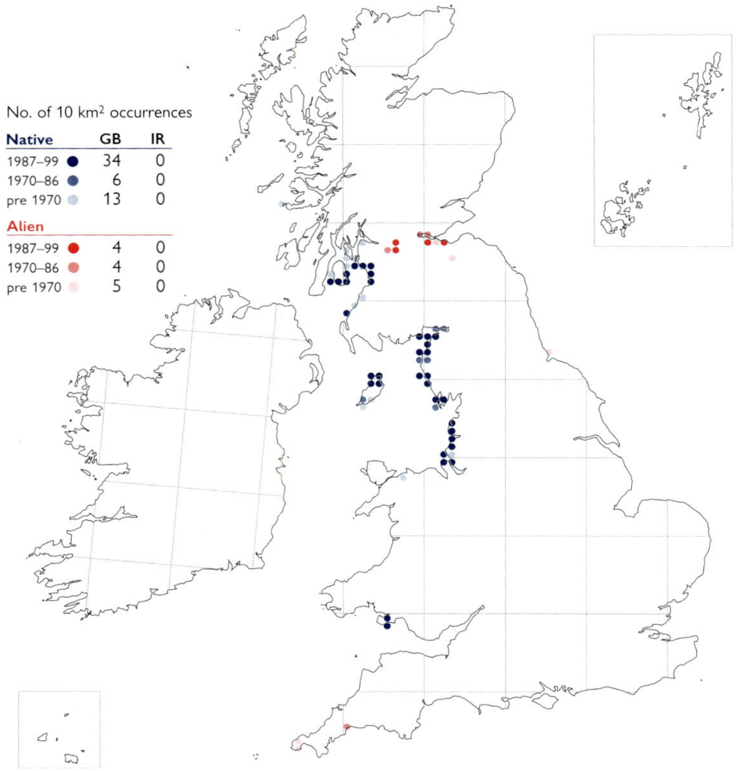

An annual or short-lived perennial herb, mainly found by the sea on open dunes and on the strand-line, and only rarely in bare fields and hedge banks near the sea. There is recent evidence to suggest that seed might be dispersed by sea (Rich, 1999b). Lowland.

Native (change for species +0.43). This subspecies is stable or even increasing in the centre of its range in N.W. England, but is slowly declining elsewhere.

Endemic.

References: Atlas (34c), Jalas & Suominen (1996), Rich (1991), Stewart *et al.* (1994).

D. A. PEARMAN

Coincya monensis subsp. *cheiranthos*
Wallflower Cabbage

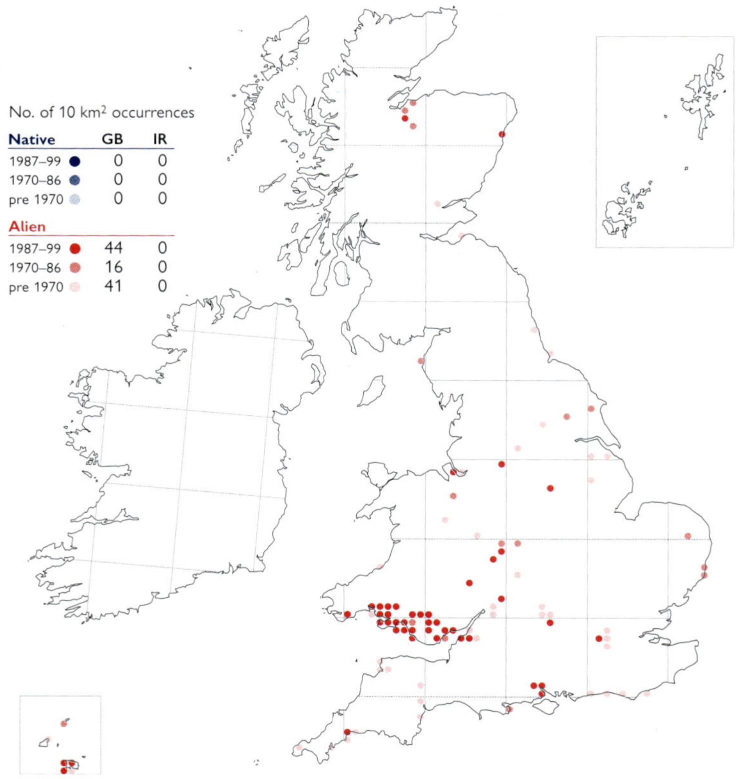

An annual or biennial herb naturalised in a few places, for example by docks, roadsides and railways, and on waste ground, but occurring mainly as a casual in a wide variety of waste places. Lowland.

Neophyte (change for species +0.43). *C. monensis* subsp. *cheiranthos* has been naturalised in Britain since 1852, and seems to be increasing, at least in S. Wales.

Suboceanic Southern-temperate element.

References: Atlas (34d), Bolòs & Vigo (1990), Jalas & Suominen (1996), Rich (1991).

D. A. PEARMAN

Coincya wrightii Lundy Cabbage

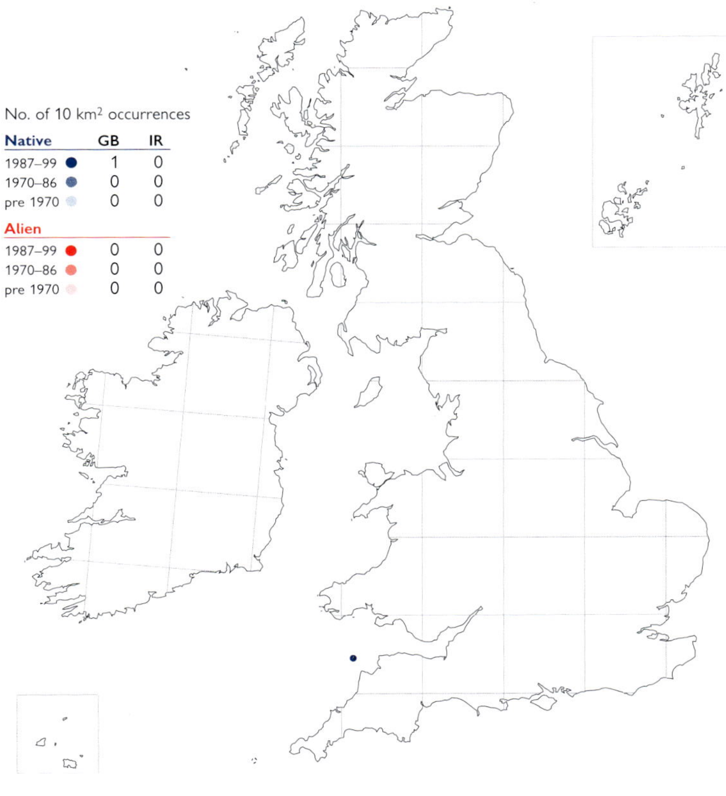

A perennial, occasionally biennial, herb, mainly found in open communities on S.-facing cliffs. It will grow on flat ground on the tops of cliffs, but only where protected from grazing animals and shielded from invading shrubs. Recent evidence suggests that seed can be dispersed in sea water. Lowland.

Native. *C. wrightii* was originally described from Lundy in 1936. It is holding its own there, but is dependent both on low levels of stocking and on the control of scrub.

Endemic.

References: Atlas (34d), Compton & Key (2000), Jalas & Suominen (1996), Marren (1971, 1972), Rich (1991, 1999b), Wigginton (1999).

D. A. PEARMAN

Hirschfeldia incana Hoary Mustard

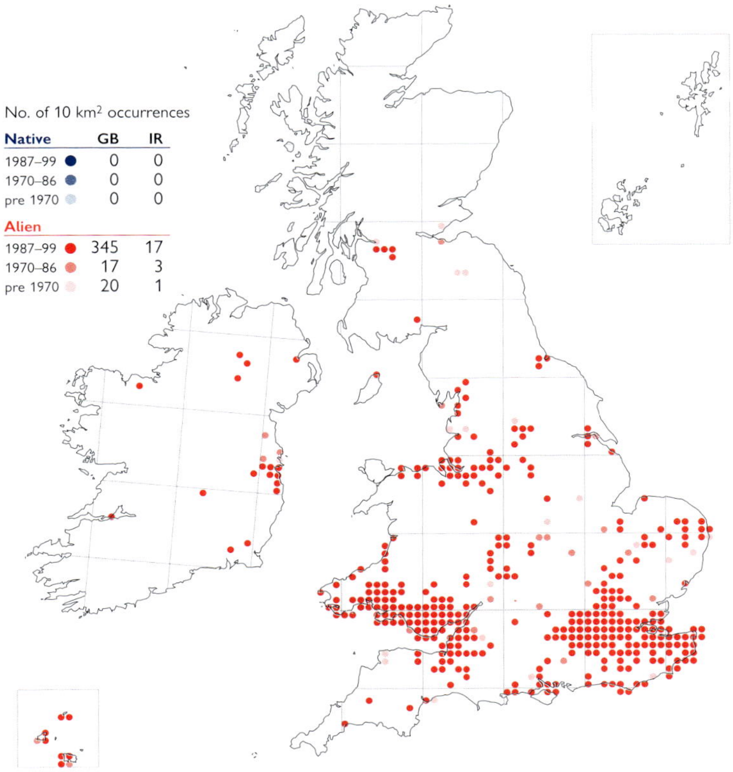

An annual or short-lived perennial herb, increasingly naturalised in a variety of waste places such as by docks, railways and roadsides, and on tips. It is often associated with grain imports and bird-seed, and frequently occurs as a casual. It was formerly introduced with wool shoddy. Lowland.

Neophyte. *H. incana* was cultivated in Britain by 1771 and was recorded from the wild by 1837, but 19th-century records are rare. It seems to be one of the few casuals that has spread from tips and survived, and it is now markedly increasing its range, having persisted at some sites for forty years.

Native of S.W. Europe, the Mediterranean region and S.W. Asia; widely naturalised further north in Europe.

References: Bolòs & Vigo (1990), Jalas & Suominen (1996), Rich (1991).

D. A. PEARMAN

Cakile maritima Sea Rocket

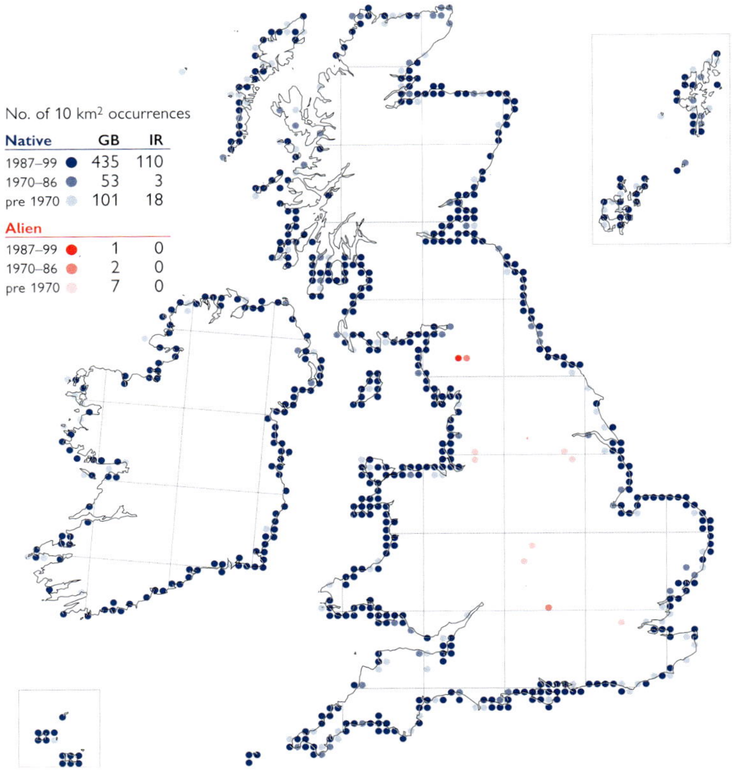

An annual, predominantly found on sandy seashores and on fore-dunes. It is often very frequent along the winter storm tide-line where there is a good source of nutrients. It is rarer on shingle beaches and is only an occasional casual elsewhere. Seeds are dispersed by tides. Lowland.

Native (change –0.38). The distribution of *C. maritima* seems stable in Britain, though numbers fluctuate enormously from year to year.

European Wide-temperate element.

References: Atlas (37a), Hocking (1982), Hultén & Fries (1986), Jalas & Suominen (1996), Meusel *et al.* (1965), Rich (1991).

D. A. PEARMAN

Rapistrum rugosum Bastard Cabbage

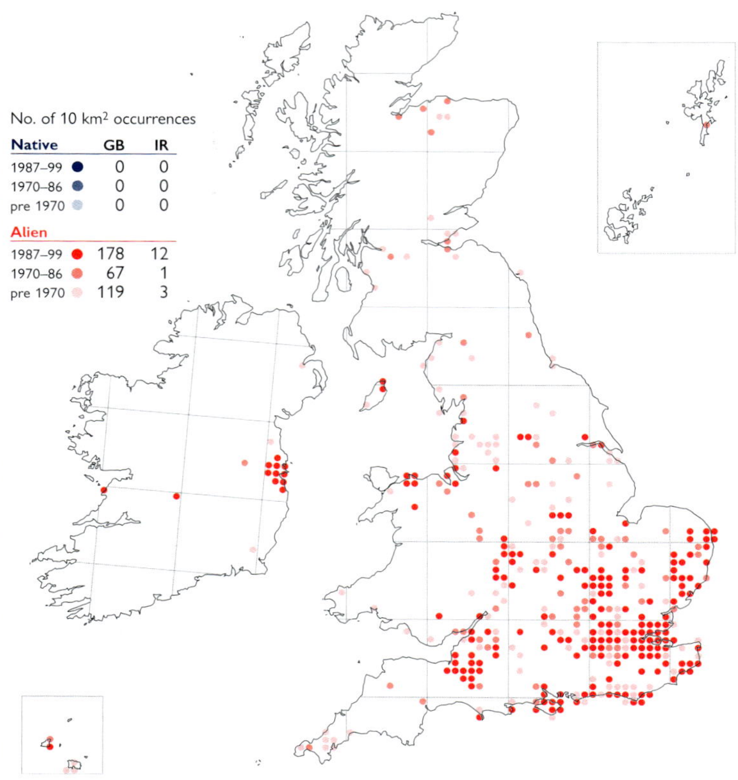

No. of 10 km² occurrences

Native	GB	IR
1987–99	0	0
1970–86	0	0
pre 1970	0	0
Alien		
1987–99	178	12
1970–86	67	1
pre 1970	119	3

An annual or short-lived perennial herb, found mainly as a casual of waste ground, but now becoming naturalised in a variety of habitats where it is sometimes invasive, such as in open grassland. It is introduced with grain and bird-seed. Lowland.

Neophyte (change +0.19). *R. rugosum* was introduced to cultivation in Britain by 1739, and was known from the wild by at least 1863. Its range has increased since the 1962 *Atlas*.

Native of the Mediterranean region and S.W. Asia; widely naturalised in temperate Europe and in the S. hemisphere.

References: Atlas (36d), Jalas & Suominen (1996), Meusel *et al.* (1965), Rich (1991).

D. A. PEARMAN

Crambe maritima Sea-kale

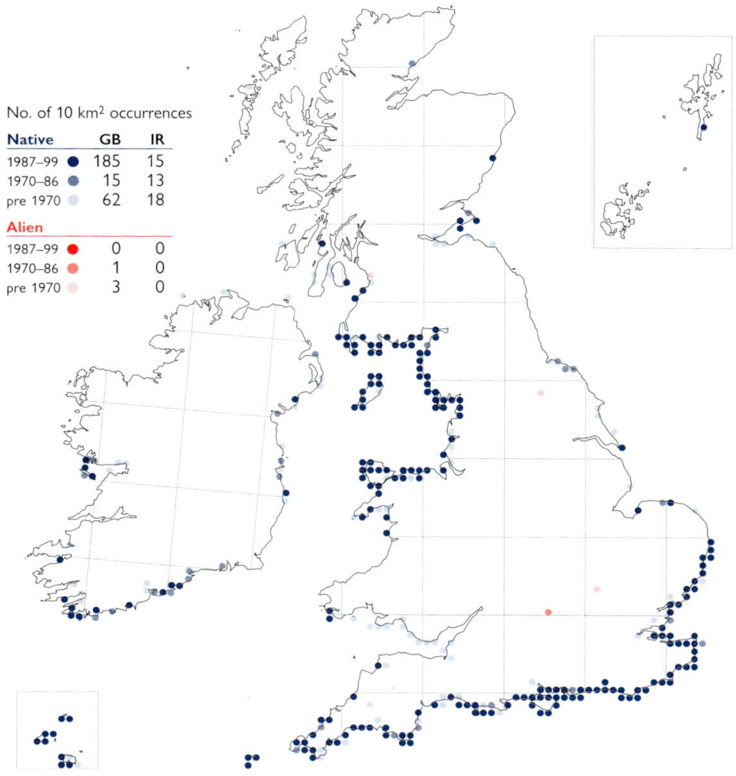

No. of 10 km² occurrences

Native	GB	IR
1987–99	185	15
1970–86	15	13
pre 1970	62	18
Alien		
1987–99	0	0
1970–86	1	0
pre 1970	3	0

A long-lived perennial herb of shingle and boulder beaches, very occasionally found on dunes (but only where these overlay shingle) and on cliffs. It reproduces by seed and from detached pieces of root. Lowland.

Native (change +0.29). *C. maritima* has declined in parts of its British range, probably because of sea-defence works which have destroyed its shingle habitats. On the other hand it has increased elsewhere, perhaps because it is now rarely gathered as a vegetable and its habitats are usually ungrazed. It has also increased in Ireland since the 1960s.

European Temperate element.

References: Atlas (36c), Curtis & McGough (1988), Hultén & Fries (1986), Jalas & Suominen (1996), Meusel *et al.* (1965), Scott & Randall (1976), Stewart *et al.* (1994).

D. A. PEARMAN

Raphanus raphanistrum subsp. *raphanistrum* Wild Radish

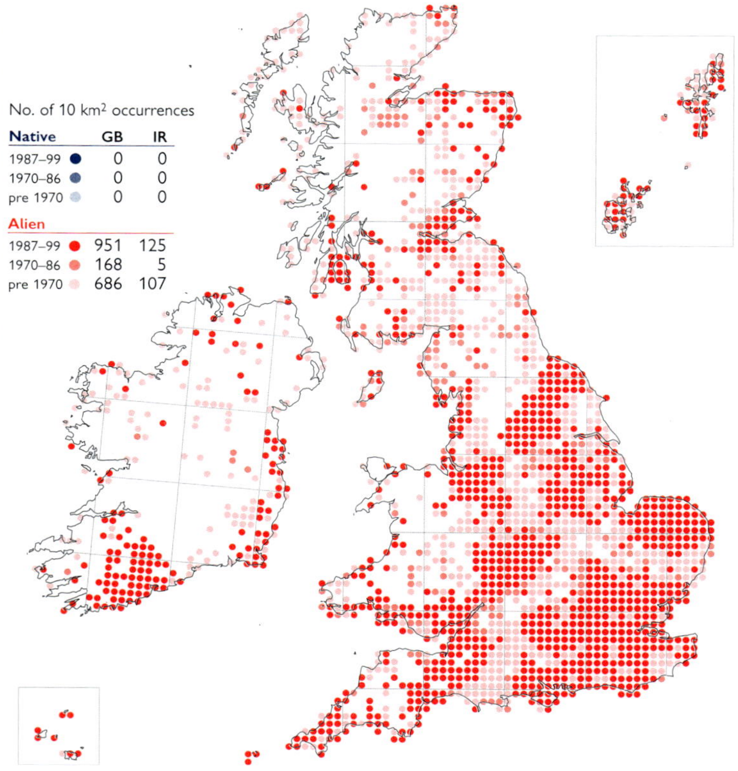

No. of 10 km² occurrences

Native	GB	IR
1987–99	0	0
1970–86	0	0
pre 1970	0	0
Alien		
1987–99	951	125
1970–86	168	5
pre 1970	686	107

An annual found as a casual or persistent weed in cultivated fields and on roadsides and waste ground. 0–380 m (Langdon Beck, Co. Durham).

Archaeophyte (change for species −1.39). *R. raphanistrum* subsp. *raphanistrum* was a noxious weed, but is easily controlled by selective herbicides and now appears to be much less common than in the past. It was mapped as 'all records' in the 1962 *Atlas*. Analysis of the database reveals that most losses have occurred since 1950.

As an archaeophyte, this subspecies has a European Southern-temperate distribution but is widely naturalised outside this range. It presumably evolved from other subspecies and spread as a weed of cultivation.

References: Atlas (36a), Atlas Supp. (8a, b), Hultén & Fries (1986), Jalas & Suominen (1996), Rich (1991).

D. A. PEARMAN

Raphanus raphanistrum subsp. *maritimus* Sea Radish

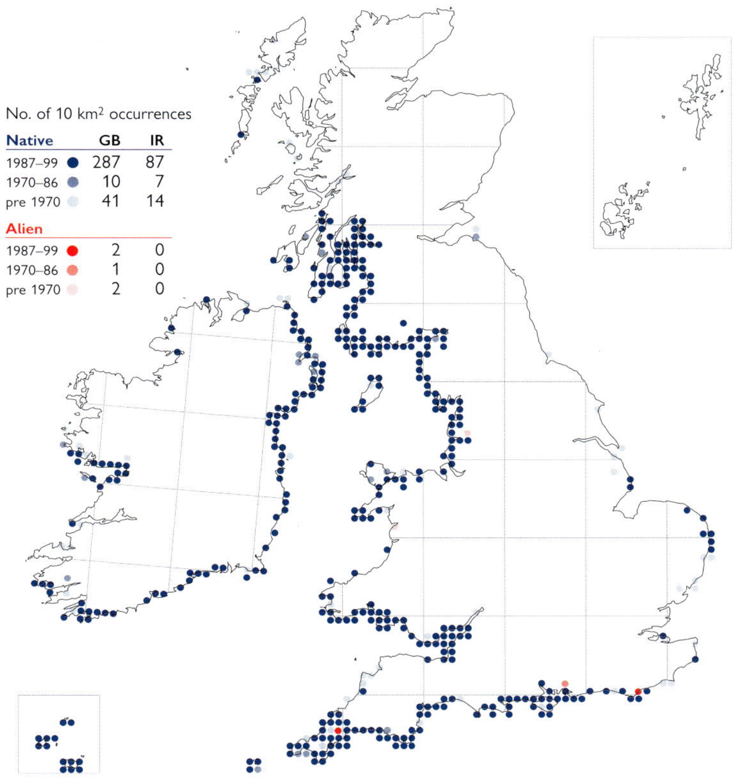

A biennial or perennial herb found in open coastal grassland, sand dunes, shingle, cliffs and disturbed ground by the sea. On parts of the east coast of Britain it grows on muddy shores. Lowland.

Native. Authors of many local floras have commented that *R. raphanistrum* subsp. *maritimus* has become more frequent in the last hundred years (e.g. Brewis *et al.*, 1996; Margetts & David, 1981). The results of the current survey provide additional support for this view.

Mediterranean-Atlantic element; it reaches its northern limit in Scotland.

References: Atlas (36b), Jalas & Suominen (1996), Rich (1991).

D. A. PEARMAN

Raphanus sativus Garden Radish

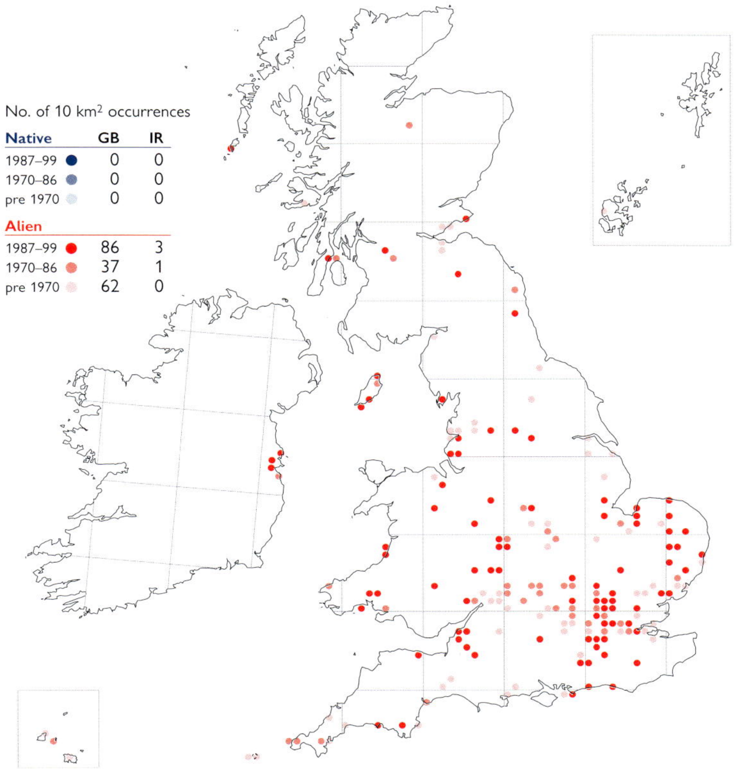

An annual, occurring casually on tips and waste places, as a garden escape or bird-seed alien. Lowland.

Casual. *R. sativus* has been cultivated in gardens since at least 995 (Harvey, 1981); the white-rooted variety had been introduced from China by 1548. A variety of cultivars is grown in Britain, all of which are included on the map. The species was recorded from the wild by 1893.

Not known as a native wild plant; it doubtless developed in cultivation from the wild *Raphanus* taxa found in the Mediterranean region.

References: Rich (1991), Smart & Simmonds (1995).

D. A. PEARMAN

Reseda luteola Weld

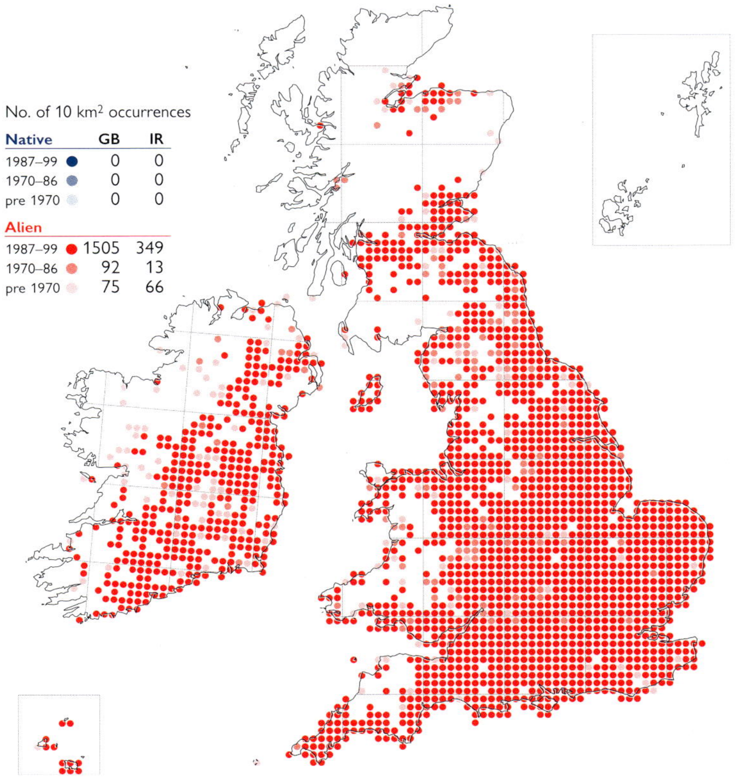

A robust biennial herb which typically occurs on neutral or base-rich soils on roadsides, waste ground and marginal land, in brick yards, gravel-pits and urban demolition sites, and, less commonly, arable or grassy areas. Generally lowland, but reaching 400 m on a roadside N. of Shap Summit (Westmorland).

Archaeophyte (change +0.69). The range of this species has been stable over many years, and has not changed appreciably since the 1962 *Atlas*. However, it now appears to be more frequent within its range.

As an archaeophyte *R. luteola* has a Eurosiberian Southern-temperate distribution; it is widely naturalised outside this range.

References: Atlas (52c), Davison (1970), Hultén & Fries (1986), Jalas & Suominen (1999), Meusel *et al.* (1965), Zohary & Hopf (2000).

G. T. D. WILMORE

Reseda alba White Mignonette

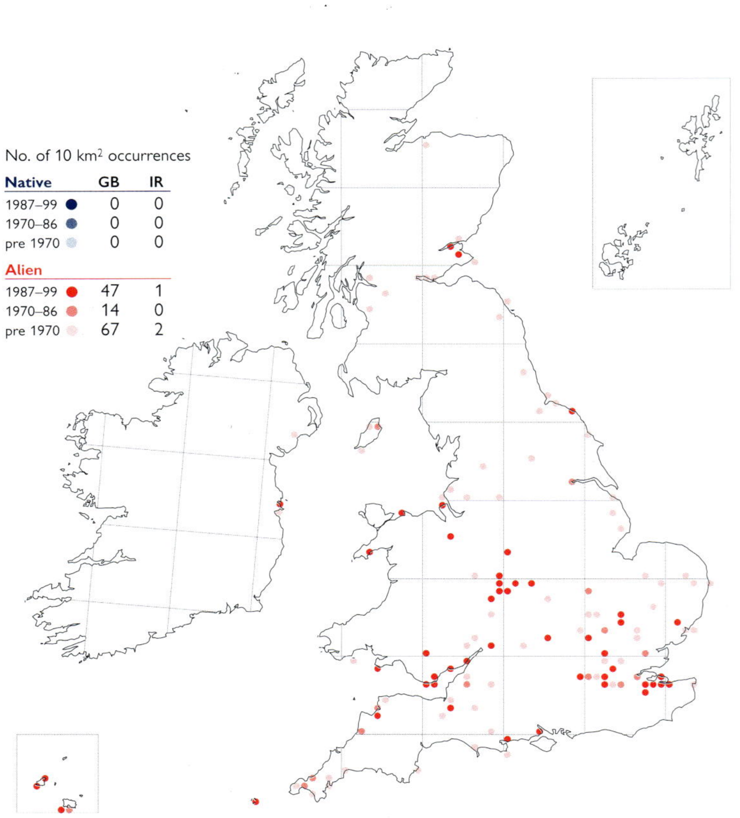

An annual or perennial herb, usually a grain, wool or bird-seed alien, but sometimes a garden escape, found in a range of man-made habitats including roadsides, railway banks, cultivated land, docklands and waste ground. Lowland.

Neophyte. Originally introduced into cultivation by 1596, *R. alba* has been known since 1826 on St Mary's (Isles of Scilly), and has been naturalised in Monmouthshire since 1968.

Native of the Mediterranean region; sparingly naturalised in N. & W. Europe.

References: Bolòs & Vigo (1990), Jalas & Suominen (1999).

G. T. D. WILMORE

Reseda lutea Wild Mignonette

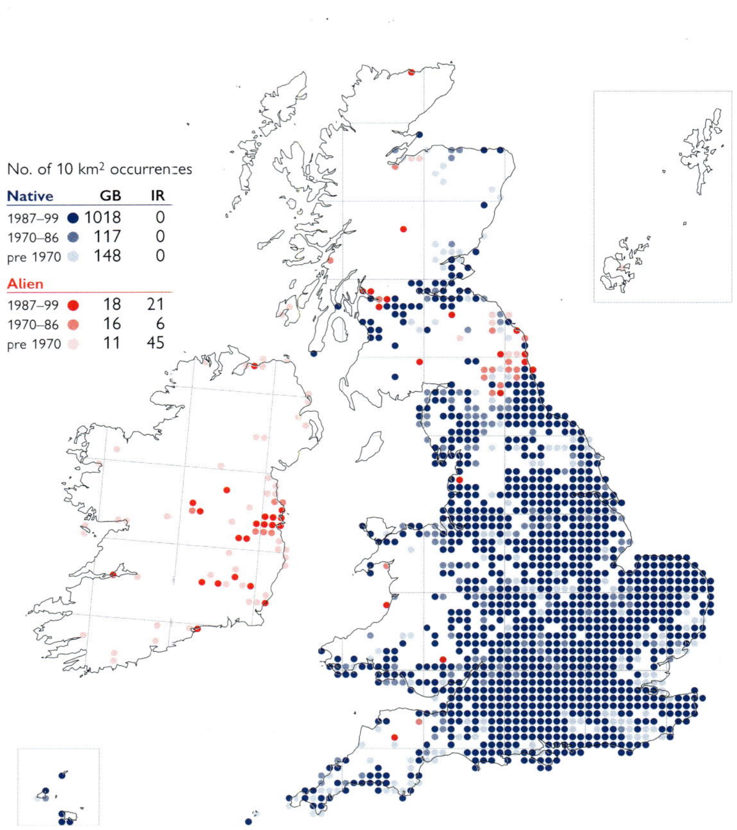

A biennial or perennial herb of well-drained soils in open habitats, occurring on waste ground and roadside verges, in marginal grassland, disused railway land, quarries and arable land, in disturbed chalk and limestone grassland and on fixed sand dunes. Generally lowland, but reaching 440 m near Stainmore (Westmorland).

Native or alien (change +0.39). *R. lutea* was recorded in Britain by 1597. There has been no significant change in its distribution since the 1962 *Atlas*.

European Southern-temperate element; widely naturalised outside its native range.

References: Atlas (52d), Hultén & Fries (1986), Jalas & Suominen (1999), Meusel *et al.* (1965).

G. T. D. WILMORE

Empetrum nigrum Crowberry

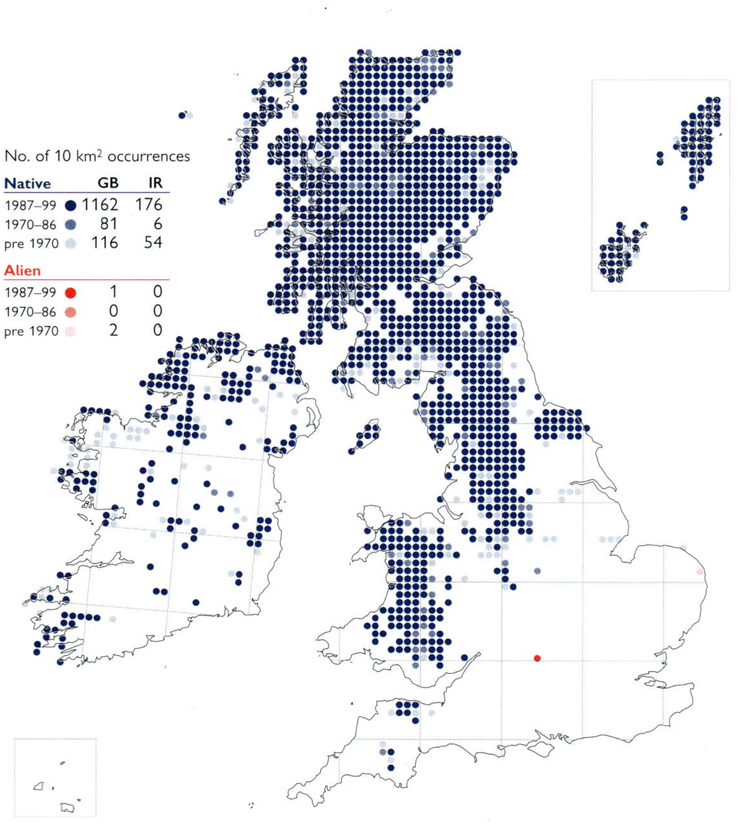

A low-growing evergreen shrub of well-drained acidic soils. It is found on moorlands and mountains, and on blanket mire where it can increase greatly after burning or where dry surfaces have been bared by erosion (Rodwell, 1991b). 0–1270 m (Cairngorms, S. Aberdeen).

Native (change −0.29). The distribution of *E. nigrum* is similar to that shown in the 1962 *Atlas*, where subsp. *hermaphroditum* and subsp. *nigrum* were mapped separately as species. Subsp. *nigrum* occurs throughout the range of the species, subsp. *hermaphroditum* is mapped separately.

Circumpolar Boreo-arctic Montane element.

References: Atlas (198c, d), Bell & Tallis (1973), Grime *et al.* (1988), Hultén & Fries (1986), Meusel *et al.* (1978).

G. T. D. WILMORE & D. A. PEARMAN

Empetrum nigrum subsp. *hermaphroditum*

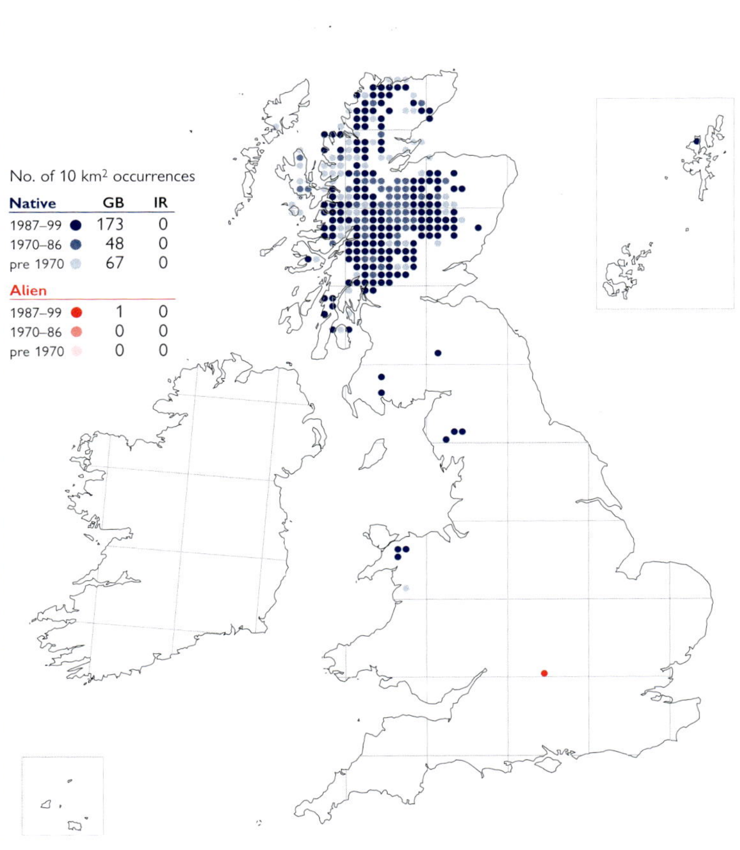

A low-growing evergreen shrub, favouring moderately well-drained acidic soils and occurring in dwarf shrub and *Racomitrium* heaths, rock crevices and dry heather moor, including areas of late snow-lie. Usually montane, occurring from 610 m to 1130 m in the Cairngorm (Easterness), but formerly on coastal dunes at Tain (E. Ross).

Native. The two subspecies can occur together, but subsp. *hermaphroditum* replaces subsp. *nigrum* at the highest altitudes. The current distribution is similar to that in the 1962 *Atlas*, and it is probably still present in many of the squares for which there are no recent records.

Circumpolar Boreo-arctic Montane element; its distribution is more northerly than the Boreal-montane subsp. *nigrum*.

References: Atlas (198d), Hultén & Fries (1986), Meusel *et al.* (1978).

G. T. D. WILMORE

Rhododendron ponticum Rhododendron

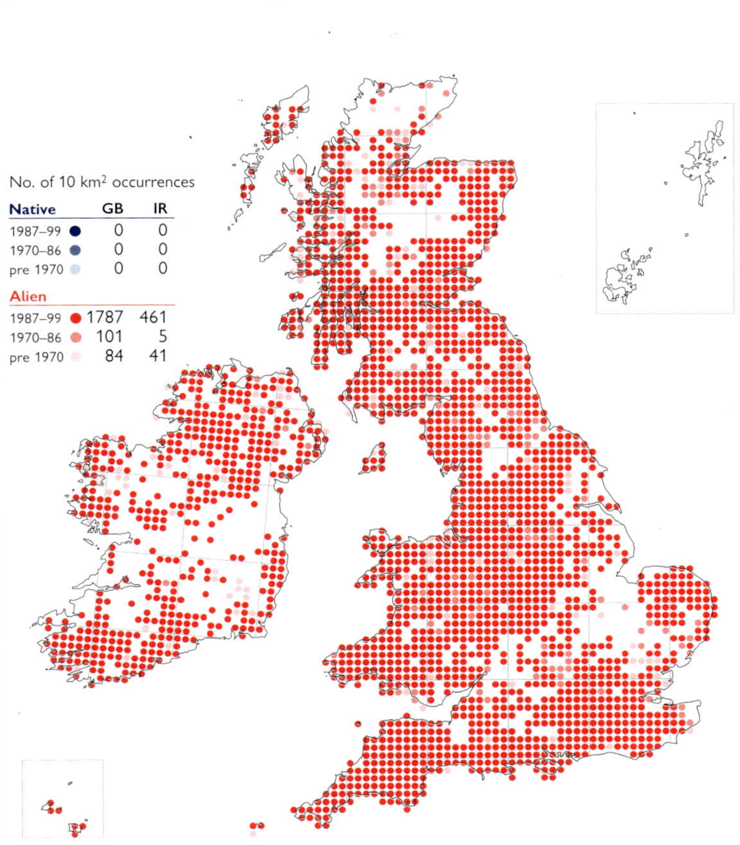

An evergreen shrub naturalised on heathy and rocky hillsides, rocky stream banks and ravines, and as an understorey in woodland on acid soils. It regenerates from seed freely and can form dense thickets. 0–600 m (Eel Crags, Cumberland).

Neophyte (change +1.83). The pollen record shows that *R. ponticum* was native to Ireland in the Hoxnian interglacial. In the current interglacial it was introduced to cultivation in 1763 and most, if not all, our plants derive from Spanish stock. It was known in the wild by at least 1894 and spread widely in the 20th century, but its initial expansion is poorly documented. Its distribution is now stable.

Native of two, disjunct areas: the Iberian peninsula and S.E. Europe, Lebanon, Turkey and the Caucasus.

References: Atlas (192c), Cross (1975), Meusel *et al.* (1978).

M. C. F. PROCTOR

Rhododendron luteum Yellow Azalea

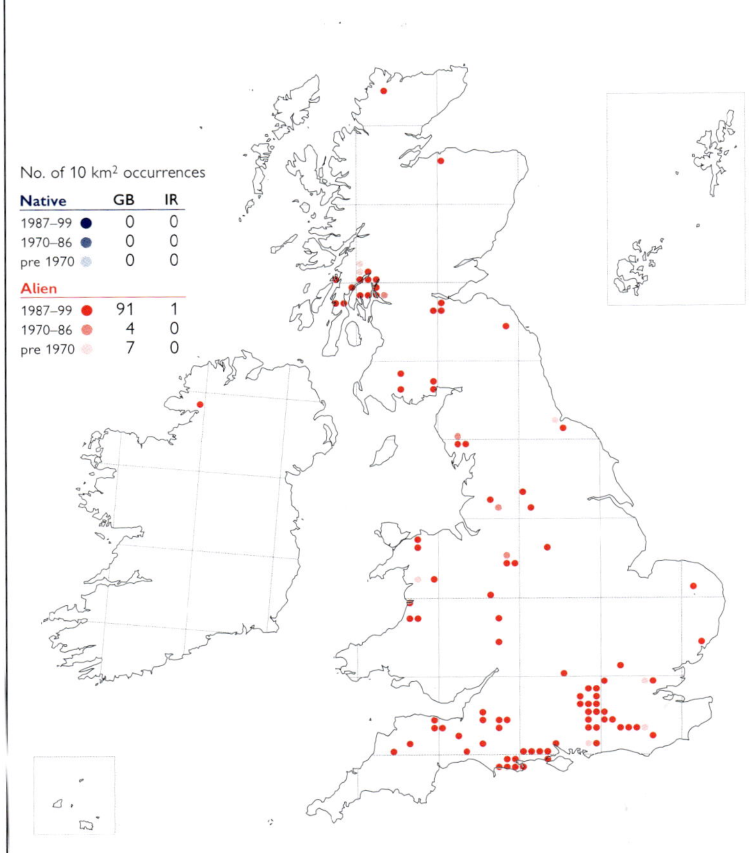

A deciduous shrub, locally naturalised on acidic soils on heathland, moorland and in open woodland. Reproduction is by seed. Lowland.

Neophyte. Introduced into cultivation in 1793, *R. luteum* is commonly grown in gardens for its beautiful, fragrant yellow flowers. It was recorded from the wild in 1939 (Bucks.).

Native of E. Europe, Turkey and the Caucasus.

References: Bean (1976), Meusel *et al.* (1978).

M. C. F. PROCTOR

Loiseleuria procumbens Trailing Azalea

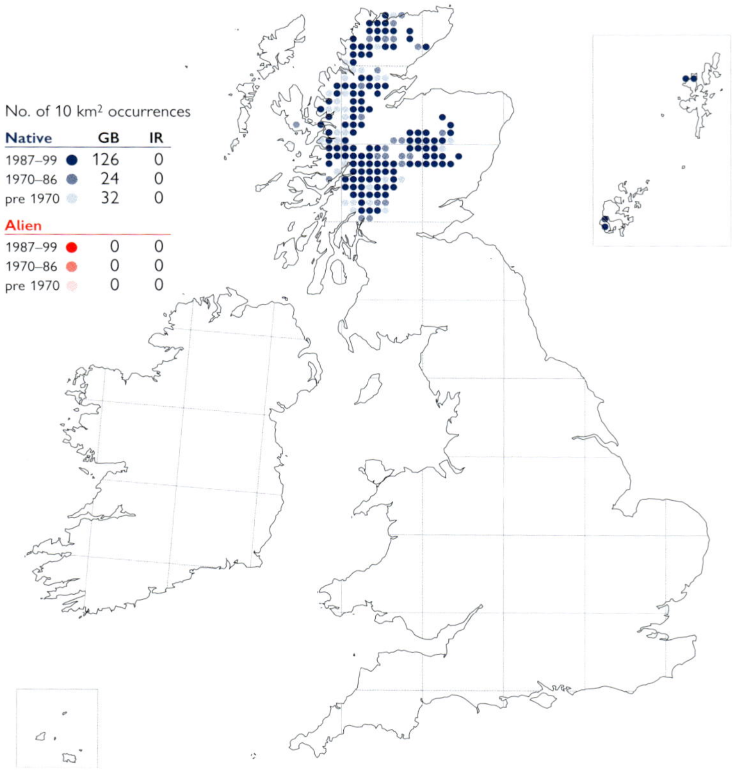

This procumbent, calcifugous dwarf shrub is found on exposed, stony mountain heaths on dry ridges and plateaux. Reproduction is mainly by seed, but it also spreads by rooting of the procumbent stems. Generally found from 500 to 900 m, but descends to 240 m on Ronas Hill (Shetland) and reaches 1100 m on Ben Macdui (S. Aberdeen).

Native (change −0.58). The distribution of *L. procumbens* is essentially stable, and it is probably still extant in many of the 10-km squares for which there are no recent records.

Circumpolar Arctic-montane element, with a disjunct distribution.

References: Atlas (192d), Hultén & Fries (1986), Meusel *et al.* (1978).

M. C. F. PROCTOR

Phyllodoce caerulea Blue Heath

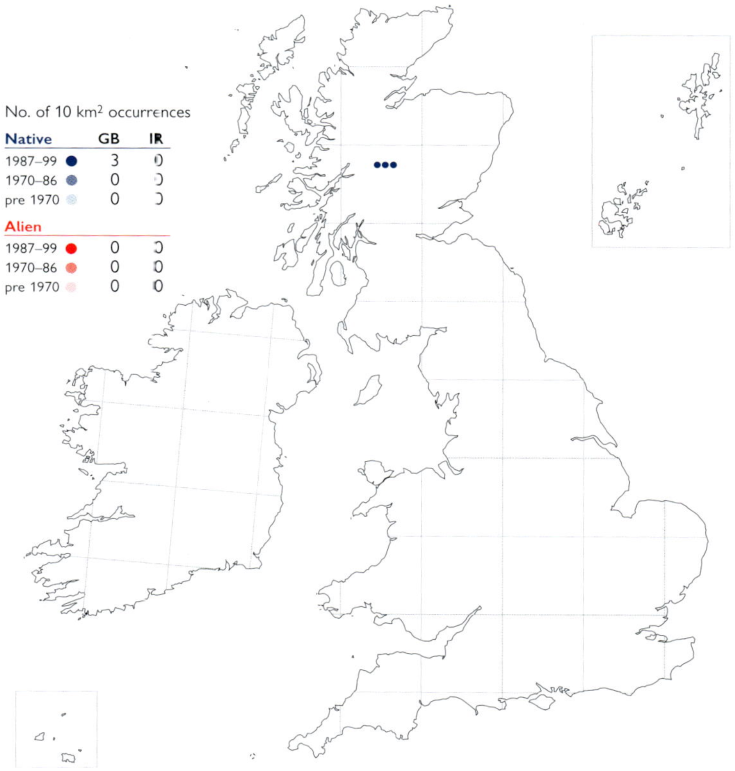

This low shrub occurs in acidic, free-draining sites on steep, usually N.- to E.-facing rocky mountain slopes. It is usually found in dwarf shrub communities, though it sometimes occurs in herb-rich grassland. All sites have a prolonged snow-lie. Flowering is irregular, and seed production generally poor. 670–800 m (Ben Alder Forest, Westerness).

Native. *P. caerulea* was discovered in Britain in 1810. The 10-km distribution is stable, but negligible recruitment, together with losses from erosion, means that populations have somewhat declined since detailed studies began in 1985.

European Arctic-montane element; also in C. and E. Asia and N. America.

References: Atlas (193a), Coker & Coker (1973), Hultén & Fries (1986), Nelson (1977), Wigginton (1999).

M. C. F. PROCTOR

Daboecia cantabrica St Dabeoc's Heath

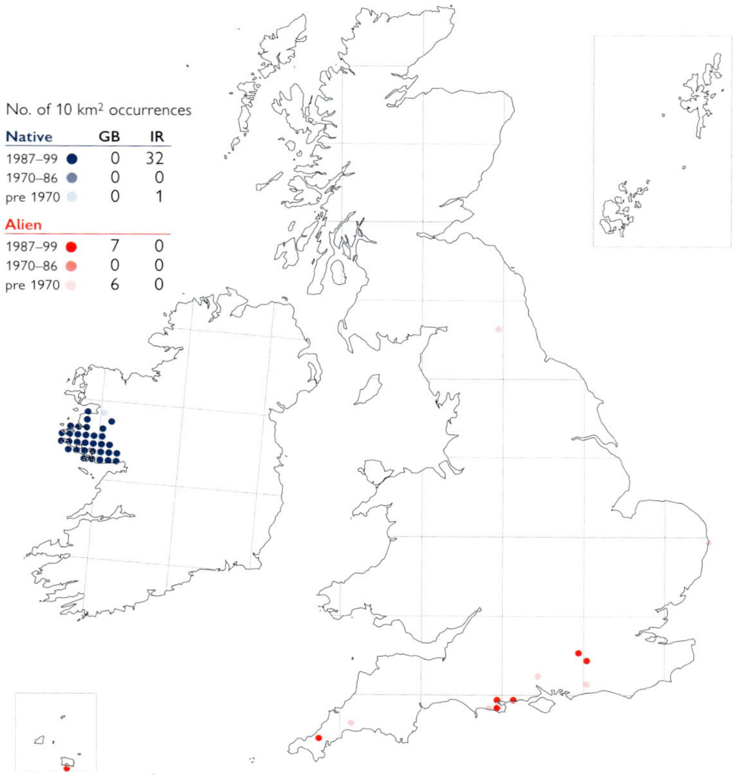

This straggling, low shrub occurs on heathland and moorland, often on rocky terrain, usually with other shrubs including *Calluna vulgaris, Erica cinerea* and *Ulex gallii*. It grows in thin acidic soils over quartzites or mica-schists, avoiding peat. 0–580 m (Ben Lettery, W. Galway, and Maamtrasna, W. Mayo).

Native (change +0.12). The distribution of *D. cantabrica* is probably little changed, if at all, since 1950. It is naturalised in a few sites in Britain.

Oceanic Southern-temperate element.

References: Atlas (193a), Dupont (1962), Nelson (2000b), Woodell (1958).

M. C. F. PROCTOR

Andromeda polifolia Bog-rosemary

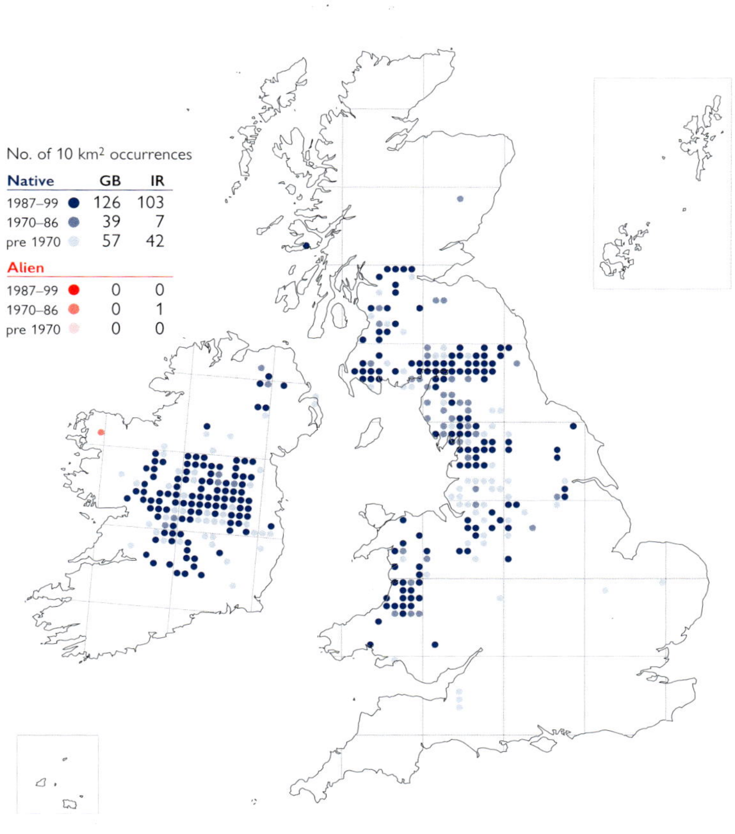

A straggling dwarf shrub of moist to wet acidic peaty ground, most abundant in lowland raised bogs but with scattered occurrences on upland peats. Fruits seldom develop. From sea level to *c.* 530 m in Wales and England, with an outlying site at 735 m on Mt Keen (S. Aberdeen).

Native (change +0.09). *A. polifolia* was lost from some sites before 1930 and its habitat has declined greatly in the last fifty years through drainage, peat extraction and afforestation. However, it remains frequent in undrained sites. Many sites have been discovered since the 1962 *Atlas*, mainly in Wales and the Clyde Basin.

Circumpolar Boreal-montane element.

References: Atlas (193b), Curtis & McGough (1988), Hultén & Fries (1986), Jacquemart (1998), Meusel *et al.* (1978), Stewart *et al.* (1994).

M. C. F. PROCTOR

Gaultheria shallon Shallon

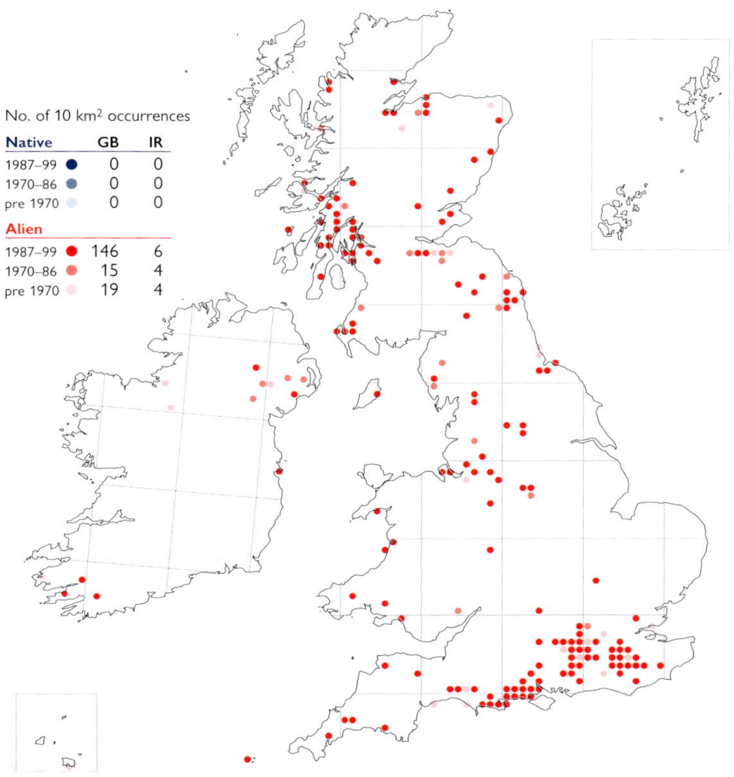

A low shrub, locally naturalised in woodland and open heathland on acidic, sandy and peaty soils. Generally lowland, but reaching 365 m at Tosson (S. Northumb.).

Neophyte. *G. shallon* was introduced into cultivation in Britain in 1826, and has been recorded in the wild since 1914. It was originally planted as cover for game birds, but has locally become a serious pest, especially on lowland heath where it regenerates rapidly after clearance. In the New Forest (S. Hants.) pigs penned in infested areas have proved effective in uprooting it.

Native of western N. America.

References: Bean (1973), Fraser *et al.* (1993), Hultén (1968), Spencer (2000).

M. C. F. PROCTOR

Gaultheria mucronata Prickly Heath

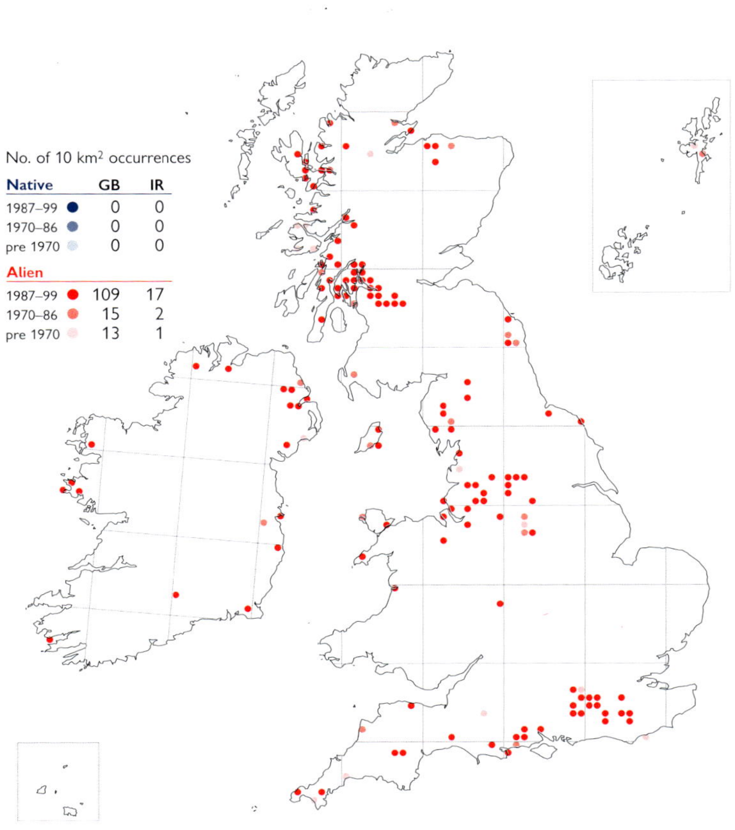

This dwarf shrub is found in open woods, roadsides and scrub on acidic, sandy soils. Its seeds are believed to be spread by birds. Lowland.

Neophyte. *G. mucronata* was introduced in 1828 as *Pernettya mucronata*, and was known in the wild by 1903. It is often grown in gardens and, although occasionally naturalised, is not such an invasive pest as *G. shallon*.

Native of S. America (Argentina, Chile).

Reference: Bean (1973).

M. C. F. PROCTOR

Arbutus unedo Strawberry-tree

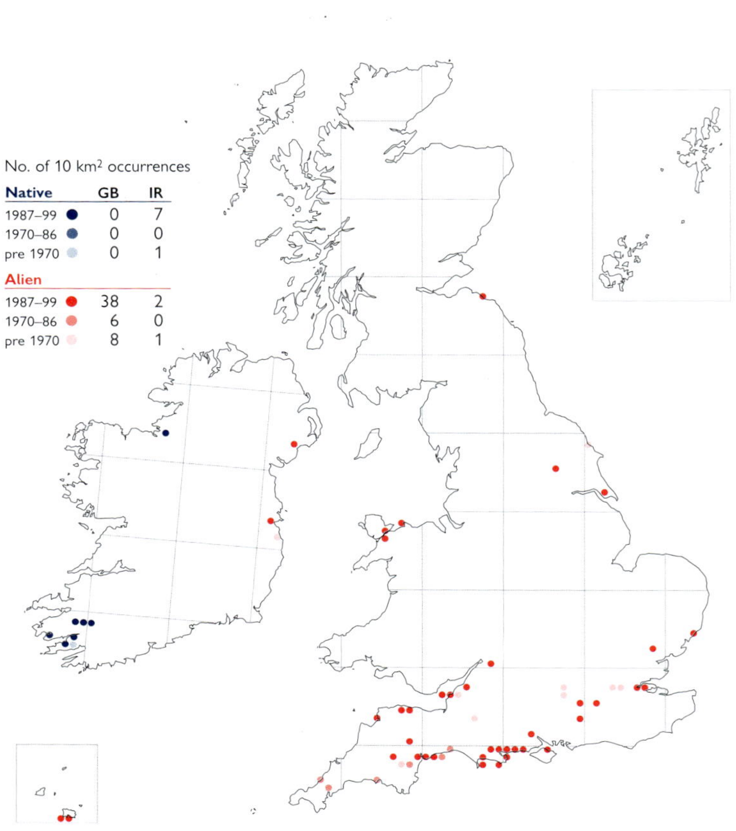

A small tree found as a native in heathy scrub and open woodland on rocky slopes and lake shores, on limestone, conglomerate, slates and sandstones; typically on very shallow soil, or rooted into rock. It also occurs as an escape from cultivation, and has become invasive at the Great Orme (Caerns.) and possibly elsewhere. Reproduction is by seed. Lowland.

Native (change +1.18). *A. unedo* was much more abundant in Ireland several centuries ago, but by the 16th century seemed to have become extinct or rare except in Co. Cork and Co. Kerry, probably largely because of its use for charcoal. There has been little change over the last hundred years.

Mediterranean-Atlantic element.

References: Atlas (193c), Meusel *et al.* (1978), Sealy & Webb (1950).

M. C. F. PROCTOR

Arctostaphylos uva-ursi Bearberry

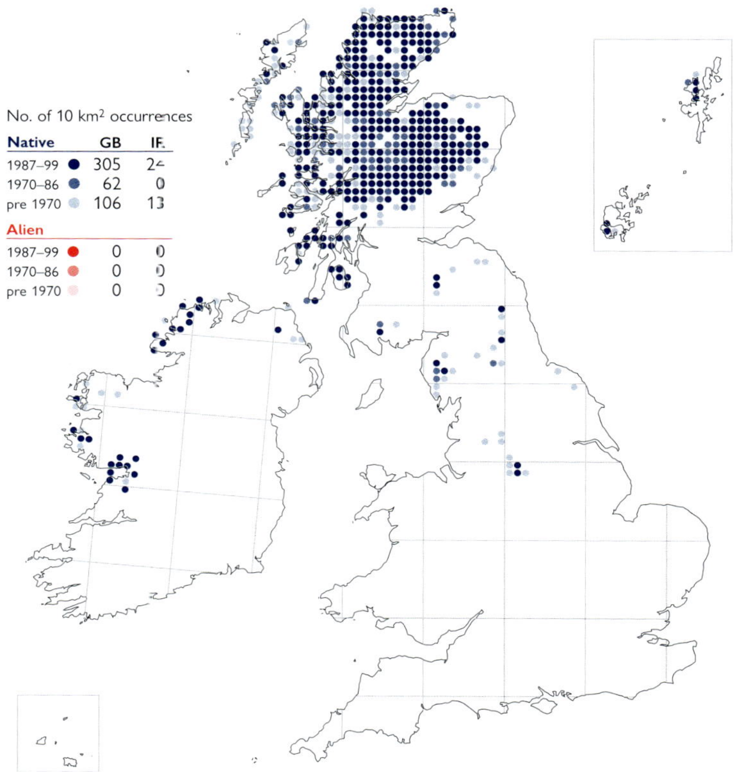

This procumbent low shrub is found on upland heaths and moorlands, often over well-drained gravelly or rocky ground, and on ravine sides. It sometimes grows in heathy grasslands on limestone, as in the Burren. From sea level in W. Ireland and Ardtoe (Westerness) to *c.* 710 m (Rannoch Moor, Mid Perth) and reportedly at 915 m in Inverness-shire.

Native (change –0.75). The range of *A. uva-ursi* is essentially unchanged since 1950, but it has suffered local declines, possibly because of moor-burning. Most losses in England took place before 1930.

Circumpolar Boreal-montane element.

References: Atlas (193d), Hultén & Fries (1986), Meusel *et al.* (1978).

M. C. F. PROCTOR

Arctostaphylos alpinus Arctic Bearberry

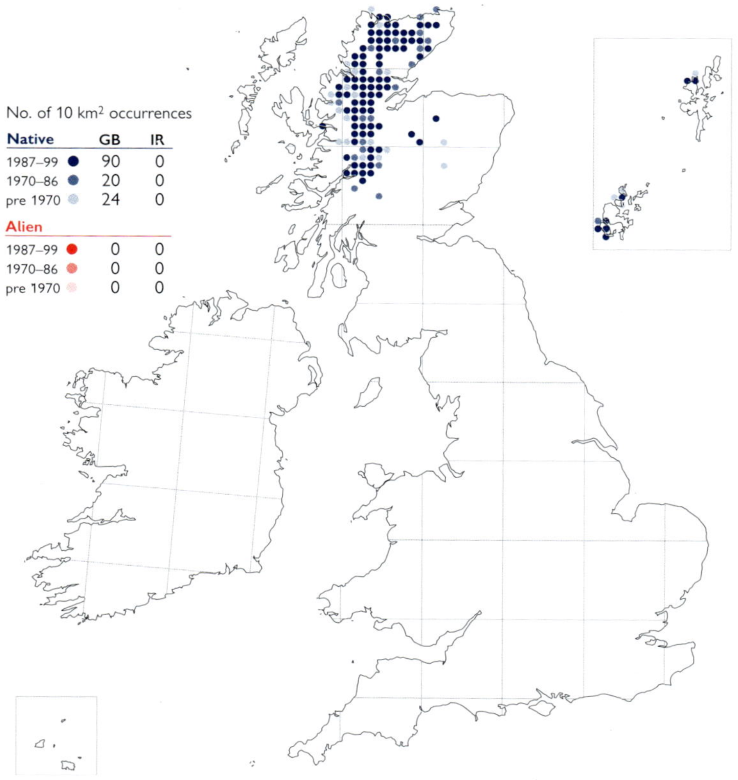

A strictly calcifugous shrub growing on acidic mineral soils or peat. It occurs on exposed upland heath, and in the northern Highlands of Scotland also on drier blanket bog. It is possibly long-lived, and fruiting is often sparse. It mostly occurs at mid-elevations, but descends to 100 m in North Roe (Shetland) and ascends to 945 m on Tom a'Choinich above Glen Affric (Easterness).

Native (change –0.22). The distribution of *A. alpina* has remained essentially unchanged over recent decades, and it probably still occurs in many of the squares for which there are only pre-1970 records.

Circumpolar Arctic-montane element.

References: Atlas (194a), Hultén & Fries (1986), Meusel *et al.* (1978), Stewart *et al.* (1994).

M. C. F. PROCTOR

Calluna vulgaris Heather

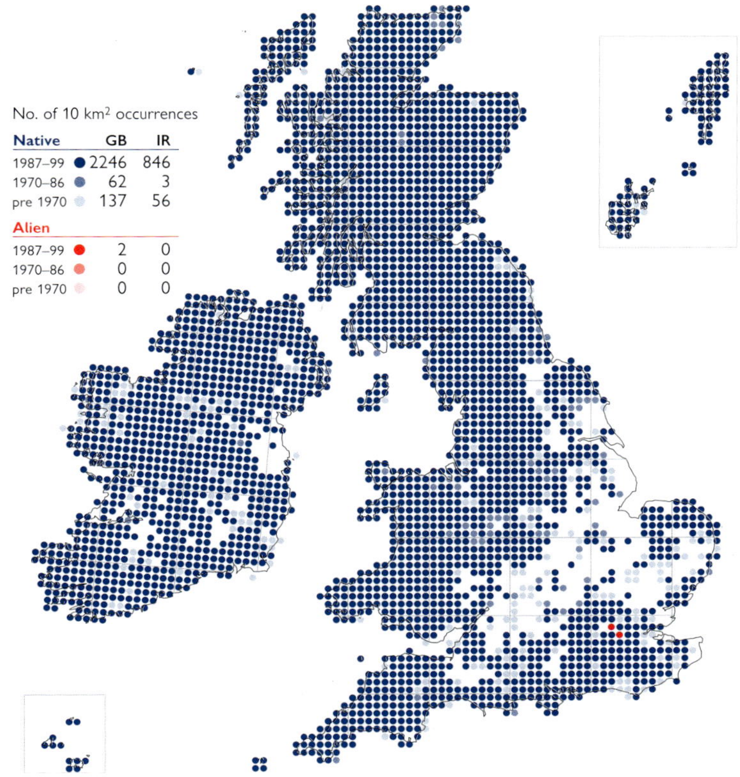

A low shrub, often dominant on heaths, moors and nutrient-poor grasslands, and in open woodland on acidic soils, ranging from dry exposed habitats to wet peat bogs. It can colonise newly available habitats. 0–1040 m (Macgillycuddy's Reeks, S. Kerry), with a single bush at 1095 m on Beinn a'Bhuird (S. Aberdeen).

Native (change –0.64). Suitable habitat for *C. vulgaris* has declined greatly, particularly in much of England, since 1950 through loss of heathland to forestry, agriculture, mineral workings and scrub. It cannot tolerate continued heavy grazing, and has declined in some upland areas for this reason.

European Boreo-temperate element; also in C. Asia and widely naturalised outside its native range.

References: Atlas (194b), Gimingham (1960), Grime *et al.* (1988), Hultén & Fries (1986), Meusel *et al.* (1978).

M. C. F. PROCTOR

Erica ciliaris Dorset Heath

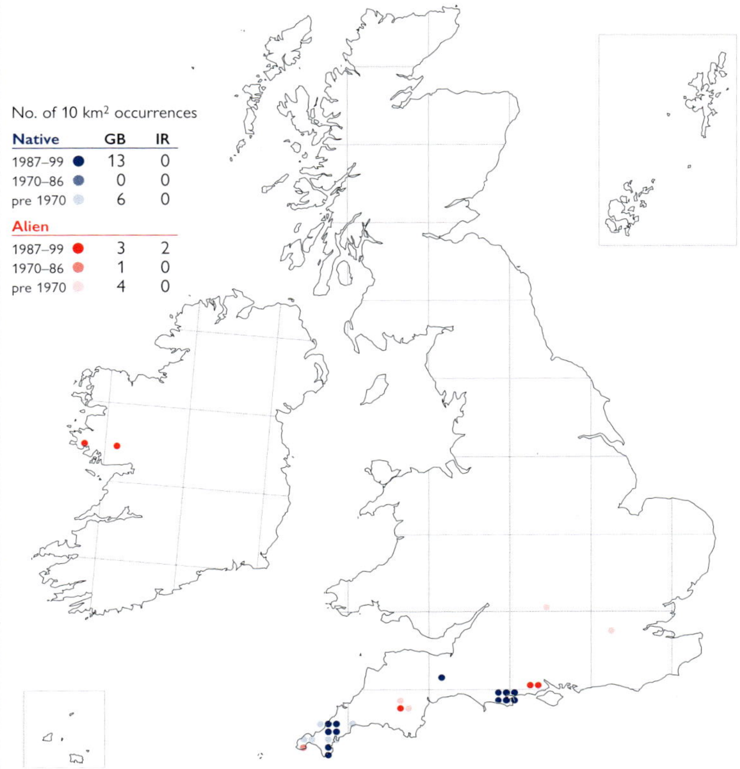

This low shrub occurs on moist heathland, extending into relatively dry heath, and also into wet valley bogs, mainly on the drier hummocks. Seedlings establish on bare ground, but in closed habitats reproduction is usually vegetative. Generally lowland, but reaching 400 m on Dartmoor (S. Devon).

Native (change –0.11). Much heathland with *E. ciliaris* in Dorset has been lost to forestry, but recently this trend has been stabilised or reversed and there is now some indication of a natural extension in its range. The Cornish sites are probably relics of larger populations. Other populations may be native or introductions to natural habitats.

Oceanic Southern-temperate element.

References: Atlas (194d), Curtis & McGough (1988), Edgington (1999), Perring & Sell (1968), Rose *et al.* (1996), Wigginton (1999).

M. C. F. PROCTOR

Erica mackaiana Mackay's Heath

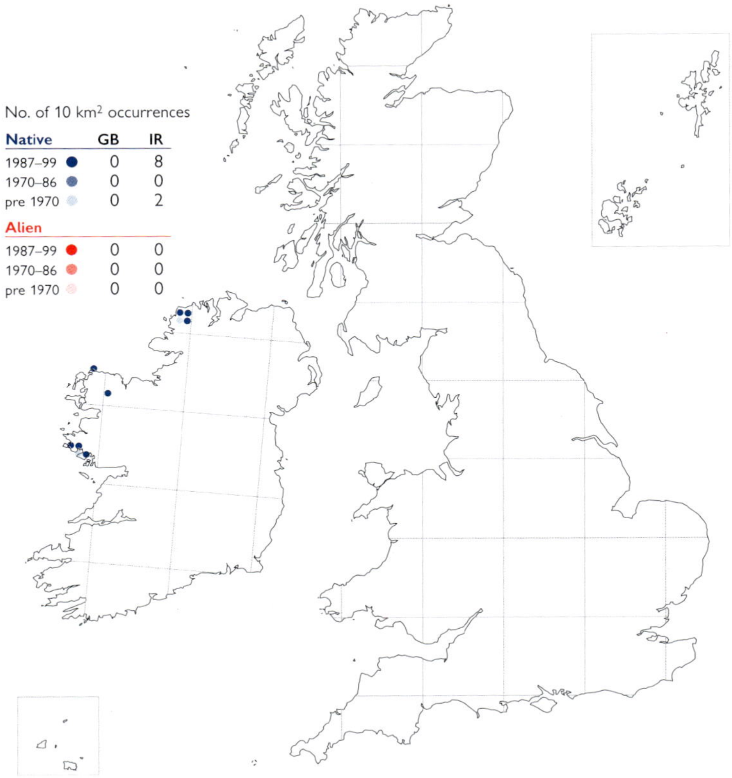

The habitats of this low shrub are blanket mire and rocky wet heath, where it occupies a somewhat narrower range of habitat than *E. tetralix*, avoiding the wettest sites. It grows on deep peat. The Irish plants of *E. mackaiana* never set seed. Their pollen fertility varies, but they produce sufficient fertile pollen to hybridise freely with *E. tetralix* to give the sterile hybrid *E.* × *stuartii*. Lowland.

Native. Populations of *E. mackaiana* appear to be stable.

Oceanic Temperate element; restricted to N.W. Spain and Ireland.

References: Atlas (194d), Curtis & McGough (1988), Perring & Sell (1968), Webb (1955b).

M. C. F. PROCTOR

Erica tetralix Cross-leaved Heath

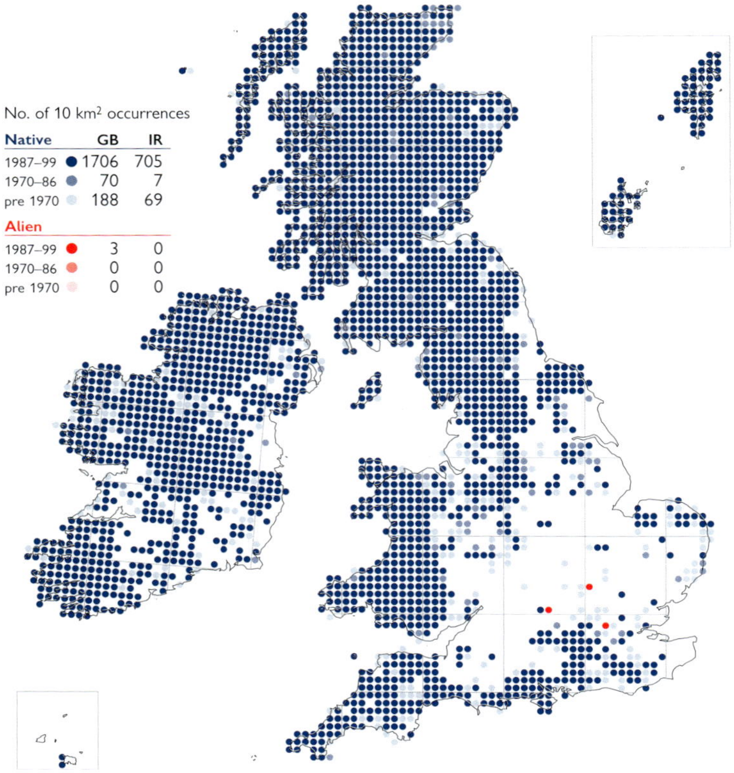

A sprawling low shrub found in a very wide range of mires and wet heaths, extending into drier heath in S.W. Britain. It is usually on wet, nutrient-poor organic soils, but can also grow in mesotrophic or eutrophic conditions. Generally from the lowlands to *c.* 670 m, but reaching 880 m in Coire Etchachan (S. Aberdeen).

Native (change –0.91). *E. tetralix* was mapped as 'all records' in the 1962 *Atlas.* Analysis of the database reveals that the widespread decline in S. England has taken place since 1950. In common with other heathland and bog species it has suffered from habitat loss, but it remains common wherever suitable conditions occur.

Suboceanic Temperate element.

References: Atlas (194c), Bannister (1966), Hultén & Fries (1986), Meusel *et al.* (1978).

M. C. F. PROCTOR

Erica cinerea Bell Heather

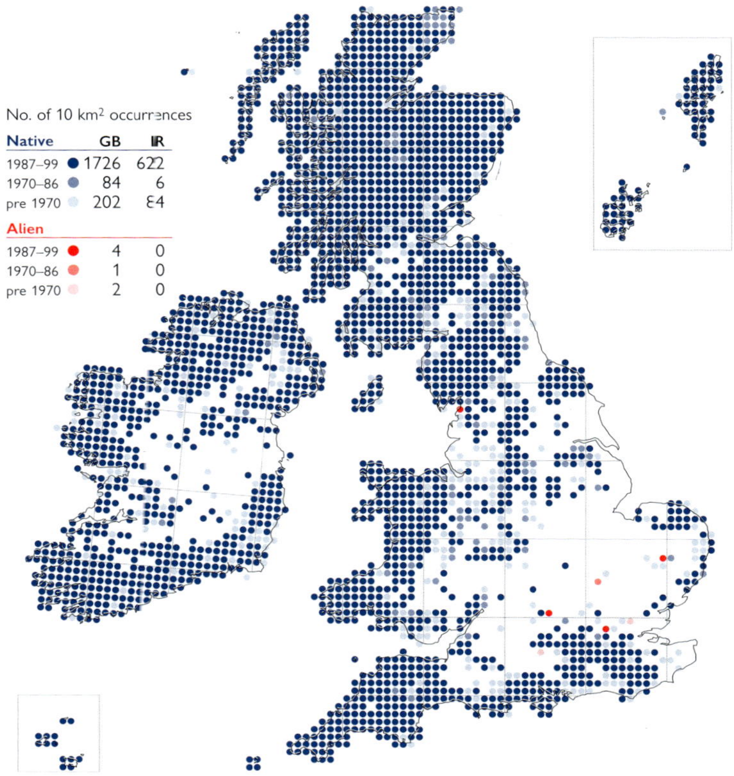

This small shrub occurs on thin, acidic, peaty or mineral soils in well-drained situations, on dry heaths, and as an occasional undershrub in open-canopy *Pinus sylvestris* or *Quercus* woodland. It is found in some calcareous grasslands that are leached and acidic at the surface ('limestone heath'). 0–790 m (Purple Mountain, S. Kerry); a report from 1220 m in the Grampians is very doubtful (Preston & Hill, 1997).

Native (change –0.94). *E. cinerea* has declined in S. England from loss of heathland habitat, and has disappeared from many former 'chalk heath' sites through encroachment of rank grass and scrub following reduction in sheep and rabbit grazing.

Oceanic Temperate element.

References: Atlas (195a), Bannister (1965), Dupont (1962), Grime *et al.* (1988), Hultén & Fries (1986), Meusel *et al.* (1978).

M. C. F. PROCTOR

Erica erigena Irish Heath

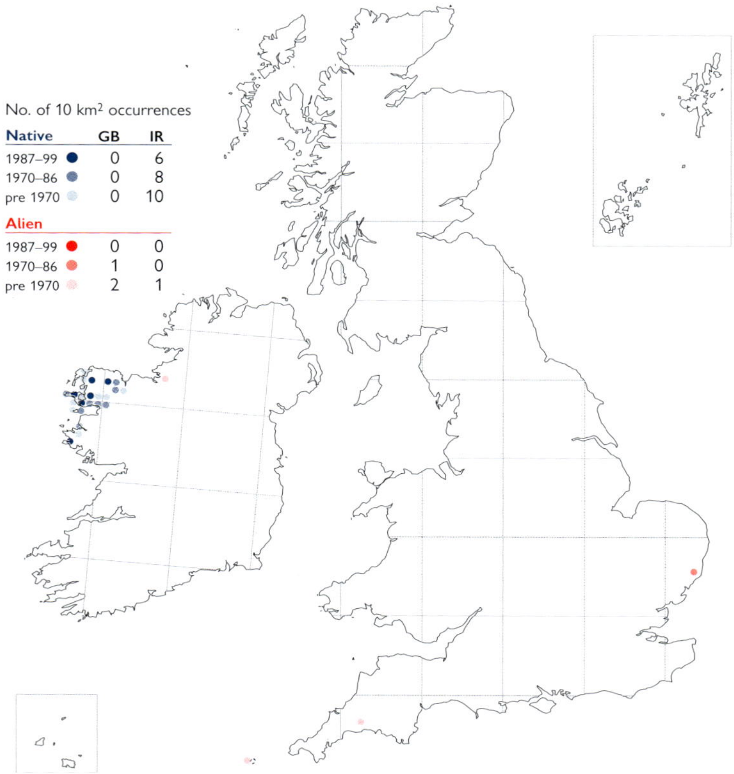

A shrub found in damp or boggy, more or less base-rich moorland, usually on slopes, often close to streams or on lake shores. Lowland.

Native. Although it is normally regarded as a native, Foss & Doyle (1988) have suggested that *E. erigena* might have been introduced to Ireland in the 15th century by pilgrims or traders. There is no indication of any appreciable change in its distribution in recent decades.

Oceanic Southern-temperate element, with a disjunct distribution.

References: Atlas (195b), Bolòs & Vigo (1995).

M. C. F. PROCTOR

Erica vagans Cornish Heath

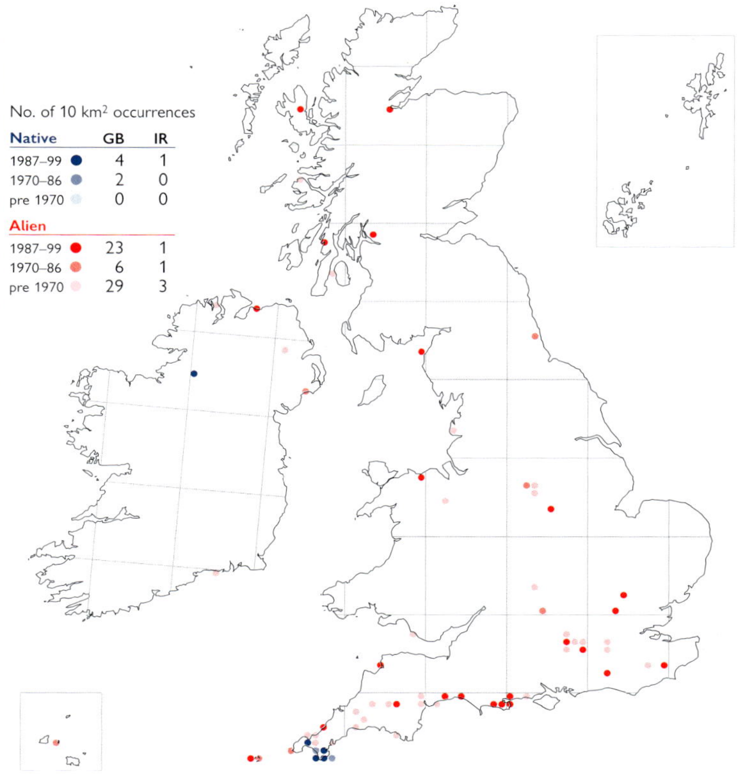

A locally abundant or co-dominant dwarf shrub in heathland with *Calluna* and *Erica cinerea*, or *Ulex* spp., often with calcicolous herbs, over ultrabasic rocks (serpentine and gabbro); also found on moist gley soils. Seedlings and plantlets can be frequent, but often die of drought; older plants regenerate from the base after winter burning. Lowland.

Native (change –0.07). In Britain, about one third of the heaths containing *E. vagans* were lost between 1908 and 1980. The remainder now have statutory protection, and the distribution seems stable. In Co. Fermanagh *E. vagans* is treated here as native but may be a prehistoric introduction.

Oceanic Southern-temperate element.

References: Atlas (195b), Bolòs & Vigo (1995), Curtis & McGough (1988), Dupont (1962), Wigginton (1999).

M. C. F. PROCTOR

Vaccinium oxycoccos Cranberry

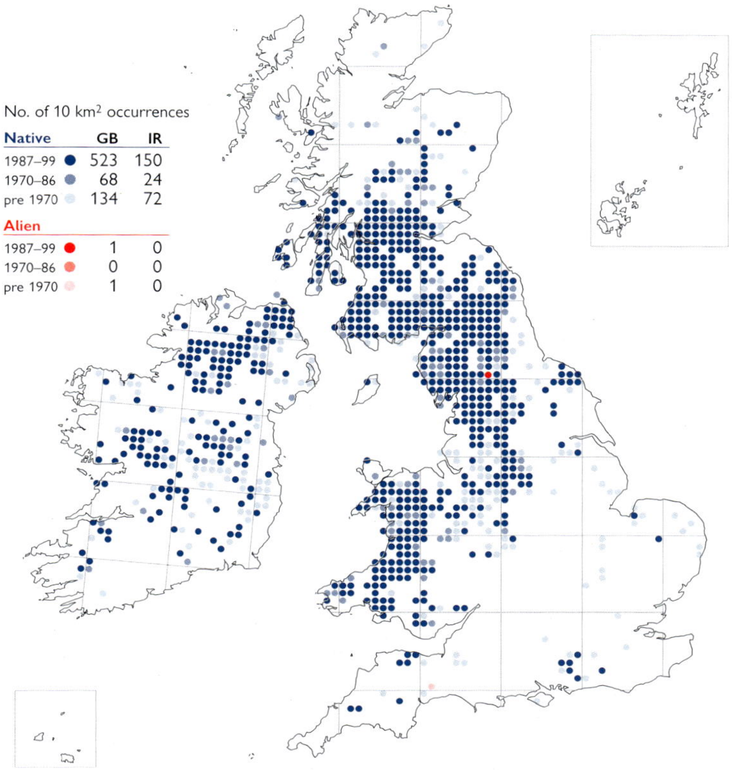

A slender, trailing dwarf shrub found in bogs and on very wet heaths, usually creeping amongst *Sphagnum*. 0–760 m (Ben Macdui, S. Aberdeen).

Native (change +0.28). Most of the losses from S. and E. England took place before 1930. Elsewhere in Britain its distribution seems generally stable, and it is much better recorded than in the 1962 *Atlas*. In Ireland there has been some decline due to peat extraction in the Midlands and afforestation in the west.

Circumpolar Boreal-montane element.

References: Atlas (196b), Hultén & Fries (1986), Jacquemart (1997), Meusel *et al.* (1978).

M. C. F. PROCTOR

Vaccinium microcarpum Small Cranberry

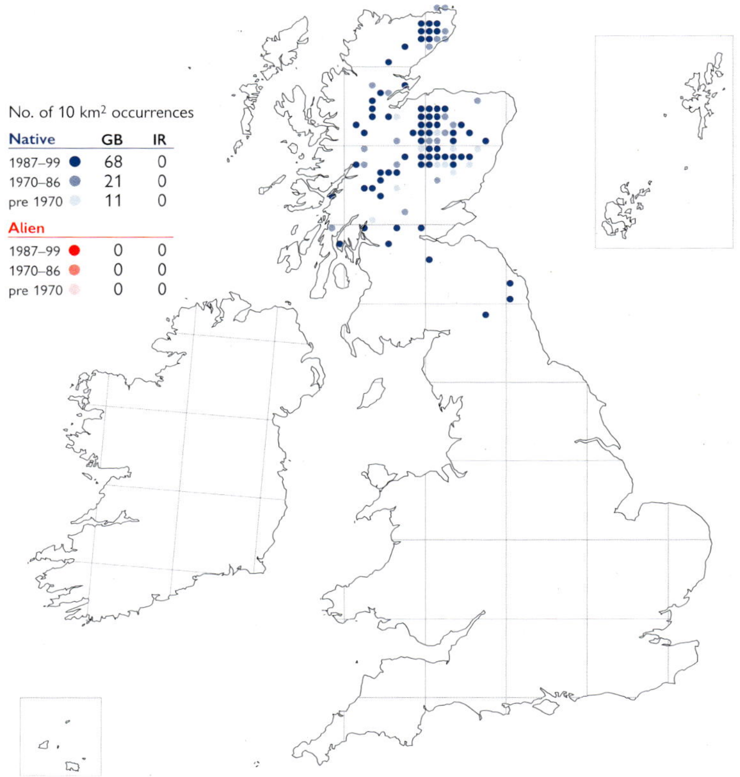

This trailing dwarf shrub occurs exclusively in *Sphagnum* mires, generally forming colonies in the drier microhabitats including the tops of hummocks. It ascends from near sea level in N.E. Scotland to 850 m on Carn nan Tri-tighearnan (Easterness).

Native (change +0.81). There has probably been no significant change in the distribution of *V. microcarpum*, the apparent widening of its range since the 1962 *Atlas* being attributed to better recording now that its taxonomy is better understood.

Circumpolar Boreal-montane element.

References: Atlas (196c), Hultén & Fries (1986), Jacquemart (1997), Meusel *et al.* (1978), Stewart *et al.* (1994).

M. C. F. PROCTOR

Vaccinium vitis-idaea Cowberry

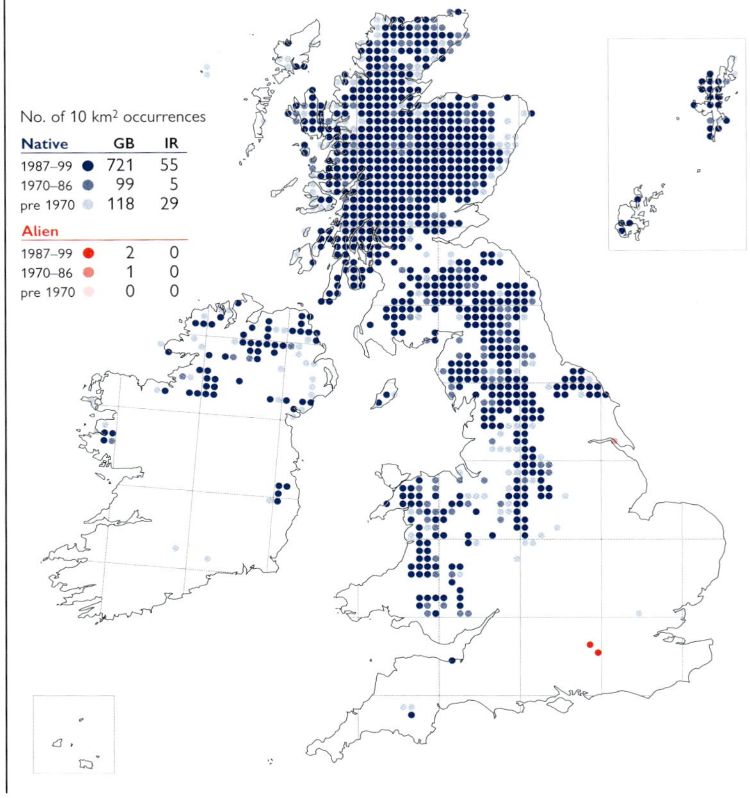

This calcifuge shrub is found on peaty heaths and moorland, in the understorey of *Quercus*, *Betula* and *Pinus* woods on acidic substrates, and on drier hummocks in blanket bogs. It ascends from 30 m by Lough Neagh (Co. Antrim) to 1095 m on Ben Lawers (Mid Perth).

Native (change –0.18). The distribution of *V. vitis-idaea* has not significantly changed since the 1962 *Atlas*.

Circumpolar Boreo-arctic Montane element.

References: Atlas (195c), Grime *et al.* (1988), Hultén & Fries (1986), Meusel *et al.* (1978), Ritchie (1955).

M. C. F. PROCTOR

Vaccinium uliginosum Bog Bilberry

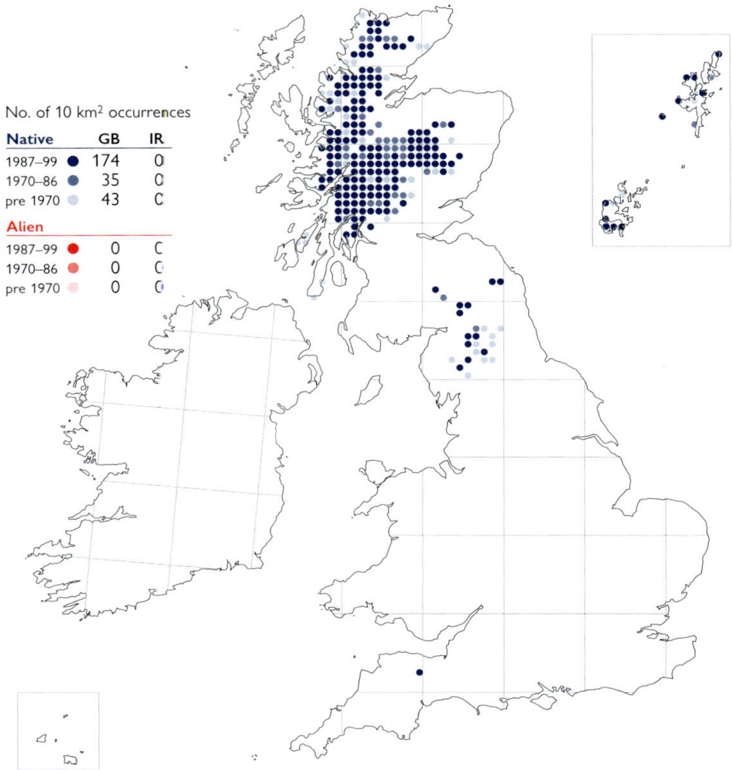

A low shrub, locally common on podsolic or peaty acidic soils in upland dwarf-shrub heaths and blanket bog, occasionally in *Nardus*-*Carex bigelowii* heath; also, rarely, in calcareous *Dryas* communities on montane ledges. From 40 m at Loch Awe (Main Argyll) to about 1130 m on Cairngorm (Easterness), but predominantly an upland plant.

Native (change –0.39). In the Highlands of Scotland *V. uliginosum* probably occurs in most squares for which there are only pre-1987 records. The losses in England mainly occurred before 1930. Its recent discovery on Exmoor (S. Somerset) is a remarkable extension of its range.

Circumpolar Boreo-arctic Montane element.

References: Atlas (196a), Hultén & Fries (1986), Jacquemart (1996), Meusel *et al.* (1978).

M. C. F. PROCTOR

Vaccinium myrtillus Bilberry

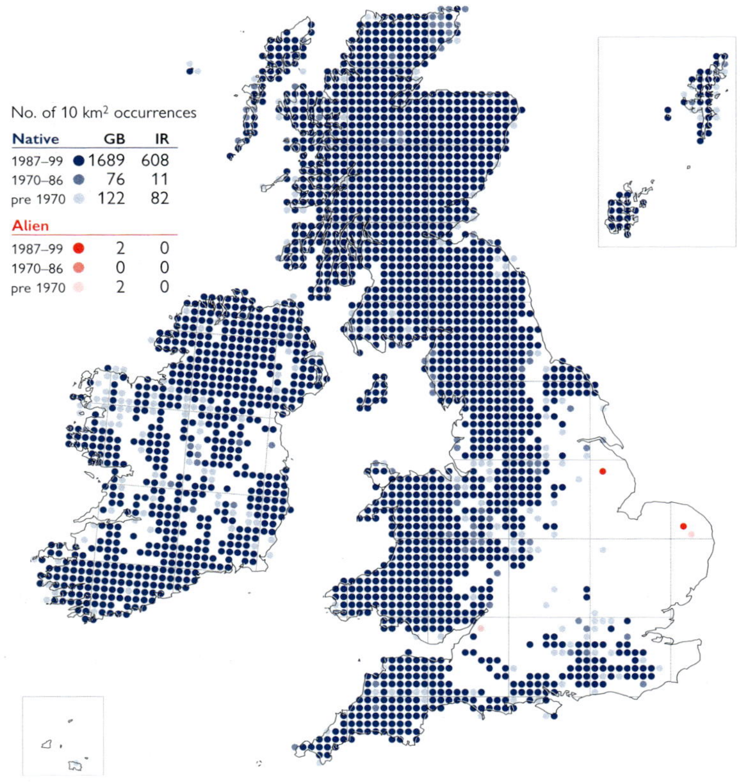

A calcifugous low shrub, common and locally dominant in well-drained heaths and moorland, especially in upland areas, and as an understorey in acid woodland of *Betula*, *Pinus* and *Quercus*; also found on hummocks in peat bogs in the north and west. It rarely regenerates from seed. 0–1300 m (Ben Macdui, S. Aberdeen).

Native (change –0.61). *V. myrtillus* was mapped as 'all records' in the 1962 *Atlas*. Analysis of the database reveals that the declines from the edge of its range in England have occurred since 1950, reflecting the loss of lowland heathland. It has also declined in C. Ireland, probably for the same reasons. Elsewhere it remains common in suitable habitats.

Eurosiberian Boreal-montane element.

References: Atlas (195d), Hultén & Fries (1986), Meusel *et al.* (1978), Ritchie (1956), Welch *et al.* (2000).

M. C. F. PROCTOR

Pyrola minor Common Wintergreen

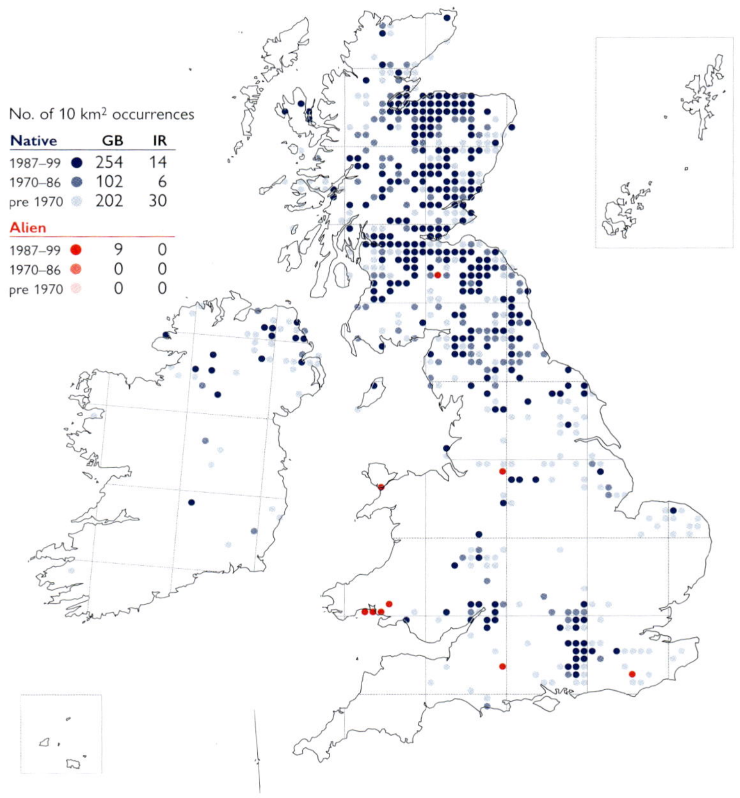

A rhizomatous, mycorrhizal, perennial evergreen herb. In S. England it is a plant of damp woodlands with deep litter, on a variety of soils; elsewhere it occurs in damp places in heaths, plantations, disused railways, on rock ledges and in sand dunes. 0–1130 m (Ben Lawers, Mid Perth).

Native (change –0.55). This species has declined throughout its scattered British range through changes in land use and management. Most losses in the south occurred before 1930. The small, southern woodland populations appear particularly vulnerable, with recent losses as sites have become too dry. It sometimes colonises plantations, where it may be introduced or arrive naturally by wind-blown seed.

Circumpolar Boreal-montane element.

References: Atlas (196d), Hultén & Fries (1986), Meusel *et al.* (1978).

F. J. RUMSEY

Pyrola media Intermediate Wintergreen

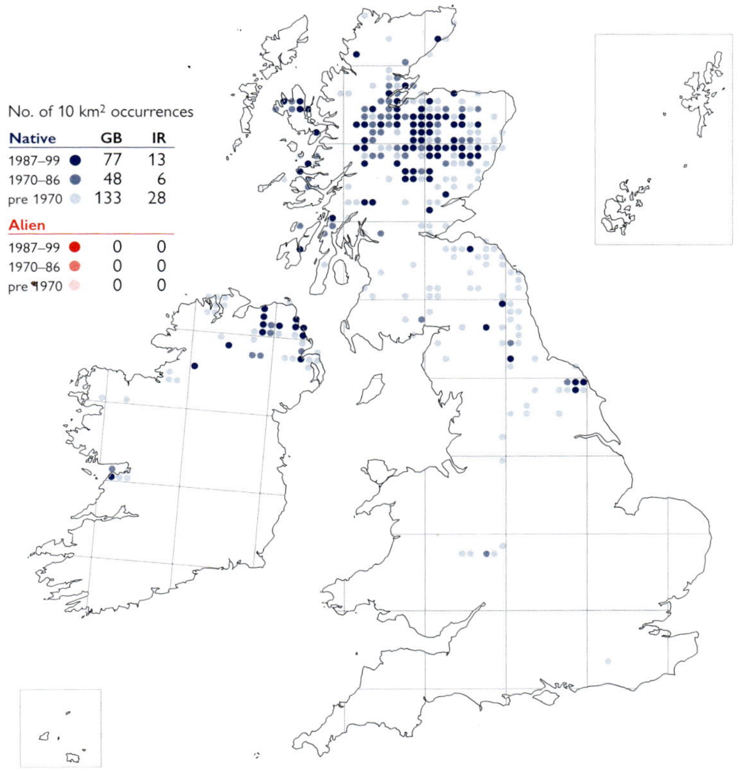

A rhizomatous, mycorrhizal, evergreen perennial herb of well-drained, mildly acidic to slightly basic soils in woods and on heaths. It is characteristic of *Arctostaphylos-Calluna* submontane heath derived from former woodland. 0–550 m (Coire Garbhlach, Easterness).

Native (change –1.09). *P. media* has been much over-recorded in the past for *P. minor*, with which it often grows. This may, in part, account for the apparently substantial historic decline. It is also very shy-flowering and may be under-recorded. Recently unfavourable woodland management and increased moorland grazing may have contributed to some losses.

Eurosiberian Boreal-montane element.

References: Atlas (197a), Curtis & McGough (1988), Hultén & Fries (1986), McVean & Ratcliffe (1962), Meusel *et al.* (1978), Stewart *et al.* (1994).

F. J. RUMSEY

Pyrola rotundifolia subsp. *rotundifolia* Round-leaved Wintergreen

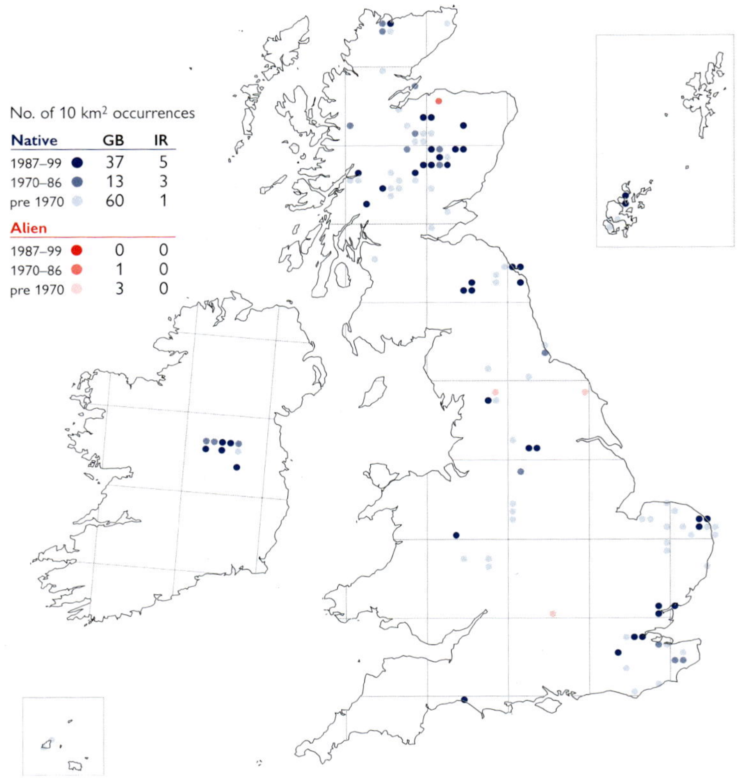

A rhizomatous, mycorrhizal, evergreen perennial herb. In England it usually grows in damp, calcareous sites including fens, disused chalk-pits and dune-slacks. In Scotland it inhabits open *Pinus sylvestris* woodland, river banks and gullies in open moorland, and montane cliff ledges. 0–760 m (Breadalbanes, Mid Perth).

Native (change for species –0.08). This subspecies has undergone a marked decline since 1930, despite some local increases in disused quarries. Reasons for the losses include afforestation and rubbish tipping, but the most serious declines, such as its near extinction in East Anglia, result from changes in fen management.

Eurosiberian Boreo-temperate element.

References: Hultén & Fries (1986), Meusel *et al.* (1978), Stewart *et al.* (1994).

F. J. RUMSEY

Pyrola rotundifolia subsp. *maritima*

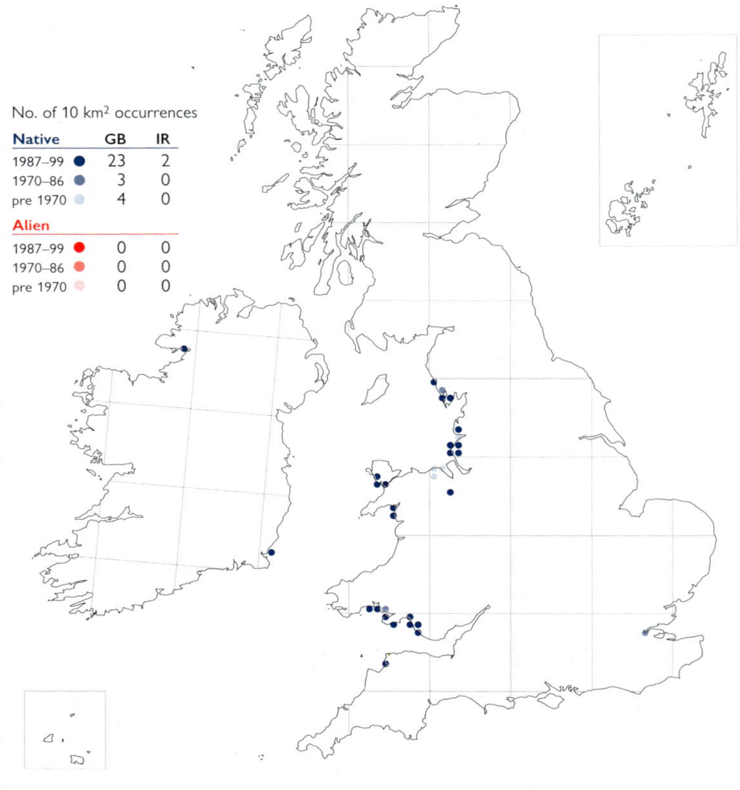

A rhizomatous, mycorrhizal, evergreen perennial herb of calcareous dune-slacks and nearby fixed dunes. It usually grows with *Salix repens* on the drier fringes of wet slacks, but also persists in conifer plantations on fixed dunes. Lowland.

Native (change for species –0.08). This plant was first described from Lancashire in 1893 and appears to have spread since then. Numbers fluctuate widely; clones appear, increase rapidly and are as likely to disappear as persist. Visiting botanists are as likely to be the inadvertent vectors of its spread as wind or wildfowl.

Suboceanic Temperate element; recorded from the Atlantic coast of Europe from N.W. France to S. Scandinavia and Germany.

References: Atlas Supp. (45a), Curtis & McGough (1988), Hultén & Fries (1986), Meusel *et al.* (1978), Stewart *et al.* (1994).

F. J. RUMSEY

Orthilia secunda Serrated Wintergreen

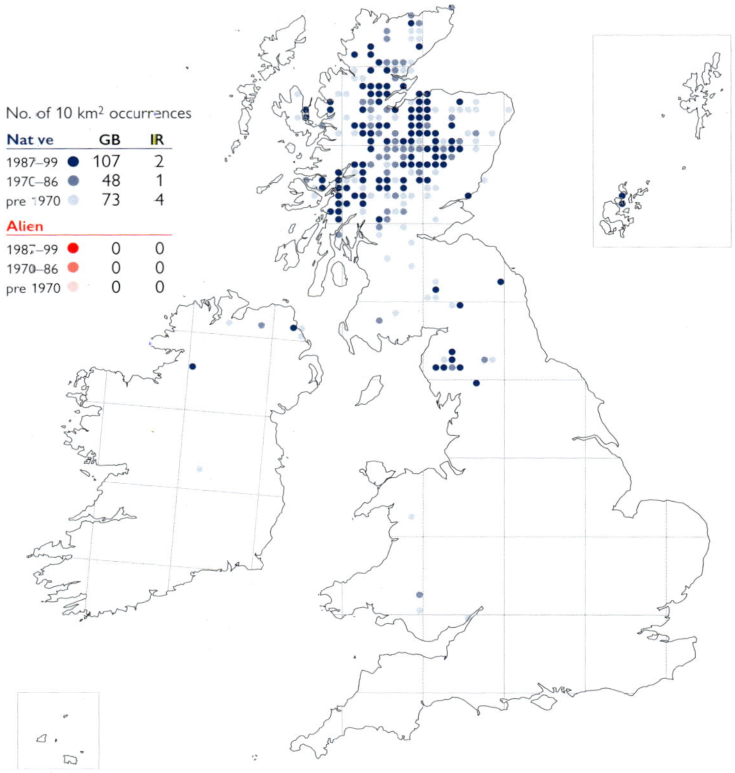

A rhizomatous, mycorrhizal, evergreen perennial herb, of damp *Calluna* and *Vaccinium*-dominated communities, mostly in *Pinus* and *Betula* woodland but also on open moorland. It also grows in clefts and on ledges in rocky gullies, and on rocky stream banks. Flowering is often erratic. From 30 m (Kirkhill, Easterness) to 690 m (Craig an Dail, S. Aberdeen).

Native (change –0.40). Poor recruitment means that the distribution has been reduced historically by fire, grazing and other moorland management practices. However, it is not clear whether it has continued to decline since 1970.

Circumpolar Boreal-montane element.

References: Atlas (197c), Curtis & McGough (1988), Hultén & Fries (1986), Meusel *et al.* (1978), Stewart *et al.* (1994).

F. J. RUMSEY

Moneses uniflora One-flowered Wintergreen

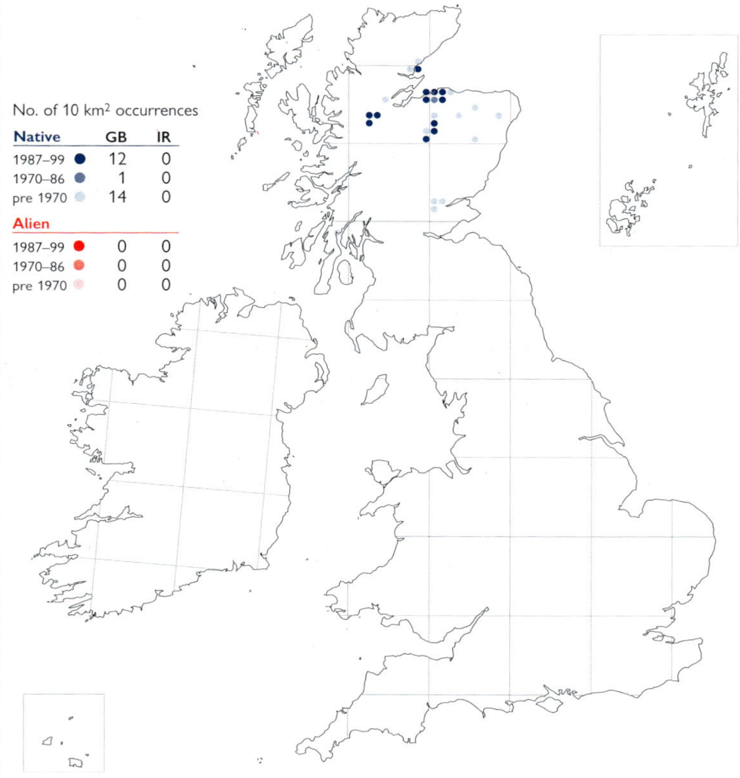

A mycorrhizal, evergreen perennial herb, spreading by rhizomes within leaf litter and bryophytes in ericaceous dwarf shrub communities of old pine plantations, rarely now in native *Pinus sylvestris* forests. The solitary flower is insect-pollinated, but recruitment from seed is rare and most propagation is vegetative. Lowland, reaching 300 m (Castle Grant, Moray).

Native (change +0.14). Diminished by past collection, this attractive species has continued to decline through changes in land use and forest management. Much of this decline took place before 1970, but its future is dependent on sympathetic management. Most remaining populations are small and vulnerable.

Circumpolar Boreal-montane element.

References: Atlas (197d), Hultén & Fries (1986), Meusel *et al.* (1978), Wigginton (1999), Wright & Lusby (1999).

F. J. RUMSEY

Monotropa hypopitys Yellow Bird's-nest

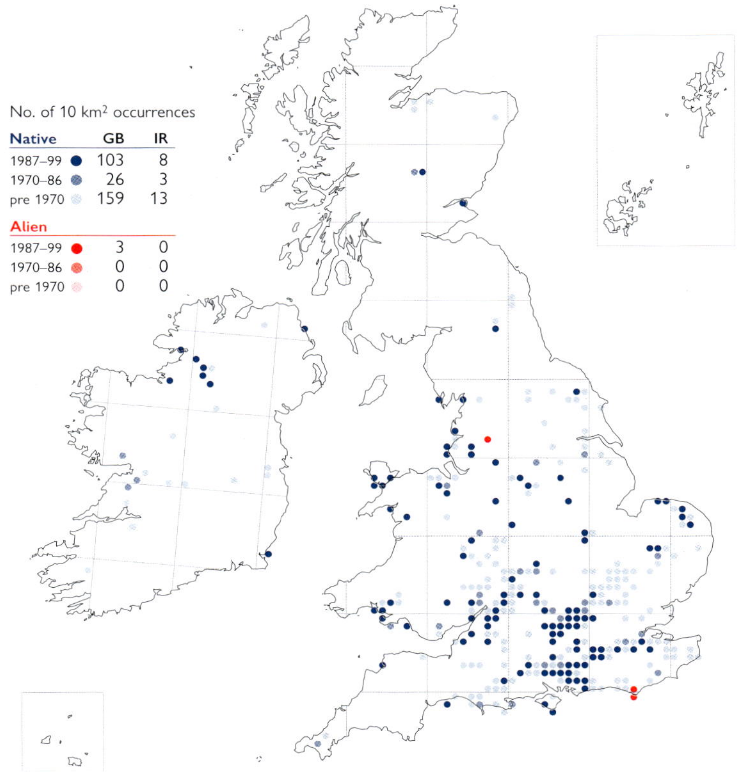

A saprophytic perennial herb of leaf litter in shaded woodlands, most frequent under *Fagus* and *Corylus* on calcareous substrates, and under *Pinus* on more acidic soils. It also grows in damp dune-slacks, where it is usually associated with *Salix repens*. 0–395 m (Buxton, Derbys.).

Native (change −1.09). A polyploid complex, with two subspecies in our area which are not easily separable morphologically. Although many sites were lost before 1930, the species has suffered a further marked decline in S. England since the 1962 *Atlas*. Though rare in Ireland, Curtis & McGough (1988) considered it to have been under-recorded in recent years.

Circumpolar Temperate element.

References: Atlas (198a), Atlas Supp. (45b, c), Hultén & Fries (1986), Meusel *et al.* (1978).

F. J. RUMSEY

Diapensia lapponica Diapensia

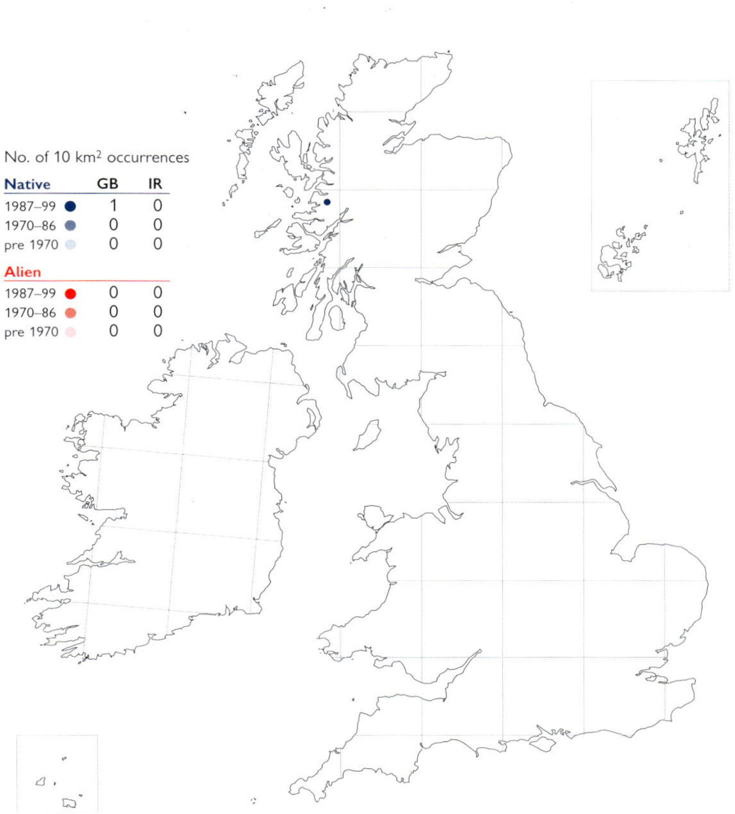

A long-lived, slow-growing, dense, cushion-forming evergreen dwarf shrub, restricted in the British Isles to a small quartz-rich rock outcrop on an exposed mountain ridge, where it grows on acidic soils in a stony fell-field. It is uncertain whether viable seed is regularly produced, and recruitment from seed at the site is certainly very limited. 780 m (near Glenfinnan, Westerness).

Native. The population (currently about 1200 clumps or mats) and the area it occupies have not changed markedly since the discovery of the species in 1951.

Circumpolar Arctic-montane element; absent from mountains of C. Europe.

References: Atlas (198b), Hultén & Fries (1986), Wigginton (1999).

F. J. RUMSEY

Primula vulgaris Primrose

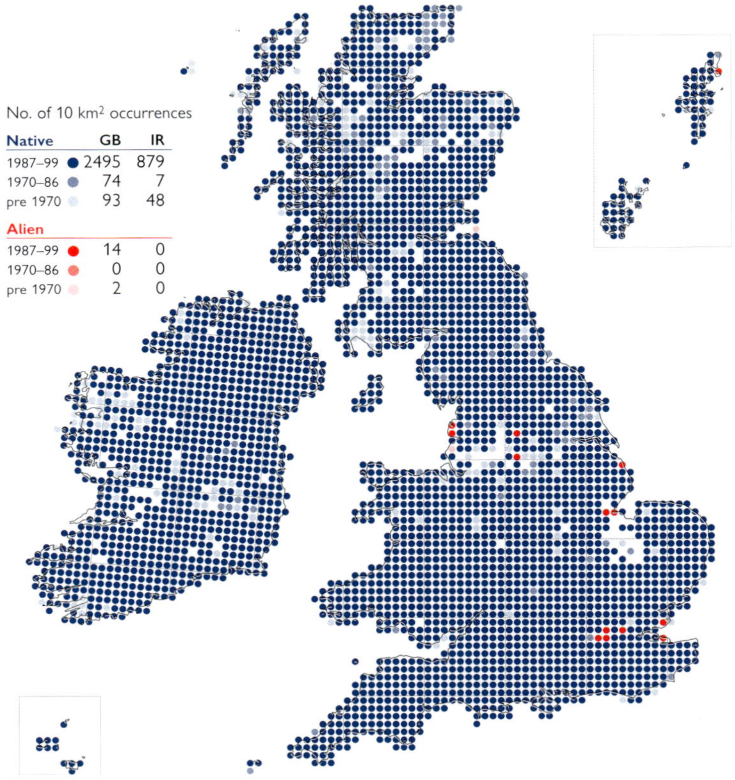

An evergreen, or sometimes aestivating, perennial herb typical of sites shaded from hot sun, found in woodland, on N.-facing banks, in hedgerows, coastal slopes and shaded montane cliffs. Reproduction is by seed, which is usually dispersed by ants. 0–850 m (Mt Brandon, S. Kerry).

Native (change +0.16). Unlike *P. veris*, populations of *P. vulgaris* have not fluctuated markedly during the last century in most parts of Britain and Ireland. However, in East Anglia woodland populations have declined greatly in response to a series of hot, dry summers from 1970 onwards (Rackham, 1999).

European Temperate element.

References: Atlas (201c), Atlas Supp. (46c), Hultén & Fries (1986), Mabey (1996), Meusel *et al.* (1978), Valverde & Silvertown (1997).

A. J. RICHARDS

Primula elatior Oxlip

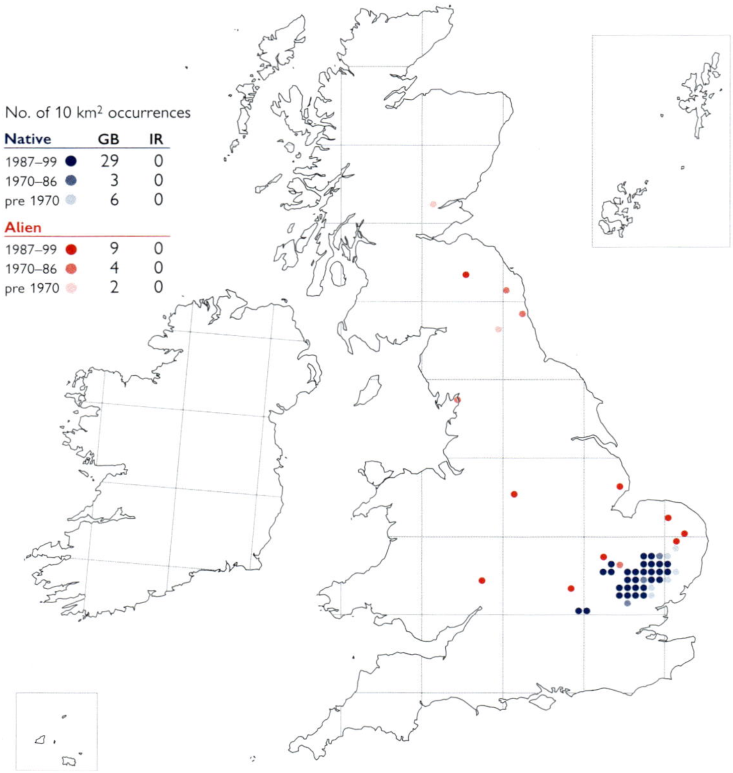

A perennial herb of woods dominated by *Acer campestre*, *Corylus*, *Fraxinus* and *Quercus robur* on damp chalky boulder-clay soils, especially where seasonal flooding occurs; rarely in wet *Alnus* woods, damp meadows and ancient hedgerows. Lowland.

Native (change +0.01). Agricultural improvement has all but eliminated this species from meadows since 1930, and in woodland numbers are reduced when coppicing ceases or conifers planted. Recently, increasing numbers of flowers have been eaten by deer. However, this has resulted in little change in the 10-km distribution.

European Temperate element, with a continental distribution in W. Europe; also in C. Asia.

References: Atlas (201b), Hultén & Fries (1986), Meusel *et al.* (1978), Preston (1993), Rackham (1999), Stewart *et al.* (1994).

A. J. RICHARDS

Primula veris Cowslip

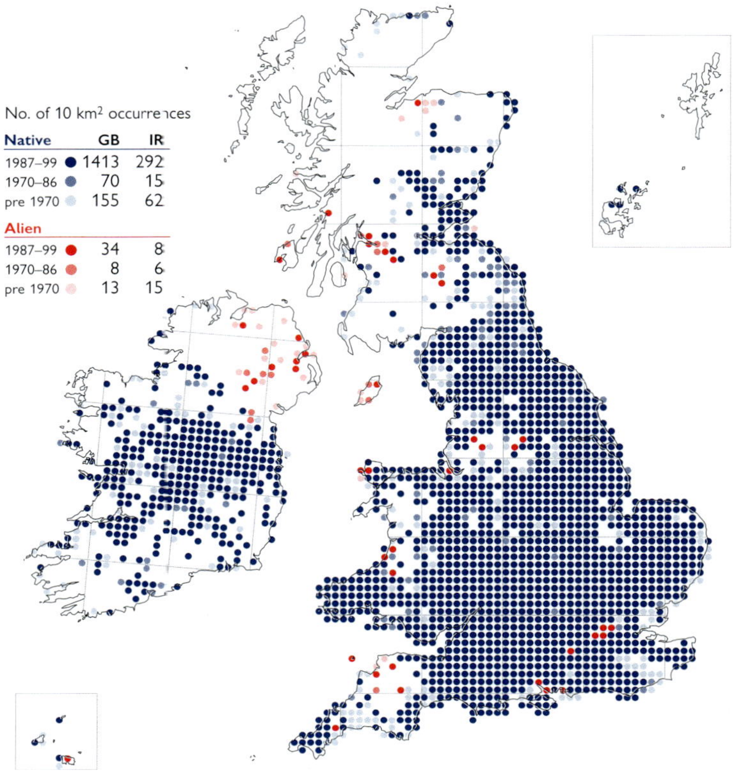

A perennial herb of well-drained, herb-rich grasslands on mesic to calcareous soils; less commonly on seasonally flooded soils, in scrub or woodland rides and edges, and on calcareous cliffs. 0–750 m (Knock Ore Gill, Westmorland), and exceptionally at 845 m (Great Dun Fell, Westmorland).

Native (change −0.32). *P. veris* suffered a marked decline between 1930 and 1980 due to the ploughing or agricultural improvement of grassland. It has, however, recently become more frequent on road verges sown with this species or with wild-flower mixtures. The current map shows little change from the 1962 *Atlas*.

Eurosiberian Temperate element.

References: Atlas (201a), Antrobus & Lack (1993), Curtis & McGough (1988), Grime *et al.* (1988), Hultén & Fries (1986), McKee & Richards (1998), Meusel *et al.* (1978), Richards (1989, 1993).

A. J. RICHARDS

Primula veris × *P. vulgaris* (*P.* × *polyantha*)

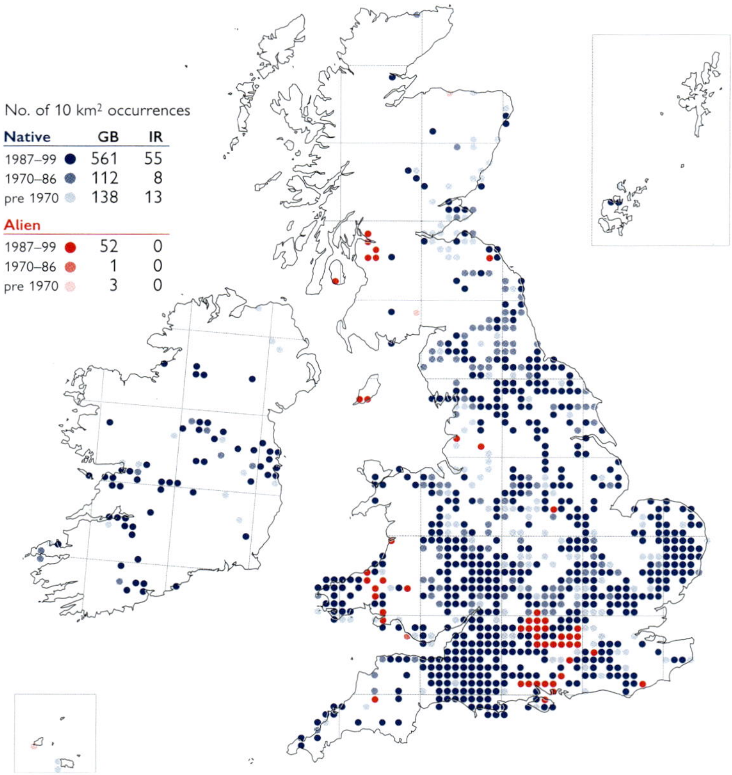

Although the parents have different ecological requirements they can be found in close proximity. Naturally occurring hybrids (including back-crosses) may be found on woodland rides and edges, in scrub, hedgerows and on roadsides verges. Garden primroses may sometimes act as the pollen parent, and plants cultivated as the garden Polyanthus also have this parentage and are occasionally planted in the wild, or found as casuals near dumped garden waste. 0–305 m (Tomintoul, Banffs.).

Native. The distribution of this hybrid is now much better known than when it was mapped by Perring & Sell (1968). Some Polyanthus records may be mapped as native in error.

Widespread in Europe.

References: Atlas Supp. (46b), Richards (1989, 1993), Stace (1975), Woodell (1965).

A. J. RICHARDS

Primula farinosa Bird's-eye Primrose

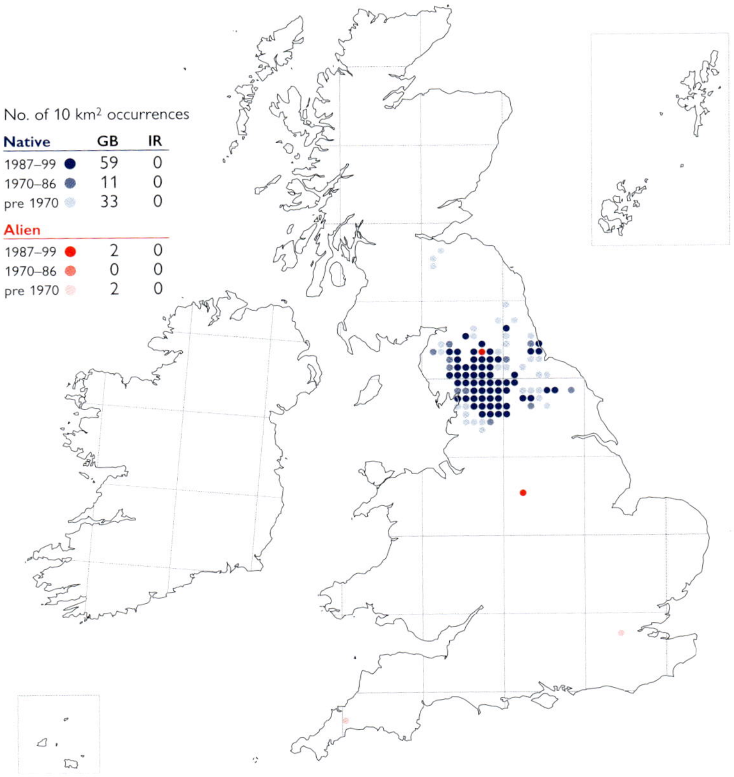

No. of 10 km² occurrences

Native	GB	IR
1987–99 ●	59	0
1970–86 ●	11	0
pre 1970 ●	33	0
Alien		
1987–99 ●	2	0
1970–86 ●	0	0
pre 1970 ●	2	0

A short-lived perennial herb typical of wet, usually spring-fed, calcareous flushes. It is often found on hummocks in springs, and on seeping banks where slippage has opened up the turf, most commonly on open marl. A few sites remain in unimproved, grazed, damp pasture. It perennates as resting-buds and reproduces by seed. 0–570 m (Jeffrey Pot, Wensleydale, N.W. Yorks.), but generally 200–400 m.

Native (change –0.46). *P. farinosa* is still frequent in suitable habitats, but local floras detail many losses due to drainage and agricultural improvement.

Eurasian Boreal-montane element, with a disjunct distribution.

References: Atlas (200c), Baker *et al.* (1994), Hultén & Fries (1986), McKee & Richards (1998), Meusel *et al.* (1978), Stewart *et al.* (1994).

A. J. RICHARDS

Primula scotica Scottish Primrose

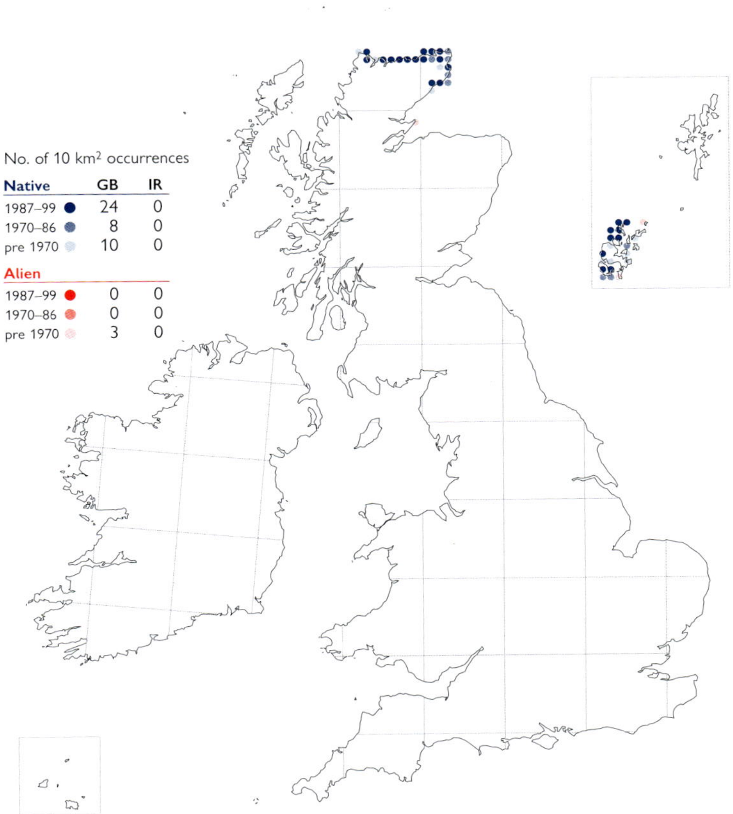

No. of 10 km² occurrences

Native	GB	IR
1987–99 ●	24	0
1970–86 ●	8	0
pre 1970 ●	10	0
Alien		
1987–99 ●	0	0
1970–86 ●	0	0
pre 1970 ●	3	0

A perennial herb, growing in a variety of moist but well-drained, usually heavily grazed, open grassland habitats that are often on calcareous substrates and sometimes liable to some sand accretion. Sites include cliff-tops, the transition zone between grassland and maritime heath, mosaics of heath and machair, and around rock outcrops. It is self-fertile, but many plants never flower. Lowland.

Native (change –0.18). This species can be adversely affected by both under- and over-grazing, most losses being attributed to one or the other. Its overall distribution is, however, unchanged since the 1962 *Atlas*.

Endemic.

References: Atlas (200d), Bullard *et al.* (1987), Hultén & Fries (1986), McKee & Richards (1998), Meusel *et al.* (1978), Ritchie (1954), Stewart *et al.* (1994), Tremayne & Richards (1997).

A. J. RICHARDS

Hottonia palustris Water-violet

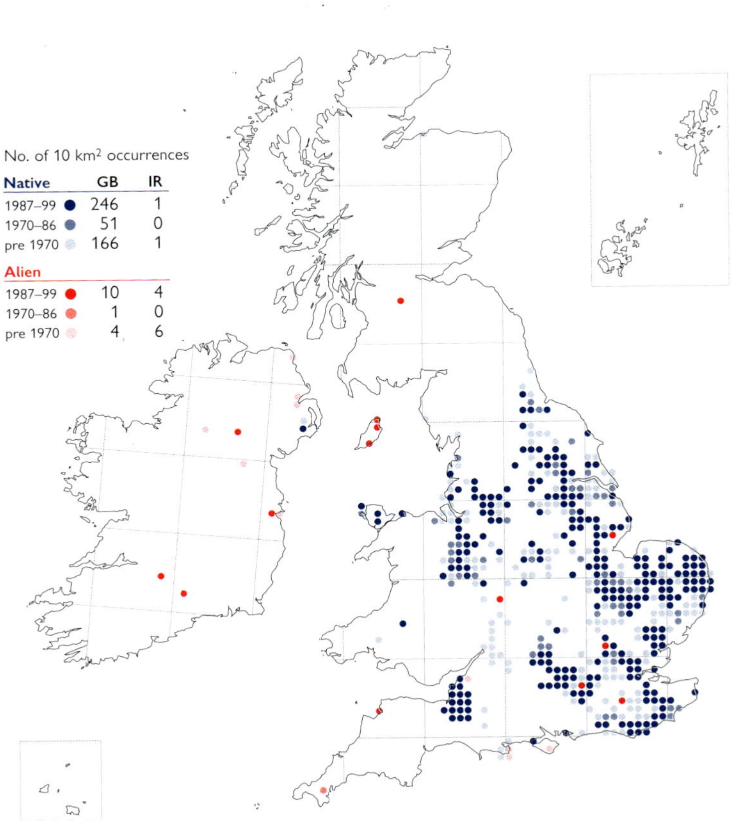

No. of 10 km² occurrences

Native	GB	IR
1987–99 ●	246	1
1970–86 ●	51	0
pre 1970 ●	166	1
Alien		
1987–99 ●	10	4
1970–86 ●	1	0
pre 1970 ●	4	6

A stoloniferous perennial of still, shallow, base-rich, clear and not eutrophicated water bodies, including ponds, ditches, ox-bows and backwaters. It can withstand shade and temporary exposure. It is heterostylous, requiring crosses between pins and thrums to set seed; vegetative growth often produces single-morph colonies that set little seed. Lowland.

Native (change –0.63). Many colonies of *H. palustris* have been lost to drainage, vegetation clearance, eutrophication, boat traffic and trampling by cattle. Most of these losses occurred before 1930, but have continued, particularly in S.E. England.

European Temperate element.

References: Atlas (201d), Brock *et al.* (1989), Curtis & McGough (1988), Hultén & Fries (1986), Meusel *et al.* (1978), Preston & Croft (1997).

A. J. RICHARDS

Cyclamen hederifolium Sowbread

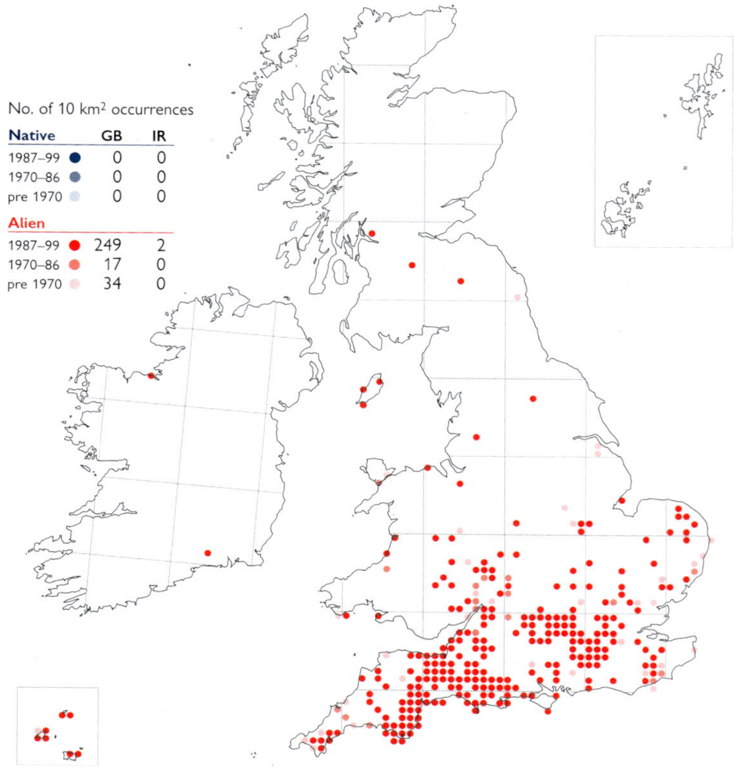

An aestivating long-lived perennial commonly cultivated in parks and gardens and reproducing by seed in dry, shaded positions, often under the canopy and against the roots of large trees. The self-fertile flowers arise laterally from initially leafless tubers at the soil surface, and seeds can be carried hundreds of metres, principally by ants. Lowland.

Neophyte (change +2.75). *C. hederifolium* was introduced into Britain before 1596, and seems to be increasingly frequent in woodlands and hedgerows near to, or as a relic of, former cultivation. It was apparently first recorded from the wild as early as 1597. It may be over-recorded for other naturalised *Cyclamen* species.

Native of S. Europe and W. Turkey.

References: Atlas (202a), Bolòs & Vigo (1995), Grey-Wilson (1997), Mabey (1996), Meusel *et al.* (1978).

A. J. RICHARDS

Lysimachia nemorum Yellow Pimpernel

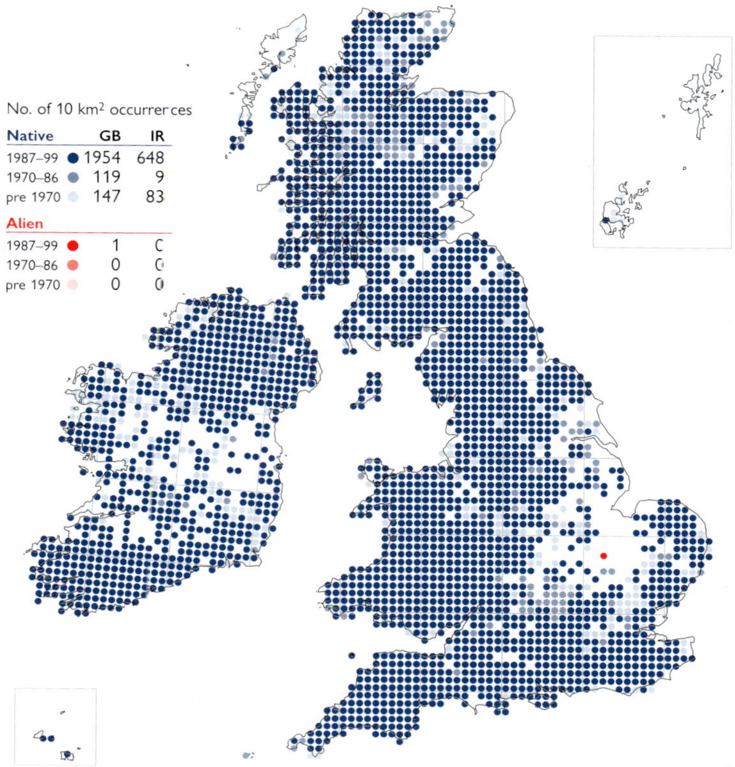

An evergreen perennial of herb-rich, usually deciduous, woodland; also in old hedges, glades, damp grassland, fens and marshes, and shaded gullies and cliffs in upland areas. It is typical of mesic brown earth soils kept relatively open by slope, shade, vernal herbs and disturbance, and avoids places where litter accumulates. 0–820 m (Breadalbanes, Mid Perth).

Native (change –0.46). Mapped as 'all records' in the 1962 *Atlas*, analysis of the database reveals that the widespread decline of *L. nemorum* in S. England has taken place since 1950. Losses have resulted from woodland destruction and replanting with conifers.

Suboceanic Temperate element.

References: Atlas (202b), Hultén & Fries (1986), Meusel *et al.* (1978).

A. J. RICHARDS

Lysimachia nummularia Creeping-Jenny

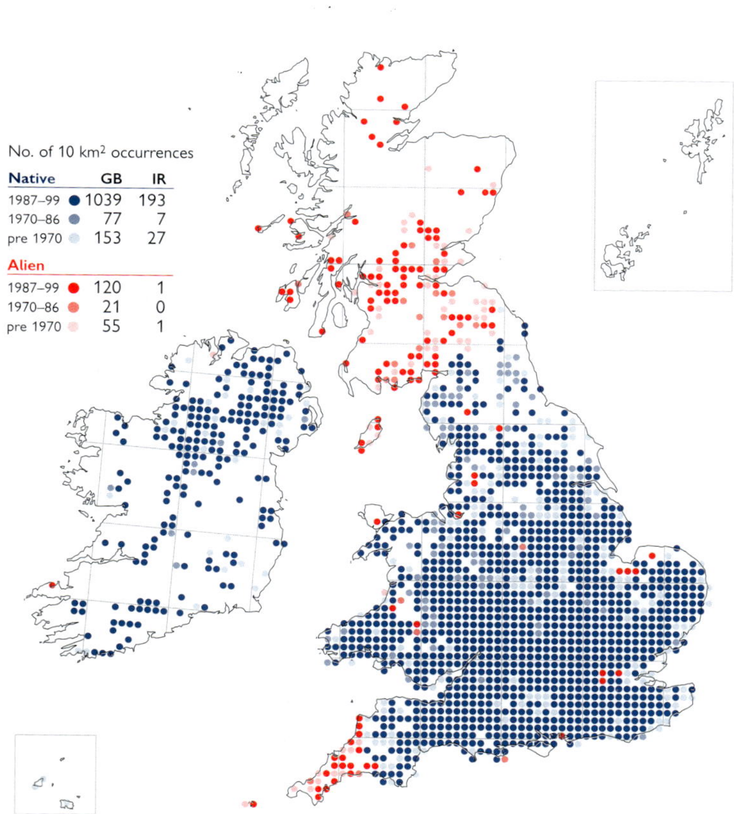

An evergreen perennial of open, damp, often clay-rich soils in shaded woodland and hedges, especially the sides of streams, and damp grassland and path-sides. It can spread rapidly from stem fragments. Seed set is rare, and some clones may be sterile. 0–365 m (S. of Buxton, Derbys.).

Native (change –0.02). *L. nummularia* is probably native in many southern sites, but is usually alien in the north. The limits of its native range are uncertain due to garden escapes, which can be well-naturalised. It was mapped as 'all records' in the 1962 *Atlas*; analysis of the database reveals that most losses have occurred since 1950.

European Temperate element; widely naturalised outside its native range.

References: Atlas (202c), Bittrich & Kadereit (1988), Hultén & Fries (1986), Meusel *et al.* (1978), Simpson *et al.* (1983).

A. J. RICHARDS

Lysimachia vulgaris Yellow Loosestrife

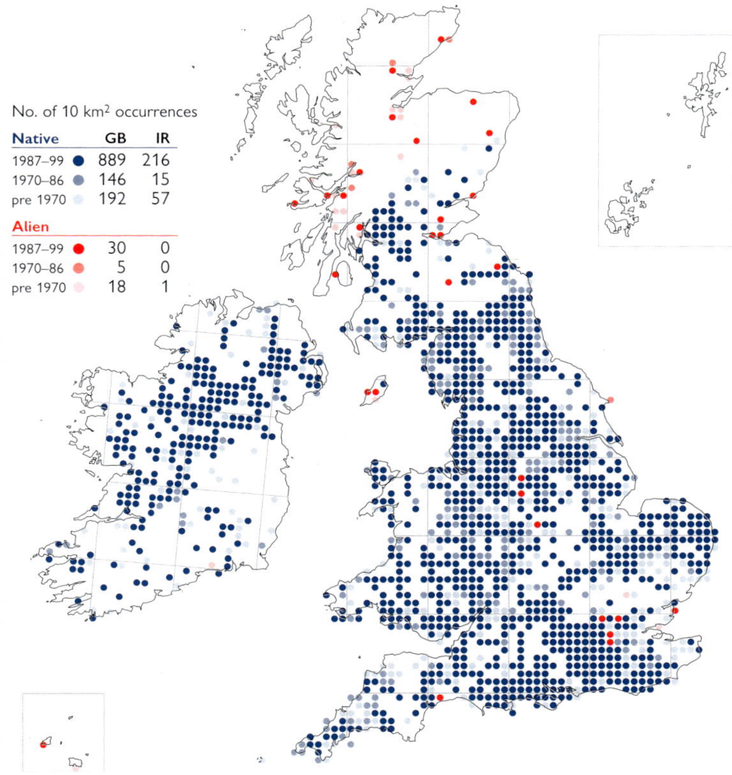

A semi-evergreen perennial spreading vegetatively by very long thin rhizomes to form large colonies in permanently wet places on organic (less commonly, mineral) soils. Habitats include river banks and streamsides, marshes, tall-herb fens, fen-carr, ponds and ditches. It withstands some shade in wet open woodland. Lowland.

Native (change +0.22). This species was mapped as 'all records' in the 1962 *Atlas*. Analysis of the database reveals that most losses, resulting from drainage and the clearance of watercourses, have taken place since 1950. It can, however, readily colonise suitable new sites, perhaps assisted by waterfowl.

Eurasian Temperate element; widely naturalised outside its native range.

References: Atlas (202d), Hultén & Fries (1986), Meusel *et al.* (1978), Proctor *et al.* (1996).

A. J. RICHARDS

Lysimachia ciliata Fringed Loosestrife

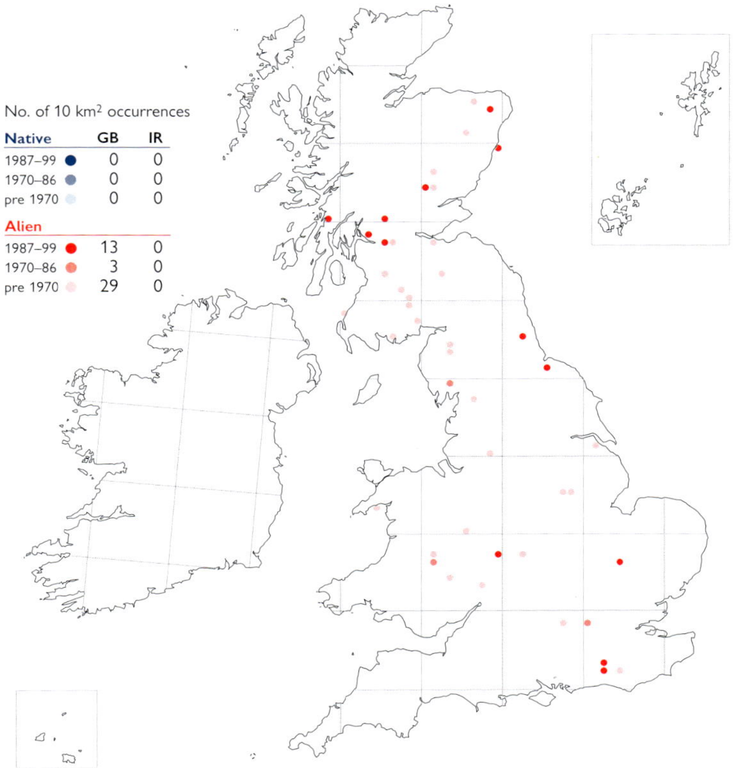

A semi-evergreen perennial herb spreading vegetatively by rhizomes to form sparse colonies, usually in continuously moist, shaded or partially shaded sites. It is found as a garden throw-out or relic of cultivation. Lowland.

Neophyte (change –0.67). *L. ciliata* was cultivated in British gardens by 1732, but is seldom grown today. It was known from the wild by 1849, but seems not to have persisted at many sites and there is little evidence of recent introductions.

Native of N. America.

Reference: Atlas (203a).

A. J. RICHARDS

Lysimachia punctata Dotted Loosestrife

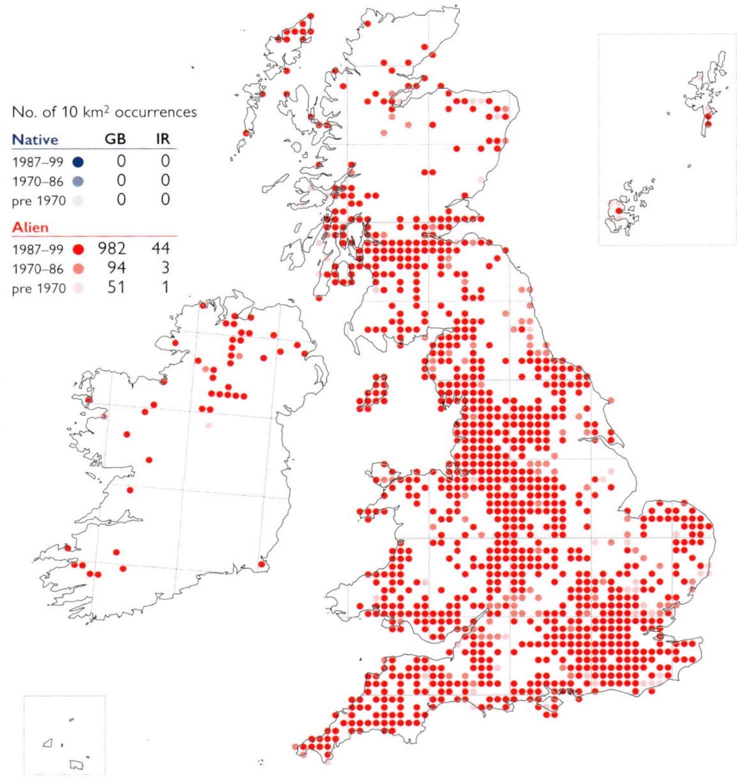

An evergreen perennial herb naturalised in rough grassland and on woodland edges, roadsides and waste ground. It spreads by short, thick rhizomes, but is apparently self-incompatible and rarely sets seed. Generally lowland, but reaching 430 m W. of Nenthead (Cumberland).

Neophyte (change +4.62). This species has been cultivated in British gardens since 1658. Its vigour means it is often thrown out with garden rubbish. It was recorded in the wild in 1853 (Angus), and is much more frequent now than shown in the 1962 *Atlas*.

Native of S.E. & E.C. Europe and W. Asia, east to Iran and the Caucasus; the western *L. punctata* and the eastern *L. verticillaris* do not appear to be specifically distinct.

References: Atlas (203b), Mabey (1996), McAllister (1999), Simpson *et al.* (1983).

A. J. RICHARDS

Lysimachia thyrsiflora Tufted Loosestrife

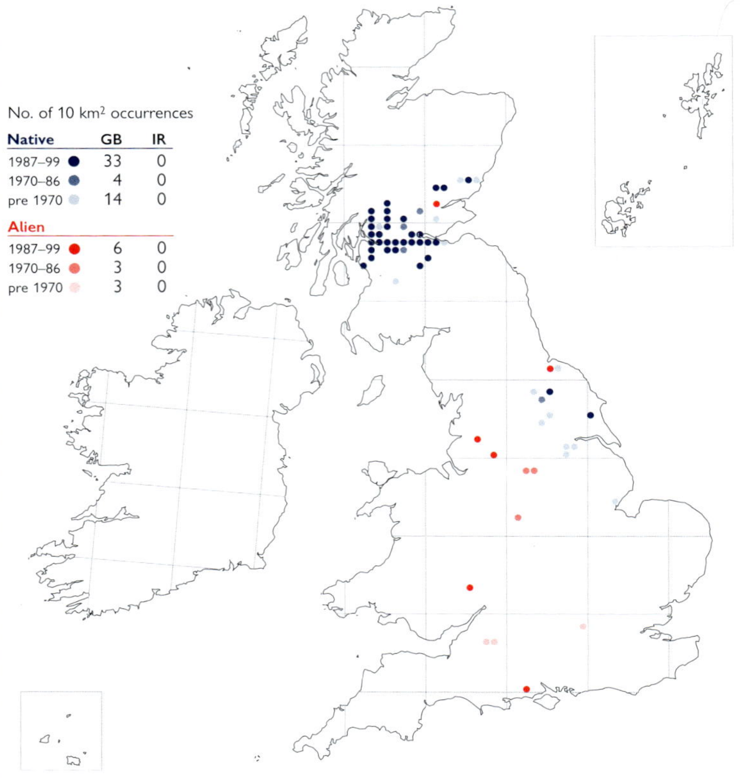

The small colonies of this perennial herb grow in shallow water in permanently wet places. Typical habitats are fens on the flood plains of rivers, lake margins, ditches, canal-sides and colliery subsidence ponds. Generally lowland, reaching 310 m at Lily Loch (Dunbarton).

Native (change +0.38). This species is maintaining its range in its Scottish strongholds, but in Yorkshire it has declined since the 1962 *Atlas*. These losses may be attributed to drainage and the degradation or loss of its habitat. It can be shy-flowering, and may be overlooked in places where it is infrequent.

Circumpolar Boreal-montane element.

References: Atlas (203d), Hultén & Fries (1986), Meusel *et al.* (1978), Stewart *et al.* (1994).

A. J. RICHARDS

Trientalis europaea Chickweed-wintergreen

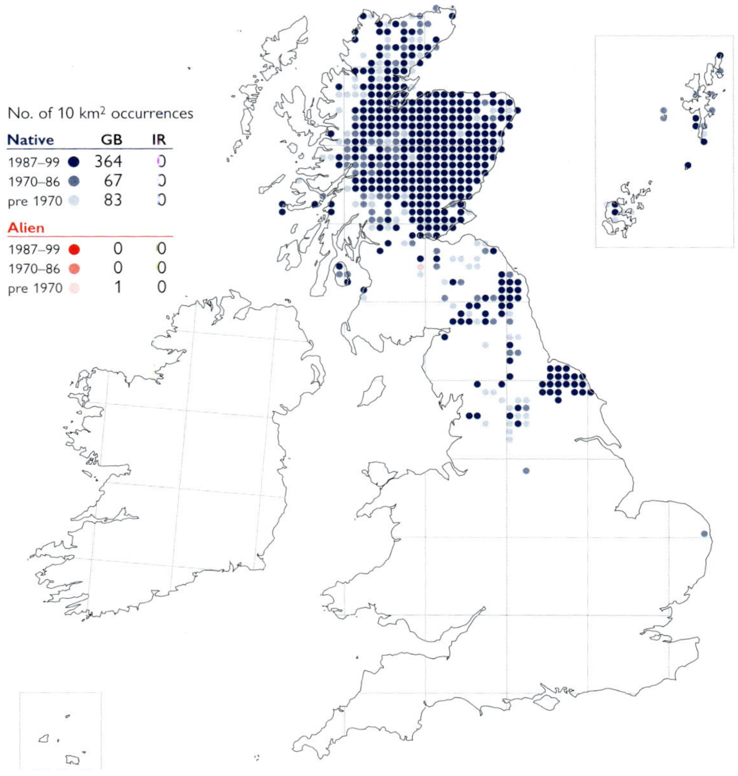

A deciduous perennial herb of moist, acidic and humus-rich, but often fertile, soils in *Betula*, *Pinus* and *Quercus* woodland and on moorland; less commonly on heaths. It is highly localised, colonies often being separated by much apparently suitable ground. Seed-set is rare. It is a good competitor, but a poor colonist. Lowland to 1100 m at Cnap Coire na Spreidhe (Easterness).

Native (change –0.27). The distribution of *T. europaea* is unchanged in Scotland, but there has been a decline in N. England since the 1962 *Atlas* because of woodland clearance and moor-burning. It is now much better recorded on the North York Moors.

Circumpolar Boreal-montane element, but absent from eastern N. America.

References: Atlas (204a), Anderson & Beare (1983), Hultén & Fries (1986), Meusel *et al.* (1978), Piqueras & Klimes (1998).

A. J. RICHARDS

Anagallis tenella Bog Pimpernel

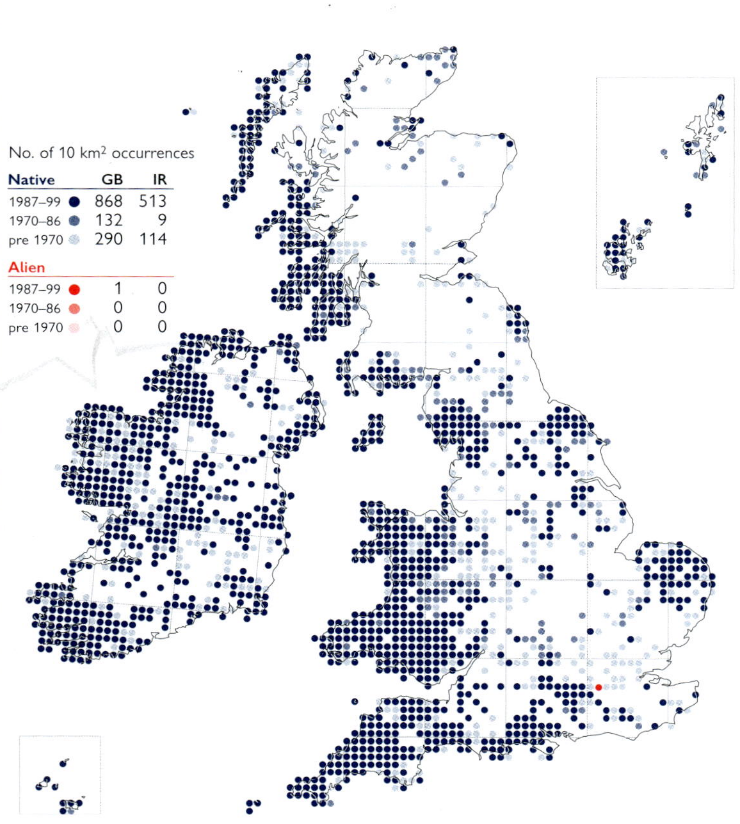

A creeping, evergreen perennial of wet open sites. In S. and E. Britain it is mostly restricted to bare soil or bryophyte mats in calcareous dune-slacks and short-sedge fens, and sometimes on acidic bogs. In the west it also occurs in a variety of soligenous and peaty mires, hillside flushes and rush-pastures, and relying on cattle, sheep or periodic flooding to keep sites open. 0–610 m (Buck of Cabrach, N. Aberdeen).

Native (change –0.54). This species was lost from many 10-km squares in S. & E. England before 1930. Since then, further sites have been lost to grassland improvement, eutrophication and drainage. The distribution is stable elsewhere.

Oceanic Southern-temperate element.

References: Atlas (204b), Dupont (1962), Hultén & Fries (1986), Meusel *et al.* (1978).

A. J. RICHARDS

Anagallis arvensis Scarlet Pimpernel

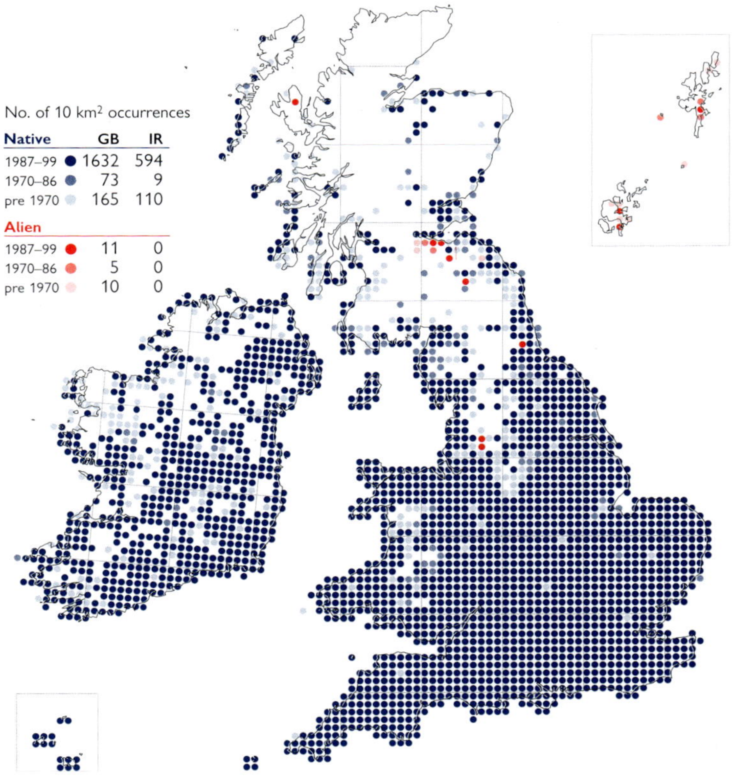

A winter-annual or, occasionally, a short-lived perennial. *A. arvensis* is common in open habitats as an arable or garden weed, and also grows around rabbit warrens and in rocky and bare sites including coastal cliffs, chalk downland, heaths and sand dunes. 0–320 m (Co. Dublin).

Native (change –0.73). *A. arvensis* was mapped as 'all records' in the 1962 *Atlas*. There have been losses in the north of its range, but it is difficult to say when they occurred. The native subsp. *arvensis* occurs throughout the range of the species. Subsp. *foemina* is mapped separately.

Eurosiberian Southern-temperate element, but widely naturalised so that distribution is now Circumpolar Southern-temperate.

References: Atlas (204c), Atlas Supp. (47a, b), Grime *et al.* (1988), Hultén & Fries (1986), Meusel *et al.* (1978).

A. J. RICHARDS

Anagallis arvensis subsp. *foemina* Blue Pimpernel

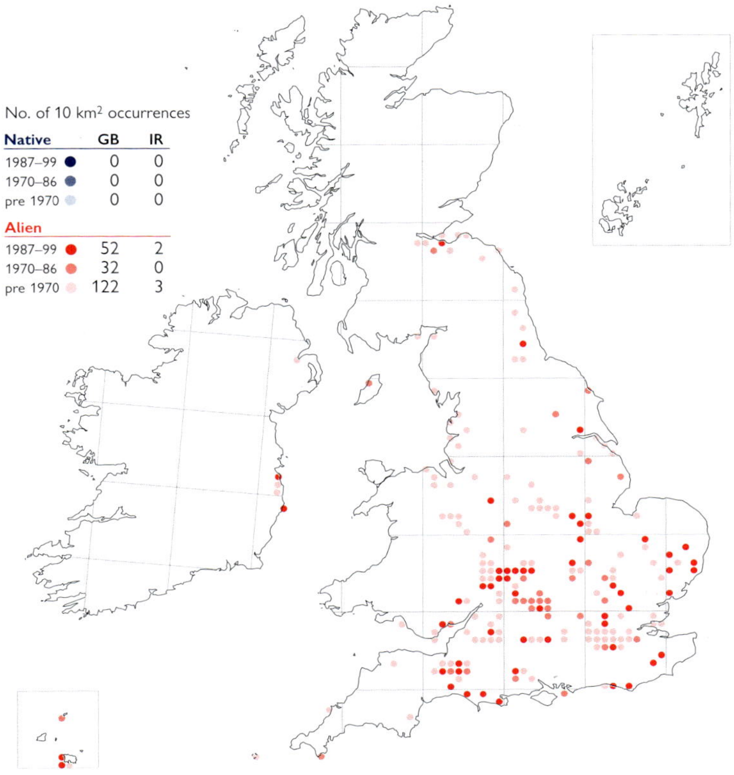

This subspecies is normally found as an arable weed and appears to be absent from the semi-natural habitats in which subsp. *arvensis* occurs. Lowland.

Archaeophyte. The blue-flowered subsp. *foemina* and the blue-flowered variant of subsp. *arvensis*, forma *azurea* (also mapped by Perring & Sell, 1968), have long been confused. Subsp. *foemina* may be declining, probably due to more intensive weed control in arable fields, but the 1968 map only included confirmed records so it is difficult to assess changes in distribution. Some recent records might be of bird-seed aliens.

As an archaeophyte subsp. *foemina* has the Eurosiberian Southern-temperate distribution of the species, but its distribution is more southerly than subsp. *arvensis*.

References: Atlas Supp. (47b), Meusel *et al.* (1978).

A. J. RICHARDS

Anagallis minima Chaffweed

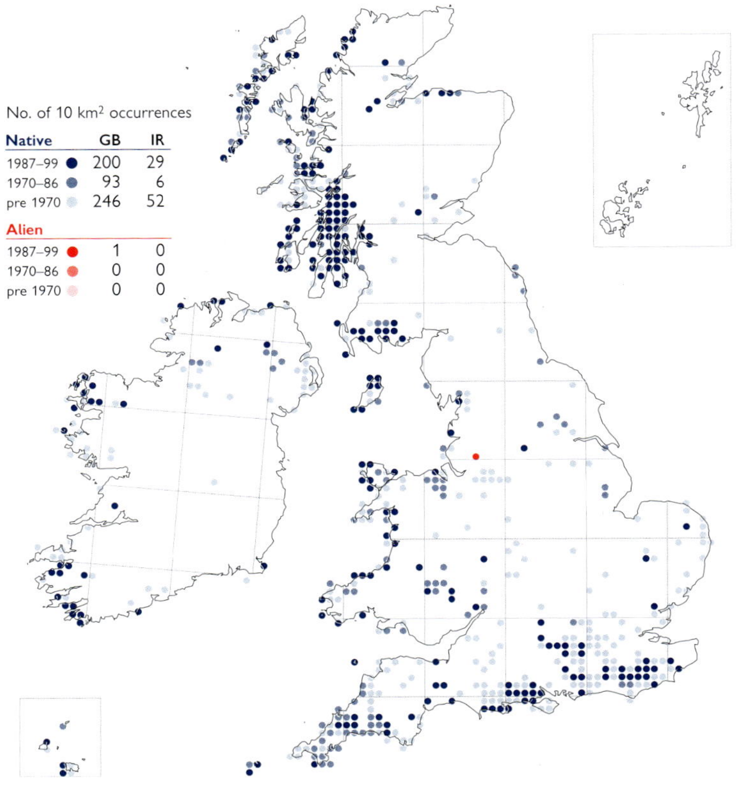

An annual of open places on damp, sandy sites, often near the sea and usually on acidic soils. Habitats include sand dunes, sandy cliffs, along paths and tracks on heathland, and in forest rides. It is a poor competitor where grazing is relaxed and general disturbance ceases. Lowland.

Native (change –1.16). Many sites, especially inland ones, were lost before 1930. In England these losses have continued or accelerated, possibly through changes in heathland management, but the distribution is stable elsewhere. It is also very easily overlooked, and is probably under-recorded.

European Temperate element; also in N. America and widely naturalised outside its native range.

References: Atlas (204d), Hultén & Fries (1986), Meusel *et al.* (1978), Salisbury (1969b).

A. J. RICHARDS

Glaux maritima Sea-milkwort

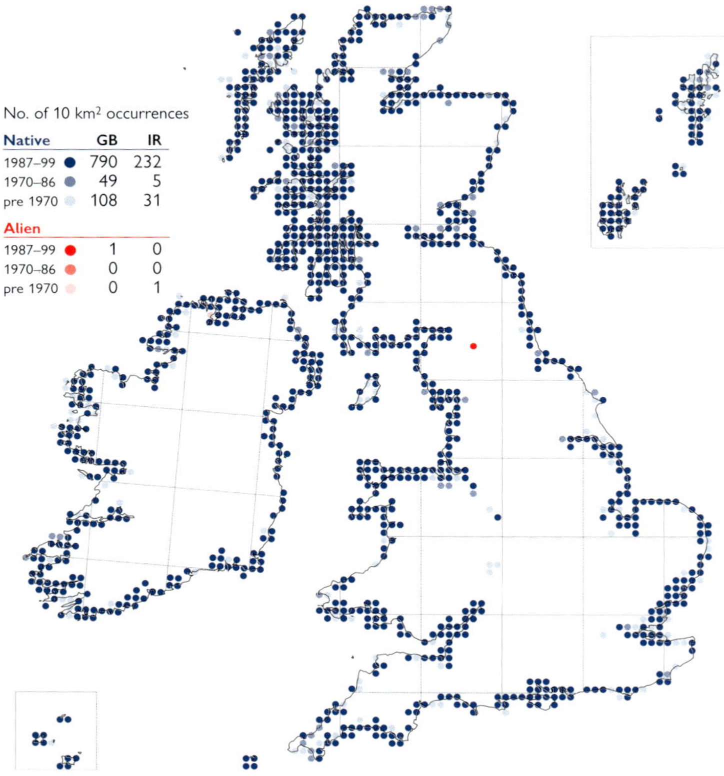

A perennial herb typically forming dense colonies on moist saline soils. Habitats include saltmarshes, strandlines, damp shingle, wet sand, brackish dune-slacks, aerobic mud and spray-drenched rock crevices. It is a poor competitor which often forms pure stands, but also grows in open communities with other halophytes. Lowland, but persisting as an introduction at 570 m (Hartside, Cumberland).

Native (change –0.41). There has been no change in the overall distribution of this species since the 1962 *Atlas*.

Circumpolar Boreo-temperate element.

References: Atlas (205a), Hultén & Fries (1986), Jerling (1988), Jerling & Elmgren (1996), Meusel *et al.* (1978).

A. J. RICHARDS

Samolus valerandi Brookweed

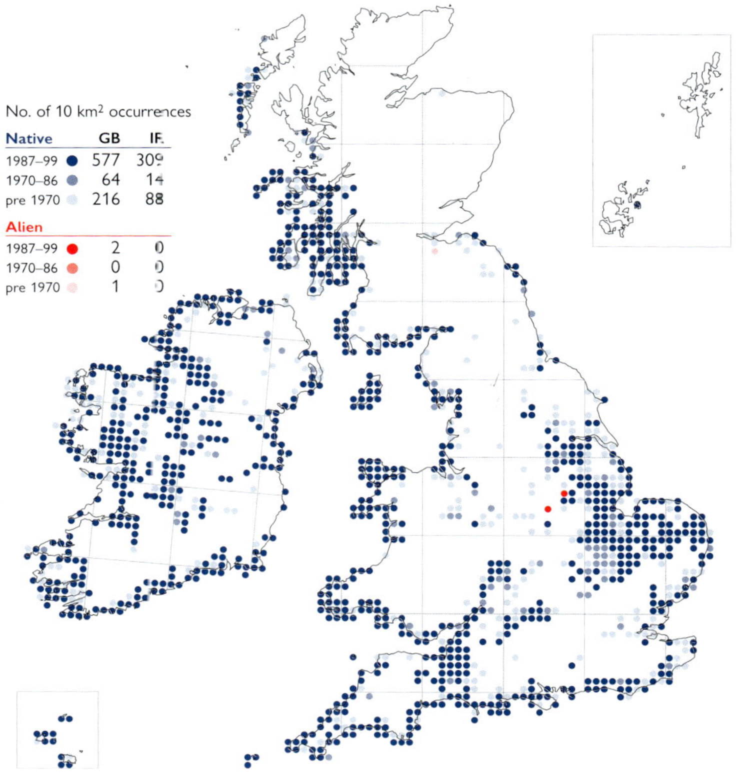

A deciduous short-lived perennial found in small colonies by permanently wet and often seasonally flooded springs, flushed sea-cliffs, puddles, ditches, lagoons and lake shores. It is limited to very open mesic, often calcareous or somewhat saline soils. Populations are often impermanent, colonising the small and transient specialised habitats which change as spring lines and water levels shift. Individuals are usually monocarpic, but may take several years to flower. Lowland.

Native (change –0.42). This species was mapped as 'all records' in the 1962 *Atlas*. Analysis of the database reveals that most of the losses, which tend to be inland and may result from drainage, have taken place since 1950.

Circumpolar Southern-temperate element.

References: Atlas (205b), Hultén & Fries (1986), Meusel *et al.* (1978).

A. J. RICHARDS

Philadelphus coronarius Mock-orange

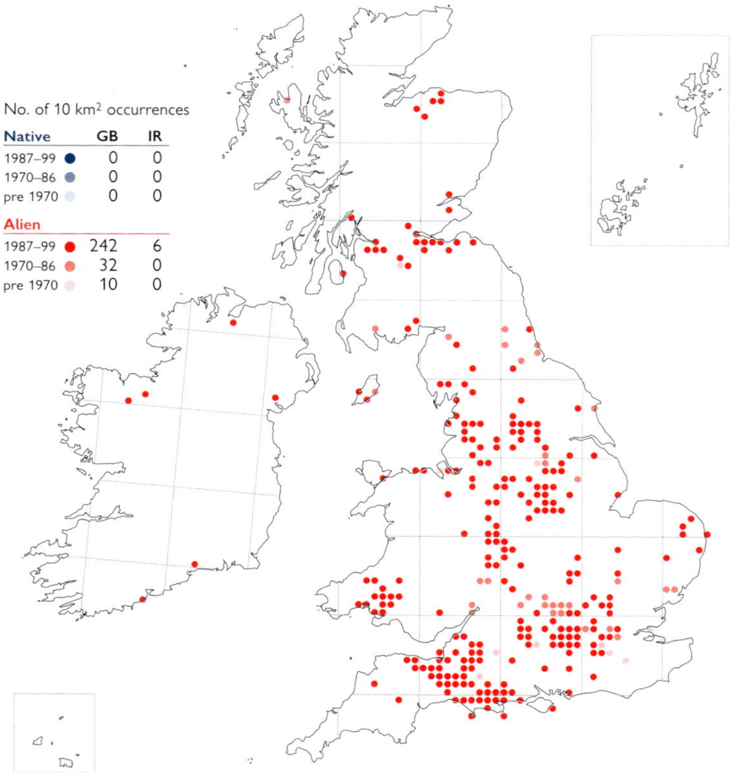

A large deciduous shrub which is occasionally planted in woodland, but is more frequent as a garden throw-out or relic of cultivation in hedges, and on roadsides and waste ground. Reproduction by seed is rare. Lowland.

Neophyte. *P. coronarius* was cultivated in our area by 1596 and is commonly grown in gardens for its scented flowers. It was known in the wild by 1919. Several other species and hybrids are frequently grown, and identification is not straightforward; some records mapped here may be erroneous (Wurzell, 1991).

Apparently native in a few localities in Austria, N. Italy and the Caucasus.

References: Bean (1976), Jalas & Suominen (1999).

T. D. DINES

Philadelphus × *virginalis* (? *P. coronarius* × *P. microphyllus* × *P. pubescens*)

Hairy Mock-orange

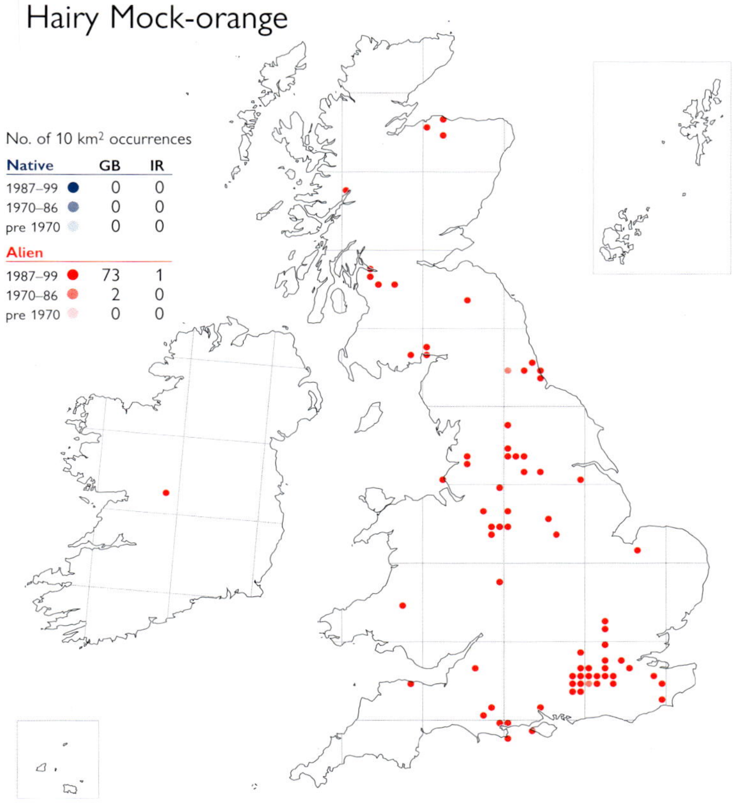

A medium to large deciduous shrub occurring as a garden throw-out or relic of cultivation in hedges and on roadsides and waste ground. Rarely, it is planted in woodland. Lowland.

Neophyte. *P.* × *virginalis*, which was raised in about 1909, is now more commonly grown in gardens for its scented flowers than *P. coronarius*, but this is not reflected in the number of records, suggesting that the hybrid has not yet become as frequent in the wild, or has been under-recorded.

A hybrid of garden origin.

References: Bean (1976), Coats (1963).

T. D. DINES

Escallonia macrantha Escallonia

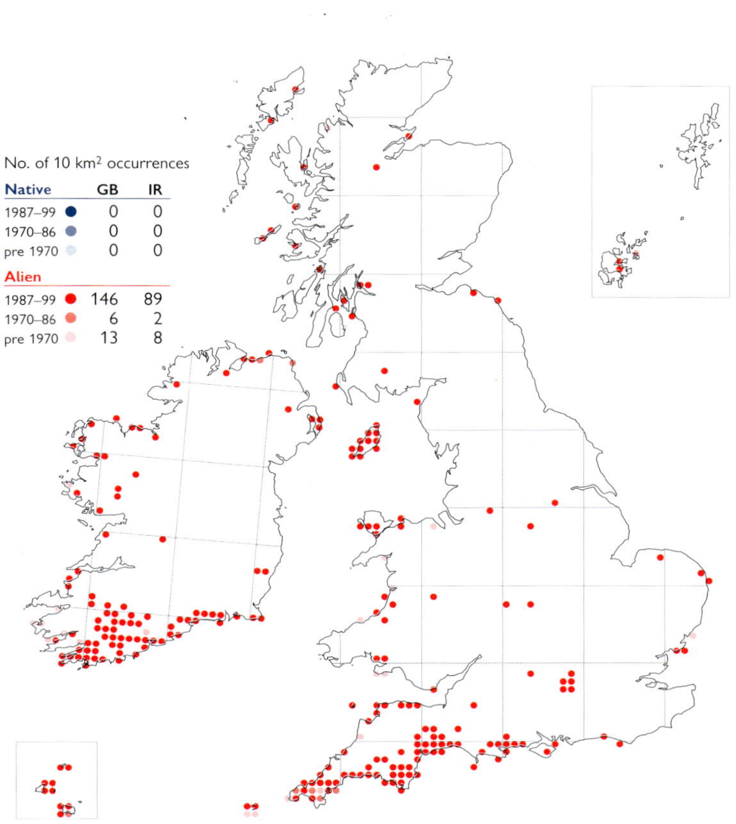

An evergreen shrub found naturalised in woodland and on cliffs, banks, roadsides and waste ground, and as a persistent relic of cultivation. Reproduction is by seed. Lowland.

Neophyte (change +2.09). *E. macrantha* was introduced to Britain in about 1847 and is often planted as a hedge, wind-break or wall covering, especially in coastal areas. It was known from the wild by at least 1905, and has now spread considerably in S.W. Britain and Ireland, showing a marked increase since the 1962 *Atlas*.

Native of S. America (Argentina, Chile).

References: Atlas (140b), Bean (1973), Coats (1963).

J. M. CROFT

Ribes rubrum Red Currant

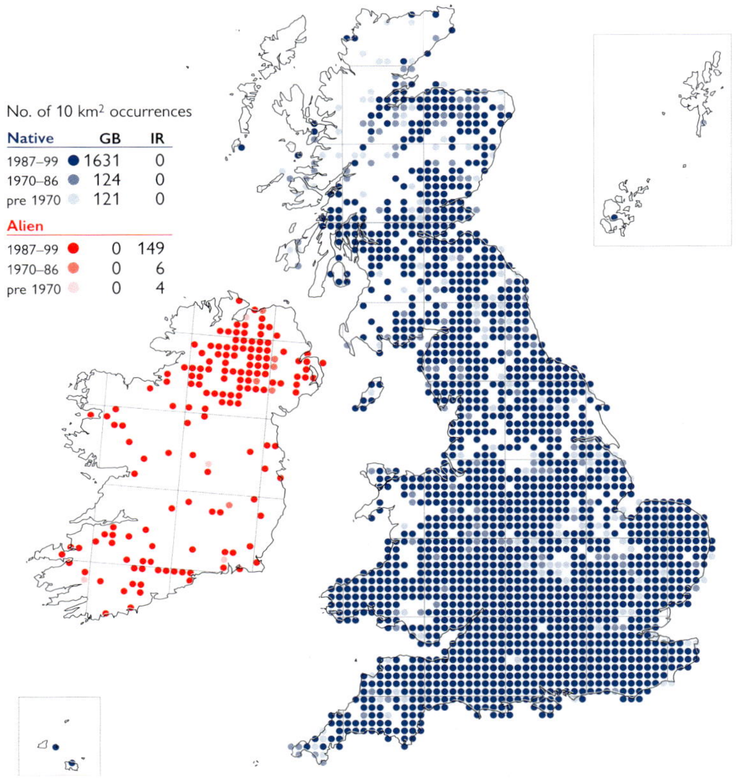

A shrub of woods, hedges and fen-carr. Generally lowland, but reaching 455 m in Teesdale (N.W. Yorks.).

Native or alien (change +1.79). This species was recorded in Britain by 1568. Small fruited plants in fen-carr and by streams in woods may be native. However, this species is widely naturalised; large fruited plants near old habitations are clearly relics of cultivation. The distinction between alien and native records is now so confused that all records in Britain are mapped as if they are native; it is alien in Ireland. The increase since the 1962 *Atlas* may partly be due to improved recording.

Suboceanic Temperate element; widely naturalised outside its native range.

References: Atlas (140c), Hultén & Fries (1986), Jalas & Suominen (1999), Mabey (1996), Roach (1985), Simpson (1982).

J. M. CROFT

Ribes spicatum Downy Currant

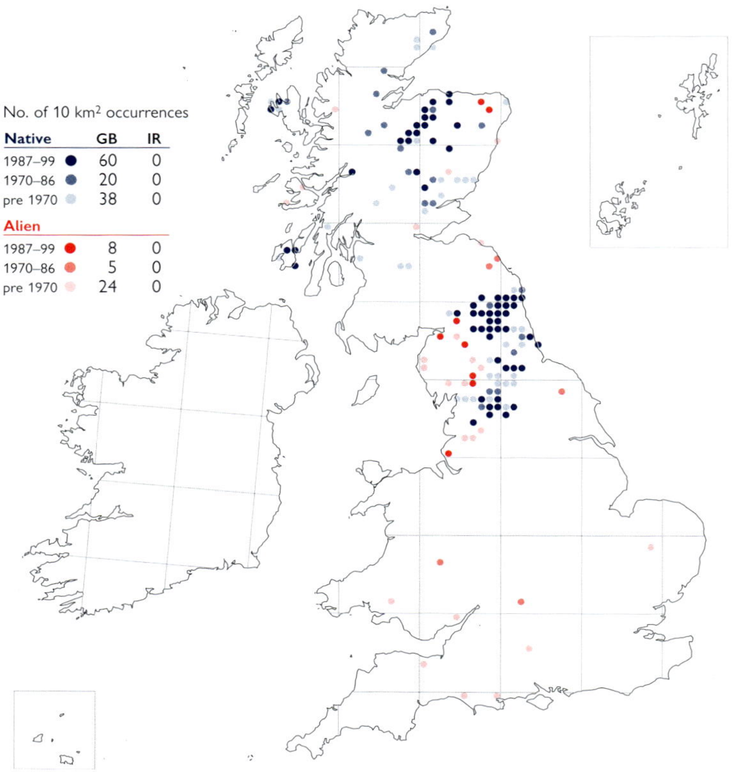

No. of 10 km² occurrences

Native		GB	IR
1987–99	●	60	0
1970–86	●	20	0
pre 1970	●	38	0
Alien			
1987–99	●	8	0
1970–86	●	5	0
pre 1970	●	24	0

A shrub of northern limestone woods, streamsides, ravines and deep grikes in limestone pavement. Generally lowland, but reaching 465 m above Malham (Mid-W. Yorks.).

Native (change −0.12). The native distribution of this species is difficult to ascertain, as cast-outs from cultivation have occasionally become established in the wild. Because identification of this species can only be certain when flowers are present, it has been under-recorded in the past. This is probably still the case, as the flowering season is early.

European Boreo-temperate element, with a continental distribution in W. Europe.

References: Atlas (140d), Jalas & Suominen (1999), Stewart *et al.* (1994).

J. M. CROFT

Ribes nigrum Black Currant

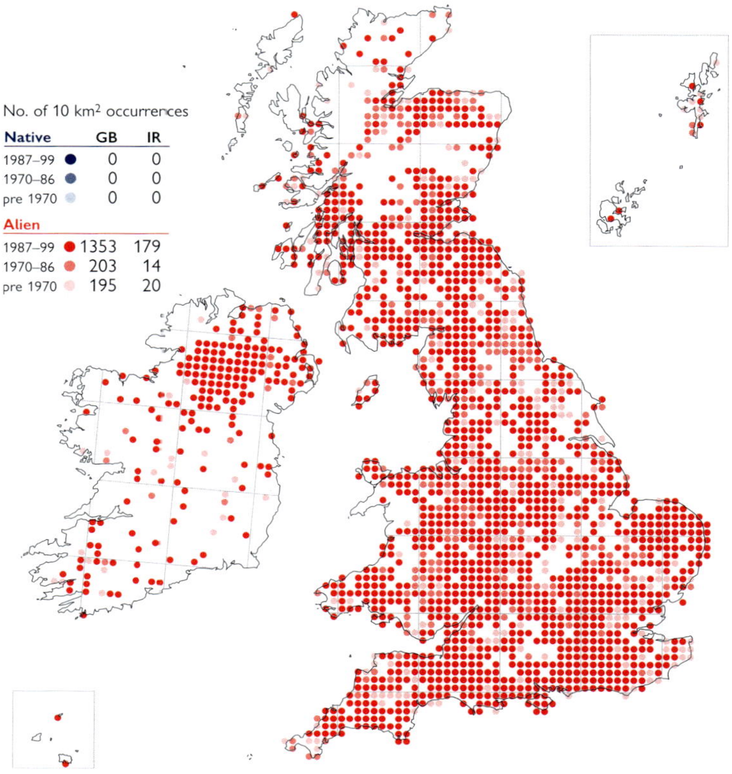

No. of 10 km² occurrences

Native		GB	IR
1987–99	●	0	0
1970–86	●	0	0
pre 1970	●	0	0
Alien			
1987–99	●	1353	179
1970–86	●	203	14
pre 1970	●	195	20

A shrub of damp woods, hedgerows and shaded streamsides, often near habitation. 0–390 m (Ingleborough, Mid-W. Yorks.).

Neophyte (change +1.76). *R. nigrum* does not appear to have been cultivated until shortly after 1600 when plants were imported from Holland. It was first recorded in the wild in 1660 and occurs as a naturalised escape throughout Britain, although some have considered it to be native in fen-carr and wet woodlands in East Anglia (Beckett *et al.*, 1999).

Eurosiberian Boreo-temperate element; widely naturalised outside its native range.

References: Atlas (141a), Halliday (1997), Hultén & Fries (1986), Jalas & Suominen (1999), Meusel *et al.* (1965), Roach (1985).

J. M. CROFT

Ribes sanguineum Flowering Currant

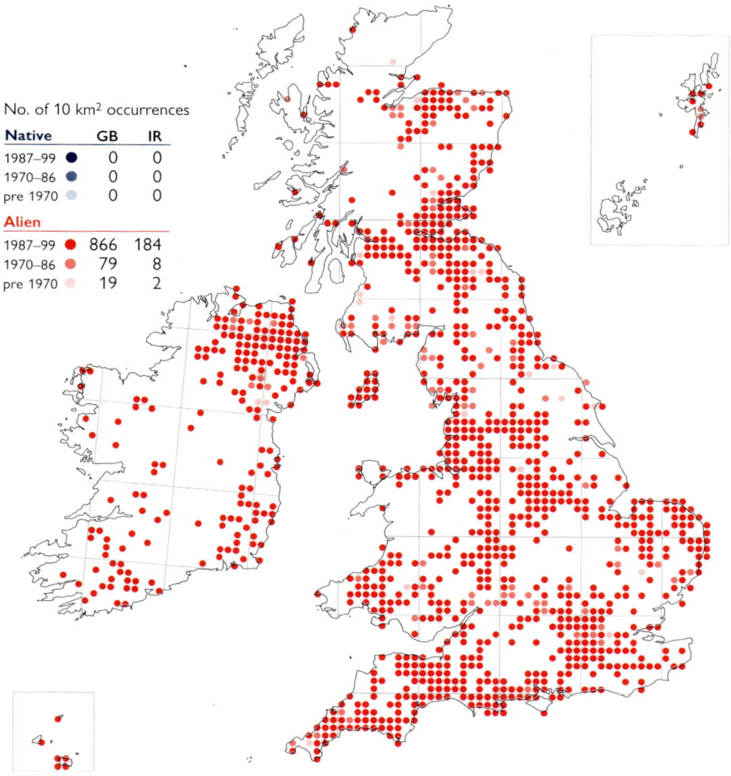

No. of 10 km² occurrences

Native		GB	IR
1987–99	●	0	0
1970–86	●	0	0
pre 1970	●	0	0
Alien			
1987–99	●	866	184
1970–86	●	79	8
pre 1970	●	19	2

This shrub is naturalised in a variety of habitats, including woods, verges, hedges and waste ground. Lowland, reaching 300 m at Brough (Westmorland).

Neophyte. *R. sanguineum* was introduced to Britain in 1826, and is widely grown in gardens and for ornamental hedging. It was known from the wild by 1916. Although changes in its distribution cannot easily be assessed, it is probably increasing, but is undoubtedly under-recorded in some areas.

Native of western N. America.

Reference: Bean (1980).

J. M. CROFT

Ribes alpinum Mountain Currant

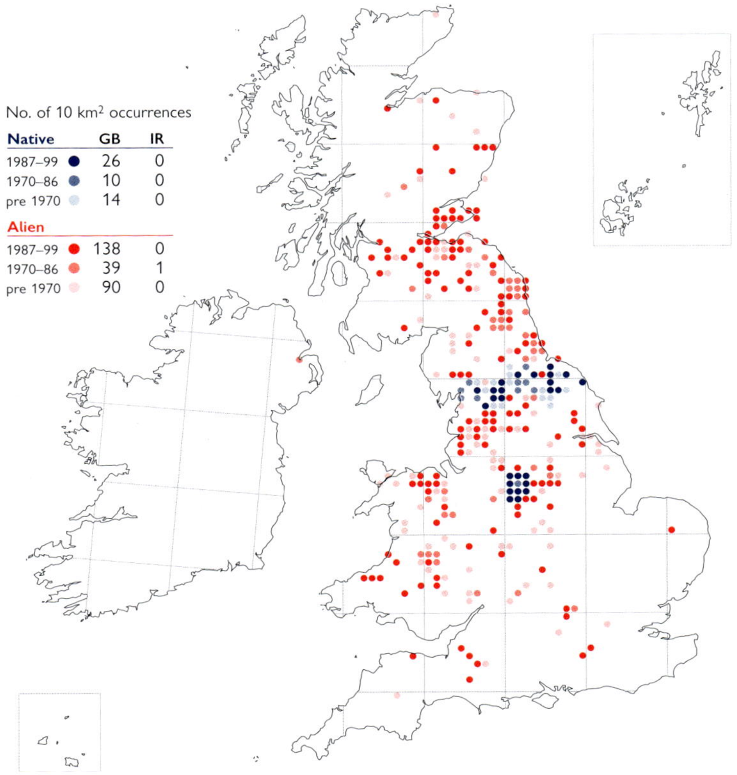
No. of 10 km² occurrences
Native GB IR
1987–99 26 0
1970–86 10 0
pre 1970 14 0
Alien
1987–99 138 0
1970–86 39 1
pre 1970 90 0

A dioecious shrub of limestone woods, rocky hedgerows and streamsides, often trailing over small cliffs and steep rocks in shaded places. It is also grown in gardens, and is found naturalised on roadsides, waste land and as a relic of cultivation. Generally lowland, reaching 365 m at Wormhill (Derbys.).

Native (change +0.45). Even within the areas where it is considered native, the distribution of this species has been obscured by escapes from cultivation as the species is widely planted for hedging or ornament. *R. alpinum* is better recorded than for the 1962 *Atlas*, and the distribution is stable.

European Boreal-montane element, but absent from the Boreal zonobiome.

References: Atlas (141b), Hultén & Fries (1986), Jalas & Suominen (1999), Meusel *et al.* (1965), Stewart *et al.* (1994).

J. M. CROFT

Ribes uva-crispa Gooseberry

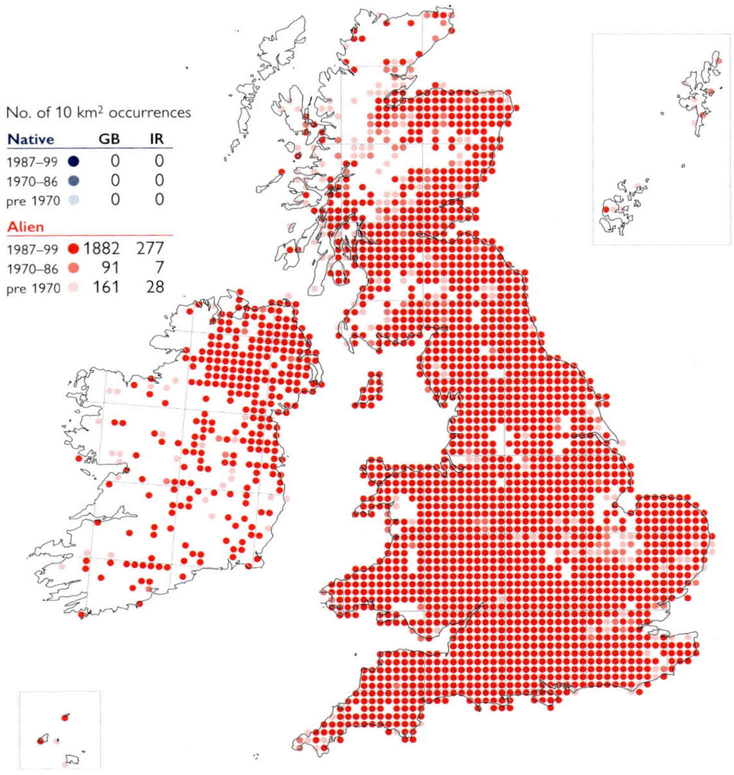
No. of 10 km² occurrences
Native GB IR
1987–99 0 0
1970–86 0 0
pre 1970 0 0
Alien
1987–99 1882 277
1970–86 91 7
pre 1970 161 28

A spiny, much-branched shrub, found as a garden escape and throw-out in deciduous woodland, hedges and scrub. It also occurs as a relic of cultivation. It is readily dispersed by birds. 0–380 m at Ingleborough (Mid-W. Yorks.).

Neophyte (change +0.72). *R. uva-crispa* has been grown in British gardens since the 13th century but was not recorded in the wild until 1763 (Cambs.). Its distribution has increased since the 1962 *Atlas*, probably due to continued escape of plants from gardens.

European Temperate element; widely naturalised outside its native range.

References: Atlas (141c), Hultén & Fries (1986), Jalas & Suominen (1999), Mabey (1996), Roach (1985).

J. M. CROFT

Crassula tillaea Mossy Stonecrop

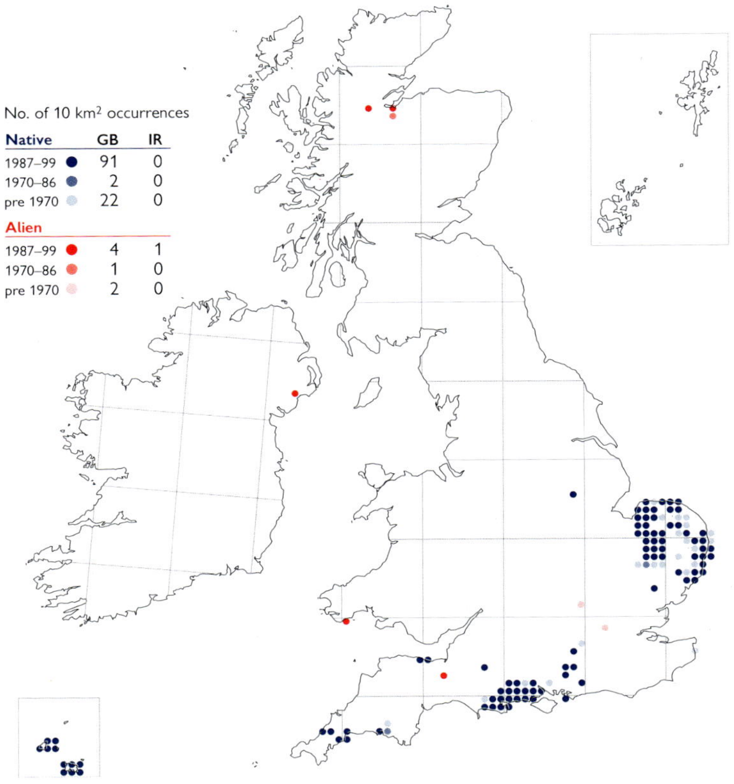
No. of 10 km² occurrences
Native GB IR
1987–99 91 0
1970–86 2 0
pre 1970 22 0
Alien
1987–99 4 1
1970–86 1 0
pre 1970 2 0

A tiny annual, growing on bare, often compacted, sandy or gravelly ground. It is often found on rutted tracks, paths and other areas where the ground is kept open by disturbance and periodic flooding. It can withstand only minimal competition from other vegetation. Lowland.

Native (change +0.86). The British range of this species appears to be expanding. Since the 1962 *Atlas* it has become much more frequent in S.W. England and has colonised forestry rides in N.E. Scotland. In Cornwall, where it was not known until 1988, it is known from several car- and caravan-parks on sandy sites.

Submediterranean-Subatlantic element.

References: Atlas (135c), Bolòs & Vigo (1984), French *et al.* (1999), Jalas & Suominen (1999), Stewart *et al.* (1994).

J. M. CROFT

Crassula aquatica Pigmyweed

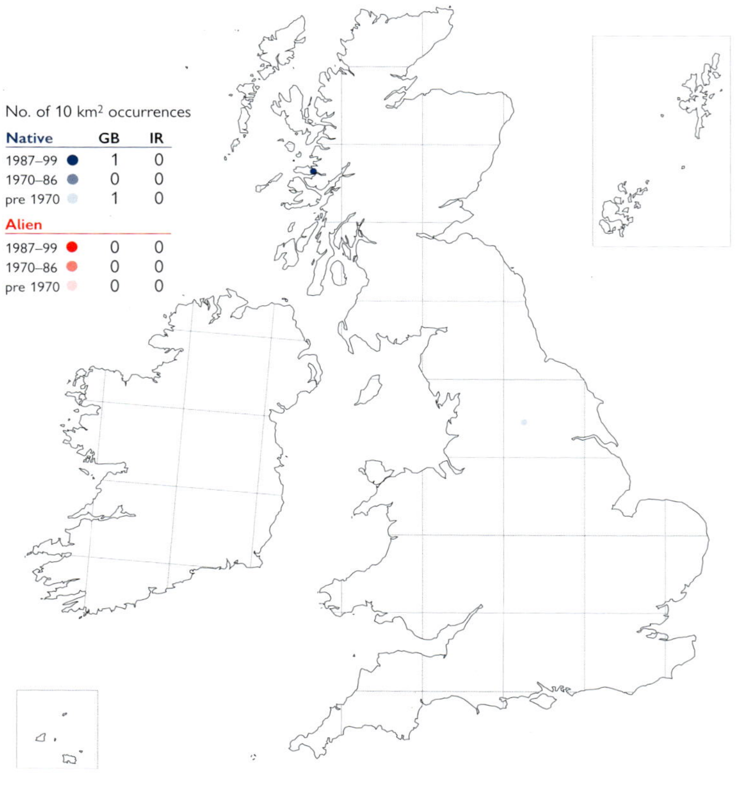

A small, slender annual which is currently known from only one locality. It grows in shallow water or on wet mud exposed by fluctuating water levels at the side of the River Shiel (Westerness), and on damp peaty soil where the riverside vegetation has been disturbed. The population is well-established but varies in size annually. Lowland.

Native or alien. *C. aquatica* formerly occurred at Adel Dam (Mid-W. Yorks.) where it was discovered in 1921, was last seen in 1938 and had disappeared by 1949. It was discovered by the River Shiel in 1969.

Circumpolar Temperate element, with a continental distribution in W. Europe and a disjunct distribution elsewhere.

References: Atlas (135c), Hultén & Fries (1986), Jalas & Suominen (1999), Meusel *et al.* (1965), Preston & Croft (1997), Wigginton (1999).

J. M. CROFT

Crassula helmsii New Zealand Pigmyweed

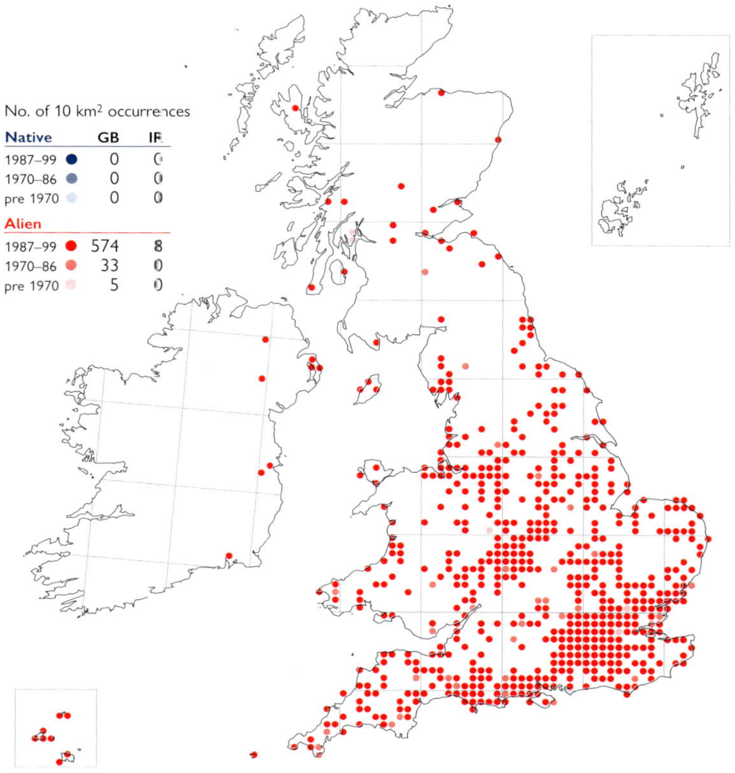

This perennial herb grows submerged in sheltered waters up to 3 metres deep or as an emergent on damp ground. It grows on soft substrates in a variety of habitats, including ponds, lakes, reservoirs, canals and ditches and can tolerate a wide range of water chemistry. It can form dense, virtually pure stands. Lowland.

Neophyte. This aggressively colonising species was first cultivated in Britain in 1927 and was discovered in the wild in 1956 (Greensted, Essex). Since the late 1970s it has spread rapidly north and west, and was first recorded in Ireland in 1994 at Gosford Castle (Co. Armagh).

Native of Australia and New Zealand.

References: Jalas & Suominen (1999), Preston & Croft (1997).

J. M. CROFT

Umbilicus rupestris Navelwort

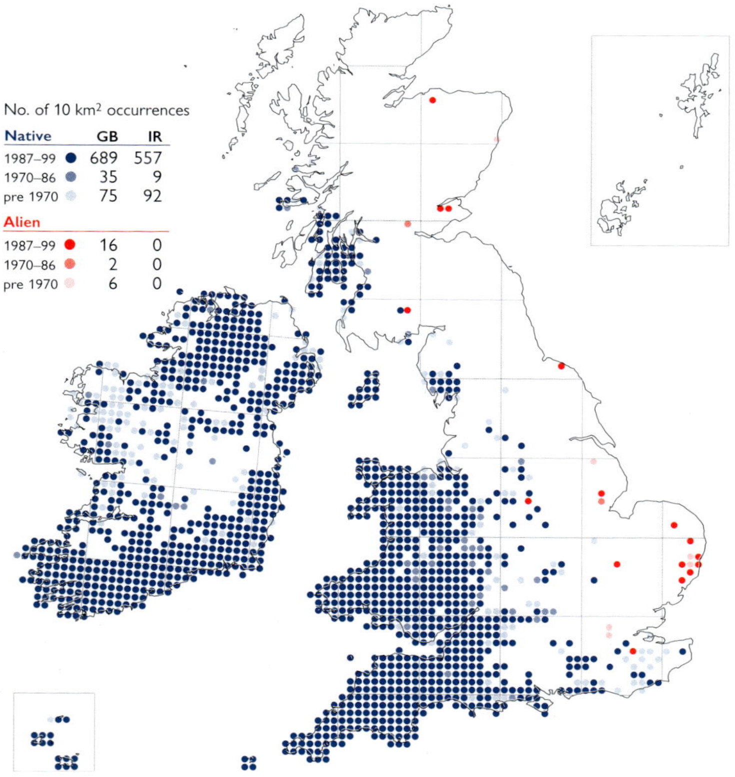

A perennial herb, growing on walls, in rock crevices and on stony hedge banks, mainly on acidic substrates. In Cornwall it has even been seen growing as an epiphyte on the boughs of large trees. 0–550 m (Berwyn Mountains near Pistyll Rhaiadr, Monts.).

Native (change –0.12). This species is still predominantly western and shows little change in distribution since the 1962 *Atlas*, in which it was mapped as 'all records'. Some records in E. England are introductions, as for example in Norfolk where it is assumed to have been planted.

Mediterranean-Atlantic element.

References: Atlas (135d), Bolòs & Vigo (1984), French *et al.* (1999), Jalas & Suominen (1999).

J. M. CROFT

Sempervivum tectorum House-leek

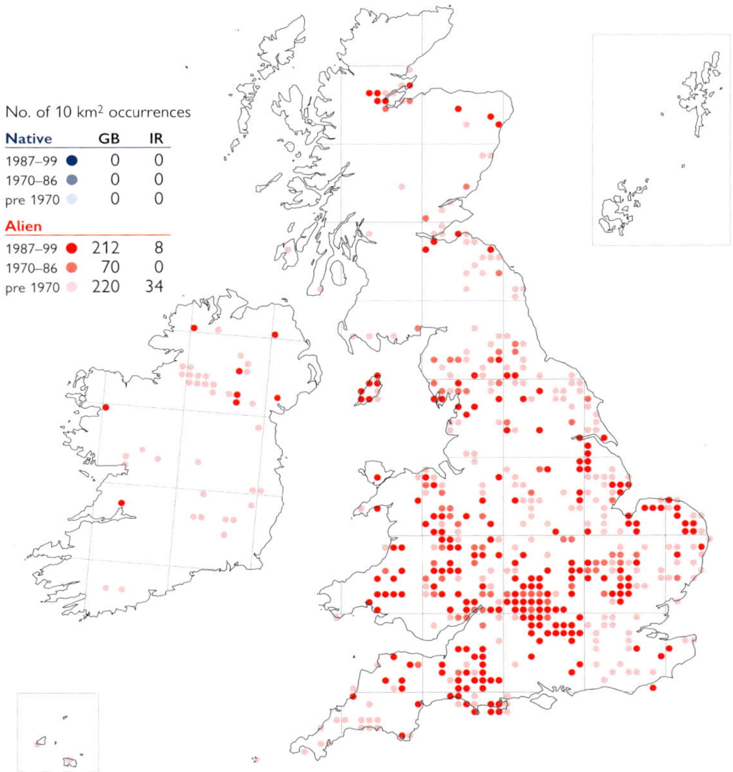

A long-lived, evergreen perennial, planted and more or less naturalised on tiled and thatched roofs, old walls, gate pillars and porches, and in churchyards. It is also occasionally found on stabilised sand dunes. Lowland.

Neophyte. *S. tectorum* has been grown in gardens since at least 1200 (Harvey, 1981), and was often planted on porches and roofs as a supposed protection against fire, lightning and thunderbolts. It was known in the wild by 1629. Some county floras mention a marked decline since the 19th century, especially where old cottages and walls have been pulled down and thatch has been replaced by slate.

Native of the mountains of C. & S. Europe.

References: French *et al.* (1999), Grigson (1955), Jalas & Suominen (1999), Le Sueur (1984).

J. M. CROFT

Sedum rosea Roseroot

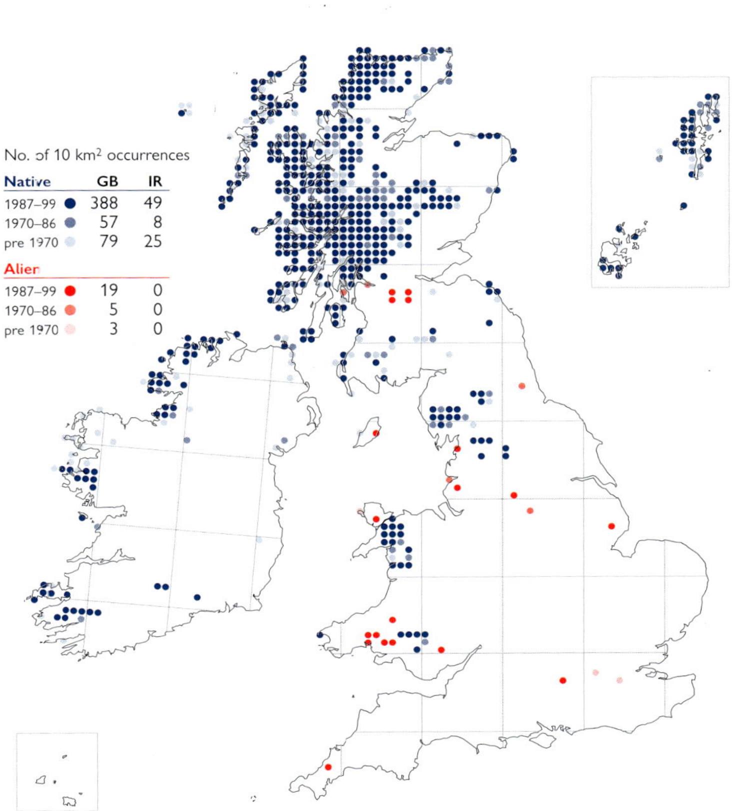

A rhizomatous perennial herb which grows on sea-cliffs and in mountains in rock crevices and on moist rock ledges. Very rarely, in W. Ireland, it occurs on coastal limestone pavement. In montane habitats it usually occupies sites which are at least slightly base-enriched. Although descending to sea level in N.W. Britain and Ireland, it is usually found above 300 m, reaching 1160 m on Ben Lawers (Mid Perth).

Native (change –0.41). There has been little change in distribution of *S. rosea* since the 1962 *Atlas*. It was previously grown as a source of perfume extracted from the roots, and is occasionally recorded as a garden escape.

Circumpolar Arctic-montane element.

References: Atlas (133a), Hultén & Fries (1986), Jalas & Suominen (1999), Meusel *et al.* (1965), Woodward (1975), Woodward & Pigott (1975).

J. M. CROFT

Sedum spectabile Butterfly Stonecrop

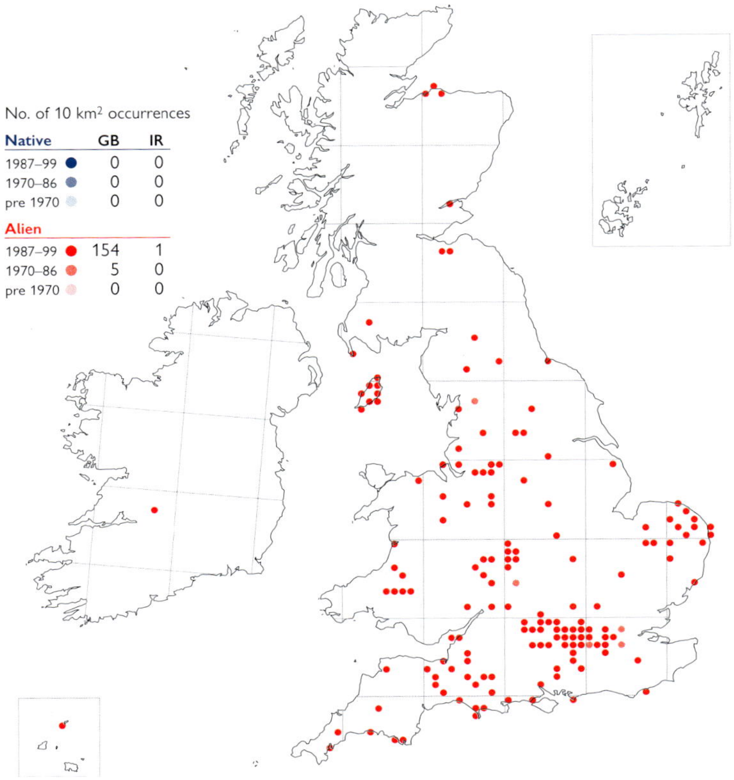

A perennial herb found well-naturalised in woodland and on roadsides, waste ground and rubbish tips, having originated as a garden throw-out. Lowland.

Neophyte. *S. spectabile* was first cultivated in Britain in 1868, and it has been known at Nettleton (N. Wilts.) since 1930. It is probably increasing in the wild as it is widely grown in gardens to attract butterflies but can become large and is frequently discarded when it outgrows the available space.

Native of China and Japan.

J. M. CROFT

Sedum telephium Orpine

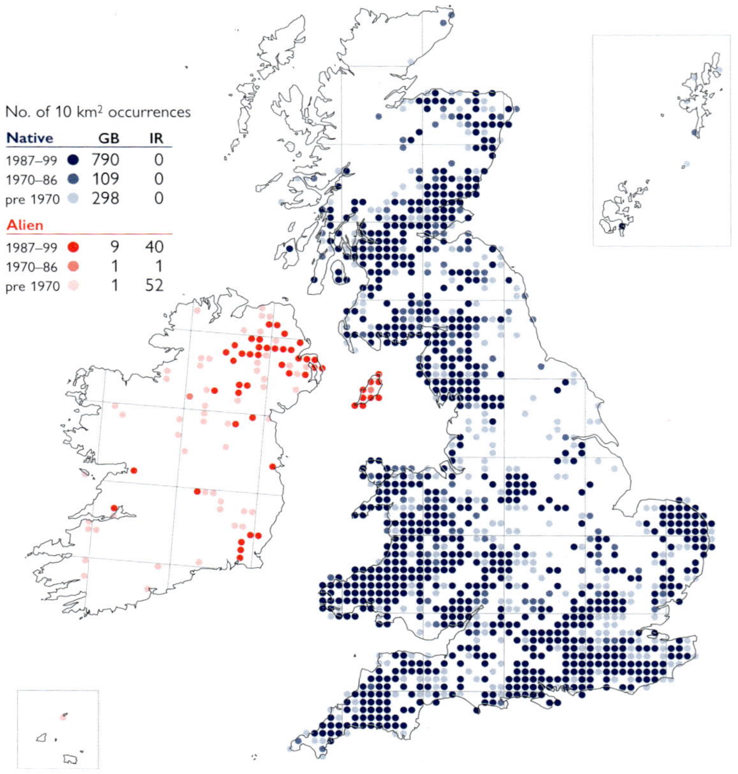

A perennial herb, found on wood-borders, hedge banks, roadsides, rocky banks and in limestone pavement, often in very small but very persistent colonies. It also occurs as an uncommon ancient woodland plant, but sometimes fails to flower in this habitat. 0–455 m (Falcon Clints, Co. Durham).

Native (change −0.34). Although native in some habitats, many colonies have become naturalised near houses as this species is grown in gardens, and the native range is now hopelessly obscured by such escapes; all British records are mapped as if they are native.

Eurasian Temperate element; widely naturalised outside its native range.

References: Atlas (133b), Hultén & Fries (1986), Jalas & Suominen (1999), Mabey (1996), Meusel *et al.* (1965), Rackham (1975, 1980), Woodward (1975), Woodward & Pigott (1975).

J. M. CROFT

Sedum spurium Caucasian-stonecrop

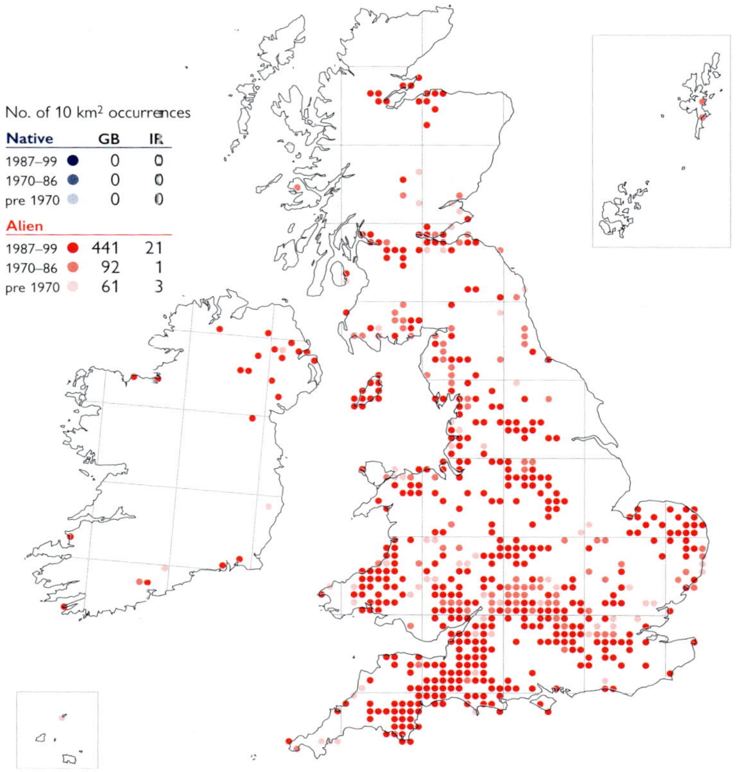

A mat-forming evergreen perennial herb, growing on walls, rocks and roadside banks. Lowland.

Neophyte (change +2.94). *S. spurium* was introduced to cultivation in 1816. It is often grown in gardens and frequently escapes to form very persistent colonies; it was known from the wild by 1910. The many new records since the 1962 *Atlas* probably indicate a real spread rather than better recording of aliens.

Native of the Caucasus; widely naturalised in temperate W. Europe.

References: Atlas (133c), Jalas & Suominen (1999).

J. M. CROFT

Sedum rupestre Reflexed Stonecrop

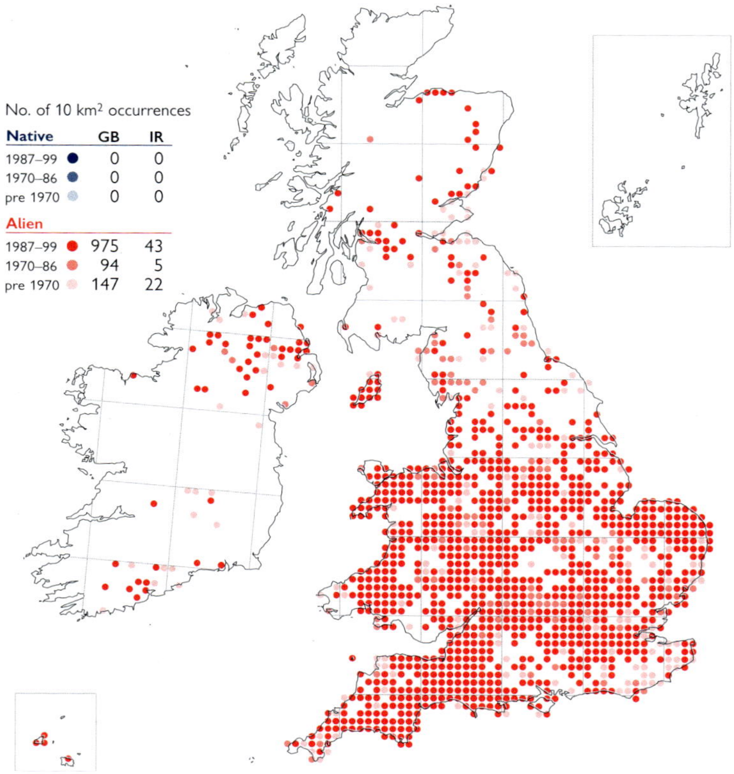

A perennial herb growing on old walls, rock outcrops, roadside banks and waste ground. Generally lowland, but reaching 365 m in N. Wales, and recorded at 845 m on Great Dun Fell (Westmorland).

Neophyte (change +2.20). *S. rupestre* has been cultivated since the 17th century for the leaves, which were eaten as a spring salad. It was known in the wild by 1666. Since the 1962 *Atlas* its distribution has increased, but it is unclear whether this is due to better recording or a genuine spread.

A European Temperate species, extending from N. Spain to S. Scandinavia and the northern half of the Balkan peninsula.

References: Atlas (135a), Grigson (1955), Hultén & Fries (1986), Jalas & Suominen (1999), Meusel *et al.* (1965).

J. M. CROFT

Sedum forsterianum Rock Stonecrop

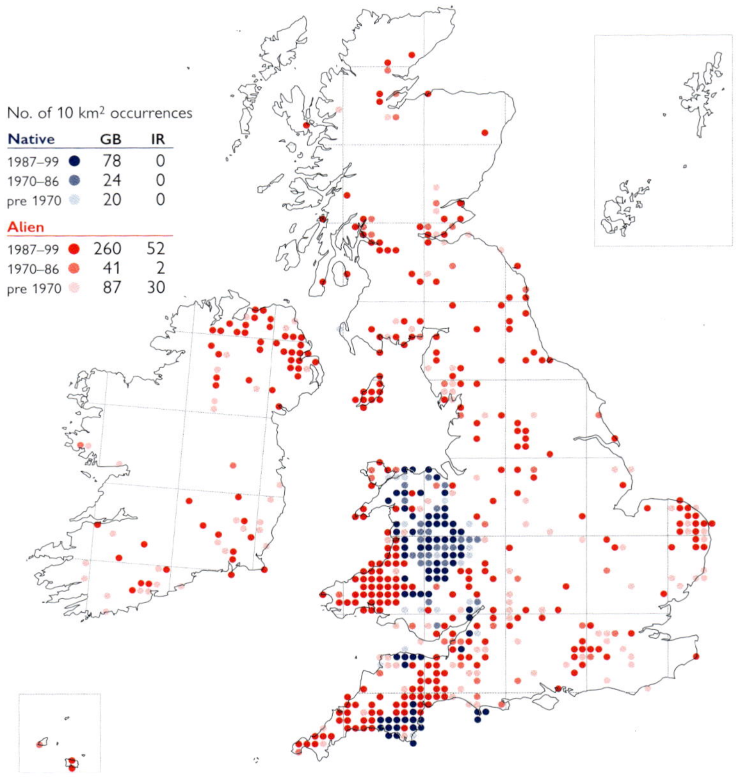

A mat-forming perennial herb of open, dry, well-drained habitats, including rocks and screes, wooded cliffs and gullies. Naturalised colonies are found in churchyards, on waste ground, mine waste, walls and railway land. Lowland to 600 m as native (Llyn y Fan Fach, Carms.).

Native (change +1.54). This species is grown in rockeries and on graves, and has become naturalised in many areas outside its native range. There has been a considerable increase in records since the 1962 *Atlas*. The distribution of native and introduced plants, even in its core western areas, is now hopelessly muddled.

Oceanic Southern-temperate element.

References: Atlas (134d), Beckett *et al.* (1999), Hultén & Fries (1986), Jalas & Suominen (1999), Meusel *et al.* (1965), Stewart *et al.* (1994).

J. M. CROFT

Sedum acre Biting Stonecrop

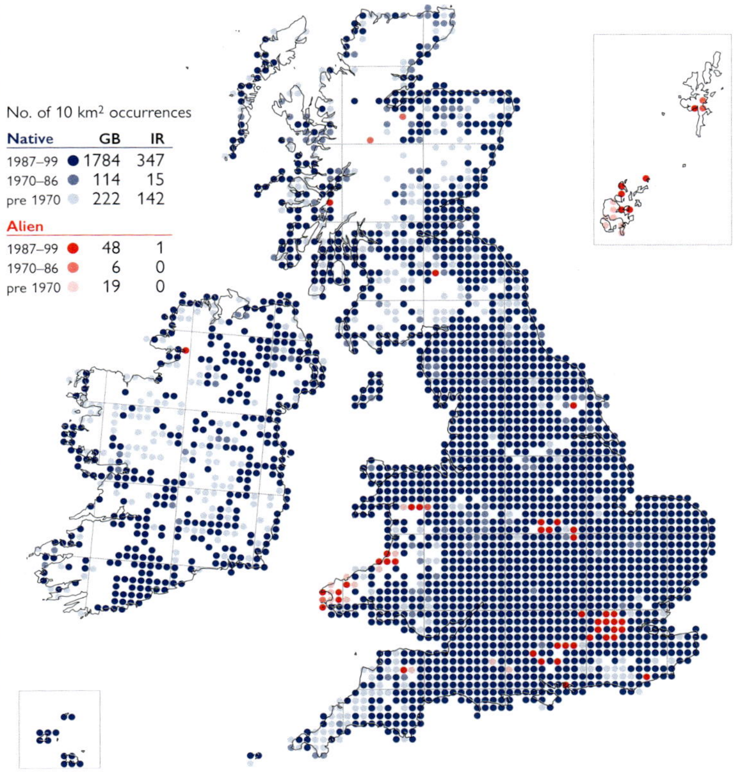

A perennial herb of dry, undisturbed and open habitats on skeletal, or virtually non-existent, acidic or basic soils. Typical natural habitats include shingle, sand dunes, cliffs and steeply sloping, S.-facing rocks. It is also frequent on walls, roofs, gravel tracks, pavements and road verges. It is resistant to grazing. 0–500 m (Arncliffe, Mid-W. Yorks.), with an exceptional record at 845 m on Great Dun Fell (Westmorland).

Native (change –0.24). The distribution of *S. acre* is stable, except in Cornwall and Ireland where it is now only locally plentiful.

European Temperate element; widely naturalised outside its native range.

References: Atlas (134c), French *et al.* (1999), Grime *et al.* (1988), Hultén & Fries (1986), Jalas & Suominen (1999), Meusel *et al.* (1965).

J. M. CROFT

Sedum sexangulare Tasteless Stonecrop

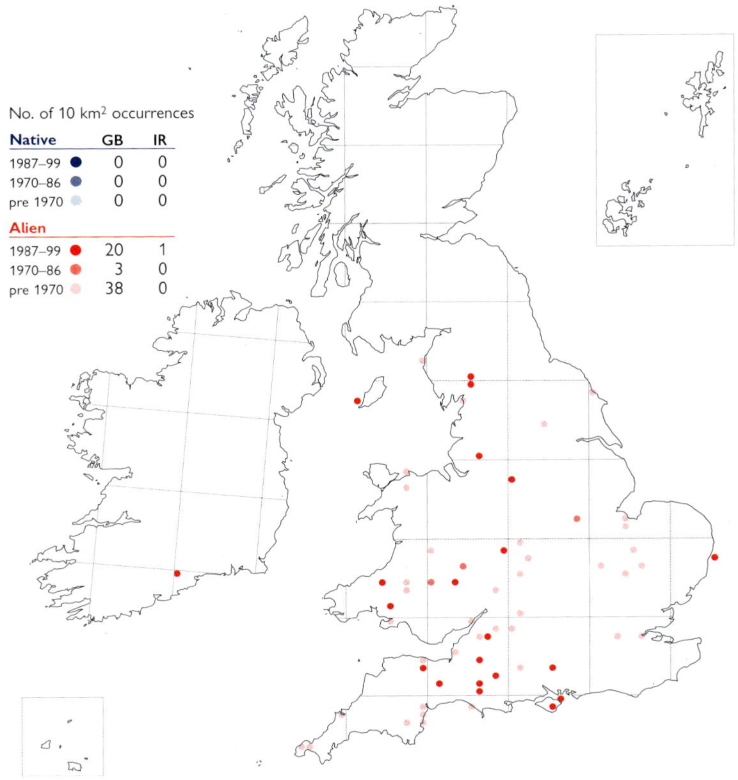

A perennial herb, found as a garden escape which can become established on lane banks, cliffs, walls, rocky places, stone steps and in churchyards. Lowland.

Neophyte. *S. sexangulare* was recorded in the wild by 1763 and has been known from Wick Rocks, Bristol (W. Gloucs.) since at least 1869. It is now only infrequently grown in gardens.

A European Temperate species, with a distribution centred in C. Europe but extending locally east to Ukraine.

References: Hultén & Fries (1986), Jalas & Suominen (1999), Meusel *et al.* (1965), White (1912).

J. M. CROFT

Sedum album White Stonecrop

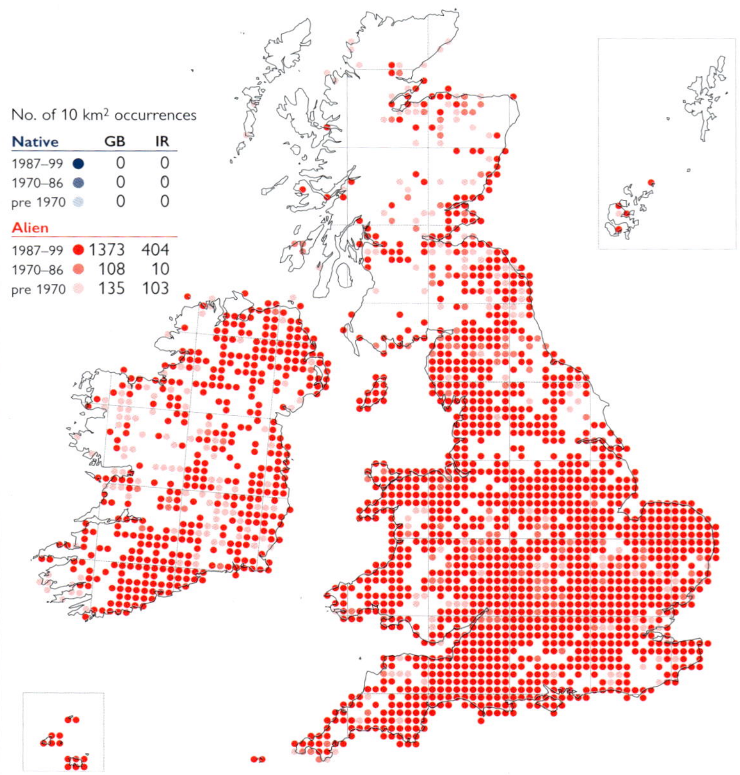

This creeping perennial herb grows on open, dry sites, such as limestone rocks, walls, roofs, the concrete of old airfield runways, maritime shingle, paths and gravel on graves. 0–570 m (Hartside, Cumberland).

Archaeophyte (change +2.41). *S. album* is treated here as an alien, though some authors have suggested that it may possibly occur as a native on limestone rocks at the eastern end of the Mendips and in S. Devon. There has been a marked increase in the frequency of this species since the 1962 *Atlas*.

As an archaeophyte *S. album* has a Submediterranean-Subatlantic distribution; it is widely naturalised outside this range.

References: Atlas (134b), Green *et al.* (1997), Hultén & Fries (1986), Jalas & Suominen (1999), Meusel *et al.* (1965).

J. M. CROFT

Sedum anglicum English Stonecrop

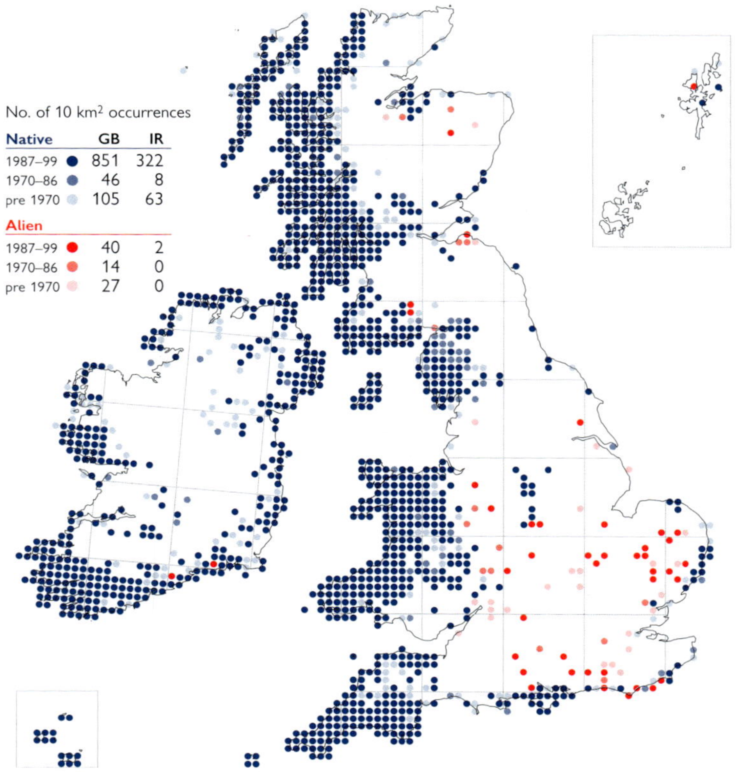

This creeping perennial herb of base-poor substrates occurs on rocks, dunes and shingle, and is also known from dry grassland. It is a characteristic plant of open ground on acidic rock outcrops near the sea. It also grows on old walls, rocky hedge banks, and quarries and mine spoil on acidic substrates. 0–1080 m (Snowdon, Caerns.).

Native (change –0.21). The only appreciable change in the distribution of this species since the 1962 *Atlas* is an increase of inland records, representing escapes from cultivation.

Oceanic Temperate element.

References: Atlas (134a), Hultén & Fries (1986), Jalas & Suominen (1999).

J. M. CROFT

Sedum dasyphyllum Thick-leaved Stonecrop

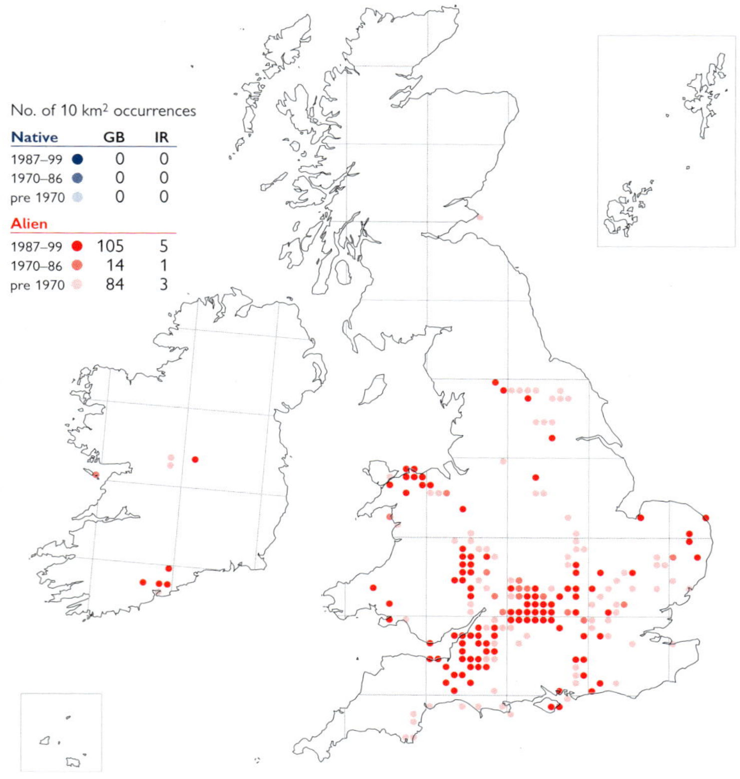

A small perennial herb with ascending stems readily rooting at the base which provide an anchorage to the crumbling surfaces of old walls on which it characteristically grows. Other habitats include quarries, cemeteries and limestone rocks. Lowland.

Neophyte (change +0.32). *S. dasyphyllum* is a garden plant which has become naturalised, and persists on many church and castle walls in S. and C. Britain. It was known from the wild by 1724, but there is little evidence of an expansion in its range since the 1962 *Atlas*.

Native of S. Europe, N. Africa and S.W. Turkey; locally naturalised in N.W. Europe.

References: Atlas (133d), Bolòs & Vigo (1984), Jalas & Suominen (1999).

J. M. CROFT

Sedum villosum Hairy Stonecrop

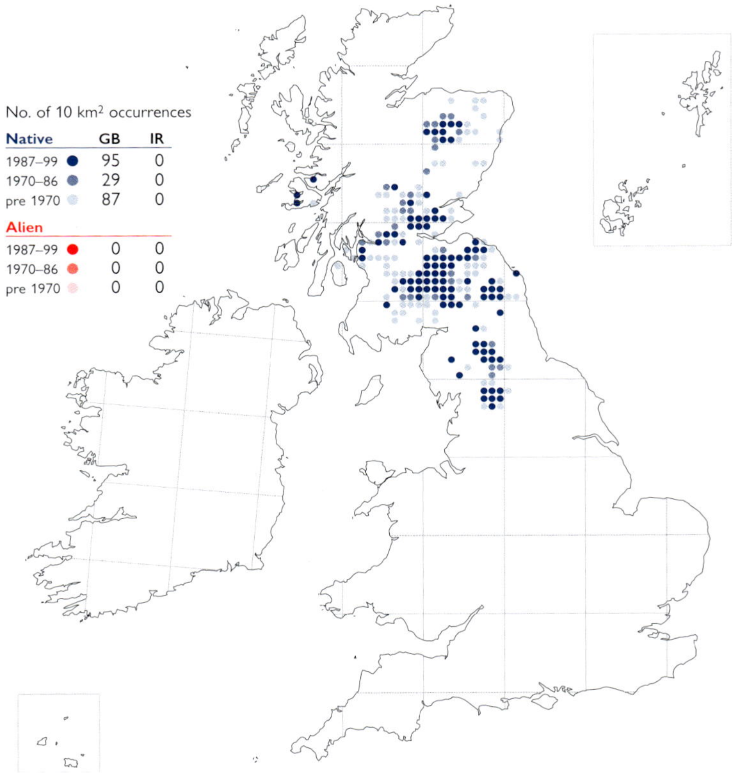

A small biennial or perennial herb which grows in at least slightly base-enriched, wet, stony ground and on streamsides in hilly areas, and in montane, often bryophyte-rich, flushes. From near sea level to 1100 m (Breadalbanes, Mid Perth), but mostly between 250 m and 500 m.

Native (change –0.76). Some losses occurred before 1930, and these have continued, especially in lowland areas, due to drainage and forestry. It is probably still present in many upland squares for which there are only pre-1987 records. Its range has, however, been extended westwards since the 1962 *Atlas* by records from Westerness and Mid Ebudes.

European Boreo-arctic Montane element; also in N. America.

References: Atlas (135b), Hultén & Fries (1986), Jalas & Suominen (1999), Meusel *et al.* (1965), Stewart *et al.* (1994).

J. M. CROFT

Bergenia crassifolia Elephant-ears

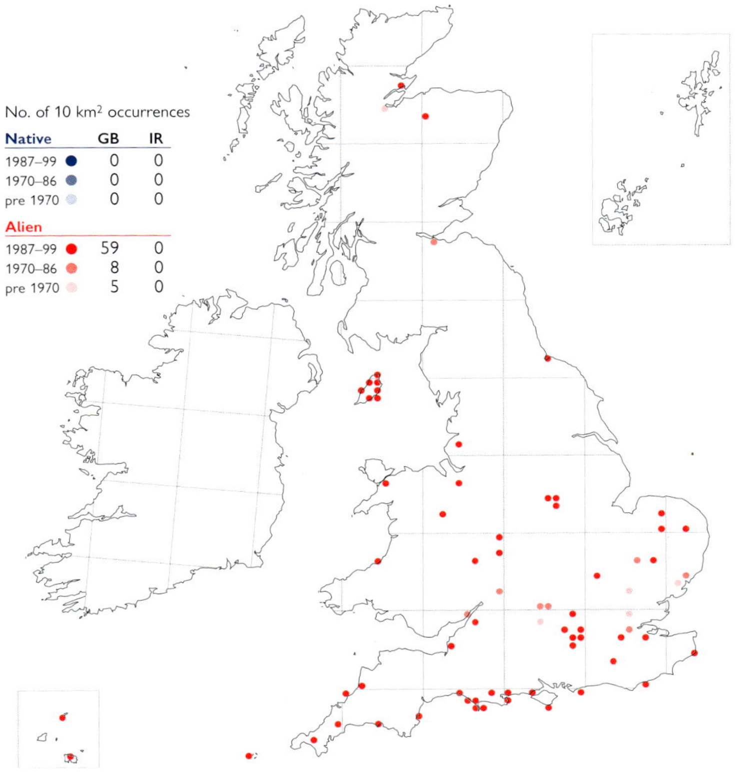

A robust, rhizomatous clump-forming perennial herb occurring as an established garden escape in woodland, copses and hedgerows and on roadside verges, railway banks, waste ground, rubbish tips and in chalk-pits. It prefers some shade but can grow in the open. It does not normally set seed owing to self-incompatibility. Lowland.

Neophyte. *B. crassifolia* was introduced into cultivation in Britain in 1765. It was not recorded from the wild until 1962, and seems to be spreading in S. Britain, especially in milder areas by the coast.

Native of Russia (Altai) eastwards to Mongolia, Korea and China.

References: Pan (1988), Yeo (1966).

R. J. GORNALL

Darmera peltata Indian-rhubarb

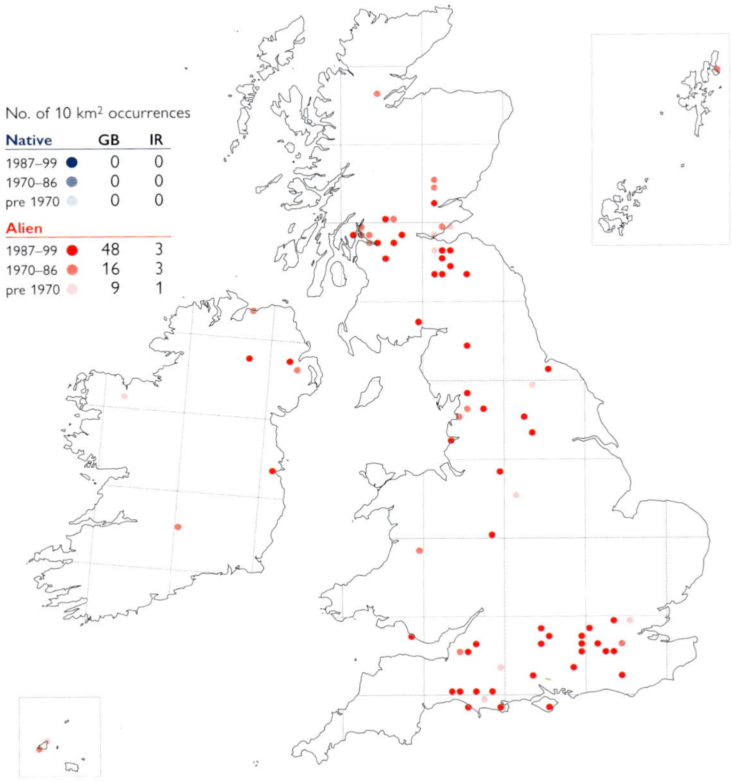

A robust, rhizomatous perennial herb with distinctive peltate leaves that appear after the flowers. It occurs in damp places as an established garden escape, but does not normally appear to set seed. Lowland.

Neophyte. *D. peltata* was introduced into cultivation in Britain in 1873 and was known in the wild by about 1920. It may be spreading in some areas.

Native of western N. America.

Reference: Jalas & Suominen (1999).

R. J. GORNALL

Saxifraga hirculus Marsh Saxifrage

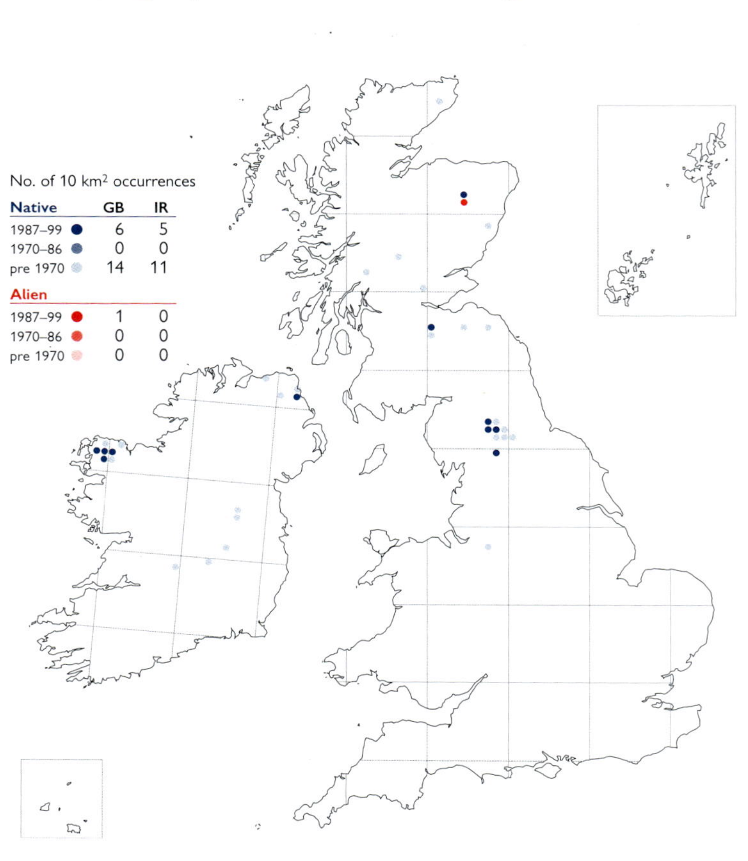

A stoloniferous perennial herb found in wet, base-rich flushes and mires, especially where the vegetation is checked by grazing. Formerly near sea level, it survives now only in its upland localities usually at 300–650 m, but reaching 750 m (Mickle Fell, N.W. Yorks.).

Native (change –0.30). The 1962 *Atlas* showed a decline up to about 1930, with lowland populations (mostly in Scotland) disappearing owing to afforestation, drainage and other demands of agriculture. Since then, however, there has been only a slight reduction in the number of 10-km squares.

Circumpolar Boreo-arctic Montane element.

References: Atlas (136c), Curtis & McGough (1988), Hedberg (1992), Hultén & Fries (1986), Jalas & Suominen (1999), Meusel *et al.* (1965), Webb & Gornall (1989), Wigginton (1999).

R. J. GORNALL

Saxifraga cymbalaria Celandine Saxifrage

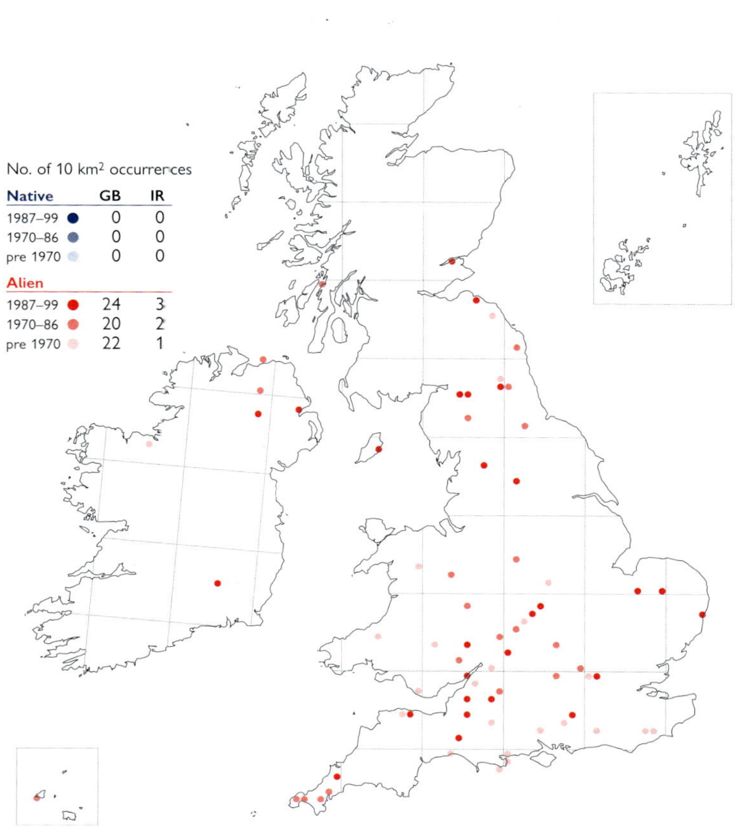

An annual, possibly sometimes biennial, herb, occurring in damp, usually shaded habitats such as at the foot of walls and on old rockeries. Setting abundant seed, the plant reproduces readily and can become a weed in gardens and nurseries. Lowland.

Neophyte. This species was introduced into cultivation in 1880 and was known from the wild by 1908. It appears to be increasing, at least in England. Most records are of var. *huetiana*, though var. *cymbalaria* has also been claimed.

Native of S.W. Asia (N. Turkey, Caucasus and N.W. Iran) with outlying localities in N. Africa and Europe, where it is only native in one locality, in Romania.

References: Jalas & Suominen (1999), Webb & Gornall (1989).

R. J. GORNALL

Saxifraga nivalis Alpine Saxifrage

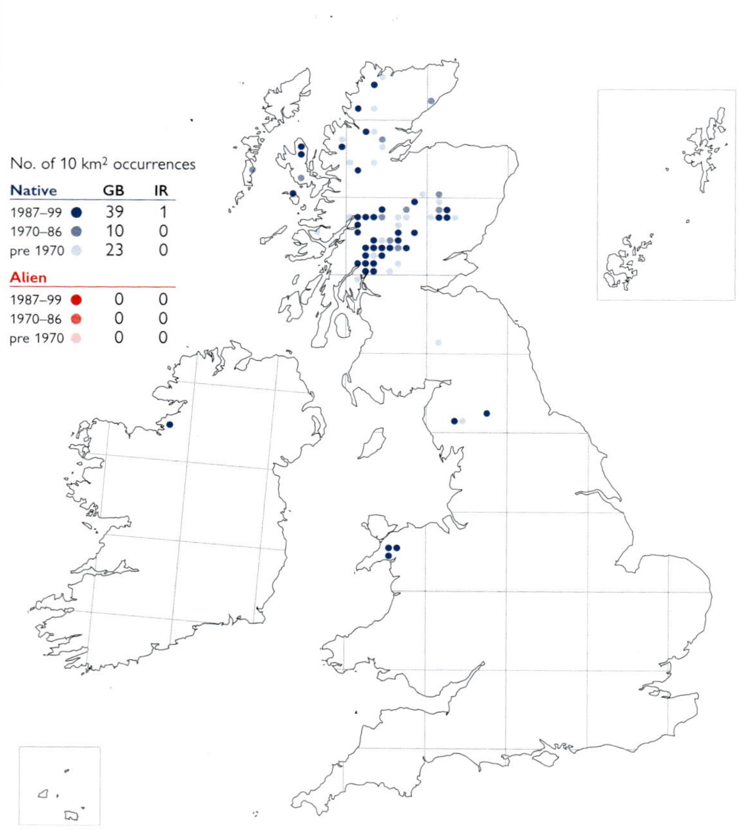

A perennial, rhizomatous herb growing on damp, shady, base-rich rocks and cliffs. It is usually found in crevices and on ledges where competing vegetation does not overtop it. From 365 m at Quiraing (N. Ebudes) to 1210 m on Ben Lawers (Mid Perth), with unlocalised records at 1300 m in the Cairngorms.

Native (change –0.50). The distribution of *S. nivalis* has remained stable since the 1962 *Atlas*.

Circumpolar Arctic-montane element.

References: Atlas (136a), Curtis & McGough (1988), Hultén & Fries (1986), Jalas & Suominen (1999), Meusel *et al.* (1965), Stewart *et al.* (1994), Webb & Gornall (1989).

R. J. GORNALL

Saxifraga stellaris Starry Saxifrage

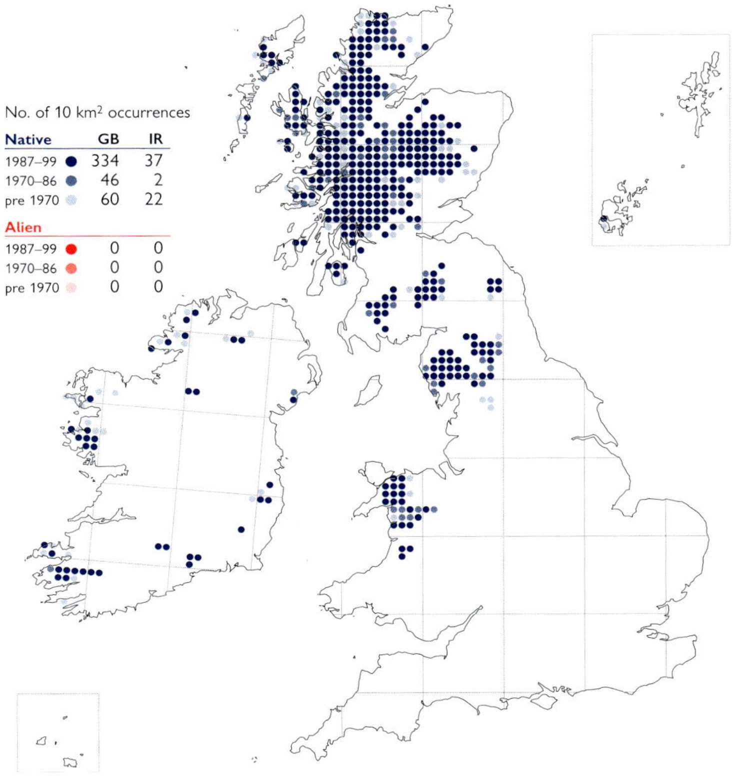

A perennial, stoloniferous herb found in open (rarely shaded), wet flushes, growing by mountain streams or on wet rock ledges and cliff-faces, usually in base-poor soil. It is commonly found from about 200–1000 m, although it has been recorded at 1340 m on the top of Ben Nevis (Westerness).

Native (change –0.58). There has been no appreciable change in the distribution of *S. stellaris* since the 1962 *Atlas*.

European Arctic-montane element; also in N. America.

References: Atlas (136b), Hultén & Fries (1986), Jalas & Suominen (1999), Meusel *et al.* (1965), Webb & Gornall (1989).

R. J. GORNALL

Saxifraga spathularis St Patrick's-cabbage

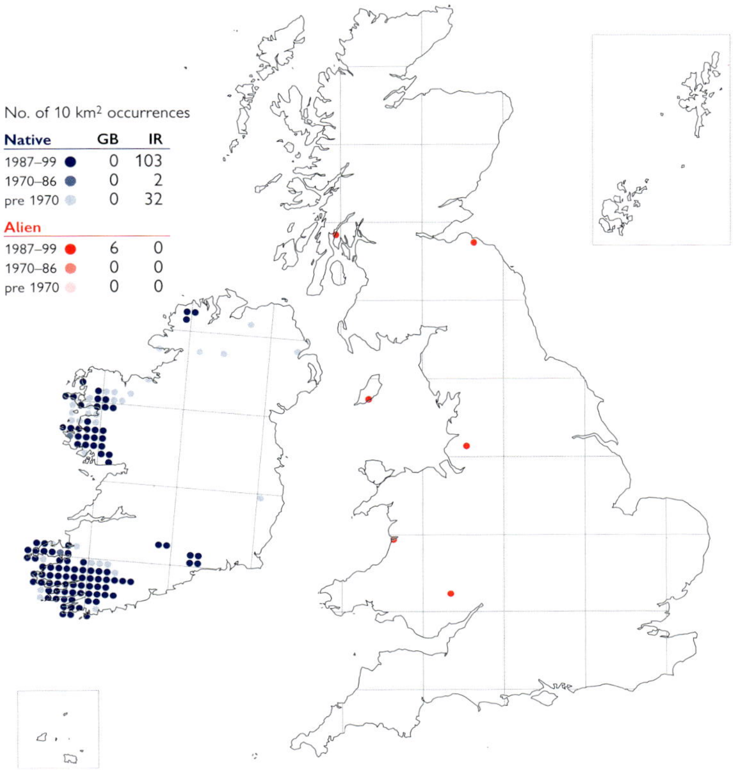

A perennial, stoloniferous herb usually found in acid conditions in areas of high rainfall (exceeding one metre per year over much of its range), where it grows in humid, rocky woods, on shady mountain cliffs and relatively unshaded S.-facing slopes. 0–1040 m on Carrantuohill (S. Kerry).

Native. There has been no appreciable change in the distribution of *S. spathularis* since the 1962 *Atlas*.

Oceanic Temperate element; restricted as a native to N. Portugal, N.W. Spain and Ireland.

References: Atlas (136d), Dupont (1962), Hultén & Fries (1986), Jalas & Suominen (1999), Waldren & Scally (1993), Webb & Gornall (1989).

R. J. GORNALL

Saxifraga spathularis × *S. umbrosa* (*S.* × *urbium*) Londonpride

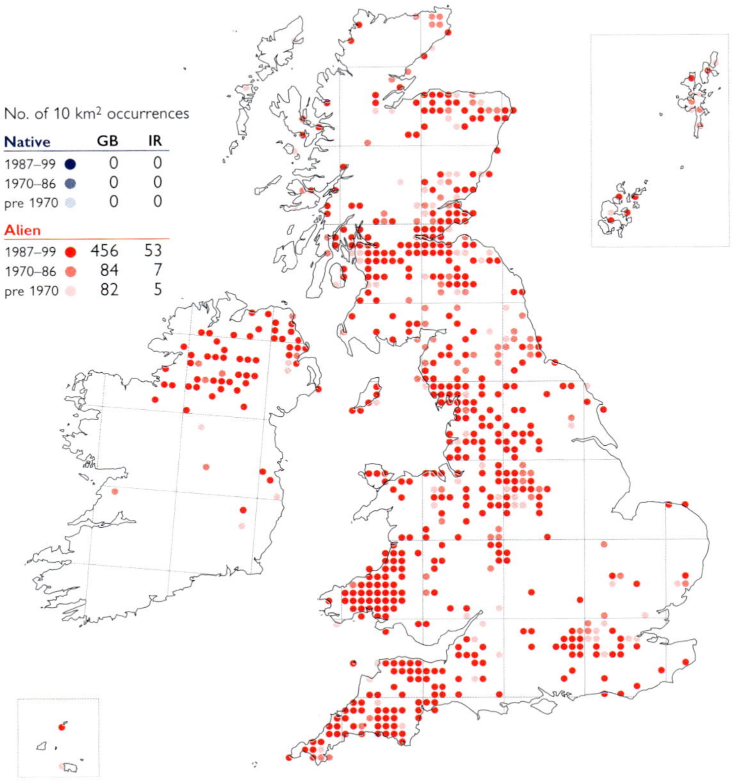

A perennial, stoloniferous herb, usually growing in shaded or at least damp places in woods, by streams, on banks and walls, and among rocks. Generally lowland, but reaching 335 m at Parsley Hay (Derbys.).

Neophyte. *S.* × *urbium* is known to have been grown in Britain before 1700. It has escaped and become established in suitable places throughout our area, despite being shy-fruiting owing to sterility in the majority of specimens. It was recorded from the wild in 1837 (Staffs.).

A hybrid of garden origin.

References: Pugsley (1936), Stace (1975).

R. J. GORNALL

Saxifraga hirsuta Kidney Saxifrage

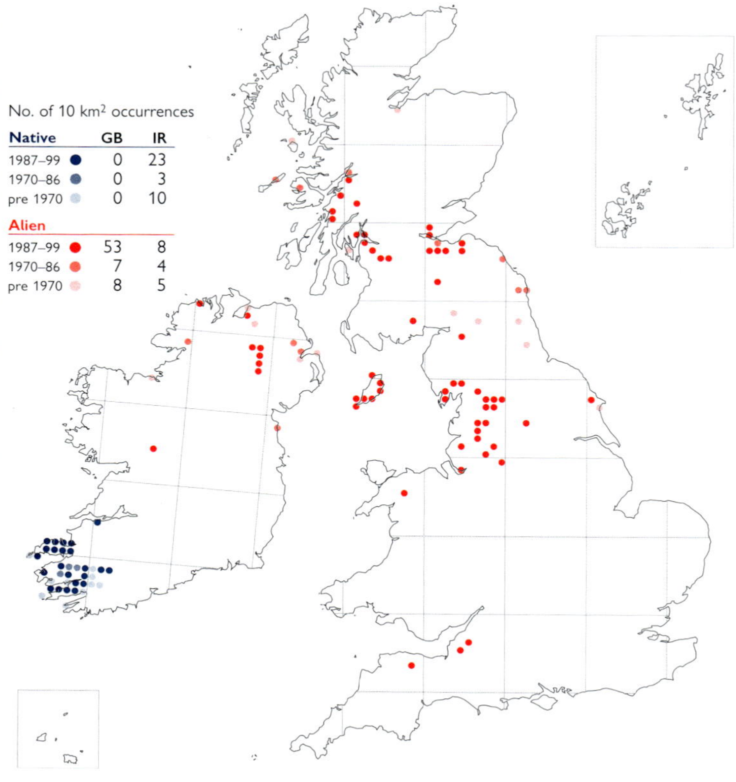

A perennial, stoloniferous herb, found growing only in damp, shaded places, such as woods, N.-facing cliffs and banks, and by streams and on rocks in the mountains. Naturalised populations in Britain, often derived from garden escapes, are often found on limestone, whereas in its native Ireland the species occurs only on siliceous rock. 0–915 m (Mt Brandon, S. Kerry).

Native (change +1.39). There has been no appreciable change in the native range of *S. hirsuta* since the 1962 *Atlas*, although naturalised populations in Britain have increased considerably, especially in the north and west.

Oceanic Temperate element.

References: Atlas (137a), Jalas & Suominen (1999), Meusel *et al.* (1965), Waldren & Scally (1993), Webb & Gornall (1989).

R. J. GORNALL

Saxifraga hirsuta × *S. spathularis* (*S.* × *polita*) False Londonpride

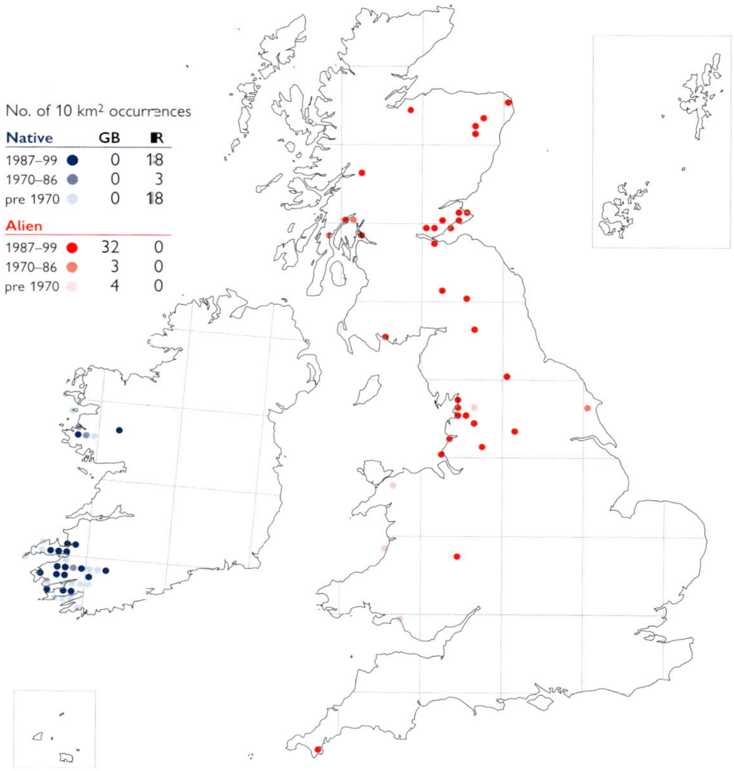

A perennial, stoloniferous herb that is naturalised in shaded or at least damp places. It is usually found in woods, by streams, on banks and amongst rocks. Lowland.

Native. The hybrid is variable and fertile, with backcrossing to both parents where they co-occur, although neither of them is found in Co. Galway and Mayo. *S.* × *polita* has also escaped from gardens and become naturalised in Britain. There is some evidence of spread in Britain since it was mapped by Perring & Sell (1968).

S. × *polita* is known elsewhere only from N.W. Spain, where it is much rarer than in its Irish range.

References: Atlas Supp. (40a), Pugsley (1936), Stace (1975).

R. J. GORNALL

Saxifraga oppositifolia Purple Saxifrage

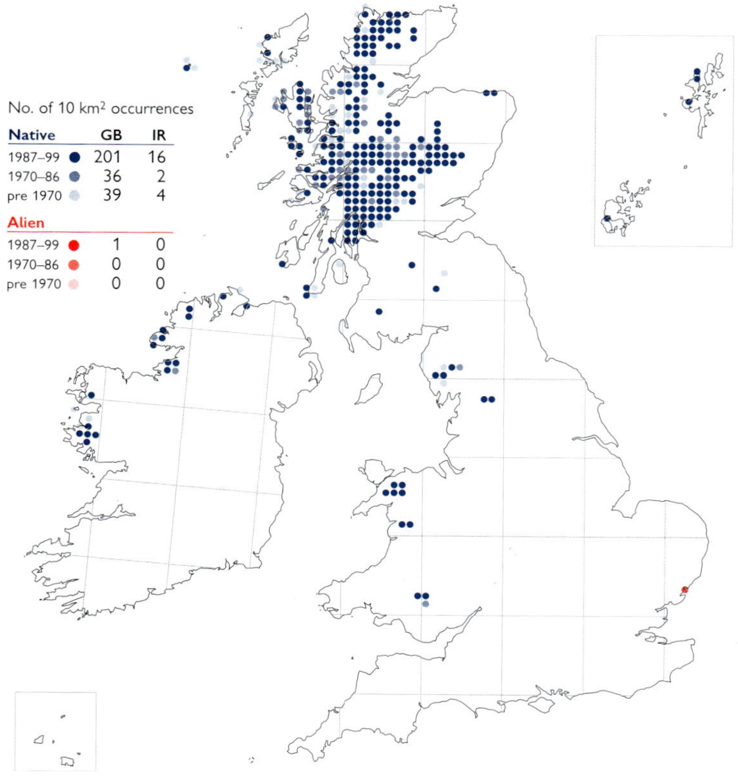

A prostrate to more or less densely caespitose perennial herb, growing on open, moist but well-drained, base-rich rocks and stony ground, mainly on cliff-faces, ledges, stony flushes and scree slopes, the southern sites having a northerly aspect. From near sea level to 1210 m (Ben Lawers, Mid Perth), but usually between 300 and 1000 m.

Native (change −0.45). The distribution of *S. oppositifolia* is stable. There is little evidence to support the formal taxonomic recognition of prostrate as opposed to caespitose ecotypes.

Circumpolar Arctic-montane element.

References: Atlas (139b), Brysting *et al.* (1996), Curtis & McGough (1988), Hultén & Fries (1986), Jalas & Suominen (1999), Jones & Richards (1956), Meusel *et al.* (1965), Webb & Gornall (1989).

R. J. GORNALL

Saxifraga aizoides Yellow Saxifrage

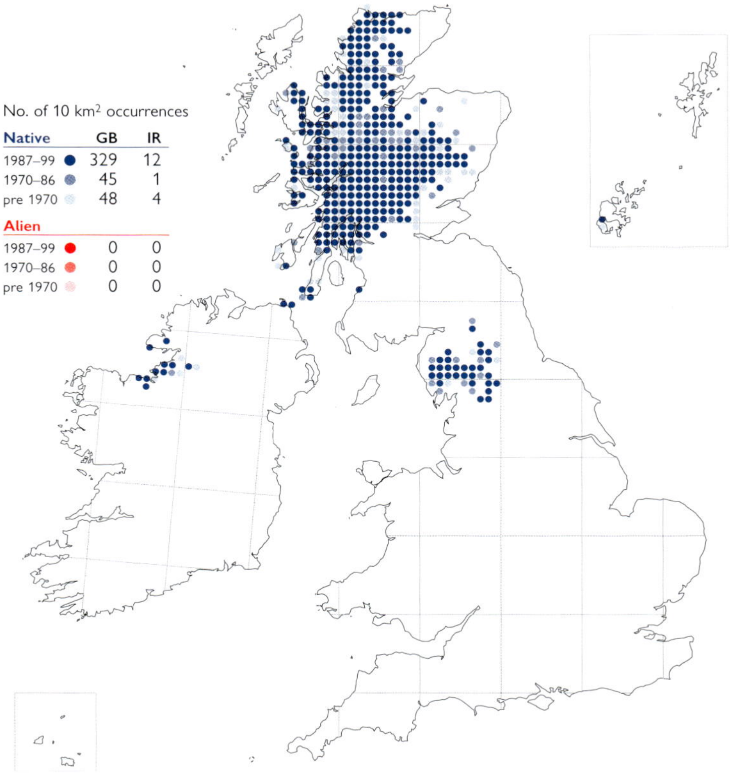

No. of 10 km² occurrences

Native	GB	IR
1987–99 ●	329	12
1970–86 ●	45	1
pre 1970 ●	48	4
Alien		
1987–99 ●	0	0
1970–86 ●	0	0
pre 1970 ●	0	0

A perennial herb usually found by the side of mountain streams, in open stony flushes on gently- or steeply-sloping ground, or sprawling over wet rocks; less commonly, on wet screes and sand dunes. It may rapidly colonise disturbed soil, such as newly dug cuttings on forestry tracks, in the vicinity of natural habitats. Whilst not strictly a calcicole, it avoids the most base-poor substrates. 0–1175 m (Breadalbanes, Mid Perth).

Native (change –0.61). There has been no appreciable change in the distribution of *S. aizoides* since the 1962 *Atlas*.

European Arctic-montane element; also in N. America.

References: Atlas (139a), Curtis & McGough (1988), Hultén & Fries (1986), Jalas & Suominen (1999), Meusel *et al.* (1965), Webb & Gornall (1989).

R. J. GORNALL

Saxifraga rivularis Highland Saxifrage

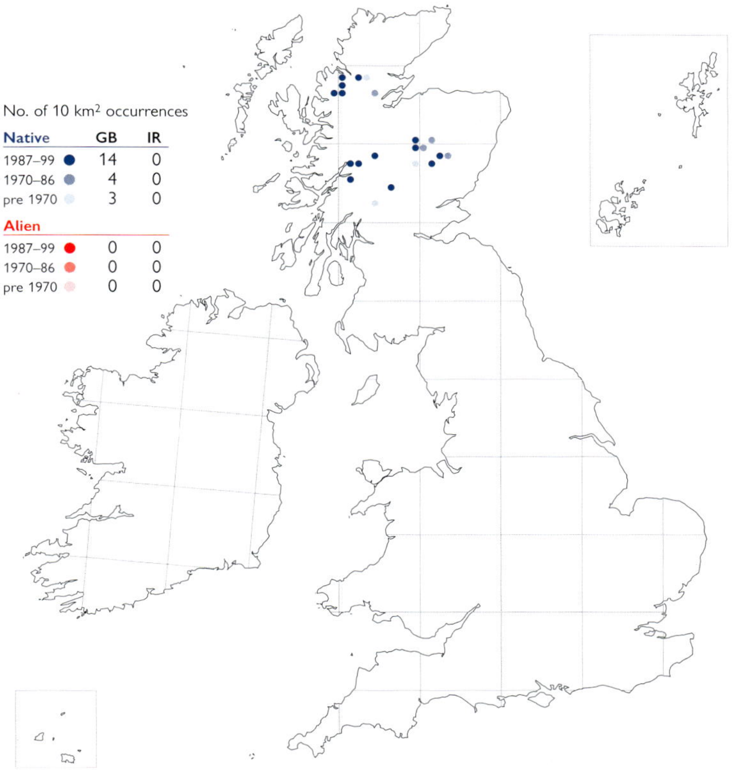

No. of 10 km² occurrences

Native	GB	IR
1987–99 ●	14	0
1970–86 ●	4	0
pre 1970 ●	3	0
Alien		
1987–99 ●	0	0
1970–86 ●	0	0
pre 1970 ●	0	0

A perennial, bulbiliferous and stoloniferous herb of basic rocks. *S. rivularis* grows on ledges or under overhangs, in damp, steep, N.- to E.-facing gullies, or more rarely in bryophyte-rich flushes on exposed scree. From 795 m in the Lairig Ghru (Easterness) to 1200 m on Ben Nevis (Westerness).

Native (change +0.19). The distribution of *S. rivularis* has been stable since the 1962 *Atlas*, though a few old records have been confirmed relatively recently.

Circumpolar Arctic-montane element; absent from mountains of C. Europe.

References: Atlas (138a), Hollingsworth *et al.* (1998), Hultén & Fries (1986), Jalas & Suominen (1999), Webb & Gornall (1989), Wigginton (1999).

R. J. GORNALL

Saxifraga cernua Drooping Saxifrage

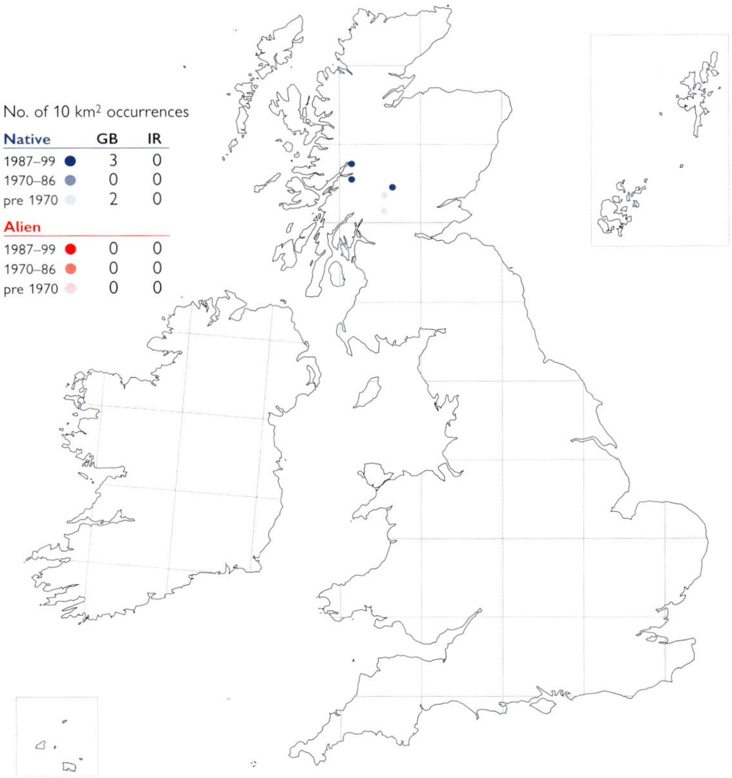

No. of 10 km² occurrences

Native	GB	IR
1987–99 ●	3	0
1970–86 ●	0	0
pre 1970 ●	2	0
Alien		
1987–99 ●	0	0
1970–86 ●	0	0
pre 1970 ●	0	0

A perennial, bulbiliferous herb of basic rocks. It grows in crevices and beneath overhangs where exposure is limited and humidity is high, usually in sites with late snow-lie. *S. cernua* flowers only infrequently, and, apparently, seed is not set in Britain; vegetative reproduction is by axillary bulbils. From 920 m (Bidean nam Bian, Main Argyll) to 1170 m (Ben Lawers, Mid Perth).

Native. Although additional populations of *S. cernua* have been discovered since the 1962 *Atlas*, there has been no change in the number of 10-km squares in which it occurs.

Circumpolar Arctic-montane element.

References: Atlas (137d), Hultén & Fries (1986), Jalas & Suominen (1999), Webb & Gornall (1989), Wigginton (1999).

R. J. GORNALL

Saxifraga granulata Meadow Saxifrage

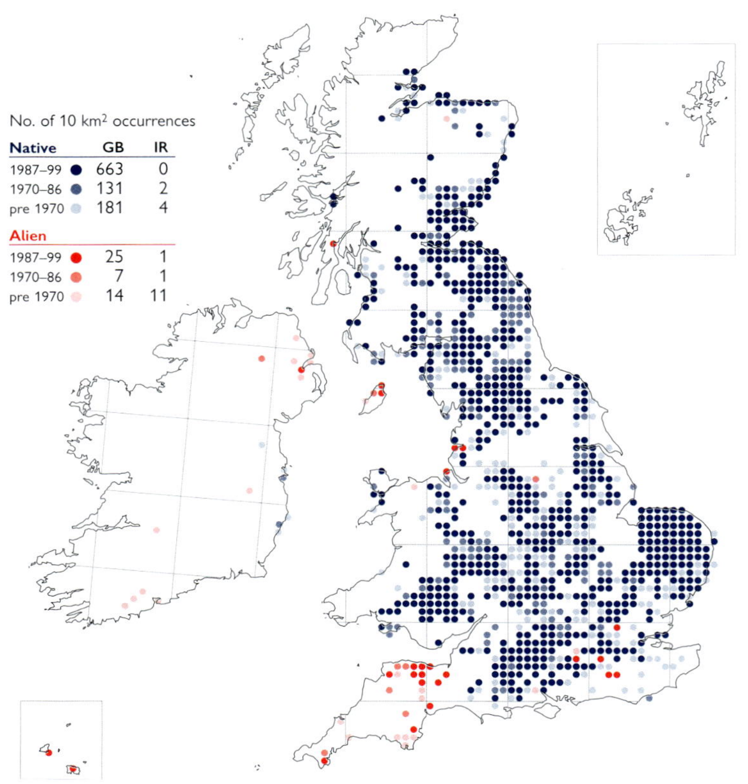

A perennial herb with a bulbiliferous rhizome, growing in moist but well-drained, often lightly grazed, base-rich and neutral grassland, in unimproved pastures and hay meadows, and on grassy banks. More rarely, it occurs on shaded river banks and in damp woodland. It is also locally naturalised near houses and in churchyards. Generally lowland, but reaching 580 m N.E. of Helbeck Fell (Westmorland).

Native (change –0.26). *S. granulata* has been lost from many sites in S. England since the 1962 *Atlas*, as grasslands have been improved. Naturalised populations include a double-flowered cultivar.

European Temperate element.

References: Atlas (137c), Curtis & McGough (1988), Hultén & Fries (1986), Jalas & Suominen (1999), Meusel *et al.* (1965), Webb & Gornall (1989).

R. J. GORNALL

Saxifraga hypnoides Mossy Saxifrage

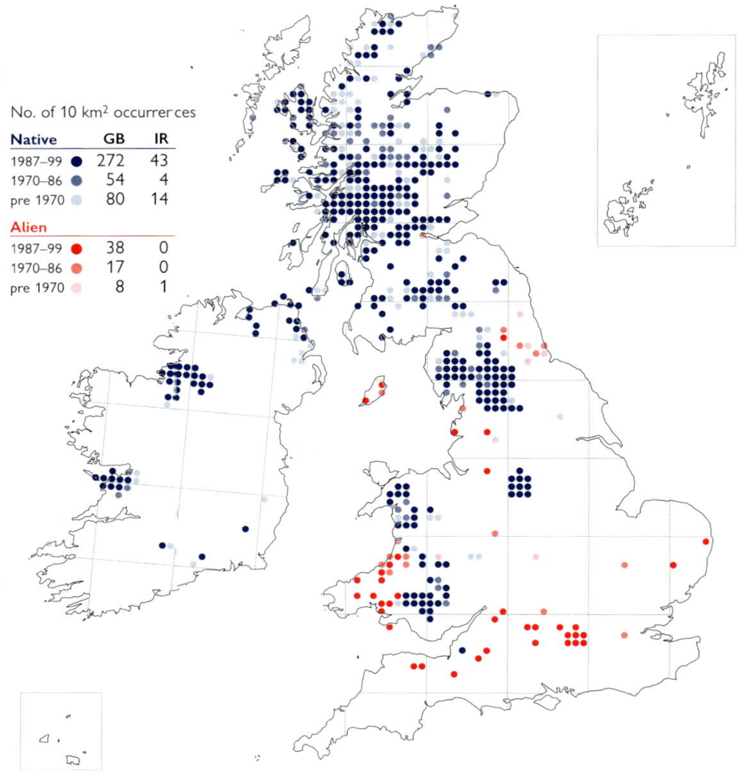

A perennial, stoloniferous herb growing on moist rocks, screes, cliffs and by mountain streams, rarely on sand dunes, often in partial shade. Substrates are frequently base-rich, although it can grow on acidic rocks. It is also cultivated and sometimes escapes. 0–1215 m (Ben Lawers, Mid Perth), but generally from 200–760 m.

Native (change –0.54). The distribution of *S. hypnoides* is generally similar to that in the 1962 *Atlas*, although it may have declined in N. Scotland. Plants from W. Ireland and Wales are diploid, whereas those from N. Ireland, N. England and Scotland are tetraploid.

Oceanic Boreal-montane element.

References: Atlas (138d), Dupont (1962), Hultén & Fries (1986), Jalas & Suominen (1999), Meusel *et al.* (1965), Parker (1979), Webb (1950), Webb & Gornall (1989).

R. J. GORNALL

Saxifraga rosacea subsp. *rosacea* Irish Saxifrage

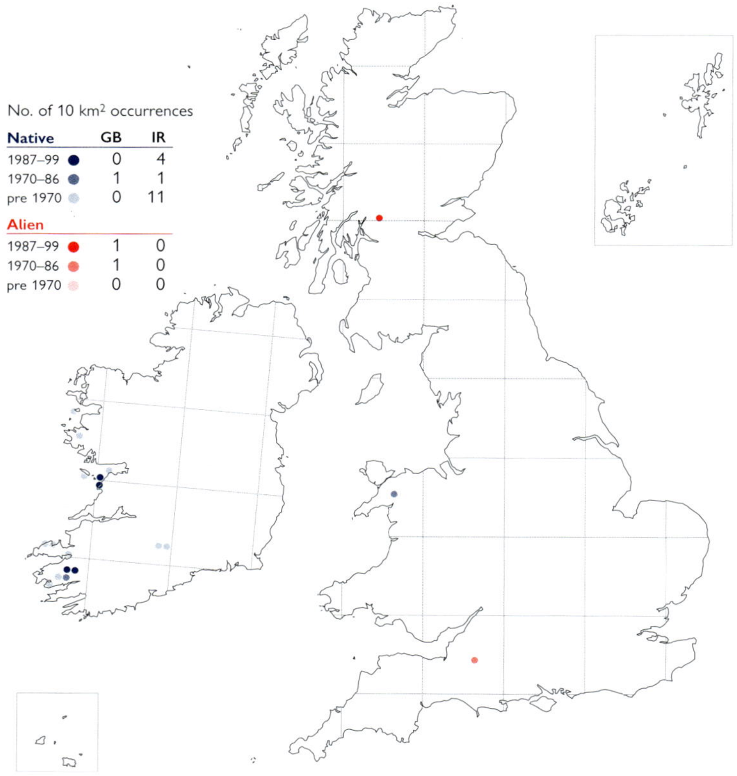

A stoloniferous perennial herb of well-drained but moist, rocky or stony substrates with little or no soil cover. It is especially common among rocks in and by mountain streams, but also occurs on cliff ledges, in rocky gullies, on scree slopes and sea-cliffs, often in open, N.- to E.-facing sites. 0–960 m (Macgillycuddy's Reeks, S. Kerry), but rarely below 500 m.

Native. It is not clear whether the apparent losses of this taxon in Ireland since 1930 are the result of under-recording. It was last recorded in N. Wales in 1978. Herbarium specimens from apparently native habitats in Scotland are of dubious provenance.

Suboceanic Boreal-montane element.

References: Atlas (138c), Hultén & Fries (1986), Jalas & Suominen (1999), Parker (1979), Webb & Gornall (1989).

R. J. GORNALL

Saxifraga rosacea subsp. *hartii*

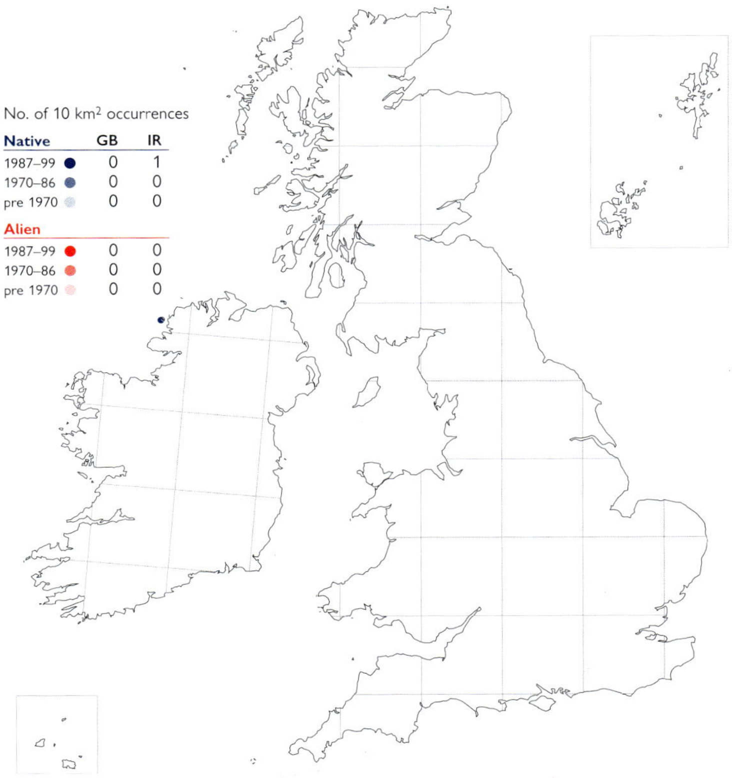

No. of 10 km² occurrences

Native	GB	IR
1987–99 ●	0	1
1970–86 ●	0	0
pre 1970 ●	0	0
Alien		
1987–99 ●	0	0
1970–86 ●	0	0
pre 1970 ●	0	0

S. rosacea subsp. *hartii* grows on sea-cliffs, where it has apparently been recorded from a number of stations on the W. and N. parts of Aranmore Island (W. Donegal). The altitude is reported to be about 90 m.

Native. Although still recorded in one 10-km square, as in the 1962 *Atlas*, the known populations have declined severely due to overgrazing and the dumping of rubbish.

Endemic.

References: Atlas (138b), Curtis & McGough (1988), Jalas & Suominen (1999), Parker (1979), Webb (1950), Webb & Gornall (1989).

R. J. GORNALL

Saxifraga cespitosa Tufted Saxifrage

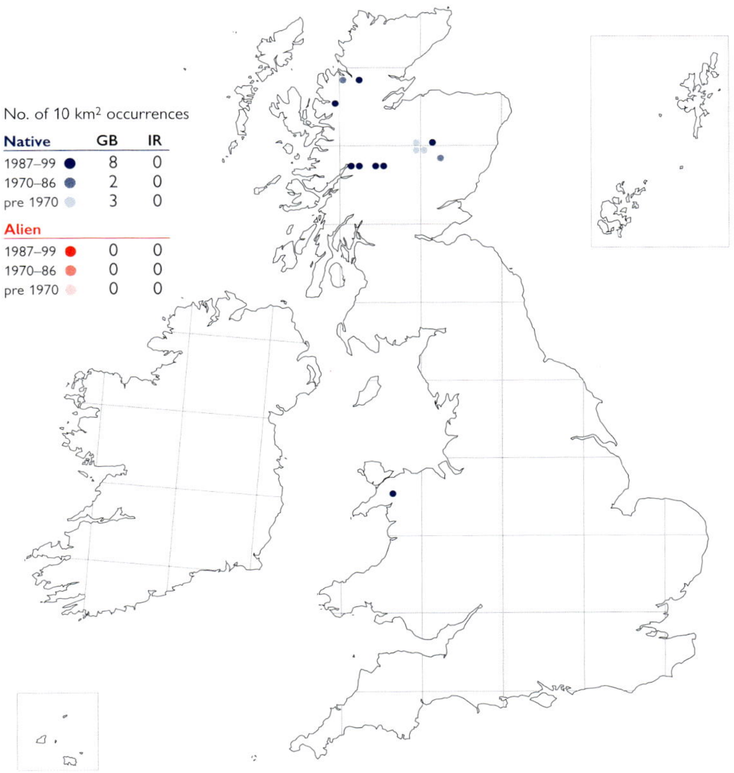

No. of 10 km² occurrences

Native	GB	IR
1987–99 ●	8	0
1970–86 ●	2	0
pre 1970 ●	3	0
Alien		
1987–99 ●	0	0
1970–86 ●	0	0
pre 1970 ●	0	0

A cushion-forming, perennial herb of well-drained base-rich rocks. It is found on mossy ledges, in crevices and on boulder-scree slopes. It appears to be highly susceptible to drought. From 520 m (Cwm Idwal, Caerns.) to 1180 m (Ben Nevis, Westerness).

Native (change –0.10). There has been a slight decline since the 1962 *Atlas*, especially in the Cairngorms. A re-stocking programme at the Welsh site, begun in 1978 and using local seed, appears to have been largely successful, notwithstanding considerable fluctuations in population size.

Circumpolar Arctic-montane element; absent from mountains of C. Europe.

References: Atlas (138b), Hultén & Fries (1986), Jalas & Suominen (1999), Meusel *et al.* (1965), Parker (1981, 1996), Webb (1950), Webb & Gornall (1989), Wigginton (1999).

R. J. GORNALL

Saxifraga tridactylites Rue-leaved Saxifrage

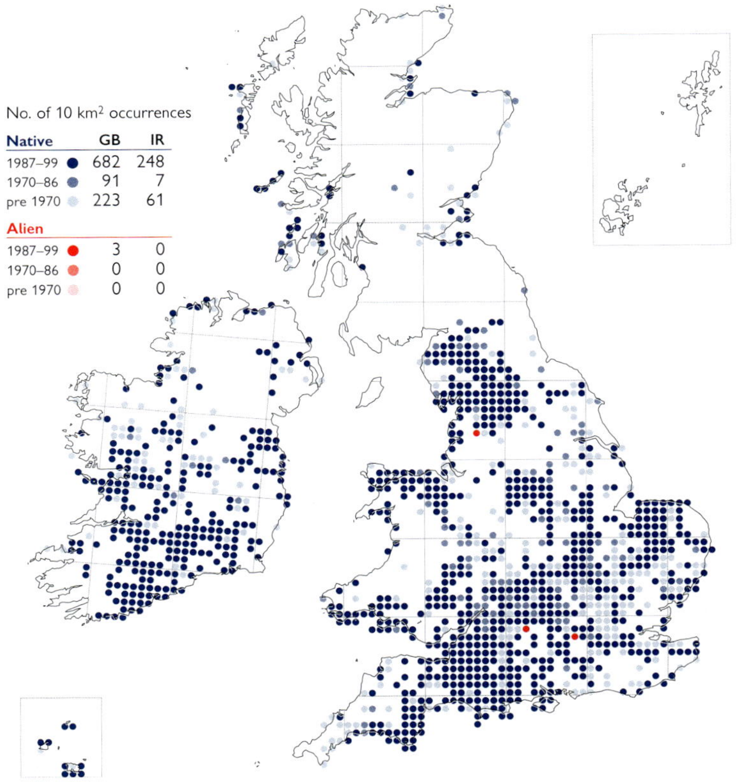

No. of 10 km² occurrences

Native	GB	IR
1987–99 ●	682	248
1970–86 ●	91	7
pre 1970 ●	223	61
Alien		
1987–99 ●	3	0
1970–86 ●	0	0
pre 1970 ●	0	0

A winter-annual that occurs in dry, open habitats such as sandy grassland, limestone pavement and rock ledges, cliffs and screes, and on man-made structures like mortared walls, pavements and railway tracks. It is most commonly found on base-rich substrates, often on skeletal soils or virtually bare rock. Usually lowland, but ascending to 595 m on Widdale Fell (N.W. Yorks.).

Native (change –0.12). *S. tridactylites* was mapped as 'all records' in the 1962 *Atlas*. It is now much better recorded but there has been a widespread decline in S. and E. England. Analysis of the database reveals that these losses have occurred since 1950.

European Southern-temperate element.

References: Atlas (137b), Grime *et al.* (1988), Hultén & Fries (1986), Jalas & Suominen (1999), Meusel *et al.* (1965), Webb & Gornall (1989).

R. J. GORNALL

Tolmiea menziesii Pick-a-back-plant

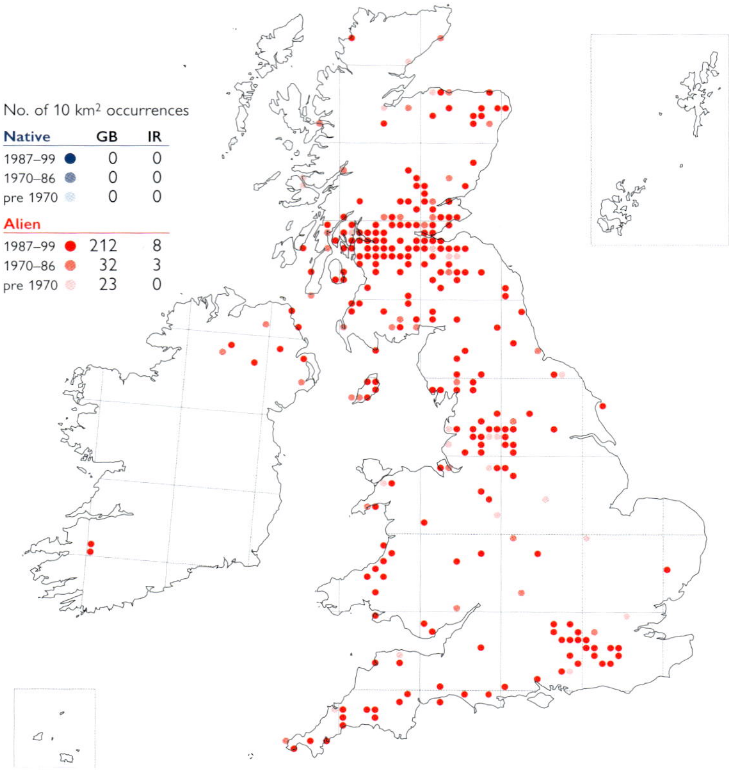

No. of 10 km² occurrences

		GB	IR
Native			
1987–99	●	0	0
1970–86	●	0	0
pre 1970	●	0	0
Alien			
1987–99	●	212	8
1970–86	●	32	3
pre 1970	●	23	0

A very shortly rhizomatous perennial, growing in nitrogen-rich soils in damp woods, by streams and other moist, shady places, occasionally on tips and waste ground. Some, or all, plants are self-incompatible and, especially if seed is lacking, the foliar embryos are likely to play an important role in their spread. Lowland.

Neophyte. Introduced in 1812 and grown as a house plant as well as in gardens, *T. menziesii* was recorded from the wild by 1928 and has become locally naturalised. It is increasing within its core areas. The species has two cytotypes, but it is not known whether our plants are diploid or tetraploid.

Native of western N. America.

References: Hultén (1968), Jalas & Suominen (1999), Soltis (1984).

R. J. GORNALL

Tellima grandiflora Fringecups

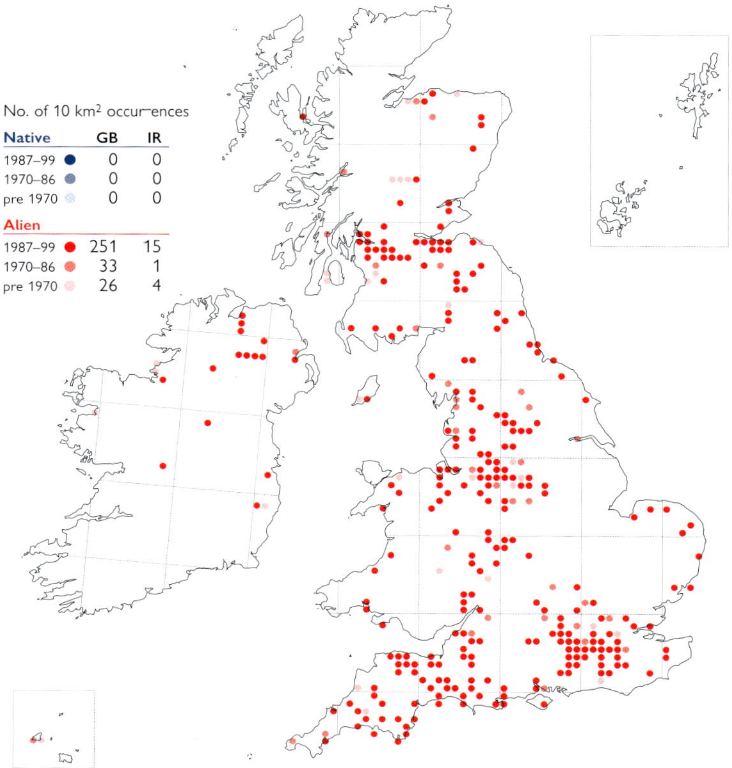

No. of 10 km² occurrences

		GB	IR
Native			
1987–99	●	0	0
1970–86	●	0	0
pre 1970	●	0	0
Alien			
1987–99	●	251	15
1970–86	●	33	1
pre 1970	●	26	4

A very shortly rhizomatous perennial, growing in nitrogen-rich soils in damp woods and hedgerows. It is an established garden escape, setting copious seed. Lowland.

Neophyte. *T. grandiflora* was introduced to cultivation in Britain in 1826. It was recorded from the wild by 1908, and seems to be spreading.

Native of western N. America.

References: Hultén (1968), Jalas & Suominen (1999).

R. J. GORNALL

Chrysosplenium oppositifolium Opposite-leaved Golden-saxifrage

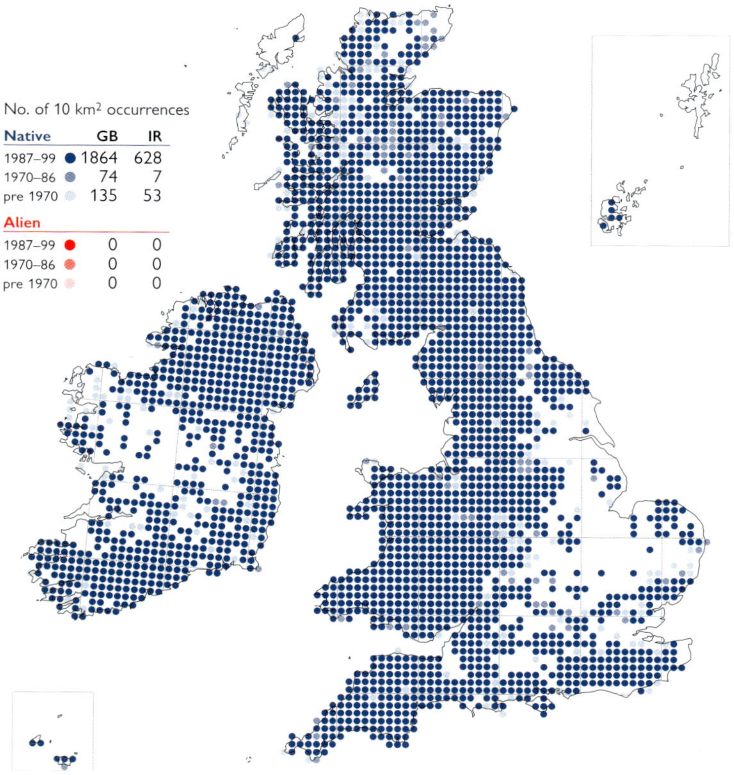

No. of 10 km² occurrences

		GB	IR
Native			
1987–99	●	1864	628
1970–86	●	74	7
pre 1970	●	135	53
Alien			
1987–99	●	0	0
1970–86	●	0	0
pre 1970	●	0	0

A perennial, stoloniferous herb found in boggy ground and seepages in woods, on streamsides, wet rocks and mountain ledges, usually in shade; also in grikes and sink-holes in limestone in the N. Pennines. It is generally found in more acidic habitats than *C. alternifolium*, but the two are often found together. 0–1100 m (Ben Nevis, Westerness).

Native (change −0.36). There has been no appreciable change in the distribution of *C. oppositifolium* since the 1962 *Atlas*.

Suboceanic Temperate element.

References: Atlas (139c), Grime *et al.* (1988), Hultén & Fries (1986), Jalas & Suominen (1999), Meusel *et al.* (1965).

R. J. GORNALL

Chrysosplenium alternifolium
Alternate-leaved Golden-saxifrage

Native	GB	IR
1987–99 ●	541	0
1970–86 ●	110	0
pre 1970 ●	139	0
Alien		
1987–99 ●	0	0
1970–86 ●	0	0
pre 1970 ●	0	0

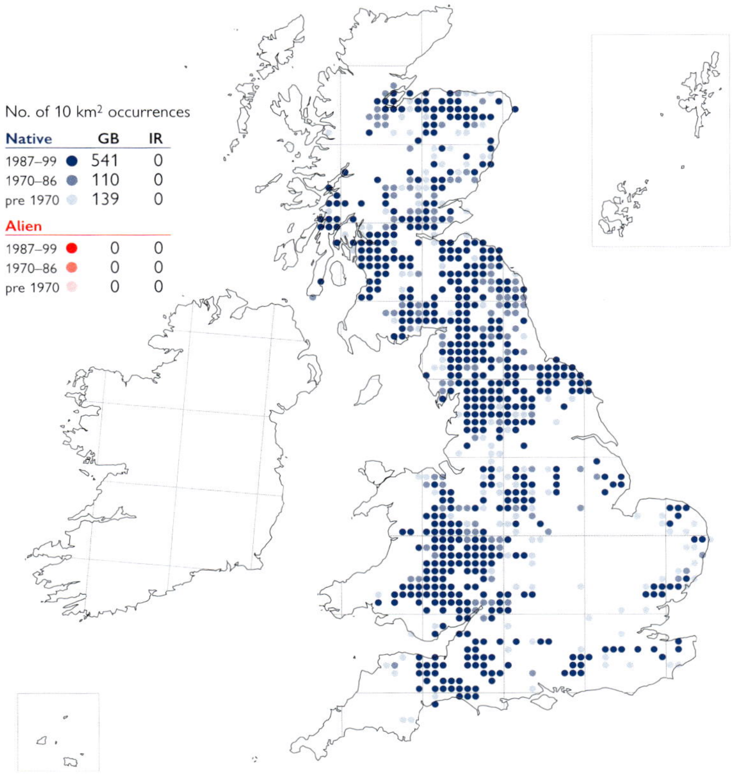

A perennial, stoloniferous herb found in boggy ground in woods, by streamsides, on wet rocks and mountain ledges, cliffs and gullies, usually in shade and usually with a supply of alkaline water. It is also found in montane, bryophyte-dominated flushes. 0–915 m (Ben Lawers, Mid Perth).

Native (change +0.62). This species is much better recorded than in the 1962 *Atlas*. The decline in East Anglia and the English S.W. Midlands had taken place before 1930.

Circumpolar Wide-boreal element.

References: Atlas (139d), Hultén & Fries (1986), Jalas & Suominen (1999), Meusel *et al.* (1965).

R. J. GORNALL

Parnassia palustris Grass-of-Parnassus

Native	GB	IR
1987–99 ●	677	198
1970–86 ●	110	19
pre 1970 ●	304	85
Alien		
1987–99 ●	1	0
1970–86 ●	0	0
pre 1970 ●	0	0

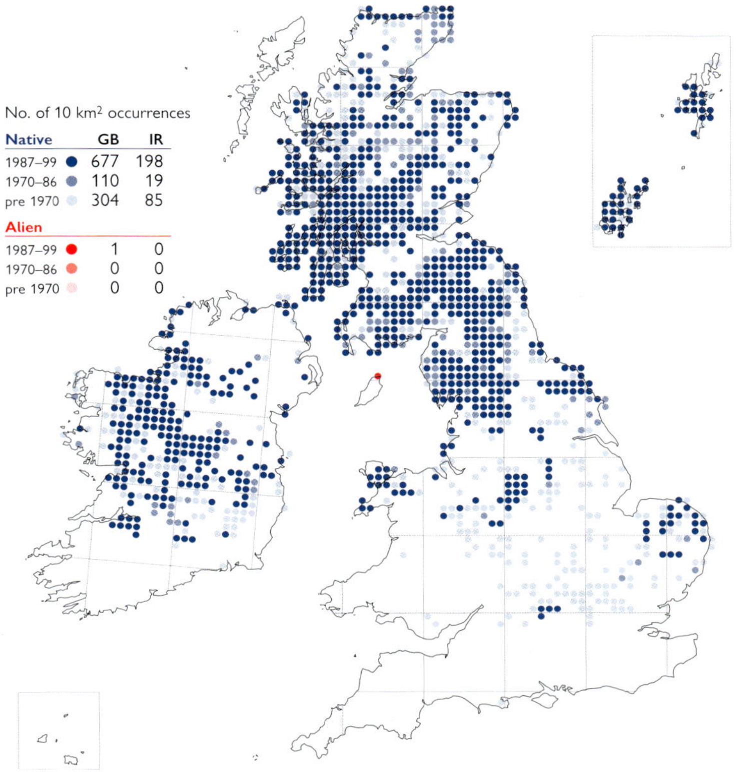

A perennial, shortly rhizomatous herb of base-rich flushes in short grassland, mires, fens, dune-slacks and machair. The last two habitats support the coastal ecotype, var. *condensata*. 0–1005 m on Ben Lawers (Mid Perth).

Native (change –0.84). The marked decline in the southern part of its range, mainly due to land drainage, was evident in the 1962 *Atlas*, and appears to be continuing. This has chiefly affected diploid populations. Its distribution in N. and N.W. Britain and Ireland (tetraploid populations) appears to be stable, though the reasons for the paucity of recent records in N. Scotland are unclear.

Circumpolar Boreo-temperate element.

References: Atlas (140a), Gornall (1988), Hultén & Fries (1986), Hultgård (1987), Jalas & Suominen (1999), Meusel *et al.* (1965), Wentworth & Gornall (1996).

R. J. GORNALL

Spiraea agg. Brideworts

Native	GB	IR
1987–99 ●	0	0
1970–86 ●	0	0
pre 1970 ●	0	0
Alien		
1987–99 ●	804	135
1970–86 ●	115	3
pre 1970 ●	152	15

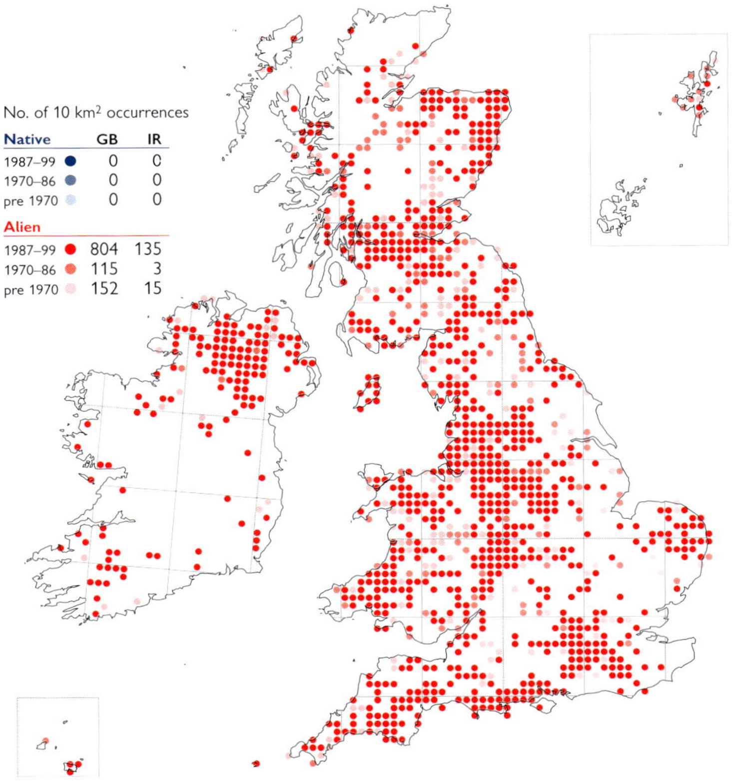

A group of deciduous perennial shrubs found naturalised in hedges, and on roadsides and river banks. All are derived from ornamental plantings or garden throw-outs. Lowland, to 435 m at Nenthead (Cumberland).

Neophyte (change +2.14). *Spiraea* taxa were being cultivated in Britain by 1640. Until recently the taxonomy of the genus has been obscure, and many recorders are still unable to distinguish the segregates, which include *S. alba, S. canescens, S. douglasii, S. media, S. japonica, S. salicifolia, S. tomentosa* and their hybrids. Records of the aggregate have increased significantly since the 1962 *Atlas*, where it was mapped as *S. salicifolia*.

An aggregate of alien species and their hybrids, the former natives of the N. hemisphere.

References: Atlas (118c), Rich & Jermy (1998).

D. J. McCOSH

Spiraea salicifolia Bridewort

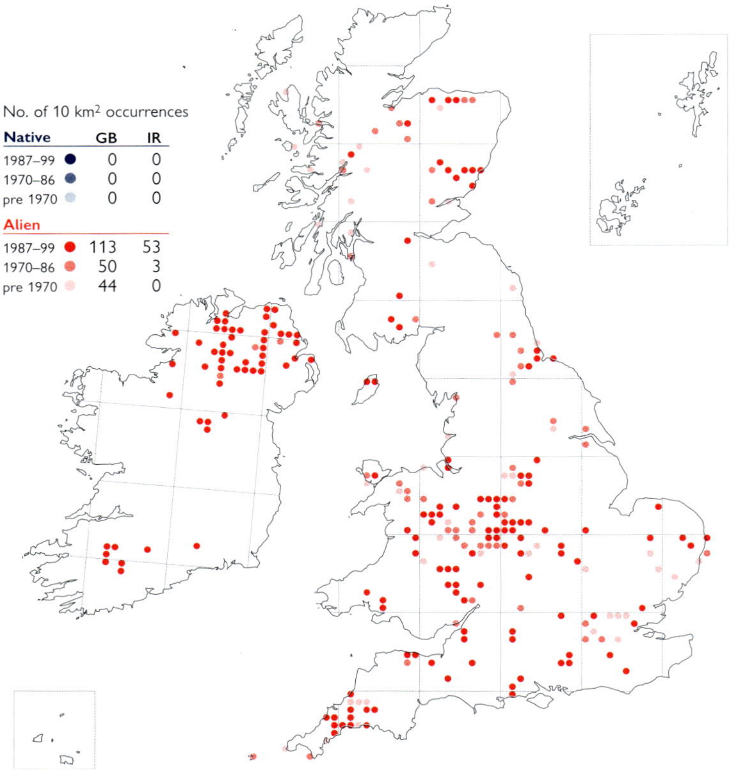

No. of 10 km² occurrences

Native		GB	IR
1987–99	●	0	0
1970–86	●	0	0
pre 1970	●	0	0
Alien			
1987–99	●	113	53
1970–86	●	50	3
pre 1970	●	44	0

A deciduous shrub, planted for ornament and occasionally naturalised in woodland and on roadsides, river banks and waste places. Lowland.

Neophyte. Although a map of *S. salicifolia* appears in the 1962 *Atlas*, the genus was not understood fully at that time and that map represents *Spiraea* agg. True *S. salicifolia*, which was introduced to cultivation in Britain by 1665, was known from the wild by 1805. It is still under-recorded.

Native of Eurasia, from C. & E. Europe eastwards to Japan.

References: Bean (1980), Coats (1963), Rich & Jermy (1998).

D. J. McCOSH

Spiraea alba × *S. douglasii* (*S.* × *billardii*) Billard's Bridewort

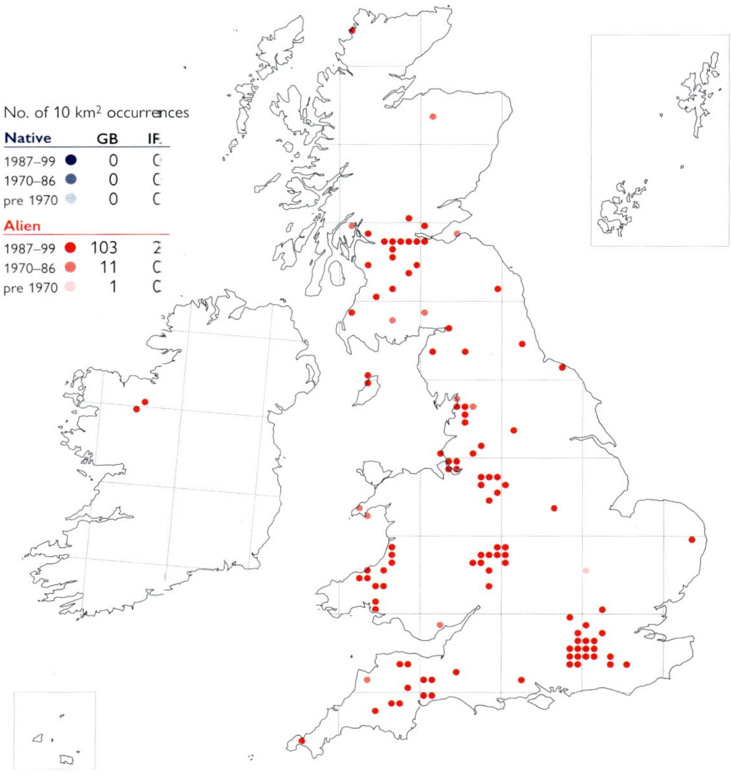

No. of 10 km² occurrences

Native		GB	IR
1987–99	●	0	0
1970–86	●	0	0
pre 1970	●	0	0
Alien			
1987–99	●	103	2
1970–86	●	11	0
pre 1970	●	1	0

A deciduous shrub, naturalised in hedgerows and on roadsides, railway banks and waste ground. Lowland.

Neophyte. *S.* × *billardii* has been in cultivation since at least 1854. It was known from the wild by 1964 and is probably increasing. It is, however, over-recorded for *S.* × *pseudosalicifolia*.

A hybrid of garden origin.

References: Rich & Jermy (1998), Silverside (1990a), Stace (1975).

D. J. McCOSH

Spiraea alba × *S. salicifolia* (*S.* × *rosalba*) Intermediate Bridewort

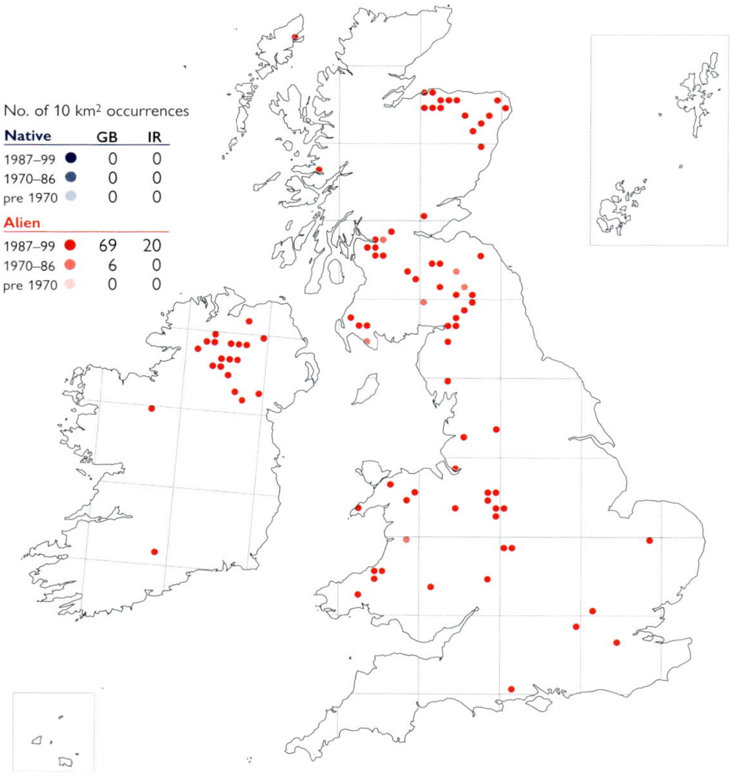

No. of 10 km² occurrences

Native		GB	IR
1987–99	●	0	0
1970–86	●	0	0
pre 1970	●	0	0
Alien			
1987–99	●	69	20
1970–86	●	6	0
pre 1970	●	0	0

A deciduous shrub, widely grown in gardens and becoming naturalised on roadsides, river banks and waste ground. It usually originates from gardens or from discarded surplus garden material, or as a relic of cultivation. Suckers are freely produced, enabling dense thickets to form. Lowland.

Neophyte. The first dates of cultivation and discovery in the wild of *S.* × *rosalba* are unknown. It is likely to be increasing.

A hybrid of garden origin.

References: Rich & Jermy (1998), Silverside (1990a).

D. J. McCOSH

Spiraea douglasii Steeple-bush

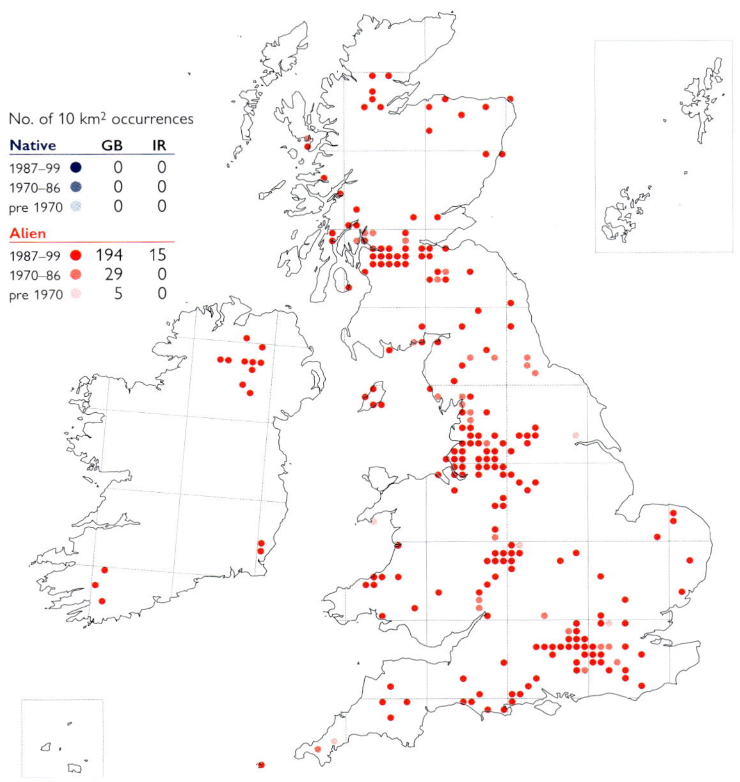

A deciduous shrub, naturalised by roads, in hedges or waste places, sometimes forming large patches. It also occurs as a relic of cultivation. Lowland.

Neophyte. *S. douglasii* has been cultivated in Britain since 1827 and it was known from the wild by 1910. It may have been over-recorded in the past due to confusion with its hybrids.

Native of western N. America.

References: Bean (1980), Coats (1963), Hultén (1968), Rich & Jermy (1998).

D. J. McCOSH

Spiraea douglasii × *S. salicifolia* (*S.* × *pseudosalicifolia*) Confused Bridewort

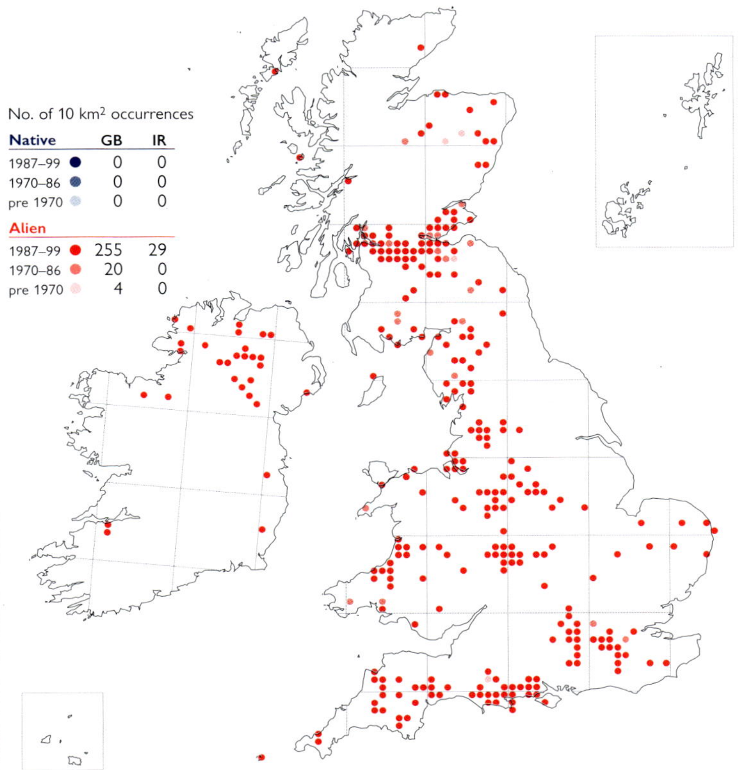

A perennial shrub, naturalised on roadsides and river banks where it forms dense thickets. Usually lowland, reaching 435 m at Nenthead (Cumberland).

Neophyte. This hybrid has been cultivated in Britain since about 1850 but was not noted in the wild until 1984. It is probably under-recorded.

A hybrid of garden origin.

References: Rich & Jermy (1998), Silverside (1990a).

D. J. McCOSH

Aruncus dioicus Buck's-beard

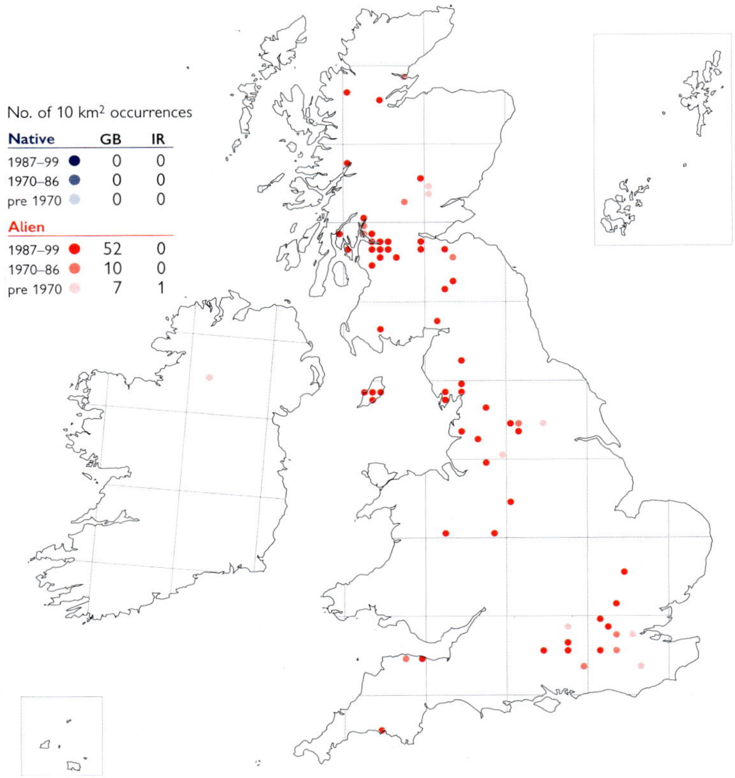

A very large dioecious perennial herb, widely grown in gardens and also planted in estate woodland. Rarely, it becomes abundantly naturalised, reproducing by seed when male and female plants exist together. More frequently it occurs as single-sex clones that spread only by means of rhizomatous growth. Lowland.

Neophyte. *A. dioicus* was cultivated in Britain by 1633. It was not recorded from the wild, however, until 1950.

A. dioicus has a disjunct, circumboreal distribution.

References: Hultén (1971), Meusel *et al.* (1965).

D. J. McCOSH

Filipendula vulgaris Dropwort

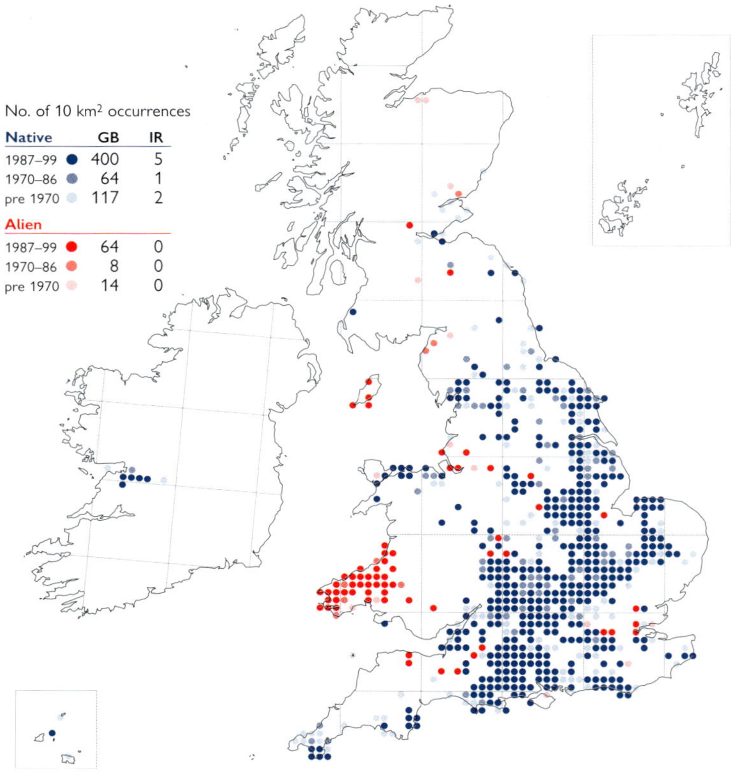

A perennial herb, mainly occurring in calcareous grassland on chalk and limestone downs, and in rough pasture; also found on coastal and inland heaths over limestone, chalk and other basic rocks, including serpentine. It is frequently planted in churchyards in W. Wales. Generally lowland, but reaching 365 m in N.E. Yorks.

Native (change –0.07). *F. vulgaris* has declined in its chalk habitats in S. England because of the lack of grazing or conversion of grassland to arable, but the distribution seems unchanged elsewhere. It is grown in gardens, sometimes escaping and becoming naturalised. Deciding whether populations are native or alien is difficult in some areas.

Eurosiberian Temperate element.

References: Atlas (118d), Curtis & McGough (1988), Hultén & Fries (1986), Meusel *et al.* (1965).

D. J. McCOSH

Filipendula ulmaria Meadowsweet

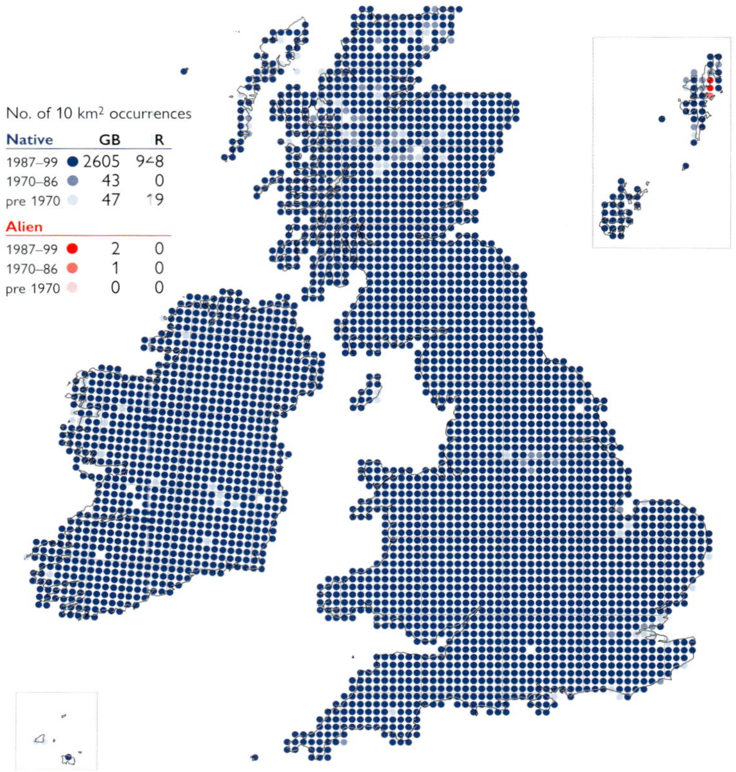

A perennial herb of damp or wet habitats, on moderately fertile, neutral or calcareous substrates; it is characteristic of sites where water levels fluctuate and is absent from permanently waterlogged ground. Typical habitats include wet woodland, damp meadows, swamps and tall-herb fens, damp roadsides, ditches and railway banks, and montane tall-herb communities. It also occasionally grows in somewhat drier conditions, such as N.-facing chalk grassland. It is intolerant of grazing and rarely flowers in shade. 0–880 m (Breadalbanes, Mid Perth).

Native (change –0.10). The distribution of *F. ulmaria* is stable.

Eurasian Boreo-temperate element; widely naturalised outside its native range.

References: Atlas (119a), Grime *et al.* (1988), Grose (1957), Hultén & Fries (1986), Meusel *et al.* (1965).

D. J. McCOSH

Kerria japonica Kerria

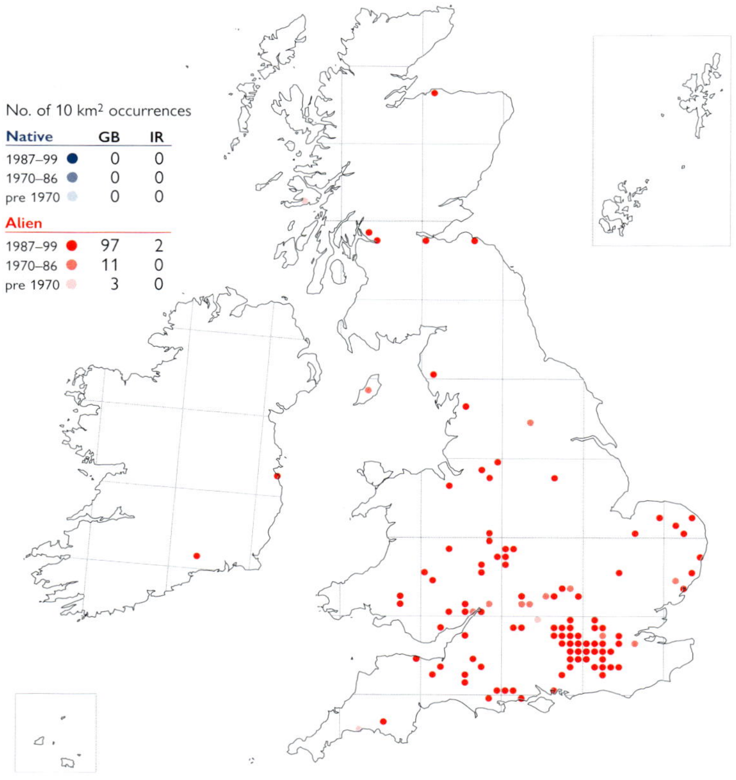

A deciduous shrub which occurs as a garden throw-out in woodland and hedgerows, and on river banks, roadsides and rubbish tips, and as a relic of cultivation. Reproduction by seed is rare, but it suckers readily and is often naturalised. Lowland.

Neophyte. *K. japonica* was introduced to cultivation in 1804 for its yellow flowers and, being easy to grow, is now extremely common in gardens. However, it was not formally recorded from the wild until 1965 (Berks.). It is probably increasing as garden material is continually discarded, but may have been ignored by some recorders in the past.

Native of China.

References: Bean (1973), Coats (1963).

D. J. McCOSH

Rubus chamaemorus Cloudberry

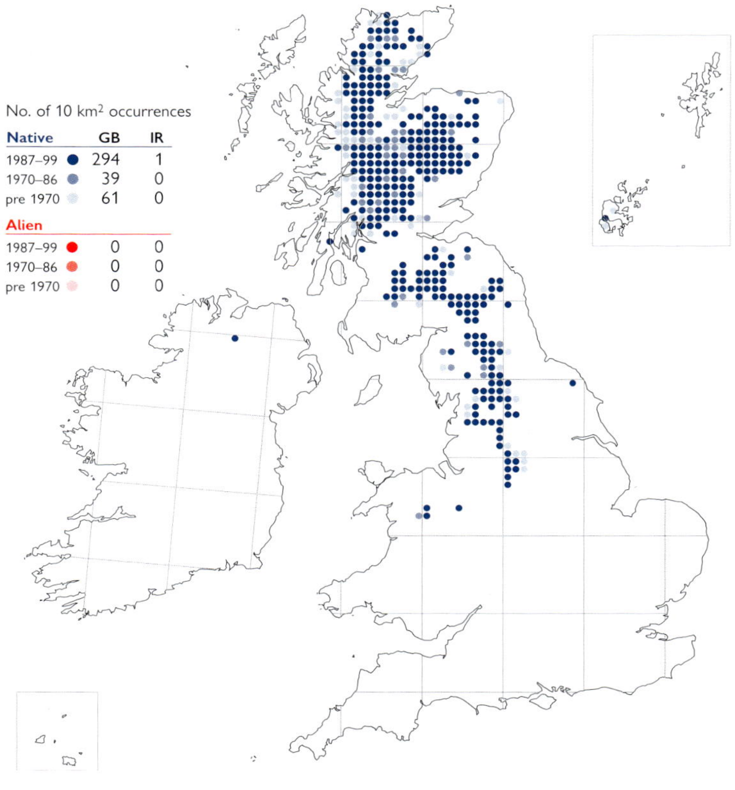

A dioecious herb of wet, base-poor peats on moorland and blanket mire, spreading by extensively creeping rhizomes and by seed. Usually above 600 m and reaches at least 1160 m (Beinn a' Bhuird, S. Aberdeen, and Carn Eige, E. Ross), but descends to *c.* 90 m in Denbighshire.

Native (change –0.47). The overall distribution of *R. chamaemorus* since the 1962 *Atlas* is stable, although there have been losses on the edges of its range, particularly in W. Scotland. These losses may be due to drainage, afforestation and moor-burning.

Circumpolar Boreal-montane element.

References: Atlas (119b), Curtis & McGough (1988), Edees & Newton (1988), Hultén & Fries (1986), Meusel *et al.* (1965), Taylor (1971).

D. J. McCOSH

Rubus tricolor Chinese Bramble

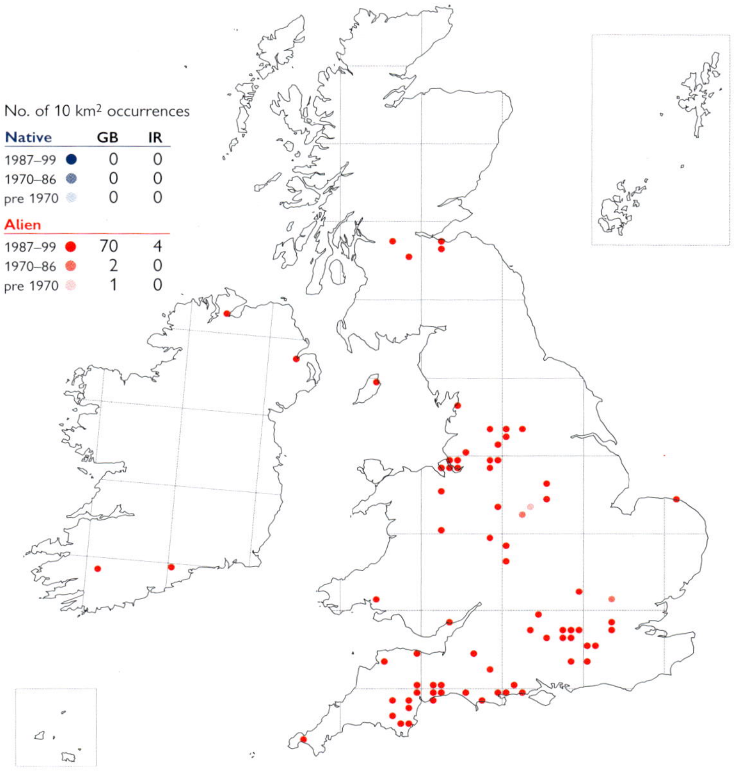

A procumbent semi-evergreen shrub, producing trailing stems several metres long. It is grown in gardens and is increasingly being mass-planted for ground-cover in parks and amenity plantings. It is found in hedges and woodland, and on roadsides, banks and waste ground. Reproduction is by rooting stem-tips and also by seed. Lowland.

Neophyte. This species was first cultivated in Britain in 1908. It was reported to be well-naturalised in the wild in 1976 (Derbys.). It is probably increasing.

Native of China.

Reference: Bean (1980).

D. J. McCOSH

Rubus saxatilis Stone Bramble

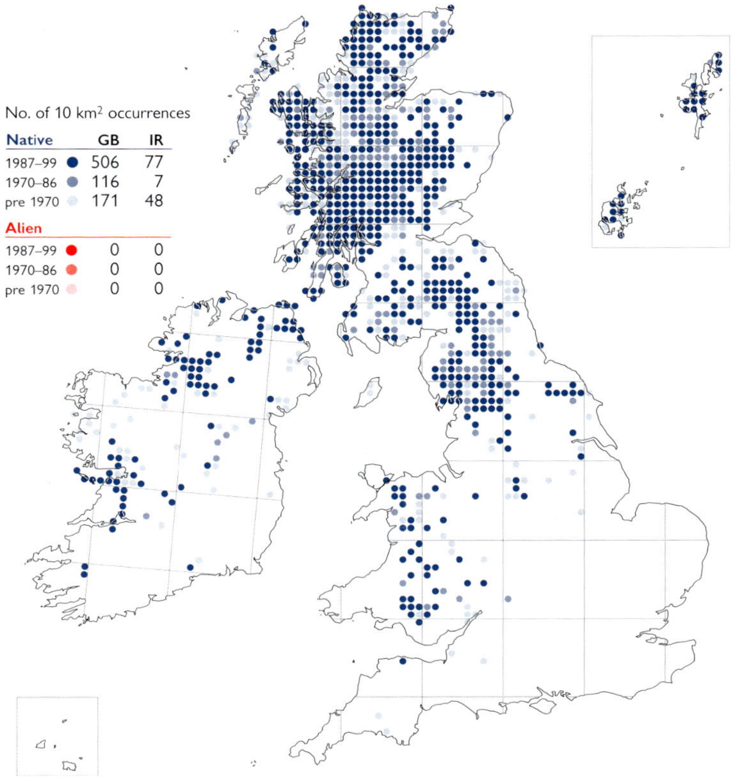

A perennial, stoloniferous deciduous herb, usually occurring on basic soils on crags, in ravines and in rocky woodland, occasionally in less acidic rocky heathland and on river-side shingle. 0–975 m (Breadalbanes, Mid Perth).

Native (change –0.27). *R. saxatilis* has been lost from some lowland areas, probably through habitat destruction and changes in woodland management. Many of these losses occurred before 1930, and the distribution is now largely stable. It is rarely frequent, and may be under-recorded in some areas.

Eurasian Boreo-temperate element; also in Greenland.

References: Atlas (119c), Edees & Newton (1988), Hultén & Fries (1986), Meusel *et al.* (1965).

D. J. McCOSH

Rubus arcticus Arctic Bramble

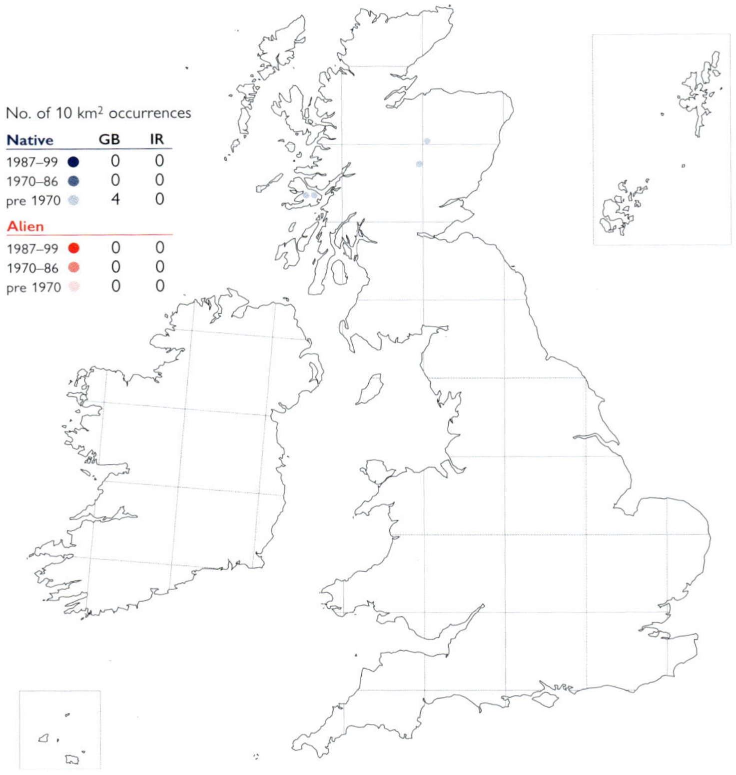

A herb with a creeping rootstock, from which arise annual stems. In Scotland it grew on mountains, but details of its habitats, associates and altitude are not known. In Finland it grows in damp, open *Betula*, *Pinus* or *Abies* woods, commonly near damp meadows, at fairly low altitudes.

Native. First recorded from Mull in 1768, there are herbarium specimens from a few scattered localities in Scotland, the last dating from 1850. The plant may be sterile in Britain, and it might have been brought here as seeds in the droppings of wintering birds.

Circumpolar Boreal-montane element; absent from mountains of C. Europe.

References: Edees & Newton (1988), Harley (1956), Hultén & Fries (1986), Marren (1999), Welch (1995).

D. A. PEARMAN

Rubus idaeus Raspberry

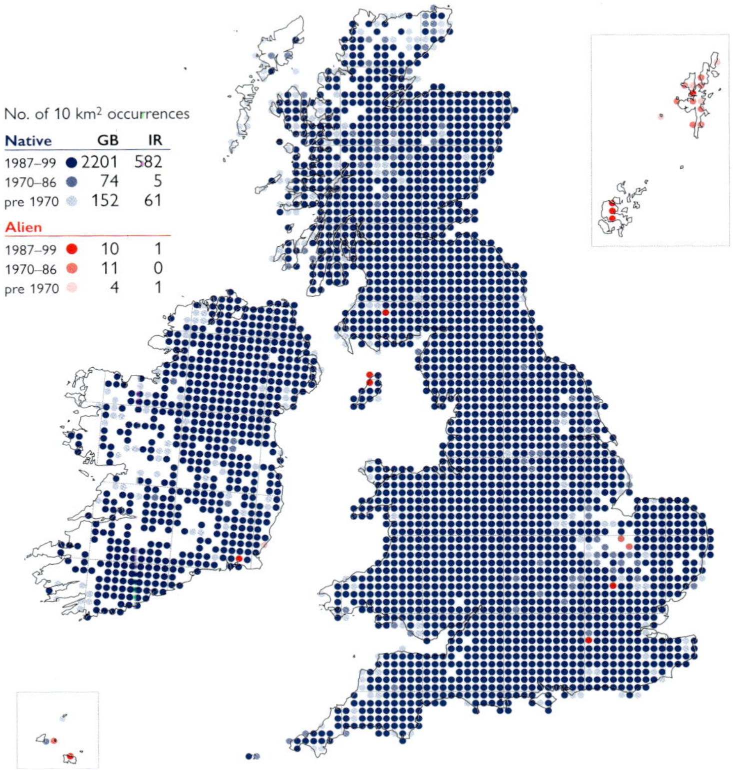

This shrub is found in open woodland, downland scrub, heathland, and sometimes in hedgerows. It also occurs on rough and waste ground as an escape from cultivation. In the uplands, it grows on the drier ledges of basic crags and ravines, and below base-rich cliffs. It spreads by bird-dispersed fruit, and by suckering, often forming thickets. 0–745 m (Atholl, E. Perth).

Native (change –0.09). *R. idaeus* has remained widespread, with no change in its distribution. It is very widely cultivated, and it can be difficult to separate native and alien populations.

Circumpolar Boreo-temperate element; widely naturalised outside its native range.

References: Atlas (119d), Edees & Newton (1988), Haskell (1960), Hultén & Fries (1986), Meusel *et al.* (1965), Roach (1985).

D. J. McCOSH

Rubus spectabilis Salmonberry

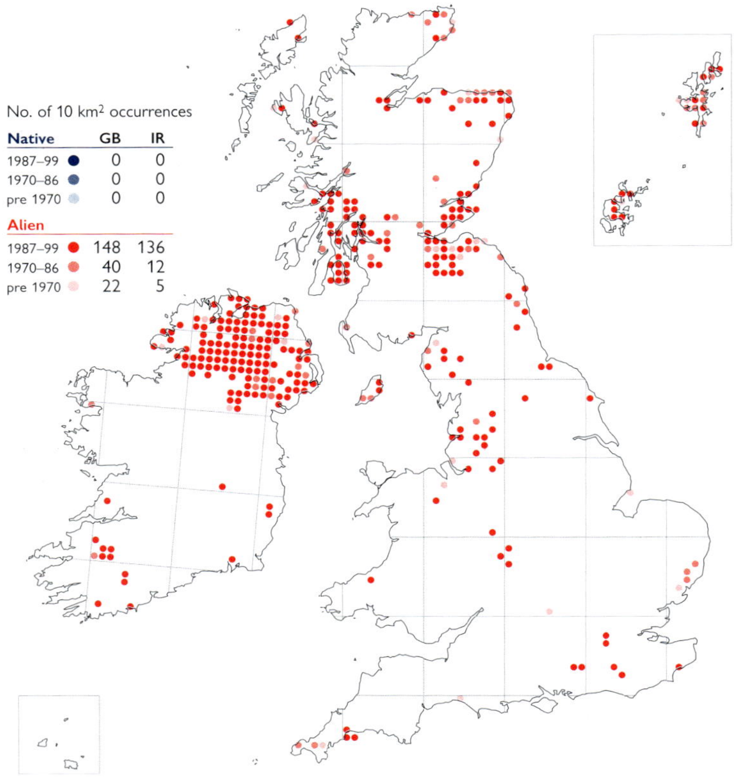

A deciduous, vigorous, suckering shrub, thoroughly naturalised in woods and hedges, often forming large thickets. Whilst most populations arise from plantings, it is also bird-sown. Lowland.

Neophyte. Introduced into cultivation in Britain in 1827, this species is planted in gardens for ornament and as game cover in estate woodland. It was known from the wild by at least 1899 and is probably increasing in some areas.

Native of western N. America.

References: Bean (1980), Edees & Newton (1988), Hultén (1968), Oleskevich *et al.* (1996).

D. J. McCOSH

Rubus fruticosus agg. Brambles

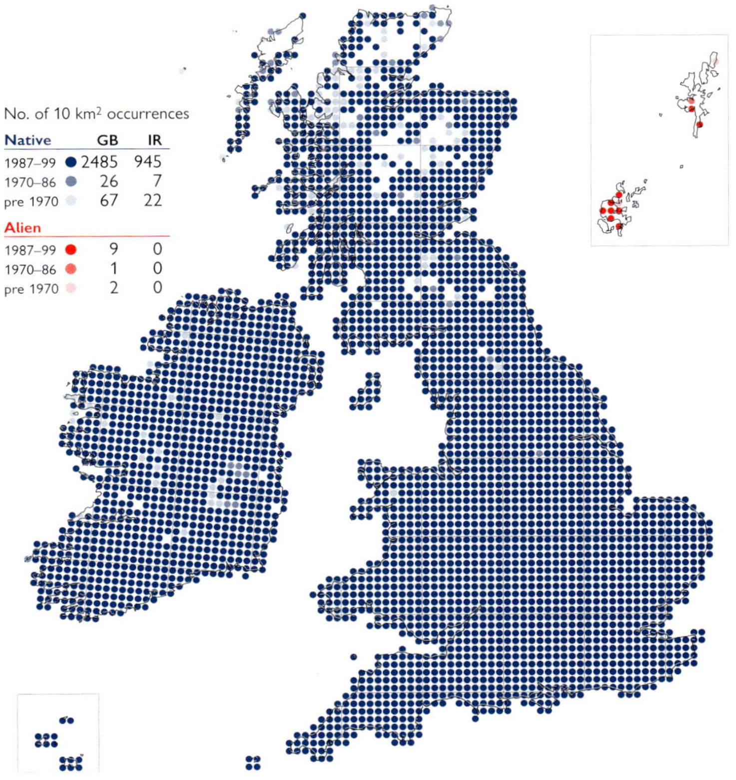

Deciduous or semi-evergreen shrubs of woods, scrub, banks, hedges, heaths and waste places. They can form dominant stands and although they have a very wide ecological tolerance they reach maximum vigour and diversity on acidic soils. They spread by bird-dispersed seeds, and by tip-rooting stems. 0–490 m (Harwood, Co. Durham).

Native (change –0.29). *R. fruticosus* is a taxonomically intractable aggregate of over 320 microspecies in *Rubus* subgenus *Rubus* (except *R. caesius*), most of which are facultatively apomictic. There is no change in the distribution of the aggregate since the 1962 *Atlas*.

European Southern-temperate element.

References: Atlas (120b), Atlas Supp. (22–27), Amor & Richardson (1980), Edees & Newton (1988), Grime *et al.* (1988), Hultén & Fries (1986), Roach (1985).

D. J. McCosh

Rubus caesius Dewberry

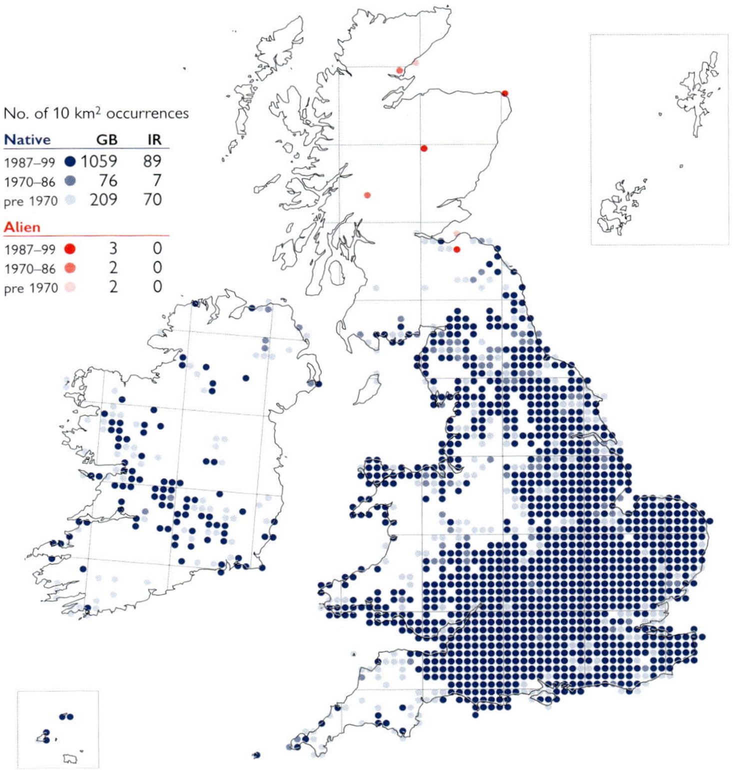

A deciduous shrub of hedges, woodland borders and rides, scrub, dry grassland and semi-stable dunes, mainly on basic soils; also in fen carr. 0–320 m (Alston, Cumberland).

Native (change –0.34). In the past *R. caesius* has been much confused with forms of *R. fruticosus* (series *Corylifolii*), which are sometimes regarded as having originated from hybrids between *R. caesius* and other microspecies of the *R. fruticosus* aggregate. The distribution was mapped as 'all records' in the 1962 *Atlas*. There seem to have been declines in the north and west of its range, and in Ireland.

Eurosiberian Temperate element; widely naturalised outside its native range.

References: Atlas (120a), Edees & Newton (1988), Hultén & Fries (1986), Meusel *et al.* (1965).

D. J. McCosh

Potentilla fruticosa Shrubby Cinquefoil

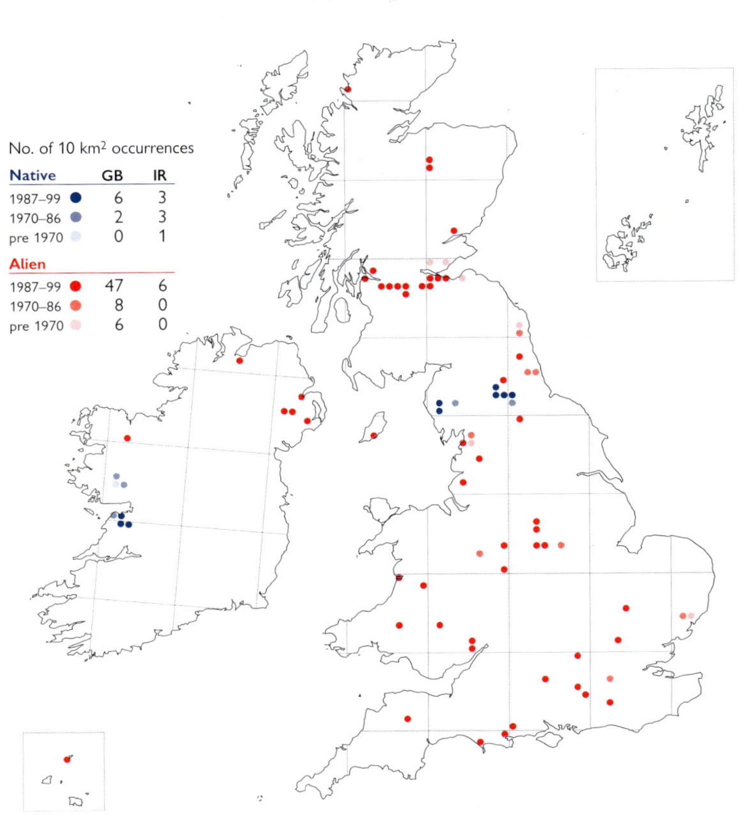

In England, this shrub is found on basic, damp rock ledges and silty, sandy or gravelly river-flats liable to inundation. In Ireland, it occurs in rocky places subject to flooding, usually around loughs and turloughs. It also occurs as a garden escape or relic in waste places. 0–700 m (Pillar, Cumberland).

Native (change +1.44). Drainage and land clearance have caused some losses of this species in Ireland, and some English populations were lost before 1940. However, the current native distribution is stable and alien plants are increasing in frequency.

Circumpolar Boreal-montane element, but absent from Scandinavia and with a disjunct distribution elsewhere.

References: Atlas (120c), Curtis & McGough (1988), Elkington & Woodell (1963), Hultén & Fries (1986), Meusel *et al.* (1965), Wigginton (1999).

D. J. McCosh

Potentilla palustris Marsh Cinquefoil

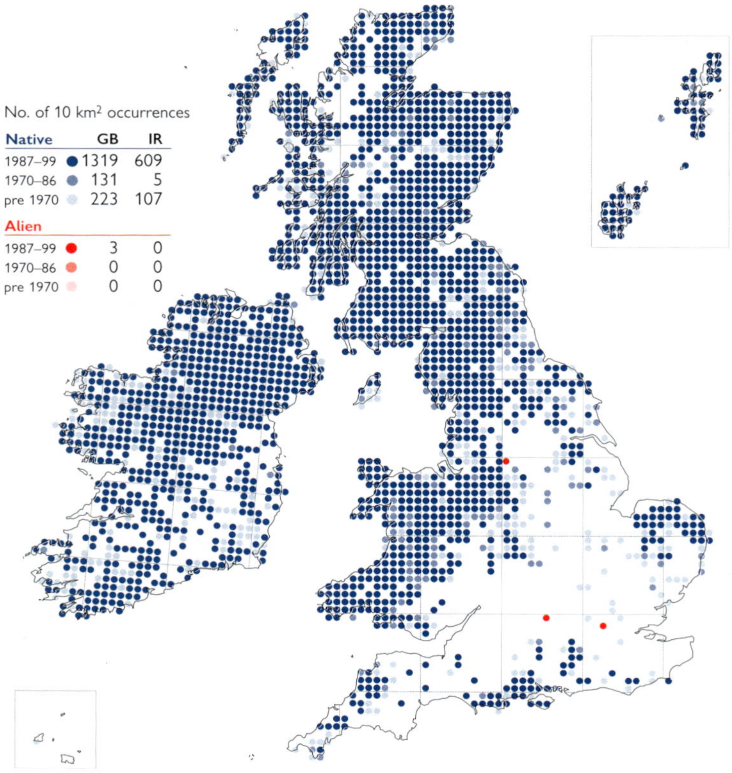

A rhizomatous perennial herb found in permanently flooded swamps, and in mires and wet meadows where the summer water table lies below the soil surface. It prefers nutrient-poor but slightly to moderately base-rich water and grows on a wide range of soils. Habitats include the edges of lakes, natural hollows, bog pools, peat cuttings and floating rafts of vegetation. 0–800 m (Tom a'Choinnich, E. Ross).

Native (change −0.21). *P. palustris* was mapped as 'all records' in the 1962 *Atlas*. It has declined since that date in S. and E. England and S. Ireland through drainage, agricultural improvement and a lack of grazing.

Circumpolar Boreo-temperate element.

References: Atlas (120d), Byfield & Pearman (1996), Hultén & Fries (1986), Meusel *et al.* (1965), Preston & Croft (1997).

D. J. McCOSH

Potentilla anserina Silverweed

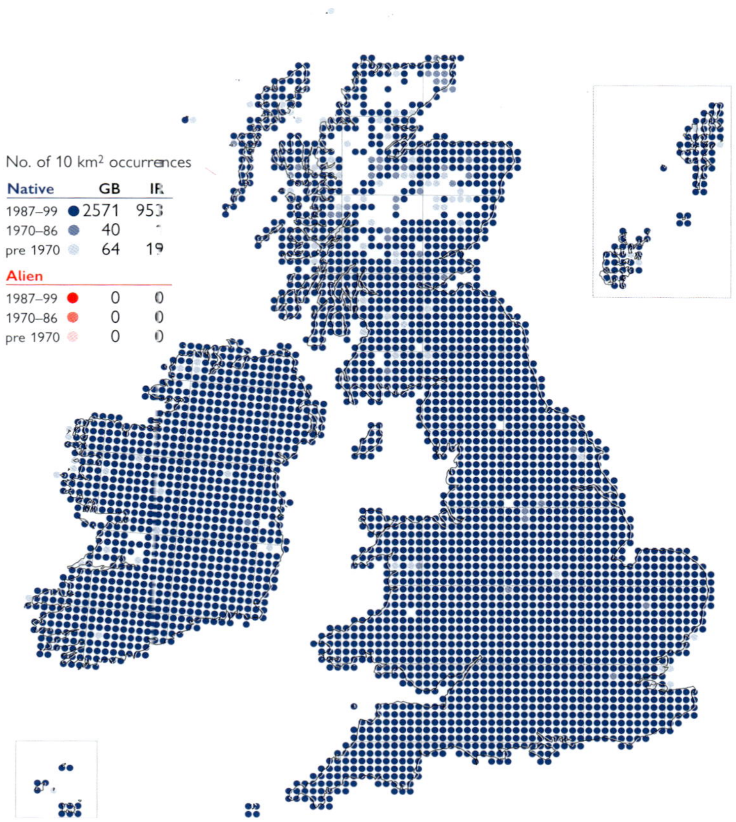

A stoloniferous perennial herb, occurring on dry, damp or periodically inundated soils in open grassy swards or on bare ground. Habitats include land subject to seasonal inundation with fresh or brackish water, upper saltmarsh, shorelines, dunes, rough ground and roadsides. It spreads vigorously by stolons, but fruiting is often poor. Generally lowland, but reaching 580 m at Hartside (Cumberland), and exceptionally at 845 m (Great Dun Fell, Westmorland).

Native (change −0.23). There has been no appreciable change in the distribution of this species since the 1962 *Atlas*.

Circumpolar Boreo-temperate element; widely naturalised outside its native range.

References: Atlas (121b), Hultén & Fries (1986), Meusel *et al.* (1965), Miyanishi *et al.* (1991), Ockendon & Walters (1970).

M. J. WIGGINTON

Potentilla rupestris Rock Cinquefoil

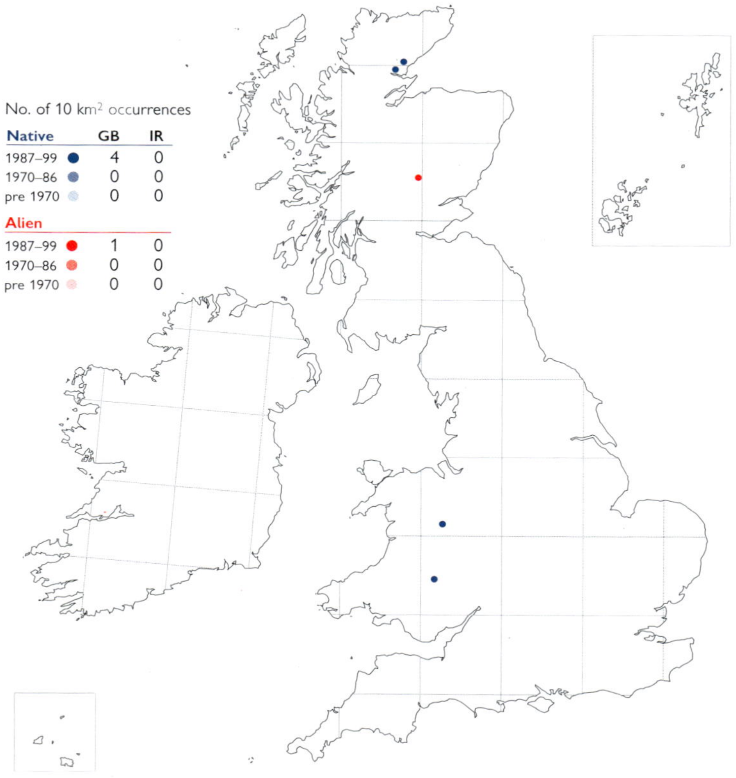

A perennial herb which, in three of its four native sites, grows in thin, dry soils in open, rocky or cliff habitats subject to summer drought. It grows over basic igneous rocks, but usually on mildly acidic soils. The southernmost site in Wales is a rock platform by a river, partially shaded by trees. Lowland.

Native. All populations are small, but currently appear to be more or less stable. The Craig Breidden (Monts.) site was much affected in the past by quarrying and the plant came close to extinction there, but the population has been bolstered by transplants.

Mediterranean-montane element; widely naturalised outside its native range.

References: Atlas (120c), Hultén & Fries (1986), Kay & John (1995), Meusel *et al.* (1965), Wigginton (1999), Wilson, Whittington *et al.* (1995).

M. J. WIGGINTON

Potentilla argentea Hoary Cinquefoil

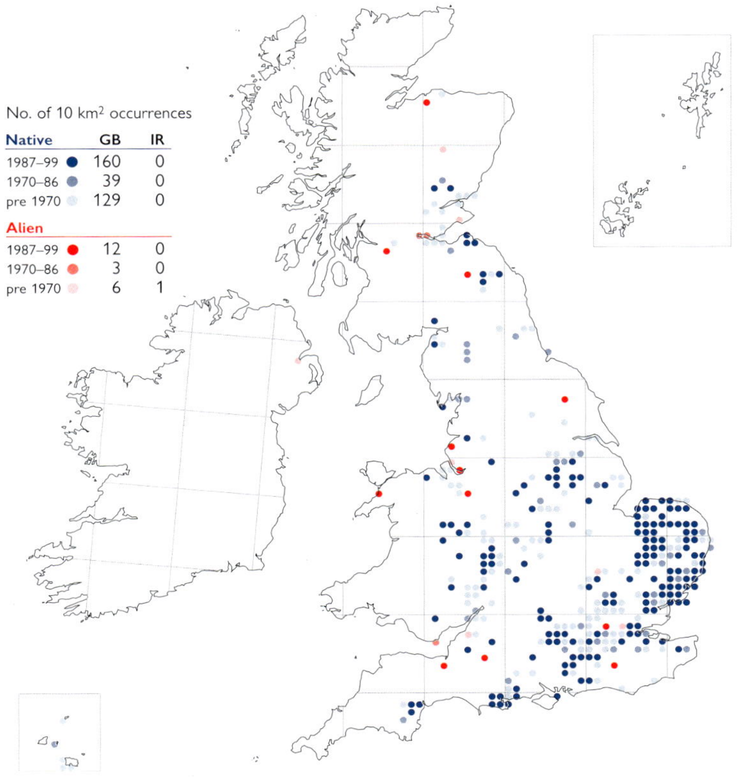

A perennial herb of dry, freely-draining, gravelly or sandy soils, found in open grassy swards on commons, in pastures, on banks, in pits and on tracks and waste ground. Reproduction is by seed, but it does not readily colonise new sites. Lowland.

Native (change –0.78). The 1962 *Atlas* indicates a decline in *P. argentea* before 1930, and more sites have been lost subsequently. It may still be in slow decline, largely due to habitat loss. However, although populations can fluctuate dramatically in size they can be extremely long-lived.

Eurosiberian Temperate element; widely naturalised outside its native range.

References: Atlas (121c), Hultén & Fries (1986), Meusel *et al.* (1965), Stewart *et al.* (1994), Werner & Soule (1976).

D. J. McCOSH

Potentilla recta Sulphur Cinquefoil

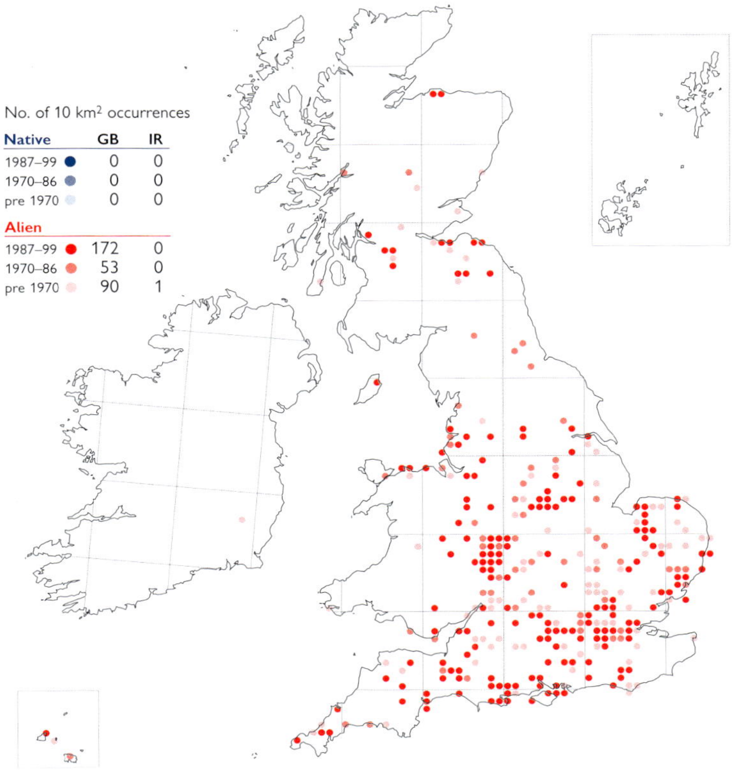

A perennial herb, originating from gardens or as a contaminant of grass seed and naturalised on waste ground, roadside banks and grassy places; rarely occurring as a casual. Lowland.

Neophyte (change +0.99). *P. recta* was introduced into Britain by 1648, and was known from the wild by 1858 (Middlesex). It seems to have become more frequent since the 1962 *Atlas*, but this may be an artifact of better recording.

A European Boreo-temperate species, naturalised in Scandinavia north of its native range, and in N. America.

References: Atlas (121d), Hultén & Fries (1986), Werner & Soule (1976).

D. J. McCOSH

Potentilla intermedia Russian Cinquefoil

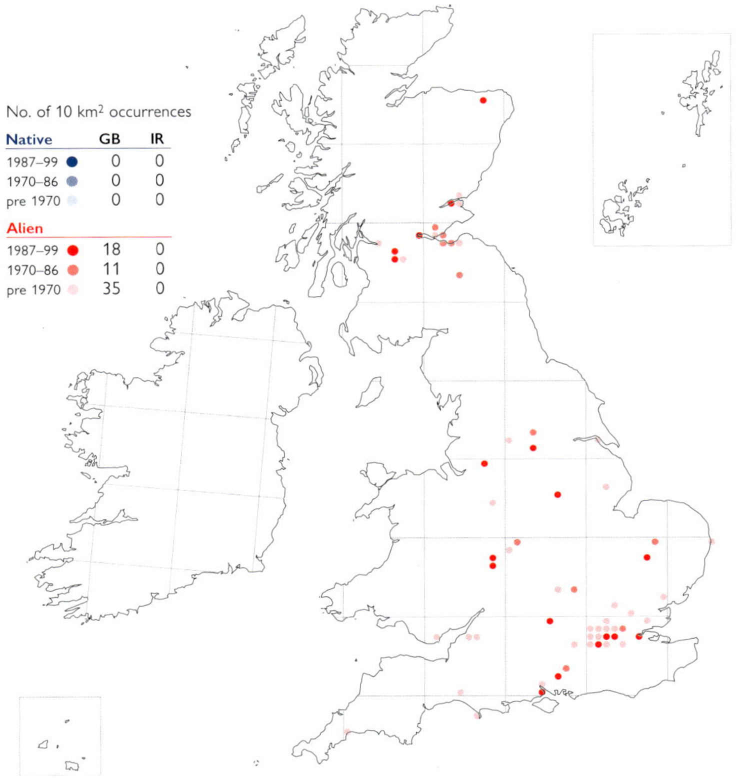

A biennial or short-lived perennial herb, found as a casual or naturalised grain alien in grassy places and waste ground. Lowland.

Neophyte. *P. intermedia* was brought into cultivation in Britain by 1786. It was first recorded in the wild in 1866 (Fife). It has not been recorded as frequently in Middlesex as it was before 1970, and appears to be declining elsewhere.

Native of European Russia; a widespread casual or naturalised plant in Europe, E. Asia and N. America.

Reference: Hultén & Fries (1986).

D. J. McCOSH

Potentilla norvegica Ternate-leaved Cinquefoil

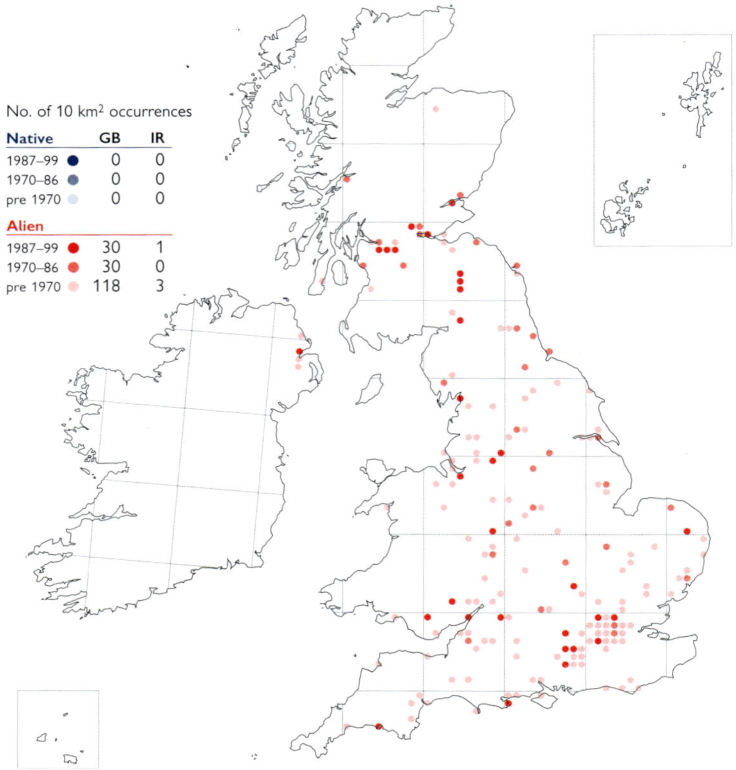

An annual or short-lived perennial herb found in waste places, quarries and on old railway lines. It is often introduced with grain or bird-seed. Populations are usually casual, but may occasionally become naturalised. Lowland.

Neophyte (change –1.68). *P. norvegica* was introduced into cultivation in Britain in 1680, but was frequently found as a grain impurity in the late 19th century, being known from the wild by 1868 (E. Gloucs.). It has declined, except in the Forth-Clyde valley where the cluster of recent records indicates that it may be increasing.

A Circumpolar Boreo-temperate species, absent as a native but naturalised or casual in W. Europe.

References: Atlas (122a), Hultén & Fries (1986), Werner & Soule (1976).

D. J. McCOSH

Potentilla crantzii Alpine Cinquefoil

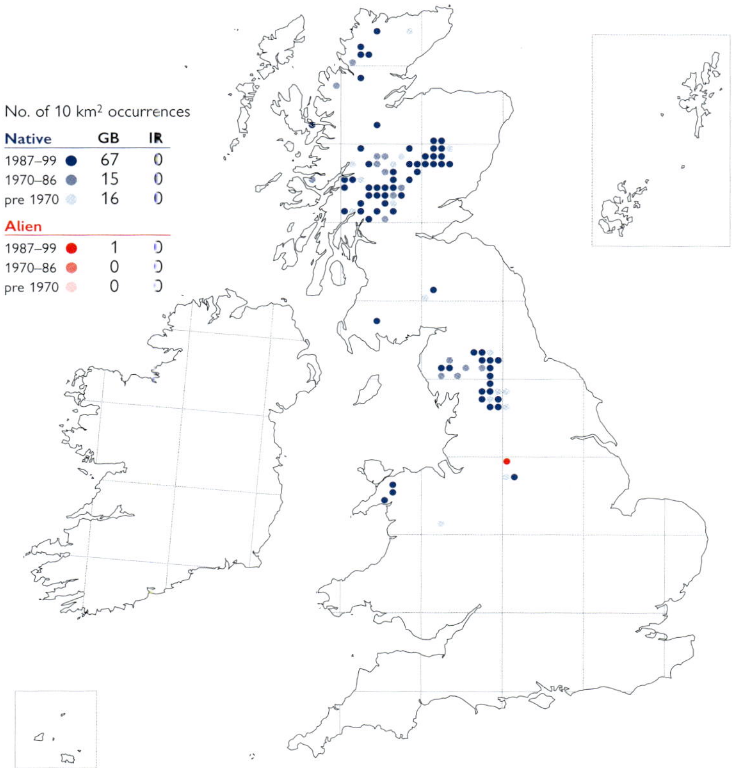

A perennial herb of dry base-rich rock faces, cliffs and ledges, close-grazed calcareous grassland and, occasionally, river shingle. It is a pseudogamous apomict, reproducing mostly by seed, with very limited vegetative spread. Generally montane, reaching 1065 m on Ben Lawers (Mid Perth), but descending to 250 m in Assynt (W. Sutherland).

Native (change –0.21). Although some sites were lost before 1930, the distribution of this species seems to have been stable in recent years. Unlike *P. neumanniana*, British populations of *P. crantzii* are relatively homogeneous.

Eurosiberian Boreo-arctic Montane element; also in N. America.

References: Atlas (122c), Hultén & Fries (1986), Meusel *et al.* (1965), Stewart *et al.* (1994).

D. J. McCOSH

Potentilla neumanniana Spring Cinquefoil

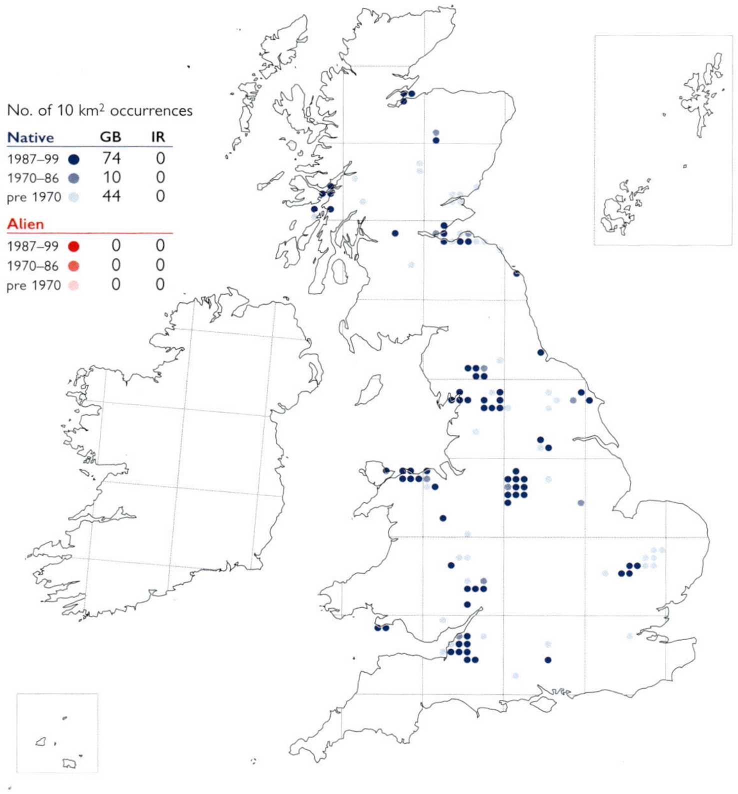

An apomictic, perennial, mat-forming herb of undisturbed, dry, basic, open habitats including coastal limestone rocks, inland rocks, crags and screes on limestones and other basic rocks, and dry chalk grassland. Reproduction is by seed and by rooting stolons. Generally lowland, but reaching 335 m at Crosby Ravensworth (Westmorland); there are old records at 610 m from Loch Loch (E. Perth).

Native (change –0.17). This species has declined or been lost from many sites in S. England, particularly those on chalk and sand (such as Breckland). Populations on limestone crags and cliffs appear to be more stable. *P. neumanniana* is morphologically and cytologically variable.

European Temperate element.

References: Atlas (122b), Hultén & Fries (1986), Meusel *et al.* (1965), Stewart *et al.* (1994).

D. J. McCOSH

Potentilla erecta Tormentil

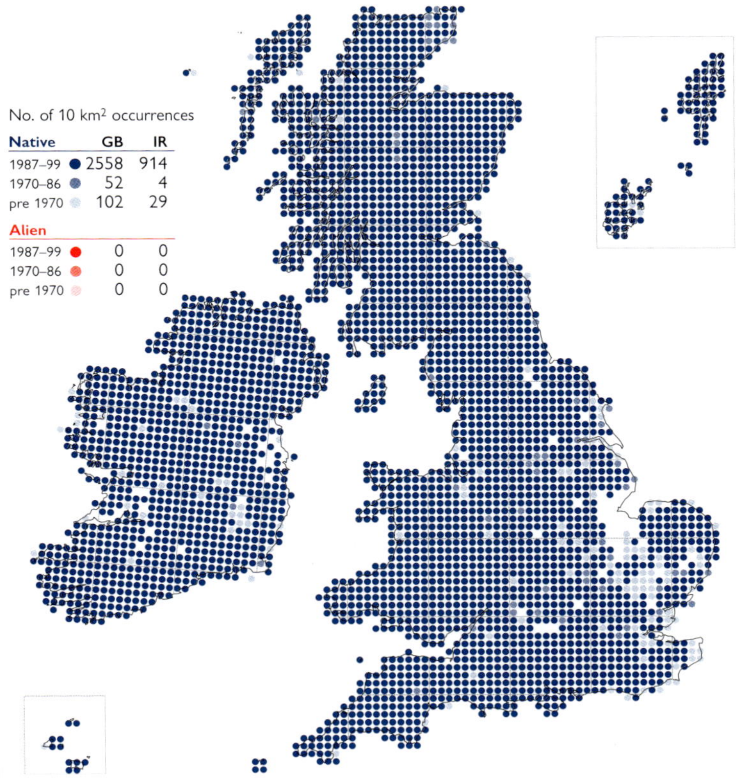

A perennial herb found in a wide variety of habitats on more or less acidic soils, including lowland, upland and montane grassland, hay- and fen-meadows, moorland and heathland, blanket and raised mires, open woodland, wood borders and hedge banks. It is unpalatable to stock. 0–1040 m (Carrantuohill, S. Kerry).

Native (change –0.50). The distribution of *P. erecta* is stable, other than in S.E. England where a decline has taken place. Local floras indicate that these losses have occurred since 1950. Two subspecies are recognised; subsp. *erecta* is found throughout the range of the species and subsp. *strictissima* is mapped separately.

Eurosiberian Boreo-temperate element.

References: Atlas (122d), Grime *et al.* (1988), Hultén & Fries (1986), Meusel *et al.* (1965).

D. J. McCOSH

Potentilla erecta subsp. *strictissima*

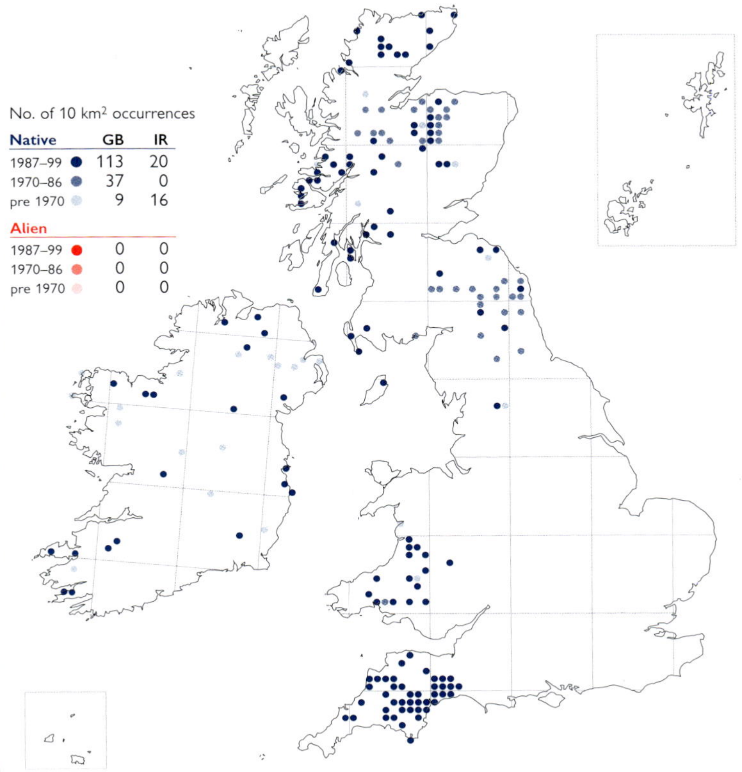

A perennial herb found in a variety of acidic habitats on peaty or skeletal mineral soils. Typical habitats include heathland, moorland, rocks and rock ledges, and occasionally grassland. Found at a range of altitudes but predominantly upland, reaching 920 m on Coire an Lochan (Easterness).

Native. Although this subspecies is under-recorded, its distribution is much better known than when mapped by Richards (1973).

European Boreal-montane element.

Reference: Rich & Jermy (1998).

T. D. DINES

Potentilla anglica Trailing Tormentil

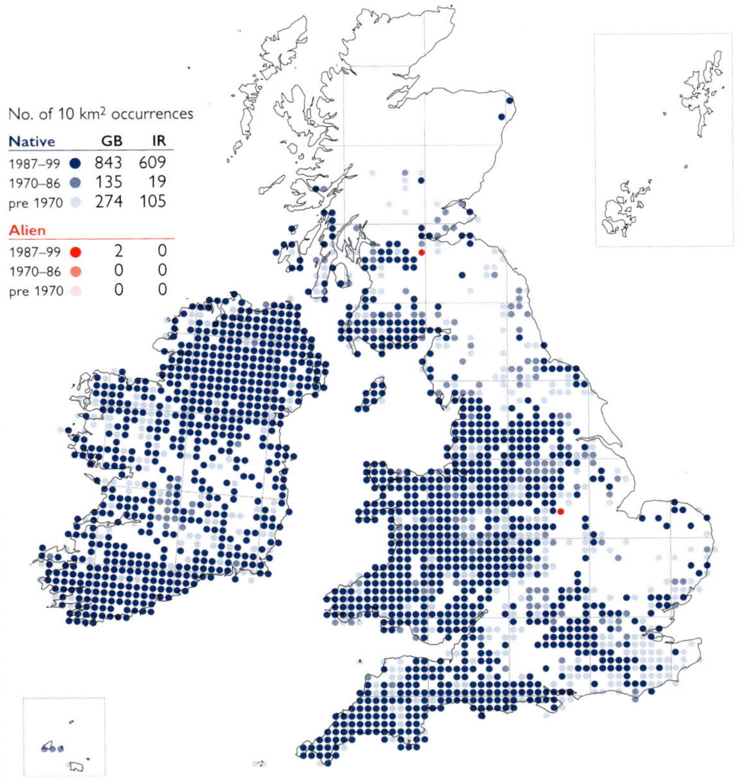

A procumbent perennial of heaths, dry banks, woodland borders and field edges, usually on well-drained acidic soils, but avoiding podsols. It can also occur on waste ground and railway banks. 0–410 m (Fron Hill, Rads.).

Native (change +0.11). *P. anglica* can be morphologically very similar to some forms of its hybrids with *P. erecta* or *P. reptans*, and the map may contain some errors. It was mapped as 'all records' in the 1962 *Atlas*. The current map suggests a decline in S. and E. England and C. Ireland. Analysis of the database reveals that many of these losses have occurred since 1950.

European Temperate element; also in N. America.

References: Atlas (123a), Hultén & Fries (1986), Meusel *et al.* (1965), Rich & Jermy (1998).

D. J. McCOSH

Potentilla anglica × *P. erecta* (*P.* × *suberecta*)

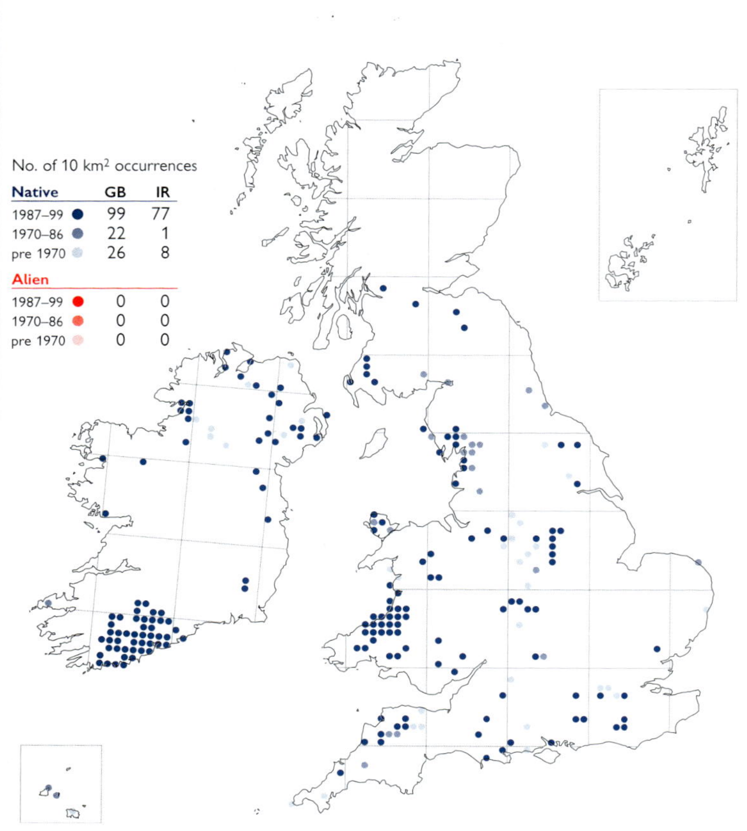

This hybrid occurs in the vicinity of the parents, rarely in isolation from them, on heaths, dry banks, woodland borders and field edges, usually on well-drained acidic soils. It is partially sterile, but a few achenes may be formed. Generally lowland, but upper altitudinal limit unknown.

Native. Seasonal and environmental variation in both *P. anglica* and *P.* × *suberecta* can make identification of this hybrid very difficult, and it may be over-recorded in some areas and under-recorded in others.

Apparently widespread in temperate regions of Europe.

References: Rich & Jermy (1998), Stace (1975).

D. J. McCOSH

Potentilla anglica × *P. reptans* & *P. erecta* × *P. reptans* (*P.* × *mixta*) Hybrid Cinquefoils

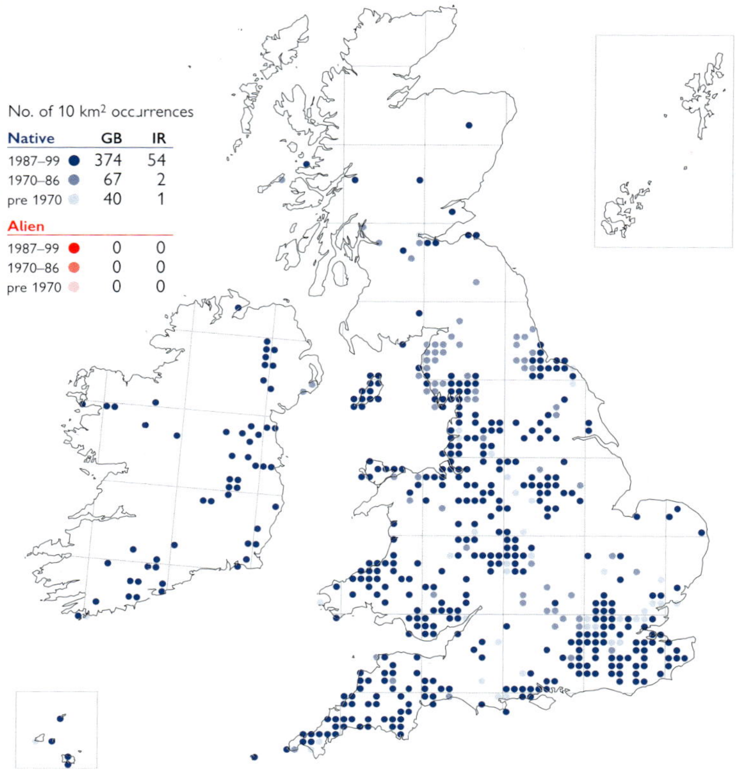

These hybrids are found on roadsides, hedge banks, in rough bare places and waste ground, and sometimes in woodland rides. They may occur with any of the parents, or in isolation from them. They are sterile, spreading by runners. Generally lowland, but upper altitudinal limit unknown.

Native. This map includes records of *P.* × *mixta sens. str.* (*P. anglica* × *P. reptans*) and *P.* × *italica* (*P. erecta* × *P. reptans*), which are impossible to separate in the field. Some forms of the parents can be very similar to their hybrids. The hybrids are probably much under-recorded.

These hybrids are common in Europe.

References: Rich & Jermy (1998), Stace (1975).

D. J. McCOSH

Potentilla reptans Creeping Cinquefoil

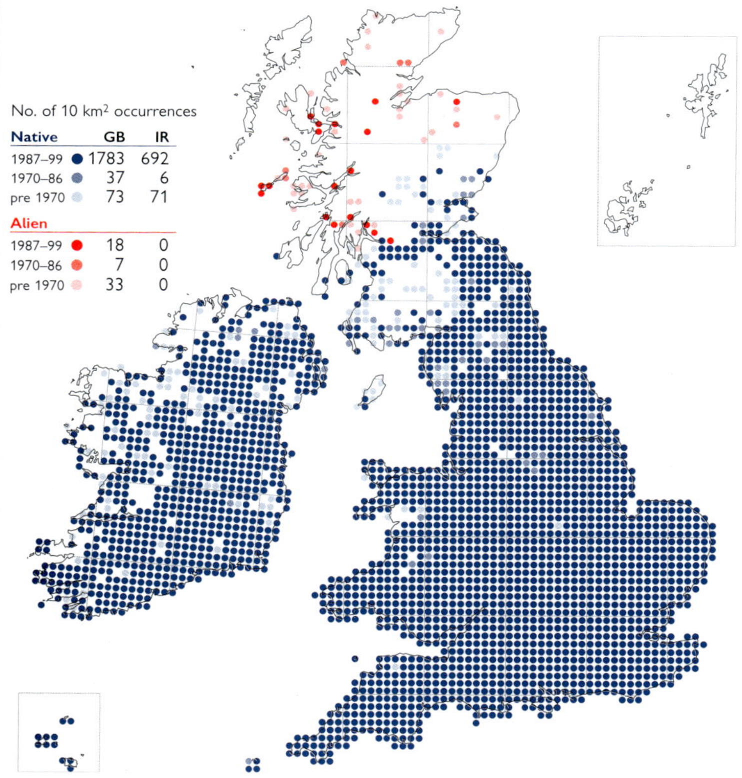

A perennial herb of woodland rides, grassland, hedgerows, banks and roadsides, and waste and cultivated ground, generally on neutral to basic soils. It reproduces by seed and spreads rapidly by runners. Generally lowland, reaching 415 m at Stainmore (Westmorland).

Native (change −0.62). The distribution of *P. reptans* is generally stable. It may only be present as an introduction at scattered localities in Scotland north of its main range.

Eurosiberian Southern-temperate element; widely naturalised outside its native range.

References: Atlas (123b), Hultén & Fries (1986), Rich & Jermy (1998).

D. J. McCOSH

Potentilla sterilis Barren Strawberry

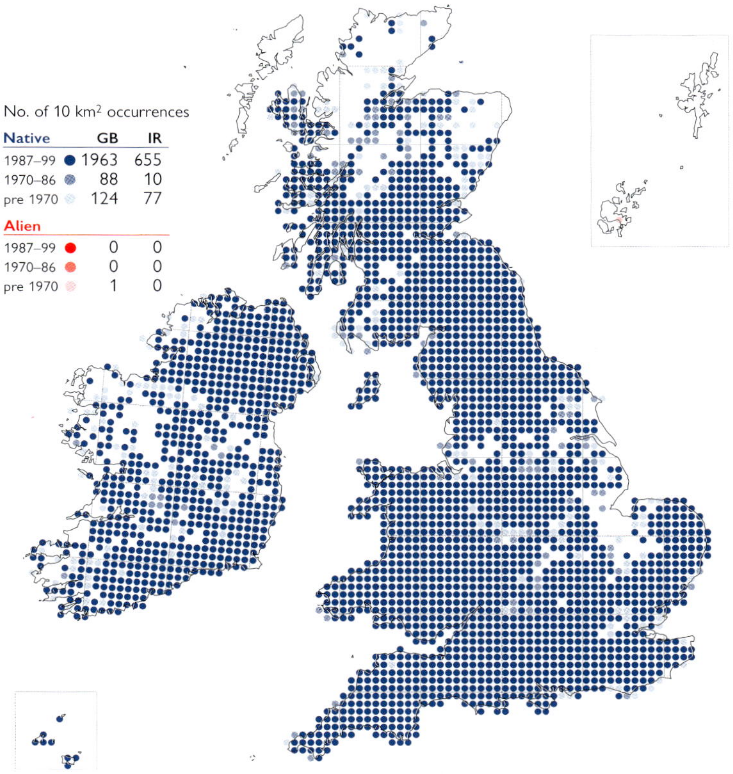

A perennial herb of relatively infertile, dry but not droughted soils in open woods, woodland margins, scrub, grassy hedge banks and rock crevices; also occasionally in meadows and on walls. In the lowlands it is usually found in partially shaded sites but it extends into open habitats in upland areas. 0–790 m (Helvellyn, Cumberland).

Native (change –0.30). There is no change in the distribution of *P. sterilis* since the 1962 *Atlas*.

Suboceanic Temperate element; also in N. America.

References: Atlas (121a), Dupont (1962), Grime *at al.* (1988), Hultén & Fries (1986), Meusel *et al.* (1965).

D. J. McCOSH

Sibbaldia procumbens Sibbaldia

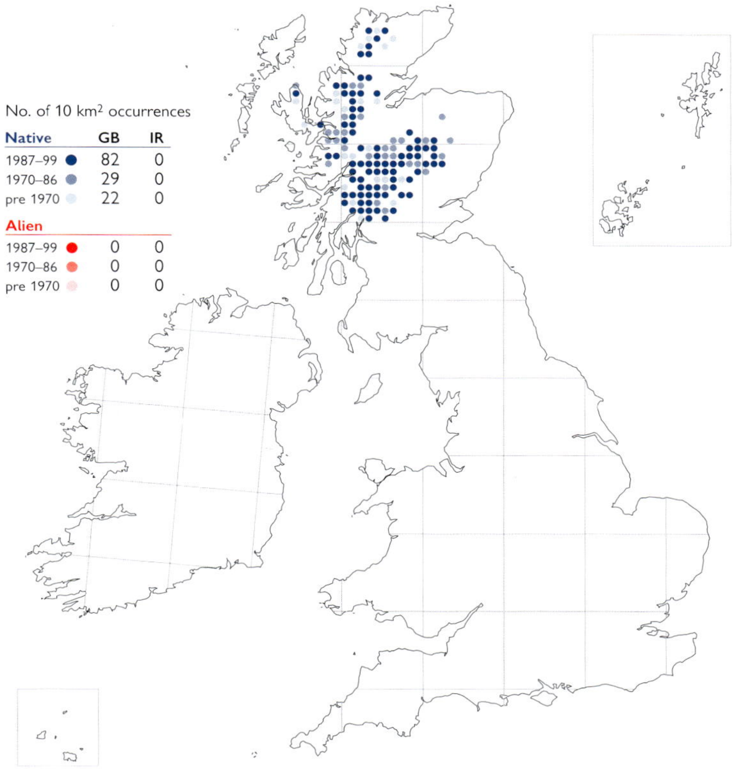

This montane perennial herb has two main habitats. It is most abundant in areas of late snow-lie, in corries and hollows and especially under cornices along a ridge. Its other main habitat is on bare, stony surfaces of high plateaux, often in areas of severe wind-scour where permanent snow does not lie. From 425 m (Sgurr na Coinnich, Skye, N. Ebudes) to 1310 m (Ben Nevis, Westerness).

Native (change –0.75). *S. procumbens* is under-recorded in some areas, though the map is more complete than in the 1962 *Atlas*. It is probably still present in 10-km squares for which there are only pre-1987 records.

Circumpolar Arctic-montane element, with a disjunct distribution.

References: Atlas (123c), Coker (1966), Hultén & Fries (1986), Meusel *et al.* (1965), Stewart *et al.* (1994).

D. J. McCOSH

Fragaria vesca Wild Strawberry

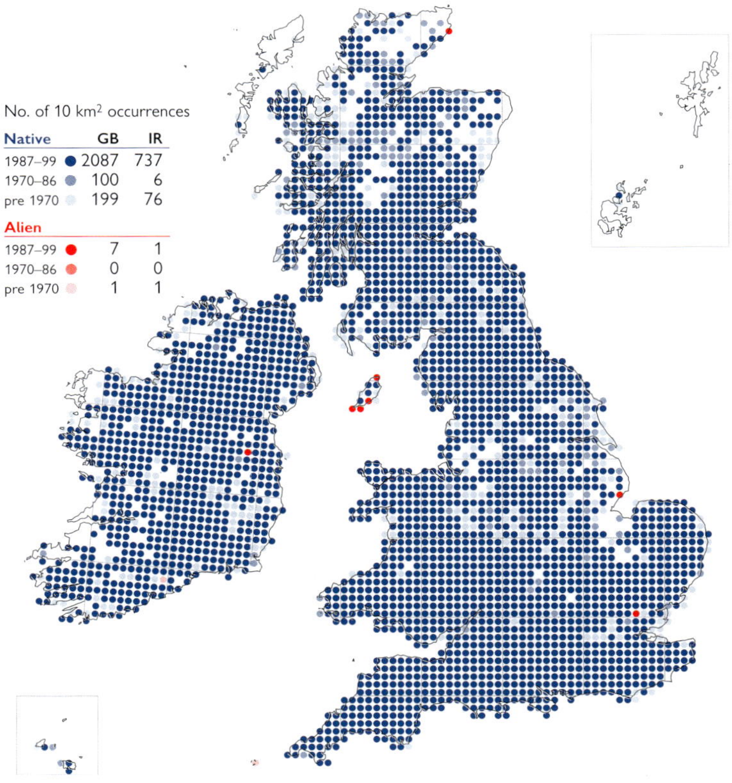

A perennial stoloniferous herb of dry, sometimes stony, soils in woodland and scrub, on hedge banks, railway banks and roadsides and on basic rock outcrops and screes in upland areas. It also colonises open ground in quarries and chalk-pits, and grows on walls. It reproduces by seed, and spreads by rooting stolons. 0–640 m (Atholl, E. Perth).

Native (change –1.09). The map suggests that there may have been a decline in the frequency of *F. vesca* since the 1962 *Atlas*.

Eurosiberian Temperate element; also in N. America and widely naturalised outside its native range.

References: Atlas (123d), Grime *et al.* (1988), Hultén & Fries (1986), Meusel *et al.* (1965), Roach (1985).

D. J. McCOSH

Fragaria moschata Hautbois Strawberry

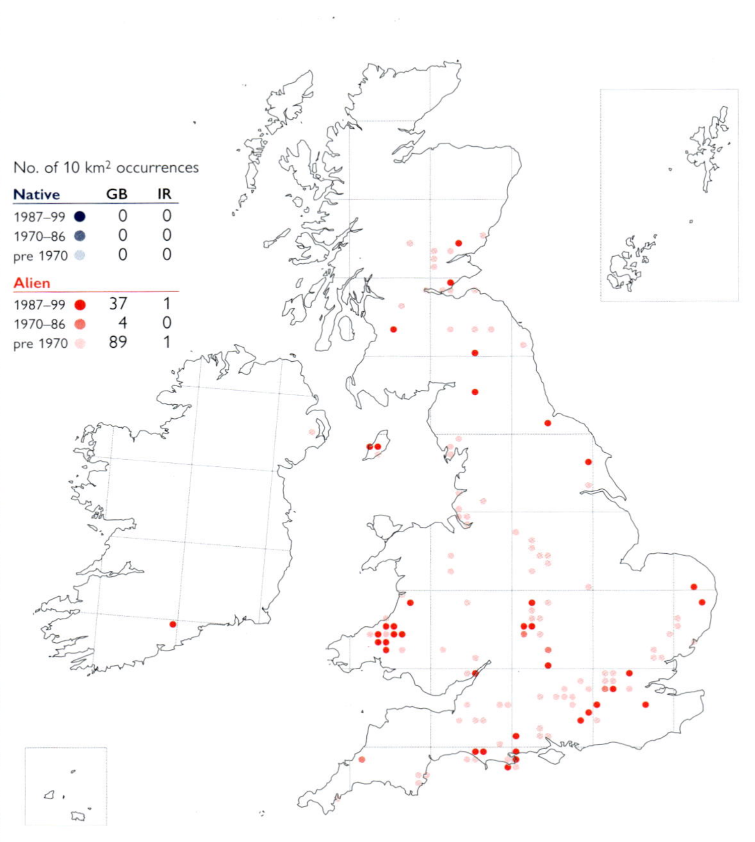

This stoloniferous perennial herb is found naturalised in woodland and hedgerows and on roadsides. Lowland.

Neophyte. This species was cultivated in Britain by 1629 and was formerly grown for its fruit. It was recorded in the wild in 1810, and has been known at two places in Surrey (Hascombe and West Horsley) since 1931. It may be over-recorded for other *Fragaria* taxa.

A European Temperate species, with a native distribution centred on C. Europe but naturalised north to Scandinavia.

References: Hultén & Fries (1986), Roach (1985).

D. J. McCOSH

Fragaria × *ananassa* Garden Strawberry

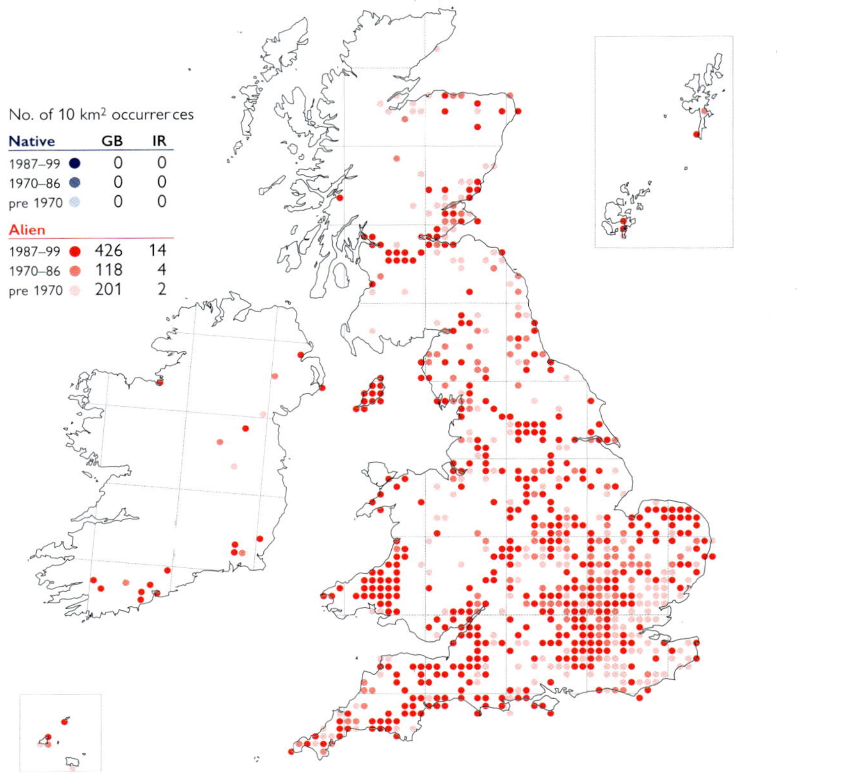

A perennial stoloniferous herb, widely naturalised on railway banks and in waste places, and a casual on rubbish tips. It usually grows as a naturalised plant on deeper soils than *F. vesca*. Lowland.

Neophyte (change +0.42). This commonly cultivated hybrid has been grown in Britain since 1806 and is a frequent escape from gardens or fruit farms, or arises from discarded or bird-sown fruits. Early records were confused with *F. moschata*, but it was certainly present in the wild by 1900. There are many more records than in the 1962 *Atlas*, probably due to both a genuine increase in the wild and improved recording of aliens.

A hybrid between two American species, developed in France in the 18th century.

References: Atlas (124a), Roach (1985), Stace (1975).

D. J. McCOSH

Duchesnea indica Yellow-flowered Strawberry

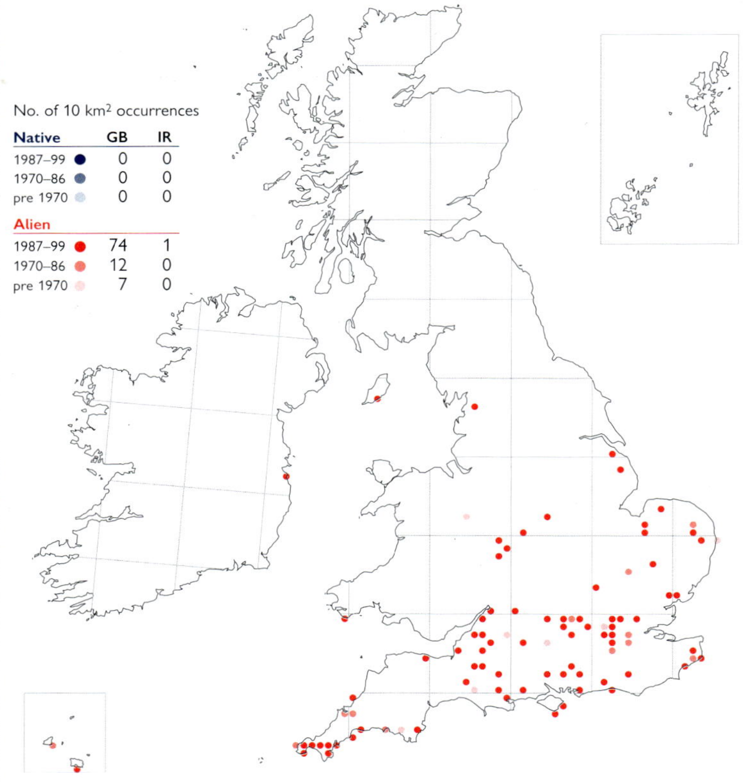

A stoloniferous perennial herb found naturalised in woodland and churchyards, and on banks, tracks and roadsides. Lowland.

Neophyte. *D. indica* has been in cultivation in Britain since 1805, and grown in gardens for its yellow flowers and strawberry-like, but inedible, fruits. It was recorded from the wild by 1879 and appears to be increasing.

Probably native in S. & E. Asia; widely naturalised elsewhere.

D. J. McCOSH

Geum rivale Water Avens

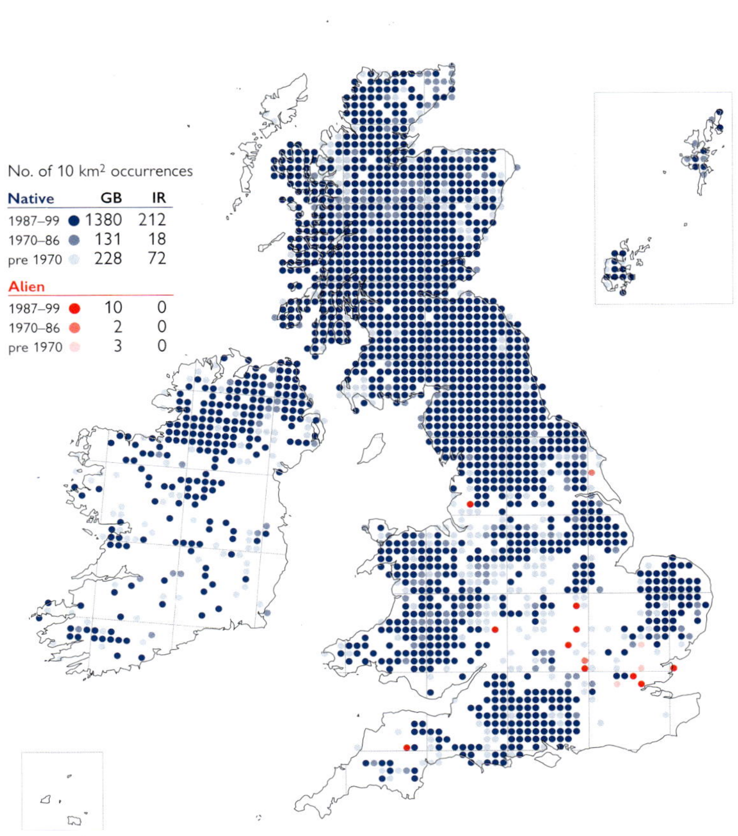

A perennial herb of mildly acidic to calcareous, slow-draining or wet soils, in shaded or open habitats, including streamsides and flushes in deciduous woodland, carr, herb-rich hay meadows and montane willow scrub and tall-herb communities on ledges. Reproduction is by seed and by rhizomatous spread. 0–975 m (Beinn a' Chaoruinn, Westerness).

Native (change –0.70). *G. rivale* was mapped as 'all records' in the 1962 *Atlas*. There have been widespread declines, particularly in C. and S. England, and analysis of the database reveals that these have mostly occurred since 1950. The species is also grown in gardens and sometimes escapes.

Eurosiberian Boreo-temperate element; also in N. America.

References: Atlas (124c), Hultén & Fries (1986), Meusel *et al.* (1965), Taylor (1997b).

D. J. McCOSH

Geum rivale × *G. urbanum* (*G.* × *intermedium*) Hybrid Avens

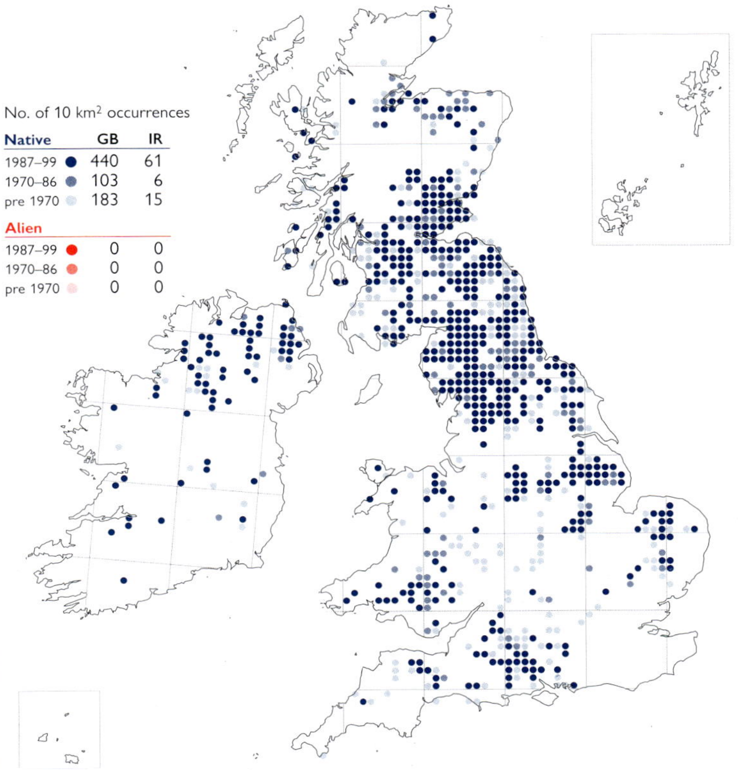

A highly fertile and variable hybrid, frequently occurring where the habitats of the parents overlap, especially where there is some disturbance. 0–450 m (Nenthead, Cumberland).

Native. The range of *G.* × *intermedium* is similar to that in Perring and Sell (1968), but is now much better recorded. As with *G. rivale*, there appear to have been some losses, particularly in C. & S. England. Hybrid swarms are often extremely variable, backcrosses to the parents being frequent.

Frequent with the parents in Europe and W. Asia.

References: Atlas Supp. (28a), Stace (1975), Taylor (1997a).

D. J. McCOSH

Geum urbanum Wood Avens

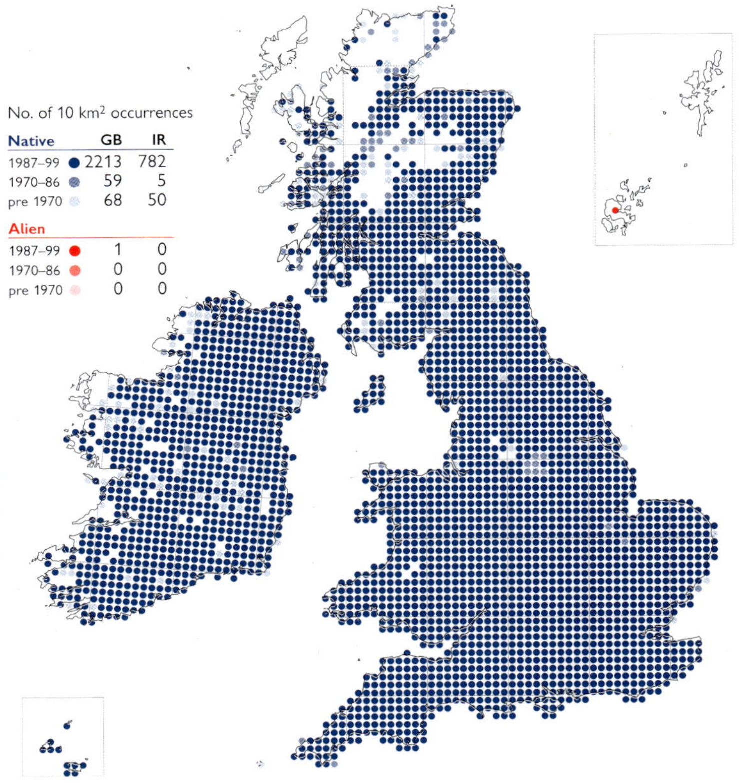

A perennial herb of freely-draining, mildly acidic to calcareous soils. It is found in moderate shade in deciduous woodland (especially in disturbed sites in secondary woodland), scrub and hedgerows and in more disturbed and open habitats, where it may grow as a street or garden weed. Generally lowland, but reaching 450 m on Alston Moor (Cumberland).

Native (change –0.53). There has been no change in the distribution of *G. urbanum* since the 1962 *Atlas*.

Eurosiberian Temperate element.

References: Atlas (124b), Grime *et al.* (1988), Hultén & Fries (1986), Meusel *et al.* (1965), Taylor (1997a).

D. J. McCOSH

Dryas octopetala Mountain Avens

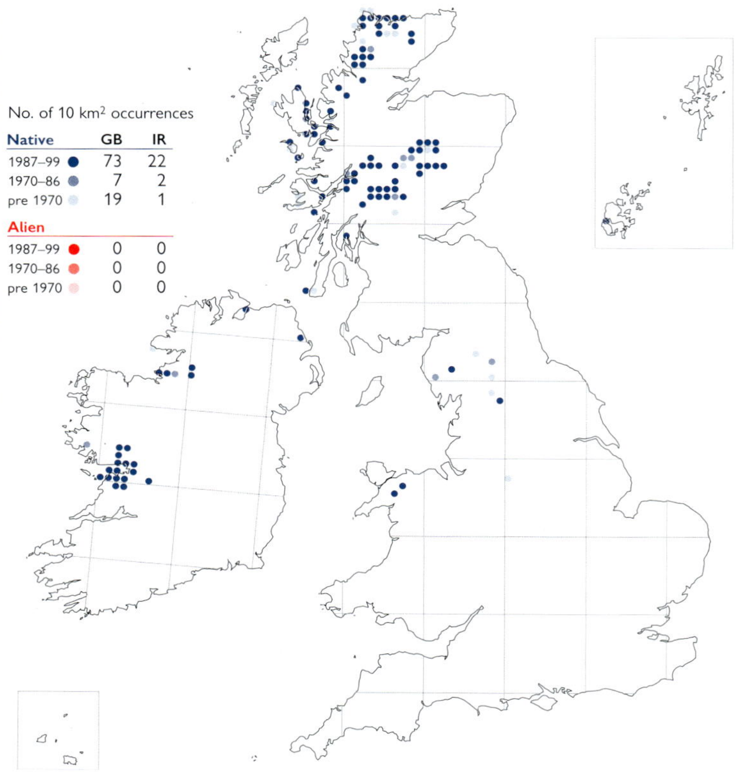

A dwarf procumbent and creeping shrub, typically found on basic ledges and rock crevices on mountains, but also occurring in upland calcareous grassland, on coastal shell-sand in N. Scotland, and limestone pavement in Ireland. From sea level in W. Sutherland and the Burren (Co. Clare) to 1035 m on Ben Avon (Banffs.).

Native (change –0.35). The distribution of this species appears to be stable overall, but populations in the Lake District are now very small. It is better recorded than in the 1962 *Atlas*.

Circumpolar Arctic-montane element, but absent from eastern N. America.

References: Atlas (124d), Curtis & McGough (1988), Elkington (1971), Hultén & Fries (1986), Meusel *et al.* (1965), Stewart *et al.* (1994).

D. J. McCOSH

Agrimonia eupatoria Agrimony

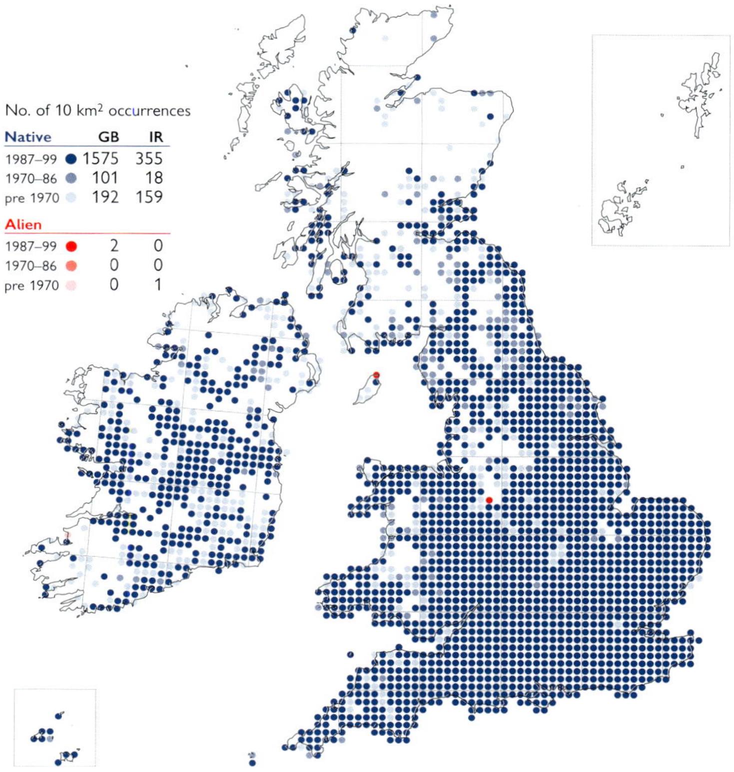

A perennial herb of basic and neutral soils, occurring in hedge banks, on woodland margins and rides, in field-borders and open grassland, on roadsides and railway banks, and sometimes in waste places. It reproduces and spreads by seed. 0–365 m (N.W. Yorks.).

Native (change –0.89). The map suggests a slight decline in the distribution of *A. eupatoria* since the 1962 *Atlas*.

Eurosiberian Southern-temperate element; widely naturalised outside its native range.

References: Atlas (125a), Hultén & Fries (1986), Meusel *et al.* (1965).

D. J. McCOSH

Agrimonia procera Fragrant Agrimony

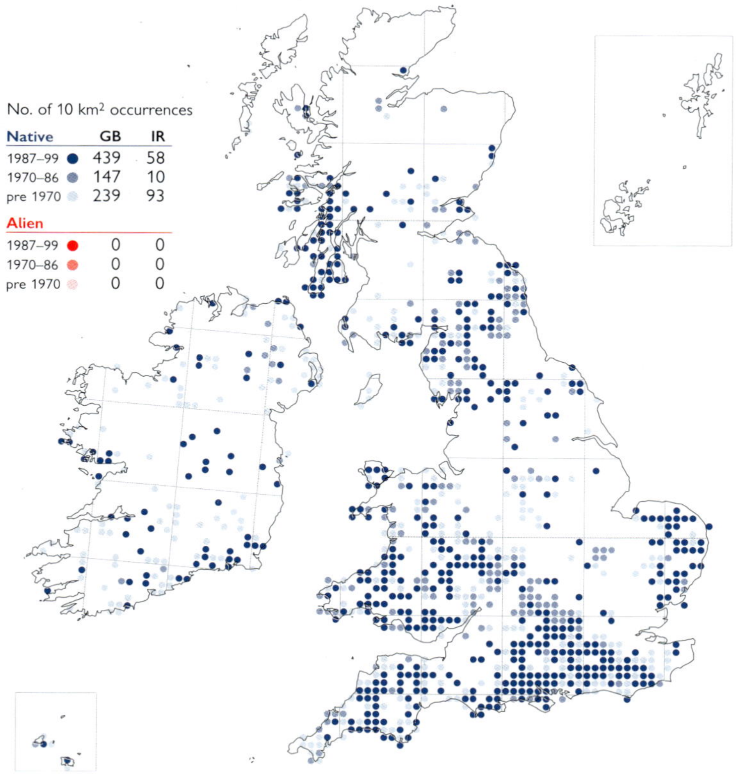

An erect perennial herb found in similar habitats to *A. eupatoria*, including hedgerows, woodland margins and roadside verges, generally on soils which are neither strongly calcareous nor strongly acidic. 0–335 m (Fortingall, Mid Perth).

Native (change –0.38). *A. procera* has been much confused in the past with *A. eupatoria* var. *sepium*, which can also be glandular and aromatic. The overall range is little changed since the 1962 *Atlas*. The apparent increases in East Anglia, W. Scotland and elsewhere are probably attributable to recorders' increased familiarity with this species.

European Temperate element.

References: Atlas (125b), Hultén & Fries (1986), Meusel *et al.* (1965).

D. J. McCOSH

Sanguisorba officinalis Great Burnet

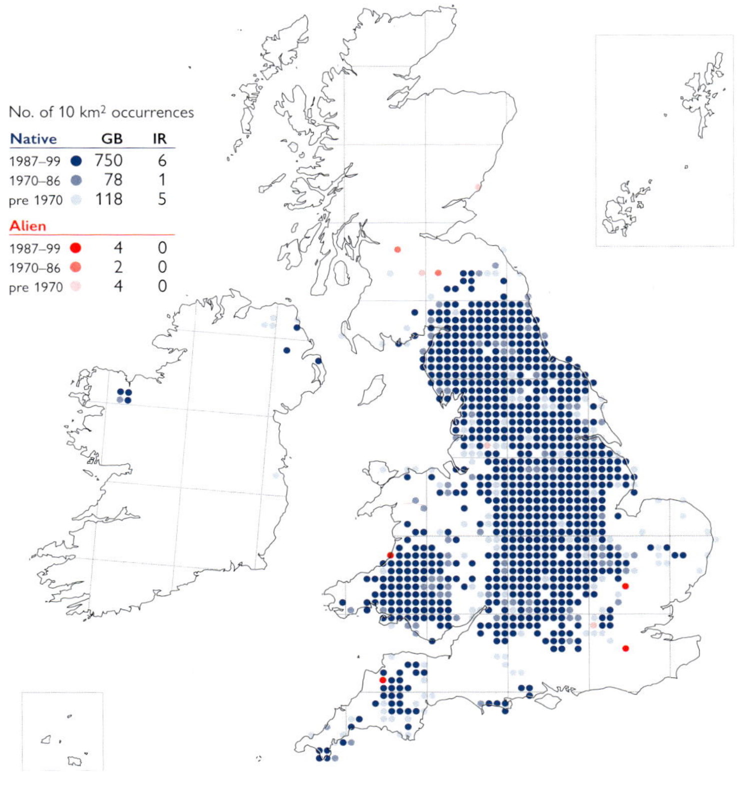

A perennial herb of neutral grassland, occurring on alluvial or peaty soils in damp or dry, unimproved pastures, hay meadows and marshy meadows, on river banks and lake shores and in base-enriched flushes on grassy heaths. 0–460 m near Cauldron Snout, Teesdale (Westmorland).

Native (change −0.23). Whilst the range of *S. officinalis* remains the same as in the 1962 *Atlas*, losses have occurred through the improvement of pastures.

Circumpolar Boreo-temperate element, but absent as a native from eastern N. America; widely naturalised outside its native range.

References: Atlas (126b), Curtis & McGough (1988), Hultén & Fries (1986), Meusel *et al.* (1965).

D. J. McCOSH

Sanguisorba minor subsp. *minor* Salad Burnet

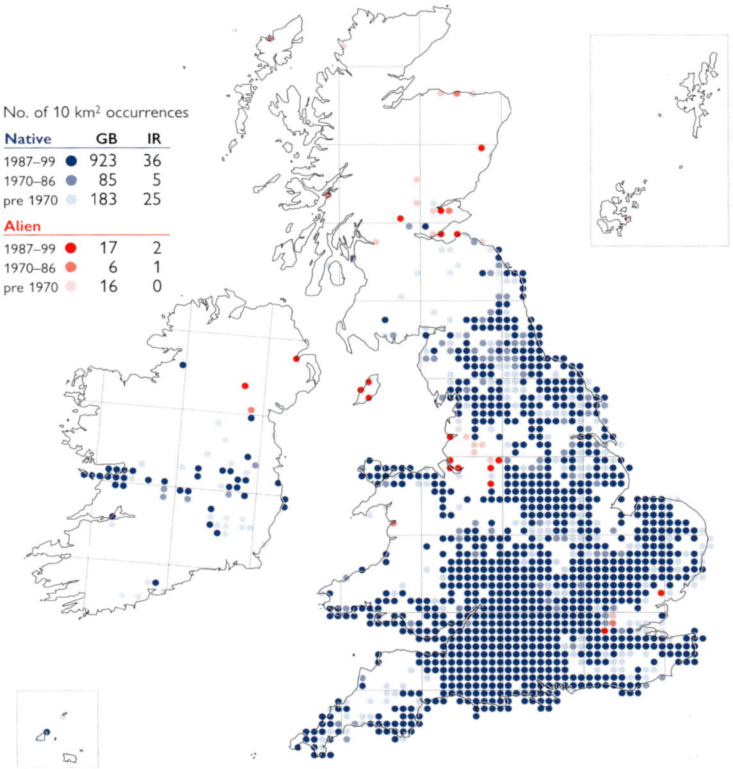

A perennial herb, almost confined to dry, infertile grassland on chalk and limestone, but also occurring on boulder-clay. It is often abundant on downland, but also grows in rock crevices, scree, quarries and on roadside banks. It is occasionally recorded with ericaceous shrubs on leached downland summits and on heathland, but only where rooted into basic horizons below. 0–500 m (N.W. Yorks.).

Native (change for species −0.16). The distribution of this species is broadly similar to that in the 1962 *Atlas*.

S. minor has a Eurosiberian Southern-temperate distribution; subsp. *minor* occurs in temperate Europe and is replaced by other taxa further south and east.

References: Atlas (126c), Grime *at al.* (1988), Hultén & Fries (1986), Meusel *et al.* (1965).

M. J. WIGGINTON

Sanguisorba minor subsp. *muricata* Fodder Burnet

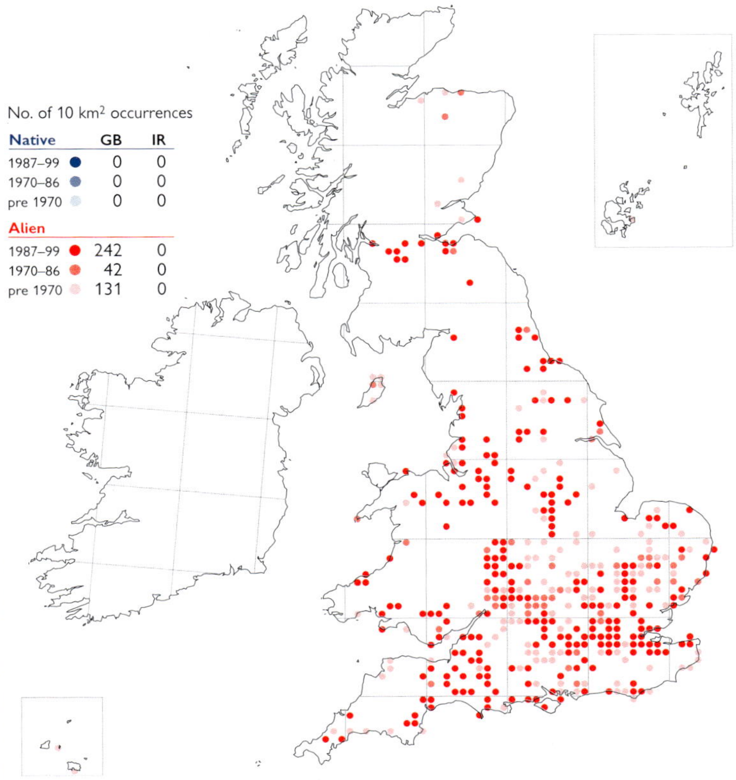

This perennial herb is a relic of cultivation which occurs as a casual, semi-established or naturalised plant in grassy places, on field edges, tracksides, banks, roadsides and railways. Lowland.

Neophyte (change for species −0.16). Subsp. *muricata*, introduced in 1803, was formerly grown for fodder. It is often grown in gardens and sometimes introduced with wildflower mixtures for calcareous soils. It was recorded from the wild by 1849 and seems to be increasing.

Native of S. Europe; widely naturalised further north.

Reference: Atlas (126d).

D. J. McCOSH

Acaena novae-zelandiae Pirri-pirri-bur

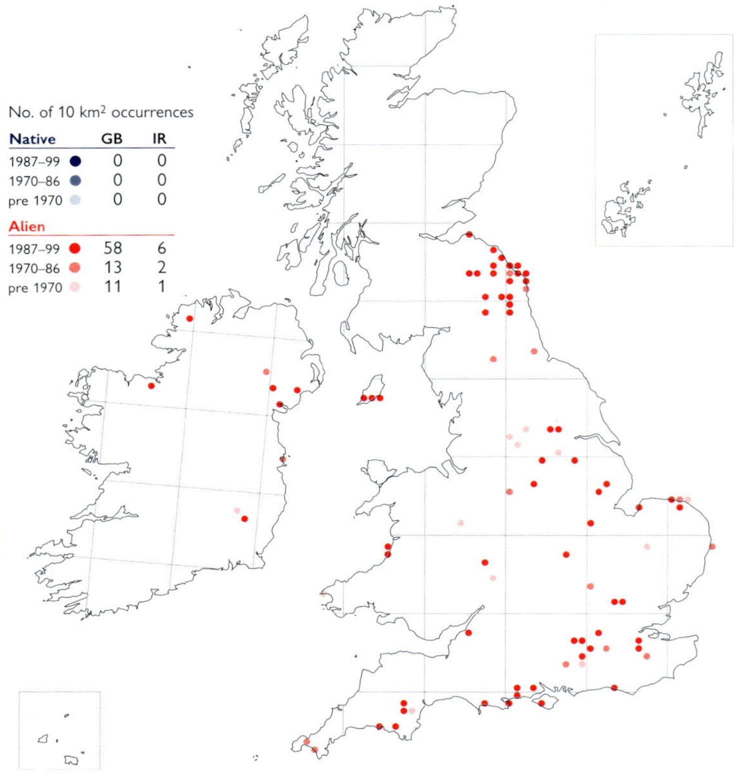

A prostrate dwarf perennial herb of freely-draining soil, naturalised in sparsely vegetated sites subject to moderate disturbance. Habitats include sand dunes, cliffs, heaths, conifer plantations on sandy soils, old gravel workings, roadsides and disused railways. Reproduction is from seed, and sometimes from pieces of rooted stolon. Lowland.

Neophyte. This species was introduced to Britain as a wool contaminant and its spread into semi-natural habitats was often from woollen mills; it was first recorded in the wild in 1901. However, some colonies appear to have resulted from the dumping of garden refuse. It can be very persistent, and its spread continues in some areas.

Native of Australia and New Zealand.

References: Gynn & Richards (1985), Yeo (1973).

D. J. McCOSH & M. J. WIGGINTON

Alchemilla alpina Alpine Lady's-mantle

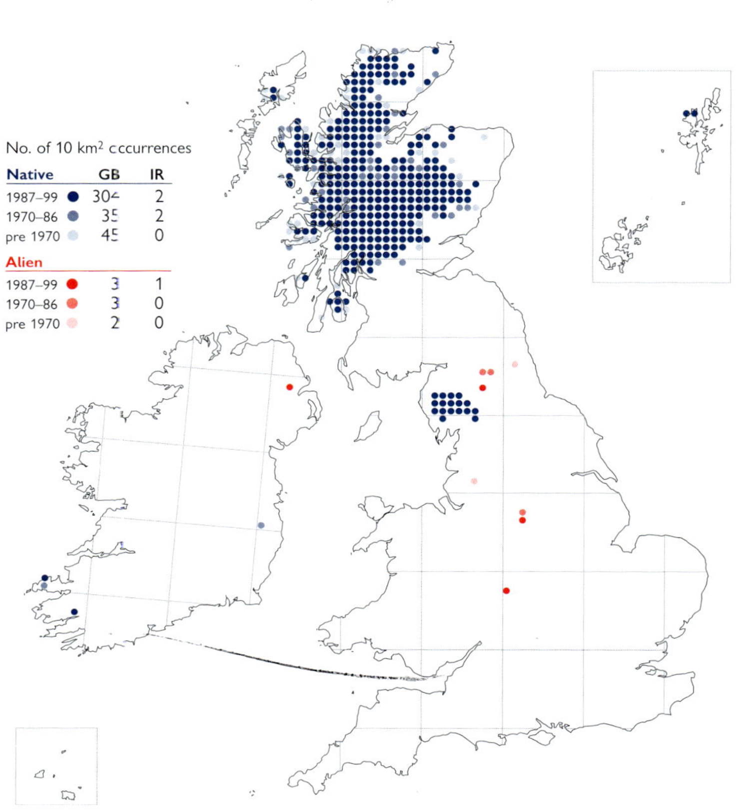

A perennial herb of montane grassland and grass-heath, scree, cliffs, rocky streamsides, rock crevices and ledges. It is found in well-drained habitats, in areas of solifluction and late snow-lie, and sometimes on mountain slopes subject to severe wind-scour. The soils range from acidic to strongly calcareous. It is frequently washed down to lower levels on river gravels. From near sea level in N.W. Scotland to 1270 m on Ben Macdui (S. Aberdeen).

Native (change –0.61). The distribution of *A. alpina* is stable.

European Arctic-montane element; also in Greenland.

References: Atlas (125c), Curtis & McGough (1988), Hultén & Fries (1986), Meusel *et al.* (1965).

D. J. McCOSH

Alchemilla conjuncta Silver Lady's-mantle

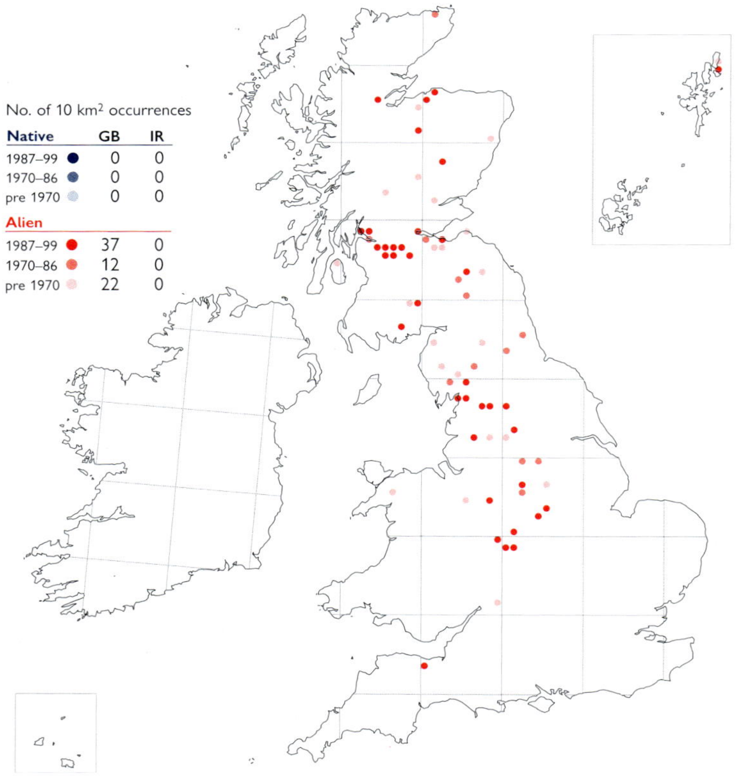

A small perennial herb found naturalised in two distinct habitats: montane grassland and streamsides, and in the lowlands on roadsides, river banks and in rough grassland. Some upland localities are very remote. Lowland to 455 in Glen Doll (Angus).

Neophyte. *A. conjuncta* was once regarded as native in its upland sites, but these are now thought to be deliberate introductions. It has been cultivated in Britain as a rockery plant since *c.* 1800, and was known from Glen Doll by 1837. It was mapped by Perring & Sell (1968), since when its range has increased slightly.

Native of the Jura and S.W. Alps.

Reference: Atlas Supp. (29a).

D. J. McCOSH

Alchemilla glaucescens

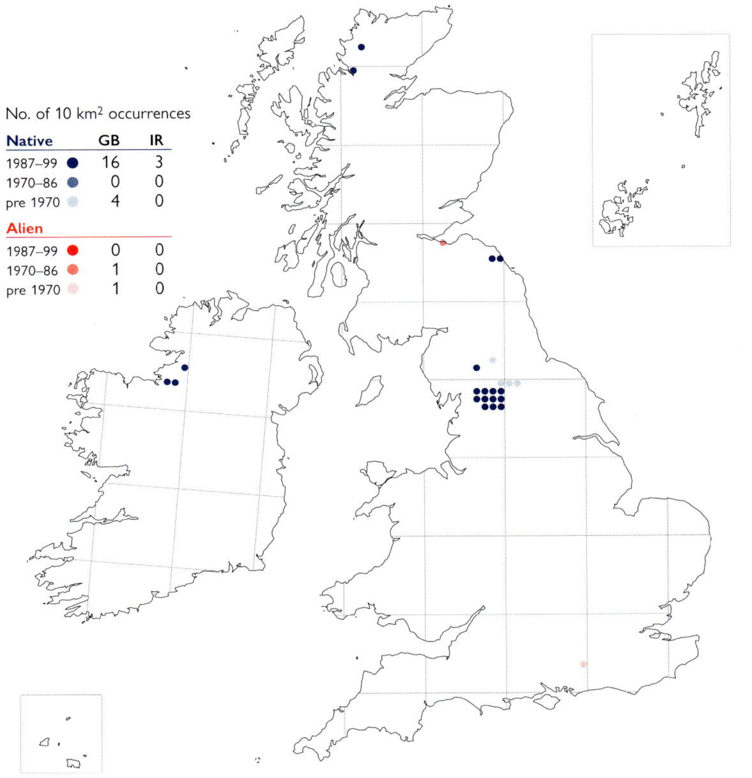

A perennial herb of limestone grassland and grassy banks by roads and rivers in N. England where it may be locally abundant; also in limestone grassland in Scotland and Ireland. Rarely elsewhere as an escape. Lowland, to 570 m on Whernside (N.W. Yorks.).

Native. There is no change in the distribution of *A. glaucescens* in England. It was discovered in Berwickshire in 1982. In Ireland the distribution is stable.

European Boreo-temperate element, with a continental distribution in W. Europe.

References: Atlas Supp. (29b), Hultén & Fries (1986), Wigginton (1999).

D. J. McCosh

Alchemilla monticola

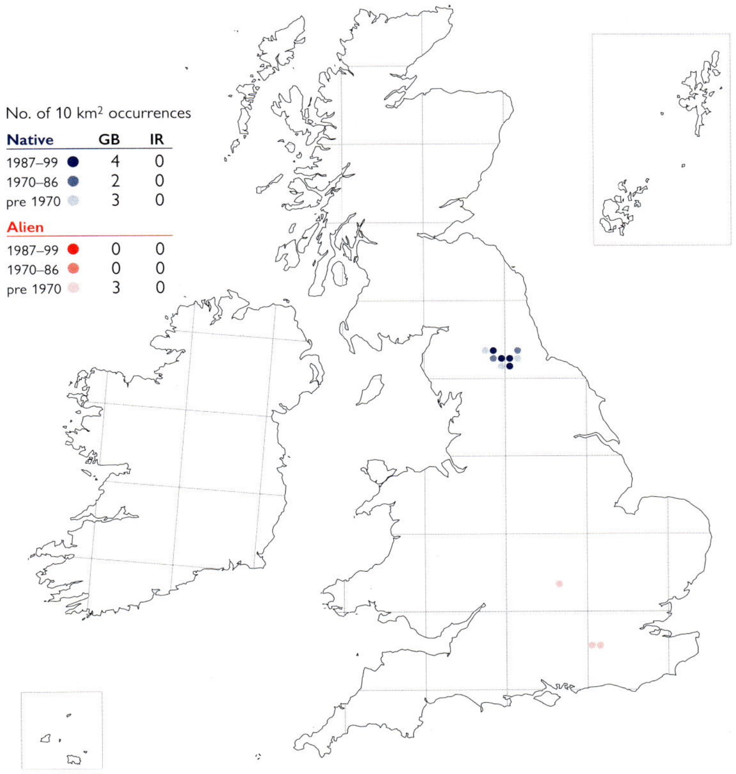

A perennial herb of neutral grassland, occurring in unimproved, species-rich hay meadows, but confined to the margins of those which are intensively managed. It is also locally frequent on road verges, particularly those which support a hay meadow flora. From 150 m (Barnard Castle, N.W. Yorks.) to 450 m (near Harwood Beck, Co. Durham).

Native. Since 1960, partial surveys have shown some losses of this species from hay fields due to the intensification of grass production, and from road verges, and it might now be lost from some 10-km squares for which there is no post-1970 record.

European Boreo-temperate element, with a continental distribution in W. Europe.

References: Atlas Supp. (30a), Bradshaw (1962), Hultén & Fries (1986), Wigginton (1999).

M. J. Wigginton

Alchemilla subcrenata

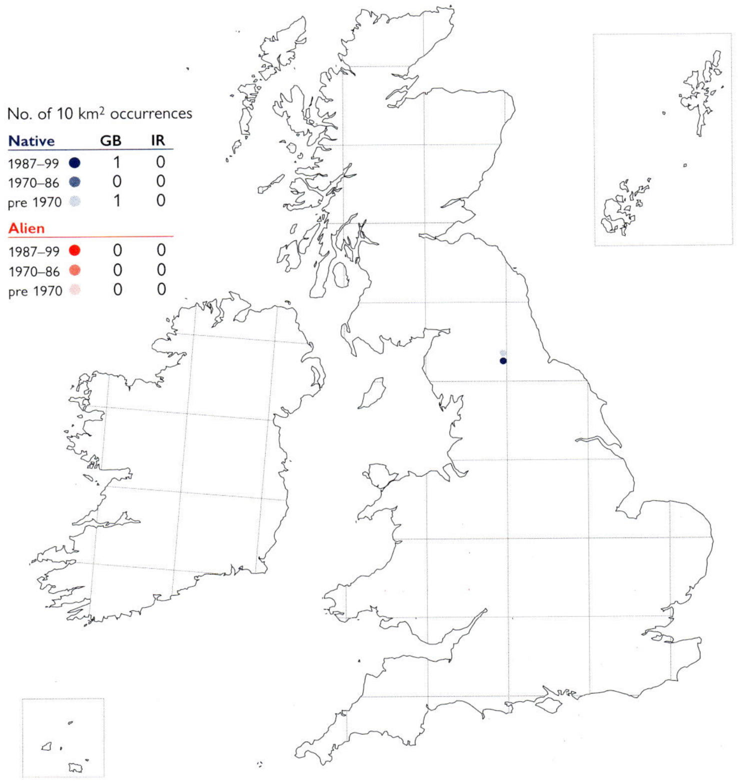

This perennial herb has been recorded in species-rich hay meadows, along the margins of those which are more intensively managed, and from unimproved pasture. The altitude range is *c.* 270–320 m in Weardale and Teesdale (Co. Durham).

Native. *A. subcrenata* was first found in 1951, and has only ever been known from a very restricted area of Teesdale, and one site to the north, in Weardale. Since 1990 it has been recorded in only two sites, and a search of three others in 1996 failed to detect it. A thorough survey is required to determine its current distribution.

European Boreo-temperate element, with a continental distribution in W. Europe.

References: Atlas Supp. (29c), Hultén & Fries (1986), Wigginton (1999).

M. J. Wigginton

Alchemilla acutiloba

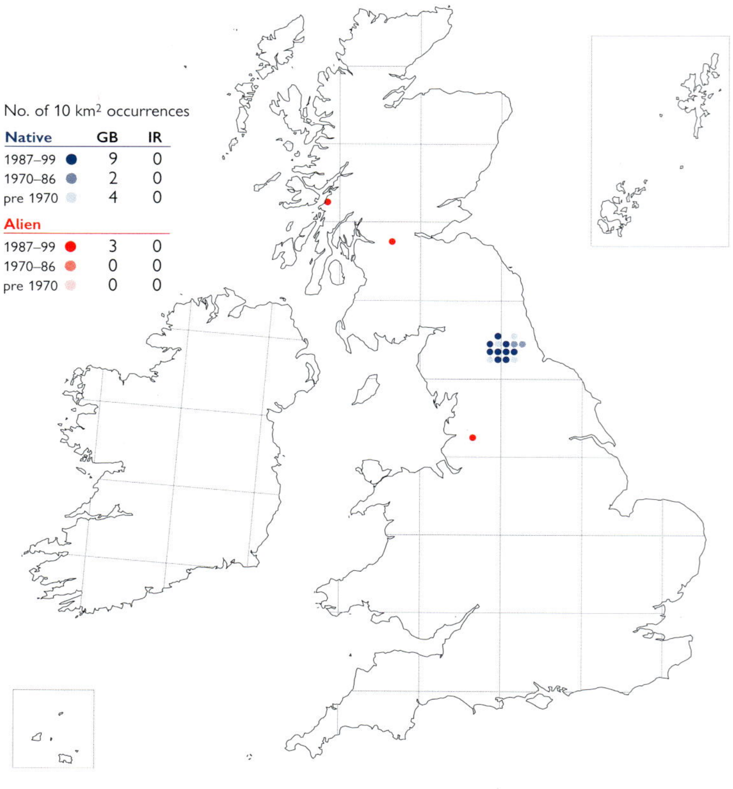

This perennial apomict of neutral grassland is found in unimproved, species-rich hay meadows, or on the margins of meadows which have been re-seeded or more intensively managed. It is also locally frequent on herb-rich road verges, and occasional on railway banks. From 140 m at Wolsingham to 450 m at Lanehead (both Co. Durham).

Native. There have been local losses of this species from hay fields due to agricultural intensification, and from roadside verges, and partial surveys since 1960 suggest it might have been lost from some of its 10-km squares. It has been naturalised in Lanarkshire since 1992.

European Boreo-temperate element, with a continental distribution in W. Europe.

References: Atlas Supp. (31a), Bradshaw (1962), Dickson *et al.* (1993), Hultén & Fries (1986), Wigginton (1999).

M. J. WIGGINTON

Alchemilla micans

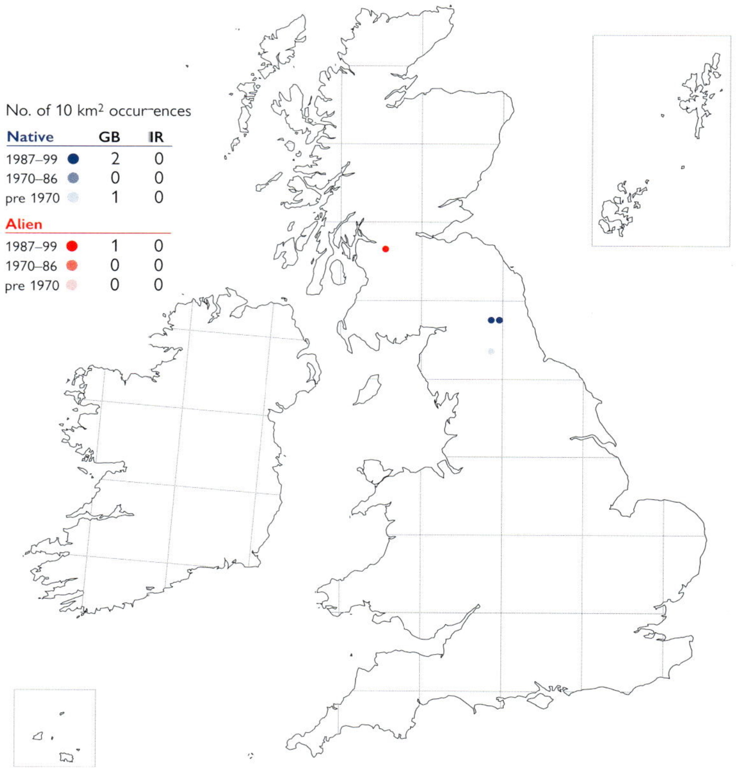

The largest population of this perennial apomict occurs in species-rich grazed pasture on shallow soil overlying Carboniferous limestone. Small populations have also been recorded from a rough pasture, from tall herbage in an ungrazed hay meadow and from a roadside verge. Lowland.

Native. This species was not recognised in Britain until 1976, and is currently known only from Northumberland. The single record from Durham was made in 1924, and casual plants were recorded in Lanarkshire in 1986 and 1992.

European Boreo-temperate element, with a continental distribution in W. Europe.

References: Hultén & Fries (1986), Wigginton (1999).

D. J. McCOSH & M. J. WIGGINTON

Alchemilla xanthochlora

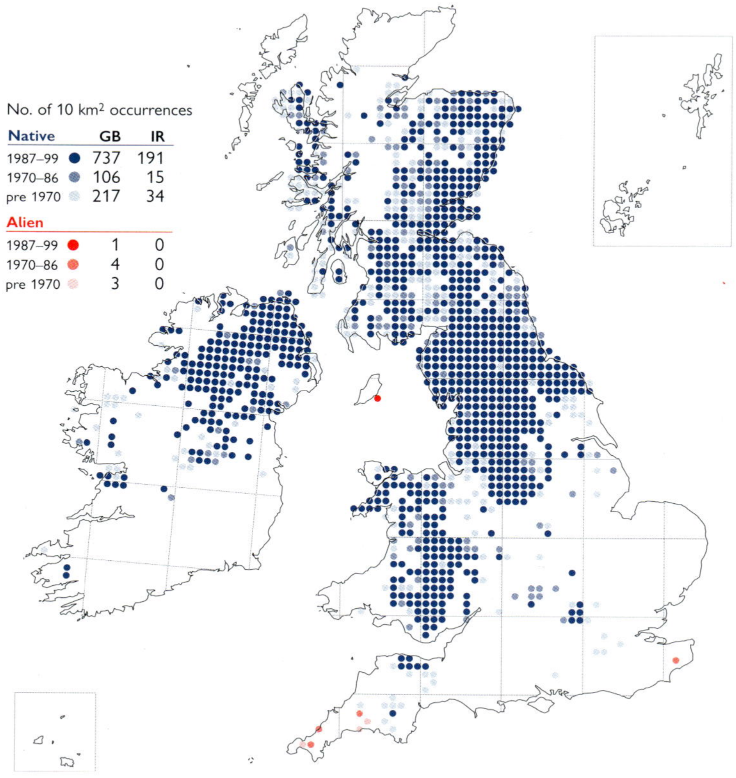

This perennial herb of neutral or mildly calcareous grassland occurs in lowland pastures and hay meadows, on upland hill-slopes, in rough grassy places, on river banks and streamsides, woodland borders and rides, and roadside verges. 0–885 m (Caenlochan, Angus).

Native. The distribution of this species is generally stable, but there may have been losses in C. and S. Scotland. These may, however, simply indicate under-recording.

European Temperate element.

References: Atlas Supp. (31b), Hultén & Fries (1986).

D. J. McCOSH

Alchemilla filicaulis subsp. *filicaulis*

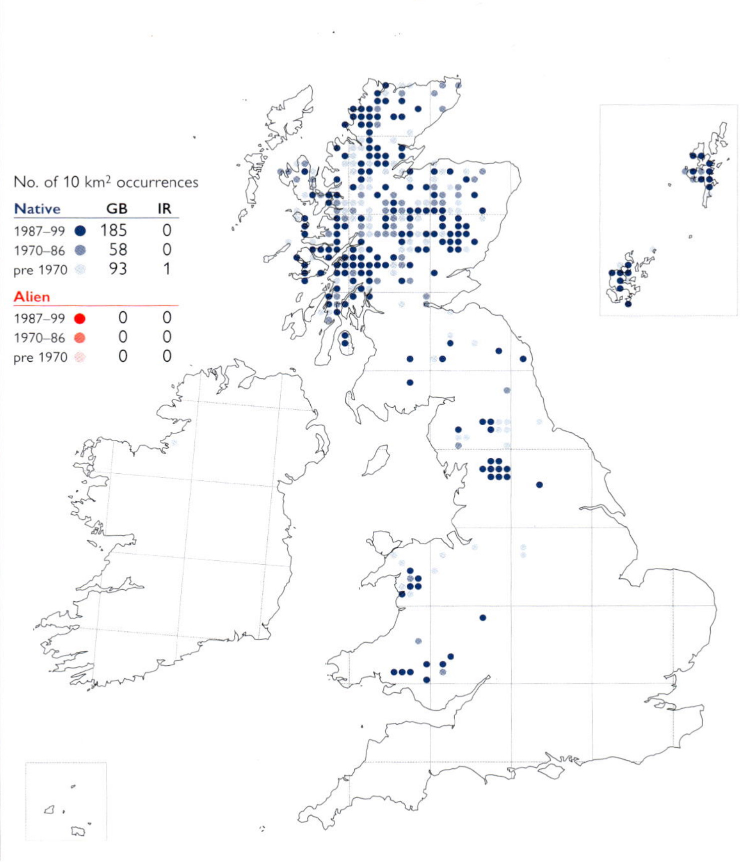

A perennial apomict of calcareous or neutral grassland and grass-heath, found in lowland pasture, on hill-slopes, on herb-rich banks, and in flushed areas; also on rock outcrops, and mountain ledges of basic rock. From sea level in Orkney and Shetland to 975 m on Beinn a' Chaoruinn (Westerness).

Native. There have been no significant changes in the distribution of *A. filicaulis* subsp. *filicaulis* since it was mapped by Perring & Sell (1968). It remains under-recorded in some areas.

European Boreal-montane element; also in N. America.

References: Atlas Supp. (30c), Hultén & Fries (1986).

D. J. McCOSH

Alchemilla filicaulis subsp. *vestita*

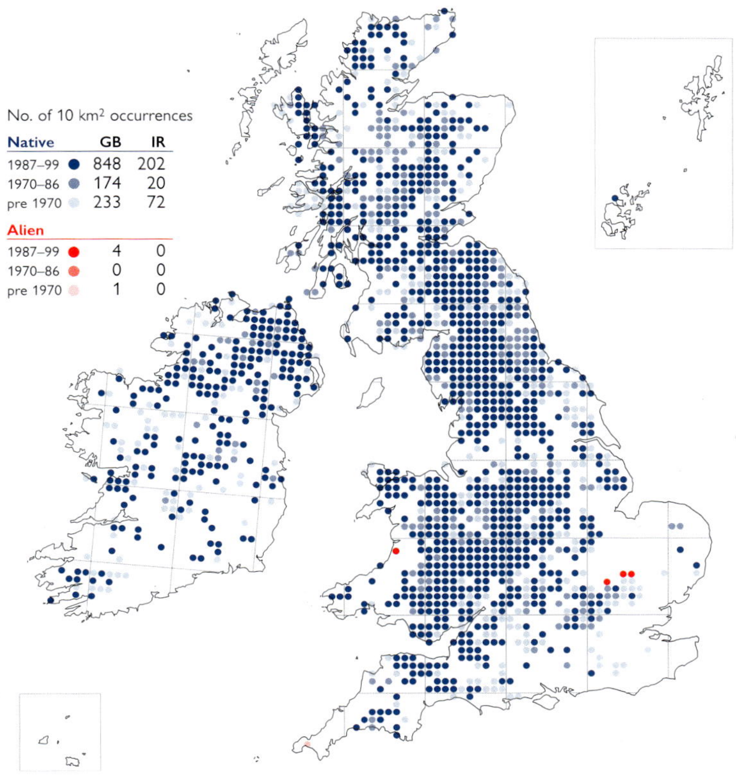

The habitats of this perennial apomict include rough pasture, grassy hill-slopes, banks and mountain flushes, woodland borders and rides, and roadside verges; it is also locally found on superficial clay-with-flints on chalk downs. 0–915 m (Ben Lui, Mid Perth).

Native. There is no significant change in the distribution of this subspecies over most of Britain and Ireland, but in lowland areas some sites have been lost, probably since 1930.

European Boreal-montane element; also in N. America.

References: Atlas Supp. (30b), Hultén & Fries (1986).

D. J. McCOSH

Alchemilla minima

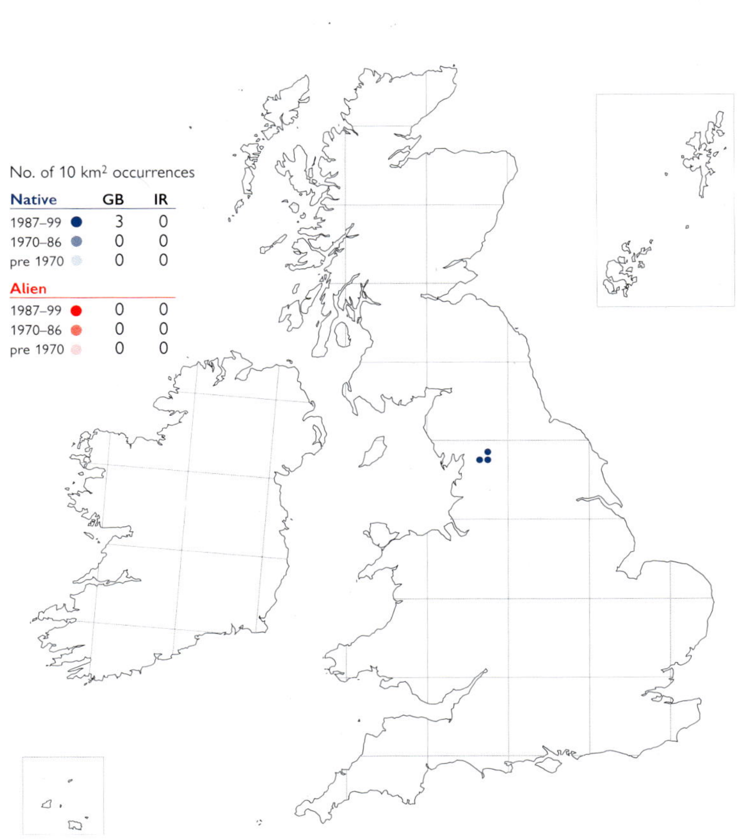

A dwarf perennial herb confined to tightly-grazed, moist, *Festuca ovina* grassland overlying Carboniferous limestone. From below 300 m to 610 m at Ingleborough (Mid-W. Yorks.).

Native. This taxon was first recognised as a separate species in 1947. It is a dwarf plant that may have been derived from *A. filicaulis* by natural selection through centuries of intensive grazing. Its overall distribution is likely to be stable, though new sites continue to be found within its very restricted range.

Endemic.

References: Atlas Supp. (29c), Wigginton (1999).

D. J. McCOSH & M. J. WIGGINTON

Alchemilla glomerulans

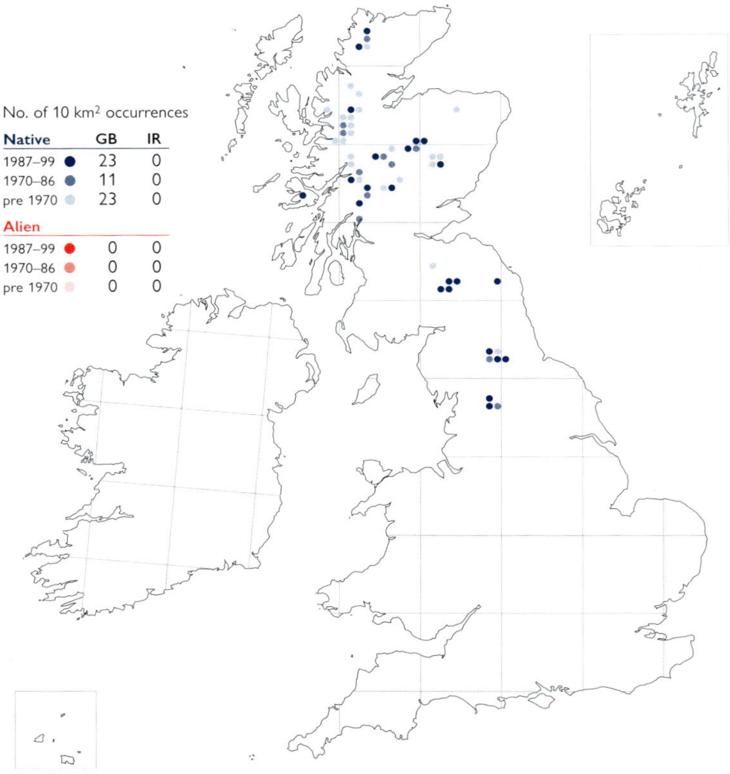

A perennial herb, typically found in ungrazed or lightly-grazed base-poor grassy habitats in the C. and N. Scottish mountains; a depauperate form grows in heavily-grazed flushes, and screes below cliffs. In Teesdale and Craven it occurs on roadsides and in species-rich hay meadows. From 145 m near Selkirk (Selkirks.) to 1030 m (Cairngorm Corries, Easterness).

Native. The current distribution of this critical species is unclear, especially in Scotland, because of the shortage of post-1970 records. However, it is unlikely to have disappeared from many of the sites for which there are only pre-1970 records, and there has probably been little change overall.

European Boreo-arctic Montane element; also in N. America.

References: Atlas Supp. (31c), Hultén & Fries (1986), Stewart *et al.* (1994).

D. J. McCOSH

Alchemilla wichurae

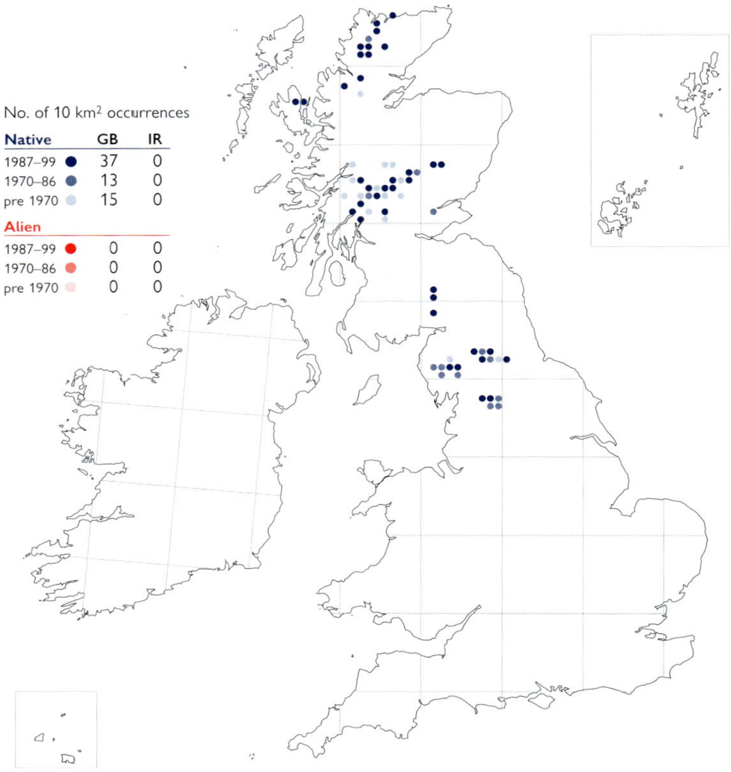

A perennial herb of tightly-grazed base-rich grassland and herb-rich rock ledges on outcrops and cliffs; sometimes colonising damp scree and bare cracks in limestone and basalt. It seems to prefer moist soils, frequently occurring near waterfalls and seepages. From sea level in N. Scotland to 990 m on Ben Lawers (Mid Perth).

Native. The distribution of *A. wichurae* appears to be more or less stable in England, but in Scotland this is less certain due to the dearth of recent records. It may, however, still be present in many squares for which there are only pre-1970 records.

European Boreal-montane element; also in Greenland.

References: Atlas Supp. (32b), Hultén & Fries (1986), Stewart *et al.* (1994).

D. J. McCOSH

Alchemilla glabra

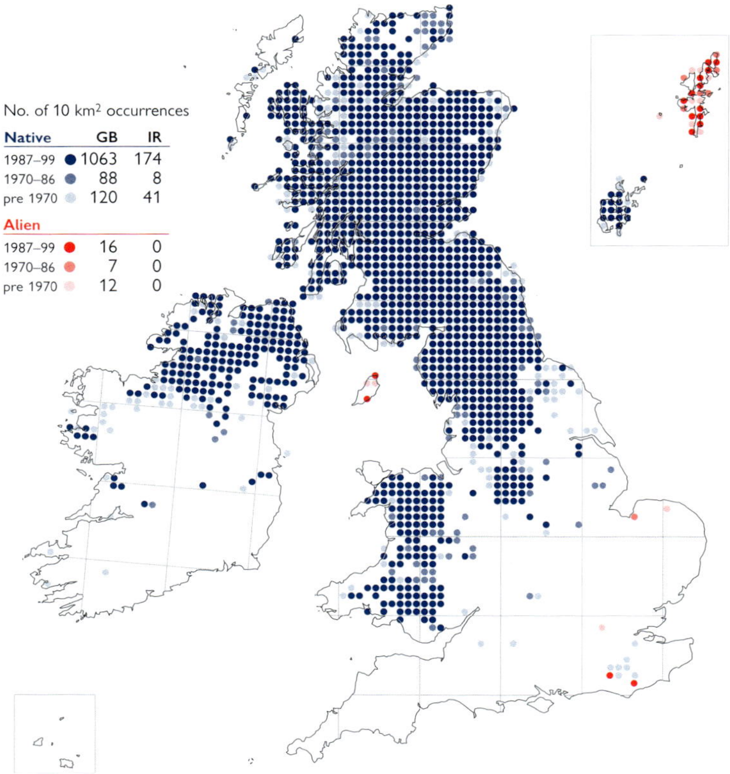

A perennial apomict of grassland habitats, occurring in lowland pasture, hay meadows, grass-heath on hillsides, roadsides and herb-rich banks kept moist by seeping water; also amongst tall vegetation on mountain ledges and in rocky river gorges. *A. glabra* seems to prefer damp soils, often occurring on stream banks, and in habitats subject to spray or temporary inundation. Lowland to 1215 m on Ben Lawers (Mid Perth).

Native. There has been no significant change in the distribution of *A. glabra* since it was mapped by Perring & Sell (1968), but there have been local losses on the southern boundaries of its range.

European Boreo-temperate element; widely naturalised outside its native range.

References: Atlas Supp. (32a), Hultén & Fries (1986), Meusel *et al.* (1965).

M. J. WIGGINTON

Alchemilla mollis

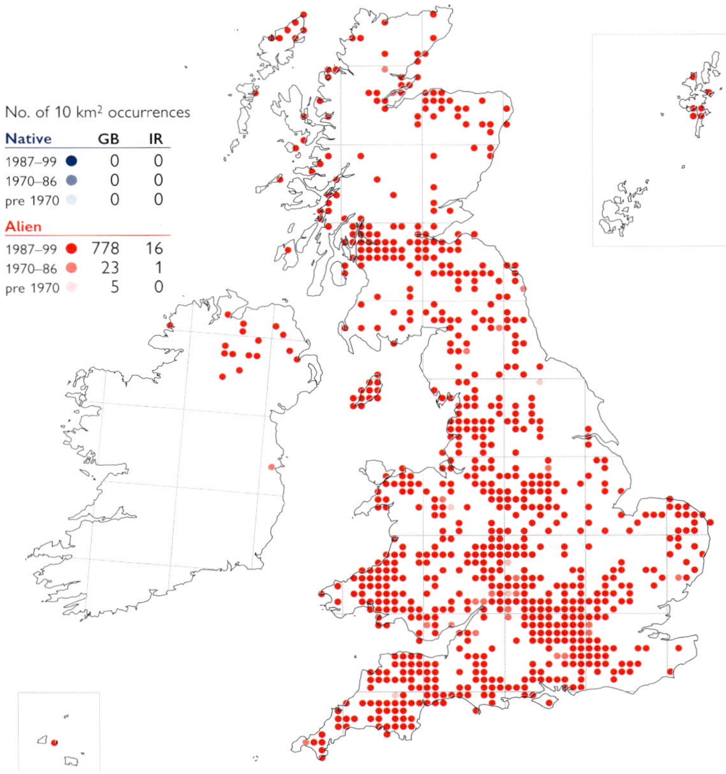

A vigorous perennial herb, frequently grown in gardens and found naturalised on roadsides, river banks, rough ground and anywhere where garden refuse is dumped. It also occurs as a casual on rubbish tips. It reproduces mainly by seed, which is produced abundantly, but it can also spread by rhizome fragments. Generally lowland, but established at 520 m below Ben Lawers (Mid Perth).

Neophyte. *A. mollis*, cultivated in Britain since 1874, was first recorded in the wild in 1948. There has been a massive increase in 10-km square records since the five mapped by Perring & Sell (1968), perhaps reflecting both its increasing popularity in gardens and better recording.

Native of S.E. Europe and S.W. Asia.

Reference: Atlas Supp. (32c).

D. J. McCOSH

Aphanes arvensis agg. Parsley-pierts

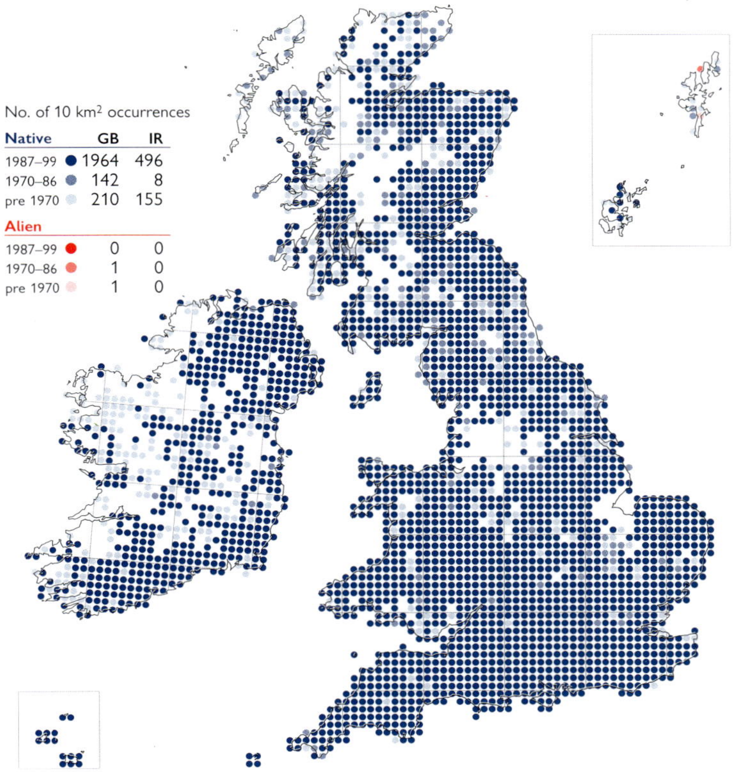

These annuals occur in arable fields, in thin and open grassland, and in bare patches on rough ground on dry, basic or acidic soils. In all these places, the cover of perennial species is restricted by disturbance or summer drought. 0–610 m (Great Dun Fell, Westmorland).

Native (change –0.32). There is no change in the overall distribution of this aggregate, which includes both *A. arvensis* and *A. australis*, since the 1962 *Atlas*. Vegetative plants cannot be assigned with confidence to the segregates, and some recorders have not distinguished fruiting material of the two species. Intermediates also occur and are also mapped here.

The distribution of the two species in this aggregate is given in their respective accounts.

References: Atlas (126a), Hultén & Fries (1986), Meusel *et al.* (1965).

D. J. McCOSH

Aphanes arvensis Parsley-piert

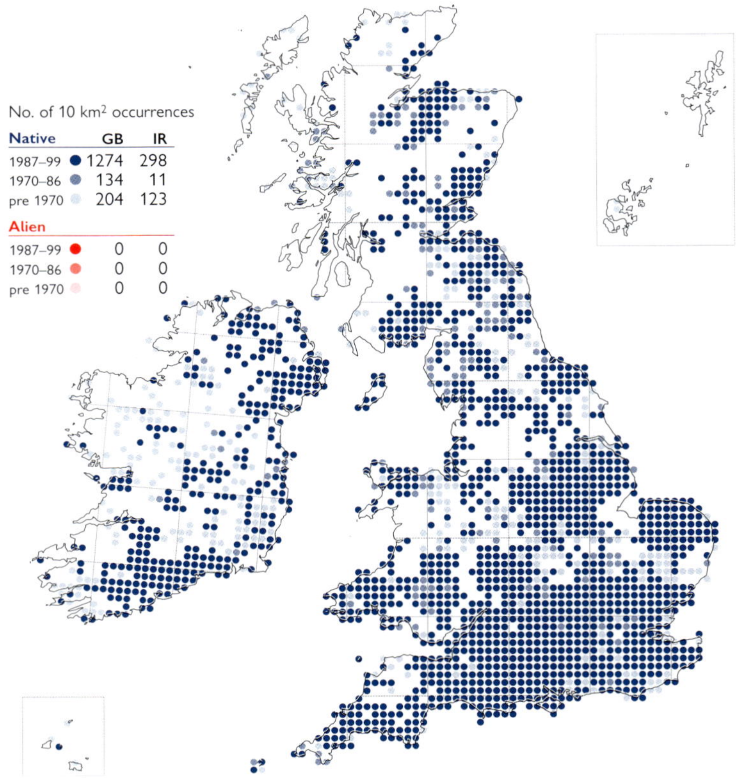

A winter- or, less frequently, spring-germinating annual of dry, basic to somewhat acidic soils in arable fields, bare patches in grassland and lawns, heaths and woodland rides, open ground in rough and waste places, gravel-pits and along railways. 0–610 m (Great Dun Fell, Westmorland).

Native. The distribution of *A. arvensis* shows little change from the map in Perring & Sell (1968), although the species is now much better recorded.

European Temperate element; widely naturalised outside its native range.

References: Atlas Supp. (33a), Grime *et al.* (1988), Hultén & Fries (1986), Meusel *et al.* (1965).

D. J. McCOSH

Aphanes australis Slender Parsley-piert

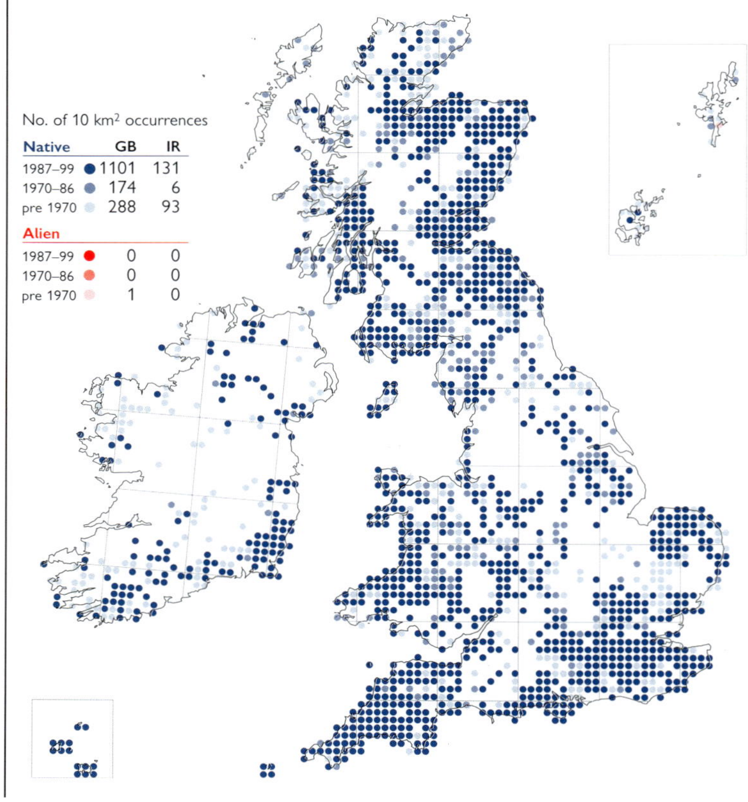

A winter- or, less frequently, spring-germinating annual of acidic, sandy or gravelly soils in woodland rides, on sand dunes, in sand- and gravel-pits, on tracks and roadside verges; also found in dry rocky places. It tends to replace *A. arvensis* on less fertile and more acidic soils. Generally lowland, but reaching 380 m at Loch Lee, Glen Esk (Angus).

Native. The distribution of this species is little changed from that shown in Perring & Sell (1968), although it is now much better recorded.

European Temperate element; widely naturalised outside its native range.

References: Atlas Supp. (33b), Hultén & Fries (1986), Meusel *et al.* (1965).

D. J. McCOSH

Rosa multiflora Many-flowered Rose

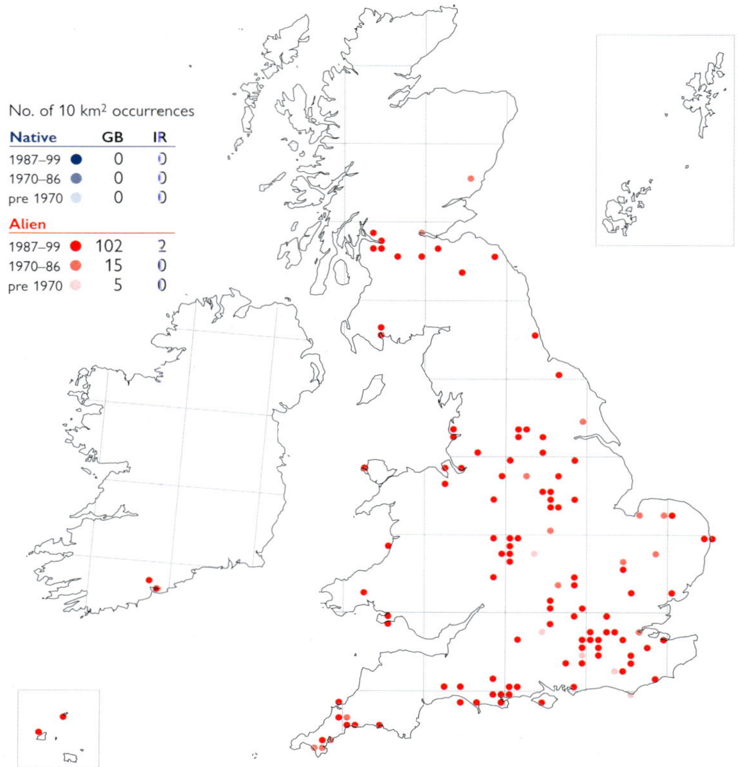

A deciduous shrub or climber found planted or as a naturalised garden escape in woodland, hedges, copses, scrub and on railway banks and waste ground. It can grow on the poorest of soils. Reproduction is by seed. Lowland.

Neophyte. *R. multiflora* has been cultivated since 1804, but only became popular in gardens and amenity planting schemes after its reintroduction in 1862. It is also used as root-stock for garden roses. It has been known in the wild since 1930 (Surrey) and its distribution is probably stable.

Native of E. Asia.

References: Bean (1980), Coats (1963), Graham & Primavesi (1993).

T. D. DINES

Rosa arvensis Field-rose

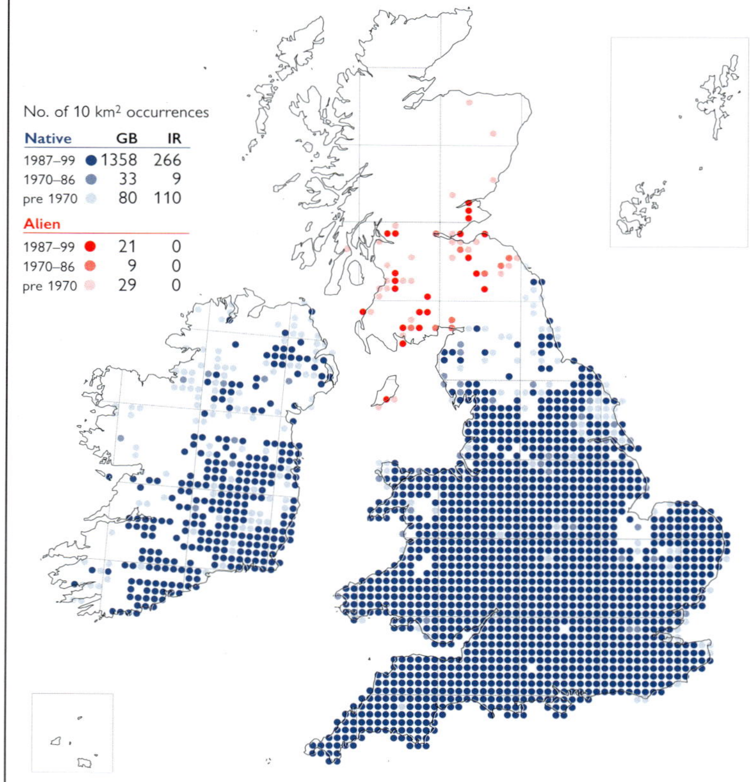

A deciduous shrub with weak flexuous stems which often climb over other vegetation, often forming dense patches. It grows on a wide variety of soils, but avoids very acidic sites, and is found on woodland edges, in clearings and along rides, on roadsides and railway banks and in scrub and hedgerows. 0–410 m (Titterstone Clee Hill, Salop).

Native (change –0.17). There is no change in the distribution of *R. arvensis* since the 1962 *Atlas*, although it is now much better recorded, especially in Wales, N.W. England and Ireland.

European Temperate element.

References: Atlas (127b), Graham & Primavesi (1993), Hultén & Fries (1986), Meusel *et al.* (1965).

R. MASKEW

Rosa arvensis × *R. canina* (*R.* × *verticillacantha*)

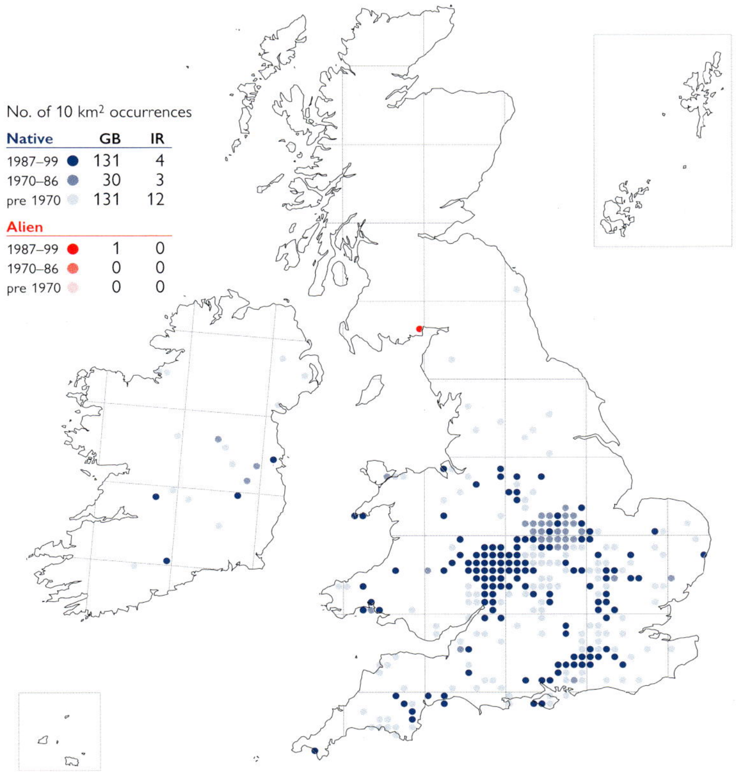

A deciduous shrub with arching stems which occurs along woodland edges, by rides and in clearings, and in hedgerows and scrub. It grows on a variety of soils, but avoids very acidic sites. It is partially fertile, usually producing a majority of well-formed hips along with a smaller number of sterile fruits. Generally lowland, but reaching 325 m at Blaentwrch (Cards.).

Native. This hybrid occurs throughout the range of *R. arvensis*, but sometimes one or both parents is absent from the immediate vicinity of hybrid bushes. When *R. arvensis* is the female parent, it is often conspicuous and the most easily recognisable of *Rosa* hybrids. Although better recorded than when mapped by Graham & Primavesi (1993), it is still severely under-recorded in many areas.

Wider distribution uncertain.

Reference: Stace (1975).

R. MASKEW

Rosa pimpinellifolia Burnet Rose

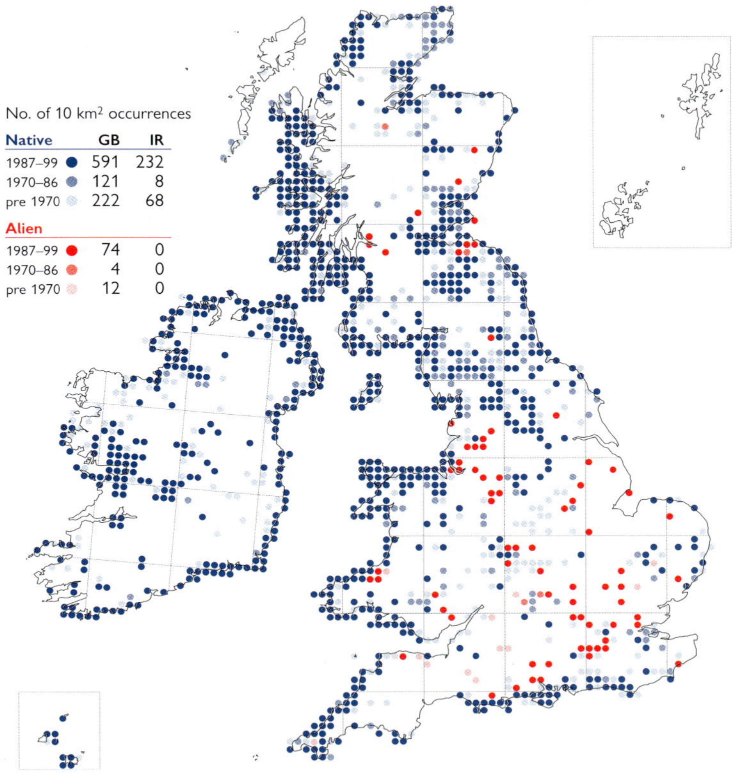

A low, suckering deciduous shrub most often found on sand dunes and sea-cliffs, but also inland on sandy and less acidic heaths, in scrub and hedgerows on chalk and limestone, and on basic cliff ledges in upland areas. Cultivars grown in gardens sometimes become naturalised. 0–610 m (Glaramara, Cumberland).

Native (change −0.05). The distribution of this species is stable on the coast. It was lost from inland sites before 1930; the habitats at these localities were not well documented but some recent losses have been from heathland. The increase in alien records since the 1962 *Atlas* has blurred the native range.

Eurasian Temperate element, with a disjunct distribution; widely naturalised outside its native range.

References: Atlas (127c), Graham & Primavesi (1993), Hultén & Fries (1986), Meusel *et al.* (1965).

R. MASKEW

Rosa pimpinellifolia × *R. sherardii* (*R.* × *involuta*)

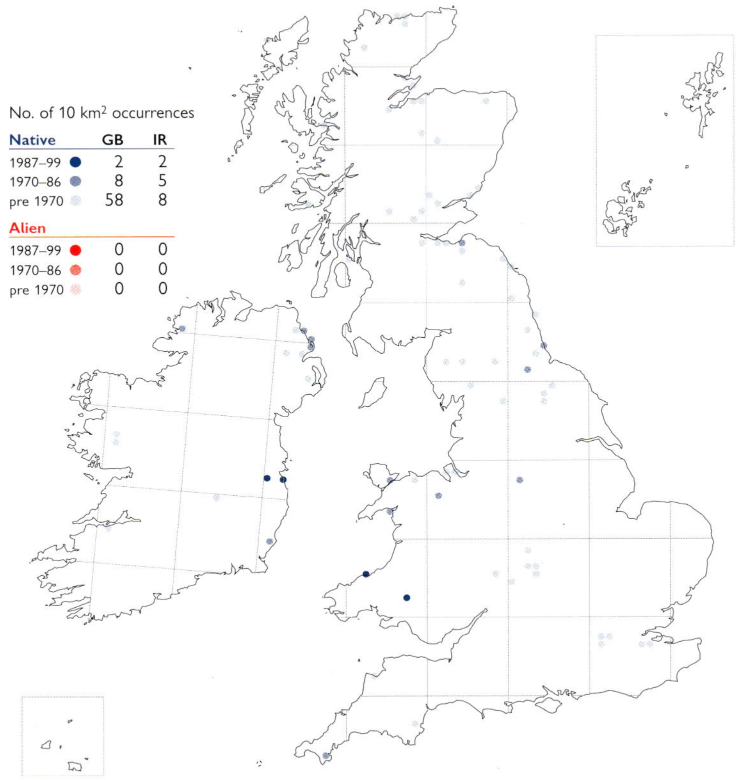

A variable, tall or low-growing, sometimes suckering, deciduous shrub. It is found in a variety of open habitats where the two species occur together, such as sand dunes, sea-cliffs, heaths, scrub and hedgerows. It is partially fertile, usually with very few well-formed hips. Lowland.

Native. As with other *Rosa* hybrids, this taxon is difficult to identify. The map is based primarily on herbarium specimens, hence the prevalence of historical records. It is, however, a genuinely rare hybrid.

Wider distribution uncertain.

References: Graham & Primavesi (1993), Stace (1975).

R. MASKEW

Rosa rugosa Japanese Rose

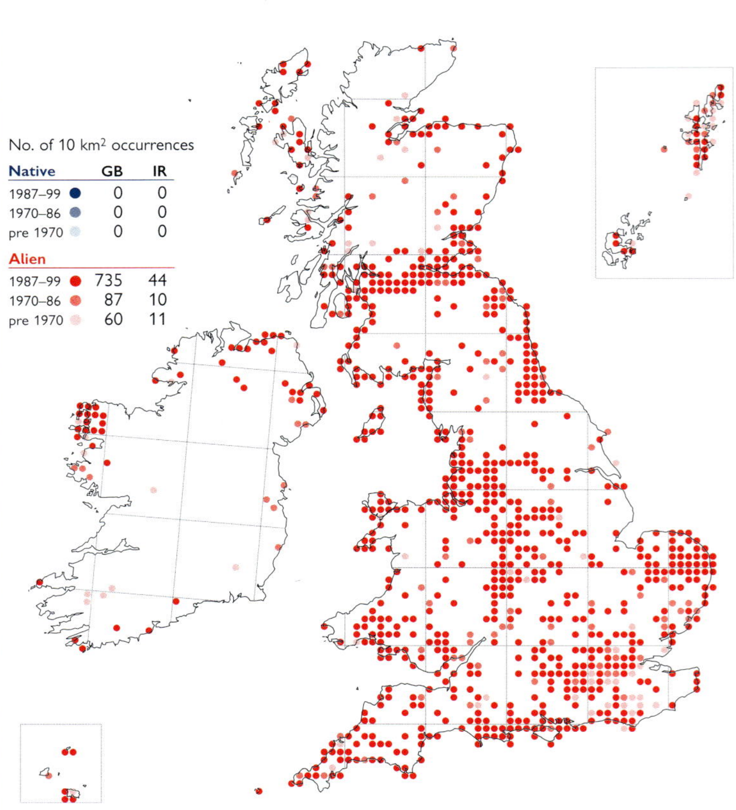

A suckering, deciduous shrub found planted or as a garden escape or throw-out in hedgerows and on sand dunes, sea-cliffs, road verges and waste ground; also occurring as a relic of cultivation. It is often well-naturalised, forming large thickets. Lowland.

Neophyte. *R. rugosa* was introduced into cultivation in 1796, but was not successfully grown until its re-introduction in 1845. It is very common in gardens, parks and amenity plantings, and was first recorded in the wild in 1927 (Cumberland). Its distribution is increasing, but the significant increase in records since it was mapped by Graham & Primavesi (1993) is probably due to better recording.

Native of E. Asia, widely naturalised in N. & C. Europe and has spread around the Baltic by sea-dispersed fruits.

Reference: Bean (1980).

R. MASKEW

Rosa 'Hollandica' Dutch Rose

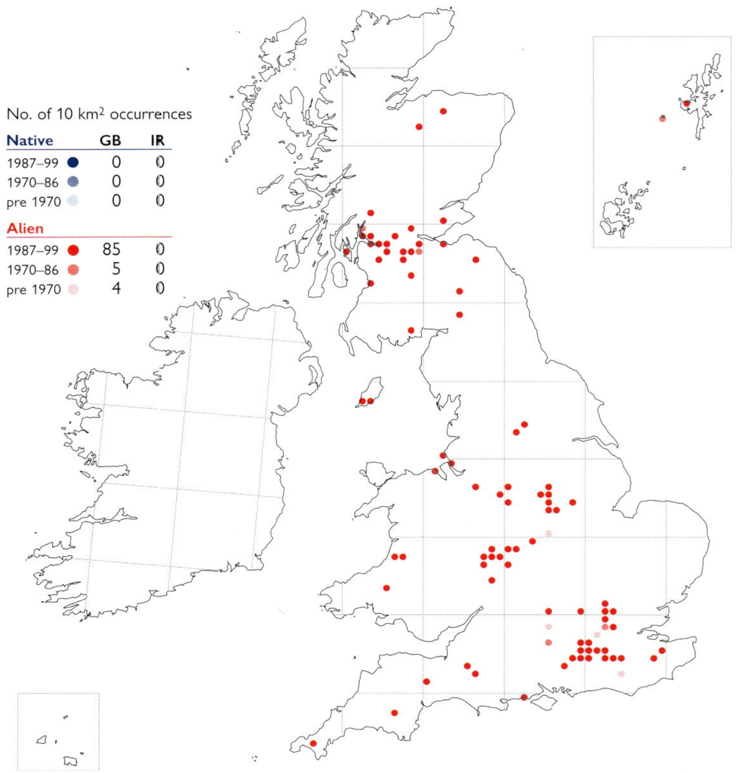

An upright deciduous shrub, suckering to form dense thickets. It is found as a garden escape or throw-out in hedgerows and on roadsides and waste ground; it also occurs as a relic of cultivation. Lowland.

Neophyte. This cultivar, first raised in 1888, is frequently used as a rootstock for garden roses. It was known in the wild by 1955 (E. Sussex). Since it was treated by Graham & Primavesi (1993) there has been a greater awareness of this plant, although it is still under-recorded.

A hybrid of garden origin; *R. rugosa* is one parent but the other is unknown.

Reference: Bean (1980).

T. D. DINES

Rosa ferruginea Red-leaved Rose

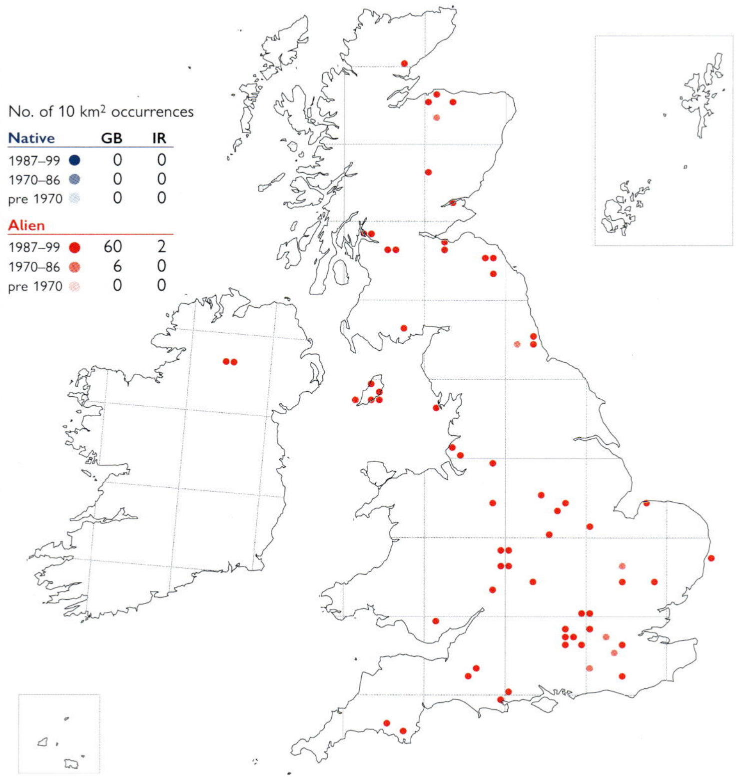

An erect, deciduous shrub, suckering freely to form thickets. It is found as a naturalised garden escape or throw-out in woodland, hedgerows and scrub, and on roadsides, railway banks, sand dunes and waste ground; it also occurs as a relic of cultivation. Reproduction is by seed, which can be bird-sown. Lowland.

Neophyte. *R. ferruginea*, which has been in cultivation since before 1830, is very popular in gardens. Like *R.* 'Hollandica', there has been an increase in records in recent years as recorders have become more familiar with it.

Native of the mountains of C. & S. Europe, from the Pyrenees to the Balkans and Carpathians.

References: Bean (1980), Bolòs & Vigo (1984), Graham & Primavesi (1993).

T. D. DINES

Rosa stylosa Short-styled Field-rose

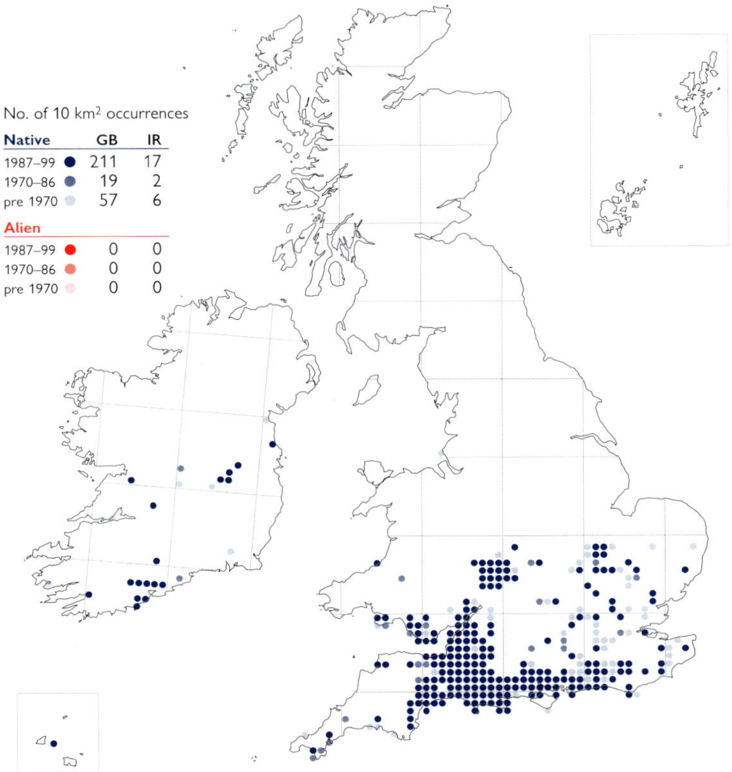

A deciduous shrub found almost exclusively on well-drained calcareous soils overlying chalk, limestone, clay and sand. It tolerates slightly shaded habitats, and is found in open woodland, hedgerows, disused quarries and scrub. Lowland.

Native. *R. stylosa* is a frequent species in parts of S.W. England but a rare plant of ancient woods and species-rich hedges at the north-eastern edge of its range, where it may have declined slightly. However, it is probably under-recorded, especially in those areas where it is rare.

European Temperate element.

Reference: Graham & Primavesi (1993).

R. MASKEW

Rosa canina Dog-rose

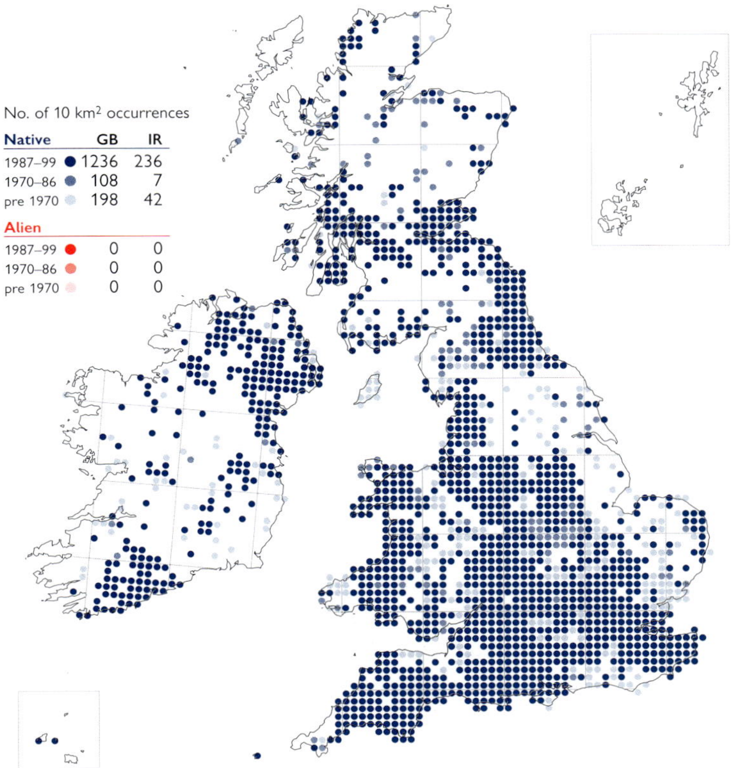

A deciduous shrub of well-drained calcareous to moderately acidic soils. Habitats include woodland, scrub, hedgerows, cliffs, river banks, rock outcrops, roadsides, railways and waste ground. It can rapidly colonise open, disturbed sites. 0–550 m (Breadalbanes, Mid Perth).

Native. This is the most widespread and common segregate of the *R. canina* aggregate. Four 'Groups' within the species were described and mapped by Graham & Primavesi (1993). The distribution of the species is probably stable. The name *R. canina* is often used for other species and hybrid 'Dog-roses'; as many such records have been rejected the true plant is under-recorded in some areas, particularly in Ireland.

European Temperate element; widely naturalised outside its native range.

Reference: Hultén & Fries (1986).

R. MASKEW

Rosa canina × *R. obtusifolia* (*R.* × *dumetorum*)

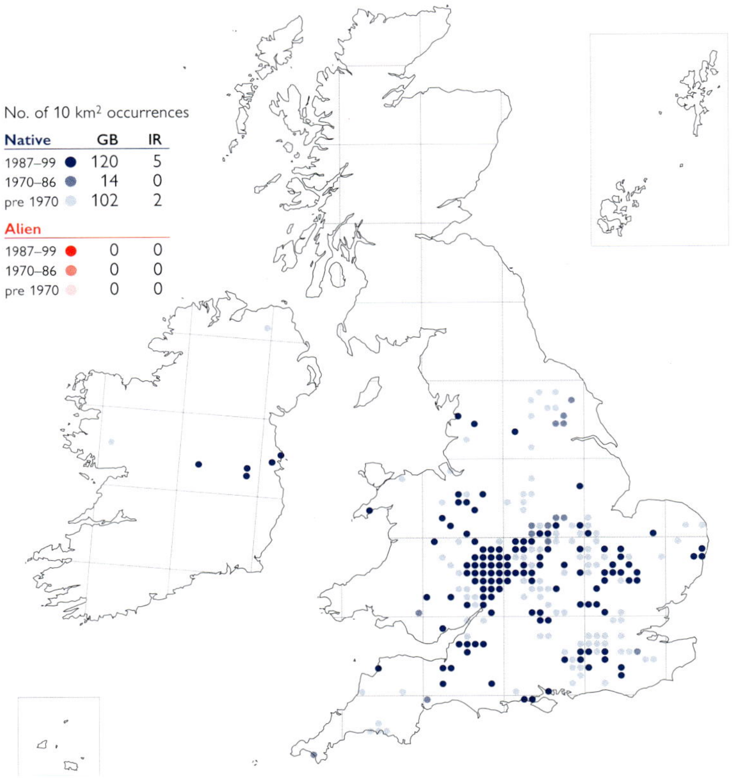

A deciduous shrub found in scrub, hedgerows, on road verges, tracksides and other open habitats. It is a fertile hybrid which usually occurs in areas where both parents are present, but is not uncommon in some areas when *R. obtusifolia* is rare or even absent. It grows mainly on calcareous soils. Lowland.

Native. This common hybrid frequently backcrosses with its parents, often producing populations displaying a complete range of intermediates between the two species. Its distribution is now better known than when mapped by Graham & Primavesi (1993), but it remains under-recorded in many areas and is often mis-identified as *R. obtusifolia* or the pubescent form of *R. canina.*

Wider distribution uncertain.

Reference: Stace (1975).

R. MASKEW

Rosa canina × *R. rubiginosa* (*R.* × *nitidula*)

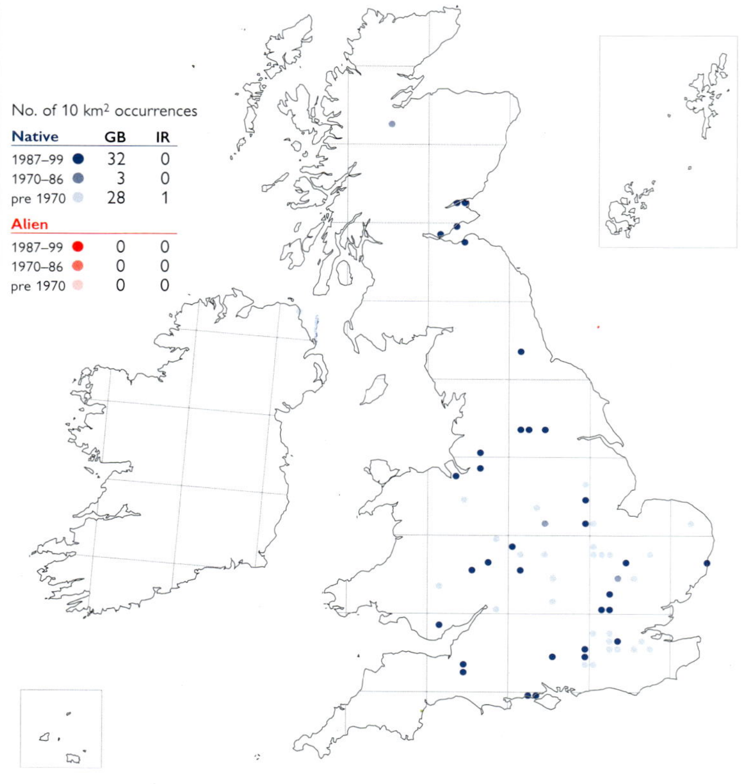

An upright deciduous shrub of woodland, hedgerows, scrub, rocky grassland, sand dunes, cliffs, disused quarries and roadsides. It prefers well-drained calcareous soils. It has recently been found in urban areas where *R. rubiginosa* has been used for landscaping. It is fertile and backcrossing is frequent. Lowland.

Native. The distribution of this hybrid is probably stable, although it is almost certainly under-recorded. Back-crosses are likely to be recorded as one or other parent.

Wider distribution uncertain.

References: Graham & Primavesi (1993), Stace (1975).

T. D. DINES

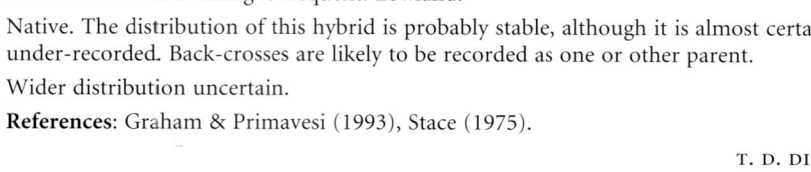

Rosa canina × *R. sherardii* (*R.* × *rothschildii*)

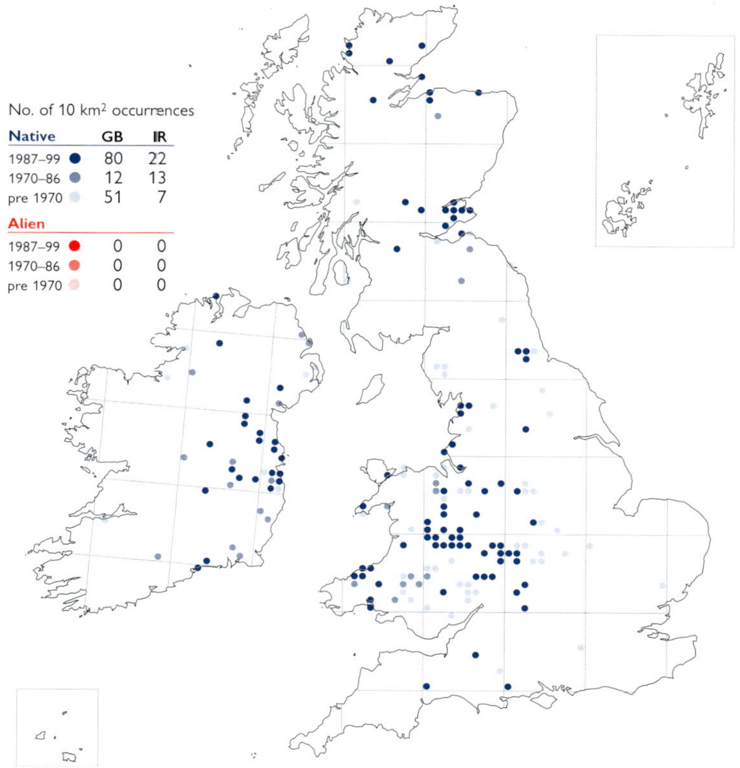

A robust deciduous shrub with flexuous or climbing stems. It occurs on woodland edges and in hedgerows, lanesides, scrub and open grassland. It grows in soils varying from calcareous to mildly acidic. It is usually fertile. Lowland.

Native. The distribution of this uncommon hybrid is now slightly better known since it was mapped by Graham & Primavesi (1993). It is, however, under-recorded, and sometimes confused with its parents.

Wider distribution uncertain.

Reference: Stace (1975).

R. MASKEW

Rosa canina × *R. stylosa* (*R.* × *andegavensis*)

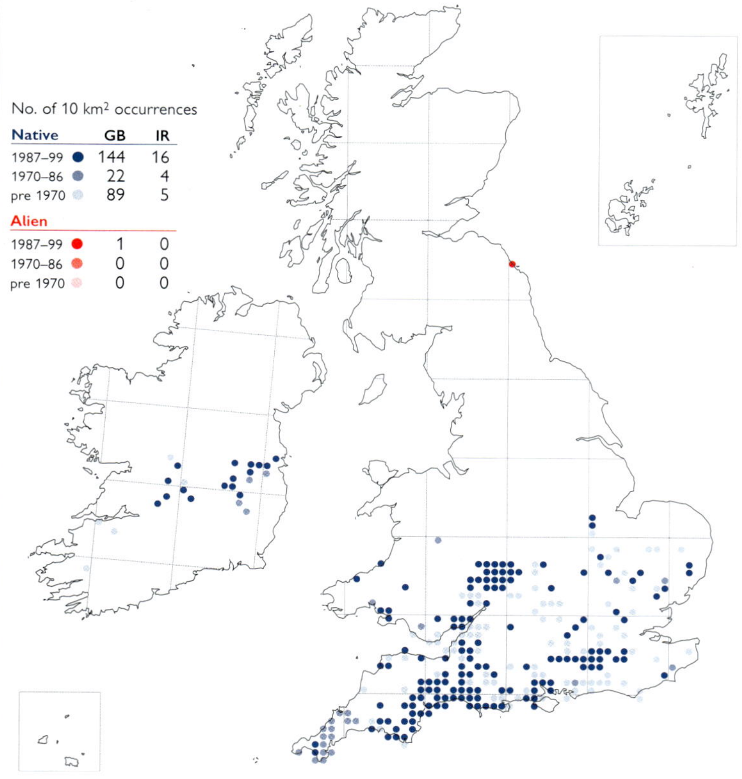

A large deciduous shrub with flexuous or climbing stems. It prefers calcareous soils and is found in hedgerows, along woodland borders, tracksides and in scrub and rough grassland. The hybrid is fertile and backcrosses are frequent; an almost continuous range of intermediates between the parents can be found in some populations. Lowland.

Native. This hybrid is frequent within the western part of the range of *R. stylosa*. Although under-recorded in some areas, its distribution is probably stable, and is much better known now than when mapped by Graham & Primavesi (1993). It is frequently misidentified as its parents.

Wider distribution uncertain.

Reference: Stace (1975).

R. MASKEW

Rosa canina × *R. tomentosa* (*R.* × *scabriuscula*)

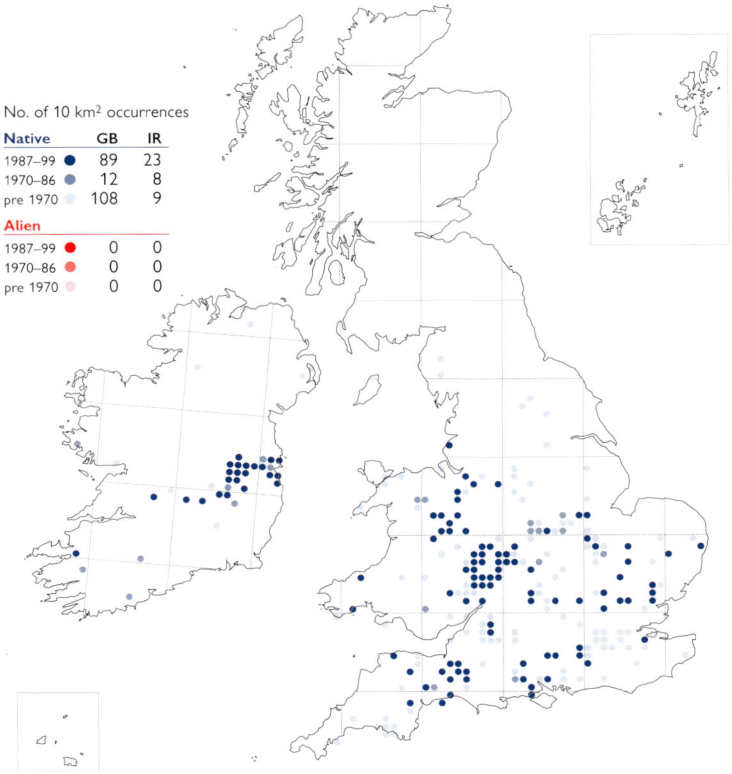

No. of 10 km^2 occurrences

Native	GB	IR
1987–99	89	23
1970–86	12	8
pre 1970	108	9
Alien		
1987–99	0	0
1970–86	0	0
pre 1970	0	0

A tall deciduous shrub with climbing or arching stems. It grows along woodland borders, in scrub and hedgerows, on road and track sides and on railway banks. It prefers well-drained calcareous soils, but is also found on basic clay substrates. It is usually fertile. Lowland.

Native. The distribution of *R.* × *scabriuscula* is probably stable. It is likely to be under-recorded, however, as it is often mistaken for its parents or for *R. canina* × *R. sherardii*.

Wider distribution uncertain.

References: Graham & Primavesi (1993), Stace (1975).

R. MASKEW

Rosa caesia subsp. *caesia* Hairy Dog-rose

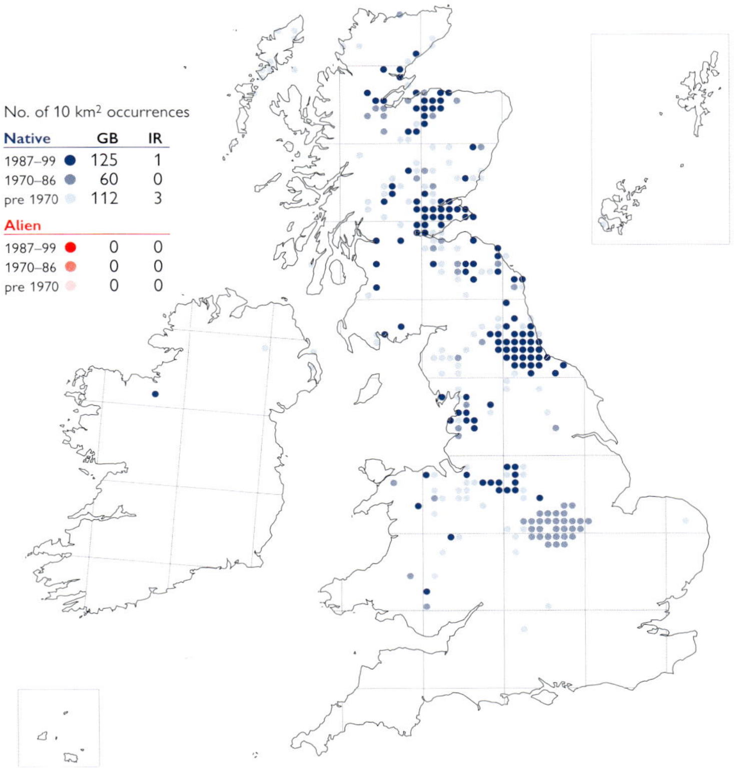

No. of 10 km^2 occurrences

Native	GB	IR
1987–99	125	1
1970–86	60	0
pre 1970	112	3
Alien		
1987–99	0	0
1970–86	0	0
pre 1970	0	0

A deciduous shrub which grows on a wide range of well-drained calcareous to mildly acidic soils. It is found on woodland edges, hedgerows, scrub, rocky grassland and river banks, cliffs and rock outcrops, road verges, railways and waste ground. It can rapidly colonise open habitats on disturbed ground. Lowland.

Native. Although now much better recorded than when mapped by Graham & Primavesi (1993), this subspecies is still under-recorded in many areas. It is often confused with other species and hybrids, particularly *R.* × *dumalis* and the pubescent form of *R. canina*. Its distribution is probably stable.

European Temperate element.

R. MASKEW

Rosa caesia subsp. *glauca* Glaucous Dog-rose

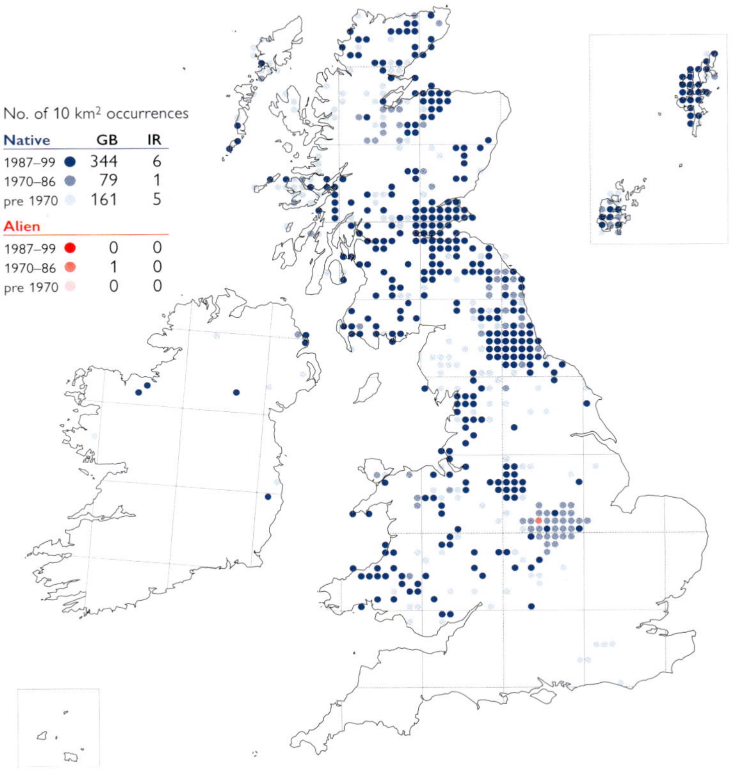

No. of 10 km^2 occurrences

Native	GB	IR
1987–99	344	6
1970–86	79	1
pre 1970	161	5
Alien		
1987–99	0	0
1970–86	1	0
pre 1970	0	0

A deciduous shrub which grows on a wide range of well-drained calcareous to mildly acidic soils. It is found on woodland edges, hedgerows, scrub, rocky grassland and river banks, cliffs and rock outcrops, sand dunes, road verges, railways and waste ground. 0–445 m (Tynehead, Cumberland).

Native. This subspecies is now much better recorded than when mapped by Graham & Primavesi (1993), especially in Scotland. It tends to replace *R. canina* in northern Scotland. It remains under-recorded in many areas, however, and it is often confused with other species and hybrids, especially *R.* × *dumalis*. Its distribution is probably stable.

European Temperate element.

R. MASKEW

Rosa caesia × *R. canina* (*R.* × *dumalis*)

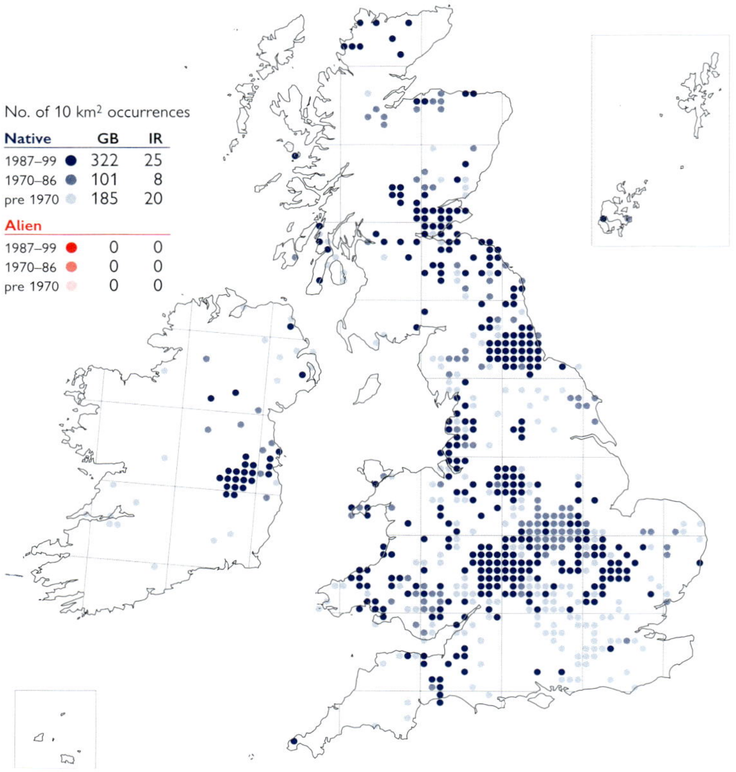

An often robust hybrid found in a wide variety of habitats, including woodland, scrub, hedgerows, rough grassland, sand dunes, tracksides and waste ground. It is fertile and back-crosses to its parents, often producing complex hybrid populations. Lowland.

Native. This is probably the most frequent *Rosa* hybrid in Britain and Ireland. Towards the south of its range it may be common in areas where *R. caesia* is absent. Although now much better recorded than when it was mapped by Graham & Primavesi (1993), it remains under-recorded in many areas. Its overall distribution is probably stable. Hybrids involving both subspecies of *R. caesia* are included on the map.

Wider distribution uncertain.

References: Preston (1997), Stace (1975).

R. MASKEW

Rosa obtusifolia Round-leaved Dog-rose

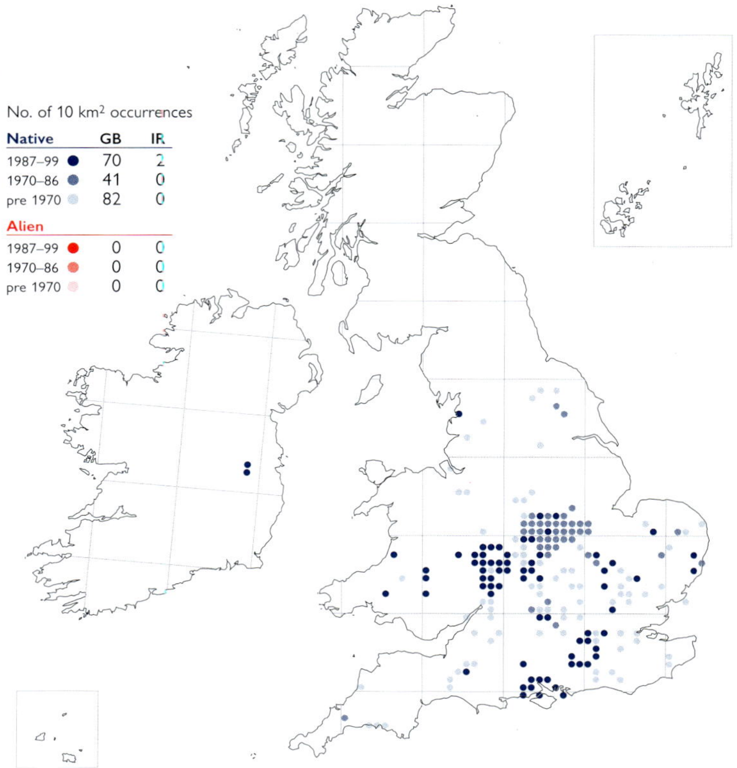

A low-growing, deciduous shrub with arching stems found along woodland borders and in hedgerows, scrub, rough grassland and road verges, gravel-pits and waste land. It prefers well-drained calcareous to mildly acidic soils. Lowland.

Native. The distribution of this species is probably stable. It is, however, an inconspicuous plant which is certainly under-recorded in many areas. It is also frequently confused with other species and hybrid 'Dog-roses', especially the pubescent form of *R. canina* and *R.* × *dumetorum*.

European Temperate element.

References: Graham & Primavesi (1993), Hultén & Fries (1986).

R. MASKEW

Rosa tomentosa Harsh Downy-rose

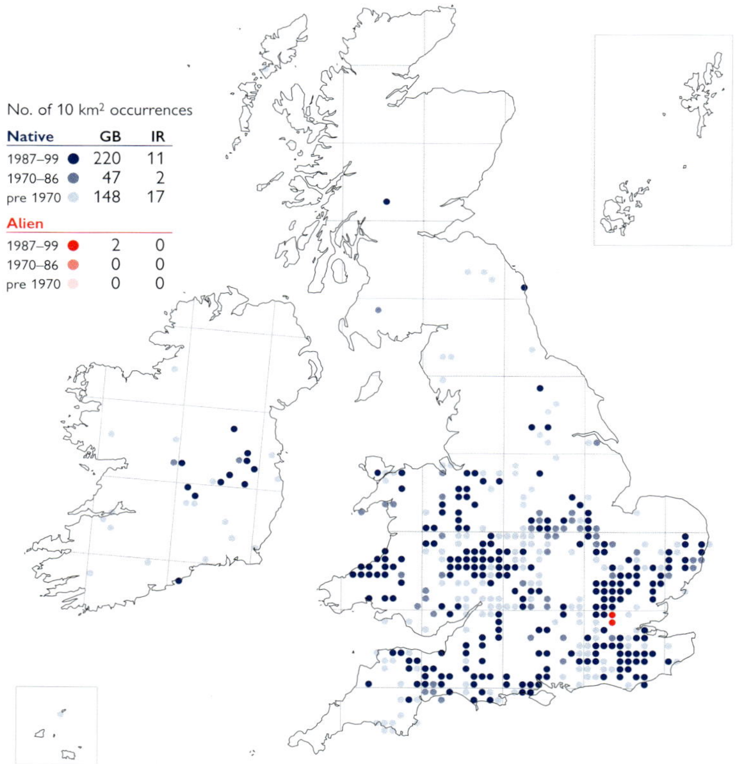

A deciduous shrub with arching or climbing stems, which prefers calcareous to mildly acidic soils and is found on woodland edges and in hedgerows, where it appears to thrive in relatively shady conditions, but also in more open habitats including scrub, rough grassland, disused quarries and less acidic heaths. Lowland.

Native. The large proportion of pre-1970 records of this species probably reflects the fact that many records are based on herbarium specimens. Like many *Rosa* species, *R. tomentosa* is probably under-recorded. It can be difficult to identify, especially in areas where hybrids with *R. canina* (*R.* × *scabriuscula*) are found.

European Temperate element.

References: Graham & Primavesi (1993), Hultén & Fries (1986), Meusel *et al.* (1965).

R. MASKEW

Rosa sherardii Sherard's Downy-rose

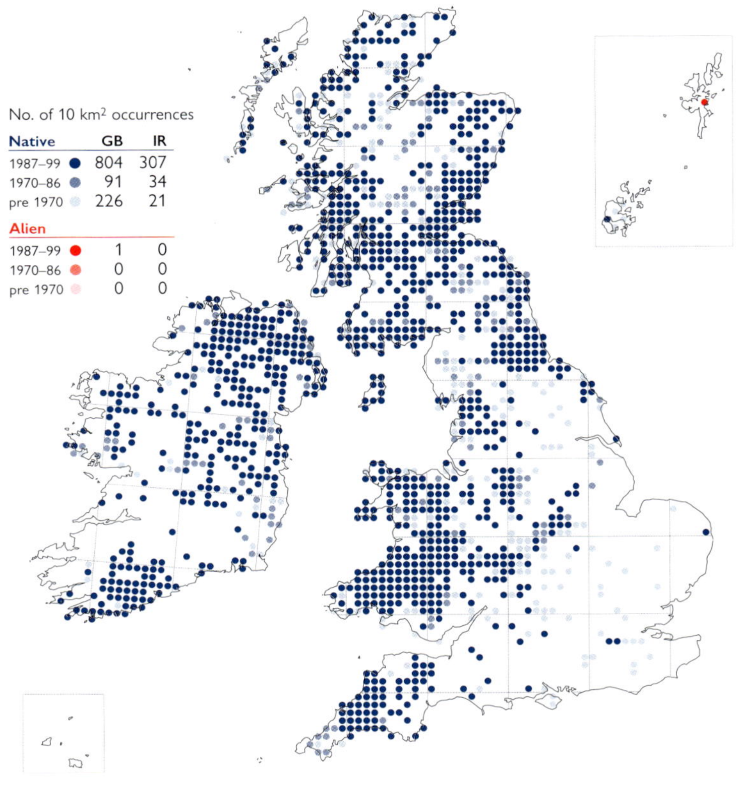

An erect, deciduous shrub of woodland edges, hedgerows, scrub, rough grassland, heathland, rock outcrops, cliffs and lane- and track-sides. It grows on a variety of soils and is able to tolerate both moderately acid and wet conditions. Lowland.

Native. This species is not too difficult to identify, and better recorded than many *Rosa* taxa, although it remains under-recorded in some areas. Its overall distribution is stable, other than in S.E. England where most of the pre-1970 records are based on herbarium specimens.

European Temperate element.

References: Graham & Primavesi (1993), Hultén & Fries (1986).

R. MASKEW

Rosa mollis Soft Downy-rose

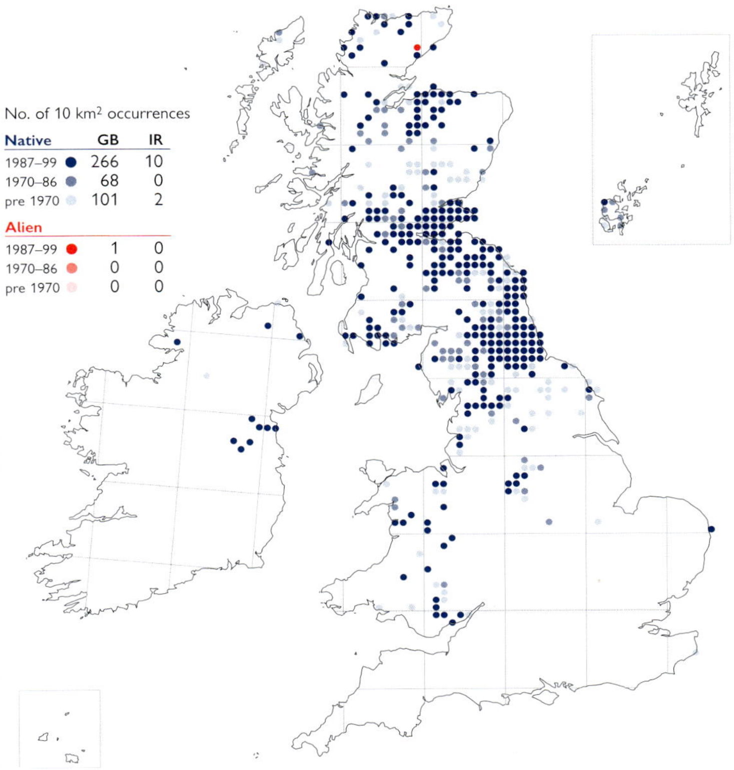

An erect, deciduous shrub, suckering freely and sometimes forming dense thickets. It occurs in woodland, hedges, scrub and rough grassland, and on rocky streamsides, rock outcrops, screes, cliffs, sand dunes, waste ground and on roadsides and pathsides. It grows in a variety of well-drained soils, but avoids very acidic conditions. 0–470 m (Pennant Dyfi, Merioneth).

Native. Like *R. sherardii*, this species is not too difficult to identify and the map is probably a fairly accurate reflection of its true distribution, although it remains under-recorded in some areas, and it is often confused with *R. mollis* hybrids. Its overall distribution is probably stable.

European Boreo-temperate element.

References: Graham & Primavesi (1993), Hultén & Fries (1986).

R. MASKEW

Rosa mollis × *R. pimpinellifolia* (*R.* × *sabinii*)

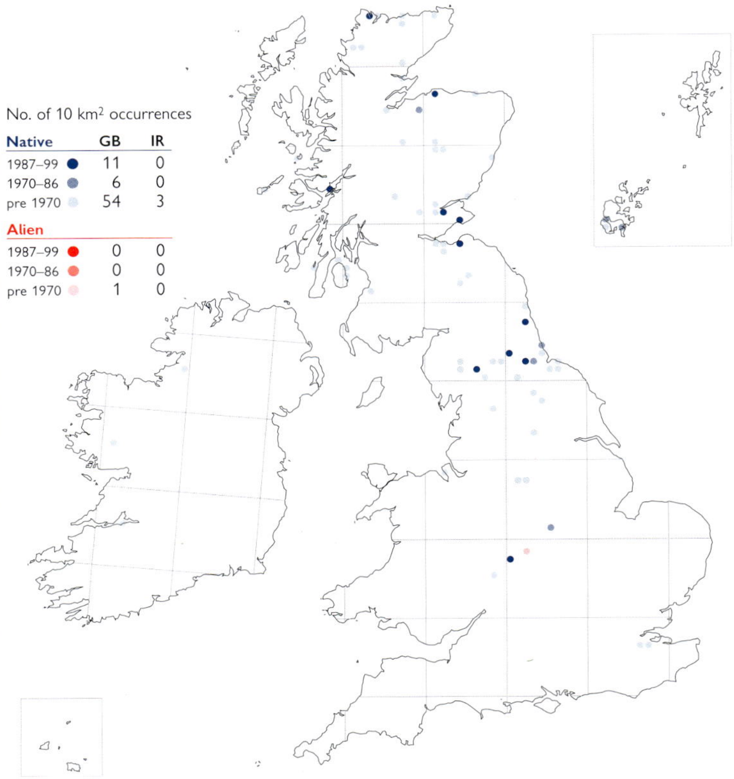

An erect, freely suckering, deciduous shrub, usually found with the parents in hedgerows, rocky grassland, and more typically in scrub on sand dunes. It is usually sterile, but occasionally produces a few well-formed hips. Lowland.

Native. The map of *R.* × *sabinii* suggests that it has declined in many areas. This is probably mainly due to under-recording in recent years, but some losses may be genuine if inland populations have been lost to habitat destruction, as is the case with *R. pimpinellifolia*. It is sometimes grown in gardens, and this may be the source of some of the records.

Wider distribution uncertain.

Reference: Stace (1975).

T. D. DINES

Rosa rubiginosa Sweet-briar

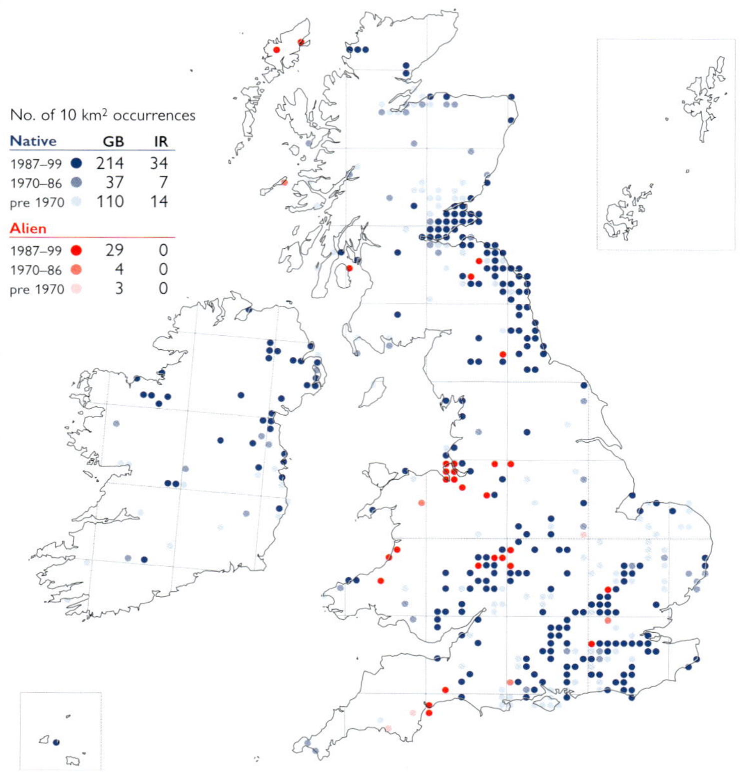

An erect deciduous shrub, characteristically found in scrub and hedgerows on chalk and limestone, but also found in quarries, on railway banks and on waste ground. It is often frequent as a colonist of under-grazed chalk grassland. Lowland.

Native. This species is not difficult to identify and is better recorded than many other *Rosa* taxa. Some populations have been lost, probably through habitat destruction. However, it is now frequently planted in parks, amenity areas and along new roadsides, and such alien occurrences are increasing.

European Temperate element; widely naturalised outside its native range.

References: Graham & Primavesi (1993), Hultén & Fries (1986).

R. MASKEW

Rosa micrantha Small-flowered Sweet-briar

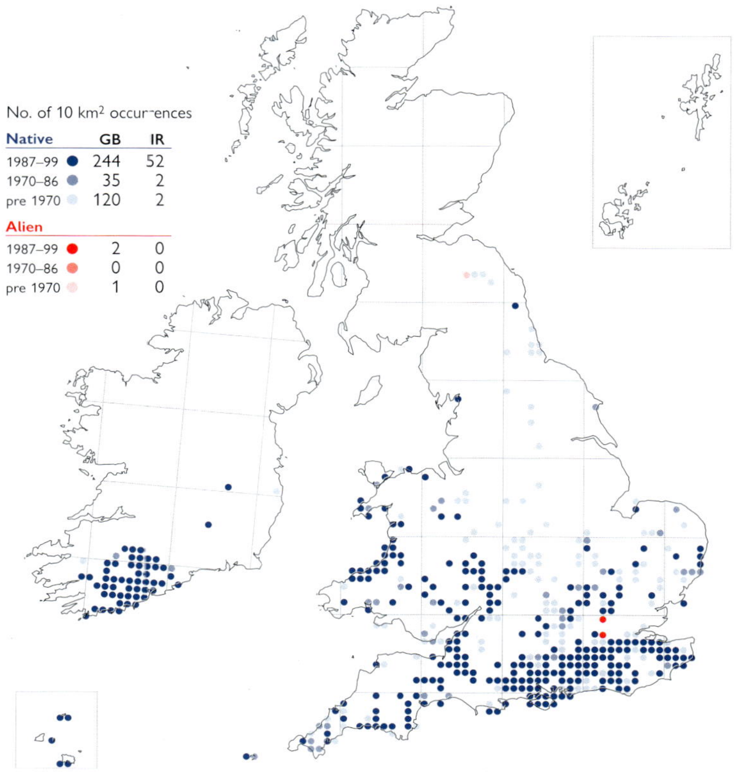

A tall, climbing, deciduous shrub, typically found in woodland, scrub and hedgerows, and also more open habitats such as chalk grassland, heathland, sea-cliffs, disused quarries and railway banks. It grows on a range of well-drained soils, but is most frequent on calcareous substrates and avoids the more acidic sites. Lowland.

Native. The distribution of *R. micrantha* is now much better known than when it was mapped by Graham & Primavesi (1993). It appears to have declined slightly in some areas, possibly through habitat destruction, but it remains under-recorded.

European Temperate element.

Reference: Hultén & Fries (1986).

R. MASKEW

Rosa agrestis Small-leaved Sweet-briar

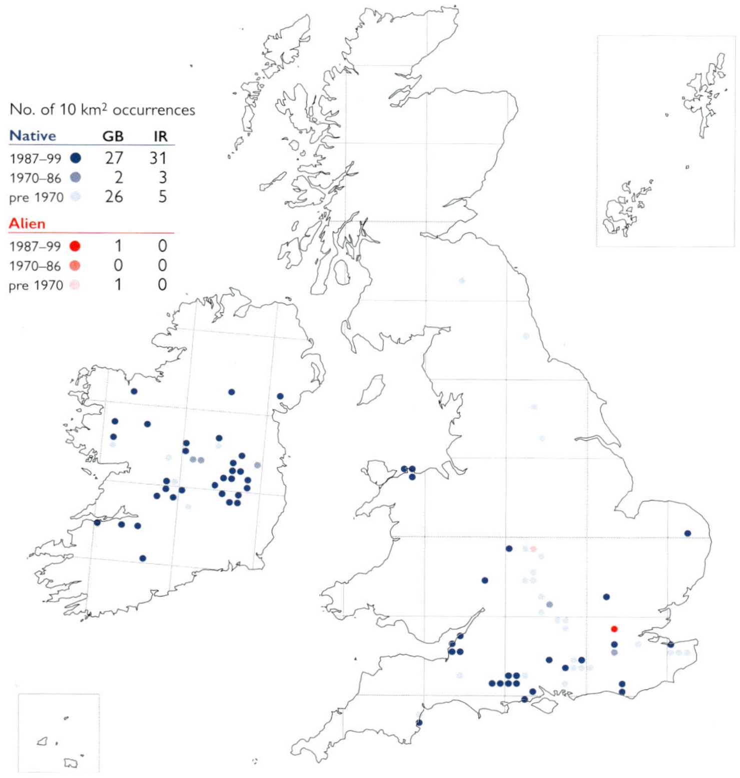

An erect, deciduous shrub of open scrub on dry, calcareous grassland overlying chalk or limestone. Populations often consist of only a few individuals. Lowland.

Native. This species was neglected for many years; recorders have only recently become familiar with it and its distribution is not yet fully documented. It has been lost from some sites due to ploughing, and from others through a reduction in grazing leading to over-shading. However, new sites continue to be found (there were recent records from 16 10-km squares in Britain in Wigginton, 1999), and more doubtless await discovery. In Ireland, its distribution appears to be better known and more stable, but even there it is probably under-recorded.

European Temperate element.

References: Graham & Primavesi (1993), Hultén & Fries (1986).

R. MASKEW

Prunus cerasifera Cherry Plum

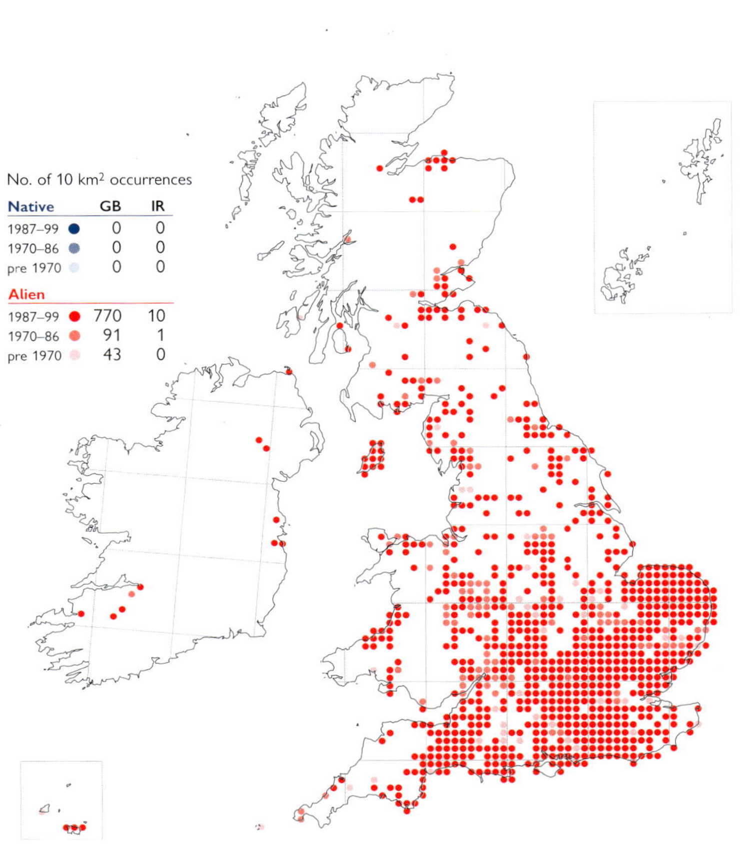

A shrub or small tree of roadsides, hedges, woods and copses and ornamental plantings. It rarely if ever suckers, and only occasionally sets fruit. Lowland.

Neophyte (change +3.43). There is considerable doubt about the history of this species in Britain. Known in cultivation in the 16th century, it was not recorded until the 20th century in many areas where it is now frequent. It may have been overlooked by recorders unaware of its early flowering, and there has been a vast increase in 10-km square records since the 1962 *Atlas* as this character has become more widely known. It may also be confused with varieties of *P. domestica* and *P. spinosa*.

Native of S.E. Europe, S.W. & C. Asia.

References: Atlas (129a), Bean (1976), Roach (1985), Sell (1991).

D. J. McCOSH

Prunus spinosa Blackthorn

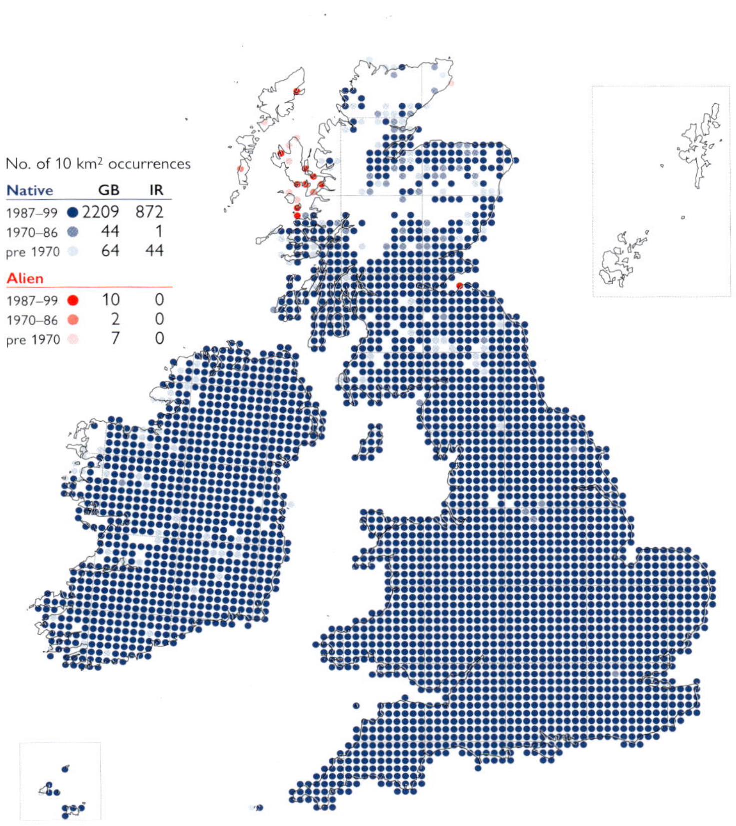

A deciduous shrub or small tree of open woodlands, scrub, hedgerows, screes and cliff-slopes; a prostrate form also occurs on shingle beaches. It grows on a wide variety of soils. It reproduces by seed, and spreads vegetatively by suckers, often forming dense thickets. In many areas, native populations have been augmented by deliberate planting in hedgerows and copses. 0–500 m (Cross Fell, Cumberland).

Native (change +0.40). The distribution of *P. spinosa* is stable.

European Temperate element; widely naturalised outside its native range.

References: Atlas (128c), Hopkins (1996), Hultén & Fries (1986), Meusel *et al.* (1965), Rackham (1980).

D. J. McCOSH

Prunus domestica Wild Plum

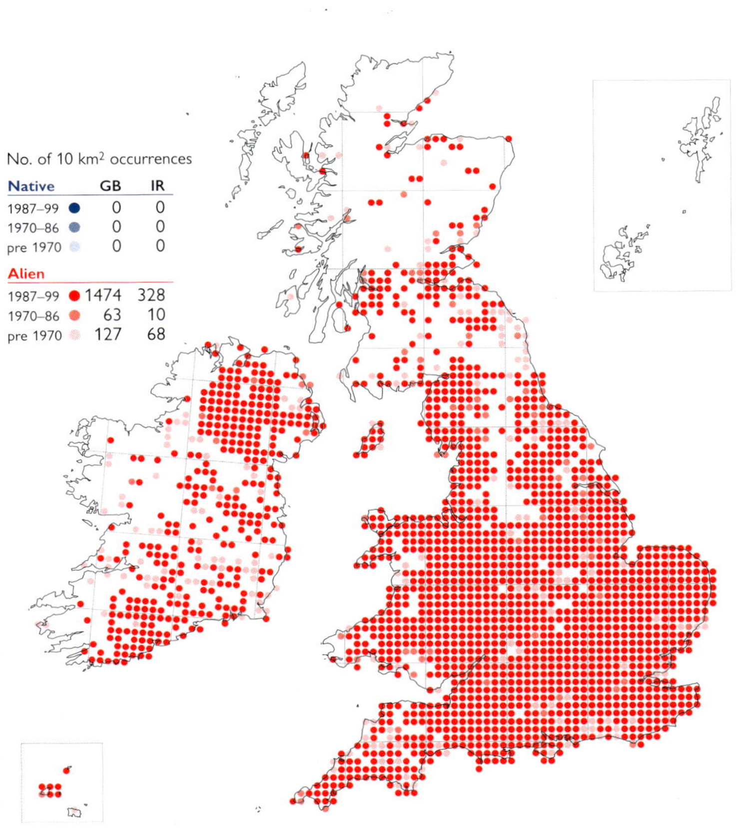

A shrub or tree, naturalised in hedges, wood-borders, scrub and waste places. Lowland.

Archaeophyte (change +2.19). This species has been grown in gardens since at least 995 (Harvey, 1981) and known from the wild since 1777. Three subspecies, subsp. *domestica* (plum), subsp. *insititia* (bullace) and subsp. *italica* (greengage) are often recognised. Many plants are relics of cultivation, but subsp. *domestica* is still being introduced from discarded plum stones. Although its overall range is similar to that in the 1962 *Atlas*, it is much better recorded.

Apparently derived from hybrids of *P. cerasifera* and *P. spinosa*; such hybrids occur in the Caucasus but *P. domestica* has been much modified in cultivation.

References: Atlas (128d), Bolòs & Vigo (1984), Roach (1985), Sell (1991), Zohary & Hopf (2000).

D. J. McCOSH

Prunus domestica × *P. spinosa* (*P.* × *fruticans*)

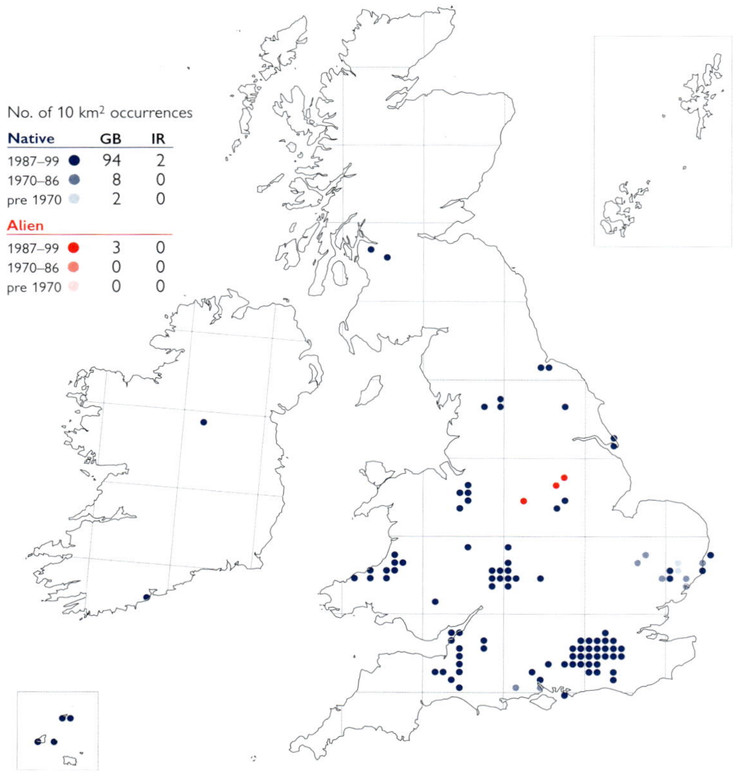

A shrub or small tree of hedgerows, scrub and railway banks. The hybrid is fertile, and can spread vigorously by suckers. Lowland.

A spontaneous hybrid between native and alien parents. This hybrid has not previously been mapped at a national scale, so trends in its distribution are difficult to assess. It is, however, likely to be under-recorded, especially as few recorders would know where to draw the line between the very variable *P. domestica* and this variable hybrid.

Apparently widespread in Europe.

Reference: Stace (1975).

D. J. McCOSH

Prunus avium Wild Cherry

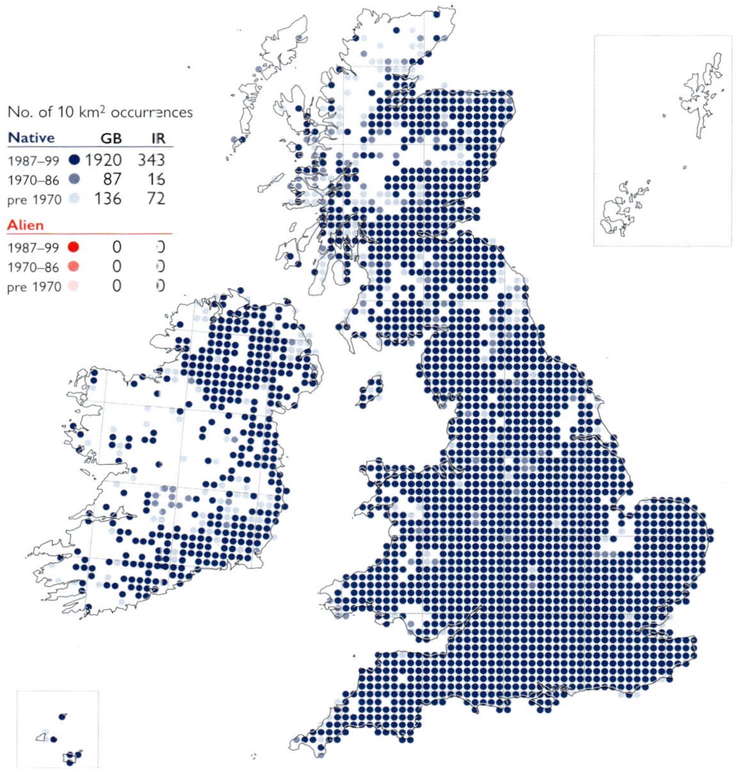

The natural habitats of this small to medium-sized tree include deciduous woodland and hedges on fertile soils, but it is widely planted as an ornamental or fruit tree in parks and gardens. It spreads by fruit and suckers. 0–400 m (N. of Garrigill, Cumberland).

Native (change +1.29). The distribution of *P. avium* has not changed since the 1962 *Atlas*. It is frequently difficult to decide whether populations are native or alien, and all records have been mapped here as if they are native.

European Temperate element; widely naturalised outside its native range.

References: Atlas (129b), Hultén & Fries (1986), Meusel *et al.* (1965), Rackham (1980), Roach (1985), Sell (1991, 1992), Zohary & Hopf (2000).

D. J. McCOSH

Prunus cerasus Dwarf Cherry

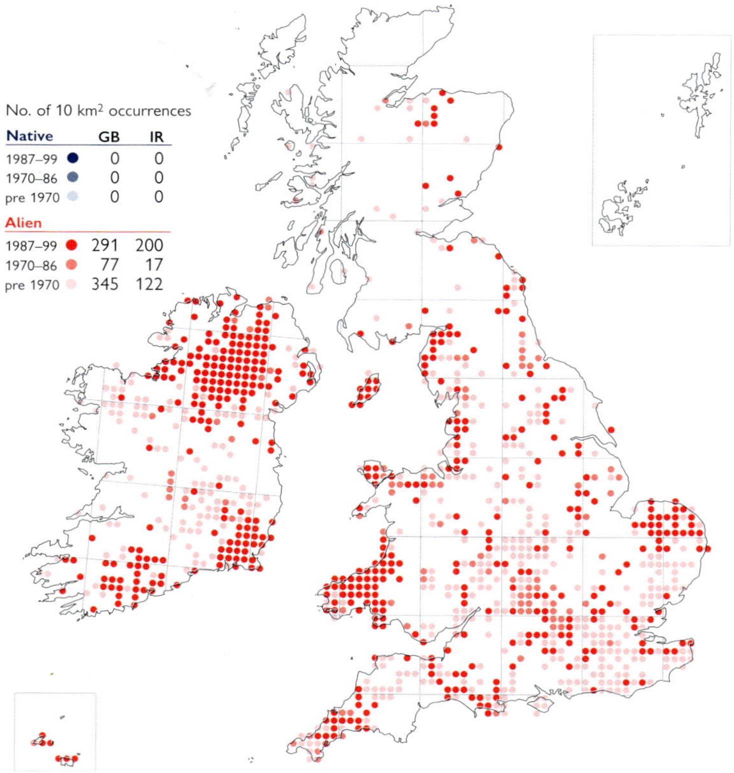

A shrub or small tree of hedgerows, copses and wood-borders. It spreads by fruit or suckers, and can sometimes form dense thickets. Lowland.

Archaeophyte (change –0.90). Cultivars of this species were being grown in the 16th century, and it is one of the parents of the Morello Cherry. Most populations have been deliberately planted, rather than bird-sown. The 1962 *Atlas* indicated a strong decline in England prior to 1930, but *P. cerasus* is much confused with *P. avium* and its past and present distribution is to that extent uncertain. There has been a small decline since the 1962 *Atlas*.

Native of S.W. Asia.

References: Atlas (129c), Roach (1985), Sell (1991), Zohary & Hopf (2000).

D. J. McCOSH

Prunus padus Bird Cherry

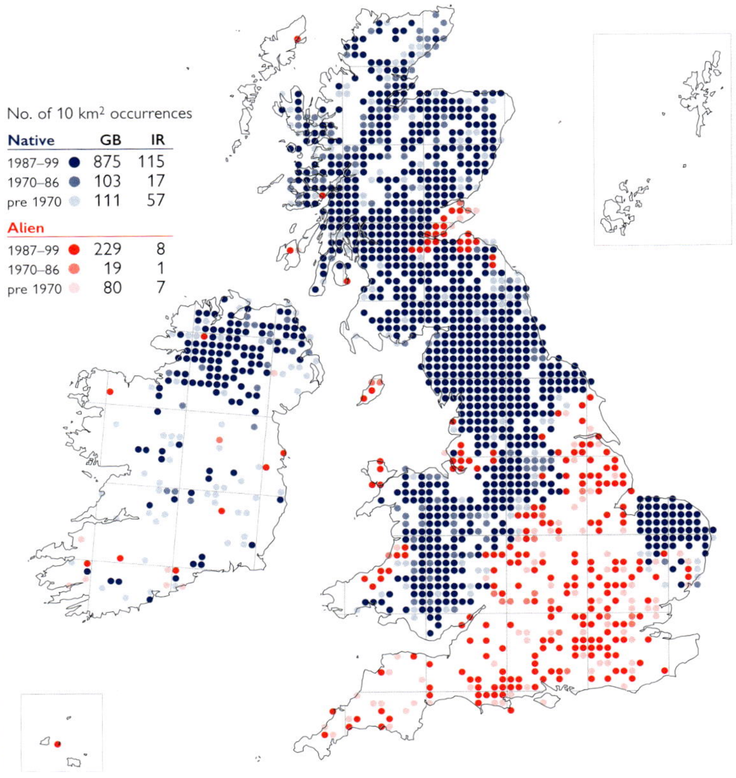

A deciduous shrub or small tree of moist woodland and scrub, streamsides and shaded rocky places; also in fen-carr in East Anglia. It occurs on a wide variety of soil types, but is most frequent on damp calcareous or base-rich substrates, and avoids very dry or very acidic conditions. It spreads by fruit and suckers, often forming thickets. 0–650 m (Dove Crag, Westmorland).

Native (change +0.58). The distribution of *P. padus* is largely unchanged since the 1962 *Atlas*. It is widely planted both within, and to the south of, its native range. The distinction between native and alien populations is sometimes unclear.

Eurasian Boreo-temperate element.

References: Atlas (129d), Curtis & McGough (1988), Hultén & Fries (1986), Leather (1996), Meusel *et al.* (1965).

D. J. McCosh

Prunus serotina Rum Cherry

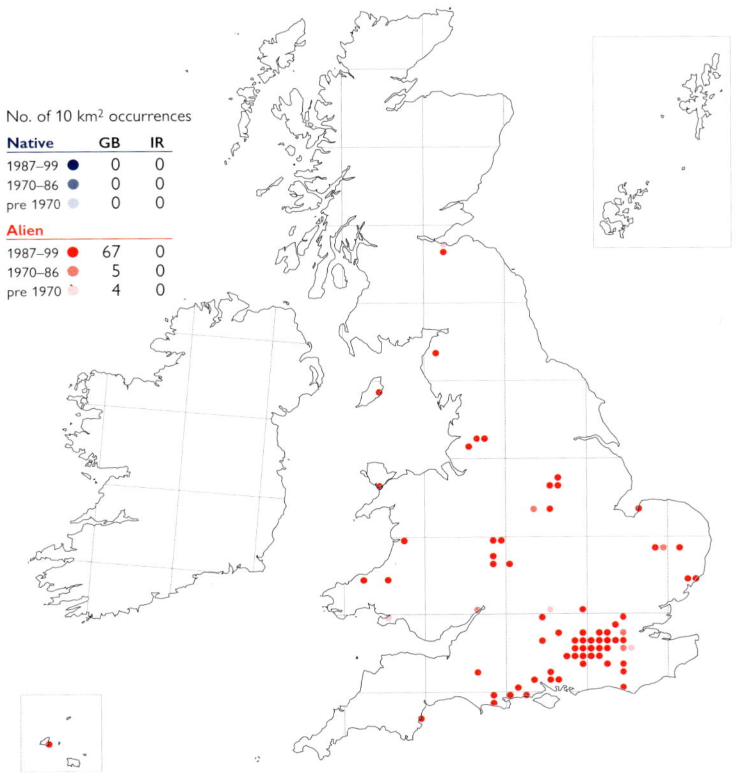

A small or medium-sized tree, commonly planted in gardens and found naturalised in woodland and hedgerows, and on roadside verges, riversides and heaths. Reproduction is by seed and the species can sometimes become invasive. Lowland.

Neophyte. This species, which was being cultivated in Britain by 1629, was first recorded from the wild in 1853 (Surrey). It is likely to be increasing.

Native of eastern N. America.

Reference: Bean (1976).

D. J. McCosh

Prunus lusitanica Portugal Laurel

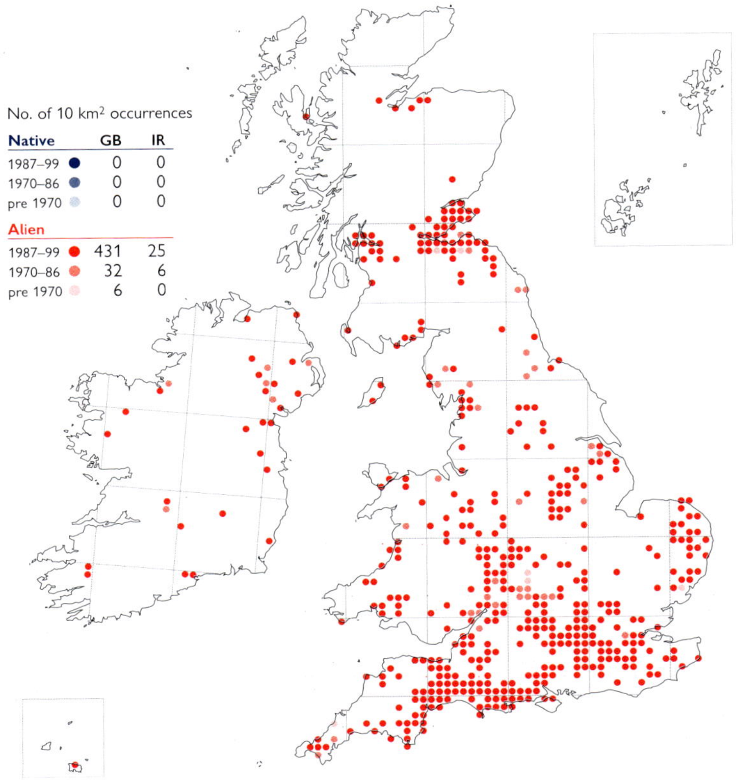

An evergreen shrub, frequently planted in parks and gardens, and now well-established in woods, scrub and on waste ground. It is frequently self-sown. Lowland.

Neophyte. *P. lusitanica*, cultivated in Britain by 1648, is less commonly planted than *P. laurocerasus*. It was noted in the wild by 1927, and may be unevenly recorded.

Native of the Iberian peninsula.

References: Bean (1976), Bolòs & Vigo (1984).

D. J. McCosh

Prunus laurocerasus Cherry Laurel

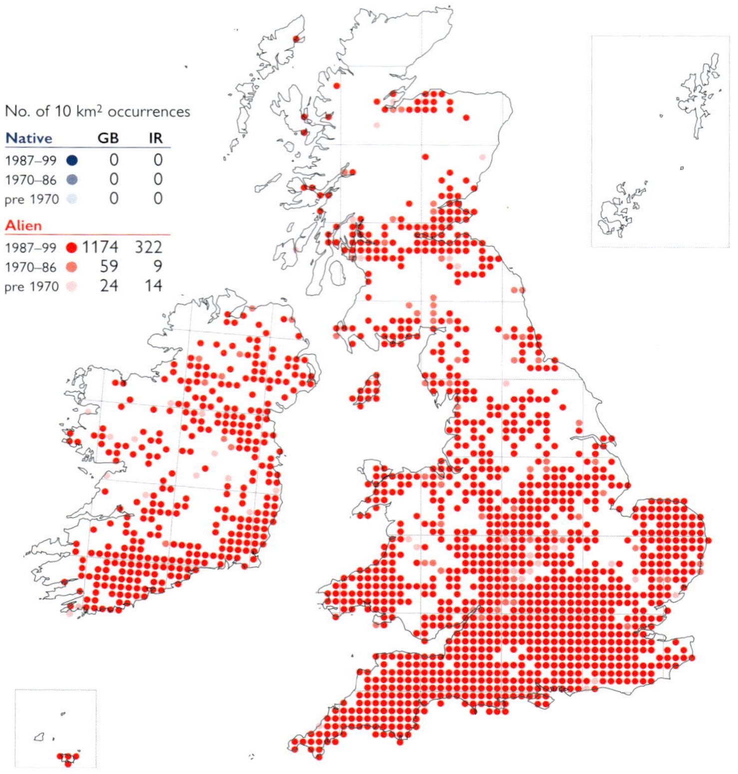

A glossy-leaved evergreen shrub or small tree, naturalised in woods and scrub, and sometimes self-sown. Layering can occur when the branches of old trees lie on the ground. Lowland.

Neophyte (change +4.70). *P. laurocerasus* was being cultivated in Britain by 1629 and was known from the wild by 1886. It is now commonly planted for amenity, and the current map shows a much greater frequency than in the 1962 *Atlas*.

Native of the Balkan peninsula.

References: Atlas (130a), Bean (1976).

D. J. McCOSH

Chaenomeles speciosa Chinese Quince

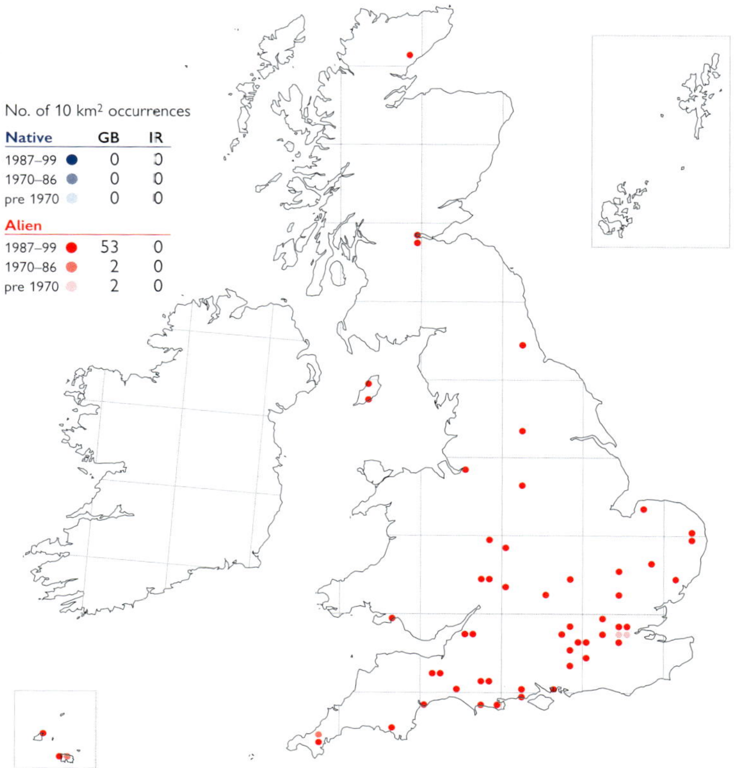

A small, deciduous shrub found in woodland, hedgerows, scrub, rough grassland and on waste ground. It occurs as a garden escape, being bird-sown, and as a relic of cultivation. Plants may be long-lived and persistent, but rarely reproduce by seed or become naturalised. Lowland.

Neophyte. This species has been cultivated in Britain since around 1796, with a second introduction in 1830, and is extremely popular in gardens. It was first recorded in the wild in 1963. Its distribution is likely to be stable, although it is probably under-recorded and may be confused with *C. japonica* and the hybrid *C. japonica* × *C. speciosa*.

Native of China.

Reference: Bean (1970).

T. D. DINES

Pyrus cordata Plymouth Pear

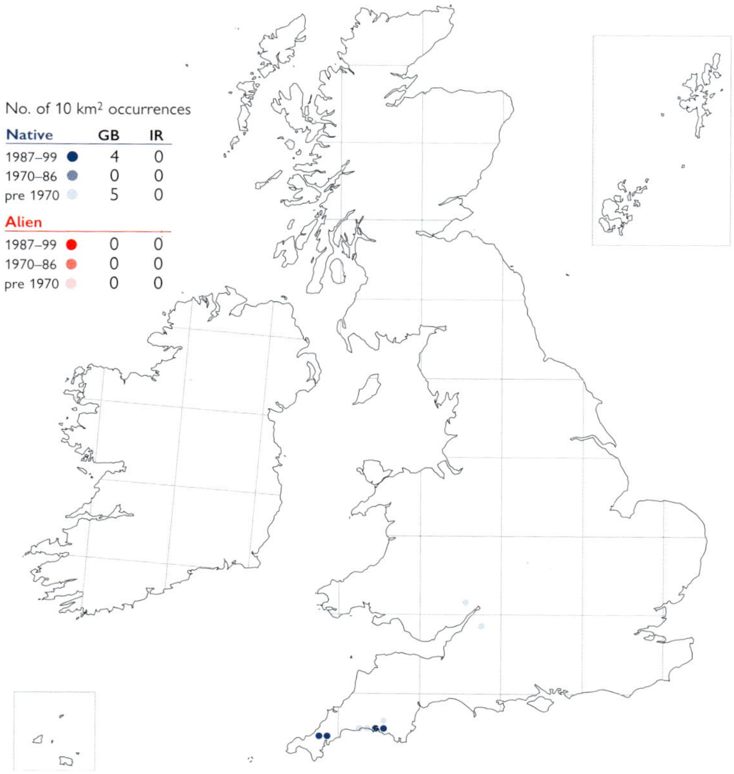

In Britain, this small deciduous tree is restricted to just a few hedgerows in Devon and Cornwall. Reproduction is by suckering; fruit-set varies greatly from year to year, and production of fertile seed is negligible. Lowland.

Native or alien. *P. cordata*, which was first recorded in 1870, has always been very rare in Britain, and its low seed-fertility offers little scope for it to spread naturally to new sites. However, the surviving populations now are being bolstered through carefully documented breeding and restocking programmes (Jackson, 1995).

Oceanic Temperate element.

References: Atlas (132c), Wigginton (1999).

D. J. McCOSH & S. J. LEACH

Pyrus communis sens. lat. Pears

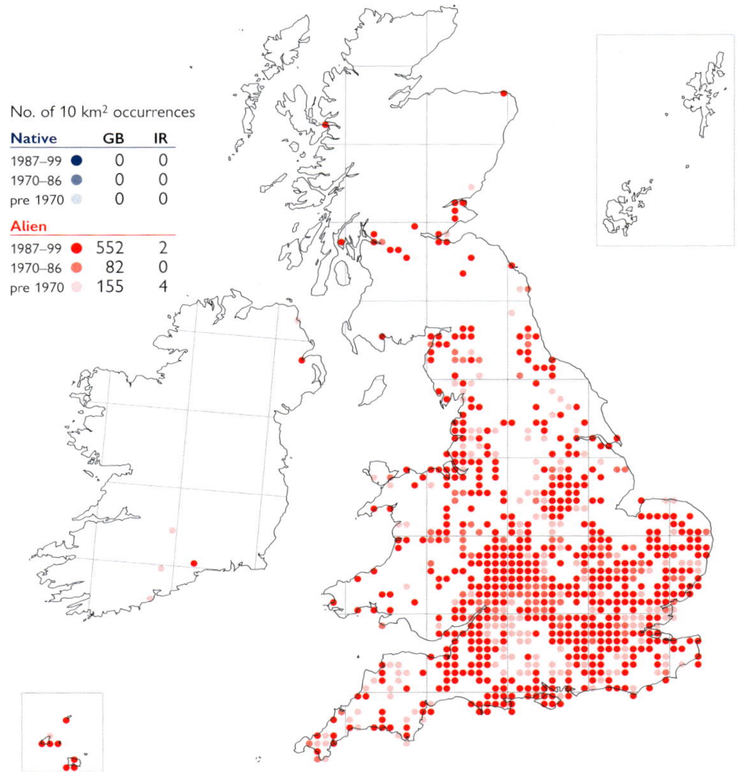

Deciduous trees or shrubs found in hedges, woodland margins and old gardens, and on railway banks and waste ground. They spread easily by seed and by discarded cores. Lowland.

Archaeophyte (change +1.49). *P. communis sens. lat.* has been grown in gardens since at least 995 (Harvey, 1981). Records have increased since the 1962 *Atlas*, but this is almost certainly due to better recording. The cultivated pear (*P. communis*) and wild pear (*P. pyraster*) are rarely distinguished by recorders.

P. communis sens. lat. includes *P. pyraster*, which may be native to C. & S. Europe and S.W. & C. Asia, and *P. communis*, which has a hybrid origin in cultivation.

References: Atlas (132b), Bolòs & Vigo (1984), Meusel *et al.* (1965), Roach (1985), Spinage (2000), Vaughan & Geissler (1997), Zohary & Hopf (2000).

D. A. PEARMAN

Malus sylvestris sens. lat. Apples

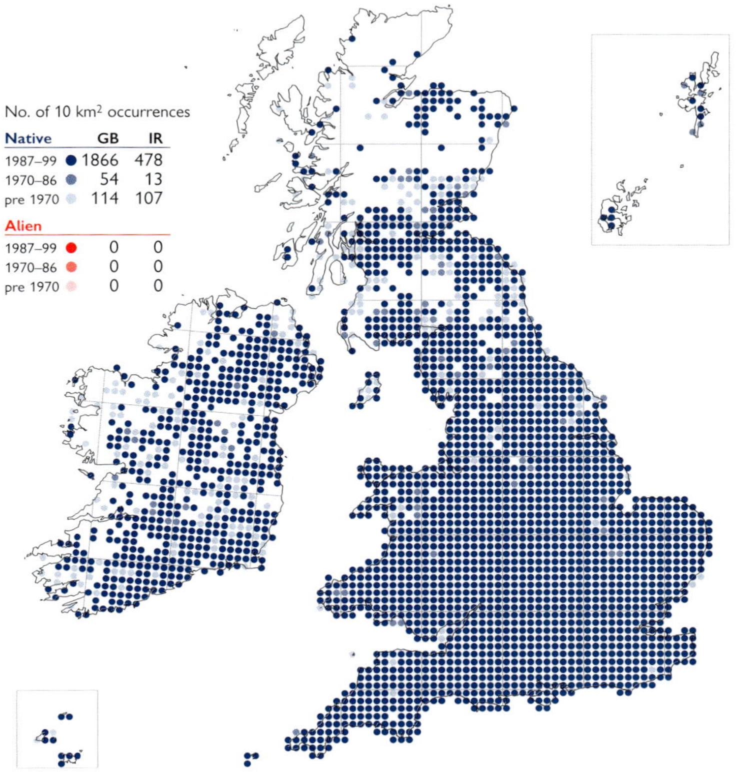

These are small trees of hedgerows, scrub, copses, roadsides and rough ground, usually occurring as single trees. 0–380 m in Swindale (Westmorland).

Native (change +0.57). This map covers both the native *M. sylvestris sens. str.* (Crab Apple) and the alien *M. domestica* (Apple) and therefore equates to the map in the 1962 *Atlas*. Confirmed records of both taxa were mapped by Perring & Sell (1968). The two species hybridise and seem to be connected by a range of intermediates, which are included on the map. It can be difficult to separate native and alien populations, and all records are mapped as if they are native.

European Temperate element; widely naturalised outside its native range.

References: Atlas (132d), Atlas Supp. (39a, b), Roach (1985), Hultén & Fries (1986).

D. J. McCOSH & D. A. PEARMAN

Sorbus domestica Service-tree

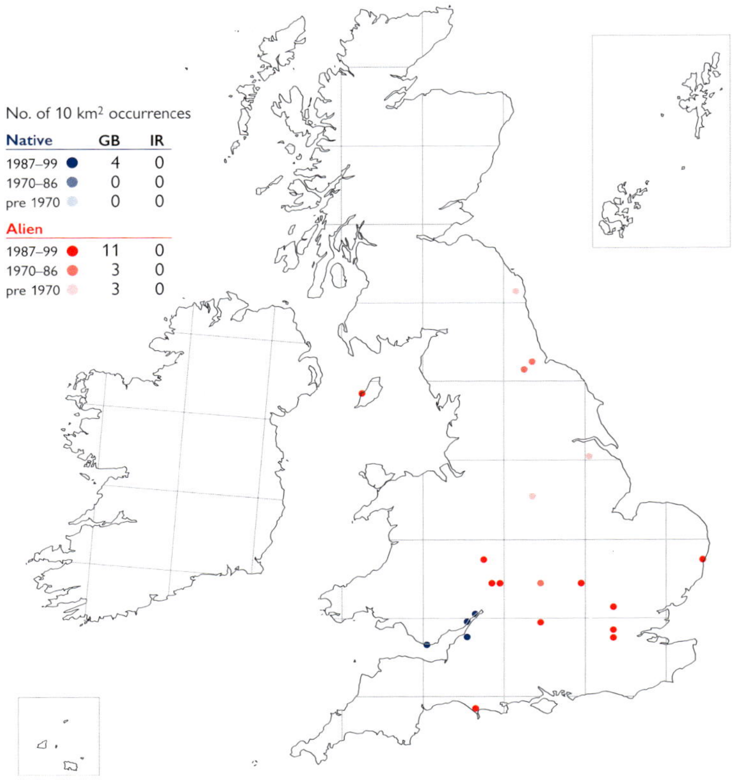

A shrub or small tree found on S.-facing coastal cliff ledges and in cliff scrub and gorge woodland, mostly on limestone, but occasionally on mudstone or shale. Although a sexual species, flowering is often poor and fruit rarely produced. Suckering is frequent. Lowland.

Native. This species was originally known as a single tree in the Wyre Forest (Worcs.), first described in 1678. Five trees now exist near there, these probably being cuttings from the original. However, its wild coastal-cliff habitat was discovered in 1973, and 22 trees are now known in Wales and 8 in England. It is also rarely grown in gardens and parks.

European Southern-temperate element.

References: Q. O. N. Kay (1998), Meusel *et al.* (1965), Paton (1967), Wigginton (1999).

M. C. F. PROCTOR & M. J. WIGGINTON

Sorbus aucuparia Rowan

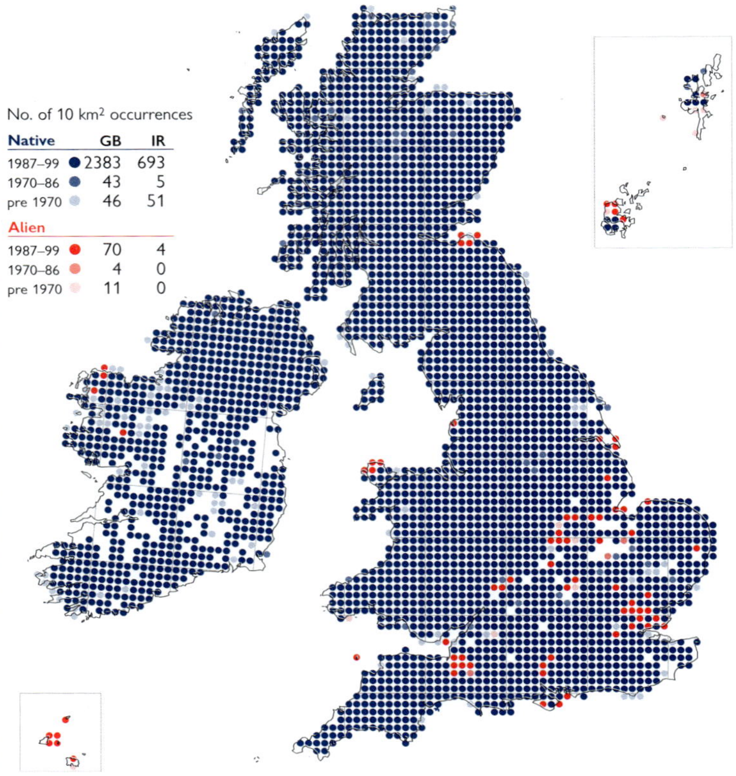

A small to medium-sized tree of woods, cliffs, rock outcrops and rocky riversides. It can also be bird-sown from planted trees on waste ground and by railways. It avoids calcareous and heavy soils and dense shade. A sexual species, it flowers and fruits freely. 0–870 m (Helvellyn, Cumberland, and in the Rannoch area, Mid Perth).

Native (change +0.86). There has been no appreciable change in the native distribution of *S. aucuparia* since the 1962 *Atlas*. The apparent increase probably results from increased recording of planted trees, some of which may be erroneously mapped as native.

Eurasian Boreo-temperate element.

References: Atlas (131c), Grime *et al.* (1988), Hultén & Fries (1986), Meusel *et al.* (1965), Raspé *et al.* (2000).

M. C. F. PROCTOR

Sorbus pseudofennica Arran Service-tree

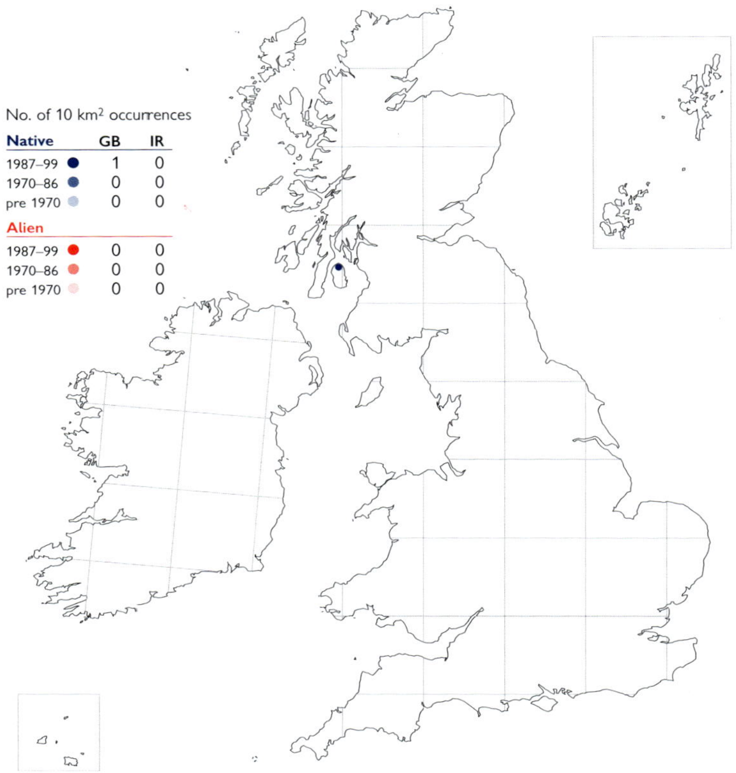

This shrub or small tree grows on the rocky sides of stream ravines and granite crags, and on rocky moorland, and is occasionally found on flatter, well-drained ground by the stream bed. It is apomictic and fully fertile. Lowland.

Native. There is no evidence of any change in the population of this species since 1970. Most of the population, totalling about 400 trees, is in a single site.

Endemic.

References: Atlas Supp. (34a), Boyd *et al.* (1988), Hull & Smart (1984), Wigginton (1999).

M. C. F. PROCTOR

Sorbus arranensis

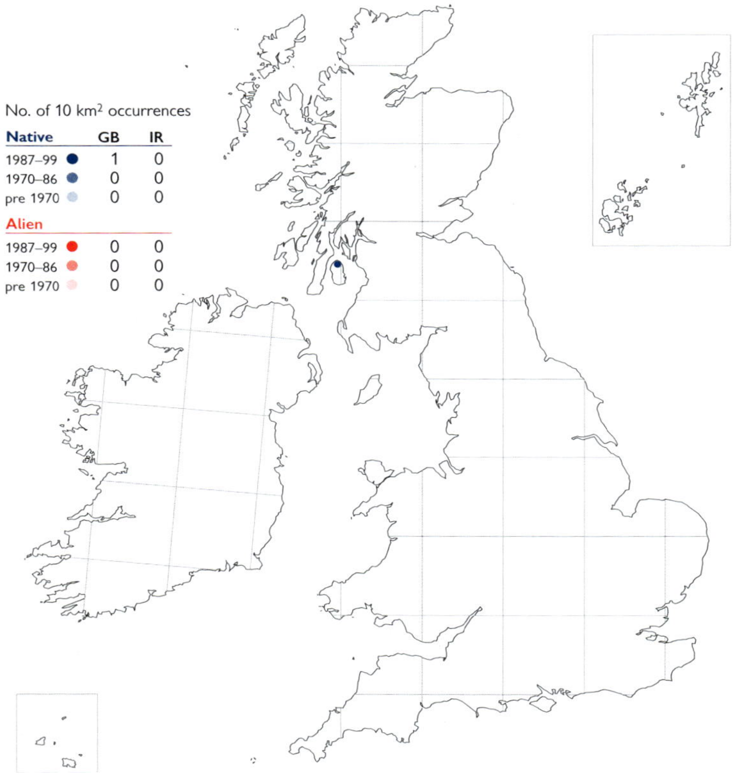

A shrub or small tree of rocky streamsides and outcrops on granite. It is a fully fertile apomictic hybrid derived from *S. aucuparia* and *S. rupicola*, and natural regeneration occurs. Lowland.

Native. The population of about 400 trees is probably about the same size as that of *S. pseudofennica*, but is somewhat more widely distributed within its restricted range. There is no evidence of any significant change in its distribution or population size since 1970.

Endemic.

References: Atlas Supp. (35a), Boyd *et al.* (1988), Hull & Smart (1984), Wigginton (1999).

M. C. F. PROCTOR

Sorbus leyana

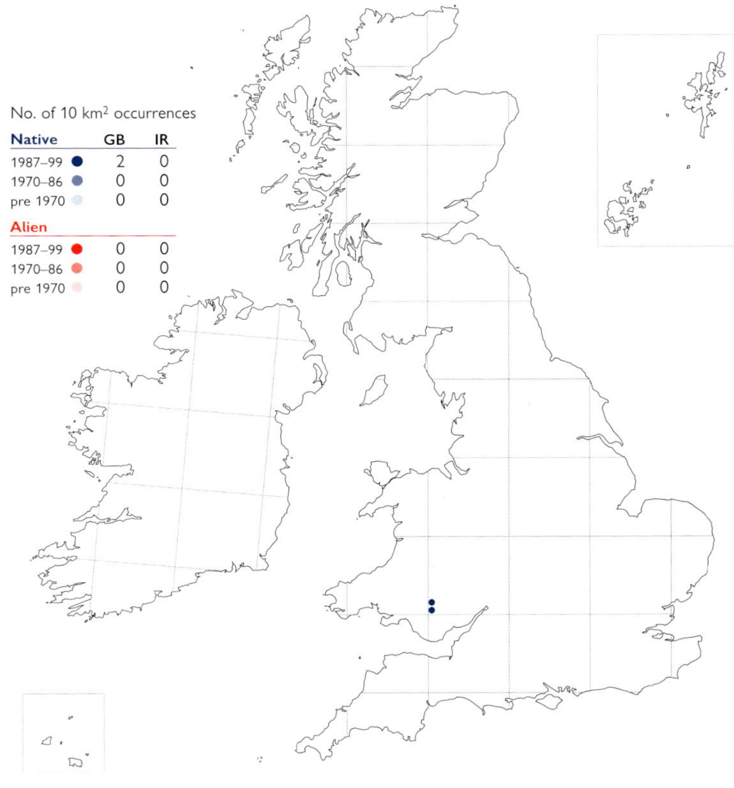

A shrub or small tree, found in scrub or open woodland on Carboniferous limestone crags. This apomictic species usually produces fruit only sparsely, and germination is poor, with little natural regeneration occurring. Lowland, reaching *c.* 300 m on Darren Fach (Brecs.).

Native. Approximately twenty trees of this species are known, in two very localised populations about a kilometre apart. Saplings are vigorous when grown in cultivation from wild seed, and some of them have been planted at one of the native sites (Penmoelallt, Brecs.) to bolster the natural population of only three trees.

Endemic.

References: Atlas Supp. (35a), Q. O. N. Kay (1998), Proctor & Groenhof (1992), Wigginton (1999), Woods (1998).

M. C. F. PROCTOR

Sorbus minima

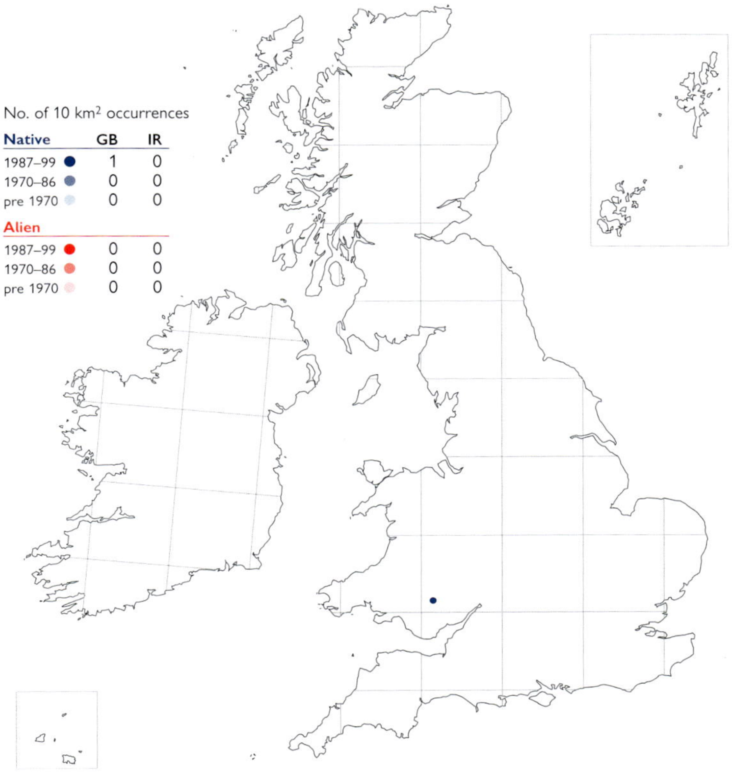

An apomictic shrub or small tree which grows on crags and steep rocky slopes on Carboniferous limestone near Llangattock, Breconshire. Fruit is abundant in good years. 360–480 m on Craig-y-Cilau (Brecs.).

Native. Though very localised, this species occurs in considerable quantity in its main area, and the population is probably stable.

Endemic.

References: Atlas Supp. (35a), Q. O. N. Kay (1998), Proctor & Groenhof (1992), Wigginton (1999), Woods (1998).

M. C. F. PROCTOR

Sorbus intermedia

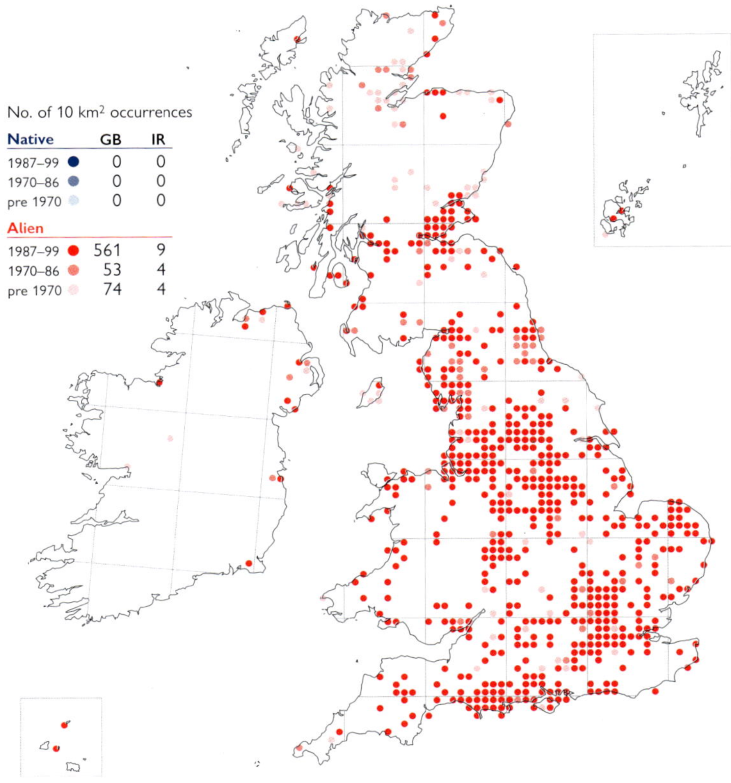

This tree is frequently self-sown in copses and on waste ground. Generally lowland, but reaching 380 m near Alston (Cumberland).

Neophyte. *S. intermedia* was introduced to cultivation in 1789, and is now widely planted as an ornamental tree, especially in town streets and parks. It was known from the wild by 1908, and is much more frequent than when it was mapped by Perring & Sell (1968).

Native of S. Sweden and the Baltic region.

References: Atlas Supp. (34b), Challice & Kovanda (1978), Hultén & Fries (1986), Liljefors (1955), Meusel *et al.* (1965).

M. C. F. PROCTOR

Sorbus anglica

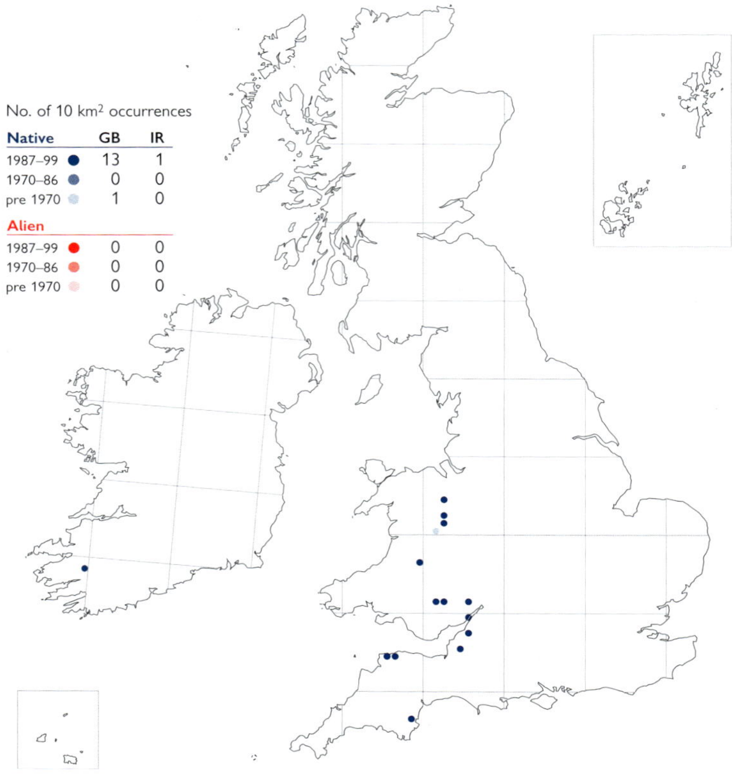

A shrub or small tree, occurring in open rocky woods and scrub, and on more exposed cliffs and stony slopes, usually on limestone, but locally also on other more or less base-rich rocks. An apomict, varying somewhat in different parts of its range. It seems to regenerate freely from seed. Generally lowland, but reaching 395 m at Llangattock Quarries (Brecs.).

Native. The distribution and populations of *S. anglica* appear to be stable.

Endemic.

References: Atlas Supp. (35b), Proctor & Groenhof (1992), Wigginton (1999), Woods (1998).

M. C. F. PROCTOR

Sorbus aria

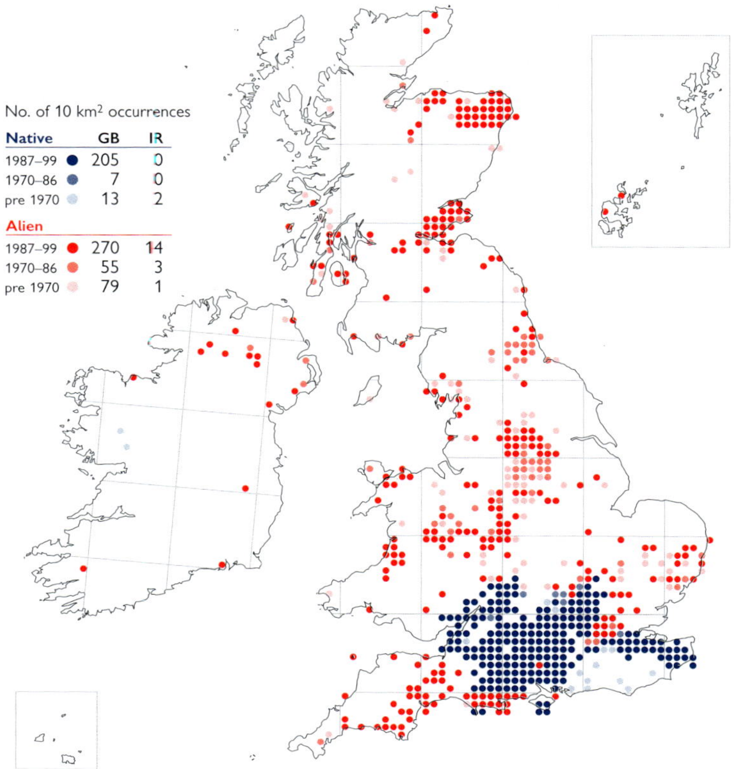

A small to medium-sized tree, native in scrub and open woodland on well-drained soils over chalk, limestone, and occasionally more acid substrates, and an introduction in parks, gardens and streets. It establishes freely from seed. Generally lowland, but reaching 455 m in Teesdale (N.W. Yorks.).

Native (change +0.82). *S. aria* is widespread and variable, and one of the three common sexually-reproducing species of the genus in Britain. It is probably native only in S. England, and perhaps in Co. Galway, and its native distribution is stable. It is widely planted as an ornamental tree, often as one of its cultivars, and widely naturalised.

European Temperate element.

References: Atlas Supp. (36a), Hultén & Fries (1986), Meusel *et al.* (1965).

M. C. F. PROCTOR

Sorbus aria × S. aucuparia (S. × thuringiaca)

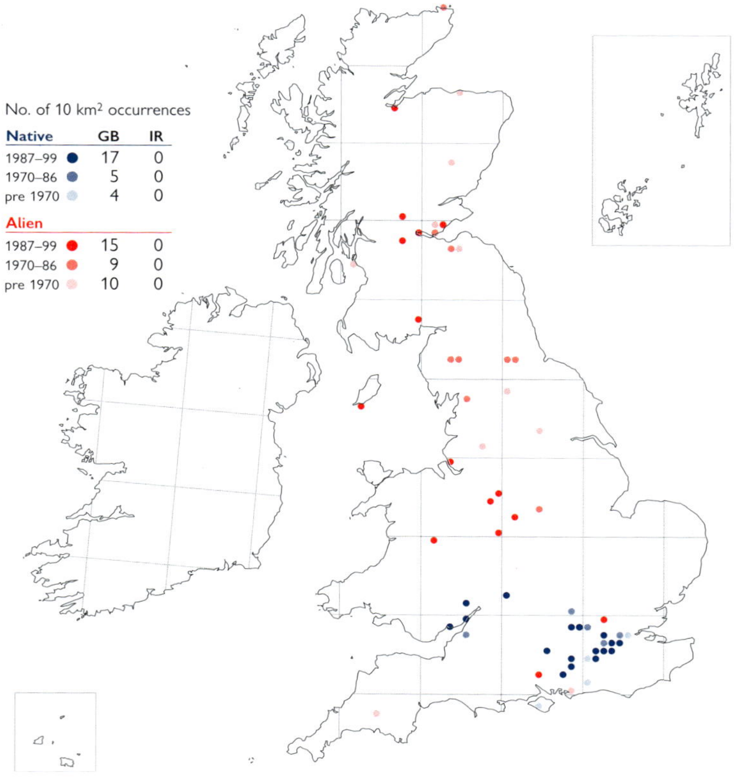

A small tree occurring sporadically where the parent species grow together, and as a planted tree in woodland and parks. It is partially fertile and can set viable seed. Lowland.

Native. This hybrid may be under-recorded, but appears to be genuinely rare as a native tree.

Widespread in Europe and perhaps more frequent as a planted tree than as a spontaneous hybrid.

Reference: Stace (1975).

M. C. F. PROCTOR

Sorbus leptophylla

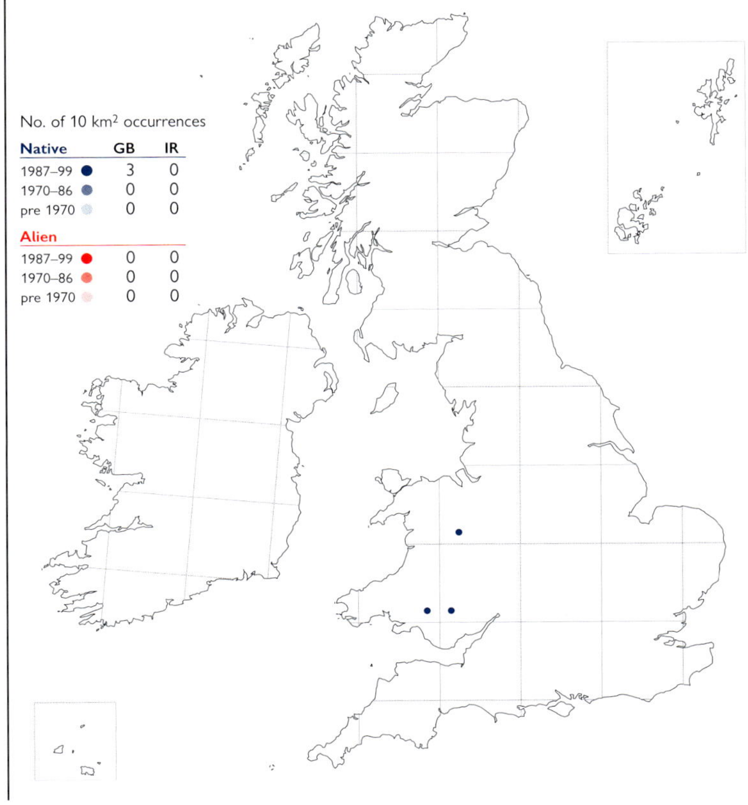

No. of 10 km² occurrences

Native		GB	IR
1987–99	●	3	0
1970–86	●	0	0
pre 1970	●	0	0
Alien			
1987–99	●	0	0
1970–86	●	0	0
pre 1970	●	0	0

An apomictic small tree, typically forming a loose sprawling mass on vertical Carboniferous limestone cliffs, with *Fraxinus excelsior* and other trees, occasionally as more upright individuals on steep rocky slopes. From 290 m (Craig-y-Rhiwarth) to about 380 m (Craig-y-Cilau, Brecs.).

Native. The two small populations in S. Wales are probably stable. A few small trees on Craig Breidden (Monts.) with similar leaf and fruit shape and peroxidase phenotype appear to belong to this species.

Endemic.

References: Atlas Supp. (36b), Proctor & Groenhof (1992), Wigginton (1999), Woods (1998).

M. C. F. PROCTOR

Sorbus wilmottiana

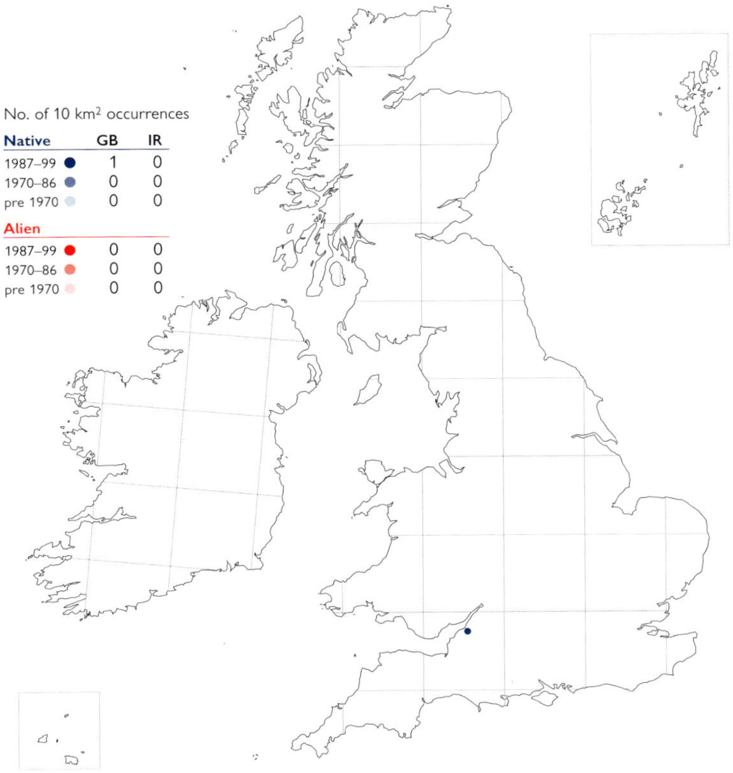

No. of 10 km² occurrences

Native		GB	IR
1987–99	●	1	0
1970–86	●	0	0
pre 1970	●	0	0
Alien			
1987–99	●	0	0
1970–86	●	0	0
pre 1970	●	0	0

This apomictic small tree or scrub occurs only on steep, rocky, Carboniferous limestone slopes in the Avon Gorge (N. Somerset and W. Gloucs.). Lowland.

Native. This species was first described in 1967. It has always been rare, and the population currently comprises only a few trees.

Endemic.

References: Atlas Supp. (36b), Nethercott (1998), Proctor & Groenhof (1992), Wigginton (1999).

M. C. F. PROCTOR

Sorbus eminens

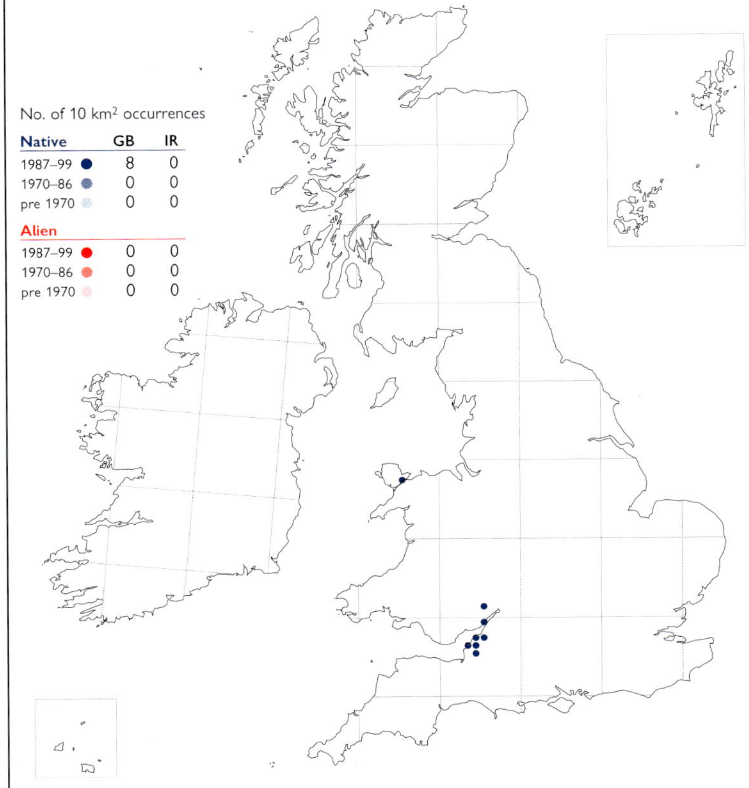

No. of 10 km² occurrences

Native		GB	IR
1987–99	●	8	0
1970–86	●	0	0
pre 1970	●	0	0
Alien			
1987–99	●	0	0
1970–86	●	0	0
pre 1970	●	0	0

A small to medium-sized tree of rocky grassland, scrub and open woodland, usually on Carboniferous limestone. Fruiting tends to be erratic, but plants regenerate from seed where conditions are suitable. Lowland.

Native. This species may include several taxa. The type plant is from the Wye valley, and populations there differ from populations elsewhere and are probably not conspecific with them.

Endemic.

References: Atlas Supp. (36b), Nethercott (1998), Proctor & Groenhof (1992), Wigginton (1999).

M. C. F. PROCTOR

Sorbus hibernica

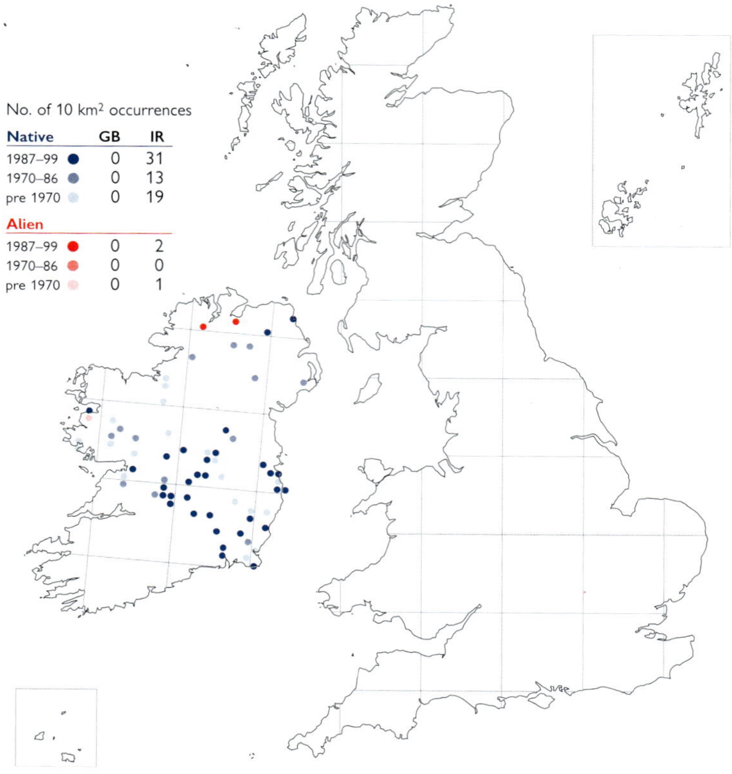

This small to medium-sized tree occurs in rocky grassland, hedges, scrub and open woodland, most often on Carboniferous limestone. It is apomictic and reproduces by seed. Lowland.

Native. This species is apparently somewhat variable. It is morphologically close to *S. eminens*, and has the same peroxidase phenotype. The distribution appears to be stable.

Endemic.

References: Atlas Supp. (37a), Parnell & Needham (1998).

M. C. F. PROCTOR

Sorbus porrigentiformis

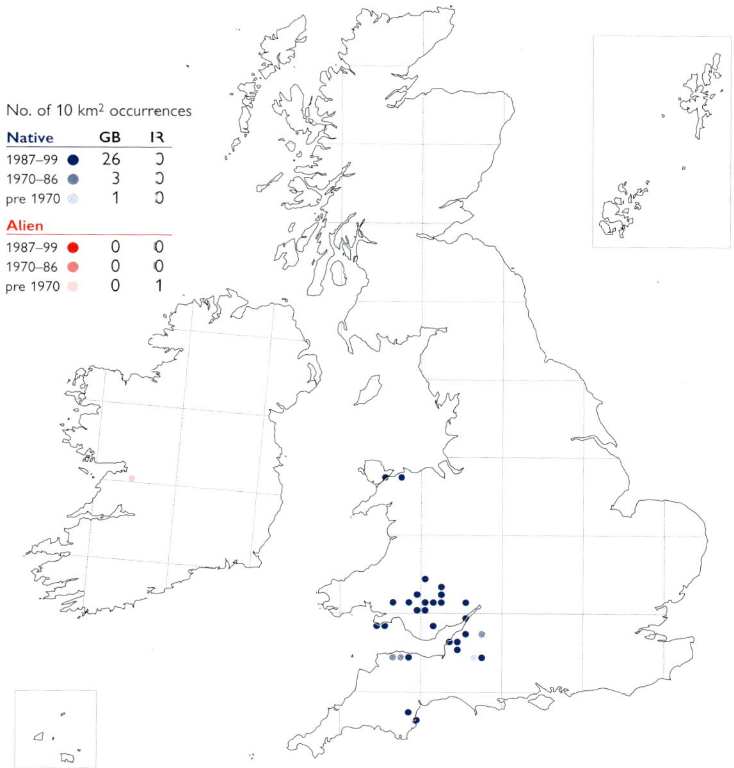

A shrub or small tree of cliffs and rock outcrops, occasionally found in rocky grassland or persisting into woodland, usually on Carboniferous or other hard limestones, very locally on other rocks. It is apomictic, and readily sets fruit, though the amount varies from year to year. Lowland to about 530 m on Craig Cerrig-gleisiad (Brecs.).

Native. This species is well circumscribed, and is one of our more widespread apomictic *Sorbus* microspecies. The distribution appears to be stable.

Endemic.

References: Atlas Supp. (37a), Proctor & Groenhof (1992), Stewart *et al.* (1994), Woods (1998).

M. C. F. PROCTOR

Sorbus lancastriensis

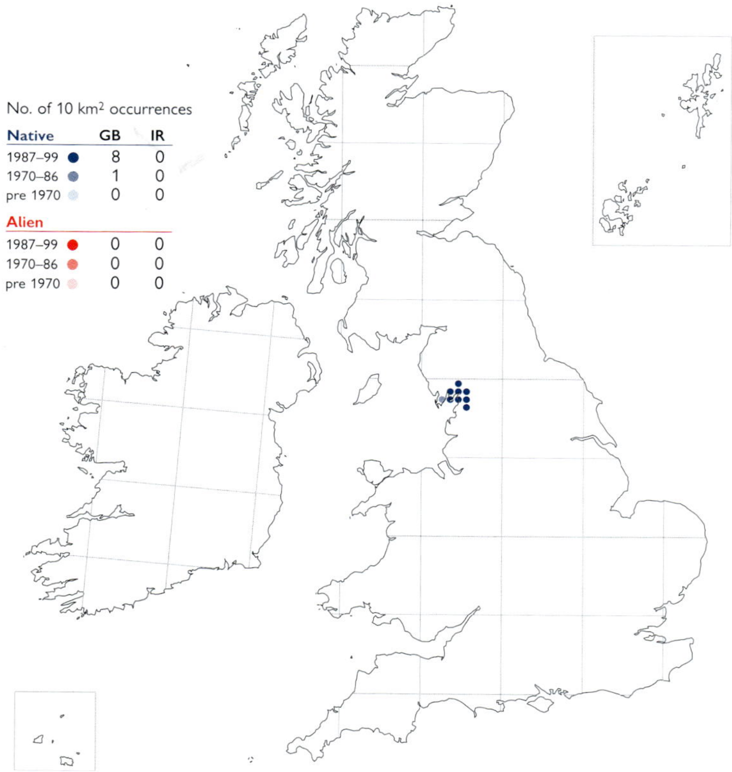

This shrub or small tree occurs on cliffs, rock outcrops and rocky scrub or open woodland, usually on Carboniferous limestone. It is apomictic, sets fruit freely, and regeneration occurs where grazing allows. Lowland.

Native. *S. lancastriensis* is related to *S. porrigentiformis* and *S. rupicola*, and occupies a similar range of habitats. The distribution and population size appear to be stable.

Endemic.

References: Atlas Supp. (37b), Rich & Baeker (1998), Wigginton (1999).

M. C. F. PROCTOR

Sorbus rupicola

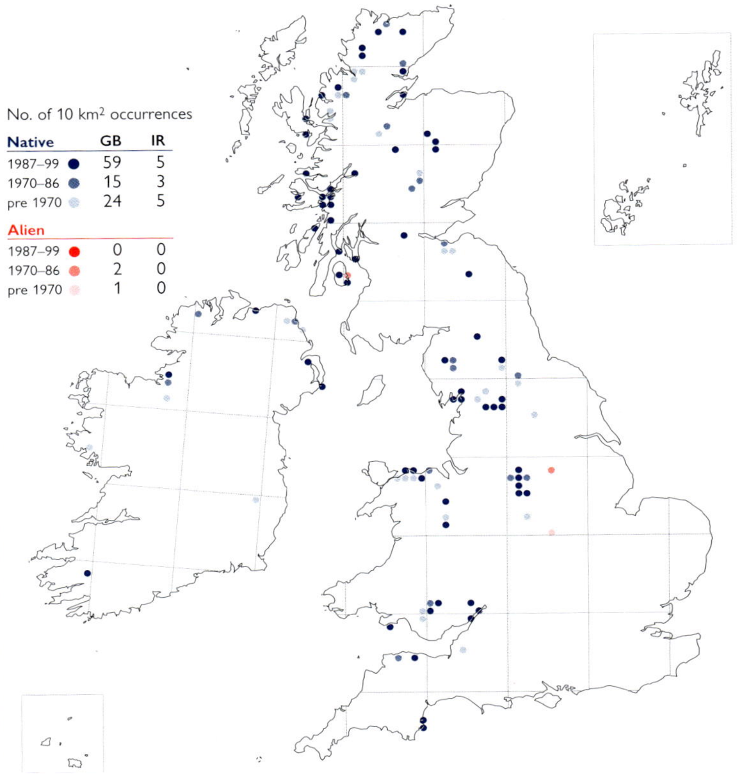

A shrub or small tree of cliffs and rock outcrops, generally found on hard limestones or other basic rocks, most often in exposed sites but also in less exposed rocky woodland. From near sea level in S. Devon and the Gower to 500 m at Creag Bhuilg in Glen Avon (Banffs.).

Native. This is the most widespread apomictic *Sorbus* species in Britain, but many populations are small, sometimes consisting of only a single tree. The distribution appears to be stable.

Suboceanic Boreal-montane element; it has a widely disjunct distribution in Britain, Estonia, Norway and Sweden.

References: Atlas Supp. (37b), Hultén & Fries (1986), Stewart *et al.* (1994).

M. C. F. PROCTOR

Sorbus vexans

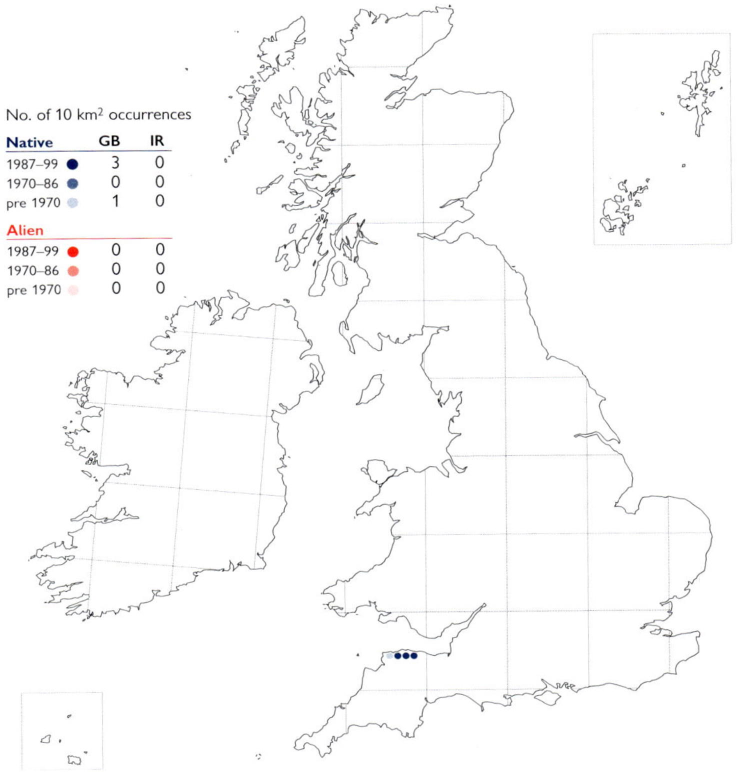

This small tree is found in rocky woodland on Old Red Sandstone, occasionally in open cliff scrub close to the shore but more usually 0.2–0.6 km inland. It is unusual among the *aria*-group of species in occurring entirely on acid soils. It is an apomictic species in which fruiting is erratic from year to year, but after favourable seasons saplings are frequent. Lowland.

Native. The total population probably numbers a hundred or more trees, and has undergone no substantial change since the species was described in 1957.

Endemic.

References: Atlas Supp. (37a), Wigginton (1999).

M. C. F. PROCTOR

Sorbus subcuneata

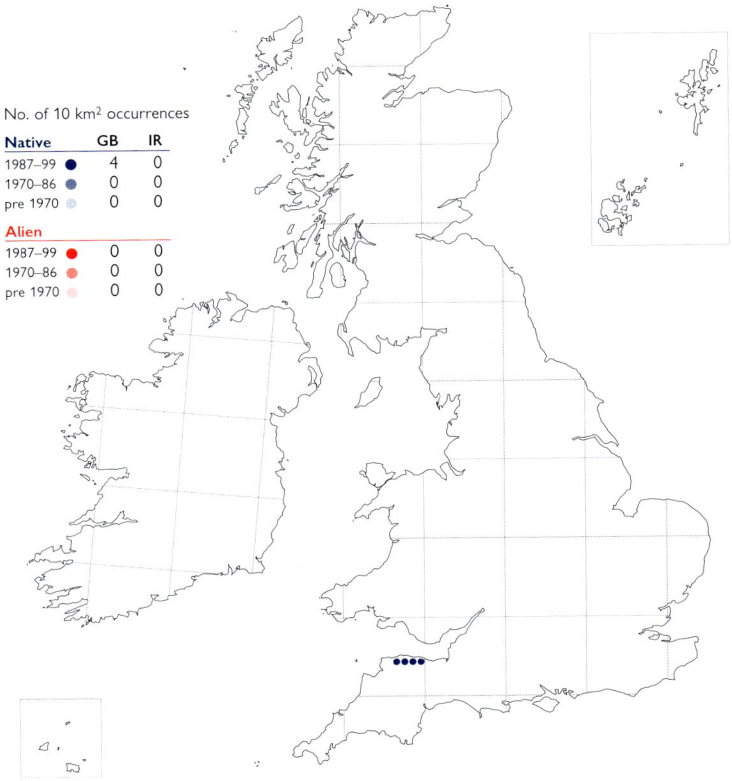

This apomictic, small to medium-sized tree, is generally found in open, rocky *Quercus petraea* woodland on Old Red Sandstone. In favourable seasons, abundant fruit is produced. Young trees are frequent. Lowland.

Native. This species is mainly concentrated in the East Lyn valley near Watersmeet (N. Devon), with small outlying populations to the east and west. The population appears to be generally stable.

Endemic.

References: Atlas Supp. (38a), Sell (1989), Wigginton (1999).

M. C. F. PROCTOR

Sorbus devoniensis

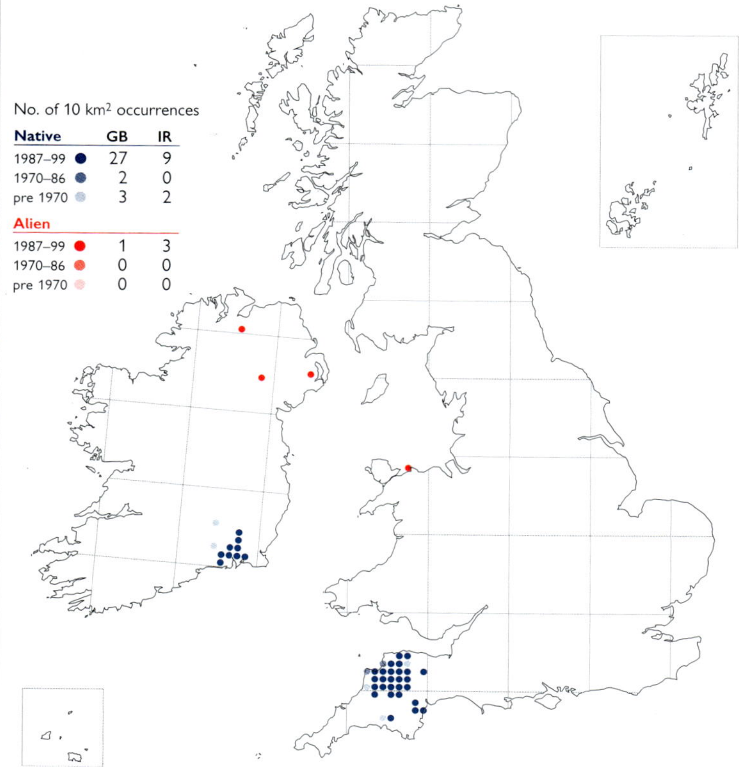

A large shrub to medium-sized tree found in hedgerows, rocky (often coastal) woodland, and occasionally in more open moorland situations, mainly on non-calcareous shale, slates and grit. It is apomictic. Fruit is abundant in favourable years, and the species readily establishes from seed, with young trees frequent in suitable habitats. Lowland.

Native. *S. devoniensis* is better recorded now than when mapped by Perring & Sell (1968), and its overall distribution since then appears to be stable. It is mainly a hedgerow plant in its core area in N. Devon, but is found in undoubted natural habitats on the coast. Its range in both Devon and Ireland may have been extended by past planting for fruit.

Endemic.

References: Atlas Supp. (38b), Stewart *et al.* (1994).

M. C. F. PROCTOR

Sorbus croceocarpa

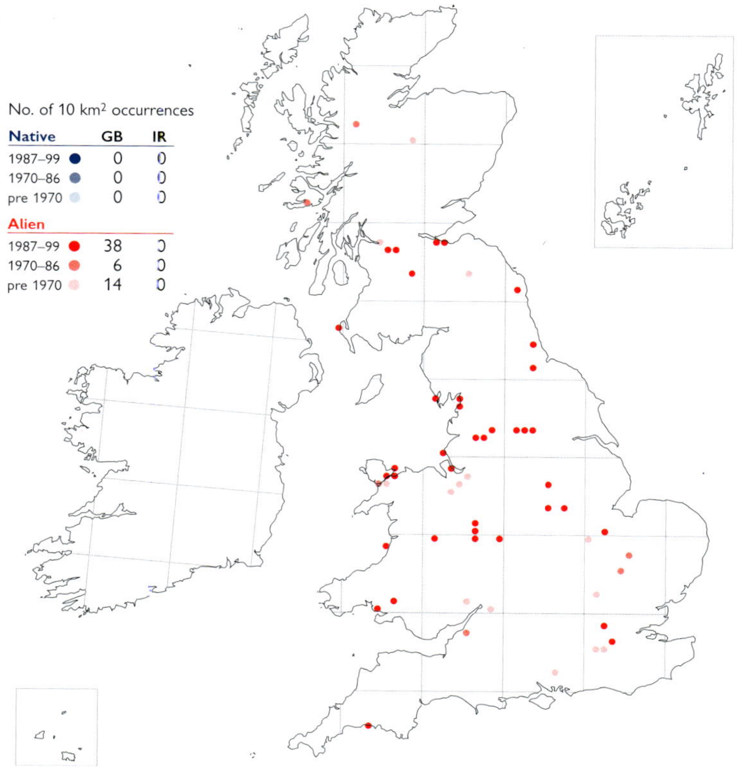

This medium-sized tree is found in scrub and open woodland, probably on mainly clayey, base-rich soils. It is freely naturalised on broken limestone and boulder-clay slopes along the Menai Strait (Caerns.). The uniform morphology and peroxidase phenotype suggests that *S. croceocarpa* is apomictic. It is occasionally self-sown. Lowland.

Neophyte. This species has possibly been cultivated in Britain since 1874, and is now widely planted as an ornamental. It had been collected from the wild by 1909 (Easterness), but it was not described as a distinct species until 1989. Trends in its distribution are uncertain.

Known only as a cultivated or naturalised plant; native range unknown.

References: Bean (1980), Sell (1989).

M. C. F. PROCTOR

Sorbus bristoliensis

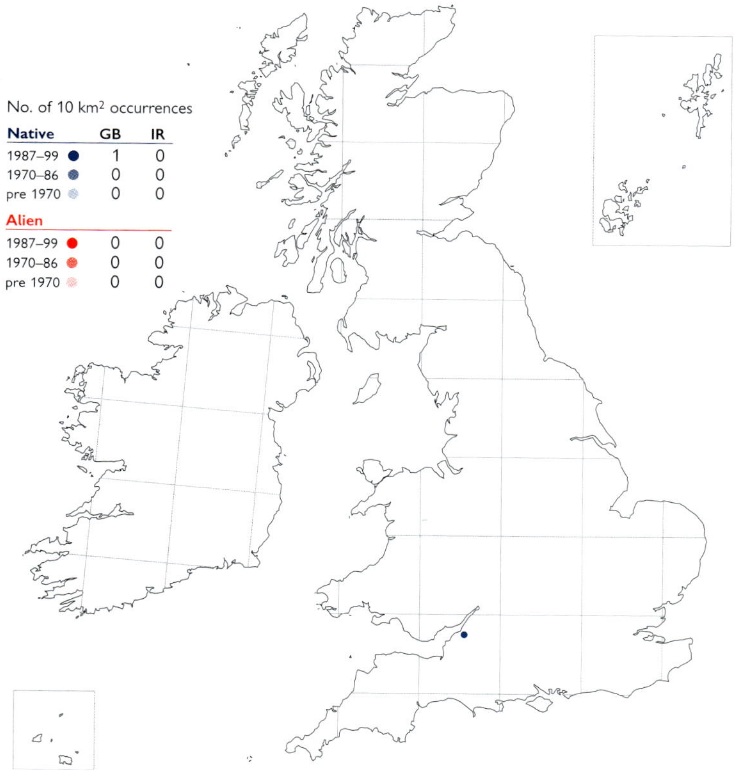

A medium-sized tree of rocky scrub and woodland on Carboniferous limestone. It is apomictic, and fruiting can be erratic. Lowland.

Native. This species is very localised, occurring on both sides of the Avon Gorge where the total population of about a hundred trees may be increasing in response to conservation management.

Endemic.

References: Atlas Supp. (38a), Nethercott (1998), Sell (1989), Wigginton (1999).

M. C. F. PROCTOR

Sorbus latifolia

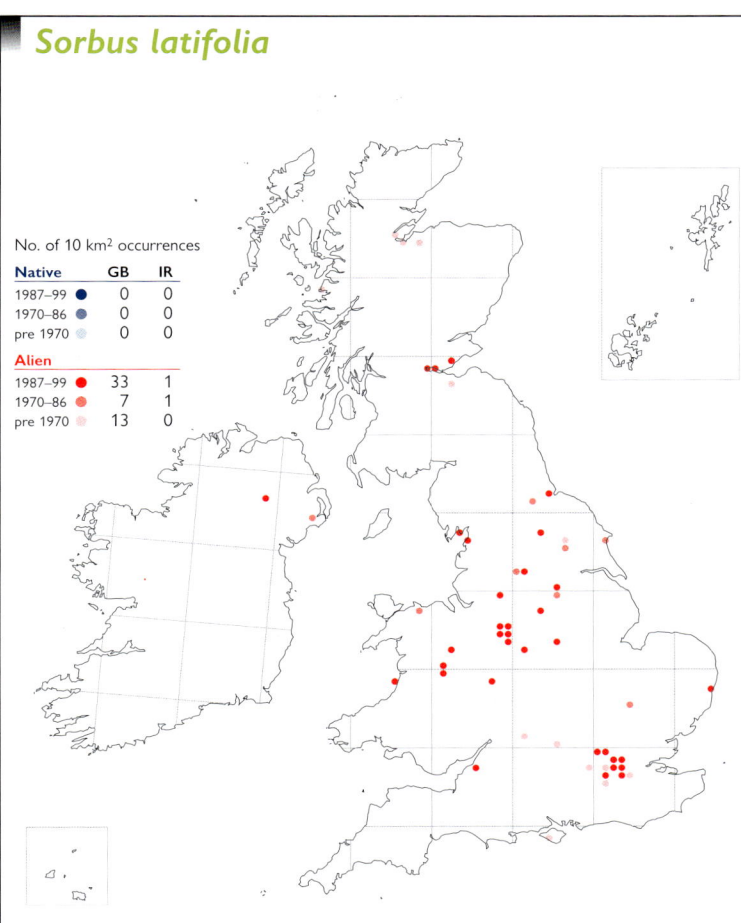

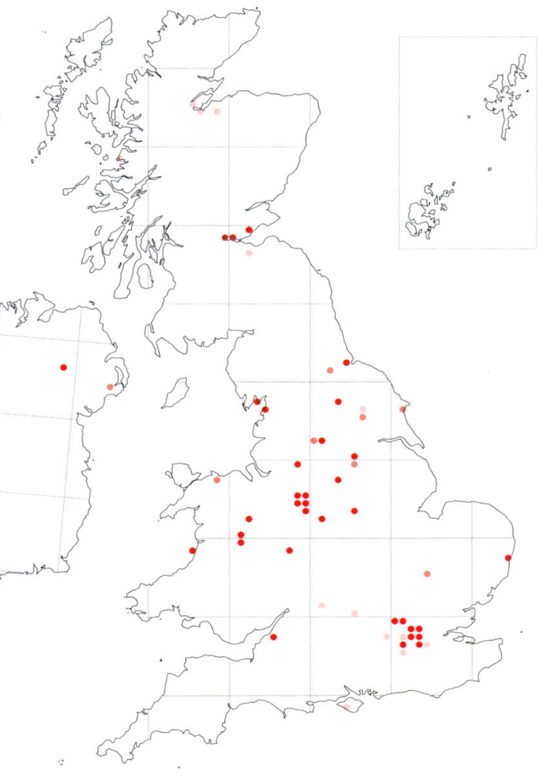

No. of 10 km² occurrences

		GB	IR
Native			
1987–99	●	0	0
1970–86	●	0	0
pre 1970	●	0	0
Alien			
1987–99	●	33	1
1970–86	●	7	1
pre 1970	●	13	0

A medium-sized tree which is sometimes planted in parks and large gardens and occasionally naturalised in various habitats including woodland, scrub, rocky slopes and river banks. Generally lowland, but reaching 450 m at Carreg Cennen (Carms.).

Neophyte. *S. latifolia* was recorded in Britain by 1866 (Midlothian). It is the least frequently cultivated species of the *S. latifolia* aggregate, and can be confused with *S. croceocarpa*.

Native of the Iberian Peninsula, France and S.W. Germany.

References: Bean (1980), Sell (1989).

M. C. F. PROCTOR

Sorbus torminalis Wild Service-tree

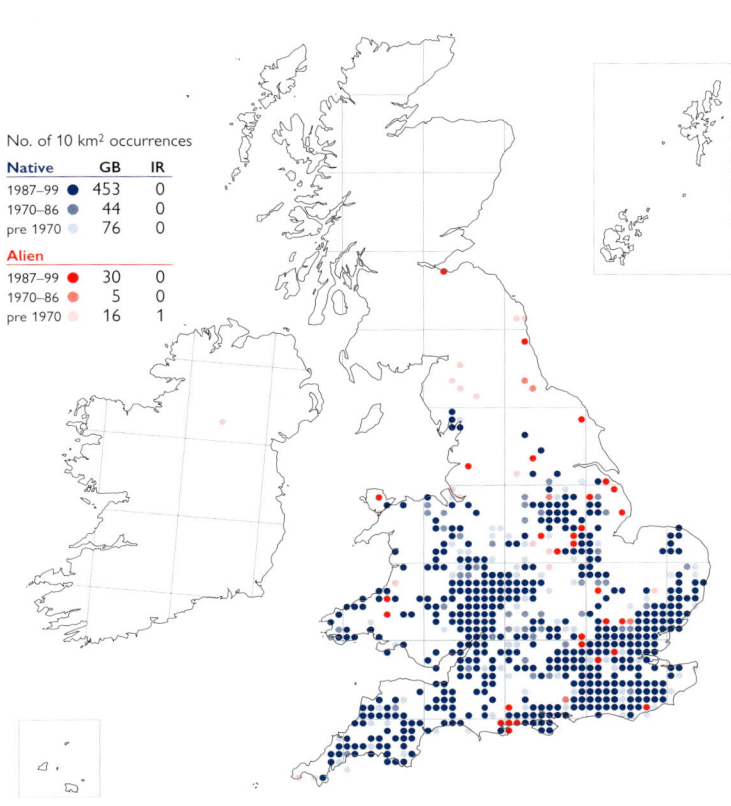

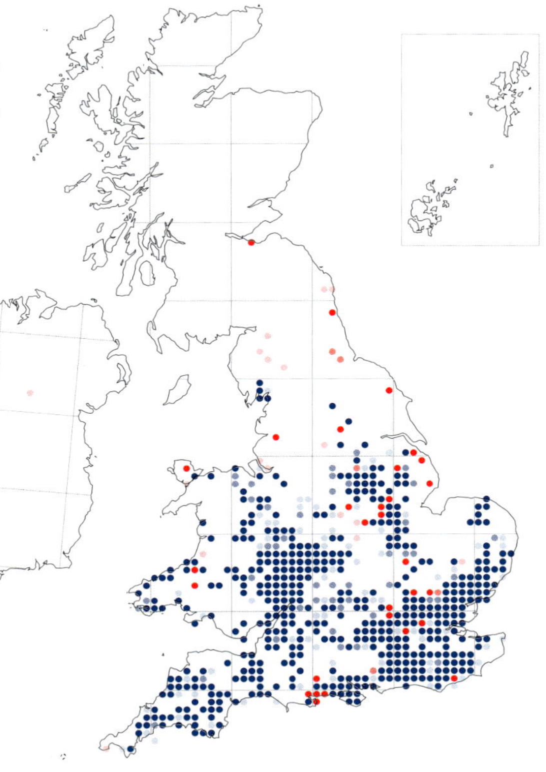

No. of 10 km² occurrences

		GB	IR
Native			
1987–99	●	453	0
1970–86	●	44	0
pre 1970	●	76	0
Alien			
1987–99	●	30	0
1970–86	●	5	0
pre 1970	●	16	1

A medium-sized tree on a wide variety of soils, often clayey, or over limestone, found in old woods and hedgerows and occasionally in scrub. Planted trees occur in parks and plantations. This is one of our three widespread sexually-reproducing species of *Sorbus*, and it may also form small stands through root-layering. Lowland.

Native (change +0.22). *S. torminalis* is often regarded as an indicator of ancient woodland. It is easily overlooked and is more frequent than was thought at the time of the 1962 *Atlas*.

European Temperate element.

References: Atlas (132a), Hultén & Fries (1986), Meusel *et al.* (1965), Roper (1993).

M. C. F. PROCTOR

Amelanchier lamarckii Juneberry

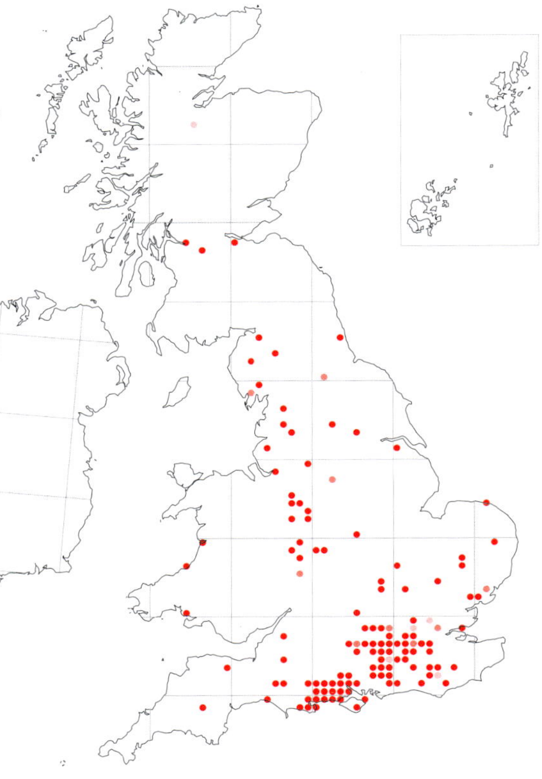

No. of 10 km² occurrences

		GB	IR
Native			
1987–99	●	0	0
1970–86	●	0	0
pre 1970	●	0	0
Alien			
1987–99	●	118	1
1970–86	●	9	0
pre 1970	●	5	0

A shrub or small tree mostly growing on acidic, usually sandy soils and naturalised in open woodland (often *Quercus* or *Betula*), wood borders and scrub, and on dry heaths and roadsides. Lowland.

Neophyte. Introduced into cultivation in 1746, *A. lamarckii* is widely planted for ornament, and is considered to be the only *Amelanchier* naturalised in Britain (Schroeder, 1970). It was first recorded from the wild in 1887 (Middlesex); on the New Forest (S. Hants.) and Surrey heaths it was initially planted and is now spreading from bird-sown fruits (Brewis *et al.*, 1996).

Native of N. America; naturalised in N.W. Europe.

References: Bean (1970), Coats (1963).

D. J. McCOSH

Cotoneaster frigidus Tree Cotoneaster

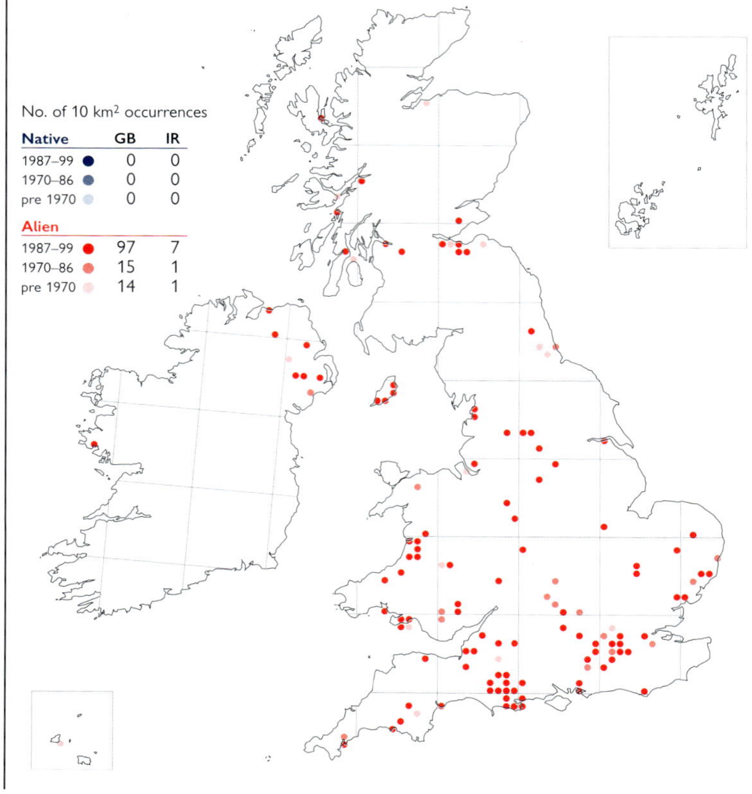

A deciduous or semi-evergreen large shrub or small tree, found as a garden escape, relic of cultivation or planted in woodland, hedgerows, rough grassland, quarries, marl- and sand-pits, along paths and on railway banks. It sets abundant seed, which is often bird-sown, and is frequently naturalised. Lowland.

Neophyte. *C. frigidus* has been cultivated in Britain since 1824, but was not recorded from the wild until 1918 (Glam.). The true species is rarely cultivated now, being superseded by *C.* × *watereri* (*C. frigidus* × *C. salicifolius*). Some of the mapped records may be misidentifications of this hybrid.

Native of the Himalayas.

Reference: Bean (1973).

T. D. DINES

Cotoneaster frigidus × *C. salicifolius* (*C.* × *watereri*) Waterer's Cotoneaster

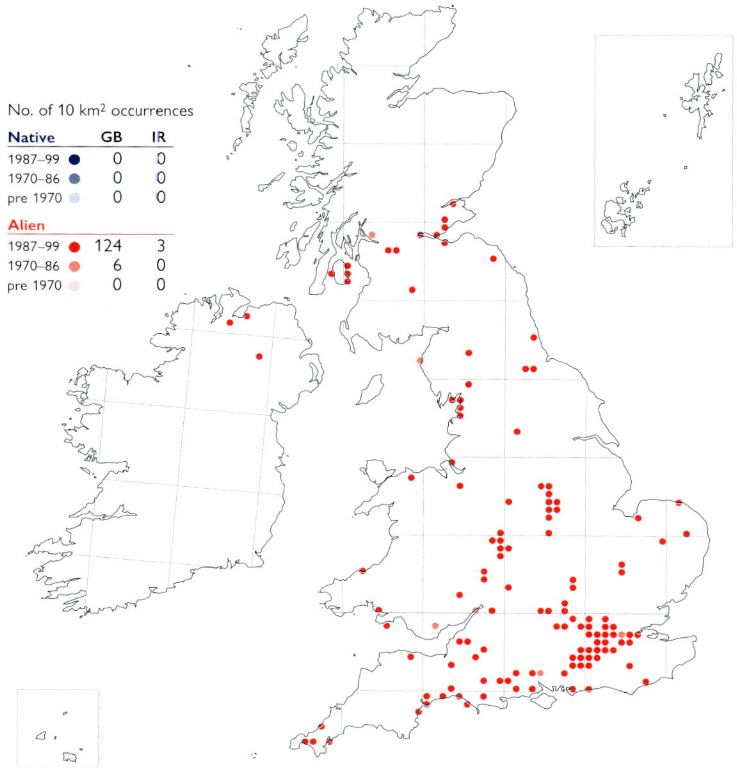

A large, semi-evergreen arching shrub found as a garden escape, relic of cultivation and planted shrub in woodland, scrub, hedgerows, rough grassland, amenity areas, churchyards, and on roadsides, railway banks, walls and waste ground. The hybrid is fertile and populations are often naturalised. Lowland.

Neophyte. This is an extremely variable hybrid, first raised in 1928; many cultivars are available and are commonly grown in gardens. Cotoneasters are a difficult group taxonomically and this hybrid is one of the most problematic; it is certainly under-recorded and confused with other taxa, particularly *C. salicifolius*. It does not seem to have been recorded in the wild until 1968 (Oxon), and is probably increasing.

A hybrid of garden origin.

References: Bean (1973), Coats (1963).

T. D. DINES

Cotoneaster salicifolius Willow-leaved Cotoneaster

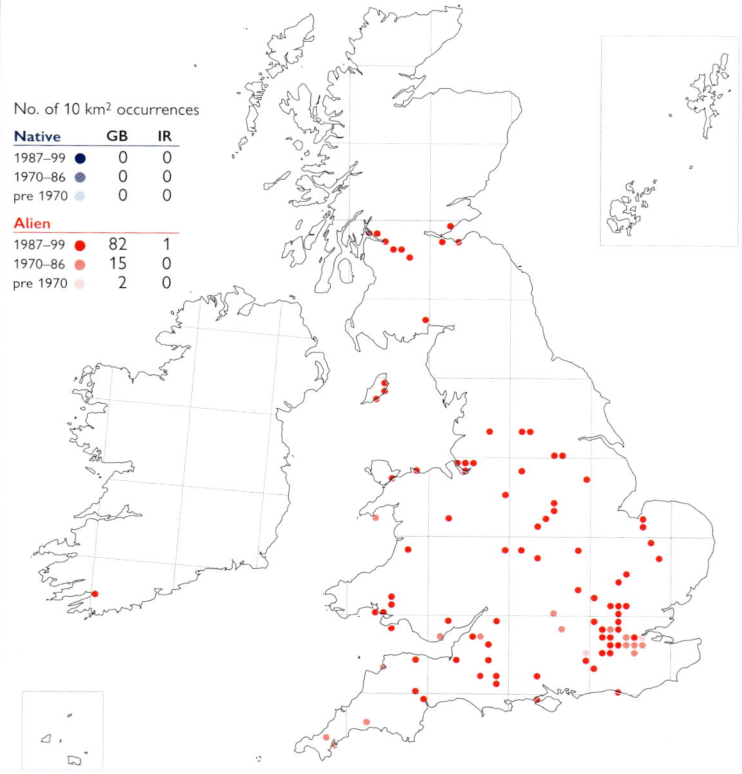

A large, evergreen arching shrub found in woodland, scrub, hedgerows, rough grassland and amenity areas, and on roadsides, railway banks, walls and waste ground. It is sometimes planted but more frequently occurs as a bird-sown garden escape; abundant seed is produced and populations are often naturalised. Lowland.

Neophyte. *C. salicifolius* has been cultivated in Britain since 1908 and is very popular in gardens. It is probably under-recorded but may also be mistaken for other taxa, especially *C.* × *watereri*, its hybrid with *C. frigidus*. It was recorded in the wild in 1966 (Surrey), and is likely to be increasing.

Native of W. China.

Reference: Bean (1973).

T. D. DINES

Cotoneaster lacteus Late Cotoneaster

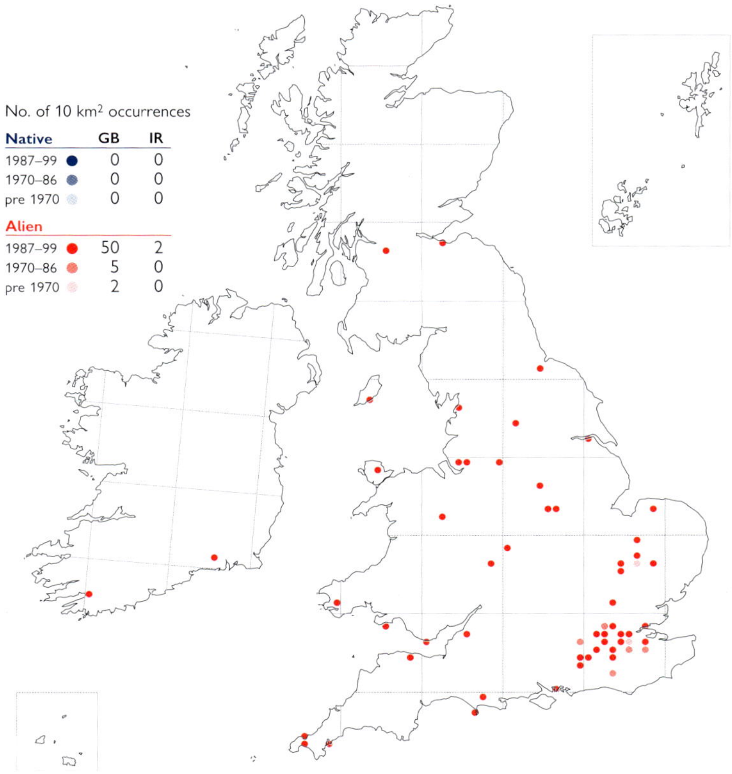

No. of 10 km² occurrences

		GB	IR
Native			
1987–99	●	0	0
1970–86	●	0	0
pre 1970	●	0	0
Alien			
1987–99	●	50	2
1970–86	●	5	0
pre 1970	●	2	0

This large, spreading, evergreen shrub is found as a bird-sown garden escape or is frequently planted in woodland, hedgerows and rough grassland, and on roadsides, railway banks, walls and waste ground. Lowland.

Neophyte. This attractive species has been cultivated in Britain since 1913. It is popular in gardens, and was first recorded in the wild in 1976 (Surrey). It is probably under-recorded, but is unlikely to be increasing at the same rate as *C.* × *watereri.*

Native of S.W. China.

Reference: Bean (1973).

T. D. DINES

Cotoneaster microphyllus agg. Small-leaved Cotoneasters

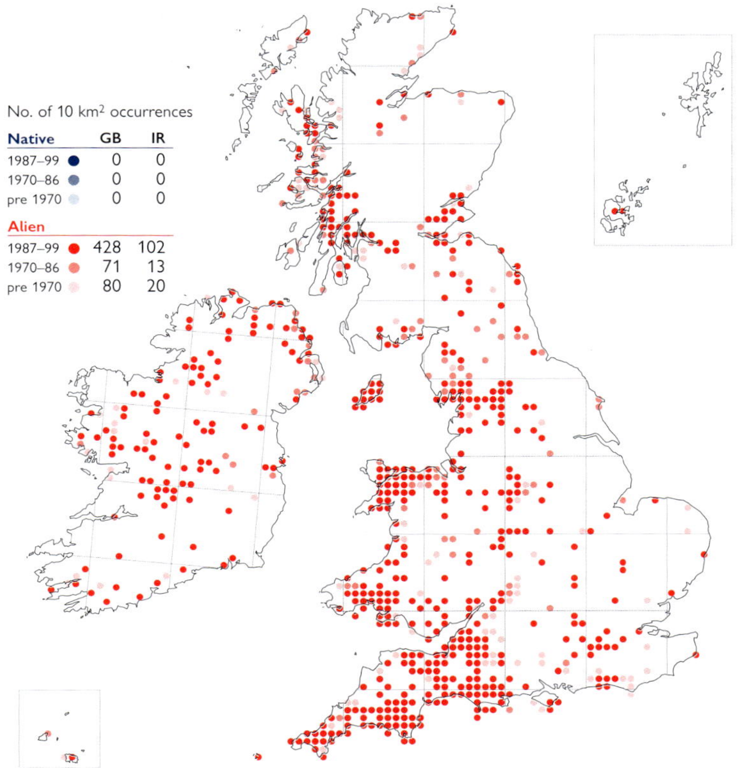

No. of 10 km² occurrences

		GB	IR
Native			
1987–99	●	0	0
1970–86	●	0	0
pre 1970	●	0	0
Alien			
1987–99	●	428	102
1970–86	●	71	13
pre 1970	●	80	20

These evergreen shrubs form carpeting mounds in rocky grassland and quarries, and on roadsides, railway banks, cliffs, scree, walls and pavements. They are often well-naturalised, especially on well-drained, calcareous substrates. Lowland.

Neophyte (change +1.54). The twelve species in this aggregate (species 21–32, Stace, 1997), are difficult to identify. They have been cultivated since at least 1824, and one (*C. integrifolius*) was recorded in the wild in 1892, at a site in N. Somerset where it still occurs. The vast majority of records are of *C. integrifolius*, and the map of this species in the 1962 *Atlas* is equivalent to this aggregate. It has increased since then, and is better recorded.

The species in this aggregate are natives of N. India, the Himalayas and China.

References: Atlas (130d), Bean (1973).

T. D. DINES

Cotoneaster horizontalis Wall Cotoneaster

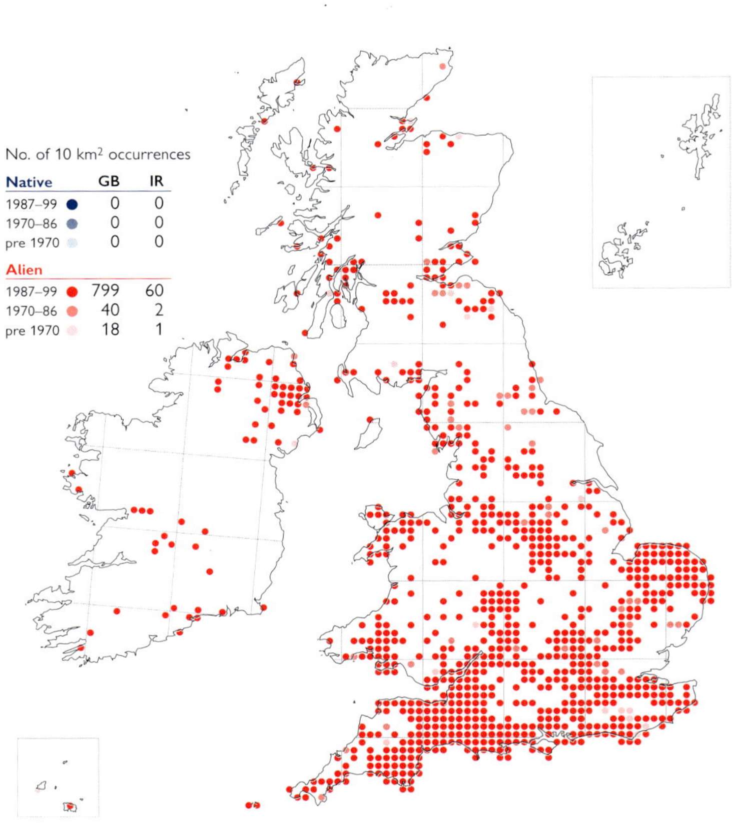

No. of 10 km² occurrences

		GB	IR
Native			
1987–99	●	0	0
1970–86	●	0	0
pre 1970	●	0	0
Alien			
1987–99	●	799	60
1970–86	●	40	2
pre 1970	●	18	1

A deciduous arching or prostrate shrub, found in rocky grassland and on railway banks, cliffs, limestone pavement, quarries, chalk pits, pavements and walls. It occurs as a garden escape or throw-out, and is also frequently bird-sown. It reproduces readily from seed and is usually very well-naturalised, sometimes even posing a threat to native vegetation. Lowland.

Neophyte. *C. horizontalis* was introduced into cultivation around 1879 and is extremely popular in gardens. It was first recorded from the wild in 1940. It appears to be increasing rapidly.

Native of W. China.

Reference: Bean (1973).

D. J. McCOSH

Cotoneaster divaricatus Spreading Cotoneaster

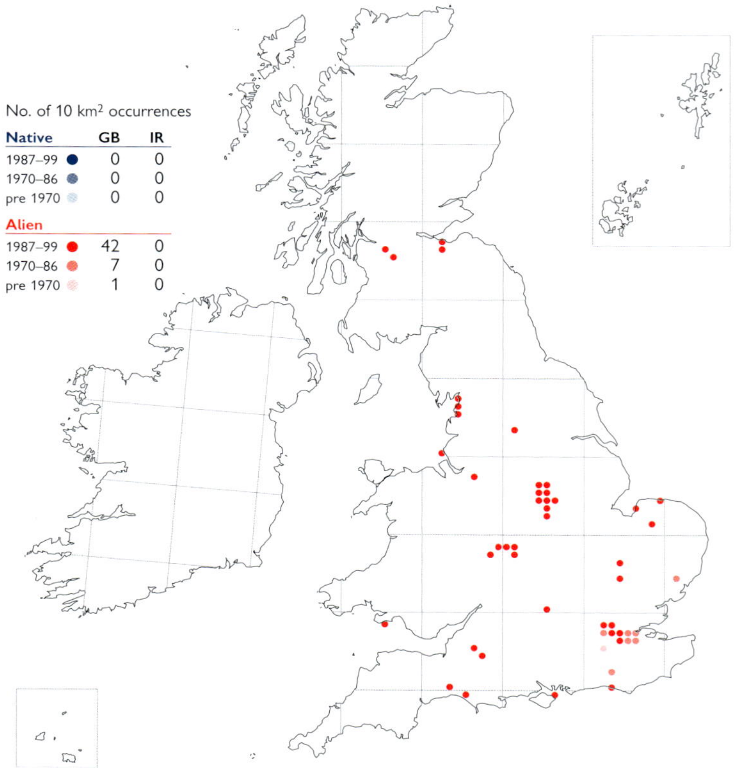

A medium-sized, spreading deciduous shrub found in scrub and rough grassland, and on roadsides, railway banks, walls and pavements. It is sometimes planted but is more frequent as a bird-sown garden escape, populations often becoming well-naturalised. Lowland.

Neophyte. This species was introduced to cultivation in 1904 and was first recorded in the wild in 1983 (Kent). It is probably under-recorded and may be spreading.

Native of W. China.

Reference: Bean (1973).

T. D. DINES

Cotoneaster cambricus Wild Cotoneaster

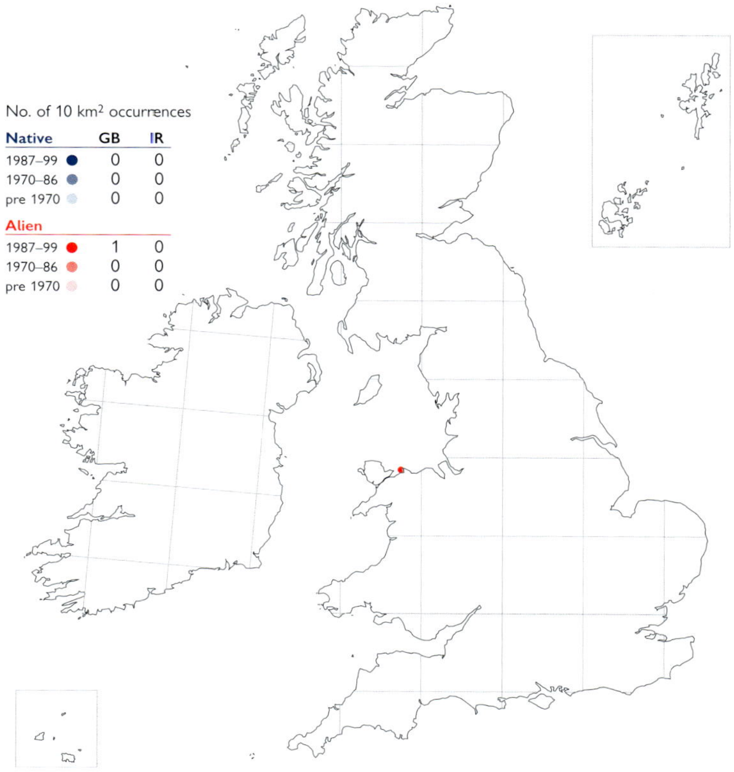

A spreading apomictic deciduous shrub confined to limestone rocks and ledges on the Great Orme (Caerns.). The few plants which remain from a larger population do not appear to regenerate either by seed or vegetatively. Lowland.

Neophyte. Although described by Fryer & Hylmö (1994) as a native endemic species, Kay & John (1995) consider it to be more likely of garden origin and have cast doubt on its distinction from *C. integerrimus*, a species which was in cultivation by 1656. It has declined since its discovery in about 1825 due to poor fruiting, grazing and scrub encroachment.

A member of the *C. integerrimus* complex, which has a European Boreal-montane distribution but occurs as a garden escape outside its native range.

References: Atlas (130b), Hultén & Fries (1986), Meusel *et al.* (1965), Wigginton (1999).

M. J. WIGGINTON

Cotoneaster simonsii Himalayan Cotoneaster

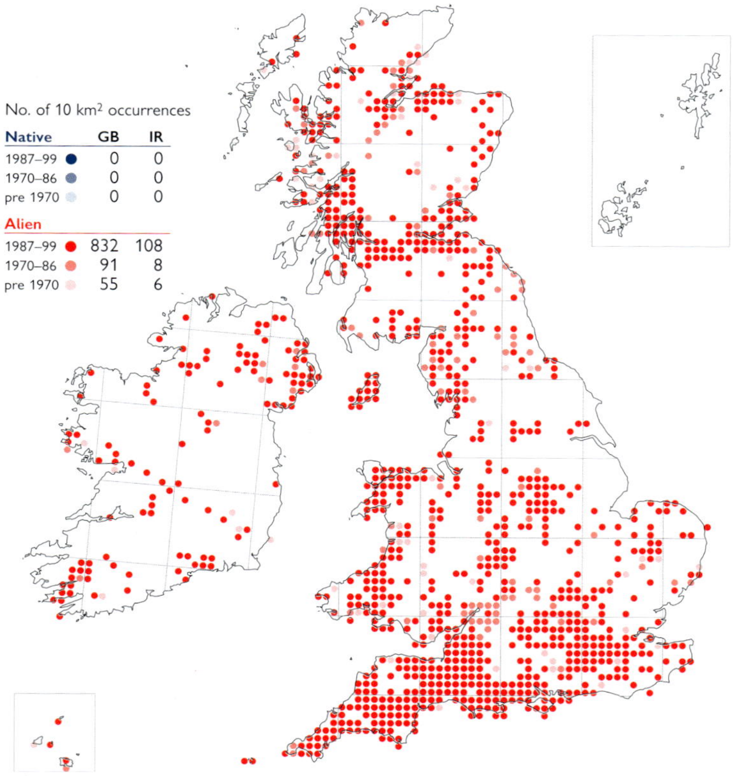

An erect, deciduous shrub, found in woodland, hedgerows, rocky grassland and scrub and on heathland, rock outcrops, walls, pavements, quarries and waste ground. It is frequently naturalised, and often originates from bird-sown seed. Generally lowland, but reaching 335 m near Watermillock (Cumberland).

Neophyte (change +3.55). This species, introduced from the Himalayas in 1865, is very popular in gardens and in amenity planting schemes. It was first recorded in the wild in 1910. Its distribution has increased considerably since it was mapped in the 1962 *Atlas*, due to both better recording and a genuine spread.

Native of the Himalayas.

References: Atlas (130c), Bean (1973).

D. J. McCOSH

Cotoneaster bullatus Hollyberry Cotoneaster

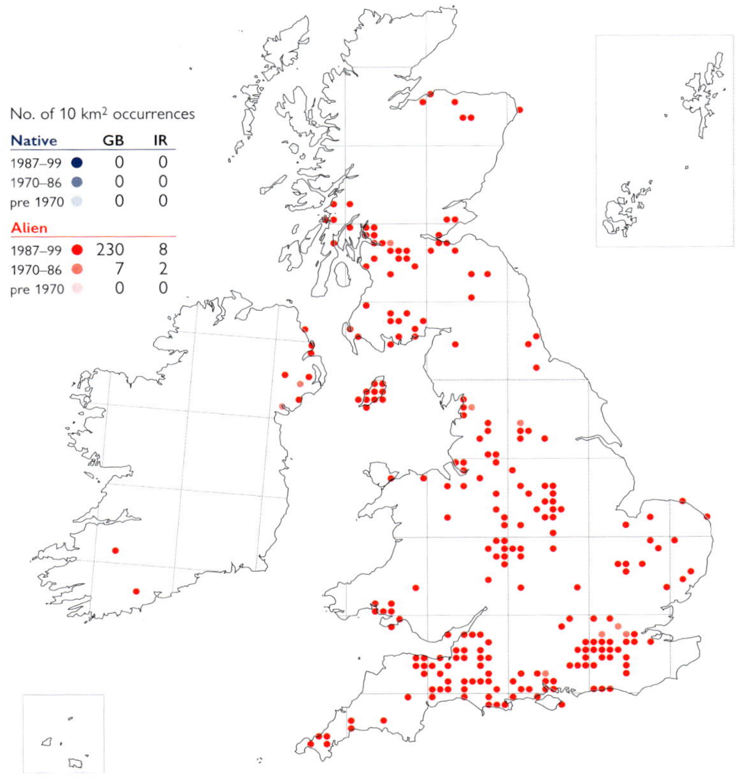

A large deciduous shrub, found naturalised in woodland, scrub, hedgerows and quarries, and on railway banks, roadsides, sand dunes, cliffs, walls and waste ground. Reproduction is by seed, which is frequently bird-sown. Lowland.

Neophyte. *C. bullatus* was introduced into cultivation in 1898, and is now widely planted in gardens and in roadside amenity schemes. It was known from the wild in Britain by at least 1957 (S. Lancs.), and is probably increasing, but it may have been over-recorded for *C. rehderi.*

Native of W. China.

Reference: Bean (1973).

D. J. McCOSH

Cotoneaster rehderi Bullate Cotoneaster

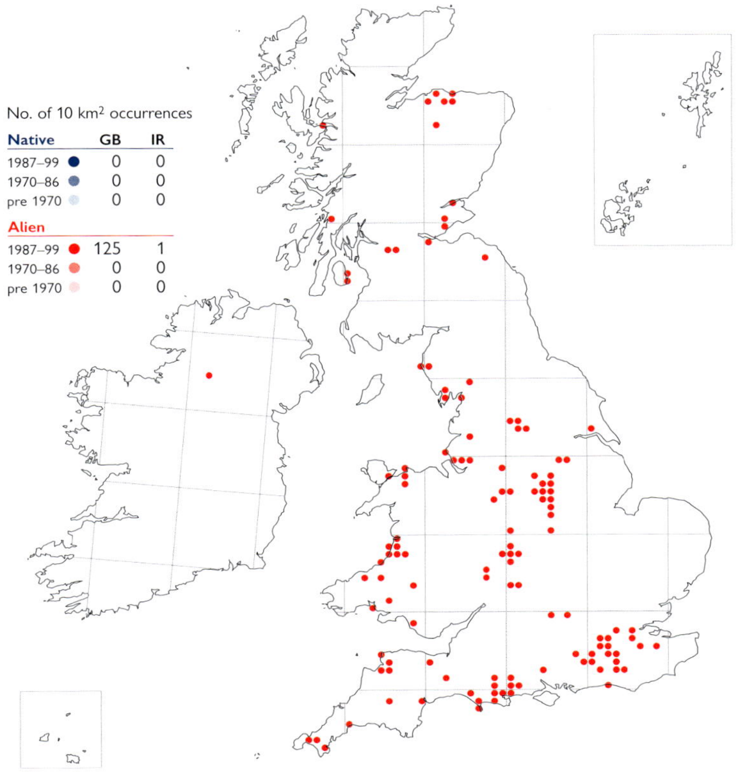

A medium to large deciduous shrub found in woodland, hedgerows, scrub, rough grassland and quarries. It is frequently naturalised, readily reproducing by seed which is often bird-sown. Lowland.

Neophyte. This species, which has been in cultivation in Britain since 1908, is popular in gardens but does not appear to have been recorded from the wild until 1986 (N. Hants.). It is probably under-recorded, and is often confused with or overlooked as *C. bullatus.*

Native of W. China.

Reference: Bean (1973).

T. D. DINES

Cotoneaster franchetii Franchet's Cotoneaster

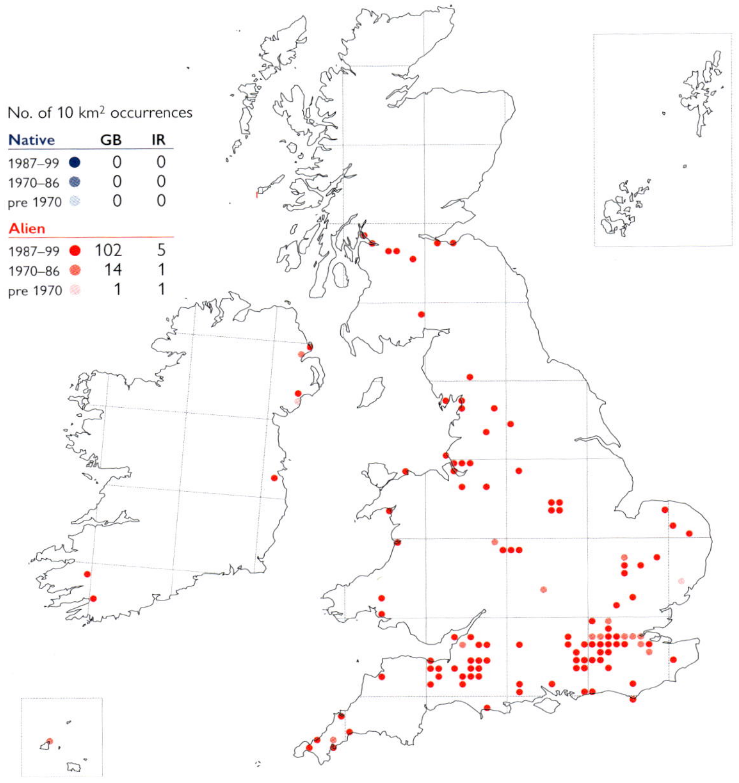

A medium-sized, arching evergreen shrub found as a garden escape in woodland, scrub, rocky grassland, quarries and sand- and chalk-pits, and on railway banks, cliffs, walls, pavements and waste ground. It is frequently bird-sown, and populations are often well-naturalised. Lowland.

Neophyte. *C. franchetii* has been cultivated in Britain since about 1895, and is very popular in gardens. It was first recorded in the wild in Britain in 1977 (Westmorland) and in Ireland in 1986 (Co. Kerry). It is probably under-recorded, and may be mistaken for *C. sternianus* and *C. dielsianus*, but is likely to be increasing.

Native of S.W. China.

Reference: Bean (1973).

T. D. DINES

Cotoneaster sternianus Stern's Cotoneaster

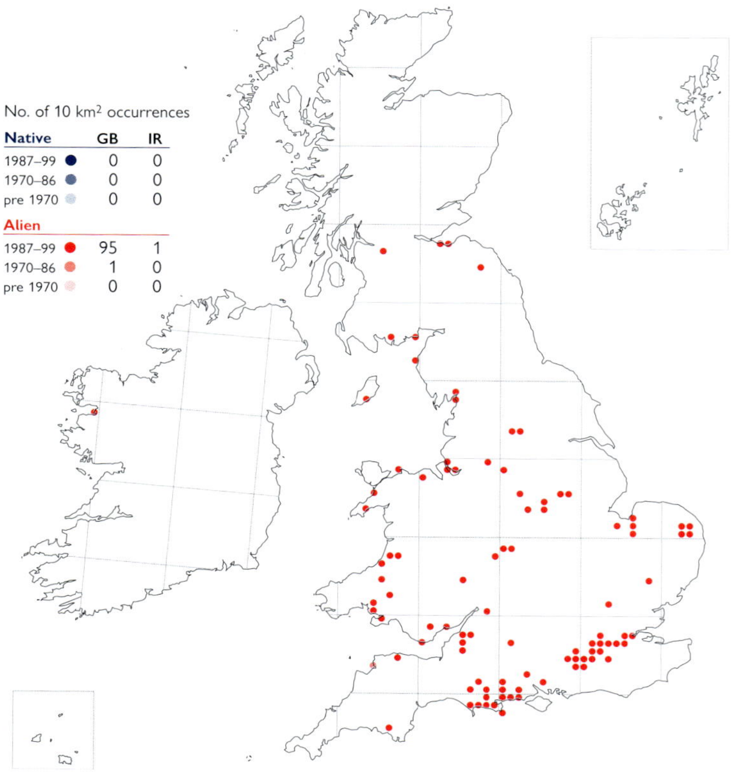

No. of 10 km² occurrences

Native	GB	IR
1987–99 ●	0	0
1970–86 ●	0	0
pre 1970 ●	0	0
Alien		
1987–99 ●	95	1
1970–86 ●	1	0
pre 1970 ●	0	0

This medium-sized evergreen shrub occurs as a garden escape in woodland, scrub and rough grassland, and on roadsides, railways, walls and pavements. It is frequently bird-sown, and can become well-naturalised. Lowland.

Neophyte. *C. sternianus* was introduced to cultivation in Britain in 1919, and is now popular in gardens. It is often mistaken for other species, particularly *C. franchetii.* It was recorded from the wild in 1981 (Middlesex and E. Kent) and may be increasing.

Native of S.W. China.

Reference: Bean (1973).

T. D. DINES

Cotoneaster dielsianus Diel's Cotoneaster

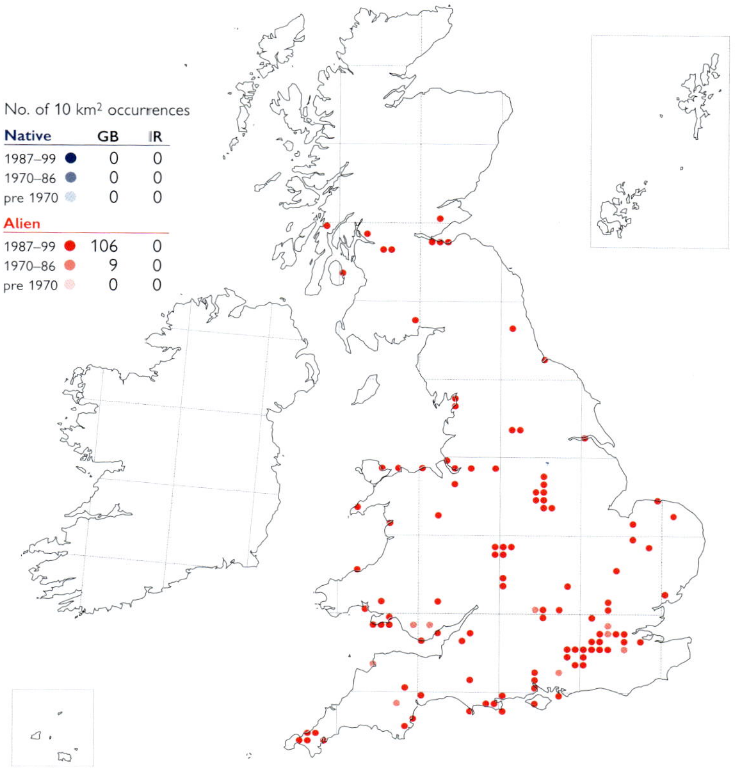

No. of 10 km² occurrences

Native	GB	IR
1987–99 ●	0	0
1970–86 ●	0	0
pre 1970 ●	0	0
Alien		
1987–99 ●	106	0
1970–86 ●	9	0
pre 1970 ●	0	0

An erect, arching deciduous shrub found as a garden escape in woodland, scrub, rough grassland and quarries, and on roadsides, railways, cliffs, walls and pavements. Lowland.

Neophyte. This species, popular in gardens, is much confused with *C. franchetii.* It has been cultivated in Britain since 1900, and was recorded in the wild in 1965. It may be increasing, but is probably under-recorded.

Native of China.

Reference: Bean (1973).

T. D. DINES

Pyracantha coccinea Firethorn

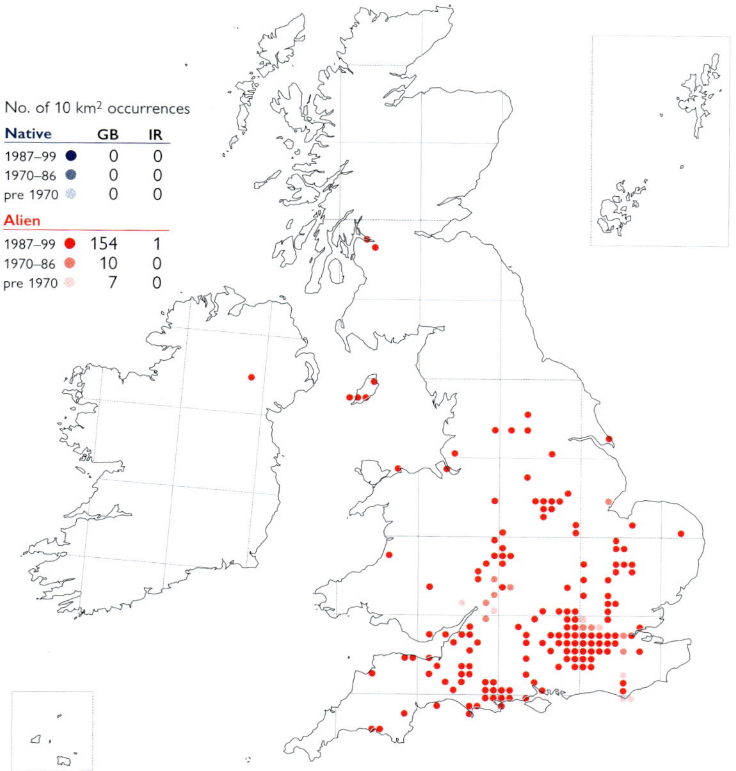

No. of 10 km² occurrences

Native	GB	IR
1987–99 ●	0	0
1970–86 ●	0	0
pre 1970 ●	0	0
Alien		
1987–99 ●	154	1
1970–86 ●	10	0
pre 1970 ●	7	0

An evergreen shrub found deliberately planted or as a garden escape, throw-out or relic of cultivation in hedgerows and amenity areas and on roadsides, banks, railways and waste ground, and in walls, pavement cracks and quarries. It reproduces readily from seed, and is often naturalised. Lowland.

Neophyte. This popular garden plant was cultivated in Britain by 1629. Dunn (1905) commented that it is 'occasionally noticed in a semi-wild state', and it is likely to be increasing.

Native of S. Europe and S.W. Asia.

References: Bean (1976), Bolòs & Vigo (1984).

T. D. DINES

Mespilus germanica Medlar

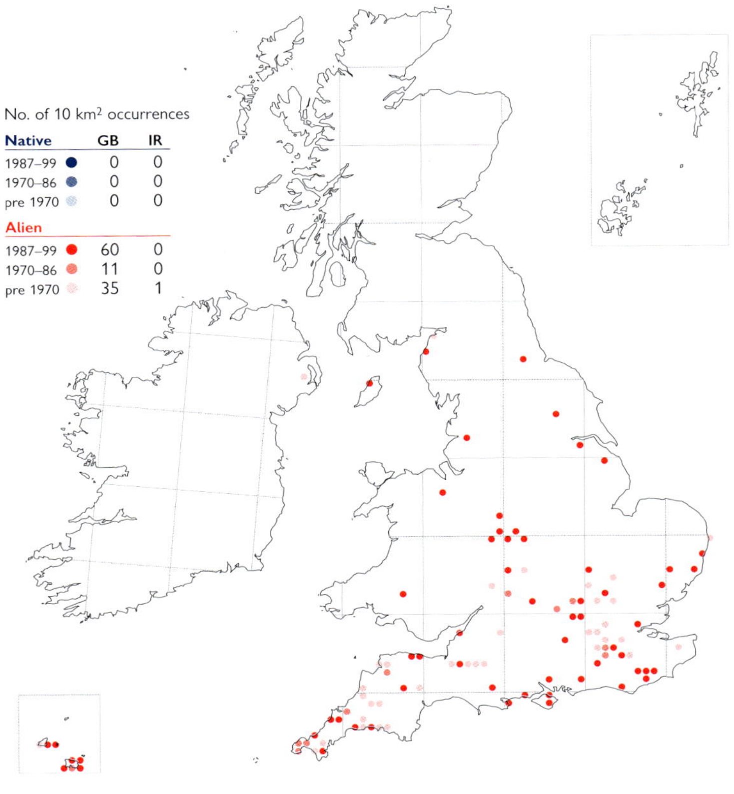

A long-lived shrub or small tree occasionally found in hedges or woods, and as a relic of cultivation. In Britain, regeneration from seed appears to be very rare, but vegetative spread by suckering has been recorded. Lowland.

Archaeophyte. *M. germanica* has been grown in Britain since 995 (Harvey, 1981) and was later much planted in gardens and orchards for its fruits. It has apparently been less often planted in recent years, but trends in its distribution are difficult to discern.

Native of S.W. Asia and perhaps S.E. Europe; widely naturalised in C. Europe.

References: Bolòs & Vigo (1984), Pyne (1997), Roach (1985).

D. J. McCOSH

Crataegus persimilis Broad-leaved Cockspurthorn

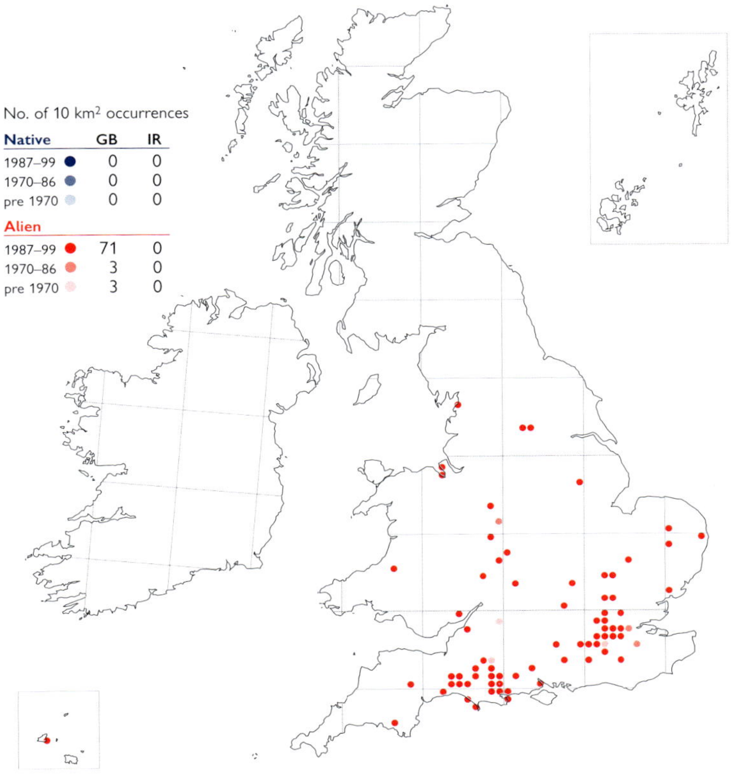

A large deciduous shrub or small tree found in woodland, plantations, hedgerows, roadsides, parks and amenity plantings. It is frequently planted, but can also be bird-sown or a relic of cultivation. Lowland.

Neophyte. *C. persimilis* has been cultivated in Britain since 1791; it is popular in gardens and is being increasingly planted on roadsides and in amenity plantings. It does not appear to have been recorded in the wild until 1934. It is probably under-recorded, and often overlooked as *C. crus-galli.*

Origin unknown; possibly a hybrid between two N. American species.

References: Bean (1970), Wurzell (1992b).

T. D. DINES

Crataegus monogyna Hawthorn

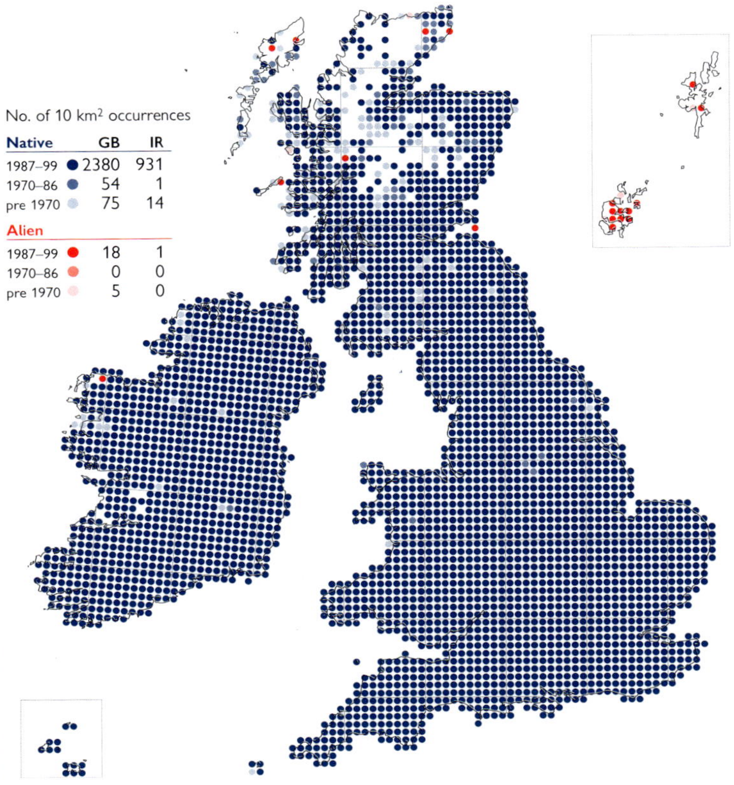

A deciduous shrub or tree of hedgerows, scrub and wood-borders, and as an understorey in open woodland on a wide range of soils. It can persist as scattered bushes in grazed sites, spreading rapidly when grazing declines or ceases. Its prolifically produced fruits are an important winter food for birds. 0–610 m (Melmerby High Scar, Cumberland).

Native (change –0.76). This species has been widely planted as a hedging plant for many centuries, and the limits of its native distribution are unclear. In N. Scotland it is often confined to the vicinity of habitation, and in some areas is certainly introduced. Its distribution is stable.

European Temperate element; widely naturalised outside its native range.

References: Atlas (131b), Grime *et al.* (1988), Hultén & Fries (1986), Meusel *et al.* (1965).

D. J. McCOSH

Crataegus laevigata Midland Hawthorn

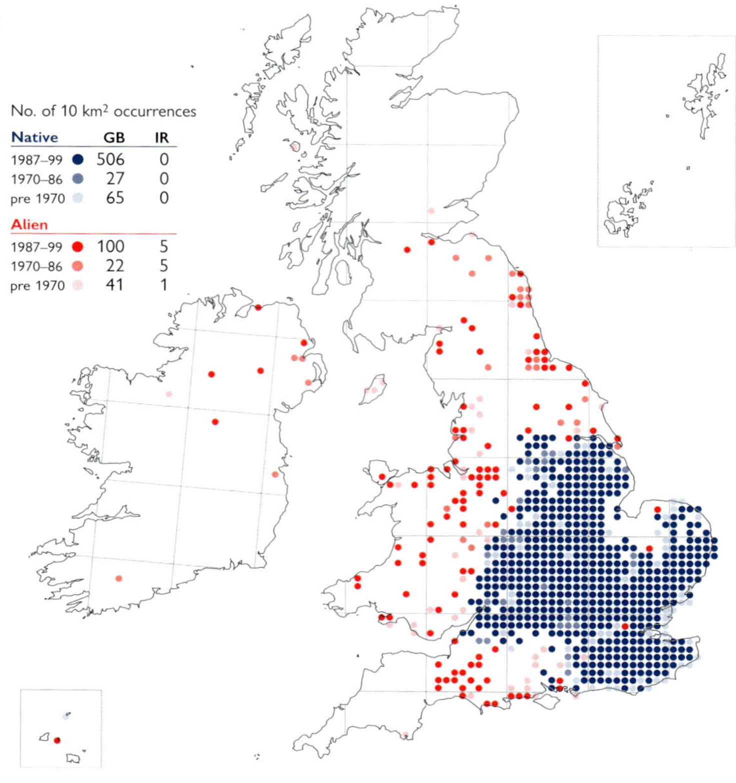

A shrub or small tree typically occurring in ancient woods, wood-borders, old hedgerows and boundary banks on clay soils. It appears to be more tolerant of shade than *C. monogyna*. Lowland.

Native (change +0.32). The frequency of *C. laevigata* has changed little in its core range since the 1962 *Atlas*, but there are now many scattered new records beyond that area. It is widely planted both inside and outside its core area, and most of the new records must relate to planted populations. *C. laevigata* is not always easy to distinguish from *C.* × *media*, and some records may be errors for the hybrid.

European Temperate element.

References: Atlas (131a), Hultén & Fries (1986), Meusel *et al.* (1965).

D. J. McCOSH

Crataegus laevigata × *C. monogyna* (*C.* × *media*)

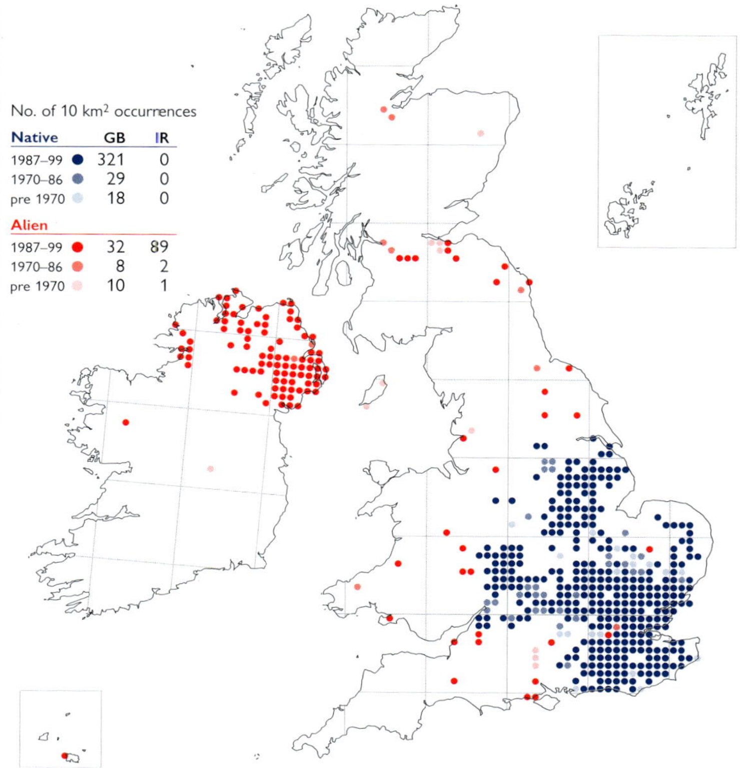

A shrub or small tree of deciduous woods, scrub, and hedgerows on a wide range of soil types. Crosses between the species are as fertile as crosses between two plants of the same species. The hybrid is fully fertile and very variable, partly because of its ability to back-cross to the parents. Lowland.

Native. This hybrid can occur in the absence of *C. laevigata*, but presumably as a native only in sites where the latter species formerly occurred. The distribution closely matches that of *C. laevigata*, except in N.E. Ireland where the hybrid was widely planted in the 19th and 20th centuries from material derived from English stock.

Frequent in those areas of Europe where both parents occur.

References: Byatt (1975), Hackney (1992), Stace (1975).

D. J. McCOSH

Robinia pseudoacacia False-acacia

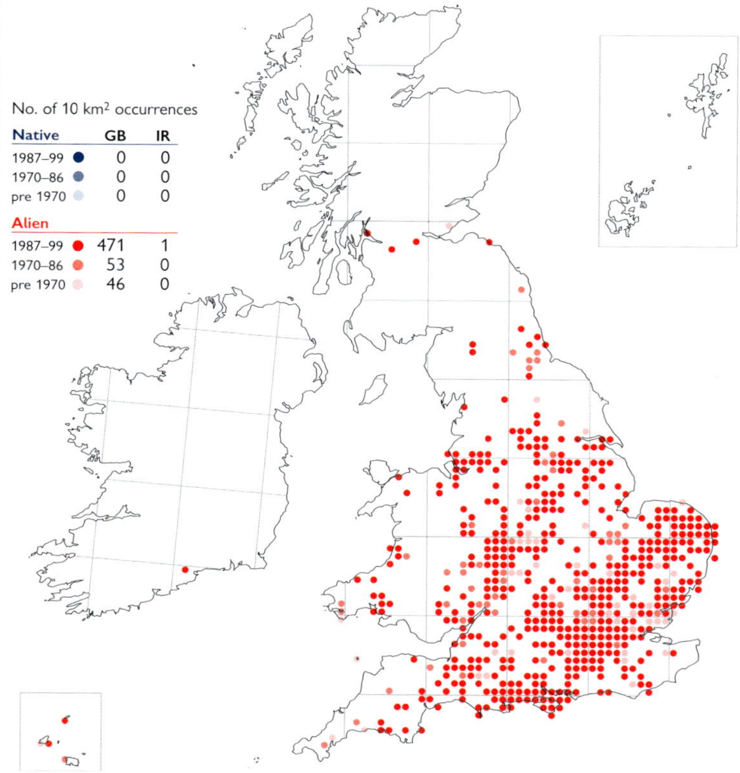

A deciduous tree, extensively planted, and spreading by suckering and, less often, by seed. Its seedlings arise on roadsides, in pavement cracks and on waste ground in urban areas. It also occurs in woods and on scrubby banks, but only does well on light soils. Lowland.

Neophyte. *R. pseudoacacia* was introduced in the 1630s and was known from the wild by at least 1888.

Native of eastern N. America; widely naturalised from planted trees in C. & S. Europe, where its suckering can become a serious problem.

References: Bean (1980), Bolòs & Vigo (1984), Mitchell (1996).

D. A. PEARMAN

Galega officinalis Goat's-rue

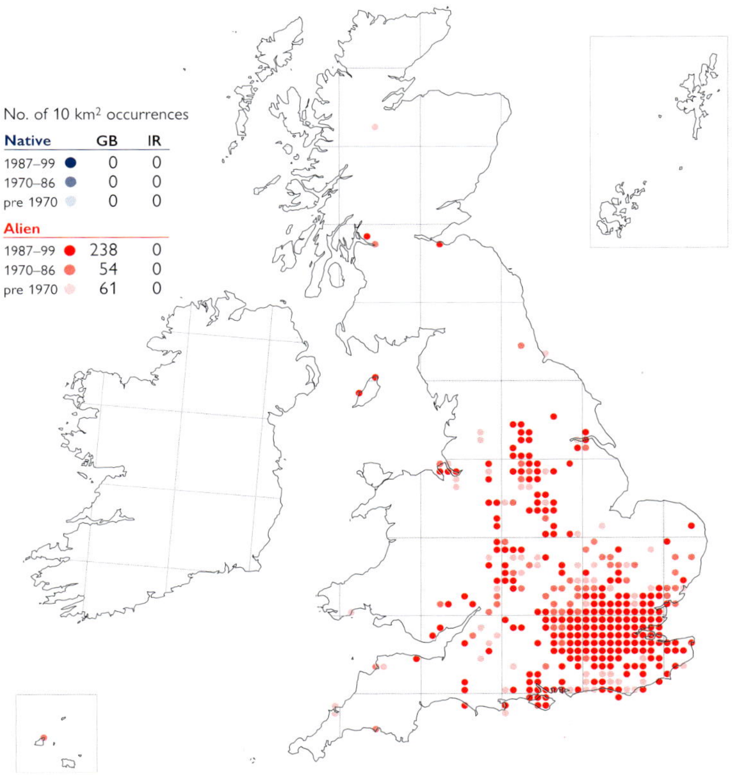

No. of 10 km² occurrences

Native		GB	IR
1987–99	●	0	0
1970–86	●	0	0
pre 1970	●	0	0
Alien			
1987–99	●	238	0
1970–86	●	54	0
pre 1970	●	61	0

A perennial herb of waste places, gravel pits, roadsides, railway banks and rubbish tips. Lowland.

Neophyte. Though introduced into cultivation by 1568, and first recorded in the wild in 1640, the spread of this species is very recent. It was not mapped in the 1962 *Atlas*, yet in parts of London is now found in 75% of the tetrads in each 10-km square. To date, it has predominantly been an urban plant, but it is now spreading into rural areas. Many garden forms are hybrids between *G. officinalis* and *G. patula* (*G.* × *hartlandii*) and it is probable that many records refer to this taxon.

A European Temperate species, absent as a native from much of W. Europe.

Reference: Burton (1983).

D. A. PEARMAN

Colutea arborescens Bladder-senna

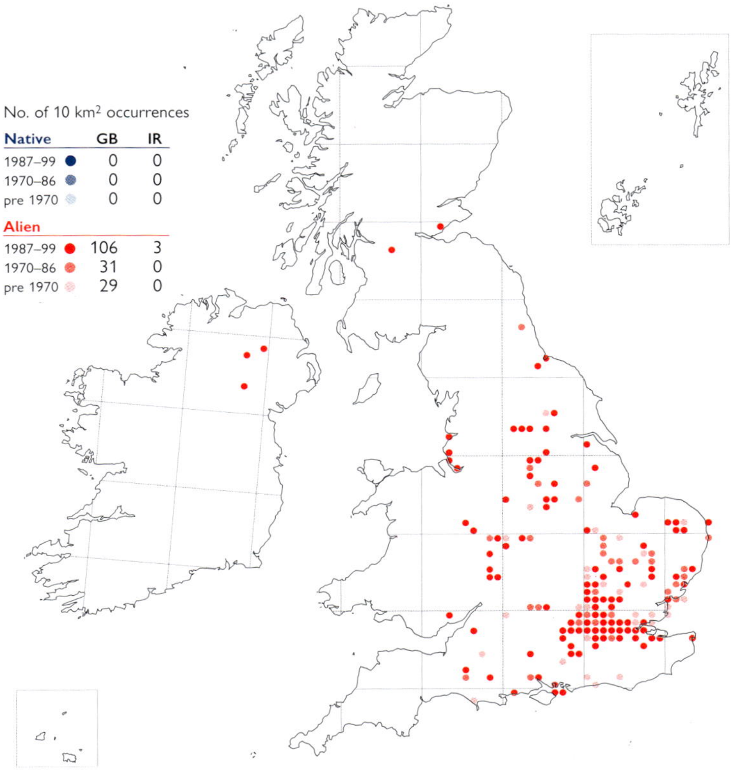

No. of 10 km² occurrences

Native		GB	IR
1987–99	●	0	0
1970–86	●	0	0
pre 1970	●	0	0
Alien			
1987–99	●	106	3
1970–86	●	31	0
pre 1970	●	29	0

A deciduous shrub now well-established in waste places and rubbish tips, and particularly in rough grassland and scrub on railway banks. Lowland.

Neophyte. This species was introduced into Britain by 1568, but has only become naturalised extensively since 1900, spreading initially along railways but then becoming more widespread.

Native of S. Europe.

References: Bean (1970), Bolòs & Vigo (1984).

D. A. PEARMAN

Astragalus danicus Purple Milk-vetch

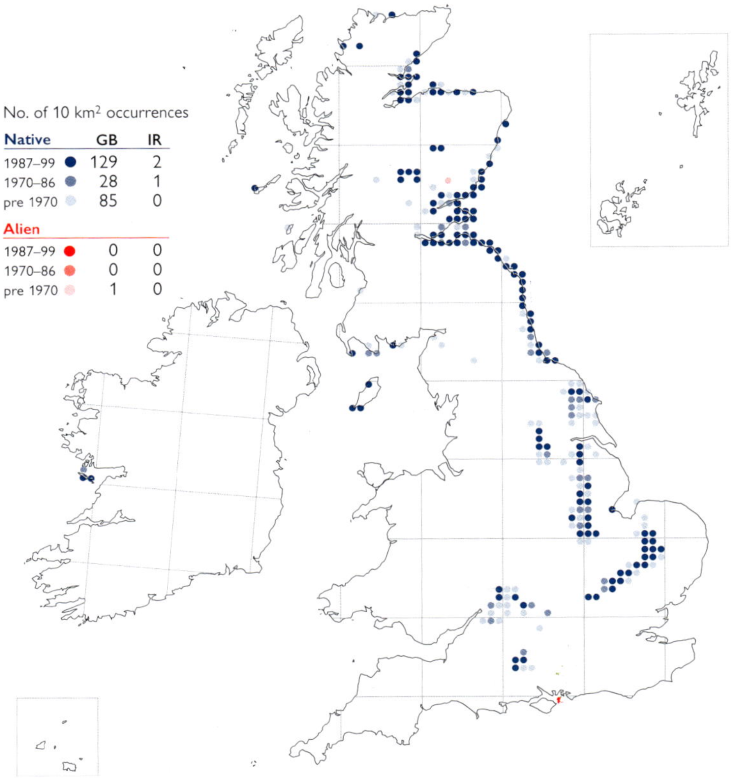

No. of 10 km² occurrences

Native		GB	IR
1987–99	●	129	2
1970–86	●	28	1
pre 1970	●	85	0
Alien			
1987–99	●	0	0
1970–86	●	0	0
pre 1970	●	1	0

A perennial herb of short unimproved turf on well-drained calcareous soils, predominantly on chalk and limestone, but also on sand dunes and machair. In Scotland, it also grows on Old Red Sandstone sea-cliffs and on mica-schist. 0–710 m (Meall an Daimh, E. Perth).

Native (change −0.88). *A. danicus* showed only a modest loss, mainly in Gloucestershire and Yorkshire, before 1930. Since then it has declined substantially on the chalk in S. England and limestone in N.E. England, largely due to agricultural improvement or lack of grazing. It is stable elsewhere.

Circumpolar Temperate element, with a continental distribution in W. Europe and absent from eastern N. America.

References: Atlas (111c), Curtis & McGough (1988), Hultén & Fries (1986), Meusel *et al.* (1965).

D. A. PEARMAN

Astragalus alpinus Alpine Milk-vetch

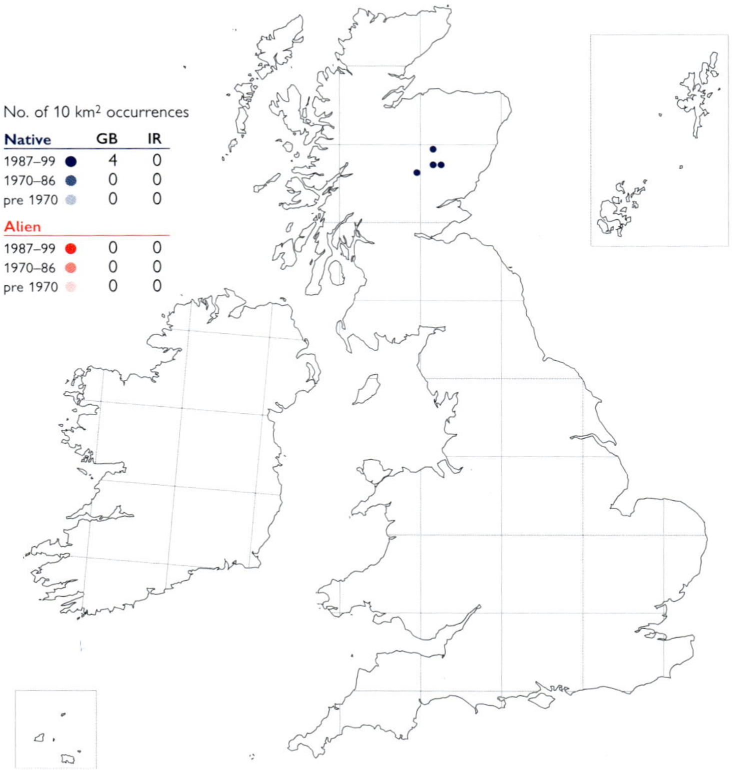

A perennial herb of species-rich, locally flushed calcareous grassland, and on base-rich ledges and rocky outcrops. Montane, from 650 m (Ben Vrackie, E. Perth) to 770 m (Creig an Dail Bheag, S. Aberdeen).

Native. *A. alpinus* is still extant in all four of its known localities, though populations on unstable rocks and crags are threatened by trampling and erosion, whilst colonies in grassland are suffering reduced flowering and seed set due to intensive grazing by sheep and deer.

Circumpolar Arctic-montane element.

References: Atlas (111b), Hultén & Fries (1986), Wigginton (1999).

D. A. PEARMAN

Astragalus glycyphyllos Wild Liquorice

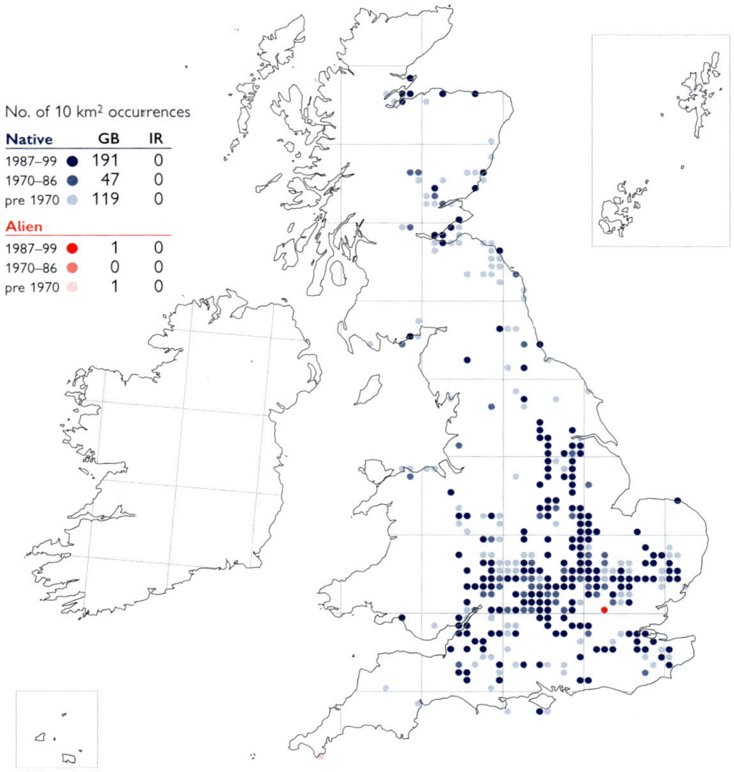

A straggling perennial herb of cliffs, wood-borders, chalk pits and scrubby grassland on railway banks and road verges; mainly on calcareous soils and thriving on warm, sheltered banks and hollows without too much grazing. Generally lowland, but reaching 365 m at Winskill Stones (Mid-W. Yorks.).

Native (change −0.36). This species is now much better recorded than it was for the 1962 *Atlas*, but county floras indicate a steady decline in sites because of the loss of rough grasslands and the tidying up of roadside banks and verges. It is often abundant where it occurs, but populations are becoming increasingly isolated.

European Temperate element; also in C. Asia.

References: Atlas (111d), Hultén & Fries (1986), Meusel *et al.* (1965).

D. A. PEARMAN

Oxytropis halleri Purple Oxytropis

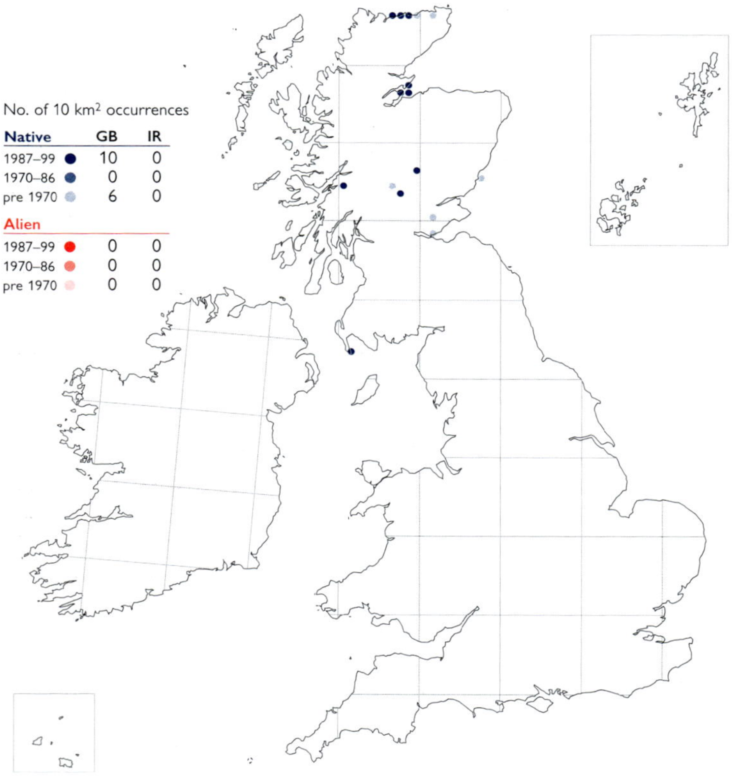

A perennial herb found on mountain rock ledges and grassy slopes on Dalradian limestone and schists, and on base-rich sandstone sea-cliffs and calcareous sand dunes. The larger populations occur at coastal sites in Sutherland. From sea level at Bettyhill (W. Sutherland) to 760 m on Ben Vrackie (E. Perth).

Native (change +0.16). Apart from losses in the 19th century, the distribution of this species appears to be stable, although some sites may be suffering from over-grazing and others from scrub encroachment.

European Boreal-montane element, but absent from the Boreal zonobiome. In mainland Europe *O. halleri* is strictly montane.

References: Atlas (112a), Bolòs & Vigo (1984), Wigginton (1999).

D. A. PEARMAN

Oxytropis campestris Yellow Oxytropis

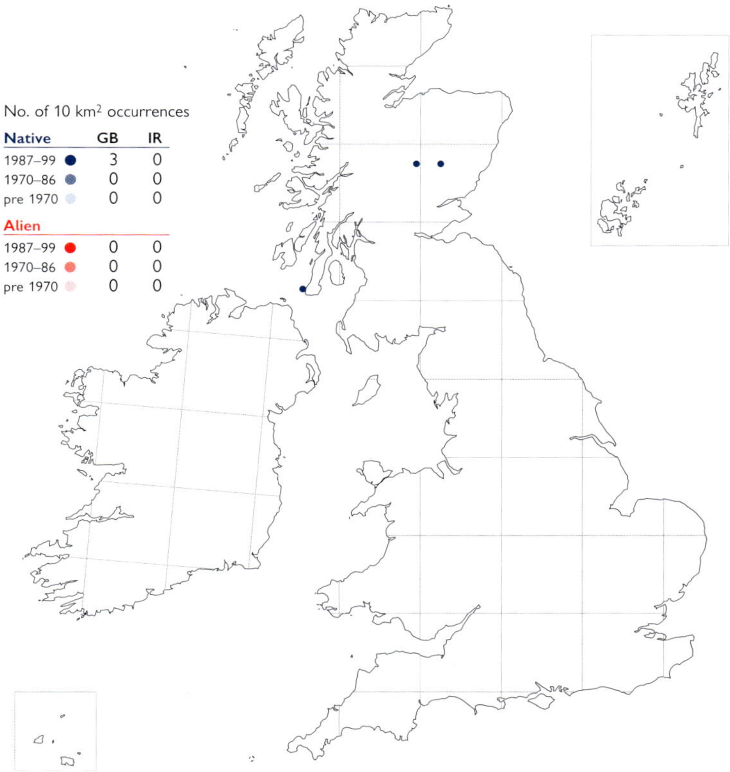

No. of 10 km² occurrences

Native	GB	IR
1987–99 ●	3	0
1970–86 ●	0	0
pre 1970 ●	0	0
Alien		
1987–99 ●	0	0
1970–86 ●	0	0
pre 1970 ●	0	0

A tufted perennial herb, confined in Britain to limestone and calcareous schists, where it grows in open communities of a southerly to south-westerly aspect. Its two inland sites are upland, between 500 and 640 m at Clova (Angus), but its third site is on sea-cliffs between 25 and 180 m.

Native. Populations of this relict species seem quite stable, despite past collecting and current grazing regimes.

Eurasian Arctic-montane element, with a continental distribution in W. Europe.

References: Atlas (112b), Hultén & Fries (1986), Meusel *et al.* (1965), Wigginton (1999).

D. A. PEARMAN

Onobrychis viciifolia Sainfoin

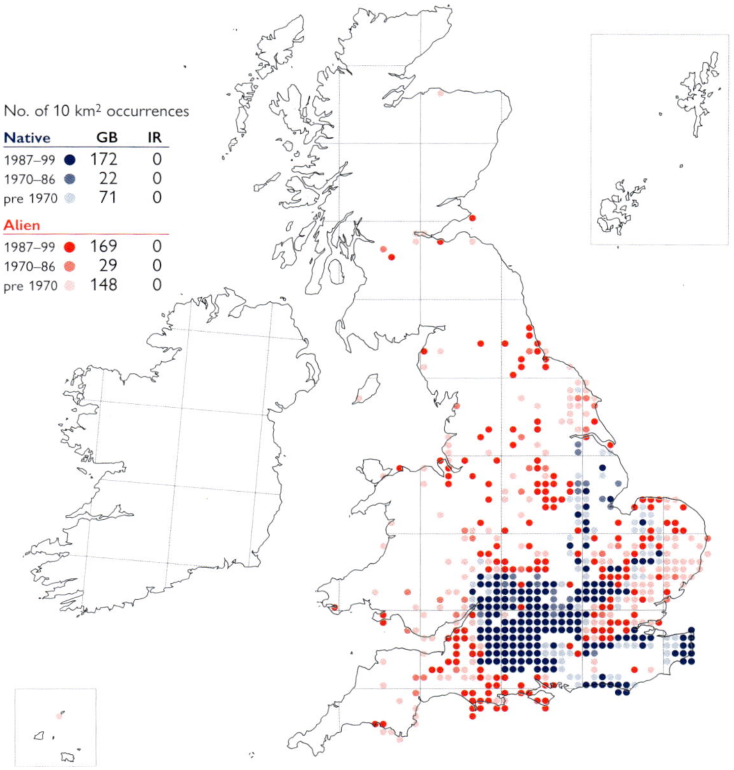

No. of 10 km² occurrences

Native	GB	IR
1987–99 ●	172	0
1970–86 ●	22	0
pre 1970 ●	71	0
Alien		
1987–99 ●	169	0
1970–86 ●	29	0
pre 1970 ●	148	0

A perennial herb which occurs in a dwarf form in unimproved chalk grassland. Robust alien variants are found on grassy banks, roadsides and by tracks on chalk and less often on other calcareous soils. They can be abundant on newly sown roadsides. Generally lowland, reaching 335 m near Taddington (Derbys.).

Native or alien (change −0.76). The chalk grassland form of *O. viciifolia* may be native, but the limits of its native distribution have been obscured by aliens. Agricultural variants were introduced in the 17th century and widely cultivated for fodder until the 19th century. The species is increasing as a constituent of wild-flower mixtures and as a contaminant of grass-seed.

Eurosiberian Temperate element; widely naturalised outside its native range.

References: Atlas (113c), Fearn (1987), Salisbury (1964).

D. A. PEARMAN

Anthyllis vulneraria Kidney Vetch

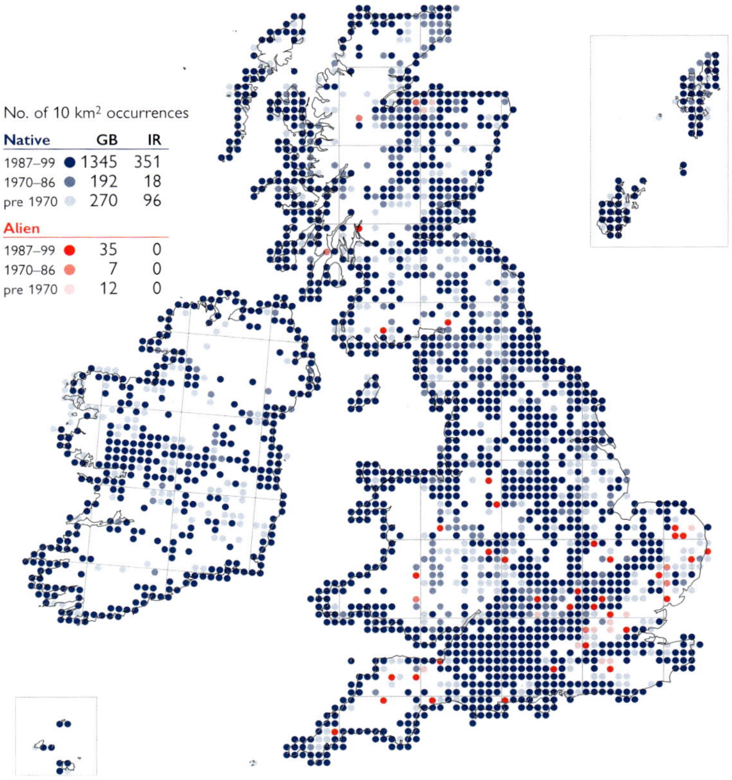

No. of 10 km² occurrences

Native	GB	IR
1987–99 ●	1345	351
1970–86 ●	192	18
pre 1970 ●	270	96
Alien		
1987–99 ●	35	0
1970–86 ●	7	0
pre 1970 ●	12	0

A perennial herb of rock outcrops and open turf on S.-facing slopes, on free-draining neutral to base-rich, often calcareous, soils. On the coast it is found on sea-cliffs, shingle and sand dunes. It is increasing as an alien on roadsides. Generally lowland, but reaching 945 m (Ben Lawers, Mid Perth).

Native (change +0.45). The native distribution is stable. Five subspecies are known in our area; subsp. *vulneraria* is found throughout the range of the species, subsp. *polyphylla* is a sporadic alien, subsp. *corbierei* is found on a few sea-cliffs and subsp. *carpatica* and subsp. *lapponica* are mapped separately.

European Boreo-temperate element; widely naturalised outside its native range.

References: Atlas (109d), Atlas Supp. (21a-d), Cullen (1986), Hultén & Fries (1986), Meusel *et al.* (1965), Salisbury (1964).

D. A. PEARMAN

Anthyllis vulneraria subsp. *carpatica*

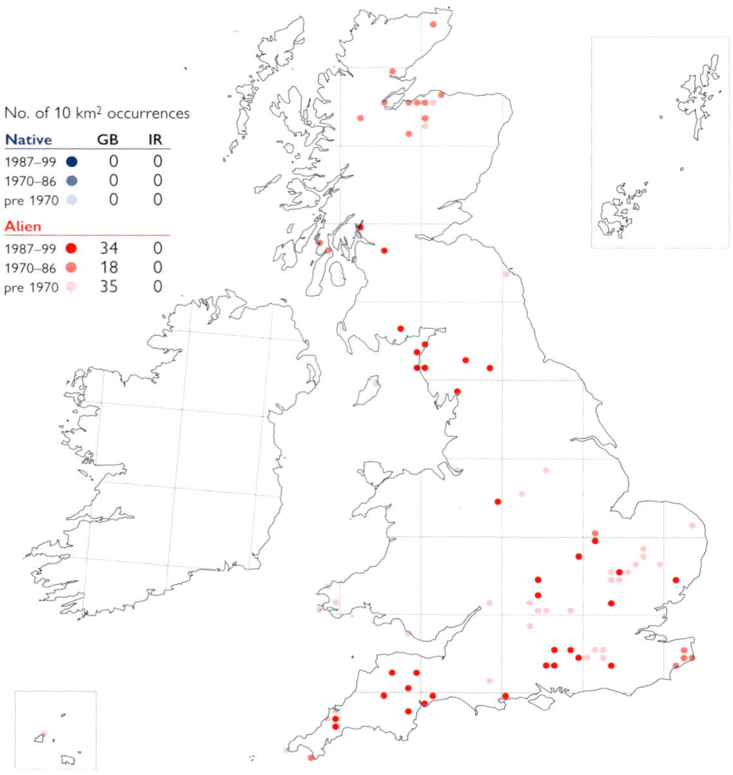

A short-lived perennial introduced in grass-seed mixtures, and appearing on roadsides and in waste places on a range of soils. Lowland.

Neophyte. This subspecies, which is represented in our area by var. *pseudovulneraria*, was collected in the wild in Britain as early as 1895 (Guernsey). It is increasingly sown in new amenity plantings. Many recorders have not identified *A. vulneraria* to subspecies, and this taxon is probably more widespread than the map suggests.

Native of N.W. & C. Europe.

References: Atlas Supp. (21c), Cullen (1986).

D. A. PEARMAN

Anthyllis vulneraria subsp. *lapponica*

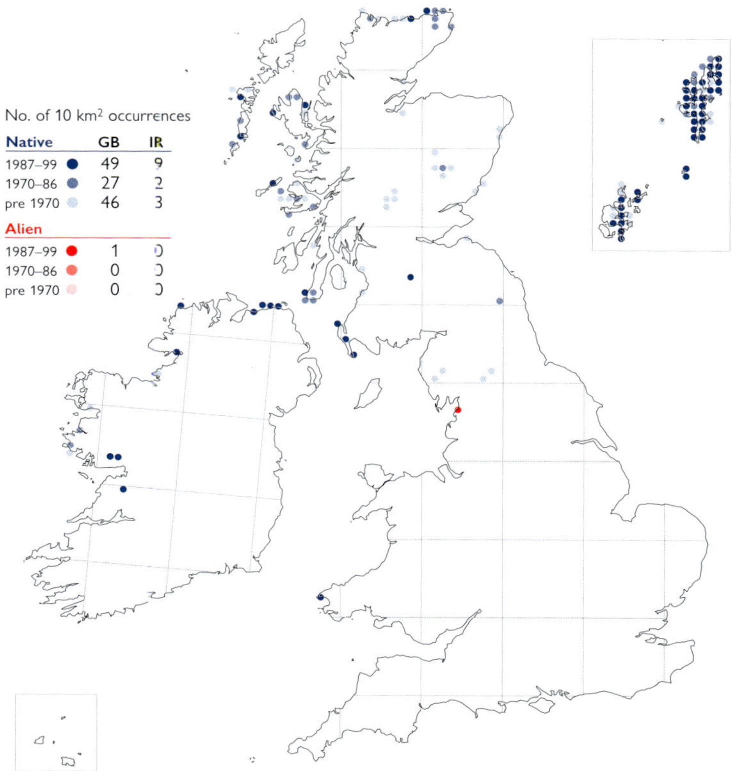

A perennial herb of sand dunes, machair, coastal and inland cliffs and rock ledges. It is also recorded from more ruderal habitats, including railway banks and forest tracks. It grows over a range of base-rich substrates, including basalt, limestones, mica-schist and serpentine. 0–945 m (Ben Lawers, Mid Perth).

Native. This subspecies is better recorded around the Scottish coast than when it was mapped by Perring & Sell (1968), but shows no change in overall distribution.

European Boreal-montane element; absent from mountains of central Europe.

References: Atlas Supp. (21d), Cullen (1986).

D. A. PEARMAN

Lotus glaber Narrow-leaved Bird's-foot-trefoil

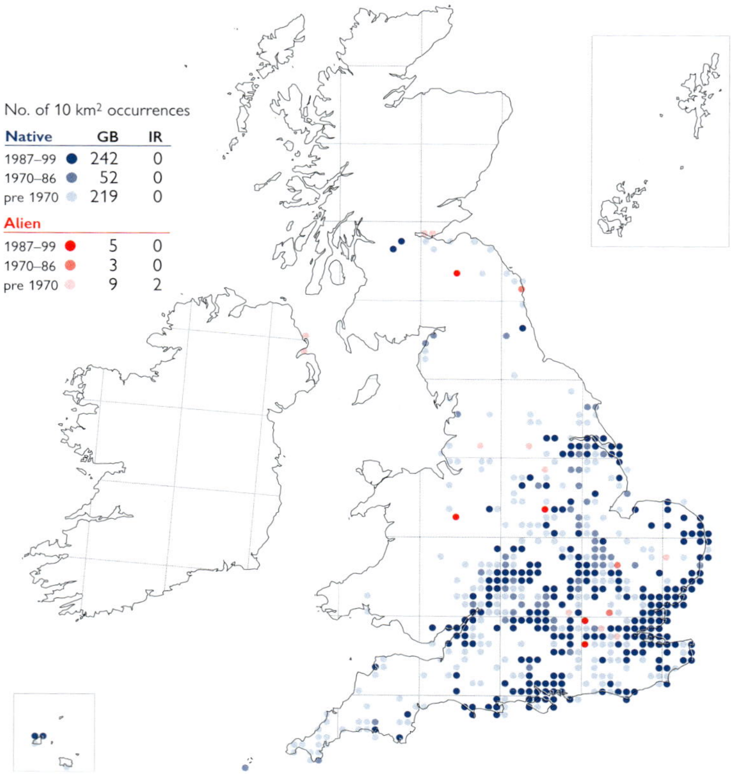

A perennial herb of coastal grazing marshes and sea-walls; inland it occurs in rough grassland, gravel-, sand-, chalk- and clay pits, and on railway banks and roadside verges. It is found on a range of neutral to calcareous soils, but shows a preference for estuarine silts and heavy clays. Lowland.

Native (change –0.55). On the coast the distribution of *L. glaber* appears to be stable, apart from losses in Devon and Cornwall where it may only ever have occurred as a casual. In many areas inland it is thought to be a casual or recent colonist of disturbed habitats, and the limits of its native range are not clear.

European Southern-temperate element; widely naturalised outside its native range.

References: Atlas (110b), Hultén & Fries (1986).

D. A. PEARMAN

Lotus corniculatus Common Bird's-foot-trefoil

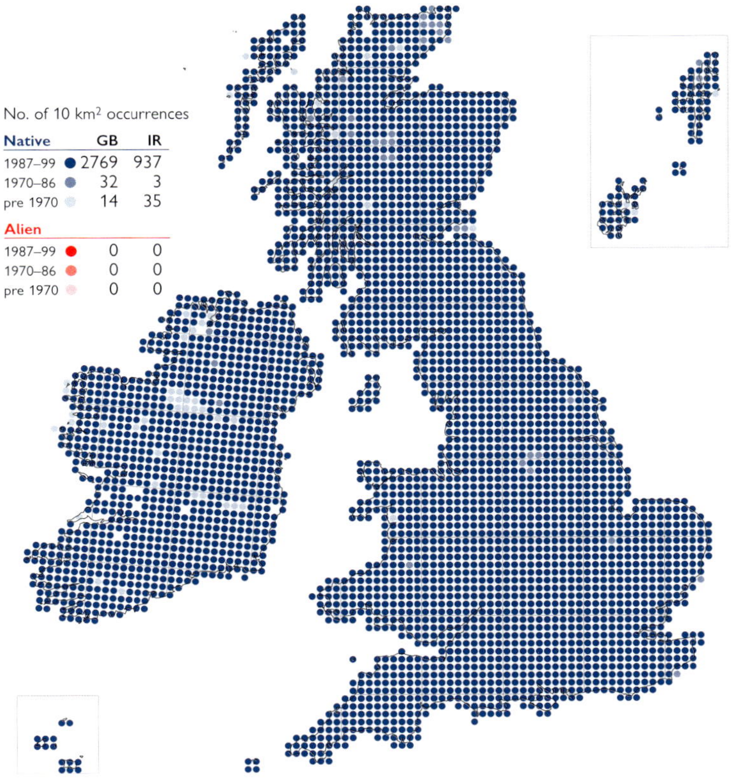

No. of 10 km² occurrences

Native		GB	IR
1987–99	●	2769	937
1970–86	●	32	3
pre 1970	●	14	35
Alien			
1987–99	●	0	0
1970–86	●	0	0
pre 1970	●	0	0

A perennial herb of grasslands, including well-drained meadows, chalk and limestone downs, hill pastures and montane rock ledges; also on coastal cliff-tops, shingle and sand dunes. It is absent from only the most acidic and infertile soils. Alien genotypes, introduced from seed mixtures, occur on roadsides. 0–915 m (Stuich an Lochan, Glen Lyon, Mid Perth).

Native (change +1.09). The overall distribution of this species is unchanged since the 1962 *Atlas*, despite the fact that it is suppressed in improved pastures and is possibly a poor competitor where grazing ceases.

Eurasian Southern-temperate element; widely naturalised outside its native range.

References: Atlas (110a), Grime *et al.* (1988), Hultén & Fries (1986), Jones & Turkington (1986), Meusel *et al.* (1965), Turkington & Franco (1980).

D. A. PEARMAN

Lotus pedunculatus Greater Bird's-foot-trefoil

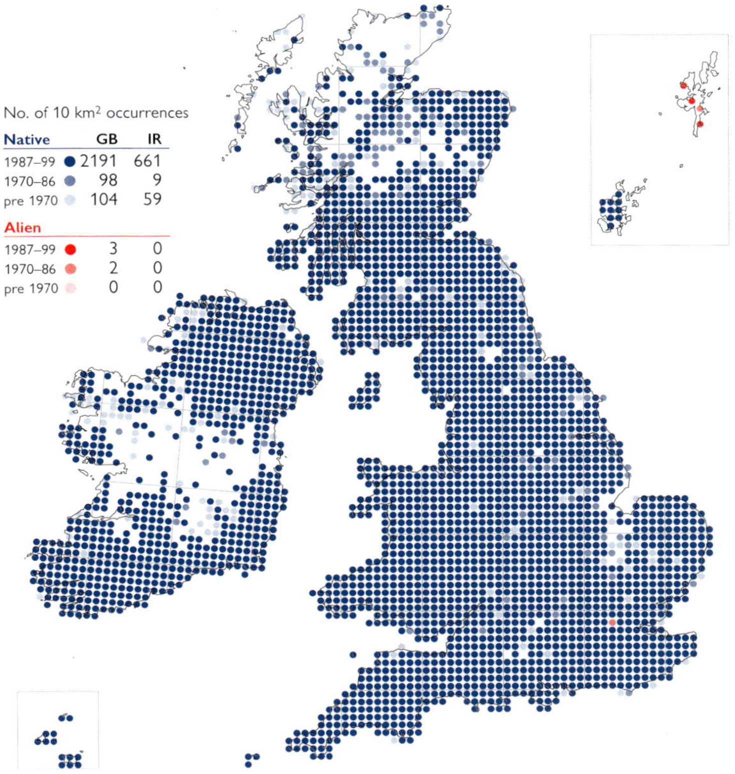

No. of 10 km² occurrences

Native		GB	IR
1987–99	●	2191	661
1970–86	●	98	9
pre 1970	●	104	59
Alien			
1987–99	●	3	0
1970–86	●	2	0
pre 1970	●	0	0

A perennial herb of rushy pastures, wet meadows, marshes, ditches, the margins of lakes, ponds and rivers, wet road verges and other marshy places. It is probably more frequent on acidic soils than calcareous ones. Generally lowland, but reaching 490 m on Dartmoor (S. Devon).

Native (change –0.06). The distribution of *L. pedunculatus* shows no change since the 1962 *Atlas*.

European Temperate element; widely naturalised outside its native range.

References: Atlas (110c), Grime *et al.* (1988), Hultén & Fries (1986), Meusel *et al.* (1965).

D. A. PEARMAN

Lotus subbiflorus Hairy Bird's-foot-trefoil

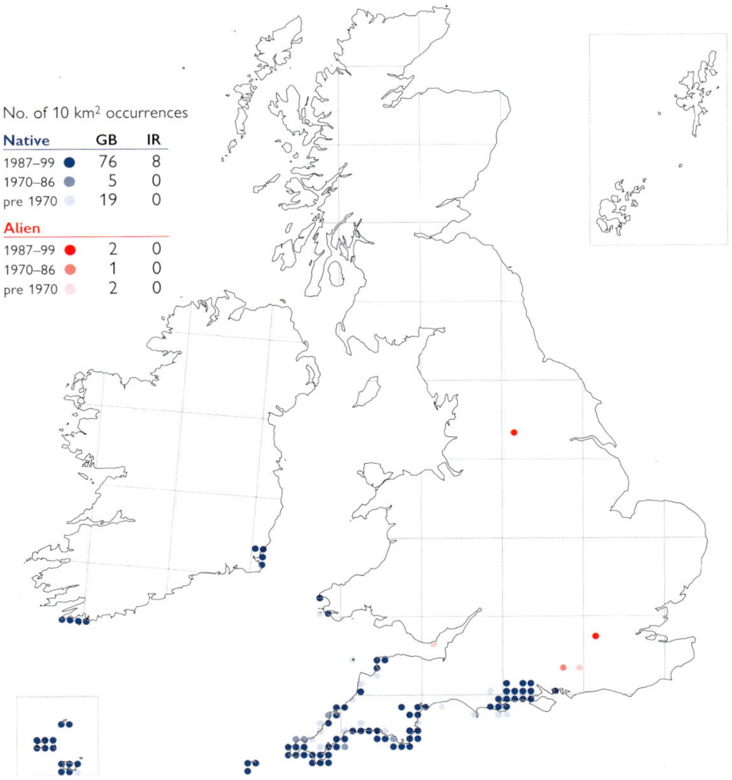

No. of 10 km² occurrences

Native		GB	IR
1987–99	●	76	8
1970–86	●	5	0
pre 1970	●	19	0
Alien			
1987–99	●	2	0
1970–86	●	1	0
pre 1970	●	2	0

An annual or short-lived perennial herb typically occurring around rock outcrops and in dry, open grassland on coastal cliffs; often on relatively sheltered, sunny banks alongside cliff-top paths and trackways, and thriving on shallow, drought-prone neutral to moderately acidic soils. It often occurs with *Lotus angustissimus*. In Dorset, Hampshire and Wexford it occurs inland in open, sandy grassland and on verges. Lowland.

Native (change +0.22). This species is much better recorded than for the 1962 *Atlas*. Increasing scrub is restricting its habitats in Devon and Cornwall, but the seed seems viable for long periods and colonies can reappear when conditions improve.

Suboceanic Southern-temperate element.

References: Atlas (110d), Curtis & McGough (1988), Kay & John (1995), Stewart *et al.* (1994).

D. A. PEARMAN

Lotus angustissimus Slender Bird's-foot-trefoil

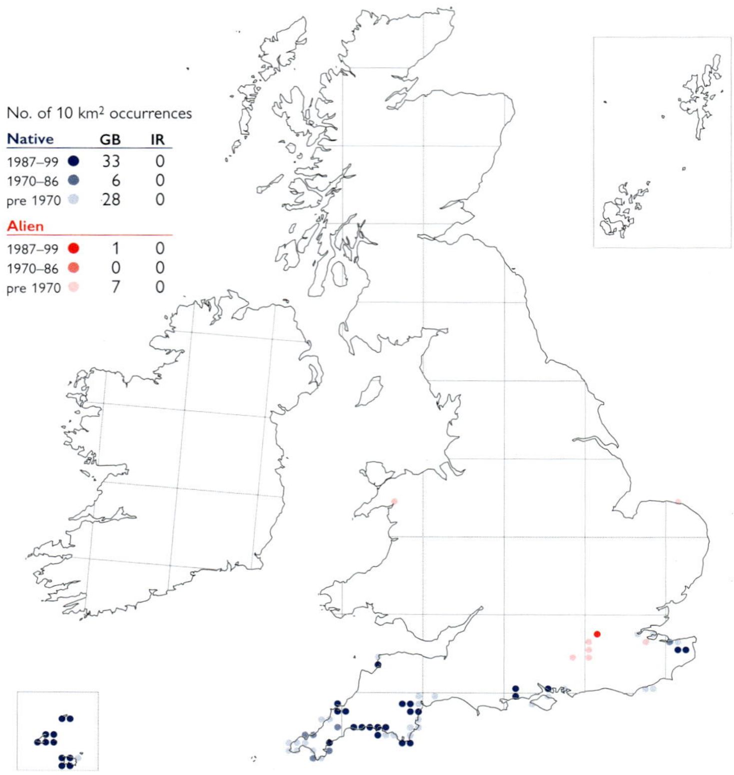

An annual of thin, drought-prone soils on sea-cliffs, growing around rock outcrops, on sunny banks by tracks and footpaths, and in open areas amongst scrub. In Devon, it occurs in similar habitats inland. The outlying sites in Hampshire and Kent are associated with sand- and gravel-workings. It often grows with *L. subbiflorus*. Lowland.

Native (change –0.23). Recent surveys have shown that *L. angustissimus* is more widespread in Devon and Cornwall than was previously thought. However, many colonies are small and at risk from scrub encroachment. The persistent seed bank may enable it to respond to a resumption of grazing or disturbance.

European Southern-temperate element.

References: Atlas (111a), Wigginton (1999).

D. A. PEARMAN

Ornithopus perpusillus Bird's-foot

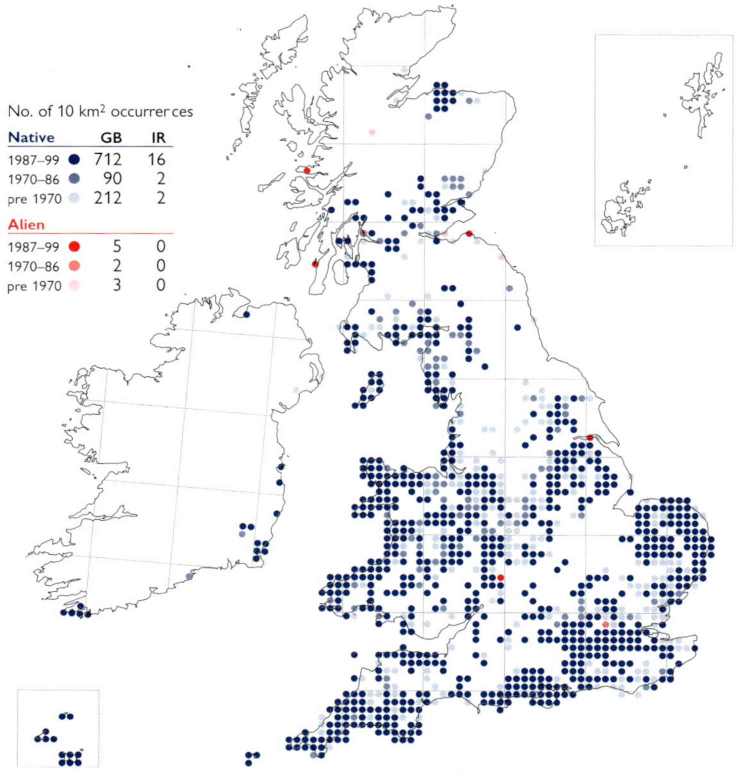

A winter-annual of short, open grassland on free-draining acidic sands and gravels; also around rock outcrops, on sand dunes and in bare patches on dry heathland. It is frequently found on drought-prone, sunny banks beside tracks and paths, and can quickly colonise bare areas after fires and other disturbances. Generally lowland, but reaching 380 m at Ettrick (Selkirks.).

Native (change –0.18). There has been no significant change in the distribution of *O. perpusillus* in England since the 1962 *Atlas*, but it appears to be more widespread in Wales, Scotland and Ireland than previously thought.

Suboceanic Temperate element; widely naturalised outside its native range.

References: Atlas (112c), Curtis & McGough (1988), Hultén & Fries (1986), Meusel *et al.* (1965).

D. A. PEARMAN

Ornithopus pinnatus Orange Bird's-foot

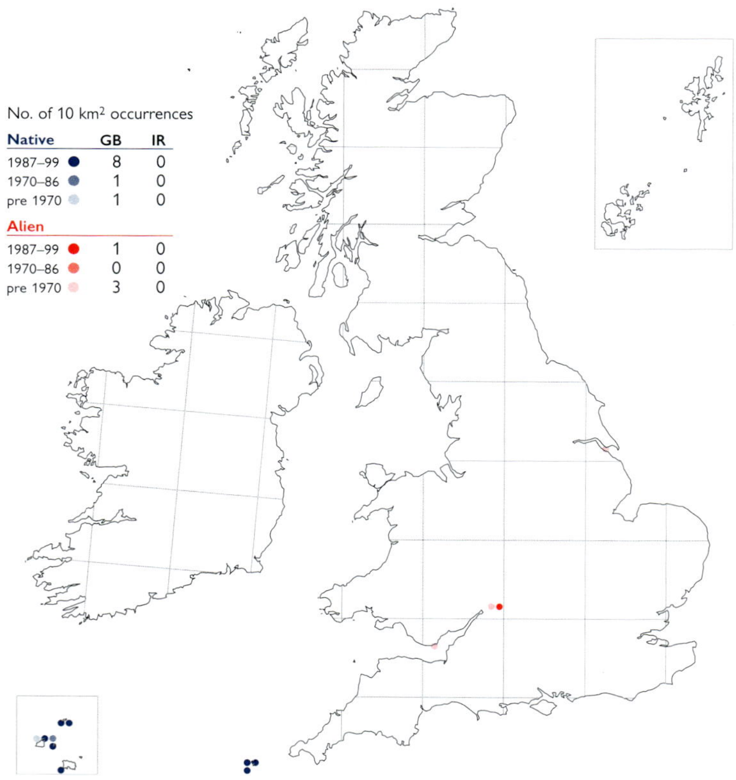

An annual found on dry heaths and heathy grassland, around granite cairns, in consolidated dune turf and in disturbed areas such as gardens and bulb-fields with sandy soil. Inland it is a rare casual of docks and waste ground. Lowland.

Native. Populations of *O. pinnatus* in the Channel Islands and the Isles of Scilly appear to be stable. Reports of it growing on the Cornish mainland remain unconfirmed (French *et al.*, 1999).

Suboceanic Southern-temperate element.

References: Atlas (112d), Bolòs & Vigo (1984), Wigginton (1999).

D. A. PEARMAN

Hippocrepis comosa Horseshoe Vetch

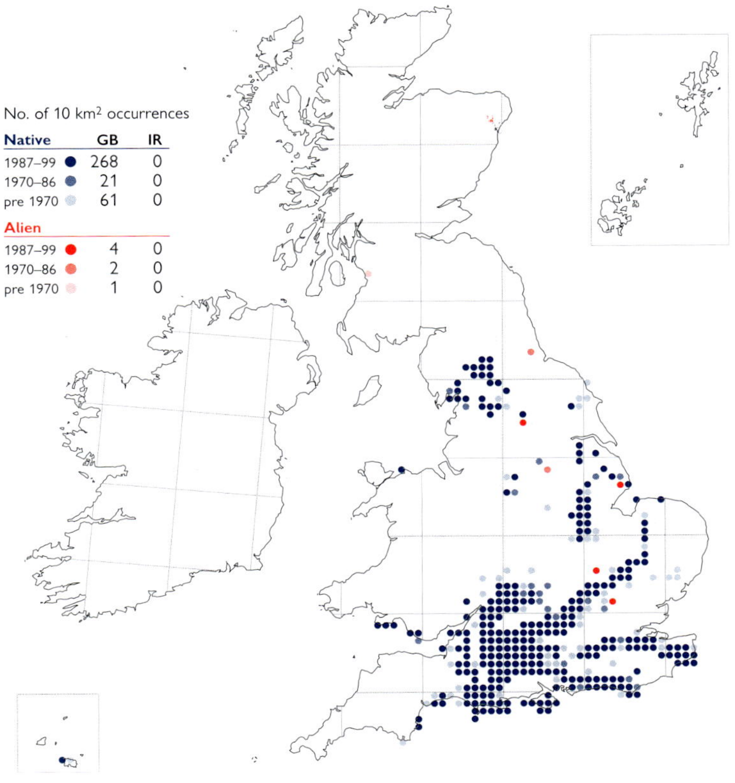

A perennial herb occurring in dry, sunny pastures on chalk and limestone, and on rock ledges on limestone cliffs. It prefers a S.-facing aspect, an open situation, and is able to colonise bare chalk, such as that in new road cuttings. Generally lowland, but reaching 600 m on Long Fell, Warcop (Westmorland).

Native (change −0.54). The map suggests little change in this species, other than a few losses from outlying squares and around the fringes of its core areas. However, county floras report declines, even within its main strongholds, due to grassland improvement, conversion to arable and scrub encroachment following the cessation of grazing.

European Temperate element.

References: Atlas (113b), Fearn (1973), Meusel *et al.* (1965).

D. A. PEARMAN

Securigera varia Crown Vetch

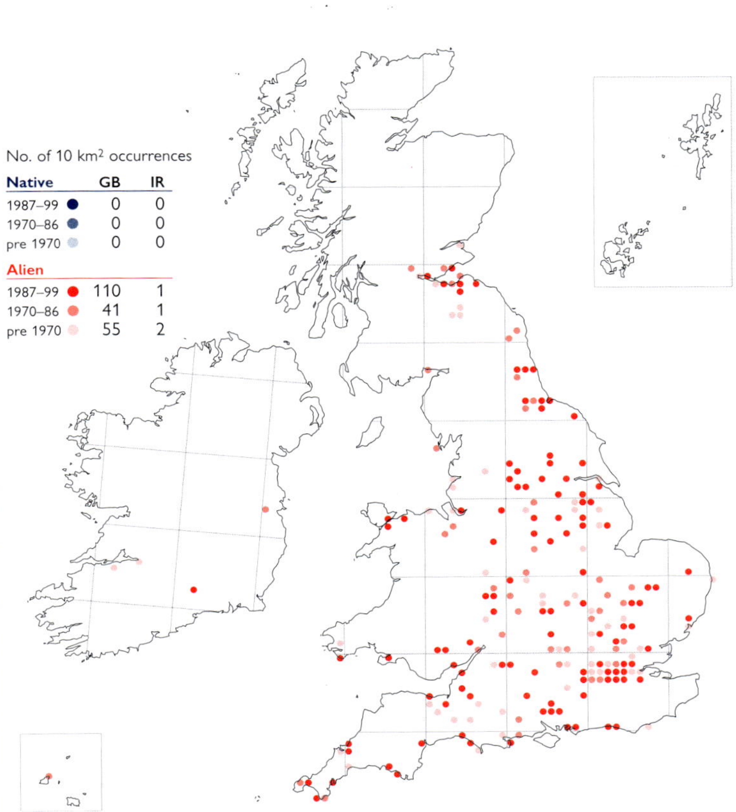

A sprawling perennial, naturalised in grassy habitats on roadsides, banks, in quarries and waste places. Lowland.

Neophyte (change +1.15). *S. varia* was introduced into gardens by 1597 but not recorded from the wild until 1843. Most occurrences are likely to be escapes from gardens, but it is occasionally found as a contaminant of clover seed. Populations are often transient, but it is a deep-rooted plant and it sometimes lives for many years.

Native of C. & S. Europe and W. Asia; naturalised further north in Europe, in N. America and elsewhere.

References: Atlas (113a), Meusel *et al.* (1965).

D. A. PEARMAN

Vicia orobus Wood Bitter-vetch

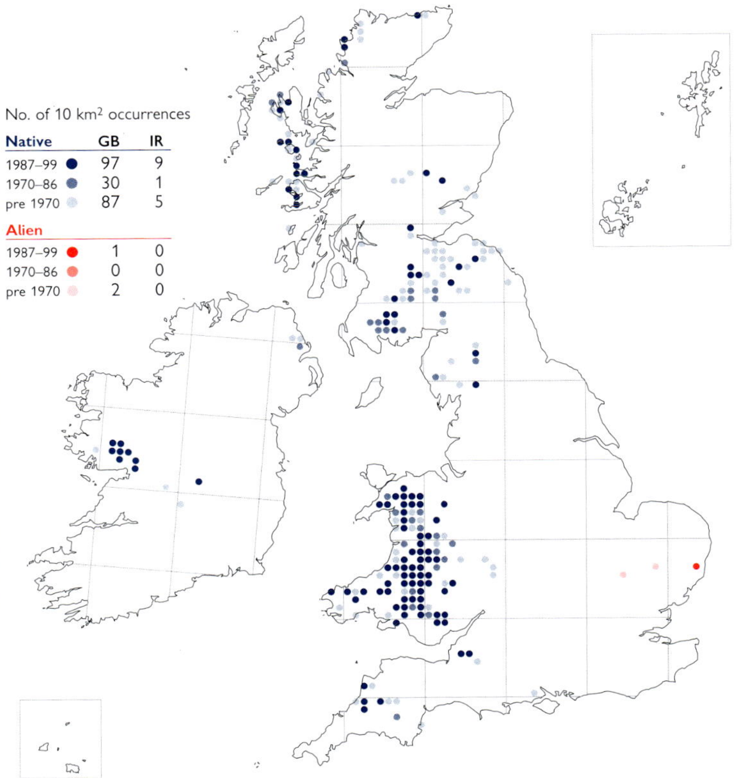

A perennial herb of grassy, often slightly base-enriched habitats on banks and the edges of fields, particularly amongst stones, boulders or bushes. Usually between 200 and 300 m, but down to sea level in Sutherland, and reaching 455 m on the Wast Water screes (Cumberland).

Native (change −0.34). This species is adversely affected by overgrazing and undergrazing, both of which have contributed to its decline, though losses have also resulted from grassland improvement and land reclamation. Britain has a significant proportion of the world population of this species.

Suboceanic Temperate element.

References: Atlas (114d), Curtis & McGough (1988), Hultén & Fries (1986), Kay & John (1994), Meusel *et al.* (1965), Stewart *et al.* (1994).

D. A. PEARMAN

Vicia cracca Tufted Vetch

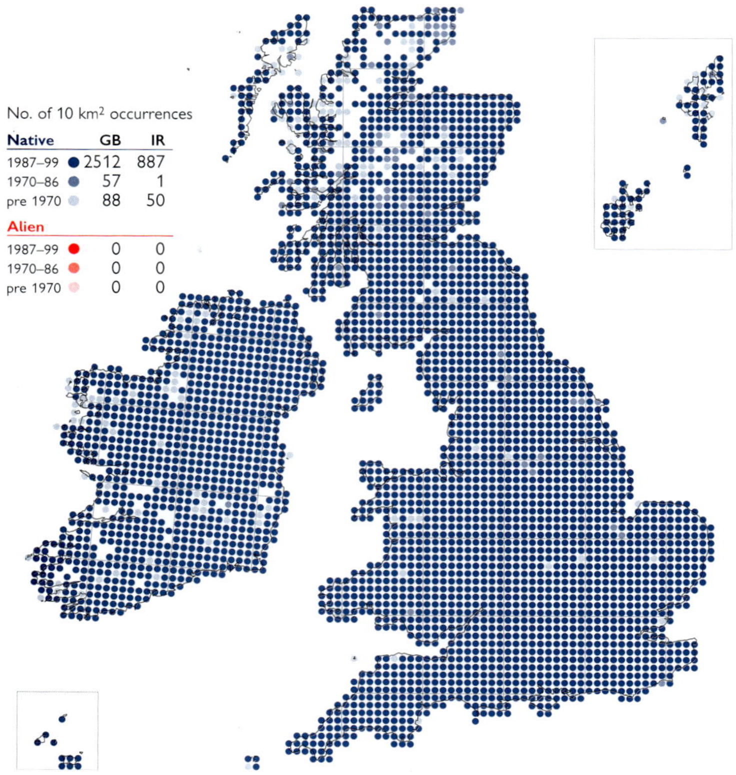

A scrambling perennial herb of hedgerows, waysides, wood-borders, scrubby grassland and river and canal banks. It also occurs in permanent pastures and hay meadows, and in marshes and tall-herb fens, but avoids permanently wet sites. Generally lowland, but reaching 550 m at Moor House, Teesdale (Westmorland).

Native (change –0.37). The distribution of this species has remained unchanged since the 1962 *Atlas*.

Eurasian Boreo-temperate element, but naturalised in N. America so distribution is now Circumpolar Boreo-temperate.

References: Atlas (114c), Aarssen *et al.* (1986), Grime *et al.* (1988), Hultén & Fries (1986), Meusel *et al.* (1965).

D. A. PEARMAN

Vicia tenuifolia Fine-leaved Vetch

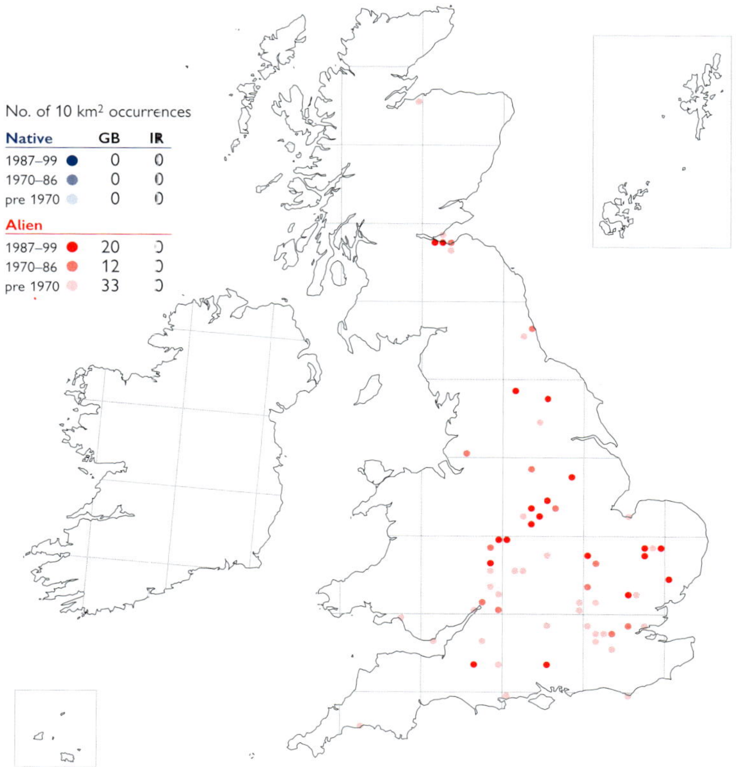

A perennial herb occasionally found naturalised on grassy banks, verges and waste ground, especially by railways. It is often introduced as a contaminant of grain. Lowland.

Neophyte. *V. tenuifolia* was grown in Britain by 1799, but not recorded from the wild until it was reported as a casual arising from imported grain in 1859. Some European botanists treat it as a subspecies of *V. cracca*.

A variable, Eurosiberian Southern-temperate species which is absent as a native from N.W. Europe.

References: Hultén & Fries (1986), Meusel *et al.* (1965).

D. A. PEARMAN

Vicia sylvatica Wood Vetch

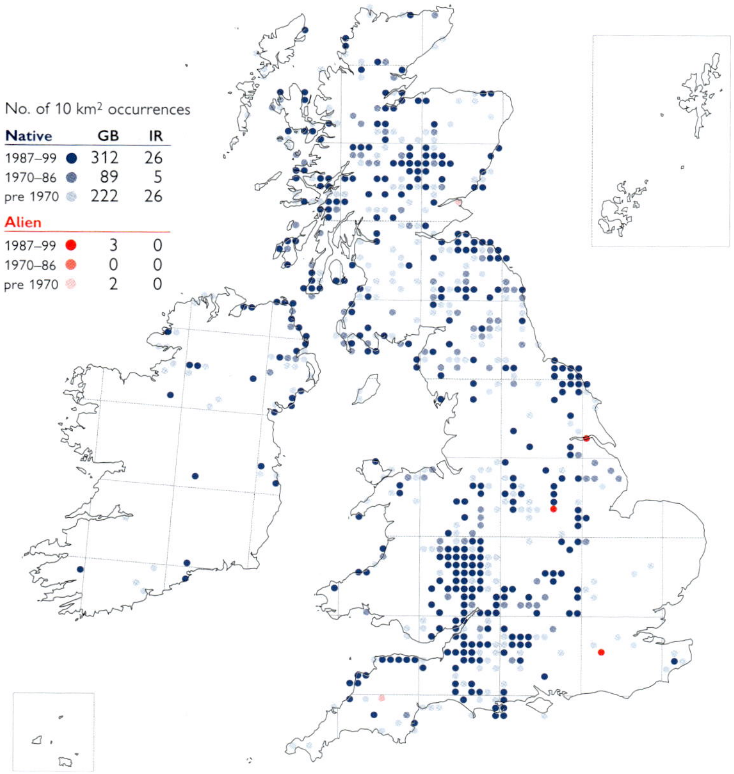

A climbing or scrambling perennial of hedges, wood-borders and clearings, scrub, rough ungrazed grassland on cliffs, wooded gorges and also on shingle, screes and railway banks. A genetically dwarf variant (var. *condensata*) occurs on shingle and coastal cliffs in N. & W. Britain and Ireland. 0–675 m (Breadalbanes, Mid Perth).

Native (change –0.71). *V. sylvatica* is much better recorded than in the 1962 *Atlas*, but there have been losses across much of its range and many county floras report local declines. It needs light in its woodland edge habitat so a decline in coppicing might be affecting its abundance.

Eurosiberian Boreo-temperate element.

References: Atlas (115a), Akeroyd (1996b), Hultén & Fries (1986), Meusel *et al.* (1965).

D. A. PEARMAN

Vicia villosa Fodder Vetch

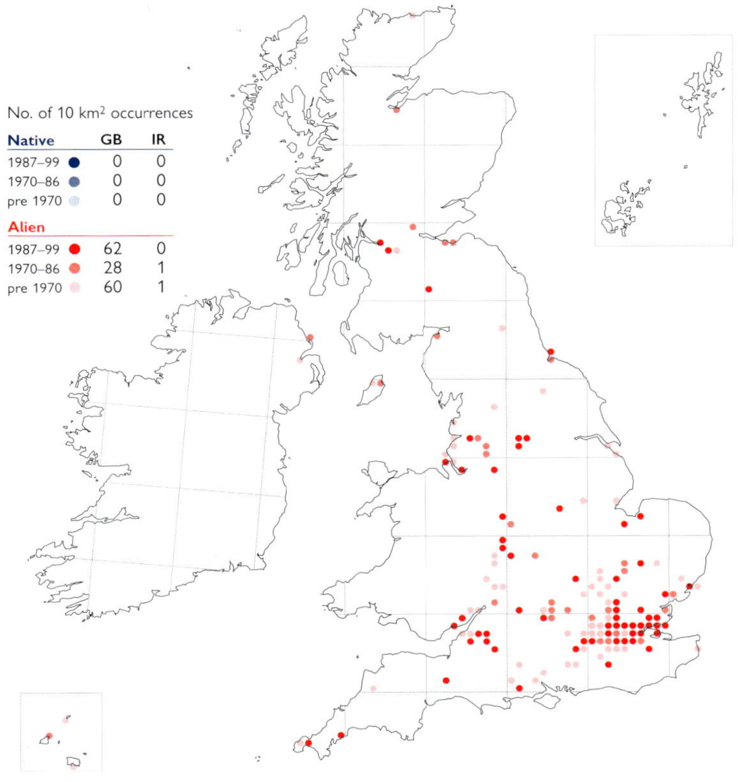

A scrambling annual, usually derived from grain, bird-seed or wool shoddy imports, and occurring as a casual on waste ground, tips and in arable fields. It sometimes survives for a few years on grassy banks. Lowland.

Neophyte. *V. villosa* was first grown in Britain in 1815, but arose in the wild as a grain contaminant in 1857. Trends in its distribution are unclear.

A variable, European Southern-temperate species which is absent as a native from much of N.W. Europe.

Reference: Hultén & Fries (1986).

D. A. PEARMAN

Vicia hirsuta Hairy Tare

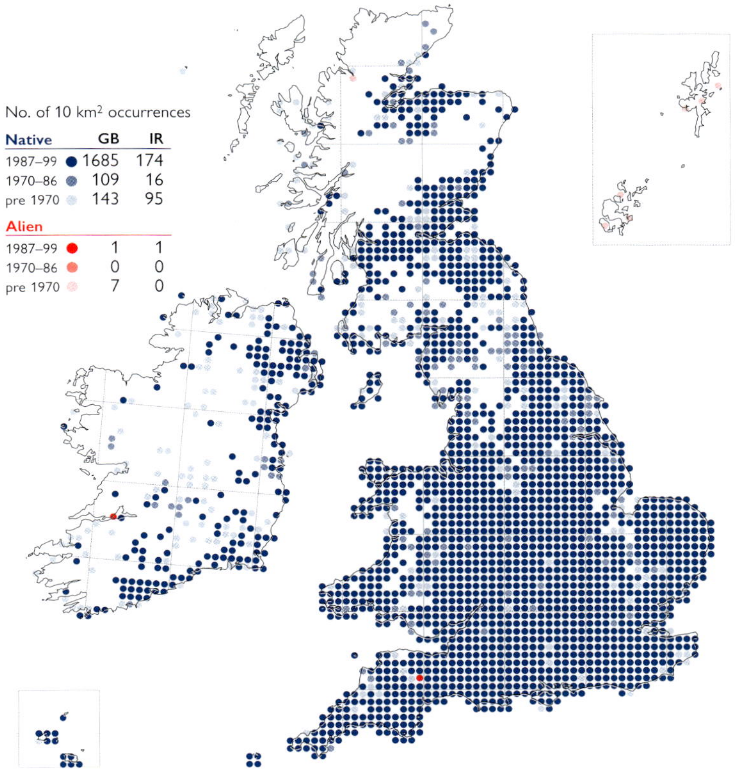

A scrambling annual of rough and disturbed ground, including road and railway banks, scrubby grassland, hedgerows, sheltered sea-cliffs and consolidated shingle beaches; also along the edges of arable fields, and on rubbish tips and waste ground. 0–335 m (Dartmoor, S. Devon).

Native (change +0.05). There has been no discernible change in the distribution of *V. hirsuta* since the 1962 *Atlas*. It was a troublesome cornfield weed in the 19th century, but is now much less frequent in that habitat.

European Temperate element, but widely naturalised so distribution is now Circumpolar Temperate.

References: Atlas (113d), Hultén & Fries (1986).

D. A. PEARMAN

Vicia parviflora Slender Tare

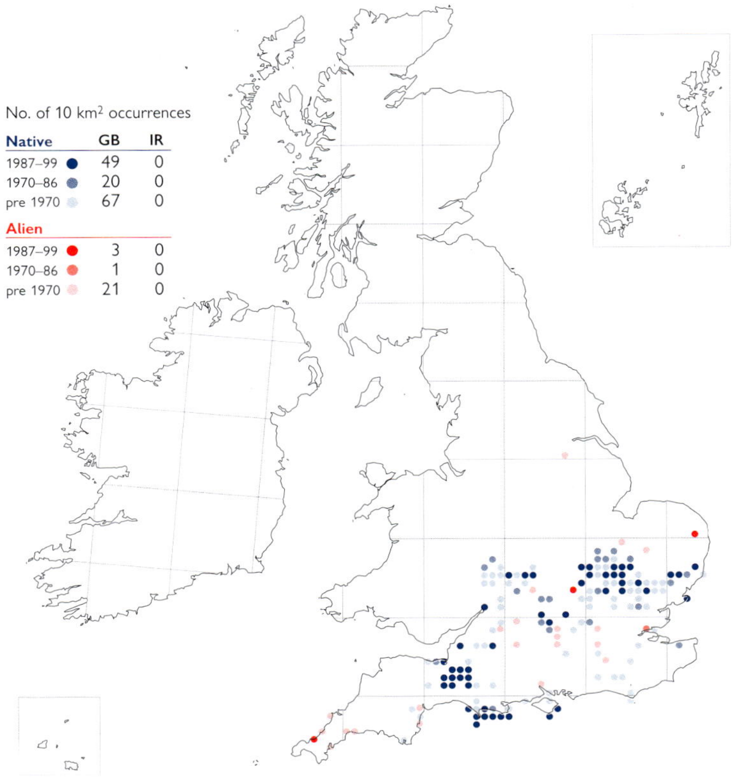

A scrambling annual of sticky calcareous clay soils which are frequently wet in winter but baked dry in summer. *V. parviflora* occurs in hedgerows, on tracks and verges, grassy banks, coastal cliffs and the edges of arable fields; also, less frequently, on urban waste ground, in municipal flower beds and as a casual of legume crops. Lowland.

Native (change −1.05). There is evidence of a widespread decline of this species in its arable habitats, commencing before 1930 but still continuing. It may have been over-looked or over-recorded previously due to confusion with *V. tetrasperma*. Its distribution in other habitats remains stable.

Submediterranean-Subatlantic element.

References: Atlas (114b), Stewart *et al.* (1994).

D. A. PEARMAN

Vicia tetrasperma Smooth Tare

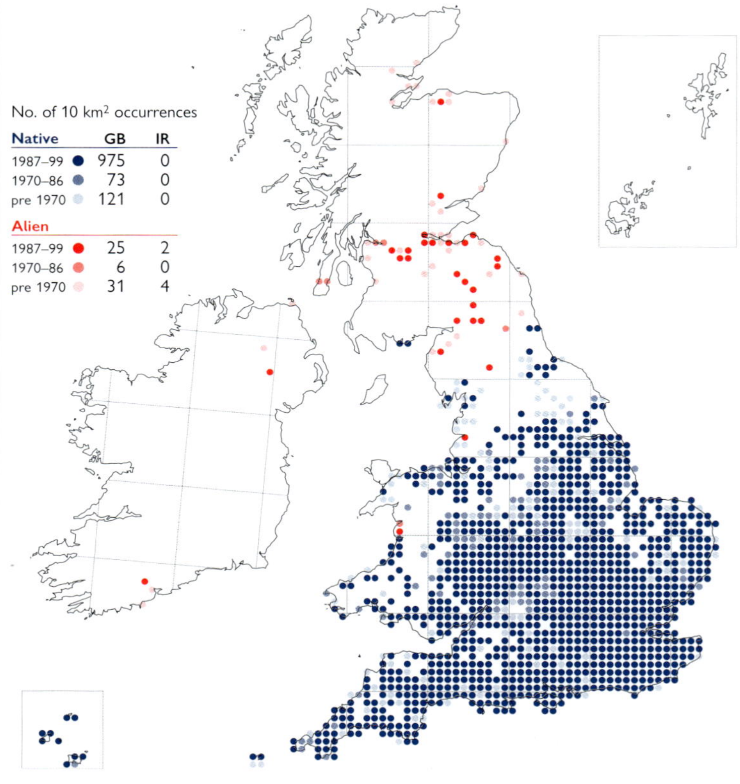

A scrambling annual of hedgerows, scrub and wood-borders, and of rough grassland on roadsides, railway banks and coastal cliffs; also found in disturbed places, including urban waste ground and arable field margins. Lowland.

Native (change +0.45). There appears to be little change in the distribution of *V. tetrasperma* since the 1962 *Atlas*, but county floras report declines throughout the range. The status of some records in N.W. England and Scotland is uncertain.

European Temperate element, but widely naturalised so distribution is now Circumpolar Temperate.

References: Atlas (114a), Aarssen *et al.* (1986), Hultén & Fries (1986).

D. A. PEARMAN

Vicia sepium Bush Vetch

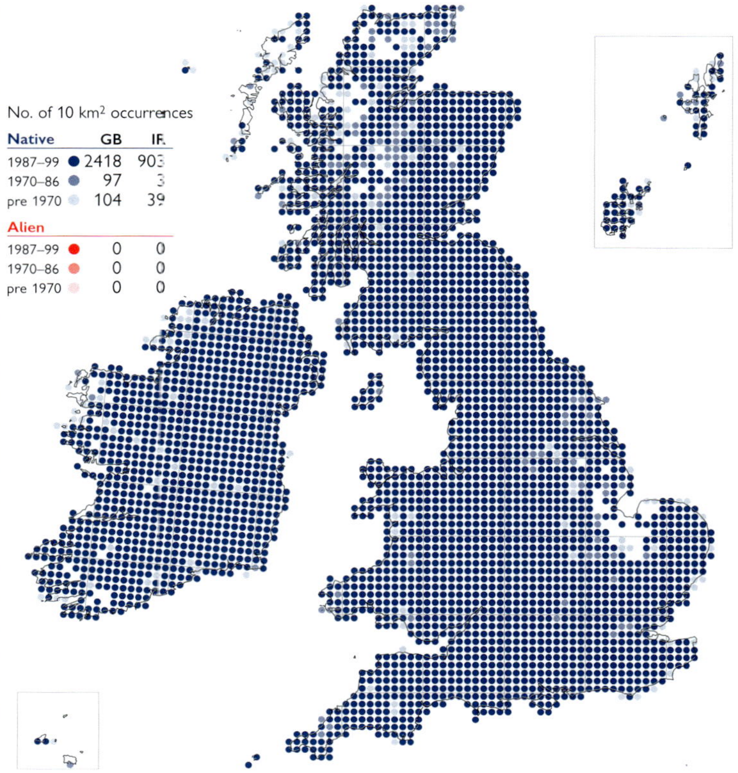

A climbing or scrambling perennial herb of hedge banks, waysides, wood-borders and lightly grazed grasslands. In upland areas it favours more open, ungrazed situations but it is most frequently found on neutral or basic soils. Generally lowland, but reaching 820 m in Caenlochan (Angus).

Native (change –0.43). *V. sepium* shows no change in its overall distribution since the 1962 *Atlas*.

Eurosiberian Boreo-temperate element; widely naturalised outside its native range.

References: Atlas (115b), Grime *et al.* (1988), Hultén & Fries (1986), Meusel *et al.* (1965).

D. A. PEARMAN

Vicia sativa Common Vetch

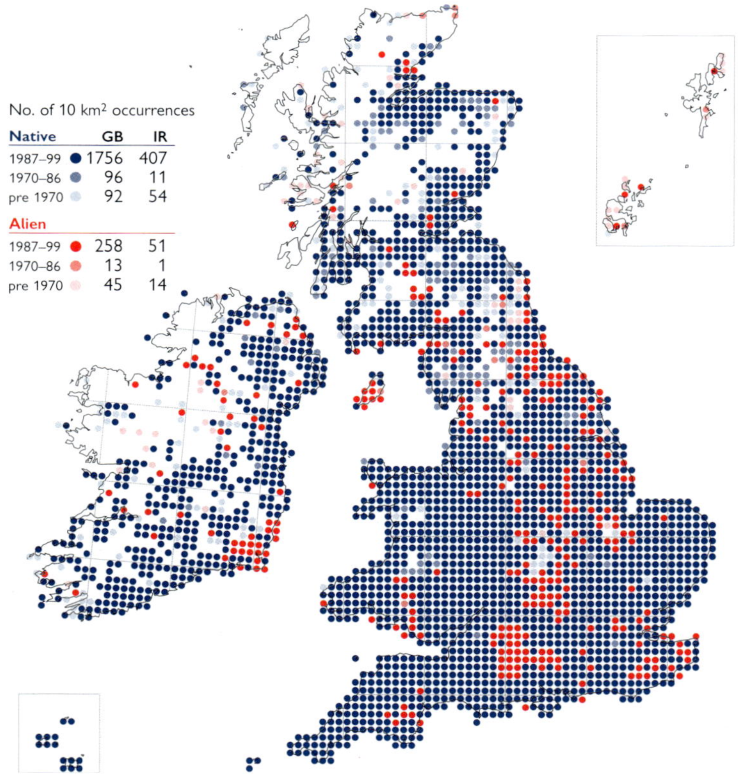

An annual of grassy and wayside places, particularly on dry and sandy sites. It was also grown as a fodder crop, and has widely escaped and become naturalised in many ruderal habitats. Lowland.

Native (change +0.19). There is little change in the distribution of *V. sativa* since the 1962 *Atlas*. Three subspecies are found in our area (subsp. *nigra*, subsp. *segetalis* and subsp. *sativa*) and these are mapped separately. The map overestimates the native distribution as subsp. *nigra* has been over-recorded as a native at inland sites.

European Southern-temperate element; widely naturalised outside its native range.

References: Bolòs & Vigo (1984), Hollings & Stace (1978).

D. A. PEARMAN

Vicia sativa subsp. *nigra*

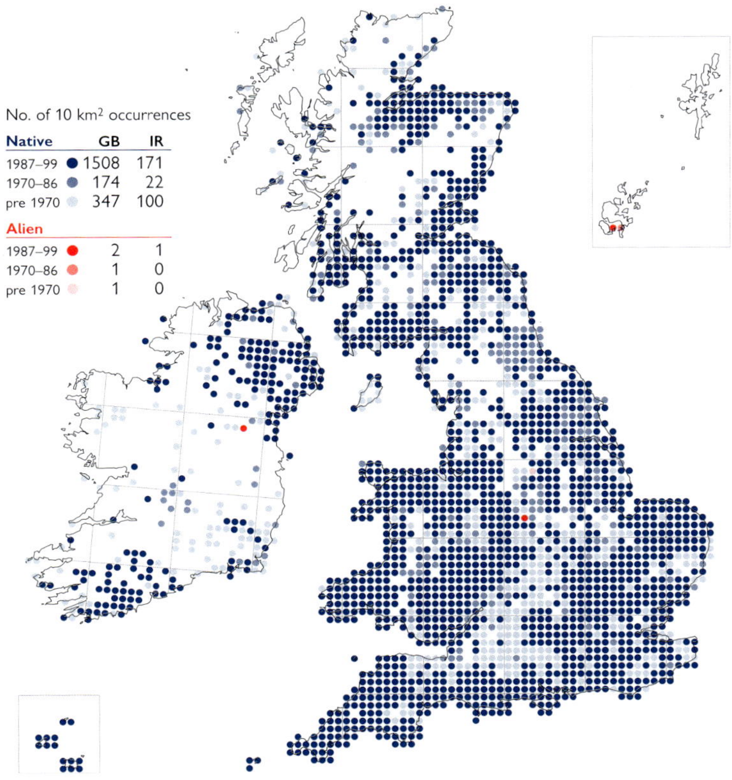

A procumbent or ascending annual, found as a native in many grassy and waste places, particularly on dry and sandy sites such as dunes, shingle, sea-cliffs and heathland. Inland, it is often found as an introduction in grassy sites, but these are rarely recorded as alien. Generally lowland, reaching 330 m at Dent Head (N.W. Yorks.).

Native. Some inland records of this subspecies are probably errors for subsp. *segetalis* as these two subspecies were not separated in national Floras until Stace (1991). Other inland records probably represent introduced rather than native plants.

Wider distribution uncertain.

References: Atlas (115d, also includes subsp. *segetalis*), Aarssen *et al.* (1986), Hollings & Stace (1978), Hultén & Fries (1986).

D. A. PEARMAN

Vicia sativa subsp. *segetalis*

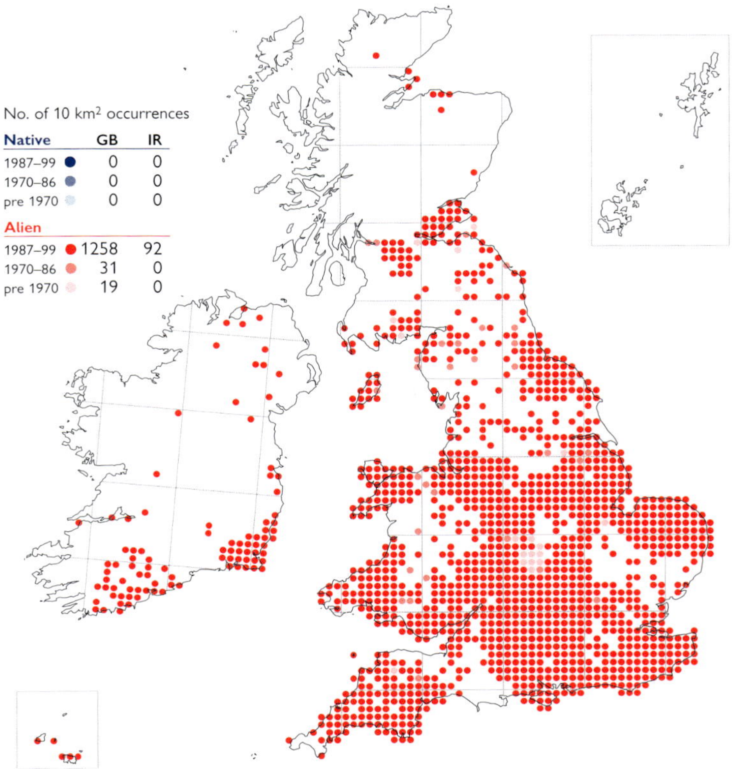

A robust annual of grassy places, including field-borders, roadsides and waste ground. Lowland.

Archaeophyte. Subsp. *segetalis* has long been grown as a fodder crop, but it is rarely cultivated now. It has not been systematically recorded, and the map is certainly incomplete; for example, all Dorset records have been ascribed to subsp. *nigra*. Nevertheless, it has become the most frequent of the *V. sativa* subspecies in our area.

Wider distribution uncertain.

References: Atlas (115d, also includes subsp. *nigra*), Aarssen *et al.* (1986), Hollings & Stace (1978).

D. A. PEARMAN

Vicia sativa subsp. *sativa*

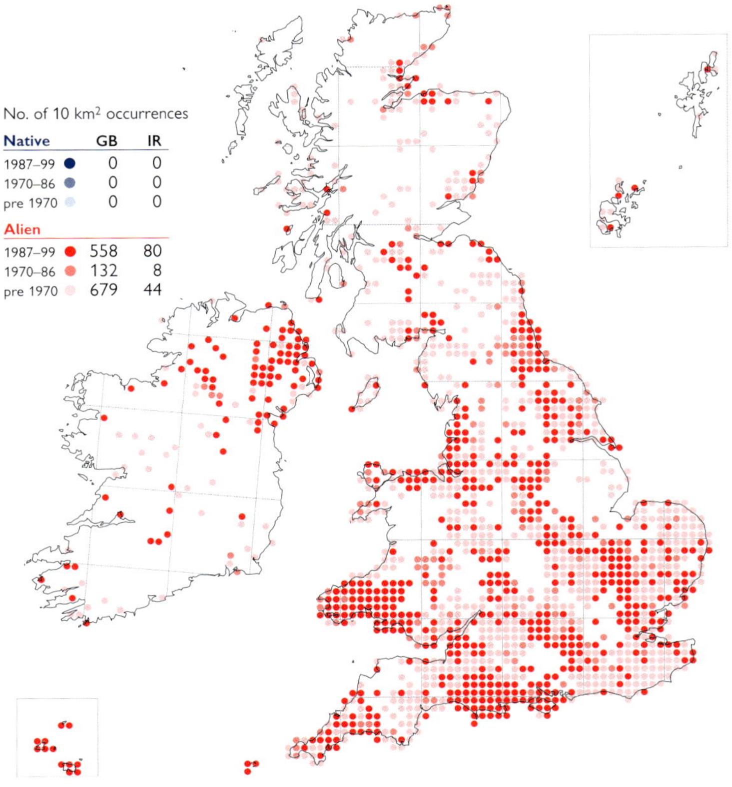

A robust annual of waste and cultivated ground and field edges. 0–385 m (Ingleborough, Mid-W. Yorks.).

Archaeophyte. This subspecies, which was recorded in the wild by 1660, was formerly much grown for fodder, and it was then regularly recorded as a casual. However, the acreage of farmland sown to 'vetches and tares' decreased by 91% between 1891 and 1958, and the incidence of *V. sativa* subsp. *sativa* has decreased accordingly. It is, however, still used as a 'green manure' by gardeners. Until recent years this taxon was misunderstood and many records may be errors for other subspecies.

Probably derived from cultivation of wild forms of *V. sativa*; now very extensively naturalised in Europe and other continents.

References: Aarssen *et al.* (1986), Hollings & Stace (1978), Killick (1975), Killick *et al.* (1998), Zohary & Hopf (2000).

D. A. PEARMAN

Vicia lathyroides Spring Vetch

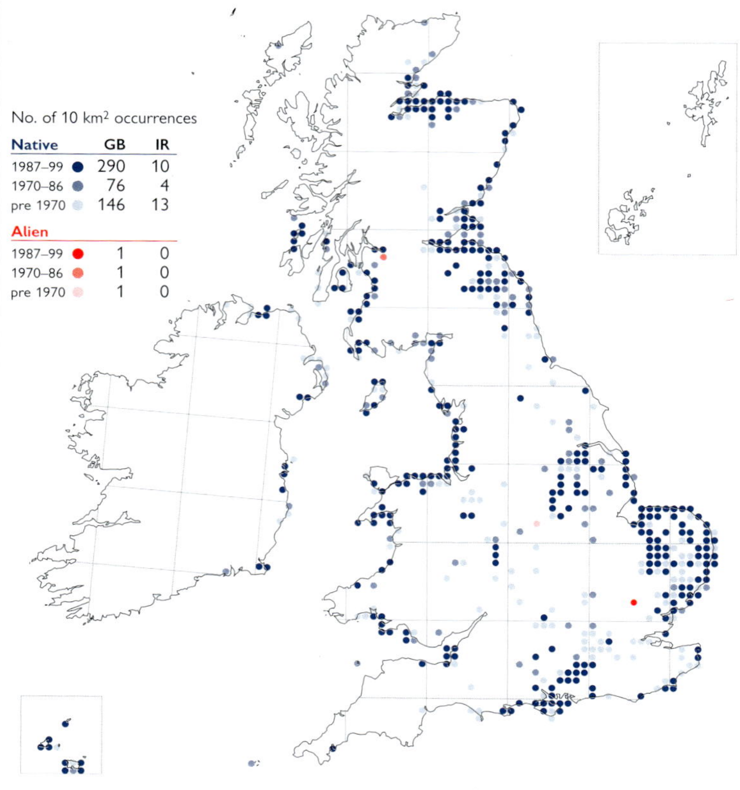

An annual of sand dunes and short, summer-parched grasslands on sandy soils by the coast; also on disturbed ground, old walls, and in dry heathland on sands and gravels inland. Lowland.

Native (change –0.36). *V. lathyroides* is easily overlooked because of its small size and early-flowering, but in some areas may also have been over-recorded for dwarf forms of *V. sativa* subsp. *nigra*, with which it often grows. However, it is very much better recorded than it was for the 1962 *Atlas*.

European Temperate element.

References: Atlas (116a), Curtis & McGough (1988), Hollings & Stace (1978), Hultén & Fries (1986), Meusel *et al.* (1965).

D. A. PEARMAN

Vicia lutea Yellow-vetch

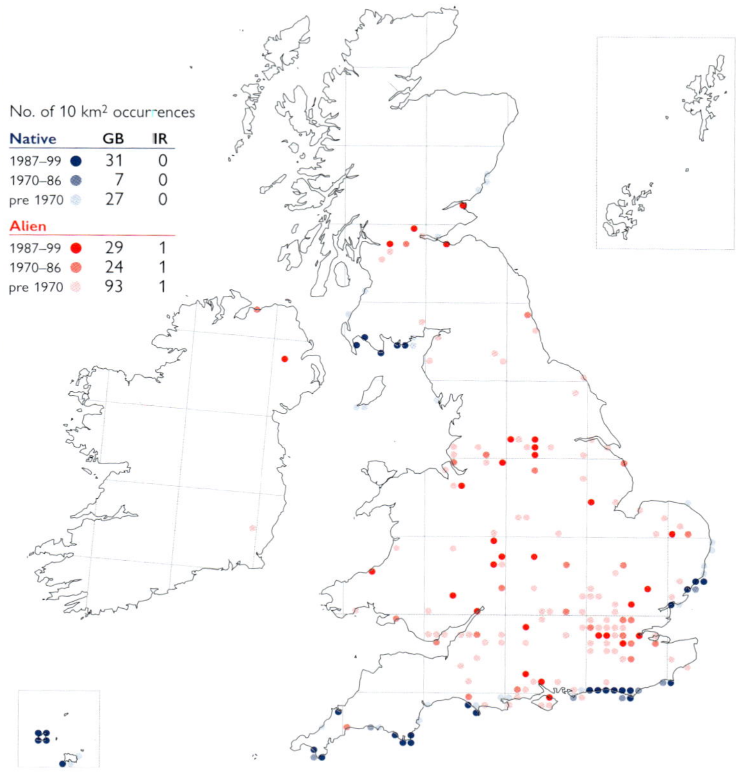

An annual found as a native in a variety of coastal habitats, including scrubby grassland and cliffs, and on open yet consolidated shingle. In S. Scotland it is confined to sheltered sea-cliffs. Inland it is found as a casual, or sometimes in persistent populations, on road-sides, quarries and railway banks. Lowland.

Native (change –0.85). The native distribution of *V. lutea* is probably stable. However, it can be difficult to separate alien and native records on the coast, which may mask any changes in the distribution of native populations. The current map indicates an overall decline in alien occurrences.

Submediterranean-Subatlantic element; widely naturalised outside its native range.

References: Atlas (115c), Bolòs & Vigo (1984), Stewart *et al.* (1994).

D. A. PEARMAN

Vicia bithynica Bithynian Vetch

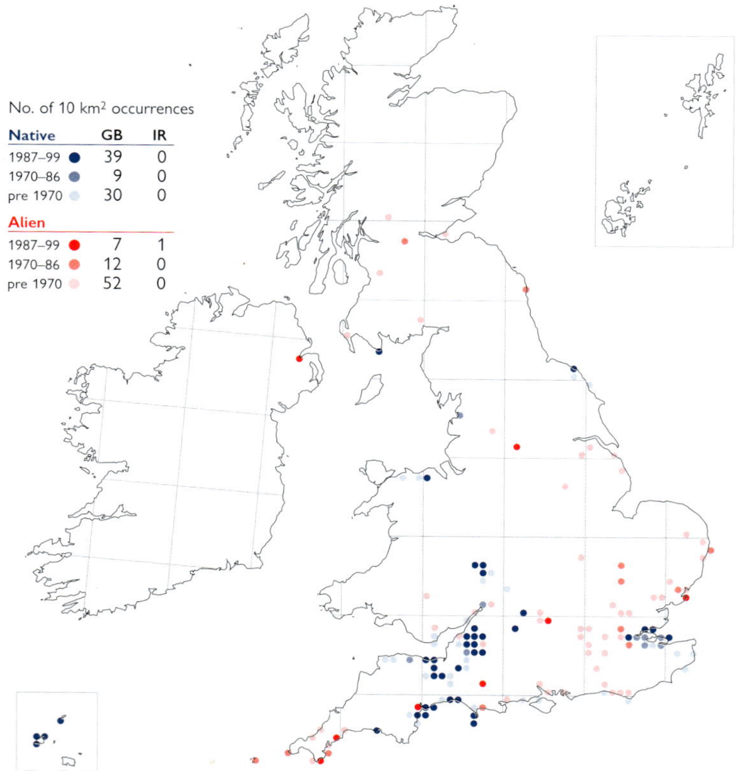

A scrambling annual found in rough grassland on coastal undercliffs, and inland in open hedges, scrubby grassland and on railway banks. At many inland sites it was probably introduced as a contaminant of legume crops, but can be persistent along hedges and tracksides. Lowland.

Native (change –0.52). *V. bithynica* appears to have declined in several of its coastal sites, which are now much more overgrown than previously. The status of some inland populations is uncertain, and it is possible that more of these could be native than are shown on the map.

Mediterranean-Atlantic element.

References: Atlas (116b), Bolòs & Vigo (1984), Stewart *et al.* (1994).

D. A. PEARMAN

Vicia faba Broad Bean

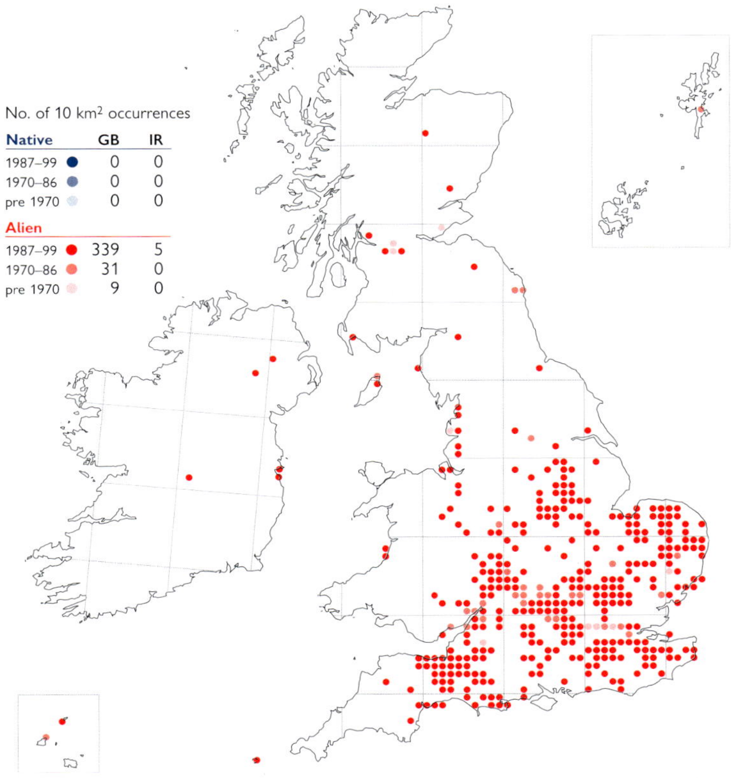

A robust annual occurring on waste ground, set-aside fields and rubbish tips. Most populations are casual. Lowland.

Neophyte. *V. faba*, cultivated in the Middle East for eight thousand years, spread to W. Europe by the second millennium BC. Seeds from the Iron Age have been found in deposits at Glastonbury and it has been grown in British gardens since 1200 (Harvey, 1981). It is widely grown as a vegetable and increasingly as a fodder crop, and is frequently found as an escape from cultivation. Small-seeded variants are introduced with bird-seed.

As *V. faba* is not known as a wild plant, it presumably originated by selection in cultivation.

References: Vaughan & Geissler (1997), Zohary & Hopf (2000).

D. A. PEARMAN

Lathyrus japonicus Sea Pea

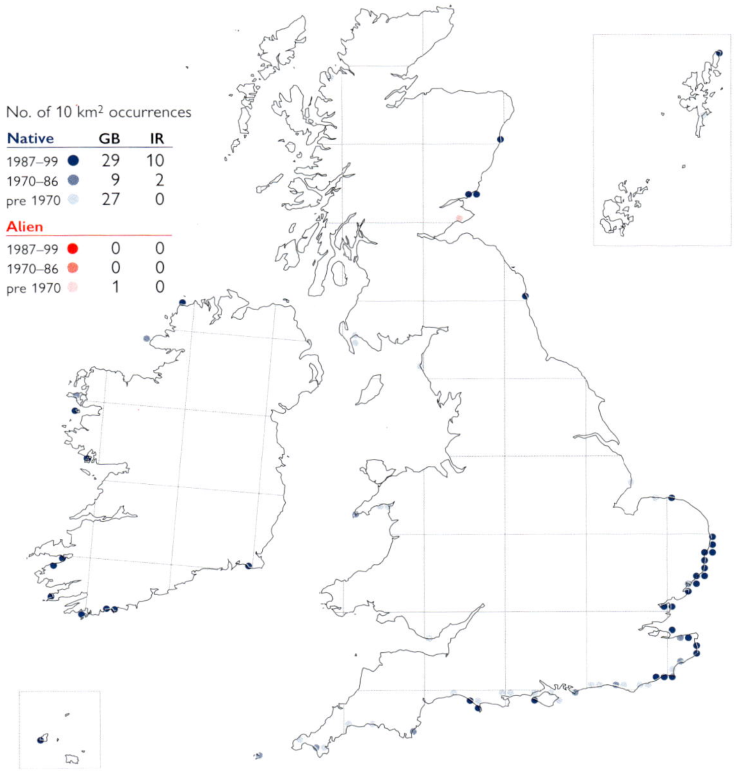

A long-lived perennial herb, forming large and conspicuous patches on shingle beaches, or rarely, in smaller quantities on blown sand. Lowland.

Native (change –0.32). The distribution of *L. japonicus* is stable within its core areas, although some populations in the north and west are transient, perhaps arriving as drift seeds from America, or reappearing after long intervals from buried seed. It has declined at some sites due to trampling. There are many more records from Ireland than in the 1962 *Atlas*.

European Boreo-arctic Montane element; a coastal species also known in E. Asia and N. America.

References: Atlas (118a), Brightmore & White (1963), Curtis & McGough (1988), Hultén & Fries (1986), Meusel *et al.* (1965), Randall (1977), Stewart *et al.* (1994).

D. A. PEARMAN

Lathyrus linifolius Bitter-vetch

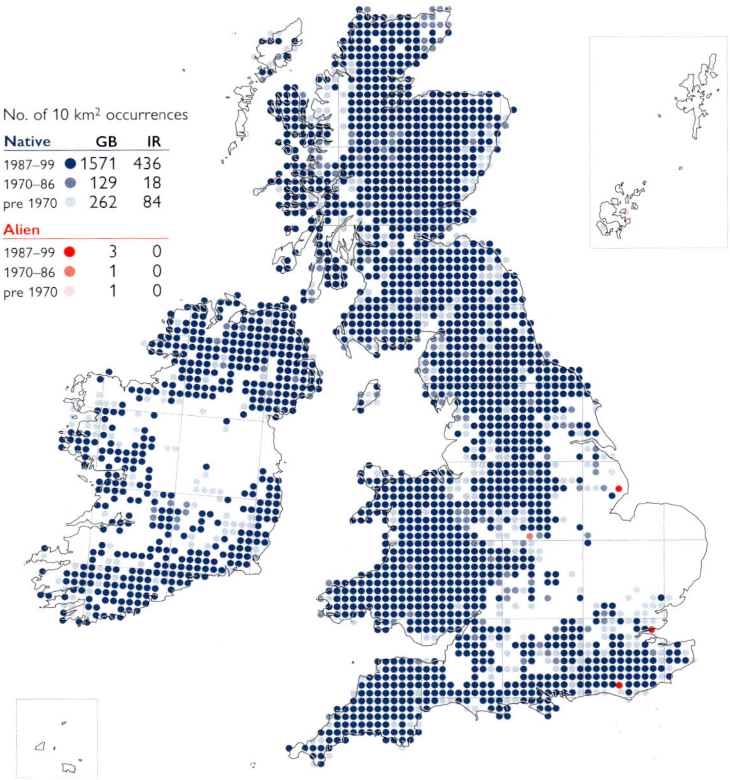

A perennial herb of moist, infertile neutral and acidic soils in heathy meadows, lightly grazed pastures, grassy banks and open woodlands; also on stream banks and rock ledges in the uplands. Generally lowland, but reaching 760 m in Mid Perth.

Native (change –0.93). *L. linifolius* was mapped as 'all records' in the 1962 *Atlas*. Analysis of the database reveals that most losses have occurred since 1950. In lowland habitats it has declined with the loss of meadows and unimproved pasture.

European Temperate element.

References: Atlas (118b), Grime *et al.* (1988), Hultén & Fries (1986), Meusel *et al.* (1965).

D. A. PEARMAN

Lathyrus pratensis Meadow Vetchling

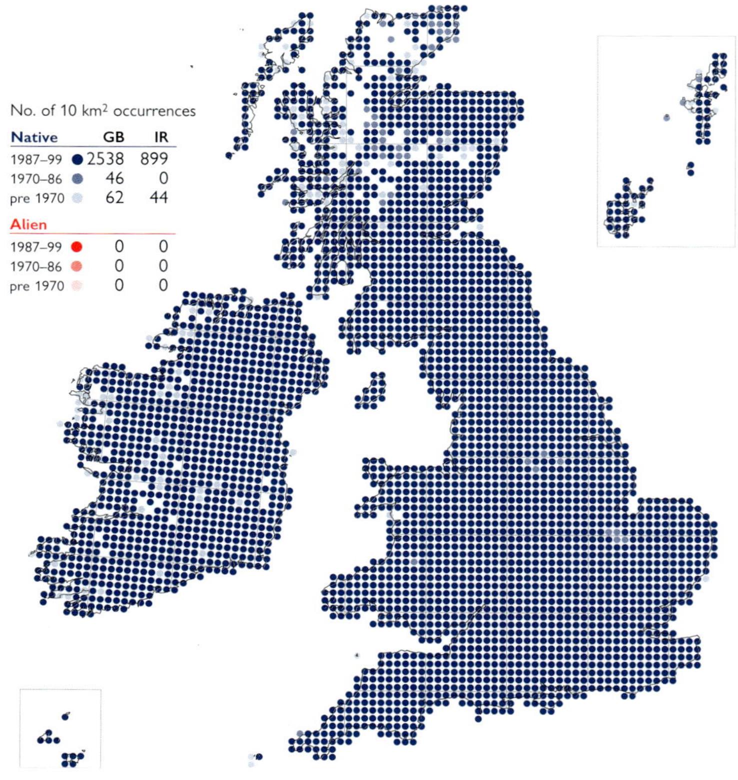

A rhizomatous perennial herb of moderately fertile soils on roadside and railway banks, hedges, unimproved pastures, hay meadows and other grassy habitats. Seed production is not abundant, and vegetative propagation from the rhizome is an important mechanism of reproduction, particularly in disturbed sites. Generally lowland, reaching 450 m in Co. Durham, and with an exceptional record at 845 m on Great Dun Fell (Westmorland).

Native (change –0.17). The distribution of *L. pratensis* appears to be stable.

Eurosiberian Boreo-temperate element; widely naturalised outside its native range.

References: Atlas (117a), Grime *et al.* (1988), Hultén & Fries (1986), Meusel *et al.* (1965).

D. A. PEARMAN

Lathyrus palustris Marsh Pea

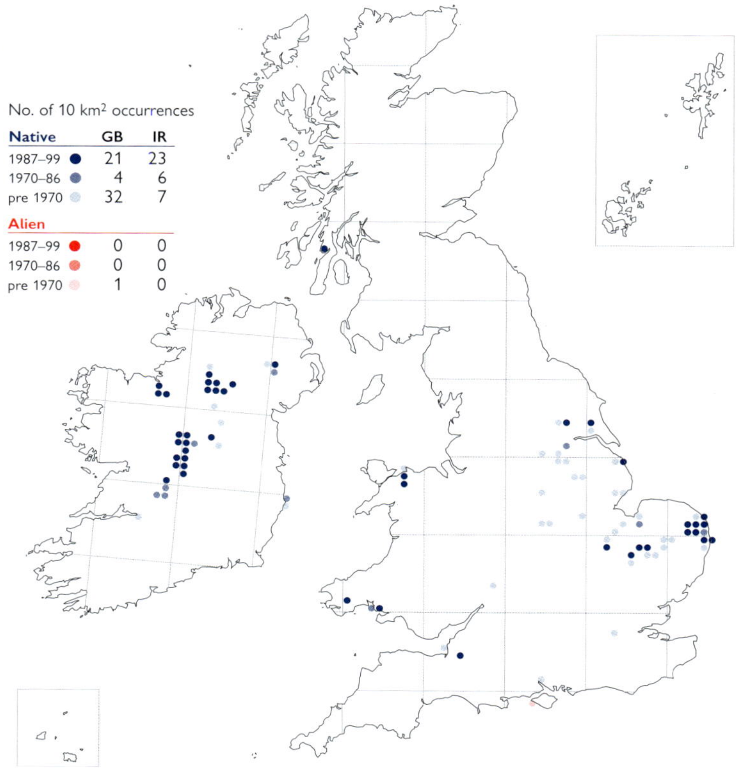

A perennial herb of base-rich fens, reed-beds and fen-meadows; also, rarely, on marshy ground by rivers. Lowland.

Native (change +0.23). *L. palustris* was lost from many sites in E. England before the end of the 19th century, and is still declining there due to drainage and lack of management. New populations have been discovered in Wales and Kintyre since 1970. One of these, which grew at Tywyn Burrows (Carms.) but has since disappeared, was the American var. *pilosus*, which probably originated as a drift seed. *L. palustris* is thought to be increasing in Ireland and is certainly better recorded there than in the 1962 *Atlas*.

Circumpolar Boreo-temperate element.

References: Atlas (117d), Curtis & McGough (1988), Hultén & Fries (1986), Meusel *et al.* (1965), Stewart *et al.* (1994).

D. A. PEARMAN

Lathyrus tuberosus Tuberous Pea

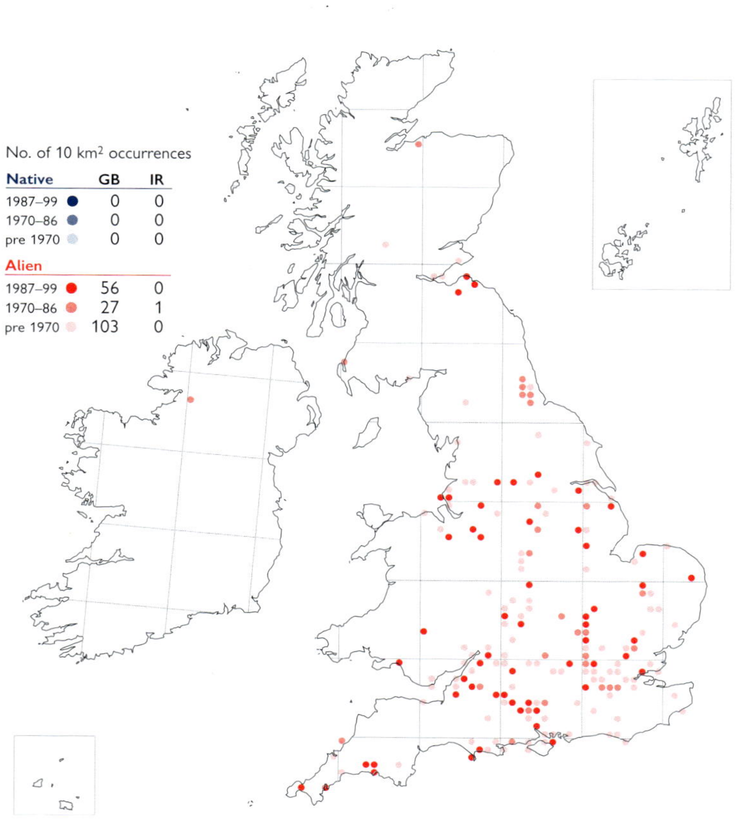

A scrambling perennial herb of hedgerows, rough grassland and waste places; also occasionally found in arable field margins where its tubers allow it to persist. Lowland.

Neophyte (change –0.99). *L. tuberosus* was being cultivated in gardens by 1596, but it is a common cornfield weed in Europe and may have been introduced before this date as a seed contaminant. It was first recorded in the wild in 1708. It is also introduced with chicken food and is probably dispersed by birds.

A Eurosiberian Temperate species, absent as a native from much of W. Europe.

References: Atlas (117b), Hultén & Fries (1986), Meusel *et al.* (1965), Salisbury (1964).

D. A. PEARMAN

Lathyrus grandiflorus
Two-flowered Everlasting-pea

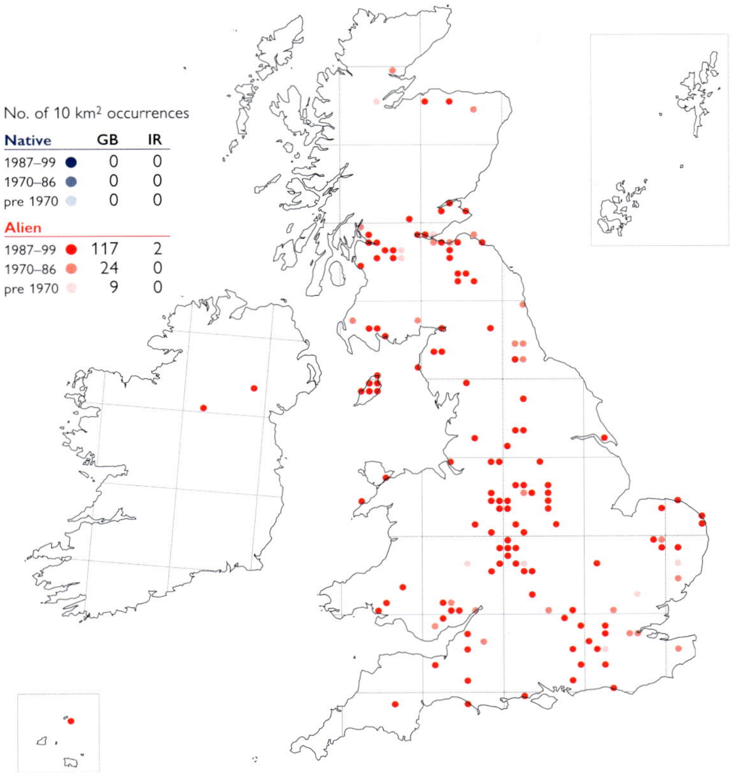

A scrambling perennial herb found as a garden escape or relic of cultivation in hedges and on waste ground and railway banks. It is a persistent species which produces a deep-rooted tuber, but reproduction by seed is rare. Lowland.

Neophyte. Introduced to cultivation in 1814, *L. grandiflorus* makes a magnificent garden plant. It was known from the wild by at least 1908, and is clearly increasing.

Native of Italy, Sicily and the Balkans.

D. A. PEARMAN

Lathyrus sylvestris Narrow-leaved Everlasting-pea

Native	GB	IR
1987–99	251	0
1970–86	52	0
pre 1970	147	0
Alien		
1987–99	23	0
1970–86	6	1
pre 1970	26	0

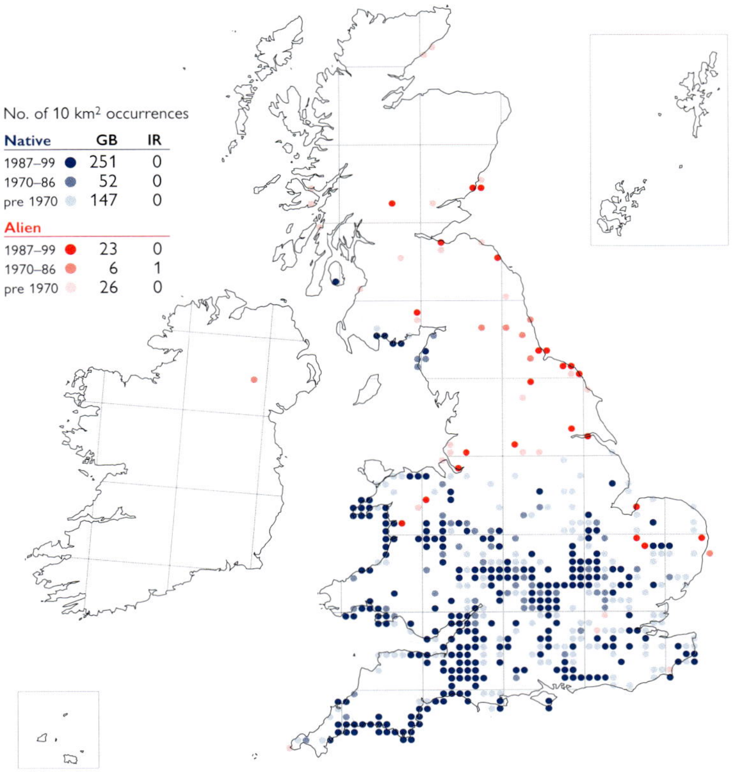

A scrambling perennial herb found in hedges, wood-borders, scrub, and on rough banks and sheltered sea-cliffs. It is sometimes cultivated in gardens and can occur as an escape in habitats such as roadsides and railway banks. Away from the coast it seems to prefer calcareous soils. Lowland.

Native (change –0.36). The distribution of *L. sylvestris* is stable. It is often difficult to know whether inland populations are native or alien and some of those mapped in the English Midlands may only be casual.

European Temperate element; widely naturalised outside its native range.

References: Atlas (117c), Hultén & Fries (1986), Meusel *et al.* (1965).

D. A. PEARMAN

Lathyrus latifolius Broad-leaved Everlasting-pea

Native	GB	IR
1987–99	0	0
1970–86	0	0
pre 1970	0	0
Alien		
1987–99	621	8
1970–86	88	1
pre 1970	64	0

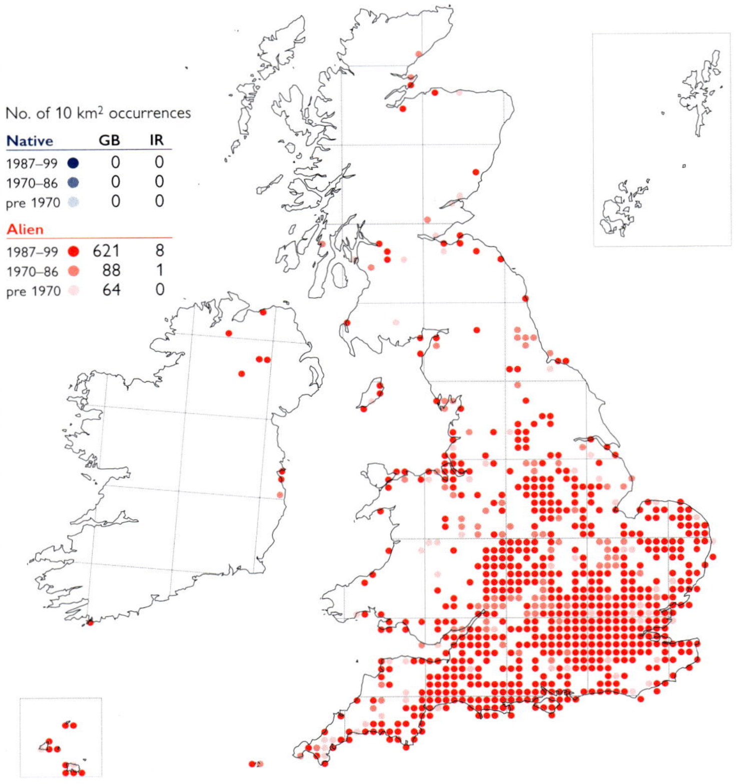

An attractive perennial herb, which has escaped from gardens and is now well-naturalised, scrambling on rough banks, sea-cliffs, railway banks and waste ground. It grows in similar habitats to *L. sylvestris* and is sometimes confused with it. Generally lowland, but reaching 340 m at Sparklow (Derbys.).

Neophyte. *L. latifolius* has been a garden plant since the 15th century, and, once established, is very persistent. It was first recorded in the wild in 1670, and appears to be spreading.

A European Southern-temperate species, native north to approximately 50°N.

Reference: Hultén & Fries (1986).

D. A. PEARMAN

Lathyrus hirsutus Hairy Vetchling

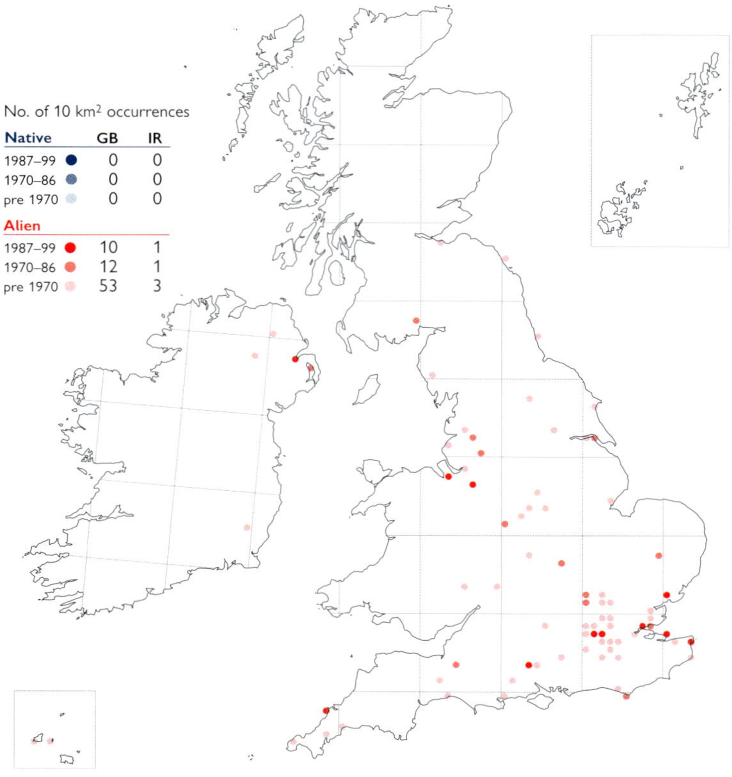

No. of 10 km² occurrences

Native	GB	IR
1987–99	0	0
1970–86	0	0
pre 1970	0	0
Alien		
1987–99	10	1
1970–86	12	1
pre 1970	53	3

A scrambling annual, found on grassy banks and waste ground as a garden escape or bird-seed or grain contaminant. Populations can persist, as in a few areas around London, but are usually casual. Lowland.

Neophyte. *L. hirsutus* was once considered to be native in S. Essex, where it was discovered at Hadleigh Castle in 1666 and survived until at least 1958, when the management of its grassland habitat changed. It has been introduced into some Essex nature reserves. Elsewhere it was not infrequent as a casual, but is now much rarer.

A European Southern-temperate species.

References: Bolòs & Vigo (1984), Jermyn (1974).

D. A. PEARMAN

Lathyrus nissolia Grass Vetchling

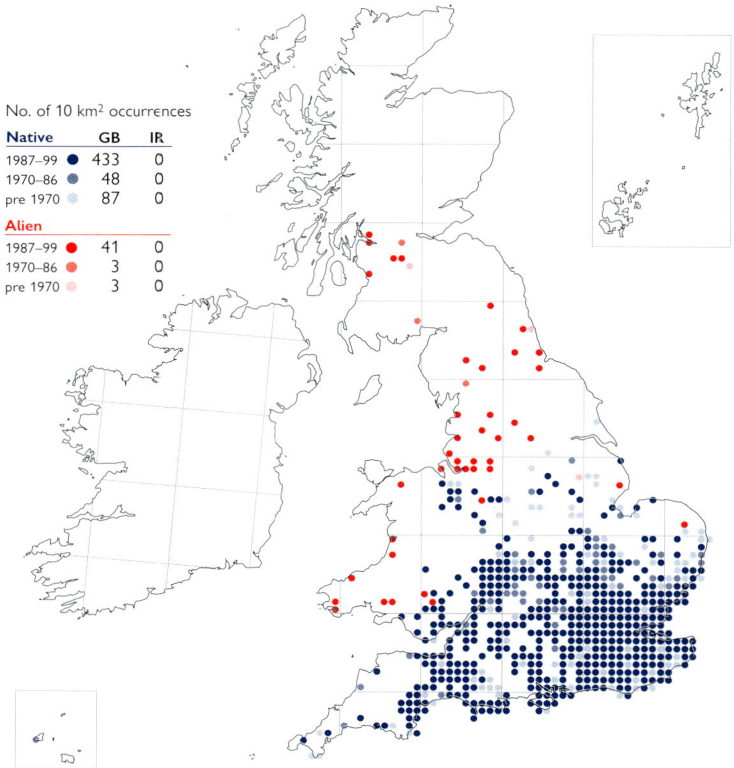

No. of 10 km² occurrences

Native	GB	IR
1987–99	433	0
1970–86	48	0
pre 1970	87	0
Alien		
1987–99	41	0
1970–86	3	0
pre 1970	3	0

An inconspicuous annual of open, often disturbed, habitats on chalk and heavy calcareous clay soils. It is found on grassy banks, verges, railway banks, woodland rides, and coastal grassland and shingle. Lowland.

Native (change +0.54). There has been a considerable increase in 10-km square records since the 1962 *Atlas*, particularly in C.S. and S.E. England. Several recent county floras suggest that it is spreading in ruderal situations, and there are also records from restored open-cast mine sites where it may have been introduced as a constituent of commercial seed mixtures. It is unclear whether recent records in N.W. England are introductions or represent a northerly expansion of its native range.

European Temperate element.

References: Atlas (116d), Cannon (1964), Meusel *et al.* (1965).

D. A. PEARMAN

Lathyrus aphaca Yellow Vetchling

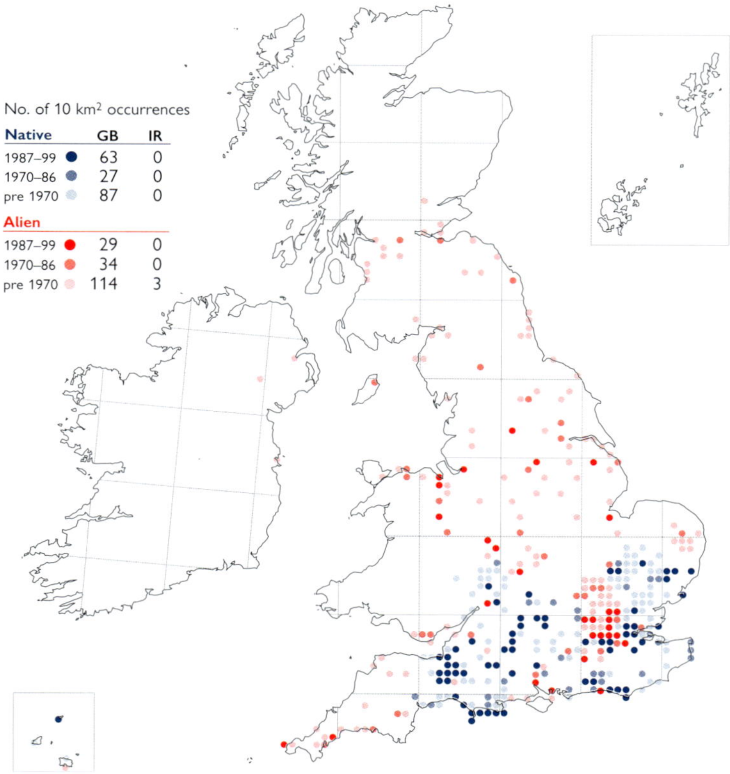

No. of 10 km² occurrences

Native	GB	IR
1987–99	63	0
1970–86	27	0
pre 1970	87	0
Alien		
1987–99	29	0
1970–86	34	0
pre 1970	114	3

The only persistent populations of this annual are in open grassy habitats on chalk, limestone and calcareous clay soils, especially near the coast. The species is possibly native in such habitats, but it also occurs as a casual in waste places, and as an arable weed where it may have been introduced as a contaminant of legume crops. Lowland.

Native or alien (change −1.38). *L. aphaca* was first recorded in 1632. It can be difficult to distinguish between possibly native and introduced populations. Its distribution is stable in its persistent sites; many losses are referable to casual records.

Submediterranean–Subatlantic element; also in C. Asia and widely naturalised outside its native range.

References: Atlas (116c), Meusel *et al.* (1965), Stewart *et al.* (1994).

D. A. PEARMAN

Pisum sativum Garden Pea

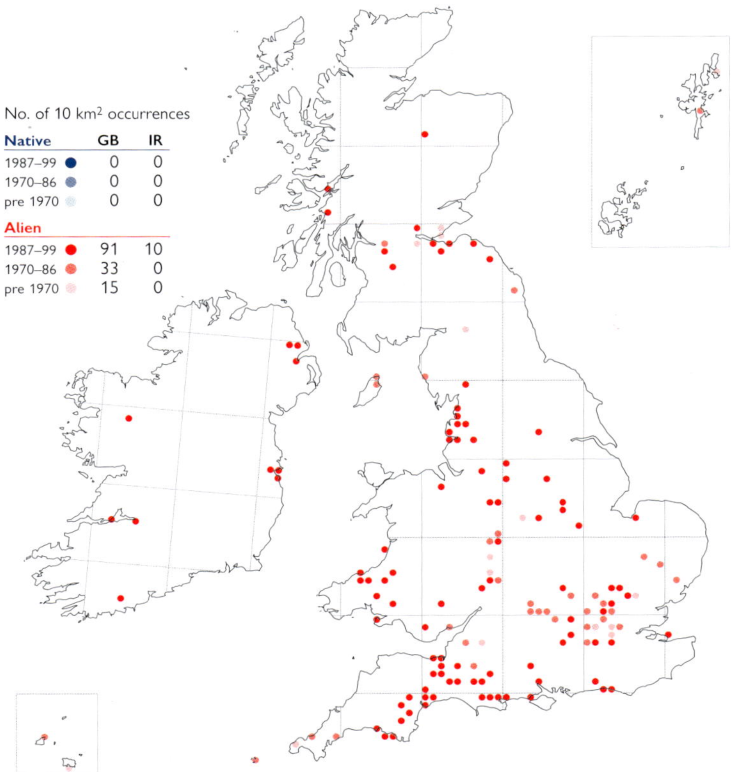

A climbing or sprawling annual, found as a casual on field margins, on tips and in waste places. Lowland.

Casual. *P. sativum* was a very early crop in the Middle East, dating from *c.* 7000 BC. It is not known when it reached Britain, but it was certainly present two thousand years ago and has been cultivated in British gardens since at least 1200 (Harvey, 1981). Commercial scale cropping is now widespread, as well as cultivation as a domestic vegetable, and it has been recorded in the wild since at least 1888.

Apparently native in the Mediterranean region and S.W. Asia, but its native range is obscured by its spread in cultivation.

References: Bolòs & Vigo (1984), Vaughan & Geissler (1997), Zohary & Hopf (2000).

D. A. PEARMAN

Ononis reclinata Small Restharrow

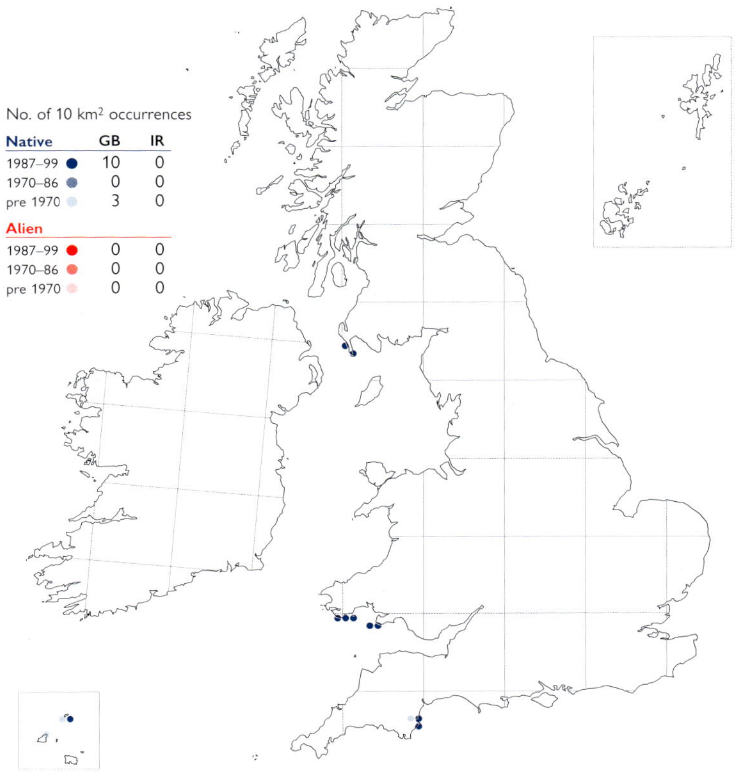

An annual of thin, dry, calcareous soils with a low organic content, especially on S.- or S.W.-facing coastal cliffs of limestone and, in Scotland, greywacke. In the Channel Islands it occurs in consolidated dune turf. Populations often fluctuate in size, sometimes very markedly. Lowland.

Native (change +0.27). The overall distribution of *O. reclinata* has not changed significantly since the 1962 *Atlas*. New sites have, however, been discovered near to known colonies, but the species was lost in Guernsey in 1956 through the growth of rank vegetation.

Mediterranean–Atlantic element.

References: Atlas (102a), Meusel *et al.* (1965), Wigginton (1999).

D. A. PEARMAN

Ononis spinosa Spiny Restharrow

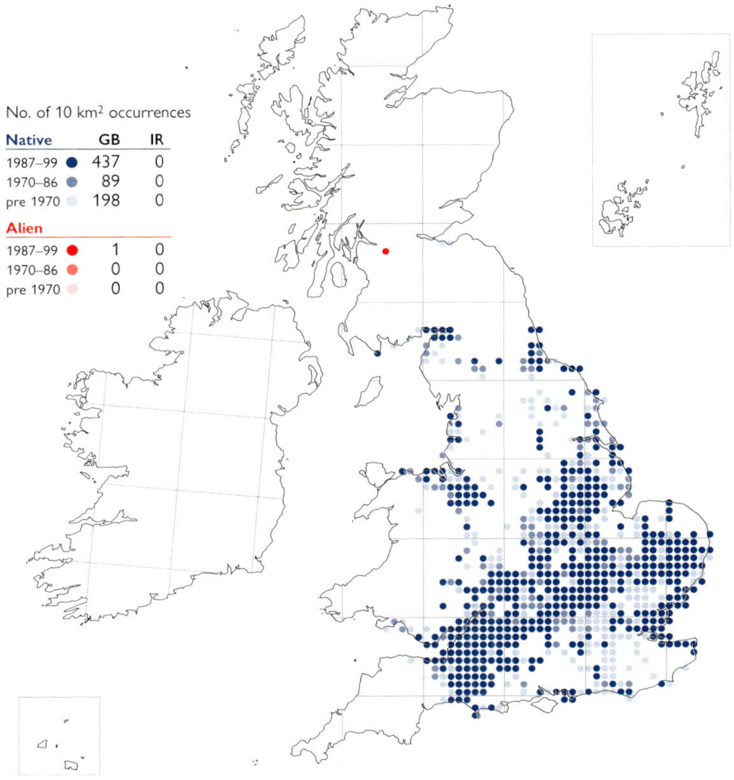

A woody perennial of infertile calcareous grasslands on chalk, limestone and heavy calcareous clay soils. It also occurs on the coast in grazing marshes and on earthen sea-walls. Lowland.

Native (change –0.82). *O. spinosa* is probably declining, especially in S.E. England, because of agricultural improvements. It is over-recorded for spiny forms of *O. repens* (var. *horrida*); some old records of *O. spinosa* in the 1962 *Atlas* are now regarded as dubious.

Eurosiberian Southern-temperate element.

References: Atlas (101d), Hultén & Fries (1986), Ivimey-Cook (1969), Meusel *et al.* (1965).

D. A. PEARMAN

Ononis repens Common Restharrow

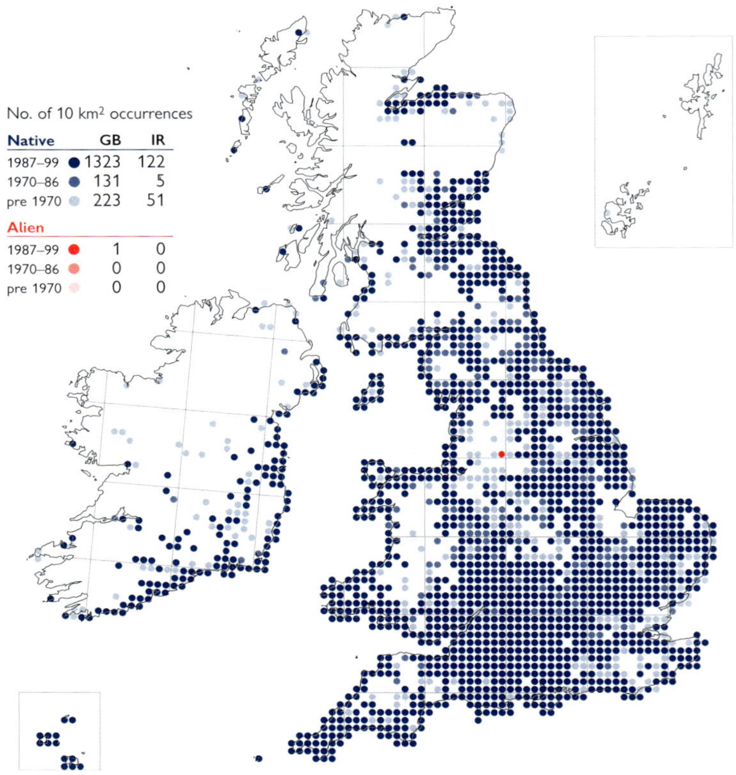

A rhizomatous perennial sub-shrub, widespread in grasslands on base-rich, well-drained, light soils, and also on calcareous boulder-clays. On the coast it occurs on sand dunes and shingle, and inland it is frequently a colonist of sandy or gravelly road verges. 0–365 m (Weardale, Co. Durham).

Native (change –0.45). There has been no appreciable change in the overall distribution of *O. repens* since the 1962 *Atlas*, where it was mapped as 'all records', although it does appear to have declined in parts of C. England and N.E. Scotland. Small, densely hairy plants from maritime sands in Devon have been referred to subsp. *maritima*, but their taxonomic status needs clarification.

European Temperate element.

References: Atlas (101c), Hultén & Fries (1986), Ivimey-Cook (1969), Meusel *et al.* (1965).

D. A. PEARMAN

Melilotus altissimus Tall Melilot

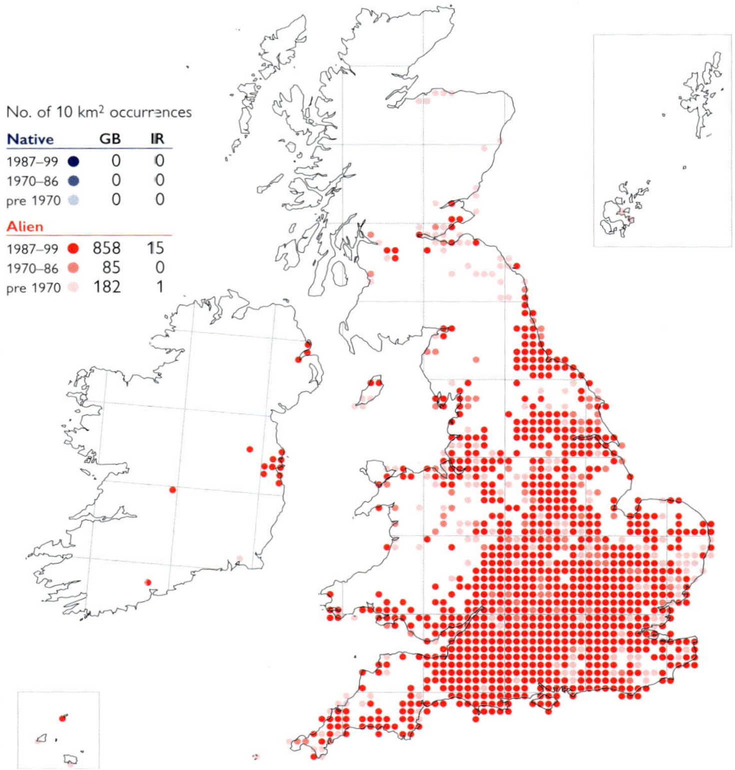

A biennial or short-lived perennial herb occurring in disturbed grassland and on roadsides, field-borders and waste places. Lowland.

Archaeophyte (change +0.73). *M. altissimus* has been known in Britain since the 16th century, and seems much more rural in its distribution than the other species of *Melilotus*. Its distribution is probably stable, but nomenclatural confusion in the past with *M. officinalis* may have obscured earlier trends.

A European Temperate species, with a native range now completely obscured by introductions.

References: Atlas (103d), Hultén & Fries (1986).

D. A. PEARMAN

Melilotus albus White Melilot

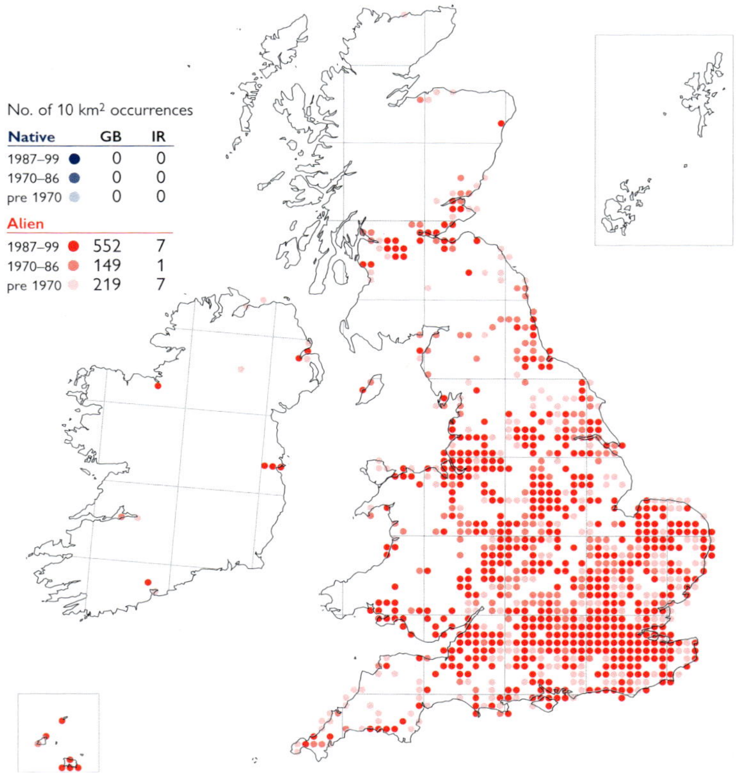

An annual or biennial herb principally occurring as a casual on waste or disturbed ground, in railway sidings and by roadsides. Generally lowland, but reaching 335 m near Brassington (Derbys.).

Neophyte (change –0.20). *M. albus*, first recorded in the wild in Britain in 1822, is occasionally grown as green fodder, but it also arrives as a contaminant of lucerne, and as a wool and bird-seed alien. The distribution is probably stable but it is difficult to generalise about trends as some populations are short-lived; it may be increasing in some areas.

A native of Eurasia (perhaps S. Europe, W. & C. Asia) which has now spread to temperate and warm-temperate regions almost worldwide.

References: Atlas (104b), Hultén & Fries (1986), Turkington *et al.* (1978).

D. A. PEARMAN

Melilotus officinalis Ribbed Melilot

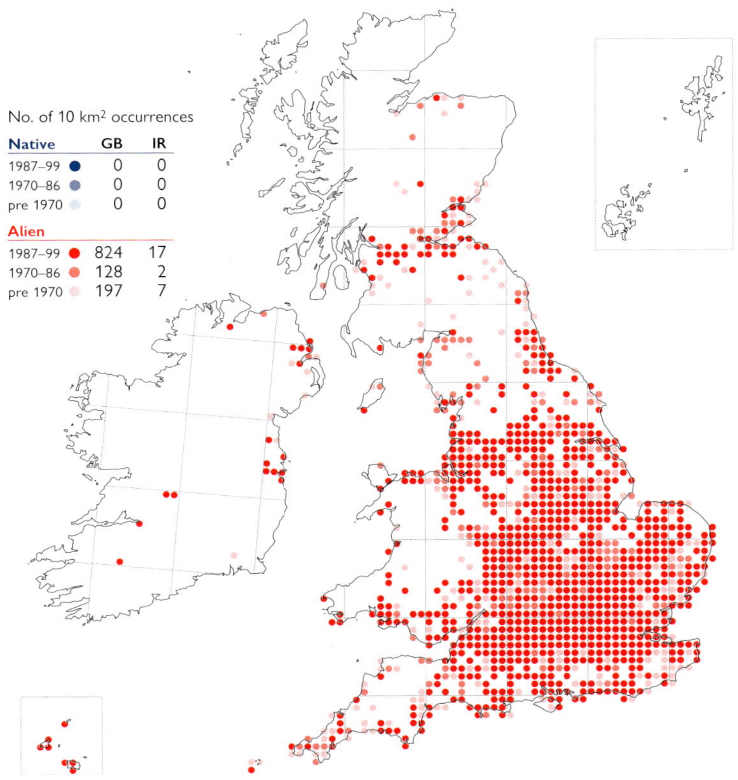

A biennial herb, widely naturalised on roadsides, field-borders and grassy banks, but also occurring as a casual on rubbish tips and waste ground. Lowland.

Neophyte (change +0.02). *M. officinalis* came here as seeds introduced with clover from America, first being recorded in 1835. It is now better recorded than it was in the 1962 *Atlas*, but its overall distribution is stable.

Probably native from C. & S. Europe eastwards to C. Asia, but this distribution is obscured by alien occurrences; a widespread alien in N. America.

References: Atlas (104a), Hultén & Fries (1986), Meusel *et al.* (1965), Turkington *et al.* (1978).

D. A. PEARMAN

Melilotus indicus Small Melilot

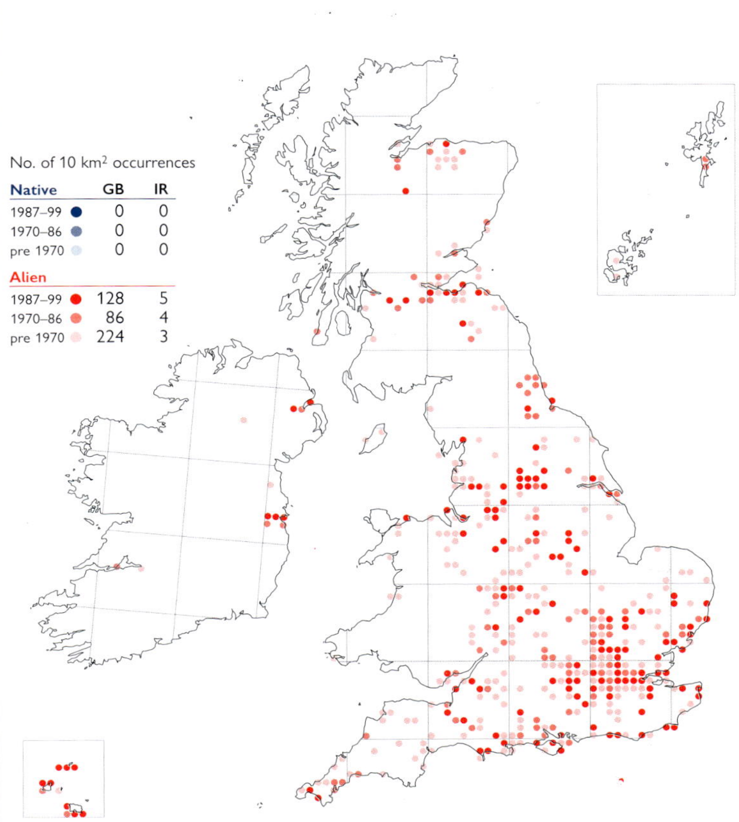

An annual usually occurring as a casual on waste ground and rubbish tips, and sometimes on cultivated ground. It is principally an alien of wool shoddy and bird-seed. Lowland.

Neophyte (change –1.59). *M. indicus* originally arrived in this country with S. American clover seed and was the latest of the four *Melilotus* species mapped to be recorded in the wild, in 1852. (It had apparently been cultivated in Britain in 1680 but did not spread from this source.) There has been a considerable decline in its distribution since the 1962 *Atlas*. Unlike the other *Melilotus* species, it does not seem to have spread far from rubbish tips.

Native of S. Europe, N. Africa, S.W. & C. Asia; widely naturalised further north in Europe and elsewhere.

References: Atlas (104c), Bolòs & Vigo (1984).

D. A. PEARMAN

Medicago lupulina Black Medick

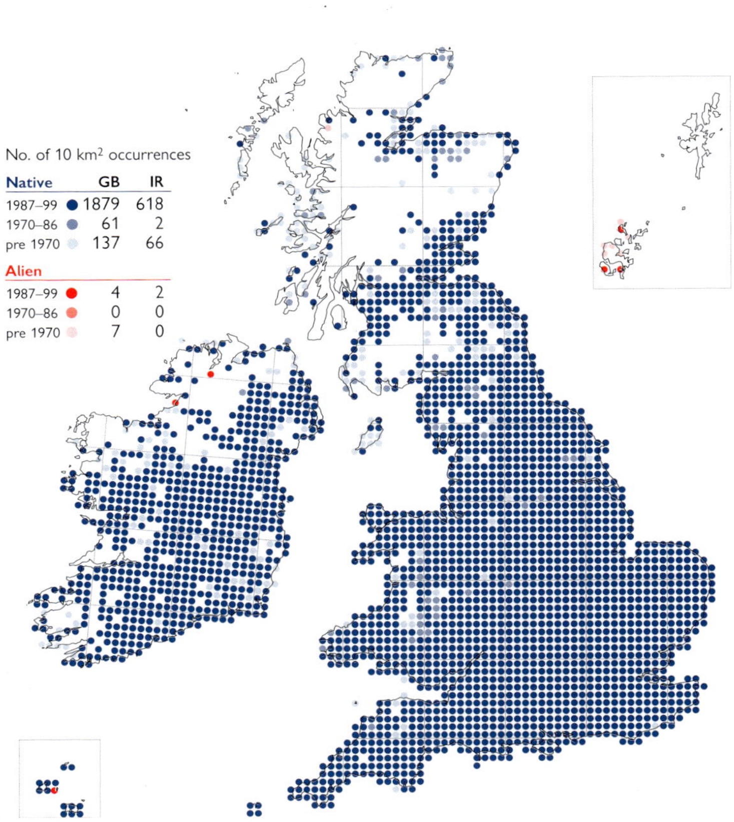

An annual or short-lived perennial herb of dry grassland and disturbed places on relatively infertile neutral or calcareous soils; often in sunny S.-facing pastures and on roadside banks, waste places and walls. Its has a long-lived seed bank. Generally lowland, but reaching 440 m at Nenthead (Cumberland).

Native (change –0.43). The overall distribution of *M. lupulina* is stable. It declined during the 20th century in some areas of W. Scotland (Pearman & Preston, 2000) but some pre-1970 records in Scotland may result from confusion with *Trifolium dubium*.

Eurosiberian Temperate element, but widely naturalised so that distribution is now Circumpolar Temperate.

References: Atlas (102d), Grime *et al.* (1988), Hultén & Fries (1986), Meusel *et al.* (1965), Turkington & Cavers (1979).

D. A. PEARMAN

Medicago sativa subsp. *falcata* Sickle Medick

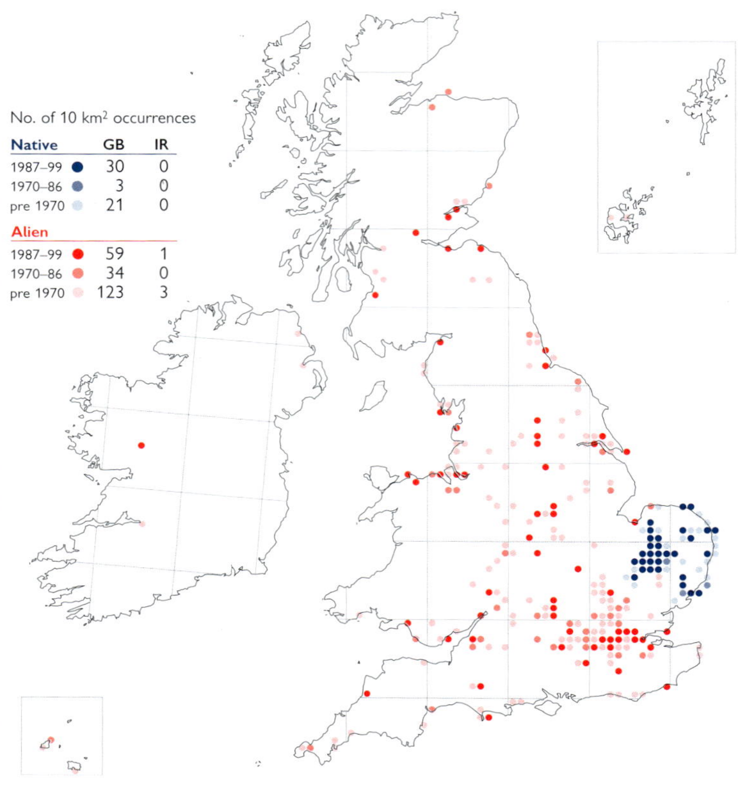

A perennial herb of grassy heaths, sea-walls, roadsides and tracks, chiefly found on calcareous soils and sands. It is often confined to the rear of roadside verges as it is sensitive to mowing. Lowland.

Native (change for species –0.56). This subspecies has declined on the coast and adjoining heaths through habitat loss. It is intolerant of livestock and rabbit grazing (Trist, 1979), and it hybridises with *M. sativa* subsp. *sativa*. Some of the casual records are probably variants of subsp. *varia* which approach subsp. *falcata*.

Eurosiberian Boreo-temperate element.

References: Atlas (102b), Hultén & Fries (1986), Stewart *et al.* (1994).

D. A. PEARMAN

Medicago sativa subsp. *varia* Sand Lucerne

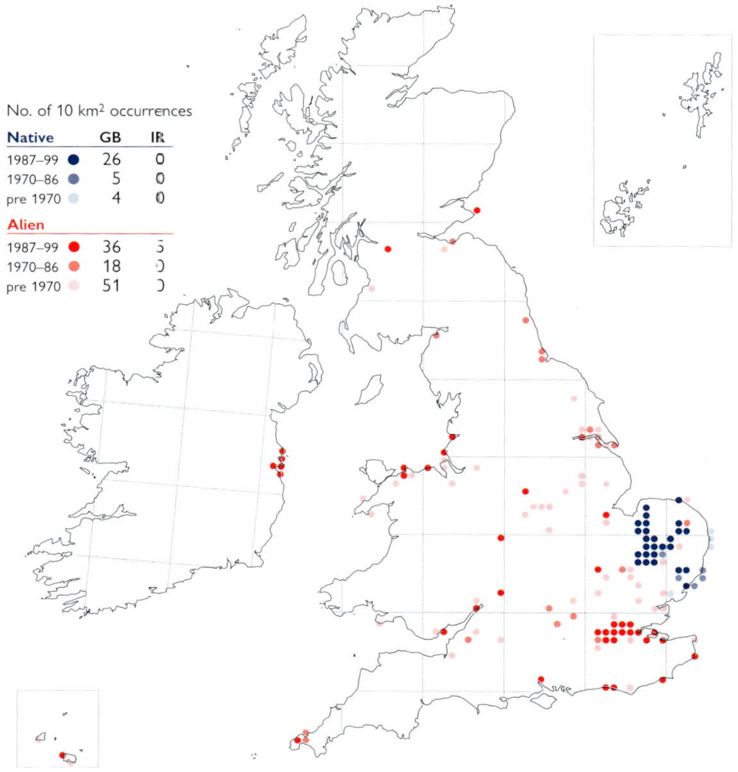

An annual or perennial herb, arising as a fertile hybrid between *M. sativa* subsp. *sativa* and *M. sativa* subsp. *falcata*, which is now established in many grassy and sandy places. Hybrid seed from Germany, called 'Grimm' Lucerne, is also grown as a fodder crop. Lowland.

A spontaneous hybrid between native and alien parents (change for species –0.56). The core distribution of this subspecies mirrors that of its native parent, *M. sativa* subsp. *falcata*, and is stable. Elsewhere it can be introduced independently of either parent, and backcrosses are occasionally recorded. It was recorded in Suffolk by 1804.

Widespread in Europe and in temperate regions elsewhere.

References: Atlas Supp. (20a), Beckett *et al.* (1999), Burton (1983).

D. A. PEARMAN

Medicago sativa subsp. *sativa* Lucerne

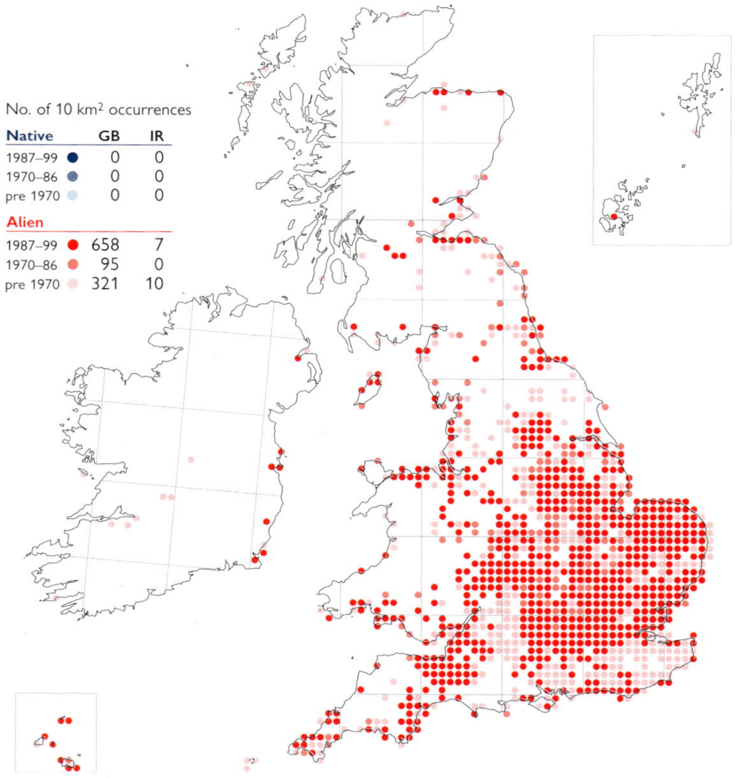

A deep-rooted perennial herb naturalised on field margins, tracksides, rough grassland and waste ground. Lowland.

Neophyte (change for species –0.56). *M. sativa* subsp. *sativa* was first cultivated as a fodder and green manure crop in Britain in the 17th century and was recorded from the wild by 1804. It was more frequently grown in a period of world protein shortage in the 1950s, especially on dry sandy soils, than it is now. Some relict populations, however, are very persistent and long-lived.

A cultivated plant of obscure origin, perhaps native to S.W. Asia but widely naturalised in both N. & S. hemispheres. A wide range of forms has been developed for different climates.

References: Atlas (102c), Smart & Simmonds (1995), Thirsk (1997), Trist (1971).

D. A. PEARMAN

Medicago minima Bur Medick

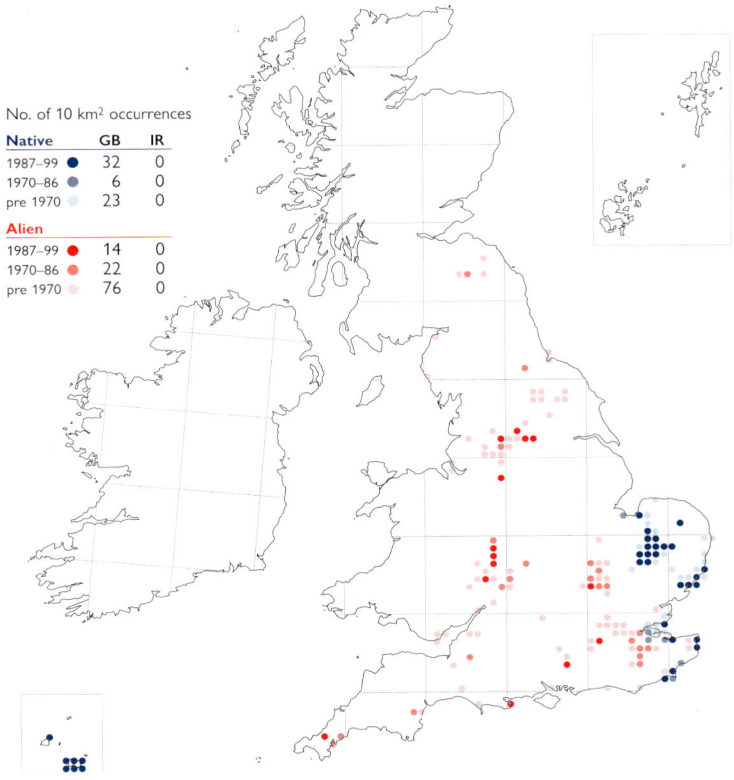

A winter-annual of dry, open, sandy or gravelly places; also occasionally found as a casual, introduced with wool shoddy. Lowland.

Native (change –1.97). *M. minima* is now less frequent on the coast due to habitat destruction and lack of grazing, allowing sites to become too rank for this low-growing species. It is now only rarely recorded as a casual.

Eurosiberian Southern-temperate element; widely naturalised outside its native range.

References: Atlas (103a), Hultén & Fries (1986), Meusel *et al.* (1965), Stewart *et al.* (1994).

D. A. PEARMAN

Medicago polymorpha Toothed Medick

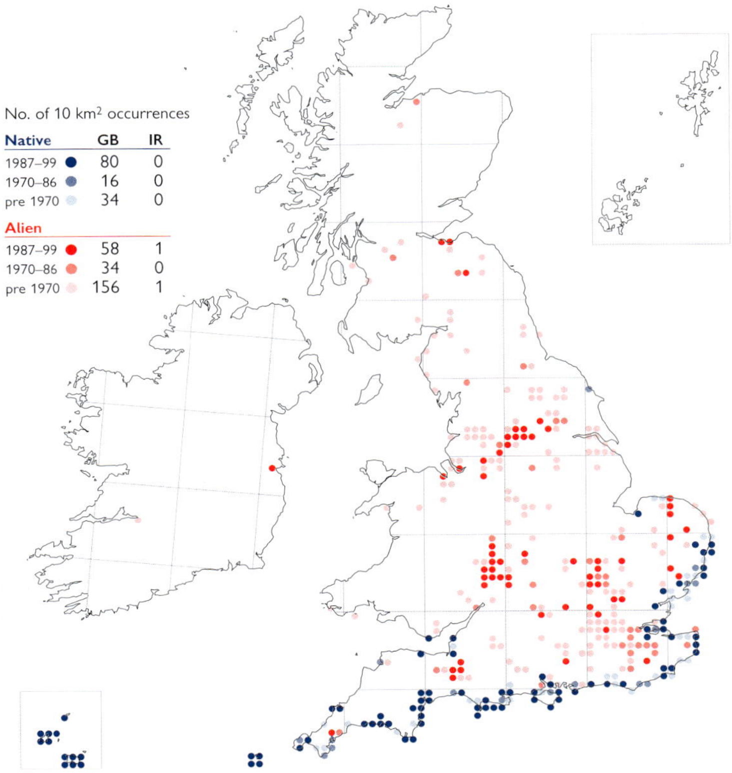

An annual found in open sandy and gravelly habitats by the coast. It occurs in short, open grassland on summer-parched banks and cliffs with other annuals, particularly in S.W. England. Inland, like other *Medicago* species, it occurs as a casual, especially with wool shoddy. Lowland.

Native (change –1.34). *M. polymorpha* has declined at its coastal stations through scrub encroachment and lack of grazing, although it may be overlooked or confused with *M. arabica*. It is much scarcer nowadays as a casual inland.

Submediterranean-Subatlantic element.

References: Atlas (103b), Bolòs & Vigo (1984), Stewart *et al.* (1994).

D. A. PEARMAN

Medicago arabica Spotted Medick

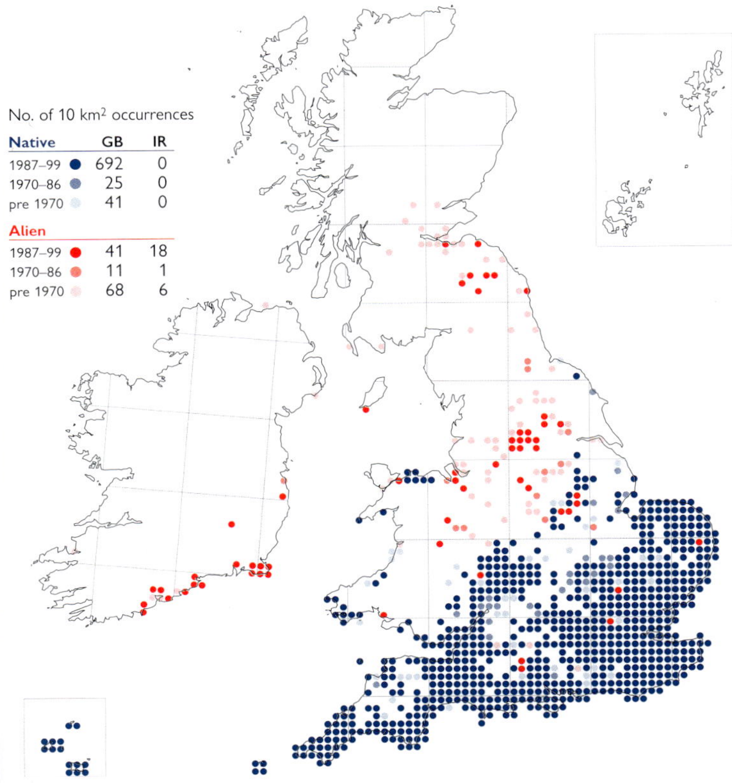

A winter-annual of grassy places, often on light, sandy and gravelly soils, particularly near the coast. It grows as a weed in lawns and frequently occurred as a casual in fields manured with wool shoddy. It is able to persist as small plants in mown lawns, and can grow as very robust plants in nutrient-rich habitats. Lowland.

Native (change +0.69). There has been a marked increase in *M. arabica* since the 1962 *Atlas*, particularly in S. England and the Midlands. Recent county floras suggest it has become more common in inland areas, for reasons that are not clear. For example, it was very rare in Oxfordshire before 1930, but is now widespread.

Submediterranean-Subatlantic element; widely naturalised outside its native range.

References: Atlas (103c), Bolòs & Vigo (1984), Killick *et al.* (1998).

D. A. PEARMAN

Trifolium ornithopodioides Bird's-foot Clover

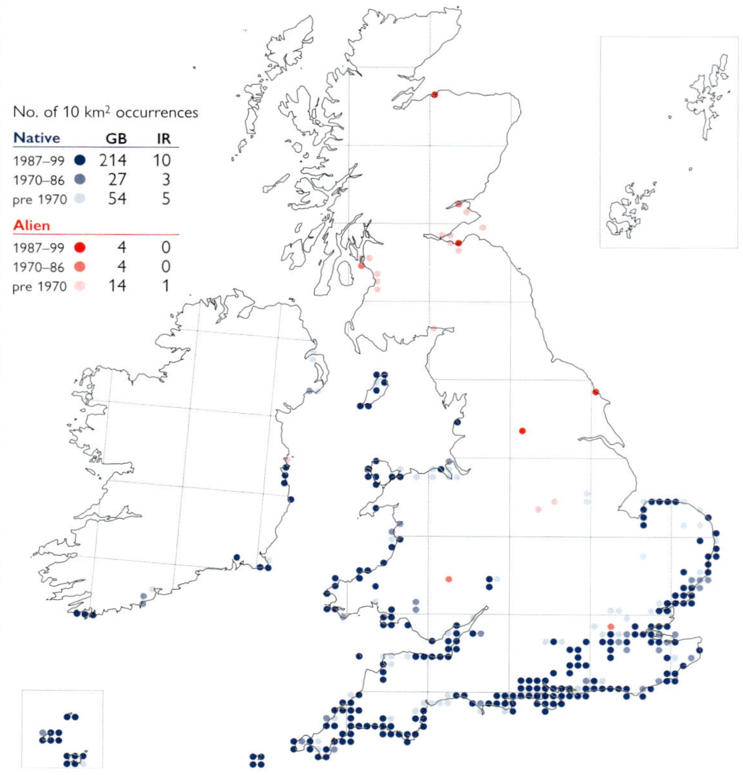

A winter-annual of acidic sands, gravels and compacted shingle, occurring on bare ground in disturbed, often much trampled places like car parks, tracks and paths, and occasionally in lawns and on heavily grazed commons. It prefers sites that are moist in winter and parched in summer. Lowland.

Native (change +0.42). This inconspicuous species has been under-recorded in the past and it is now much better recorded than in the 1962 *Atlas*. Inland, it may be declining at some sites due to lack of management or building developments.

Suboceanic Southern-temperate element.

References: Atlas (104d), Bolòs & Vigo (1984), Stewart *et al.* (1994).

D. A. PEARMAN

Trifolium repens White Clover

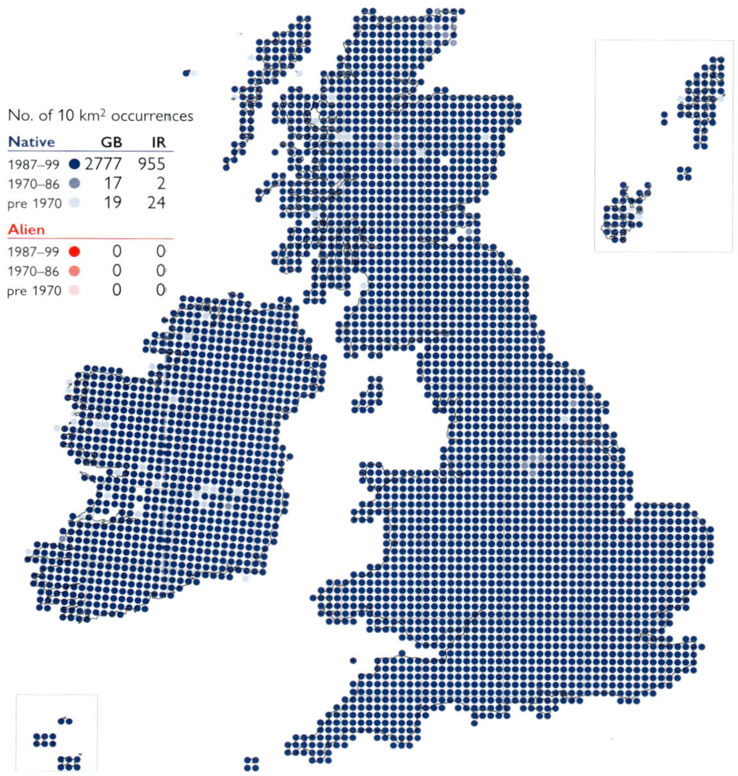

A stoloniferous perennial herb, occurring in grasslands on all but the wettest or most acidic soils; also on waste ground and in other ruderal habitats. It is very tolerant of grazing, mowing and trampling and is often scarce or absent in taller grassland. It is very widely sown as a component of short and medium term leys, and on roadsides, and many commercial cultivars are available. Generally below 400 m, but reaching 880 m in the Breadalbanes (Mid Perth).

Native (change −1.31). There has been no change in the range of this ubiquitous species.

Eurosiberian Boreo-temperate element, but widely naturalised so that distribution is now Circumpolar Boreo-temperate.

References: Atlas (108c), Burdon (1983), Davis (1992), Grime *et al.* (1988), Hultén & Fries (1986), Turkington & Burdon (1983).

D. A. PEARMAN

Trifolium occidentale Western Clover

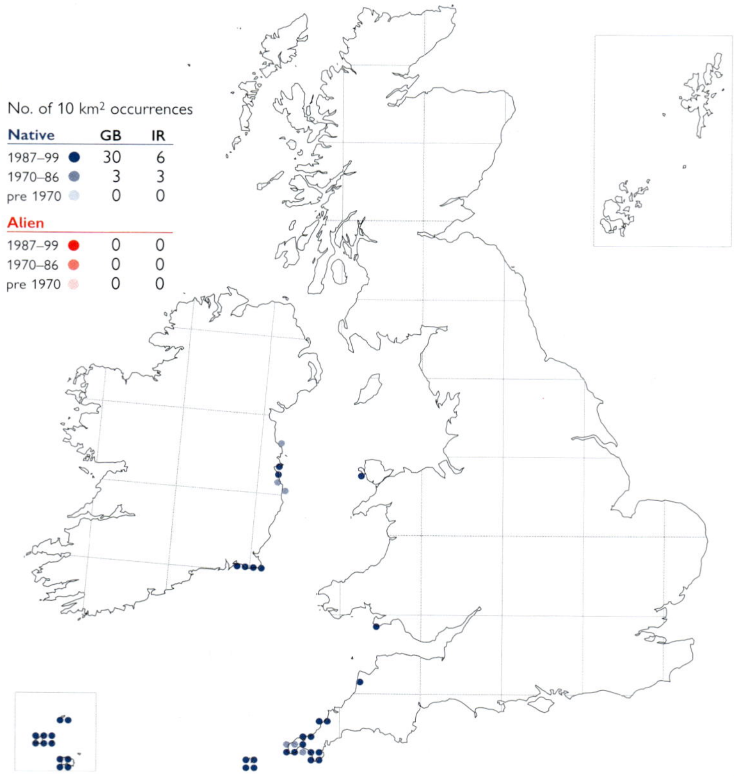

A stoloniferous perennial herb of dry, species-rich coastal grasslands, often growing on cliff-slopes around rock outcrops, or on stabilised sand. It is largely restricted to exposed sites liable to drenching by salt-laden winds, and rarely occurs more than 100 m from the sea. Lowland.

Native. *T. occidentale* was discovered in 1957, and described as a new species from Cornwall and the Channel Islands in 1961; populations there appear to be stable. It was found in Ireland in 1979 and in Wales in 1987, and may have been overlooked elsewhere. The Cornish populations suffer heavy mortality in hot summers but subsequently recover from seed.

Oceanic Temperate element; restricted to the coast of W. Europe.

References: Atlas Supp. (20b), Blackstock & Roberts (1998), Coombe (1961), Stewart *et al.* (1994).

D. A. PEARMAN

Trifolium hybridum Alsike Clover

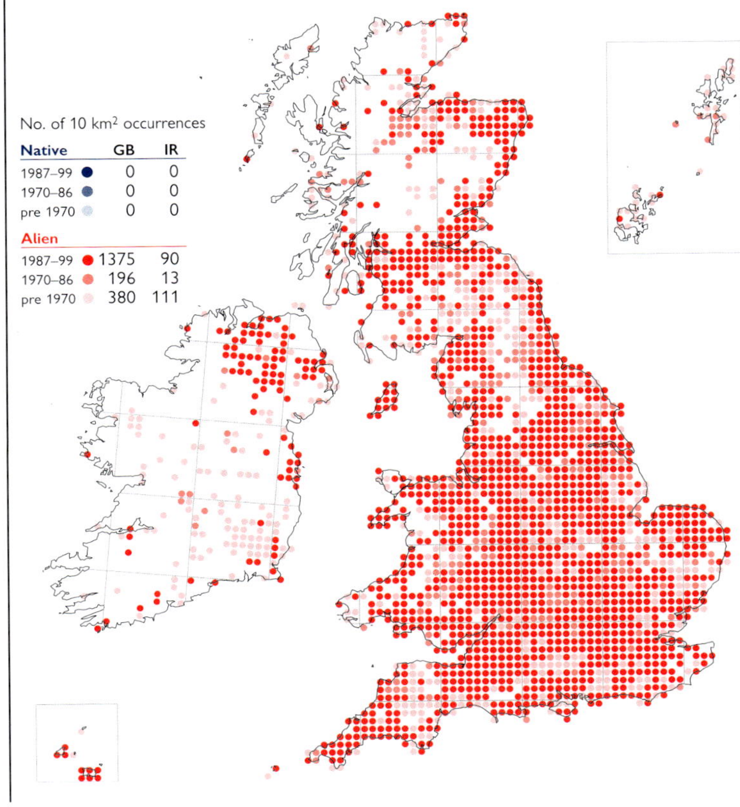

An annual of grassy banks, meadows, roadsides and waste places. It is a frequent, often unintentional, constituent of seed mixtures, but rarely persists once a closed sward has developed. Its persistence in some areas probably relies on repeated introductions. 0–350 m at Stainmore (Westmorland).

Neophyte (change –0.48). *T. hybridum* used to be much grown as a forage crop and was recorded from the wild by 1762. It is still grown on a small scale as a green manure. It was mapped as 'all records' in the 1962 *Atlas*, and its distribution is stable.

Native subspecies occur in S. Europe & S.W. Asia, especially in the mountains. Most of our plants are the cultivated subsp. *hybridum*, which is widely naturalised further north in Europe and elsewhere.

References: Atlas (108b), Hultén & Fries (1986), Meusel *et al.* (1965).

D. A. PEARMAN

Trifolium glomeratum Clustered Clover

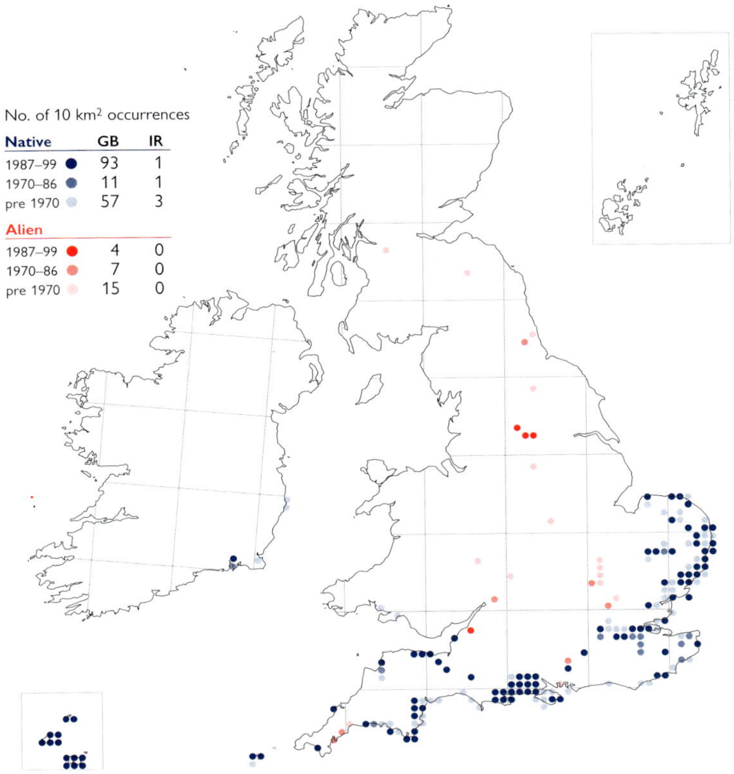

A winter-annual typically occurring in short open communities on light, drought-prone often somewhat acidic sandy or stony soils near the coast. Habitats include pathside banks, seafront lawns and cliff-slopes; also in sandy pastures, arable land and in the Isles of Scilly as a weed of bulb-fields. It is a rare casual inland. Lowland.

Native (change –0.11). The losses of *T. glomeratum* mostly date from before 1930. It now seems stable at its coastal sites, which are better recorded than for the 1962 *Atlas*, and, indeed, since it was last mapped on a national scale in 1994.

Mediterranean–Atlantic element.

References: Atlas (107d), Bolòs & Vigo (1984), Curtis & McGough (1988), Stewart *et al.* (1994).

D. A. PEARMAN

Trifolium suffocatum Suffocated Clover

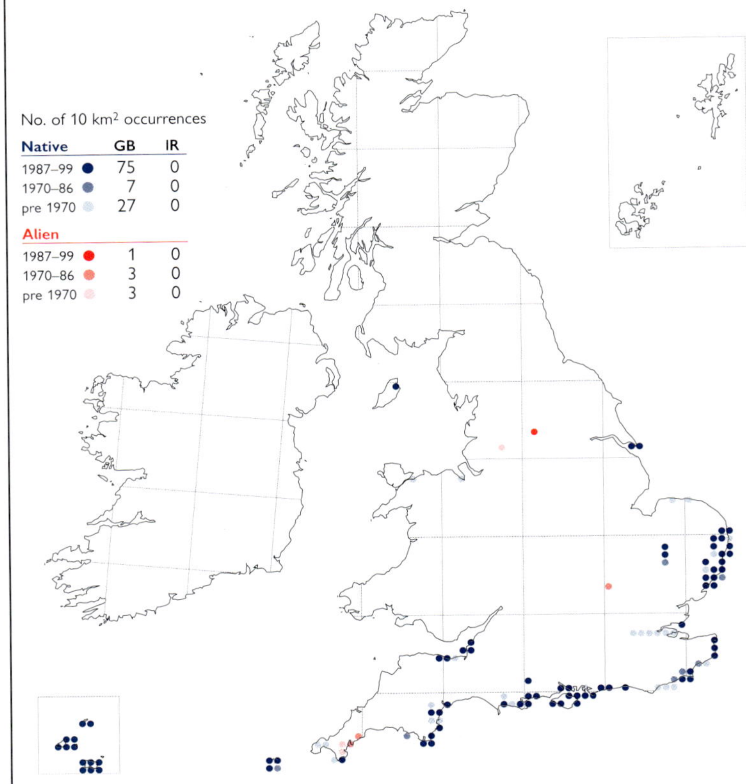

A winter-annual found on thin, dry soils on rocky coasts or on acidic compacted sand and shingle, either in open turf or on bare ground, and often part of a species-rich mosaic of annuals or bulbous plants. It occasionally grows on moister soils, but only in situations that are baked dry in summer. It was also a very rare alien of wool shoddy or spent tan. Lowland.

Native (change +0.14). Most inland populations of *T. suffocatum* were lost before 1930. There is little evidence of any change in its coastal distribution, where it was under-recorded until recently.

Mediterranean–Atlantic element.

References: Atlas (108a), Bolòs & Vigo (1984), Stewart *et al.* (1994).

D. A. PEARMAN

Trifolium strictum Upright Clover

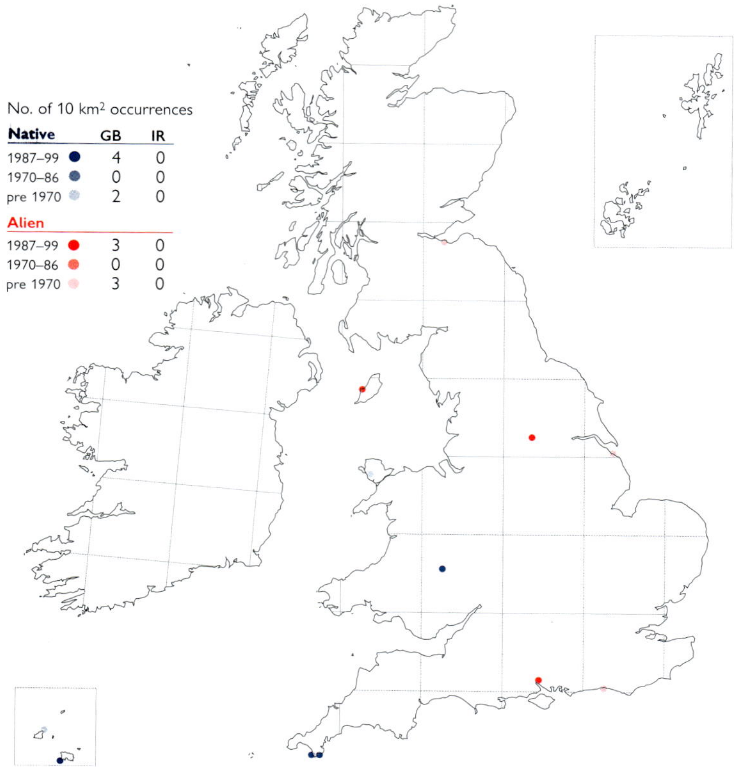

A winter-annual of shallow soils over schists, basalt and serpentine, preferring rock outcrops and S.-facing cliff-slopes kept open by grazing and drought. The number of plants in a population can fluctuate annually; the seeds are long-lived and plants are most numerous in years following summer drought. Lowland.

Native. The distribution of *T. strictum* is stable in Britain. In the Channel Islands it is now restricted to one site in Jersey, while in Guernsey it has not been seen, despite searching, since 1933.

Submediterranean–Subatlantic element.

References: Atlas (106a), Bolòs & Vigo (1984), Wigginton (1999).

D. A. PEARMAN

Trifolium fragiferum Strawberry Clover

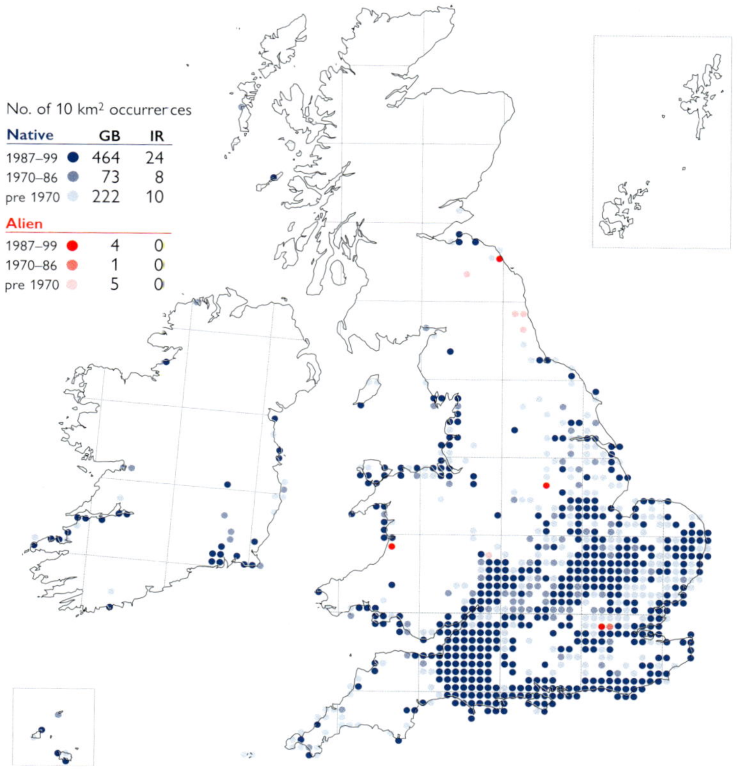

A procumbent perennial herb, rooting at the nodes. On the coast it is found behind salt-marshes, on earthen sea-walls and grazing marshes. Inland it occurs in pastures or by tracks on damp alluvial or calcareous clay soils. It sometimes grows in long-established amenity grassland or lawns. Lowland.

Native (change –0.81). The distribution of *T. fragiferum* shows declines in some areas, largely due to the improvement of old pastures or their conversion to arable, although such losses do not always show at the 10-km scale (e.g. in Dorset). It is still frequent in its coastal habitats, despite the extensive draining and ploughing of grazing marshes.

Eurosiberian Southern-temperate element; widely naturalised outside its native range.

References: Atlas (108d), Hultén & Fries (1986), Meusel *et al.* (1965).

D. A. PEARMAN

Trifolium resupinatum Reversed Clover

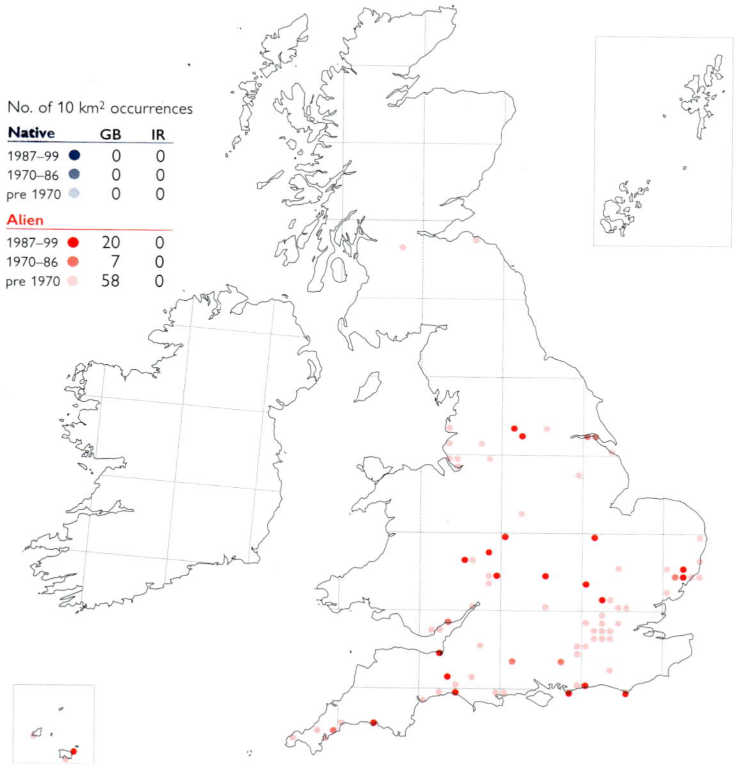

An annual found on waste ground, roadside verges, shoddy fields, as a tan-bark alien and a contaminant of grain and bird-seed. It is usually casual, but several persistent populations are known in grassland in S. England. Lowland.

Neophyte. *T. resupinatum* has been cultivated since 1713, and was first recorded in the wild in 1830. Distributional changes are difficult to assess, but it has persisted at Combwich (S. Somerset) since 1929.

Apparently a native of S.W. Asia and perhaps S.E. Europe, but long cultivated for fodder and now widely naturalised in S. Europe, Africa, S. America and elsewhere.

Reference: Bolòs & Vigo (1984).

D. A. PEARMAN

Trifolium aureum Large Trefoil

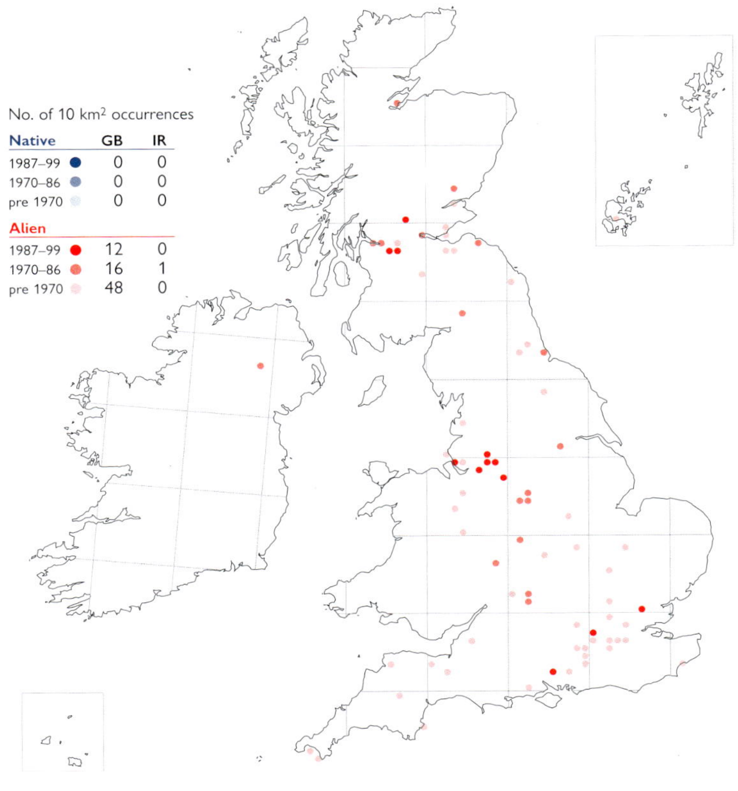

An annual found as a casual or naturalised in rough grassland and on waste ground, arising from bird-seed, wool shoddy or as a grain contaminant. Generally lowland, but reaching 320 m at Brassington (Derbys.).

Neophyte. *T. aureum* was introduced in 1815, and first recorded in the wild in 1838. It has declined as a casual, possibly through improved seed cleaning techniques and the less frequent use of wool shoddy.

A European Temperate species, absent as a native from much of W. Europe; naturalised in N. America and elsewhere.

References: Hultén & Fries (1986), Meusel *et al.* (1965).

D. A. PEARMAN

Trifolium campestre Hop Trefoil

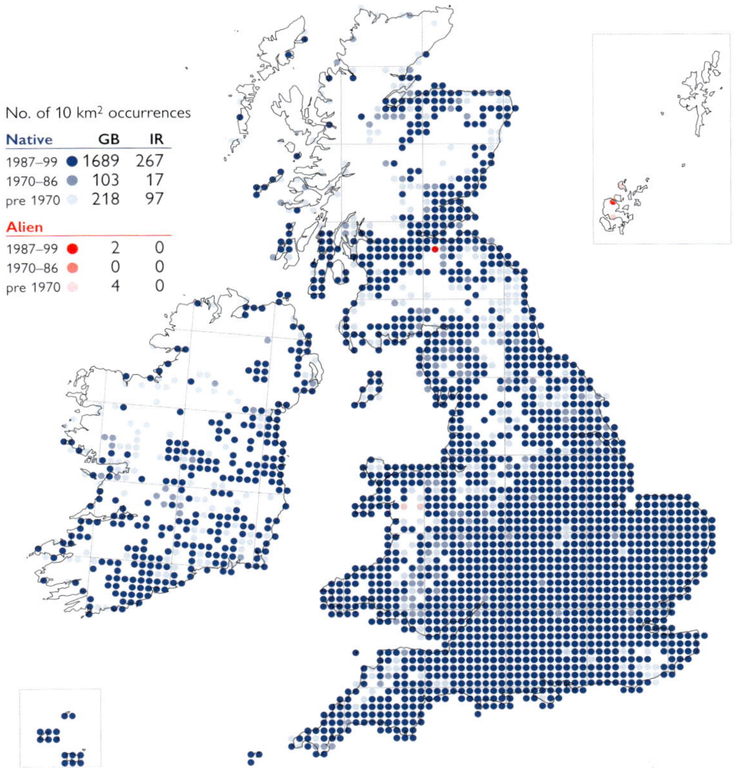

A winter-annual of grassland habitats on dry, relatively infertile neutral or base-rich soils, and thus more demanding and less frequent than *T. dubium*. It also occurs on spoil heaps from slate and limestone quarries. 0–350 m (Minninglow, Derbys.).

Native (change –0.45). *T. campestre* was mapped as 'all records' in the 1962 *Atlas*. The timing and causes of its apparent decline across parts of the north and west of its range are unclear.

Eurosiberian Southern-temperate element; widely naturalised outside its native range.

References: Atlas (109a), Hultén & Fries (1986), Meusel *et al.* (1965).

D. A. PEARMAN

Trifolium dubium Lesser Trefoil

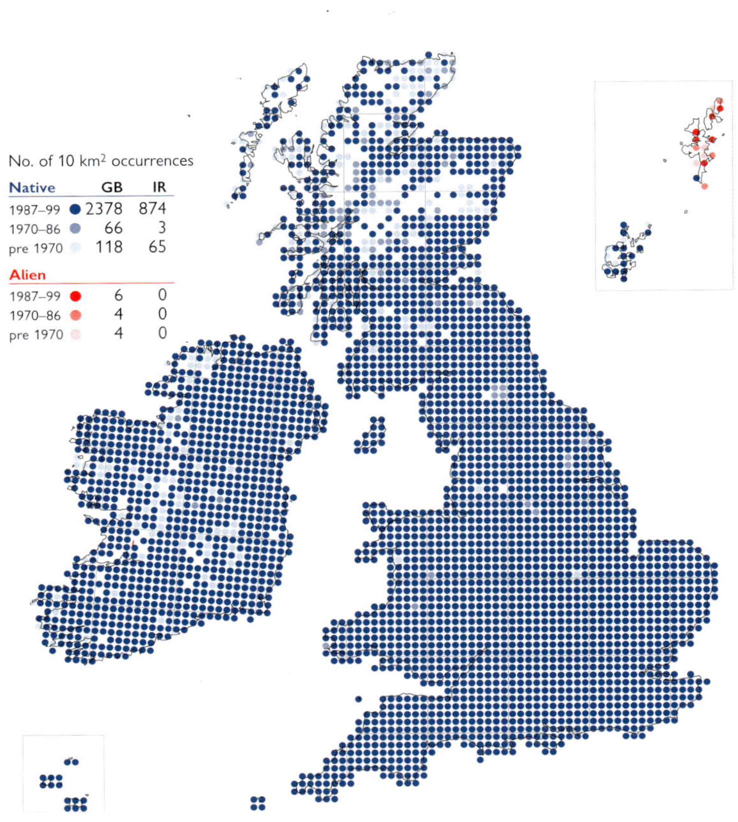

A winter-annual of hay meadows, waysides and waste places, and also frequent in lawns. Like *T. campestre*, it is most frequent in dry grasslands, but can be abundant too in winter-flooded meadows and damp pastures, and can thrive even in fairly nutrient-enriched situations. It also occurs in open habitats such as on rock outcrops, quarry spoil and railway ballast. Generally lowland, but reaching 530 m at Garrigill (Cumberland).

Native (change –0.11). There is no change in the distribution of *T. dubium*.

European Temperate element; widely naturalised outside its native range.

References: Atlas (109b), Grime *et al.* (1988), Hultén & Fries (1986).

D. A. PEARMAN

Trifolium micranthum Slender Trefoil

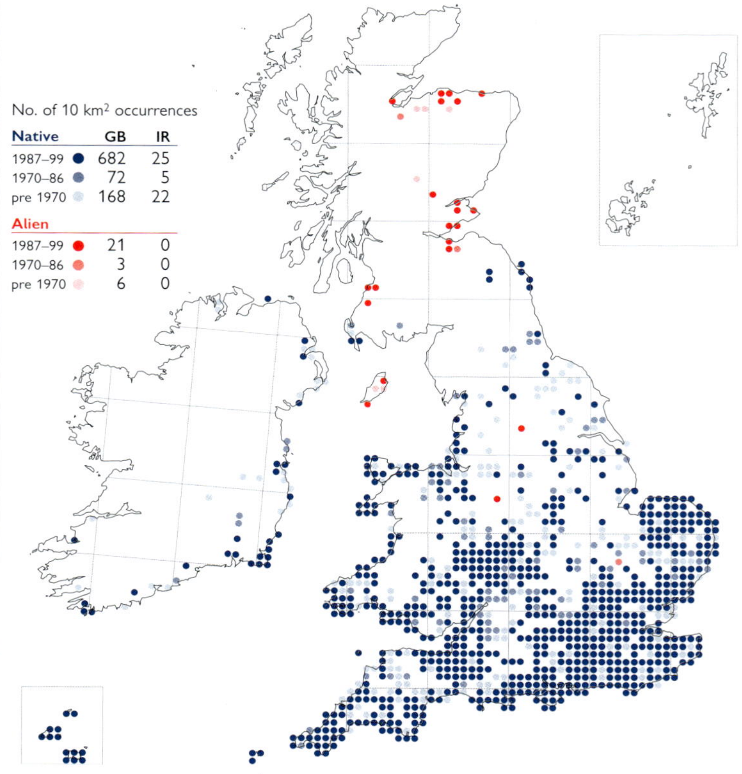

A winter-annual of neutral or moderately acidic soils, found on the coast in open, sandy or gravelly grassland rich in annuals and inland in drought-prone pastures, on paths and verges, and as a weed in lawns. It is tolerant of grazing, mowing and heavy trampling. Generally lowland, but reaching 365 m near Bowes (N.W. Yorks.).

Native (change +0.62). *T. micranthum* is frequently confused with depauperate specimens of *T. dubium*, and the map may include such errors, but it is undoubtedly much better recorded than it was for the 1962 *Atlas*. Even allowing for this, however, it is likely that it has undergone a considerable increase in England and Wales since 1950.

Submediterranean-Subatlantic element; widely naturalised outside its native range.

References: Atlas (109c), Hultén & Fries (1986).

D. A. PEARMAN

Trifolium pratense Red Clover

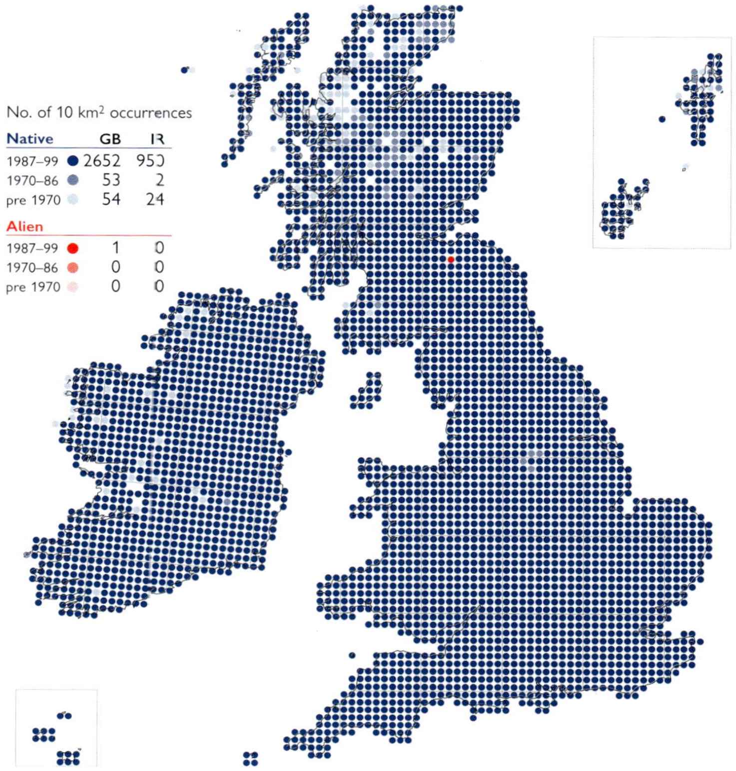

A perennial herb, native in a wide range of grasslands other than on the most acidic soils, and also common in waste places. The map includes the large, agriculturally selected variants (var. *sativum*) that are extensively sown into stubble and as components of short-term leys. 0–850 m (Scottish Highlands).

Native (change –0.18). There has been no change in the distribution of *T. pratense* since the 1962 *Atlas*.

Eurosiberian Temperate element, but widely naturalised so that distribution is now Circumpolar Temperate.

References: Atlas (105a), Grime *et al.* (1988), Hultén & Fries (1986).

D. A. PEARMAN

Trifolium medium Zigzag Clover

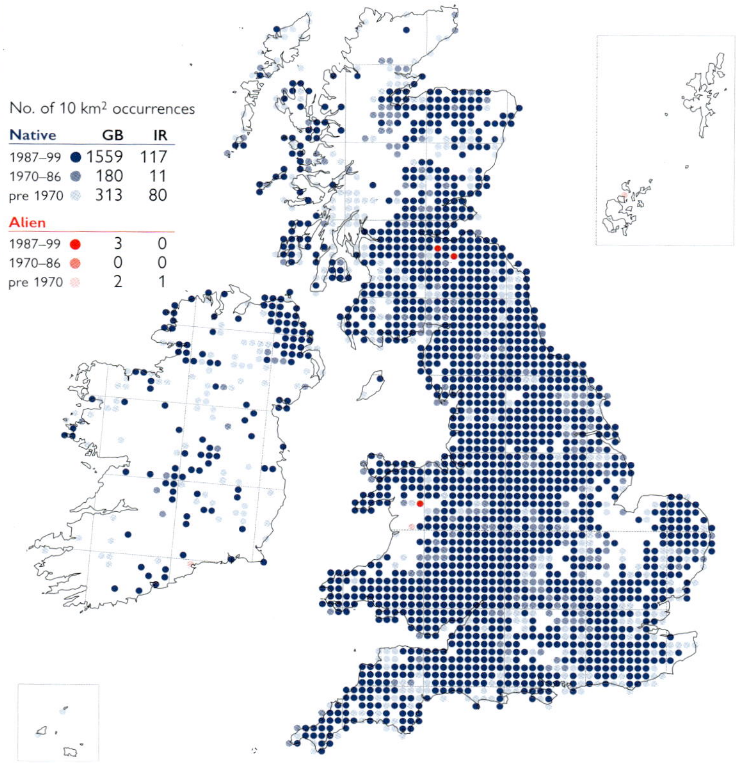

A rhizomatous perennial herb, mostly found in neutral grasslands on heavy soils, although it also occurs in hedgerows and on wood edges, and in ruderal habitats such as quarry spoil-heaps and railway banks. In upland areas it is also found on rocky stream-sides and in tall-herb communities on rock ledges, and on heaths in Ireland. Generally lowland, but reaching 610 m on Helvellyn (Cumberland).

Native (change –0.53). *T. medium* was mapped as 'all records' in the 1962 *Atlas*. Analysis of the database reveals that the marked decline in S. and E. England has occurred since 1950, the loss of permanent pasture being the main cause.

Eurosiberian Boreo-temperate element; widely naturalised outside its native range.

References: Atlas (105c), Grime *et al.* (1988), Hultén & Fries (1986), Meusel *et al.* (1965).

D. A. PEARMAN

Trifolium ochroleucon Sulphur Clover

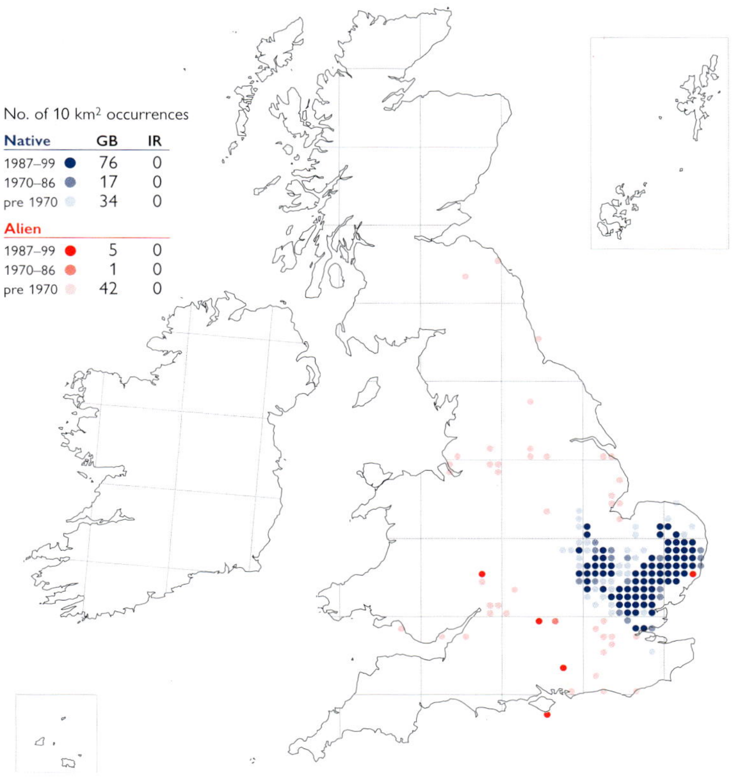

A perennial herb found on chalky boulder-clays or, more rarely, chalk, in pastures and on road verges, trackways and wood-borders. It also occurs as a casual outside its native core area. Lowland.

Native (change –0.84). *T. ochroleucon* is now rare in pastures, as large areas of suitable grassland have been converted to arable. Even in surviving grasslands there have been marked declines due to eutrophication, lack of grass cutting and encroachment of scrub. Many of its old roadside sites have been destroyed as a result of road-widening and re-routing schemes.

European Temperate element.

References: Atlas (105b), Bolòs & Vigo (1984), Stewart *et al.* (1994).

D. A. PEARMAN

Trifolium incarnatum subsp. *incarnatum* Crimson Clover

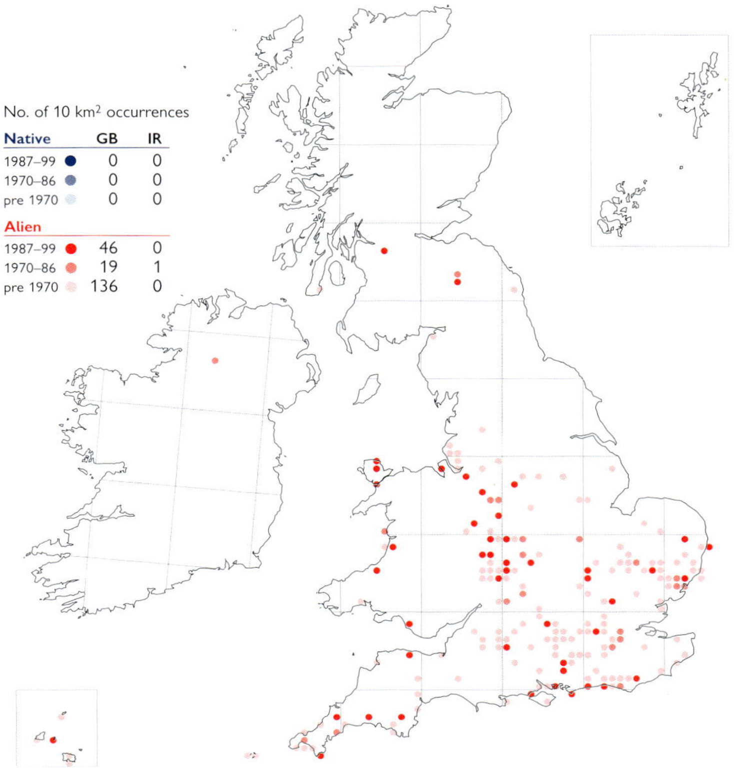

An annual of field margins and waste places. Lowland.

Neophyte (change for species –1.76). *T. incarnatum* subsp. *incarnatum* was cultivated in Britain by 1596 and was much grown as a fodder crop from the middle of the 19th century, but is less frequently cultivated now. It was recorded from the wild by 1838, but now only occurs as a sporadic casual.

Only known with certainty as a cultivated plant, although connected through intermediates to subsp. *molinerii* in S. Europe.

D. A. PEARMAN

Trifolium incarnatum subsp. *molinerii* Long-headed Clover

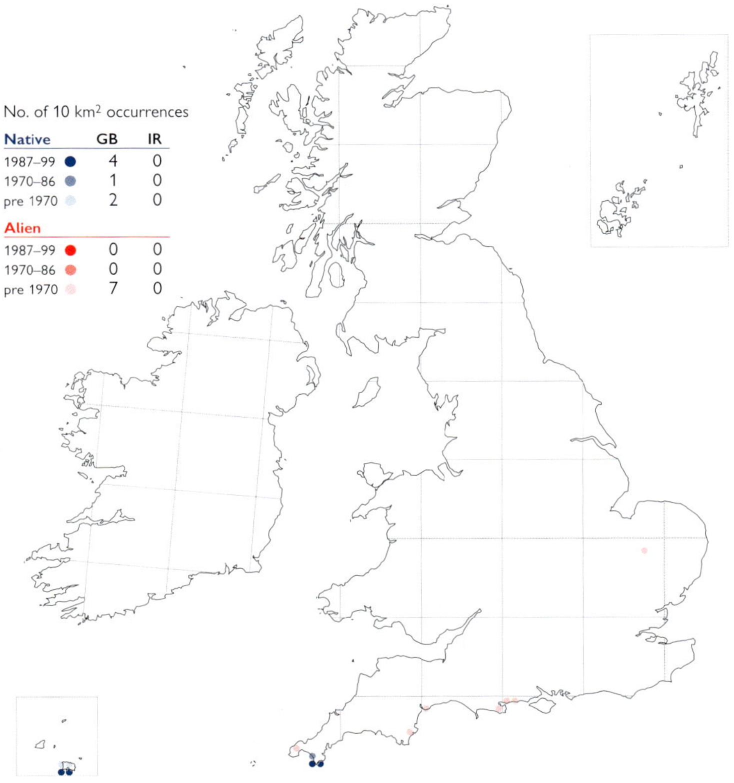

A winter-annual of schists, but not serpentine soils, on cliff-slopes in open habitats that are severely droughted in the summer. It has occurred as a casual at ports. Lowland and strictly maritime, only occurring within 200 m of the sea.

Native. In England, this subspecies still occurs at its five known localities on the Lizard peninsula (W. Cornwall), although population size varies considerably from year to year, depending on climatic conditions. In Jersey it also fluctuates in abundance in its best known site, an islet in Portelet Bay.

Mediterranean–Atlantic element.

References: Atlas (106b), Martin & Frost (1980), Wigginton (1999).

D. A. PEARMAN

Trifolium striatum Knotted Clover

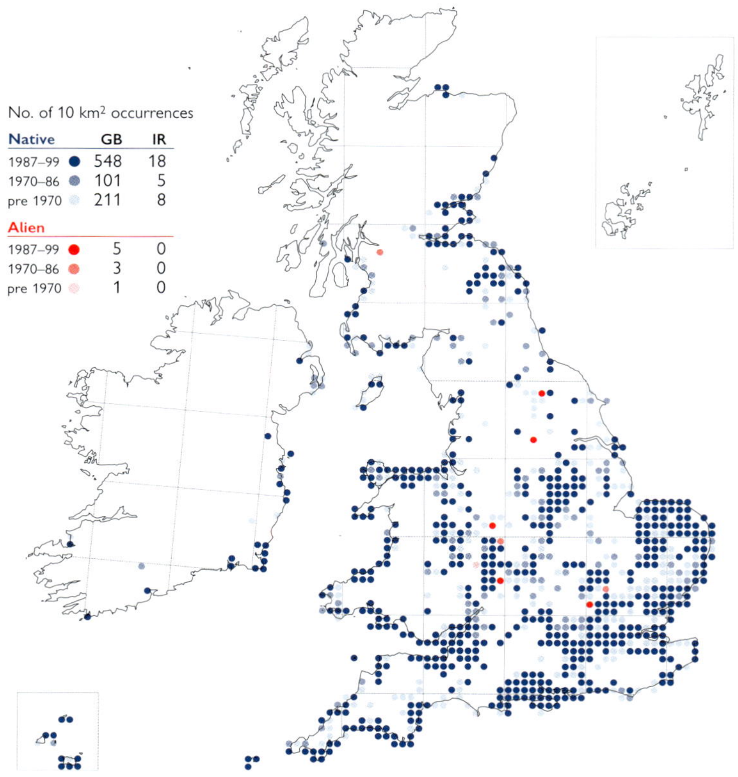

A winter-annual occurring in short, open communities around rock outcrops and on thin, relatively infertile drought-prone soils. Habitats include well-drained pastures, grassy banks and road verges. Although often growing in acidic sites, it is also found on base-rich neutral or even highly calcareous soils. 0–320 m (Derbys.).

Native (change –0.11). Since the 1962 *Atlas*, some sites of *T. striatum* have been lost owing to the improvement or abandonment of pastures and other developments.

European Southern-temperate element; widely naturalised outside its native range.

References: Atlas (106d), Coombe (1956b), Hultén & Fries (1986).

D. A. PEARMAN

Trifolium bocconei Twin-headed Clover

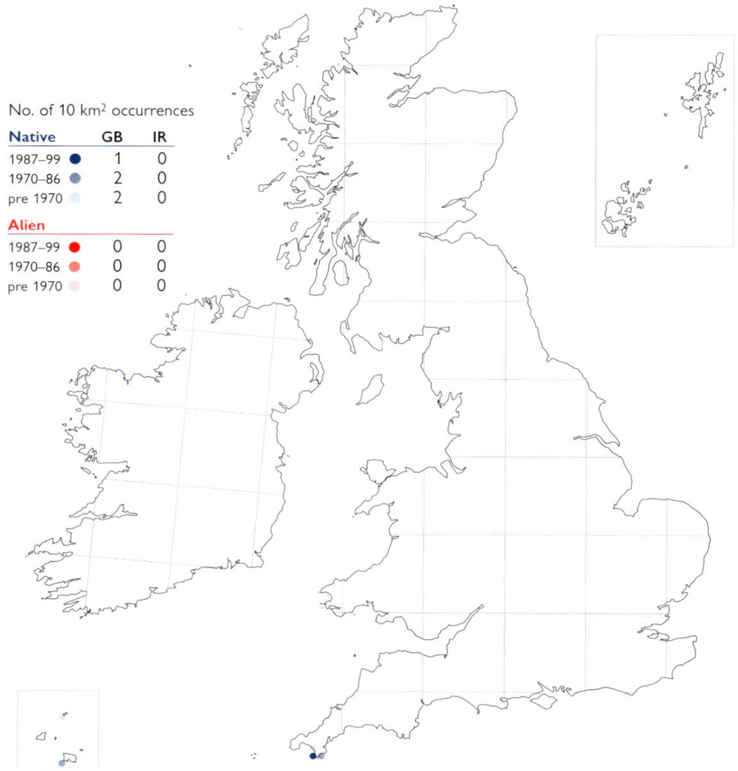

A winter-annual of shallow soils over serpentine, or rarely schist, and favouring sheltered S.-facing summer-droughted grasslands near the sea. Lowland.

Native. This is the rarest of the Lizard clovers. It fluctuates in numbers with climatic variation but, as the critical factor is the need for open habitats, it is largely dependent on adequate levels of grazing. Its distribution is almost unchanged since 1960, when long-term monitoring of populations began. It is also rare in Jersey, where it has not been found at one of its two localities since rabbit numbers were reduced by myxomatosis.

Mediterranean-Atlantic element.

References: Atlas (107b), Bolòs & Vigo (1984), Wigginton (1999).

D. A. PEARMAN

Trifolium scabrum Rough Clover

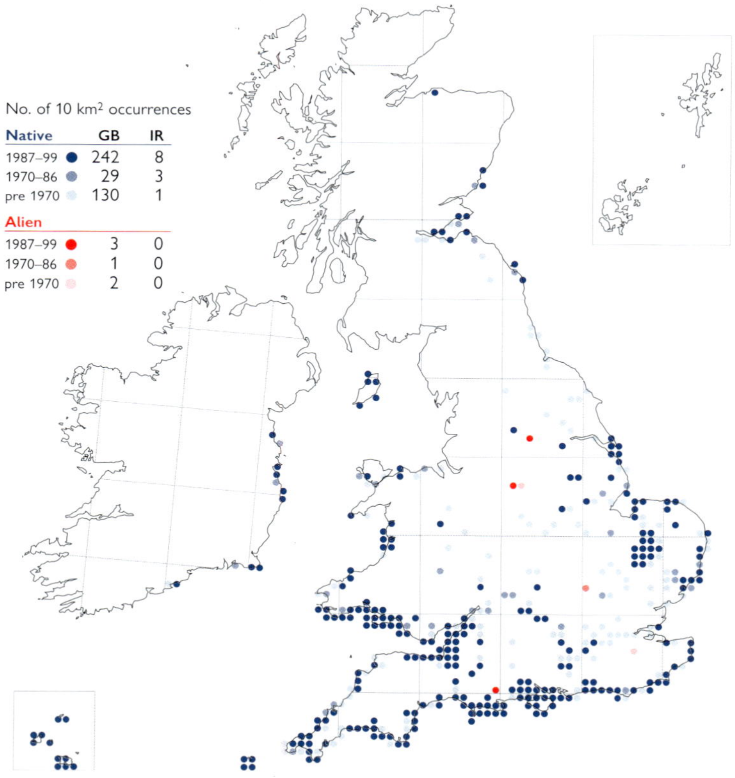

A winter-annual of thin, infertile, drought-prone soils over limestone, sand and gravel; by the sea also in summer-parched cliff-top grasslands. It grows in similar habitats to *T. striatum*, and frequently grows with it, but tends to occupy rockier, drier and more strongly leached (acidic) microsites. Inland, it is also found on road verges. Lowland.

Native (change –0.39). Coastal populations of *T. scabrum* are largely stable, while inland losses mainly occurred before 1930 and may have involved casual plants.

Submediterranean-Subatlantic element.

References: Atlas (107a), Bolòs & Vigo (1984), Coombe (1956b).

D. A. PEARMAN

Trifolium arvense Hare's-foot Clover

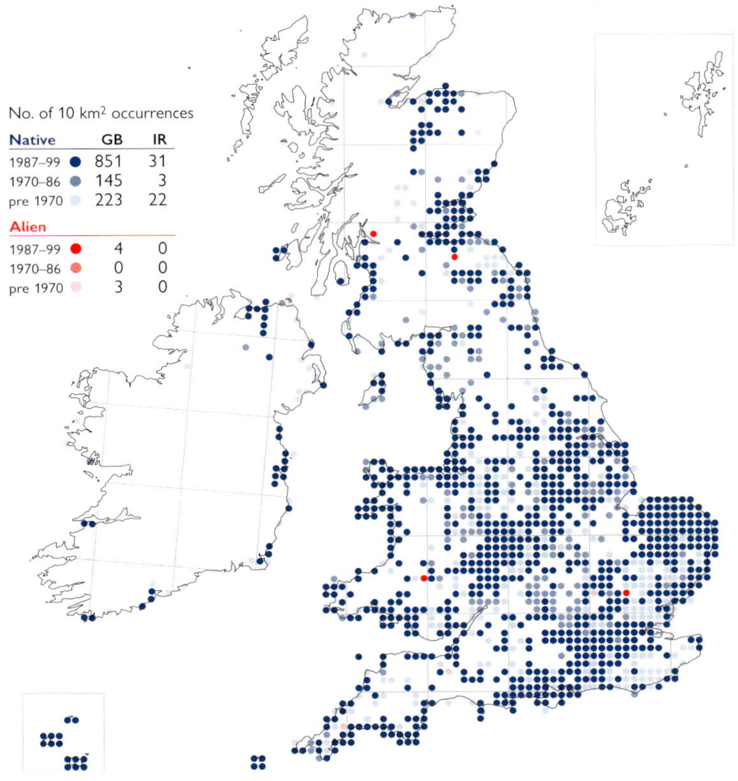

An annual of open rocky or sandy habitats, such as acidic heathlands, sea-cliffs and sand dunes; also on railway ballast and waste ground inland and in disturbed grassland and set-aside fields on light, sandy soils. Lowland.

Native (change –0.01). *T. arvense* was mapped as 'all records' in the 1962 *Atlas*. Analysis of the database reveals that the widespread decline in S.E. England has taken place since 1950.

Eurosiberian Southern-temperate element; widely naturalised outside its native range.

References: Atlas (106c), Hultén & Fries (1986), Meusel *et al.* (1965).

D. A. PEARMAN

Trifolium squamosum Sea Clover

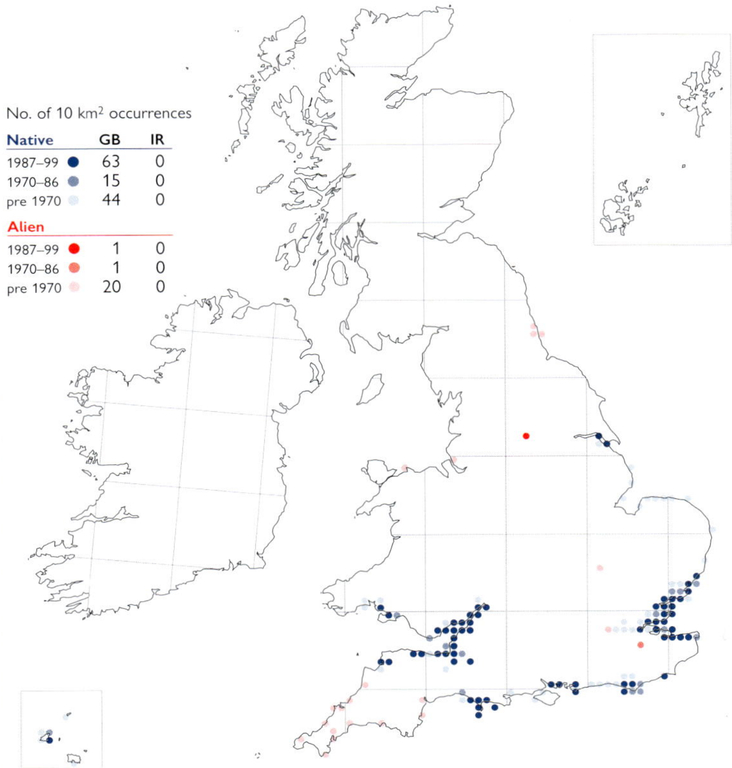

An annual found in the dry, uppermost parts of saltmarshes, brackish meadows and by tidal rivers and creeks. It is rarely found inland in open grassland on calcareous soils, and as a casual of waste places and by railways. Lowland.

Native (change –0.32). Many sites for *T. squamosum* have been lost due to coastal developments, including sea-defence schemes, and a lack of grazing on coastal grasslands. Most losses occurred before 1930, but have continued since then, although it is fairly stable in its remaining sites.

Mediterranean–Atlantic element.

References: Atlas (105d), Bolòs & Vigo (1984), Stewart *et al.* (1994).

D. A. PEARMAN

Trifolium subterraneum Subterranean Clover

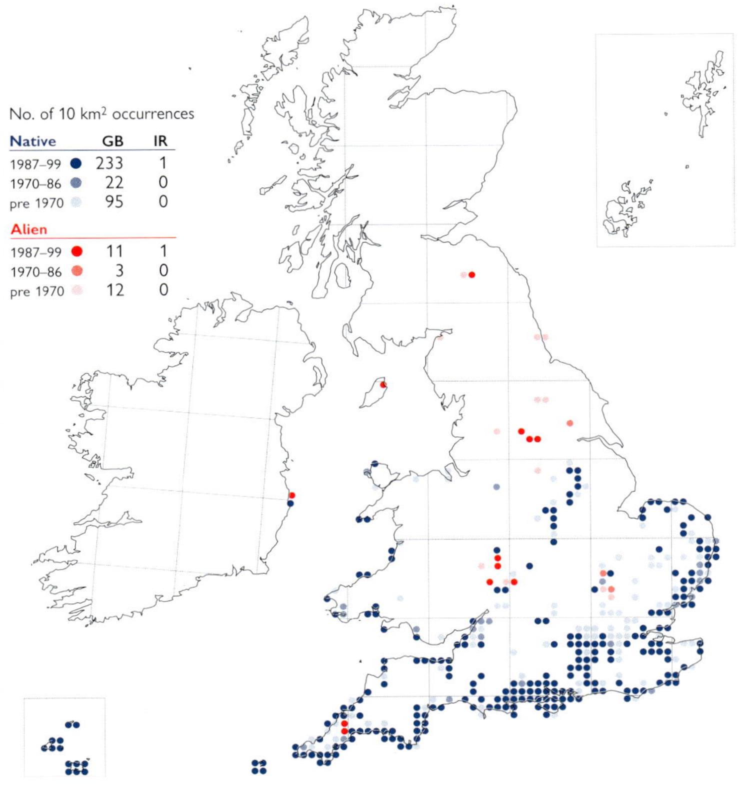

A procumbent winter-annual found near the coast in open grassland or heathland on thin, free-draining neutral to acidic sands, gravels and shingle; inland it occurs in summer-parched grasslands on chalk and limestone. Lowland.

Native (change –0.10). Many inland sites for *T. subterraneum* were lost in the 19th and early 20th centuries, but since 1930 its distribution seems to have remained fairly stable. Robust alien variants were frequently recorded as casuals introduced with wool shoddy.

Submediterranean-Subatlantic element; widely naturalised outside its native range.

References: Atlas (107c), Bolòs & Vigo (1984), Curtis & McGough (1988).

D. A. PEARMAN

Lupinus arboreus Tree Lupin

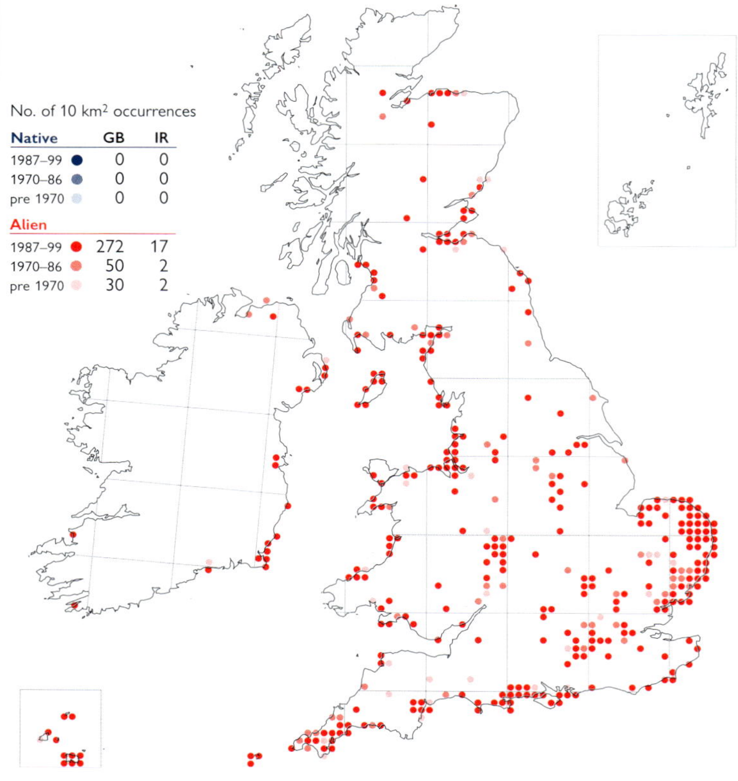

A short-lived, semi-evergreen shrub, widely planted on sand dunes and (in Cornwall) on china-clay tips, where it can cover large areas by virtue of its copiously produced seed, which is long-lived. It also occurs on roadsides and railway banks and in waste places. Lowland.

Neophyte (change +1.84). When first introduced to cultivation in 1793, this species was thought to need greenhouse protection. It was known from the wild by at least 1945, and had become well established by the time of the 1962 *Atlas*. It has considerably increased its range since then, and is considered a threat to native vegetation in some areas, for example on sand dunes in Hampshire and Norfolk.

Native of western N. America (California).

References: Atlas (99b), Bean (1973), Beckett *et al.* (1999), Brewis *et al.* (1996).

D. A. PEARMAN

Lupinus arboreus × *L. polyphyllus* (*L.* × *regalis*) Russell Lupin

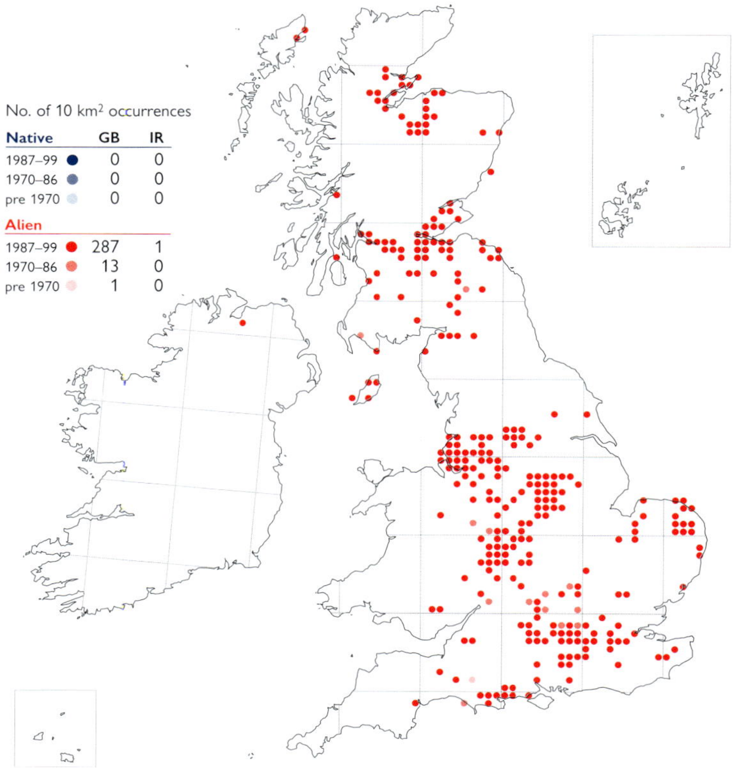

A short-lived perennial herb well-naturalised on rough ground, motorway and railway banks, and riverside shingle. Lowland.

Neophyte. This hybrid was first cultivated in Britain in 1937 after many years of experiments with other *Lupinus* species and hybrids, and is very popular as a garden plant. It was recorded from the wild by 1955 (Oxon), and appears to be increasing. Some cultivars may well contain genes of other lupin species, including *L. nootkatensis*, *L. mutabilis* and *L. hartwegii*. Backcrosses to *L. arboreus* also occur. It is unevenly recorded and sometimes erroneously recorded as *L. polyphyllus*.

A hybrid of garden origin, which also occurs spontaneously in Britain where the parents are established together.

Reference: Gorer (1970).

D. A. PEARMAN

Lupinus polyphyllus Garden Lupin

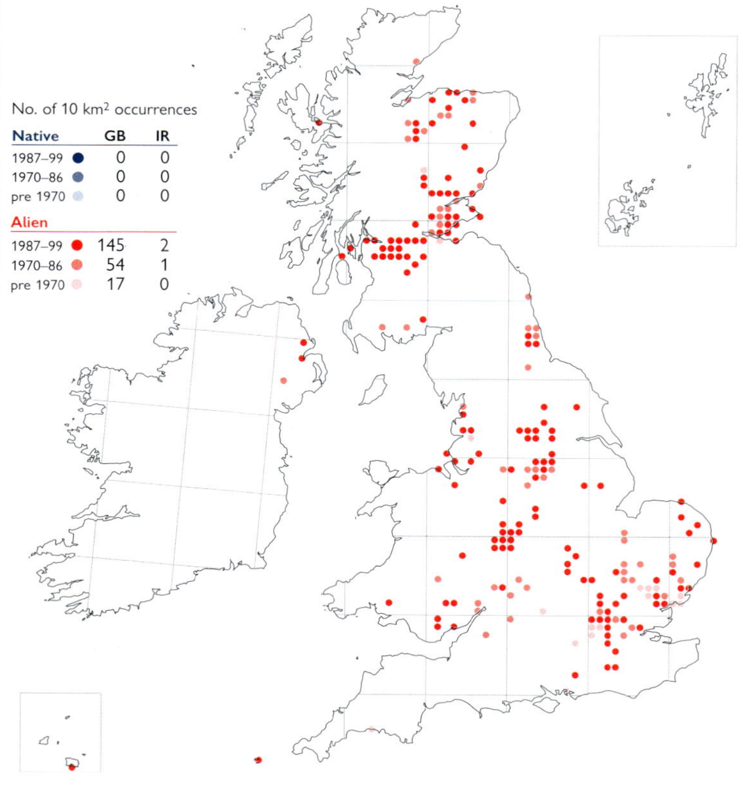

A biennial or short-lived perennial herb, found on roadside and railway banks, river shingle, rubbish tips and waste places. Populations are often naturalised. Lowland.

Neophyte. This species, introduced to cultivation in 1826, is now rarely grown, its hybrid with *L. arboreus* (*L.* × *regalis*) being the common garden plant. For this reason it is likely that a large proportion of modern records, at least south of Scotland, are misidentifications of the hybrid. It was known in the wild by 1900.

Native of western N. America.

References: Bolòs & Vigo (1984), Hultén (1968).

D. A. PEARMAN

Lupinus nootkatensis Nootka Lupin

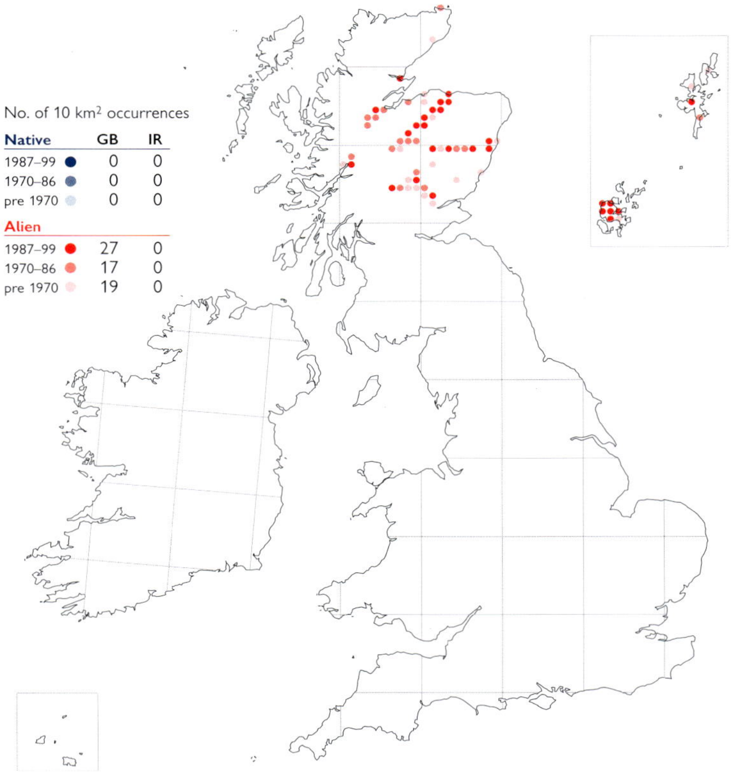

A perennial herb, found on river shingles and river banks in mainland Scotland, and streamsides in Orkney and Shetland. It has also been deliberately planted at some sites to improve nitrogen-poor upland soils. Reproduction is by seed; it is often well-naturalised and can be locally abundant. 0–340 m (Spittal of Glenshee, E. Perth).

Neophyte (change –0.83). Introduced into gardens in 1794, *L. nootkatensis* was first recorded in the wild by the River Dee in 1862, having escaped from the grounds of Balmoral Castle (S. Aberdeen). It is now rarely cultivated, having been replaced by *L.* × *regalis*, but the distribution in the wild appears to be stable. Records in England and Wales are probably errors for *L.* × *regalis*, and are not mapped.

Native of N.E. Asia and western N. America.

Reference: Atlas (99a).

D. A. PEARMAN

Laburnum anagyroides Laburnum

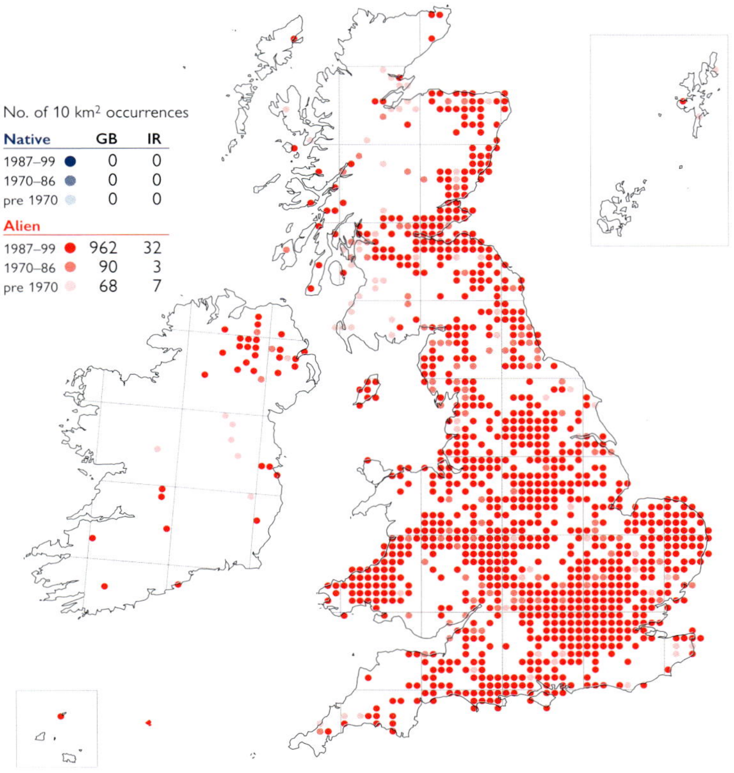

A small tree, much planted in parks, gardens and on waysides, thriving mainly on acid soils and frequently self-seeding into waste ground and by roads and railways. In W. Britain it was formerly used in hedging, and persists locally. Generally lowland, but reaching 425 m near New Radnor (Rads.).

Neophyte (change +3.71). *L. anagyroides* was introduced into cultivation in Britain before 1596 and was known from the wild by 1879. The likely explanation for the huge increase in records since the 1962 *Atlas* appears to be the increased interest in recording alien trees.

Native of the mountains of C. Europe.

References: Atlas (99c), Bean (1973), Chater (1992).

D. A. PEARMAN

Laburnum alpinum Scottish Laburnum

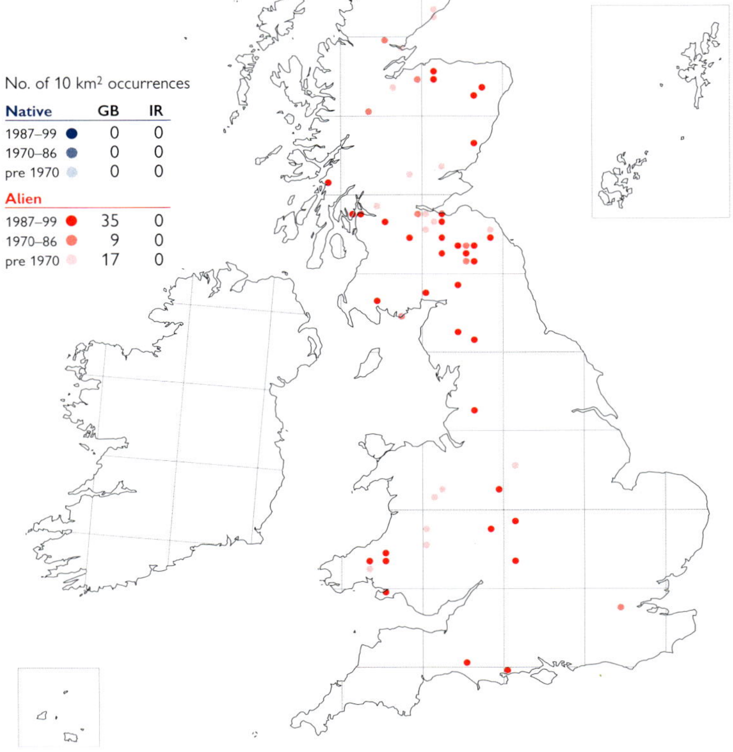

A medium-sized tree, usually found planted on roadsides and in woodland and hedges, occasionally in quantity. It rarely reproduces from seed, but does sometimes form thickets in this way. Lowland.

Neophyte. Although *L. alpinum* was in cultivation in Britain by 1596, it is still infrequent in gardens and in the wild, from where it was known by at least 1908. It is often confused with *L. anagyroides* and *L.* × *watereri* (*L. anagyroides* × *L. alpinum*), and it may be under-recorded. Most modern garden plants are the hybrid.

Native of the mountains of C. & S.E. Europe.

References: Bean (1973), Chater (1992).

D. A. PEARMAN

Cytisus striatus Hairy-fruited Broom

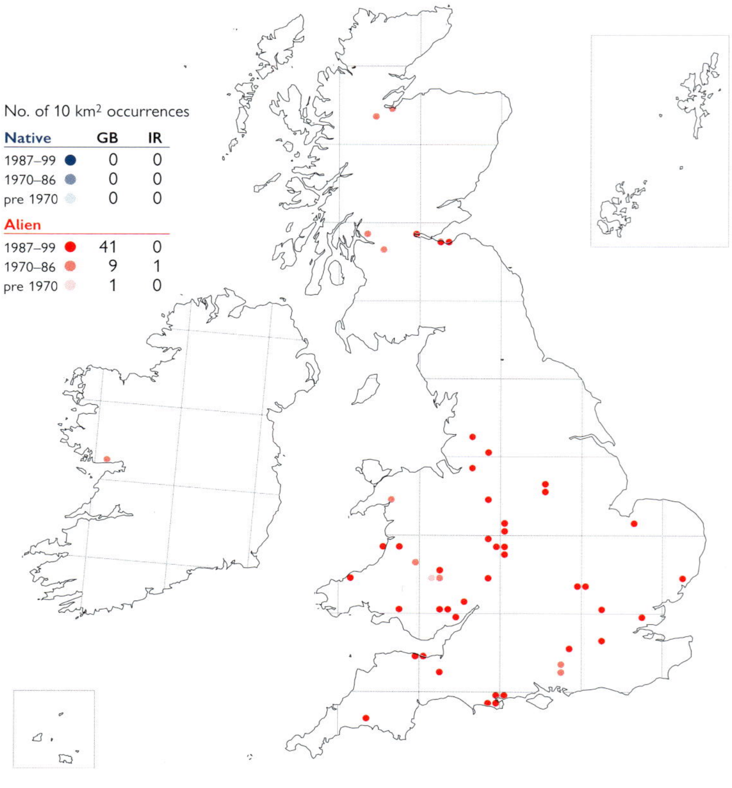

A shrub which has been used locally in roadside plantings and which reproduces freely from seed, sometimes forming extensive populations. It is also occasionally planted in parks and amenity areas. Lowland.

Neophyte. *C. striatus* was introduced to cultivation in 1816. It is very rarely grown in gardens, but seems popular in roadside planting schemes, when it was possibly first used *c.* 1959 (Clement, 1978b). It was first recorded in the wild in 1963 (Rads.), but was probably present before then.

Native of Portugal and western Spain, where possible hybrids with *C. scoparius* have been reported.

Reference: Bean (1970).

D. A. PEARMAN

Cytisus scoparius Broom

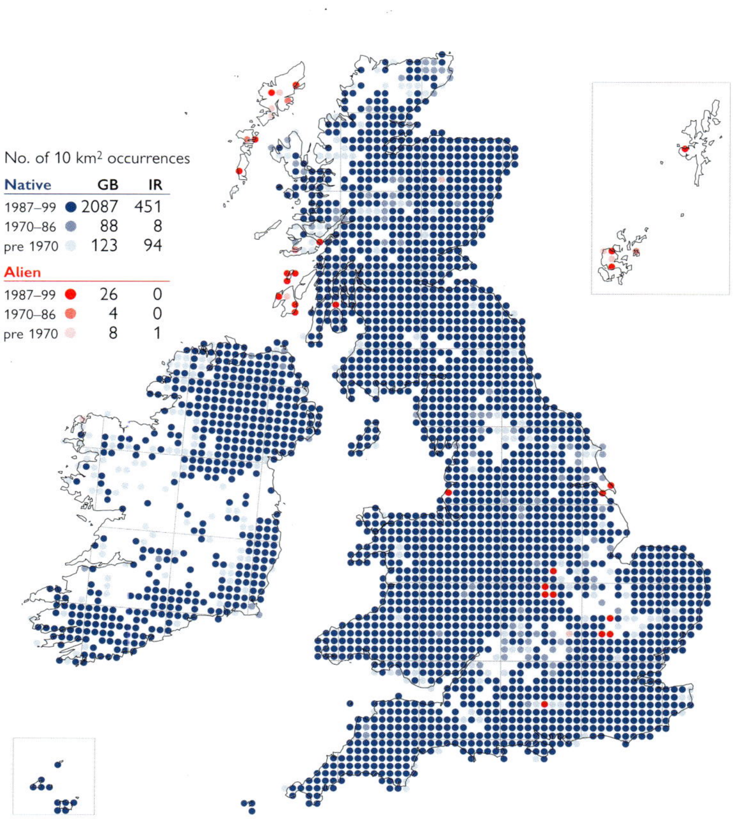

C. scoparius subsp. *scoparius* is an erect shrub of sandy acidic soils, on heaths, open woodland, railway banks, stony riversides and, particularly, on roadside banks and verges where it may often be planted. The prostrate subsp. *maritimus* grows on western sea-cliffs. Generally lowland, but reaching 640 m in Atholl (E. Perth).

Native (change 0.00). There has been little change in the distribution of *C. scoparius* since the 1962 *Atlas*. In N. Britain, entire populations may die back during very cold winters. Subsp. *scoparius* occurs throughout the range of the species; subsp. *maritimus* is mapped separately.

European Temperate element; widely naturalised outside its native range.

References: Atlas (101b), Hultén & Fries (1986), Meusel *et al.* (1965).

D. A. PEARMAN

Cytisus scoparius subsp. *maritimus* Prostrate Broom

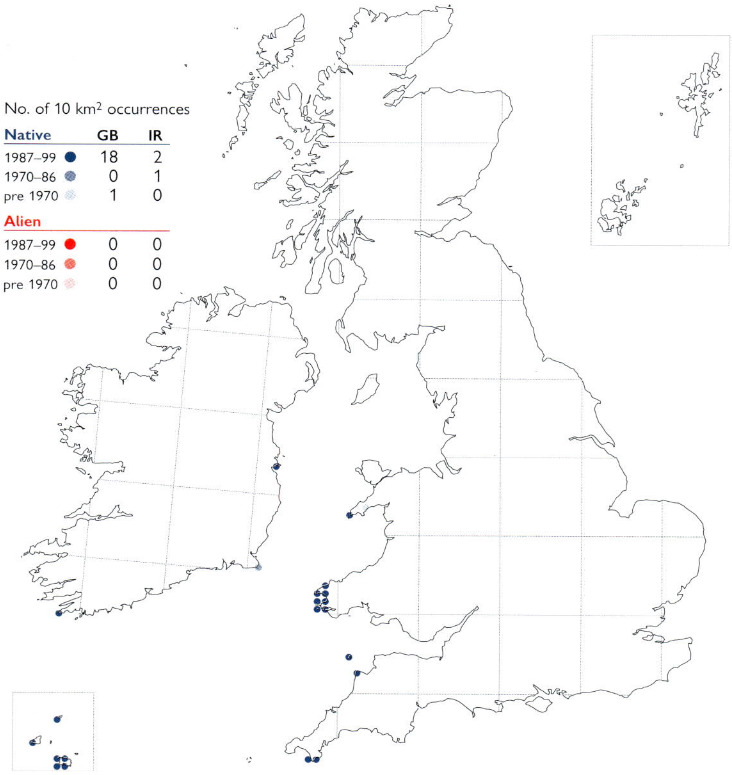

A prostrate shrub of exposed sea-cliffs, typically near the crest of the cliff in sites that are exposed to wind but not to heavy salt-spray. Very rarely, it may be found on rocky outcrops as far as one kilometre inland. It maintains its character in cultivation and breeds true. Lowland.

Native. The distribution of subsp. *maritimus* appears to be stable. Kay & John (1995) describe it as very distinct in its morphology, habitat and distribution. Some low-growing variants of subsp. *scoparius* occur, but these lack the intensely prostrate habit of subsp. *maritimus* and are only phenotypically dwarfed.

Oceanic Temperate element.

References: Atlas Supp. (19b), Wigginton (1999).

D. A. PEARMAN

Spartium junceum Spanish Broom

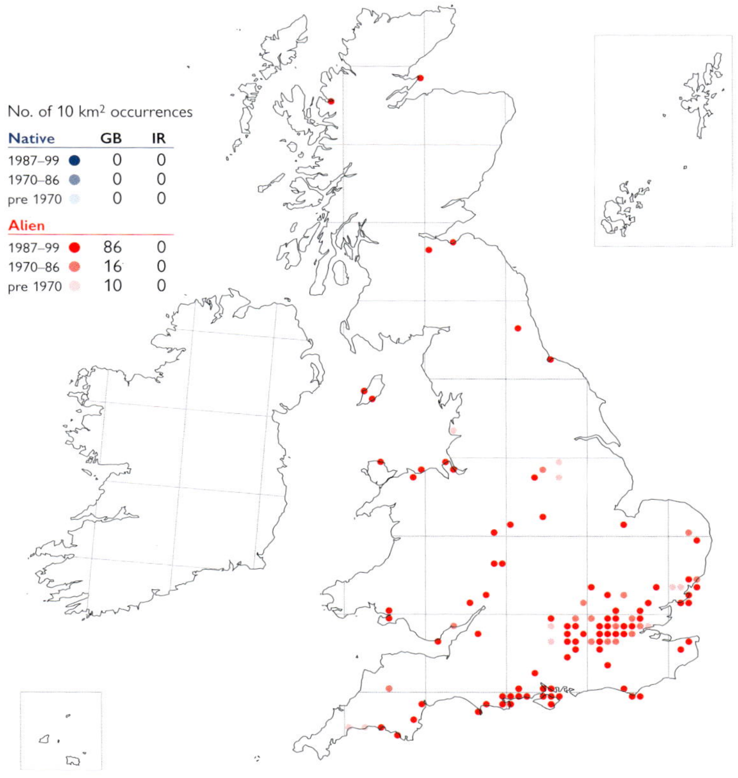

A shrub planted in amenity areas and on roadsides. It sets seed readily and can quickly become naturalised on light soils by roads, railways, and (in N. Essex) along coastal cliffs. Lowland.

Neophyte. Introduced into British gardens by 1548, *S. junceum* is frequently grown for its very sweetly scented flowers. It was known in the wild by at least 1905, but it was little noticed until the 1950s; since then plantings by new roads have led to its introduction to many new areas.

Native of the Mediterranean region.

References: Bean (1980), Burton (1983), Meusel *et al.* (1965).

D. A. PEARMAN

Genista tinctoria Dyer's Greenweed

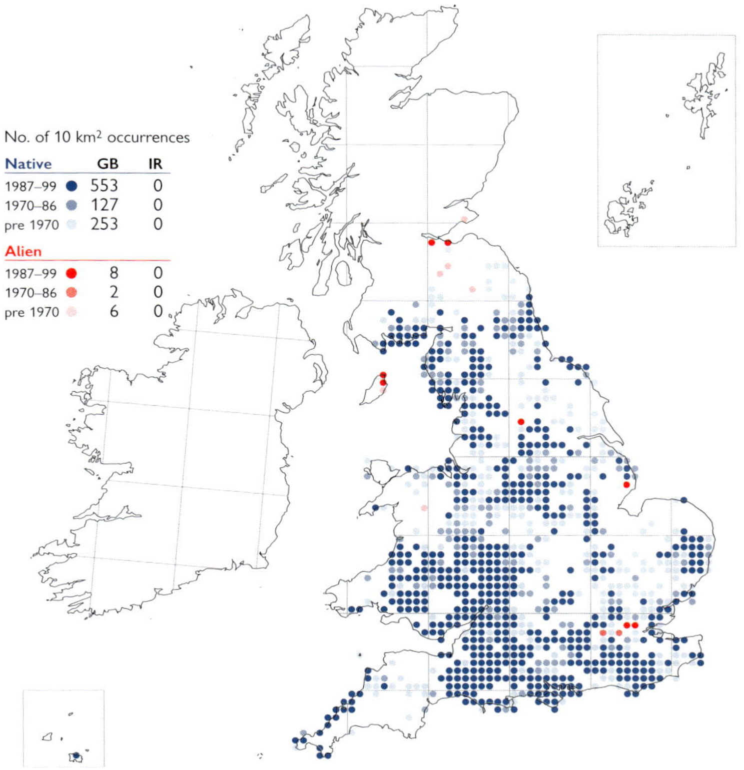

A small deciduous shrub of rough pastures, old meadows, grassy heaths, cliffs, road verges and field edges on heavy soils, usually calcareous to slightly acidic clays. Lowland.

Native (change –0.77). This species has declined considerably since the 1940s through the loss of old pastures, and is now a rare sight in some counties. Subsp. *tinctoria* occurs throughout the range of the species; subsp. *littoralis* is mapped separately.

European Temperate element.

References: Atlas (99d), Hultén & Fries (1986), Meusel *et al.* (1965).

D. A. PEARMAN

Genista tinctoria subsp. *littoralis*

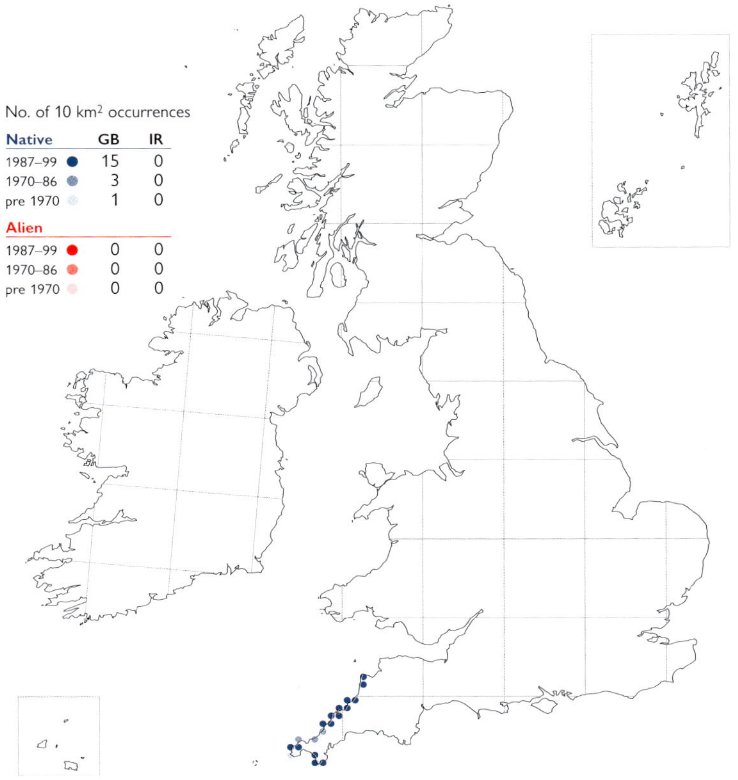

A small procumbent shrub of cliff-top grassland and maritime heaths; on the Lizard (W. Cornwall) it also occurs inland on *Erica vagans* heaths. Lowland.

Native. The distribution of this subspecies is probably stable, and there is little change from the map of prostrate forms of *G. tinctoria* in Perring & Sell (1968). All prostrate forms of *G. tinctoria* in Cornwall are mapped.

Oceanic Temperate element. Subsp. *littoralis* occurs in Britain and N.W. France (Brittany); other procumbent variants occur in exposed habitats elsewhere.

Reference: Atlas Supp. (19a).

D. A. PEARMAN

Genista pilosa Hairy Greenweed

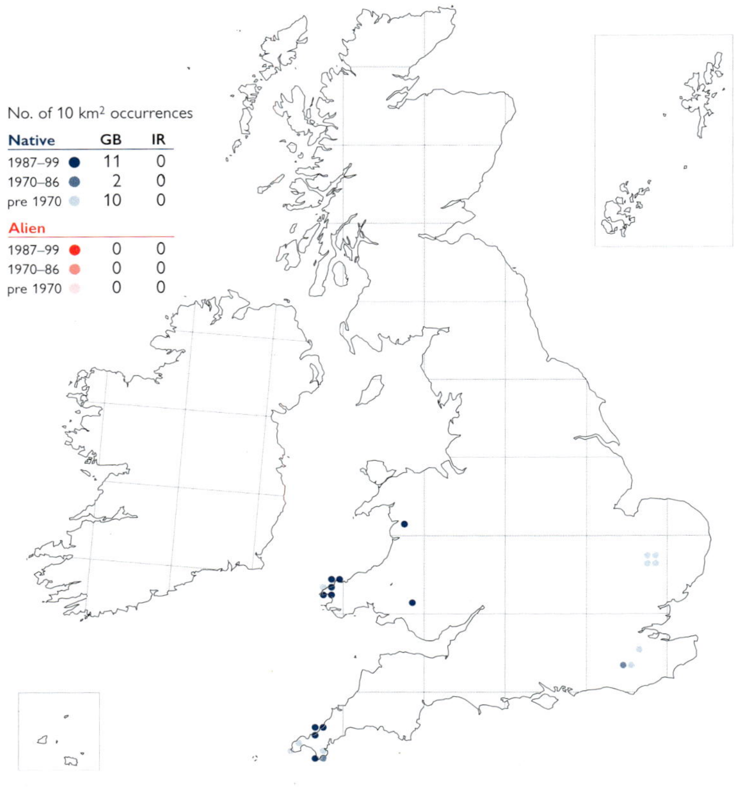

A low, usually prostrate, shrub growing mainly on rocky cliff-tops in maritime grassland or heath. It also grows in limestone grassland in Brecon, on rock ledges in the Cadair Idris range, and formerly on dry heaths in S. England. Generally lowland, but reaching 710 m on Cadair Idris (Merioneth).

Native (change –0.26). All *G. pilosa* sites on inland heaths have been lost through agricultural improvement or lack of management. The Breckland populations had become extinct by 1866 but the last inland site, in Sussex, survived until 1977 or perhaps 1980. Elsewhere, the distribution and populations seem stable.

European Temperate element.

References: Atlas (100b), Hultén & Fries (1986), Meusel *et al.* (1965), Rich *et al.* (1996), Wigginton (1999).

D. A. PEARMAN

Genista anglica Petty Whin

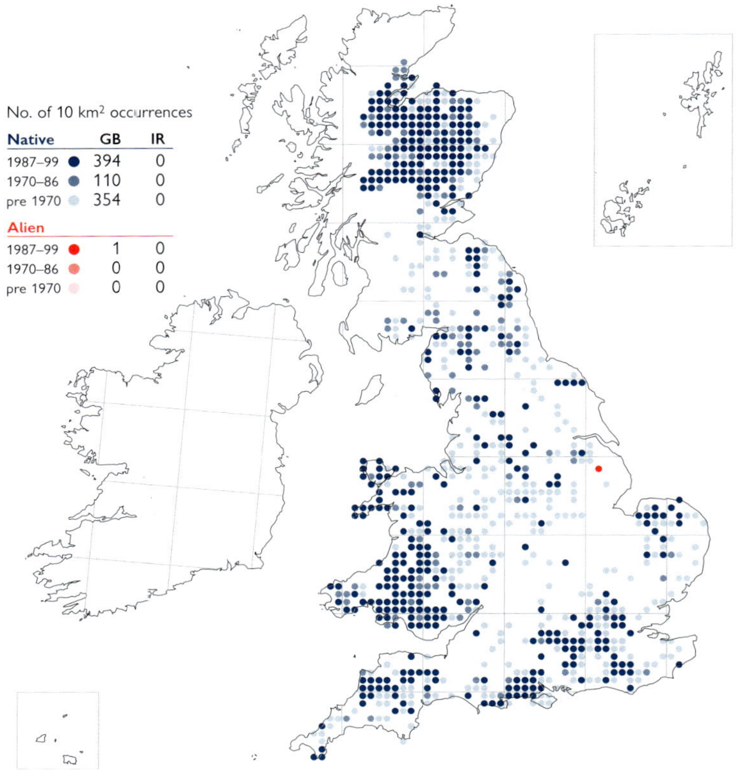

A small spiny shrub, found in the lowlands on relatively humid grass heaths and around the drier fringes of bogs. In upland areas it occurs in heathy, damp, unimproved pastures. Predominantly lowland but reaching 730 m in Atholl (E. Perth).

Native (change –1.09). Away from the New Forest (S. Hants.), with its long continuity of grazing, there has been a very substantial decline of this species in England and S. Scotland. The inherent fertility of many of its heathland sites means that they are frequently vulnerable to agricultural improvement, or else are apt to be become quickly overgrown following cessation of grazing or other management.

Oceanic Temperate element.

References: Atlas (100a), Byfield & Pearman (1996), Dupont (1962), Hultén & Fries (1986), Meusel *et al.* (1965).

D. A. PEARMAN

Genista hispanica Spanish Gorse

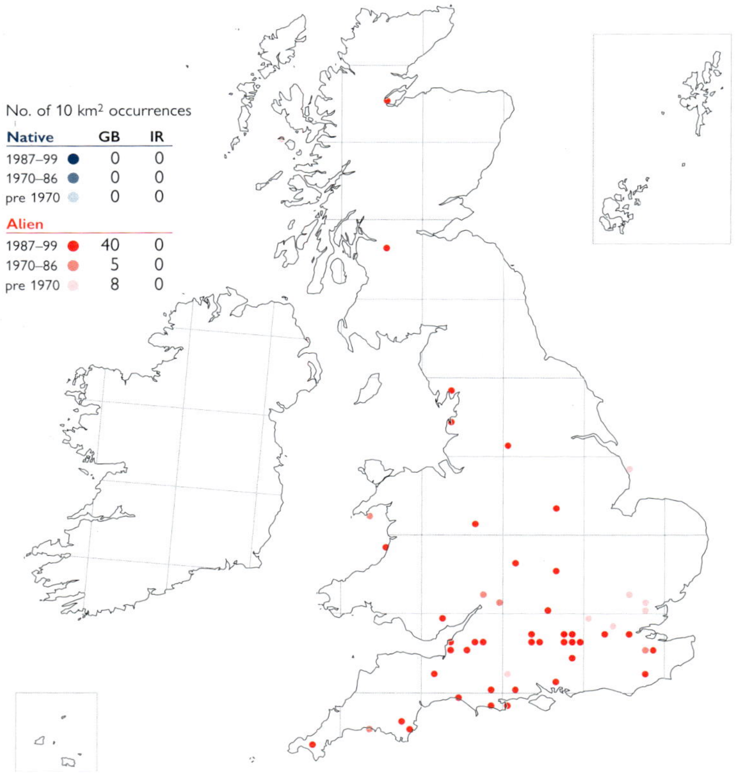

A densely spiny, small shrub which is occasionally planted on roadsides and in amenity areas, from where it has sometimes established itself by seed onto nearby sandy or rocky banks. Lowland.

Neophyte. Cultivated in Britain by 1759, *G. hispanica* is popular as a garden plant but still uncommon in the wild; it was not recorded in the wild until 1927 (Cards.) but is probably increasing.

Native of S.W. Europe (France and Spain).

References: Bean (1973), Bolòs & Vigo (1984), Meusel *et al.* (1965).

D. A. PEARMAN

Ulex europaeus Gorse

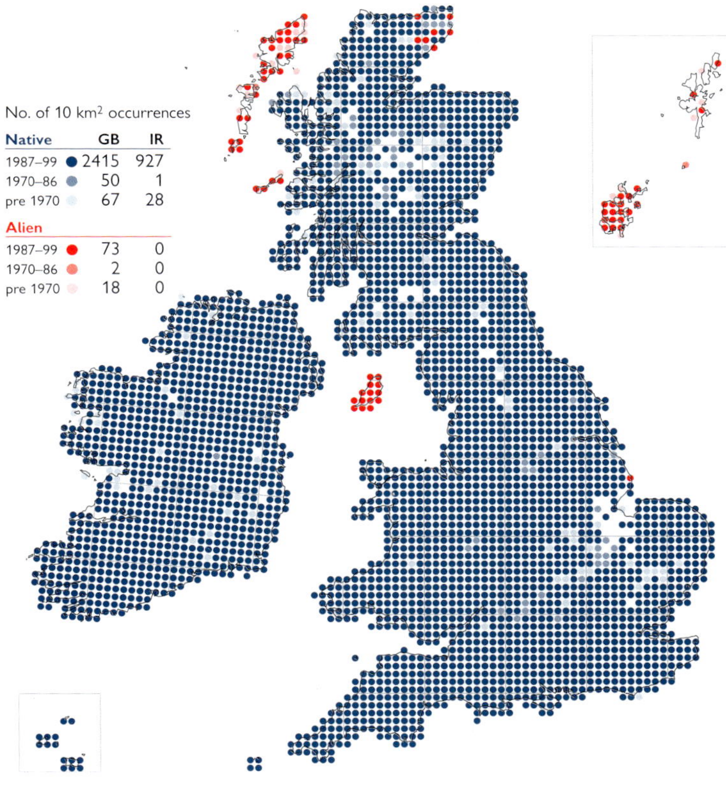

A shrub of mildly acidic soils, including leached soils on chalk and limestone, and acidic sands and gravels. It occurs in under-grazed pastures, woodland rides, on sea-cliffs and sand dunes, and on waste ground and railways. It is sometimes planted as hedges or game-shelter, and on roadsides. Generally lowland, but reaching 640 m on Carnedd Dafydd (Caerns.).

Native (change −0.34). *U. europaeus* has increased in abundance with the increase of ruderal habitats and the relaxation of grazing on lowland heaths and cliff-tops since the 1930s. It may be an introduction in some areas where it is mapped as native.

Oceanic Temperate element, but naturalised in suboceanic parts of Europe and elsewhere.

References: Atlas (100c), Grime *et al.* (1988), Hultén & Fries (1986), Meusel *et al.* (1965).

D. A. PEARMAN

Ulex europaeus × *U. gallii*

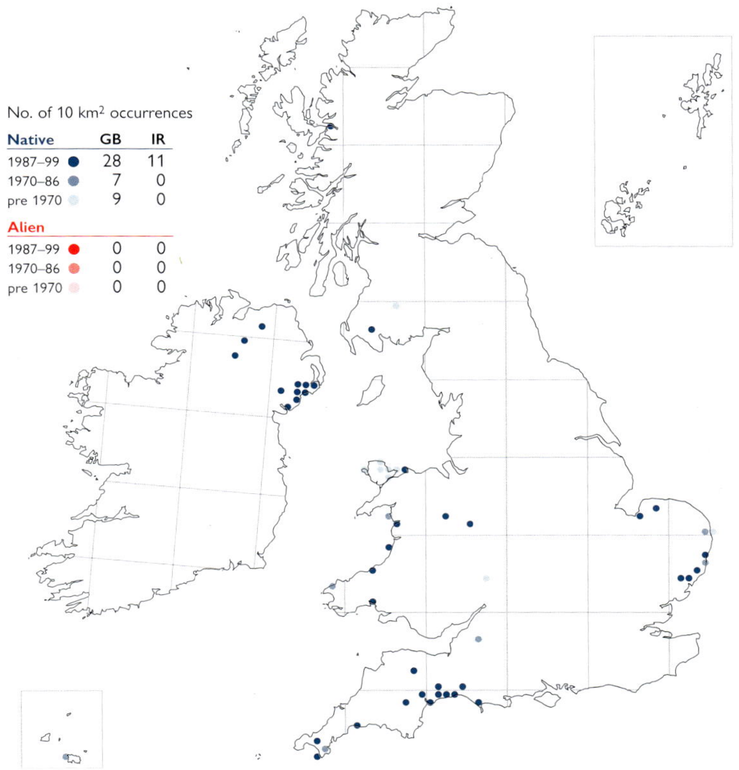

A shrub of acidic substrates, including leached soils overlying calcareous rocks, occurring with the parental species on sea-cliffs, heaths, roadside banks, waysides and pastures. It is highly fertile. Lowland.

Native. There are insufficient data to assess any changes in the distribution of this taxon. It is morphologically variable, so that some hybrids may have been overlooked as one or the other parent, and conversely pure species may have been recorded as the hybrid.

This hybrid is also reported from France (Brittany).

Reference: Stace (1975).

D. A. PEARMAN

Ulex gallii Western Gorse

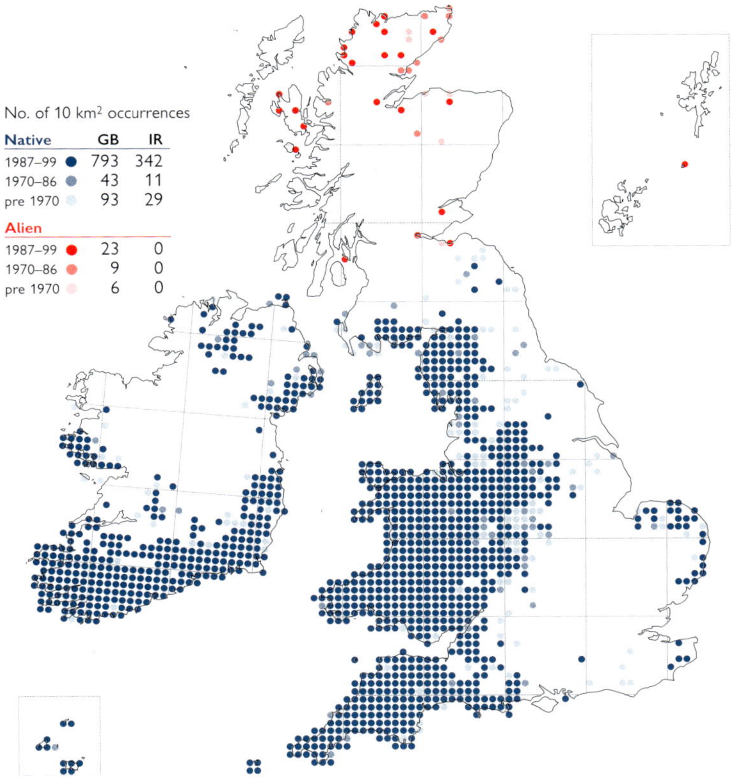

A shrub of heaths on infertile acidic soils, including leached soils overlying chalk and limestone; also found on sea-cliffs, in under-grazed or abandoned pastures, and on scrubby banks and waste ground. Generally lowland, but reaching 670 m in Macgillycuddy's Reeks (S. Kerry).

Native (change +0.20). The distribution of *U. gallii* is probably stable, despite local losses from grazing pressures. The boundary between the range of this species and that of *U. minor* is remarkably sharp, other than in Dorset, but there has been much misrecording. There are new records for N. Scotland, where it may have been sown beside newly-made roads or, flowering late in the year, may previously have been overlooked.

Oceanic Temperate element.

References: Atlas (100d), Kirchner & Bullock (1999), Proctor (1965).

D. A. PEARMAN

Ulex minor Dwarf Gorse

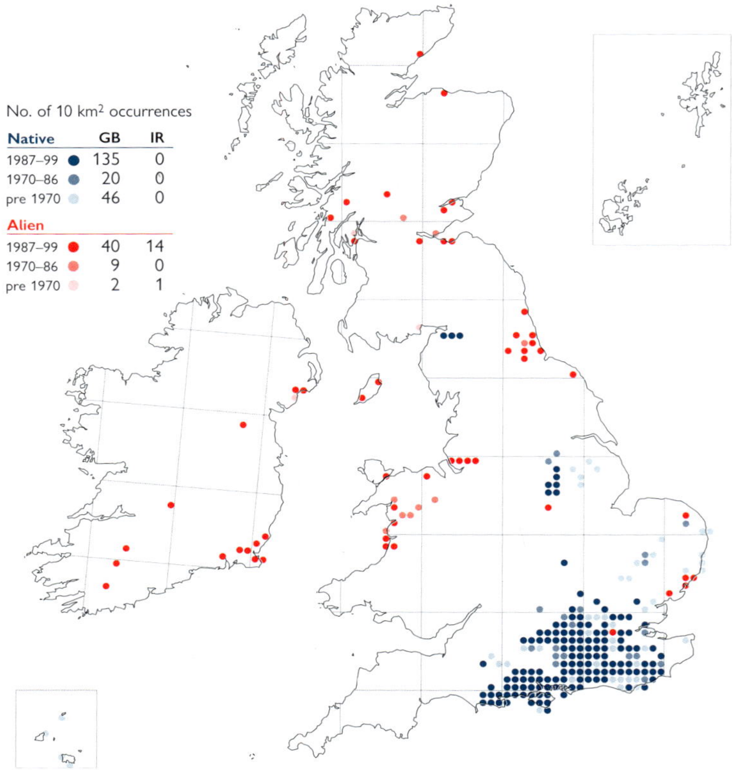

A small, sometimes procumbent shrub of heaths on free-draining acidic, nutrient-poor soils over podsolised sands and gravels, and occasionally over superficial deposits overlying chalk. In heathland areas it can persist as an under-storey shrub in scrub and secondary woodland. It also occurs on undergrazed heathy pastures and, rarely, on wet heaths. Lowland.

Native (change +0.20). Whilst there has been little overall decline, many sites for *U. minor* have been lost to development and forestry, and to increasing scrub where grazing has ceased. Alien records have increased in Britain and Ireland; some of these may result from its use in seed mixtures for roadsides.

Oceanic Southern-temperate element.

References: Atlas (101a), Proctor (1965), Stewart *et al.* (1994).

D. A. PEARMAN

Hippophae rhamnoides Sea-buckthorn

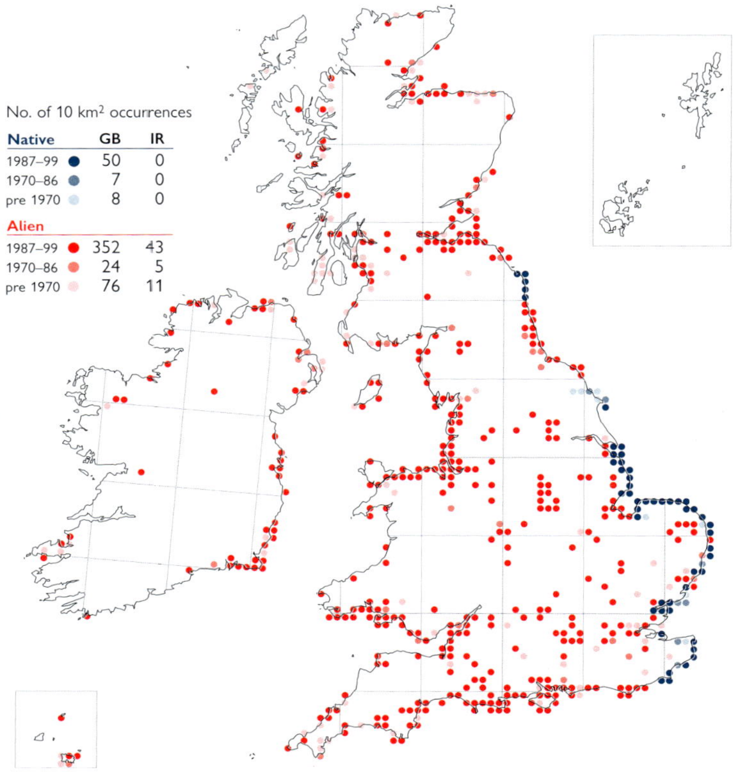

A thorny deciduous shrub or small tree of stabilised sand dunes and coastal banks, spreading by rhizomes and layering and often forming dense thickets. It is dioecious and wind-pollinated, flowering in the winter or early spring on bare wood and fruiting in autumn. Lowland.

Native (change +1.27). A dominant plant of the late glacial period and still found in montane habitats in continental Europe and the Himalayas, but in Britain native only in coastal habitats. It is planted widely as an amenity shrub within and outside its native range, and its invasive spread can pose a major threat to native vegetation.

European Boreo-temperate element; also in C. Asia and widely naturalised outside its native range.

References: Atlas (144a), Hultén & Fries (1986), Meusel *et al.* (1978), Pearson & Rogers (1962), Stewart *et al.* (1994).

A. J. RICHARDS

Myriophyllum verticillatum Whorled Water-milfoil

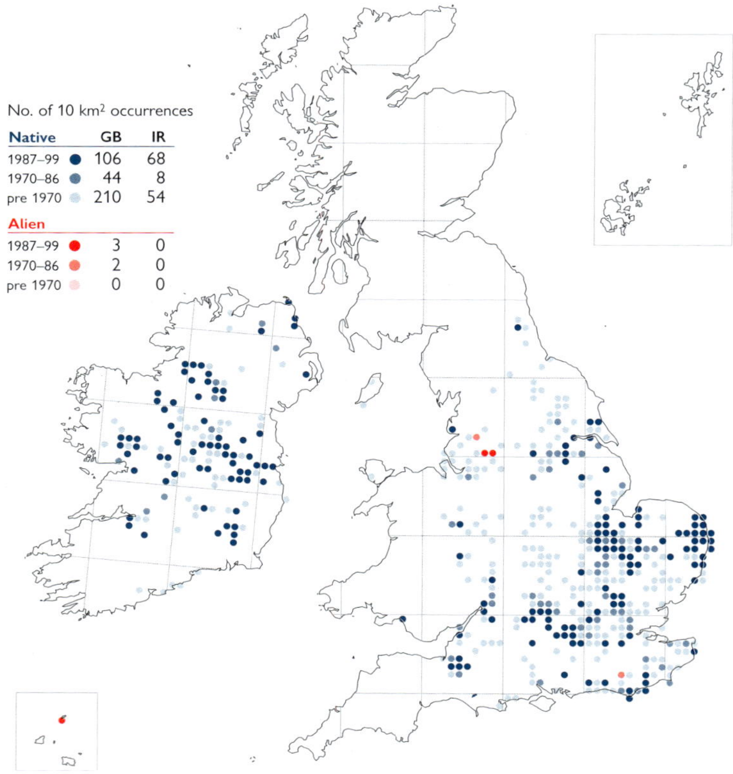

M. verticillatum is a robust, perennial plant of clear or slightly turbid, still or slowly flowing calcareous water in lakes, streams, canals and ditches. It occurs over both peaty and inorganic substrates. It flowers and sets seed, and also perennates by specialised turions which are produced in the leaf axils. Lowland.

Native (change –0.89). This species is probably more frequent in the calcareous central plain of Ireland than elsewhere in our area, and there it is probably still under-recorded. The decline in *M. verticillatum* in S. England began before the 1962 *Atlas*, and has continued since then.

Circumpolar Temperate element.

References: Atlas (149d), Hultén & Fries (1986), Meusel *et al.* (1978), Preston & Croft (1997), Stewart *et al.* (1994).

C. D. PRESTON

Myriophyllum aquaticum Parrot's-feather

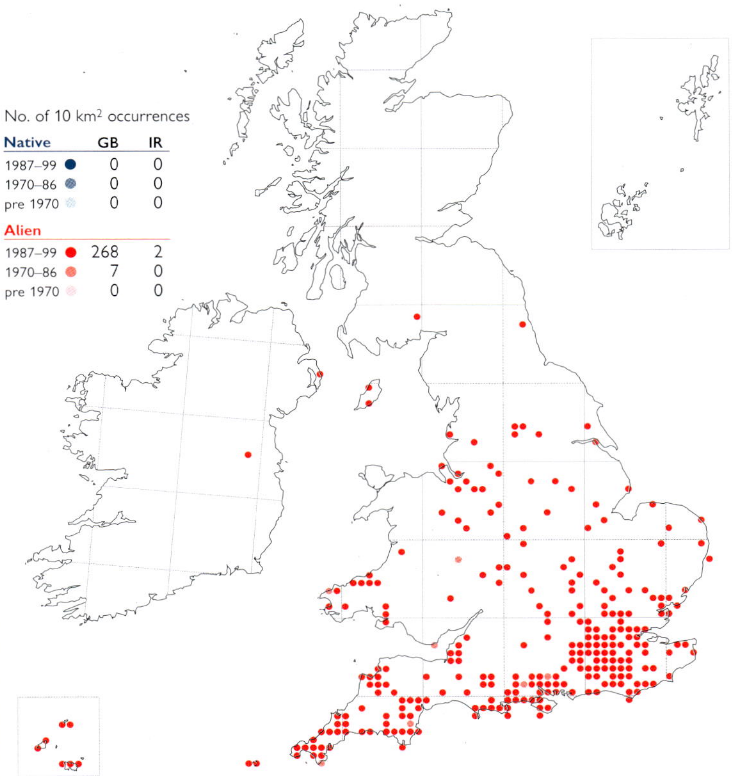

No. of 10 km² occurrences

Native	GB	IR
1987–99 ●	0	0
1970–86 ●	0	0
pre 1970 ●	0	0
Alien		
1987–99 ●	268	2
1970–86 ●	7	0
pre 1970 ●	0	0

This perennial often grows in emergent masses in small, sheltered, eutrophic water bodies, especially ponds and ditches but also reservoirs, canals and flooded mineral workings. Only female plants are known in Britain and they spread clonally by vegetative fragmentation. Lowland.

Neophyte. This species, which has been grown in water gardens in Britain since 1878, was first recorded in the wild in 1960 in Surrey and in 1969 in E. Sussex; in Ireland it was first seen in Co. Down in 1990. It is introduced when surplus garden plants are dumped in the wild and the extent to which it spreads naturally to new sites is unknown.

Native of central S. America; female plants are now widely naturalised in warm temperate and tropical areas elsewhere.

Reference: Preston & Croft (1997).

C. D. PRESTON

Myriophyllum spicatum Spiked Water-milfoil

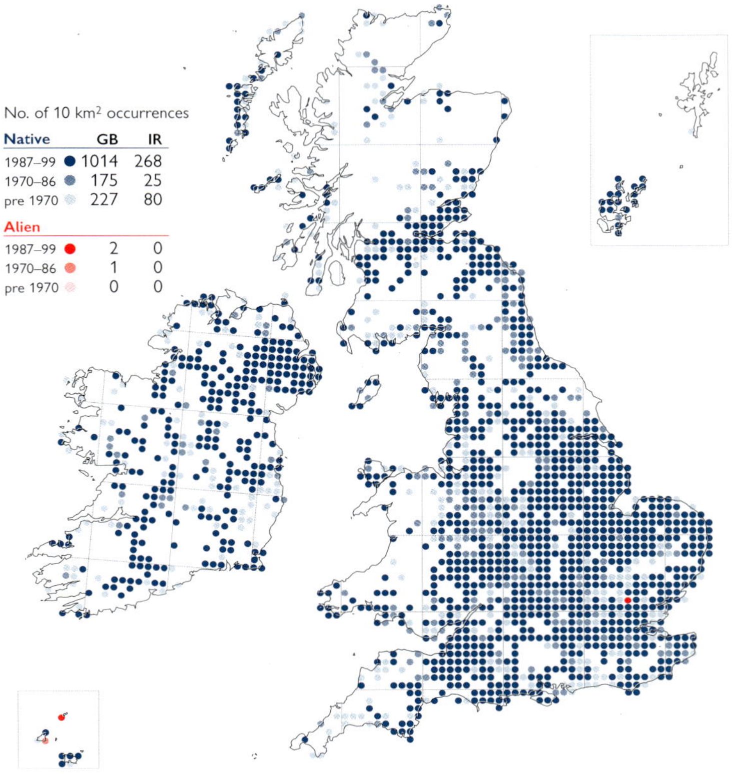

No. of 10 km² occurrences

Native	GB	IR
1987–99 ●	1014	268
1970–86 ●	175	25
pre 1970 ●	227	80
Alien		
1987–99 ●	2	0
1970–86 ●	1	0
pre 1970 ●	0	0

This species grows in a wide range of meso-eutrophic or eutrophic and often calcareous waters. These include lakes, ponds, rivers, canals and ditches. It may persist in managed ditches and rivers and colonises newly flooded mineral workings. It also grows in slightly brackish sites. 0–390 m (Drumore Loch, E. Perth).

Native (change +0.63). This is the commonest *Myriophyllum* species in the eutrophic lowlands of Britain and in some areas of Ireland. Like many aquatic plants it is better recorded now than in the 1962 *Atlas*.

Eurasian Temperate element, but naturalised in N. America so distribution is now Circumpolar Temperate.

References: Atlas (150a), Aiken *et al.* (1979), Hultén & Fries (1986), Meusel *et al.* (1978), Preston & Croft (1997).

C. D. PRESTON

Myriophyllum alterniflorum Alternate Water-milfoil

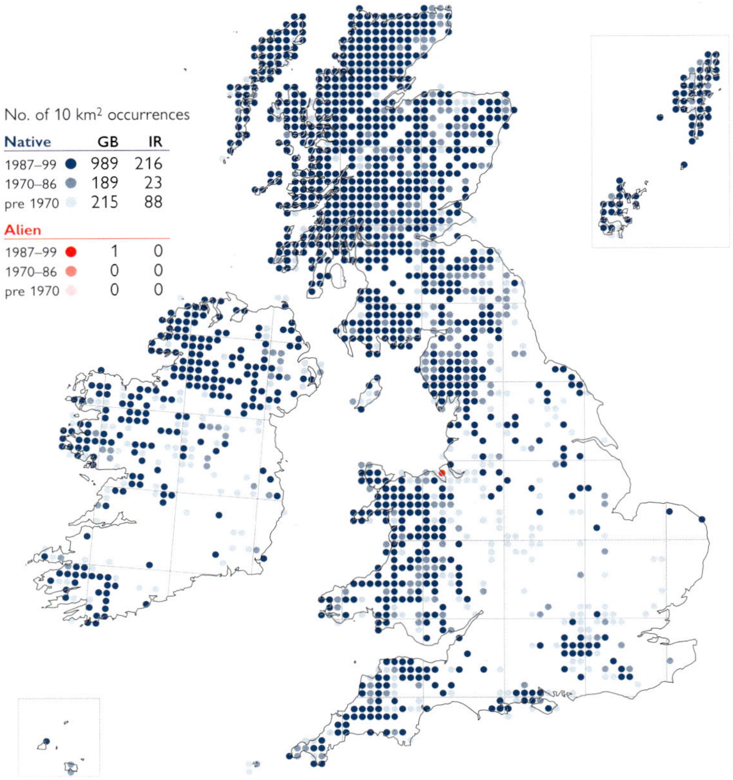

No. of 10 km² occurrences

Native	GB	IR
1987–99 ●	989	216
1970–86 ●	189	23
pre 1970 ●	215	88
Alien		
1987–99 ●	1	0
1970–86 ●	0	0
pre 1970 ●	0	0

This submerged aquatic perennial occurs in both standing and flowing waters, including rapidly flowing, peaty streams and rivers in which few other macrophytes grow. In Scotland and Ireland it occurs in a wide range of habitats (including, occasionally, highly calcareous sites) but in S.E. England it is confined to acidic, mesotrophic or oligotrophic waters. 0–780 m (Lochan an Tairbh-uisge, Mid Perth).

Native (change +1.00). The map of this species is much more complete than that in the 1962 *Atlas*, with many more 10-km square records in Wales, Scotland and Ireland. The losses in lowland England largely occurred before 1930.

Suboceanic Boreo-temperate element; also in N. America.

References: Atlas (150b), Hultén & Fries (1986), Meusel *et al.* (1978), Preston & Croft (1997).

C. D. PRESTON

Gunnera tinctoria Giant-rhubarb

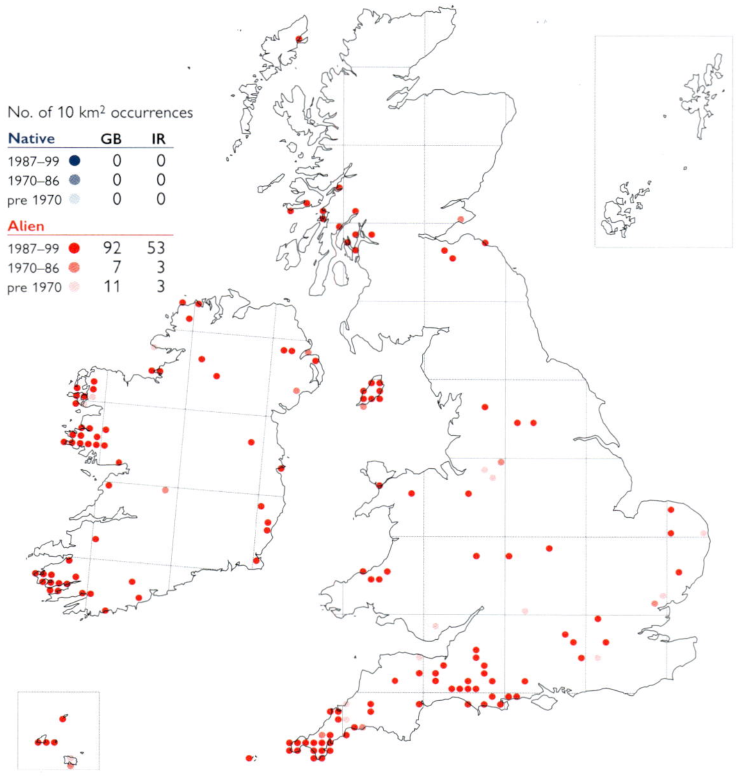

A large, robust perennial herb, capable of forming dense thickets of leaves in damp rough grassland, woodland, shaded places near lakes and rivers, and on sheltered sea-cliffs and beaches beside streams. It frequently reproduces by seed, producing long-lived and sometimes very invasive populations in suitable habitats. Lowland.

Neophyte. Introduced into cultivation in 1849, *G. tinctoria* is becoming increasingly popular in gardens, but often outgrows its space and is then discarded. It was known from the wild by at least 1908, and is probably increasing. Some mapped records may be errors for *G. manicata*, which is regarded as the more frequent species in Devon.

Native of western S. America.

T. D. DINES

Lythrum salicaria Purple-loosestrife

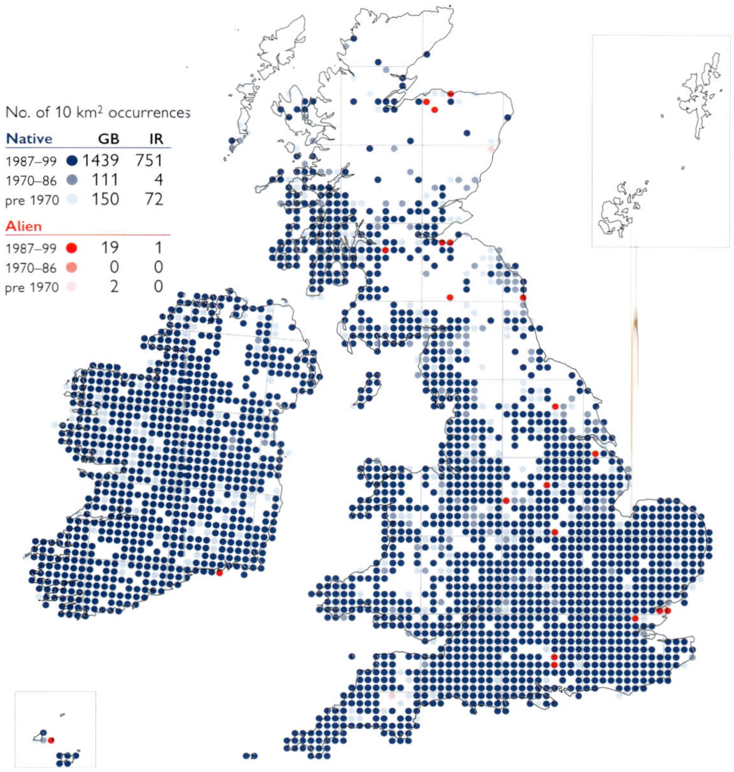

A perennial herb growing on the margins of slow-flowing rivers, canals, lakes, flooded gravel-pits, in tall-herb fens and willow carr. It thrives in permanently wet, or periodically inundated, fertile soils and tends to avoid acidic conditions. 0–440 m (Lake Ferta, S. Kerry).

Native (change −0.08). Although there have been local losses due to habitat destruction, this species is a good coloniser and can quickly become established in new sites. It is sometimes introduced to ponds and newly created wetlands. Its overall distribution is probably stable.

Eurasian Temperate element, but naturalised in N. America so distribution is now Circumpolar Temperate.

References: Atlas (142d), Hultén & Fries (1986), Mal *et al.* (1992), Meusel *et al.* (1978).

R. WILSON

Lythrum hyssopifolium Grass-poly

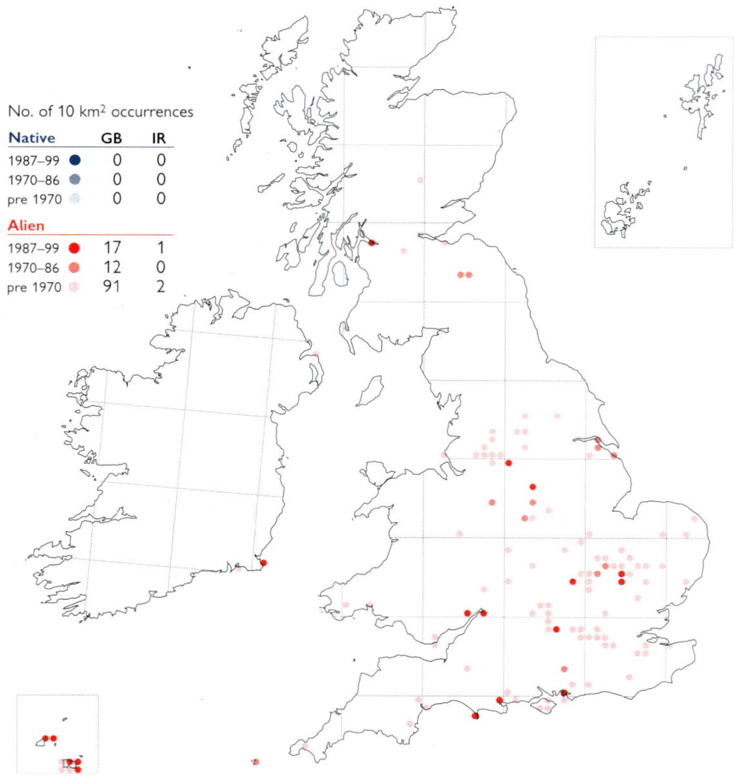

An annual of disturbed ground which is flooded in winter, including hollows and ruts in arable fields, and damp pastures disturbed in winter by numerous waterfowl. It sometimes occurs as a casual from seeds introduced with grain or from other sources. Lowland.

Archaeophyte (change −1.12). *L. hyssopifolium* has been rare since the middle of the 19th century. It may appear erratically at some sites, and can quickly colonise new ones, making trends difficult to judge. The population at Slimbridge (E. Gloucs.) was discovered in 1985 and had expanded by 1993 to an estimated 600 000 plants.

As an archaeophyte *L. hyssopifolium* has a Eurosiberian Southern-temperate distribution; it is widely naturalised outside this range.

References: Atlas (143a), Callaghan (1998), Meusel *et al.* (1978), Wigginton (1999).

R. WILSON

Lythrum portula Water-purslane

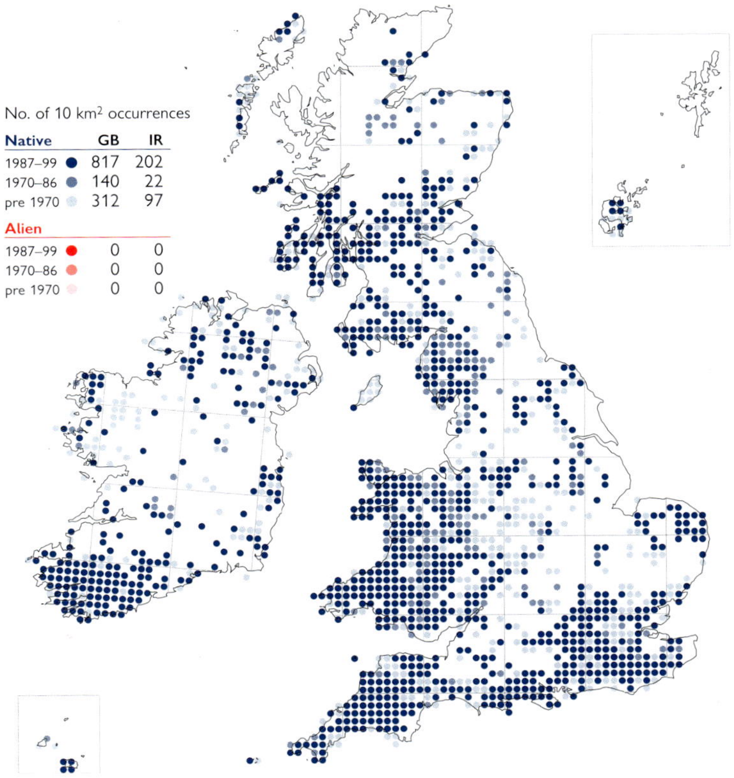

An annual of acidic or calcium-deficient silty soils at the muddy margins of pools and in temporarily flooded habitats such as rutted tracks, woodland rides, heathland pools and the draw-down zone of reservoirs. It avoids the most acidic and nutrient-poor soils and is rarely found over peat. 0–460 m (source of R. Teme, Herefs.).

Native (change +0.32). *L. portula* was mapped as 'all records' in the 1962 *Atlas*. Analysis of the database reveals that many losses have occurred since 1950. Reasons for this decline include ponds being drained or becoming overgrown, and rutted tracks being resurfaced or filled in. It is now much better recorded, particularly in W. Britain.

European Temperate element.

References: Atlas (143b), Atlas Supp. (41a, b), Hultén & Fries (1986), Meusel *et al.* (1978), Preston & Croft (1997).

R. WILSON

Daphne mezereum Mezereon

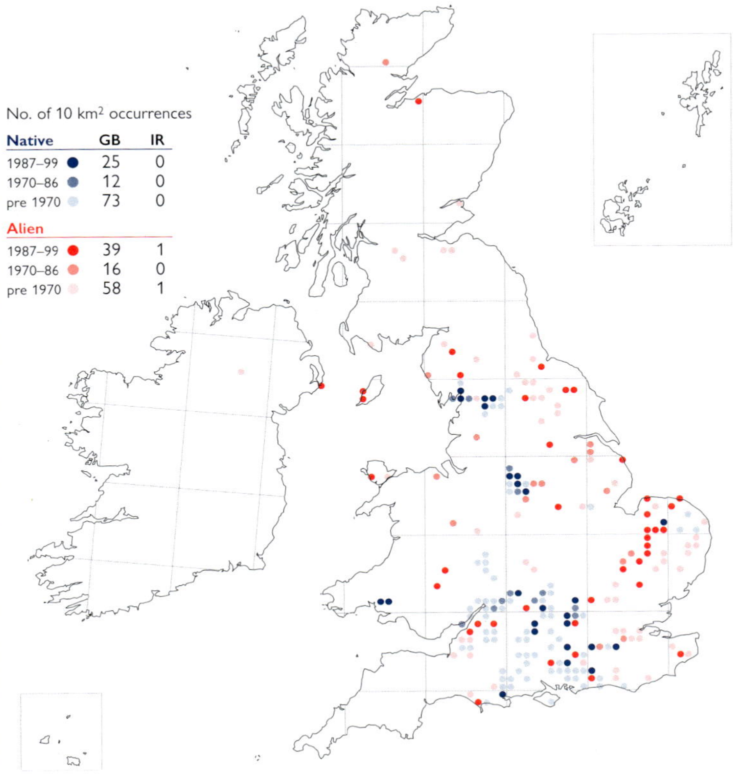

A deciduous shrub of calcareous woodland, often on steep, sometimes rocky, slopes with little ground cover, but rarely in deep shade. It also grows in chalk-pits, and in wet, species-rich fens. It reproduces by seed and is self-fertile. 0–335 m (Ling Gill, Ribblesdale, Mid-W. Yorks.).

Native or alien (change –0.06). *D. mezereum* was not recorded in the wild until 1752. It has been cultivated for centuries and escapes are common, so its native range is somewhat uncertain. Native populations have suffered from habitat loss and uprooting, but it is common in gardens, is frequently bird-sown, and is increasing as an alien.

Eurosiberian Boreo-temperate element; widely naturalised outside its native range.

References: Atlas (143c), Hultén & Fries (1986), Meusel *et al.* (1978), Stewart *et al.* (1994).

A. J. RICHARDS

Daphne laureola Spurge-laurel

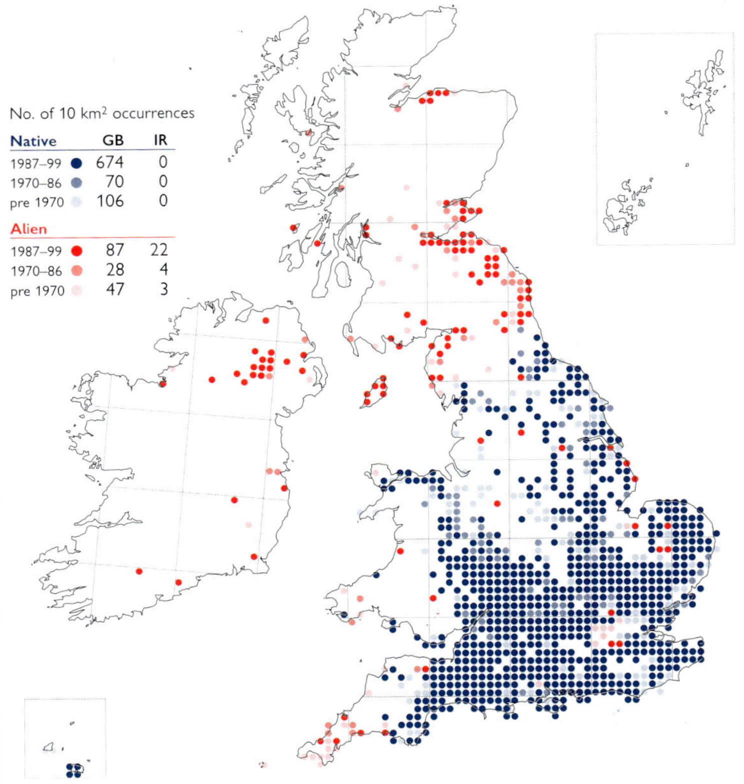

An evergreen, low-growing shrub of heavy, neutral to basic soils in deciduous woodland, often in quite deep shade. It reproduces by self-layering and by seed, but as it flowers in the early spring and requires cross-pollination by flies or moths, seed-set is often poor. Lowland.

Native (change +0.10). *D. laureola* is usually regarded as a native in England and Wales, but it is also a relic of cultivation or introduction in many sites, and across much of its British range is more typical of pheasant-rearing estates and parklands than semi-natural woodland. Many squares mapped as native are more likely to be alien.

Submediterranean-Subatlantic element.

References: Atlas (143d), Meusel *et al.* (1978).

A. J. RICHARDS

Epilobium hirsutum Great Willowherb

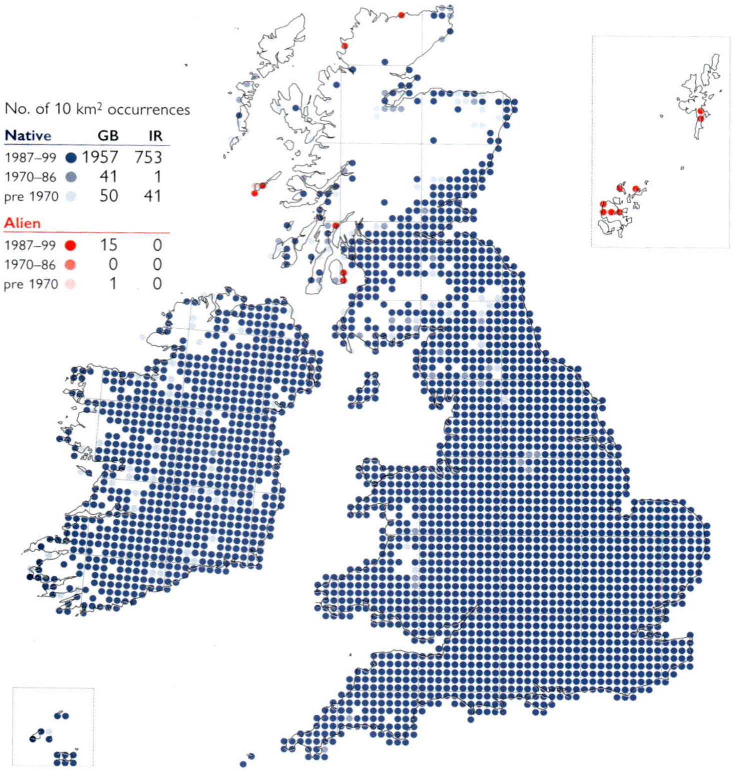

No. of 10 km² occurrences

Native		GB	IR
1987–99	●	1957	753
1970–86	●	41	1
pre 1970	●	50	41
Alien			
1987–99	●	15	0
1970–86	●	0	0
pre 1970	●	1	0

A tall perennial herb of open habitats, including ditches, marshes, the edges of streams or ponds, damp woodland margins and waste ground. It thrives in wet, fertile, neutral to basic habitats, although it can tolerate dry ground. It spreads from seed, or from branching rhizomes which may result in dense stands. 0–365 m (Derbys.), with an exceptional record at 845 m on Great Dun Fell (Westmorland).

Native (change +0.12). The distribution of *E. hirsutum* is much the same as in the 1962 *Atlas*, although increases in Wales, N. England and S.W. Scotland suggest that it may have expanded into these areas.

Eurasian Southern-temperate element; widely naturalised outside its native range.

References: Atlas (144c), Grime *et al.* (1988), Hultén & Fries (1986), Meusel *et al.* (1978), Shamsi & Whitehead (1974).

G. D. KITCHENER

Epilobium parviflorum Hoary Willowherb

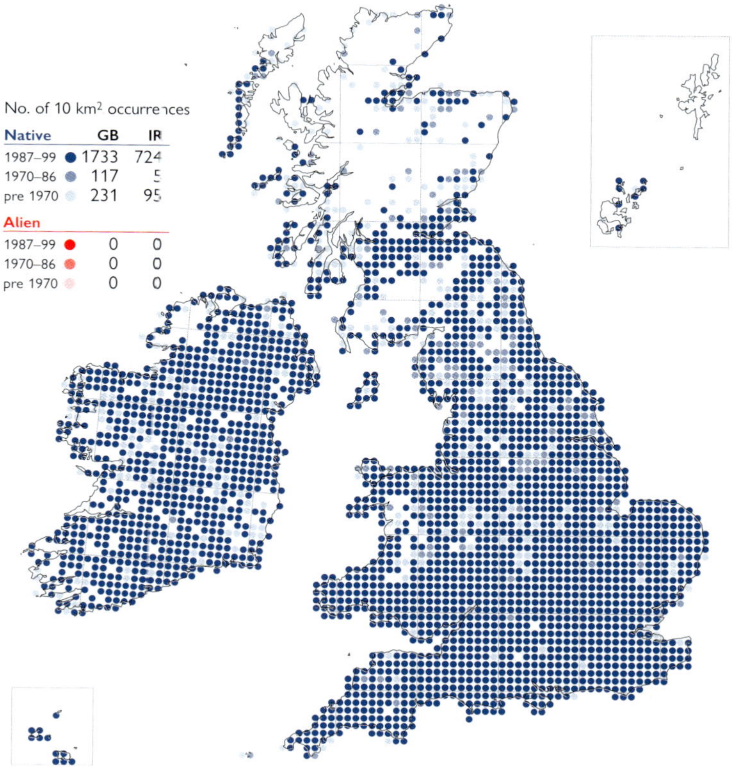

No. of 10 km² occurrences

Native		GB	IR
1987–99	●	1733	724
1970–86	●	117	5
pre 1970	●	231	95
Alien			
1987–99	●	0	0
1970–86	●	0	0
pre 1970	●	0	0

A perennial herb, most frequent in marshes, fens, streamsides and other disturbed wet places, spreading by air-borne seed. However, the species will also grow in dry habitats, including quarries, banks and waste ground and as a street weed. Generally lowland, reaching 365 m in Derbyshire.

Native (change –0.41). There is no evidence of significant change in the range of *E. parviflorum* since the 1962 *Atlas*.

European Temperate element.

References: Atlas (144d), Grime *et al.* (1988), Hultén & Fries (1986).

G. D. KITCHENER

Epilobium montanum Broad-leaved Willowherb

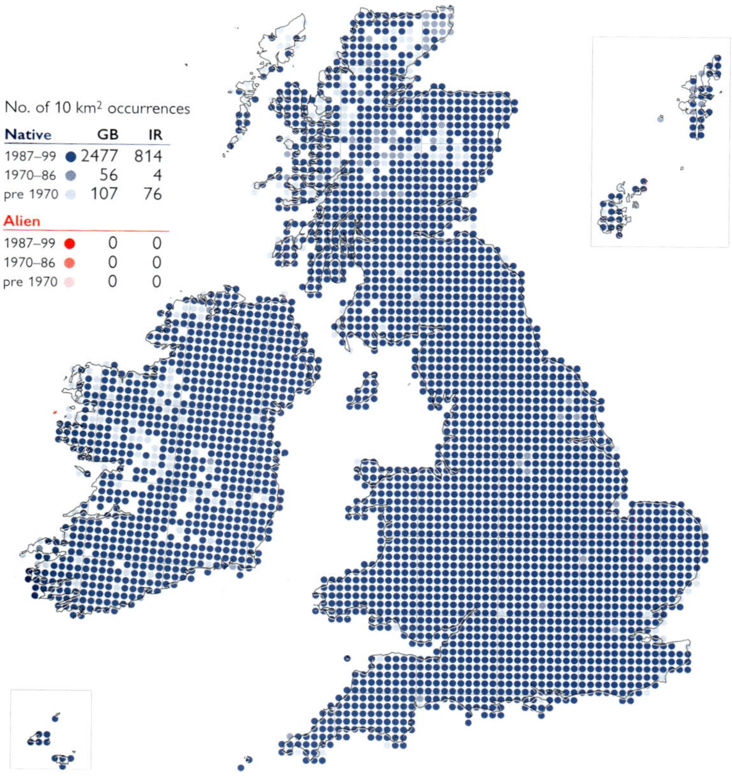

No. of 10 km² occurrences

Native		GB	IR
1987–99	●	2477	814
1970–86	●	56	4
pre 1970	●	107	76
Alien			
1987–99	●	0	0
1970–86	●	0	0
pre 1970	●	0	0

A perennial herb growing in woods, hedge banks, gardens and waste ground, and sometimes on rock ledges or gullies, often in shade. It spreads by wind-borne seed. 0–790 m (Snowdon, Caerns.), with an exceptional record at 845 m on Great Dun Fell (Westmorland).

Native (change –0.39). There is no evidence for changes in the national distribution of *E. montanum* since the 1962 *Atlas*; it is a frequent plant throughout its range.

European Temperate element; also in C. and E. Asia.

References: Atlas (145a), Grime *et al.* (1988), Hultén & Fries (1986), Meusel *et al.* (1978), Myerscough & Whitehead (1966, 1967).

G. D. KITCHENER

Epilobium montanum × E. obscurum (E. × aggregatum)

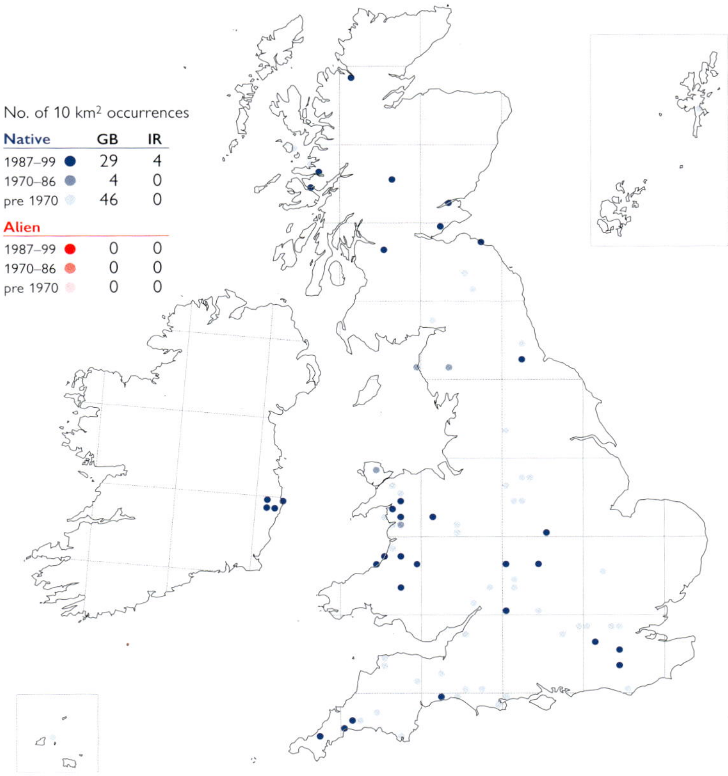

A perennial hybrid of disturbed ground, including gardens, quarries and woodland margins where the parents are found together. This hybrid is capable of producing F_2 progeny, and there is evidence of back-crossing but this is not normally found in large populations and F_2 plants are probably uncommon. Lowland.

Native. This hybrid is widespread but almost certainly under-recorded.

Widespread in Europe.

References: Seavey & Raven (1977), Stace (1975), Thakur (1965).

G. D. KITCHENER

Epilobium montanum × E. parviflorum (E. × limosum)

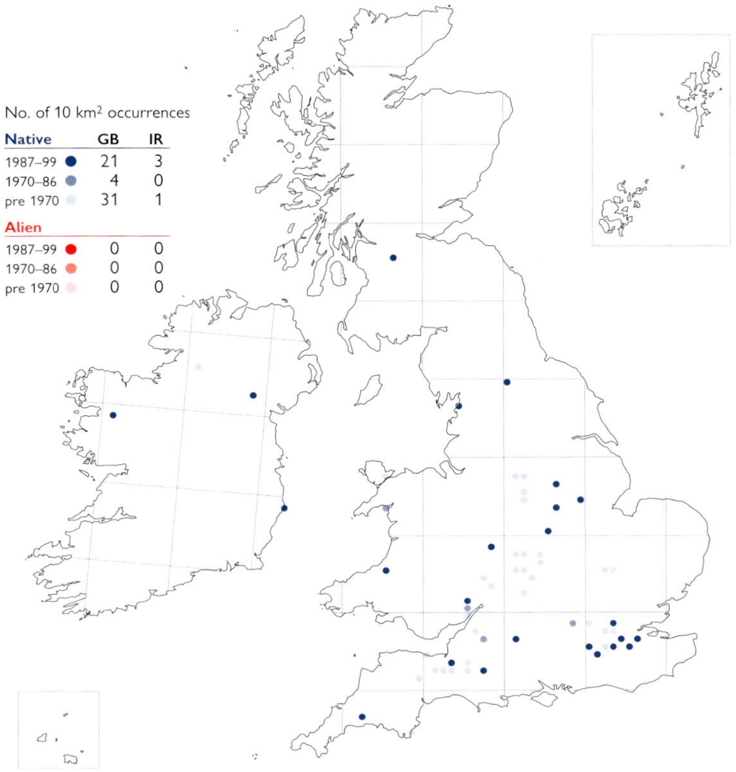

A perennial hybrid of disturbed ground where the parents are found together, for example, in quarries, felled woodland and waste ground. This hybrid is capable of producing F_2 progeny, and there is evidence of back-crossing but this is not normally found in large populations, suggesting that F_2 plants are uncommon. Lowland.

Native. This hybrid is almost certainly under-recorded, but the map reflects the more southern distribution of *E. parviflorum.*

Widespread in Europe.

References: Compton (1913), Lehmann & Schwemmle (1927), Stace (1975), Thakur (1965).

G. D. KITCHENER

Epilobium lanceolatum
Spear-leaved Willowherb

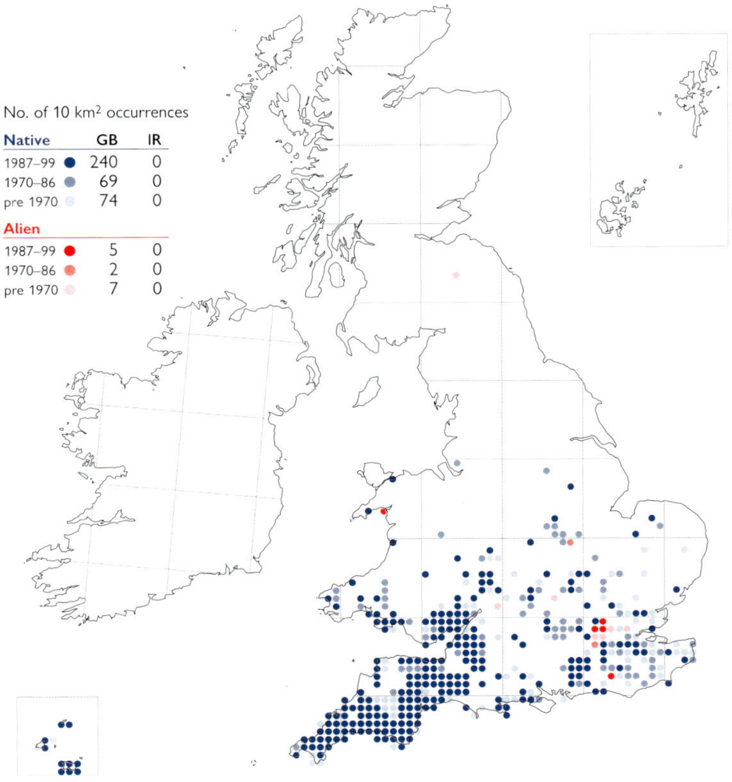

A perennial herb which grows in various dry habitats in S.W. England, including roadsides, walls, banks, quarries, streets and dunes. Elsewhere, it is more frequently a garden weed. 0–400 m (Dartmoor, S. Devon).

Native (change +0.07). There have been both apparent gains and losses of *E. lanceolatum* since the 1962 *Atlas* which are not easy to interpret, especially as it is a species which must be unfamiliar to many recorders. It has either spread, or been more comprehensively recorded, in S.W. England. Elsewhere, some garden weed occurrences may be impermanent, but nurseries act as a vector for fresh colonists.

Submediterranean-Subatlantic element.

References: Atlas (145b), Meusel *et al.* (1978), Stewart, *et al.* (1994), Walters (1979).

G. D. KITCHENER

Epilobium tetragonum
Square-stalked Willowherb

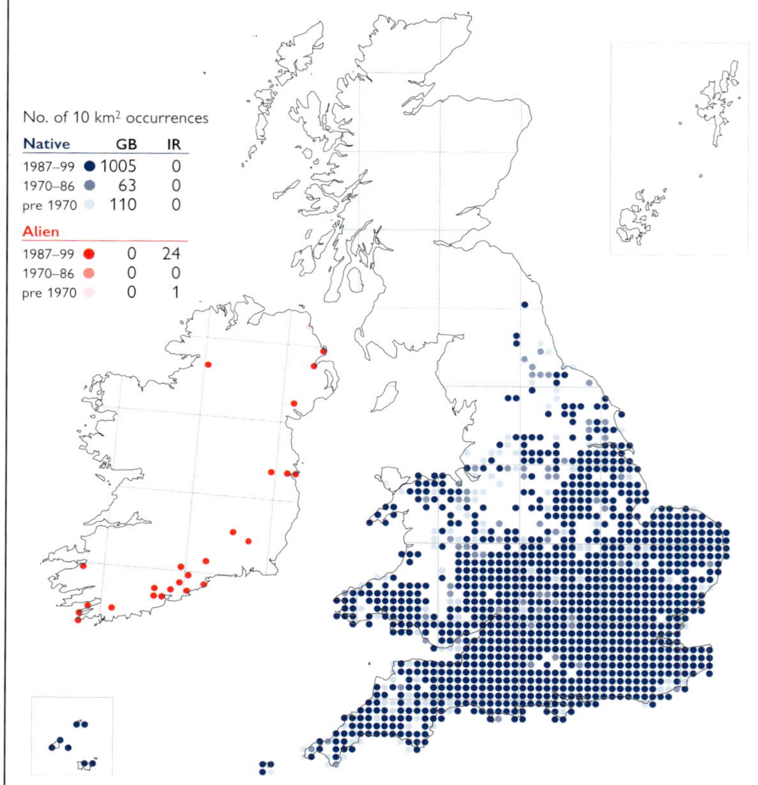

A perennial herb occurring in damp or dry habitats including cultivated or waste ground, gardens, quarries, streamsides and woodland margins. It spreads by seed. Lowland.

Native (change +1.66). The distribution of *E. tetragonum* in Britain remains stable and essentially southern, though it is now far better recorded. Many records have been made in Scotland since the 1962 *Atlas*, but none have been verified by expert examination and all have been treated as probable errors for *E. obscurum* or other species. It has increased in Ireland.

Eurosiberian Temperate element.

References: Atlas (146a, b), Hultén & Fries (1986).

G. D. KITCHENER

Epilobium obscurum Short-fruited Willowherb

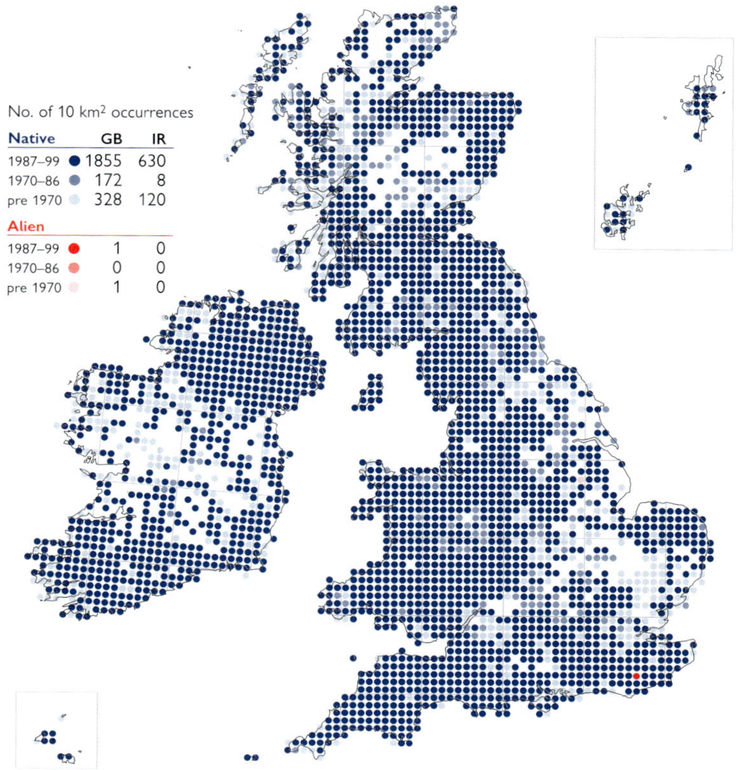

A perennial herb found on cultivated or waste ground, marshes, streamsides and woodland margins. It thrives in damp habitats, but tolerates dry ground, and spreads by wind-borne seed and long stolons. 0–775 m (Mangerton Mountain, S. Kerry) and 845 m on Great Dun Fell (Westmorland).

Native (change +0.38). This species has been more comprehensively recorded than it was in the 1962 *Atlas*, particularly in N. Ireland and W. Britain. There have, however, been local losses in E. England and W. Scotland. The reasons for these losses are unclear, unless recorders in the 1950s were less familiar with the alien *E. ciliatum* and reported it as this species.

European Temperate element.

References: Atlas (146c), Grime *et al.* (1988), Hultén & Fries (1986).

G. D. KITCHENER

Epilobium roseum Pale Willowherb

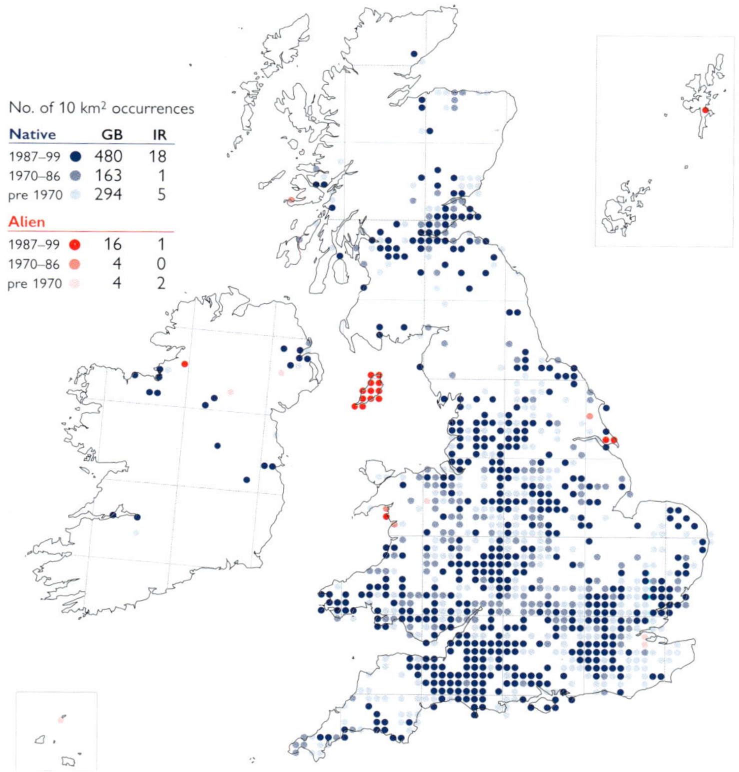

A perennial herb of damp disturbed places, found near streams and canals, in woods, on shaded banks, and in street gutters and gardens. Generally lowland, but reaching 560 m on Wold Fell (Mid-W. Yorks.).

Native (change –0.25). *E. roseum* is a distinctive species, especially once seen, which is much better recorded now than it was in the 1962 *Atlas*. The current map, however, shows many losses at the 10-km square level, and there seems to have been some retrenchment of distribution in favour of *E. ciliatum*, although the patterns are difficult to interpret.

Eurosiberian Temperate element.

References: Atlas (145c), Hultén & Fries (1986).

G. D. KITCHENER

Epilobium ciliatum American Willowherb

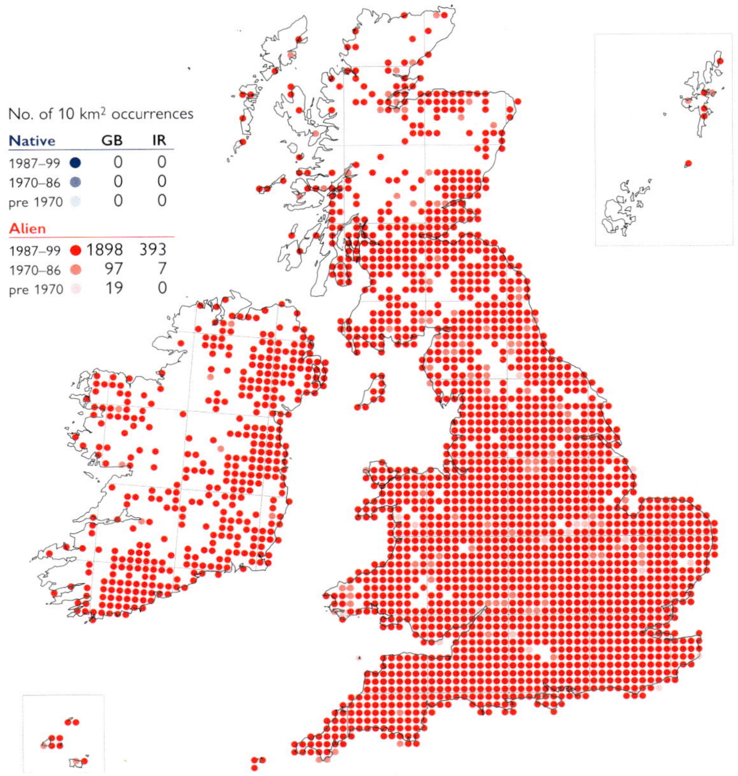

A perennial herb found on disturbed ground such as in gardens, shrubberies, felled woodland, waste ground, on walls and in pavement cracks. It spreads readily by seed, often colonising newly disturbed sites. It occasionally grows in semi-natural habitats, including marshes. 0–450 m (above Garrigill, Cumberland).

Neophyte (change +3.88). This species was first collected in 1891, but was unrecognised until the 1930s. It established itself most strongly in south-east England, and has spread rapidly north and west. The current map shows a dramatic extension from the 1962 *Atlas* and it is now one of our most frequent aliens.

Native of N. America; widely naturalised in C. & N. Europe.

References: Atlas (145d), Grime *et al.* (1988), Hultén & Fries (1986), Meusel *et al.* (1978), Myerscough & Whitehead (1966, 1967), Preston (1988).

G. D. KITCHENER

Epilobium ciliatum × *E. hirsutum* (*E.* × *novae-civitatis*)

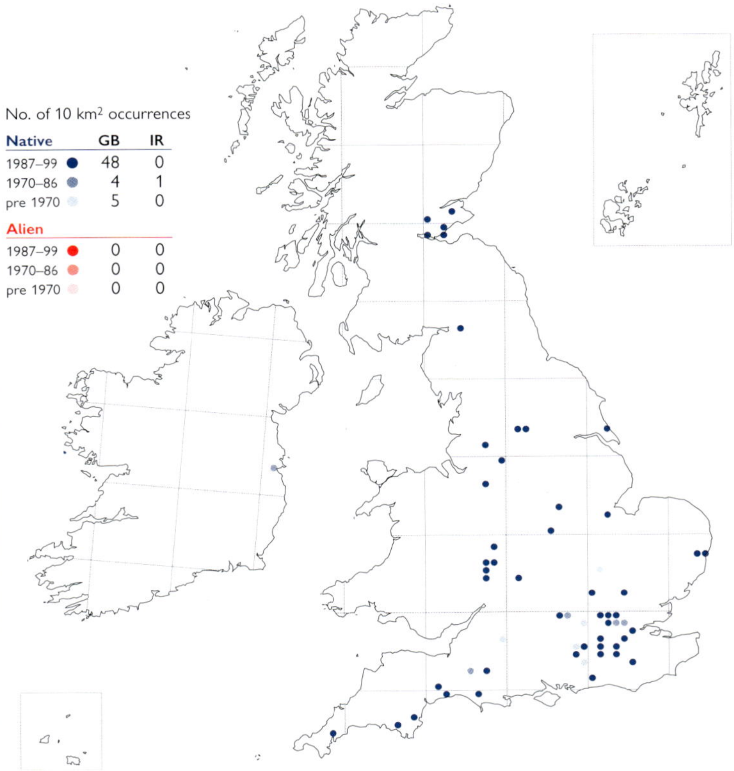

A perennial herb of disturbed ground where the parents are found together, including gardens, waste ground, ditches, felled woodland, quarries and set-aside. Some viable seed is produced, but the plant is usually found as single specimens, suggesting that F_2 progeny are rare. Lowland.

A spontaneous hybrid between native and alien parents. *E.* × *novae-civitatis* was first recorded in 1936 (N. Hants.). It may be better recorded than many *Epilobium* hybrids due to its showy flowers, but almost certainly remains under-recorded.

Recorded from C. Europe.

Reference: Stace (1975).

G. D. KITCHENER

Epilobium ciliatum × *E. montanum*

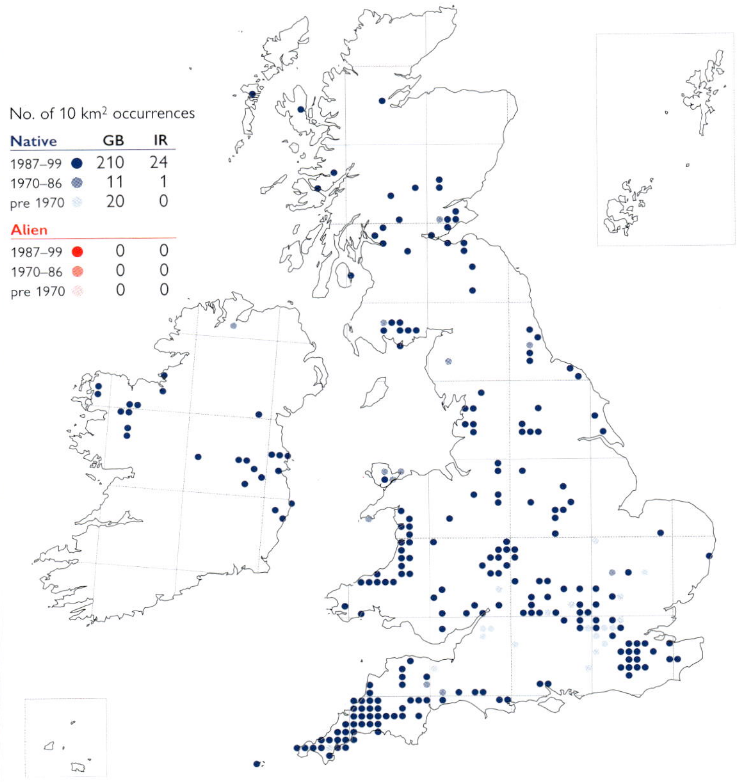

A perennial hybrid of disturbed ground where the parents may be found together, occurring on waste ground and roadsides, and in gardens. The hybrid is partly self-fertile, but populations are usually small, suggesting that F_2 plants are uncommon. Generally lowland, but reaching 400 m at Llanddewi-Brefi (Cards.).

A spontaneous hybrid between native and alien parents. This may be the commonest *Epilobium* hybrid in our area. It was first collected in 1930 (Surrey), as soon as *E. ciliatum* was recognised, but like many *Epilobium* hybrids it is almost certainly under-recorded. Localised concentrations of records indicate the activities of recorders familiar with the hybrid.

Recorded from C. Europe.

References: Brockie (1970), Stace (1975), Thakur (1965).

G. D. KITCHENER

Epilobium ciliatum × *E. obscurum*

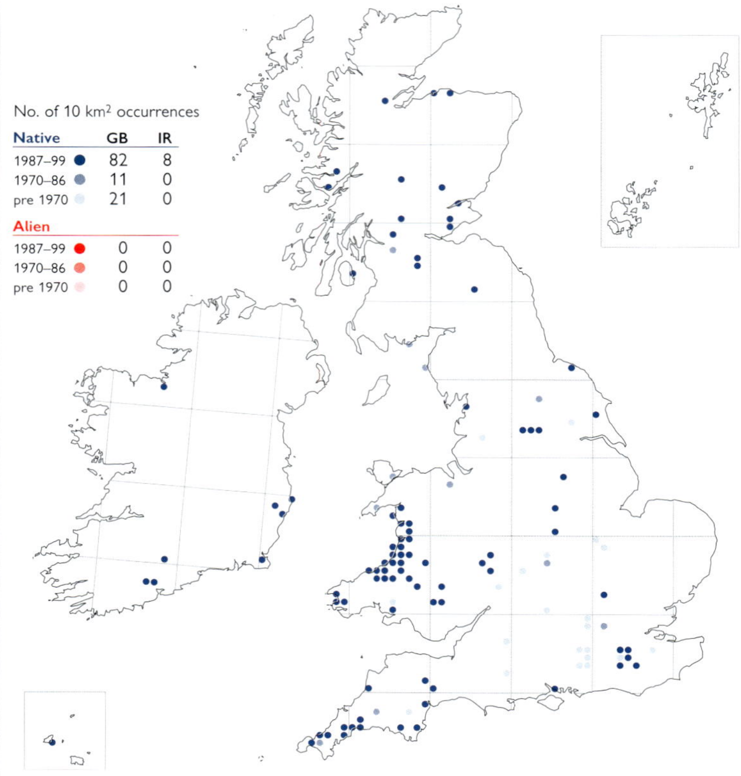

A perennial hybrid of disturbed ground, found in sites such as waste ground, gardens and roadsides where the parents may be found together. The hybrid is partly self-fertile, and is not normally found in significant numbers, suggesting that F_2 plants are uncommon. Lowland.

A spontaneous hybrid between native and alien parents. This hybrid was first recorded in 1934 (Surrey) and has not been mapped on a national level before. It is certainly under-recorded; localised concentrations of records indicate the activities of recorders familiar with the hybrid. Where both *E. obscurum* and *E. tetragonum* are present, their hybrids with *E. ciliatum* may be difficult to separate.

Recorded from C. Europe.

References: Ash (1947), Brockie (1970), Stace (1975), Thakur (1965).

G. D. KITCHENER

Epilobium ciliatum × *E. parviflorum*

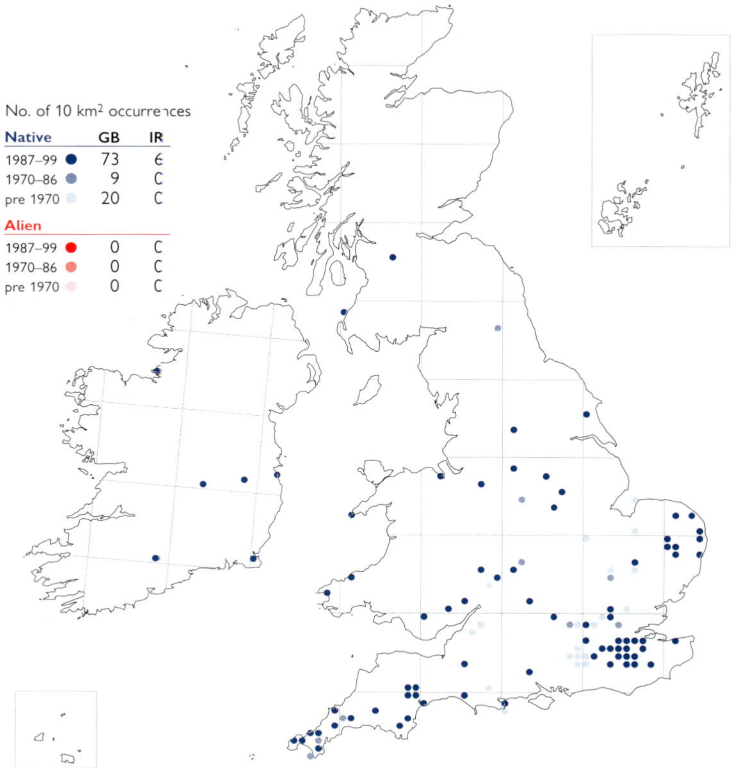

A perennial herb of disturbed places where the parents coincide, such as in quarries and felled woodland, and on waste ground and roadsides. This hybrid can produce F_2 progeny, and there are occasional large populations which may exhibit back-crossing. Generally lowland.

A spontaneous hybrid between native and alien parents. This hybrid was first recorded in 1934 in Surrey. Its distribution follows that of its parents, with the relative infrequency of records in N. Britain reflecting the more limited presence of *E. parviflorum* there. It is certainly under-recorded.

Recorded from C. Europe.

References: Stace (1975), Thakur (1965).

G. D. KITCHENER

Epilobium palustre Marsh Willowherb

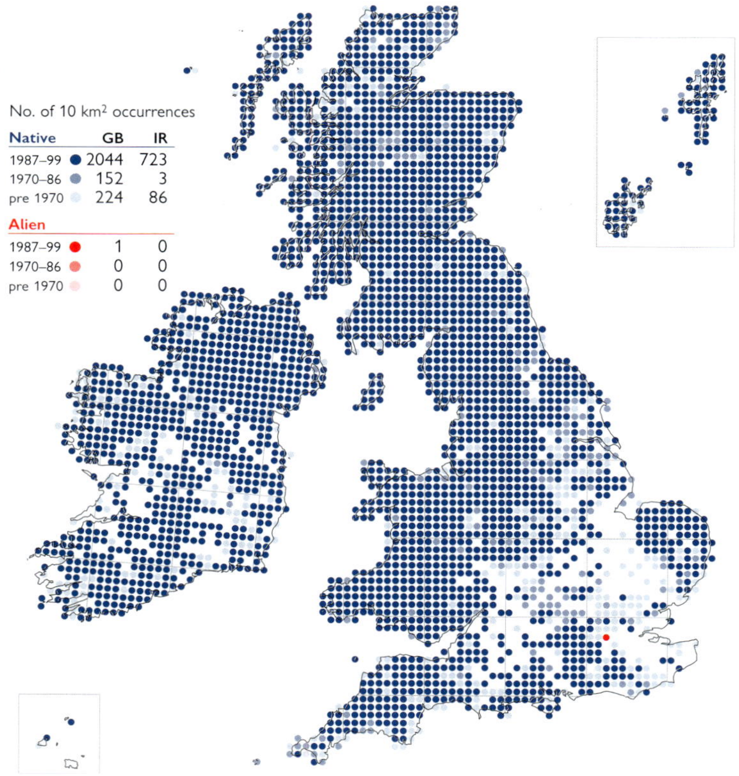

A stoloniferous perennial herb of wet acidic sites, occurring in bogs, marshes, flushes and ditches, and on stream or lake margins. It spreads by wind-borne seed and by turions which form at the ends of the stolons and can be carried off by winter floods. 0–795 m (Foel Fras, Caerns.) and 845 m on Great Dun Fell (Westmorland).

Native (change −0.18). *E. palustre* was mapped as 'all records' in the 1962 *Atlas*. The current map shows losses, particularly in S.E. England, and analysis of the database reveals that most of these have occurred since 1950. These declines are probably due to agricultural improvement, drainage and eutrophication.

Circumpolar Boreo-temperate element.

References: Atlas (146d), Grime *et al.* (1988), Hultén & Fries (1986), Kytövuori (1969), Meusel *et al.* (1978).

G. D. KITCHENER

Epilobium anagallidifolium Alpine Willowherb

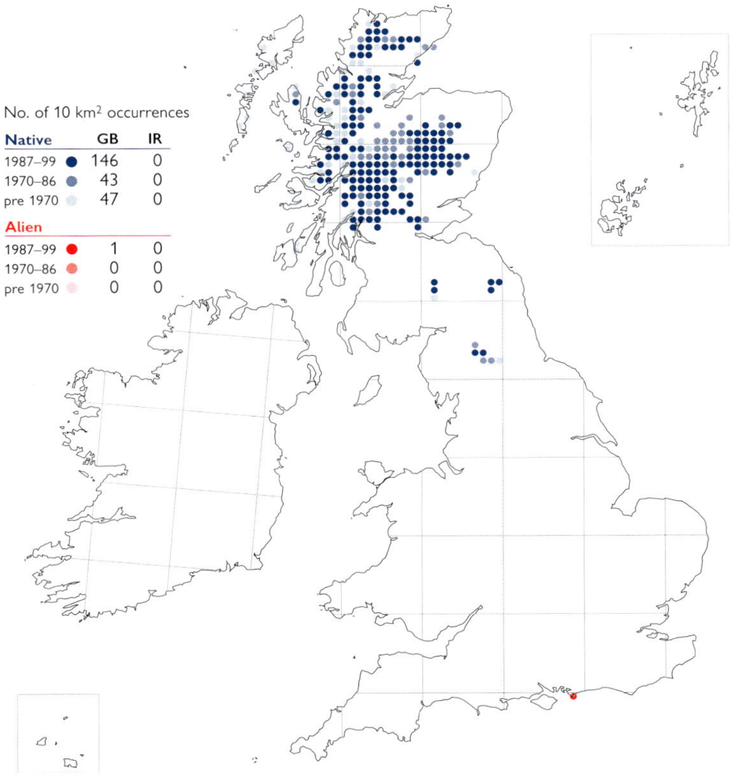

A very shortly stoloniferous perennial herb, growing in mossy mountain flushes, on steep wet slopes and by streams; sometimes washed down and persisting for a while by streams and rivers at lower altitudes. It grows from 155 m near Inchnadamph (W. Sutherland) to 1190 m on Ben Lawers (Mid Perth), but usually in the upper part of that range.

Native (change –0.76). This montane species may still be present in many of the 10-km squares for which there are only pre-1987 records.

Circumpolar Arctic-montane element, with a disjunct distribution.

References: Atlas (147a), Hultén & Fries (1986), Kytövuori (1972), Meusel *et al.* (1978).

G. D. KITCHENER

Epilobium alsinifolium Chickweed Willowherb

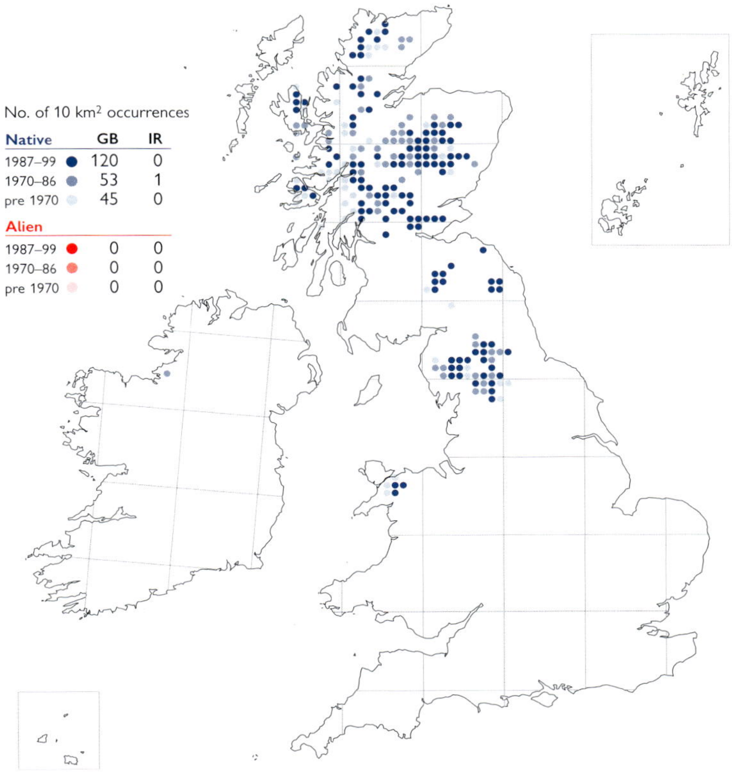

A montane perennial herb, spreading by stolons and occurring in or by mountain springs or streams, or on irrigated mountain ledges, often in moss carpets; sometimes temporarily established by streams at lower altitudes. It occurs on acidic and basic substrates, and tolerates eutrophic conditions. From 120 m on Eigg (N. Ebudes) to 1140 m on Bidean nam Bian (Main Argyll).

Native (change –0.41). The range of this species is largely unchanged since the 1962 *Atlas*, but is better recorded. It may still be present in many of the squares for which there are only pre-1987 records, but over-grazing may account for some local losses.

European Arctic-montane element; also in Greenland.

References: Atlas (147b), Curtis & McGough (1988), Hultén & Fries (1986), Kytövuori (1972), Meusel *et al.* (1978), Stewart *et al.* (1994).

G. D. KITCHENER

Epilobium brunnescens New Zealand Willowherb

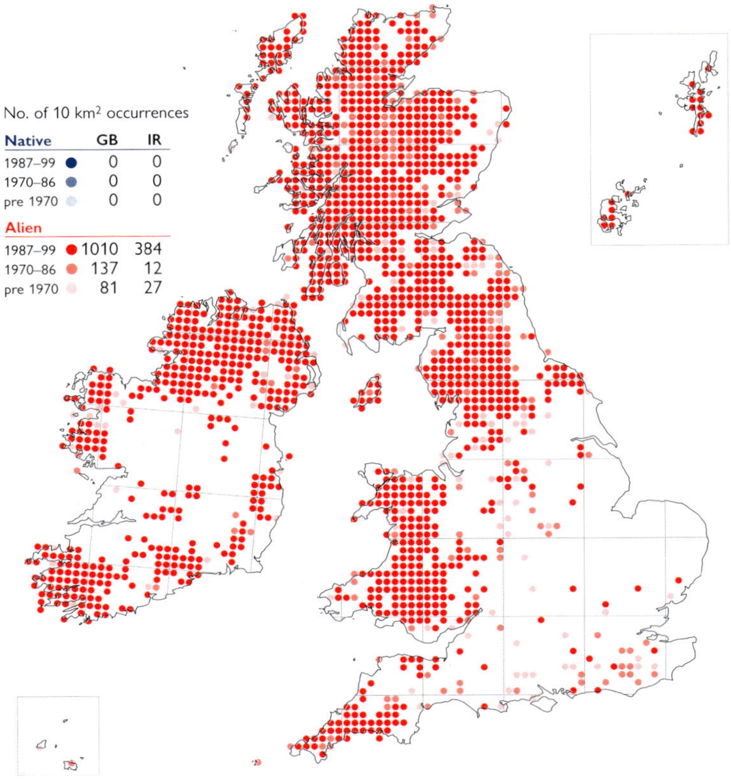

A creeping perennial herb of moist open areas, on gravel, gritty or stony soils, on substrates ranging from acidic to very base-rich. It grows on streamsides, ditches, tracks and paths, screes, quarries, damp stone walls and banks. It spreads by rooting at the nodes, and by seed, and is well-naturalised in many remote localities. 0–915 m (Y Garn, Caerns.) and reportedly higher in Ireland.

Neophyte (change +1.42). This species was first recorded in the wild in 1904 in Edinburgh, and its subsequent spread in the wild has accelerated from the 1930s. It is much more widespread than when mapped in the 1962 *Atlas*, particularly in the north and west.

Native of New Zealand.

References: Atlas (147c), Davey (1961), Harrison (1968), Hayward (1995), Kitchener & McKean (1998), Raven & Raven (1976). G. D. KITCHENER

Chamerion angustifolium Rosebay Willowherb

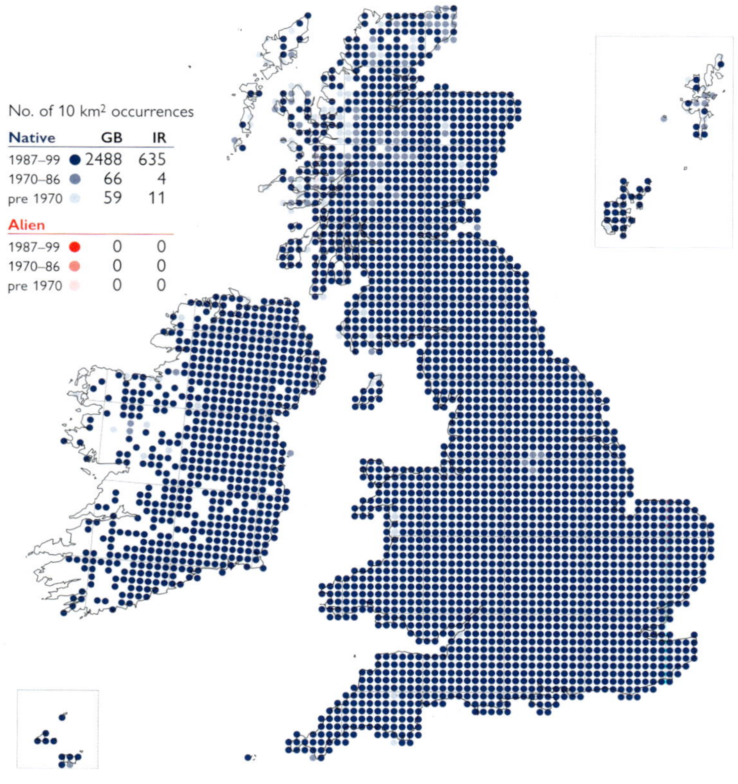

A rhizomatous perennial herb of moderately fertile soils, forming dense stands on disturbed, often burnt ground, on heaths, in woodland rides and clearings, on sand dunes, along tracksides, roadsides and railways, and in waste places; also in upland areas on rock ledges and screes. 0–975 m (Lochnagar, S. Aberdeen).

Native (change −0.01). This was a rare upland species in the early 19th century, but by the time of the 1962 *Atlas* its major extension of distribution in mainland Britain, perhaps originating from an overseas source, had already taken place. Since 1962 it has consolidated its range in Ireland. All records are mapped as if they are native.

Circumpolar Boreo-temperate element.

References: Atlas (147d), Broderick (1990), Grime *et al.* (1988), Harvey (1966), Hultén & Fries (1986), Meusel *et al.* (1978), Myerscough (1980).

G. D. KITCHENER

Ludwigia palustris Hampshire-purslane

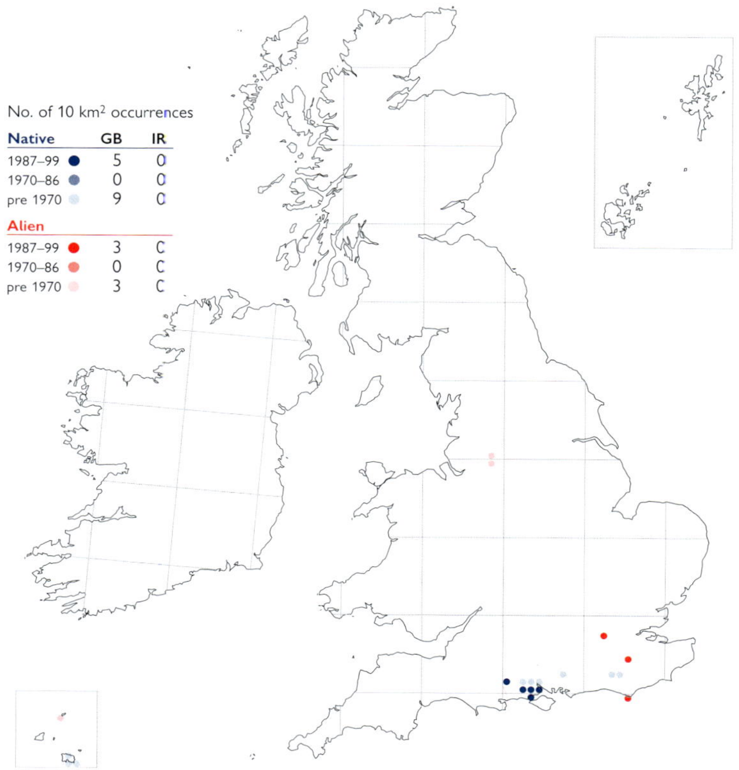

A perennial herb found by or in seasonally-inundated pits or pools overlying base-rich clays. It also occurs in runnels within valley mires and in poached areas in pasture or woodland glades. Lowland.

Native (change +0.19). The range of *L. palustris* continues to expand in the New Forest (S. Hants.), where it benefits from disturbance by livestock but soon disappears when enclosed. The Dorset colony, found in 1996, has been confirmed as this species. A recent Surrey record has been redetermined as the horticultural hybrid *L.* × *kentiana* and other alien records may require checking.

European Southern-temperate element; also in N. America and widely naturalised outside its native range.

References: Atlas (144b), Clement (2000), Cox (1997), Meusel *et al.* (1978), Preston & Croft (1997), Wigginton (1999).

D. A. PEARMAN

Oenothera agg. Evening-primroses

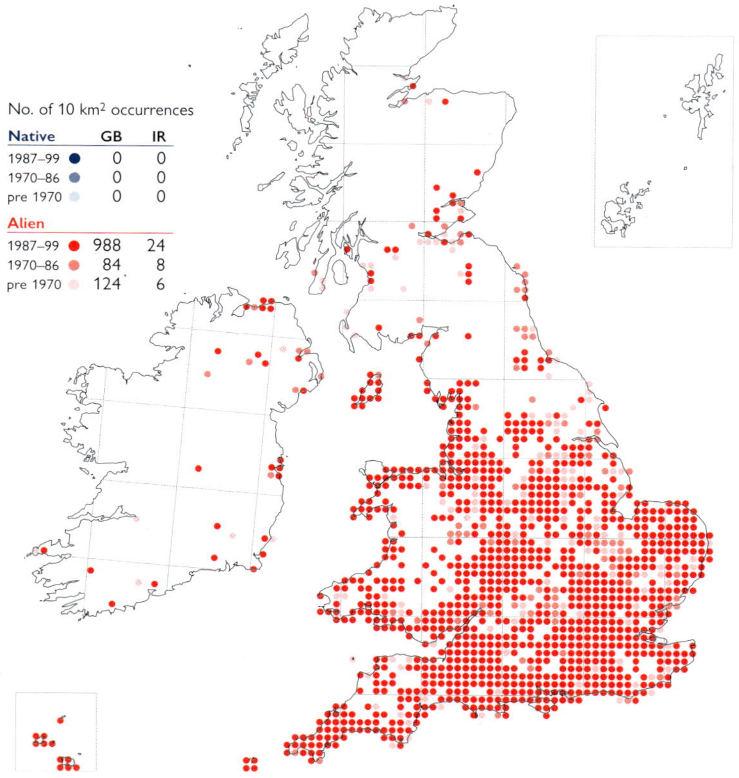

Annual or biennial herbs of open habitats on sandy or waste ground. Lowland.

Neophyte (change +1.02). *Oenothera* taxa were cultivated in Britain by 1629. The taxonomic interpretation of this genus has changed considerably over the last fifty years. Hybridisation and backcrossing are frequent, reciprocal crosses often having different characters, making identification very difficult. Some records mapped as species may therefore refer to hybrids.

An aggregate of closely related species originally from N. & S. America, and their hybrids and derivatives.

References: Atlas (148a-d), Dietrich *et al.* (1997), Rostański (1982).

C. A. STACE

Oenothera glazioviana
Large-flowered Evening-primrose

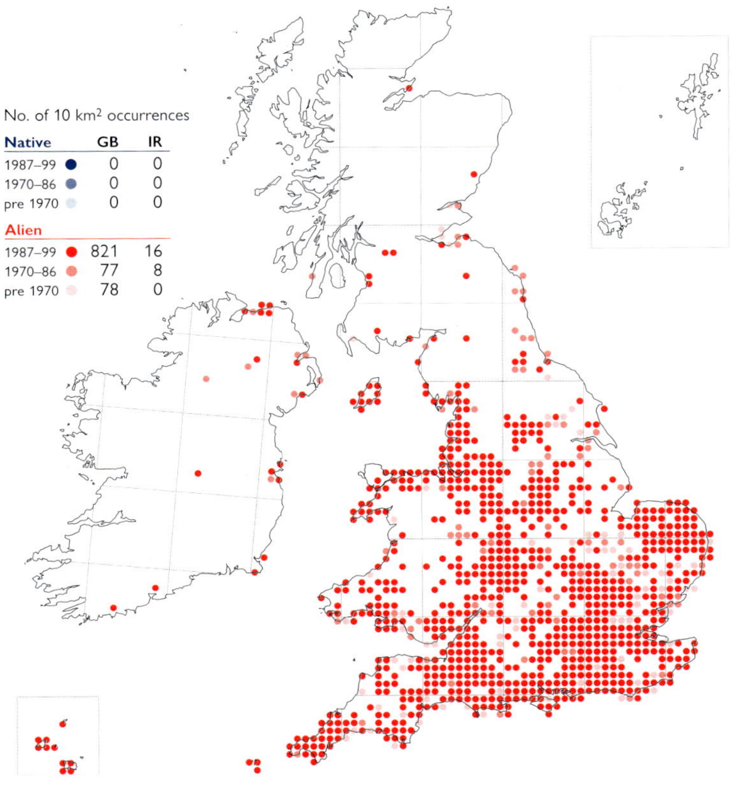

A tall, conspicuous biennial of open ground on sandy soils, often near the sea but also inland on waste ground, railway sidings and roadsides, in old cultivated areas, quarries and sand-pits. Large numbers are often found on freshly opened ground, though populations quickly decrease as the vegetation closes. Lowland.

Neophyte. *O. glazioviana* was first cultivated in Britain in about 1858, and recorded from the wild in 1866. It has spread and consolidated its range since the 1962 *Atlas*, and is now by far the commonest species of the genus in Britain. It readily hybridises with *O. biennis* and *O. cambrica.*

Native of N. America.

References: Atlas (148b), Dietrich *et al.* (1997), Rostański (1982).

C. A. STACE

Oenothera biennis Common Evening-primrose

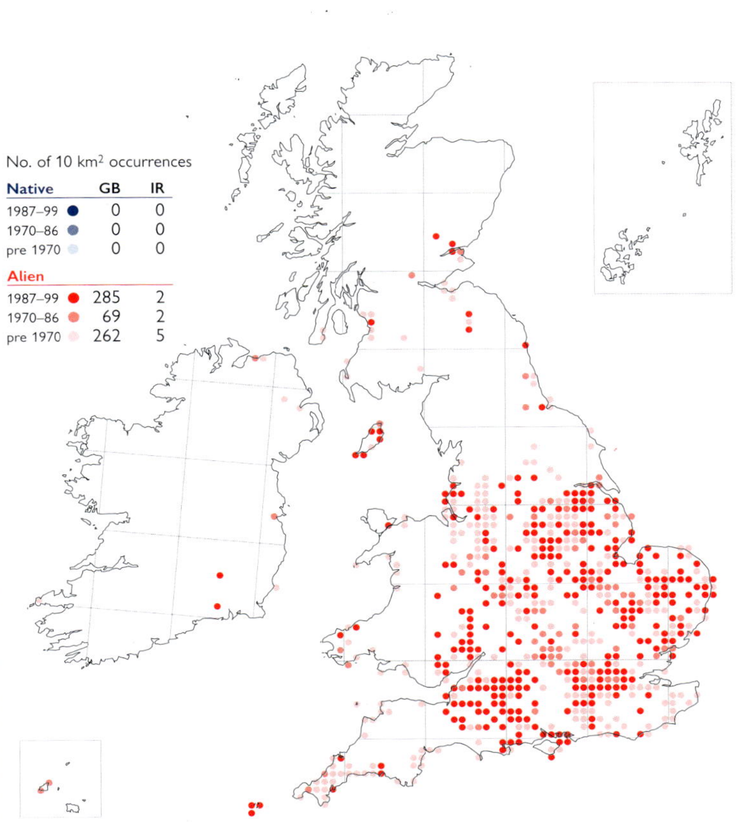

A tall biennial of open ground on sandy soils, found on sand dunes, river banks, waste ground, railway sidings and roadsides, in neglected cultivated fields, quarries, sand-pits and rubbish tips. Large numbers often occur on freshly opened ground, but they quickly decrease as the vegetation closes. Lowland.

Neophyte. *O. biennis* was cultivated in Britain by 1629 and was recorded in the wild in about 1650. It often grows with other *Oenothera* species, and hybrids may outnumber pure *O. biennis.* Losses since the 1962 *Atlas* are possibly due to the transient nature of some colonies and past misidentifications.

O. biennis was introduced either from N. America, or from Europe from N. American progenitors.

References: Atlas (148a), Dietrich *et al.* (1997), Hall *et al.* (1988), Hultén & Fries (1986), Meusel *et al.* (1978), Rostański (1982).

C. A. STACE

Oenothera biennis × *O. glazioviana* (*O.* × *fallax*) Intermediate Evening-primrose

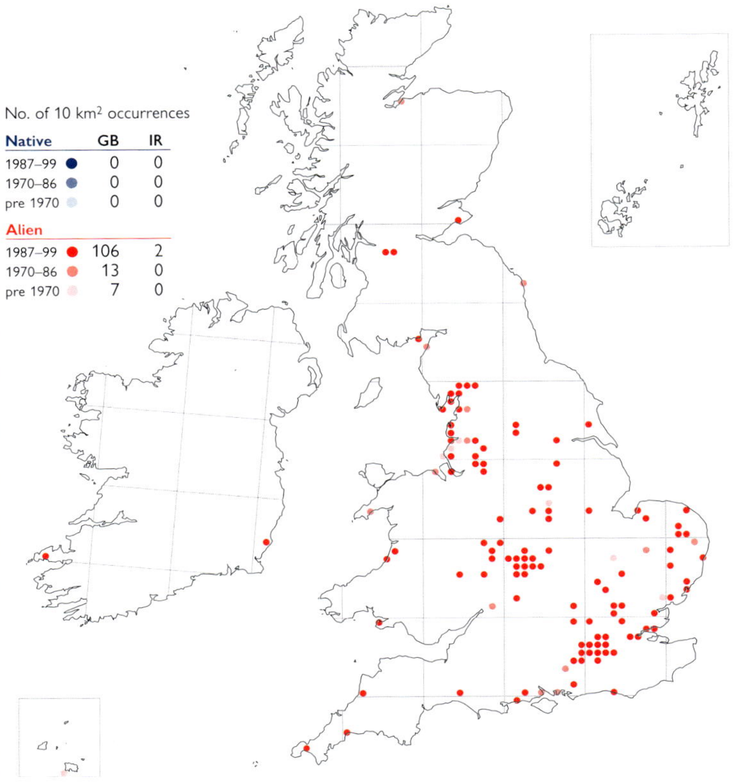

A tall biennial of open ground on light soils, especially on dunes and in sandy places, but also on waste ground, quarries, railway sidings and roadsides. Lowland.

A spontaneous hybrid between two alien parents. This hybrid often arises spontaneously, but it frequently occurs in the absence of one parent (especially *O. biennis*) or even both, its presence often indicating the former existence of the parental species at the same site. It also occurs independently as a garden escape. It was first recorded in the wild in 1978, and is probably under-recorded.

O. × *fallax* is also recorded from several other countries in C. Europe.

References: Bowra (1992b, 1997), Dietrich *et al.* (1997), Rostański (1982), Stace (1975).

C. A. STACE

Oenothera cambrica
Small-flowered Evening-primrose

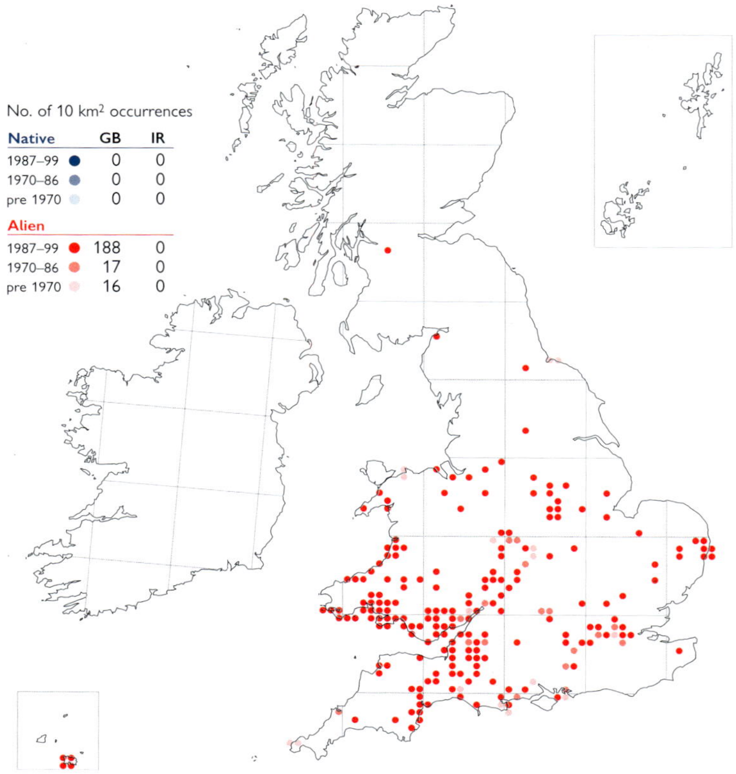

A tall biennial of open ground on light soils, found on sand dunes, sandy seashores and river banks, waste ground, tracksides, railway land and roadsides, dockland, and in old cultivated areas, quarries and sand-pits. Lowland.

Neophyte. *O. cambrica* was introduced to cultivation in Britain in 1775, and was first recorded from the wild in Cardiff (Glamorgan) in 1833. Hybrids with both *O. biennis* and *O. glazioviana* form readily and can outnumber or replace pure *O. cambrica*. It is now considerably more common than when mapped in the 1962 *Atlas*.

Probably native of N. America, or evolved in Europe from ancestors originally from N. America.

References: Atlas (148d), Bowra (1999), Dietrich (1991), Rostański (1982).

C. A. STACE

Oenothera stricta Fragrant Evening-primrose

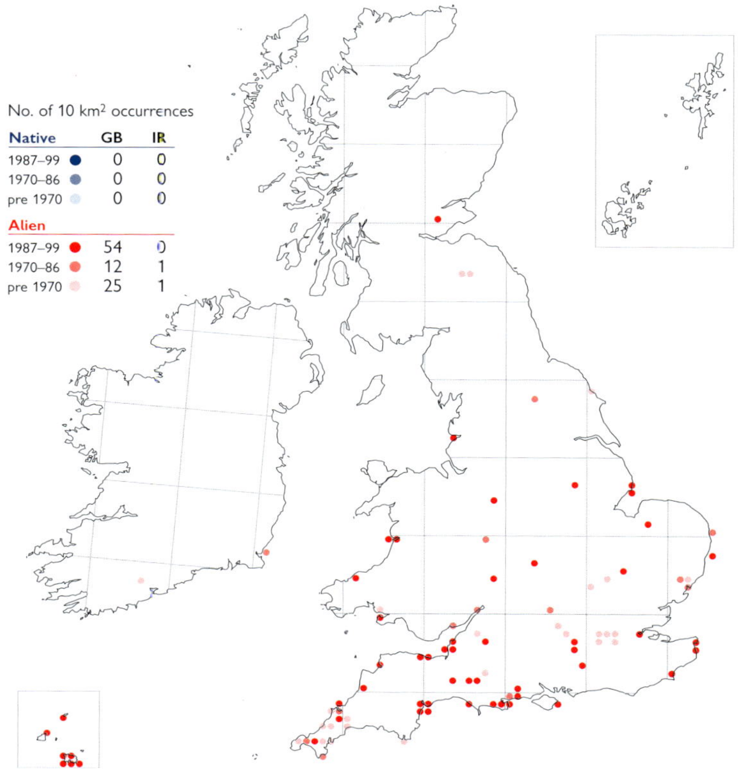

An erect to ascending biennial of open ground on light, frequently sandy soils, often in long-lasting, well-naturalised colonies near the sea but usually only casual inland. Lowland.

Neophyte. This species was first cultivated in Britain in 1790, and was found in the wild in the Channel Islands in 1847 and in Britain in 1852. It does not hybridise with the other *Oenothera* species, and remains scarce.

Native of S. America (Chile).

References: Atlas (148c), Rostański (1982).

C. A. STACE

Fuchsia magellanica Fuchsia

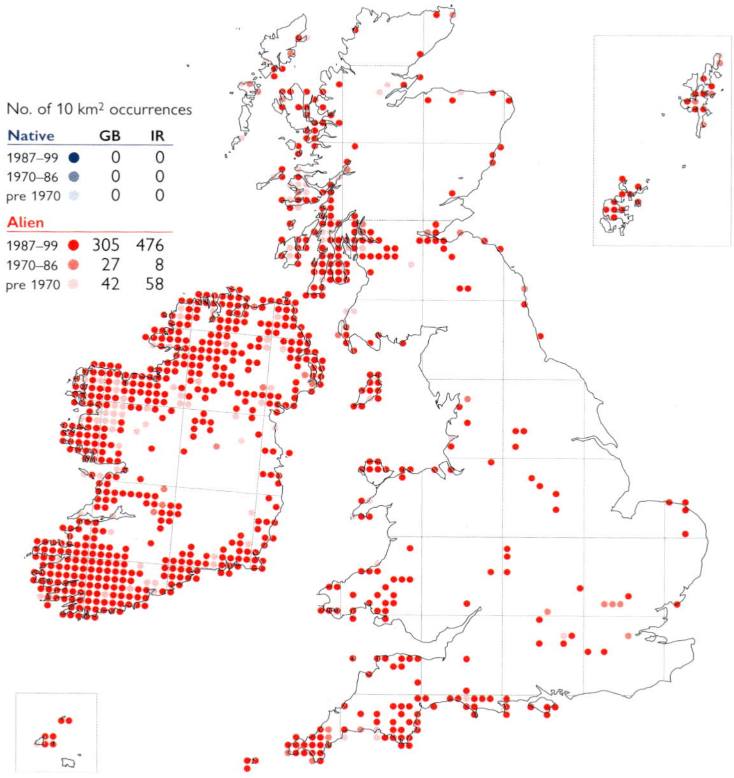

A spreading shrub, principally found as a planted hedge or in the gardens of abandoned cottages but also naturalised in hedgerows, scrub, by streams and amongst rocks and on walls. Reproduction is by suckering. Generally lowland, but upper altitudinal limit unknown.

Neophyte (change +1.85). *F. magellanica* was introduced to our area in 1788 and first recorded from the wild in 1857. Nearly all the hedges in W. Britain and Ireland are the cultivar 'Riccartonii', which apparently arose in a Scottish nursery before 1850 and is sterile. Other taxa occur, but their taxonomy has yet to be elucidated.

Native of S. America (Argentina, Chile).

References: Atlas (149a), Bean (1973).

D. A. PEARMAN

Circaea lutetiana Enchanter's-nightshade

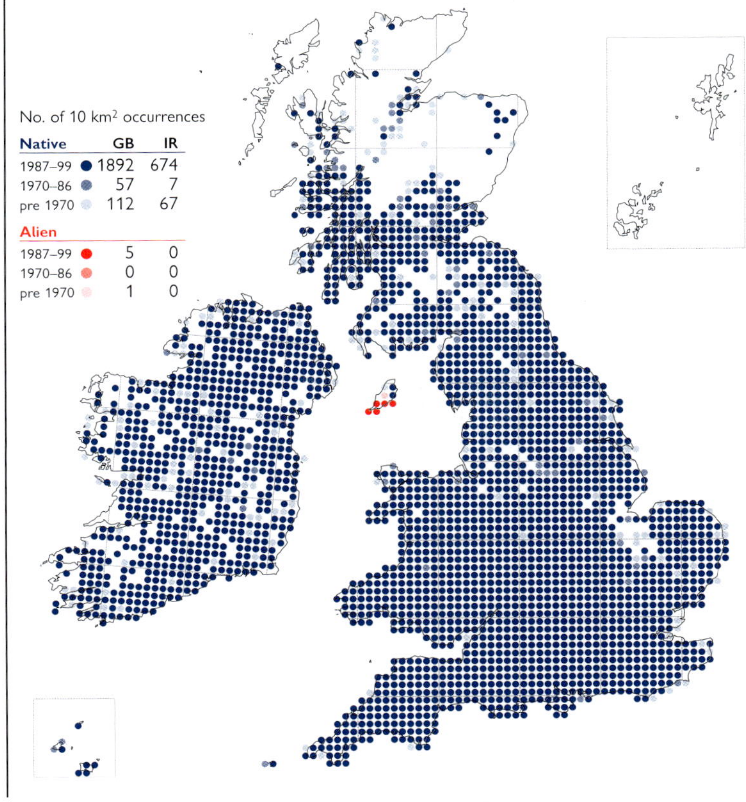

No. of 10 km² occurrences

Native		GB	IR
1987–99	●	1892	674
1970–86	●	57	7
pre 1970	●	112	67
Alien			
1987–99	●	5	0
1970–86	●	0	0
pre 1970	●	1	0

A perennial herb found in moist, usually base-rich, shaded habitats, including both ancient and secondary woodland, hedgerows, scrub, stream and river banks, and as a weed of cultivation. Its spread is aided by its long, brittle but very persistent rhizomes. Generally lowland, but reaching 465 m at Highfolds Scar (Mid-W. Yorks.).

Native (change –0.38). There has been no appreciable change in the overall range of this species since the 1962 *Atlas*.

European Temperate element; also in C. and E. Asia and N. America.

References: Atlas (149b), Grime *et al.* (1988), Hultén & Fries (1986), Meusel *et al.* (1978).

D. A. PEARMAN

Circaea alpina Alpine Enchanter's-nightshade

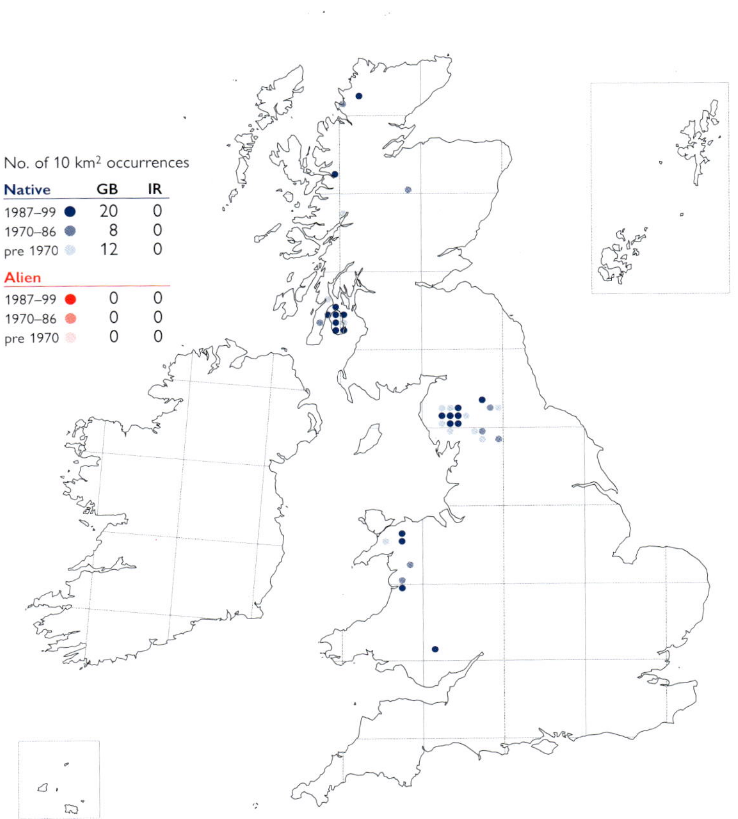

No. of 10 km² occurrences

Native		GB	IR
1987–99	●	20	0
1970–86	●	8	0
pre 1970	●	12	0
Alien			
1987–99	●	0	0
1970–86	●	0	0
pre 1970	●	0	0

A perennial herb typically associated with seepage areas within rocky, bryophyte-rich *Quercus* woodland, but also found amongst boulders and scree by the sides of streams and waterfalls, under *Pteridium*, and even amongst *Sphagnum*. It probably spreads by rhizomes and stolons, as well as by seed. Generally lowland, but reaching 755 m on Knock Fell (Westmorland).

Native. *C. alpina* was formerly confused with *C.* × *intermedia*, and its status and distribution were not clarified until 1963. The overall distribution is probably stable.

Circumpolar Boreal-montane element.

References: Atlas Supp. (42a), Hultén & Fries (1986), Meusel *et al.* (1978), Raven (1963), Stewart *et al.* (1994).

D. A. PEARMAN

Circaea alpina × *C. lutetiana* (*C.* × *intermedia*) Upland Enchanter's-nightshade

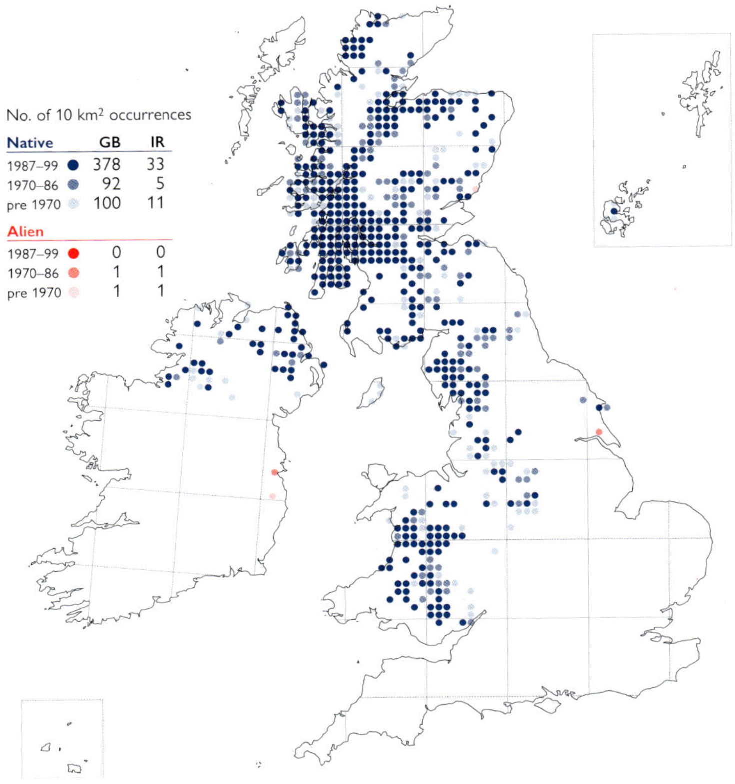

No. of 10 km² occurrences

Native		GB	IR
1987–99	●	378	33
1970–86	●	92	5
pre 1970	●	100	11
Alien			
1987–99	●	0	0
1970–86	●	1	1
pre 1970	●	1	1

A perennial herb occurring in moist wooded or shaded habitats, by streams and amongst wet rocks; also sometimes in gardens and on disturbed roadsides. It spreads by rhizomes, stolons and occasionally by seed, even though it is usually sterile. It can persist in the absence of one or both parents. Generally lowland, but reaching 560 m at Allt Creagach, Coignafearn (Easterness).

Native (change +0.48). This vigorous hybrid perhaps arose shortly after the last Glacial period when *C. alpina* was more frequent. The map of *C. alpina* agg. in the 1962 *Atlas* is equivalent to *C.* × *intermedia* here. The hybrid has apparently increased since then, probably owing to better recording.

Widespread in temperate Europe; also C. & E. Asia and N. America.

References: Atlas (149c), Hultén & Fries (1986), Meusel *et al.* (1978), Raven (1963), Stace (1975).

D. A. PEARMAN

Cornus sanguinea Dogwood

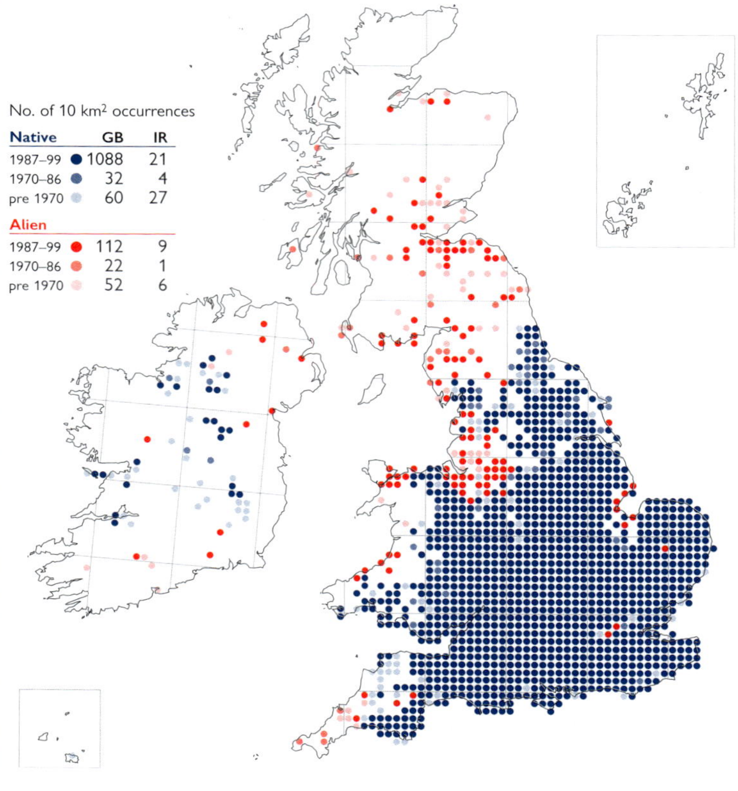

A deciduous shrub, locally frequent in woodland, scrub, hedgerows and shelter-belts on limestone soils or base-rich clays, and sometimes dominant in hedges and scrub on chalk. It is frequently planted in landscaping schemes and is introduced sporadically, or occurs as an escape, outside its native range. Lowland.

Native (change –0.06). There has been no significant change in the distribution of *C. sanguinea* since the 1962 *Atlas*, but introductions have begun to blur its native range. In Ireland the native range follows Scannell & Synnott (1987), but more work is needed to clarify the situation there.

European Temperate element.

References: Atlas (152c), Hultén & Fries (1986), Meusel *et al.* (1978).

G. T. D. WILMORE

Cornus sericea Red-osier Dogwood

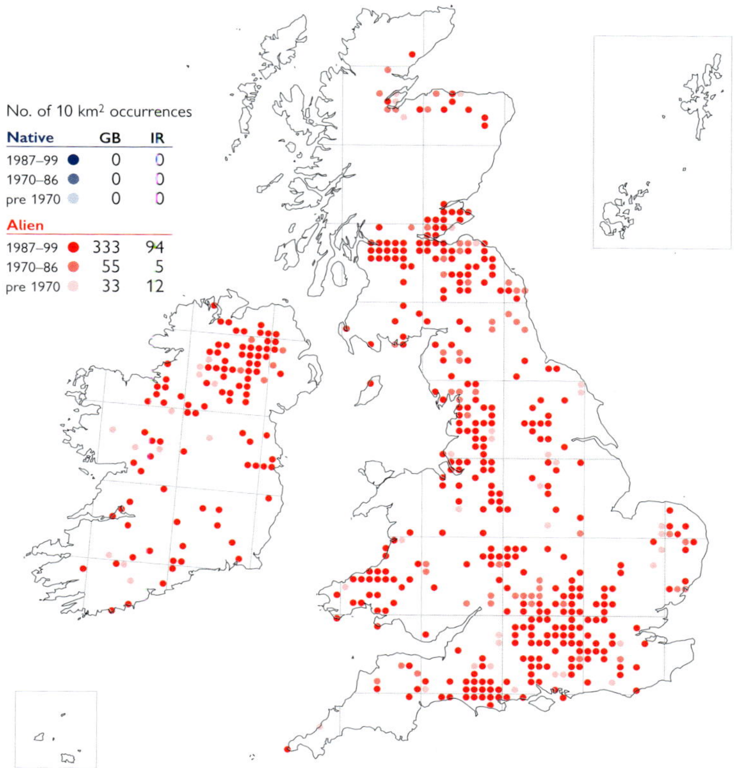

A deciduous shrub naturalised in woodland and along riversides, sometimes suckering to produce extensive thickets; also much planted in parkland, amenity plantings and on roadsides and sometimes occurs as an escape on waste ground and marginal land. Lowland.

Neophyte. *C. sericea* was cultivated in Britain by 1683. It was known from the wild by at least 1905 and is now well-naturalised in many places. It is probably increasing.

Native of N. America; closely related to the Eurasian *C. alba*.

References: Bean (1970), Hultén & Fries (1986), Kelly (1990), Meusel *et al.* (1978).

G. T. D. WILMORE

Cornus alba White Dogwood

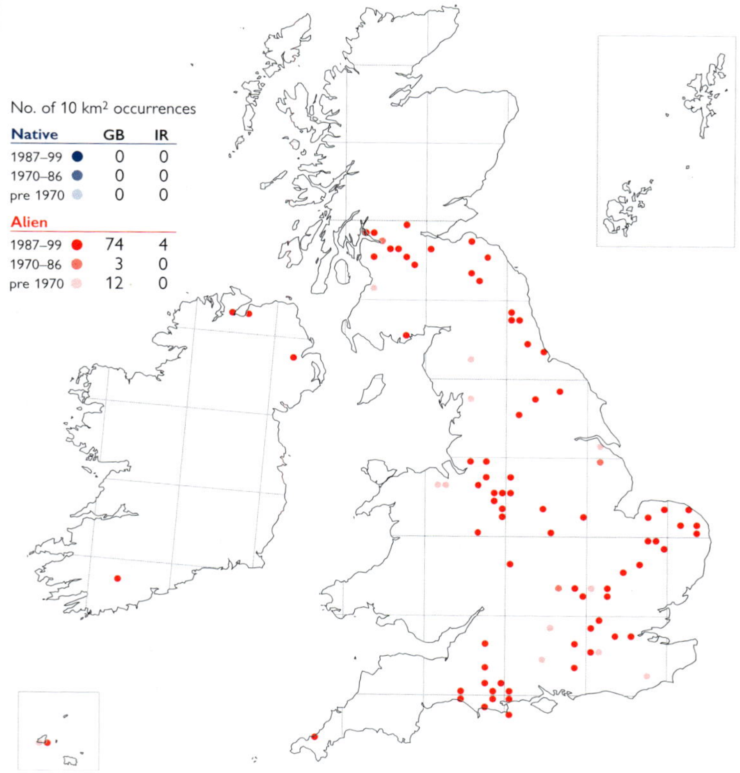

A deciduous, suckering shrub found in hedges, on roadsides, in parks and in amenity plantings. It occurs as a garden escape, and is also planted for game cover. Populations can be naturalised, but never to the same extent as *C. sericea*. Lowland.

Neophyte. This species was introduced to cultivation in 1741. The first confirmed record in the wild was in 1875. The reliability of some older records in the wild is uncertain, due to confusion with *C. sericea*, and it may be under-recorded.

Native of Eurasia, where it extends from European Russia to the Far East (150°E).

References: Bean (1970), Hultén & Fries (1986), Meusel *et al.* (1978).

H. J. KILLICK

Cornus mas Cornelian-cherry

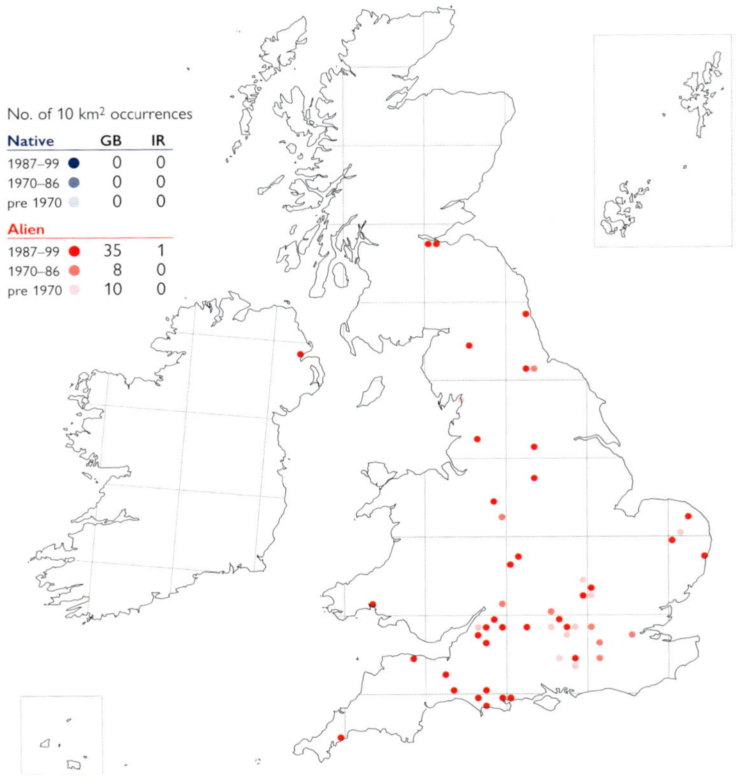

A shrub or small tree found in woodland, hedgerows and scrub and on roadsides. Plants can be long-lived, but fruit infrequently; although they can be bird-sown, they are rarely naturalised. Lowland.

Neophyte. *C. mas* was cultivated in Britain by 1596, and is popular in gardens. However, it was not formally recorded from the wild until 1927 (Oxon).

Native of C. & S.E. Europe and W. Asia.

References: Bean (1970), Meusel *et al.* (1978).

H. J. KILLICK

Cornus suecica Dwarf Cornel

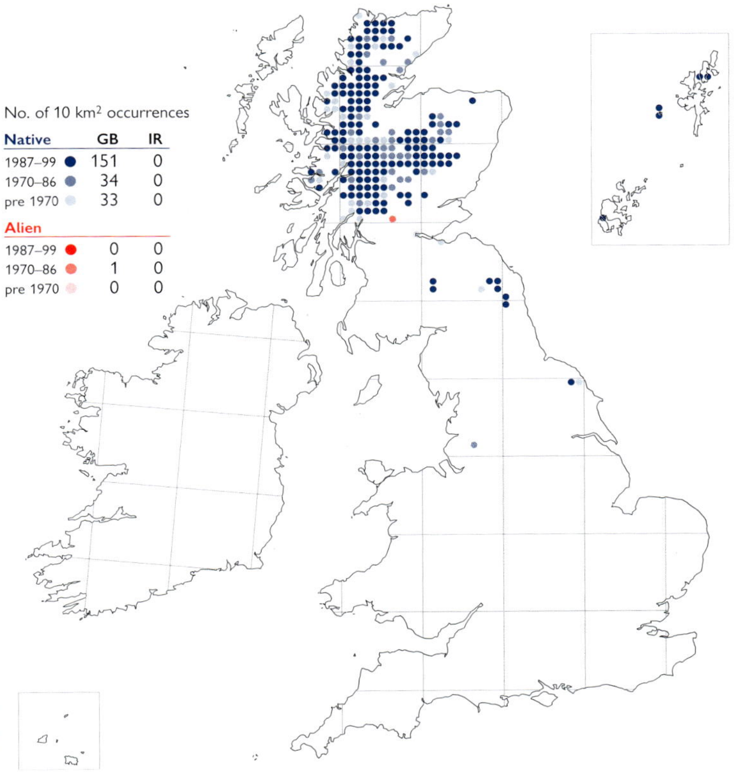

A low-growing rhizomatous perennial herb of wet, base-poor peats at moderate to high altitudes, including areas of late snow-lie. It is almost entirely confined to montane dwarf shrub heath communities, but also extends into acid montane grassland. From 135 m in N.W. Yorkshire to 915 m in Atholl (E. Perth).

Native (change –0.42). The distribution is very similar to that in the 1962 *Atlas*, and it is probably still present in remote 10-km squares for which only pre-1987 records are available. It was last seen, however, at its most southerly native locality, Turton Moor (S. Lancs.), in 1977.

European Boreo-arctic Montane element, but absent from mountains of C. Europe; also in E. Asia and N. America.

References: Atlas (152d), Hultén & Fries (1986), Meusel *et al.* (1978), Taylor (1999b).

G. T. D. WILMORE

Aucuba japonica Spotted-laurel

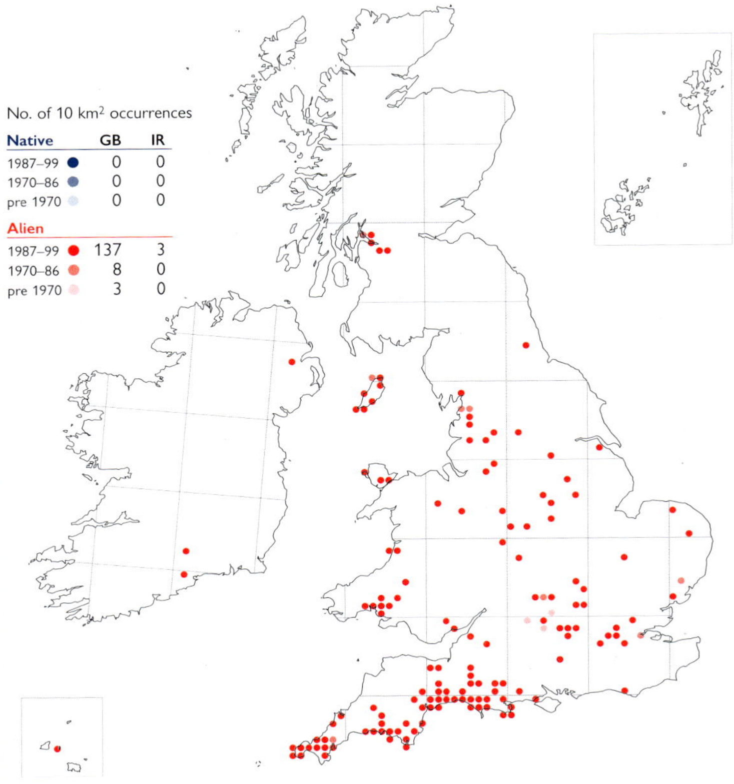

An evergreen shrub found in woodland, plantations, hedges and parks, and on roadsides; usually being deliberately planted or as a relic of cultivation. Populations can occasionally become naturalised, reproducing by seed, especially in S. Britain. Lowland.

Neophyte. *A. japonica* has been cultivated in Britain since 1783 and is now very frequent in gardens and amenity plantings. It was recorded in the wild in 1978, and it was found naturalised in woodland at Trevarrack (W. Cornwall) in 1981.

Native of E. Asia; first introduced from Japan as plants with yellow-speckled leaves, which still predominate in cultivation.

Reference: Bean (1970).

H. J. KILLICK

Griselinia littoralis New Zealand Broadleaf

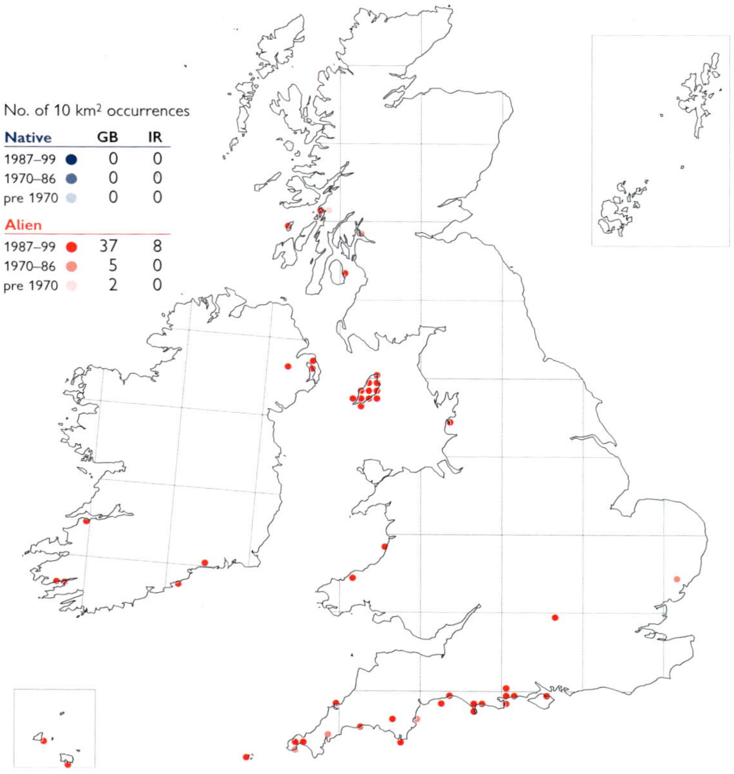

A large evergreen shrub found in woodland, hedges, shelter-belts, churchyards, parks and on waste ground and sea-cliffs. Plants are dioecious and seldom fruit; self-seeding is only very rarely reported. Lowland.

Neophyte. This species has been cultivated in Britain since 1872. It is popular in gardens, and is frequently planted in wind-breaks in coastal areas. especially in the west. It has been known from the wild since at least 1957, and its distribution may be increasing in coastal areas.

Native of New Zealand.

Reference: Bean (1973).

H. J. KILLICK

Thesium humifusum Bastard-toadflax

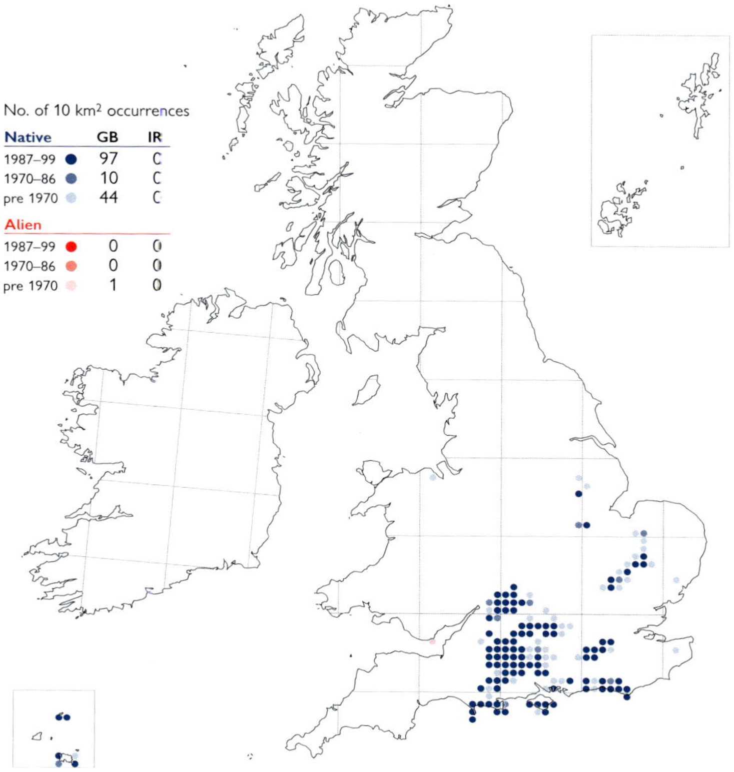

A perennial, with prostrate herbaceous shoots from a woody rootstock, which is hemiparasitic on the roots of various herbs. It is found in short, usually grazed, species-rich calcareous grassland, chiefly on chalk, less frequently on limestone, and rarely on clays or calcareous sandy soils near the coast. Little is known of its reproductive biology. Lowland.

Native (change −0.21). The distribution of this species is largely unchanged since the 1962 *Atlas*. Losses, most of which occurred before 1930, have resulted from the ploughing of downland, scrub encroachment and nutrient enrichment through fertiliser application, particularly in East Anglia. It was last recorded in Kent in 1963.

Oceanic Temperate element.

References: Atlas (152b), Jalas & Suominen (1976), Stewart *et al.* (1994).

F. J. RUMSEY

Viscum album Mistletoe

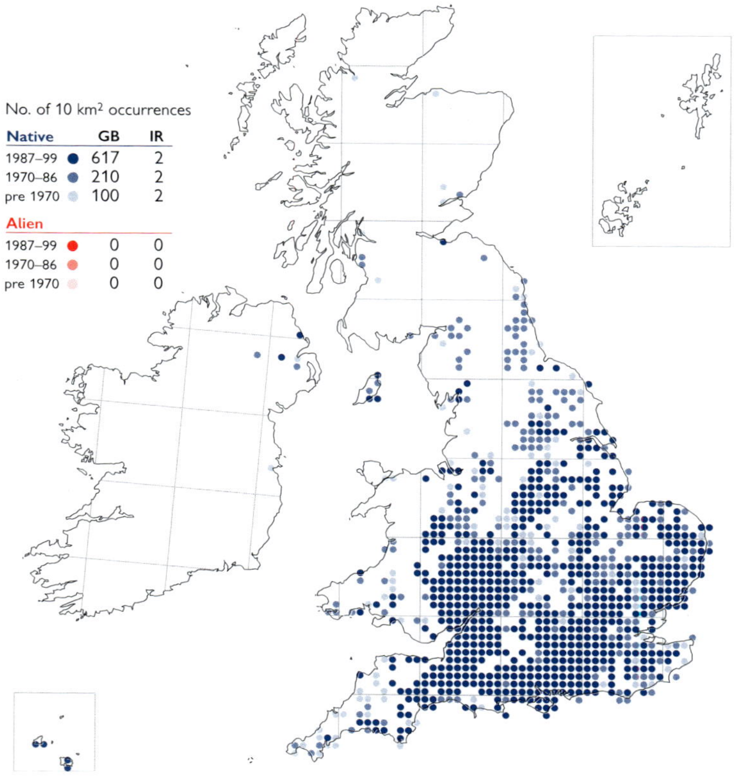

A hemiparasite on a wide range of trees in orchards, hedgerows, parklands and gardens, but rarely in dense or primary woodlands. Its most frequent hosts are *Malus* spp., followed by *Tilia* × *europaea*, *Crataegus* spp. (its commonest native hosts), *Populus* spp., *Acer* spp., *Salix* spp. and *Robinia pseudoacacia*. Lowland.

Native (change +0.97). The native range of *V. album* is so obscured by introductions that all records are mapped as if they are native. A survey in 1969–72 produced many new records, suggesting that it had previously been under-recorded. A repeat survey in 1994–8 provided no evidence for any marked changes in its distribution.

European Temperate element; also in C. and E. Asia.

References: Atlas (152a), Briggs (1999), Hultén & Fries (1986), Jalas & Suominen (1976), Meusel *et al.* (1965), Perring (1973).

F. J. RUMSEY

Euonymus europaeus Spindle

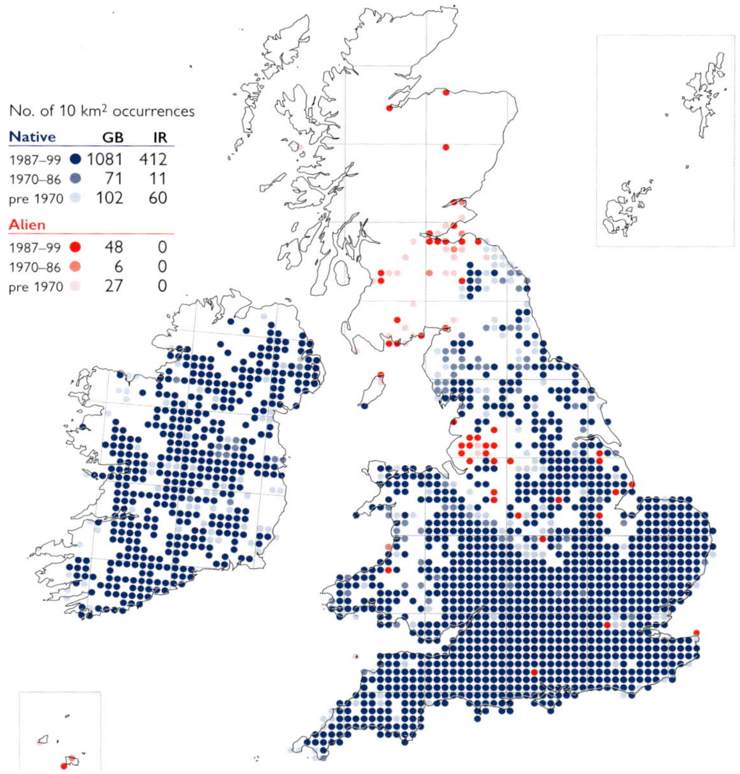

A deciduous shrub or small tree found in hedges, scrub and open deciduous woodland on free-draining base-rich soils, particularly those overlying chalk and limestone. It is also planted in woodlands, hedgerows and gardens from where it can become well-naturalised in the wild. 0–380 m (Craig-y-Benglog, Merioneth).

Native (change +0.15). The considerable increase in records of this species in N. England, Wales and Ireland since the 1962 *Atlas* probably results from better recording of a shrub which is easily overlooked when it is not fruiting. In Scotland, at the northern edge of its range, it is particularly difficult to determine whether some populations are native or alien.

European Temperate element; widely naturalised outside its native range.

References: Atlas (98a), Hultén & Fries (1986), Meusel *et al.* (1978), Rackham (1980).

T. D. DINES

Euonymus japonicus Evergreen Spindle

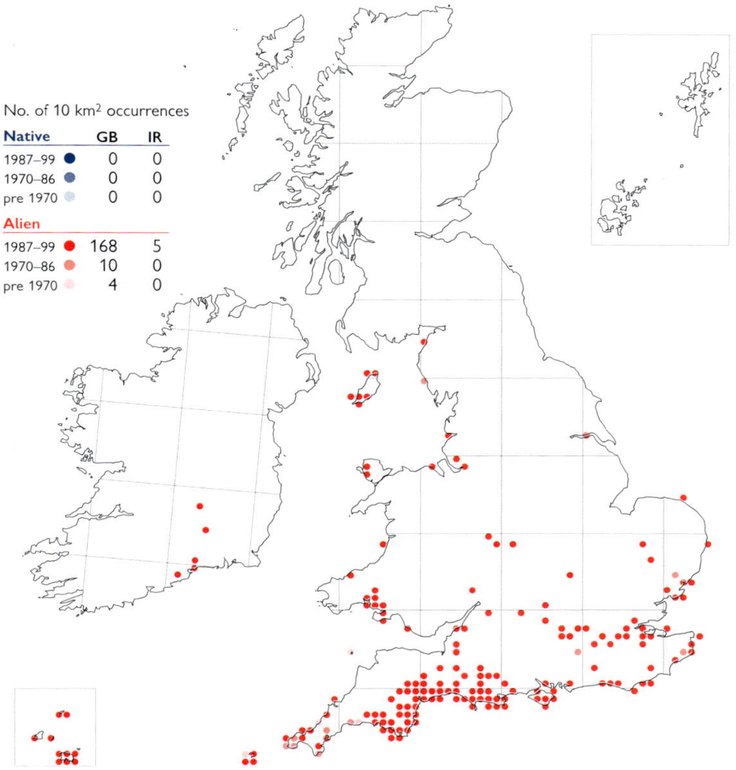

An evergreen shrub or small tree widely planted for hedging, especially in coastal areas, and also found as a garden throw-out and relic of cultivation in woodland and on sea-cliffs and roadsides. Reproduction by seed has been reported from S. Britain, and the species can be established on sheltered sea-cliffs and in woods. Lowland.

Neophyte. *E. japonicus* was introduced to cultivation in 1804 and has been known from the wild since 1897 (Isles of Scilly). It is probably increasing due to its continued popularity in gardens.

Native of Japan.

Reference: Bean (1973).

T. D. DINES

Ilex aquifolium Holly

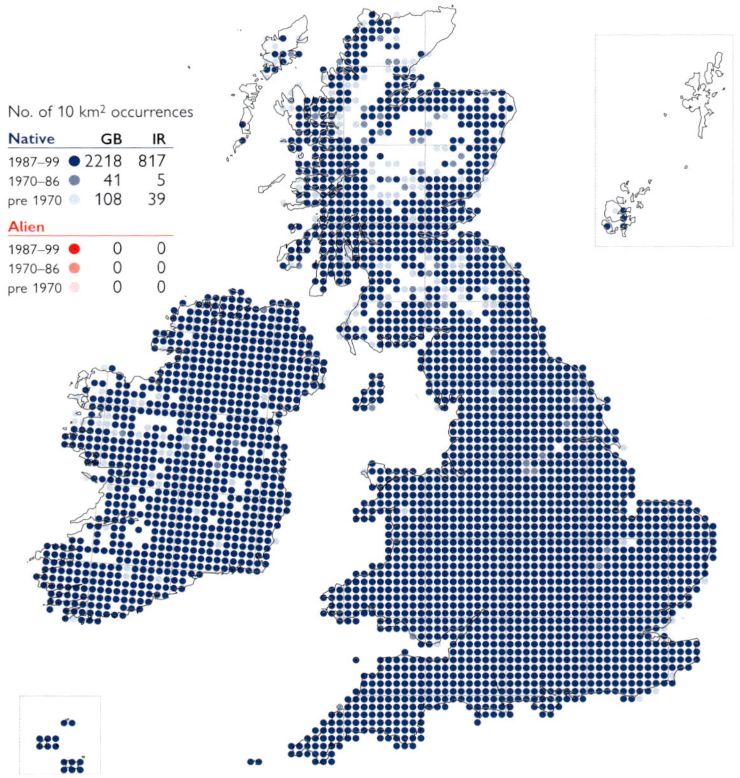

An evergreen shrub of deciduous woodlands, especially those on acidic soils in which *Fagus* and *Quercus* predominate; often a frequent or locally dominant undershrub but rarely dominating the canopy. Its susceptibility to browsing can limit regeneration. It is also found in wood-pasture, scrub and hedgerows, and on ledges of acidic cliffs, and is often planted in amenity areas and parkland. 0–600 m (Eel Crags, Cumberland).

Native (change –0.16). Widespread planting has completely obscured the native distribution of this species and all records are mapped as if they are native. There is little change in its overall distribution since the 1962 *Atlas*.

Suboceanic Southern-temperate element.

References: Atlas (97d), Hultén & Fries (1986), Meusel *et al.* (1978), Peterken (1975), Peterken & Lloyd (1967).

G. T. D. WILMORE

Ilex aquifolium × *I. perado* (*I.* × *altaclerensis*) Highclere Holly

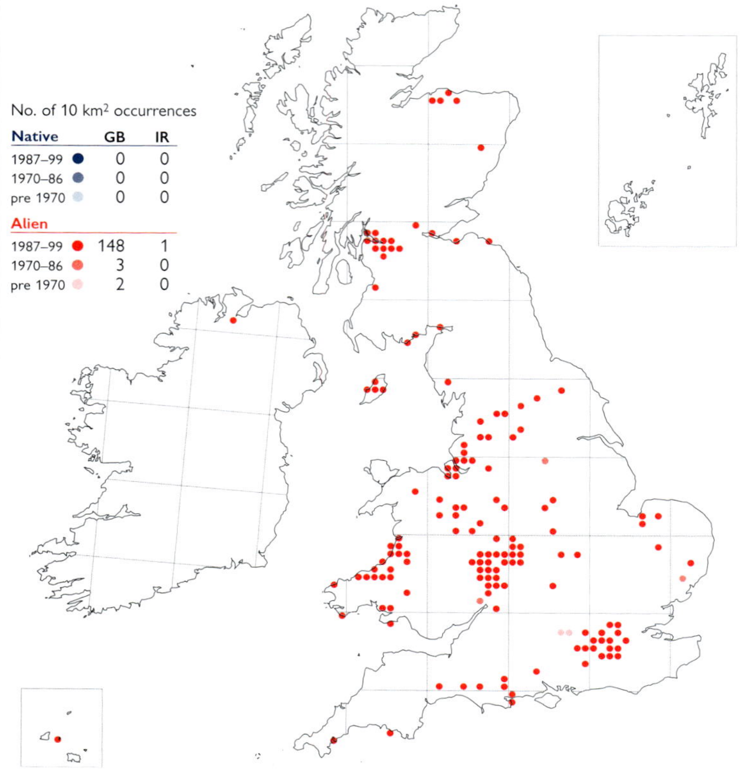

This large evergreen shrub or small tree is found as a garden escape or planted in woodland, hedges, parks and amenity areas, and on roadsides and waste ground; it also occurs as a relic of cultivation. The hybrid is fully fertile and bird- and self-sown plants are often recorded. Lowland.

Neophyte. *I.* × *altaclerensis* arose in cultivation just before 1800 and Loddiges Nursery distributed plants from Highclere Castle (N. Hants.) from *c.* 1836. Like *I. aquifolium*, it is available as numerous cultivars which are extremely popular in gardens. Separating the two taxa can be very difficult, and the hybrid is almost certainly under-recorded. It is probably increasing.

A hybrid of garden origin.

References: Bean (1973), Mitchell (1996).

T. D. DINES

Buxus sempervirens Box

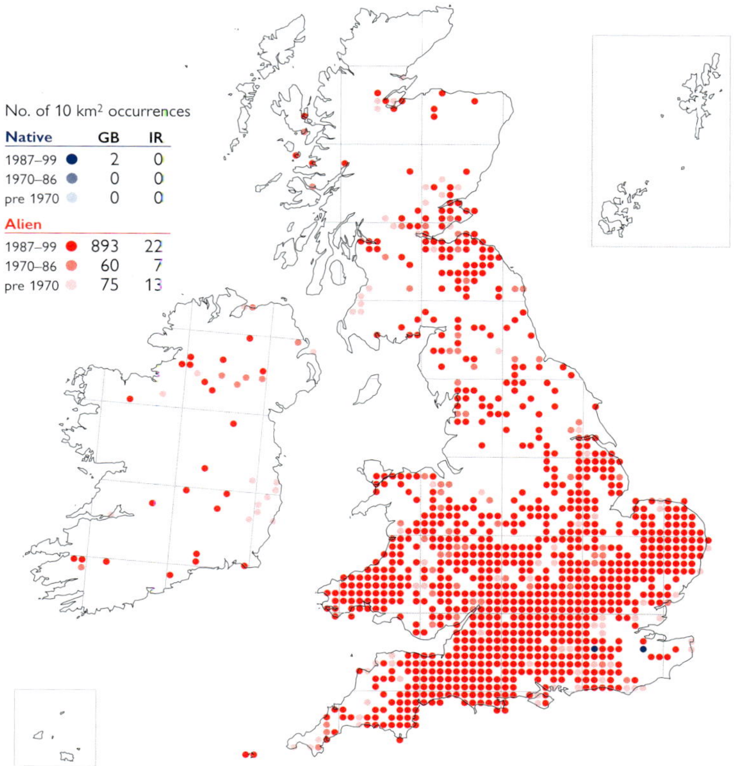

An evergreen shrub or small tree, native to woodlands and thickets on steep slopes on chalk, and in scrub on chalk downland. It is popular for hedging in gardens and is often planted in woodlands, often becoming naturalised. Lowland.

Native (change +2.54). Although *B. sempervirens* is thought to be native at some sites, including Box Hill (Surrey), Boxley (E. Kent) and possibly elsewhere, it has been widely cultivated since Roman times and the limits of its native range are uncertain. Its alien range has increased dramatically since the 1962 *Atlas* due to widespread planting and more efficient recording of alien trees and shrubs.

Submediterranean-Subatlantic element; widely naturalised outside its native range.

References: Atlas (98b), Meusel *et al.* (1978), Wigginton (1999).

T. D. DINES

Mercurialis perennis Dog's Mercury

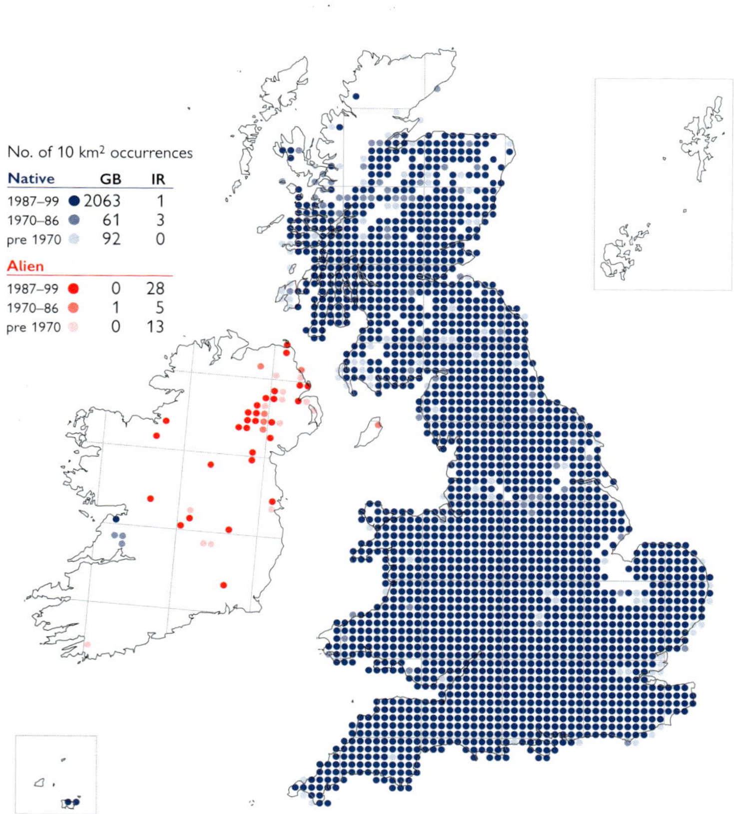

A rhizomatous, dioecious perennial herb usually growing on damp but free-draining base-rich soils. In the lowlands it is largely restricted to shaded sites, including ancient woodland, older secondary woodland, hedgerows and shaded banks, but in the uplands it occurs on unshaded basic crags, scree, cliff ledges and in ravines, particularly on moist N.-facing slopes, and it also grows in the grikes of limestone pavements. 0–1005 m (Ben Lawers, Mid Perth).

Native (change –0.65). There has been no appreciable change in the distribution of *M. perennis* since the 1962 *Atlas*. In Ireland, it is apparently native only in the Burren (Scannell & Synnott, 1987).

European Temperate element.

References: Atlas (169b), Curtis (1981), Grime *et al.* (1988), Hultén & Fries (1986), Meusel *et al.* (1978), Webb & Scannell (1983).

J. H. S. COX

Mercurialis annua Annual Mercury

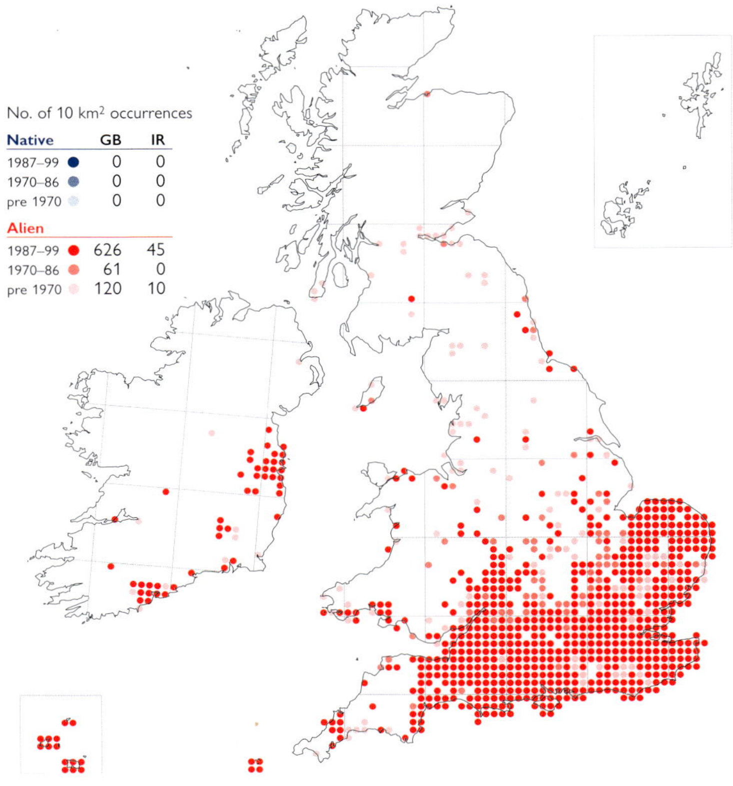

A dioecious annual of disturbed waste places, cultivated ground, particularly in allotments and gardens, rubbish tips, walls, and roadsides, thriving on light, nutrient-rich soils. It produces a long-lived seed bank. Lowland.

Archaeophyte (change +0.28). *M. annua* is known from archaeological deposits in Viking York. It has spread from ports and town gardens, and comparison of the current map with that in the 1962 *Atlas* suggests that is still increasing.

As an archaeophyte *M. annua* has a Submediterranean-Subatlantic distribution.

References: Atlas (169c), Hultén & Fries (1986), Kenward & Hall (1995), Meusel *et al.* (1978).

J. H. S. COX

Euphorbia peplis Purple Spurge

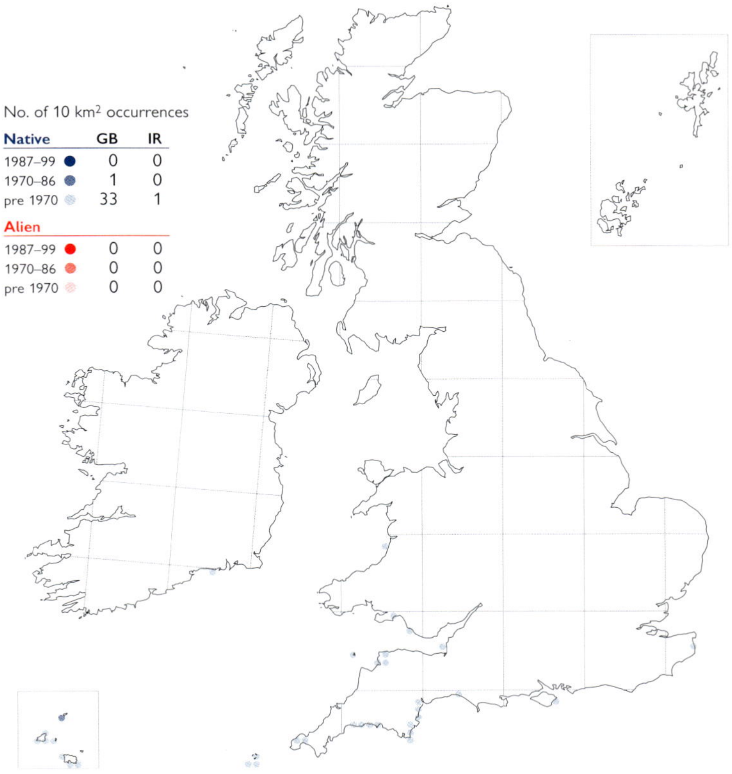

An annual which grows on fine shingle or coarse sand just above the high-water mark of spring tides. Lowland.

Native (change −1.49). *E. peplis* is assumed to be extinct in Britain, the last record being from Lundy (N. Devon) in 1965, and in Ireland, where it was last recorded at Garrais Cove (Waterford) in 1839. It may also be extinct in the Channel Islands, where it was last recorded in Alderney in 1976. It is probably vulnerable to winter storm damage on Atlantic coasts. However, it could possibly recolonise from sea-borne seed derived from populations in mainland Europe.

Mediterranean-Atlantic element.

References: Atlas (169d), Bolòs & Vigo (1990), Curtis & McGough (1988), Lousley (1971).

J. H. S. COX

Euphorbia hyberna Irish Spurge

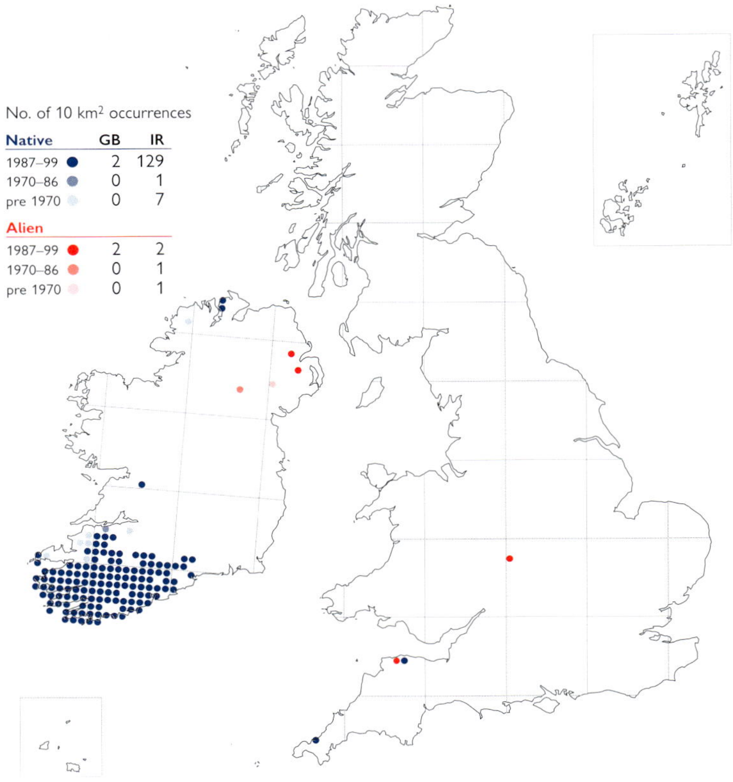

A rhizomatous perennial of woodland glades, hedgerows and shaded stream banks, growing best when receiving dappled sun for at least part of the day. Lowland, reaching 500 m in Waterville (S. Kerry) and reportedly to 550 m elsewhere in Co. Kerry.

Native. The distribution of *E. hyberna* is stable in Ireland. Most authorities consider it to be native in Britain, although there is a possibility that it could have been introduced from Ireland. At its Cornish site it is vulnerable to the spread of more vigorous species, including *Lonicera periclymenum* and *Rubus fruticosus*, following the cessation of coppicing.

Suboceanic Southern-temperate element.

References: Atlas (170b), Bolòs & Vigo (1990), Dupont (1962), Wigginton (1999).

J. H. S. COX

Euphorbia dulcis Sweet Spurge

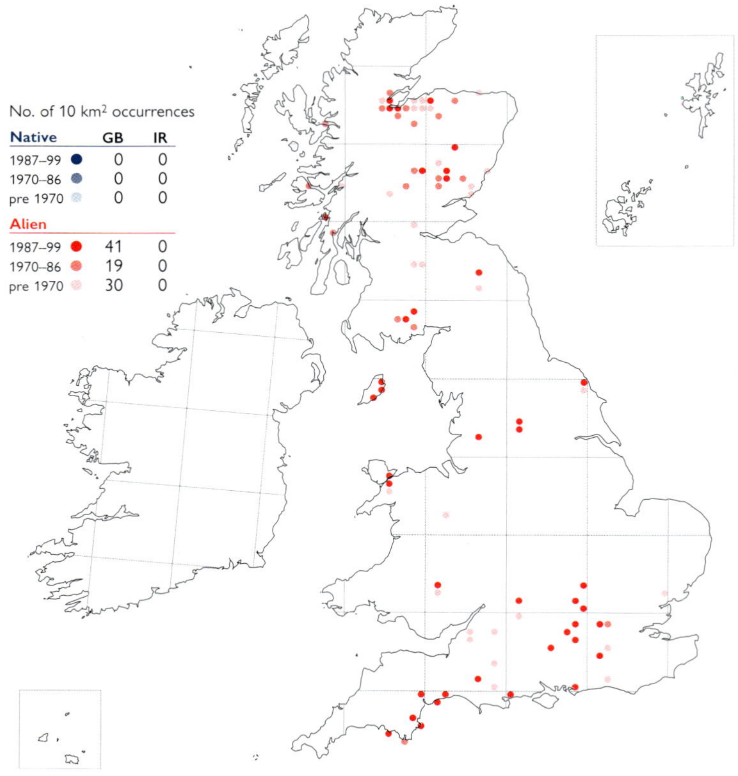

A rhizomatous perennial herb found naturalised in woodland and scrub, and on roadsides and river banks. Lowland.

Neophyte (change +0.46). This species has been cultivated in Britain since 1634, but is not now commonly grown. It was first recorded from the wild in 1849. It is now better recorded than in the 1962 *Atlas*.

A European Temperate species, reaching its eastern limit in Macedonia.

References: Atlas (170c), Meusel *et al.* (1978).

J. H. S. COX

Euphorbia platyphyllos Broad-leaved Spurge

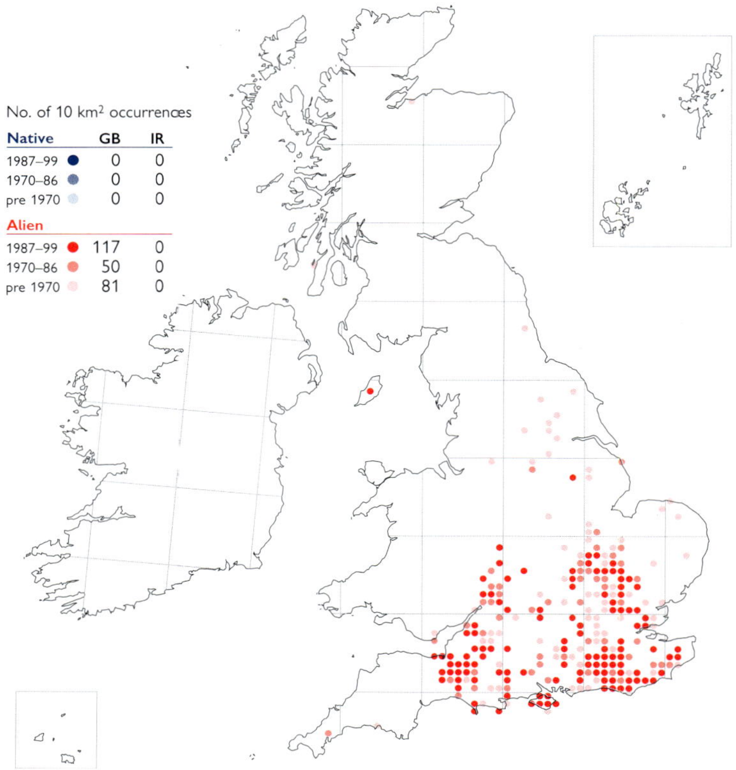

An annual of cultivated and waste ground, usually growing on calcareous clays but sometimes on lighter chalk or limestone soils. It is found most frequently at the margins of arable fields, and occasionally on roadsides. Its seed is thought to be long-lived in the soil. Lowland.

Archaeophyte (change –0.24). *E. platyphyllos* had declined considerably prior to 1930. Many of the records from new 10-km squares since the 1962 *Atlas* are of single plants or small populations, and the species is still declining due to agricultural intensification.

As an archaeophyte *E. platyphyllos* has a European Southern-temperate distribution.

References: Atlas (170d), Bolòs & Vigo (1990), Stewart *et al.* (1994).

J. H. S. COX

Euphorbia serrulata Upright Spurge

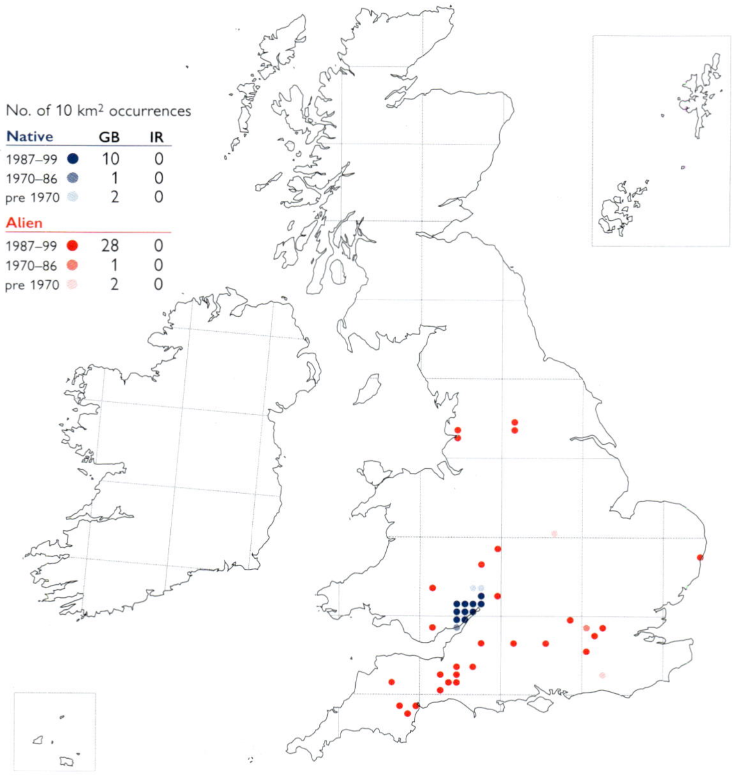

An annual or biennial herb of open deciduous woodland, tracks and hedge banks, growing on calcareous soils and occasionally on alluvial gravels and clays. Populations tend to decline as shade and competition increase, but soil disturbance stimulates the germination of even long-buried seeds. Lowland.

Native or alien (change +1.20). *E. serrulata* was not recorded in the wild until 1773. Its distribution has shown little change in those areas where it appears to be native, but populations have declined markedly in recent years and many now comprise fewer than twenty individuals. It has become more frequent as a garden escape since the 1962 *Atlas*, and these populations can be remarkably persistent.

European Southern-temperate element.

References: Atlas (171a), Wigginton (1999).

J. H. S. COX

Euphorbia helioscopia Sun Spurge

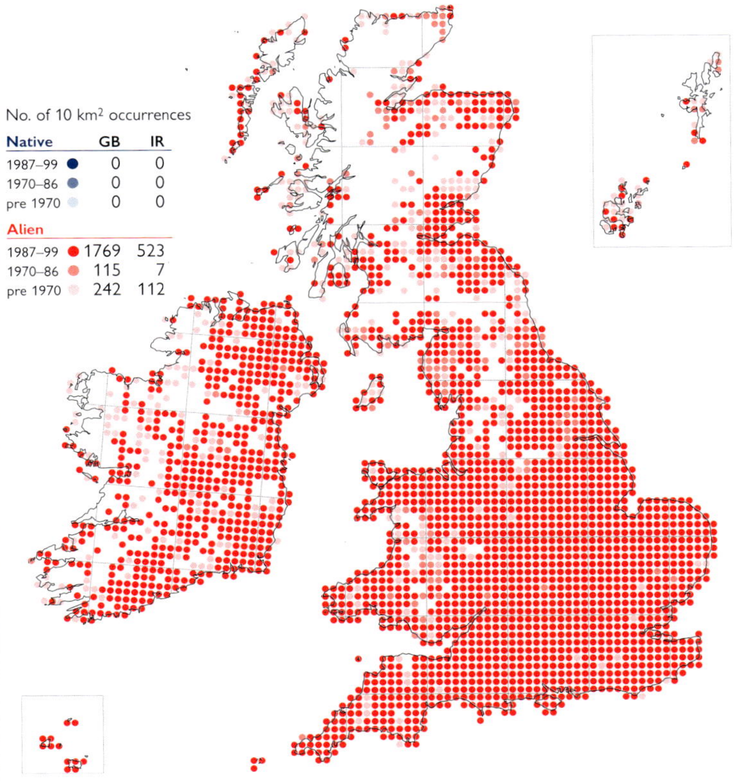

An annual growing in cultivated and disturbed ground in gardens, on waste ground and in arable fields, particularly with root and leaf crops. It thrives on dry, well-drained, neutral to base-rich soils in sun-warmed situations. The seeds may be dispersed by ants. Mainly lowland, but ascending to 450 m in Clun Forest (Salop).

Archaeophyte (change –0.77). There has been no change in the overall distribution of this species since the 1962 *Atlas*, but the map suggests a decline at the edges of its range.

As an archaeophyte *E. helioscopia* has a Eurasian Southern-temperate distribution, but it is naturalised in N. America so its distribution is now Circumpolar Southern-temperate.

References: Atlas (171b), Hultén & Fries (1986), Meusel *et al.* (1978).

J. H. S. COX

Euphorbia lathyris Caper Spurge

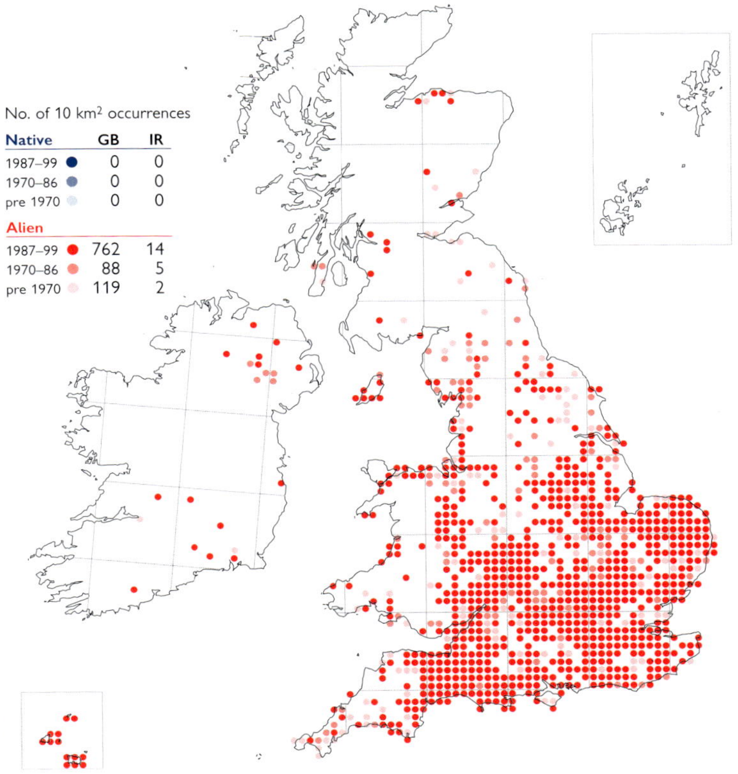

A biennial of disturbed ground and waste places, including roadsides, abandoned gardens, old quarries and rubbish tips, often near human habitation; it also occurs in open woodland. The seeds are very long-lived. Lowland.

Archaeophyte (change +2.16). There are many more records for this species than in the 1962 *Atlas*. It is persistent in a few woods in S. England, and a casual or naturalised alien in other areas, originating as an escape from cultivation or from bird-seed. It is often grown in gardens as an ornamental, and has a reputation for deterring moles.

As an archaeophyte *E. lathyris* has a European Southern-temperate distribution; it is widely naturalised outside this range.

References: Atlas (170a), Bolòs & Vigo (1990).

J. H. S. COX

Euphorbia exigua Dwarf Spurge

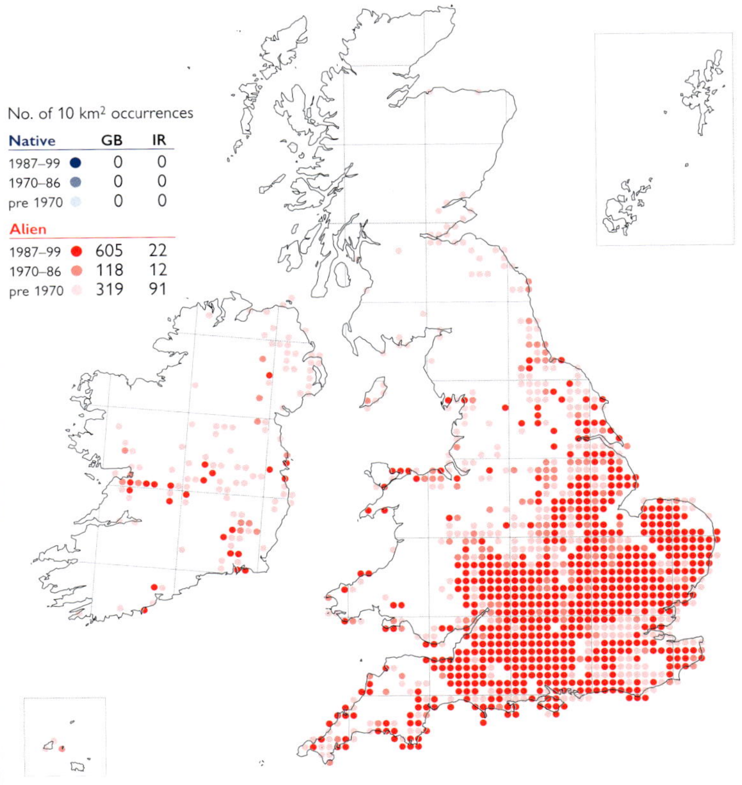

An annual of arable land, less frequently occurring in other areas of disturbed ground such as gardens, waste ground and bare patches in dry grassland. It favours dry, light and base-rich soils in sunny situations. Lowland.

Archaeophyte (change –1.18). This species was lost from many areas of Ireland and N. and W. Britain before 1930. Since the 1962 *Atlas*, it has declined further, especially in S.E. England. Most losses are due to the intensification of agriculture, particularly the increased use of herbicides.

As an archaeophyte *E. exigua* has a European Southern-temperate distribution.

References: Atlas (171d), Hultén & Fries (1986), Meusel *et al.* (1978).

J. H. S. COX

Euphorbia peplus Petty Spurge

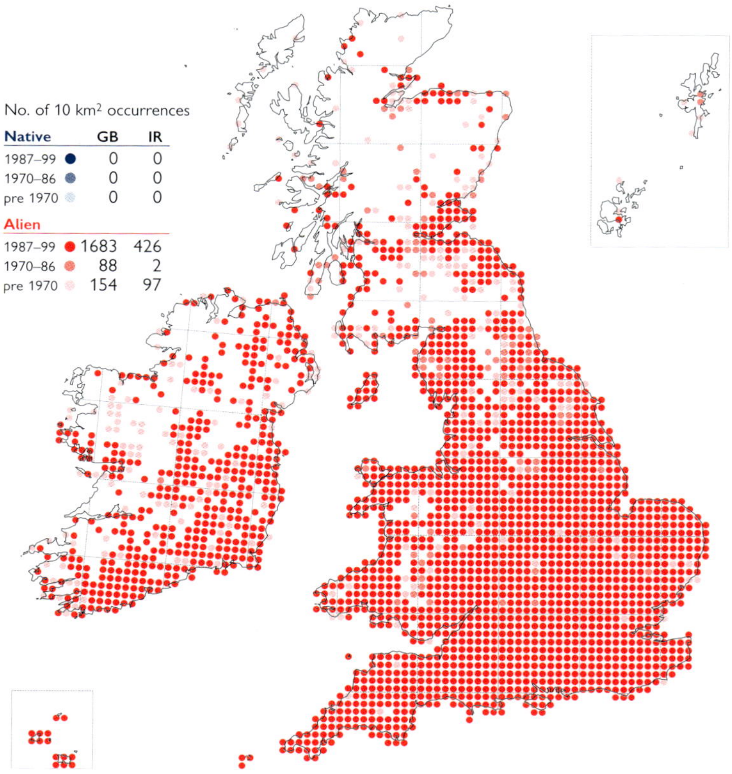

An annual of cultivated, disturbed and waste ground, frequently growing close to human habitation and favouring well-drained and nutrient-rich soils in sun-warmed situations. 0–410 m in Mid-W. Yorkshire.

Archaeophyte (change –0.17). The distribution of this species shows no change since the 1962 *Atlas*. *E. peplus* var. *minima* has been recorded as a casual garden weed, brought in with garden plants from Italy.

As an archaeophyte *E. peplus* has a European Southern-temperate distribution; it is widely naturalised outside this range.

References: Atlas (171c), Hultén & Fries (1986), Meusel *et al.* (1978).

J. H. S. COX

Euphorbia portlandica Portland Spurge

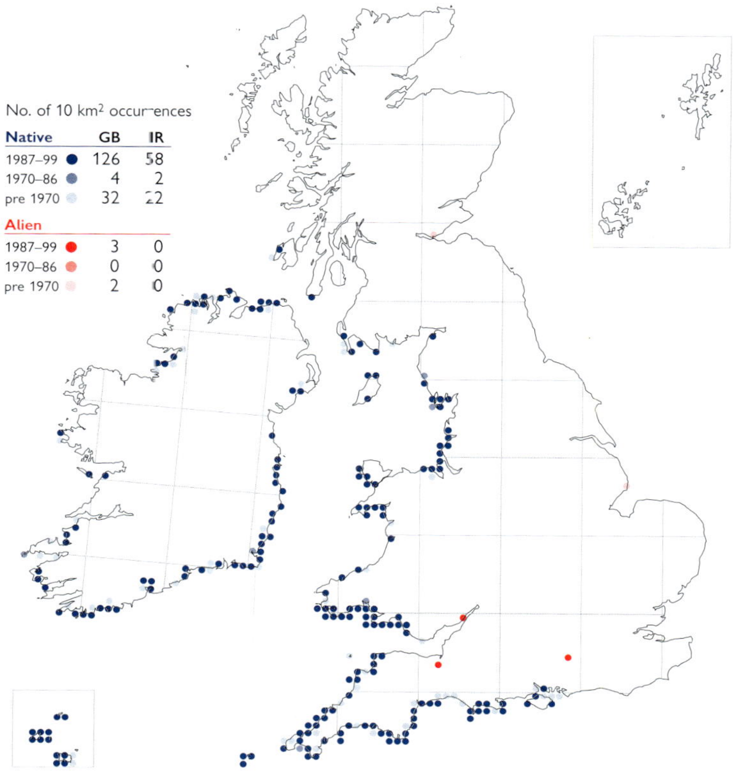

A biennial or short-lived perennial herb, growing in a wide range of coastal habitats. It occurs on cliffs, rocky slopes and steep maritime grasslands overlying many different rock types, and also on shingle and sheltered or semi-fixed sand dunes. Lowland.

Native (change –0.09). The distribution of *E. portlandica* probably changed very little during the 20th century, although it is now more comprehensively recorded than it was in the 1962 *Atlas*.

Oceanic Southern-temperate element.

References: Atlas (172a), Stewart *et al.* (1994).

J. H. S. COX

Euphorbia paralias Sea Spurge

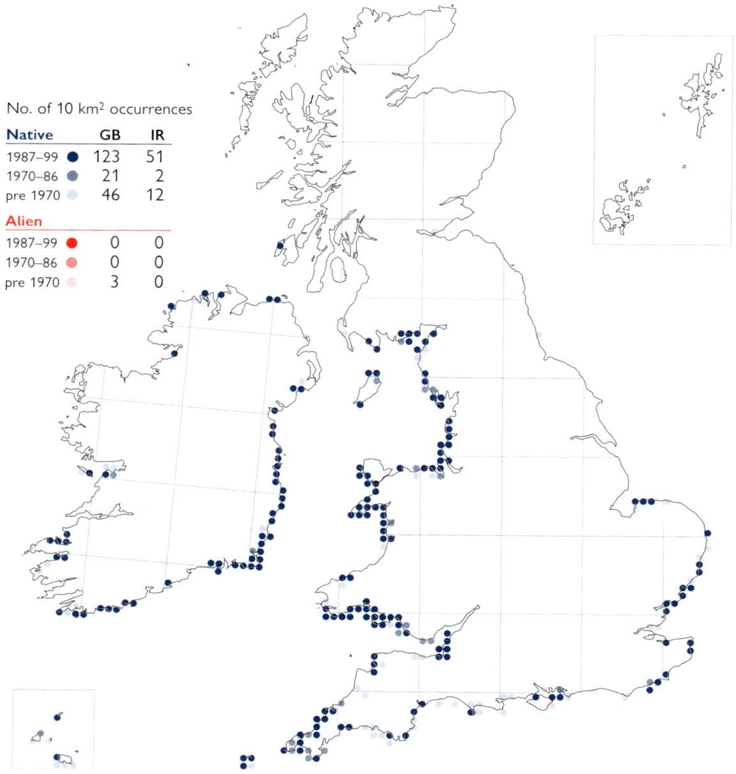

A deep-rooted perennial herb, thriving on free-draining mobile or semi-stable sand dunes, often in the company of *E. portlandica*. It also occurs along the drift-line of sandy foreshores and less frequently on shingle. Lowland.

Native (change –0.35). There has been little change in the distribution of *E. paralias* since the 1962 *Atlas*, though there have been some local losses, such as in Dorset and Devon, due to excessive disturbance of its coastal habitats.

Mediterranean-Atlantic element.

References: Atlas (172b), Bolòs & Vigo (1990), Stewart *et al.* (1994).

J. H. S. COX

Euphorbia esula Leafy Spurge

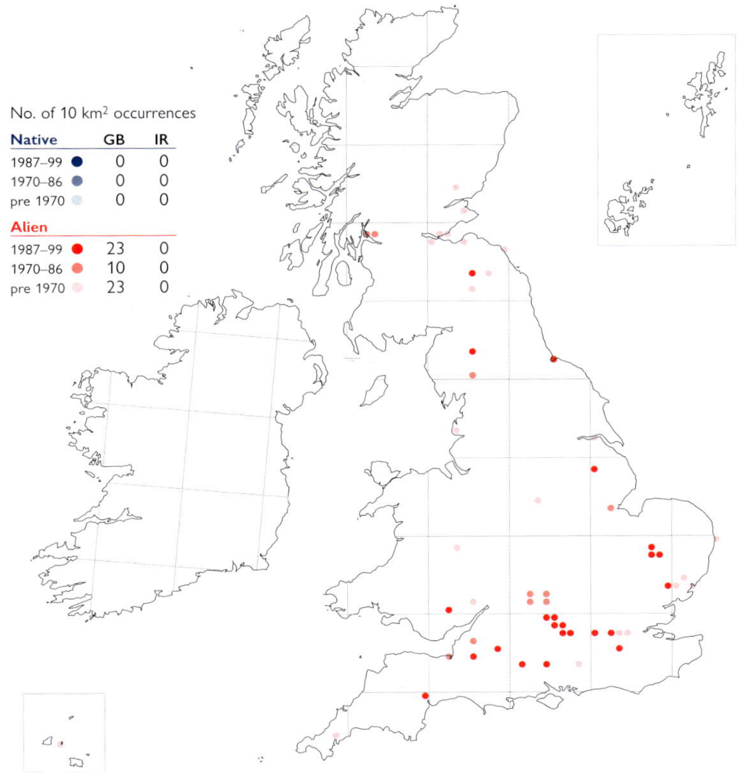

A perennial herb naturalised in similar habitats to *E.* × *pseudovirgata*, including tracks, hedgerows, waste ground and road verges. Lowland.

Neophyte. *E. esula* was cultivated in Britain by 1570. It may also have been originally introduced with grain or timber and was first recorded in the wild in 1805. The *E. esula* group is taxonomically complex, and records made before the publication of Stace (1991) should be treated with caution. Stace (1997) recognises three species and three hybrids in the British Isles, but says that at least 60 species comprise the aggregate worldwide, and that other taxa may be present here. The map in the 1962 *Atlas* was, effectively, that of *E.* × *pseudovirgata*.

Native of S. Europe; naturalised north of its native range.

References: Best *et al.* (1980), Meusel *et al.* (1978).

J. H. S. COX

Euphorbia esula × *E. waldsteinii* (*E.* × *pseudovirgata*) Twiggy Spurge

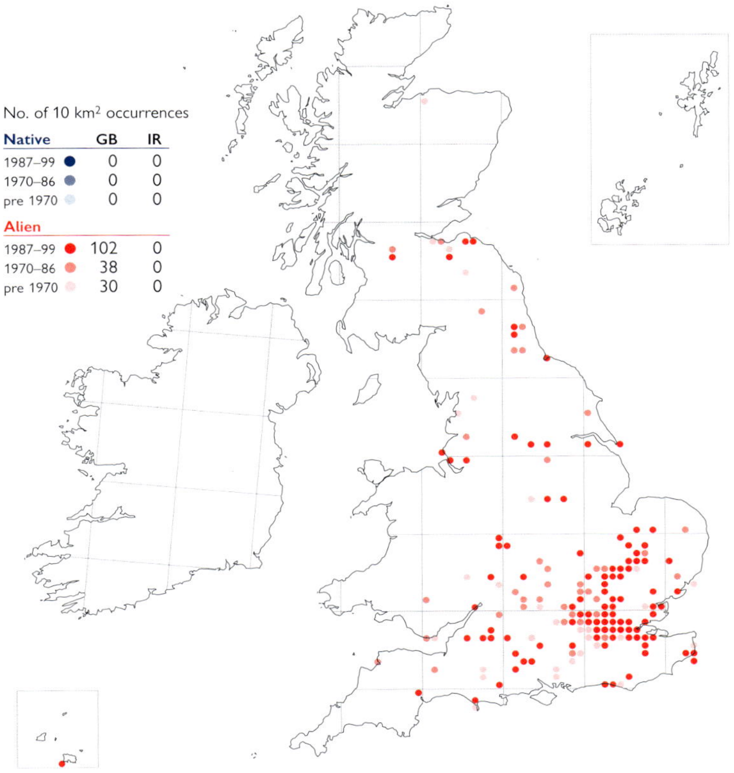

This perennial hybrid is naturalised on waste ground, grassy banks, road verges, tracks and in hedgerows. Lowland.

Neophyte. *E.* × *pseudovirgata* may have been originally introduced with grain, timber or fodder for horses, and it also occurs as a garden escape. Its dates of introduction are obscure, as it was initially confused with *E. esula*, but it was known from the wild by at least 1937. Most of the records of *E. esula* agg. in the 1962 *Atlas* represent this hybrid, and there has been little change in its distribution since then.

This taxon occurs naturally with its parents in C. & E. Europe.

Reference: Stace (1975).

J. H. S. COX

Euphorbia cyparissias Cypress Spurge

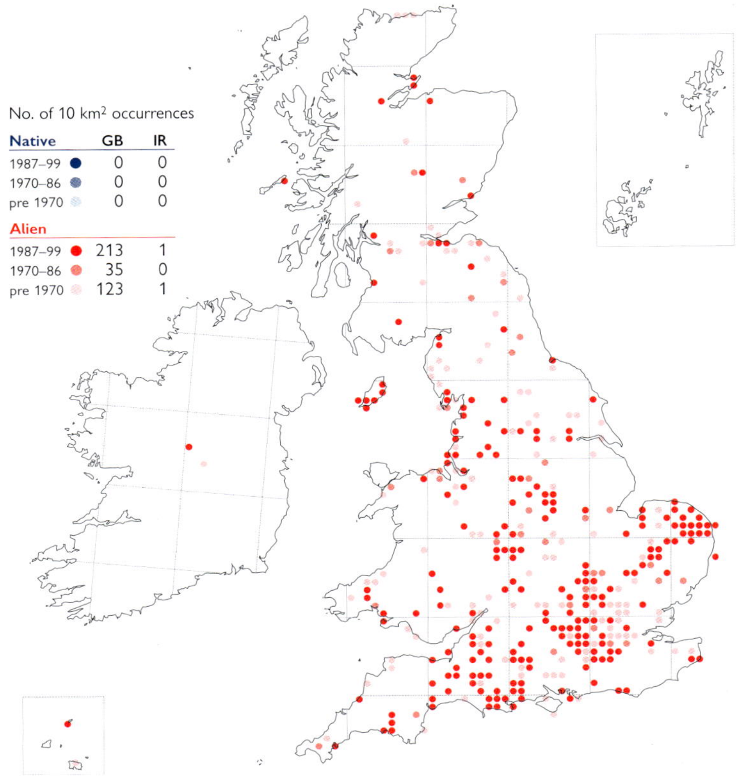

A rhizomatous perennial herb widely naturalised as a garden escape in waste places such as tracksides, roadsides, walls and sandy banks, and also in calcareous grassland and amongst scrub. It can colonise arable margins, and may become established on sand dunes. It also grows on some racecourses where it could have been introduced with horse-feed or bedding. Lowland.

Neophyte (change +0.98). *E. cyparissias* was cultivated in Britain by 1640 and was first recorded in the wild in 1799. It is sometimes regarded as a native of chalk grassland in S.E. England. There are many more records now than in the 1962 *Atlas*.

A European Temperate species, absent as a native from much of W. Europe; widely naturalised outside its native range.

References: Atlas (172d), Hultén & Fries (1986), Meusel *et al.* (1978).

J. H. S. COX

Euphorbia amygdaloides Wood Spurge

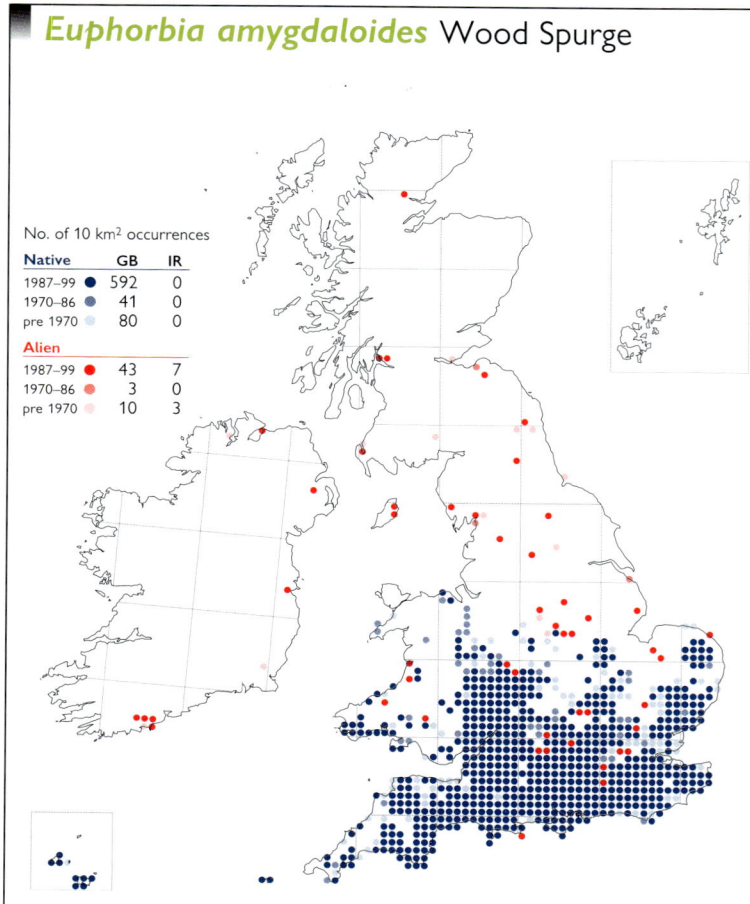

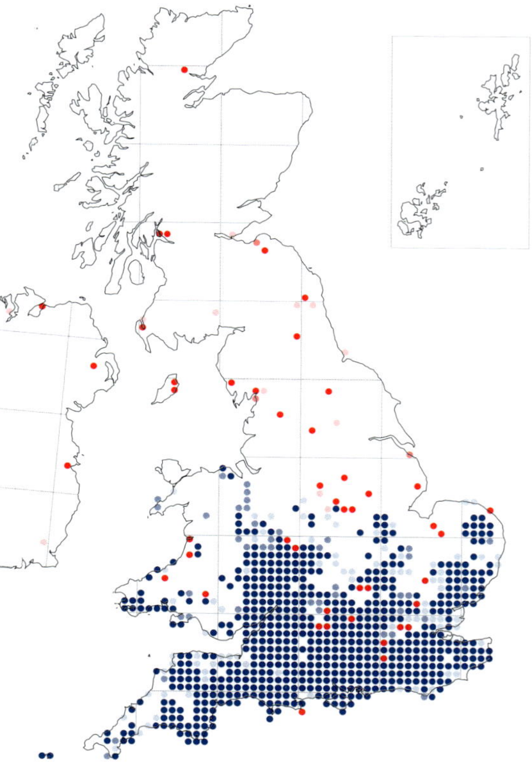

No. of 10 km^2 occurrences

Native	GB	IR
1987–99 ●	592	0
1970–86 ●	41	0
pre 1970 ●	80	0
Alien		
1987–99 ●	43	7
1970–86 ●	3	0
pre 1970 ●	10	3

A rhizomatous perennial herb of neutral or acidic soils in old woods and shaded hedge banks, more rarely found amongst scrub and around rock outcrops. In woods it is a light-demanding plant which may re-appear from buried seed after coppicing. It is also cultivated as a garden plant, where it is persistent and can be very invasive. Generally lowland, reaching 455 m at Rhydymain (Merioneth).

Native (change −0.22). There has been no change in the distribution of this species since the 1962 *Atlas*. The native subsp. *amygdaloides* occurs throughout the native range of the species; the alien subsp. *robbiae* is mapped separately.

European Temperate element.

References: Atlas (173a), Meusel *et al.* (1978), Rackham (1980).

J. H. S. COX

Euphorbia amygdaloides subsp. *robbiae*

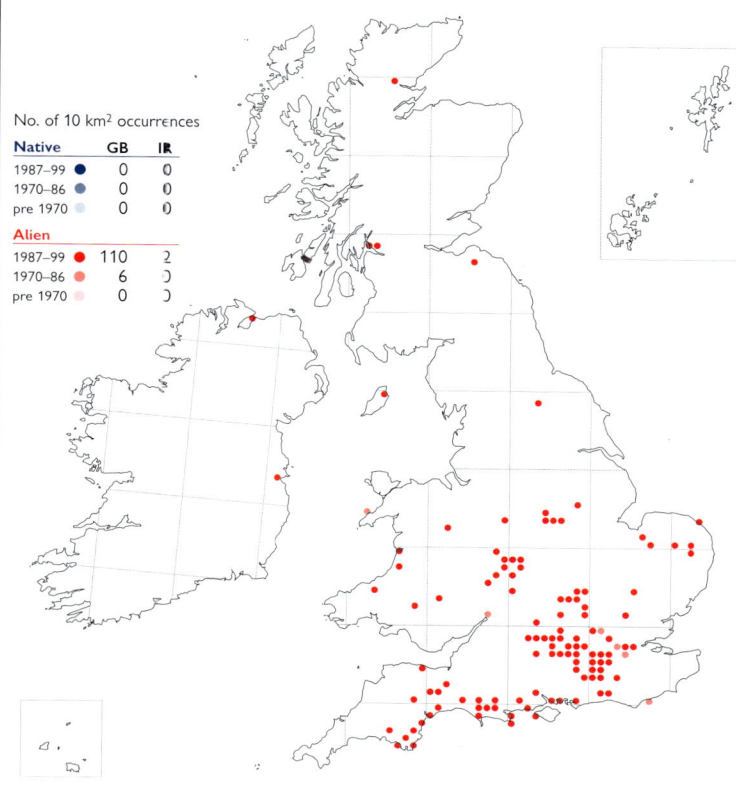

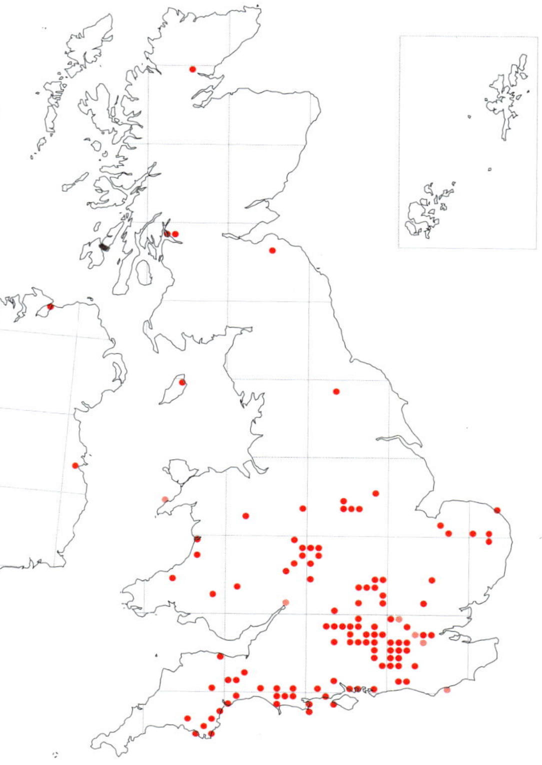

No. of 10 km^2 occurrences

Native	GB	IR
1987–99 ●	0	0
1970–86 ●	0	0
pre 1970 ●	0	0
Alien		
1987–99 ●	110	2
1970–86 ●	6	0
pre 1970 ●	0	0

A vigorous rhizomatous perennial herb found naturalised in woodland and rough grassland, and on shaded banks, roadsides and along paths. It mainly occurs as a garden escape or throw-out; some populations are large and very long-lived. Reproduction is principally by vegetative spread, but occasionally by seed. Lowland.

Neophyte. This species was first cultivated in Britain *c.* 1891 and is now frequent in gardens and amenity plantings. However, it was not formally recorded from the wild until 1977. It is probably spreading but may be under-recorded.

Native of N.W. Turkey.

Reference: Clement (1997).

J. H. S. COX

Euphorbia characias Mediterranean Spurge

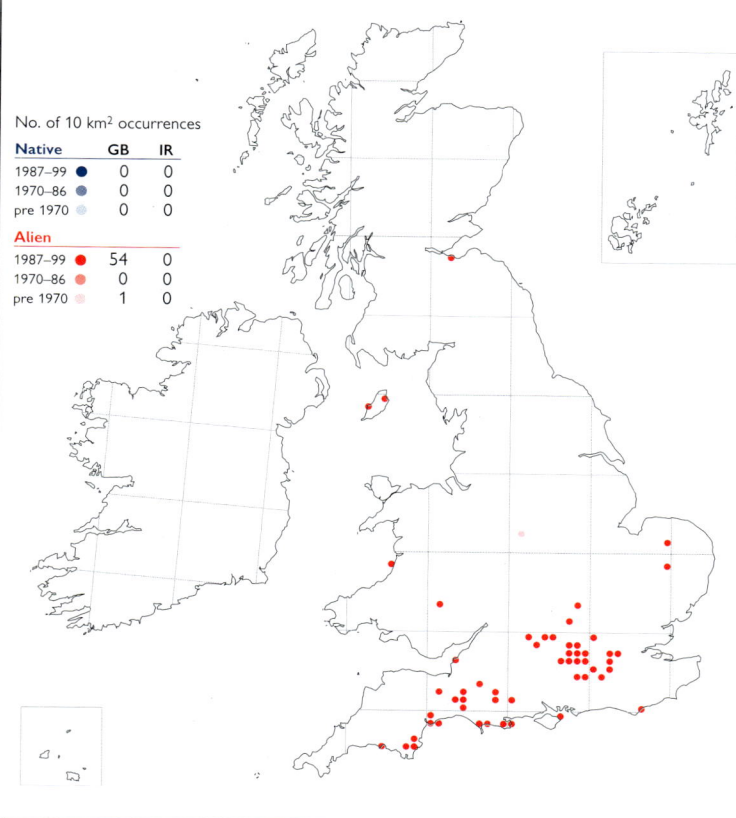

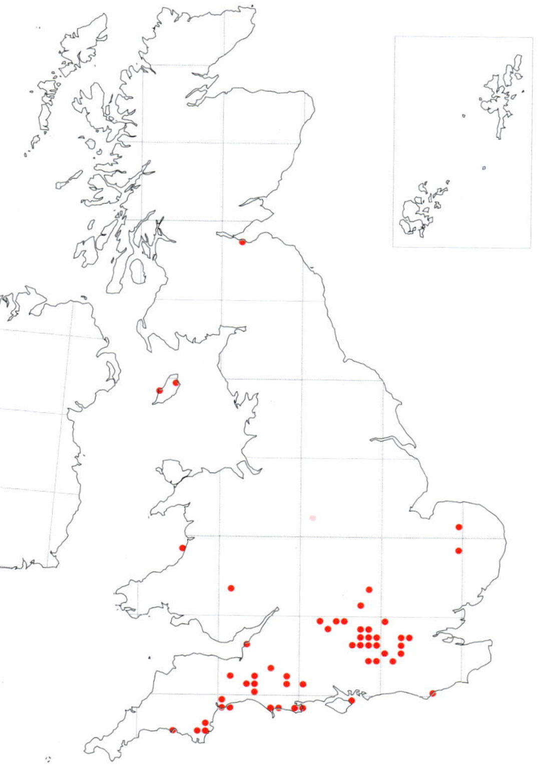

No. of 10 km^2 occurrences

Native	GB	IR
1987–99 ●	0	0
1970–86 ●	0	0
pre 1970 ●	0	0
Alien		
1987–99 ●	54	0
1970–86 ●	0	0
pre 1970 ●	1	0

An evergreen, semi-shrubby perennial found naturalised on roadsides, banks, along paths, on waste ground and in other urban habitats. It originates as a garden escape, throw-out and as a relic of cultivation. Reproduction by seed has been reported. Lowland.

Neophyte. This species was grown in Britain by 1629 and it is now a popular garden plant. It was known from the wild by 1797, when it was considered to be native. It is probably increasing but may be under-recorded. Both subsp. *characias* and subsp. *wulfenii* have been recorded, but are not mapped separately.

Native of the Mediterranean region.

Reference: Meusel *et al.* (1978).

J. H. S. COX

Rhamnus cathartica Buckthorn

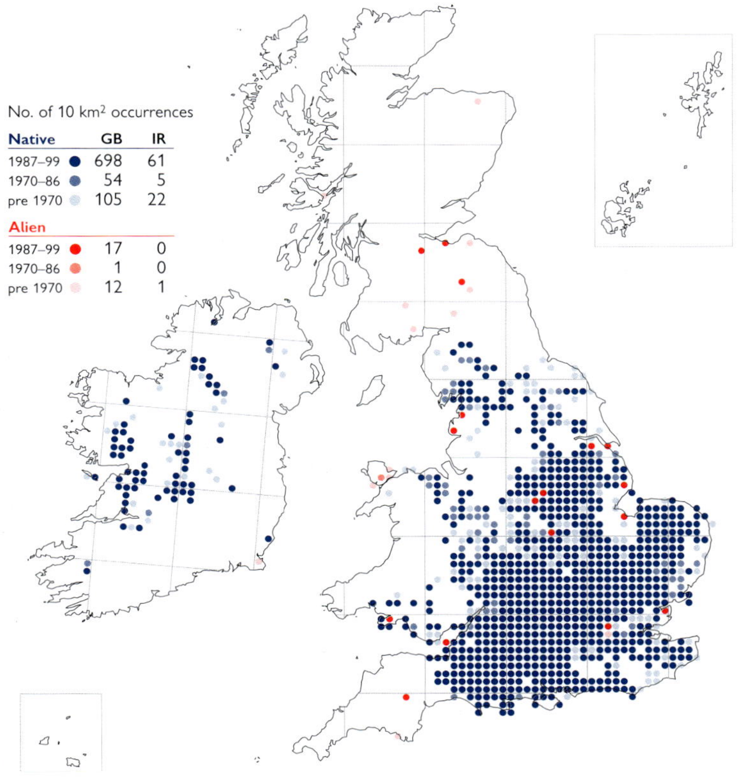

This strongly calcicolous shrub or small tree is found in a variety of habitats, including as an undershrub in woodland (usually *Fraxinus* and *Quercus*), in scrub and hedgerows and in fen-carr and damp *Alnus* woods. Reproduction is by seed, and it regenerates from cut stools and after grazing or burning. 0–380 m (near Malham Tarn, Mid-W. Yorks.).

Native (change –0.04). There is some evidence of a decline in the distribution of *R. cathartica* in W. Britain since the 1962 *Atlas*, although it is now much better recorded in Ireland.

Eurosiberian Temperate element; widely naturalised outside its native range.

References: Atlas (98c), Godwin (1943), Hultén & Fries (1986), Meusel *et al.* (1978).

G. T. D. WILMORE

Frangula alnus Alder Buckthorn

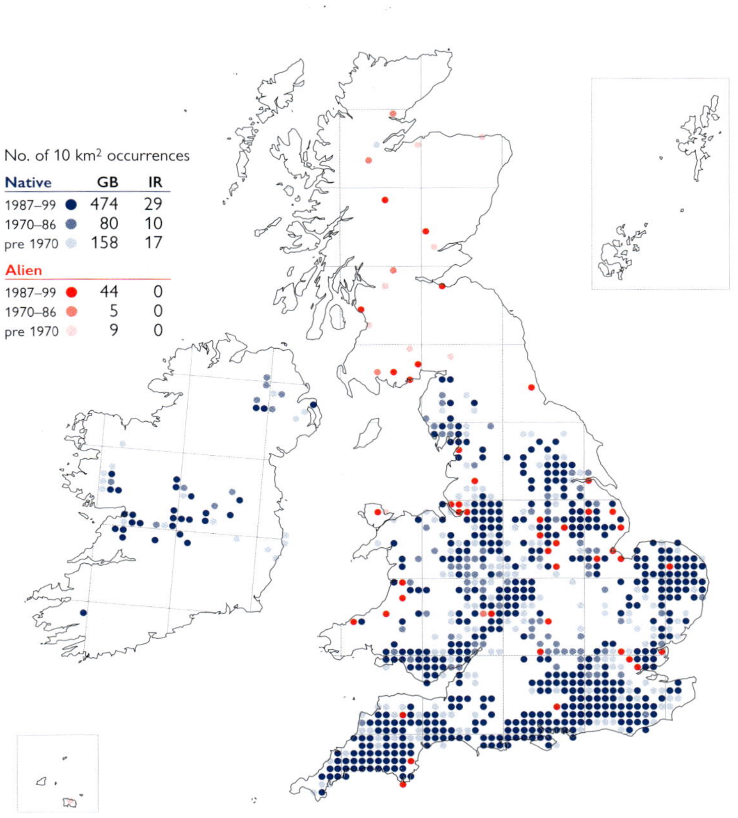

This deciduous shrub or small tree grows on a wide range of soils, but avoids drought-prone and permanently waterlogged sites. It is found in scrub on fen peat, on the edges of raised mires, on heaths and in valley mires, in scrub, hedgerows and in woodland. It regenerates strongly after cutting, burning or grazing. Generally lowland, but reaching 450 m at The Arch (Cards.).

Native (change –0.16). This species was formerly planted for charcoal production, and this may have blurred its native distribution in some areas. Its overall distribution is stable, but there have been some losses since the 1962 *Atlas*.

Eurosiberian Temperate element; widely naturalised outside its native range.

References: Atlas (98d), Curtis & McGough (1988), Godwin (1943), Hultén & Fries (1986), Meusel *et al.* (1978).

G. T. D. WILMORE

Vitis vinifera Grape-vine

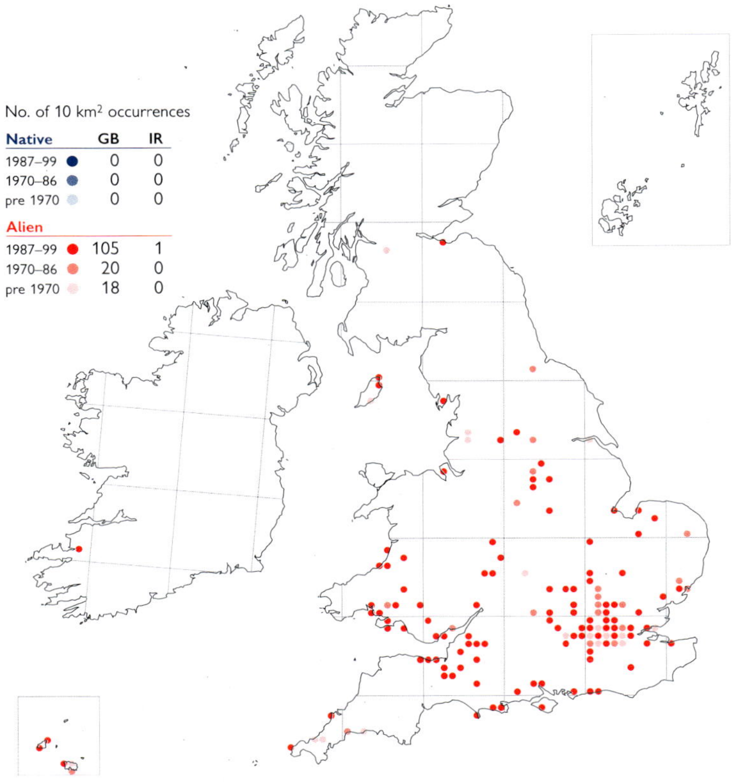

A perennial scrambling woody climber naturalised in hedges and scrub and along river banks, often as a garden escape or relic of cultivation, and occasionally as a casual on tips. Lowland.

Neophyte. *V. vinifera* was grown in Britain for wine-making in Roman times and the Dark Ages, commonly in the Medieval period and more or less continuously since then. It was recorded from the wild by at least 1908, and is now much more frequently reported, probably due to its popularity in gardens, to a renewed interest in wine production, and to an increased interest in recording aliens.

Native of S. & C. Europe, N. Africa and S.W. Asia, but limits of native range obscured by its spread in cultivation.

References: Barty-King (1997), Bolòs & Vigo (1990), Meusel *et al.* (1978), Vaughan & Geissler (1997), Zohary & Hopf (2000).

D. A. PEARMAN

Parthenocissus quinquefolia Virginia-creeper

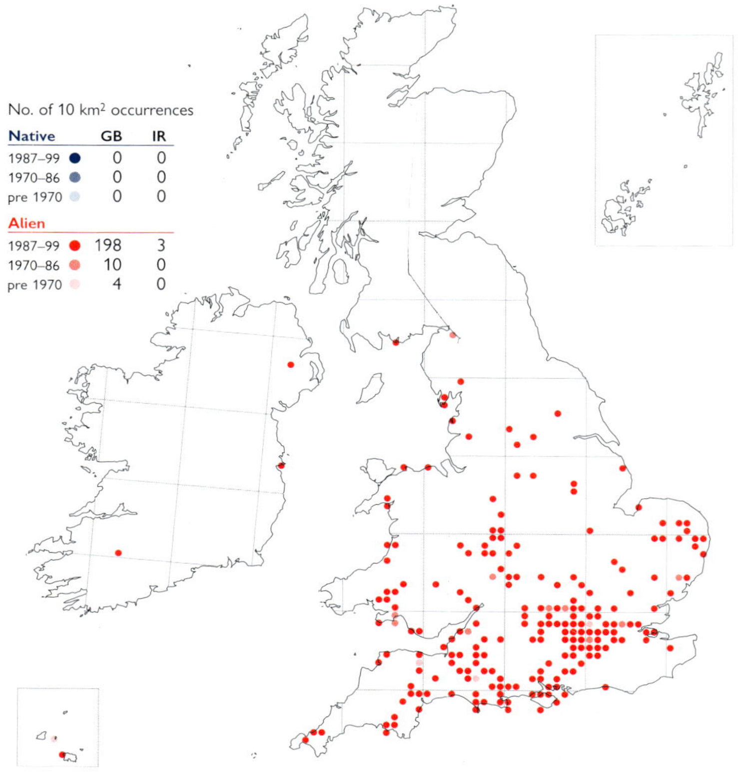

A self-clinging perennial climber well-established in hedgerows and also on railway banks and roadsides. Lowland.

Neophyte. *P. quinquefolia* was introduced into cultivation before 1629, and is very widely grown for its autumn colour. It was recorded in the wild by 1927. The name Virginia-creeper was also applied to *P. tricuspidata* (now Boston-ivy), and there may have been some confusion between these two species.

Native of N. America.

References: Bean (1976), Coats (1963).

D. A. PEARMAN

Parthenocissus inserta False Virginia-creeper

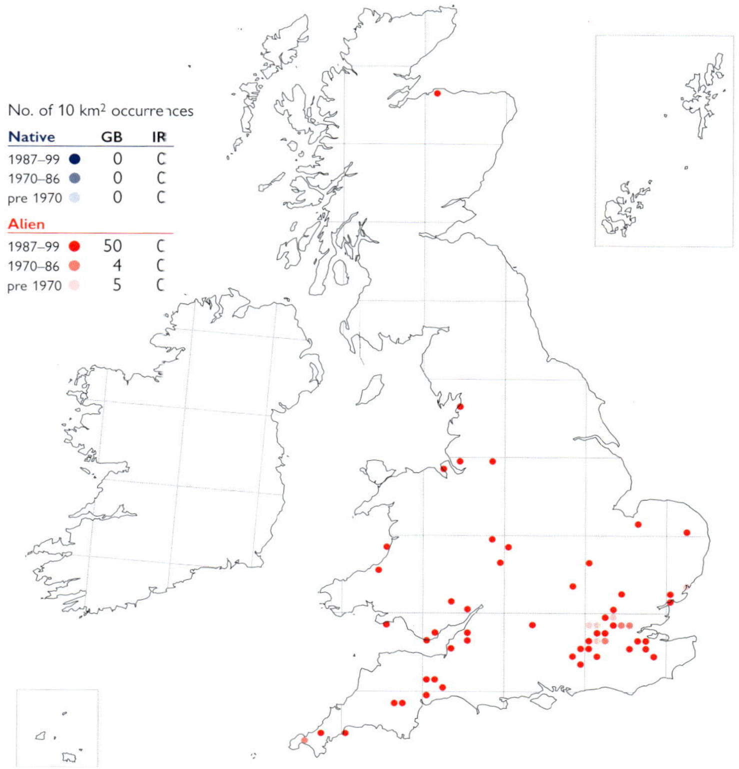

A perennial scrambling woody climber widely grown in gardens and occasionally found naturalised on rubbish tips, waste ground and old walls, and in hedges, usually as a garden escape or throw-out, or as a relic of cultivation. Lowland.

Neophyte. Introduced into cultivation before 1824, *P. inserta* has been confused with *P. quinquefolia*, and has been thought to be conspecific with it. It was recorded from the wild by 1948 and appears to be increasing, but is perhaps just better recorded now.

Native of N. America.

References: Bean (1976), Coats (1963).

D. A. PEARMAN

Linum bienne Pale Flax

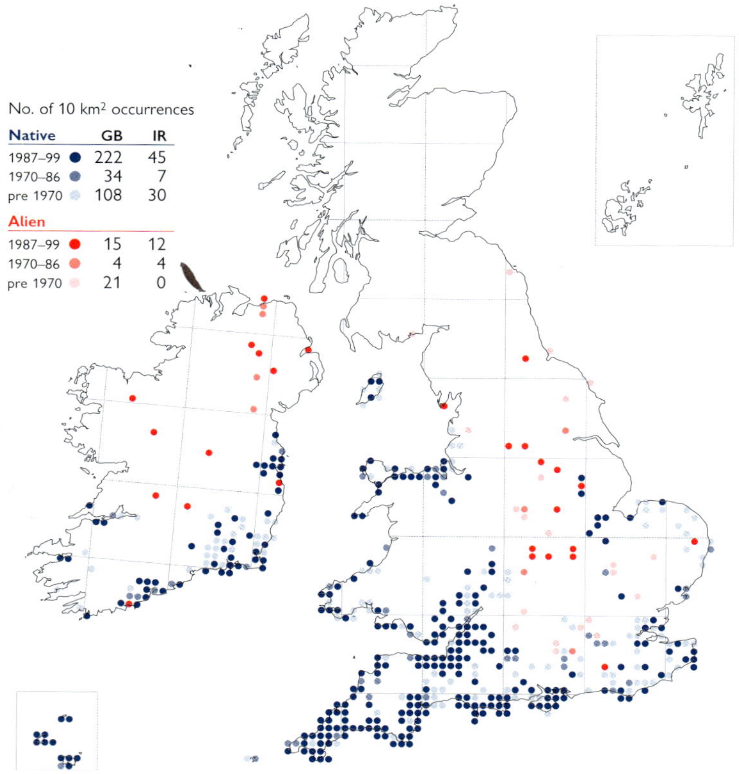

An annual, biennial or short-lived perennial herb of dry grassy places and grassland-scrub mosaics, chiefly near the sea; its habitats include cliff-slopes and coombes, path and field margins, roadsides, railway banks and old quarries. It appears to favour warm, sheltered, S.-facing slopes and relatively infertile, drought-prone soils. Lowland.

Native (change +0.06). *L. bienne* is much more frequent, or much more frequently recorded, than in the 1962 *Atlas*. The additional records are balanced by losses from some inland squares due to habitat destruction and intensification of grassland management.

Mediterranean-Atlantic element.

References: Atlas (89d), Bolòs & Vigo (1990).

S. J. LEACH

Linum usitatissimum Flax

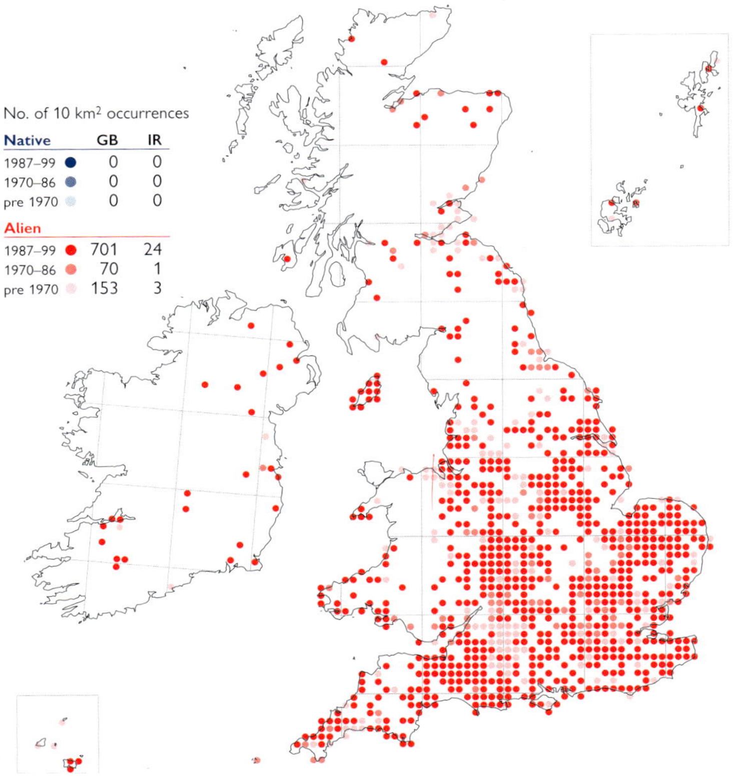

A robust annual found on road verges, rubbish tips and waste ground and locally, rather surprisingly, on stone reservoir banks. It is also a moderately frequent bird-seed alien. Lowland.

Neophyte. *L. usitatissimum* has been cultivated since at least 1240 (Harvey, 1981), and was formerly widely grown for fibre. Since the early 1990s it has been increasingly planted for linseed oil production. It was known from the wild by 1632 (Kent). Escapes from cultivation have become much more common, but are usually transient.

Originated in cultivation; perhaps derived by selection from *L. bienne.*

References: Thirsk (1997), Zohary & Hopf (2000).

G. T. D. WILMORE

Linum perenne Perennial Flax

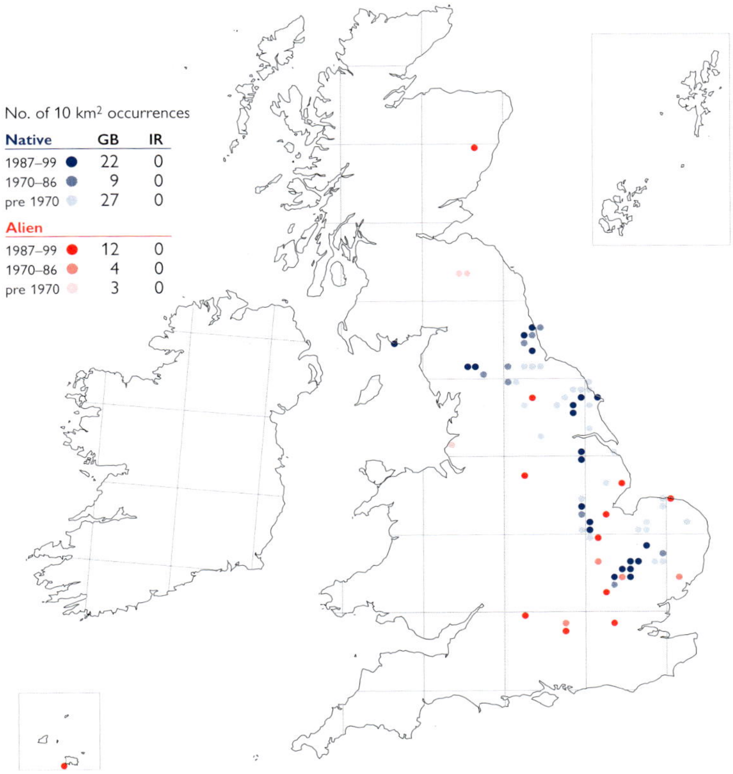

A perennial herb, growing in open, well-drained situations on base-rich substrates, including lightly-grazed grassland, dry banks and roadsides. From near sea level in Kirkcudbrightshire to 340 m near Crosby Ravensworth (Westmorland).

Native (change +0.43). This species has become extinct at several localities in the last hundred years, but its current distribution is similar to that shown in the 1962 *Atlas.* However, many surviving populations are very small and presumably vulnerable to land-use change.

Circumpolar Boreo-temperate element, with a disjunct distribution. The *L. perenne* group is taxonomically complex; our plant is subsp. *anglicum* which is a British endemic.

References: Atlas (90a), Hultén & Fries (1986), Meusel *et al.* (1978), Ockendon (1968), Stewart *et al.* (1994).

G. T. D. WILMORE

Linum catharticum Fairy Flax

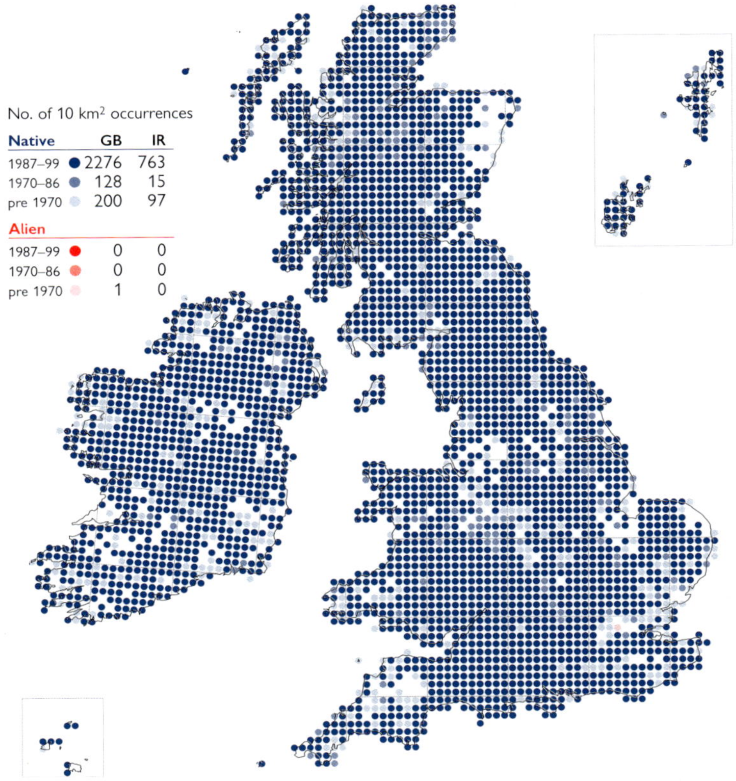

An annual or biennial herb of dry, infertile calcareous or base-rich substrates, but also found in flushed sites on neutral or mildly acidic soils. It occurs in a wide range of calcareous grasslands, mires and flushes, in short-sedge fen-meadows, on outcrops and ledges of basic rock, road cuttings, quarry spoil and lead-mine debris, and very locally on dry heaths. 0–840 m (Breadalbanes, Mid Perth).

Native (change –0.44). *L. catharticum* was mapped as 'all records' in the 1962 *Atlas.* It has probably declined in many areas, particularly in S. England where local losses are beginning to show at the 10-km square scale.

European Temperate element; also in N. America.

References: Atlas (90b), Grime *et al.* (1988), Hultén & Fries (1986), Meusel *et al.* (1978).

G. T. D. WILMORE

Radiola linoides Allseed

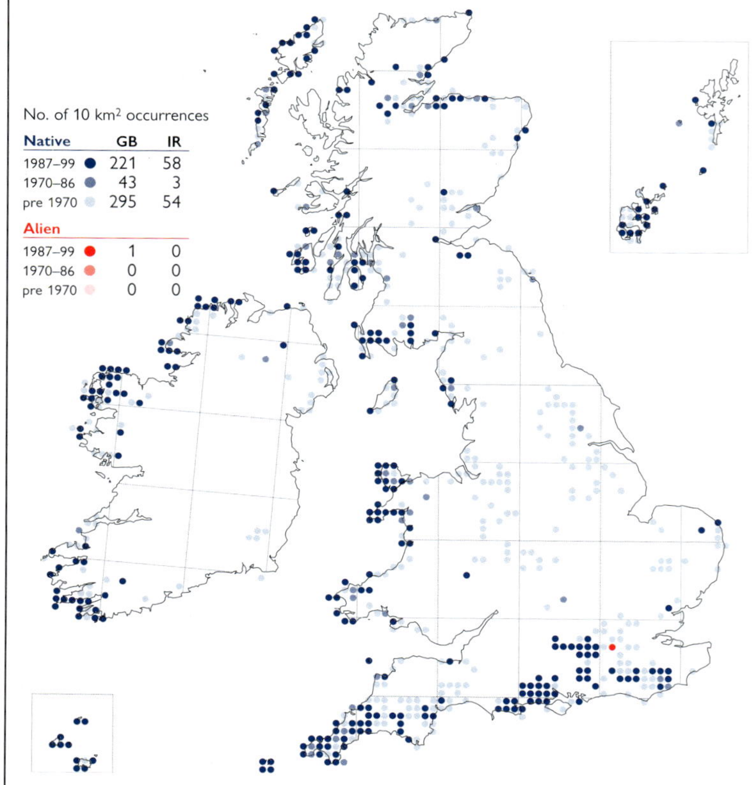

A small annual of damp, bare, infertile, peaty or sandy ground in acid grasslands and heaths, by ponds, on tracks and in woodland rides. Near the coast it occurs in dune-slacks, sandy grassland, on machair, and in soil-filled rock cracks. Lowland.

Native (change −0.87). *R. linoides* suffered a considerable decline before 1930, largely due to the loss of lowland heaths or a lack of grazing and disturbance on them. These losses have continued in England: it was lost from 95% of a sample of sites in Dorset between 1935 and 1992 and remaining populations are very small. Its distribution seems to be stable elsewhere. It is easily overlooked and may be under-recorded in some areas.

European Temperate element.

References: Atlas (90c), Byfield & Pearman (1996), Hultén & Fries (1986), Meusel *et al.* (1978).

G. T. D. WILMORE

Polygala vulgaris Common Milkwort

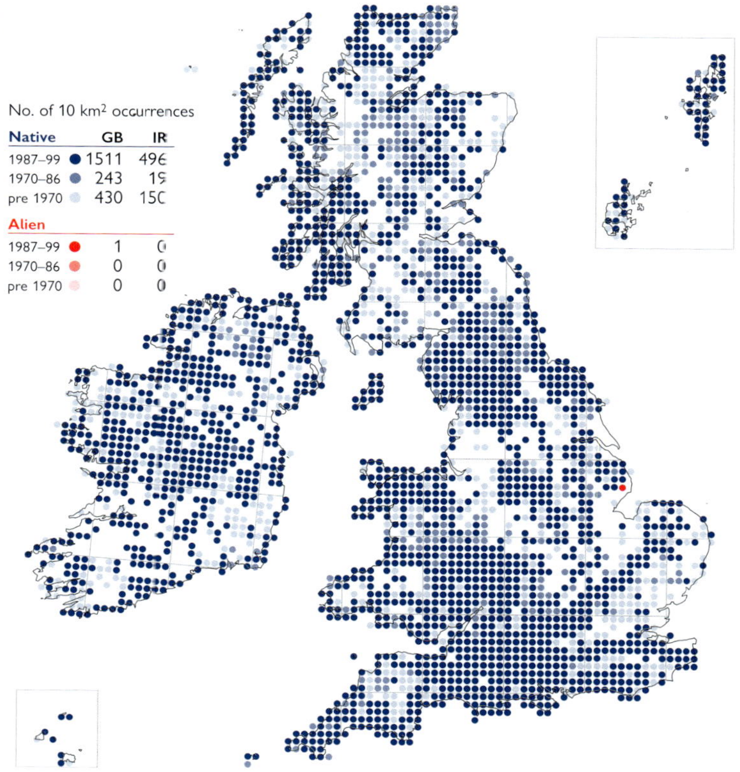

A perennial herb which usually grows in short, moderately infertile neutral to basic grassland on banks, hill-slopes crags and sand dunes. It also occurs in acid grasslands, heaths and fen-meadows. 0–730 m (Mourne Mountains, Co. Down).

Native (change −1.14). This species was mapped as 'all records' in the 1962 *Atlas*. There has been no significant change in the overall range of this species, but there have been very substantial losses within its range. Analysis of the database reveals that most of these occurred since 1950. Declines are presumably due to the intensification of grassland management and conversion to arable.

European Temperate element.

References: Atlas (55d), Grime *et al.* (1988), Hultén & Fries (1986).

M. J. Y. FOLEY

Polygala serpyllifolia Heath Milkwort

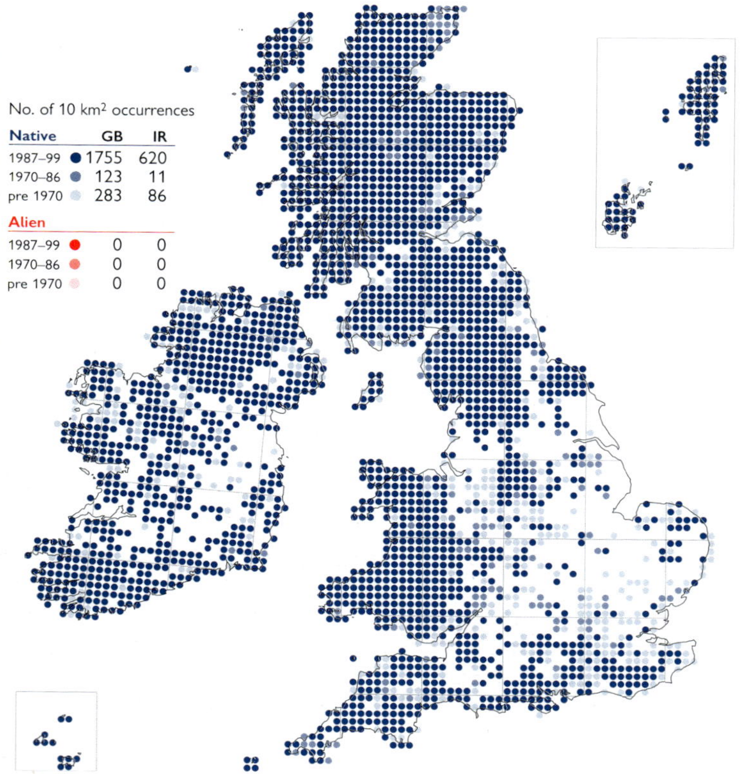

A perennial herb occurring on acidic soils in grasslands, moors, heaths and mires. 0–1035 m (Ben Lawers, Mid Perth).

Native (change −0.50). This species was mapped as 'all records' in the 1962 *Atlas*. There has been a substantial decline in S. England, and analysis of the database reveals that these declines have occurred since 1950.

Suboceanic Temperate element; also in Greenland.

References: Atlas (56a), Dupont (1962), Hultén & Fries (1986), Meusel *et al.* (1978).

M. J. Y. FOLEY

Polygala calcarea Chalk Milkwort

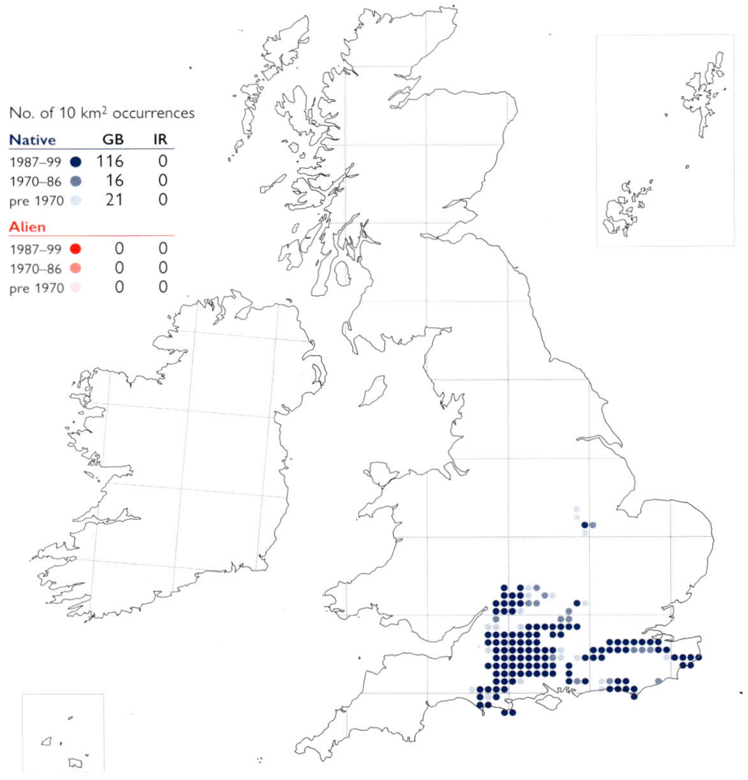

A perennial herb found in tightly-grazed chalk and limestone grassland, usually on warm S.-facing slopes. It is a poor competitor which disappears if insufficient grazing allows coarser grasses to become dominant. Lowland.

Native (change –0.37). *P. calcarea* remains locally frequent within its main area of distribution but seems to have been lost from some localities towards the periphery of its range, possibly due to a lack of grazing.

Oceanic Southern-temperate element.

References: Atlas (56b), Meusel *et al.* (1978), Stewart *et al.* (1994).

M. J. Y. FOLEY

Polygala amarella Dwarf Milkwort

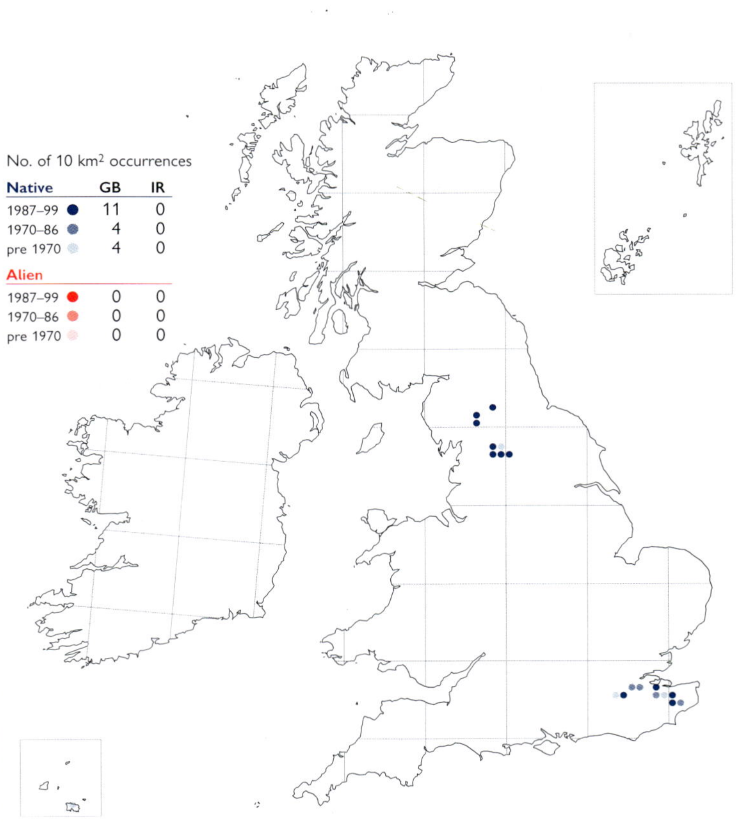

A perennial herb of base-rich substrates. On the North Downs it is usually found in close-grazed chalk grassland, and in N. England in limestone grassland, on rock ledges and in fissures in limestone scars, and sometimes on damp stream banks and in areas of eroded sugar limestone. Generally lowland, but reaching 530 m on Cronkley Fell (N.W. Yorks.).

Native (change –0.10). The overall range of this species has changed little, although it has disappeared from some sites and greatly declined in others, particularly in Kent, during the last few decades. This is probably caused by changes in grazing regimes.

European Boreo-temperate element, with a continental distribution in W. Europe.

References: Atlas (56c), Hultén & Fries (1986), Meusel *et al.* (1978), Wigginton (1999).

M. J. Y. FOLEY

Aesculus hippocastanum Horse-chestnut

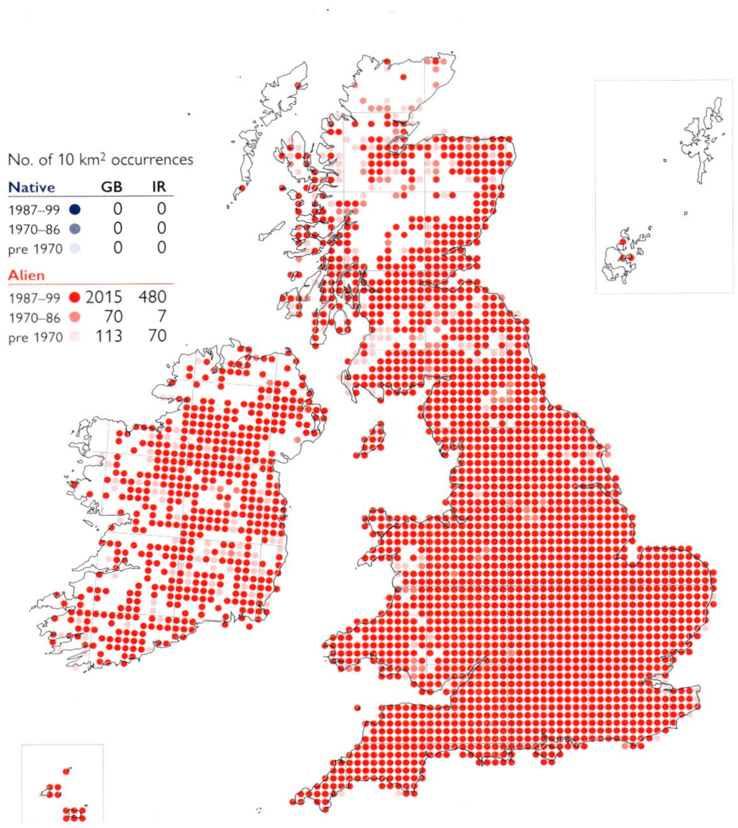

A tree of parkland, large gardens and estates, churchyards, urban streets and village greens; also a constituent of deciduous and mixed woodland. It is sometimes self-sown in scrubby areas, waste ground or rough grassland, and occasionally regenerates in woodland, but is rarely fully naturalised. Generally lowland, but reaching 505 m at Ashgill (Cumberland).

Neophyte (change +1.08). This species was introduced into cultivation in 1612 or 1615, and was recorded from the wild by 1870. It is now better recorded in S.W. England, Wales and Ireland than in the 1962 *Atlas*, but its overall range has changed little.

Native of the Balkan peninsula; other species in the family are found in S. & E. Asia and the Americas.

References: Atlas (97c), Bean (1970), Mabey (1996), Meusel *et al.* (1978), Mitchell (1996).

G. T. D. WILMORE

Aesculus carnea Red Horse-chestnut

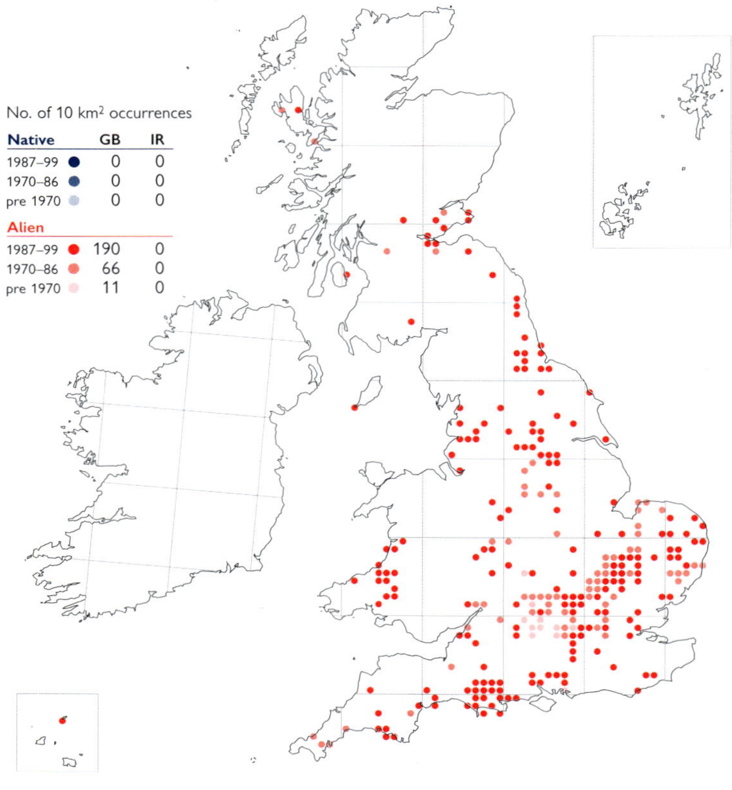

An ornamental tree planted in parkland and large gardens, and as a street tree. It also occurs, rarely, as a minor introduction in deciduous and mixed woodland. Lowland.

Neophyte. This species, which has been cultivated in Britain since 1818, was recorded in the wild by 1955. It was not mapped in the 1962 *Atlas*, so changes in its distribution are difficult to assess.

An alloploid hybrid of garden origin, derived from *A. hippocastanum* and *A. pavia* in the early 19th century.

Reference: Bean (1970).

G. T. D. WILMORE

Acer platanoides Norway Maple

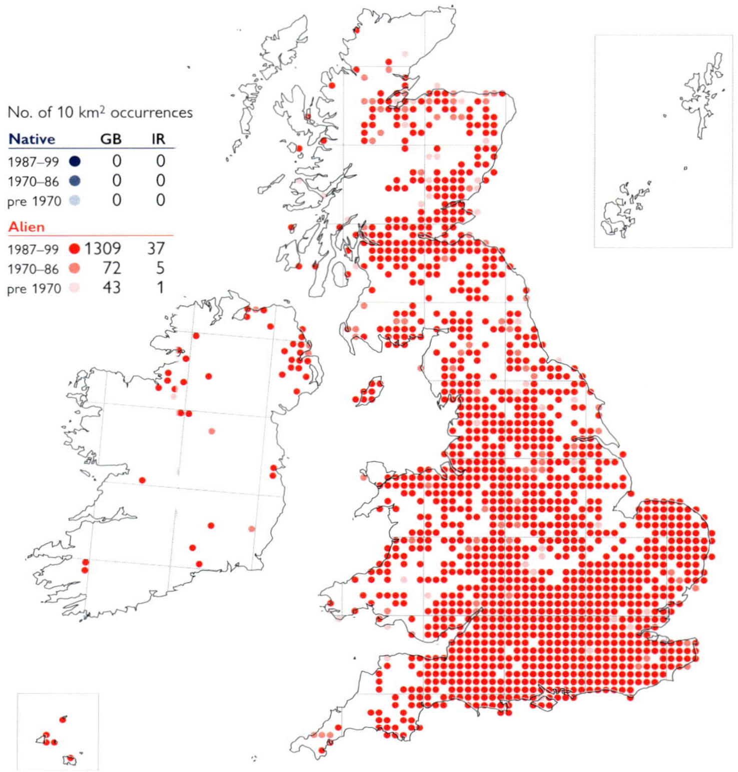

A deciduous tree planted in woodland, hedgerows, amenity areas, gardens and along roads. It tolerates a wide range of soil types and is frequently self-sown, becoming naturalised in secondary woodland, rough grassland, scrub and urban waste land. Generally lowland, but reaching 340 m at Alston (Cumberland).

Neophyte. *A. platanoides* was in cultivation in Britain by 1683, and has been known from the wild since at least 1905. Trends in its distribution are difficult to assess, but it is probably increasing due to continued planting.

A European Temperate species, absent as a native from much of western Europe.

References: Bean (1970), Hultén & Fries (1986), Jones (1945), Meusel *et al.* (1978), Mitchell (1996).

T. D. DINES

Acer cappadocicum Cappadocian Maple

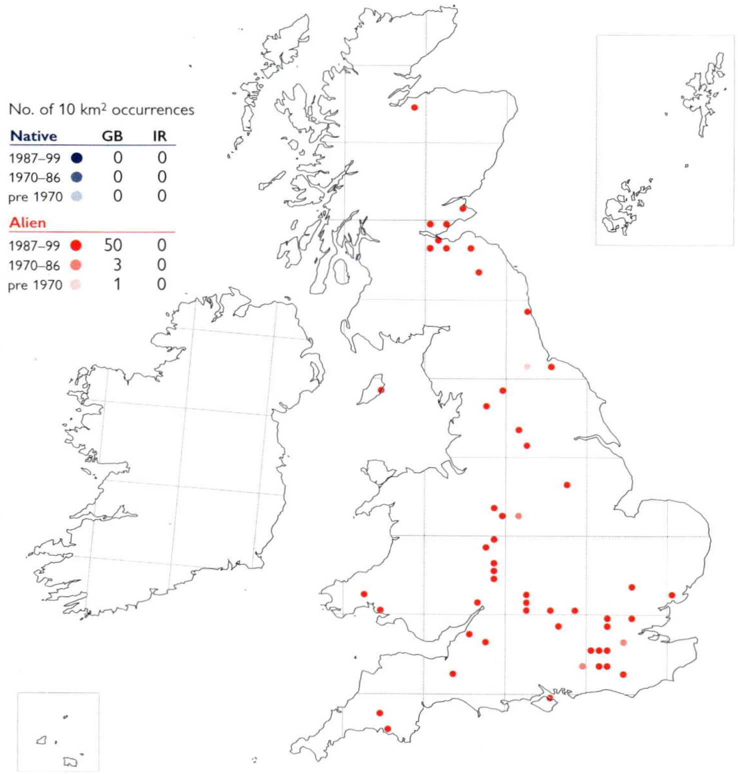

A fast-growing deciduous tree planted in parks and large gardens, and on roadsides, and self-sown in grassland, hedgerows and waste land. Extensive suckering has been reported. Lowland.

Neophyte. *A. cappadocicum* was introduced into cultivation in 1838, and is now widely grown as an amenity and garden tree. It was not recorded from the wild until 1977 and trends in its distribution are difficult to assess.

Native of S.W. & C. Asia.

References: Bean (1970), Meusel *et al.* (1978), Mitchell (1996).

T. D. DINES

Acer campestre Field Maple

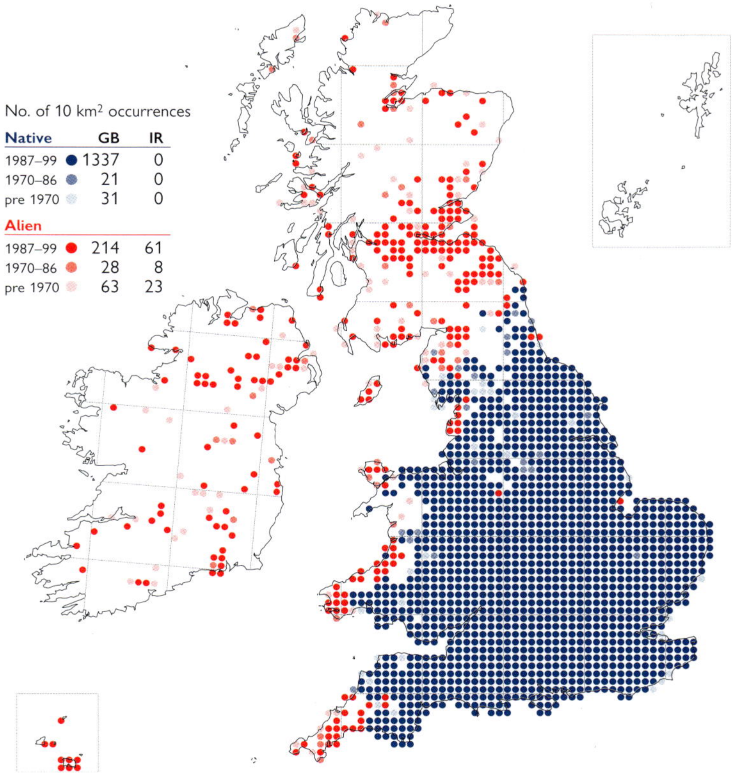

A deciduous tree, native in woodland, scrub and old hedgerows on a wide range of moist, usually base-rich, soils. It is also widespread as a planted tree in amenity areas, on farmland, along roads and in hedgerows and coppice. It fruits erratically, sometimes producing only male flowers following a year of prolific fruiting. 0–380 m (Llanthony, Brecon).

Native (change +0.35). Since the 1962 *Atlas* there have been many more records of *A. campestre* from S.W. England and S. Wales. It has also been very extensively planted both within and beyond its native range, and these introductions tend to blur the boundary of its native range.

European Temperate element.

References: Atlas (97b), Hultén & Fries (1986), Jones (1945), Meusel *et al.* (1978), Rackham (1980).

T. D. DINES

Acer pseudoplatanus Sycamore

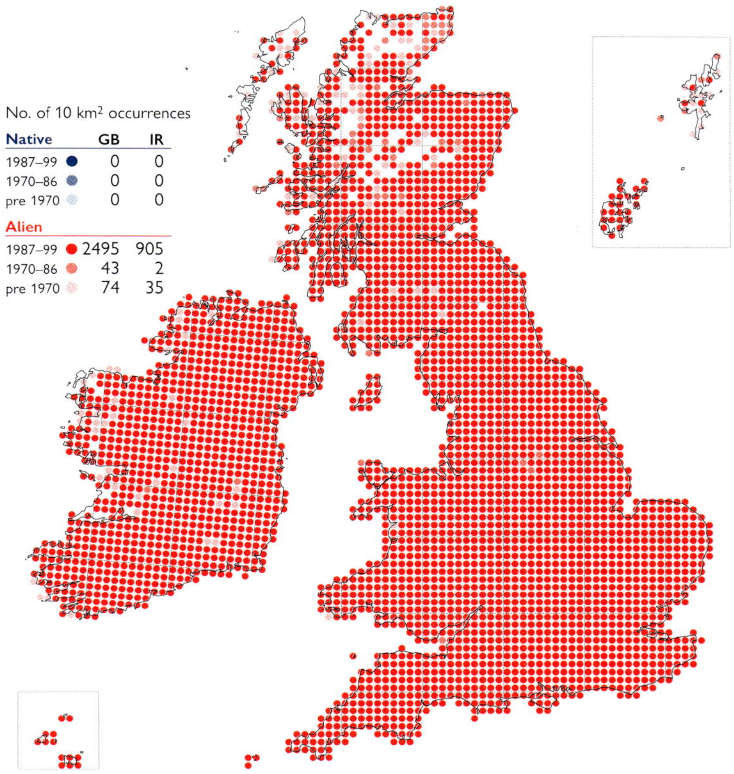

A large, rapidly growing deciduous tree of plantations, woods, parkland, estates, large gardens and roadsides, prolifically self-sowing and naturalised in a very wide range of natural, semi-natural and man-made habitats, avoiding only the most acidic and water-logged soils. In upland areas, however, it is often restricted to sites associated with habitation. 0–580 m (Dowgang Hush, Cumberland).

Neophyte (change –0.40). *A. pseudoplatanus* was introduced to Britain in the 16th century and was widely planted from the late 18th century onwards; it was first recorded from the wild in 1632. There has been little change in its distribution since the 1962 *Atlas*.

A European Temperate species, mainly found in the mountains of C. & S. Europe.

References: Atlas (97a), Grime *et al.* (1988), Jones (1945), Meusel *et al.* (1978), Mitchell (1996).

T. D. DINES

Acer saccharinum Silver Maple

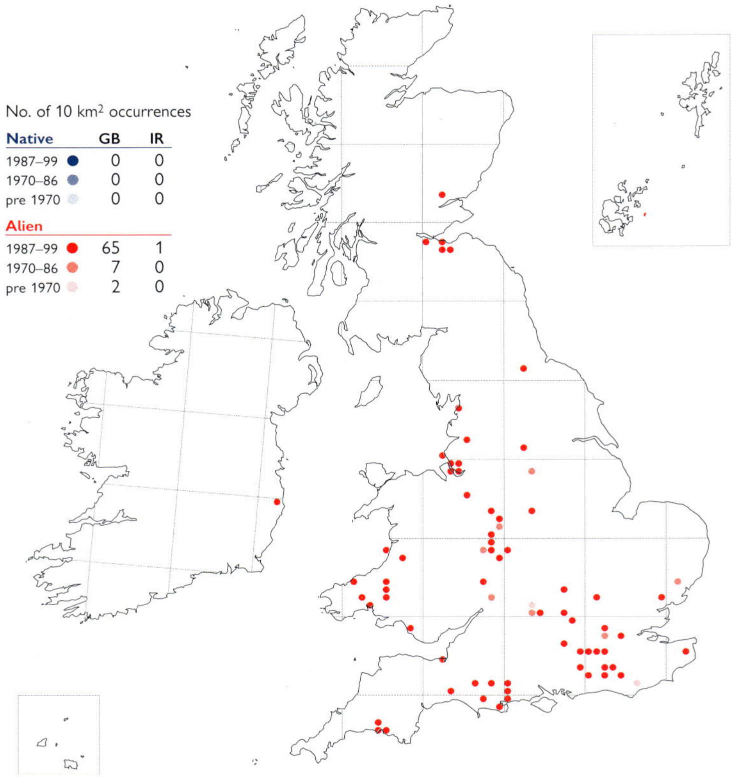

A large, fast-growing, deciduous tree of town parks, large gardens and roadsides. It rarely sets seed in Britain, although establishment from seed has been reported from the London area. Lowland.

Neophyte. *A. saccharinum* was introduced to Britain in 1725. It is widely planted, but was not recorded from the wild until 1959. It is an increasingly popular roadside tree, and may therefore be increasing in the wild.

Native of eastern N. America.

References: Bean (1970), Mitchell (1996).

T. D. DINES

Acer negundo Ashleaf Maple

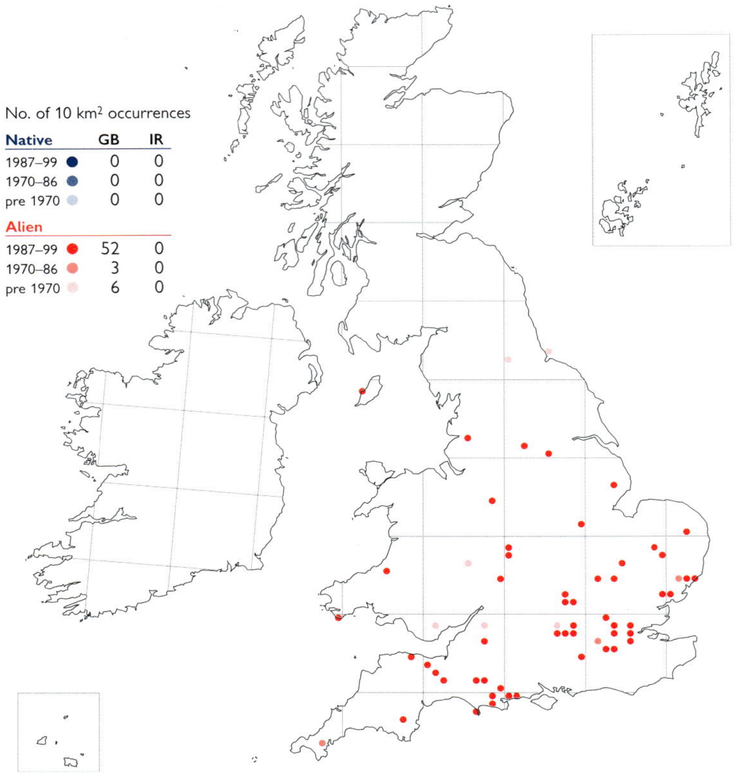

A dioecious, deciduous tree planted in town parks, gardens, car parks and streets, and sometimes also found on railway banks. Establishment from seed is rare, but natural regeneration has been reported in S.E. England. Lowland.

Neophyte. *A. negundo* was cultivated in Britain by 1688, but was not recorded from the wild until 1913. It is now commonly planted for ornament, usually as a variegated form, though fortunately this often reverts to the green-leaved form when its shoots are not pruned.

Native of eastern N. America.

Reference: Bean (1970).

T. D. DINES

Rhus typhina Stag's-horn Sumach

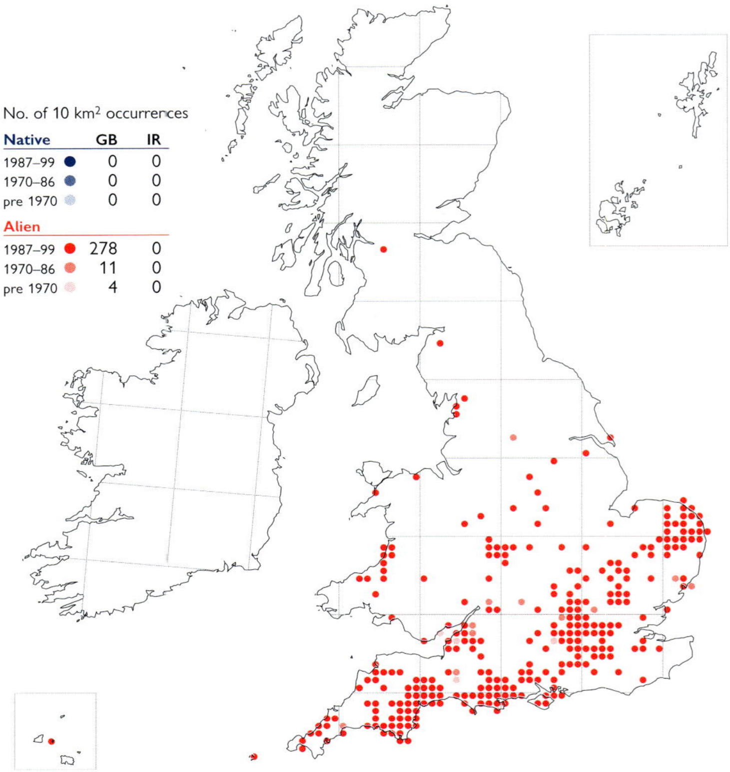

A deciduous shrub or small tree found as a garden escape or throw-out on roadsides, railway banks and, less frequently, on waste ground and rubbish tips; it also occurs as a relic of cultivation. It can become naturalised, reproducing vigorously by suckers, but seed-set is very poor as the species is dioecious and most populations consist of a single sex. Lowland.

Neophyte. This species was cultivated in Britain by 1629, and is extremely popular in gardens. Many records are questionably wild, and it was not recorded as naturalised until 1966 (Berks.). The vast majority of records are recent, probably reflecting the increased interest in recording aliens rather than a genuine increase in its distribution.

Native of eastern N. America.

Reference: Bean (1980).

H. J. KILLICK

Ailanthus altissima Tree-of-heaven

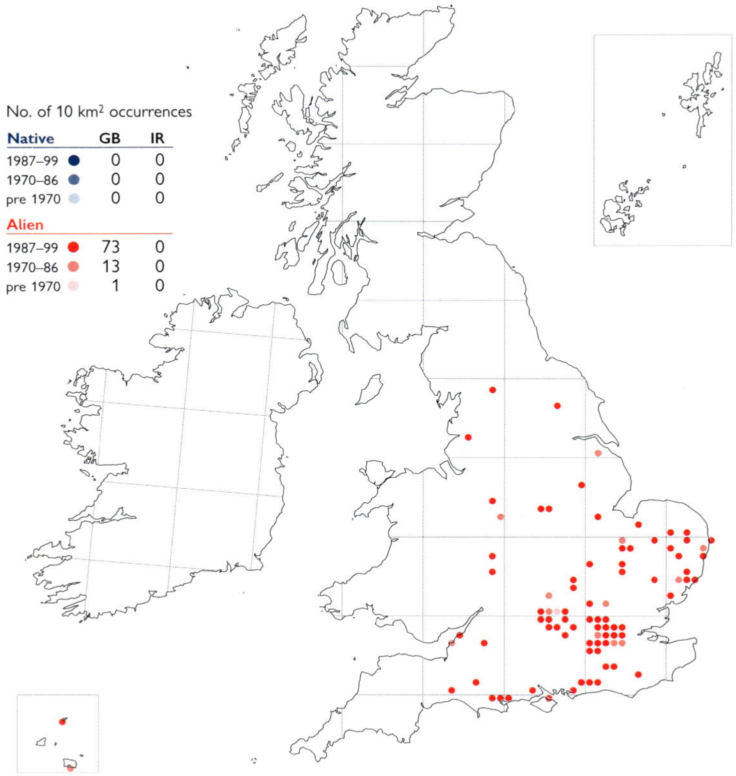

A large tree planted in streets and parks in urban areas, but rarely found elsewhere. It is extremely tolerant of atmospheric pollution. It grows rapidly, and spreads by suckering, but rarely sets seed in Britain. Lowland.

Neophyte. *A. altissima* was introduced into cultivation in 1751, but not recorded from the wild until 1935.

Native of China; extensively naturalised in C. & S. Europe where it is an extremely invasive species, once described to C. D. Preston as the 'Tree of Hell'.

Reference: Bean (1970).

D. A. PEARMAN

Oxalis corniculata Procumbent Yellow-sorrel

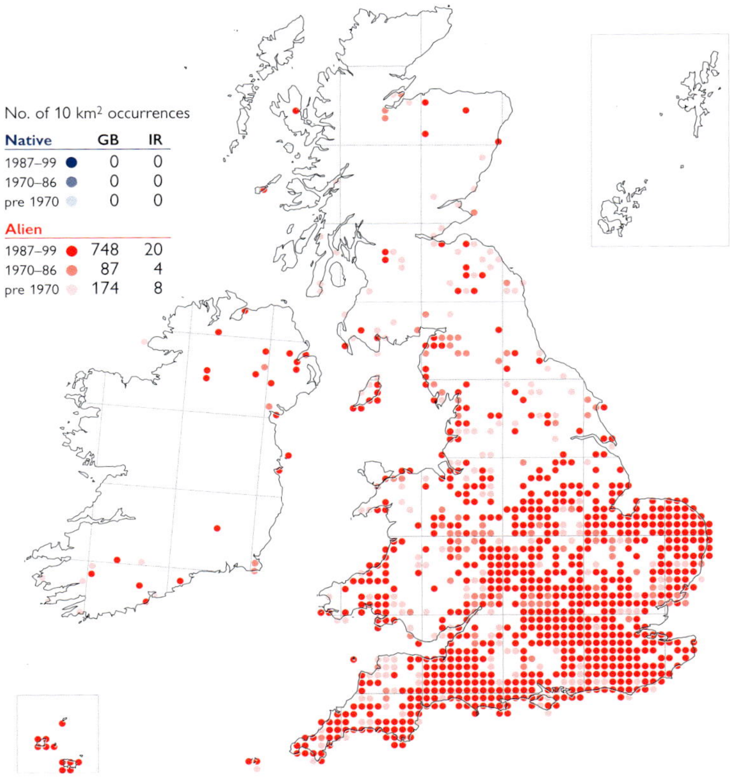

A scrambling annual or short-lived perennial herb which is often a pernicious weed of cultivated land, disturbed areas and paths. It is self-compatible and the seeds are explosively ejected up to two metres from the capsules. Its brittle stems readily root at the nodes. Lowland.

Neophyte (change +1.62). *O. corniculata* was cultivated in Britain by 1656 and was first recorded in the wild in 1770, but the main spread has been in the 20th century and this has continued since the 1962 *Atlas*. It has consolidated its range in the south, and is now more widely recorded in N. Britain and Ireland.

Native range unknown; now found in warm temperate and tropical regions throughout the world.

References: Atlas (95c), Lovett-Doust *et al.* (1985), Hultén & Fries (1986), Meusel *et al.* (1978), Rich & Jermy (1998), Watson (1997), Young (1958).

M. F. WATSON

Oxalis exilis Least Yellow-sorrel

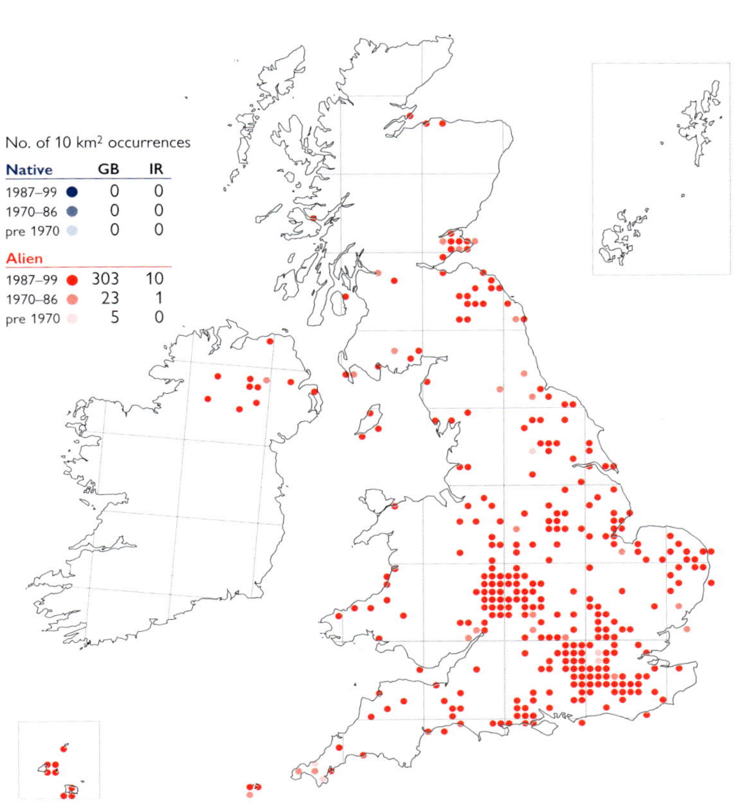

A small, prostrate annual or short-lived perennial herb grown in garden rockeries, escaping into habitats such as disturbed ground, pathsides, the base of walls, and crevices in paved areas, in shaded or open situations. Most populations are casual, but in some sites it is known to have persisted for many years. Lowland.

Neophyte. As *O. exilis* is often mistaken for small forms of *O. corniculata*, and it was not mapped in the 1962 *Atlas*, it is difficult to assess changes in its distribution. It was first recorded in the wild in 1926, and may be under-recorded.

Native of Australasia.

References: Reid (1975), Rich & Jermy (1998), Watson (1997).

M. F. WATSON

Oxalis stricta Upright Yellow-sorrel

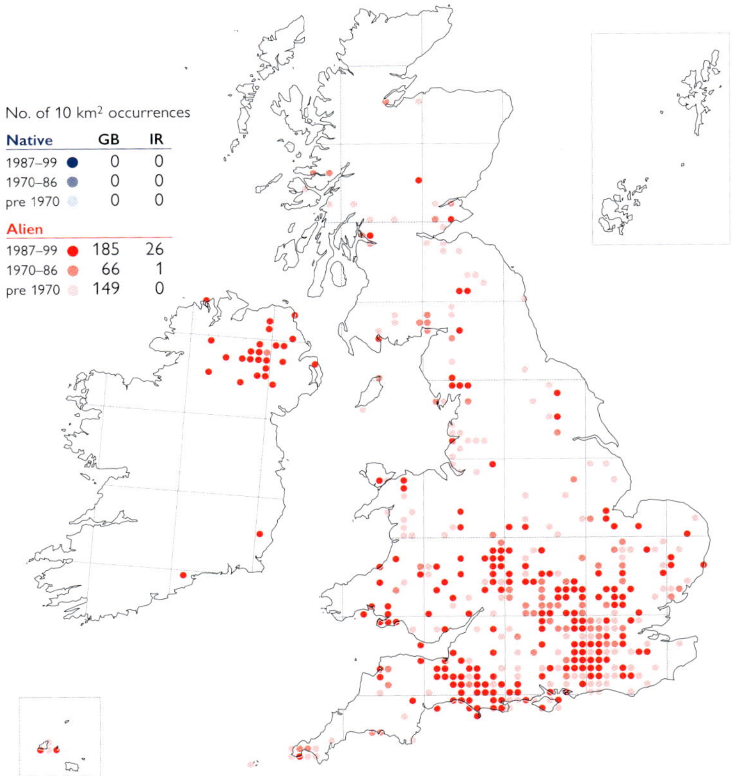

An erect annual of cultivated ground, disturbed areas and paths, in shaded or semi-shaded situations. It spreads by seed or occasionally by underground rhizomes. Seed is freely set by self-pollination, and forcibly ejected from the capsules over a distance of up to two metres. Lowland.

Neophyte (change –0.09). *O. stricta* was cultivated in Britain by 1658 and has long been frequent in S. England; it was recorded from the wild by 1823. It has spread westwards since the 1962 *Atlas*. Most populations are casual, but is has been known to persist for many years in some places.

Native of N. America and E. Asia; extensively naturalised in temperate Europe.

References: Atlas (95d), Lovett-Doust *et al.* (1985), Hultén & Fries (1986), Meusel *et al.* (1978), Rich & Jermy (1998), Watson (1997), Young (1958).

M. F. WATSON

Oxalis articulata Pink-sorrel

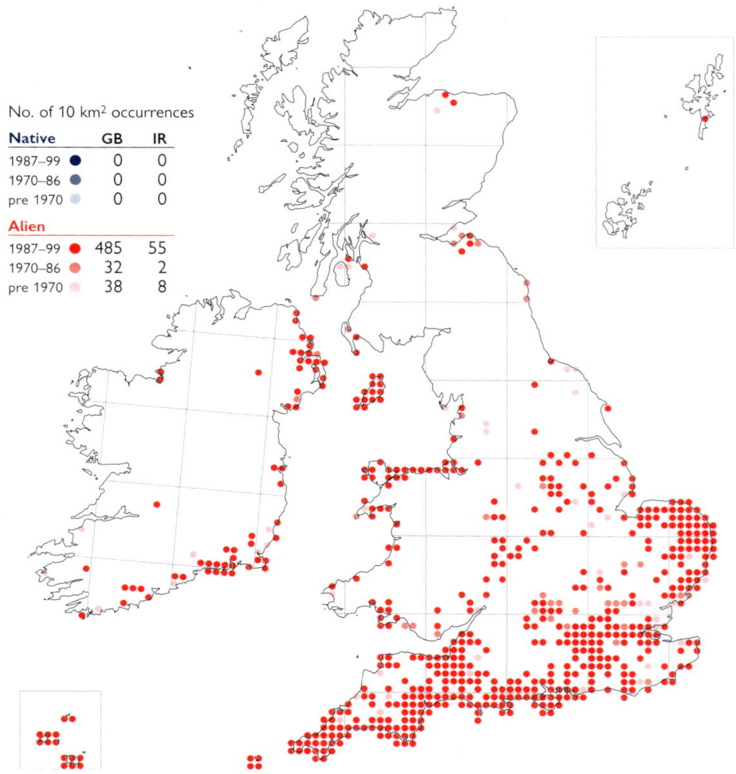

A vigorous rhizomatous perennial herb occasionally naturalised in disturbed areas, on waste ground, roadsides and seashores, but always associated with human habitation. It reproduces mainly by semi-woody rhizome fragments, though some seed-set is reported. Lowland.

Neophyte. *O. articulata* was introduced in 1870, and is now frequently grown in gardens as an ornamental. It was known from the wild by 1912, and is thought to be increasing, although such changes are difficult to assess.

Native of temperate S. America.

References: Bolòs & Vigo (1990), Watson (1997), Young (1958).

M. F. WATSON

Oxalis acetosella Wood-sorrel

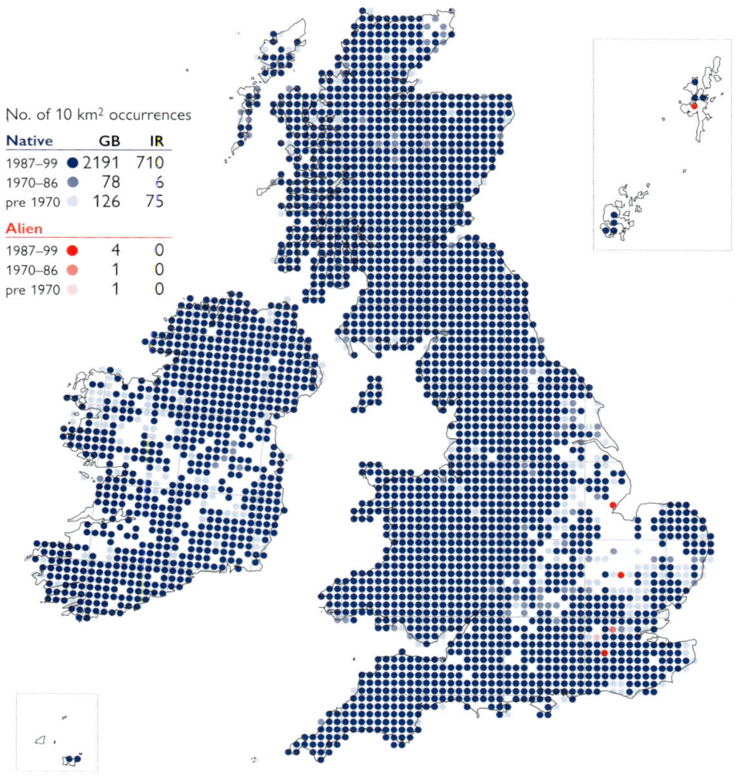

A perennial creeping herb of woodland, hedgerows, banks, and other moist, usually shaded, habitats; also in rough montane grassland, grikes in limestone pavement, *Vaccinium* communities, bryophyte-rich block screes, and rock ledges. It grows on both calcareous and non-calcareous soils, though only those which are moisture-retentive. It is one of the few species able to survive the deep shade of conifer plantations. 0–1160 m (Ben Lawers, Mid Perth).

Native (change −0.74). The distribution of *O. acetosella* has little changed since the 1962 *Atlas*, where it was mapped as 'all records'. The bulk of the losses in E. England probably occurred before 1950.

Eurasian Boreo-temperate element.

References: Atlas (95b), Grime *et al.* (1988), Hultén & Fries (1986), Meusel *et al.* (1978), Packham (1978), Young (1958).

M. F. WATSON

Oxalis debilis Large-flowered Pink-sorrel

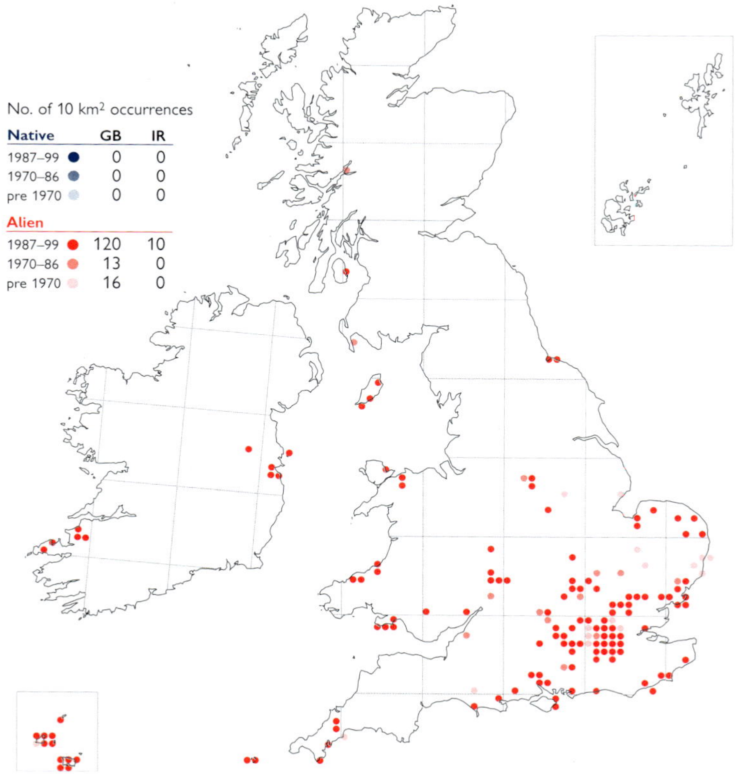

A bulbous perennial herb, formerly widely grown as an ornamental and readily becoming naturalised in gardens and on waste ground. It spreads rapidly by easily detached bulblets that are resistant to all but the strongest herbicides, and in places it has become an almost ineradicable weed. Lowland.

Neophyte. Introduced in 1826, *O. debilis* has spread slowly, despite its weedy nature and ability to become naturalised. It was recorded in the wild by 1900.

Native of temperate S. America.

References: Bolòs & Vigo (1990), Watson (1997), Young (1958).

M. F. WATSON

Oxalis latifolia Garden Pink-sorrel

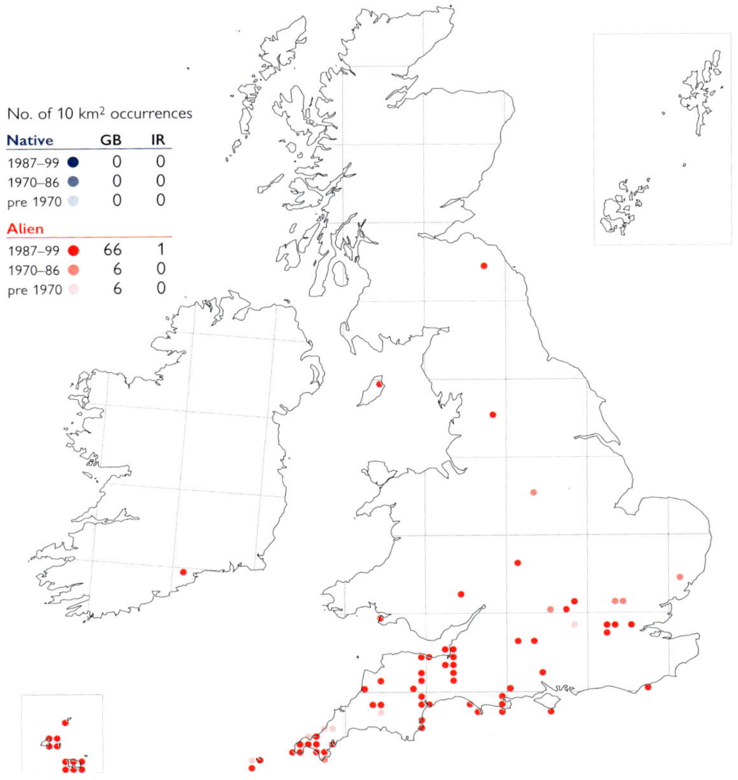

A bulbous perennial which was formerly grown as an ornamental. It is a weed of nurseries and gardens, from which it spreads to rubbish tips and other sites where garden refuse is dumped. It does not set seed, but spreads vigorously by easily detached bulblets that are resistant to all but the strongest herbicides. Lowland.

Neophyte. *O. latifolia* was recorded in the wild by 1921. It is less frost-hardy than the similar *O. debilis*, and was very considerably reduced in Guernsey in the severe 1962–3 winter, although it soon recovered (McClintock, 1975). Changes in its distribution cannot be readily assessed.

Native of S. & C. America; now a notorious agricultural weed of mild climates worldwide.

References: Bolòs & Vigo (1990), Watson (1997), Young (1958).

M. F. WATSON

Oxalis incarnata Pale Pink-sorrel

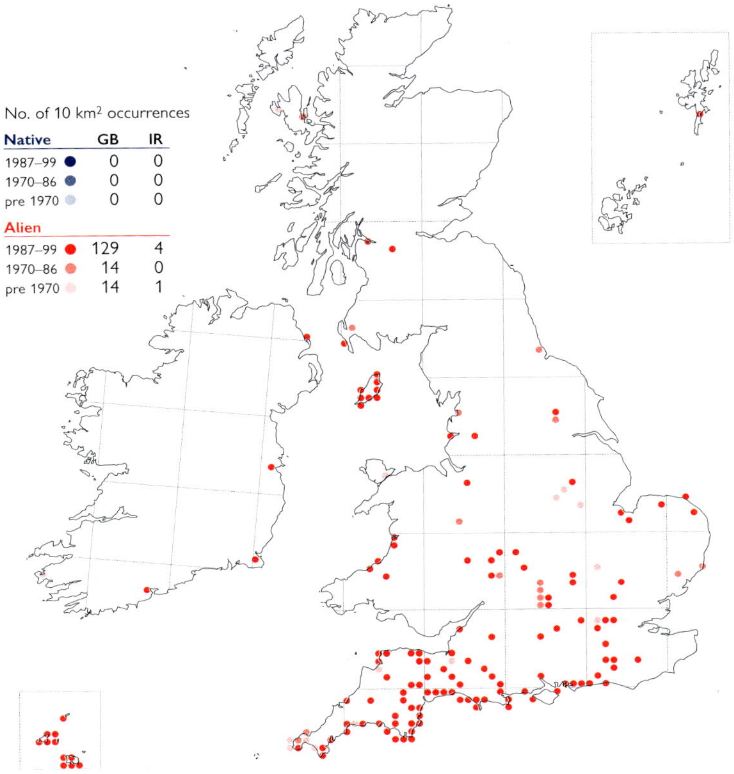

A perennial bulbous herb with annual erect branching stems. It is cultivated in gardens, occasionally escaping to nearby disturbed, shaded sites, hedge banks, stone walls and pavement cracks. It does not set seed, but spreads by bulblets produced in the axils of the aerial stems. Lowland.

Neophyte. *O. incarnata* was cultivated in Britain by 1739 and recorded in the wild by 1912 (Middlesex). Changes in its distribution are not easily assessed.

Native of S. Africa.

References: Watson (1997), Young (1958).

M. F. WATSON

Geranium endressii French Crane's-bill

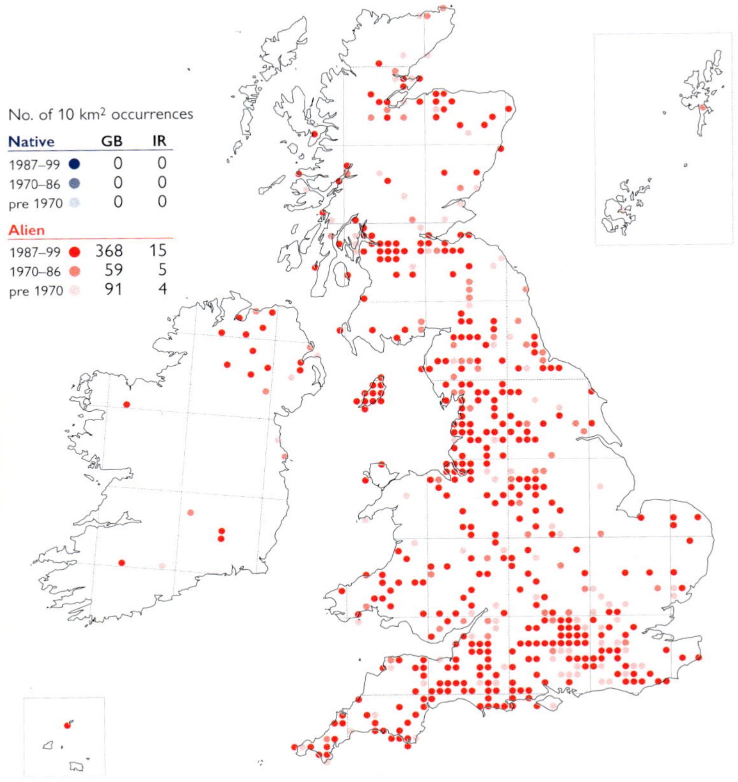

A rhizomatous perennial herb found as a garden escape on grassy or wooded banks and roadsides around habitation; also occurring as a garden throw-out on rubbish tips and waste ground. Lowland.

Neophyte (change +2.07). *G. endressii*, introduced to Britain in 1812, is common in gardens and has been recorded in the wild since 1906. It has increased since the 1962 *Atlas*, although some records may be errors for *G. × oxonianum* as several cultivars of the hybrid are very similar to this parent. The frequent cultivar 'Wargrave Pink' is probably also a form of *G. × oxonianum*, though Yeo (1985) considers it to be a cultivar of *G. endressii*.

Native of the Pyrenees; the native range is very restricted.

References: Atlas (91b), Dupont (1962), Meusel *et al.* (1978).

S. J. LEACH

Geranium endressii × *G. versicolor* (*G.* × *oxonianum*) Druce's Crane's-bill

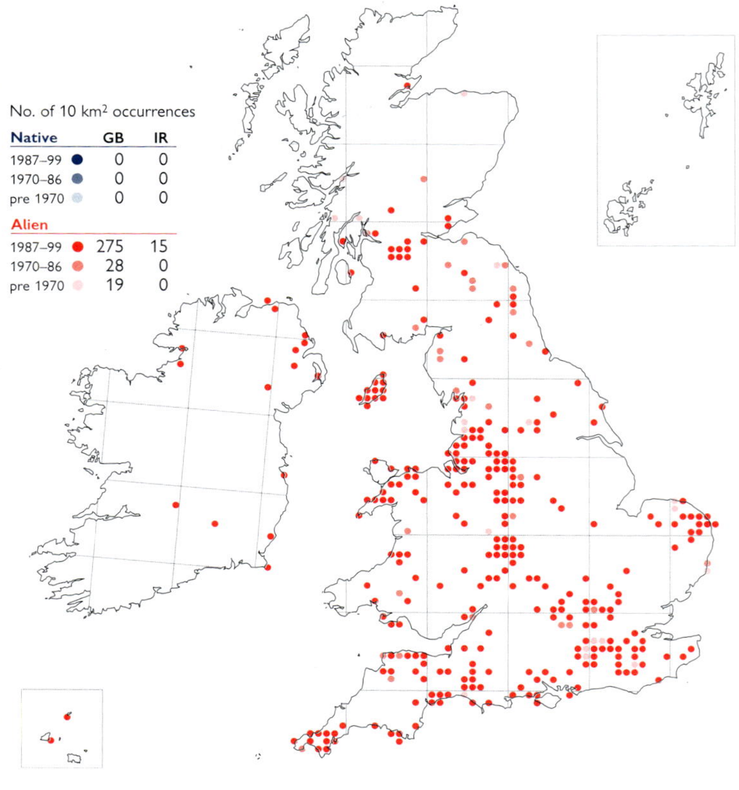

A clump-forming perennial herb, occasionally naturalised in open woodland and rough grassland on railway banks and roadsides, usually close to habitation, and also occurring as a garden throw-out on tips and waste ground. Lowland.

Neophyte. This hybrid was first cultivated in Britain in 1932 and is much grown in gardens; it has been recorded in the wild since 1954. It is an extremely variable taxon; whilst the large-flowered cultivar 'Claridge Druce' is distinctive, smaller-flowered cultivars of this hybrid such as 'A. T. Johnson' and 'Winscombe' may frequently have been recorded in error as *G. endressii.*

A hybrid of garden origin.

References: Stace (1975), Yeo (1985).

S. J. LEACH

Geranium versicolor Pencilled Crane's-bill

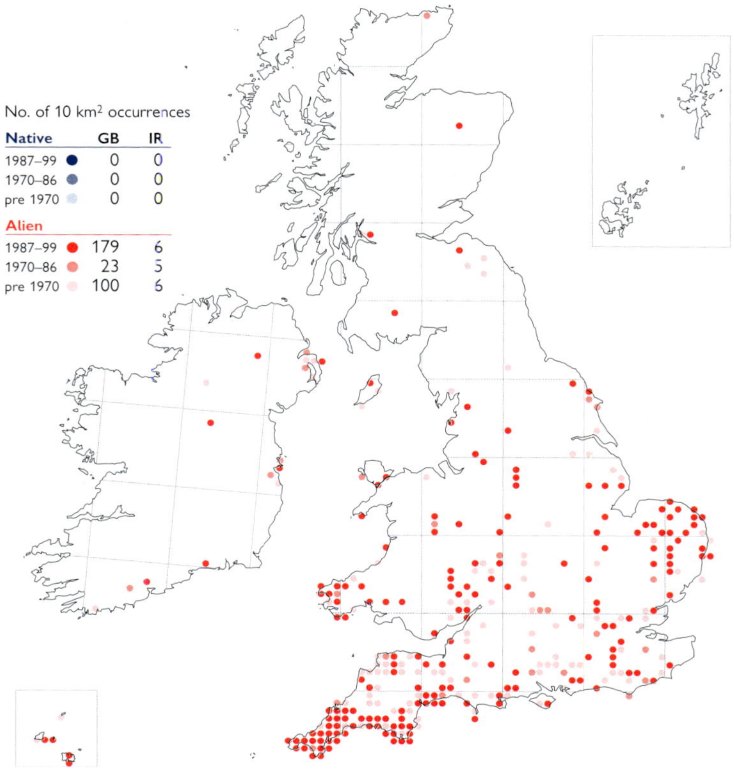

This rhizomatous perennial herb is a frequent garden escape or throw-out, occurring in grassy places on roadsides and railway banks, in hedge banks and wood-borders, usually close to habitation and showing a strong preference for warm, sheltered, often somewhat shaded situations. Lowland.

Neophyte (change +0.74). Cultivated in Britain by 1629, *G. versicolor* has been known in the wild since 1820. It is now better recorded and more widespread than when mapped in the 1962 *Atlas*.

Native of Italy, Sicily and the southern Balkans.

References: Atlas (91c), Yeo (1985).

S. J. LEACH

Geranium nodosum Knotted Crane's-bill

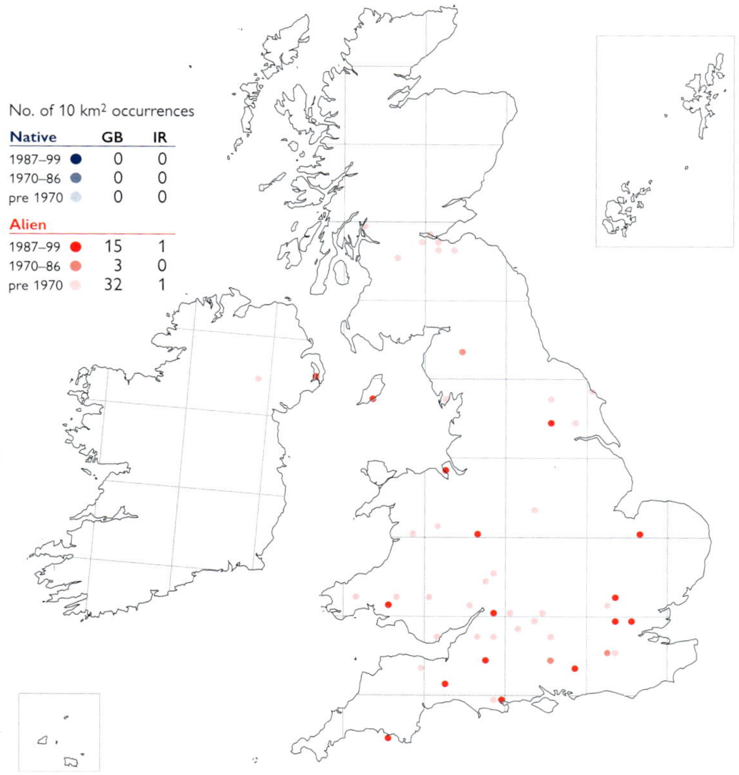

A shortly rhizomatous perennial herb occurring as a garden escape or throw-out in wood-borders, hedgerows, churchyards and in rough grassland on railway banks and roadsides, usually close to habitation. Lowland.

Neophyte. This species was introduced to cultivation in Britain in 1633 and was first recorded in the wild in 1801. Unlike several other cultivated *Geranium* taxa, there is little indication that it is increasing.

Native of the mountains of S. Europe from the Pyrenees to the Balkans.

References: Bolòs & Vigo (1990), Yeo (1985).

S. J. LEACH

Geranium rotundifolium
Round-leaved Crane's-bill

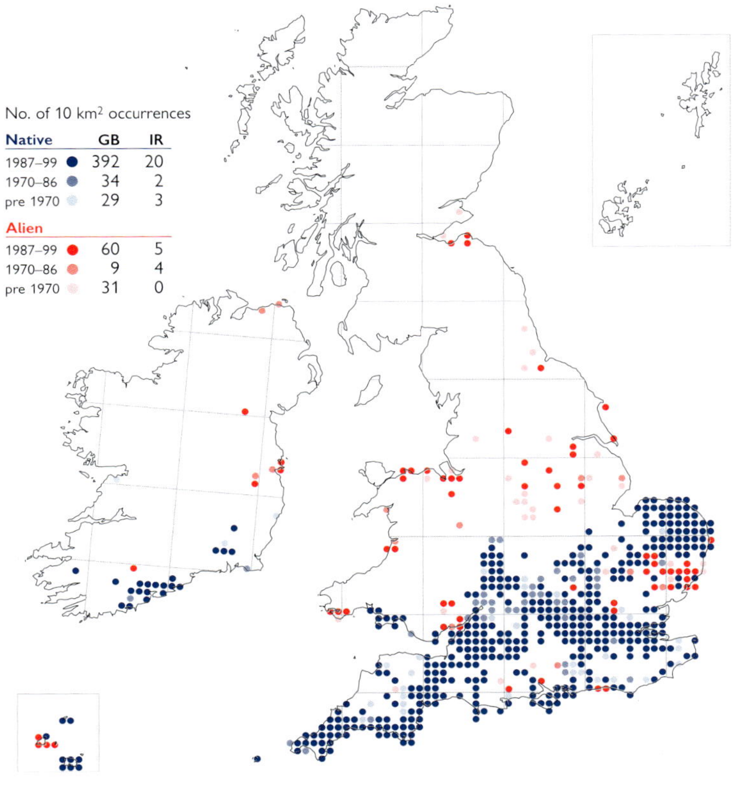

An annual of hedgerows, dry roadside-banks and wall-tops, especially close to the sea, but spreading to roadside verges, rubble heaps, railway ballast and waste ground. It is also a garden and street weed. Colonies may be very persistent, even where the species is confined to weedy habitats. Lowland.

Native (change +1.70). *G. rotundifolium* has increased markedly since the 1962 *Atlas*. As a presumed native it is restricted to S. Britain and the extreme south of Ireland. However, given its spread in parts of S. England, it is possible that some recent, supposedly alien, records to the north of this could in fact represent a natural extension of its native range.

Eurosiberian Southern-temperate element; widely naturalised outside its native range.

References: Atlas (93a), Curtis & McGough (1988), Meusel *et al.* (1978).

S. J. LEACH

Geranium sylvaticum Wood Crane's-bill

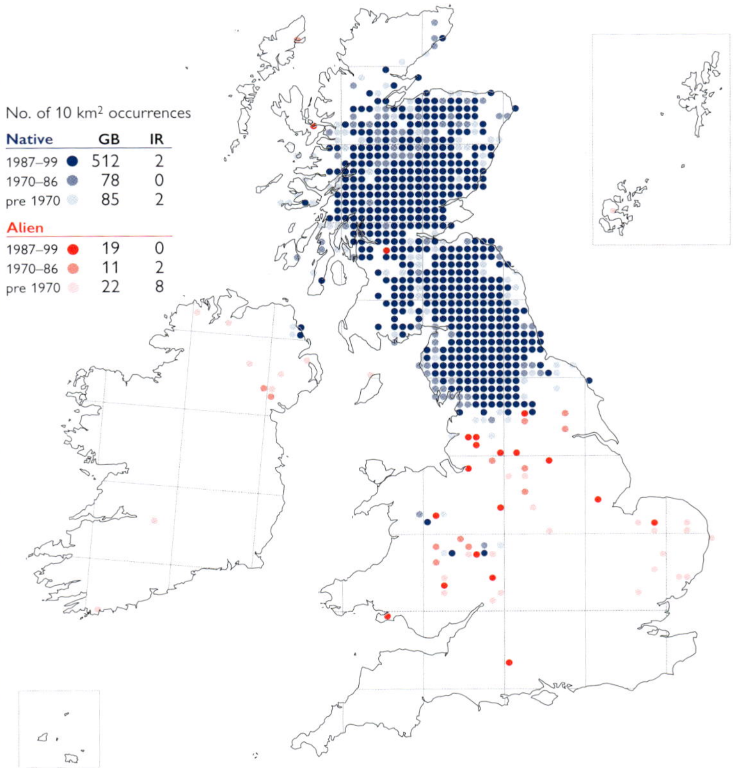

A stoutly rhizomatous perennial herb of hay meadows, ungrazed damp woodlands, streamsides and mountain rock ledges, and in many areas a characteristic feature of lane-side hedge banks and verges. Mainly upland, to 1005 m on Ben Lawers (Mid Perth).

Native (change –0.45). *G. sylvaticum* has declined locally since the 1962 *Atlas*, particularly on the edges of its range, having been lost from many hay meadows due to the increased use of fertilisers and the widespread change to silage production (Halliday, 1997). It is grown in gardens, though much less frequently than *G. pratense* and *G. sanguineum*, and sometimes escapes.

Eurosiberian Boreal-montane element.

References: Atlas (91a), Curtis & McGough (1988), Hultén & Fries (1986), Meusel *et al.* (1978), Yeo (1985).

S. J. LEACH

Geranium pratense Meadow Crane's-bill

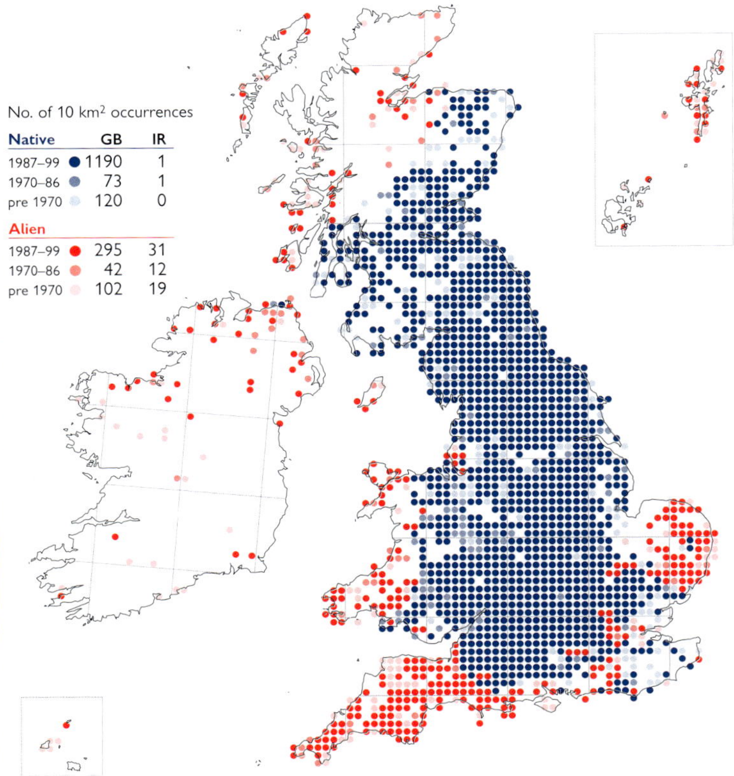

A perennial herb of rough grassland on verges, railway banks and streamsides, and in damp hay meadows and lightly grazed pastures, mainly on calcareous soils. Generally lowland, reaching 375 m at Alston (Cumberland), but exceptionally at 845 m on Great Dun Fell (Westmorland).

Native (change +0.15). *G. pratense* was formerly frequent in hay meadows, but has become increasingly restricted to roadsides due to changes in agricultural practices. Alien sites have increased markedly since the 1962 *Atlas*; separating native and alien plants within its supposed native range can be difficult.

Eurasian Boreo-temperate element; widely naturalised outside its native range.

References: Atlas (90d), Hultén & Fries (1986), Meusel *et al.* (1978), Yeo (1985).

S. J. LEACH

Geranium sanguineum Bloody Crane's-bill

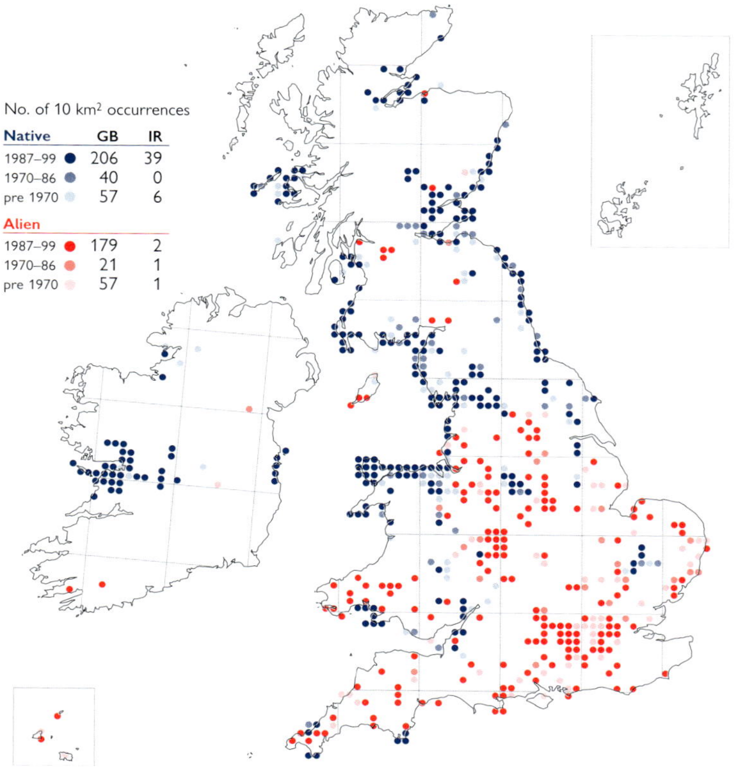

A rhizomatous, perennial herb of base-rich grasslands and scrub, open rocky woodlands, coastal cliffs and stabilised sand dunes; mainly on the coast but also inland on limestone pavements and cliff ledges, and in chalk and limestone grassland. As a native it has a curiously patchy distribution, and is often restricted to localised substrates such as dolerite and serpentine. It also occurs as a garden escape or throw-out on grassy banks, verges, tips and waste ground. 0–420 m (Ingleborough, Mid-W. Yorks.).

Native (change +0.83). The native distribution of *G. sanguineum* is similar to that shown in the 1962 *Atlas*, but it is now much more frequent as a garden escape.

European Temperate element.

References: Atlas (92a), Halliday (1997), Hibberd (1994), Hultén & Fries (1986), Meusel *et al.* (1978), Yeo (1985).

S. J. LEACH

Geranium columbinum Long-stalked Crane's-bill

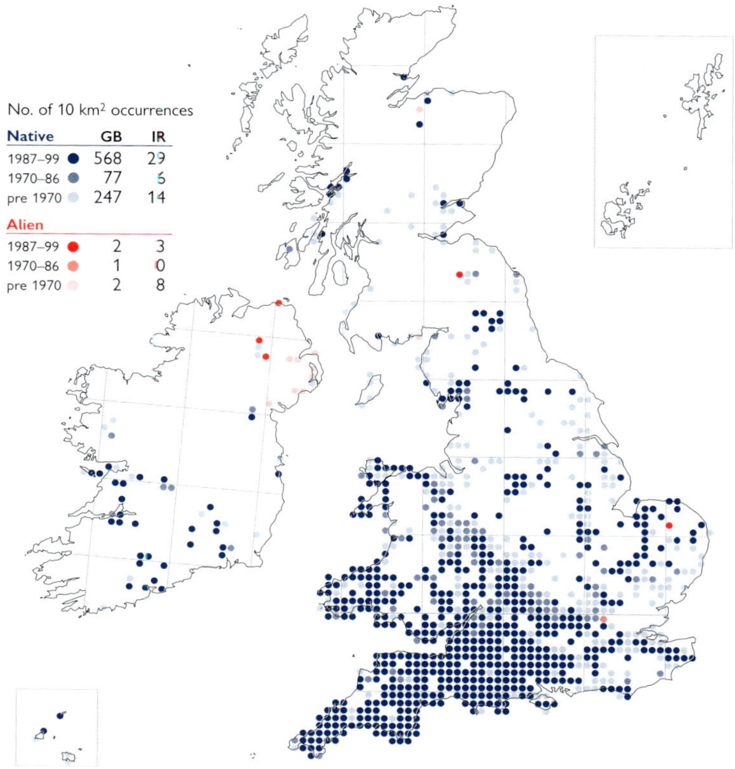

An annual of dry grasslands and grassland-scrub mosaics. Its habitats include sand dunes, scrubby cliff slopes, hedge banks, field margins, chalk and limestone downland, railway banks and old quarries. It is usually on calcareous soils, and is often a pioneer on disturbed sites. It favours warm, sheltered, often S.-facing banks and hollows. Lowland.

Native (change –0.34). This species was mapped as 'all records' in the 1962 *Atlas*. There has been a widespread decline in S.E. and N. England; analysis of the database reveals that most losses have occurred since 1950. These are probably due to habitat destruction, intensification of grassland management and scrub encroachment.

European Temperate element; widely naturalised outside its native range.

References: Atlas (92c), Hultén & Fries (1986), Meusel *et al.* (1978).

S. J. LEACH

Geranium dissectum Cut-leaved Crane's-bill

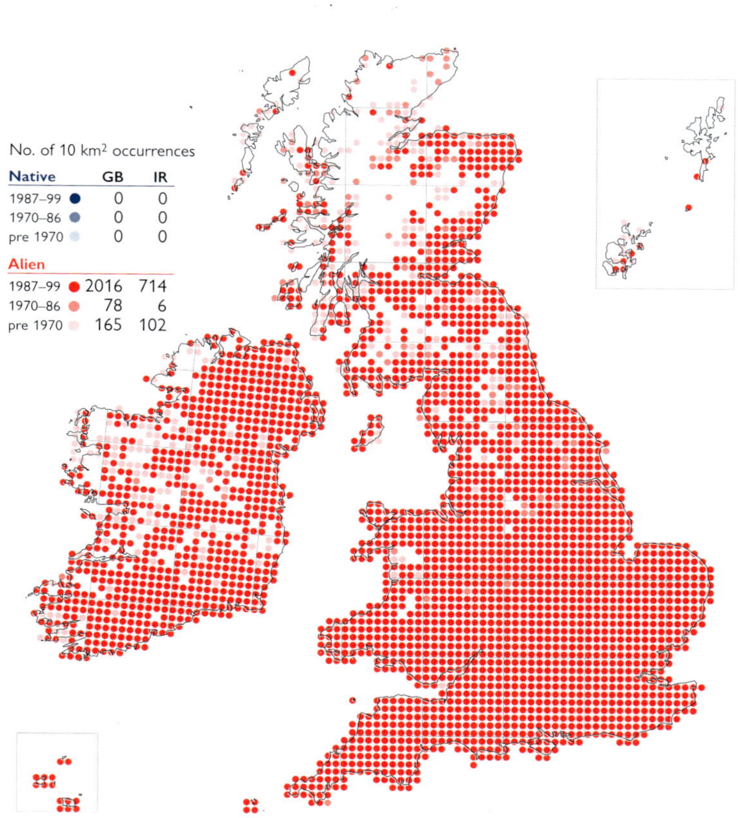

An annual of grasslands, hedge banks, waysides and waste ground, and a common weed of flower borders, allotments and arable fields. Generally lowland, but reaching 380 m at Braemar (S. Aberdeen).

Archaeophyte (change –0.09). There appears to have been little change in the range of *G. dissectum* since the 1962 *Atlas*, although it is now recorded from many more squares. Its ability to thrive in disturbed, nutrient-enriched habitats makes it likely that this species is becoming more abundant in many areas.

As an archaeophyte *G. dissectum* has a European Southern-temperate distribution; it is widely naturalised outside its native range.

References: Atlas (92d), Grime *et al.* (1988), Hultén & Fries (1986), Meusel *et al.* (1978).

S. J. LEACH

Geranium ibericum × *G. platypetalum* (*G.* × *magnificum*) Purple Crane's-bill

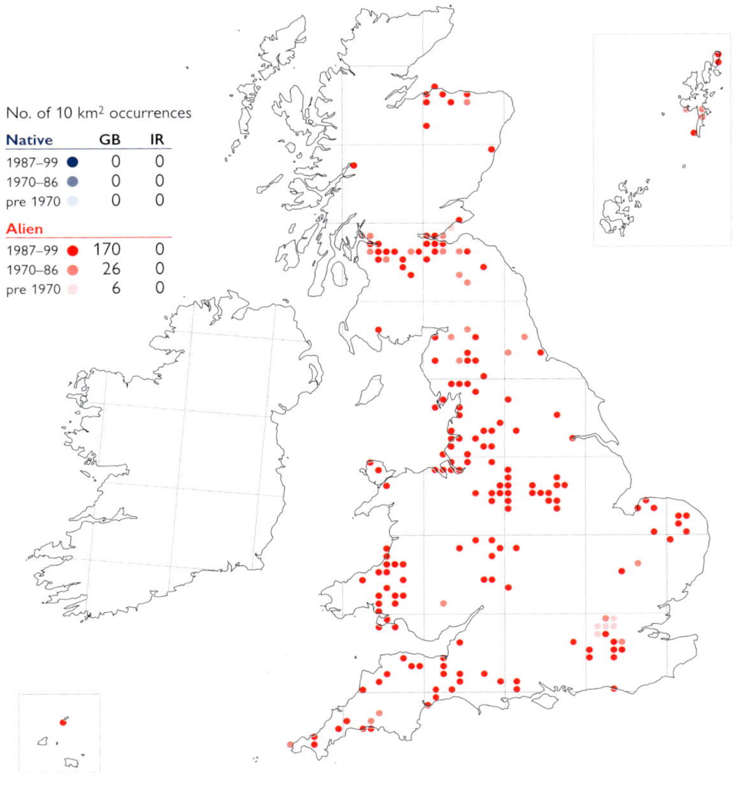

A clump-forming perennial herb, occurring in the wild as a long-persisting escape or throw-out on roadsides, railway banks and waste ground. Lowland.

Neophyte. This sterile hybrid is popular amongst gardeners for its striking blue-purple flowers and ground-covering abilities. It was first recorded in the wild in 1932 (Middlesex) and is probably increasing. The lack of records from Ireland is surprising.

A hybrid of garden origin; both its parents are found in the Caucasus.

Reference: Yeo (1985).

S. J. LEACH

Geranium pyrenaicum Hedgerow Crane's-bill

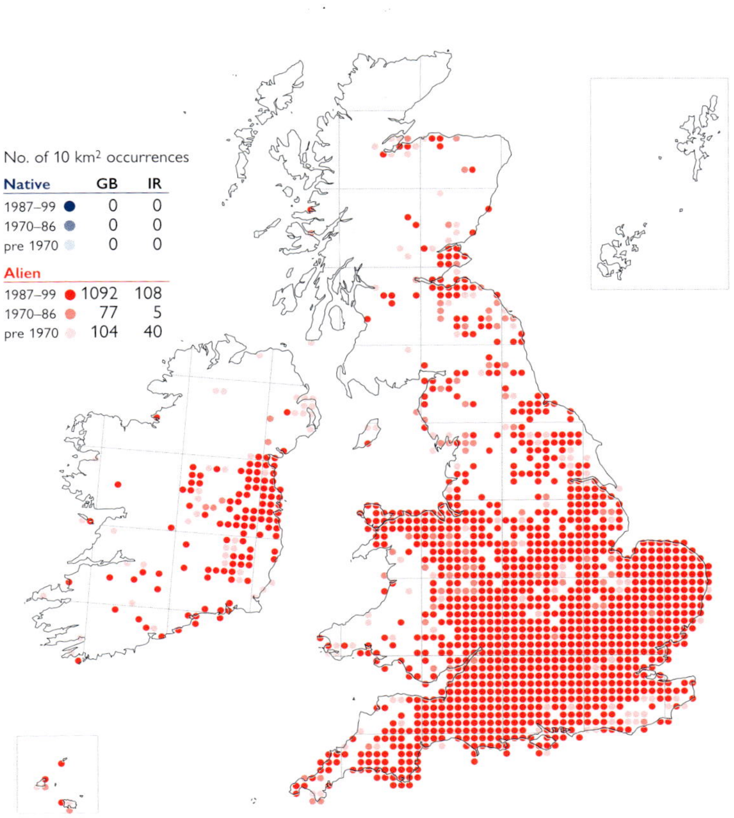

A perennial herb of hedgerows, roadsides, field margins, rough grassy banks and waste places; often found growing close to habitation, and possibly sometimes occurring as a garden escape or throw-out. Lowland.

Neophyte (change +1.14). This conspicuous species was not recorded in Britain until 1762 and, although considered possibly native by Stace (1997), it is almost certainly an alien which appears to be increasing in many areas. In Hampshire, for example, it was described as being very rare in the late 19th century but is now widespread and fairly common.

Native of the mountains of S. Europe & S.W. Asia; it spread rapidly in W. & C. Europe in the 18th and 19th centuries and now has a European Temperate distribution.

References: Atlas (92b), Brewis *et al.* (1996), Hultén & Fries (1986), Meusel *et al.* (1978).

S. J. LEACH

Geranium pusillum Small-flowered Crane's-bill

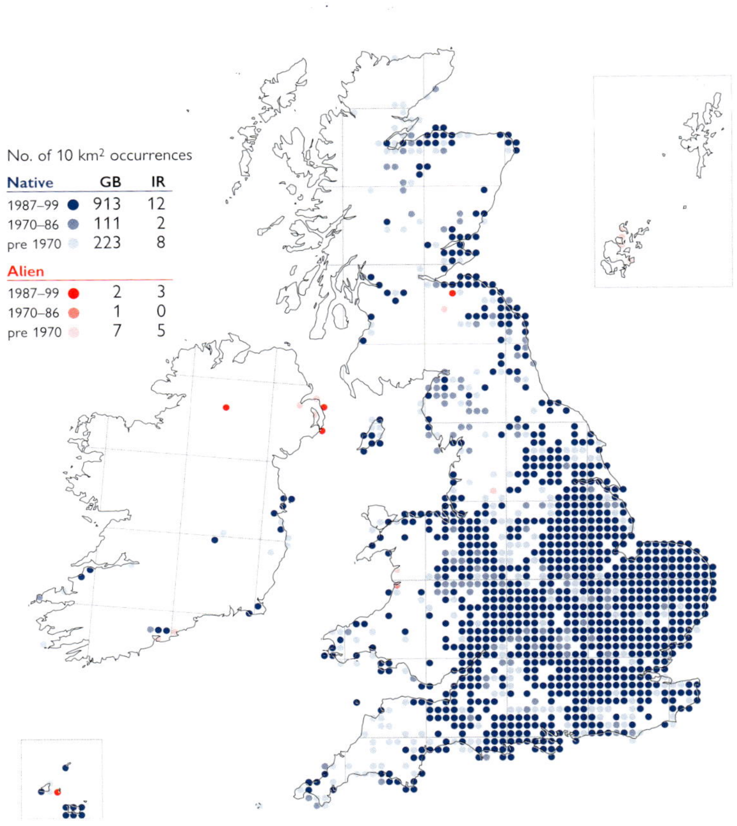

An annual of cultivated land, open summer-droughted grasslands, roadsides and waste places, thriving in well-drained, sandy soils. Lowland.

Native (change +0.16). There has been little change in the distribution of this species since the 1962 *Atlas*, although it is much better recorded now. *G. pusillum* is an elusive species and, because of its similarity with *G. molle*, it may have been overlooked in some squares.

Eurosiberian Temperate element; widely naturalised outside its native range.

References: Atlas (93c), Hultén & Fries (1986), Meusel *et al.* (1978).

S. J. LEACH

Geranium molle Dove's-foot Crane's-bill

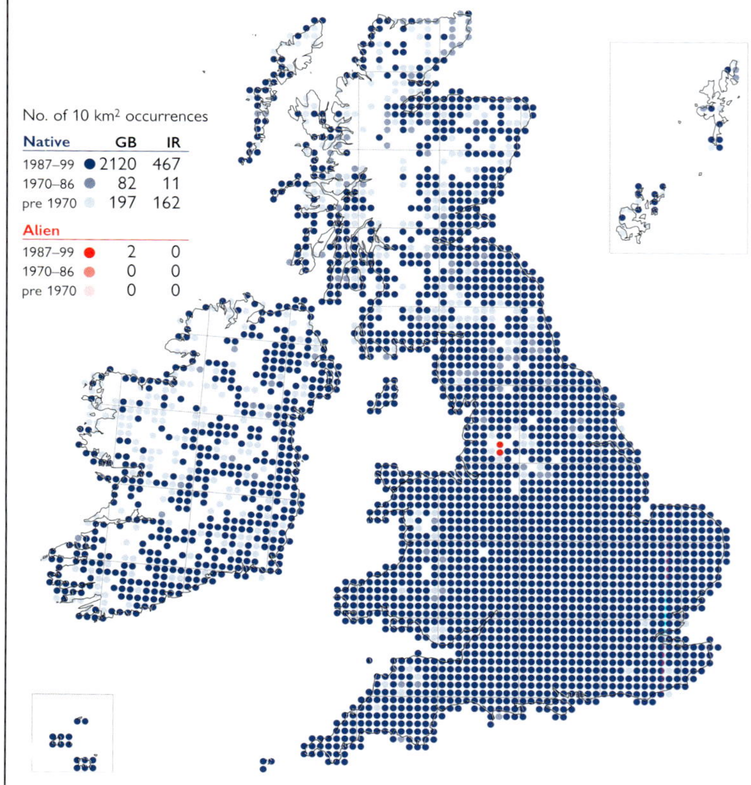

An annual found in a wide array of open habitats, including dry grasslands, rock outcrops, cultivated land, garden lawns, verges and waste ground. Generally lowland, but reaching 550 m at Moor House (Westmorland).

Native (change −0.46). There appears to have been little change in the distribution of this species since the 1962 *Atlas*.

European Southern-temperate element; widely naturalised outside its native range.

References: Atlas (93b), Grime *et al.* (1988), Hultén & Fries (1986), Meusel *et al.* (1978).

S. J. LEACH

Geranium macrorrhizum Rock Crane's-bill

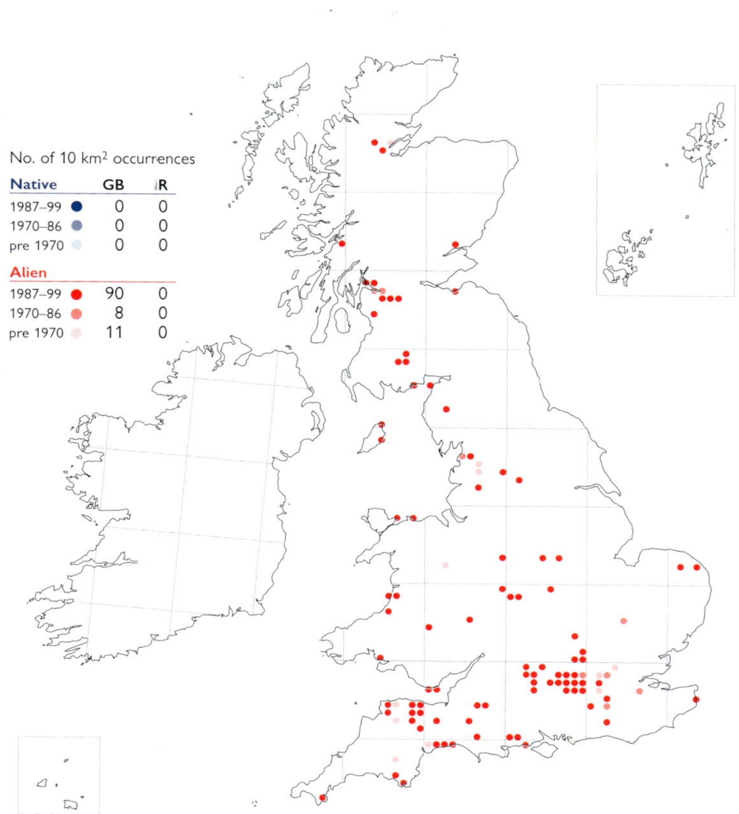

A rhizomatous perennial herb, much grown in gardens and occasionally occurring as an escape or throw-out in hedge banks, open woodland and on roadsides, usually not far from habitation. Lowland.

Neophyte. This species, introduced into Britain by 1576, is much grown in gardens, and is very persistent once established. It has been known in the wild since 1835, and is probably increasing. There are several cultivars and in some areas the white-flowered 'Album' appears to be the most frequently encountered in the wild.

Native of the mountains of S. Europe (Alps, Apennines, Balkan peninsula and Carpathians).

Reference: Yeo (1985).

S. J. LEACH

Geranium lucidum Shining Crane's-bill

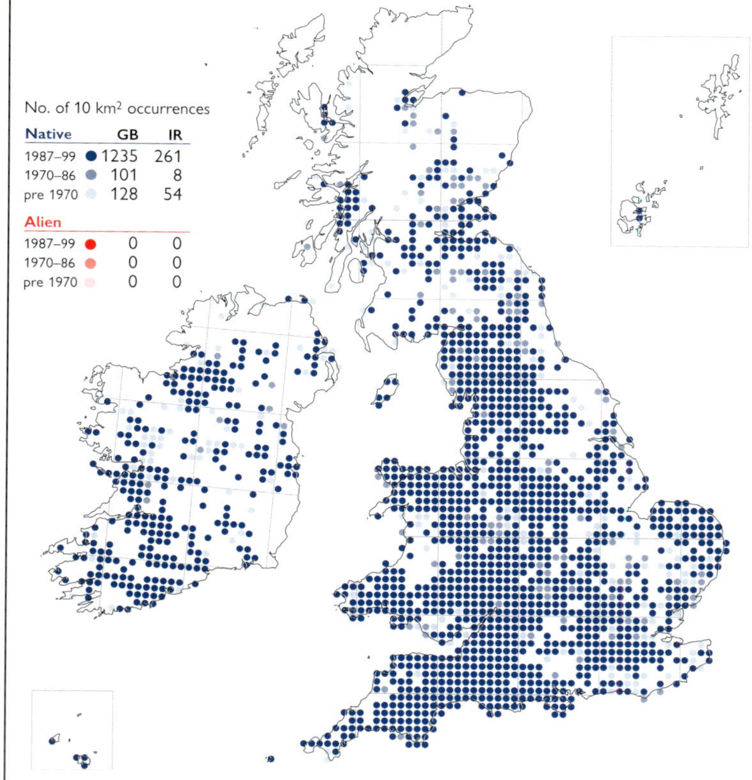

An annual of roadside-banks, rock outcrops and scree, preferring calcareous soils and characteristic of limestone districts. It is widespread in artificial habitats, including mortared walls, churchyards, roadsides, waste ground and railway ballast; also as an escape from gardens. Generally lowland, but reaching 610 m at Melmerby High Scar (Cumberland).

Native (change +1.42). *G. lucidum* can be invasive, and there has been an expansion of its range and a marked increase in the number of 10-km squares since the 1962 *Atlas*. It is often very difficult to separate alien and native occurrences, and all records are mapped as if they are native.

Submediterranean-Subatlantic element.

References: Atlas (93d), Hultén & Fries (1986), Meusel *et al.* (1978), Yeo (1985).

S. J. LEACH

Geranium robertianum Herb-Robert

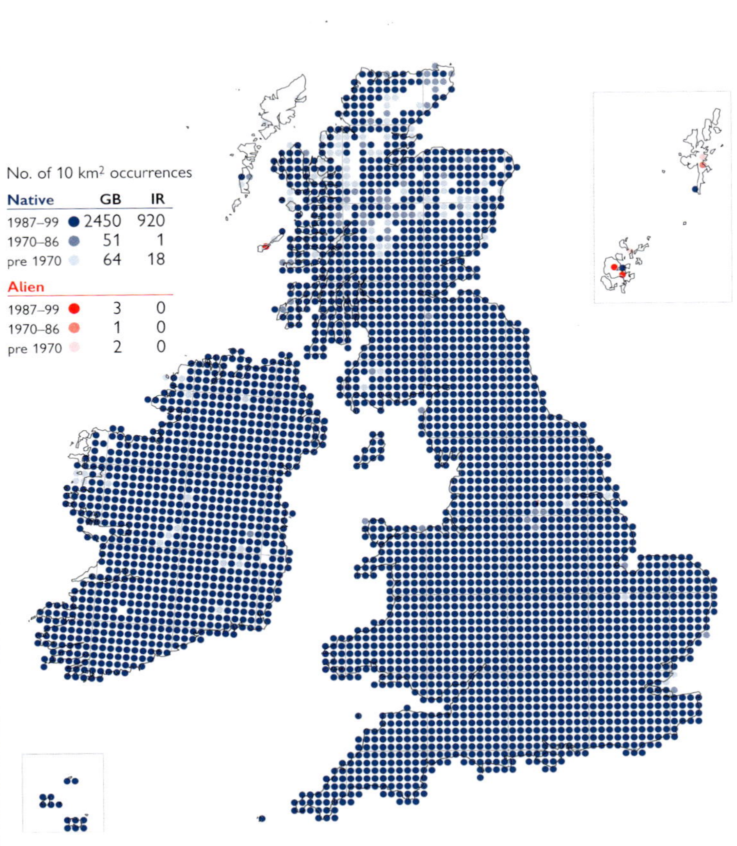

An annual or biennial shade-tolerant herb found on a wide range of soil types, except those that are strongly acidic. Its habitats include woods, hedgerows, walls, shaded banks, limestone pavements, screes and coastal shingle; also in disturbed artificial habitats. 0–700 m (Great Dun Fell, Westmorland).

Native (change −0.41). There is no evidence of any appreciable change since the 1962 *Atlas*. *G. robertianum* is very variable; subsp. *maritimum* and subsp. *celticum* are mapped by Perring & Sell (1968).

European Temperate element; also in C. and E. Asia and widely naturalised outside its native range.

References: Atlas (94a), Atlas Supp. (16a, b), Baker (1955), Grime *et al.* (1988), Hultén & Fries (1986), Meusel *et al.* (1978), Stace (1997).

S. J. LEACH

Geranium purpureum Little-Robin

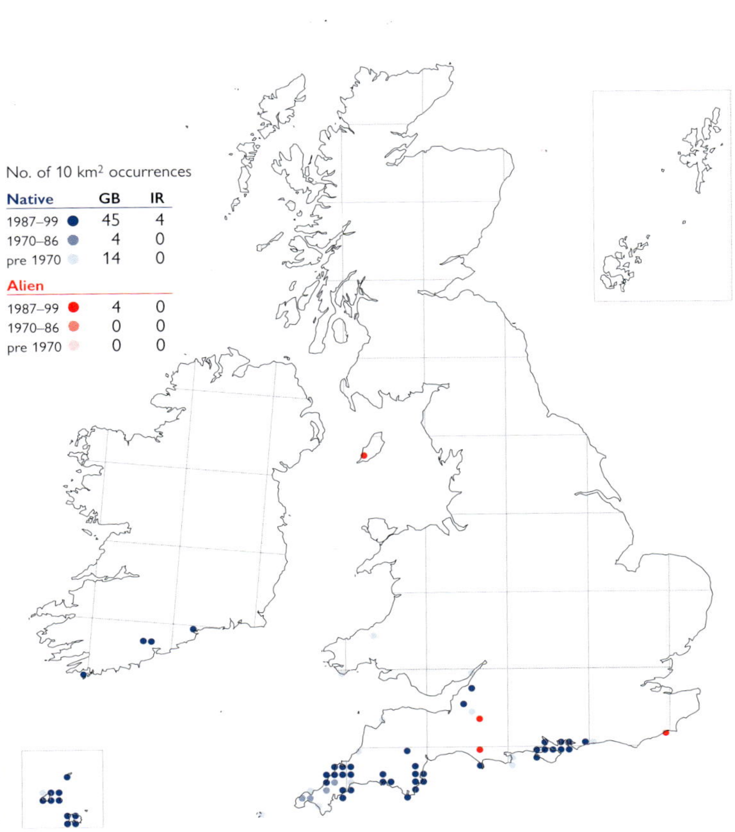

Subspecies *purpureum* is an upright annual in stony or rocky places near the sea, on sheltered cliffs, disused railway lines, and particularly by roads and fields on the earth-and-stone sides of Cornish hedge banks. Subsp. *forsteri* is a prostrate plant of stabilised areas at the top of shingle beaches. Lowland.

Native (change +0.22). Since the 1962 *Atlas*, *G. purpureum* has been lost from some outlying areas, but in Cornwall it has increased markedly since about the mid-1970s, possibly benefiting from modern road verge management and recent mild winters.

Mediterranean-Atlantic element.

References: Atlas (94b), Atlas Supp. (17a, b), Baker (1955), Curtis & McGough (1988), FitzGerald (1990), French *et al.* (1999), Hultén & Fries (1986), Meusel *et al.* (1978), O'Mahony (1985), Wigginton (1999).

S. J. LEACH

Geranium phaeum Dusky Crane's-bill

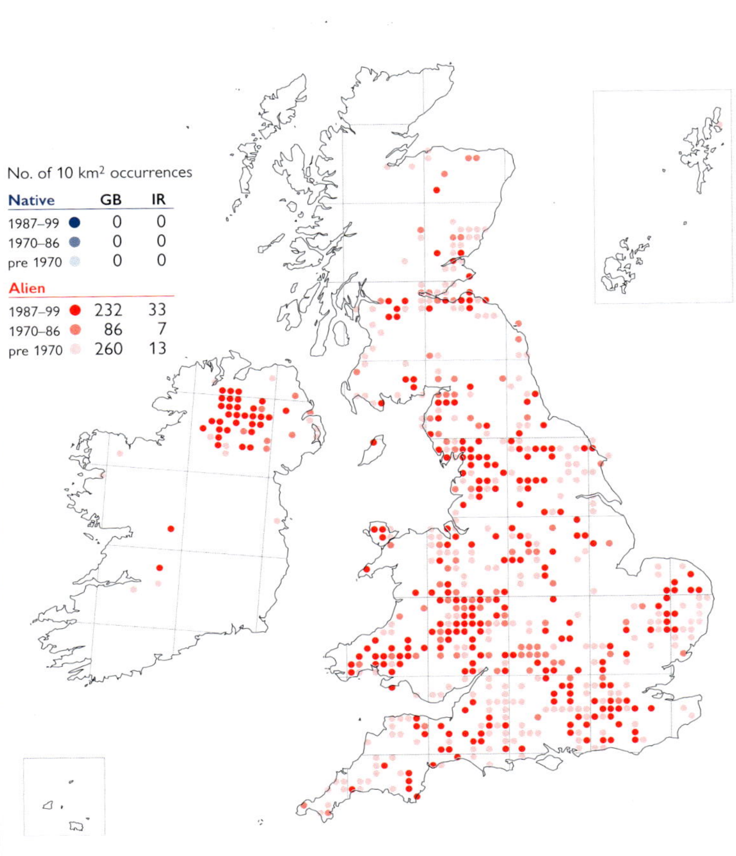

This clump-forming perennial herb is well-naturalised on roadsides and railway banks, and in churchyards and wood-borders; it usually grows close to habitation as a garden escape or throw-out but is sometimes deliberately planted in the wild. It favours shaded situations and moist, fertile soils. Lowland.

Neophyte (change −0.67). *G. phaeum* has been known in the wild since 1724. Although better recorded since the 1962 *Atlas*, there is little evidence of any appreciable change in its distribution. It is a variable taxon with many named cultivars; most records are of the purplish-black flowered forms of var. *phaeum*, though the pinkish-mauve var. *lividum* and white 'Album' have also been reported from the wild.

Native of the mountains of C. & S. Europe.

References: Atlas (91d), Bolòs & Vigo (1990), Hibberd (1994), Yeo (1985).

S. J. LEACH

Erodium maritimum Sea Stork's-bill

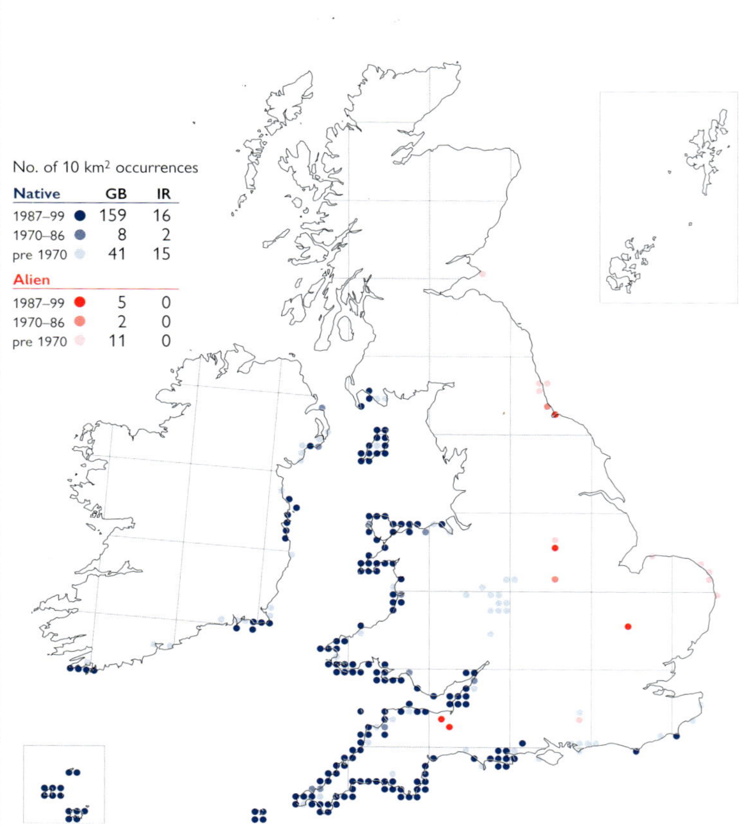

An annual of trampled or closely-grazed cliff-top grasslands, disturbed sand dunes and gull-infested sea-cliffs, and around coastal settlements on walls and pavements. Inland, it has been recorded from limestone grassland (Somerset), in heathland areas by sandy tracks and gravel workings and, rarely, as an introduction on railway ballast. Lowland.

Native (change +0.38). There has been no appreciable change in distribution since the 1962 *Atlas*, though it has gone from most of its presumed native inland sites. However, in Dorset inland records associated with quarrying have shown a distinct increase.

Suboceanic Southern-temperate element, with a disjunct distribution.

References: Atlas (94c), Bolòs & Vigo (1990), Stewart *et al.* (1994).

S. J. LEACH

Erodium moschatum Musk Stork's-bill

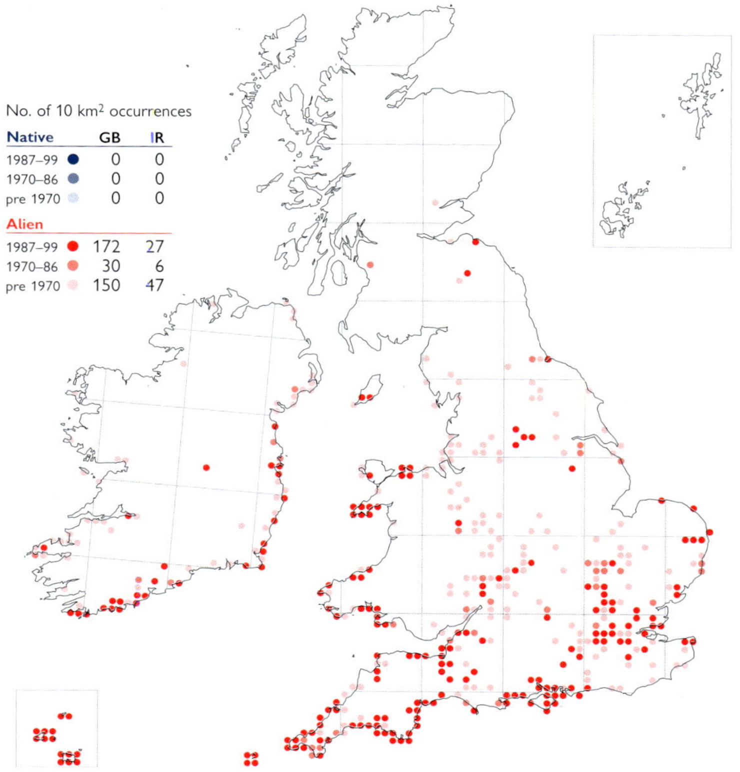

An annual of barish places near the sea, in disturbed sand dunes, on roadsides, wall-tops, field margins and waste ground. In the Isles of Scilly it is a frequent bulb-field weed. It is recorded inland as a casual, sometimes introduced with wool shoddy. Lowland.

Archaeophyte (change +0.47). This species is well-established in the coastal regions of S.W. England, Wales and Ireland, where it is often considered a native. In these areas it is now more frequent than was the case in the 1962 *Atlas*. It remains an uncommon casual elsewhere.

As an archaeophyte *E. moschatum* has a Mediterranean-Atlantic distribution; it is widely naturalised outside this range.

References: Atlas (94d), Bolòs & Vigo (1990), Stewart *et al.* (1994).

S. J. LEACH

Erodium cicutarium agg. Common Stork's-bill

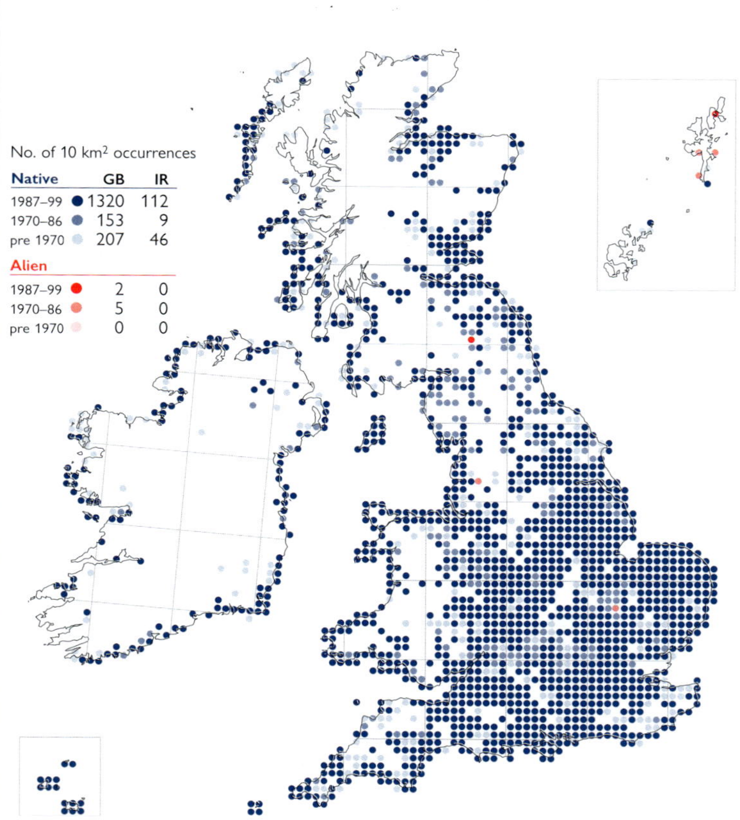

Annuals of well-drained sandy and rocky places, sand dunes, summer-parched grasslands and heaths; they are also found on roadsides, stone walls and railway ballast, and are common wool aliens. Generally lowland, but reaching 420 m on Kirkstone Pass (Westmorland).

Native (change –0.11). There is little evidence of appreciable change in the two species included here (*E. cicutarium* and *E. lebelii*) since the 1962 *Atlas*, though they may be spreading in some inland areas beside salt-treated roads. *E. cicutarium* occurs throughout range of the aggregate, *E. lebelii* is mapped separately.

Eurosiberian Southern-temperate element, but widely naturalised so that distribution is now Circumpolar Southern-temperate.

References: Atlas (95a), Atlas Supp. (18a), Hultén & Fries (1986), Meusel *et al.* (1978).

S. J. LEACH

Erodium lebelii Sticky Stork's-bill

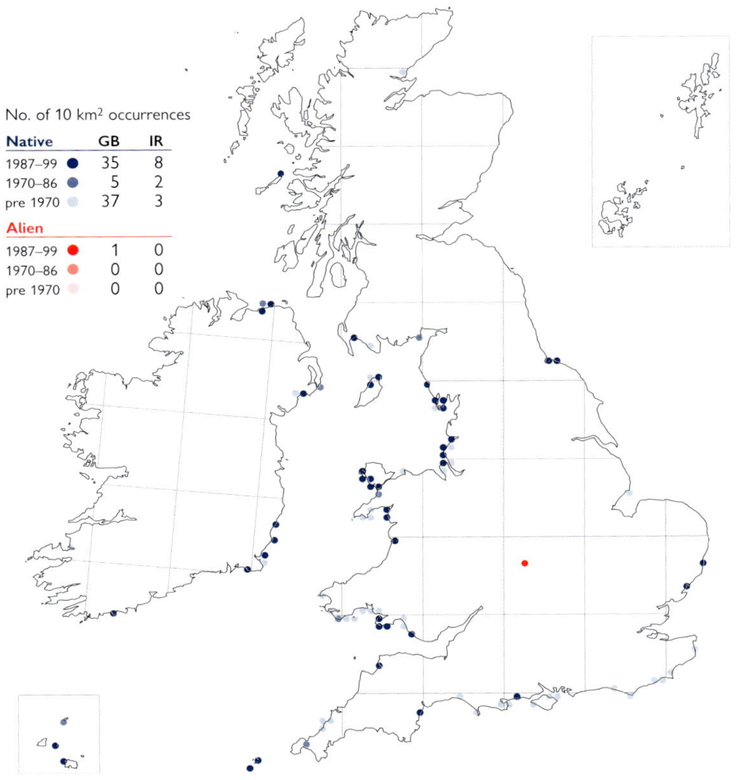

An annual of barish places in stabilised sand dunes. Lowland.

Native. The map of *E. lebelii* in Perring & Sell (1968) was based on herbarium material. The current range is broadly similar, although the map suggests that in some areas recent recorders have not distinguished this species from the closely related *E. cicutarium*. Some pre-1970 records of *E. lebelii* could be errors for the taxonomically uncertain *E. cicutarium* subsp. *dunense*.

Suboceanic Southern-temperate element.

Reference: Atlas Supp. (18b).

S. J. LEACH

Limnanthes douglasii Meadow-foam

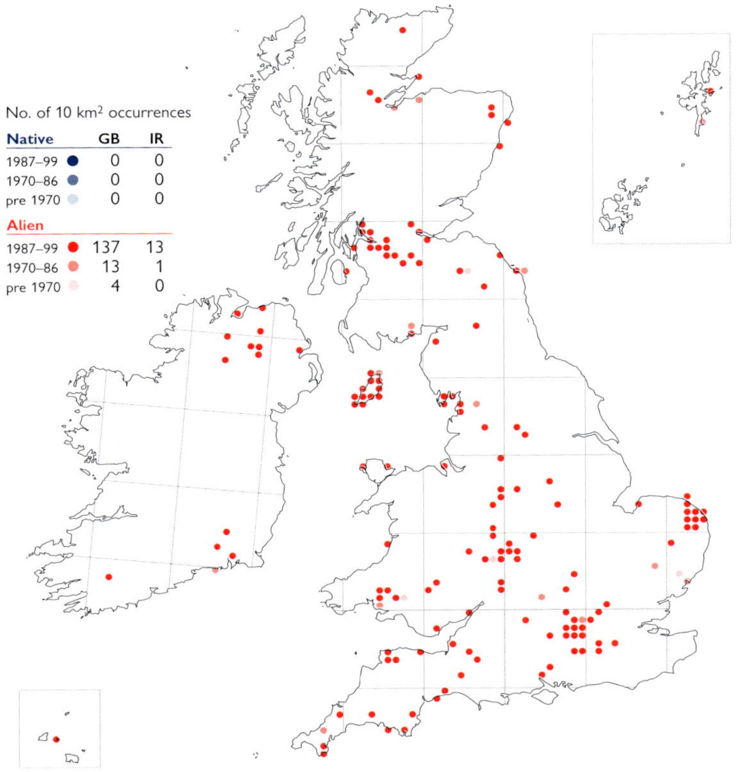

An annual grown in gardens and found in the wild by roads and on sea-cliffs, beaches, lake shores, waste places and rubbish tips. Although plants set abundant seed, populations are usually casual, but may persist for a few years. Lowland.

Casual. *L. douglasii* was first cultivated in Britain in 1833 and is very popular in gardens, almost always under the name 'Poached-egg Plant'. It was known from the wild by at least 1870, and Dunn (1905) noted that it had 'been found once or twice as an escape from cultivation'. It is probably increasing.

Native of western N. America (California).

H. J. KILLICK

Tropaeolum majus Nasturtium

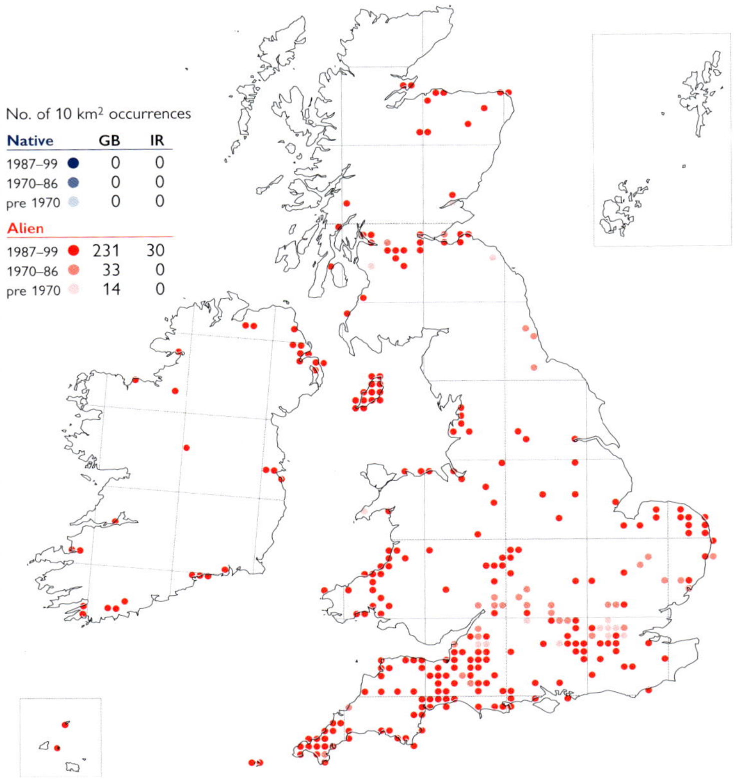

A climbing or trailing annual of cultivated land, rubbish tips and disturbed ground. The plant is usually casual, arising from seed on plants discarded from gardens. Although plants are frost-sensitive they usually set fertile seed, and, since these seeds are frost-resistant, populations can persist for a few years. Lowland.

Casual. *T. majus*, introduced to Britain in 1686, is very commonly cultivated in gardens. It was known from the wild by at least 1908, but changes in its distribution are difficult to assess.

Ancestral variants of *T. majus* probably occur as natives in S. America but the plant has been much modified in cultivation, perhaps in part through hybridisation with other species.

T. D. DINES

Impatiens noli-tangere Touch-me-not Balsam

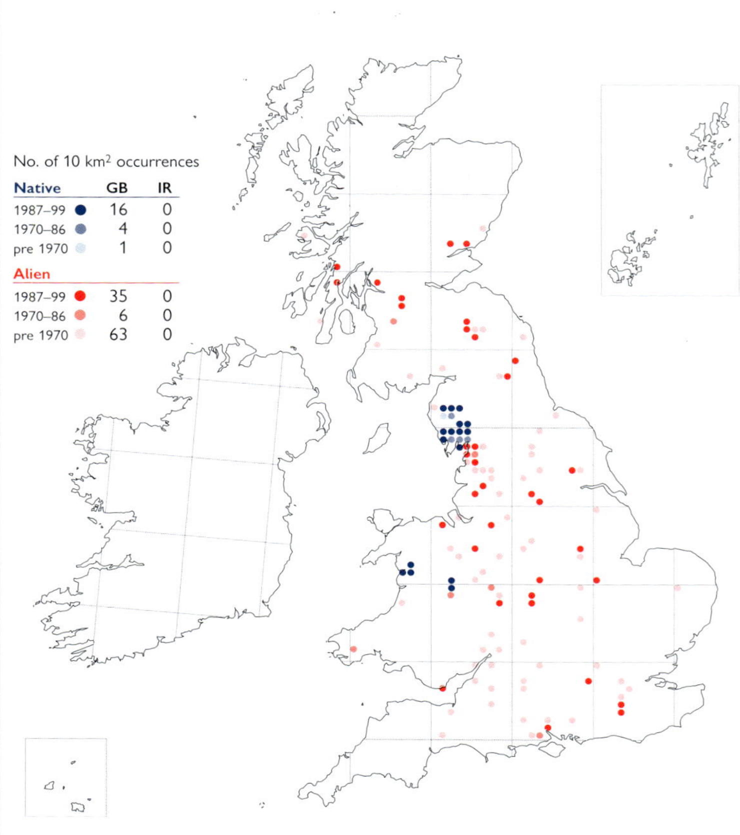

Our only native *Impatiens* is an annual plant of nutrient-rich soils in damp but not waterlogged woodland, occurring on streamsides and in valley-side seepages. Lowland.

Native (change −0.77). The species has not shown any significant change in its native distribution in Britain over a long period, and remains rare and local. Most non-native records are probably of short-lived casuals, and it appears now to be rarely cultivated.

Eurasian Temperate element; also in western N. America. The British sites for this species are amongst several scattered western outliers detached from its continuous range from C. Europe eastwards.

References: Atlas (96a), Hultén & Fries (1986), Meusel *et al.* (1978), Stewart *et al.* (1994).

R. M. BURTON

Impatiens capensis Orange Balsam

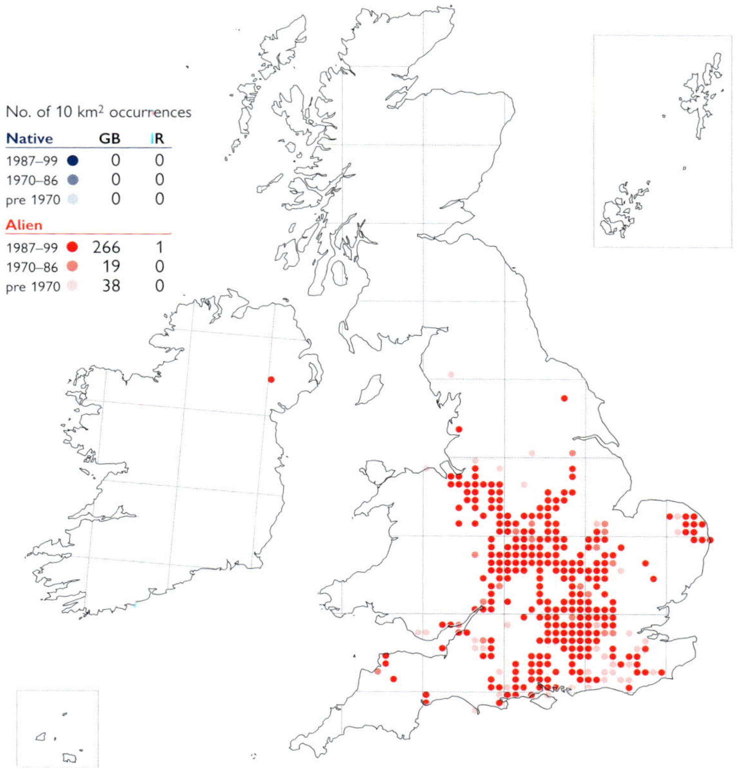

An annual, thoroughly naturalised by rivers, canals and adjacent reservoirs. The seeds can be ejected from its capsules to a distance of a few metres, and can be dispersed by water. Lowland.

Neophyte (change +0.71). Introduced probably in the very early 19th century, the first record of *I. capensis* outside cultivation was from Surrey in 1822. There has since been a steady expansion into semi-natural habitats, which still continues. The Norfolk population originated from an independent introduction to a tributary of the Bure near Aylsham in 1927 (Petch & Swann, 1968).

Native of North America; also naturalised in France and Germany, although on a smaller scale than in our area.

References: Atlas (96b), Hultén & Fries (1986), Meusel *et al.* (1978), Trewick & Wade (1986).

R. M. BURTON

Impatiens parviflora Small Balsam

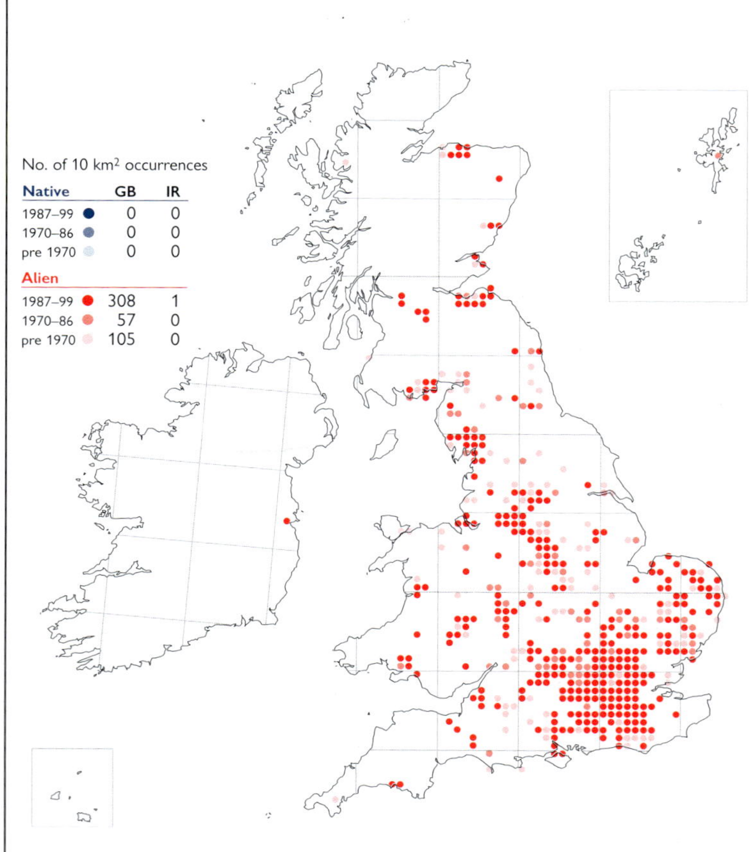

This species is usually found as a well-naturalised alien in semi-natural woodland and plantations, especially along tracks, but it also occurs in N. and W. Britain along shaded river banks, and in E. Scotland on unshaded river shingle. There are also scattered records of casual plants in timber yards. Lowland.

Neophyte (change +0.10). *I. parviflora* was introduced in 1823. It was recorded in the wild in England in 1851 (Surrey), and in Scotland in 1864. Since then it has increased steadily.

Native of C. Asia; widely naturalised in temperate Europe, all European populations possibly originating from a single introduction.

References: Atlas (96c), Coombe (1956a), Hultén & Fries (1986), Meusel *et al.* (1978).

R. M. BURTON

Impatiens glandulifera Indian Balsam

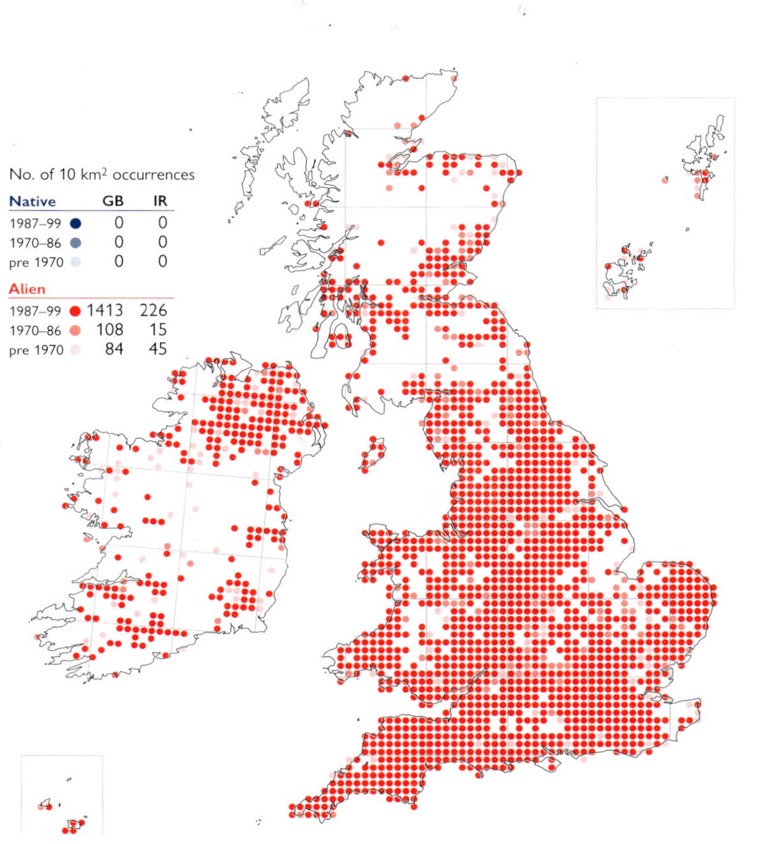

This species is most frequent on the banks of waterways, where it often forms continuous stands, but is also established in damp woodland, flushes and mires. The tallest annual in Britain, its rapid growth can shade out even *Urtica dioica*. Lowland.

Neophyte (change +1.85). *I. glandulifera* was introduced as an ornamental garden plant in 1839 and was first recorded in the wild in 1855 (Middlesex). It became naturalised independently in many different places, but did not attain its most rapid rate of increase until almost a century later. There has been a significant increase in its frequency since the 1962 *Atlas*.

Native of the Himalayas; widely naturalised in temperate Europe.

References: Atlas (96d), Beerling & Perrins (1993), Bolòs & Vigo (1990), Grime *et al.* (1988).

R. M. BURTON

Hedera colchica Persian Ivy

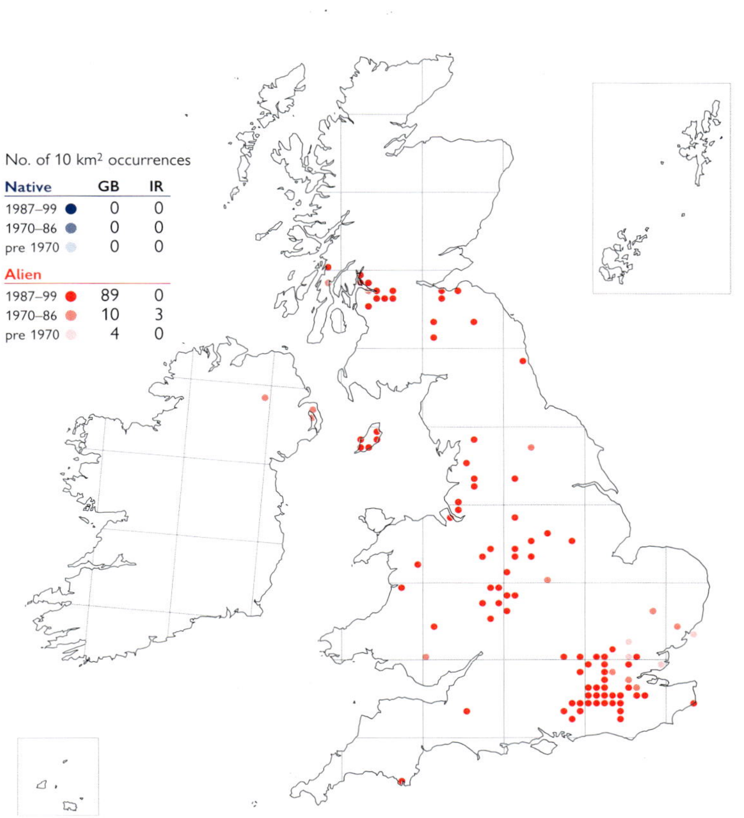

An evergreen perennial climber found as a garden escape in woodland, hedges and scrub and on roadsides, railway banks, walls and waste ground. It reproduces vegetatively and by seed, which can be bird-sown. Lowland.

Neophyte. *H. colchica* has been cultivated in Britain since 1851 and is popular in gardens and amenity plantings, where it is used for ground-cover. It was first recorded in the wild in 1959, and is certainly under-recorded; it may be increasing.

Native of N. Turkey and the Caucasus.

References: Bean (1973), Meusel *et al.* (1978).

T. D. DINES

Hedera helix Ivy

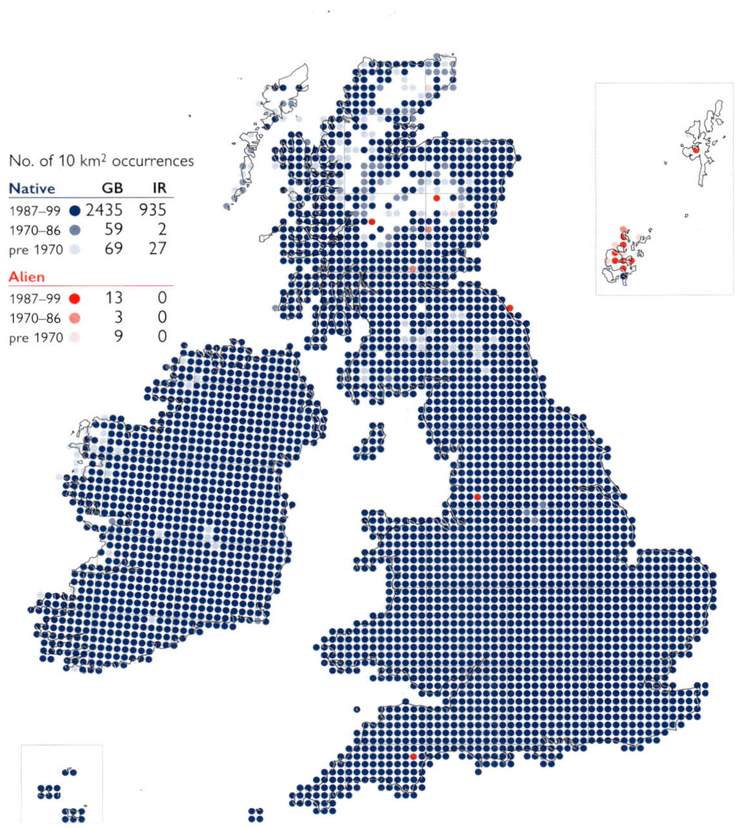

An evergreen perennial woody climber most characteristic of woodland, scrub and hedgerows, but also common on walls, rock outcrops and cliffs. It may carpet the ground in secondary woodland. It generally favours basic to moderately acidic soils. It is highly palatable to deer and stock, and in grazed upland areas becomes restricted to inaccessible rock outcrops. 0–610 m (Mourne Mountains, Co. Down).

Native (change −0.65). The range of this species is unchanged since the 1962 *Atlas*. The two native subspecies (subsp. *helix* and subsp. *hibernica*) are mapped separately.

European Southern-temperate element.

References: Atlas (153a), Grime *et al.* (1988), Hultén & Fries (1986), Meusel *et al.* (1978).

G. T. D. WILMORE

Hedera helix subsp. *helix* Common Ivy

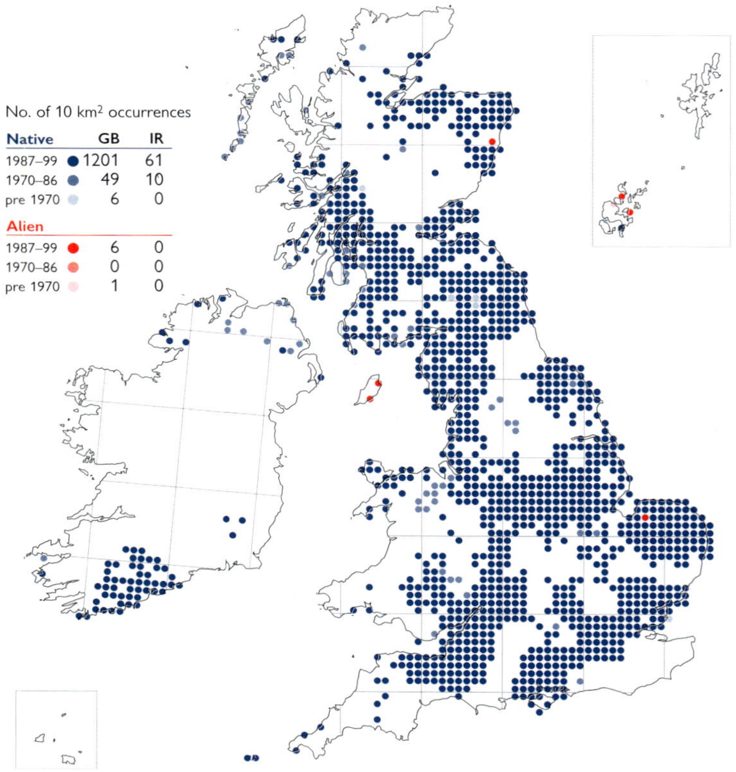

An evergreen perennial climber of woodland, scrub, hedgerows, walls, rock outcrops and cliffs, which may also sprawl over the ground. It avoids only the most acidic soils and is highly palatable to deer and stock; in grazed upland areas it becomes restricted to inaccessible rock outcrops. Altitudinal limit unknown.

Native. This diploid subspecies occurs throughout the British and Irish range of the species. However, being the nominate race, it is poorly recorded in some areas, but the map shows that it is frequent in areas where recorders take care to identify it. The two subspecies, however, are not always easy to separate.

European Southern-temperate element.

References: McAllister & Rutherford (1990), Rich & Jermy (1998).

T. D. DINES

Hedera helix subsp. *hibernica* Atlantic Ivy

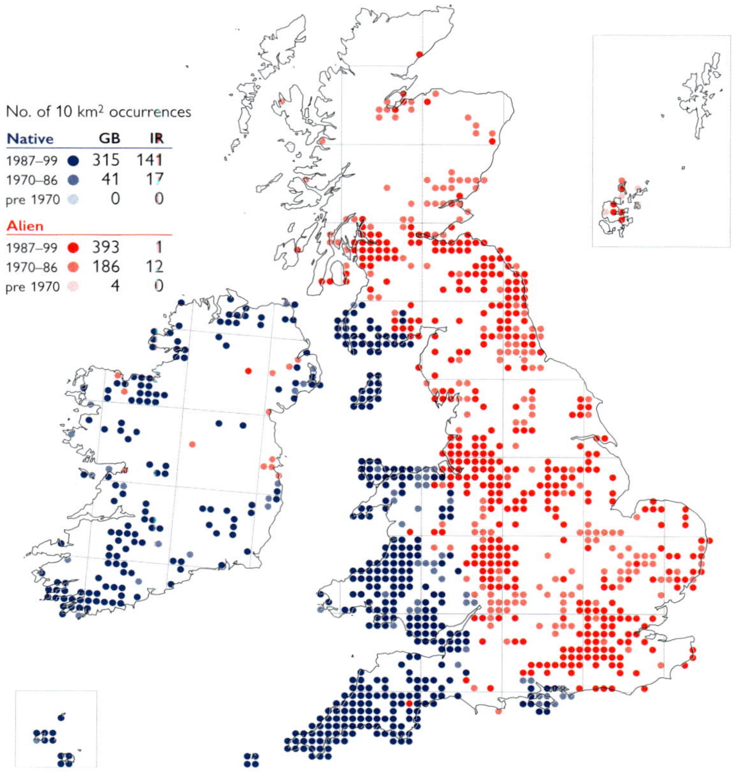

An evergreen perennial climber of woodland, scrub, hedgerows, walls, rock outcrops and cliffs, which may also sprawl over the ground. It avoids only the most acidic soils. Altitudinal limit unknown.

Native. This tetraploid subspecies is more frequent in W. Britain and Ireland than subsp. *helix*. It is much better known than when mapped by McAllister & Rutherford (1990), but remains very under-recorded in some areas, particularly in Ireland. The two subspecies can be difficult to separate. Alien records refer to the cultivar 'Hibernica', which is widely grown and has been found naturalised since at least 1838; separating alien and native populations within the native range can be difficult.

Oceanic Southern-temperate element.

Reference: Rich & Jermy (1998).

T. D. DINES

Hydrocotyle vulgaris Marsh Pennywort

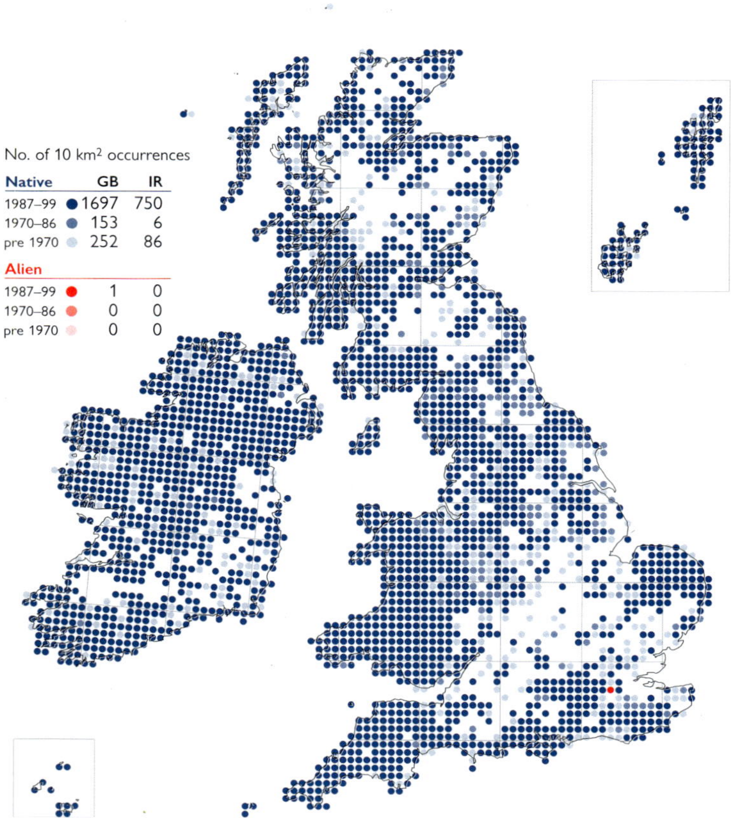

A mat-forming perennial herb found in a wide range of damp or wet habitats, including carr, mires, fens, fen-meadows, swamps, marshes, in soakways and along spring-lines, and in dune-slacks and wet hollows in stabilised shingle. In very oceanic areas it grows in drier habitats, such as turfed wall-tops. 0–530 m (Tal-y-Fan, Caerns.).

Native (change –0.53). *H. vulgaris* was mapped as 'all records' in the 1962 *Atlas*. It has declined, and analysis of the database reveals that most of the losses have occurred since 1950. Drainage and development have steadily eliminated its sites in S.E. England.

Suboceanic Southern-temperate element.

References: Atlas (153b), Grime *et al.* (1988), Hultén & Fries (1986), Tutin (1980).

M. SOUTHAM

Sanicula europaea Sanicle

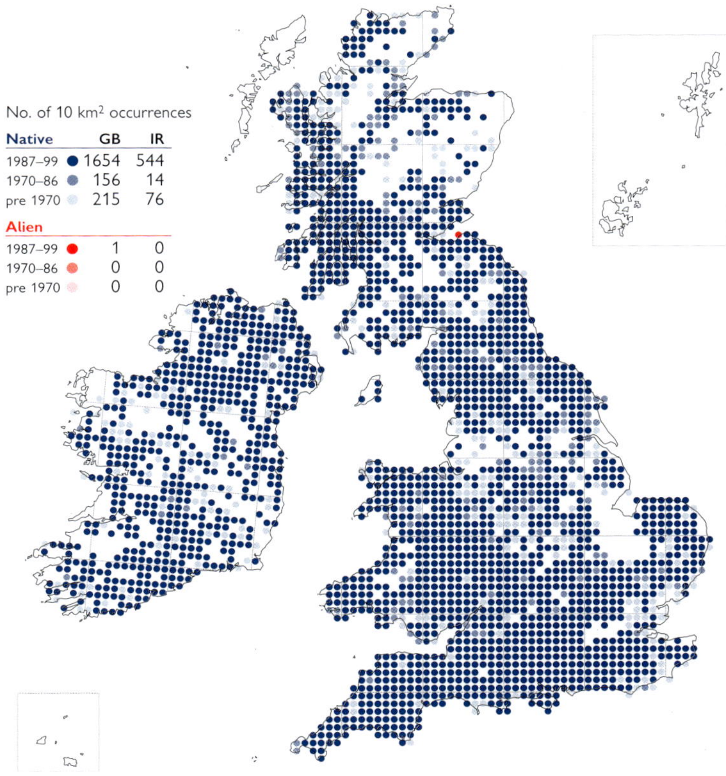

No. of 10 km² occurrences

Native		GB	IR
1987–99	●	1654	544
1970–86	●	156	14
pre 1970	●	215	76
Alien			
1987–99	●	1	0
1970–86	●	0	0
pre 1970	●	0	0

A perennial herb of moist soil in deciduous woodland, often where *Fagus*, *Fraxinus* or *Quercus* spp. predominate; also locally in substantial hedge banks and shaded roadsides. In the north and west of its range it is sometimes found in relict woodland in gorges and in sheltered stream ravines. The substrate is usually calcareous or otherwise base-rich, but can occasionally be neutral or mildly acidic. Generally lowland, but reaching 500 m above Malham (Mid-W. Yorks.).

Native (change –0.98). In some areas there has been some decline in the frequency of this species, which was mapped as 'all records' in the 1962 *Atlas*.

European Temperate element; also in C. Asia.

References: Atlas (153c), Grime *et al.* (1988), Hultén & Fries (1986), Meusel *et al.* (1978), Tutin (1980).

M. SOUTHAM

Astrantia major Astrantia

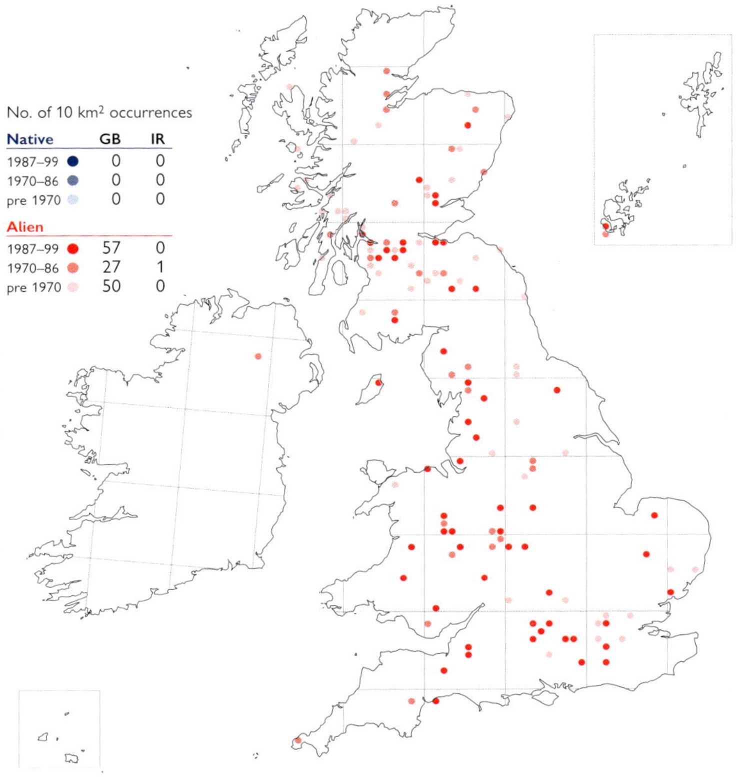

No. of 10 km² occurrences

Native		GB	IR
1987–99	●	0	0
1970–86	●	0	0
pre 1970	●	0	0
Alien			
1987–99	●	57	0
1970–86	●	27	1
pre 1970	●	50	0

A long-lived perennial herb occurring on waste ground as an escape from cultivation or as an introduction in partially shaded habitats, most often near habitation. Lowland.

Neophyte. Introduced into cultivation by 1596, *A. major* is a popular garden plant and various cultivars are grown. This is the first map showing its national distribution, so changes are difficult to assess. It has been known in the wild near Craven Arms (Salop) since at least 1841.

Native of C. Europe.

References: Meusel *et al.* (1978), Tutin (1980).

M. SOUTHAM

Eryngium maritimum Sea-holly

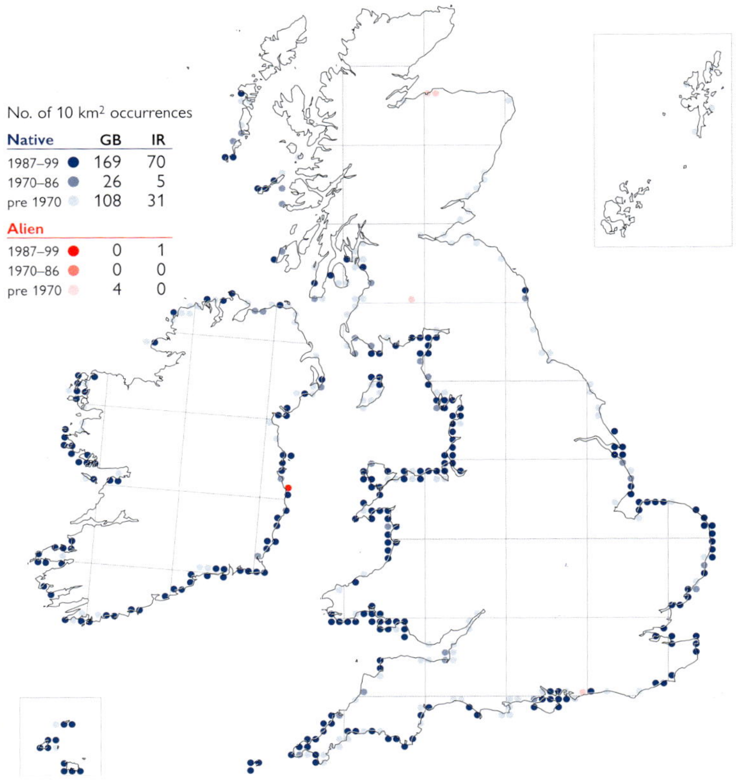

No. of 10 km² occurrences

Native		GB	IR
1987–99	●	169	70
1970–86	●	26	5
pre 1970	●	108	31
Alien			
1987–99	●	0	1
1970–86	●	0	0
pre 1970	●	4	0

A glaucous, spiny perennial herb confined to coasts, occurring mainly on incipient and mobile sand dunes and occasionally on shingle. Lowland.

Native (change –0.80). This species disappeared from most of its sites in N.E. England and E. Scotland before 1930, for reasons which are unclear. There has evidently been some further decline since then.

European Southern-temperate element.

References: Atlas (153d), Hultén & Fries (1986), Meusel *et al.* (1978), Tutin (1980).

M. SOUTHAM

Eryngium campestre Field Eryngo

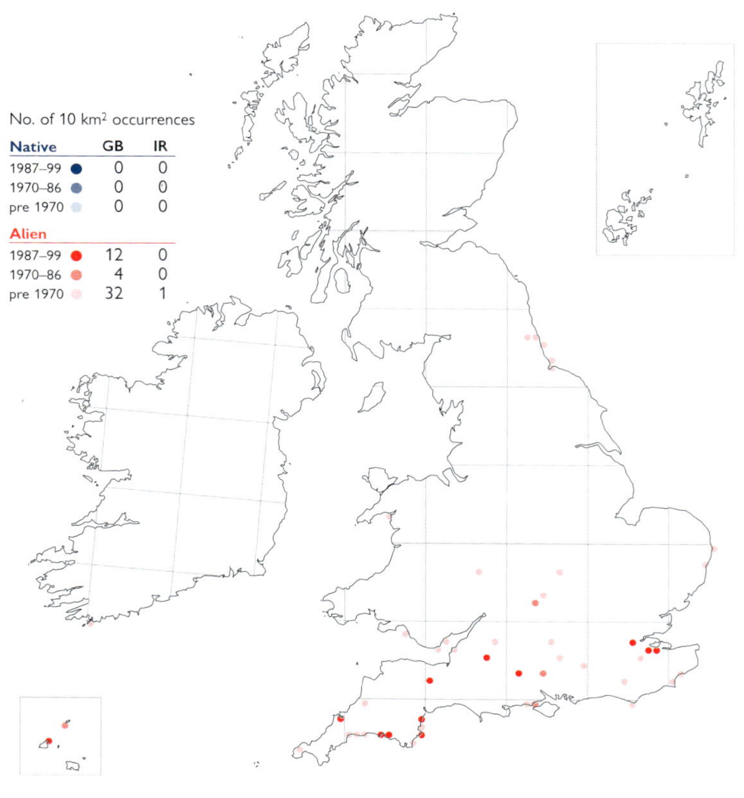

A perennial herb of well-drained neutral or calcareous soils in old pastures and coastal grassland in S.W. England, where it is very long-established and was once considered to be native. Elsewhere, short-lived or casual populations have been reported from pastures, roadsides and rough ground. Lowland.

Archaeophyte (change –0.41). *E. campestre* was first recorded in the wild in Devon, where it was seen by Ray on Monday 7th July, 1662. Sites in Devon and Somerset have statutory protection and the species is persisting there with appropriate management.

As an archaeophyte *E. campestre* has a European Southern-temperate distribution.

References: Atlas (154a), Tutin (1980), Meusel *et al.* (1978), Wigginton (1999).

M. SOUTHAM

Chaerophyllum temulum Rough Chervil

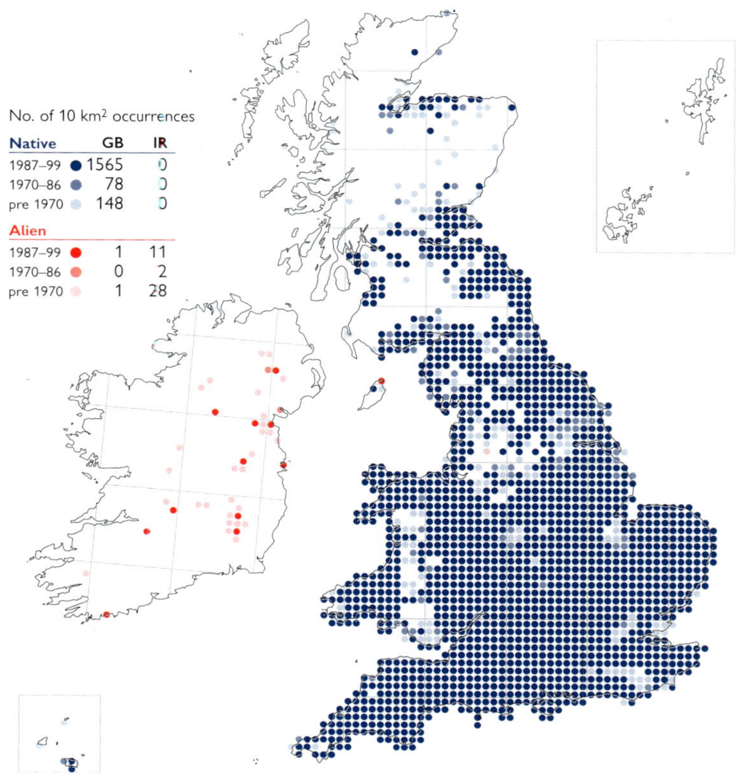

A biennial herb, especially characteristic of rank grassland on roadside verges, by hedges and along wood-borders and forest rides; also found on railway banks and in waste places. It tolerates light shade, but rarely occurs on damp and acidic soils. Reproduction is by seed. 0–365 m (Derbys. and Mid-W. Yorks.).

Native (change –0.64). There has been little appreciable change in the distribution in Britain since the 1962 *Atlas*, but it seems to have declined in Ireland, where it is probably introduced (Scannell & Synnott, 1987).

European Temperate element.

References: Atlas (154b), Hultén & Fries (1986), Meusel *et al.* (1978), Tutin (1980).

M. SOUTHAM

Anthriscus sylvestris Cow Parsley

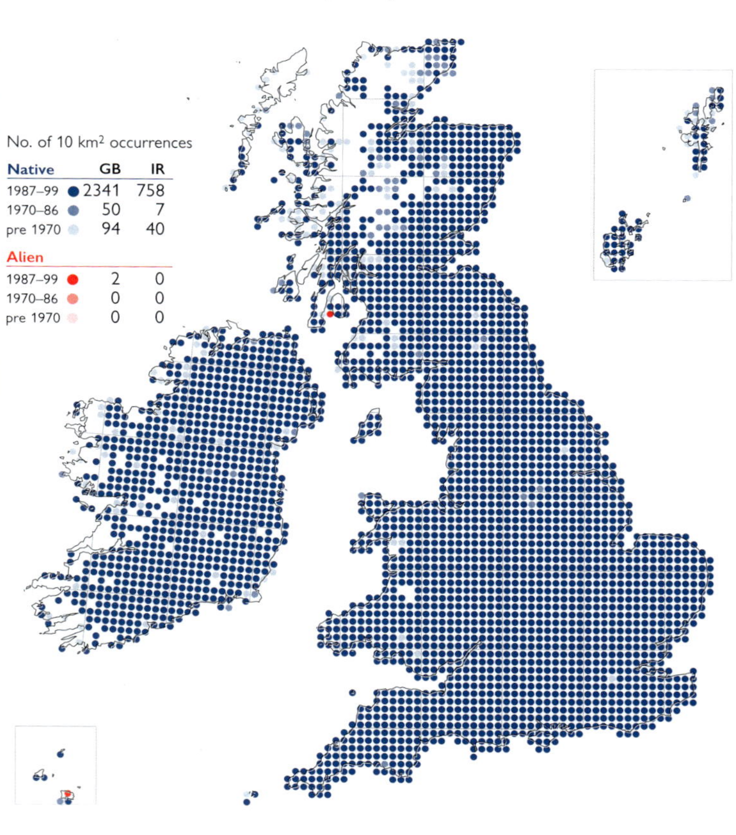

A robust perennial herb, characteristic of roadsides and hedgerows, but also occurring in abandoned pastures and under-managed hay meadows, in woodland rides and edges, on railway banks, and on waste and cultivated ground. Very wet or dry habitats are avoided. Seedlings are sometimes plentiful but extension of colonies also occurs by offsets from the main stems. 0–760 m (Mt Brandon, S. Kerry) and 845 m on Great Dun Fell (Westmorland).

Native (change –0.19). There has been no change in the distribution of *A. sylvestris* at the 10-km square scale since the 1962 *Atlas*, but it may have become more frequent on roadsides because of changes in management (e.g. Killick *et al.*, 1998).

Eurasian Boreo-temperate element.

References: Atlas (154d), Grime *et al.* (1988), Hultén & Fries (1986), Meusel *et al.* (1978), Tutin (1980).

M. SOUTHAM

Anthriscus cerefolium Garden Chervil

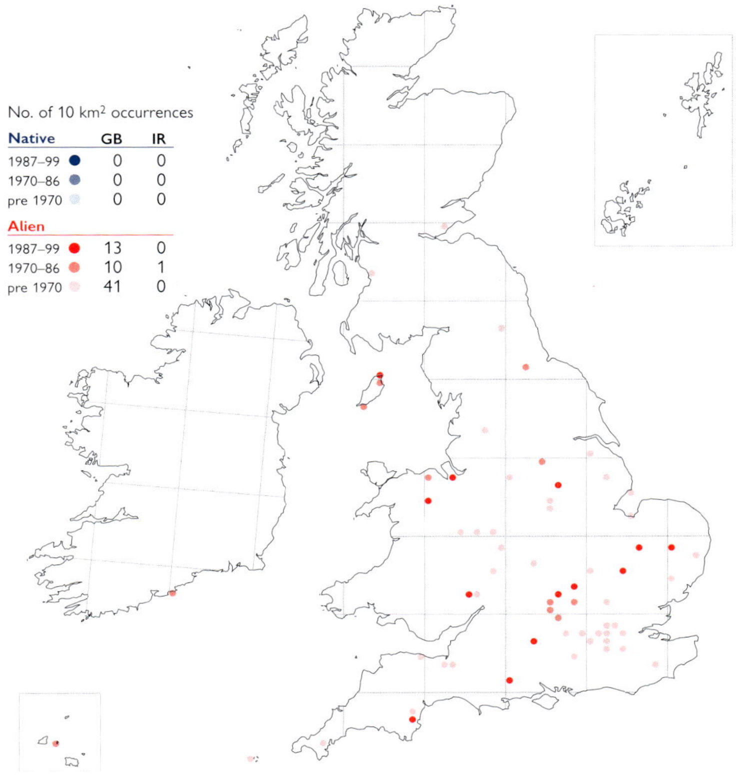

An erect or spreading perennial, cultivated as a garden herb and found naturalised in open and ruderal habitats, including a rock face in Herefordshire. It also occurs as a casual on waste ground, roadsides and rubbish tips. Lowland.

Neophyte. *A. cerefolium* was grown in British gardens by 995 (Harvey, 1981), but has decreased in popularity in recent years. It was recorded from the wild in 1633. It has been naturalised in Herefordshire since at least 1867.

Native of S.E. Europe and the eastern Mediterranean region, but widely naturalised outside this area through its spread in cultivation.

References: Bolòs & Vigo (1990), Tutin (1980).

M. SOUTHAM

Anthriscus caucalis Bur Chervil

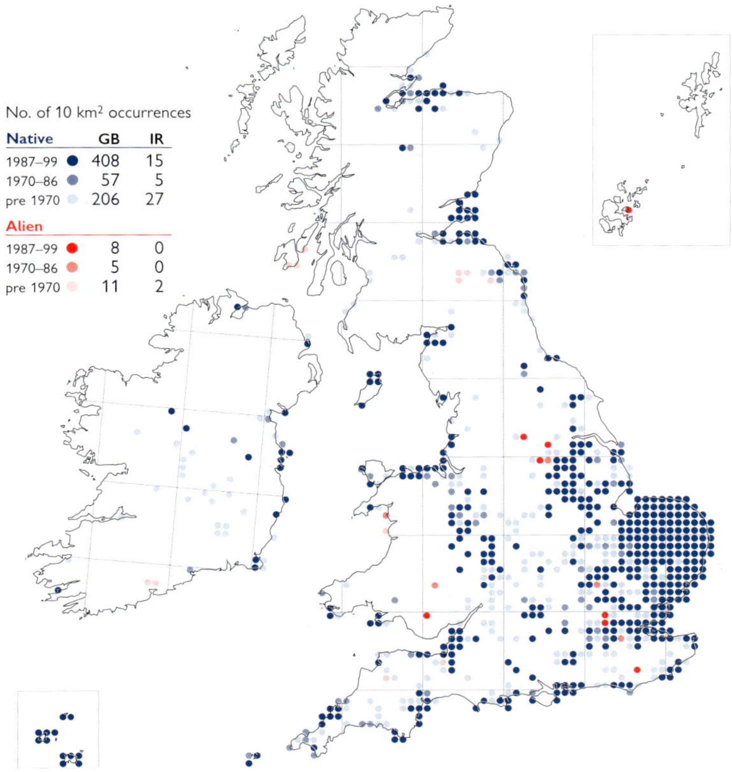

An annual of open habitats on well-drained, mainly sandy or gravelly soils, including dry grassland, hedge banks, roadsides, sea-walls, waste ground, gravel-pits and arable fields. Lowland.

Native (change –0.16). Although many sites for *A. caucalis* were lost before 1930, especially in S. England, its distribution has stabilised since the 1962 *Atlas*. Most losses were due to changes in agricultural practices and land use.

European Temperate element.

References: Atlas (154c), Hultén & Fries (1986), Meusel *et al.* (1978), Tutin (1980).

M. SOUTHAM

Scandix pecten-veneris Shepherd's-needle

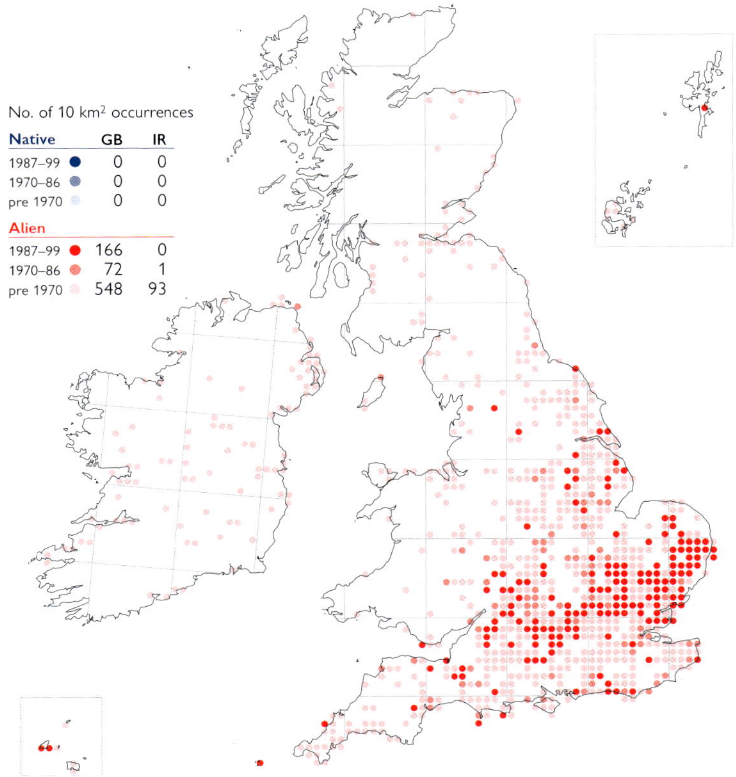

An annual of arable fields, particularly on calcareous clay soils; occasionally on paths and banks beside current or former arable sites, and rarely on waste ground, coastal cliffs, and in gardens. Generally lowland, reaching 320 m in Teesdale (Co. Durham).

Archaeophyte (change –3.65). This species has decreased very greatly since 1950 as a result of modern agricultural methods, especially herbicide treatments. It now appears to be extinct in Ireland, but is still occasionally seen in large numbers in East Anglia.

As an archaeophyte *S. pecten-veneris* has a Eurosiberian Southern-temperate element; it is widely naturalised outside this range. Its spread northwards was assisted by agriculture.

References: Atlas (155a), Curtis & McGough (1988), Hultén & Fries (1986), Meusel *et al.* (1978), Stewart *et al.* (1994), Tutin (1980).

M. SOUTHAM

Myrrhis odorata Sweet Cicely

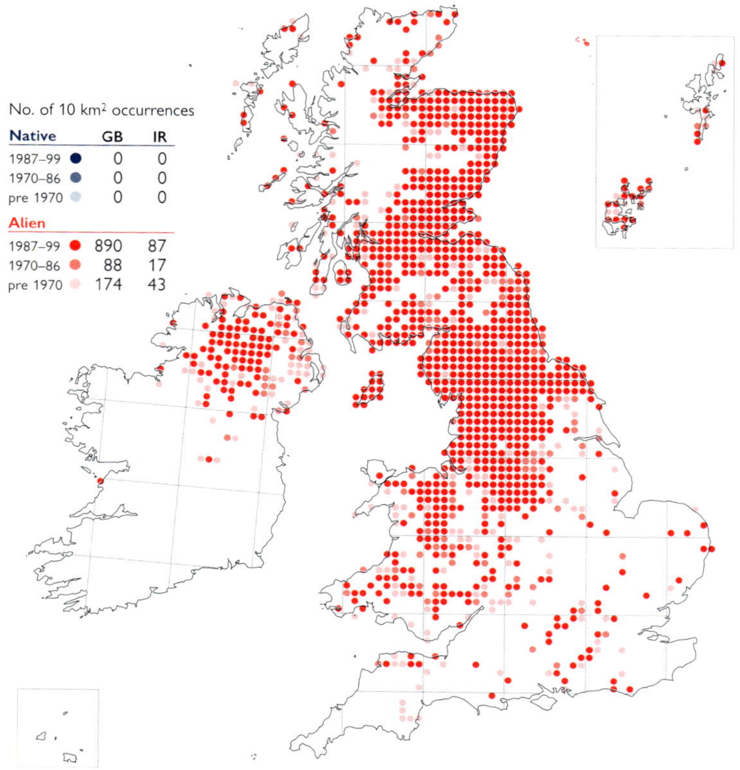

A perennial herb of hedge banks, woodland margins, roadside verges, river banks and other grassy places. Many sites are near houses or old settlements, indicating its origin in cultivation, but it is also often found in places remote from habitation. Generally lowland, but reaching 500 m in E. Allendale (S. Northumb.).

Neophyte (change –0.25). This species was first recorded from the wild in 1777. There has been little appreciable change in its distribution since the 1962 *Atlas*.

Native of the mountains of C. & S. Europe; widely naturalised elsewhere in temperate Europe.

References: Atlas (155b), Grime *et al.* (1988), Hultén & Fries (1986), Meusel *et al.* (1978), Tutin (1980).

M. F. WATSON

Coriandrum sativum Coriander

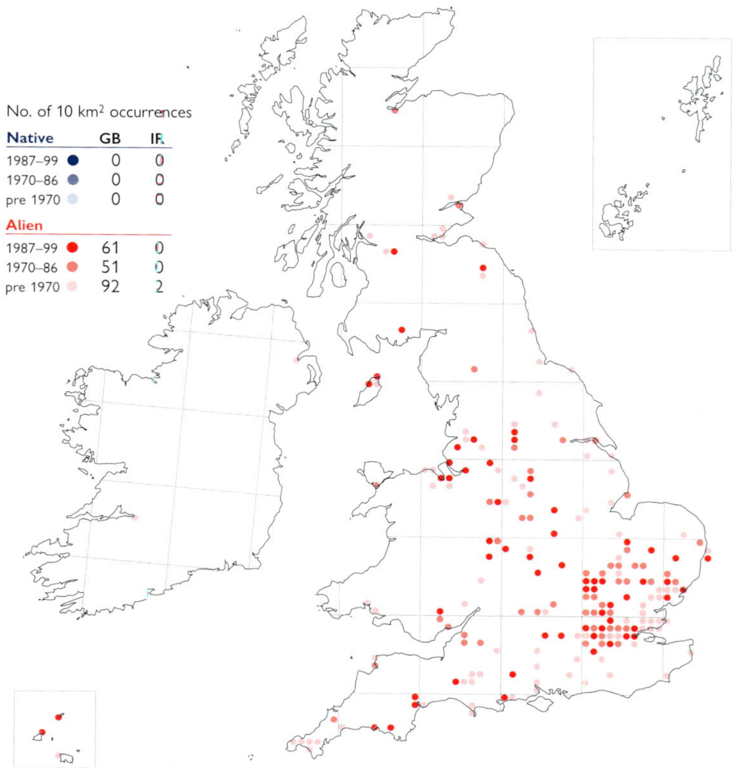

An annual, usually occurring as a casual on rubbish tips or disturbed ground, mostly from bird-seed and culinary sources; also established in a few places, such as on roadsides in N.W. Essex. Lowland.

Neophyte. *C. sativum* was grown in Britain by 995 (Harvey, 1981). Although increasingly cultivated as a culinary herb, its distribution in the wild is limited by the fact that young plants are very frost-tender. It was first recorded from the wild in 1793.

Apparently native of N. Africa and W. Asia; widely naturalised in S. Europe and elsewhere.

References: Bolòs & Vigo (1990), Tutin (1980), Vaughan & Geissler (1997), Zohary & Hopf (2000).

M. SOUTHAM

Smyrnium olusatrum Alexanders

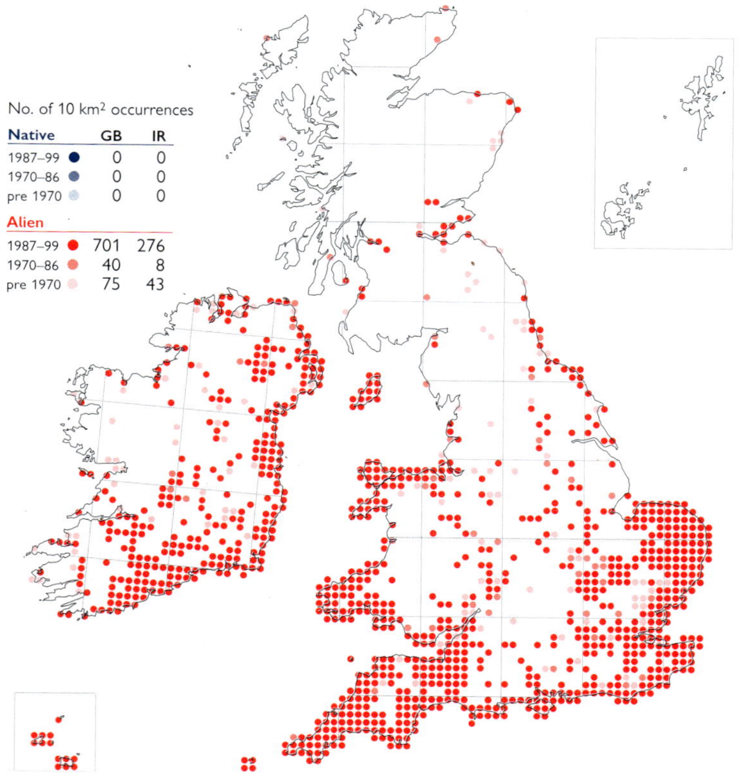

A robust perennial herb naturalised in hedge banks, on cliffs, at the base of walls, and on grassy roadsides, pathsides and waste ground, mainly near the sea. Lowland.

Archaeophyte (change +0.66). *S. olusatrum* was introduced in Roman times, and was widely cultivated until displaced by celery in the 15th century. The distribution is largely unchanged since the 1962 *Atlas*, although it appears to be increasing in some inland areas. There is no satisfactory explanation for its predominantly coastal distribution as some inland populations have persisted for many decades.

Native of the Mediterranean region and S. Europe, north to N.W. France.

References: Atlas (156d), Bolòs & Vigo (1990), Salisbury (1964), Tutin (1980).

M. F. WATSON

Bunium bulbocastanum Great Pignut

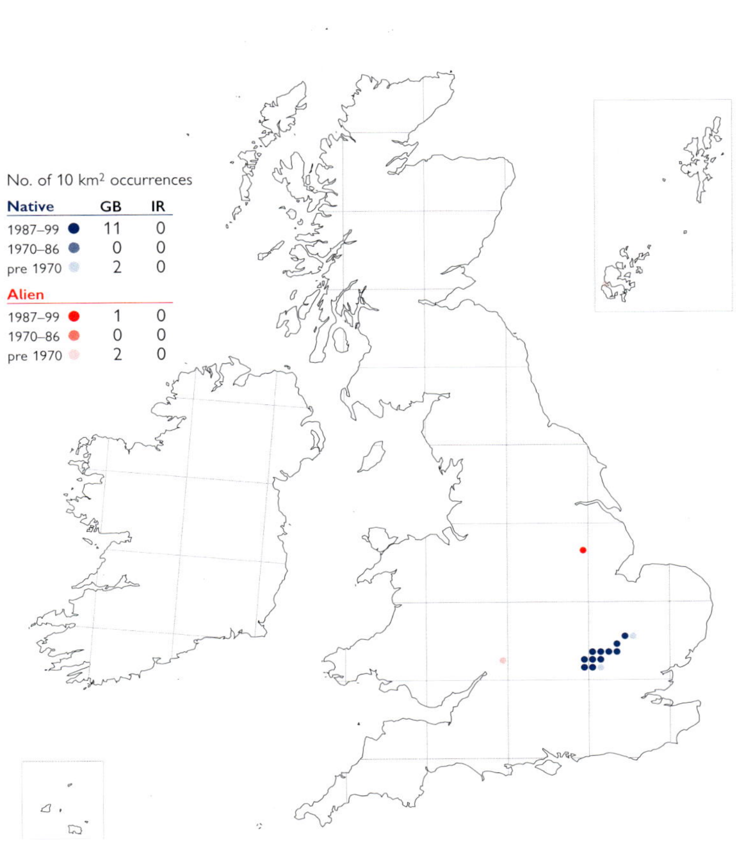

A tuberous perennial herb of dry chalk soils, most frequent in arable fields, especially where cultivation has ceased, and sometimes dominant in arable reverting to pasture. It also grows in rough or broken turf on chalk downs, field edges, in hedgerows and scrub, on roadside verges and in quarries. Reproduction is by seed. Lowland.

Native (change +0.14). There has been little change in the range of this species since the 1962 *Atlas*. It requires open soil for seedling establishment, but mature plants can thrive in closed swards and the tubers survive shallow-ploughing. Its absence from the North and South Downs is difficult to explain in view of its abundance in similar habitat at Boulogne (France).

Suboceanic Southern-temperate element.

References: Atlas (161a), Bolòs & Vigo (1990), Tutin (1980), Wigginton (1999).

M. SOUTHAM

Conopodium majus Pignut

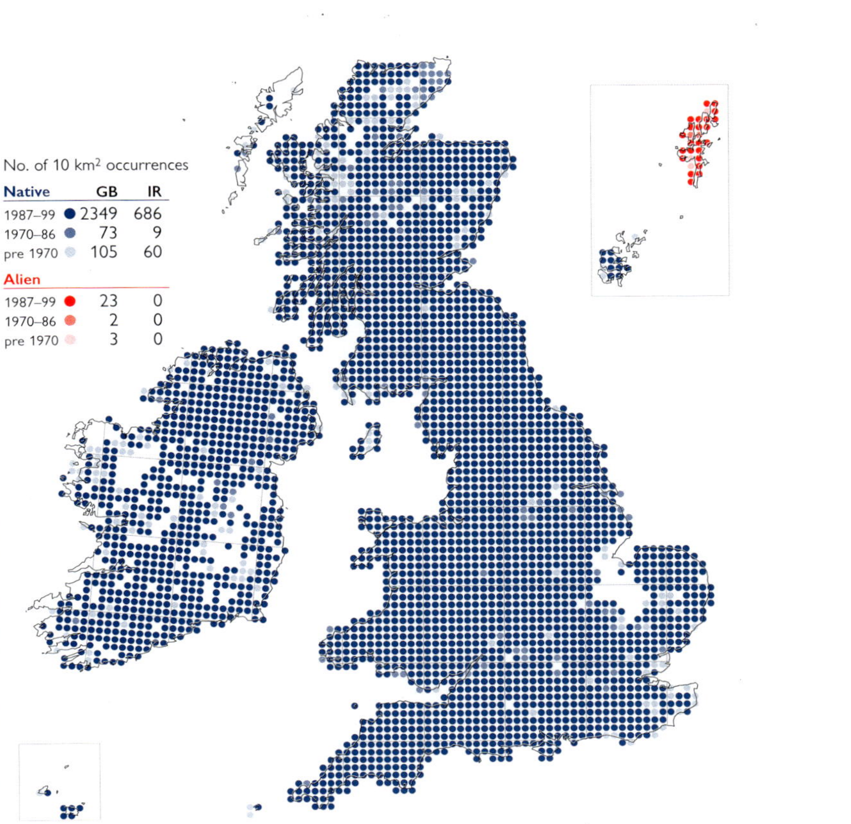

A perennial herb, found in damp or shaded meadows and pastures, hedgerows, roadside verges, copses and woodlands; especially characteristic of some types of northern hay meadow (Rodwell, 1992). It grows on a wide range of acidic and base-rich soils. 0–700 m (Breadalbanes, Mid Perth), and 845 m on Great Dun Fell (Westmorland).

Native (change –0.19). *C. majus* has declined locally as its grassland habitats have been improved (Brewis *et al.*, 1996), but its national distribution at a 10-km scale is little changed. It was introduced in Shetland in the 19th century, but its status in Orkney is unclear.

Oceanic Temperate element.

References: Atlas (161b), Grime *et al.* (1988), Hultén & Fries (1986), Tutin (1980).

M. SOUTHAM

Pimpinella major Greater Burnet-saxifrage

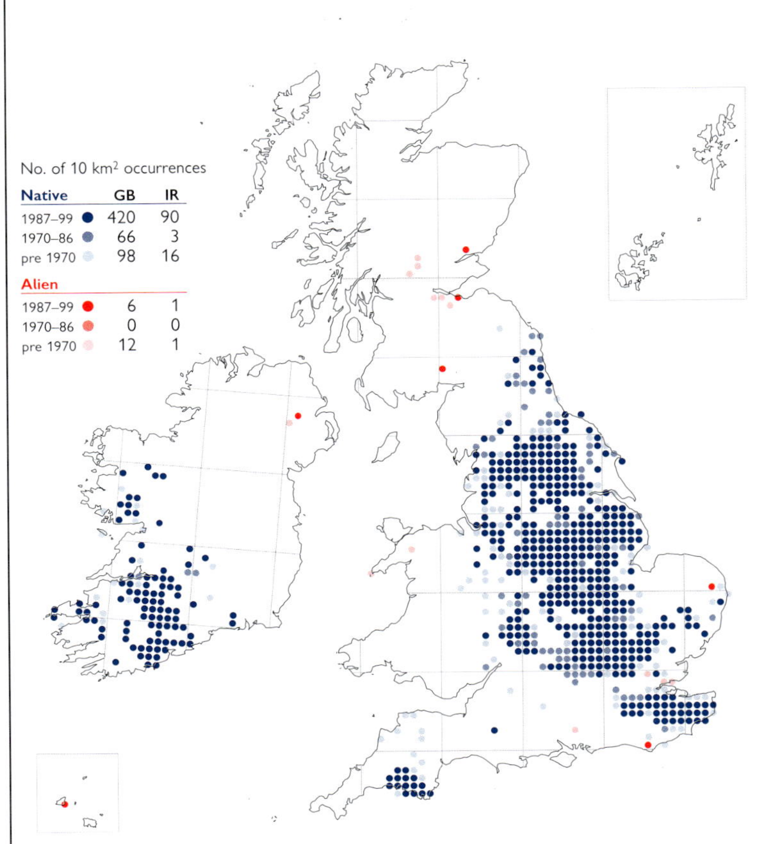

A perennial herb, mainly on basic soils derived from chalk and limestone, but also on clay, and most often found on roadsides, hedge banks, railway banks and wood edges, sometimes persisting on roadsides when neighbouring woods have been removed. Generally lowland, reaching 320 m in Derbyshire.

Native (change –0.16). The 1962 *Atlas* showed a few losses of *P. major* before 1930, but there is little evidence of any further decline since then; indeed, it is now much better recorded in some areas. Its absence from large areas of S. England is difficult to explain.

European Temperate element. Its scattered, mainly lowland, distribution in our area contrasts with its widespread presence in mountains in mainland Europe.

References: Atlas (161d), Hultén & Fries (1986), Meusel *et al.* (1978), Rackham (1980), Tutin (1980).

M. SOUTHAM

Pimpinella saxifraga Burnet-saxifrage

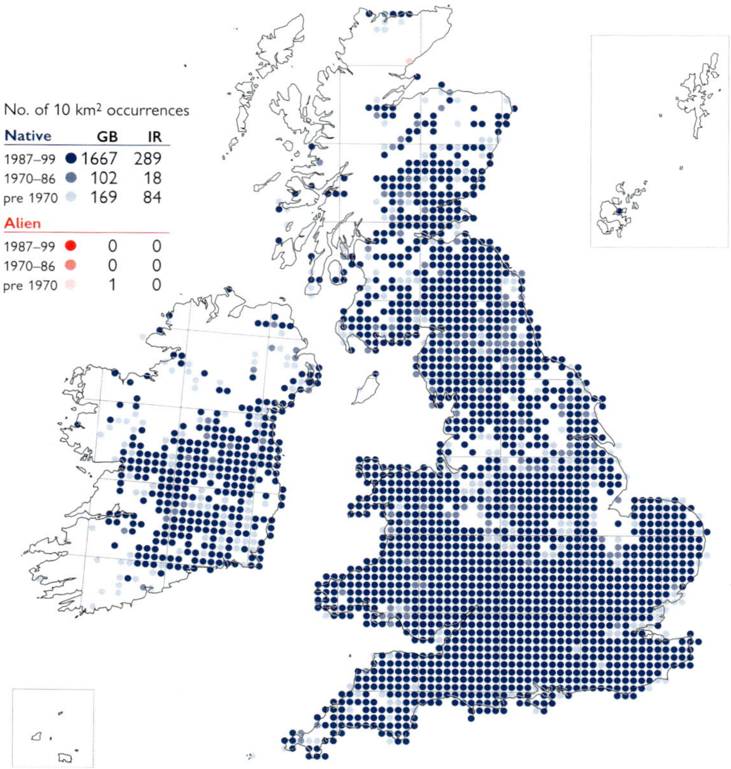

No. of 10 km² occurrences

		GB	IR
Native			
1987–99	●	1667	289
1970–86	●	102	18
pre 1970	●	169	84
Alien			
1987–99	●	0	0
1970–86	●	0	0
pre 1970	●	1	0

A perennial herb of grassy habitats on well-drained soils, favouring those which are calcareous or otherwise base-rich, but also on acidic sands. It occurs on grazed and ungrazed chalk and limestone downs, in rough pasture and other grassland, in woodland edges and open rides; less frequently on roadsides and rough ground. Generally lowland, but reaching 810 m on Snowdon (Caerns.) and Dollywagon Pike (Westmorland).

Native (change –0.31). There has been no appreciable change in the distribution of *P. saxifraga* since the 1962 *Atlas*, where it was mapped as 'all records'.

Eurosiberian Temperate element.

References: Atlas (161c), Grime *et al.* (1988), Hultén & Fries (1986), Meusel *et al.* (1978), Tutin (1980).

M. SOUTHAM

Aegopodium podagraria Ground-elder

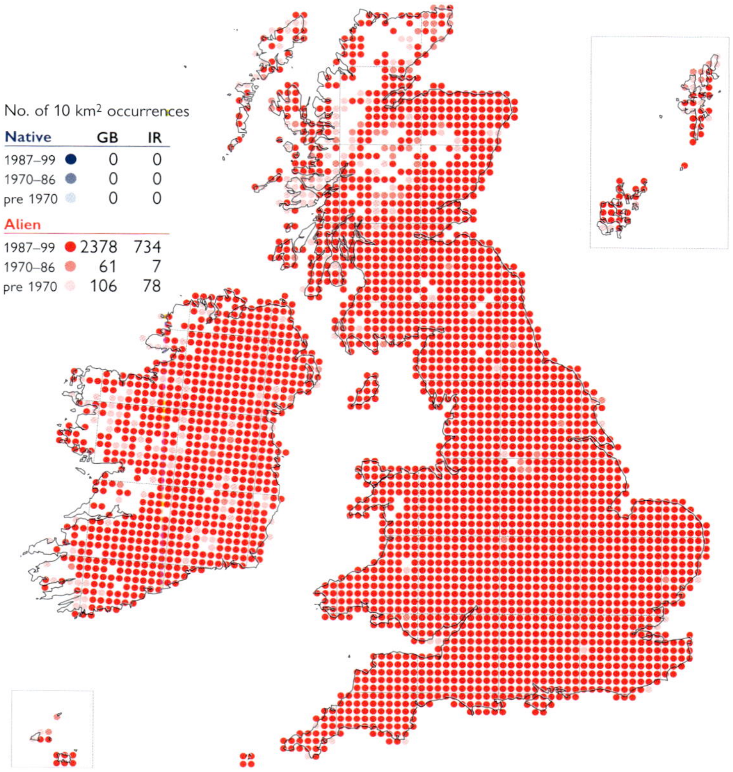

No. of 10 km² occurrences

		GB	IR
Native			
1987–99	●	0	0
1970–86	●	0	0
pre 1970	●	0	0
Alien			
1987–99	●	2378	734
1970–86	●	61	7
pre 1970	●	106	78

A rhizomatous perennial herb found in a wide variety of disturbed habitats, especially hedgerows, road verges, churchyards, neglected gardens and waste ground. It typically occurs near habitation or in the vicinity of abandoned settlements, such as in woodlands. It reproduces by seed and by its aggressive rhizome system. 0–450 m (Nenthead, Cumberland).

Archaeophyte (change –0.45). Archaeological evidence suggests that *A. podagraria* was introduced in Roman times, probably for medicinal and culinary use. There is little change in its range since the 1962 *Atlas*.

A Eurosiberian Temperate species, certainly native in C. Europe but native limits obscured by its spread in cultivation and as a weed.

References: Atlas (162a), Hultén & Fries (1986), Meusel *et al.* (1978), Tutin (1980).

M. SOUTHAM

Sium latifolium Greater Water-parsnip

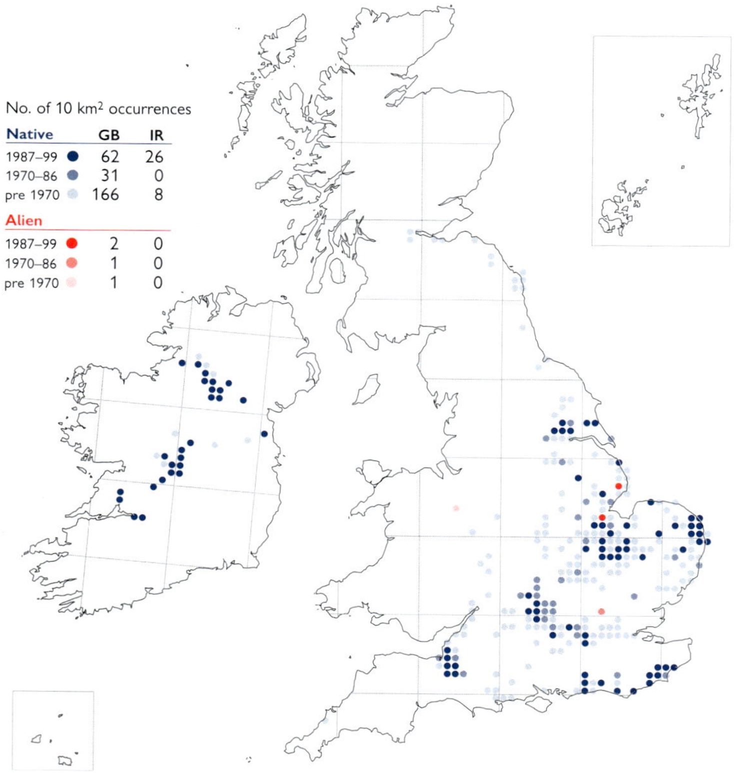

No. of 10 km² occurrences

		GB	IR
Native			
1987–99	●	62	26
1970–86	●	31	0
pre 1970	●	166	8
Alien			
1987–99	●	2	0
1970–86	●	1	0
pre 1970	●	1	0

This perennial herb was once typical of very wet, species-rich, tall-herb fen, which often developed as floating mats at the margins of lakes and large rivers; now it is generally found in ditches, growing amongst other emergent species or in reedswamp. It prefers alkaline conditions. It is readily grazed by stock, and often restricted to inaccessible ditch-banks. Lowland.

Native (change –1.83). The substantial decline of this species was already apparent in the 1962 *Atlas* and has continued since then, largely due to the effects of habitat destruction, drainage, reclamation and changes in the management of water levels.

Eurosiberian Temperate element.

References: Atlas (162b), Hultén & Fries (1986), Meusel *et al.* (1978), Stewart *et al.* (1994), Tutin (1980).

M. SOUTHAM

Berula erecta Lesser Water-parsnip

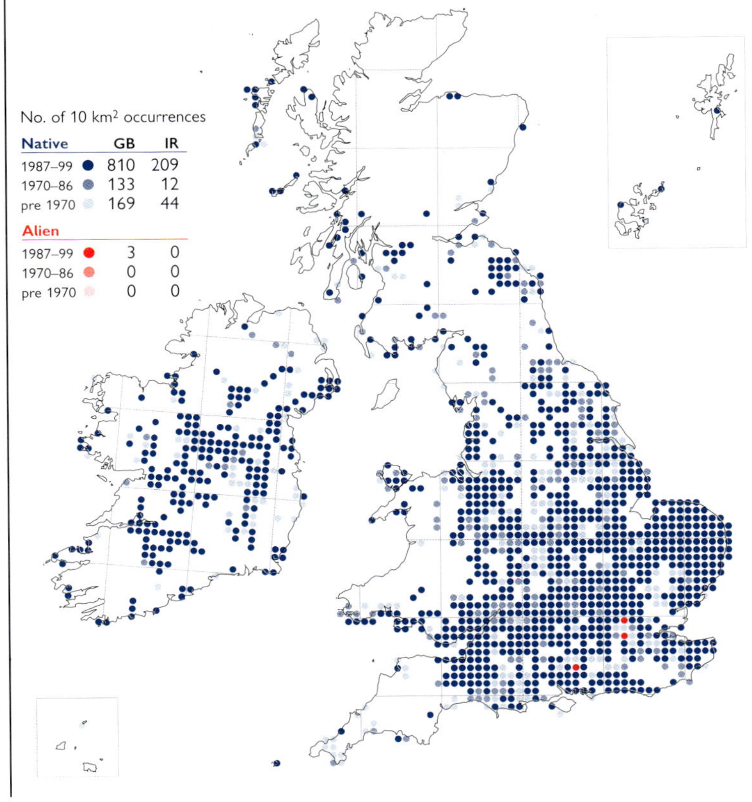

This stoloniferous perennial occurs as a submerged aquatic in rivers and streams, and as an emergent species at the edges of lakes, ponds, rivers, ditches and canals, and in marshes. It is also found on seasonally-flooded wet ground, and usually roots into fine silt or mud. Vegetative spread is by short-lived stolons or rhizomes, but little is known of the frequency of reproduction by seed. Lowland.

Native (change –0.02). This species was mapped as 'all records' in the 1962 *Atlas*. Declines have occurred, and analysis of the database reveals that most of these losses have occurred since 1950. They are probably caused by drainage and habitat destruction.

European Temperate element; also in C. Asia and N. America.

References: Atlas (162c), Hultén & Fries (1986), Preston & Croft (1997), Tutin (1980).

M. SOUTHAM

Crithmum maritimum Rock Samphire

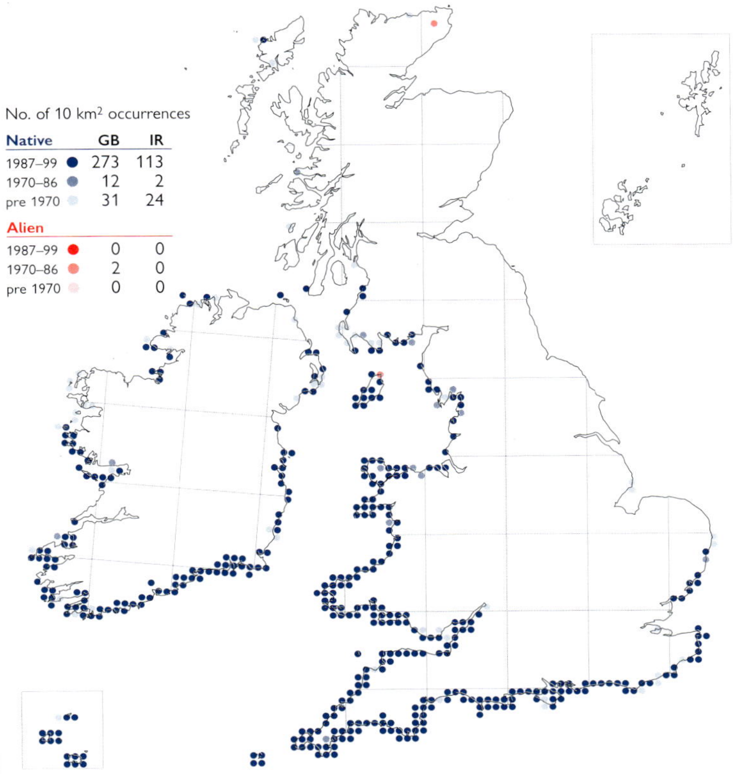

A fleshy perennial herb of spray-drenched rock crevices and ledges on sea-cliffs, coastal rocks and on stabilised shingle; also in maritime grassland and artificial habitats like harbour walls and stone sea defences. It appears indifferent to soil reaction, being found on many rock types from chalk and limestone to granite. Lowland.

Native (change +0.23). There has been no appreciable change in the distribution of *C. maritimum* since the 1962 *Atlas*.

Mediterranean-Atlantic element.

References: Atlas (162d), Bolòs & Vigo (1990), Malloch & Okusanya (1979), Okusanya (1979a, b, c), Tutin (1980).

M. J. WIGGINTON

Seseli libanotis Moon Carrot

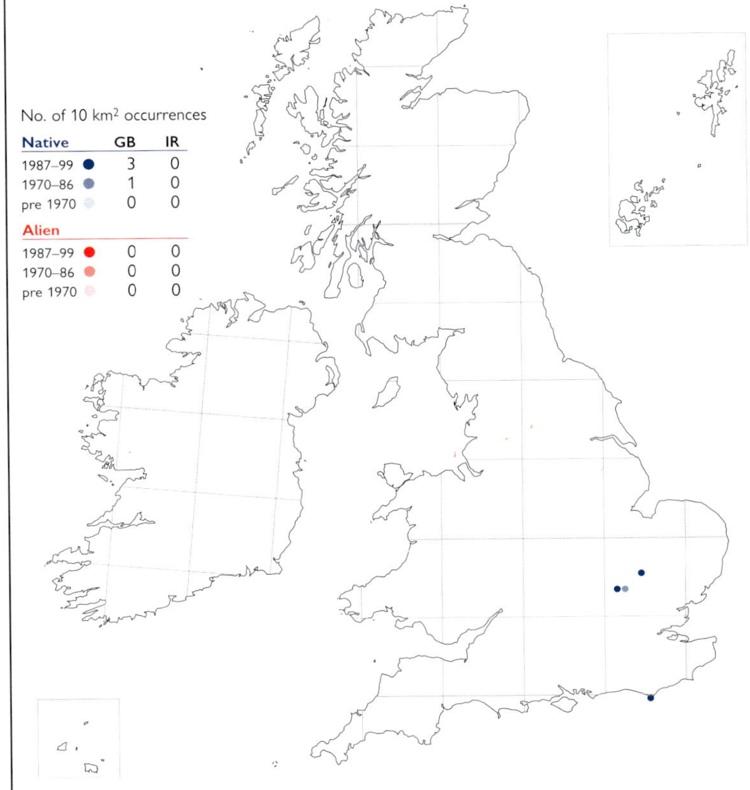

This herb is usually biennial, though it is sometimes a short-lived monocarpic perennial. It is mainly a plant of chalk grassland, but in Cambridgeshire is also found on chalky roadside banks and on ledges in an abandoned chalk quarry. Lowland.

Native. The national distribution of *S. libanotis* is stable, and all populations lie within SSSIs. However, colonies in Cambridgeshire are small and vulnerable, and one has declined significantly in recent years, presumably because of poor grassland management.

Eurasian Temperate element, with a continental distribution in W. Europe.

References: Atlas (163a), Hultén & Fries (1986), Meusel *et al.* (1978), Tutin (1980), Wigginton (1999).

M. SOUTHAM

Oenanthe fistulosa Tubular Water-dropwort

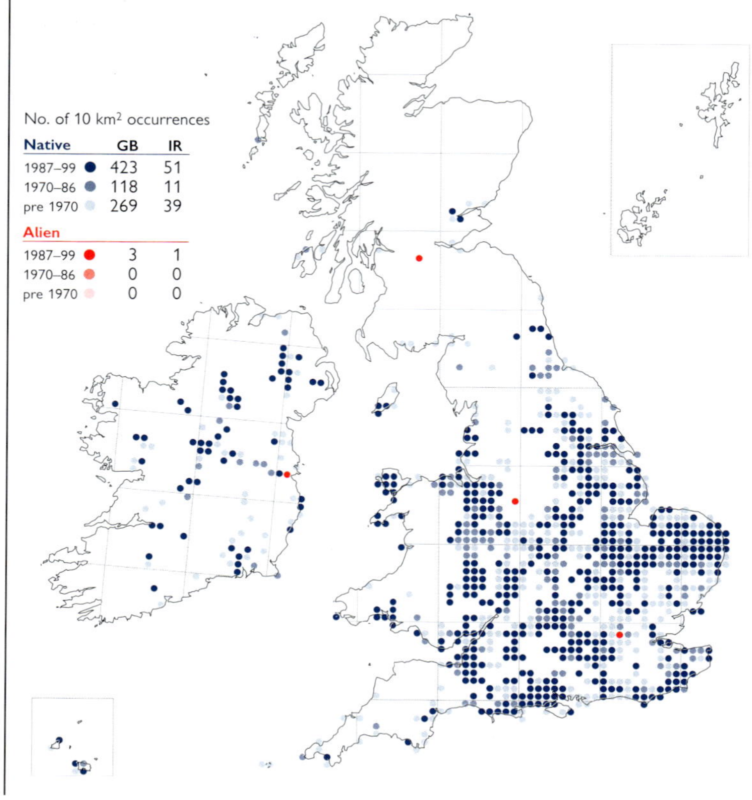

A perennial herb of damp or wet habitats, usually in areas of winter flooding. It occurs in meadows and pastures in the flood plains of rivers, in marshes and fens, and in emergent and fringing vegetation by rivers, streams, canals, ditches, lakes and ponds. It reproduces by seed, and spreads by stolons. In heavily grazed swards seed is produced virtually at ground level from secondary growth. Lowland.

Native (change −1.18). This species, mapped as 'all records' in the 1962 *Atlas*, has appreciably declined because of drainage and the re-seeding of old grassland or its conversion to arable. Most losses appear to have occurred since 1950.

European Temperate element.

References: Atlas (163b), Hultén & Fries (1986), Meusel *et al.* (1978), Preston & Croft (1997), Tutin (1980).

M. SOUTHAM & M. J. WIGGINTON

Oenanthe silaifolia Narrow-leaved Water-dropwort

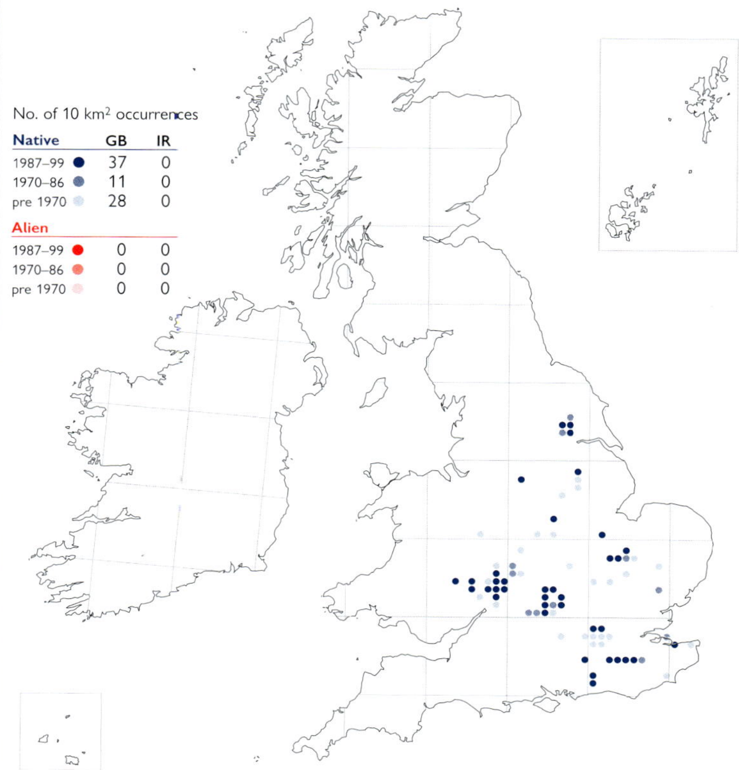

This perennial herb is found in damp grassland which receives calcareous flood-water in winter. It normally grows in hay meadows and may occur abundantly in lammas meadows, but only as depauperate individuals in more intensively farmed land. It also occurs on damp streamsides. Lowland.

Native (change +0.37). Many sites were lost before 1930, but new sites for this species have been discovered since the 1962 *Atlas*, and its known range has been extended to E. Yorkshire. It is eliminated by even quite modest agricultural intensification. All records from Dorset, mapped in both the 1962 *Atlas* and in Stewart *et al.* (1994), are now regarded as errors for *O. lachenalii*.

European Southern-temperate element.

References: Atlas (163d), Tutin (1980).

M. SOUTHAM

Oenanthe pimpinelloides Corky-fruited Water-dropwort

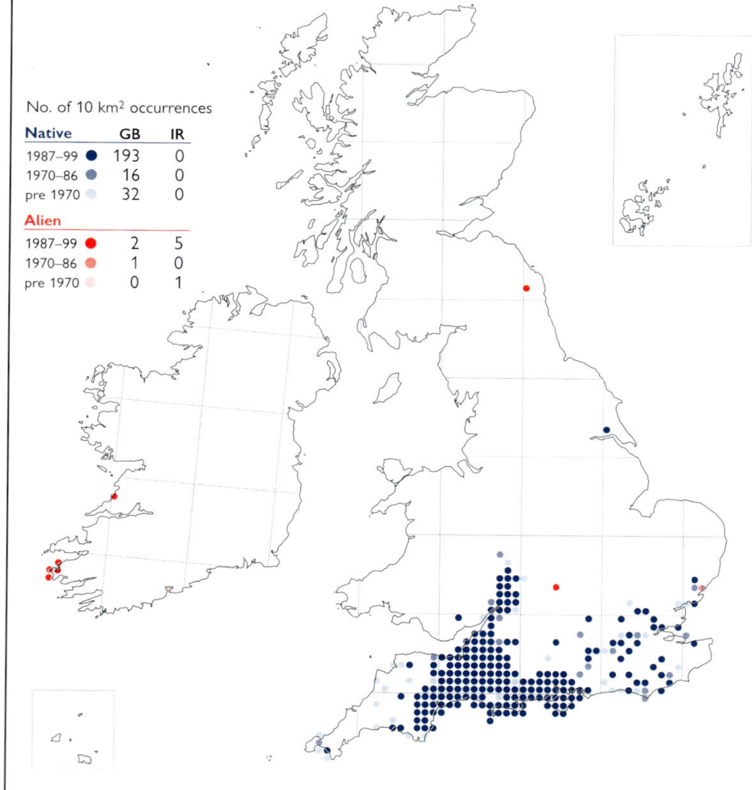

A tuberous perennial herb, found in hay meadows and pastures, especially those which are horse-grazed, and on roadsides. It grows in both damp and dry grassland, being the only *Oenanthe* which grows in dry habitats in our area. Lowland.

Native (change +0.48). Despite many losses, this species is still present in very large numbers on some roadsides and in unimproved fields. Many of its outlying sites have been discovered since 1960, due to an expansion of its range or the existence of previously overlooked sites. It was recently discovered in Co. Clare, where it may have been introduced in a seed mixture.

Mediterranean-Atlantic element.

References: Atlas (163c), Bolòs & Vigo (1990), Curtis & McGough (1988), Stewart *et al.* (1994), Tutin (1980).

M. SOUTHAM

Oenanthe lachenalii Parsley Water-dropwort

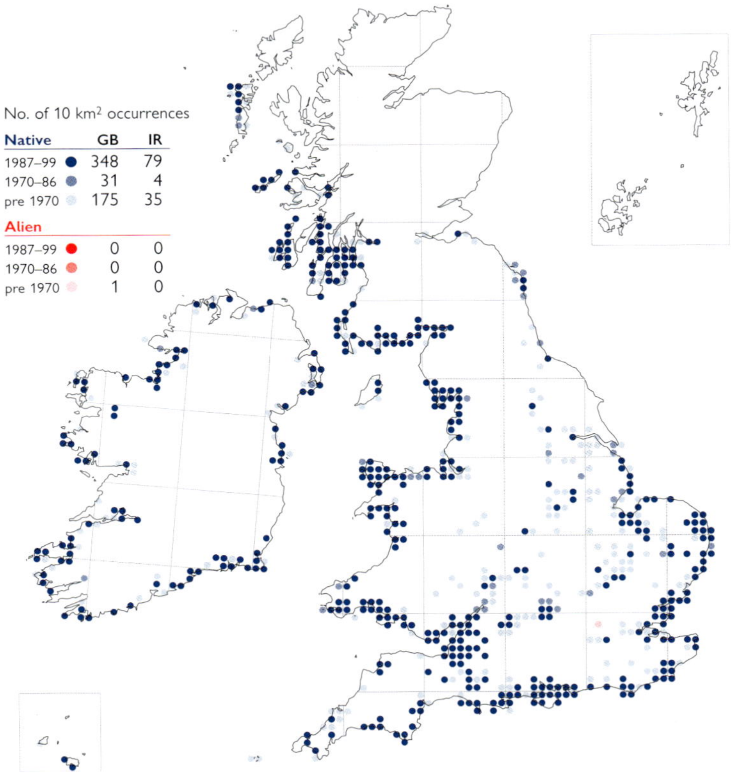

In coastal areas this perennial herb occurs in the uppermost parts of saltmarshes, in rough grassland in drained estuarine marshes, by brackish dykes and the lower reaches of tidal rivers. Inland, it is found in base-enriched habitats, including marshes, fen-meadows and tall-herb fen. Lowland.

Native (change −0.36). *O. lachenalii* is still widespread and locally plentiful, especially around the coast, but inland sites have been lost to drainage and land-fill. It was mapped as 'all records' in the 1962 *Atlas*.

Suboceanic Southern-temperate element.

References: Atlas (164a), Hultén & Fries (1986), Tutin (1980).

M. SOUTHAM

Oenanthe crocata Hemlock Water-dropwort

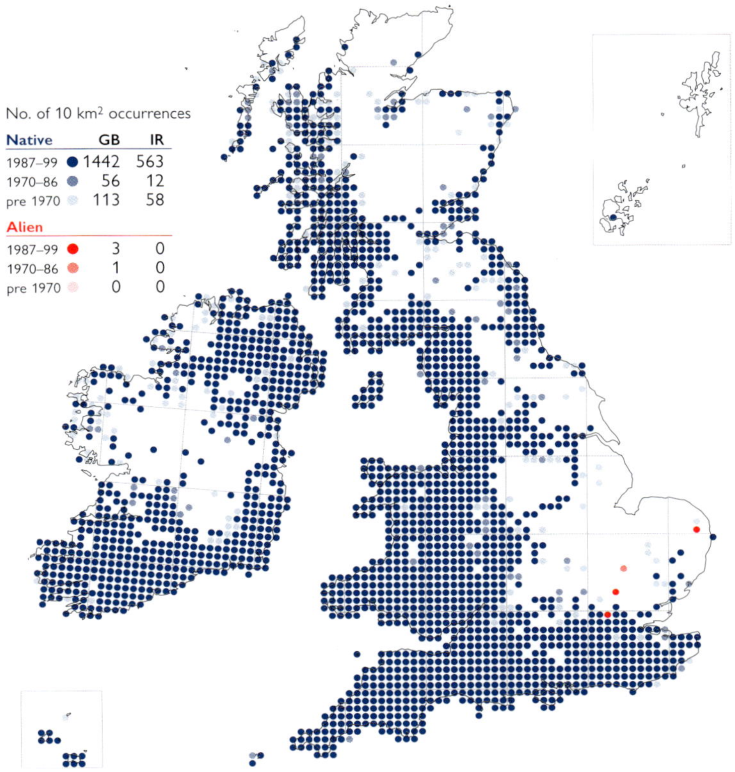

A tuberous perennial herb of shallow water in ditches, the banks of streams, rivers, canals, lakes and ponds, roadside culverts, marshes and wet woodland, among boulders at the top of beaches and on dripping or flushed sea-cliffs. Reproduction is usually by seed, and it perhaps spreads when detached tuberous roots are washed downstream. Generally lowland, but reaching 320 m in The Paps (N. Kerry).

Native (change −0.04). Though highly poisonous to man and animals, this species is rarely eradicated from sites, and its national distribution is largely stable.

Suboceanic Southern-temperate element.

References: Atlas (164b), Preston & Croft (1997), Tutin (1980).

M. SOUTHAM & M. J. WIGGINTON

Oenanthe fluviatilis River Water-dropwort

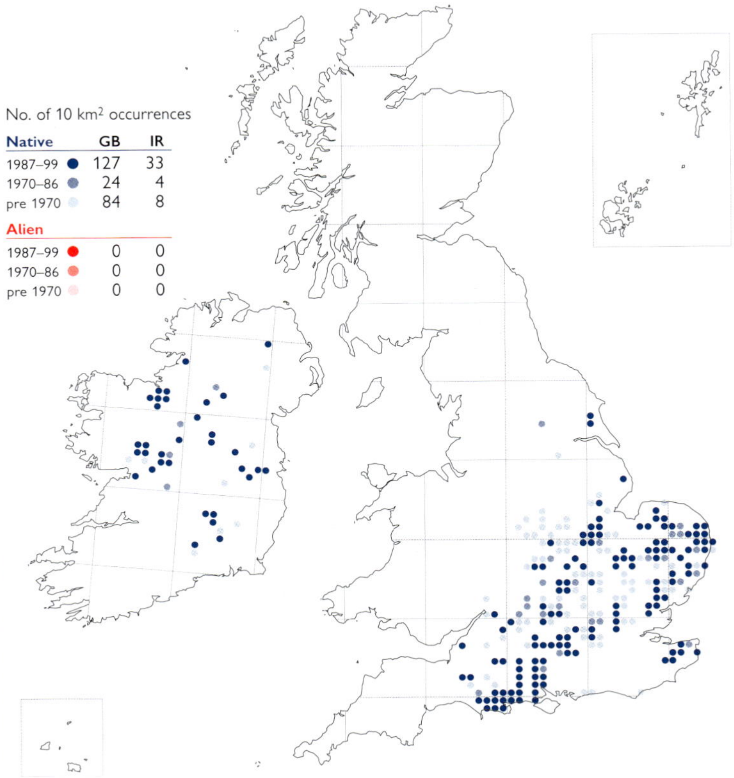

An aquatic perennial herb, most frequent in clear, meso-eutrophic water of calcareous streams and rivers; also found in canals and ditches, but rarely in ponds. In flowing water, propagation is usually by plants rooting at nodes, or by vegetative fragmentation. Flowering is more frequent in still or sluggish water, but the frequency of reproduction from seed is unknown. Lowland.

Native (change +0.19). The decline in this species was evident in the 1962 *Atlas* and has continued. The main reasons are eutrophication from agricultural run-off, dredging, and canalisation of rivers. However, it was clearly under-recorded in 1962 and may still be overlooked in some areas.

Oceanic Temperate element.

References: Atlas (164d), Hultén & Fries (1986), Preston & Croft (1997), Stewart *et al.* (1994), Tutin (1980).

M. SOUTHAM & M. J. WIGGINTON

Oenanthe aquatica Fine-leaved Water-dropwort

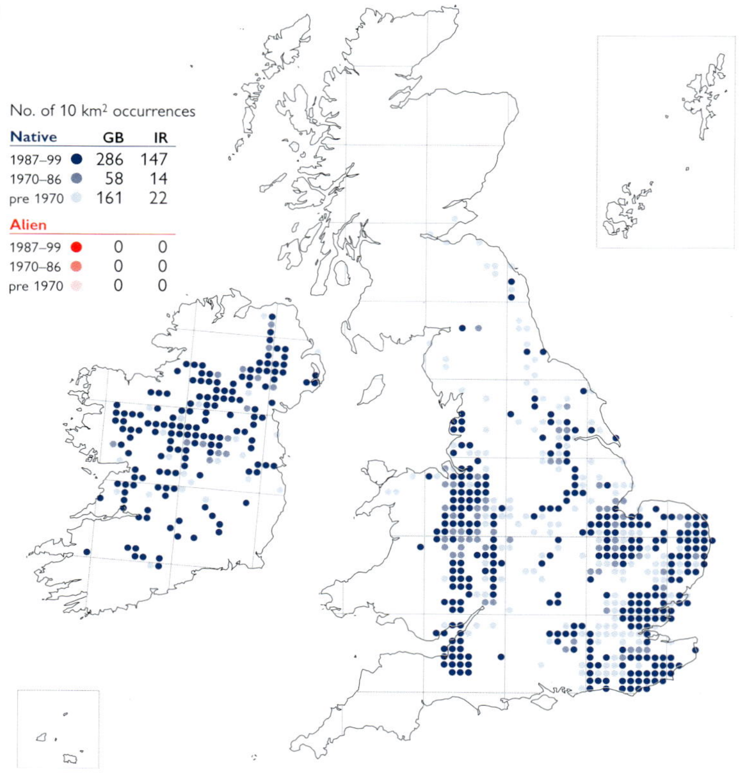

A tuberous perennial herb of still or slow-moving water, usually occurring on deep, silty, often eutrophic, substrates in shallow ponds and ditches, often where water fluctuates in depth. It also grows in open vegetation by sheltered lakes, reservoirs, canals, streams and rivers, and in marshes and seasonally flooded depressions. Lowland.

Native (change −0.35). *O. aquatica* has declined throughout its British range but particularly in the east. It is, however, better recorded now in its core areas than it was in the 1962 *Atlas*.

Eurosiberian Temperate element; also in E. Asia.

References: Atlas (164c), Hultén & Fries (1986), Meusel *et al.* (1978), Preston & Croft (1997), Tutin (1980).

M. SOUTHAM & M. J. WIGGINTON

Aethusa cynapium Fool's Parsley

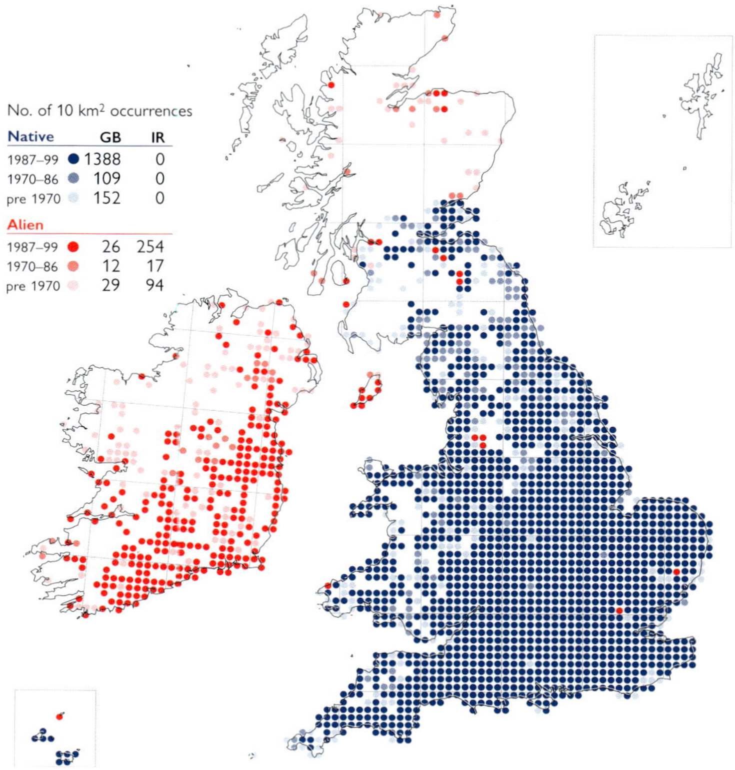

An annual of hedge banks, waste places, arable fields and other cultivated ground. Lowland.

Native or alien (change −0.41). *A. cynapium* was first recorded in Britain by 1597. It was mapped as 'all records' in the 1962 *Atlas*. The current map indicates losses in N. & W. Britain, and in Ireland, where it is thought to be alien (Scannell & Synnott, 1987). The reasons for these losses are unclear, but analysis of the database reveals that most have taken place since 1950. Two subspecies occur in our area: subsp. *cynapium* occurs throughout the range of the species; subsp. *agrestis* is probably an archaeophyte and is most frequent on arable land in S. Britain.

European Temperate element.

References: Atlas (165a), Hultén & Fries (1986), Tutin (1980).

M. SOUTHAM

Foeniculum vulgare Fennel

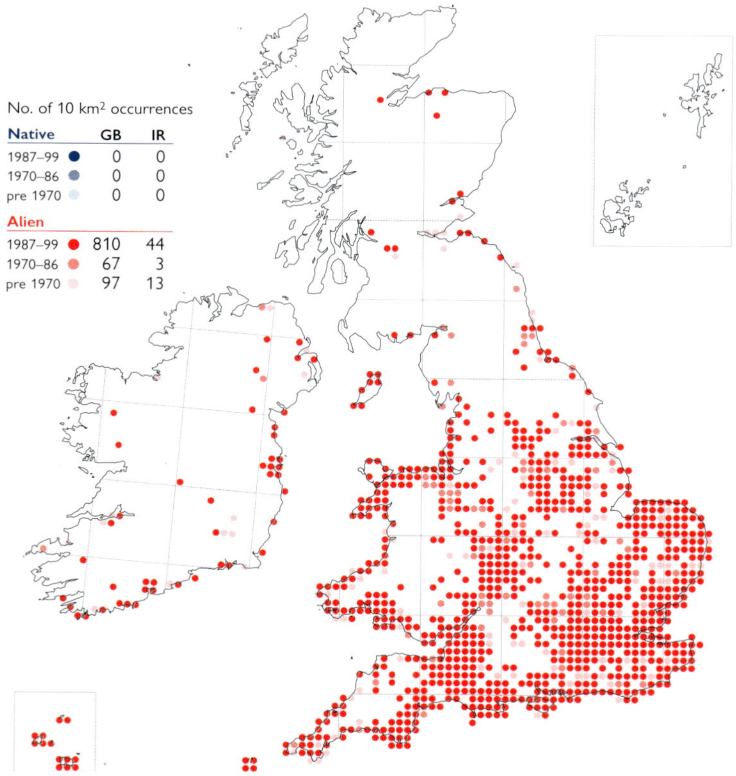

A perennial herb found in marshes, on sea-walls, in gravel-pits, on roadsides and waste ground and on rubbish tips. It has a deep tap-root, allowing it to survive in droughted habitats and making it very persistent. It reproduces freely from seed. Lowland.

Archaeophyte (change +1.17). Archaeological evidence suggests that *F. vulgare* has been used as a culinary herb since Roman times. It is common in gardens, but rarely cultivated on a field scale. It is much more frequent now than when mapped in the 1962 *Atlas*, especially inland.

Subsp. *piperitum* is native to the Mediterranean region; our plant is subsp. *vulgare* which was derived from it in cultivation and is widely naturalised in Europe and elsewhere.

References: Atlas (165b), Bolòs & Vigo (1990), Tutin (1980).

M. SOUTHAM

Anethum graveolens Dill

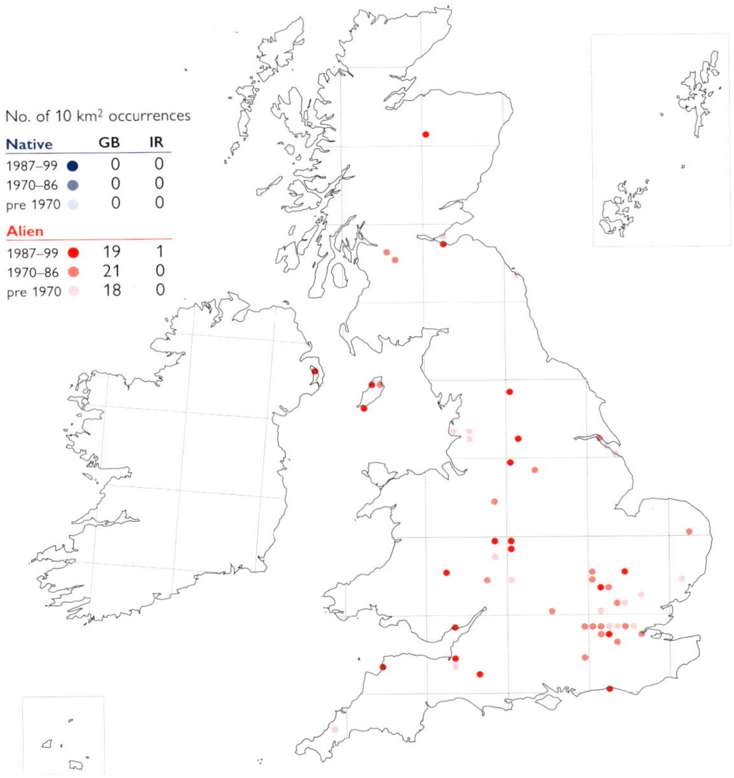

An aromatic annual thriving on light, well-drained soils, occurring as a casual in habitats associated with man, including in and near gardens, waste places and rubbish dumps. It is frost-sensitive. Lowland.

Casual. *A. graveolens* was grown in British gardens by 995 (Harvey, 1981), but was not recorded from the wild until 1863. Like other flavouring agents, it is increasingly grown as culinary influences from Europe and Asia become more prevalent.

Native range obscure; the species perhaps originates in warm-temperate Asia but is widespread as a casual in Europe and more or less naturalised in the Mediterranean region.

References: Meusel *et al.* (1978), Tutin (1980), Zohary & Hopf (2000).

M. SOUTHAM

Silaum silaus Pepper-saxifrage

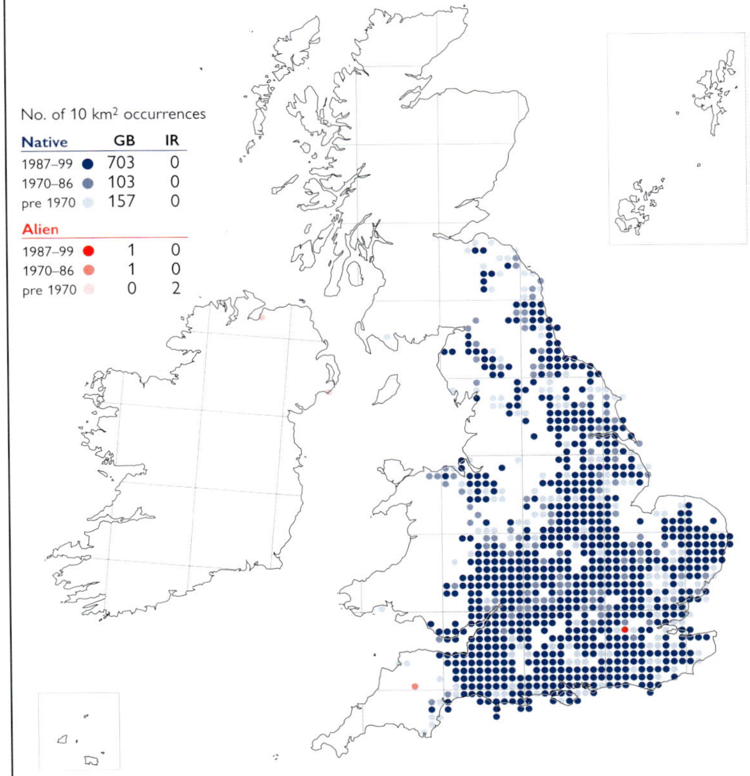

S. silaus is found in damp, unimproved neutral grassland, usually on clay soils. Its habitats include hay- and water-meadows, species-rich pastures and roadsides; it is occasionally found on chalk downs, railway banks and vegetated shingle. Lowland.

Native (change –0.42). The 1962 *Atlas* indicated little change in the distribution of *S. silaus* before 1930. Since the 1962 *Atlas* there is evidence of a small but distinct decline throughout its range, e.g. in Somerset (Green *et al.*, 1997) and Oxfordshire (Killick *et al.*, 1998).

Eurosiberian Temperate element.

References: Atlas (165c), Hultén & Fries (1986), Meusel *et al.* (1978), Tutin (1980).

M. SOUTHAM

Meum athamanticum Spignel

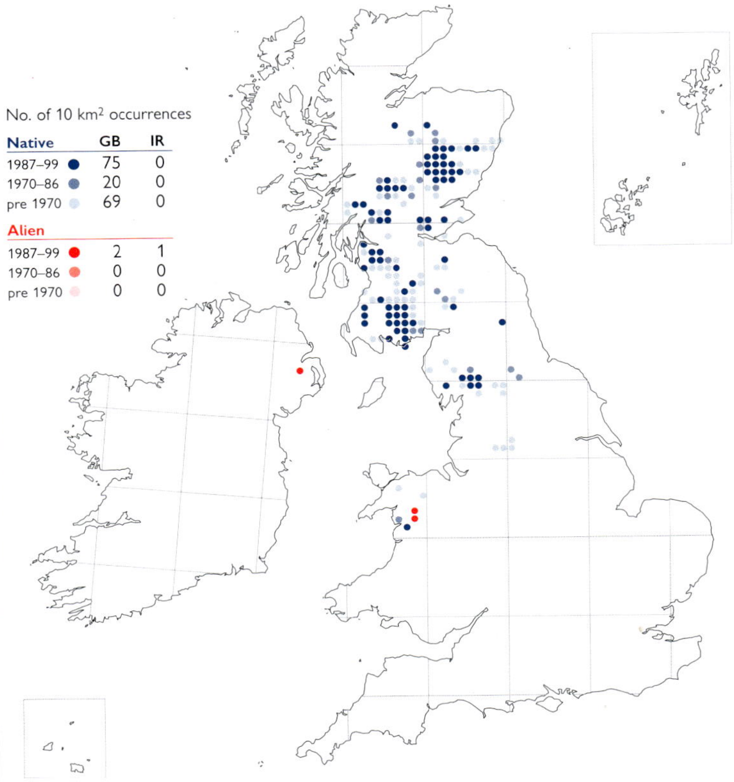

A perennial herb of deep brown-earth neutral or mildly acidic soils occurring in dry, unimproved grassland in pastures, hay meadows and on roadside-banks. Mostly found below 300 m, though there are populations at 610 m at White Coombe (Dumfriess.) and Fealar (E. Perth).

Native (change –0.40). Many of the pre-1970 sites for *M. athamanticum* were lost before 1930 as a result of agricultural improvement of grassland and probably also through deliberate destruction because it can taint cows' milk. There have been further losses since the 1962 *Atlas*, though new sites have also been discovered, notably in N. Wales and Cumbria.

European Boreal-montane element, but absent from the Boreal zonobiome.

References: Atlas (165d), Hultén & Fries (1986), Meusel *et al.* (1978), Stewart *et al.* (1994), Tutin (1980).

M. F. WATSON

Physospermum cornubiense Bladderseed

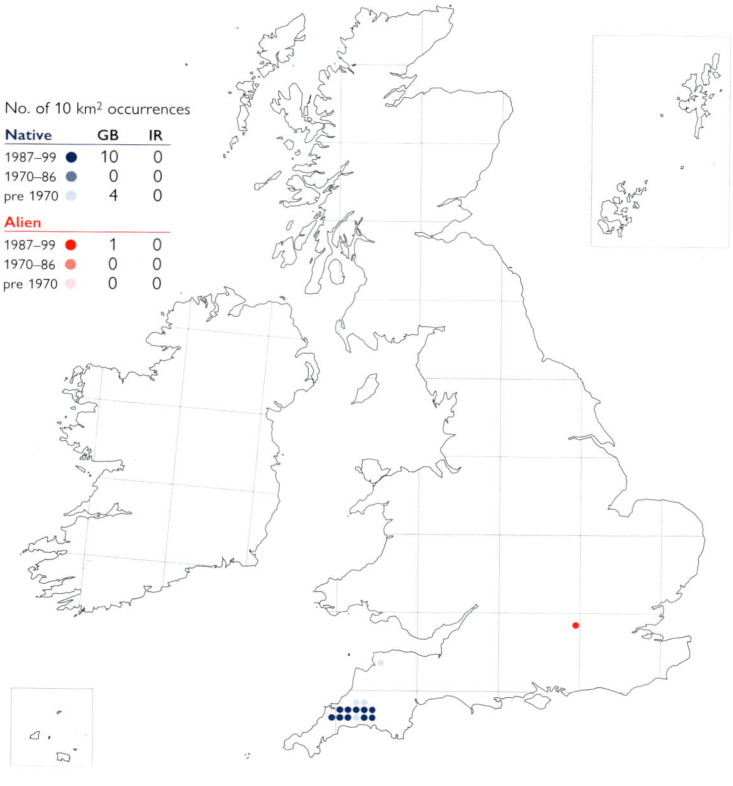

This rhizomatous perennial herb is often found in substantial, loose colonies in open woodland, in *Ulex* scrub on heaths, on rough grassy slopes (often in stream valleys), in *Molinia* grassland, and on shaded roadside banks. Reproduction is by seed, and the plant regenerates strongly after burning or clearance. Lowland.

Native (change +0.07). Since the late 1970s the number of sites of *P. cornubiense* has fallen from nearly fifty to about twenty, and populations have also strongly declined. This may be attributed to factors rendering its habitats more densely shaded or destroying them altogether, including the lack of woodland management, afforestation, scrub clearance and the loss of grazing.

European Temperate element.

References: Atlas (157a), Tutin (1980), Wigginton (1999).

M. J. WIGGINTON

Conium maculatum Hemlock

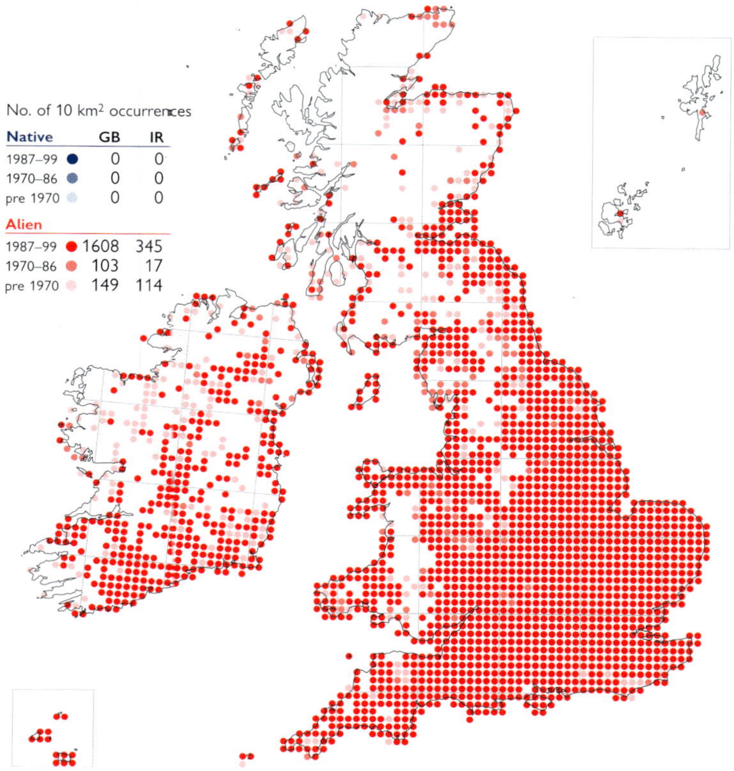

A biennial of damp places, such as ditches and river banks, and of drier habitats, including rough grassland, waste ground, rubbish tips and roadsides. It is a colonist of disturbed areas, particularly on dredged mud, sometimes forming large stands. Generally lowland, but reaching 305 m at Llynheillyn (Rads.).

Archaeophyte (change –0.02). The distribution of *C. maculatum* is little changed since the 1962 *Atlas*, although the map suggests declines on the edges of its range. Conversely, local increases in its core areas have been reported, particularly in open habitats (Killick *et al.*, 1998).

As an archaeophyte *C. maculatum* has a Eurosiberian Southern-temperate distribution; it is widely naturalised outside this range.

References: Atlas (157b), Hultén & Fries (1986), Meusel *et al.* (1978), Tutin (1980).

M. SOUTHAM

Bupleurum falcatum Sickle-leaved Hare's-ear

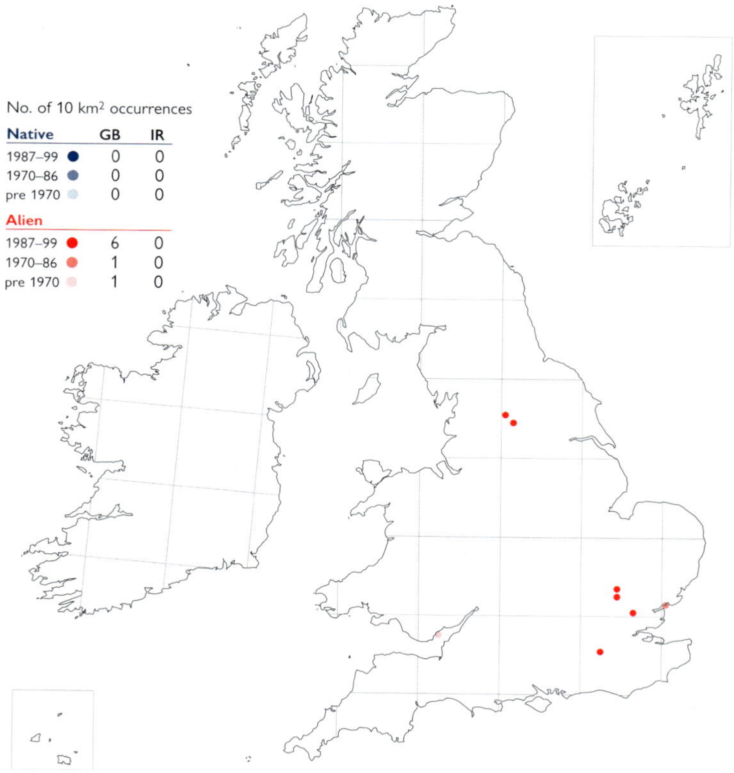

This biennial or short-lived perennial herb has been recorded in hedge banks and field-borders, on ditch banks and on roadside verges, but only recently in the latter habitat. Plants reproduce by seed, which appears to remain viable for only one year. Lowland.

Neophyte. This species, which was cultivated in gardens by 1739, has only been naturalised in a single locality, at Norton Heath in S. Essex, where it was first recorded in 1831 and last seen in 1962. The present population there was established from seed taken from cultivated plants derived from the original colony. It is quite widely grown in gardens.

A Eurasian Southern-temperate species.

References: Atlas (157d), Field (1994), Meusel *et al.* (1978), Tutin (1980), Wigginton (1999).

M. SOUTHAM

Bupleurum tenuissimum Slender Hare's-ear

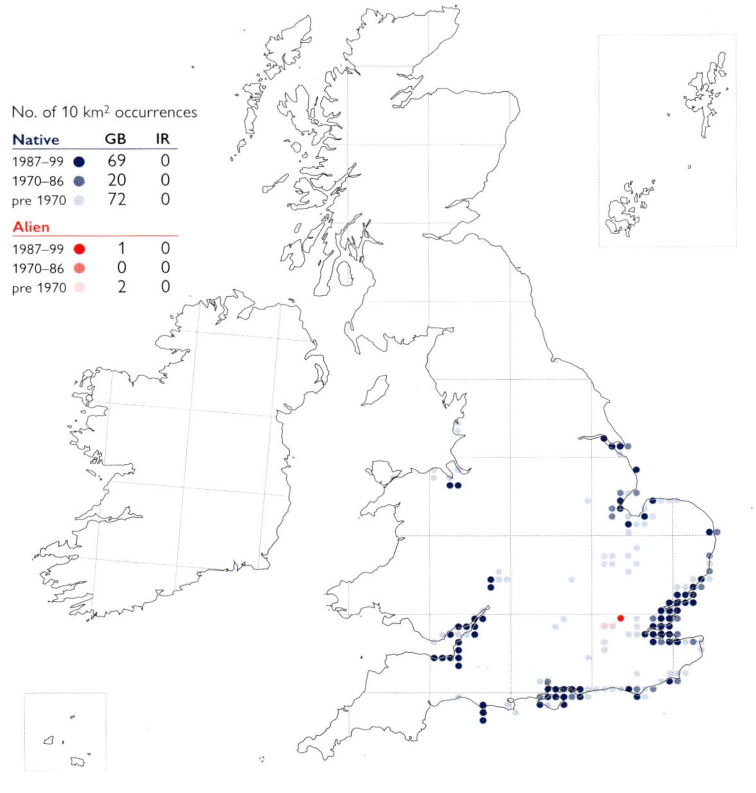

No. of 10 km² occurrences

Native	GB	IR
1987–99 ●	69	0
1970–86 ●	20	0
pre 1970 ●	72	0
Alien		
1987–99 ●	1	0
1970–86 ●	0	0
pre 1970 ●	2	0

This slender, often diminutive, annual is primarily a colonist of thinly vegetated or disturbed coastal sites, including coastal banks, sea walls, drained estuarine marshes and the margins of brackish ditches. Inland populations formerly grew on commons and roadsides; it still grows on commons near Malvern (Worcs.). Lowland.

Native (change –0.97). Most of the inland sites for this species were lost before 1930, and the distribution seems to have been largely stable since the 1962 *Atlas*. It was recorded from the central reservation of the A2 near Dartford (W. Kent) in 1982 but has otherwise shown little tendency to colonise salted roadsides.

European Southern-temperate element.

References: Atlas (158a), Coombe (1994), Hultén & Fries (1986), Meusel *et al.* (1978), Stewart *et al.* (1994), Tutin (1980).

M. SOUTHAM

Bupleurum baldense Small Hare's-ear

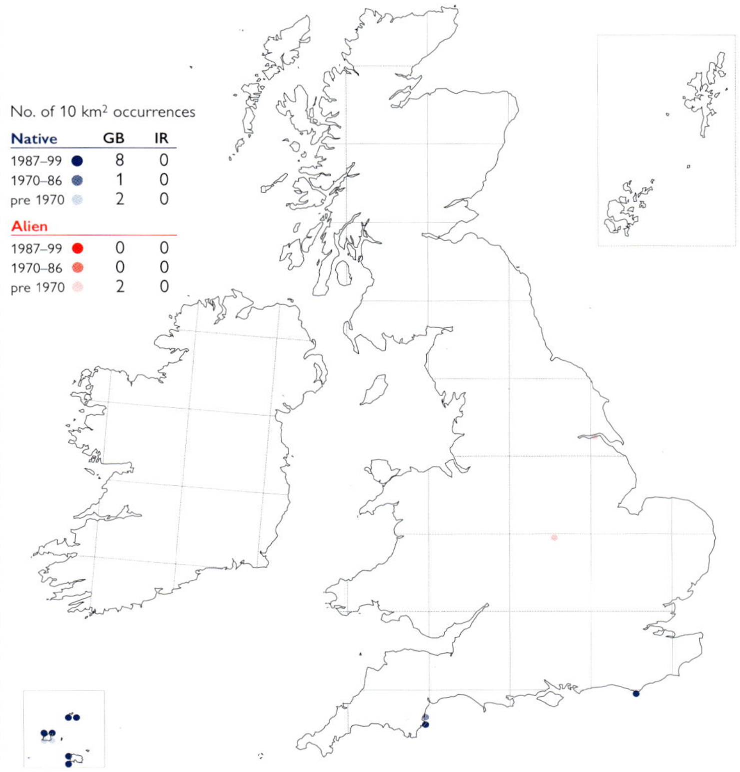

No. of 10 km² occurrences

Native	GB	IR
1987–99 ●	8	0
1970–86 ●	1	0
pre 1970 ●	2	0
Alien		
1987–99 ●	0	0
1970–86 ●	0	0
pre 1970 ●	2	0

This diminutive annual is found in rabbit-grazed coastal grassland over calcareous substrates. Near Beachy Head (E. Sussex), it is found where the turf breaks to bare chalk near the eroding and overhanging cliff-edge, and in Devon on made-ground as well as in cliff-top turf. In the Channel Islands it grows in open turf on consolidated sand dunes and more rarely on cliffs. Lowland.

Native. *B. baldense* has been lost from some sites in Devon, but the populations at Beachy Head and those in the Channel Islands are stable.

Mediterranean-Atlantic element.

References: Atlas (157d), Bolòs & Vigo (1990), Tutin (1980), Wigginton (1999).

M. SOUTHAM

Bupleurum rotundifolium Thorow-wax

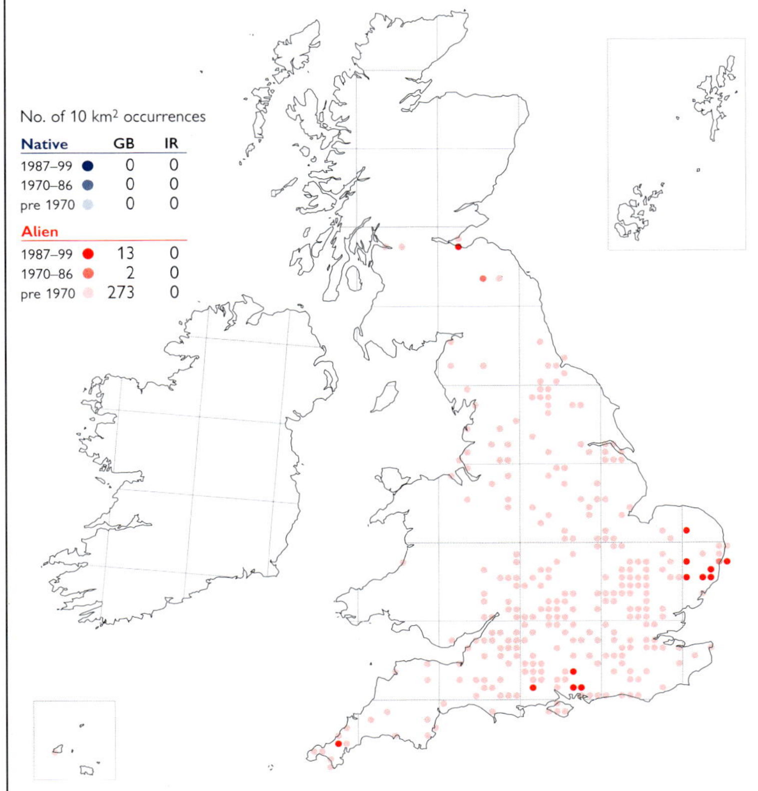

No. of 10 km² occurrences

Native	GB	IR
1987–99 ●	0	0
1970–86 ●	0	0
pre 1970 ●	0	0
Alien		
1987–99 ●	13	0
1970–86 ●	2	0
pre 1970 ●	273	0

This annual was formerly an arable weed of chalk and limestone soils, but it is now a rare bird-seed casual. Lowland.

Archaeophyte (change –4.58). *B. rotundifolium* was frequent in the mid 19th century but, with seed screening, became rare by the 20th century, and has been extinct in arable habitats since the 1960s. Germination of its seed is affected by late frosts, and its persistence depended on repeated re-introductions, often with clover seed, from S.W. Europe. It is often confused with *B. subovatum*, and some erroneous records may be mapped. Most of the East Anglian sites are deliberate introductions.

B. rotundifolium may originate in S.W. Asia; it is widespread as an archaeophyte in C. & S. Europe and has been introduced to many other areas worldwide.

References: Atlas (157c), Meusel *et al.* (1978), Tutin (1980).

M. F. WATSON

Bupleurum subovatum False Thorow-wax

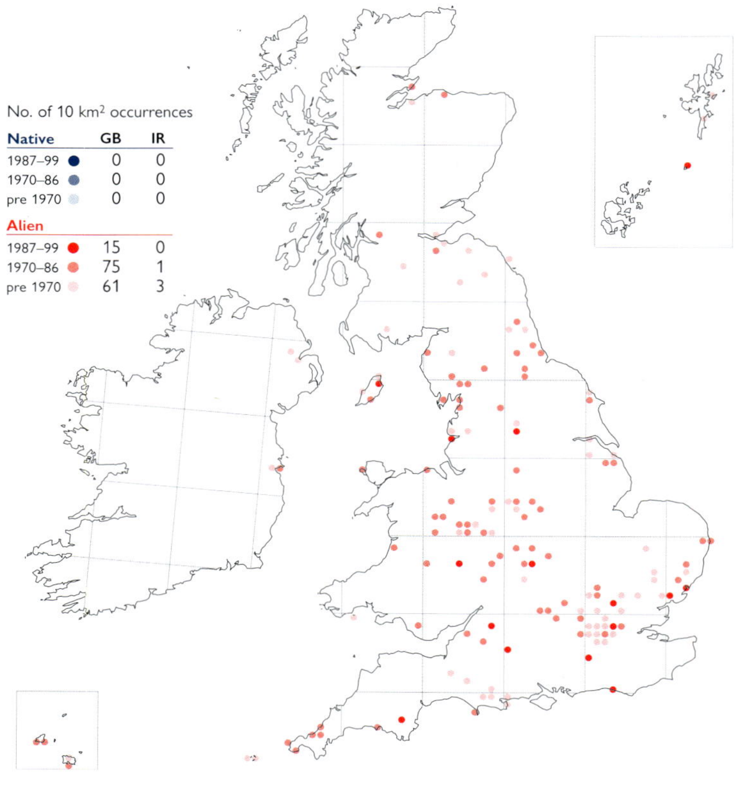

An annual occurring casually on disturbed and waste ground, on rubbish tips and in gardens. Lowland.

Neophyte. The first records of *B. subovatum* in Britain date from 1859. It showed a marked increase in frequency from about 1950 onwards, perhaps because of the increased availability and use of wild-bird seed, but is now much less frequent. It has often been confused with *B. rotundifolium*.

Native of the Mediterranean region.

Reference: Tutin (1980).

M. F. WATSON

Trinia glauca Honewort

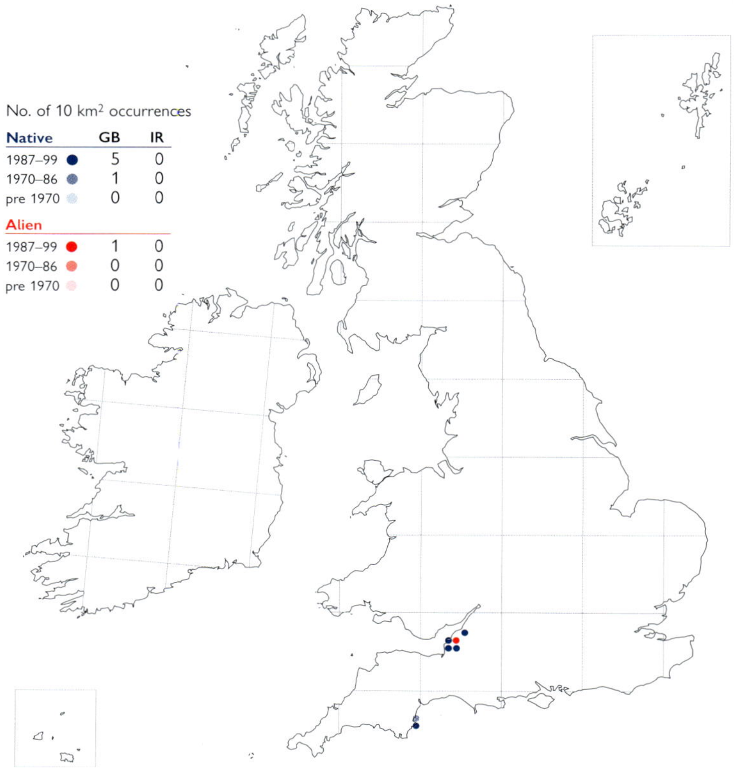

This monocarpic, dioecious perennial herb is restricted to dry limestone sites, typically occurring in short-grazed, open, species-rich turf on S.-facing slopes. In heavily grazed turf the plant can be perennial until the opportunity arises to flower. Reproduction is by seed. Lowland.

Native (change +0.12). The distribution of this species has not changed since the 1962 *Atlas*, and the national population appears to be stable. It persists at one alien site in N. Somerset.

European Southern-temperate element.

References: Atlas (158b), Meusel *et al.* (1978), Tutin (1980), Wigginton (1999).

M. SOUTHAM

Apium graveolens Wild Celery

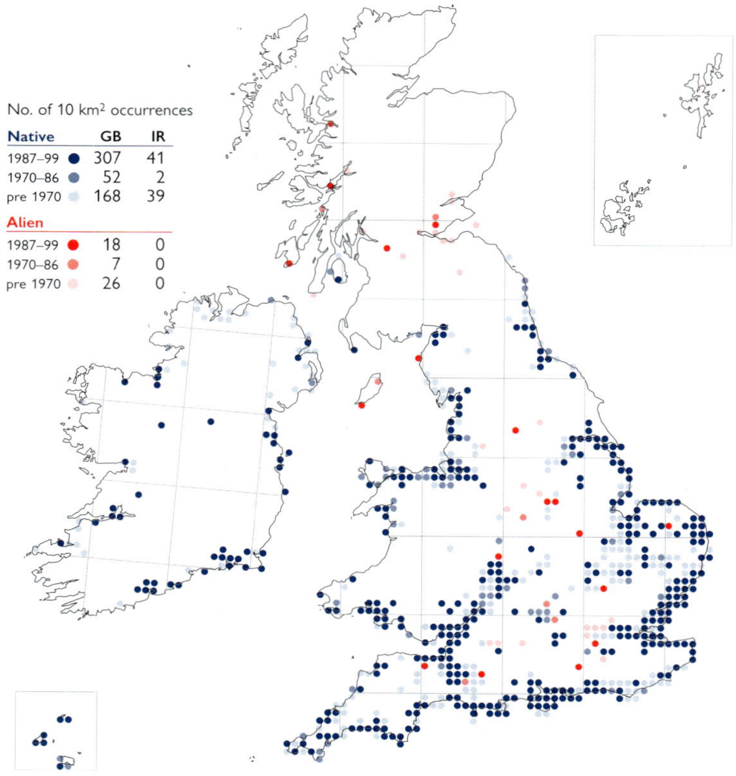

A biennial or monocarpic perennial herb found on sea-walls, beside brackish ditches, on tidal river banks and drift lines, and the uppermost parts of saltmarshes. Inland it occurs on disturbed ground in marshes, by ponds and ditches and occasionally in gravel-pits. Lowland.

Native (change –0.63). Many inland sites for *A. graveolens* were lost before 1930. The coastal distribution shows a slight decline since the 1962 *Atlas*, especially in Sussex.

Eurosiberian Southern-temperate element; widely naturalised outside its native range.

References: Atlas (158c), Hultén & Fries (1986), Meusel *et al.* (1978), Tutin (1980), Zohary & Hopf (2000).

M. SOUTHAM

Apium nodiflorum Fool's-water-cress

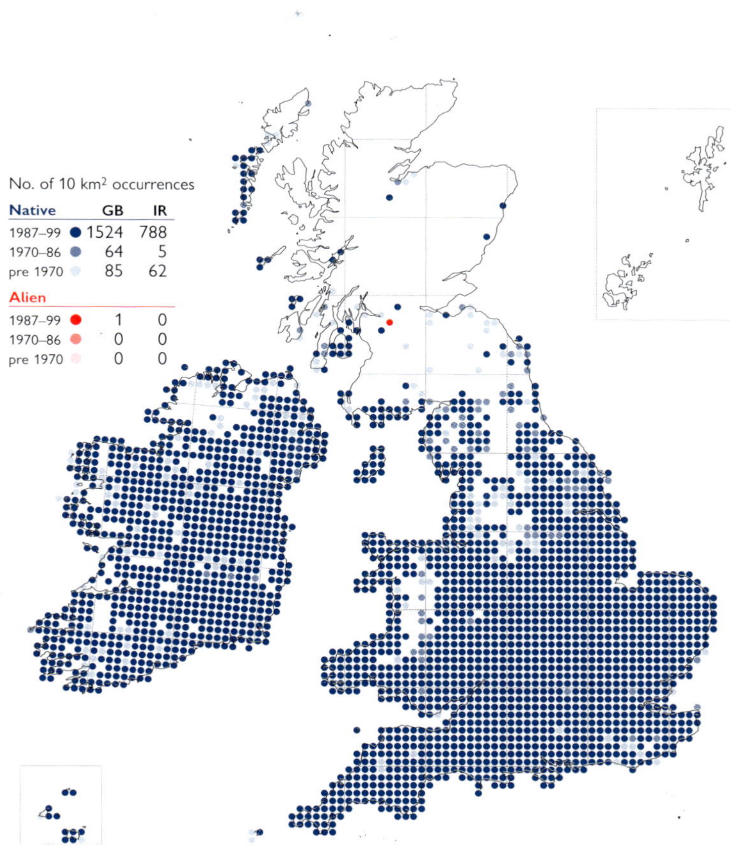

Native	GB	IR
1987–99	1524	788
1970–86	64	5
pre 1970	85	62
Alien		
1987–99	1	0
1970–86	0	0
pre 1970	0	0

A perennial herb, found in shallow water in streams, ditches, swamps and marshes, and on seasonally exposed mud at the edges of ponds, lakes, rivers and canals, sometimes scrambling into nearby vegetation. It is characteristic of nutrient-enriched sites. Generally lowland, but reaching 335 m E. of Shap (Westmorland).

Native (change –0.31). The distribution of this species is stable. It was mapped as 'all records' in the 1962 *Atlas*. There has been confusion between this species and *Berula erecta*, particularly in S. Scotland where some of the older records may be erroneous.

Eurosiberian Southern-temperate element; widely naturalised outside its native range.

References: Atlas (158d), Grime *et al.* (1988), Preston & Croft (1997), Tutin (1980).

M. SOUTHAM

Apium repens Creeping Marshwort

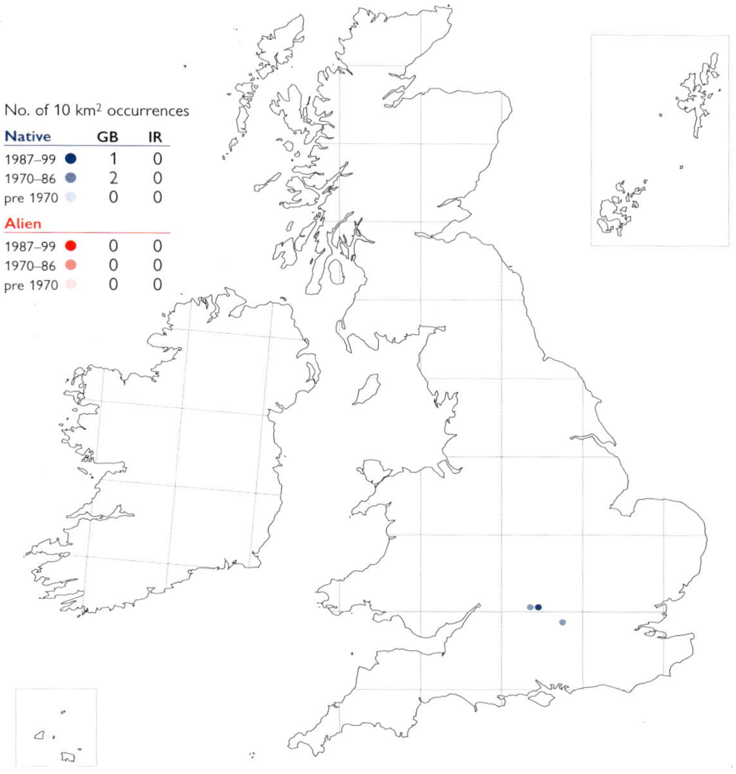

A creeping perennial herb of damp meadows and shallow water in ditches and ponds, generally at sites subject to winter flooding. At Port Meadow (Oxon) it grows in disturbed, shortly-grazed neutral grassland and adjacent open soil, where trampling by horses and cattle keeps the habitat open. Seed is produced by plants that escape grazing. Lowland.

Native. The distribution of *A. repens* has long been uncertain because of confusion with morphologically similar forms of *A. nodiflorum* and hybrids between them. Genetic studies in the 1990s have confirmed its presence at Port Meadow. All records outside Oxfordshire require confirmation.

European Temperate element.

References: Atlas (159a), Grassly *et al.* (1996), Hultén & Fries (1986), Tutin (1980), Wigginton (1999).

M. F. WATSON

Apium inundatum Lesser Marshwort

Native	GB	IR
1987–99	431	251
1970–86	142	12
pre 1970	360	88
Alien		
1987–99	1	0
1970–86	0	0
pre 1970	0	0

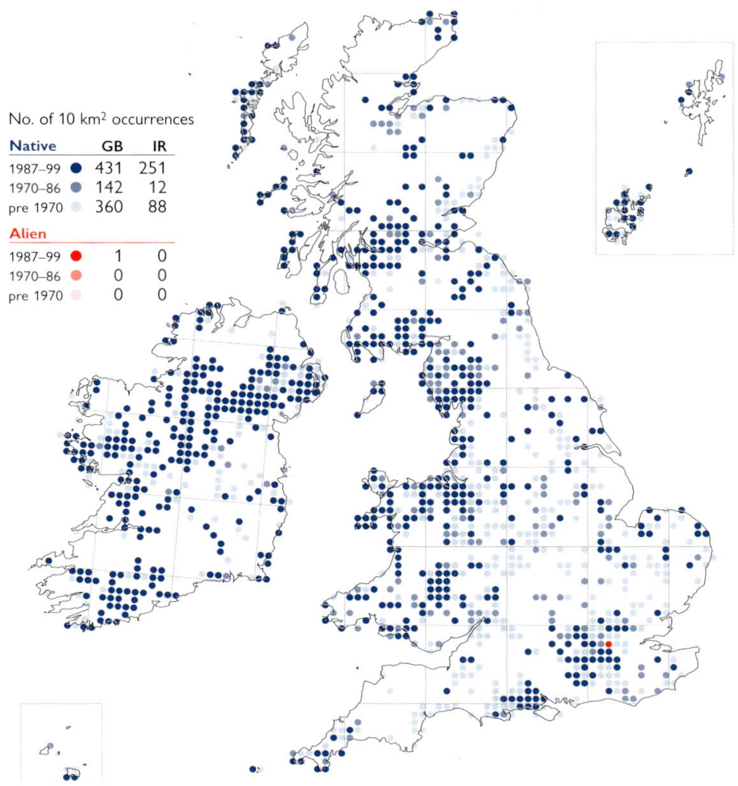

This perennial herb occurs in permanent shallow water in streams, ditches, ponds, canals and backwaters, and in sites which are subject to periodic desiccation, such as the edges of lakes, pools, reservoirs and dune-slacks. It is confined to oligotrophic or mesotrophic habitats, and most sites are base-poor. Reproduction is by seed. 0–500 m (Cronkley Fell, N.W. Yorks.).

Native (change –0.54). The considerable decline of *A. inundatum* in Britain, which has continued since the 1962 *Atlas*, is the result of the destruction of shallow water bodies, drainage and eutrophication. Trends in Ireland are difficult to assess.

Suboceanic Temperate element.

References: Atlas (159b), Hultén & Fries (1986), Meusel *et al.* (1978), Preston & Croft (1997), Tutin (1980).

M. SOUTHAM

Apium inundatum × *A. nodiflorum* (*A.* × *moorei*)

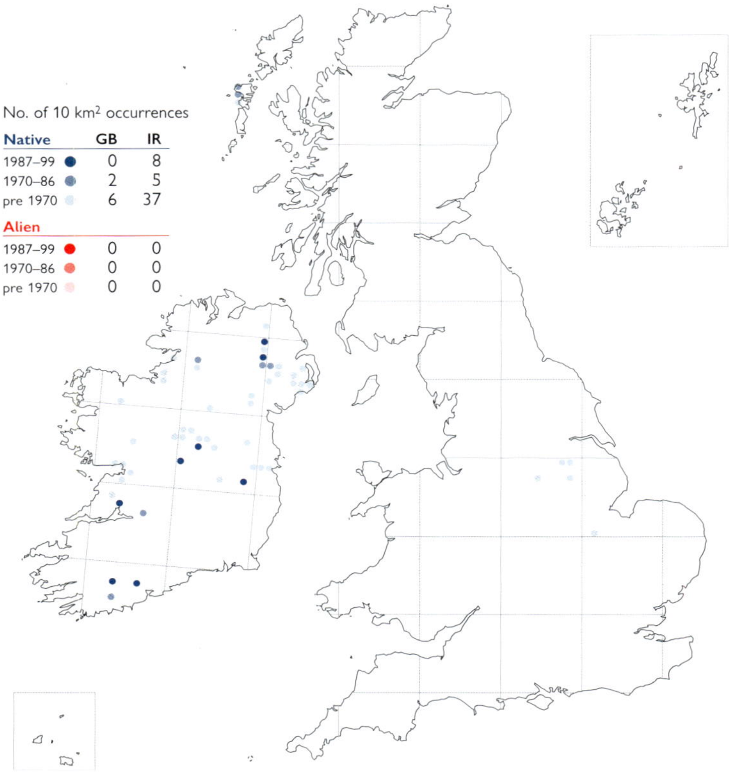

No. of 10 km² occurrences

Native	GB	IR
1987–99 ●	0	8
1970–86 ●	2	5
pre 1970 ●	6	37
Alien		
1987–99 ●	0	0
1970–86 ●	0	0
pre 1970 ●	0	0

This perennial sterile hybrid is found in marshes and at the sides of lakes, rivers, streams and canals. Lowland.

Native. *A.* × *moorei* is apparently very rare in Britain, but widely scattered in Ireland. It is easily overlooked and is probably under-recorded.

Apparently only known from Britain and Ireland.

References: Atlas Supp. (42b), Preston & Croft (1997), Stace (1975), Tutin (1980).

M. SOUTHAM

Petroselinum crispum Garden Parsley

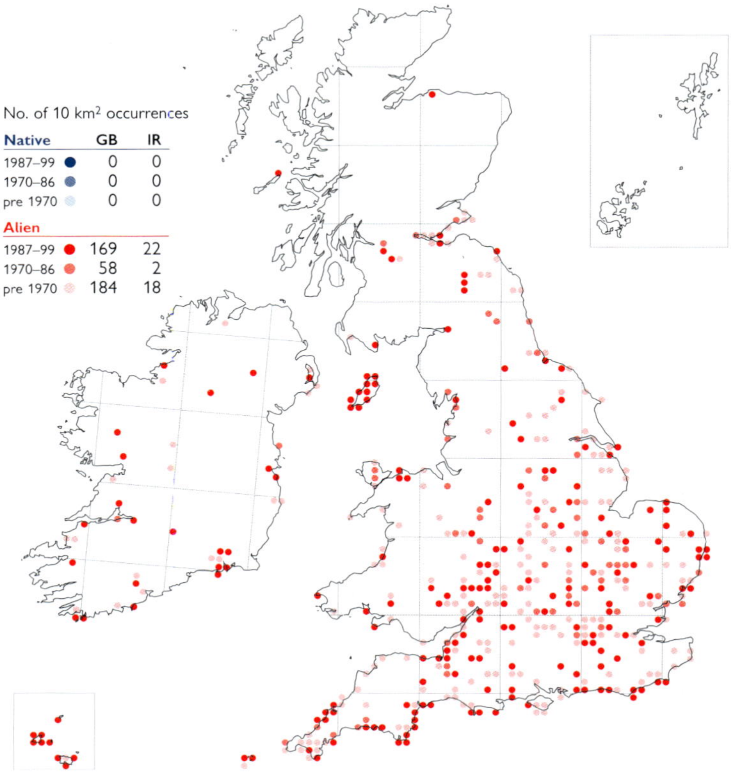

No. of 10 km² occurrences

Native	GB	IR
1987–99 ●	0	0
1970–86 ●	0	0
pre 1970 ●	0	0
Alien		
1987–99 ●	169	22
1970–86 ●	58	2
pre 1970 ●	184	18

A biennial herb which forms small but persistent colonies on cliffs, banks and waste ground in coastal areas. These colonies usually have uncrisped leaves. The crisped form which is commonly grown in gardens occurs as a casual close to habitation both near the sea and inland. Lowland.

Archaeophyte (change –0.34). *P. crispum* has been cultivated in British gardens since at least 995 (Harvey, 1981). Although most records are casual, it is very persistent in many of its coastal sites. It is difficult to assess any changes in its distribution.

A cultivated species of uncertain origin which is now widely naturalised in Europe and other continents.

References: Atlas (159c), Bolòs & Vigo (1990), Tutin (1980).

M. SOUTHAM

Petroselinum segetum Corn Parsley

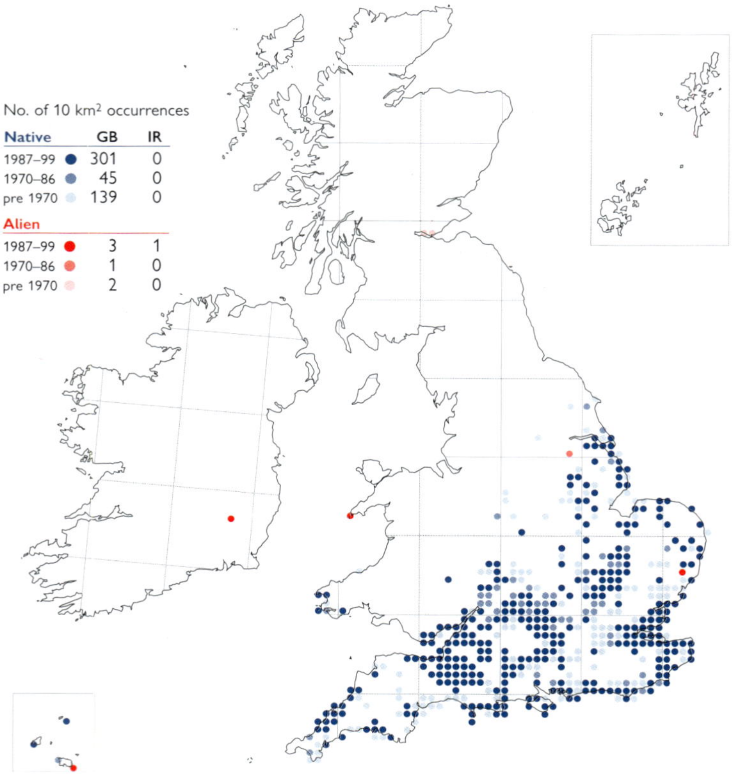

No. of 10 km² occurrences

Native	GB	IR
1987–99 ●	301	0
1970–86 ●	45	0
pre 1970 ●	139	0
Alien		
1987–99 ●	3	1
1970–86 ●	1	0
pre 1970 ●	2	0

This slender biennial of well-drained calcareous soils on clay or chalk is found on arable fields margins, on grassy banks, roadsides, railway banks, river banks, by sea walls, in drained estuarine marshes, on rough waste ground and occasionally as a garden weed. Lowland.

Native (change +0.12). *P. segetum* was thought to be declining at the time of the 1962 *Atlas*. Whilst it has undoubtedly been lost from some 10-km squares, many more new records have been added since 1962.

Suboceanic Southern-temperate element.

References: Atlas (159d), Bolòs & Vigo (1990), Tutin (1980).

M. SOUTHAM

Sison amomum Stone Parsley

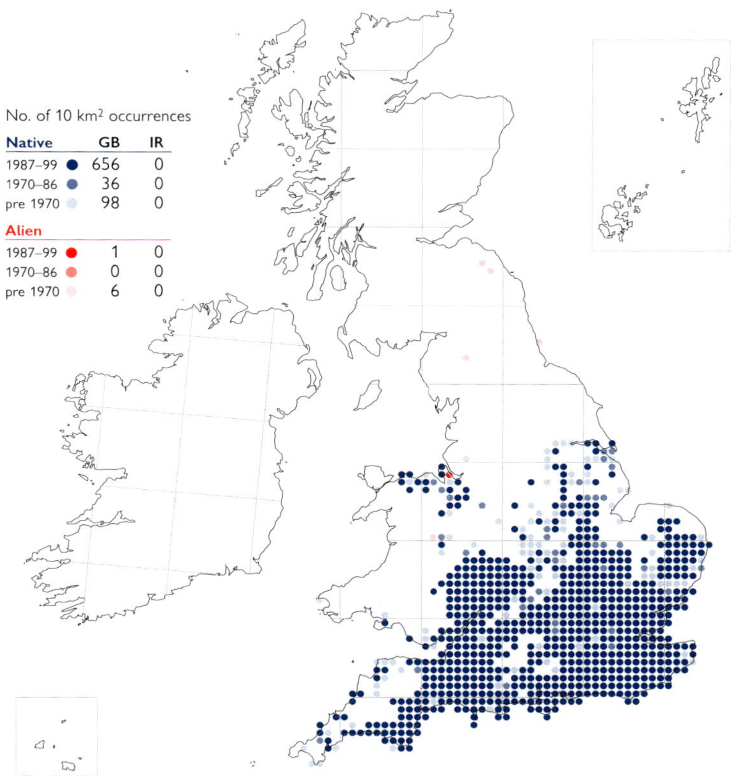

A biennial herb of sticky clay and better drained neutral to calcareous soils, found mainly in hedgerows, on banks, rough scrubby grassland, waysides and disturbed waste ground. Lowland.

Native (change –0.19). There has been no appreciable change in the 10-km distribution of *S. amomum* since the 1962 *Atlas.*

Submediterranean-Subatlantic element.

References: Atlas (160a), Bolòs & Vigo (1990), Tutin (1980).

M. SOUTHAM

Cicuta virosa Cowbane

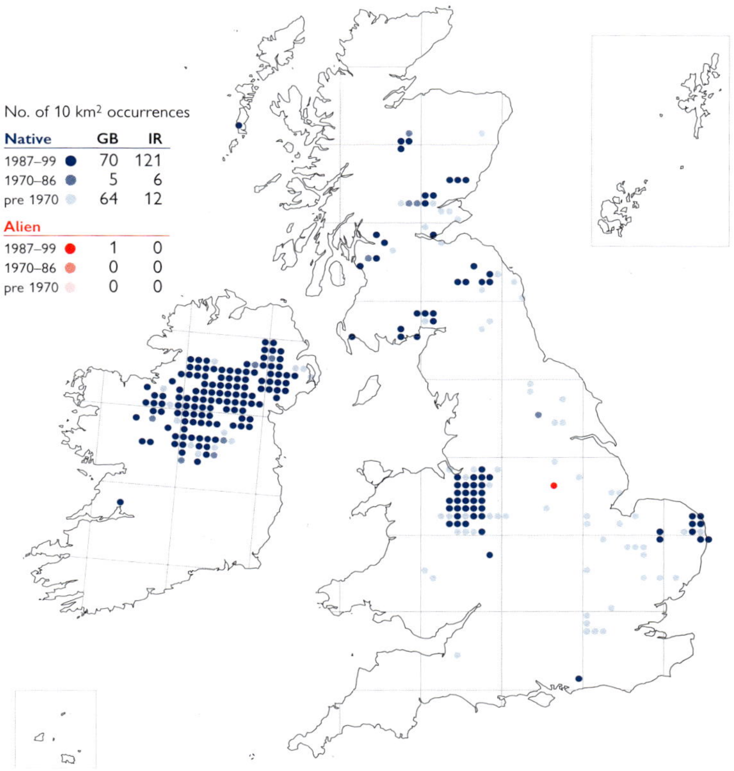

This perennial herb grows in shallow water on the margins of standing or slowly flowing water, including lakes, ponds, rivers, streams, ditches and canals, or in deeper water on floating mats of vegetation. It also occurs in tall-herb fen, in marshy pasture and on damp mud. Lowland.

Native (change +0.55). Many losses of *C. virosa* are probably due to drainage, but it may have been eradicated from some pastures because of its toxicity to cattle. Some losses were very early: the last record in Cambridgeshire, for example, was made in the 18th century. However, the distribution now seems largely stable.

Eurasian Boreo-temperate element, with a continental distribution in W. Europe.

References: Atlas (160b), Hultén & Fries (1986), Meusel *et al.* (1978), Mulligan & Munro (1981), Stewart *et al.* (1994), Tutin (1980).

M. F. WATSON

Ammi majus Bullwort

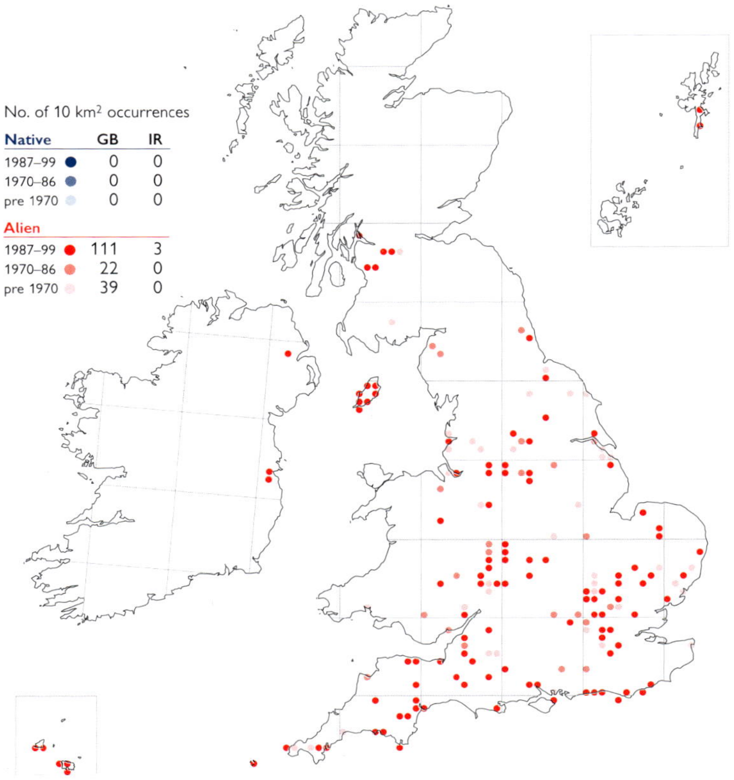

An annual, found in parkland and neglected gardens, on spoil-tips and roadsides. It is a wool-shoddy or bird-seed alien which is usually casual. Lowland.

Neophyte. Introduced by 1551 from S. Europe, *A. majus* was first recorded in the wild in 1845 (Sussex). Records of it are increasing, but only as a casual. It was notably frequent in the London area in 1993 (Wurzell, 1994b).

Native of S. Europe, N. Africa and S.W. Asia.

References: Bolòs & Vigo (1990), Tutin (1980).

M. SOUTHAM

Falcaria vulgaris Longleaf

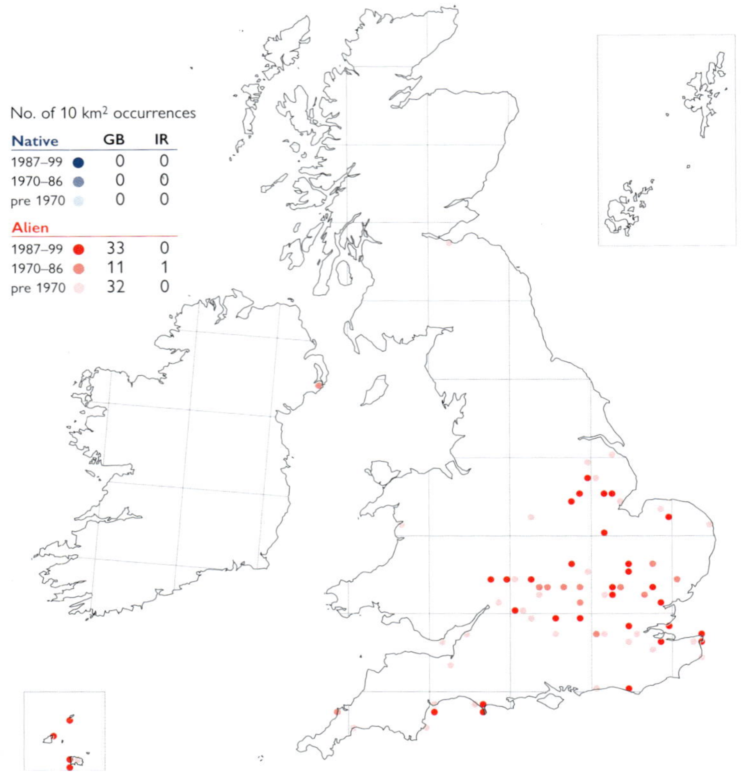

This rhizomatous perennial herb has been recorded from a wide variety of habitats, including pasture, arable fields, roadsides, cliff-top grassland, scrub, chalk quarries, gravel-pits, railway ballast, river banks and waste ground. Viable seed is produced in Britain only in warm summers, and large colonies probably result from rhizomatous growth. Lowland.

Neophyte. *F. vulgaris* was introduced as a garden plant in 1726 and was first recorded from the wild in 1858. Some populations are well-naturalised, but others are only transient. The distribution of naturalised populations appears to be stable.

A Eurosiberian Temperate species, absent as a native from most of W. Europe.

References: Hultén & Fries (1986), Meusel *et al.* (1978), Tutin (1980).

M. SOUTHAM

Carum carvi Caraway

Native	GB	IR
1987–99	0	0
1970–86	0	0
pre 1970	0	0
Alien		
1987–99	46	4
1970–86	31	1
pre 1970	228	25

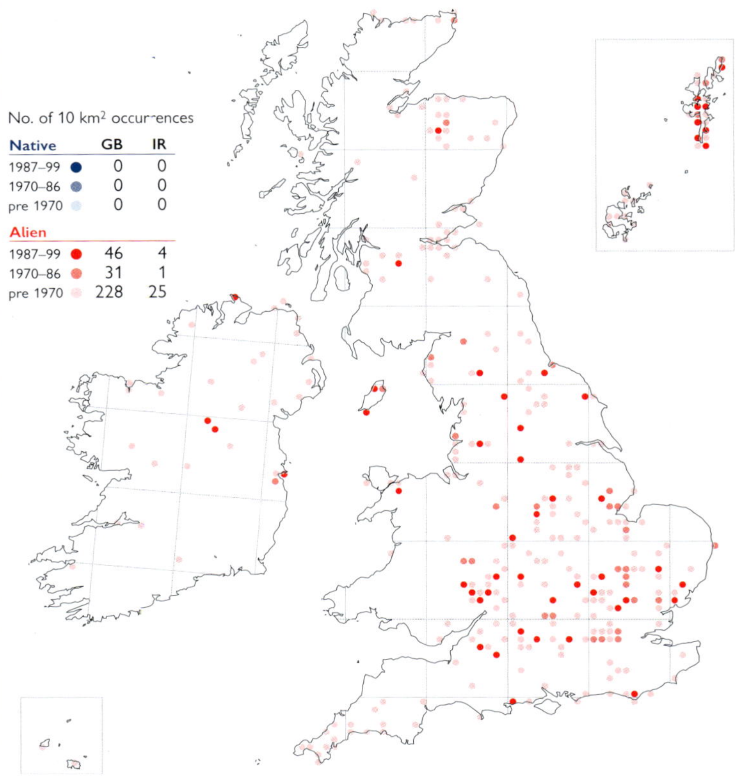

A monocarpic perennial herb, found naturalised in meadows, on sand dunes, roadsides and railway banks, and as a casual in waste places and on rubbish tips. Generally lowland, but formerly reaching 425 m at Blair Atholl (E. Perth).

Archaeophyte (change –2.22). *C. carvi* was introduced from Europe before 1375 (Harvey, 1981). It is well-naturalised in Shetland, but is uncommon elsewhere, being much less frequently cultivated than formerly. Many occurrences are casual, presumably arising from fruits imported as flavouring agents.

Apparently native of Europe, W. & C. Asia but native range obscured by spread in cultivation; now widespread in boreal and temperate zones in the N. hemisphere.

References: Atlas (160d), Hultén & Fries (1986), Meusel *et al.* (1978), Tutin (1980).

M. SOUTHAM

Carum verticillatum Whorled Caraway

Native	GB	IR
1987–99	246	27
1970–86	22	0
pre 1970	29	15
Alien		
1987–99	1	0
1970–86	0	0
pre 1970	1	0

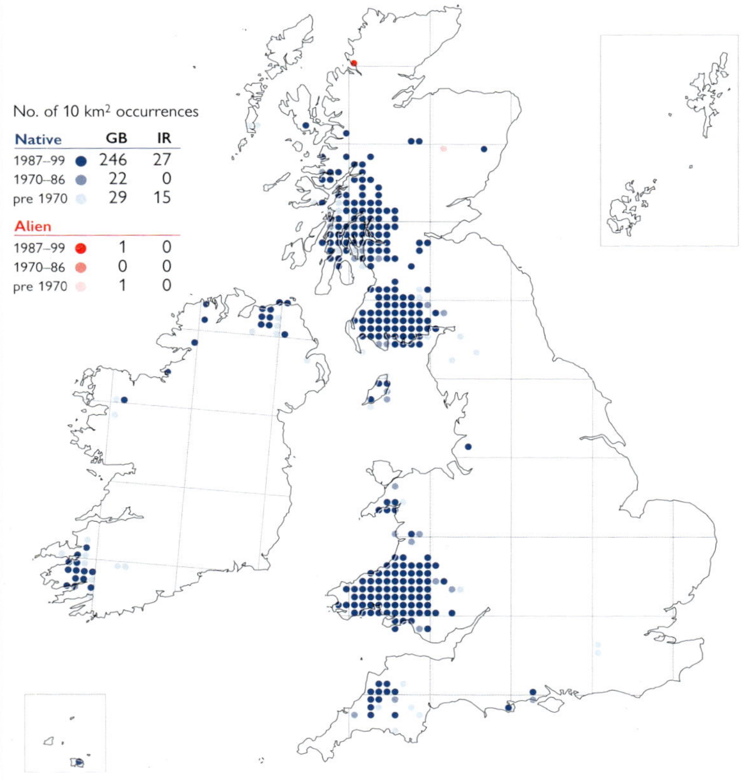

A perennial, calcifuge herb of marshes, streamsides, damp meadows, rushy pastures and on wet hillsides with a pronounced soligenous influence. 0–425 m (Llyn Berwyn, Cards.) and 440 m in S. Kerry.

Native (change +0.22). The distribution of this species is generally stable. Since the 1962 *Atlas*, some sites have been lost, including the remaining Surrey site in 1967. However, new sites have been discovered, particularly in S. Wales, N. & W. Scotland and Ireland. In W. Scotland, and perhaps elsewhere, it may be becoming more frequent in pastures where drainage is increasingly neglected.

Oceanic Southern-temperate element.

References: Atlas (160c), Blackstock *et al.* (1991), Kay & John (1994), Marren (1981), Meusel *et al.* (1978), Tutin (1980).

M. F. WATSON

Selinum carvifolia Cambridge Milk-parsley

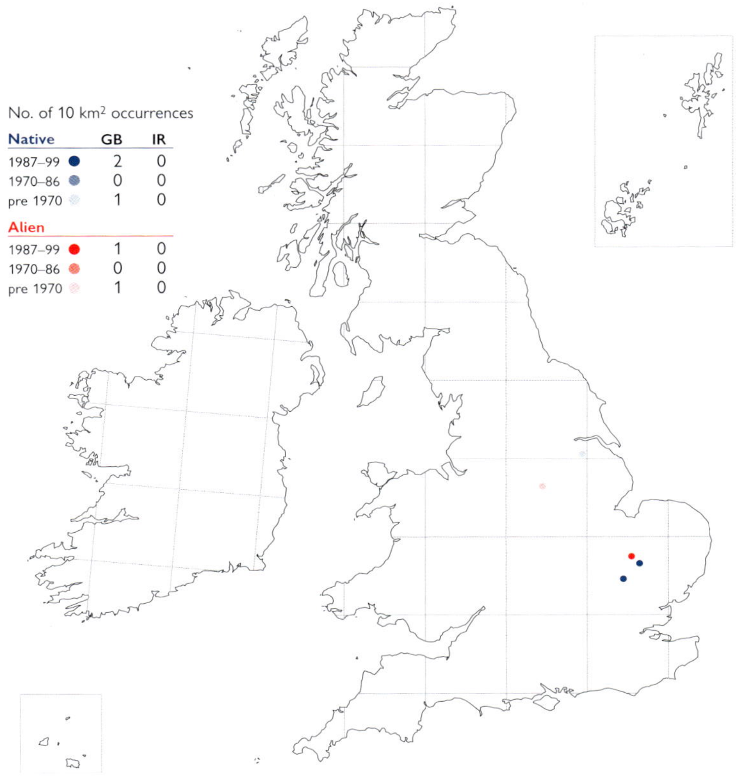

No. of 10 km² occurrences

Native	GB	IR
1987–99 ●	2	0
1970–86 ●	0	0
pre 1970	1	0
Alien		
1987–99 ●	1	0
1970–86 ●	0	0
pre 1970	1	0

A perennial herb of fens, damp meadows and rough-grazed marshy pasture on calcareous peaty soils or fen peat overlying chalk. It does not grow on the wettest ground in fens, but prefers slightly better-drained fringe areas and low banks. Lowland.

Native. *S. carvifolia* was last recorded in Lincolnshire in 1931 and in Nottinghamshire by 1952. The species is now confined to three statutorily protected sites in Cambridgeshire, where the plant is thriving.

European Temperate element.

References: Atlas (166a), Hultén & Fries (1986), Meade (1989), Meusel *et al.* (1978), O'Leary (1989), Tutin (1980), Wigginton (1999).

M. F. WATSON

Ligusticum scoticum Scots Lovage

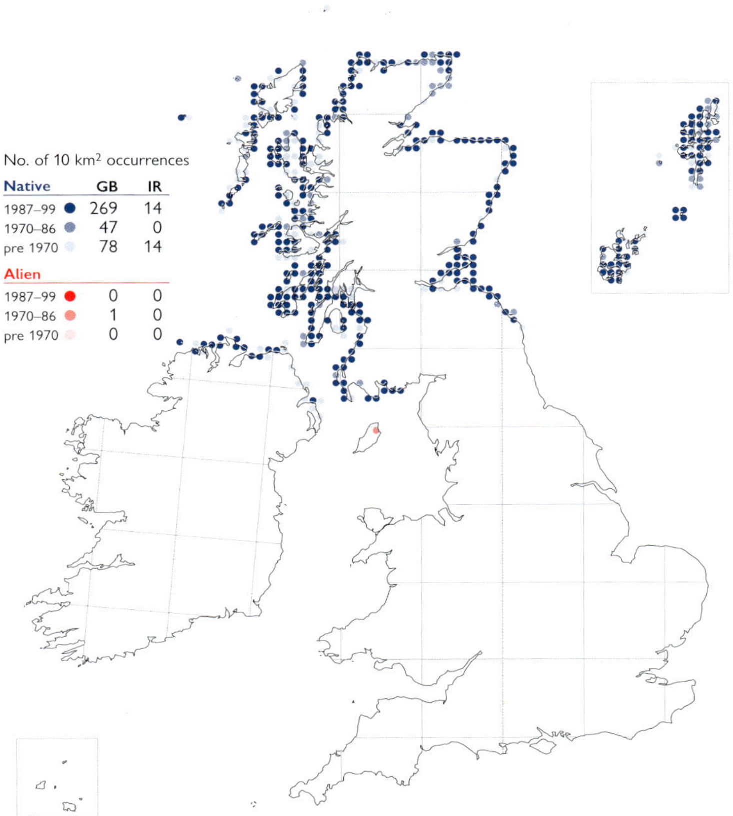

No. of 10 km² occurrences

Native	GB	IR
1987–99 ●	269	14
1970–86 ●	47	0
pre 1970	78	14
Alien		
1987–99 ●	0	0
1970–86 ●	1	0
pre 1970	0	0

A perennial herb of coastal rock crevices and free-draining skeletal soils by the sea. Habitats include cliffs, rocky shores and platforms, spray-drenched shingle, stabilised sand dunes and stone sea-defence walls. The seeds float and retain some viability even after a year in sea water. Lowland.

Native (change –0.29). The distribution of *L. scoticum* in Scotland is little changed, but there are losses in N. Ireland, some since the 1962 *Atlas*. Drought sensitivity and the requirement for cold, wet conditions for germination may be important in limiting the southern range of this species in our area.

European Boreo-arctic Montane element; a coastal species also found in E. Asia and N. America.

References: Atlas (166b), Curtis & McGough (1988), Hultén & Fries (1986), Palin (1988), Tutin (1980).

M. F. WATSON

Angelica sylvestris Wild Angelica

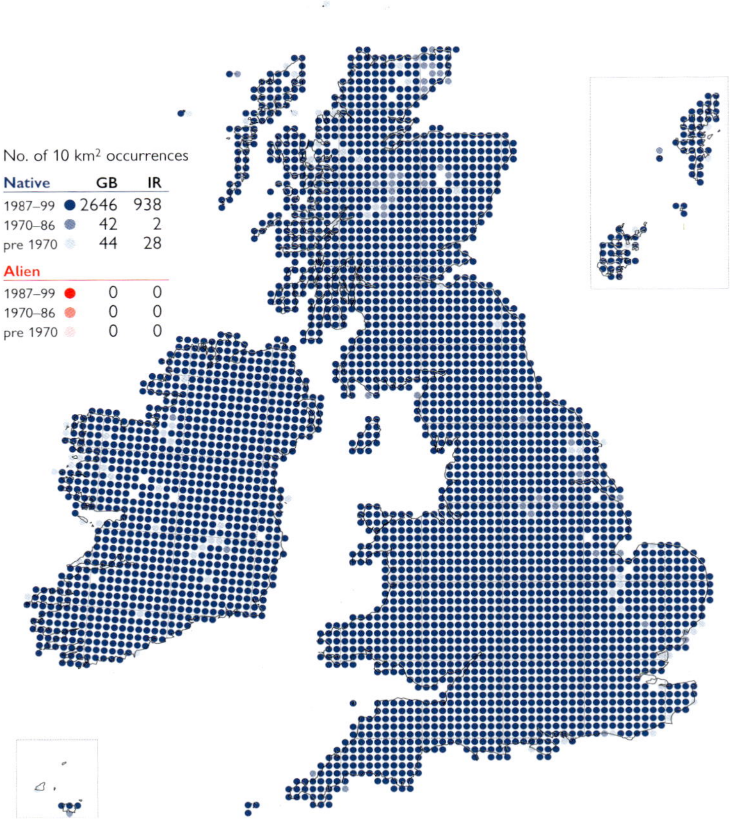

No. of 10 km² occurrences

Native	GB	IR
1987–99 ●	2646	938
1970–86 ●	42	2
pre 1970	44	28
Alien		
1987–99 ●	0	0
1970–86 ●	0	0
pre 1970	0	0

A perennial herb, occurring on base-enriched soils in a wide variety of habitats, including damp woods and carr, damp neutral grassland, marshes, mires, swamps and tall-herb fens, sea-cliffs, ungrazed montane grassland and mountain ledges. Reproduction is by seed. 0–855 m (Helvellyn, Cumberland).

Native (change +0.12). The distribution of *A. sylvestris* is stable.

Eurosiberian Boreo-temperate element.

References: Atlas (166c), Grime *et al.* (1988), Hultén & Fries (1986), Meusel *et al.* (1978), Tutin (1980).

M. SOUTHAM

Angelica archangelica Garden Angelica

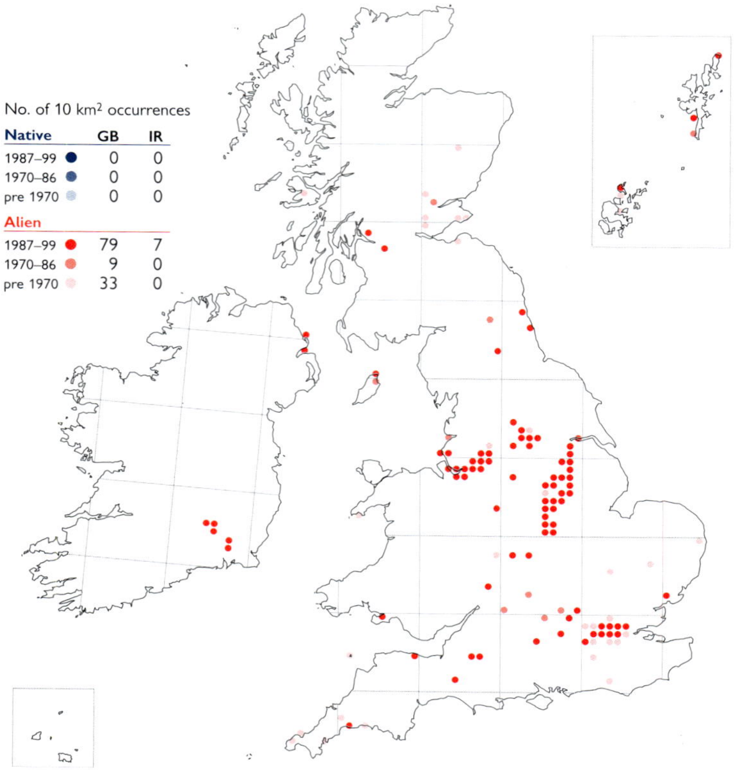

A robust monocarpic perennial herb, naturalised on riversides, roadsides and in waste places. It is otherwise seen only in small numbers in gardens or as an escaped casual. Lowland.

Neophyte (change +1.34). *A. archangelica* was cultivated in Britain by 1568 and was first recorded from the wild in *c.* 1700 (Middlesex). It is now rarely grown and its range has not spread significantly since the 1962 *Atlas*, other than as a casual, but it has increased in frequency within its core areas.

A Eurasian Boreal-montane species; a rare example of a northern species which has spread south in Europe as a result of cultivation.

References: Atlas (166d), Hultén & Fries (1986), Meusel *et al.* (1978), Tutin (1980), Vaughan & Geissler (1997).

M. SOUTHAM

Levisticum officinale Lovage

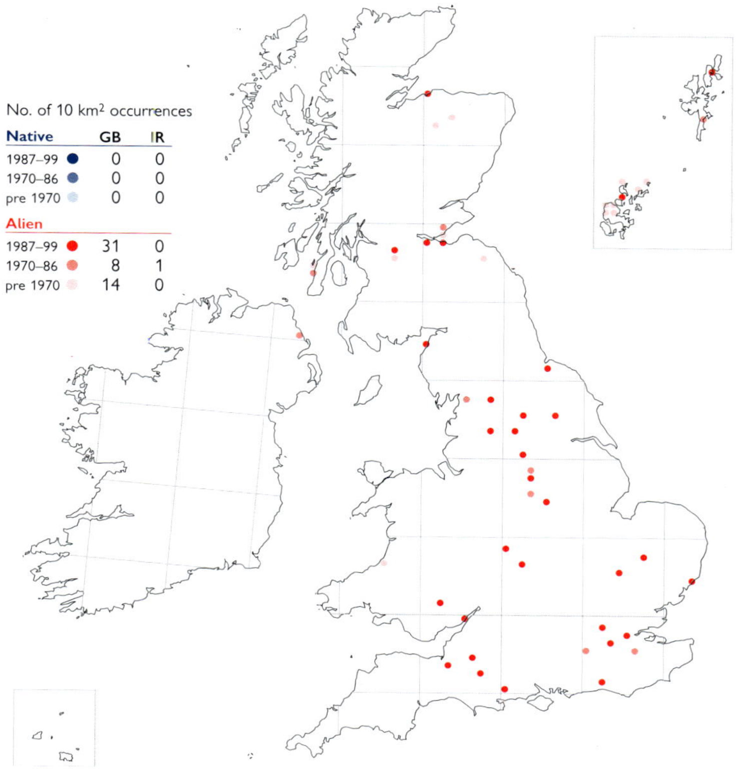

A clump-forming perennial herb, much grown as a culinary herb and found naturalised on rough ground and by walls and paths. It also occurs as a relic of cultivation. Lowland.

Neophyte. *L. officinale* was being grown in gardens by 995 (Harvey, 1981) and was first recorded from the wild in 1883 (Orkney). Its overall frequency is probably stable.

Native of Iran and Afghanistan; widely naturalised in Europe, N. America and elsewhere through its use as a herb.

References: Hultén & Fries (1986), Tutin (1980).

M. SOUTHAM

Peucedanum officinale Hog's Fennel

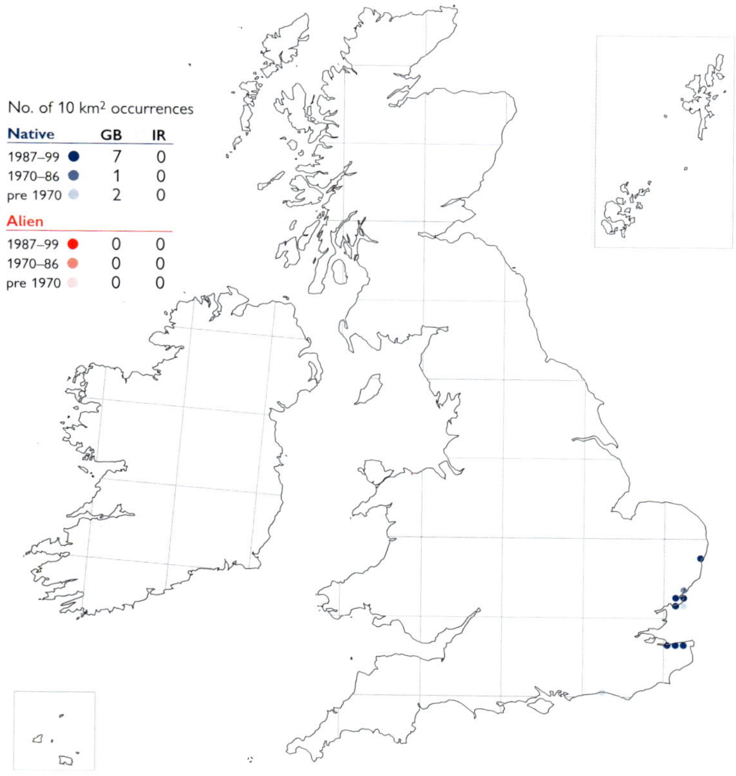

This perennial herb of coastal grassland occurs in rough grassland and scrubby places adjoining saltmarsh or brackish grazing marsh, on creek sides and on sea walls; also on waste ground and, rarely, on roadsides. Plants spread by rhizomatous growth, but seed ripens only in warm years. Lowland.

Native (change +0.29). This species was recorded in 1666 in Sussex, but has not been seen there subsequently. It appears to have become more frequent in Essex since 1950, though some outlying populations have been lost. Its discovery in 1990 at Southwold, E. Suffolk, is a significant extension of its range.

European Southern-temperate element.

References: Atlas (167a), Meusel *et al.* (1978), Randall & Thornton (1996), Tutin (1980), Wigginton (1999).

M. SOUTHAM

Peucedanum palustre Milk-parsley

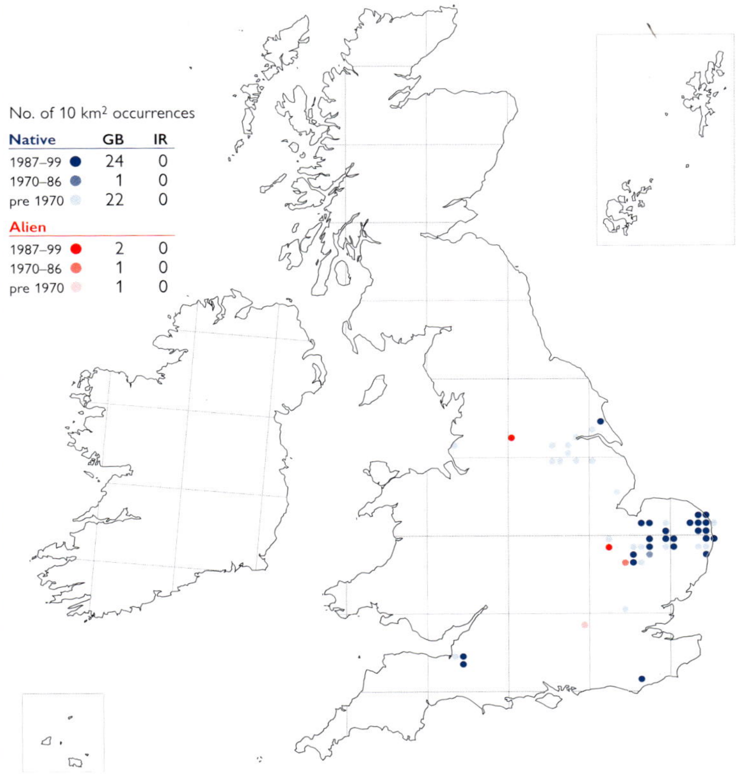

A biennial or short-lived perennial herb, mainly growing on permanently damp peat, often in sites flooded in winter. It is most characteristic of tall-herb fen, being found in both cut and uncut stands. It can survive in fen scrub and alder-carr, and it occurs rarely in marshes and damp pasture. Lowland.

Native (change –0.07). Many sites of *P. palustre* were lost in the 19th century, following drainage and reclamation. Since 1930, further sites have been lost in East Anglia, principally to scrub invasion. Most extant sites are in nature reserves.

Eurosiberian Boreo-temperate element, with a continental distribution in W. Europe.

References: Atlas (167a), Harvey & Meredith (1981), Hultén & Fries (1986), Meredith & Grubb (1993), Meusel *et al.* (1978), Stewart *et al.* (1994), Tutin (1980).

M. F. WATSON

Peucedanum ostruthium Masterwort

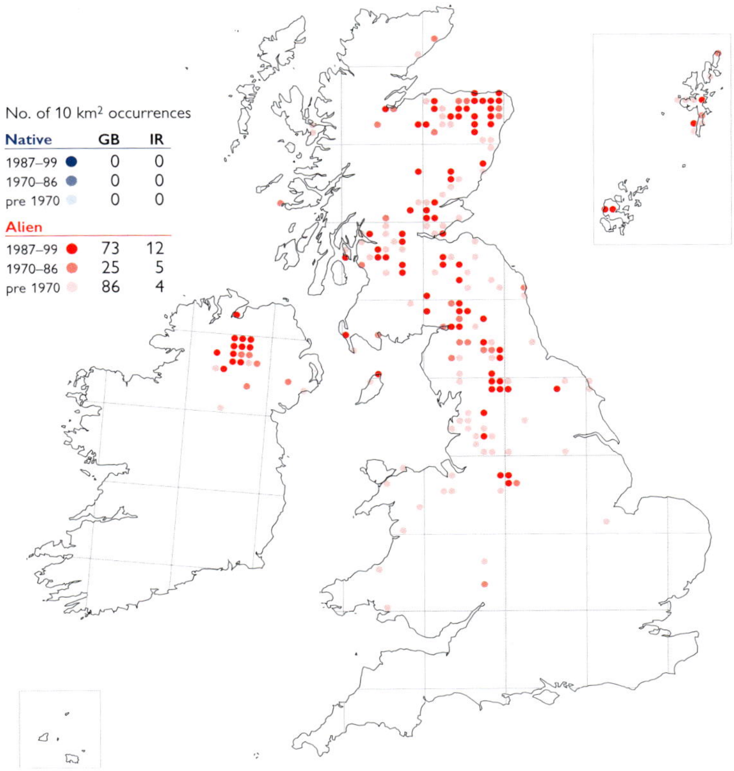

A perennial herb naturalised in moist or damp grassy areas, including marshy pasture, on hillsides and by streams and rivers, and sometimes established around farm buildings. Generally lowland, but reaching 385 m at Langdon Beck (Co. Durham).

Archaeophyte (change +0.03). *P. ostruthium* was formerly cultivated as a pot-herb or for veterinary purposes. It is very persistent and shows no sign of a change in its distribution since the 1962 *Atlas*.

Native of the mountains of C. & S.W. Europe.

References: Atlas (167b), Hultén & Fries (1986), Tutin (1980).

M. F. WATSON

Pastinaca sativa Wild Parsnip

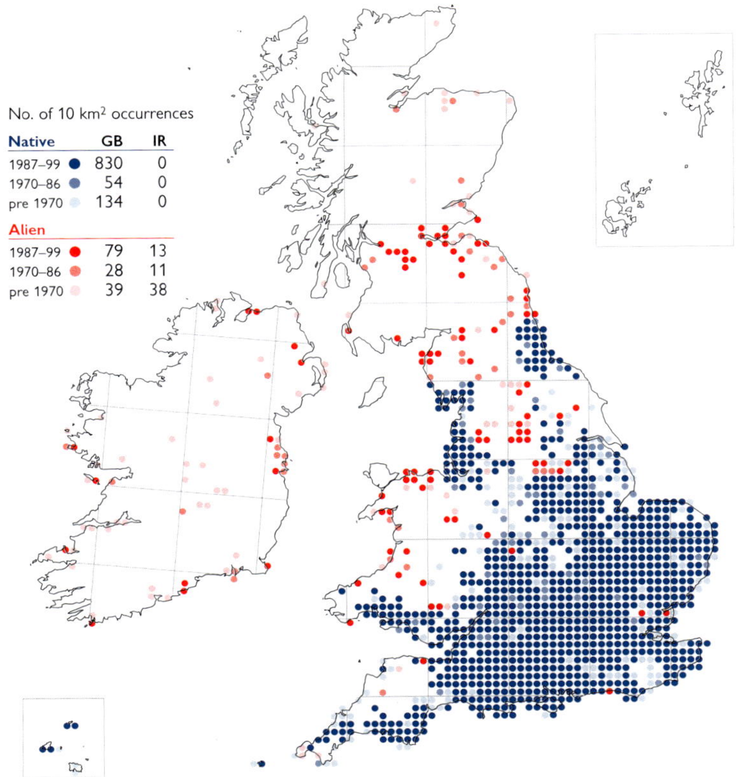

This biennial herb occurs in neutral and calcareous grassland, especially in chalk and limestone districts. It is found in rank swards on downland, on roadsides, railway banks, and rough and uncultivated land. 0–380 m (Stainmore, Westmorland).

Native (change –0.39). There has been little appreciable change in the distribution of *P. sativa* since the 1962 *Atlas*, although there have been losses at the edges of its range. It may be increasing in some areas along new roads (Halliday, 1997; French *et al.*, 1999).

Eurosiberian Temperate element; widely naturalised outside its native range.

References: Atlas (167c), Hultén & Fries (1986), Meusel *et al.* (1978), Tutin (1980), Zohary & Hopf (2000).

M. SOUTHAM

Heracleum sphondylium Hogweed

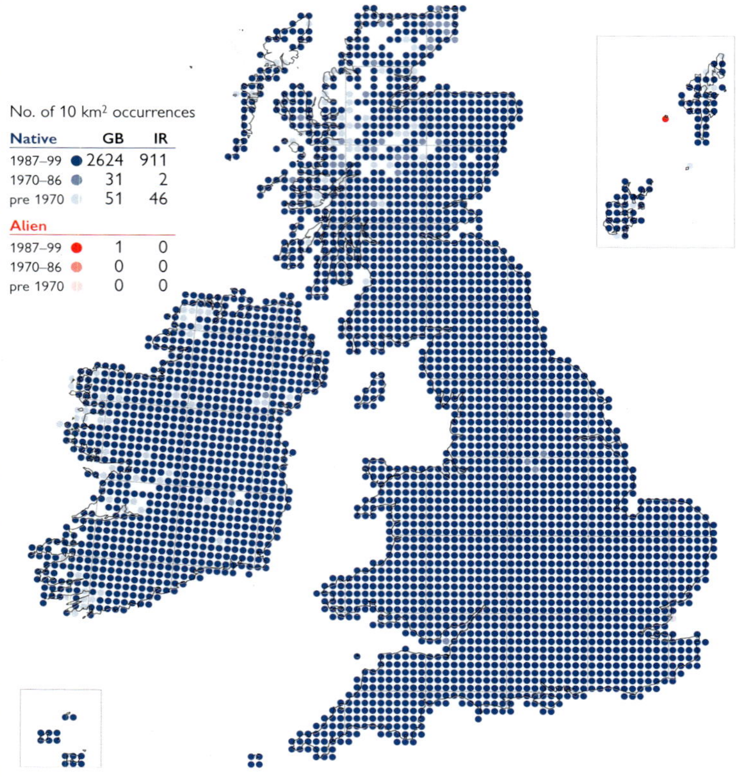

A robust perennial herb of dry or moist, neutral to calcareous soils. It has a wide habitat range, including rough and disturbed grassland, especially on roadsides and trackways, woodland rides, scrub, river banks, stabilised dunes, coastal cliffs, montane tall-herb vegetation and waste ground. 0–1005 m (Breadalbanes, Mid Perth).

Native (change +0.08). There has been no change in the distribution of *H. sphondylium* since the 1962 *Atlas*.

Eurasian Boreo-temperate element; widely naturalised outside its native range.

References: Atlas (167d), Grime *et al.* (1988), Hultén & Fries (1986), Meusel *et al.* (1978), Sheppard (1991), Tutin (1980).

M. F. WATSON

Heracleum mantegazzianum Giant Hogweed

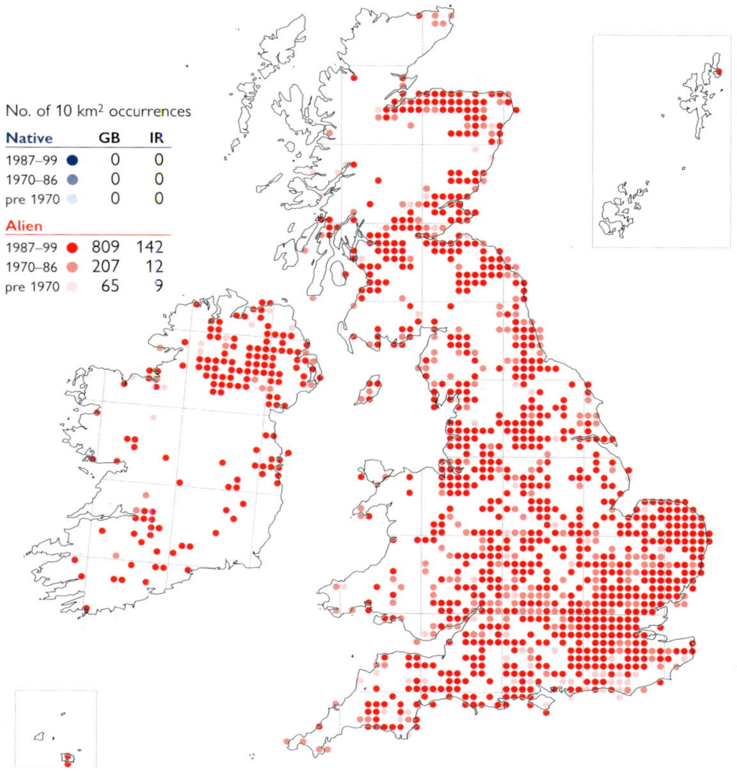

A very large perennial herb occurring in derelict gardens, neglected urban places and waste ground, on rubbish tips, roadsides and by streams and rivers, forming large colonies if allowed to prosper. It spreads by seed, which is prolifically produced. Lowland.

Neophyte (change +2.09). *H. mantegazzianum* was introduced to gardens as a monumental curiosity by 1820. It was deliberately planted by rivers and ponds and was first recorded in the wild in 1828 (Cambs.). Its spread has been rapid since the 1962 *Atlas*, despite the fact that the plant has been the subject of control measures now that its capacity to cause dermatitis has become known.

Native of S.W. Asia.

References: Atlas (168a), Tiley *et al.* (1996), Tutin (1980).

M. SOUTHAM

Tordylium maximum Hartwort

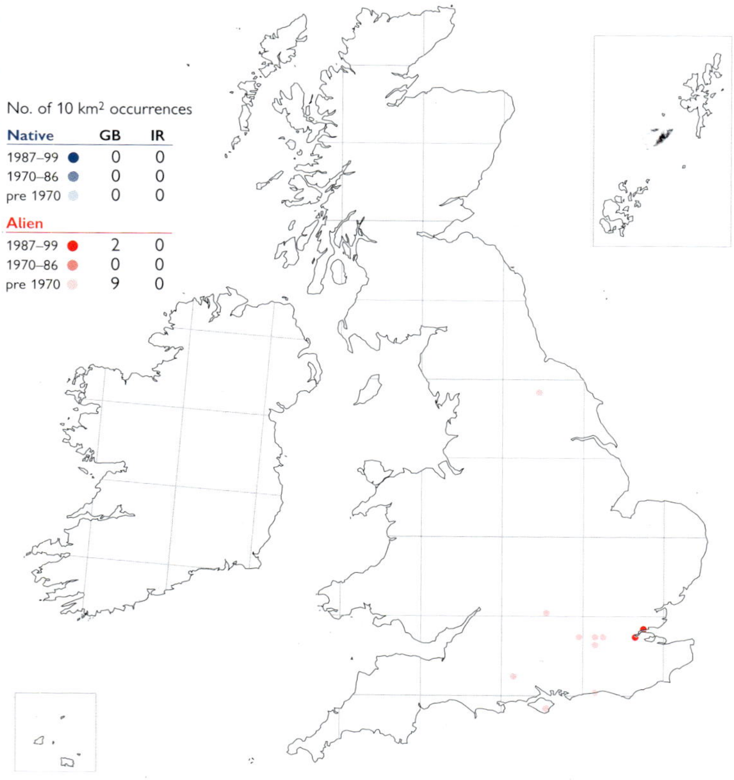

An annual or biennial herb of neutral grassland and in grassy thorn scrub, on clayey or alluvial soils. The sites are sheltered and S.-facing. Its seed is apparently short-lived. Lowland.

Neophyte. *T. maximum* was first recorded in Britain around 1670. It persisted in Middlesex until 1837, and was recorded at Tilbury (S. Essex), between 1875 and 1984. The current localities at Benfleet (S. Essex) were discovered in 1949 and 1966, where some populations have survived development and encroaching scrub, aided by seed collection and sowing *in situ*. *T. maximum* is treated here as a rare alien, but regarded as a possible native by Wigginton (1999).

Native from S. & S.C. Europe and N. Turkey to the Caucasus and N. Iran.

References: Bolòs & Vigo (1990), Tutin (1980).

D. A. PEARMAN

Torilis japonica Upright Hedge-parsley

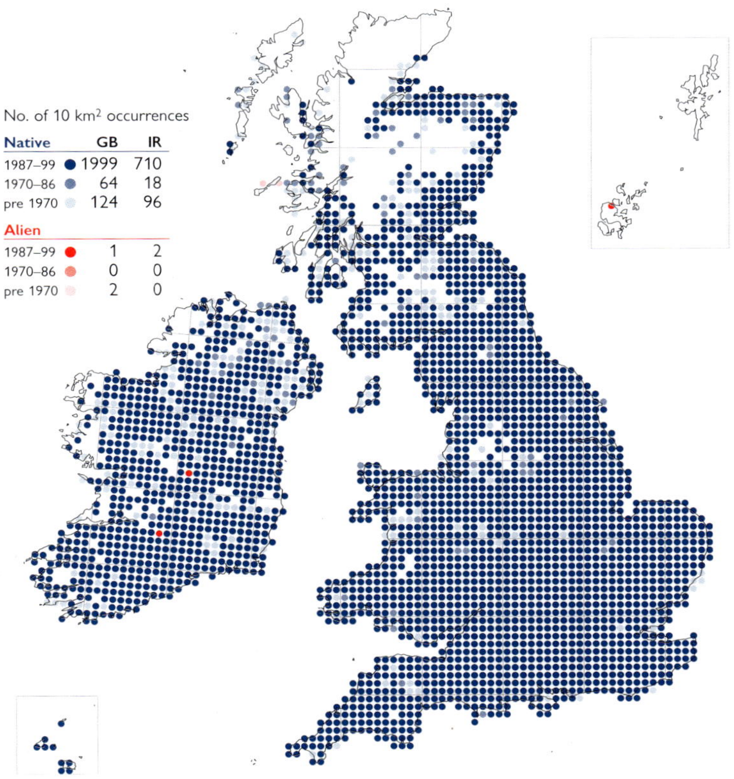

An annual, or rarely biennial, herb of dry neutral and basic soil, found in woodland margins, hedgerows, rough and rank grassland, and on roadside verges. 0–410 m (Craven Pennines, Mid-W. Yorks.).

Native (change –0.48). The distribution of *T. japonica* is little changed since the 1962 *Atlas*.

Eurasian Temperate element; widely naturalised outside its native range.

References: Atlas (155c), Grime *et al.* (1988), Hultén & Fries (1986), Meusel *et al.* (1978), Tutin (1980).

M. F. WATSON

Torilis arvensis Spreading Hedge-parsley

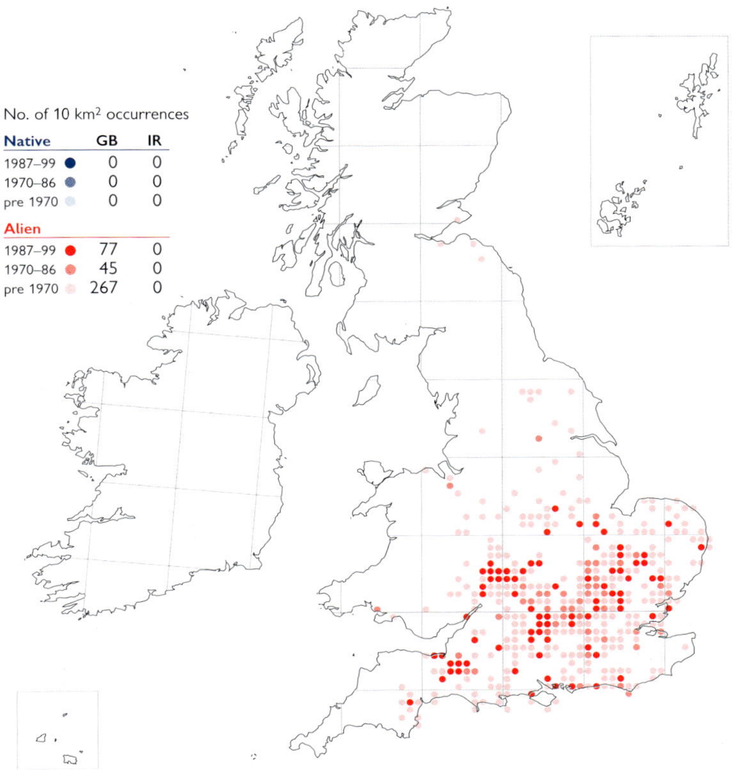

An annual, rarely biennial, herb, almost exclusively found on arable land in autumn-sown cereals, but sometimes in other arable crops; also on waste and disturbed ground. It is perhaps most frequent on calcareous clays, but is found on a wide range of soils, including sands and gravels. Lowland.

Archaeophyte (change –2.56). Once frequent, this species had already lost nearly half its sites by 1930, and since then its accelerating decline has been one of the most dramatic shown by any arable weed. It is a victim of intensive crop management, being vulnerable to herbicides and unable to compete in dense crop swards.

As an archaeophyte *T. arvensis* has a Eurosiberian Southern-temperate distribution.

References: Atlas (155d), Bolòs & Vigo (1990), Stewart *et al.* (1994), Tutin (1980).

M. F. WATSON

Torilis nodosa Knotted Hedge-parsley

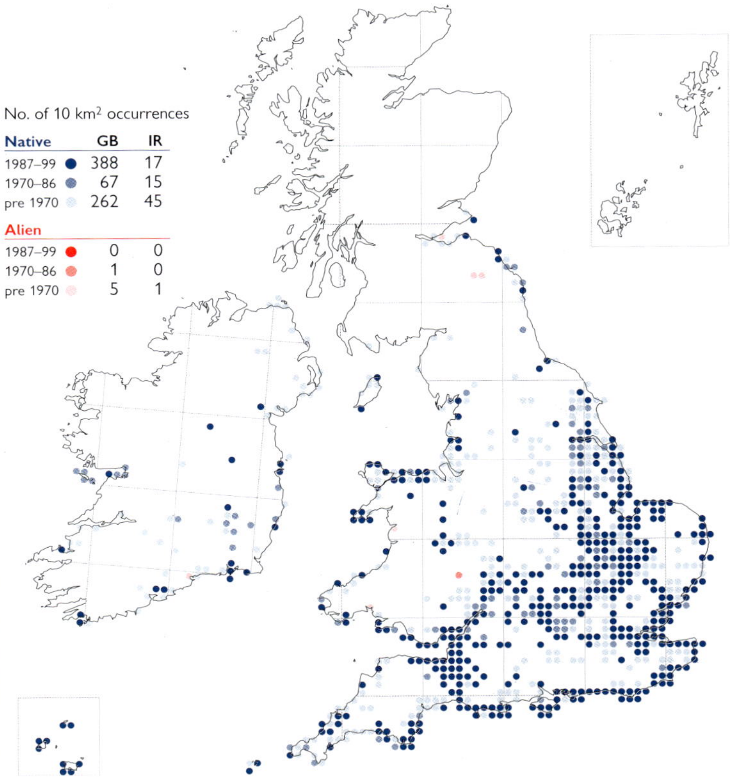

An annual found in a wide range of dry, sparsely vegetated habitats, including open grassland, sunny banks, sea walls, cliff-tops, arable fields, tracks and waste ground; occasionally in disused sand- and gravel-pits, and on rubbish tips. Lowland.

Native (change –0.36). *T. nodosa* has declined since the 1962 *Atlas* in many inland areas, although it can be locally very abundant (Green *et al.*, 1997). It is stable at most of its coastal sites, or even increasing, as in Dorset.

Mediterranean-Atlantic element, but naturalised north of its native range so that distribution is now Submediterranean-Subatlantic.

References: Atlas (156a), Meusel *et al.* (1978), Tutin (1980).

M. F. WATSON

Daucus carota Wild Carrot

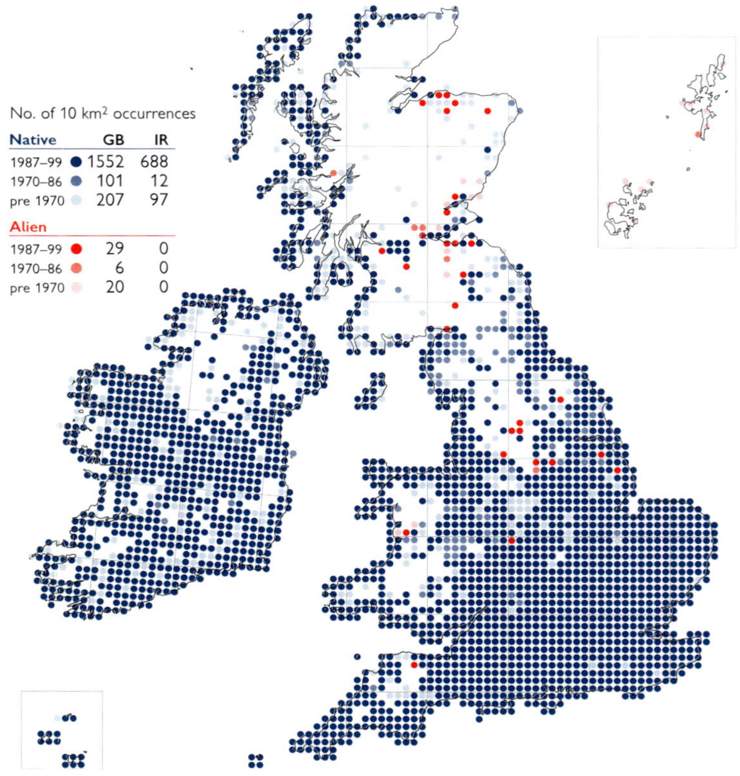

No. of 10 km² occurrences

Native		GB	IR
1987–99	●	1552	688
1970–86	●	101	12
pre 1970	●	207	97
Alien			
1987–99	●	29	0
1970–86	●	6	0
pre 1970	●	20	0

A biennial herb of fairly infertile, well-drained, often calcareous, soils. Habitats include disturbed or open turf on chalk downs, rough grassland on roadsides, waysides and railway banks, quarries, chalk- and gravel-pits, and waste ground. 0–400 m (Connor Hill, S. Kerry).

Native (change –0.59). The map suggests a decline in *D. carota* at the northern fringes of its range. Three subspecies are recognised in our area: subsp. *carota* (Wild Carrot) occurs throughout the range of the species, the cultivated subsp. *sativus* (Carrot) is a casual on tips and a relic of cultivation, and subsp. *gummifer* is mapped separately.

Eurosiberian Southern-temperate element, but widely naturalised so that distribution is now Circumpolar Southern-temperate.

References: Atlas (168b), Dale (1974), Grime *et al.* (1988), Hultén & Fries (1986), Meusel *et al.* (1978), Tutin (1980).

M. SOUTHAM

Daucus carota subsp. *gummifer* Sea Carrot

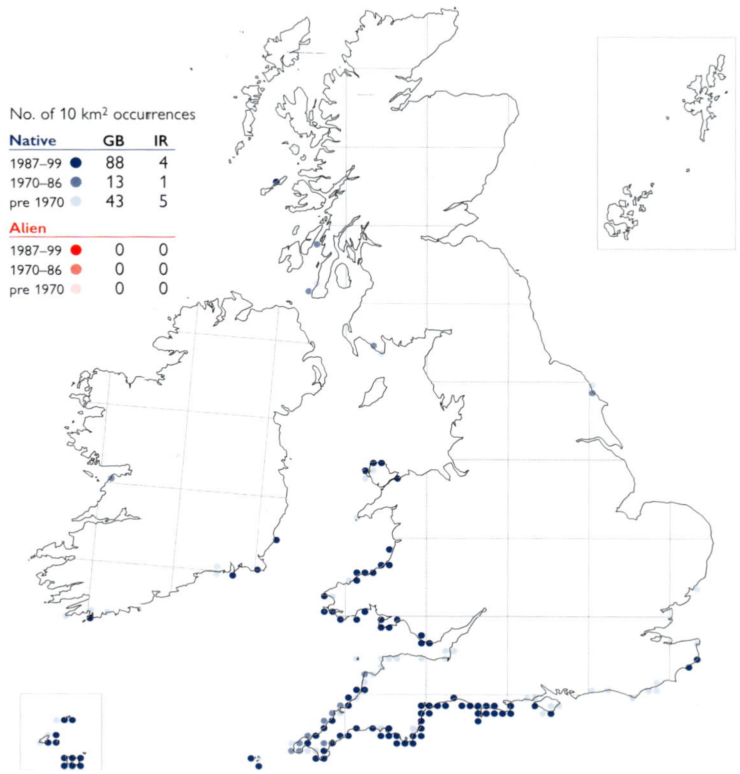

No. of 10 km² occurrences

Native		GB	IR
1987–99	●	88	4
1970–86	●	13	1
pre 1970	●	43	5
Alien			
1987–99	●	0	0
1970–86	●	0	0
pre 1970	●	0	0

This biennial herb is entirely coastal, occurring in both open and closed grassland on cliffs and on stable sand dunes. Lowland.

Native. There has been no appreciable change it the distribution of subsp. *gummifer* since it was mapped by Perring & Sell (1968), but it is now much better recorded. A range of morphological intermediates exist between subsp. *gummifer* and subsp. *carota*. The most extreme form occurs in Cornwall and Devon, with the features differentiating it from subsp. *carota* becoming less obvious further east and north (Perring & Sell, 1968).

Oceanic Temperate element; known only from the Atlantic coast of Europe.

References: Atlas Supp. (43a), Malloch & Okusanya (1979), Okusanya (1979a, b, c), Rich & Jermy (1998), Tutin (1980).

M. SOUTHAM

Cicendia filiformis Yellow Centaury

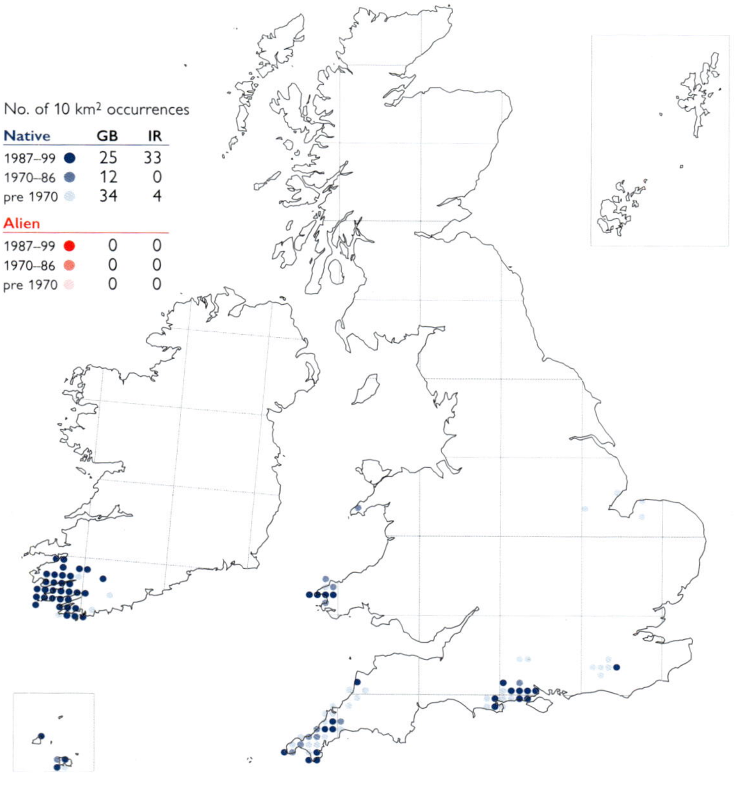

No. of 10 km² occurrences

Native		GB	IR
1987–99	●	25	33
1970–86	●	12	0
pre 1970	●	34	4
Alien			
1987–99	●	0	0
1970–86	●	0	0
pre 1970	●	0	0

An annual of open heathland habitats, growing on sandy and peaty soils of relatively high base-status which are damp in winter and spring; it is also found in damp pasture, woodland rides, dune-slacks and on cliffs. Reduced competition, caused by winter-flooding, grazing and disturbance, is essential. Lowland.

Native (change –0.70). The 1962 *Atlas* showed major losses before 1930, especially in Cornwall, and this species has declined further in areas such as Dorset. It is threatened at many sites because of the lack of grazing and the upgrading of heathland tracks with hardcore. In west Britain and Ireland, however, local losses may have been balanced by newly discovered sites.

Submediterranean-Subatlantic element.

References: Atlas (207a), Hultén & Fries (1986), Meusel *et al.* (1978), Stewart *et al.* (1994).

R. D. PORLEY

Exaculum pusillum Guernsey Centaury

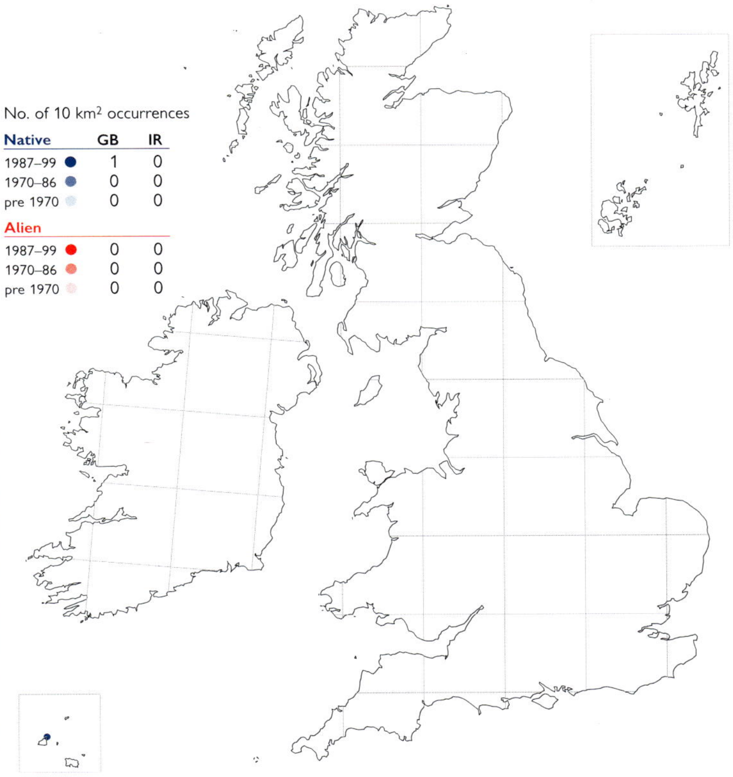

No. of 10 km² occurrences

Native	GB	IR
1987–99	1	0
1970–86	0	0
pre 1970	0	0
Alien		
1987–99	0	0
1970–86	0	0
pre 1970	0	0

This tiny procumbent to ascending annual grows in moist, open, short turf in coastal dune-slacks. Winter-flooding and rabbit-grazing maintain the open conditions which this species requires. Lowland.

Native. This species was discovered in 1850, since when very few populations have been found. Some of the populations have apparently disappeared, but the plant persists at other sites, though its abundance varies from year to year.

Suboceanic Southern-temperate element.

References: Atlas (207a), Bolòs & Vigo (1995), McClintock (1975).

R. D. PORLEY

Centaurium scilloides Perennial Centaury

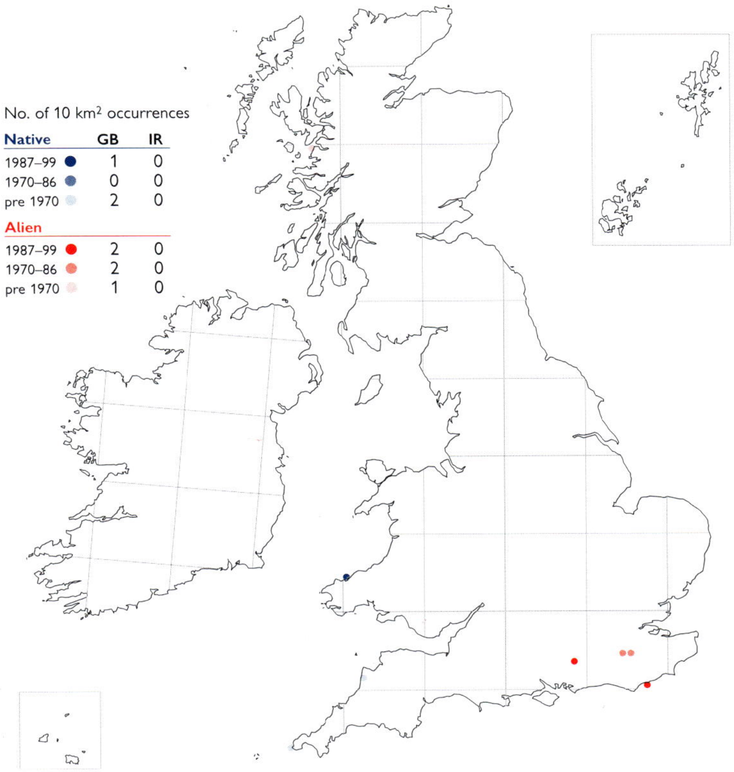

No. of 10 km² occurrences

Native	GB	IR
1987–99	1	0
1970–86	0	0
pre 1970	2	0
Alien		
1987–99	2	0
1970–86	2	0
pre 1970	1	0

This perennial herb occurs on freely draining soils on the slopes of coastal cliffs. Most native populations are in grassland and maritime dwarf-shrub heath, often along eroded and trampled edges, with some extending into dune grassland. In S.E. England it occurs in lawns at both coastal and inland sites. Lowland.

Native. The decline of grazing and invasion of scrub are likely causes for the disappearance of this species from Cornwall, where it was first seen in the 1950s. The first record from lawns in S.E. England was made in 1974. In these localities it may be an escape from cultivation.

Oceanic Southern-temperate element; restricted to disjunct localities along the coast of W. Europe.

References: Atlas (207c), Dupont (1962), Wigginton (1999).

R. D. PORLEY

Centaurium erythraea Common Centaury

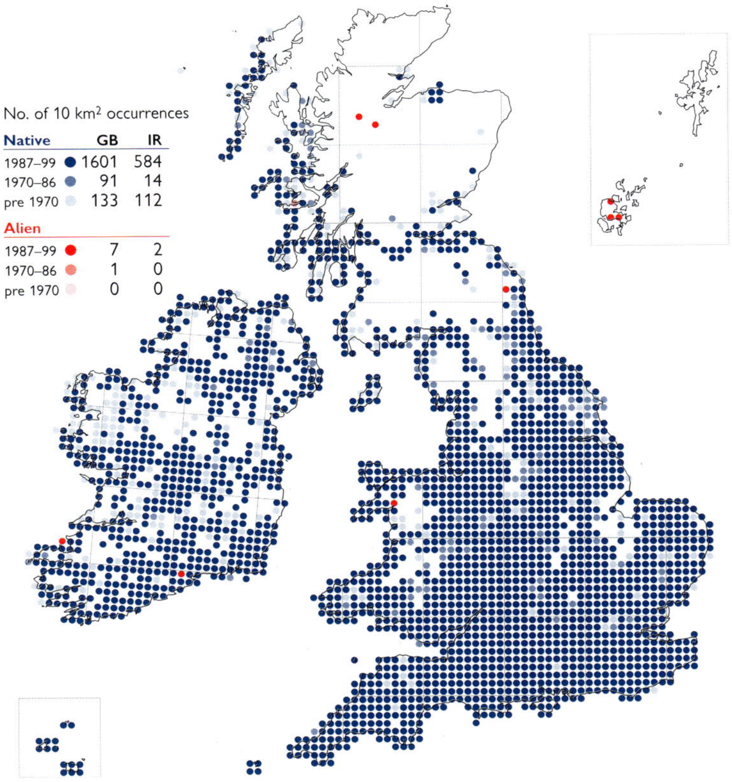

No. of 10 km² occurrences

Native	GB	IR
1987–99	1601	584
1970–86	91	14
pre 1970	133	112
Alien		
1987–99	7	2
1970–86	1	0
pre 1970	0	0

A biennial, rarely annual, herb of mildly acidic to calcareous, well-drained, often disturbed, soils, occurring in a wide range of habitats including chalk and limestone grassland, heathland, woodland rides and open scrub, dune grassland, quarries, spoil-heaps and road verges. 0–330 m (Wauchope, Roxburghs.).

Native (change +0.03). The distribution of this ecologically wide-ranging and mobile species shows little change from the 1962 *Atlas*.

European Southern-temperate element; widely naturalised outside its native range.

References: Atlas (207d, 208a), Grime *et al.* (1988), Hultén & Fries (1986), Meusel *et al.* (1978).

R. D. PORLEY

Centaurium littorale Seaside Centaury

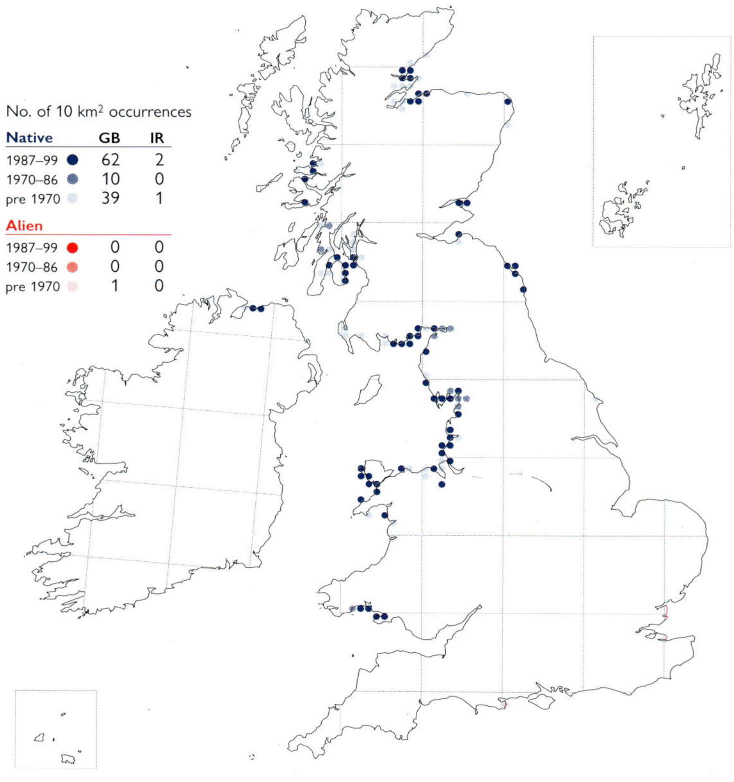

No. of 10 km² occurrences

Native		GB	IR
1987–99	●	62	2
1970–86	●	10	0
pre 1970	●	39	1
Alien			
1987–99	●	0	0
1970–86	●	0	0
pre 1970	●	1	0

This biennial herb is confined to coastal dunes, the uppermost levels of saltmarshes and calcareous, humus-rich turf near the sea where competing vegetation is checked and the habitat kept open by grazing or trampling. Lowland.

Native (change +0.03). There appears to be little change in the overall distribution of *C. littorale* since the 1962 *Atlas*; although some populations have been lost, new ones have been discovered.

European Temperate element.

References: Atlas (208b), Curtis & McGough (1988), Hultén & Fries (1986), Meusel *et al.* (1978), Preston & Pearman (2000), Stewart *et al.* (1994).

R. D. PORLEY

Centaurium pulchellum Lesser Centaury

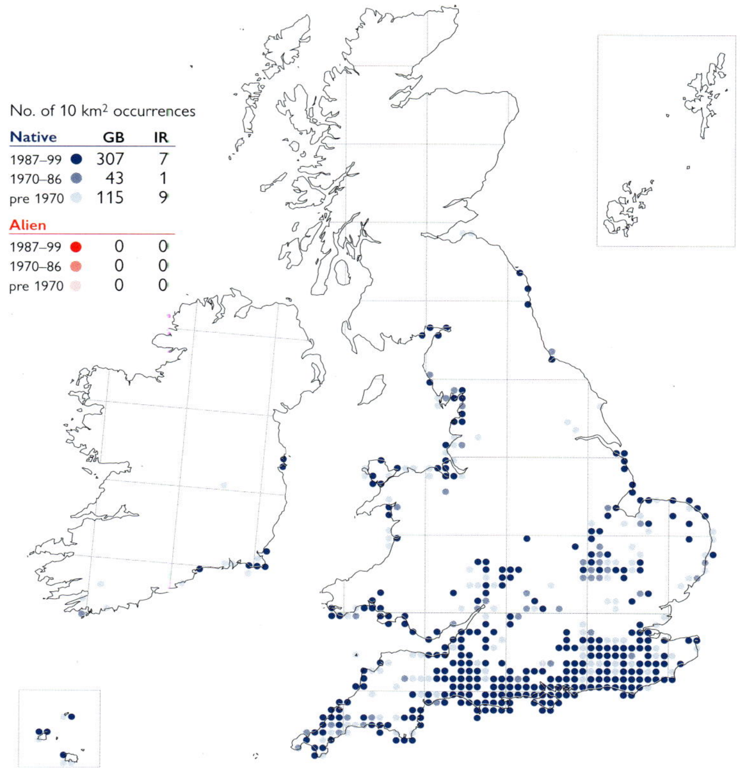

No. of 10 km² occurrences

Native		GB	IR
1987–99	●	307	7
1970–86	●	43	1
pre 1970	●	115	9
Alien			
1987–99	●	0	0
1970–86	●	0	0
pre 1970	●	0	0

An erect annual of mildly acidic to calcareous soils. Inland it is found in dry, open grasslands and heaths, in woodland rides, marl pits and other open, disturbed ground. On the coast it is a plant of open sandy and muddy grassy places, often by estuaries, sand dunes and in upper saltmarsh. Lowland.

Native (change +0.10). The overall distribution of this species has remained stable although there has been some decline on heathland due to lack of management (particularly grazing), and coastal development has led to the loss of some of its former sites. It is much better recorded than in the 1962 *Atlas*.

Eurosiberian Southern-temperate element.

References: Atlas (207b), Curtis & McGough (1988), Hultén & Fries (1986), Meusel *et al.* (1978).

R. D. PORLEY

Centaurium tenuiflorum Slender Centaury

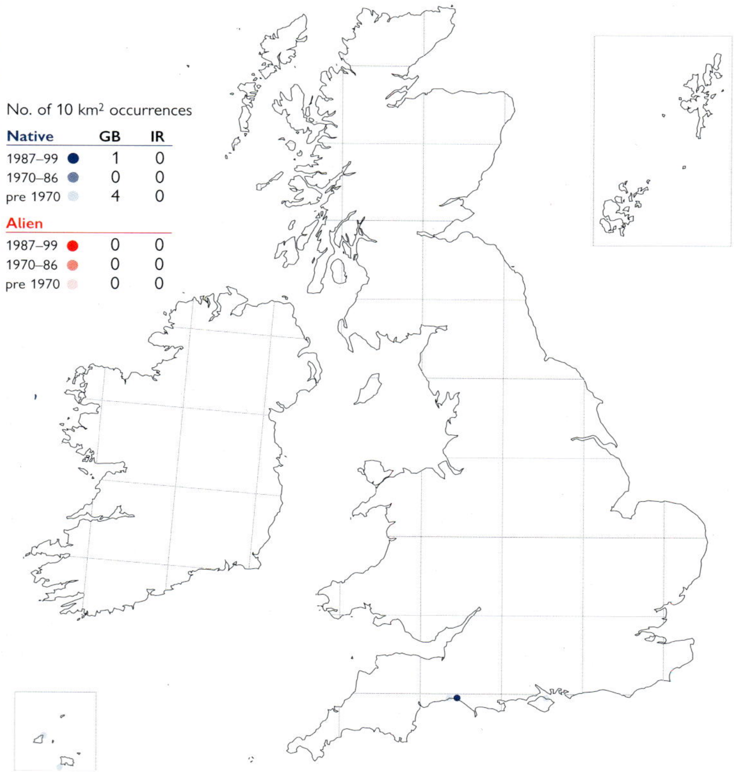

No. of 10 km² occurrences

Native		GB	IR
1987–99	●	1	0
1970–86	●	0	0
pre 1970	●	4	0
Alien			
1987–99	●	0	0
1970–86	●	0	0
pre 1970	●	0	0

This annual occurs on open, poorly drained sandy or clayey soils on slumping coastal cliffs. It will not persist in a closed sward, tending to appear a few years after the habitat is opened up, and disappear after about ten years unless open conditions are re-created. Lowland.

Native. Though this species is rare and very localised, the unstable nature of its habitat should ensure its survival in Dorset. It was last recorded in the Isle of Wight in 1953; later records from here are errors. It was recorded in the Channel Islands in 1837, but not subsequently.

Mediterranean-Atlantic element.

References: Atlas (207c), Meusel *et al.* (1978), Wigginton (1999).

R. D. PORLEY

Blackstonia perfoliata Yellow-wort

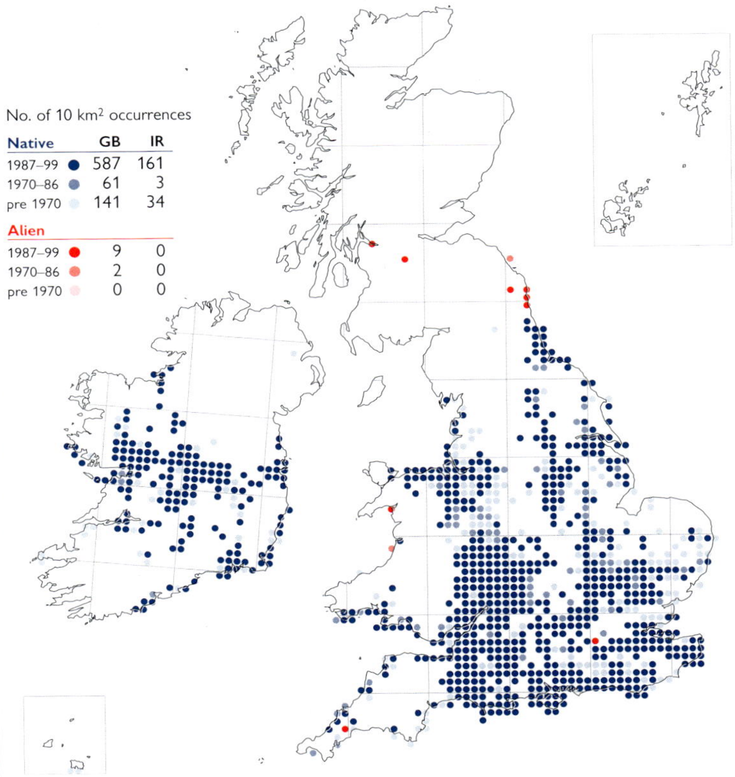

An annual or biennial herb of open dry (but frequently winter-wet), often stony, shallow basic soils. Its main habitats are calcareous grasslands and fixed sand dunes, but it can be an abundant colonist of disturbed ground, including quarries and railway cuttings, and on road verges and pathsides. Lowland.

Native (change +0.12). There has been little change in the overall distribution of this species since the 1962 *Atlas*, although it has spread into Northumberland since then (Swan, 1993).

Submediterranean-Subatlantic element.

References: Atlas (208c), Meusel *et al.* (1978).

R. D. PORLEY

Gentianella ciliata Fringed Gentian

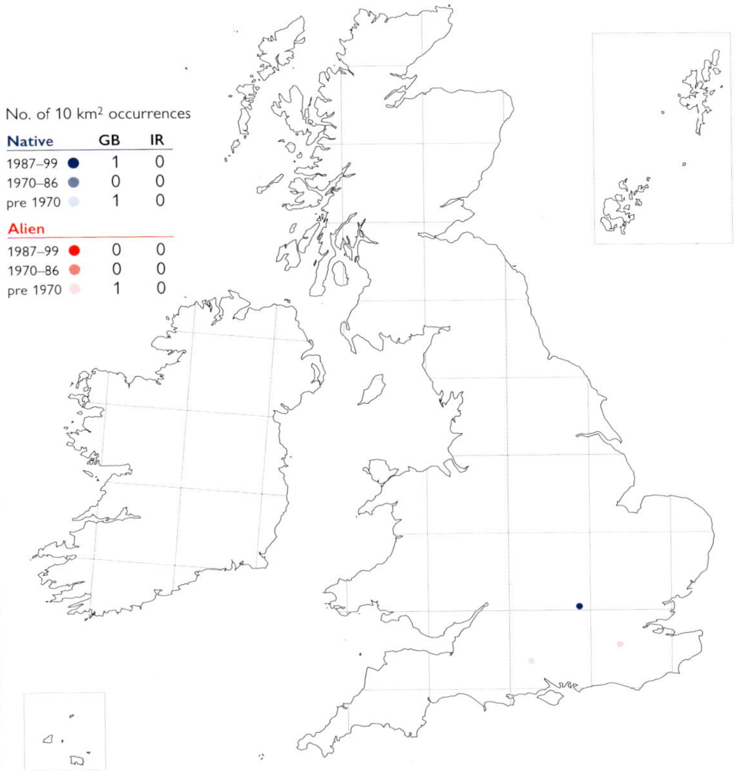

A biennial, or perhaps annual, herb of chalk grassland, growing in shortly-grazed, herb-rich turf on a steep W.-facing slope. Lowland.

Native or alien. Although first reported as *G. ciliata* from Wendover (Bucks.) in 1875, these plants were subsequently dismissed as *Campanula glomerata* until the species was re-discovered, in perhaps the same field, in 1982. This, the only known extant population, is variable in size, reaching a peak in the late 1980s but since declining to almost none. Recent study of herbarium specimens has revealed previously unknown collections from Wiltshire (1892) and Surrey (1910), the latter being considered an alien.

Eurosiberian Temperate element, with a continental distribution in W. Europe.

References: Hultén & Fries (1986), Meusel *et al.* (1978), Wigginton (1999).

R. D. PORLEY

Gentianella campestris Field Gentian

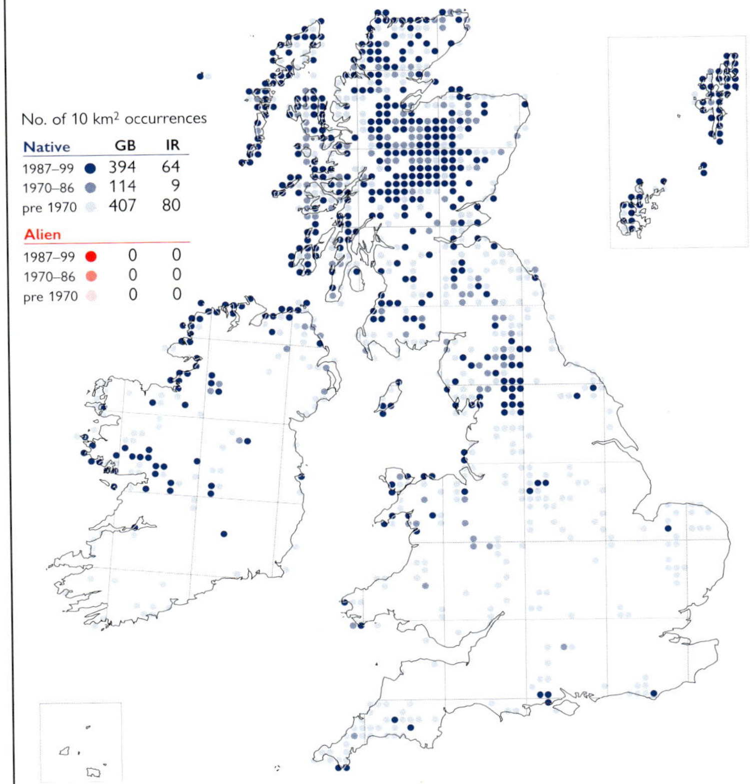

A biennial, occasionally annual, herb of mildly acidic to neutral soils in a variety of open habitats, including pastures, hill grassland, grassy heaths, sand dunes, machair and road verges. On limestone it probably indicates surface leaching or the presence of non-calcareous superficial deposits. Generally lowland, but reaching 915 m at Cairnwell (E. Perth, S. Aberdeen).

Native (change −1.28). *G. campestris* had already suffered a marked decline before 1930 and sites are still being lost through overgrazing in the uplands and the neglect of lowland pastures. In its English stronghold, Cumbria, it has disappeared from half the 10-km squares for which there are post-1930 records.

European Boreo-temperate element.

References: Atlas (209b), Halliday (1997), Hultén & Fries (1986), Meusel *et al.* (1978).

R. D. PORLEY

Gentianella germanica Chiltern Gentian

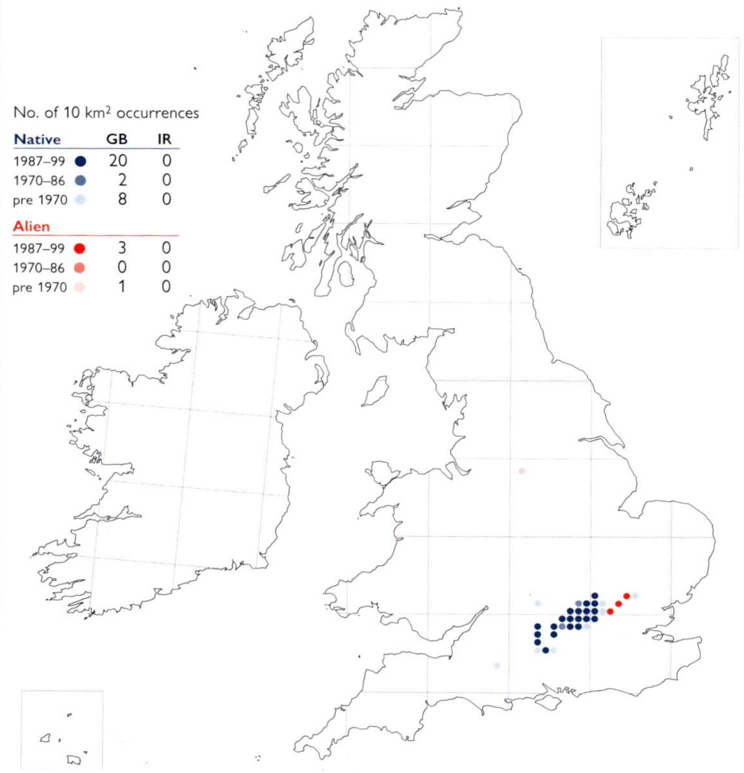

An annual or biennial herb of shallow chalk soils, occurring in chalk grassland (frequently near tracks) and chalk-pits, and sometimes in open scrub and on woodland margins. Lowland.

Native (change –0.17). The distribution of this species has not changed significantly since the 1962 *Atlas*. There have, however, been some local losses, and populations vary in size from year to year depending on weather and disturbance. *G. amarella* often accompanies this species, and hybrids (*G.* × *pamplinii*) frequently occur, with back-crossing to *G. amarella*. Hybridisation has been implicated in the loss of some populations of *G. germanica*.

European Temperate element, with a continental distribution in W. Europe.

References: Atlas (209c), Meusel *et al.* (1978), Stewart *et al.* (1994), Killick *et al.* (1998).

R. D. PORLEY

Gentianella amarella Autumn Gentian

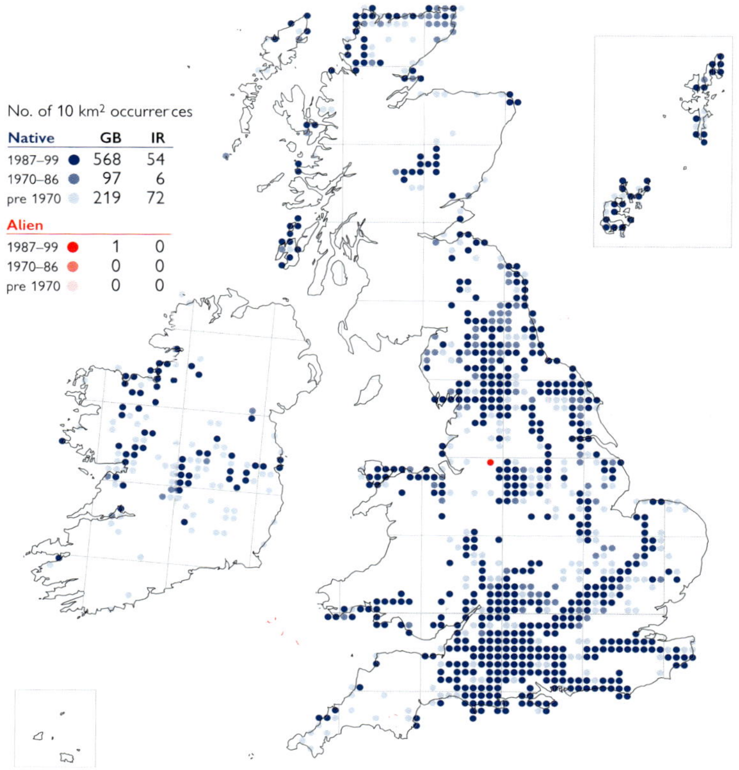

An annual or biennial herb of well-drained basic soils, typically occurring in grazed chalk and limestone grassland, on calcareous dunes and machair, on spoil-tips and in cuttings and quarries. Mainly lowland, ascending to 750 m (Knock Fell, Westmorland).

Native (change –0.75). There is some evidence of a decline in this species since the 1962 *Atlas*. In Ireland only subsp. *hibernica* has been recorded; subsp. *amarella* is widespread in Britain north to C. Scotland, and subsp. *septentrionalis* is mapped separately. *G. amarella* hybridises freely with *G. anglica* and *G. uliginosa*. However, recent molecular research suggests that the boundaries between these taxa are not clear-cut.

Circumpolar Boreo-temperate element.

References: Atlas (209d), Hultén & Fries (1986), Meusel *et al.* (1978).

R. D. PORLEY

Gentianella amarella subsp. *septentrionalis*

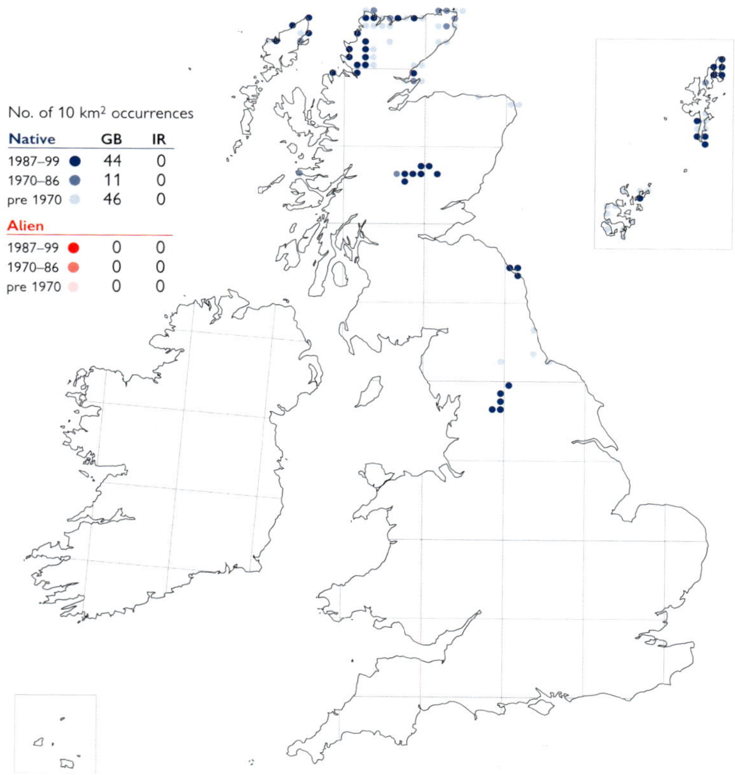

An annual or biennial herb growing on well-drained calcareous soils on machair and sand dunes, and in the Grampians on grassy mica-schist slopes and ledges and in limestone grassland. Generally lowland, but upper altitudinal limit unknown.

Native. It is difficult to assess trends in this taxon since it has not been consistently recorded and is poorly understood. Subsp. *druceana*, mapped separately by Perring & Sell (1968), is doubtfully distinct from subsp. *septentrionalis*, and all records are mapped here as the latter.

Endemic.

References: Atlas Supp. (48a, b), Pritchard (1960).

R. D. PORLEY

Gentianella anglica Early Gentian

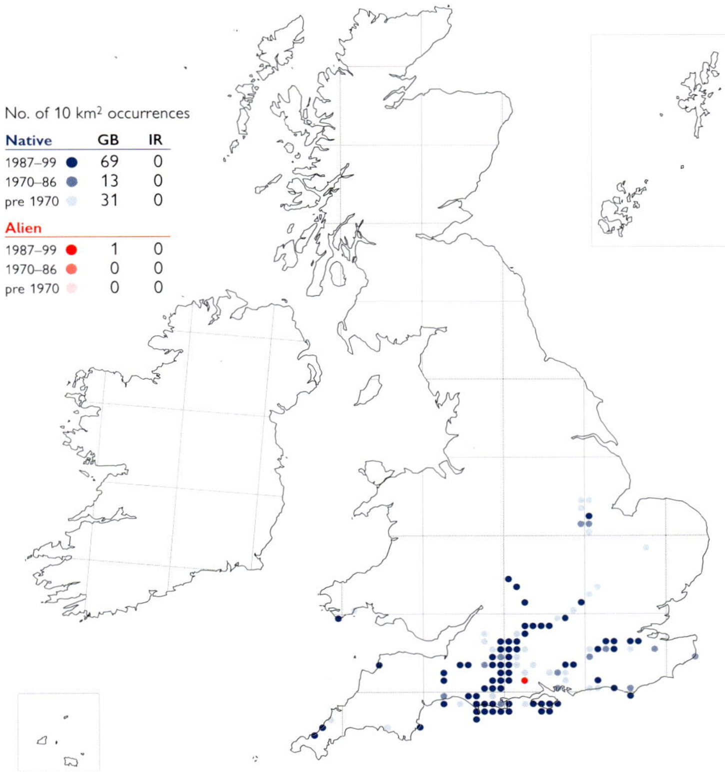

An annual or biennial herb of shallow calcareous soils, especially on the chalk, in closely grazed calçareous grassland, quarries, on cliff-tops and sand dunes. Lowland.

Native (change –0.32). Since the 1962 *Atlas*, many populations of *G. anglica* have declined or been lost, largely due to reduced grazing. Recently, however, some new sites have been found as a result of intensive surveys. The hybrid between *G. anglica* and *G. amarella* (*G.* × *davidiana*), previously described as *G. anglica* subsp. *cornubiensis*, has been reported from S. England. However, preliminary genetic analysis suggests that distinctions between *G. anglica*, *G. amarella* and the hybrid can no longer be supported (Winfield & Parker, 2000).

Endemic.

References: Atlas (210a), Rich (1997a), Rich *et al.* (1997), Stewart *et al.* (1994).

R. D. PORLEY

Gentianella uliginosa Dune Gentian

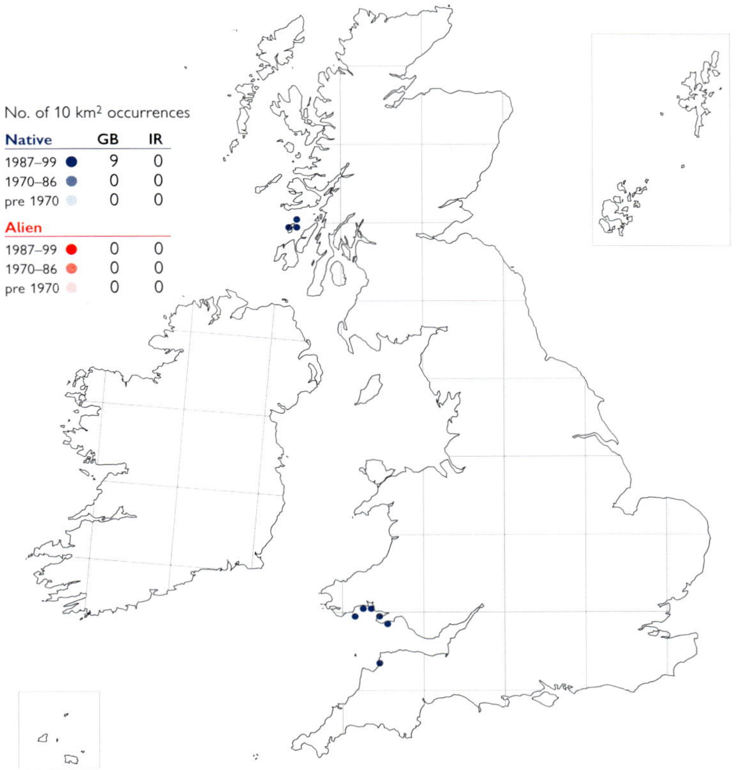

An annual of coastal dunes, dune-slacks and machair, usually in open ground or short vegetation maintained by grazing, disturbance or winter flooding. Lowland.

Native. Since the 1962 *Atlas*, populations have been discovered in Scotland, and the species re-discovered in 1998 at Braunton Burrows (N. Devon) after a gap of about seventy years. It is threatened by the lack of grazing at some Welsh sites, and perhaps also by hybridisation with *G. amarella*, although the latter view is not supported by Kay & John (1995).

European Temperate element.

References: Atlas (210a), Gulliver (1998), Holyoak (1999), Hultén & Fries (1986), Meusel *et al.* (1978), Rich (1996), Wigginton (1999).

R. D. PORLEY

Gentiana pneumonanthe Marsh Gentian

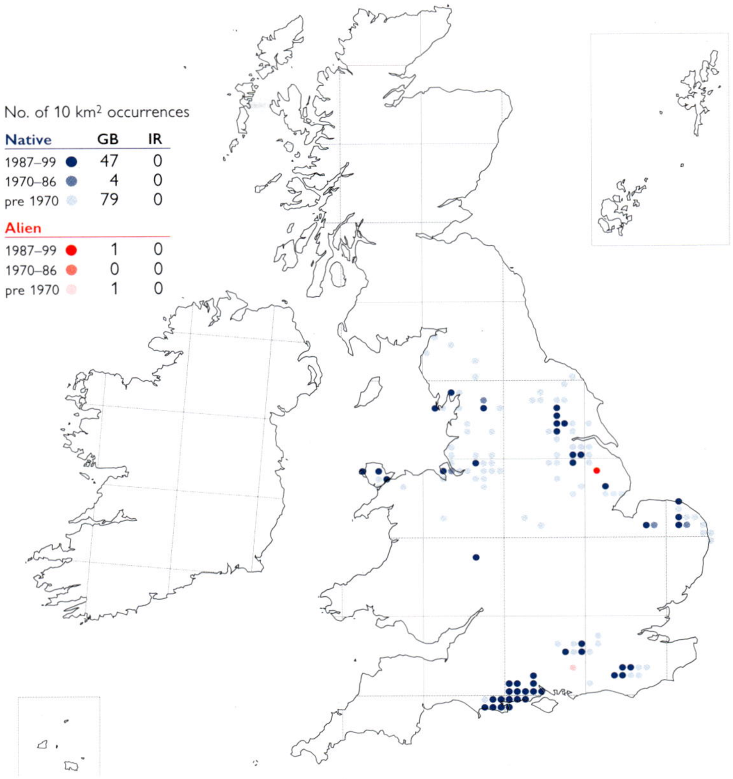

A long-lived perennial herb of damp acidic grassland and wet heaths, usually on relatively enriched soils, and often where there is seasonal movement of surface water. The opening up of the habitat by grazing or occasional light burning favours this species by promoting flowering. Lowland.

Native (change –0.31). This species was already declining by the 1930s and the map in the 1962 *Atlas* included many local extinctions, due largely to drainage, development or neglect of its sites. This decline has continued to the present day.

Eurosiberian Temperate element.

References: Atlas (208d), Hultén & Fries (1986), Meusel *et al.* (1978), Simmonds (1946), Stewart *et al.* (1994).

R. D. PORLEY

Gentiana verna Spring Gentian

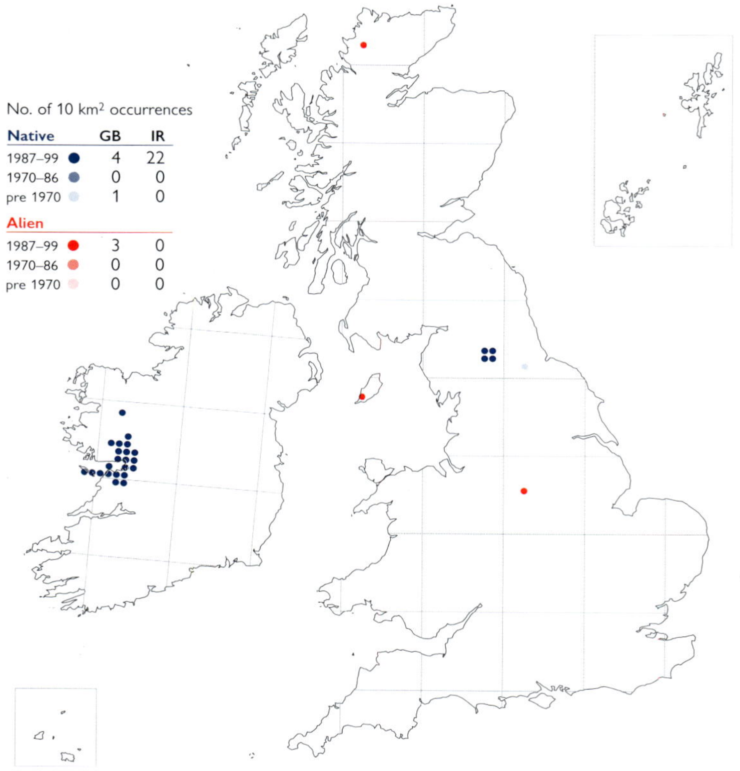

A perennial herb of open, often stony, limestone grassland and calcareous glacial drift. It is also found on hummocks in calcareous flush communities, and in Ireland also on limestone pavement and fixed dunes. Upland in England, ascending from about 350 m to *c.* 730 m on Little Fell (Westmorland), but lowland in Ireland on the Burren (Co. Clare) down to near sea level.

Native (change +0.21). There has been no appreciable change in distribution of this species since the 1962 *Atlas*, but numbers have declined, particularly in outlying populations. Overgrazing in the uplands is a continuing threat to this species in Britain.

European Arctic-montane element; also in C. Asia.

References: Atlas (209a), Elkington (1963), Halliday (1997), Hultén & Fries (1986), Meusel *et al.* (1978), Wigginton (1999).

R. D. PORLEY

Gentiana nivalis Alpine Gentian

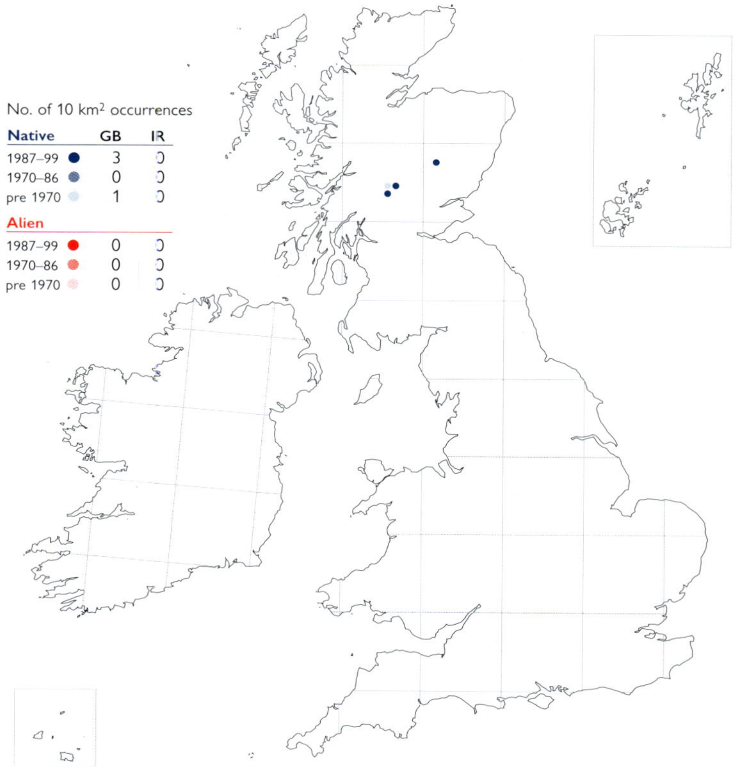

G. nivalis is an annual or biennial herb of calcareous soils, most populations occurring in grazed herb-rich grassland. It is found on rock ledges, vegetated screes and adjacent slopes. Montane, from 730 m to 1095 m on Ben Lawers (Mid Perth).

Native. Populations of this species vary greatly in numbers from year to year, but seem stable in the longer term. Current high levels of grazing cause the loss of many plants, but this may mostly be offset by the corresponding control of competitors and the creation of new ground for seedlings. This species is one of very few annuals in the British Arctic-montane flora.

European Arctic-montane element; also in N. America.

References: Atlas (209a), Hultén & Fries (1986), Meusel *et al.* (1978), Wigginton (1999).

R. D. PORLEY

Vinca minor Lesser Periwinkle

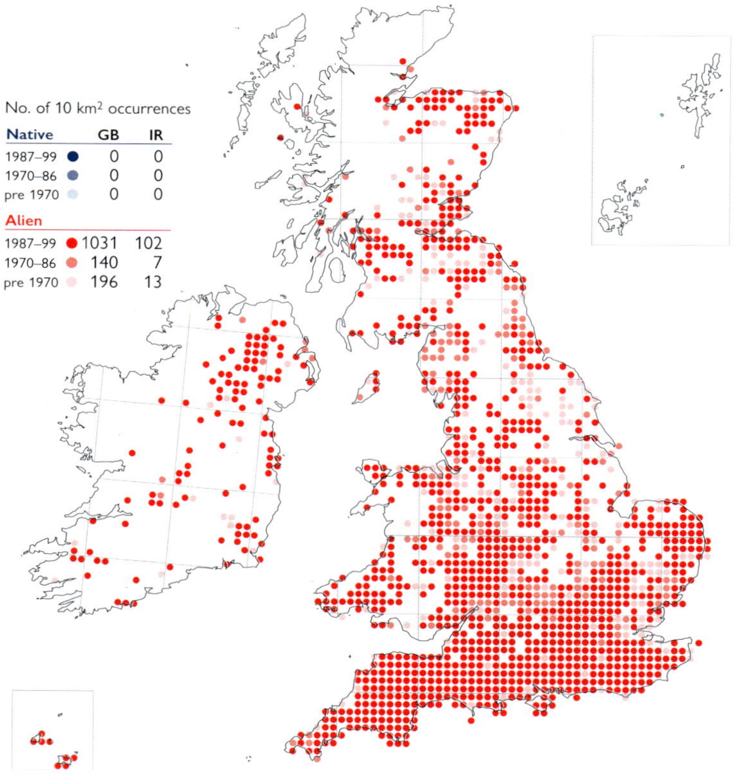

A perennial herb of woodland, roadside banks and verges, waste places and rubbish tips. Generally lowland, but reaching 380 m at Glenlivet (Banffs.).

Archaeophyte (change +0.48). This species is similar to *V. major* but, being smaller and less invasive, is more widely cultivated and therefore more frequent as an escape. It was grown in gardens in Britain by 995 (Harvey, 1981). There has been an increase in records since the 1962 *Atlas*, probably due to a genuine increase as well as better recording.

Apparently native to S., W. & C. Europe, C. & S. Russia and the Caucasus, but the limits of the native range have been obscured by its spread in cultivation.

References: Atlas (206c), Meusel *et al.* (1978).

T. D. DINES

Vinca major Greater Periwinkle

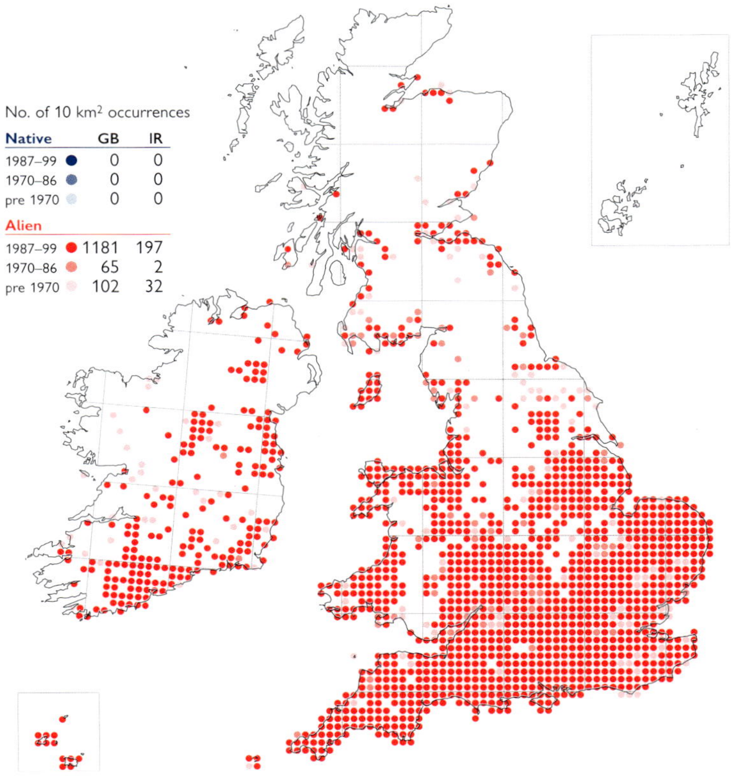

A perennial herb of waste places, rubbish tips, roadside verges, shaded banks and woodland. Lowland.

Neophyte (change +1.49). *V. major* was being cultivated in Britain by 1597. It is widely grown as a ground-cover species, but its robust habit and vigorous growth means that it is frequently discarded, and often becomes well-established in suitable habitats. It was recorded from the wild by 1650 (Middlesex). It has increased since the 1962 *Atlas*, probably mostly due to continued dumping of garden material, although it is also likely to be better recorded.

The naturalised plant, subsp. *major*, is a native of the European Mediterranean region; subsp. *hirsuta* is native to N. Turkey and the Caucasus.

References: Atlas (206d), Meusel *et al.* (1978).

T. D. DINES

Nicandra physalodes Apple-of-Peru

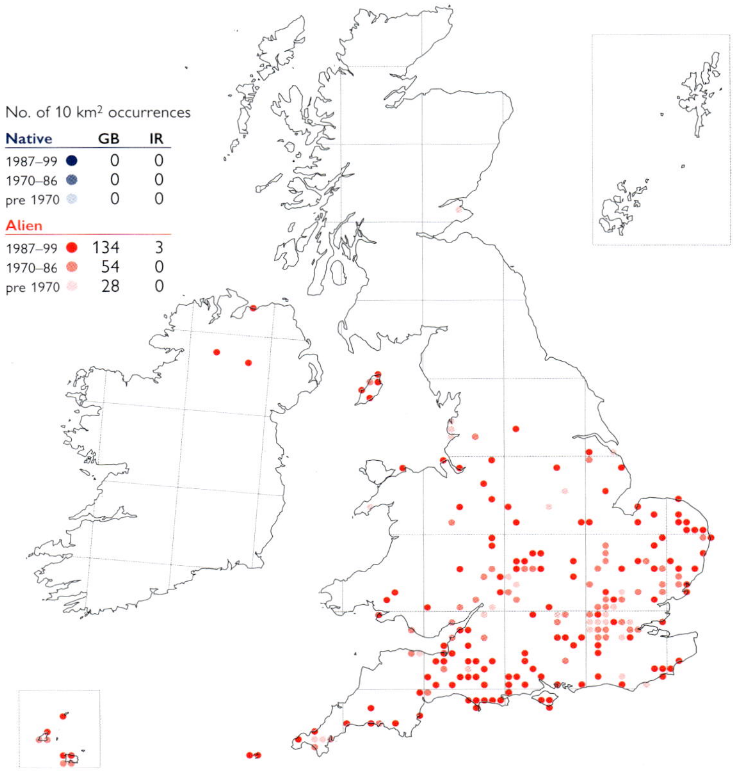

A large annual found on rubbish tips, waste land and as a garden escape. *N. physalodes* is usually a casual, but seeds can overwinter, resulting in established populations in milder areas. Lowland.

Neophyte. *N. physalodes* was grown in Britain by 1759. It is occasionally cultivated and sometimes occurs as a bird-seed alien. It was recorded in the wild in 1860 (Middlesex). Its size and appearance mean that it is unlikely to be overlooked.

Native of S. America (Peru).

T. D. DINES

Lycium agg. Teaplants

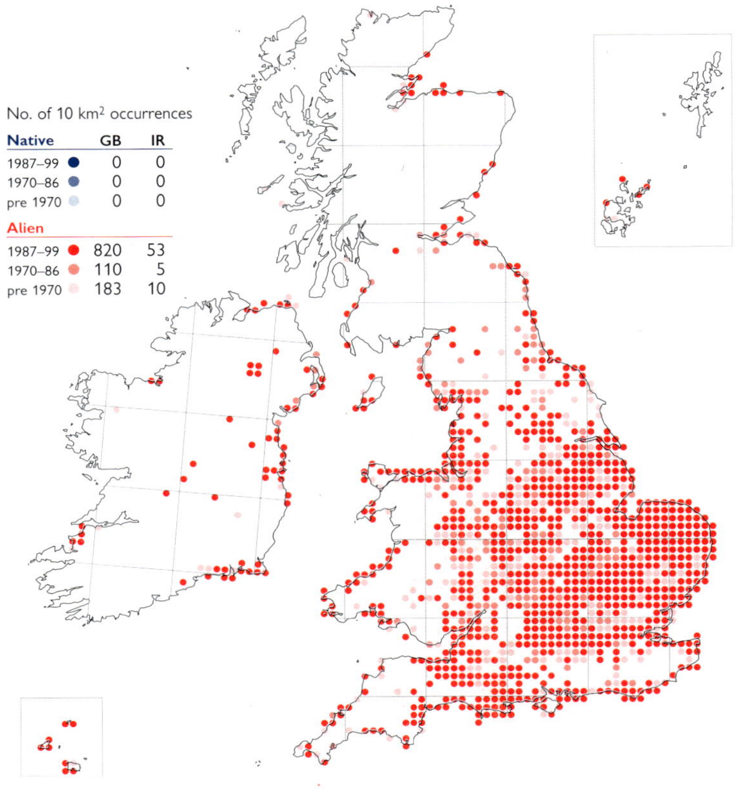

The two supposed species of *Lycium* in our area (*L. barbarum* and *L. chinense*) are suckering, deciduous, spiny shrubs which are grown as hedges, particularly in coastal areas. Both grow readily from bird-distributed seed and become established on shingle, waste land, hedge banks and walls. 0–350 m (near Brassington, Derbys.).

Neophyte (change +0.13). The species were introduced before 1696, and they have always been confused in our area. Since many records are unreliable, both species are mapped here together. They have been recorded from the wild since 1848. There has been no change in the distribution since the 1962 *Atlas*.

Both *L. barbarum* and *L. chinense* are natives of China.

References: Atlas (219a), Atlas Supp. (50a), Bean (1973), Coats (1963), Meusel *et al.* (1978).

T. D. DINES

Atropa belladonna Deadly Nightshade

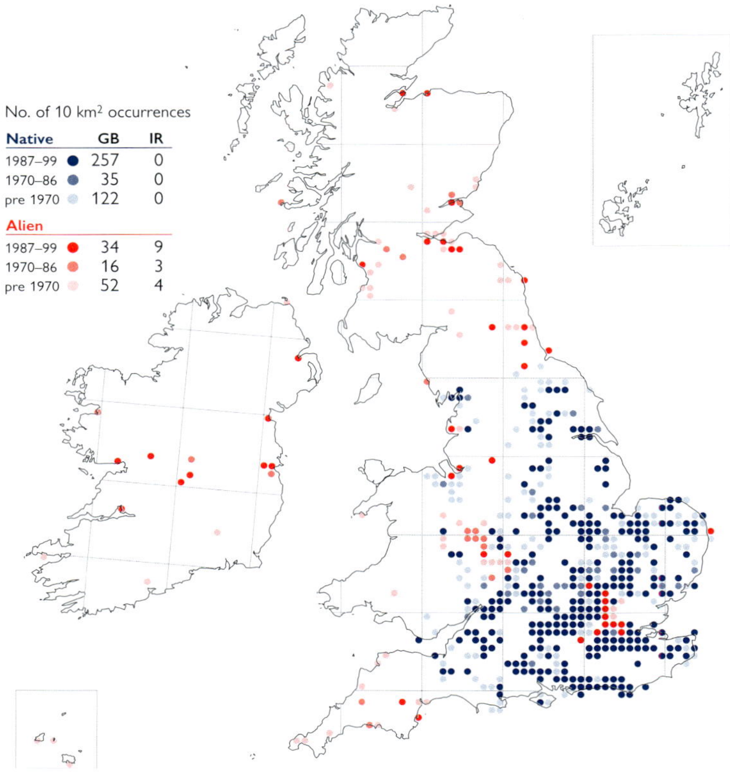

A robust rhizomatous perennial herb of dry disturbed ground, field margins, hedgerows and open woodland. It is native only on calcareous soils, particularly those overlying chalk, but it occurs on a wider range of soils as an alien, where it is often a relic of cultivation as a medicinal herb. Reproduction is mainly by bird-distributed seed. Lowland.

Native (change –0.33). Although highly persistent, *A. belladonna* has declined due to agricultural improvement and, in some cases, specific eradication. Some populations may well be ancient introductions and it is difficult to determine the status of many populations, even within its core area.

European Temperate element.

References: Atlas (219b), Butcher (1947), Meusel *et al.* (1978).

T. D. DINES

Hyoscyamus niger Henbane

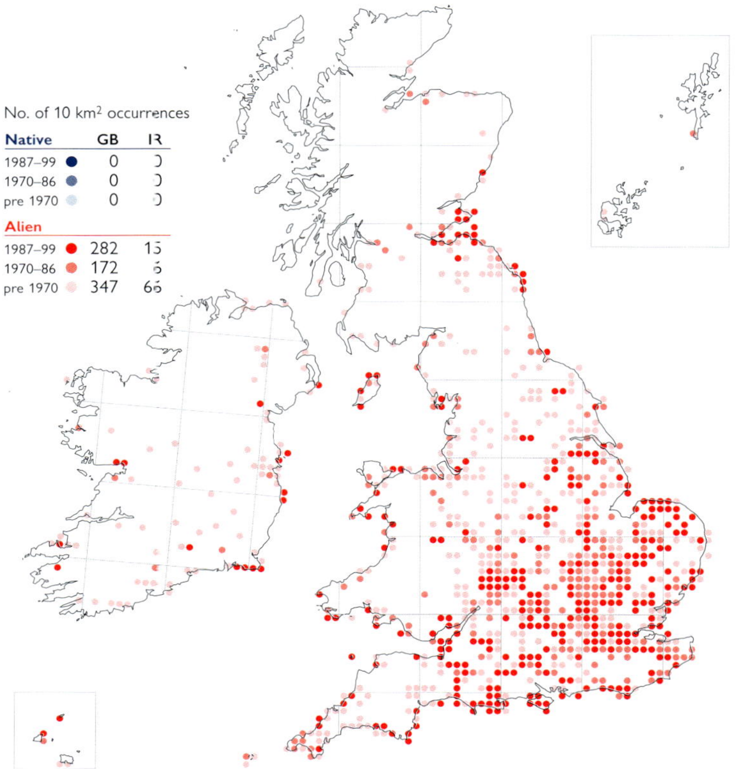

A biennial herb of dry, calcareous soils, particularly those overlying chalk, and on coastal sandhills, sandy open areas and waste ground. It prefers disturbed ground, including rabbit warrens and building sites. Lowland.

Archaeophyte (change –1.38). There is a continuous archaeological record of *H. niger* in Britain from the Bronze Age onwards. Declines before 1930, particularly in Ireland, were evident in the 1962 *Atlas* and have continued markedly since then, mostly through the increased use of herbicides.

As an archaeophyte *H. niger* has a Eurosiberian Southern-temperate distribution, but it is widely naturalised so that it is now Circumpolar Southern-temperate.

References: Atlas (219c), Curtis & McGough (1988), Hultén & Fries (1986), Meusel *et al.* (1978), Stewart *et al.* (1994).

T. D. DINES

Physalis alkekengi Japanese-lantern

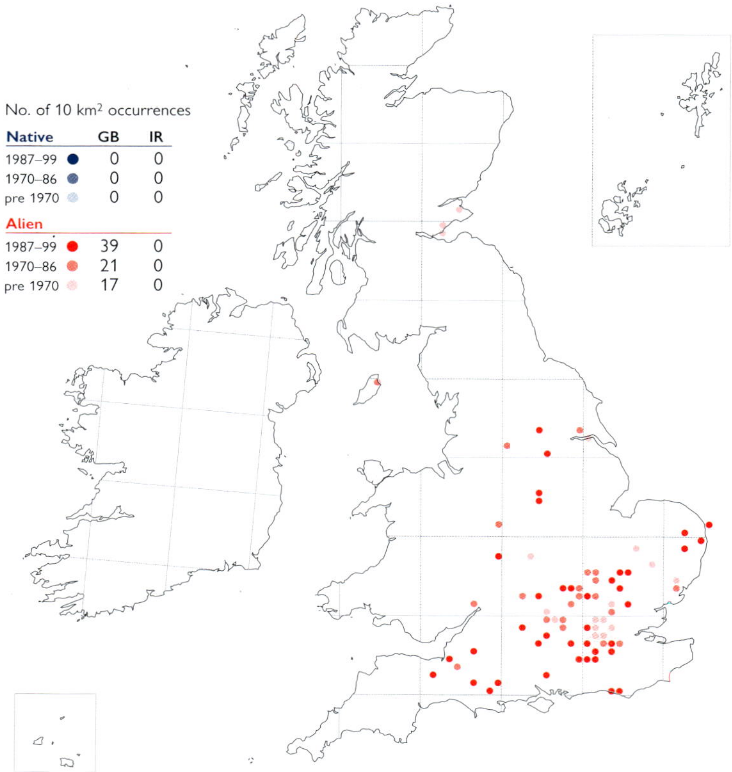

P. alkekengi is a rhizomatous perennial herb found as a garden escape on rubbish tips, railway banks and waste land. Most records are of established colonies; dispersal by seed is limited and the majority of populations are probably derived from garden throw-outs. It is very persistent when established. Lowland.

Neophyte. *P. alkekengi* was introduced from S. Europe by 1548, and is often cultivated for its ornamental fruits. It was known from the wild by 1650.

Native of C. & S. Europe and S.W. & C. Asia.

Reference: Meusel *et al.* (1978).

T. D. DINES

Lycopersicon esculentum Tomato

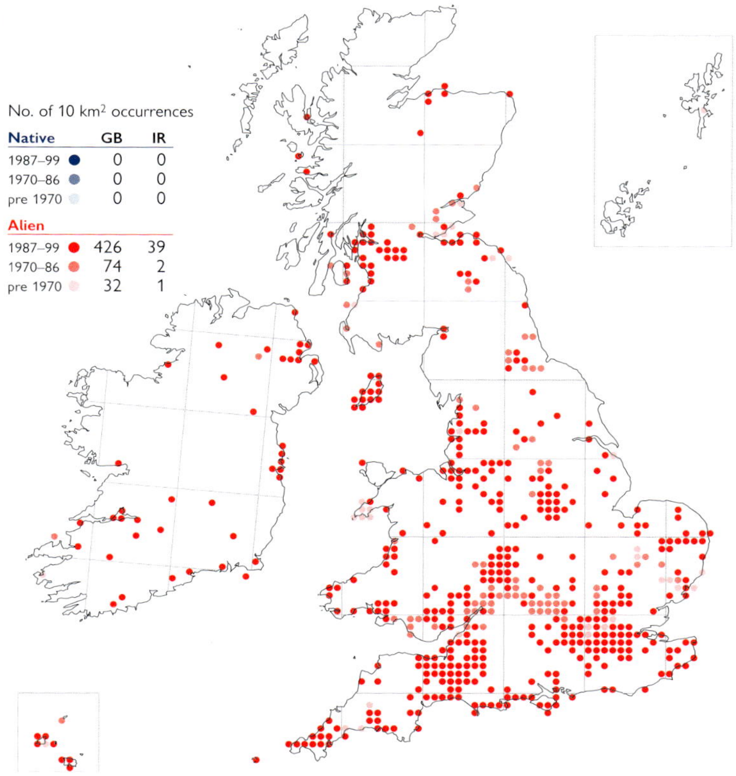

A scrambling annual of waste ground, rubbish tips and sewage works. The species behaves as a perennial in its native range, but both the plant and, to a lesser extent, its seeds are frost-sensitive. Most populations are therefore killed each year, with new plants arising from seeds discarded as fresh fruit or from human sewage. Lowland.

Neophyte. *L. esculentum* was introduced into Britain by 1595, and is very widely cultivated. It was known from the wild by at least 1905. Casual plants have not been recorded systematically until recently and are still ignored by many recorders.

Native of C. & S. America; a frequent casual in Europe but not truly naturalised here.

References: Bolòs & Vigo (1995), Smart & Simmonds (1995), Vaughan & Geissler (1997).

T. D. DINES

Solanum nigrum Black Nightshade

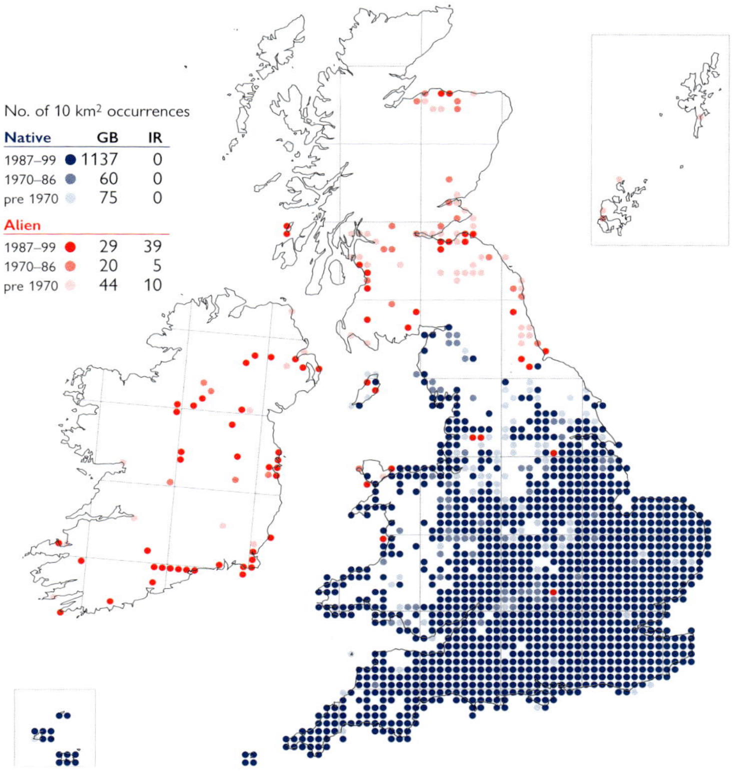

An annual weed of cultivated and waste land, especially where the soil is nutrient-rich. Lowland.

Native or alien (change +0.44). *S. nigrum* may be native in S.E. England, but it is certainly introduced and usually casual in N. England, Wales, Scotland and Ireland. It was recorded in the wild by 1597. It appears to have declined at the edge of its native distribution, but these populations may have been only casual. Some alien records may refer to subsp. *schultesii* from S. Europe.

Eurasian Southern-temperate element; widely naturalised outside its native range.

References: Atlas (220a), Basset & Monro (1985), Hultén & Fries (1986), Meusel *et al.* (1978).

T. D. DINES

Solanum physalifolium Green Nightshade

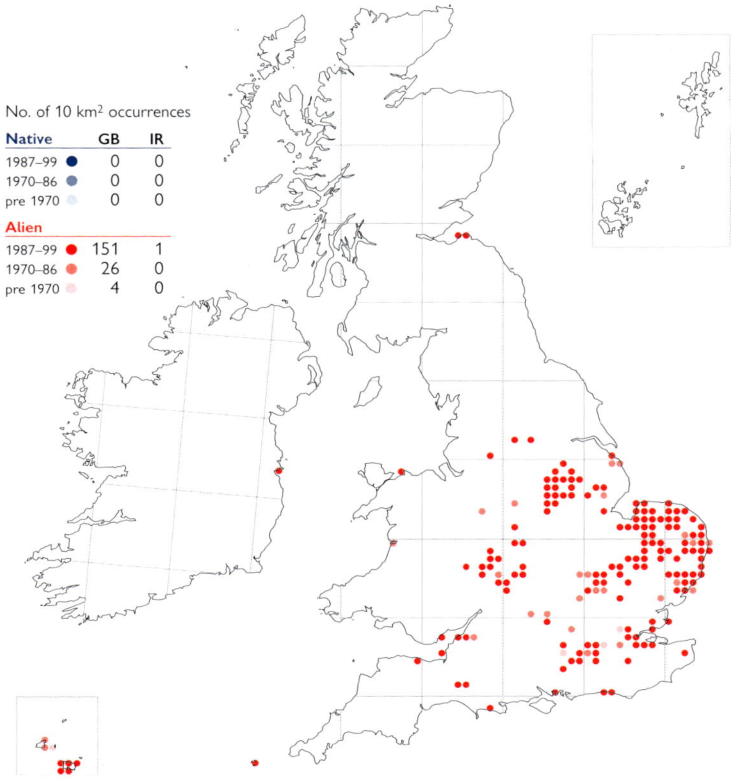

An often prostrate annual of cultivated ground, waste land and refuse tips. The majority of populations are casual, but the species can occasionally become established. Lowland.

Neophyte. *S. physalifolium* was recorded in the wild by 1949 (Norfolk). It is increasing rapidly, especially as a seed contaminant, and is sometimes found in abundance, particularly in crops of sugar-beet. It may be under-recorded in some areas.

Native of S. America.

T. D. DINES

Solanum sarachoides Leafy-fruited Nightshade

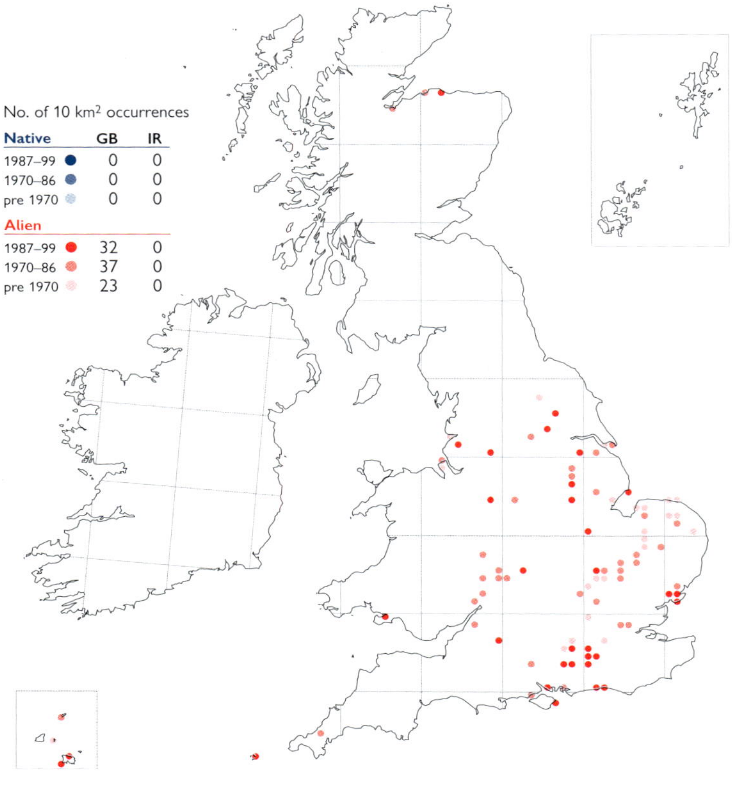

No. of 10 km² occurrences

Native	GB	IR
1987–99	0	0
1970–86	0	0
pre 1970	0	0
Alien		
1987–99	32	0
1970–86	37	0
pre 1970	23	0

An erect or decumbent annual, found as a casual on cultivated ground, rubbish tips and waste ground. Rarely, it becomes naturalised. Lowland.

Neophyte. *S. sarachoides* was recorded in the wild in 1897 (S. Lincs.). In common with several other casual American aliens, it is often associated with carrot seed, and was imported by this means during the Second World War. It appears to be increasing in Britain, but some records may result from confusion with the more widespread *S. physalifolium*.

Native of S. America.

Reference: Lousley (1953).

T. D. DINES

Solanum dulcamara Bittersweet

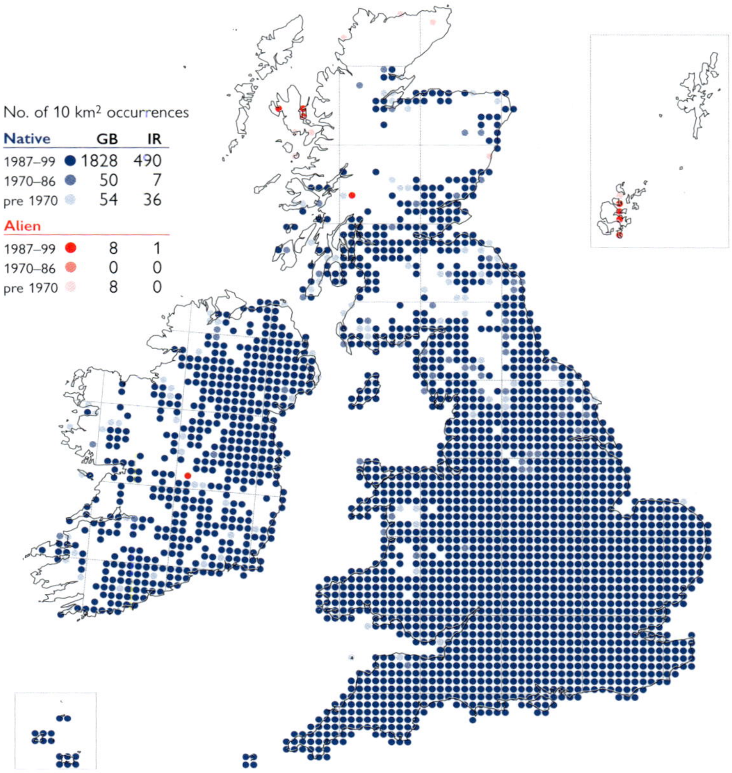

No. of 10 km² occurrences

Native	GB	IR
1987–99	1828	490
1970–86	50	7
pre 1970	54	36
Alien		
1987–99	8	1
1970–86	0	0
pre 1970	8	0

A scrambling, woody perennial growing in woodland, thickets, hedgerows, ditches, and, as var. *marinum*, on shingle beaches. It often grows in moist habitats and is common in swamps and tall-herb fens, and beside rivers and lakes, where it can even grow in shallow water. Lowland.

Native (change –0.11). There has been no appreciable change in the distribution of *S. dulcamara* since the 1962 *Atlas*.

Eurasian Southern-temperate element, but naturalised in N. America so distribution is now Circumpolar Southern-temperate.

References: Atlas (219d), Grime *et al.* (1988), Hultén & Fries (1986), Meusel *et al.* (1978).

T. D. DINES

Solanum tuberosum Potato

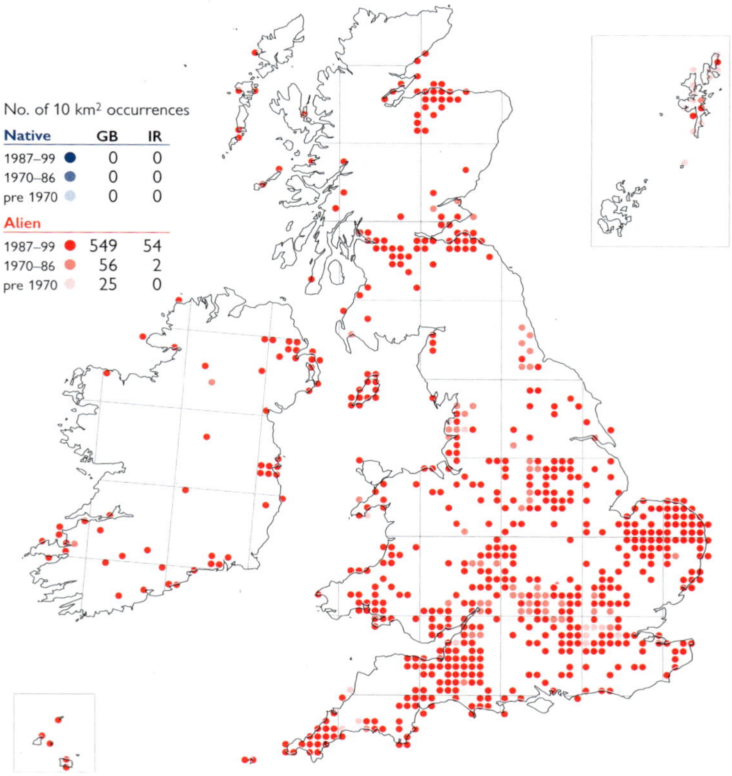

No. of 10 km² occurrences

Native	GB	IR
1987–99	0	0
1970–86	0	0
pre 1970	0	0
Alien		
1987–99	549	54
1970–86	56	2
pre 1970	25	0

A rhizomatous perennial herb of cultivated and waste land, rubbish tips and on coastal sand and shingle where domestic waste has been dumped. In disturbed habitats it is usually casual, but the production of tubers allows some populations in more stable sites to become established. Lowland.

Neophyte. Potatoes have been cultivated in the Andes for at least 7000 years. *S. tuberosum* was introduced to England in about 1590, becoming a staple food of the poor, especially in Scotland and Ireland, by 1800. It was known from the wild by at least 1908, and its distribution is likely to be stable, but like many crop casuals, the species is under-recorded in some areas.

Native of S. America.

References: Bolòs & Vigo (1995), Hawkes (1990), Jackson (1986), Phillips & Rix (1993).

T. D. DINES

Solanum rostratum Buffalo-bur

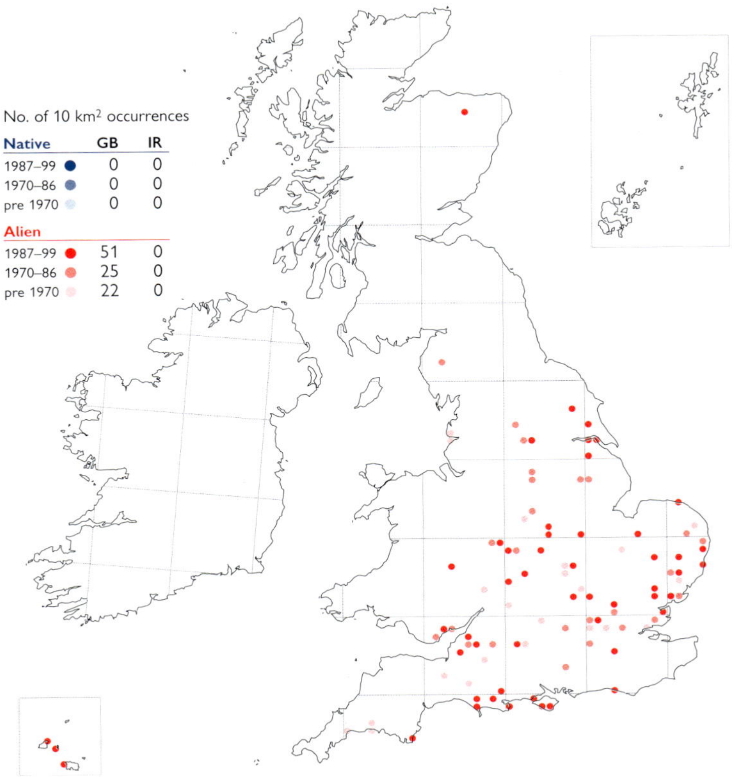

An erect annual, found as a casual on roadsides, arable fields, rubbish tips and waste places where it is introduced with grain, wool shoddy and bird-seed. Lowland.

Casual. *S. rostratum* was first grown in Britain in 1823. It was recorded in the wild in 1886, where it arose as a seed impurity, and it may be increasing.

Native of south-east N. America and Mexico.

Reference: Bolòs & Vigo (1995).

T. D. DINES

Datura stramonium Thorn-apple

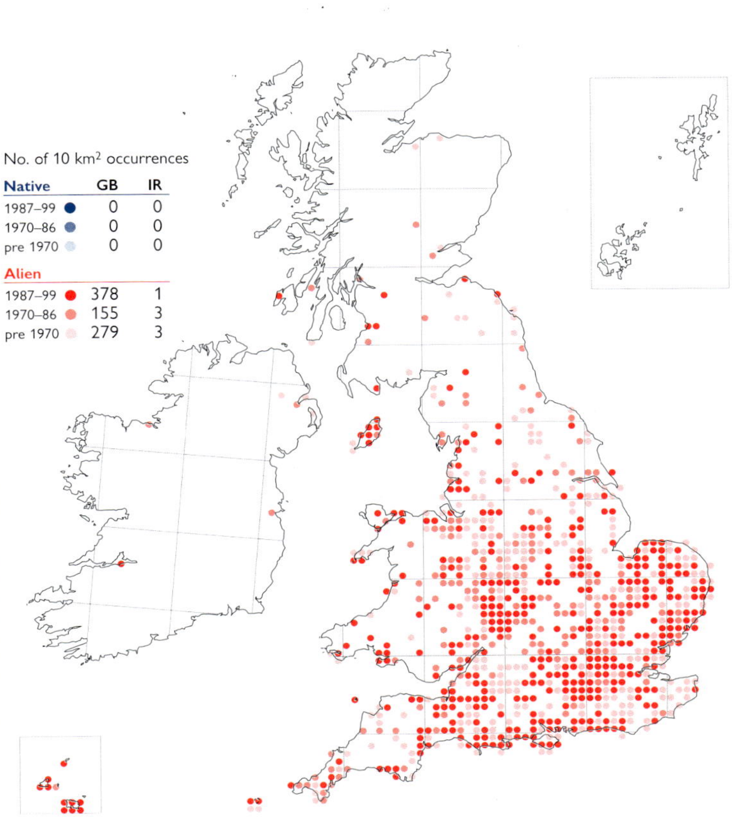

An annual of waste ground, rubbish tips, cultivated and disturbed ground. Most populations are casual, but the species can become naturalised, or reappear after long periods from dormant seeds. It most frequently arises as a garden escape or from bird-seed, but also from oil-seed and grain. Lowland.

Neophyte (change –0.71). *D. stramonium* was cultivated in Britain by 1597 and was grown commercially for alkaloids used to treat asthma. It was first recorded in the wild in 1777. Casual populations make changes in its distribution difficult to assess, but it may have declined in recent years.

Native range unknown, possibly America or the Black Sea region; now widespread in temperate and sub-tropical regions.

References: Atlas (220b), Hultén & Fries (1986), Mabey (1996), Meusel *et al.* (1978), Weaver & Warwick (1984).

T. D. DINES

Nicotiana alata Sweet Tobacco

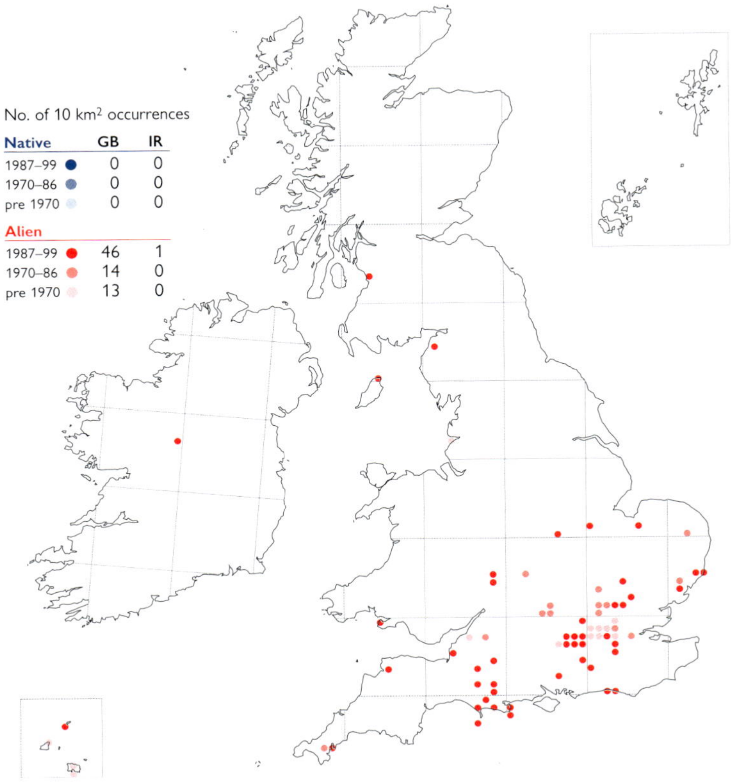

A potentially large, erect annual, occurring as a casual on rubbish tips and waste and cultivated ground, sometimes persisting for a few years. Reproduction is by seed. Lowland.

Casual. *N. alata* was first cultivated in Britain in 1829, and was recorded from the wild by 1928. It is frequently grown in gardens for its scented flowers, and is surprisingly uncommon in the wild given its popularity in gardens. Hybrids with *N. forgetiana* (*N.* × *sanderae*) are common in gardens and may have been overlooked for this species.

Native of S. America.

T. D. DINES

Petunia axillaris × P. integrifolia (P. × hybrida) Petunia

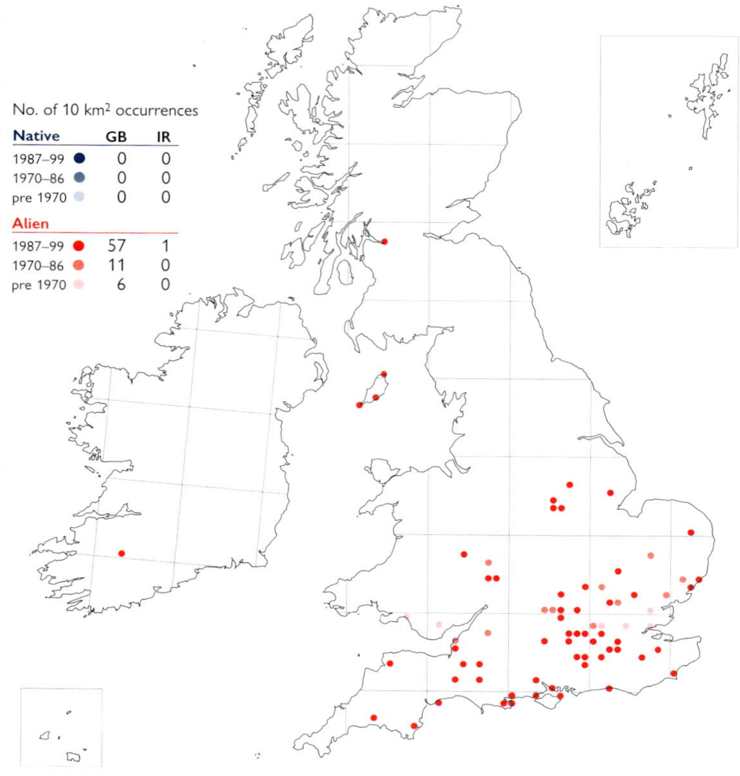

A procumbent to erect annual, found on rubbish tips and waste ground. It is usually casual, but sometimes persists for a few years. It is fertile, and sets abundant, but frost-sensitive, seed. Lowland.

Neophyte. *P. × hybrida* was raised before 1837 and is extremely popular in cultivation, but surprisingly rare in the wild. It was recorded as a casual by 1948.

A hybrid of garden origin.

T. D. DINES

Convolvulus arvensis Field Bindweed

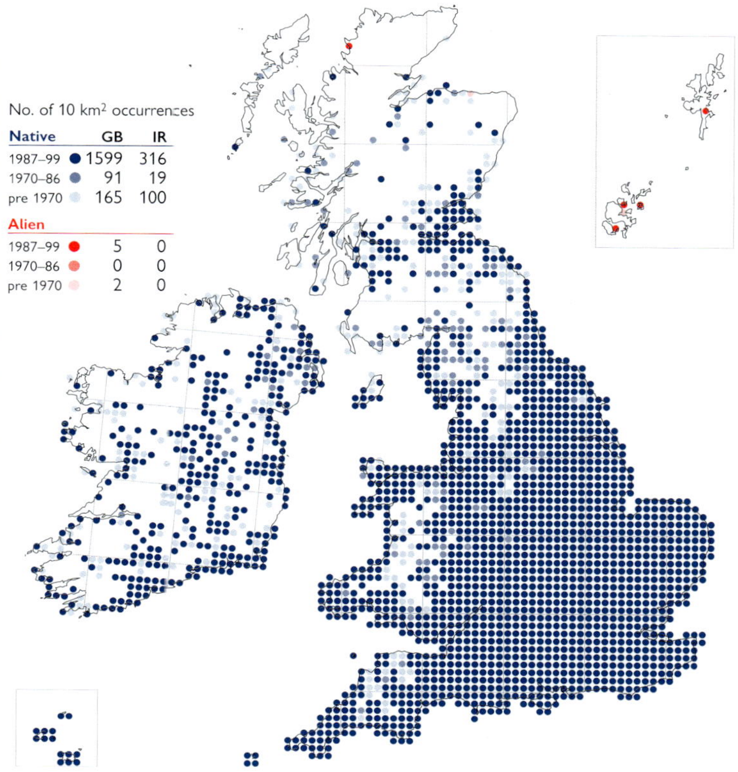

A trailing or climbing perennial herb, found on waste or cultivated ground, waysides, railway banks, open scrub and rough or short grassland, including disturbed chalk downland. Lowland.

Native (change –0.70). *C. arvensis*, which was mapped as 'all records' in the 1962 *Atlas*, shows local losses in the north and west of its range.

Eurosiberian Southern-temperate element, but widely naturalised so that distribution is now Circumpolar Southern-temperate.

References: Atlas (217b), Grime *et al.* (1988), Hultén & Fries (1986), Meusel *et al.* (1978), Weaver & Riley (1982).

G. M. KAY

Calystegia soldanella Sea Bindweed

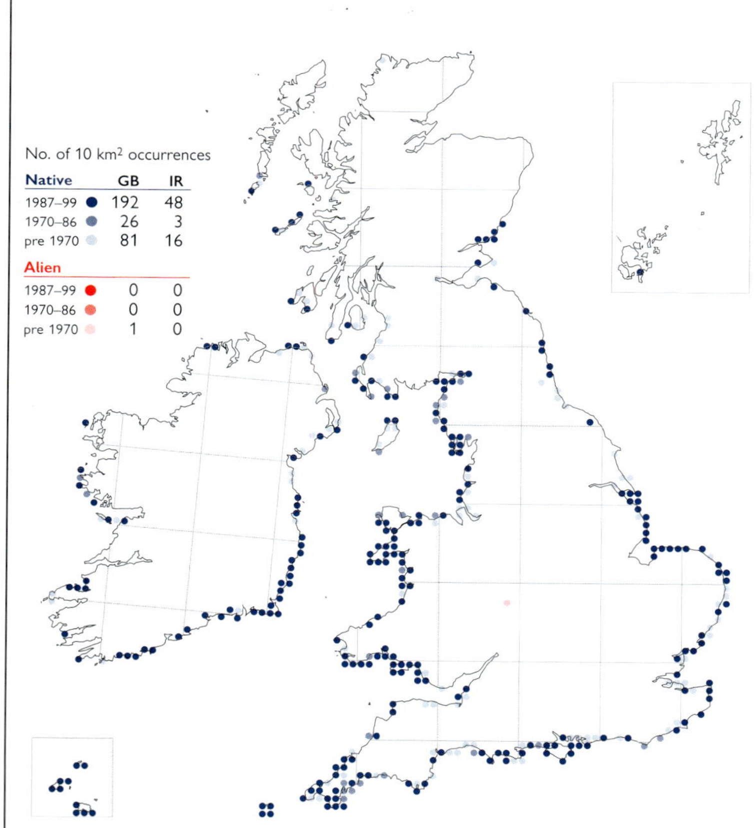

A trailing perennial herb found on sand dunes, and above the strand-line on sand and shingle beaches, often with *Eryngium maritimum*. Lowland.

Native (change –0.58). The disturbance and loss of its habitat has led to many losses of this species, particularly in S. and E. England, though it seems to be relatively resistant to trampling.

Mediterranean-Atlantic element; also in western N. America and widely naturalised outside its native range.

References: Atlas (218b), Hultén & Fries (1986), Meusel *et al.* (1978).

G. M. KAY

Calystegia sepium Hedge Bindweed

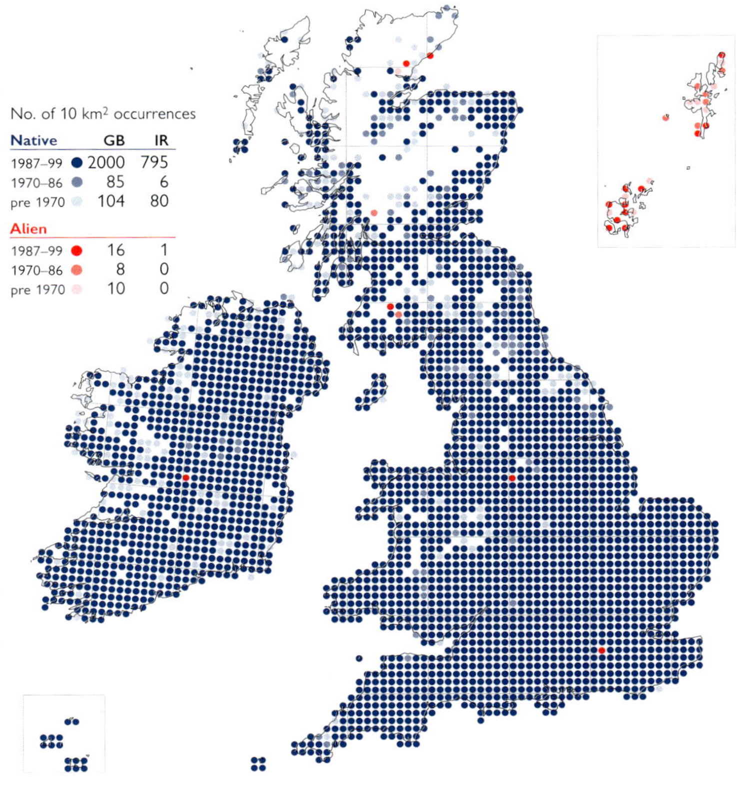

A perennial climber, occurring in hedges, scrub, woodland edges, tall-herb fens, in open *Salix* and *Alnus* carr, and on railway banks and waste ground. It also occurs in artificial habitats in built-up areas and near habitation. 0–365 m (near Buxton, Derbys.).

Native (change +0.69). There is no change in the range of this species since the 1962 *Atlas*. Two subspecies occur in our area: subsp *sepium* is found throughout the range of the species and subsp. *roseata* is mapped separately. A third, subsp. *spectabilis*, was formerly recorded as a garden relic in Merionethshire.

Circumpolar Temperate element.

References: Atlas (217c), Grime *et al.* (1988), Hultén & Fries (1986), Meusel *et al.* (1978).

G. M. KAY

Calystegia sepium subsp. *roseata*

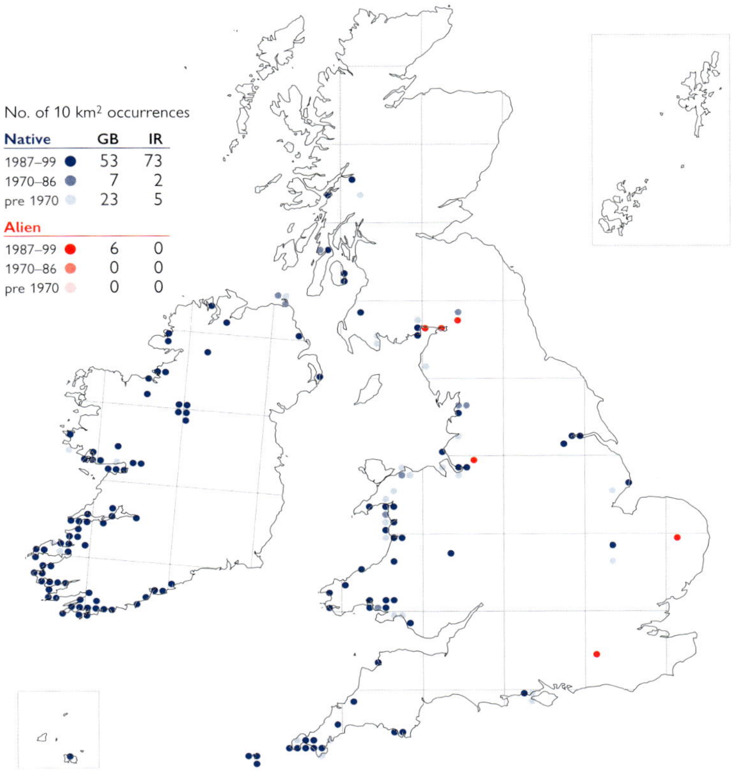

This subspecies typically occurs in brackish habitats at the upper edges of saltmarshes, in reedbeds, and in grassy waste places in coastal regions. It is also found inland on river-sides, in fens and on rough ground. Lowland.

Native. *C. sepium* subsp. *roseata* was first described in 1967 and mapped by Perring & Sell (1968). It is now better recorded, and the distribution appears to be largely stable.

Oceanic Temperate element; also known from temperate S. America, Easter Island and Australasia.

References: Atlas Supp. (49b), Brummitt & Chater (2000), Sell (1967).

G. M. KAY

Calystegia sepium × *C. silvatica* (*C.* × *lucana*)

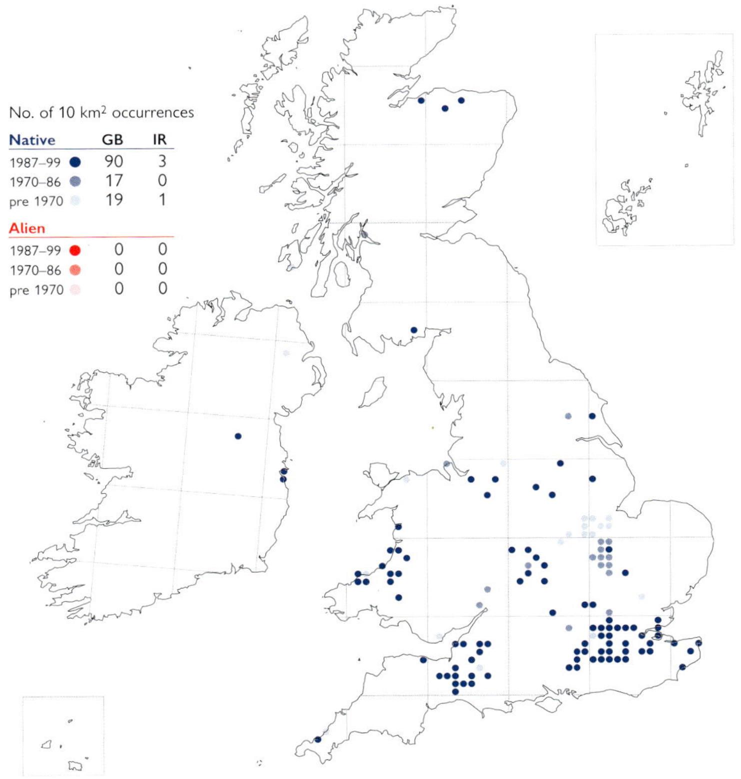

This hybrid occupies similar habitats to the parents, including hedges, fences and waste ground. Clones vary in fertility, from highly fertile to nearly sterile. Lowland.

A spontaneous hybrid between native and alien parents. Trends in the distribution of this hybrid are difficult to ascertain. It is still very poorly recorded.

Widespread in S. Europe and N. Africa.

Reference: Stace (1975).

G. M. KAY

Calystegia pulchra Hairy Bindweed

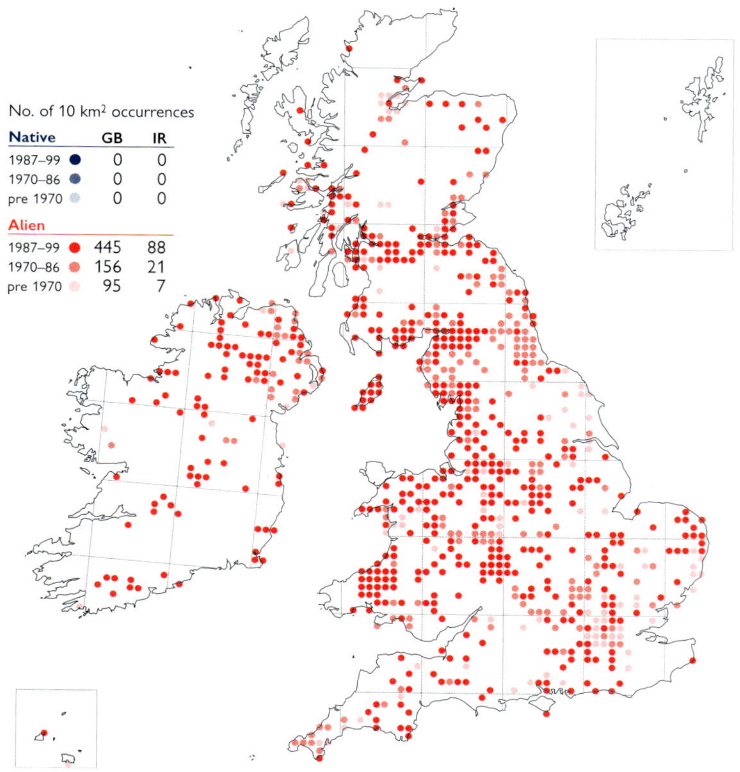

A climbing perennial found naturalised in hedges and on waste ground, usually close to habitation. Reproduction is mainly vegetative and seed rarely ripens in our area. Lowland.

Neophyte (change +2.78). *C. pulchra* has been cultivated in Britain since 1823. The earliest specimens from our area were collected in 1867 (from London, though perhaps a garden plant) and 1884 (Edinburgh). It was not recognised in the wild until 1956, and described as a distinct species by Brummitt & Heywood (1960). It is much better recorded now than it was in the 1962 *Atlas*.

Origin unknown – either originating in cultivation or a native of N.E. Asia; widely naturalised in N. & C. Europe.

References: Atlas (217d), Atlas Supp. (49c), Brummitt & Chater (2000), Meusel *et al.* (1978).

G. M. KAY

Calystegia silvatica Large Bindweed

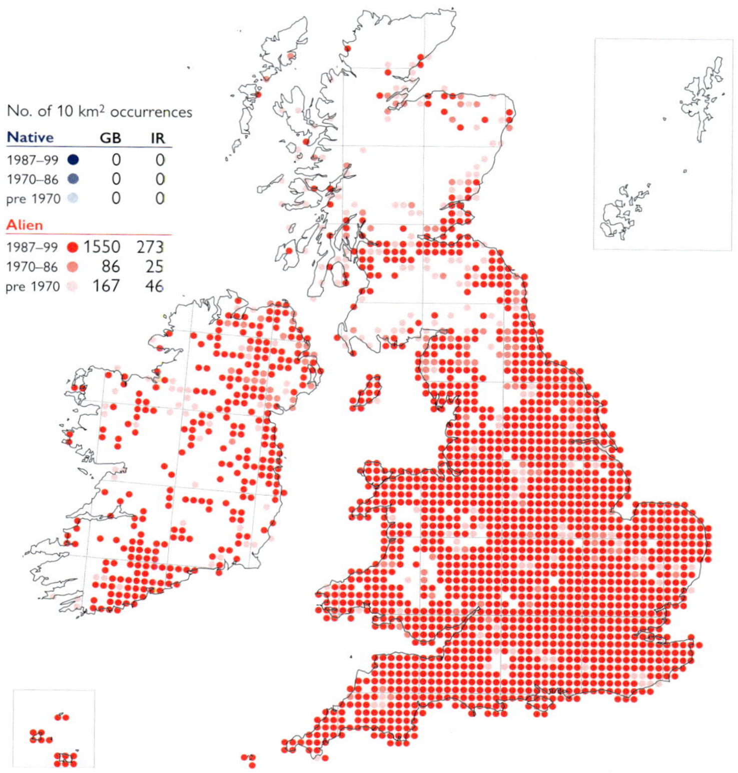

A climbing or scrambling perennial of hedges, fences, gardens and waste ground. It also spreads into semi-natural habitats away from habitation. Seed-set is often very poor. 0–350 m (Ravenstonedale, Westmorland).

Neophyte (change +0.47). *C. silvatica* has been cultivated in Britain since 1815. It was collected from the wild in 1863 but not recognised as distinct from *C. sepium* until 1948, by which time it was widespread. The distribution is now stable. Two subspecies occur in Britain; subsp. *disjuncta* is common, but subsp. *silvatica* may be more frequent in W. Britain and Ireland.

Native of the Mediterranean region and S.W. Asia.

References: Atlas (218a), Brummitt (1996), Brummitt & Chater (2000), Hultén & Fries (1986), Lousley (1948a), Meusel *et al.* (1978).

G. M. KAY

Cuscuta europaea Greater Dodder

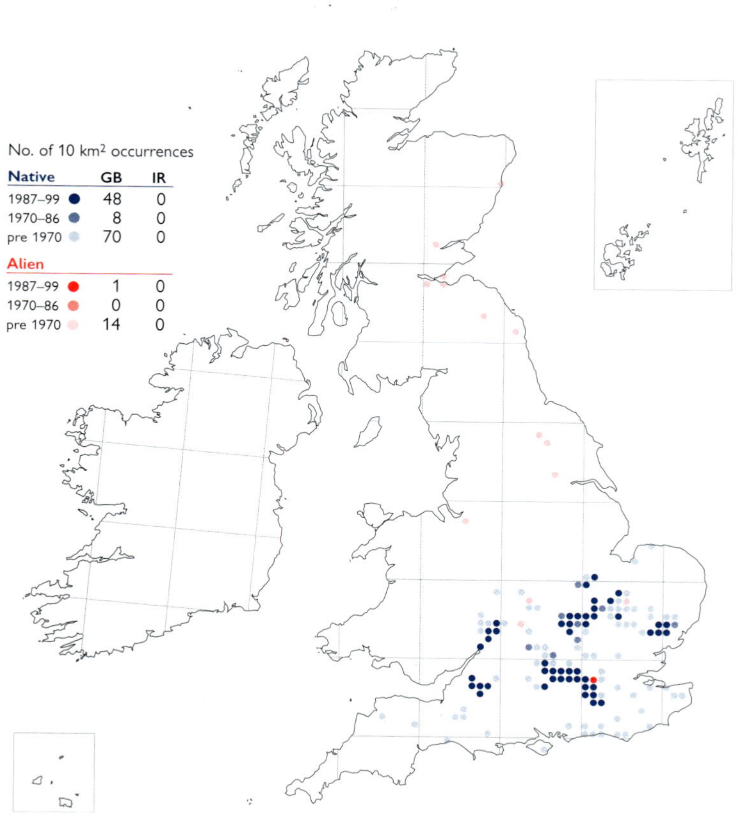

An annual, rarely perennial, rootless twining holoparasite of damp nitrophilous places, especially the banks of rivers, but also hedges and ditches. Its primary host is usually *Urtica dioica*, rarely *Humulus lupulus* or other species, whence it can spread to a wide spectrum of secondary hosts. It often grows close to flowing water, which may disperse the seeds. Lowland.

Native (change +0.04). *C. europaea* has been found in many new sites in the E. Midlands since 1970, presumably due to better recording. However, it appears to be in decline in some areas, though is probably still overlooked in others.

Eurosiberian Temperate element; widely naturalised outside its native range.

References: Atlas (218c), Hultén & Fries (1986), Meusel *et al.* (1978), Stewart *et al.* (1994), Verdcourt (1948).

F. J. RUMSEY

Cuscuta epithymum Dodder

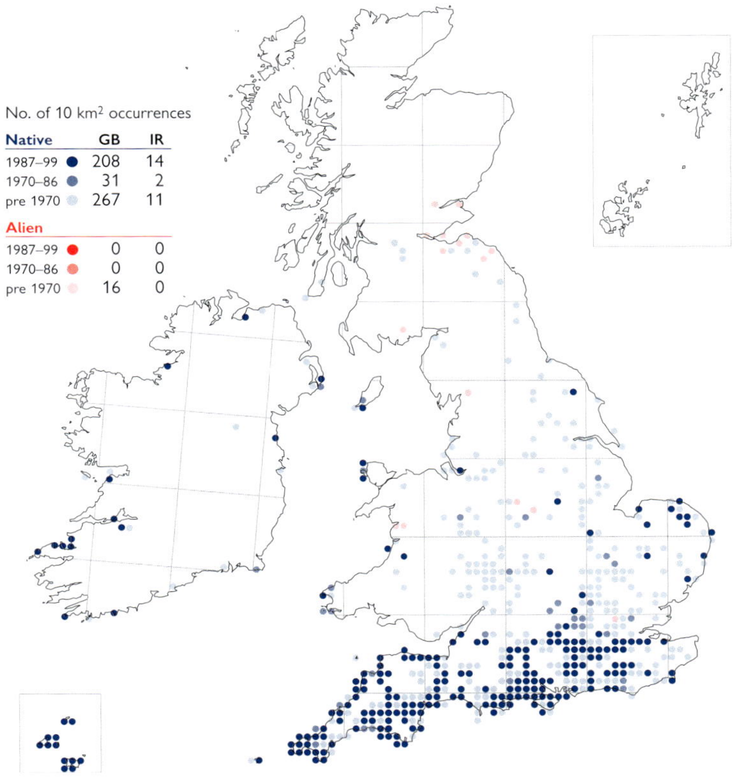

An annual, rarely perennial, rootless twining herb, parasitic on the stems of a wide variety of small shrubs and herbs (most frequently *Calluna vulgaris*, *Thymus polytrichus*, *Ulex gallii* and *U. minor*) on heathland, chalk downland and fixed dune grasslands. It is also casual on field crops and in arable field-borders at the northern and western extent of its range. Lowland.

Native (change −1.28). The loss of lowland heath, ploughing of chalk downlands, and an increase in scrub have caused a decline in this species since 1930. Whilst it is still locally abundant, the map suggests a continuing substantial decline.

Eurosiberian Southern-temperate element; widely naturalised outside its native range.

References: Atlas (218d), Hultén & Fries (1986), Meusel *et al.* (1978).

F. J. RUMSEY

Menyanthes trifoliata Bogbean

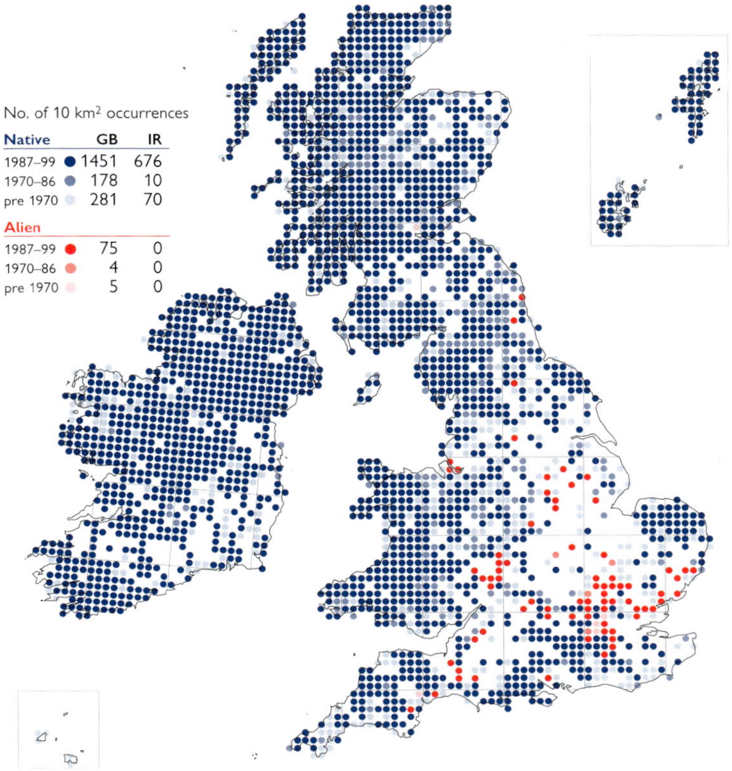

M. trifoliata is a rhizomatous perennial which grows as an emergent at the shallow edge of lakes, pools or slow-flowing rivers, or in swamps, flushes or dune-slacks. It tolerates a wide range of water chemistry, but is intolerant of shade. 0–1005 m (Beinn Heasgarnich, Mid Perth).

Native (change −0.04). Although this species remains frequent over much of our area, it has decreased in S.E. England because of the drainage of wetlands in both historic and recent times. In some of these areas it is now more frequent as a planted ornamental than as a native.

Circumpolar Boreo-temperate element.

References: Atlas (210b), Hewett (1964), Hultén & Fries (1986), Meusel *et al.* (1978), Preston & Croft (1997).

C. D. PRESTON

Nymphoides peltata Fringed Water-lily

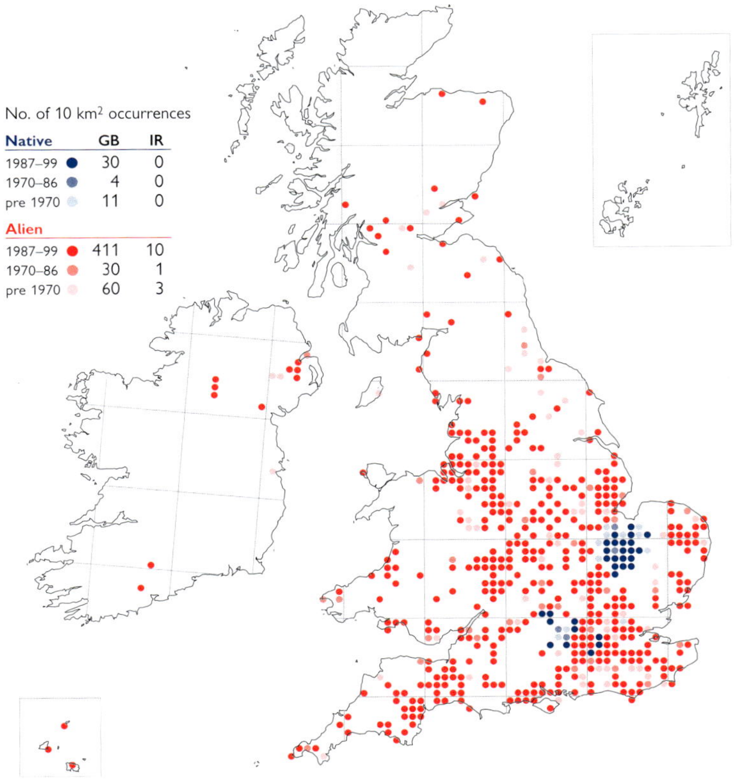

A rhizomatous perennial which grows in water 0.5–2 metres deep in lakes, ponds, slowly flowing rivers, canals and large fenland ditches. As a native it is a plant of calcareous and eutrophic water. Lowland.

Native (change +2.81). *N. peltata* had declined as a native in the Thames valley by 1960, but remains frequent in the fenlands of East Anglia. It is popularly grown as an ornamental, and has become widely naturalised from material deliberately planted in the wild or discarded as surplus stock. These records obscure some of the older native occurrences.

Eurasian Temperate element; widely naturalised outside its native range.

References: Atlas (210c), Atlas Supp. (48c), Hultén & Fries (1986), Meusel *et al.* (1978), Preston & Croft (1997), Stewart *et al.* (1994).

C. D. PRESTON

Polemonium caeruleum Jacob's-ladder

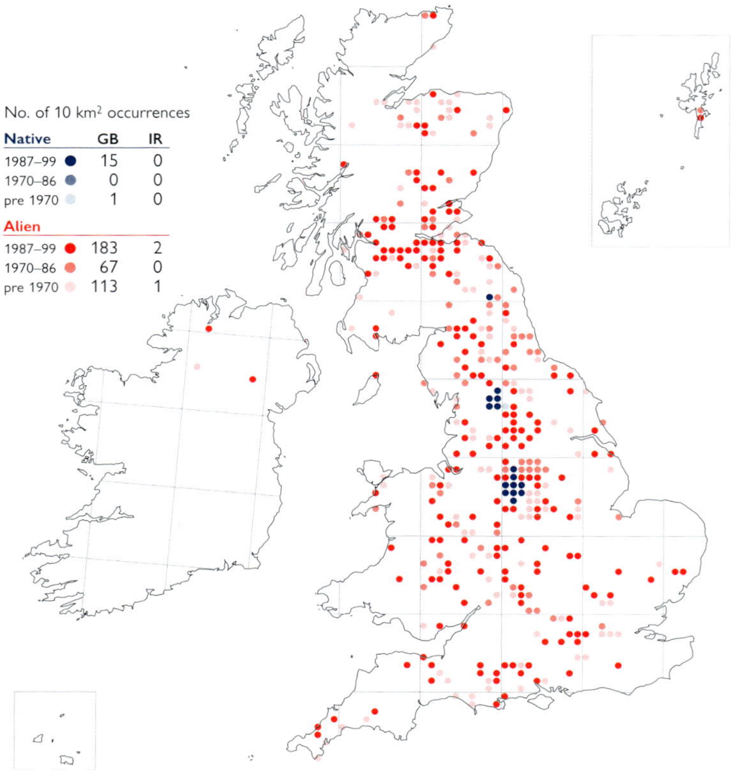

A clump-forming perennial herb, largely restricted as a native to steep but stabilised limestone screes, usually in partial shade, but also found on andesite debris and river-cliffs in Northumberland. It is confined to sites where the soil remains moist. Alien populations occur along hedgerows, on river banks and in other places near habitation. 200–580 m (Pen-y-ghent, Mid-W. Yorks.) as native, but down to sea level as an alien.

Native (change +1.17). The native range seems generally stable. *P. caeruleum* is popular in gardens, but confusion of garden and native forms is unlikely (Pigott, 1958).

Eurosiberian Boreal-montane element, with a continental distribution in W. Europe; widely naturalised outside its native range.

References: Atlas (210d), Hultén & Fries (1986), Meusel *et al.* (1978), Wigginton (1999).

A. J. RICHARDS

Phacelia tanacetifolia Phacelia

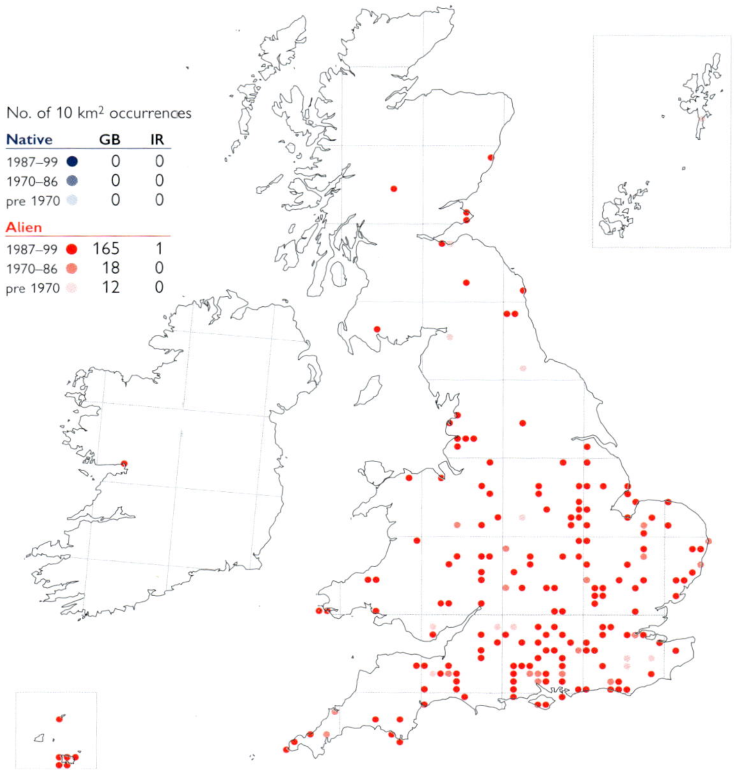

An annual grown in gardens for ornament, in allotments as a 'green manure', and in fields as a source of nectar for bees and for hoverflies that prey on aphids. It is found in cultivated ground (including the edges of arable fields), rubbish tips, waste ground, pavement cracks, and in newly sown grass leys, where it originates as a seed contaminant. It also occurs as a bird-seed and pheasant feed alien. Most populations are casual. Lowland.

Neophyte. This species has been cultivated in Britain since 1832. It was first recorded in the wild in 1885 (S.W. Yorks.), and appears to be increasing.

Native of western N. America (California).

References: Briggs (1997a), Leach (1997).

H. J. KILLICK

Lithospermum purpureocaeruleum Purple Gromwell

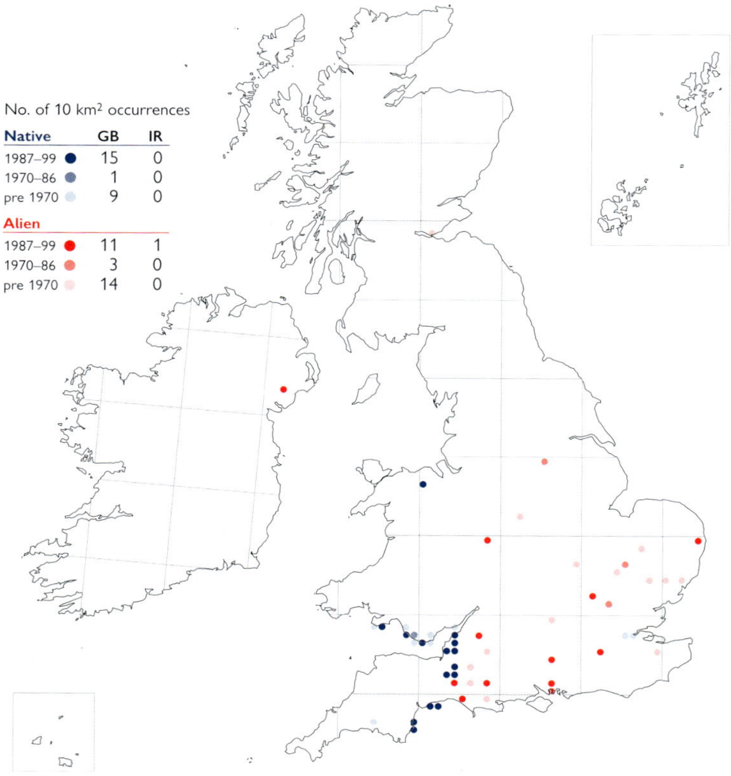

A perennial herb with creeping woody stems occurring in chalk and limestone districts in two distinct habitats. Inland, it grows in woodland edges and rides, and on lanesides and banks in partial shade. On the coast, it is found amongst naturally dwarfed, open scrub on slumped cliffs, slopes and crags. It spreads by seed and from the stems rooting at nodes. It also occurs as a garden escape on roadsides and waste ground. Lowland.

Native (change –0.33). The native distribution of this species seems to have been more or less stable since 1930, though some populations have been shaded out in neglected or unmanaged woodland. It has increased as an alien since the 1962 *Atlas*.

European Southern-temperate element.

References: Atlas (215d), Meusel *et al.* (1978), Wigginton (1999).

D. WELCH

Lithospermum officinale Common Gromwell

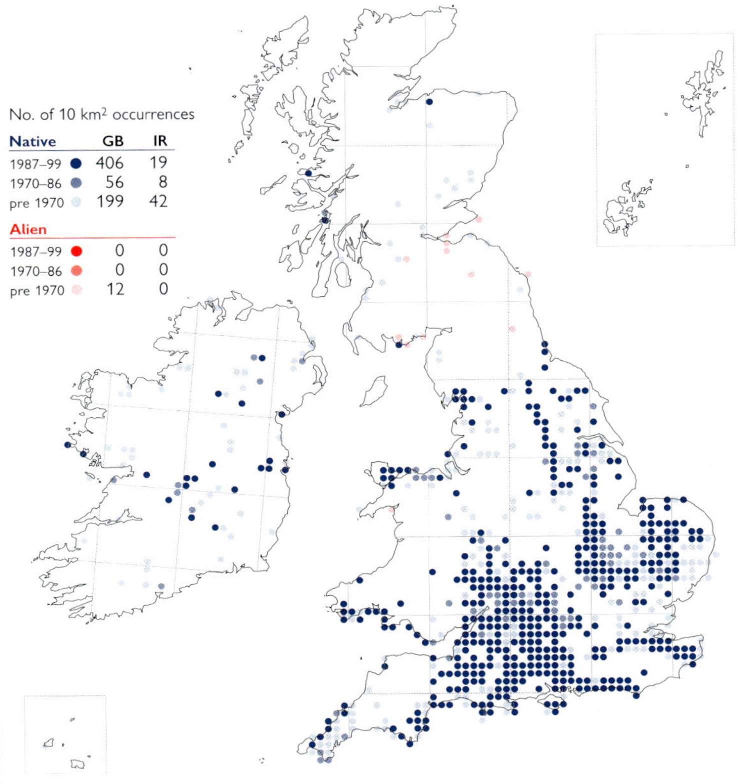

A shortly rhizomatous perennial herb which grows in grassland, hedgerows and wood margins, mostly on base-rich soils. Lowland.

Native (change –0.59). There seems to have been something of a decline in this species in England since the 1962 *Atlas*. Most of the losses in Scotland and N. England occurred before 1930. It has been considered alien in Scotland, but its long occurrence (from 1764 in Angus) at some sites and its semi-natural, base-rich habitats (for example, rocky riversides and limestone scree) suggest otherwise.

Eurosiberian Temperate element; widely naturalised outside its native range.

References: Atlas (216a), Hultén & Fries (1986), Meusel *et al.* (1978).

D. WELCH

Lithospermum arvense Field Gromwell

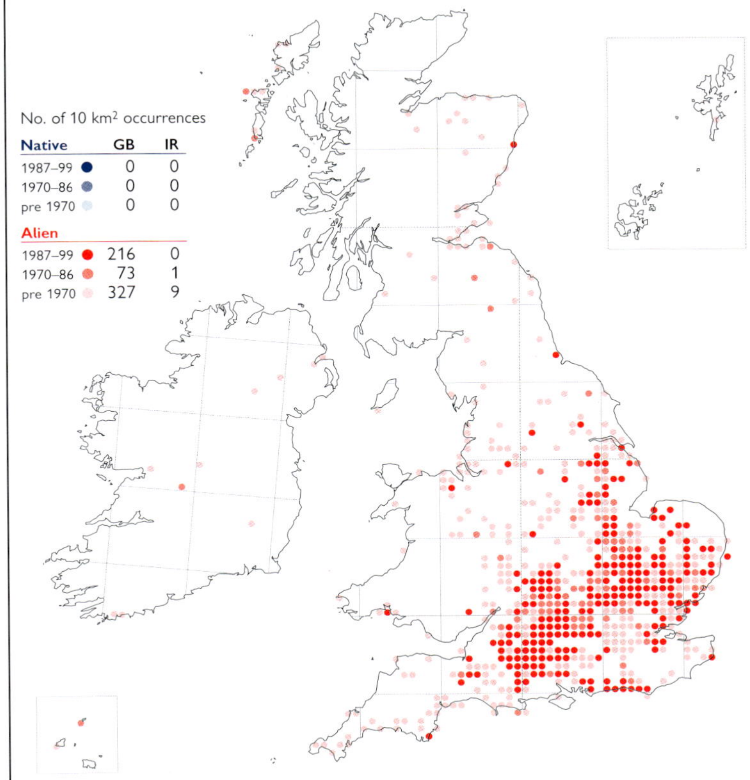

An annual of arable fields, occasionally found on waste ground and in other disturbed habitats, favouring light, dry, calcareous soils. Seed is short-lived and populations depend upon regular disturbance for survival. Lowland.

Archaeophyte (change –1.91). Archaeological evidence suggests that this species has been an arable weed in Britain since the Bronze Age. It has declined substantially since the 1950s because of agricultural intensification, and in many areas it is now uncommon in arable fields. Seed can be transported with grain, resulting in casual populations outside its core range.

As an archaeophyte *L. arvense* has a Eurosiberian Southern-temperate distribution; it is widely naturalised outside this range.

References: Atlas (216b), Hultén & Fries (1986), Meusel *et al.* (1978), Svensson & Wigren (1986).

D. WELCH

Echium vulgare Viper's-bugloss

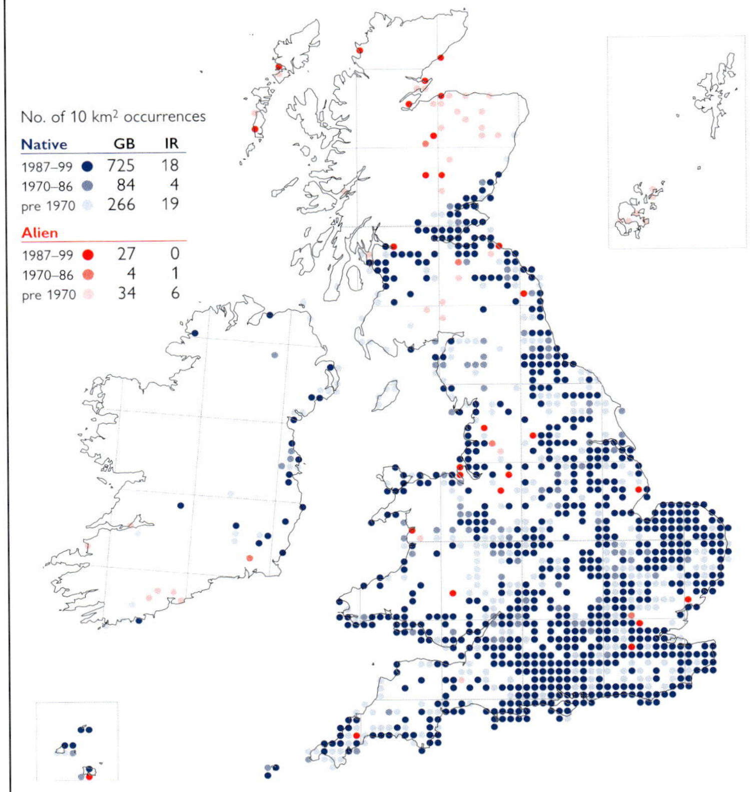

A biennial of grassy and disturbed habitats on well-drained soils. It is found in bare places on chalk and limestone downs, on heaths, in quarries and chalk-pits, in cultivated and waste land, along railways and roadsides, and by the coast on cliffs, sand dunes and shingle. Generally lowland, but formerly reaching 365 m as an alien at Braemar (S. Aberdeen).

Native (change –0.24). Since 1930 *E. vulgare* has declined somewhat in frequency, due to agricultural intensification, reclamation, and the development of neglected ground.

Eurosiberian Temperate element; widely naturalised outside its native range.

References: Atlas (216d), Hultén & Fries (1986), Meusel *et al.* (1978).

D. WELCH

Echium plantagineum Purple Viper's-bugloss

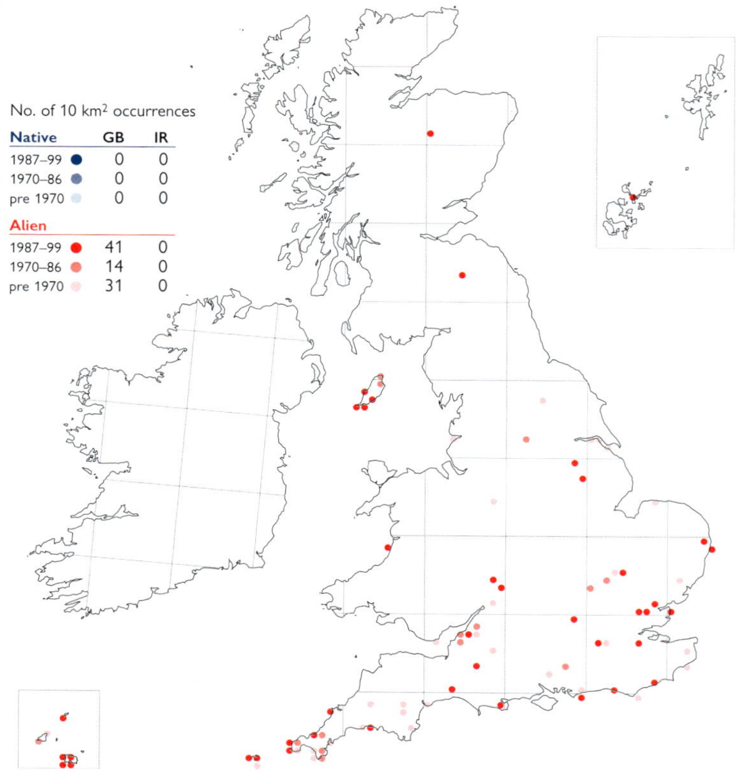

An annual or biennial herb growing as a weed in arable fields, on cliffs and in open sandy habitats by the coast. It has a long-lived seed bank, and populations vary greatly in size from year to year. It also occurs casually as a rare garden escape or outcast. Lowland.

Archaeophyte (change +0.36). *E. plantagineum* has been known from Jersey since 1690 but was first recorded in Cornwall in 1856. It is well-established in Jersey, West Penwith (W. Cornwall) and the Isles of Scilly, where it has been considered native by many authors. Elsewhere it is a casual.

As an archaeophyte *E. plantagineum* has a Mediterranean-Atlantic distribution; it is widely naturalised outside this range, and a notorious weed in Australia.

References: Atlas (217a), Bolòs & Vigo (1995), Butterfield (1999), Wigginton (1999).

D. WELCH

Pulmonaria officinalis Lungwort

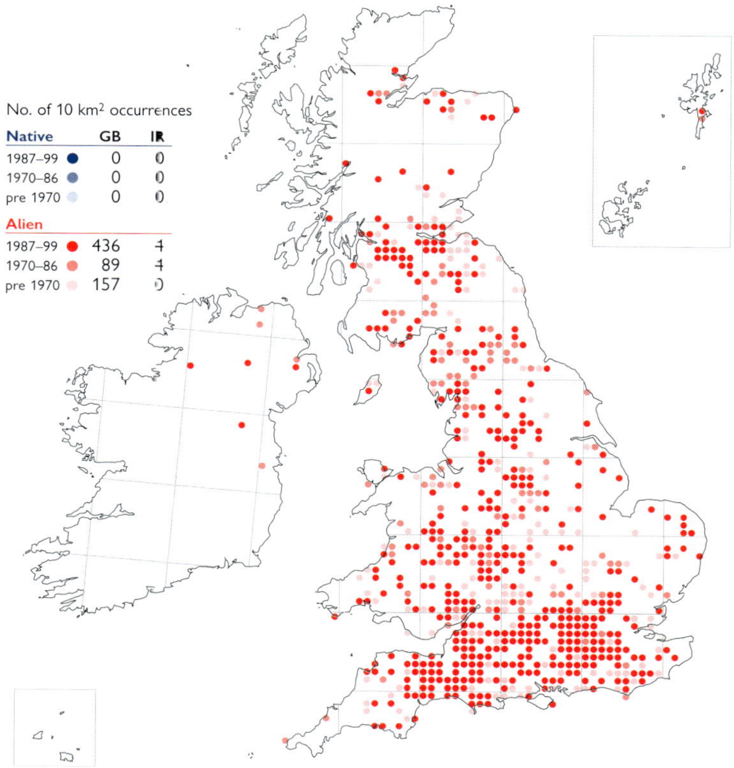

A perennial herb, naturalised in woodlands and scrub, on banks and rough ground, and also occurring on rubbish tips and waste ground. Generally lowland, but reaching 385 m (Forest-in-Teesdale, Co. Durham).

Neophyte (change +1.77). This species was cultivated in Britain before 1597, and is now commonly grown in gardens. Though some occurrences were treated as possibly native in the 1962 *Atlas*, it is now regarded as an introduction at all its British sites; it was recorded from the wild by 1793. The number of records has increased since the 1962 *Atlas*, particularly in southern England, reflecting both increased abundance and more intensive recording of aliens.

A European Temperate species, absent as a native from much of W. Europe.

References: Atlas (213b), Hultén & Fries (1986), Meusel *et al.* (1978).

D. WELCH

Pulmonaria obscura Suffolk Lungwort

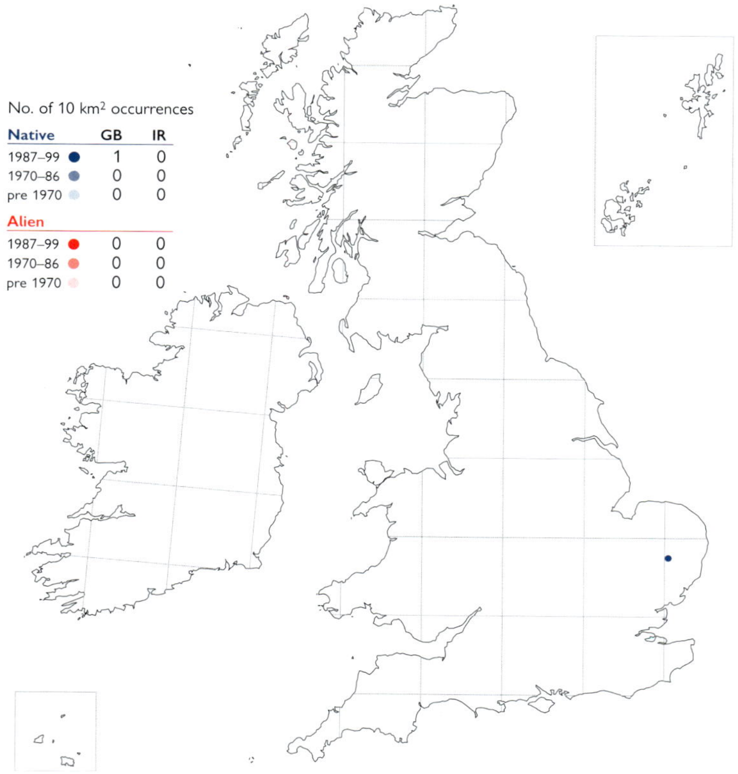

This perennial herb grows on poorly drained chalky boulder clay in three ancient woods with a long history of coppice management; *Acer campestre*, *Corylus* and *Fraxinus* are the usual dominant trees. It reproduces by seed and by rhizomatous spread. Lowland.

Native. *P. obscura* is considered native in Britain because its habitat and associated species are similar to those in its native N. Europe, and because it is rarely cultivated here. It has declined in two woods since the 1930s due to the cessation of coppicing, and has been lost from another through habitat destruction.

European Temperate element, with a continental distribution in W. Europe.

References: Birkinshaw & Sanford (1996), Hultén & Fries (1986), Wigginton (1999).

D. WELCH

Pulmonaria longifolia
Narrow-leaved Lungwort

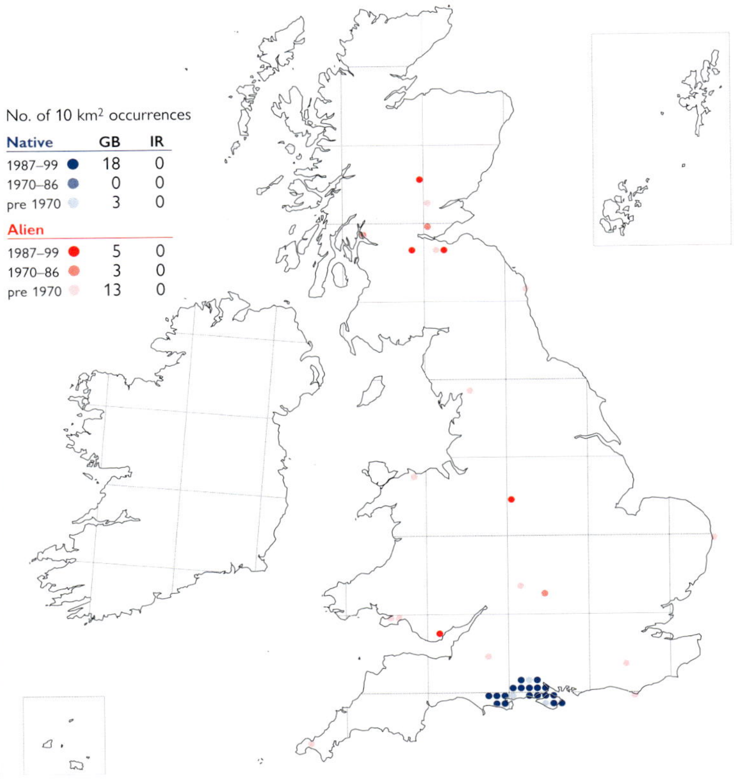

A perennial herb of lightly shaded habitats, mostly on base-rich clay soils, in coppiced woodland, wood-pasture and *Pteridium aquilinum* heathland; it also grows in hedge banks and marl-pits. Though it seeds freely and reproduces vegetatively, even vigorous colonies seldom spread into apparently suitable contiguous ground. Lowland.

Native (change –0.01). Although *P. longifolia* is mapped here in a few more 10-km squares than in the 1962 *Atlas*, this is mainly due to more intensive recording, and some colonies have been lost, particularly from hedge banks. It is sometimes grown in gardens, and occasional escapes have been noted.

Oceanic Temperate element.

References: Atlas (213a), Hultén & Fries (1986), Meusel *et al.* (1978), Stewart *et al.* (1994).

D. WELCH

Symphytum officinale Common Comfrey

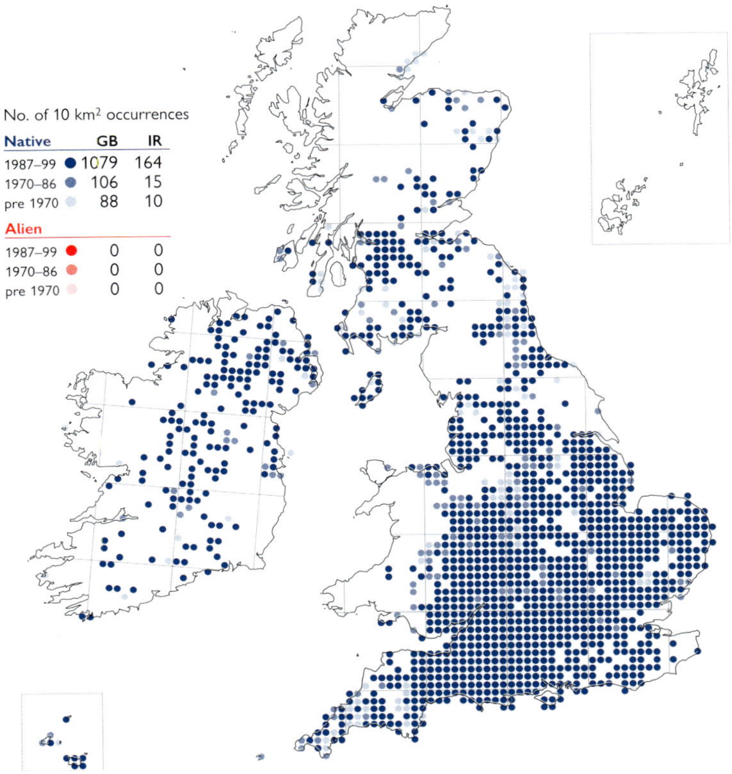

This tall perennial herb occurs on the banks of streams and rivers, in ditches, fens and marshes, and on damp road verges. Generally lowland, reaching 320 m near Buxton (Derbys.).

Native (change +0.34). *S. officinale* was over-recorded for *S.* × *uplandicum* in the 1962 *Atlas* and this confusion still obscures its true distribution. It may well be alien in much of N. and W. Britain and in Ireland but it is impossible to determine the native range with any certainty and all records are mapped as if they are native. Both diploid and tetraploid cytotypes occur and are morphologically separable. *S.* × *uplandicum* sometimes back-crosses with this parent (Perring, 1994).

European Temperate element; also in C. Asia and widely naturalised outside its native range.

References: Atlas (211c), Hultén & Fries (1986), Meusel *et al.* (1978). D. WELCH

Symphytum asperum × *S. officinale* (*S.* × *uplandicum*) Russian Comfrey

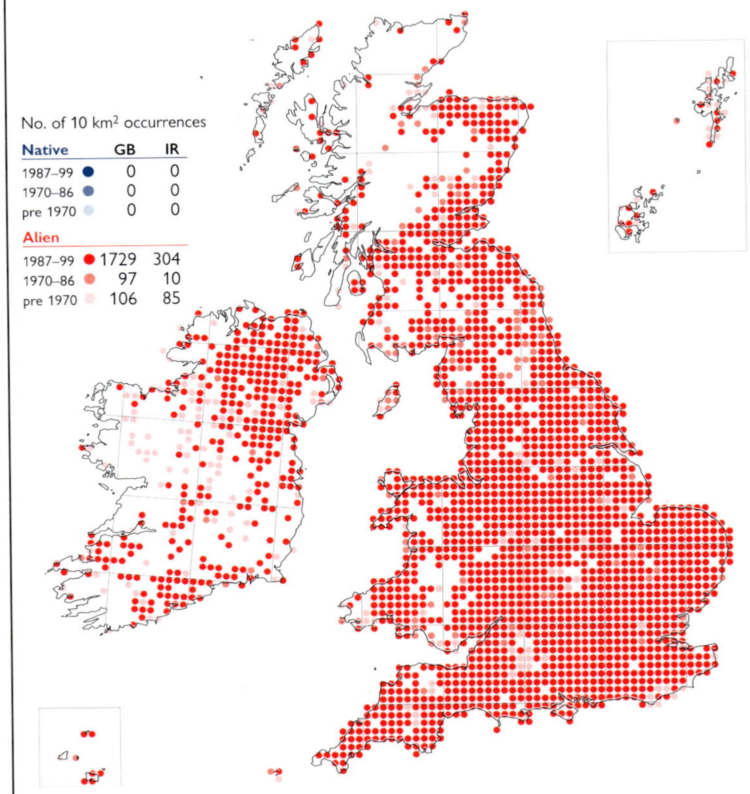

The habitats of this perennial herb include rough and waste ground, railway banks, road-sides, hedge banks and woodland margins. Generally lowland, but reaching 365 m at Alston (Cumberland).

Neophyte. This hybrid was introduced as a forage plant in 1870. It was widely cultivated in Britain in the late 19th and early 20th centuries, and known from the wild by 1884. Two cytotypes occur, with 2n=36 and 2n=40. The map in Perring & Sell (1968) was only provisional; it is therefore difficult to assess changes in distribution, though the plant is clearly well-established and probably increasing.

This hybrid is known from the Caucasus; it has been spread in cultivation and is naturalised in temperate Europe.

References: Atlas Supp. (49a), Perring (1994), Stace (1975). D. WELCH

Symphytum tuberosum Tuberous Comfrey

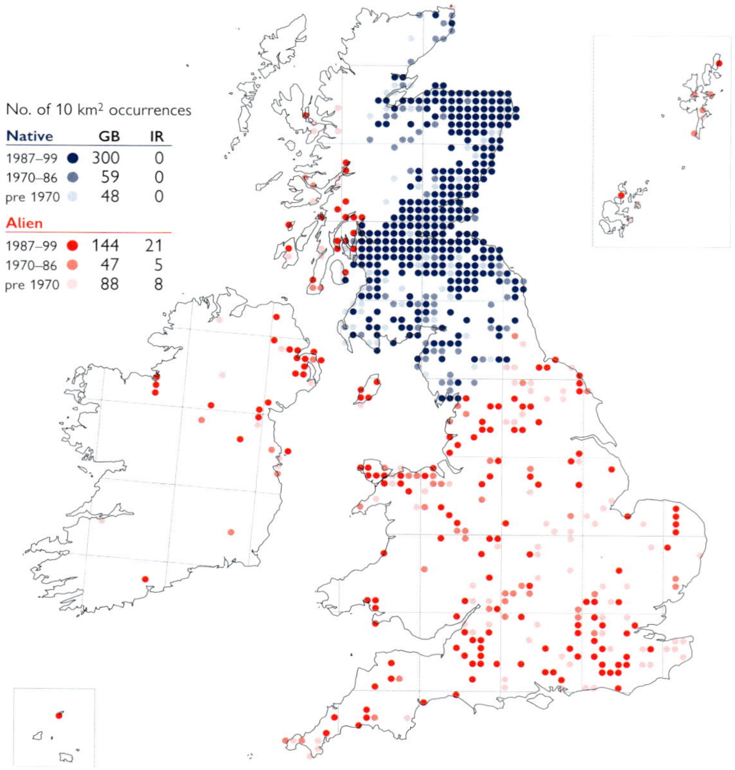

The native habitats of this perennial herb are damp woodland, ditches, stream and river banks, where it occurs in both shaded and open situations. As an alien, it occurs on roadside verges, waste ground and other disturbed sites. Generally lowland, but reaching 335 m in Mid Perth.

Native (change +0.11). This species has long been considered introduced to Ireland, Wales and most of England. In Scotland too its native status is sometimes questioned, partly in view of its late discovery (1777), and the limit of any native range has been obscured by escapes. There is no significant change in its presumed native distribution since the 1962 *Atlas*, though alien records are much more frequent in England.

European Temperate element.

References: Atlas (212b), Meusel *et al.* (1978).

D. WELCH

Symphytum 'Hidcote Blue' *(S. grandiflorum* × *?S.* × *uplandicum)* Hidcote Comfrey

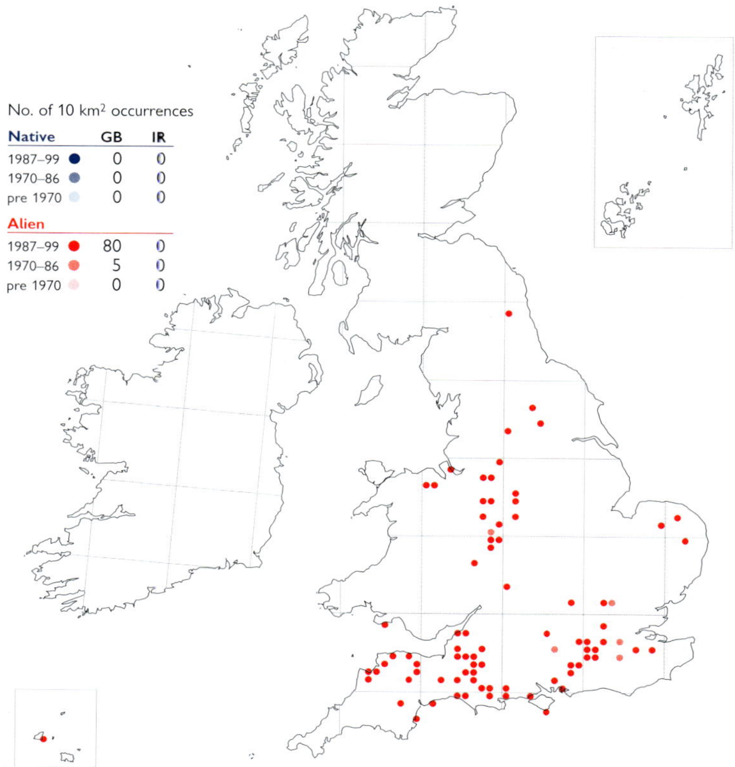

This perennial herb occurs on roadsides and in hedges and woodland. It is a garden escape or throw-out which can occasionally become naturalised. Lowland.

Neophyte. This plant, raised at Hidcote (E. Gloucs.) not long before 1930, grows aggressively in gardens and is often discarded. It was first noticed in the wild in 1979, and is evidently increasing.

A hybrid of garden origin.

Reference: Leslie (1982).

D. WELCH

Symphytum grandiflorum Creeping Comfrey

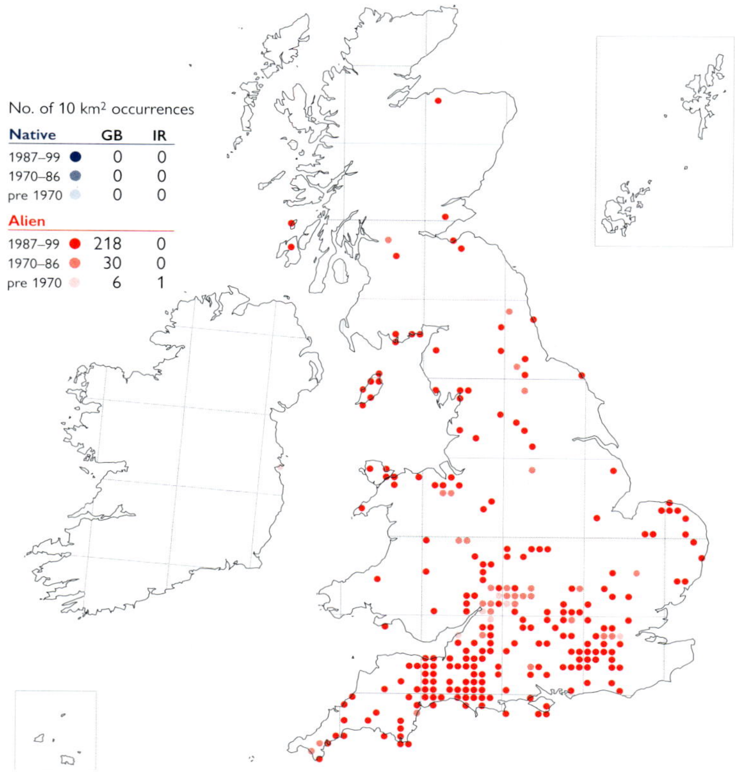

A perennial herb widely grown under trees or in the open in parks, gardens and churchyards and found naturalised in woods, hedges and on shaded banks and riversides, often spreading vigorously by stoloniferous and rhizomatous growth. Lowland.

Neophyte. *S. grandiflorum* was introduced to cultivation in Britain not long before 1900; it was recorded from the wild in 1898. It is popular in gardens, especially for ground cover, but can be extremely aggressive and is often discarded. It has doubtless increased in frequency in recent years.

Native of the Caucasus.

D. WELCH

Symphytum orientale White Comfrey

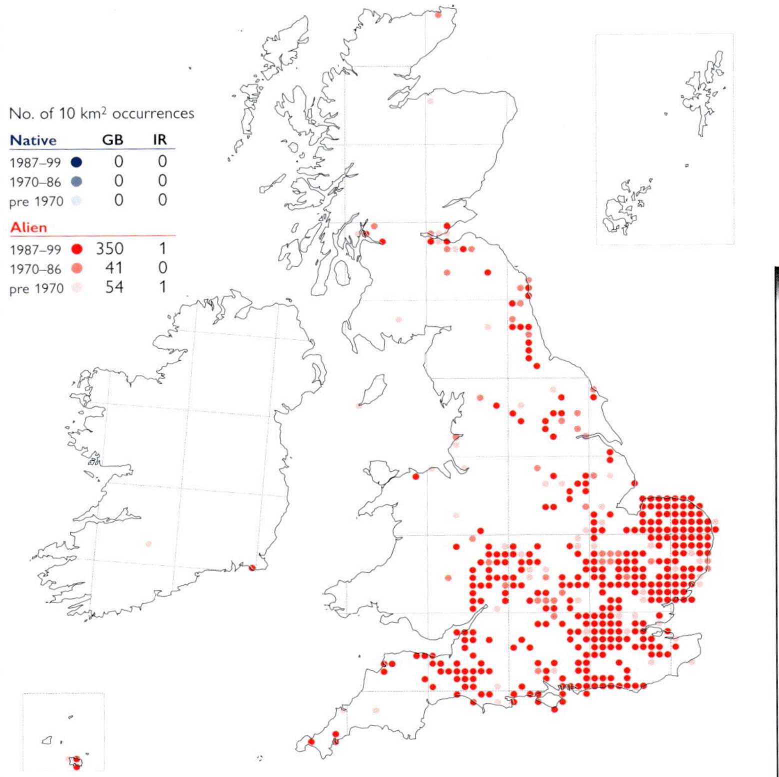

This perennial herb is found as an escape or outcast in hedgerows and copses, on lanesides, by roads and railways, and on waste ground. It is often naturalised, and sometimes regenerates from seed. Lowland.

Neophyte (change +1.83). *S. orientale* was introduced to gardens by 1752, when it was known to have been grown in Cambridge, and was known from the wild by 1849. Its distribution has increased greatly since the 1962 *Atlas* owing to a genuine spread, although part of the increase is also due to better recording of alien species.

Native of S. Russia, N.W. Turkey and the Caucasus.

Reference: Atlas (212a).

D. WELCH

Brunnera macrophylla Great Forget-me-not

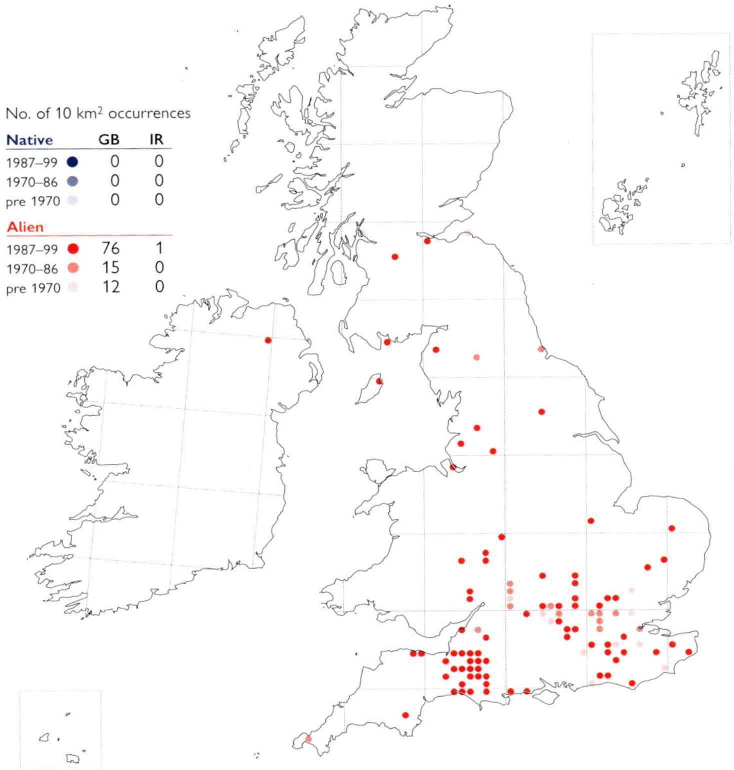

A shortly-rhizomatous clump-forming perennial herb found as a persistent garden escape or throw-out. It occurs in rough grassland, woodland and on rubbish tips. Lowland.

Neophyte. This species has been cultivated in British gardens since 1830 and is very popular. It was recorded in the wild by 1928. It may previously have been under-recorded, having sometimes been identified as *Omphalodes verna*, but even so it is evidently increasing.

Native of the Caucasus and N. Asia.

D. WELCH

Anchusa officinalis Alkanet

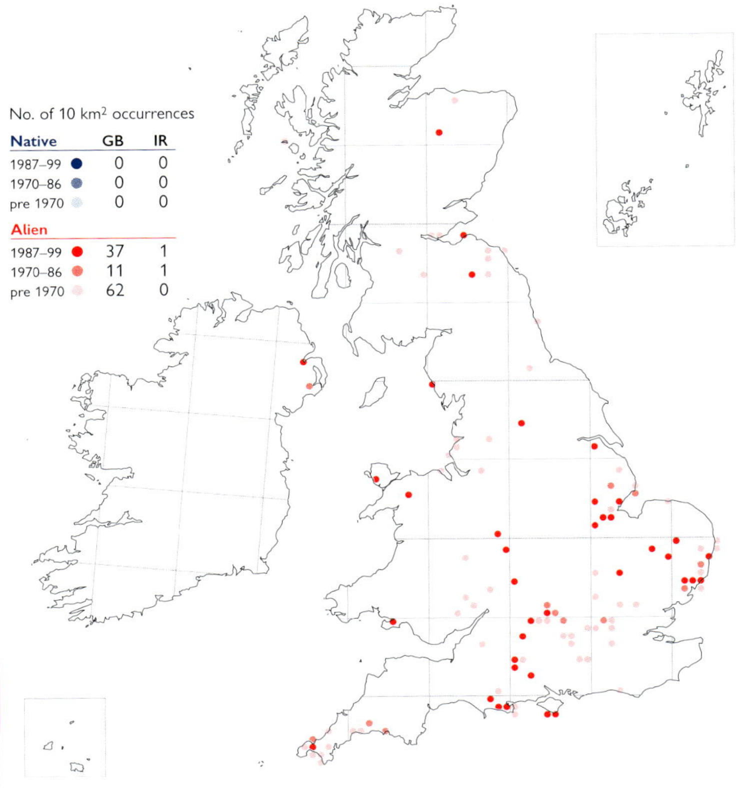

A tall perennial herb of rough and waste ground, hedgerows, railway banks, and tips, originating from gardens or from bird-seed. Populations are generally impermanent, but are occasionally naturalised, as, for example, on sand dunes at Phillack Towans (W. Cornwall). Lowland.

Neophyte. *A. officinalis* has been grown in British gardens since at least 1200 (Harvey, 1981), and was formerly also cultivated for fodder. It was first formally recorded in the wild in 1799.

A European Temperate species.

Reference: Hultén & Fries (1986).

D. WELCH

Anchusa azurea Garden Anchusa

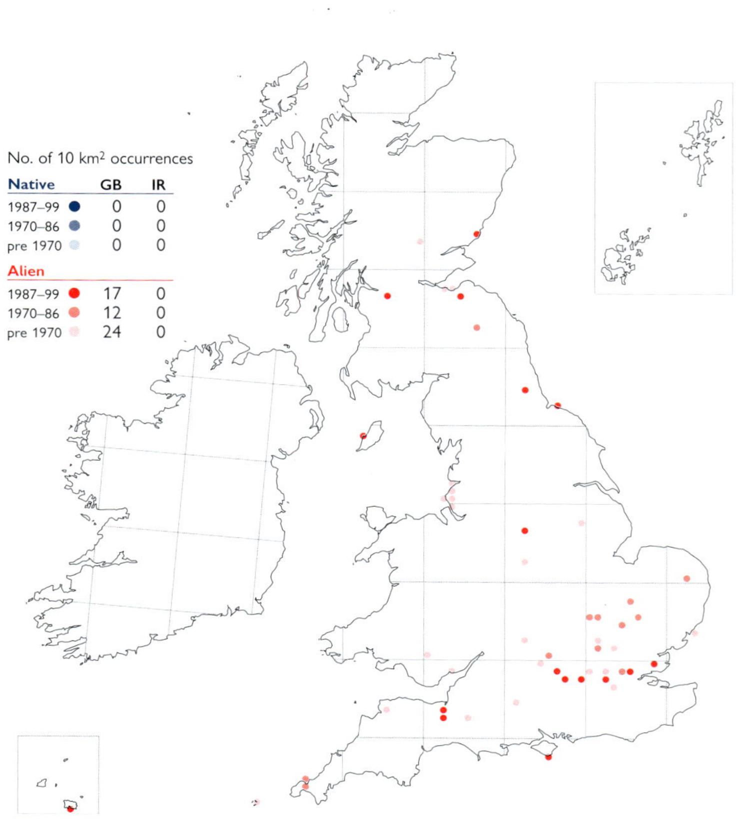

A tall perennial herb of rough and waste ground, and rubbish tips. It occurs as a garden escape, but also arises from grain and bird-seed. It is rarely naturalised. Lowland.

Neophyte. *A. azurea* was once grown as a fodder plant and was also introduced as an impurity in grain seed. It is often grown in gardens, introduced by 1597, being commoner in cultivation than *A. officinalis*. It was first recorded in the wild in 1866 (Sussex), and has been naturalised at Phillack Towans (W. Cornwall) since 1914.

Native of S. Europe, the Mediterranean region and W. Asia.

Reference: Bolòs & Vigo (1995).

D. WELCH

Anchusa arvensis Bugloss

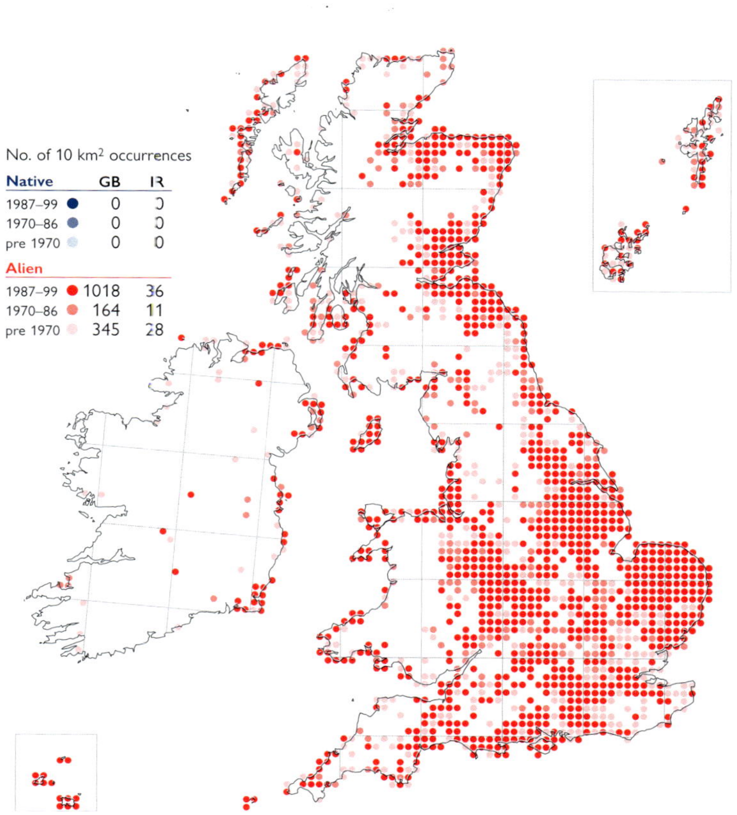

This annual weed is mostly found on well-drained soils in arable fields, but it also occurs near the sea on sandy heaths, in disturbed dunes and on waste ground. Lowland, but with a casual record at 420 m near Ballater (S. Aberdeen).

Archaeophyte (change –0.70). *A. arvensis* was mapped as 'all records' in the 1962 *Atlas*, and analysis of the database reveals that most of the losses have occurred since 1950. The main cause has been agricultural intensification, including the increased use of herbicides.

As an archaeophyte *A. arvensis* has a Eurosiberian Temperate distribution; it is widely naturalised outside this range.

References: Atlas (212d), Hultén & Fries (1986), Meusel *et al.* (1978).

D. WELCH

Pentaglottis sempervirens Green Alkanet

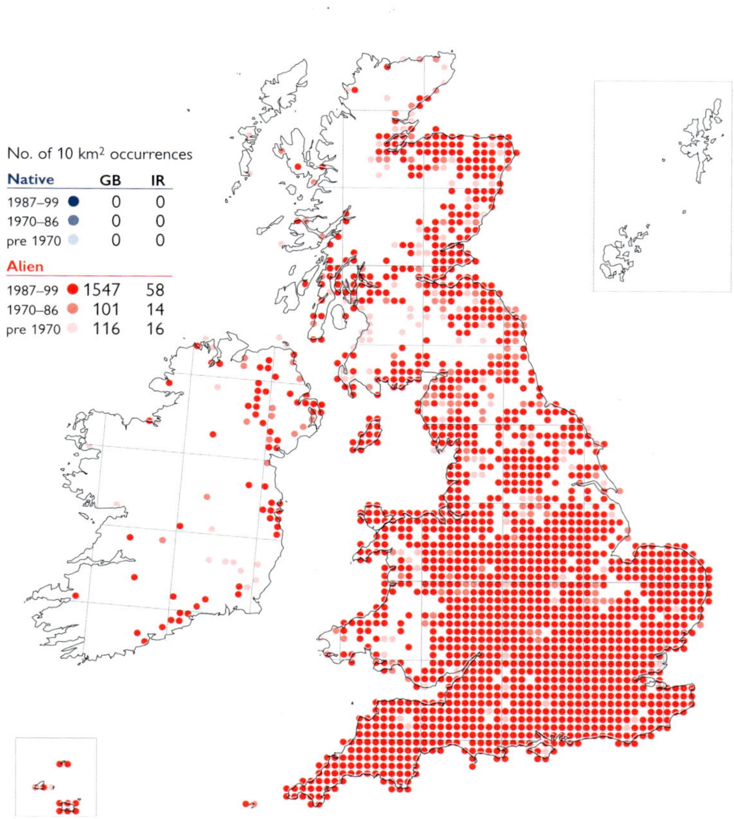

This erect perennial herb is mostly found near habitation in lightly shaded habitats, including waste ground, roadside-banks, hedgerows, scrub and woodland, but it also grows on riversides. It reproduces prolifically from seed and can be very invasive. Generally lowland, but reaching 380 m at Upper Glenlivet (Banffs.).

Neophyte (change +1.81). *P. sempervirens*, introduced to British gardens before 1700, is frequently grown in gardens. It was known from the wild by 1724. Already widespread at the time of the 1962 *Atlas*, it has increased further during the past forty years.

Native of S.W. Europe, but naturalised in W. Europe north of its native range.

Reference: Atlas (212c).

D. WELCH

Borago officinalis Borage

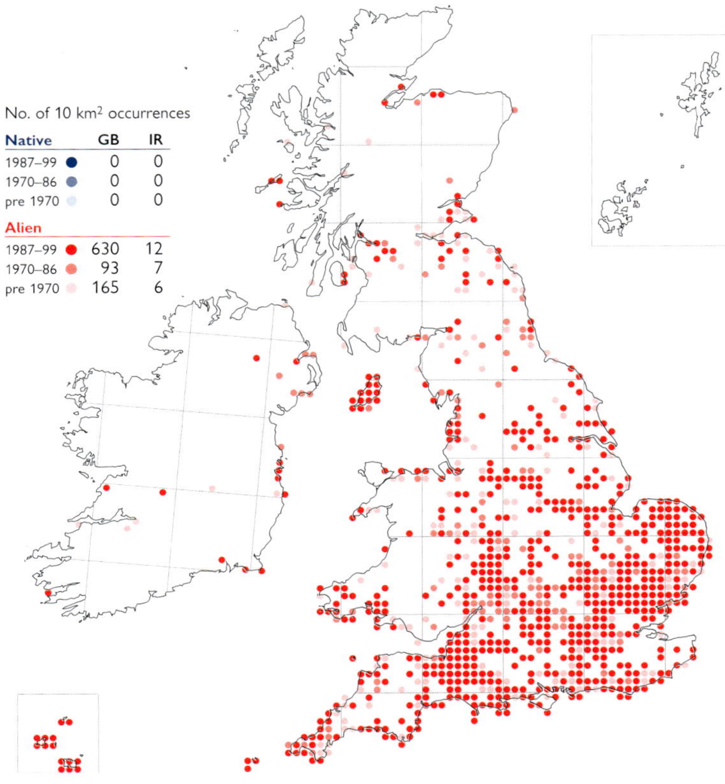

An annual occurring as a casual garden escape on roadsides and waste ground. It also arises from bird-seed and as a relic of cultivation as a minor crop. It is rarely naturalised. Lowland, with an exceptional record at 425 m (Alston, Cumberland).

Neophyte. *B. officinalis* has been grown in gardens since at least 1200 (Harvey, 1981). It was first recorded from the wild in 1777, and Dunn (1905) described it as a frequent escape from gardens. It is increasingly planted as a nectar-source for bees.

Native of the Mediterranean area; the limits of its native range have been obscured by its spread in cultivation.

Reference: Bolòs & Vigo (1995).

D. WELCH

Trachystemon orientalis Abraham-Isaac-Jacob

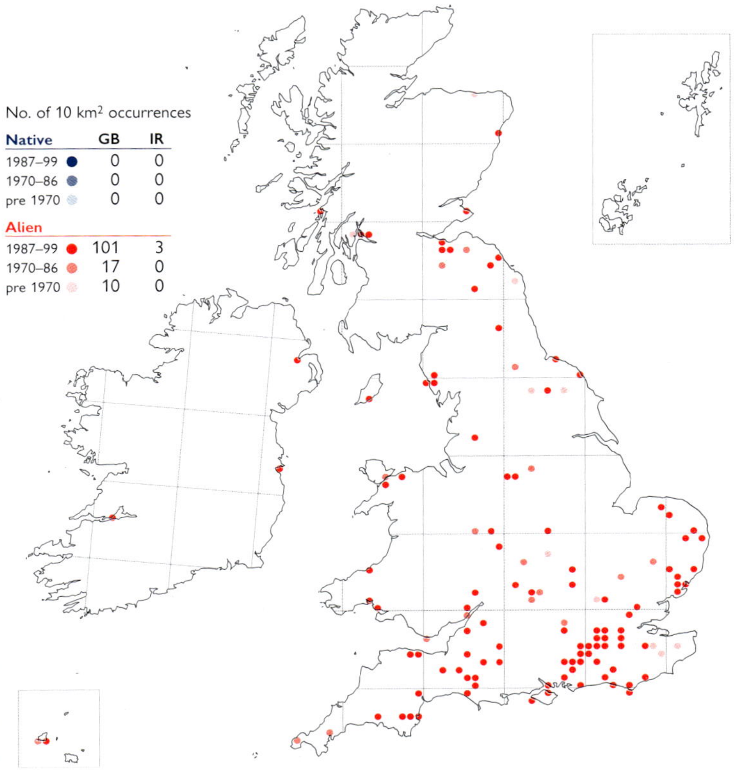

A rhizomatous perennial grown in gardens and occasionally naturalised in damp woods, laneside banks and other shaded places. Lowland.

Neophyte. Cultivated in Britain by 1752, this species was recorded from the wild by 1844 (Cumberland). Trends in its distribution are difficult to assess.

Native of E. Bulgaria, N. Turkey and the western Caucasus.

D. WELCH

Mertensia maritima Oysterplant

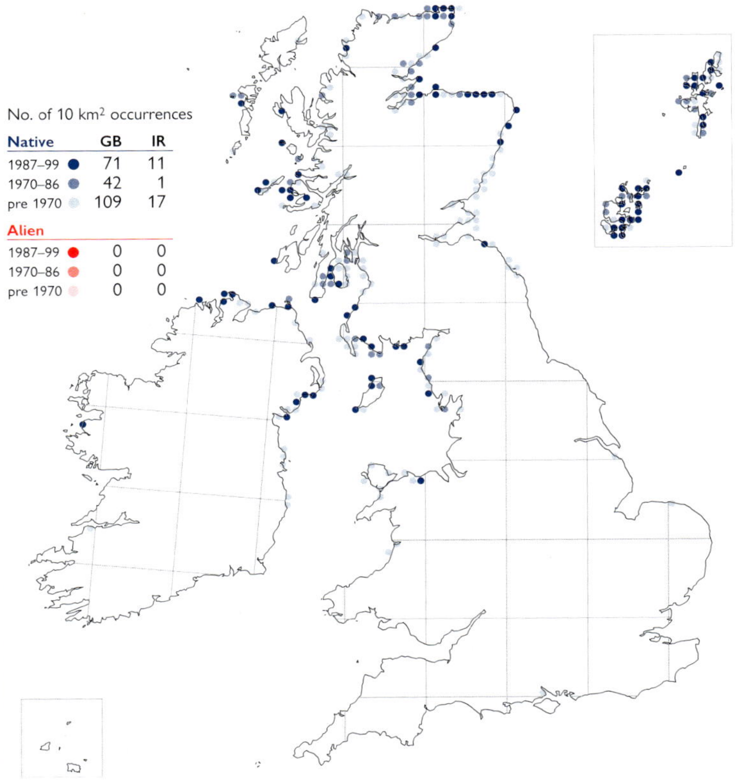

A perennial herb, usually found on gravelly beaches and shingle but sometimes on sand. It can also colonise earth and rocks tipped at the coast (Randall, 1988). Seeds can survive prolonged immersion in sea water, and dispersion in sea currents enables colonisation of new, but sometimes transient, sites. Lowland.

Native (change –0.53). The distribution of *M. maritima* has varied markedly since 1800. In Britain expansion in the far north has been balanced by contraction in the south. In Ireland it declined during the 19th century, but is now increasing. Losses result from storms, recreational pressures, shingle removal and grazing.

European Boreo-arctic Montane element; a coastal species also found in E. Asia and N. America.

References: Atlas (216c), Curtis & McGough (1988), Hultén & Fries (1986), Scott (1963a), Stewart *et al.* (1994), Welch & Innes (1999).

D. WELCH

Amsinckia micrantha Common Fiddleneck

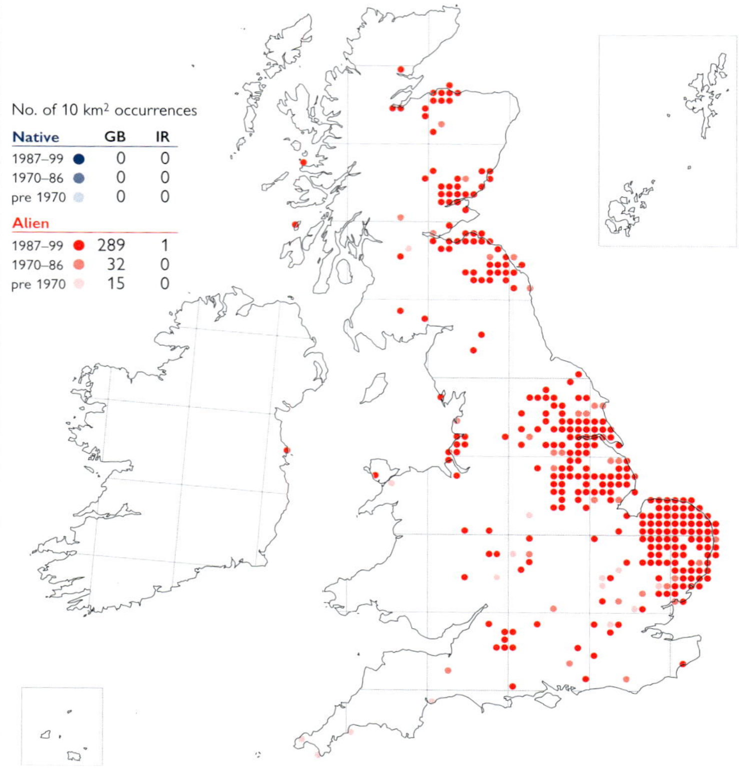

An annual weed of arable land and waste ground. It is sometimes an abundant or even pernicious weed on light, sandy soils. It arises as a contaminant of grain and from wool shoddy. Lowland.

Neophyte. *A. micrantha* was first cultivated in Britain in 1836, but was not recorded from the wild until later, in 1910, when it was introduced as a seed impurity. It has increased steadily since the 1950s, and is clearly much more frequent now than formerly. It has been confused with *A. lycopsoides*.

Native of western N. America.

References: Beckett *et al.* (1999), Clement (1999), Simpson (1982).

D. WELCH

Asperugo procumbens Madwort

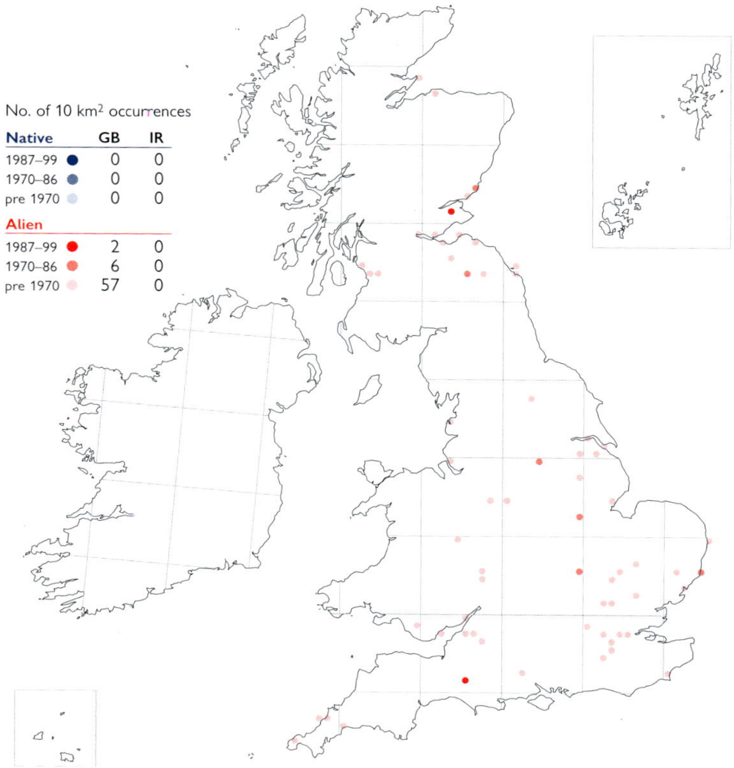

A small to medium-sized herb found in arable fields, and on rough and waste ground (especially near ports). It is introduced in grain or with wool shoddy. Most occurrences are casual but it persisted on waste ground at Auchmithie (Angus) for almost forty years. Lowland.

Neophyte. This species was first recorded in the wild in 1660. It has declined and is now very rare, improved cleaning of grain and a decline in the use of wool shoddy being the likely causes.

Native range completely obscured by its spread with cultivation; now widespread in Europe and W. & C. Asia.

Reference: Hultén & Fries (1986).

D. WELCH

Myosotis scorpioides Water Forget-me-not

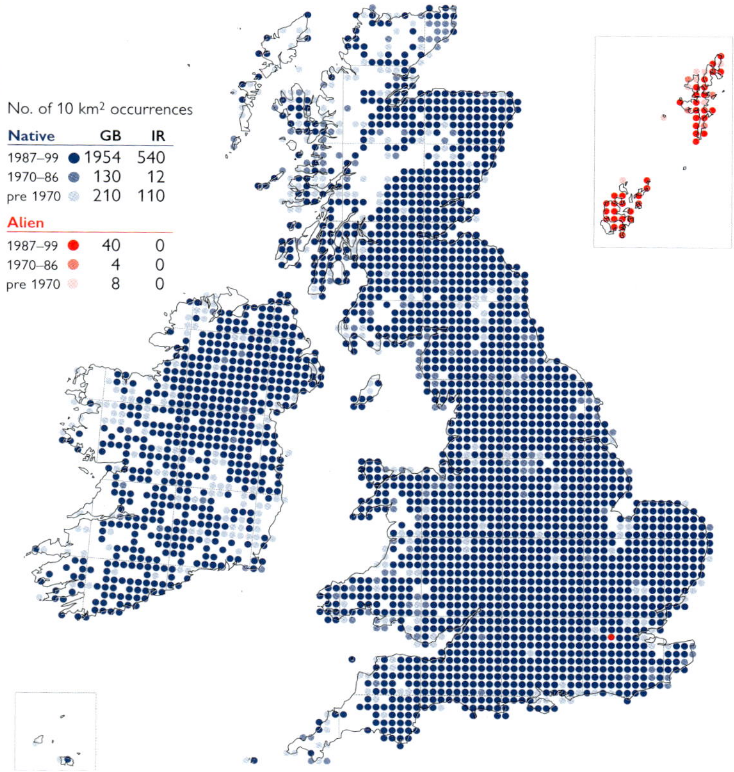

A stoloniferous or rhizomatous perennial herb found in damp or wet habitats, usually in fertile, calcareous to mildly acidic soils. It is usually terrestrial, occurring by lakes, ponds, rivers and streams, in marshes and in fens, but may sometimes be aquatic, forming submerged patches or floating rafts. 0–600 m (Moor House, Westmorland).

Native (change –0.77). The distribution of *M. scorpioides* has been generally stable over many decades. It has sometimes been confused with *M. secunda*, and some errors may be included on the map.

Eurosiberian Temperate element; widely naturalised outside its native range.

References: Atlas (213c), Grime *et al.* (1988), Hultén & Fries (1986), Preston & Croft (1997), Welch (1967).

D. WELCH

Myosotis secunda Creeping Forget-me-not

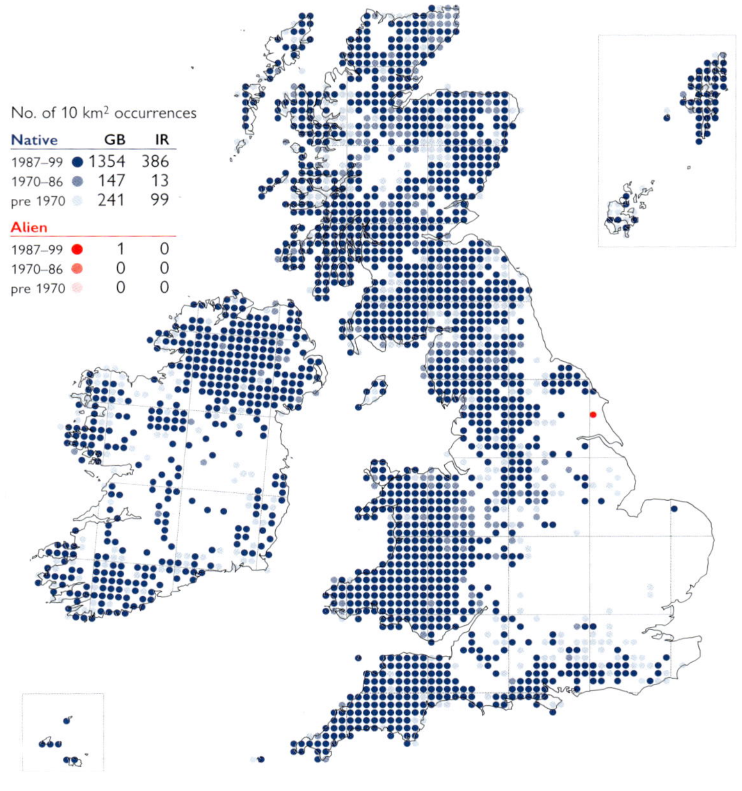

A stoloniferous annual to perennial herb found by streams and pools, in marshy pasture, moorland flushes and springs. It prefers acid peaty soils, and usually avoids calcareous soils. 0–805 m (Carnedd Llewelyn, Caerns.).

Native (change +0.52). This species has sometimes been confused with *M. scorpioides* and *M. laxa*, and the map in the 1962 *Atlas* certainly contained some errors. It is now more comprehensively recorded, and there appears to have been little change in its distribution, other than declines at the edge of its range in C. & S. England. It was mapped as 'all records' in the 1962 *Atlas*.

Oceanic Temperate element.

References: Atlas (213d), Welch (1967).

D. WELCH

Myosotis stolonifera Pale Forget-me-not

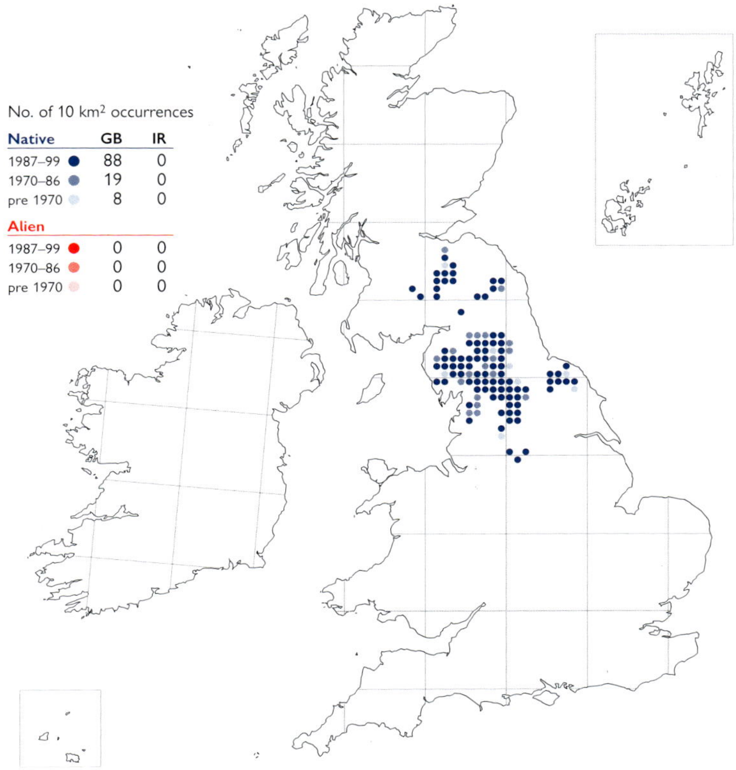

A perennial herb growing by rills and along base-rich spring-lines and flushes. Mainly upland, reaching 820 m on Cross Fell (Cumberland), and down to 130 m in the Lune valley (Cumberland).

Native (change +0.77). This species was first collected in Britain in 1919, from the N. Pennines, and initially described as an endemic, *M. brevifolia* (Salmon, 1926). It was inadequately recorded for the 1962 *Atlas*, the plant not being sought in areas distant from the original sites. However, following a study of its characteristic features, *M. stolonifera* became better known and it has been found more widely, notably in S. Scotland. It may still be overlooked elsewhere.

Oceanic Boreal-montane element; confined to Portugal, Spain and Britain.

References: Atlas (214a), Stewart *et al.* (1994), Welch (1967).

D. WELCH

Myosotis laxa Tufted Forget-me-not

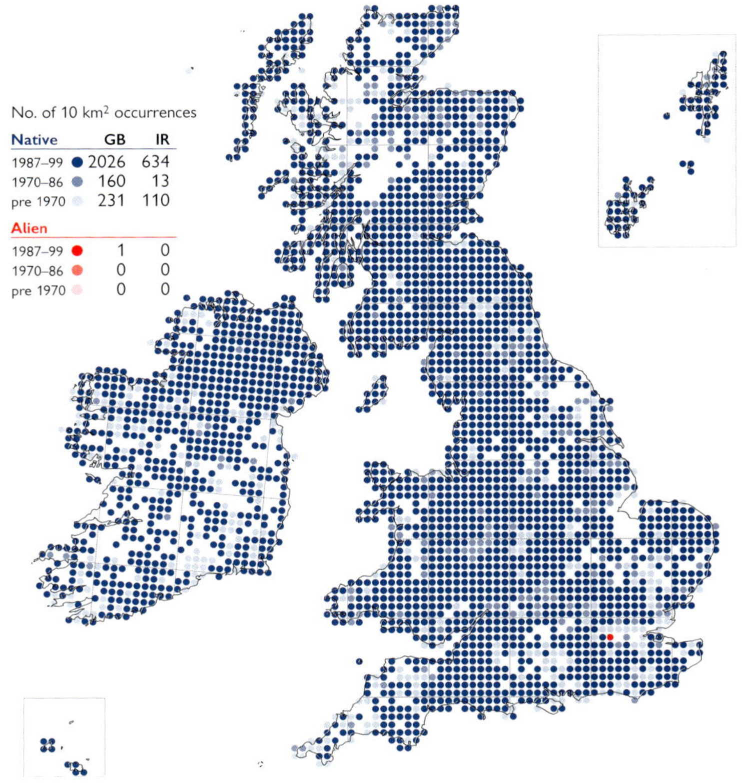

An annual or biennial herb of wet ground, often growing in open places trampled by livestock or where there has been other disturbance. It occurs in marshes, fen-meadows, rush-pastures, and by lakes, ponds, canals, rivers and streams. Lowland to 530 m in Atholl (E. Perth) and 550 m in Co. Londonderry.

Native (change +0.65). *M. laxa* was mapped as 'all records' in the 1962 *Atlas*, and apart from local losses in S.E. England, seems little changed in distribution.

Circumpolar Boreo-temperate element, with a disjunct distribution.

References: Atlas (214b), Hultén & Fries (1986).

D. WELCH

Myosotis sicula Jersey Forget-me-not

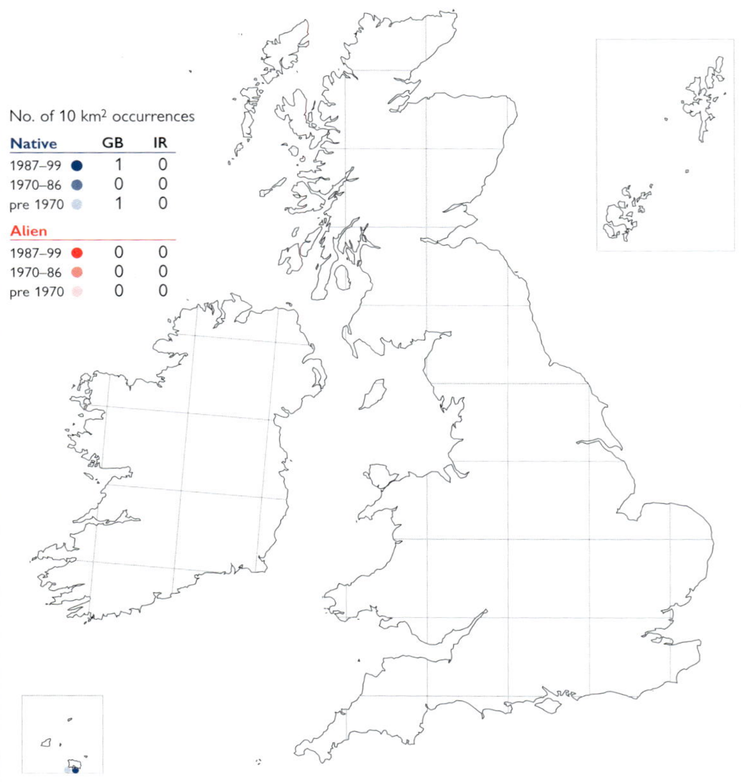

An annual only ever known from two sites on Jersey, in damp places on Ouaisné Common and by a small pool near the coast at Noirmont. Lowland.

Native. *M. sicula* was discovered at both its Jersey localities by A. J. Wilmott in 1922. It was last seen at Ouaisné Common in 1957. The Noirmont population disappeared in the 1980s when the pool became shaded by *Salix* species, but these were cleared in 1983 and it was rediscovered at the site in 1992. The very small population there maintains a precarious existence.

Mediterranean-Atlantic element.

References: Atlas (214c), Bolòs & Vigo (1995), Le Sueur (1984).

D. WELCH

Myosotis alpestris Alpine Forget-me-not

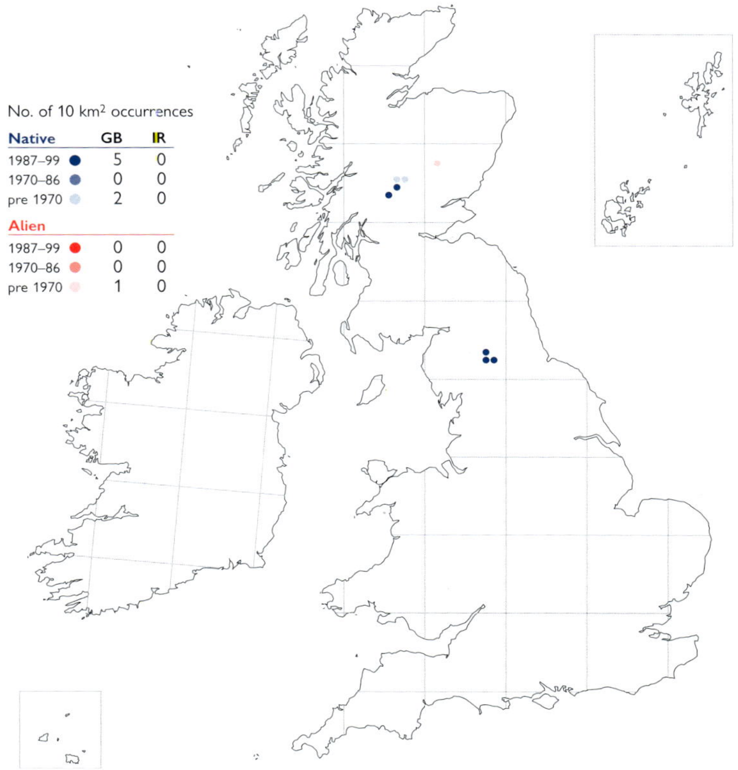

A perennial herb found in two contrasting habitats: heavily-grazed limestone grassland on base-rich well-drained soils in the Pennines, and both on and below mica-schist ledges on ungrazed cliffs in Perthshire, often in open communities. Reproduction is by seed. From 685 m in Teesdale (N.W. Yorks. and Westmorland) to 1180 m on Ben Lawers (Mid Perth).

Native (change –0.22). The distribution of *M. alpestris* is generally stable, though some populations in the Pennines appear to be in decline. The English and Scottish populations are ecotypically distinct (Elkington, 1964).

Circumpolar Arctic-montane element, with a disjunct distribution.

References: Atlas (214c), Hultén & Fries (1986), Meusel *et al.* (1978), Wigginton (1999).

D. WELCH

Myosotis sylvatica Wood Forget-me-not

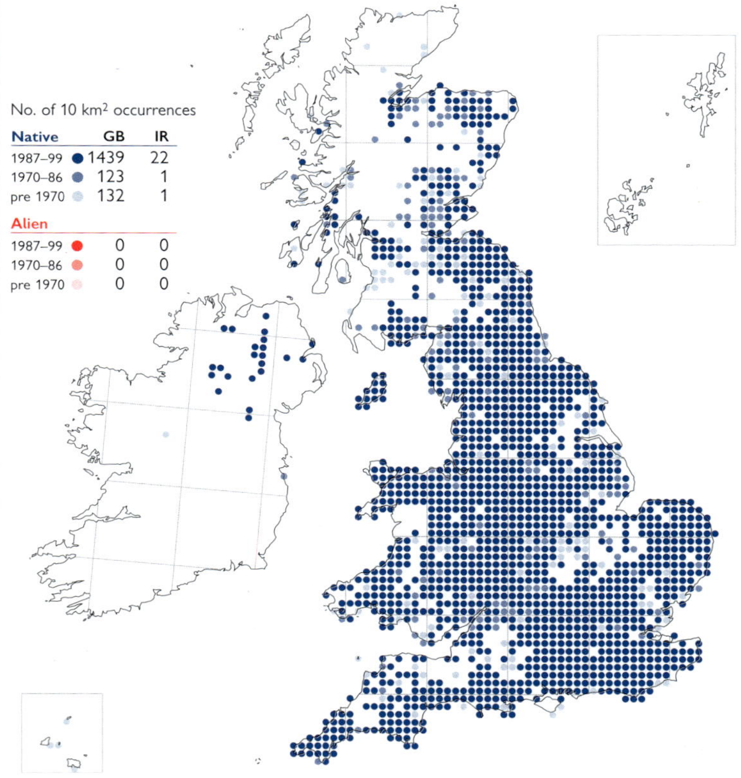

An erect biennial or perennial herb growing as a native, at least in England, on damp, fertile soils in woodland and rocky grassland. It is much more widespread in a wider range of habitats as a garden escape. 0–485 m (East Stone Gill, N.W. Yorks.).

Native (change +2.18). The native range of *M. sylvatica* has been totally obscured by escapes, and all records are mapped as if they are native. Alien plants were once similar to the natives but are now often recognisable by their more varied, brighter flower colours. Such escapes were excluded from the map in the 1962 *Atlas*. Large-flowered variants of *M. arvensis* may have been reported as this species.

Eurasian Temperate element; widely naturalised outside its native range.

References: Atlas (214d), Hultén & Fries (1986), Meusel *et al.* (1978), Rich & Jermy (1998).

D. WELCH

Myosotis arvensis Field Forget-me-not

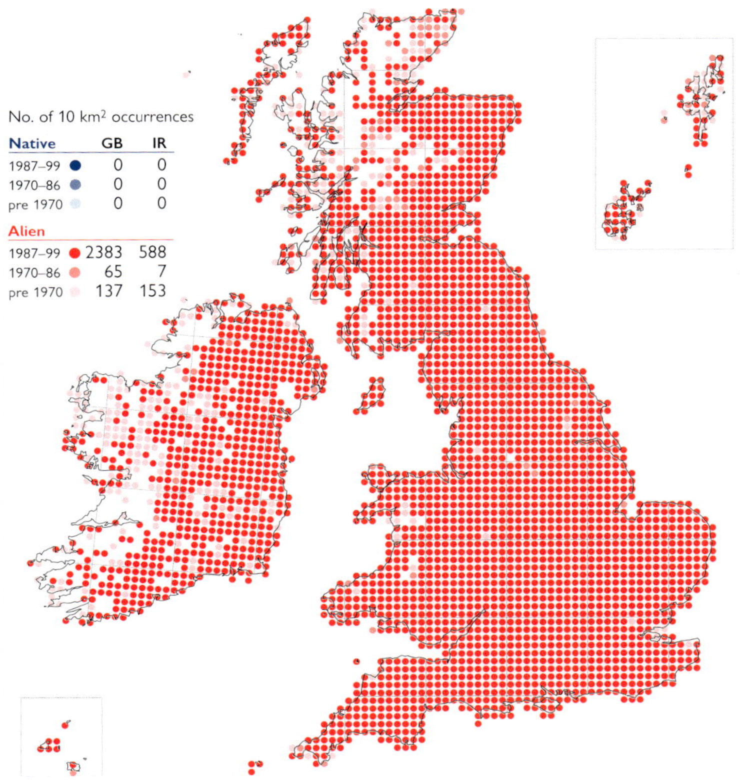

An annual or biennial herb of open or disturbed ground, especially cultivated fields. Other habitats include woodland edges, open grassland, hedges, scrub, roadsides, walls and quarries. 0–610 m (Nenthead, Cumberland), and exceptionally at 845 m on Great Dun Fell (Westmorland).

Archaeophyte (change –0.34). Despite changes in agricultural practice, the distribution of *M. arvensis* has remained stable since 1900, due in part to its flexible life history and seed longevity. The larger-flowered var. *sylvestris* is often mistaken for *M. sylvatica*, but such errors are unlikely to have affected the map.

As an archaeophyte *M. arvensis* has a Eurosiberian Boreo-temperate distribution; it is widely naturalised outside this range.

References: Atlas (215a), Grime *et al.* (1988), Hultén & Fries (1986), Rich & Jermy (1998).

D. WELCH

Myosotis ramosissima Early Forget-me-not

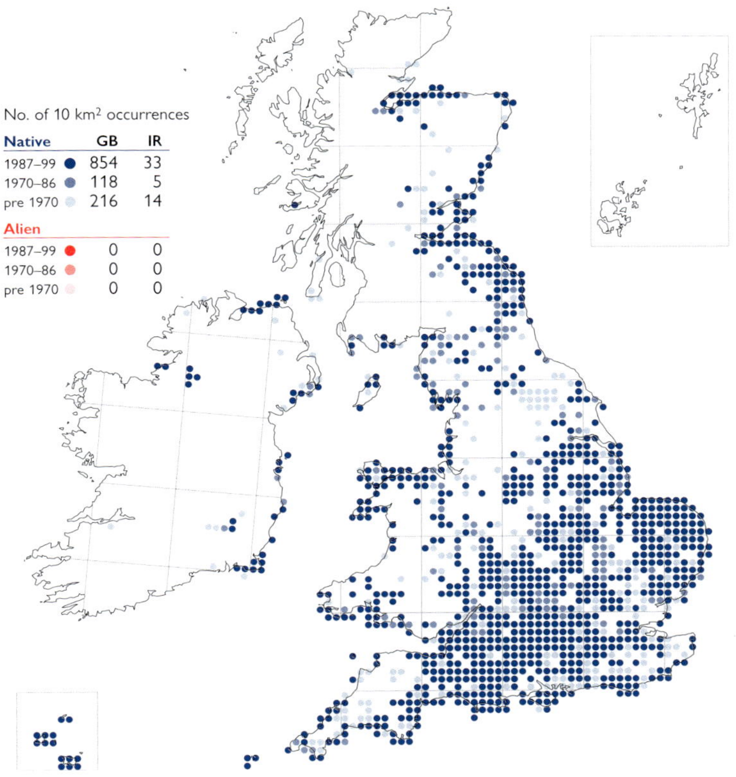

An annual of open habitats or bare ground on dry, relatively infertile soils. It is found in chalk and limestone grassland, on sandy heaths and banks, stabilised dunes, the borders of sandy cultivated fields, railway tracks, rocks, walls, gravel-pits, quarry spoil and waste ground. 0–430 m (above Swindale, Brough, Westmorland).

Native (change +0.11). *M. ramosissima* was mapped as 'all records' in the 1962 *Atlas*. There have been some local losses since 1950, possibly caused by the loss of close-grazed semi-natural communities in inland areas.

European Southern-temperate element.

References: Atlas (215c), Grime *et al.* (1988), Hultén & Fries (1986).

D. WELCH

Myosotis discolor Changing Forget-me-not

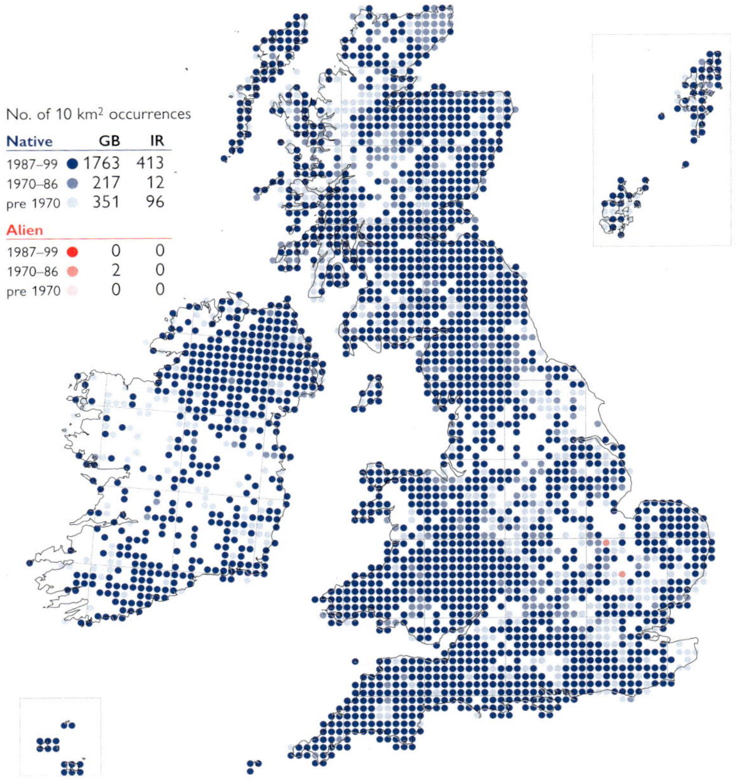

An annual of open grassland and disturbed ground occurring in a wide range of habitats, including fen- and hay-meadows, pastures, moorland edges, marshes, dune-slacks, arable field margins, road verges, railway tracks, chalk- and gravel-pits, rocks and walls. Generally lowland, but reaching 610 m on Little Fell (Westmorland) and 845 m on Great Dun Fell (Westmorland).

Native (change +0.14). *M. discolor* was mapped as 'all records' in the 1962 *Atlas*. Analysis of the database reveals that the widespread decline in S. and E. England has taken place since 1950. It is, however, much more comprehensively recorded in the south and west.

European Temperate element.

References: Atlas (215b), Hultén & Fries (1986), Meusel *et al.* (1978).

D. WELCH

Lappula squarrosa Bur Forget-me-not

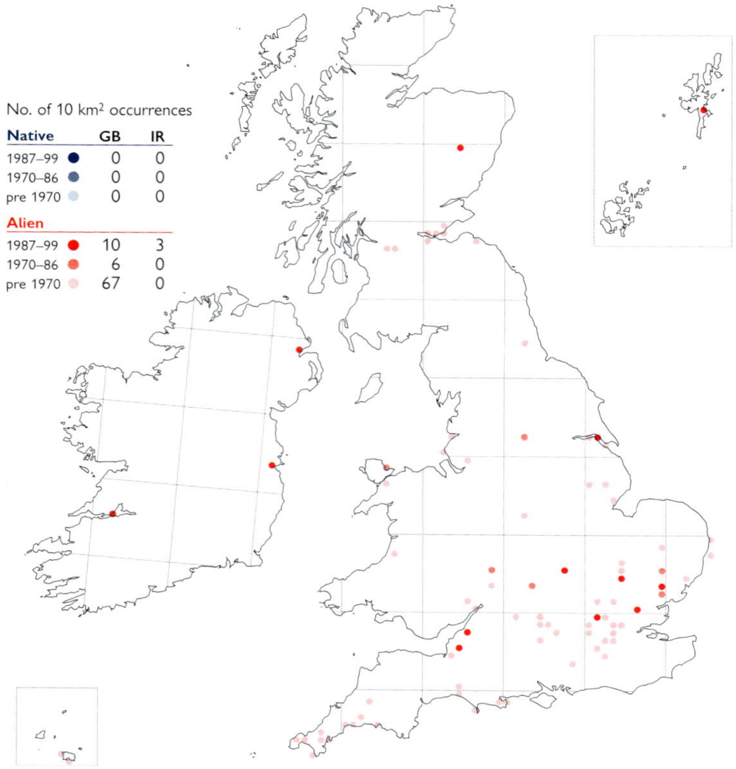

An annual introduced in bird-seed, grass-seed, grain and wool shoddy and occurring as a casual on waste ground and rubbish tips. Lowland.

Casual. This species was cultivated in Britain by 1683 and has been recorded from the wild since 1871. The present map suggests that it may be less frequent than formerly. European grain imports were probably the main source of introduction earlier in the 20th century, but recently grass-seed imports from N. America may have been more important.

L. squarrosa has a Eurosiberian Temperate distribution, but it is rare in W. Europe; it is widely naturalised in N. America and elsewhere.

References: Hultén & Fries (1986), Meusel *et al.* (1978).

D. WELCH

Omphalodes verna Blue-eyed-Mary

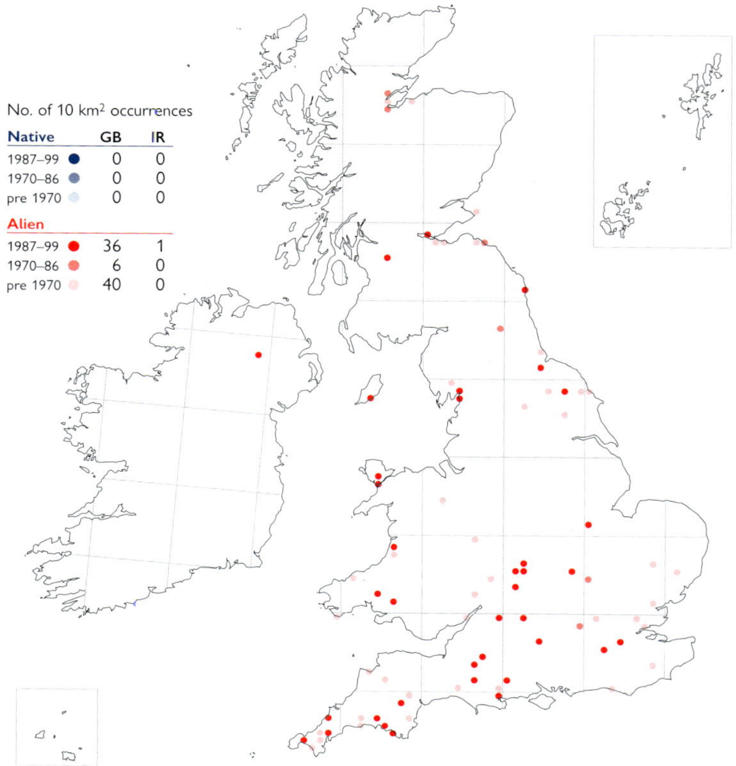

This creeping perennial is a garden escape or outcast which has become naturalised in woodland and along lanes. Lowland.

Neophyte. *O. verna* was cultivated in Britain by 1633, but appears to be less frequent in gardens than formerly. It was known from the wild by at least 1840, and Dunn (1905) described it as 'occasionally recorded in England as naturalised near gardens'. It can be very persistent.

Native of the mountains of S.C. & S.E. Europe.

Reference: Meusel *et al.* (1978).

D. WELCH

Cynoglossum officinale Hound's-tongue

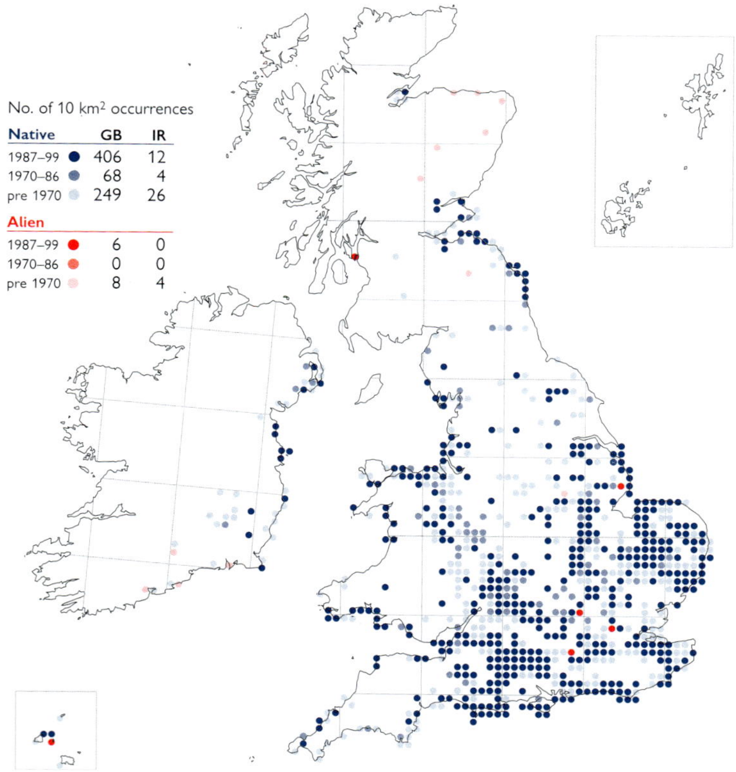

A biennial herb of disturbed ground, growing mostly on dry, often base-rich soils. Habitats include coastal dunes, shingle, open grassland, woodland margins and clearings, field edges, cleared land and gravelly waste. It is unpalatable to grazing animals and is often frequent on disturbed ground by rabbit warrens. Generally lowland, but reaching 400 m on Eglwyseg Rocks, Llangollen (Denbs.).

Native (change −1.09). *C. officinale* has declined sharply since the 1950s, loss of habitat and herbicide spraying doubtless being major factors. Some old records in N. Scotland and Ireland are considered to be relics of cultivation by herbalists.

Eurosiberian Temperate element; widely naturalised outside its native range.

References: Atlas (211a), Hultén & Fries (1986), Jong *et al.* (1990), Upadhyaya *et al.* (1988).

D. WELCH

Cynoglossum germanicum
Green Hound's-tongue

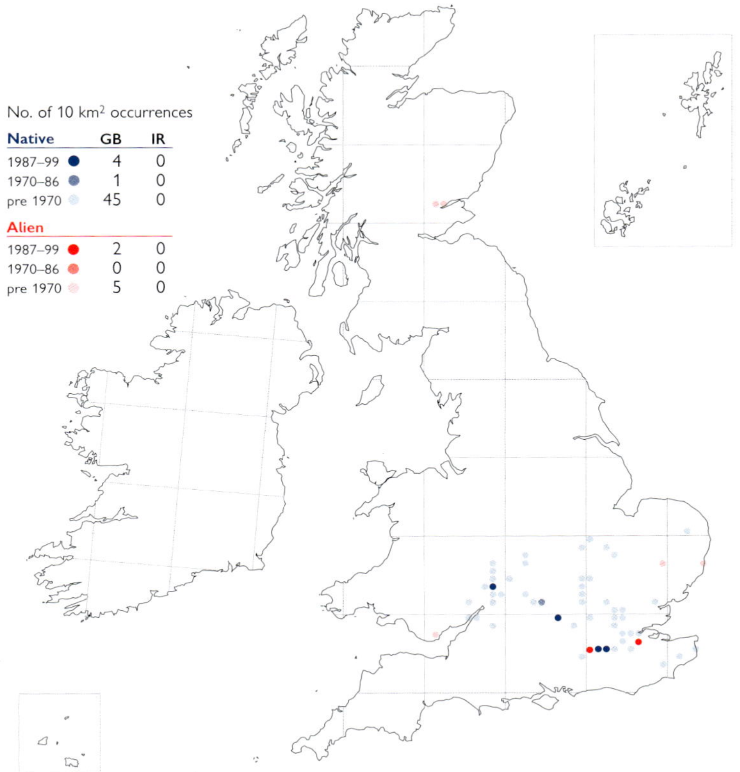

A biennial or short-lived perennial herb of glades in or margins of deciduous woodland, usually *Fagus* or *Quercus*; sometimes also occurring in hedge banks. It is found mainly on calcareous, freely-draining, loamy soils. Lowland.

Native (change −0.52). The marked decline of this species before 1930 was shown in the 1962 *Atlas*. The reasons for it are unclear, but it may be more apparent than real as many old records appear to have been of transient populations. Further sites have been lost since 1930, but the great storm of 1987 opened up its woodland habitats, and since then huge populations have been recorded in Surrey.

European Temperate element, with a continental distribution in W. Europe.

References: Atlas (211b), Meusel *et al.* (1978), Wigginton (1999).

D. WELCH

Verbena officinalis Vervain

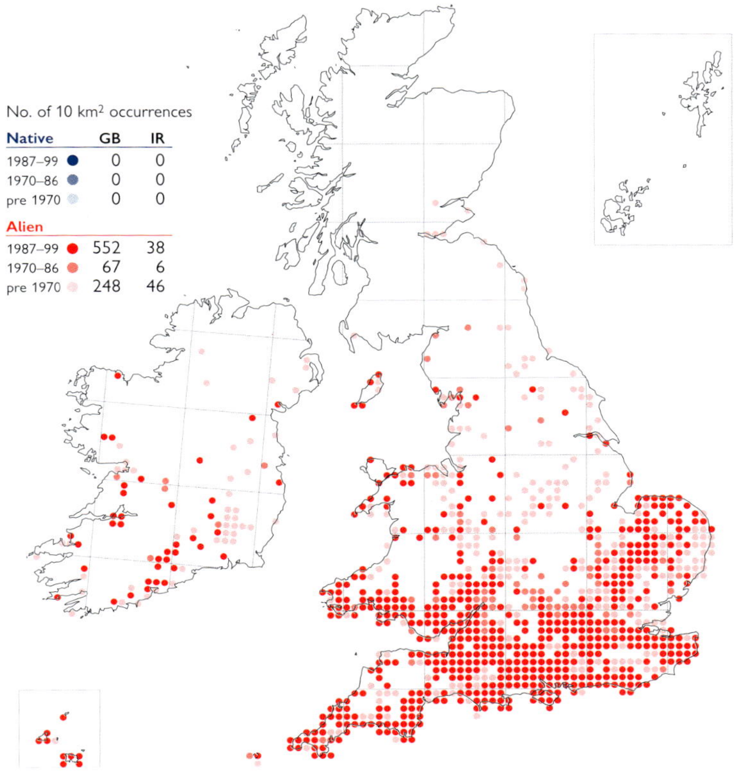

A perennial herb, usually of open habitats or bare ground on freely-draining, often calcareous soils. It is most frequent in rough grassland and scrub, on roadsides, and on sheltered coastal cliffs and rock outcrops; less often in quarries and gravel-pits, and on streamsides, wood-borders and walls. Lowland.

Archaeophyte (change −0.43). *V. officinalis* has been growing around human settlements since the Neolithic, and it was widely cultivated in medieval gardens. Since the 1962 *Atlas* there have been substantial losses in East Anglia, and some elsewhere.

As an archaeophyte *V. officinalis* has a Eurasian Southern-temperate distribution; it is widely naturalised outside this range.

References: Atlas (240c), Hultén & Fries (1986), Meusel *et al.* (1978), Woodward (1997).

R. M. BURTON

Stachys officinalis Betony

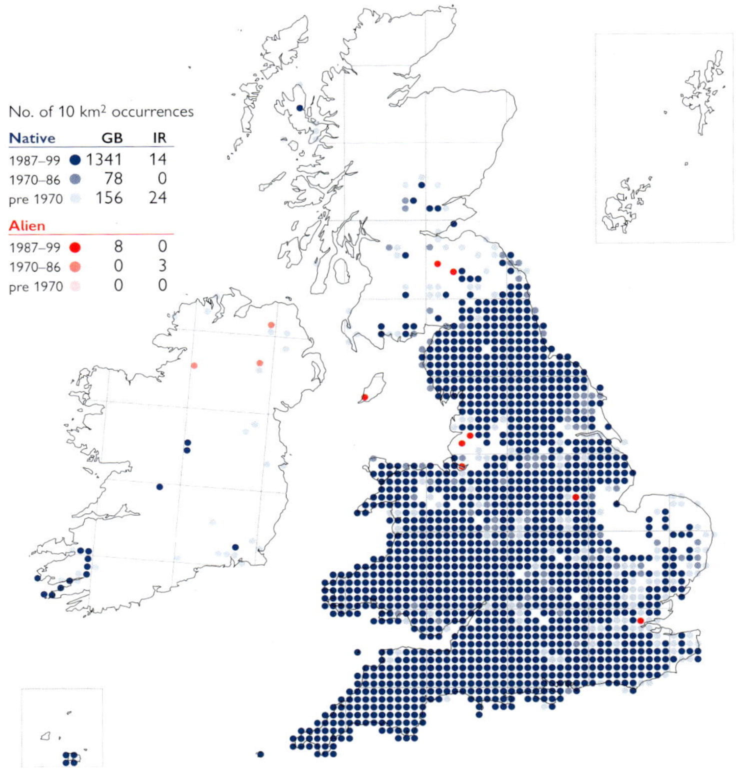

A perennial herb of hedge banks, grassland, heaths, open woods and woodland rides and margins. It is occasionally found in cliff-top grassland, sometimes as the genetically dwarf var. *nana*. It favours mildly acidic soils, but is also found on those that are neutral or somewhat calcareous. 0–460 m (Teesdale, Westmorland).

Native (change −0.62). *S. officinalis* has suffered local losses in England and Ireland as a result of the loss and improvement of permanent pastures, the ploughing of fields to the edge of woods with consequent loss of the marginal flora and the shading of woodland grassland following a decline in coppicing.

European Temperate element; also in C. Asia.

References: Atlas (246d), Curtis & McGough (1988), Grime *et al.* (1988), Hultén & Fries (1986), Meusel *et al.* (1978), Rackham (1980).

K. WALKER

Stachys byzantina Lamb's-ear

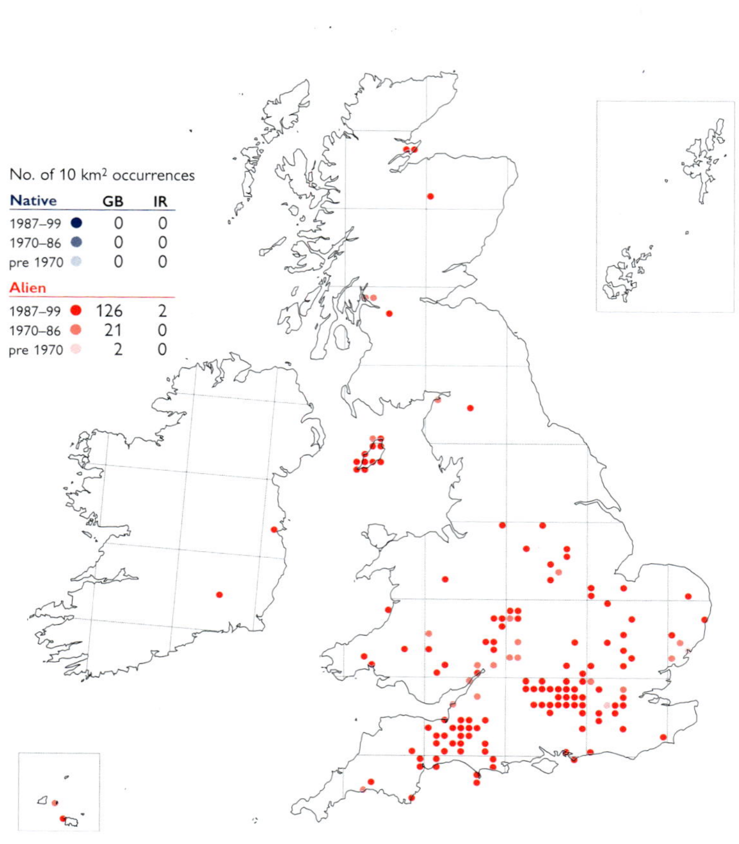

A stoloniferous perennial herb found as a garden escape, throw-out or, occasionally, a relic of cultivation. It is often persistent on roadsides, rubbish tips, waste ground and in quarries. Lowland.

Neophyte. This species, introduced into cultivation in 1782, has long been extremely popular in gardens. It was first recorded in the wild in 1858. It may be increasing, but is unevenly recorded.

Native of S.W. Asia.

Reference: Bolòs & Vigo (1995).

K. WALKER

Stachys germanica Downy Woundwort

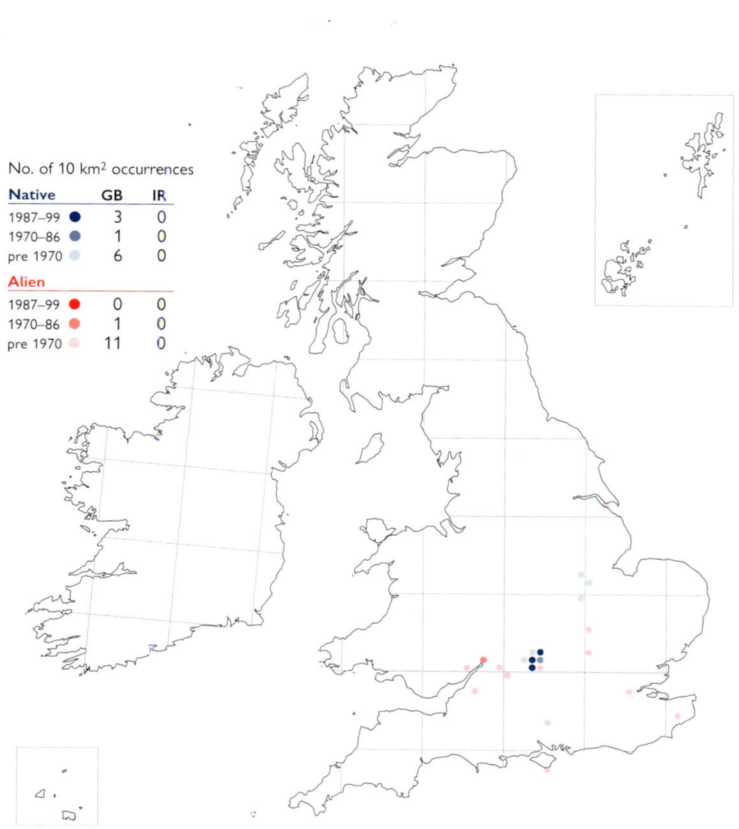

A biennial or short-lived perennial herb of woodland margins, grassy banks, ancient hedgerows and green lanes overlying oolitic limestone; occasionally recorded from stony fields and quarries. The reappearance of plants following disturbance and scrub clearance suggests that seeds remain viable over long periods. Lowland.

Native (change –0.27). *S. germanica* may have been more widespread than the current map suggests, but it declined markedly in the 19th century. It became confined to the present small area of Oxfordshire before 1930, and, though it is still restricted to this area, recent conservation work has bolstered the population to some extent.

European Temperate element.

References: Atlas (247b), Bolòs & Vigo (1995), Dunn (1997), Marren (1988), Wigginton (1999).

K. WALKER

Stachys alpina Limestone Woundwort

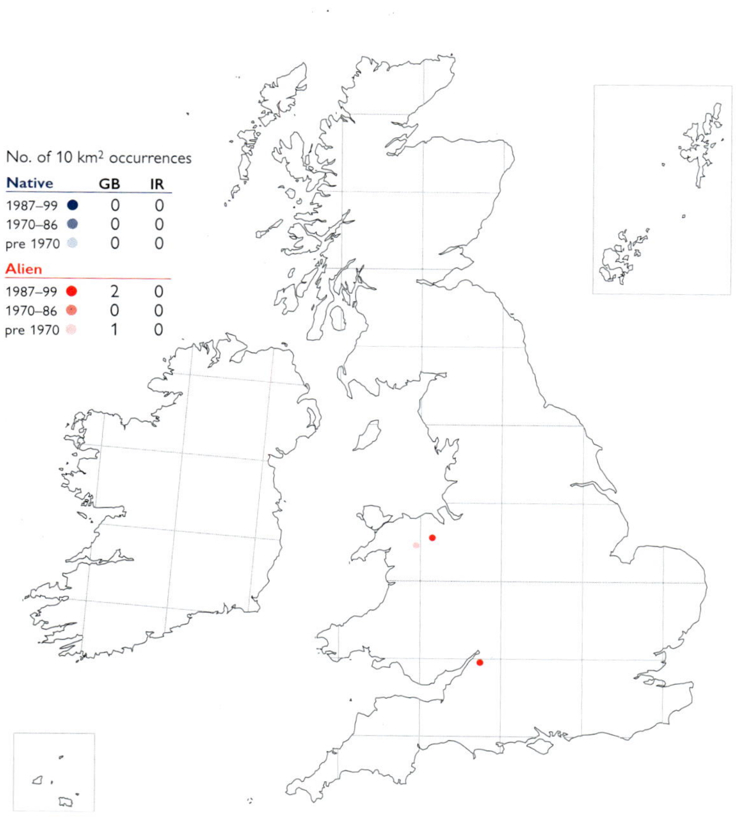

A perennial herb of open woodlands, wood-borders, hedge banks and trackways on thin soils overlying calcareous rock. Lowland.

Neophyte. *S. alpina* was cultivated in Britain by 1597. Although it has long been considered to be native, this is extremely doubtful; Kay & John (1995) conclude that it is a relatively recent introduction. The Gloucestershire population may have spread shortly after its discovery in 1897, but is now apparently stable, albeit maintained artificially by sowing seed. It was found in Denbighshire in 1927, but colonies there are slowly declining; remaining plants are often re-introductions. It can reappear after disturbance, however, and buried seed may exist in former sites.

S. alpina has a European Temperate distribution.

References: Atlas (247b), Bolòs & Vigo (1995), Wigginton (1999).

K. WALKER

Stachys sylvatica Hedge Woundwort

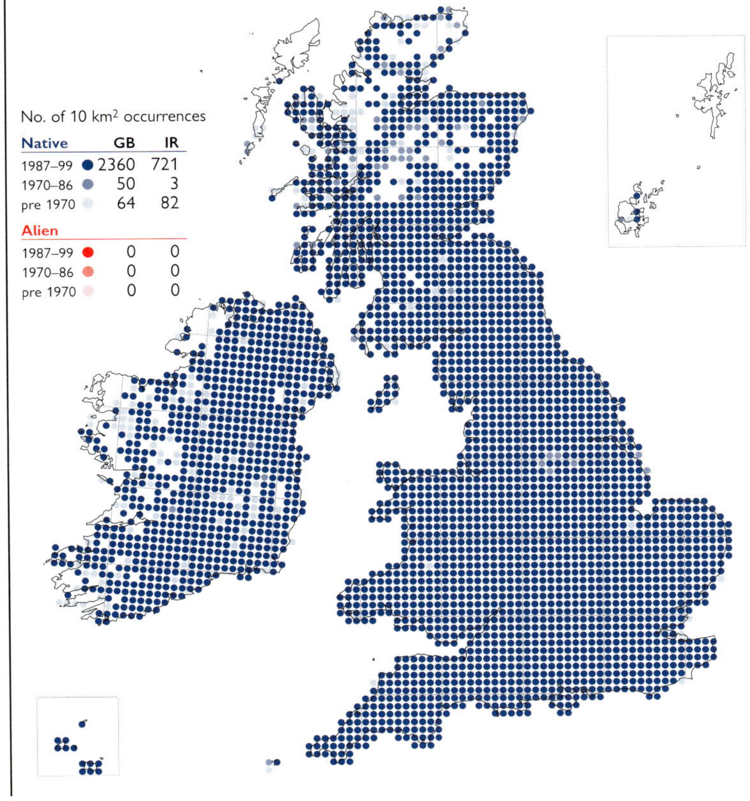

A rhizomatous perennial of woods, hedgerows, the banks of rivers and streams, rough grassland and waste places, and, locally, a persistent garden weed. It characteristically grows in moist, fertile, mildly acidic to basic soils in disturbed or lightly to moderately shaded sites. It spreads by vigorous rhizomatous extension, and reproduces by seed and by rhizome fragments. 0–500 m (above Malham, Mid-W. Yorks.), with an exceptional record at 845 m on Great Dun Fell (Westmorland).

Native (change –0.49). There is no evidence for a change in the overall distribution of *S. sylvatica.*

Eurosiberian Temperate element.

References: Atlas (247d), Grime *et al.* (1988), Hultén & Fries (1986), Meusel *et al.* (1978).

K. WALKER

Stachys palustris Marsh Woundwort

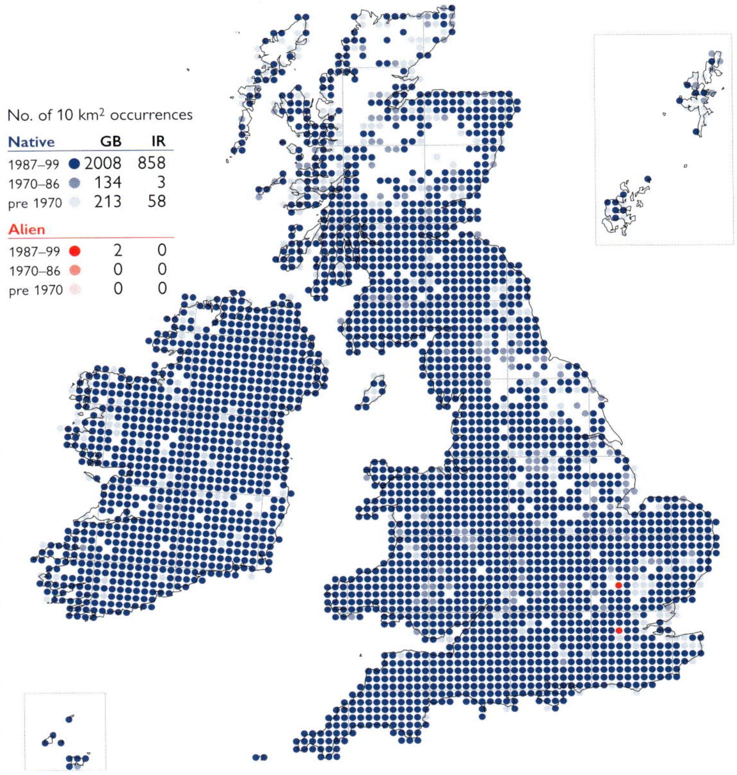

A perennial herb of damp places. It grows by streams, rivers, ditches and ponds, in fens, marshes and swamps, on rough ground and occasionally in cultivated fields. It is typically found on intermittently flooded and poorly drained soils. 0–540 m (Moor House, Teesdale, Westmorland).

Native (change +0.01). There is little evidence of any change in the overall distribution of *S. palustris.* It was mapped in the 1962 *Atlas* as 'all records'. Broad-leaved forms of *S.* × *ambigua* have formerly been mistaken for *S. palustris* by visitors to Scotland, and some such errors may still be mapped.

Circumpolar Boreo-temperate element.

References: Atlas (247c), Hultén & Fries (1986), Meusel *et al.* (1978), Morton (1973), Wilcock & Jones (1974).

K. WALKER

Stachys palustris × *S. sylvatica* (*S.* × *ambigua*) Hybrid Woundwort

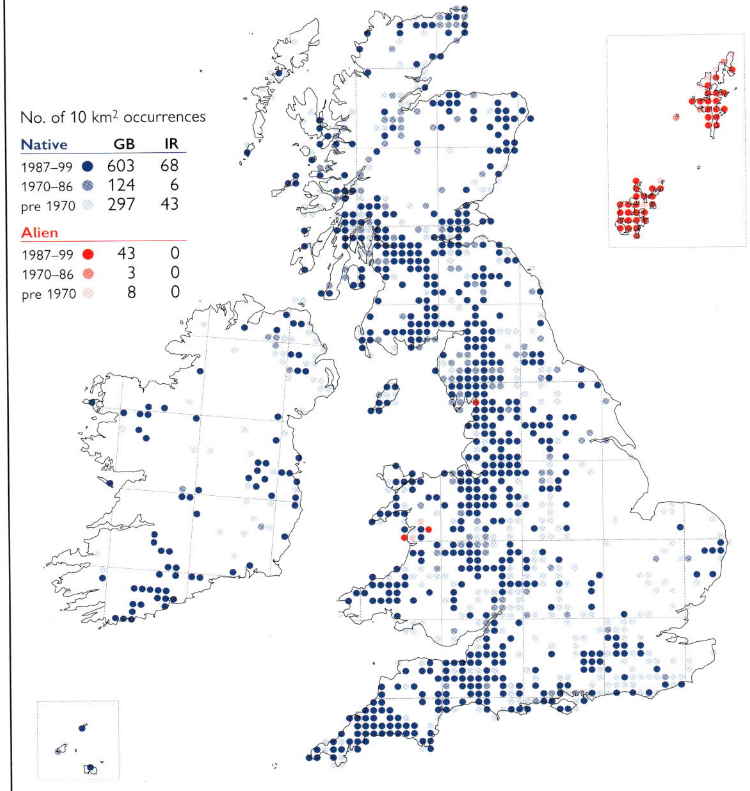

A sterile rhizomatous perennial herb occurring in the habitats of either parent, but frequently in the absence of both. It grows besides streams, on river banks and roadsides, on the edge of woodland, on waste ground and as a weed of cultivated ground, particularly in the north and west where it occurs in the absence of *S. sylvatica.* It reproduces by rhizome fragments. 0–300 m (Pentre Llyncwmmer, Cards.).

Native. Although localised declines have been reported (e.g. Mountford, 1994), and the map shows some loss in S.E. England, there is no evidence for an appreciable change in the distribution of *S.* × *ambigua*, and it may be more frequent than the current map suggests.

Widespread in Europe.

References: Atlas Supp. (69b), Grime *et al.* (1988), Morton (1973), Stace (1975), Wilcock & Jones (1974).

K. WALKER

Stachys annua Annual Yellow-woundwort

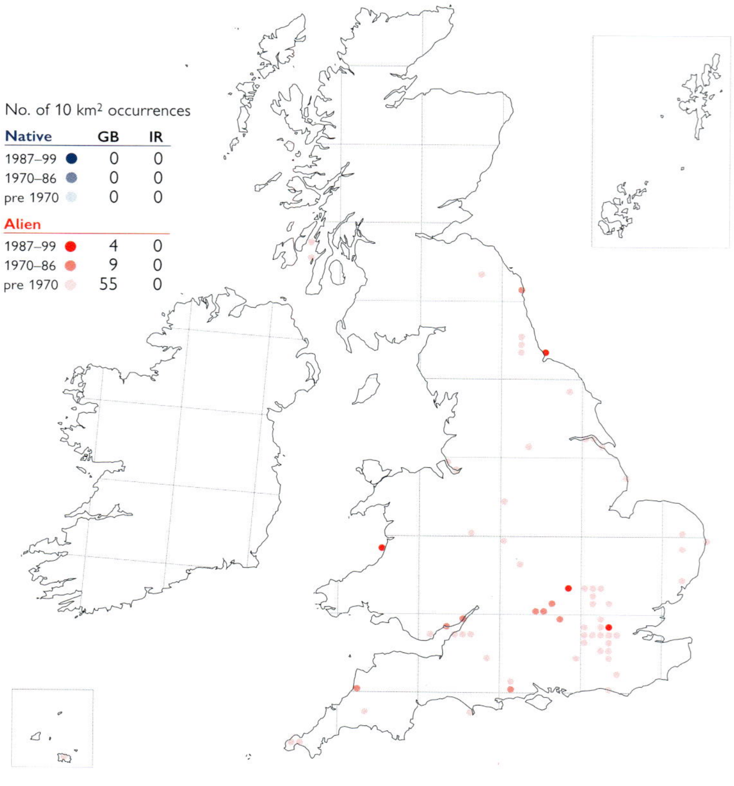

An annual found in arable fields, on waste ground and occasionally in gardens. It is usually casual but sometimes persists for several years, and originates as a contaminant of grain and oil-seed. Lowland.

Casual. This species was cultivated in Britain by 1713. It was recorded in the wild by 1830 (Kent) and was at one time so abundant in parts of S.E. England that it was considered to be native. It has declined dramatically, probably due to improvements in seed cleaning techniques.

A variable species with a European Temperate distribution; absent as a native from W. Europe.

References: Meusel *et al.* (1978), Styles (1976).

K. WALKER

Stachys arvensis Field Woundwort

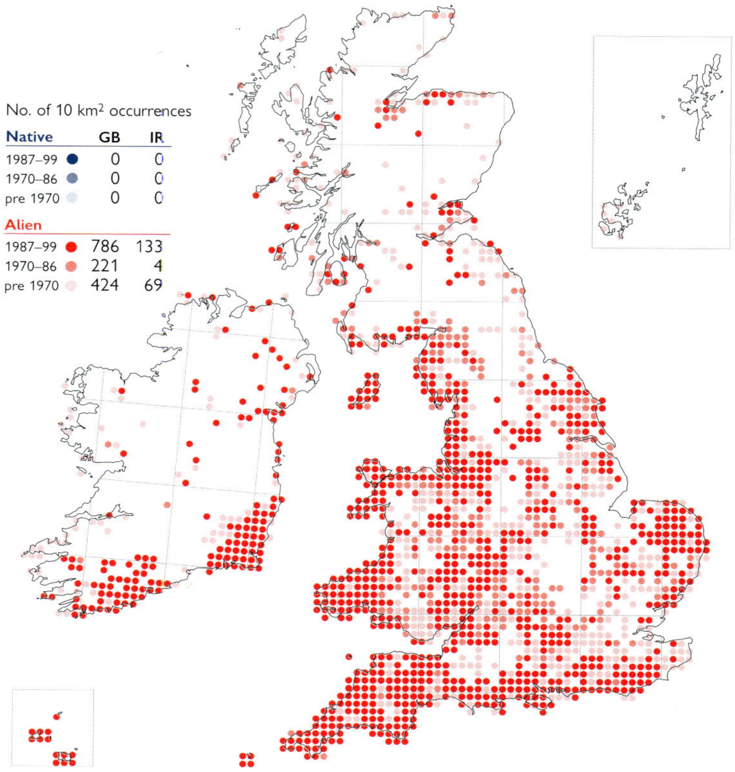

A summer- or winter-annual of arable fields, allotments and gardens, waste ground and road verges, usually on non-calcareous soils. It occurs on limestone outcrops in W. Ireland. Generally lowland, but reaching 380 m near Simonsbath (S. Somerset).

Archaeophyte (change –1.17). *S. arvensis* was formerly a frequent weed of arable land. Mapped as 'all records' in the 1962 *Atlas*, analysis of the database reveals that many losses took place before 1950, but its decline has accelerated since then.

As an archaeophyte *S. arvensis* has a Suboceanic Southern-temperate distribution; it is widely naturalised outside this range.

References: Atlas (247a), Hultén & Fries (1986), Meusel *et al.* (1978).

K. WALKER

Ballota nigra Black Horehound

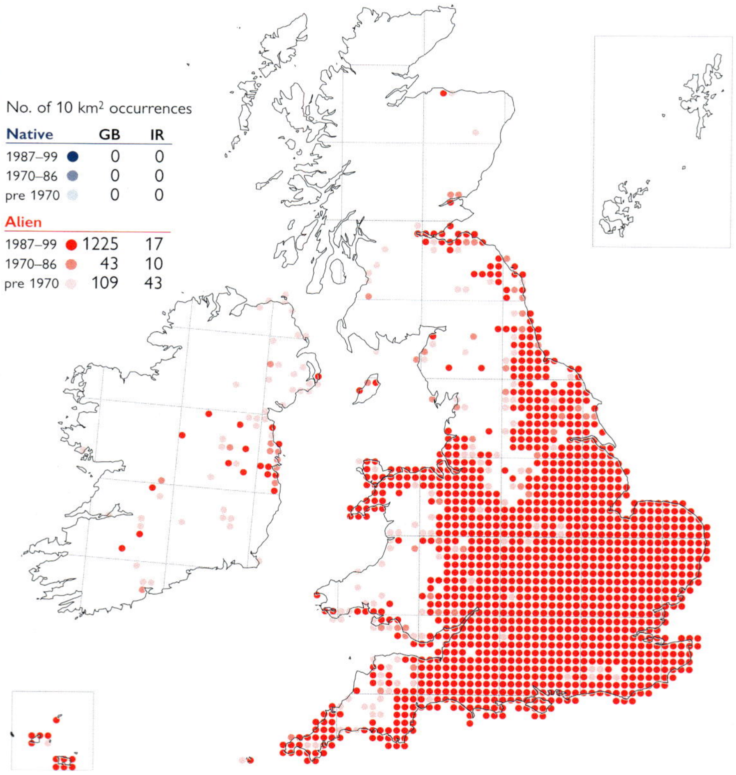

A foetid perennial herb of hedgerows, field-borders, walls, waysides and waste ground, often on disturbed nutrient-rich soils near habitations. Lowland, though it has been recorded as a casual at 480 m on Helvellyn (Cumberland).

Archaeophyte (change –0.37). Archaeological evidence suggests that *B. nigra* has been associated with human settlements since the Iron Age. It shows some small declines in the north and west, but many of these populations may only have been casual.

As an archaeophyte *B. nigra* has a European Southern-temperate distribution.

References: Atlas (248a), Halliday (1997), Hultén & Fries (1986), Meusel *et al.* (1978).

K. WALKER

Leonurus cardiaca Motherwort

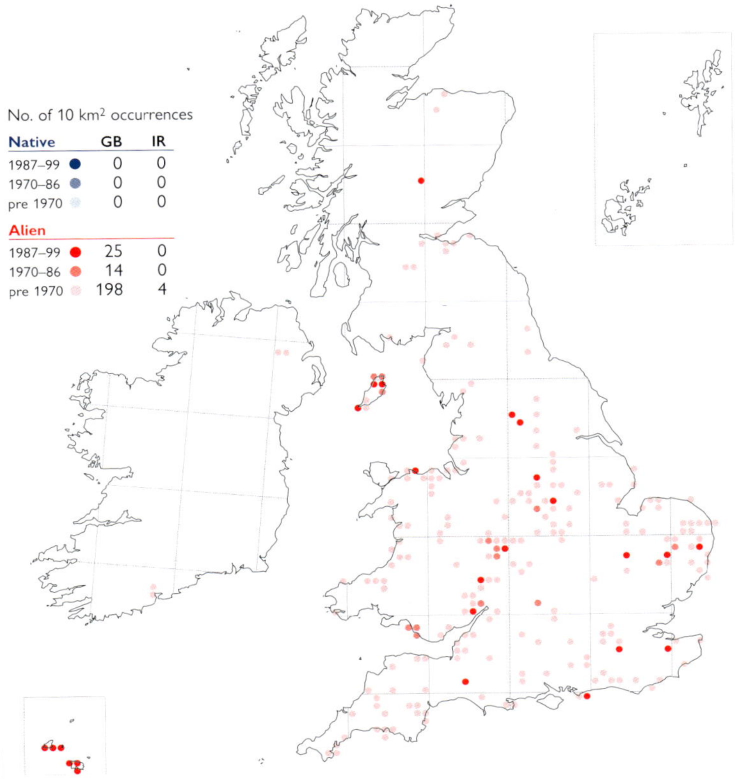

A rhizomatous perennial herb naturalised on waysides and in waste places, often near habitation, and often persisting for long periods. Lowland.

Neophyte (change −3.05). *L. cardiaca* was first introduced from continental Europe in the Middle Ages as a medicinal herb and later as an impurity of imported grain. It was recorded in the wild by 1597. The marked contraction in range shown in the 1962 *Atlas* has continued as a result of the gradual extinction of established populations and a decreasing rate of introduction, the latter due in part to more effective seed cleaning.

L. cardiaca sens. lat. is a Eurosiberian Temperate species, but the limits of its native range have been obscured by its spread in cultivation.

References: Atlas (249d), Hultén & Fries (1986), Mabey (1996).

K. WALKER

Lamiastrum galeobdolon subsp. *galeobdolon*

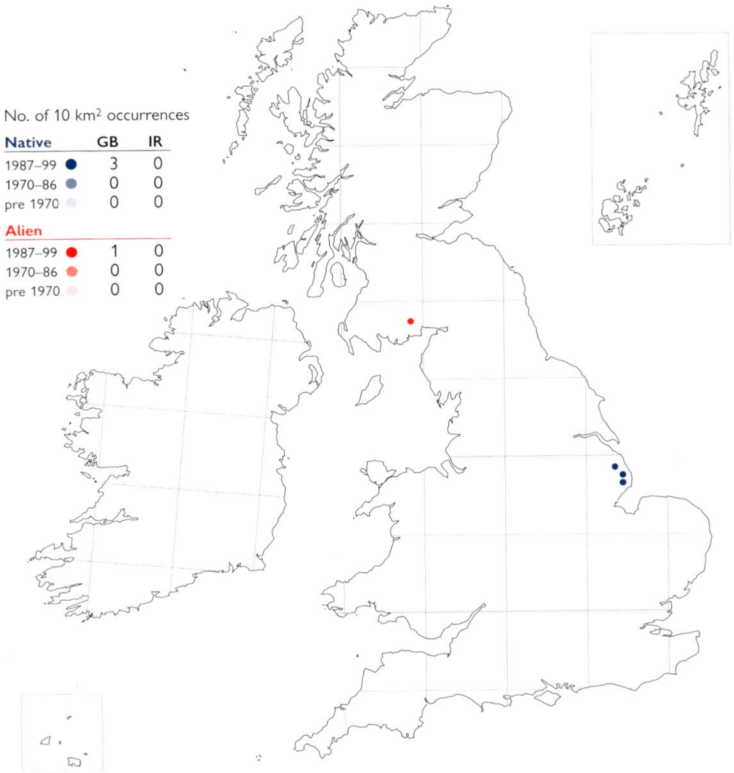

A stoloniferous herbaceous perennial herb of deciduous woods, woodland edges and hedgerows. Lowland.

Native. *L. galeobdolon* subsp. *galeobdolon* is diploid and was first recorded in Britain by Wegmüller (1971). It has been confirmed cytologically only from N. Lincolnshire and Kirkcudbrightshire, although at the latter site it is probably introduced. It may be overlooked elsewhere because of its similarity to the widespread subsp. *montanum*.

European Temperate element; this subspecies extends further north and east than subsp. *montanum*.

References: Hultén & Fries (1986), Meusel *et al.* (1978), Packham (1983), Rich & Jermy (1998).

K. WALKER

Lamiastrum galeobdolon subsp. *montanum* Yellow Archangel

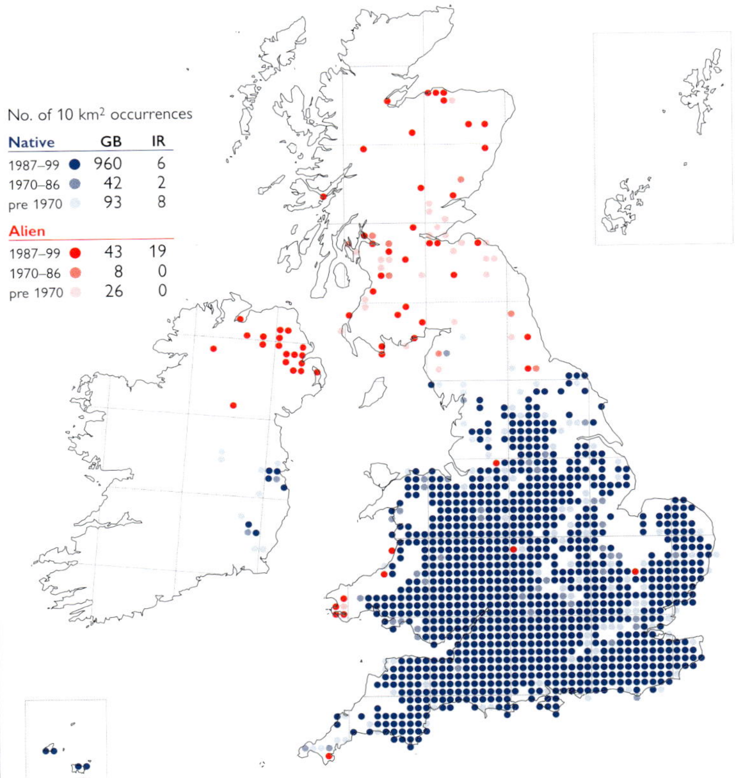

A stoloniferous perennial herb of moist woodland, hedges, roadsides and grikes of limestone pavement, usually on heavy soils. It is often associated with ancient woods and wood-relic hedges (Rackham, 1980), particularly in the east. 0–425 m (Ystradfellte, Brecs.).

Native (change for species +1.07). *L. galeobdolon* subsp. *montanum* has declined since the 1960s in Ireland as a result of road improvements and construction schemes. There is no evidence for a significant change in distribution elsewhere since the 1962 *Atlas*. In Scotland variegated forms of garden origin occur as casuals.

European Temperate element; subsp. *montanum* occurs in the southern and western part of the species' range.

References: Curtis & McGough (1988), Grime *et al.* (1988), Hultén & Fries (1986), Meusel *et al.* (1978), Packham (1983), Rich & Jermy (1998).

K. WALKER

Lamiastrum galeobdolon subsp. *argentatum*

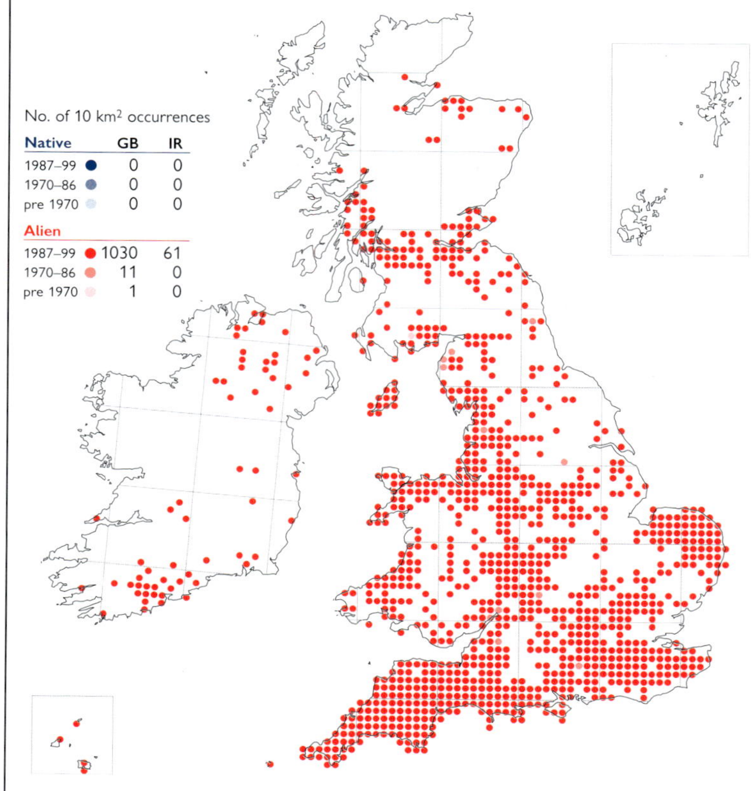

A stoloniferous herbaceous perennial growing on a variety of soils, and naturalised at the edge of woodland, on roadside verges and on tracksides, sometimes in unshaded places and often in places where garden rubbish is dumped. Lowland.

Neophyte (change for species +1.07). The history of this plant in our area is obscure. It was not recognised taxonomically until Smejkal described it as a new taxon in 1975 and Rutherford & Stirling (1987) brought it to the attention of British and Irish botanists. It was probably introduced in the late 1960s; the first record in the wild was in 1974. It is increasing rapidly, but seems to have been unevenly recorded and may be more common than the map suggests.

Origin uncertain; perhaps a cultivated variant of subsp. *montanum*.

References: Clement & Foster (1994), Rich & Jermy (1998).

K. WALKER

Lamium album White Dead-nettle

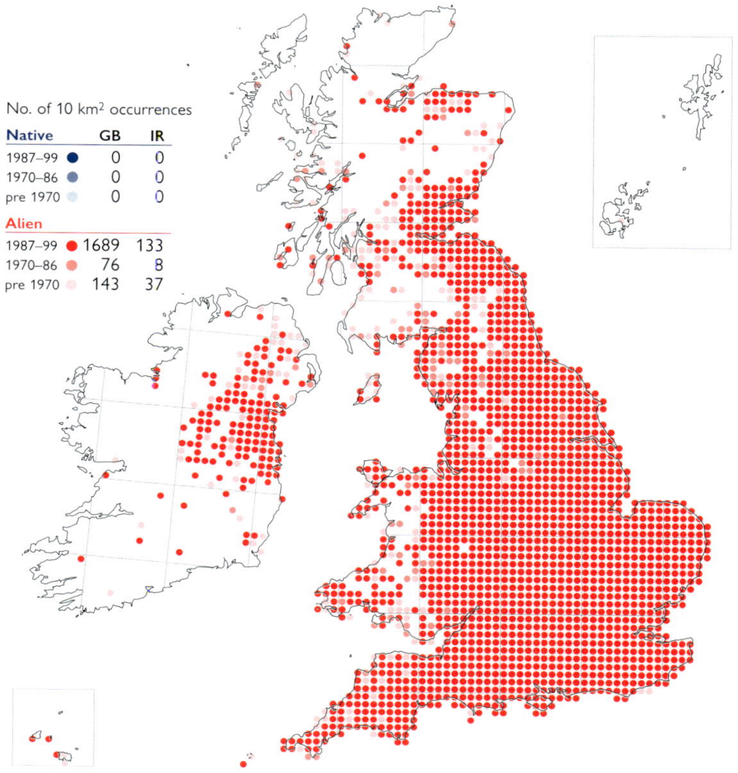

A rhizomatous or sometimes stoloniferous perennial herb of secondary woodland, hedge banks, waysides and rough ground, often growing on fertile soils close to habitation. 0–345 m (Garrigill, Cumberland).

Archaeophyte (change –0.65). There is no evidence for an overall change in the range of *L. album* since the 1962 *Atlas*.

As an archaeophyte *L. album* has a Eurasian Boreo-temperate distribution; it is widely naturalised outside this range.

References: Atlas (249c), Grime *et al.* (1988), Hultén & Fries (1986), Meusel *et al.* (1978).

K. WALKER

Lamium maculatum Spotted Dead-nettle

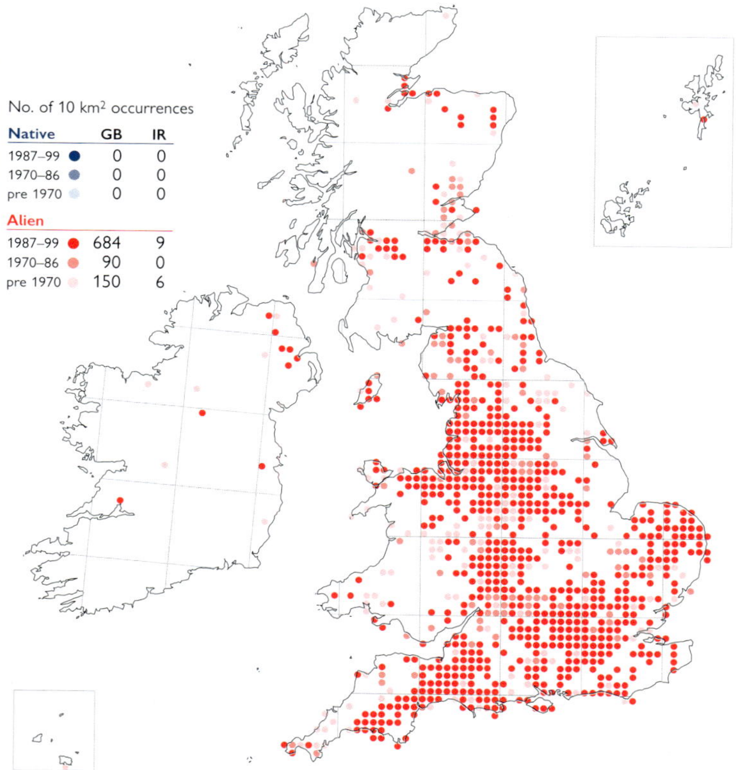

This rhizomatous or stoloniferous perennial herb is naturalised on rough ground, rubbish tips, roadsides and waste places, usually close to habitation. Generally lowland, but recorded at 490 m above Garrigill (Cumberland).

Neophyte. This species was introduced into cultivation from Italy in 1683, and recorded in the wild by about 1730. It is still widely grown as a garden plant. It is certainly increasing in many areas.

A European Temperate species, absent as a native from much of W. Europe.

References: Hultén & Fries (1986), Meusel *et al.* (1978).

K. WALKER

Lamium purpureum Red Dead-nettle

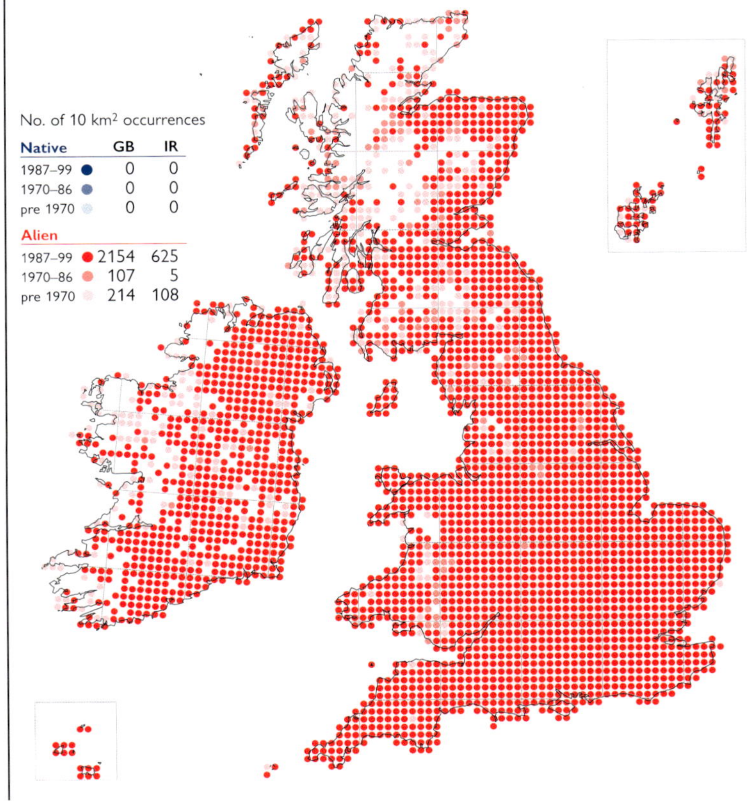

This annual is a frequent colonist of fertile and disturbed soils, and is found in cultivated and waste ground, gardens, hedgerows, on roadside verges, along railways, around rock outcrops and in rough grassland. 0–610 m (Grasshill, Co. Durham).

Archaeophyte (change –1.09). As in the 1962 *Atlas*, *L. purpureum* is present in nearly all lowland areas but the map suggests a decline in Scotland, possibly due to the abandonment of marginal arable land.

As an archaeophyte *L. purpureum* has a European Temperate distribution; it is widely naturalised outside this range.

References: Atlas (249b), Grime *et al.* (1988), Hultén & Fries (1986).

K. WALKER

Lamium hybridum Cut-leaved Dead-nettle

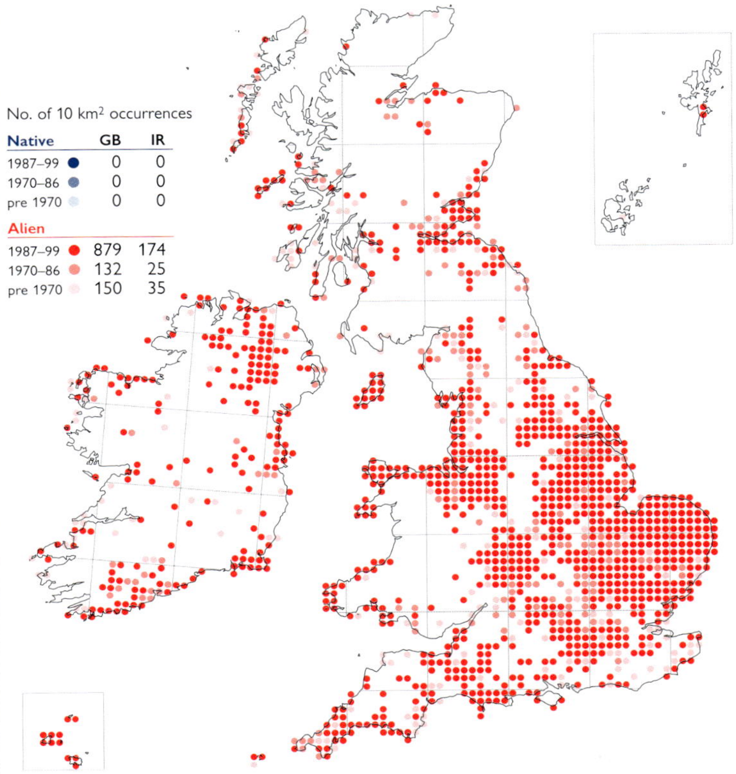

An annual of cultivated, waste and disturbed ground on dry soils, often occurring as a weed of heavily fertilised, broad-leaved crops. 0–320 m (Derbyshire).

Archaeophyte (change +1.57). *L. hybridum* shows a considerable increase in distribution compared with the 1962 *Atlas*, presumably due to its ability to exploit conditions of high fertility. However, it is often confused with *L. purpureum*, so may previously have been under-recorded. It is possibly still overlooked in some areas, and may be more common than the current map suggests.

As an archaeophyte *L. hybridum* has a European Temperate distribution.

References: Atlas (249a), Hultén & Fries (1986).

K. WALKER

Lamium confertum Northern Dead-nettle

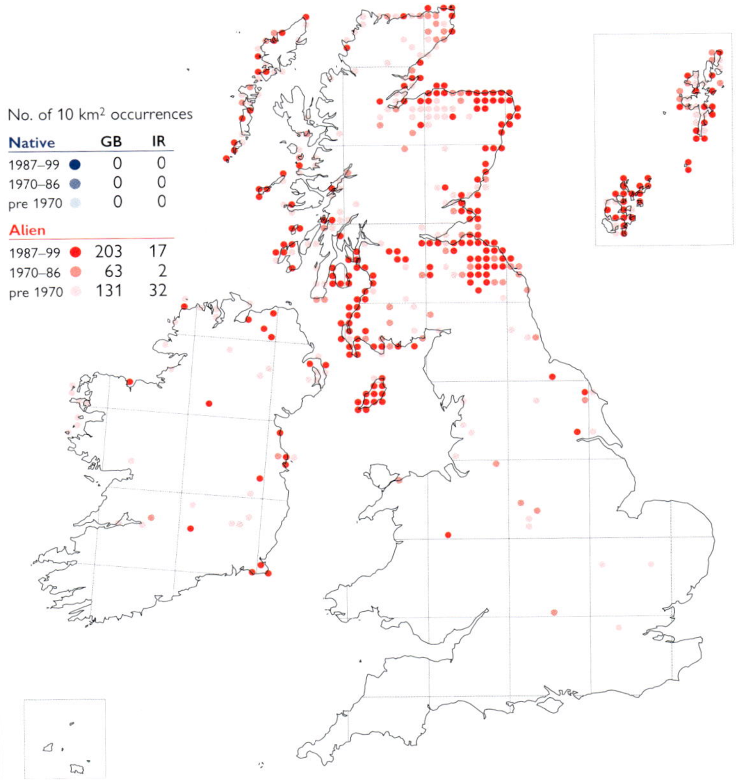

An annual of cultivated and waste ground. Generally lowland, but formerly recorded as a casual at 320 m in Derbyshire.

Archaeophyte (change –0.40). The map suggests a decline in the frequency of *L. confertum* in Scotland. Elsewhere, the decline apparent in the 1962 *Atlas* has continued; naturalised populations no longer exist in England, where it is now present only as a casual.

As an archaeophyte *L. confertum* has a European Boreal-montane distribution; it also occurs in Greenland.

References: Atlas (248d), Hultén & Fries (1986).

K. WALKER

Lamium amplexicaule Henbit Dead-nettle

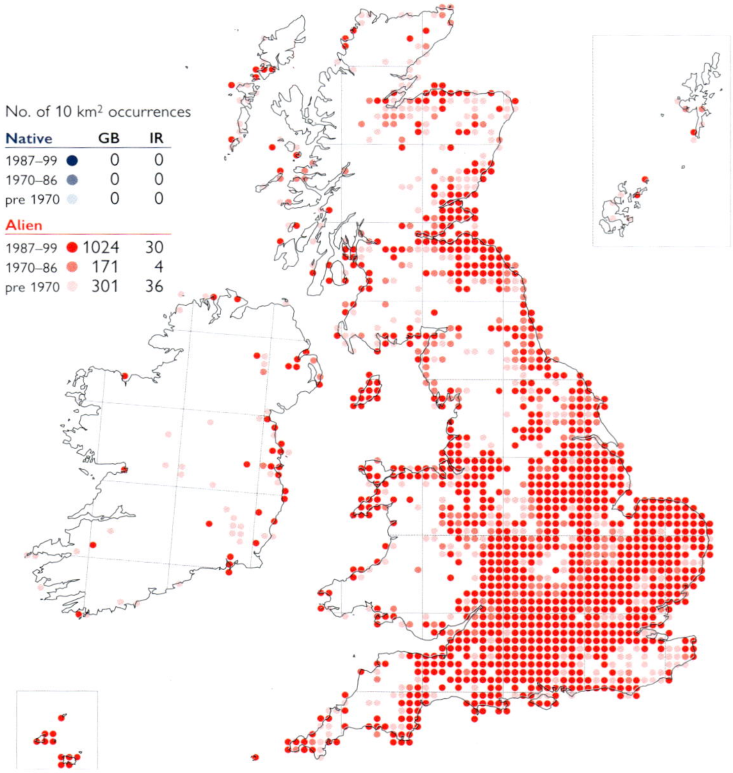

An annual of open, cultivated and waste ground, usually found on light, dry soils. It also occurs on walls, by railways and in cracks in pavements. 0–455 m (Atholl, E. Perth).

Archaeophyte (change –0.22). Although now much better recorded than it was in the 1962 *Atlas*, the map suggests a slight decline in this species, especially in S. England and N.E. Scotland. It is decreasing as a weed of cultivated land because of the increased use of herbicides (Brewis *et al.*, 1996).

As an archaeophyte *L. amplexicaule* has a Eurosiberian Southern-temperate distribution, but it is widely naturalised so that distribution is now Circumpolar Southern-temperate.

References: Atlas (248c), Hultén & Fries (1986), Meusel *et al.* (1978).

K. WALKER

Galeopsis segetum Downy Hemp-nettle

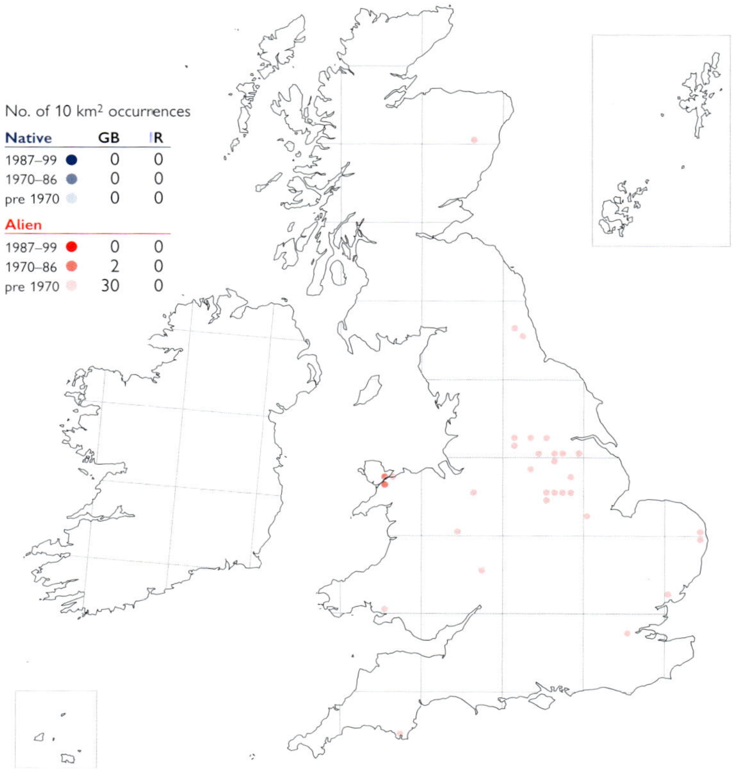

An annual weed of arable and waste ground, most frequently found as a casual in root-crops. Lowland.

Archaeophyte. This species was formerly established at one site near Bangor (Caerns.). Although it used to appear annually after its discovery there in 1802, it has been seen only once since the 1962 *Atlas* (in 1975) and is now presumed to be extinct. It has often been regarded as a native at this locality. Elsewhere it is a rare and declining casual.

As an archaeophyte *G. segetum* has a Suboceanic Temperate distribution.

References: Atlas (250b), Meusel *et al.* (1978).

K. WALKER

Galeopsis angustifolia Red Hemp-nettle

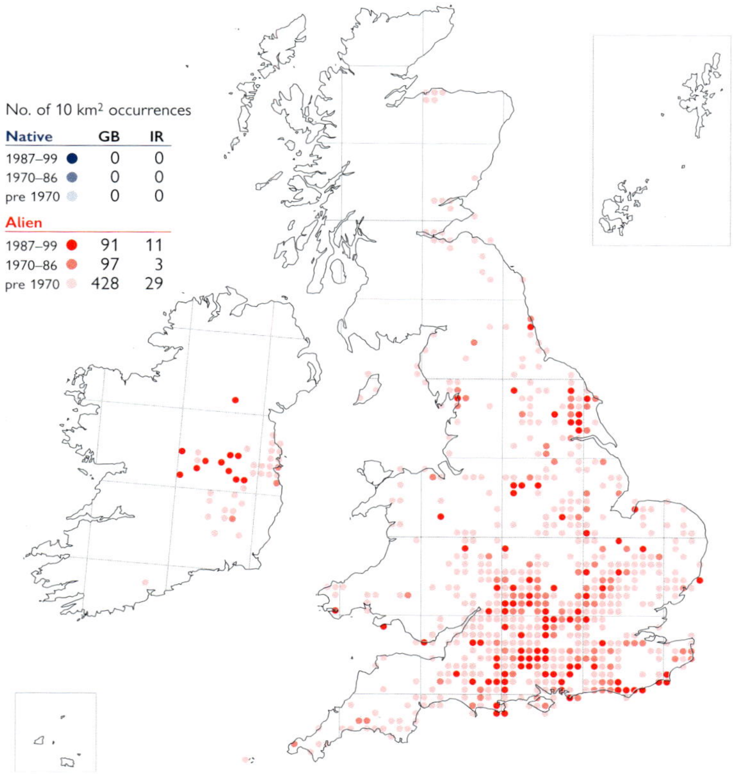

An annual of arable land, waste places and open ground on calcareous substrates, including limestone pavements and scree; also found on eskers and on coastal sand and shingle. This late-flowering species often fails to set seed within winter-sown crops. 0–320 m (Derbys.).

Archaeophyte (change –3.31). *G. angustifolia* was formerly a common cornfield weed in some areas (Druce, 1927), but the contraction in range shown in the 1962 *Atlas* has accelerated following a shift from spring- to winter-sown crops and cleaner crop husbandry. It is increasing on ground disturbed by gravel extraction in Ireland.

European Temperate element.

References: Atlas (250a), Curtis & McGough (1988), Hultén & Fries (1986), Meusel *et al.* (1978), Stewart *et al.* (1994), Townsend (1962).

K. WALKER

Galeopsis speciosa Large-flowered Hemp-nettle

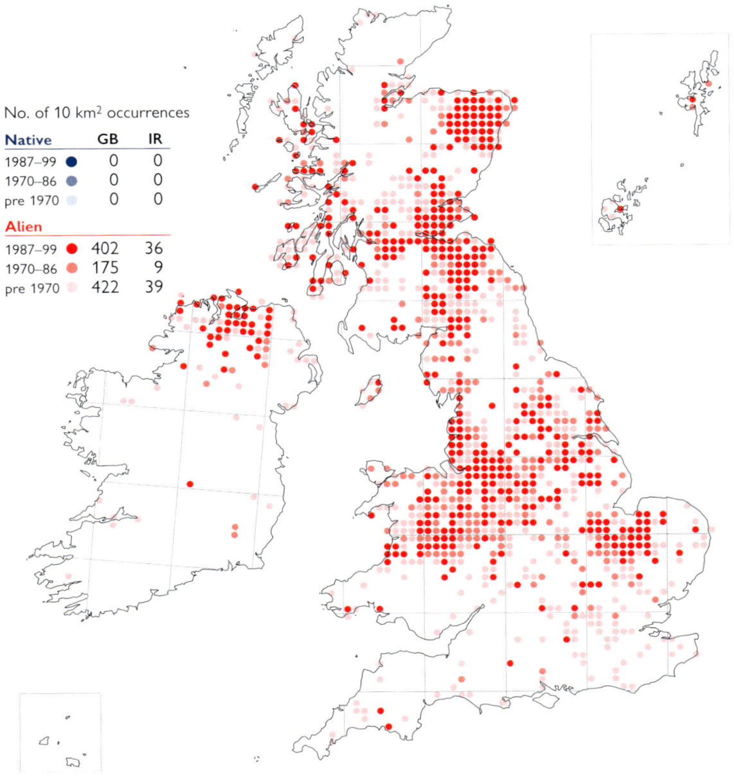

An annual weed of cultivated, marginal and waste ground, often within root-crops (especially potatoes) on peaty soils. 0–445 m (Clun Forest, Salop).

Archaeophyte (change −1.82). This species has declined markedly since the 1962 *Atlas*. It is often associated with traditional arable farming and has suffered where modern methods of cultivation and weed control have been introduced.

As an archaeophyte *G. speciosa* has a Eurosiberian Boreo-temperate distribution; it is widely naturalised outside this range.

References: Atlas (250d), Hultén & Fries (1986), Meusel *et al.* (1978).

K. WALKER

Galeopsis tetrahit sens. lat. Common Hemp-nettles

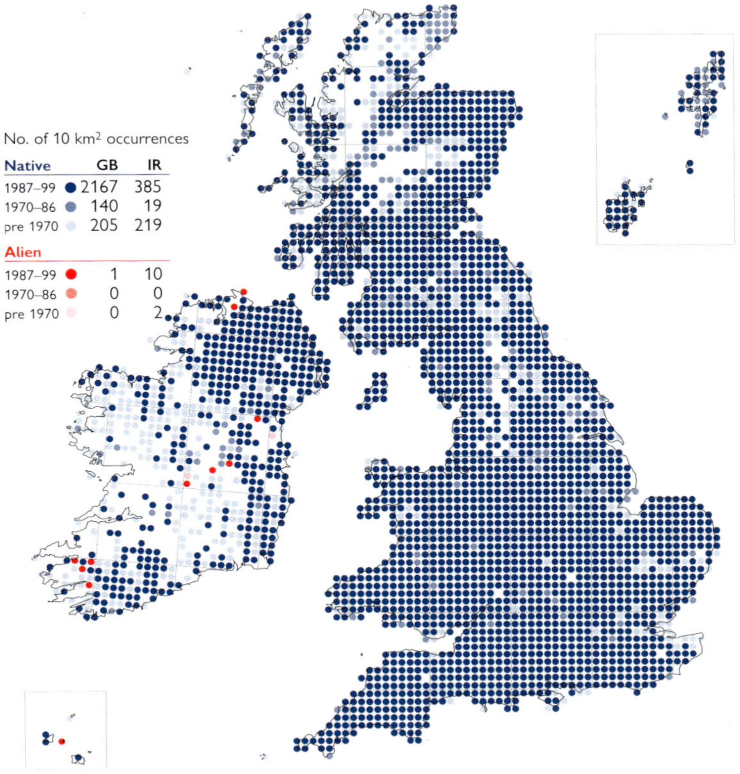

These annuals grow in moist, moderately shaded semi-natural habitats such as woodland clearings, ditches, fens, river banks, roadside verges and wet heaths, but perhaps more frequently in disturbed arable, waste and cultivated ground. 0–445 m (Clun Forest, Salop).

Native (change −0.61). Records of *G. bifida* and *G. tetrahit* are included on this map, which is equivalent to the map of *G. tetrahit sens. lat.* in the 1962 *Atlas*.

The world distribution of the two segregates is given in the following accounts.

References: Atlas (250c), Grime *et al.* (1988), Hultén & Fries (1986), O'Donovan & Sharma (1987).

K. WALKER

Galeopsis tetrahit Common Hemp-nettle

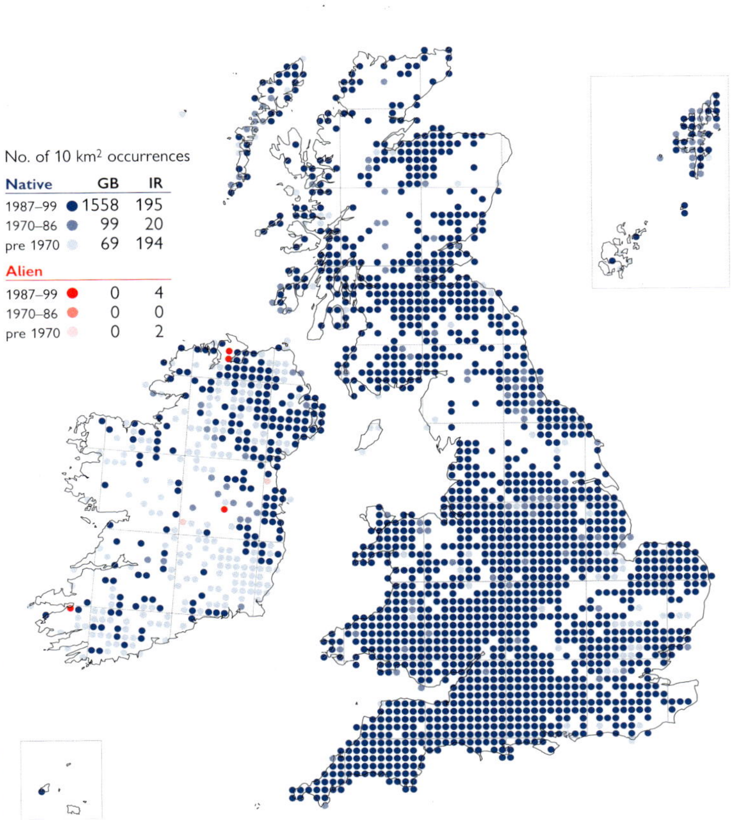

An annual of habitats such as woodland clearings, ditches, fens, river banks, roadside verges and wet heaths, but perhaps more frequent in disturbed arable, waste and cultivated ground. 0–450 m (Clun Forest, Salop).

Native. *G. tetrahit sens. str.* has not previously been mapped at a national level. It is clear from the map that in some areas recorders have not distinguished it from *G. bifida*.

European Boreo-temperate element.

References: Grime *et al.* (1988), Hultén & Fries (1986), Meusel *et al.* (1978), O'Donovan & Sharma (1987).

K. WALKER

Galeopsis bifida Bifid Hemp-nettle

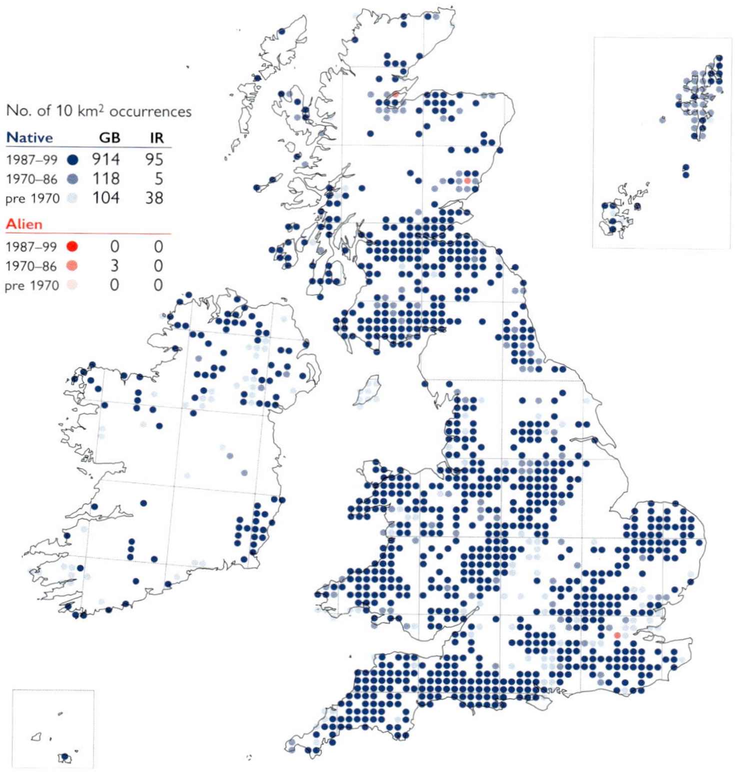

No. of 10 km² occurrences

Native	GB	IR
1987–99 ●	914	95
1970–86 ●	118	5
pre 1970 ●	104	38
Alien		
1987–99 ●	0	0
1970–86 ●	3	0
pre 1970 ●	0	0

An annual of arable, waste and cultivated ground, and less often woodland clearings and ditch-sides. It is found in similar places to *G. tetrahit*, and often grows with it, although more strictly an arable weed. Lowland.

Native. *G. bifida* was mapped within *G. tetrahit sens. lat.* in the 1962 *Atlas*. As a consequence there is no basis from which to assess any changes in distribution. It is almost certainly under-recorded.

Eurasian Boreo-temperate element, but naturalised in N. America so distribution is now Circumpolar Boreo-temperate.

References: Grime *et al.* (1988), Hultén & Fries (1986), Meusel *et al.* (1978), O'Donovan & Sharma (1987).

K. WALKER

Melittis melissophyllum Bastard Balm

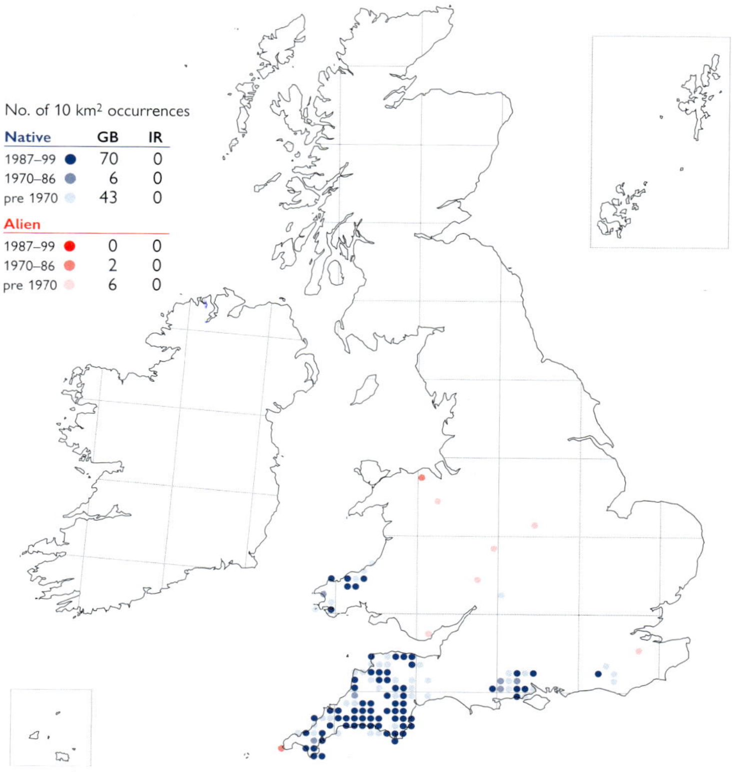

No. of 10 km² occurrences

Native	GB	IR
1987–99 ●	70	0
1970–86 ●	6	0
pre 1970 ●	43	0
Alien		
1987–99 ●	0	0
1970–86 ●	2	0
pre 1970 ●	6	0

A strong-smelling perennial herb of woodland, wood-borders, hedge banks and scrub on base-rich soils. In the New Forest (S. Hants.), it is a plant of ancient woodland. It favours light shade and can be abundant in cleared or coppiced woodland. It is intolerant of grazing. Lowland.

Native (change –0.47). The distribution of *M. melissophyllum* in Devon and Cornwall is apparently stable. However, it has declined markedly in the New Forest and Dorset over the past twenty years as a result of overshading and pony grazing, although at some sites it has reappeared after scrub clearance and coppicing.

European Temperate element.

References: Atlas (246a), Brewis *et al.* (1996), Kay & John (1995), Meusel *et al.* (1978), Stewart *et al.* (1994).

K. WALKER

Marrubium vulgare White Horehound

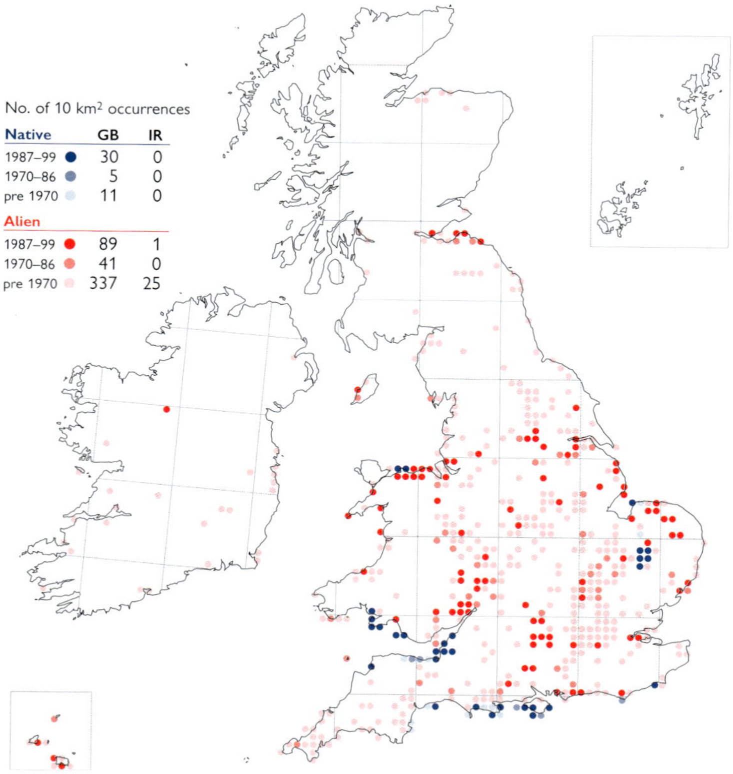

No. of 10 km² occurrences

Native	GB	IR
1987–99 ●	30	0
1970–86 ●	5	0
pre 1970 ●	11	0
Alien		
1987–99 ●	89	1
1970–86 ●	41	0
pre 1970 ●	337	25

A perennial herb, probably native only near the sea on open, exposed cliff-top grasslands and slopes overlying limestone and chalk, and on sandy banks and verges in Breckland. It is cultivated for tea and its medicinal properties, and is naturalised in rough and waste places; it also occurs as a wool-shoddy alien. Lowland.

Native (change –2.02). The distinction between native and alien populations, particularly in coastal areas, can be difficult. However, native sites seem to have declined since the 1962 *Atlas* due to lack of grazing, and its decline as an alien, already apparent by 1962, has also continued.

Eurosiberian Southern-temperate element; widely naturalised outside its native range.

References: Atlas (251c), Hultén & Fries (1986), Mabey (1996), Meusel *et al.* (1978), Stewart *et al.* (1994).

K. WALKER

Scutellaria galericulata Skullcap

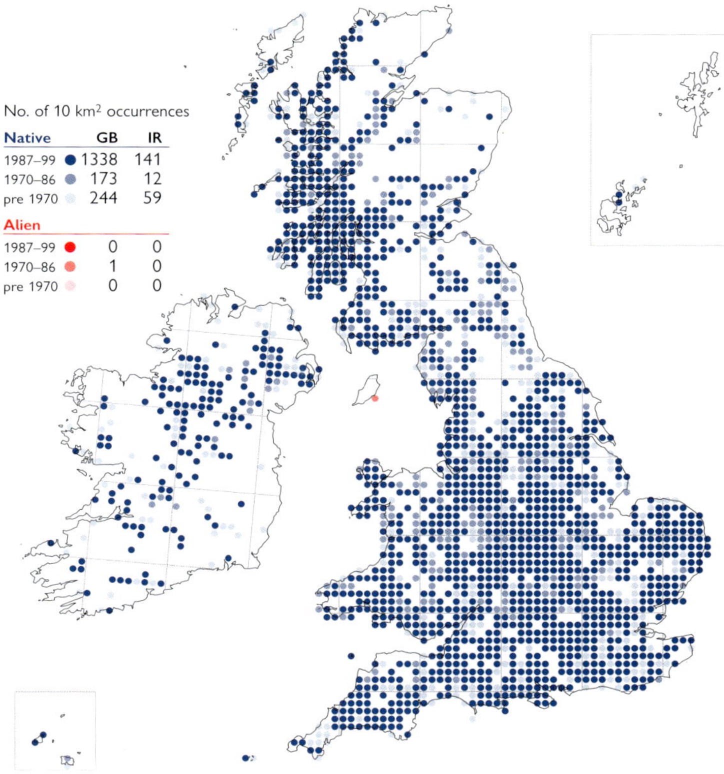

A perennial herb associated with a variety of wetland habitats including ponds, rivers, canals, marshes, fens, fen-meadows, wet woodland and dune-slacks. It also grows on coastal boulder beaches in Scotland. Generally lowland, but reaching 365 m at High Cup Gill (Westmorland).

Native (change –0.39). *S. galericulata* was mapped as 'all records' in the 1962 *Atlas*. It has declined in England as a result of drainage and habitat destruction (Mountford, 1994). Elsewhere there is no evidence for a change in distribution.

Eurosiberian Boreo-temperate element; also in N. America.

References: Atlas (251d), Hultén & Fries (1986), Meusel *et al.* (1978).

K. WALKER

Scutellaria minor Lesser Skullcap

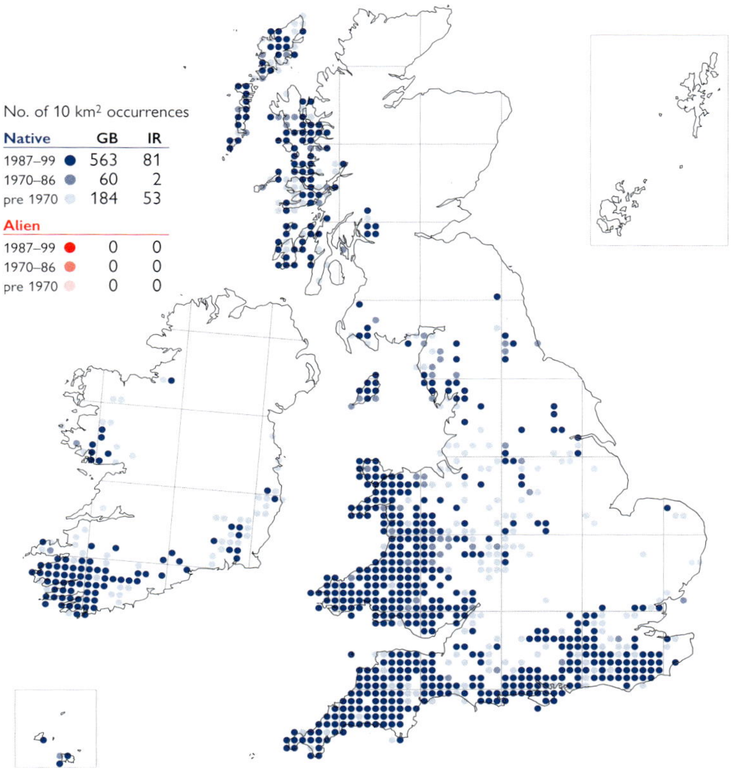

A perennial herb of wet heaths, bogs, marshes and moist, heathy woodlands on acidic, oligotrophic or slightly mesotrophic soils. 0–440 m (Woodhead, Cheshire).

Native (change +0.03). Many populations of *S. minor* were lost before 1930, and the range shown in the 1962 *Atlas* has continued to contract as a result of drainage and habitat loss.

Suboceanic Southern-temperate element.

References: Atlas (252a), Dupont (1962), Hultén & Fries (1986), Meusel *et al.* (1978).

K. WALKER

Teucrium scorodonia Wood Sage

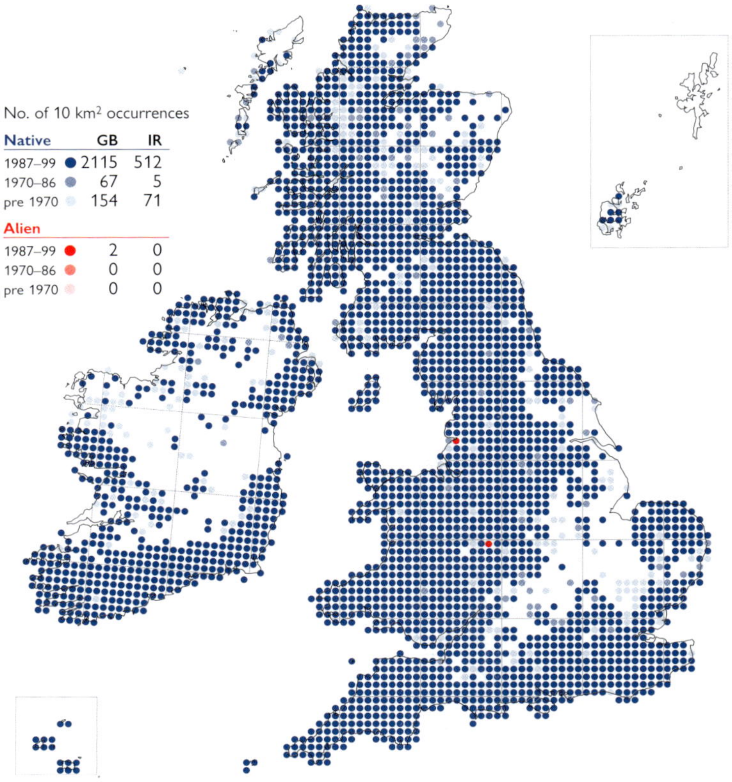

A rhizomatous perennial herb of well-drained, acidic to mildly calcareous mineral soils, occurring in a wide range of habitats including woodland, hedgerows, scrub, heaths, limestone grassland and pavement, mountain ledges, dunes and shingle, and amongst *Pteridium*. 0–550 m (Pistyll Rhaeadr, Monts.).

Native (change –0.69). There is no evidence for a change in the range of *T. scorodonia* since the 1962 *Atlas*.

Suboceanic Southern-temperate element.

References: Atlas (253a), Grime *et al.* (1988), Hultén & Fries (1986), Hutchinson (1968), Meusel *et al.* (1978).

K. WALKER

Teucrium chamaedrys Wall Germander

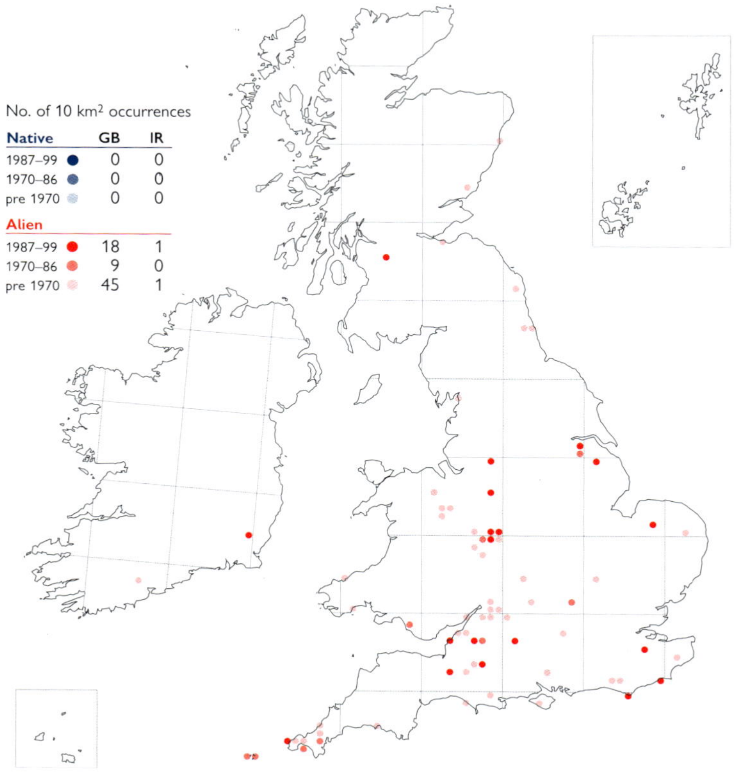

This perennial herb is usually found on walls, rocks and dry banks. A population of small, almost prostrate, plants has been known in cliff-top chalk grassland at Cuckmere Haven (E. Sussex) since 1945, where it may be native (Rose, 1988). Lowland.

Neophyte (change –0.41). This species, first recorded in the wild in 1710, appears to be declining as a garden escape. The identity of extant populations needs to be checked as the normal garden plant is probably the hybrid *T. chamaedrys* × *T. lucidum*. Some populations are very long-lived; one has been known on a wall at Curry Mallet (S. Somerset) since 1922.

European Southern-temperate element, with a continental distribution in W. Europe.

References: Atlas (252b), Clement & Foster (1994), Meusel *et al.* (1978), Wigginton (1999).

K. WALKER

Teucrium scordium Water Germander

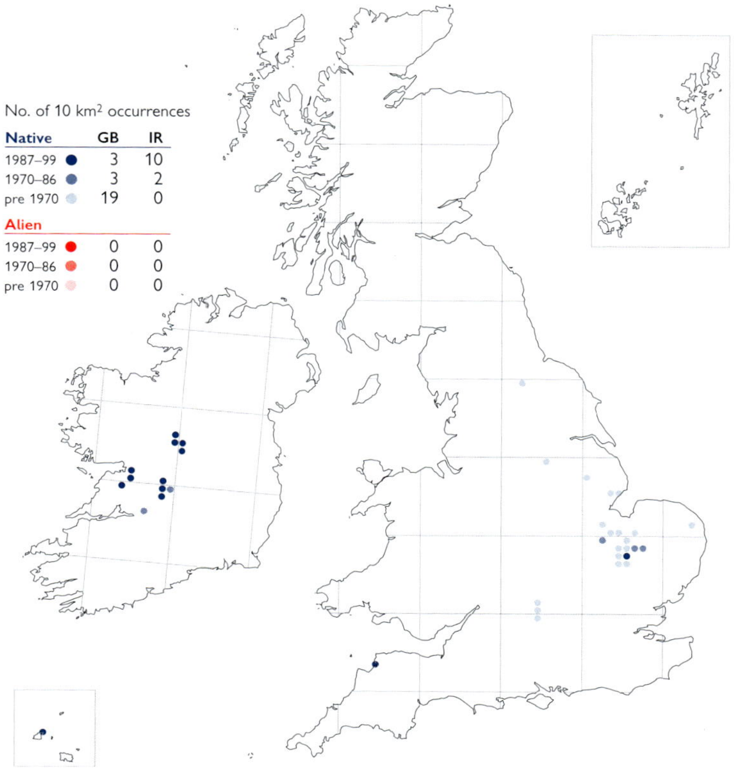

This stoloniferous perennial herb has been recorded in a variety of wetland habitats with fluctuating water levels, including the margins of dune-slack pools, reed-fen, clay-pits and the banks of rivers, ponds and ditches. In Ireland it is often recorded from turloughs. Flowering and seed production can be poor. Lowland.

Native (change –0.64). The long-term decline of *T. scordium* in England, apparent from the 1962 *Atlas*, has continued as a result of drainage, reclamation and eutrophication of its wetland habitats. Although remaining populations are apparently stable, they are threatened by lack of management, scrub encroachment and shading. It is stable in Ireland.

Eurosiberian Southern-temperate element.

References: Atlas (252c), Hultén & Fries (1986), Wigginton (1999).

K. WALKER

Teucrium botrys Cut-leaved Germander

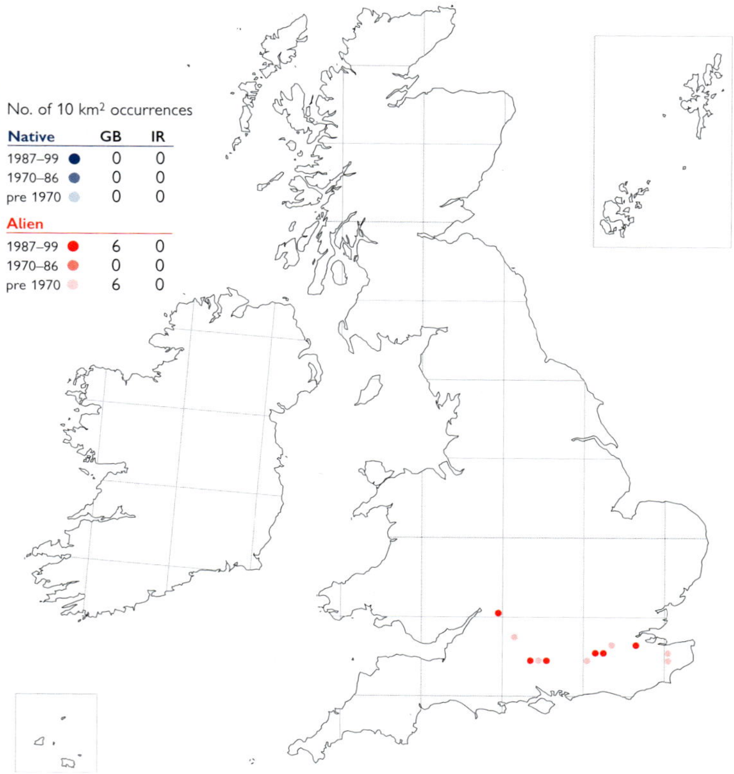

A biennial herb of bare ground within open grassland, arable field margins, and open fallow overlying chalk and limestone; occasionally recorded on spoil-tips and in disused quarries. Lowland.

Neophyte (change –0.42). *T. botrys* was cultivated in Britain by 1633 and first recorded in 1844 at Box Hill (Surrey), and has sometimes been considered to be native. It has declined since 1930 due to agricultural intensification, scrub encroachment and lack of grazing. However, it benefits from disturbance and at some sites thousands of plants have been recorded following cultivation or conservation management such as harrowing and turf cutting.

European Temperate element, but absent as a native from much of W. Europe.

References: Atlas (252d), Meusel *et al.* (1978), Rich (1997b), Wigginton (1999).

K. WALKER

Ajuga reptans Bugle

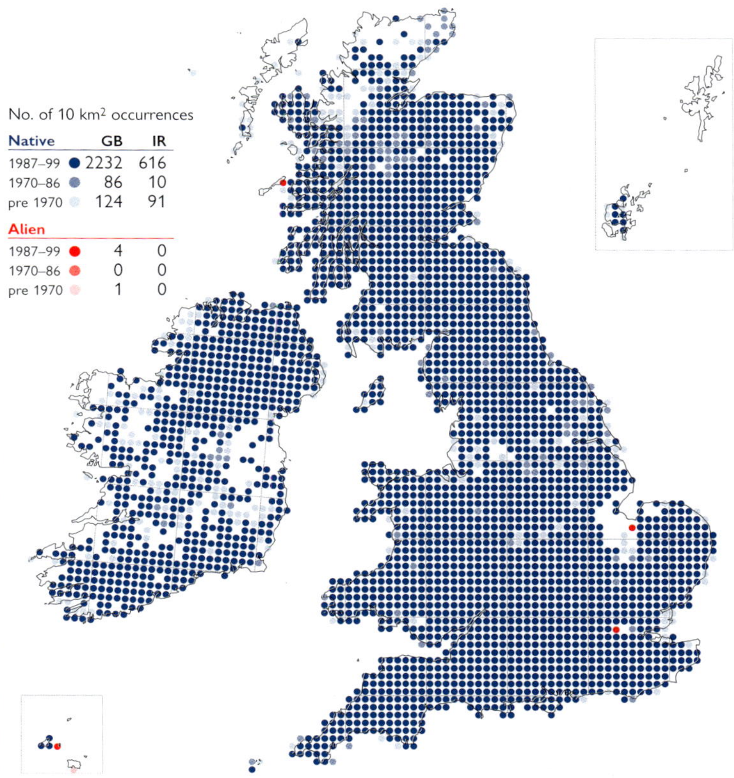

A rhizomatous perennial herb of damp deciduous woods and woodland rides, shaded places and unimproved grassland on neutral or acidic soils, sometimes occurring in flushed ground. 0–760 m on Y Foel-fras (Caerns.).

Native (change –0.56). There is no evidence of an appreciable change in the distribution of *A. reptans* since the 1962 *Atlas.*

European Temperate element.

References: Atlas (253c), Hultén & Fries (1986), Meusel *et al.* (1978).

K. WALKER

Ajuga pyramidalis Pyramidal Bugle

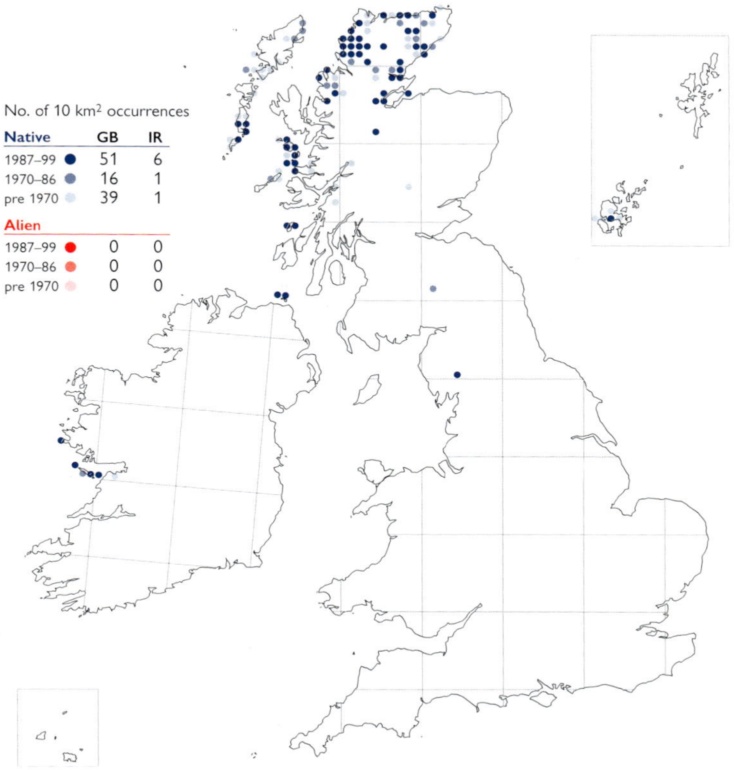

A perennial herb of free-draining slopes, rock crevices and shallow peat in open heathland and grassland overlying moderately acidic, or occasionally neutral or basic, soils. Reproduction is mainly from seed, which is long-lived and often germinates after disturbance. 0–650 m on Ill Bell (Westmorland).

Native (change –0.34). This species has been found in many new sites since the 1962 *Atlas*, particularly in N. and W. Ireland and remoter areas of Scotland where further populations probably await discovery. Conversely, some sites, such as on Coll (Mid Ebudes), have apparently been lost through more intensive grazing.

European Boreal-montane element.

References: Atlas (253d), Curtis & McGough (1988), Gulliver (1997), Hultén & Fries (1986), Meusel *et al.* (1978), Rich, Kay & Sydes (1999), Stewart *et al.* (1994).

K. WALKER

Ajuga chamaepitys Ground-pine

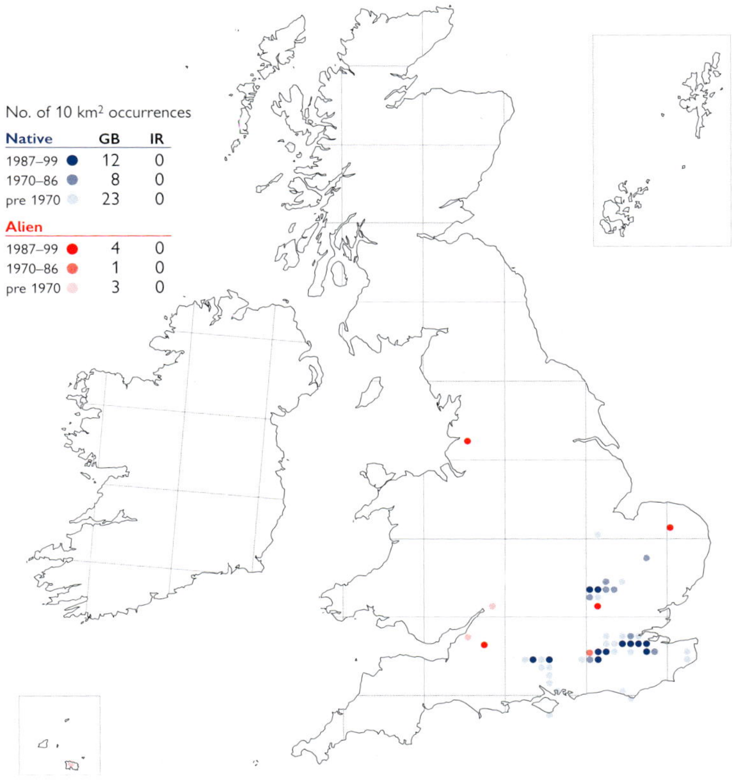

An annual or biennial herb of arable field margins and bare tracks on calcareous soils, and on open chalk downland. Its seeds are long-lived and this has led to its reappearance following disturbance at some sites. Lowland.

Native or archaeophyte (change –0.62). Known as a British plant since 1551, *A. chamaepitys* has declined considerably over the past fifty years due to herbicide spraying, abandonment of fallow land and succession to coarse grassland, scrub and woodland on chalk slopes. It has benefited from conservation management at some sites.

European Southern-temperate element.

References: Atlas (253b), Meusel *et al.* (1978), Stewart *et al.* (1994), Wigginton (1999).

K. WALKER

Nepeta cataria Cat-mint

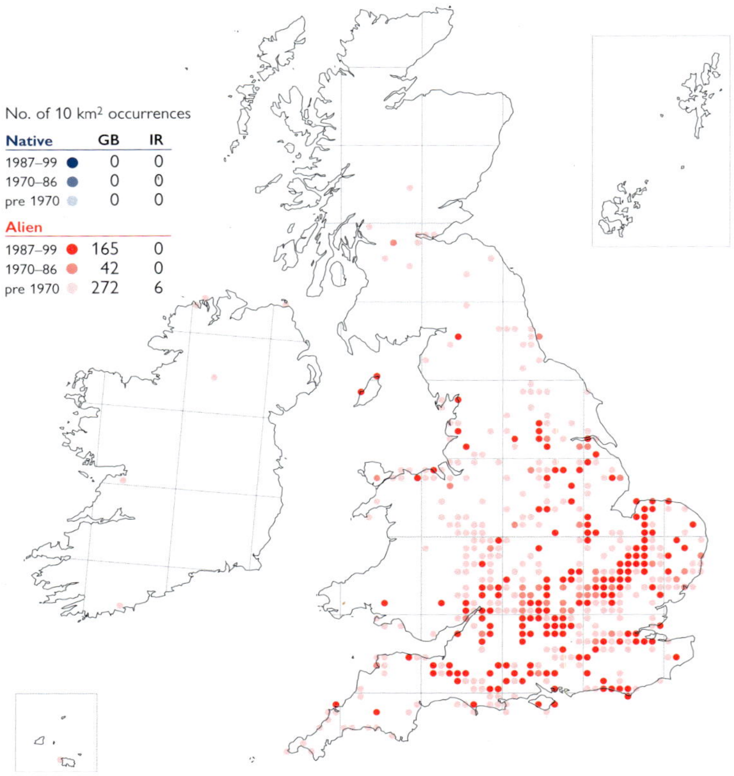

A perennial herb of open grassland, waysides, hedge banks, roadsides and rough ground on calcareous soils. In addition to such long-established populations, plants occasionally escape from gardens and give rise to casual populations which rarely persist for more than a few seasons. The seed is long-lived. Lowland.

Archaeophyte (change –1.23). The marked contraction in range of *N. cataria* shown in the 1962 *Atlas* has continued. In Northamptonshire it has declined as a result of hedgerow removal (Gent *et al.*, 1995). Elsewhere, the intensification of agriculture and the growth of scrub have claimed many sites.

As an archaeophyte *N. cataria* has a Eurosiberian Temperate distribution; it is widely naturalised outside this range.

References: Atlas (251a), Hultén & Fries (1986), Meusel *et al.* (1978).

K. WALKER

Nepeta nepetella × *N. racemosa* (*N.* × *faassenii*) Garden Cat-mint

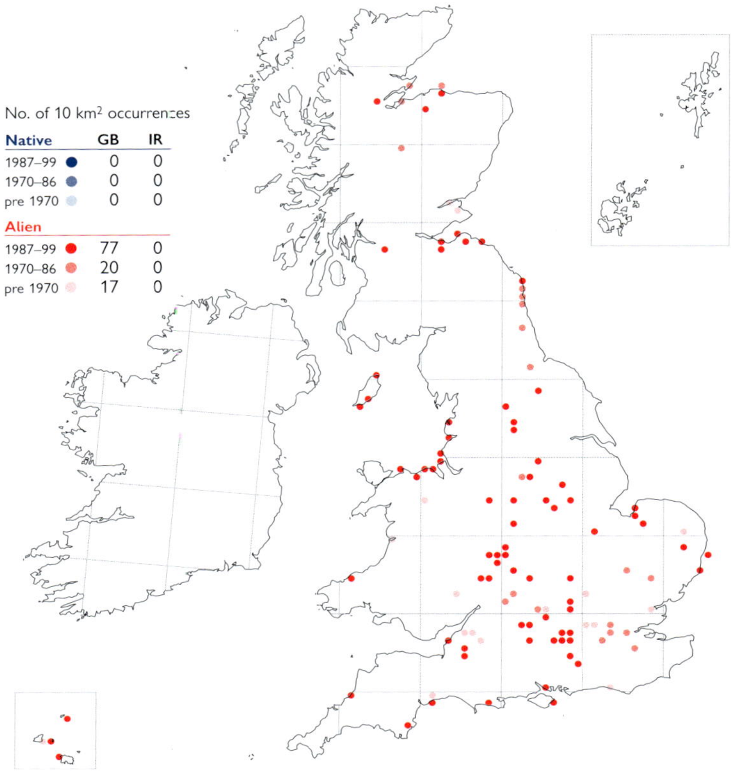

A rhizomatous perennial herb found as a garden throw-out or escape on roadsides, rubbish tips and waste ground. It can persist and spread by means of its extensive rhizomes. The hybrid is sterile; fertile plants have been recorded but these may be referable to *N. racemosa*. Lowland.

Neophyte. This hybrid was raised in 1784 and is extremely popular in gardens. It may be increasing in the wild, where it was not recorded until 1928.

A hybrid of garden origin.

K. WALKER

Glechoma hederacea Ground-ivy

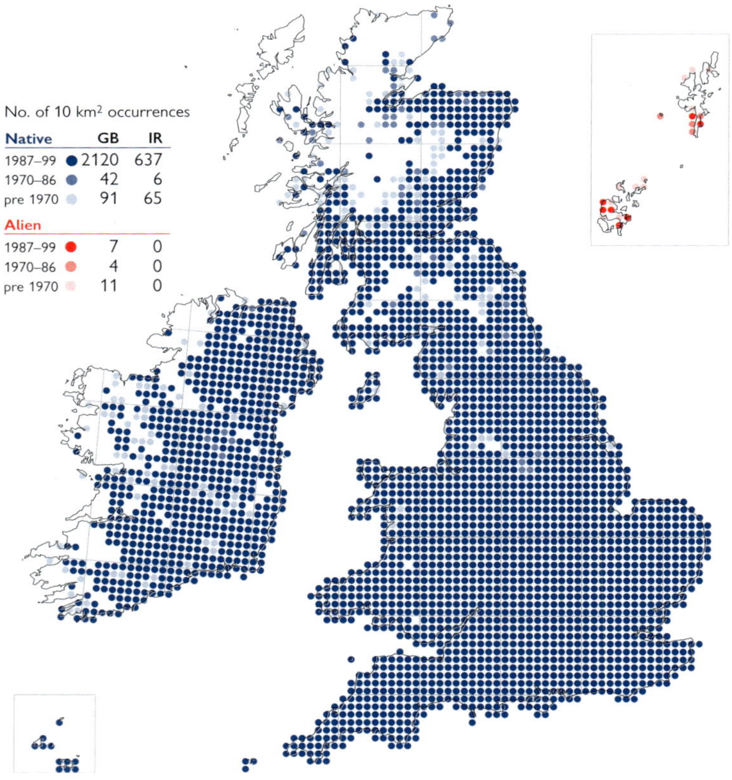

A carpet-forming stoloniferous perennial herb of woods, grassland, hedgerows and waste places, usually on fertile soils. It usually spreads vegetatively by rapid growth of its creeping stems, and seed-set is often very low. 0–465 m (Hawkswick Clowder, Mid-W. Yorks.).

Native (change –0.56). Although there is no evidence for a major change in distribution at the national scale, *G. hederacea* is apparently increasing within lowland woods, particularly in S.E. England, where excessive deer grazing has led to a decline in more palatable woodland ground-flora species (Cooke, 1994).

Eurasian Boreo-temperate element, but naturalised in N. America so distribution is now Circumpolar Boreo-temperate.

References: Atlas (251b), Grime *et al.* (1988), Hultén & Fries (1986), Hutchins & Price (1999), Meusel *et al.* (1978).

K. WALKER

Prunella vulgaris Selfheal

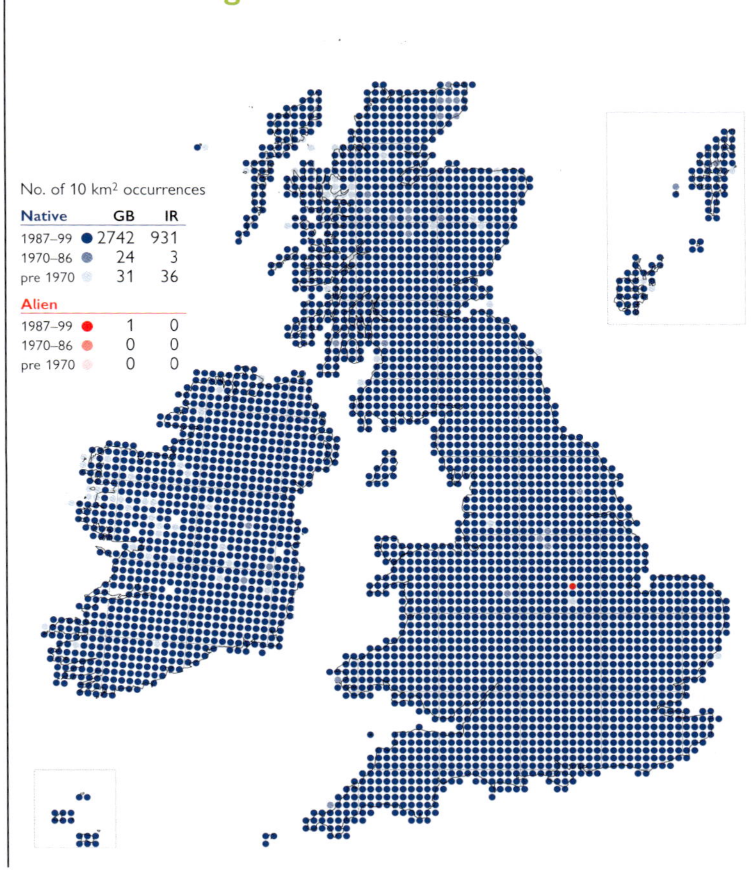

A patch-forming perennial herb of neutral and calcareous grassland, growing in clearings in woods, in meadows, pastures and lawns, on roadsides and waste ground, typically associated with moist, fertile soils. It forms clonal patches in short-grazed turf and spreads by the detachment of daughter ramets. 0–755 m (Knock Fell, Westmorland), and 845 m on Great Dun Fell (Westmorland).

Native (change +0.60). *P. vulgaris* is very frequent throughout Britain and Ireland, and in areas where it has been mapped in detail it has usually proved to be present in almost all tetrads. There is no evidence for a change in overall distribution.

Circumpolar Wide-temperate element; widely naturalised outside its native range.

References: Atlas (246b), Grime *et al.* (1988), Hultén & Fries (1986), Meusel *et al.* (1978).

K. WALKER

Prunella laciniata Cut-leaved Selfheal

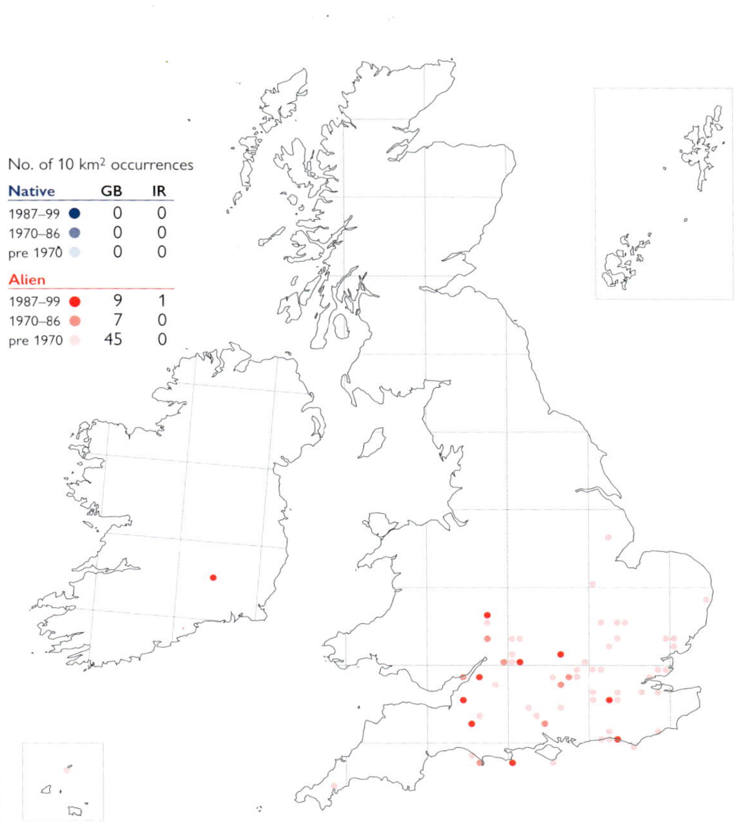

A rhizomatous perennial herb found naturalised in dry calcareous soils in grassland and on roadsides, waste ground and along woodland rides. Lowland.

Neophyte (change −2.60). *P. laciniata* has been regarded as native in some grassland sites. However, it was being cultivated in Britain by 1713 and was not recorded in the wild until 1886. At some sites it is known to have been introduced with clover seed. A widespread decline was evident by 1930, and losses have increased since the 1962 *Atlas*. It is now rare, but new sites are occasionally found; it was first recorded in Ireland in 1999. Its hybrid with *P. vulgaris* is present at many of its sites.

P. laciniata has a European Temperate distribution.

References: Atlas (246c), Clement (1985b), Hultén & Fries (1986), Meusel *et al.* (1978), Morton (1973).

K. WALKER

Melissa officinalis Balm

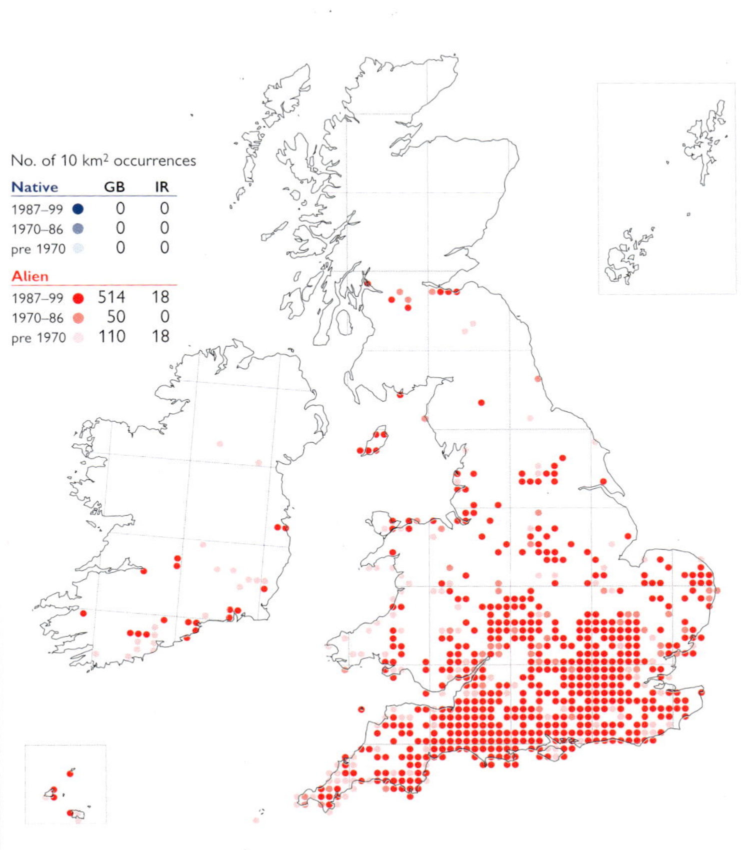

A lemon-scented perennial herb, found on banks and road verges close to habitation, and on rubbish dumps and waste ground. This species is much grown in gardens where it seeds prolifically, often becoming a weed and escaping into nearby habitats. Lowland.

Neophyte (change +1.73). *M. officinalis* was being cultivated in British gardens by 995 (Harvey, 1981) and was recorded in the wild by 1763. It is a culinary herb which seems to be increasing as a garden escape.

Native of S. Europe, the Mediterranean region and S.W. Asia, but its native range is obscured by naturalised populations further north.

References: Atlas (244d), Bolòs & Vigo (1995), Mabey (1996).

K. WALKER

Clinopodium menthifolium Wood Calamint

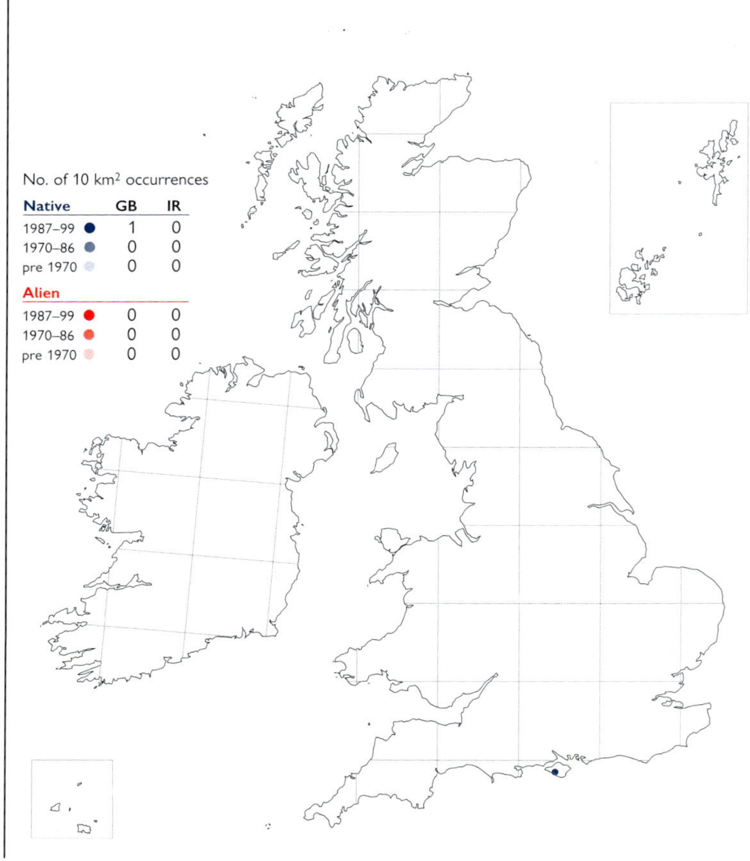

A rhizomatous perennial herb of lightly-shaded woodland edges and scrub overlying chalk. Lowland.

Native. This species has been known since 1843 from a single dry chalk valley in the Isle of Wight. Although once abundant, the cessation of coppicing in the 1940s led to a marked decline, but this has been stemmed by the resumption of coppicing and the clearance of invasive ground-cover and more nutrient demanding tall herbs such as *Eupatorium cannabinum* and *Urtica dioica*.

European Temperate element.

References: Atlas (243c), Wigginton (1999).

K. WALKER

Clinopodium ascendens Common Calamint

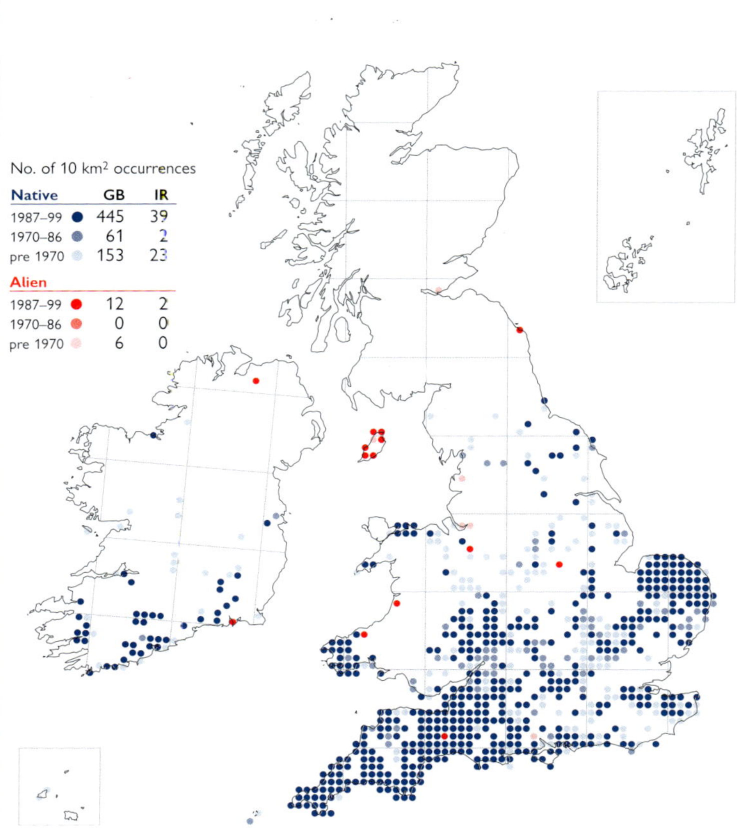

A rhizomatous perennial of hedge banks, road verges, rough scrubby grassland and rocky outcrops, usually on dry base-rich soils. It may occur as a relic of cultivation. Generally lowland, but reaching 380 m at Conistone (Mid-W. Yorks.).

Native (change +0.04). Some sites for *C. ascendens* were lost before 1930, particularly in the north of its range. There have been further losses since the 1962 *Atlas*, the reasons for which include the lack of scrub control in its grassland habitats, and changes in road verge management. However, it is now much better recorded in some areas, particularly East Anglia and Somerset.

European Temperate element.

Reference: Atlas (243d).

K. WALKER

Clinopodium calamintha Lesser Calamint

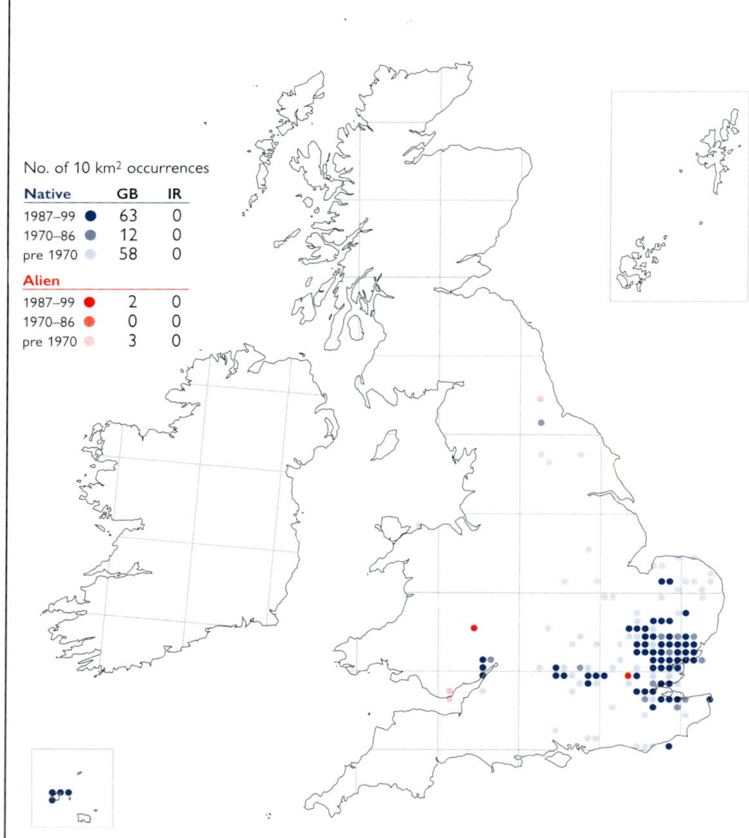

A short-lived perennial herb of dry, S.-facing banks and rough grassland on calcareous, sandy or gravelly soils. Formerly a pasture plant, it is now largely confined to roadsides, railway banks, churchyards and waste ground. Lowland.

Native (change –0.31). The native range of *C. calamintha* is uncertain because of former confusion with *C. ascendens* and the occurrence of garden escapes. Despite these uncertainties it is clear that its range has contracted, and many colonies have been lost following habitat destruction and changes in cutting regimes in its grassland habitats.

Submediterranean-Subatlantic element.

References: Atlas (244a), Stewart *et al.* (1994).

K. WALKER

Clinopodium vulgare Wild Basil

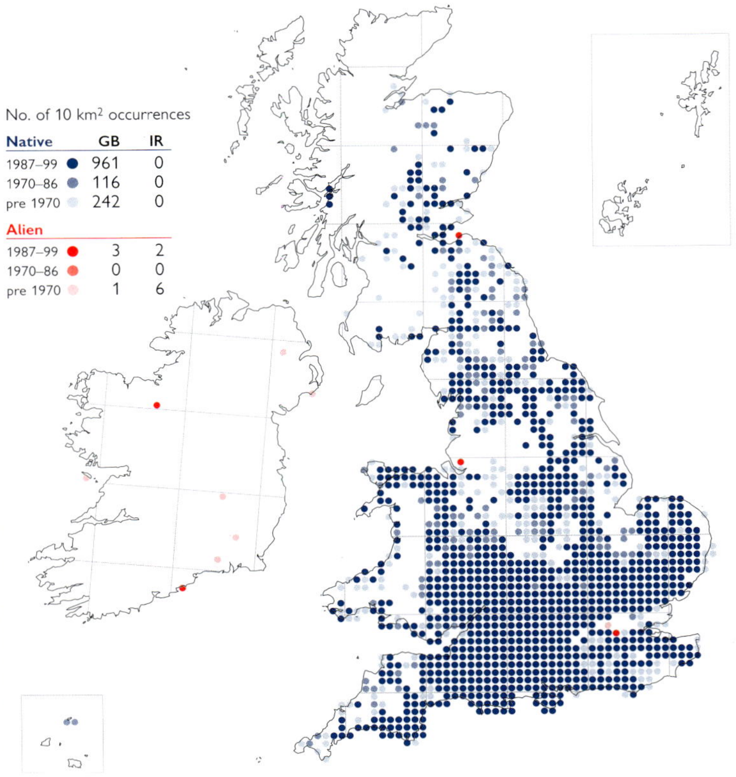

A rhizomatous perennial herb of hedges, woodland margins, coarse scrubby grassland, coastal cliffs and sand dunes, typically on dry calcareous soils. It is also found on waste ground, old quarries, and railway-sides. 0–395 m (W. Perth).

Native (change –0.67). There is no evidence of a change in the distribution of *C. vulgare* in S. England since the 1962 *Atlas* (where it was mapped as 'all records'), although it is apparently decreasing in the northern half of its British range. Its decline in Cumbria has been noted by Halliday (1997).

Circumpolar Temperate element, with a disjunct distribution.

References: Atlas (244c), Grime *et al.* (1988), Hultén & Fries (1986), Meusel *et al.* (1978).

K. WALKER

Clinopodium acinos Basil Thyme

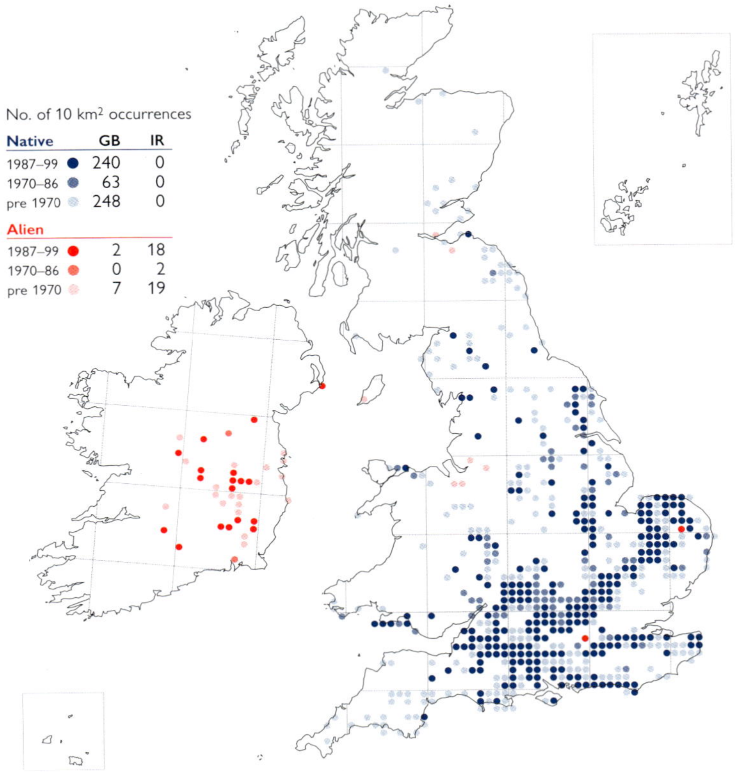

A usually annual herb of open habitats in dry grassland, rocky ground or arable fields. In Britain it usually grows on calcareous soils, whereas in Ireland it occurs on sandy and gravelly sites, including eskers. It is also a rare casual of waste ground, quarries and banks by roads and railways. Lowland.

Native (change –1.59). *C. acinos* has substantially declined as a result of more efficient methods of weed control and, in Ireland, gravel extraction. In many areas it is no longer found in arable fields, surviving only in less intensively managed habitats. It is considered to be alien in Ireland (Scannell & Synnott, 1987).

European Temperate element; also in C. Asia.

References: Atlas (244b), Curtis & McGough (1988), Hultén & Fries (1986), Meusel *et al.* (1978).

K. WALKER

Origanum vulgare Wild Marjoram

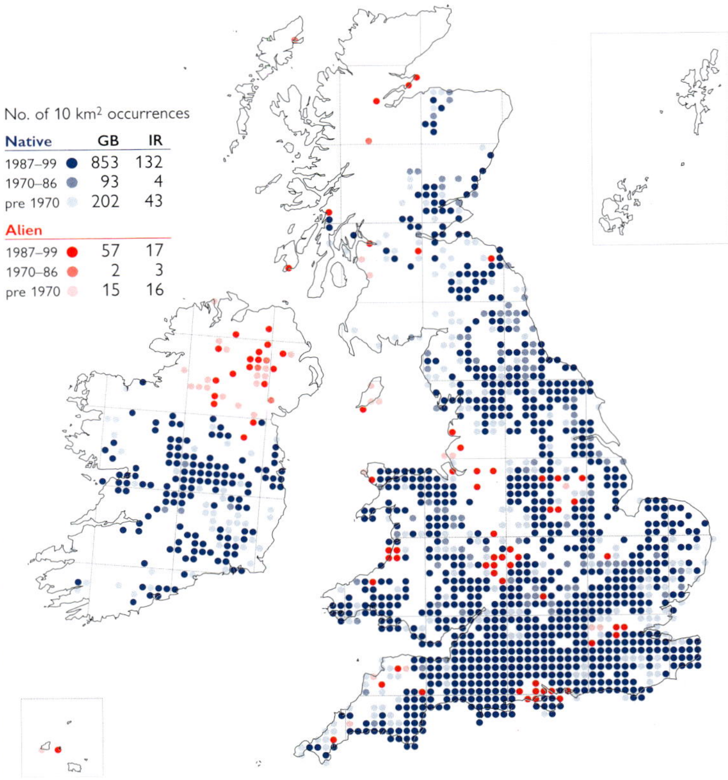

This herbaceous perennial herb of dry, infertile, calcareous soils is found in grassland, hedge banks, and scrub, and is a colonist of bare or sparsely vegetated ground, including quarries and road verges. It is occasionally naturalised from gardens. It is intolerant of heavy grazing. 0–410 m (N.W. Yorks.).

Native (change –0.10). *O. vulgare* was mapped as 'all records' in the 1962 *Atlas*. It has declined slightly except in the main areas of chalk and limestone soils.

Eurasian Southern-temperate element; widely naturalised outside its native range.

References: Atlas (242c), Grime *et al.* (1988), Hultén & Fries (1986), Meusel *et al.* (1978).

K. WALKER

Thymus pulegioides Large Thyme

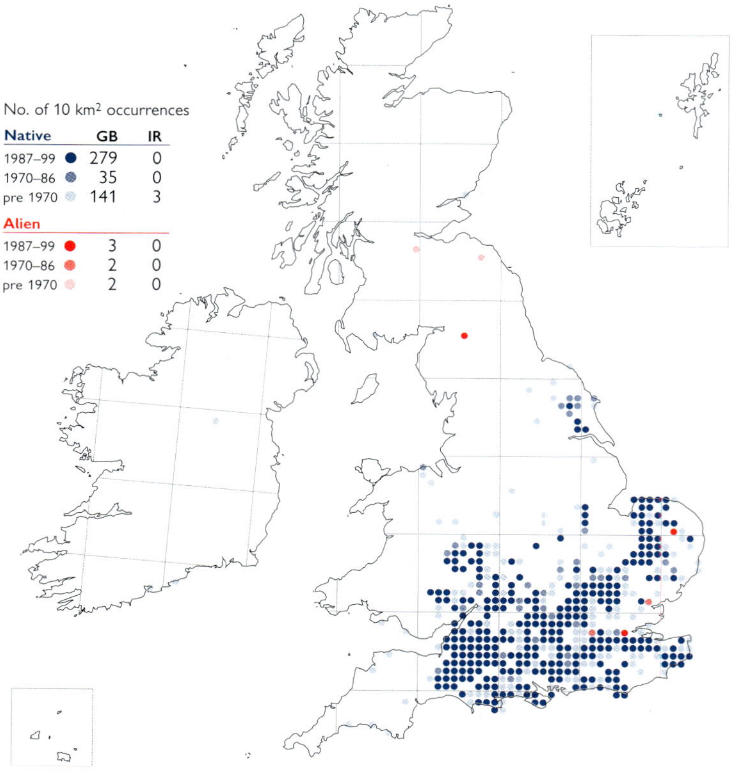

No. of 10 km² occurrences

Native		GB	IR
1987–99	●	279	0
1970–86	●	35	0
pre 1970	●	141	3
Alien			
1987–99	●	3	0
1970–86	●	2	0
pre 1970	●	2	0

A prostrate perennial herb of bare ground, short turf or coarse grassland on chalk, more rarely on sands and gravels on heaths and fixed dunes. It is more tolerant of competition than *T. polytrichus*. Lowland.

Native (change –0.38). This species was formerly confused with *T. polytrichus*, and the editors of the 1962 *Atlas* were cautious in accepting records that had not been verified by experts. There is less confusion now, and the species is better recorded. It has clearly declined at the edges of its range.

European Temperate element.

References: Atlas (242d), Hultén & Fries (1986), Meusel *et al.* (1978), Pigott (1955).

K. WALKER

Thymus polytrichus Wild Thyme

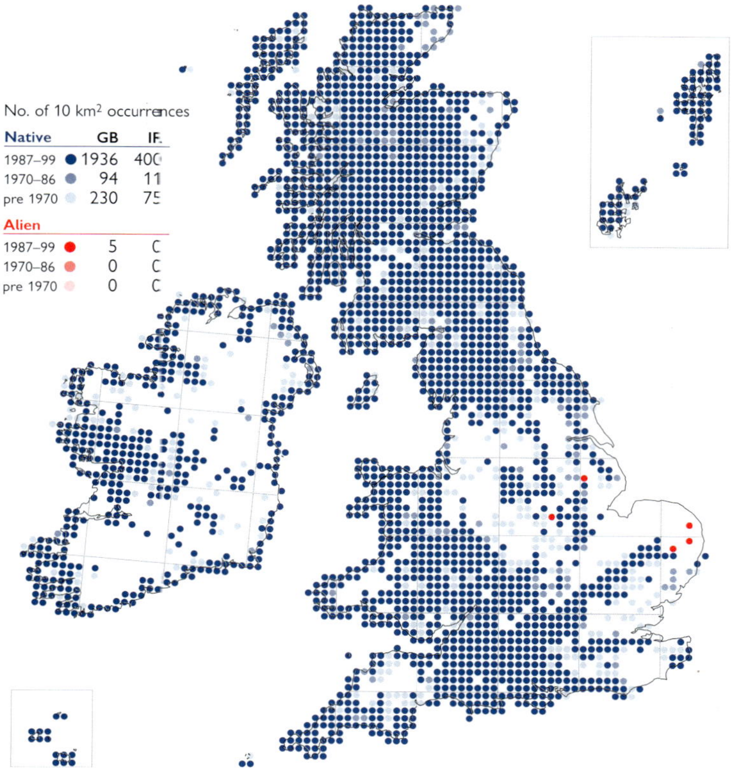

No. of 10 km² occurrences

Native		GB	IR
1987–99	●	1936	400
1970–86	●	94	11
pre 1970	●	230	75
Alien			
1987–99	●	5	0
1970–86	●	0	0
pre 1970	●	0	0

A perennial herb of free-draining, calcareous or base-rich substrates, including chalk, limestone, sands and gravels. It occurs in short grassland on heaths, downland, sea-cliffs and sand dunes, and around rock outcrops and hummocks in calcareous mires. It is also frequent in upland grassland and on montane cliffs, rocks and ledges. 0–1125 m (Ben Lawers, Mid Perth).

Native (change –0.64). *T. polytrichus* is still very common in suitable habitats. There is evidence for some losses in the southern part of its range since the 1962 *Atlas*.

European Boreo-temperate element.

References: Atlas (243b), Grime *et al.* (1988), Hultén & Fries (1986), Meusel *et al.* (1978), Pigott (1955).

K. WALKER

Thymus serpyllum Breckland Thyme

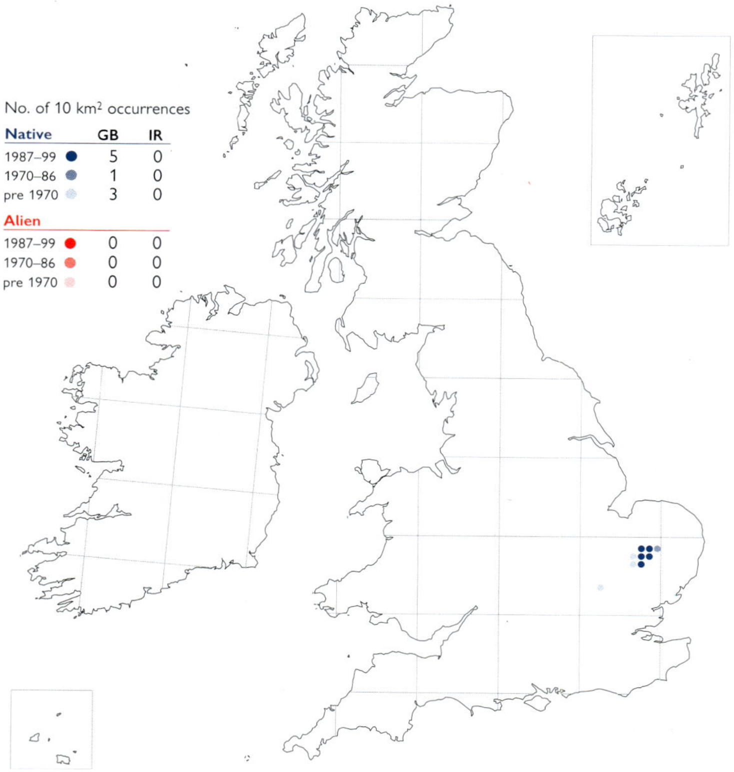

No. of 10 km² occurrences

Native		GB	IR
1987–99	●	5	0
1970–86	●	1	0
pre 1970	●	3	0
Alien			
1987–99	●	0	0
1970–86	●	0	0
pre 1970	●	0	0

A small prostrate perennial herb confined to dry sandy heaths and grasslands overlying chalk drift, and on inland dunes, especially in areas disturbed by rabbits or sheep. Lowland.

Native (change –0.11). *T. serpyllum* was first recorded in 1773, but since then, many sites have been lost to forestry and cultivation. However, the recovery of rabbit populations and increase in sheep grazing has ensured the maintenance of suitable conditions at its remaining sites.

European Boreo-temperate element, with a continental distribution in W. Europe; widely naturalised outside its native range.

References: Atlas (243a), Hultén & Fries (1986), Meusel *et al.* (1978), Pigott (1955), Trist (1979), Wigginton (1999).

K. WALKER

Lycopus europaeus Gipsywort

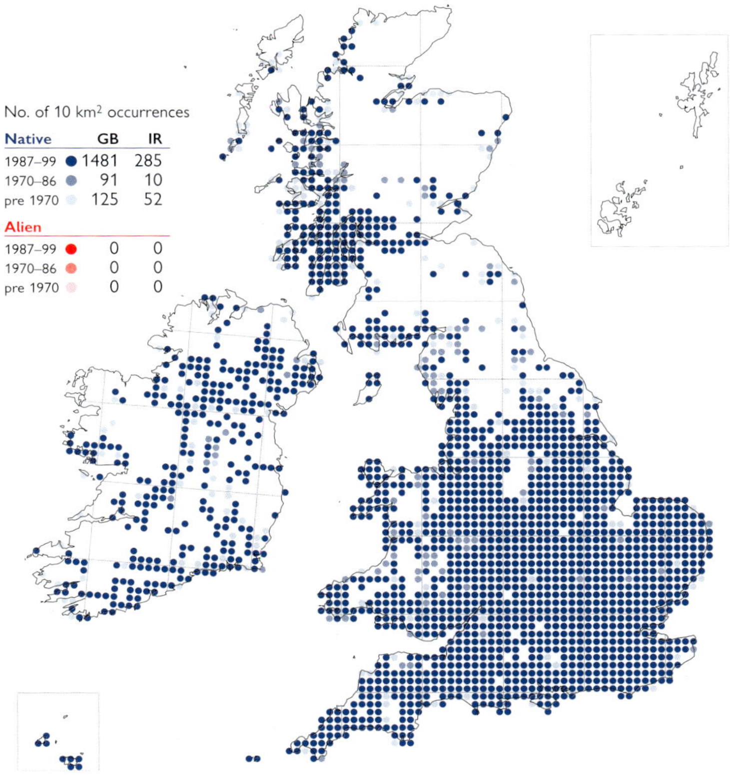

A rhizomatous perennial herb of wet habitats on organic and mineral soils, including the banks of rivers, streams, lakes and ditches, fens, fen carr, the top of beaches and dune-slacks. It is tolerant of temporary flooding, and is often an early colonist of exposed mud and shallow standing water in newly created wetlands. 0–485 m (Lochan Learg nan Lunn, Mid Perth).

Native (change –0.01). The current distribution of *L. europaeus* is apparently stable. Though the species may have disappeared from some sites as a result of drainage and clearance of watercourses, these losses are likely to have been balanced by colonisation of new sites, particularly gravel-pits.

Eurosiberian Temperate element; widely naturalised outside its native range.

References: Atlas (242b), Hultén & Fries (1986), Meusel *et al.* (1978).

K. WALKER

Mentha arvensis Corn Mint

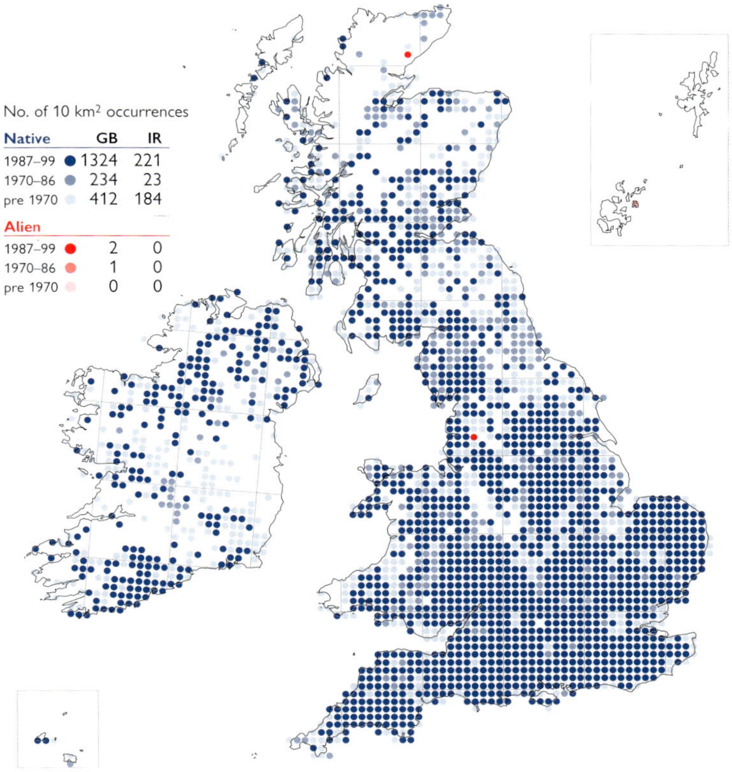

A rhizomatous perennial, rarely annual, herb of arable fields, woodland rides, marshy pastures and waste places; overlapping in habitat with *M. aquatica* but typically replacing it in drier habitats or where water levels fluctuate markedly. 0–390 m (Drumore Loch, E. Perth).

Native (change –1.30). *M. arvensis* was mapped as 'all records' in the 1962 *Atlas*. Analysis of the database reveals that a substantial decline has occurred since 1950. It has declined as a weed of cultivated land, and this may partly account for the significant loss from the English Midlands northwards. Some early records may refer to *M.* × *verticillata*.

Circumpolar Boreo-temperate element; widely naturalised outside its native range.

References: Atlas (241a), Hultén & Fries (1986), Meusel *et al.* (1978).

K. WALKER

Mentha arvensis × *M. spicata* (*M.* × *gracilis*) Bushy Mint

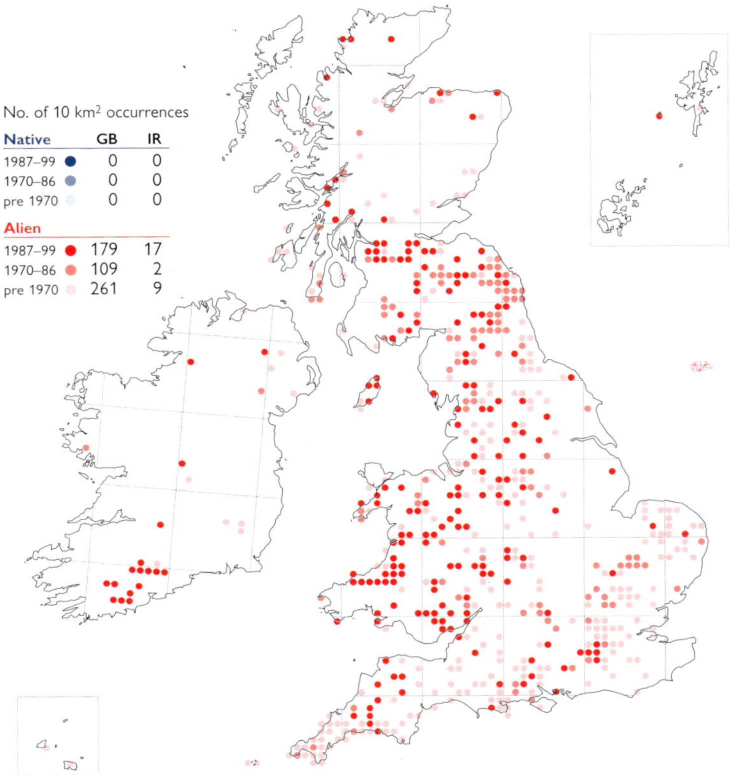

A rhizomatous perennial herb of damp places and waste ground, often in the absence of either parent. 0–360 m (near Tregaron, Cards.).

A spontaneous hybrid between native and alien parents. Plants arising spontaneously are pubescent, but glabrous clones are more commonly established as garden escapes or throw-outs. As these variants have not been recorded separately, all records are mapped as if they are alien. The map is similar to that in Perring & Sell (1968), and there is no evidence for a change in range, but there are few recent records. It has been mis-identified as *M.* × *verticillata* and as the non-British *M.* × *dalmatica*.

This hybrid is frequently cultivated and occurs widely in Europe.

References: Atlas Supp. (65b), Graham (1950), Stace (1975).

K. WALKER

Mentha aquatica Water Mint

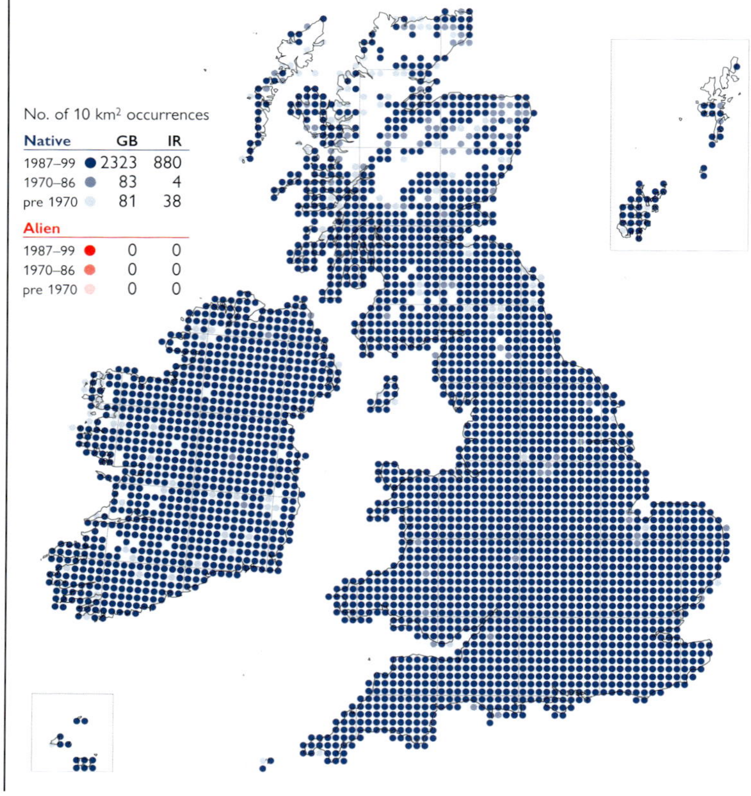

A rhizomatous perennial herb, typically associated with permanently wet habitats adjacent to open water, often partially or wholly submerged. It grows by ditches, ponds and rivers, in marshes, wet pastures, dune-slacks and fens, and in wet woods. It spreads clonally by extensive rhizomes, and by detached rhizome fragments, which are often dispersed by water. 0–455 m (N.W. Yorks.).

Native (change –0.11). The current map for *M. aquatica* is similar to that in the 1962 *Atlas*. There is no evidence for a change in its overall distribution.

European Temperate element; widely naturalised outside its native range.

References: Atlas (241b), Grime *et al.* (1988), Hultén & Fries (1986).

K. WALKER

Mentha aquatica × *M. arvensis* (*M.* × *verticillata*) Whorled Mint

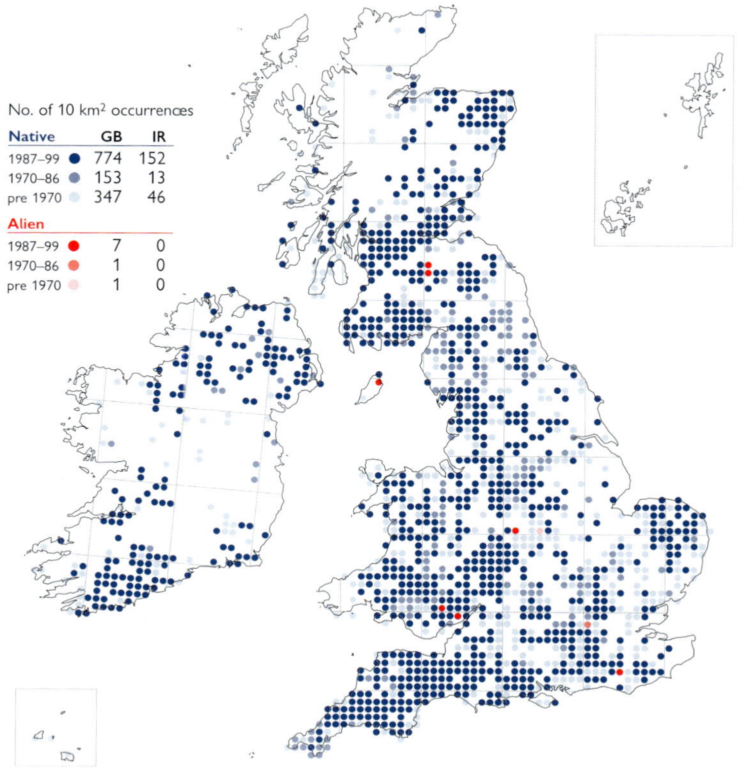

A rhizomatous perennial herb of damp places, including arable fields, wet grassland, track-sides, woodland rides, marshes, river banks, lake-shores, pond-sides and disturbed ground. It often grows in slightly drier habitats than *M. aquatica*. It is usually sterile; highly fertile plants are sometimes found and may be back-crosses. Generally lowland, but reaching 365 m at Beguildy (Rads.).

Native. This naturally occurring hybrid has been lost from some wetland sites (Mountford, 1994) and, presumably, from some arable sites as well.

Frequent in mainland Europe, although rare in the Mediterranean region.

References: Atlas Supp. (66a), Stace (1975).

K. WALKER

Mentha aquatica × *M. arvensis* × *M. spicata* (*M.* × *smithiana*) Tall Mint

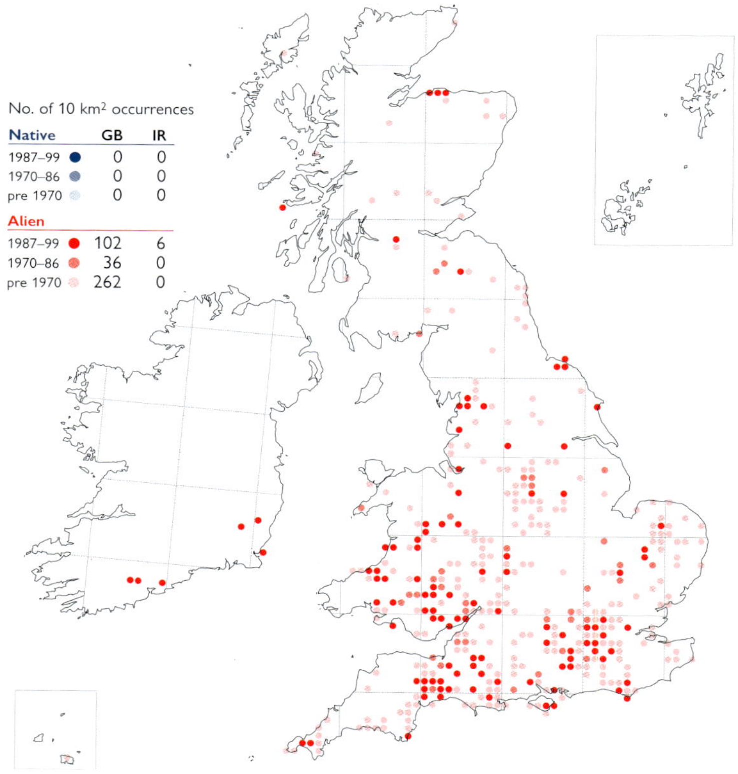

A rhizomatous perennial herb of damp places and on waste ground, occasionally arising spontaneously where *M. spicata* and *M.* × *verticillata* occur together, but more typically a garden throw-out. It is usually sterile. Lowland.

Neophyte. This hybrid was first recorded in the wild in 1724. Many of the records mapped were compiled by Perring & Sell (1968). Declines have been reported but it is unclear whether these are due to genuine losses, to under-recording, or to re-assessment of the characters separating this hybrid from *M.* × *gracilis*. In Surrey, garden mints have become rarer in wild places and more frequent on rubbish tips due to the more organised collection of garden waste (Lousley, 1976a).

This hybrid occurs widely in C. Europe.

References: Atlas Supp. (66b), Mountford (1994), Stace (1975), Wade *et al.* (1994).

K. WALKER

Mentha aquatica × *M. spicata* (*M.* × *piperita*) Peppermint

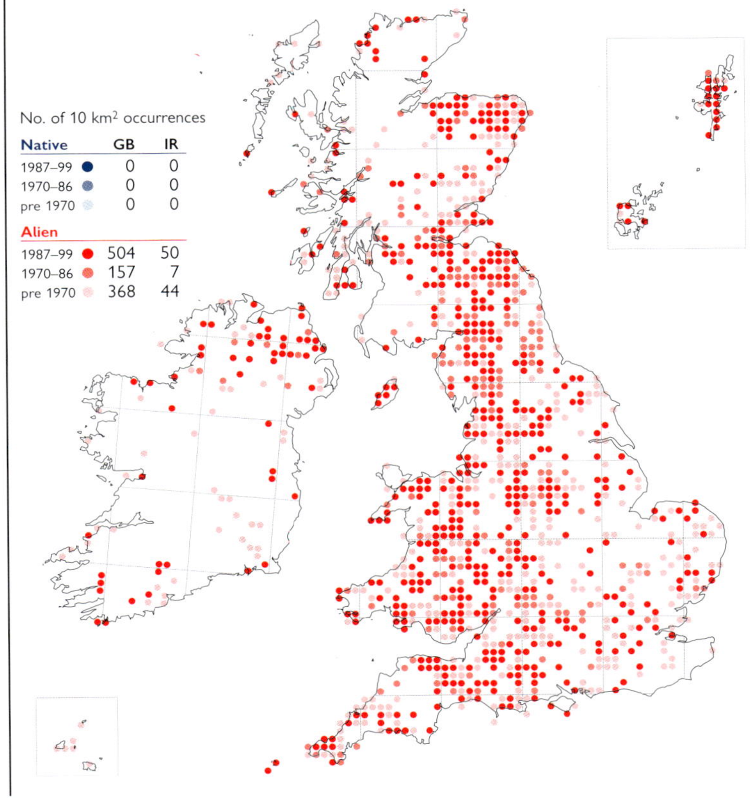

A rhizomatous perennial herb of damp ground and waste places. Glabrous plants are thought to be garden escapes or throw-outs, while pubescent forms are likely to have arisen spontaneously. 0–450 m (Grwyne Fawr, Mons.).

A spontaneous hybrid between native and alien parents. The distribution of this hybrid has not changed appreciably since it was mapped by Perring & Sell (1968), who included pubescent plants formerly mis-determined as *M.* × *dumetorum* (*M. aquatica* × *M. longifolia*). The extent of spontaneous hybridisation is unclear because glabrous and hairy forms are rarely recorded separately, and all records are mapped here as if they are alien.

Widespread in Europe, both as a spontaneous hybrid and as a garden escape.

References: Atlas Supp. (68b, c), Graham (1951), Stace (1975).

K. WALKER

Mentha spicata Spear Mint

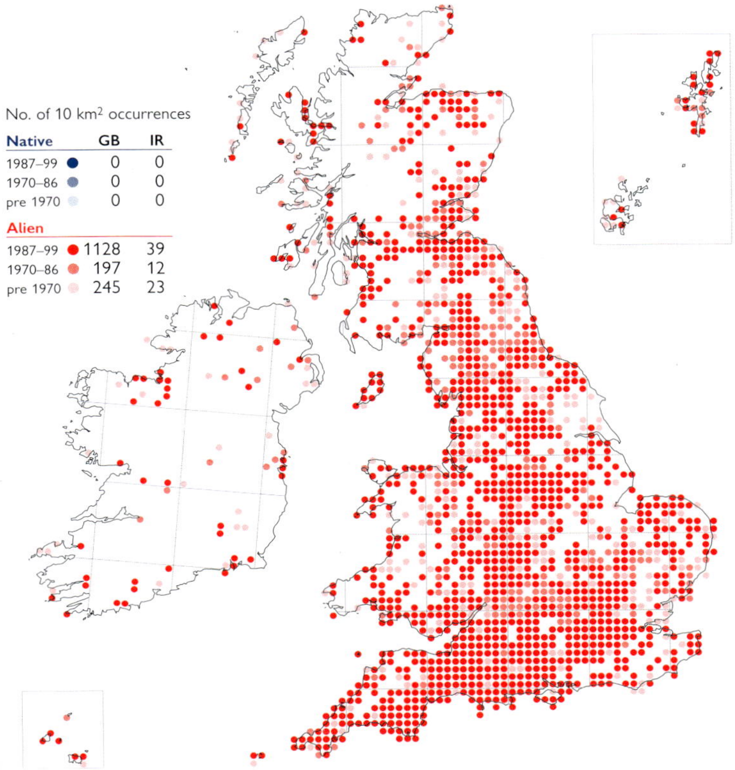

A rhizomatous perennial herb naturalised in a variety of damp or wet habitats, and on rough and waste ground, usually close to habitation. Generally lowland, but reaching 350 m at Alston (Cumberland).

Archaeophyte (change +1.69). *M. spicata* was probably under-recorded in the 1962 *Atlas* because of confusion with *M. longifolia*, *M. suaveolens* and their hybrids. It is very commonly cultivated and spreads vigorously, so it is often discarded with garden rubbish. In some counties it is now the most common naturalised garden mint.

Origin unknown; derived from *M. longifolia* and *M. suaveolens* by hybridisation and chromosome doubling. It probably arose in cultivation.

References: Atlas (241c, d), Atlas Supp. (67d), Graham (1958), Hultén & Fries (1986).

K. WALKER

Mentha spicata × *M. suaveolens* (*M.* × *villosa*) Apple-mint

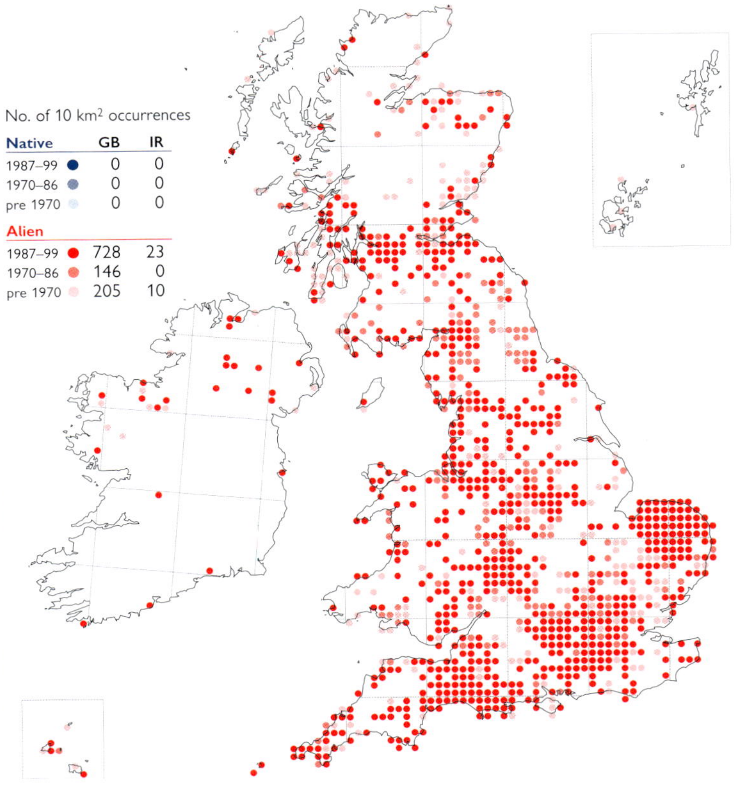

A rhizomatous perennial herb, naturalised in damp, rough and waste ground, and on roadsides. Lowland.

Neophyte. *M.* × *villosa* was known from the wild by at least 1882 and is an increasing garden escape. It was mapped as *M.* × *niliaca* by Perring & Sell (1968), but at that stage it was thought to be under-recorded. It may still be more frequent than the current map suggests. *M. scotica*, formerly considered as a separate species or a variant of *M.* × *villosa*, has now been placed within *M. spicata*. *Mentha* species and hybrids are taxonomically difficult and rather unfashionable with recorders; there is no illustrated handbook so recording effort has been decidedly uneven.

Widespread in Europe.

References: Atlas Supp. (67a, b), Stace (1975).

K. WALKER

Mentha longifolia × *M. spicata* (*M.* × *villosonervata*) Sharp-toothed Mint

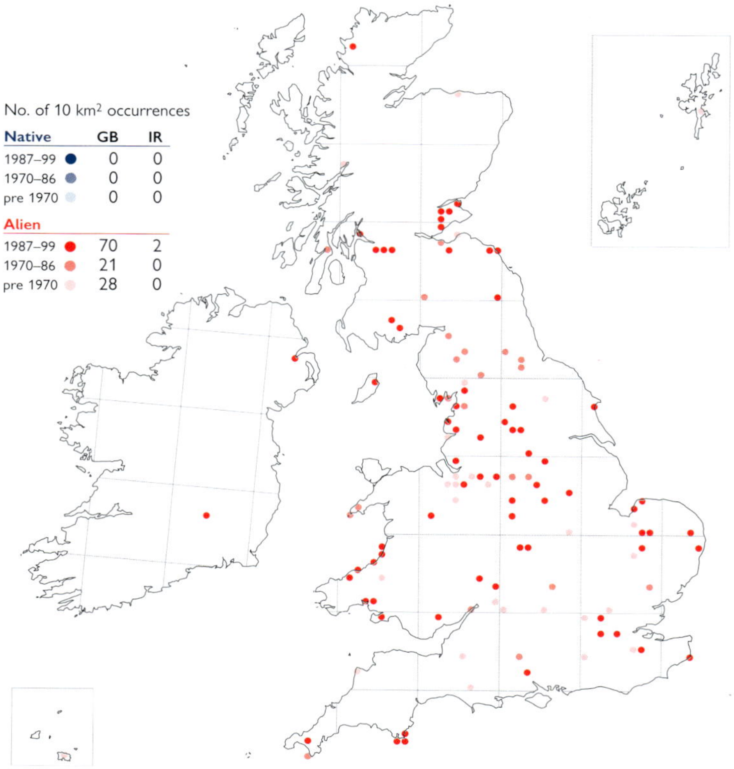

No. of 10 km² occurrences

	GB	IR
Native		
1987–99	0	0
1970–86	0	0
pre 1970	0	0
Alien		
1987–99	70	2
1970–86	21	0
pre 1970	28	0

A perennial herb found established in rough grassland, waste ground and on rubbish tips. It occurs as a garden escape and throw-out, and as a relic of cultivation. Lowland.

Neophyte. This hybrid is commonly cultivated in gardens and was known from the wild by 1934. The apparent increase shown on the map may be due to former taxonomic difficulties, particularly confusion with *M. spicata* and *M.* × *villosa*, taxa from which it is not readily separable. It is still probably under-recorded.

Probably widespread in Europe, but the distribution is uncertain due to confusion with *M.* × *villosa*.

Reference: Stace (1975).

K. WALKER

Mentha longifolia × *M. suaveolens* (*M.* × *rotundifolia*) False Apple-mint

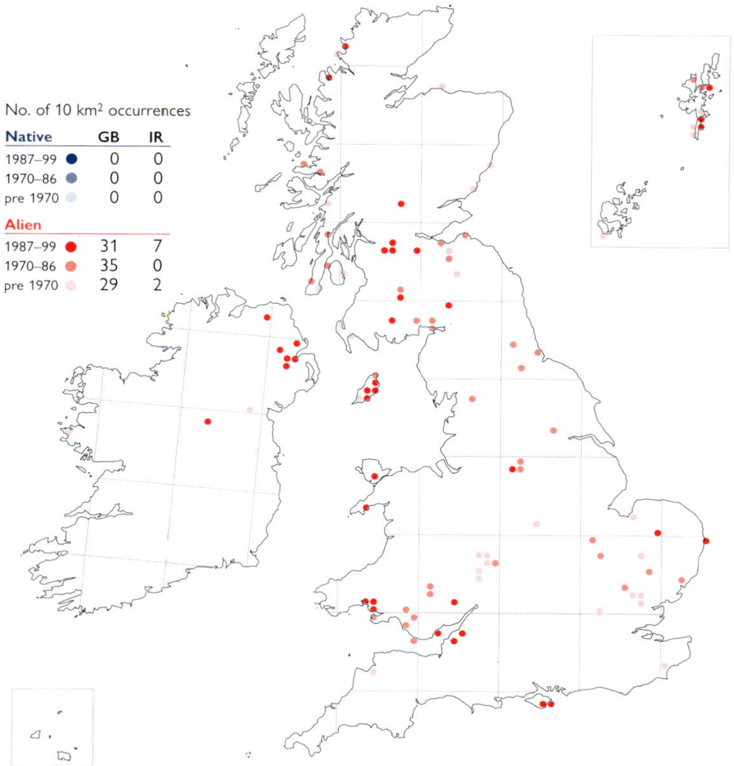

No. of 10 km² occurrences

	GB	IR
Native		
1987–99	0	0
1970–86	0	0
pre 1970	0	0
Alien		
1987–99	31	7
1970–86	35	0
pre 1970	29	2

A perennial herb found as a garden escape or throw-out on rubbish tips, waste ground and roadsides. Lowland.

Neophyte. As with all hybrid *Mentha* taxa, uncertainties in identification mean that this hybrid is under-recorded, so it is difficult to assess changes in distribution. It was known to occur in the wild by 1900. It is often confused with its parents and with *M.* × *villosa*.

Frequent with the parents in Europe and S.W. Asia.

References: Atlas Supp. (67c), Stace (1975).

K. WALKER

Mentha suaveolens Round-leaved Mint

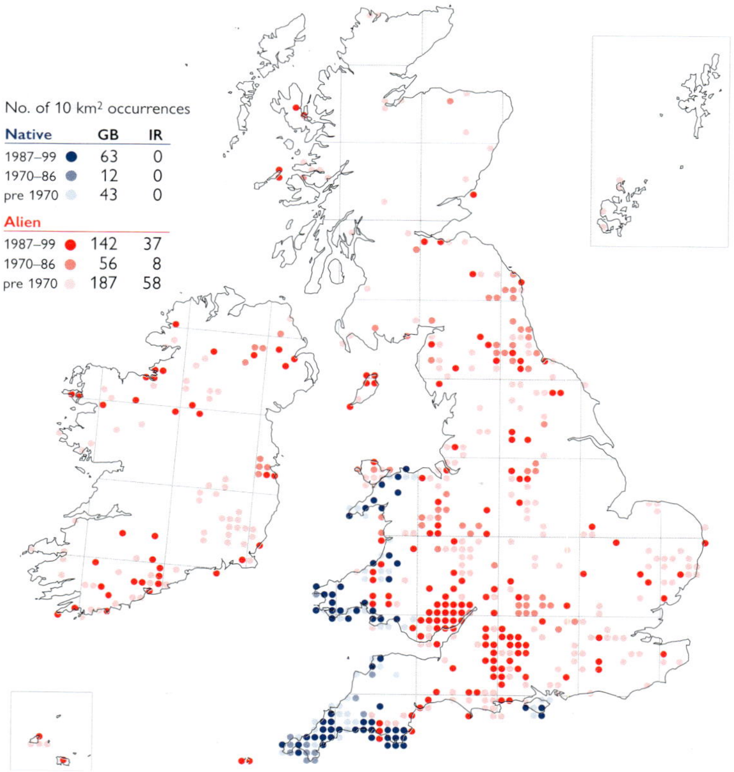

No. of 10 km² occurrences

	GB	IR
Native		
1987–99	63	0
1970–86	12	0
pre 1970	43	0
Alien		
1987–99	142	37
1970–86	56	8
pre 1970	187	58

A rhizomatous perennial herb of damp places. It is probably native only in S.W. England and Wales, and elsewhere occurs as a garden escape, often forming extensive colonies on roadsides and waste ground. Lowland.

Native (change –0.32). The apparent decline of *M. suaveolens* shown in the 1962 *Atlas* may have continued. However, its former abundance may be in part due to over-recording, as many early records of *M. suaveolens* (*M. rotundifolia*) are probably errors for *M.* × *villosa* (cf. Brewis *et al.*, 1996; Halliday, 1997). It is possibly native as mapped, but its native range is obscured by garden escapes.

Submediterranean-Subatlantic element.

References: Atlas (242a), Hultén & Fries (1986).

K. WALKER

Mentha pulegium Pennyroyal

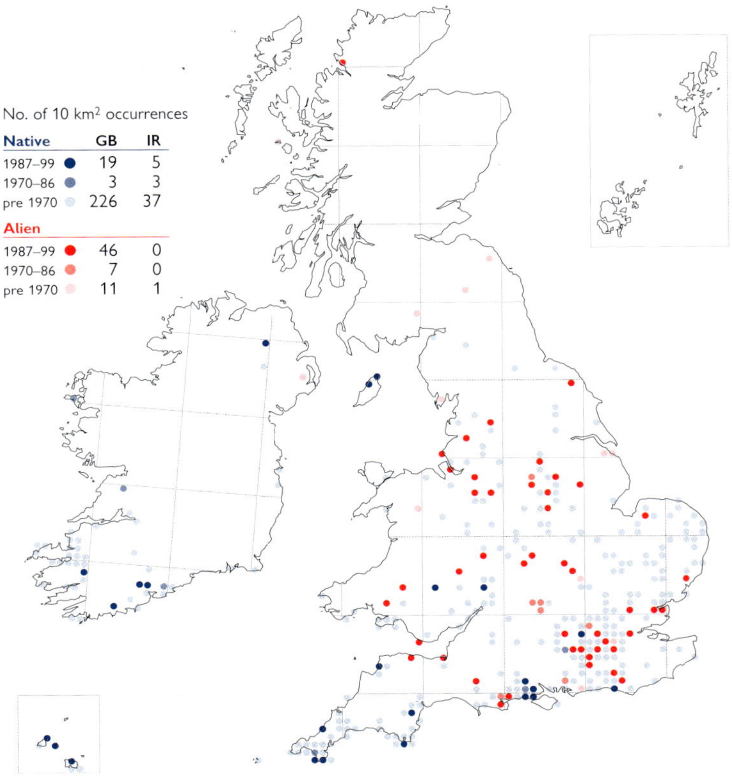

A short-lived perennial herb of seasonally inundated grassland overlying silt and clay. The majority of native populations are now confined to pools, runnels, ruts and poached areas on heavily grazed village greens, but habitats also include damp heathy pastures, lake shores and coastal grassland. Lowland.

Native (change –0.70). *M. pulegium* was lost from many sites before 1930, but has declined further since then due to habitat destruction and the loss of traditional grazing on village greens. A robust variety has been introduced with N. American seed mixtures to a number of sites (Briggs, 1997b), and the species certainly appears to be increasing as an alien.

European Southern-temperate element.

References: Atlas (240d), Bolòs & Vigo (1995), Curtis & McGough (1988), Kay & John (1995), Stewart *et al.* (1994), Wigginton (1999).

K. WALKER

Mentha requienii Corsican Mint

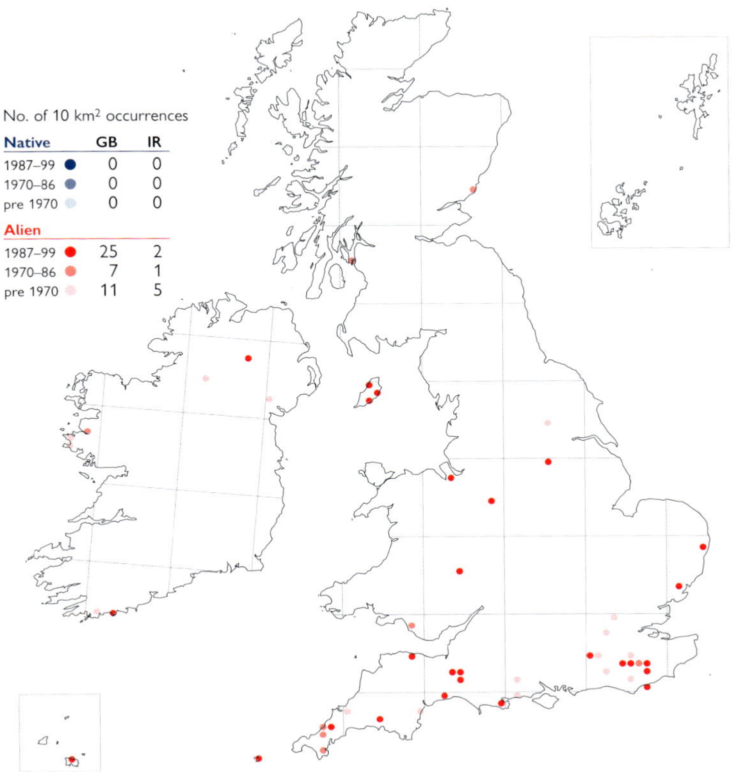

A perennial herb which occurs as a weed in cultivated ground and occasionally as well-established populations in damp grassy and rocky places, in woodlands, along tracks, paths and pavements and on rubbish tips. Generally lowland, but reaching 305 m on Slieve Gullion (Co. Armagh).

Neophyte. This species was introduced in 1829 and is frequent in cultivation. It was first recorded in the wild in 1890 (Cumbria) and was mapped by Perring & Sell (1968). There are more records since then, probably due to improved recording.

Native of Corsica, Sardinia and the nearby Italian island of Montecristo; possibly naturalised in Portugal, but rare in mainland Europe.

Reference: Atlas Supp. (65a).

K. WALKER

Lavandula angustifolia × *L. latifolia* (*L.* × *intermedia*) Garden Lavender

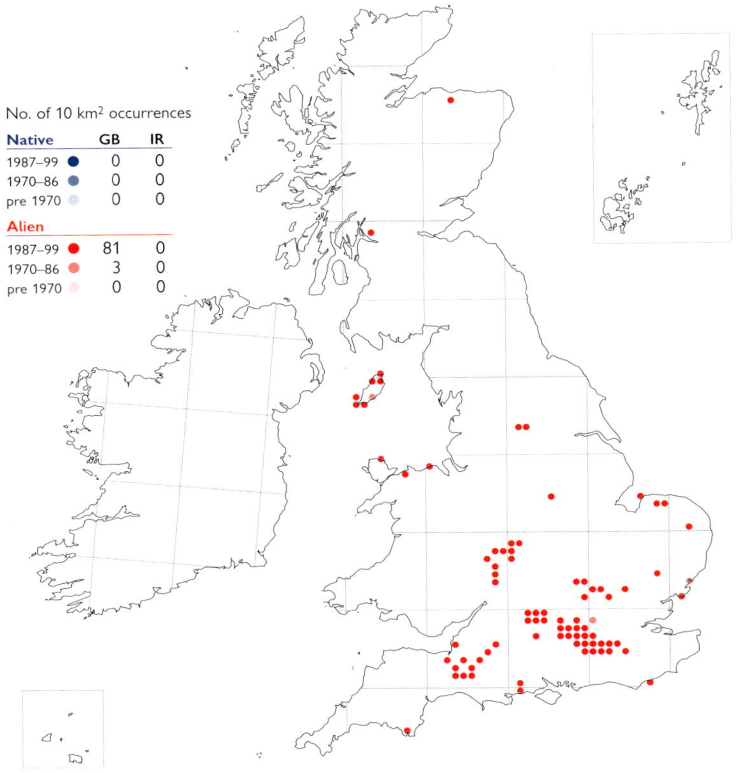

A small evergreen shrub, occurring as a garden escape, throw-out or relic of cultivation, almost always in urban habitats. It is cultivated as a commercial crop in East Anglia. Lowland.

Neophyte. This hybrid was introduced into cultivation in the 17th century and is extremely popular in gardens. It was not recorded from the wild, however, until 1984 and the map indicates that it may be unevenly recorded. The hybrid is not very fertile, and most records of self-sown or bird-sown plants from walls, pavement cracks, banks and chalk pits, many of which are mapped here, are probably referable to *L. angustifolia*.

A spontaneous hybrid in the W. Mediterranean region.

Reference: Bean (1973).

K. WALKER

Rosmarinus officinalis Rosemary

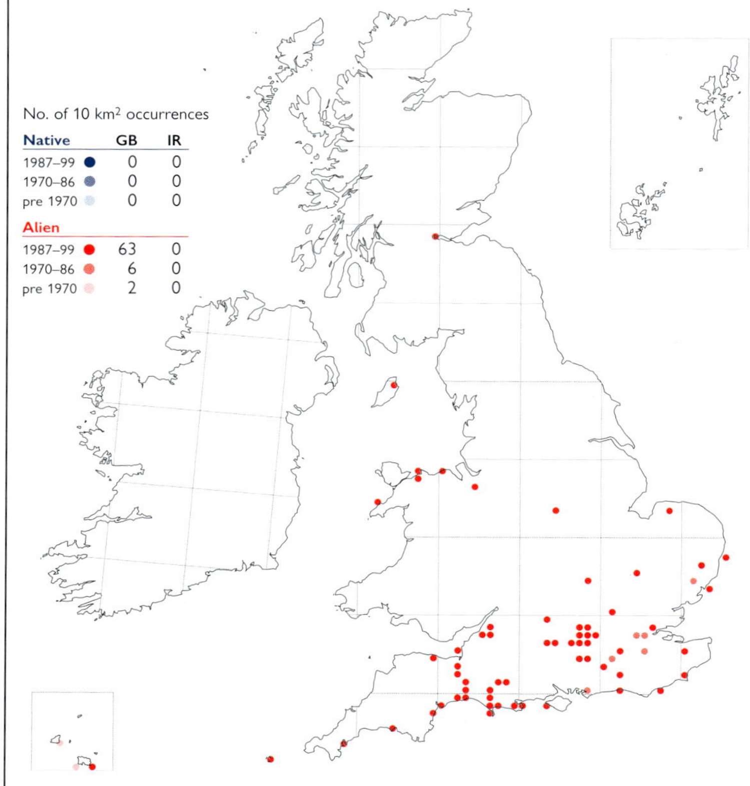

An evergreen shrub, usually found as a throw-out or relic of cultivation, almost always close to habitation. Self- and bird-sown plants have been recorded from walls, pavement cracks and waste ground. Lowland.

Neophyte. *R. officinalis* has been cultivated as a herb in Britain since at least 1375 (Harvey, 1981) and is extremely popular in gardens. It was recorded from the wild by 1969. A number of cultivars are available, and have been recorded from the wild, including the white-flowered 'Alba'.

Native of the S.W. Europe, the Mediterranean region and the Caucasus.

References: Bean (1980), Bolòs & Vigo (1995).

K. WALKER

Salvia pratensis Meadow Clary

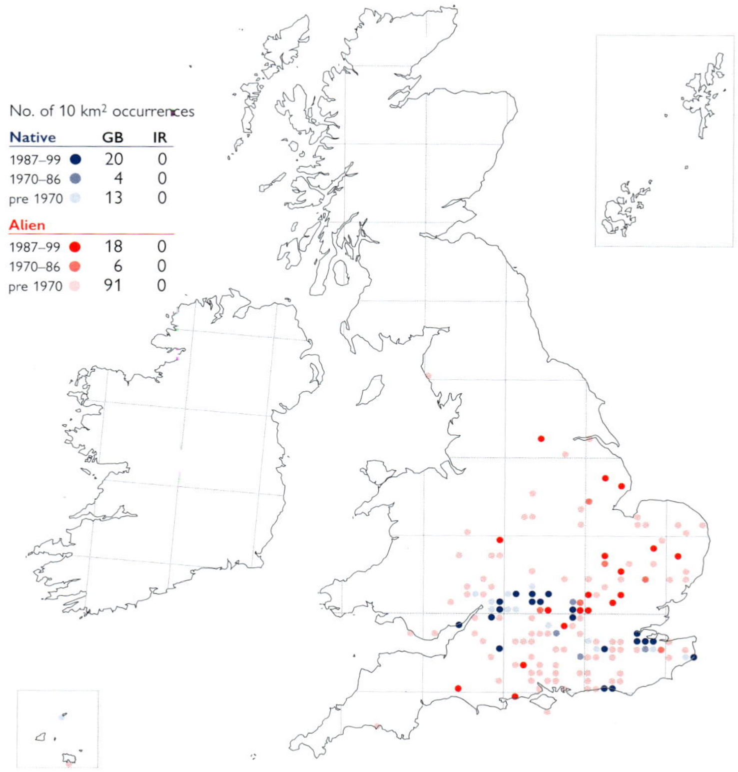

A long-lived perennial herb of unimproved grassland, lane-sides, road verges and disturbed ground on well-drained soils overlying chalk and limestone. It is occasionally established from gardens or as a casual in waste places. Lowland.

Native or alien (change –0.75). *S. pratensis* was not recorded from the wild until 1699, but was known to Elizabethan gardeners, and its native status has often been questioned. Most of the losses of native sites seem to have taken place before 1950, and there is little evidence for a significant decline in recent years. Introductions also appear to be decreasing.

European Temperate element.

References: Atlas (245b), Hultén & Fries (1986), Kay & John (1995), Meusel *et al.* (1978), Rich *et al.* (1999b), Wigginton (1999).

K. WALKER

Salvia verbenaca Wild Clary

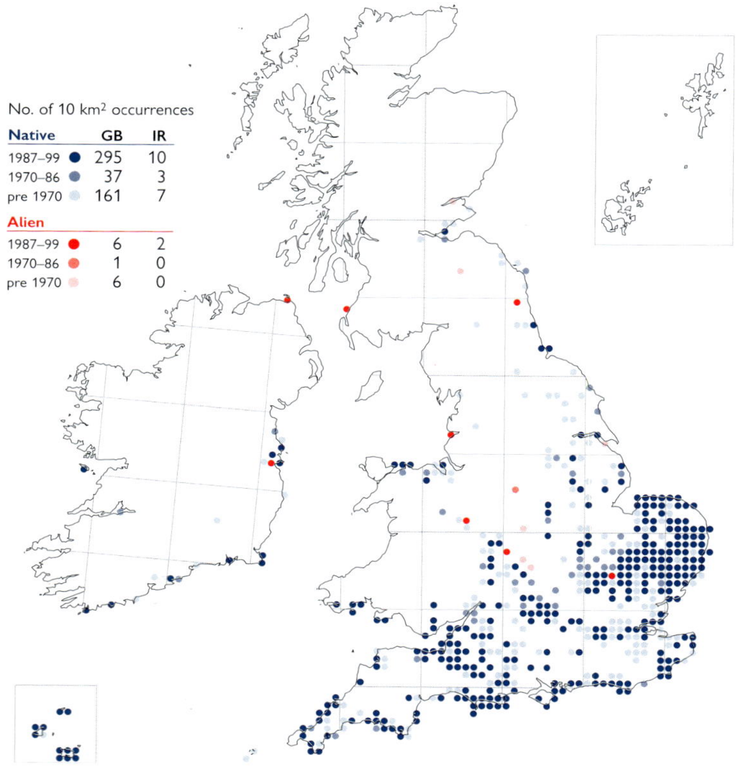

An aromatic perennial herb of open grassland on sunny banks, sand dunes and roadsides; usually on well-drained, base-rich soils, including sticky calcareous clays that are wet in winter and baked dry in summer. In S.E. England, it is often associated with churchyards because of the medieval practice of sowing it on graves (Sturt, 1995). In Ireland, it is almost exclusively coastal. Lowland.

Native (change –0.51). The decline of *S. verbenaca* was already apparent in the 1962 *Atlas* and it has continued, particularly inland and in the north of its range. Most losses are probably due to changes in land use.

Mediterranean-Atlantic element.

References: Atlas (245c, d), Bolòs & Vigo (1995), Curtis & McGough (1988), Mabey (1996).

K. WALKER

Salvia viridis Annual Clary

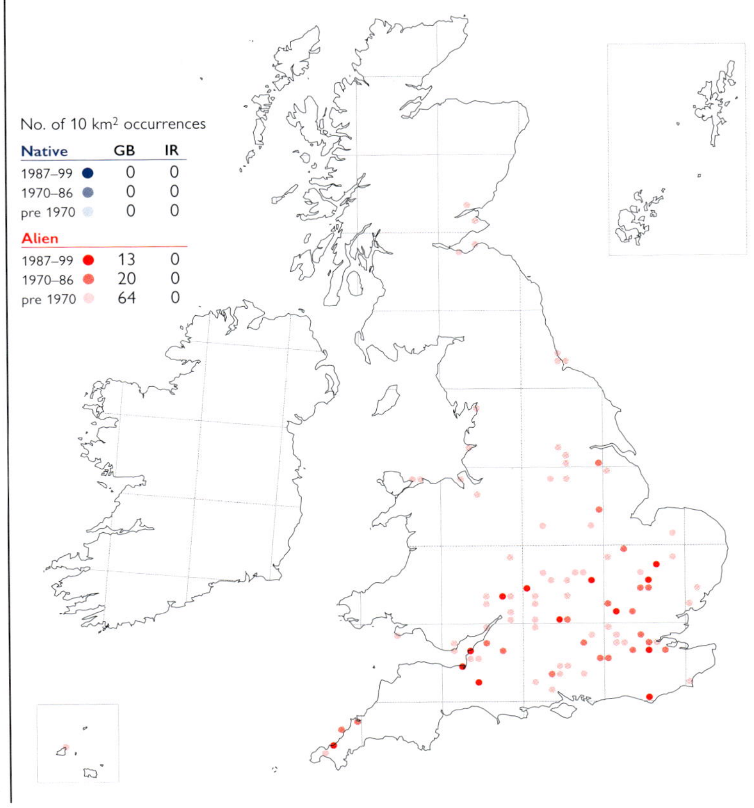

An annual which occurs as a garden escape or bird-seed alien on roadsides, rubbish tips and waste ground. It is usually casual, but some populations persist for a few years. Lowland.

Neophyte. This species was introduced to cultivation in Britain by 1596 and was recorded from the wild by 1859. Given its popularity in gardens, the apparent decline in records since 1970 is difficult to explain; it may be an artefact caused by the accumulation of casual records over many years up to 1970.

Native of the Mediterranean region and S.W. Asia.

Reference: Bolòs & Vigo (1995).

K. WALKER

Salvia verticillata Whorled Clary

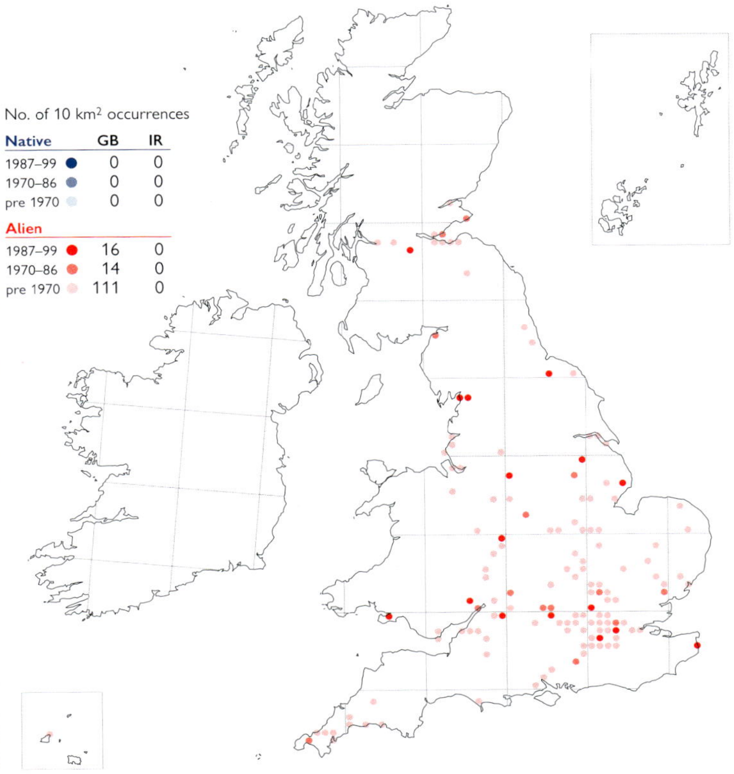

A foetid perennial herb, usually casual, but sometimes naturalised, on road verges, waste ground and by railways. It arises as a garden escape or a grain contaminant. Lowland.

Neophyte (change –2.18). *S. verticillata* was introduced into cultivation in Britain by 1594, and was recorded from the wild by 1857. The apparent decline shown in the 1962 *Atlas* has continued. Most populations are short-lived and there are few recent records, although it has persisted at Phillack Towans (W. Cornwall) since 1918 together with several other garden plants.

A European Southern-temperate species, absent as a native from N.W. Europe.

References: Atlas (245a), Hultén & Fries (1986).

K. WALKER

Hippuris vulgaris Mare's-tail

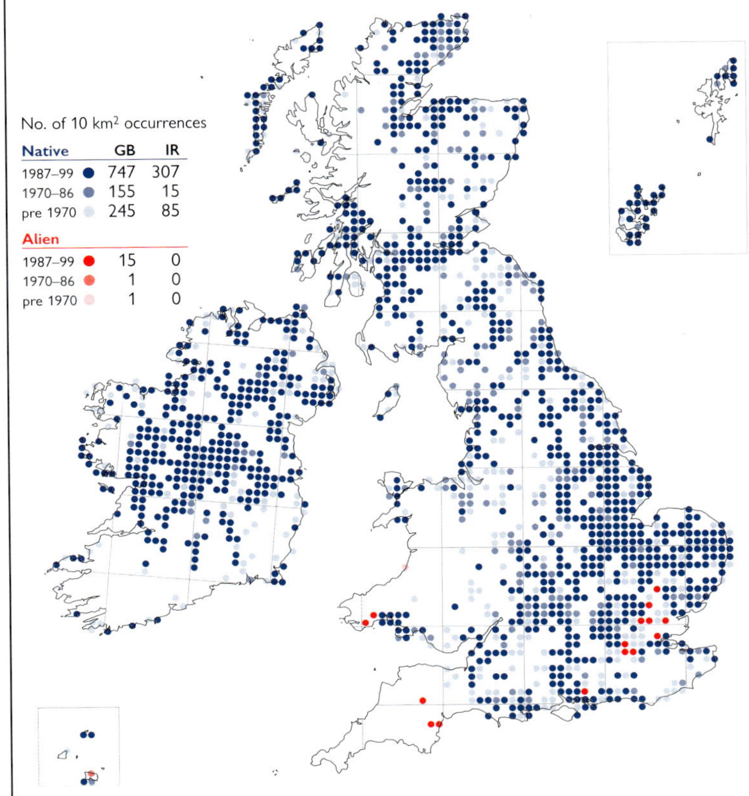

This herbaceous perennial occurs in two growth forms. Plants with long, flaccid stems grow as submerged aquatics, and are sometimes abundant in clear calcareous water. More rigid, stiffly erect shoots grow as emergents at the edge of lakes and ponds, in swamps or in upland flushes. These may be very robust when growing on deep, eutrophic mud. 0–900 m (Moine Mhor, Easterness), but rare above 400 m.

Native (change –0.05). There is little evidence for any marked change in the distribution of this species, although it has spread as an introduction in some counties.

Circumpolar Boreo-temperate element.

References: Atlas (150c), Hultén & Fries (1986), Meusel *et al.* (1978), Preston & Croft (1997).

C. D. PRESTON

Callitriche hermaphroditica
Autumnal Water-starwort

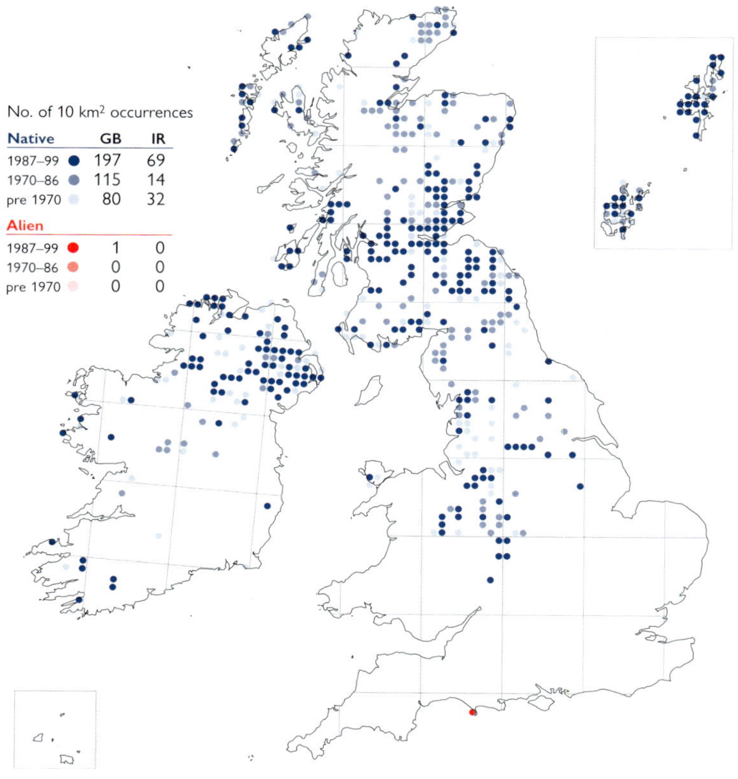

This species occurs in mesotrophic lakes, canals and gravel-pits. It is usually annual, although some populations may perennate. 0–390 m at Drumore Loch (E. Perth).

Native (change −0.21). *C. hermaphroditica* was under-recorded in the 1962 *Atlas*. There is no evidence of a general decline, though it has been lost from some sites, possibly because of deteriorating water quality. However, recent records from gravel-pits in Lincolnshire and canals in the Midlands suggest that it may be spreading in those areas. It is morphologically variable; populations in our area include at least three forms.

Circumpolar Boreal-montane element.

References: Atlas (151c), Hultén & Fries (1986), Martinsson (1991b), Meusel *et al.* (1978), Preston & Croft (1997), Schotsman (1967), Stewart *et al.* (1994).

R. V. LANSDOWN

Callitriche truncata Short-leaved Water-starwort

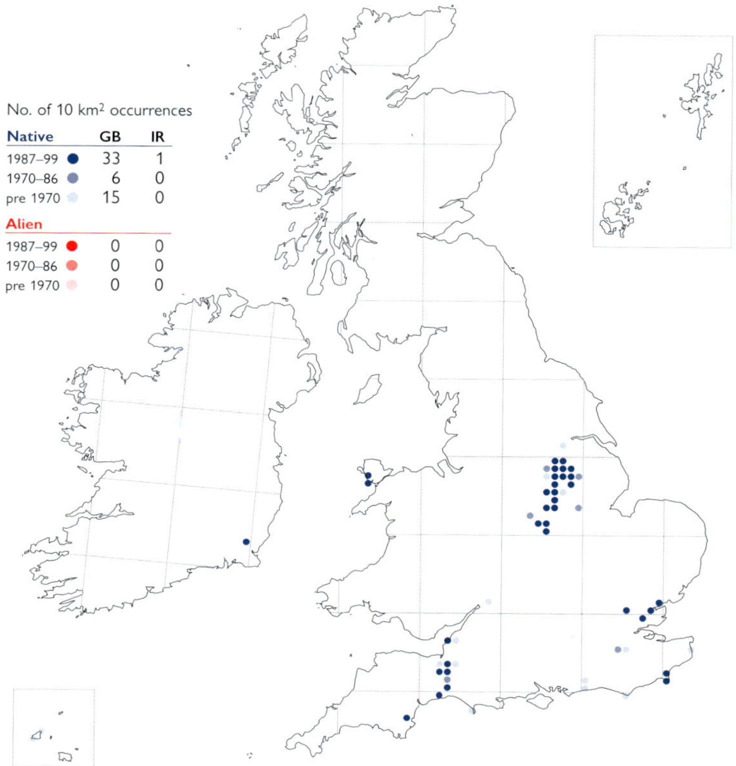

An annual or occasionally perennial herb, growing in rivers, canals, ditches, lakes and gravel-pits, typically in base-rich mesotrophic or eutrophic waters, and rarely as a terrestrial plant on wet mud. Lowland.

Native (change +0.47). *C. truncata* appears to be increasing and spreading northwards in Britain. It has been discovered since the 1962 *Atlas* in Essex, Lincolnshire and Anglesey, and its range now overlaps with that of *C. hermaphroditica.*

Mediterranean-Atlantic element. *C. truncata* is also spreading northwards in mainland Europe.

References: Atlas (151d), Barry & Wade (1986), Curtis & McGough (1988), Hultén & Fries (1986), Lansdown (1999), Meusel *et al.* (1978), Preston & Croft (1997), Schotsman (1967), Stewart *et al.* (1994).

R. V. LANSDOWN

Callitriche stagnalis sens. lat.
Common Water-starwort

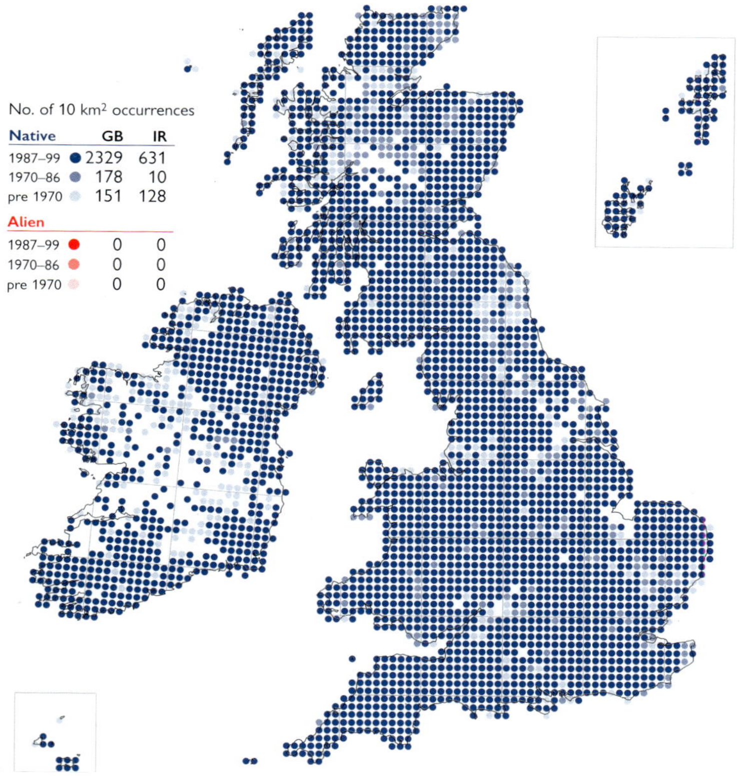

Annual or perennial herbs found on rutted tracks, in ephemeral pools and at the margins of ditches and rivers. They also grow in shallow permanent water. 0–610 m on Great Dun Fell (Westmorland), and possibly to 890 m on Foel Grach (Caerns.).

Native (change +1.51). As in the 1962 *Atlas*, records of *C. platycarpa* are included here with *C. stagnalis*. Due to difficulties of identification, many cannot be assigned to either species. *C. stagnalis sens. str.* is believed to be the most widespread *Callitriche* species in our area, but the name is often applied to any *Callitriche* species with floating leaves and it is not possible to produce a reliable map of this segregate.

C. stagnalis sens. lat. is a member of the European Temperate element and is widely naturalised elsewhere.

References: Atlas (150d), Grime *et al.* (1988), Hultén & Fries (1986), Meusel *et al.* (1978), Preston & Croft (1997), Schotsman (1967).

R. V. LANSDOWN

Callitriche platycarpa
Various-leaved Water-starwort

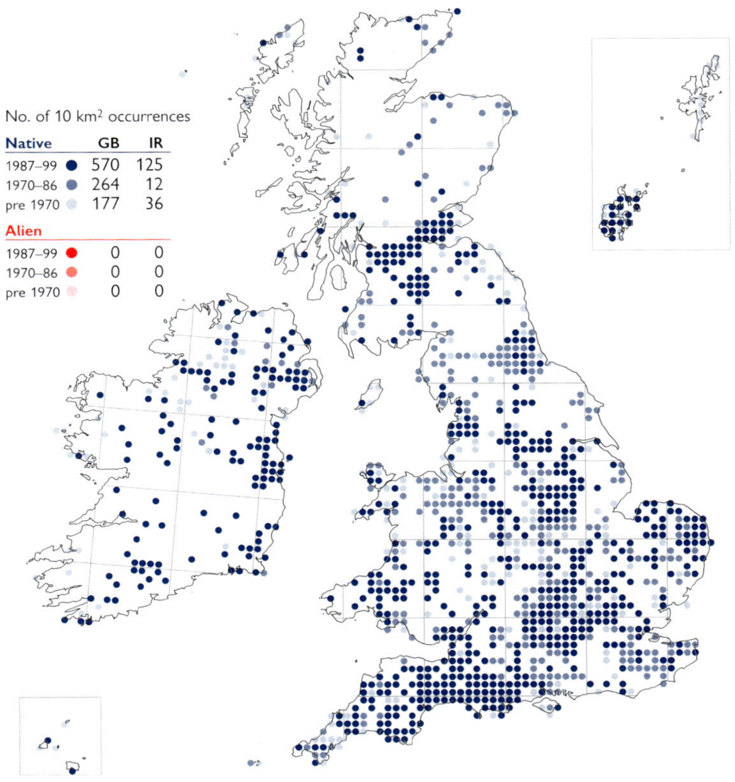

A perennial herb, occurring in most types of water body and as a terrestrial form on wet mud. However, it appears to be most frequent in eutrophic waters, particularly ditches and canals. Generally lowland, but reaching 520 m at Llynnoedd Ieuan (Cards.).

Native. Records for *C. platycarpa* were mapped with *C. stagnalis* in the 1962 *Atlas*, so no direct comparison is possible. It is still under-recorded, and it is possible that it is the most abundant species of *Callitriche* in lowland Britain. It may be spreading in Britain because of eutrophication, as appears to have happened in Sweden.

European Temperate element.

References: Bolòs & Vigo (1995), Martinsson (1991a), Preston & Croft (1997), Schotsman (1967).

R. V. LANSDOWN

Callitriche obtusangula Blunt-fruited Water-starwort

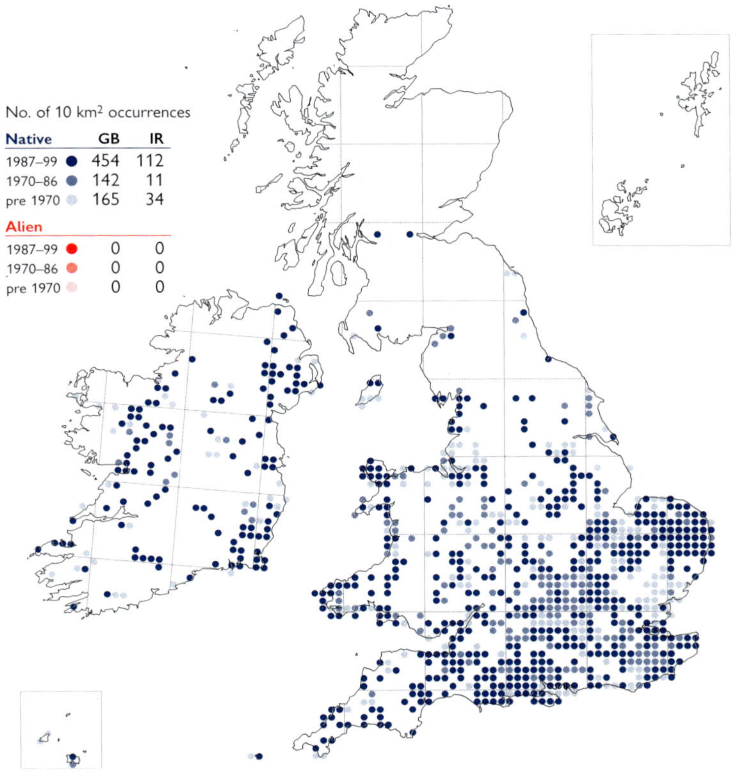

A perennial herb typical of permanent, still or slow-flowing mesotrophic to eutrophic waters, extending into brackish water in coastal grazing marshes. It also grows in a terrestrial form on wet mud as water levels drop. Lowland.

Native (change +1.35). *C. obtusangula* was under-recorded for the 1962 *Atlas*. There is no evidence of any genuine change in its distribution.

Suboceanic Southern-temperate element.

References: Atlas (151a), Meusel *et al.* (1978), Preston & Croft (1997), Schotsman (1967).

R. V. LANSDOWN

Callitriche palustris

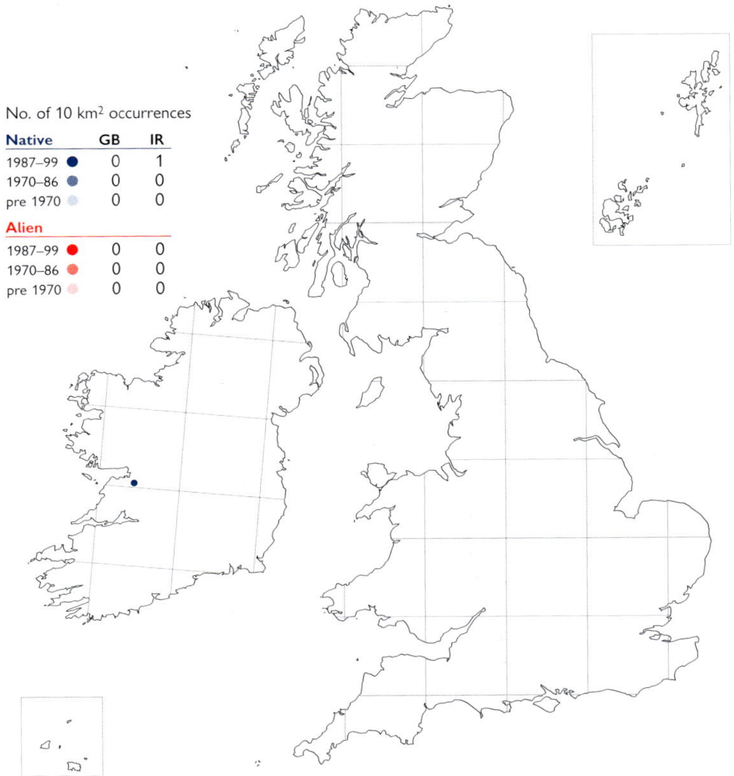

An annual or possibly perennial herb, which was discovered growing on clay in the dry bed of a turlough in Co. Galway in 1999. In Europe it is recorded from ephemeral water bodies, particularly lakes with marked water level fluctuations, temporary pools and rutted tracks, and is most frequent as a terrestrial form on wet mud. Lowland.

Native. The occurrence of *C. palustris* in our area has only recently been confirmed. All previous records have been refuted or remain unconfirmed. It was discovered in Dunbartonshire in 2000, after the close of recording for this *Atlas*.

Circumpolar Boreal-temperate element.

References: Hultén & Fries (1986), Lansdown & Bruinsma (2000), Meusel *et al.* (1978), Schotsman (1967).

R. V. LANSDOWN

Callitriche hamulata sens. lat.
Intermediate Water-starwort

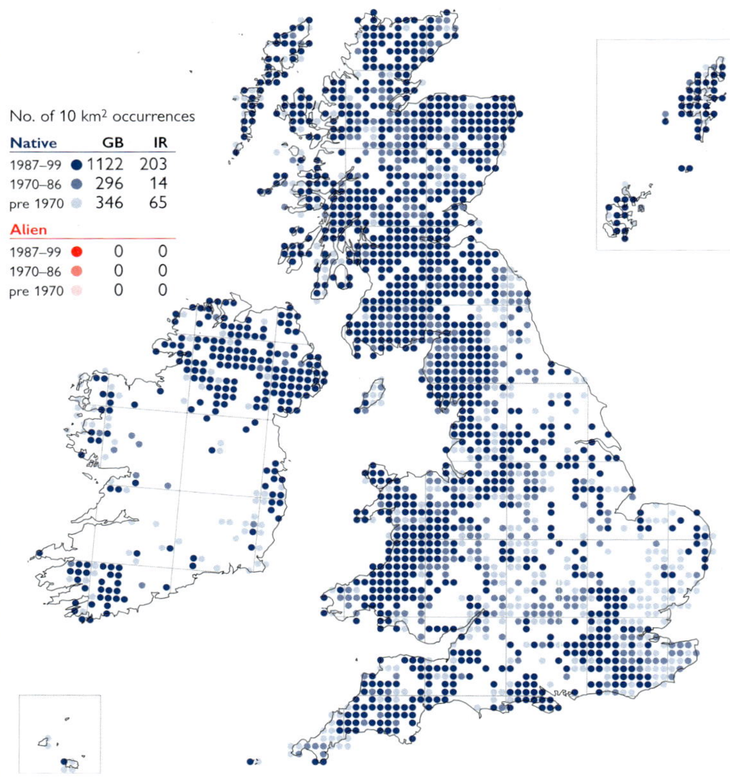

These perennial herbs grow in both deep, still water and fast-flowing rivers, particularly in acidic, oligotrophic water. They may be annual in temporary water bodies, such as pools, ditches and reservoir edges. 0–915 m (Sgurr na Lapaich, Easterness).

Native (change +1.12). As in the 1962 *Atlas*, this map combines records for *C. hamulata* and *C. brutia*, but the broadly defined species is much better recorded now. So many records are referable to *C. hamulata sens. lat.* that a reliable map of *C. hamulata sens. str.* cannot be produced. There is also doubt over the taxonomic separation of these two species.

C. hamulata sens. lat. has a European Wide-temperate distribution; it also occurs in Greenland and as an alien in Australasia.

References: Atlas (151b), Hultén & Fries (1986), Meusel *et al.* (1978), Preston & Croft (1997), Schotsman (1967).

R. V. LANSDOWN

Callitriche brutia Pedunculate Water-starwort

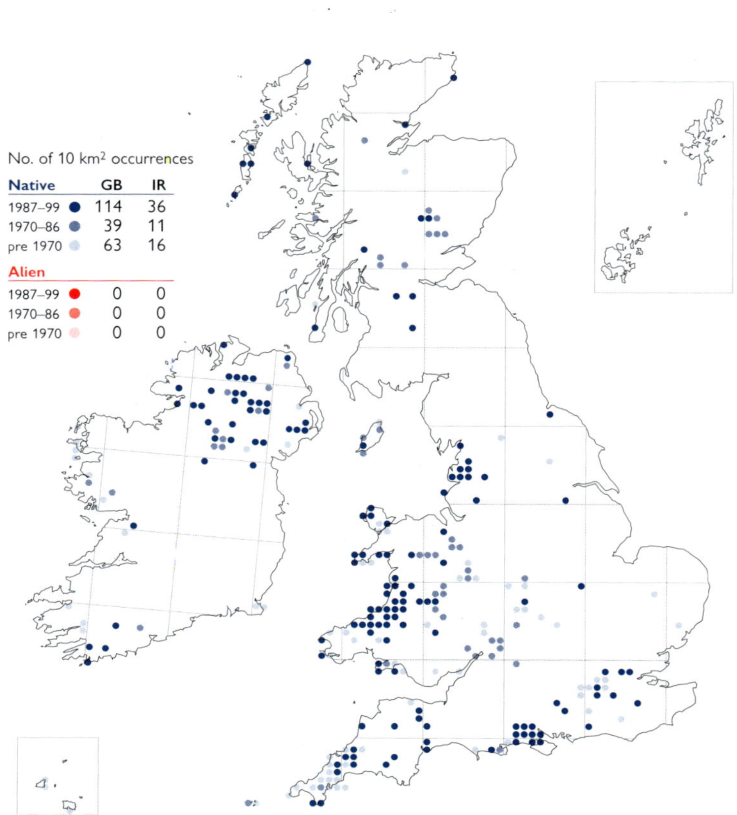

An annual or perennial herb, growing in ephemeral pools, ruts and poached muddy ground, although it may also occur in permanent water. Generally lowland, but discovered in 2000 at 950 m on Ben Lawers (Mid Perth).

Native. This appears to be the most abundant *Callitriche* in some areas, such as the New Forest (S. Hants.) and the Lizard (W. Cornwall), and is certainly more widespread than previously thought. However, it was not differentiated from *C. hamulata* in the 1962 *Atlas*, and no direct comparison is possible. Only forms with pedunculate fruit can be separated from *C. hamulata* without chromosome counts.

European Southern-temperate element.

References: Hultén & Fries (1986), Meusel *et al.* (1978), Preston & Croft (1997).

R. V. LANSDOWN

Plantago coronopus Buck's-horn Plantain

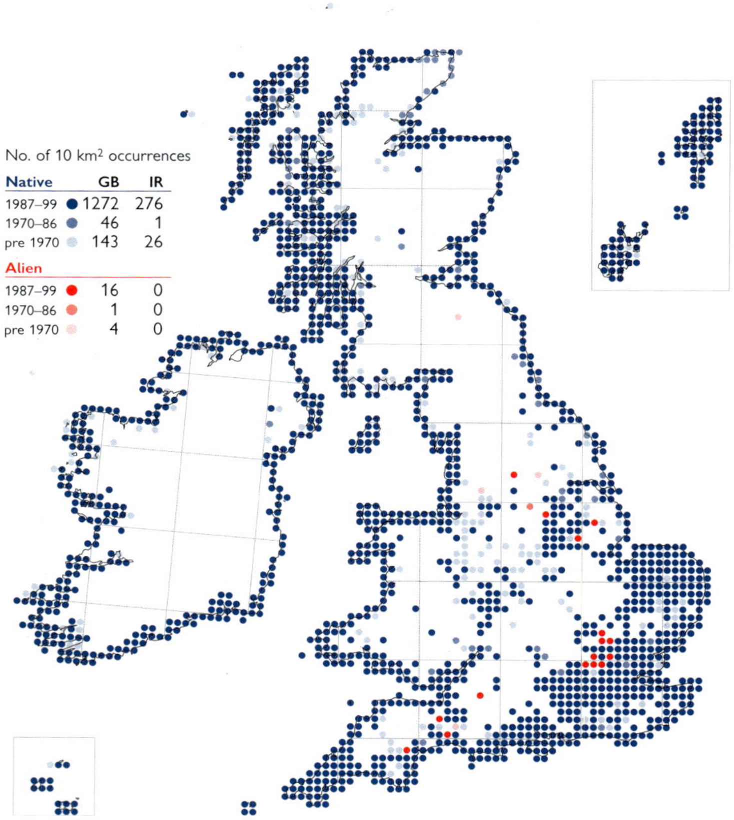

A perennial herb of dry, open, often heavily trampled, habitats on acidic to basic stony or sandy soils, and rock crevices. It occurs in open grassland, on heaths, sand dunes and shingle, sea-cliffs and sea-walls, waste ground and by paths. Always known inland in S. and E. England, plants increasingly occur beside salt-treated roads. 0–340 m (Chagford, S. Devon).

Native (change +0.16). Unlike many halophytes, there has been no major change in the distribution of this species since the 1962 *Atlas*, although many sites in the Midlands have been lost since 1950 as heathland has been improved.

Eurosiberian Southern-temperate element; widely naturalised outside its native range.

References: Atlas (255a), Dodds (1953), Hultén & Fries (1986), Meusel *et al.* (1978).

G. M. KAY

Plantago maritima Sea Plantain

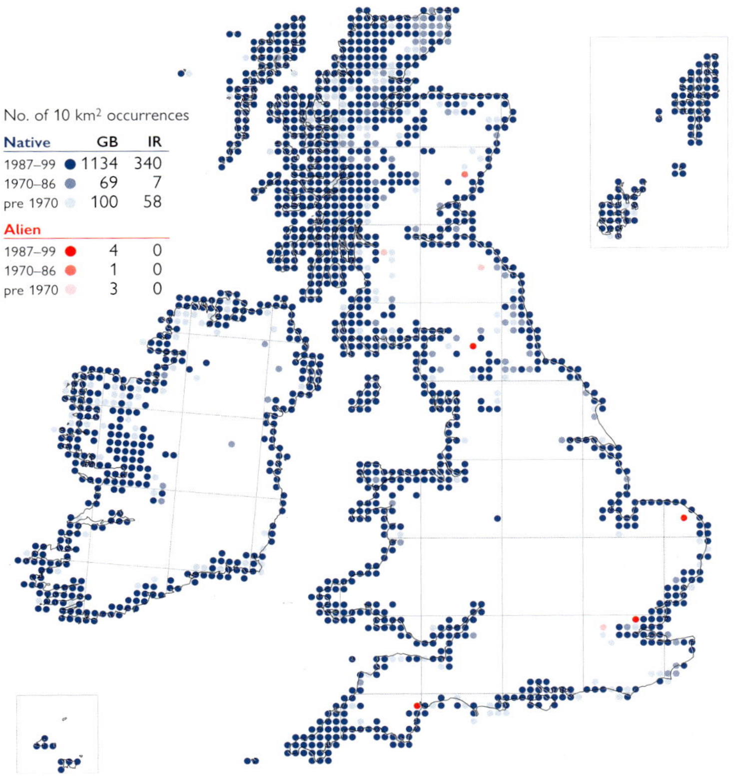

A perennial herb of the middle and upper zones of saltmarshes, coastal turf, rocks and cliffs, on coastal heaths and occasionally on shingle beaches and inland saltmarshes. In the uplands it is found in species-rich pastures, on stream banks, rock ledges and scree, and in stony flushes. It occasionally colonises inland road verges. 0–790 m (Caerns., Mid Perth and Co. Mayo).

Native (change −0.28). The distribution of *P. maritima* in semi-natural habitats shows little change since the 1962 *Atlas*. It has spread on the verges of salt-treated roads, especially in Scotland, but this is barely evident at the 10-km scale.

Eurosiberian Wide-boreal element; also in N. America.

References: Atlas (254d), Hultén & Fries (1986), Meusel *et al.* (1978).

G. M. KAY

Plantago major Greater Plantain

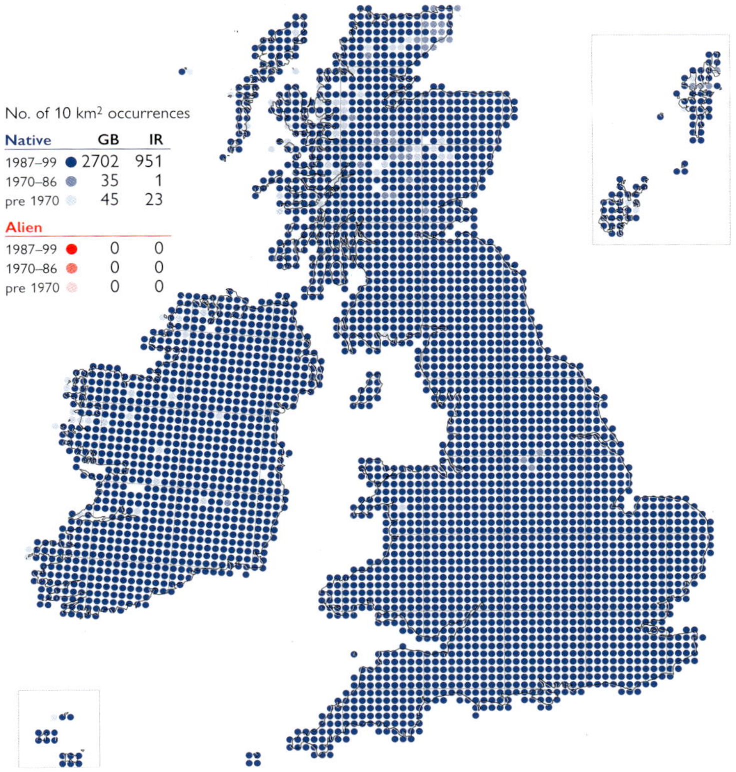

A perennial herb of open habitats; it is most frequent on trampled paths and tracks, disturbed field edges and roadsides, and in gardens, but it also occurs in some closed grasslands. It grows in a wide range of soils, avoiding only very acidic sites, and can produce a large and persistent seed bank. 0–625 m (Knock Fell), with an exceptional record at 845 m on Great Dun Fell (both Westmorland).

Native (change +0.09). The distribution of this ubiquitous species is stable. Subsp. *major* occurs throughout the range of the species, subsp. *intermedia* is mapped separately.

Eurasian Wide-temperate element, but naturalised in N. America so distribution is now Circumpolar Wide-temperate.

References: Atlas (254a), Grime *et al.* (1988), Hawthorn (1974), Hultén & Fries (1986), Meusel *et al.* (1978), Sagar & Harper (1964).

G. M. KAY

Plantago major subsp. *intermedia*

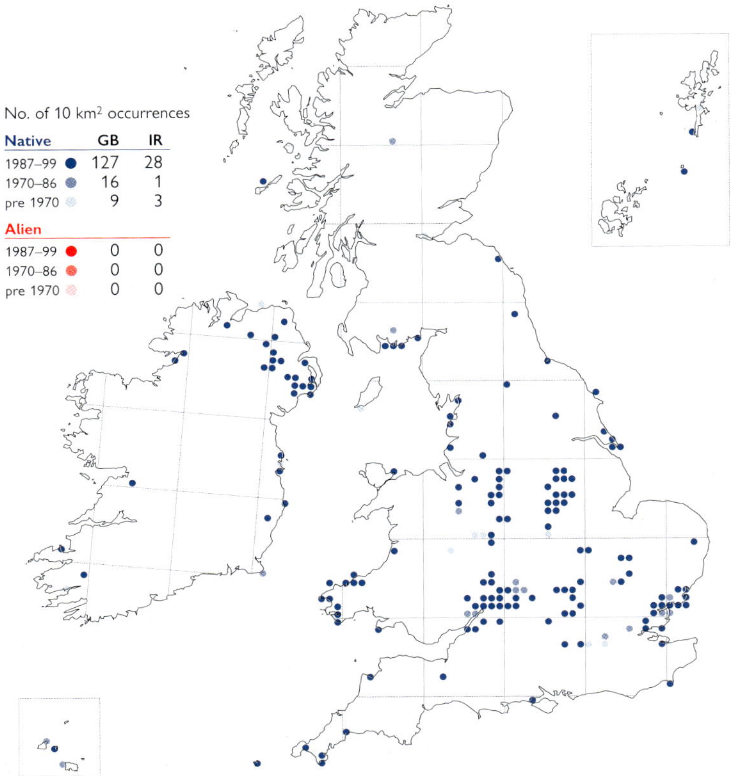

An annual or perennial herb of damp, open habitats, including somewhat saline soils by coastal creeks and the upper end of saltmarshes, the edge of rivers and streams, damp mud exposed in summer by ponds and reservoirs, and winter-flooded hollows in arable fields. It has also been recorded beside salt-treated roads. Lowland.

Native. This subspecies was first recognised in Britain by Lousley (1958), but not described in standard floras until Clapham *et al.* (1987). Wolff & Morgan-Richards (1999) have suggested that it is specifically distinct from subsp. *major*, but the morphological separation in our area is not clear-cut. It is under-recorded.

Subsp. *intermedia* is widespread in Europe; its wider distribution is uncertain.

References: Akeroyd & Doogue (1988), Preston & Whitehouse (1986).

G. M. KAY

Plantago media Hoary Plantain

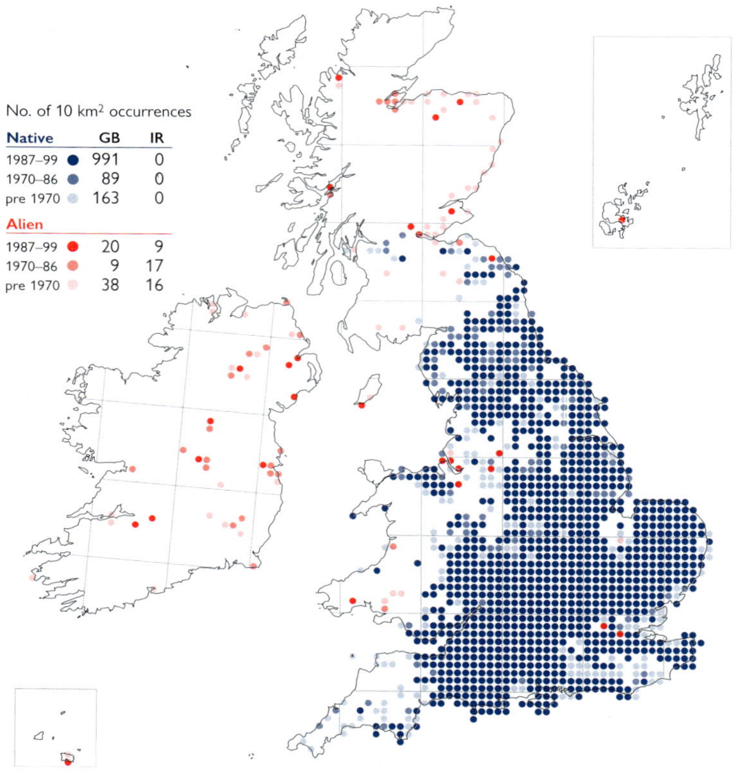

A perennial herb, characteristic of chalk and limestone soils but also occurring on heavy clay soils. The main habitats are downland grassland and tracks, calcareous pasture and mown grassland (such as churchyards); it is less frequent in hay meadows and on fixed dunes, and is sometimes found in water-meadows which receive calcareous water (Grose, 1957). Seed appears to be short-lived. 0–520 m (S. Northumb.).

Native (change –0.79). *P. media* shows declines since the 1962 *Atlas* around the limits of its core area, mainly due to improvements to permanent pastures.

Eurasian Temperate element; widely naturalised outside its native range.

References: Atlas (254b), Hultén & Fries (1986), Meusel *et al.* (1978), Sagar & Harper (1964).

G. M. KAY

Plantago lanceolata Ribwort Plantain

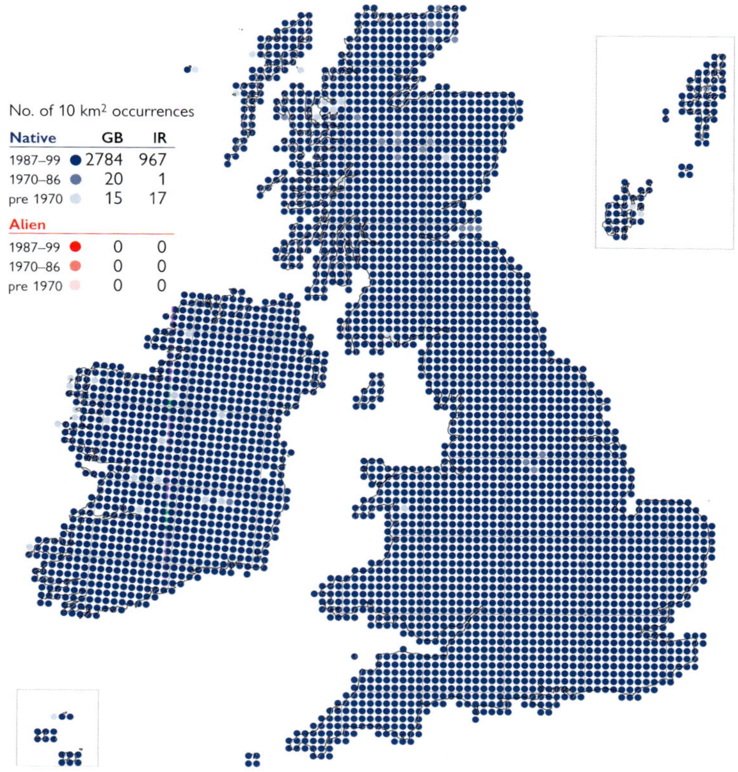

A perennial herb, found in a wide range of habitats over all but the most acidic soils. It occurs in meadows and pastures, in upland grasslands, on rock ledges and crevices, sand dunes and cliffs (including sites subject to sea-spray), on roadsides and river banks, in cultivated and waste ground, in lawns and on walls. Seed is moderately long-lived. 0–790 m in Atholl (E. Perth), and 845 m on Great Dun Fell (Westmorland).

Native (change +1.35). There has been no change in the distribution of this species since the 1962 *Atlas*. It is genetically and phenotypically variable.

Eurosiberian Southern-temperate element, but widely naturalised so that distribution is now Circumpolar Southern-temperate.

References: Atlas (254c), Cavers *et al.* (1980), Grime *et al.* (1988), Hultén & Fries (1986), Meusel *et al.* (1978), Sagar & Harper (1964).

G. M. KAY

Plantago arenaria Branched Plantain

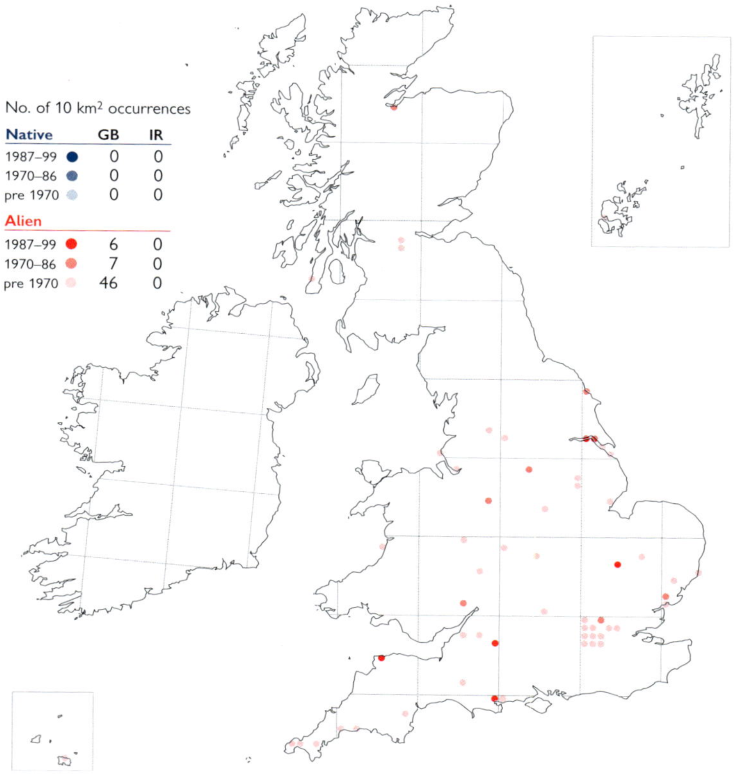

An annual of open sandy places, including waste ground, and in docklands, usually casual but sometimes naturalised. Lowland.

Neophyte. Although *P. arenaria* was cultivated in Britain in 1804, it arrived as a grain impurity and was first recorded in the wild around 1860, and is now of only rare occurrence. There has been much nomenclatural confusion between this species and the allied *P. afra*; some mapped records may be errors for the latter, which Stace (1997) suggests is now commoner than *P. arenaria* in Britain.

P. arenaria has a Eurosiberian Southern-temperate distribution; its native range has been obscured by its spread as a weed and it is widely naturalised in other continents.

References: Hultén & Fries (1986), Meusel *et al.* (1978).

G. M. KAY

Littorella uniflora Shoreweed

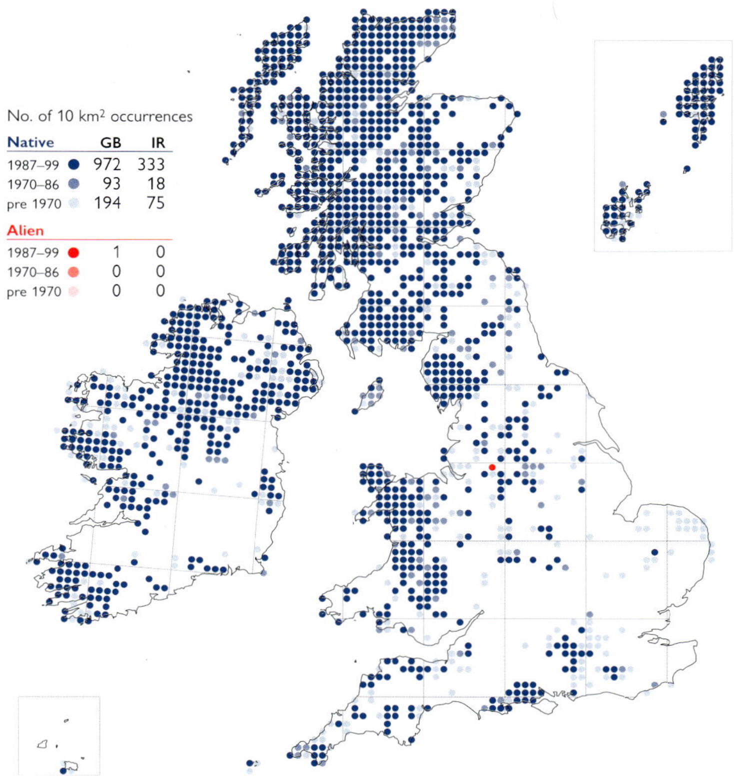

A perennial herb of oligotrophic or mesotrophic waters, found in lakes, reservoirs, rivers, streams, ponds and winter-flooded dune-slacks, growing on stones, gravel, sand, peat, marl or soft mud. It grows to a depth of about 4 metres and can form a dense band in the draw-down zone around lakes and reservoirs. It reproduces by seed and vegetatively by rooting stolons. Seeds remain viable for decades. 0–825 m (Ffynnon Llyffant, Caerns.).

Native (change +0.40). Drainage had resulted in the loss of many lowland sites for *L. uniflora* before 1930, especially in S. England, and further losses have occurred since then. It is now much better recorded that it was at the time of the 1962 *Atlas*.

Suboceanic Temperate element.

References: Atlas (255b), Hultén & Fries (1986), Meusel *et al.* (1978), Preston & Croft (1997).

M. J. WIGGINTON

Buddleja davidii Butterfly-bush

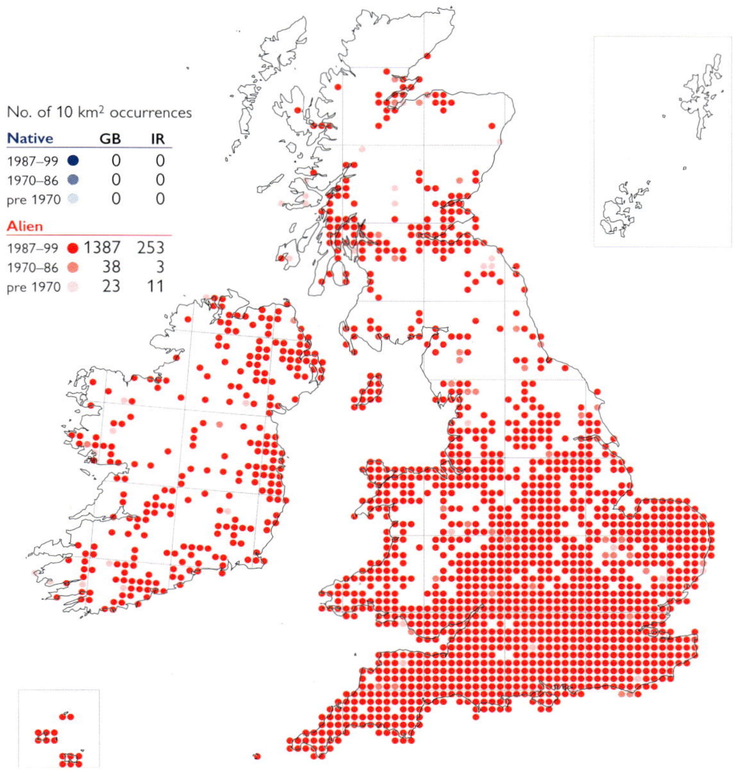

A large deciduous shrub, now very well established on waste ground, by railways, in quarries, on roadsides and generally in urban habitats, where it often grows on walls and neglected buildings. It prefers dry, disturbed sites where large populations can develop from its wind-dispersed seed. Lowland.

Neophyte (change +3.73). *B. davidii* was introduced into cultivation in the 1890s, and quickly became very popular in gardens. It was known to be naturalised in the wild in Merioneth by 1922 and Middlesex by 1927, and shown as locally well established in S. England in the 1962 *Atlas*. In recent decades it has spread rapidly throughout lowland Britain and, to a lesser extent, Ireland.

Native of China.

References: Atlas (205c), Bolòs & Vigo (1995).

T. D. DINES

Buddleja globosa Orange-ball-tree

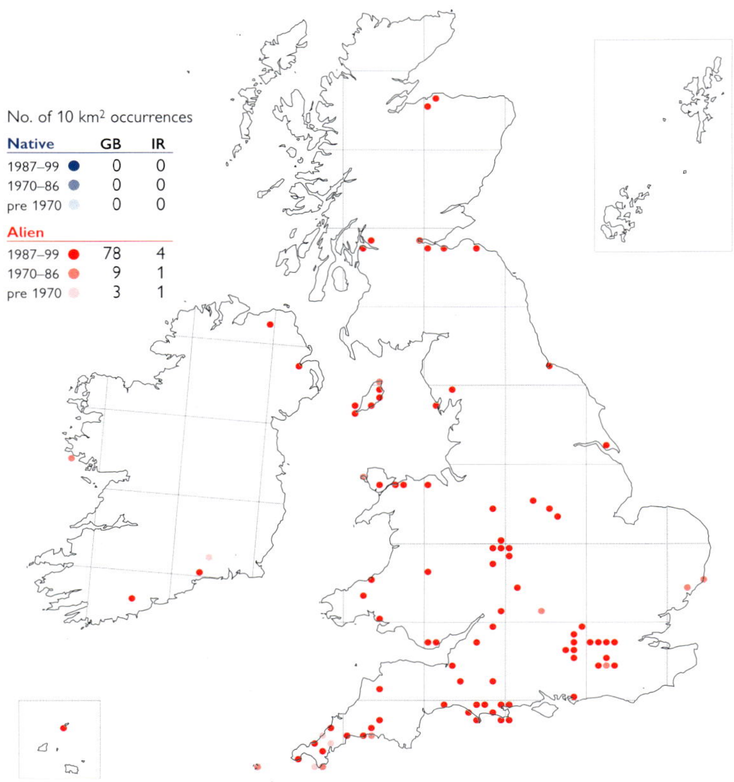

A large deciduous shrub, found as a garden escape or throw-out on roadsides and waste ground, and as a relic of cultivation. Reproduction is by seed but the species is rarely naturalised. Lowland.

Neophyte. Introduced into cultivation in 1774, *B. globosa* has recently become more popular in gardens. It remains, however, a rare escape, and was not recorded from the wild until 1964.

Native of S. America (Chile, Peru).

References: Bean (1970), Coats (1963).

T. D. DINES

Forsythia suspensa × *F. viridissima* (*F.* × *intermedia*) Forsythia

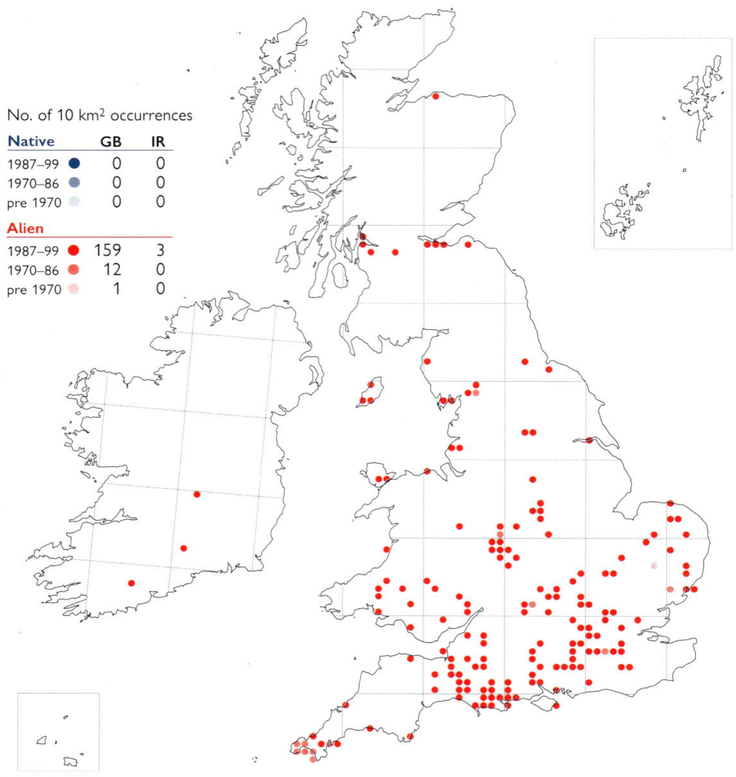

A deciduous shrub found as a relic of cultivation at sites of former habitation and sometimes arising from outcast material in hedgerows, on waste ground, rubbish tips, roadsides and river banks. All plants in the wild seem to be derived from vegetative fragments and it does not appear to set viable seed. Lowland.

Neophyte. *F.* × *intermedia* was raised in *c.* 1880 in Germany and was first grown in Britain ten years later. It has been known from the wild since 1971. It is a very common and vigorous garden plant, which is possibly increasing in the wild with continued dumping of garden rubbish.

A hybrid of garden origin.

References: Bean (1973), Coats (1963).

T. D. DINES

Fraxinus excelsior Ash

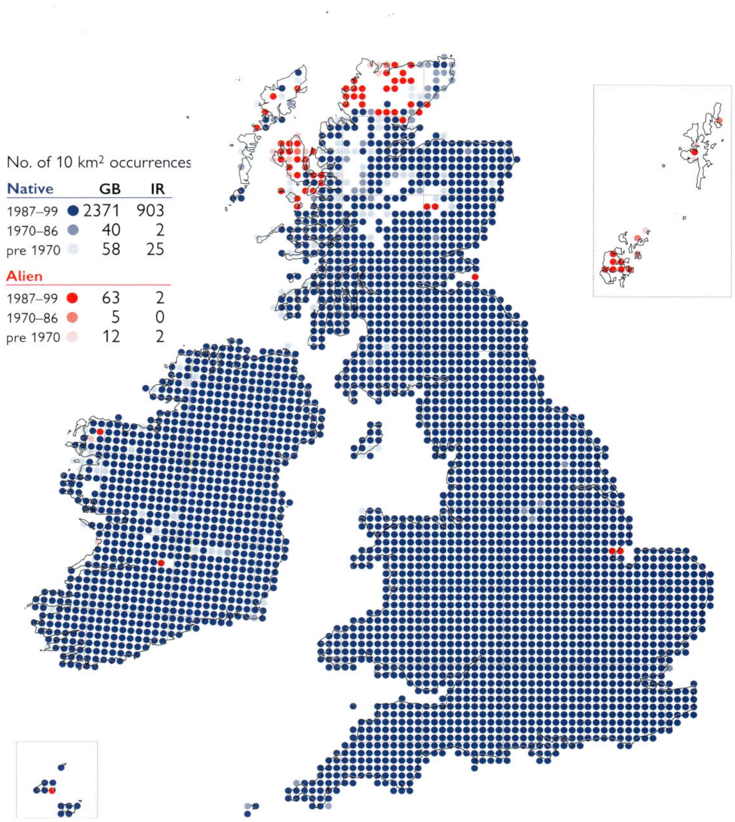

A deciduous tree of woodland, scrub and hedgerows, especially on moist, basic soils, but also frequent on rock scars and cliffs, stabilised scree and the grikes of limestone pavement. It can tolerate periodically waterlogged soils, being found around springs and in *Alnus* and *Salix* carr. In managed woodland it may be grown as a timber tree or coppice. It is a rapid coloniser of waste ground, disused quarries and railway banks. 0–585 m (Cwm Idwal, Caerns.).

Native (change –0.73). The range of *F. excelsior* is stable. In N. Scotland it is native on limestone and widely planted elsewhere; differentiating native from alien populations can be difficult.

European Temperate element.

References: Atlas (205d), Grime *et al.* (1988), Hultén & Fries (1986), Meusel *et al.* (1978), Rackham (1980), Wardle (1961).

T. D. DINES

Syringa vulgaris Lilac

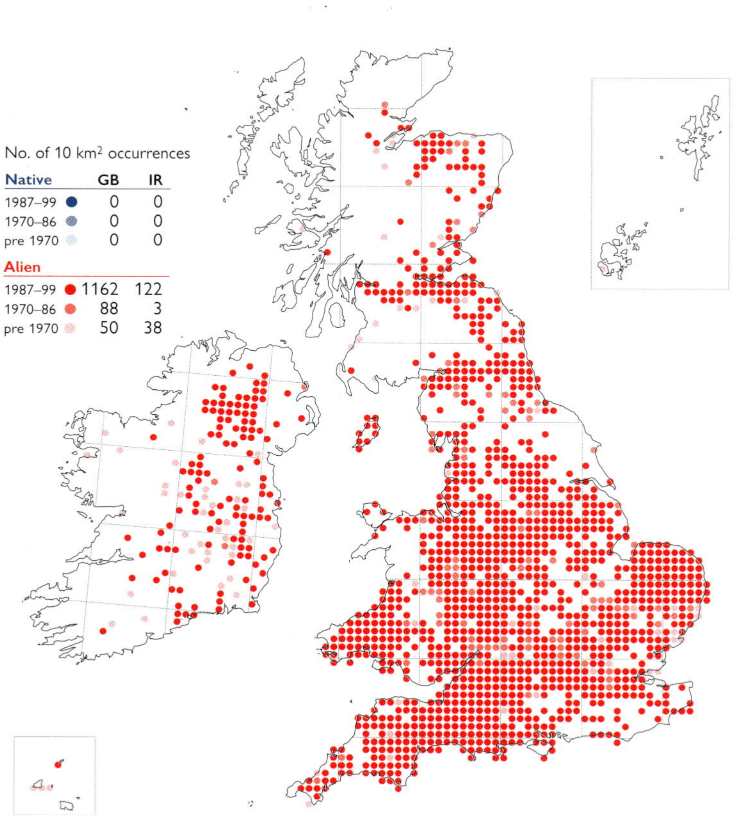

A strongly suckering deciduous shrub, occurring as a relic of cultivation in sites of former habitation, or more or less naturalised in hedges, on roadsides, railway banks, tips and waste ground. It is occasionally planted for hedging well away from habitation. Reproduction is mostly vegetative, and establishment from seed is rare. Lowland.

Neophyte (change +4.48). *S. vulgaris* was cultivated in Britain by 1597, becoming extremely popular in parks and gardens in the 19th century. It was known from the wild by at least 1879. The astonishing increase in records since the 1962 *Atlas* is probably due to both better recording of aliens and a genuine increase.

Native of S.E. Europe, but modified by selection and cultivation.

References: Atlas (206a), Bean (1980), Fiala (1988), Meusel *et al.* (1978).

T. D. DINES

Ligustrum vulgare Wild Privet

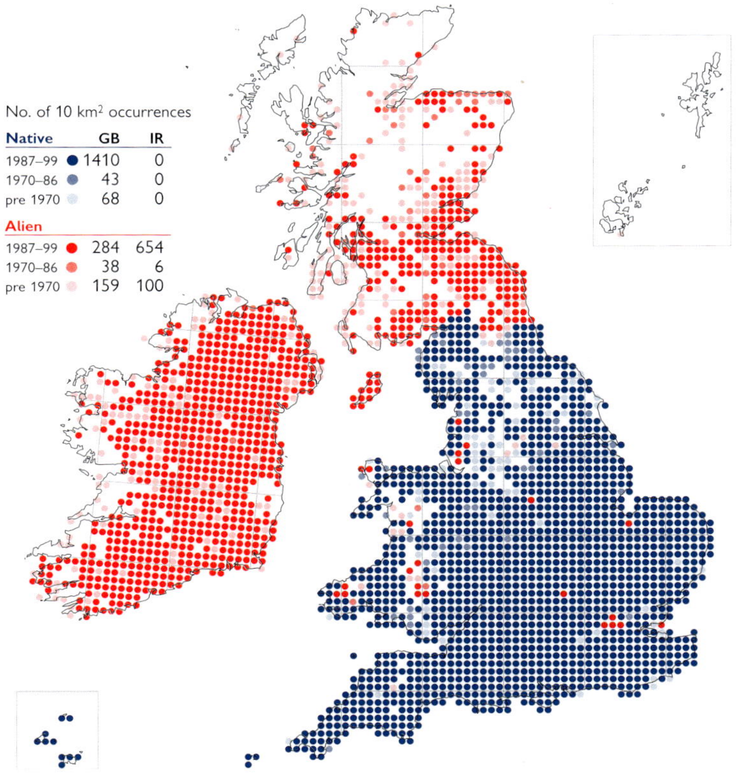

A deciduous to semi-evergreen shrub found as a native in hedgerows, woodland and scrub, preferring well-drained, calcareous or base-rich soils. It is also often planted, particularly in hedges and woodland, and occurs as a garden escape and a relic of cultivation. 0–490 m (Craig y Cilau, Brecs.).

Native (change –0.69). There has been no significant change in the native range of *L. vulgare* since the 1962 *Atlas*, although alien records are mapped here for the first time. It can be difficult to decide if some populations are alien or native; in Ireland it is mapped as alien but may be native at some coastal sites. Some records may be errors for *L. ovalifolium.*

European Temperate element; widely naturalised outside its native range.

References: Atlas (206b), Hultén & Fries (1986), Meusel *et al.* (1978).

T. D. DINES

Ligustrum ovalifolium Garden Privet

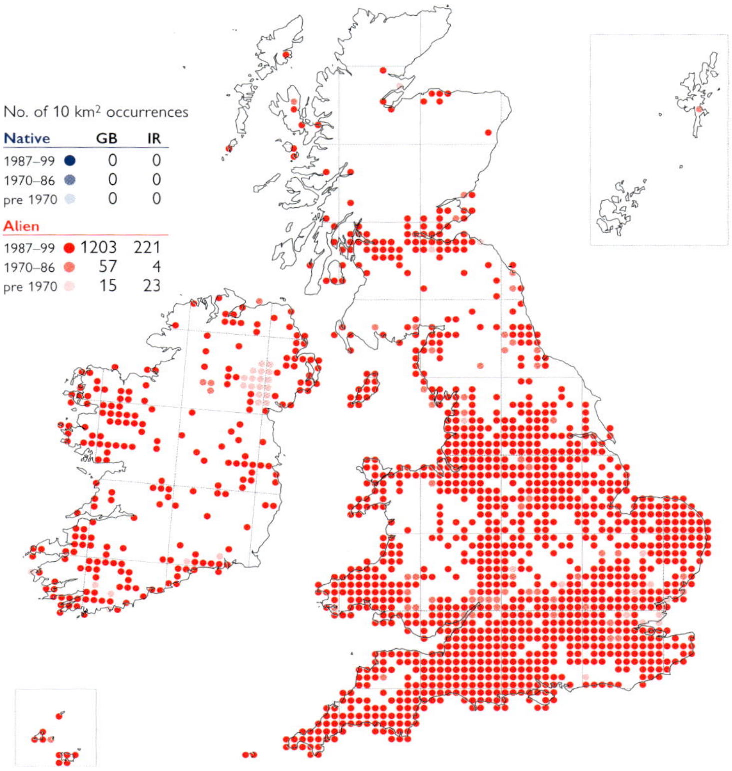

A semi-evergreen or evergreen shrub, abundantly planted for hedging and found as a persistent relic in old gardens, or a garden throw-out in hedges, rubbish tips, waste land and on railway banks. Plants very occasionally arise from bird-sown seed. Lowland.

Neophyte. *L. ovalifolium* has been cultivated in Britain since 1842, and was first recorded from the wild in 1939. Trends in its distribution are difficult to assess, but it is probably increasing through the continued disposal of garden rubbish. It is unevenly recorded.

Native of Japan.

Reference: Bean (1973).

T. D. DINES

Verbascum blattaria Moth Mullein

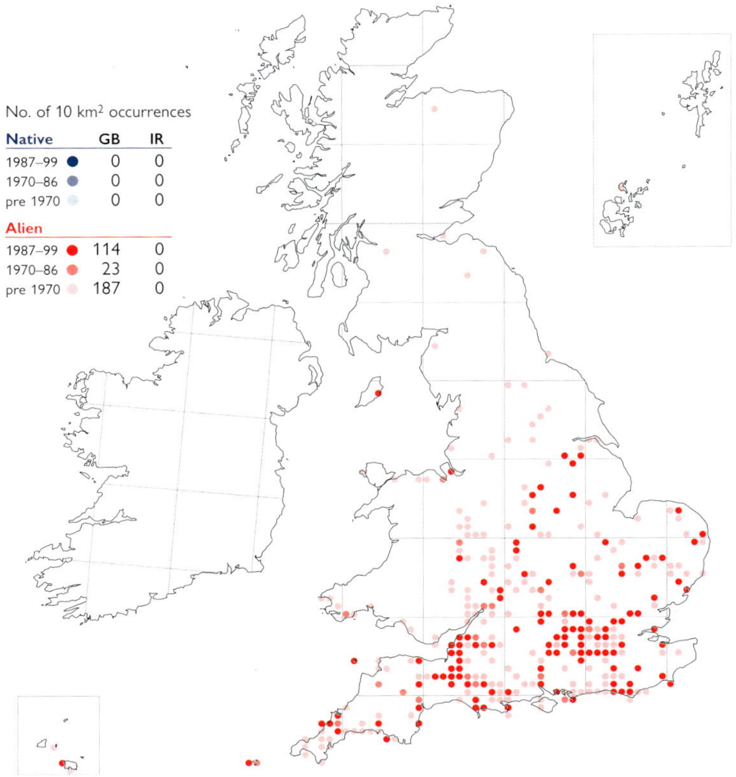

A biennial, occasionally annual, herb of waste ground, rough grassland, field-borders and waysides; also in old quarries and gravel-pits. Occurrences are generally casual, but some populations appear to be persistent. Lowland.

Neophyte (change –0.39). *V. blattaria* was being cultivated in Britain by 1596 and has been known from the wild since at least 1629. It was described as 'scattered irregularly . . . over the central and southern counties of England' by Dunn (1905) and this still summarises its overall distribution.

V. blattaria has a Eurosiberian Southern-temperate distribution; it is naturalised in Europe north of its native range.

References: Atlas (221d), Gross & Werner (1978), Meusel *et al.* (1978).

A. HORSFALL

Verbascum virgatum Twiggy Mullein

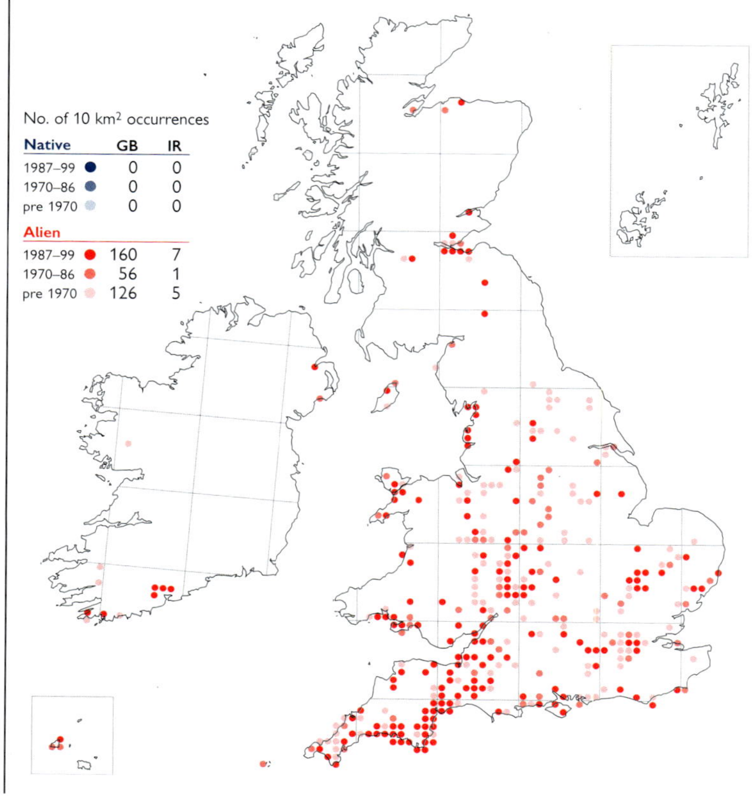

A biennial herb, naturalised on dry banks, walls, field margins, rough grassland, pastures and sheltered sea-cliffs in S.W. England; elsewhere a casual of waste ground, rubbish tips, re-seeded road verges, sand-pits, tracks and disturbed coastal dunes. It reproduces by seed, easily colonising open habitats, but does not survive much competition. Lowland.

Neophyte (change +0.35). *V. virgatum* was first recorded in the wild in Britain in 1787. Its distribution is probably stable in S.W. England, where it has often been considered to be native or probably native.

A Suboceanic Southern-temperate species.

References: Atlas (222a), Meusel *et al.* (1978), Stewart *et al.* (1994).

A. HORSFALL

Verbascum phlomoides Orange Mullein

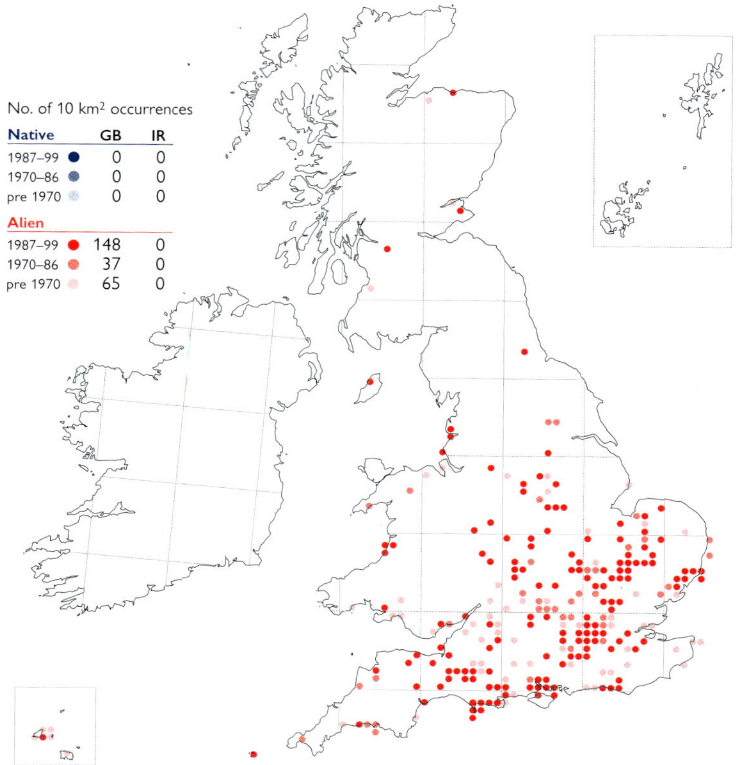

This biennial herb occurs in hedge banks, by waysides, on open rough ground and rubbish tips, generally on sandy or stony soils. It is sometimes naturalised in C. and S. Britain, but occurrences are usually casual. Seed is freely produced. Lowland.

Neophyte (change +0.87). *V. phlomoides* was cultivated in Britain by 1739 and was known from the wild by 1838. Although now rarely cultivated, it seems to be more frequent in the wild than it was at the time of the 1962 *Atlas*.

V. phlomoides has a European Temperate distribution.

Reference: Atlas (220d).

A. HORSFALL

Verbascum densiflorum Dense-flowered Mullein

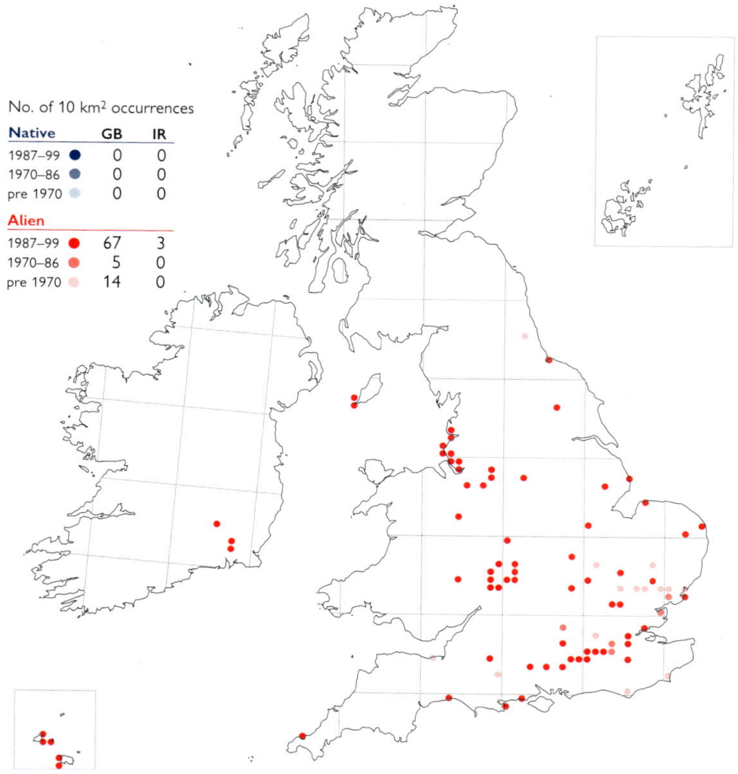

A large, biennial herb found as a garden escape on waste ground, roadsides and rubbish tips. Populations are usually casual, but may persist for a few years. Lowland.

Neophyte. This species is popular in gardens, where it has been grown since about 1825. The earliest report from the wild, in 1838, is an error but it was 'occasionally recorded as a garden escape in England' according to Dunn (1905). It is probably over-recorded for *V. thapsus*, and for *V. phlomoides* and its hybrids.

A European Temperate species, but rare or absent as a native from westernmost Europe.

Reference: Hultén & Fries (1986).

T. D. DINES

Verbascum thapsus Great Mullein

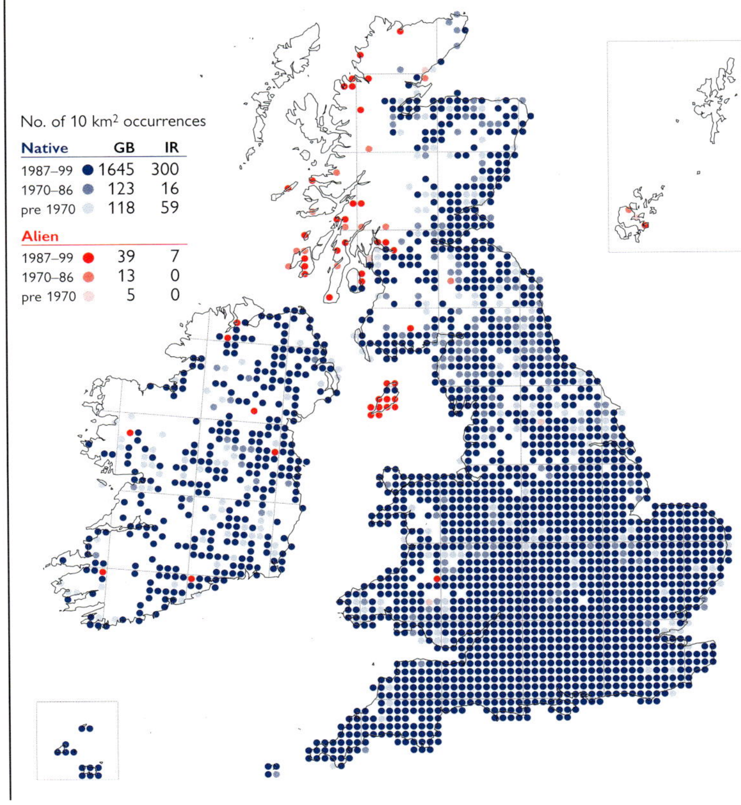

A biennial herb of open scrub and hedge banks, waysides, railway banks and sidings, rough grassy places, waste ground and quarries. It prefers well-drained soils, especially those over a sand, gravel or chalk substrate. It produces copious long-lived seed and may become an abundant colonist. It is a frequent garden escape, becoming established on rubbish tips when thrown out. 0–370 m (White Scars, Ingleborough, Mid-W. Yorks.).

Native (change +0.27). There has been little change in the distribution of *V. thapsus* since the 1962 *Atlas*. The map may overestimate its occurrence as a native and underestimate the extent to which it occurs as an alien.

Eurosiberian Temperate element; widely naturalised outside its native range.

References: Atlas (220c), Gross & Werner (1978), Hultén & Fries (1986), Meusel *et al.* (1978).

A. HORSFALL

Verbascum nigrum Dark Mullein

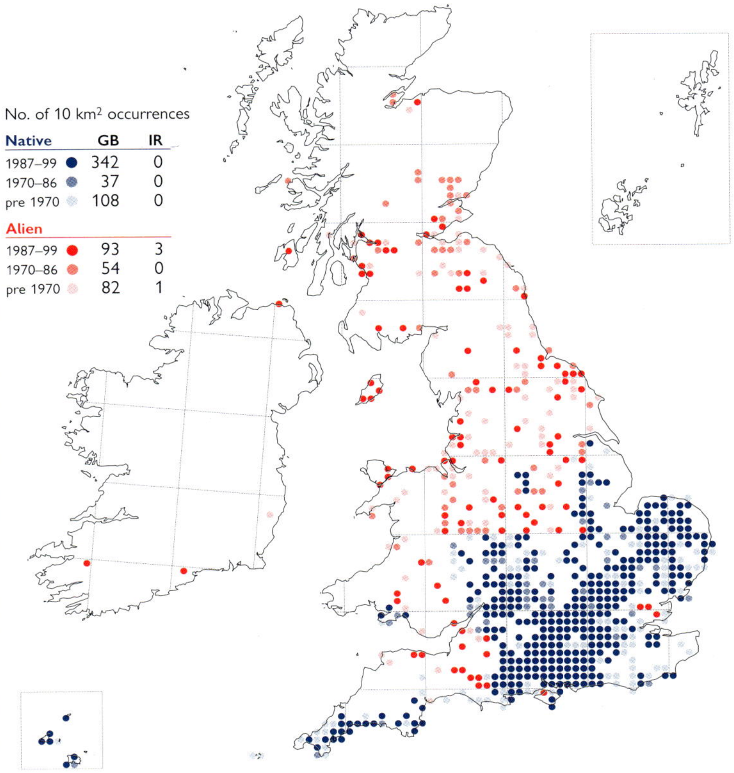

This biennial or short-lived perennial herb is found on road verges and banks, in hedge banks and other grassy places, on walls and in cultivated ground, including arable field margins. It prefers well-drained calcareous soils. 0–335 m (Brassington, Derbys.).

Native (change –0.12). *V. nigrum* is a mobile species; changes in the management of hedge banks and roadsides may be responsible for local losses, but can also be of benefit in assisting the dispersal of seed to new areas. This species is often grown in gardens, and the distinction between native and alien populations has become more blurred as garden escapes become more frequent.

Eurosiberian Temperate element.

References: Atlas (221c), Hultén & Fries (1986).

A. HORSFALL

Verbascum pulverulentum Hoary Mullein

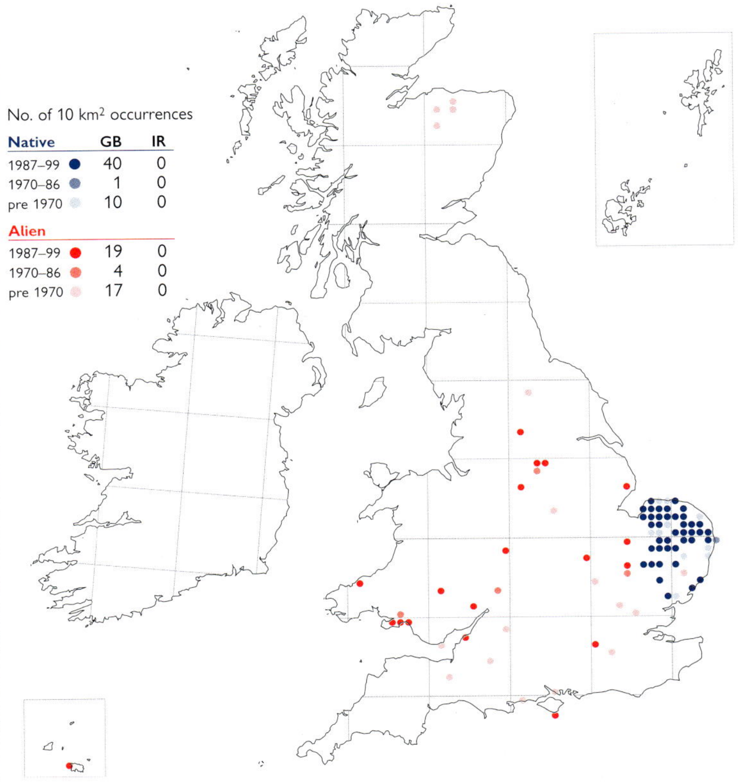

This monocarpic perennial is found on roadside verges and railway banks, in old quarries and gravel-pits, in hedge banks, rough ground, and locally on coastal shingle (its only 'natural' habitat). Outside its core area it is usually a casual of waste ground. Seed remains viable for many years and new populations can appear after soil disturbance. Lowland.

Native or alien (change +0.94). This species was first recorded in Britain in 1670. Since the 1962 *Atlas* it has been recorded in many additional 10-km squares, perhaps because of more systematic recording. Losses have been reported from its core areas in East Anglia, but other populations of this spectacular species are safeguarded by admirers.

Submediterranean-Subatlantic element.

References: Atlas (221b), Bolòs & Vigo (1995), Stewart *et al.* (1994).

A. HORSFALL

Verbascum lychnitis White Mullein

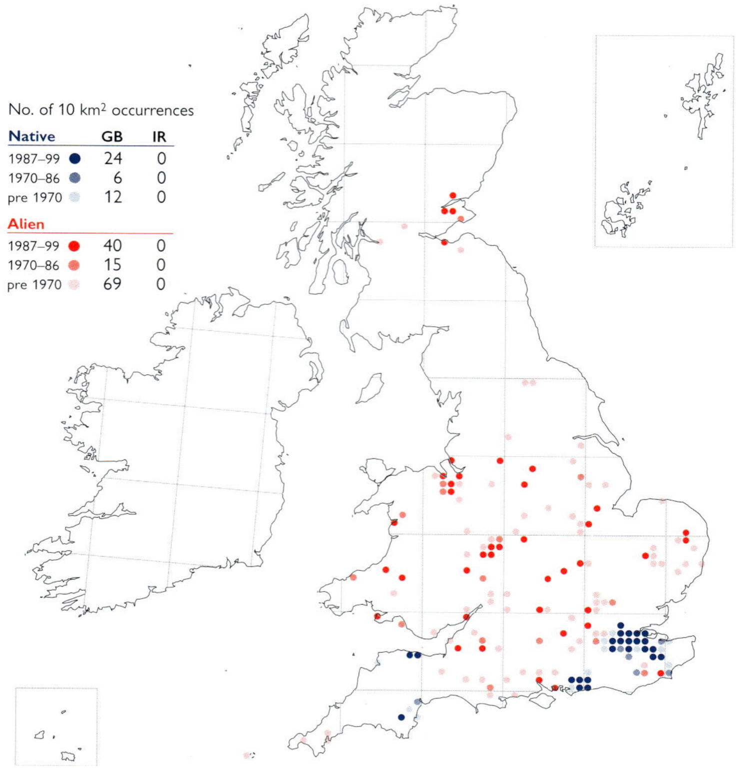

A biennial, or occasionally short-lived perennial, herb of dry, usually calcareous soil, occurring in rough pastures, recently cleared woodland, on railway banks, tracksides and road verges and in quarries and waste places. Seed is copiously produced, and remains viable for many years. It freely hybridises with other *Verbascum* species. Lowland.

Native (change –0.23). There has been little change in the overall distribution of *V. lychnitis*, but because it depends on periodic disturbance its abundance can vary markedly from year to year. Large populations can arise where woodland is cleared in forestry operations or as a result of storm damage.

European Temperate element.

References: Atlas (221a), Meusel *et al.* (1978), Stewart *et al.* (1994).

A. HORSFALL

Scrophularia nodosa Common Figwort

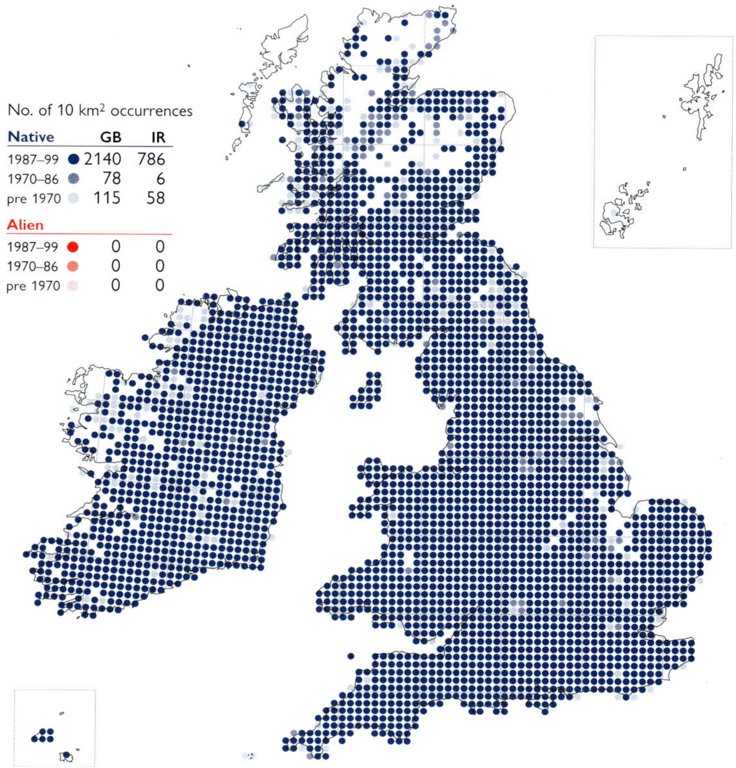

A perennial herb of open or shaded habitats, preferring fertile soils and found in damp woodland, woodland rides, hedge banks, ditches and riversides, and sometimes in drier sites on waste ground. Generally lowland, but reaching 480 m at Angle Tarn, Patterdale (Westmorland).

Native (change –0.37). The distribution of *S. nodosa* is stable.

Eurosiberian Temperate element.

References: Atlas (224d), Hultén & Fries (1986), Meusel *et al.* (1978).

A. HORSFALL

Scrophularia auriculata Water Figwort

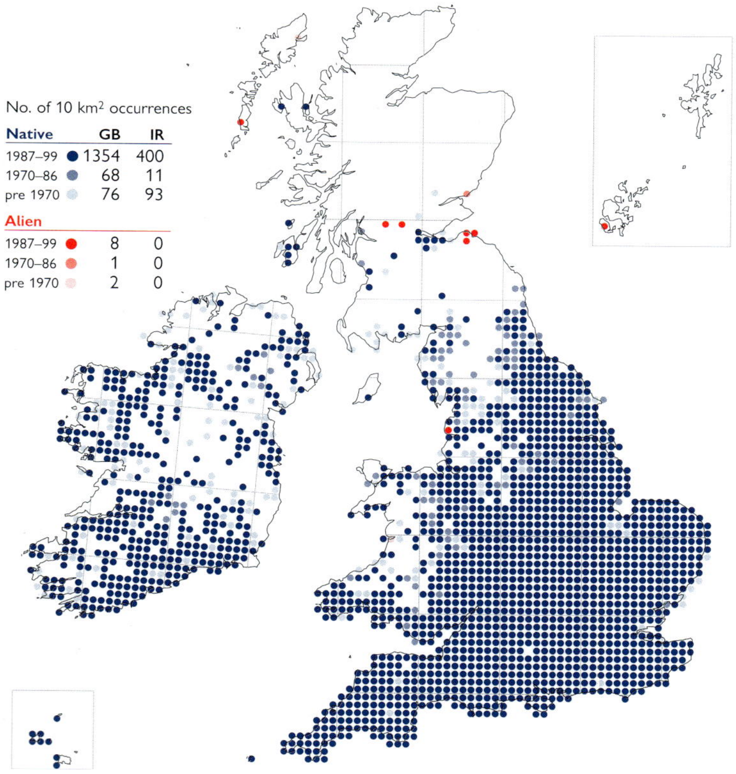

A perennial herb of wet places on the margins of lakes, rivers, streams and canals, and in ditches, marshes and wet woodlands. Lowland.

Native (change –0.21). The distribution of *S. auriculata* is broadly similar to that in the 1962 *Atlas*. There are rather more records at the northern edge of its contiguous range, probably because of more thorough recording in previously under-worked areas. Many records from N. Scotland have been discounted as errors for *S. nodosa*.

Suboceanic Southern-temperate element.

References: Atlas (225a), Meusel *et al.* (1978).

A. HORSFALL

Scrophularia umbrosa Green Figwort

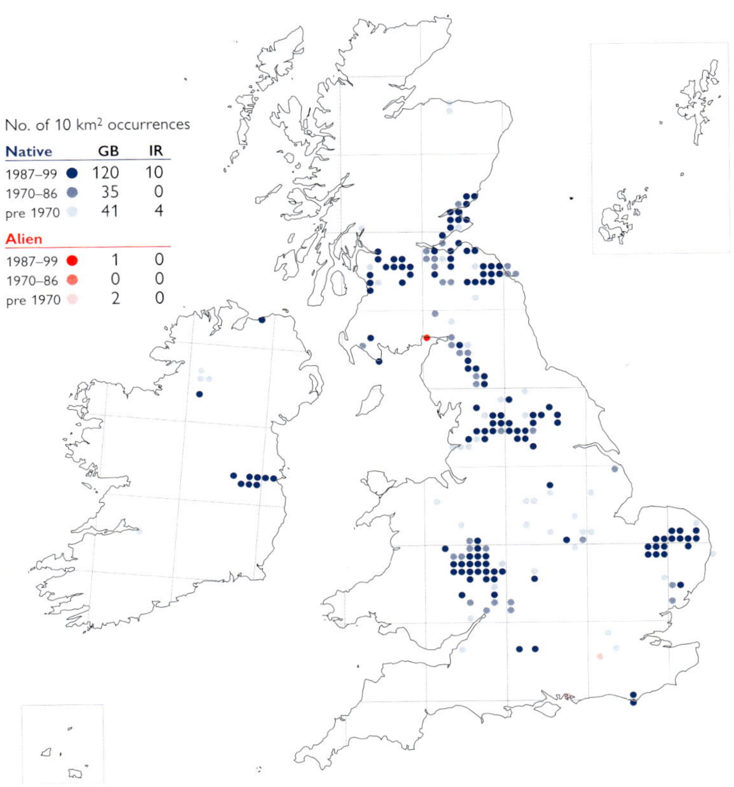

A rhizomatous perennial herb which grows on fertile soils by streams and rivers, and in damp woodland, in both open and shaded places. Generally lowland.

Native (change +0.72). Although *S. umbrosa* is treated as a native species it may be a relatively recent colonist. It was first recorded in Britain in 1840 (Salop) and in Ireland in 1895 (Co. Dublin). It was first recorded relatively recently in counties such as Berwickshire (1852), Norfolk (1904), Angus (1910) and Cumbria (1924), and it has a puzzlingly patchy distribution. It continues to increase, especially within its current strongholds.

Eurosiberian Temperate element, with a continental distribution in W. Europe.

References: Atlas (225b), Curtis & McGough (1988), Hultén & Fries (1986), Meusel *et al.* (1978), Stewart *et al.* (1994).

A. HORSFALL

Scrophularia scorodonia Balm-leaved Figwort

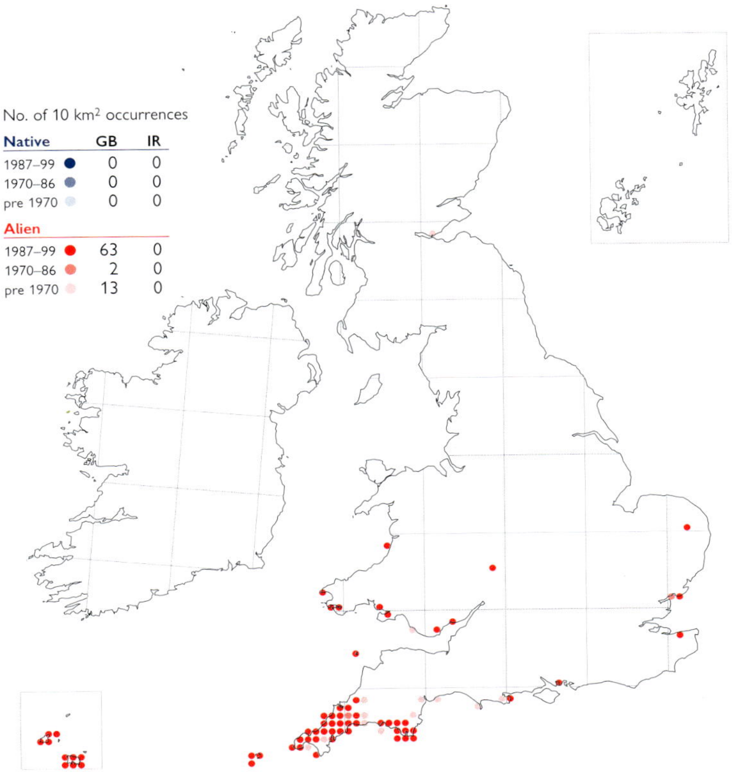

A long-lived perennial herb of hedge banks, scrubby field-borders, rough ground on cliff-tops, waste places, disused quarries and old walls. Lowland.

Neophyte (change +0.75). *S. scorodonia* has been known in Jersey since 1689 and in Cornwall since 1712. As the mainland sites are almost entirely near ports and its habitats are ruderal, it may have been accidentally introduced. It is unclear whether the increase in recent records since the 1962 *Atlas* indicates a real extension of range, or merely better recording, although even within its core areas it appears to be increasing.

Oceanic Southern-temperate element.

References: Atlas (225c), Dupont (1962), Meredith (1994), Wigginton (1999).

A. HORSFALL

Scrophularia vernalis Yellow Figwort

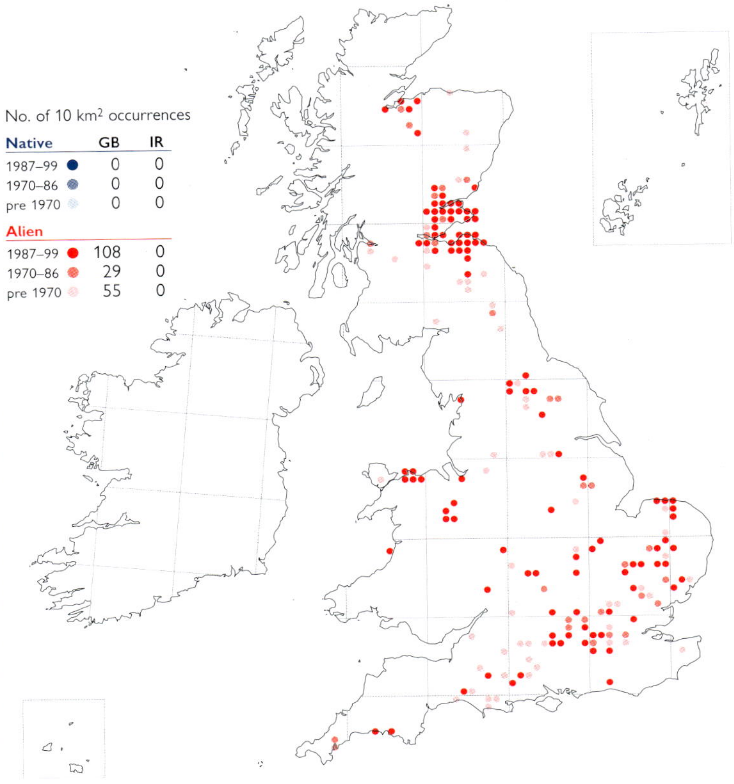

A biennial, sometimes perennial, herb found in woodland clearings, plantations, hedge banks and rough waste ground, usually in shade. Lowland.

Neophyte (change +0.54). This species was first recorded in the wild in 1633. By the 18th century it was known from several localities in, for example, Surrey, Sussex and N. Wales, and it had become naturalised throughout Britain by 1930. The number of records has increased since the 1962 *Atlas*, particularly in S.E. Scotland.

Native of the mountains of C. & S. Europe and the Caucasus.

References: Atlas (225d), Salisbury (1964).

A. HORSFALL

Mimulus agg. Monkeyflowers

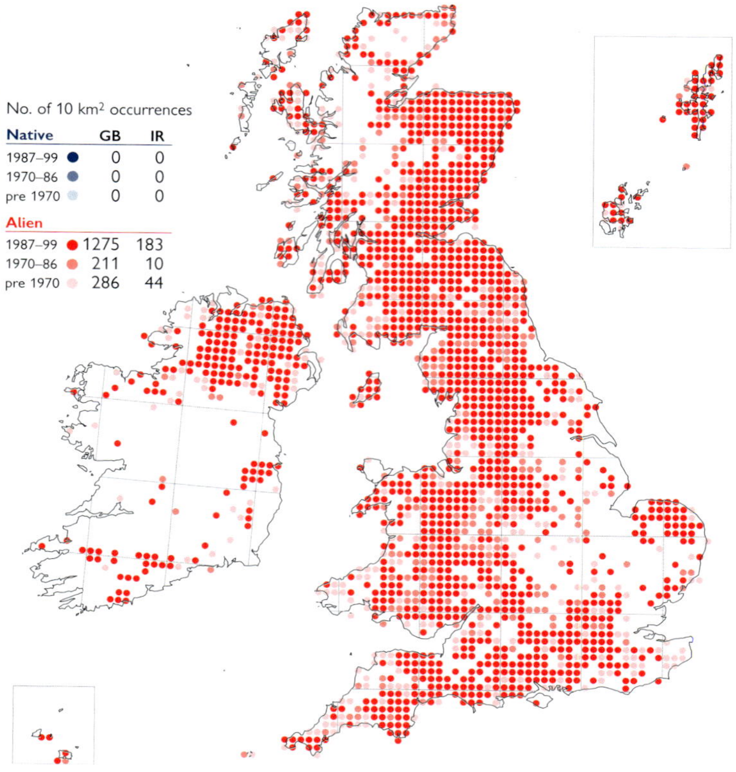

A genus of vigorous, stoloniferous perennials which have become widely naturalised in damp places. 0–440 m (above Garrigill, Cumberland).

Neophyte (change –0.47). *Mimulus* taxa were cultivated as waterside garden plants by 1759. They have been known to be established in the wild since at least 1830 and some have hybridised both in cultivation and in the wild to form a critical complex of closely related taxa. Because of identification difficulties, the individual taxa are probably under-recorded. Many species and cultivars of *Mimulus* are now commercially available and further garden escapes can be expected.

An aggregate comprising alien species from western N. & S. America, and their hybrids.

References: Atlas (226a-c), Rich & Jermy (1998), Silverside (1994).

A. HORSFALL

Mimulus moschatus Musk

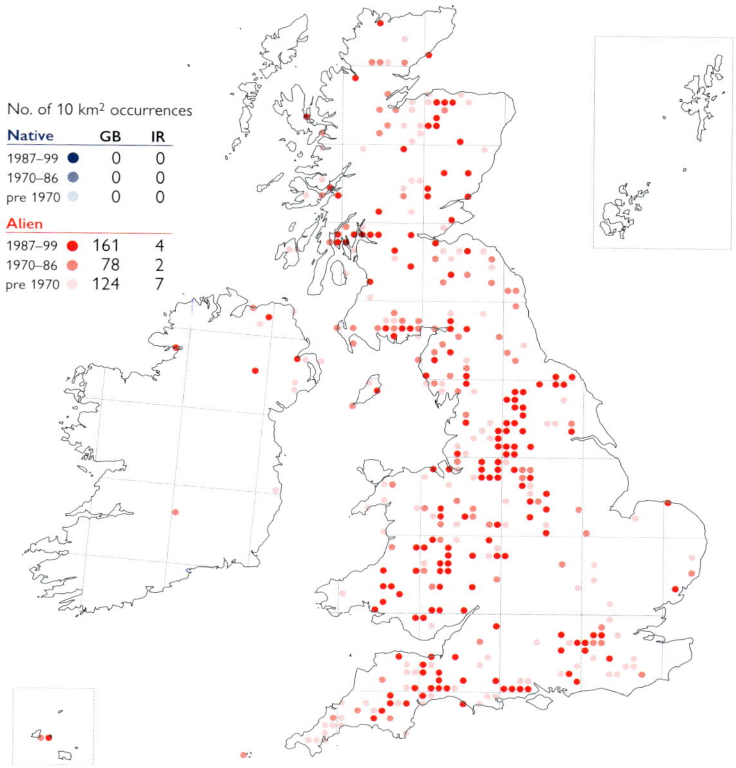

A decumbent perennial herb, casual or naturalised in damp, often shaded places, including the muddy edges of ditches, wooded swamps, damp woodland rides, by ponds and in damp pasture. Seed has considerable longevity in the soil. Lowland.

Neophyte. *M. moschatus* was introduced to gardens in 1826, spreading from there into semi-natural habitats; it was recorded from the wild by 1866. It is a very distinct species which has not hybridised with the other *Mimulus* taxa, and ought to be well recorded. Its distribution is stable.

Native of western N. America.

References: Atlas (226c), Rich & Jermy (1998).

A. HORSFALL

Mimulus guttatus Monkeyflower

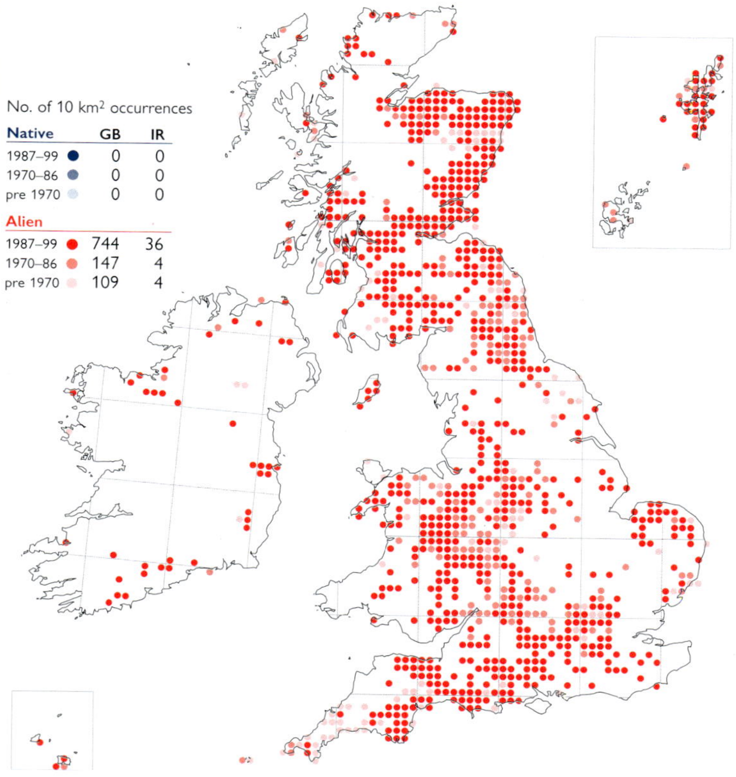

This vigorous perennial herb is found in wet places by streams, rivers and ponds, in damp meadows, marshy ground and open woodland. It spreads both by seed and vegetatively, rooting from the nodes in wet mud or gravel. Generally lowland, but upper altitudinal limit unknown.

Neophyte. *M. guttatus* is thought to have been introduced to cultivation in 1812. It soon became established in semi-natural habitats, being known in the wild by 1824. It continues to be a popular garden plant, but its distribution is probably stable. It has been much over-recorded for other taxa in the uplands.

Native of western N. America.

References: Atlas (226a), Hultén & Fries (1986), Meusel *et al.* (1978), Rich & Jermy (1998), Silverside (1994).

A. HORSFALL

Mimulus guttatus × *M. luteus* (*M.* × *robertsii*)
Hybrid Monkeyflower

Native	GB	IR
1987–99 ●	0	0
1970–86 ●	0	0
pre 1970	0	0
Alien		
1987–99 ●	333	115
1970–86 ●	99	7
pre 1970	41	3

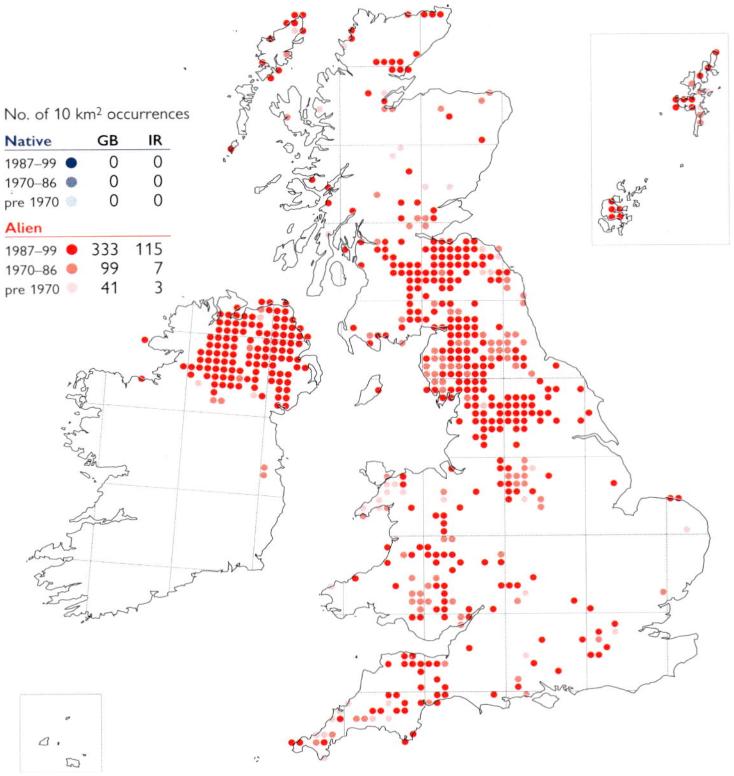

A perennial hybrid of damp and wet places, including on mud and shingle by streams and rivers, in flushes and in marshy ground. This hybrid is partially fertile, and can spread vegetatively from stem fragments. It ascends to over 440 m above Garrigill (Cumberland) and probably to 610 m in the Ochils (W. Perth).

Neophyte. *M.* × *robertsii* was first recognised in Britain by Roberts (1964) but not named as *M.* × *robertsii* until 1990. The earliest record traced dates to 1872 (Berwicks.). It is the commonest taxon of high ground, and many records of *Mimulus* agg., *M. luteus*, and even *M. guttatus* are probably referable to this hybrid, which is therefore likely to be under-recorded.

A hybrid of garden origin.

References: Rich & Jermy (1998), Silverside (1990b, 1994), Stace (1975).

A. HORSFALL

Mimulus cupreus × *M. guttatus* (*M.* × *burnetii*)
Coppery Monkeyflower

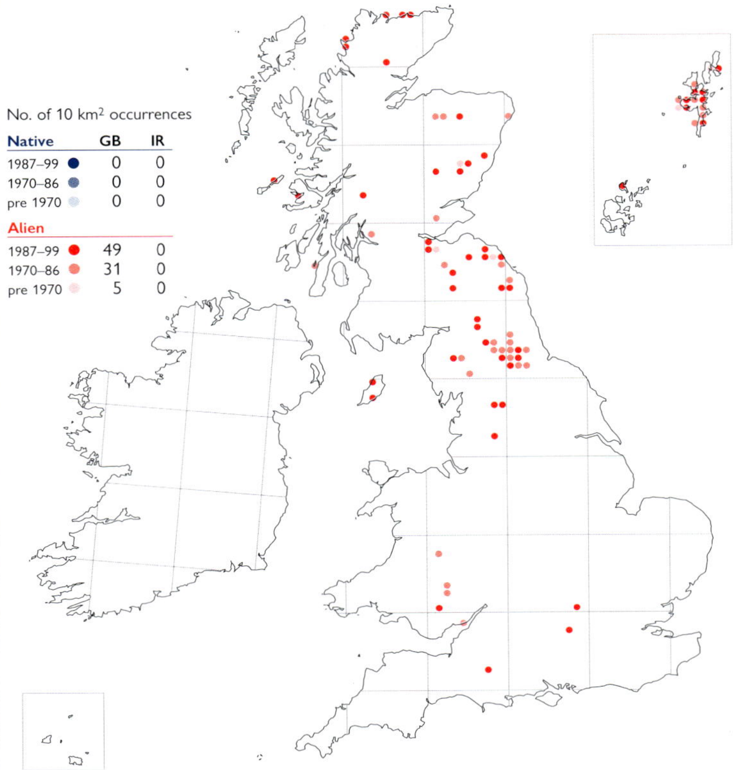

In common with *M. guttatus* and *M. luteus*, this plant of wet places is found in marshy ground by watercourses, in flushes and on river shingle. It is usually sterile, but rarely sets a few viable seeds. Generally lowland, but reaching 415 m above Garrigill (Cumberland).

Neophyte. This hybrid was initially identified as *M. cupreus*, but this species is not naturalised in Britain. The hybrid was raised by Dr Burnet of Aberdeen *c.* 1901. It was first collected in 1931 (Cumberland) and was apparently well-established by 1957. Roberts (1964, 1968) established the hybrid nature of the British populations.

A hybrid of garden origin; the parents *M. cupreus* (S. America) and *M. guttatus* (N. America) are allopatric.

References: Silverside (1994), Stace (1975), Rich & Jermy (1998).

M. J. WIGGINTON

Mimulus luteus Blood-drop-emlets

Native	GB	IR
1987–99 ●	0	0
1970–86 ●	0	0
pre 1970	0	0
Alien		
1987–99 ●	29	2
1970–86 ●	53	0
pre 1970	104	3

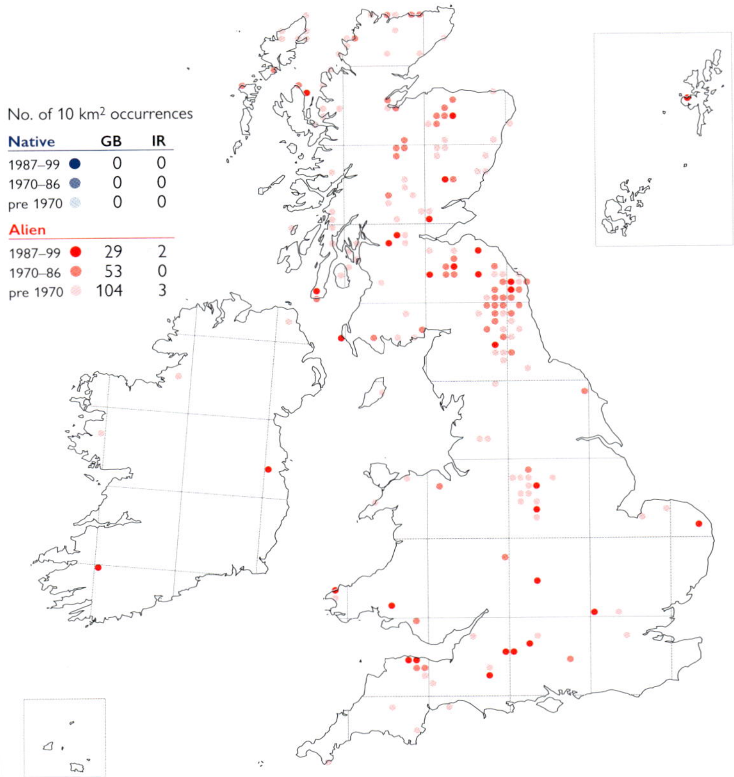

A creeping mat-forming perennial herb naturalised in damp or wet places such as marshes and flushes, and on riversides and river shingle. It reproduces by seed and vegetatively from stem fragments. Generally lowland, but upper altitudinal limit unknown.

Neophyte. *M. luteus* was introduced into cultivation *c.* 1826. It is very local and has been much over-recorded for *M.* × *robertsii* (*M. guttatus* × *M. luteus*). The map may contain some errors, particularly amongst the pre-1987 records.

Native of S. America (Andes).

References: Atlas (226b), Silverside (1994), Rich & Jermy (1998).

A. HORSFALL

Limosella aquatica Mudwort

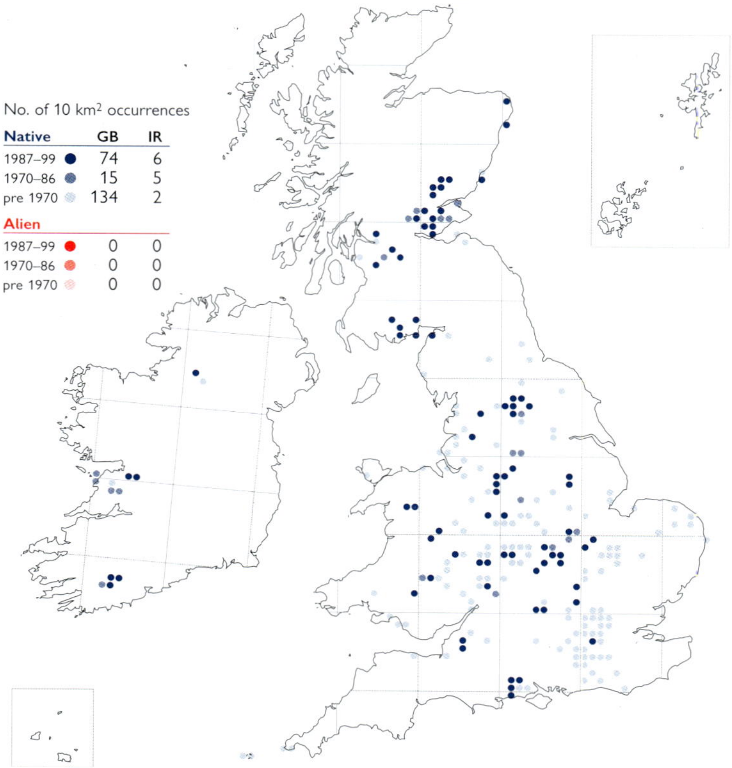

No. of 10 km^2 occurrences

Native	GB	IR
1987–99 ●	74	6
1970–86 ●	15	5
pre 1970 ●	134	2
Alien		
1987–99 ●	0	0
1970–86 ●	0	0
pre 1970 ●	0	0

An annual of the muddy edges of rivers, lakes, reservoirs, pools, ditches, rutted tracks and roadsides. In the Burren (Co. Clare) it also occurs in limestone solution hollows. It may prefer mildly acidic, nutrient enriched soils. Plants reproduce by seed and also spread by stolons. 0–455 m (Malham Moor, Mid-W. Yorks.).

Native (change +1.00). Though this species is erratic in appearance, there have been many losses, mainly before 1930 but still continuing. Reasons for this include drainage, the infilling of ponds, and lack of grazing. The remarkable increase in Scottish records since the 1962 *Atlas* may result from better recording or a genuine spread.

Circumpolar Boreo-temperate element.

References: Atlas (226d), Curtis & McGough (1988), Hultén & Fries (1986), Meusel *et al.* (1978), Stewart *et al.* (1994).

A. HORSFALL

Limosella australis Welsh Mudwort

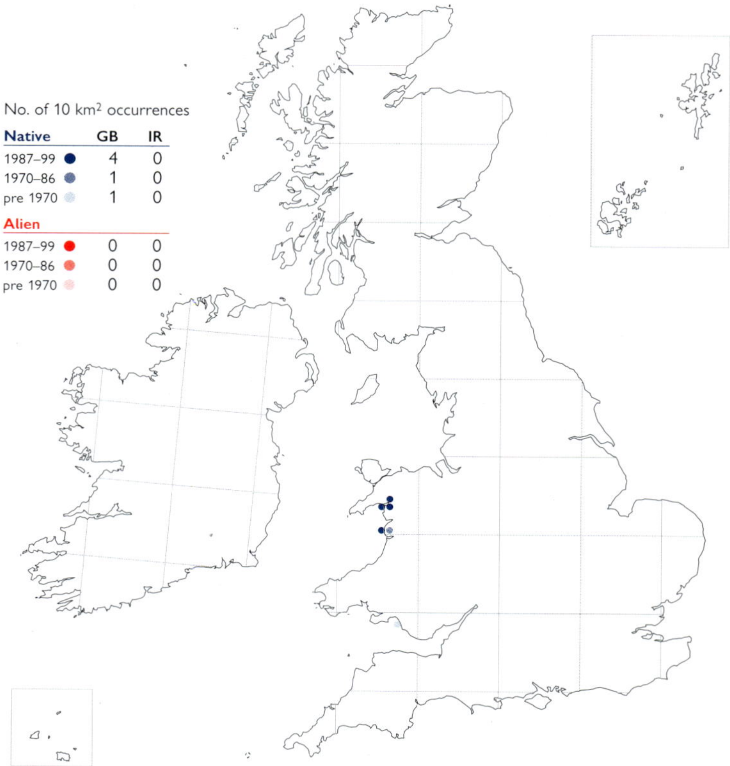

No. of 10 km^2 occurrences

Native	GB	IR
1987–99 ●	4	0
1970–86 ●	1	0
pre 1970 ●	1	0
Alien		
1987–99 ●	0	0
1970–86 ●	0	0
pre 1970 ●	0	0

An annual to short-lived perennial of mudflats and saltmarsh pools, occurring as scattered plants or in dense swards. Reproduction by seed occurs, but plants spread mostly by stolons. Lowland.

Native. This species was first recorded in 1897 in Glamorgan, but colonies in S. Wales had probably died out by the 1940s. It was not found in N. Wales until 1921. It can be locally abundant, but numbers fluctuate markedly annually. It has been suggested that it might be a recent arrival, or even a ballast-alien (Jones, 1991), but the evidence for this is inconclusive.

Oceanic Boreo-temperate element; in Europe restricted to Wales, but widespread in coastal regions of eastern N. America.

References: Atlas (227a), Hultén & Fries (1986), Kay & John (1995), Meusel *et al.* (1978), Wigginton (1999).

A. HORSFALL

Antirrhinum majus Snapdragon

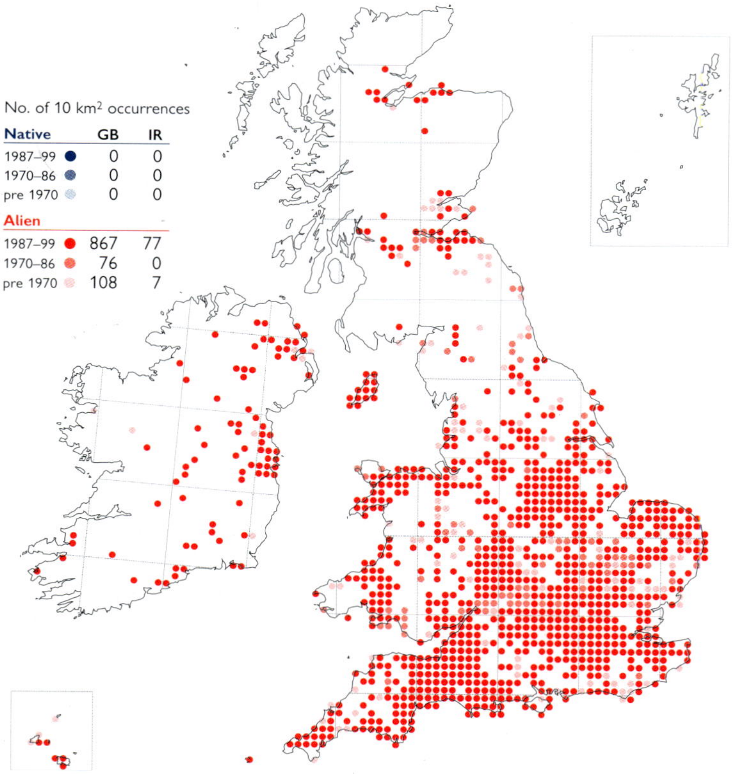

No. of 10 km^2 occurrences

Native	GB	IR
1987–99 ●	0	0
1970–86 ●	0	0
pre 1970 ●	0	0
Alien		
1987–99 ●	867	77
1970–86 ●	76	0
pre 1970 ●	108	7

An annual or short-lived perennial herb, widely naturalised on old walls, waysides, pavement cracks, waste ground and rubbish tips. Populations can be long-lived, and the species reproduces readily from seed. Lowland.

Neophyte (change +2.84). *A. majus* has been cultivated here since Elizabethan times and is very popular in gardens. It was first recorded in the wild in 1762. Its distribution has increased significantly since the 1962 *Atlas* and this is probably due to a genuine spread as well as the better recording of aliens.

Native of the S.W. Europe and the W. Mediterranean region.

References: Atlas (222c), Bolòs & Vigo (1995).

A. HORSFALL

Chaenorhinum minus Small Toadflax

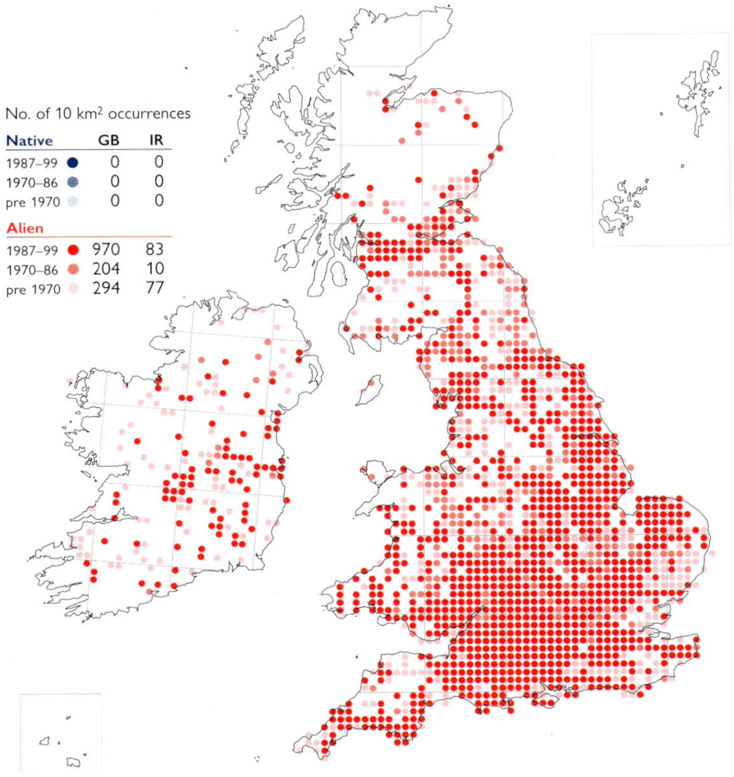

A spring-germinating annual of open habitats on well-drained, often calcareous, soils, including cultivated fields, forestry tracks, on rough waste ground, old walls, quarries, and especially along railways. 0–425 m (Kirkstone Pass, Westmorland).

Archaeophyte (change –0.63). *C. minus* was mapped as 'all records' in the 1962 *Atlas*. It has declined in many areas, particularly in Ireland. It was once a familiar weed of cultivated farmland, but agricultural intensification has now rendered it much rarer in this habitat and it is now more likely to be found along railways and in railway yards.

As an archaeophyte *C. minus* has a European Temperate distribution; it is widely naturalised outside this range.

References: Atlas (223d), Grime *et al.* (1988), Hultén & Fries (1986).

A. HORSFALL

Misopates orontium Weasel's-snout

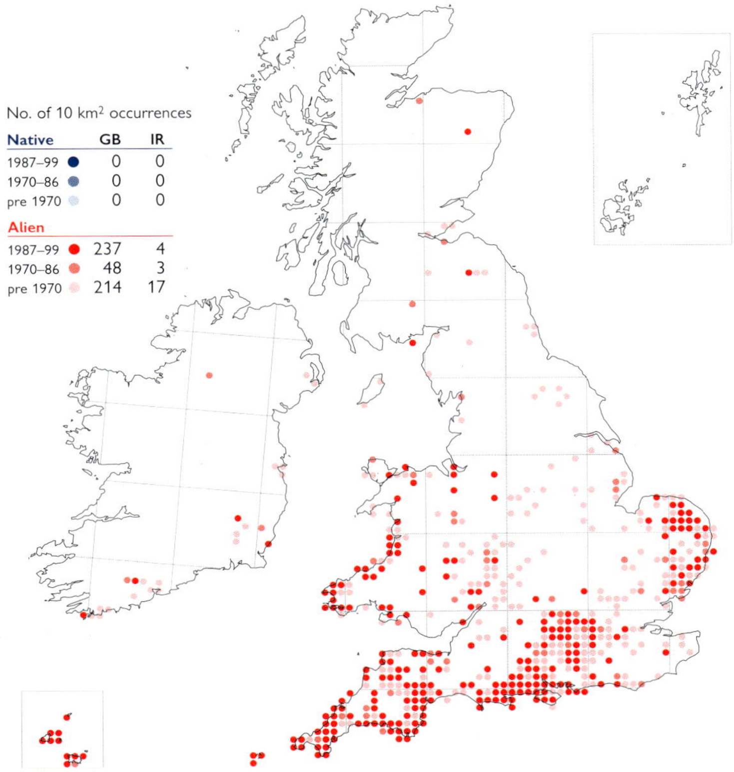

A spring-germinating annual of light soils, found in arable and other cultivated ground including among horticultural crops, and in gardens and waste places. It reproduces by seed, but cold, wet summers inhibit its germination and growth. Lowland.

Archaeophyte (change –0.89). *M. orontium* has declined because of agricultural intensification and the more widespread autumn sowing of crops. Salisbury (1961) noted that it appeared to be less frequent than formerly, and since then it has declined sharply, disappearing from almost half of its 10-km squares in S.E. England in the last forty years.

As an archaeophyte *M. orontium* has a Eurosiberian Southern-temperate distribution; it is widely naturalised outside this range.

References: Atlas (222b), Curtis & McGough (1988), Hultén & Fries (1986), Meusel *et al.* (1978), Wilson (1991).

A. HORSFALL

Cymbalaria muralis Ivy-leaved Toadflax

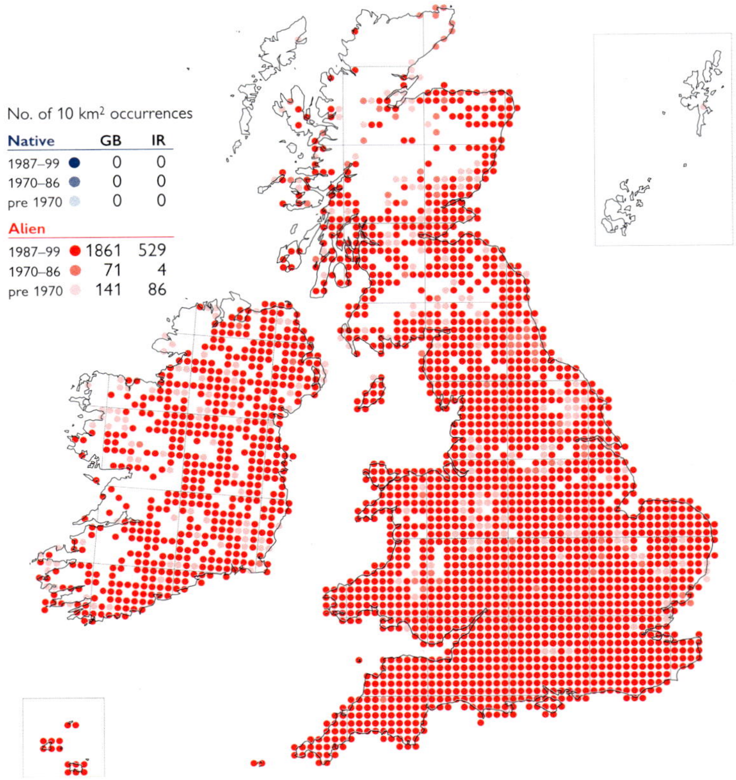

This perennial herb is well-established on old walls and bridges, pavements, and in other well-drained rocky and stony places, often near habitation. It is also found as large, prostrate patches on shingle beaches. It can root from fragments or from nodes, and its seeds germinate readily in brick and stone mortar. 0–450 m (Garrigill, Cumberland).

Neophyte (change –0.10). *C. muralis* was introduced into gardens before 1602, and records from the wild date from 1640 (Herts.). Comparison of the current map with that in the 1962 *Atlas* suggests that its distribution is now stable.

Native of the mountains of S.C. & S.E. Europe; widely naturalised through much of temperate and southern Europe.

References: Atlas (224c), Meusel *et al.* (1978), Salisbury (1964).

A. HORSFALL

Cymbalaria pallida Italian Toadflax

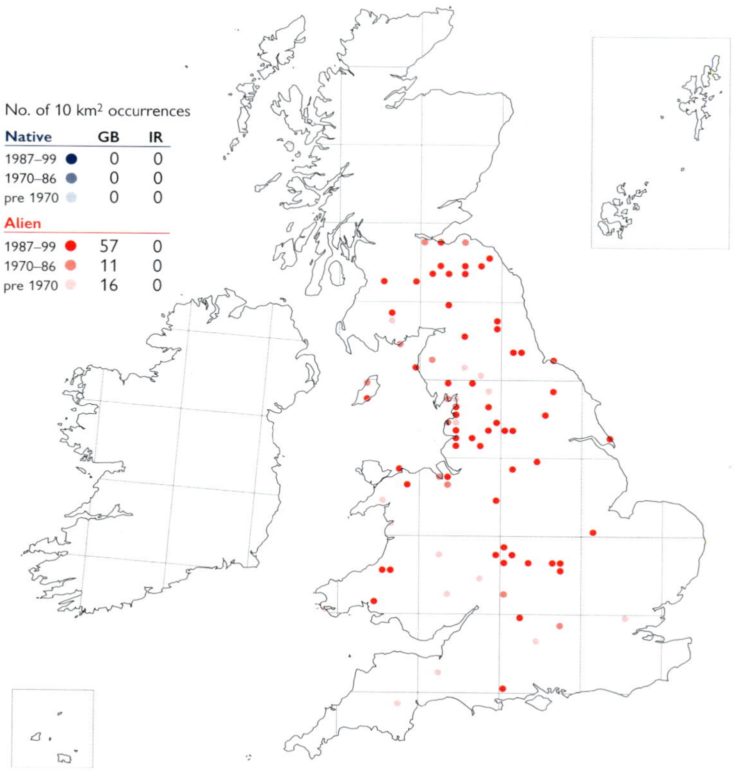

A perennial herb found as a garden escape on walls, pavements, shingle, rocky banks and waste ground. Populations are often naturalised, spreading readily by seed. Lowland.

Neophyte. This species was introduced into cultivation in Britain by 1882 and was known from the wild by 1924 (Westmorland). It was naturalised on a shingle beach at Bardsea (Westmorland) from 1952 to 1977. It is difficult to assess any changes in its distribution, especially as some populations may have been overlooked but others may have been recorded in error for pale-flowered variants of *C. muralis*.

Native of the mountains of C. Italy.

Reference: Meusel *et al.* (1978).

T. D. DINES

Kickxia elatine Sharp-leaved Fluellen

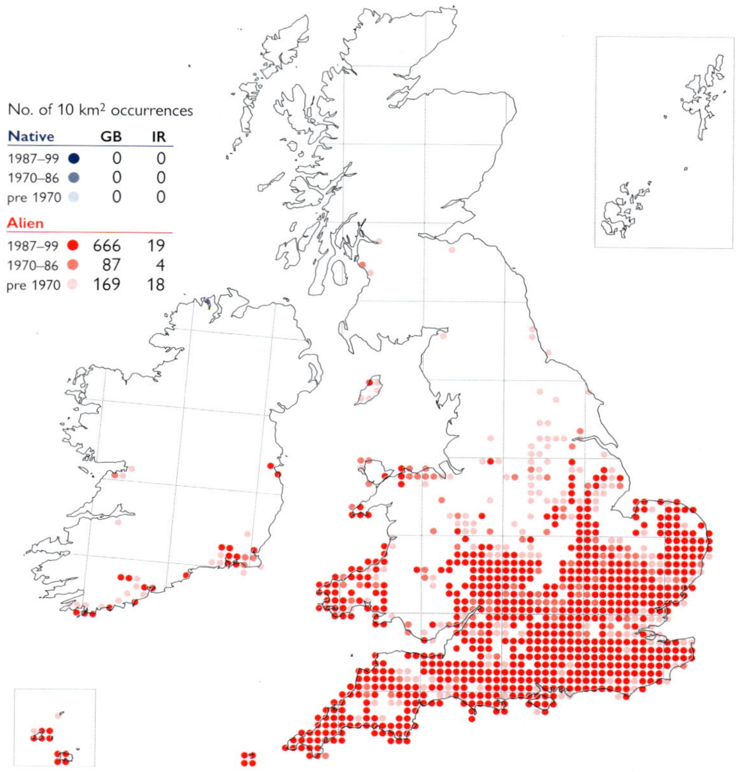

An annual of basic soils, including light soils over chalk and calcareous boulder-clay, found on the headlands and margins of arable fields (particularly cornfields), and less commonly on tracks, waste ground and in gardens. It is also found on sandy soils, and has been recorded on open peaty ground (Co. Cork). Lowland.

Archaeophyte (change –0.18). Although this species is often only present in small and scattered populations, its overall distribution shows only limited losses. Simpson (1982) suggests that some *Kickxia* seeds germinate in late summer and may therefore escape herbicides.

As an archaeophyte *K. elatine* has a European Southern-temperate distribution; it is widely naturalised outside this range.

References: Atlas (224b), Curtis & McGough (1988), Hultén & Fries (1986), Meusel *et al.* (1978).

A. HORSFALL

Kickxia spuria Round-leaved Fluellen

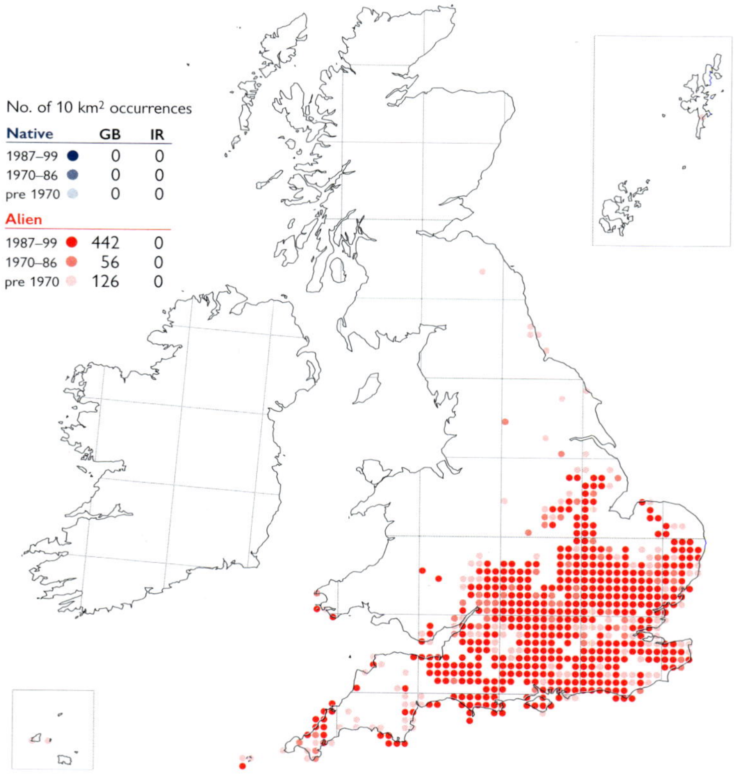

An annual of arable land and open waste ground, usually on calcareous soils, in similar habitats to *K. elatine* and often growing with it. Lowland.

Archaeophyte (change –0.07). Like *K. elatine*, this species has a similar range to that mapped in the 1962 *Atlas*. Although both species are often thought to be declining, they have never been abundant weeds and the distribution may be more stable than is often supposed.

As an archaeophyte *K. spuria* has a European Southern-temperate distribution.

References: Atlas (224a), Meusel *et al.* (1978).

A. HORSFALL

Linaria vulgaris Common Toadflax

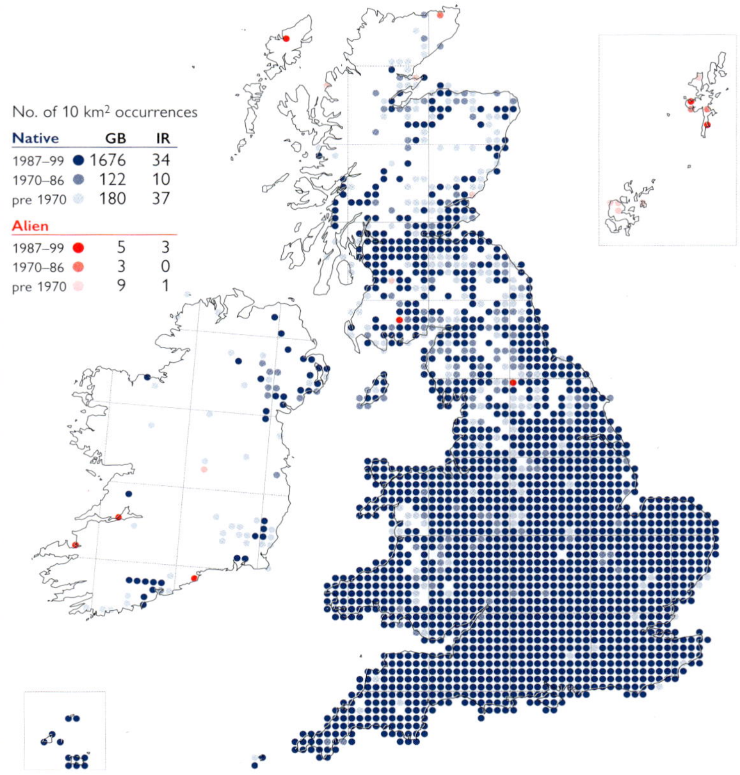

This perennial herb is found in open grassy places, on stony and waste ground, hedge banks, road verges, railway banks and cultivated land, especially on calcareous soils. It reproduces by seed and also spreads by creeping rhizomes. 0–360 m near Alston (Cumberland).

Native (change –0.80). The overall range of *L. vulgaris* is stable, although there have been local losses in the north of its British range and in some areas of Ireland. It was mapped as 'all records' in the 1962 *Atlas*.

Eurasian Boreo-temperate element, but naturalised in N. America so distribution is now Circumpolar Boreo-temperate.

References: Atlas (223c), Grime *et al.* (1988), Hultén & Fries (1986), Meusel *et al.* (1978), Saber *et al.* (1995).

A. HORSFALL

Linaria purpurea Purple Toadflax

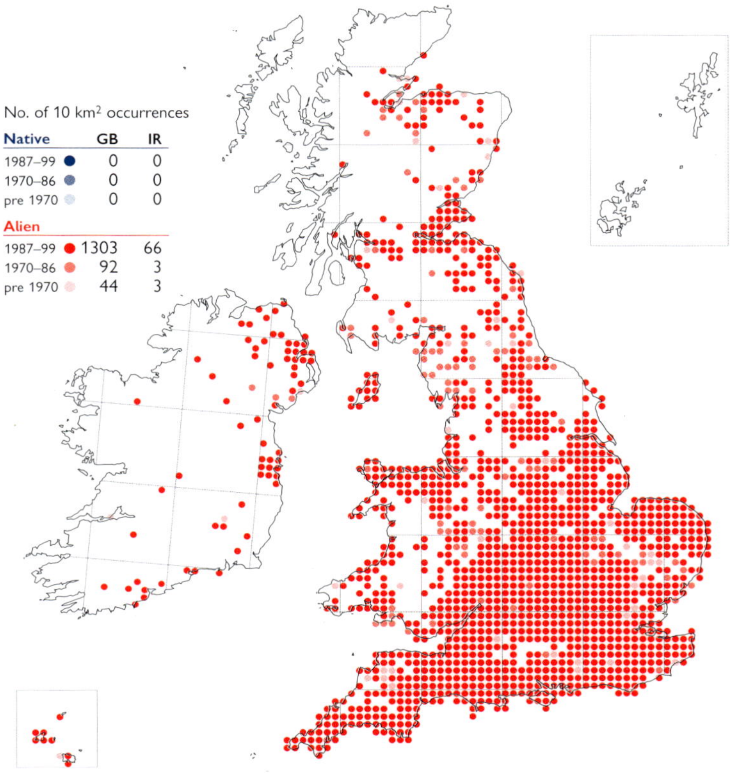

A perennial herb, occurring as a garden escape or outcast on waste ground, roadsides and banks, along railways, on pavements, walls and rubbish tips, and in quarries. Lowland.

Neophyte (change +3.66). *L. purpurea* was introduced into cultivation by 1648 and is a popular garden plant with several named cultivars. It was known from the wild by *c.* 1830 (Middlesex), and its distribution has greatly increased since the 1962 *Atlas*, probably due to both better recording and a genuine spread.

Native of C. & S. Italy and Sicily.

Reference: Atlas (223a).

A. HORSFALL

Linaria repens Pale Toadflax

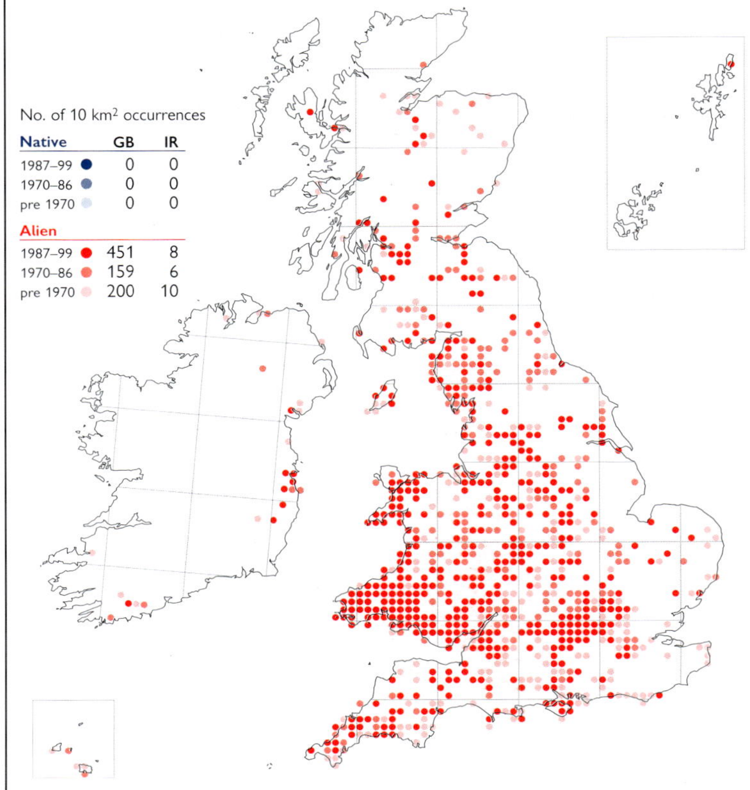

This rhizomatous perennial herb is found on rough and waste ground, stony and cultivated land, grassy banks and along railway tracks, usually on dry, calcareous or base-rich soils. 0–335 m (near Brassington, Derbys.).

Archaeophyte (change +0.30). There are many more records now of *L. repens* than in the 1962 *Atlas*, but the overall range is unchanged. It has gone from some 10-km squares, perhaps reflecting the tendency for it to behave as a transient colonist of open habitats.

As an archaeophyte *L. repens* has a Suboceanic Temperate distribution; it is widely naturalised outside this range.

References: Atlas (223b), Hultén & Fries (1986), Meusel *et al.* (1978).

A. HORSFALL

Linaria repens × *L. vulgaris* (*L.* × *sepium*)

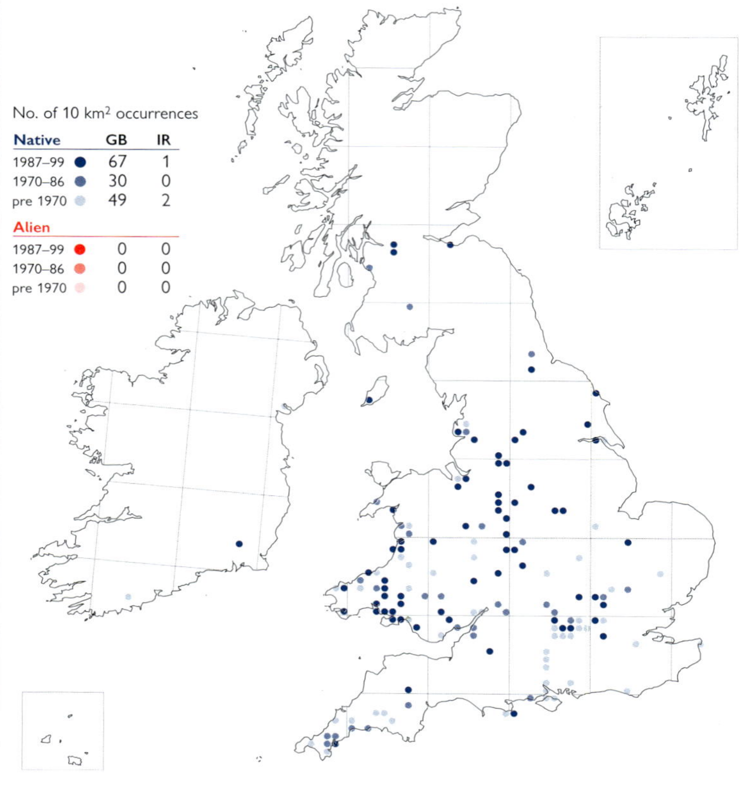

This perennial hybrid typically occurs on dry, calcareous soils on banks, walls and railway cuttings, and on rough or stony waste ground. It often occurs in the absence of one or both parents; it is fertile and backcrosses readily to form swarms of intermediates. Lowland.

A spontaneous hybrid between native and alien parents. *L.* × *sepium* is better recorded than when mapped by Perring & Sell (1968), but is probably still frequently overlooked.

Widespread in W. & C. Europe.

References: Atlas Supp. (50b), Stace (1975).

A. HORSFALL

Linaria supina Prostrate Toadflax

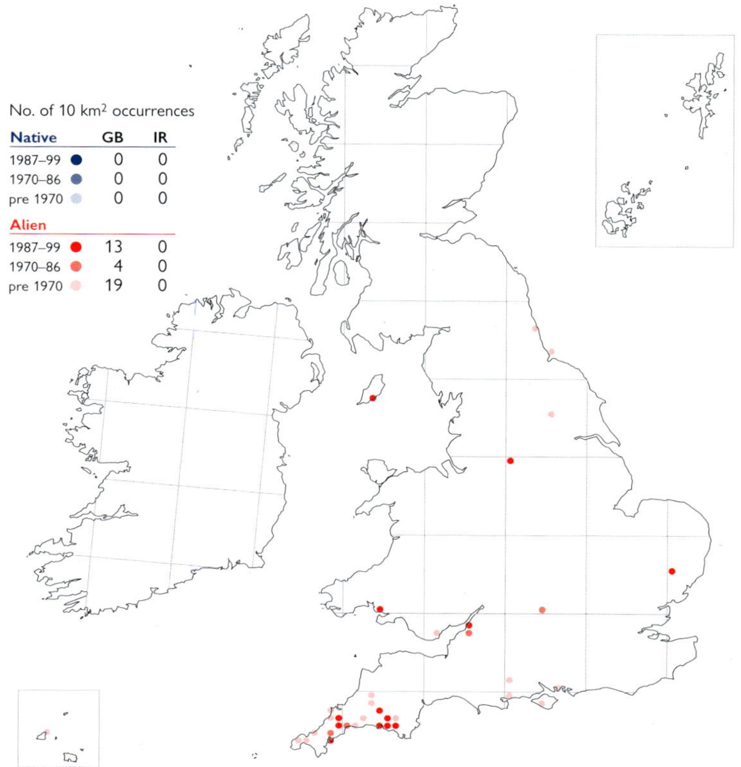

An annual found on sandy sea shores, waste places and walls near the sea, and on roadside banks and beside railways. It is naturalised in some parts of S.W. Britain, but only occurs as a rare casual elsewhere. Lowland.

Neophyte (change +0.13). This species has been cultivated in Britain since at least 1728. It was first recorded in the wild in 1848 at Par (E. Cornwall) and Hayle (W. Cornwall), where it was formerly considered to be native. There are no published records of it at Hayle after 1909, and it was last seen at Par in 1987. It is now a rare casual.

Native of S.W. Europe, extending east to N. Italy.

References: Atlas (222d), Bolòs & Vigo (1995), FitzGerald (1990), French *et al.* (1999).

T. D. DINES

Linaria pelisseriana Jersey Toadflax

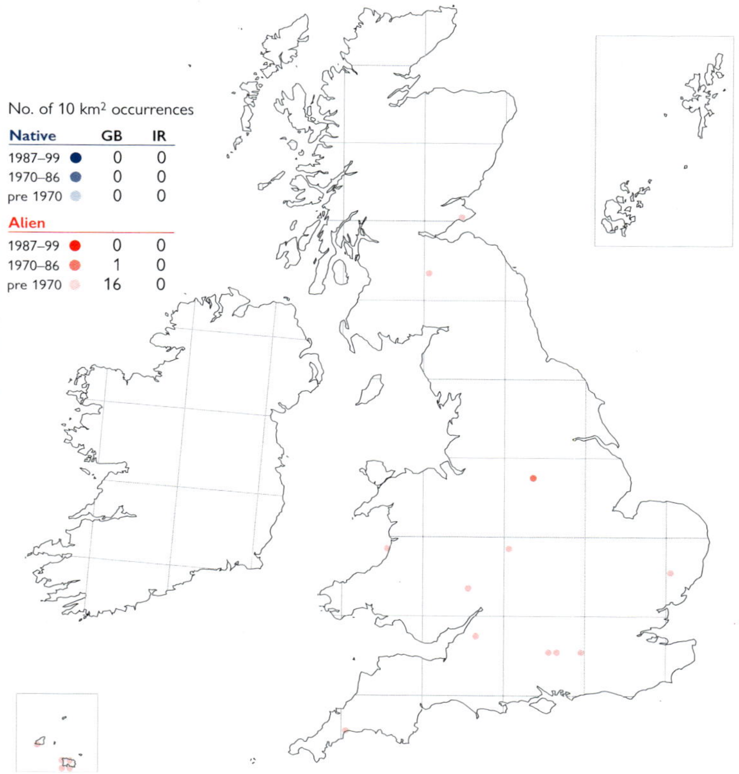

An annual of hedge banks, rough ground and rocky places. Lowland.

Neophyte. *L. pelisseriana* was first cultivated in Britain in 1640. It was recorded from the wild in Jersey in 1837, where it was seen sporadically at scattered localities until 1955. It is often regarded as a native to the island, but its history suggests that it is more likely to be an alien. It is only casual elsewhere.

A Mediterranean-Atlantic species.

References: Atlas (222d), Bolòs & Vigo (1995).

A. HORSFALL

Linaria maroccana Annual Toadflax

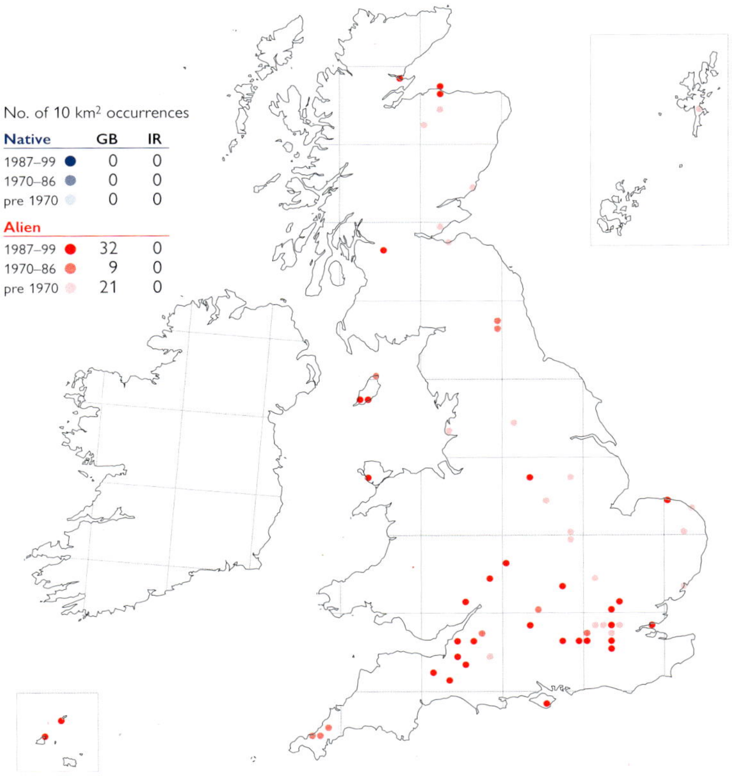

No. of 10 km² occurrences

Native	GB	IR
1987–99	0	0
1970–86	0	0
pre 1970	0	0
Alien		
1987–99	32	0
1970–86	9	0
pre 1970	21	0

An annual occurring as a casual garden escape on rubbish tips, waysides and in waste places, usually near habitation. It has also been recorded on sea-cliffs. The seeds germinate readily. Lowland.

Casual. *L. maroccana* was introduced into cultivation in Britain in 1872 and was recorded from the wild by 1928. Its overall distribution is probably stable.

Native of N. Africa (Morocco).

A. HORSFALL

Digitalis purpurea Foxglove

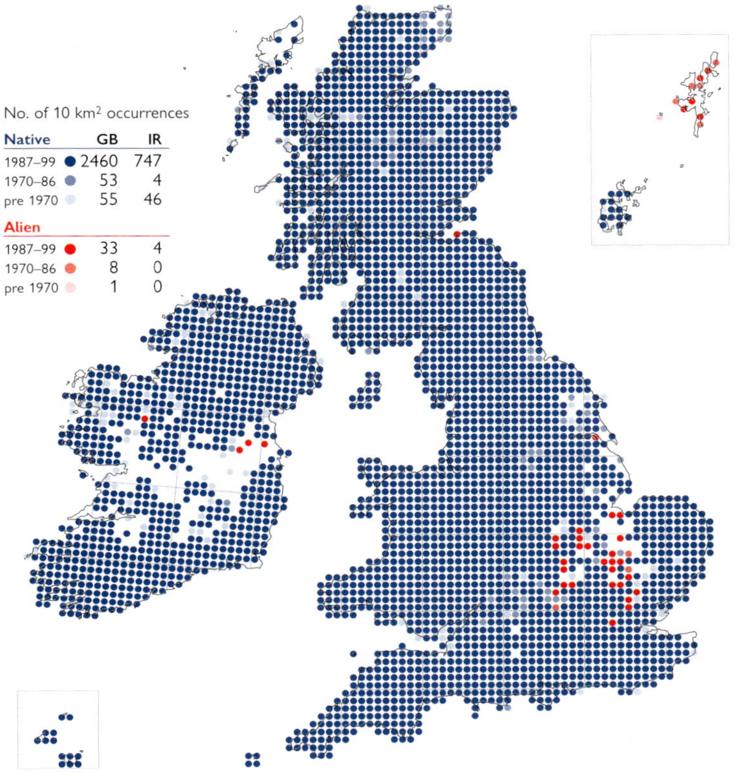

No. of 10 km² occurrences

Native	GB	IR
1987–99	2460	747
1970–86	53	4
pre 1970	55	46
Alien		
1987–99	33	4
1970–86	8	0
pre 1970	1	0

A biennial or short-lived perennial, common on acidic soils in hedge banks, open woods and woodland clearings, on heath and moorland margins, river banks, montane rocky slopes, sea-cliffs, walls and waste land. It is often found in great abundance in disturbed or burnt areas, such as recently felled forestry plantations. 0–880 m (Mt Brandon, S. Kerry).

Native (change +0.72). The distribution of *D. purpurea* is stable, a major factor in its success being its prolific seed production and persistent seed bank. It is a common garden plant and can occur as an escape outside its native range and also on calcareous soils.

Suboceanic Southern-temperate element; widely naturalised outside its native range.

References: Atlas (227d), Grime *et al.* (1988), Hultén & Fries (1986), Meusel *et al.* (1978).

A. HORSFALL

Erinus alpinus Fairy Foxglove

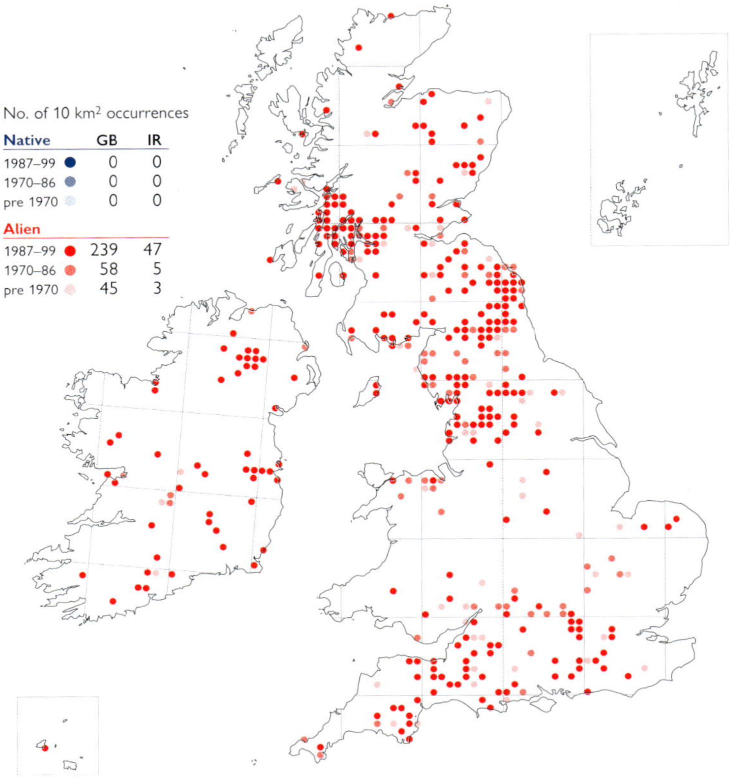

No. of 10 km² occurrences

Native	GB	IR
1987–99	0	0
1970–86	0	0
pre 1970	0	0
Alien		
1987–99	239	47
1970–86	58	5
pre 1970	45	3

This short-lived, semi-evergreen perennial herb occurs in the crevices of old walls and in other stony places, often on limestone or bricks with lime mortar. It seeds freely and thrives in full sun. Generally lowland, but reaching 350 m at Alston (Cumberland).

Neophyte (change +1.52). *E. alpinus* is a popular garden plant, cultivated by 1739, which seeds prolifically and becomes well-naturalised in new sites. It was known from the wild by 1867. There seems to have been a considerable spread since the 1962 *Atlas*, perhaps in part a result of the greater propensity of botanists to record plants on garden walls.

Native of the mountains of S.W. and S.C. Europe.

References: Atlas (227c), Meusel *et al.* (1978).

A. HORSFALL

Veronica serpyllifolia Thyme-leaved Speedwell

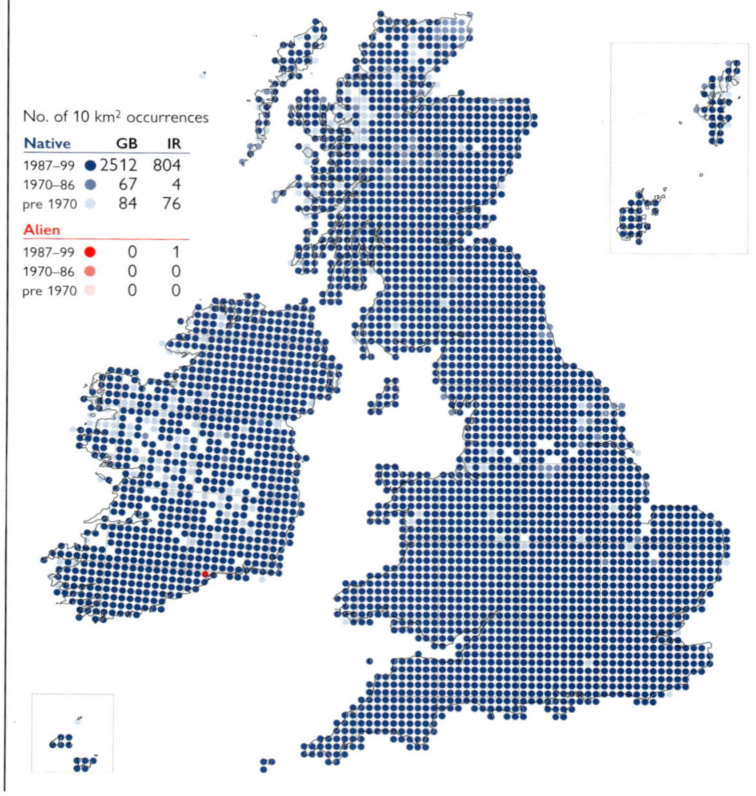

A low perennial herb with creeping and rooting stems. It is widespread in both natural and artificial habitats, including woodland rides, grassland, heaths, flushes, damp rock ledges, cultivated land, lawns, waste ground and damp paths. 0–1160 m (Braeriach, S. Aberdeen).

Native (change +0.80). The range of *V. serpyllifolia* is stable. Subsp. *serpyllifolia* occurs throughout the range of the species up to altitudes of 825 m (Cross Fell, Cumberland); subsp. *humifusa* is mapped separately.

Circumpolar Boreo-temperate element; widely naturalised outside its native range.

References: Atlas (230c), Hultén & Fries (1986), Meusel *et al.* (1978).

A. HORSFALL

Veronica serpyllifolia subsp. *humifusa*

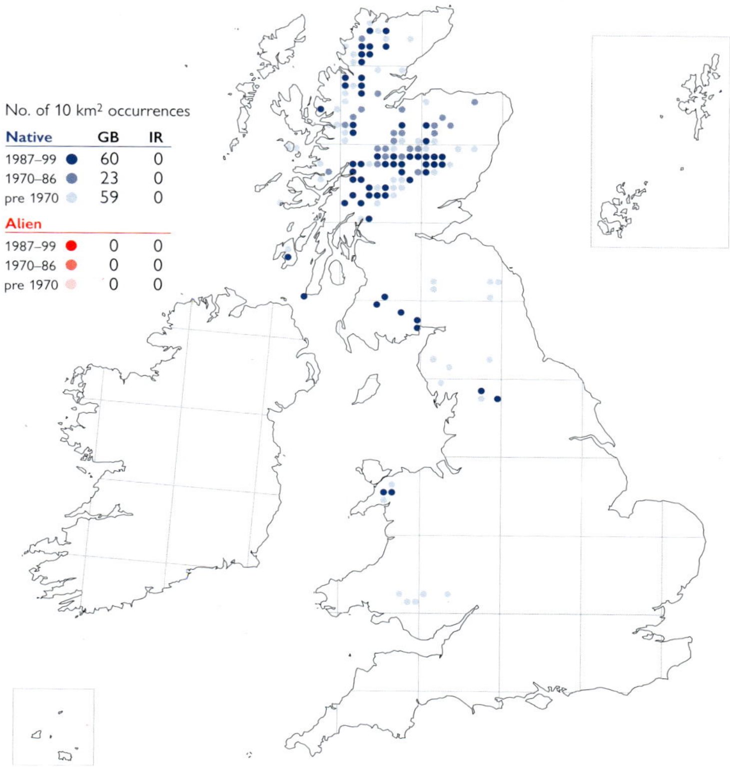

A creeping, perennial upland and montane herb of rock ledges, flushes and wet gravel, from 120 m (N. of Daviot, Easterness) to 1160 m on Braeriach (S. Aberdeen).

Native. It seems unlikely there have been any significant changes in the distribution of this subspecies since it was mapped by Perring & Sell (1968), though it is still probably under-recorded.

Circumpolar Arctic-montane element, with a disjunct distribution.

References: Atlas Supp. (51c), Hultén & Fries (1986), Meusel *et al.* (1978).

A. HORSFALL

Veronica alpina Alpine Speedwell

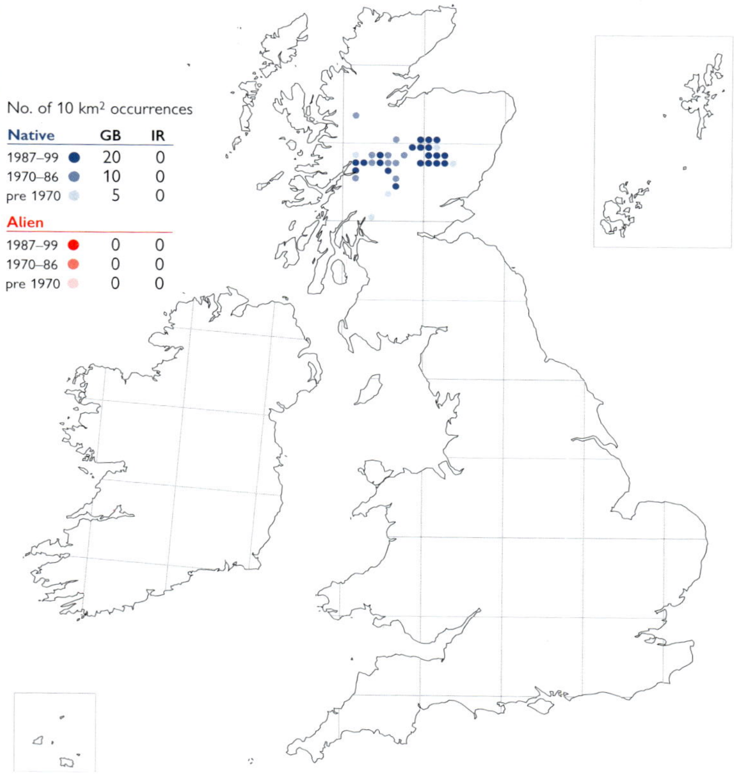

This small montane perennial herb typically occurs in areas of late snow-lie in open, often rocky, places on well-drained but slightly moist ground. It grows on both acidic and calcareous substrates, but most of its sites are subject to some base-enrichment from flushing. From 760 m above Loch Callater (S. Aberdeen) to 1190 m on Aonach Beag (Westerness).

Native (change –0.29). There seems to be little appreciable change in the distribution of *V. alpina* since the 1962 *Atlas*, and it is probably extant in most of the 10-km squares for which there are only pre-1987 records.

Eurosiberian Arctic-montane element; also in C. Asia and N. America.

References: Atlas (230b), Hultén & Fries (1986), Meusel *et al.* (1978), Stewart *et al.* (1994).

M. J. WIGGINTON

Veronica fruticans Rock Speedwell

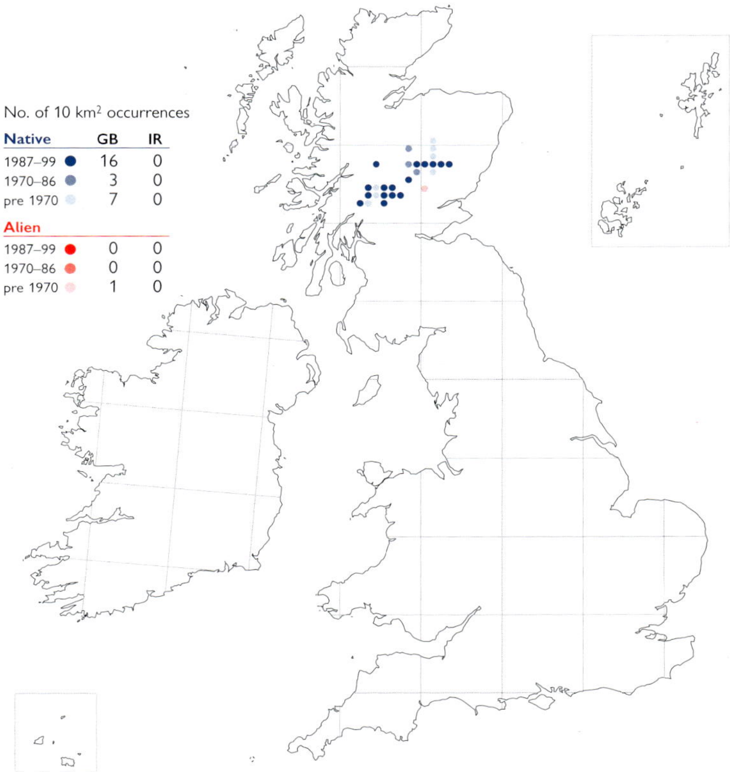

No. of 10 km² occurrences

Native		GB	IR
1987–99	●	16	0
1970–86	●	3	0
pre 1970	●	7	0
Alien			
1987–99	●	0	0
1970–86	●	0	0
pre 1970	●	1	0

A small, rather woody perennial, restricted to calcareous substrates and occurring on dry open slopes and rock ledges on crags, in sites which are usually S.-facing and inaccessible to grazing animals. Montane, from 540 m (Meal an Fhiodhain, Mid Perth) to 1100 m (Ben Lawers, Mid Perth).

Native (change +0.11). It is uncertain whether the distribution of *V. fruticans* is stable. Some populations are known to have been lost, for reasons which are uncertain, and the numbers of plants at extant sites are small. The additional 10-km square records since the 1962 *Atlas* can probably be ascribed to better recent recording.

European Arctic-montane element; also in Greenland.

References: Atlas (230a), Hultén & Fries (1986), Meusel *et al.* (1978), Wigginton (1999).

A. HORSFALL

Veronica officinalis Heath Speedwell

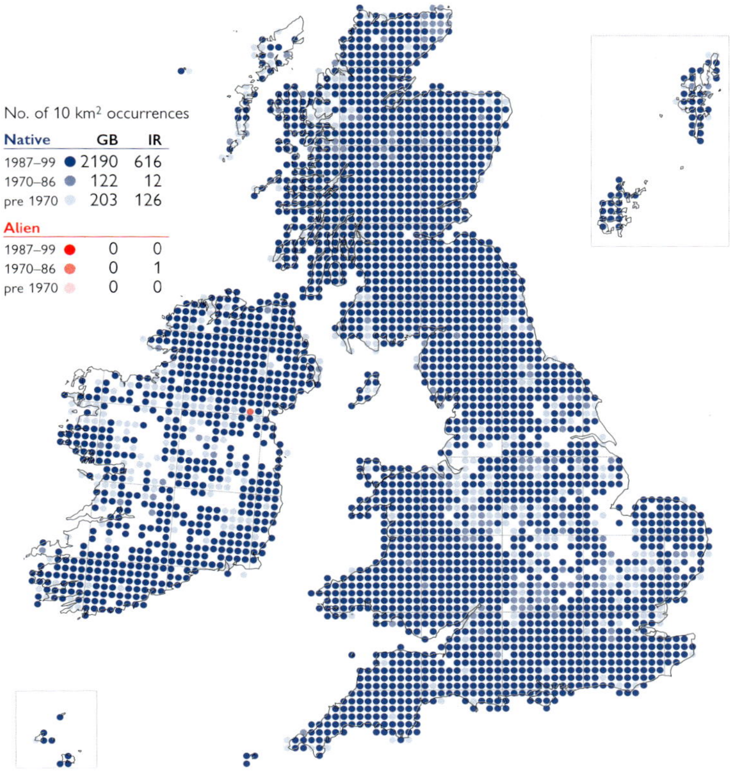

No. of 10 km² occurrences

Native		GB	IR
1987–99	●	2190	616
1970–86	●	122	12
pre 1970	●	203	126
Alien			
1987–99	●	0	0
1970–86	●	0	1
pre 1970	●	0	0

This perennial herb is found in open woods and woodland rides, on banks, in grassland and on heathland. It grows on well-drained, often moderately acidic or leached soils, and in some grasslands is confined to raised ground or anthills. 0–880 m (Cadair Idris, Merioneth).

Native (change –0.84). The broad distribution of *V. officinalis* is little changed, though there have clearly been local declines where its habitats have been lost, particularly in the English Midlands. It was mapped as 'all records' in the 1962 *Atlas*.

European Boreo-temperate element; widely naturalised outside its native range.

References: Atlas (229a), Hultén & Fries (1986), Meusel *et al.* (1978).

A. HORSFALL

Veronica chamaedrys Germander Speedwell

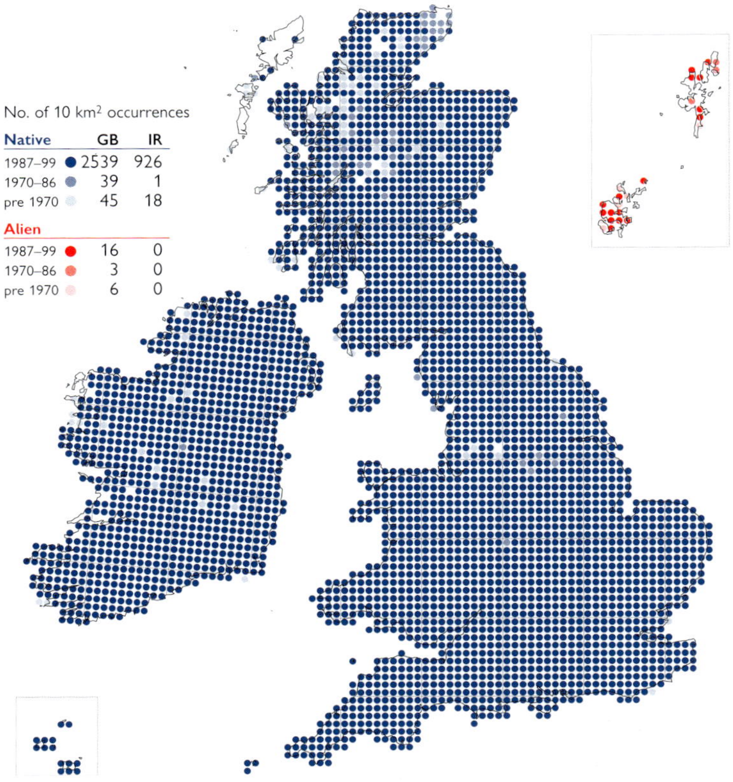

No. of 10 km² occurrences

Native		GB	IR
1987–99	●	2539	926
1970–86	●	39	1
pre 1970	●	45	18
Alien			
1987–99	●	16	0
1970–86	●	3	0
pre 1970	●	6	0

A stoloniferous perennial herb of woods, hedge banks, grassland, rock outcrops, upland screes, road verges, railway banks and waste ground, found on most soil types except the most impoverished. It also occurs on anthills on chalk downland. It spreads vegetatively by prostrate stems which root at the nodes; reproduction from seed appears to be comparatively rare. 0–750 m (Meal na Teanga, Loch Lochy, Westerness), with an unlocalised record of 820 m elsewhere in the Scottish Highlands.

Native (change –0.50). There has been no change in the distribution of *V. chamaedrys* since the 1962 *Atlas*.

Eurosiberian Boreo-temperate element; widely naturalised outside its native range.

References: Atlas (229c), Grime *et al.* (1988), Hultén & Fries (1986), Meusel *et al.* (1978).

A. HORSFALL

Veronica montana Wood Speedwell

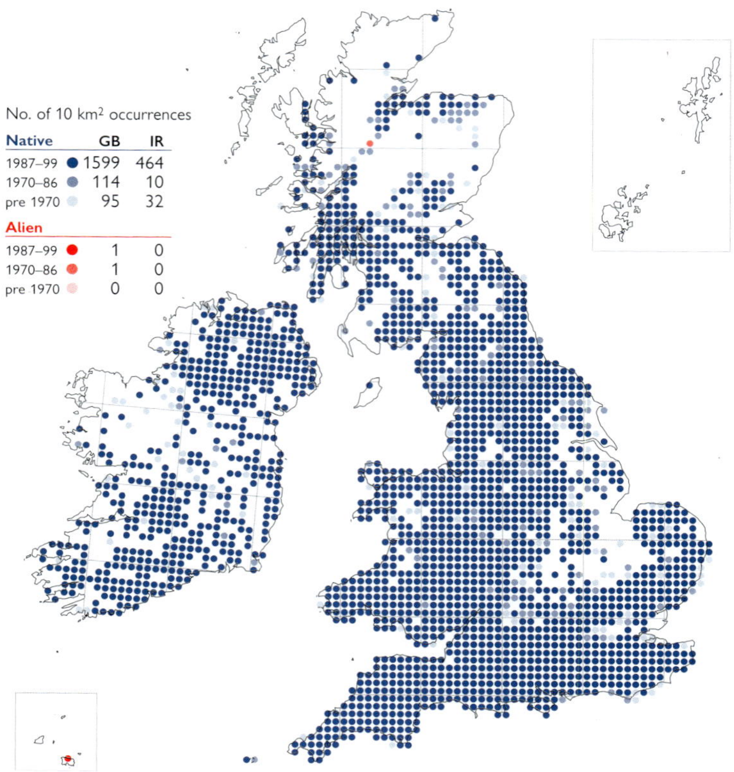

A perennial herb of damp basic to mildly acidic soils in long-established, mixed deciduous woodland, scrub and shaded hedge banks. It is found on loamy and sandy soils and on heavy clay. Generally lowland, but reaching 435 m at Pont y Daf (Brecs.).

Native (change +0.48). The distribution of *V. montana* is little changed since the 1962 *Atlas*, although it is now better recorded in W. Scotland.

European Temperate element.

References: Atlas (229b), Grime *et al.* (1988), Hultén & Fries (1986), Meusel *et al.* (1978).

A. HORSFALL

Veronica scutellata Marsh Speedwell

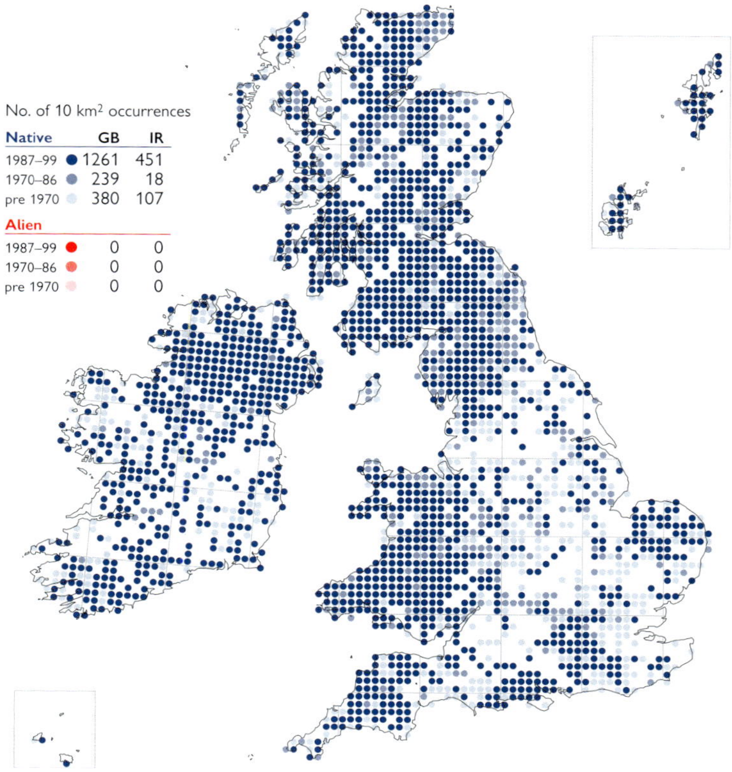

This perennial herb is found in a wide range of wetland habitats, including pond and lake margins, marshes, fens and fen-meadows, wet grassland, hillside flushes, bogs and wet heath, often on acidic soils. It occurs in both open habitats and amongst tall vegetation. 0–780 m on Cross Fell (Cumberland).

Native (change –0.06). *V. scutellata* is now much better recorded than in the 1962 *Atlas*, where it was mapped as 'all records'. The current map shows widespread declines, especially in C. & S. England and C. Ireland. Analysis of the database reveals that these losses have occurred since 1950.

Eurosiberian Boreo-temperate element; also in N. America and widely naturalised outside its native range.

References: Atlas (228d), Atlas Supp. (51a), Hultén & Fries (1986), Meusel *et al.* (1978).

A. HORSFALL

Veronica beccabunga Brooklime

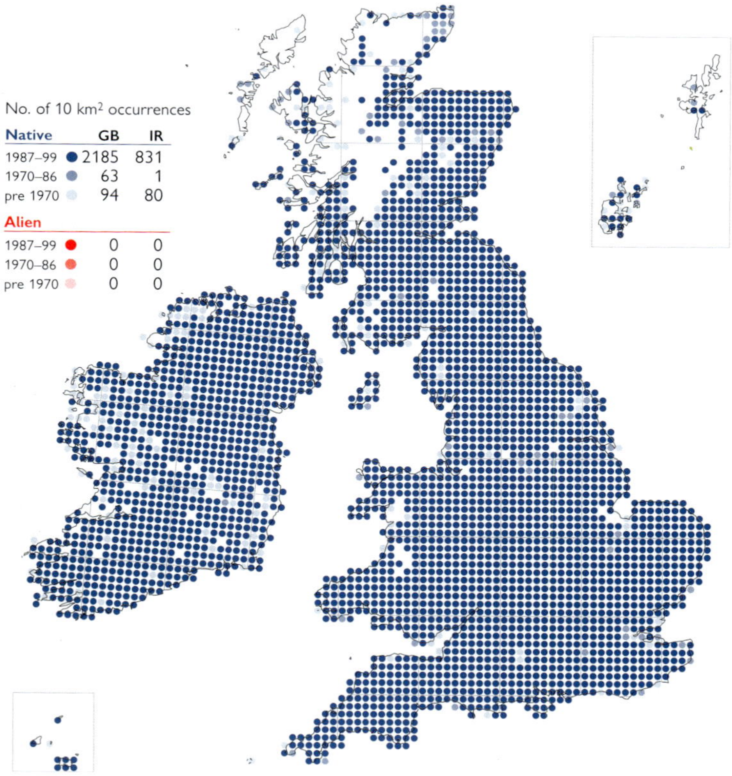

This robust perennial herb occurs on all but the most infertile substrates in a wide range of wetland habitats: in shallow water, by rivers, streams and ponds, in ditches, marshy hollows in pastures, flushes, wet woodland rides and rutted tracks. It thrives in fairly open habitats, competing poorly in dense stands of taller plants. Propagation is by seed and vegetatively from rooted stems. 0–845 m (Great Dun Fell, Westmorland).

Native (change –0.31). There is no evidence of any change in the distribution of *V. beccabunga* since the 1962 *Atlas*.

Eurosiberian Temperate element; widely naturalised outside its native range.

References: Atlas (228a), Grime *et al.* (1988), Hultén & Fries (1986), Meusel *et al.* (1978), Preston & Croft (1997).

A. HORSFALL

Veronica anagallis-aquatica
Blue Water-speedwell

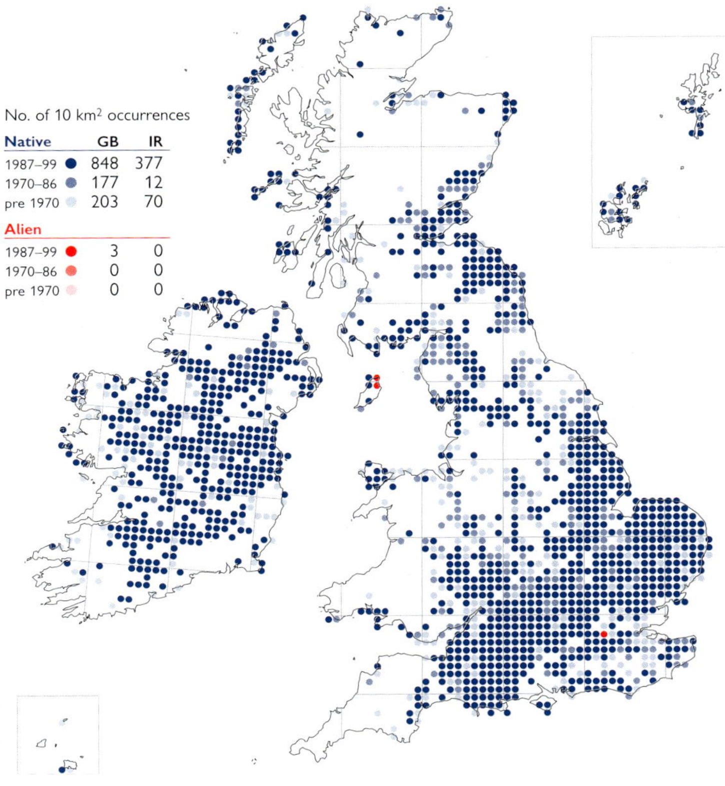

An annual found on fertile substrates by rivers, streams and ponds, in ditches and in flooded clay- and gravel-pits. It grows as a vegetative plant submerged in shallow water, or as a flowering emergent, or as a terrestrial plant in marshy habitats and disturbed ground at the water's edge. Reproduction is by seed and by rooted stem fragments. 0–380 m (Malham Tarn, Mid-W. Yorks.).

Native (change +0.05). This species was mapped as 'all records' in the 1962 *Atlas*. Most of the losses have occurred since 1950 but the distribution is probably stable now, though the species may be over-recorded for its hybrid with *V. catenata* in some areas.

Eurasian Southern-temperate element; widely naturalised outside its native range.

References: Atlas (228b), Burnett (1997), Hultén & Fries (1986), Meusel *et al.* (1978), Preston & Croft (1997).

A. HORSFALL

Veronica anagallis-aquatica × *V. catenata* (*V.* × *lackschewitzii*)

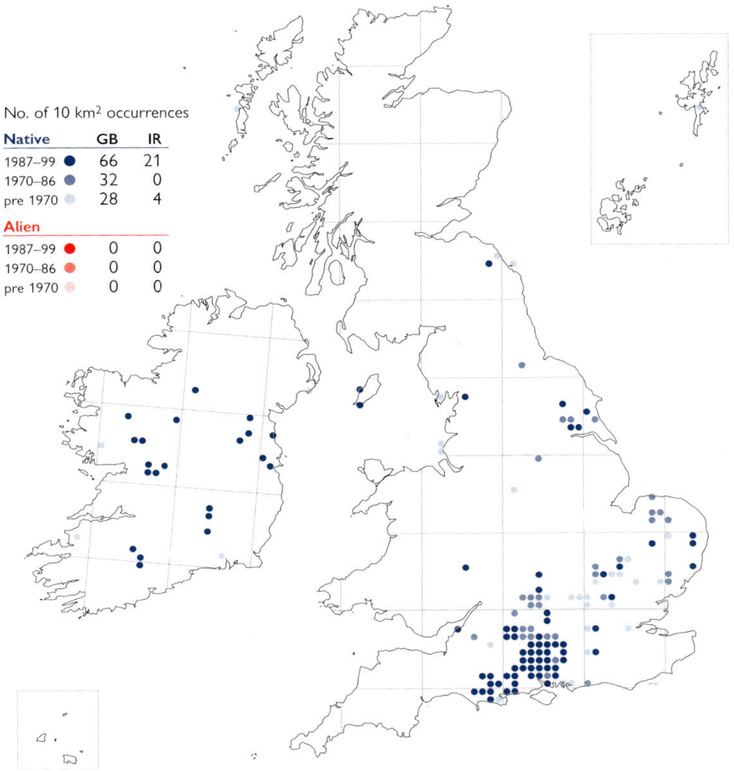

This annual or sometimes perennial hybrid is found in similar habitats to its parents, including the edges of ponds, lakes, ditches, streams and rivers. It can often be found with them, but may occur in the absence of one or both, and in Hampshire replaces the parents over a large area (Brewis *et al.*, 1996). The F_1 hybrid is usually highly sterile, but F_2 and later generations can be more fertile. It may perennate, but its life cycle needs further study. Lowland.

Native. Records of this hybrid have gradually accumulated since it was first described in 1929 but it is almost certainly under-recorded, especially in Ireland.

Widespread in temperate Europe, and also recorded from N. America where *V. anagallis-aquatica* is an introduction.

References: Burnett (1997), Preston & Croft (1997), Stace (1975).

A. HORSFALL

Veronica catenata Pink Water-speedwell

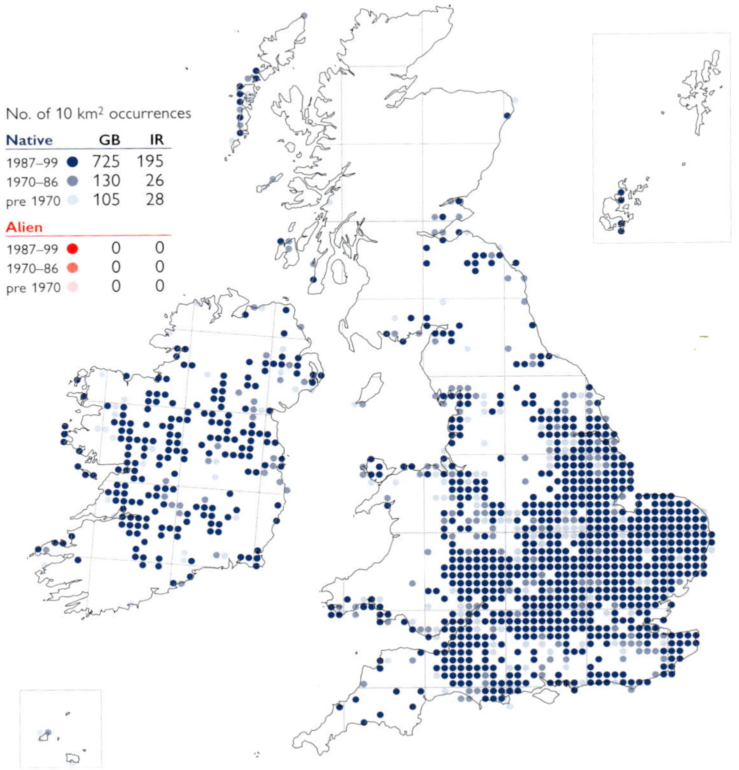

This usually annual species is found in shallow water and on the muddy edges of rivers, streams, ponds and lakes, in dune-slacks, and in clay-, gravel- and chalk-pits. Although often found with *V. anagallis-aquatica*, its habitats are more restricted, being more frequent on the muddy edges of standing waters. Lowland.

Native (change +0.37). The distribution of *V. catenata* has changed little since the 1962 *Atlas*, but it is now much better recorded.

Circumpolar Temperate element, with a disjunct distribution.

References: Atlas (228c), Burnett (1997), Hultén & Fries (1986), Meusel *et al.* (1978), Preston & Croft (1997).

A. HORSFALL

Veronica praecox Breckland Speedwell

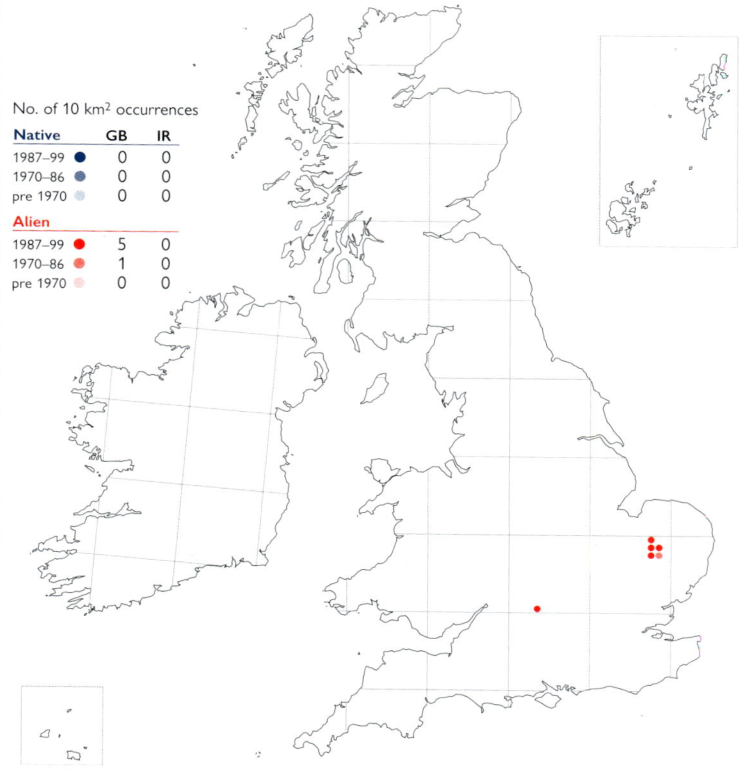

An annual found naturalised on free-draining sandy soils, usually where there is regular disturbance. Habitats include the edges of arable fields, on tracks, sandy banks, and open rough grassland. Lowland.

Neophyte. This species was grown in Britain in 1775. It was not recorded from the wild until 1933, and has only ever been known at a few sites. Numbers in individual populations can fluctuate greatly, and it is vulnerable to herbicides. It is less common in Breckland than at the time of the 1962 *Atlas*, although it has been successfully introduced to new sites.

A European Southern-temperate species, absent as a native from much of W. Europe.

References: Atlas (230a), Hultén & Fries (1986), Meusel *et al.* (1978), Trist (1979).

M. J. WIGGINTON

Veronica triphyllos Fingered Speedwell

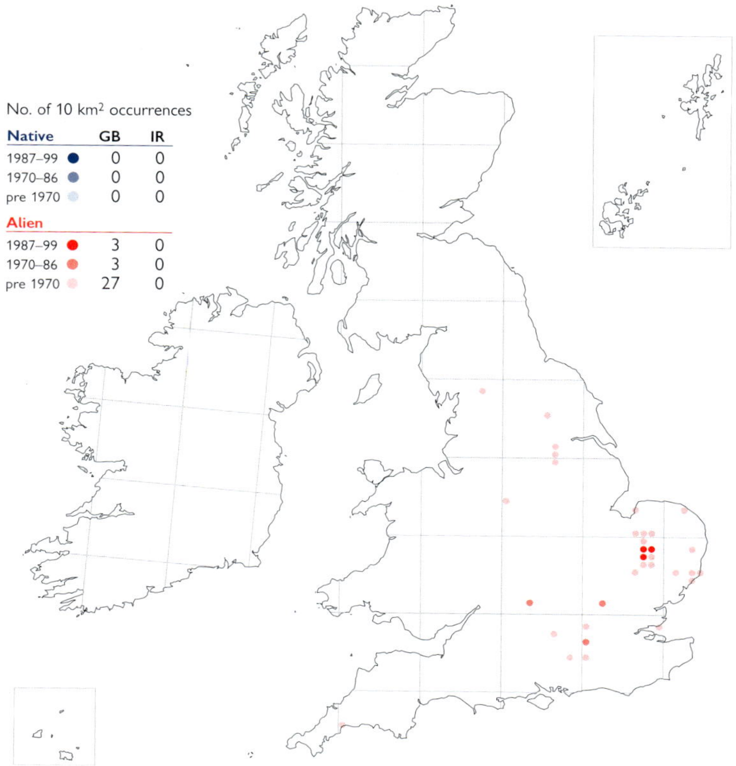

Recently, this annual of sandy calcareous or slightly acidic soils has been found on the margins of arable fields and on sandy banks, but it was formerly also known from tracks, fallow fields, gravel-pits and waste ground. Regular disturbance is needed to maintain sufficient open ground for germination. Lowland.

Archaeophyte (change –0.82). *V. triphyllos* was first recorded in Britain in 1670, and it has been long-established in Breckland. Its main decline took place before 1930, and since the 1962 *Atlas* it has been confined to very few sites. Conservation management is ensuring its survival at remaining sites.

As an archaeophyte *V. triphyllos* has a European Temperate distribution, with a continental distribution in W. Europe.

References: Atlas (231c), Hultén & Fries (1986), Meusel *et al.* (1978), Wigginton (1999).

A. HORSFALL

Veronica arvensis Wall Speedwell

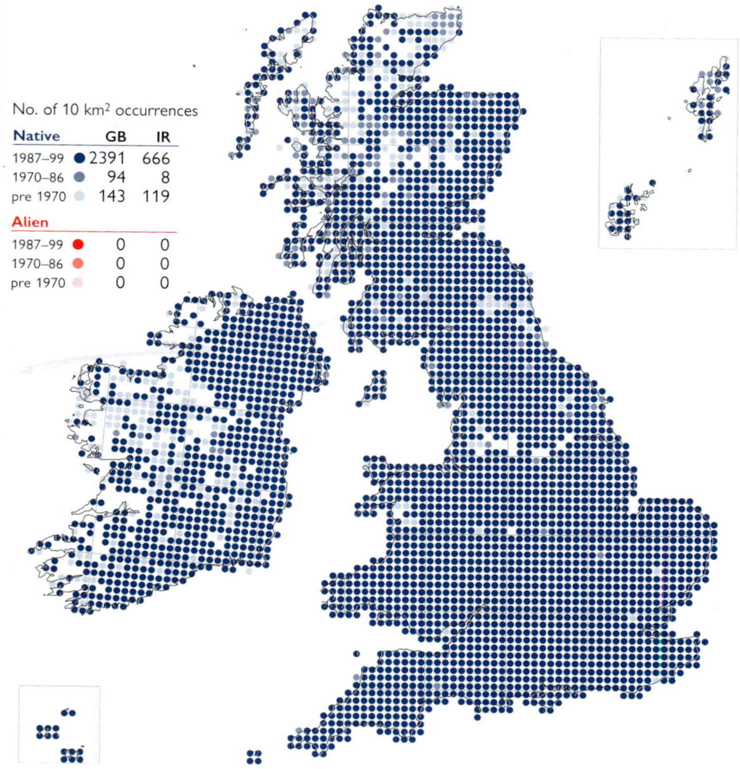

An annual of cultivated land, open grassland, heaths, sand dunes, gravelled paths and tracks, waste ground, banks, walls and pavements, usually on dry soils. In closed grassland it may be restricted to anthills. Seed remains viable in the soil for several years. Generally lowland, but reaching at least 820 m in the Breadalbanes (Mid Perth) and exceptionally at 845 m on Great Dun Fell (Westmorland).

Native (change +0.48). There is no evidence of any significant change in the distribution of *V. arvensis* since the 1962 *Atlas*, other than in W. Ireland where it may have been under-recorded in recent years.

European Southern-temperate element, but widely naturalised so that distribution is now Circumpolar Southern-temperate.

References: Atlas (231a), Grime *et al.* (1988), Hultén & Fries (1986), Meusel *et al.* (1978).

A. HORSFALL

Veronica verna Spring Speedwell

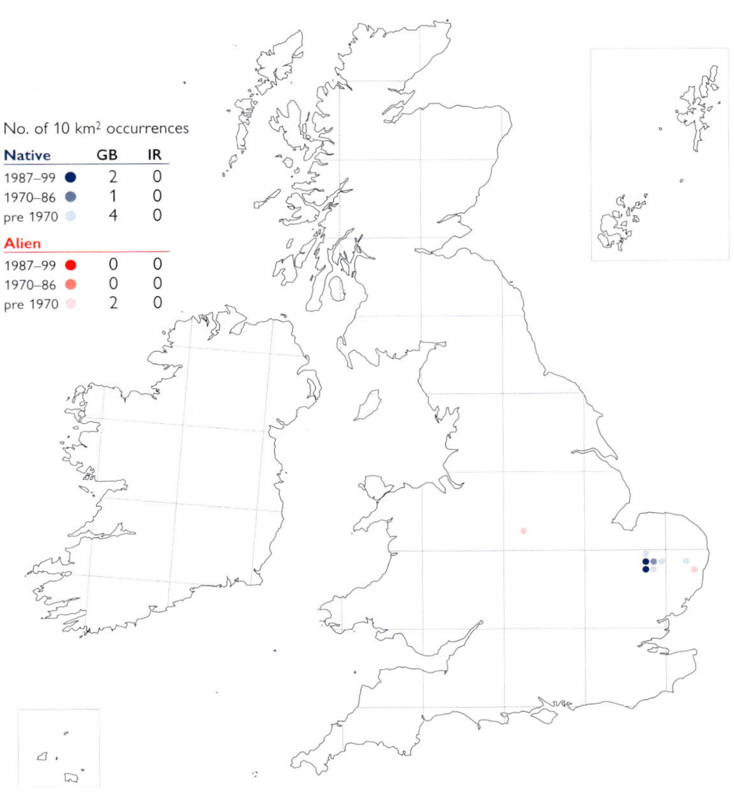

An annual of infertile sandy soils, occurring in short grassland and uncultivated, sometimes stony, places including rabbit warrens. *V. verna* does not occur on cultivated land, but depends on intensive grazing by sheep or rabbits to keep its habitat open. Lowland.

Native (change –0.64). Unlike *V. praecox* and *V. triphyllos*, *V. verna* is a plant of semi-natural habitats in Breckland and is therefore considered native. It has declined considerably, and is now found in only twelve sites in two 10-km squares. It has been introduced to two reserves, where it is flourishing.

Eurosiberian Temperate element, with a continental distribution in W. Europe.

References: Atlas (231b), Hultén & Fries (1986), Meusel *et al.* (1978), Wigginton (1999).

A. HORSFALL

Veronica peregrina American Speedwell

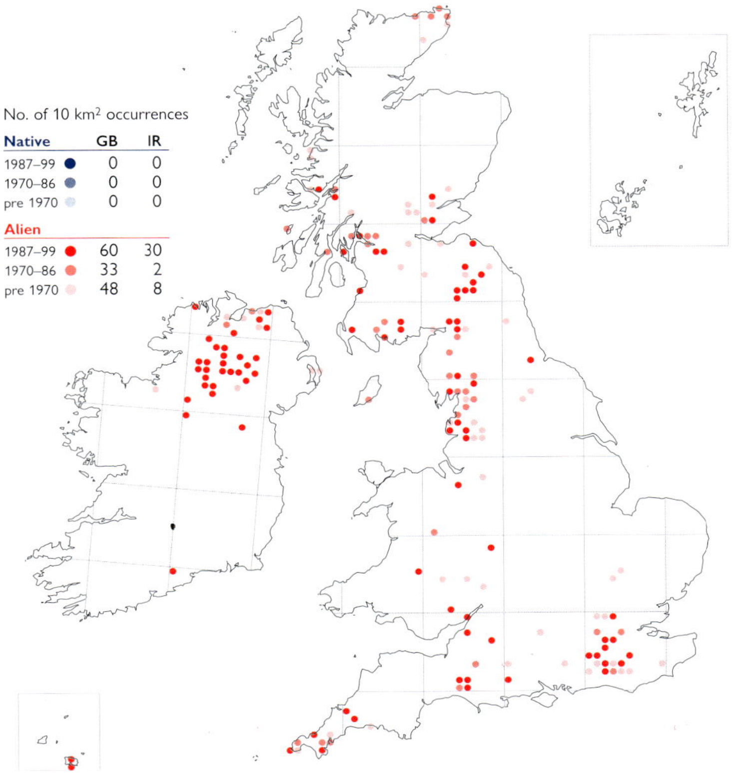

An annual occurring as a naturalised or casual weed of parks, gardens, garden centres, allotments and other cultivated ground, and in damp waste places and by streams. Lowland.

Neophyte (change +0.15). *V. peregrina* was introduced by 1680, apparently via mainland Europe. It was first recorded in the wild in 1836, and has shown a slow spread since the 1962 *Atlas*. It has increased markedly in N. Ireland since 1970.

Native of N. & S. America.

References: Atlas (230d), Bangerter (1964, 1966), Hultén & Fries (1986), Meusel *et al.* (1978).

A. HORSFALL

Veronica agrestis Green Field-speedwell

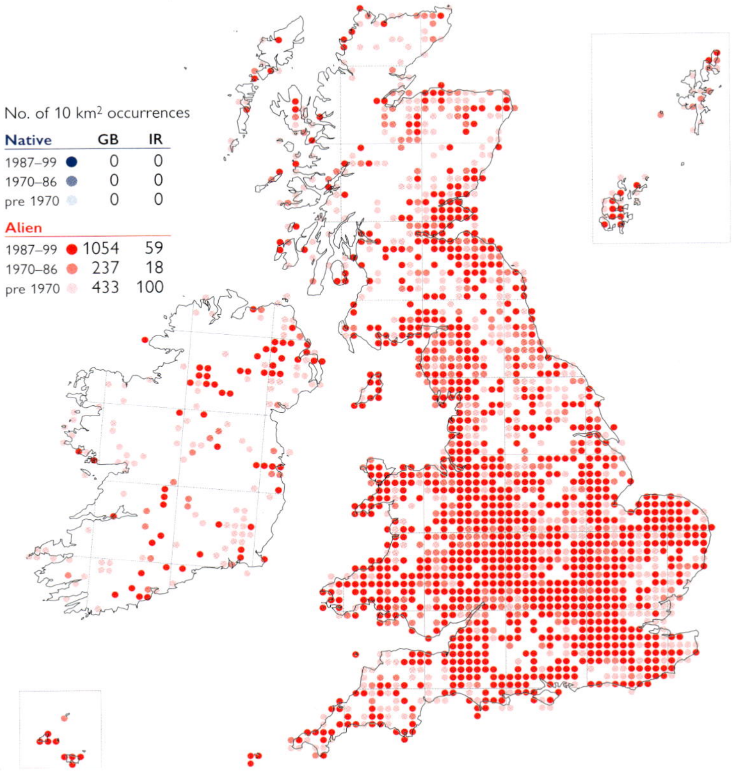

This spring-germinating annual is a colonist of cultivated land, waysides, gardens and allotments. It prefers soils which are well-drained and acidic, occurring on calcareous substrates only when there is surface leaching. 0–410 m on Malham Moor (Mid-W. Yorks.) and 455 m in E. Perth.

Archaeophyte (change –0.38). *V. agrestis* was mapped as 'all records' in the 1962 *Atlas*, but even then was considered to be diminishing (Salisbury, 1961). It has continued to decrease, mainly due to changing agricultural practices, and is no longer a familiar cornfield weed.

As an archaeophyte *V. agrestis* has a European Temperate distribution; it is widely naturalised outside this range.

References: Atlas (232c), Hultén & Fries (1986), Meusel *et al.* (1978).

A. HORSFALL

Veronica polita Grey Field-speedwell

Native	GB	IR
1987–99 ●	0	0
1970–86 ●	0	0
pre 1970 ●	0	0
Alien		
1987–99 ●	793	39
1970–86 ●	157	5
pre 1970 ●	295	63

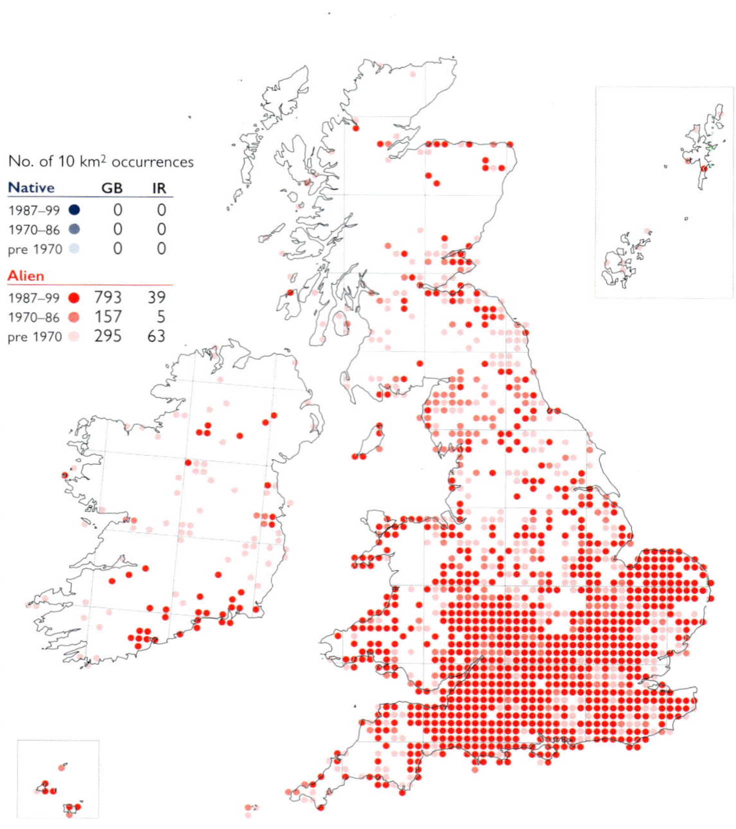

An annual of cultivated fields and gardens, typically growing on light, sandy, often calcareous soils. Generally lowland, but reaching 350 m near Kaber (Westmorland).

Neophyte (change +0.07). *V. polita* was first recorded from the wild in Britain in 1777. Its distribution now seems stable in S. & E. England and Wales, but appears to be declining in N. England, Scotland and Ireland. It was mapped as 'all records' in the 1962 *Atlas*.

A Eurosiberian Southern-temperate species.

References: Atlas (232b), Hultén & Fries (1986).

A. HORSFALL

Veronica persica Common Field-speedwell

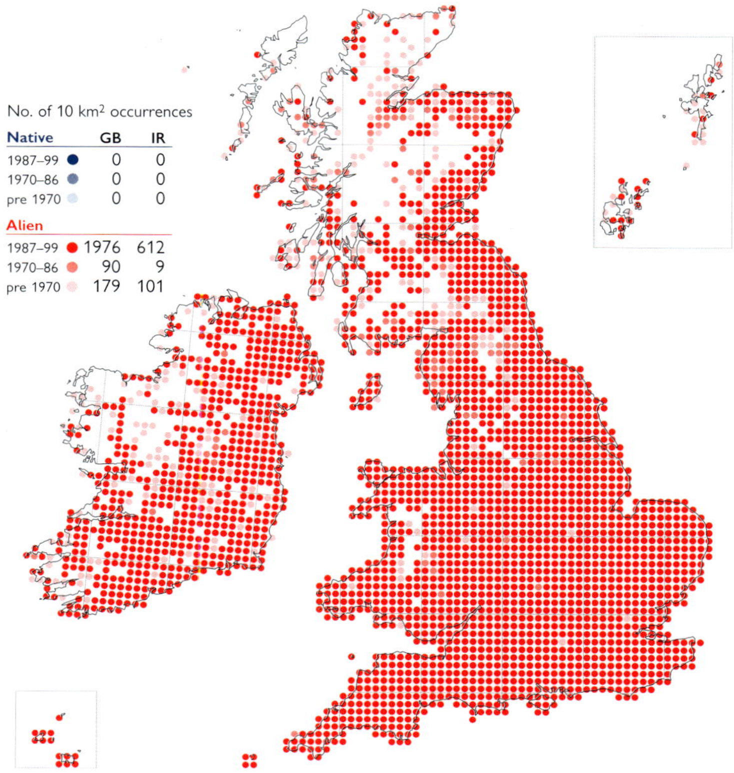

Native	GB	IR
1987–99 ●	0	0
1970–86 ●	0	0
pre 1970 ●	0	0
Alien		
1987–99 ●	1976	612
1970–86 ●	90	9
pre 1970 ●	179	101

An annual of arable fields, other cultivated areas and waste ground, found on a wide range of fertile soils. It is self-fertile and seeds prolifically, the seeds forming a persistent seed bank and germinating throughout the year. It also spreads vegetatively from stem fragments. 0–350 m (Alston, Cumberland).

Neophyte (change –0.37). This species was first recorded in the wild in 1826 and rapidly extended its range during the 19th century. It remains frequent even in arable fields.

Probably native to mountains of the Caucasus & N. Iran, where it may have originated by the hybridisation of *V. ceratocarpa* and *V. polita*. Widespread as a weed in Europe, N. Africa, N. America, Japan and New Zealand.

References: Atlas (232a), Grime *et al.* (1988), Fischer (1987), Hultén & Fries (1986), Meusel *et al.* (1978).

A. HORSFALL

Veronica filiformis Slender Speedwell

Native	GB	IR
1987–99 ●	0	0
1970–86 ●	0	0
pre 1970 ●	0	0
Alien		
1987–99 ●	1832	352
1970–86 ●	112	9
pre 1970 ●	78	36

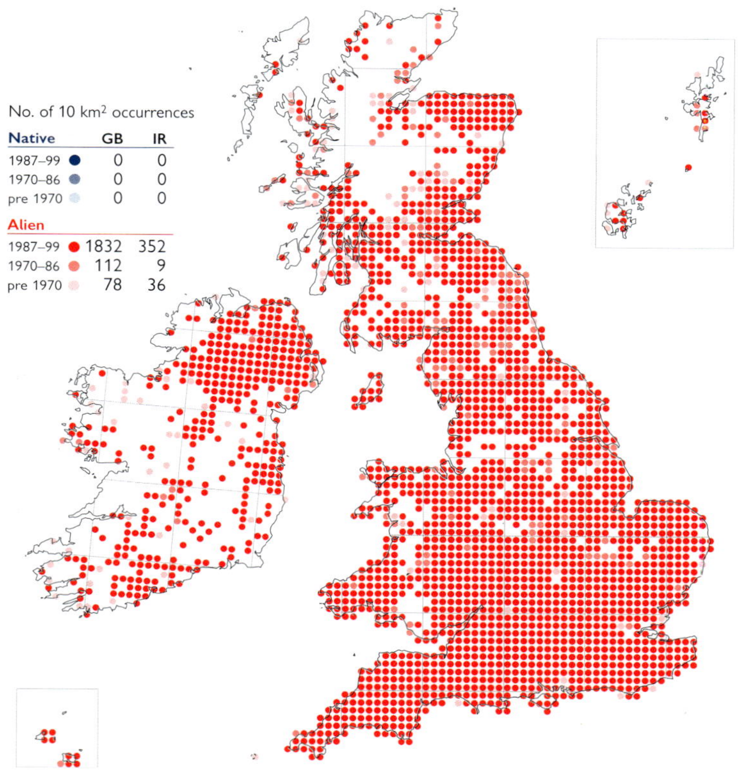

A perennial herb of lawns and churchyards, and also found on roadsides, paths, grassy banks and streamsides. It is self-incompatible, rarely setting seed in our area but often spreads from fragments after mowing. Generally lowland, but reaching 450 m at Nenthead (Cumberland).

Neophyte (change +2.69). This species was certainly cultivated in Britain by 1808 but was not widely grown until the 20th century. It was first recorded in the wild in 1838 but not reported again until 1927. Thereafter it spread rapidly, and was widespread at the time of the 1962 *Atlas*. Since then it has further increased and greatly consolidated its range.

Native of N. Turkey and the Caucasus.

References: Atlas (232d), Bangerter & Kent (1957, 1962, 1965), Harris & Lovell (1980), Hultén & Fries (1986), Meusel *et al.* (1978).

A. HORSFALL

Veronica hederifolia Ivy-leaved Speedwell

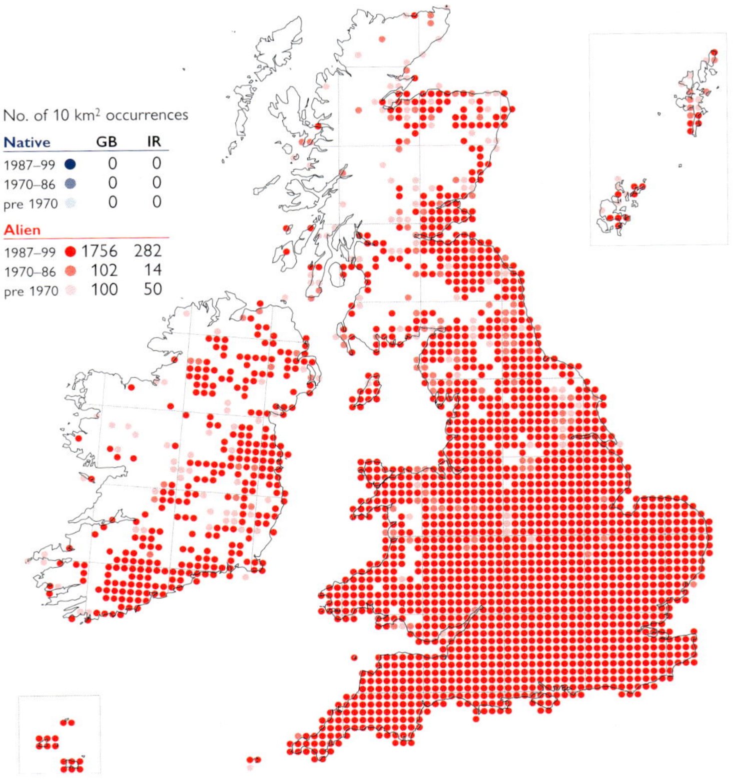

No. of 10 km² occurrences

Native		GB	IR
1987–99	●	0	0
1970–86	●	0	0
pre 1970	●	0	0
Alien			
1987–99	●	1756	282
1970–86	●	102	14
pre 1970	●	100	50

An annual of cultivated and waste ground, woodland rides, hedge banks, walls, banks and gardens, found on sandy, loam or clay soils. *V. hederifolia* seeds freely, with germination in spring or autumn. 0–380 m (Malham Moor, Mid-W. Yorks.).

Archaeophyte (change +0.57). The distribution and frequency of *V. hederifolia* remains stable. Two cytologically distinct subspecies occur in our area and are mapped separately.

As an archaeophyte *V. hederifolia* has a European Southern-temperate distribution; it is widely naturalised outside this range.

References: Atlas (231d), Hultén & Fries (1986), Meusel *et al.* (1978), Salisbury (1964).

A. HORSFALL

Veronica hederifolia subsp. *hederifolia*

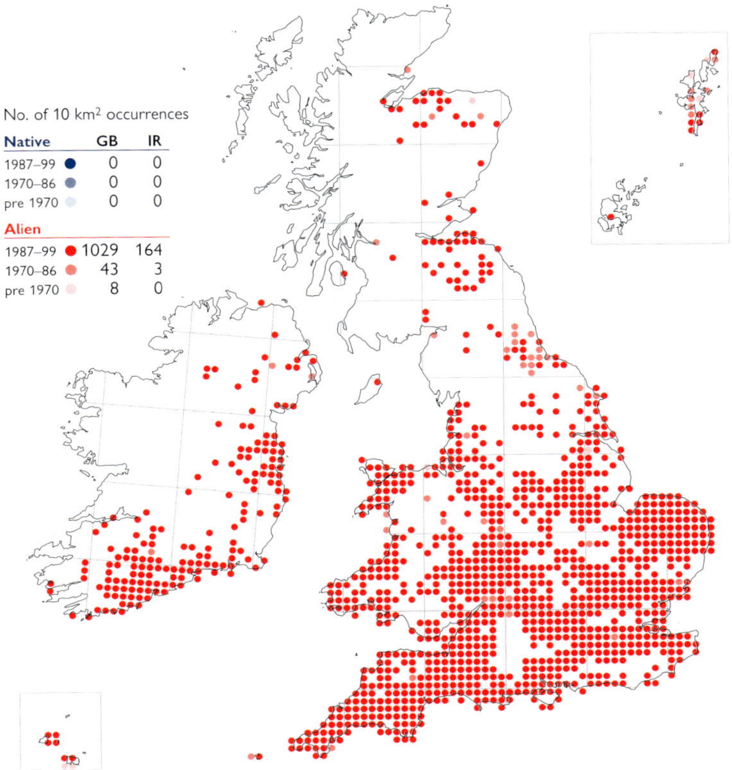

No. of 10 km² occurrences

Native		GB	IR
1987–99	●	0	0
1970–86	●	0	0
pre 1970	●	0	0
Alien			
1987–99	●	1029	164
1970–86	●	43	3
pre 1970	●	8	0

This annual is most frequent in open, unshaded habitats, often in disturbed ground where soil has been recently turned. Typical habitats include arable fields, gardens, allotments, waste ground and roadside banks. Generally lowland, but upper altitudinal limit unknown.

Archaeophyte. Although typical forms of subsp. *hederifolia* are clearly distinct from subsp. *lucorum*, both subspecies are morphologically variable and it is often difficult to name plants with certainty. They have therefore been under-recorded and some mapping errors are likely.

Subsp. *hederifolia* occurs throughout the range of the species.

References: Fischer (1975), Meusel *et al.* (1978).

M. J. WIGGINTON

Veronica hederifolia subsp. *lucorum*

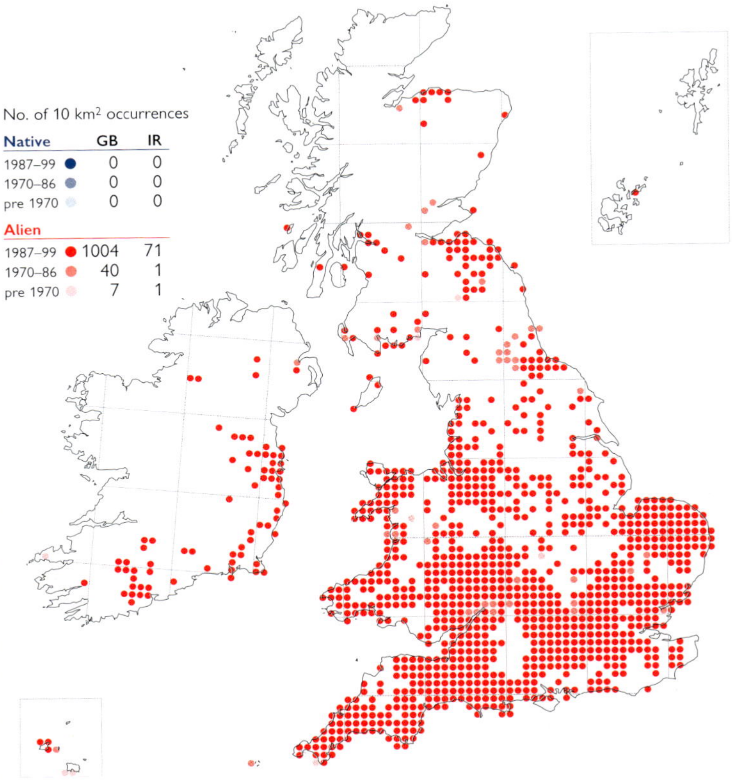

No. of 10 km² occurrences

Native		GB	IR
1987–99	●	0	0
1970–86	●	0	0
pre 1970	●	0	0
Alien			
1987–99	●	1004	71
1970–86	●	40	1
pre 1970	●	7	1

In contrast to subsp. *hederifolia*, subsp. *lucorum* is usually found in shady places, including woodland rides and hedgerows, though plants may also occur in open sites. Generally lowland, but upper altitudinal limit unknown.

Archaeophyte. Although subsp. *lucorum* differs from subsp. *hederifolia* in chromosome number, the morphological separation of the two taxa is often difficult. They have therefore been under-recorded and some mapping errors are likely.

Subsp. *lucorum* has a European Temperate distribution; it extends less far south in Europe than subsp. *hederifolia*.

References: Fischer (1975), Meusel *et al.* (1978).

M. J. WIGGINTON

Veronica longifolia Garden Speedwell

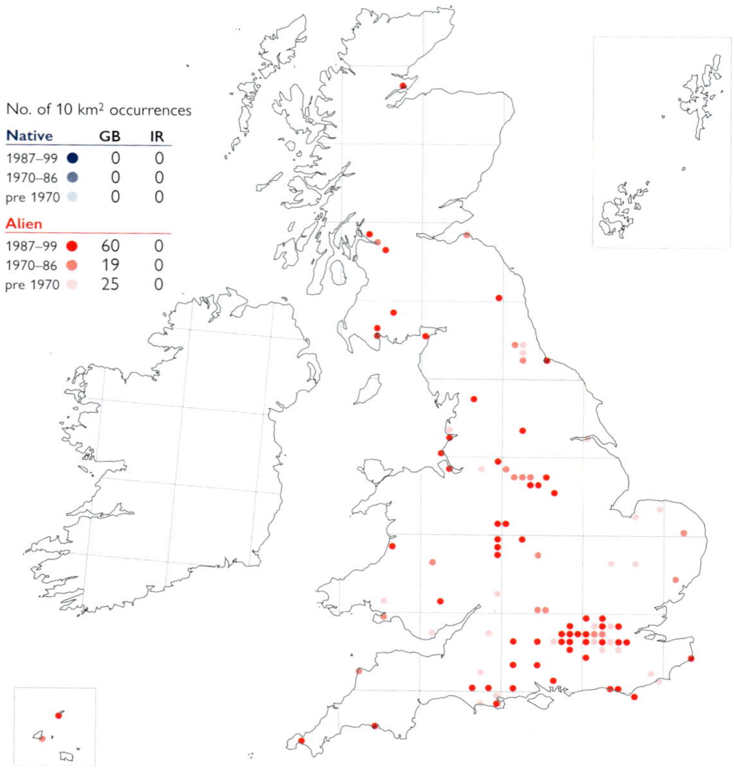

This woody-based perennial herb is found as a naturalised garden escape or casual in rough grassland and hedgerows and on roadsides, waste ground and rubbish tips. Lowland.

Neophyte. *V. longifolia* was cultivated in Britain by 1731, and is popular in gardens. It was first recorded in the wild in 1928. It may be increasing, but has been recorded in error for garden forms of *V. spicata* and, especially, the fertile hybrid *V. longifolia* × *V. spicata*.

A Eurasian Boreo-temperate species, which is absent as a native from W. Europe.

Reference: Hultén & Fries (1986).

T. D. DINES

Veronica spicata Spiked Speedwell

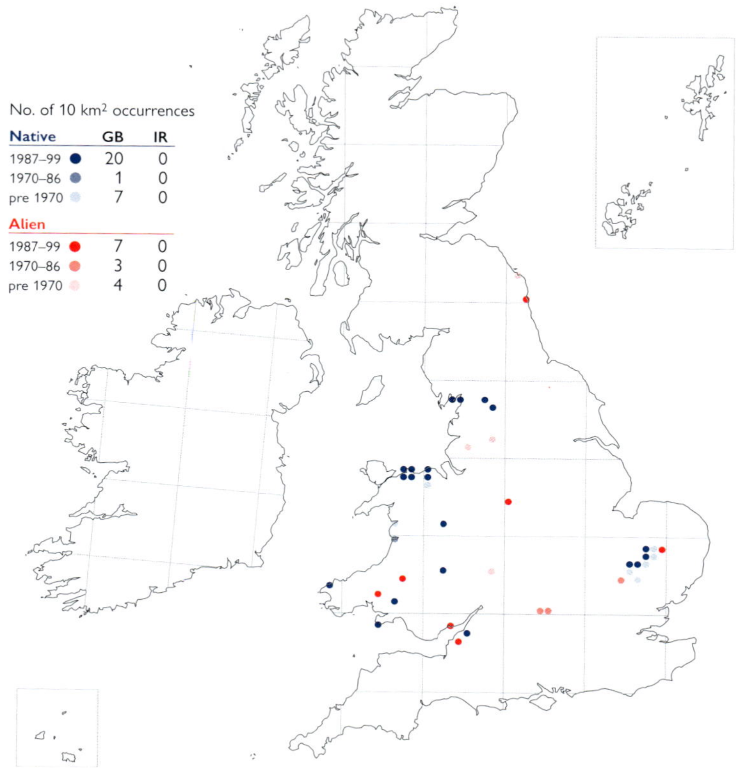

A perennial herb of well-drained, nutrient-poor soils. In East Anglia, subsp. *spicata* usually grows on acidic to base-rich sandy soils in open, shortly-grazed grassland. Elsewhere, subsp. *hybrida* grows in thin soils on base-rich cliffs, grassland and rocks. Generally lowland, but reaching 400 m in Ribblesdale (Mid-W. Yorks.).

Native (change +0.13). The distribution of *V. spicata* is stable, though there have been losses of subsp. *spicata*, especially before 1930. The two subspecies may not be distinct and are not formally distinguished by Stace (1997) or recognised in *Flora Europaea* (Tutin *et al.*, 1972).

Eurosiberian Temperate element, with a continental distribution in W. Europe.

References: Atlas (229d), Atlas Supp. (51b), Hultén & Fries (1986), Meusel *et al.* (1978), Kay & John (1995), Stewart *et al.* (1994), Wigginton (1999), Wilson *et al.* (2000).

A. HORSFALL

Hebe salicifolia Koromiko

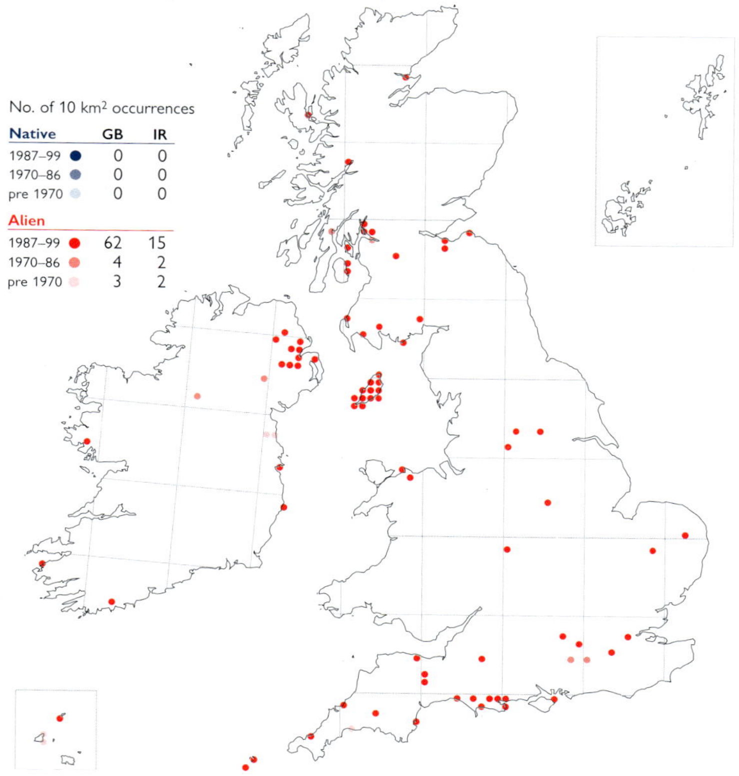

An evergreen shrub found planted or as a garden escape on roadside and in hedgerows, sea-cliffs, quarries, pavement cracks and on walls. It reproduces freely by seed, and can become well-established, especially in coastal areas. Lowland.

Neophyte. This species has been cultivated in Britain since 1843, and is extremely popular in gardens. It was first recorded in the wild in 1913, but trends in distribution are difficult to assess. *H. salicifolia* is the parent of many cultivated hybrids, and some records may be in error for these.

Native of S. America (Chile) and New Zealand.

Reference: Bean (1973).

T. D. DINES

Hebe elliptica × H. speciosa (H. × franciscana) Hedge Veronica

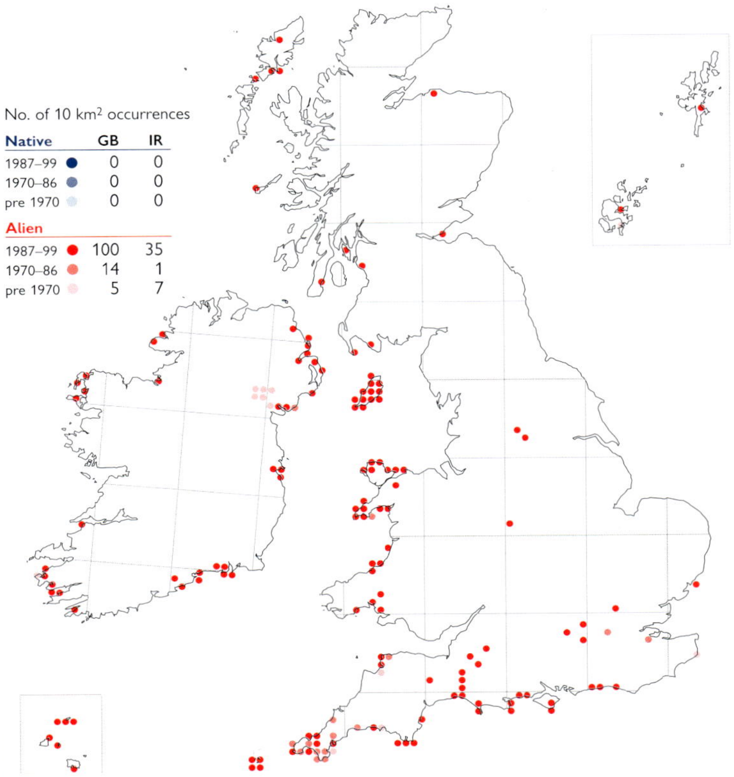

An evergreen shrub found as a garden escape or planted in coastal hedge banks and on sea-cliffs, where it is frequently well-naturalised, and also on waste ground, walls, pavement cracks and in amenity plantings. The hybrid is fully fertile and reproduces abundantly by seed. Lowland.

Neophyte. This hybrid was first raised in 1859, with the most frequent cultivar, 'Blue Gem', being raised in 1868. It is extremely popular in gardens, especially in coastal areas. It was known from the wild in Britain by 1931 and in Ireland by 1972, but it was probably present in both countries well before these dates. It appears to be increasing.

A hybrid of garden origin; both parents are native of New Zealand.

References: Chalk (1986), Clement (1985a), Green (1973), Stace (1975).

T. D. DINES

Sibthorpia europaea Cornish Moneywort

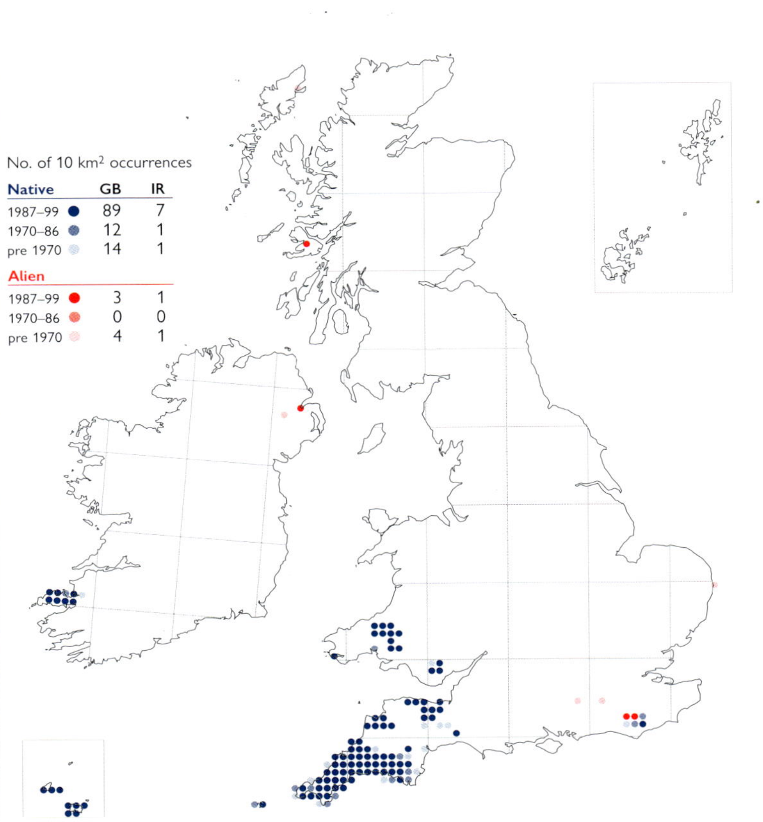

A procumbent perennial herb of acidic soils in damp, shady places, including woodland, banks by small streams and ditches (often creeping over a carpet of mosses), on wet heathland, on thin soil over granite walls or other masonry, on shaded paths and lawns, and occasionally in other damp habitats. 0–515 m (Connor Hill, S. Kerry).

Native (change –0.14). Many more recent records of *S. europaea* are available now compared with the 1962 *Atlas*, presumably reflecting the improved recording of this inconspicuous species rather than any increase in range or frequency. In Ireland, it appears to be threatened by the aggressive spread of *Epilobium brunnescens*.

Oceanic Temperate element.

References: Atlas (227b), Curtis & McGough (1988), Stewart *et al.* (1994).

A. HORSFALL

Melampyrum cristatum Crested Cow-wheat

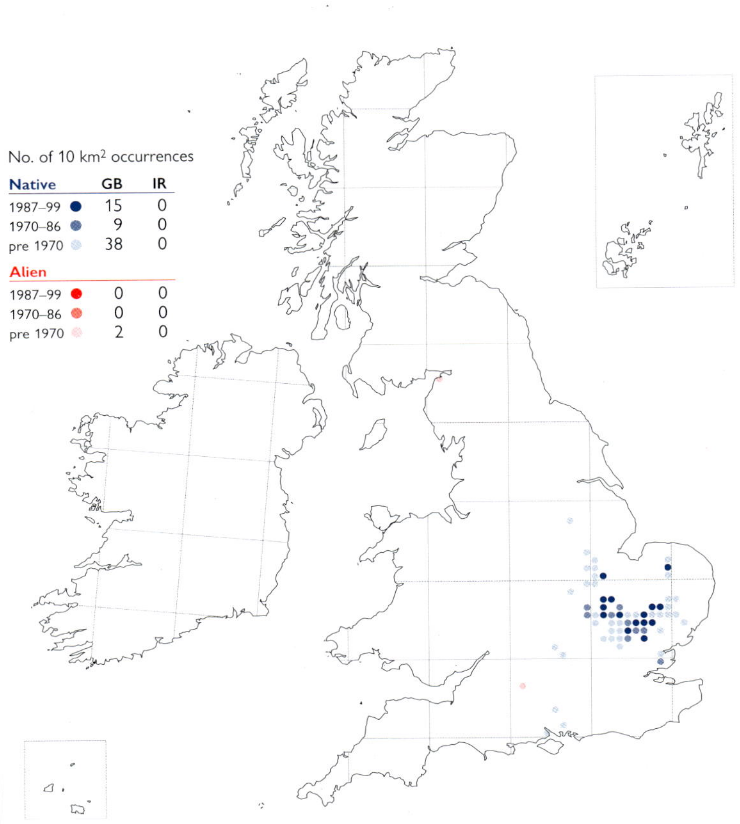

An annual hemiparasite of various woody and herbaceous species; mostly found on the margins of ancient *Quercus robur* woodlands, their clearings and rides and in associated field hedge banks on chalky boulder-clay soils. It is very rarely found in open grassland. Lowland.

Native (change –0.88). Much of the loss of *M. cristatum* had taken place before 1930, but the range has further contracted since then, with losses occurring through the cessation of traditional woodland management, and herbicide spraying of verges and field-borders. It has apparently been lost from its sole grassland site, in S. Essex, through trampling by horses.

Eurosiberian Temperate element, with a continental distribution in W. Europe.

References: Atlas (234a), Horrill (1972), Hultén & Fries (1986), Meusel *et al.* (1978), Stewart *et al.* (1994).

F. J. RUMSEY

Melampyrum arvense Field Cow-wheat

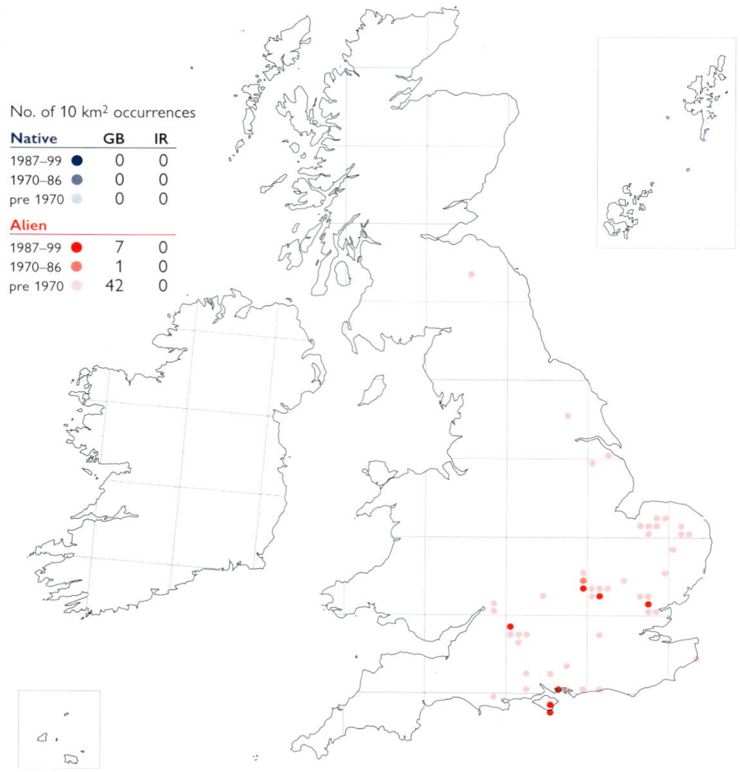

An annual hemiparasite, mainly on the roots of grasses. Formerly an arable weed, it now occurs in open grassland beside hedges and ditch-banks, field-borders, in a disused brick-pit and on slumping chalk cliff-faces. Lowland.

Neophyte (change –0.49). Although an archaeophyte in parts of N. Europe, this conspicuous species was not recorded in Britain until 1724. Originally introduced with crop-seed, it has declined due to improved seed cleaning, agricultural intensification and a lack of disturbance at some sites. It was rare by 1930, and has continued to decline since then. It has been deliberately planted at some sites.

M. arvense has a European Temperate distribution, although it is absent from much of W. Europe.

References: Atlas (234b), Hultén & Fries (1986), Meusel *et al.* (1978), Wigginton (1999).

F. J. RUMSEY

Melampyrum pratense Common Cow-wheat

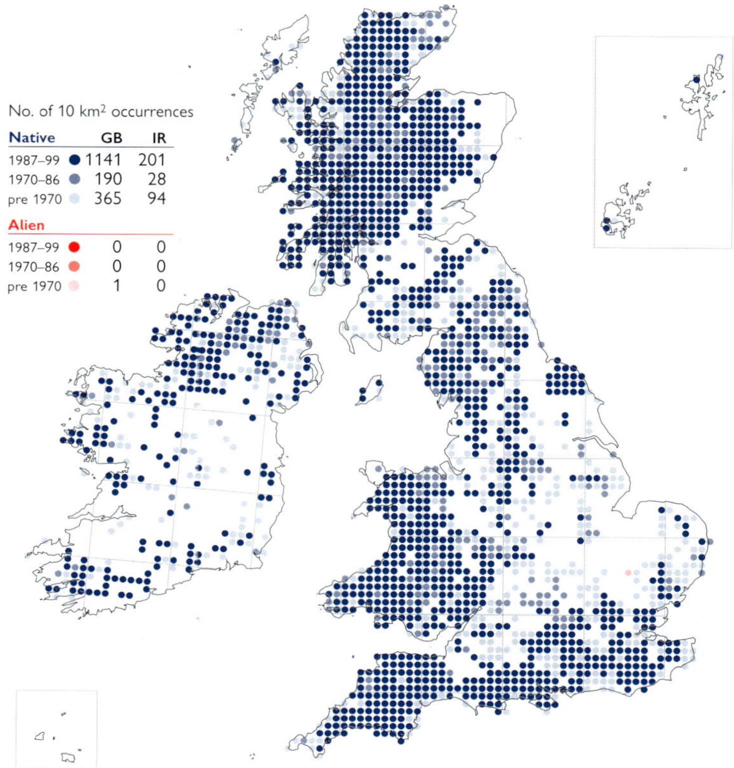

An annual hemiparasite of woods, scrub, heaths and upland moorlands on well-drained, nutrient-poor acidic soils; more rarely in scrub, hedgerows and deciduous woodland on chalk and limestone. The large seeds are distributed by ants. Lowland to 960 m (Macgillycuddy's Reeks, S. Kerry).

Native (change –0.88). *M. pratense* was mapped as 'all records' in the 1962 *Atlas*. Many of the losses apparent on the map occurred before 1930, but the decline has accelerated since then. This is probably due to habitat loss and the cessation of traditional woodland management.

Eurosiberian Boreo-temperate element.

References: Atlas (234c), Atlas Supp. (54a, b), Hultén & Fries (1986), Smith (1963).

F. J. RUMSEY

Melampyrum sylvaticum Small Cow-wheat

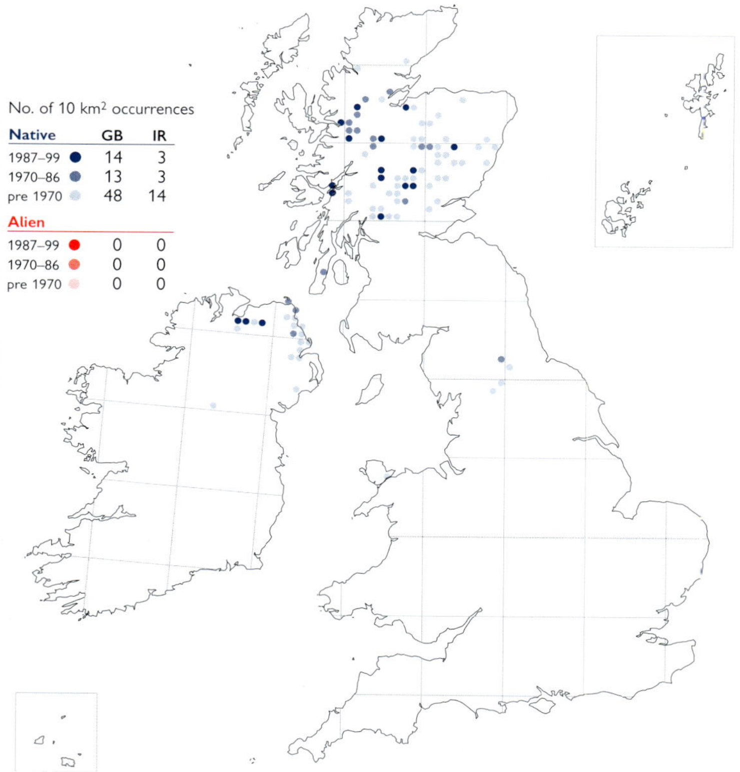

An annual hemiparasite found in humid, lightly shaded situations on damp, usually somewhat enriched, acidic soils; in wooded ravines, in grassy hollows and on banks in woodlands and on upland cliff ledges. Near sea level to 760 m on Aonach air Chrith (W. Ross).

Native (change –0.58). This species has declined most in the southern and lowland parts of its range, and especially in N. Ireland where many losses occurred before 1930. Afforestation, nutrient enrichment, grazing and trampling by livestock have contributed to its decline. However, it is almost certainly under-recorded. It has been confused with some forms of *M. pratense*.

European Boreal-montane element.

References: Atlas (234d), Curtis & McGough (1988), Hultén & Fries (1986), Meusel *et al.* (1978), Rich, FitzGerald & Sydes (1998), Stewart *et al.* (1994).

F. J. RUMSEY

Euphrasia officinalis agg. Eyebrights

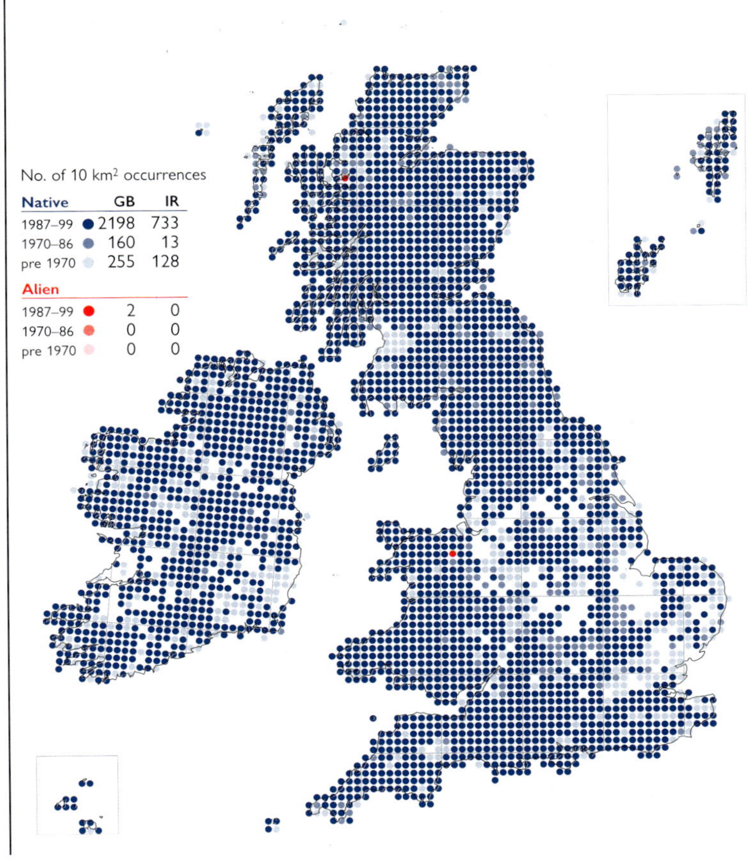

A group of small, hemiparasitic annuals on the roots of diverse herbs and small shrubs, mainly found in permanent or semi-permanent grasslands. 0–1215 m (Ben Lawers, Mid Perth).

Native (change −1.61). *E. officinalis* agg. includes all the *Euphrasia* taxa in our area. Hybridisation (sometimes involving backcrossing and the formation of locally distinct populations) and morphological plasticity mean that identification can be difficult, and comments on distributional trends for many taxa are tentative. The aggregate has declined due to habitat loss and agricultural improvement. Most taxa are under-recorded.

For the distribution of the component taxa, see the following accounts.

References: Atlas (235a), Grime *et al.* (1988), Meusel *et al.* (1978), Perring & Sell (1968), Silverside (1998), Stace (1975), Yeo (1978).

F. J. RUMSEY

Euphrasia rostkoviana subsp. *rostkoviana*

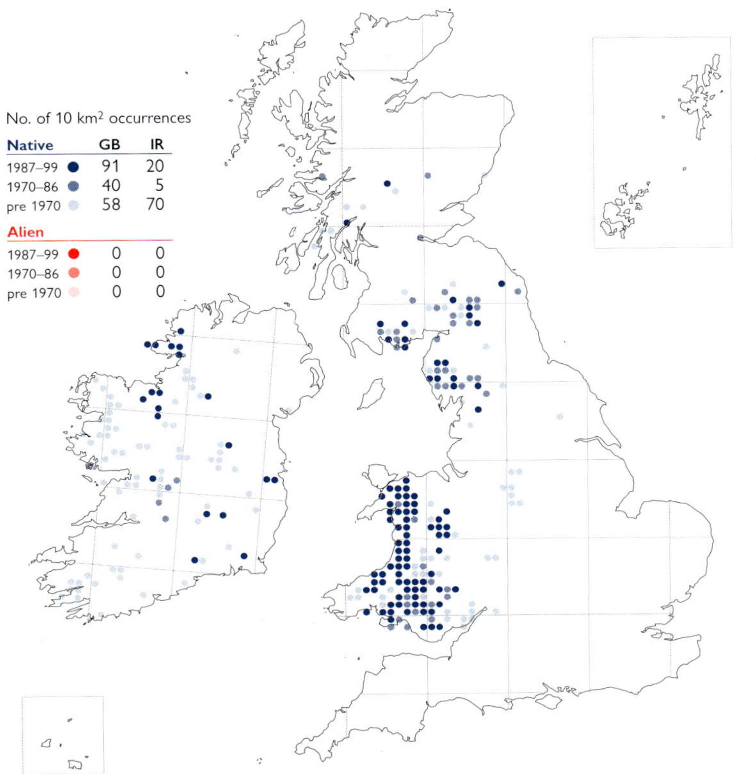

An annual of rather damp, herb-rich hay meadows, riverside grasslands, lightly grazed pastures and grassy roadsides. It is replaced in shorter turf and more acidic sites by *E. anglica*, with which it completely intergrades. Generally lowland, but reaching 420 m at Llyn Gorast (Cards.).

Native. This diploid taxon is probably declining through agricultural improvement and loss of habitat. It is regarded as a subspecies of *E. officinalis sens. str.* by Silverside (1991, 1998); in his interpretation *E. officinalis* subsp. *officinalis* is a plant of N.E. Europe. It is under-recorded, especially in Ireland.

European Boreo-temperate element.

References: Atlas Supp. (61a), Stewart *et al.* (1994).

F. J. RUMSEY

Euphrasia rostkoviana subsp. *montana*

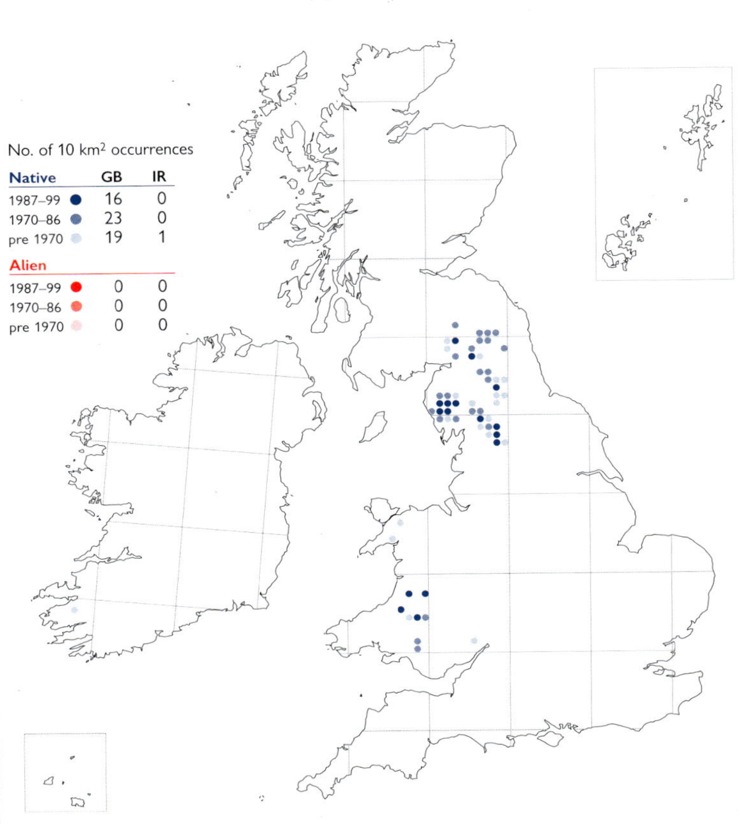

An annual of upland hay meadows, primarily of relatively dry sites but also in wet meadows and upland fens, rarely if ever colonising improved grasslands. Most sites are at 200–455 m (Smallburn, N. Northumb.), but it has been recorded at 100 m.

Native. This diploid taxon is difficult to separate from subsp. *rostkoviana*, with which it may hybridise. Its distribution is uncertain through difficulties of identification, but it is undoubtedly declining as hay meadows are lost or agriculturally improved. It is regarded as a subspecies (subsp. *monticola*) of *E. officinalis sens. str.* by Silverside (1991, 1998).

European Boreo-temperate element.

References: Atlas Supp. (61b), Stewart *et al.* (1994).

F. J. RUMSEY

Euphrasia rivularis

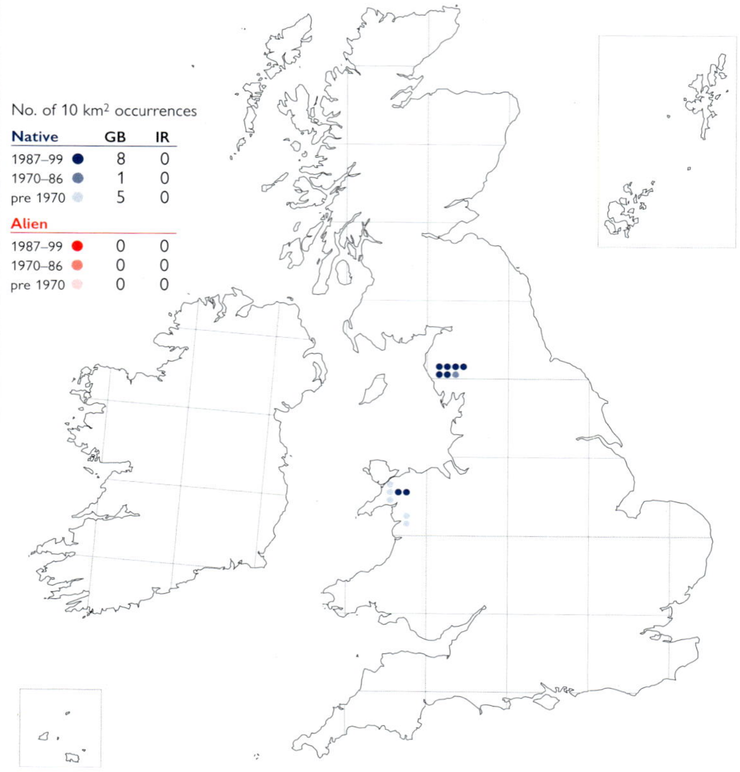

An annual of upland rocky flushes, seepage areas and wet rock ledges on montane cliffs. It grows from 380 m at Langstrath (Westmorland) to *c.* 750 m on Snowdon (Caerns.).

Native. *E. rivularis* is diploid and is closely allied to *E. rostkoviana*, from which it has probably arisen through hybridisation with *E. micrantha*. Its distribution appears to be stable in the Lake District, but the species seems to have declined in Wales where recent searches have failed to re-find it in many known sites.

Endemic.

References: Atlas Supp. (61c), Wigginton (1999).

F. J. RUMSEY

Euphrasia anglica

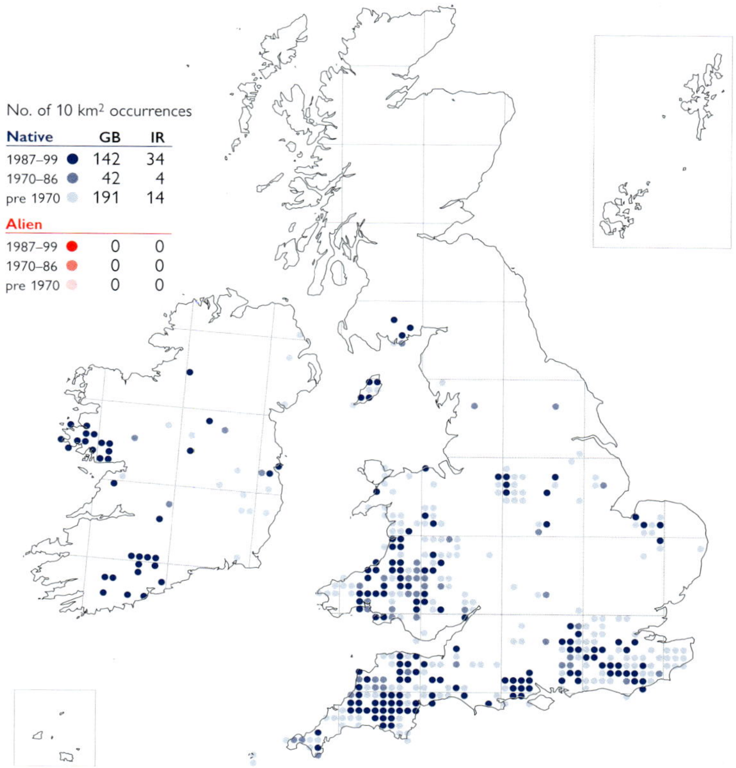

A diploid annual of tightly grazed habitats on damp acidic substrates. It grows in old pastures, heaths, moorlands, and disused quarries. Generally lowland, but reaching 395 m at Ystrad-ffin (Carms.).

Native. *E. anglica* is most distinct in S.W. England; elsewhere material can be difficult to name (Silverside, 1991). It frequently occurs in mixed populations showing introgression with a range of tetraploid species, and some of these local forms have become stabilised and show distinct ecological preferences. Because of these difficulties, the distribution is imperfectly known. It is regarded as a subspecies of *E. officinalis sens. str.* by Silverside (1991, 1998).

Endemic; the identification of similar plants in N. France requires confirmation (Yeo, 1978).

Reference: Atlas Supp. (62a).

F. J. RUMSEY

Euphrasia vigursii

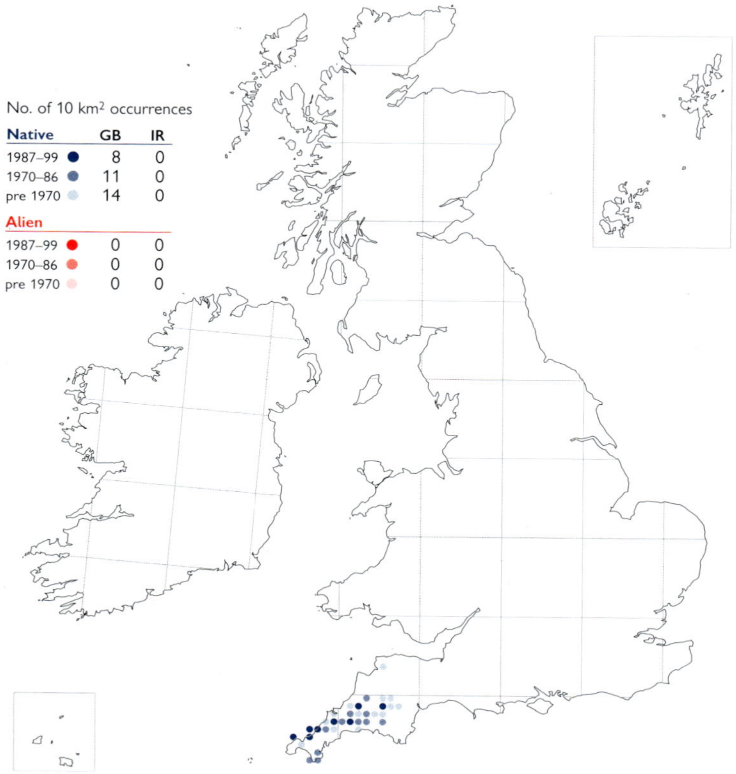

An annual that is characteristic of *Agrostis curtisii-Ulex gallii* heaths. In coastal areas it occurs mainly on cliff-tops in patches of short, species-rich turf around rock outcrops, and by tracks and paths, where scrub is suppressed but the vegetation is not too open. Inland it occurs on lightly grazed damp heaths and open moorland. Generally lowland, but reaching 330 m on Kit Hill (E. Cornwall).

Native. This diploid species has substantially declined since 1950 at its inland sites through habitat dereliction and destruction, including the decline of grazing. *E. vigursii* is believed to have arisen as a stabilised segregate following introgression between *E. anglica* and *E. micrantha*.

Endemic.

References: Atlas Supp. (62b), French *et al.* (1999), Wigginton (1999).

F. J. RUMSEY

Euphrasia arctica subsp. *arctica*

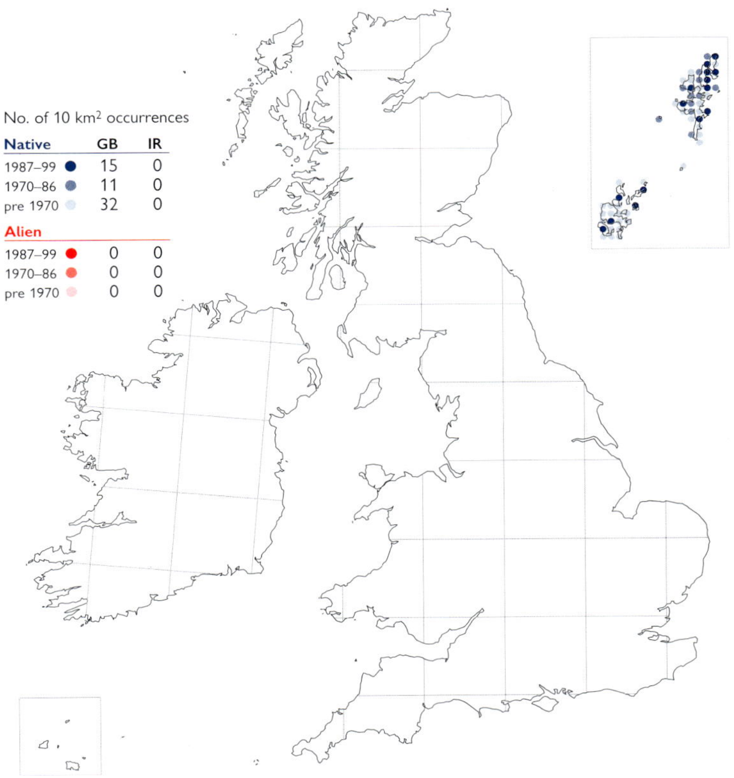

This tetraploid annual is found in damp meadows and marshes. Lowland.

Native. The current taxonomic treatment of *E. arctica* is based on Yeo's (1978) revision of *Euphrasia* in Europe; the plants treated here as *E. arctica* were mapped as *E. borealis* and *E. brevipila* by Perring & Sell (1968). It is difficult to assess changes in the distribution of populations now referred to subsp. *arctica*. It is possible that some mainland populations in Scotland and Cumbria should be referred to this subspecies.

Oceanic Boreal-montane Element; restricted to Orkney, Shetland and the Faeroes.

References: Hultén & Fries (1986), Rich & Jermy (1998), Scott & Palmer (1987).

F. J. RUMSEY

Euphrasia arctica subsp. *borealis*

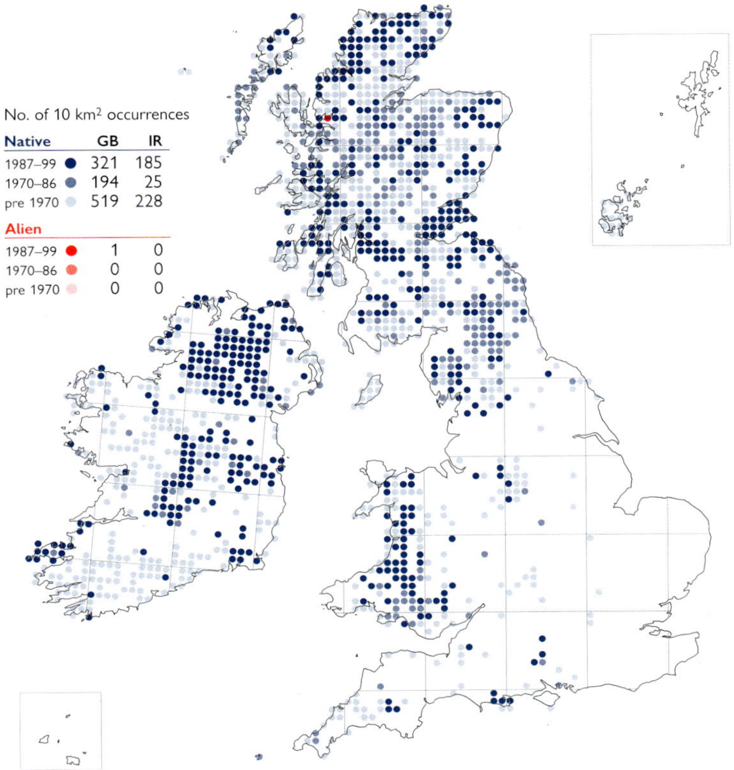

An annual, locally common in damp, rough grasslands, pastures, hay meadows, river banks and on roadsides, often in damper sites than *E. nemorosa* which it replaces in Scotland. Generally lowland, but reaching 760 m (Lochan na Lairige, Mid Perth).

Native. *E. arctica* subsp. *borealis* is a very variable tetraploid taxon, and is under-recorded both because of recent changes in the taxonomic treatment of *E. arctica sens. lat.* and through confusion with *E. nemorosa* and hybrids with that species. All records of *E. arctica* outside Orkney and Shetland have been mapped as subsp. *borealis*, but some Scottish and Lake District populations approach subsp. *arctica* (Halliday, 1997).

Oceanic Boreal-temperate element.

References: Hultén & Fries (1986), Yeo (1978).

F. J. RUMSEY

Euphrasia arctica × *E. confusa*

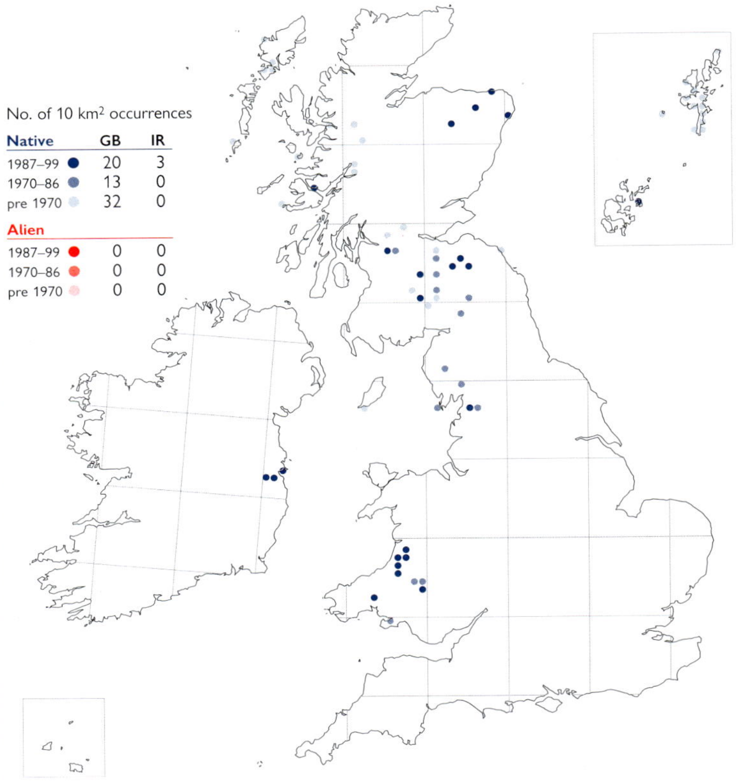

A fertile hybrid of pastures, stabilised dunes, roadsides and upland grasslands. Both subspecies of *E. arctica* are involved, although the hybrid with subsp. *arctica* is restricted to Orkney and Shetland. Generally lowland, but upper altitudinal limit unknown.

Native. This hybrid may occur in uniform colonies which replace its parents, but also forms hybrid swarms where its parents meet. It is not always separable from depauperate forms of *E. arctica* and is certainly under-recorded.

Wider distribution uncertain.

References: Rich & Jermy (1998), Stace (1975).

F. J. RUMSEY

Euphrasia tetraquetra

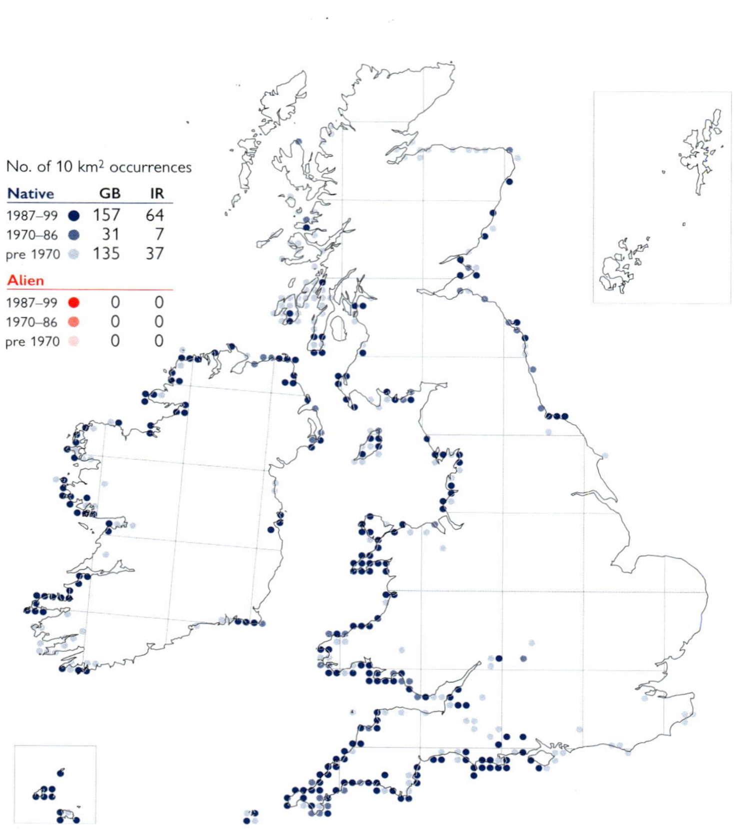

An annual of short turf on exposed coastal cliffs and sand dunes, but also locally inland on chalk and limestone pastures. Lowland.

Native. Whilst usually distinct, this tetraploid species is often confused with other *Euphrasia* taxa, particularly in N. Scotland where dwarf forms of *E. nemorosa* and *E. foulaensis* can be very similar. Its distribution has not changed markedly since it was mapped by Perring & Sell (1968), although it is now better recorded in some areas. Records north of W. Ross have been omitted in the absence of recently determined specimens.

Oceanic Temperate element.

References: Atlas Supp. (58b), Yeo (1978).

F. J. RUMSEY

Euphrasia nemorosa

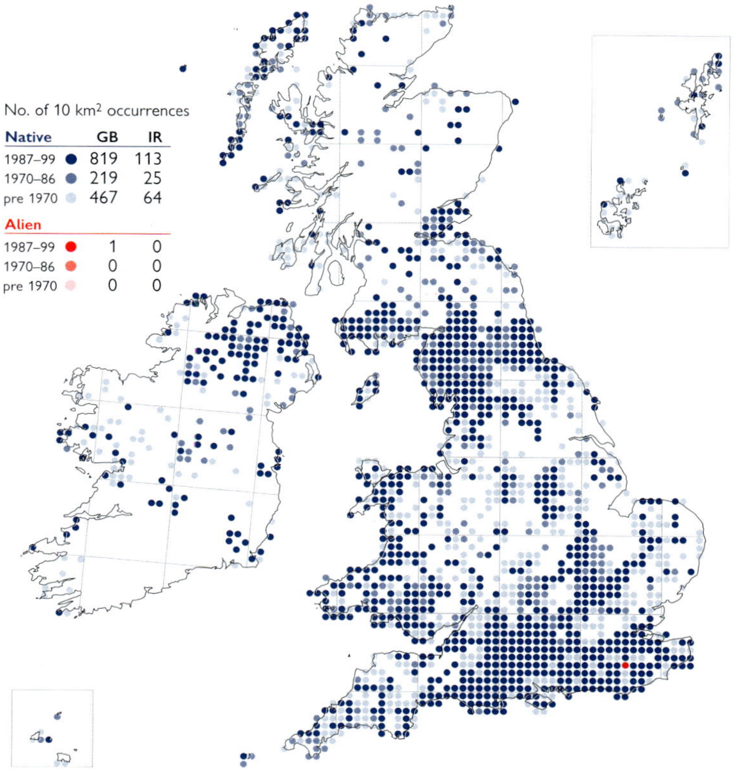

An annual occurring in short grasslands, on heaths, downs and dunes, in open scrub, woodland rides and upland moorlands. It is absent from agriculturally improved land. 0–825 m (Cross Fell, Cumberland).

Native. This tetraploid species has not shown any marked change in its range since it was mapped by Perring & Sell (1968). It is the most common and ecologically diverse of our eyebrights, becoming more restricted to calcareous soils at low altitudes in the north. However, it forms hybrids with many other *Euphrasia* species, and introgressed populations can be locally abundant, making identification difficult.

European Temperate element.

References: Atlas Supp. (58c), Hultén & Fries (1986).

F. J. RUMSEY

Euphrasia pseudokerneri

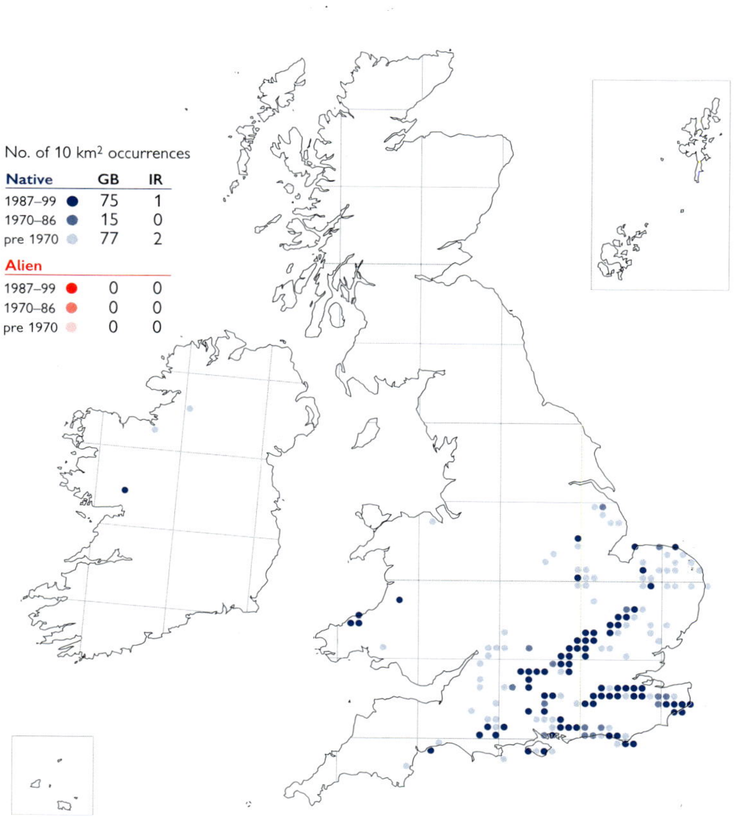

An annual of herb-rich downland turf on chalk and soft limestones, rarely found on harder limestones in Ireland or as forma *elongata* in damp fens, and recently discovered in calcareous flushes, a lead mine and coastal grassland in Cardiganshire. Lowland.

Native. *E. pseudokerneri*, a tetraploid, is decreasing through the ploughing up of its habitat and agricultural improvement of downland pastures. Changes in land management have favoured other *Euphrasia* species, particularly *E. nemorosa*, which may, in turn, have increased the incidence of hybridisation.

Endemic; it is replaced by *E. stricta* in Europe.

References: Atlas Supp. (59b), Stewart *et al.* (1994).

F. J. RUMSEY

Euphrasia confusa

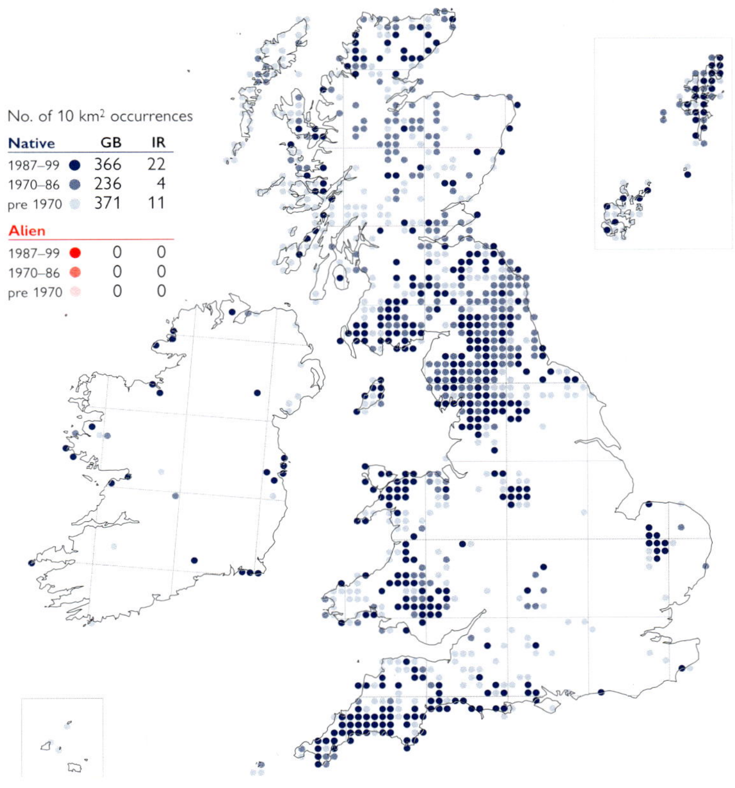

An annual of grazed pasture and grassy heathland on free-drained, acidic or calcareous soils. It is especially characteristic of hill pastures in N. & W. Britain, but is occasionally found in open vegetation on sandy soils elsewhere. It is rare and mainly coastal in Ireland. It can withstand intense grazing pressure. 0–660 m (Cross Fell, Cumberland).

Native. *E. confusa* is a widespread and frequent tetraploid species which has not changed appreciably in its distribution since it was mapped by Perring & Sell (1968), although it is now better recorded. As with other *Euphrasia* taxa, however, extensive hybridisation makes identification difficult and it is likely to be under-recorded.

Oceanic Boreo-temperate element.

Reference: Atlas Supp. (59a).

F. J. RUMSEY

Euphrasia confusa × *E. nemorosa*

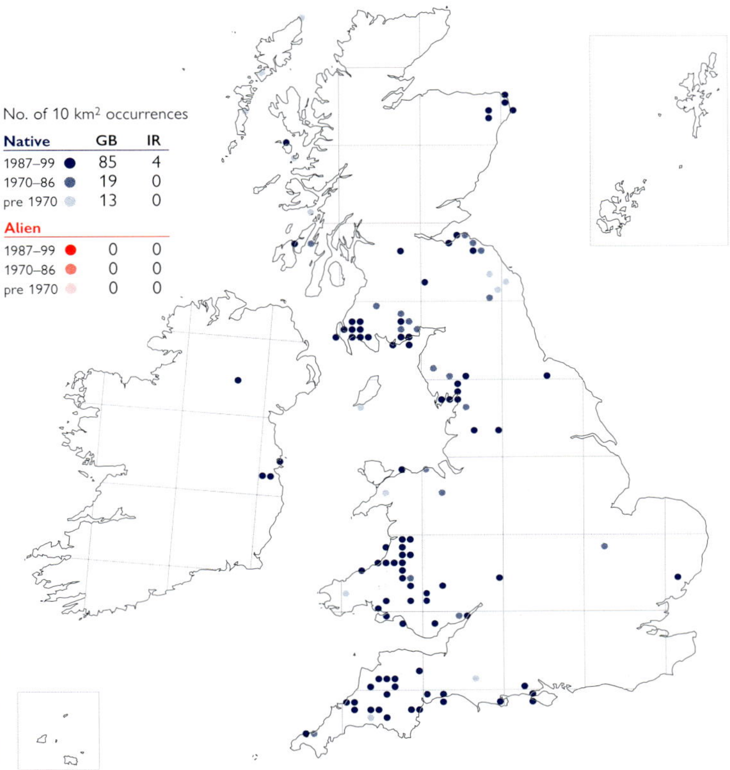

A widespread fertile hybrid, occurring in pastures, on roadsides, in disturbed ground and as a natural colonist of formerly ploughed land. Generally lowland, but reaching 570 m at Llyn Llygad Rheidol (Cards.).

Native. This hybrid may occur in large uniform colonies in the absence of one or both parents, and may entirely replace them in some districts. It tends to become restricted to coastal areas in the northern part of its range, as does *E. nemorosa*. Hybrid plants are variable and form a complete continuum between the parental taxa; they are certainly under-recorded.

Wider distribution uncertain.

Reference: Stace (1975).

F. J. RUMSEY

Euphrasia confusa × *E. scottica*

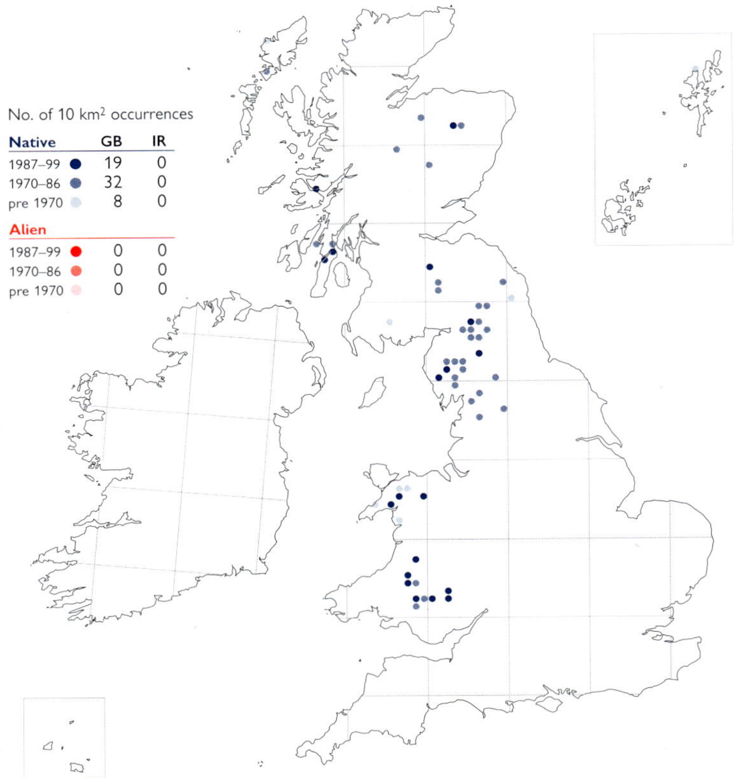

A fertile hybrid of flushes and damp stream banks in upland areas, occurring throughout much of the range of *E. scottica* and absent only where *E. confusa* is rare. It is, however, often found in the absence of one or other parent. The altitudinal range is unknown.

Native. This hybrid is certainly under-recorded. It has not been previously mapped on a national scale.

Wider distribution uncertain.

References: Rich & Jermy (1998), Stace (1975).

F. J. RUMSEY

Euphrasia frigida

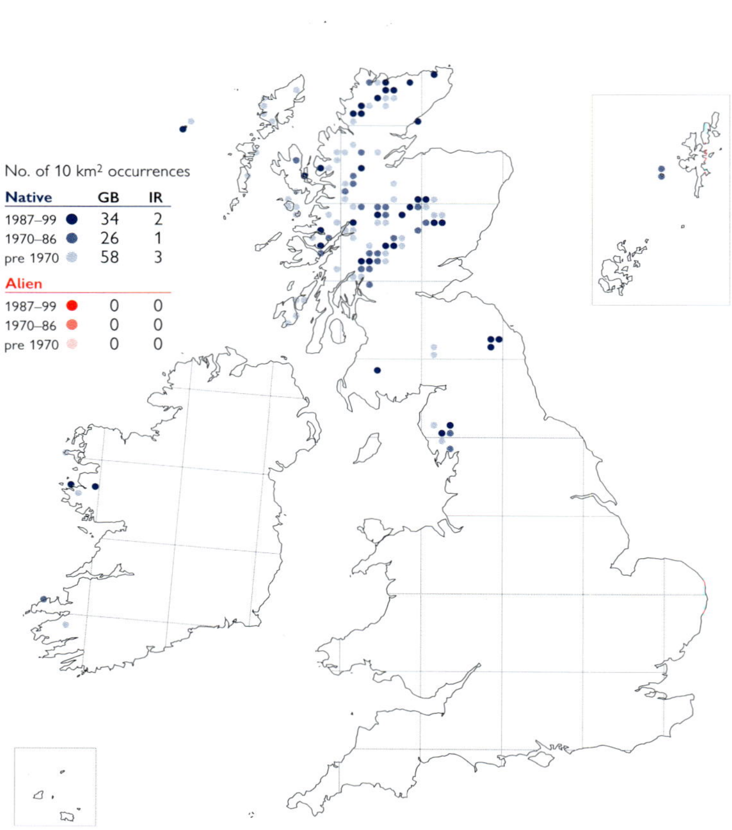

An annual of damp or wet, usually rather basic, cliff ledges. It occurs at 200 m on Foula (Shetland), but is usually found above 400 m, reaching 1190 m on Aonach Beag (Westerness).

Native. There is no evidence of any change in the distribution of this tetraploid species, although it is much better recorded than it was when mapped by Perring & Sell (1968). It is, however, almost certainly under-recorded, and is likely still to be present in many of the 10-km squares for which there are only pre-1987 records.

Eurosiberian Arctic-montane element; also in N. America.

References: Atlas Supp. (56a), Hultén & Fries (1986), Meusel *et al.* (1978), Stewart *et al.* (1994).

F. J. RUMSEY

Euphrasia foulaensis

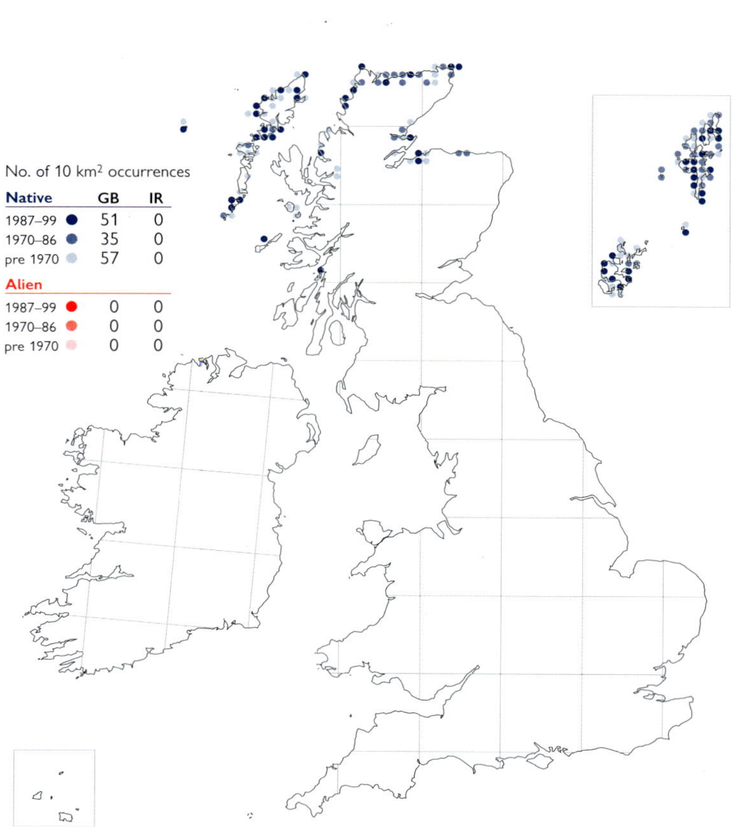

An annual occurring in damp, open turf on coastal cliff-tops, and at the upper fringe of saltmarshes. Its cliff-top sites are subject to sea spray, but it avoids the most exposed sites. *E. foulaensis* seems unable to survive in rank turf, and grazing by sheep or rabbits is essential for its survival. Lowland.

Native. This tetraploid species can be locally abundant. Some sites have been lost through agricultural improvement or the cessation of grazing, but an apparent decline may rather reflect the lack of recent critical recording. However, it is much better recorded now than when mapped by Stewart *et al.* (1994).

Oceanic Boreal-montane element.

References: Atlas Supp. (56b), Meusel *et al.* (1978).

F. J. RUMSEY

Euphrasia cambrica

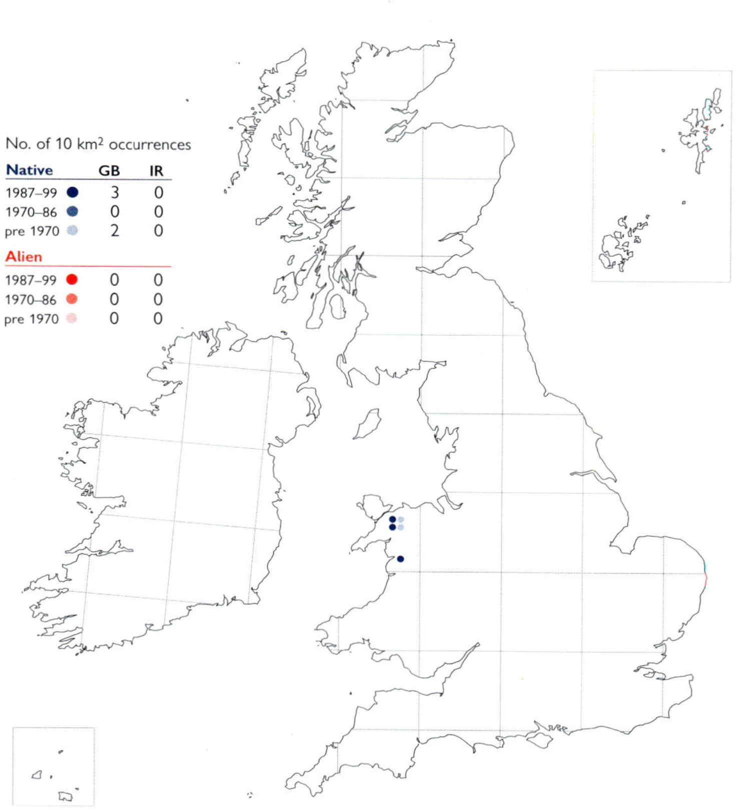

An annual of well-drained, basic, sheep-grazed grassland on mountain slopes. More rarely it grows in wetter, base-enriched flushes where it can occur with *E. rivularis*. An upland species, reaching 880 m on Cadair Idris (Merioneth).

Native. *E. cambrica*, a tetraploid species, is easily overlooked because of its small size and inconspicuous flowers. It has been confused in the past with *E. ostenfeldii*, with which it forms hybrids, as at Cwm Idwal (Caerns.). Its distribution is better known than when mapped by Perring & Sell (1968).

Endemic.

References: Atlas Supp. (58a), Wigginton (1999).

F. J. RUMSEY

Euphrasia ostenfeldii

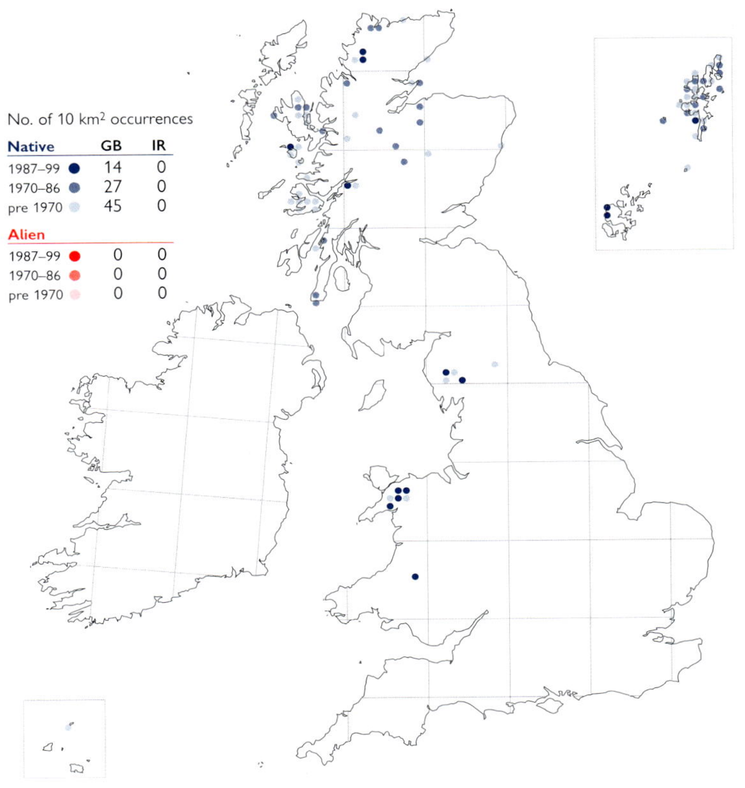

This annual occurs in sparsely vegetated areas in very well-drained, exposed habitats, including dry limestone rock ledges, eroding sea-cliffs, fine-gravel screes, bare serpentine debris and sandy coastal turf. 0–760 m (Cul Mor, W. Ross, and Helvellyn, Cumberland).

Native. Although many plants now recognised as this tetraploid species were previously referred to *E. curta*, the current taxonomic treatment only dates from 1971 and *E. ostenfeldii* is therefore under-recorded. It has also been confused with *E. marshallii* and *E. rotundifolia*, and hybrids are known with *E. foulaensis*. Populations from the southern end of its range (var. *rupestris*) may merit subspecific status.

Oceanic Boreo-arctic Montane element.

Reference: Stewart *et al.* (1994).

F. J. RUMSEY

Euphrasia marshallii

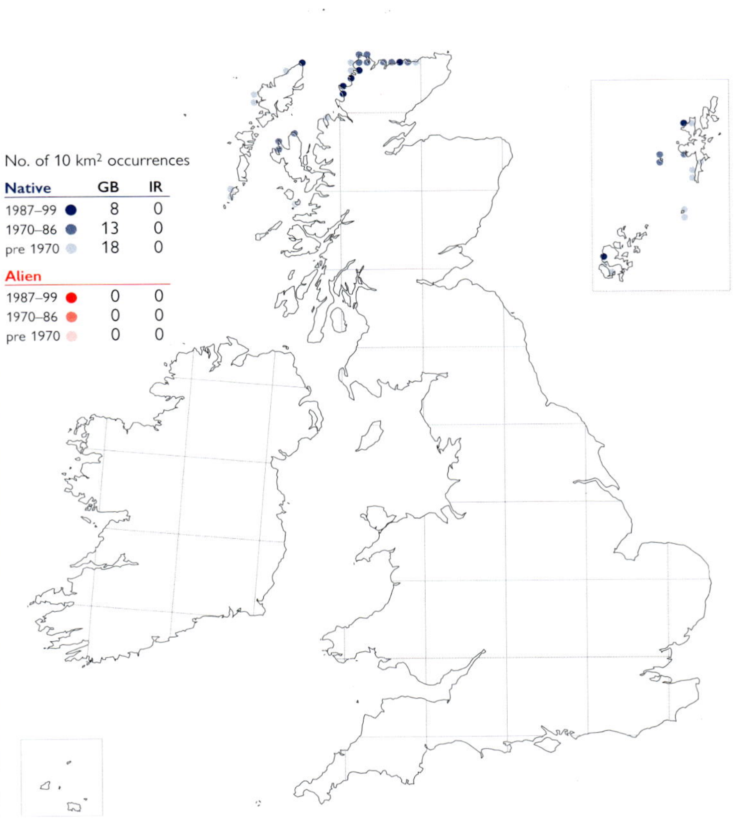

An annual of coastal rocks and eroding sea-cliff edges below maritime *Calluna vulgaris*-*Empetrum nigrum* heath. *Plantago* spp., particularly *P. maritima*, are constant associates and may act as hosts. Lowland.

Native. This tetraploid species was first described in 1929, and despite its relative distinctiveness, past confusion with other species and current under-recording make it difficult to assess changes in distribution. However, populations appear to be relatively stable with only limited losses through cultivation of cliff-tops. In more basic situations, *E. marshallii* is often replaced by its hybrid with *E. nemorosa*. There are many more recent records than when mapped by Wigginton (1999).

Endemic.

References: Atlas Supp. (57b), Stewart *et al.* (1994).

F. J. RUMSEY

Euphrasia rotundifolia

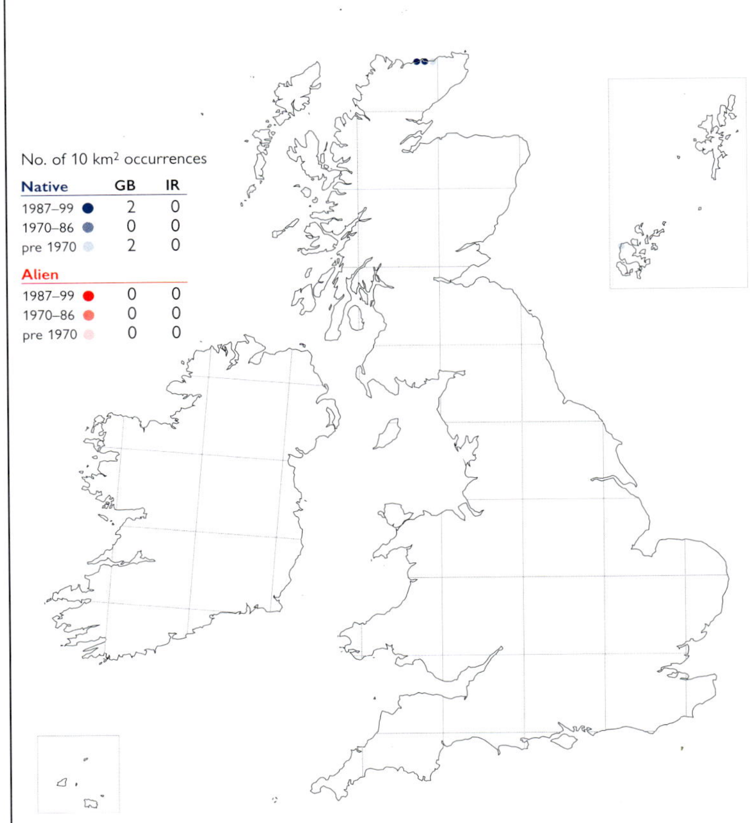

An annual of flushed basic turf on sea-cliffs. At its currently known sites, *Primula scotica* is a constant associate. Lowland.

Native. Though a re-interpretation of herbarium material has resulted in a more restrictive view of this tetraploid taxon than that adopted in Perring & Sell (1968), its taxonomic status remains uncertain. It is apparently derived through hybridisation between *E. foulaensis*, *E. ostenfeldii* and *E. marshallii* and is sometimes hardly distinguishable from the last, with which it regularly grows, when both are dwarfed.

Endemic.

References: Atlas Supp. (56c), Wigginton (1999).

F. J. RUMSEY

Euphrasia campbelliae

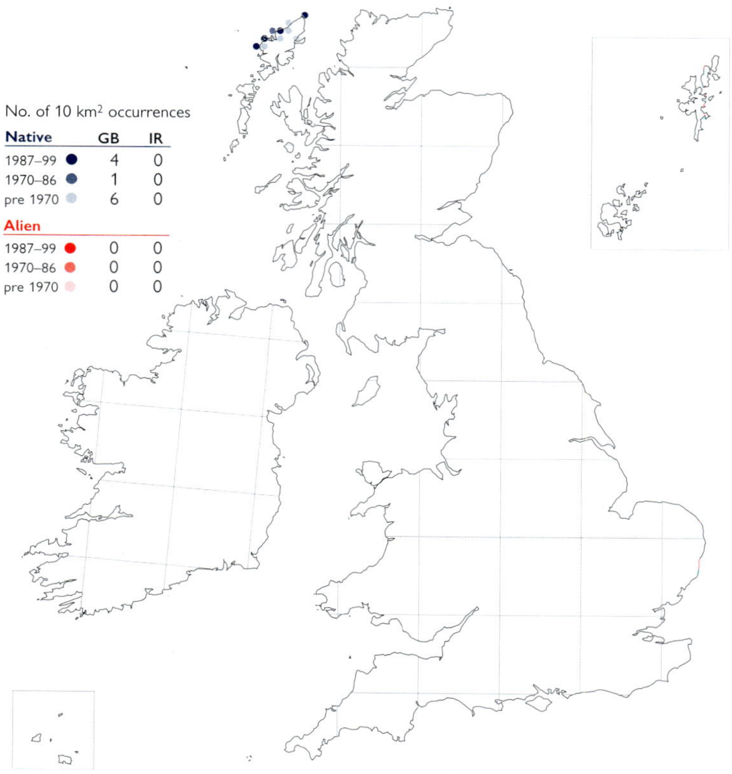

An annual of damp, mossy, grazed heathy turf near the sea, in sedge-rich communities or in grassy dwarf-shrub heath, especially where *Calluna vulgaris* and *Erica tetralix* are kept short by grazing. Lowland.

Native. This tetraploid taxon was first described as a species in 1940, and is probably of complex hybrid origin. Hybrids are formed with *E. confusa*, *E. micrantha* and *E. nemorosa* without apparent threat to the integrity of the few known populations of the species. There is no overall change in its distribution, although there are more recent records now than when mapped by Wigginton (1999).

Endemic.

Reference: Atlas Supp. (57a).

F. J. RUMSEY

Euphrasia micrantha

Native	**GB**	**IR**
1987–99	376	86
1970–86	184	13
pre 1970	381	91
Alien		
1987–99	0	0
1970–86	0	0
pre 1970	0	0

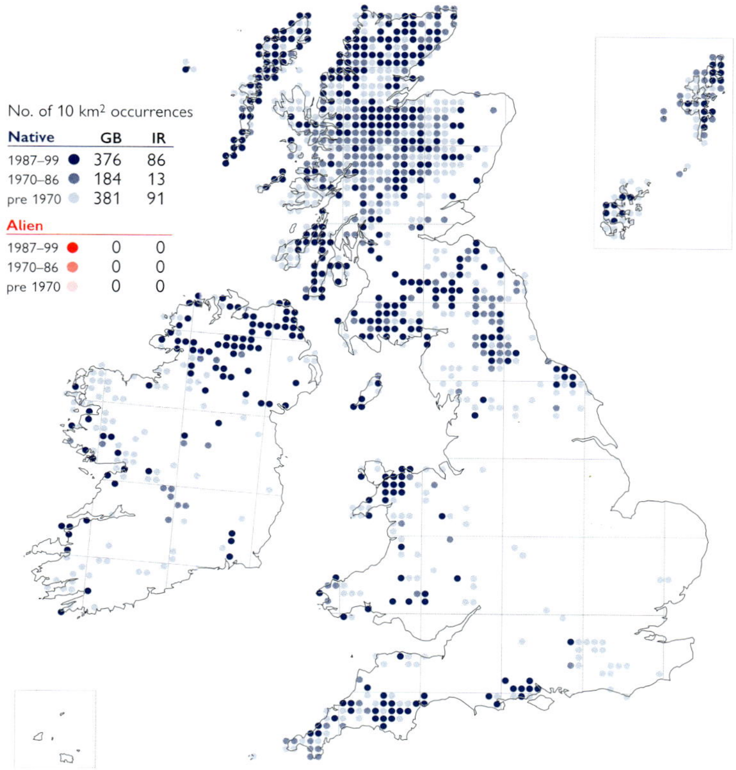

An annual of dry or damp places on acid heaths and moorland. It also grows on open clay and sandy substrates in disturbed habitats, such as disused gravel-pits and old airfields. 0–530 m (Redburn Common, Co. Durham).

Native. This tetraploid species has declined in S.E. England, where it may now be almost extinct. Elsewhere, many apparent losses are due to under-recording or former confusion with other taxa, particularly *E. scottica*, and it is likely to be largely stable. Many populations show signs of hybridisation with *E. nemorosa*. In N.W. Scotland it is supposedly largely replaced by *E. micrantha* × *E. scottica*, but there are very few records of this hybrid.

European Temperate element.

References: Atlas Supp. (55a), Hultén & Fries (1986), Meusel *et al.* (1978), Rich & Jermy (1998), Yeo (1978).

F. J. RUMSEY

Euphrasia scottica

Native	**GB**	**IR**
1987–99	298	36
1970–86	152	7
pre 1970	151	32
Alien		
1987–99	0	0
1970–86	0	0
pre 1970	0	0

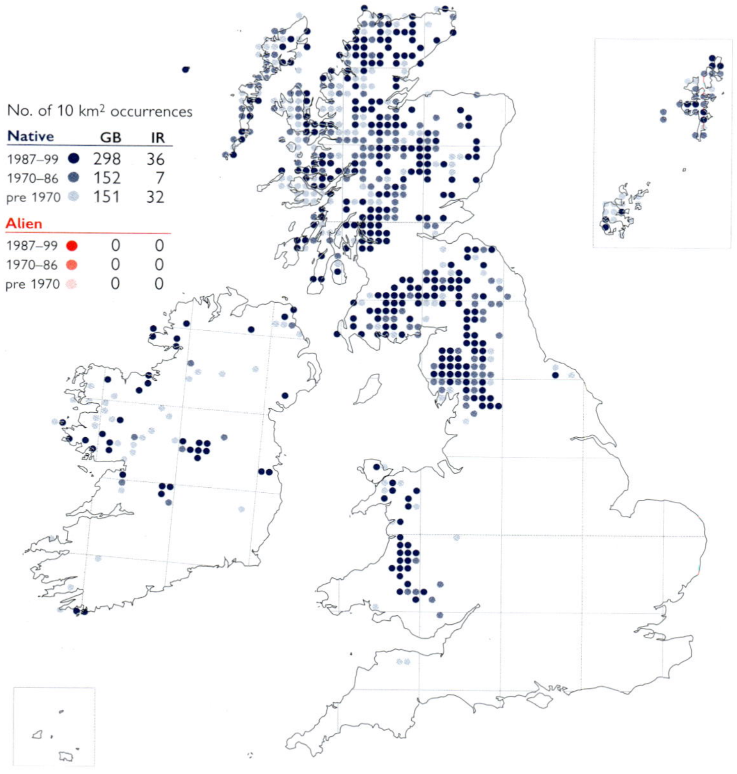

An annual which is particularly associated with flush communities and wet moorland in upland areas. It is similar to *E. micrantha* and perhaps not specifically distinct from it (Stace, 1997). 0–915 m (Sgurr na Lappaich, Glen Strathfarrar, Easterness).

Native. The distribution of this tetraploid species is somewhat uncertain because of confusion with other taxa and hybridisation. It is certainly under-recorded. In N.W. Scotland it is supposedly largely replaced by *E. micrantha* × *E. scottica*, but there are very few records of this hybrid.

European Boreal-montane element.

References: Atlas Supp. (55b), Hultén & Fries (1986), Meusel *et al.* (1978).

F. J. RUMSEY

Euphrasia heslop-harrisonii

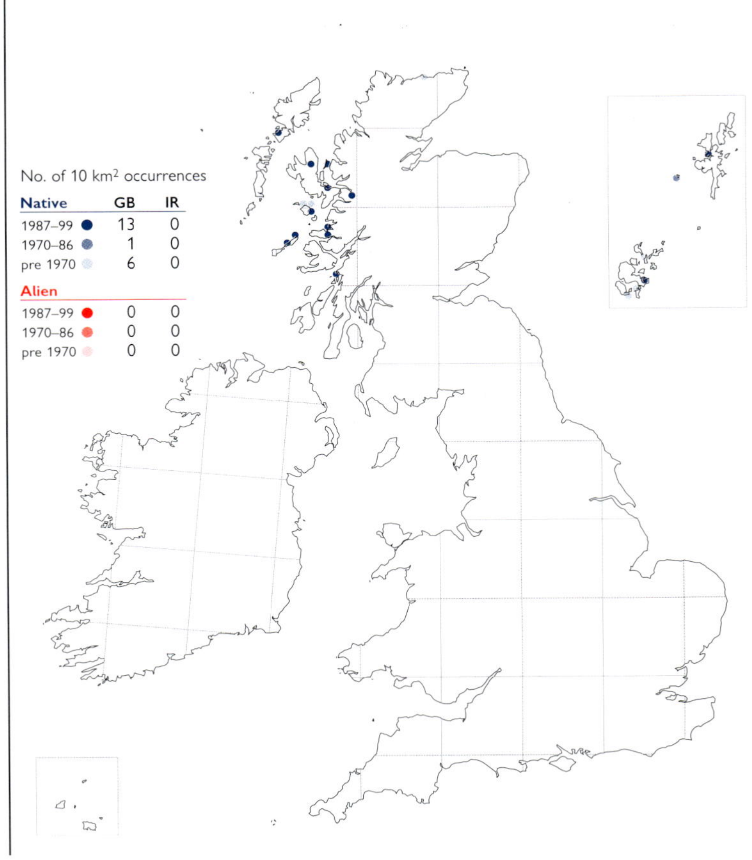

No. of 10 km² occurrences

Native		GB	IR
1987–99	●	13	0
1970–86	●	1	0
pre 1970	●	6	0
Alien			
1987–99	●	0	0
1970–86	●	0	0
pre 1970	●	0	0

An annual largely restricted to turfy areas in saltmarshes immediately above the high water mark, where it is associated with *Plantago* spp. More rarely it occurs on grassy banks in the spray zone. Lowland.

Native. Though first described in 1945, this tetraploid species was poorly known until clarified by Yeo (1978). Since it was mapped by Perring & Sell (1968) it has been recorded at more sites, but may still be under-recorded. *E. heslop-harrisonii* is dependent on the continuation of grazing of its saltmarsh habitat.

Endemic.

References: Atlas Supp. (58a), Wigginton (1999).

F. J. RUMSEY

Euphrasia salisburgensis

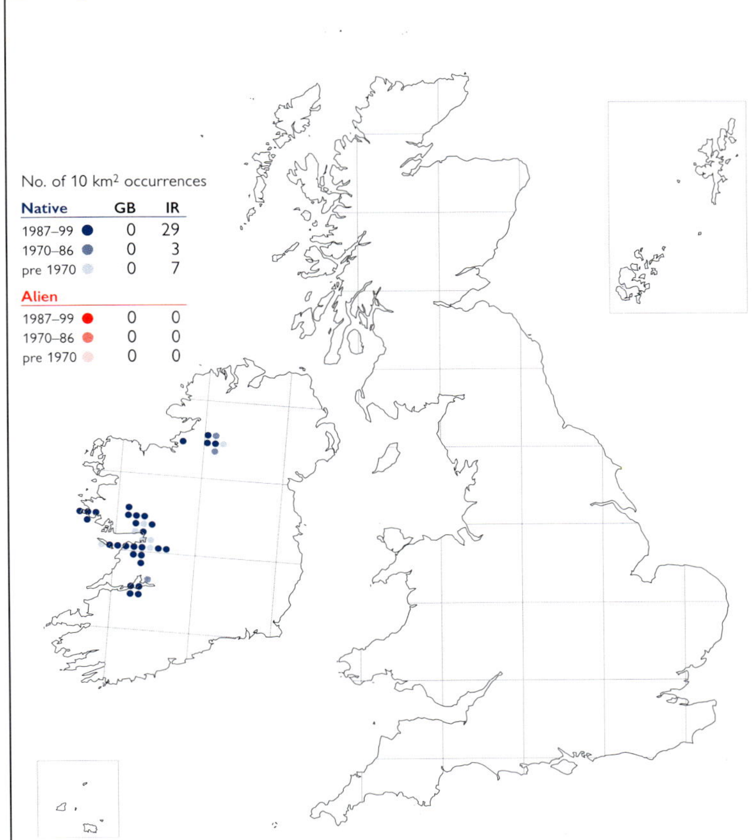

No. of 10 km² occurrences

Native		GB	IR
1987–99	●	0	29
1970–86	●	0	3
pre 1970	●	0	7
Alien			
1987–99	●	0	0
1970–86	●	0	0
pre 1970	●	0	0

An annual of calcareous grasslands in karstic areas, on maritime dunes and montane limestone cliffs. 0–550 m (Ben Bulben, Co. Sligo).

Native. The distribution of this tetraploid species appears to be stable. Plants from cliffs on Ben Bulben approach the boreal species *E. lapponica*, but may alternatively be derived through introgression with *E. frigida*. Hybrids between *E. salisburgensis* and other *Euphrasia* spp. are rare, and are largely but not completely sterile. A specimen from Yorkshire collected in 1885/6 is now considered to have resulted from confusion or mislabelling in the herbarium.

European Boreo-arctic Montane element; Irish material is referable to the endemic var. *hibernica*.

References: Atlas (235b), Hultén & Fries (1986), Meusel *et al.* (1978), Sledge (1975), Yeo (1975, 1978).

F. J. RUMSEY

Odontites vernus Red Bartsia

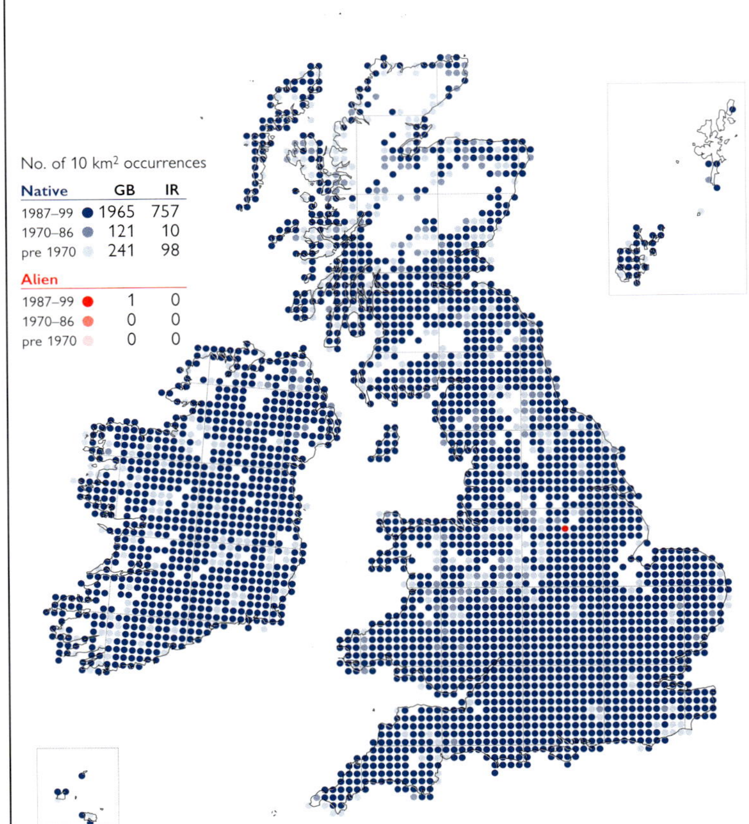

No. of 10 km² occurrences

Native		GB	IR
1987–99	●	1965	757
1970–86	●	121	10
pre 1970	●	241	98
Alien			
1987–99	●	1	0
1970–86	●	0	0
pre 1970	●	0	0

An annual root-hemiparasite of short, often trampled grasslands, tracks, waste places, the edges of arable fields, gravelly and rocky seashores and saltmarshes. 0–540 m (Nenthead, Cumberland).

Native (change –0.46). The distribution of this species has not altered significantly since the 1962 *Atlas*. Three subspecies are represented in our area: subsp. *vernus* is widespread, subsp. *serotinus* is concentrated in S. Britain and Ireland, and subsp. *litoralis* is mapped separately.

Eurasian Temperate element; widely naturalised outside its native range.

References: Atlas (235c), Atlas Supp. (63a-c), Hultén & Fries (1986).

F. J. RUMSEY

Odontites vernus subsp. *litoralis*

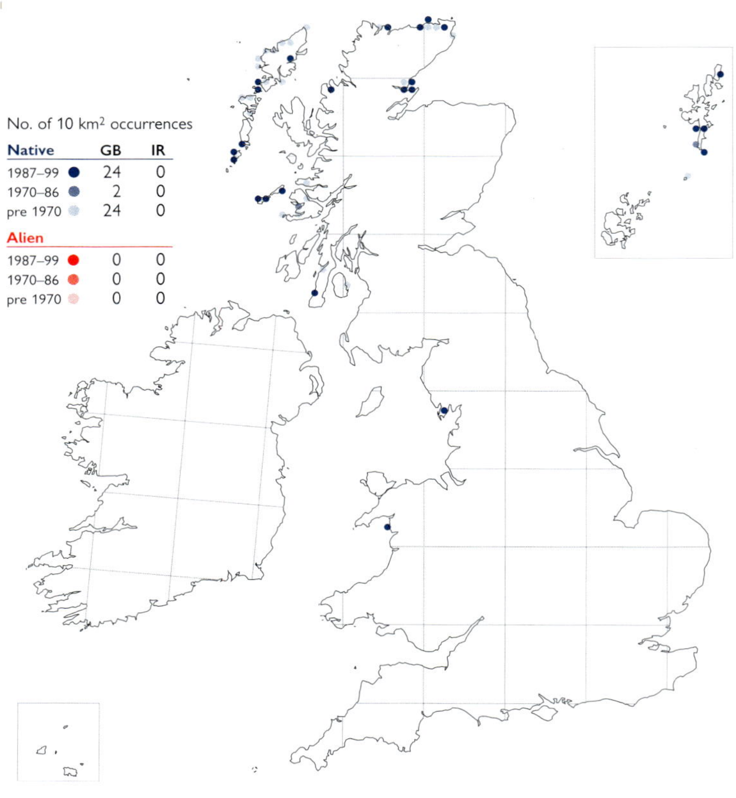

An annual root-hemiparasite of coastal grasslands on stabilised sand dunes, gravelly and rocky seashores, the upper margins of saltmarshes, and of waste ground near the sea. Lowland.

Native. Subsp. *litoralis* has been confused with depauperate specimens of the two weedy widespread subspecies, which may also occur in its coastal habitats. It was mapped by Perring & Sell (1968), but is still under-recorded.

European Boreo-temperate element. Subsp. *litoralis* was described from the Baltic; plants from S. Sweden southwards may be a distinct taxon, subsp. *pumilus*.

References: Atlas Supp. (63c), Hultén & Fries (1986), Sell (1967), Snogerup (1982).

F. J. RUMSEY

Bartsia alpina Alpine Bartsia

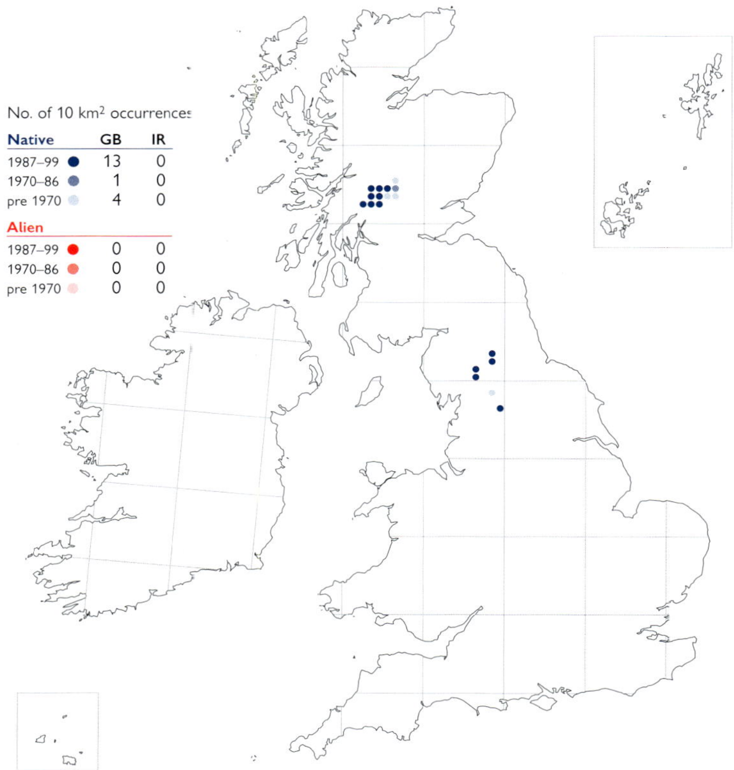

A shortly rhizomatous perennial of base-rich soils. In England it is a plant of the drier hummocks in basic flushes and runnels in damp upland pastures, and of steep, flushed, species-rich banks. In the Breadalbanes (Mid Perth) it grows on the periodically inundated ledges of mica-schist crags. Seed-set is poor. From 245 m at Orton (Westmorland) to 950 m on Beinn Heasgarnich (Mid Perth).

Native (change –0.10). *B. alpina* sites have been lost from pastures through overgrazing, trampling by cattle and drainage, but ledge communities are largely unaffected. The overall distribution is unchanged from that shown in the 1962 *Atlas*.

European Arctic-montane element; also in N. America.

References: Atlas (236a), Hultén & Fries (1986), Meusel *et al.* (1978), Wigginton (1999).

F. J. RUMSEY

Parentucellia viscosa Yellow Bartsia

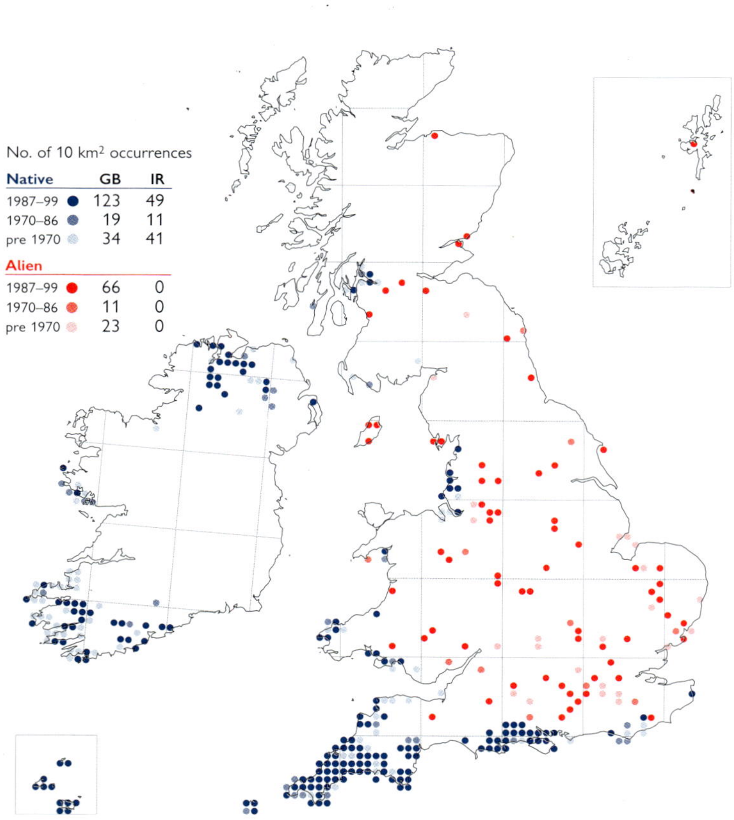

A hemiparasitic annual of damp, open grassy places on sandy soils, often by tracks. It normally occurs in drier dune-slacks and in reclaimed heath-pasture, but is also found on pathsides, rough and scrubby grassland and field-borders, and increasingly in re-seeded amenity grasslands and waste places. It thrives on disturbance. Lowland.

Native (change +0.64). This species has increased northwards and eastwards in Britain, largely through introductions from seed mixtures. Conversely, the re-seeding of old pasture has led to some decline over the same period at inland sites in S.W. England. In Ireland, it appears to be relatively stable in the north, but it has declined significantly in the south-west.

Mediterranean-Atlantic element; widely naturalised outside its native range.

References: Atlas (235d), Bolòs & Vigo (1995), Stewart *et al.* (1994).

F. J. RUMSEY

Rhinanthus angustifolius Greater Yellow-rattle

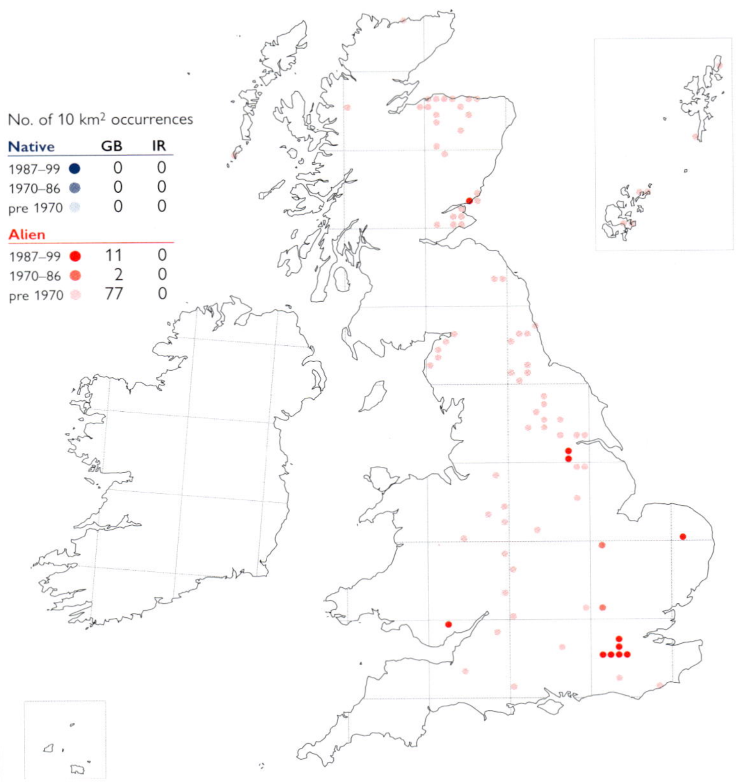

This annual root-parasite was formerly a widespread weed of arable land in E. Britain. However, most of the remaining sites are on the North Downs, in grassland and open scrub on chalk. In Lincolnshire, it occurs on peat in an area of cleared *Pteridium* and on railway ballast. In Angus, a tiny colony survives in sandy coastal grassland. Lowland.

Neophyte (change –0.10). This species was first recorded in the wild in 1724. Nearly all its pre-1970 sites were lost before 1930. It was not recognised on the North Downs until 1966 (Lousley, 1976a). Hay cutting after flowering at some of its North Downs sites may have assisted its spread. Some old records may be errors for the variable *R. minor*.

Eurosiberian Boreo-temperate element.

References: Atlas (233c), Hultén & Fries (1986), Meusel *et al.* (1978), Wigginton (1999).

F. J. RUMSEY

Rhinanthus minor Yellow-rattle

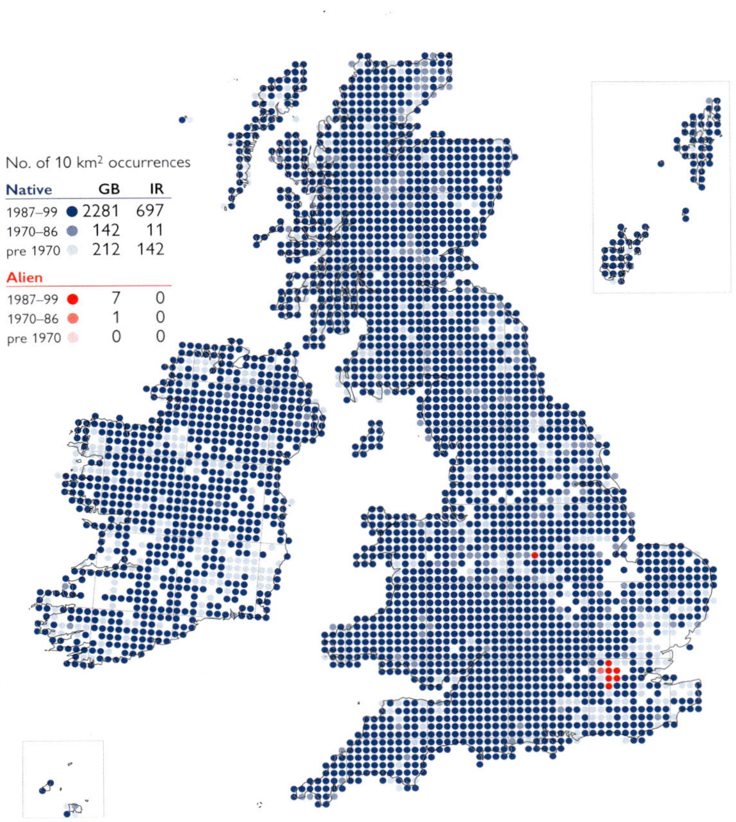

An annual root-hemiparasite of nutrient-poor grasslands, including permanent pastures, hay meadows, the drier parts of fens, flushes in lowland and upland grasslands, and on montane ledges; also on roadsides and waste ground. 0–1065 m (Ben Lawers, Mid Perth).

Native (change –0.49). *R. minor* declined throughout the 20th century in semi-natural grasslands. It is now included in many wild-flower seed mixtures. It is very variable and six intergrading and apparently inter-fertile subspecies have been recognised in our area. However, whilst they show broad geographic and ecological distinctions, some local races cannot be clearly assigned to any of them.

European Boreo-temperate element; widely naturalised outside its native range.

References: Atlas (233d), Grime *et al.* (1988), Hultén & Fries (1986), Meusel *et al.* (1978).

F. J. RUMSEY

Rhinanthus minor subsp. *minor*

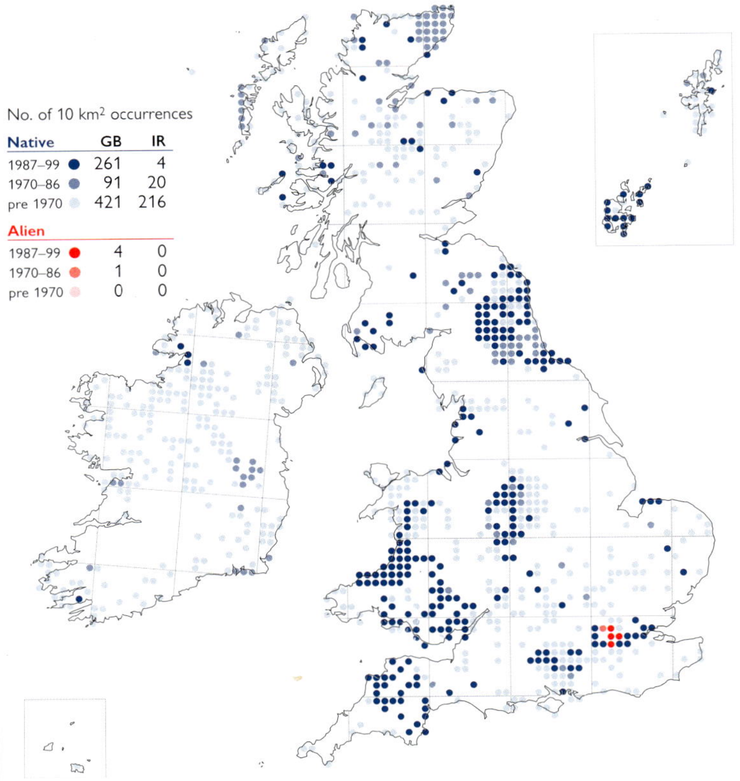

A little-branched, early-summer flowering, ecotype found in a range of lowland grassy, well-drained, nutrient-poor, basic habitats, including unimproved pasture, hay meadows, road verges and rough grassland. 0–570 m (Moor House, Westmorland).

Native. The poor coverage of this subspecies in many areas reflects the fact that most recorders have not distinguished the type subspecies from the others. The remarks on the trends in the distribution of *R. minor* also apply to this subspecies.

European Boreo-temperate element; occurring almost throughout the range of the species but rarer in the boreal zone than subsp. *stenophyllus*.

Reference: Atlas Supp. (53a).

F. J. RUMSEY

Rhinanthus minor subsp. *stenophyllus*

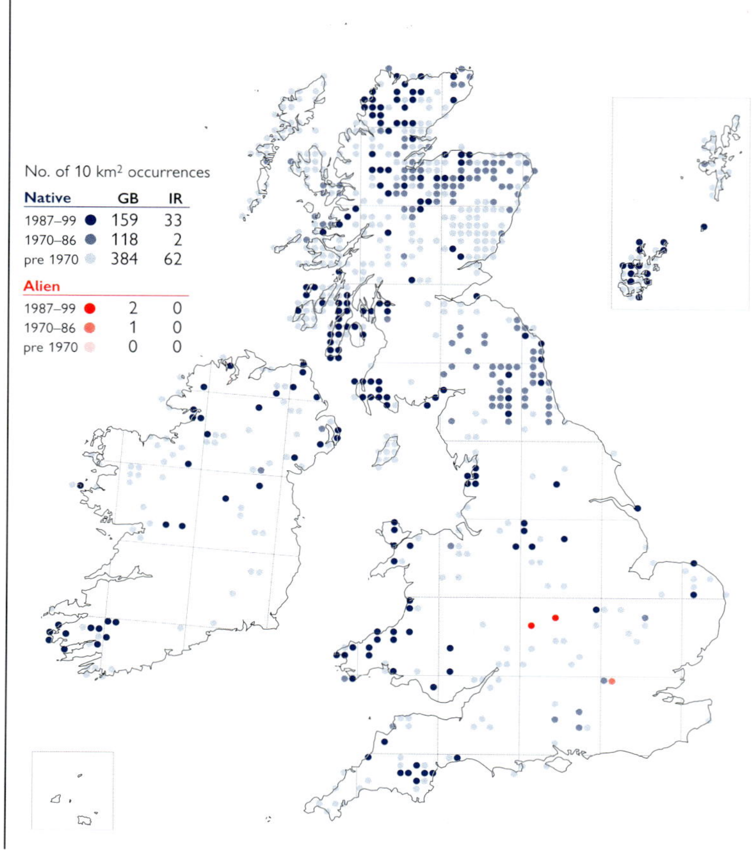

A much-branched, floriferous, late-flowering ecotype which occurs in fens and damp grasslands in S. England, but largely replaces subsp. *minor* in a wide variety of habitats over much of N. England and Scotland. 0–500 m (Loch Callater, S. Aberdeen).

Native. A widespread taxon which is certainly under-recorded in northern and western Britain and in Ireland. The losses in S.E. England are probably real as much of its habitat there has been improved.

European Boreo-montane element.

Reference: Atlas Supp. (53b).

F. J. RUMSEY

Rhinanthus minor subsp. *monticola*

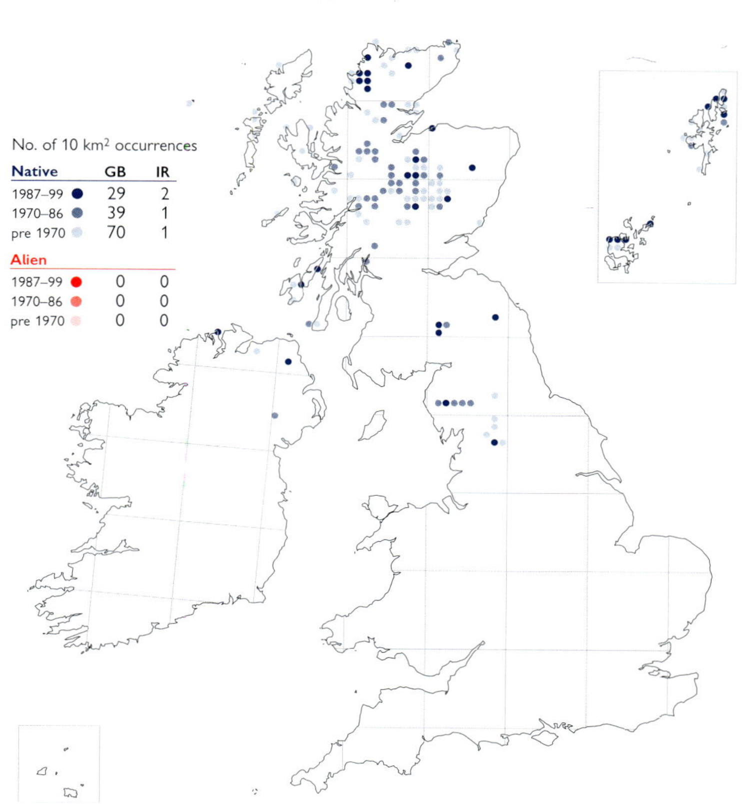

A squat, branched, but few-flowered, autumnal-flowering ecotype of base-rich grassland, serpentine heath and basic rock ledges. It grows at sea level in N. & W. Scotland and W. Ireland, is perhaps most frequent above 200 m and ascends to at least 855 m on Helvellyn (Cumberland).

Native. This often heavily pigmented, dull-flowered plant is not unduly difficult to identify but it is probably greatly under-recorded, as it suffers from the same neglect as the other segregates of this difficult complex. It is unlikely to have changed greatly in distribution in the recent past.

European Boreo-montane element.

Reference: Atlas Supp. (53c).

F. J. RUMSEY

Rhinanthus minor subsp. *calcareus*

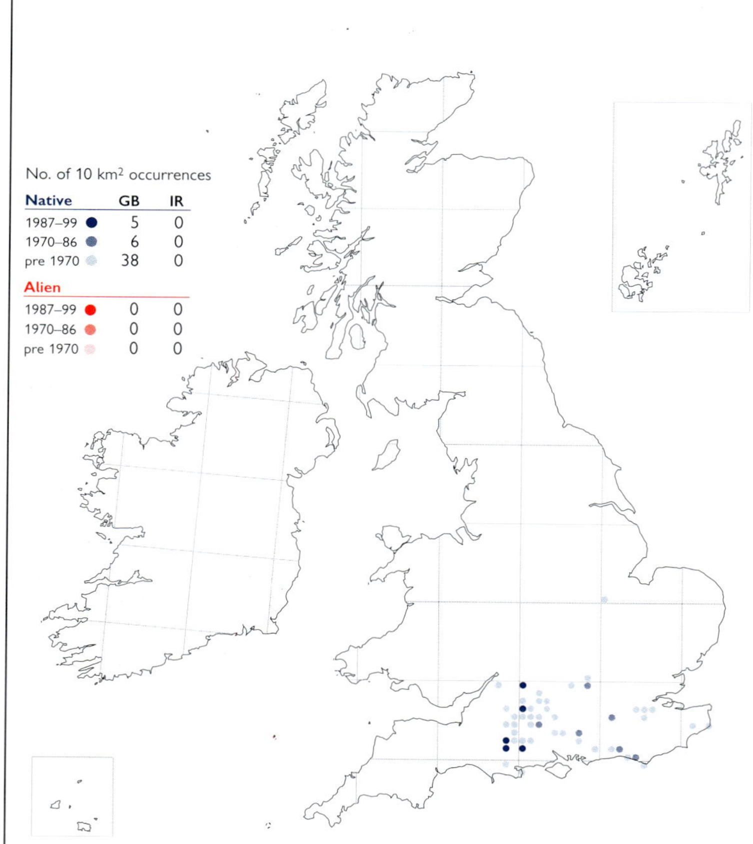

A much-branched, many noded, late-flowering ecotype of dry, basic downland turf. Lowland.

Native. Almost all of the records mapped are those collected from herbaria by Perring & Sell (1968). This taxon is possibly declining with the loss of semi-natural grassland, but it is much too poorly recorded to be sure. Plants from the Burren resemble this subspecies but British experts do not regard them as sufficiently similar to be included within it (Webb & Scannell, 1983).

A little-known taxon, not recorded outside our area.

Reference: Atlas Supp. (52b).

F. J. RUMSEY

Rhinanthus minor subsp. *lintonii*

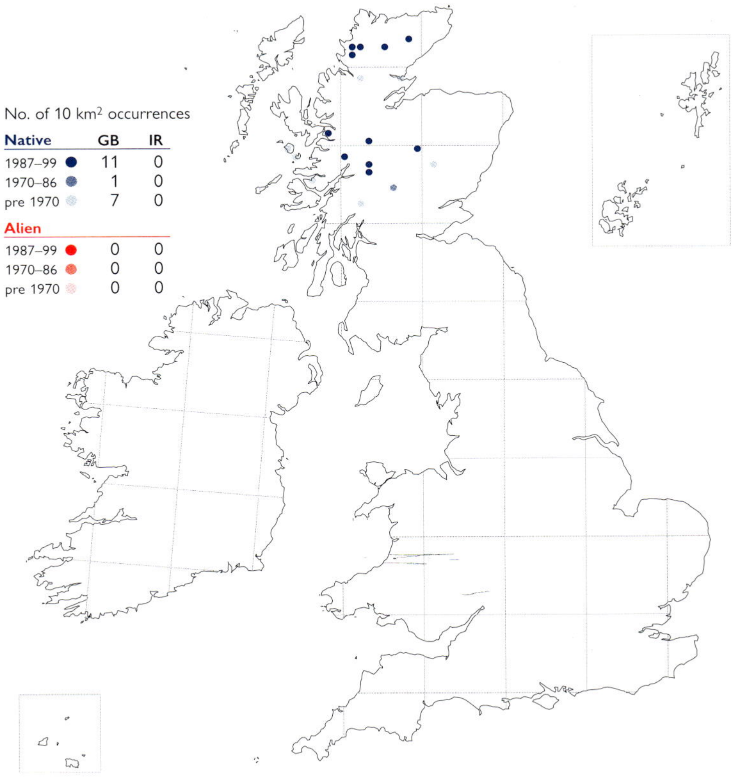

A late-flowering ecotype of montane grassland, flushes and damp rock ledges. It has been recorded from 550 m (Coire Chailein, Main Argyll) to 915 m or more (Ben Avon, Banffs.).

Native. This subspecies is probably derived from hybridisation between subsp. *borealis* and subspp. *monticola* and *stenophyllus* (Stace, 1997). It has not been previously mapped and is probably under-recorded.

Not recorded outside Britain.

Reference: Sell (1967).

F. J. RUMSEY

Rhinanthus minor subsp. *borealis*

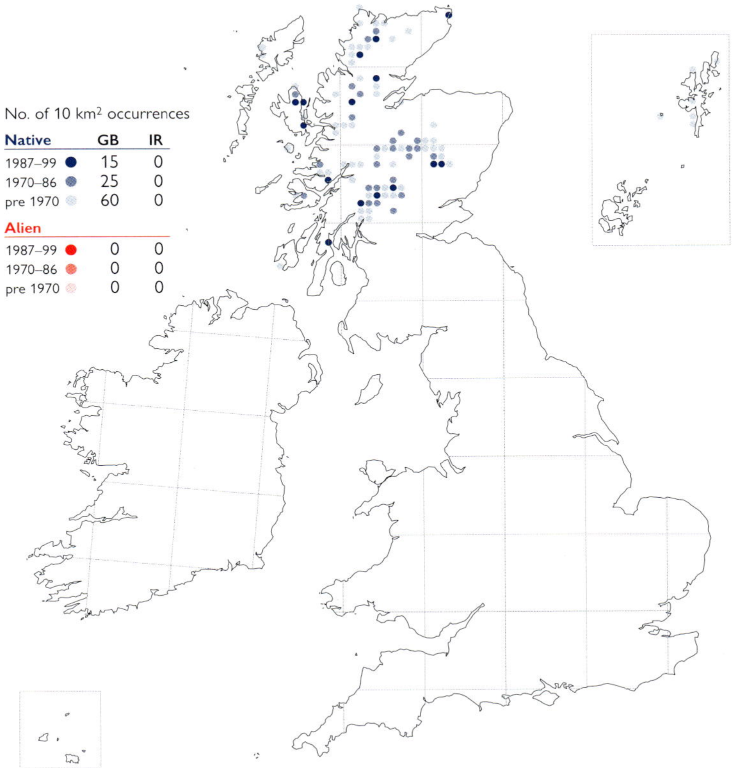

A small, unbranched, late-flowering ecotype of grassy places, cliffs and flushes on mountains. It is also recorded from sea-cliffs in Shetland. An upland plant ascending to 1065 m on Ben Lawers (Mid Perth).

Native. Like the other subspecies of *R. minor*, this taxon is probably under-recorded. However, some records may be misidentifications of the less well-known subsp. *lintonii*.

European Boreo-arctic Montane element; Hultén & Fries (1986) treat this taxon as conspecific with the N. American *R. groenlandicus*.

References: Atlas Supp. (53d), Meusel *et al.* (1978).

F. J. RUMSEY

Pedicularis palustris Marsh Lousewort

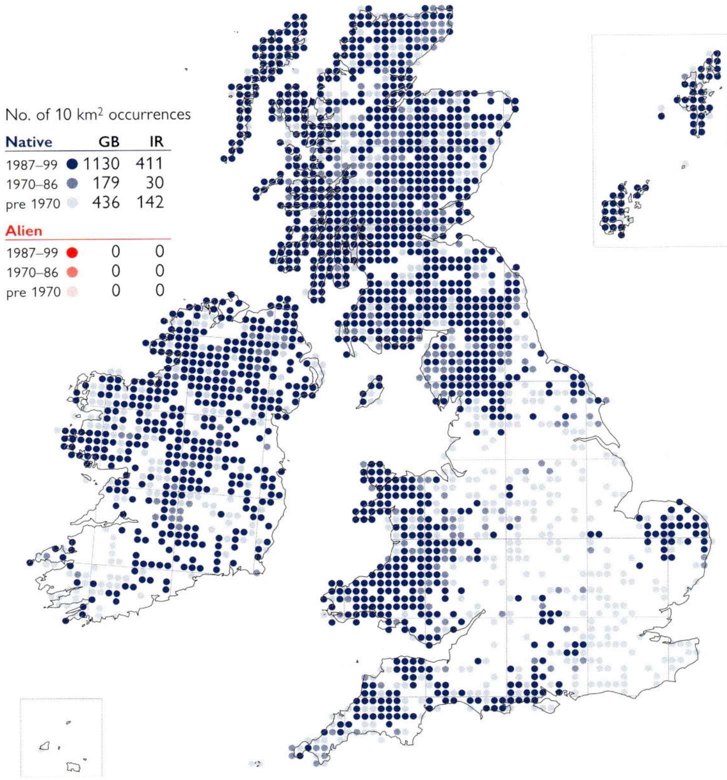

An annual to biennial root-hemiparasitic herb of a wide range of base-rich to acidic, moist habitats, including wet heaths, valley bogs, wet meadows, ditches, fens and hillside flushes. Its sites are usually more enriched than those preferred by *P. sylvatica*. 0–550 m (E. Highlands of Scotland).

Native (change –0.88). The decline of *P. palustris* in C. & S. England and many parts of Ireland was already apparent in the 1962 *Atlas* and the species has suffered a further substantial decline since then, mostly through habitat loss due to drainage and agricultural improvement. Elsewhere, there are only local losses.

European Boreo-temperate element.

References: Atlas (233a), Hultén & Fries (1986), Meusel *et al.* (1978).

F. J. RUMSEY

Pedicularis sylvatica Lousewort

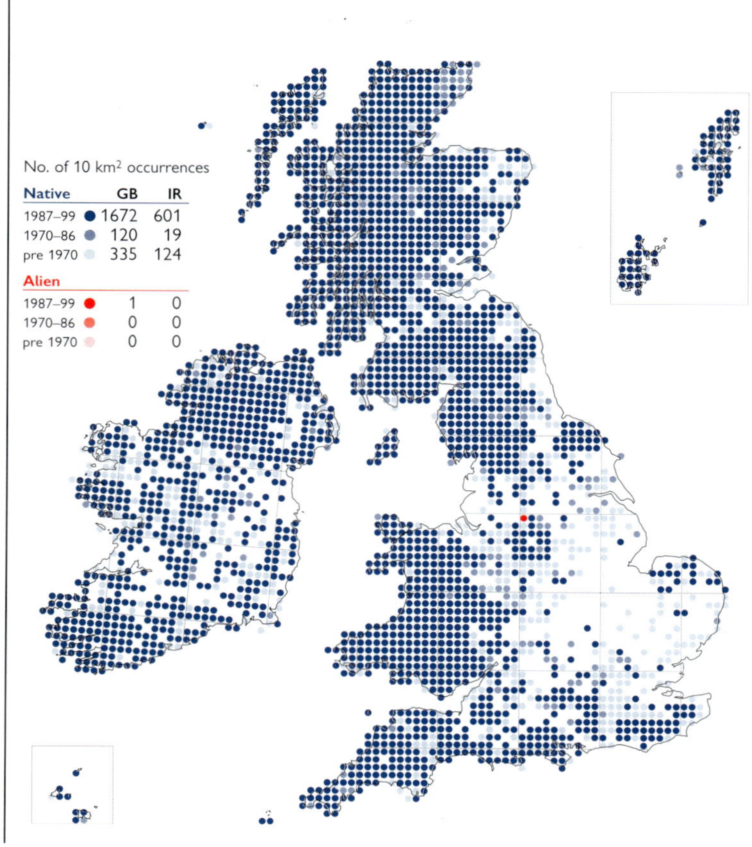

A perennial, rarely biennial, root-hemiparasitic of acidic soils, found on damp grassy heaths, moorlands, upland flushed grasslands and the drier parts of bogs and marshes. 0–915 m (Macgillycuddy's Reeks, S. Kerry).

Native (change –1.28). *P. sylvatica* was mapped as 'all records' in the 1962 *Atlas*. There has been a widespread decline in S. and E. England, which can be attributed to the loss of heathlands and unimproved grasslands. Analysis of the database reveals that most losses have occurred since 1950. Two subspecies occur in our area: subsp. *sylvatica* occurs throughout the range of the species; subsp. *hibernica* is mapped separately.

European Temperate element.

References: Atlas (233b), Hultén & Fries (1986), Meusel *et al.* (1978).

F. J. RUMSEY

Pedicularis sylvatica subsp. *hibernica*

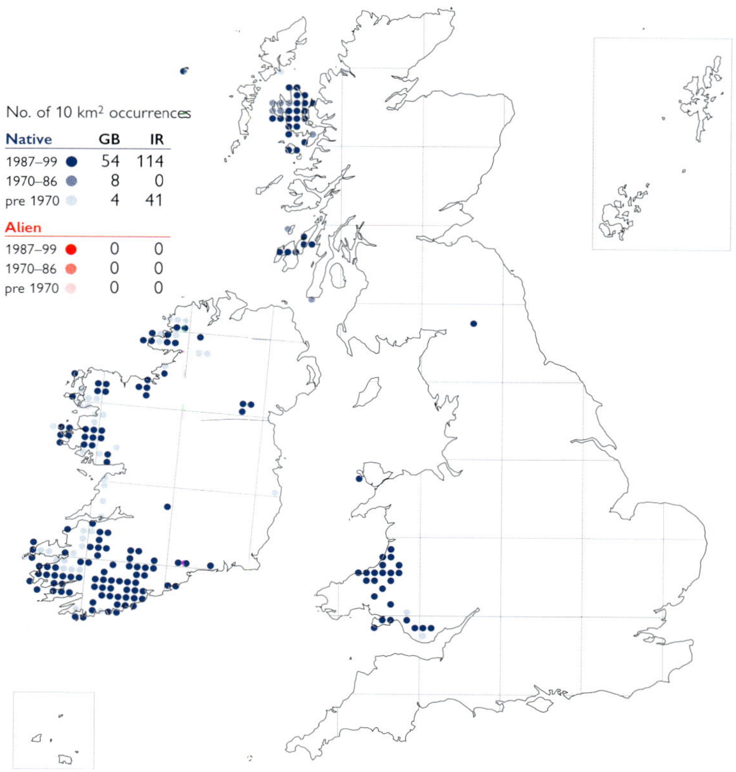

This subspecies largely replaces subsp. *sylvatica* in W. Ireland, where it occurs on damp moorlands and blanket bogs. 0–330 m (E. of Tregaron, Cards.).

Native. This taxon was first described and mapped by Webb (1956) from Ireland and the Outer Hebrides. It has only recently been recorded in England and Wales and may be more widespread in W. Britain than the map suggests. Intermediates with subsp. *sylvatica* occur, especially in the eastern portion of its Irish range.

Oceanic Boreal-montane element; its relationship to the pubescent subsp. *lusitanica* of the Iberian peninsula requires clarification.

References: Atlas Supp. (52a), Meusel *et al.* (1978), Rich (1994b).

F. J. RUMSEY

Lathraea squamaria Toothwort

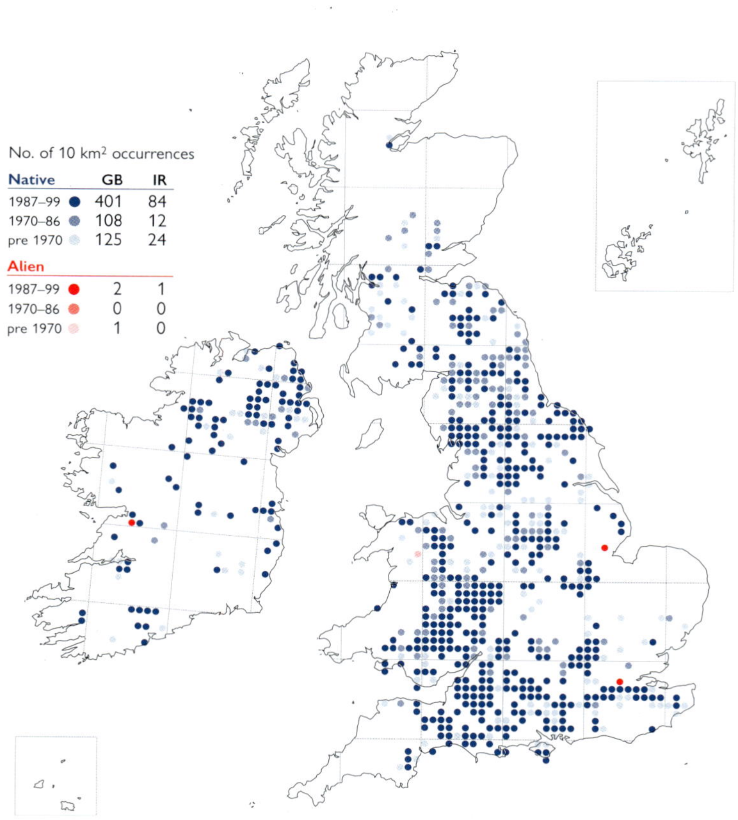

An annual or perennial herb, parasitic on the roots of a range of woody plants, especially *Corylus, Fraxinus* and *Ulmus glabra*. Its typical habitats include deciduous woodland, hedgerows, and river and stream banks. Generally lowland, but reaching 350 m at Alston (Cumberland).

Native (change –0.36). The distribution of *L. squamaria* is similar to that shown in the 1962 *Atlas*. There have been 10-km square losses throughout its range, the reasons for which are unclear.

European Temperate element; also in C. Asia.

References: Atlas (236b), Hultén & Fries (1986), Meusel *et al.* (1978).

M. J. Y. FOLEY

Lathraea clandestina Purple Toothwort

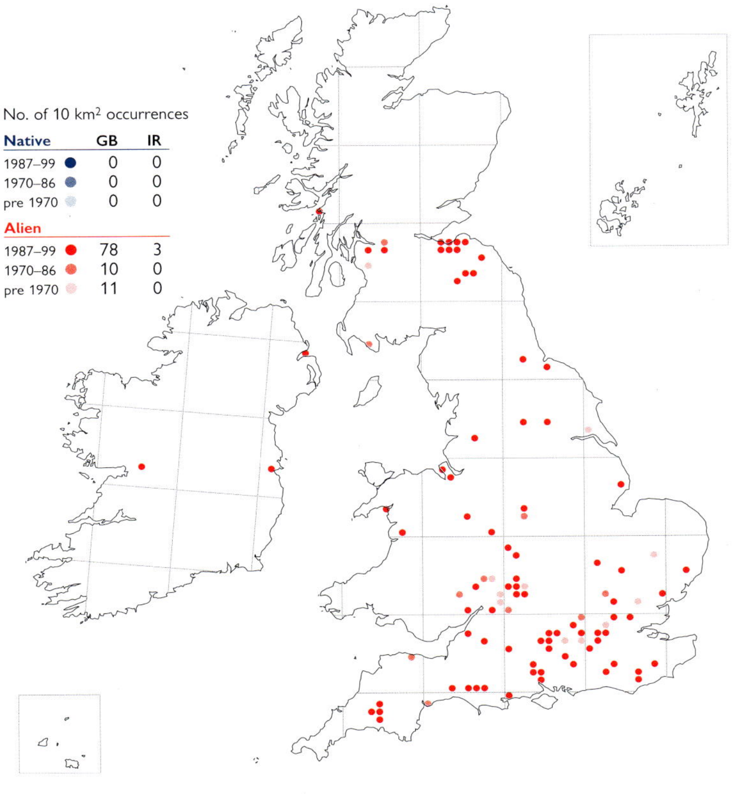

A root parasite mainly found on species of *Alnus*, *Salix* and *Populus*. It usually occurs in damp, shaded places in open woodland, coppices and along hedgerows, and especially near stream and river margins where dispersal may be effected by flood water. Lowland.

Neophyte. *L. clandestina* was introduced to cultivation in Britain around 1888. It is grown in gardens as an attractive curiosity, from where many populations are likely to have originated. It was first reported from the wild in 1908 at Coe Fen (Cambs.), where it was probably deliberately planted and where it still survives.

A Suboceanic Temperate species, native to Belgium, France, Spain and Italy.

References: Atkinson (1996), Bolòs & Vigo (1995), Meusel *et al.* (1978).

M. J. Y. FOLEY

Orobanche purpurea Yarrow Broomrape

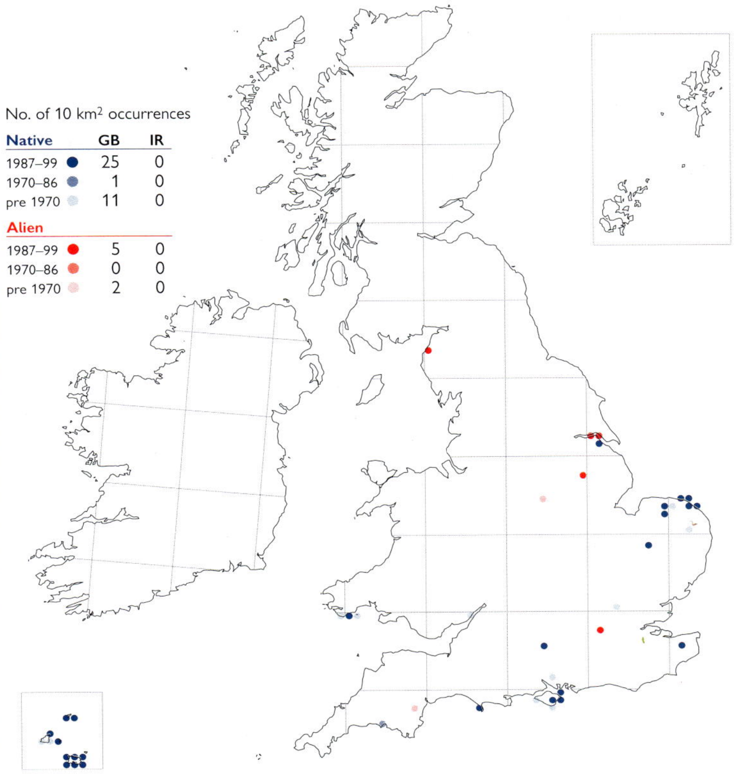

An annual or possibly perennial herb, parasitic on *Achillea millefolium*. It typically occurs on dry, somewhat basic soils in cliff-top grassland and on roadsides and grassy banks, usually near the sea. More rarely, it occurs in disturbed artificial habitats. Flowers can reappear after decades of absence, suggesting that the seeds are long-lived or that plants can persist without flowering for many years. Lowland.

Native (change +0.50). In recent years the loss or decline of some populations of this species has to some extent been balanced by the appearance, or re-appearance, of others. The population at Maryport docks (Cumberland) has increased rapidly since it was first found in 1983.

European Temperate element.

References: Atlas (236c), Hultén & Fries (1986), Rumsey & Jury (1991), Wigginton (1999).

M. J. Y. FOLEY

Orobanche rapum-genistae Greater Broomrape

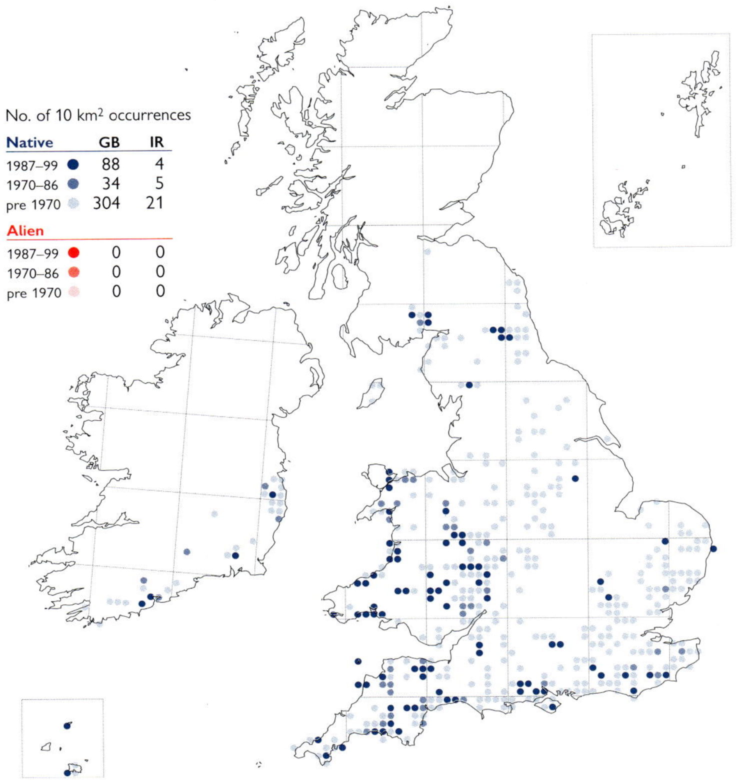

A perennial root parasite of leguminous shrubs, especially *Ulex europaeus* and *Cytisus scoparius*, but also known to occur occasionally on *Genista tinctoria*. Its habitat, governed by that of its hosts, is mainly scrub, but hedge banks and track-sides are also favoured. Lowland.

Native (change –0.35). This species suffered a dramatic decline in the 19th and early 20th century. This is largely unexplained, although changes in land-use were probably at least partly responsible. The decline appears to have halted by 1950, and subsequently populations have been found at a greater rate than they have been lost.

Suboceanic Southern-temperate element.

References: Atlas (236d), Curtis & McGough (1988), Meusel *et al.* (1978), Rumsey & Headley (1998), Rumsey & Jury (1991), Stewart *et al.* (1994).

M. J. Y. FOLEY

Orobanche caryophyllacea Bedstraw Broomrape

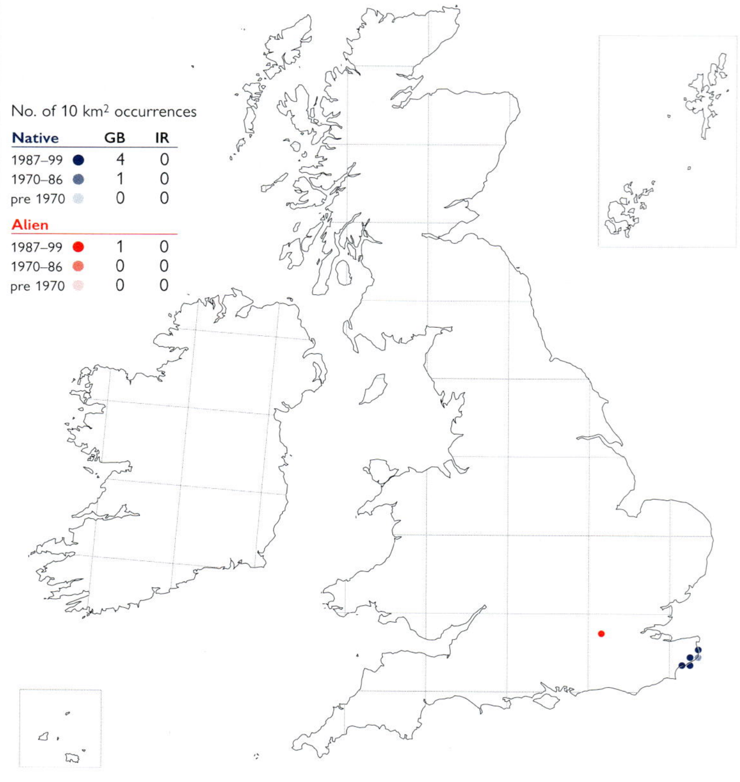

A root parasite mainly of *Galium mollugo* and *G. verum*, probably perennial and sometimes long-lived. It occurs in stabilised dune grassland, and in scrub and hedge banks on chalk downs and undercliffs. Most populations are small, but some of those on dunes are of a considerable size. Lowland.

Native (change +0.01). Populations of this species on the North Downs (E. Kent) are vulnerable, and may be declining because of habitat change, but those on dunes seem reasonably secure. Native records from other parts of Britain, including Scotland, are now considered to be errors.

European Temperate element, with a continental distribution in W. Europe.

References: Atlas (237b), Meusel *et al.* (1978), Rumsey & Jury (1991), Wigginton (1999).

M. J. Y. FOLEY

Orobanche elatior Knapweed Broomrape

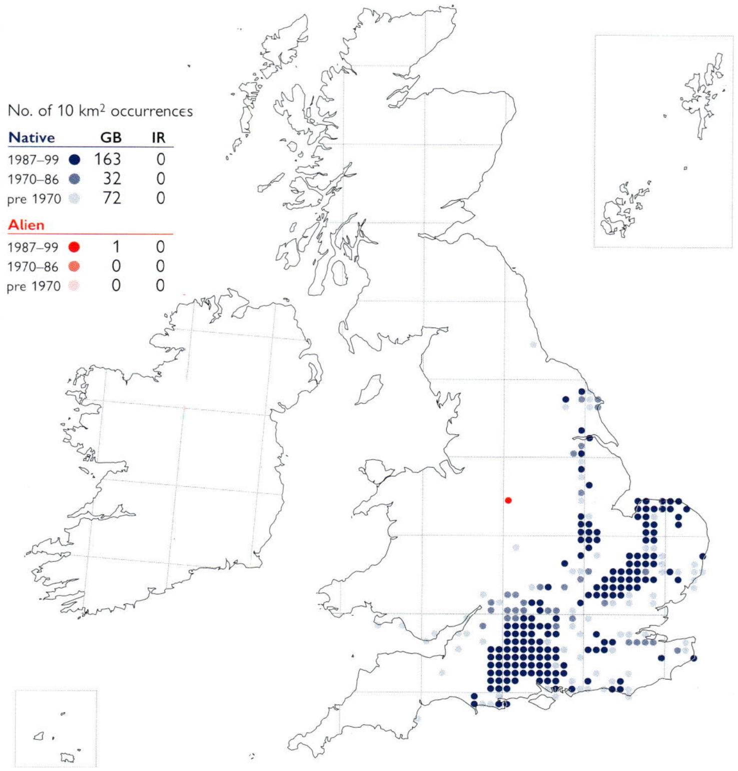

Apparently perennial, this species is almost exclusively parasitic upon *Centaurea scabiosa*, and is mainly found in chalk and limestone grassland. It may also form substantial populations in man-made habitats such as road verges, railway banks and quarries. Lowland.

Native (change –0.33). There appears to have been a gradual contraction in the range of *O. elatior*, especially in areas outside its core range where populations tend to be small. Most losses are probably due to habitat destruction. The distribution is stable in its core areas.

Eurosiberian Temperate element, with a continental distribution in W. Europe.

References: Atlas (237c), Hultén & Fries (1986), Meusel *et al.* (1978), Rumsey & Jury (1991).

M. J. Y. FOLEY

Orobanche alba Thyme Broomrape

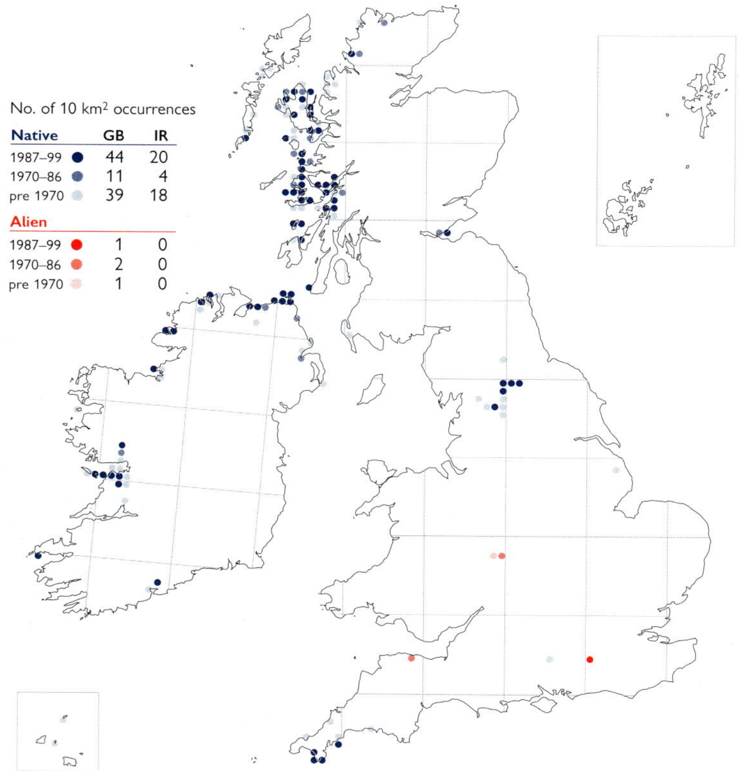

An annual, or possibly perennial, root parasite of *Thymus polytrichus*. Its principal habitat is base-rich rocky coastal slopes, but it also occurs inland on stabilised scree below limestone outcrops in N. England. Generally lowland, but reaching *c.* 490 m at Nappa Scar, Wensleydale (N.W. Yorks.).

Native (change –0.38). Populations of *O. alba* can vary greatly in size from year to year, but the overall range of the species appears to be stable. It is probably still present in many of the Scottish squares where it has not been recorded since 1987.

European Temperate element; also in C. Asia.

References: Atlas (237a), Ballantyne (1992), Hultén & Fries (1986), Rumsey & Jury (1991), Stewart *et al.* (1994).

M. J. Y. FOLEY

Orobanche reticulata Thistle Broomrape

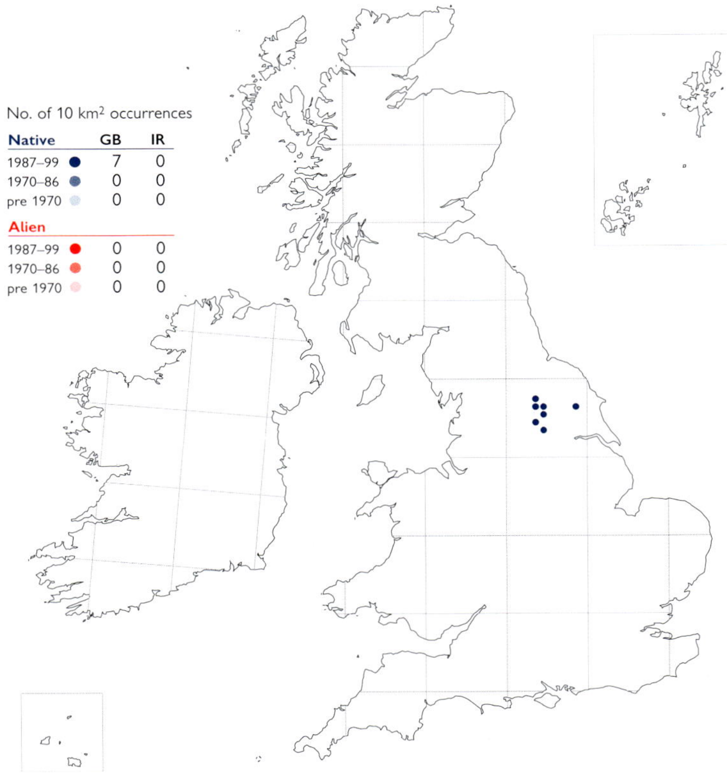

Probably perennial, this root parasite of thistles occurs in rough grassland, on road verges, and especially on river margins and flood plains where its frequent, sometimes transient, occurrence suggests that seed is readily dispersed by water. It is almost restricted to Magnesian limestone districts. Lowland.

Native. The distribution of *O. reticulata* appears to be stable, although many new populations have recently been found.

European Temperate element, with a continental distribution in W. Europe; also in C. Asia. Our plant is subsp. *procera*, which has a lowland distribution in Europe and is quite distinct from the montane subsp. *reticulata*.

References: Atlas (237b), Foley (1993, 2000), Hultén & Fries (1986), Meusel *et al.* (1978), Newlands & Smith (1998), Rumsey & Jury (1991), Wigginton (1999).

M. J. Y. FOLEY

Orobanche hederae Ivy Broomrape

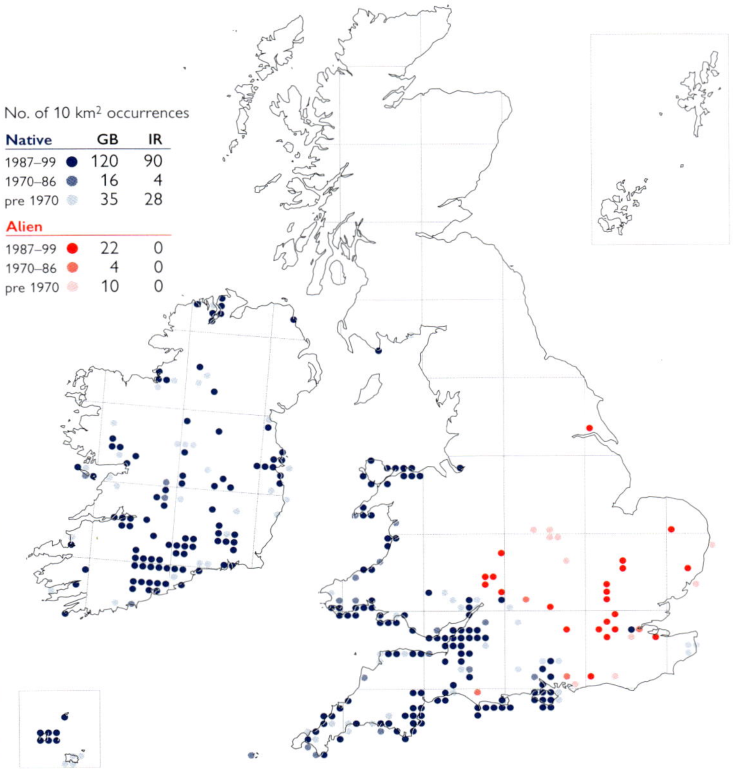

An annual or perennial parasite which grows on the roots of *Hedera helix*, especially subsp. *hibernica*, and, rarely, on other cultivated Araliaceae. Its habitat is that of its host and includes coastal cliffs, open rocky woodland, quarries, hedge banks and other similar habitats. Lowland.

Native (change +0.20). Western populations of *O. hederae* in coastal habitats appear to be stable. In S.E. England it has been recorded with increasing frequency in artificial habitats, including gardens, where it is probably introduced, although the possibility of long-range dispersal of the light seeds cannot always be ruled out.

Submediterranean-Subatlantic element.

References: Atlas (238b), Curtis & McGough (1988), Jones (1987), Meusel *et al.* (1978), Rumsey & Jury (1991), Stewart *et al.* (1994).

M. J. Y. FOLEY

Orobanche artemisiae-campestris Oxtongue Broomrape

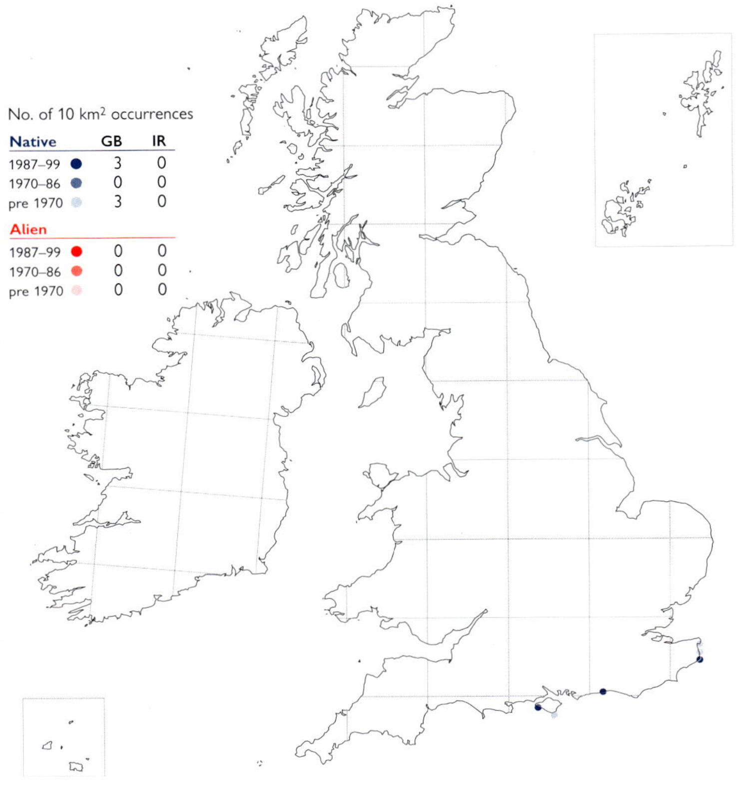

An annual or perennial occurring mainly on the ledges and fine calcareous debris of coastal chalk cliffs where it parasitises species of Asteraceae, especially *Picris hieracioides*. Lowland.

Native. This species is probably decreasing. Populations are usually quite small and some have been lost as a result of cliff falls and erosion. The inland records mapped in the 1962 *Atlas* are now thought to be erroneous (Stace, 1997).

European Southern-temperate element.

References: Atlas (238a), Hultén & Fries (1986), Meusel *et al.* (1978), Rumsey & Jury (1991), Wigginton (1999).

M. J. Y. FOLEY

Orobanche minor Common Broomrape

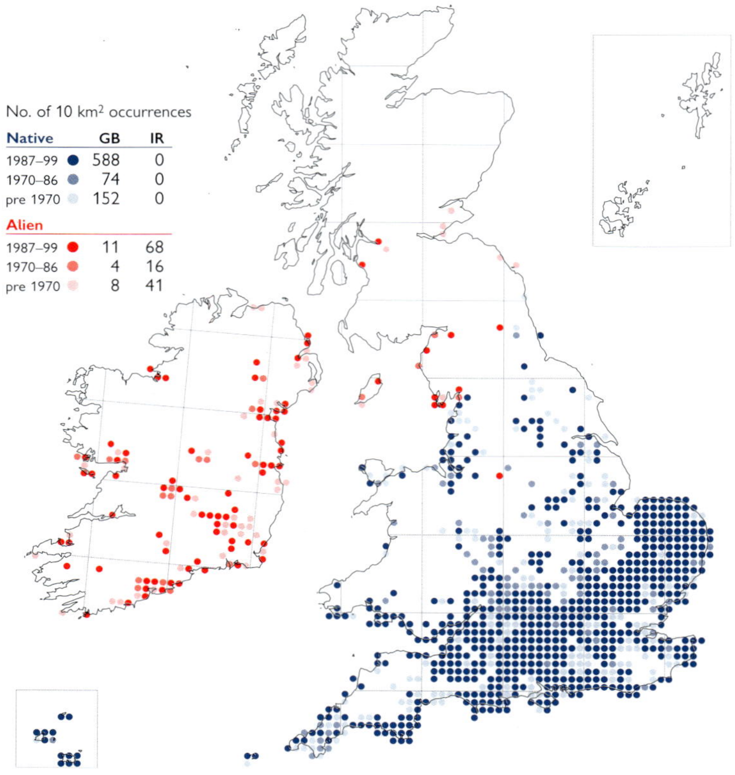

This usually annual root parasite occurs on a wide range of hosts, but mainly on species of Fabaceae and Asteraceae. Var. *maritima* is native on sand dunes and cliffs on the south coast (Stewart *et al.*, 1994), with *Daucus carota* subsp. *gummifer* as its main host. Var. *minor*, the most frequent of several poorly differentiated varieties, is probably alien. It is usually found on cultivated land (often introduced with grass seed) and other disturbed ground. Lowland.

Native (change –0.20). The frequent introductions of this species give it a dynamic distribution, so that long-term trends in frequency are difficult to assess.

European Southern-temperate element; widely naturalised outside its native range.

References: Atlas (237d, 238c), Hultén & Fries (1986), Meusel *et al.* (1978), Rumsey & Jury (1991).

M. J. Y. FOLEY

Acanthus mollis Bear's-breech

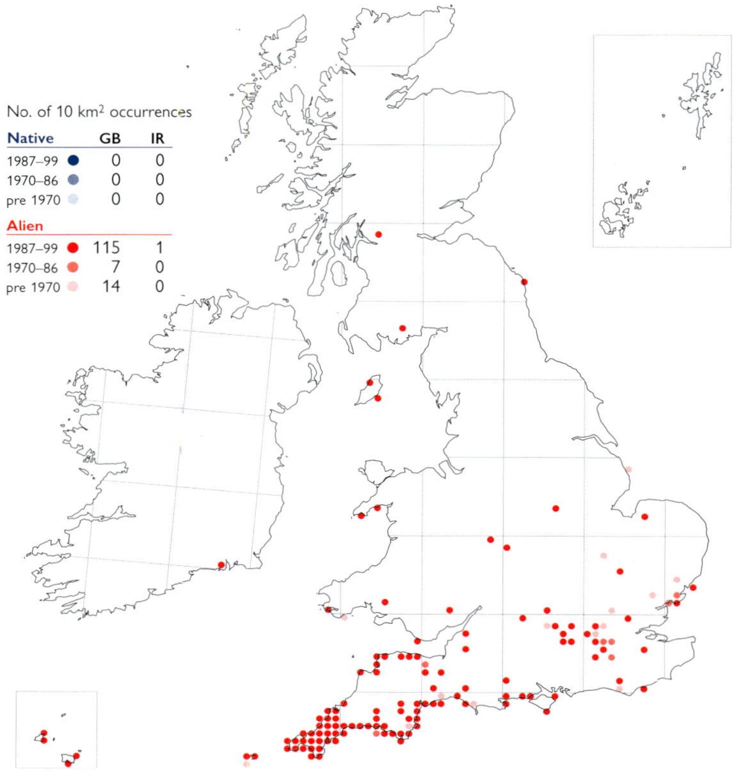

A robust, clump-forming perennial herb, persisting where it is dumped from gardens in sites such as roadsides, railway banks, waste places and in woodlands. Flowering is irregular, particularly in shaded sites. Although seed is produced, spread is almost exclusively through dispersal of root fragments. Lowland.

Neophyte (change +1.55). This plant is popular in gardens for its attractive foliage, and has been cultivated in our area since at least 1548. It was first recorded in the wild in 1820, and is increasingly found naturalised near gardens and where garden refuse is tipped. It has spread since the 1962 *Atlas*, probably due to a genuine increase in the wild and also to better recording.

Native of the Mediterranean region.

References: Atlas (240b), Bolòs & Vigo (1995).

F. J. RUMSEY

Pinguicula lusitanica Pale Butterwort

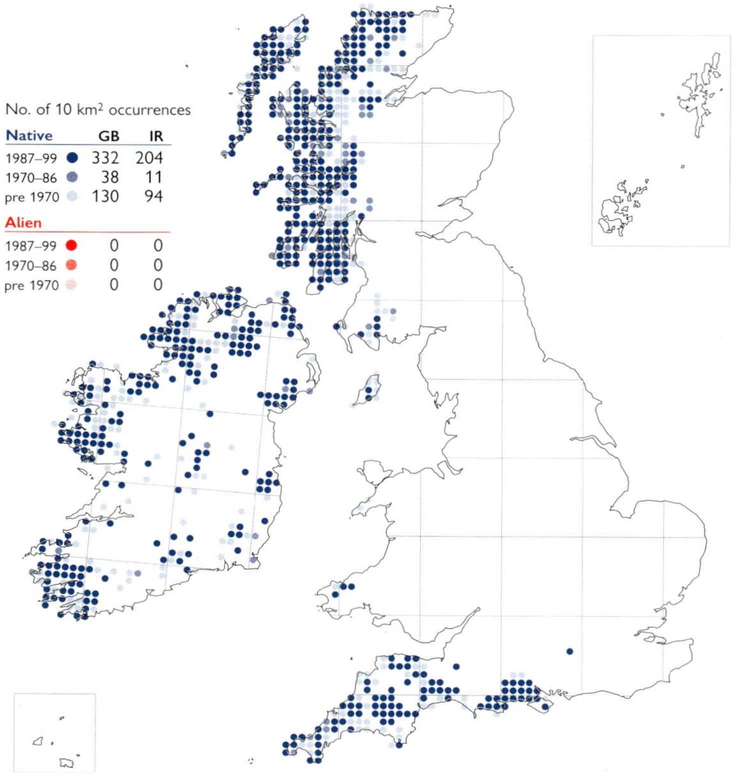

An insectivorous perennial herb which retains its insect-trapping leaves through the winter. It grows on damp bare peat and at the bases of grass, rush or sedge tussocks beside moorland rills, drainage ditches on former bogs, acidic flushes and wet heaths, often in places trampled by livestock or deer. 0–490 m (Dartmoor, S. Devon, and the Mourne Mountains, Co. Down).

Native (change –0.83). *P. lusitanica* is an easily overlooked species which has declined in some areas since the 1962 *Atlas*, largely through loss of habitat, changes in management and scrub encroachment.

Oceanic Temperate element.

References: Atlas (238d), Dupont (1962).

F. J. RUMSEY

Pinguicula alpina Alpine Butterwort

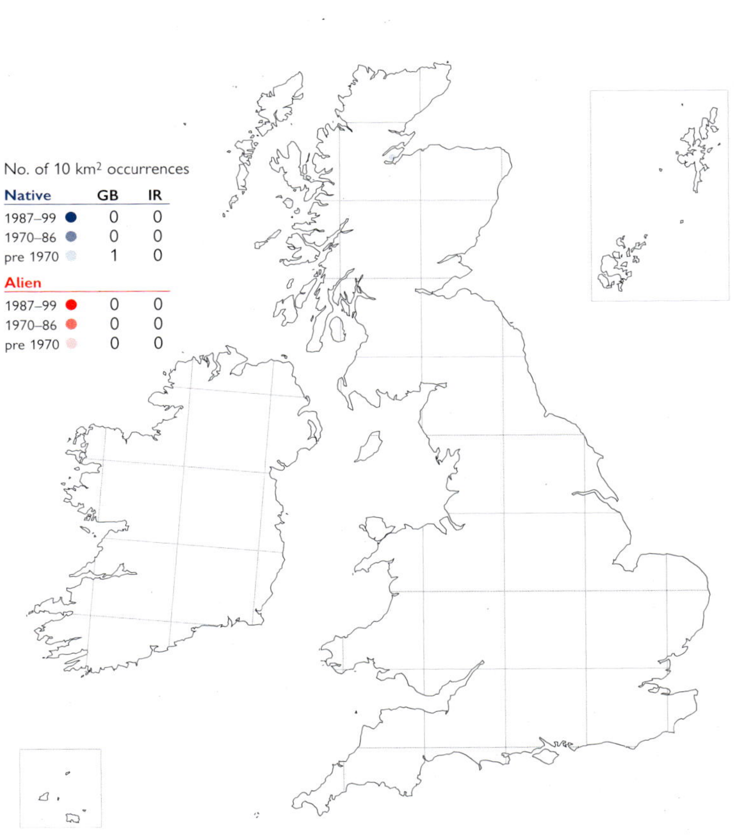

A small, rosette-forming, insectivorous perennial herb formerly known from one site, described as a bog or moor. In mainland Europe, it grows in base-rich flushes, stony mires and the drier parts of open boggy heath, often at high altitude. Lowland.

Native or alien. *P. alpina* was known from a single site near Avoch (E. Ross), where it was first reported in 1831 and last recorded before 1919. It gradually declined as its site was encroached upon by cultivation and colonised by seedling conifers; it was also heavily collected. Its restriction to this single, otherwise unremarkable, lowland mire has led to suspicions that it may have been planted.

Eurosiberian Arctic-montane element.

References: Atlas (239a), Hultén & Fries (1986), Lusby (1998), Marren (1999), Meusel *et al.* (1978).

F. J. RUMSEY

Pinguicula vulgaris Common Butterwort

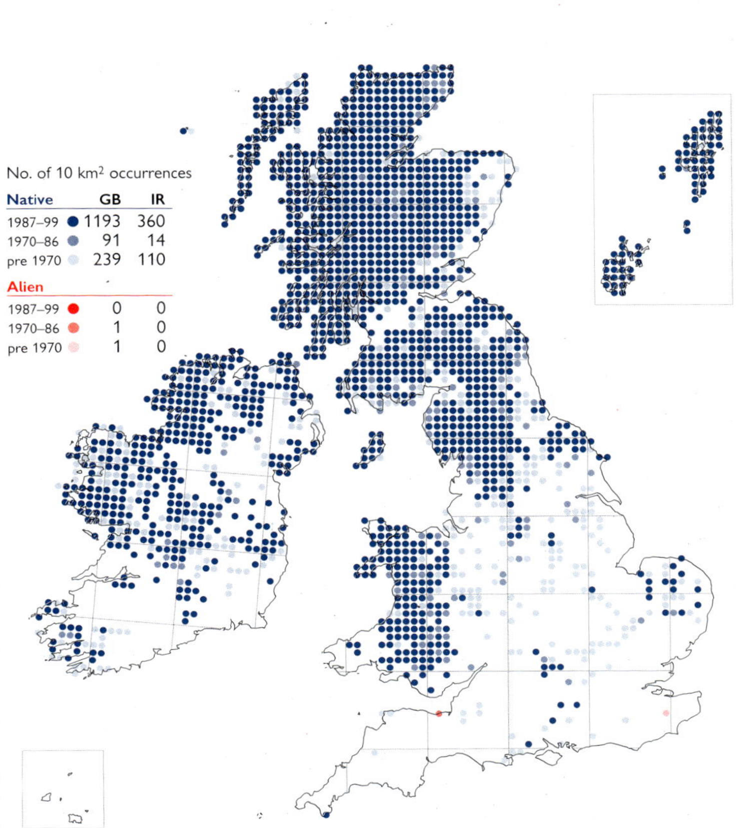

A rosette-forming, insectivorous perennial herb of damp, nutrient-poor habitats, overwintering as a rootless bud. It is found in bogs, in crevices of irrigated rocks and rock ledges, in base-poor as well as base-rich open flushes, and in open bryophyte-dominated communities in fens. 0–970 m (Beinn Heasgarnich, Mid Perth).

Native (change –0.76). Many lowland sites of *P. vulgaris* were lost before the end of the 19th century due to drainage and agricultural intensification. This loss has continued, with over half the 10-km squares which remained in C. & S. England after 1930 now lost, including many in Norfolk. The distribution elsewhere is stable.

Circumpolar Boreal-montane element, with a disjunct distribution.

References: Atlas (239b), Hultén & Fries (1986), Meusel *et al.* (1978).

F. J. RUMSEY

Pinguicula grandiflora Large-flowered Butterwort

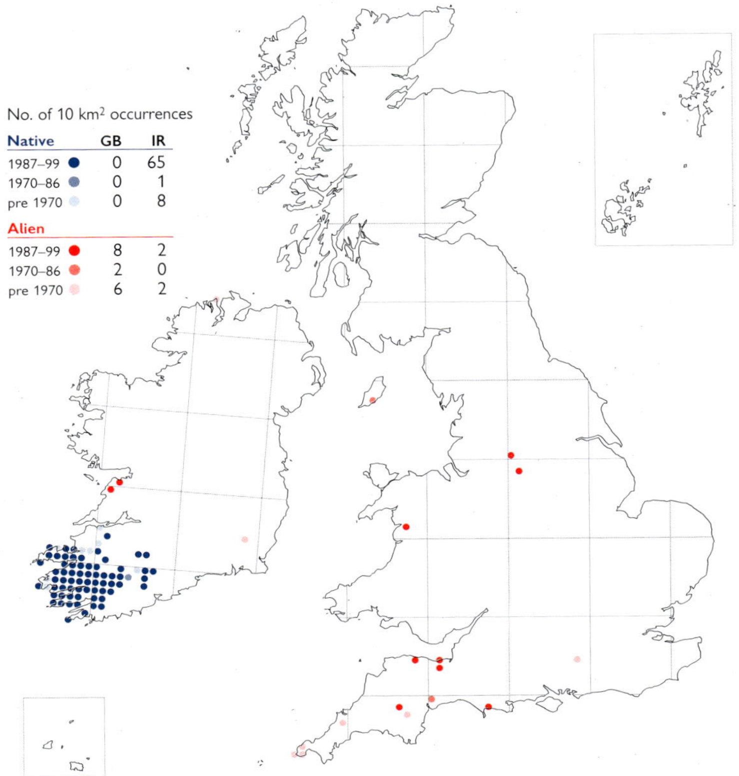

A rosette-forming, insectivorous perennial herb, overwintering as a rootless bud which also functions as a vegetative propagule. It is found on wet rocks, flushed moorland and acidic bogs. 0–855 m (Macgillycuddy's Reeks, S. Kerry).

Native. There has been little change in the native Irish range of *P. grandiflora*, where this species may be locally abundant. It has been planted in damp places and is occasionally naturalised in Britain.

Oceanic Temperate element.

References: Atlas (239a), Meusel *et al.* (1978), Scully (1916).

F. J. RUMSEY

Utricularia vulgaris sens. lat.

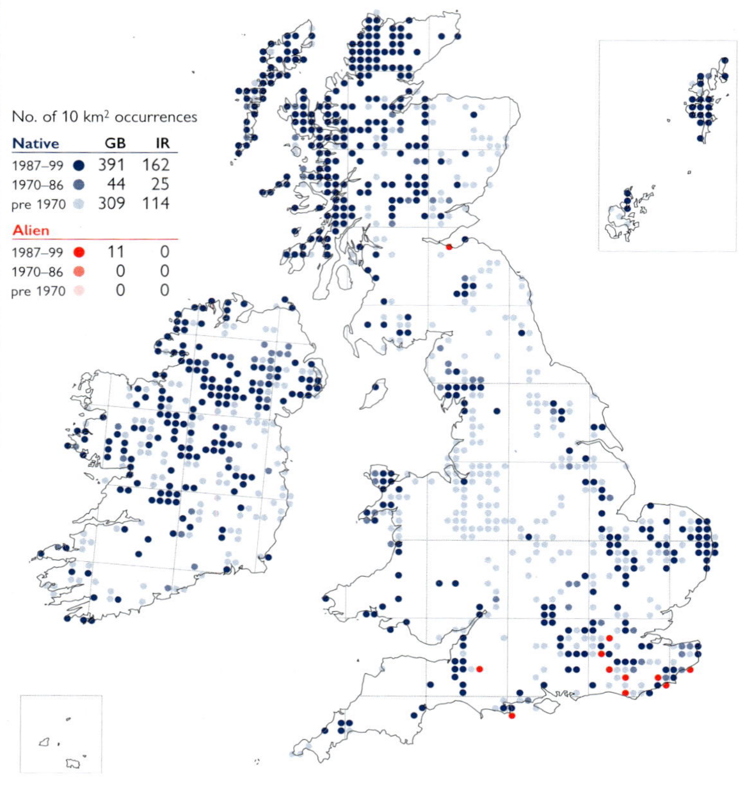

Free-floating, insectivorous perennials found in still or sluggish, acidic to basic, nutrient-poor waters over inorganic or peaty substrates. Flowering is irregular, and reproduction is usually by turions. Generally lowland, but reaching 640 m in Atholl (E. Perth).

Native (change +0.47). This aggregate, which comprises *U. australis* and *U. vulgaris*, is much better recorded in Scotland and N. Ireland than it was in the 1962 *Atlas*. It has declined in many areas, mainly due to drainage, peatland destruction and eutrophication. The two species can only be identified with certainty when flowering.

The distribution of the two species is given in their respective accounts.

References: Atlas (239c), Hultén & Fries (1986), Meusel *et al.* (1978), Preston & Croft (1997), Taylor (1989).

F. J. RUMSEY

Utricularia vulgaris sens. str. Greater Bladderwort

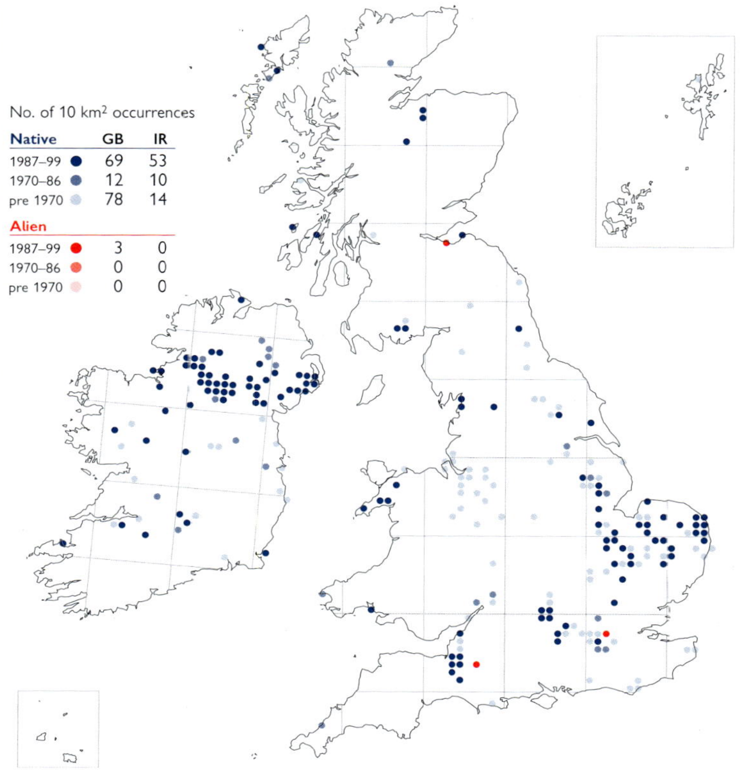

U. vulgaris is found in oligotrophic and mesotrophic, base-rich waters. Habitats include sheltered bays in limestone lakes, ponds, ditches and pools in calcareous fens and grazing marshes, and flooded clay-, marl-, and gravel-pits. Flowering is temperature dependent, variable annually, and less frequent in the north of its range. Generally lowland, but upper altitudinal limit unknown.

Native. The map is based largely on flowering plants together with records of vegetative plants identified by Taylor, and the range of the species is greatly underestimated. Distributional change is difficult to determine, but the species has declined in lowland England.

Eurosiberian Temperate element.

References: Atlas Supp. (64a), Hultén & Fries (1986), Meusel *et al.* (1978), Preston & Croft (1997), Taylor (1989).

F. J. RUMSEY

Utricularia australis Bladderwort

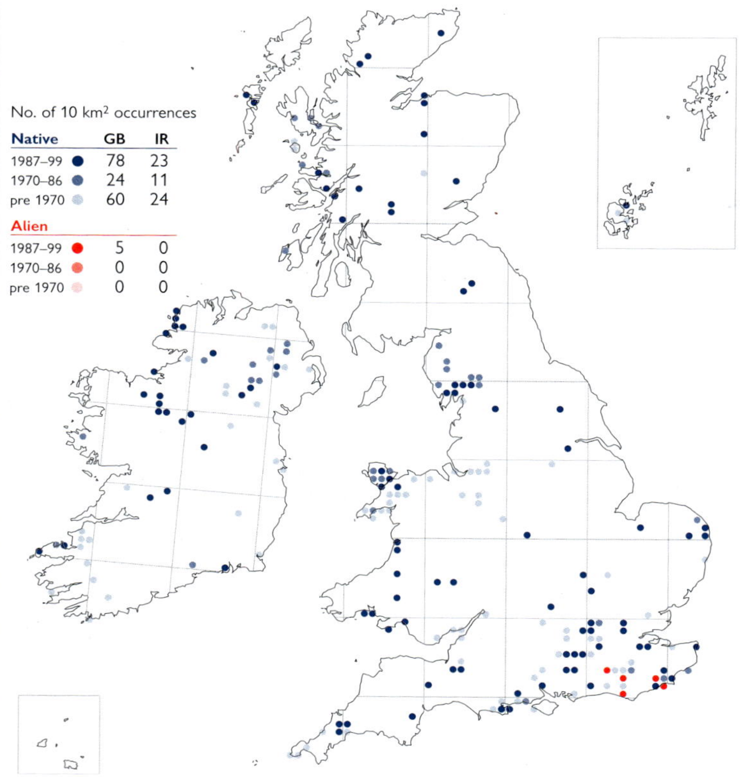

A species mainly of acidic water, found in lakes, ponds, reservoirs, slow-flowing streams, ditches, canals, and swampy ground over mineral or peaty soil. It also occurs, however, in moderately calcareous sites. Generally lowland, but reaching 335 m on Lambrigg Fell (Westmorland), and probably at 500 m in Llyn Anafon (Caerns.).

Native. The map is based mainly on flowering plants, which, since most flowering is in the south, must give it a southern bias and greatly underestimate the range. Distributional change is difficult to assess, though it has decreased in lowland England.

Eurasian Boreo-temperate element.

References: Hultén & Fries (1986), Meusel *et al.* (1978), Preston & Croft (1997), Thor (1979).

F. J. RUMSEY

Utricularia intermedia sens. lat.

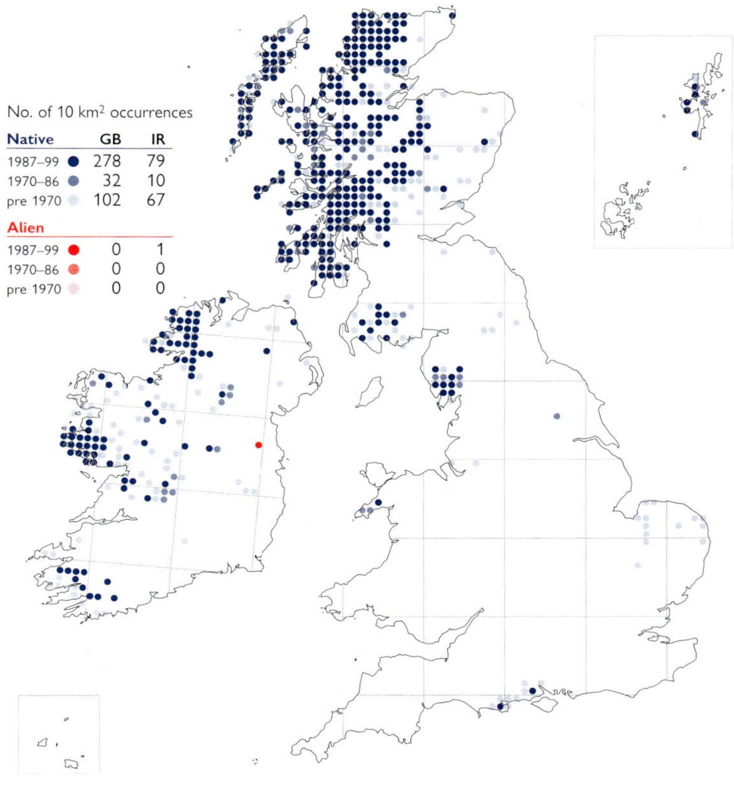

Perennial, insectivorous herbs, most frequent in shallow, oligotrophic water in acidic and peaty sites, though also occurring in calcareous sites. They rarely flower, and reproduction is mainly by turions. 0–650 m (Loch Achlarich, Mid Perth).

Native (change +0.40). A poorly-known complex of three species, *U. intermedia sens. str.*, *U. ochroleuca* and *U. stygia*, for which there are rather few reliable records. The aggregate has declined, especially in the southern part of its range, due to habitat destruction and eutrophication. It is, however, much better recorded in Scotland and Ireland than in the 1962 *Atlas*.

Circumpolar Boreal-montane element.

References: Atlas (239d), Hultén & Fries (1986), Meusel *et al.* (1978), Preston & Croft (1997), Taylor (1989), Thor (1979, 1987, 1988).

F. J. RUMSEY

Utricularia minor Lesser Bladderwort

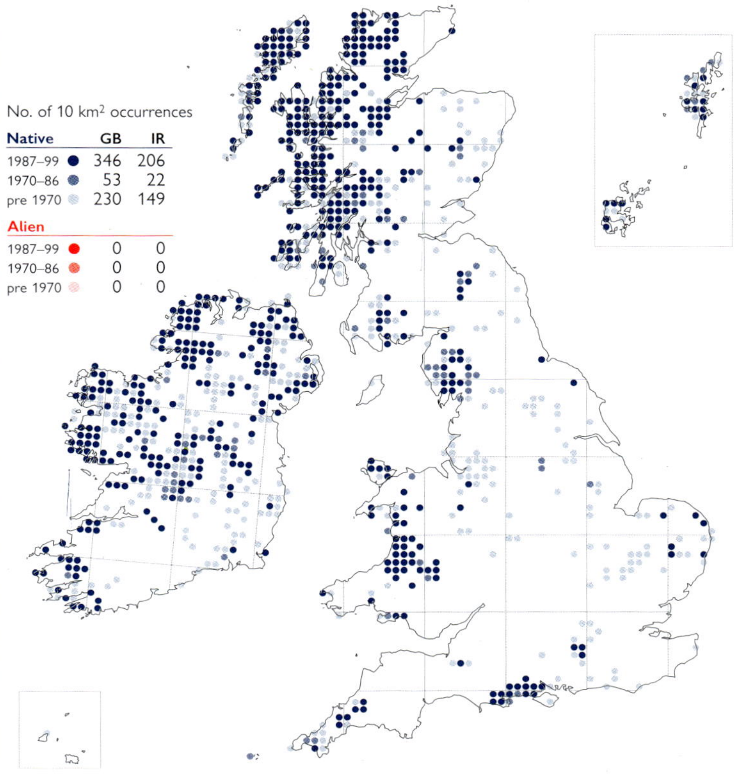

A perennial herb of nutrient-poor, acidic, or sometimes base-rich, shallow water in bog pools and abandoned peat cuttings, at the edges of lakes amongst emergent vegetation, in ditches and small ponds, and in fens. 0–600 m (Haystacks Tarn, Cumberland), and possibly to 685 m in Scotland.

Native (change +0.20). *U. minor* is free-flowering, and can be reliably identified on both flowering and vegetative characters. A decline was apparent in the 1962 *Atlas*, and this has continued in S. and E. England, and perhaps also in S.E. Ireland, due to habitat destruction and eutrophication. It may still be under-recorded in the northern and western parts of its range.

Circumpolar Boreo-temperate element.

References: Atlas (240a), Hultén & Fries (1986), Meusel *et al.* (1978), Preston & Croft (1997).

F. J. RUMSEY

Campanula patula Spreading Bellflower

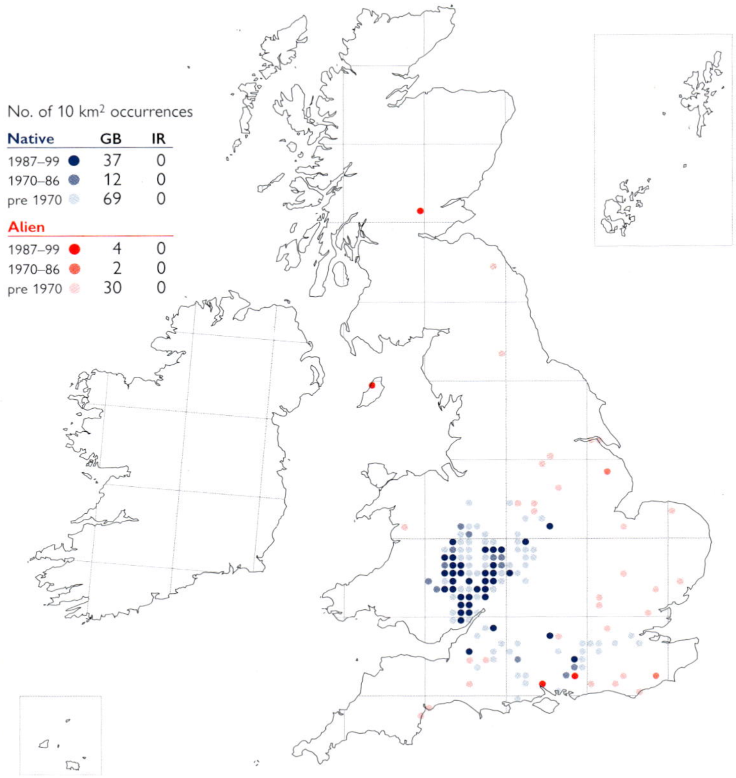

A biennial herb of dry, well-drained, sunny sites on fairly infertile sandy or gravelly soils. It is found in open woodland, on banks and rock outcrops. Reproduction is by seed, which needs disturbed sites for germination, but which is long-lived, allowing the plant to reappear after long absences. Lowland.

Native (change –0.77). The decline of *C. patula* was already apparent in the 1962 *Atlas*. It has since disappeared from many sites through the cessation of coppicing and other disturbance in woodland, and the increased use of herbicides on roadsides and railway banks.

European Temperate element, with a continental distribution in W. Europe; widely naturalised outside its native range.

References: Atlas (257b), Hultén & Fries (1986), Meusel & Jäger (1992), Stewart *et al.* (1994).

T. D. DINES

Campanula rapunculus Rampion Bellflower

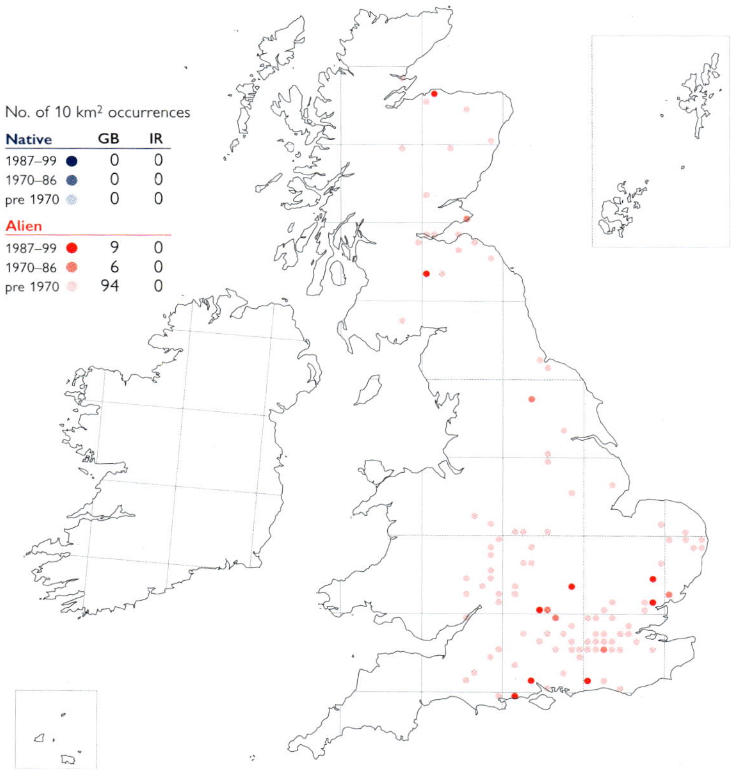

A perennial herb found naturalised in rough grassland and on roadsides, railway banks and in quarries. It also occurs as a relic of cultivation. Reproduction is from seed and rhizome fragments. Lowland.

Archaeophyte (change −2.16). *C. rapunculus* was once frequently grown in gardens in our area for ornament and its edible roots. It was recorded from the wild as early as 1597, but fell out of favour as a vegetable around 1700 and has consequently declined seriously. It is now rarely encountered, either in cultivation or in the wild.

C. rapunculus is a variable species with a European Southern-temperate distribution; it is naturalised in Europe north of its native range.

References: Atlas (257c), Meusel & Jäger (1992).

T. D. DINES

Campanula persicifolia Peach-leaved Bellflower

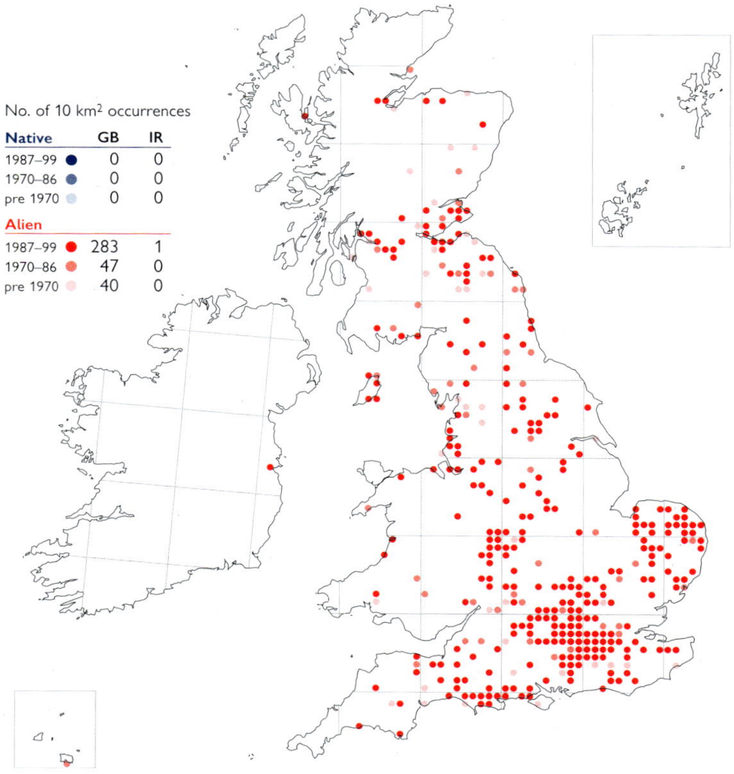

A rhizomatous perennial herb found on waste ground, waysides and banks, and in hedgerows and woods. It is commonly cultivated, and occurs as a casual or, more frequently, as established and persistent populations. Lowland.

Neophyte (change +2.80). *C. persicifolia*, introduced into cultivation before 1596, was claimed as native when it was found in the wild by Druce (1903), but the evidence for this is very unconvincing. The large increase in records since the 1962 *Atlas* is probably due to continued escapes from cultivation, as well as to better recording of aliens.

A European Temperate species, absent as a native from much of W. Europe.

References: Atlas (256c), Hultén & Fries (1986), Meusel & Jäger (1992).

T. D. DINES

Campanula medium Canterbury-bells

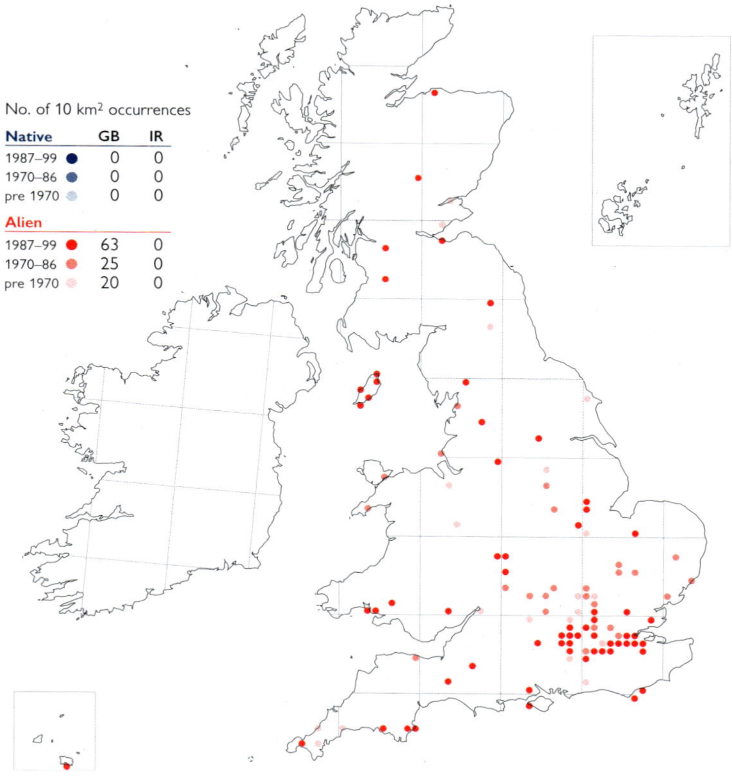

This large biennial herb occurs on disturbed waste ground, road verges and banks. It is usually casual but can become established, particularly on railway banks and on chalk soils. Lowland.

Neophyte. *C. medium* was introduced to Britain from Europe by 1597 and is now common in cultivation. It was first recorded in the wild in 1870. It was not mapped in the 1962 *Atlas*, so changes in its distribution are difficult to assess.

Native of S.E. France and N. & C. Italy.

T. D. DINES

Campanula glomerata Clustered Bellflower

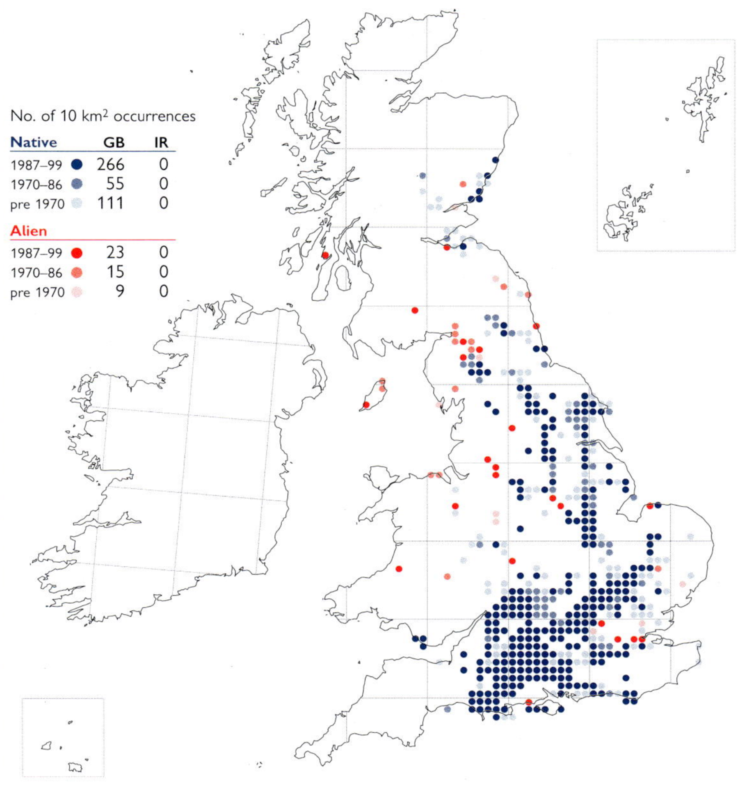

A perennial herb of calcareous grassland, scrub, open woodland, cliffs and sand dunes. It is most frequent on chalk and oolite, and curiously absent from apparently suitable habitat on other limestones and base-rich substrates. It also occurs as a garden escape on roadsides and waste ground. 0–355 m (Oddendale, Westmorland).

Native (change –0.51). This species has seen a gradual decline outside its core areas, due to habitat loss and a decline in grazing. Many populations are small, often consisting of only a few plants (Green *et al.*, 1997). *C. glomerata* is widely grown in gardens and several cultivars are available.

Eurasian Temperate element, with a continental distribution in W. Europe.

References: Atlas (256d), Hultén & Fries (1986), Meusel & Jäger (1992).

T. D. DINES

Campanula portenschlagiana Adria Bellflower

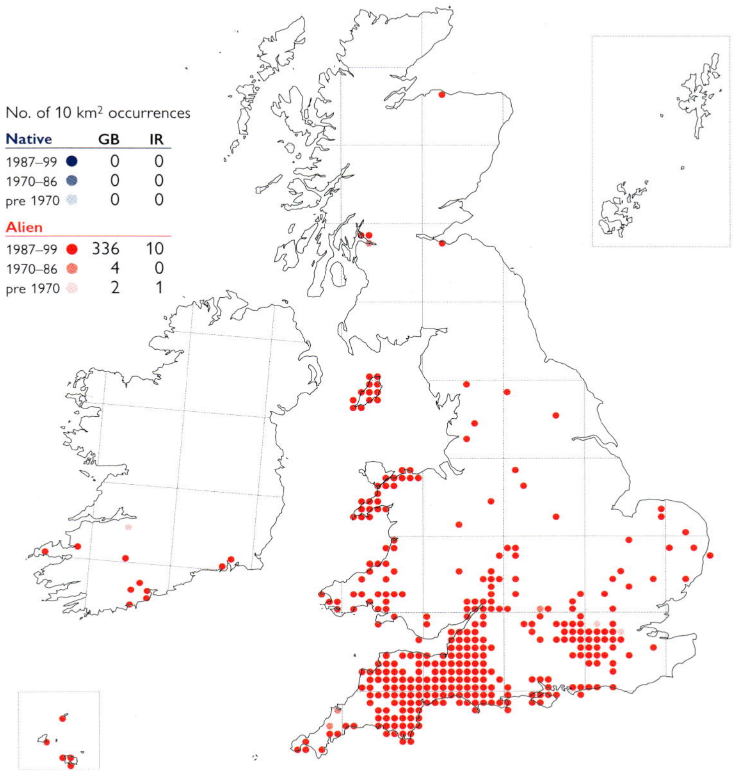

A low-growing perennial herb, found established on walls, in pavement cracks and on rocky banks and waste ground. Populations arise from seed and stem fragments, which root readily, and from garden throw-outs. Lowland.

Neophyte. *C. portenschlagiana* was introduced into cultivation in 1835, and is now very popular in gardens. It was first recorded in the wild in 1922. It was brought to the attention of botanists in our area by Grenfell (1982) and Stace (1991); it is unevenly recorded, although the western range shown by the map is probably genuine. It is increasing rapidly in some areas.

Native of W. Jugoslavia.

T. D. DINES

Campanula poscharskyana Trailing Bellflower

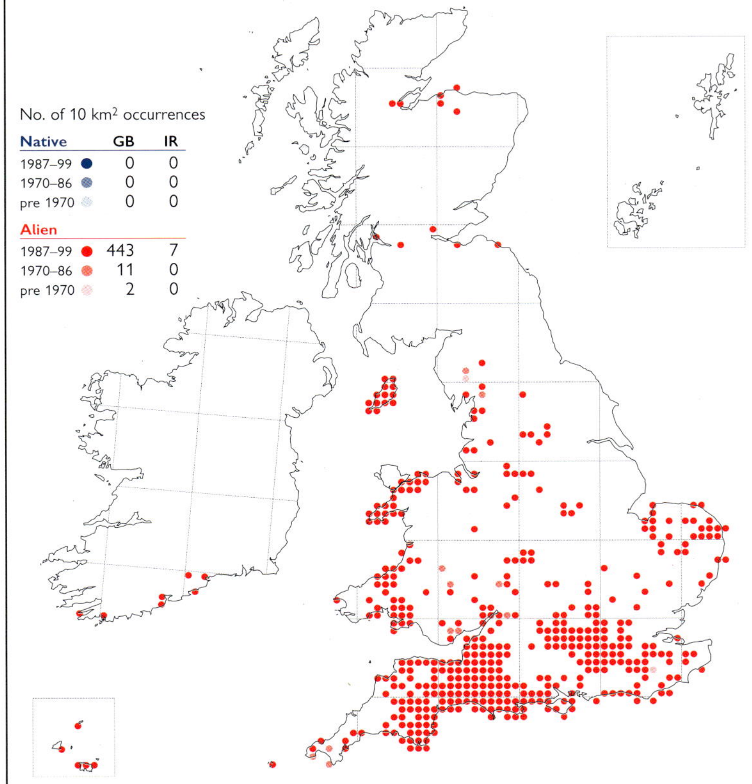

A low-growing perennial herb, widely grown in gardens and becoming established in the wild on walls, in pavement cracks and on rocky banks and waste ground. Populations arise from seed and stem fragments, which root readily, and from garden throw-outs. Lowland.

Neophyte. *C. poscharskyana*, introduced into cultivation in 1931, is frequently grown in gardens. Clement (1978a) described it as 'often seen questionably wild about gardens', giving only one locality. Like *C. portenschlagiana*, it was described and illustrated by Grenfell (1982) but not included in a national flora until Stace (1991). It was first recorded in the wild in 1957 (Worcs.) and is increasing rapidly in some areas, but the species is unevenly recorded.

Native of W. Jugoslavia.

T. D. DINES

Campanula latifolia Giant Bellflower

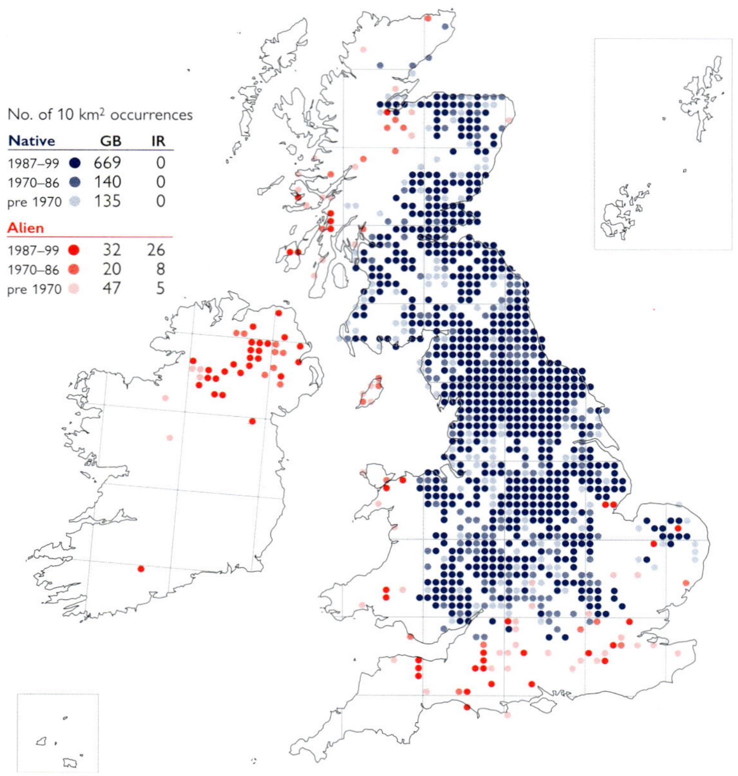

A large perennial herb of damp woodland, wooded riversides and hedgerows, usually on fertile, neutral or calcareous soils. It is also grown in gardens, occurring as an established alien on waste ground, roadsides and hedge banks. Generally lowland, but reaching 390 m at Ingleborough (Mid-W. Yorks.).

Native (change –0.23). Within its native range, this species has shown little change in overall distribution since the 1962 *Atlas*. It is, however, becoming more frequent as an alien. It is regarded as an alien in Ireland (Scannell & Synnott, 1987).

European Temperate element; also in C. Asia.

References: Atlas (255d), Hultén & Fries (1986), Meusel & Jäger (1992).

T. D. DINES

Campanula trachelium Nettle-leaved Bellflower

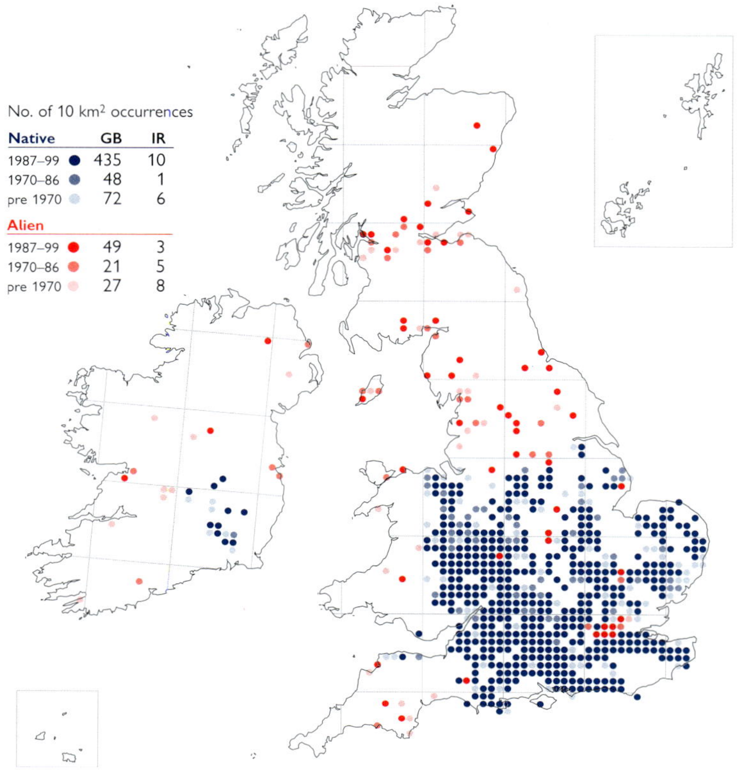

A large perennial herb, found as a native on dry, base-rich, usually calcareous soils in woodland, scrubby grassland and hedge banks; in Ireland it is also reported from river banks and swamp woodland. It is also grown in gardens, and occurs as a naturalised alien on a wider range of soils and habitats. Generally lowland, but reaching 320 m in Monk's Dale (Derbys.).

Native (change +0.14). This species has declined slightly through habitat loss arising from the cessation of coppicing and the removal of hedgerows. Alien records were not mapped in the 1962 *Atlas*.

European Temperate element; also in C. Asia.

References: Atlas (256a), Curtis & McGough (1988), Goodwillie (1999a), Hultén & Fries (1986), Meusel & Jäger (1992).

T. D. DINES

Campanula rapunculoides Creeping Bellflower

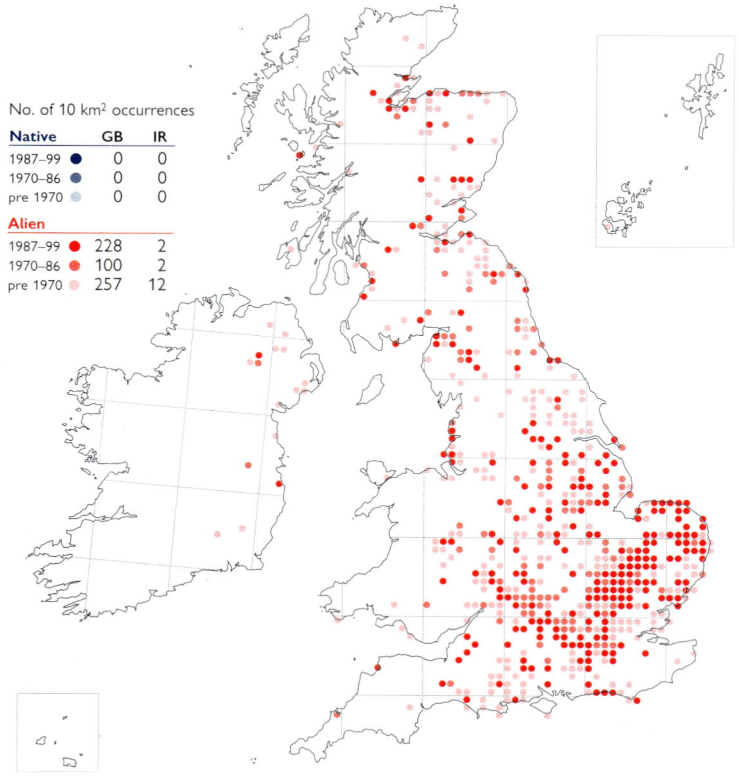

A rhizomatous perennial herb of roadside verges, grassy banks, railway banks and occasionally woodland and field-borders. Generally lowland, but reaching 365 m at Felindre (Rads.).

Neophyte (change –1.24). This species was cultivated in Britain by 1568 and was first recorded from the wild around 1708. It is frequently grown in gardens where it can become invasive and is often discarded. Despite this, populations in the wild are often short-lived and this may help to explain the decline, which was already apparent in the 1962 *Atlas*.

A European Temperate species, apparently native north to the Netherlands, but naturalised to S. Fennoscandia.

References: Atlas (256b), Hultén & Fries (1986), Meusel & Jäger (1992).

T. D. DINES

Campanula rotundifolia Harebell

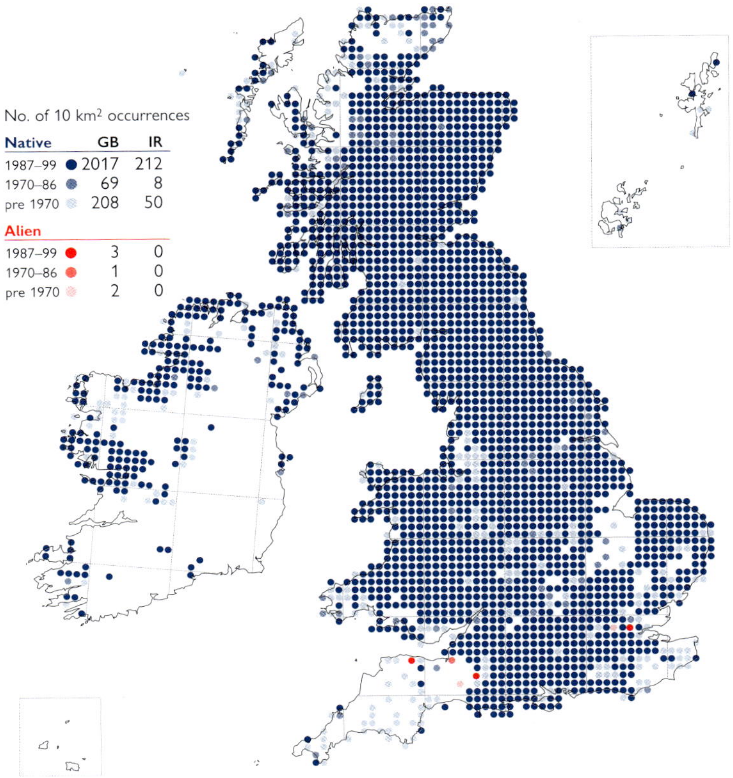

A rhizomatous perennial herb of dry, open, infertile habitats including grassland, fixed dunes, rock ledges, roadsides and railway banks. It tolerates a wide range of soil pH, being found on both mildly acidic and calcareous substrates, and heavy-metal tolerant races are known. 0–1160 m (Breadalbanes, Mid Perth).

Native (change –0.92). There has been no significant change in the range of this species, other than some local declines at the edges of its range. It was mapped as 'all records' in the 1962 *Atlas*. Its absence from some apparently suitable areas, especially in S.W. England, is very difficult to explain.

Circumpolar Boreo-temperate element.

References: Atlas (257a), Grime *et al.* (1988), Hultén & Fries (1986), Meusel & Jäger (1992).

T. D. DINES

Legousia hybrida Venus's-looking-glass

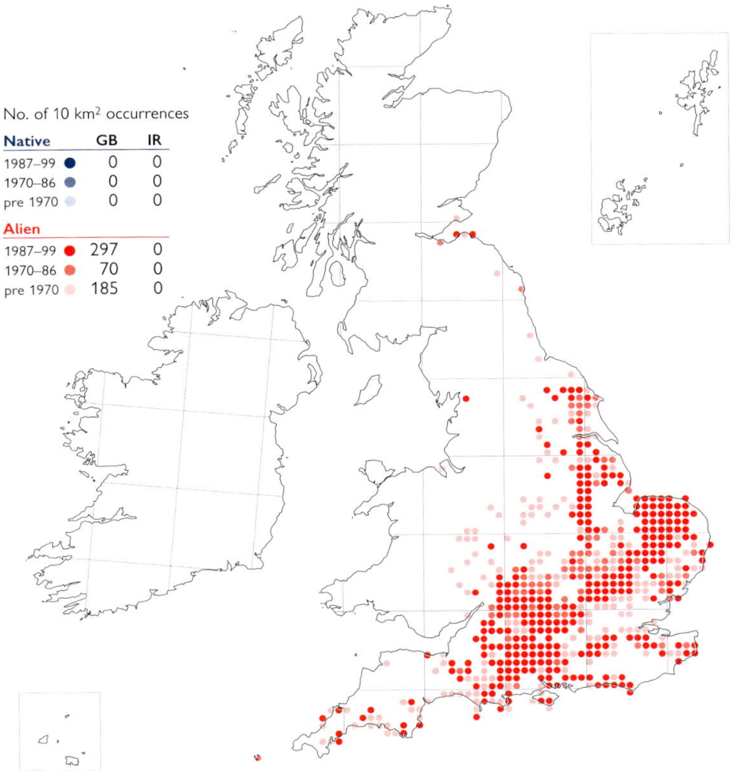

This annual of arable fields is usually found on calcareous soils, especially on chalk. Outside its core areas, it occurs as a casual in disturbed sites such as motorway banks. Lowland.

Archaeophyte (change –0.60). *L. hybrida* has significantly declined since the 1940s because of the use of herbicides and changing methods of arable cultivation. However, the seed is long-lived and populations can reappear after long periods of absence.

As an archaeophyte *L. hybrida* has a European Southern-temperate distribution.

References: Atlas (257d), Hultén & Fries (1986), Meusel & Jäger (1992).

T. D. DINES

Wahlenbergia hederacea Ivy-leaved Bellflower

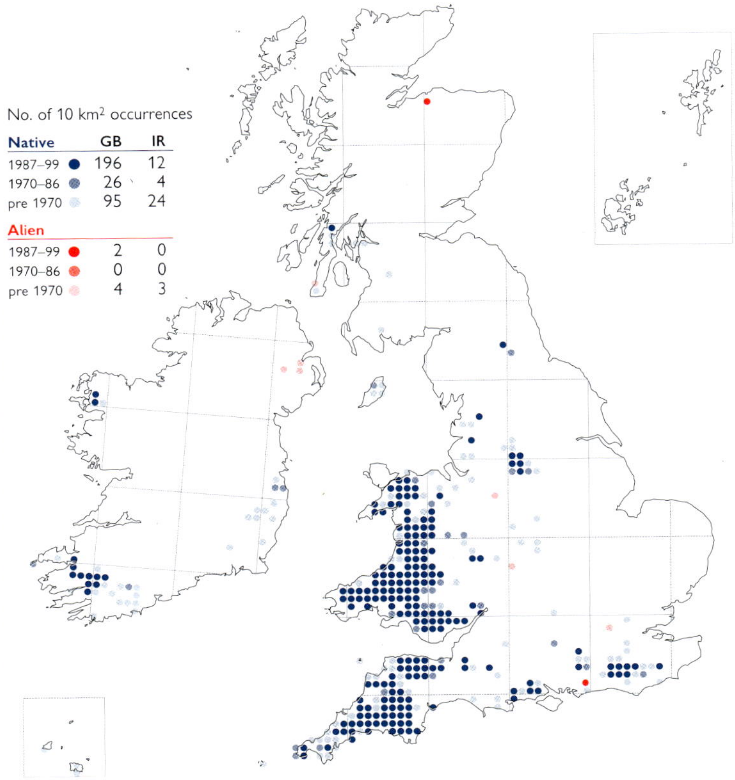

A small, low-growing perennial herb found in damp, wet or boggy places on acidic soils, occurring on heaths, heathy pastures, moors, open woodland and *Salix* carr, and by streams and in flushes. In Ireland, it is most frequent beside streams and is absent from pastures. It prefers areas with moving, rather than standing, water. 0–485 m (Killakee, Co. Dublin).

Native (change –0.30). The decline in this species from the edges of its range was already apparent by the 1962 *Atlas*, and has continued since then. Reasons for losses include habitat destruction, improvement of pastures, increased grazing and peat extraction.

Oceanic Southern-temperate element.

References: Atlas (255c), Meusel & Jäger (1992).

T. D. DINES

Phyteuma spicatum Spiked Rampion

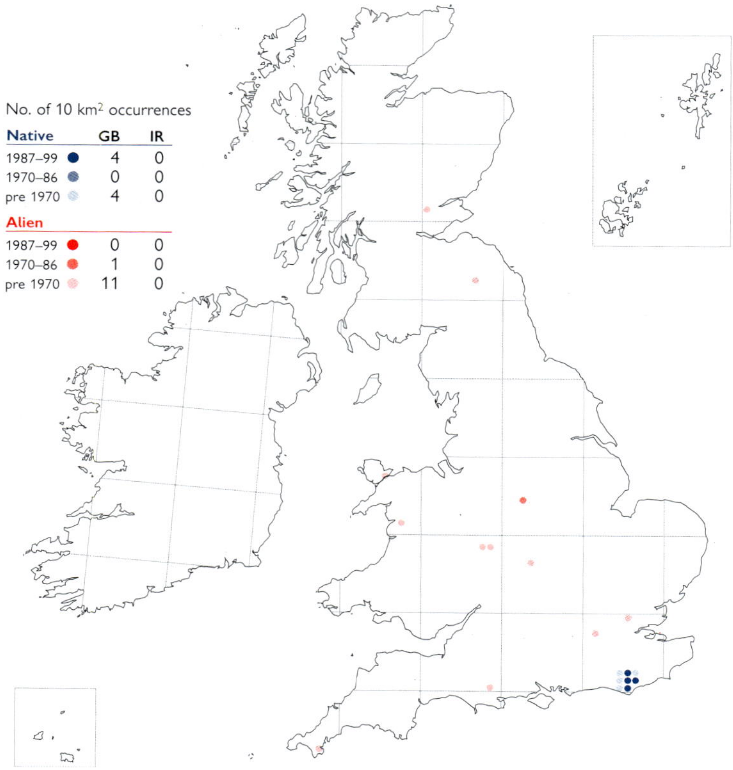

A long-lived perennial herb of damp, fertile, acid soils on road verges, streamsides and in coppiced woodland. Reproduction is by seed, which is long-lived, but recruitment at some sites is negligible. Lowland.

Native or alien (change –0.73). *P. spicatum* has been grown for centuries as a medicinal plant, and was first recorded in the wild in 1640. Plants in Sussex were first recorded in 1824, suggesting that although traditionally regarded as native, it might be an introduction there. It was formerly more widespread and more abundant within this stronghold. It cannot tolerate shade and has disappeared from many sites through a lack of coppicing.

European Temperate element.

References: Atlas (258b), Hultén & Fries (1986), Meusel & Jäger (1992), Wheeler & Hutchings (1999), Wigginton (1999).

T. D. DINES

Phyteuma orbiculare Round-headed Rampion

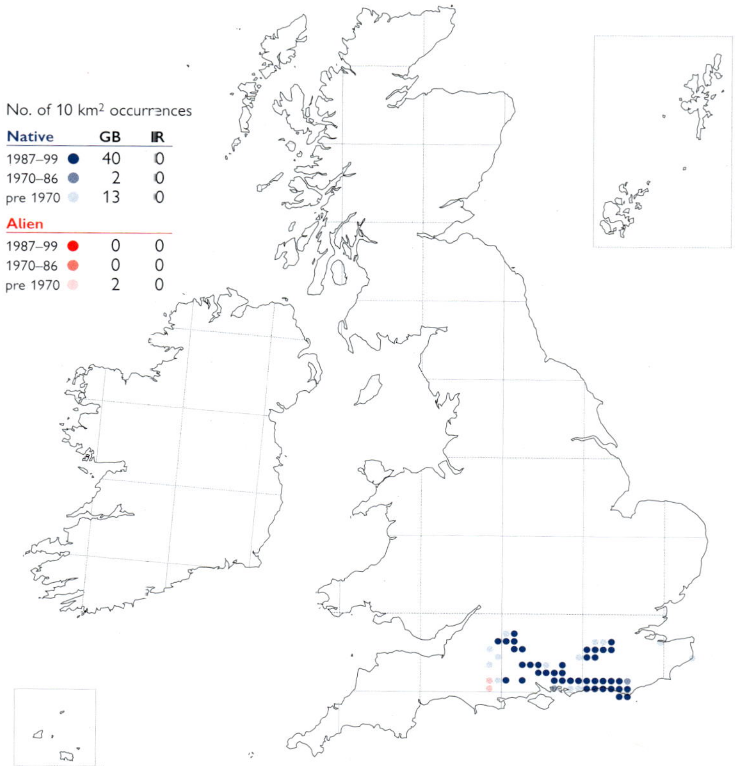

A perennial herb of species-rich chalk grassland, open scrub, earthworks and verges. It is tolerant of grazing, and seems to prefer grazed areas, but also grows in neighbouring ungrazed grassland. Propagation is mostly by seed but it also spreads by stoloniferous growth. Lowland.

Native (change –0.16). This species was last recorded before 1930 in most of the squares for which there is no recent record. Its 10-km square distribution now appears to be stable, although some sites have been lost through ploughing of chalk grassland or its agricultural improvement.

European Boreo-temperate element; at low altitudes in W. Europe and in mountains of C. Europe.

References: Atlas (258a), Meusel & Jäger (1992), Stewart *et al.* (1994).

T. D. DINES

Jasione montana Sheep's-bit

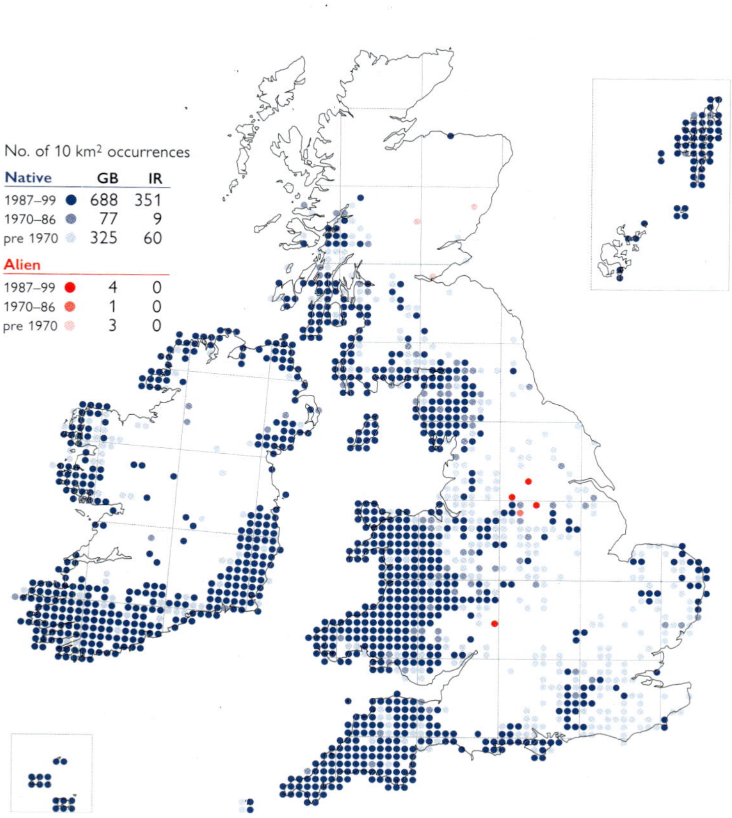

A biennial herb of acidic, shallow, well-drained soils. It occurs on sea-cliffs, in maritime grasslands and heaths and on stabilised sand dunes, and inland on heathland, stone walls, hedge banks and railway cuttings. Propagation is by seed and disturbed, open sites and recently burnt ground are frequently colonised. 0–955 m (Mt Brandon, S. Kerry).

Native (change –1.08). Many sites in C. and E. England were lost before 1930, and *J. montana* has further declined in the English Midlands and S.E. England since the 1962 *Atlas*. Reasons for losses include the loss of lowland heaths and the growth of coarser vegetation following a decline in rabbit grazing. Its distribution is more stable in Ireland.

European Temperate element.

References: Atlas (258c), Hultén & Fries (1986), Meusel & Jäger (1992), Parnell (1985).

T. D. DINES

Lobelia urens Heath Lobelia

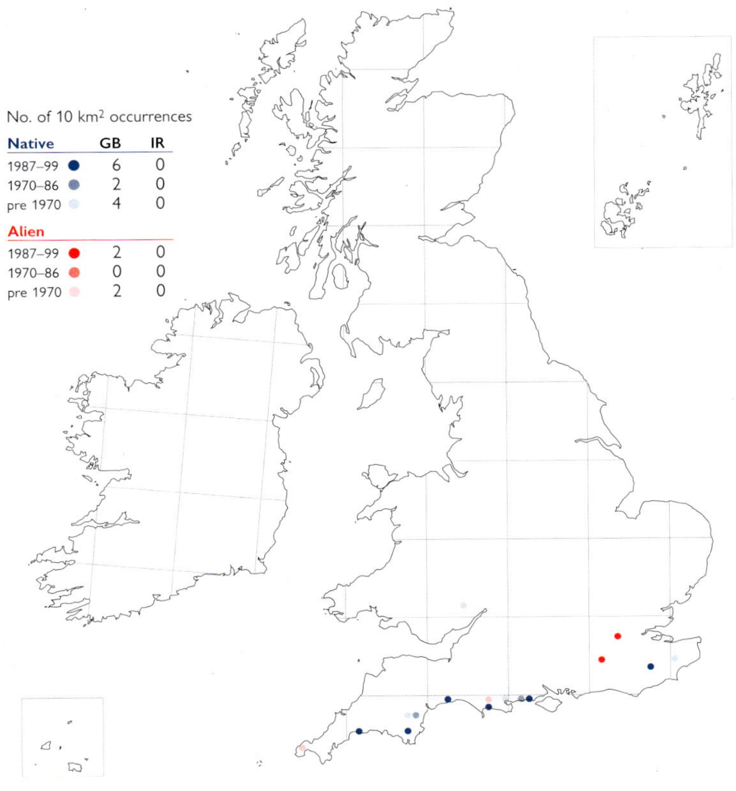

A rhizomatous perennial herb of rough pastures and grassy heaths. It is confined to infertile acidic soils that are often seasonally waterlogged. Reproduction is by seed, which may be long-lived, and germination seems to be stimulated by disturbance. Most populations fluctuate erratically in size. Lowland.

Native (change –0.19). *L. urens* has decreased gradually and it is now known from only six localities. Populations have been lost to afforestation and agricultural improvement, with continued threats from lack of appropriate habitat management, especially coppicing.

Oceanic Southern-temperate element.

References: Atlas (258d), Brightmore (1968), Meusel & Jäger (1992), Wigginton (1999).

T. D. DINES

Lobelia erinus Garden Lobelia

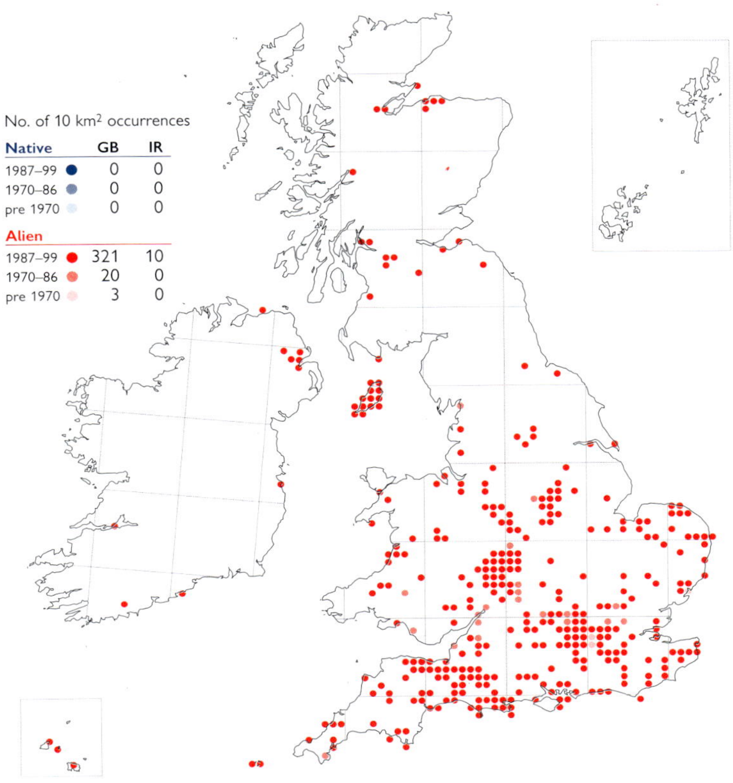

This species, capable of growing as a perennial but usually annual in our area, is found on waste ground, rubbish tips and pavements. It is usually casual, often arising from seed which drops from window boxes or hanging baskets, but some populations persist in sheltered urban sites. Lowland.

Neophyte. *L. erinus* was cultivated in Britain by 1752 and is extremely widely grown as an ornamental, but was not recorded from the wild until 1913. It is unevenly recorded and changes in distribution are difficult to assess.

Native of S. Africa.

T. D. DINES

Lobelia dortmanna Water Lobelia

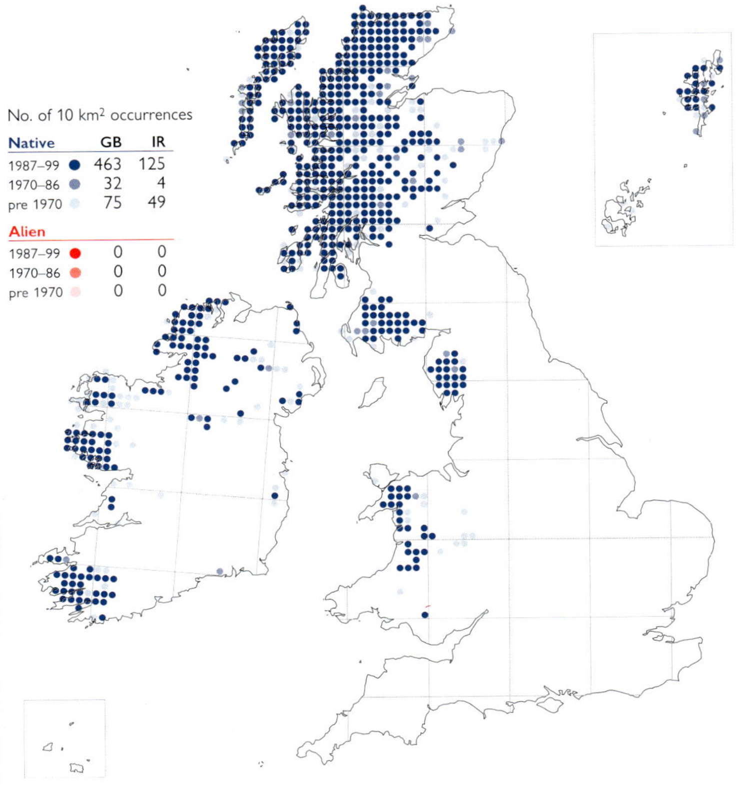

A small, rosette-forming perennial herb of oligotrophic lakes with acidic substrates. It is slow-growing, with little ability to withstand shade or competition and is therefore confined to shallow water less than 2 metres deep. Reproduction and dispersal is by seed, which can remain viable for thirty years. 0–745 m (Llyn Bâch, Caerns.).

Native (change –0.05). *L. dortmanna* is still frequent in the north and west but has declined, largely before 1930, from the eastern edge of its range in both Britain and Ireland through eutrophication.

European Boreal-montane element; also in N. America.

References: Atlas (259a), Farmer (1989), Hultén & Fries (1986), Meusel & Jäger (1992), Preston & Croft (1997), Woodhead (1951b).

T. D. DINES

Sherardia arvensis Field Madder

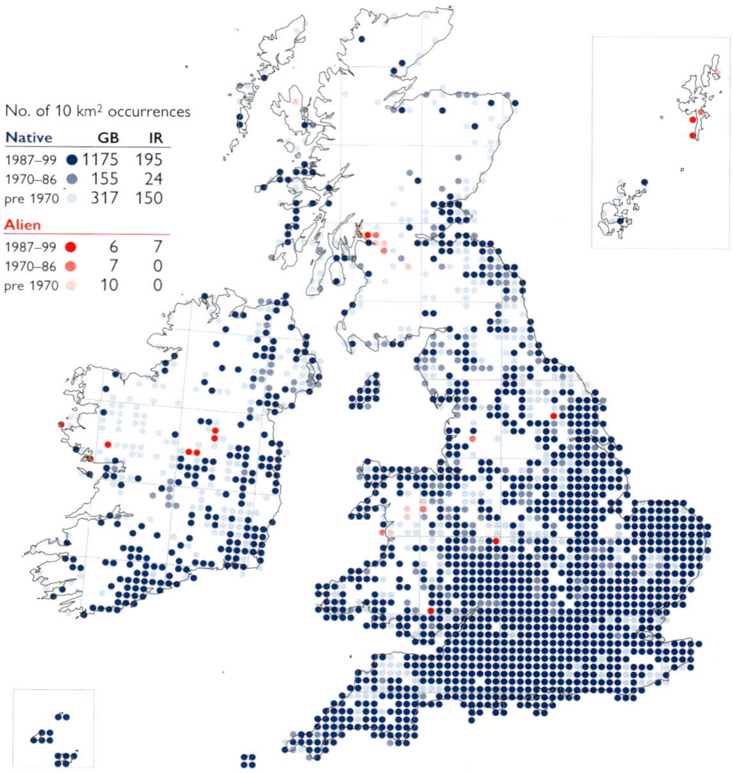

An annual of open, droughted grasslands, sheltered cliffs, sand dunes, arable fields, waste ground, waysides and verges. Mainly lowland, but reaching 365 m in Aberdeenshire.

Native (change –0.94). Formerly frequent, this species is now much decreased due to agricultural intensification. It was mapped as 'all records' in the 1962 *Atlas*. Analysis of the database reveals that most of these losses have occurred since 1950. Whilst remaining common in some coastal localities and locally inland, it is now no more than a rare casual over parts of its range. It may well be native only in some western coastal habitats and an archaeophyte elsewhere.

European Southern-temperate element; it is widely naturalised outside its native range.

References: Atlas (259b), Hultén & Fries (1986), Meusel & Jäger (1992), Rich & Woodruff (1996).

W. R. MEEK

Asperula cynanchica Squinancywort

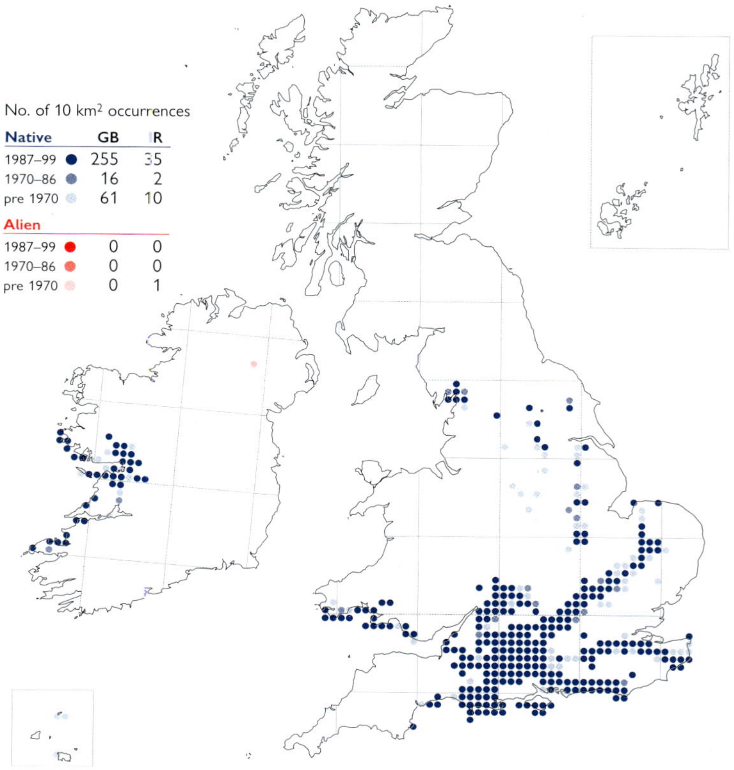

A rhizomatous perennial herb of dry, calcareous grasslands and sand dunes. 0–305 m (Burren, Co. Clare).

Native (change –0.47). This species has undergone a substantial decline in some eastern parts of its range since the 1962 *Atlas* as a result of the ploughing or improvement of its calcareous grassland habitat, and the neglect of what remains. Populations elsewhere are probably relatively stable, although the species is still declining in some limestone areas as, for example, in Oxfordshire (Killick *et al.*, 1998). There are two subspecies in our area: subsp. *cynanchica* is found throughout the range of the species; subsp. *occidentalis* has been recorded from Alderney, Dorset, S. Wales and W. Ireland.

European Temperate element.

References: Atlas (259c), Bolòs & Vigo (1995), Tutin & Chater (1974).

W. R. MEEK

Asperula arvensis Blue Woodruff

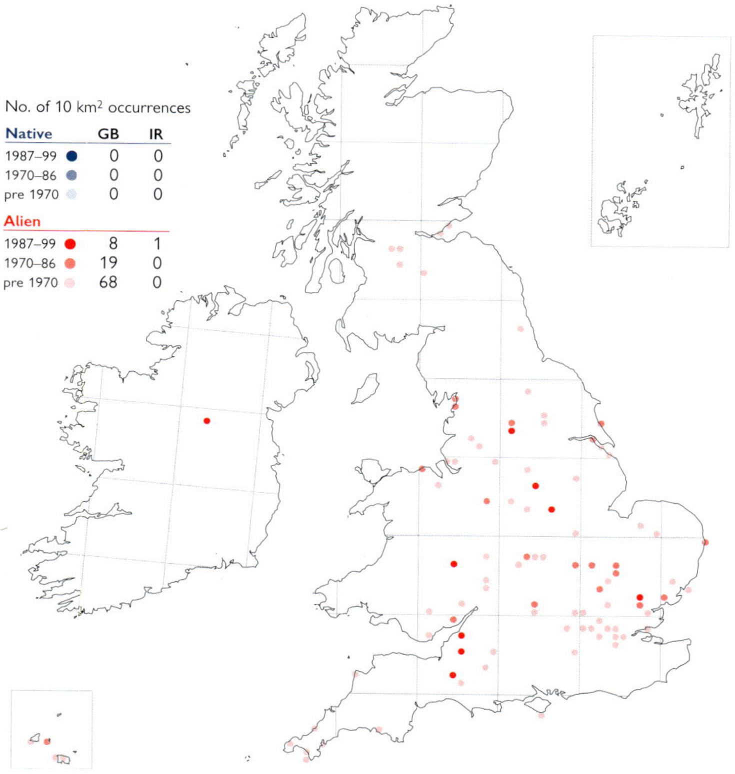

An annual of arable land, waste places and rubbish tips. Lowland.

Casual. *A. arvensis* was introduced into cultivation in Britain by the 16th century, and was first recorded from the wild in 1541. Recent records may be from bird-seed.

Probably native of S.W. Asia and perhaps S. Europe, but it has spread as an archaeophyte to C. Europe and occurs further north as a casual.

Reference: Meusel & Jäger (1992).

W. R. MEEK

Galium boreale Northern Bedstraw

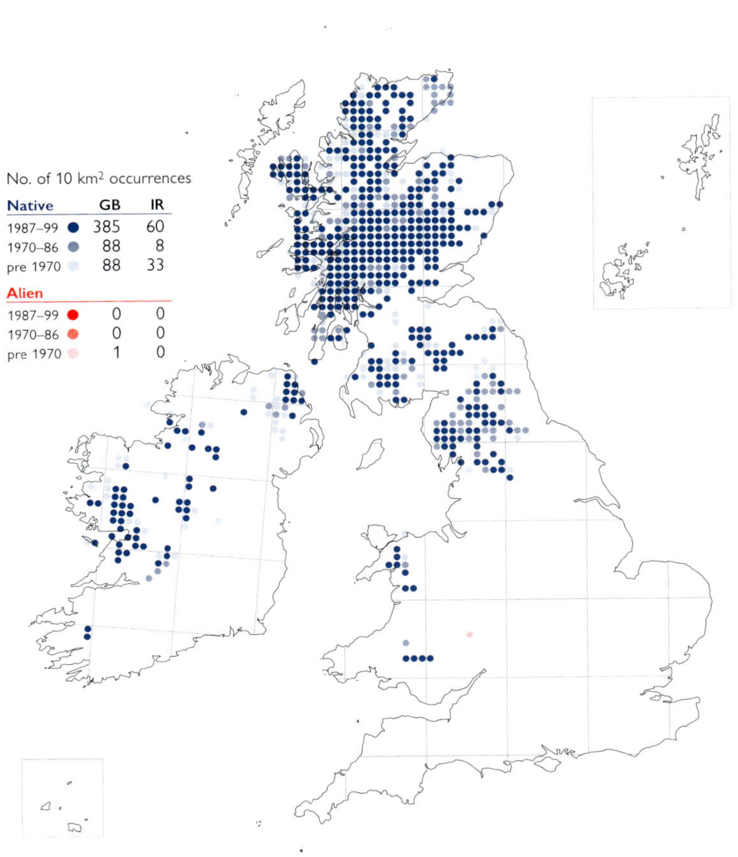

A perennial herb of damp, usually base-enriched substrates, occurring in rocky places, on mountain ledges and screes, in base-rich flushes in montane grassland, on river shingle and stony lake shores, and on stabilised sand dunes. 0–1065 m (Ben Lawers, Mid Perth).

Native (change –0.52). The distribution of this species is little changed in Scotland, but it seems to have declined in England, particularly since 1970, where increased grazing pressure may be causing losses (Halliday, 1997). It is better recorded now than it was for the 1962 *Atlas*, especially in N. Ireland.

Circumpolar Boreo-temperate element.

References: Atlas (260b), Hultén & Fries (1986), Meusel & Jäger (1992).

W. R. MEEK

Galium odoratum Woodruff

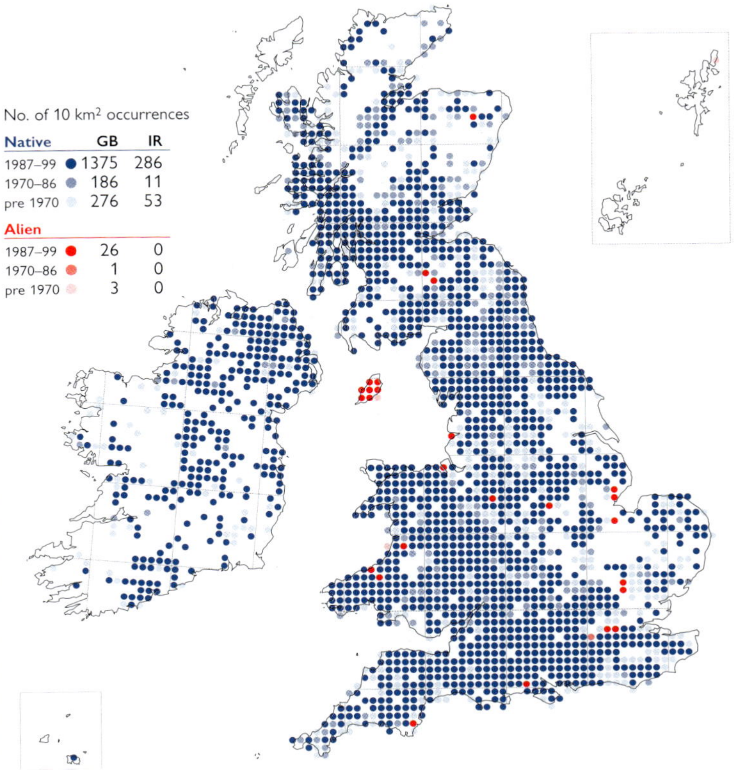

A rhizomatous perennial herb which grows in deciduous woodland, scrub and shaded hedge banks on base-rich or neutral, often damp, soils. This species is thought to be a good indicator of ancient woodlands in lowland England. 0–640 m (Atholl, E. Perth).

Native (change –0.62). *G. odoratum* was mapped as 'all records' in the 1962 *Atlas*. It has declined as older woodlands have been removed or replanted with conifers, especially in East Anglia and S.E. England, but it remains frequent and locally abundant across much of its range. It is sometimes grown in gardens and some records near habitation may be escapes.

European Temperate element; also in C. and E. Asia.

References: Atlas (260a), Hultén & Fries (1986), Meusel & Jäger (1992), Peterken (1974), Rackham (1980).

W. R. MEEK

Galium uliginosum Fen Bedstraw

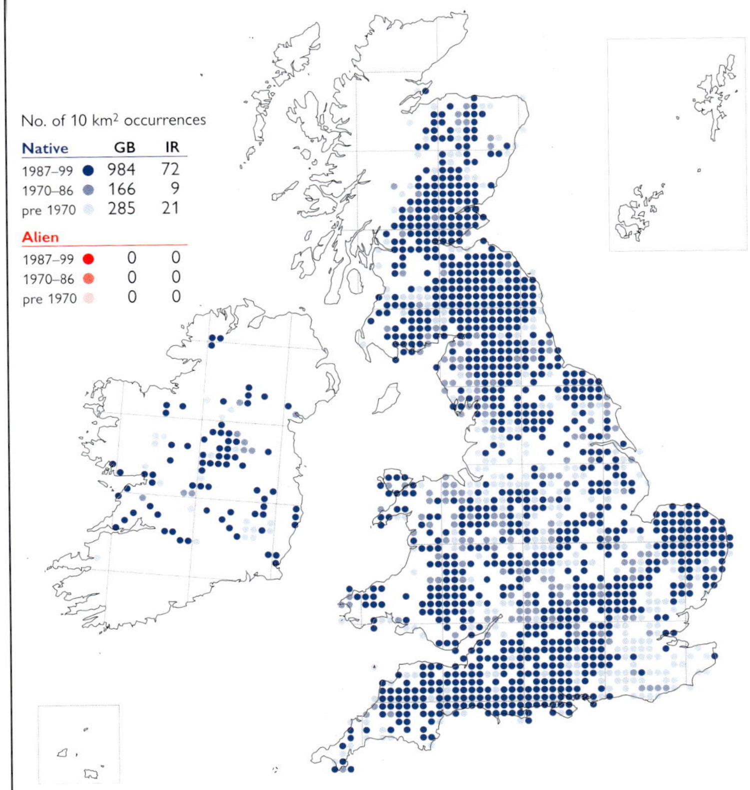

A perennial herb of base-rich marshes and fens, usually in drier and more calcareous habitats than *G. palustre*, although the two may sometimes be found growing together. 0–750 m (Cross Fell, Cumberland).

Native (change –0.14). *G. uliginosum* has tended to be over-recorded in the past; in particular all records from N.W. Scotland are treated as doubtful and are not mapped. Some other early records may also be errors for *G. palustre*. Even allowing for this, analysis of the database reveals a widespread decline in this species since 1950. It was mapped as 'all records' in the 1962 *Atlas*.

Eurasian Boreo-temperate element.

References: Atlas (262b), Hultén & Fries (1986), Meusel & Jäger (1992), Mountford (1994), Rich & Woodruff (1996).

W. R. MEEK

Galium constrictum Slender Marsh-bedstraw

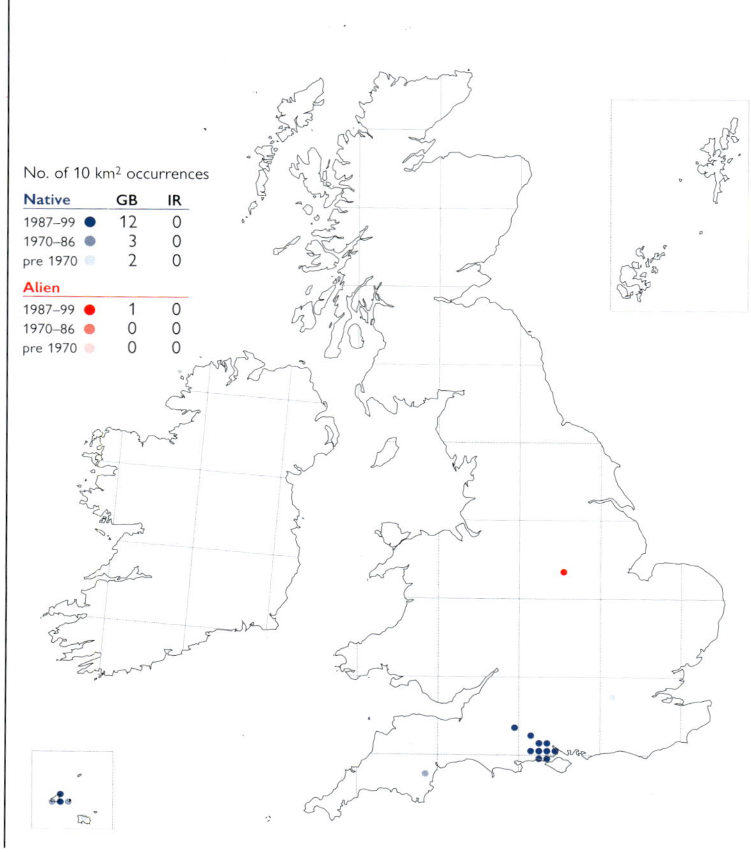

A perennial herb found around the margins of ponds which dry out in summer, on New Forest 'lawns' and in track ruts, and locally in marl-pits and ditches in water-meadows. Lowland.

Native. The many new records for *G. constrictum* in recent years probably result from a genuine increase in the number of sites as well as greater recording effort. New Forest (S. Hants.) populations seem reasonably secure as long as current management practices continue. It was last seen in Devon in 1972. The record from Yorkshire shown in the 1962 *Atlas* is considered to be an error.

Mediterranean-Atlantic element.

References: Atlas (262a), Brewis *et al.* (1996), Meusel & Jäger (1992), Wigginton (1999).

W. R. MEEK

Galium palustre Common Marsh-bedstraw

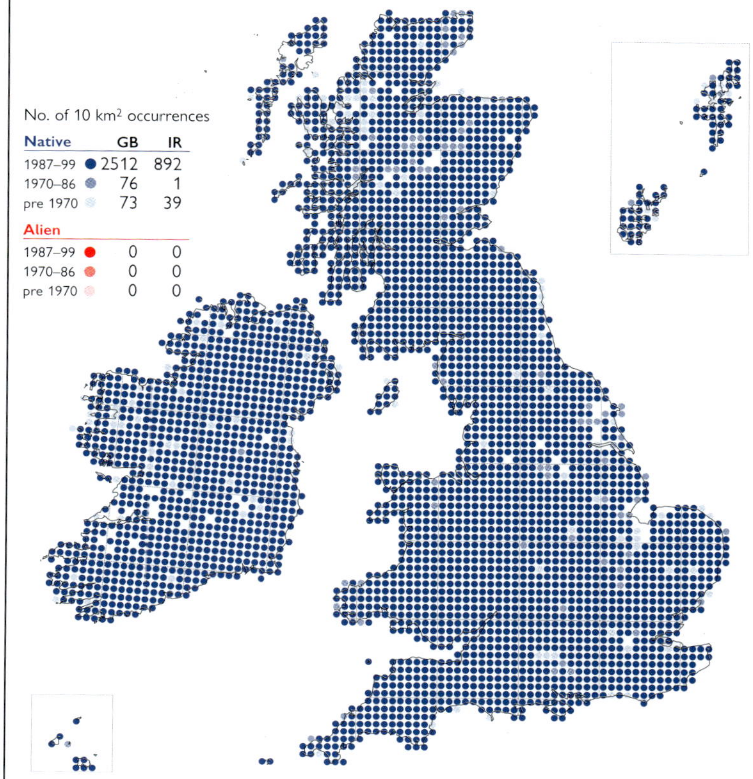

A perennial herb of seasonal or permanently wetland habitats, including wet meadows, marshes, fens, ditches, ponds and lakesides. It grows on a wide range of soil types, but has a preference for non-calcareous substrates. 0–825 m (Cross Fell, Cumberland).

Native (change +0.07). Despite widespread destruction of wetlands, there is no evidence to suggest a change in the distribution of this species since the 1962 *Atlas*, perhaps due to its ability to colonise new standing water habitats. Two subspecies (subsp. *palustre* and subsp. *elongatum*) occur in our area throughout the range of the species, but have not been recorded evenly.

Eurosiberian Boreo-temperate element; also in N. America.

References: Atlas (261d), Grime *et al.* (1988), Hultén & Fries (1986), Meusel & Jäger (1992), Mountford (1994).

W. R. MEEK

Galium verum Lady's Bedstraw

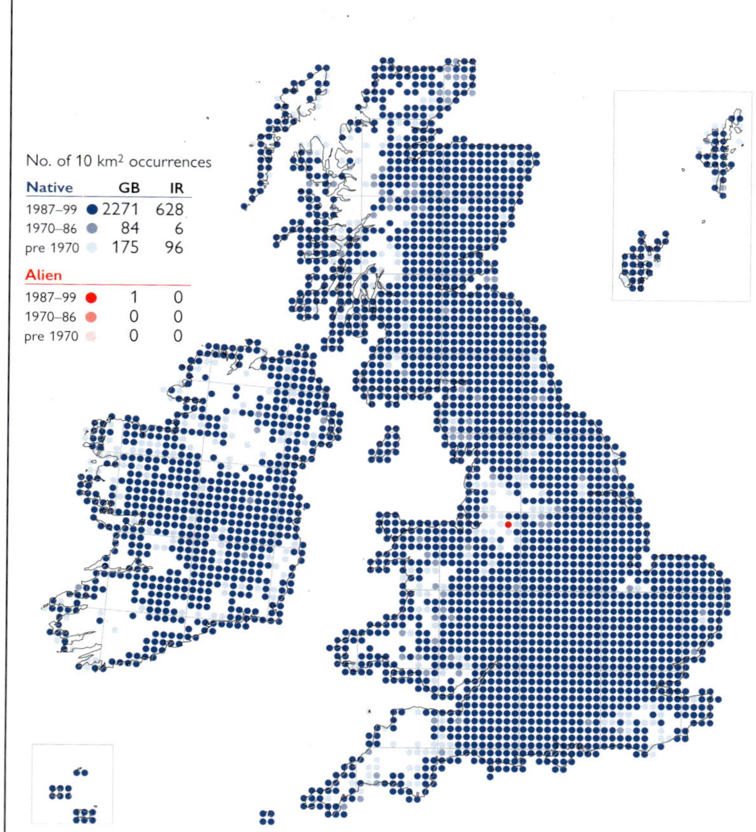

A stoloniferous perennial herb of well-drained, relatively infertile neutral or calcareous soils. Habitats include hay meadows, pastures, chalk and limestone downland, rock outcrops, quarries, coastal cliff-tops, dune grasslands and machair, roadsides and railway banks. Procumbent plants (var. *maritimum*) occur widely in coastal habitats. 0–780 m (Coire na Gabhar, Mid Perth).

Native (change –0.85). Although it has declined locally as a result of grassland improvement, *G. verum* remains common across much of its range. It is a frequent constituent of wild-flower seed mixtures.

Eurasian Boreo-temperate element; widely naturalised outside its native range.

References: Atlas (260d), Grime *et al.* (1988), Hultén & Fries (1986), Meusel & Jäger (1992).

W. R. MEEK

Galium mollugo Hedge Bedstraw

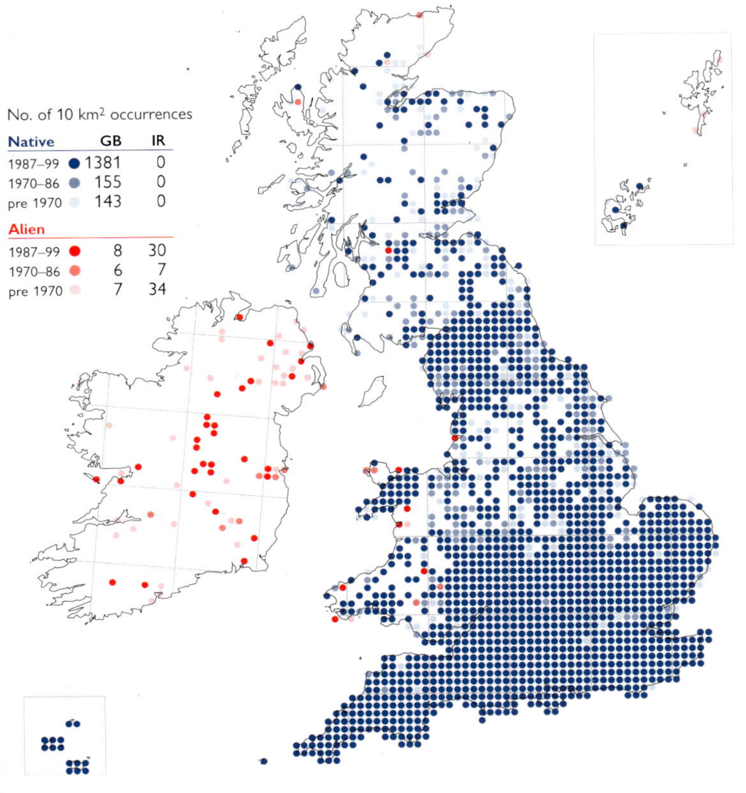

A stoloniferous perennial herb of well-drained, calcareous or base-rich soils. Habitats include rough and permanent grassland, waysides, railway banks, roadsides, hedge banks, woodland edges, scrub and waste ground. 0–845 m on Great Dun Fell (Westmorland).

Native (change +0.04). *G. mollugo* shows little change in distribution since the 1962 *Atlas*, although it is now much better recorded, especially in Wales. It is common in wild-flower seed mixtures, and may be increasing on roadsides. It is regarded as alien in Ireland (Scannell & Synnott, 1987). Two subspecies may occur in our area (subsp. *mollugo* and subsp. *erectum*) but their taxonomic status is uncertain.

European Boreo-temperate element; widely naturalised outside its native range.

References: Atlas (260c), Hultén & Fries (1986), Meusel & Jäger (1992).

W. R. MEEK

Galium mollugo × *G. verum* (*G.* × *pomeranicum*)

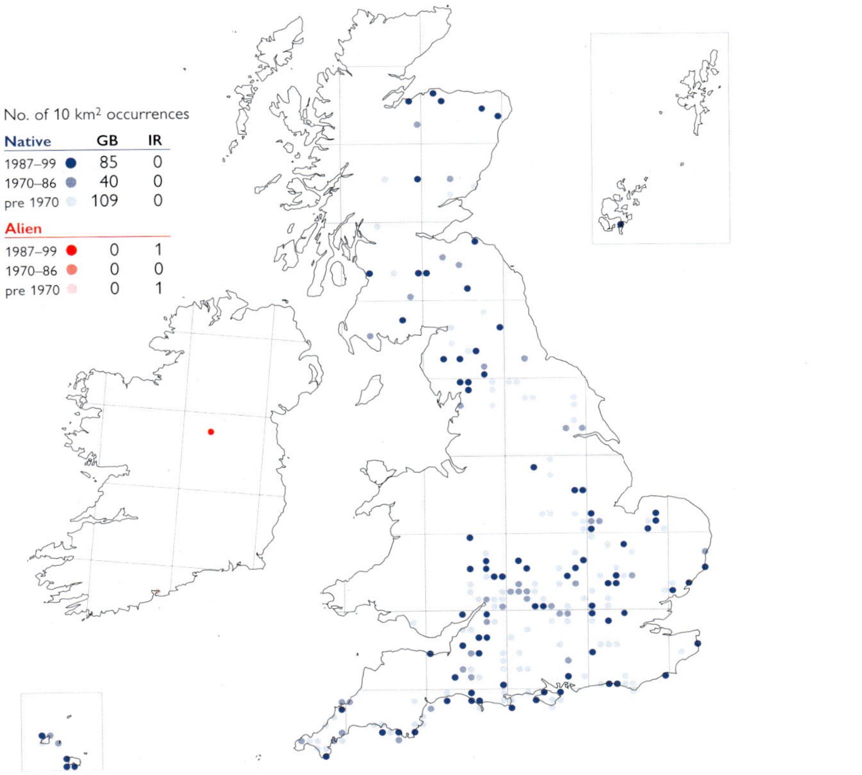

This vigorous hybrid, which is intermediate in all characters between its parents, occasionally arises where the two species occur together. Lowland.

Native. Although this hybrid is described by Stace (1997) as being frequent, there seem to be relatively few recent records. It is much better recorded than when mapped by Perring & Sell (1968), but is probably still under-recorded. A robust alien variety is reported to be planted on roadside verges (Rich & Jermy, 1998).

Widespread in Europe.

References: Atlas Supp. (70b), Stace (1975).

W. R. MEEK

Galium pumilum Slender Bedstraw

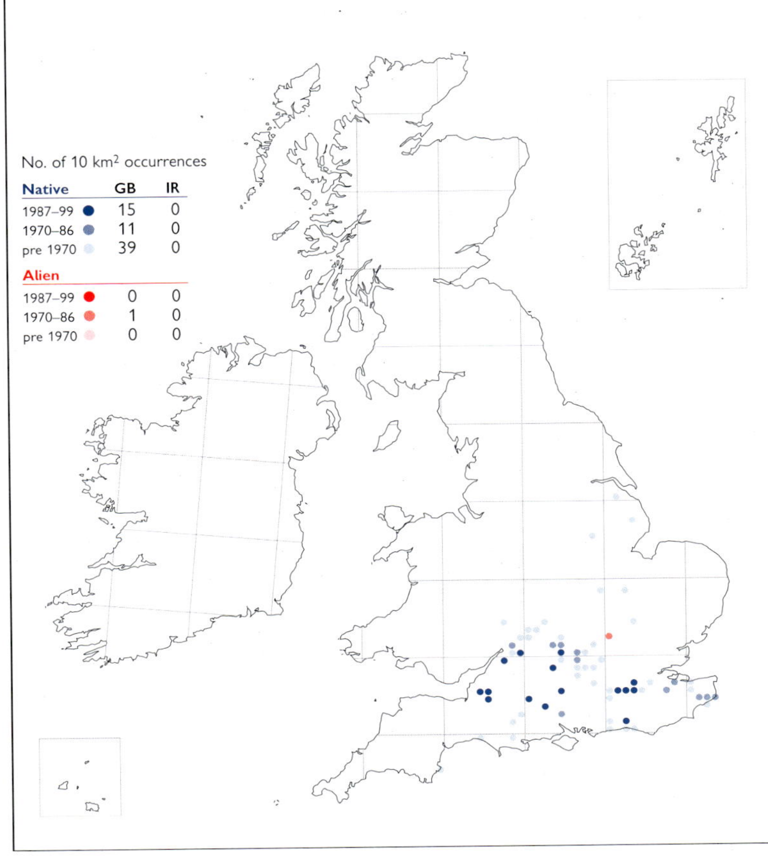

A perennial herb of species-rich chalk and limestone grassland where the sward is kept short by grazing, mowing, exposure or disturbance, and on anthills. It also grows on limestone spoil-heaps. 0–375 m (near Pitlochry, E. Perth).

Native (change −1.32). Old records of *G. pumilum* are unreliable as the species was previously confused with *G. sterneri*, so changes in its distribution are difficult to assess. However, confirmed specimens exist from many sites where this species now appears to be absent, the loss of chalk grassland and reduced grazing probably accounting for any decline. Plants at Cheddar (N. Somerset) formerly referred to *G. fleurotii* are now included within *G. pumilum*.

European Temperate element.

References: Atlas (261b), Hultén & Fries (1986), Meusel & Jäger (1992), Stewart *et al.* (1994).

W. R. MEEK

Galium sterneri Limestone Bedstraw

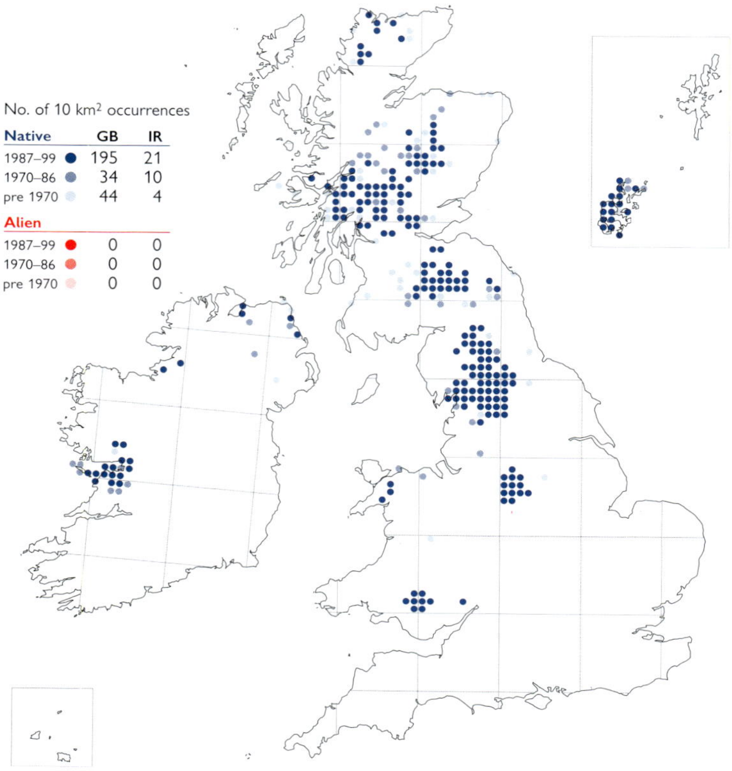

No. of 10 km² occurrences

Native		GB	IR
1987–99	●	195	21
1970–86	●	34	10
pre 1970	●	44	4
Alien			
1987–99	●	0	0
1970–86	●	0	0
pre 1970	●	0	0

A perennial herb of short grassland overlying limestone and basic igneous rocks and mica-schist, also growing on vegetated scree slopes, rock ledges and limestone pavement. Lowland to 975 m on Creag Mhor (Mid Perth).

Native (change +0.69). *G. sterneri* is much better recorded now than for the 1962 *Atlas*. Populations in rocky, upland habitats are probably stable, whilst it may have declined in limestone grassland because of agricultural improvement or the lack of grazing.

Suboceanic Boreal-montane element.

References: Atlas (261c), Grime *et al.* (1988), Meusel & Jäger (1992), Stewart *et al.* (1994).

W. R. MEEK

Galium saxatile Heath Bedstraw

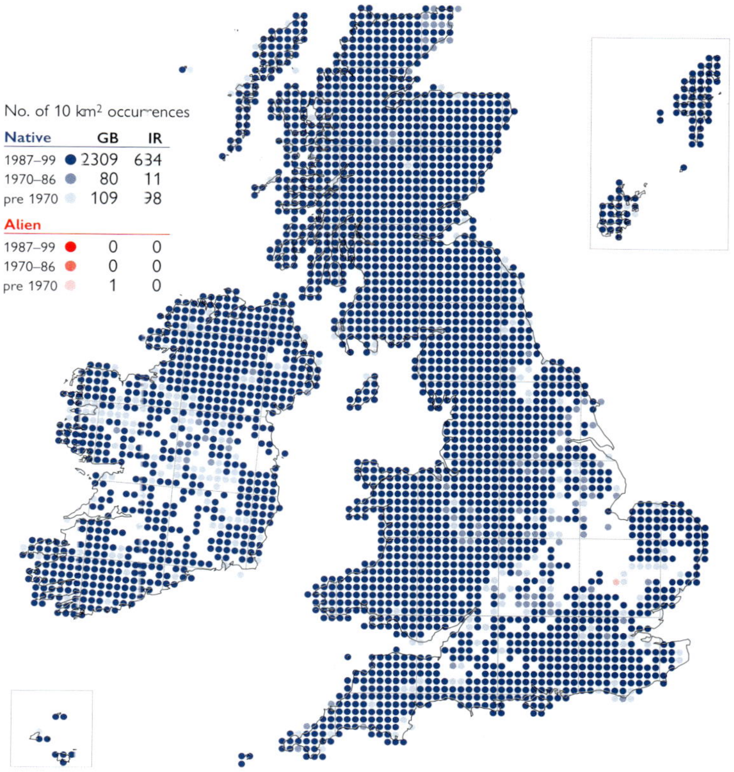

No. of 10 km² occurrences

Native		GB	IR
1987–99	●	2309	634
1970–86	●	80	11
pre 1970	●	109	98
Alien			
1987–99	●	0	0
1970–86	●	0	0
pre 1970	●	1	0

A low-growing perennial herb of infertile acidic soils, occurring in grassland, on heaths, in rocky places and open woods, and locally on disturbed or derelict ground. 0–1215 m (Ben Lawers, Mid Perth).

Native (change –0.15). This species remains abundant in upland areas, and is a useful indicator of unimproved hill grassland. However, it has declined locally in the lowlands as a result of habitat destruction.

Suboceanic Temperate element.

References: Atlas (261a), Dupont (1962), Grime *et al.* (1988), Hultén & Fries (1986), Meusel & Jäger (1992).

W. R. MEEK

Galium aparine Cleavers

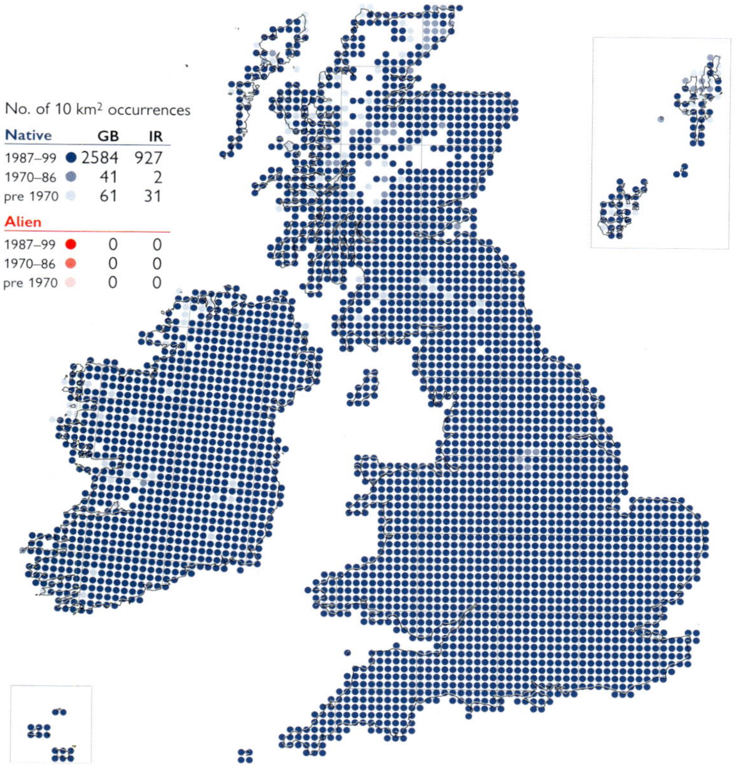

No. of 10 km² occurrences

Native		GB	IR
1987–99	●	2584	927
1970–86	●	41	2
pre 1970	●	61	31
Alien			
1987–99	●	0	0
1970–86	●	0	0
pre 1970	●	0	0

A scrambling annual of cultivated land, hedges, river banks, waysides, soil heaps and waste places, also growing in more natural habitats such as scree slopes and shingle. It inhabits both tall-herb and ruderal communities, thriving on highly fertile soil. Seed is dispersed by mammals. 0–440 m (Clun Forest, Salop).

Native (change –0.09). The overall distribution of *G. aparine* is stable. It is a highly successful and abundant weed, especially in arable crops, and is probably increasing in abundance despite the use of species-specific agricultural herbicides.

European Temperate element, but widely naturalised so distribution is now Circumpolar Temperate.

References: Atlas (262d), Grime *et al.* (1988), Hultén & Fries (1986), Malik & Vanden Born (1988), Meusel & Jäger (1992), Taylor (1999a).

W. R. MEEK

Galium spurium False Cleavers

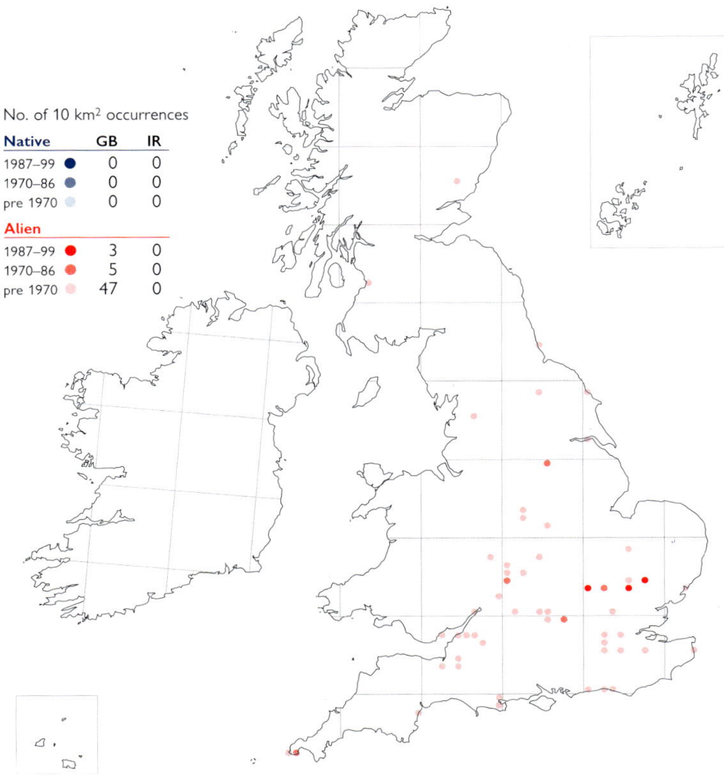

An annual, formerly occurring as an arable weed on autumn-cultivated land, particularly in cereal and potato crops, but in recent years known only from a few allotments and nearby roadside verges. It may occasionally arise from a residual seed bank during road-works. It flowers in autumn. Lowland.

Neophyte (change −1.87). This species was first recorded in the wild in 1806 and has been known from around Saffron Walden (N. Essex) since 1844, although it is now much rarer than it was when first discovered. It has only been recorded as a casual elsewhere.

Circumpolar Temperate element, with a continental distribution in W. Europe; the limits of its native range have been obscured by its spread in cultivation.

References: Atlas (263a), Hultén & Fries (1986), Jermyn (1974), Malik & Vanden Born (1988).

W. R. MEEK

Galium tricornutum Corn Cleavers

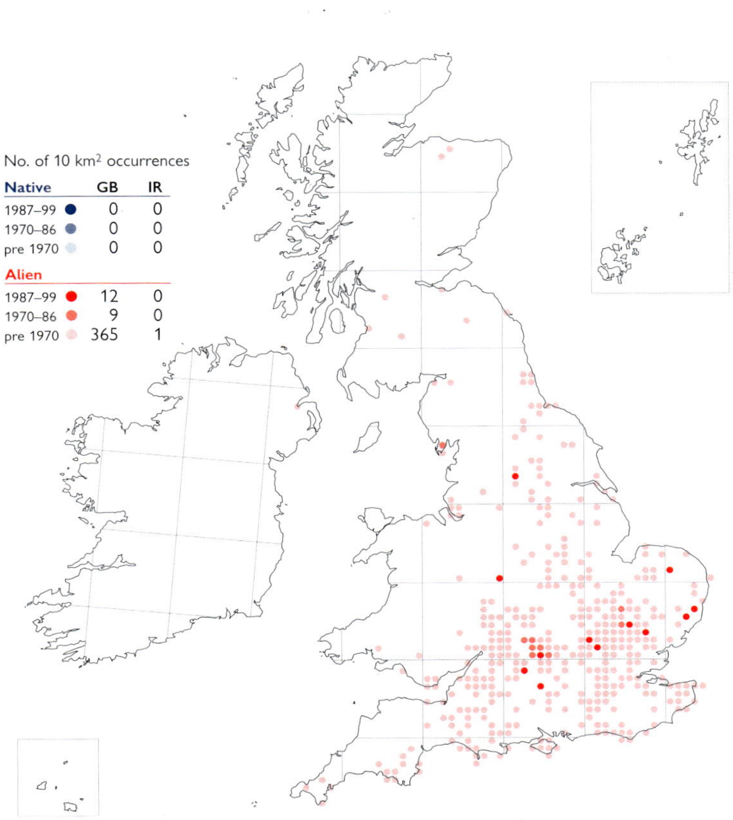

An annual of cereal fields and disturbed ground, chiefly on dry calcareous soils. Rarely, it can arise as a casual from the seed bank during earth-moving. Lowland.

Archaeophyte (change −4.78). The very substantial decline, already apparent by 1930, has continued since the 1962 *Atlas*, due to the intensification of arable farming. Populations are usually very small, and large ones tend to be transient. Re-introduction into protected sites is occasionally attempted using seed of English origin.

As an archaeophyte *G. tricornutum* has a Eurosiberian Southern-temperate distribution.

References: Atlas (262c), Bolòs & Vigo (1995), Wigginton (1999).

W. R. MEEK

Galium parisiense Wall Bedstraw

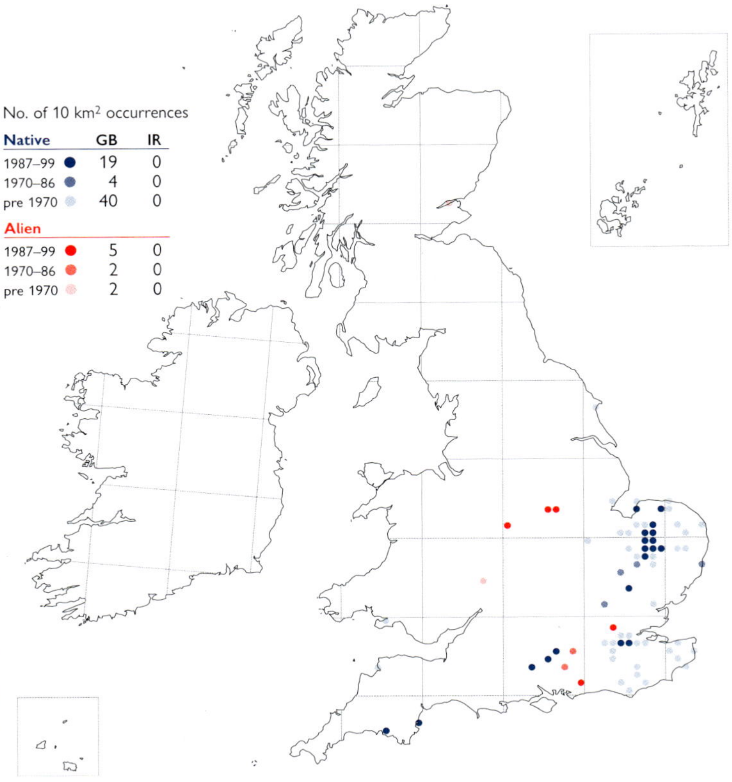

An annual of old walls and bare ground on calcareous or neutral substrates. It is intolerant of competition, and is susceptible to nutrient-enrichment. It may occasionally occur as a casual or short-term introduction well outside its normal range. Lowland.

Native or alien (change −0.57). *G. parisiense* was first recorded in Britain in 1690. Always scarce, and lost from many 10-km squares before 1930, it has undergone further decline since the 1962 *Atlas* as many of the open, infertile soils on which it grew have undergone agricultural improvement and ancient walls have been cleaned, rebuilt or demolished.

Submediterranean-Subatlantic element.

References: Atlas (263b), Bolòs & Vigo (1995), Stewart *et al.* (1994).

W. R. MEEK

Cruciata laevipes Crosswort

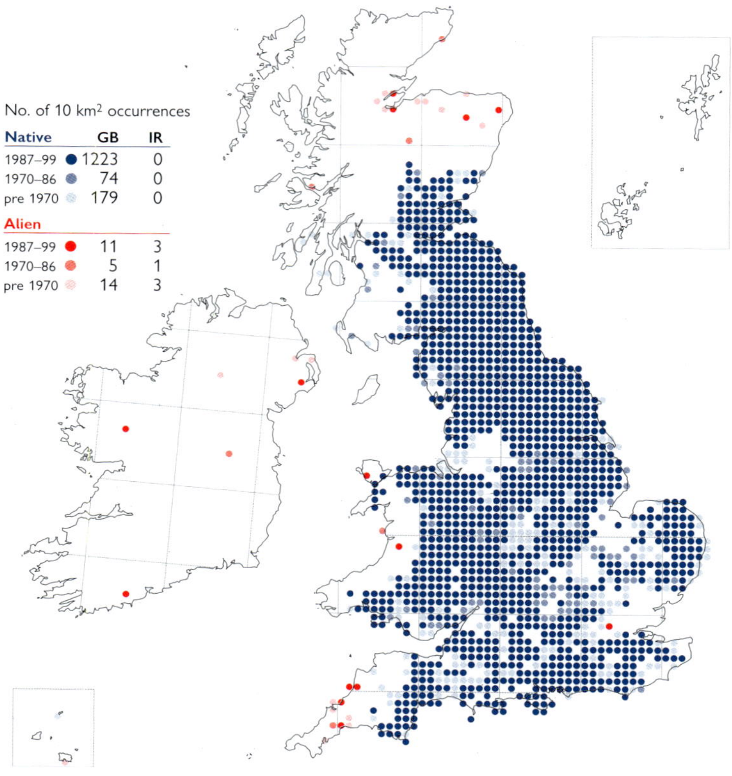

A perennial herb of deep, well-drained neutral or calcareous soils, typically occurring in ungrazed grassland, open scrub, hedge banks, woodland rides and edges, and on waysides. 0–550 m (Garrigill, Cumberland).

Native (change –0.77). *C. laevipes* has shown some losses since the 1962 *Atlas*. Although characteristic of older and relatively undisturbed habitats, it is able to colonise roadsides, river- and railway banks, and remains fairly frequent across its range. It is considered alien in Ireland (Scannell & Synnott, 1987), where it remains rare.

Eurosiberian Temperate element.

References: Atlas (259d), Grime *et al.* (1988), Meusel & Jäger (1992).

W. R. MEEK

Rubia peregrina Wild Madder

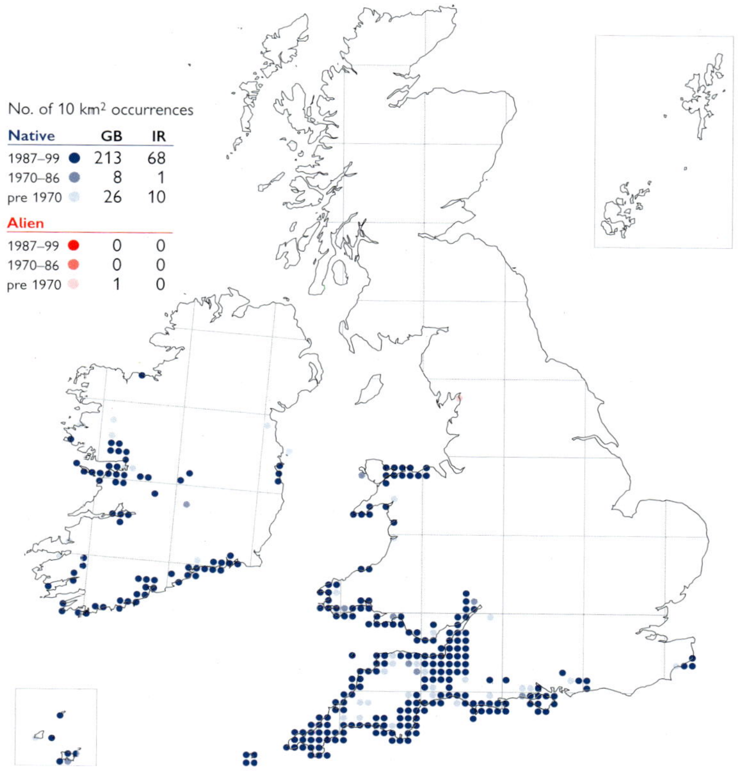

A scrambling, evergreen perennial of hedge banks, scrub, walls, cliffs and other rocky places near the coast, or very locally on calcareous soils further inland. Lowland.

Native (change +0.17). Although geographically restricted, this resilient species can be frequent where it occurs and shows no change in its distribution since the 1962 *Atlas*, although this is now much better recorded along the coast of S. Ireland. Relaxation of grazing on coastal cliff-tops may have led to an increase in available habitat.

Mediterranean-Atlantic element.

References: Atlas (263c), Bolòs & Vigo (1995).

W. R. MEEK

Sambucus racemosa Red-berried Elder

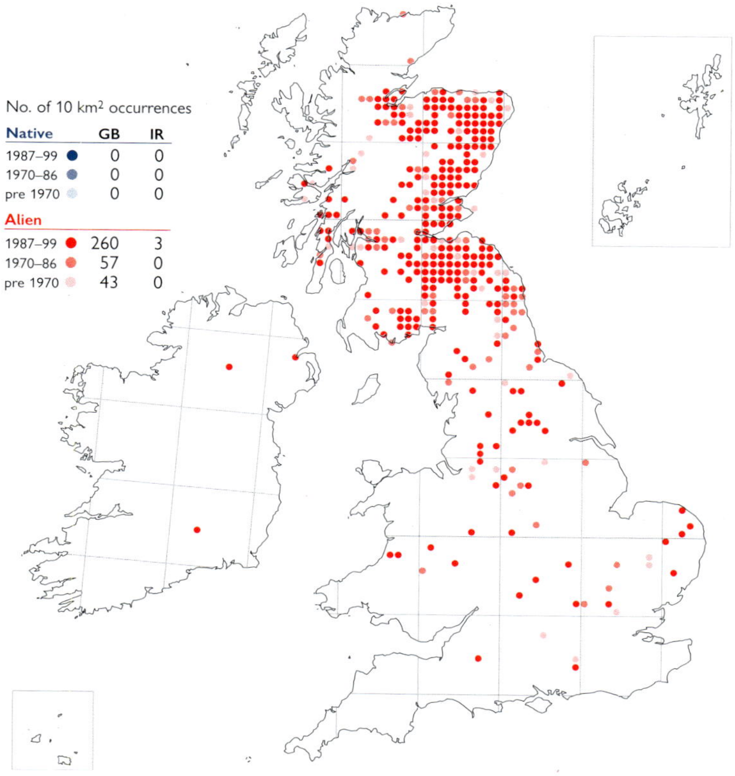

A deciduous shrub established in woodland, shrubberies, hedges and waste ground, and planted as game cover in parts of N. England and Scotland. Lowland.

Neophyte (change +0.79). This species was introduced in the 16th century and is quite widely cultivated. Dunn (1905) described it as 'occasionally noticed as a semi-wild plant near gardens and shrubberies'. Most of its spread as a garden escape took place in the 20th century, with much consolidation since the 1962 *Atlas*.

A very variable Circumpolar Boreo-temperate species, comprising several subspecies. It is absent as a native from much of N. & W. Europe.

References: Atlas (264b), Bean (1980), Hultén & Fries (1986), Meusel & Jäger (1992).

G. T. D. WILMORE

Sambucus nigra Elder

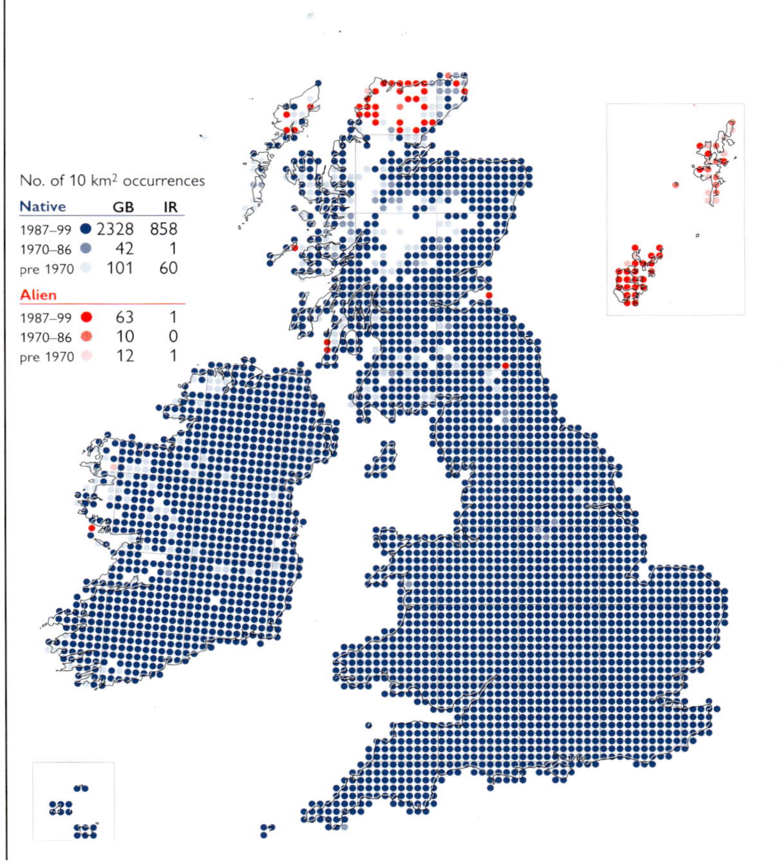

A deciduous shrub or small tree of fertile soils in a wide range of habitats including woodland, hedgerows, grassland, scrub, waste ground, roadsides and railway banks. It is resistant to rabbit grazing and often occurs around warrens. 0–470 m (above Nenthead, Cumberland).

Native (change –0.75). The current range of *S. nigra* is similar to that shown in the 1962 *Atlas*. It may have been originally introduced at some sites, but it is often impossible to distinguish native and alien occurrences, especially as it has often spread naturally by bird-sown seeds. In N. Scotland it is often restricted to the vicinity of human settlements.

European Temperate element; widely naturalised outside its native range.

References: Atlas (264a), Grime *et al.* (1988), Hultén & Fries (1986), Meusel & Jäger (1992).

G. T. D. WILMORE

Sambucus ebulus Dwarf Elder

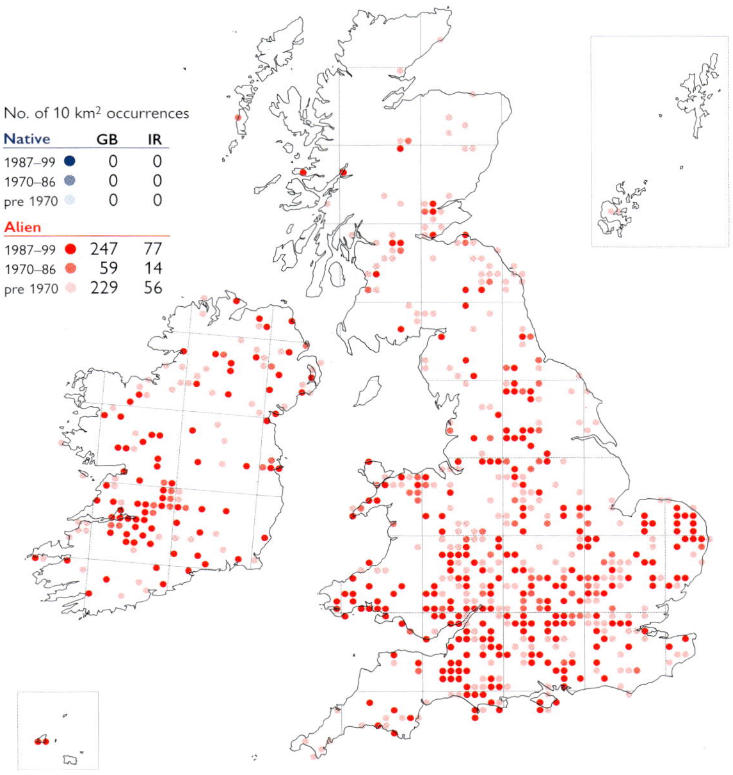

A robust perennial rhizomatous herb occurring infrequently in hedgerows, on roadsides and waste ground, usually in small numbers but locally forming thickets (Lavin & Wilmore, 1994). Lowland.

Archaeophyte (change –0.17). The distribution of *S. ebulus* is little changed since the 1962 *Atlas*.

As an archaeophyte *S. ebulus* has a European Southern-temperate distribution.

References: Atlas (263d), Meusel & Jäger (1992).

G. T. D. WILMORE

Viburnum opulus Guelder-rose

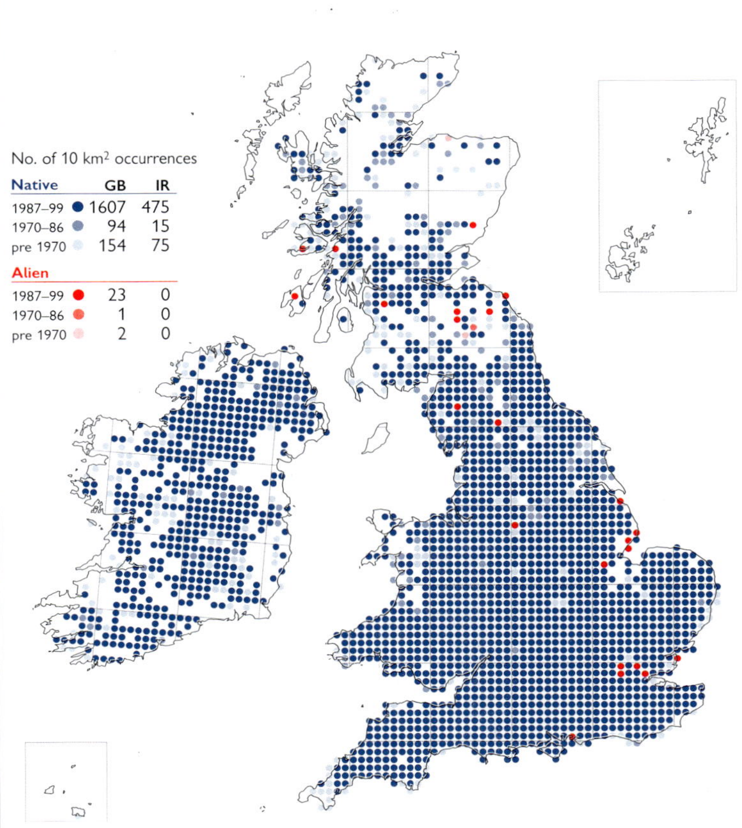

This deciduous shrub of neutral or calcareous soils is found in woodland, scrub and hedgerows, in fen carr and *Alnus* and *Salix* thickets, and on stream banks, favouring damp places, but also found in dry habitats. It is planted in parks and gardens and plants which spread from these sites to the wild sometimes include yellow-fruited cultivars (Lousley, 1976a). 0–400 m (south of Garrigill, Cumberland).

Native (change –0.15). The distribution of *V. opulus* is similar to that shown in the 1962 *Atlas*.

Circumpolar Temperate element.

References: Atlas (264d), Hultén & Fries (1986), Meusel & Jäger (1992).

G. T. D. WILMORE

Viburnum lantana Wayfaring-tree

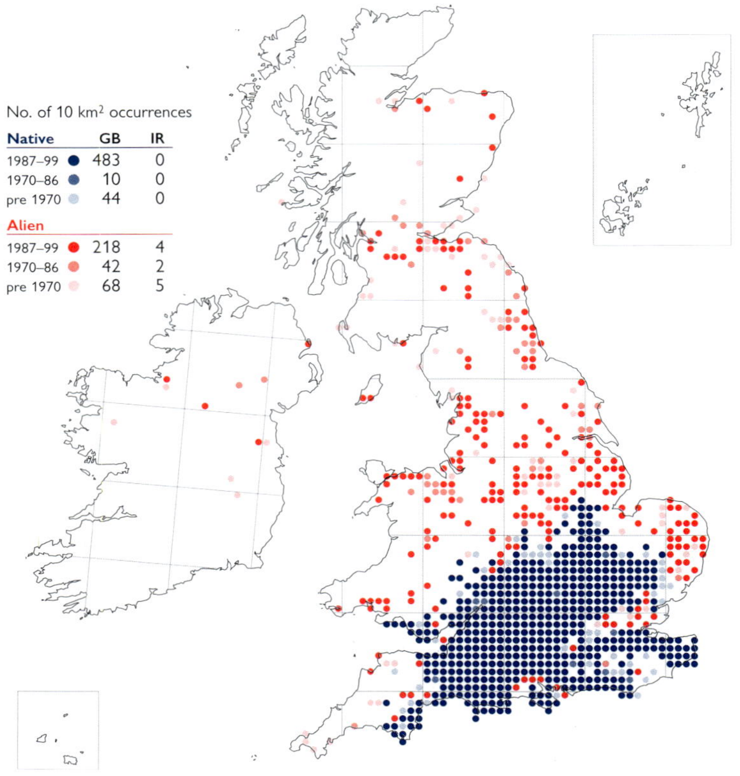

A deciduous shrub, found in woodland, scrub and hedgerows, especially on base-rich soils. It is particularly characteristic of chalk and limestone districts. It is now frequently planted on roadsides and may also appear when used as a stock for other cultivated species of *Viburnum* (Stace, 1997). Lowland.

Native (change +0.37). The native distribution of *V. lanata* is little changed since the 1962 *Atlas*, though local losses have occurred. There are now many more alien records, and the northern limits of its range have become blurred due to these. The status of some populations can be difficult to determine.

European Temperate element.

References: Atlas (264c), Meusel & Jäger (1992).

G. T. D. WILMORE

Viburnum tinus Laurustinus

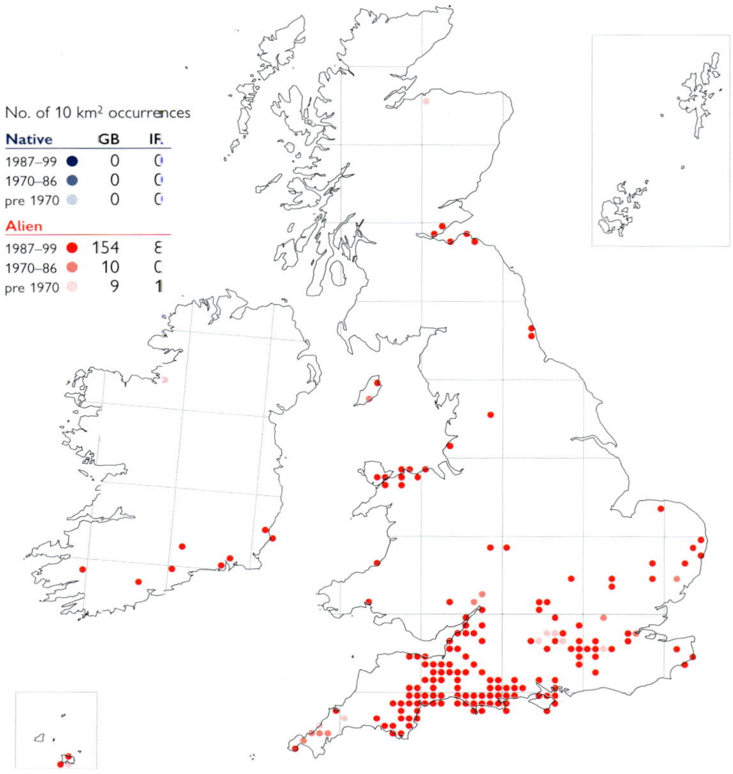

A large, evergreen shrub found as a garden escape or planted on sea-cliffs, banks, woodland, rough grassland, roadsides, railway banks, amenity plantings and waste ground; also a relic of cultivation. Reproduction is by seed and populations can become well-established, especially those on calcareous soils overlying chalk and limestone in coastal localities in S. England and N. Wales. Lowland.

Neophyte. *V. tinus* has been valued in gardens for its winter flowers since its introduction before 1596. It was recorded from the wild by 1941, and appears to be increasing.

Native of the S.W. Europe and the Mediterranean region.

References: Bean (1980), Bolòs & Vigo (1995).

T. D. DINES

Symphoricarpos albus Snowberry

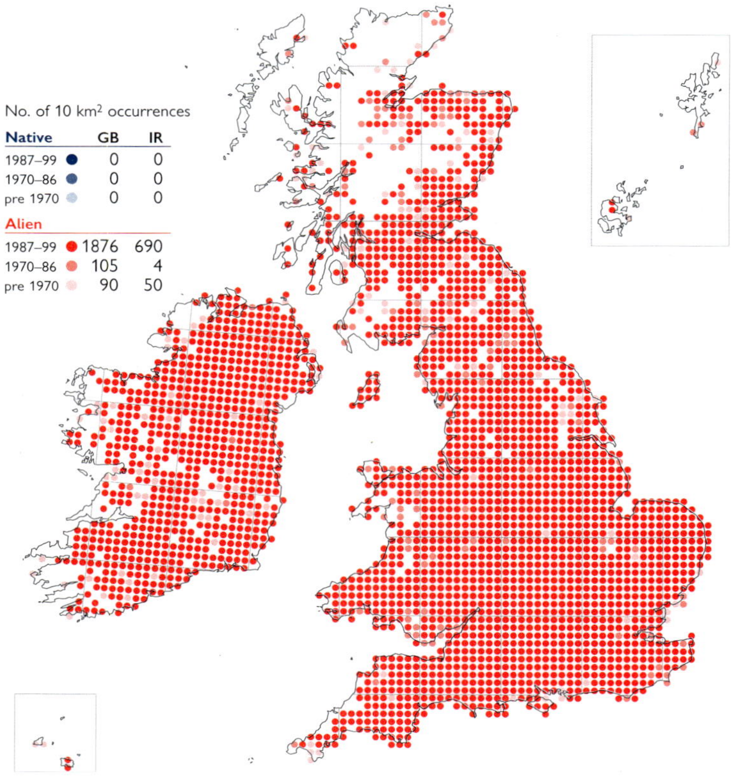

A bushy rhizomatous shrub widely naturalised in woodland, scrub, hedgerows and on waste ground; also formerly quite widely planted as cover for game in woodland. It reproduces by suckering and fruits freely, but rarely regenerates from seed. It spreads very slowly; dense thickets are normally the result of close initial planting. 0–385 m (Forest-in-Teesdale, Co. Durham).

Neophyte (change +1.74). *S. albus* was introduced into cultivation in Britain in 1817, and was known from the wild by 1863. Since the 1962 *Atlas* it has consolidated its position almost everywhere.

Native of N. America; the plant naturalised here is the western var. *laevigatus*.

References: Atlas (265a), Bean (1980), Coats (1963), Gilbert (1995).

G. T. D. WILMORE

Symphoricarpos microphyllus × *S. orbiculatus* (*S.* × *chenaultii*) Hybrid Coralberry

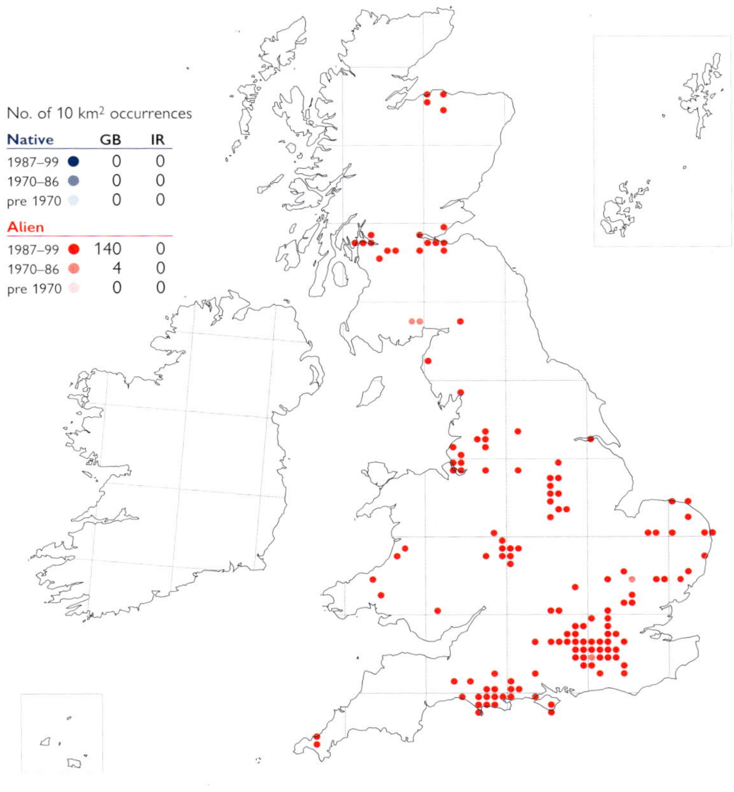

A deciduous shrub found naturalised in woodland, scrub, hedgerows and on waste ground, but not to the same extent as *S. albus*. Like *S. albus*, it reproduces by suckering and fruits freely, but rarely regenerates from seed. Lowland.

Neophyte. This hybrid arose in 1910, and it is often grown in gardens or planted for game cover. It was not recorded in the wild until 1974 (Cambs.), and is likely to be increasing. It has been mistaken in the past for *S. orbiculatus*.

A hybrid of garden origin.

Reference: Bean (1980).

T. D. DINES

Linnaea borealis Twinflower

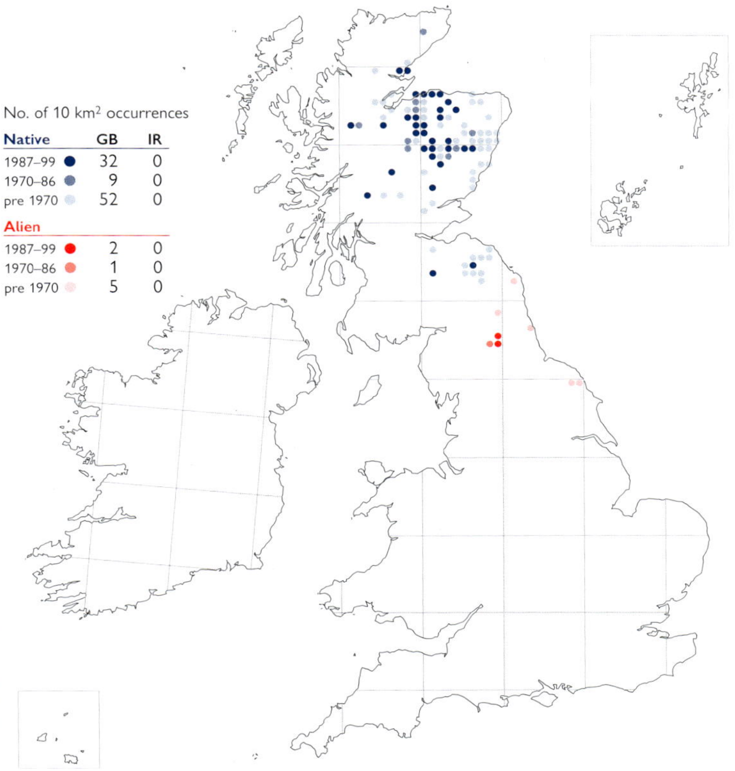

A creeping perennial, woody at the base, of both native and planted *Pinus sylvestris* woodland, where it occurs in slight to moderate shade, on barish ground or leaf litter, sometimes with an acidic heathy herb flora. It spreads vegetatively and by seed, though seedling establishment seems largely restricted to disturbed ground. 0–730 m (Easterness).

Native (change +0.07). There were severe losses of *L. borealis* before 1930, mostly through the clearance of native woodland. It may still be decreasing, although it has been better recorded recently. All English records are probably introductions, perhaps with tree seedlings.

Circumpolar Boreal-montane element.

References: Atlas (265b), Hultén & Fries (1986), Meusel & Jäger (1992), Stewart *et al.* (1994), Swan (1993).

G. T. D. WILMORE

Leycesteria formosa Himalayan Honeysuckle

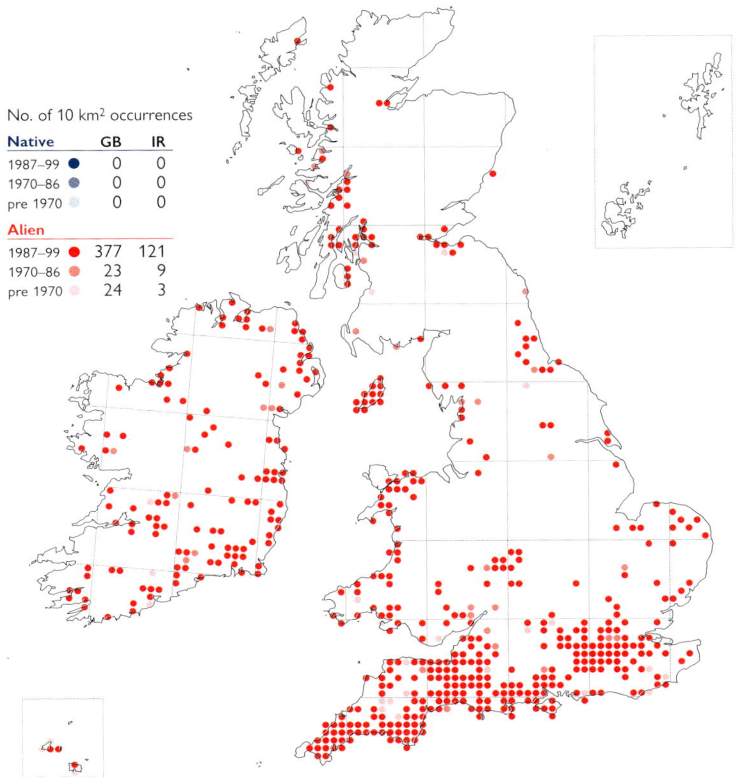

A deciduous shrub, locally established as a garden escape in woodland, hedgerows, shrubberies and on waste ground. It is sometimes planted as cover for pheasants, and also arises from bird-sown seed. Lowland.

Neophyte. This species was introduced into cultivation in Britain in 1824, and was known from the wild by at least 1905. Its range is said to be increasing nationally (Clement & Foster, 1994) and locally (Bowen, 2000; French *et al.*, 1999), but the species was not mapped for the 1962 *Atlas*, so such trends are difficult to assess.

Native of the Himalayas.

Reference: Bean (1973).

G. T. D. WILMORE

Lonicera pileata Box-leaved Honeysuckle

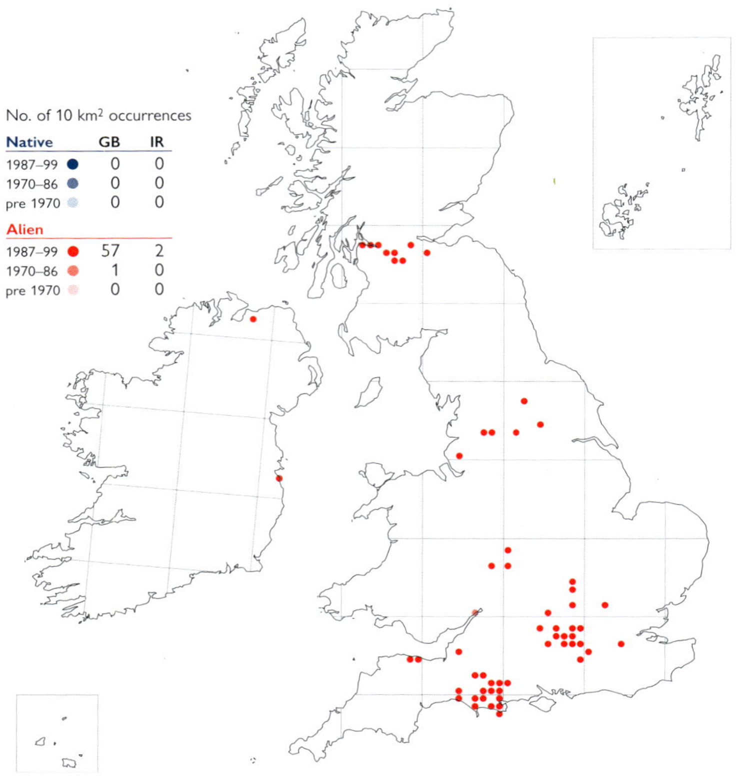

A small to medium-sized evergreen shrub, found planted or as a garden escape on roadsides and in parks, woodland, hedges and amenity areas. It can become well-established and is sometimes self- or bird-sown. Seedlings have also been found in pavement cracks and on walls. Lowland.

Neophyte. *L. pileata* was introduced to cultivation in Britain in 1900. It was first recorded from the wild in 1959 (Ketton, Leics.), and is increasingly planted in roadside shrubberies and plantations for ground-cover. It may be increasing in the wild.

Native of China.

Reference: Bean (1973).

T. D. DINES

Lonicera nitida Wilson's Honeysuckle

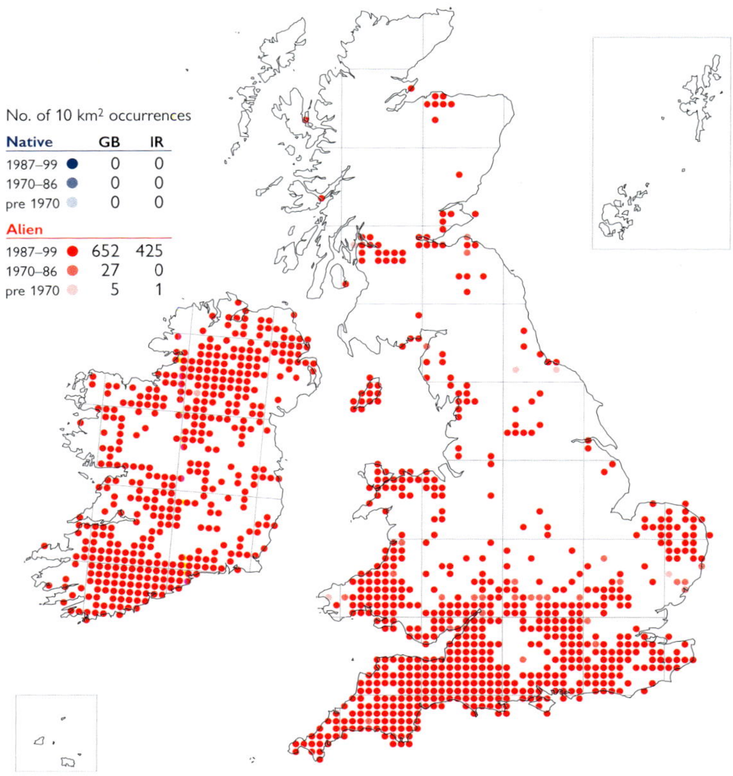

An evergreen shrub, grown extensively for hedging, but then rarely flowering; also occurring locally as a well-established garden escape, self-sown into woodland, scrub, hedgerows and waste ground, where it can persist for many years. Lowland.

Neophyte. This species has been introduced into cultivation twice; the first clone, which arrived in 1908, never flowers, but the second, available since 1939, flowers freely. The species is widely established and is likely to be spreading, especially in the south and west. It was first recorded from the wild in 1955.

Native of China.

References: Bean (1973), Coats (1963).

G. T. D. WILMORE

Lonicera xylosteum Fly Honeysuckle

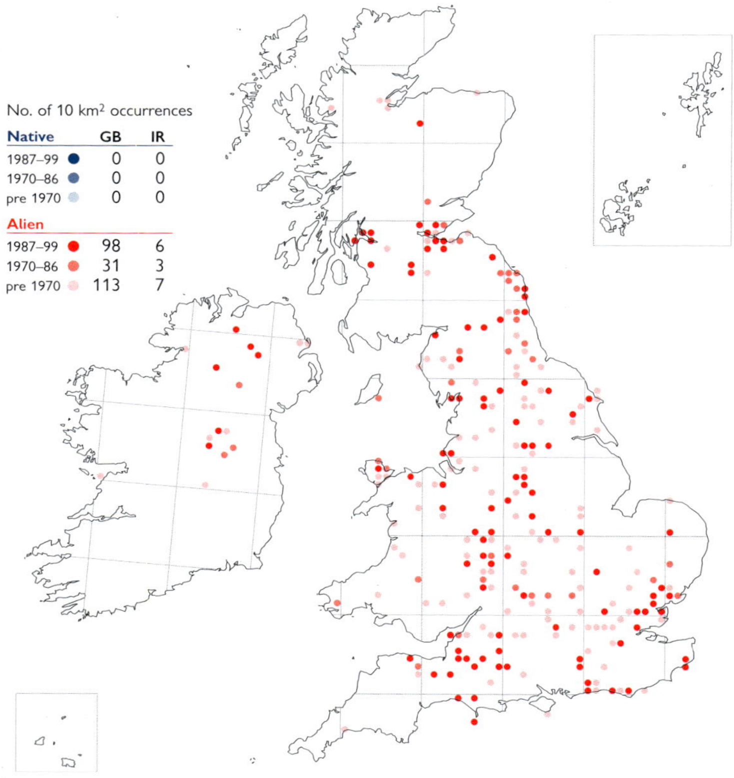

A deciduous shrub, growing on basic or neutral soils in woodland, hedgerows and scrub, generally occurring in small populations. Lowland.

Neophyte (change +0.58). *L. xylosteum* was being cultivated in Britain by 1683. First recorded in the wild as a casual in 1770, it has been known from the chalk scarp of the South Downs near Amberley (W. Sussex) since 1801, where it grows in ancient woodland, hedgerows or scrub. Although arguably native at these localities, it is more likely to be an introduction (Webb, 1985). It was lost before 1930 from many sites where it is an undoubted garden escape.

A Eurosiberian Temperate species; widely naturalised outside its native range.

References: Atlas (265c), Hultén & Fries (1986), Meusel & Jäger (1992), Wigginton (1999).

G. T. D. WILMORE

Lonicera japonica Japanese Honeysuckle

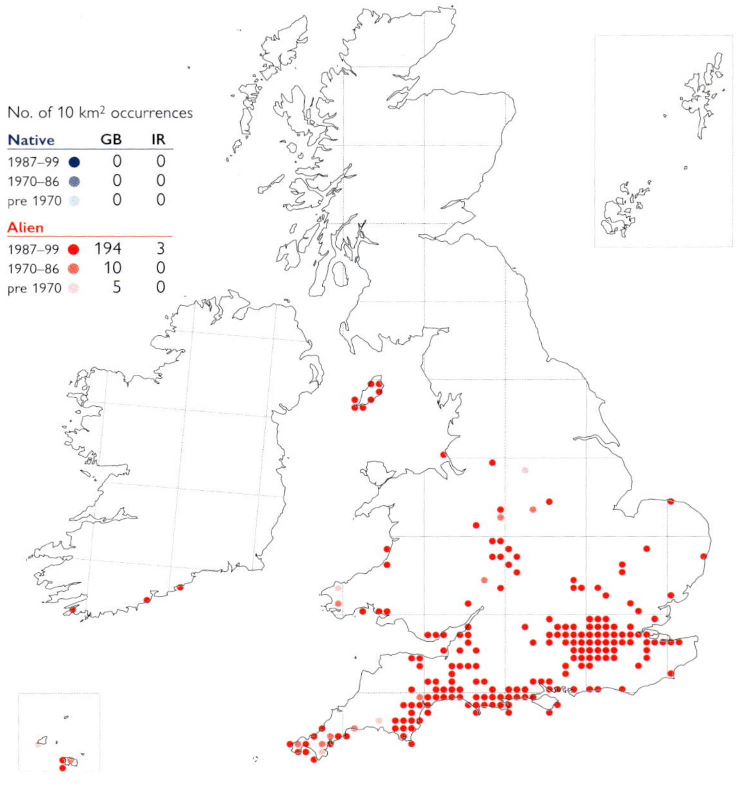

A semi-evergreen twining shrub infrequently occurring in woodland, scrub, hedgerows and on waste ground, sometimes forming extensive thickets in S. England. Lowland.

Neophyte. This species was introduced into Britain in 1806. Some populations are long-established, including that at Bere Ferrers in S. Devon where it has been known since 1937. Its distribution seems to be increasing.

Native of E. Asia; naturalised in N. America, where it can be a pernicious weed.

References: Bean (1973), Bolòs & Vigo (1995).

G. T. D. WILMORE

Lonicera periclymenum Honeysuckle

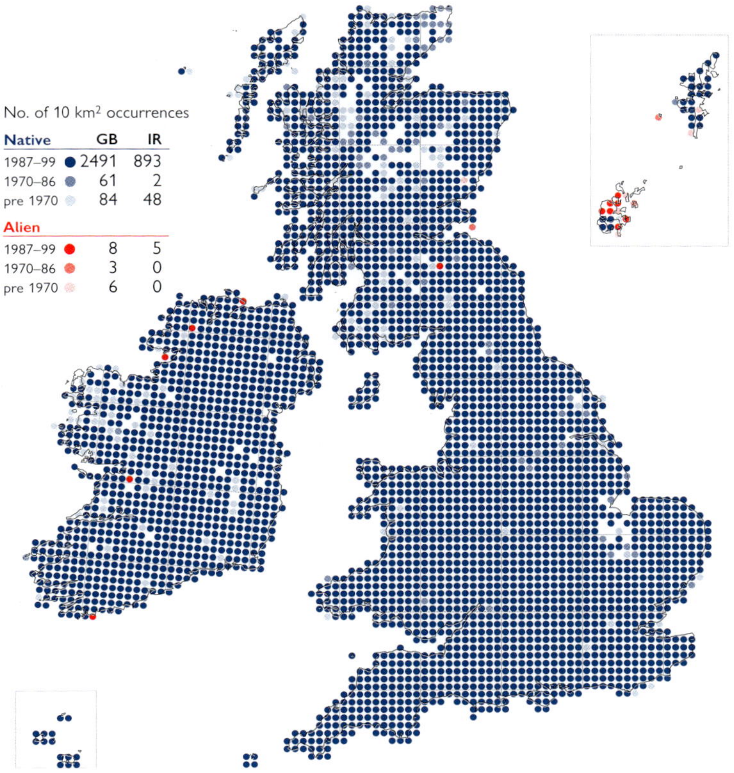

A perennial deciduous twining shrub or woody climber, found in woodland, scrub and hedgerows, and on shaded rocks. It prefers freely-drained, moderately basic to acidic soils, but also grows on poorly-drained base-rich clays. 0–610 m (Fairfield, Westmorland, and in Co. Down).

Native (change –0.11). The current distribution of *L. periclymenum* is similar to that shown in the 1962 *Atlas*.

Suboceanic Southern-temperate element.

References: Atlas (265d), Grime *et al.* (1988), Hultén & Fries (1986), Meusel & Jäger (1992).

G. T. D. WILMORE

Lonicera caprifolium Perfoliate Honeysuckle

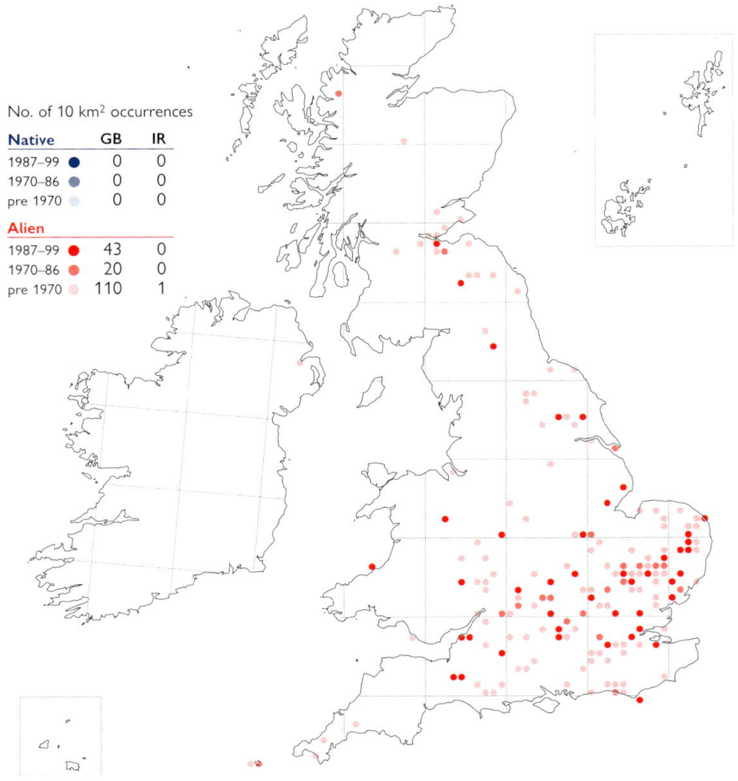

A twining, glaucous deciduous shrub, naturalised in woodland, scrub and hedgerows, perhaps establishing mainly from bird-sown seed from garden plants. Lowland.

Neophyte (change –0.73). *L. caprifolium* has been known in cultivation in Britain since at least 1596, and was recorded in the wild by 1763 (Cambs.). As with *L. xylosteum*, it was lost from most of its sites before 1930. However, it is less frequently cultivated now and most modern garden plants are its hybrid with *L. etrusca*, *L.* × *italica*. This may explain the paucity of modern records of *L. caprifolium* and even some of those mapped may represent misidentifications of the hybrid.

Native of E.C. and S.E. Europe from Italy eastwards, and of S.W. Asia; widely naturalised elsewhere in Europe.

References: Atlas (266a), Bean (1973), Coats (1963).

G. T. D. WILMORE

Adoxa moschatellina Moschatel

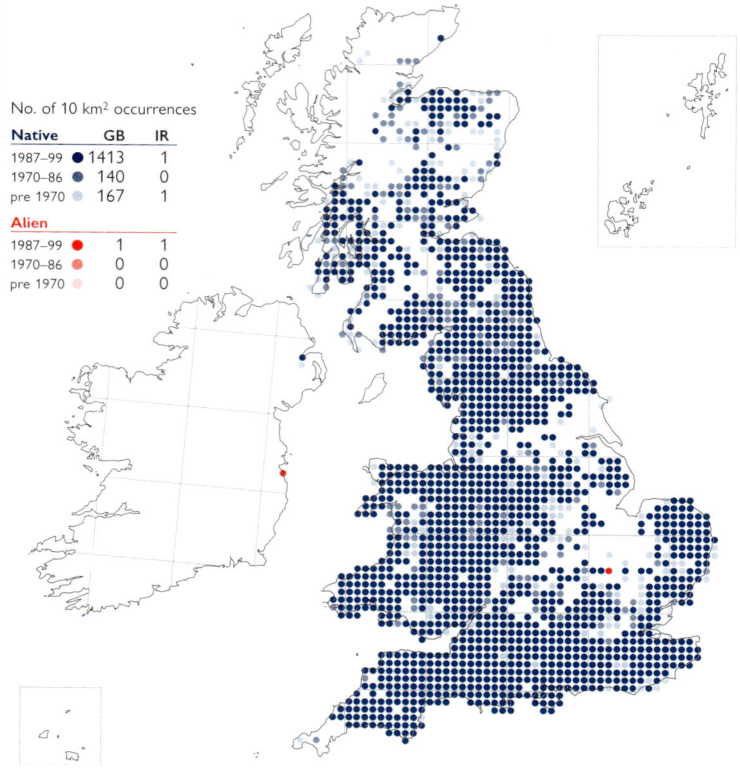

A perennial rhizomatous herb of mesic brown earth soils on the shaded banks of rivers and streams, in deciduous woodlands and shaded hedge banks; also occasionally in shaded base-rich sites in mountains. This is a vernal species which disappears by May or June in the lowlands. It is self-fertile, reproducing by seed and vegetatively. 0–1065 m (Ben Lawers, Mid Perth).

Native (change –0.05). Populations of *A. moschatellina* within old woodlands have remained stable. In some areas *A. moschatellina* is possibly under-recorded due to its early season. Its native status in Ireland has been questioned (Hackney, 1992).

Circumpolar Boreo-temperate element, with a disjunct distribution.

References: Atlas (266b), Curtis & McGough (1988), Hultén & Fries (1986), Meusel & Jäger (1992).

A. J. RICHARDS

Valerianella locusta Common Cornsalad

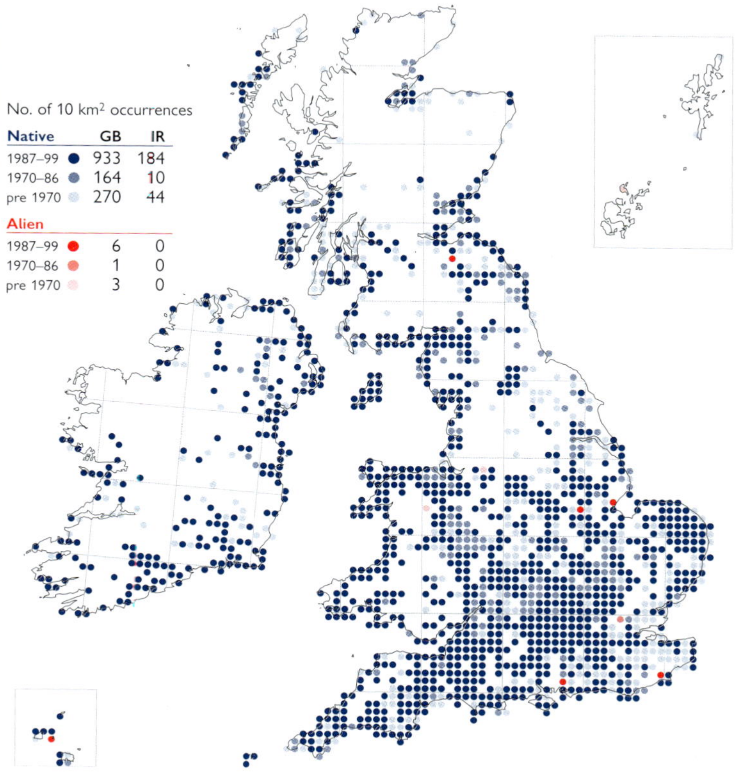

A winter-annual which occurs with other annuals on thin soils around rock outcrops and on scree, and on sand dunes and coastal shingle. It also grows in a wide range of disturbed habitats, including walls, gravel paths, railway tracks, in paving, gardens and, rarely, on arable land. Coastal populations are often the dwarf var. *dunensis*. The species is also commercially grown as a winter salad crop. 0–365 m (Wormhill, Derbys.).

Native (change –0.11). The distribution of *V. locusta* is broadly stable, with some local losses. Mapped as 'all records' in the 1962 *Atlas*, analysis of the database reveals that most of the losses have occurred since 1950.

European Temperate element; widely naturalised outside its native range.

References: Atlas (266c), Atlas Supp. (71a), Hultén & Fries (1986).

P. J. WILSON

Valerianella carinata Keeled-fruited Cornsalad

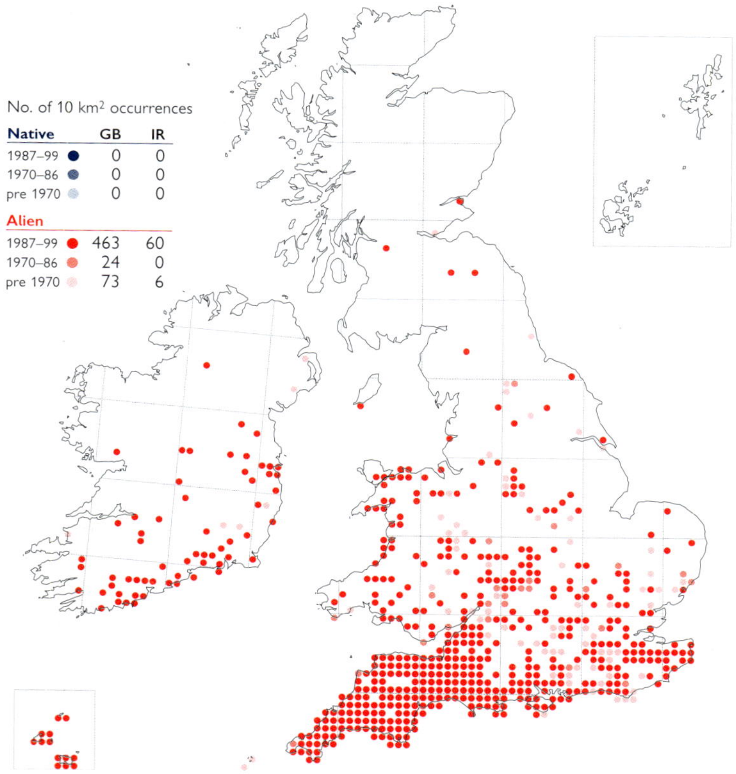

This autumn-germinating annual is found mainly in sites associated with human activity. Habitats include walls, gravel paths, paving, railway tracks and gardens. It does not occur in arable land. Lowland.

Archaeophyte (change +2.15). *V. carinata* has greatly increased since the 1962 *Atlas*, and it is now more common than *V. locusta* in some areas, especially S.W. England. It is difficult to account for such a dramatic increase, which cannot be entirely due to the fact that the differentiating characters are now better known.

As an archaeophyte *V. carinata* has a European Southern-temperate distribution.

References: Atlas (266d), Meusel & Jäger (1992).

P. J. WILSON

Valerianella rimosa Broad-fruited Cornsalad

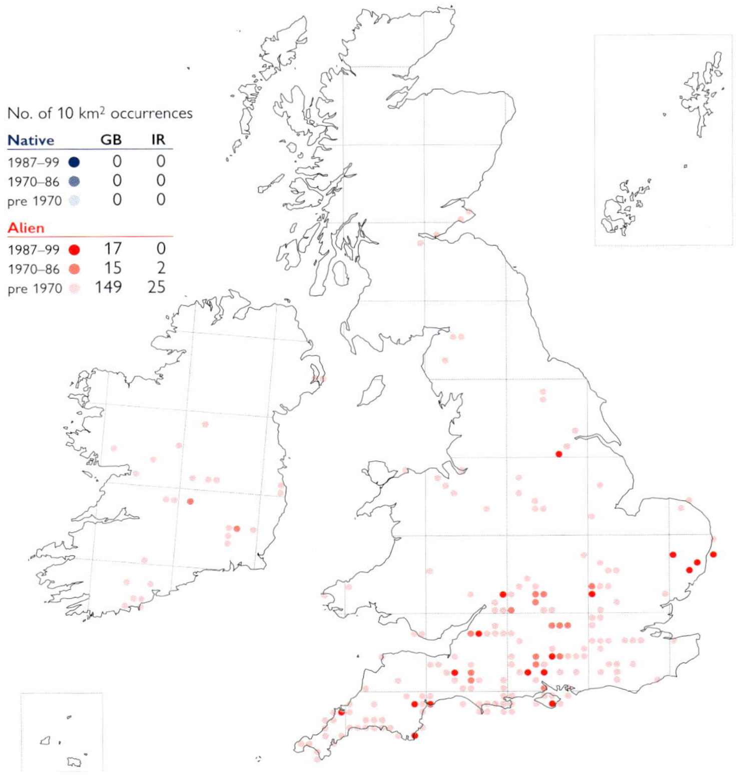

An annual of arable land, generally found on fields margins which have escaped intensive management. It occurs on sand, calcareous clay and chalk soils, often as part of a species-rich annual community. Other habitats include quarry edges and spoil-tips. Seed may be relatively short-lived, and germination occurs in spring and autumn. Lowland.

Archaeophyte (change –2.55). *V. rimosa* is known from archaeological evidence to have been present in Britain from the Iron Age. A decline was already apparent by the 1962 *Atlas*, and has continued since then, largely due to agricultural intensification. Recent records in Suffolk result from deliberate introductions.

As an archaeophyte *V. rimosa* has a European Temperate distribution.

References: Atlas (267a), Hultén & Fries (1986), Wilson (1990), Wigginton (1999).

P. J. WILSON

Valerianella dentata Narrow-fruited Cornsalad

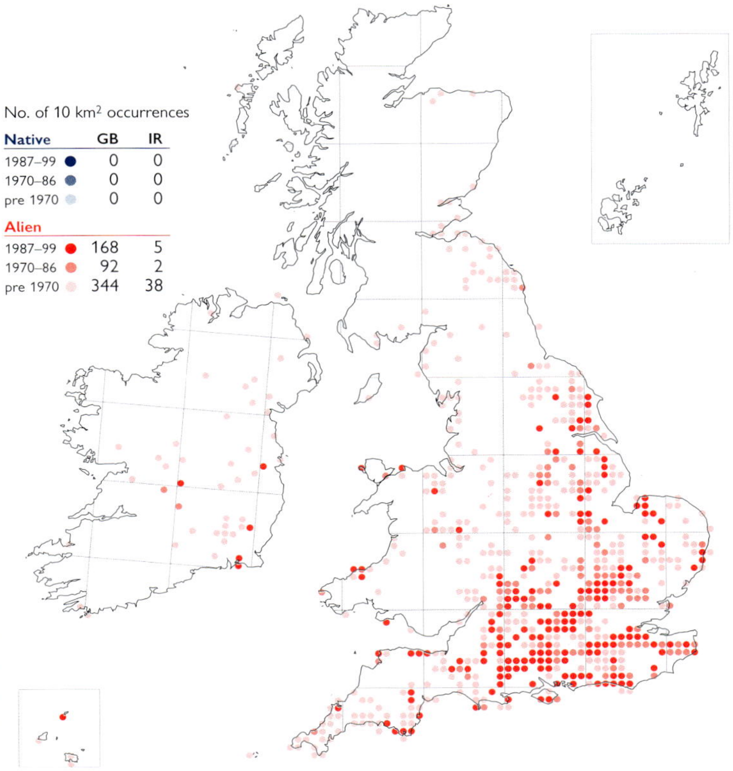

An annual of arable land, especially on chalky soils, but locally also on sand and calcareous clay. It can be found most frequently in the corners of fields and along field edges which have escaped intensive management. It can germinate both in spring and autumn, but is most frequently found in spring-sown crops. It has moderately long-lived seeds. Lowland.

Archaeophyte (change –1.86). There is a continuous archaeological record of this species from the Bronze Age onwards. It has decreased substantially since the 1962 *Atlas* with the intensification of arable farming.

As an archaeophyte *V. dentata* has a European Temperate distribution.

References: Atlas (267c), Hultén & Fries (1986), Meusel & Jäger (1992), Stewart et al. (1994), Wilson & Aebischer (1995).

P. J. WILSON

Valerianella eriocarpa Hairy-fruited Cornsalad

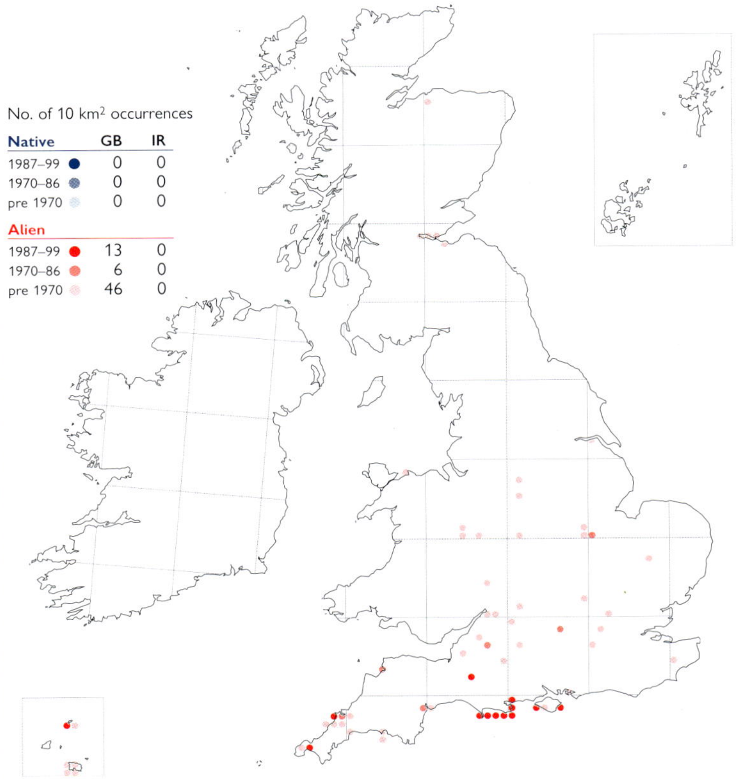

A winter-annual of drought-prone, stony substrates, found on cliff edges, calcareous banks, walls, quarries and other dry open habitats. It has been considered as native on bare and disturbed limestone and hard chalk in Dorset and the Isle of Wight (Wigginton, 1999). Some older records are from arable land, where it was probably casual. Lowland.

Neophyte (change –0.69). *V. eriocarpa* was recorded as cultivated in Britain in 1821. Although it has been known in the wild as a casual since 1845, the extent of the established populations in Dorset and the Isle of Wight was only discovered in the 1990s.

V. eriocarpa has a Submediterranean-Subatlantic distribution; it has perhaps spread northwards from the Mediterranean region as a cultivated plant or weed.

References: Atlas (267b), Bolòs & Vigo (1995).

P. J. WILSON

Valeriana officinalis Common Valerian

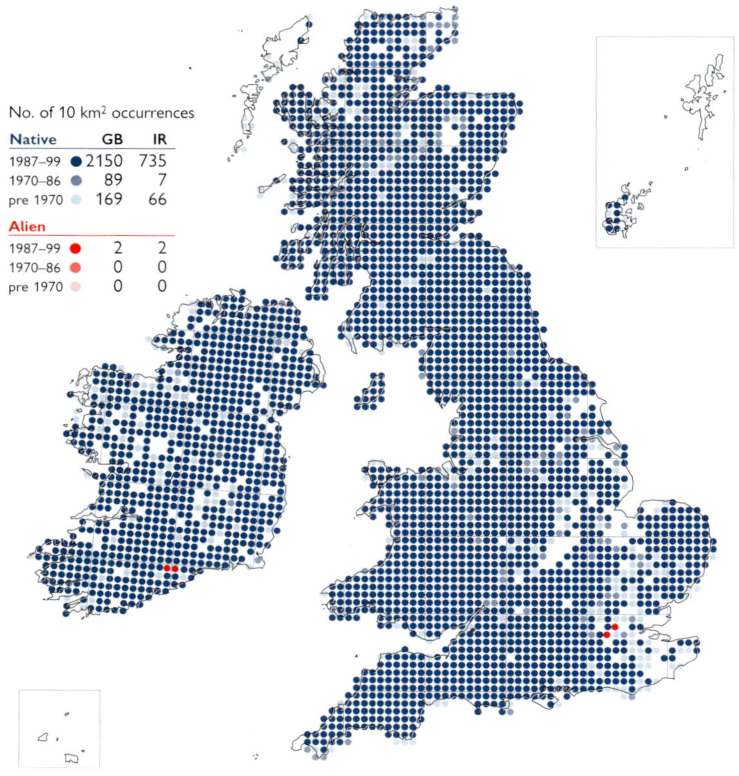

A perennial herb found in a wide range of habitats. Subsp. *sambucifolia* occurs in damp grassland, marshes, fens, water margins and ditches, and wet woods throughout the country; subsp. *collina* in dry calcareous grasslands, hedge banks and woodland rides in S. and C. Britain. Lowland to 805 m on Mt Brandon (S. Kerry).

Native (change –0.64). *V. officinalis* was mapped as 'all records' in the 1962 *Atlas*. A widespread decline has taken place in S.E. England, and analysis of the database reveals that these losses have occurred since 1950.

Eurasian Boreo-temperate element; widely naturalised outside its native range.

References: Atlas (267d), Grime *et al.* (1988), Hultén & Fries (1986), Meusel & Jäger (1992).

P. J. WILSON

Valeriana pyrenaica Pyrenean Valerian

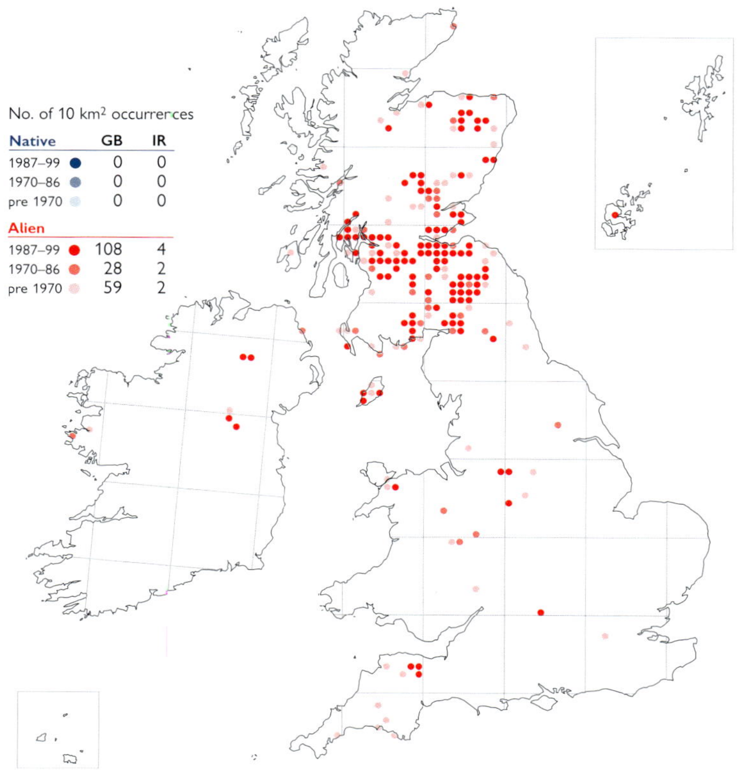

A perennial herb, naturalised in damp woods and shady hedge banks. Generally lowland, but reaching 305 m near Bridgend, Dufftown (Banffs.).

Neophyte (change –0.35). This species has been grown in Britain since at least 1692, and was first recorded in the wild in 1782 as a supposed native. Populations are often long-lived, and its distribution has changed little since the 1962 *Atlas*.

Native of the Pyrenees and the Cordillera Cantabrica.

References: Atlas (268a), Bolòs & Vigo (1995).

P. J. WILSON

Valeriana dioica Marsh Valerian

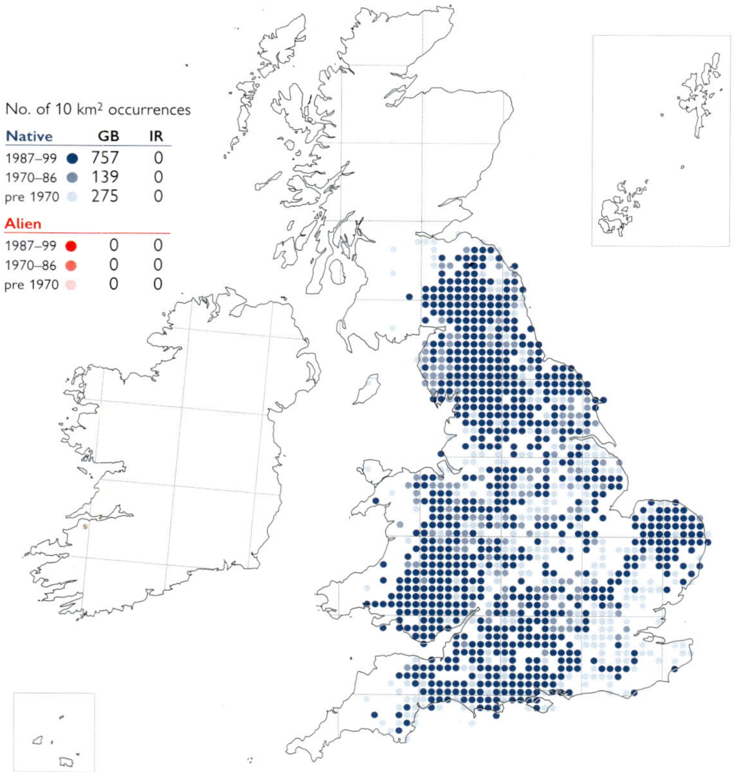

A perennial herb of calcareous mires, marshy grassland, water-meadows, flushes, *Salix* fen-carr and *Alnus* woods. 0–780 m (Cross Fell, Cumberland).

Native (change –0.67). This species showed some losses before 1930 in the 1962 *Atlas*, and this decline has accelerated since then. This has followed the degradation or loss of its wetland habitats.

European Temperate element; also in C. Asia and N. America.

References: Atlas (268b), Hultén & Fries (1986), Meusel & Jäger (1992).

P. J. WILSON

Centranthus ruber Red Valerian

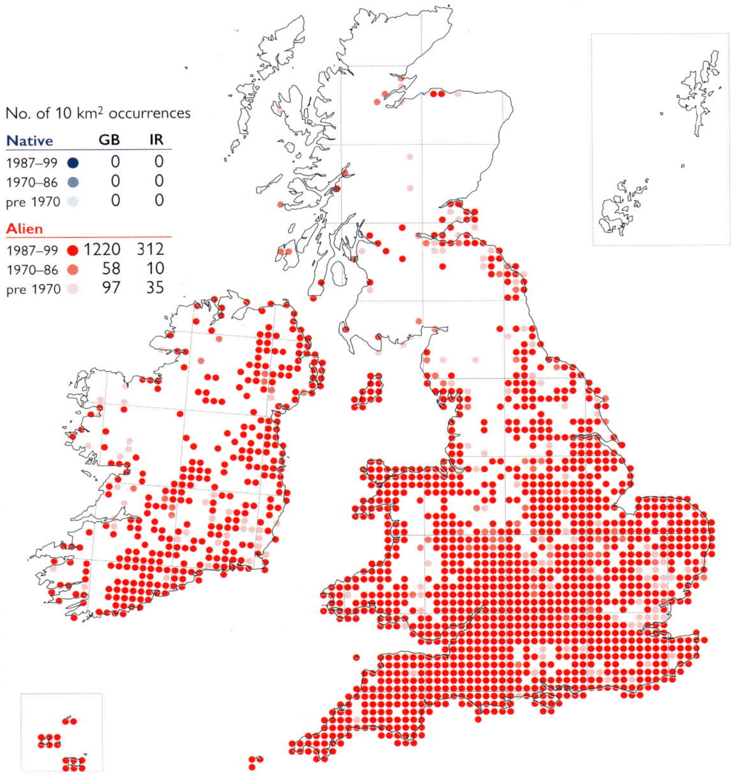

No. of 10 km² occurrences

Native		GB	IR
1987–99	●	0	0
1970–86	●	0	0
pre 1970	●	0	0
Alien			
1987–99	●	1220	312
1970–86	●	58	10
pre 1970	●	97	35

A perennial herb thoroughly naturalised on sea-cliffs, limestone rock outcrops and pavements, rocky waste ground, in quarries, on railway banks, on old walls and buildings, and in other well-drained, disturbed and open habitats. Lowland.

Neophyte (change +1.15). *C. ruber* was grown in Britain by 1597 and is a popular garden plant. It was first recorded in the wild in 1763 (Cambs.). There are many more records for this species than when it was mapped in the 1962 *Atlas*, although its range has not changed significantly.

Native of S.W. Europe and the Mediterranean region; widely naturalised further north in W. & C. Europe.

References: Atlas (268c), Bolòs & Vigo (1995).

P. J. WILSON

Dipsacus fullonum Wild Teasel

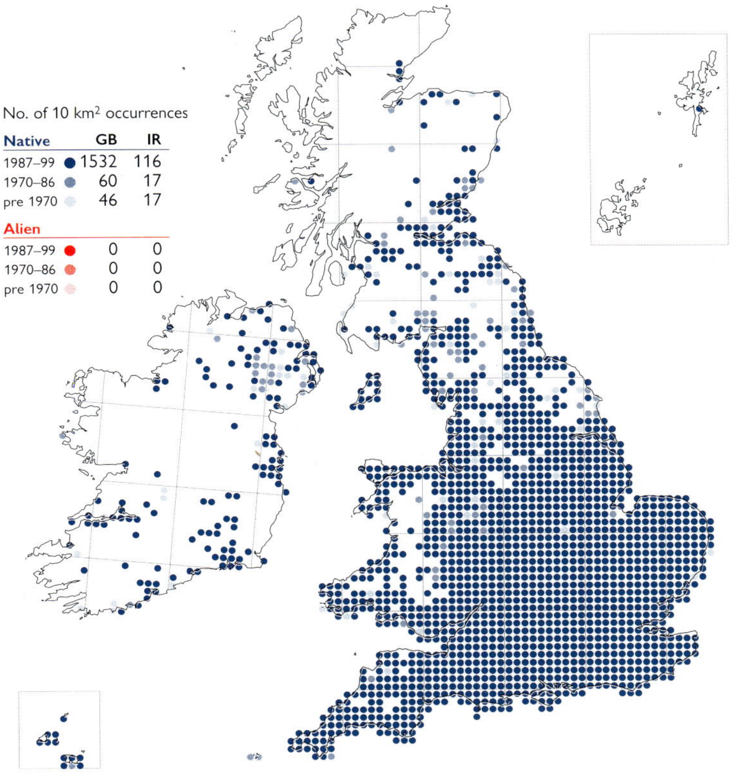

No. of 10 km² occurrences

Native		GB	IR
1987–99	●	1532	116
1970–86	●	60	17
pre 1970	●	46	17
Alien			
1987–99	●	0	0
1970–86	●	0	0
pre 1970	●	0	0

A robust biennial herb, frequent in rough grassland, wood margins, thickets and hedgerows, and on roadsides and waste ground on a very wide range of soil types. It fruits prolifically, and often colonises bare ground after disturbance. 0–365 m (Garrigill, Cumberland).

Native or alien (change +0.82). It is impossible to distinguish native and introduced populations of this ruderal species and all records are mapped as if they are native. The distribution of *D. fullonum* is broadly similar to that shown in the 1962 *Atlas*, but there are now many more records, perhaps often of casuals, outside its former core area.

European Temperate element.

References: Atlas (268d), Bolòs & Vigo (1995).

G. T. D. WILMORE

Dipsacus sativus Fuller's Teasel

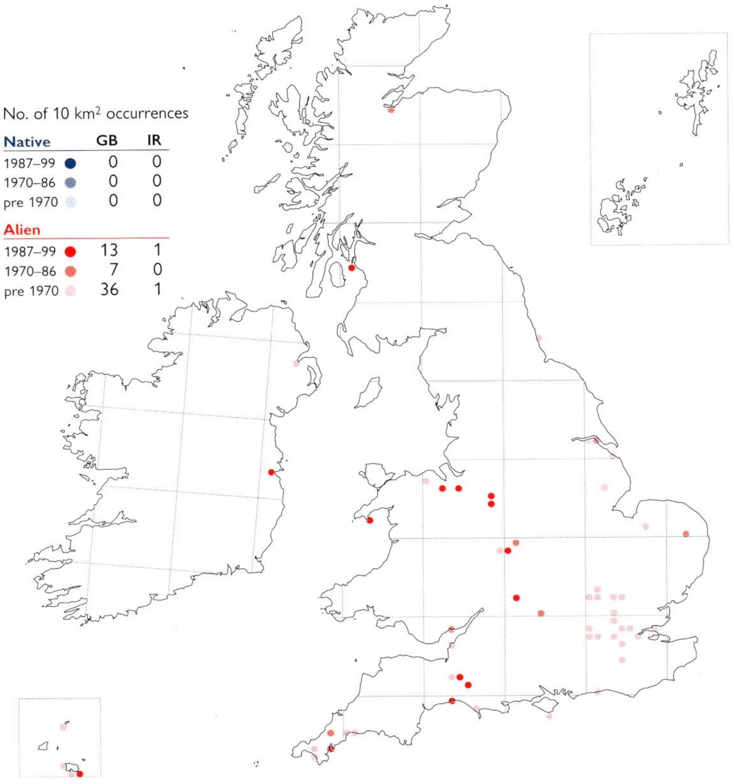

No. of 10 km² occurrences

Native		GB	IR
1987–99	●	0	0
1970–86	●	0	0
pre 1970	●	0	0
Alien			
1987–99	●	13	1
1970–86	●	7	0
pre 1970	●	36	1

A robust biennial herb, occasionally found as an escape from cultivation or as a bird-seed alien on waste ground, railway sidings and tips. Lowland.

Neophyte. *D. sativus* was formerly cultivated for use in raising the nap of woollen cloth after fulling, and is still grown for that purpose in Somerset (Green *et al.*, 1997). It was first recorded from the wild in 1762.

Origin uncertain; presumably derived by selection from a wild *Dipsacus* species.

G. T. D. WILMORE

Dipsacus pilosus Small Teasel

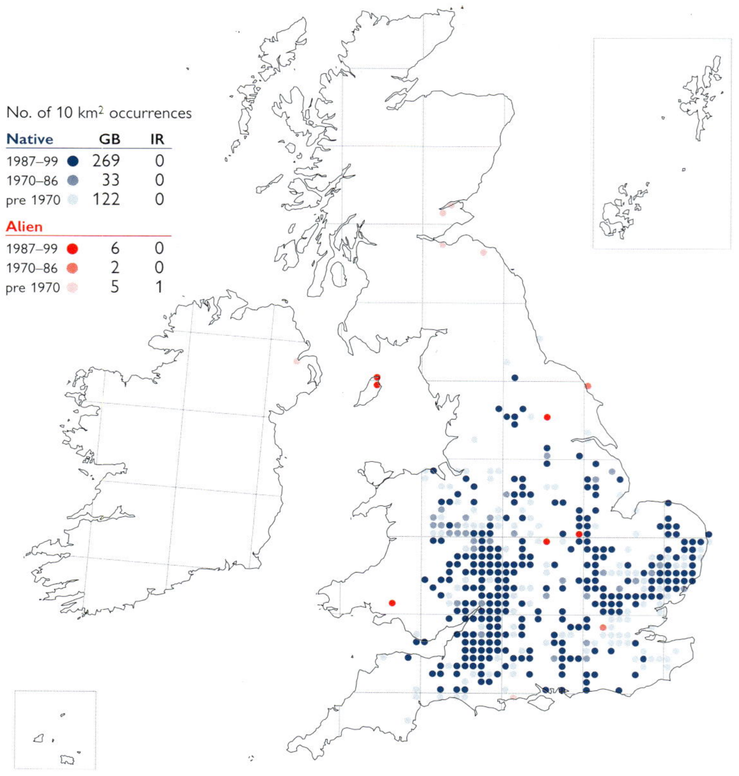

A biennial herb occurring locally on woodland edges, rides and in clearings, in scrub and hedgerows, on ditch-sides and stream and river banks. It also grows in quarries and on waste ground. It prefers damp, calcareous soils. Lowland.

Native (change +0.06). *D. pilosus* has been lost from some squares since the 1962 *Atlas*. It needs disturbance for germination, and agricultural improvement or a lack of woodland management may account for the losses.

European Temperate element.

References: Atlas (269a), Meusel & Jäger (1992).

G. T. D. WILMORE

Cephalaria gigantea Giant Scabious

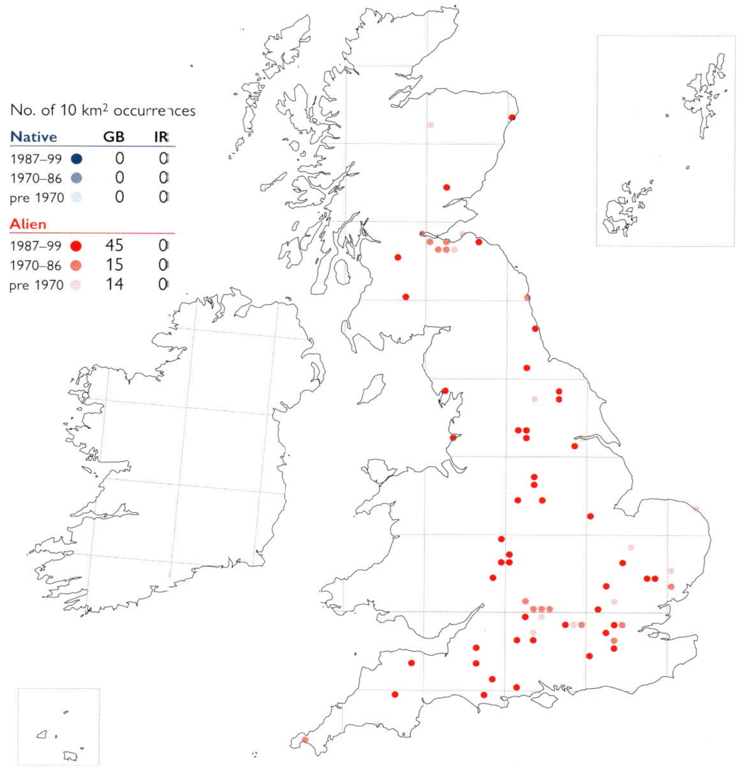

A large perennial herb, found as a garden escape or throw-out in rough grassland and on roadsides, river banks, railway banks, and waste ground. It can become well-established through rhizomatous growth. Lowland.

Neophyte. *C. gigantea* was being cultivated in Britain by 1759 and is now popular in gardens. It can easily outgrow its space, however, and is frequently discarded, being recorded in the wild by 1920. It appears to be increasing.

Native of the Caucasus.

T. D. DINES

Knautia arvensis Field Scabious

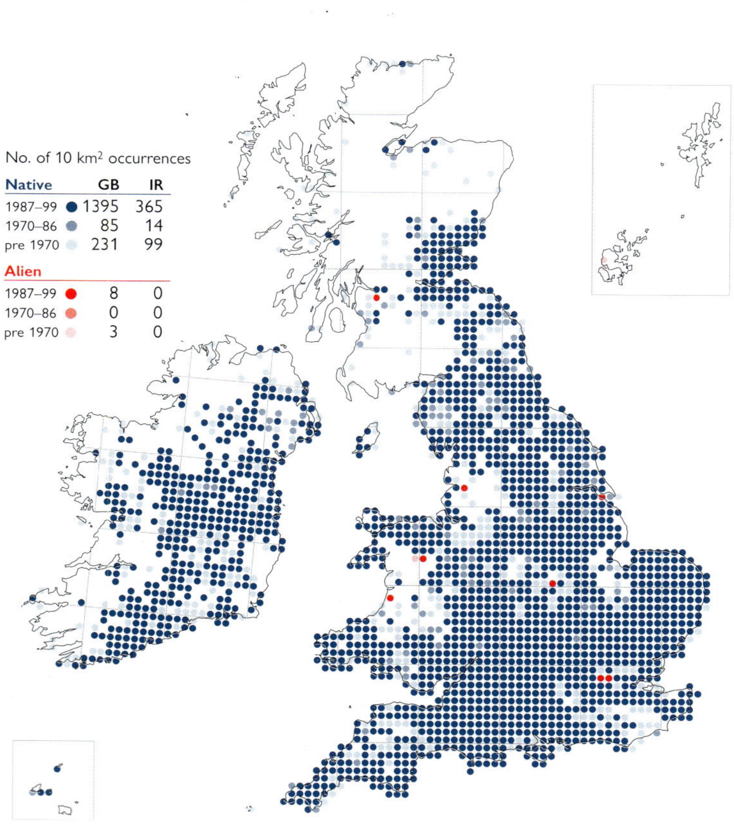

A perennial herb of calcareous and neutral grassland on well-drained, especially basic soils. It is found in chalk and limestone grassland, in rough pasture, open hedgerows and wood borders, and as a colonist on roadside verges, railway banks and grassy waste ground. It is also a locally common weed of cultivation, especially in field-borders on the chalk. 0–365 m (Derbys.).

Native (change −0.88). *K. arvensis* was mapped as 'all records' in the 1962 *Atlas*. Analysis of the database reveals that most of the losses have occurred since 1950. It is included in some wild-flower seed mixtures.

Eurosiberian Temperate element; widely naturalised outside its native range.

References: Atlas (269b), Hultén & Fries (1986), Meusel & Jäger (1992).

G. T. D. WILMORE

Succisa pratensis Devil's-bit Scabious

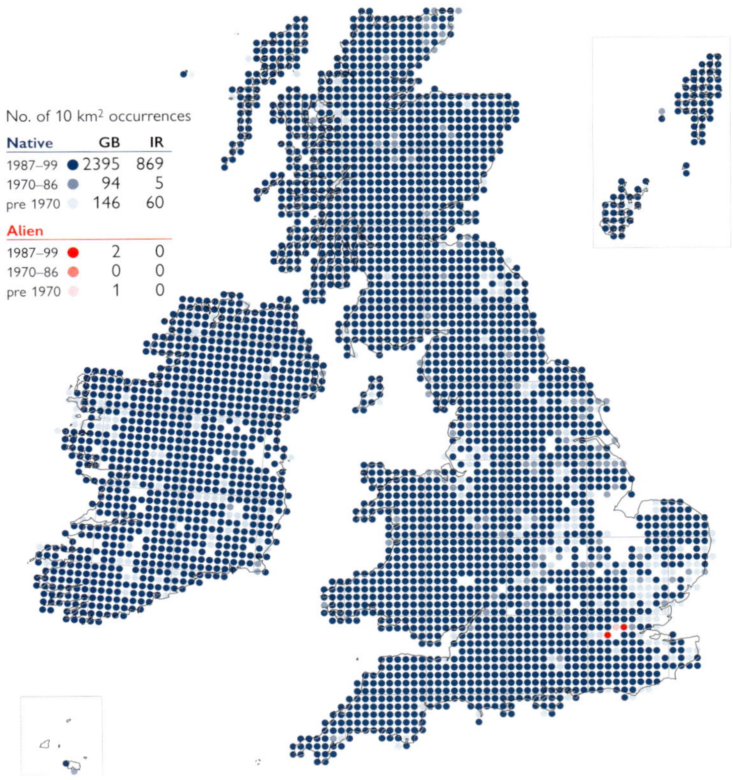

A perennial herb, growing in a wide range of moist to moderately free-draining habitats, and favouring mildly acidic soils. It occurs in woodland rides, in heathland and grassland and in mires, and in the uplands on cliff ledges and in ravines. 0–970 m (Carnedd Dafydd, Caerns.).

Native (change –0.57). *S. pratensis* was mapped as 'all records' in the 1962 *Atlas*. A widespread decline in S. and E. England has taken place, and analysis of the database reveals that most losses have occurred since 1950. These losses are largely attributable to the improvement of grassland and heaths.

Eurosiberian Temperate element.

References: Atlas (269d), Adams (1955), Grime *et al.*, (1988), Hultén & Fries (1986), Meusel & Jäger (1992).

G. T. D. WILMORE

Scabiosa columbaria Small Scabious

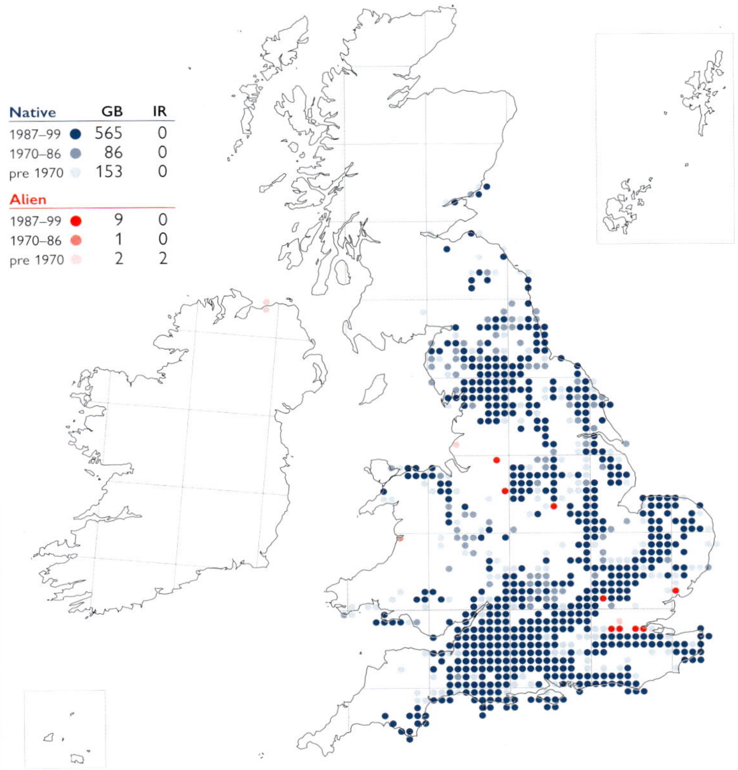

A perennial herb, usually found on dry, relatively infertile soils. Habitats include calcareous pastures on downs, hill-slopes and banks, and it is occasionally found on cliffs and rock outcrops and in disused chalk and limestone quarries. Generally lowland, reaching 640 m on Cross Fell (Cumberland).

Native (change –0.71). Very few losses of *S. columbaria* were evident in the 1962 *Atlas*. Since then, however, it has declined throughout its range, but particularly outside its core areas, due to the loss of permanent grassland and a lack of grazing leading to coarser swards.

European Temperate element.

References: Atlas (269c), Grime *et al.* (1988), Hultén & Fries (1986), Meusel & Jäger (1992).

G. T. D. WILMORE

Echinops sphaerocephalus Glandular Globe-thistle

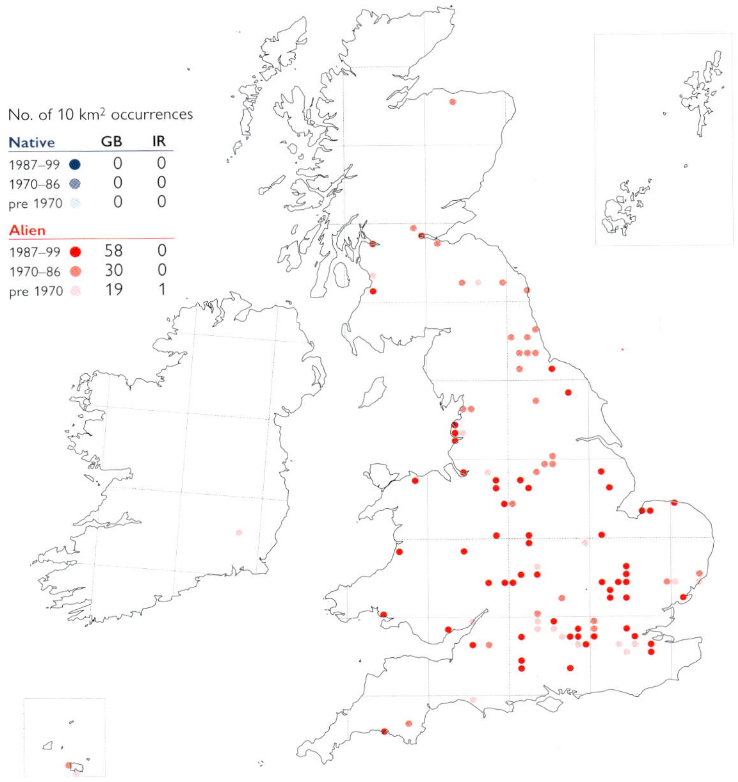

A biennial to perennial herb grown in gardens and found as an escape on roadsides, railway banks, waste ground and in disused quarries. It is usually casual but sometimes persists. Lowland.

Neophyte. *E. sphaerocephalus* was cultivated in Britain by 1596, but was not recorded from the wild until 1908. It is less common in gardens than *E. bannaticus* or *E. exaltatus* and possibly over-recorded because of confusion with these more widely grown species.

E. sphaerocephalus has a Eurosiberian Southern-temperate distribution, extending north into W. Europe to C. France.

References: Kent (1968), Meusel & Jäger (1992).

F. H. PERRING

Echinops exaltatus Globe-thistle

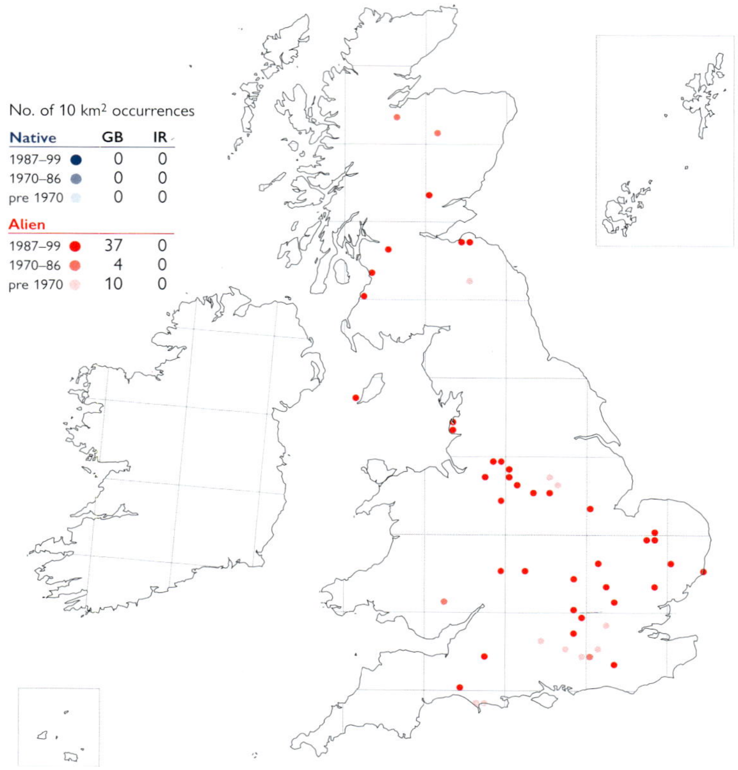

A biennial to perennial herb widely grown in gardens and found on roadsides and waste ground. It is occasionally naturalised, but is more often only casual. Lowland.

Neophyte. This species, first cultivated in Britain in 1822, is less frequent in gardens than *E. bannaticus* and is consequently much rarer in the wild. It was first recorded in 1931 (Middlesex).

Native of E.C. and S. Europe.

References: Kent (1968), Meusel & Jäger (1992).

F. H. PERRING

Echinops bannaticus Blue Globe-thistle

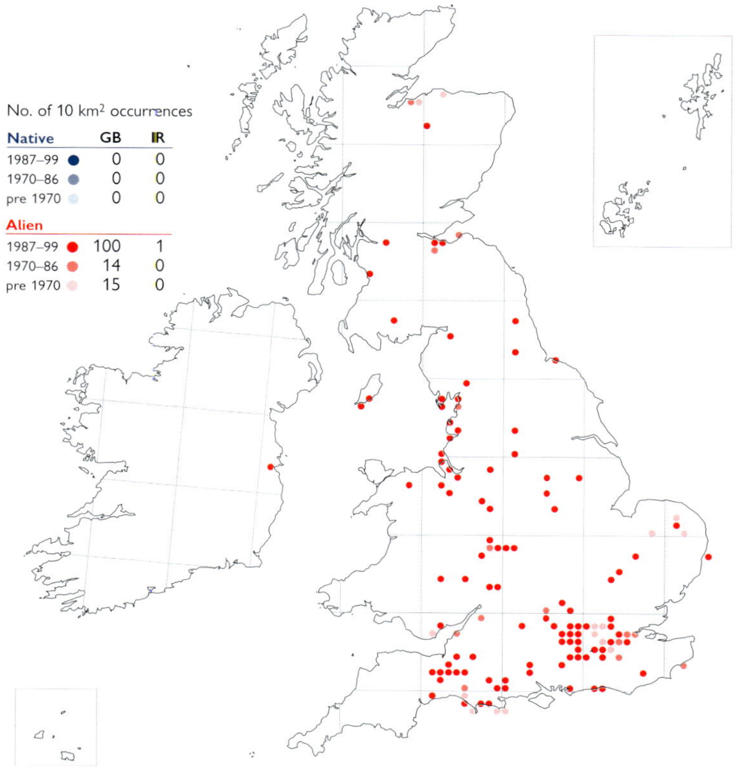

A biennial to perennial herb, very commonly grown in gardens and found naturalised on roadsides, railway banks and waste ground. Lowland.

Neophyte. First brought into cultivation in 1832, *E. bannaticus* is now widely available to gardeners, especially the variety 'Taplow Blue'. It is now the most frequent *Echinops* in the wild, where it was first recorded in 1938. Most records of *E. ritro* refer to this species.

Native of S.E. Europe east to the Crimea.

Reference: Kent (1968).

F. H. PERRING

Carlina vulgaris Carline Thistle

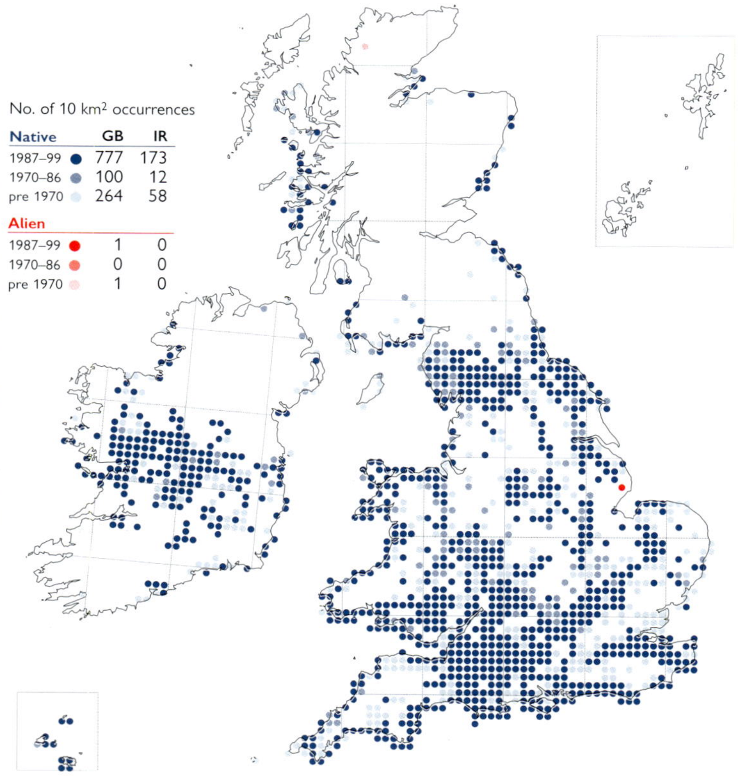

A monocarpic perennial herb, typically occurring in well-grazed grassland on dry, infertile calcareous or base-rich soils, but also in more open habitats, including dry rock ledges, screes, quarry floors, coastal cliffs and sand dunes. 0–455 m (Haweswater, Westmorland).

Native (change –0.85). *C. vulgaris* was mapped as 'all records' in the 1962 *Atlas*. A widespread decline has taken place, and analysis of the database reveals that most losses have occurred since 1950. Losses are partly due to habitat destruction and a lack of grazing. It has declined on dunes in Co. Down through their destruction and acidification.

Eurosiberian Temperate element.

References: Atlas (286d), Greig-Smith & Sagar (1981), Grime *et al.* (1988), Hackney (1992), Hultén & Fries (1986), Meusel & Jäger (1992).

F. H. PERRING

Arctium lappa Greater Burdock

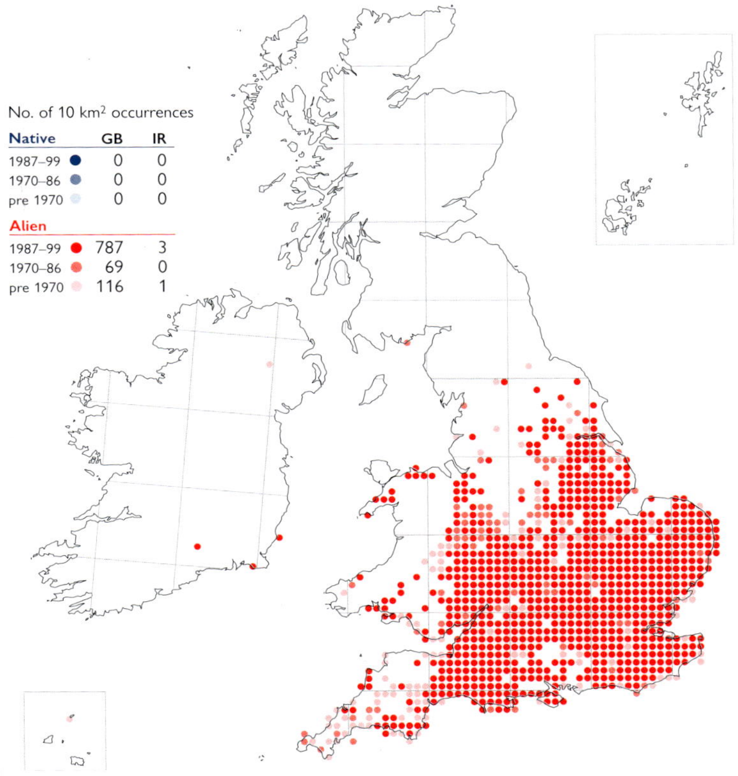

A monocarpic perennial herb of streamsides and river banks, roadside verges, tracks and waysides, field-borders, waste land and other disturbed places. Lowland.

Archaeophyte (change +0.51). *A. lappa* was probably over-recorded in the 1962 *Atlas* because '*Arctium lappa*' was used as the name of the *Arctium* aggregate at that time. The distribution seems stable. It may only be a recent colonist outside its core area in England.

As an archaeophyte *A. lappa* has a Eurosiberian Temperate distribution, but it is widely naturalised so that its distribution is now Circumpolar Temperate.

References: Atlas (287a), Gross *et al.* (1980), Hultén & Fries (1986), Meusel & Jäger (1992), Perring (1960).

F. H. PERRING

Arctium minus sens. lat. Lesser Burdock

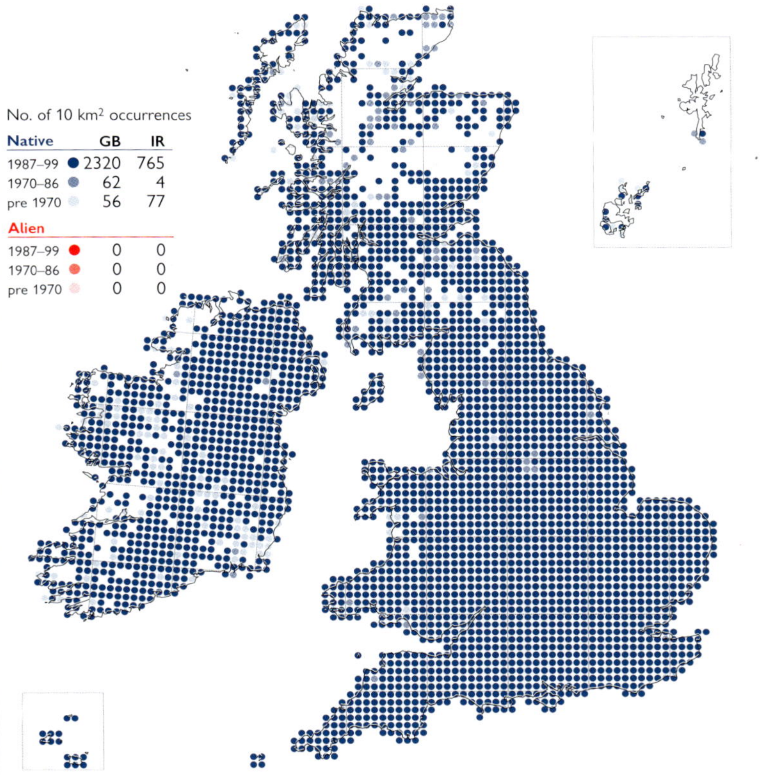

A monocarpic perennial of woodlands, scrub, hedgerows, roadsides, railway banks, rough pastures, sand dunes and waste ground. It is autogamous, but outbreeding can produce fertile variants which, by constant inbreeding, produce a great variety of almost pure lines. 0–390 m (near Nenthead, Cumberland).

Native (change –0.41). A variable taxon which has received differing taxonomic treatments in recent years. The species mapped here is the *A. minus* of Stace (1991), which was split into *A. minus* and *A. nemorosum* in Stace (1997). Its distribution is stable.

Eurasian Temperate element, but naturalised in N. America so distribution is now Circumpolar Temperate.

References: Atlas (287b), Atlas Supp. (73a, b), Grime *et al.* (1988), Gross *et al.* (1980), Hultén & Fries (1986), Meusel & Jäger (1992), Perring (1960).

F. H. PERRING

Saussurea alpina Alpine Saw-wort

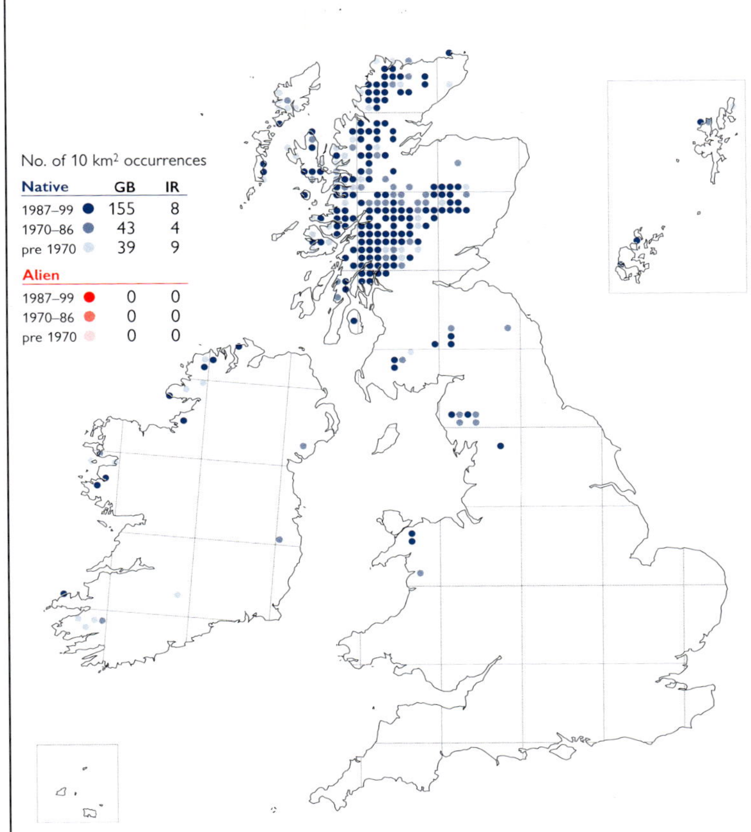

A perennial herb of damp, base-rich cliffs, screes and other open ground, occasionally found in flushed areas and sometimes washed down mountain streams to become established on rocky stream banks and riverside shingle. It reproduces by seed, and spreads vegetatively by new rosettes arising from stolons. From near sea level in Caithness to 1170 m on Ben Lawers (Mid Perth).

Native (change –0.51). Though the map of *S. alpina* shows no discernible change since the 1962 *Atlas*, there is evidence of local declines since the 1950s in the Lake District (Halliday, 1997) and on Irish mountains (Curtis & McGough, 1988).

Eurasian Arctic-montane element.

References: Atlas (290d), Godwin (1975), Hultén & Fries (1986), Meusel & Jäger (1992).

F. H. PERRING

Carduus tenuiflorus Slender Thistle

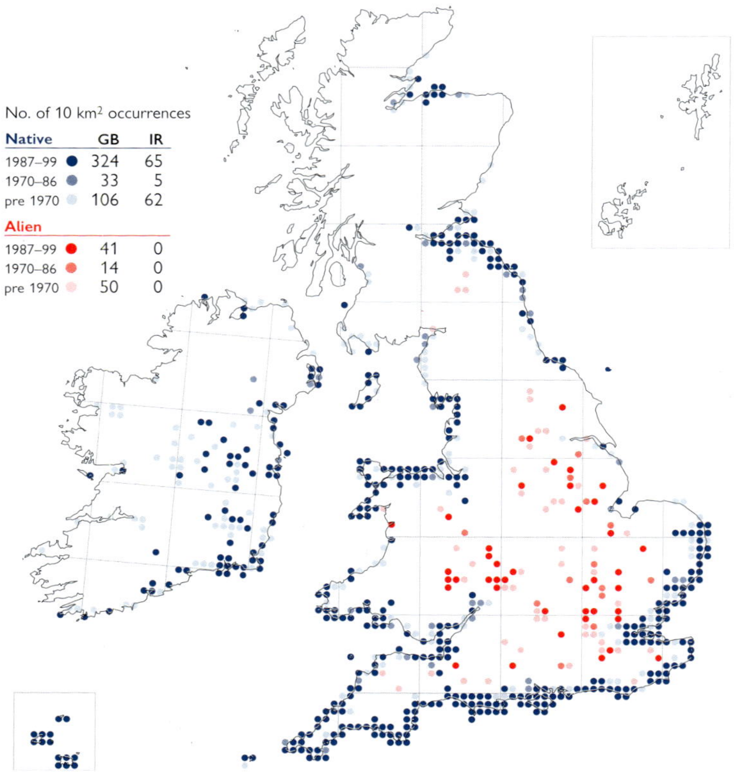

An annual or biennial herb of dry, coastal grasslands, sea-bird colonies, sea-walls, upper edges of beaches, sandy waste ground and roadsides. Inland, it occurs on well-drained soils, often, but not always, as an alien. Lowland.

Native (change –0.14). The current distribution of *C. tenuiflorus* is similar to that mapped in the 1962 *Atlas*. It is at its northern limit in E. Sutherland and Caithness, but in continental Europe it reaches only the southern tip of Holland. The inland records in England frequently represent wool aliens, but these can be difficult to distinguish from native occurrences.

Suboceanic Southern-temperate element.

References: Atlas (287c), Meusel & Jäger (1992).

F. H. PERRING

Carduus crispus Welted Thistle

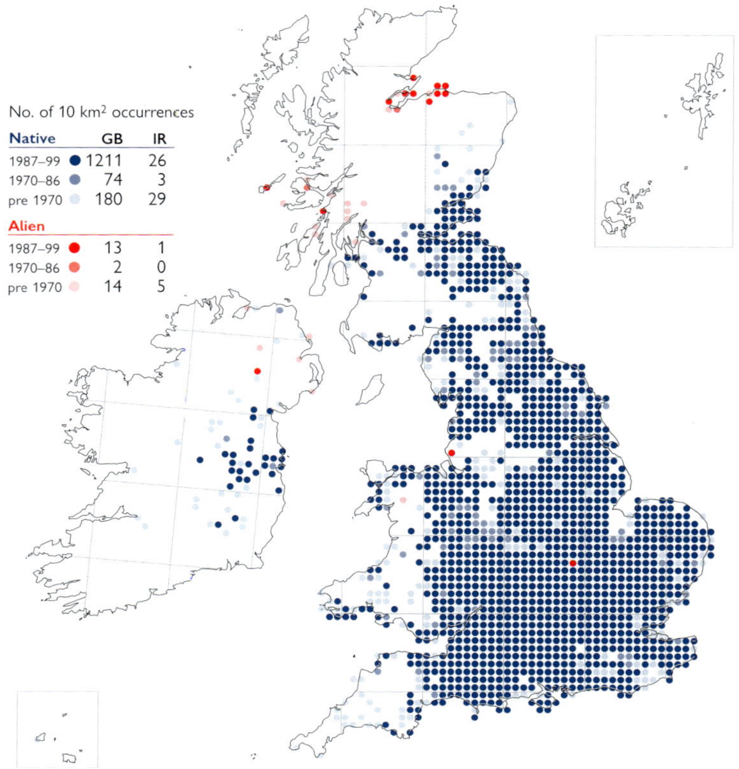

A biennial herb of woodland margins, ditch-banks, damp hedge bottoms, streamsides, tall grassland, roadsides, railway banks and waste places, especially on clay soils with a high nutrient status. In Ireland, it is confined as a native to dry banks and waste places. Generally lowland, but reaching 365 m above Castleton (Derbys.).

Native (change –0.18). The core distribution of *C. crispus* is stable. It may only have been present as a casual in squares at the periphery of its range from which it has not been recorded recently. In N. Scotland it is only present as an introduction in waste places.

Eurosiberian Temperate element, but widely naturalised so that distribution is now Circumpolar Temperate.

References: Atlas (288a), Desrochers *et al.* (1988), Hultén & Fries (1986), Meusel & Jäger (1992).

F. H. PERRING

Carduus crispus × *C. nutans* (*C.* × *stangii*)

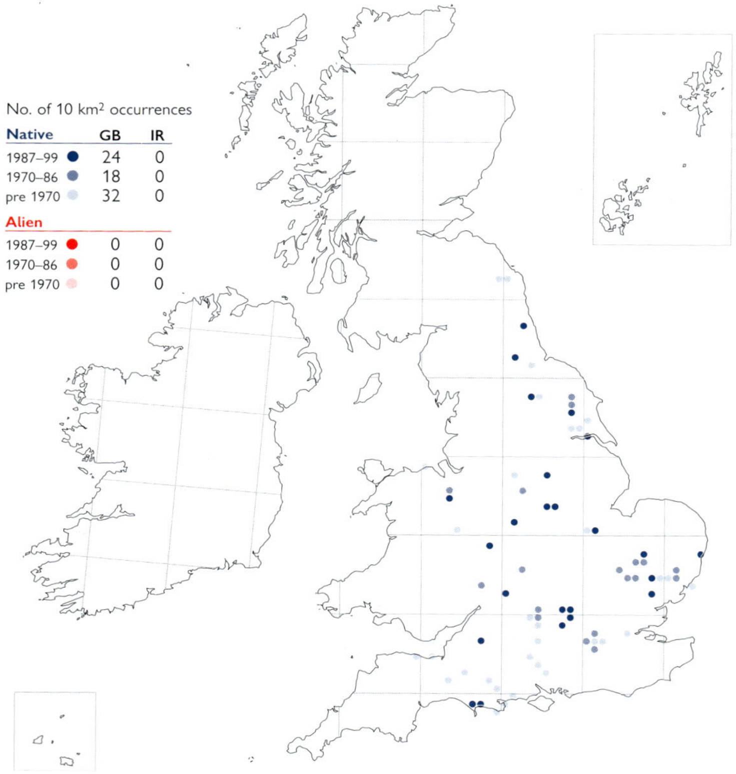

This hybrid is often found where the parents grow close to one another, in habitats such as grassland, roadside verges and waste ground. Lowland.

Native. Trends in the distribution of this hybrid are difficult to discern. It is probably under-recorded.

Widespread in Europe.

Reference: Stace (1975).

F. H. PERRING

Carduus nutans Musk Thistle

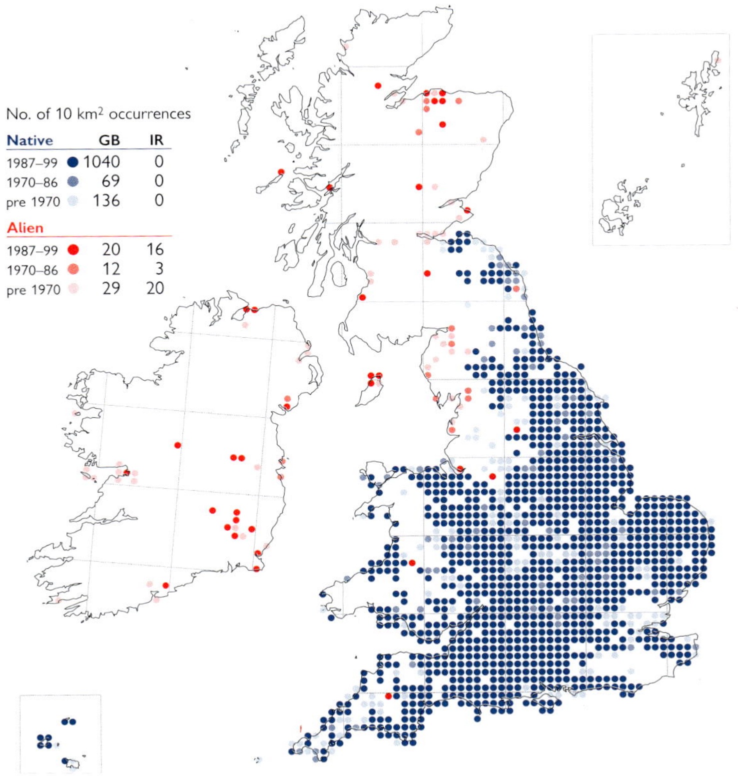

This biennial, or sometimes perennial, herb is mainly found on chalk, limestone or lime-enriched soils, but also occurs on sandy or shingly ground. It is found in rough, often overgrazed or recently established pastures, on roadsides and in disturbed places. Generally lowland, but reaching *c.* 530 m at High Cup Nick (Westmorland).

Native (change –0.15). The overall distribution of *C. nutans* is similar to that in the 1962 *Atlas*, but it is now better recorded in many areas. It has always been rare in Ireland, where it is generally considered to have been introduced as a seed contaminant.

Eurosiberian Temperate element; widely naturalised outside its native range.

References: Atlas (287d), Curtis & McGough (1988), Desrochers *et al.* (1988), Hultén & Fries (1986), Meusel & Jäger (1992).

F. H. PERRING

Cirsium eriophorum Woolly Thistle

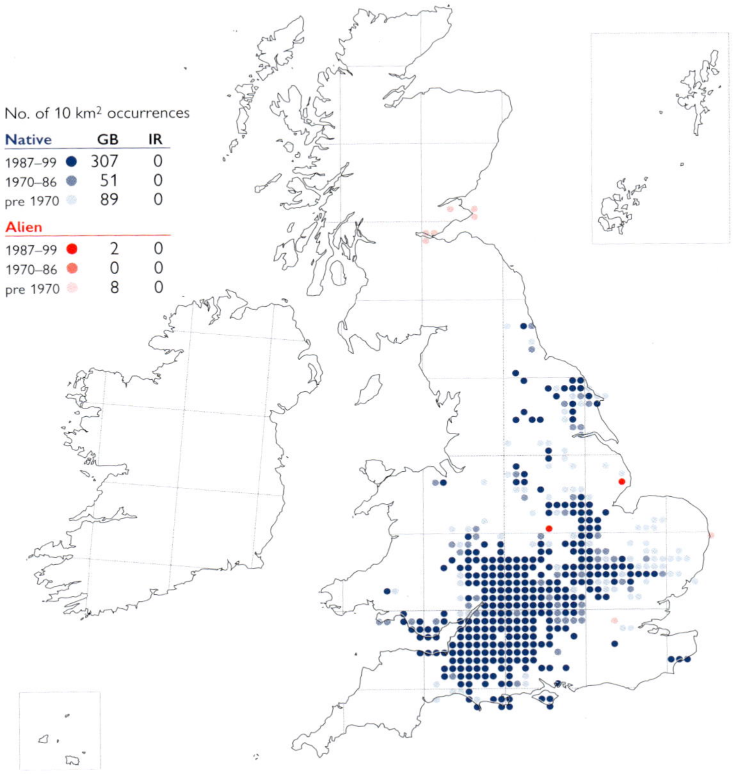

A robust monocarpic perennial herb occurring in dry, often ungrazed, grasslands, open scrub and woods on limestone, chalk and lime-rich clay. It also grows in disturbed habitats created by quarrying. 0–310 m (Hassop Mines, Derbys.).

Native (change –0.08). There has been little overall change in the distribution of *C. eriophorum* since the 1962 *Atlas*. It has declined in some areas on the edges of its range, particularly in East Anglia where populations may have been only casual.

European Temperate element.

References: Atlas (288b), Bolòs & Vigo (1995), Meusel & Jäger (1992), Tofts (1999).

F. H. PERRING

Cirsium vulgare Spear Thistle

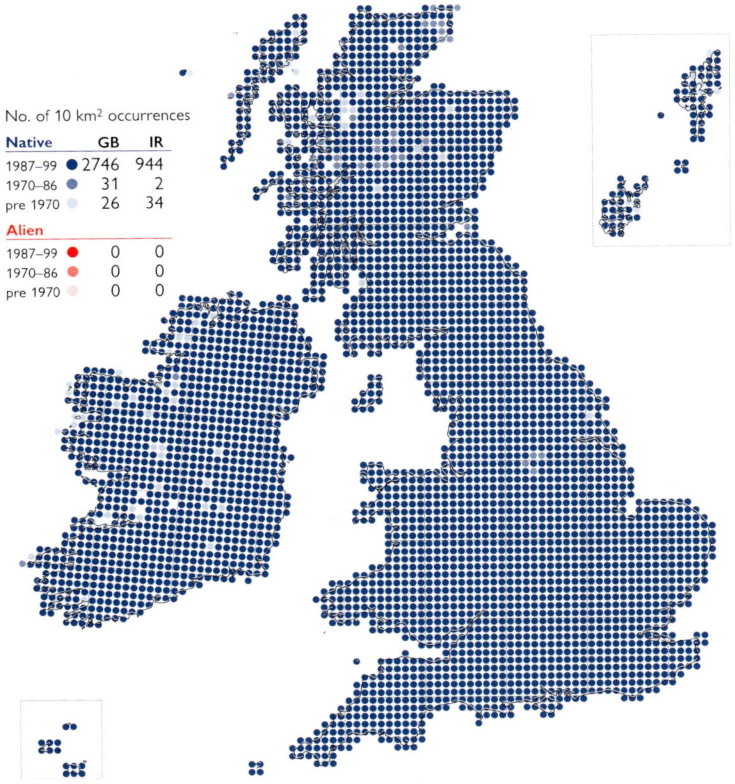

A monocarpic perennial occurring in a wide array of habitats, including overgrazed pastures and rough grassland, sea-cliffs, dunes, drift lines and well-drained, fertile, disturbed habitats including arable fields, spoil heaps, waste ground and burnt areas in woodland. 0–685 (Breadalbanes, Mid Perth), and exceptionally at 845 m on Great Dun Fell (Westmorland).

Native (change +0.80). Though statutorily listed as a noxious weed in Britain and subject to control by landowners, there has been no change in the overall range since the 1962 *Atlas*, and it is probably increasing in man-made habitats.

Eurosiberian Temperate element; widely naturalised outside its native range.

References: Atlas (288c), Grime *et al.* (1988), Hultén & Fries (1986), Klinkhamer & Jong (1993), Meusel & Jäger (1992).

F. H. PERRING

Cirsium dissectum Meadow Thistle

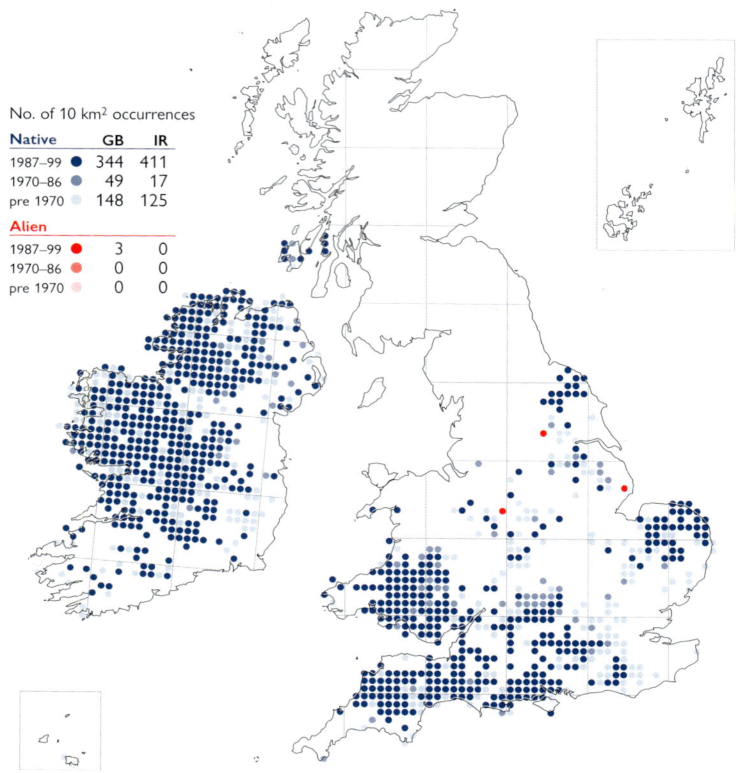

A shortly stoloniferous perennial herb of fens, fen-meadows, flood-pastures, bog margins and poorly-drained meadows on acid to neutral, usually peaty, soils. It often grows in sites subject to marked vertical or lateral movement of water. 0–500 m (Co. Sligo).

Native (change –0.14). *C. dissectum* has been lost from many localities in S. and E. England since the 1962 *Atlas*, following drainage or shading as its sites undergo succession to scrub or woodland. It is now much better recorded elsewhere. The contrast between the distribution of this species in Britain and Ireland, and between this species and *C. heterophyllum*, are unresolved biogeographical problems.

Oceanic Temperate element.

References: Atlas (289d), Kay & John (1994), Meusel & Jäger (1992).

F. H. PERRING

Cirsium dissectum × *C. palustre* (*C.* × *forsteri*)

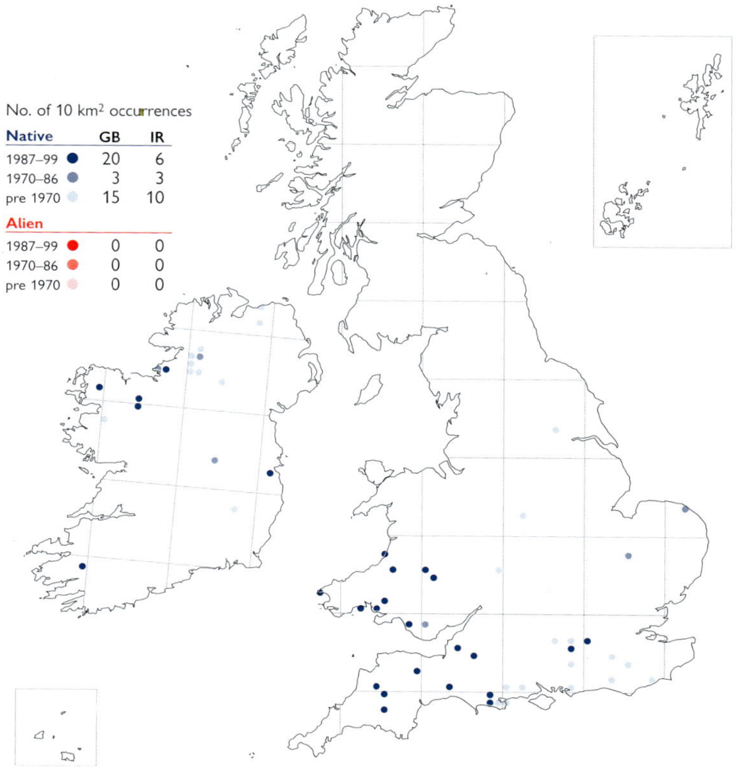

This hybrid occurs, sometimes abundantly, with the parents, in fens, bogs, marshes and wet grassland. Generally lowland, but reaching over 300 m at Kings Nympton (N. Devon).

Native. Though this hybrid has not previously been mapped on a national scale, there is unlikely to have been much change in its distribution. It is probably under-recorded.

C. × *forsteri* is recorded from the mainland of W. Europe (France and the Netherlands).

Reference: Stace (1975).

F. H. PERRING

Cirsium tuberosum Tuberous Thistle

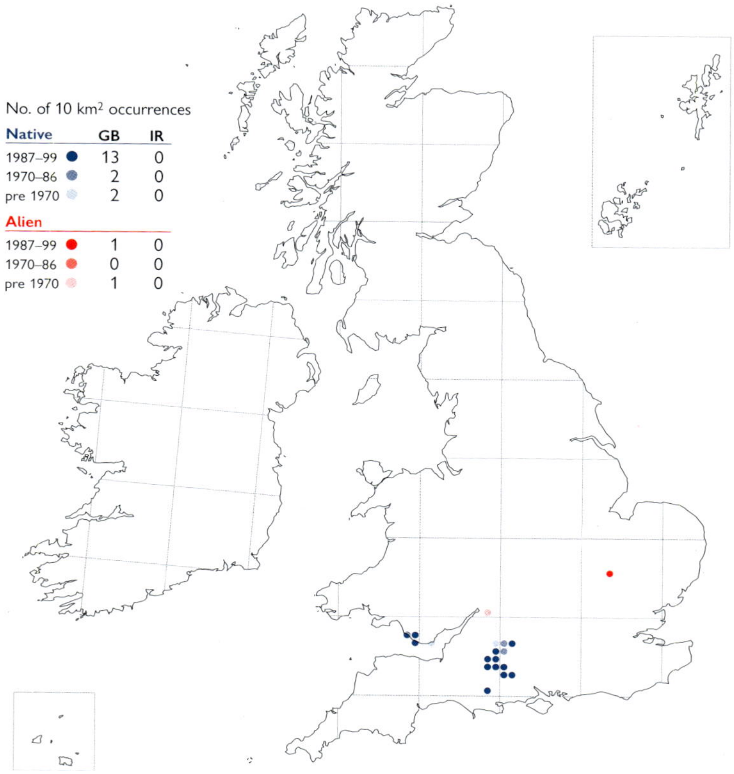

A perennial herb of old chalk and limestone grassland, often on slopes with a N. or N.W. aspect, and sometimes occurring in rank swards. It spreads by producing axillary basal rosettes to form clonal patches, and also reproduces by seed. Lowland.

Native (change +0.41). This species is declining in its two main areas in Wiltshire and Glamorgan. The reasons for the decline include habitat destruction by ploughing, coastal erosion, changes in land management and, particularly in Wiltshire, hybridisation with *Cirsium acaule*. It became extinct as a native in its sole Cambridgeshire locality in 1973, but has subsequently been re-introduced there.

Suboceanic Southern-temperate element.

References: Atlas (290a), Kay & John (1994), Meusel & Jäger (1992), Wigginton (1999).

F. H. PERRING

Cirsium heterophyllum Melancholy Thistle

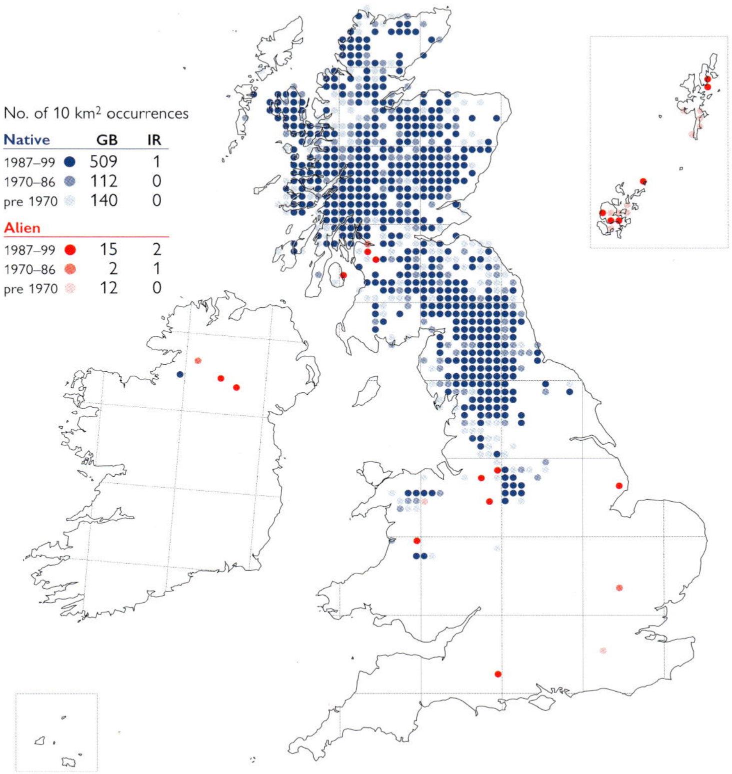

No. of 10 km² occurrences

Native		GB	IR
1987–99	●	509	1
1970–86	●	112	0
pre 1970	●	140	0
Alien			
1987–99	●	15	2
1970–86	●	2	1
pre 1970	●	12	0

A perennial herb of stream banks, hay meadows, damp roadside verges and moist woodland margins. Mostly upland, reaching 760 m in the Breadalbanes (Mid Perth) and possibly to *c.* 975 m elsewhere in Scotland.

Native (change –0.44). *C. heterophyllum* is declining in some areas due to the shift from traditionally managed hay meadows to silage making, and as a result of unsympathetic management of roadside verges. Its restriction to a few localities in northern Ireland is characteristic of several other species in the Eurosiberian Boreal-montane element in our flora.

Eurosiberian Boreal-montane element.

References: Atlas (289c), Curtis & McGough (1988), Hultén & Fries (1986), Meusel & Jäger (1992).

F. H. PERRING

Cirsium acaule Dwarf Thistle

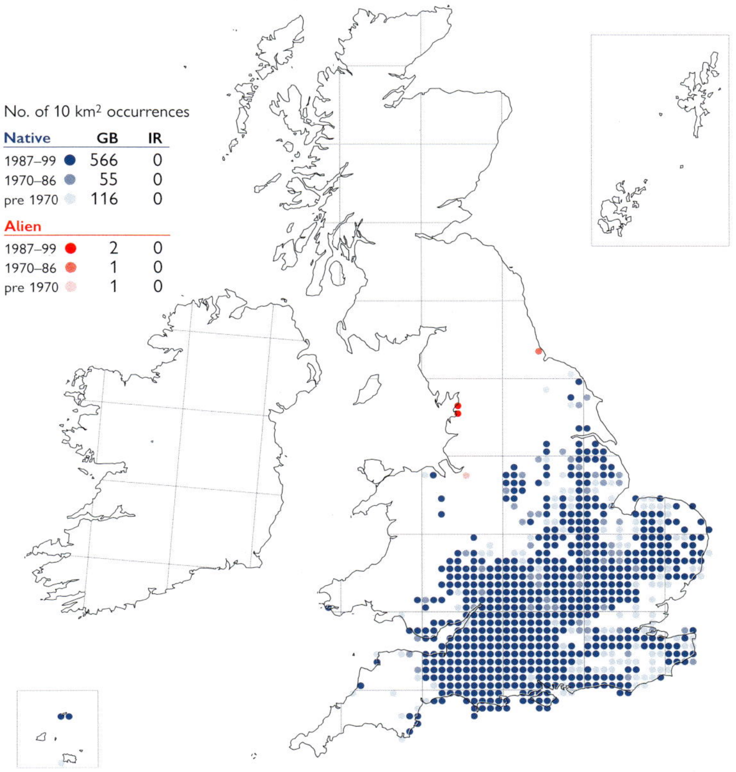

No. of 10 km² occurrences

Native		GB	IR
1987–99	●	566	0
1970–86	●	55	0
pre 1970	●	116	0
Alien			
1987–99	●	2	0
1970–86	●	1	0
pre 1970	●	1	0

A rosette-forming perennial herb of short swards on base-rich soils, particularly on chalk and limestone. The northerly and westerly limits appear to be determined by summer warmth and in areas such as the Yorkshire Wolds and Derbyshire it is almost wholly confined to S.W.-facing slopes. It benefits from the sward being grazed to less than 10–15 cm, or frequent mowing, but is destroyed by heavy trampling. Generally lowland, but reaching 425 m at Trefil (Mons.).

Native (change –0.52). *C. acaule* shows very little change since the 1962 *Atlas*, other than in S.E. England where there have been losses, probably due to habitat loss.

European Temperate element.

References: Atlas (289b), Hultén & Fries (1986), Meusel & Jäger (1992), Pigott (1968).

F. H. PERRING

Cirsium palustre Marsh Thistle

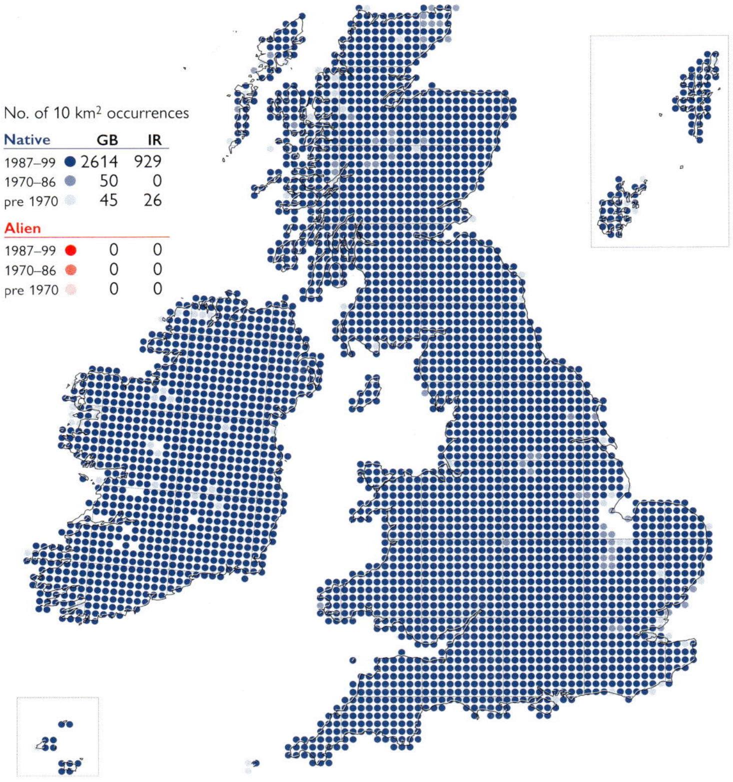

No. of 10 km² occurrences

Native		GB	IR
1987–99	●	2614	929
1970–86	●	50	0
pre 1970	●	45	26
Alien			
1987–99	●	0	0
1970–86	●	0	0
pre 1970	●	0	0

A monocarpic perennial herb of mires, fens, marshes, damp grassland, rush-pastures, wet woodland, montane springs and flushes, and tall-herb vegetation on mountain ledges. It reproduces by seed, which may persist for many years, as, for example, during the dark phase of a coppice cycle. 0–760 m (Cross Fell, Cumberland), and 845 m on Great Dun Fell (Westmorland).

Native (change +0.15). There has been no overall change in the distribution of *C. palustre* since the 1962 *Atlas*.

Eurosiberian Boreo-temperate element; widely naturalised outside its native range.

References: Atlas (288d), Grime *et al.* (1988), Hultén & Fries (1986), Meusel & Jäger (1992).

F. H. PERRING

Cirsium arvense Creeping Thistle

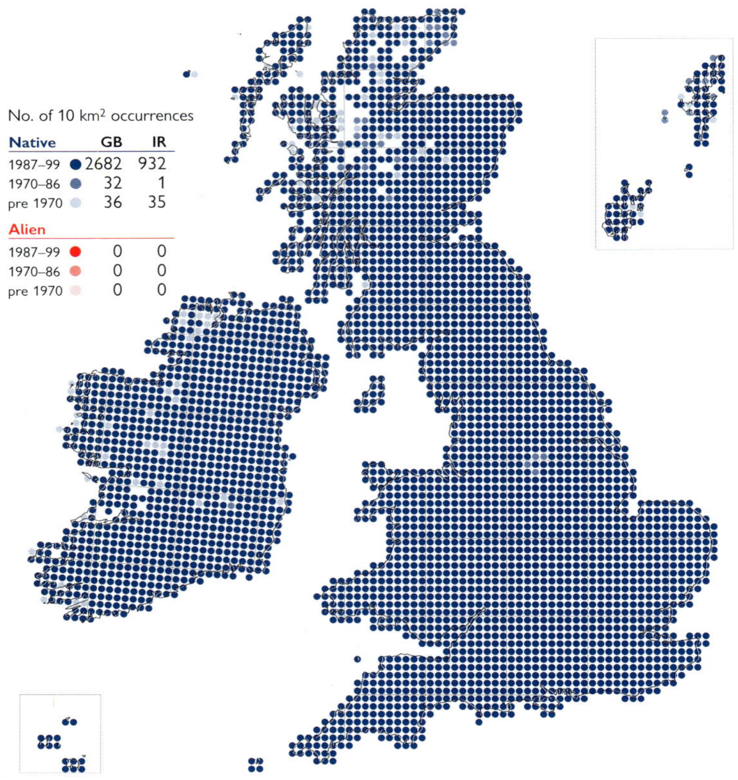

Native	**GB**	**IR**
1987–99	2682	932
1970–86	32	1
pre 1970	36	35
Alien		
1987–99	0	0
1970–86	0	0
pre 1970	0	0

A perennial herb of over-grazed pastures, hay meadows and rough grassy places, roadsides, arable fields and other cultivated land, and in urban habitats and waste ground. Plants regenerate freely from rhizome fragments which are broken up by ploughing or other disturbance. 0–700 m (Breadalbanes, Mid Perth), and at 845 m on Great Dun Fell (Westmorland).

Native (change +0.47). Though listed as a noxious weed in Britain under the 1959 Weeds Act, there has been no change in distribution since the 1962 *Atlas* and probably little over many decades before that.

Eurasian Temperate element, but naturalised in N. America so distribution is now Circumpolar Temperate.

References: Atlas (289a), Grime *et al.* (1988), Hultén & Fries (1986), Kay (1985), Meusel & Jäger (1992), Moore (1975), Werner (1975).

F. H. PERRING

Onopordum acanthium Cotton Thistle

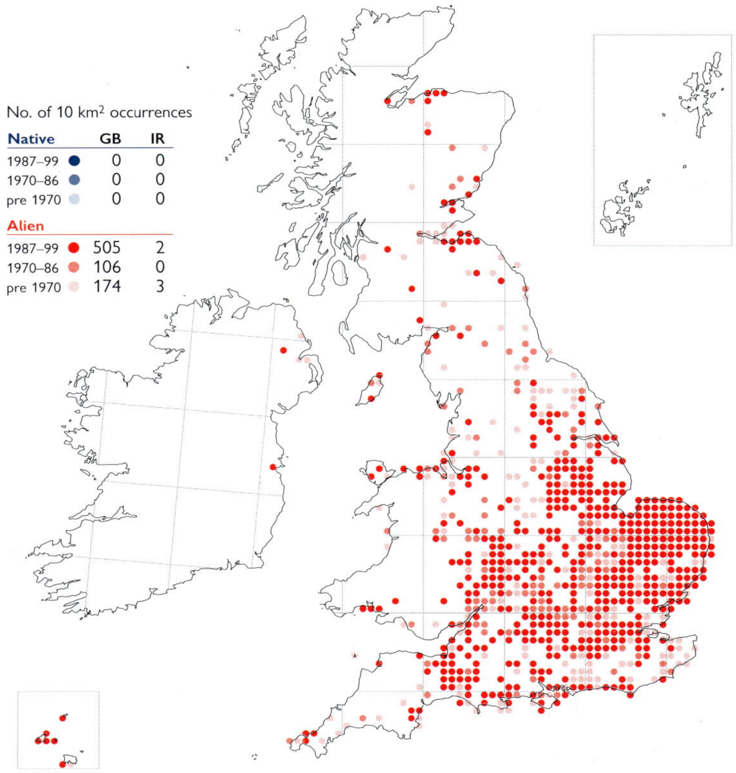

A tall biennial herb of fields, hedgerows, rubbish tips and other waste places, often near market gardens and farm buildings, and perhaps dispersed to new sites with manure or contaminated straw. 0–330 m (near Alston, Cumberland).

Archaeophyte (change +0.66). There is archaeological evidence for the presence of *O. acanthium* in Britain from the Iron Age onwards. It appears to have increased in frequency since the 1962 *Atlas*, possibly as an escape from gardens where it is frequently grown for ornament.

As an archaeophyte *O. acanthium* has a Eurosiberian Temperate distribution; it is widely naturalised outside this range.

References: Atlas (290c), Hultén & Fries (1986), Meusel & Jäger (1992).

F. H. PERRING

Silybum marianum Milk Thistle

Native	**GB**	**IR**
1987–99	0	0
1970–86	0	0
pre 1970	0	0
Alien		
1987–99	238	13
1970–86	77	13
pre 1970	243	38

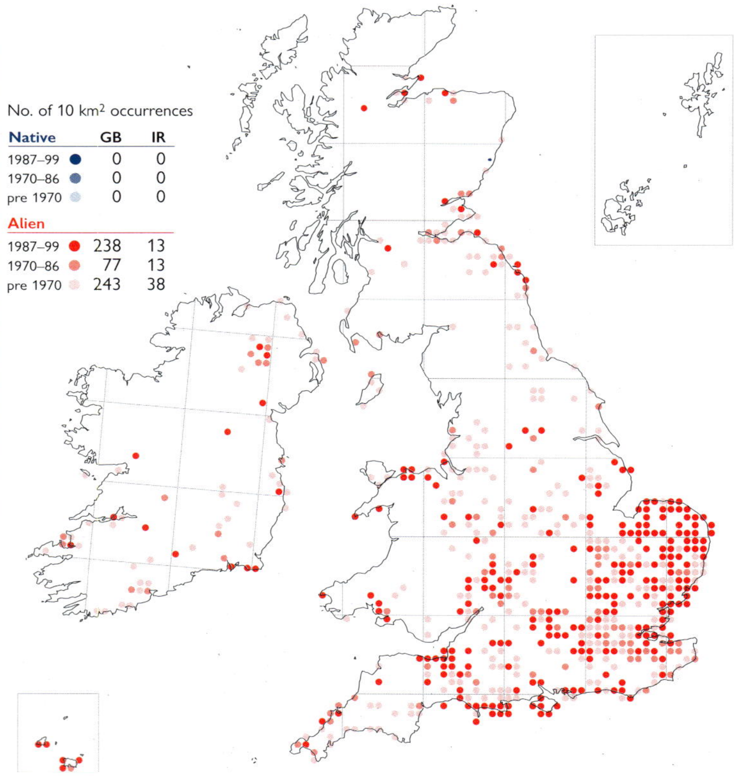

An annual or biennial, found in rough pasture, on grassy banks, in hedgerows and on waste ground. It is locally well-established and persistent, especially in coastal habitats in S. England, but is also a widespread casual. Lowland.

Archaeophyte (change –0.07). *S. marianum* occurs as an introduction with wool shoddy, bird-, grass- and oil-seed, and as a garden escape. The large number of pre-1987 records is unlikely to reflect a decline in this species, but rather it represents the accumulation of casual records over many years.

Native of the Mediterranean region; naturalised or casual throughout much of Europe and in N. America and Australia.

References: Atlas (290b), Bolòs & Vigo (1995), Doogue *et al.* (1998).

F. H. PERRING

Serratula tinctoria Saw-wort

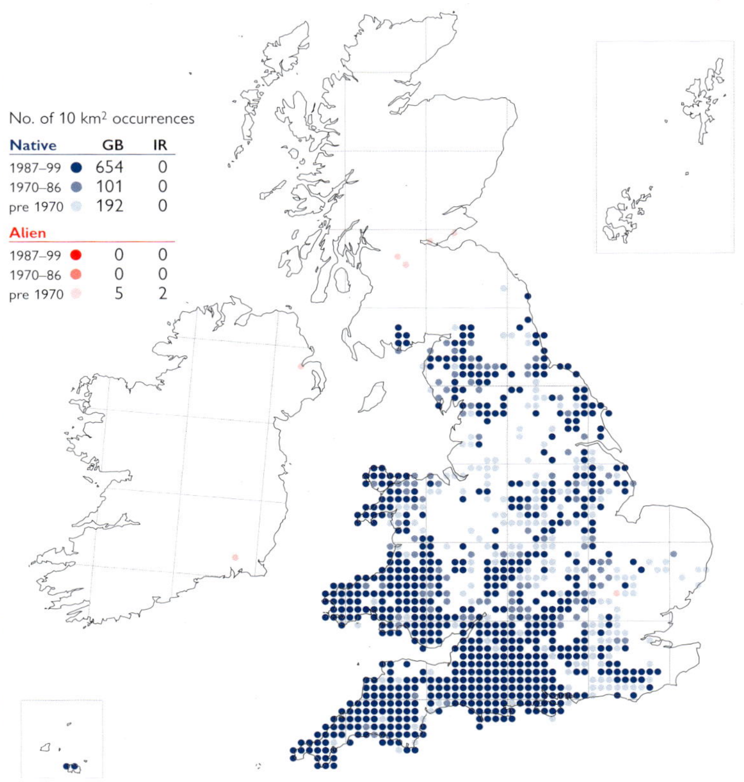

A perennial herb of calcareous grassland, hay and fen-meadows, wet heaths and heathy mires, open scrub and woodland, rocky lake shores and cliff-tops; also in artificial habitats including roadside verges and railway banks. 0–560 m (Fur Tor, Dartmoor, S. Devon).

Native (change –0.21). *S. tinctoria* has declined in England since the 1962 *Atlas*, due to drainage, the improvement of pastures and the loss of grasslands in woods and on wood-margins. It was first recorded in Ireland as a casual in 1893. It was found in Co. Wexford in 1925, but has not been seen at its sole locality there since the 1950s and was probably destroyed during the building of a marina.

European Temperate element.

References: Atlas (292d), Curtis & McGough (1988), Hultén & Fries (1986), Meusel & Jäger (1992), Rackham (1980).

F. H. PERRING

Centaurea scabiosa Greater Knapweed

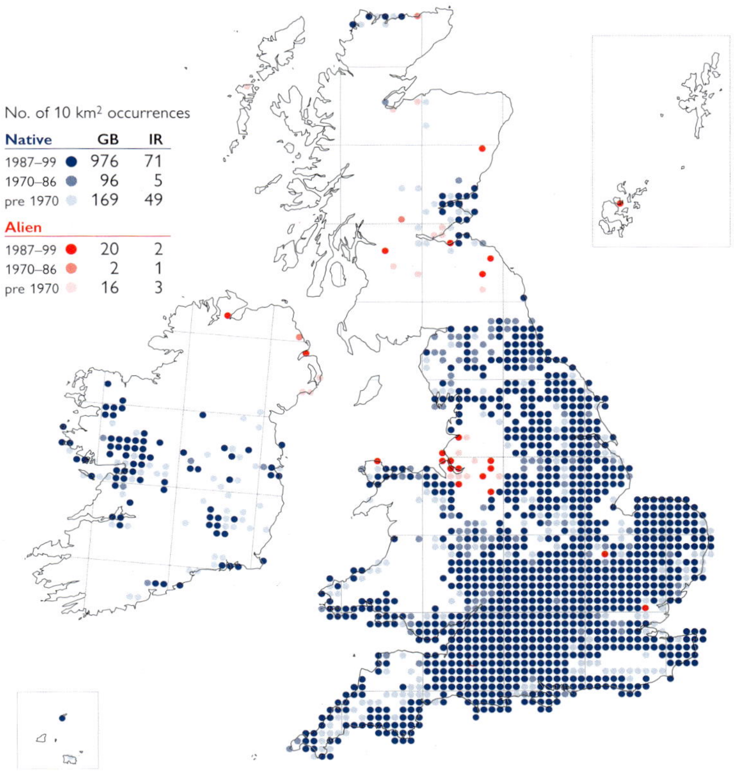

A tufted, winter-green, perennial herb of dry, usually calcareous soils, found in grassland, scrub and woodland edges, on cliffs, roadsides, railway banks, quarries and waste ground. 0–320 m (Matlock, Derbys.).

Native (change –0.49). *C. scabiosa* has apparently declined due to the destruction of old grasslands, although it is able to colonise disturbed habitats. Plants with entire basal leaves, var. *succisifolia*, are found in coastal populations in Wales and Scotland, but their taxonomic status is uncertain (Valentine, 1980).

Eurosiberian Temperate element; widely naturalised outside its native range.

References: Atlas (291a), Grime *et al.* (1988), Grose (1957), Hultén & Fries (1986), Meusel & Jäger (1992).

F. H. PERRING

Centaurea montana Perennial Cornflower

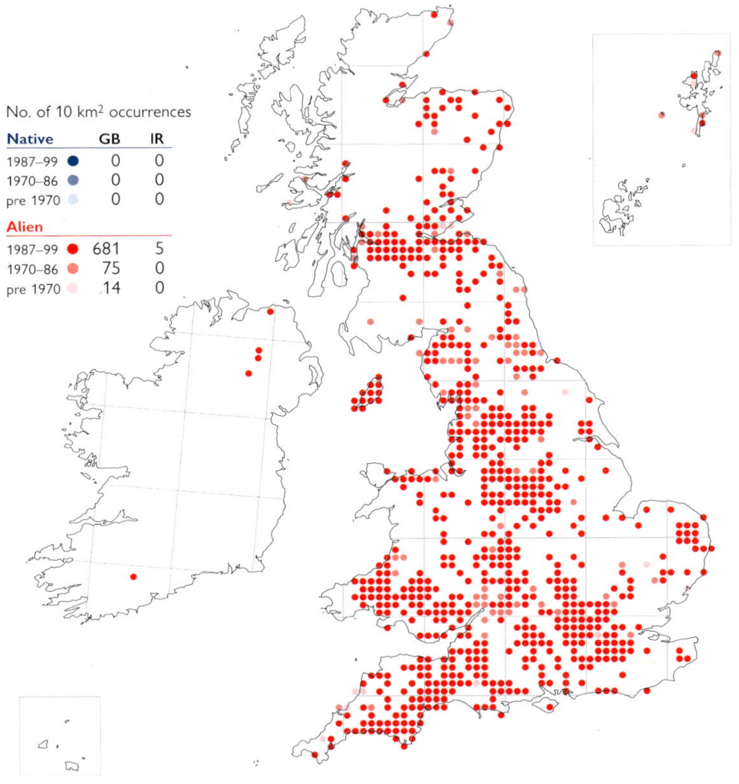

A vigorous perennial herb, spreading by rhizomes and frequently naturalised on roadsides, railway banks and in other waste places. Mainly lowland, reaching 380 m at Garrigill (Cumberland).

Neophyte. This species was introduced to cultivation in Britain before 1596; it is widely grown in gardens and was known from the wild by 1888. It was not mapped in the 1962 *Atlas*, but it is probably spreading. It is very persistent, and many populations are long-established.

Native of the mountains of C. and S. Europe.

Reference: Meusel & Jäger (1992).

F. H. PERRING

Centaurea cyanus Cornflower

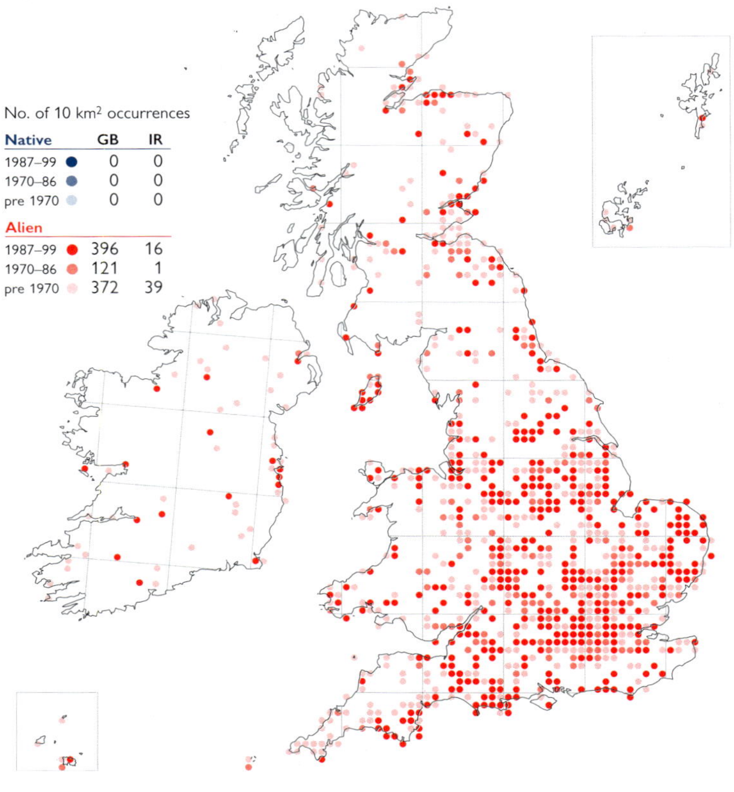

No. of 10 km² occurrences

	GB	IR
Native		
1987–99	0	0
1970–86	0	0
pre 1970	0	0
Alien		
1987–99	396	16
1970–86	121	1
pre 1970	372	39

C. cyanus formerly occurred as an annual weed of arable habitats. Since 1986 it has been recorded in very few arable fields, but it is now frequent in waste places, on roadsides and on rubbish tips as a casual arising from gardens and wild-flower seed mixtures. 0–350 m (Blackwell, Derbys.).

Archaeophyte (change –0.39). Known to have been present in Britain from the Iron Age onwards, *C. cyanus* remained a serious weed until seed cleaning began a rapid decline, accelerated by the use of herbicides. Since the 1980s it has increased through wild-flower seed.

As an archaeophyte *C. cyanus* has a distribution centred on the European Temperate region; it has been widely naturalised outside this range.

References: Atlas (291b), Curtis & McGough (1988), Hultén & Fries (1986), Meusel & Jäger (1992), Stewart *et al.* (1994), Wigginton (1999).

F. H. PERRING

Centaurea calcitrapa Red Star-thistle

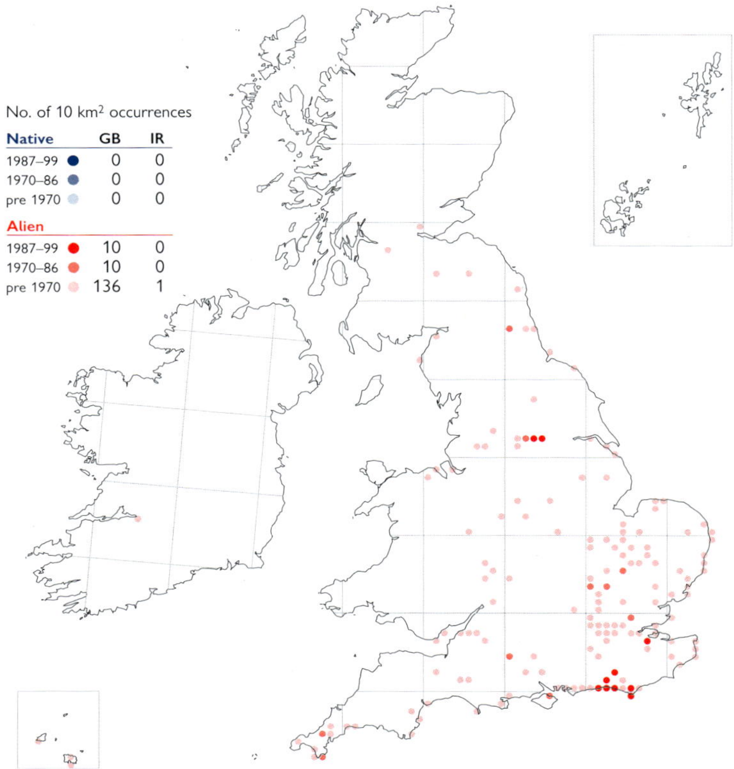

No. of 10 km² occurrences

	GB	IR
Native		
1987–99	0	0
1970–86	0	0
pre 1970	0	0
Alien		
1987–99	10	0
1970–86	10	0
pre 1970	136	1

A biennial herb of waste ground and tracksides in dry grassland, and on banks on well-drained sandy, gravelly or light chalky soils. Lowland.

Archaeophyte (change –2.34). There is a single archaeological record of this species from a Roman site, and it was recorded in the historic period by 1597. It is sometimes regarded as native in Sussex, where it grows in habitats similar to those near the Somme estuary in N. France (Wigginton, 1999). It is also established in Kent, where it has been known since 1839 at Chatham. Elsewhere it is a declining casual from wool, bird-seed, lucerne seed and esparto.

Native of the Mediterranean region; widespread as an introduction further north but now declining outside its native range.

References: Atlas (292b), Meusel & Jäger (1992).

F. H. PERRING

Centaurea solstitialis Yellow Star-thistle

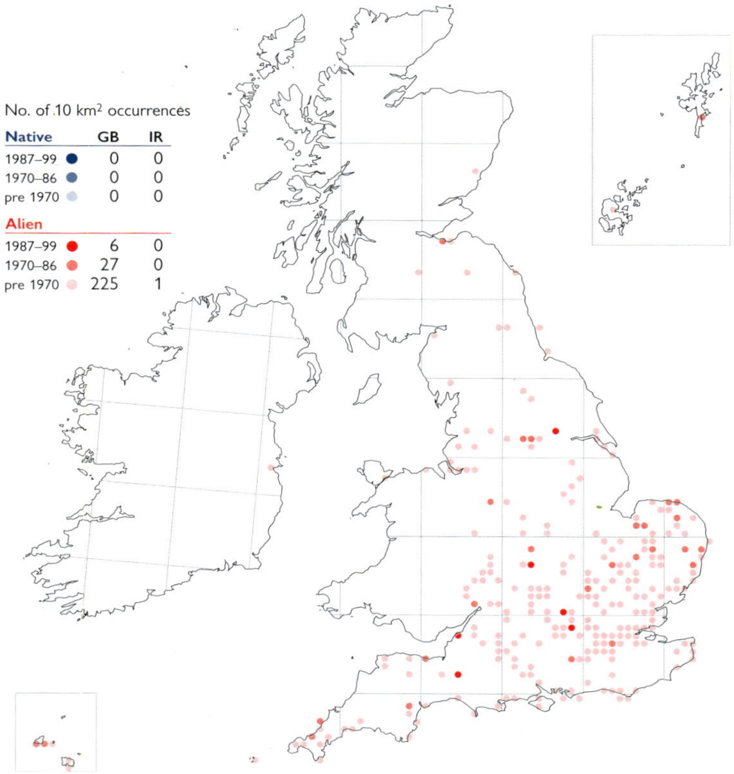

No. of 10 km² occurrences

	GB	IR
Native		
1987–99	0	0
1970–86	0	0
pre 1970	0	0
Alien		
1987–99	6	0
1970–86	27	0
pre 1970	225	1

An annual or, rarely, biennial herb introduced with grain, bird-seed, lucerne or sainfoin seed, or wool, and found most frequently in arable fields and on waste ground. Lowland.

Neophyte (change –5.62). *C. solstitialis* was known from the wild by at least 1778, and formerly persisted in lucerne and sainfoin fields. Dunn (1905) described it as 'of frequent occurrence in Southern England', but it is much less common than it was before 1930, and now only known as a casual.

Native of S. Europe and S.W. Asia; widely naturalised elsewhere.

References: Atlas (292c), Meusel & Jäger (1992).

F. H. PERRING

Centaurea diluta Lesser Star-thistle

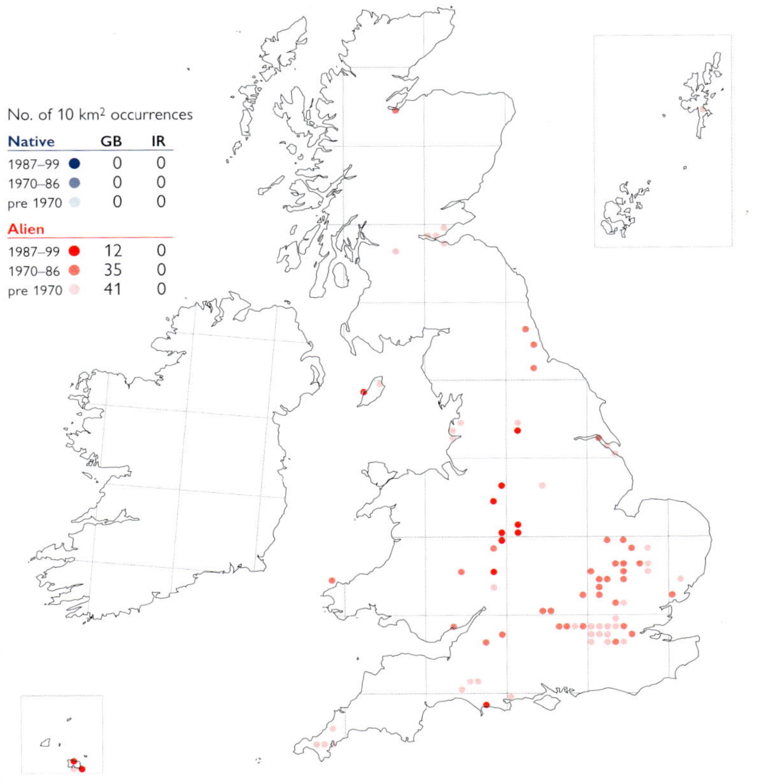

This perennial herb, often behaving as an annual, is a casual of waste places and rubbish tips, mainly from bird-seed but also from wool and grain. Lowland.

Casual. *C. diluta* was introduced into cultivation in 1781. It was first recorded in the wild in 1904, and the species is now a very rare casual.

A native of S.W. Spain, N.W. Africa and the Atlantic Islands.

Reference: Bolòs & Vigo (1995).

F. H. PERRING

Centaurea nigra Common Knapweed

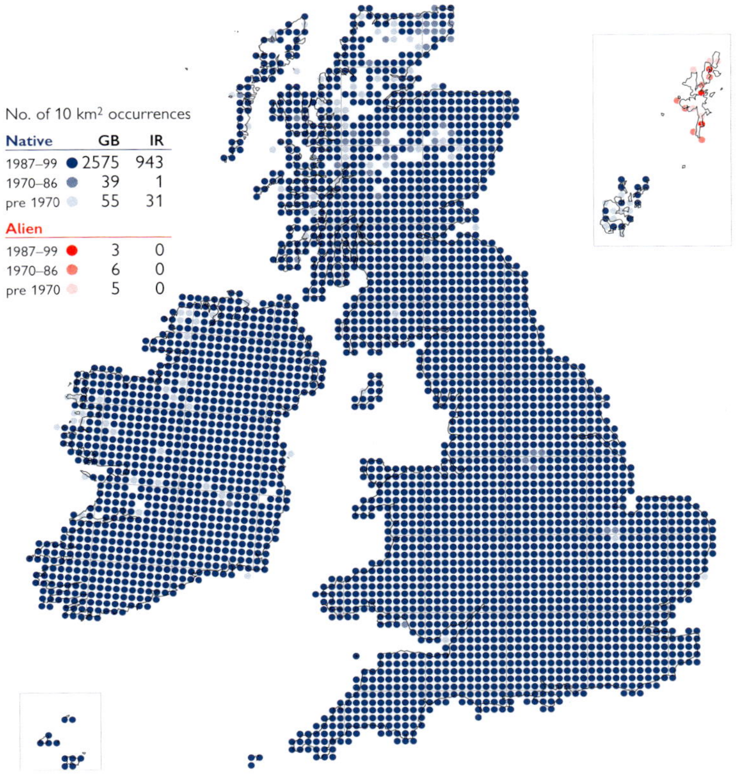

A perennial herb of meadows and pastures, sea-cliffs, roadsides, railway banks, scrub, woodland edges, field borders and waste ground, on a wide range of soils. 0–580 m (Cadair Idris, Merioneth) and exceptionally at 845 m on Great Dun Fell (Westmorland).

Native (change –0.25). The distribution of *C. nigra* is stable. Forms on light soils in England and Wales are sometimes recognised as subsp. *nemoralis*. Elsewhere, subsp. *nigra* predominates. Intermediates occur where both subspecies grow, and also in the absence of subsp. *nemoralis*.

Suboceanic Temperate element; widely naturalised outside its native range.

References: Atlas (291d), Atlas Supp. (74a-c), Grime *et al.* (1988), Grose (1957), Hultén & Fries (1986), Marsden-Jones & Turrill (1954), Meusel & Jäger (1992), Ockendon *et al.* (1969).

F. H. PERRING

Carthamus tinctorius Safflower

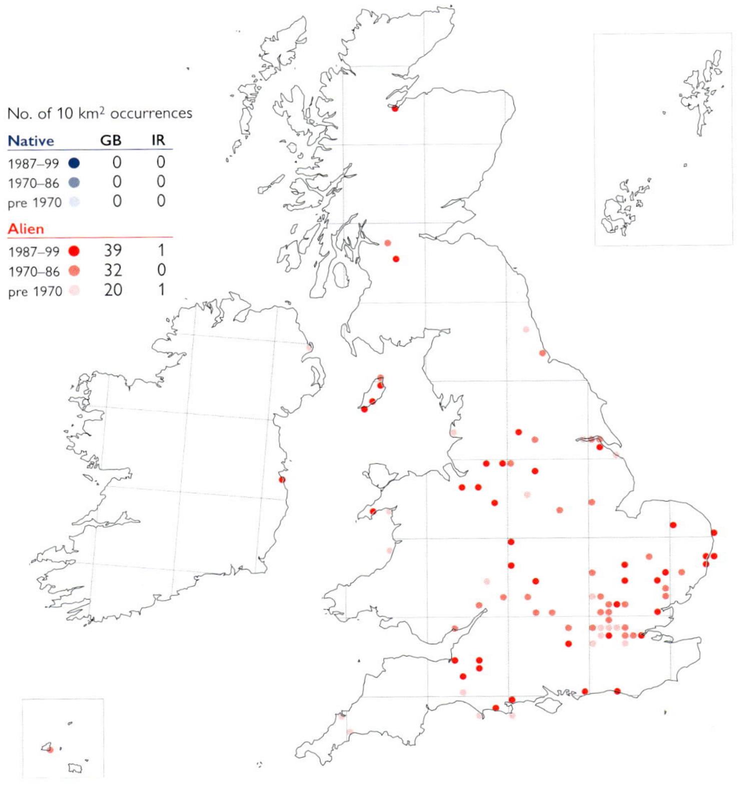

An annual or biennial herb used in bird-seed mixtures and appearing as a casual on rubbish tips, where farmyard manure has been dumped, and in other waste places. Lowland.

Casual. This species has been in cultivation in Britain since at least 1551, when it was grown for the oil extracted from its achenes and the red and saffron dyes from its flowers. It was not formally recorded from the wild, however, until 1899 (Dorset).

C. tinctorius is not known as a wild plant, but it has long been cultivated in S. Europe, S.W. & S.C. Asia and elsewhere.

References: Meusel & Jäger (1992), Rougemont (1989), Zohary & Hopf (2000).

F. H. PERRING

Cichorium intybus Chicory

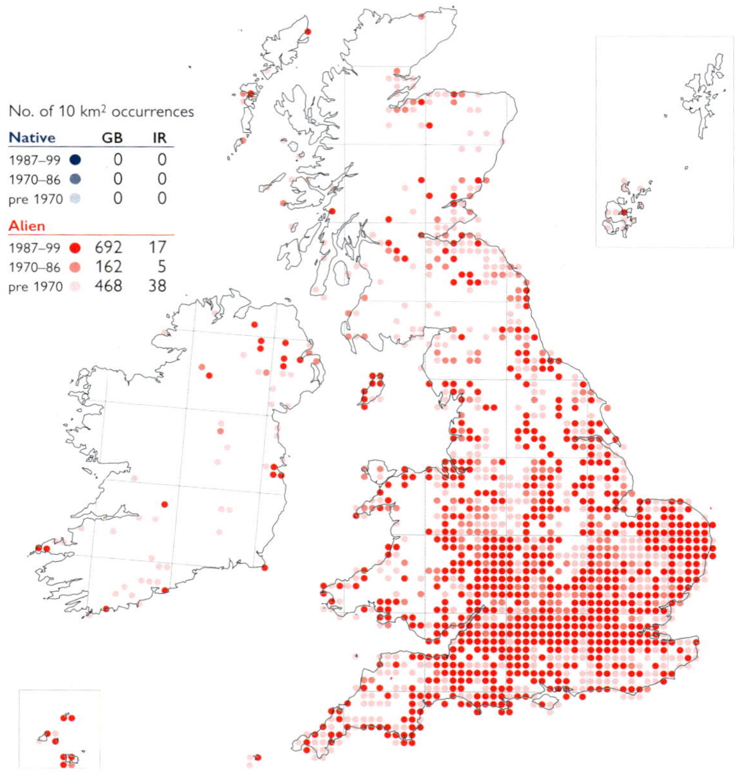

A perennial herb of roadsides, field margins and rough grassland on a wide range of soils. Lowland.

Archaeophyte (change –1.27). Though *C. intybus* was formerly regarded as a native, at least in England and Wales, doubt is now cast on that status by most modern local Floras, which suggest it is always a relic of its former cultivation as a fodder crop. This species has declined as it is now rarely cultivated.

As an archaeophyte *C. intybus* has a Eurosiberian Southern-temperate distribution, but it is widely naturalised so that its distribution is now Circumpolar Southern-temperate.

References: Atlas (293a), Hultén & Fries (1986), Meusel & Jäger (1992).

F. H. PERRING

Arnoseris minima Lamb's Succory

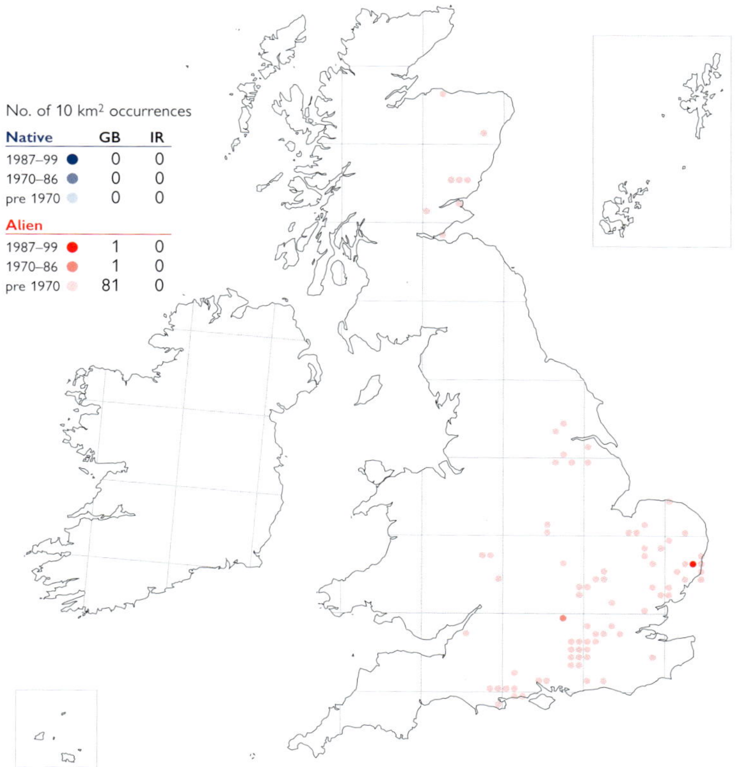

An annual weed of cornfields or fallow ground on the most infertile, acidic, sandy soils. It was particularly associated with soils over Lower Greensand. Lowland.

Archaeophyte (change –3.72). Although widely scattered in the 18th and 19th centuries, *A. minima* was probably persistent only in E. & S.E. England. It declined from the start of the 20th century and by 1953 was limited to about twelve localities. Except for a deliberate introduction in E. Suffolk, it has not been recorded since 1971. Its decline was due to the increased use of fertilisers and herbicides, and improved seed screening methods.

As an archaeophyte *A. minima* has a European Temperate distribution.

References: Atlas (293c), Hultén & Fries (1986), Meusel & Jäger (1992), Silverside (1990c).

F. H. PERRING

Lapsana communis Nipplewort

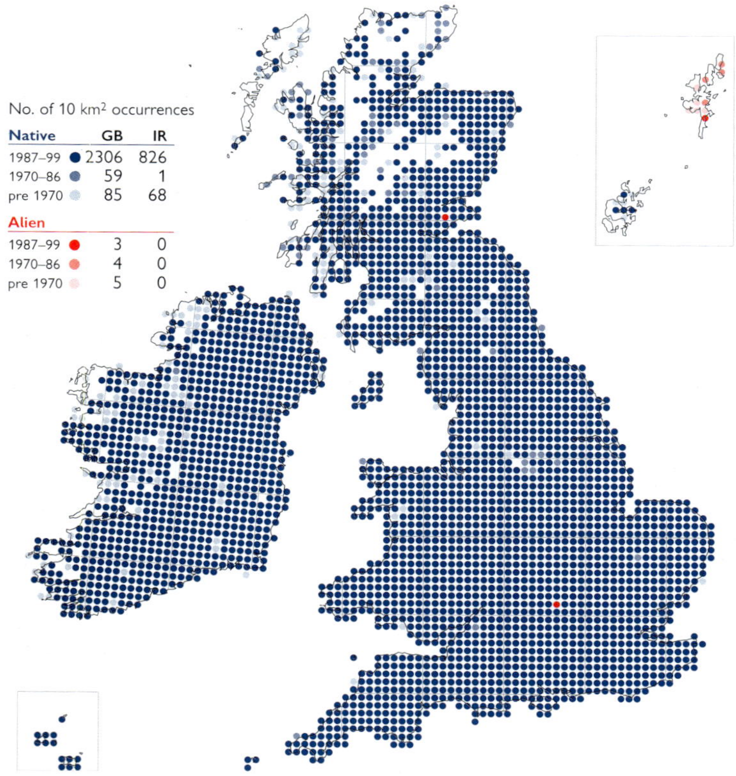

An annual to perennial herb, typically occurring in disturbed and shaded places, and thriving over a wide range of soil acidity and moisture. Habitats include open woodland, scrub, hedgerows, waste and cultivated ground, railway banks, roadsides and old walls. 0–440 m (Clun Forest, Salop).

Native or alien (change –0.47). There has been no discernible change in the range of *L. communis* since the 1962 *Atlas*. Almost all our plants are subsp. *communis*, but the normally perennial subsp. *intermedia*, a well-established introduction, has been recorded from a few sites.

European Temperate element; also in C. Asia and widely naturalised outside its native range.

References: Atlas (293b), Burtt (1950), Grime *et al.* (1988), Hultén & Fries (1986), Meusel & Jäger (1992), Sell (1981).

F. H. PERRING

Hypochaeris radicata Cat's-ear

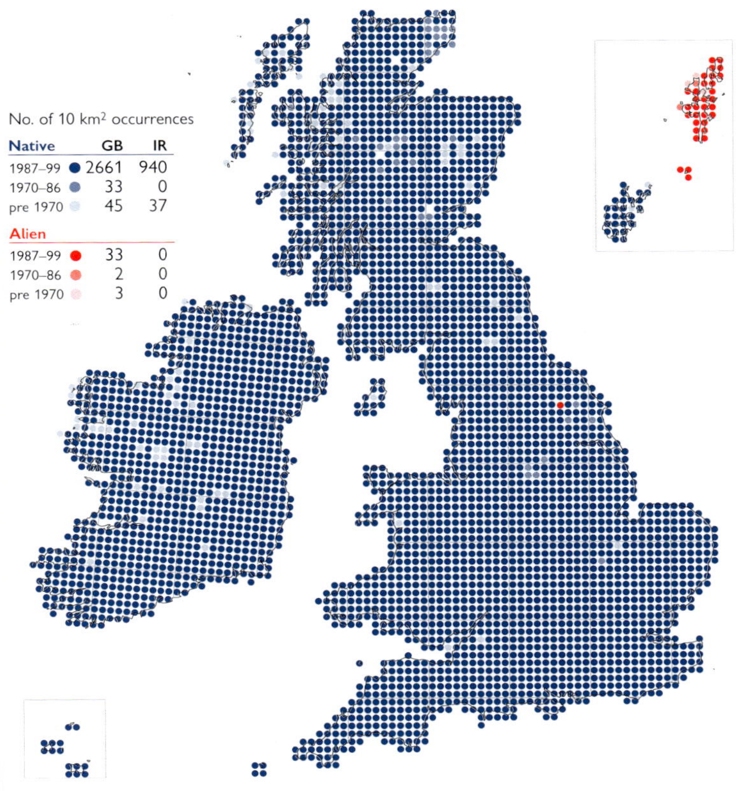

No. of 10 km^2 occurrences

Native	GB	IR
1987–99 ●	2661	940
1970–86 ●	33	0
pre 1970 ●	45	37
Alien		
1987–99 ●	33	0
1970–86 ●	2	0
pre 1970 ●	3	0

A perennial herb of meadows, pastures, lawns, heathland, cliff-tops, sand dunes, roadsides, railway banks and waste ground, on slightly acidic, usually free-draining soils. It is very tolerant of drought, and is absent from sites subject to prolonged waterlogging. 0–530 m (S. of Garrigill, Cumberland), and to 610 m on Macgillycuddy's Reeks (S. Kerry).

Native (change +0.61). There has been no discernible change in the range of *H. radicata* since the 1962 *Atlas*. It was first recorded in Shetland in 1889.

European Southern-temperate element; widely naturalised outside its native range.

References: Atlas (293d), Aarssen (1981), Grime *et al.* (1988), Hultén & Fries (1986), Meusel & Jäger (1992), Scott & Palmer (1987), Turkington & Aarssen (1983).

F. H. PERRING

Hypochaeris glabra Smooth Cat's-ear

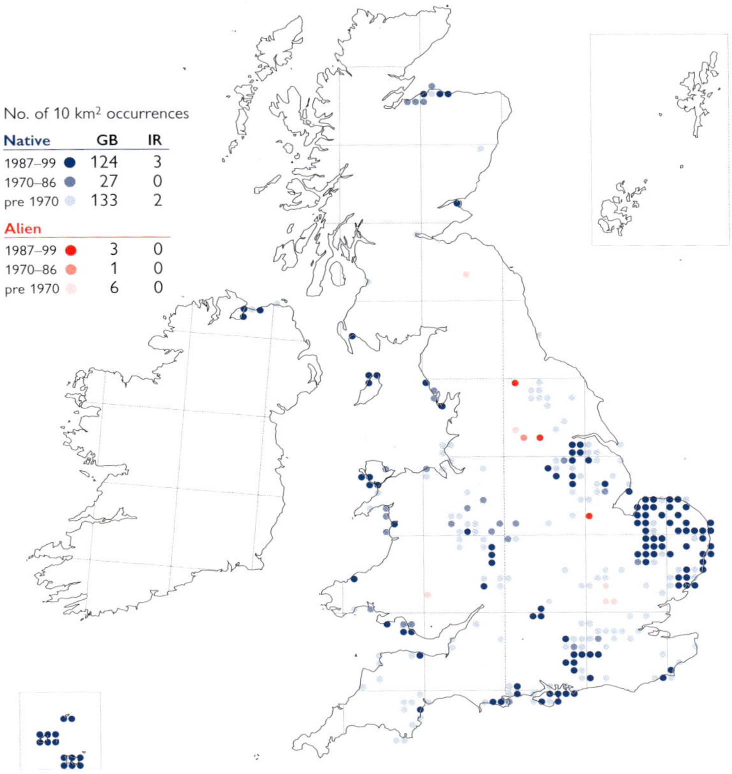

No. of 10 km^2 occurrences

Native	GB	IR
1987–99 ●	124	3
1970–86 ●	27	0
pre 1970 ●	133	2
Alien		
1987–99 ●	3	0
1970–86 ●	1	0
pre 1970 ●	6	0

An annual of open summer-parched grasslands and heathy pastures, on usually acidic, nutrient-poor, sandy or gravelly soils; also occurring in dune grassland and on sandy shingle. It was formerly widespread as a weed of arable fields, and as a wool-shoddy alien. Lowland.

Native (change −1.01). *H. glabra* is declining in semi-natural habitats, even though there have been many new records since it was mapped by Stewart *et al.* (1994). It has been lost from some squares as a result of agricultural improvement or loss of grazing. It can easily be overlooked, however, particularly as the flowers close in the afternoon.

European Southern-temperate element; widely naturalised outside its native range.

References: Atlas (294a), Curtis & McGough (1988), Fone (1989), Hultén & Fries (1986), Meusel & Jäger (1992).

F. H. PERRING

Hypochaeris maculata Spotted Cat's-ear

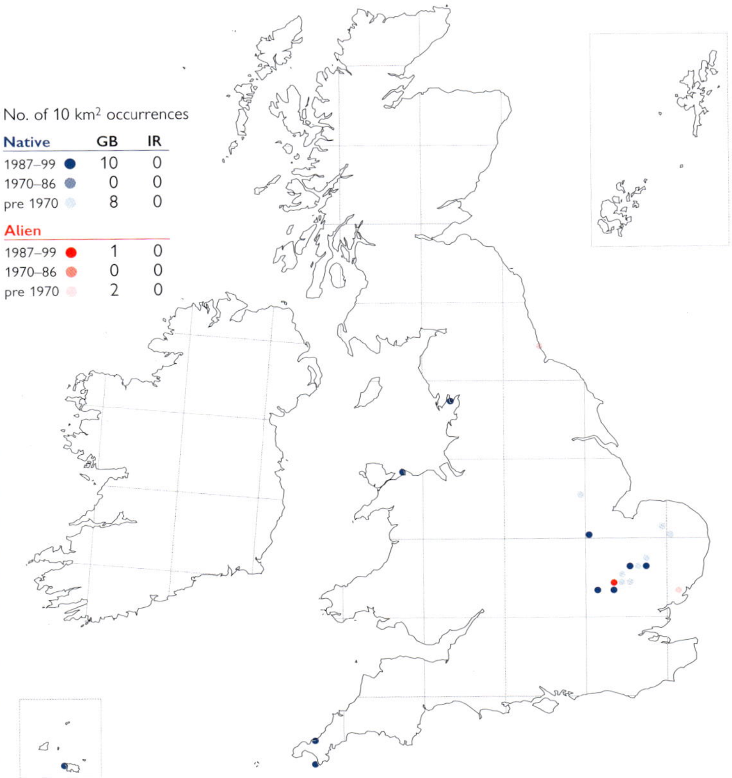

No. of 10 km^2 occurrences

Native	GB	IR
1987–99 ●	10	0
1970–86 ●	0	0
pre 1970 ●	8	0
Alien		
1987–99 ●	1	0
1970–86 ●	0	0
pre 1970 ●	2	0

A perennial herb confined to free-draining, usually base-rich substrates. It occurs on chalk and limestone downland, on coastal cliffs over limestone and serpentine, and on wind-blown calcareous sand on the N. Cornwall coast. In Jersey, it grows on exposed granite cliffs. Lowland.

Native (change −0.10). This species is now known from only half the sites recorded in the 19th century. These losses, all inland, have been due to habitat destruction by ploughing, or lack of grazing followed by invasion of coarse grasses and scrub. Whilst two of the remaining sites each have over a thousand plants, three sites have fewer than ten.

Eurosiberian Temperate element, with a continental distribution in W. Europe.

References: Atlas (294b), Hultén & Fries (1986), Meusel & Jäger (1992), Wells (1976), Wigginton (1999).

F. H. PERRING

Leontodon autumnalis Autumn Hawkbit

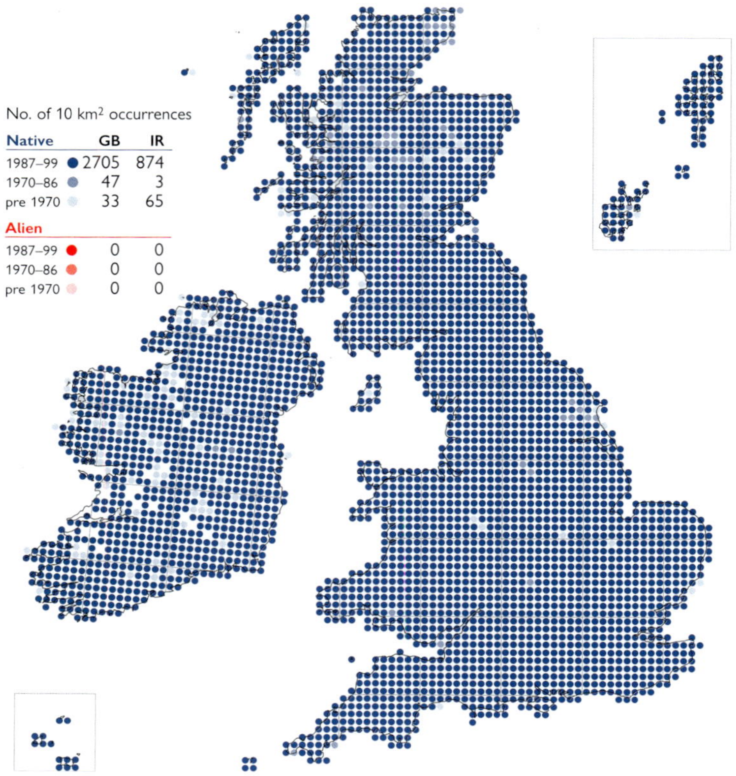

A perennial herb of meadows and pastures, open scrub, heaths, moorland, saltmarshes, fixed dunes and roadsides in the lowlands, and also of screes, flushes and lake margins in the uplands. 0–975 m (Macgillycuddy's Reeks, S. Kerry).

Native (change +1.33). *L. autumnalis* is tolerant of trampling and competes well in taller vegetation. It has increased on roadsides, but its overall national range is stable. It is very variable, with several ecotypes ranging from tall, hairy plants on montane cliff ledges to small, virtually glabrous plants in saltmarshes.

European Boreo-temperate element; widely naturalised outside its native range.

References: Atlas (294c), Allen (1957), Grime *et al.* (1988), Hultén & Fries (1986), Meusel & Jäger (1992).

F. H. PERRING

Leontodon hispidus Rough Hawkbit

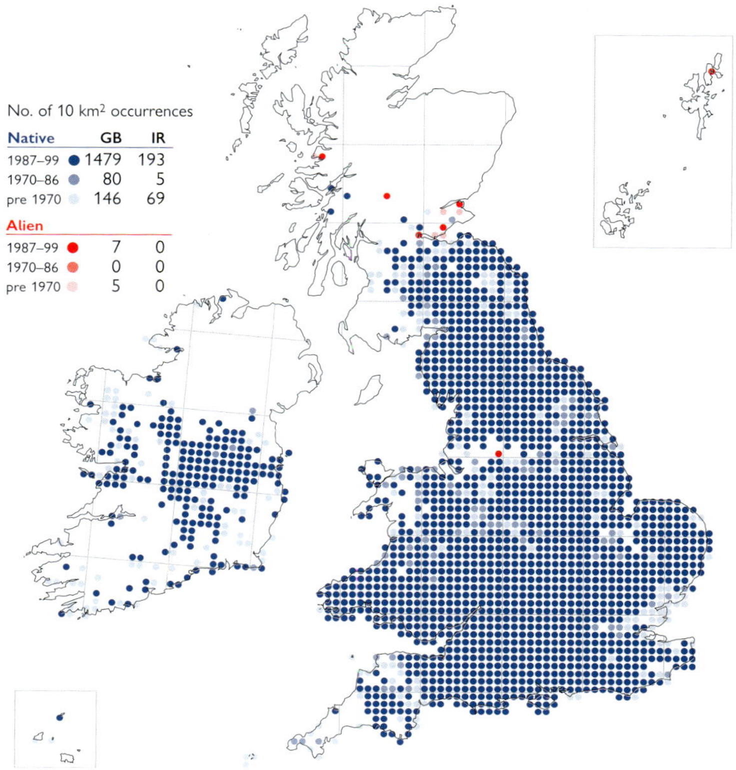

A perennial herb of dry, neutral or calcareous soils, occurring in hay meadows, pastures and other grasslands, on roadside verges, railway banks, rock ledges and in quarries. It readily spreads by wind-dispersed seeds into open habitats. 0–575 m (Harwood Dale, Cheviot) and exceptionally at 845 m on Great Dun Fell (Westmorland).

Native (change –0.59). There has been no discernible change in the distribution of *L. hispidus* since the 1962 *Atlas*. It is possibly over-recorded in some areas in error for *L. saxatilis*; almost all Scottish Highland records have been treated as dubious. Almost or wholly glabrous plants (var. *glabratus*) occasionally occur.

European Temperate element.

References: Atlas (294d), Grime *et al.* (1988), Hultén & Fries (1986), Meusel & Jäger (1992).

F. H. PERRING

Leontodon saxatilis Lesser Hawkbit

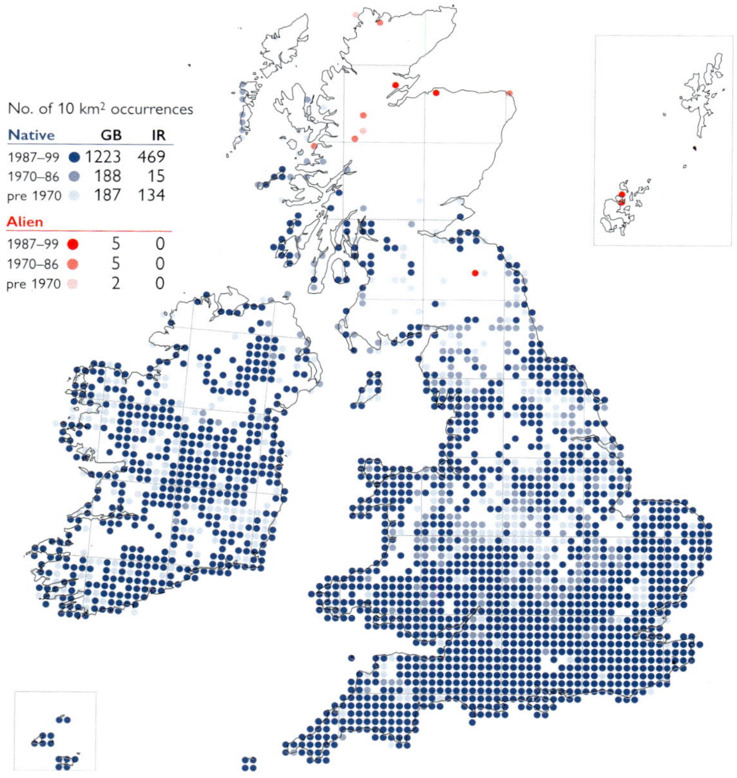

A perennial or biennial herb of often heavily grazed and trampled grassland, dry moorland, stony banks, limestone and other basic rock outcrops, fixed dunes, tracks, and sand- and gravel-pits. It prefers well-drained, calcareous to mildly acidic soils, but also occurs in periodically wet habitats such as dune-slacks, rutted tracks and pond margins on sandy soils. 0–500 m (Brown Clee Hill, Salop).

Native (change +0.21). *L. saxatilis* was mapped as 'all records' in the 1962 *Atlas*. It is now much better recorded but has apparently declined in some areas since 1950. It can be confused with *L. autumnalis* and *L. hispidus*.

Suboceanic Southern-temperate element; widely naturalised outside its native range.

References: Atlas (295a), Hultén & Fries (1986), Meusel & Jäger (1992).

F. H. PERRING

Picris echioides Bristly Oxtongue

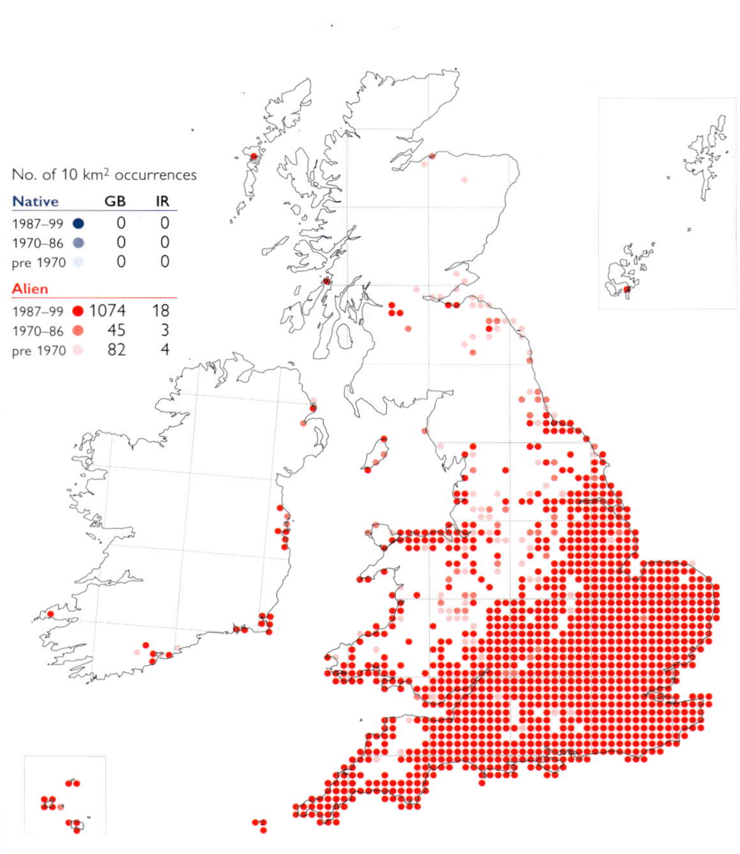

An annual or biennial herb of open grassland, roadsides, field margins, cliffs and scree slopes, river banks, sea walls and waste places, especially on lime-rich clay soils. Lowland.

Archaeophyte (change +0.77). *P. echioides* was mapped with a somewhat patchy distribution in the 1962 *Atlas*, being found mainly south of a line from the Severn to the Humber. Since then it has consolidated within that range. It is only casual in Scotland.

As an archaeophyte *P. echioides* has a European Southern-temperate distribution; it is widely naturalised outside this range.

References: Atlas (295b), Meusel & Jäger (1992).

F. H. PERRING

Picris hieracioides Hawkweed Oxtongue

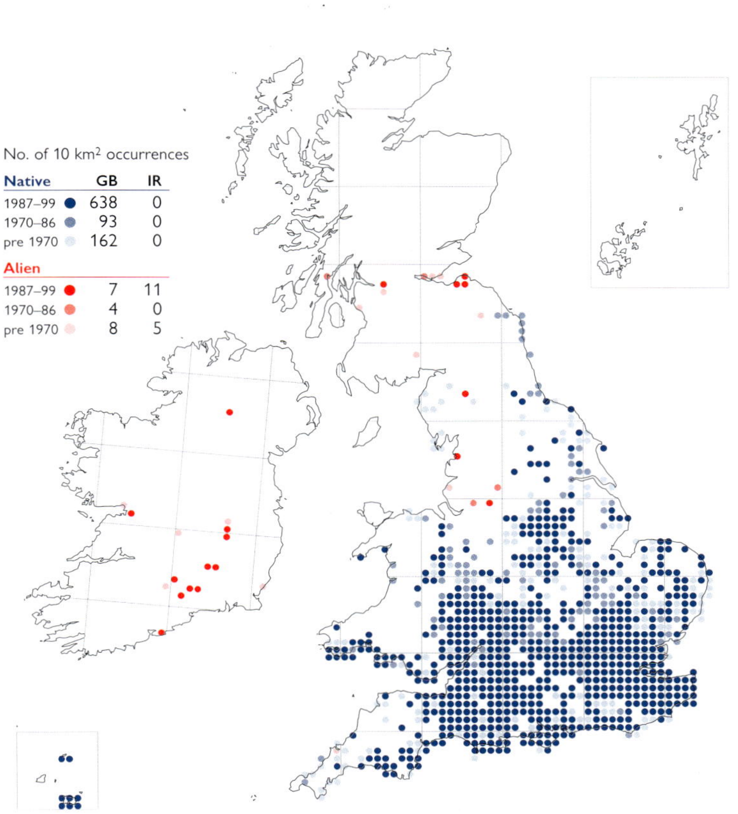

A biennial or perennial herb, mainly of calcareous soils, occurring in the less heavily grazed swards in chalk and limestone grassland, on roadsides and railway banks, and in quarries and lime-pits. It is intolerant of heavy grazing and is a poor competitor in dense vegetation. Lowland.

Native (change –0.06). *P. hieracioides* has shown little change in distribution since the 1962 *Atlas*, though there have been some losses in the northern and eastern parts of its range. It has been introduced to Ireland, where it has spread towards the south and west along railways.

Eurasian Temperate element.

References: Atlas (295c), Hultén & Fries (1986), Meusel & Jäger (1992).

F. H. PERRING

Scorzonera humilis Viper's-grass

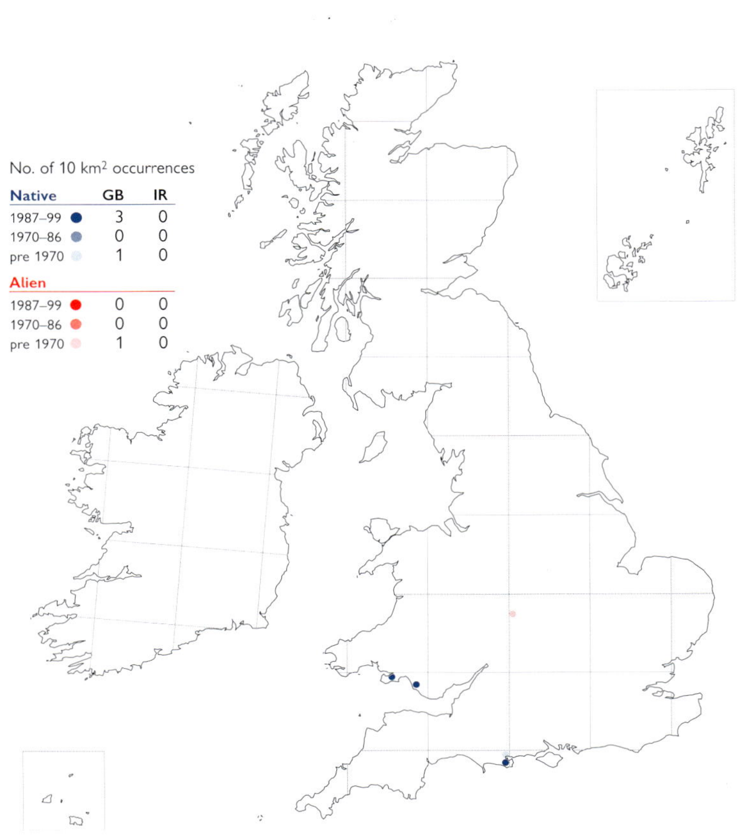

A perennial herb of damp, unimproved grasslands and fen-meadows on relatively infertile, neutral or mildly acidic soils. Lowland.

Native or alien. This species was first recorded in Britain at one site in Dorset in 1914. That site still supports a large population, though the field is now very overgrown. It was recorded once at another site in Dorset, in 1927. The small Warwickshire population, discovered in 1954, had gone by 1967. The two Glamorgan sites were discovered in 1996 and 1997. The native status of this species is sometimes questioned, but as its habitats are similar to those in continental Europe, most authorities accept it as native in Britain.

European Temperate element.

References: Atlas (296b), Hultén & Fries (1986), Meusel & Jäger (1992), Wigginton (1999).

F. H. PERRING

Tragopogon pratensis Goat's-beard

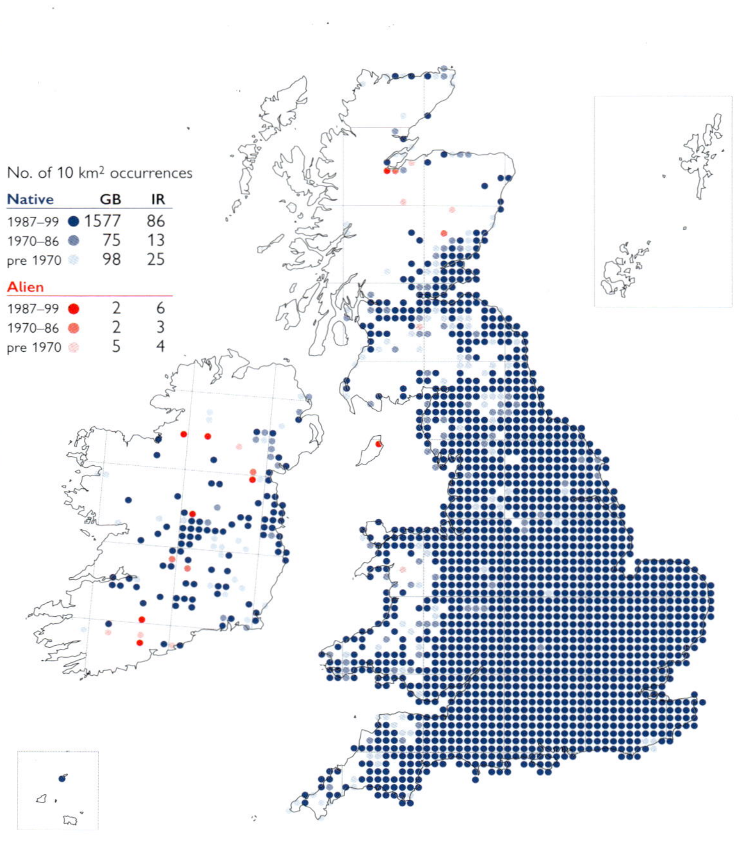

An annual to perennial herb of tall grassland in meadows and pastures, on field margins, sand dunes, roadsides, railway banks and waste ground. 0–365 m (Buxton, Derbys.).

Native (change –0.30). The distribution of *T. pratensis* is almost unchanged since the 1962 *Atlas*. It tolerates occasional mowing, but with the decline of traditional hay meadows has become increasingly restricted to more disturbed habitats. The native subsp. *minor* occurs throughout the range of the species, the alien subsp. *pratensis* is mapped separately.

Eurosiberian Temperate element; widely naturalised outside its native range.

References: Atlas (295d), Barling (1955), Hultén & Fries (1986), Meusel & Jäger (1992), Rich & Jermy (1998).

F. H. PERRING

Tragopogon pratensis subsp. *pratensis*

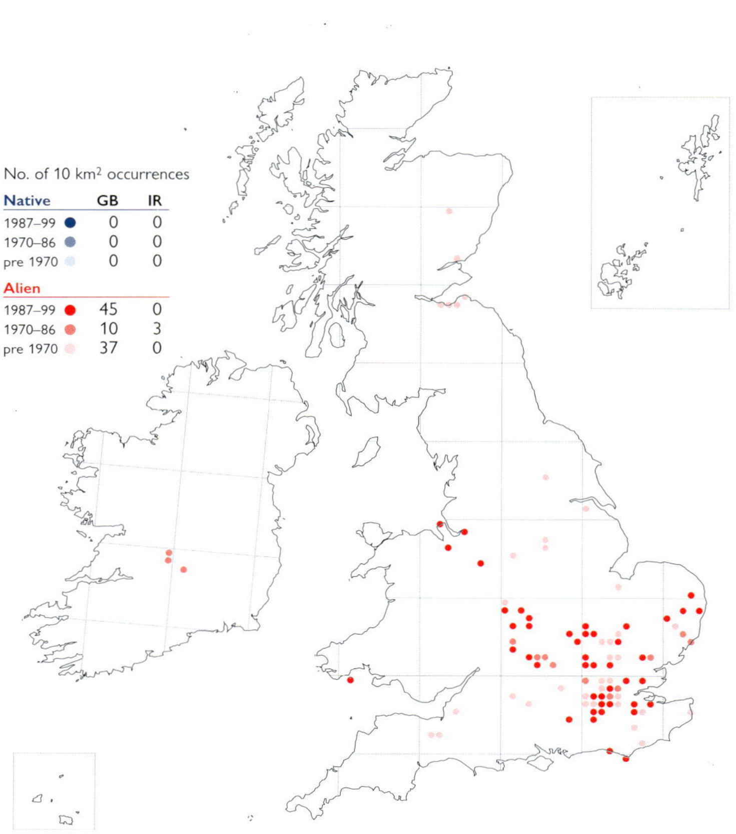

An annual to perennial herb of dry grassland, roadsides, rubbish tips and waste ground. It is usually casual in the north and west of its range. Lowland.

Neophyte. This subspecies was first recorded in Britain in 1836 (Middlesex). It may be slightly over-recorded; only records based on freshly opened flowers are reliable because the characters separating the subspecies become obscure very soon after fertilisation (Rich & Jermy, 1998).

Subsp. *pratensis* occurs throughout much of the European range of the species.

References: Barling (1955), Meusel & Jäger (1992).

F. H. PERRING

Tragopogon porrifolius Salsify

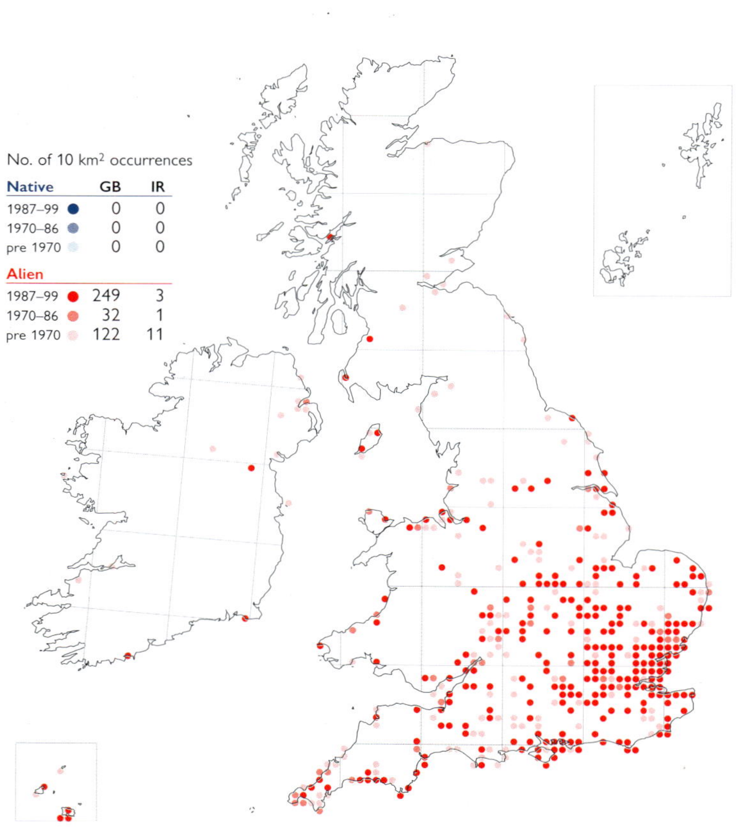

An annual or biennial herb, occasionally escaping from cultivation and naturalised on sea-walls, cliffs, rough grassland and road verges, especially in S.E. England. Elsewhere it is usually only casual. Lowland.

Neophyte (change +1.08). This species was first recorded in the wild in Britain in 1695 (Middlesex). Its range is similar to that shown in the 1962 *Atlas*, but it is now better recorded.

Native of the Mediterranean region; the native range of the cultivated subsp. *porrifolius* is uncertain but it is widely naturalised in temperate Europe.

References: Atlas (296a), Bolòs & Vigo (1995).

F. H. PERRING

Sonchus palustris Marsh Sow-thistle

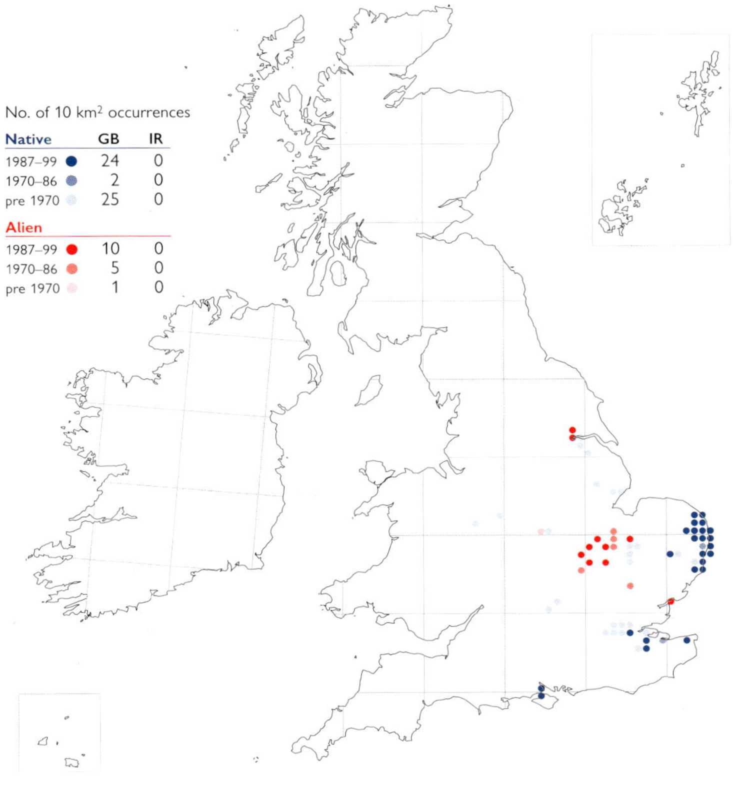

A perennial herb of tall vegetation beside rivers on damp peaty or silty soils rich in nitrogen. It is also moderately tolerant of saline conditions, and can grow near tidal river mouths. Lowland.

Native (change +0.18). Whilst urban developments have caused a decline in *S. palustris* in the Thames Valley and in Kent there is some evidence of an increase in Broadland and E. Suffolk. It became extinct in Cambridgeshire through drainage long before 1930, but has spread from a colony introduced to Woodwalton Fen (Hunts.). The Hampshire population, though first found in 1959, appears to be native. The Yorkshire plants are thought to have been introduced with *Salix* from East Anglia.

Eurosiberian Temperate element.

References: Atlas (297c), Hultén & Fries (1986), Meusel & Jäger (1992), Stewart *et al.* (1994).

F. H. PERRING

Sonchus arvensis Perennial Sow-thistle

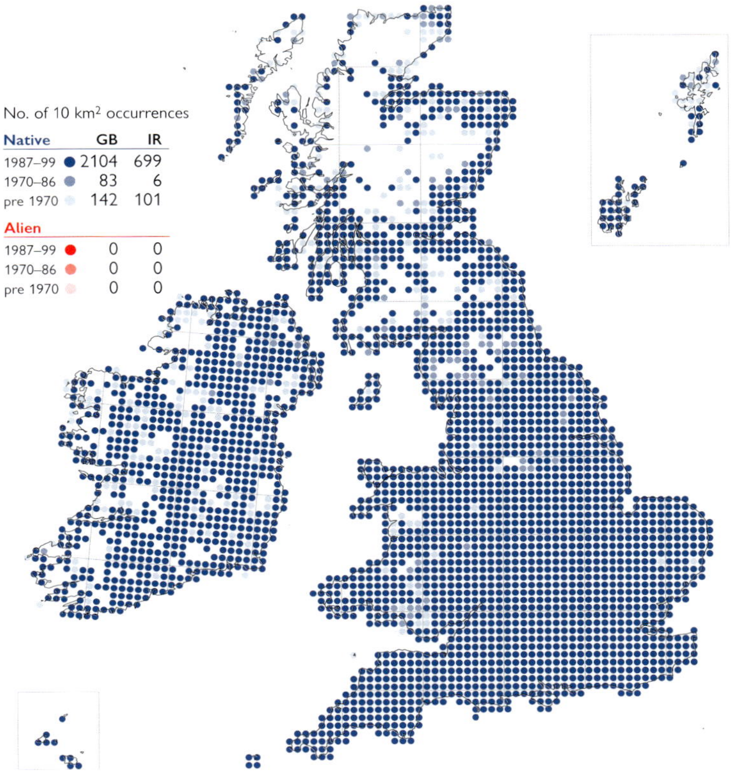

A creeping perennial herb of roadside verges, ditch and river banks, sea-walls and the upper parts of beaches and saltmarshes, particularly along strand lines; it is also frequent in arable fields, where it can be a weed, and on waste ground. It prefers disturbed, nutrient-enriched soils. 0–445 m (near Bishop's Castle, Salop).

Native (change −0.12). There has been no change in the distribution of *S. arvensis* since the 1962 *Atlas*. A glabrous form, subsp. *uliginosus*, has been recorded throughout Britain, usually from wetter habitats. Its taxonomic status and ecology require further study.

Eurosiberian Temperate element, but widely naturalised so that distribution is now Circumpolar Temperate.

References: Atlas (297d), Hultén & Fries (1986), Lemna & Messersmith (1990), Lousley (1968), Meusel & Jäger (1992).

F. H. PERRING

Sonchus oleraceus Smooth Sow-thistle

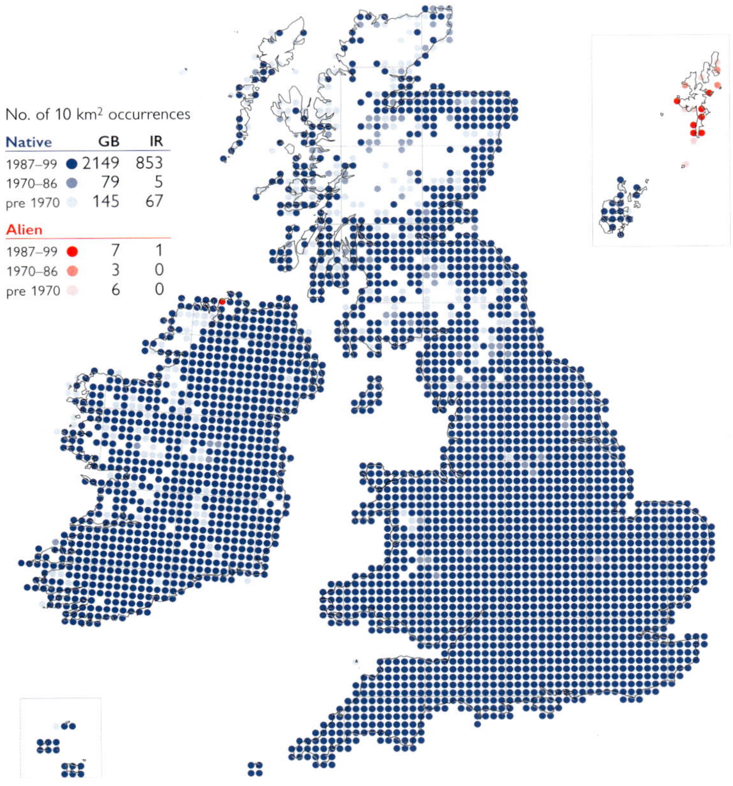

An overwintering annual of disturbed or trampled grasslands, coastal cliff-slopes, roadside verges, arable fields, manure heaps, walls, pavement cracks, gardens and waste places. It is intolerant of grazing but an invasive weed of bare ground. It is more frequent in coastal habitats than *S. asper* but less frequent in the uplands. 0–365 m (Kisdon, N.W. Yorks.).

Native (change −0.42). There has been no discernible change in the distribution of *S. oleraceus* since the 1962 *Atlas*.

European Southern-temperate element, but widely naturalised so that distribution is now Circumpolar Southern-temperate.

References: Atlas (298a), Grime *et al.* (1988), Hultén & Fries (1986), Hutchinson *et al.* (1984), Lewin (1948), Meusel & Jäger (1992).

F. H. PERRING

Sonchus asper Prickly Sow-thistle

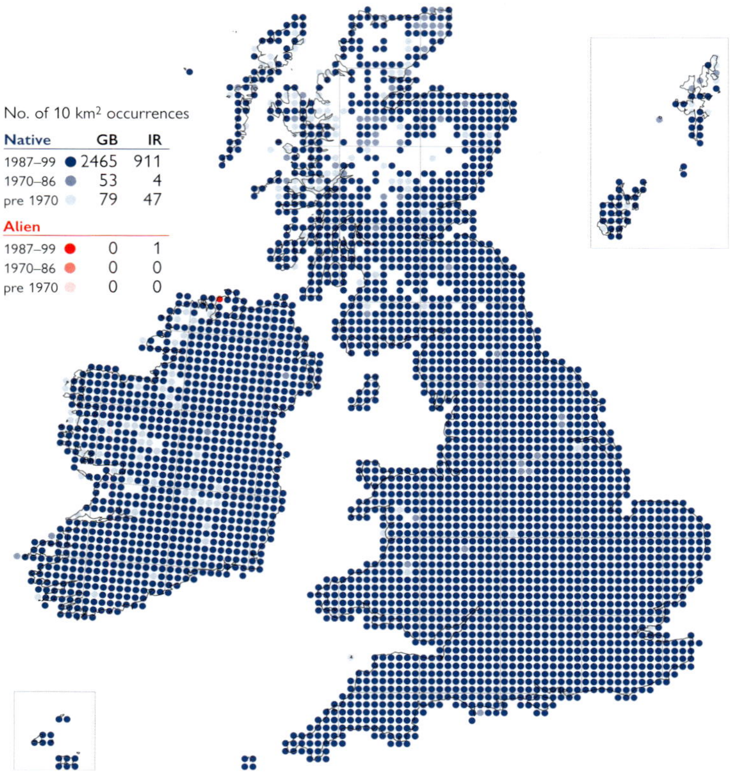

An overwintering annual of rough grassland, scrub, roadside verges, quarries, rock outcrops, railway lines, arable fields, manure heaps, gardens and waste places. It prefers dry, disturbed, sandy soils and is intolerant of grazing, but can be an invasive weed of bare ground. *S. asper* tolerates rather wetter conditions than *S. oleraceus* and is more frequent in the uplands. 0–395 m (Weardale, Co. Durham).

Native (change +0.78). There has been no discernible change in the distribution of *S. asper* since the 1962 *Atlas.*

European Southern-temperate element, but widely naturalised so that distribution is now Circumpolar Southern-temperate.

References: Atlas (298b), Grime *et al.* (1988), Hultén & Fries (1986), Hutchinson *et al.* (1984), Lewin (1948), Meusel & Jäger (1992).

F. H. PERRING

Lactuca serriola Prickly Lettuce

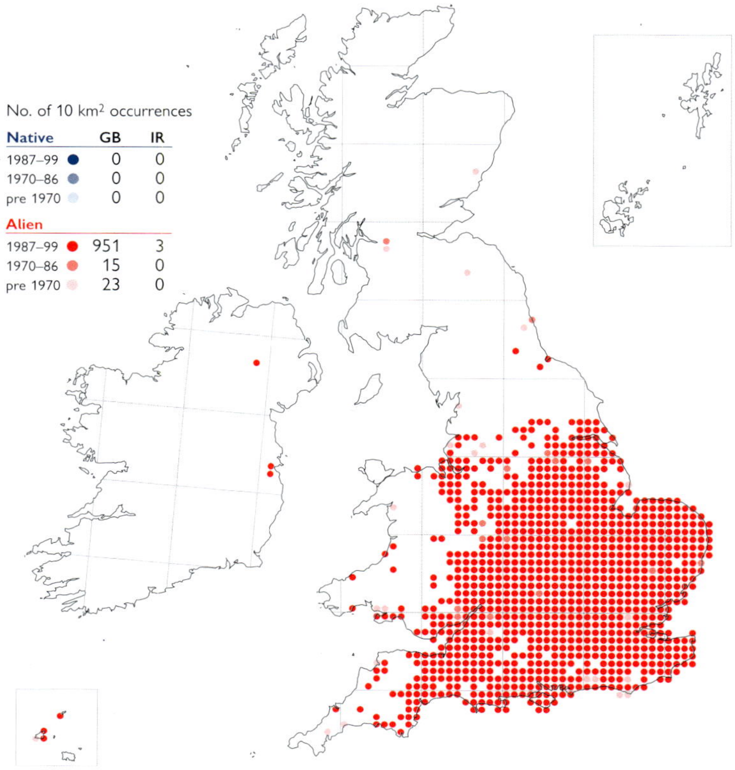

An annual or biennial herb of roadsides, waste ground, gravel-pits and sea walls, often rapidly colonising newly turned soil. Also occasionally in semi-natural habitats, such as shingle banks and sand dunes. Lowland.

Archaeophyte (change +2.70). *L. serriola*, first recorded in Britain in 1632, spread rapidly during the 20th century, often with road development. Since the 1962 *Atlas* it has spread northwards and westwards, and greatly consolidated its range. It was first recorded in Ireland in 1996.

As an archaeophyte *L. serriola* has a Eurosiberian Southern-temperate distribution; it is widely naturalised outside this range.

References: Atlas (296c), Bowra (1992a), Carter & Prince (1985), Hultén & Fries (1986), Meusel & Jäger (1992), Reynolds (1999).

F. H. PERRING

Lactuca virosa Great Lettuce

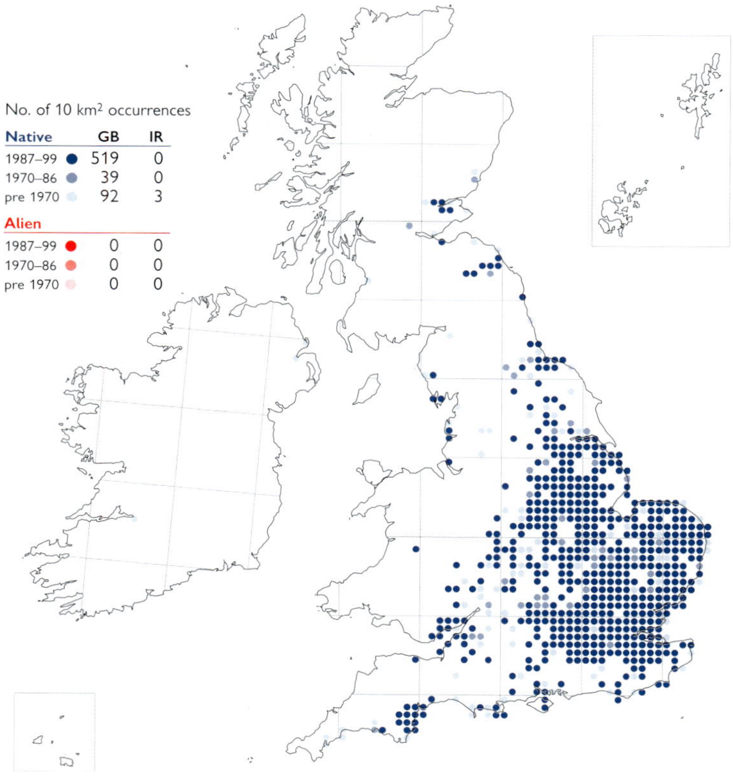

An annual or biennial herb, sensitive to grazing, occurring as a native on coastal cliffs, inland rock outcrops and perhaps sand dunes, but much more widespread as a plant of rank calcareous grassland, woodland margins, road-banks, quarries, tracks and rough ground. Lowland.

Native (change +1.16). *L. virosa* was first recorded in Britain in 1570 but it was often recorded in error for *L. serriola* forma *integrifolia* before 1930, so it may have been rarer then than records suggest. Road development has greatly assisted its spread since 1980. Recorders have not distinguished the alien sites and all records are therefore mapped as if they are native.

Suboceanic Southern-temperate element.

References: Atlas (296d), Boorman & Fuller (1984), Meusel & Jäger (1992), Oswald (2000), Salisbury (1953).

F. H. PERRING

Lactuca saligna Least Lettuce

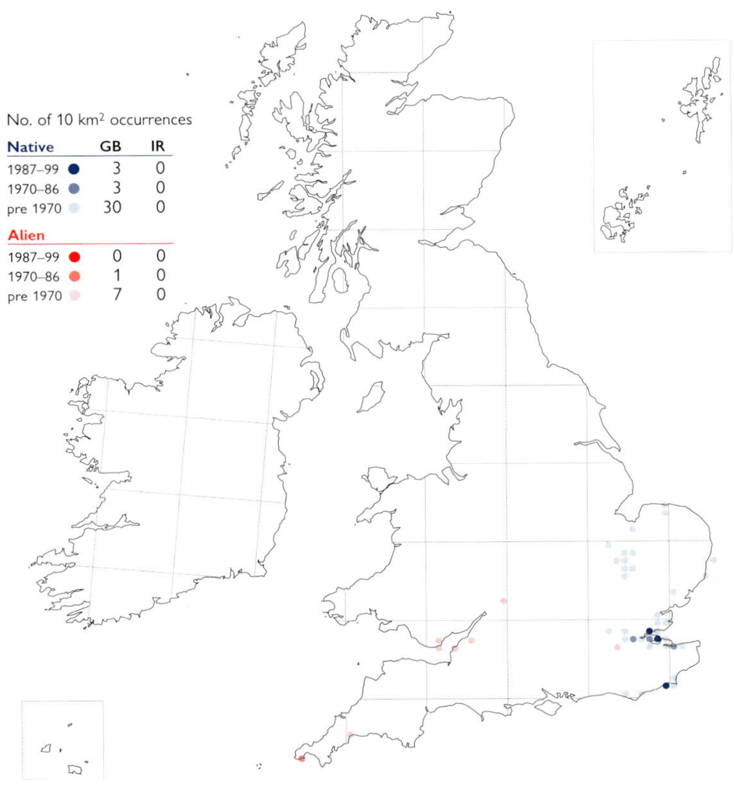

This autumn- or spring-germinating annual occurs on disturbed, sandy shingle and old sea walls, growing on sparsely vegetated ground; also formerly on the banks of rivers and ditches in East Anglia, and, rarely, on paths and cliffs in Essex. Lowland.

Native (change –1.51). *L. saligna* has suffered a severe decline owing to sea-wall refurbishment and river engineering. It was extinct at about half its known sites by 1930, and survived in East Anglia until 1953. Inundation by sea water caused a dramatic decline in the present Sussex population in the 1990s, but the Essex population is thriving, and benefits from cattle grazing. It was re-found on the Isle of Grain (W. Kent) in 1999.

European Southern-temperate element.

References: Atlas (297a), Meusel & Jäger (1992), Wigginton (1999).

F. H. PERRING

Cicerbita alpina Alpine Blue-sow-thistle

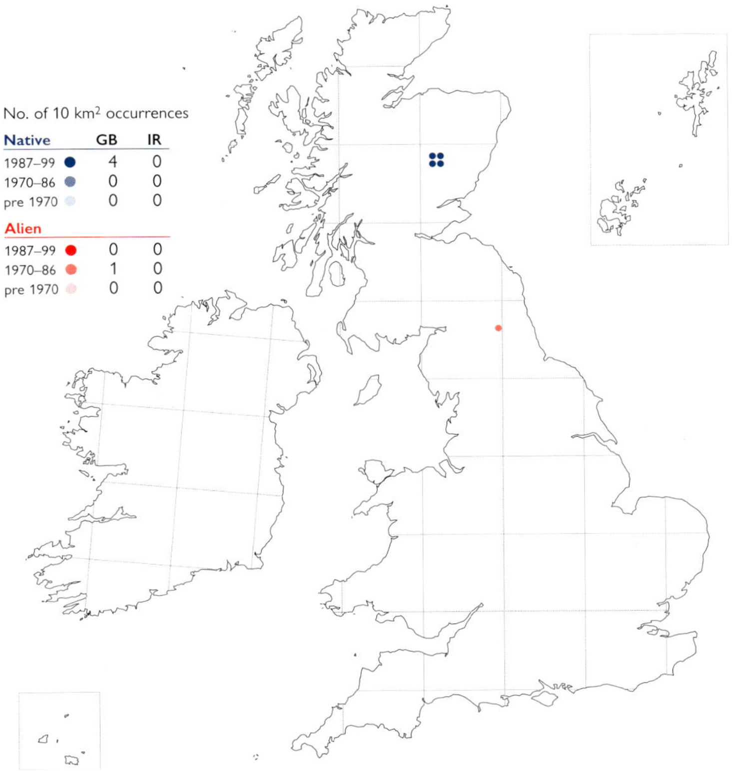

A tall perennial of ledges inaccessible to grazing animals on moist, predominantly N.-facing acidic rocks, often where there is late snow-lie. From 700 m in Glen Doll (Angus) to 1090 m on Lochnagar (S. Aberdeen), but formerly at 530 m in Glen Canness (Angus).

Native. This is one of our rarest mountain plants and only four sites are extant. The perennial threat is from trampling or grazing by deer and other animals. It is an outbreeding species with several colonies perhaps comprising single clones with poor seed production.

European Boreal-montane element; a woodland and sub-montane meadow species across much of its European range.

References: Atlas (298c), Hultén & Fries (1986), Marren *et al.* (1986), Meusel & Jäger (1992), Wigginton (1999).

F. H. PERRING

Cicerbita macrophylla Common Blue-sow-thistle

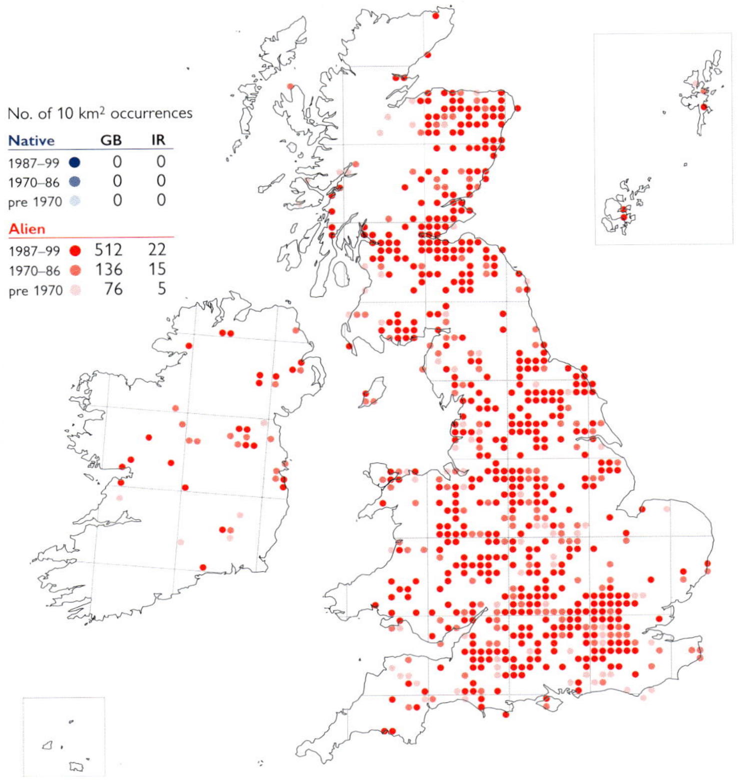

A tall perennial herb which has escaped from gardens and become established on roadsides, pond margins and river banks where, with its long rhizomes, it forms large clonal patches. Generally lowland, but reaching 320 m at Tomintoul (Banffs.).

Neophyte. This species was introduced into British gardens in 1823, and was first recorded as naturalised by a roadside at Glenridding, Ullswater (Cumberland) in 1915. It is still spreading in Britain and Ireland.

Our plant is subsp. *uralensis*, from the Urals; subsp. *macrophylla* is found in the Caucasus.

References: Meusel & Jäger (1992), Sell (1986).

F. H. PERRING

Mycelis muralis Wall Lettuce

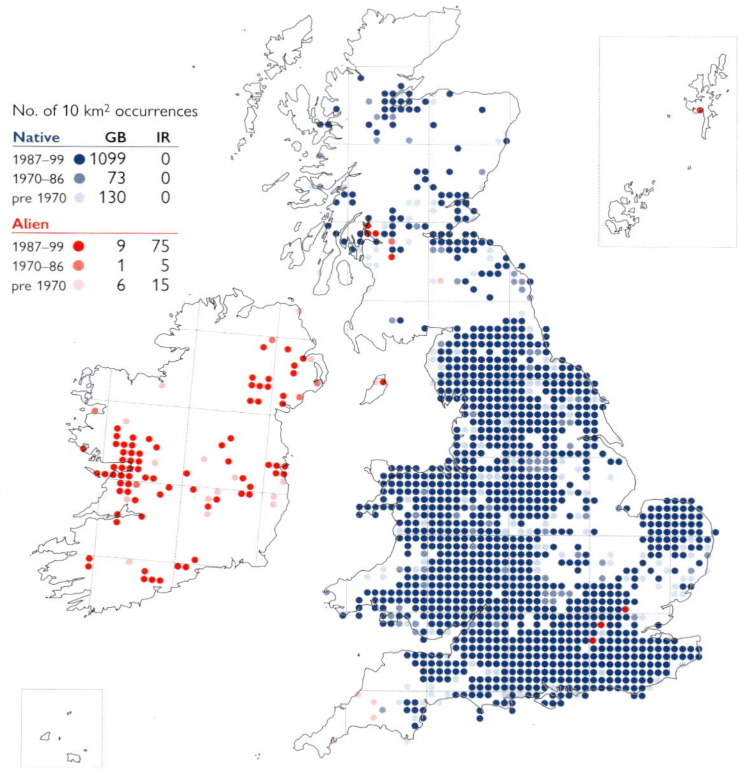

A winter-green perennial herb of shaded walls, rock outcrops and hedge banks, and in woodland, wood margins and scrub, especially on chalk and limestone but also on acidic rocks in some areas. In the Burren (Co. Clare) it occurs on open limestone pavement. 0–500 m (Alston Moor, Cumberland).

Native (change +0.01). The distribution of *M. muralis* has remained largely unchanged since the 1962 *Atlas*, though there are now more records from Scotland and Ireland. It is almost certainly an introduction in Ireland, where it is now frequent in semi-natural habitats in the Burren, where it was not recorded until 1939, and southern Co. Down.

European Temperate element.

References: Atlas (297b), Clabby & Osborne (1999), Grime *et al.* (1988), Hultén & Fries (1986), Meusel & Jäger (1992), Webb & Scannell (1983).

F. H. PERRING

Taraxacum agg. Dandelions

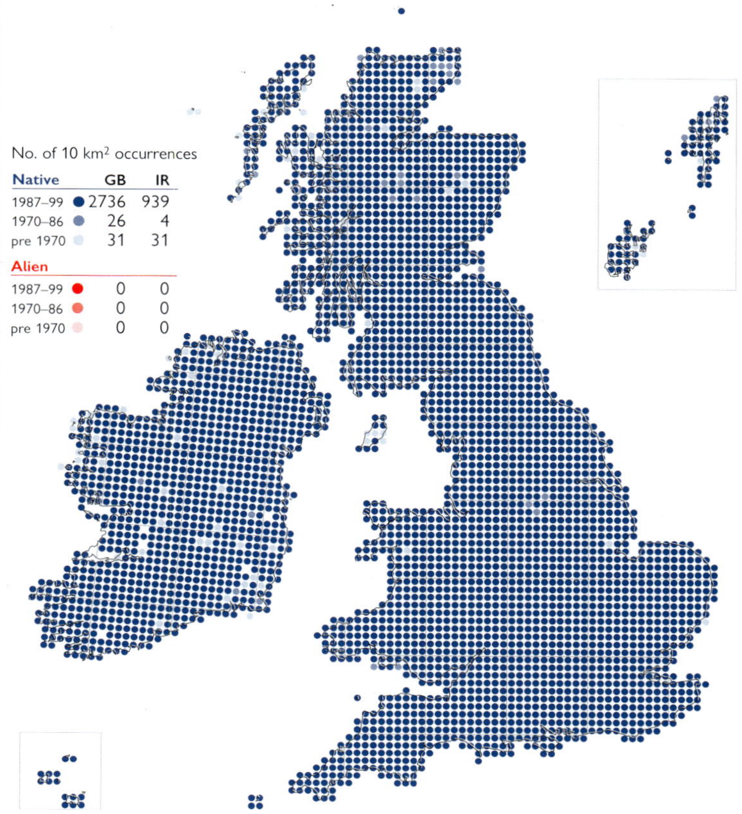

These tap-rooted perennial herbs occur in a very wide range of habitats, but prefer disturbed sites. Some microspecies are associated with natural or semi-natural habitats, including sand dunes, chalk grassland, fens, flushes and cliffs, but most occur in disturbed habitats such as pastures, roadside verges, lawns, tracks, paths and waste ground. 0–1220 m (Scotland).

Native (change +0.43). The distribution of *Taraxacum* agg. is stable. This taxonomically difficult genus comprises 229 apomictic microspecies in our area, of which over 40 are probably endemic and about 100 are alien. The introduced taxa are rarely recorded as such, and all records are mapped here as native.

Circumpolar Wide-temperate element.

References: Atlas (300c), Dudman & Richards (1997), Grime *et al.* (1988), Hultén & Fries (1986), Meusel & Jäger (1992).

T. D. DINES

Crepis paludosa Marsh Hawk's-beard

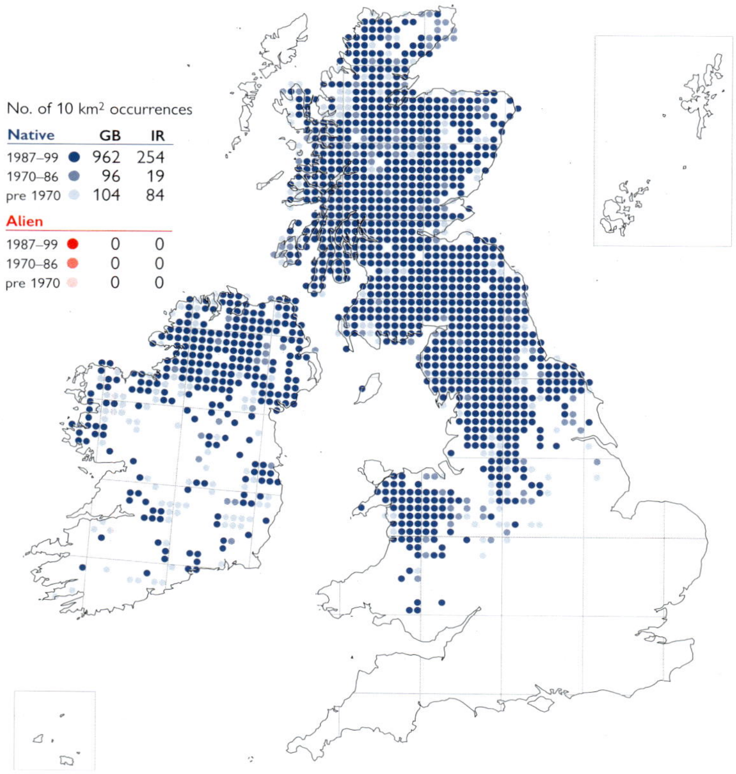

In the uplands, this perennial herb is found on rocky, wooded streamsides, where it often grows on water-splashed rocks, and also in sheltered gullies and flushed banks. At lower altitudes it occurs in fens, damp hay meadows, ditches and on roadside verges. Lowland to 915 m (Easterness).

Native (change –0.27). There has been no overall change in the distribution of *C. paludosa* since the 1962 *Atlas*, but it has declined at the southern edge of its range in both England and Ireland following habitat destruction and the drainage of lowland marshes.

European Boreo-temperate element.

References: Atlas (300b), Hultén & Fries (1986), Meusel & Jäger (1992).

F. H. PERRING

Crepis mollis Northern Hawk's-beard

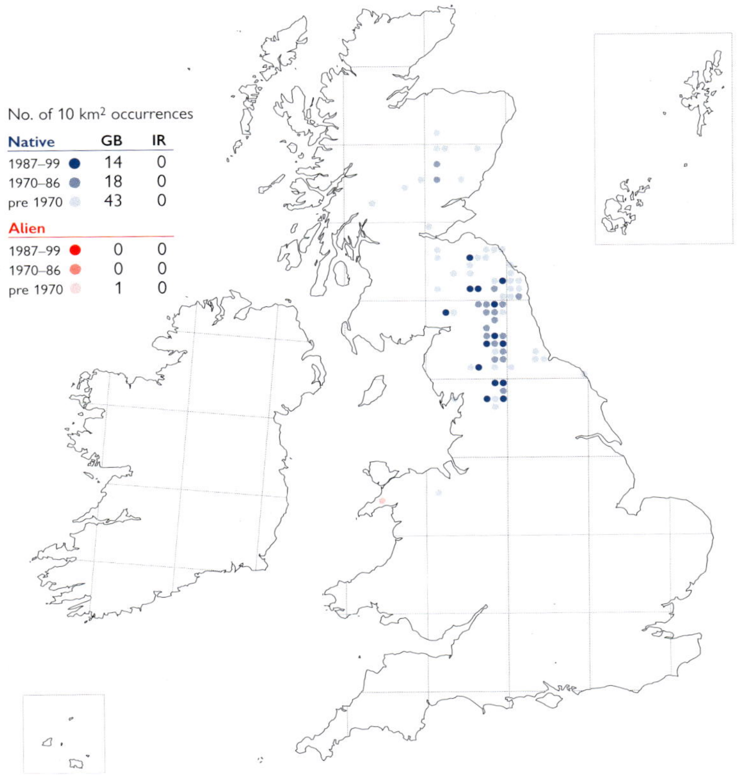

A winter-green perennial of herb-rich grassland and wood-pasture on shallow base-rich soils. From 90 m beside the Swarland Burn (Cheviot) to 670 m in Caenlochan (Angus).

Native (change –1.20). This species has declined and is no longer present in many of its late 19th century localities. It has not been seen since 1986 in over half the 10-km squares for which there are post-1970 records, indicating a substantial decline in the last thirty years. Conversely, however, it was under-recorded in the past and has been discovered in new localities in N. England and S. Scotland since 1960.

European Temperate element, with a continental distribution in W. Europe.

References: Atlas (299c), Meusel & Jäger (1992), Stewart *et al.* (1994).

F. H. PERRING

Crepis biennis Rough Hawk's-beard

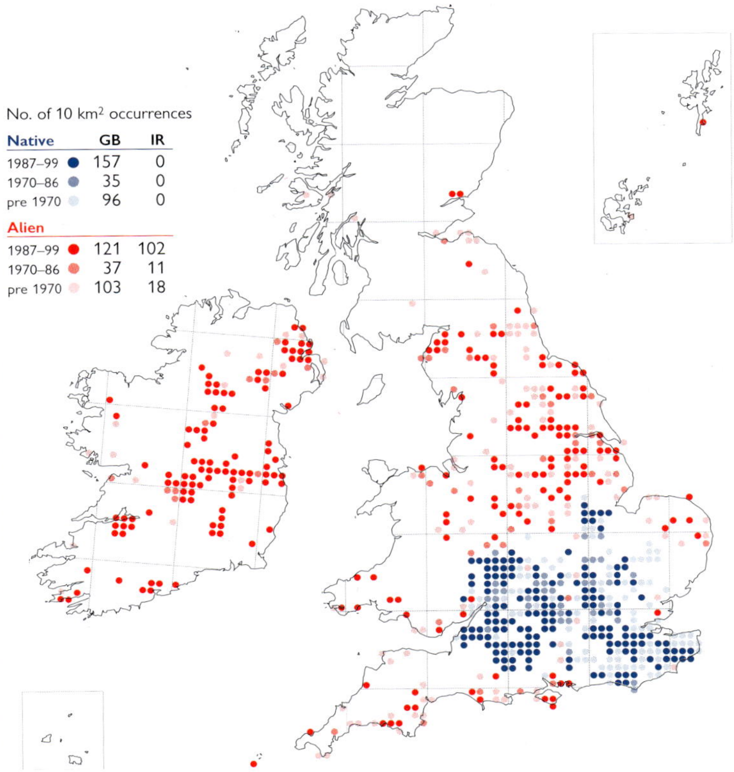

A stout, biennial herb of rough grassland and woodland margins on chalk soils in S.E. England, but elsewhere introduced, often with grass seed, and persisting locally in pastures, in arable fields and on field margins, roadsides, dry banks and waste ground. Lowland.

Native (change –0.02). Since the 1962 *Atlas C. biennis* has consolidated its native range, and spread as an alien elsewhere, especially in Ireland; it was first recorded in Shetland in 1999. In N. and W. Britain it is usually only casual. There have, however, been losses from native sites in Kent and Sussex since the 1962 *Atlas*, the reasons for which are unclear.

European Temperate element.

References: Atlas (299d), Hultén & Fries (1986), Meusel & Jäger (1992).

F. H. PERRING

Crepis capillaris Smooth Hawk's-beard

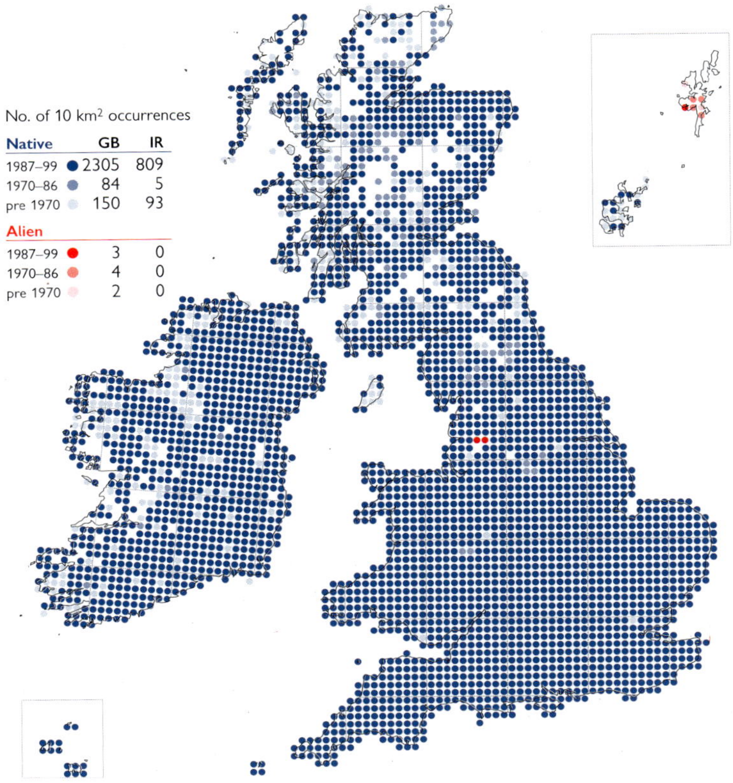

This morphologically variable winter-green annual is an early colonist of open ground, found in thin grassland, on road verges, lawns, spoil heaps, rocky banks and other open habitats. It is a poor competitor in closed vegetation. 0–445 m (Clun Forest, Salop).

Native (change –0.17). The distribution of *C. capillaris* appears to be stable, but it may be increasing locally in man-made habitats.

European Temperate element; widely naturalised outside its native range.

References: Atlas (300a), Grime *et al.* (1988), Hultén & Fries (1986), Meusel & Jäger (1992).

F. H. PERRING

Crepis vesicaria Beaked Hawk's-beard

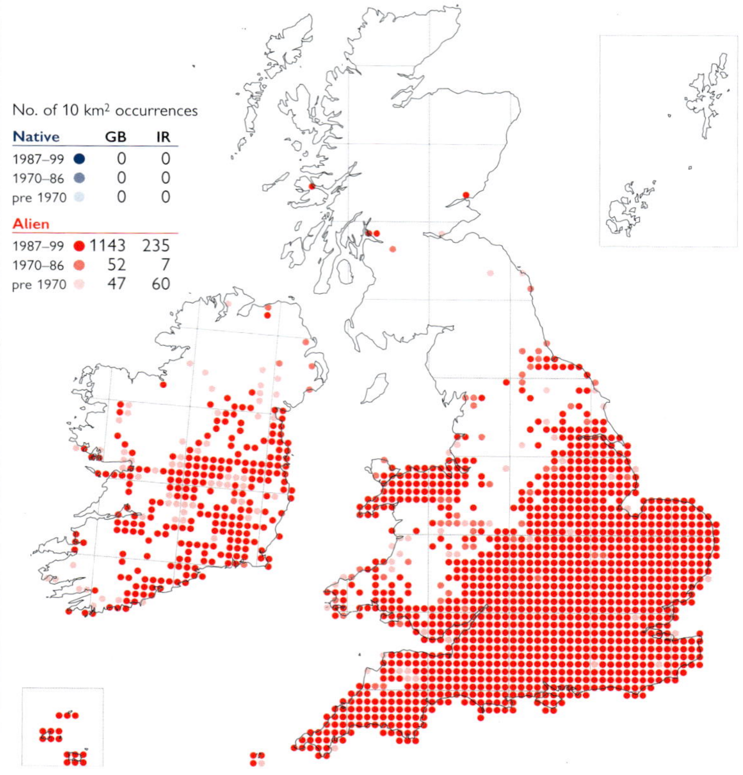

A usually biennial herb, sometimes annual or perennial, of lightly mown or grazed grassland on roadsides, lawns, railway banks and in waste places. Lowland.

Neophyte (change +0.60). In Britain, this species was first recorded in 1713 in Kent. It spread rapidly, reaching the west coast of Ireland in 1896. It is now the commonest yellow composite in flower on roadsides in S. Britain and S. Ireland in May. It has failed to spread far into N. England, and the first authentic record from Co. Durham was not made until 1951. There are many more records in W. England, Wales and Ireland than in the 1962 *Atlas*.

Native of the Mediterranean region and S.W. Asia.

References: Atlas (299b), Meusel & Jäger (1992).

F. H. PERRING

Crepis setosa Bristly Hawk's-beard

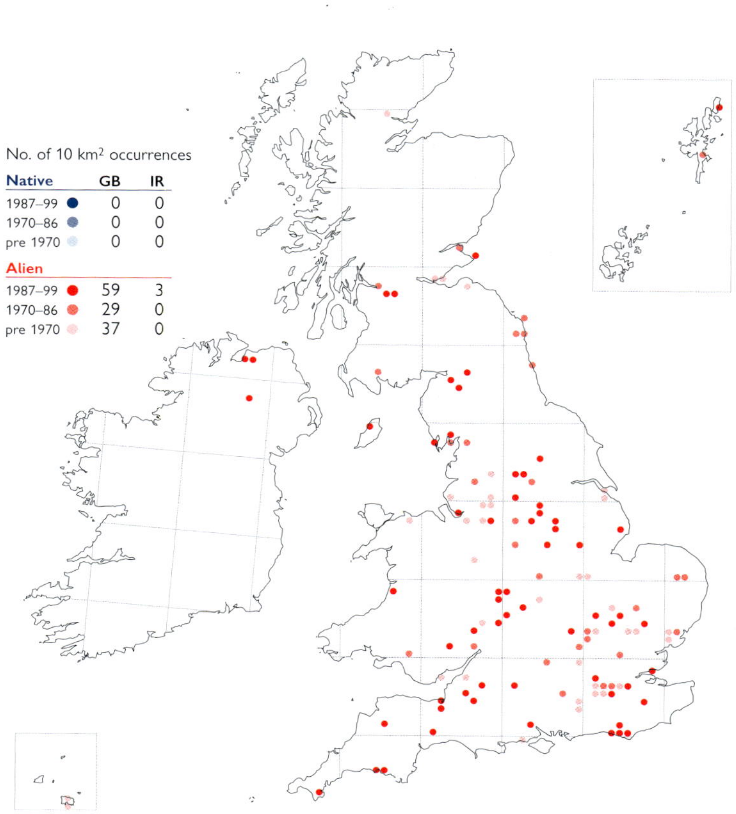

An annual which is most frequent in newly sown grass-clover leys, but is also found in arable fields and on waste ground; it sometimes persists but is usually casual. Lowland.

Casual. This species was first recorded from N. Essex in 1843, and from N. Yorkshire in 1857. It is probably more frequent now than formerly.

C. setosa has a European Southern-temperate distribution.

Reference: Bolòs & Vigo (1995).

F. H. PERRING

Crepis foetida Stinking Hawk's-beard

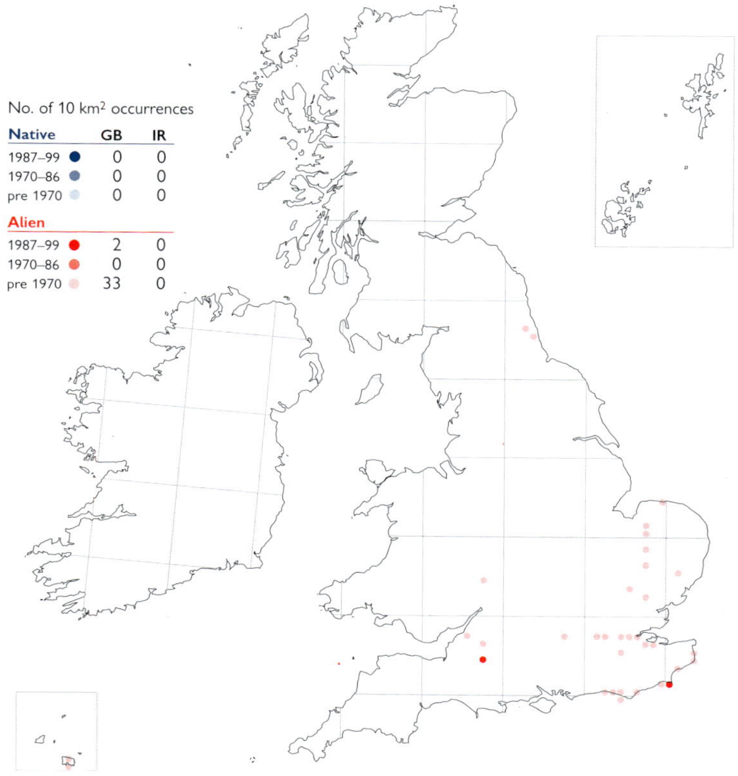

An annual or biennial herb now found only on disturbed coastal shingle, but formerly also in open sandy or chalky habitats inland. Lowland.

Archaeophyte. *C. foetida* is now only known as an established population from Dungeness (E. Kent). It became extinct there in 1980, but following its re-introduction in 1992 a new population has been established in a shingle-heath community. The reasons for its extinction are unclear, though biotic factors may have been significant as rabbits seem partial to it. The Dungeness plant is subsp. *foetida*; the rare casual records may be referable to the C. and S.E. European subsp. *commutata* or to subsp. *rhoeadifolia*.

As an archaeophyte *C. foetida* has a Eurosiberian Southern-temperate distribution.

References: Atlas (299a), Meusel & Jäger (1992), Wigginton (1999).

F. H. PERRING

Crepis praemorsa Leafless Hawk's-beard

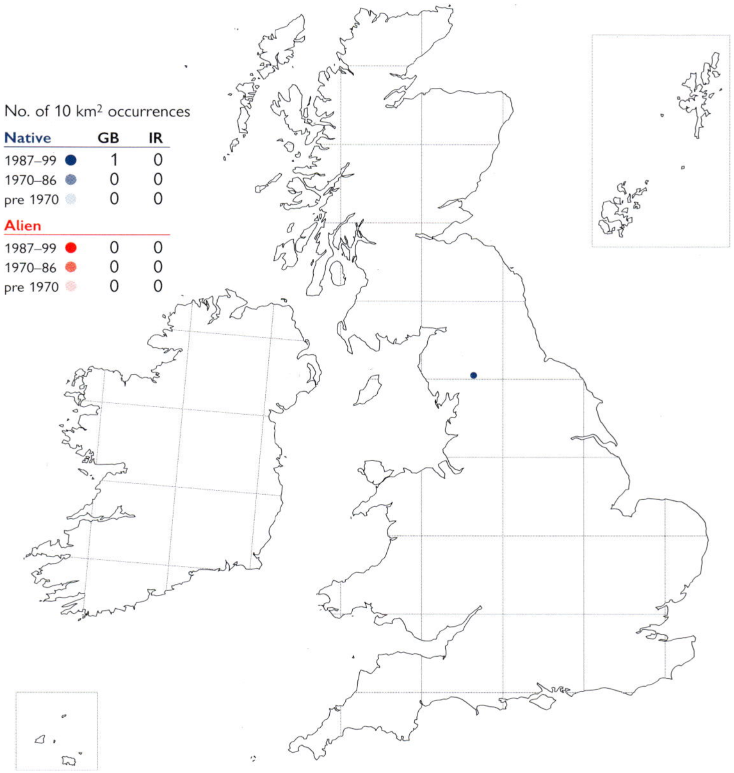

No. of 10 km² occurrences

Native	GB	IR
1987–99	1	0
1970–86	0	0
pre 1970	0	0
Alien		
1987–99	0	0
1970–86	0	0
pre 1970	0	0

A perennial herb, confined to low banks of limestone drift at the edges of a hay meadow and grazed pasture, at an altitude of 240 m (Westmorland).

Native. This species was found for the first time in Britain (and W. Europe) in Westmorland in 1988. In 1996, the population comprised about two hundred individuals. Doubt has been expressed about its status, and it is regarded as an archaeophyte in some European countries. However, the similarity of its British and Scandinavian habitats and the undisturbed nature of the site suggest that it may be native in Britain.

Eurosiberian Temperate element, with a continental distribution in W. Europe.

References: Halliday (1997), Hultén & Fries (1986), Meusel & Jäger (1992), Wigginton (1999).

F. H. PERRING

Pilosella peleteriana Shaggy Mouse-ear-hawkweed

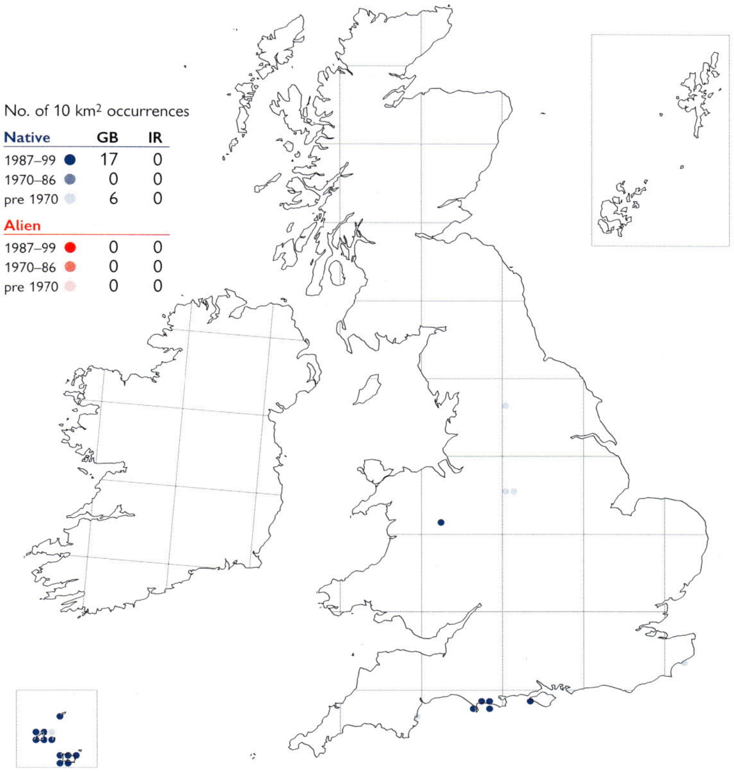

No. of 10 km² occurrences

Native	GB	IR
1987–99	17	0
1970–86	0	0
pre 1970	6	0
Alien		
1987–99	0	0
1970–86	0	0
pre 1970	0	0

A perennial herb, spreading by short, thick stolons, occurring on steep, well-drained slopes in chalk and limestone grassland (especially on the edges of paths, on small terraces and at the edge of sea-cliffs), on dry dolerite rock shelves and quarry waste, on shallow soils overlying granite, and on dunes. Lowland.

Native. *P. peleteriana* was first mapped by Perring & Sell (1968). The current map shows some losses before 1970, but the species is now better recorded in Dorset and the overall distribution is stable. There is no specimen supporting the Merionethshire record mapped in Perring & Sell (1968) and the plant has not been re-found.

Suboceanic Boreo-temperate element.

References: Atlas Supp. (132a), Hultén & Fries (1986), Meusel & Jäger (1992), Wigginton (1999).

F. H. PERRING

Pilosella officinarum Mouse-ear-hawkweed

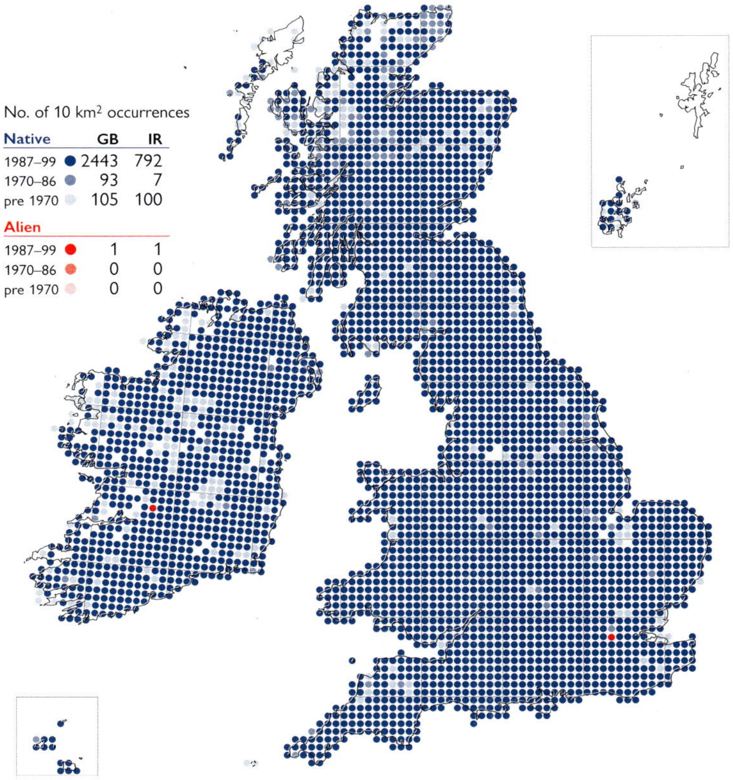

No. of 10 km² occurrences

Native	GB	IR
1987–99	2443	792
1970–86	93	7
pre 1970	105	100
Alien		
1987–99	1	1
1970–86	0	0
pre 1970	0	0

A stoloniferous, perennial herb of dry habitats, including short grassland, heaths, sand dunes, screes, rock outcrops, quarries and cliffs. It grows on both base-rich and acidic substrates. 0–915 m (Ben Macdui, S. Aberdeen).

Native (change –0.59). There has been no discernible change in the distribution of *P. officinarum* since the 1962 *Atlas*. It is variable, and seven subspecies have been recognised. However, all are more or less connected by intermediates, and are not or only partially discrete geographically and ecologically; Stace (1997) considers they are no more than varieties.

European Temperate element; widely naturalised outside its native range.

References: Atlas (298d), Bishop & Davy (1994), Grime *et al.* (1988), Hultén & Fries (1986), Meusel & Jäger (1992), Rich & Jermy (1998).

F. H. PERRING

Pilosella flagellaris subsp. *flagellaris*

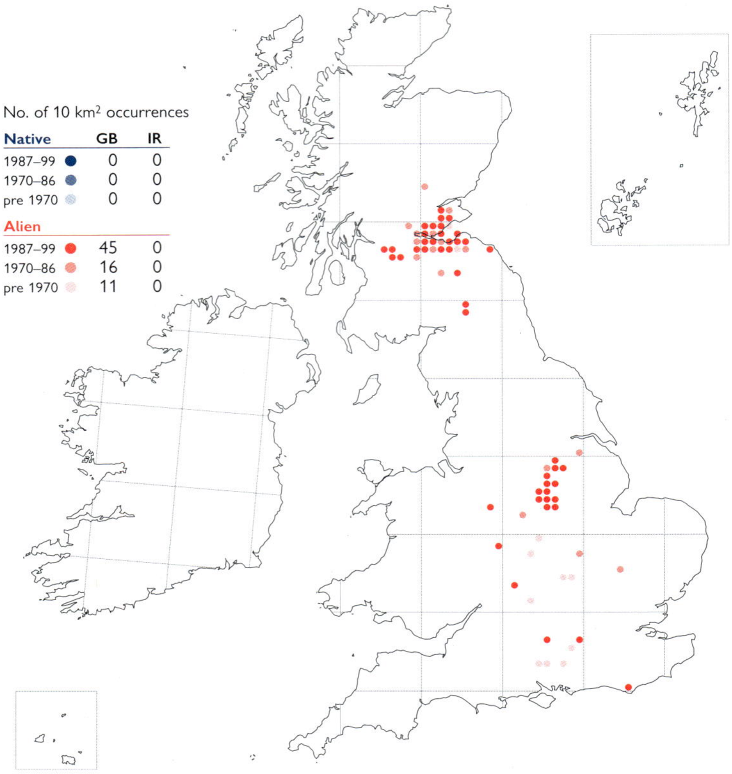

No. of 10 km² occurrences

Native	GB	IR
1987–99 ●	0	0
1970–86 ●	0	0
pre 1970 ●	0	0
Alien		
1987–99 ●	45	0
1970–86 ●	16	0
pre 1970 ●	11	0

A perennial, stoloniferous herb, sometimes grown in gardens and found naturalised on roadsides and railway banks. Reproduction is by seed and vegetative spread. Lowland.

Neophyte. This subspecies has been cultivated in Britain since 1816. It was first recorded in the wild in 1869 on railway banks at Granton, Edinburgh (Midlothian), where it may have escaped from the Botanic Garden. Trends in its distribution are difficult to assess.

P. flagellaris subsp. *flagellaris* has a European Boreo-temperate range, with a continental distribution in W. Europe.

Reference: Atlas Supp. (132b).

F. H. PERRING

Pilosella flagellaris subsp. *bicapitata* Shetland Mouse-ear-hawkweed

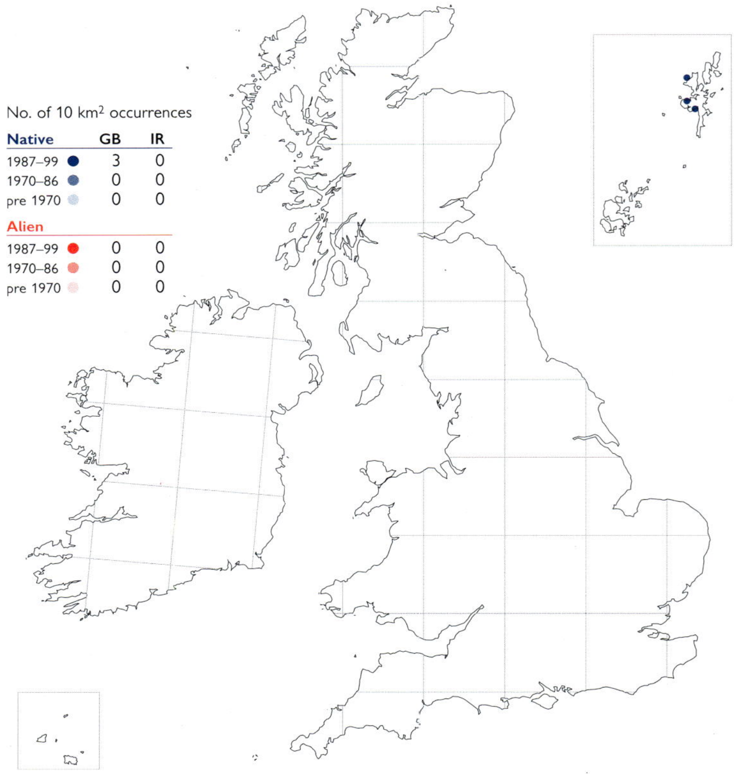

No. of 10 km² occurrences

Native	GB	IR
1987–99 ●	3	0
1970–86 ●	0	0
pre 1970 ●	0	0
Alien		
1987–99 ●	0	0
1970–86 ●	0	0
pre 1970 ●	0	0

An endemic perennial herb of grassy limestone rocky outcrops, heathy granulitic gneiss and feldspathic rocky sea-banks in three localities in Shetland, where it was first described in 1962. Lowland.

Native. There has been no change in the distribution of this subspecies at the 10-km scale. However, two of its three localities are accessible to sheep, and the increased grazing in recent years has suppressed flowering and might have reduced the populations. The occurrence of this endemic subspecies in Shetland at such a distance from native populations of subsp. *flagellaris* in C. & S. Europe presents a fascinating problem in plant distribution.

Endemic.

References: Atlas Supp. (132b), Scott (1968), Wigginton (1999).

F. H. PERRING

Pilosella aurantiaca Fox-and-cubs

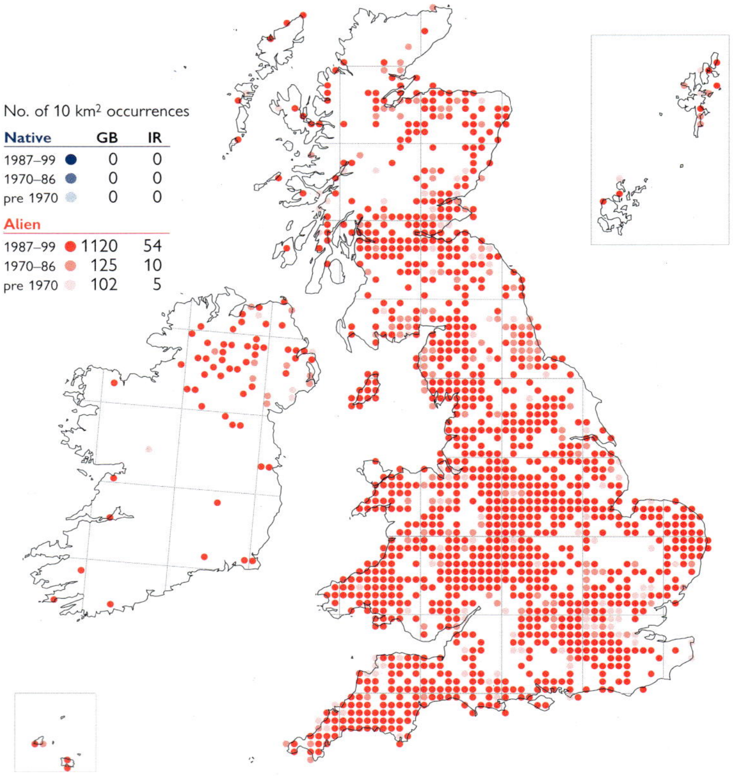

No. of 10 km² occurrences

Native	GB	IR
1987–99 ●	0	0
1970–86 ●	0	0
pre 1970 ●	0	0
Alien		
1987–99 ●	1120	54
1970–86 ●	125	10
pre 1970 ●	102	5

A stoloniferous or rhizomatous perennial herb, which has escaped from gardens and become naturalised on railway and roadside banks, on walls and in churchyards and other grassy and waste places. 0–445 m (Nenthead, Cumberland).

Neophyte. *P. aurantiaca* was grown in gardens by 1629 and recorded from the wild by 1793. Two subspecies are recognised in our area, subsp. *aurantiaca* and subsp. *carpathicola*. Although expertly determined records of these were mapped by Perring & Sell (1968), they have not been comprehensively recorded since then and are not mapped separately here.

P. aurantiaca has a European Boreal-montane distribution; it is naturalised outside this range in Europe, N. America and elsewhere.

References: Atlas Supp. (134a, b), Hultén & Fries (1986), Meusel & Jäger (1992).

F. H. PERRING

Hieracium agg. Hawkweeds

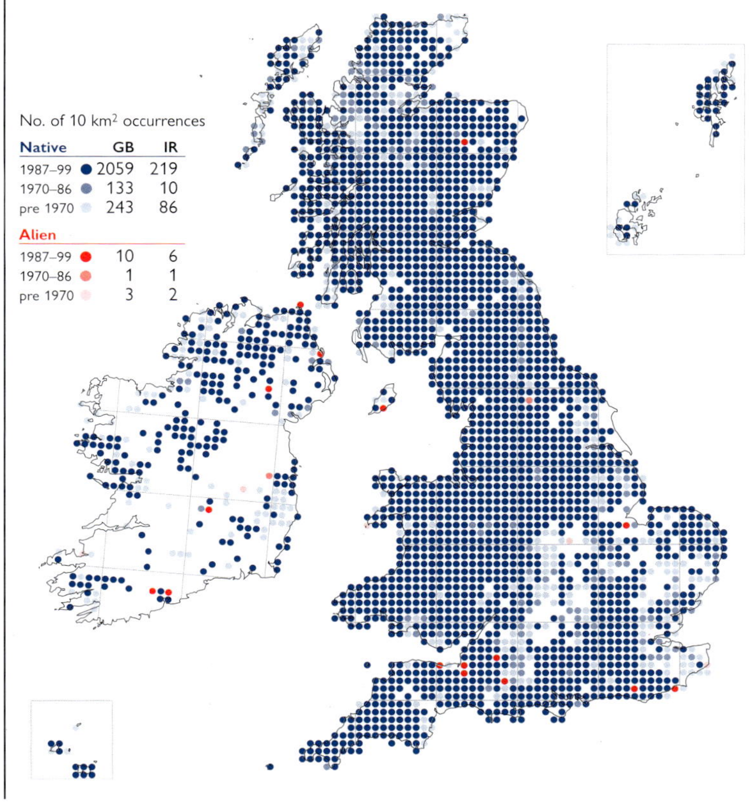

A group of tap-rooted perennial herbs growing in a wide range of habitats, but with a preference for infertile, rocky substrates. Habitats include rocky grassland, limestone pavement, cliffs, upland rock ledges, rocky river banks, roadsides, railway banks, walls and quarries. In grazed areas plants are often restricted to inaccessible sites. 0–1220 m (Cairngorms, S. Aberdeen).

Native. *Hieracium* agg. currently includes 262 microspecies in our area, and reliable records of 221 were mapped by Perring & Sell (1968). The distribution of the aggregate is probably stable. The map of *Hieracium pilosella sens. lat.* in the 1962 *Atlas* comprises taxa now placed in *Pilosella.*

Circumpolar Wide-boreal element.

References: Atlas Supp. (75–131), Bolòs & Vigo (1995), Grime *et al.* (1988), Hultén & Fries (1986), Meusel & Jäger (1992).

T. D. DINES

Filago vulgaris Common Cudweed

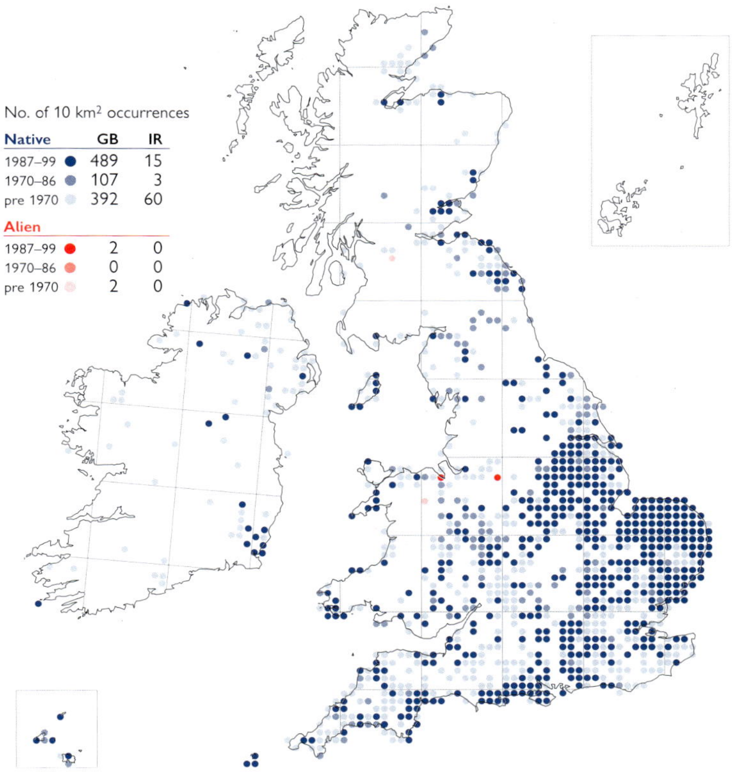

An autumn- or spring-germinating annual of dry, open, acidic to neutral and occasionally calcareous habitats including open grassland, quarries and rocky ledges, sand-pits and dunes, sandy heaths and tracks, and arable and other cultivated ground. Lowland.

Native (change –1.20). Populations of *F. vulgaris* may fluctuate in size in response to climatic conditions and disturbance, but overall there has been a progressive decline, very marked in S. and W. England and Ireland, largely as a result of changing agricultural practices and the cultivation of marginal land.

European Southern-temperate element; also recorded in C. Asia. It reaches its northern limit in Scotland and Jutland.

References: Atlas (277a), Hultén & Fries (1986), Meusel & Jäger (1992).

G. HALLIDAY

Filago lutescens Red-tipped Cudweed

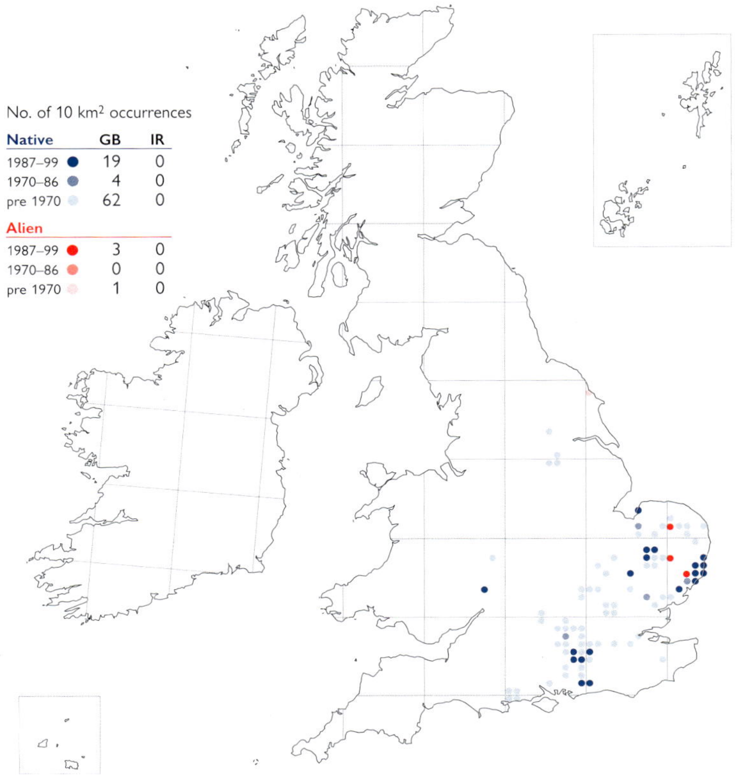

A winter- or spring- annual of dry, open, sandy or gravelly acidic to neutral soils such as the edges of arable fields, tracks, sand-pits, heaths and commons, and particularly characteristic of rabbit scrapes. Populations can vary greatly in size annually. Lowland.

Native or alien (change –0.34). First recorded in 1846, there was a dramatic decline in *F. lutescens* after the 1950s, attributable to habitat loss, the reduction of the rabbit population and changing agricultural practices. Since 1993 there has been a sharp increase in the number of records due to special surveys and conservation management.

As a possible native, *F. lutescens* has a European Temperate distribution.

References: Atlas (277b), Hultén & Fries (1986), Meusel & Jäger (1992), Rich (1999a), Wigginton (1999).

G. HALLIDAY

Filago pyramidata Broad-leaved Cudweed

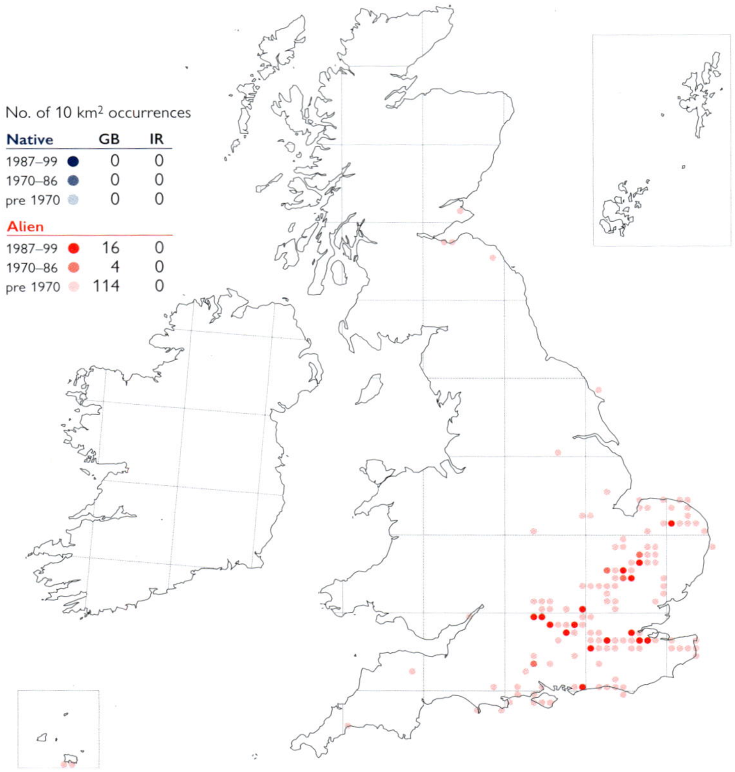

An annual of well-drained soils usually kept open through drought or disturbance. Formerly, it was most frequent as a weed of arable land on calcareous or acidic sandy soils, but most remaining sites are in chalk quarries or on chalk spoil. Populations vary greatly in size annually. Lowland.

Archaeophyte (change −1.14). This species has suffered a major decline which is most probably attributable to changing agricultural practices, such as autumn cereal growing and the increasing use of herbicides.

As an archaeophyte *F. pyramidata* has a Submediterranean-Subatlantic distribution. It reaches its northern limit in England, and is declining in the eastern part of its European range.

References: Atlas (277c), Meusel & Jäger (1992), Rich (1999c), Wigginton (1999).

G. HALLIDAY

Filago minima Small Cudweed

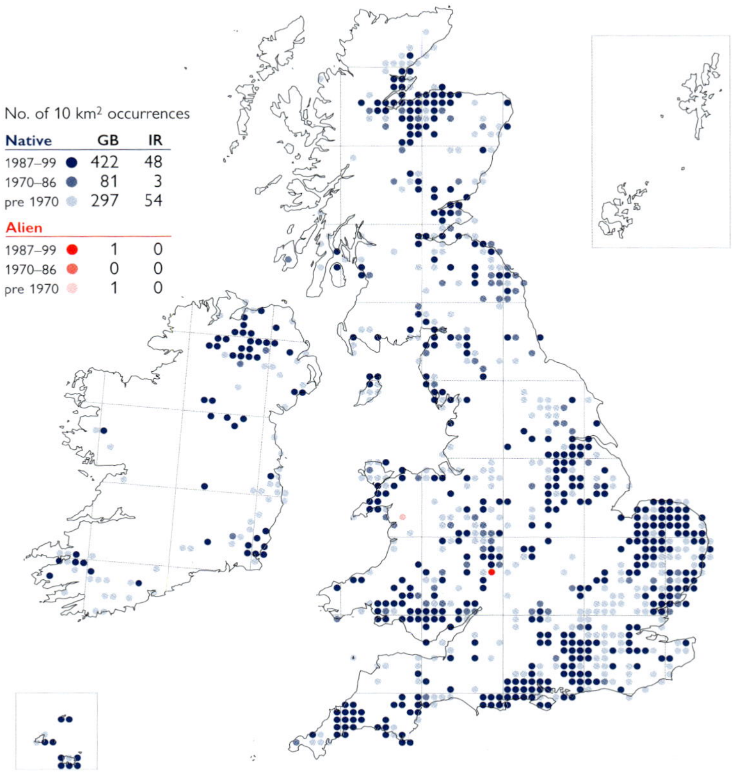

An annual of dry, open, infertile, acidic to neutral soils in a wide range of habitats, including arable fields, open grassland, quarries and mine spoil, woodland tracks, sandy heaths, sand-pits and dunes. 0–365 m (Rannoch, Mid Perth).

Native (change −0.91). *F. minima* is nowhere particularly frequent, and like all species of *Filago*, is in continuing slow decline across much of its British and Irish range.

European Temperate element; it reaches its northern limit in Scotland.

References: Atlas (278a), Curtis & McGough (1988), Hultén & Fries (1986), Meusel & Jäger (1992).

G. HALLIDAY

Filago gallica Narrow-leaved Cudweed

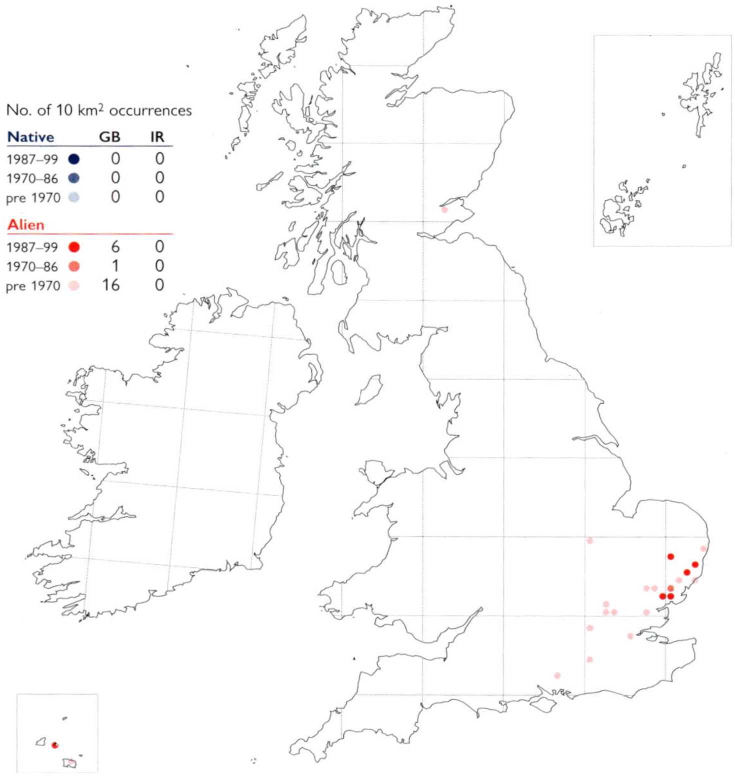

An annual of well-drained, sandy and gravelly soils in open, disturbed sites such as arable field margins, grassy banks, gravel-pits and quarries, tracks and roadsides. Lowland.

Archaeophyte (change +0.01). *F. gallica*, first recorded in Britain in 1696, was known from at least thirty sites in S.E. England. In the late 19th and 20th centuries changing agricultural practices and reduced rabbit disturbance led to its extinction in 1955. It was re-introduced to its last known site in Essex in 1994 using local seed, and has been deliberately planted in Suffolk. It survives on Sark (Channel Islands).

F. gallica is native of the Mediterranean region; it is probably introduced in the northern part of its Submediterranean-Subatlantic range.

References: Atlas (277d), Meusel & Jäger (1992), Rich, Gibson & Marsden (1999), Wigginton (1999).

G. HALLIDAY

Antennaria dioica Mountain Everlasting

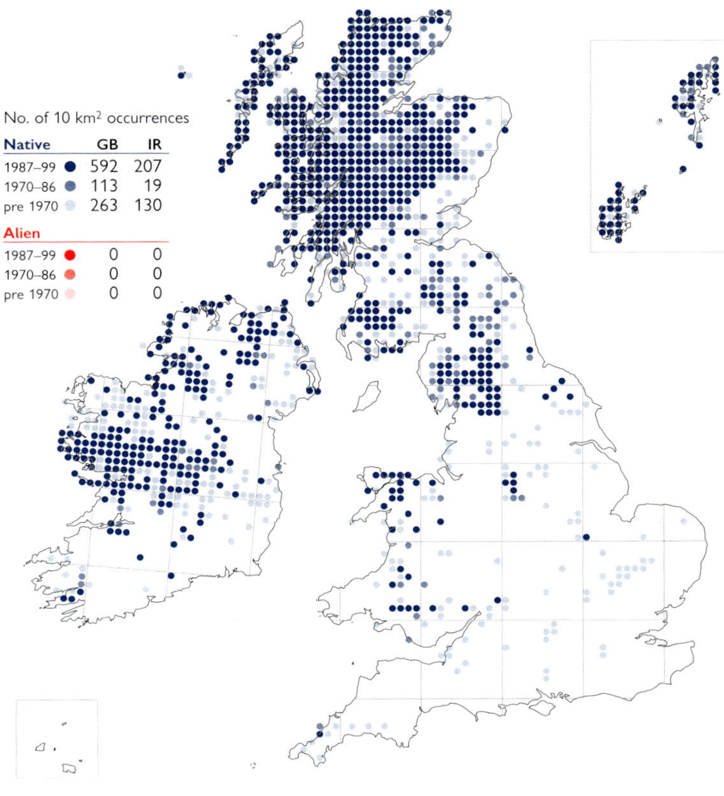

A shortly stoloniferous perennial herb of thin, basic to mildly acidic soils. Its lowland habitats include chalk and limestone grassland, heathland, coastal cliff-tops, sand dunes and machair. In upland areas, habitats include rock ledges, crags, streamsides, screes, well-drained acidic grasslands, heathy pastures and dwarf-shrub heaths. 0–885 m (Macgillycuddy's Reeks, S. Kerry).

Native (change –0.88). *A. dioica* declined markedly in the century before the 1962 *Atlas*, and there have been further losses since then. Most losses are due to the ploughing of its habitats, or to intensification of grassland management. It is now a mostly coastal and upland species in Britain.

Eurasian Boreo-temperate element.

References: Atlas (279d), Hultén & Fries (1986), Meusel & Jäger (1992).

G. HALLIDAY

Anaphalis margaritacea Pearly Everlasting

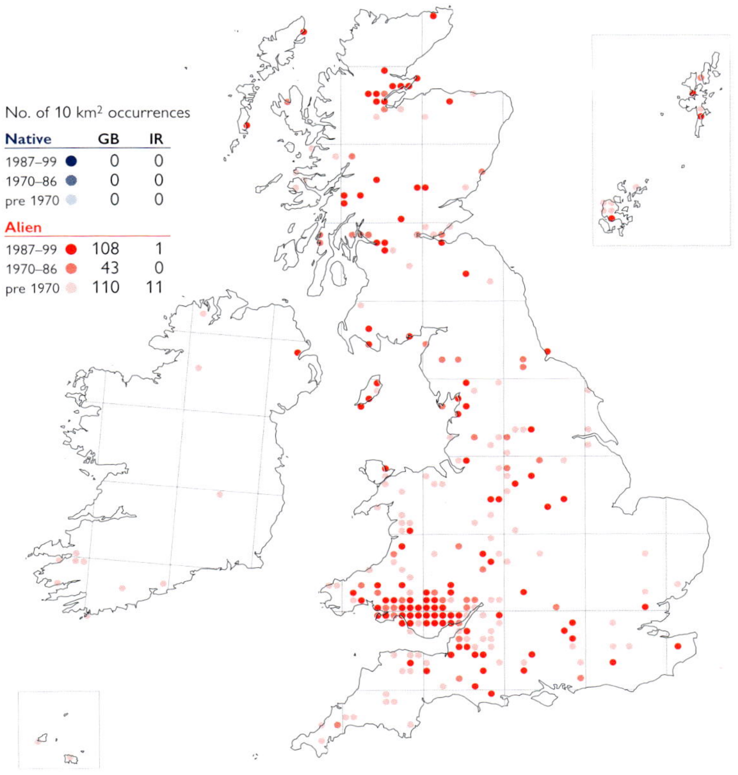

A rhizomatous perennial herb established on waste ground and short grassland, especially roadside verges, lane- and railway-sides, coal mine slag-heaps and in open areas in woodland. Generally lowland, but reaching 410 m at Eisteddfa Gurig (Cards.).

Neophyte (change +0.07). *A. margaritacea* was being grown in British gardens by 1596. It was first recorded in the wild before 1709, when it was already well-naturalised along the Rhymney River, Glamorgan. There has been little change in overall distribution since the 1962 *Atlas*, although there are more additional scattered records and it has consolidated its range in S. Wales.

Native of N. America and N.E. Asia; widely naturalised in N. & C. Europe.

References: Atlas (279c), Wade *et al.* (1994).

G. HALLIDAY

Gnaphalium norvegicum Highland Cudweed

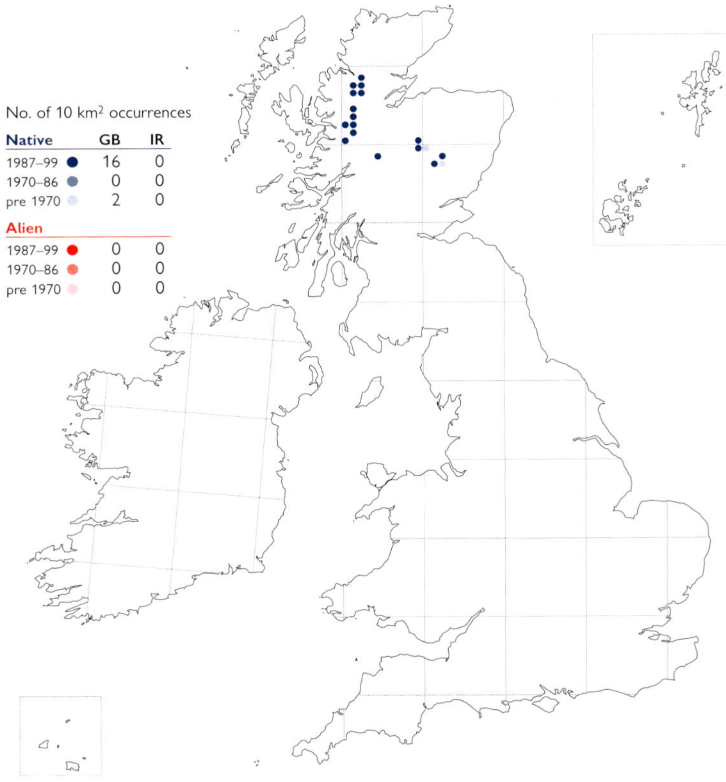

This perennial herb occurs on ungrazed rock ledges, crags, river gorges, screes and in gullies, preferring a southerly or easterly aspect and an acidic, well-drained mineral soil. From 600 m (Aonach air Chrith, W. Ross) to 980 m (Sgurr na Lapaich, Easterness).

Native (change +0.58). Although there is little evidence of a significant decline in the number of sites (currently about thirty), populations of *G. norvegicum* are small and vulnerable to rock falls and avalanches, though new habitat is also created this way. There are now twice as many records as in the 1962 *Atlas*, largely due to a special survey carried out in the 1990s.

European Arctic-montane element; also in C. Asia and N. America.

References: Atlas (278c), Hultén & Fries (1986), Meusel & Jäger (1992), Wigginton (1999).

G. HALLIDAY

Gnaphalium sylvaticum Heath Cudweed

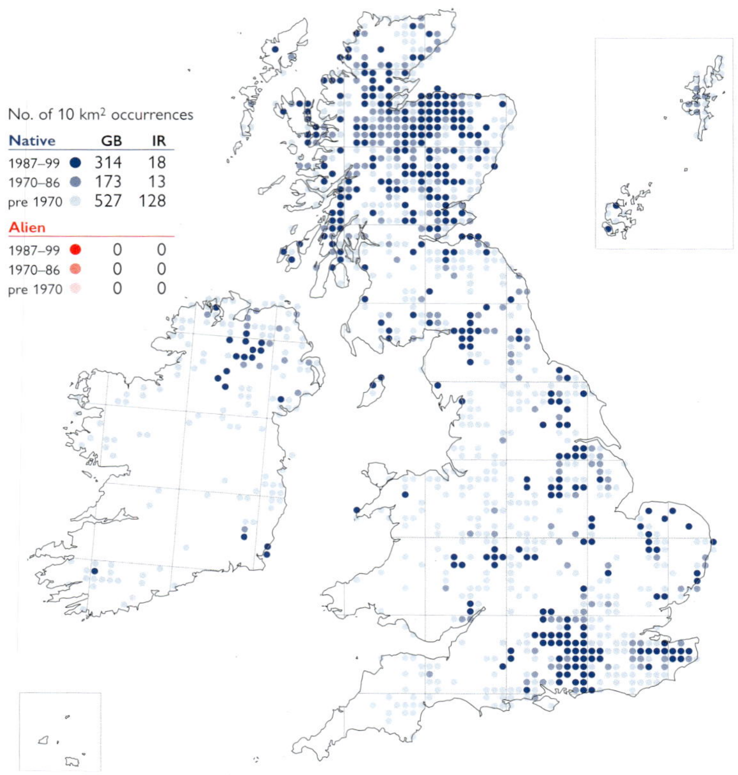

A short-lived perennial herb of open communities on dry, acidic, often sandy or gravelly soils. Habitats include heaths and heathy pastures, sand-pits, dunes, tracks and, especially, open woodland and forestry rides in areas of former heathland. 0–850 m (Breadalbanes, Mid Perth).

Native (change –2.65). There was little indication in the 1962 *Atlas* that *G. sylvaticum* was decreasing outside Ireland, Wales and S.W. England. However, it now appears to be declining throughout its range, despite probably being one of the few native vascular plant species to benefit from the extensive afforestation programmes of the 20th century.

Eurosiberian Boreo-temperate element; also in N. America.

References: Atlas (278b), Curtis & McGough (1988), Hultén & Fries (1986), Meusel & Jäger (1992), Stewart *et al.* (1994).

G. HALLIDAY

Gnaphalium supinum Dwarf Cudweed

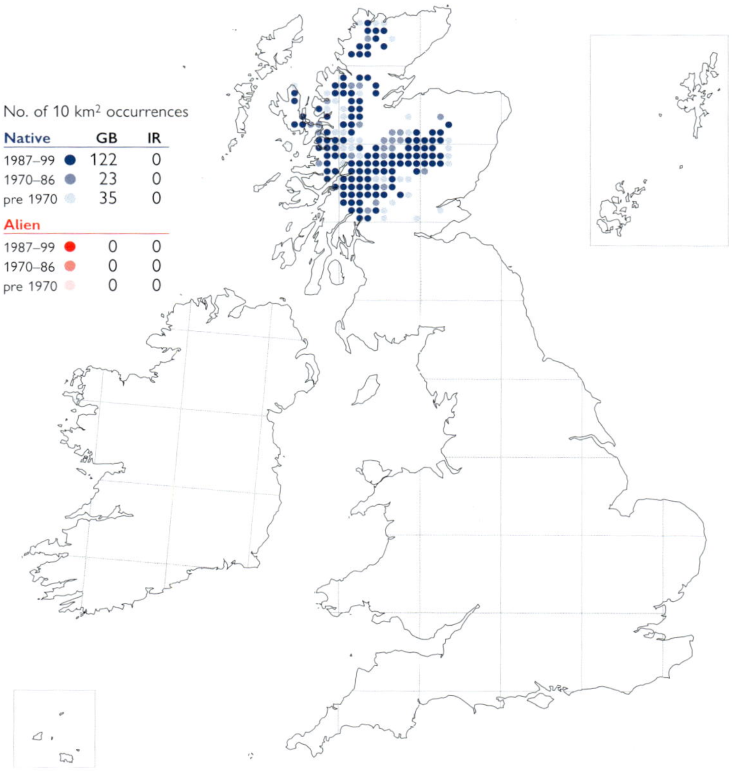

A dwarf, perennial herb, found on mountain-top fell-field communities, wet grassy slopes, cliffs, moraines and late snow-patches, where it grows in sites which are relatively well-drained and stony and dry out in summer. Usually from 455 to 1305 m (Ben Macdui, S. Aberdeen), but descending to 300 m on river gravels at Killin (Mid Perth).

Native (change –0.68). The distribution of *G. supinum* is probably stable. It may still be present in those 10-km squares for which there are only pre-1987 records.

European Arctic-montane element; also in C. Asia and N. America.

References: Atlas (278d), Hultén & Fries (1986), Meusel & Jäger (1992).

G. HALLIDAY

Gnaphalium uliginosum Marsh Cudweed

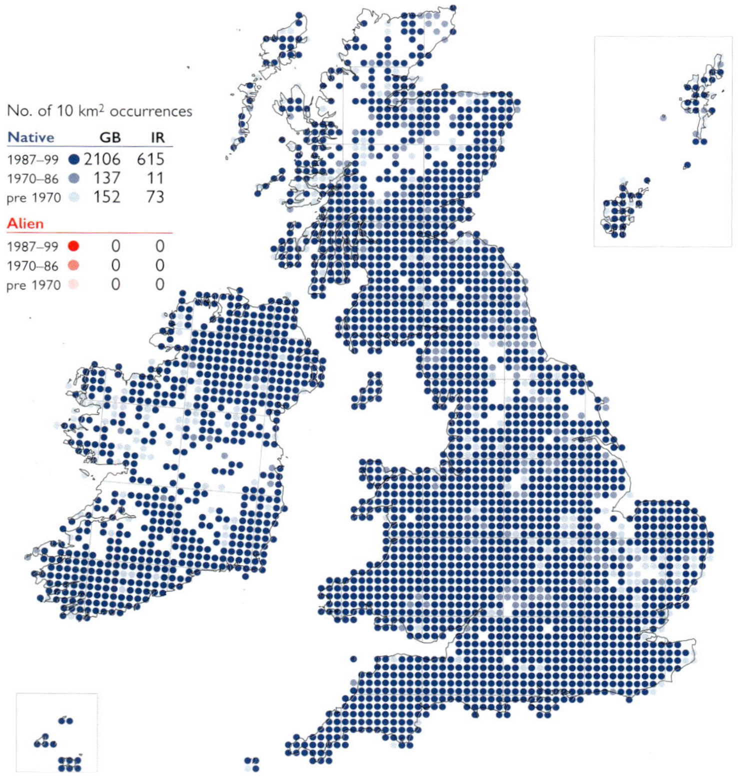

An annual of open, muddy ground, usually subject to waterlogging during the winter. It is characteristic of trampled field entrances, compacted arable and cultivated land, the margins of reservoirs and the edges of summer-dry ponds trampled by cattle. It is also found on rutted tracks on heaths and wet rides in woodland. It prefers mildly to quite strongly acidic soils. 0–470 m (Glen Kyllachy, Easterness).

Native (change +0.80). The distribution of *G. uliginosum* has increased since the 1962 *Atlas*, probably due to its preference for disturbed habitats. The species is also better recorded now.

Eurasian Boreo-temperate element, but naturalised in N. America so distribution is now Circumpolar Boreo-temperate.

References: Atlas (279a), Hultén & Fries (1986), Meusel & Jäger (1992).

G. HALLIDAY

Gnaphalium luteoalbum Jersey Cudweed

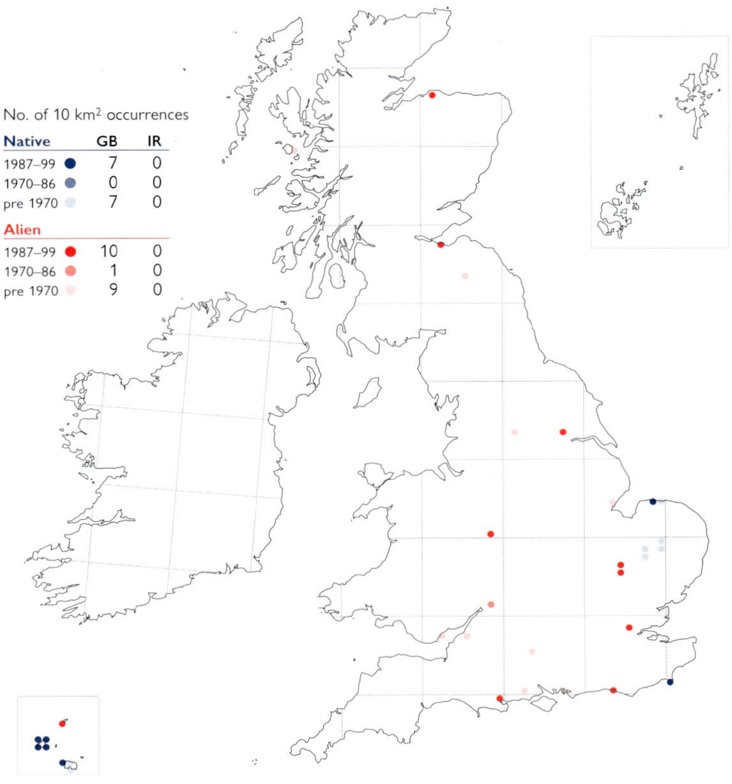

An annual or biennial of sandy fields, dune-slacks and waste ground; now restricted as an apparent native to the Channel Islands, the margins of two recently created pools in Norfolk and an area of excavated shingle in Kent. Recent records elsewhere are mainly of casual plants on waste ground, although it is thriving on tracks and a rubbish tip in Dorset. Lowland.

Native or alien (change +0.23). *G. luteoalbum* was first recorded in 1690 and became extinct at all its presumed native Breckland sites in the early 20th century. It was discovered in E. Kent in 1996.

Eurosiberian Southern-temperate element, reaching its northern limit in S. Sweden. It is widely naturalised outside its native range.

References: Atlas (279b), Hultén & Fries (1986), Meusel & Jäger (1992), Wigginton (1999).

G. HALLIDAY

Inula helenium Elecampane

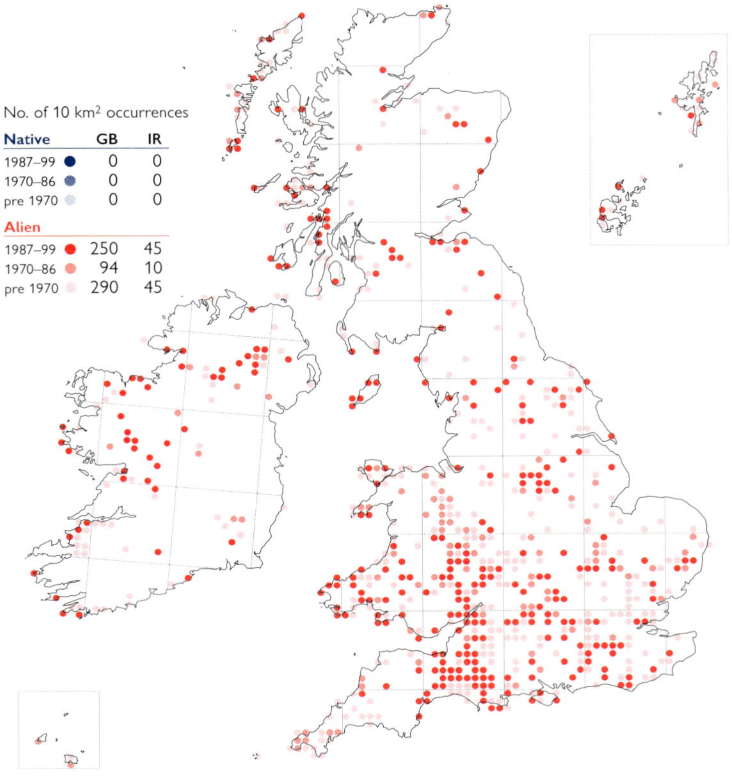

A rather robust, conspicuous and persistent perennial herb, widely if sparsely established from garden outcasts on road- and lane-sides and by woodland margins, but seldom far from habitation. Lowland.

Archaeophyte (change –0.80). *I. helenium* has been grown in gardens for its medicinal and ornamental value since at least 995 (Harvey, 1981). It is very persistent but it may be in gradual decline.

Native of W. & C. Asia; widely naturalised in temperate Europe and elsewhere.

References: Atlas (275d), Meusel & Jäger (1992).

G. HALLIDAY

Inula salicina Irish Fleabane

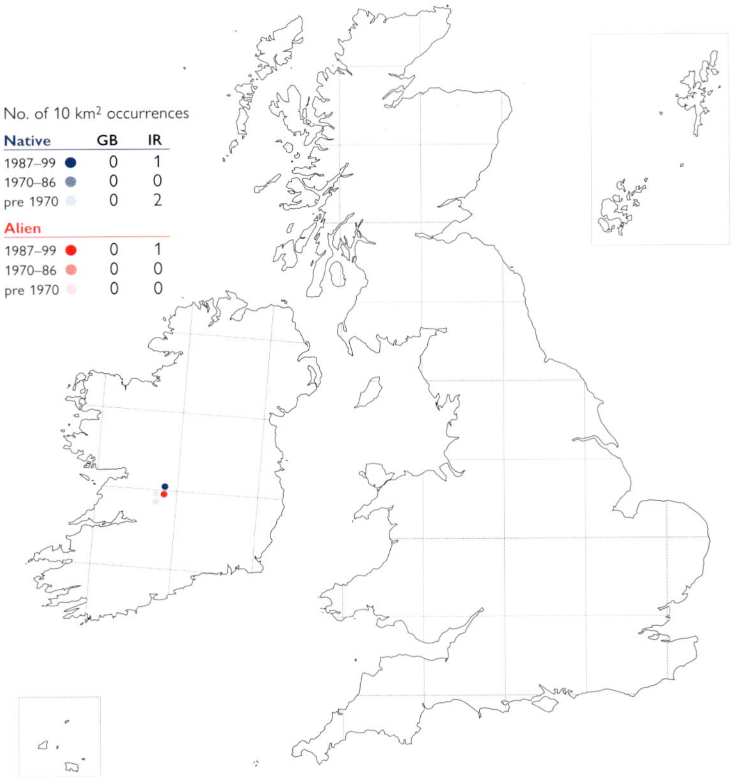

This perennial herb is known only from the northern half of Lough Derg, where it has been recorded along the limestone shoreline and on the islands, occupying an intermediate, stony habitat between the flood level and the surrounding scrub. Lowland.

Native. Populations of *I. salicina* have progressively declined. It disappeared from the S.E. Galway side of Lough Derg in the 1960s and there is now only one population in N. Tipperary. Its decline has been tentatively attributed to increased eutrophication of the lake and lake shore. It has been transplanted to one former native site.

Eurasian Temperate element, with a continental distribution in W. Europe.

References: Atlas (276a), Curtis & McGough (1988), Hultén & Fries (1986), Meusel & Jäger (1992).

G. HALLIDAY

Inula conyzae Ploughman's-spikenard

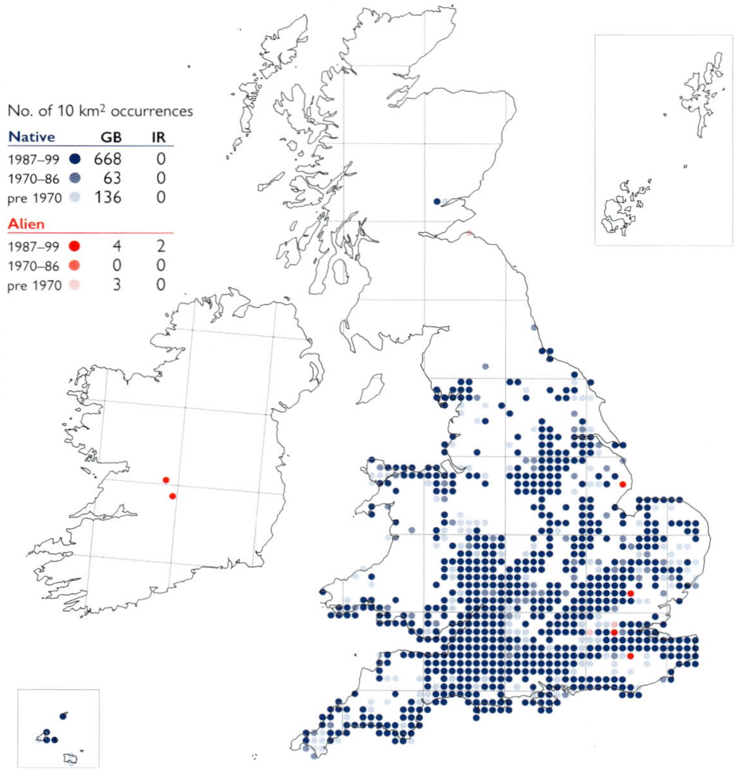

A biennial or perennial herb of dry sites, mainly on chalk or limestone, less frequently on sands and gravels, typically in places where the vegetation cover is broken or in areas of open soil or stony ground. It occurs in dry grassland, on banks, woodland margins, rides and scrub, in quarries and pits, screes (but rarely on cliffs), on the more vegetated parts of sand dunes, on roadsides and rough ground. Most sites have a southerly aspect and are unshaded. 0–305 m (Kingsdale, Mid-W. Yorks.).

Native (change –0.15). There has been no appreciable change in the distribution of *I. conyzae* since the 1962 *Atlas*.

European Temperate element; it reaches its northern limit in Jutland.

References: Atlas (276b), Grime *et al.* (1988), Hultén & Fries (1986), Meusel & Jäger (1992).

G. HALLIDAY

Inula crithmoides Golden-samphire

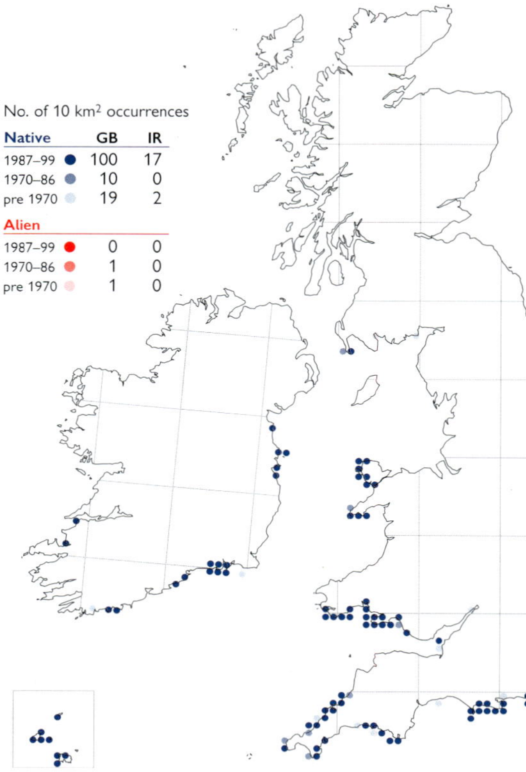

A perennial herb of two distinct types of habitat. On sea-cliffs it grows on ledges, in crevices and in open turf on calcareous or base-rich rocks, where it is often rooted in soil enriched with calcareous shell sand. In S.E. England it also occurs in saltmarshes, growing in low-marsh sites on coarse sand and above this on moderately organic soils, frequently where drift-litter accumulates. Lowland.

Native (change +0.09). This species is much better recorded now than in the 1962 *Atlas*. Its distribution is stable, with most losses from outlying squares having occurred in the 19th century.

Mediterranean-Atlantic element; it reaches its northern limit in Scotland.

References: Atlas (276a), Bolòs & Vigo (1995), Malloch & Okusanya (1979), Okusanya (1979a, b, c), Stewart *et al.* (1994).

G. HALLIDAY

Pulicaria dysenterica Common Fleabane

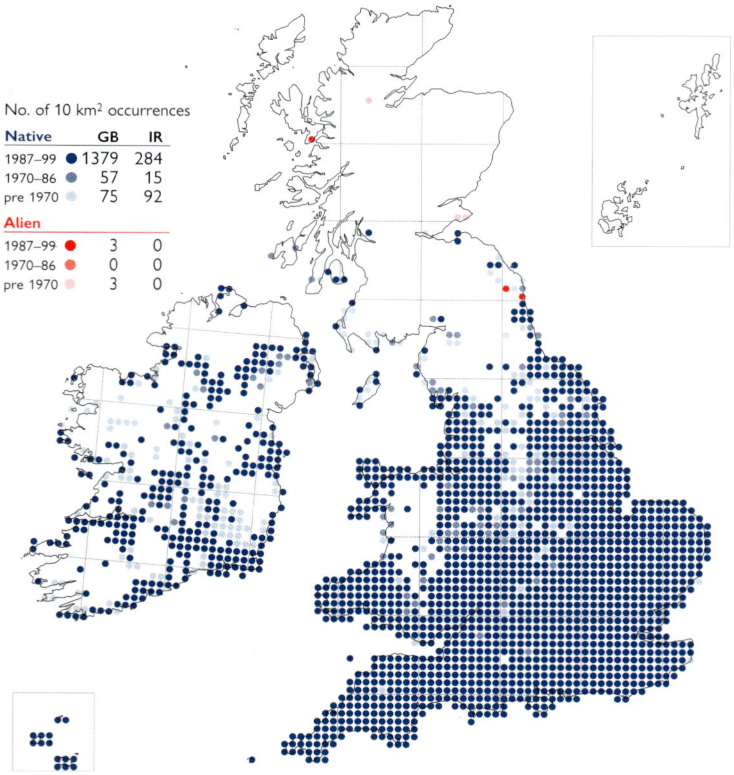

A perennial, rhizomatous herb of damp or wet, open habitats including marshy places, water- and fen-meadows, tall-herb fens, by rivers, streams, canals and ditches, in dune-slacks, wet hollows and seepages on sea-cliffs, damp woodland rides and roadside verges. It is found on a wide range of acidic and base-rich soil types. 0–325 m (Llyn Brianne, Cards.).

Native (change –0.08). The distribution of *P. dysenterica* is stable.

Eurosiberian Southern-temperate element; it reaches its northern limit in eastern Denmark.

References: Atlas (276c), Hultén & Fries (1986), Meusel & Jäger (1992).

G. HALLIDAY

Pulicaria vulgaris Small Fleabane

Native	**GB**	**IR**
1987–99	10	0
1970–86	1	0
pre 1970	115	0
Alien		
1987–99	2	0
1970–86	0	0
pre 1970	0	0

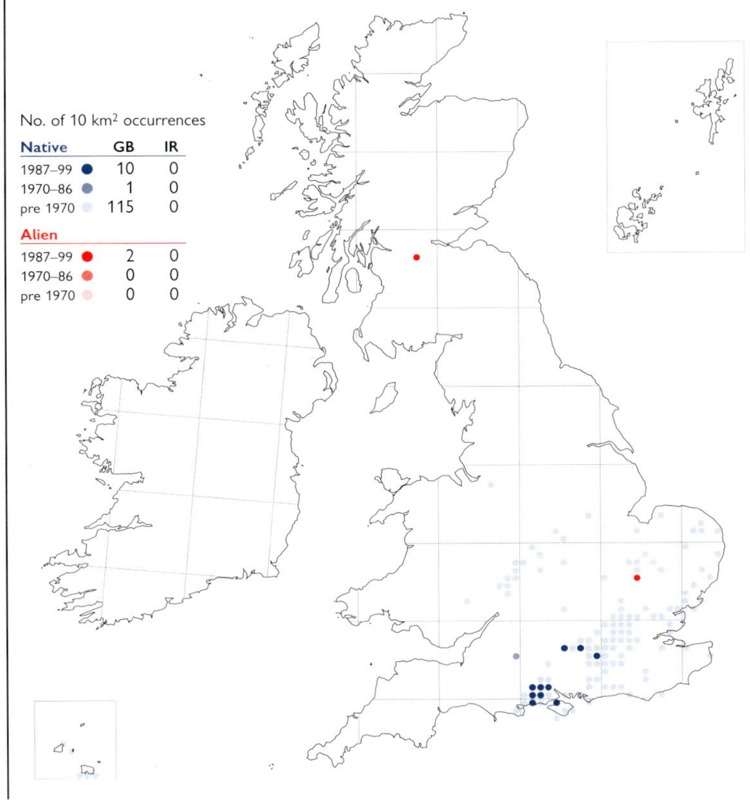

An annual of damp, winter-flooded hollows in somewhat acidic, unimproved grassland on New Forest 'lawns', on commons and village greens, and on rutted tracks. Extant sites are usually pony-grazed, this disturbance creating the open conditions needed for seedling survival. Lowland.

Native (change –0.55). Most sites for this species were lost before 1930. More recently, outside the New Forest (S. Hants.), it was lost from Wiltshire in the 1970s and the remaining sites in N. Hampshire and Surrey are very small. Losses are largely due to lack of grazing on commons, the infilling of ponds and drainage.

Eurosiberian Temperate element; it is declining throughout Europe and now reaches its northern native limit in Surrey.

References: Atlas (276d), Hultén & Fries (1986), Meusel & Jäger (1992), Prince & Hare (1981), Wigginton (1999).

G. HALLIDAY

Telekia speciosa Yellow Oxeye

Native	**GB**	**IR**
1987–99	0	0
1970–86	0	0
pre 1970	0	0
Alien		
1987–99	34	1
1970–86	7	0
pre 1970	15	0

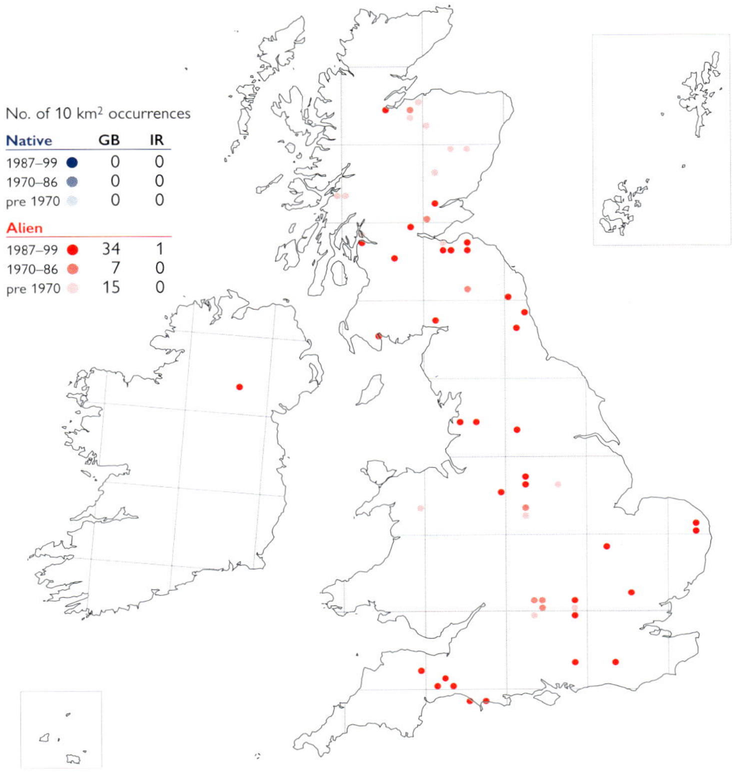

A robust perennial herb, occurring as a garden escape or occasionally as a deliberate introduction in rough grassland and by lakes and rivers. It seems to prefer damp soils and is sometimes well-naturalised. Lowland.

Neophyte. *T. speciosa* has been cultivated since 1739 and is frequently grown in gardens. It was first recorded in the wild near Lovat Bridge (Easterness) in 1914, and still persists there.

Native of the mountains of E.C. Europe, the Balkan peninsula, N. Turkey and the Caucasus; widely naturalised in C. Europe.

Reference: Meusel & Jäger (1992).

G. HALLIDAY

Solidago virgaurea Goldenrod

Native	**GB**	**IR**
1987–99	1607	390
1970–86	147	10
pre 1970	294	92
Alien		
1987–99	3	0
1970–86	1	0
pre 1970	2	0

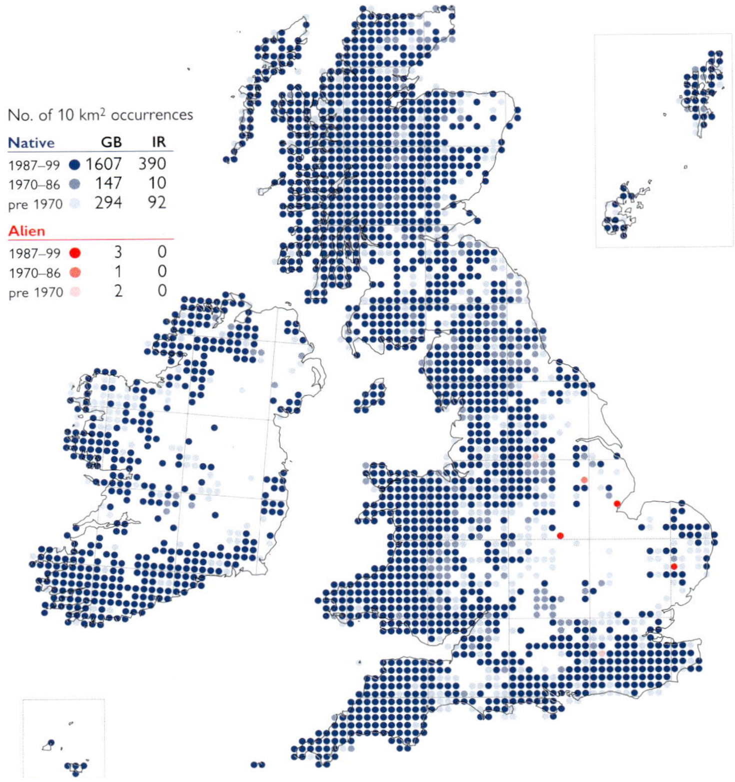

A perennial herb of free-draining, usually acidic (occasionally basic) substrates in a wide range of habitats. In the lowlands these include woods, hedge banks, heaths, banks and coastal cliff-tops; in the uplands, cliff ledges, rocks by waterfalls, rocky streamsides, tall-herb communities in gullies, montane grass-heath and fell-field. 0–1095 m (Ben Dearg, E. Ross).

Native (change –0.89). *S. virgaurea* was mapped as 'all records' in the 1962 *Atlas*. Populations in lowland Britain are disappearing due to habitat loss and analysis of the database reveals that much of this decline has occurred since 1950. It is a very variable species, with many different ecotypes.

Eurasian Boreo-temperate element.

References: Atlas (280a), Grime *et al.* (1988), Halliday (1997), Hultén & Fries (1986), Meusel & Jäger (1992).

G. HALLIDAY

Solidago canadensis Canadian Goldenrod

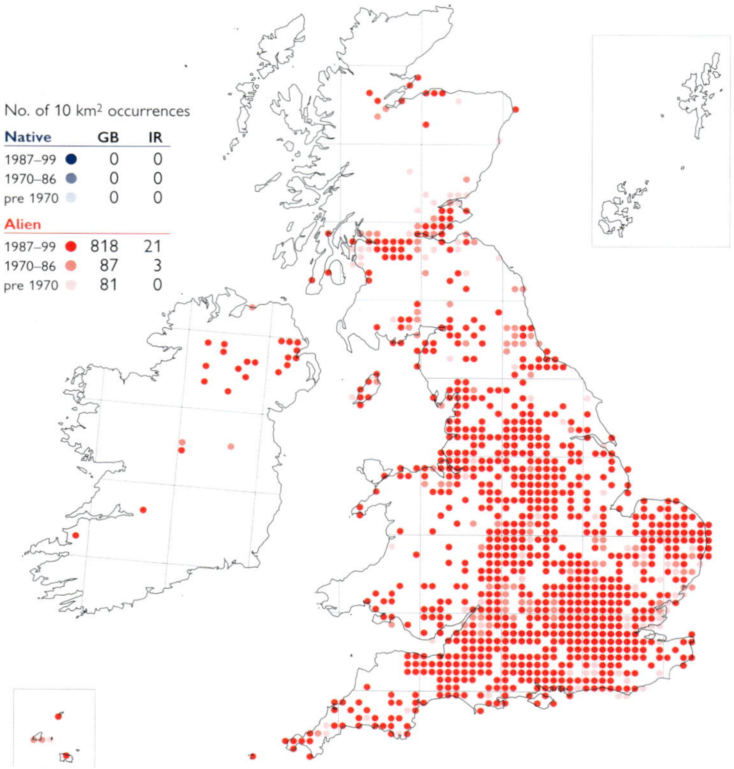

A tall, rhizomatous perennial herb naturalised on roadsides, by railways and on river banks, waste ground and spoil heaps on a wide range of soil types. Garden throw-outs can be very persistent; the plants are fertile and spread by seed. Lowland.

Neophyte. *S. canadensis* was introduced into cultivation in Britain in 1648, and it is now very popular in gardens. It was recorded in the wild in 1888 in Oxfordshire, but did not become widespread until 1930. Although formerly confused with *S. gigantea*, the maps are probably reliable in indicating that *S. canadensis* is the commoner of the two in S. Britain. It is likely to be increasing.

Native of N. America; widely naturalised in Europe.

References: Meusel & Jäger (1992), Werner *et al.* (1980).

G. HALLIDAY

Solidago gigantea Early Goldenrod

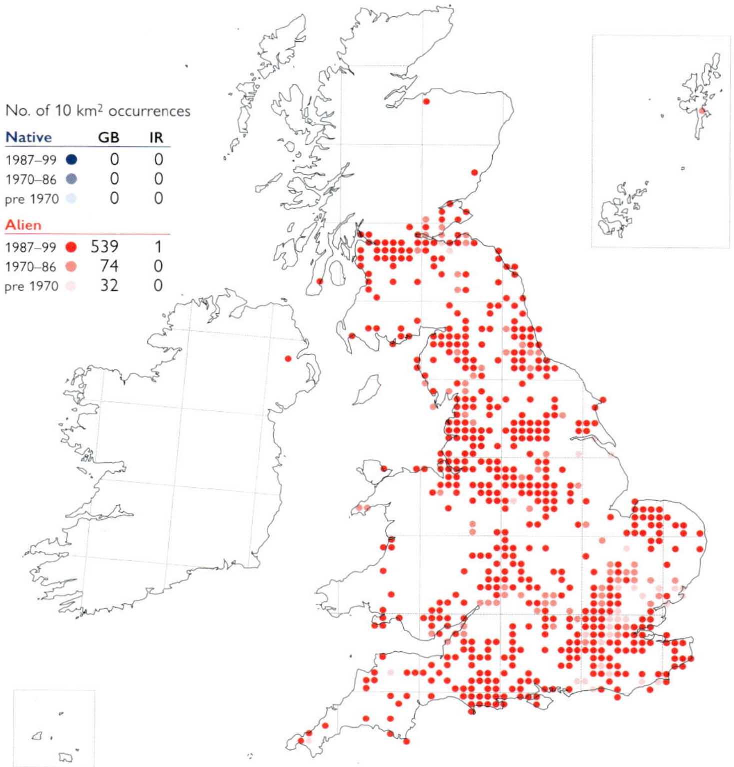

A tall, rhizomatous perennial herb naturalised in waste places, by railways, on roadside verges, river banks and rubbish tips on a wide range of soil types. Most plants in the wild originate from the dumping of garden rubbish and probably spread further by seed. Lowland.

Neophyte. *S. gigantea* was cultivated in Britain by 1758, and it is now commonly grown in gardens. It was known from the wild by 1916 but, like *S. canadensis*, it did not become widespread until after 1930. It is likely to be spreading.

Native of N. America; widely naturalised in Europe.

Reference: Meusel & Jäger (1992).

G. HALLIDAY

Aster (alien N. American taxa) Michaelmas-daisies

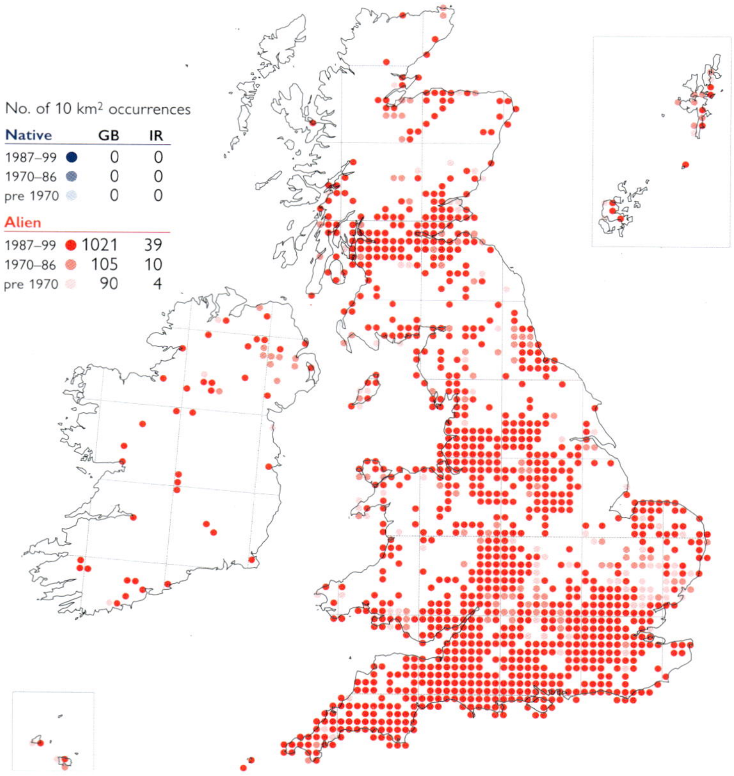

These perennial herbs are often naturalised on hedge banks, roadsides, railways, rubbish tips, waste ground and also river banks, lakesides, saltmarshes and in fens. Most colonies arise from discarded garden material, followed by rhizomatous spread or regeneration by seed. 0–345 m (Garrigill, Cumberland).

Neophyte. This map includes the species of *Aster* of N. American origin, and their hybrids. These have been grown in Britain since 1710 and are very difficult to separate. Plants are often recorded as *Aster* agg., and these and more precisely determined records are mapped here. *A. schreberi* was known from one site in Renfrewshire and *A. laevis* is a rare escape; the other three species and two hybrids are mapped separately. The maps must be regarded as provisional.

Natives of N. America.

Reference: Oliver (1998).

T. D. DINES

Aster novae-angliae Hairy Michaelmas-daisy

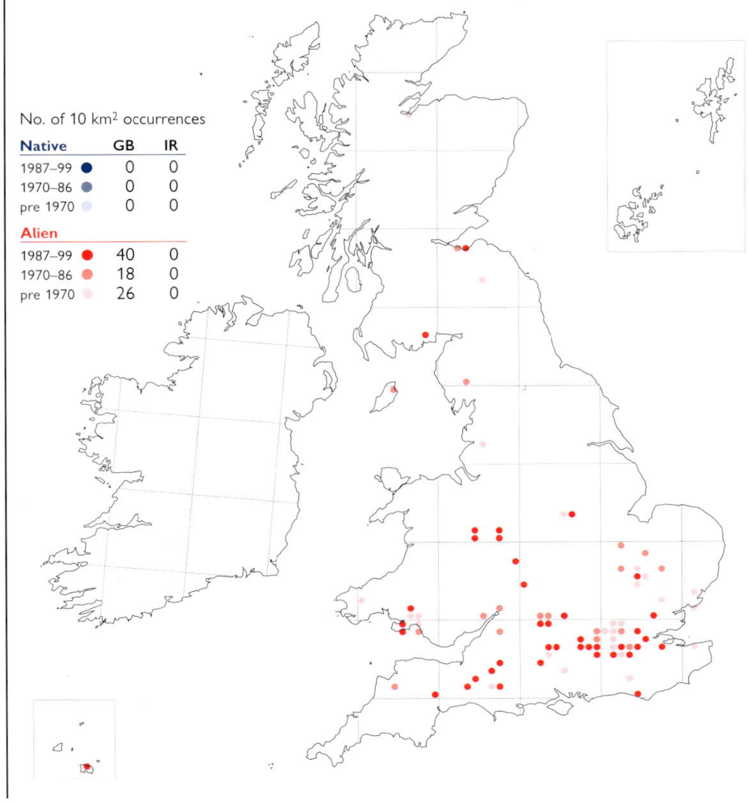

A robust perennial herb naturalised on waste ground, by railways and on roadside verges. Lowland.

Neophyte. Introduced into cultivation in 1710, *A. novae-angliae* is commonly grown in gardens, but being vigorous and long-lived, any excess garden material tends to be dumped surreptitiously and it readily becomes established. It is likely that further spread by seed from these initial colonies takes place. It was first recorded in the wild in 1915.

Native of eastern N. America; widely naturalised in C. Europe.

G. HALLIDAY

Aster laevis × *A. novi-belgii* (*A.* × *versicolor*) Late Michaelmas-daisy

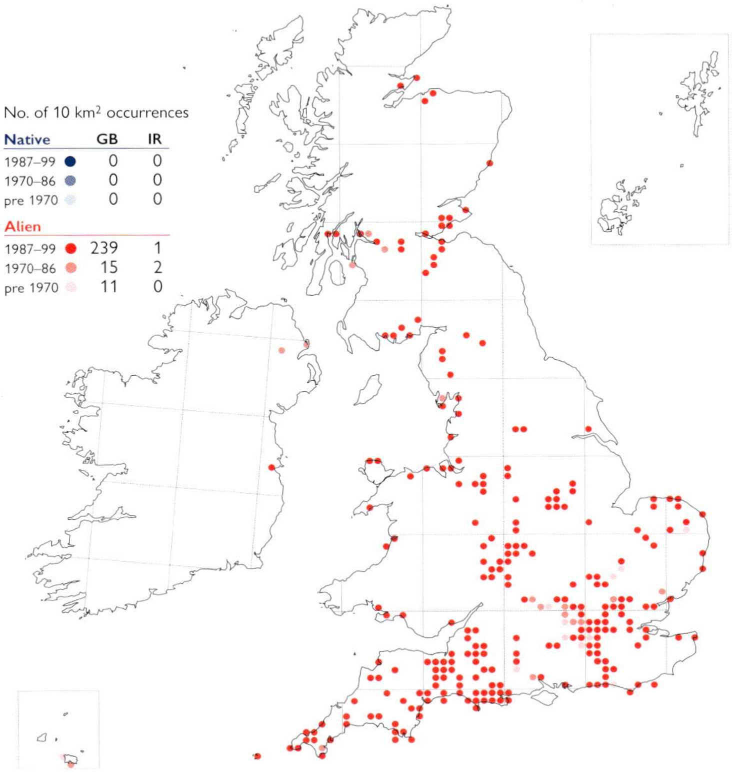

A perennial herb established in waste places, on rubbish tips, by railways and on roadsides from outcast plants from gardens. 0–345 m (Garrigill, Cumberland).

Neophyte. This vigorous hybrid originated in cultivation in Britain in 1790, and was known in the wild by 1800. As with most alien *Aster* taxa, opinions differ markedly on its frequency, reflecting the taxonomic problems posed by these plants where hybrids appear to form a complete spectrum between the parents, presumably as a result of back-crossing. However, actual proof of this, and of the fertility of the primary hybrids, is lacking.

A hybrid of garden origin; it is not known in N. America but is widely naturalised in C. Europe.

References: Oliver (1998), Stace (1975).

G. HALLIDAY

Aster novi-belgii Confused Michaelmas-daisy

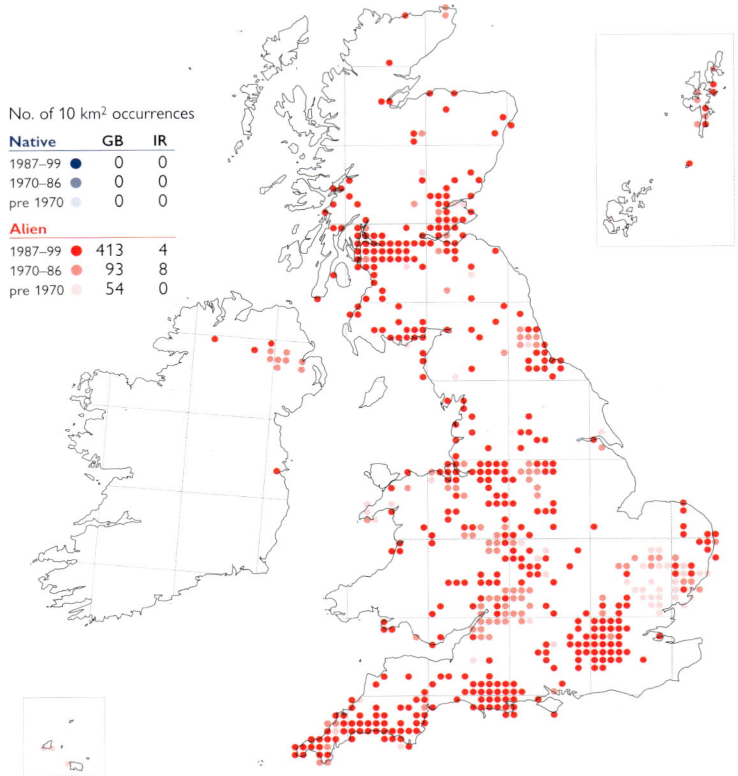

A. novi-belgii is naturalised on hedge banks, railway banks, roadsides, rubbish tips and waste ground. In addition, it sometimes occurs on river banks, lakesides and in fen vegetation. Lowland.

Neophyte. *A. novi-belgii* was introduced as a garden plant in 1710. It was recorded in the wild by 1860 and the extant population at Wicken Fen (Cambs.) was found in 1864. Stace (1997) and other authors suggest that it has been much over-recorded in the past for *A.* × *salignus* and, no doubt, also for *A.* × *versicolor*.

Native of eastern N. America; widely naturalised in N.W. & C. Europe.

References: Briggs *et al.* (1989), Meusel & Jäger (1992).

G. HALLIDAY

Aster lanceolatus
Narrow-leaved Michaelmas-daisy

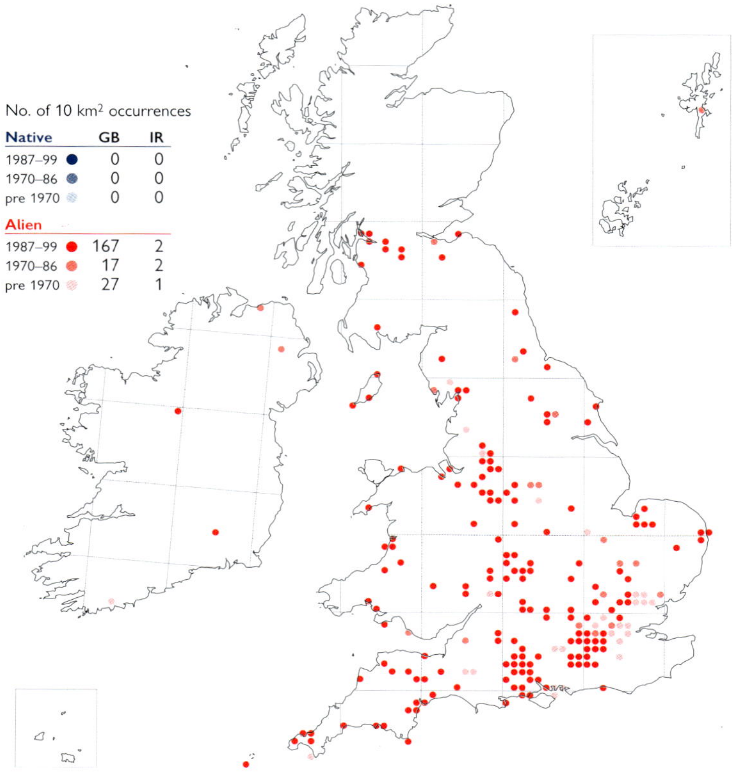

A. lanceolatus has become established along railways and river banks, on roadsides and by car parks and on waste ground and tips. Lowland.

Neophyte. *A. lanceolatus* has been cultivated since 1811 and was recorded from the wild by 1865. According to Stace (1997) it is second only to *A.* × *salignus* in frequency in Britain, yet several county Floras record it as rare. This may indicate regional variation, but is more likely to reflect the difficulties experienced in separating *A. lanceolatus* from *A.* × *salignus*. Most occurrences have arisen from garden outcasts, but it may have been accidentally introduced by N. American servicemen in Wiltshire (Oliver, 1998).

Native of eastern N. America; widely naturalised in W. & C. Europe.

Reference: Meusel & Jäger (1992).

G. HALLIDAY

Aster lanceolatus × *A. novi-belgii (A.* × *salignus)*
Common Michaelmas-daisy

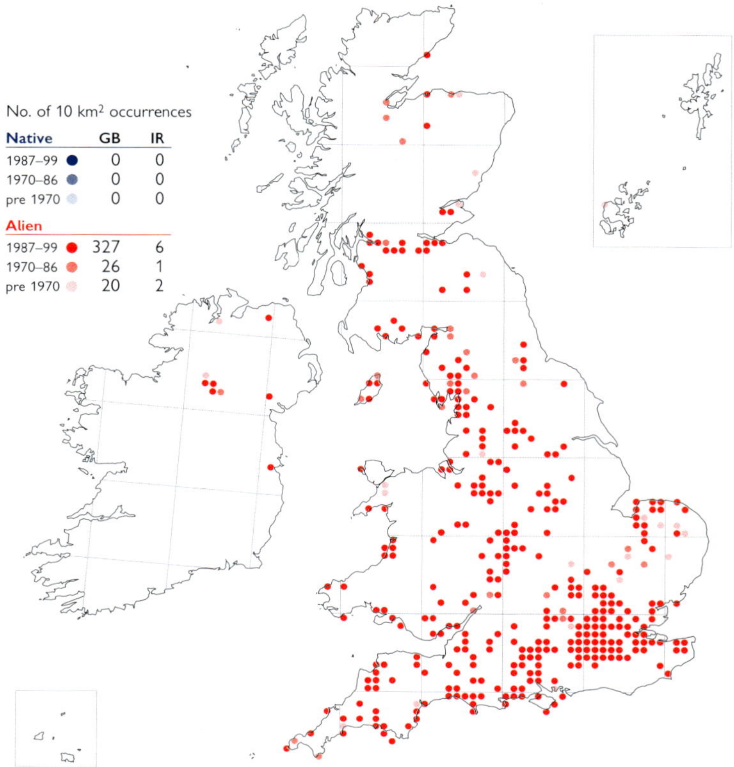

A. × *salignus* is a vigorous hybrid which has become widely established in river and lakeside habitats and in fen vegetation, as well as along roadsides and railways, and on waste ground. 0–305 m (Hardendale, Westmorland).

Neophyte. This hybrid is thought to have been introduced into cultivation in Britain in 1815, and was first recorded in the wild in 1867. It is much confused with both its parents, particularly *A. lanceolatus*. Some of the taxonomic problems might arise from back-crossing as the hybrid is reportedly somewhat fertile. It is probably under-recorded.

A hybrid of garden origin; widely naturalised in N. & C. Europe.

References: Meusel & Jäger (1992), Oliver (1998), Stace (1975).

G. HALLIDAY

Aster tripolium Sea Aster

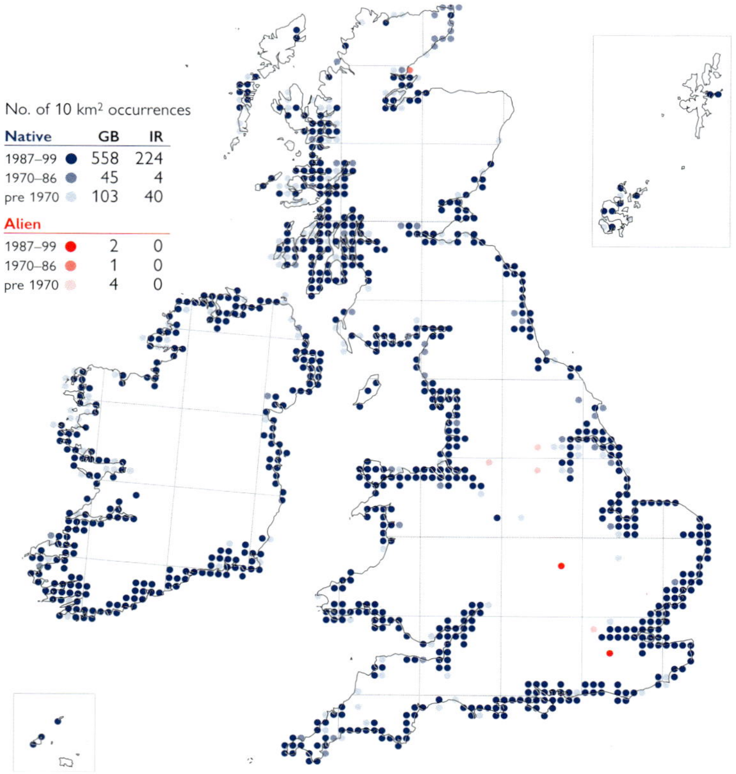

A short-lived perennial herb occurring at low elevations in ungrazed or lightly grazed saltmarshes, especially along creeksides, and also on muddy sea-banks, tidal river banks and in brackish ditches. In W. Britain and Ireland it also grows amongst rocks and on exposed sea-cliffs. It also occurs very locally in inland saltmarshes and recently it has been recorded beside salt-treated roads. Lowland.

Native (change –0.44). The distribution of *A. tripolium* has not appreciably changed since the 1962 *Atlas*. The rayless form, var. *discoideus*, was mapped by Perring & Sell (1968).

Eurasian Temperate element.

References: Atlas (280b), Atlas Supp. (72b), Clapham *et al.* (1942), Hultén & Fries (1986), Lee (1977), Meusel & Jäger (1992).

G. HALLIDAY

Aster linosyris Goldilocks Aster

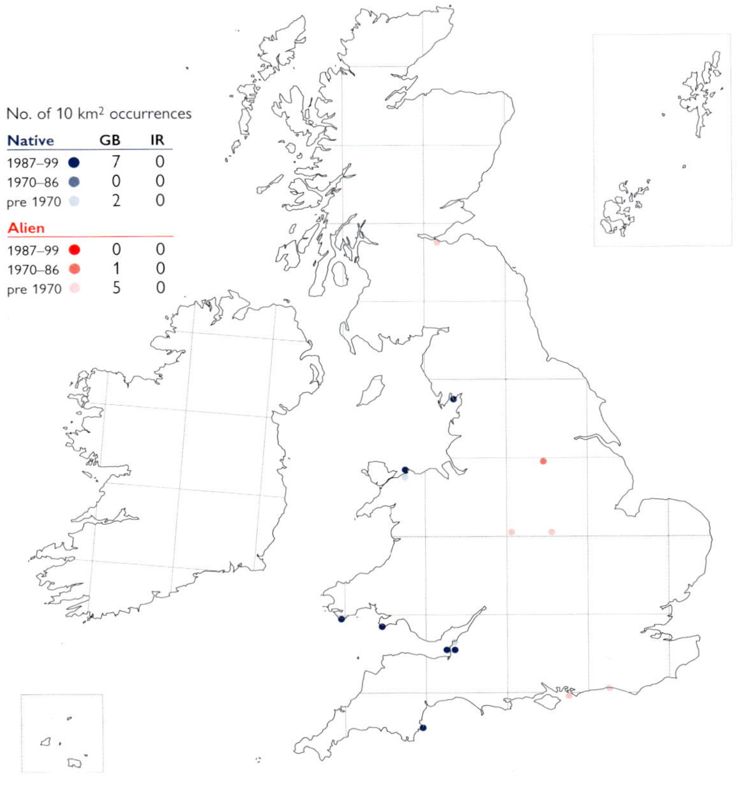

A perennial herb of shallow soil in open, grassy habitats on limestone sea-cliffs and rocky slopes, cliff-top grassland and wind-pruned heath overlying limestone. It is a poor competitor, and is usually intolerant of heavy grazing, although in Pembrokeshire it is found in low-growing, sheep-grazed, cliff-top grassland and heath. It seems to be self-incompatible and some small populations appear to represent single, self-sterile clones. Lowland.

Native (change –0.10). The distribution is of *A. linosyris* stable, but some populations are very small.

European Temperate element, with a continental distribution in W. Europe; it reaches its northern limit on Öland and Gotland.

References: Atlas (280c), Hultén & Fries (1986), Meusel & Jäger (1992), Wigginton (1999).

G. HALLIDAY & M. J. WIGGINTON

Erigeron glaucus Seaside Daisy

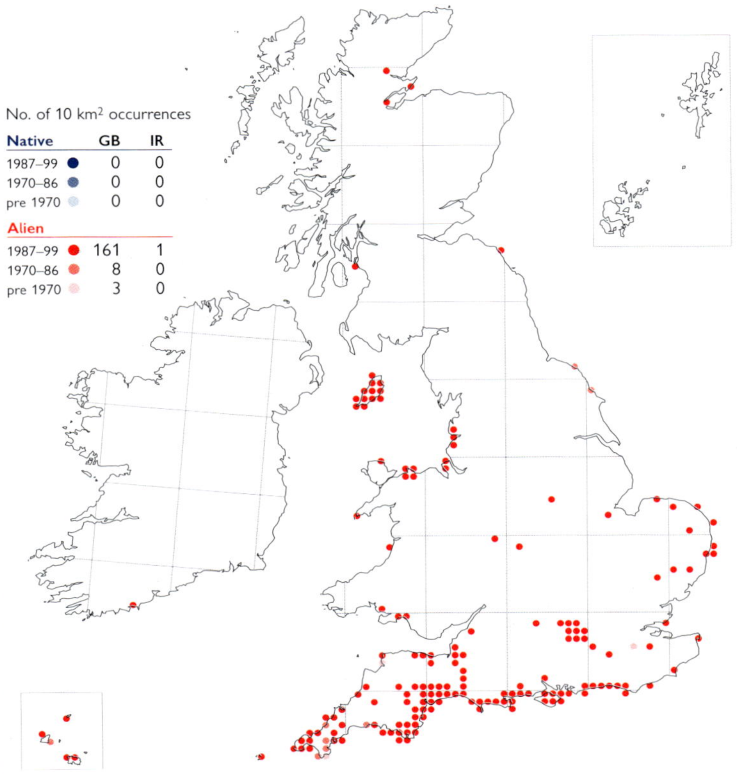

A decumbent perennial herb found as a garden escape in pavement cracks and on walls, banks, sea-cliffs and shingle. Lowland.

Neophyte. This species has been cultivated in Britain since 1812 and is now commonly grown in gardens, particularly near the coast. It was first recorded from the wild in 1942 on sea-cliffs at Bournemouth (S. Hants.). Trends in distribution are difficult to determine, but is reported to be increasing in Somerset (Green *et al.*, 1997).

Native of N. America.

G. HALLIDAY

Erigeron borealis Alpine Fleabane

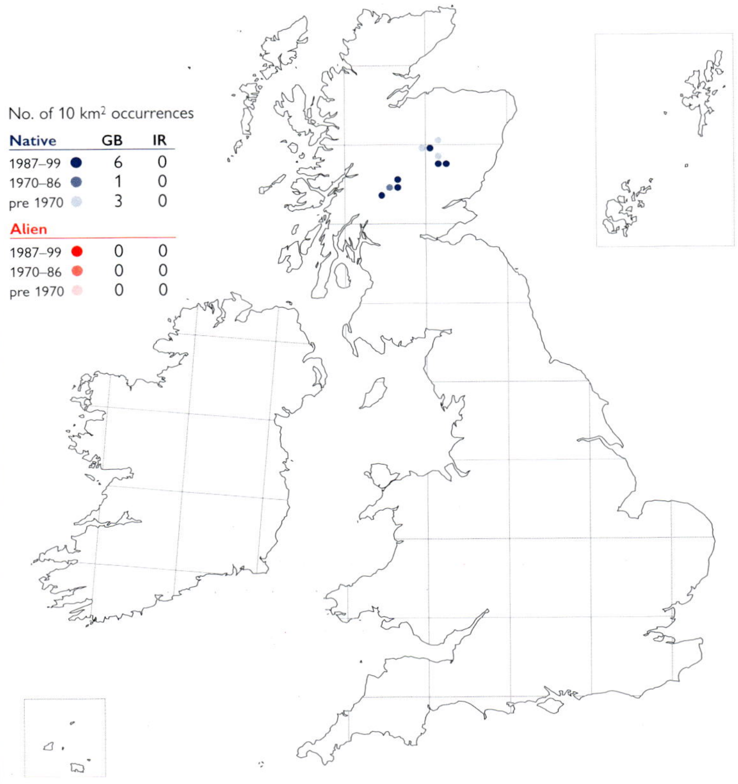

A perennial rhizomatous herb found on unstable, basic, mostly S.-facing cliff ledges of mica-schist, usually adjacent to grazed, herb-rich grassland. The surviving sites are inaccessible to grazing by sheep and deer. From 640 m on Craig Maud (Angus) to 1100 m on Creag an Fhitich (Mid Perth).

Native (change –0.11). Pre-1970 sites for *E. borealis* in the eastern Highlands of Scotland were lost before 1930, and the current distribution of the species may be stable. However, the size of some populations varies greatly from year to year, and long-term trends are unclear.

Eurosiberian Arctic-montane element; it reaches its southern limit in Scotland, being replaced by the very similar *E. neglectus* in the Alps.

References: Atlas (281a), Hultén & Fries (1986), Meusel & Jäger (1992), Wigginton (1999).

G. HALLIDAY & M. J. WIGGINTON

Erigeron karvinskianus Mexican Fleabane

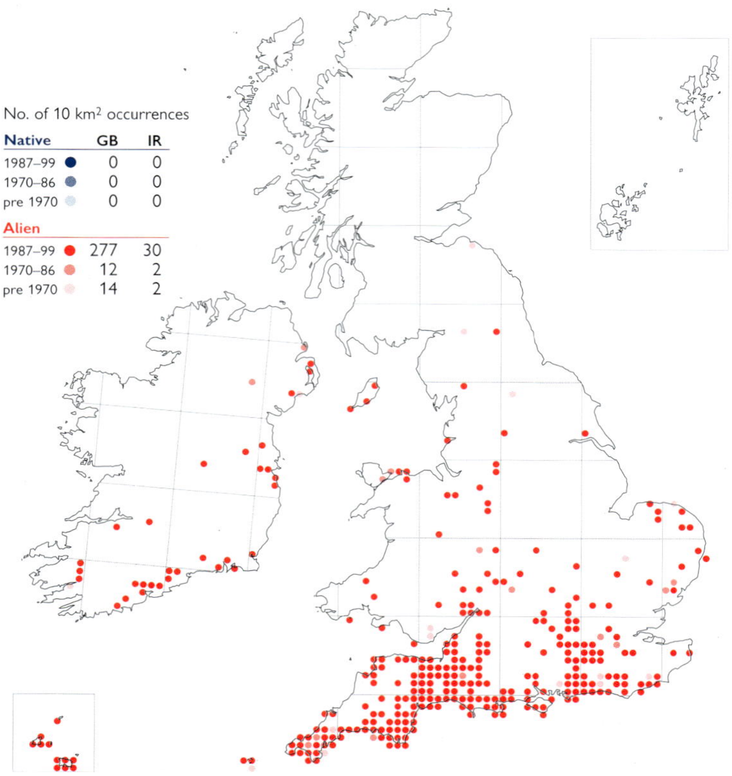

A perennial herb, well-established on walls, rock outcrops and cliffs, in cracks in pavements and on stony banks, to which it has usually spread by seed from nearby gardens. Lowland.

Neophyte (change +2.37). *E. karvinskianus* has been cultivated in Britain since 1836 and is widely grown in frost-free areas, particularly in the Channel Islands, in coastal areas of S.W. England and S. Wales and in sheltered urban sites. It was first recorded as naturalised in 1860, at St Peter Port, Guernsey, where it was locally abundant by the 1890s. It has shown a marked increase since the 1962 *Atlas*.

Native of C. America (Mexico); naturalised in W. & S. Europe.

References: Atlas (281b), McClintock (1975), Meusel & Jäger (1992).

G. HALLIDAY

Erigeron acer Blue Fleabane

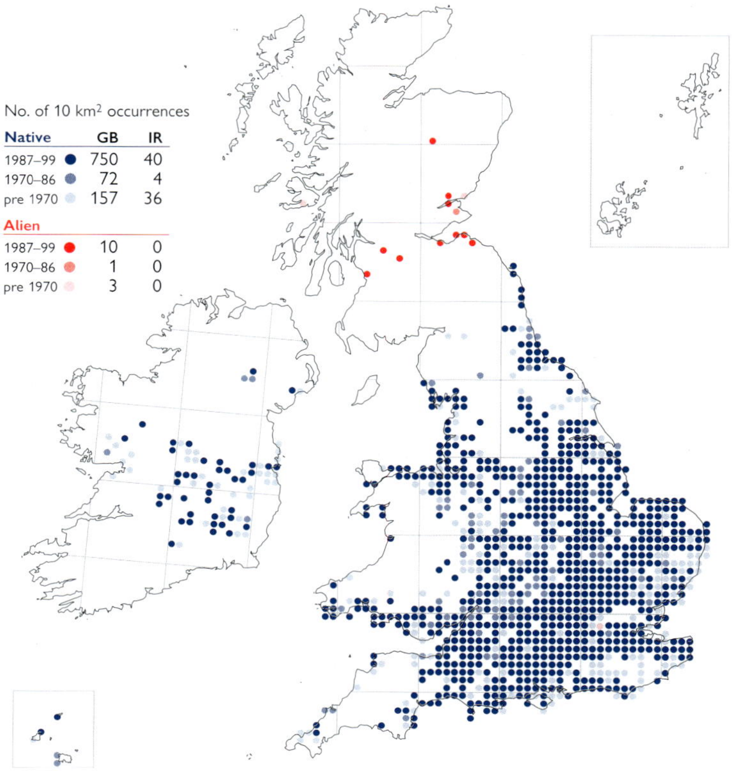

An annual or perennial herb of open, well-drained, skeletal neutral or calcareous soils, often on warm, S.-facing slopes. Habitats include sand dunes, sand-pits, spoil and waste heaps from quarries, railway ballast, industrial waste and cinders. It also grows on rock outcrops, especially of chalk and limestone and on mortared walls. 0–430 m (Banffs.).

Native (change +0.33). The overall range of *E. acer* is stable, although there are now many more records from the English Midlands than there were in the 1962 *Atlas*, and it appears to have declined locally in some areas, particularly parts of S.E. and E. England.

Circumpolar Boreo-temperate element.

References: Atlas (280d), Curtis & McGough (1988), Grime *et al.* (1988), Hultén & Fries (1986), Meusel & Jäger (1992).

G. HALLIDAY

Conyza canadensis Canadian Fleabane

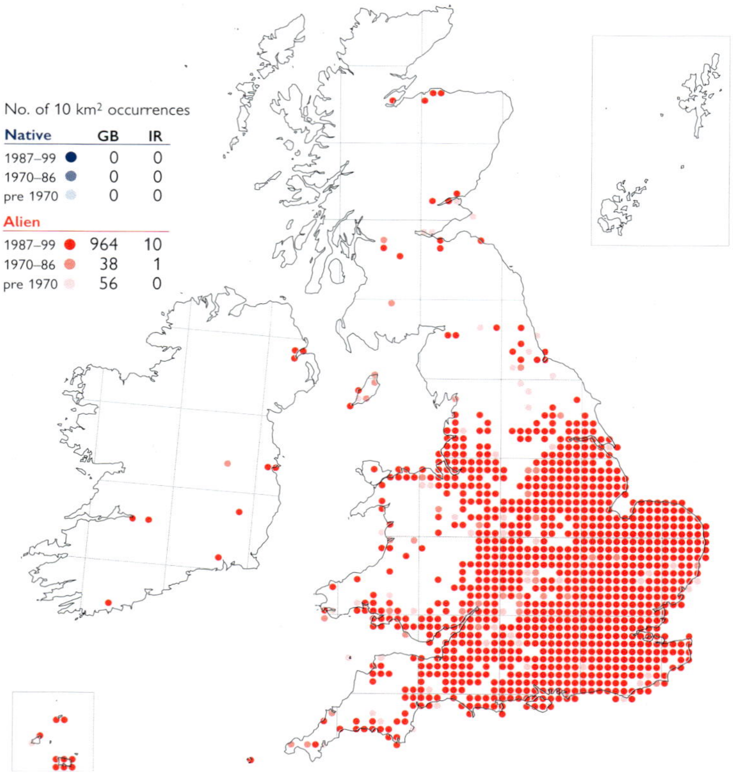

An erect annual of well-drained, open habitats such as pavements, waste places, walls, railway ballast and as a weed of cultivated ground. It is particularly characteristic of urban areas of S. and E. England. It also occurs occasionally on sand dunes and on sandy ground inland. Lowland.

Neophyte (change +1.12). This species has been known in the London area since 1690. It has shown a marked expansion of range in Britain since the 1962 *Atlas*, and was first recorded in Ireland in 1983.

Native of N. America; naturalised throughout Europe and in similar climatic regions almost worldwide.

References: Atlas (281c), Hultén & Fries (1986), Meusel & Jäger (1992).

G. HALLIDAY

Conyza sumatrensis Guernsey Fleabane

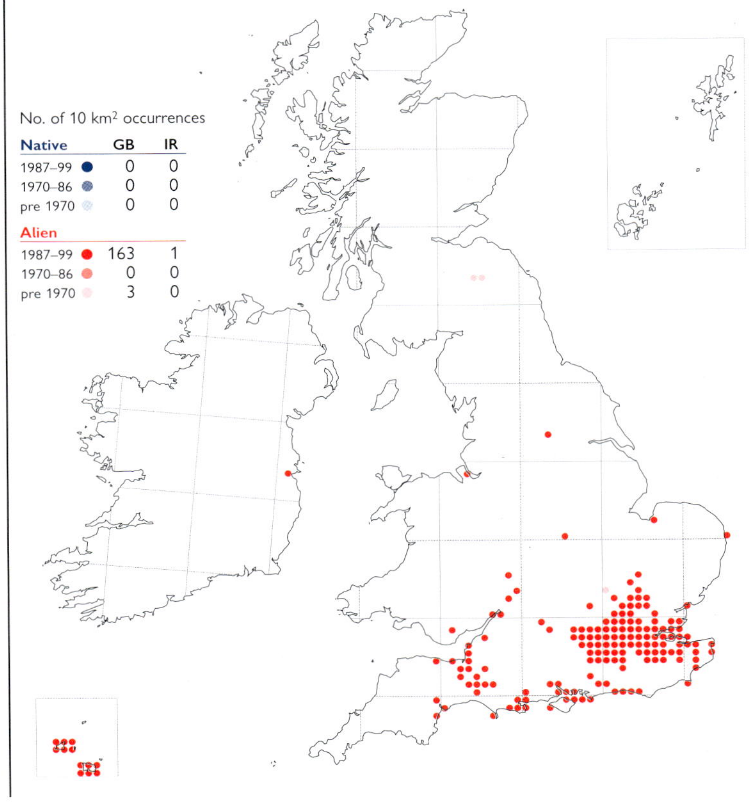

A tall and very conspicuous annual of well-drained, open and disturbed ground, such as waste land, railway-sides and docks, chiefly around towns. It also occurs as a wool casual. Lowland.

Neophyte. The first record for *C. sumatrensis* in our area was from Guernsey in 1961. It was found naturalised in S. Essex in 1974 and was well established in the London area by 1984. It appears to be spreading rapidly, and was first recorded in Ireland in 1990.

Native of S. America; widely naturalised in Europe.

References: Bolòs & Vigo (1995), Wurzell (1988, 1994a).

G. HALLIDAY

Olearia macrodonta New Zealand Holly

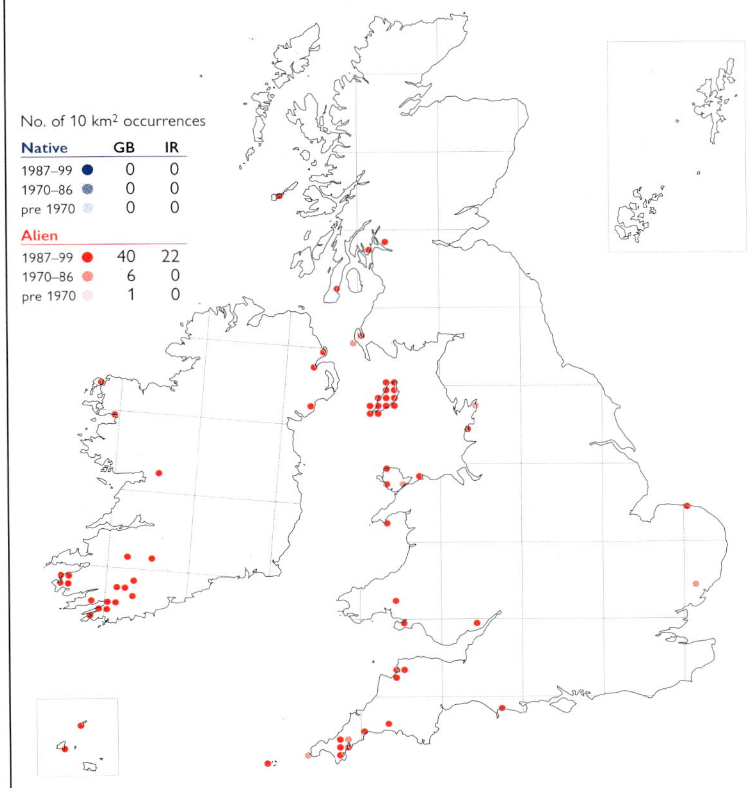

An evergreen shrub, found in hedges and scrub, and on roadsides, banks, sea-cliffs, sand dunes and waste ground. It is usually found as a garden escape or throw-out, and can become well-established in suitable habitats. Reproduction is by seed. Lowland.

Neophyte. This species was first introduced to British gardens in 1886; it is now extremely popular, especially in coastal areas. It was recorded from the wild by 1957.

Native of New Zealand.

Reference: Bean (1976).

G. HALLIDAY

Bellis perennis Daisy

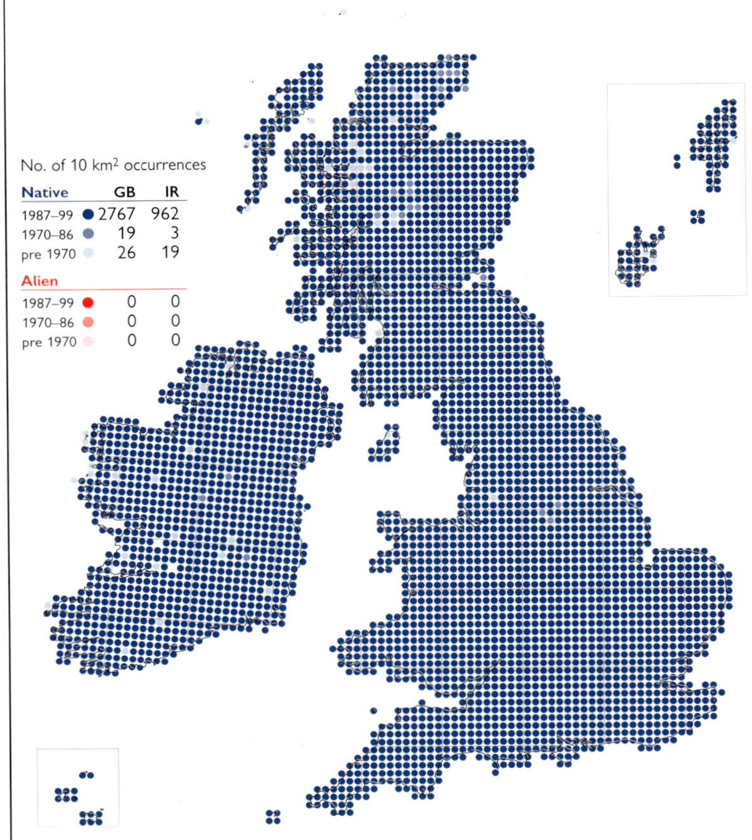

A rosette-forming, winter-green, shortly stoloniferous perennial which grows in mown or heavily grazed or trampled grassland. It occurs in practically all types of neutral and calcareous grassland but it does best in those that are relatively wet for at least part of the year. It is most familiar as a weed of lawns and recreational areas, roadside verges and pastures, but more natural habitats include stream banks, lake margins, dune-slacks and the margins of upland flushes. 0–915 m (Caenlochan, Angus).

Native (change +0.89). The range of *B. perennis* is stable.

European Temperate element; widely naturalised outside its native range.

References: Atlas (281d), Grime *et al.* (1988), Hultén & Fries (1986), Meusel & Jäger (1992).

G. HALLIDAY

Tanacetum parthenium Feverfew

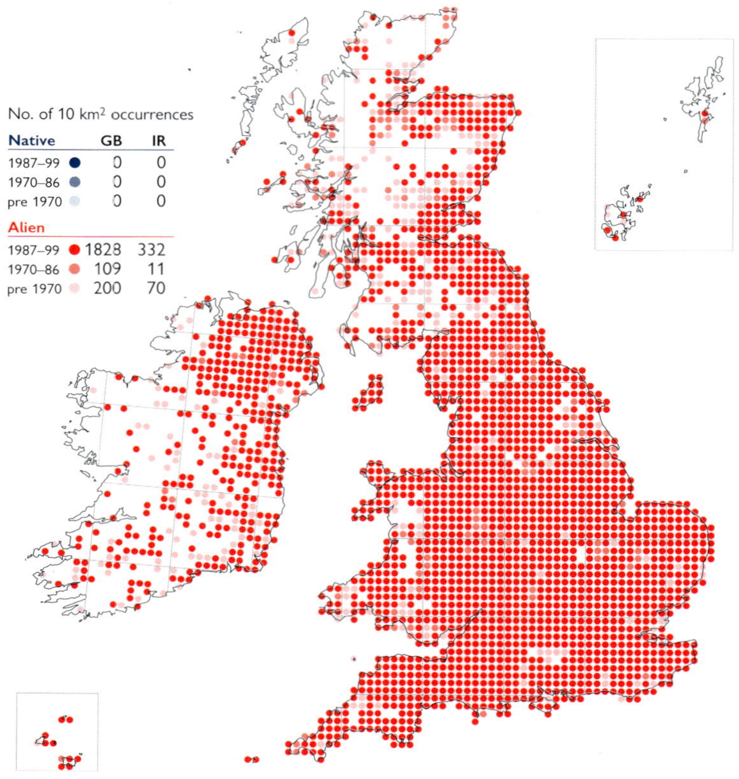

An aromatic perennial herb commonly cultivated as an ornamental plant. It is widely naturalised in gardens, on walls, waysides, tips and waste ground. It seeds freely, but has a poorly formed pappus, and disperses for only short distances; it is therefore most frequent near habitation. 0–380 m (above Alston, Cumberland).

Archaeophyte (change +0.23). *T. parthenium* was being grown in gardens for medicinal use by 995 (Harvey, 1981). Its distribution has not changed significantly since the 1962 *Atlas*.

Apparently native to the Balkan peninsula; now widespread in temperate regions throughout the world.

References: Atlas (285a), Meusel & Jäger (1992).

H. J. KILLICK

Tanacetum vulgare Tansy

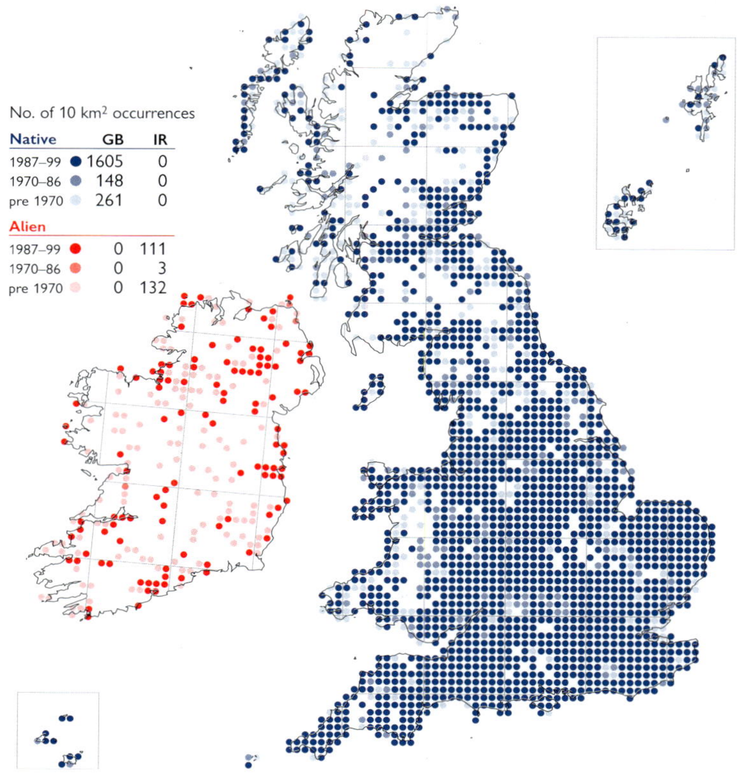

An aromatic, rhizomatous perennial herb found in grassy places by rivers, roads and railways, and on waste ground. 0–380 m (Glenlivet, Banffs.).

Native (change –0.23). This species was grown in medieval gardens as a medicinal or culinary herb, and escapes from cultivation are widely naturalised. For this reason it is often impossible to differentiate native plants from those of garden origin, and all those in Britain are mapped as if they are native; it is considered to be alien in Ireland (Scannell & Synnott, 1987). There is no significant change in its distribution since the 1962 *Atlas*.

Eurasian Boreo-temperate element, but naturalised in N. America so distribution is now Circumpolar Boreo-temperate.

References: Atlas (285b), Hultén & Fries (1986), Mabey (1996), Meusel & Jäger (1992).

H. J. KILLICK

Seriphidium maritimum Sea Wormwood

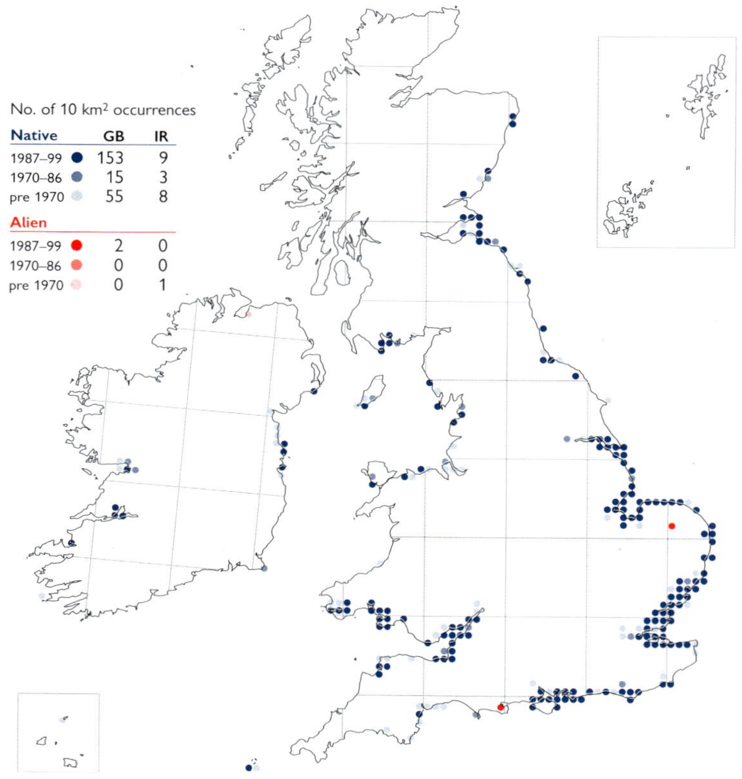

An aromatic perennial herb occurring in the upper, drier parts of saltmarshes; also found on shingle, sea-cliffs, waste ground and walls close to the sea, by brackish dykes of drained estuarine marshes and on the banks of tidal rivers. Lowland.

Native (change –0.42). *S. maritimum* was lost from many sites in N. and W. Britain (and some elsewhere) before 1930, with further losses since. It is stable elsewhere and can be locally common, and is now better recorded, or perhaps increasing, in Ireland.

Suboceanic Temperate element.

References: Atlas (286b), Hultén & Fries (1986), Leach (1984), Meusel & Jäger (1992).

H. J. KILLICK

Artemisia vulgaris Mugwort

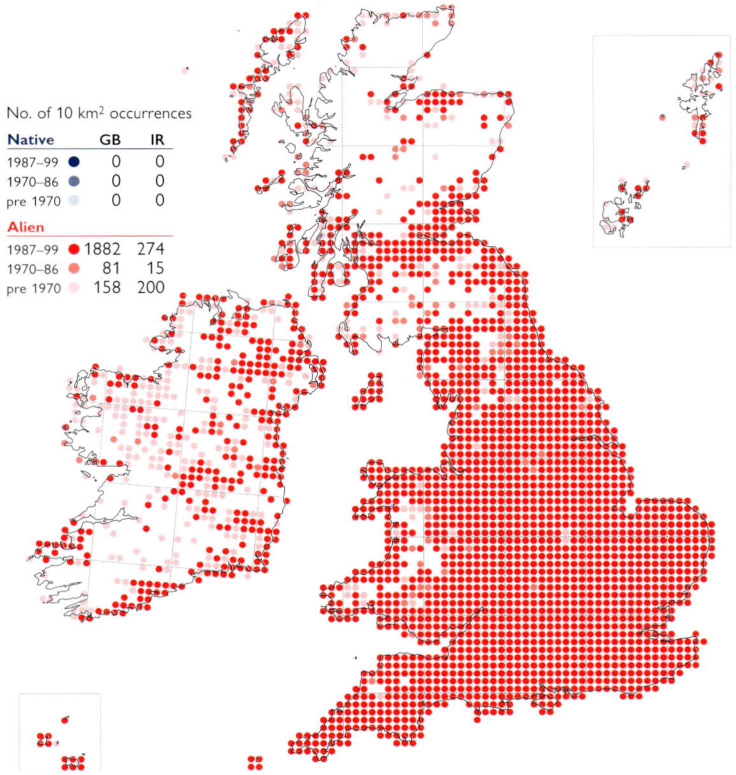

A tall aromatic perennial herb of waste places, tips, rough ground, roadside verges and waysides, usually on relatively fertile soils. The seeds, lacking a pappus, are often distributed by human activities, especially in urban areas and along road and rail systems. Generally lowland, but reaching 350 m north of Shap summit (Westmorland), and with an unlocalised record of 420 m in Aberdeenshire.

Archaeophyte (change –0.20). The distribution of this species appears to be stable.

As an archaeophyte *A. vulgaris* has a Eurosiberian Temperate distribution; it is widely naturalised outside this range.

References: Atlas (285c), Grime *et al.* (1988), Hultén & Fries (1986), Meusel & Jäger (1992).

H. J. KILLICK

Artemisia verlotiorum Chinese Mugwort

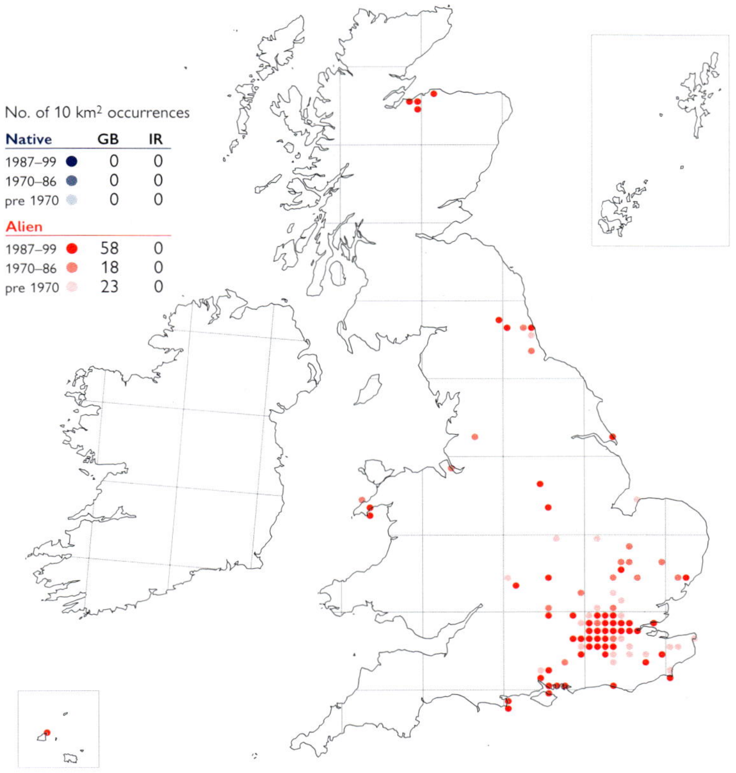

A tall rhizomatous perennial herb, naturalised and locally abundant on waysides, and in waste places and rough ground. As it flowers in October and November, it rarely sets seed, and its spread is often by detached pieces of rhizome. Lowland.

Neophyte (change +0.10). *A. verlotiorum* has gradually extended its range in Europe. It was first collected from the wild in Britain in 1908 in Middlesex but not recognised as a distinct species until 1938–9. By 1950 it was well-established in Surrey and Middlesex; its subsequent spread has been relatively slow.

Native of S.W. China; widely naturalised in W. & C. Europe and other continents.

References: Atlas (285d), Brenan (1950), Burton (1983), Meusel & Jäger (1992).

H. J. KILLICK

Artemisia absinthium Wormwood

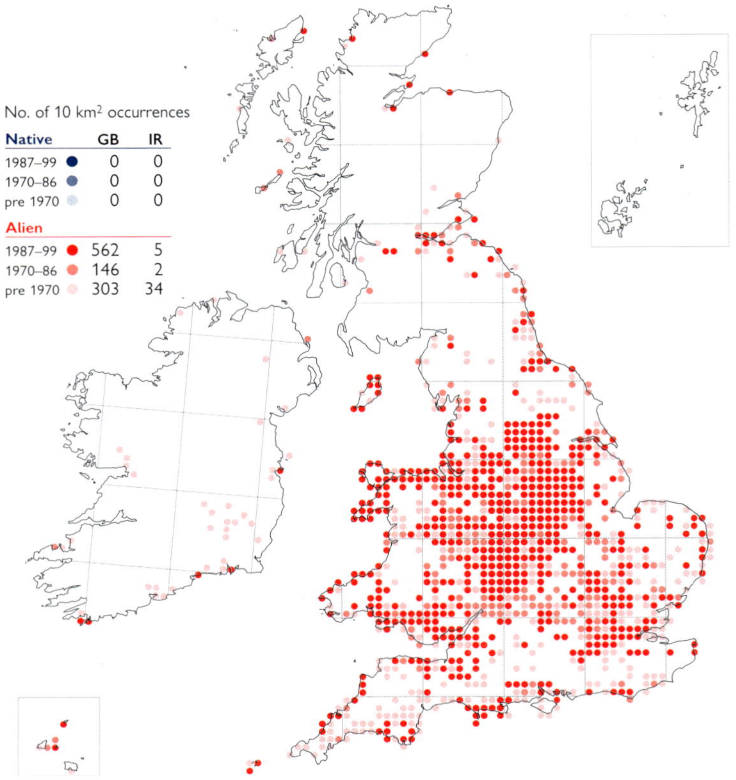

An aromatic perennial herb of waste and rough ground, waysides, railway sidings, rubbish tips, gravel-pits, quarries and other anthropogenic habitats. 0–370 m (Teesdale, Co. Durham).

Archaeophyte (change –0.46). This species, which was being grown in British gardens by 1200 (Harvey, 1981), was formerly cultivated for medicine and flavouring. It is often persistent, especially in urban and maritime locations where it is less at risk from frost damage, and the distribution is more or less stable.

As an archaeophyte *A. absinthium* has a Eurosiberian Temperate distribution; it is widely naturalised outside this range.

References: Atlas (286a), Grime *et al.* (1988), Hultén & Fries (1986), Maw *et al.* (1985), Meusel & Jäger (1992).

H. J. KILLICK

Artemisia norvegica Norwegian Mugwort

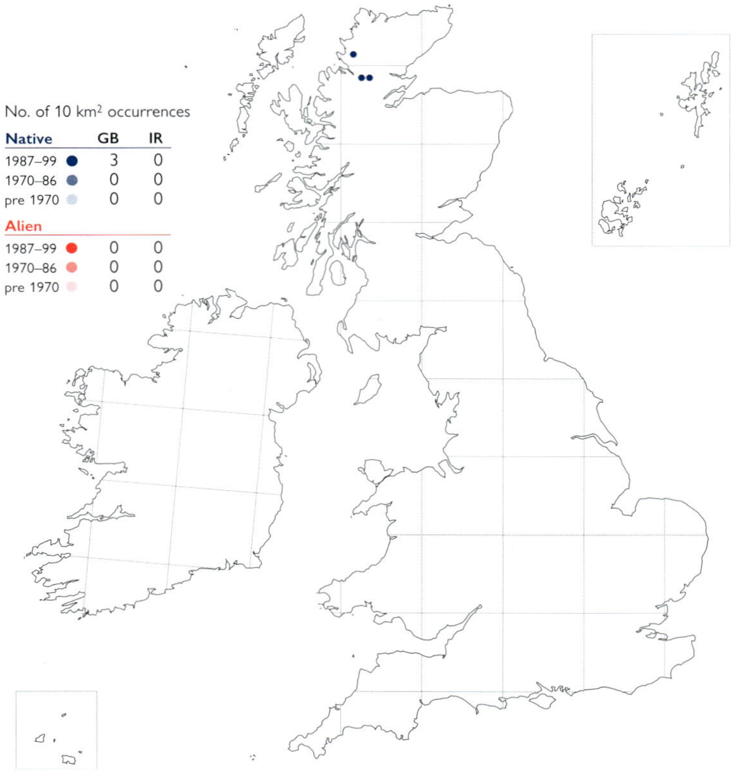

A small rhizomatous perennial of mountain tops, usually occurring in exposed situations on or near the summit ridge. Habitats include bare stony ground, *Racomitrium* heath, bouldery crests of solifluction terraces, and sometimes hollows between rocks. The relative importance of sexual and vegetative reproduction in British populations is uncertain. 700–870 m (Seana Bhraigh, E. Ross).

Native. This species was discovered in 1950 in W. Ross, and subsequently at two additional sites in the same vice-county. Its populations fluctuate in size, but appear to be stable in the longer term.

European Arctic-montane element; rare in the Arctic zonobiome and absent from mountains of C. Europe.

References: Atlas (285d), Hultén & Fries (1986), Wigginton (1999).

M. J. WIGGINTON

Artemisia campestris Field Wormwood

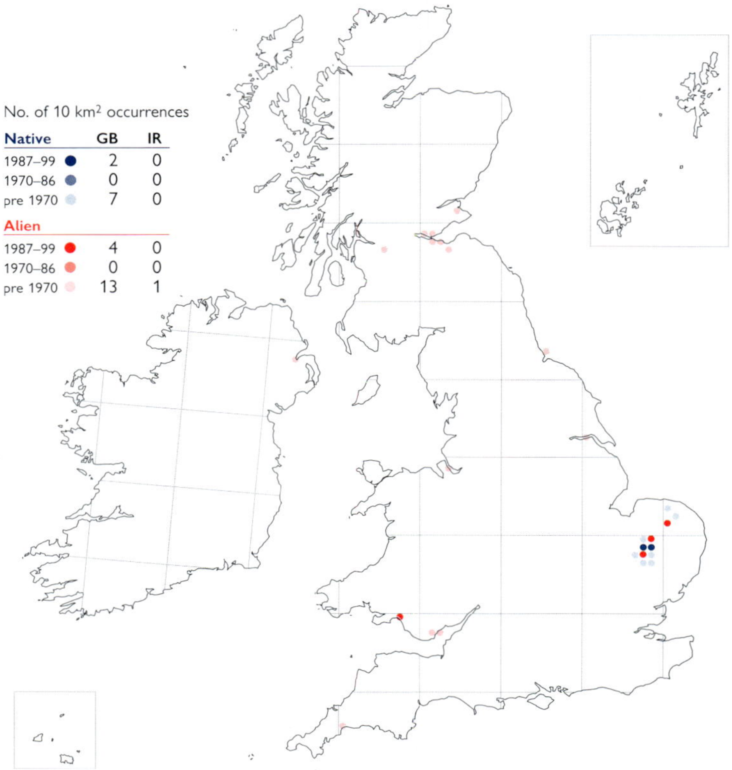

A perennial herb found in the Breckland in short open grassland, grass-heath, on forest rides and tracks, in abandoned arable fields, and on roadsides. It does not persist in tall, closed turf but sometimes reappears following disturbance. Lowland.

Native (change –0.42). Many sites of *A. campestris* have been lost to agriculture, forestry or building development. It is extant at only three native sites, but alien populations have been established using native seed. It is vulnerable to grazing, and native populations survive only in rabbit exclosures. The naturalised population on sand dunes in Glamorgan appears to be in decline.

Eurosiberian Temperate element, with a continental distribution in W. Europe.

References: Atlas (286c), Hultén & Fries (1986), Meusel & Jäger (1992), Wigginton (1999).

H. J. KILLICK

Santolina chamaecyparissus Lavender-cotton

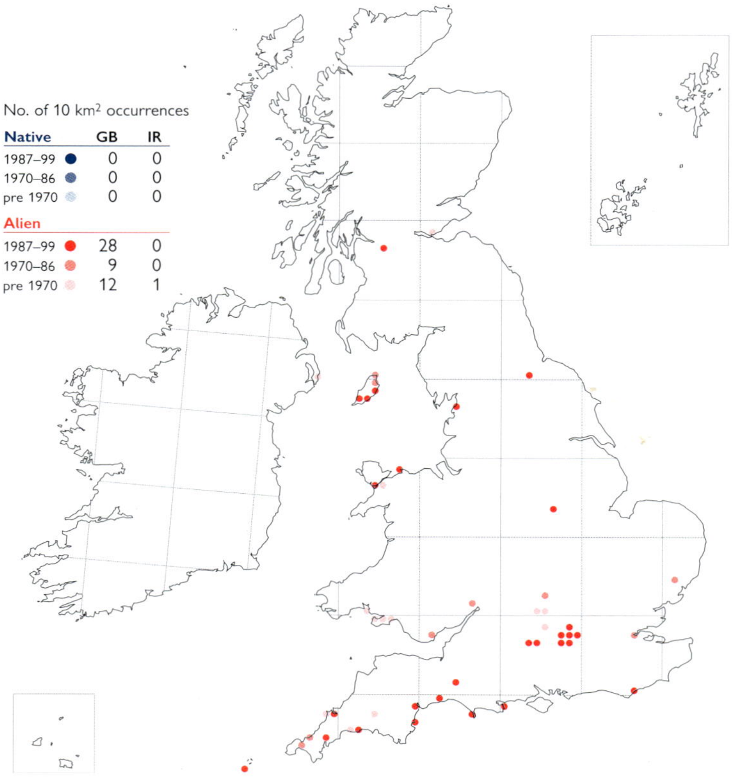

A small evergreen shrub persistent on rubbish tips, rough ground, and as a relic of cultivation, and more recently established on some sandy shores. Lowland.

Neophyte. This species, introduced to cultivation in Britain by 1548, is frequently grown in gardens. As a garden escape it was known by 1905.

Native of the Mediterranean region.

References: Bolòs & Vigo (1995), Dunn (1905).

H. J. KILLICK

Otanthus maritimus Cottonweed

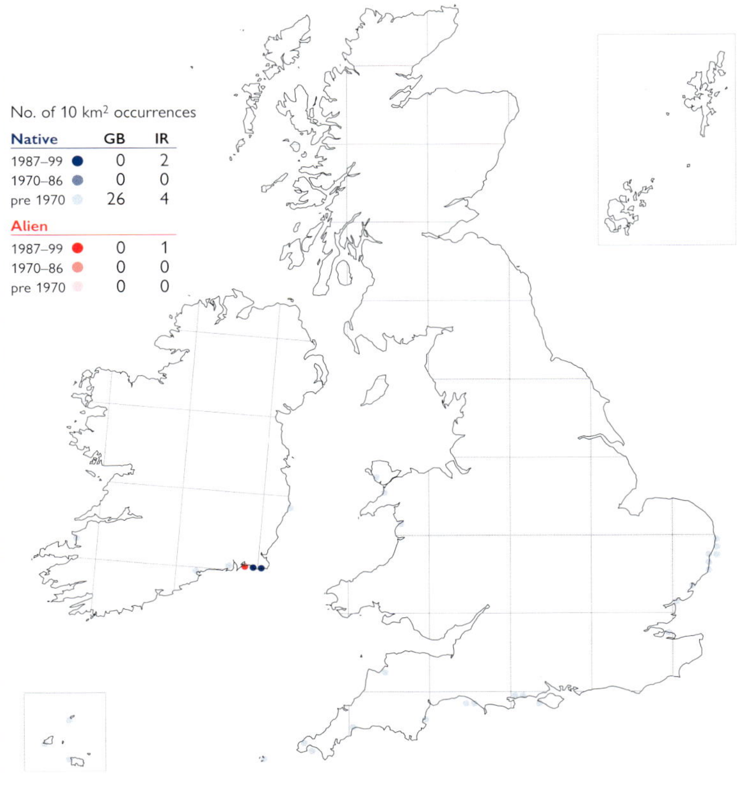

A perennial herb of sand dunes and stabilised shingle. Lowland.

Native (change −1.49). First recorded in our area in the early 17th century, this species has undergone a major decline since 1850. It is now restricted as a native to just two sites in Co. Wexford, where populations appear to be stable; it has been introduced to a third locality. It is extinct in Britain, the last records being from Cornwall in 1933 and the Isles of Scilly in 1936. It was last recorded in the Channel Islands (on Jersey) in 1926.

Mediterranean-Atlantic element; more frequent on stable Mediterranean beaches than on the stormy Atlantic coast.

References: Atlas (283c), Bolòs & Vigo (1995), Curtis & McGough (1988), Hurst (1901), Lousley (1971), Marren (1999).

H. J. KILLICK

Achillea ptarmica Sneezewort

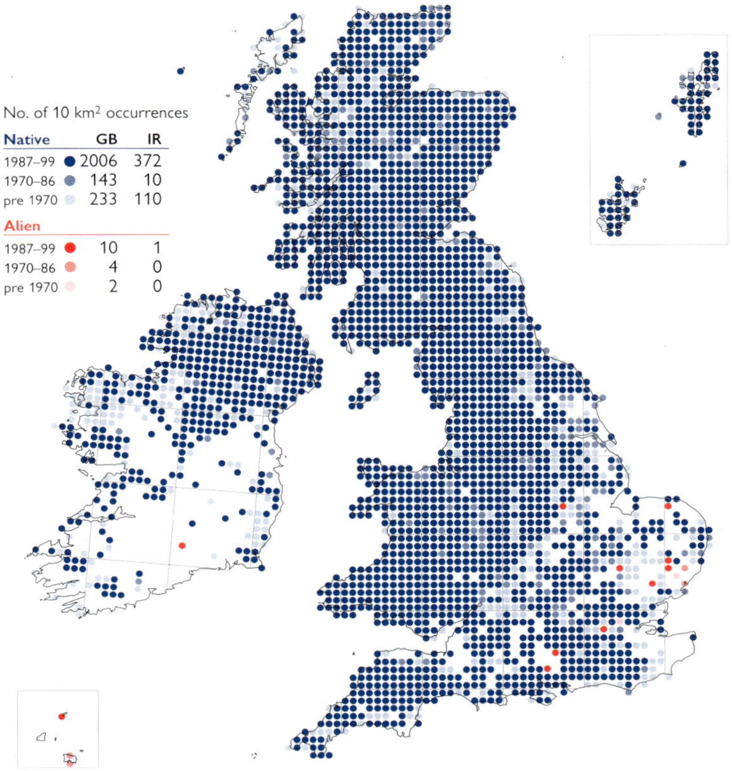

A perennial herb of damp or wet habitats on a wide range of soils, including fen- and water-meadows, rush-pasture, marshes, streamsides, wet heath, springs and flushes on hill slopes and occasionally in wet woodland. It is also established from cultivation in churchyards, and on roadsides and waste ground. 0–770 m (Cross Fell, Cumberland).

Native (change −0.65). *A. ptarmica* was mapped as 'all records' in the 1962 *Atlas*. It is now better recorded in some areas, but has declined in others. Analysis of the database reveals that while many losses occurred before 1950, they have accelerated since then; most are due to drainage and habitat destruction.

Eurasian Boreo-temperate element; widely naturalised outside its native range.

References: Atlas (283b), Hultén & Fries (1986), Meusel & Jäger (1992).

H. J. KILLICK

Achillea millefolium Yarrow

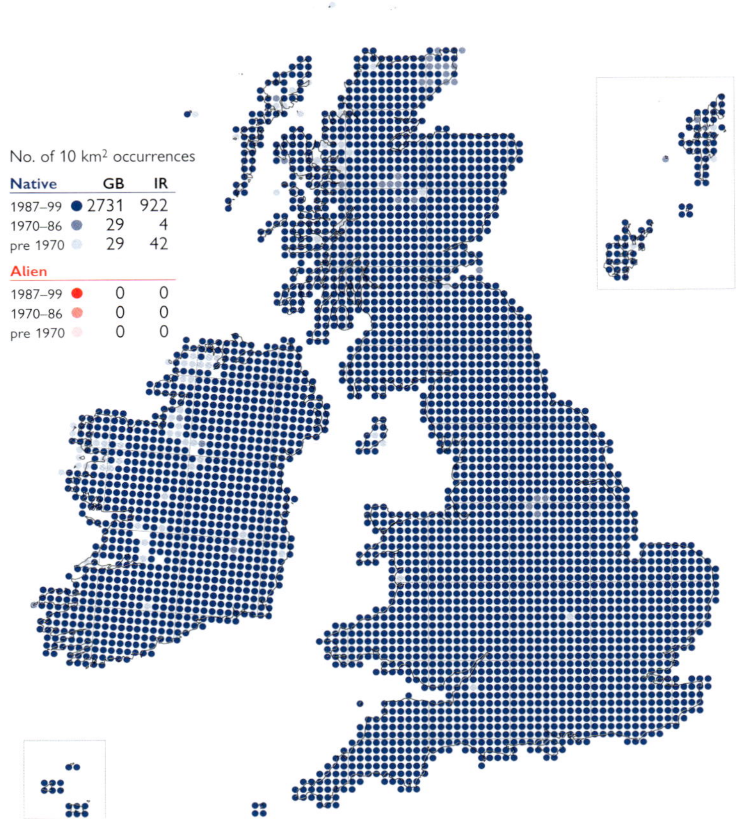

A perennial herb found in all kinds of grassland habitats, ranging from lawns to montane communities irrigated by melting snow-beds; also found on coastal sand dunes and stabilised shingle, waysides and waste ground. It tolerates drought, and grows in most soils except the most nutrient-poor, permanently waterlogged or strongly acidic. 0–1210 m (Ben Lawers, Mid Perth).

Native (change +0.29). There has been no significant change in the distribution of *A. millefolium* since the 1962 *Atlas*.

Eurasian Boreo-temperate element, but naturalised in N. America so distribution is now Circumpolar Boreo-temperate.

References: Atlas (283a), Grime *et al.* (1988), Hultén & Fries (1986), Meusel & Jäger (1992), Warwick & Black (1982).

H. J. KILLICK

Chamaemelum nobile Chamomile

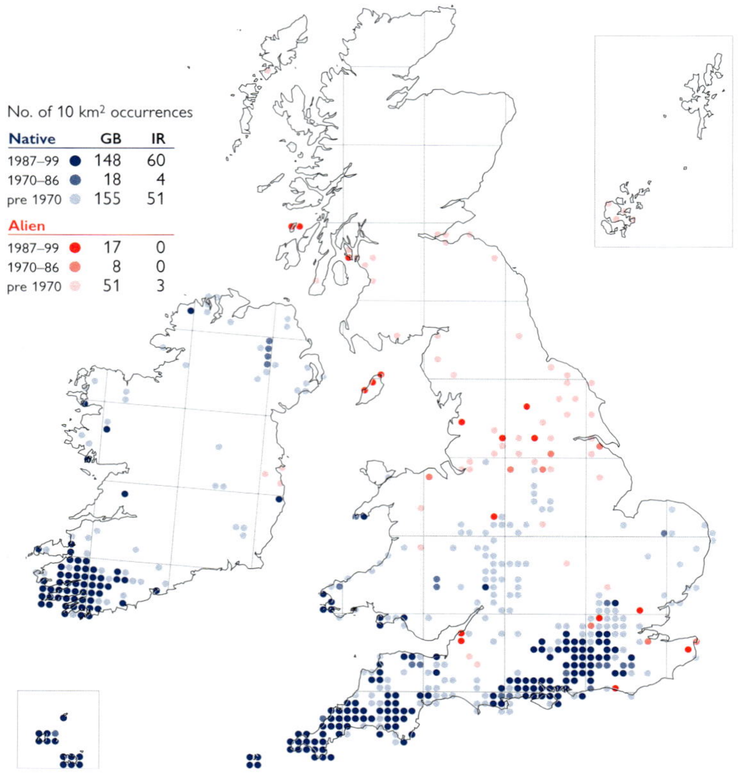

A perennial herb found in moderately acidic, seasonally wet grassland, especially on sandy commons and pastures where mowing, trampling or grazing discourages competitors; also in coastal grassland and on cliffs, where exposure and trampling maintain a short sward. 0–465 m (near Priestleap, S. Kerry).

Native (change –0.92). *C. nobile* had decreased considerably by 1930, and this decline continues due to drainage and the cessation of grazing. In Dorset, for example, over 90% of the sites recorded in the 1930s had gone by 1993 (Byfield & Pearman, 1996). However, it remains stable in its core areas, and is better recorded now than when mapped by Stewart *et al.* (1994).

Suboceanic Southern-temperate element.

References: Atlas (282d), Bolòs & Vigo (1995), Kay & John (1994), Westerhoff & Clark (1992).

H. J. KILLICK

Anthemis arvensis Corn Chamomile

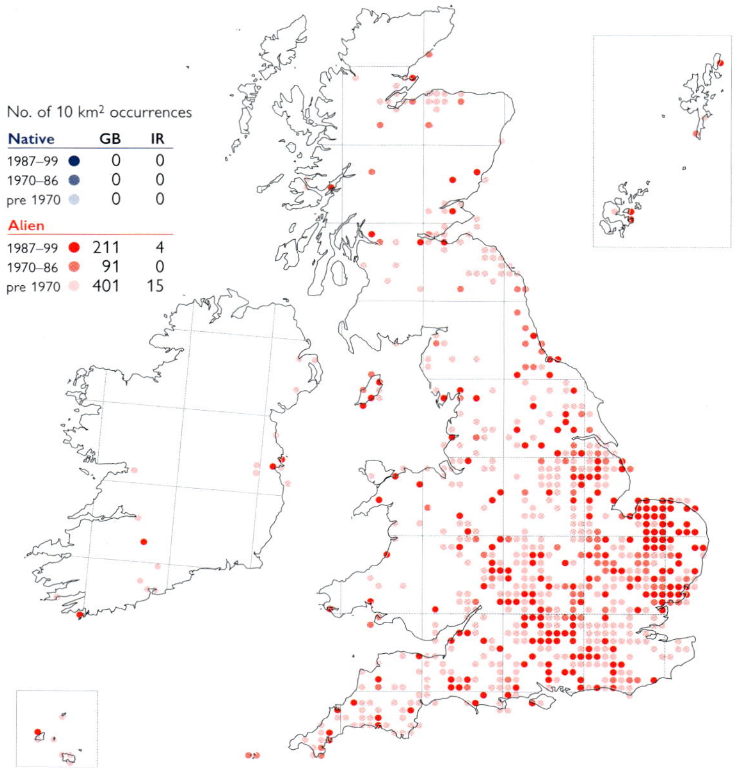

An aromatic annual of light calcareous or sandy soils, growing in arable fields, especially cereals; also in leys, field-borders and waste places, and on roadsides and disturbed ground near the sea. It is occasionally introduced as a contaminant of grass-seed or in wild-flower seed mixtures. Lowland.

Archaeophyte (change –1.79). This species declined substantially in the 20th century, and especially since the 1962 *Atlas*. It was fairly resistant to the first phenoxy herbicides but is more susceptible to other, more recently developed, compounds. Many northern occurrences are casual.

As an archaeophyte *A. arvensis* has a European Southern-temperate distribution; it is widely naturalised outside this range.

References: Atlas (282c), Curtis & McGough (1988), Hultén & Fries (1986), Kay (1971b), Meusel & Jäger (1992).

H. J. KILLICK

Anthemis cotula Stinking Chamomile

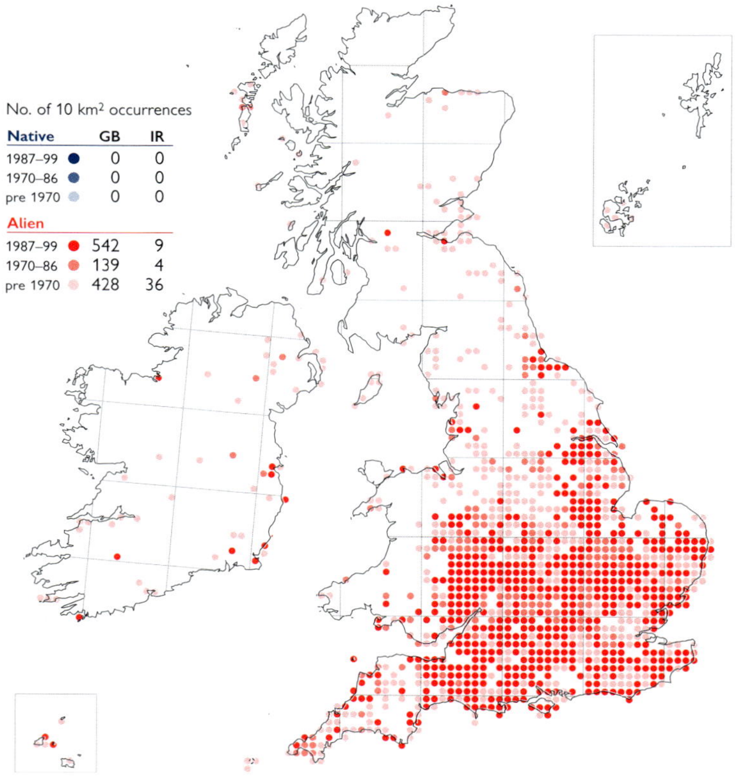

A foetid annual of cereals and other arable crops. In some areas it favours heavy soils, including clay, clay-loam and marl, being replaced by *A. arvensis* on lighter soils, but it can grow on light soils, including those over chalk. Lowland.

Archaeophyte (change –1.60). *A. cotula* has probably been a serious weed of crops since the Iron Age. Although fairly resistant to the first phenoxy herbicides, it has been much reduced by more recent ones. It has declined in many areas since the 1962 *Atlas*. It is a grain-seed casual in Ireland.

As an archaeophyte *A. cotula* has a European Southern-temperate distribution; it is widely naturalised outside this range.

References: Atlas (282b), Hultén & Fries (1986), Kay (1971a), Meusel & Jäger (1992).

H. J. KILLICK

Anthemis tinctoria Yellow Chamomile

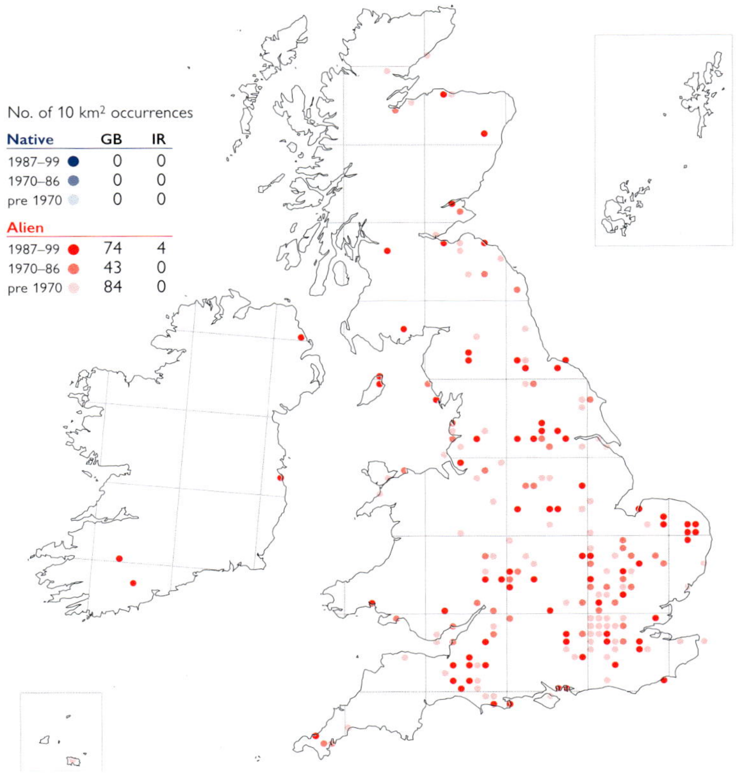

A biennial or perennial herb of waste, rough and marginal land, usually on dry soils. Populations are usually small. Lowland.

Neophyte. *A. tinctoria*, introduced by 1561, was formerly cultivated for a yellow dye, and is still grown in gardens as an ornamental. It was first noted in the wild in 1690 (Durham). It is a casual or an established escape, and perhaps also a bird-seed alien.

A Eurosiberian Temperate species, which has spread northwards in Europe as an escape from cultivation.

References: Hultén & Fries (1986), Meusel & Jäger (1992).

H. J. KILLICK

Chrysanthemum segetum Corn Marigold

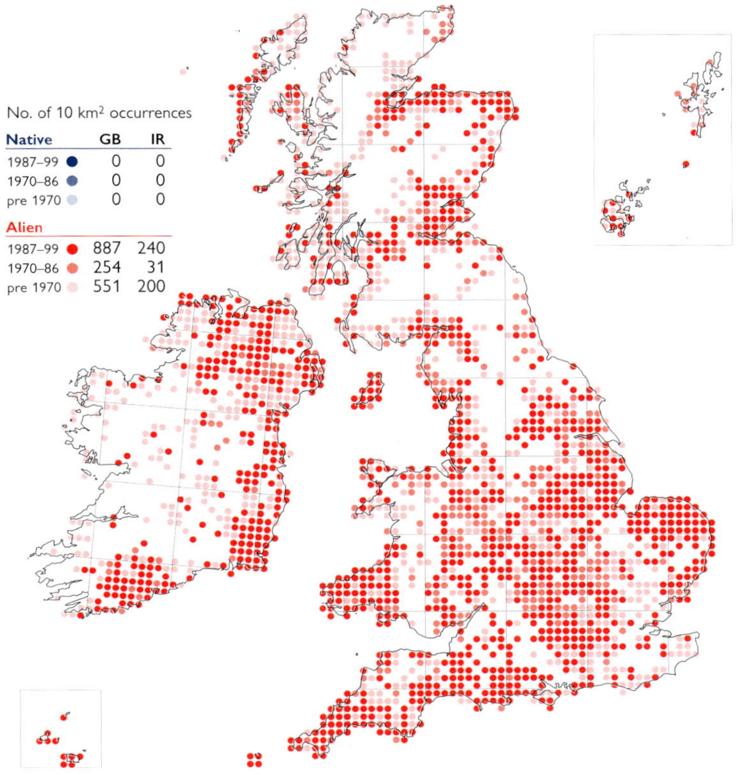

A mainly spring-germinating annual of light, sandy or loamy soils deficient in calcium, found in arable fields and other disturbed habitats, on roadsides, waste ground and rubbish tips. 0–410 m (Eisteddfa Gurig, Cards.).

Archaeophyte (change −1.80). There is a continuous archaeological record of *C. segetum* in Britain from the Iron Age onwards. It was a serious weed in Victorian times, but is now much reduced due to improved seed cleaning, liming, herbicides and the shift to autumn-sown crops. Much of this decline has taken place since 1930.

As an archaeophyte *C. segetum* has a European Southern-temperate distribution; it is widely naturalised outside this range.

References: Atlas (284c), Howarth & Williams (1972), Hultén & Fries (1986), Meusel & Jäger (1992), Wilson (1991).

H. J. KILLICK

Leucanthemum vulgare Oxeye Daisy

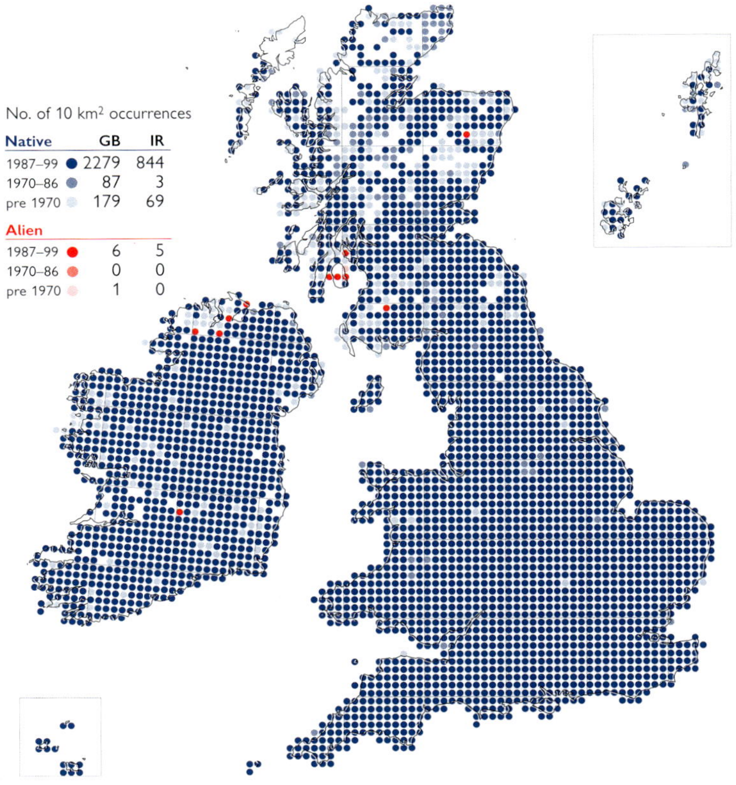

A perennial herb found in many grassy habitats, especially meadows and pastures which are cut or moderately grazed, preferring well-drained, neutral to base-rich soils; also on coastal cliffs, stabilised dunes, waste ground, by railways and newly sown roadsides. It can quickly colonise open ground. 0–845 m (Great Dun Fell, Westmorland).

Native (change −1.14). Throughout much of its range this species has been spread by human activities, sometimes being sown in grass mixtures. However, the map suggests a decline in Scotland, presumably in semi-natural habitats.

Eurosiberian Boreo-temperate element, but widely naturalised so that distribution is now Circumpolar Boreo-temperate.

References: Atlas (284d), Grime *et al.* (1988), Howarth & Williams (1968), Hultén & Fries (1986), Meusel & Jäger (1992).

H. J. KILLICK

Leucanthemum lacustre × *L. maximum* (*L.* × *superbum*) Shasta Daisy

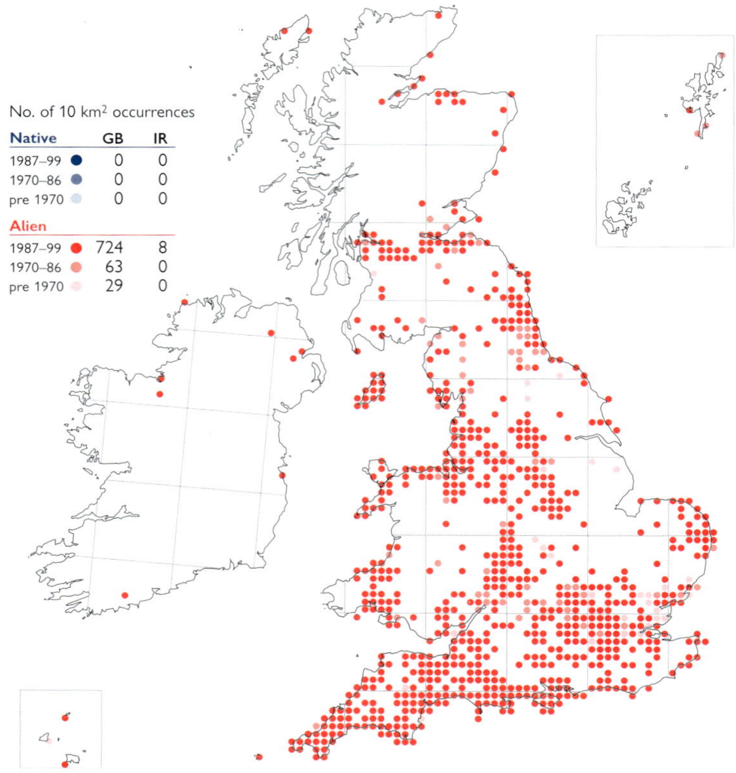

A perennial herb which is well-established in disturbed habitats, including waste and rough ground, quarries, roadsides and railway banks. Although a hybrid, it is fully fertile. Lowland.

Neophyte. *L.* × *superbum*, first introduced to cultivation in Britain in 1816, is now commonly planted in amenity schemes and in grassland on roadside verges, sometimes to the exclusion of *L. vulgare*. It was first recorded from the wild in 1913. It is unevenly recorded.

A hybrid of garden origin.

H. J. KILLICK

Matricaria recutita Scented Mayweed

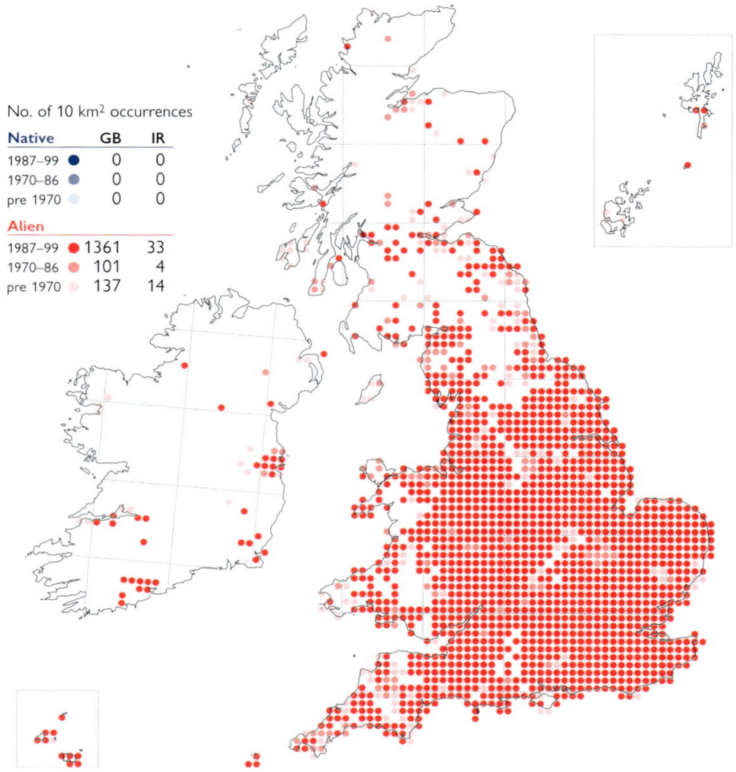

An aromatic annual of arable land, especially in cereal crops, and waste places. It usually occurs on light soils, but is sometimes found on loams and heavy clays. 0–365 m (Kirkstone, Westmorland).

Archaeophyte (change +0.92). It is difficult to account for the apparent increase in this species. In arable fields it has decreased because of herbicides, and is now more frequent in field gateways and margins than actually within the crop. However, it may now be better recorded. It is only casual in Scotland and Ireland.

As an archaeophyte *M. recutita* has a European Southern-temperate distribution, but it is widely naturalised so that its distribution is now Circumpolar Southern-temperate.

References: Atlas (284a), Hultén & Fries (1986), Meusel & Jäger (1992).

H. J. KILLICK

Matricaria discoidea Pineappleweed

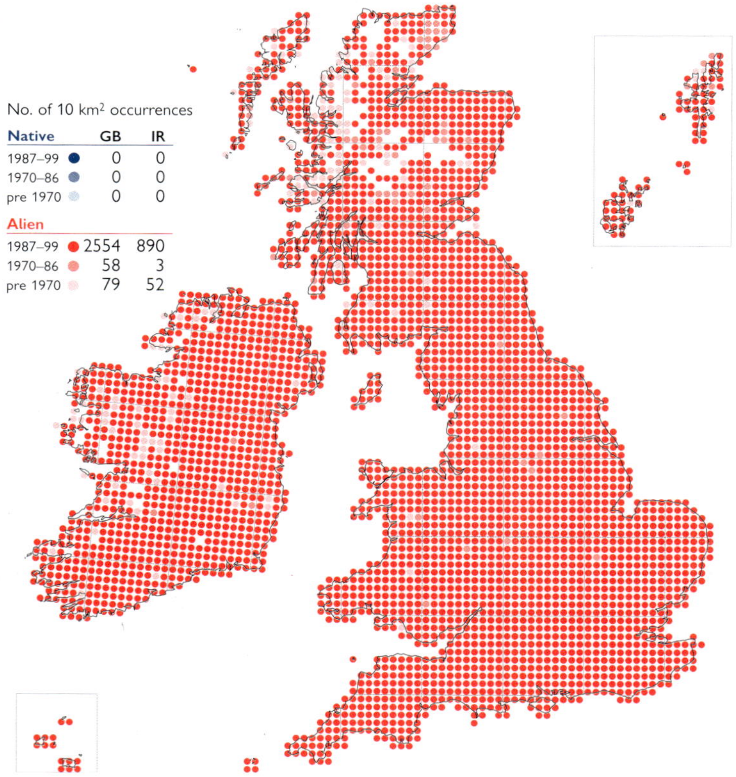

This annual of disturbed, usually fertile ground is found on roadsides, waste ground, tracks and in field gateways, and in arable crops. Generally lowland, but reaching 845 m on Great Dun Fell (Westmorland).

Neophyte (change –0.49). Although cultivated in 1781, this species was first recorded in the wild in Britain as an escape from Kew Gardens in 1871 and in Ireland in 1894, and it became one of the fastest spreading plants in the 20th century. Its spread was aided by the transport of seeds on tyres and footwear. The distribution is now stable.

Probably native of N. & S. Asia and perhaps adjacent parts of N. America; in the 19th and 20th centuries it spread to boreal and temperate zones throughout the N. hemisphere.

References: Atlas (284b), Grime *et al.* (1988), Hultén & Fries (1986), Meusel & Jäger (1992), Salisbury (1964).

H. J. KILLICK

Tripleurospermum maritimum Sea Mayweed

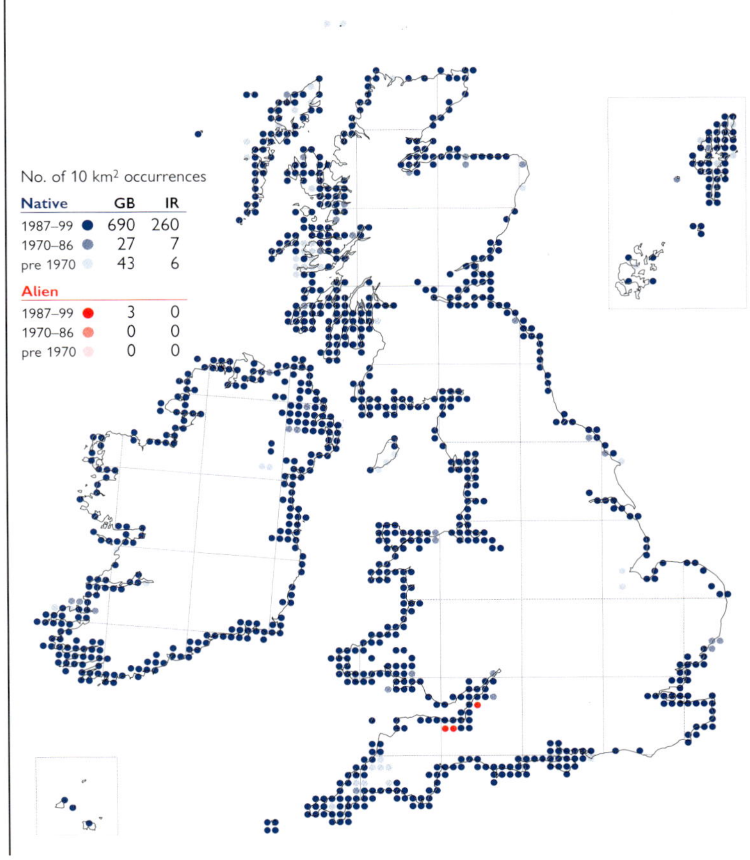

A perennial, sometimes biennial, herb occurring in a wide range of coastal habitats, including open sand, shingle, cliffs, walls and waste ground; also rarely inland on road verges. Lowland.

Native (change for species *sens. lat.* +0.31). *T. maritimum* and *T. inodorum* were treated as conspecific in the 1962 *Atlas*. Many older records are, therefore, only referable to the aggregate, but *T. maritimum sens. str.* now seems to be well recorded and its distribution is probably stable.

Circumpolar Wide-boreal element.

References: Hultén & Fries (1986), Kay (1972), Meusel & Jäger (1992).

H. J. KILLICK

Tripleurospermum inodorum Scentless Mayweed

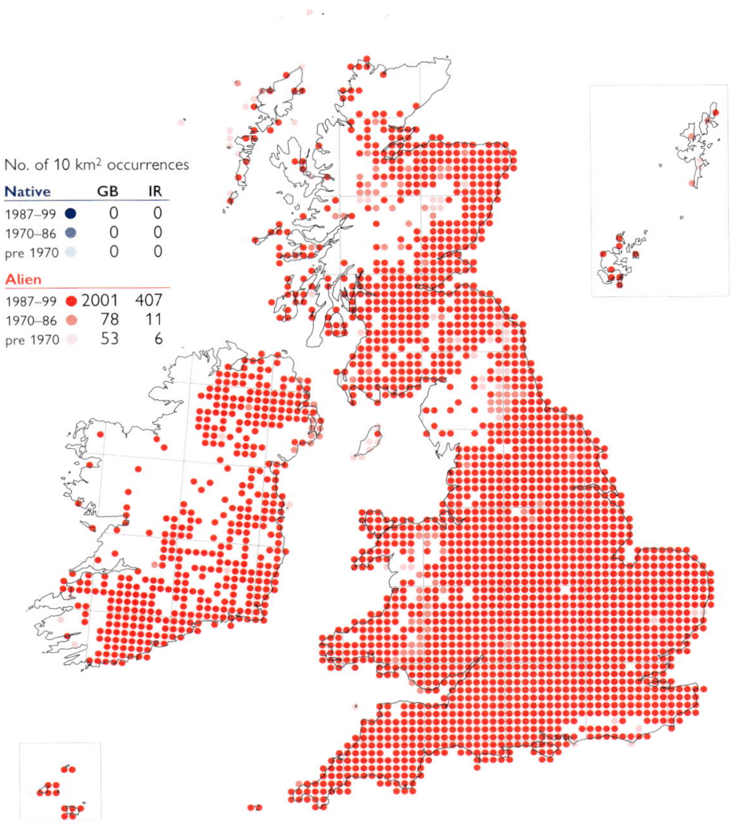

An annual weed of arable fields, farm tracks and gateways, and on waste ground on a wide range of disturbed, fertile soils; also found on roadsides, railway ballast and spoil heaps. Generally lowland, but reaching 530 m in Atholl (E. Perth).

Archaeophyte (change for species *sens. lat.* +0.31). There is a continuous archaeological record of *T. inodorum* from British sites from the Bronze Age onwards. It was not recognised as a full species distinct from *T. maritimum* until 1969 and the two taxa have not been distinguished by recorders in some areas. Its distribution is probably stable.

As an archaeophyte *T. inodorum* has a Eurosiberian Temperate distribution, but it is widely naturalised so that its distribution is now Circumpolar Temperate.

References: Grime *et al.* (1988), Hultén & Fries (1986), Kay (1969, 1994), Meusel & Jäger (1992).

H. J. KILLICK

Cotula coronopifolia Buttonweed

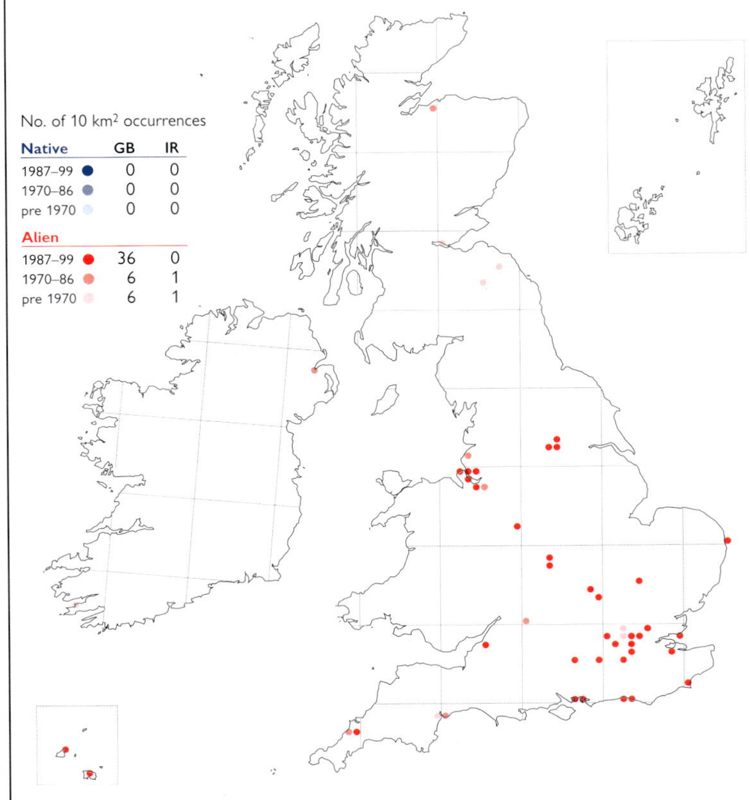

An annual to perennial herb, widely grown in gardens and also found as a wool alien. It has become naturalised in wet and marshy places, in greatest abundance on open tidal saline mud, but also in inland sites, often in areas of mining subsidence that are flooded in winter. Lowland.

Neophyte. This species, which was cultivated in Britain by 1683, has been known in the wild since 1869 (Middlesex). It has increased, especially recently, and extended its range into non-saline areas.

Native of S. Africa, naturalised (especially in coastal regions) in W. Europe, western N. America, S. America, Australasia and elsewhere.

References: Clement (1993b), Hultén & Fries (1986), Martin (1993), Meusel & Jäger (1992).

H. J. KILLICK

Senecio cineraria Silver Ragwort

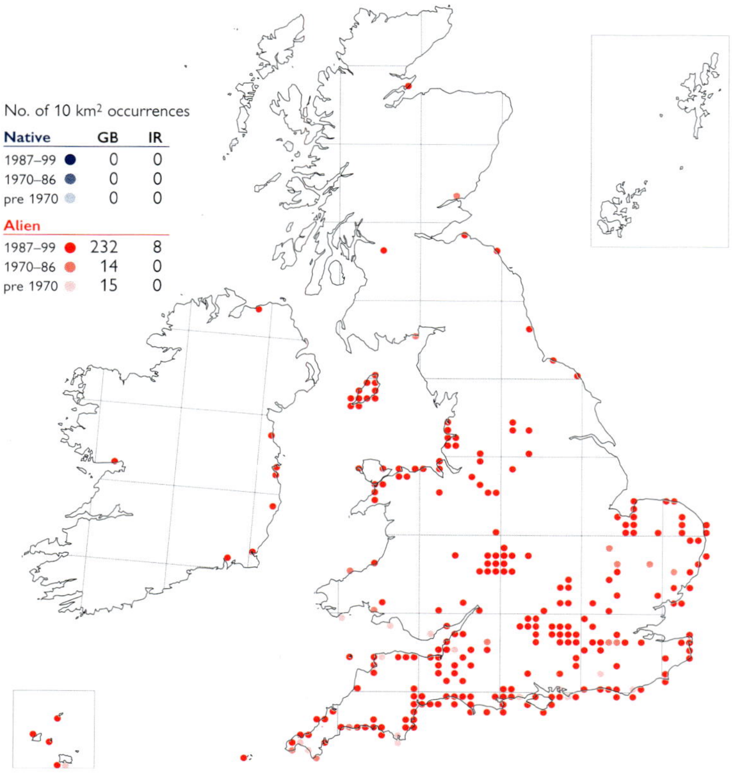

No. of 10 km² occurrences

Native	GB	IR
1987–99 ●	0	0
1970–86 ●	0	0
pre 1970 ●	0	0
Alien		
1987–99 ●	232	8
1970–86 ●	14	0
pre 1970 ●	15	0

An evergreen shrub, sometimes well-established and abundant on cliffs and rough ground near the sea and occurring as a casual inland on rubbish tips, roadside verges and waste ground. Lowland.

Neophyte (change +2.73). *S. cineraria*, introduced into Britain by 1633, was recorded in the wild by 1893. It is increasingly grown as an ornamental plant in gardens, and increasingly escapes. The map shows a substantial increase since the 1962 *Atlas*, which cannot be entirely due to the better recording of aliens.

Native of the W. & C. Mediterranean region.

References: Atlas (273d), Meusel & Jäger (1992).

H. J. KILLICK

Senecio cineraria × *S. jacobaea* (*S.* × *albescens*)

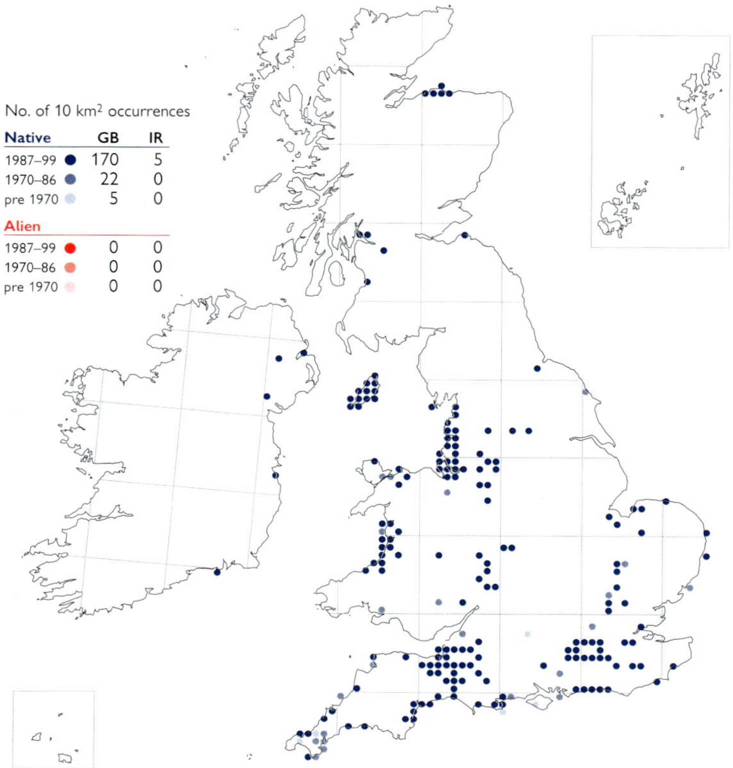

No. of 10 km² occurrences

Native	GB	IR
1987–99 ●	170	5
1970–86 ●	22	0
pre 1970 ●	5	0
Alien		
1987–99 ●	0	0
1970–86 ●	0	0
pre 1970 ●	0	0

A perennial herb of walls, pavement cracks, waste places and other urban habitats, and in coastal areas on sand dunes and shingle. Some populations may be persistent, but it also occurs as a casual. It is fertile, variable, and backcrosses with the parents. Lowland.

A spontaneous hybrid between native and alien parents. This hybrid reportedly first arose in Shropshire in 1836. It then appeared in Killiney Bay, Dublin in 1902, where it persists, and then in Cornwall in 1906. It is recorded in many places where *S. cineraria* is not found, and appears to be spreading. It arises both spontaneously and as a garden escape, but these have not been differentiated and all records are mapped as if they are native.

Wider distribution uncertain.

References: Burbidge & Colgan (1902), Colgan (1904), Murphy (1981), Stace (1975).

H. J. KILLICK

Senecio fluviatilis Broad-leaved Ragwort

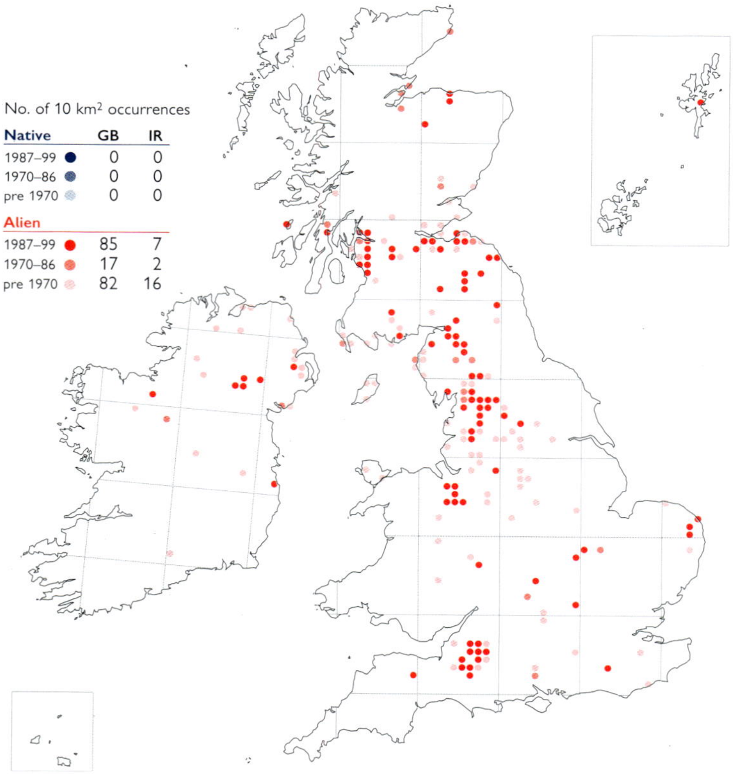

No. of 10 km² occurrences

Native	GB	IR
1987–99 ●	0	0
1970–86 ●	0	0
pre 1970 ●	0	0
Alien		
1987–99 ●	85	7
1970–86 ●	17	2
pre 1970 ●	82	16

A tall perennial herb, naturalised by streams and rivers, and in fens, fen-woodland, swamps and marshy grassland. Lowland.

Neophyte (change +0.03). Introduced before 1600 and originally grown for medicinal use, the first record of *S. fluviatilis* in the wild was in 1633. The somewhat clustered distribution of records suggest independent spread from several different sites; it is now grown in gardens, so many recent records are probably garden escapes. It has persisted in many places, especially in the north, although some pre-1930 colonies have now gone.

S. fluviatilis has a Eurosiberian Temperate distribution; it is naturalised in N.W. Europe beyond its native range.

References: Atlas (273a), Halliday (1997), Meusel & Jäger (1992).

H. J. KILLICK

Senecio paludosus Fen Ragwort

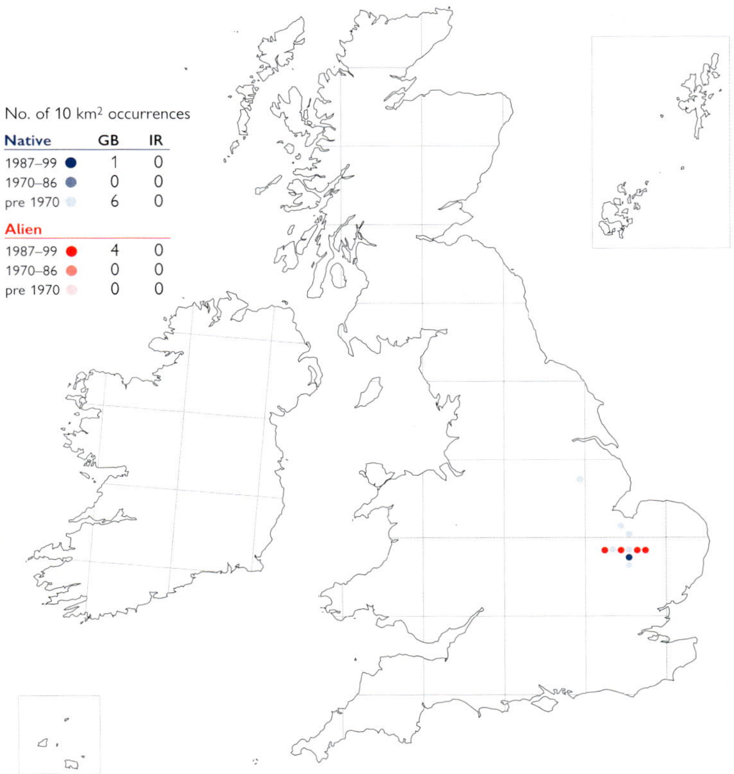

A long-lived perennial herb of tall-herb fens and ditches. The roadside ditch which is now its only native site is usually flooded in winter but dry in summer. Seed-set is poor there, but plants cultivated from material from this site show improved seed-set both in cultivation and at transplantation sites. Lowland.

Native. Many sites for this species were drained in the 18th and 19th centuries. There was no substantiated record between 1857 and 1972, when it was rediscovered near Ely (Cambs.). The species has been reintroduced in or near some of its historic sites (e.g. Wicken Fen, Cambs.).

Eurosiberian Temperate element, with a continental distribution in W. Europe.

References: Atlas (272d), Hultén & Fries (1986), Meusel & Jäger (1992), Wigginton (1999).

M. J. WIGGINTON

Senecio smithii Magellan Ragwort

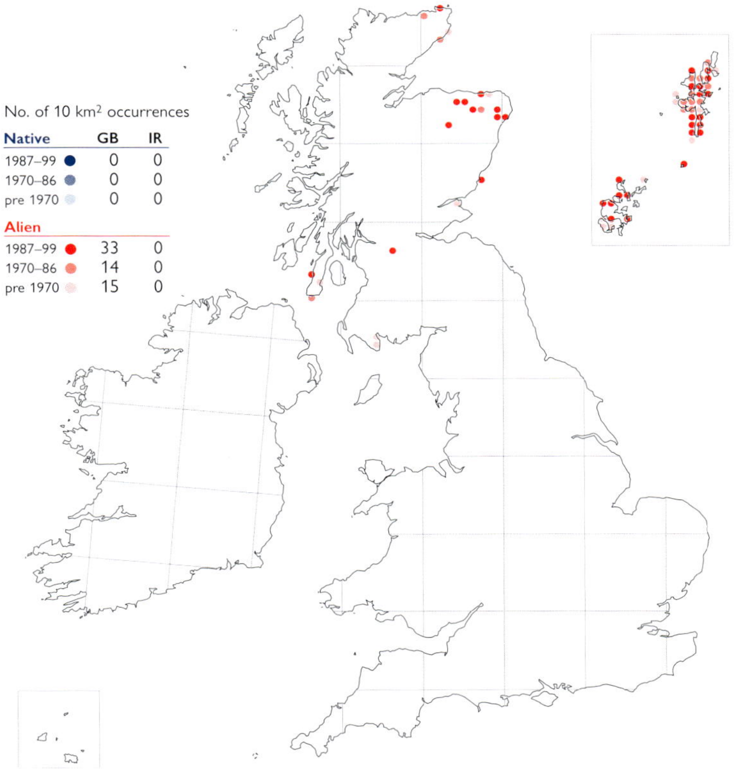

A tall perennial herb found naturalised as a garden escape or relic of cultivation in grassy meadows and pastures, on roadsides, by lakes and streams and in ditches. Lowland.

Neophyte. This species was introduced to Britain in 1895. There is a widely held belief that it was introduced by whalers who brought it back from sailing trips to Patagonia and Chile, and it is known locally as 'Falkland Islands Daisy'. However, Scott & Palmer (1987) conclude that most of the Shetland material arrived through normal horticultural channels. It was known from the wild by the 1920s and its distribution is probably stable.

Native of temperate S. America.

Reference: Druce (1928).

H. J. KILLICK

Senecio jacobaea Common Ragwort

Native	**GB**	**IR**
1987–99	2662	949
1970–86	41	1
pre 1970	36	32
Alien		
1987–99	0	0
1970–86	0	0
pre 1970	0	0

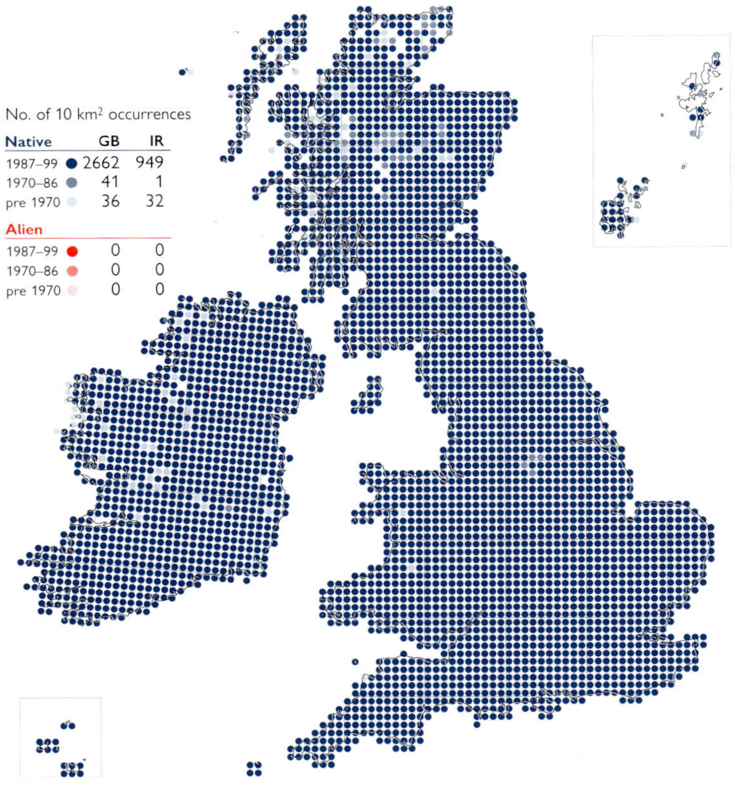

A biennial or perennial herb, widespread in grassland and especially abundant in neglected, rabbit-infested or overgrazed pastures; it also grows on sand dunes, in scrub, open woods and along woodland rides, waste ground, road verges and waysides, and on rocks, screes and walls. 0–670 m (Atholl, E. Perth, and on Mangerton, S. Kerry).

Native (change +0.11). The distribution of *S. jacobaea* is unchanged from the map in the 1962 *Atlas*. It is a notifiable weed, subject to statutory control, but this has clearly had little, if any, effect on its distribution or abundance.

Eurosiberian Temperate element.

References: Atlas (271a), Bain (1991), Grime *et al.* (1988), Harper & Wood (1957), Hultén & Fries (1986), Meusel & Jäger (1992), Wardle (1987).

H. J. KILLICK

Senecio aquaticus Marsh Ragwort

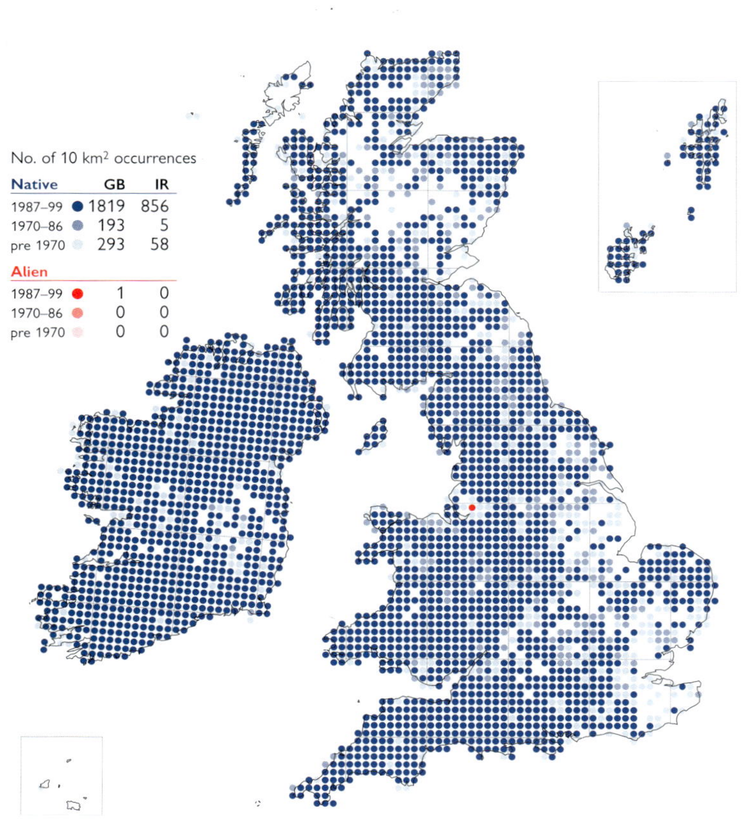

A biennial, sometimes perennial, herb occurring in marshes, wet meadows, rush-pastures, and by streams, ponds and ditches. Generally lowland, but reaching 460 m in Weardale (Co. Durham) and in Co. Wicklow.

Native (change –0.92). This species was mapped as 'all records' in the 1962 *Atlas*. There has been widespread decline in S. and E. England, and analysis of the database reveals that most losses have occurred since 1950. The decline is probably due to drainage of wet meadows and agricultural intensification.

European Temperate element.

References: Atlas (271b), Hultén & Fries (1986), Meusel & Jäger (1992).

H. J. KILLICK

Senecio aquaticus × *S. jacobaea* (*S.* × *ostenfeldii*)

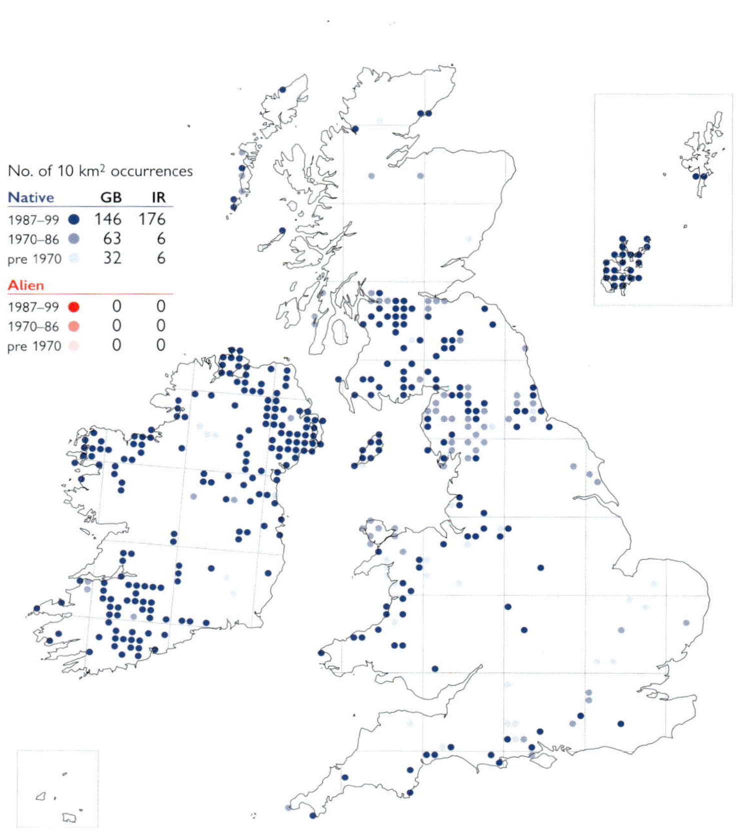

A biennial or perennial herb of damp grassland, riversides and waste places, usually growing with its parents. Generally lowland, but reaching *c.* 350 m at Leadgate (Cumberland).

Native. *S.* × *ostenfeldii* is probably under-recorded, having been overlooked or recorded as *S. jacobaea*. Less than 15% of the achenes ripen, but it backcrosses and very large and variable hybrid swarms can occur, particularly in the north and west.

Widespread in W. & C. Europe.

References: Harper & Wood (1957), Stace (1975).

H. J. KILLICK

Senecio erucifolius Hoary Ragwort

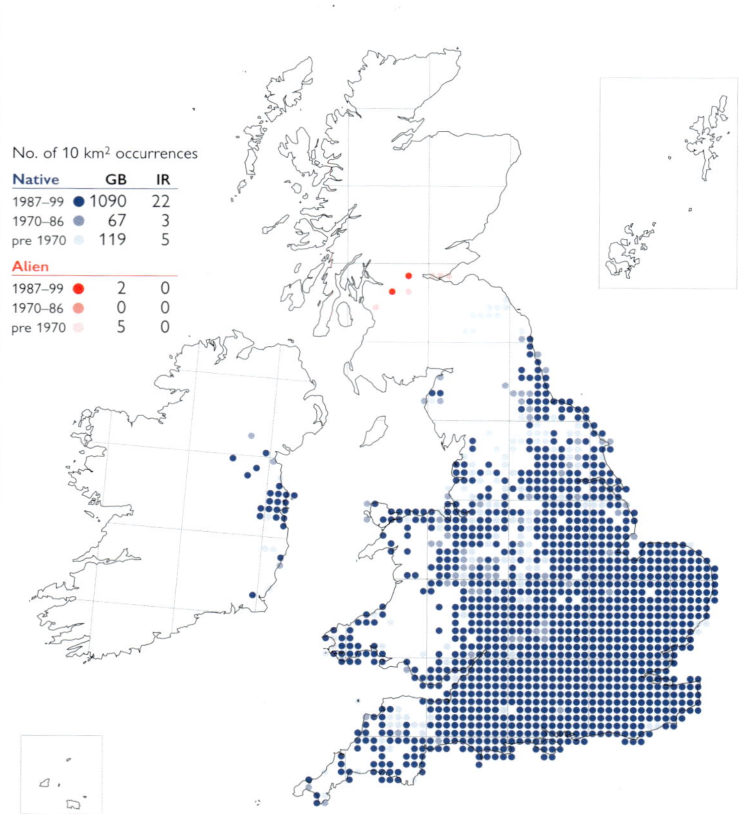

A perennial herb of grassland and disturbed habitats, including hay meadows and pastures, chalk and limestone downland, field-borders, railway banks, roadsides, waste places, shingle banks and fixed sand dunes; it is usually found on neutral or calcareous soils, especially clays that are wet in winter but baked dry in summer. Lowland.

Native (change –0.03). The distribution of *S. erucifolius* shows little change since the 1962 *Atlas*, but it may have declined locally, particularly in the English Midlands and at the northern edges of its range.

Eurosiberian Temperate element; also in E. Asia.

References: Atlas (271c), Hultén & Fries (1986), Meusel & Jäger (1992).

H. J. KILLICK

Senecio squalidus Oxford Ragwort

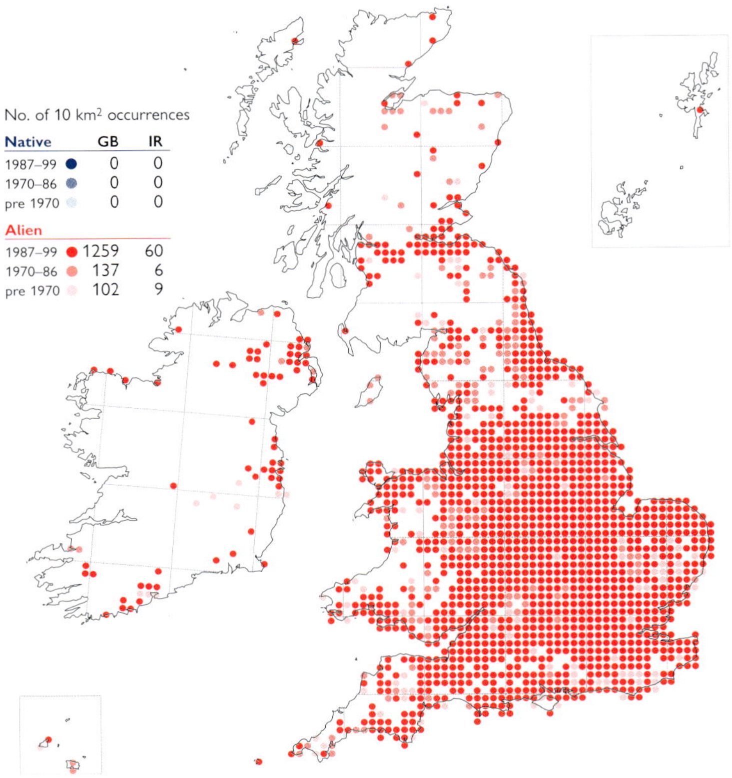

A short-lived perennial herb of waste places, walls, railways, cinders, roadsides and gardens, where it is often thoroughly established on well-drained soils. Lowland.

Neophyte (change +0.77). This species was first recorded in 1794 as an escape from Oxford Botanic Garden. It was recorded at scattered localities until the 1850s, but spread rapidly after reaching the railway in Oxford in *c.* 1879. Since the 1962 *Atlas* its range has increased and it has spread off railways into other habitats. In Ireland it was naturalised in Cork city by 1845 but its spread from there has been relatively slow.

S. squalidus appears to have arisen in cultivation in the Oxford Botanic Garden (Abbott *et al.*, 2000).

References: Atlas (271d), Grime *et al.* (1988), Kent (1956, 1957, 1960, 1964a, b).

H. J. KILLICK

Senecio squalidus × *S. viscosus* (*S.* × *subnebrodensis*)

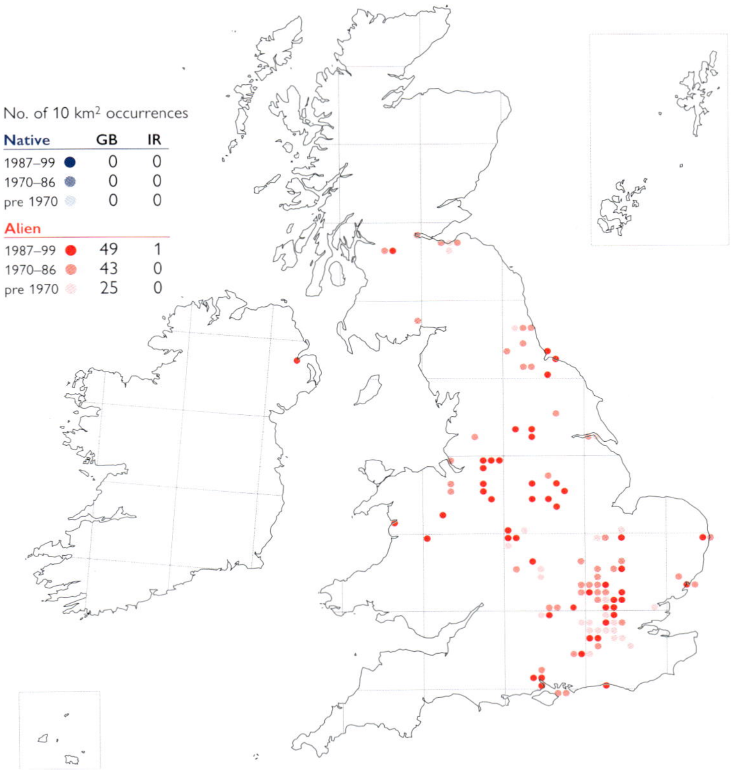

An annual or biennial herb of waste and disturbed ground, especially railways, rubble and clay-gravel paths. It is usually found where the parents grow together. Lowland.

A spontaneous hybrid between two alien parents. This hybrid was discovered in the London area in 1944, when it was particularly frequent on bombed sites. It is completely sterile, with infertile pollen and no ripe fruits, and there is no evidence of introgression.

This hybrid was originally described from Romania, but its distribution in mainland Europe is not well-documented.

References: Lousley (1946), Stace (1975).

H. J. KILLICK

Senecio squalidus × *S. vulgaris* (*S.* × *baxteri*)

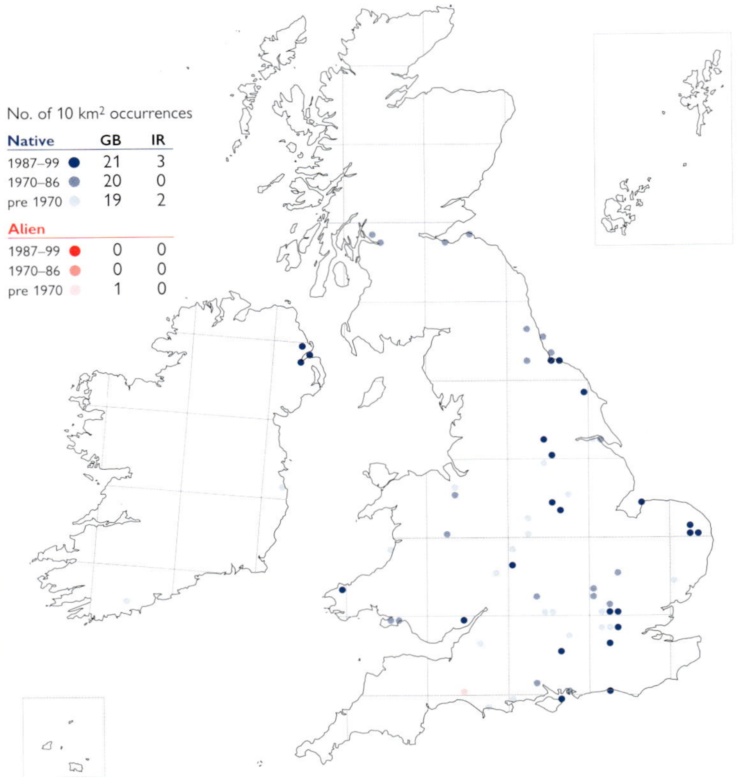

An annual or biennial herb of waste and disturbed ground, especially by railways. It is a sterile tetraploid which is unable to reproduce vegetatively or by seed. Lowland.

A spontaneous hybrid between native and alien parents. This hybrid has been known since 1892 (Oxon) and 1906 (Glamorgan) and appears sporadically.

Wider distribution uncertain.

References: Brenan (1948), Crisp (1972), Stace (1975).

H. J. KILLICK

Senecio cambrensis Welsh Groundsel

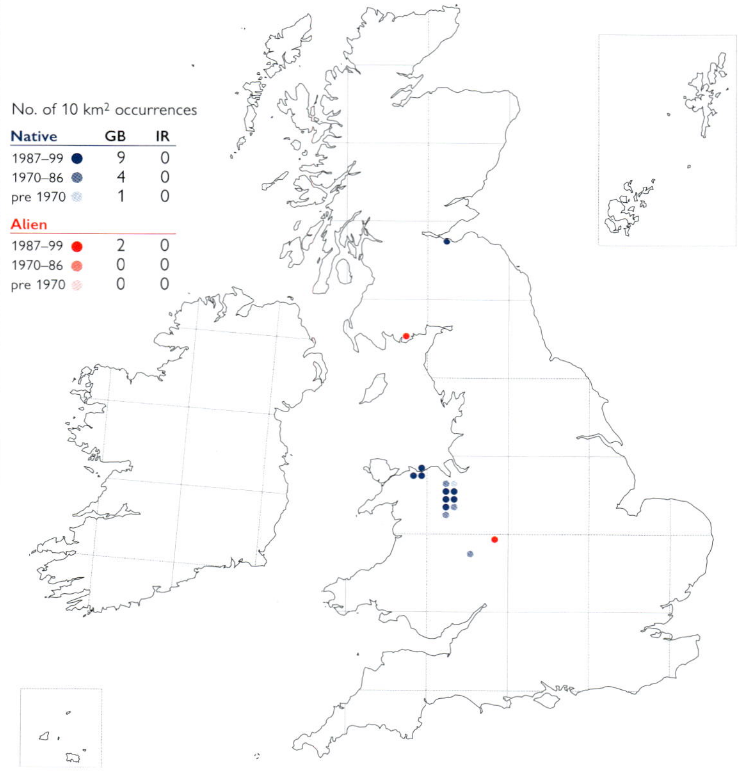

Typically an annual, though sometimes a short-lived perennial, *S. cambrensis* grows in open or disturbed sites, including waste and rough ground, on roadsides and footpaths, and in cracks in walls. Lowland.

Native. This species was first collected in Denbighshire in 1925, although the specimen was not identified as *S. cambrensis* until 1957. The first confirmed record came from Flintshire in 1948. *S. cambrensis* is believed to have arisen from the hybrid between *S. vulgaris* and *S. squalidus* by chromosome doubling. The Welsh and Scottish populations arose independently of each other; it was first recorded in Midlothian in 1974.

Endemic.

References: Atlas (271c), Ingram & Noltie (1995), Wigginton (1999), Wynne (1993).

H. J. KILLICK

Senecio vulgaris Groundsel

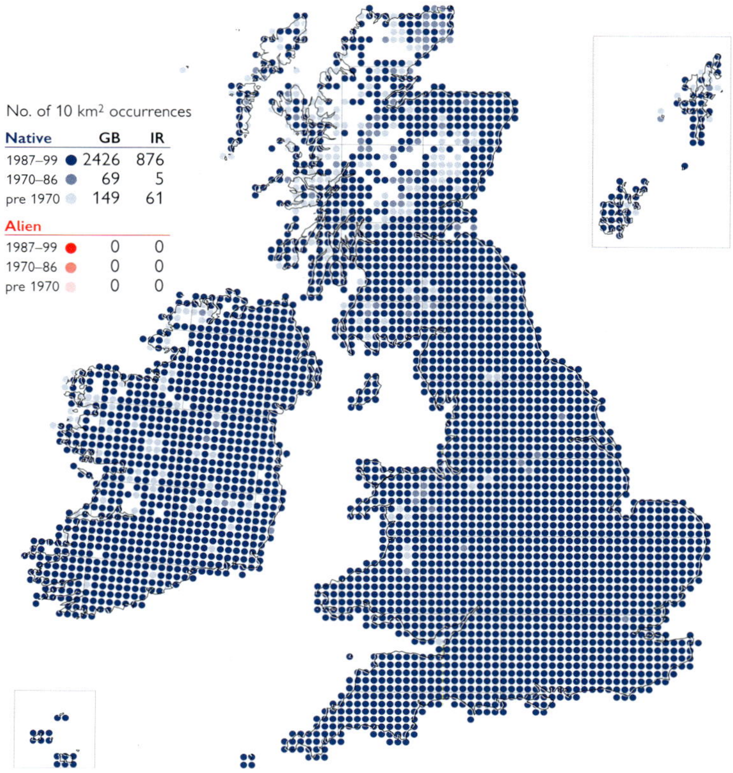

An annual of open and disturbed ground, occurring in semi-natural habitats on sand dunes and coastal cliffs, and as a weed in waste places, gardens, arable fields and other open habitats. 0–550 m (Moor House, Westmorland).

Native (change −1.08). There has been no significant change in the distribution of *S. vulgaris* since the 1962 *Atlas*, although there appears to have been a decline in the Scottish Highlands, possibly due to abandoning of marginal cultivations. The two rayed variants (var. *denticulatus* and var. *hibernicus*) were mapped by Perring & Sell (1968).

European Southern-temperate element, but widely naturalised so that distribution is now Circumpolar Southern-temperate.

References: Atlas (272c), Atlas Supp. (71b), Grime *et al.* (1988), Hultén & Fries (1986), Meusel & Jäger (1992).

H. J. KILLICK

Senecio sylvaticus Heath Groundsel

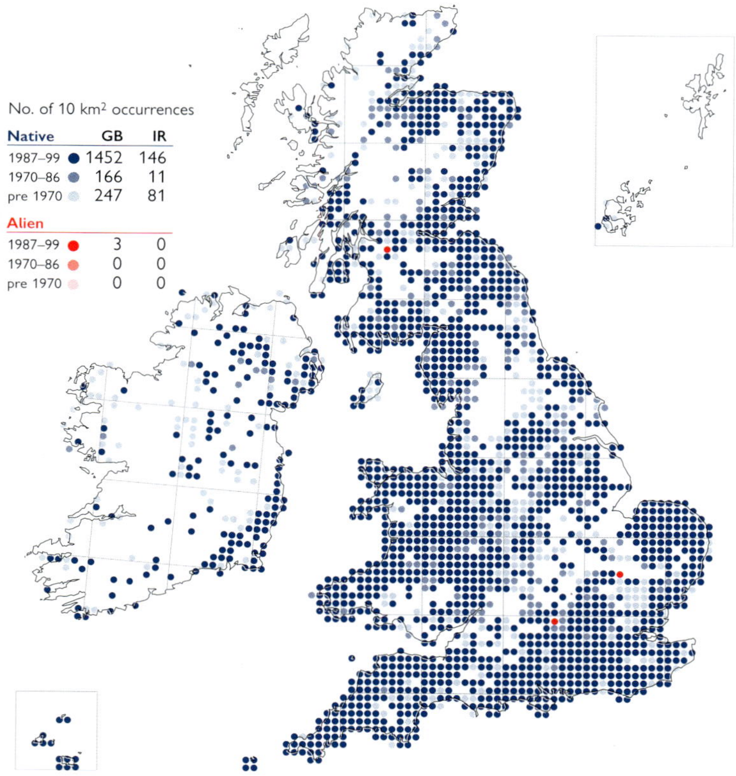

An annual of open habitats on heaths, in cleared and burnt woodland, on bushy commons, banks and sea-cliffs, usually growing on sandy, non-calcareous soils. 0–365 m (Rinns of Kells, Kirkcudbrights.).

Native (change +0.09). There has been little change in the distribution of *S. sylvaticus* since the 1962 *Atlas*.

European Temperate element.

References: Atlas (272a), Hultén & Fries (1986), Meusel & Jäger (1992).

H. J. KILLICK

Senecio viscosus Sticky Groundsel

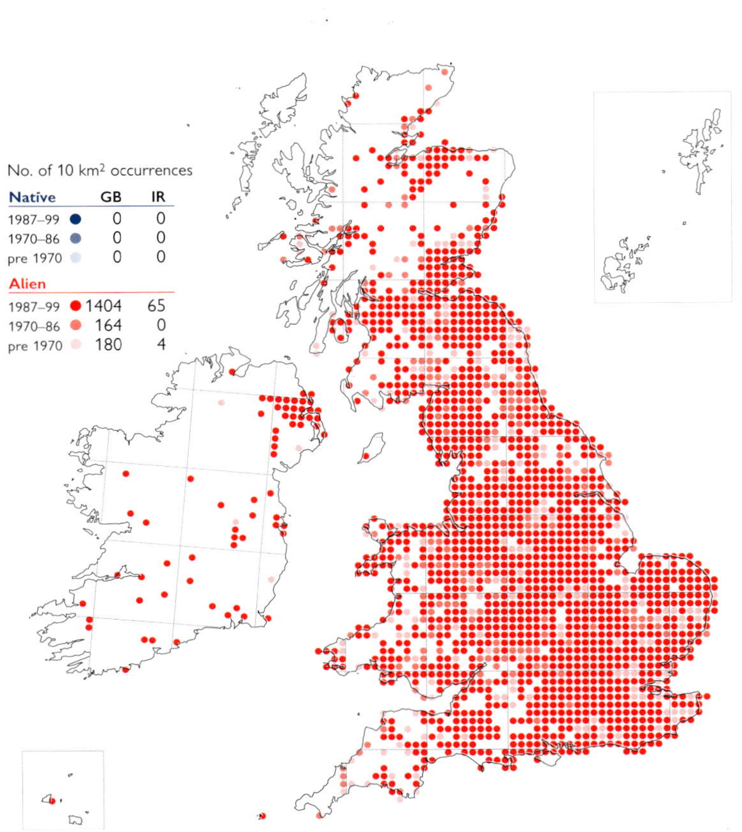

An annual of free-draining disturbed substrates including sands, gravels and cinders, found on roadsides, banks, wall-tops, pavements, railway ballast, coastal shingle and dunes, in gravel-pits, and on open rough and waste ground. 0–430 m (Nenthead, Cumberland).

Neophyte (change +0.63). This species, first recorded in 1660, has spread greatly since 1900, especially along roads and railways. Since the 1962 *Atlas* it has spread further into S.W. England and Ireland, and consolidated its range elsewhere. Genetically dwarf variants occur on maritime shingle, where the species could conceivably be native.

A European Temperate species which has spread markedly in W. & N. Europe in recent centuries.

References: Atlas (272b), Akeroyd *et al.* (1978), Grime *et al.* (1988), Hultén & Fries (1986), Meusel & Jäger (1992), Salisbury (1964). H. J. KILLICK

Tephroseris integrifolia subsp. *integrifolia* Field Fleawort

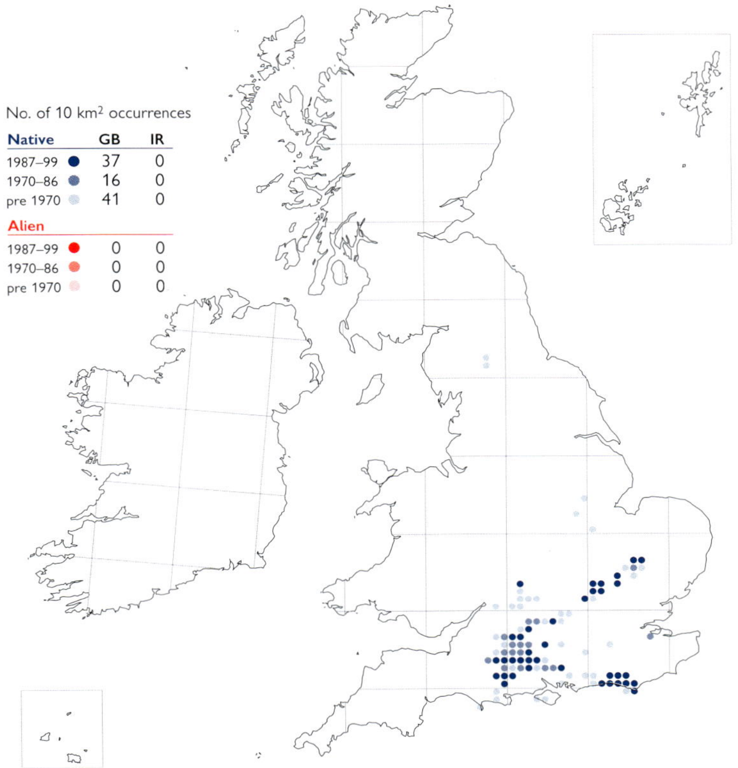

A biennial or perennial herb, growing on shallow soils on chalk and, rarely, on oolitic limestone. It occurs in short grassland on downland, ancient earthworks and tracks, favouring warm, dry, S.-facing sites. Lowland, but formerly at 550 m above Brough (Westmorland).

Native (change for species −0.79). This subspecies has decreased since 1960, and continues to do so, possibly due to under-grazing, agricultural improvement and scrub encroachment. The two Westmorland records probably refer to an undescribed, and now presumably extinct, taxon.

T. integrifolia sens. lat. is a complex of taxa with a disjunct Circumpolar Wide-boreal distribution (Hultén & Fries, 1986; Meusel & Jäger, 1992). Subsp. *integrifolia* is widespread in Europe but its eastern limit is unclear.

References: Halliday (1997), Smith (1979), Stewart *et al.* (1994). H. J. KILLICK

Tephroseris integrifolia subsp. *maritima*

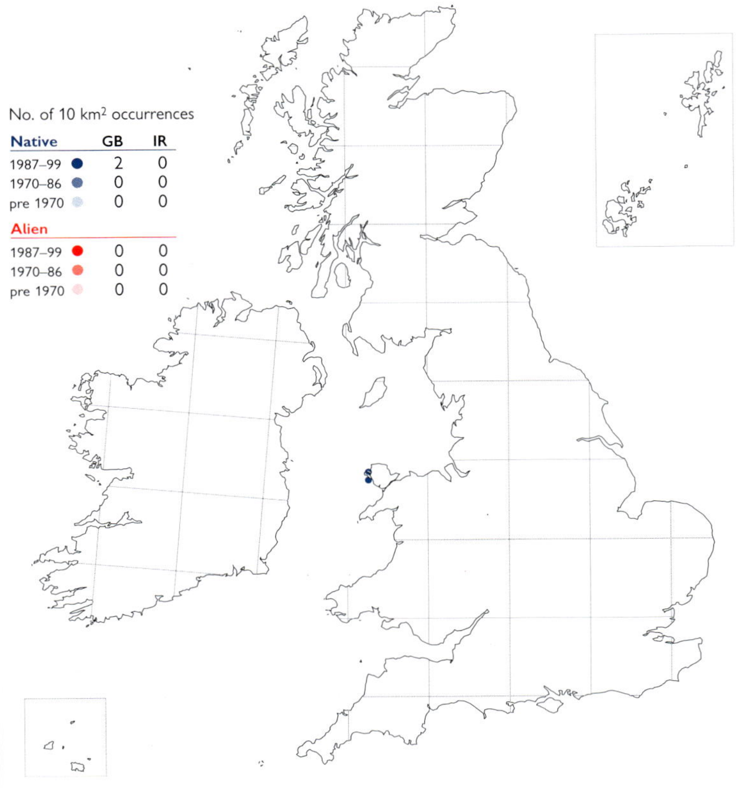

A biennial or short-lived perennial herb occurring on mildly acidic to neutral soils on grassy coastal cliff-slopes, and on ledges and in crevices on the cliff-face. Most populations occur on exposed slopes with a S.W. to N.W. aspect. Reproduction is mainly by seed. Lowland.

Native. This distinctive subspecies has only been recorded on Anglesey. It was discovered in 1813 and named as a variety by Syme in 1866. Populations appear to be stable. The taxonomic position of subsp. *maritima* perhaps requires re-assessment in the context of the complex variation shown in Europe by *T. integrifolia* and related species (cf Kay & John, 1995).

Endemic.

References: Smith (1979), Wigginton (1999).

H. J. KILLICK

Tephroseris palustris Marsh Fleawort

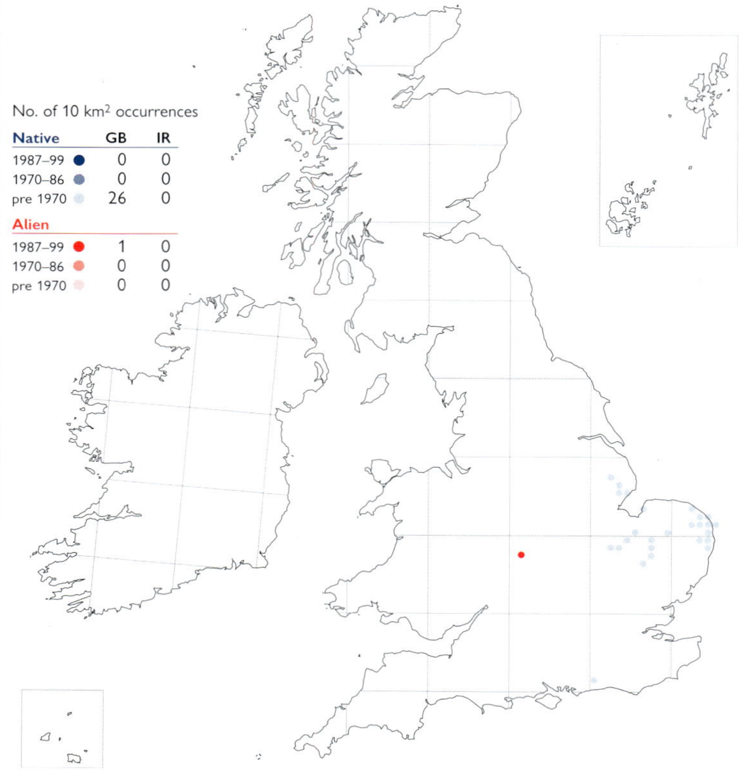

A biennial or short-lived perennial herb of pond margins and fen ditches; in Holland, it is known to be an early colonist of the bare mud on land newly reclaimed from the sea (polders). Lowland.

Native. *T. palustris* was first recorded in Britain in 1650 but had become extinct by the end of the 19th century. It was lost from Sussex in 1725, from the Cambridgeshire and Lincolnshire fens by the early 1800s, and from the Norfolk Broads by the 1890s. Its last native record was from Dersingham (Norfolk) in 1899. Drainage and agricultural changes probably caused its demise at most sites.

Circumpolar Wide-boreal element.

References: Atlas (273b), Hultén & Fries (1986), Marren (1999).

H. J. KILLICK & S. J. LEACH

Brachyglottis 'Sunshine' (*? B. compacta × B. laxifolia*) Shrub Ragwort

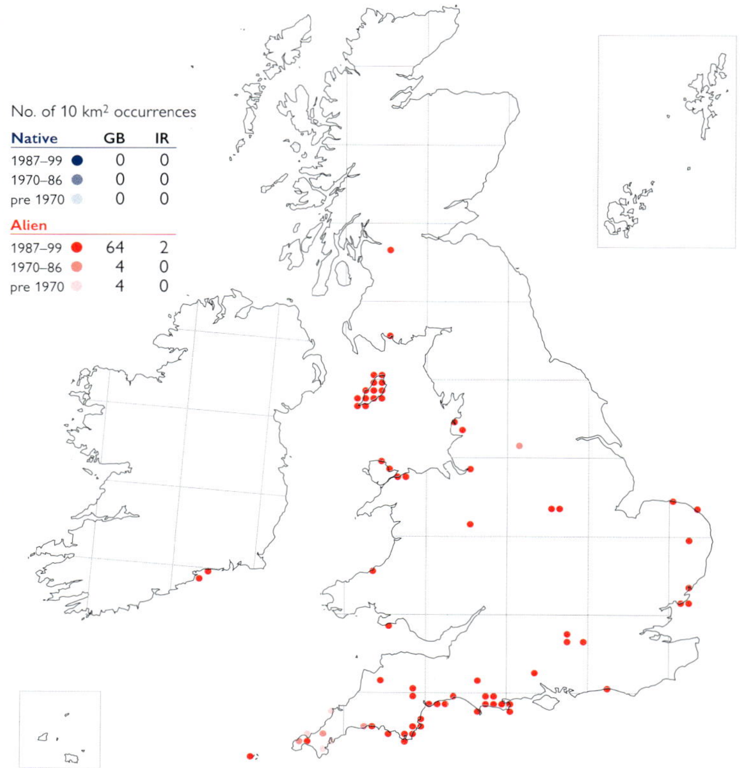

A spreading shrub, much grown in parks, gardens and on roadsides in urban areas, and readily persisting in the wild in grassy places and on rough ground and sand dunes. Lowland.

Neophyte. This plant (known as '*Senecio greyi*' in horticulture) originated in cultivation around 1910 and is becoming increasingly well-established in the wild. Most occurrences are, however, very recent and it was not recorded from the wild until 1981. Some undated records, mapped as pre-1970, may precede this.

A hybrid of garden origin and uncertain parentage.

Reference: Bean (1980).

H. J. KILLICK

Doronicum pardalianches Leopard's-bane

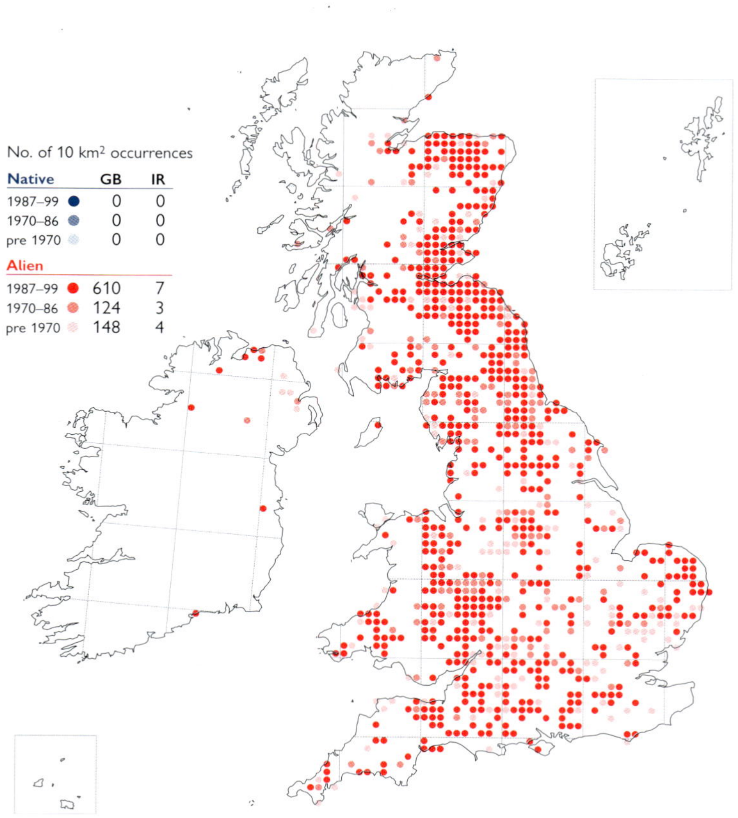

A rhizomatous perennial herb, well-naturalised in woods, plantations, and on roadsides and other shaded places. Lowland.

Neophyte (change +0.89). This species has been cultivated in Britain since the 16th century and is widely grown for ornament and formerly for medicinal purposes. It was first recorded in the wild in 1633 from Northumberland. It has consolidated and slightly expanded its range since the 1962 *Atlas*. Some mapped records may represent the hybrids *D.* × *willdenowii* and *D.* × *excelsum*.

Native of W. Europe, east to Germany and Italy.

References: Atlas (274a), Meusel & Jäger (1992).

H. J. KILLICK

Doronicum plantagineum
Plantain-leaved Leopard's-bane

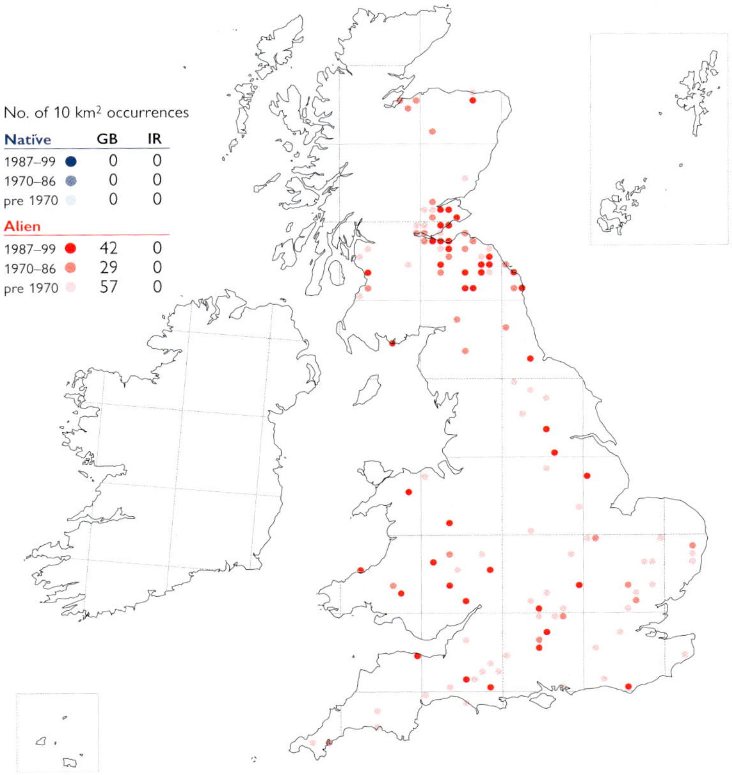

A rhizomatous perennial, occasionally naturalised in woods and waste places. Lowland.

Neophyte (change −0.57). *D. plantagineum* is a popular garden plant, introduced into cultivation in Britain by 1570 and locally naturalised since 1799. Its distribution shows no change from the 1962 *Atlas*, although it is likely that some mapped records are the hybrids *D.* × *willdenowii* or *D.* × *excelsum*.

Native of W. Europe (Iberian peninsula, Italy, France), north to N. France.

References: Atlas (274b), Bolòs & Vigo (1995), Leslie (1981a).

H. J. KILLICK

Tussilago farfara Colt's-foot

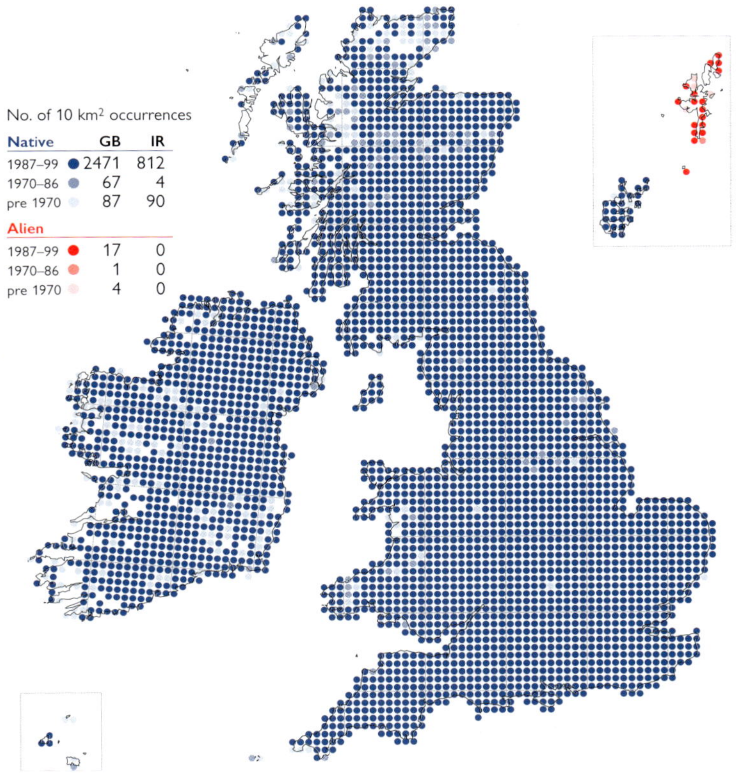

A rhizomatous perennial herb, occurring, often as a pioneer, in a wide range of moist or dry, often disturbed habitats, which include sand dunes and shingle, slumping cliff-slopes, landslides, spoil heaps, seepage areas, rough grassland, crumbling river banks, waste places and roadside verges. It can also be a troublesome arable weed. 0–1065 m in the Breadalbanes (Mid Perth).

Native (change −0.65). The distribution has shown little overall change since the 1962 *Atlas*. It is alien in Shetland, where it was first recorded in 1841.

Eurosiberian Boreo-temperate element; widely naturalised outside its native range.

References: Atlas (274c), Grime *et al.* (1988), Hultén & Fries (1986), Meusel & Jäger (1992).

H. J. KILLICK

Petasites hybridus Butterbur

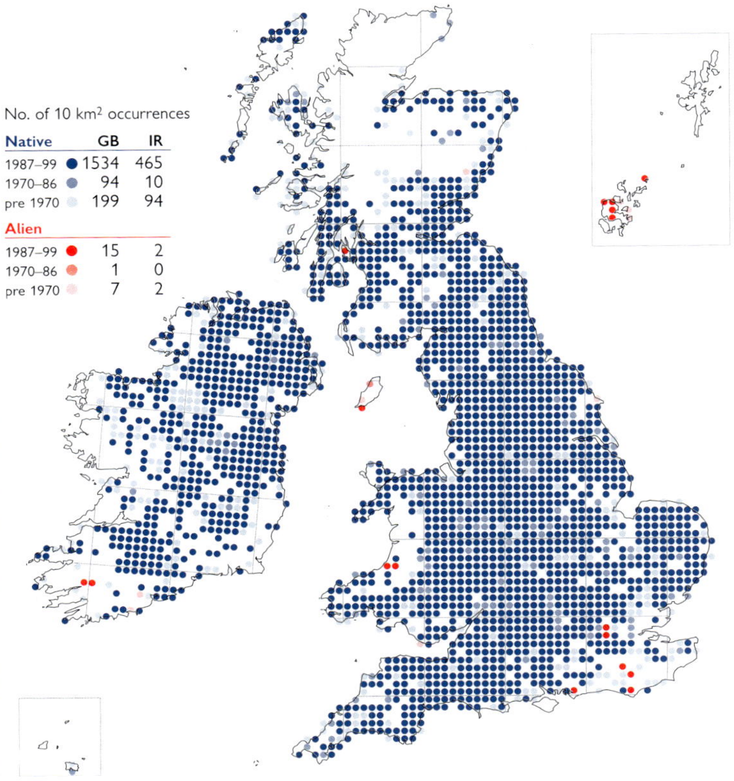

A dioecious rhizomatous perennial herb of moist, fertile, often alluvial, soils by water-courses, in wet meadows, marshes, flood plains and copses, and on roadsides. It spreads mostly vegetatively from rhizome fragments. Female plants are frequent only in N. and C. England. Male-only colonies are probably single clones, many perhaps from deliberate plantings for a source of pollen and nectar for hive bees (Stevens, 1990). 0–380 m (near Garrigill, Cumberland).

Native (change −0.15). The distribution of *P. hybridus* is little changed since the 1962 *Atlas*. Although many populations may be alien, they are rarely recorded as such.

European Temperate element.

References: Atlas (274d), Atlas Supp. (72a), Grime *et al.* (1988), Hultén & Fries (1986), Meusel & Jäger (1992), Perring & Sell (1968), Valentine (1947).

H. J. KILLICK

Petasites japonicus Giant Butterbur

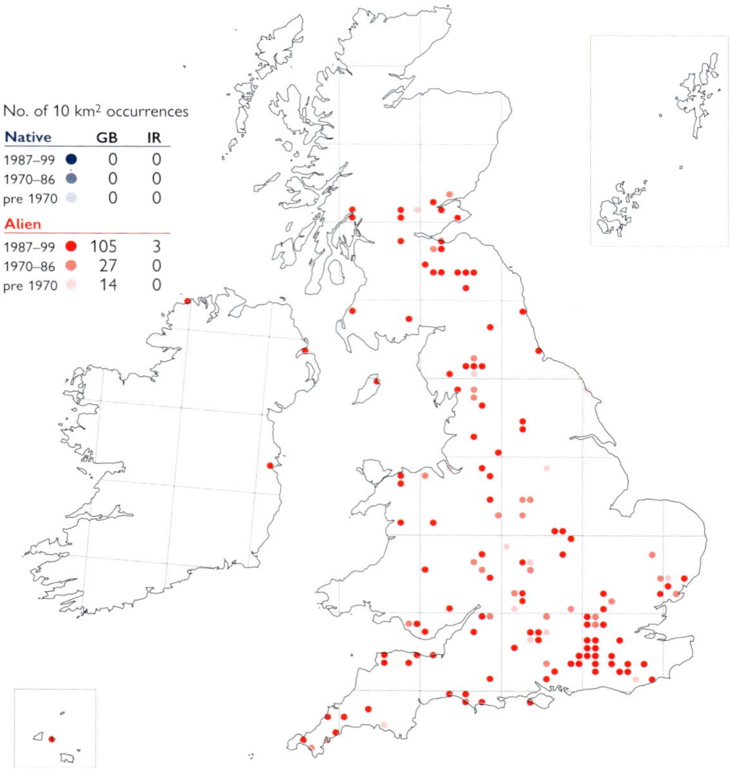

No. of 10 km² occurrences

Native	GB	IR
1987–99	0	0
1970–86	0	0
pre 1970	0	0
Alien		
1987–99	105	3
1970–86	27	0
pre 1970	14	0

A dioecious rhizomatous perennial herb, established in shaded or open sites by rivers and other damp places, in plantations and on roadsides. Lowland.

Neophyte. This species was first grown in Britain in 1897. Female plants are rarely if ever naturalised, and the male-only colonies, which are sometimes extensive, are probably single clones maintained by vegetative spread. Naturalised colonies may have originated from garden escapes or throw-outs, or may have been deliberately planted. The first record in the wild appears to be in 1924 (Bucks.). It is possibly spreading.

Native of Japan and Sakhalin.

H. J. KILLICK

Petasites albus White Butterbur

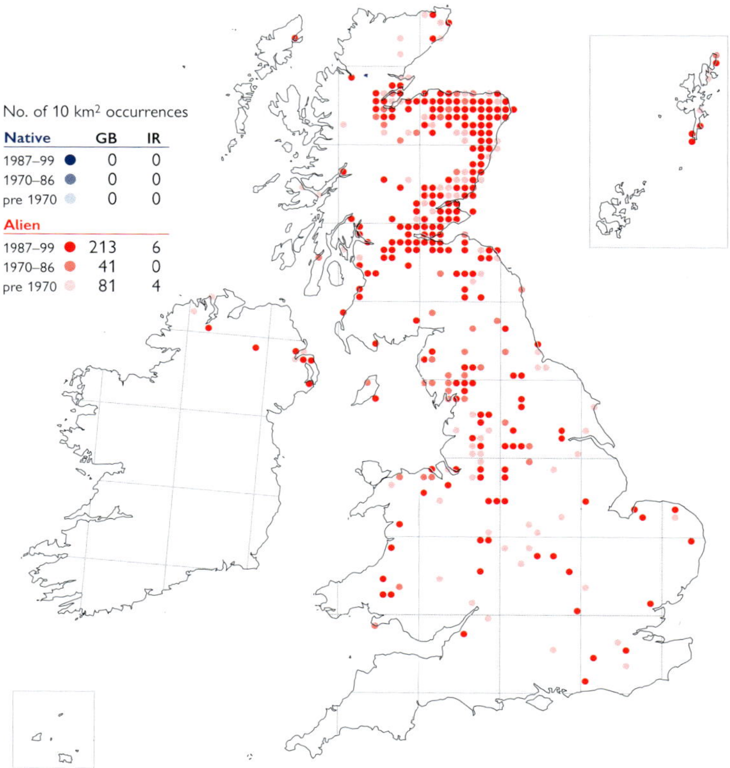

No. of 10 km² occurrences

Native	GB	IR
1987–99	0	0
1970–86	0	0
pre 1970	0	0
Alien		
1987–99	213	6
1970–86	41	0
pre 1970	81	4

A dioecious rhizomatous perennial herb found established in woods and waste places, on waysides and shady riversides, occasionally forming large stands. The male plant is much more common than the female in our area. Lowland.

Neophyte (change +0.01). *P. albus* was introduced to Britain by 1683, and naturalised by at least 1843 (W. Yorks.). It shows little overall change in distribution since the 1962 *Atlas*.

Native of the mountains of Europe and S.W. Asia.

References: Atlas (275a), Hultén & Fries (1986), Meusel & Jäger (1992).

H. J. KILLICK

Petasites fragrans Winter Heliotrope

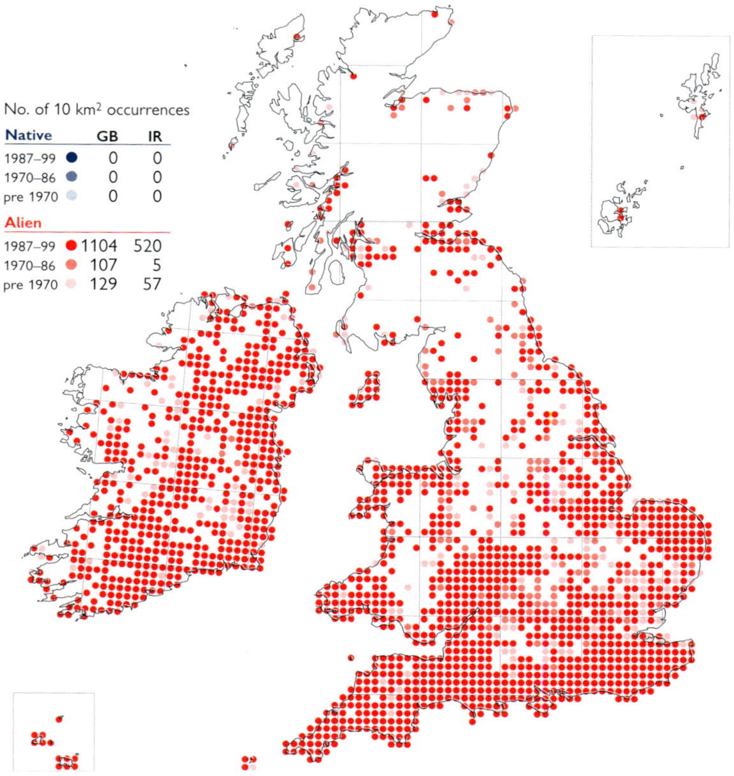

No. of 10 km² occurrences

Native	GB	IR
1987–99	0	0
1970–86	0	0
pre 1970	0	0
Alien		
1987–99	1104	520
1970–86	107	5
pre 1970	129	57

This dioecious rhizomatous perennial herb is naturalised on streamsides, banks, rough ground and roadsides, where it sometimes forms large, very persistent stands. Lowland.

Neophyte (change +0.80). *P. fragrans* was introduced in 1806, and the male plant is grown as an ornamental in gardens and some churchyards. It was known in the wild by at least 1835 (Middlesex), and was well established by the start of the 20th century (Dunn, 1905). It appears to be still spreading in Britain and Ireland. Female plants are unknown in our area.

Native of the C. Mediterranean region in Europe (Italy, Sicily, Sardinia) and of N. Africa.

References: Atlas (275b), Meusel & Jäger (1992).

H. J. KILLICK

Homogyne alpina Purple Colt's-foot

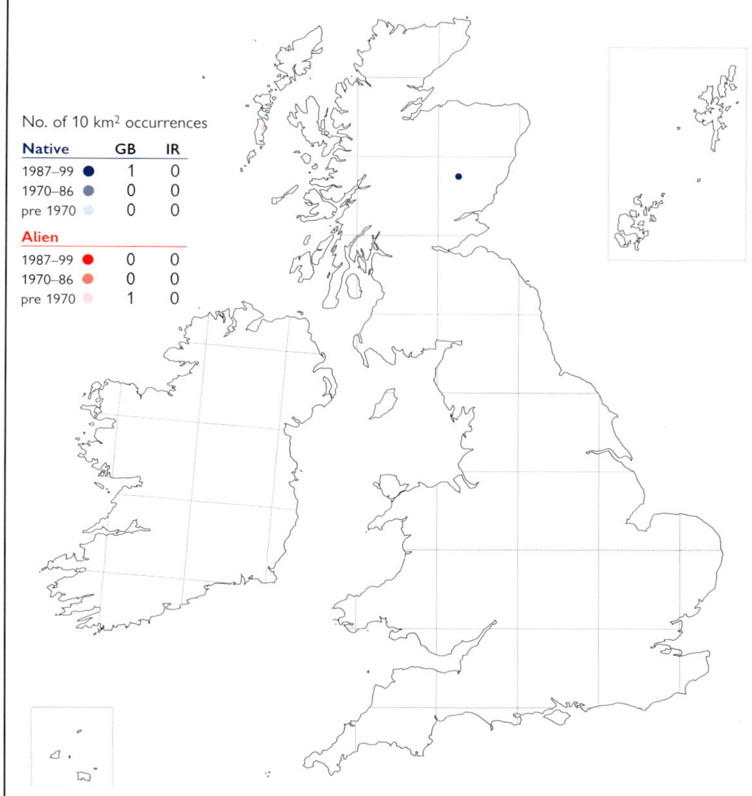

This shortly-creeping perennial herb grows as a possible native on a single, broad, cliff ledge on a wet hillside in the Clova Mountains, Angus. Reproduction appears to be only by rhizomatous spread. The altitude is *c.* 600 m.

Native or alien. This species was discovered in Clova in 1813, but not refound until 1951. There is a long-standing and unresolved debate as to whether it was deliberately planted there. Plants were translocated to nearby sites in the 1990s. *H. alpina* was also recorded from South Uist in 1955, where it was deliberately planted and is now extinct.

A European Boreo-arctic Montane species, endemic to the mountains of W., C. & E. Europe from Spain to Bulgaria.

References: Atlas (275c), Meusel & Jäger (1992), Marren (1999), Pankhurst & Mullin (1991), Wigginton (1999).

H. J. KILLICK

Calendula officinalis Pot Marigold

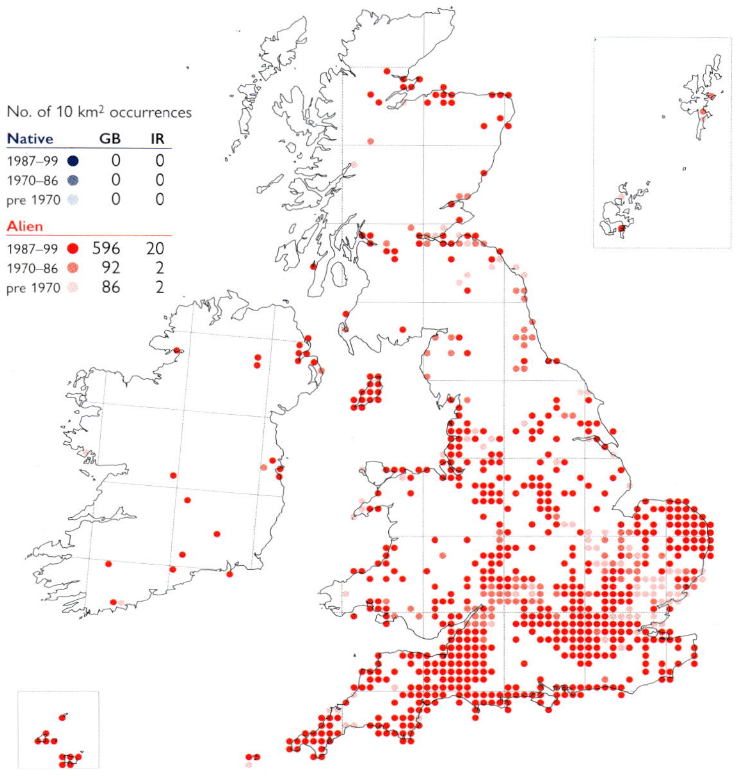

An annual typically found in or near gardens, and on waste ground and rubbish tips. Lowland.

Neophyte. This species was being grown in British gardens by 995 (Harvey, 1981) as a pot-herb, and as a medicinal and ornamental plant. It also occurs as a bird-seed and esparto alien. It readily escapes, being known from the wild by 1872 (Fife). It is increasing in the Isles of Scilly, where it was first recorded in 1939, the Channel Islands, and in sheltered places in S. Britain, but it seldom persists elsewhere.

A garden plant of unknown origin.

References: Bolòs & Vigo (1995), Lousley (1971).

H. J. KILLICK

Ambrosia artemisiifolia Ragweed

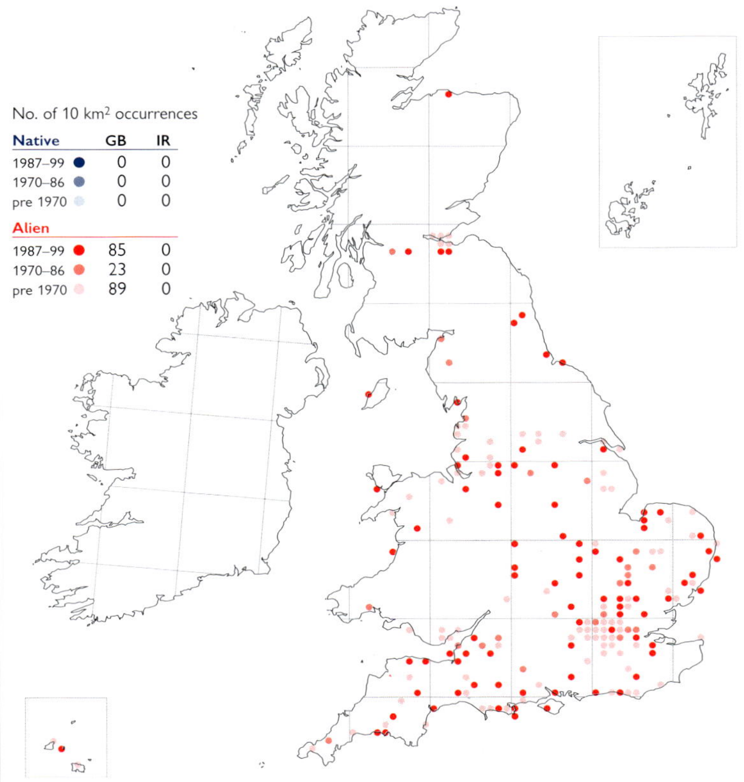

An annual occurring as a casual (rarely persisting) on rubbish tips, in dockyards, arable fields and on waste ground and in places where bird-seed is scattered. Lowland.

Neophyte. *A. artemisiifolia* was cultivated in Britain by 1759 and has been recorded as a casual since 1836. It is introduced with oil-seed, grain and other agricultural seed, and with animal and bird feed. It was not mapped in the 1962 *Atlas*.

Native of N. America; widely naturalised elsewhere.

References: Bassett & Crompton (1975), Meusel & Jäger (1992), Rich (1994c).

H. J. KILLICK

Ambrosia trifida Giant Ragweed

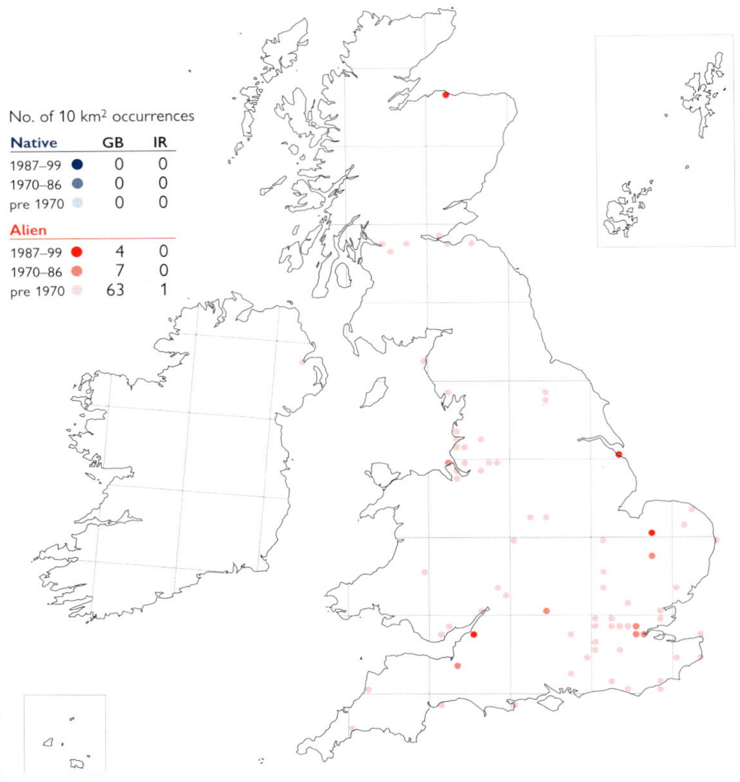

A tall annual of waste ground, rubbish tips and docks, found as a casual or sometimes persistent alien arriving with bird-seed, oil-seed, soya beans or grain. Lowland.

Casual. *A. trifida* was cultivated in Britain by 1699, but arose principally as a grain contaminant and was first recorded from the wild in 1897. It underwent a dramatic decline before 1970, as importation of loose cargoes at docks ceased and seed cleaning techniques improved.

Native of N. America.

References: Rich (1994c), Wade *et al.* (1994).

H. J. KILLICK

Xanthium strumarium Rough Cocklebur

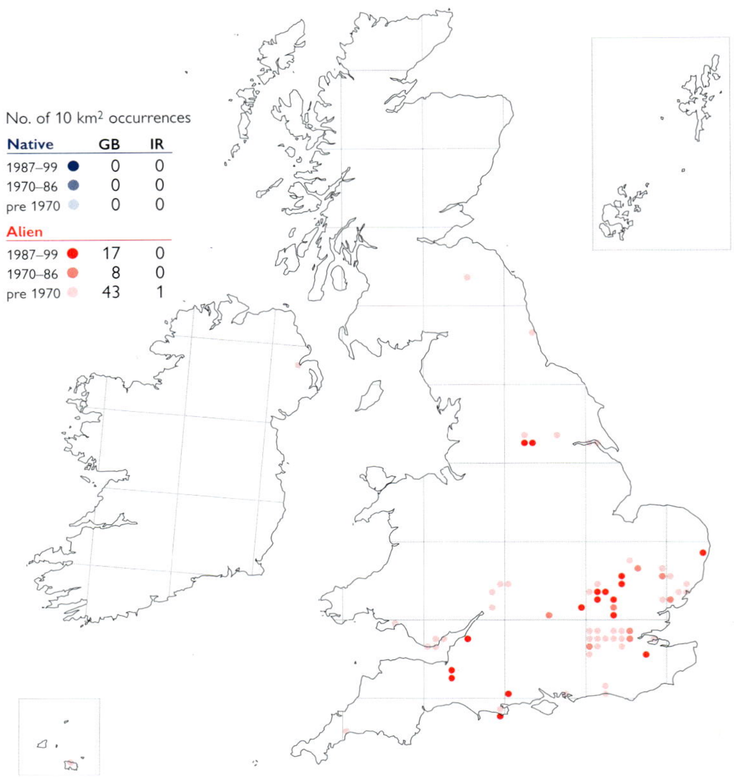

A tall annual of estuarine shores, rubbish tips, docks and waste ground, originating mainly from grain, wool, oil-seed and soya bean waste. It is usually casual, sometimes persisting for a few years. Lowland.

Neophyte. This species was known in the wild in Britain by 1597, but has declined since 1970. It is very variable and the whole '*strumarium* group' is mapped here.

The native range of *X. strumarium* is obscure; it is frequent in C. & S. Europe and a widespread weed in warmer climates throughout the world.

References: Clement (1981b), Lousley (1961), Meusel & Jäger (1992), Wade *et al.* (1994), Weaver & Lechowicz (1983).

H. J. KILLICK

Xanthium spinosum Spiny Cocklebur

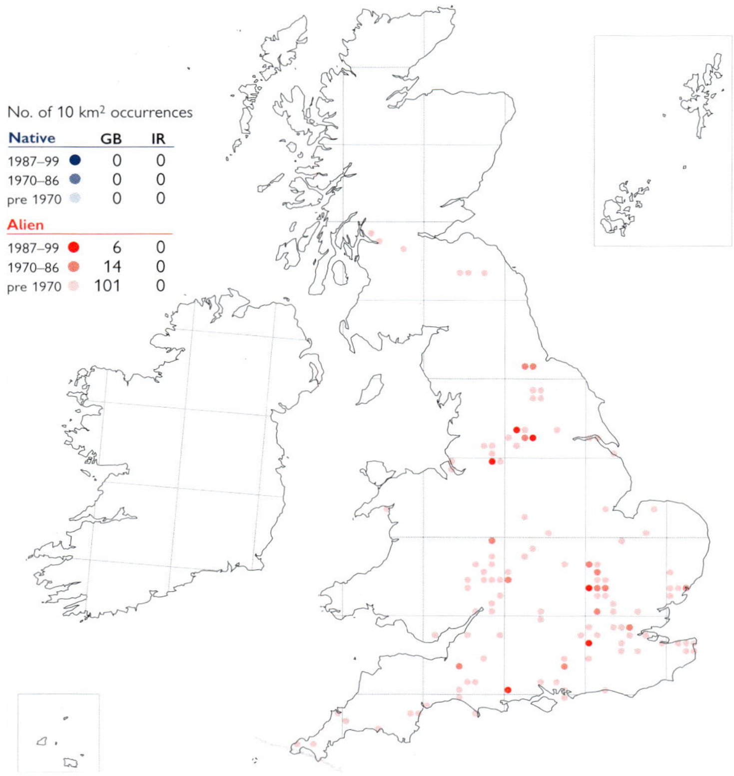

An annual occurring on waste ground, rubbish tips, in sewage works and railway sidings. It is usually casual, sometimes persisting for a few years. It generally arises from wool shoddy or bird-seed. Lowland.

Neophyte. *X. spinosum*, introduced from S. America via Europe by 1713, was known in the wild in Britain by 1846 (Herts.). It appears to be declining. *

Native of S. America; widely naturalised elsewhere.

References: Clement (1981b), Meusel & Jäger (1992).

* Along Gala Water 1911. Abundant. 1st record 1868

H. J. KILLICK

Guizotia abyssinica Niger

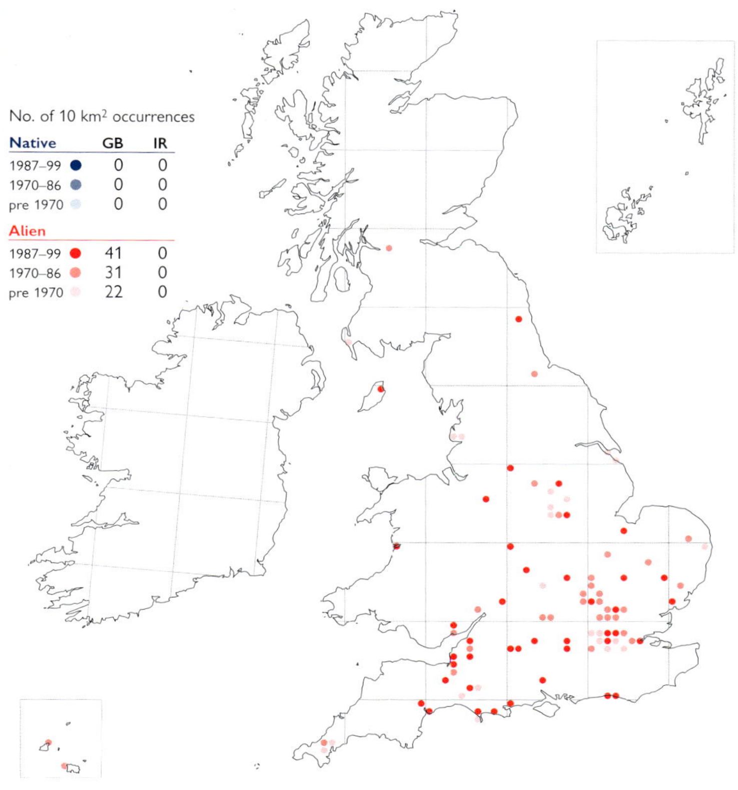

A tall annual of rubbish tips, sewage farms, roadsides and waste places, where it is usually found as a casual from bird-seed, grain, oil-seed and wool shoddy. Lowland.

Casual. *G. abyssinica* was introduced to British gardens by 1806. It has been known in the wild since 1876 (Glamorgan), and may be increasing.

Native of E. Africa.

References: Dunn (1905), Wade *et al.* (1994).

H. J. KILLICK

Helianthus annuus Sunflower

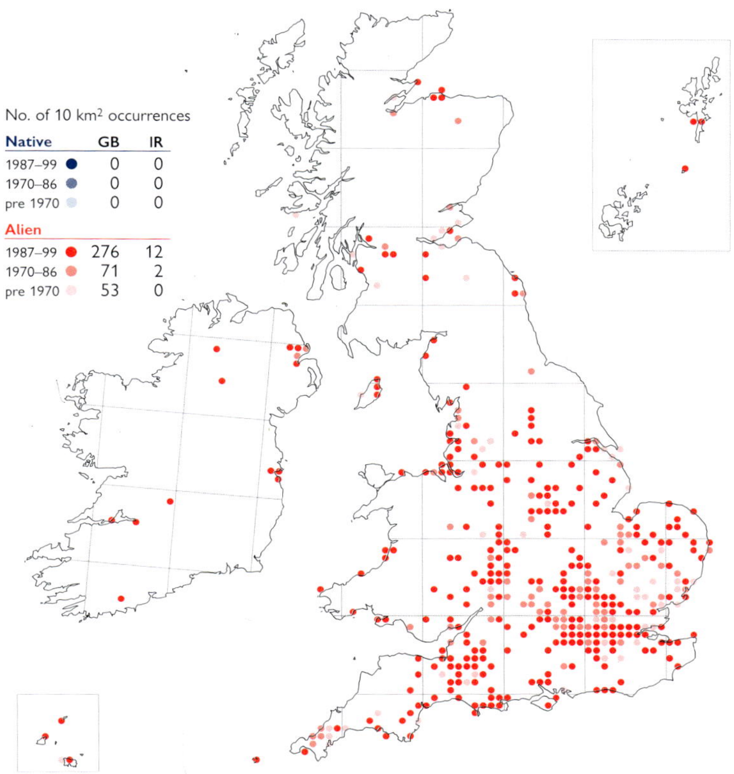

A tall, showy annual, found as a casual, or sometimes persisting, on waste ground, rubbish tips and roadsides. Lowland.

Neophyte. *H. annuus* was introduced to Britain by 1596. It is widely grown in gardens and on allotments, but is sometimes cultivated on a field scale for its oil-bearing seeds. It also arises from bird-seed, food for small mammals and from wool shoddy. It was first recorded in the wild in 1902, and is an increasingly frequent relic of cultivation.

Native of N. America.

Reference: Vaughan & Geissler (1997).

H. J. KILLICK

Helianthus pauciflorus × *H. tuberosus* (*H.* × *laetiflorus*) Perennial Sunflower

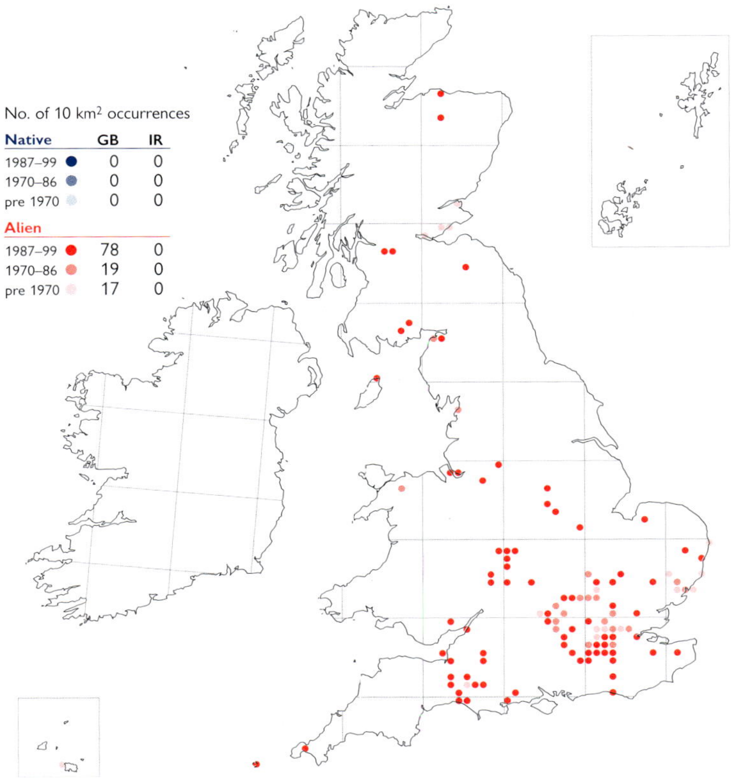

A tall perennial herb found on rubbish tips, in waste places, along field margins, tracks and verges, and on sand dunes. It occurs as a naturalised, or sometimes casual, garden escape or throw-out. Lowland.

Neophyte. This hybrid was introduced to gardens in 1815 and reported from the wild (as *H. rigidus*) by 1902. It is now the commonest perennial sunflower in gardens, and appears to be increasing in the wild. *H.* × *laetiflorus* comprises more than one taxon, probably derived in cultivation from N. American parents. Some mapped records may be misidentifications of *H. pauciflorus*.

Probably a hybrid of garden origin.

H. J. KILLICK

Helianthus tuberosus Jerusalem Artichoke

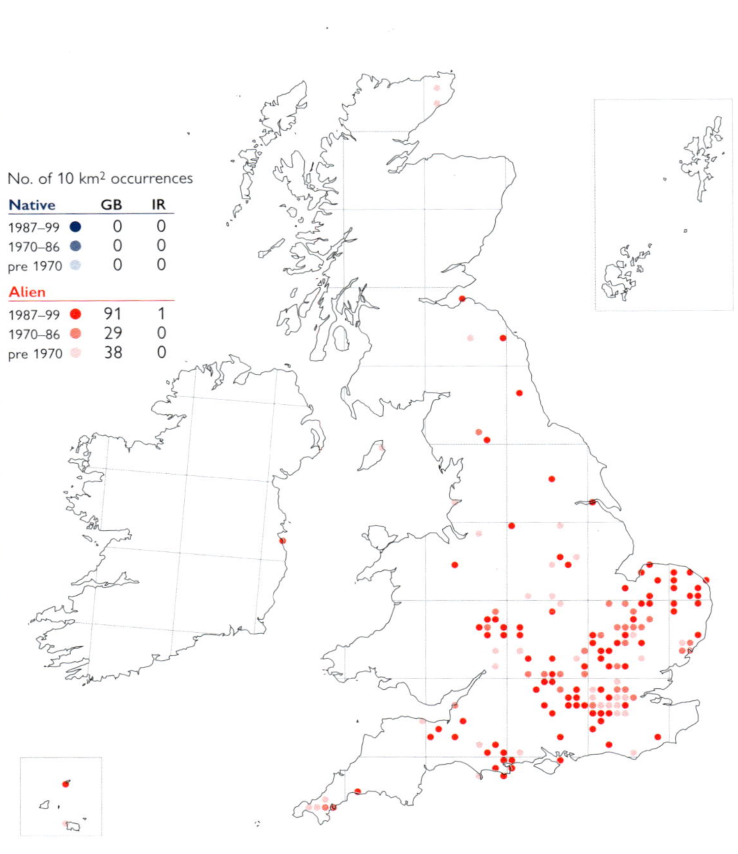

This perennial herb is occasionally recorded on rubbish tips, waste ground and as a relic of cultivation. Lowland.

Neophyte. *H. tuberosus* was cultivated in Britain by 1617 for its edible tubers, then known as 'potatoes of Canada', and is now grown in gardens. Although it flowers only after long hot summers, its tubers can survive as garden throw-outs, or plants can arise from bird-seed. It was known to occur in the wild by 1897, and is probably increasing.

Native of N. America.

References: Meusel & Jäger (1992), Swanton *et al.* (1992).

H. J. KILLICK

Galinsoga parviflora Gallant-soldier

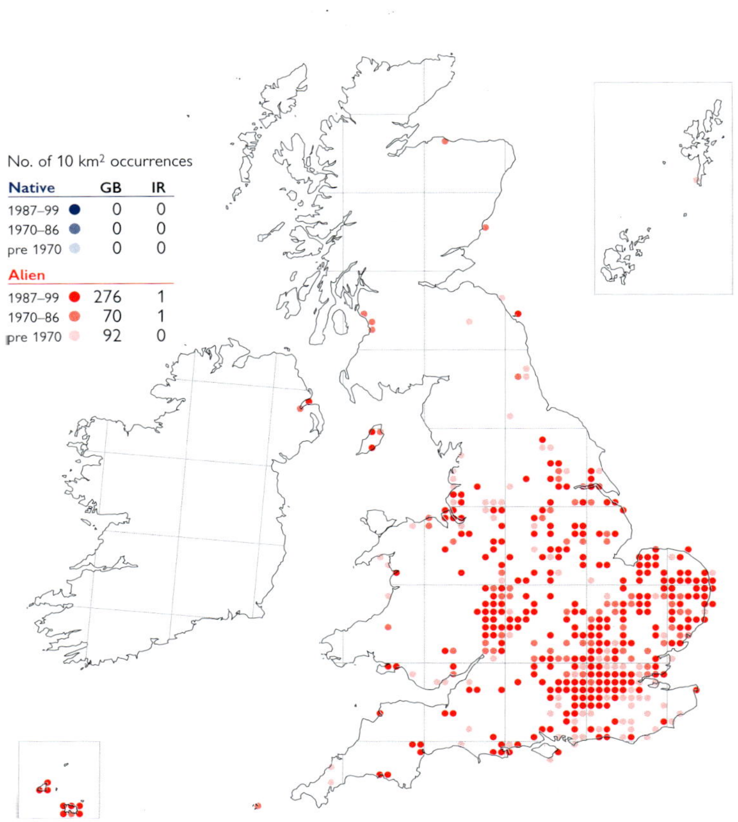

An annual weed of light soils in cultivated fields, nursery plots, garden centres and gardens, and in waste ground in urban areas. Lowland.

Neophyte (change +0.63). This species was introduced as a garden plant to Kew Gardens by 1796, from where it had escaped by 1860 (and was known as 'Kew Weed'). It has spread steadily since then, aided by transport from nursery gardens and occurring also as a wool alien. It is widespread as a casual, sometimes persisting, and has certainly undergone a considerable expansion of range and frequency since the 1962 *Atlas*.

Native of S. America; widely naturalised in other countries.

References: Atlas (270c), Hultén & Fries (1986), Meusel & Jäger (1992), Warwick & Sweet (1983).

H. J. KILLICK

Galinsoga quadriradiata Shaggy-soldier

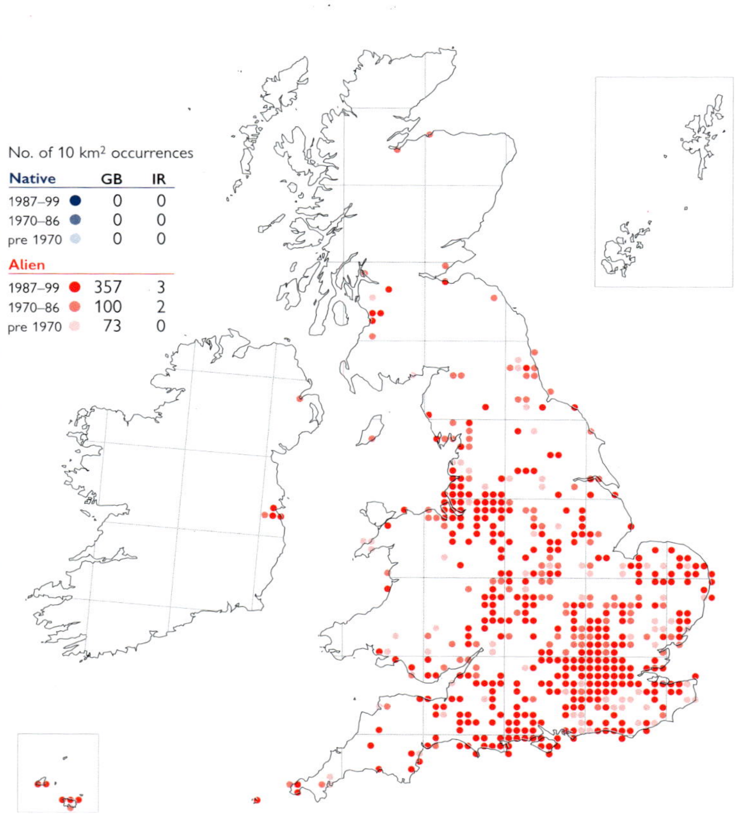

An annual of arable fields, waste ground, roadsides and rubbish tips, derelict urban sites and cracks in pavements, most often occurring in the larger conurbations. Lowland.

Neophyte (change +1.07). The species was first recorded from the wild in 1909 (Middlesex), perhaps being originally introduced with ornamental garden plants, though it is also a bird-seed alien. Early on, it may have been mistaken for *G. parviflora*, and there were records from only nine 10-km squares by 1940. In the 1962 *Atlas* it was present in 98 10-km squares, and it has increased considerably since then. It was first recorded in Ireland in 1980.

Native of C. & S. America; widely naturalised in Europe and elsewhere.

References: Atlas (270d), Burton (1983), Hultén & Fries (1986), Lousley (1971), Meusel & Jäger (1992), Warwick & Sweet (1983).

H. J. KILLICK

Bidens cernua Nodding Bur-marigold

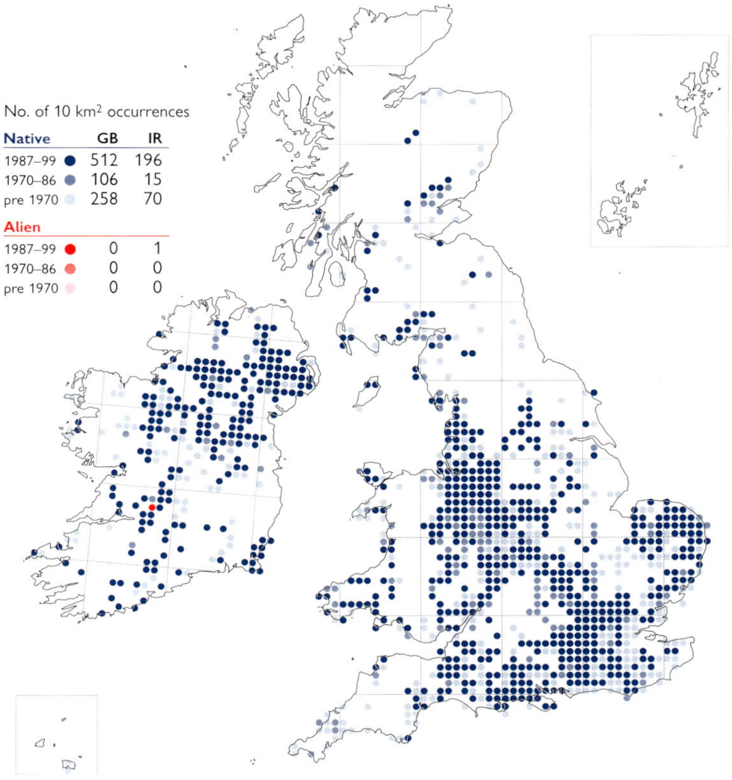

An annual, growing on a wide range of damp or wet substrates on the margins of slow-flowing rivers and streams, by ponds and meres, often in places subject to winter flooding; also in ditches and marshes. 0–310 m (Heathcote, Derbys.).

Native (change –0.54). *B. cernua* declined markedly in some areas before 1930. This decline, associated with drainage and habitat destruction, has continued, especially in S.E. England. It is now better recorded, particularly in Wales, Scotland and Ireland.

Circumpolar Temperate element.

References: Atlas (270a), Green *et al.* (1997), Grose (1957), Hultén & Fries (1986), Meusel & Jäger (1992).

H. J. KILLICK

Bidens tripartita Trifid Bur-marigold

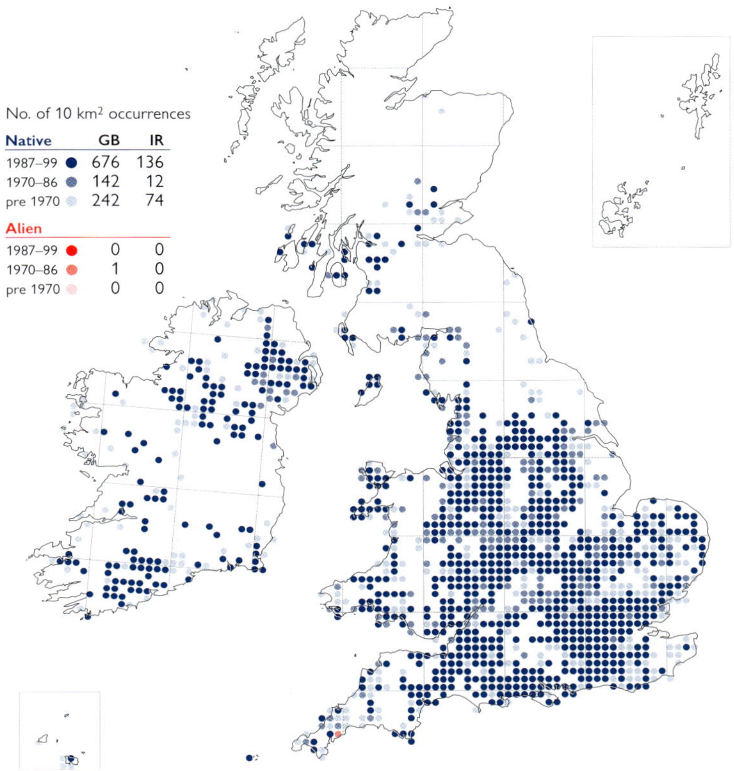

An annual of nutrient-rich mud or gravel by ponds, occurring in wet pits, by slow rivers and streams, often in areas wet in winter but exposed in summer; also found in ditches, peat workings and other damp places. It prefers less acidic and drier substrates than *B. cernua*. Generally lowland, but reaching 365 m at Pilleth (Rads.).

Native (change –0.43). This species is decreasing, particularly in S.E. England. Drainage, the infilling of ponds and ditches, and the canalisation of watercourses have all contributed to its decline. However, it is now much better recorded, especially in N. Ireland.

Eurasian Temperate element; widely naturalised outside its native range.

References: Atlas (270b), Clapham (1969), Green *et al.* (1997), Grose (1957), Hultén & Fries (1986), Meusel & Jäger (1992).

H. J. KILLICK

Bidens frondosa Beggarticks

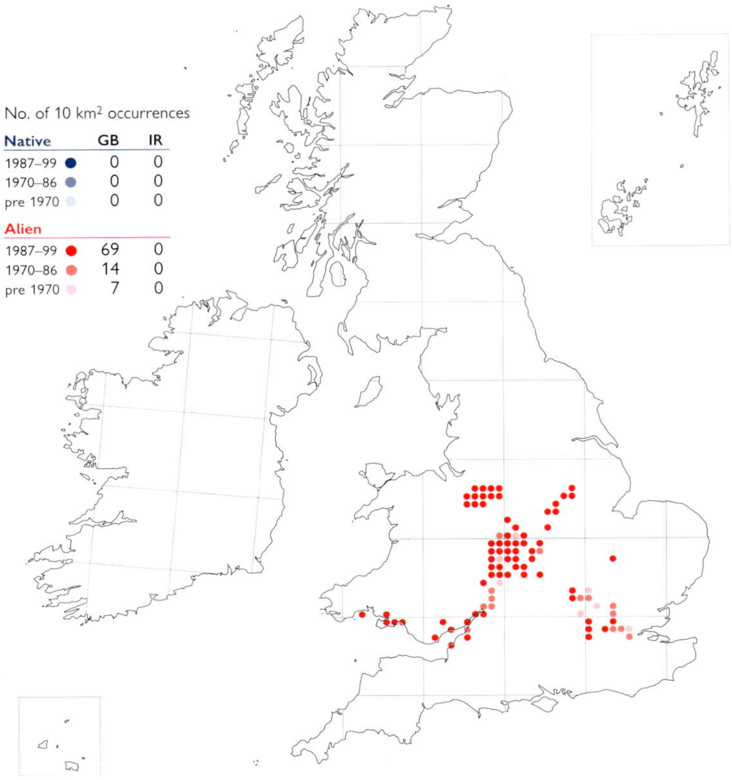

This annual is found by rivers and canals, on damp and waste ground, and on tips in urban areas and ports. Lowland.

Neophyte. *B. frondosa* was first cultivated in Britain in 1710. It was recorded as a casual in 1918 but not known as a naturalised plant until it was found on a canal towpath near Birmingham in 1952 (Cadbury *et al.*, 1971). Its vector is unknown, but it may be a wool alien. It is now frequently naturalised by canals in the English Midlands.

Native of N. & S. America; widely naturalised in temperate Europe.

References: G. M. Kay (1998), Meusel & Jäger (1992).

H. J. KILLICK

Tagetes patula French Marigold

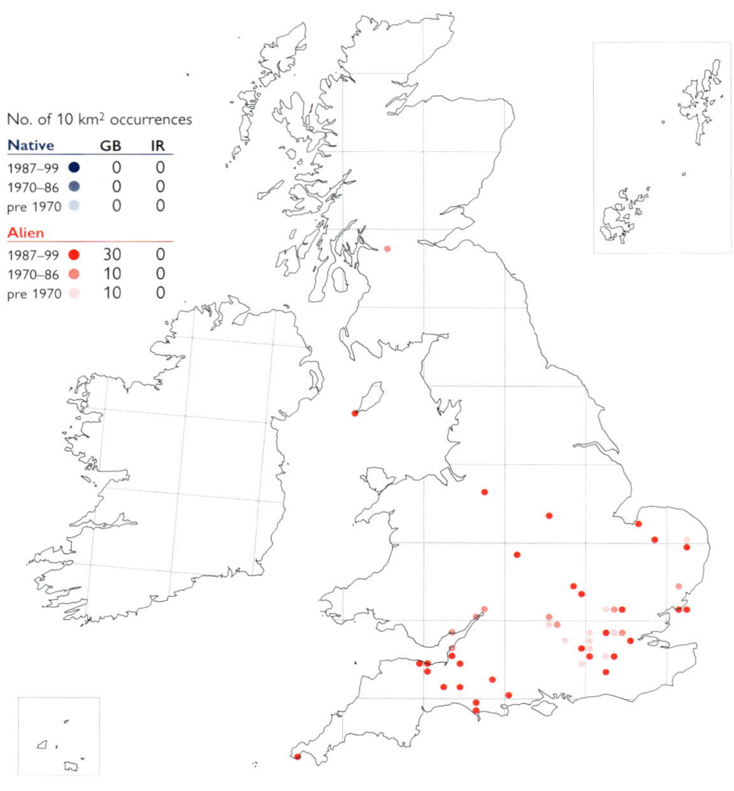

An annual of waste ground and rubbish tips, where it occurs as a casual escape or throw-out from gardens. It also occurs in parks and other cultivated ground as a relic of cultivation. Lowland.

Casual. *T. patula* was cultivated in Britain by 1573, and is extremely common in gardens, where it is grown as a bedding plant and as an insect deterrent amongst crop plants. It was recorded in the wild in 1933 (Herts.). Some authorities consider it to be conspecific with *T. erecta* (African Marigold).

Native of Mexico.

T. D. DINES

Eupatorium cannabinum Hemp-agrimony

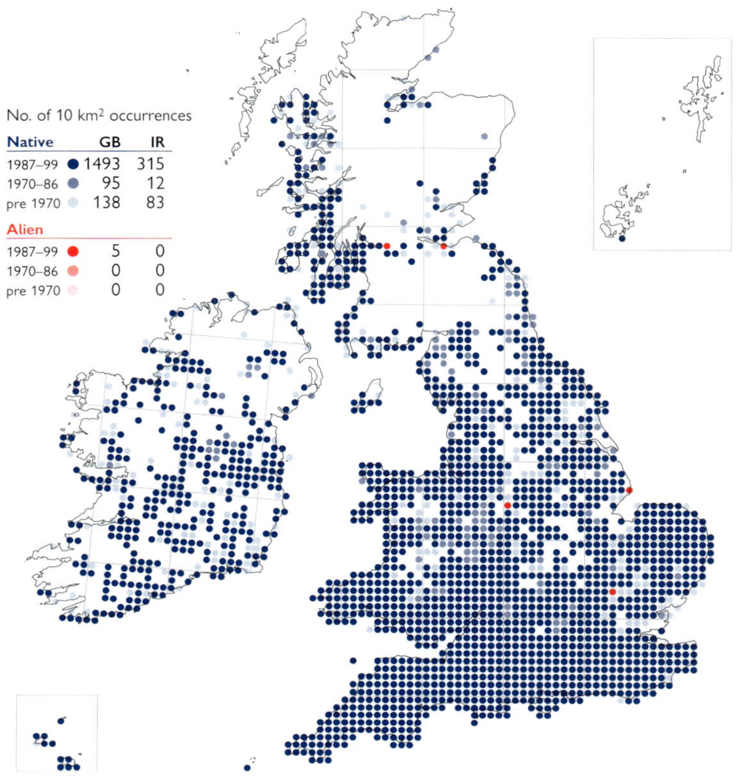

A perennial herb found on base-enriched soils in a wide range of damp or wet habitats, including marginal vegetation by ponds, lakes, rivers and canals, tall-herb fen, fen-meadows, marshes, wet woodland, mires and wet heath; also flushed areas on sea-cliffs and in dune-slacks. It is infrequent in dry habitats, but is found in dry woods and on hedge banks, on waste ground, and even on dry chalk banks. 0–350 m (Ystradfellte, Brecs.).

Native (change –0.15). The distribution of *E. cannabinum* has shown little change since the 1962 *Atlas*.

European Temperate element; also in C. Asia.

References: Atlas (282a), Hultén & Fries (1986), Meusel & Jäger (1992).

M. J. WIGGINTON

Butomus umbellatus Flowering-rush

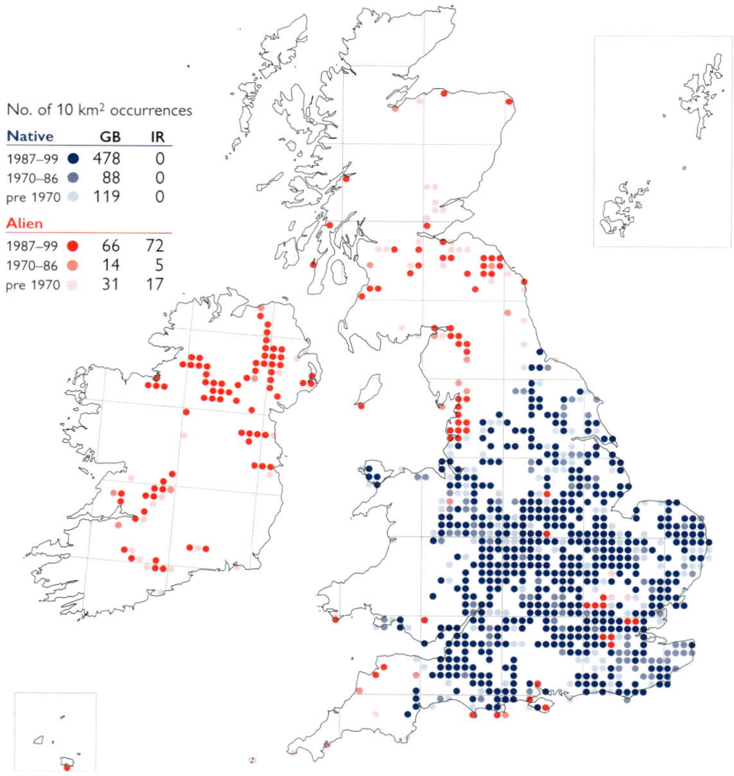

A submerged or emergent rhizomatous perennial which grows in calcareous, often eutrophic, water at the edges of rivers, lakes, canals, ditches and in swamps. It rarely sets seed but reproduces by lateral buds on the rhizome. Lowland.

Native (change –0.04). *B. umbellatus* has maintained its core distribution since the 1962 *Atlas*. It has spread in the Tweed catchment, where it was first recorded in 1956, and along the River Eden (Cumbria). In Ireland Webb & Scannell (1983) suggest that it is native in some areas, including Co. Clare, but it is treated here as alien; it spread rapidly around Lough Neagh in the 20th century.

Eurosiberian Temperate element; widely naturalised outside its native range.

References: Atlas (302b), Hultén & Fries (1986), Meusel *et al.* (1965), Preston & Croft (1997).

C. D. PRESTON

Sagittaria sagittifolia Arrowhead

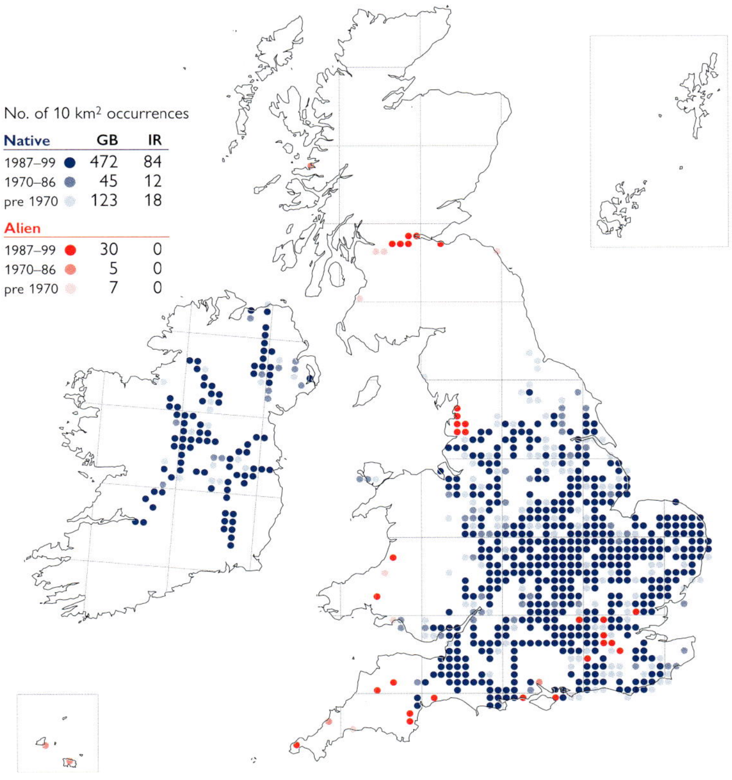

A perennial herb of shallow, still or slowly flowing, calcareous and eutrophic water. In major rivers it may be present only as submerged leaves, but in ditches, lakes, ponds and canals it often produces emergent leaves and flowers. It perennates as tubers produced on stolons in the leaf axils. Lowland.

Native (change –0.44). The overall British range of *S. sagittifolia* is stable, though in some areas the plant has been lost from ditches and is now restricted to larger rivers (Tarpey & Heath, 1990). In Ireland it is now better recorded than in the 1962 *Atlas*, and has spread through canals. It rarely colonises new habitats, but is sometimes deliberately introduced to ponds.

Eurosiberian Boreo-temperate element.

References: Atlas (302a), Hultén & Fries (1986), Preston & Croft (1997).

C. D. PRESTON

Baldellia ranunculoides Lesser Water-plantain

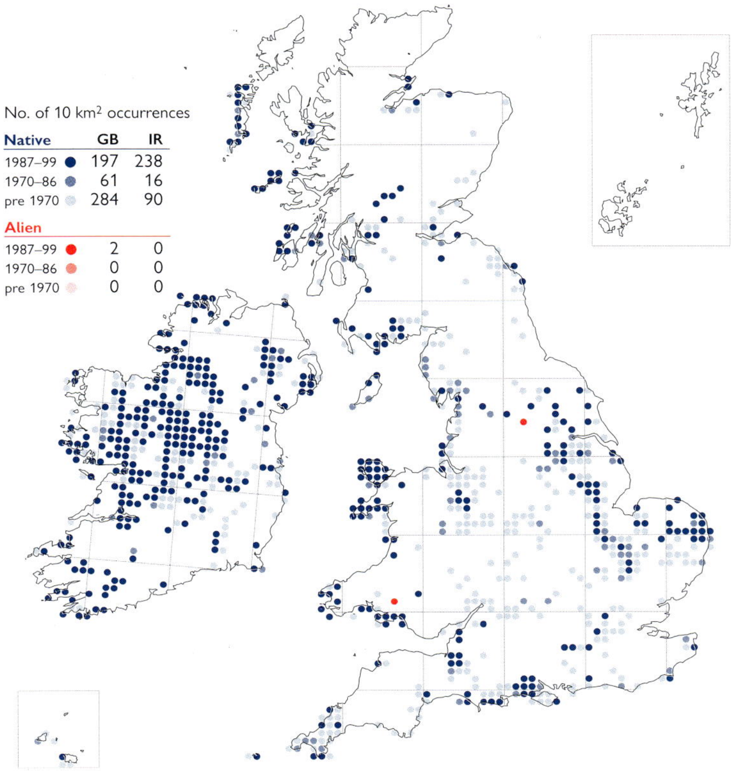

This perennial herb is restricted to habitats at the water's edge, where potential competitors are restricted by fluctuating water levels, disturbance or moderate exposure. It usually grows in mildly to strongly calcareous or brackish waters, over a range of organic or inorganic substrates. 0–320 m (W. of Libanus, Brecs.).

Native (change –1.08). The decline of *B. ranunculoides* was already apparent in the 1962 *Atlas*, and can be attributed to the destruction of small waters and the colonisation of others by more competitive species once grazing ceases. Since then it has declined further in England but seems stable in the west of its range.

Suboceanic Southern-temperate element.

References: Atlas (300d), Hultén & Fries (1986), Meusel *et al.* (1965), Preston & Croft (1997).

C. D. PRESTON

Luronium natans Floating Water-plantain

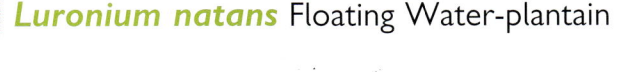

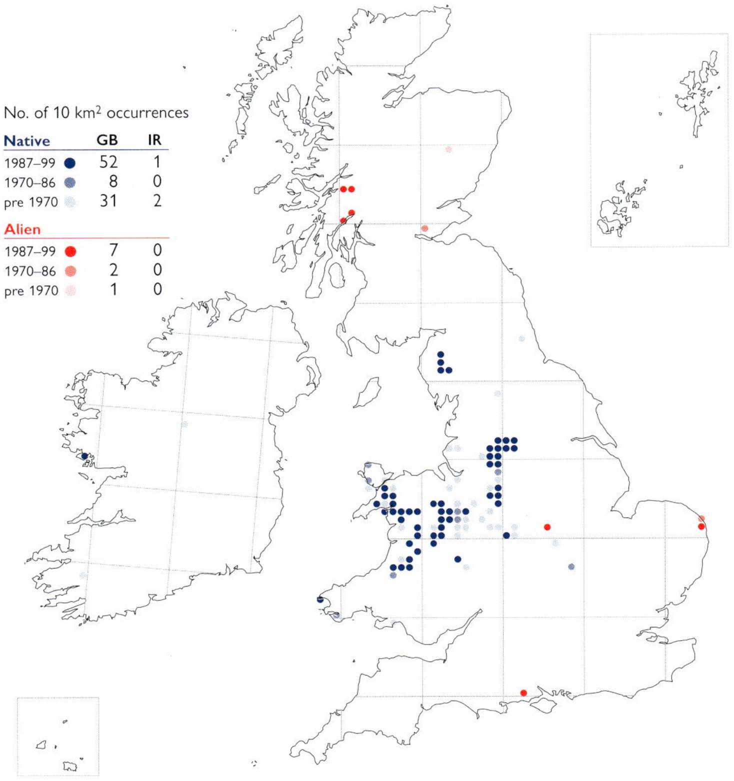

A stoloniferous perennial of mesotrophic or oligotrophic lakes, pools and slow-flowing rivers, and abandoned or little-used canals. In deep water it persists as rosettes of submerged leaves, sometimes with cleistogamous flowers, but it produces floating leaves and flowers freely in shallower water or on exposed mud. 0–450 m (Bugeilyn, Monts.).

Native (change +0.24). In Britain, *L. natans* has been lost from many of its lowland sites because of eutrophication. However, some populations, particularly of the submerged form, have been overlooked until recently in both Britain and Ireland; it may still be under-recorded.

Suboceanic Temperate element.

References: Atlas (301a), Hultén & Fries (1986), Kay *et al.* (1999), Meusel *et al.* (1965), Preston & Croft (1997), Stewart *et al.* (1994).

C. D. PRESTON

Alisma plantago-aquatica Water-plantain

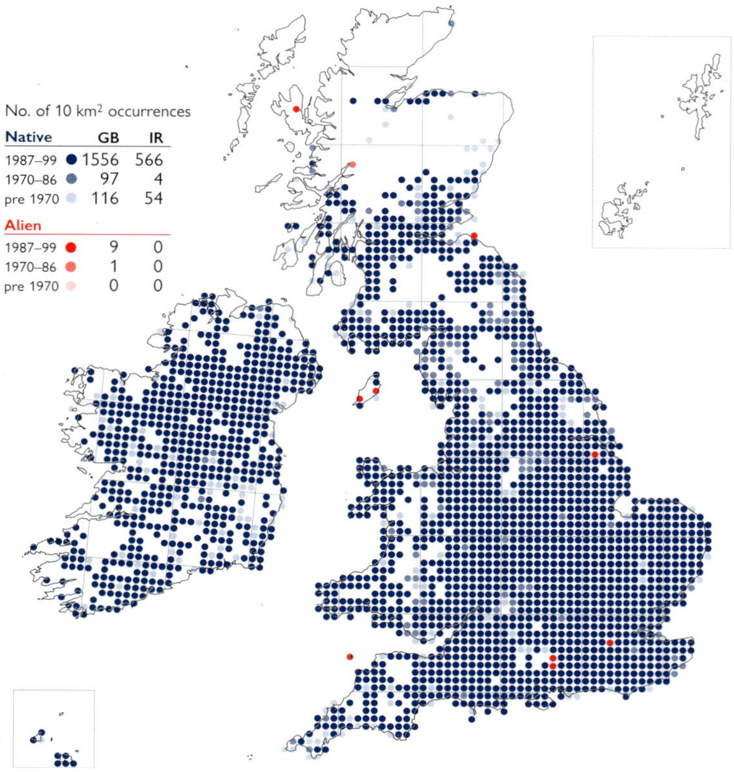

A perennial herb growing on exposed mud at the shallow edge of still or slow-flowing waters, or in marshes and swamps. It is confined to mesotrophic or eutrophic habitats. Plants fruit prolifically, and the species is a frequent colonist of newly cleaned ditches and recently flooded mineral workings. Generally lowland, but reaching 405 m at Dock Tarn (Cumberland).

Native (change –0.19). A well-recorded species with a distribution which is similar to that in the 1962 *Atlas*.

Circumpolar Wide-temperate element; widely naturalised outside its native range.

References: Atlas (301b), Grime *et al.* (1988), Hultén & Fries (1986), Meusel *et al.* (1965), Preston & Croft (1997).

C. D. PRESTON

Alisma lanceolatum Narrow-leaved Water-plantain

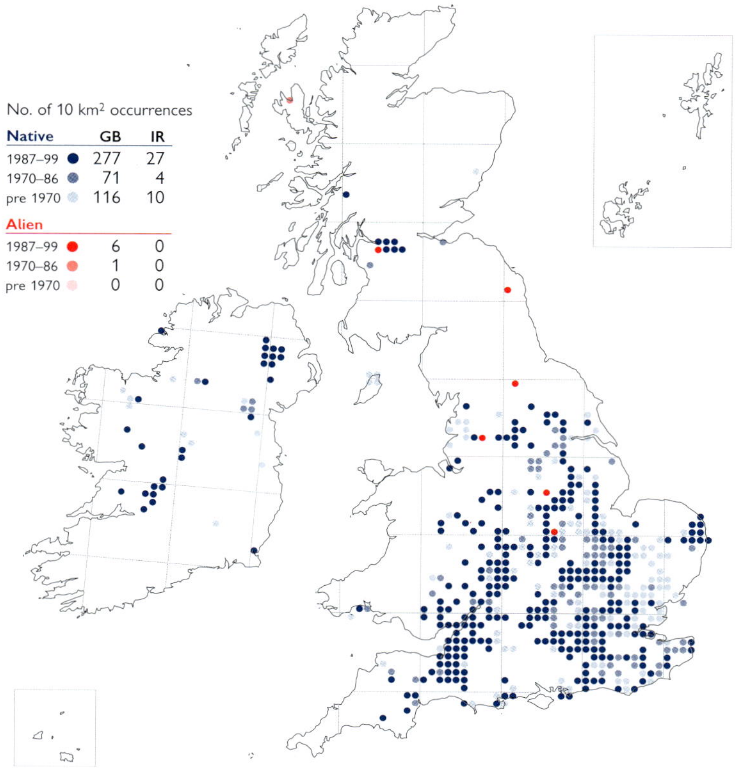

An emergent perennial herb, found in shallow water or on exposed mud at the edge of a wide range of water bodies, although in many areas it is particularly frequent in canals. It is most frequent in eutrophic, calcareous water and rooted in a fine substrate. Lowland.

Native (change +0.38). *A. lanceolatum* was not consistently recognised by British botanists as a distinct species until 1952, so distributional trends are difficult to assess. It might still be overlooked as *A. plantago-aquatica*, but the map may also include erroneous records as narrow-leaved plants of that species are sometimes mis-identified as *A. lanceolatum.*

Eurosiberian Southern-temperate element.

References: Atlas (301c), Hultén & Fries (1986), Meusel *et al.* (1965), Preston & Croft (1997).

C. D. PRESTON

Alisma gramineum Ribbon-leaved Water-plantain

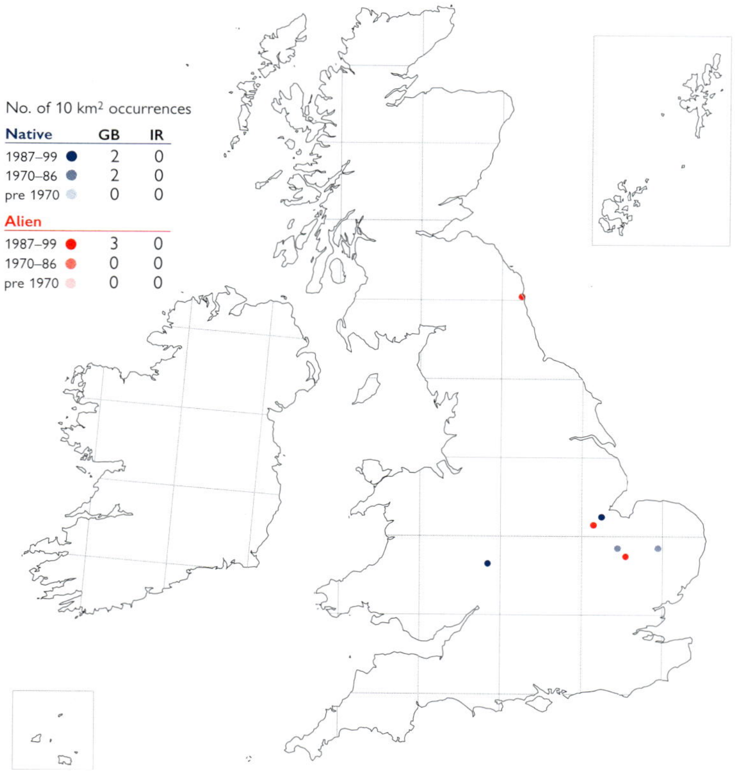

An annual or short-lived perennial which grows in shallow, eutrophic water at the edge of lakes, rivers and fenland drains. Populations may arise from buried seed after disturbance. Lowland.

Native. This species maintains a precarious presence in Britain. It was first found at Westwood Great Pool (Worcs.) in 1920 and the River Glen (S. Lincs.) in 1955; at both sites populations vary annually but are usually small and the plant sometimes fails to appear. At the other two sites the species has been found only once, in 1970. Native material has been introduced to new sites.

Circumpolar Temperate element, with a continental distribution in W. Europe and a disjunct distribution elsewhere.

References: Atlas (301c), Hultén & Fries (1986), Meusel *et al.* (1965), Preston & Croft (1997), Wigginton (1999).

C. D. PRESTON

Damasonium alisma Starfruit

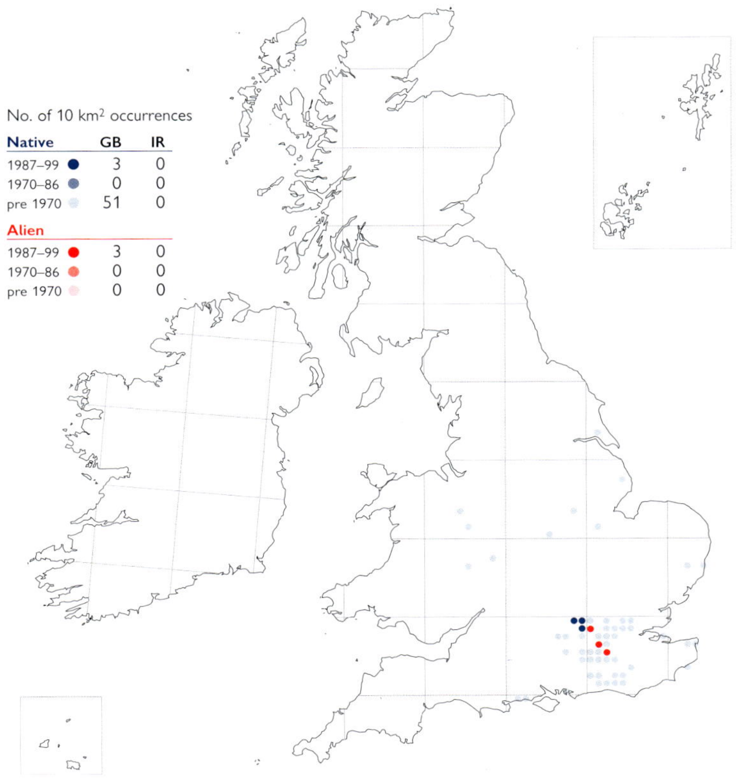

This annual grows on acidic mud by small ponds where the habitat is kept open by fluctuating water levels and disturbance by grazing animals. Its sites are usually flooded in winter but may be wet or dry in summer. The seeds appear to persist in a long-lived seed bank. Lowland.

Native (change –0.52). *D. alisma* declined catastrophically in the 19th and early 20th centuries following the decline of traditional grazing regimes. Most sites have become filled in, overgrown or reduced to suburban duck-ponds. Only one site was known in the 1980s, but the species has since responded to conservation management at some sites and has been planted at others.

Mediterranean-Atlantic element.

References: Atlas (301d), Marren (1999), Preston & Croft (1997), Wigginton (1999).

C. D. PRESTON

Hydrocharis morsus-ranae Frogbit

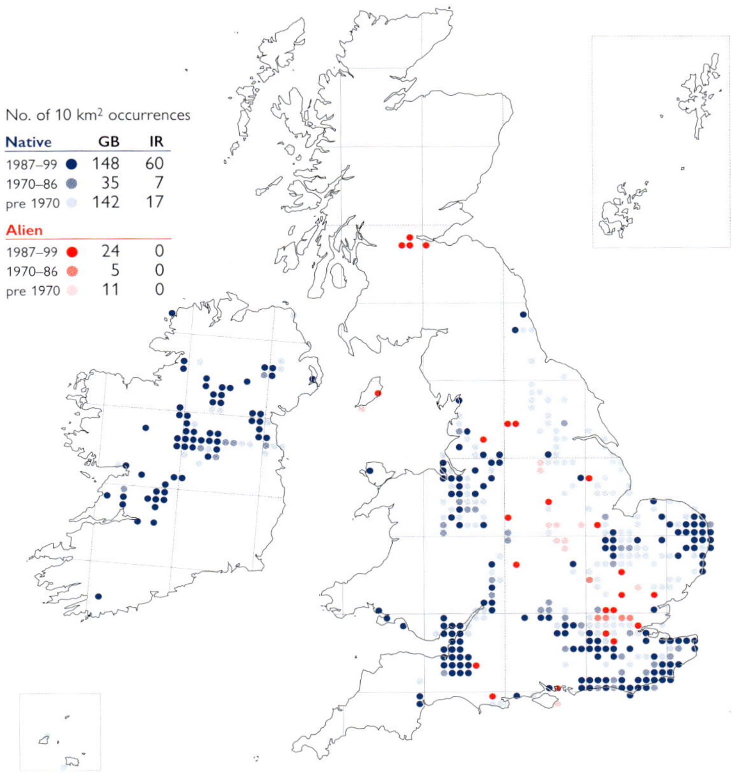

The floating rosettes of this perennial are found in shallow, calcareous, mesotrophic or meso-eutrophic water in the sheltered bays of lakes or in ponds, canals and ditches. Reproduction is primarily vegetative; viable seed is sometimes set, but seedlings are probably rare. Lowland.

Native (change –0.89). This species has declined in Britain, having suffered from the conversion of grazing marshes to arable, and from eutrophication. It is more stable in Ireland, where it is more frequent in lakes. It is increasingly planted or discarded in the wild.

Eurosiberian Temperate element; widely naturalised outside its native range.

References: Atlas (302c), Hultén & Fries (1986), Meusel *et al.* (1965), Preston & Croft (1997), Stewart *et al.* (1994).

C. D. PRESTON

Stratiotes aloides Water-soldier

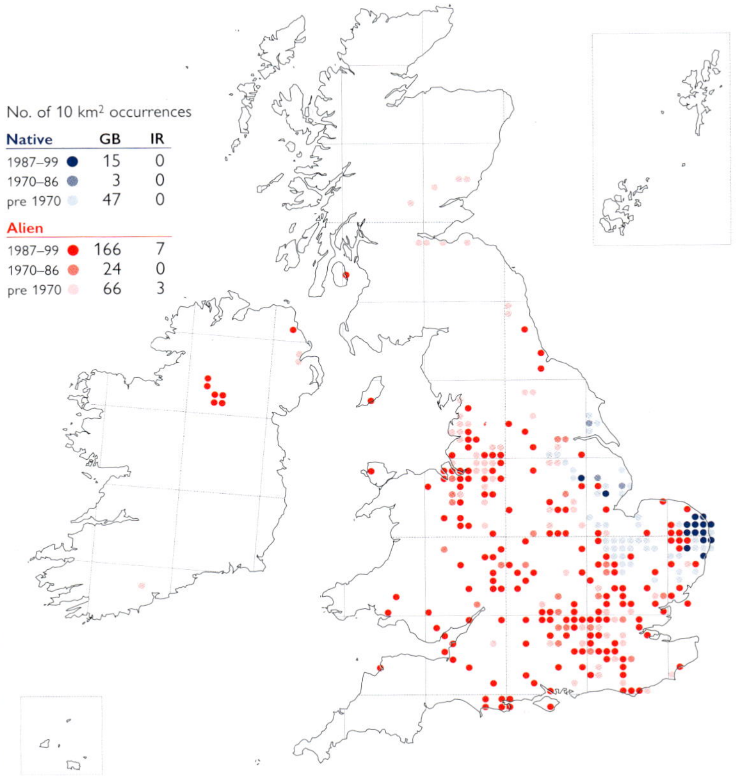

Native populations of this perennial herb are found in calcareous, meso-eutrophic lakes, ponds and ditches. Alien colonies occur in a range of other habitats, including canals. All our plants are female, reproducing vegetatively. Lowland.

Native or alien (change +1.65). This species was first recorded in 1633, and the presence of only one sex suggests that it may not be native (Cook & Urmi-König, 1983). In Britain apparently native populations have been in long-term decline, probably due to eutrophication. Alien populations are often short-lived. Forbes (2000) argues that it may be native in Co. Fermanagh.

Eurosiberian Boreo-temperate element; widely naturalised outside its native range.

References: Atlas (302d), Hultén & Fries (1986), Meusel *et al.* (1965), Preston & Croft (1997), Stewart *et al.* (1994).

C. D. PRESTON

Elodea canadensis Canadian Waterweed

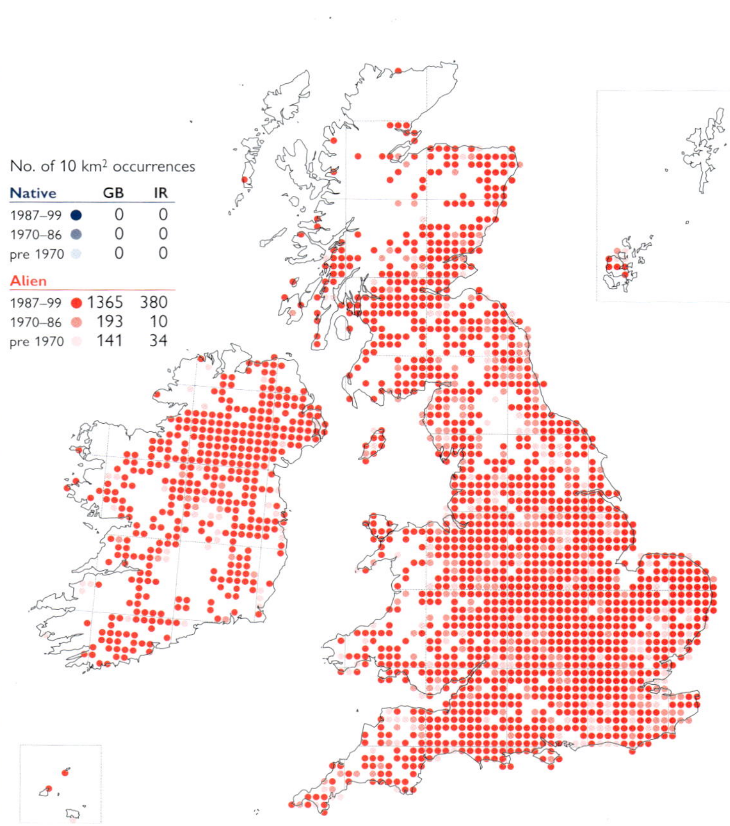

This aquatic perennial herb has a broad habitat range, growing in mesotrophic to eutrophic waters from the shallows to depths of 3 metres or more. It favours still or slowly flowing sites where silt accumulates. All plants in our area are female and reproduce vegetatively. 0–440 m (Loch Loch, E. Perth).

Neophyte (change +0.37). This species was first recorded in Ireland in 1836 and in Britain in 1842, subsequently spreading rapidly. Since the 1962 *Atlas*, where it was mapped as 'all records', it has disappeared from some areas, often being replaced by *E. nuttallii.*

Native of temperate N. America; widely naturalised in Europe and Australasia.

References: Atlas (303a), FNAEC (2000), Grime *et al.* (1988), Hultén & Fries (1986), Preston & Croft (1997), Spicer & Catling (1988).

C. D. PRESTON

Elodea nuttallii Nuttall's Waterweed

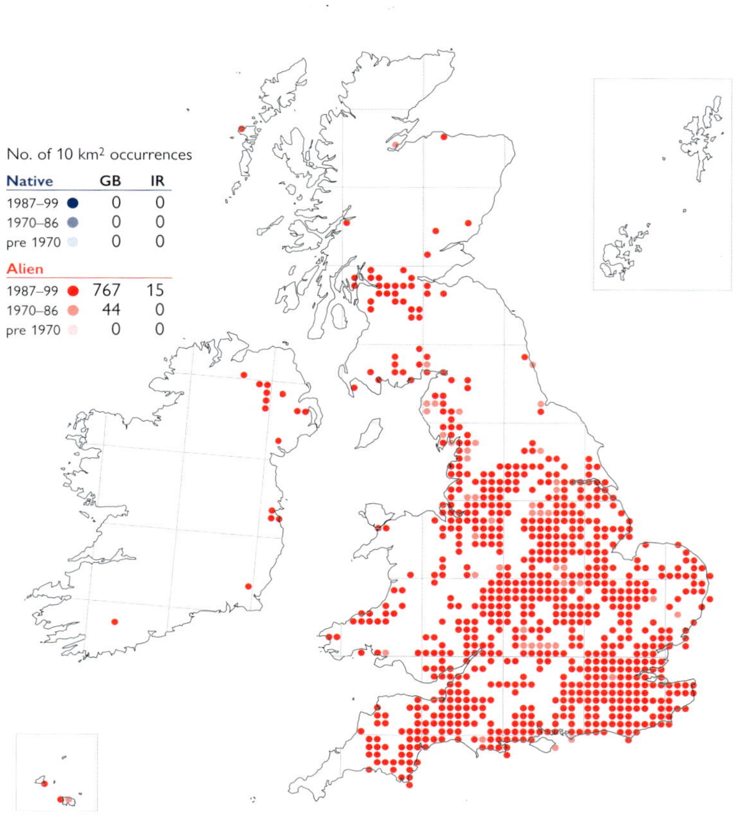

Like *E. canadensis,* this perennial herb is found in still or slowly flowing, shallow or deep water. It appears to be more restricted to eutrophic water than its congener, and it may be frequent even in highly disturbed canals and rivers. It is an effective coloniser of new habitats, despite the fact that it must spread vegetatively as all plants in our area are female. 0–315 m (Bethania, Cards.).

Neophyte. This species is grown by aquarists and pondkeepers. It was first recorded as naturalised in Britain in 1966 (Oxon), and has since spread rapidly. It was first recorded in Ireland in 1984, at Lough Neagh.

Native of temperate N. America; naturalised in Europe since 1939 and in Japan since the 1960s.

References: FNAEC (2000), Preston & Croft (1997).

C. D. PRESTON

Hydrilla verticillata Esthwaite Waterweed

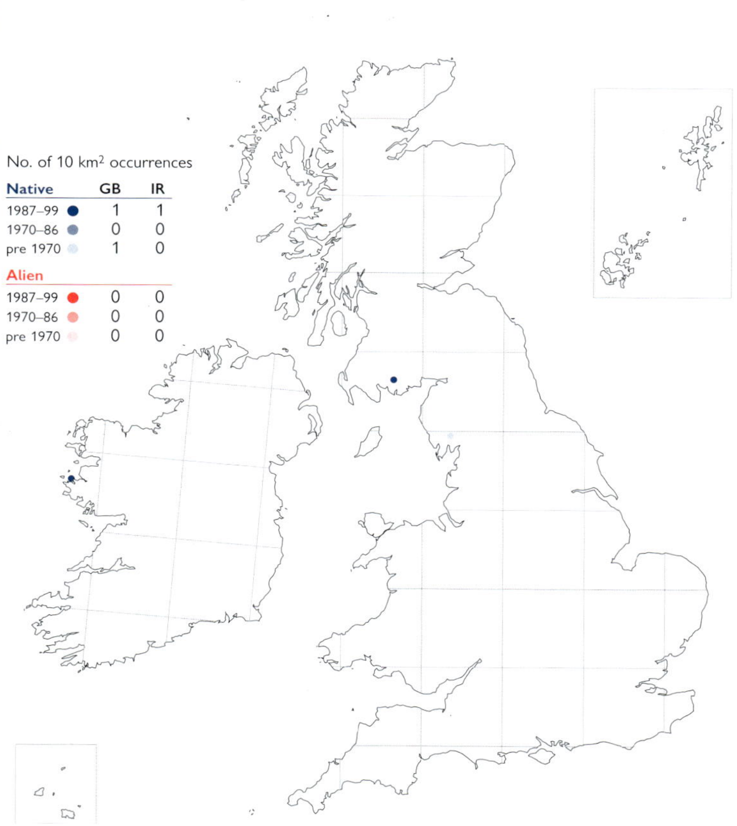

In Britain and Ireland, *H. verticillata* grows in relatively deep water in mesotrophic lakes, only occasionally extending into the shallows. The Irish population is known to be female, but flowers have never been seen in Britain. Reproduction is by turions. Lowland.

Native. Although this species was first discovered in Britain at Esthwaite Water (Westmorland) in 1914 it has not been seen at this site since 1941, and may have been eliminated by eutrophication. It survives at Lough Rusheenduff (W. Galway), where it was discovered in 1935, and it was found in Kirkcudbrightshire in 1999.

Eurasian Southern-temperate element, with a continental distribution in W. Europe; widely naturalised outside its native range.

References: Atlas (303a), Curtis & McGough (1988), Preston & Croft (1997), Scannell & Webb (1976).

C. D. PRESTON

Lagarosiphon major Curly Waterweed

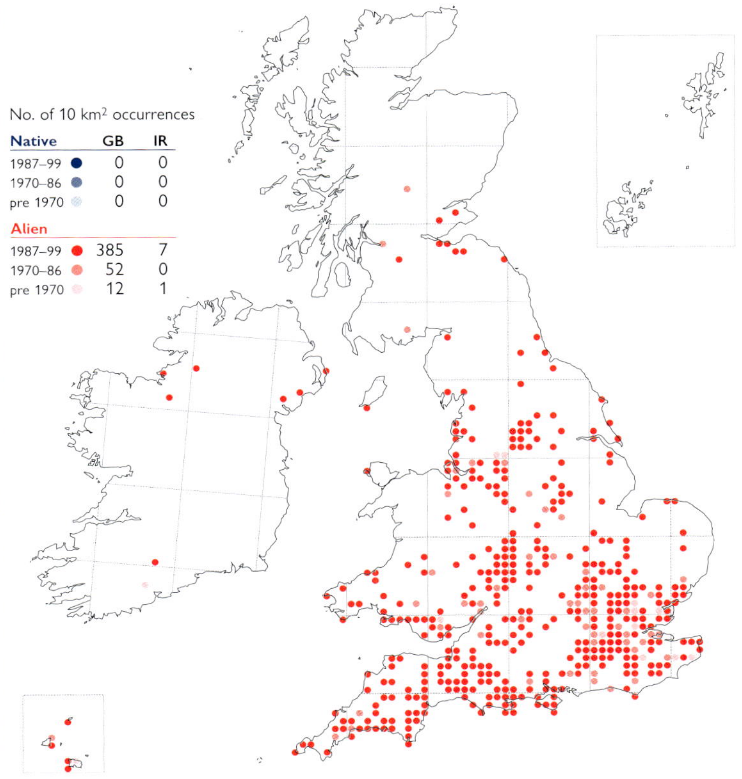

L. major is a submerged aquatic which grows in standing waters, including lakes, ponds and flooded mineral workings, and in canals. It may be abundant in small ponds. Only female plants have been recorded, and reproduction is by vegetative fragmentation. Lowland.

Neophyte. This commonly cultivated aquatic was first recorded as naturalised in Britain in 1944. Although it was not common enough to map in the 1962 *Atlas*, it has spread in recent decades, probably from the initial release of discarded material from garden ponds or aquaria into the wild and subsequent vegetative spread.

Native of southern Africa; locally established in Europe and extensively naturalised in New Zealand.

Reference: Preston & Croft (1997).

C. D. PRESTON

Aponogeton distachyos Cape-pondweed

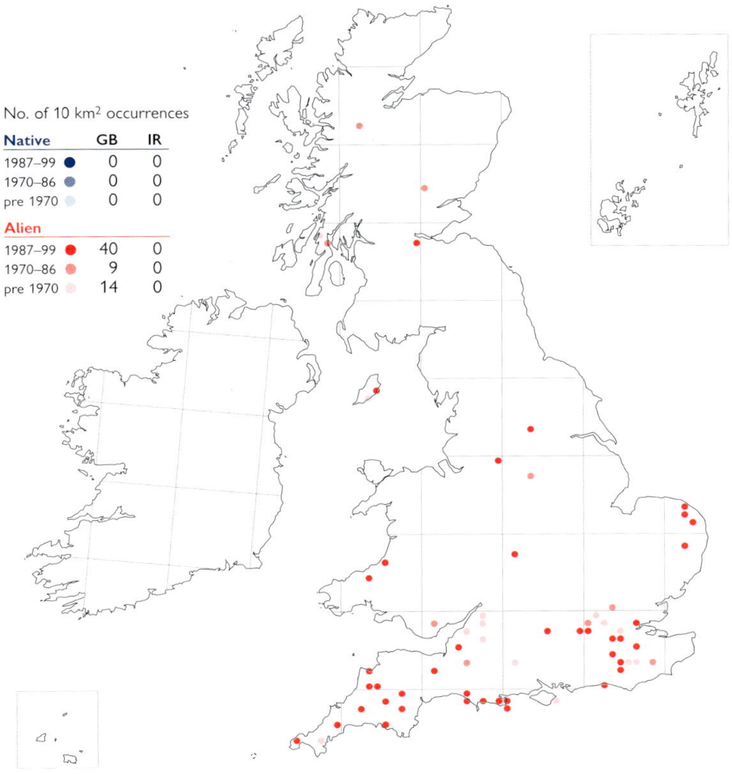

A tuberous perennial which grows as an emergent aquatic in water up to 2 m deep. It may persist for many years in lakes and ponds as a relic of cultivation, or at sites where it is introduced into the wild. It has been known to reproduce by seed in Britain, but shows little sign of spreading to new sites without human assistance. Lowland.

Neophyte. This species was first cultivated in Britain in 1788 and is now popular in water gardens for its hawthorn-scented white flowers. It was recorded from the wild by 1889 (Middlesex) and now appears to be recorded with increasing frequency.

Native of S. Africa (Cape Province); established as an alien elsewhere in Africa and in France, California, S. America, Australia and New Zealand.

References: FNAEC (2000), Preston & Croft (1997).

C. D. PRESTON

Scheuchzeria palustris Rannoch-rush

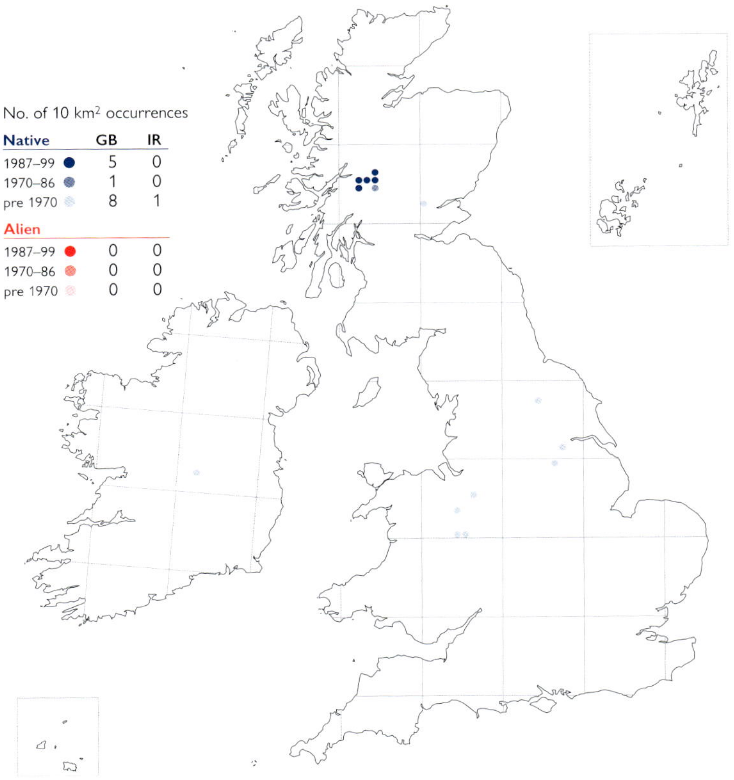

A rhizomatous perennial herb of base-poor, wet habitats, typically found in acid runnels, pools or semi-submerged *Sphagnum* lawns at pool edges. Formerly also lowland, its extant sites on Rannoch Moor are at *c.* 300 m (Mid Perth and Main Argyll).

Native. All English sites for *S. palustris* were lost before 1900 due to drainage and eutrophication. The distribution in Scotland is stable; new sites have been found on Rannoch Moor recently due to detailed recording. It was discovered in Co. Offaly in 1951, but the original site was lost to peat extraction by 1960, and a nearby transplant did not survive.

Circumpolar Boreal-montane element, with a continental distribution in W. Europe.

References: Atlas (303c), Curtis & McGough (1988), Godwin (1975), Hultén & Fries (1986), Meusel *et al.* (1965), Wigginton (1999).

F. J. RUMSEY

Triglochin palustre Marsh Arrowgrass

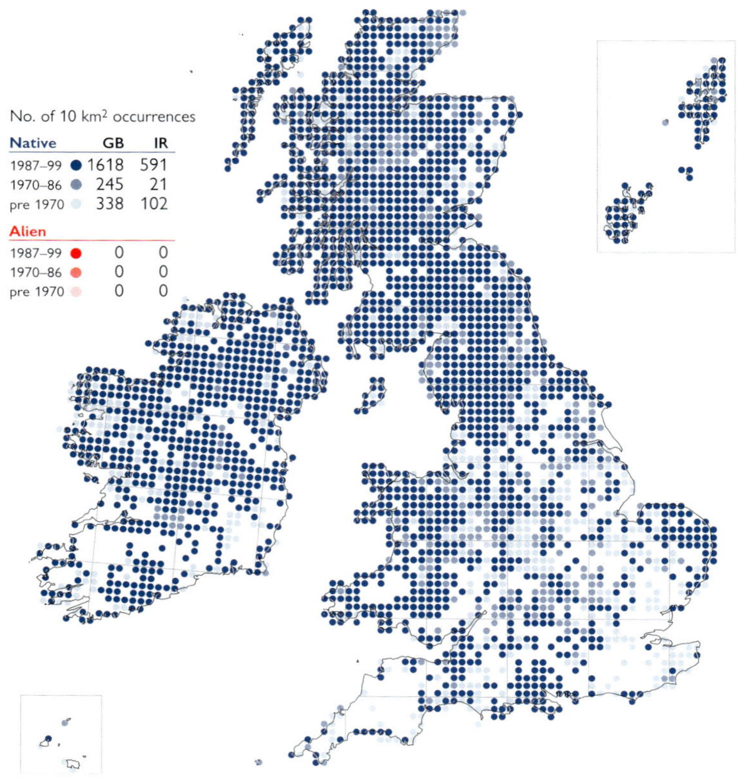

This slender, perennial, rhizomatous herb occurs in open, damp, grassy or marshy places, often on calcareous substrates. Habitats include wet meadows and rush-pastures, heaths, fens, springs and flushes, saltmarsh fringes flushed with fresh water, and river shingle in upland areas. 0–970 m (Beinn Heasgarnich, Mid Perth).

Native (change –0.22). *T. palustre* is inconspicuous, and may sometimes be overlooked. It was mapped as 'all records' in the 1962 *Atlas*. Analysis of the database reveals that the decline in S. England has taken place since 1950, owing to the loss of its habitat through drainage, agricultural intensification and a lack of grazing.

Circumpolar Boreo-temperate element.

References: Atlas (303d), Hultén & Fries (1986).

F. J. RUMSEY

Triglochin maritimum Sea Arrowgrass

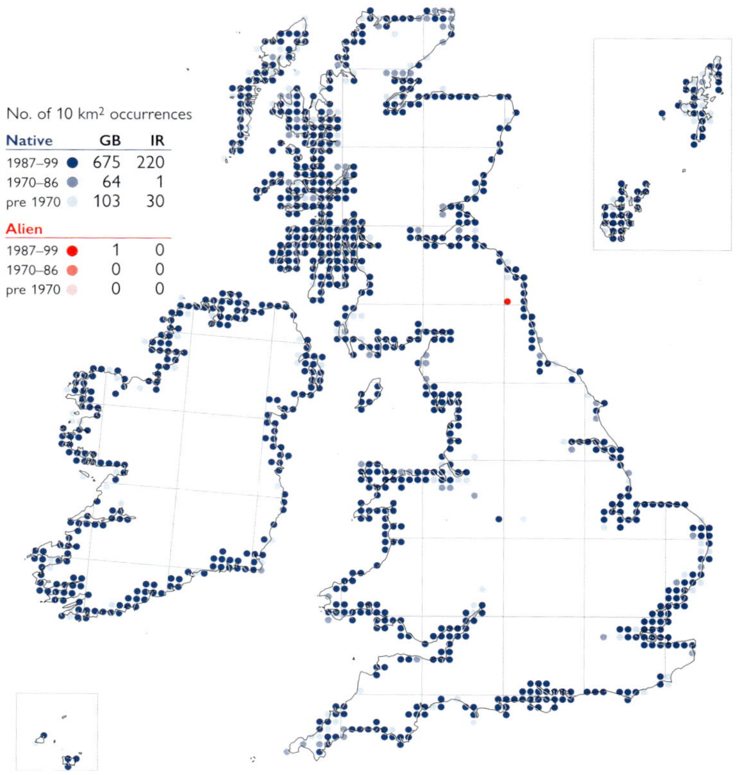

A rhizomatous perennial herb of saline habitats. It is abundant in coastal and estuarine saltmarshes, flushed coastal rocks and cliff edges subject to sea spray, and the banks of tidal rivers. Inland, it occurs in brackish pastures as, for example, over saline Keuper beds in Cheshire, and at one site in Hampshire, in flushed turf on calcareous clay. Very rarely, it grows alongside salt-treated roads. Lowland.

Native (change –0.44). This species shows no appreciable change to its coastal range since the 1962 *Atlas* but has been lost from some of its few inland sites through habitat destruction.

Circumpolar Boreo-temperate element.

References: Atlas (304a), Davy & Bishop (1991), Hultén & Fries (1986), Lee (1977).

F. J. RUMSEY

Potamogeton natans Broad-leaved Pondweed

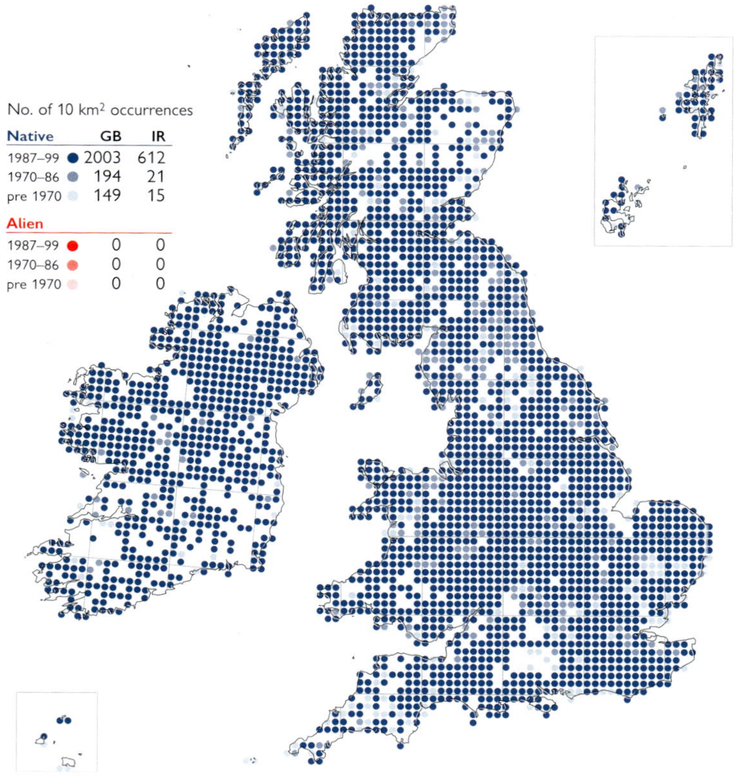

This rhizomatous perennial herb is frequent as a floating-leaved aquatic in still or slowly flowing waters, and more rarely found as plants with submerged phyllodes in more rapid streams and rivers. It has a very wide ecological tolerance, growing in oligotrophic to eutrophic and base-poor to base-rich water over a wide range of substrates. Although it may be found in shallow swamps, or in water over 5 m deep, it is most frequent at moderate depths of 1–2 metres. 0–760 m (below Stob Ban, Westerness).

Native. There is little evidence for any change in the distribution of *P. natans* since the 1962 *Atlas*.

Circumpolar Boreo-temperate element.

References: Atlas (305a), Grime *et al.* (1988), Hultén & Fries (1986), Meusel *et al.* (1965), Preston (1995), Preston & Croft (1997).

C. D. PRESTON

Potamogeton polygonifolius Bog Pondweed

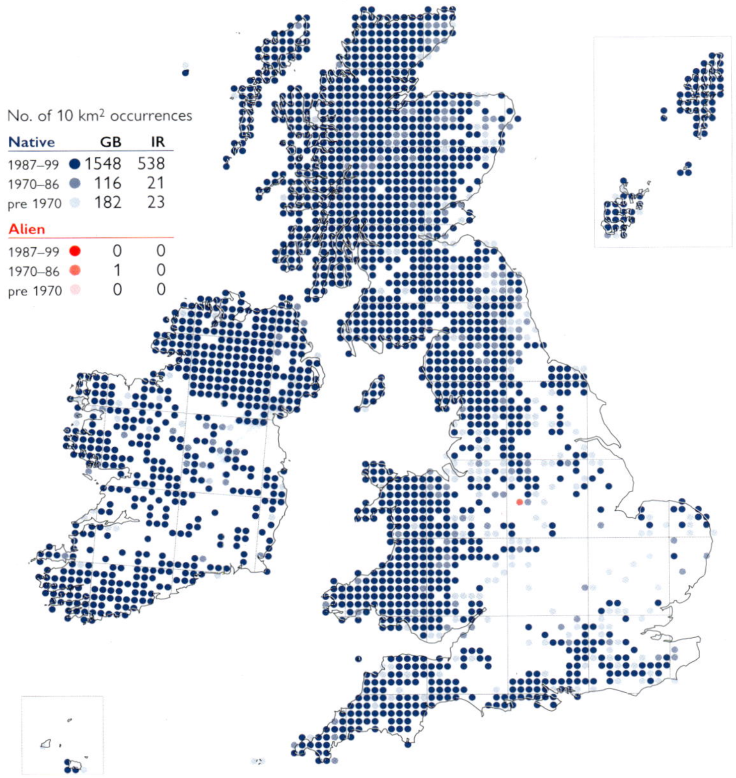

This rhizomatous perennial herb may grow as an aquatic in shallow water in lakes, pools, the backwaters of rivers, streams and ditches, or in a dwarf, subterrestrial state in wet *Sphagnum* lawns or 'brown moss' communities. It is usually restricted to acidic water, only rarely occurring in highly calcareous but nutrient-poor sites. 0–780 m (Lochan an Tairbh-uisge, Mid Perth).

Native. This species has declined in parts of S.E. England where suitable habitats have always been rare, and have been gradually reduced by habitat destruction, drainage or falling water-tables.

Suboceanic Temperate element; also in N. America.

References: Atlas (305b), Hultén & Fries (1986), Meusel *et al.* (1965), Preston (1995), Preston & Croft (1997).

C. D. PRESTON

Potamogeton coloratus Fen Pondweed

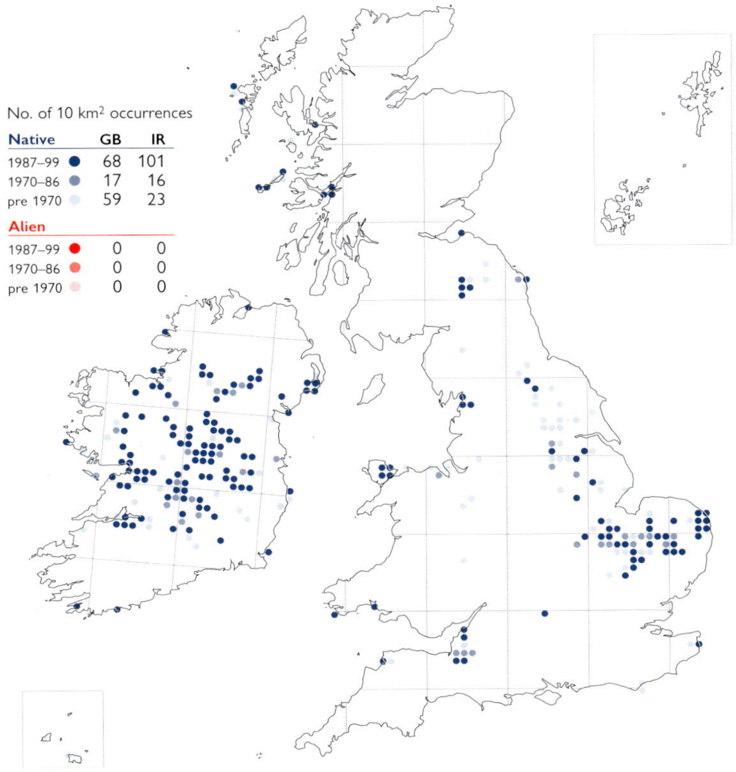

This rhizomatous perennial herb is found in shallow, calcium-rich but nutrient-poor waters in lakes, pools, clay-pits, shallow streams and ditches. It grows over a range of substrates, including peat, marl, sand and clay. Lowland.

Native (change +0.03). *P. coloratus* has declined over much of its British range, having been lost from many sites because of drainage or eutrophication. Many remaining localities are in nature reserves. It has, however, also colonised some new habitats since 1930, and the largest populations are currently those in abandoned brickfields in Peterborough. In Ireland it is still relatively frequent in the central plain.

European Southern-temperate element.

References: Atlas (305c), Hultén & Fries (1986), Preston (1995), Preston & Croft (1997), Stewart *et al.* (1994).

C. D. PRESTON

Potamogeton nodosus Loddon Pondweed

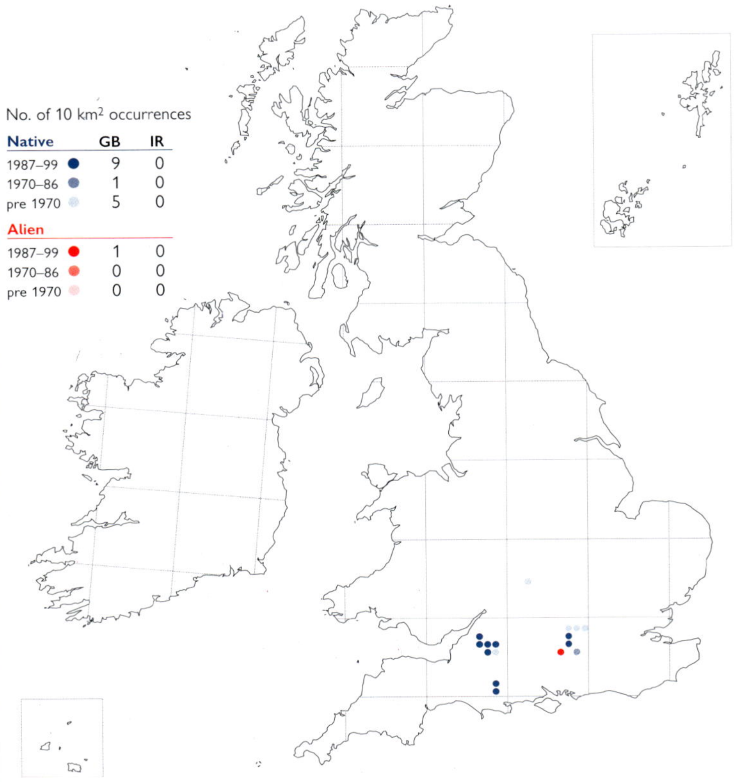

This rhizomatous perennial herb is confined to a few calcareous and moderately eutrophic rivers. It grows in shallow or fairly deep water, apparently preferring gravelly substrates and avoiding soft clays. It reproduces vegetatively but does not fruit in Britain. Lowland.

Native (change –0.18). *P. nodosus* has a relatively stable distribution in the Bristol Avon, the Dorset Stour and the R. Loddon (Berks.). It is apparently extinct in the Thames, perhaps eliminated in the 1950s by eutrophication and increasing pleasure-boat traffic. The record from the Warwickshire Stour is based on a single, 19th century specimen. Recently, it has been planted at additional sites.

Circumpolar Southern-temperate element.

References: Atlas (305d), Hultén & Fries (1986), Preston (1995), Preston & Croft (1997), Wigginton (1999).

C. D. PRESTON

Potamogeton lucens Shining Pondweed

No. of 10 km² occurrences

Native		GB	IR
1987–99	●	270	123
1970–86	●	47	20
pre 1970	●	140	28
Alien			
1987–99	●	0	0
1970–86	●	0	0
pre 1970	●	0	0

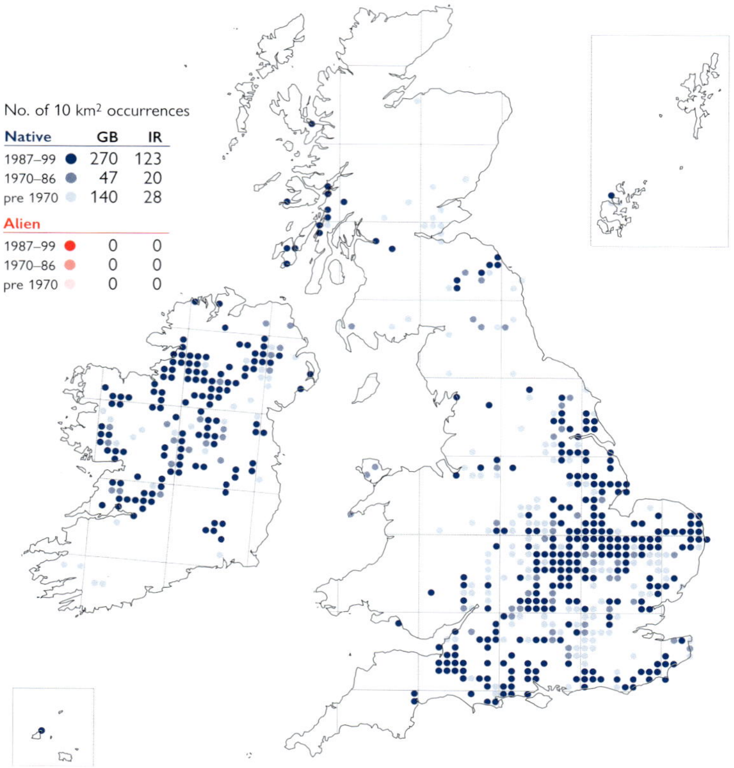

A rhizomatous perennial herb with submerged but no floating leaves, *P. lucens* grows in relatively deep, calcareous water in lakes, larger rivers, canals, flooded chalk- and gravel-pits and major fenland drains. It is found in clear, nutrient-poor, unpolluted waters as well as more eutrophic and turbid sites. 0–380 m (Malham Tarn, Mid-W. Yorks.).

Native (change +0.25). *P. lucens* remains frequent in some areas of England and Ireland, although there is evidence for decline in Surrey since 1930 and at the northern edge of its English range. It has declined more markedly in E. Scotland, apparently being more susceptible to eutrophication there.

Eurosiberian Temperate element.

References: Atlas (306a), Hultén & Fries (1986), Meusel *et al.* (1965), Preston (1995), Preston & Croft (1997).

C. D. PRESTON

Potamogeton lucens × *P. perfoliatus* (*P.* × *salicifolius*) Willow-leaved Pondweed

No. of 10 km² occurrences

Native		GB	IR
1987–99	●	23	13
1970–86	●	7	0
pre 1970	●	38	2
Alien			
1987–99	●	0	0
1970–86	●	0	0
pre 1970	●	0	0

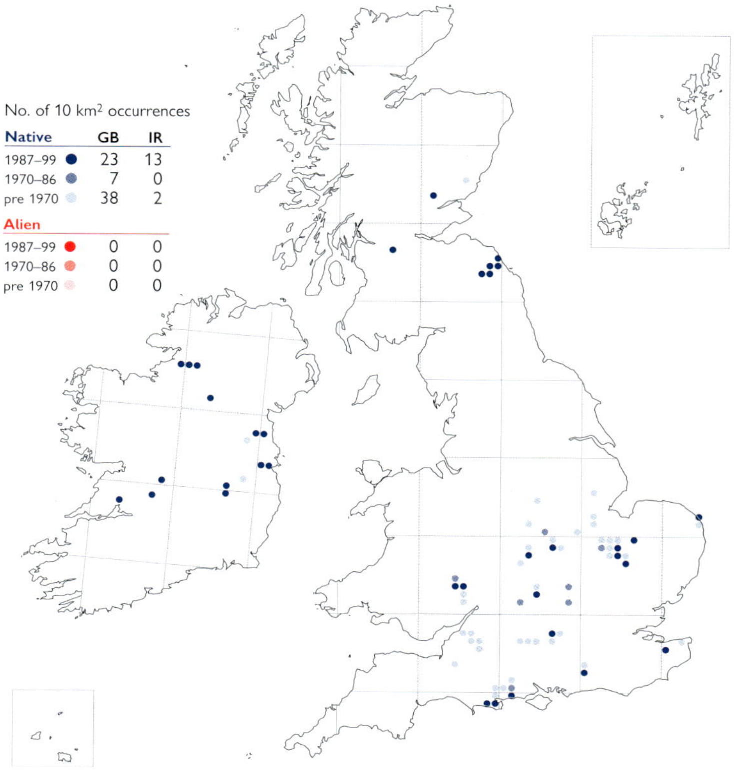

Sites where this rhizomatous perennial have become well-established and locally frequent include slow-flowing rivers, canals and large fenland drains; it is also recorded from a few lakes. It requires relatively deep, moderately or strongly calcareous, meso-eutrophic or eutrophic water. Lowland.

Native. *P.* × *salicifolius* has become extinct in some lakes because of eutrophication, and it appears to be less frequent than formerly in canals. However, it is similar in morphology to *P. lucens* and is probably under-recorded.

Widespread in temperate Europe.

References: Atlas Supp. (136b), Preston (1995), Preston & Croft (1997), Stace (1975).

C. D. PRESTON

Potamogeton gramineus Various-leaved Pondweed

No. of 10 km² occurrences

Native		GB	IR
1987–99	●	257	108
1970–86	●	74	23
pre 1970	●	142	32
Alien			
1987–99	●	0	0
1970–86	●	0	0
pre 1970	●	0	0

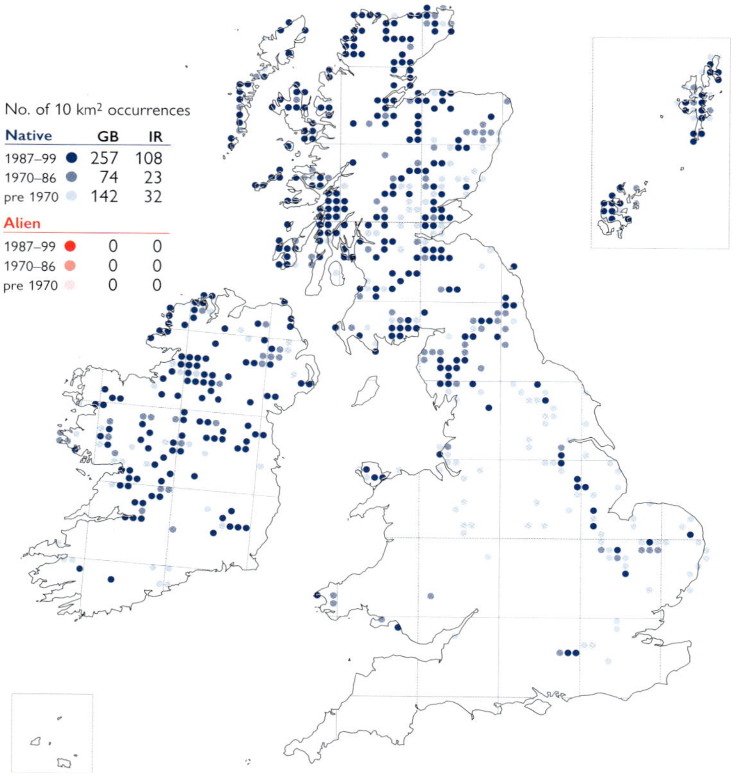

This variable perennial is found in relatively shallow water in a variety of water bodies, including lakes, reservoirs, rivers, streams, canals and ditches. It tolerates a wide range of water quality, although it is absent both from the most acidic and oligotrophic sites and from the most eutrophic. It survives occasional, short-term desiccation. 0–915 m (Meall nan Tarmachan, Mid Perth).

Native (change +0.67). Like many aquatics, *P. gramineus* is better recorded than it was in the 1962 *Atlas*. However, it has declined in England since 1930 in areas where eutrophication and the conversion of grazing land to arable have reduced the number of suitable ditches.

Circumpolar Boreo-temperate element.

References: Atlas (306b), Hultén & Fries (1986), Preston (1995), Preston & Croft (1997).

C. D. PRESTON

Potamogeton gramineus × *P. lucens* (*P.* × *zizii*) Long-leaved Pondweed

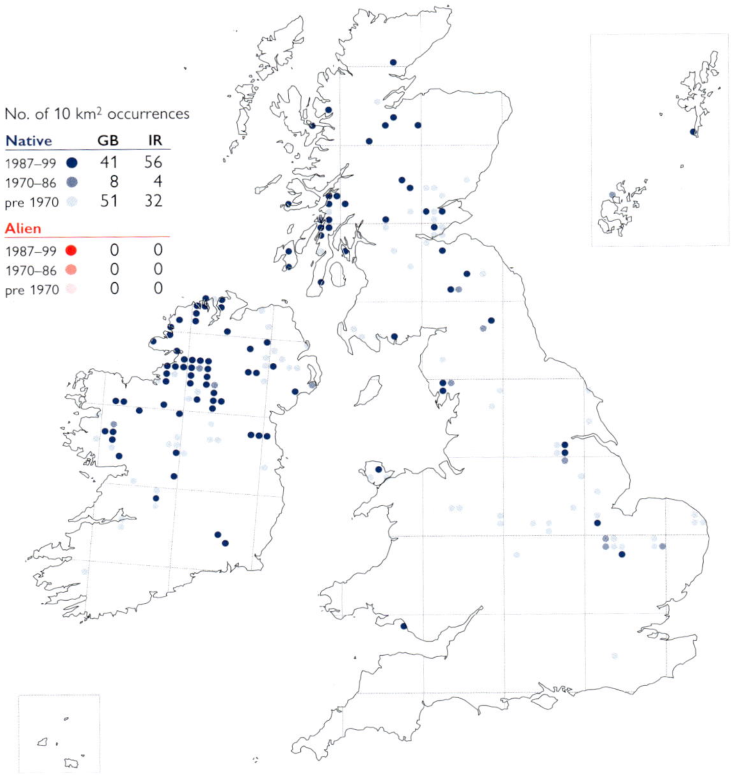

This rhizomatous perennial grows in a range of mesotrophic, somewhat base-enriched waters including lakes, rivers, streams, fenland lodes and ditches. In N. Scotland and N.W. Ireland it often occurs in the absence of *P. lucens*. It is morphologically variable and some populations are sterile whereas others set well-formed fruit. 0–395 m (Loch na Craige, Mid Perth).

Native. There is little evidence for the decline of this hybrid in N.W. Ireland and W. Scotland, where it is probably still under-recorded. In E. Ireland, E. Scotland and England there is a clear decline, which matches that of *P. lucens* in Scotland and *P. gramineus* in England.

Widespread in temperate Europe.

References: Atlas Supp. (137b), Preston (1995), Preston & Croft (1997), Stace (1975).

C. D. PRESTON

Potamogeton gramineus × *P. perfoliatus* (*P.* × *nitens*) Bright-leaved Pondweed

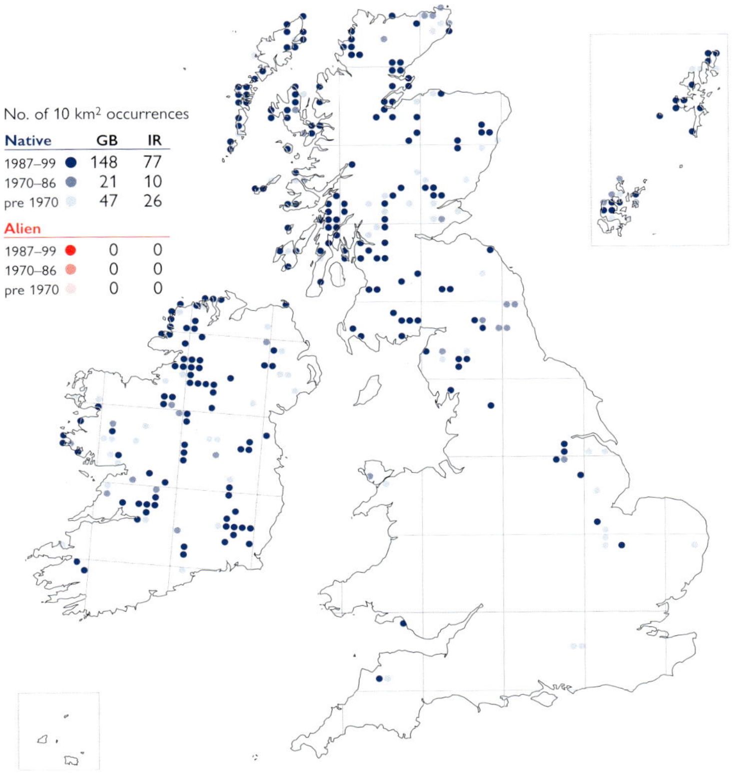

This rhizomatous perennial is found in more or less mesotrophic water in lakes, reservoirs, rivers, streams, canals and ditches. In some sites it can be the most abundant macrophyte, sometimes occurring in the absence of one or even both parents. It is favoured by moderate fluctuations in water level, and can withstand short periods of desiccation. 0–640 m (Loch Oss, W. Perth).

Native. *P.* × *nitens* is a sterile hybrid which sometimes resembles *P. gramineus* very closely, and is almost certainly under-recorded. It has been lost from some sites in East Anglia, and was last recorded from the Basingstoke Canal (N. Hants.) in 1931.

Widespread in boreal and temperate regions of Europe and also recorded elsewhere in the N. hemisphere.

References: Atlas Supp. (137c), Preston (1995), Preston & Croft (1997), Stace (1975).

C. D. PRESTON

Potamogeton alpinus Red Pondweed

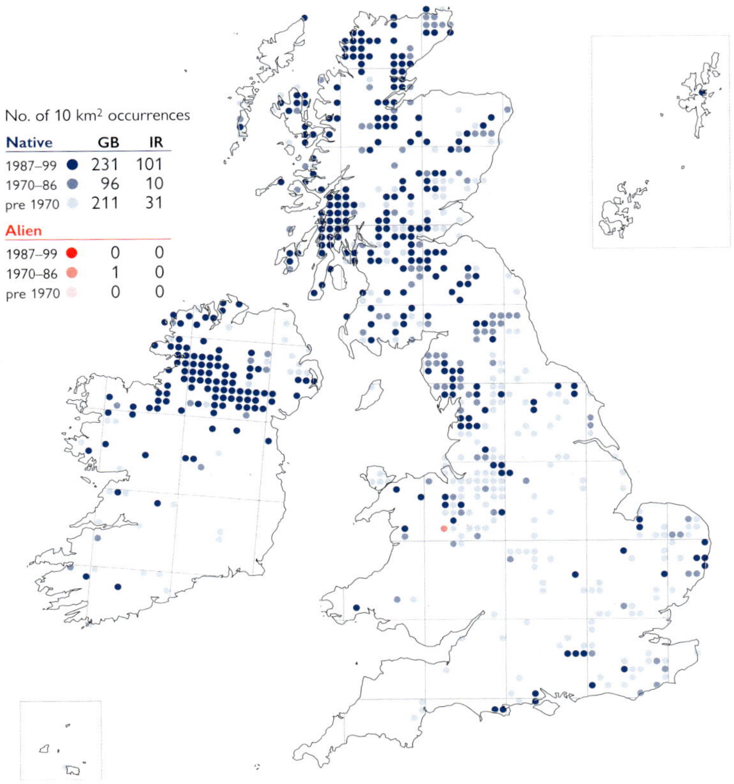

A rhizomatous perennial of still or slow-flowing water in lakes, rivers, canals, ditches and flooded mineral workings. *P. alpinus* is often found in sites where silt accumulates, such as lake inflows or backwaters in rivers. It characteristically grows in mesotrophic, often neutral or mildly acidic water. 0–945 m (Meall nan Tarmachan, Mid Perth).

Native (change +0.30). Many of the sites from which *P. alpinus* has been lost in England are small ponds, pits and ditches where it was last seen in the 19th century. There have been some further losses in England and S.E. Scotland from eutrophication, but the distribution elsewhere appears to be stable.

Circumpolar Boreal-montane element.

References: Atlas (306c), Hultén & Fries (1986), Preston (1995), Preston & Croft (1997).

C. D. PRESTON

Potamogeton praelongus
Long-stalked Pondweed

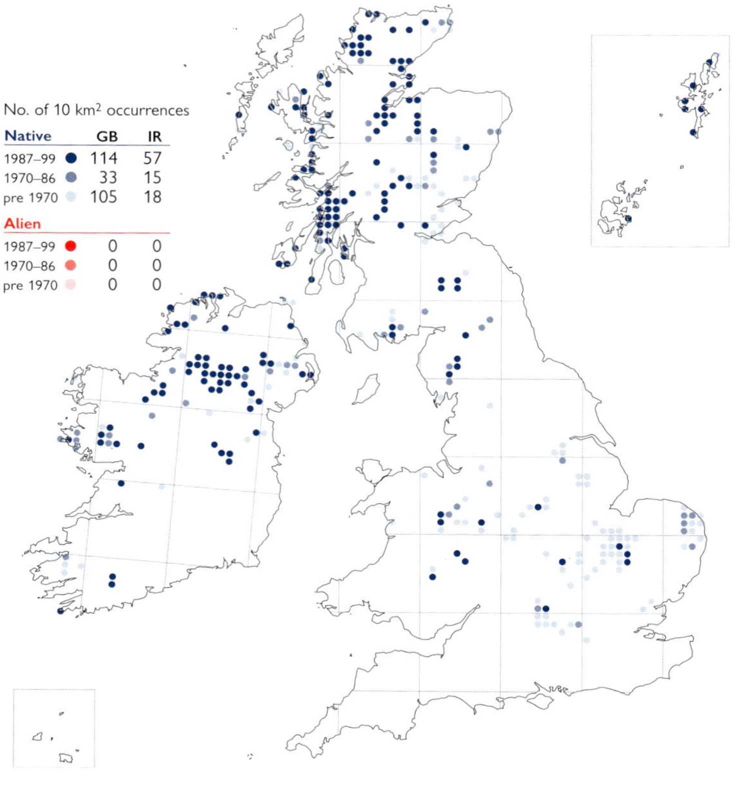

No *Potamogeton* species is more characteristic of deep water than *P. praelongus*. It is a rhizomatous perennial that usually grows at depths greater than 1 m in clear, mesotrophic water in lakes, rivers, canals and major drains. It has only rarely been recorded from shallow water. 0–800 m (Loch Coire Cheap, Mid Perth).

Native (change −0.26). *P. praelongus* appears to have been lost from many waters in the southern half of its range since 1930. The most likely cause of this decline is eutrophication. As a species of deeper water it can be inconspicuous, and it may be under-recorded in Scotland and Ireland.

Circumpolar Boreal-montane element.

References: Atlas (306d), Hultén & Fries (1986), Preston (1995), Preston & Croft (1997), Stewart *et al.* (1994).

C. D. PRESTON

Potamogeton perfoliatus Perfoliate Pondweed

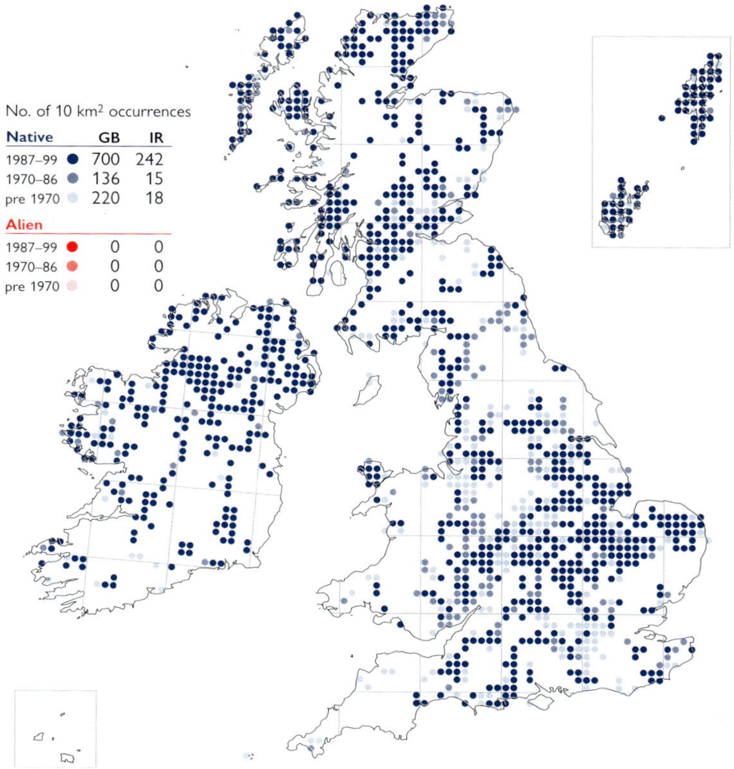

A frequent macrophyte in larger water bodies, *P. perfoliatus* occasionally grows in oligotrophic sites but is more often found in mesotrophic or eutrophic conditions. It is a rhizomatous perennial that grows in shallow water in sites which are not prone to occasional desiccation, but is most vigorous at depths of 1 m or more. 0–780 m (Loch an Tairbh-uisge, Mid Perth).

Native. There is evidence for a decline of *P. perfoliatus* in some areas, such as Surrey and Cheshire. However it remains frequent in many regions, including some where more exacting *Potamogeton* species have declined.

Circumpolar Boreo-temperate element, with a disjunct distribution.

References: Atlas (307a), Hultén & Fries (1986), Preston (1995), Preston & Croft (1997).

C. D. PRESTON

Potamogeton epihydrus American Pondweed

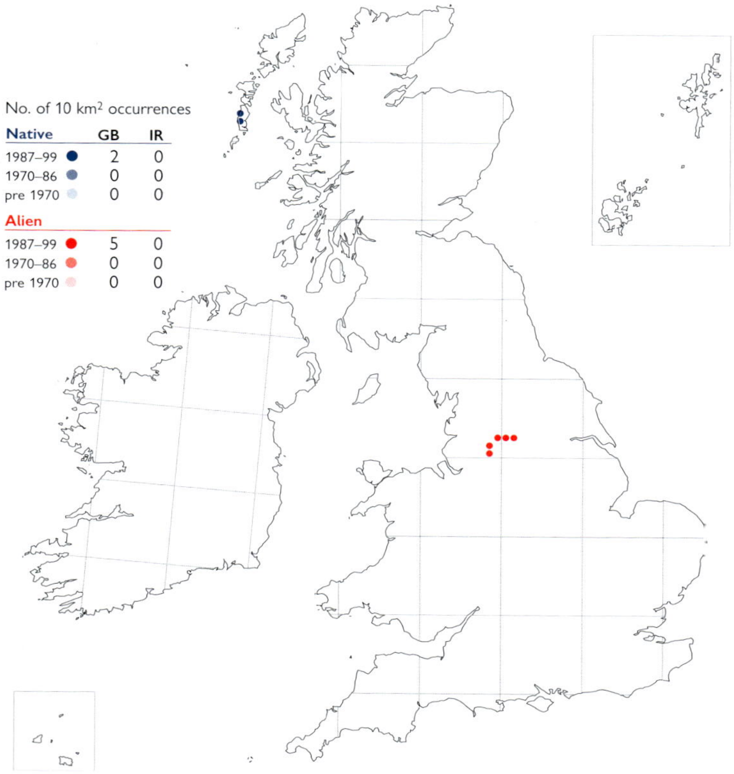

In the Outer Hebrides, this rhizomatous perennial grows in a few peaty lochans, in oligotrophic and base-poor water less than 1 m deep. It is also established in the mesotrophic Rochdale Canal and Calder & Hebble Navigation in N. England. Lowland.

Native (change +0.11). This species was first discovered in England near Halifax (S.W. Yorks.) in 1907; although it must be an introduction there, the manner of its arrival from N. America is unclear. It was not found in the Outer Hebrides, where it is native, until 1943, and there is no evidence for any subsequent change in its distribution there.

Oceanic Boreo-temperate element; in Europe restricted to Britain but widespread in N. America.

References: Atlas (307b), FNAEC (2000), Preston (1995), Preston & Croft (1997), Wigginton (1999).

C. D. PRESTON

Potamogeton friesii Flat-stalked Pondweed

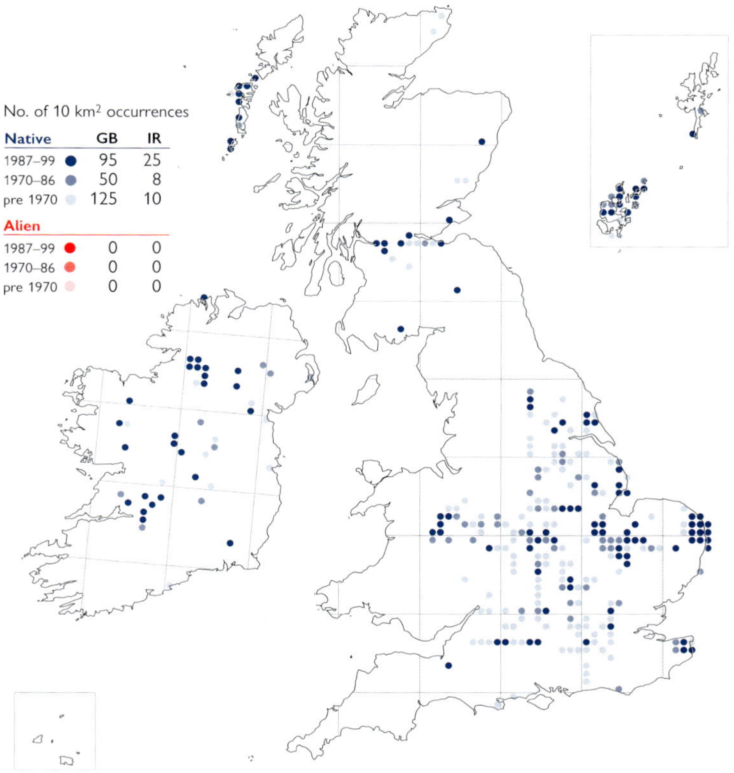

No. of 10 km² occurrences

Native	GB	IR
1987–99	95	25
1970–86	50	8
pre 1970	125	10
Alien		
1987–99	0	0
1970–86	0	0
pre 1970	0	0

A rhizomatous perennial of calcareous and often rather eutrophic, still or very slowly flowing waters. These include lakes, sluggish rivers and streams, canals, fenland lodes and flooded mineral workings. Lowland.

Native (change −1.06). *P. friesii* expanded through the canal network, and subsequently declined as canals became disused or dominated by pleasure boat traffic. Like other linear-leaved pondweeds it is probably under-recorded. The distinction between *P. friesii* and *P. pusillus* is straightforward in England but becomes difficult in N. Scotland and some records may require reassessment.

Circumpolar Boreo-temperate element, with a disjunct distribution.

References: Atlas (307c), Hultén & Fries (1986), Meusel *et al.* (1965), Preston (1995), Preston & Croft (1997), Stewart *et al.* (1994).

C. D. PRESTON

Potamogeton rutilus Shetland Pondweed

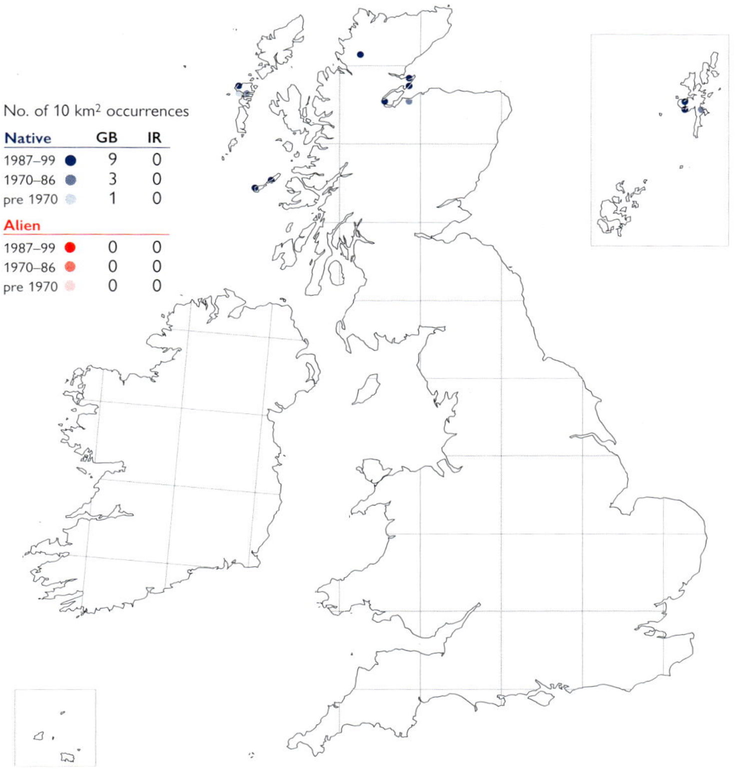

No. of 10 km² occurrences

Native	GB	IR
1987–99	9	0
1970–86	3	0
pre 1970	1	0
Alien		
1987–99	0	0
1970–86	0	0
pre 1970	0	0

This rhizomatous perennial grows submerged in unpolluted, mesotrophic or eutrophic lochs and adjoining streams, usually where there is some base-enrichment. It normally reproduces by turions, though it has recently been found fruiting in Scotland. Lowland.

Native (change +0.18). In 1962 *P. rutilus* was regarded as a very rare species, known only from the Outer Hebrides and Shetland. Intensive surveys of aquatic habitats have gradually revealed more colonies, and others doubtless remain to be discovered. It has apparently been lost from one site, Loch Flemington (Easterness), because of eutrophication.

European Boreal-montane element.

References: Atlas (305d), Hultén & Fries (1986), Preston (1995), Preston & Croft (1997), Wigginton (1999).

C. D. PRESTON

Potamogeton pusillus Lesser Pondweed

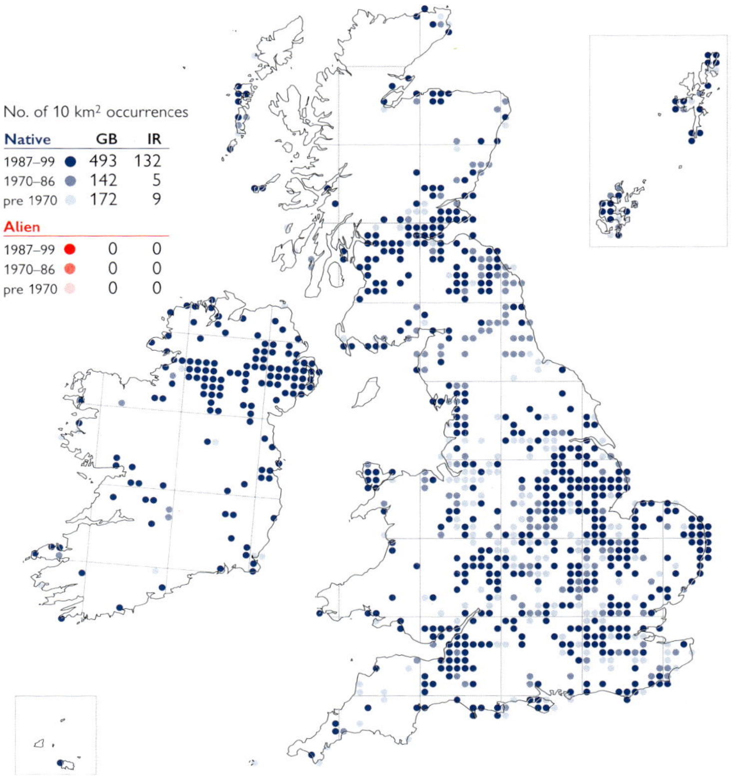

No. of 10 km² occurrences

Native	GB	IR
1987–99	493	132
1970–86	142	5
pre 1970	172	9
Alien		
1987–99	0	0
1970–86	0	0
pre 1970	0	0

A rhizomatous perennial found in standing or slowly flowing water in sheltered lakes and reservoirs, ponds, rivers, canals, ditches and flooded mineral workings. It favours mesotrophic to eutrophic water and tolerates slightly brackish conditions. 0–320 m (Stony Middleton, Derbys.).

Native (change +0.77). *P. pusillus* and *P. berchtoldii* were confused in our area until 1938. Any assessment of changes in their distribution must therefore be based on very limited evidence, especially as they may still be under-recorded. However, comparison of pre-1930 and recent records suggests that *P. pusillus* may have increased in some areas, perhaps in response to eutrophication.

Circumpolar Southern-temperate element.

References: Atlas (307d), Hultén & Fries (1986), Preston (1995), Preston & Croft (1997).

C. D. PRESTON

Potamogeton obtusifolius
Blunt-leaved Pondweed

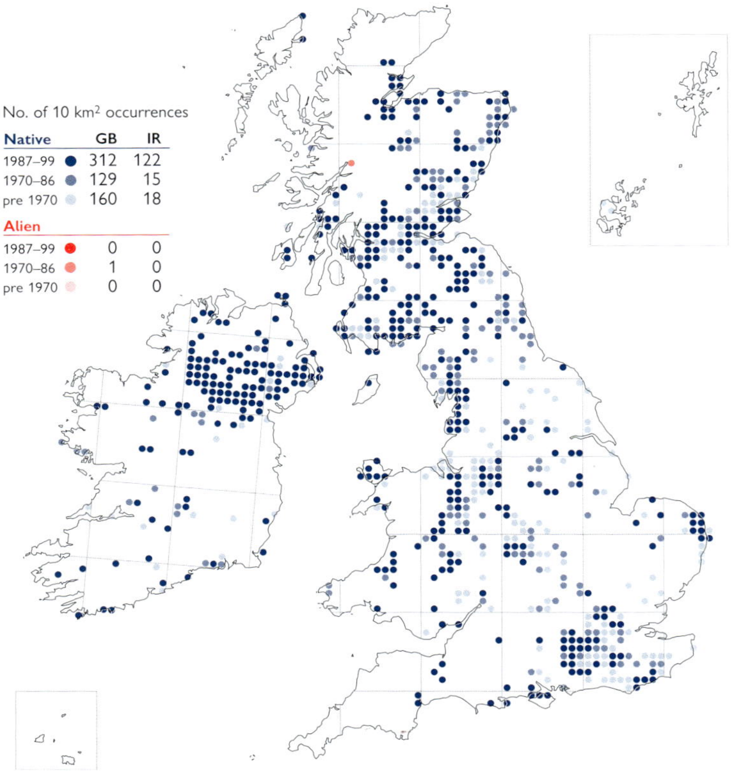

No. of 10 km² occurrences

Native		GB	IR
1987–99	●	312	122
1970–86	●	129	15
pre 1970	●	160	18
Alien			
1987–99	●	0	0
1970–86	●	1	0
pre 1970	●	0	0

A rhizomatous perennial characteristically found in mesotrophic or meso-eutrophic, acidic or neutral standing waters in lakes, ponds and flooded mineral workings, or in canals and the backwaters of rivers. It may, however, occupy a broader habitat range in areas such as N. Ireland where it is especially frequent. It fruits freely, and also reproduces by turions. 0–480 m (Angle Tarn, Patterdale, Westmorland).

Native (change +0.96). Although *P. obtusifolius* is better recorded than in the 1962 *Atlas*, the map suggests that it has declined throughout much of its British range. There is less evidence of decline in Ireland.

Circumpolar Boreo-temperate element, with a disjunct distribution.

References: Atlas (308a), Hultén & Fries (1986), Meusel *et al.* (1965), Preston (1995), Preston & Croft (1997).

C. D. PRESTON

Potamogeton berchtoldii Small Pondweed

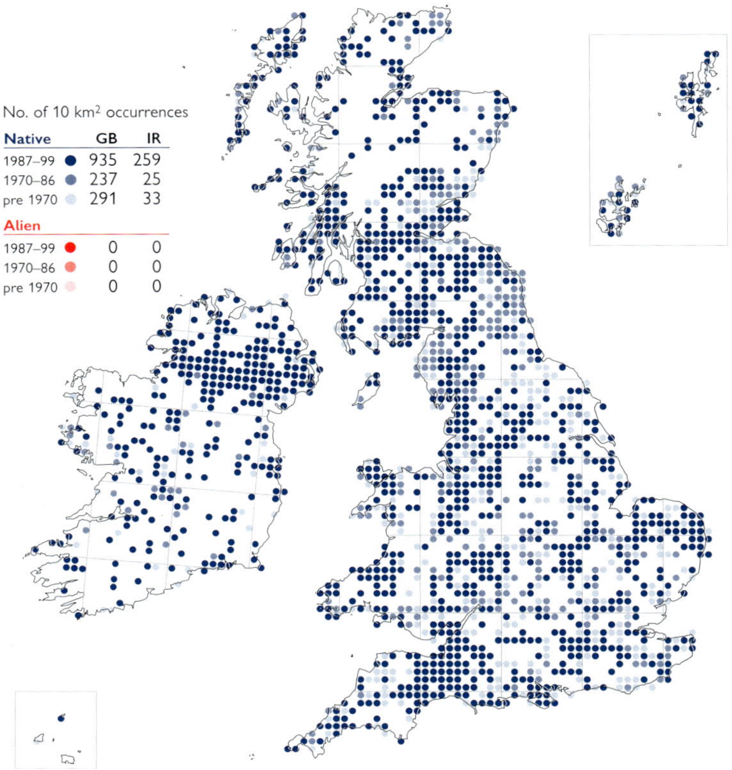

No. of 10 km² occurrences

Native		GB	IR
1987–99	●	935	259
1970–86	●	237	25
pre 1970	●	291	33
Alien			
1987–99	●	0	0
1970–86	●	0	0
pre 1970	●	0	0

A variable species which occurs in a wide range of still or slowly flowing waters, which may be base-rich or base-poor, oligotrophic, mesotrophic or eutrophic, and exposed or sheltered. *P. berchtoldii* is sometimes found in brackish sites, but is usually replaced in such habitats by *P. pusillus*. 0–710 m (Lochan nan Cat, Mid Perth).

Native (change +1.66). Until 1938, *P. berchtoldii* was confused with *P. pusillus* in our area. It is under-recorded by botanists who neglect linear-leaved pondweeds but may be over-recorded by others who fail to distinguish *P. pusillus*. Changes in its distribution are difficult to assess.

Circumpolar Boreo-temperate element.

References: Atlas (308b), Hultén & Fries (1986), Preston (1995), Preston & Croft (1997).

C. D. PRESTON

Potamogeton trichoides Hairlike Pondweed

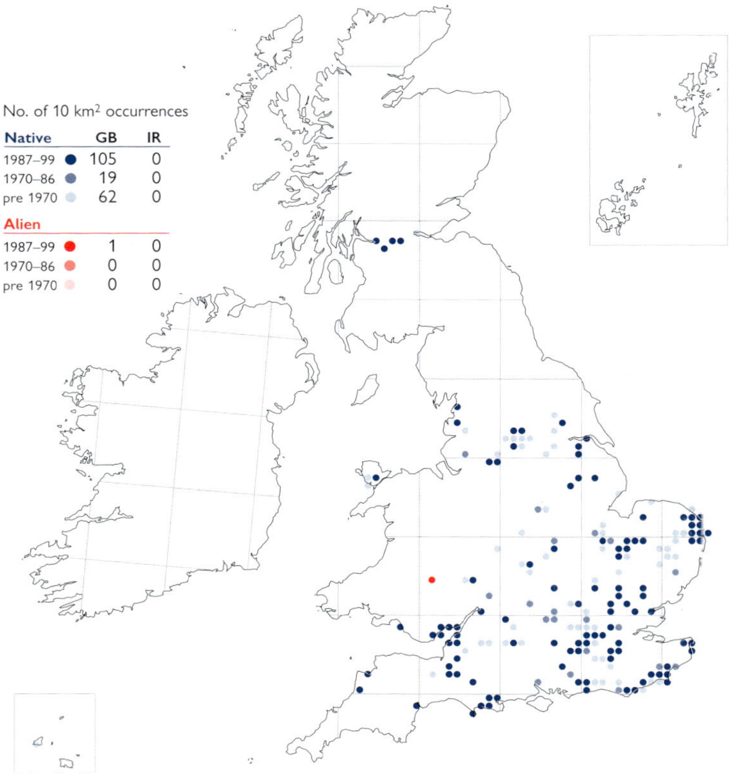

No. of 10 km² occurrences

Native		GB	IR
1987–99	●	105	0
1970–86	●	19	0
pre 1970	●	62	0
Alien			
1987–99	●	1	0
1970–86	●	0	0
pre 1970	●	0	0

P. trichoides is found in a range of still or slowly flowing, mesotrophic or eutrophic waters including lakes, ponds, rivers, canals, ditches and flooded mineral workings. It often colonises disturbed sites such as recently cleared canals and ditches. Lowland.

Native (change +0.57). Although the distribution of this elusive linear-leaved pondweed is better known now than at the time of the 1962 *Atlas*, it is probably still under-recorded as it is difficult to recognise, especially when growing with *P. pusillus*. There is evidence of a genuine increase in frequency since 1960 in the Somerset Levels, perhaps because of eutrophication.

Eurosiberian Southern-temperate element.

References: Atlas (308c), Hultén & Fries (1986), Meusel *et al.* (1965), Preston (1995), Preston & Croft (1997), Stewart *et al.* (1994).

C. D. PRESTON

Potamogeton compressus
Grass-wrack Pondweed

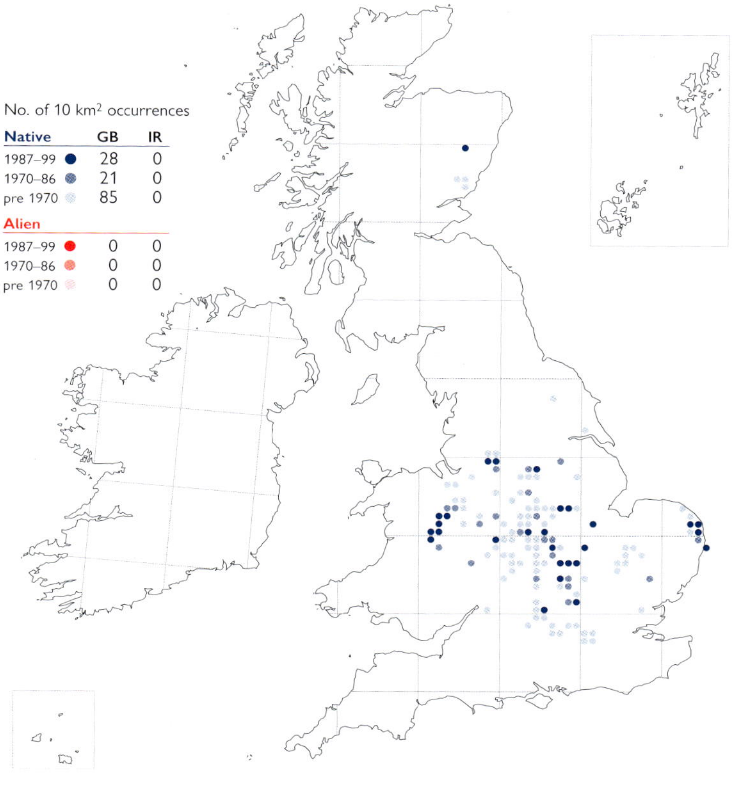

P. compressus, a rhizomatous perennial, has been recorded from a wide range of habitats: lakes, sluggish rivers, ditches, canals and flooded mineral workings. Its sites share a tendency to be still or slowly flowing, mesotrophic and slightly to moderately base-rich. Lowland.

Native (change –1.68). This species appears to have been in gradual decline for over a hundred and fifty years. It is almost extinct in lakes and rivers, and is declining in grazing marsh ditches. In Broadland populations have been lost since 1970. Some of the most vigorous surviving populations are in canals, especially the Montgomery branch of the Shropshire Union Canal.

Eurasian Boreo-temperate element.

References: Atlas (308d), Hultén & Fries (1986), Preston (1995), Preston & Croft (1997), Stewart *et al.* (1994).

C. D. PRESTON

Potamogeton acutifolius Sharp-leaved Pondweed

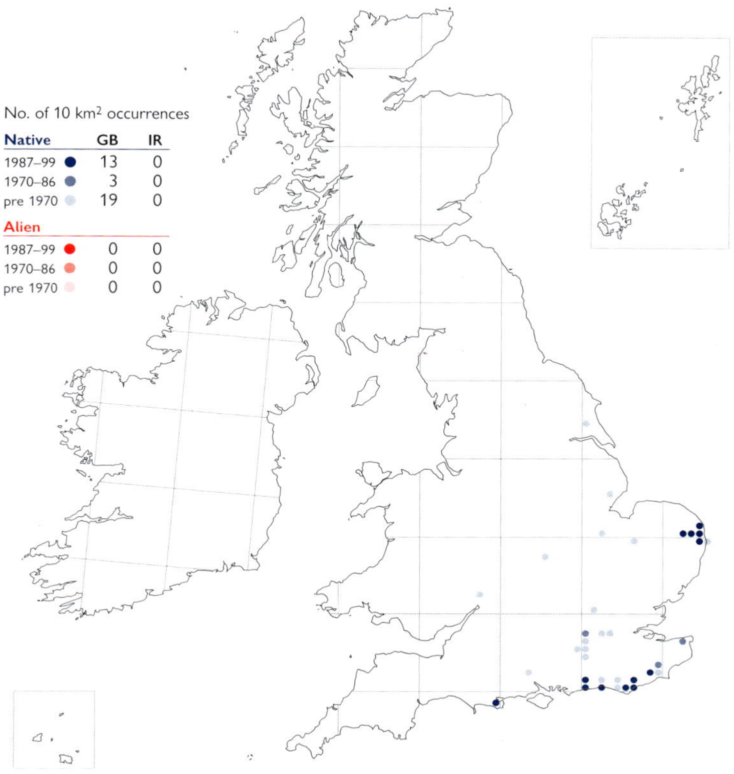

This rhizomatous perennial has a very narrow habitat range, being confined to shallow, species-rich drainage ditches in lowland grazing marshes, where it typically grows in calcareous, mesotrophic or meso-eutrophic water. Although it fruits relatively freely and also produces turions, it shows little or no propensity to colonise new habitats. Lowland.

Native (change +0.05). *P. acutifolius* appears to be in gradual, long-term decline. Since 1960 it has decreased in Norfolk and become extinct in the London area and critically endangered in Dorset. However, several vigorous populations survive in Sussex.

European Temperate element.

References: Atlas (309a), Hultén & Fries (1986), Preston (1995), Preston & Croft (1997), Preston & Pearman (1998b), Wigginton (1999).

C. D. PRESTON

Potamogeton crispus Curled Pondweed

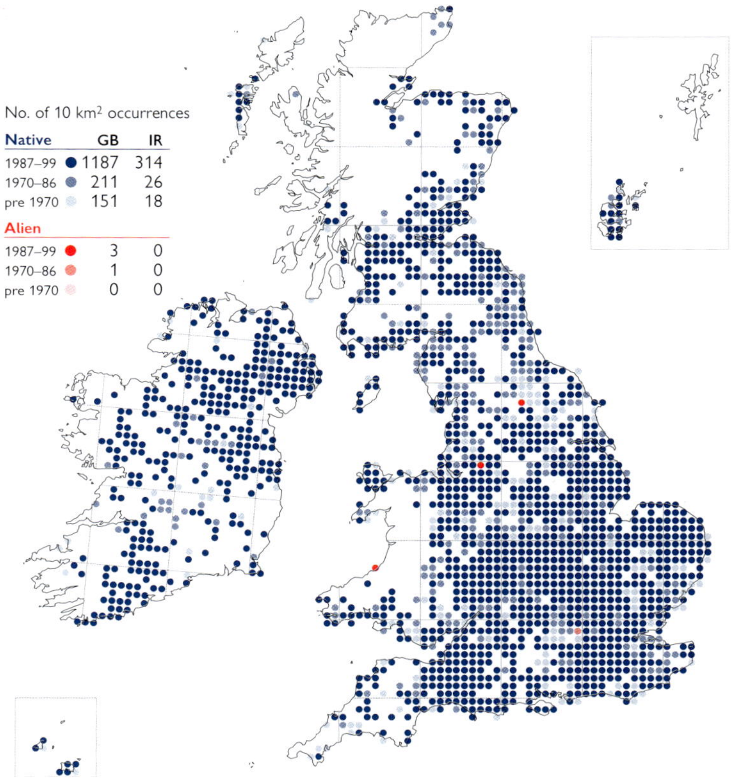

P. crispus is a rhizomatous perennial which grows in a wide range of mesotrophic or eutrophic waters. These include lakes, ponds, rivers, streams, canals, ditches and disused mineral workings. It is more tolerant of eutrophication than most British *Potamogeton* species. Generally lowland, but reaching 350 m at Drumore Loch (E. Perth).

Native. This is one of the most distinctive *Potamogeton* species, and it is therefore relatively well recorded. There is little evidence for any change in its British or Irish distribution.

Eurasian Southern-temperate element, but naturalised in N. America so distribution is now Circumpolar Southern-temperate.

References: Atlas (309b), Catling & Dobson (1985), Grime *et al.* (1988), Hultén & Fries (1986), Meusel *et al.* (1965), Preston (1995), Preston & Croft (1997).

C. D. PRESTON

Potamogeton filiformis
Slender-leaved Pondweed

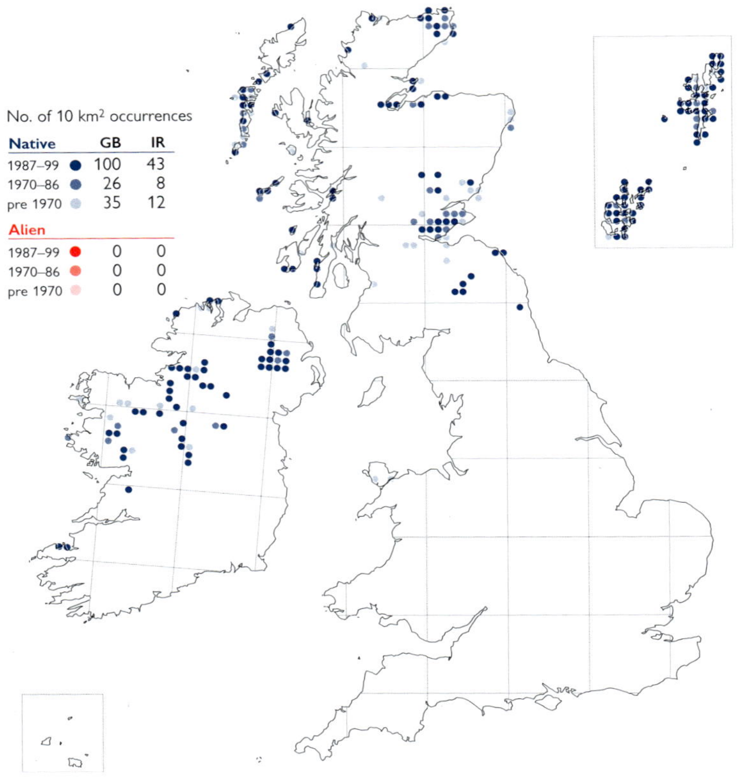

This rhizomatous herb usually grows in open vegetation in the shallow edges of lakes. It is typically found over gravel, sand, silt or mud in sites where the water is base-rich, eutrophic or slightly brackish. It also occasionally grows in rivers, streams and ditches. 0–390 m at Drumore Loch (E. Perth), formerly at 735 m at Coire Dhubh-chlair (Mid Perth).

Native (change +0.63). *P. filiformis* was last collected in Anglesey in 1826, and it may have decreased in E. Scotland since 1960, as it can no longer be found at some sites. However, there is little evidence for any decline in N. Scotland or Ireland, where it may be under-recorded.

Circumpolar Boreal-montane element.

References: Atlas (309c), Hultén & Fries (1986), Meusel *et al.* (1965), Preston (1995), Preston & Croft (1997), Stewart *et al.* (1994).

C. D. PRESTON

Potamogeton pectinatus Fennel Pondweed

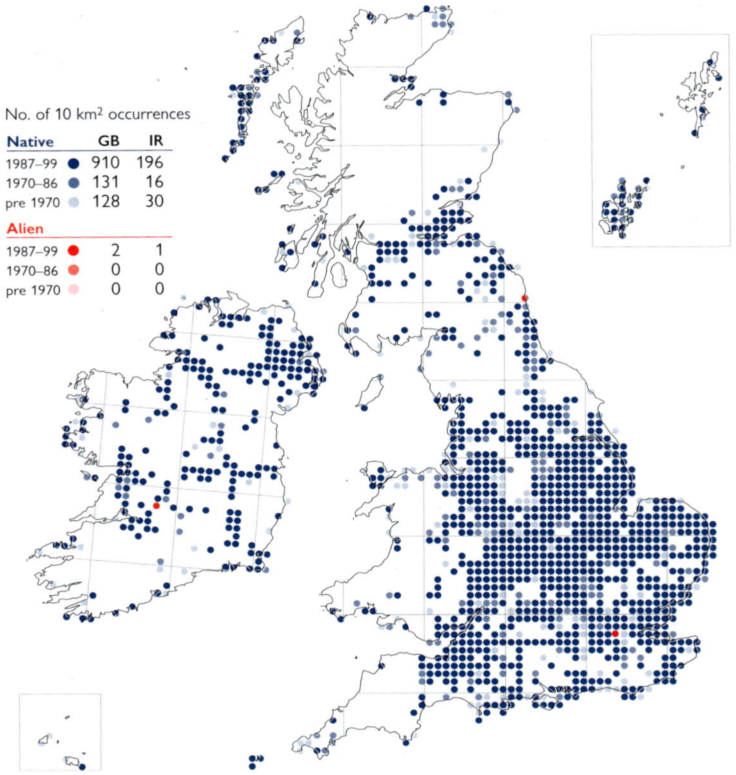

This rhizomatous, linear-leaved aquatic is a characteristic species of eutrophic or brackish waters, where it may form dense stands in lakes, reservoirs, rivers, streams, canals, ditches, ponds and flooded mineral workings. It is tolerant of disturbance in canals and navigable rivers. It is occasionally found in highly calcareous but nutrient-poor lakes. Lowland.

Native. There is little evidence for any change in the distribution of this species. Any losses from habitat destruction have been compensated for by the ability of the species to colonise newly available habitats, such as flooded sand- and gravel-pits.

Circumpolar Wide-temperate element.

References: Atlas (309d), Hultén & Fries (1986), Meusel *et al.* (1965), Preston (1995), Preston & Croft (1997).

C. D. PRESTON

Groenlandia densa Opposite-leaved Pondweed

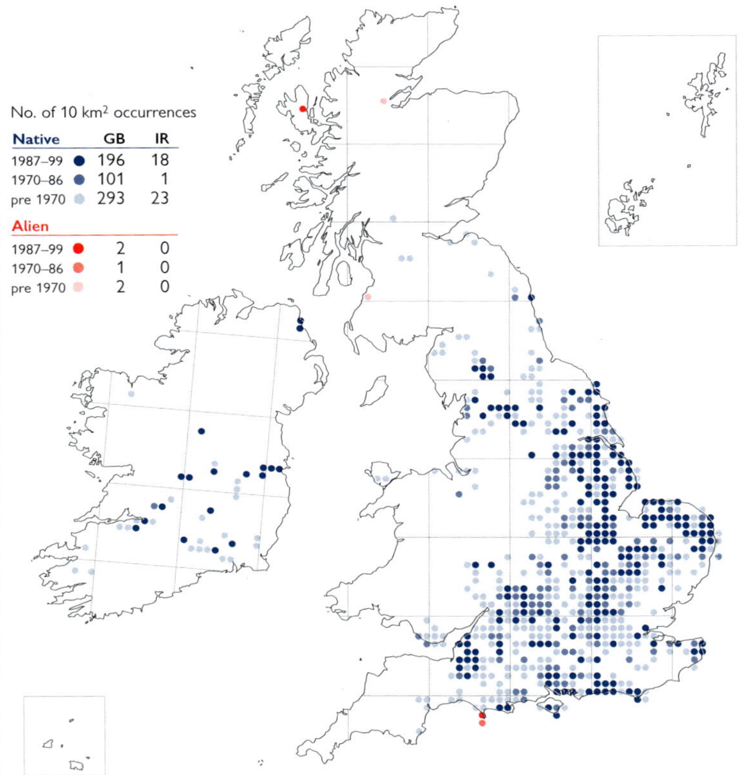

A perennial herb of shallow, clear, base-rich water which may grow in lakes and rivers, but is more frequent in smaller waters such as streams, canals, ditches and ponds. It rarely colonises newly available habitats, although it is sometimes found as an introduction in ponds. Generally lowland, but reaching 380 m at Malham Tarn (Mid-W. Yorks.).

Native (change −1.23). *G. densa* has declined in Britain due to urbanisation, which has led to its loss from the London area, eutrophication and the loss of spring-fed streams and ditches because of falling water tables. This decline began before 1930, but is still continuing.

European Temperate element.

References: Atlas (310a), Curtis & McGough (1988), Hultén & Fries (1986), Meusel *et al.* (1965), Preston (1995), Preston & Croft (1997).

C. D. PRESTON

Ruppia maritima Beaked Tasselweed

Native	GB	IR
1987–99	169	39
1970–86	72	4
pre 1970	125	37

Alien		
1987–99	1	0
1970–86	0	0
pre 1970	1	0

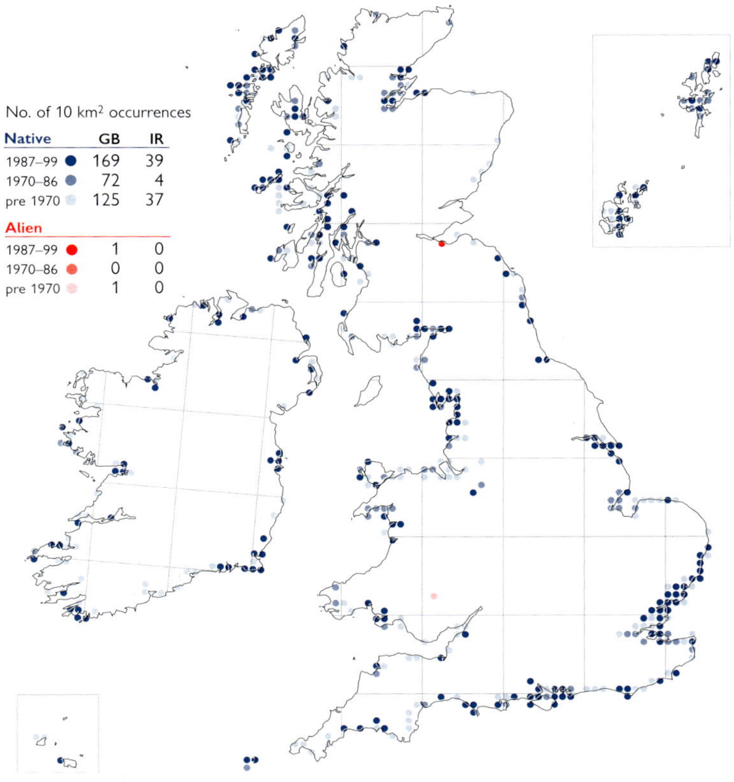

A submerged, annual or perennial aquatic of brackish waters, *R. maritima* grows in shallow water in coastal lakes, pools on saltmarshes, rock pools, creeks and ditches near the sea. It is also found as a dwarf variant on tidal mud-flats, especially in N.E. Scotland. The recent inland record in Cheshire is in an area of natural salt deposits. Lowland.

Native (change –0.34). *R. maritima* had been lost from many sites before 1930, and this loss has continued in the southern part of its range around coastal developments.

Circumpolar Wide-temperate element.

References: Atlas (310c), Hultén & Fries (1986), Preston & Croft (1997).

C. D. PRESTON

Ruppia cirrhosa Spiral Tasselweed

Native	GB	IR
1987–99	48	9
1970–86	17	2
pre 1970	57	11

Alien		
1987–99	0	0
1970–86	0	0
pre 1970	0	0

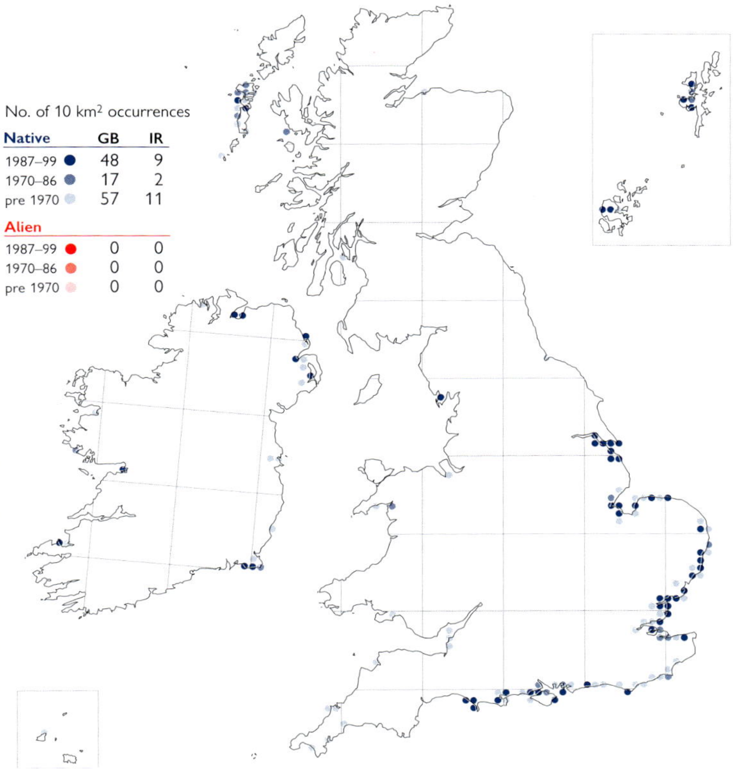

R. cirrhosa is a perennial aquatic which occurs in similar habitats to *R. maritima*, including coastal lakes, tidal inlets, creeks and brackish ditches. It usually grows in deeper water than *R. maritima* and tolerates more saline conditions, even growing with *Zostera* species. Lowland.

Native (change –0.41). *R. cirrhosa* is now much better recorded than it was for the 1962 *Atlas*. Most of the losses took place before 1930.

Circumpolar Wide-temperate element, but absent from eastern N. America.

References: Atlas (310b), Hultén & Fries (1986), Preston & Croft (1997), Stewart *et al.* (1994).

C. D. PRESTON

Najas flexilis Slender Naiad

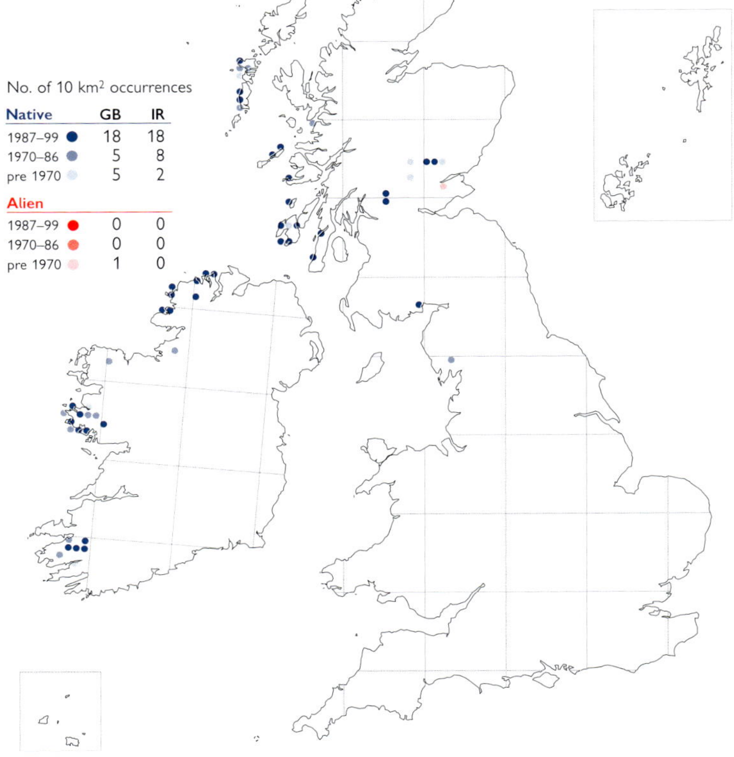

This aquatic plant is an annual which is usually found in deep, clear, mesotrophic lakes where the water receives some base-enrichment from nearby basalt, limestone or calcareous dune-sand. Lowland.

Native (change +0.48). This is an elusive species which has been recorded at an increasing number of British and Irish sites in recent years as a result of lake surveys. It has, however, apparently been lost from its only English site, Esthwaite Water (Westmorland), and some lakes in E. Scotland because of eutrophication.

Circumpolar Boreal-montane element, with a disjunct distribution.

References: Atlas (311a), Curtis & McGough (1988), Hultén & Fries (1986), Meusel *et al.* (1965), Preston & Croft (1997), Stewart *et al.* (1994).

C. D. PRESTON

Najas marina Holly-leaved Naiad

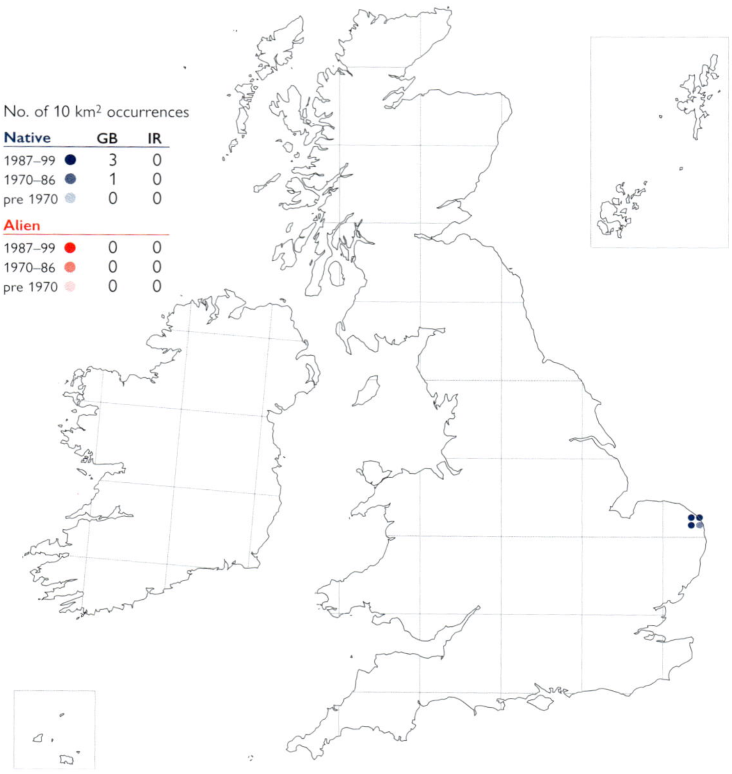

No. of 10 km² occurrences

Native	GB	IR
1987–99 ●	3	0
1970–86 ●	1	0
pre 1970 ●	0	0
Alien		
1987–99 ●	0	0
1970–86 ●	0	0
pre 1970 ●	0	0

N. marina grows in meso-eutrophic water over deep substrates of peat or silty mud in the Norfolk broads. It is a dioecious annual, with both male and female plants in Britain. Lowland.

Native. First discovered in Britain in 1883 (Hickling Broad, E. Norfolk), *N. marina* decreased in the late 1960s as a result of pollution, but has since responded to action which has been taken to reduce nutrient levels in the Norfolk Broads.

Circumpolar Southern-temperate element, with a continental distribution in W. Europe.

References: Atlas (311b), Handley & Davy (2000), Hultén & Fries (1986), Preston & Croft (1997), Wigginton (1999).

C. D. PRESTON

Zannichellia palustris Horned Pondweed

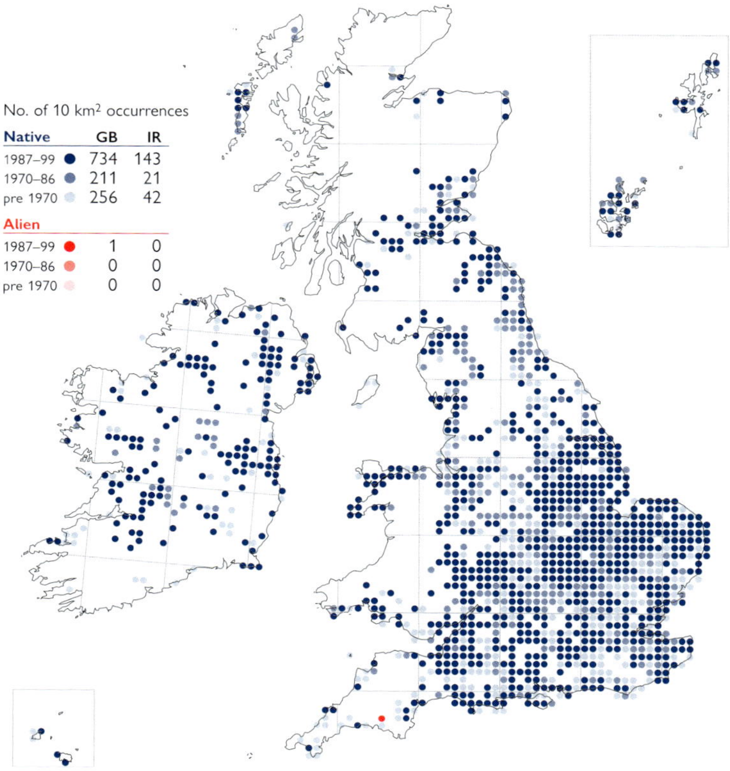

No. of 10 km² occurrences

Native	GB	IR
1987–99 ●	734	143
1970–86 ●	211	21
pre 1970 ●	256	42
Alien		
1987–99 ●	1	0
1970–86 ●	0	0
pre 1970 ●	0	0

This submerged, perennial aquatic grows in a range of shallow-water habitats. The most characteristic include clear chalk streams, eutrophic lakes and ponds, and brackish lagoons, ponds and ditches. It is a frequent colonist of disused mineral workings. 0–380 m (Llynheilyn, Rads.).

Native (change +0.17). *Z. palustris* is much more frequent within its British and Irish range than was shown in the 1962 *Atlas*. This probably represents more effective recording, but there is evidence for the expansion of this species in S. Scotland, in response to eutrophication. Conversely, the current map shows a decline in England, presumably from drainage and ditch improvements.

Circumpolar Southern-temperate element.

References: Atlas (310d), Hultén & Fries (1986), Preston & Croft (1997).

C. D. PRESTON

Zostera marina Eelgrass

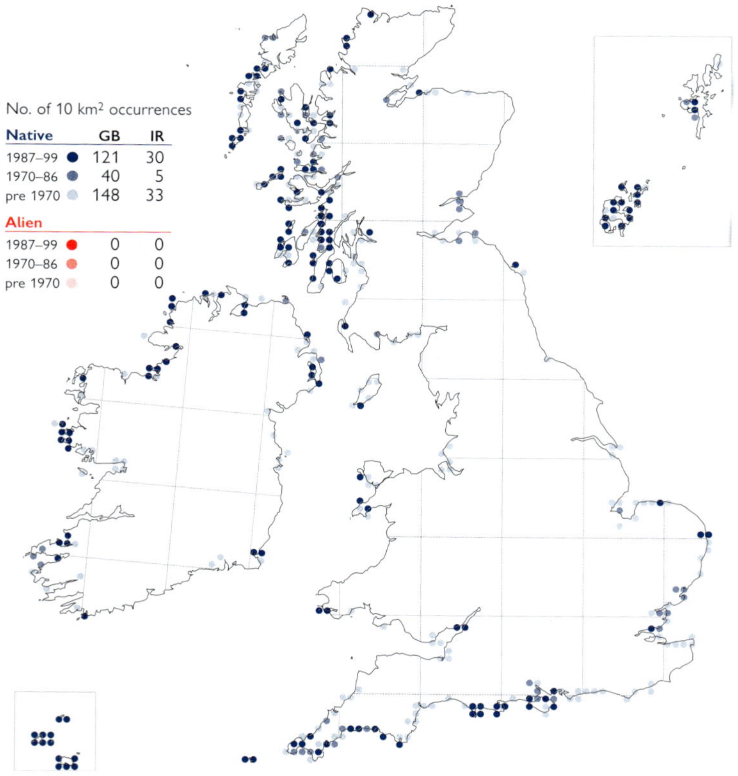

No. of 10 km² occurrences

Native	GB	IR
1987–99 ●	121	30
1970–86 ●	40	5
pre 1970 ●	148	33
Alien		
1987–99 ●	0	0
1970–86 ●	0	0
pre 1970 ●	0	0

Z. marina is a perennial which grows in the subtidal zone, on substrates of gravel, sand or sandy mud in areas which are protected from full exposure. It descends to depths of about 4 metres. Lowland.

Native (change –0.86). This species declined throughout its European range after a major outbreak of wasting disease in the 1930s. It has never fully recovered, and a further outbreak of disease was noted in the 1980s. However, there are obvious difficulties in recording this marine species and many old and recent records are based on stranded plants. The extent to which the map reflects the distribution of rooted plants is difficult to assess.

Circumpolar Wide-temperate element.

References: Atlas (304b), Hartog (1970), Hultén & Fries (1986), Meusel *et al.* (1965), Stewart *et al.* (1994), Tutin (1942).

C. D. PRESTON

Zostera angustifolia Narrow-leaved Eelgrass

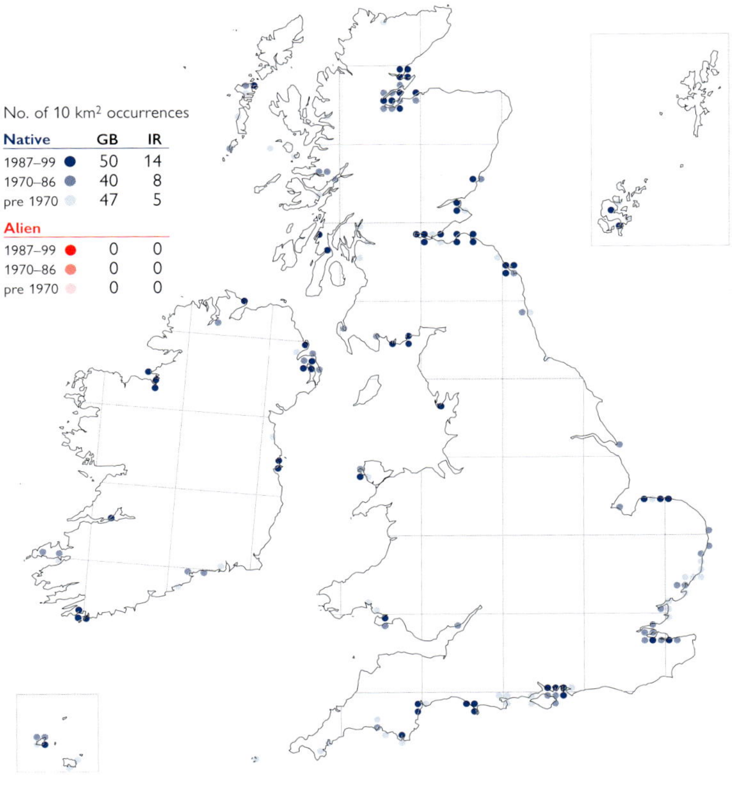

No. of 10 km² occurrences

Native		GB	IR
1987–99	●	50	14
1970–86	●	40	8
pre 1970	●	47	5
Alien			
1987–99	●	0	0
1970–86	●	0	0
pre 1970	●	0	0

Z. angustifolia is a perennial which grows on sheltered tidal mudflats, in estuaries and in coastal lagoons, usually in shallower, more turbid water than *Z. marina*. It is usually found on mud or muddy sands, between the half-tide and low-tide marks. Lowland.

Native (change –0.68). *Z. angustifolia* was not recognised as a distinct species by British botanists until described as *Z. hornemanniana* by Tutin (1936). It was under-recorded in the 1962 *Atlas*, but its distribution is now better known.

Z. angustifolia is also recorded from Denmark and Sweden; it is difficult to distinguish from narrow-leaved variants of *Z. marina* and is not regarded as specifically distinct by Hartog (1970).

References: Atlas (304c), Stewart *et al.* (1994), Tutin (1942).

C. D. PRESTON

Zostera noltei Dwarf Eelgrass

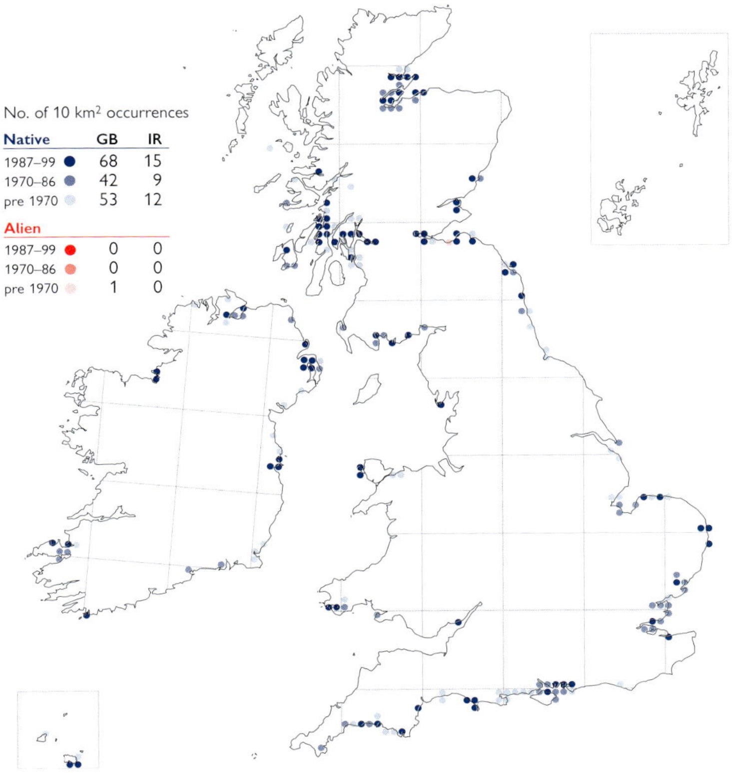

No. of 10 km² occurrences

Native		GB	IR
1987–99	●	68	15
1970–86	●	42	9
pre 1970	●	53	12
Alien			
1987–99	●	0	0
1970–86	●	0	0
pre 1970	●	1	0

Although a coastal species, this perennial is found at higher levels of the shore than other *Zostera* species. It grows in sheltered estuaries and harbours, where it is found on mixed substrates of sand and mud. Plants are often concentrated in pools or runnels on the shore. Lowland.

Native (change –0.51). The distribution of *Z. noltei* appears to be stable, and it even persists in relatively polluted waters such as the Thames estuary. The reason for the decrease in Hampshire is unclear. It may have been present only as strand-line plants in other squares where it was last recorded before 1970.

Eurasian Southern-temperate element.

References: Atlas (304d), Hartog (1970), Hultén & Fries (1986), Meusel *et al.* (1965), Stewart *et al.* (1994).

C. D. PRESTON

Acorus calamus Sweet-flag

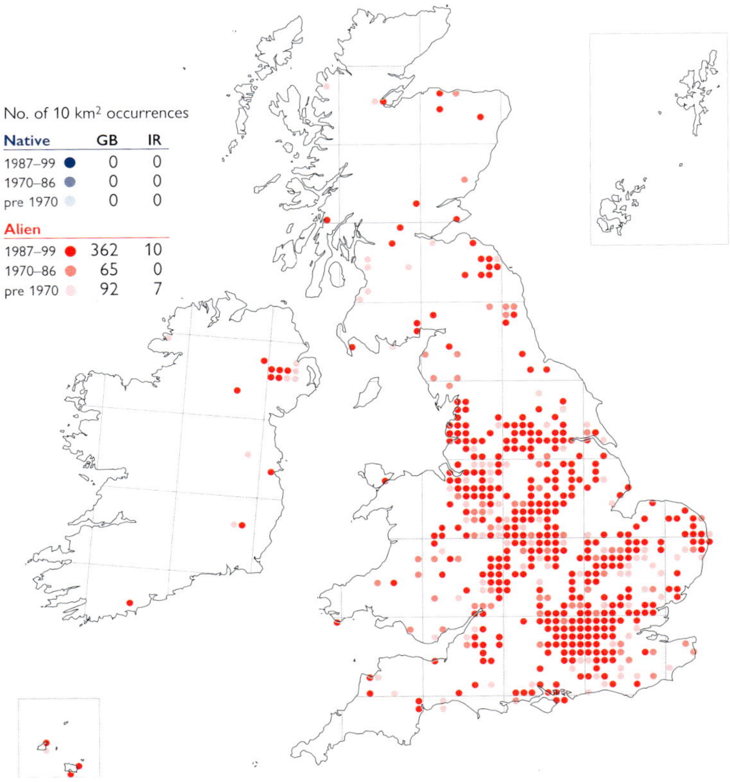

No. of 10 km² occurrences

Native		GB	IR
1987–99	●	0	0
1970–86	●	0	0
pre 1970	●	0	0
Alien			
1987–99	●	362	10
1970–86	●	65	0
pre 1970	●	92	7

A rhizomatous perennial herb growing at the margins of streams, canals, ponds and lakes in shallow, nutrient-rich calcareous water. The European plant is a sterile triploid. Lowland.

Neophyte (change +0.69). *A. calamus* was introduced into England in the 16th century and by 1668 was established in the wild. It appears to be more widespread than when mapped in the 1962 *Atlas*, probably due to better recording as the plant flowers shyly in our area and may well have been overlooked in the past.

Native plants are diploid (Siberia, N. America) or tetraploid (S. & E. Asia); the origin of the triploid is unknown but it is naturalised in Europe & W. Asia, the Himalayas and eastern N. America.

References: Atlas (343a), Hultén & Fries (1986), Preston & Croft (1997).

R. WILSON

Lysichiton americanus American Skunk-cabbage

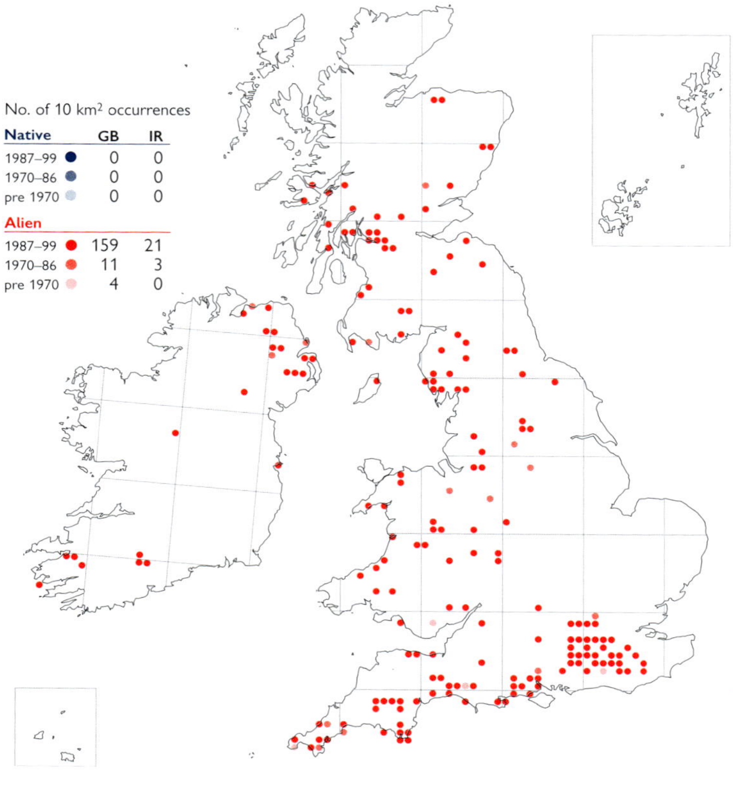

A perennial herb that is often planted beside ponds and streams in parks and gardens. It reproduces from seed and can quickly become established as an escape in carr and on swampy ground. Lowland.

Neophyte. *L. americanus* was introduced into cultivation in 1901 and was known in the wild by 1947 (Surrey). It is difficult to assess changes in distribution, but the species is likely to be increasing.

Native of western N. America.

Reference: FNAEC (2000).

R. WILSON

Arum maculatum Lords-and-Ladies

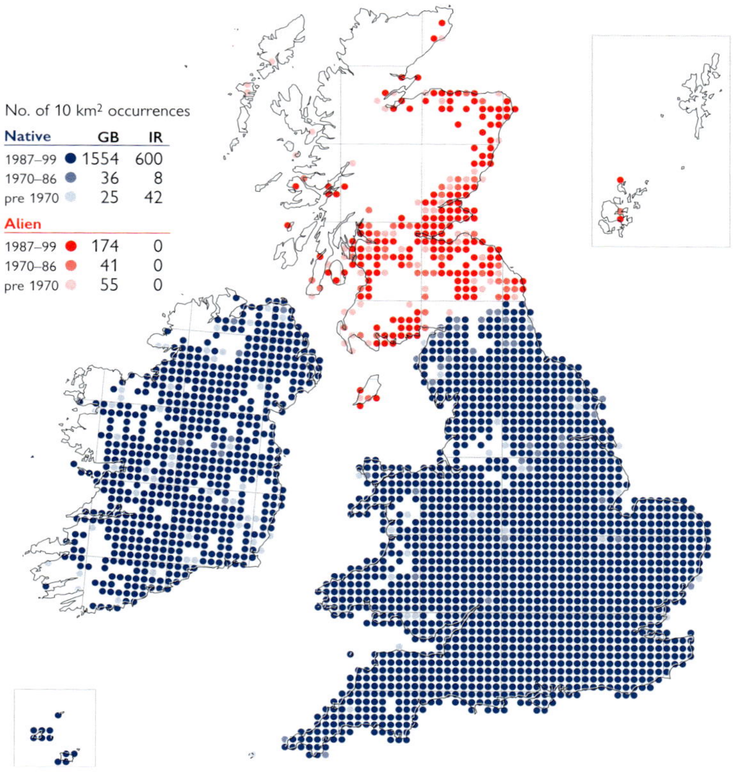

A rhizomatous perennial herb of woodlands, hedgerows and other shaded areas on moist, well-drained and reasonably fertile soils. Generally lowland, but reaching 425 m at Ystradfellte (Brecs.).

Native (change –0.28). The distribution of *A. maculatum* shows little change since the 1962 *Atlas*. There has undoubtedly been some loss caused by hedgerow removal but this has been compensated for by its ability to establish itself quickly in new areas of suitable habitat such as scrub and plantations. Local floras suggest that *A. maculatum* is alien north of S. Northumberland and Cumbria.

European Temperate element.

References: Atlas (343b), Boyce (1993), Grime *et al.* (1988), Hultén & Fries (1986), Meusel *et al.* (1965), Prime (1954, 1960), Sowter (1949).

R. WILSON

Arum italicum Italian Lords-and-Ladies

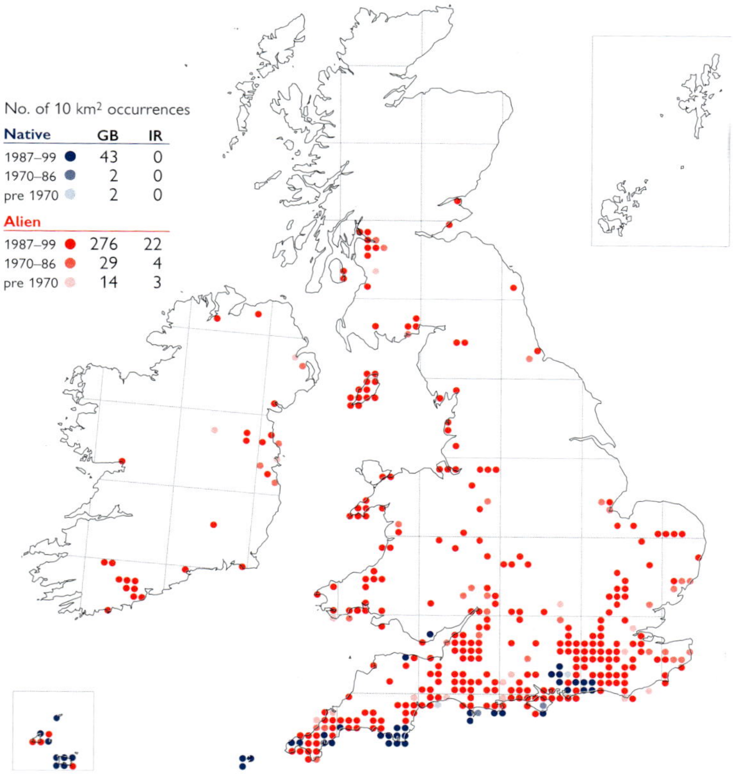

A rhizomatous perennial found in woodlands, hedge banks, scrub and field-borders. It prefers shaded, humid environments and a deep, well-drained soil reasonably rich in nutrients. Reproduction is by division of the rootstock; only subsp. *italicum* sets abundant seed. It also occurs as a garden escape. Lowland.

Native (change +2.09). The native distribution of *A. italicum* is probably stable, although alien occurrences may be increasing. The two subspecies found in our area, the native subsp. *neglectum* and the alien subsp. *italicum*, are mapped separately.

Mediterranean-Atlantic element.

References: Atlas (343c), Boyce (1993), Meusel *et al.* (1965).

R. WILSON

Arum italicum subsp. *neglectum*

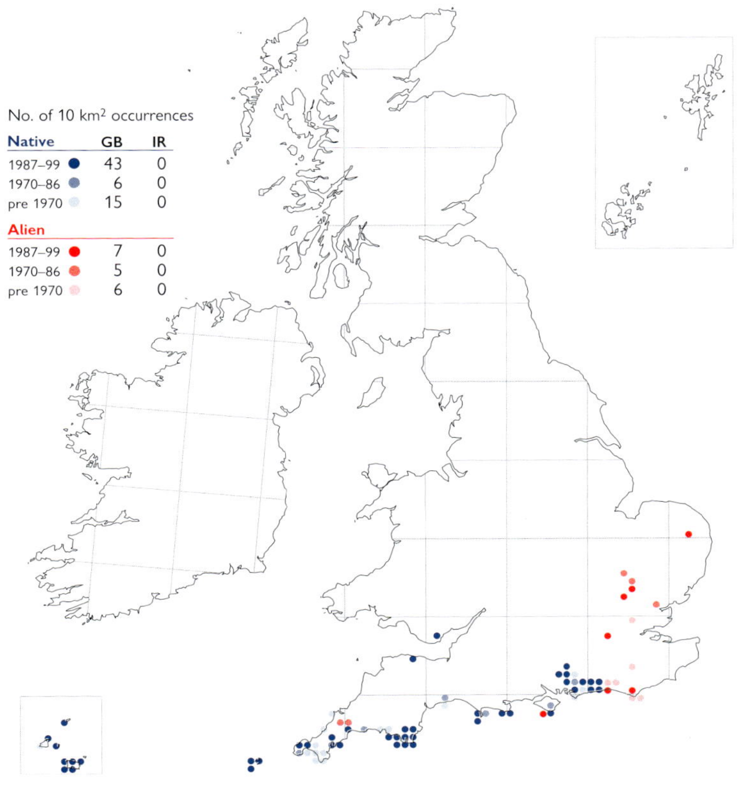

A rhizomatous perennial found in hedge banks, scrub and stony field-borders close to the coast and in hanger woodlands at inland sites. It thrives at sites where there is a shaded, humid environment (often growing on the north side of walls), and a deep, well-drained soil reasonably rich in nutrients. It reproduces by natural division of the rootstock, but little seed is set because the flowers are often eaten by animals. Lowland.

Native. The distribution of this subspecies as a native plant is probably stable. It may be increasing as a garden throw-out or escape, but it is much less frequently grown than subsp. *italicum*.

Oceanic Temperate element; found in W. Europe from N. Spain to Britain.

References: Boyce (1993), Prime (1954, 1960), Stewart *et al.* (1994).

R. WILSON

Arum italicum subsp. *italicum*

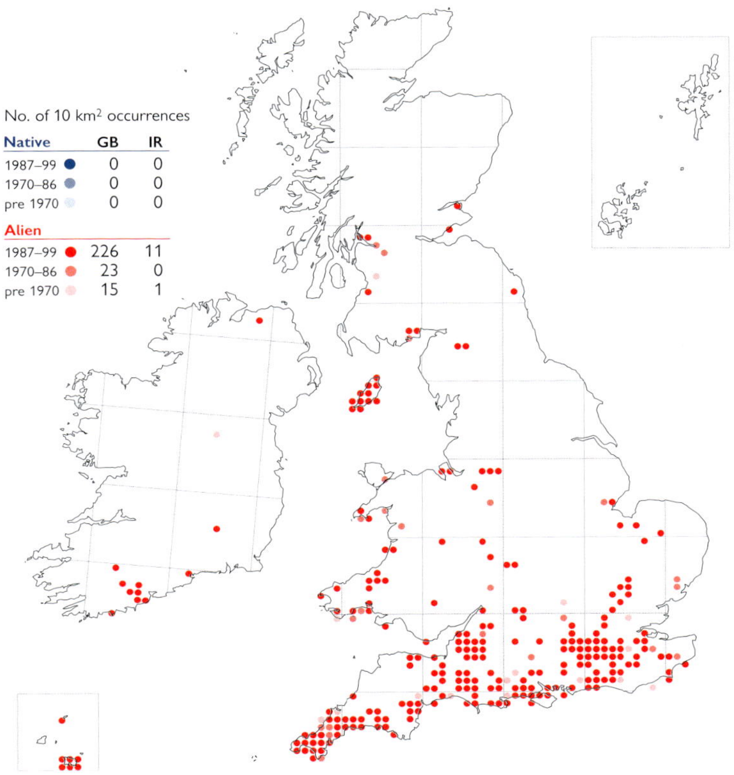

A rhizomatous perennial herb found as a garden escape in scrub and shaded hedge banks, often in places where garden waste is dumped. It sets abundant seed which is often bird-sown. Lowland.

Neophyte. This subspecies was cultivated in Britain by 1683 and was known from the wild by at least 1905. It is popular in gardens, and its distribution is probably increasing.

A. italicum subsp. *italicum* has a Mediterranean-Atlantic distribution.

References: Boyce (1993), Prime (1954, 1960).

R. WILSON

Spirodela polyrhiza Greater Duckweed

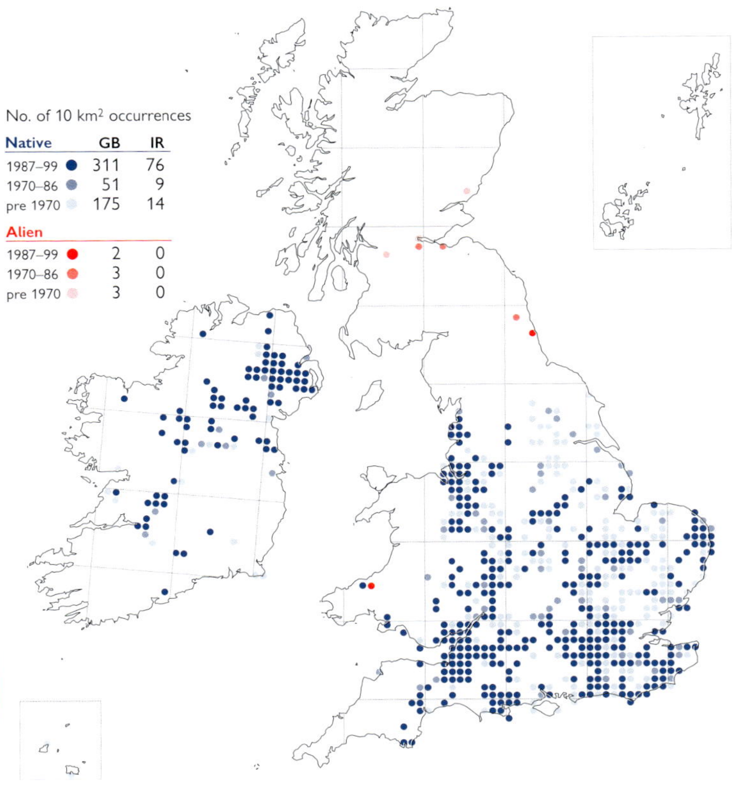

S. polyrhiza grows in base-rich water in ponds, ditches, canals and slowly flowing rivers. It is particularly frequent in grazing marshes. Flowers have only been seen once in our area (in 1906) and reproduction is by vegetative budding. Lowland.

Native (change –0.18). The loss of populations of *S. polyrhiza* in the northern part of its British range was already apparent before 1930. There have been local losses since then as ponds have disappeared or grazing marshes have been converted to arable land, but the species remains frequent within its core areas, and is much better recorded than in the 1962 *Atlas*.

Circumpolar Southern-temperate element.

References: Atlas (343d), Hultén & Fries (1986), Preston & Croft (1997).

C. D. PRESTON

Lemna gibba Fat Duckweed

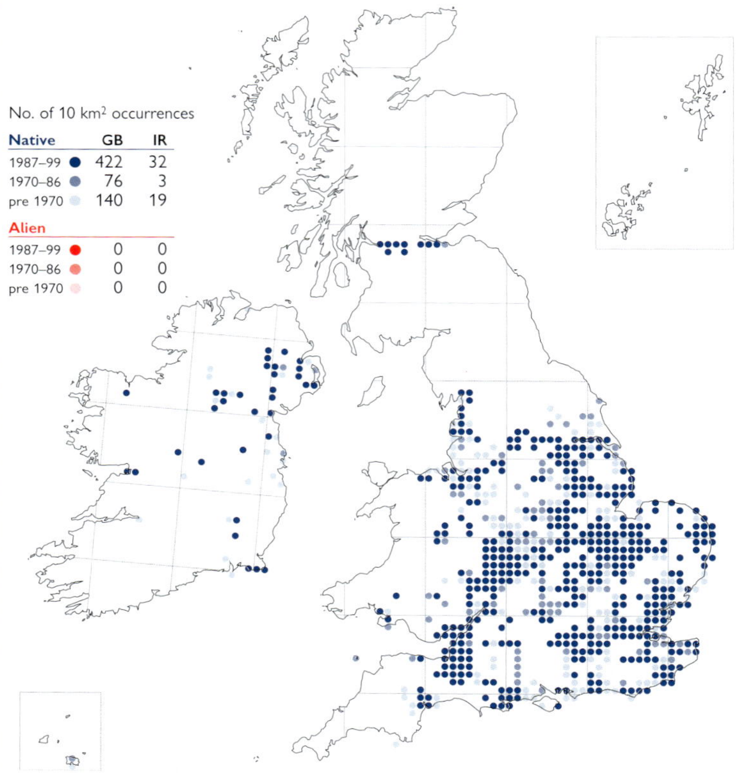

This buoyant duckweed is a plant of still or slowly flowing, eutrophic water in ponds, canals, ditches or the quiet backwaters of rivers; it can also grow in brackish water. In very eutrophic sites it may form dense masses which exclude other aquatics. It reproduces by vegetative budding, though it flowers slightly more freely than our other Lemnaceae. Lowland.

Native (change +0.07). The distribution of this species is similar to that shown in the 1962 *Atlas*. It was then described as 'casual only in Scotland' but it is now frequent in the Forth & Clyde and Union Canals, where it may have arrived by natural dispersal.

European Southern-temperate element; also in C. Asia and N. America.

References: Atlas (344c), Hultén & Fries (1986), Meusel *et al.* (1965), Preston & Croft (1997).

C. D. PRESTON

Lemna minor Common Duckweed

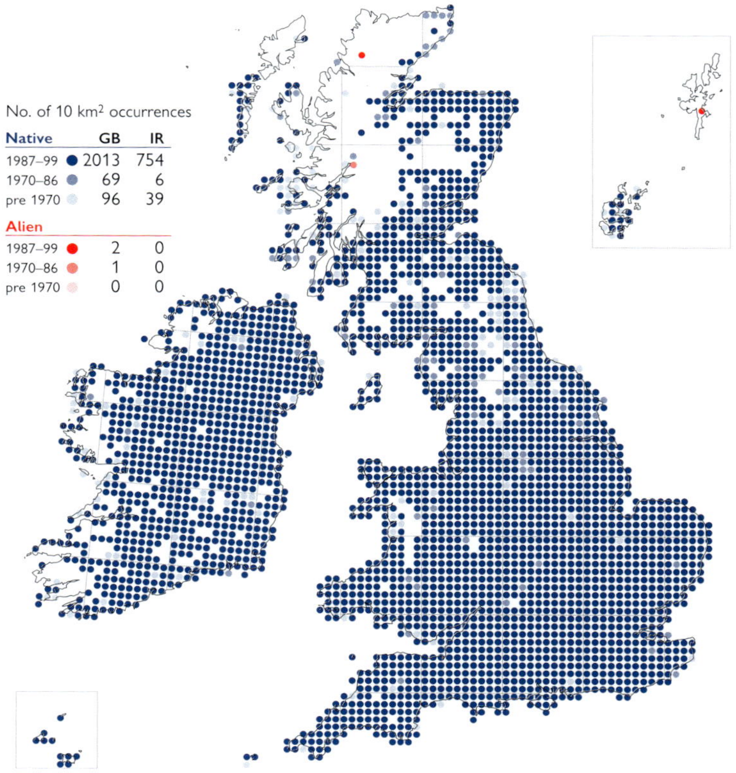

This is our most widespread and frequent floating aquatic plant, often abundant on a wide variety of still or slowly flowing, mesotrophic or eutrophic waters. It also occurs terrestrially on exposed mud, or damp stonework and rocks. Plants rarely flower and reproduction is by vegetative budding. 0–500 m (Brown Clee Hill, Salop).

Native (change +0.60). There is no evidence that the distribution of this species has changed since the 1962 *Atlas*, as the overall range is similar and the extra squares in which the species is known have almost certainly resulted from more detailed recording.

Circumpolar Southern-temperate element; widely naturalised outside its native range.

References: Atlas (344b), Grime *et al.* (1988), Hultén & Fries (1986), Meusel *et al.* (1965), Preston & Croft (1997).

C. D. PRESTON

Lemna trisulca Ivy-leaved Duckweed

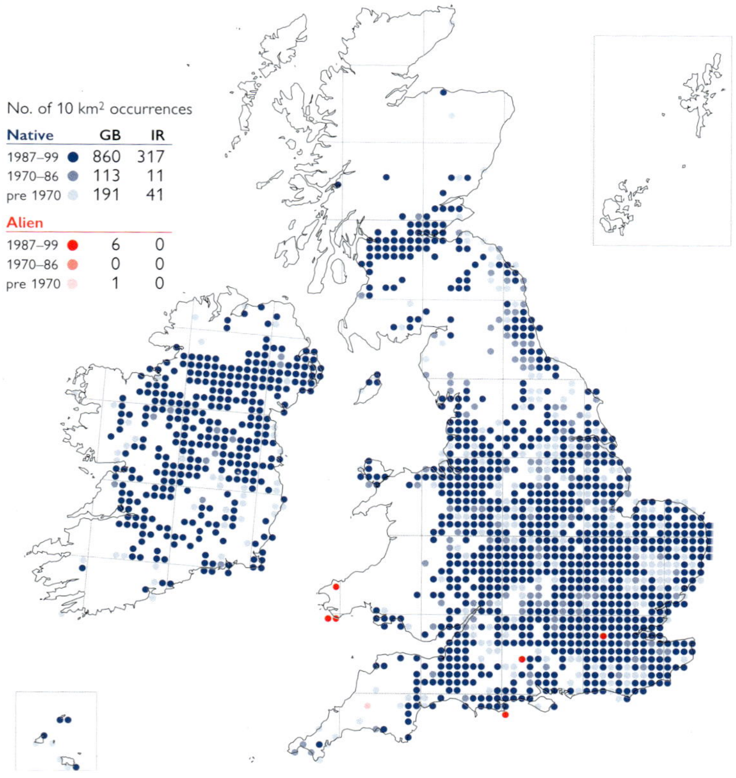

The only submerged species of *Lemna* in our area, *L. trisulca* is frequent in mesotrophic to eutrophic, still to slowly flowing waters where low nutrient levels or exposure prevent the development of a dense blanket of floating *Lemna* species. Reproduction is by vegetative budding; flowering is very rare. Generally lowland, but reaching 340 m at Kingside Loch (Selkirks.).

Native (change –0.21). There have been some local losses, but the distribution of *L. trisulca* is otherwise stable.

Circumpolar Temperate element.

References: Atlas (344a), Hultén & Fries (1986), Preston & Croft (1997).

C. D. PRESTON

Lemna minuta Least Duckweed

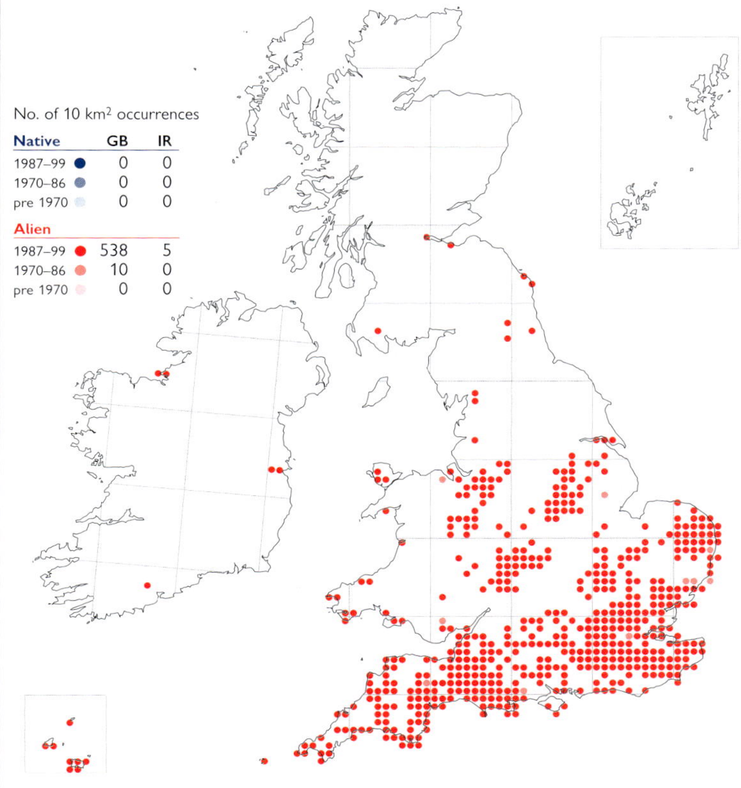

L. minuta occurs, often in abundance, as a floating aquatic on the surface of lakes, ponds, slowly flowing rivers, streams, canals and ditches. It is sufficiently shade-tolerant to occur on ponds shaded by marginal trees or woodland. Reproduction is by vegetative budding. Lowland.

Neophyte. This species was first recorded in Britain in 1977, when it was discovered in Cambridge, and it has spread rapidly especially since the late 1980s. It is probably under-recorded, as it is easily overlooked as *L. minor*. It was first found in Ireland at Blarney Castle (E. Cork) in 1993 and Ballyconnell (Co. Sligo) in 1995.

Native to temperate and subtropical N. & S. America; now widely naturalised in Europe where it was first discovered in France in 1965.

References: Cotton (1999), FNAEC (2000), Preston & Croft (1997).

C. D. PRESTON

Wolffia arrhiza Rootless Duckweed

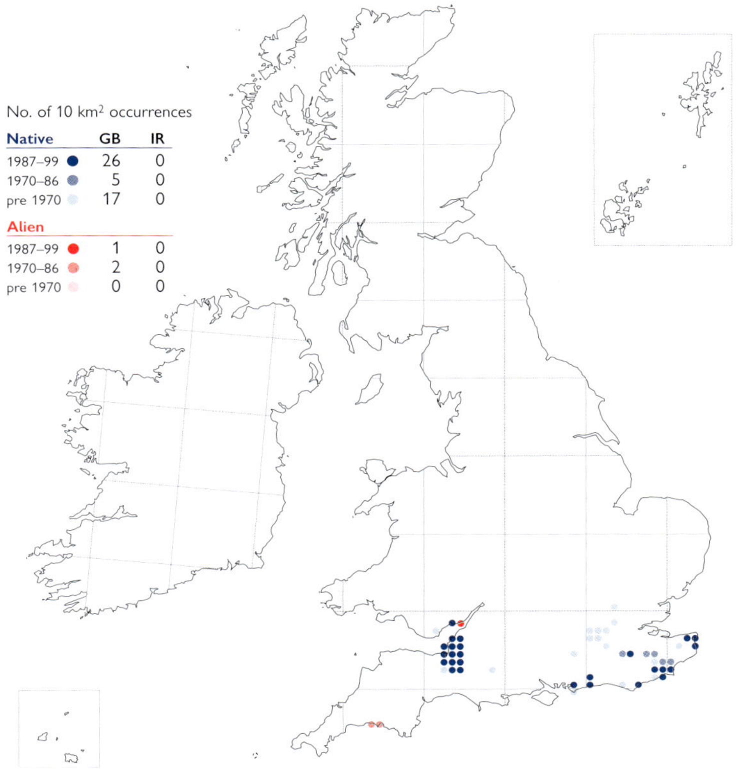

This tiny rootless plant floats on ponds and ditches as small, pure patches or scattered amongst other floating Lemnaceae. Reproduction is by vegetative budding; flowers have never been seen in our area. Lowland.

Native (change –0.03). *W. arrhiza* remains frequent in the Somerset Levels, but it has clearly declined since 1930 in the eastern part of its British range. It is now extinct in the London area.

Eurosiberian Southern-temperate element.

References: Atlas (344d), Preston & Croft (1997), Stewart *et al.* (1994).

C. D. PRESTON

Eriocaulon aquaticum Pipewort

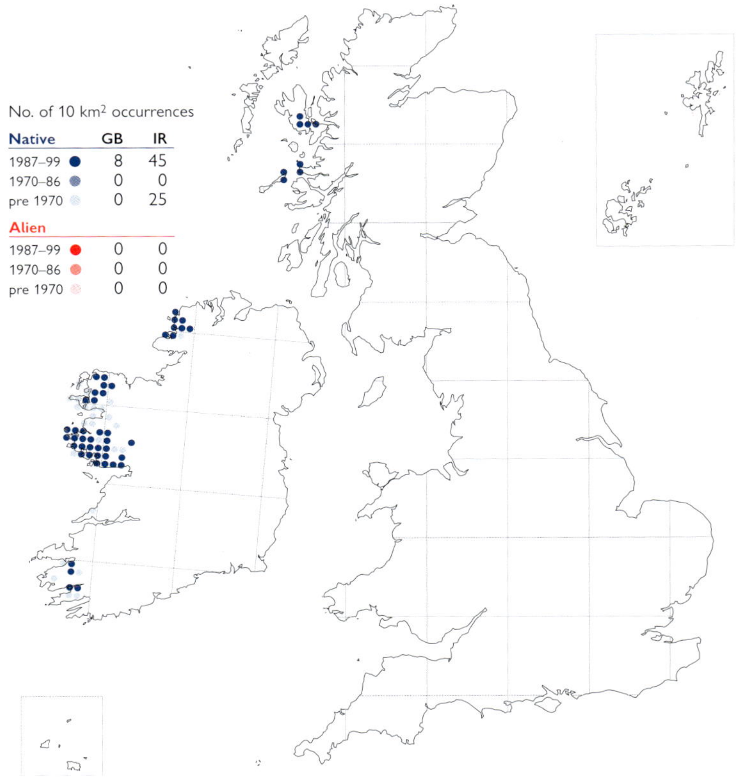

The perennial rosettes of *E. aquaticum* grow on peat or on inorganic substrates at the edge of oligotrophic lakes and pools. It ranges from levels which are often exposed above the water to those which are permanently submerged. Little is known about its reproductive ecology. Generally lowland, but reaching 300 m in Glen Lough (W. Cork).

Native (change +0.18). This species still has the restricted distribution mapped in the 1962 *Atlas*, the only significant change being its discovery in mainland Scotland in 1967. Its habitat is so widespread in W. Scotland that it is difficult to explain its limited range there.

Oceanic Boreal-montane element; in Europe restricted to Britain and Ireland but widespread in N. America.

References: Atlas (311c), FNAEC (2000), Preston & Croft (1997), Wigginton (1999).

C. D. PRESTON

Juncus squarrosus Heath Rush

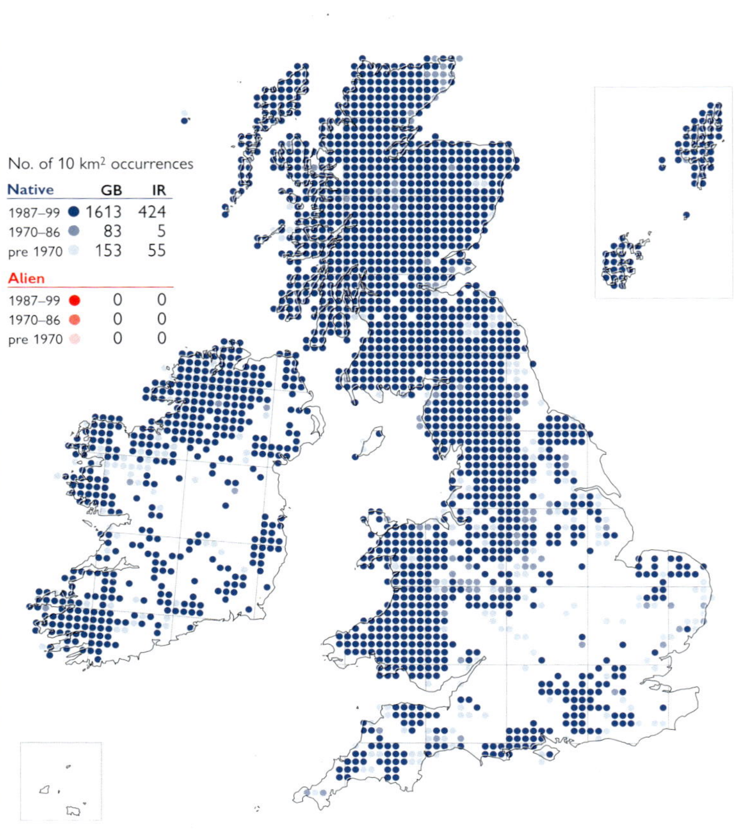

A wiry, tufted perennial herb which is characteristic of wet peaty heath and moorland, raised and valley mires and upland flushes on acidic substrates. 0–1040 m (Carrantuohill, S. Kerry).

Native. *J. squarrosus* has been lost from many sites in S. and E. England since the 1962 *Atlas*, due to drainage and agricultural improvement. The distribution is stable elsewhere.

Suboceanic Temperate element; also in Greenland.

References: Atlas (317c), Dupont (1962), Grime *et al.* (1988), Hultén & Fries (1986), Meusel *et al.* (1965), Welch (1966).

C. A. STACE

Juncus tenuis Slender Rush

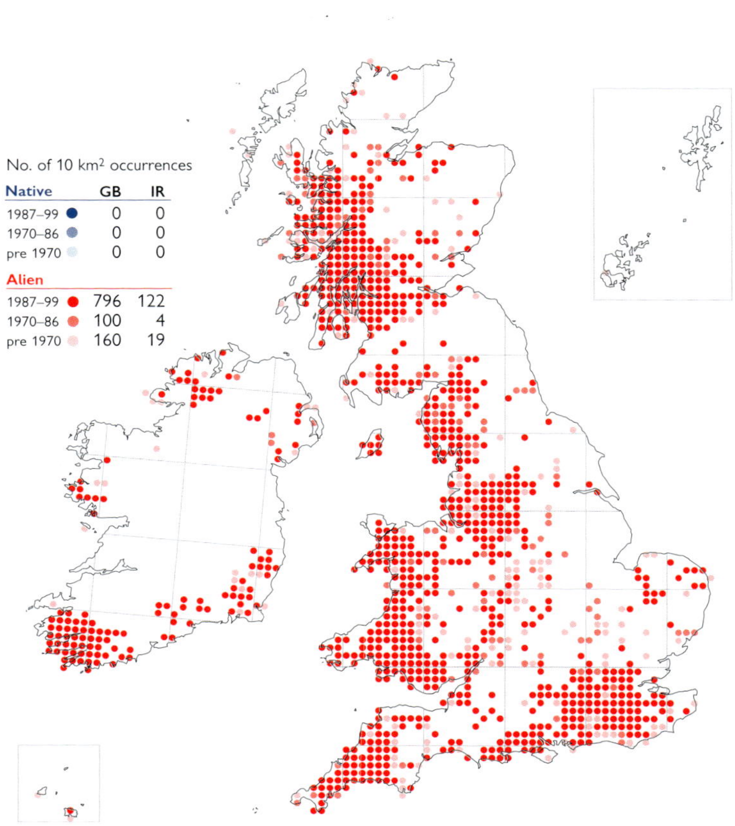

A slender, tufted perennial herb found in damp open ground by roads and lakes, on paths and in woodland and forest rides. 0–430 m (above Dent station, N.W. Yorks., and in the Caulderbeck Fells, Cumberland).

Neophyte (change +0.83). This species was first recorded from Angus in the 1790s, but did not begin to spread widely until the late 19th century. Claims for native status in Ireland are unproven. It has consolidated rather than extended its range since the 1962 *Atlas*.

Native of N. & S. America.

References: Atlas (317d), FNAEC (2000), Hultén & Fries (1986), Meusel *et al.* (1965), Richards (1943a), Salisbury (1974).

C. A. STACE

Juncus compressus Round-fruited Rush

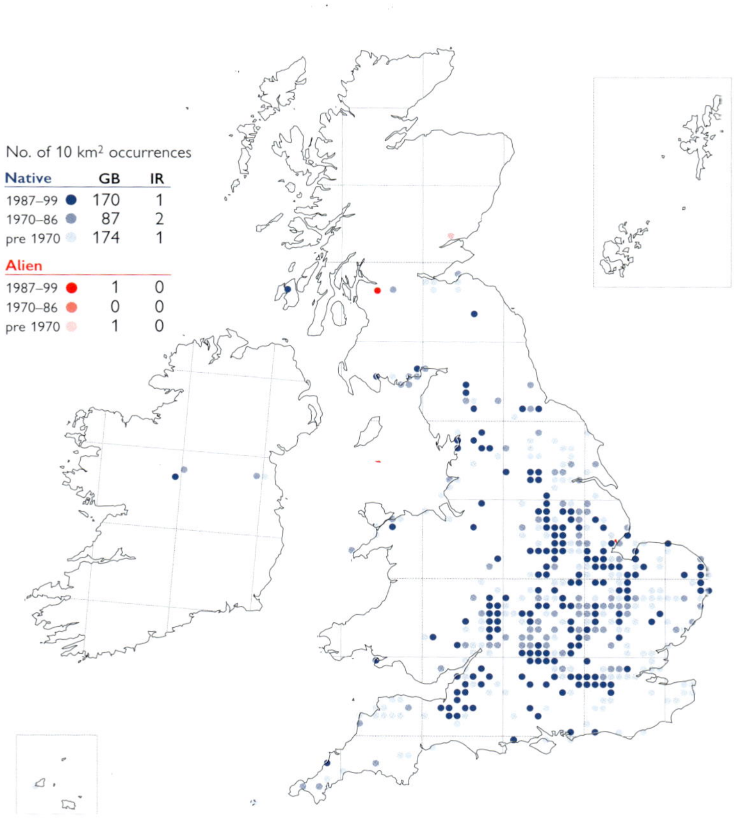

A compact to spreading, rhizomatous perennial herb found in marshes, wet meadows and pastures, often near the sea and sometimes in brackish conditions, occasionally with *J. gerardii*. 0–370 m (above Dent station, N.W. Yorks.).

Native (change –1.09). *J. compressus* has always been confused with *J. gerardii*, especially on the coast. It could well be under- or over-recorded. Many recent county floras record a decline because of drainage and the loss of permanent pastures. Conversely, it has appeared in new sites on the edges of reservoirs. It is mapped as native in Ireland, but its status there is uncertain (Scannell & Synnott, 1987).

Eurosiberian Temperate element; widely naturalised outside its native range.

References: Atlas (318a), Curtis & McGough (1988), Hultén & Fries (1986), Meusel *et al.* (1965).

C. A. STACE

Juncus gerardii Saltmarsh Rush

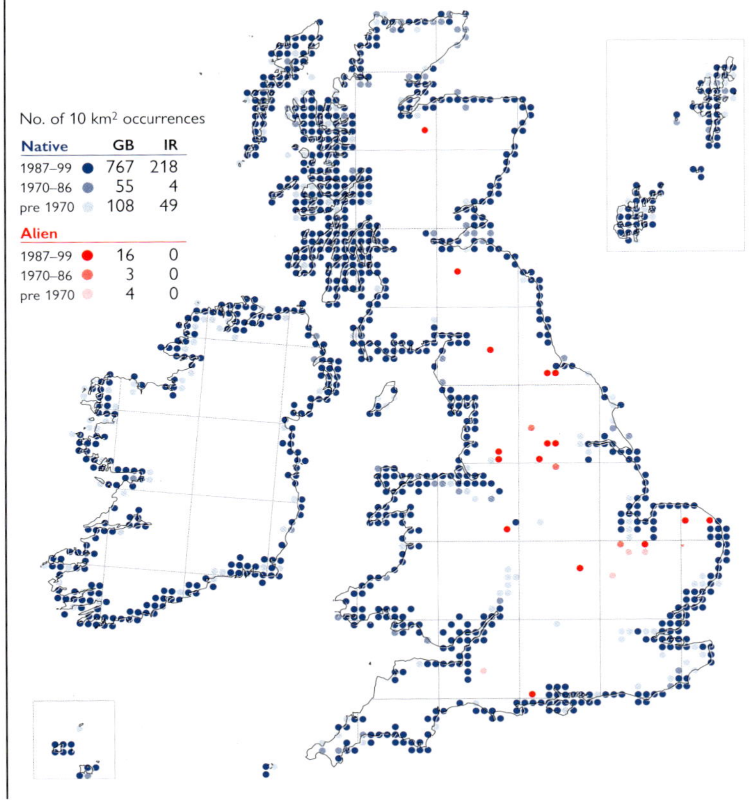

This species is apparently confined to saline habitats, mostly in the uppermost parts of coastal saltmarshes, but also around coastal rock pools, in spray-drenched cliff-top turf and at saline sites inland. Lowland, although unintentionally but successfully introduced with other halophytes on a bank below a roadside car park at 575 m at the Hartside Cafe, Cumberland (Corner, 1997).

Native (change –0.13). There is little change in the overall distribution of *J. gerardii* since the 1962 *Atlas*.

Circumpolar Wide-temperate element.

References: Atlas (318b), Hultén & Fries (1986), Meusel *et al.* (1965).

C. A. STACE

Juncus trifidus Three-leaved Rush

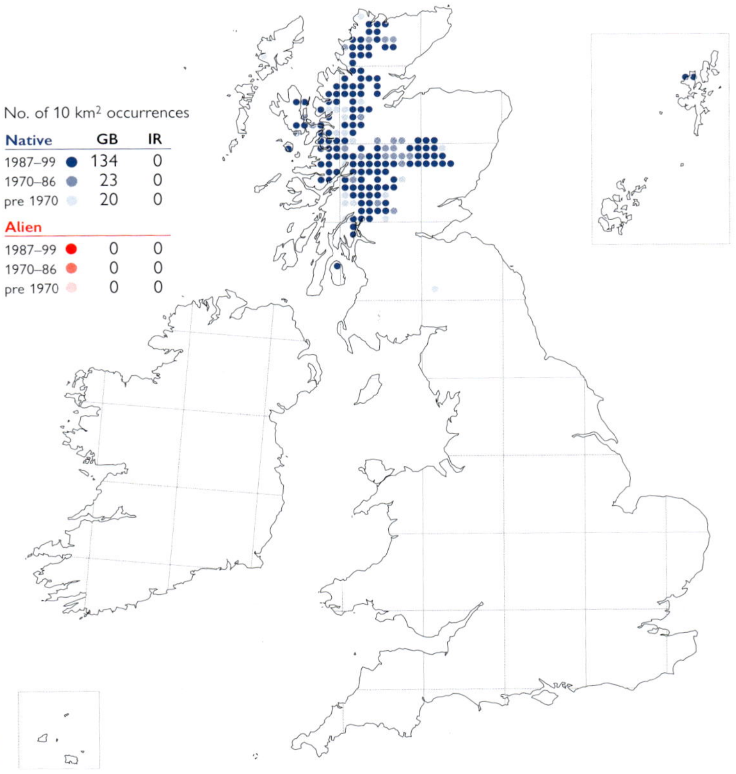

A small, densely tufted perennial herb found in bare or bryophyte- or lichen-rich places on mountains on shallow soil or in rock crevices, on both acidic and calcareous substrata. *J. trifidus* is one of the principal angiosperms of wind-swept, often almost snow-free plateau edges over *c.* 1000 m, but it also occupies sites that are snow-covered for several months. From 240 m on Ronas Hill (Shetland) to 1310 m in the Cairngorms.

Native (change –0.38). The distribution of *J. trifidus* is stable.

Eurosiberian Arctic-montane element; also in N. America.

References: Atlas (318c), Hultén & Fries (1986), Ingram (1958), McVean & Ratcliffe (1962), Meusel *et al.* (1965).

C. A. STACE

Juncus bufonius sens. lat. Toad Rushes

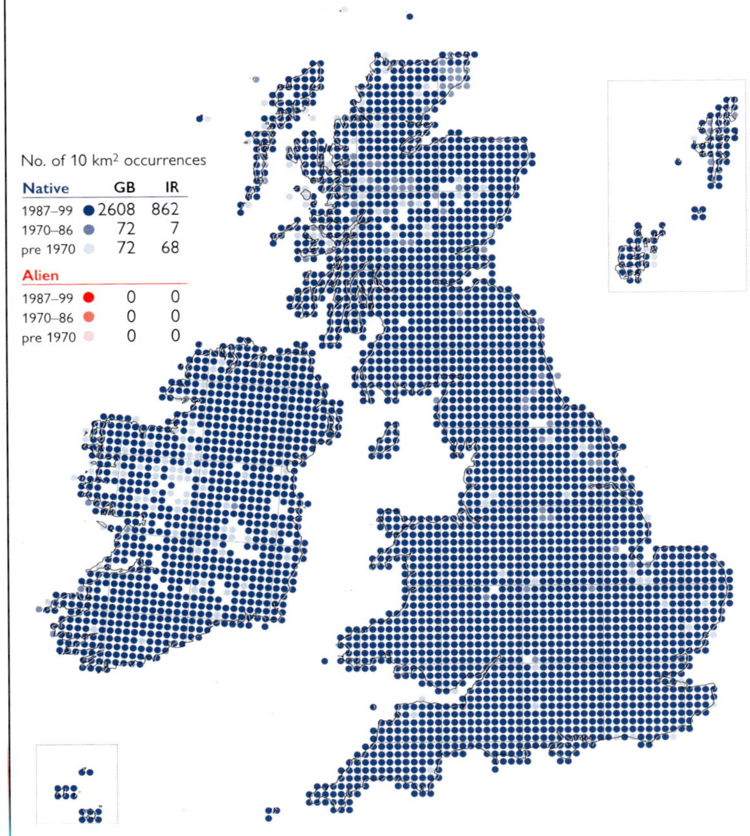

These annuals are found in a wide variety of habitats which are moist or flooded in winter, and where there is some disturbance and little or no competition. This includes the edges of ponds, lakes and marshes, dune-slacks, estuaries, saltmarshes, sandy seashores, and tracks and gateways. 0–595 m (Mangerton, S. Kerry).

Native (change +1.13). The range of *J. bufonius sens. lat.* is stable. Although the taxonomy of the British and Irish species included in the *J. bufonius* complex (*J. ambiguus, J. bufonius sens. str.* and *J. foliosus*) was elucidated in the 1970s and 1980s, they remain under-recorded as some botanists only record the aggregate.

The distribution of the segregate species is given in the following accounts.

References: Atlas (318d), Cope & Stace (1978, 1983, 1985), Grime *et al.* (1988), Hultén & Fries (1986).

C. A. STACE

Juncus foliosus Leafy Rush

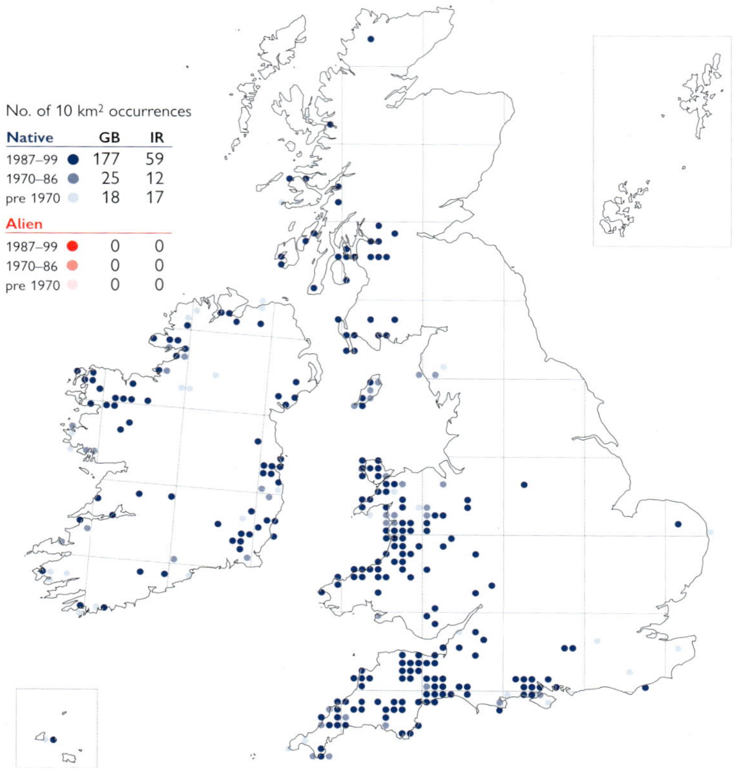

No. of 10 km² occurrences

Native		GB	IR
1987–99	●	177	59
1970–86	●	25	12
pre 1970	●	18	17
Alien			
1987–99	●	0	0
1970–86	●	0	0
pre 1970	●	0	0

This spring-germinating annual is the most robust member of the *J. bufonius* aggregate, occurring in wet fields, marshes and ditches and on the muddy margins of lakes and ponds, sometimes with *J. bufonius sens. str.* Although often found near the coast, this species seems to shun brackish water. 0–365 m (Corndon Hill, Monts.).

Native. *J. foliosus* was recognised in Britain as a subspecies in 1959 but did not become widely known until Cope & Stace (1978) treated it as a species. It is almost certainly under-recorded. However, the current map indicates a predominantly south-western distribution which fits in with its wider European range and is almost certainly a true reflection of reality.

Suboceanic Southern-temperate element.

References: Cope & Stace (1983, 1985), Grime *et al.* (1988).

C. A. STACE

Juncus bufonius sens. str. Toad Rush

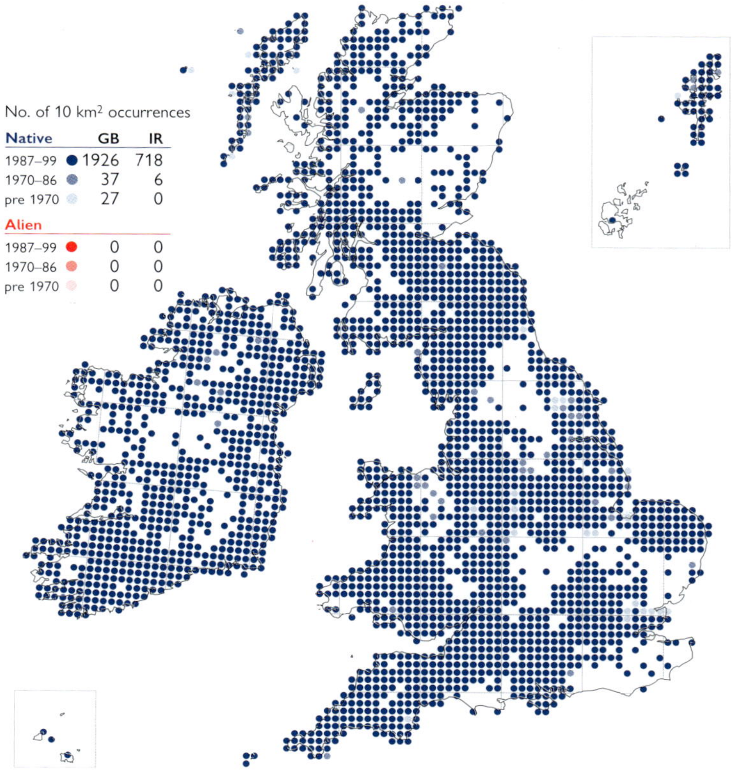

No. of 10 km² occurrences

Native		GB	IR
1987–99	●	1926	718
1970–86	●	37	6
pre 1970	●	27	0
Alien			
1987–99	●	0	0
1970–86	●	0	0
pre 1970	●	0	0

An annual of habitats where the water-table is at least seasonally high and there is little competition, including the margins of ponds, lakes, streams and rivers, marshes and dune-slacks, and rarely acid bogs. It also grows around brackish lakes and on estuarine mud- and sand-flats, and is often a weed of disturbed ground, including tracks and road-sides. 0–595 m (Mangerton, S. Kerry).

Native. *J. bufonius sens. str.* was not separated from other members of the *J. bufonius* aggregate until 1978, and it is under-recorded in some areas. It is, however, the most frequent species of the aggregate, and probably occurs throughout the range of *J. bufonius sens. lat.*

Circumpolar Wide-temperate element; widely naturalised outside its native range.

References: Cope & Stace (1978, 1983, 1985), Grime *et al.* (1988).

C. A. STACE

Juncus ambiguus Frog Rush

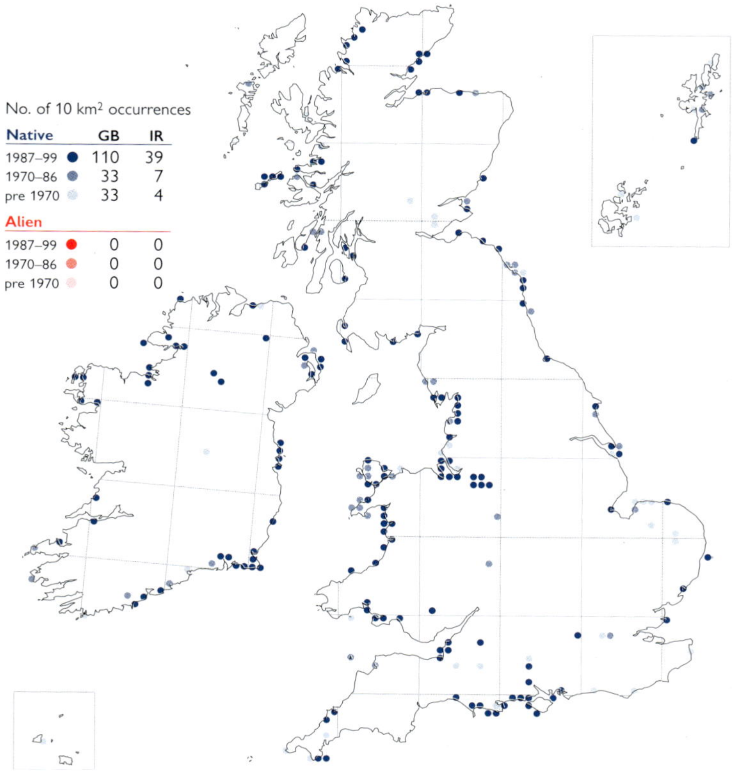

No. of 10 km² occurrences

Native		GB	IR
1987–99	●	110	39
1970–86	●	33	7
pre 1970	●	33	4
Alien			
1987–99	●	0	0
1970–86	●	0	0
pre 1970	●	0	0

This dwarf, annual, spring-germinating member of the *J. bufonius* aggregate occurs in bare damp brackish places near the coast and sometimes inland, often with *J. bufonius*. It is typical of coastal mud- and sand-flats above high-water mark and of the margins of saline and brackish lakes, and is also found on bare mud and waste ground associated with inland salt-flashes and salt-workings, and on highly basic lime-waste tips. Lowland.

Native. *J. ambiguus* was not separated from *J. bufonius* in Britain until 1978 and is often confused with variants of *J. bufonius*. It is almost certainly still under-recorded, especially in Scotland and W. Ireland.

European Southern-temperate element; also in N. America.

References: Cope & Stace (1978, 1983, 1985), Grime *et al.* (1988).

C. A. STACE

Juncus capitatus Dwarf Rush

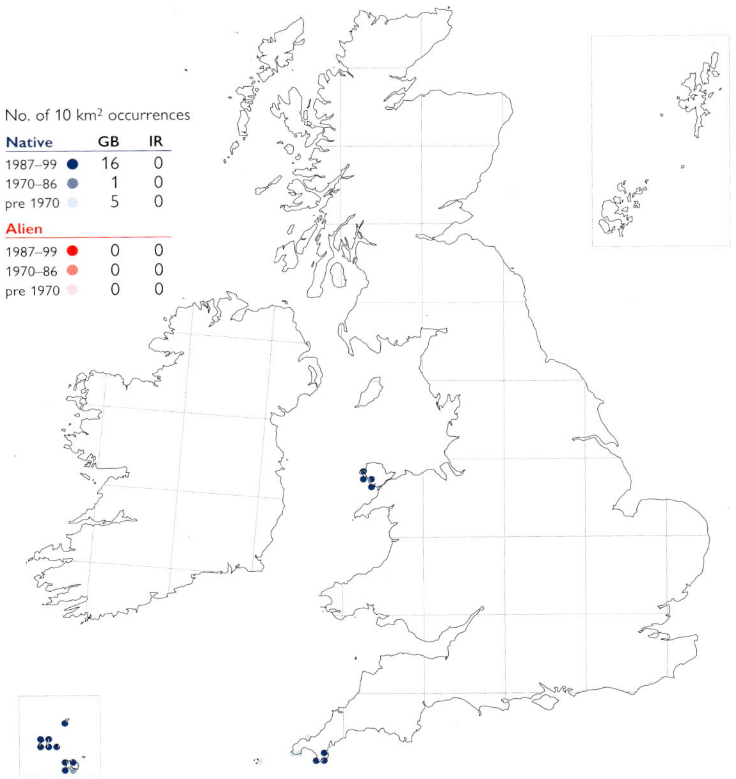

A diminutive, autumn-germinating annual of barish ground often kept open by standing water in winter and always droughted in summer. It grows around serpentine rock outcrops, on ledges of granite sea-cliffs, in dune-slacks and sometimes in quarries. Lowland.

Native. Populations vary greatly in size from year to year, and reports of extinction have sometimes proved to be premature, notably on Anglesey where it was re-discovered in 1995. The main threat to the Cornish sites is the cessation of grazing.

European Southern-temperate element; widely naturalised outside its native range.

References: Atlas (320d), Hultén & Fries (1986), Meusel *et al.* (1965), Wigginton (1999).

C. A. STACE

Juncus subnodulosus Blunt-flowered Rush

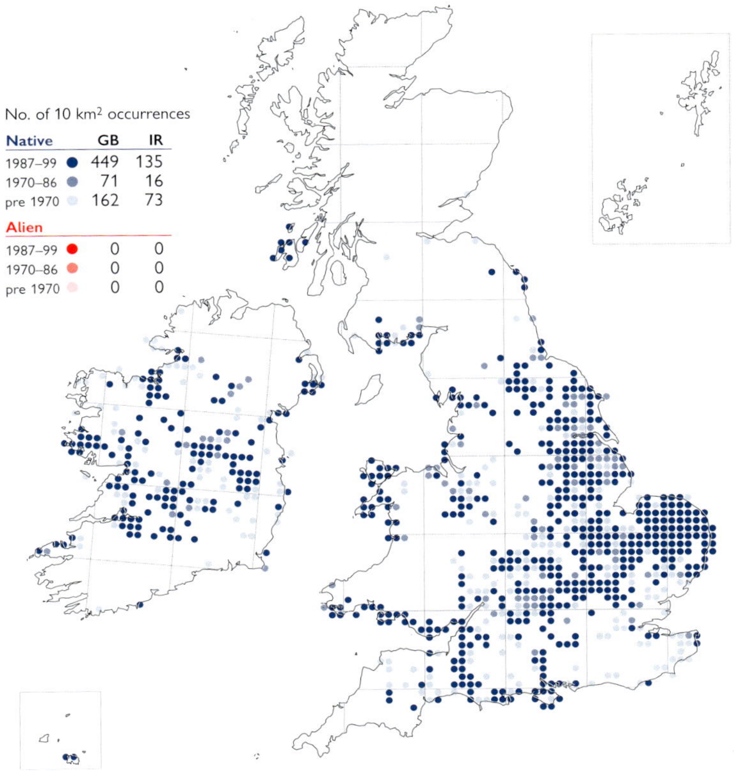

A strong, rhizomatous perennial herb growing in dense stands in fens, marshes, wet meadows, ditches and by water, usually in more base-rich conditions than any of the other jointed-rushes (sect. *Septati*); it also sometimes occurs in brackish water. Lowland.

Native (change +0.15). *J. subnodulosus* has been lost from many sites in S. England both before 1930 and subsequently, because of drainage of its habitat. However, it is now known to be more frequent in Wales than was suspected. This species is very distinct from other jointed-rushes and the map should indicate its true distribution.

European Southern-temperate element.

References: Atlas (321a), Hultén & Fries (1986), Meusel *et al.* (1965), Richards & Clapham (1941e).

C. A. STACE

Juncus alpinoarticulatus Alpine Rush

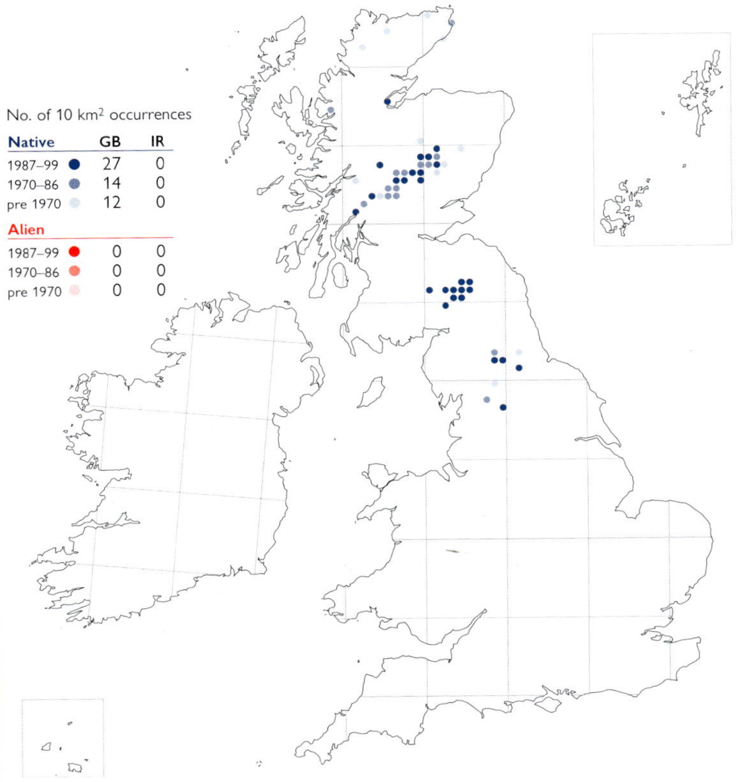

This montane, rhizomatous herb occurs in rather open wet turf in marshes and flushes and by lakes and streams, usually on base-rich soil and often over limestone. It is often found with a range of other less common montane calcicoles in bryophyte-rich habitats. From 150 m near Pitlochry (E. Perth) to 880 m on the Ben Alder range (Westerness).

Native (change –0.12). *J. alpinoarticulatus* has been much confused with variants of *J. articulatus*, and sometimes grows and hybridises with them. New finds in recent years are due to better identification skills rather than to natural range extension.

Circumpolar Boreal-montane element.

References: Atlas (321d), Hultén & Fries (1986), Stewart *et al.* (1994).

C. A. STACE

Juncus articulatus Jointed Rush

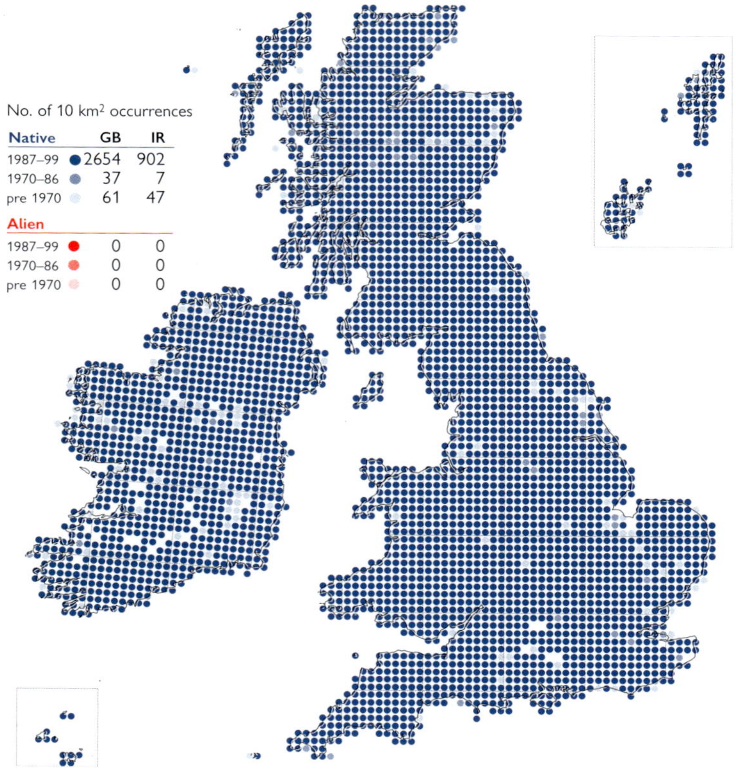

This very variable, erect to decumbent, clumped to extensively rhizomatous herb is found in a wide range of wet or damp habitats, both freshwater and brackish. It is characteristic of damp fields, marshes, ditches, flushes, rutted woodland rides, margins of ponds, lakes and streams and dune-slacks, avoiding only the most acid soils. 0–810 m (Great Dun Fell, Westmorland).

Native (change +1.26). There is no change in the overall range of *J. articulatus* since the 1962 *Atlas*.

Eurosiberian Southern-temperate element, but widely naturalised so that distribution is now Circumpolar Southern-temperate.

References: Atlas (321c), Grime *et al.* (1988), Hultén & Fries (1986), Meusel *et al.* (1965).

C. A. STACE

Juncus acutiflorus Sharp-flowered Rush

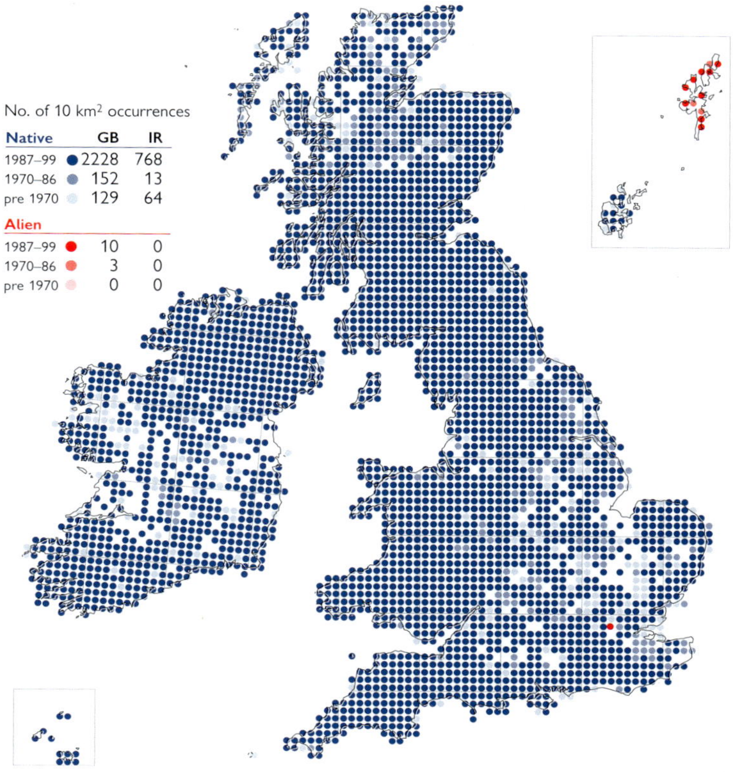

This species is usually tall and erect, and more extensively rhizomatous than *J. articulatus*. It occurs in wet or damp habitats on acidic soils, frequently at a lower pH than is tolerated by *J. articulatus*, particularly in damp meadows and pastures, marshes, bogs, wet heathland, and by ditches and ponds. 0–685 m (Mangerton, S. Kerry).

Native (change +1.16). *J. acutiflorus* was under-recorded in the 1962 *Atlas*. It often closely resembles the more robust growth-forms of *J. articulatus* and this might have produced some inaccuracies in the records of both species. However, they are both so frequent that this would be unlikely to affect the distributions shown on the maps.

European Temperate element; also in N. America.

References: Atlas (321b), Hultén & Fries (1986), McVean & Ratcliffe (1962), Meusel *et al.* (1965).

C. A. STACE

Juncus acutiflorus × *J. articulatus* (*J.* × *surrejanus*)

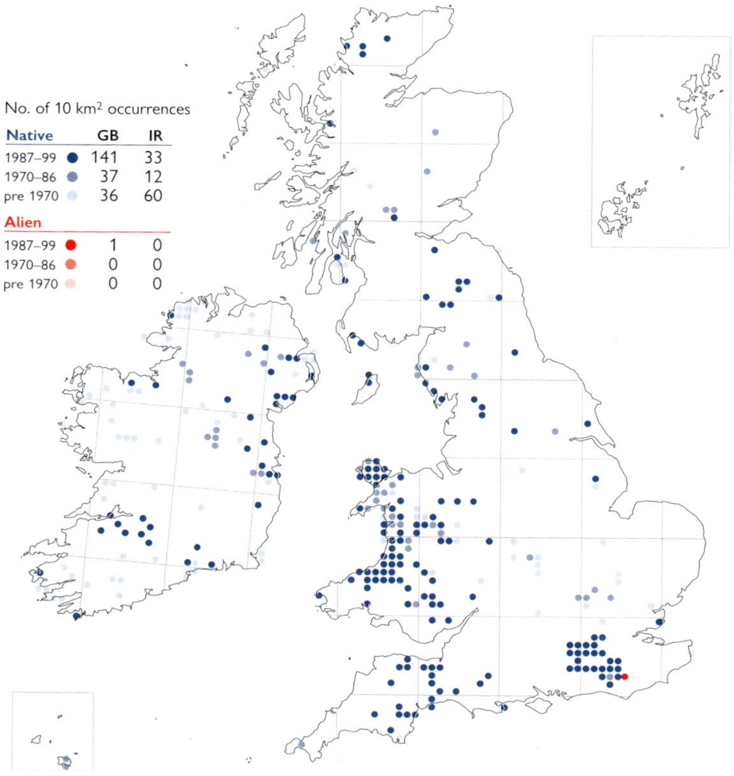

The two parents frequently cohabit in areas of medium acidity, and hybrids can occur in varying quantity in a wide range of damp and wet habitats. The hybrid is sometimes commoner than either parent (as in parts of C. Wales). It is usually highly sterile, but a small proportion of flowers can produce viable seed. Lowland to 425 m at Pont Crugnant (Monts.).

Native. This hybrid is frequently mis-identified as one or other parent, and is almost certainly under-recorded. Areas of concentrated records in S.E. England and Wales reflect the activities of recorders familiar with the hybrid.

This hybrid is also recorded from W. & C. Europe.

References: Blackstock & Roberts (1986), Stace (1975).

C. A. STACE

Juncus bulbosus Bulbous Rush

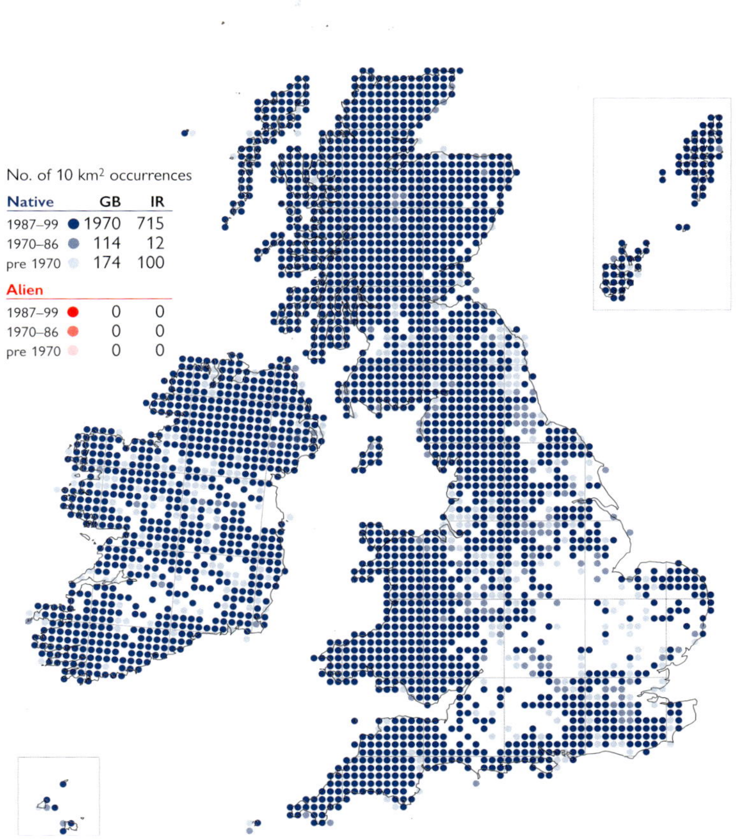

A very variable herb, ranging from tufted, terrestrial plants to submerged, floating aquatics, often rooting at the nodes and with proliferating flowers. It occurs in or by water and in open, often seasonally wet habitats, in acidic to neutral soils. Unusually, it grows in some calcareous turloughs in the Burren. 0–960 m (Caenlochan, Angus).

Native (change +0.34). The distribution of *J. bulbosus* is stable, with slight declines at the edges of its range. Some authorities recognise two segregates (*J. bulbosus* and *J. kochii*), but opinions differ widely on their taxonomic status and distribution.

European Boreo-temperate element; also in N. America.

References: Atlas (322a), Grime *et al.* (1988), Hultén & Fries (1986), Meusel *et al.* (1965), Preston & Croft (1997), Webb & Scannell (1983).

C. A. STACE

Juncus pygmaeus Pigmy Rush

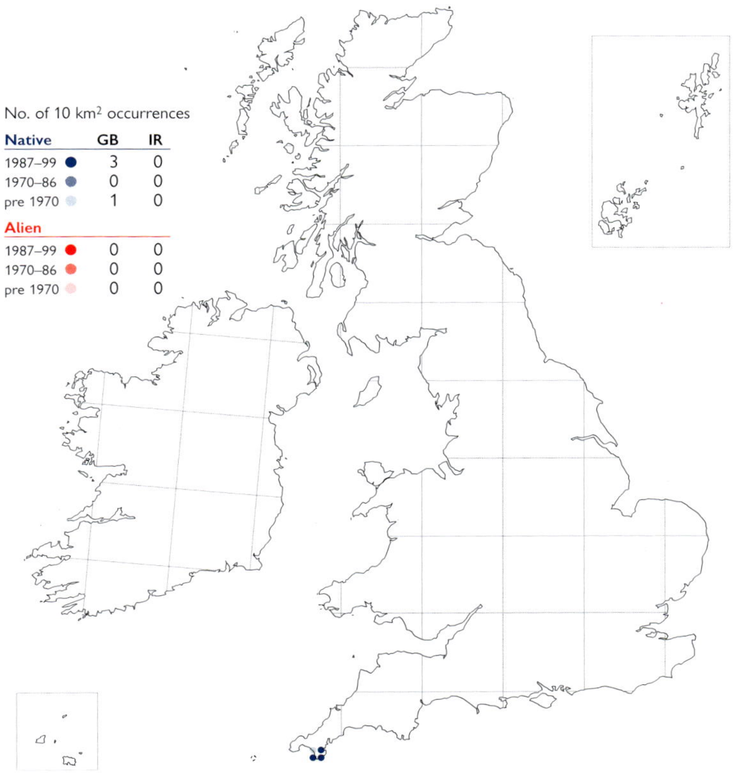

This diminutive, loosely tufted, spring-germinating annual occurs in seasonally wet, compacted open ground such as in gateways and on wheel tracks, less often in natural areas of erosion and in quarries, on serpentine heathland. Lowland.

Native. Populations of *J. pygmaeus* vary greatly in size from year to year, but there has been a strong decline since 1950, mainly due to tracks becoming abandoned and vegetated over, or infilled with hardcore.

Mediterranean-Atlantic element.

References: Atlas (322b), Hultén & Fries (1986), Wigginton (1999).

C. A. STACE

Juncus biglumis Two-flowered Rush

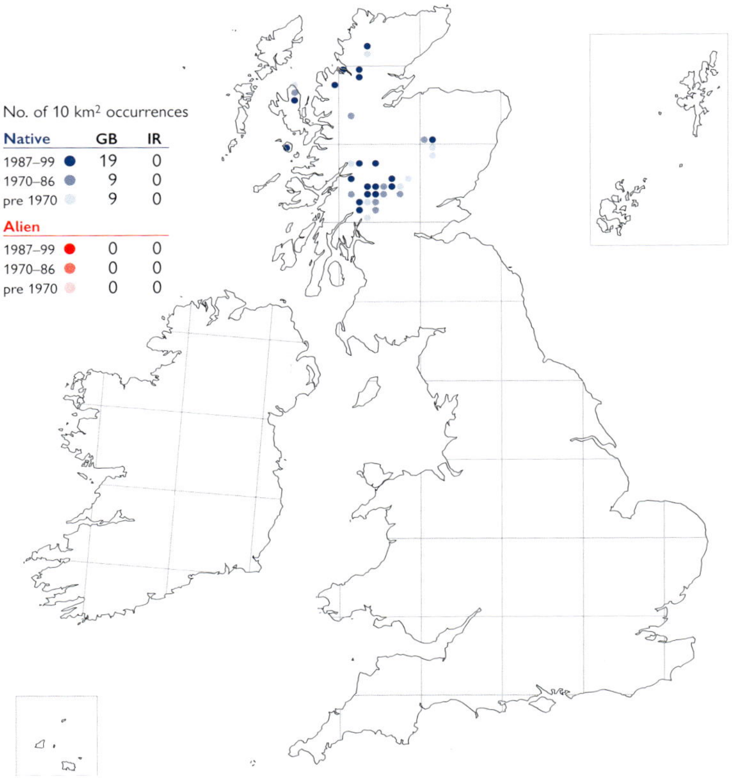

This short, tufted perennial herb occurs in damp rocky or gravelly places, ranging from well-watered rock faces and flushes to marshes with short open vegetation. It is confined to base-rich, but relatively competition-free, habitats in species-rich localities. From 460 m on Rum (N. Ebudes) to 1100 m in the Breadalbanes (Mid Perth) and Aonach Beag (Westerness).

Native (change –0.17). *J. biglumis* is probably present in most of its pre-1987 sites, and new sites have been found in N.W. Scotland. Its distribution is stable. Old records for Upper Teesdale are considered doubtful in the absence of convincing herbarium material.

Circumpolar Arctic-montane element.

References: Atlas (322c), Hultén & Fries (1986), Stewart *et al.* (1994).

C. A. STACE

Juncus triglumis Three-flowered Rush

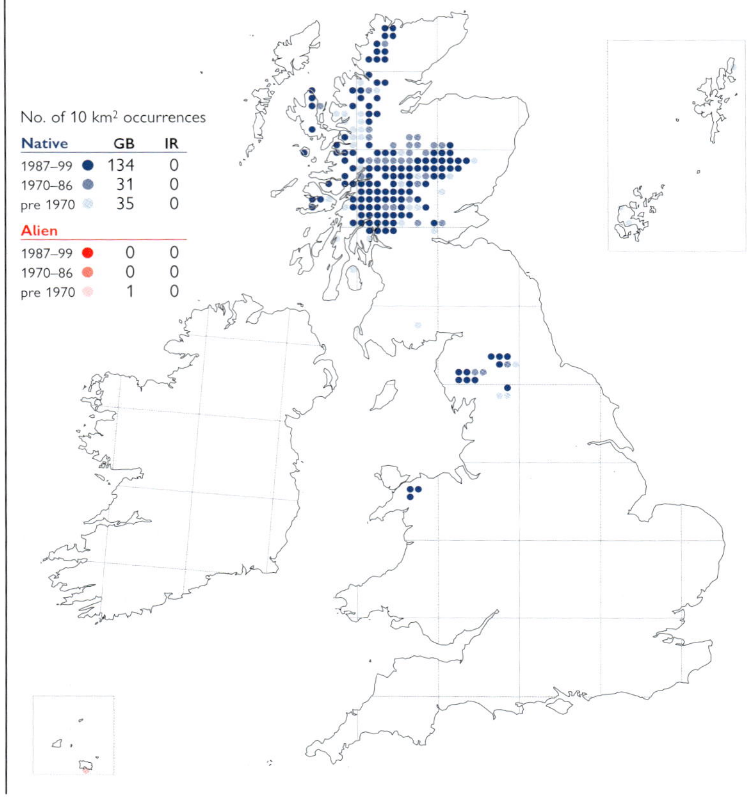

A short, tufted, montane perennial herb of base-rich damp rocky or gravelly places, flushes and small marshes with little competing vegetation. It occupies the same habitats as the rarer *J. biglumis*, and often grows near or with it, but is less restricted to high altitudes, descending from 1065 m on Snowdon (Caerns.) to, formerly, 60 m in Shetland.

Native (change –0.38). *J. triglumis* was lost from Orkney and Shetland before 1930. The overall distribution elsewhere has changed little since the 1962 *Atlas*, and it is probably still present in those squares where it has not been seen since 1987.

Circumpolar Arctic-montane element.

References: Atlas (322d), Hultén & Fries (1986).

C. A. STACE

Juncus castaneus Chestnut Rush

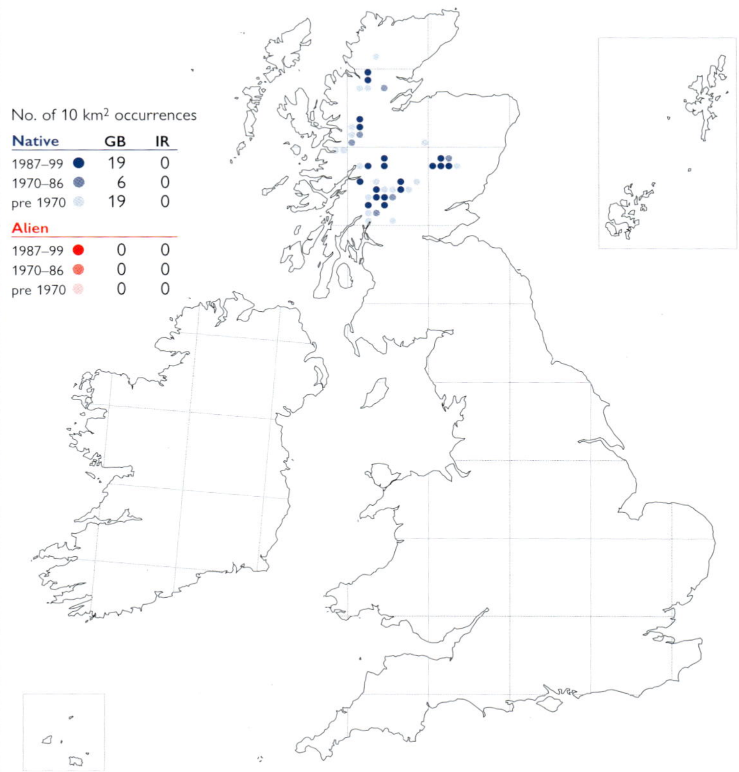

This short, tufted perennial herb often occurs in species-rich localities with one or both of *J. biglumis* and *J. triglumis*, although it is more characteristic of wetter and more calcareous habitats, and can better withstand competition from grassy vegetation. From 610 m on Sgurr na Lapaich (Easterness) to 990 m on Ben Lawers (Mid Perth).

Native (change –0.40). *J. castaneus* might still be present in some of its pre-1987 sites, but few of these have been updated since it was mapped by Stewart *et al.* (1994). Since many populations are small and often in sites where there is heavy grazing, there is some cause for concern.

Circumpolar Arctic-montane element.

References: Atlas (322b), Hultén & Fries (1986).

C. A. STACE

Juncus maritimus Sea Rush

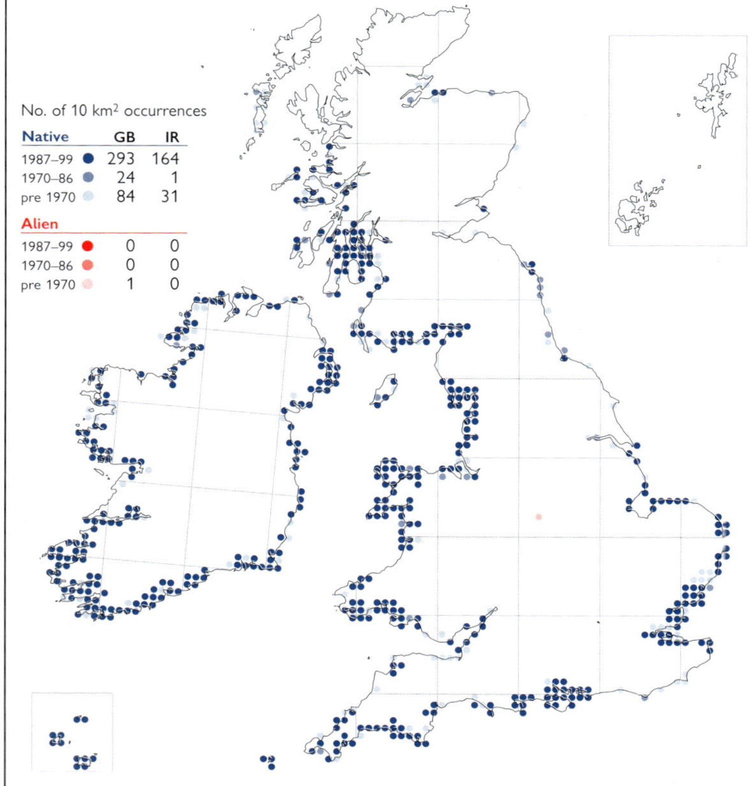

A rhizomatous, clump-forming perennial herb of saltmarshes and saline dune-slacks. It also occurs in areas subject to freshwater seepage on low, exposed rocky cliff-tops and stony sea-loch shores. It is tolerant of a wide range of salinities and soil moisture, occurring at all levels in saltmarshes and in both silty and sandy substrates. Lowland.

Native (change –0.26). *J. maritimus* has been lost since the 1962 *Atlas* from a number of sites on the S. and E. coasts of England, but the distribution is stable elsewhere.

European Southern-temperate element; widely naturalised outside its native range.

References: Atlas (320b), Hultén & Fries (1986), Meusel *et al.* (1965), Snogerup (1993).

C. A. STACE

Juncus acutus Sharp Rush

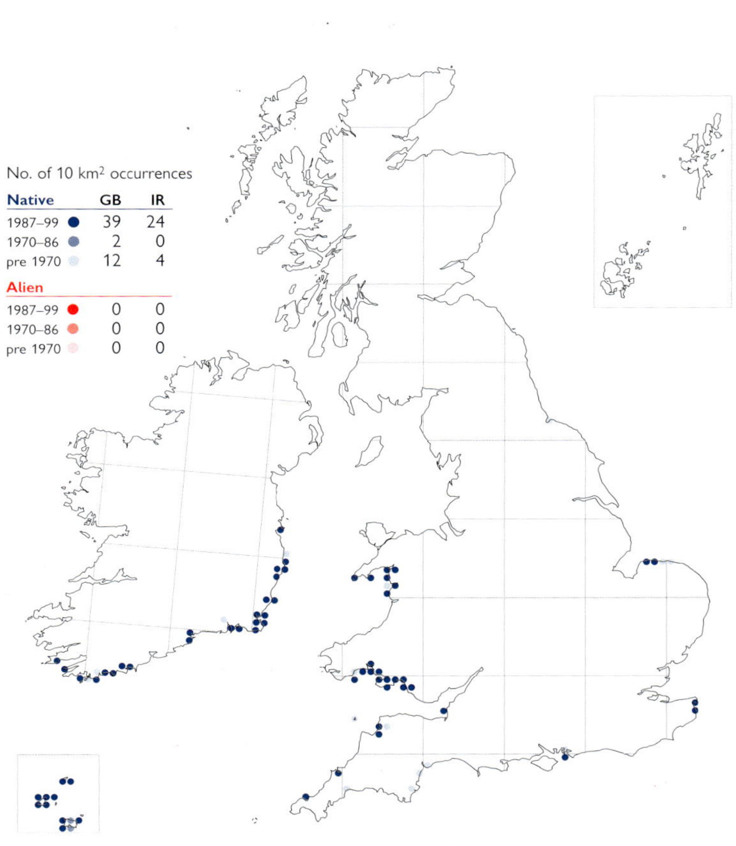

A tall, tussock-forming perennial herb typically occurring in saline or brackish dune-slacks, in the uppermost levels of dry saltmarsh and on shingle banks. There is often little competing vegetation. Lowland.

Native (change +0.01). *J. acutus* seems to be an efficient coloniser and, in Britain, sometimes quickly appears in newly available sites. There have been very few 10-km square losses since the 1962 *Atlas*. Local losses have resulted from sea-defence works or changing coastlines.

Mediterranean-Atlantic element; also in western N. America.

References: Atlas (320c), Hultén & Fries (1986), Jones & Richards (1954), Snogerup (1993), Stewart *et al.* (1994).

C. A. STACE

Juncus subulatus Somerset Rush

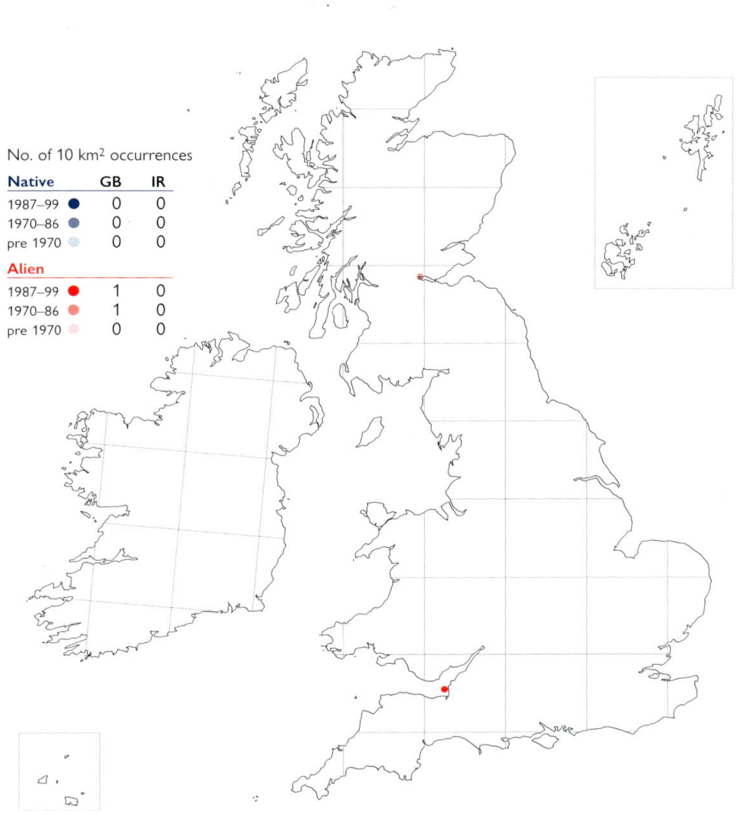

A rhizomatous plant forming large dominant patches. In N. Somerset the plant grows in brackish reed-swamp in a dune system; in Stirlingshire, it occurs in a pool on reclaimed dockland. Lowland.

Neophyte. This species was first recorded in Britain in 1957 (N. Somerset) and its mode of introduction is uncertain. It may have arrived with shipping, and is therefore treated here as a neophyte, but could have been introduced by birds (its closest native localities are on the coast of N. Spain), in which case it would be native to Britain (Willis & Davies, 1960). A second site was found in Scotland in 1983 (Stirlingshire). It shows no sign of spreading beyond its immediate points of entry.

A Mediterranean-Atlantic species.

References: Atlas (320a), Stewart (1987).

C. A. STACE

Juncus balticus Baltic Rush

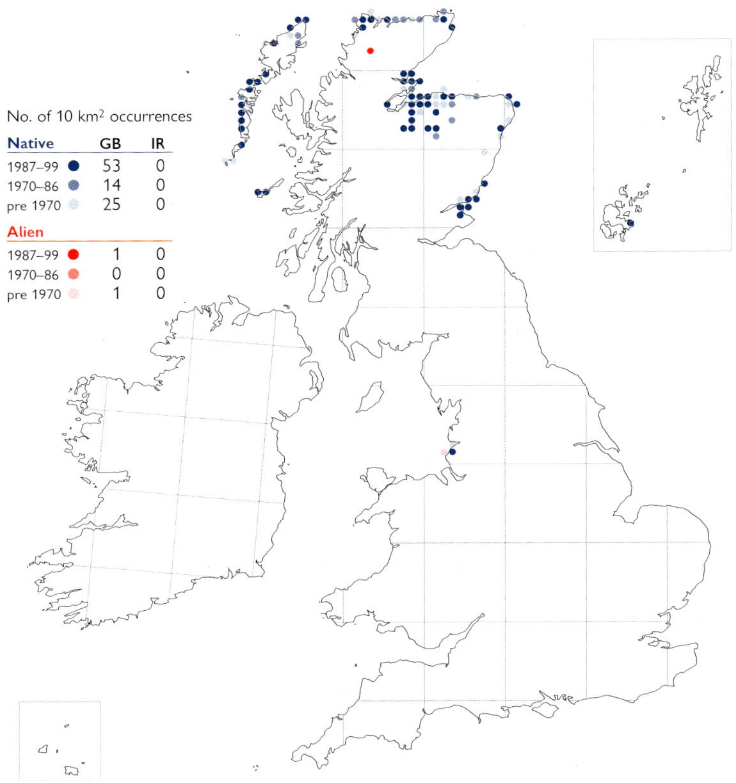

J. balticus usually grows in dune-slacks and other damp areas in maritime sand, mud or peat, frequently beside river estuaries, in open or closed vegetation. The plant is rhizomatous and rarely forms dense patches. It also occurs inland in N.E. Scotland on river-terraces or flood plains or in marshes. Generally sea level, but reaching 405 m on Slochd Mor (Easterness).

Native (change –0.34). The distribution of *J. balticus* seems stable, and it is probably still present in many of its Scottish sites that have no post-1986 records. There has been an increase in abundance in its very restricted sites in S. Lancashire over the past thirty years.

Circumpolar Boreo-arctic Montane element.

References: Atlas (320a), Hultén & Fries (1986), Smith (1984), Stace (1972), Stewart *et al.* (1994).

C. A. STACE

Juncus filiformis Thread Rush

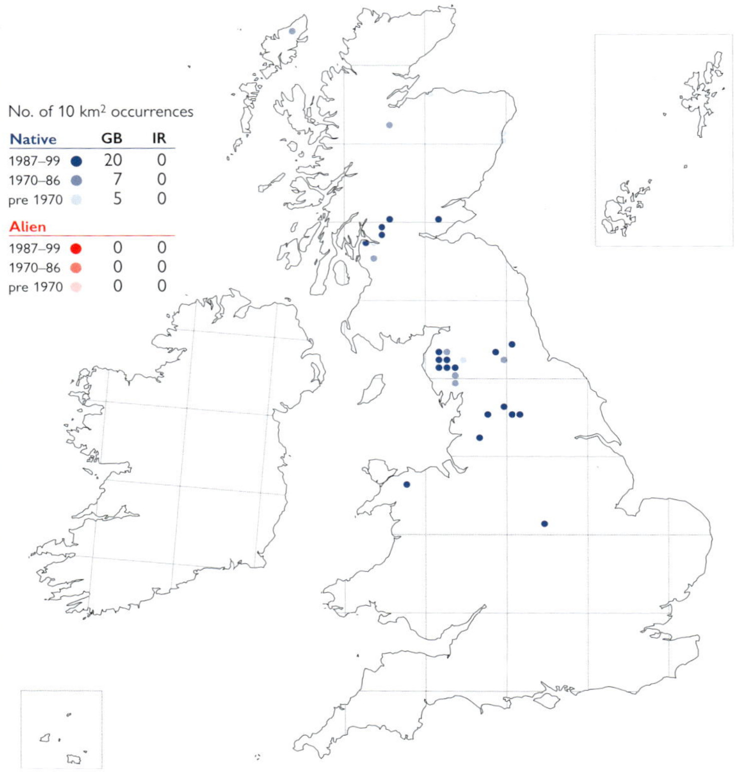

A rhizomatous perennial herb, restricted in Britain to the edges of lakes or reservoirs, mostly in a narrow fringing zone of periodically flooded wet marshy pasture or more open ground. Lowland.

Native (change +0.79). *J. filiformis* is evidently effectively dispersed, as it can appear in newly available habitats far from known sites. It is, however, very easily overlooked due to its thin, often short (and heavily grazed) stems rather sparsely distributed on extended rhizomes, and for this reason may be somewhat under-recorded. It was known from only five squares outside the Lake District in the 1962 *Atlas*.

Circumpolar Boreal-montane element.

References: Atlas (319d), Blackstock (1981), Hultén & Fries (1986), Richards (1943b), Stewart *et al.* (1994).

C. A. STACE

Juncus inflexus Hard Rush

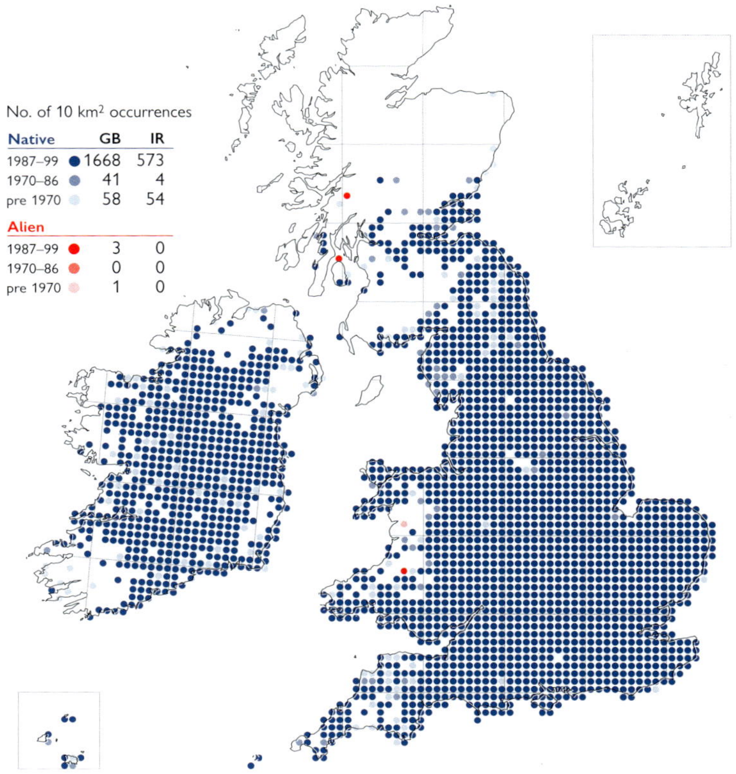

A clump-forming perennial herb of wet places by rivers, ponds and lakes, and in marshes, wet fields, ditches and occasionally dune-slacks and fens. It is almost always on base-rich soils, frequently on heavy clays, where it replaces *J. effusus*. 0–550 m (Mattergill Sike, Westmorland).

Native (change +0.04). There is no change in the distribution of this species since the 1962 *Atlas*. The absence of *J. inflexus* from some areas of England, Wales and Ireland indicates a lack of sufficient base-richness, whereas most of Scotland lies beyond its northern limit in Europe.

Eurosiberian Southern-temperate element; widely naturalised outside its native range.

References: Atlas (319a), Hultén & Fries (1986), Meusel *et al.* (1965), Richards & Clapham (1941b).

C. A. STACE

Juncus effusus Soft-rush

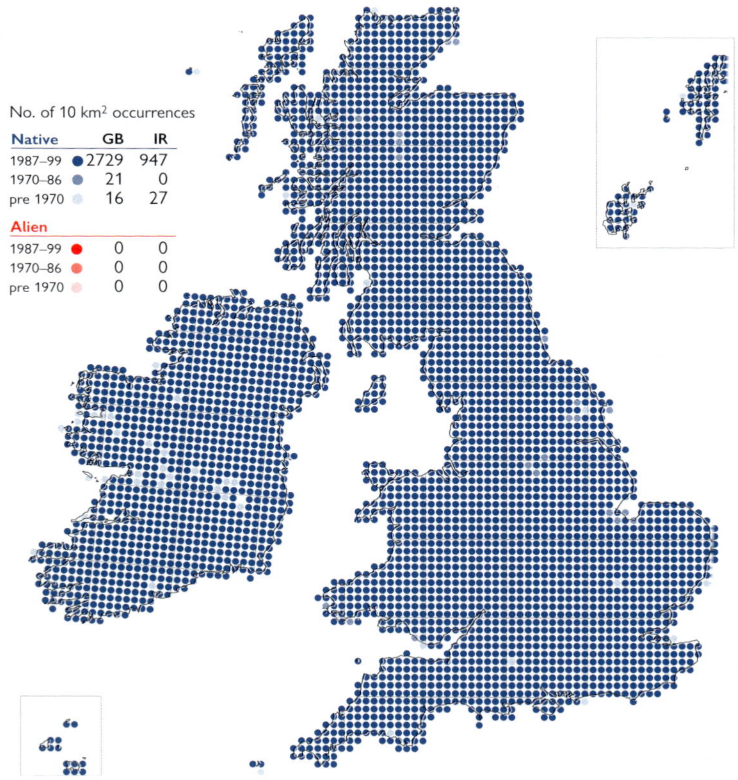

J. effusus occurs by rivers, ponds and lakes, and in marshes, wet fields, ditches and open wet woodland. It avoids base-rich soils and is most characteristic of sandy and peaty substrates, especially open heaths and moors, where it can be dominant. 0–845 m (Great Dun Fell, Westmorland) and 850 m Co. Wicklow.

Native (change +1.06). The range of *J. effusus* is unchanged since the 1962 *Atlas*.

European Southern-temperate element; also disjunctly in C. and E. Asia and N. America and widely naturalised outside its native range.

References: Atlas (319b), Agnew (1961, 1968), Grime *et al.* (1988), Hultén & Fries (1986), Meusel *et al.* (1965), Richards & Clapham (1941c).

C. A. STACE

Juncus effusus × *J. inflexus* (*J.* × *diffusus*)

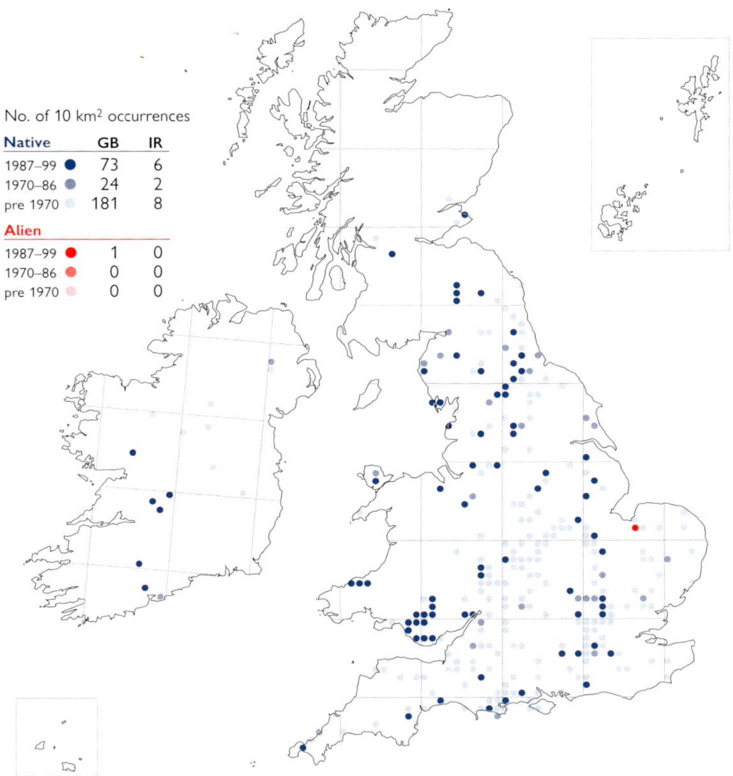

J. effusus and *J. inflexus* often occur together in lowland habitats, but their hybrid is uncommon and usually only present as one or few plants in a population. *J. inflexus* normally flowers later than *J. effusus*, and this appears to be a largely effective isolating mechanism. The hybrid is, however, not fully sterile and second or backcross generations might occur. Lowland to 305 m (Gilcambon Beck, Cumberland).

Native. The majority of pre-1970 records shown here are those mapped by Perring & Sell (1968). The range of the hybrid has not changed since then, but it is easily overlooked and the lack of recent records probably indicates under-recording.

Widespread in Europe.

References: Atlas Supp. (140a), Clifford (1959), Stace (1972, 1975).

C. A. STACE

Juncus conglomeratus Compact Rush

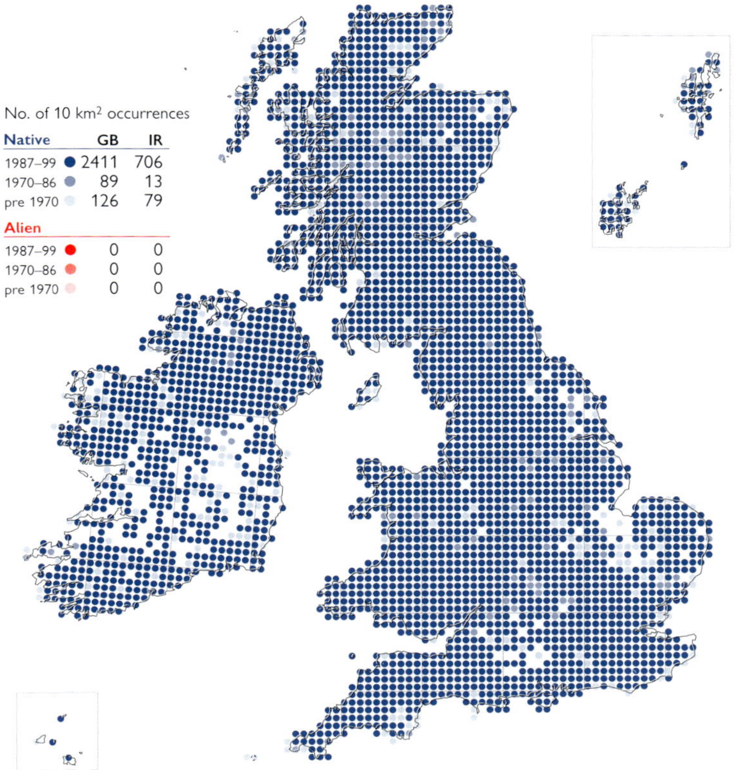

J. conglomeratus tends to occur in slightly drier habitats than *J. effusus*, avoiding the wetter places and the more acidic soils, being characteristic of damp fields, ditches, open woodland and margins of still or running water. 0–840 (Breadalbanes, Mid Perth), and 845 m on Great Dun Fell (Westmorland).

Native (change +0.84). Despite the well-documented differences between *J. effusus* and *J. conglomeratus*, they are still frequently confused. The latter species might be over-recorded, but this is unlikely to affect the map which indicates a stable distribution.

European Temperate element; widely naturalised outside its native range.

References: Atlas (319c), Agnew (1968), Hultén & Fries (1986), Meusel *et al.* (1965), Richards & Clapham (1941d).

C. A. STACE

Luzula forsteri Southern Wood-rush

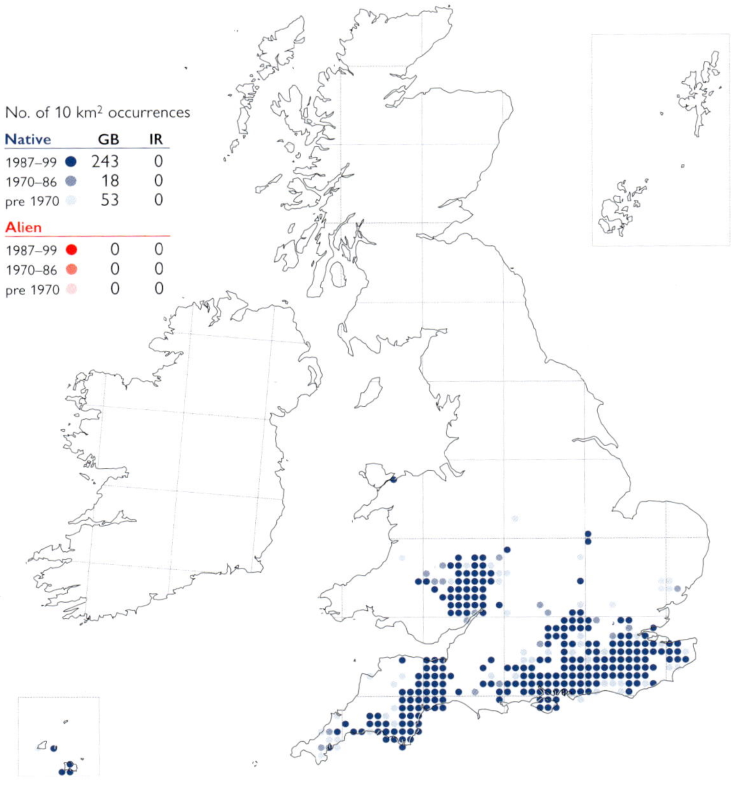

A tufted, grass-like perennial herb of woods and other moist but well-drained shaded places, often on roadside banks and in hedgerows. It is most frequent on acidic soils, but avoids the most acidic. Competition is rarely tolerated, and plants usually occur in leaf-litter or moss-dominated sites. It often grows with *L. pilosa*, and appears to have very similar ecological requirements in S. England. Lowland.

Native (change +0.25). There have been some local losses of *L. forsteri*, particularly in S.E. England, but the overall distribution is stable.

Submediterranean-Subatlantic element.

References: Atlas (323b), Meusel *et al.* (1965).

C. A. STACE

Luzula forsteri × *L. pilosa* (*L.* × *borreri*)

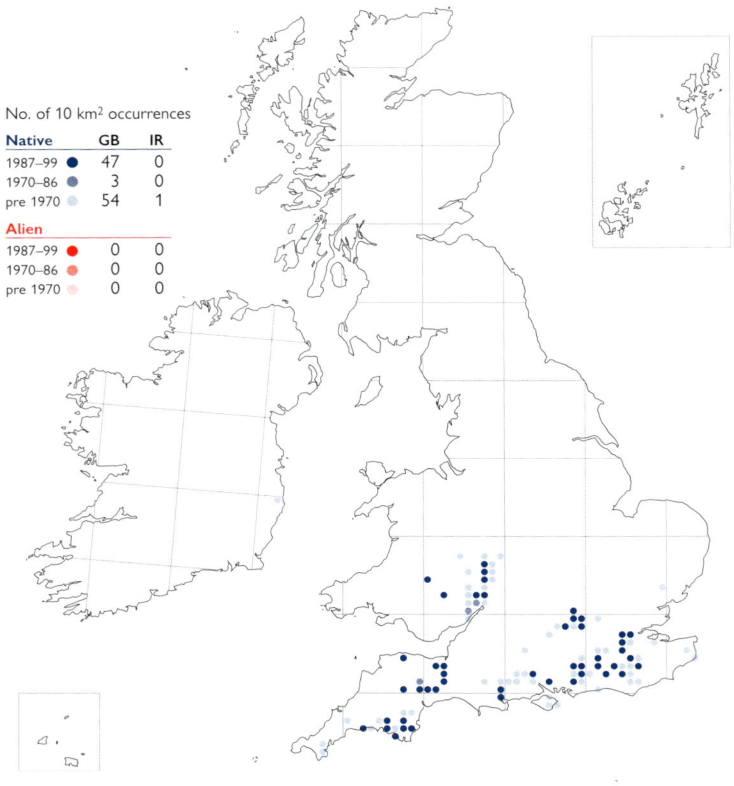

Throughout its range *L. forsteri* is usually found near *L. pilosa*, and hybrids are not infrequent since there appear to be no ecological differences between the parents. The hybrids are usually sterile, but some seed is formed and back-crossing is suspected. Lowland.

Native. The current distribution of this hybrid is similar to that mapped by Perring & Sell (1968). It is easily overlooked, and may still be present in squares for which there are only pre-1987 records. Its former presence in Ireland is remarkable, given that *L. forsteri* is absent from there.

An uncommon hybrid, recorded in mainland Europe from France and Germany.

References: Atlas Supp. (140b), Ebinger (1962), Stace (1975).

C. A. STACE

Luzula pilosa Hairy Wood-rush

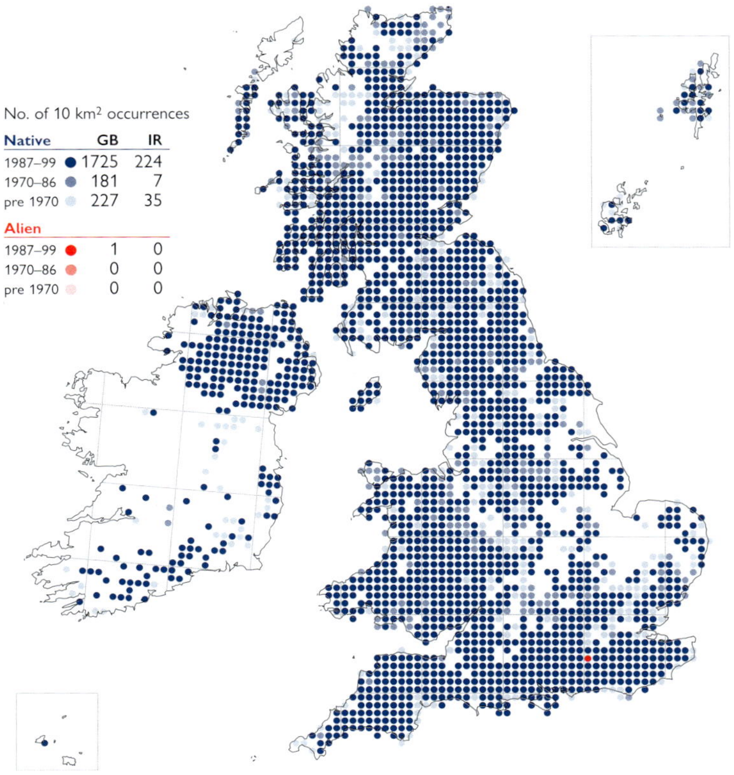

A tufted, grass-like perennial herb of woods and other moist but well-drained shaded places, often on roadside-banks and in hedgerows, generally on fairly acidic soils but not confined to them. Plants usually occur in leaf-litter or moss-dominated sites, and competition is rarely tolerated. In upland areas, more exposed sites such as quarries, pastures and rough ground are occupied. 0–670 m (Ben Lawers, Mid Perth).

Native (change –0.35). This species was mapped as 'all records' in the 1962 *Atlas*. A decline, especially in C. & E. England, has taken place and analysis of the database reveals that most of these losses have occurred since 1950.

Eurosiberian Boreo-temperate element.

References: Atlas (323a), Grime *et al.* (1988), Hultén & Fries (1986), Meusel *et al.* (1965).

C. A. STACE

Luzula sylvatica Great Wood-rush

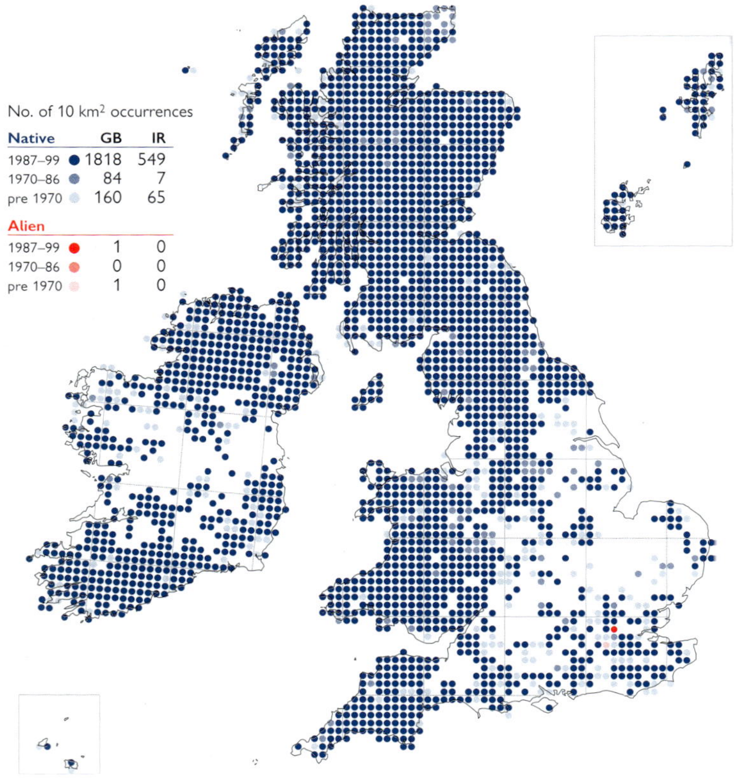

Characteristic of a range of damp, acidic, usually shaded habitats, this rhizomatous patch-former is found in lowland woods, often beside streams, on peaty heaths and moors, and on rock ledges and rocky streamsides in mountainous regions. It is intolerant of grazing and it is often confined to woods or, in the uplands, rocks. 0–1040 m (Carrantuohil, S. Kerry).

Native (change –0.02). The distribution of *L. sylvatica* is stable, except in C. & S.E. England where it appears to have been lost from many sites before 1970.

European Temperate element.

References: Atlas (323c), Hultén & Fries (1986), McVean & Ratcliffe (1962), Meusel *et al.* (1965).

C. A. STACE

Luzula luzuloides White Wood-rush

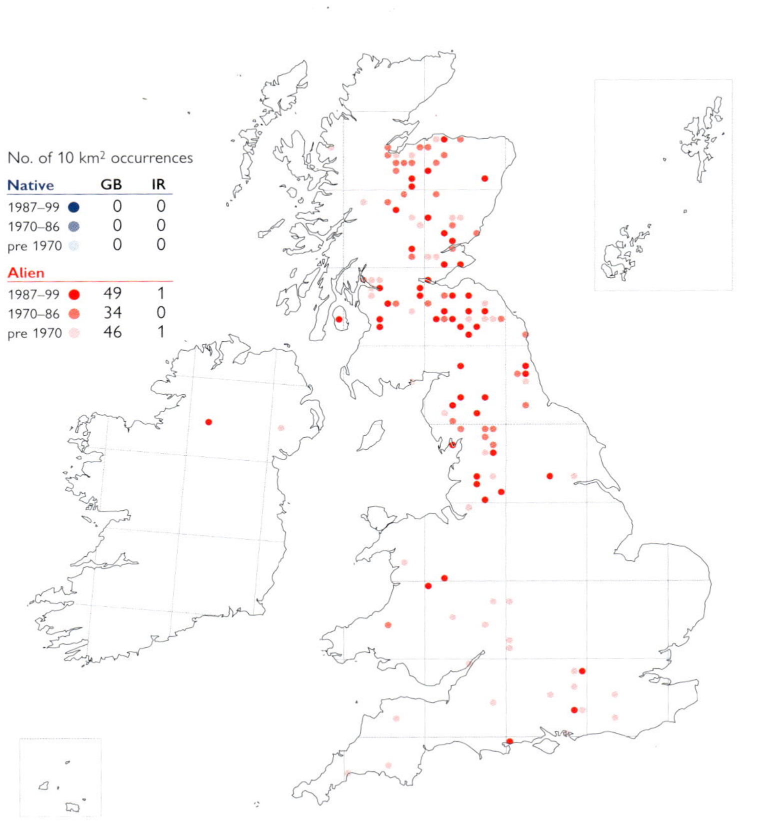

This tufted, rhizomatous herb occurs as a garden escape in woods and other moist shady places, often by streams, but sometimes in open peaty places in upland areas. It is usually found on acidic soils. Generally lowland, but reaching 365 m at Garsdale Head (Westmorland), and exceptionally at 845 m on Great Dun Fell (Westmorland).

Neophyte (change +0.18). *L. luzuloides* was introduced into Britain before 1800 and was recorded from the wild by 1871. Dunn (1905) described it as 'recorded two or three times in England'. There are more records than in the 1962 *Atlas*, which may reflect better recording or a continued spread.

Native of C. Europe; naturalised further north, especially in the Nordic countries, and in N. America.

References: Atlas (323d), Hultén & Fries (1986), Meusel *et al.* (1965).

C. A. STACE

Luzula campestris Field Wood-rush

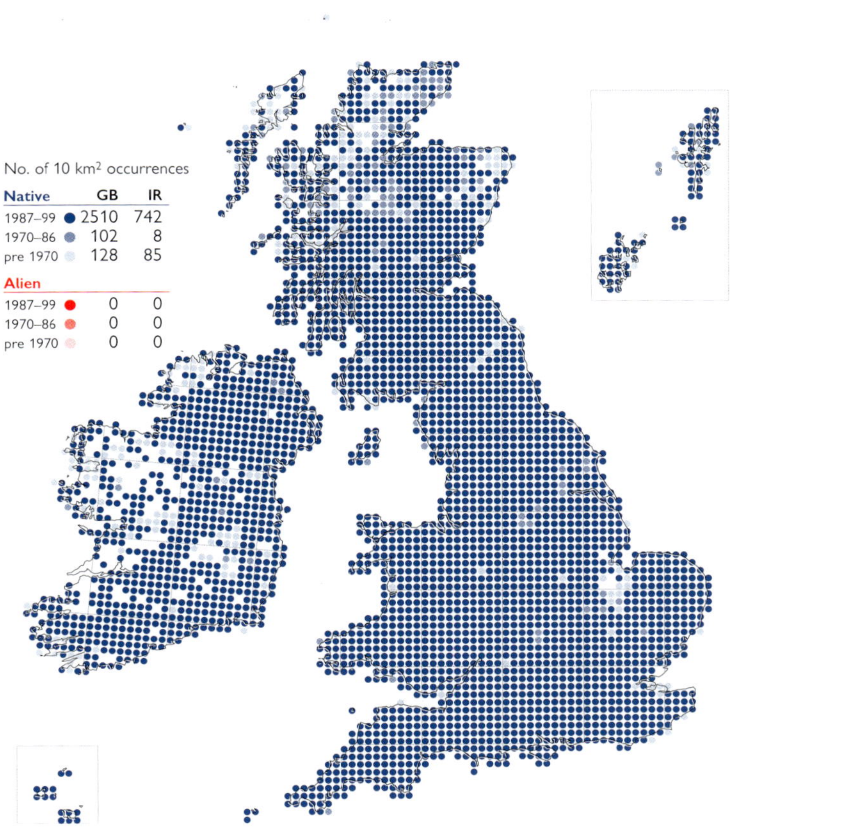

An early-flowering, low-growing, tufted but shortly rhizomatous herb characteristic of short, unshaded, relatively infertile grassland. It is found in a range of grazed or mown, often trampled, grassy places, particularly pastures, meadows, grassy verges and lawns which are moderately acidic to slightly alkaline. Taller grassland, rocky ledges and slopes, and quarries and spoil heaps are also colonised. 0–1005 m (Carnedd Dafydd, Caerns.).

Native (change −0.18). There has been little change in the distribution of this species since the 1962 *Atlas*.

European Temperate element.

References: Atlas (324c), Grime *et al.* (1988), Hultén & Fries (1986), Meusel *et al.* (1965).

C. A. STACE

Luzula multiflora Heath Wood-rush

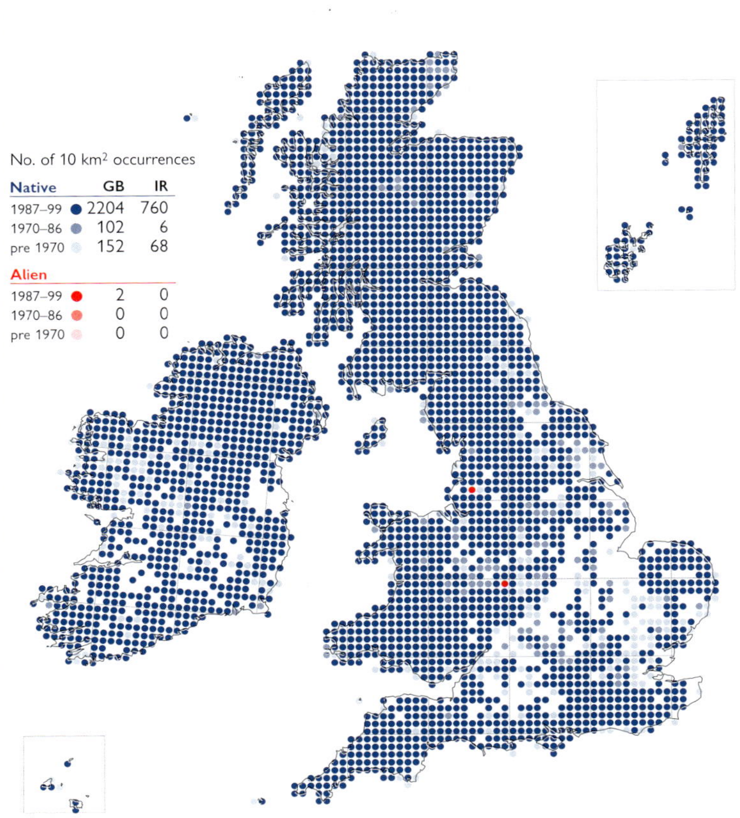

An early-flowering tufted but shortly rhizomatous herb of various grassy habitats, preferring more acidic soils and more shaded places than *L. campestris*. It occurs on heaths and moors and in the drier parts of bogs, in meadows and other grassy places, and in open woods and wood margins. 0–1020 m (Breadalbanes, Mid Perth).

Native (change +0.28). Because of past and present confusion with *L. campestris*, it is difficult to assess the validity of the apparent decline of this species in C. &. E. England. Subsp. *multiflora* and subsp. *congesta* are difficult to distinguish and are unevenly recorded; they are not mapped here.

Circumpolar Wide-boreal element; widely naturalised outside its native range.

References: Atlas (324d), Hultén & Fries (1986), Kirschner & Rich (1996).

C. A. STACE

Luzula pallidula Fen Wood-rush

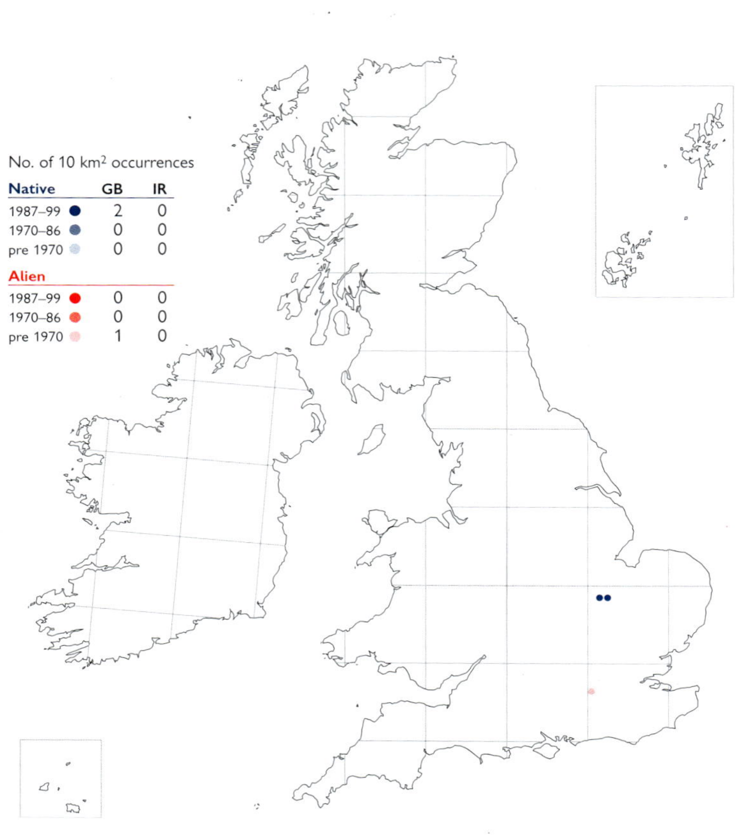

A tufted but shortly rhizomatous plant of fens, open peat, and banks, rides and glades in damp peaty woodland, particularly in disturbed ground. Lowland.

Native. *L. pallidula* is possibly now extinct, having been last recorded in 1992 at Woodwalton Fen (Hunts.). However, it is likely that a soil seed bank still exists and this might give rise to populations following renewed peat disturbance. A record from Co. Antrim requires confirmation (Hackney, 1992).

Eurasian Boreo-temperate element, with a continental distribution in W. Europe; also in N. America and widely naturalised outside its native range.

References: Atlas (324b), Hultén & Fries (1986), Meusel *et al.* (1965), Rich (1994a), Wigginton (1999).

C. A. STACE

Luzula arcuata Curved Wood-rush

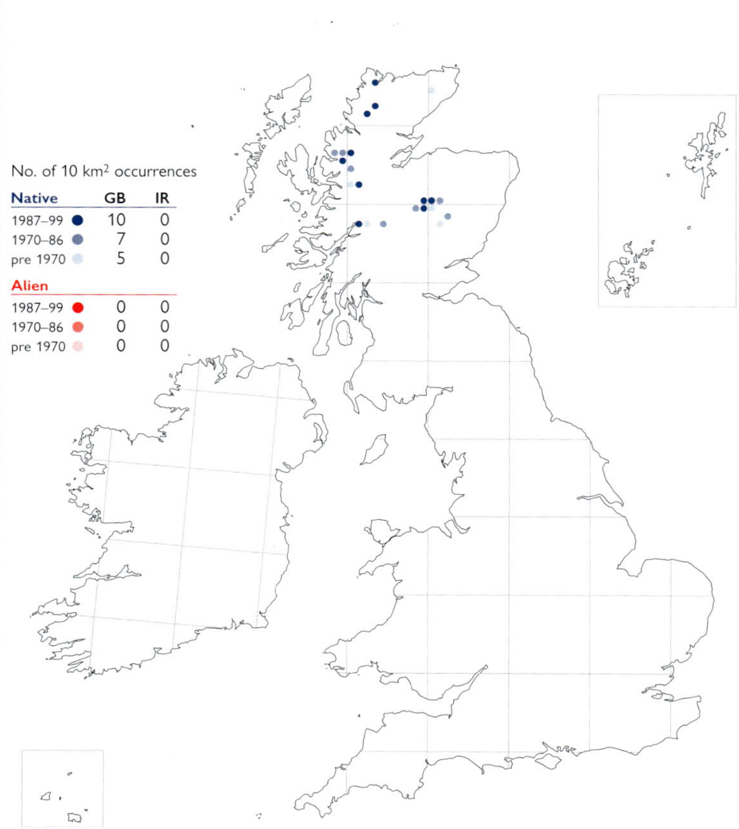

A dwarf, tufted, shortly rhizomatous and stoloniferous herb of bare windswept rocky summit ridges and plateaux that are mostly kept free of winter snow, where it often grows with *Juncus trifidus*. In high altitude corries it can occur in areas of snow-lie with greater vegetation cover. From 760 m (Slioch, W. Ross) to 1290 m (Cairn Toul, S. Aberdeen).

Native (change –0.43). *L. arcuata* has been found at new sites since the 1962 *Atlas*, but it seems likely there has been little change in its distribution, and it is probably still present in 10-km squares for which there are only pre-1987 records.

European Arctic-montane element, but absent from mountains of C. Europe; also in E. Asia and western N. America.

References: Atlas (324b), Hultén & Fries (1986), Stewart *et al.* (1994), Wigginton (1999).

C. A. STACE

Luzula spicata Spiked Wood-rush

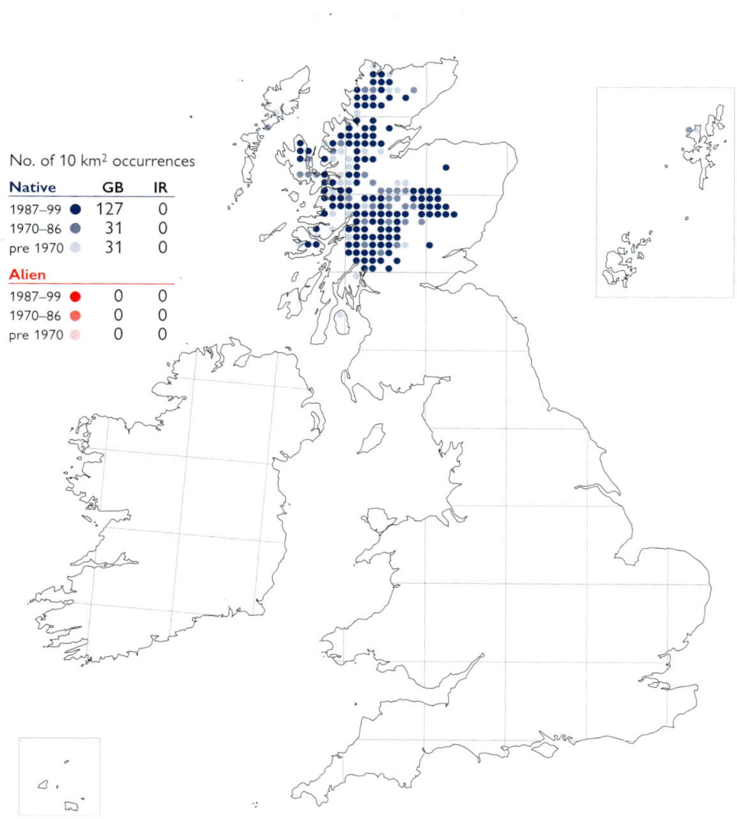

A dwarf, tufted, shortly rhizomatous and stoloniferous calcifuge of barish, open stony ground on mountains, both on flat areas and on cliffs and ledges, occasionally washed down rivers and on bulldozed hill tracks. It usually occurs in species-poor localities. From 275 m (Ronas Hill, Shetland) to 1220 m (Ben Macdui, Banffs.).

Native (change –0.72). There is virtually no change in the distribution of *L. spicata*. Two pre-1900 records from the Lake District shown in the 1962 *Atlas* are not supported by any herbarium material and are now considered dubious.

European Arctic-montane element; also in C. Asia and N. America.

References: Atlas (324a), Hultén & Fries (1986), Meusel *et al.* (1965).

C. A. STACE

Eriophorum angustifolium
Common Cottongrass

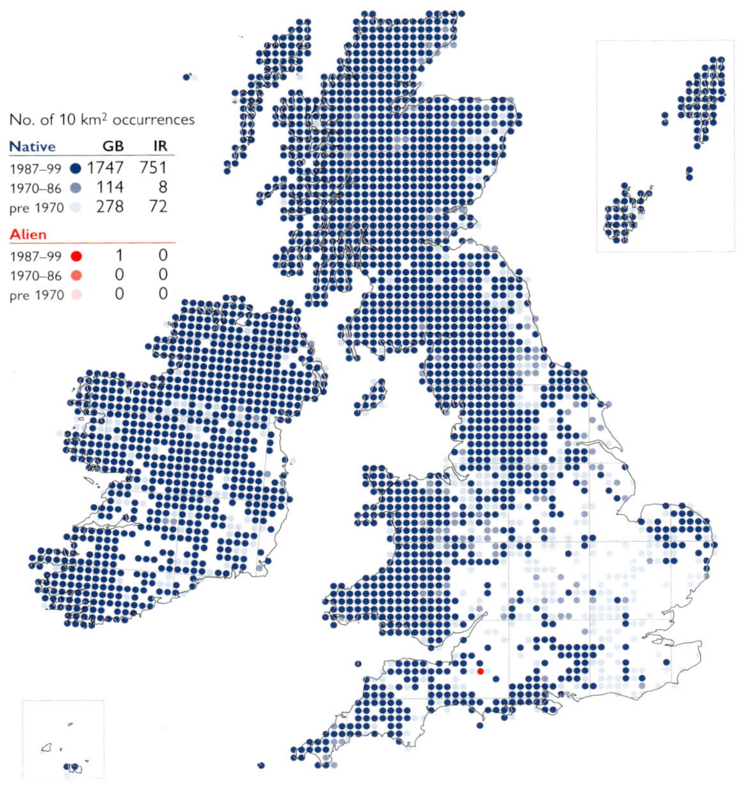

A rhizomatous perennial of open, wet, peaty ground, both calcareous and acidic, sometimes colonising peat-cuttings and often growing in standing water. Its habitats range from upland blanket bogs and hillside flushes to wet heaths and marshy meadows in the lowlands. 0–1100 m (Ben Alder, Westerness).

Native (change –0.79). Drainage, groundwater extraction and the cessation of grazing at some sites have considerably reduced the abundance of this species in the lowlands. Analysis of the database reveals that many losses took place before 1950, but have continued since then. In Hertfordshire, most sites were lost by 1900, and all by 1979 (James & Goldsmith, 1993).

Circumpolar Wide-boreal element.

References: Atlas (346c), Grime *et al.* (1988), Hultén & Fries (1986), Meusel *et al.* (1965), Phillips (1954).

M. S. PORTER & M. J. Y. FOLEY

Eriophorum latifolium Broad-leaved Cottongrass

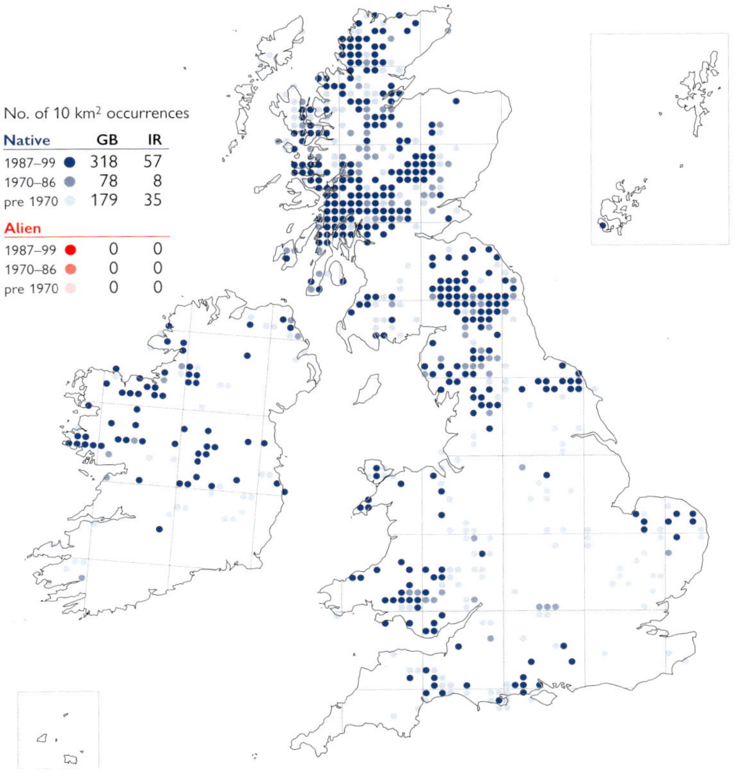

A rhizomatous perennial herb of open sites, growing in wet, base-rich lowland meadows and mires, and in fens and calcareous flushes in the uplands. 0–670 m (Breadalbanes, Mid Perth).

Native (change +0.36). This species was lost from some of its pre-1970 squares in the 19th century, and from most before 1930. These losses have continued in lowland areas as a result of afforestation and agricultural intensification. The species is much better recorded in W. Scotland, the Scottish Borders and in Ireland than it was for the 1962 *Atlas*.

European Boreo-temperate element.

References: Atlas (347a), Hultén & Fries (1986), Meusel *et al.* (1965).

M. S. PORTER & M. J. Y. FOLEY

Eriophorum gracile Slender Cottongrass

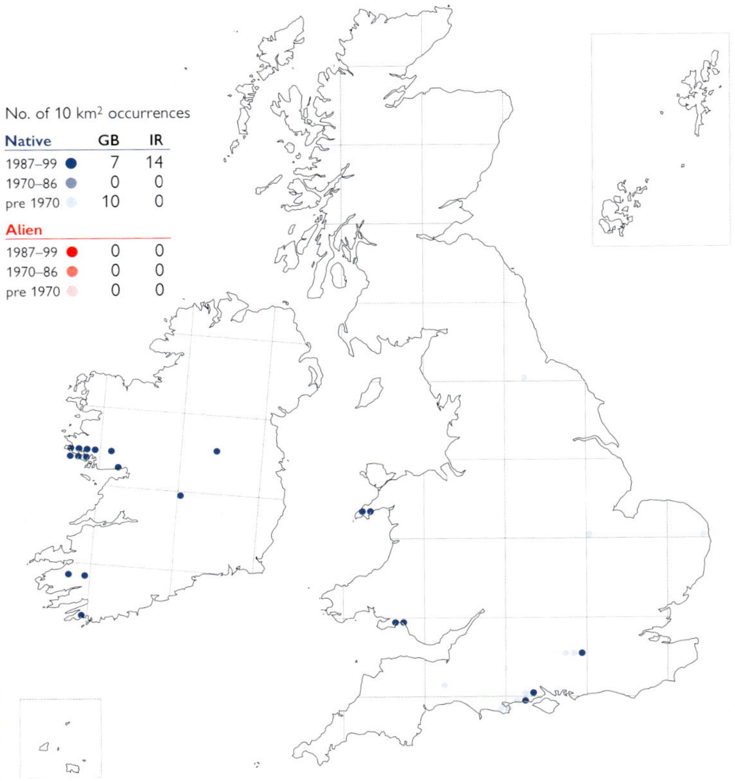

A rhizomatous perennial found in the wettest parts of bogs, transitional mires, poor fens and on the edge of *Alnus* carr, typically over liquid peats. Its sites are calcareous or moderately acidic, and have some water movement. Lowland.

Native (change –0.20). Since 1930, *E. gracile* has become extinct in Norfolk and Dorset and declined to one site in Surrey and two in Hampshire. However, it appears to be stable in Wales and Ireland, where it has been discovered since the 1962 *Atlas*. Losses have been due to drainage, afforestation and infilling.

Circumpolar Boreo-temperate element, with a continental distribution in W. Europe; rare and declining in Europe and Fennoscandia.

References: Atlas (346d), Curtis & McGough (1988), Hultén & Fries (1986), Kay & John (1995), Wigginton (1999).

M. S. PORTER & M. J. Y. FOLEY

Eriophorum vaginatum Hare's-tail Cottongrass

Native	**GB**	**IR**
1987–99	1256	460
1970–86	85	12
pre 1970	175	88
Alien		
1987–99	0	0
1970–86	1	0
pre 1970	0	0

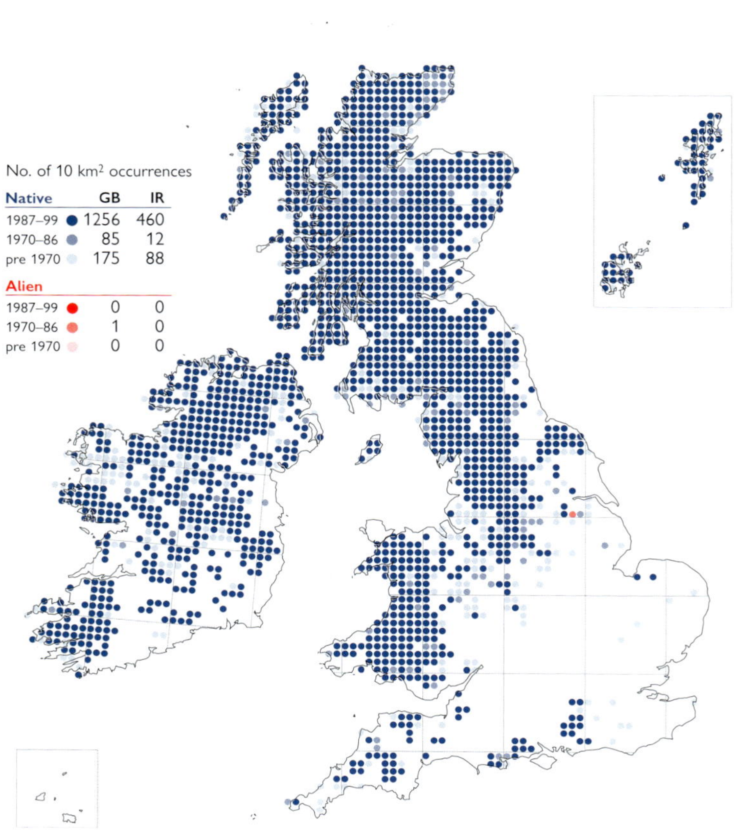

A tussock-forming rhizomatous perennial herb of wet heaths and mires, including blanket- and raised bogs. It is characteristic of wet peaty moorlands, often dominant or co-dominant with *Calluna vulgaris*, where it survives, or even increases, after burning. Its sites are always open and almost always acidic. 0–945 m (Ben Lawers, Mid Perth).

Native (change –0.36). Many lowland sites for *E. vaginatum* had been lost by 1930. Since then, further losses have occurred in the N. Midlands, but it remains abundant in many areas of N. & W. Britain and in Ireland.

Circumpolar Boreo-arctic Montane element.

References: Atlas (347b), Grime *et al.* (1988), Hultén & Fries (1986), Meusel *et al.* (1965), Wein (1973).

M. S. PORTER & M. J. Y. FOLEY

Trichophorum alpinum Cotton Deergrass

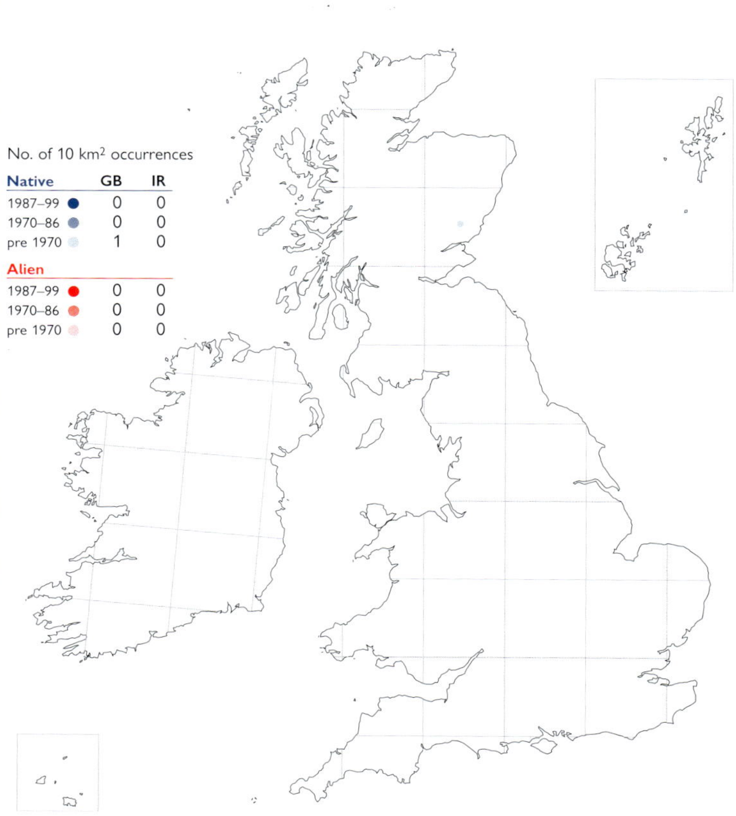

A perennial herb, formerly known from the drier parts of a single bog in Angus. Lowland.

Native. In Britain, *T. alpinum* has only been recorded from the Moss of Restennet, where it was first discovered in 1791. It was extinct by about 1804, probably as a result of the extraction of marl from the site.

Circumpolar Boreal-montane element, with a continental distribution in W. Europe.

References: Atlas (347c), Hultén & Fries (1986), Ingram & Noltie (1981), Lusby (1998), Marren (1999).

M. J. Y. FOLEY & M. S. PORTER

Trichophorum cespitosum Deergrass

Native	**GB**	**IR**
1987–99	1295	531
1970–86	73	14
pre 1970	185	84
Alien		
1987–99	0	0
1970–86	0	0
pre 1970	0	0

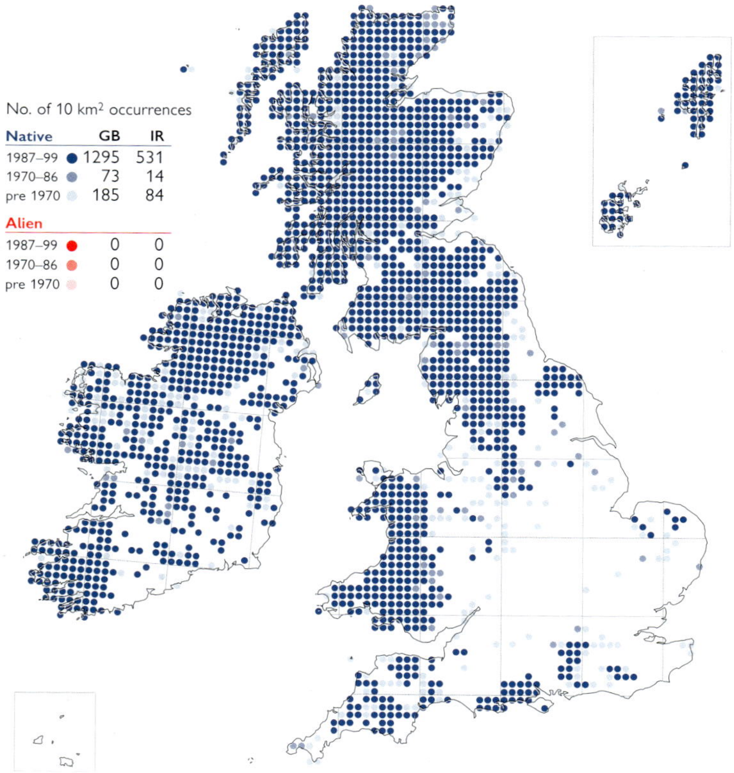

A densely tufted perennial herb occurring on peaty moors and bogs over acidic soils, persisting even in burnt and heavily deer-grazed areas. It is also local on open ground on wet lowland heaths in S. & E. England, where it often avoids the wettest sites and favours grazed, burnt and trampled areas. 0–1190 m (above Caenlochan, Angus).

Native (change –0.31). The decline of *T. cespitosum* in S. & E. England was already apparent in the 1962 *Atlas*, and has continued, but the distribution appears to be stable elsewhere. Two subspecies and their hybrid occur in our area; subsp. *germanicum* is found throughout the range of the species, subsp. *cespitosum* is mapped separately.

Circumpolar Boreal-montane element.

References: Atlas (347c), Hollingsworth & Swan (1999), Hultén & Fries (1986), Meusel *et al.* (1965), Swan (1999).

M. J. Y. FOLEY & M. S. PORTER

Trichophorum cespitosum subsp. *cespitosum*

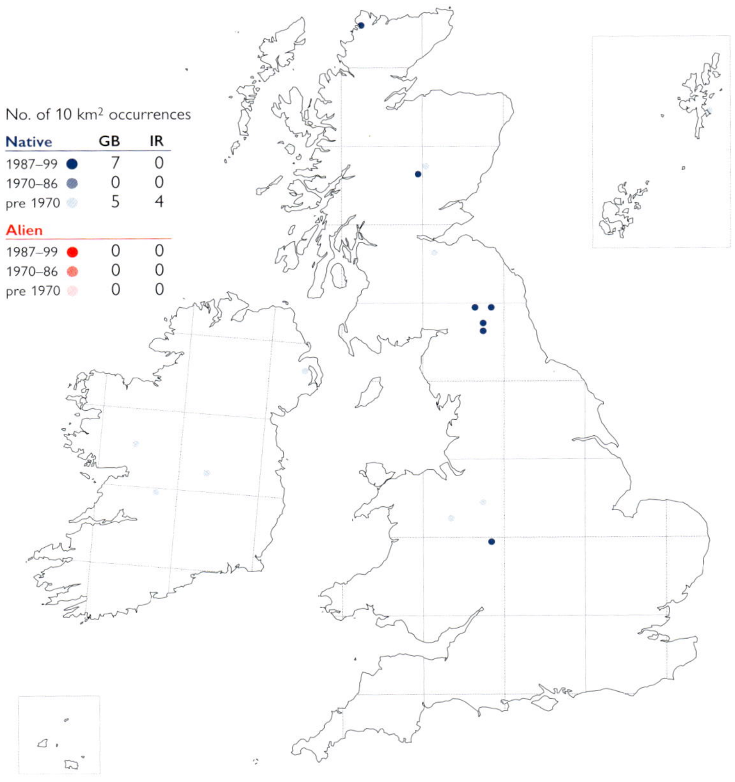

A perennial herb known only on acidic, peaty moorland where it occurs mainly on the margins of *Sphagnum* mires in areas subject to some base-enrichment. From 75 m to 620 m (Scotsman's Knowe, Cheviot).

Native. Although this taxon has traditionally been included in British Floras, its presence was only recently confirmed (Swan, 1999). Herbarium specimens date back to 1840 (Salop). It appears to be rare, but its distribution requires further study. The hybrid with subsp. *germanicum* often occurs near subsp. *cespitosum* but is more widespread.

Subsp. *cespitosum* occurs throughout the Circumpolar Boreal-montane range of the species, except in those areas of N.W. Europe where it is replaced by subsp. *germanicum*.

References: Hollingsworth & Swan (1999), Hultén & Fries (1986).

M. J. Y. FOLEY & M. S. PORTER

Eleocharis palustris Common Spike-rush

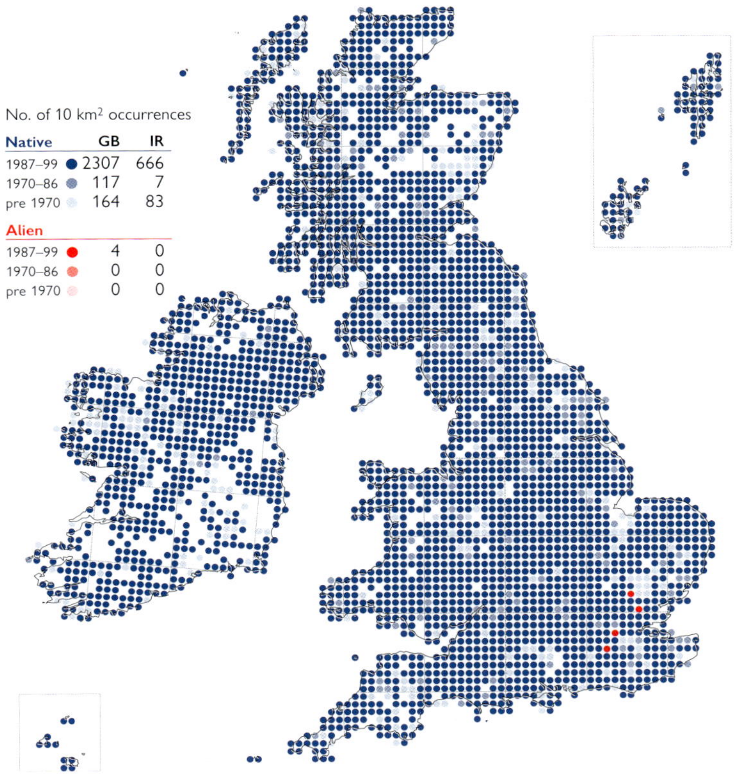

An emergent rhizomatous perennial herb, found on the margins of ponds, lakes, slow-flowing rivers and streams, in fens, marshes, swamps and wet meadows, and in ditches, dune-slacks and saltmarshes. It grows in a wide range of organic and mineral soils, but rarely on acidic peat. It spreads by rhizomes and reproduces by seed. 0–550 m (Tyne Head, Cumberland).

Native (change +0.91). Some sites have been lost since the 1962 *Atlas*, where *E. palustris* was mapped as 'all records'. Subsp. *vulgaris* is found throughout Britain and Ireland; subsp. *palustris* is mapped separately.

Eurasian Wide-temperate element.

References: Atlas (351a), Grime *et al.* (1988), Hultén & Fries (1986), Perring & Sell (1968), Preston & Croft (1997).

M. S. PORTER & M. J. Y. FOLEY

Eleocharis palustris subsp. *palustris*

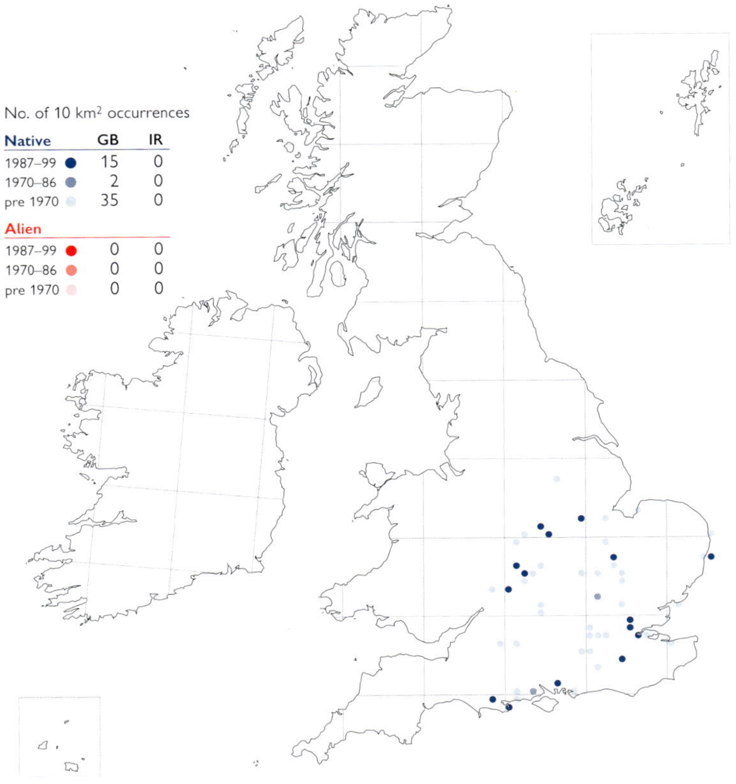

A rhizomatous perennial herb found on the margins of ponds, lakes and rivers, and in fens, marshes, wet meadows and ditches. It does not appear to be ecologically distinct from subsp. *vulgaris* and, indeed, is often found growing with it. Lowland.

Native. Both subspecies of *E. palustris* are very poorly recorded. Subsp. *palustris* was mapped by Perring & Sell (1968) using records collected by S.M. Walters, and there have been very few new records since then. It has apparently been lost from some sites where it was known to Walters, and a detailed survey is required to establish its current distribution.

Subsp. *palustris* is more widespread than subsp. *vulgaris* in Europe, extending from northern Scandinavia to the Mediterranean region.

References: Atlas Supp. (146a), Preston & Croft (1997), Walters (1949).

D. A. PEARMAN

Eleocharis austriaca Northern Spike-rush

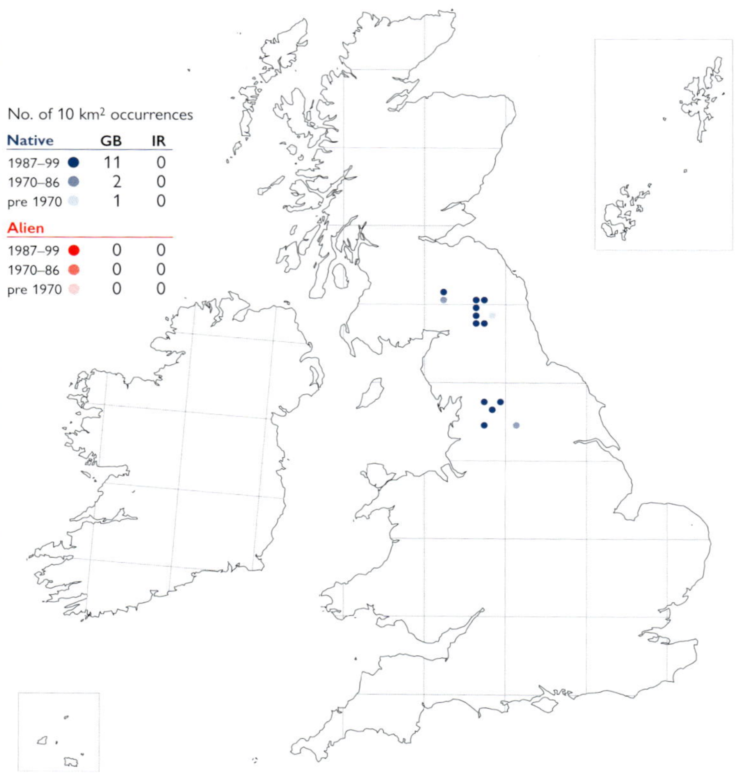

A rhizomatous perennial herb found in the middle reaches of upland rivers, usually in slacker water in places which are to some extent protected from spates, such as in shallow bays. It also grows in ditches, pools, runnels and springs. The substrate is usually gravel with some silt deposition. 60–340 m (Ribblehead, Mid-W. Yorks.).

Native. *E. austriaca* was first found in 1947, but not recognised until 1960. Since then a number of new sites have been discovered, some only to be lost again as the plant responded to changed conditions. The greater frequency of spates due to altered land-use threatens its survival in its riverine habitat.

European Boreal-montane element.

References: Atlas Supp. (146b), Hultén & Fries (1986), Walters (1963), Wigginton (1999).

M. S. PORTER & M. J. Y. FOLEY

Eleocharis uniglumis Slender Spike-rush

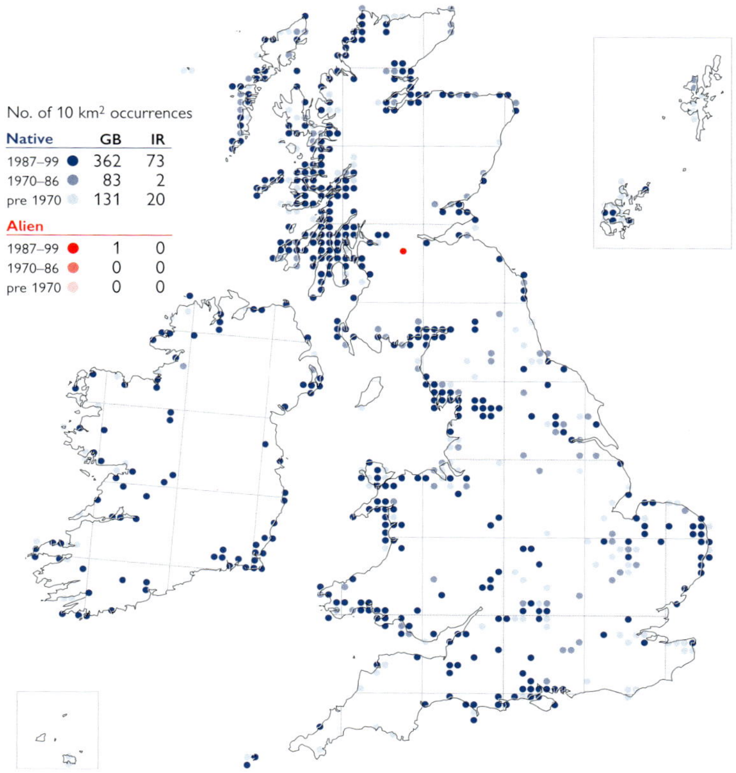

A rhizomatous perennial herb, predominantly of coastal habitats, growing in damp dune-slacks, saltmarshes, short, brackish grassland and pools in the spray zone. It also occurs inland in base-rich, wet meadows and calcareous marshes, and locally, as in Oxfordshire, by springs with higher than normal sodium content. Generally lowland, but reaching 325 m at Ponterwyd (Cards.).

Native (change +0.60). The coastal distribution of *E. uniglumis* has probably remained stable in recent years, although it is much better recorded now than in the 1962 *Atlas*. There are many more records from inland areas, where it may previously have been overlooked.

Circumpolar Temperate element.

References: Atlas (351b), Hultén & Fries (1986), Walters (1949).

M. S. PORTER & M. J. Y. FOLEY

Eleocharis multicaulis Many-stalked Spike-rush

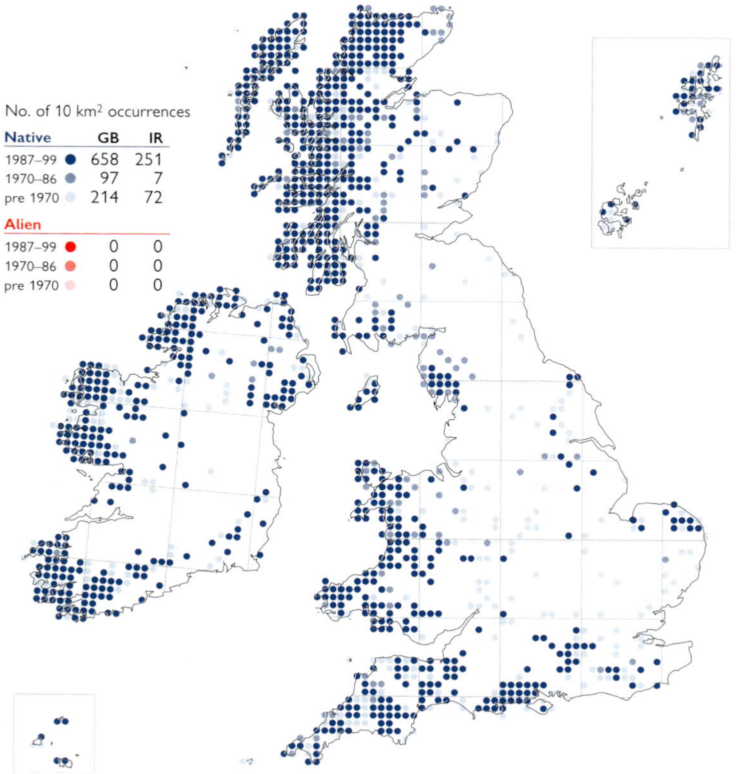

A densely tufted perennial herb mainly of acid bogs, wet heath, valley mires, pools and wet hollows over peat, and at the edge of acidic lakes; also occurring in coastal dune-slacks. 0–610 m (Macgillycuddy's Reeks, S. Kerry).

Native (change +0.47). Many sites for *E. multicaulis* were lost in the 19th century, due to drainage and changes in land use. Since 1962, it has been lost from many more sites in E. England and E. Scotland. Conversely, this rather inconspicuous species is now much better recorded in the uplands, and it is known from over six times as many 10-km squares in Wales as in the 1962 *Atlas*.

Suboceanic Temperate element.

References: Atlas (350d), Hultén & Fries (1986), Meusel *et al.* (1965).

M. S. PORTER & M. J. Y. FOLEY

Eleocharis quinqueflora Few-flowered Spike-rush

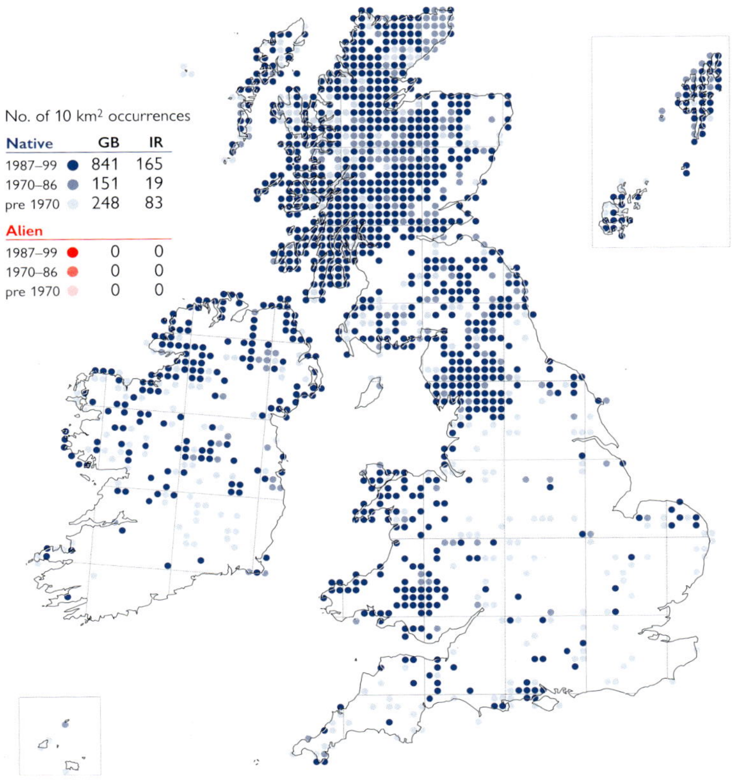

A perennial herb of base-rich marshes and fens, calcareous flushes on peaty soils, stony and muddy areas with seeping water, wet paths and short turf on banks; also in coastal cliff-flushes, dune-slacks and in the upper parts of saltmarshes. It requires open sites and is often dependent on grazing, cutting or disturbance. 0–915 m (Atholl, E. Perth).

Native (change +0.02). Many lowland sites for this species were lost in the 19th century. Since the 1962 *Atlas* it has been lost from half the extant 10-km squares in lowland England, because of drainage and lack of grazing. It is easily overlooked and is now better recorded elsewhere.

European Boreo-temperate element; also in C. and E. Asia and N. America.

References: Atlas (350c), Hultén & Fries (1986), Meusel *et al.* (1965).

M. S. PORTER & M. J. Y. FOLEY

Eleocharis acicularis Needle Spike-rush

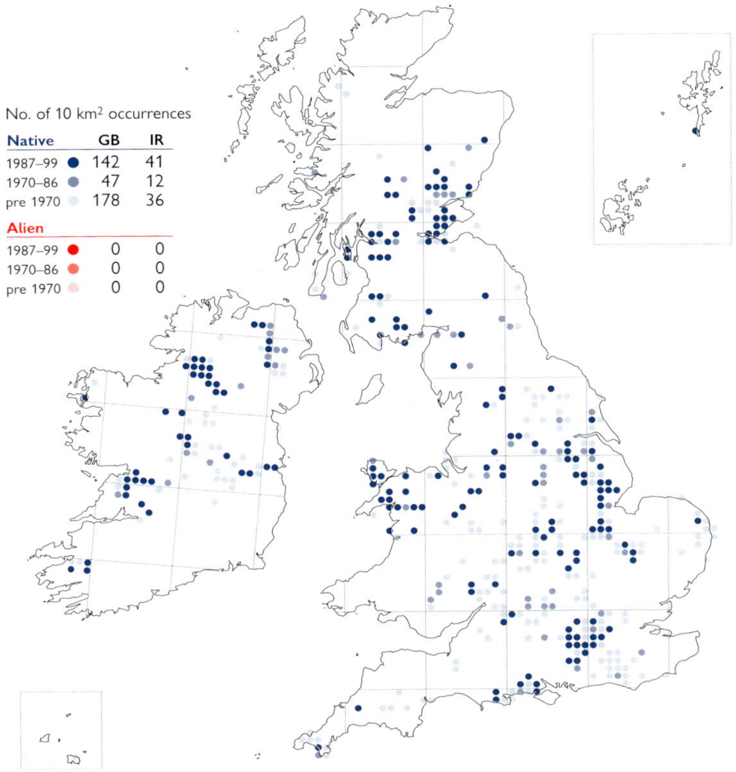

A rhizomatous perennial herb growing on the margins of lakes, ponds, reservoirs and rivers, in sites subject to winter flooding, and fully aquatic in shallow, still or slow-moving mesotrophic to eutrophic water. It roots into sand, gravel, mud or silt, often forming extensive lawns, but flowering only when exposed by falling water levels. 0–390 m (Drumore Loch, E. Perth).

Native (change –0.11). Submerged populations of *E. acicularis* are often overlooked. Furthermore, this species can reappear in former sites when conditions become suitable, and can colonise newly-flooded sand- and gravel-pits.

Circumpolar Boreo-temperate element.

References: Atlas (350b), Hultén & Fries (1986), Preston & Croft (1997), Stewart *et al.* (1994).

M. S. PORTER & M. J. Y. FOLEY

Eleocharis parvula Dwarf Spike-rush

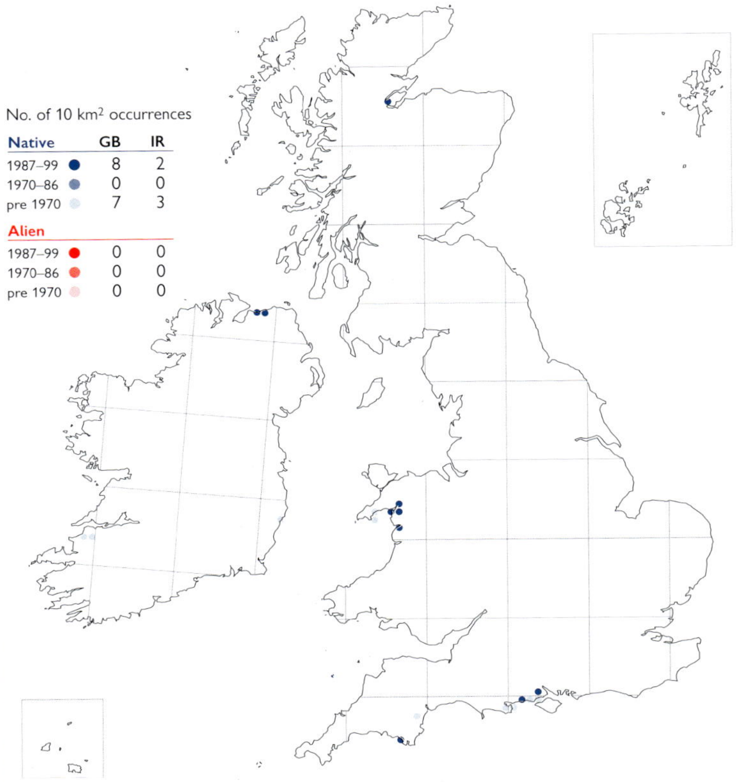

A diminutive rhizomatous perennial growing on firm estuarine mud by tidal rivers, and in tidal pans in brackish grazing marshes. It occurs close to the upper limit of tidal influence, avoiding strongly saline areas. It reproduces vegetatively by turions, and by seed, but flowering and fruiting is very poor in many localities. Lowland.

Native (change 0.00). Colonies of *E. parvula* have been lost due to dredging and the cessation of grazing, which allows the development of taller vegetation. It is easily overlooked, and its discovery in E. Ross in 1999 suggests that it may be found in other northern sites.

European Temperate element; also in E. Asia and N. America.

References: Atlas (350a), Curtis & McGough (1988), Dines & Preston (2000), Kay & John (1995), Hultén & Fries (1986), Wigginton (1999).

M. S. PORTER & M. J. Y. FOLEY

Bolboschoenus maritimus Sea Club-rush

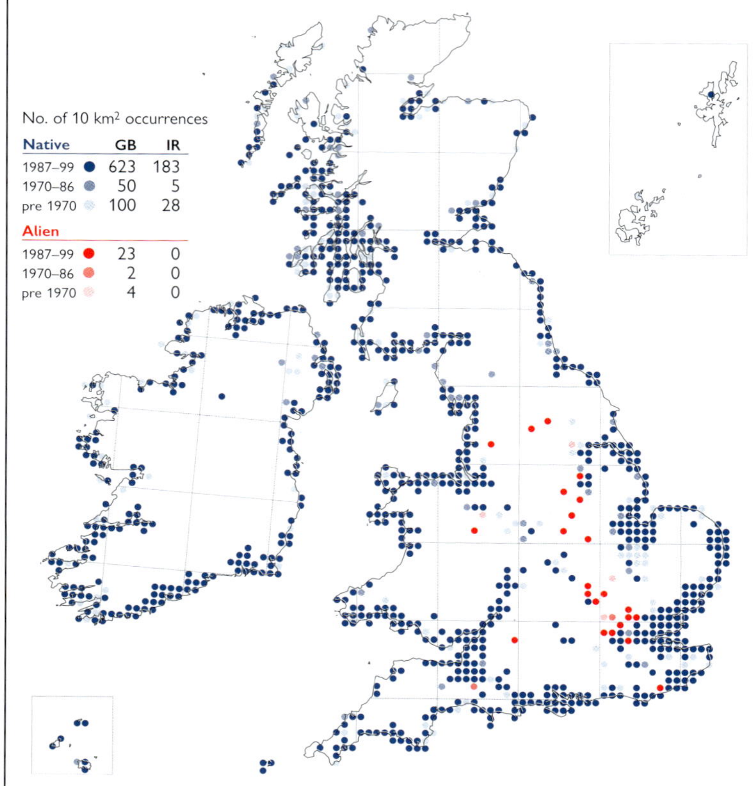

A rhizomatous perennial mainly of saline ground or in shallow brackish water, usually rooted in mud but sometimes in gravel and shingle. Coastal habitats include saltmarshes, tidal river banks, creeks, ditches, lakes, ponds, borrow-pits, and also marshes and damp pastures. It is occasionally found in freshwater habitats inland, including flooded riversides and clay- and gravel-pits. *B. maritimus* reproduces by rhizomatous spread, tubers and seed. Lowland.

Native (change 0.00). The overall distribution of *B. maritimus* is stable. It is sometimes deliberately planted in lakes and ponds.

Eurosiberian Southern-temperate element; widely naturalised outside its native range.

References: Atlas (347d), Hultén & Fries (1986), Preston & Croft (1997).

M. J. Y. FOLEY & M. S. PORTER

Scirpus sylvaticus Wood Club-rush

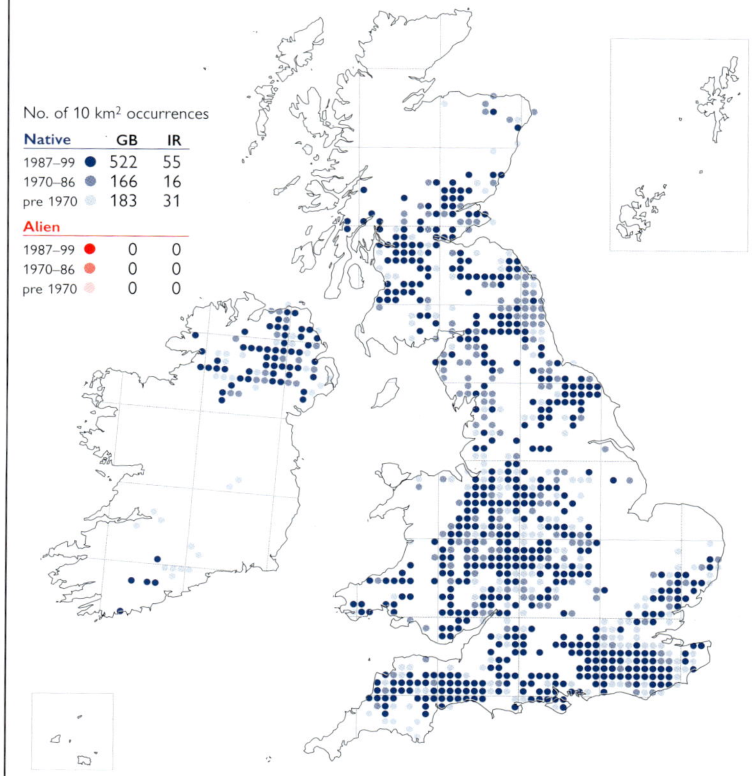

A robust rhizomatous perennial herb which may form extensive stands in swampy valley woodlands and similar shady places; also in wet pastures bordering woods and streams, and on the margins of rivers, streams, lakes and ponds. It typically grows over thick, rather eutrophic silts which are often iron-enriched (Rodwell, 1991a). Lowland, reaching 300 m at Clearburn Loch (Selkirks.).

Native (change +0.02). There have been local losses of *S. sylvaticus* in S.E. England. It is, however, much better recorded since the 1962 *Atlas*, especially in N. England, Scotland and N. Ireland.

Eurosiberian Temperate element.

References: Atlas (348a), Hultén & Fries (1986), Meusel *et al.* (1965).

M. J. Y. FOLEY & M. S. PORTER

Scirpoides holoschoenus Round-headed Club-rush

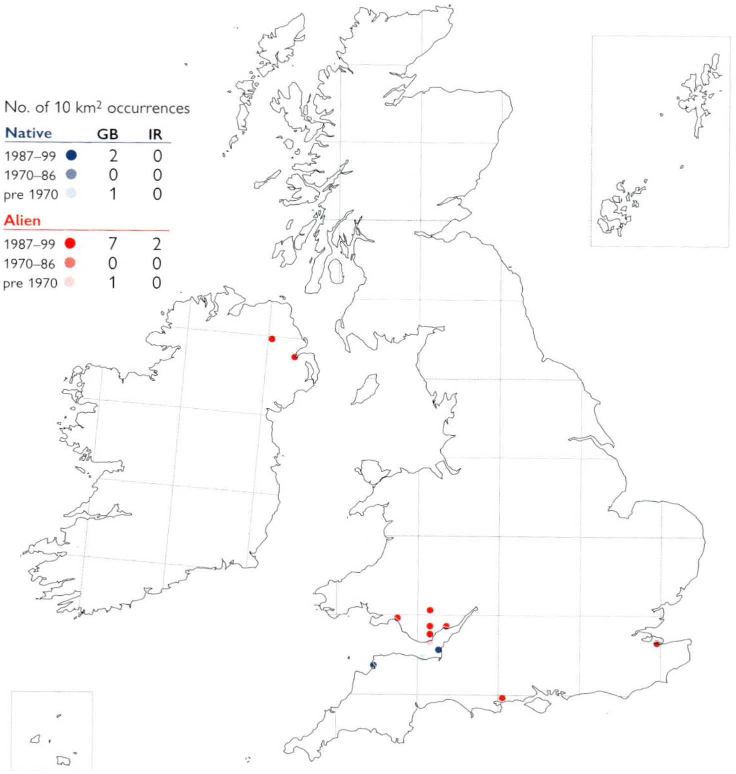

In Devon, this rhizomatous perennial herb occurs in damp dune-slacks and on adjacent low dunes, and in Somerset in a damp sandy hollow on a coastal golf course. Elsewhere, it occurs as an alien, especially in industrial areas. Substantial ripening of fruit and seed set appear to occur only after a long, hot summer. Lowland.

Native (change +0.21). The distribution of this species is stable, though native populations are at risk from scrub encroachment, which requires careful management, and hydrological changes.

Eurosiberian Southern-temperate element.

References: Atlas (348b), Wigginton (1999).

M. J. Y. FOLEY & M. S. PORTER

Schoenoplectus lacustris Common Club-rush

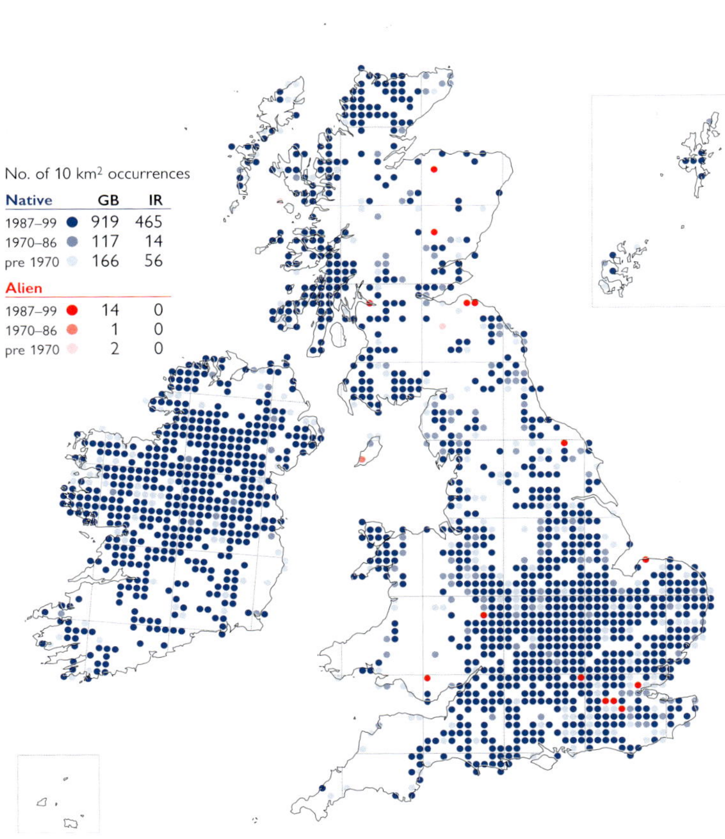

A tall rhizomatous perennial herb of standing or flowing fresh water, in conditions ranging from eutrophic and base-rich to oligotrophic and base-poor. Substrates include silt, clay, peat or gravel. It occurs in ponds, lakes, canals, dykes and slow moving rivers, usually in water 0.3–1.5 m deep, but can also be found in deeper water. Generally lowland, but reaching 405 m at Dock Tarn (Cumberland).

Native (change +0.47). The distribution of *S. lacustris* is very similar to that shown in the 1962 *Atlas*, where it was mapped as 'all records', but there have been local losses in S.E. England.

Eurosiberian Wide-temperate element.

References: Atlas (348d), Hultén & Fries (1986), Preston & Croft (1997).

M. J. Y. FOLEY & M. S. PORTER

Schoenoplectus tabernaemontani Grey Club-rush

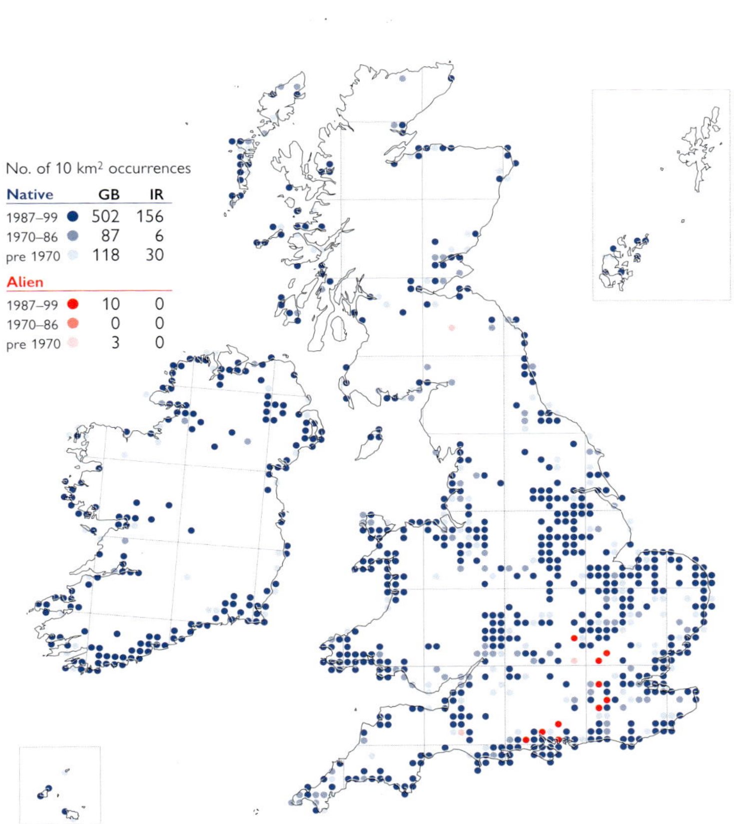

A rhizomatous perennial herb, most frequent in coastal sites where it grows in brackish water in rivers, dykes, tidal channels, lagoons and dune-slacks; also in depressions in salt-marsh and in wet pasture. Inland, it occurs by lakes, ponds, slow-flowing rivers, streams and canals, and in flooded quarries and pits. Lowland.

Native (change +0.67). The map indicates a marked increase in inland sites for this species since the 1962 *Atlas*. This might, in part, be due to better recording now that its ability to grow inland is appreciated. It has also colonised newly created gravel-pits, and may have been planted at some of its inland sites.

Eurasian Southern-temperate element.

References: Atlas (349a), Hultén & Fries (1986), Meusel *et al.* (1965), Preston & Croft (1997).

M. J. Y. FOLEY & M. S. PORTER

Schoenoplectus triqueter Triangular Club-rush

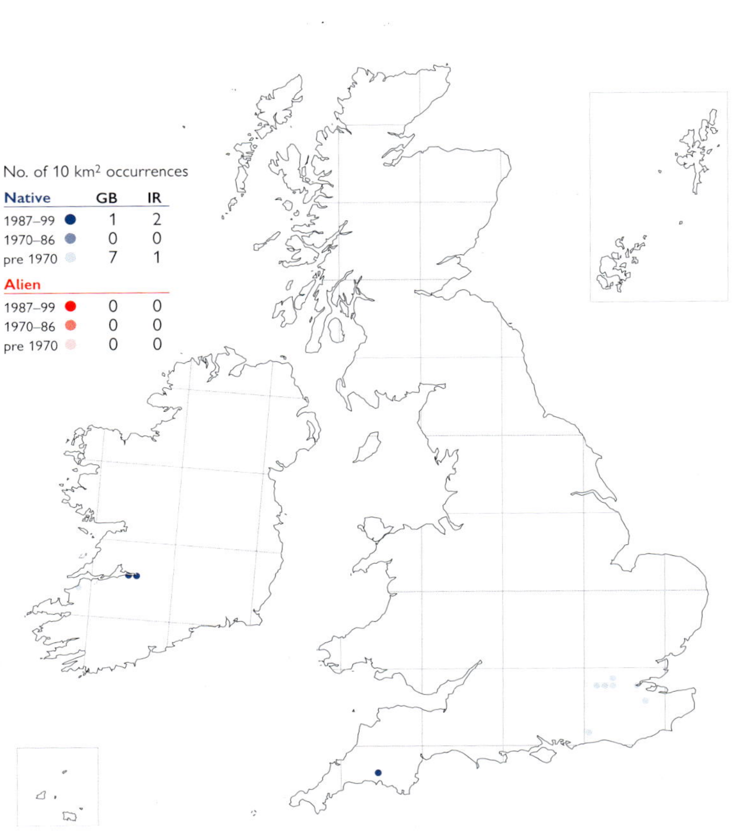

This tussock-forming, rhizomatous perennial herb occurs on mud-banks along the lower reaches of tidal rivers, where it may become submerged at the highest tides. Lowland.

Native. Populations of *S. triqueter* have been lost to land reclamation and as a result of bank construction. The surviving population on the River Tamar (S. Devon) is now extremely small, although the hybrid with *S. tabernaemontani* is more frequent there. *S. triqueter* is still present in Ireland by the River Shannon (Co. Limerick) and its tributaries, where populations are larger than in Britain but some are threatened by development.

Eurasian Temperate element.

References: Atlas (348c), Curtis & McGough (1988), Preston & Croft (1997), Wigginton (1999).

M. J. Y. FOLEY & M. S. PORTER

Schoenoplectus pungens Sharp Club-rush

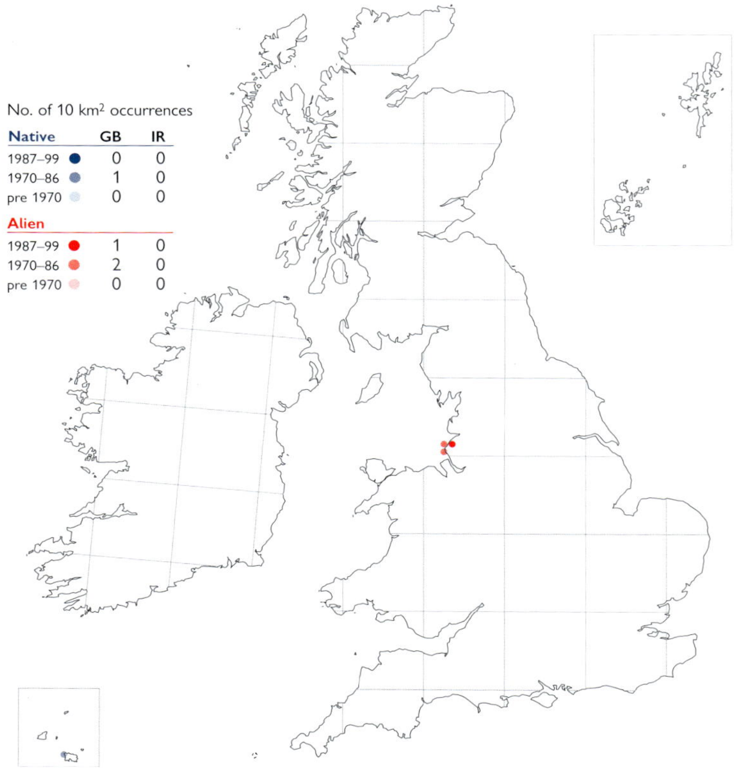

This rhizomatous perennial was formerly known from the margin of a coastal lake in Jersey, and from a wet, coastal dune-slack near Ainsdale (S. Lancs.). Lowland.

Native or alien. *S. pungens* was first recorded in our area in 1724. It declined in Jersey, where it is possibly native, during the 1940s and 1950s; it was last seen there in the early 1970s, apparently having been replaced by *Carex riparia*. It was first collected at Ainsdale in 1909, and in 1928 a large patch occurred in a dune-slack. Its origin there is obscure and it became extinct by 1978. Ainsdale stock survived in cultivation, however, and has been planted at several sites nearby.

European Temperate element; widespread in N. America.

References: Atlas (348b), Meusel *et al.* (1965), Preston & Croft (1997).

M. J. Y. FOLEY & M. S. PORTER

Isolepis setacea Bristle Club-rush

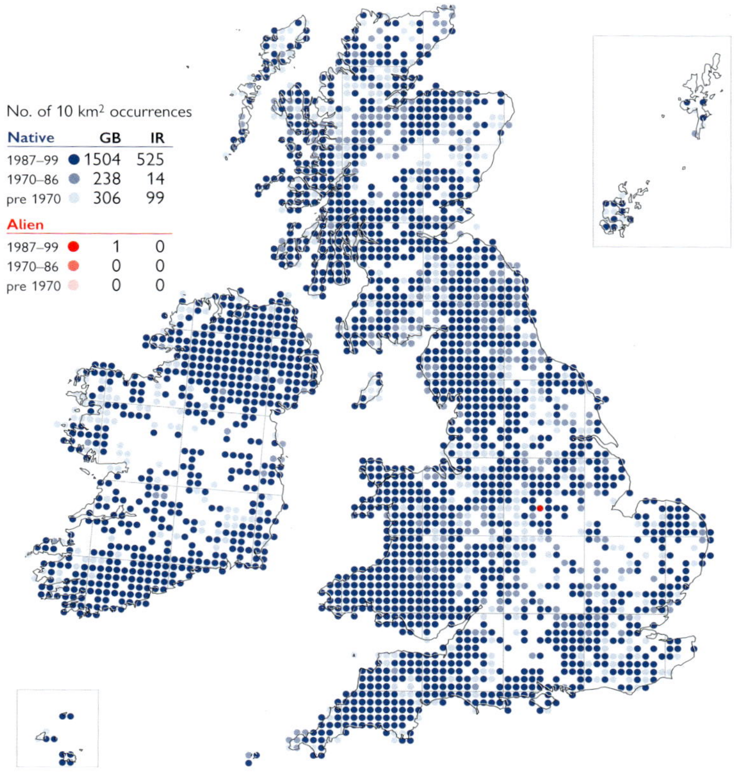

A perennial herb of open, damp, generally acidic sites, especially those subject to winter flooding. It occurs on sandy or gravelly tracks, on the shores of lakes or ponds, in short grassland, on eroding streamsides and occasionally on the coast in sand dunes or in turf in the upper zones of saltmarshes. 0–590 m (High Cup Nick, Westmorland).

Native (change +0.53). *I. setacea* was mapped as 'all records' in the 1962 *Atlas*. Analysis of the database reveals that most of the lowland losses have occurred since 1950, probably as a result of drainage. It is, however, a pioneer species of disturbed ground and can be easily overlooked.

Eurosiberian Temperate element; widely naturalised outside its native range.

References: Atlas (349b), Hultén & Fries (1986), Meusel *et al.* (1965).

M. S. PORTER & M. J. Y. FOLEY

Isolepis cernua Slender Club-rush

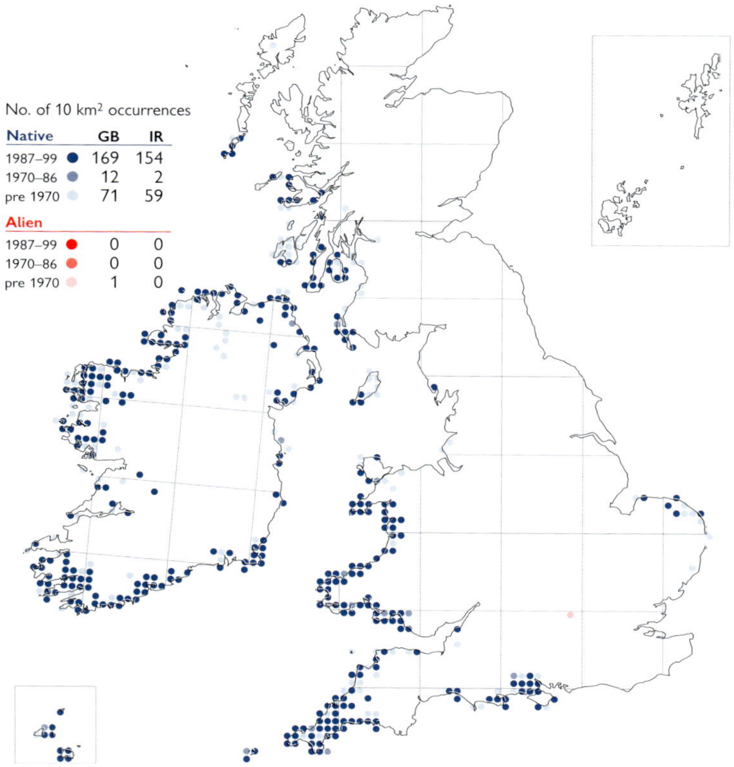

A perennial found in wet, coastal grassland, in bare or open sites over damp sand, peat and mud, in short turf and sometimes in flushes and trickles on rocky cliffs. In the New Forest (S. Hants.), where it is locally common, it occurs inland in flushed acidic or base-rich turf and in old marl-pits. Lowland.

Native (change +0.23). Some sites for *I. cernua* were lost before 1930, but the distribution appears to be little changed since the 1962 *Atlas*, though the recording of this inconspicuous species has improved.

Mediterranean-Atlantic element; also in N. America.

Reference: Atlas (349c).

M. S. PORTER & M. J. Y. FOLEY

Eleogiton fluitans Floating Club-rush

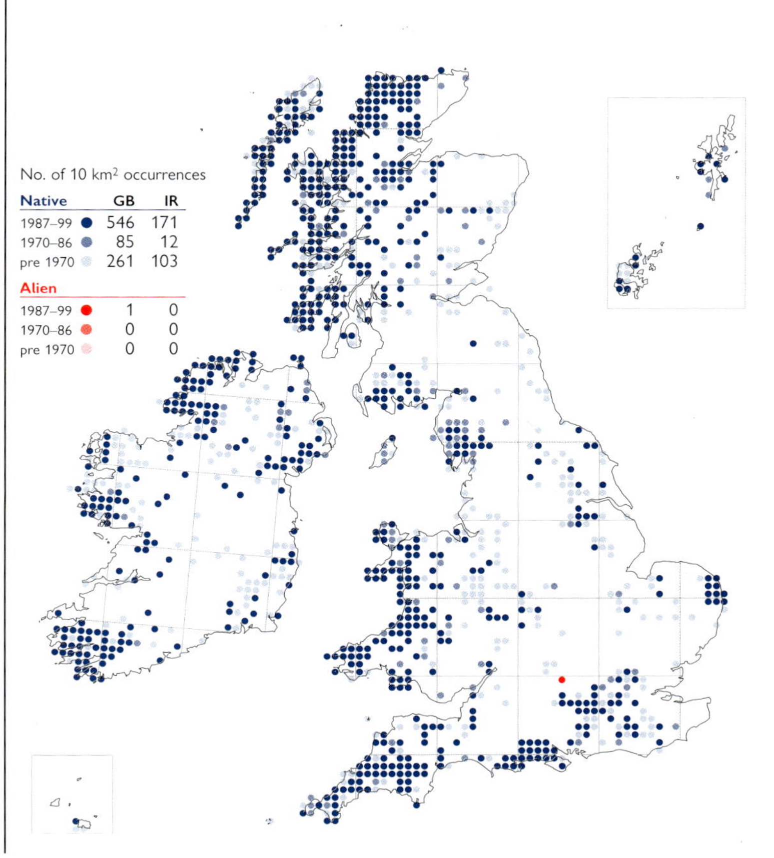
No. of 10 km² occurrences
Native GB IR
1987–99 546 171
1970–86 85 12
pre 1970 261 103
Alien
1987–99 1 0
1970–86 0 0
pre 1970 0 0

A perennial herb found mainly on peaty, acidic substrates in, or on the margins of, slow-flowing streams, ditches, pools and the sheltered shores of some larger lakes and tarns, often in seasonally flooded sites. It also occurs in muddy hollows in grasslands and heaths, and the wet floors of disused quarries, sand- and gravel-pits. 0–435 m (Llyn Gwngu, Cards., and Styhead Tarn, Cumberland).

Native (change +0.37). Many of the losses of *E. fluitans* occurred before 1930, following drainage or the loss of lowland heathland, and this decline has continued in England since the 1962 *Atlas*. It is better recorded now elsewhere.

Oceanic Southern-temperate element, with a disjunct distribution.

References: Atlas (349d), Hultén & Fries (1986), Meusel *et al.* (1965), Preston & Croft (1997).

M. J. Y. FOLEY & M. S. PORTER

Blysmus compressus Flat-sedge

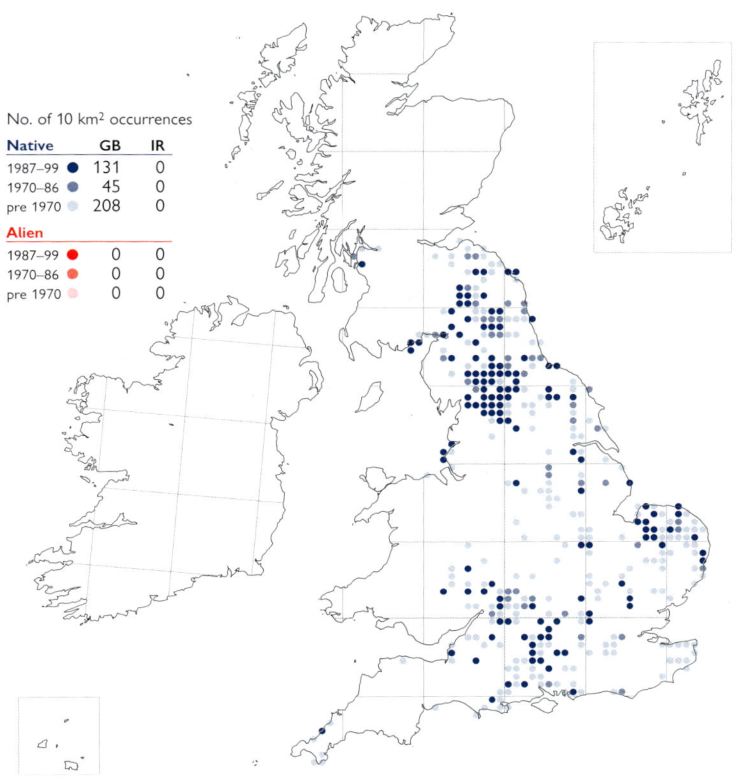
No. of 10 km² occurrences
Native GB IR
1987–99 131 0
1970–86 45 0
pre 1970 208 0
Alien
1987–99 0 0
1970–86 0 0
pre 1970 0 0

A rhizomatous perennial of open areas in marshes and fens, and in short, sedge-rich, damp grassland, calcareous flushes and stream borders which are subject to flooding. 0–490 m (Co. Durham).

Native (change –1.28). *B. compressus* has suffered a severe decline throughout its range. In the 1962 *Atlas*, more than half the squares from which it was recorded had been lost by 1930. This decline has continued, with 40% of its post-1930 squares now gone. Much of the decline has been due to drainage, the loss of unimproved damp grasslands, falling water-tables, eutrophication and the cessation of grazing.

European Temperate element, with a continental distribution in W. Europe; also in C. Asia.

References: Atlas (351c), Hultén & Fries (1986), Meusel *et al.* (1965).

M. J. Y. FOLEY & M. S. PORTER

Blysmus rufus Saltmarsh Flat-sedge

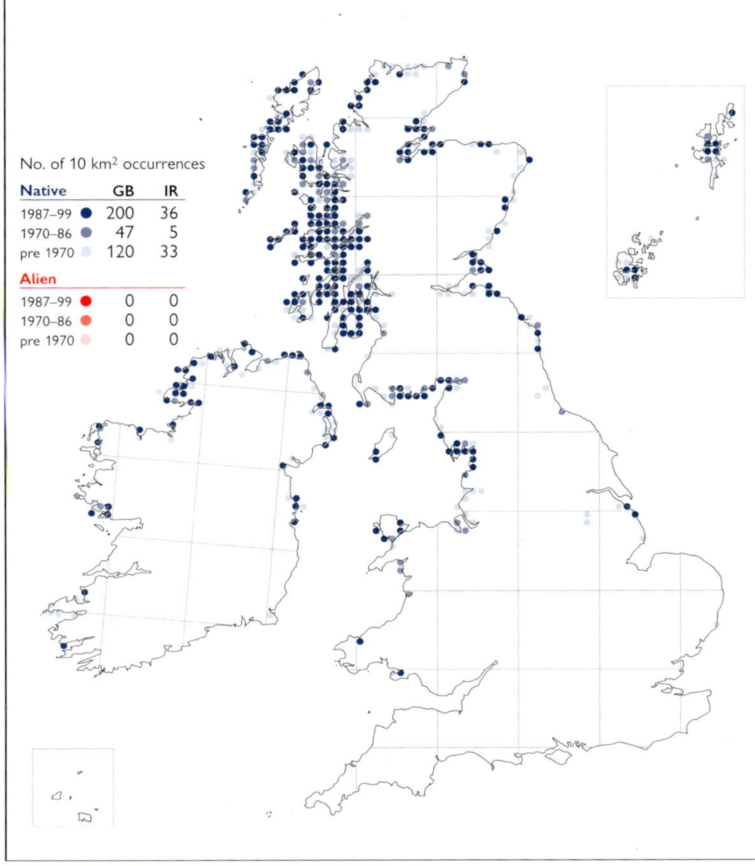
No. of 10 km² occurrences
Native GB IR
1987–99 200 36
1970–86 47 5
pre 1970 120 33
Alien
1987–99 0 0
1970–86 0 0
pre 1970 0 0

A rhizomatous perennial herb, found in sandy or gravelly wet runnels and depressions in saltmarshes, and in brackish ditches and dune-slacks. It also occurs on rocky shores, in freshwater seepages and beside streams where they debouch onto the beach. Lowland.

Native (change –0.53). Most of the losses of *B. rufus* around the Irish Sea and along the E. coast of Scotland took place before 1930, and the distribution now appears to be generally stable. It may still be present in some of the more remote 10-km squares in which it has not been seen since 1970.

European Boreal-montane element; also in C. Asia and N. America.

References: Atlas (351d), Hultén & Fries (1986), Meusel *et al.* (1965).

M. J. Y. FOLEY & M. S. PORTER

Cyperus longus Galingale

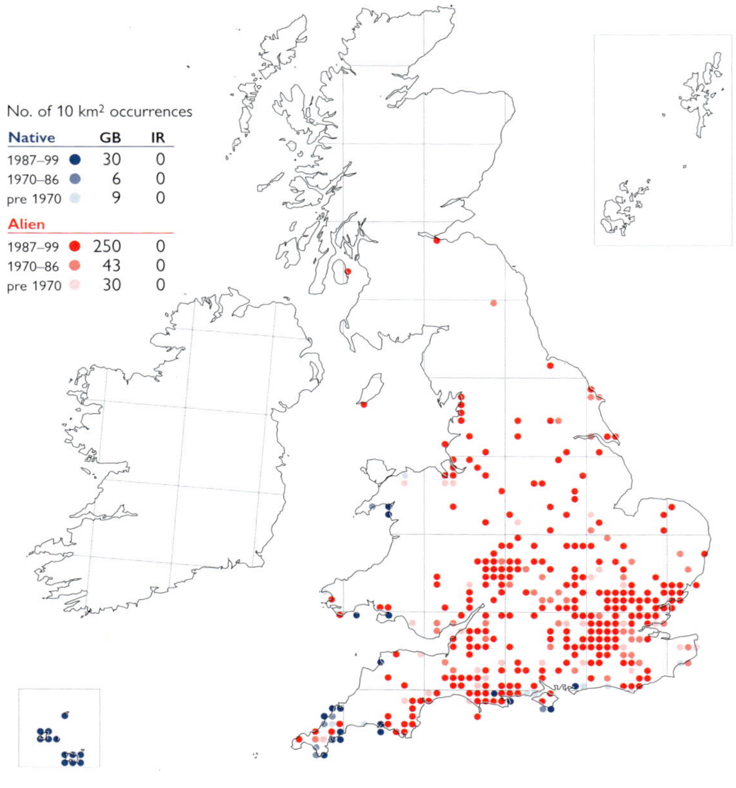

A rhizomatous perennial herb of marshes and wet pastures near the coast, and sometimes in base-rich flushes on sea-cliffs. It also occurs on pond margins and in ditches inland, where it is usually planted. Reproduction is through vigorous rhizomatous spread, and it may not set seed in Britain. Lowland.

Native (change +2.22). Since the 1962 *Atlas* this species has apparently been lost from some native sites because of agricultural change, including the cessation of grazing. However, new native sites have been discovered since 1960, notably in N. Wales, and the number of introduced sites has increased very considerably.

European Southern-temperate element.

References: Atlas (352a), Stewart *et al.* (1994).

M. J. Y. FOLEY & M. S. PORTER

Cyperus eragrostis Pale Galingale

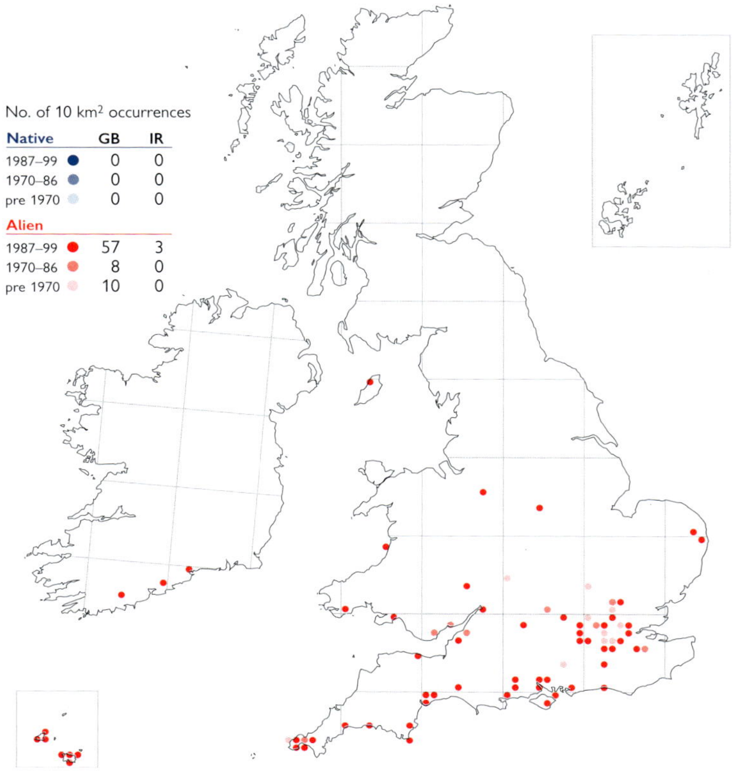

A perennial herb occurring on roadsides, river banks and pond margins, in ditches and on ballast and rough ground, where it occurs as a grass-seed or wool-shoddy alien and sometimes as a garden escape. It has become well-naturalised in some places, especially in the Channel Islands, but is more often only casual. Lowland.

Neophyte. *C. eragrostis* was possibly introduced to cultivation in Britain in 1790. It was first recorded in the wild in 1909 (Guernsey). Although some populations have been lost in the London area, the overall distribution is likely to be stable or increasing.

Native of tropical America.

M. S. PORTER & M. J. Y. FOLEY

Cyperus fuscus Brown Galingale

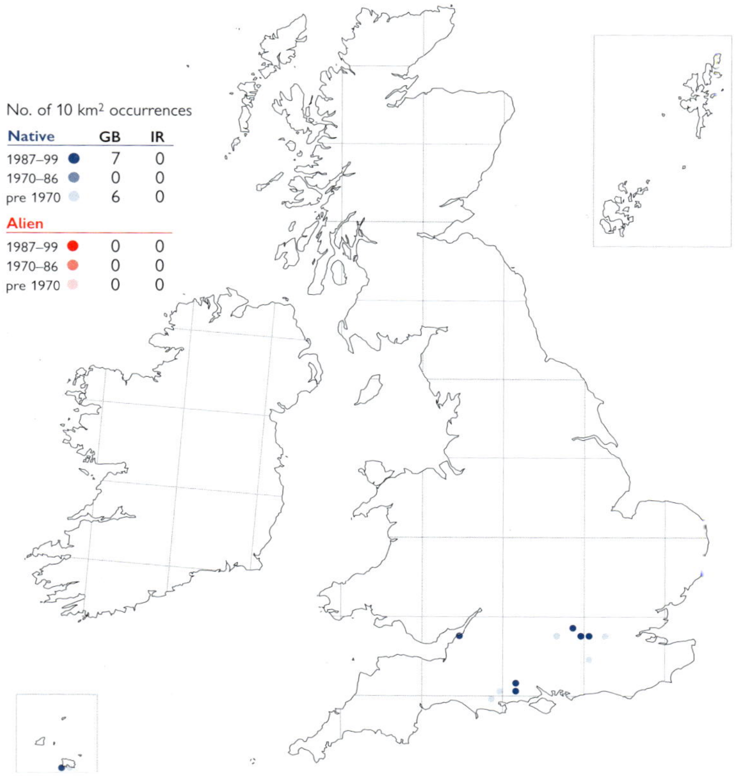

An annual of moist, open disturbed ground around the margins of ponds and by ditches, often on ground subject to winter-flooding. The substrate may be peaty, muddy or stony but humus-rich. Seed may not be set in cool summers. Lowland.

Native (change −0.32). The conditions favoured by *C. fuscus* were traditionally maintained by grazing animals, but cessation of grazing, encroachment by scrub and lowering of the water-tables have all contributed to an appreciable decline which began before 1930. However, current sites now benefit from statutory protection, and since seed appears to be long-lived, populations may be revived with suitable conservation management.

Eurosiberian Southern-temperate element.

References: Atlas (352b), Hultén & Fries (1986), Rich *et al.* (1999a), Wigginton (1999).

M. J. Y. FOLEY & M. S. PORTER

Schoenus nigricans Black Bog-rush

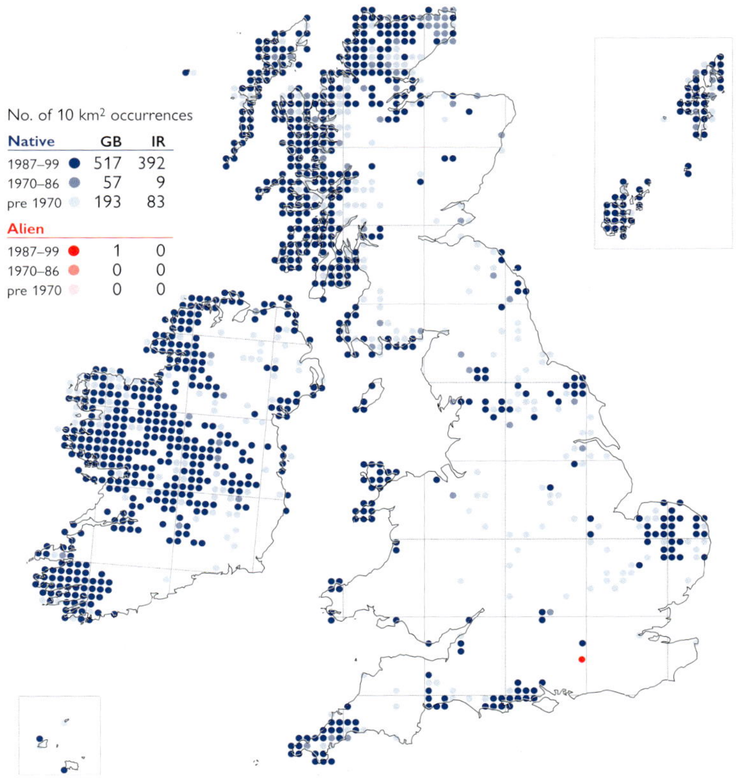

A tussock-forming perennial of calcareous and other base-rich fens (especially near springs), and of peaty flushes, marshes, bogs, dune-slacks, sea-cliff flushes, and the upper fringes of saltmarshes where there is base-rich flushing. In W. Ireland, it is frequent on acid blanket-bog. Generally lowland, but reaching 550 m in the Mourne Mountains (Co. Down).

Native (change –0.53). *S. nigricans* had widely declined in lowland sites before 1930, and this loss has continued outside its core areas. Its distribution is stable near the western coasts of Britain and in Ireland.

Eurosiberian Southern-temperate element; also in N. America.

References: Atlas (352c), Hultén & Fries (1986), Meusel *et al.* (1965), Sparling (1968).

M. J. Y. FOLEY & M. S. PORTER

Schoenus ferrugineus Brown Bog-rush

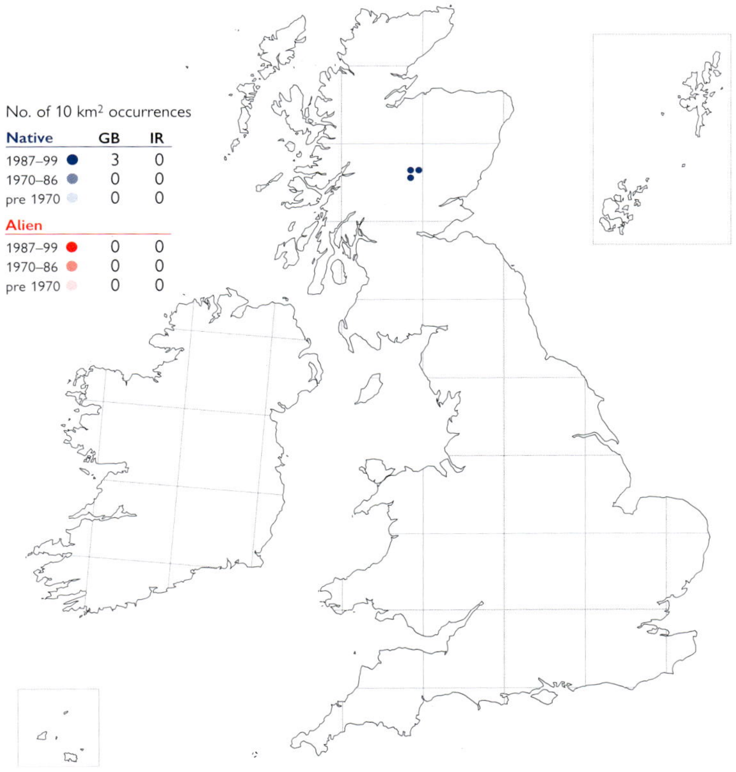

A tussock-forming perennial occurring in base-rich flushes within calcareous grassland, usually adjacent to unimproved heathland. Seed production is low, and dispersal is restricted as the seed is retained in the inflorescence for up to a year. From 200 m above Loch Tummel (Mid Perth) to 390 m as an alien on Ben Vrackie (E. Perth).

Native. *S. ferrugineus* was thought to have become extinct in Scotland when the water-levels were raised at its only known site, Loch Tummel, in 1950. At least one transplant from here survived, however, and since 1970 ten new sites have been discovered. All have some form of statutory protection.

European Boreal-montane element.

References: Atlas (352c), Hultén & Fries (1986), Meusel *et al.* (1965), Wigginton (1999).

M. J. Y. FOLEY & M. S. PORTER

Rhynchospora alba White Beak-sedge

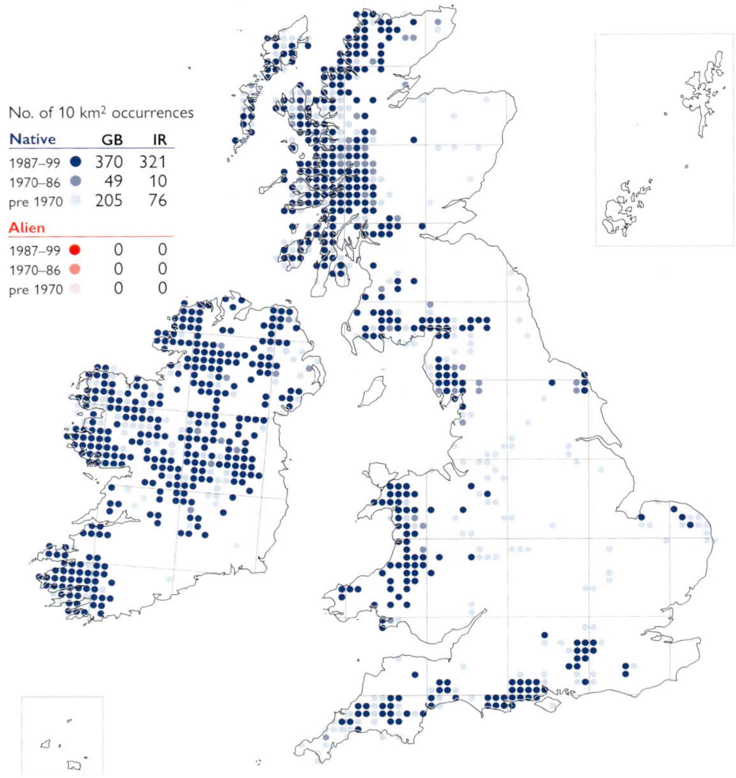

A perennial herb of base-poor acidic bogs, wet heaths and mires, often in association with *Sphagnum* species. It is intolerant of competition, preferring open sites, and is frequently found on bare wet peat, sometimes in shallow standing water. Generally lowland, but reaching 850 m (Clogwyn y Garnedd, Caerns.).

Native (change –0.43). Many former lowland sites of *R. alba* were lost in the 19th century as a result of drainage, peat extraction and afforestation. Since the 1962 *Atlas* more sites have been lost in S. England, but its distribution elsewhere has been largely stable.

Circumpolar Boreo-temperate element, with a disjunct distribution.

References: Atlas (352d), Hultén & Fries (1986), Meusel *et al.* (1965).

M. S. PORTER & M. J. Y. FOLEY

Rhynchospora fusca Brown Beak-sedge

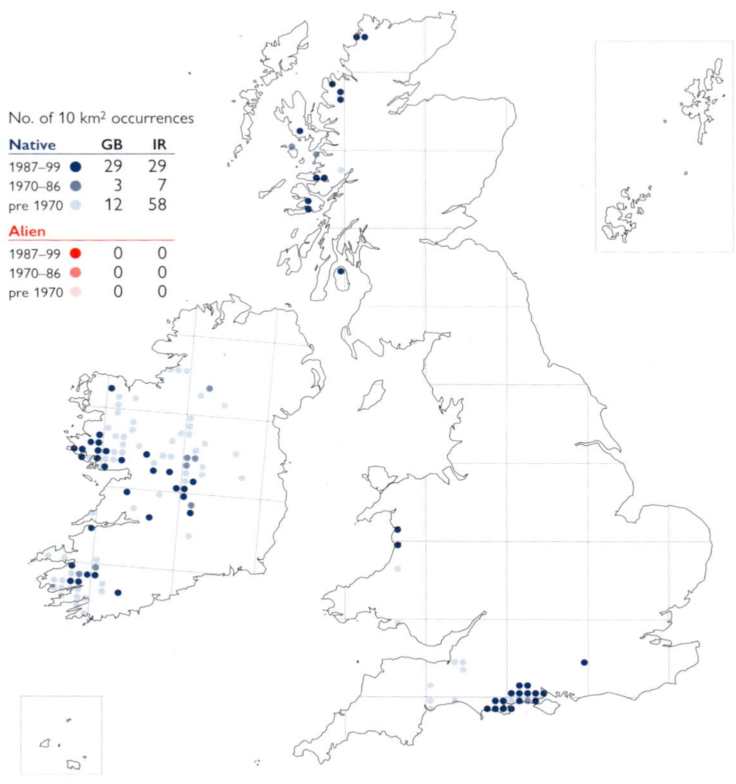

A rhizomatous perennial of wet heaths, and the margins of acidic mires, favouring bare peat where competition is limited but preferring somewhat drier sites than *R. alba* in the south of its range. It spreads vegetatively, and may reproduce by seed. Lowland.

Native (change +0.02). Although most English sites for *R. fusca* outside its core area had been lost before 1930, more have now gone, sometimes as a result of drainage but more frequently because of invading carr following the reduction or cessation of grazing. However, it is much better recorded in Scotland since the 1962 *Atlas*. In Ireland, there has been a widespread decline.

Suboceanic Boreo-temperate element; also in N. America.

References: Atlas (353a), Hultén & Fries (1986), Meusel *et al.* (1965), Stewart *et al.* (1994).

M. S. PORTER & M. J. Y. FOLEY

Cladium mariscus Great Fen-sedge

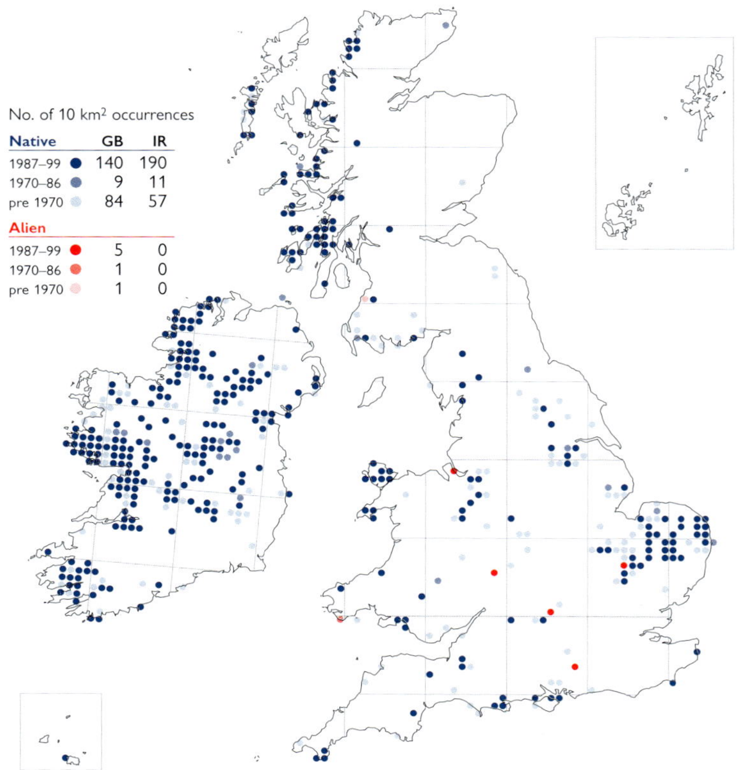

A rhizomatous perennial of oligotrophic to mesotrophic habitats, usually growing on peat. It is found in swamps at the margins of lakes and ponds and along streams, and in tall-herb fens and open fen carr. In England and Wales it is largely restricted to calcareous sites, whereas in the Hebrides and W. Ireland it also occurs in acidic areas. Lowland.

Native (change +0.11). *C. mariscus* has declined because of drainage, and some lowland sites are currently threatened by eutrophication and scrub invasion. However, it was under-recorded in Scotland in the 1962 *Atlas*, and new sites may await discovery there.

Eurosiberian Southern-temperate element; also in N. America.

References: Atlas (353b), Conway (1942), Hultén & Fries (1986), Meusel *et al.* (1965), Preston & Croft (1997).

M. S. PORTER & M. J. Y. FOLEY

Kobresia simpliciuscula False Sedge

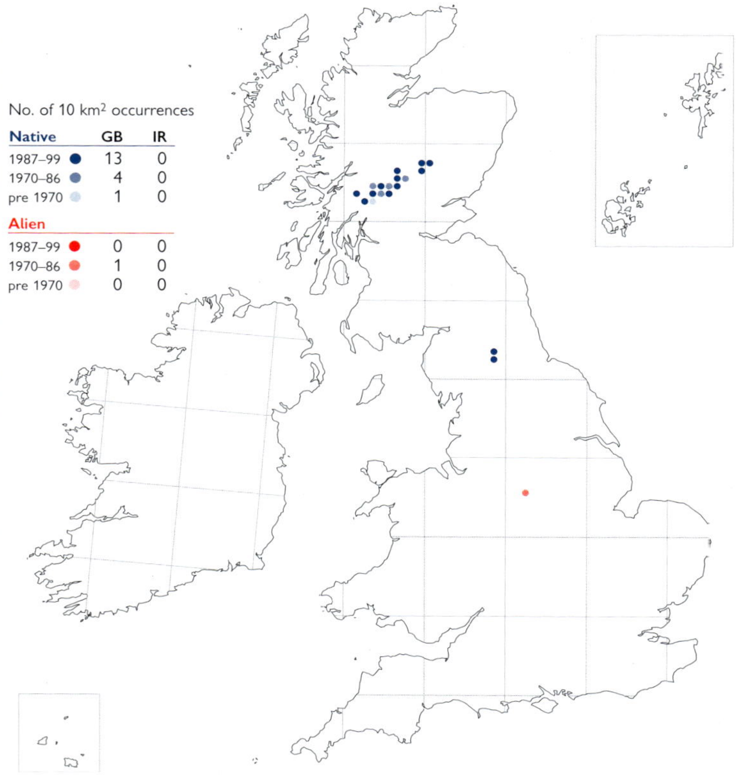

A densely tufted perennial herb of open situations, occurring in stony flushes, base-rich mires and wet, grassy or sedge-rich turf on limestone or calcareous mica-schist. Generally montane, occurring as low as 360 m in Teesdale (N.W. Yorks.) but reaching 1065 m on Meall Garbh (Mid Perth).

Native (change +0.58). The distribution of *K. simpliciuscula* is stable. Even when common, however, it is easy to overlook and it probably still occurs in most sites for which there are only pre-1987 records. Some populations are very extensive with tens of thousands of plants.

Circumpolar Arctic-montane element.

References: Atlas (353c), Hultén & Fries (1986), Meusel *et al.* (1965), Wigginton (1999).

M. S. PORTER & M. J. Y. FOLEY

Carex paniculata Greater Tussock-sedge

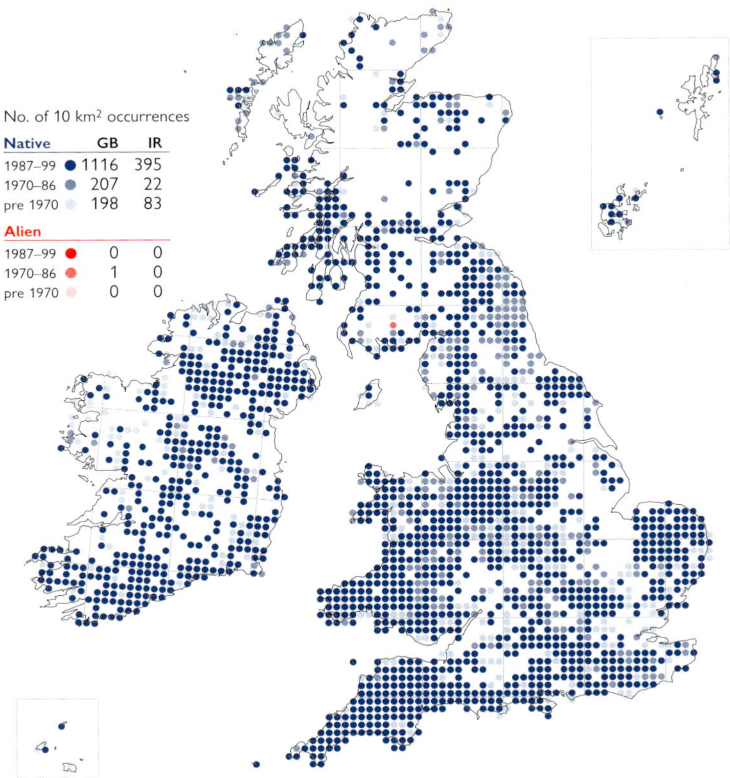

No. of 10 km² occurrences

Native	GB	IR
1987–99	1116	395
1970–86	207	22
pre 1970	198	83
Alien		
1987–99	0	0
1970–86	1	0
pre 1970	0	0

This tussock-forming perennial herb occurs in a wide range of habitats, usually somewhat base-enriched, including swamps and fens, the edges of lakes, ponds, canals and ditches, open fen-carr and swampy woodland. It usually grows in the open, where it fruits freely, but tolerates moderate shade, although it can become smaller and less vigorous, flowering only sparsely. Generally lowland, but to over 600 m on Gylchedd (Denbs.).

Native (change –0.11). The range of *C. paniculata* is stable. However, its frequency has decreased and it has been lost from many sites and is threatened in others, such as in Broadland where 'tussock-fens' have ceased to develop.

European Temperate element.

References: Atlas (365b), Hultén & Fries (1986), Jermy *et al.* (1982), Meusel *et al.* (1965), Preston & Croft (1997).

M. S. PORTER & M. J. Y. FOLEY

Carex paniculata × *C. remota* (*C.* × *boenninghausiana*)

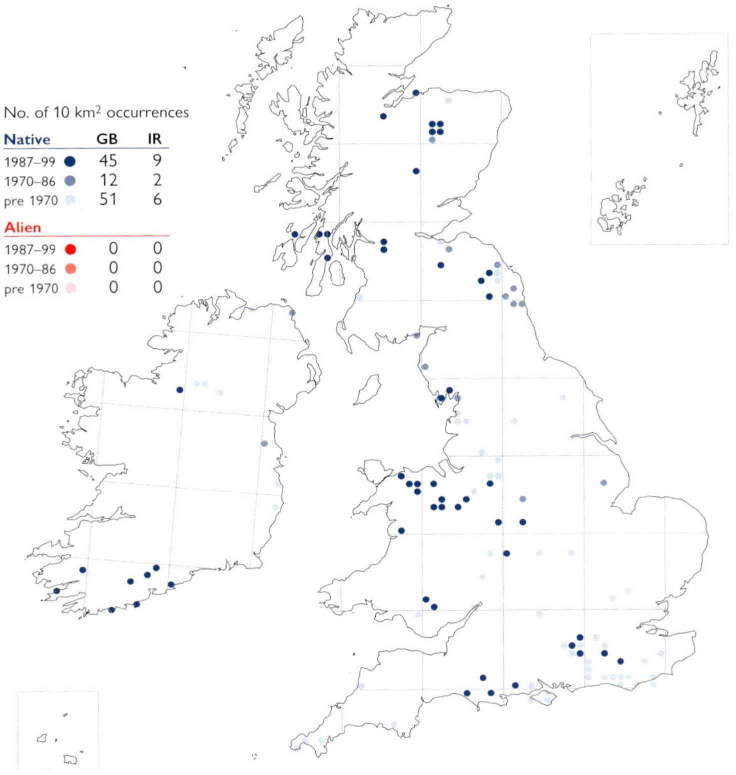

No. of 10 km² occurrences

Native	GB	IR
1987–99	45	9
1970–86	12	2
pre 1970	51	6
Alien		
1987–99	0	0
1970–86	0	0
pre 1970	0	0

A herb of wet fen-woodland and carr, especially where there is seasonal inundation, often growing in close association with *C. paniculata*. It is sterile and tolerates shade. Lowland.

Native. There is no appreciable change in the distribution of this hybrid. It is probably under-recorded.

Widespread in temperate Europe.

References: Atlas Supp. (147a), Jermy *et al.* (1982), Stace (1975).

M. J. Y. FOLEY & M. S. PORTER

Carex appropinquata Fibrous Tussock-sedge

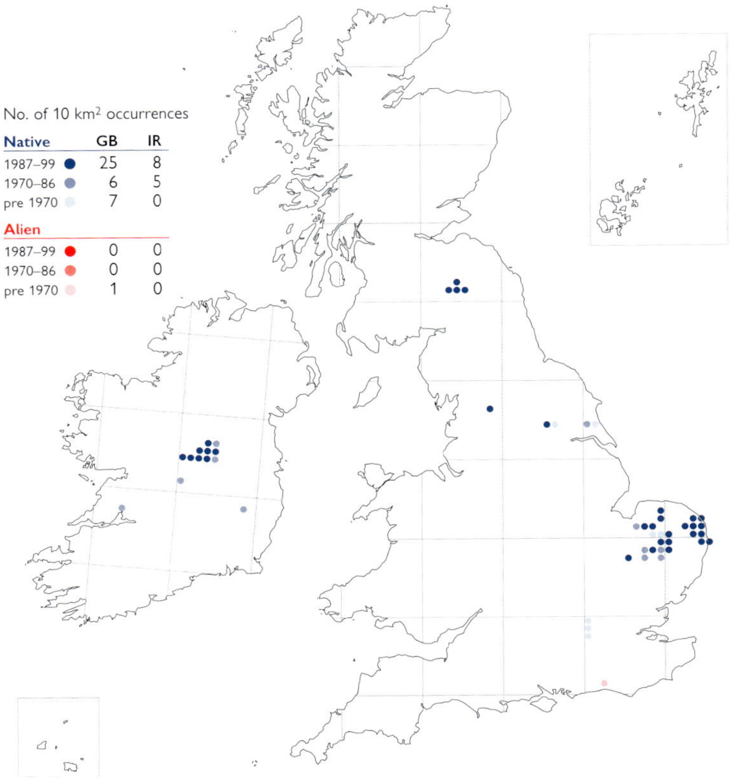

No. of 10 km² occurrences

Native	GB	IR
1987–99	25	8
1970–86	6	5
pre 1970	7	0
Alien		
1987–99	0	0
1970–86	0	0
pre 1970	1	0

A tussock-forming perennial herb, mainly occurring in open fenland but also in *Salix*-carr where, however, its numbers may sometimes be reduced by shading and drying out. Generally lowland, but reaches 380 m at Malham Tarn (Mid-W. Yorks.).

Native (change –0.17). *C. appropinquata* has declined in its East Anglian stronghold and in its Yorkshire lowland sites as a result of drainage and recent dry summers. However, the sites in the Scottish Borders, all discovered since the 1962 *Atlas*, appear to be in good heart, and the distribution in Ireland seems stable.

Eurosiberian Boreo-temperate element, with a continental distribution in W. Europe.

References: Atlas (365c), Corner (1969), Hultén & Fries (1986), Jermy *et al.* (1982), Stewart *et al.* (1994).

M. S. PORTER & M. J. Y. FOLEY

Carex diandra Lesser Tussock-sedge

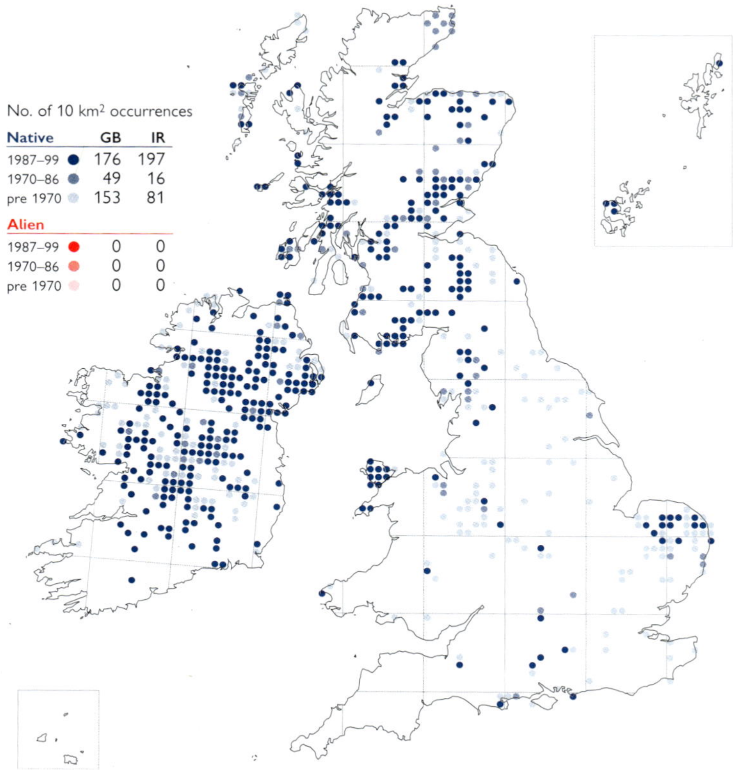

A perennial herb of wet, peaty areas, tolerating both acidic soils and those flushed by calcareous springs. It is often found on the margins of pools and in swamps, though it can also thrive on the edges of wet woods, among scattered trees or in fen-carr. Generally lowland, but reaching 370 m near Malham Tarn (Mid-W. Yorks.).

Native (change +0.22). *C. diandra* has been lost as a result of drainage and scrub encroachment from many lowland sites, especially in S. and E. England, during the 20th century. Most losses occurred before 1930, but are still continuing in England. It is now much better recorded in Scotland and N. Ireland.

Circumpolar Boreo-temperate element.

References: Atlas (365d), Hultén & Fries (1986), Jermy *et al.* (1982).

M. S. PORTER & M. J. Y. FOLEY

Carex vulpina True Fox-sedge

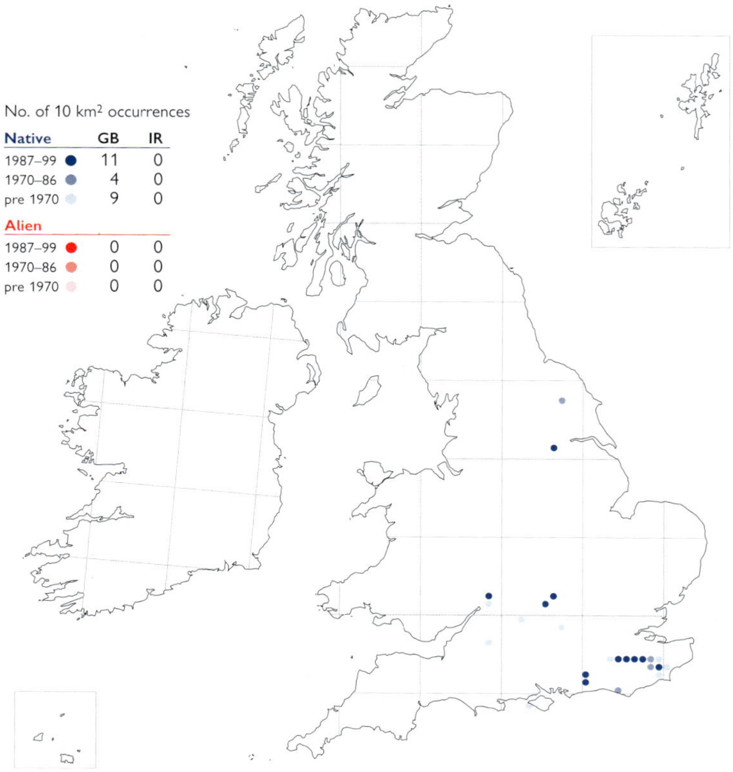

A perennial herb of wet, open or shaded habitats, usually on heavy clay soils that are flooded in winter and dry in summer. It occurs by ditches and rivers, in meadows and in a *Quercus-Crataegus* thicket. It sometimes grows in standing water. Plants fruit freely. Lowland.

Native (change –0.57). There is much confusion regarding the separation of this species from *C. otrubae*. Despite apparent differences in leaf anatomy, preliminary molecular analysis has revealed that the situation is not clear-cut, and taxonomic work continues. Distributional changes are, therefore, not easily assessed.

Eurosiberian Temperate element, with a continental distribution in W. Europe.

References: Atlas (366b), Hultén & Fries (1986), Jermy *et al.* (1982), Porley (1999), Stewart *et al.* (1994), Wigginton (1999).

R. D. PORLEY

Carex otrubae False Fox-sedge

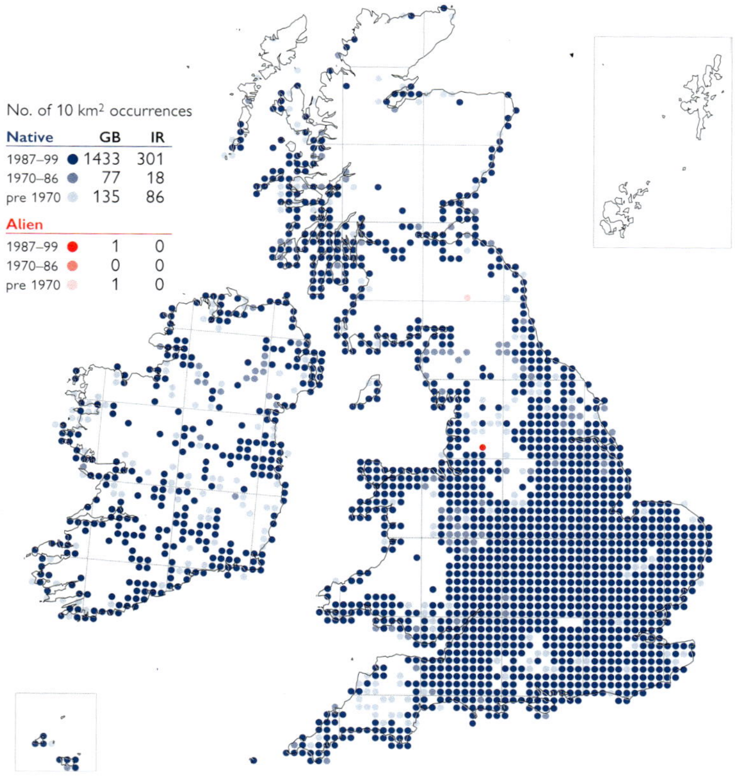

A perennial of wet habitats, usually on heavy soils. It is found on the sides of streams and ponds, in ditches, swamps, wet lowland meadows and pastures, at the upper edge of salt-marshes, and, less commonly, on damp roadsides and hedge banks. In N. England and Scotland it is essentially a coastal plant. It generally grows in slightly drier conditions than *C. vulpina*, tending to avoid standing water. Lowland.

Native (change –0.14). The overall distribution of *C. otrubae* is little changed since the 1962 *Atlas*, though loss of habitat through drainage has caused some local declines, for example in Dorset (Pearman, 1994).

Eurosiberian Southern-temperate element.

References: Atlas (366a), Hultén & Fries (1986), Jermy *et al.* (1982).

M. J. Y. FOLEY & M. S. PORTER

Carex otrubae × *C. remota* (*C.* × *pseudoaxillaris*)

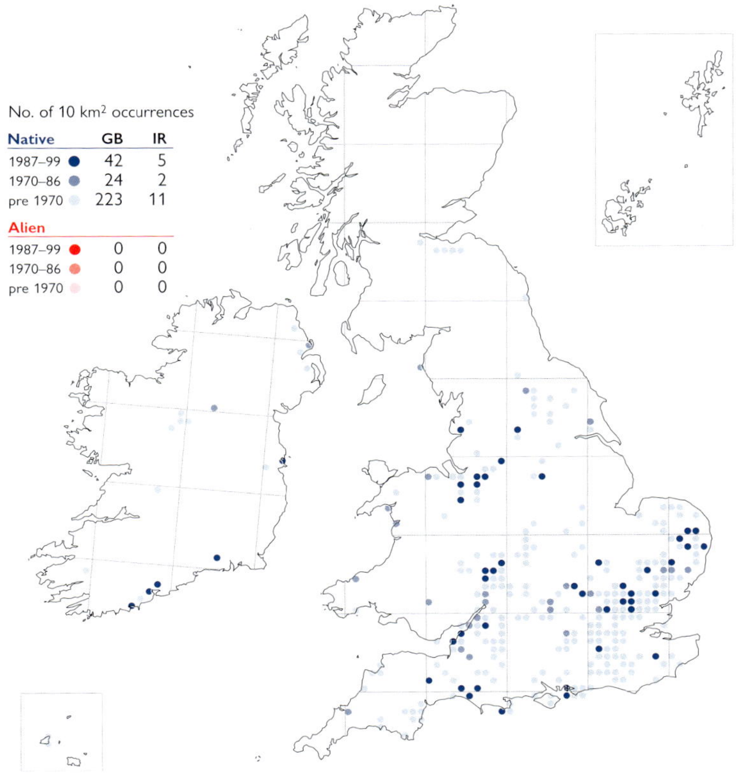

A generally sterile hybrid, found near both parents in a range of habitats, including woodland rides and clearings, ditch-sides, roadside banks and the damp ledges of sea-cliffs. Lowland.

Native. The mapped distribution of this hybrid is similar to that shown by Perring & Sell (1968), but the fact that recent records are much less numerous than older ones suggests that it is under-recorded.

Outside our area *C.* × *pseudoaxillaris* is recorded from Germany and the Netherlands.

References: Atlas Supp. (147b), Jermy *et al.* (1982), Stace (1975).

M. J. Y. FOLEY & M. S. PORTER

Carex spicata Spiked Sedge

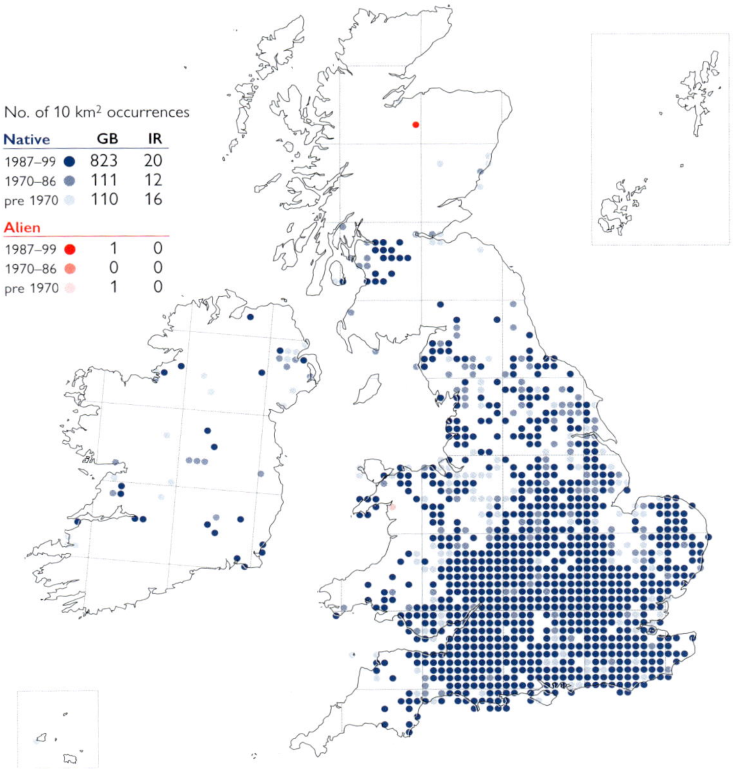

A perennial herb of rough grasslands, roadsides, railway banks, hedge banks, woodland rides and clearings, open scrub and waste ground. It is a plant of moist, neutral or slightly base-rich, heavy soils, and cannot withstand much competition. Lowland.

Native. This species was previously confused with *C. muricata* subsp. *lamprocarpa*, but a clear account of the differences was provided by Jermy *et al.* (1982) and it is now known to be more common in N.W. England and W. Scotland than indicated in the 1962 *Atlas*.

European Temperate element; widely naturalised outside its native range.

References: Atlas (368a), David & Kelcey (1985), Hultén & Fries (1986).

M. S. PORTER & M. J. Y. FOLEY

Carex muricata subsp. *muricata*

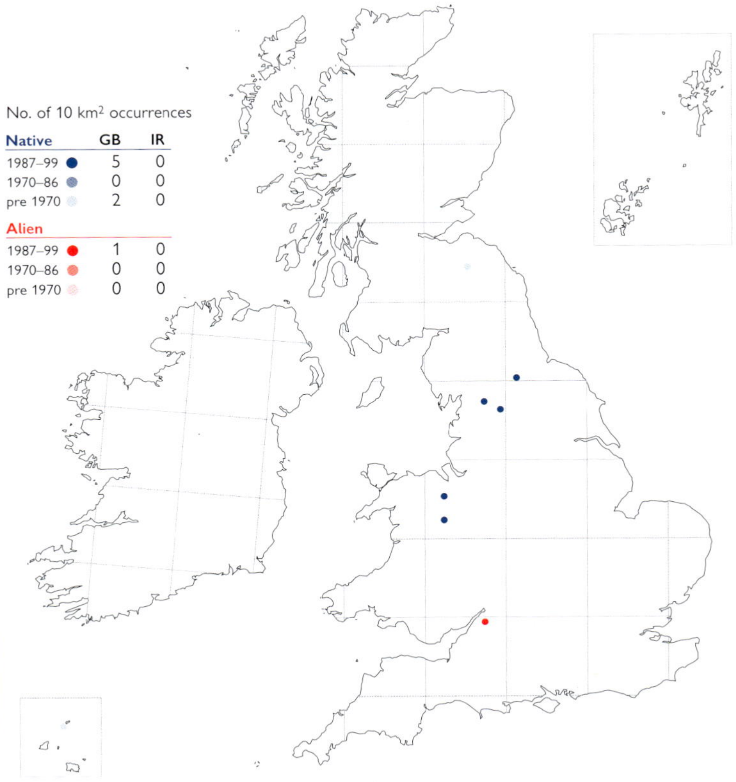

This perennial herb is confined to dry habitats over limestone, being found on limestone pavements, ledges, grassy slopes and scree; some sites are shaded. Generally lowland, but reaching 340 m in Ribblesdale (Mid-W. Yorks.).

Native. The native Gloucestershire site was lost in 1983 through overgrowth of surrounding vegetation. Plants were re-introduced, but these died out by 1989. A large Shropshire population was discovered in 1999 and an extension to it confirmed in 2000. It may be overlooked or mis-recorded elsewhere.

European Boreo-temperate element; introduced to N. America.

References: David & Kelcey (1985), Foley & Porter (2000), Jermy *et al.* (1982), Wigginton (1999).

M. S. PORTER & M. J. Y. FOLEY

Carex muricata subsp. lamprocarpa

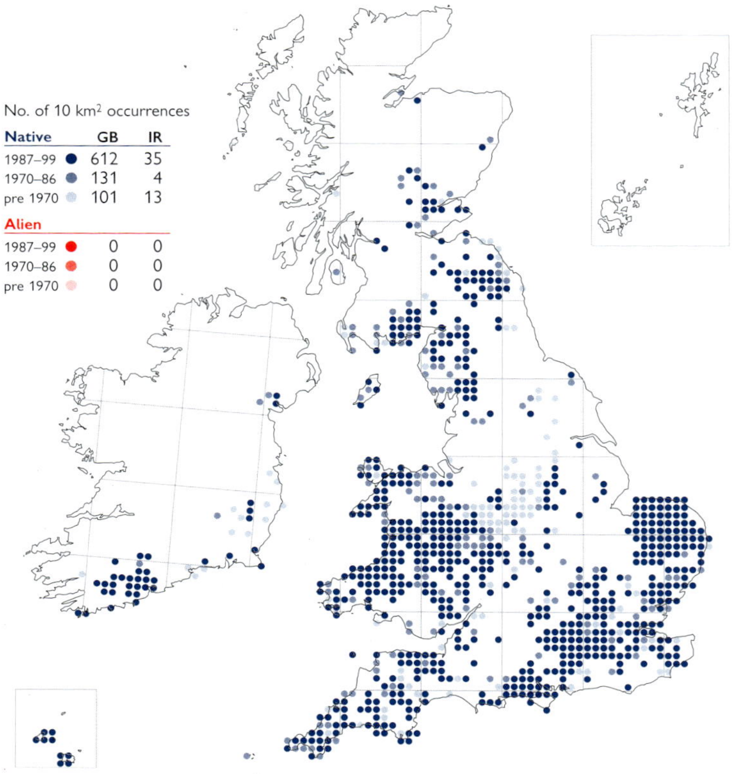

A tufted perennial herb of well-drained, light, sometimes sandy, soils, which is able to tolerate rather more acidic substrates than *C. spicata*. It is found on hedge banks, earth-filled walls and roadsides, in rough meadows, heathland and on rocky slopes, and is somewhat intolerant of shade. Generally lowland, but reaching 335 m at Cae Gaer (Cards.).

Native. The current taxonomic treatment of *C. muricata* was set out by Jermy *et al.* (1982). *C. muricata* subsp. *lamprocarpa* is probably still under-recorded in some areas; it is sometimes confused with *C. spicata*.

European Southern-temperate element.

Reference: David & Kelcey (1985).

M. S. PORTER & M. J. Y. FOLEY

Carex divulsa subsp. divulsa

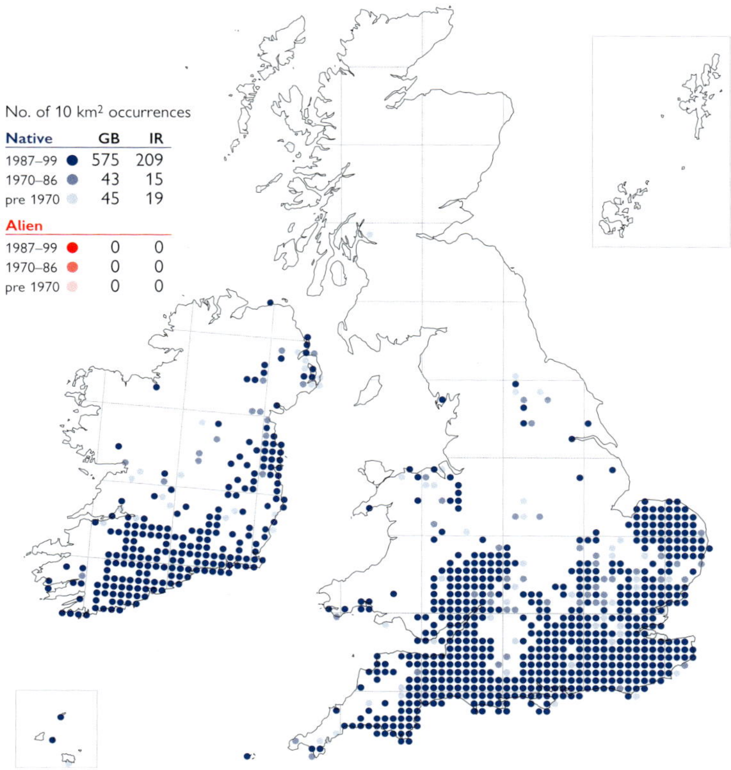

A tufted perennial herb, tolerating a wide range of soils, except those which are markedly acidic, and found in hedge banks, scrub, along woodland borders and paths, on roadsides and in rough, open grassland. Lowland.

Native. Subsp. *divulsa* was mapped as *C. polyphylla* in the 1962 *Atlas*. The overall range seems to be stable. It is now better recorded in many areas but not mapped in some vice-counties where recorders have failed to distinguish the two subspecies of *C. divulsa*.

European Southern-temperate element; the species is mapped by Hultén & Fries (1986).

References: Atlas (367b), David & Kelcey (1985), Jermy *et al.* (1982).

M. S. PORTER & M. J. Y. FOLEY

Carex divulsa subsp. leersii

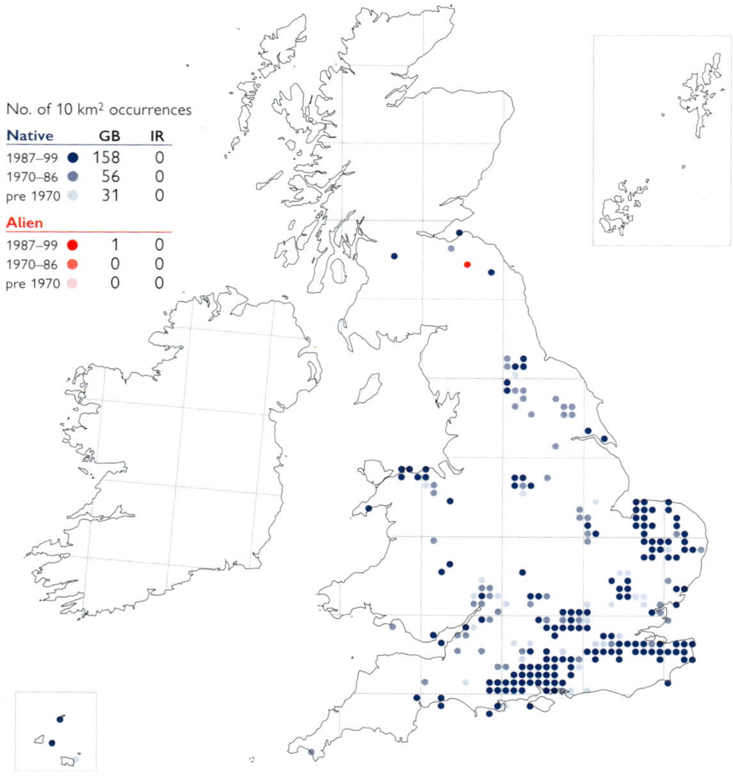

This tufted perennial herb occurs on roadsides, in woodland rides, in hedge banks and open grassland, always on chalk or limestone. Lowland.

Native. The current map of this subspecies shows a similar range to that given for *C. divulsa* in the 1962 *Atlas*, but there are many additional records, including new sites in Scotland. This reflects improved knowledge of the taxonomy of the species resulting from the work of R.W. David. However, the two subspecies of *C. divulsa* are linked by a range of intermediates which can sometimes make certain identification problematical.

Eurosiberian Southern-temperate element; this subspecies extends further north in Europe than subsp. *divulsa*.

References: Atlas (367d), David & Kelcey (1985), Jermy *et al.* (1982).

M. S. PORTER & M. J. Y. FOLEY

Carex arenaria Sand Sedge

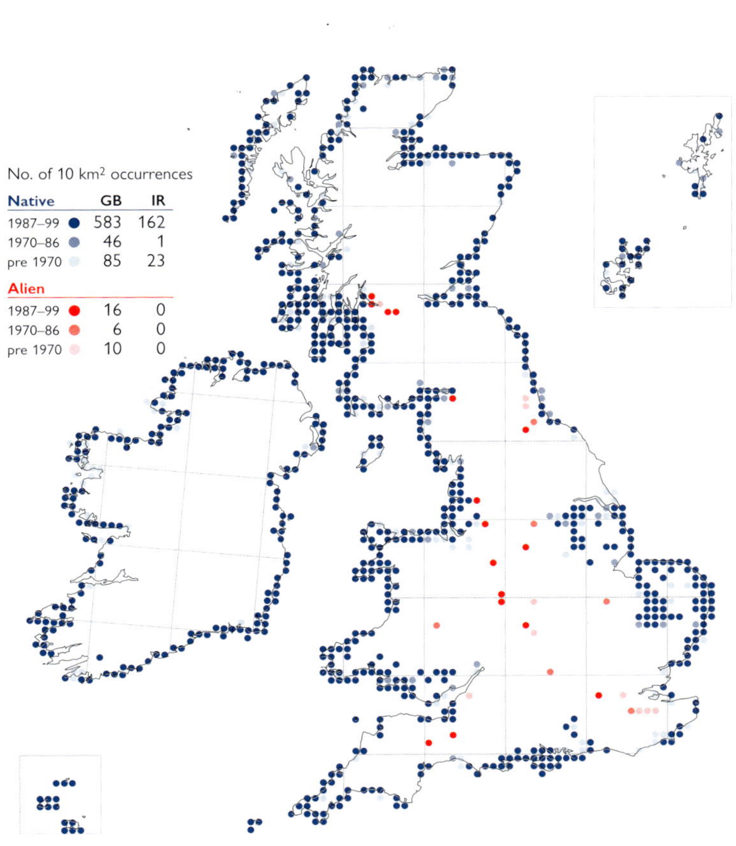

A rhizomatous perennial herb of sandy habitats, where it can be a dominant plant of fixed dunes, dune-slacks, sandy flats and on tracksides or other ruderal habitats. Though predominantly coastal, it can be locally common on dunes and heaths inland, particularly on the Lincolnshire coversands and in Breckland. It can be a rapid coloniser, spreading by far-creeping rhizomes. It is occasionally found as an alien on railway clinker. Lowland.

Native (change –0.27). There has been little change in the overall distribution of this species since the 1962 *Atlas*.

European Temperate element.

References: Atlas (366d), Hultén & Fries (1986), Jermy *et al.* (1982), Meusel *et al.* (1965), Noble (1982).

M. J. Y. FOLEY & M. S. PORTER

Carex disticha Brown Sedge

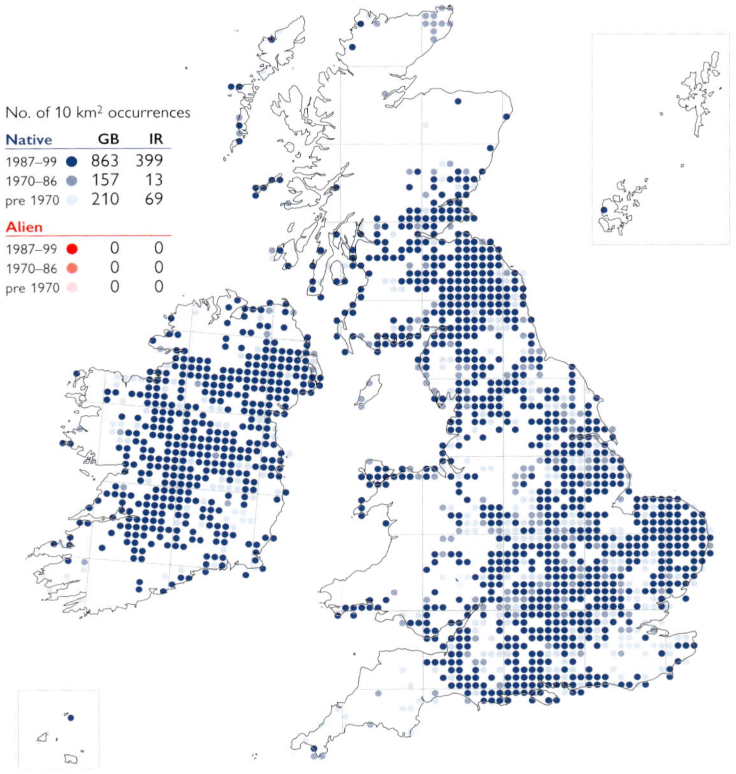

This rhizomatous species is found in wet meadows and marshes, fens, by ditches and streams, on the margins of lakes and ponds, and occasionally in dune-slacks. It almost invariably grows in well-illuminated sites, particularly on base-enriched substrates, and favours areas with a fluctuating water table. Generally lowland, but reaching 455 m on Windburgh Hill (Roxburghs.).

Native (change –0.03). *C. disticha* has declined in many areas of England since the 1962 *Atlas* because of drainage, although it can be easily overlooked, especially when it is present only as scattered non-flowering shoots.

Eurosiberian Temperate element.

References: Atlas (366c), Hultén & Fries (1986), Jermy *et al.* (1982), Meusel *et al.* (1965).

M. S. PORTER & M. J. Y. FOLEY

Carex chordorrhiza String Sedge

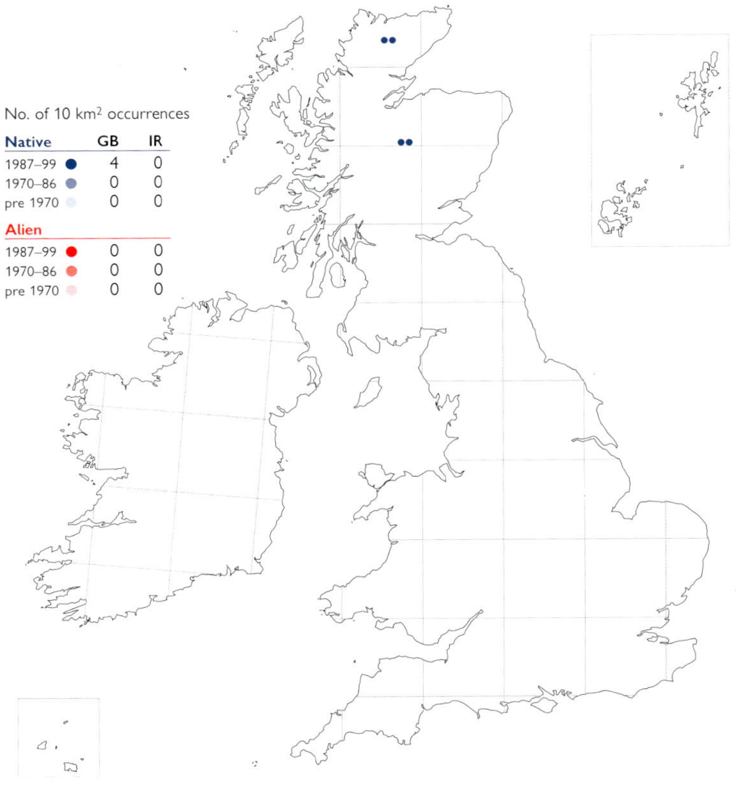

This rhizomatous perennial herb is found in fens and base-poor mires, where it usually grows in standing water. Propagation appears to be mainly vegetative by long trailing runners, although genetic variation within its populations suggests that it also reproduces by seed. Lowland.

Native. *C. chordorrhiza* may be a late glacial relic in our area. It was discovered in Westerness in 1978, and is now known to be much more frequent there than was first supposed. It is susceptible to both drainage and, apparently, to prolonged submergence.

Circumpolar Boreo-arctic Montane element, with a continental distribution in W. Europe.

References: Atlas (367a), Hultén & Fries (1986), Jermy *et al.* (1982), Meusel *et al.* (1965), Wigginton (1999).

M. J. Y. FOLEY & M. S. PORTER

Carex divisa Divided Sedge

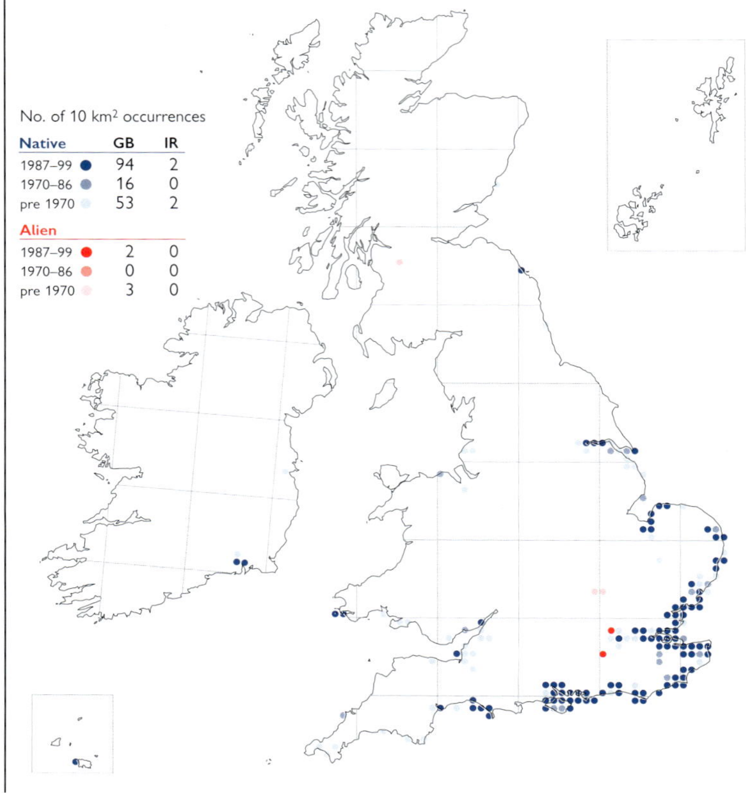

A rhizomatous perennial herb of brackish ditches, dune-slacks and damp grasslands near the sea. It avoids areas of standing water. Lowland.

Native (change –0.35). *C. divisa* suffered losses, particularly in S.W. England, before 1930. It remains frequent in suitable areas in S. and E. England, and there has been little change in its distribution since the 1962 *Atlas* despite local losses due to coastal development and the widespread conversion of grazing marshes to arable. In Ireland, the populations in Co. Dublin have been lost to building and it was considered to be extinct in Co. Kilkenny and Co. Wexford until re-discovered in 1990.

Submediterranean-Subatlantic element.

References: Atlas (367a), Curtis & FitzGerald (1994), Curtis & McGough (1988), Jermy *et al.* (1982), Stewart *et al.* (1994).

M. J. Y. FOLEY & M. S. PORTER

Carex maritima Curved Sedge

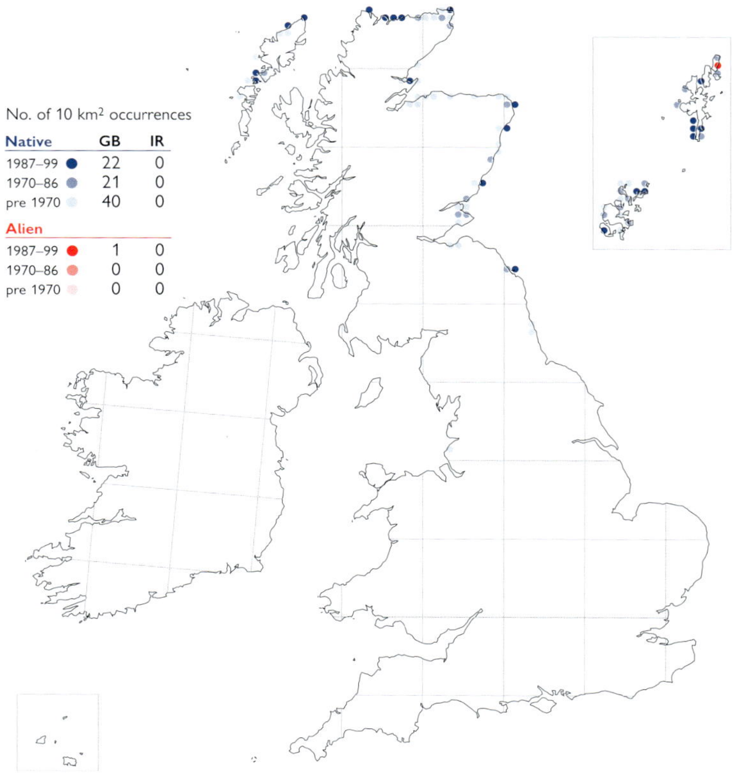

A rhizomatous perennial herb found in short vegetation in damp dune-slacks and on open sand, often close to freshwater seepages or where streams debouch onto the shore. It is mobile and can colonise new sites with suitable habitat. Populations can be very large. Lowland.

Native (change –1.34). *C. maritima* suffered a considerable decline before 1930. Assessments of more recent trends are difficult because it is a very inconspicuous and under-recorded species. Further fieldwork is required to establish its current distribution. It is probably lost from St. Andrews golf course (Fife) where it was abundant in 1984.

Circumpolar Arctic-montane element.

References: Atlas (367b), Hultén & Fries (1986), Jermy *et al.* (1982), Leach (1986), Meusel *et al.* (1965), Stewart *et al.* (1997).

M. J. Y. FOLEY & M. S. PORTER

Carex remota Remote Sedge

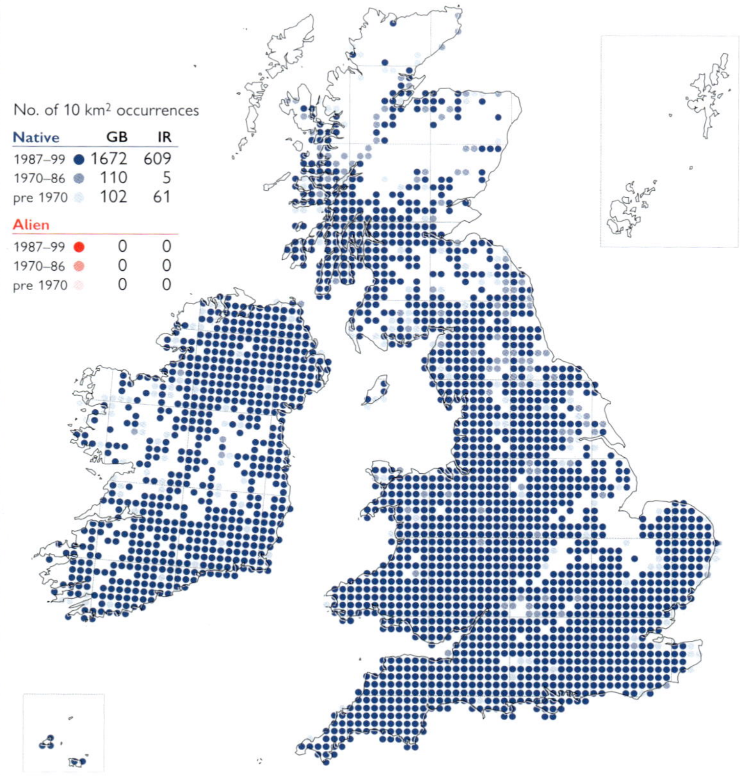

A tufted perennial herb of damp woodland and woodland rides, often growing in considerable shade. It can become dominant in favoured habitats, such as in woods that are seasonally flooded. Generally lowland, but reaching 320 m in Swaledale (N.W. Yorks.).

Native (change +0.04). There has been no significant change in the overall distribution of this species since the 1962 *Atlas*.

European Temperate element; also in C. and E. Asia.

References: Atlas (369a), Hultén & Fries (1986), Jermy *et al.* (1982), Meusel *et al.* (1965).

M. S. PORTER & M. J. Y. FOLEY

Carex ovalis Oval Sedge

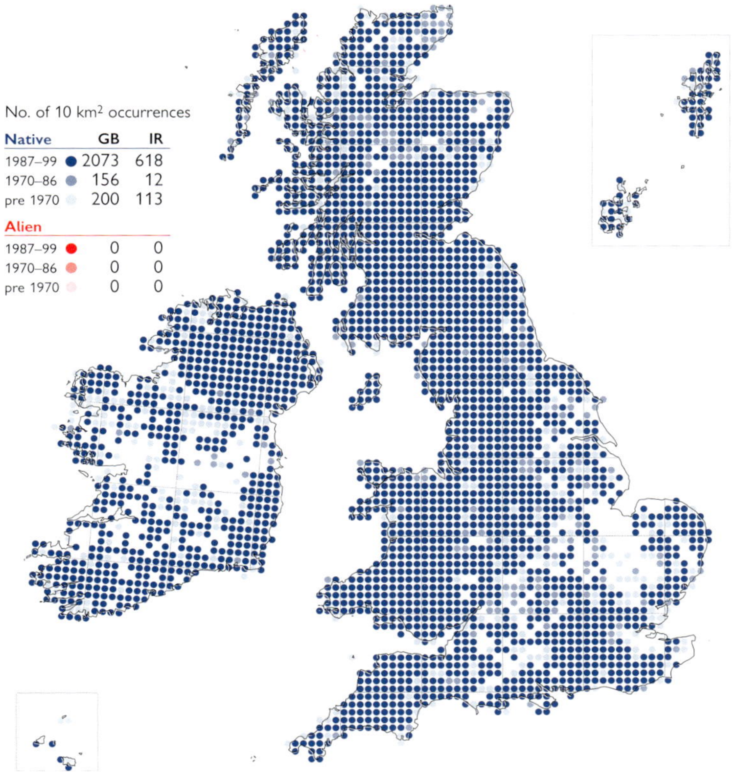

A tufted perennial herb, especially of acidic grassland, frequent in upland pastures, *Nardus* grassland and *Calluna* moorland. It is also common on lowland heaths, in damp meadows, woodland rides and edges, in ruderal habitats, and sometimes in moderately base-rich flushes and dune-slacks. It can be frequent in meadows which have been improved, and have then regressed. 0–1005 m (Lochnagar, S. Aberdeen).

Native (change –0.21). *C. ovalis* was mapped as 'all records' in the 1962 *Atlas*. It has remained widespread and common, other than in S. and E. England where there have been many local losses.

Eurosiberian Boreo-temperate element; widely naturalised outside its native range.

References: Atlas (369d), Hultén & Fries (1986), Jermy *et al.* (1982), Meusel *et al.* (1965).

M. J. Y. FOLEY & M. S. PORTER

Carex echinata Star Sedge

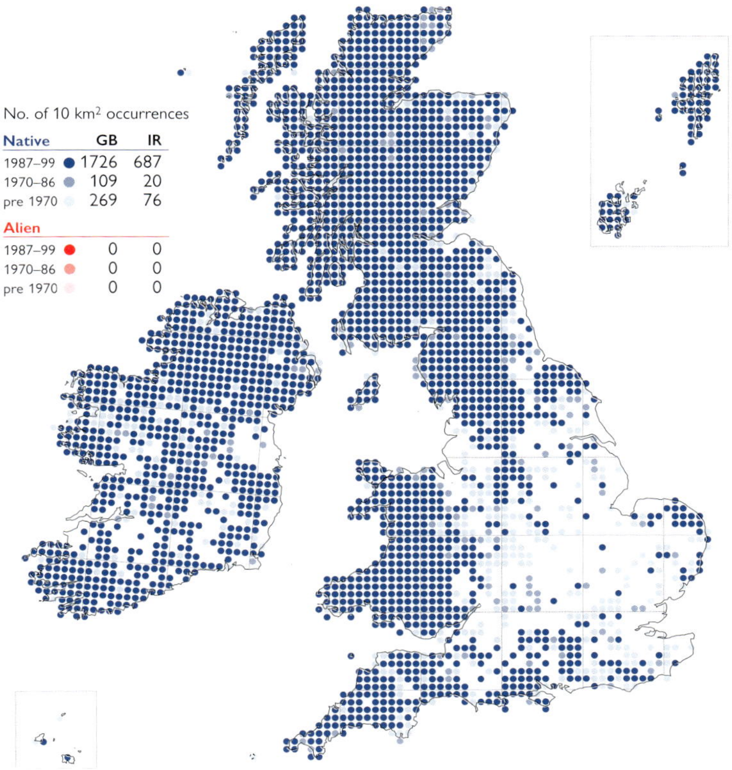

A perennial of seasonally or permanently waterlogged habitats on acidic to base-rich substrates. It is found in a wide range of mires, wet heaths, upland flushes and springs; also in rush-pastures, wet meadows, flushed grassland on hill-slopes and, rarely, in wet woodland rides. 0–1005 m (Macgillycuddy's Reeks, S. Kerry).

Native (change –0.75). *C. echinata*, which was mapped as 'all records' in the 1962 *Atlas*, was lost from many lowland squares in England in the 19th century. Since then, and since 1962, there have been further losses through drainage and agricultural improvements. The distribution is stable elsewhere.

European Boreo-temperate element; also in N. America.

References: Atlas (368d), Hultén & Fries (1986), Jermy *et al.* (1982), Meusel *et al.* (1965).

M. J. Y. FOLEY & M. S. PORTER

Carex dioica Dioecious Sedge

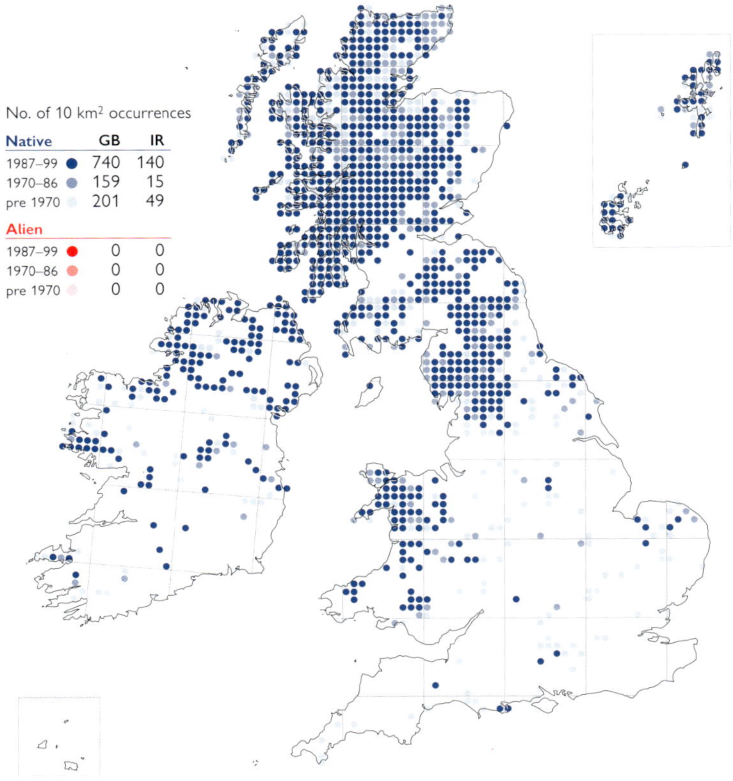

A shortly rhizomatous perennial herb of very wet, neutral to base-rich mires. It grows particularly well in silty, calcareous mud and by the edges of lime-rich flushes and springs. Although it occurs at sea level, many sites are at moderate or high altitudes, ascending to about 1000 m on Ben Lawers (Mid Perth).

Native (change –0.35). *C. dioica* declined dramatically in lowland England before 1930, due to drainage and agricultural improvement. There have been further losses there (half the remaining sites south of the Severn-Humber line), but there are now a very substantial number of new records from N. and W. Britain, and from Ireland.

Circumpolar Boreo-arctic Montane element.

References: Atlas (371a), Hultén & Fries (1986), Jermy *et al.* (1982).

M. S. PORTER & M. J. Y. FOLEY

Carex davalliana Davall's Sedge

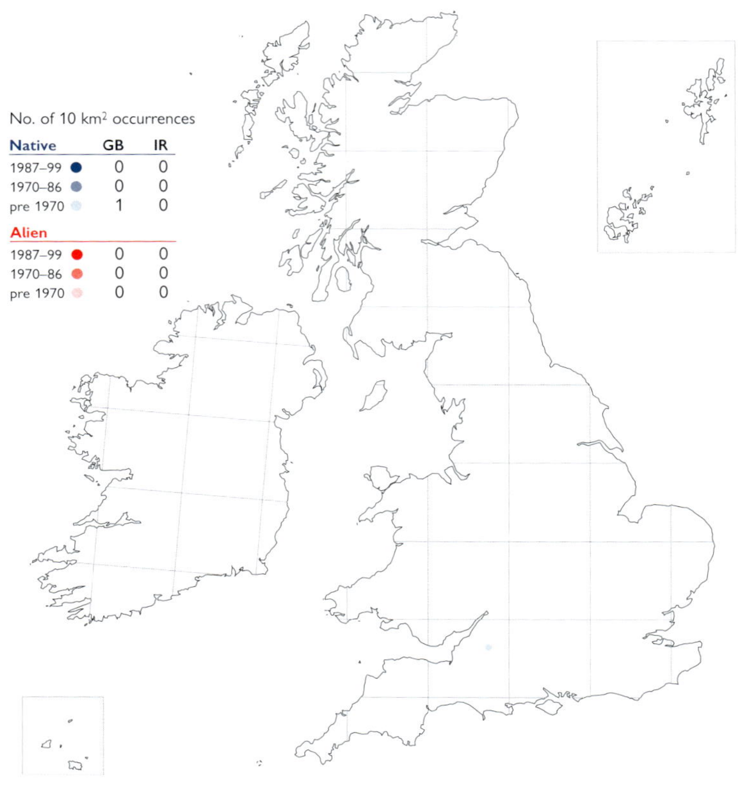

This species has been recorded in Britain only in a calcareous mire at Lansdown near Bath (N. Somerset). Lowland.

Native. *C. davalliana* is extinct. It was discovered before 1809, but Babington (1834) noted that it 'has not been found for some years'. The site was drained and subsequently built on, and there was no trace of the plant in 1852.

European Temperate element, with a continental distribution in W. Europe.

References: Atlas (370b), Hultén & Fries (1986), Jermy *et al.* (1982), Marren (1999), Meusel *et al.* (1965), White (1912).

M. J. Y. FOLEY & M. S. PORTER

Carex elongata Elongated Sedge

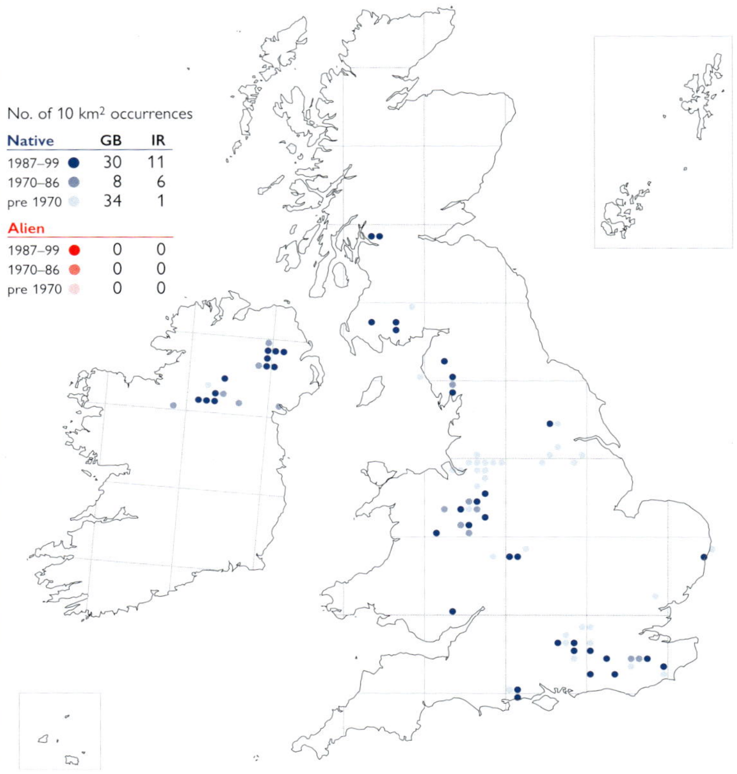

A perennial herb of wet woodlands, especially those dominated by *Alnus* and those on lake shores, but also found on pond margins, in ditches and seasonally flooded areas, and in wet meadows. In favourable conditions it can form large, loose tussocks and it sets seed freely in more open situations. Lowland.

Native (change +0.06). Some populations of *C. elongata* which were recorded before 1930 have been lost because of drainage and other forms of habitat destruction. The distribution is otherwise stable, and the species is better recorded in Ireland than in the 1962 *Atlas*.

Eurosiberian Boreal-montane element, with a continental distribution in W. Europe.

References: Atlas (368c), Curtis & McGough (1988), David (1978b), Hultén & Fries (1986), Jermy *et al.* (1982), Meusel *et al.* (1965), Stewart *et al.* (1994).

M. J. Y. FOLEY & M. S. PORTER

Carex lachenalii Hare's-foot Sedge

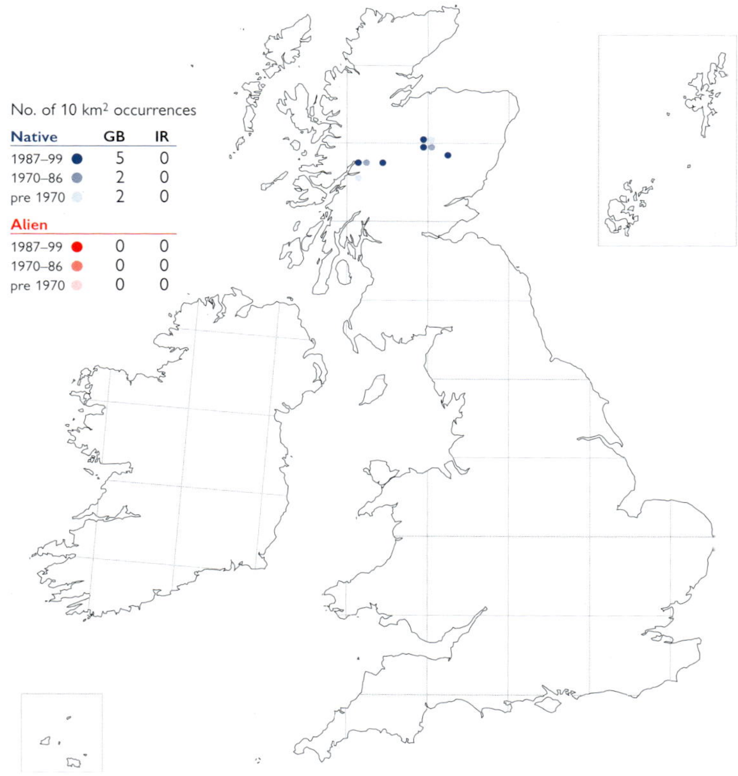

A perennial herb occurring on wet, acidic, N.-facing slopes and rock ledges, and in nutrient-poor flushes, especially in areas of late snow-lie. Montane, from 950 m on Cairn Toul (S. Aberdeen), but formerly lower in Glen Coe, to 1150 m on Ben Macdui (S. Aberdeen).

Native (change −0.22). The distribution of *C. lachenalii* seems to be stable, although some sites are very remote; it may still be present in some squares for which there are only pre-1987 records.

Circumpolar Arctic-montane element.

References: Atlas (369c), Hultén & Fries (1986), Jermy *et al.* (1982), Wigginton (1999).

M. J. Y. FOLEY & M. S. PORTER

Carex curta White Sedge

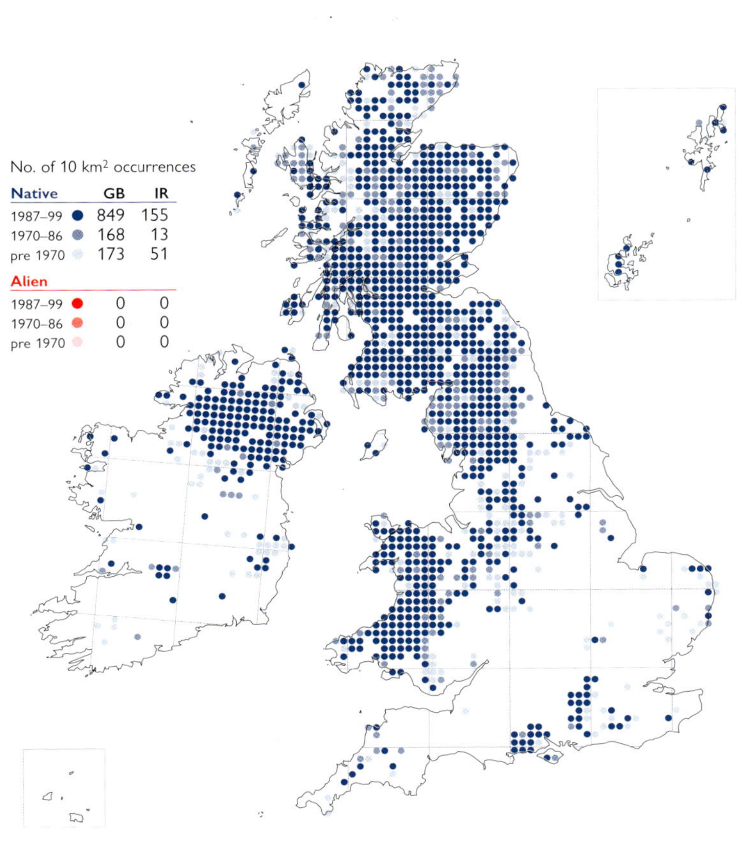

A perennial herb of lowland bogs, floating *Sphagnum* rafts in lowland basin mires, nutrient-poor mires in the mountains, and wet, acidic, occasionally sandy heaths. 0–1100 m (Ben Alder, Westerness).

Native (change +0.17). The 1962 *Atlas* shows a decline in this species since 1930, especially in England. Since then, a few more sites in the lowlands have been lost to drainage, but many more have been discovered in the west, particularly in Wales and N. Ireland.

Circumpolar Boreal-montane element.

References: Atlas (369b), Hultén & Fries (1986), Jermy *et al.* (1982), Meusel *et al.* (1965).

M. J. Y. FOLEY & M. S. PORTER

Carex hirta Hairy Sedge

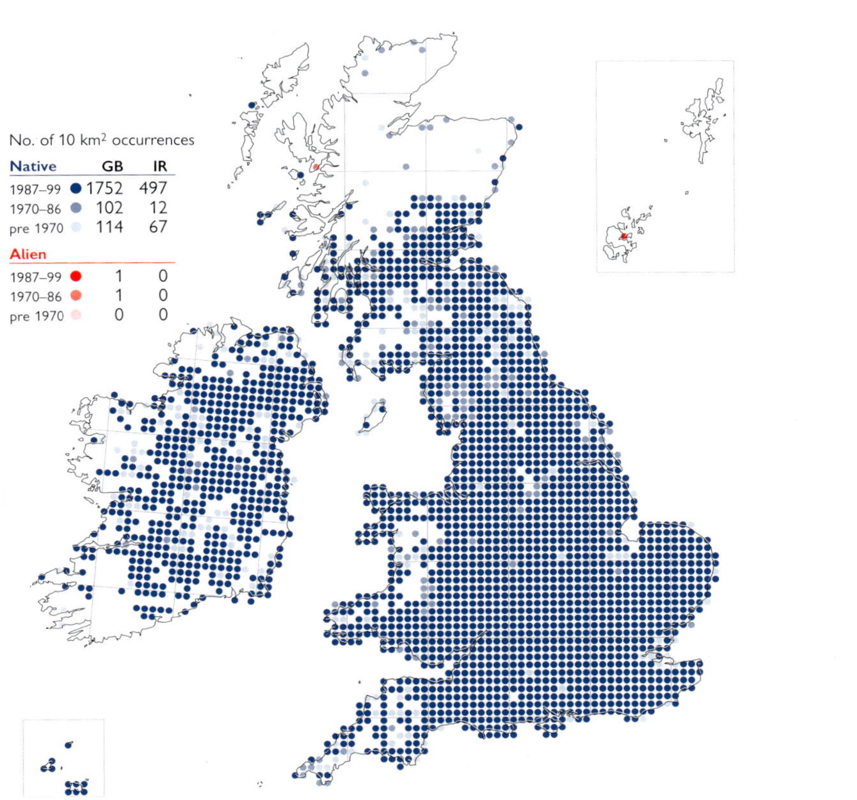

Found in a wide variety of damp, grassy habitats, this rhizomatous perennial herb is particularly common in areas influenced by man: waste ground, tracksides, roadside verges and hedge banks. It is also frequent, however, in rough grassland, hayfields, and occasionally sand dunes, marshes and damp, open woods. It avoids highly acidic, infertile soils. 0–470 m (Hartside, Cumberland).

Native (change +0.17). The range and frequency of *C. hirta* are likely to have been stable over many years, and the distribution is similar to that mapped in the 1962 *Atlas*.

European Temperate element; widely naturalised outside its native range.

References: Atlas (361b), Hultén & Fries (1986), Jermy *et al.* (1982).

M. S. PORTER & M. J. Y. FOLEY

Carex lasiocarpa Slender Sedge

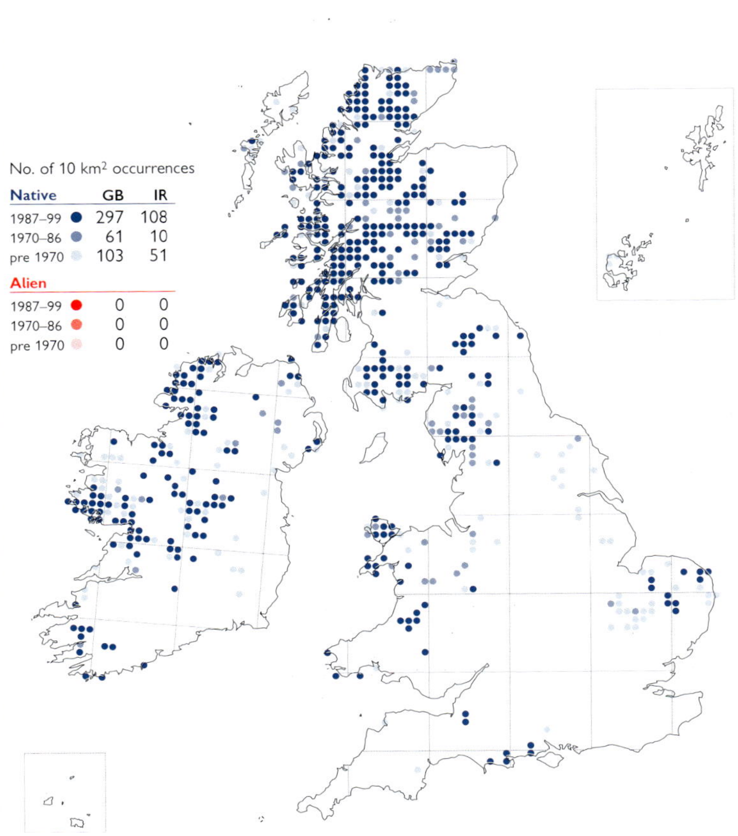

This rhizomatous perennial grows in reed-swamps and other vegetation at the edges of lakes, pools and slow-flowing streams and rivers, and in flushes and wet hollows in fens; it can also colonise old peat workings and drainage channels. It generally occurs in nutrient-poor water, which may be base-rich or base-poor. 0–650 m (E. of Beinn Heasgarnich, Mid Perth).

Native (change +0.73). Some *C. lasiocarpa* localities were lost to drainage before 1930, and since the 1962 *Atlas* there have been further declines in East Anglia and in other areas of lowland England and E. Ireland. Conversely, more intensive fieldwork has revealed many new 10-km squares elsewhere.

Circumpolar Boreal-montane element.

References: Atlas (361c), Hultén & Fries (1986), Jermy *et al.* (1982), Preston & Croft (1997).

M. J. Y. FOLEY & M. S. PORTER

Carex acutiformis Lesser Pond-sedge

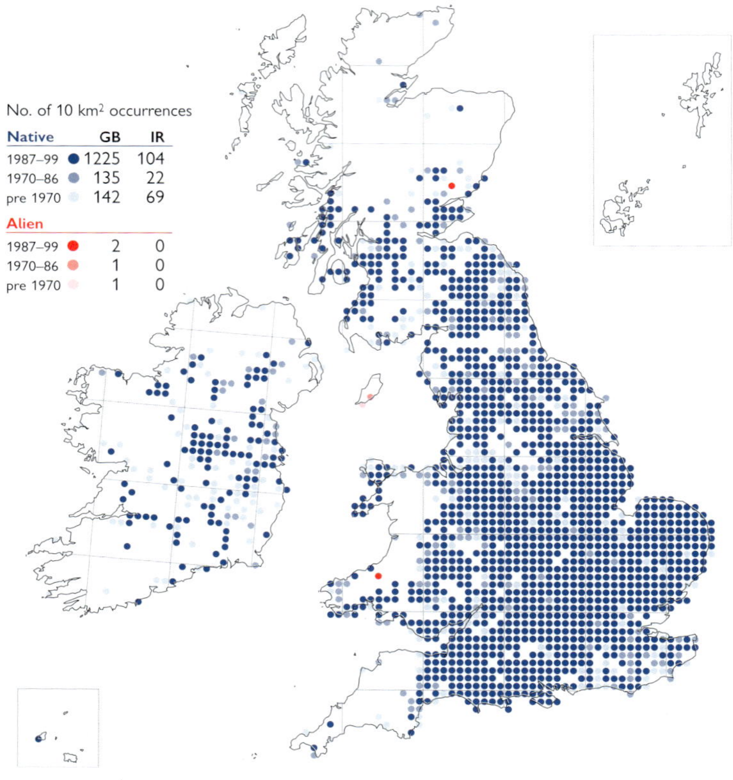

This rhizomatous perennial herb grows in base-rich, mesotrophic and eutrophic water, often occurring in extensive stands. Its habitats include river banks, lakes and pond margins, marshy areas near streams, fen-meadows and tall-herb fen, and carr; it is notably shade-tolerant. It avoids sites subject to marked fluctuations in water level. Reproduction by seed is poor. 0–370 m (near Garrigill, Cumberland).

Native (change +0.16). *C. acutiformis* was mapped as 'all records' in the 1962 *Atlas*. Apart from some local declines, its distribution seems generally stable.

Eurosiberian Temperate element.

References: Atlas (358c), Grime *et al.* (1988), Hultén & Fries (1986), Jermy *et al.* (1982), Meusel *et al.* (1965), Preston & Croft (1997).

M. J. Y. FOLEY & M. S. PORTER

Carex riparia Greater Pond-sedge

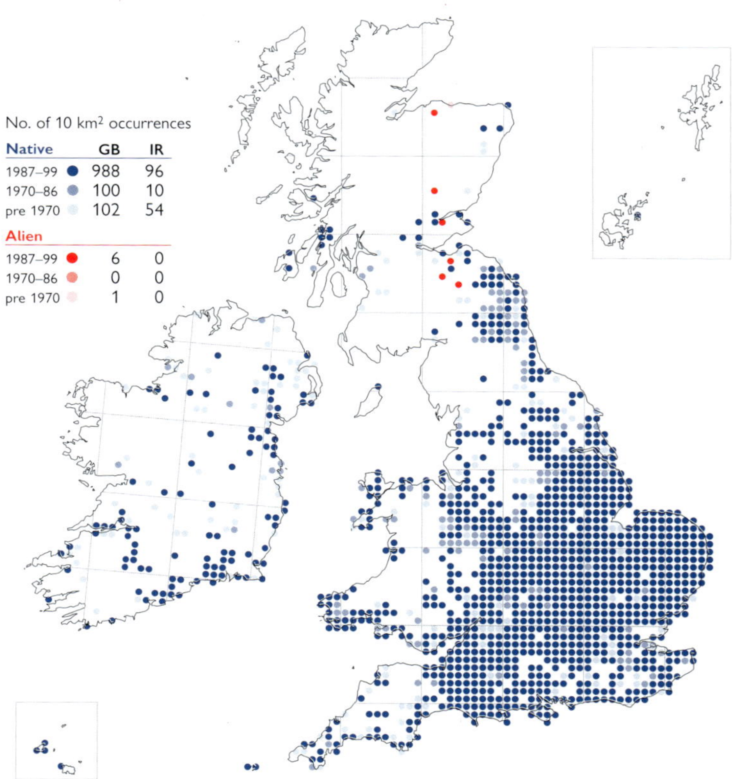

This rhizomatous perennial is found in reed-swamps, on the edges of pools and lakes, in marshy areas and wet woodland, and along the banks of slow-flowing rivers and canals. It is a species of base-rich, mesotrophic or eutrophic sites. Reproduction is mainly vegetative, but new colonies can arise from seed. Lowland.

Native (change +0.18). The distribution of *C. riparia* is broadly similar to that shown in the 1962 *Atlas*, and the plant is much better recorded in Ireland. Records from N.W. Scotland have been treated as errors. It is sometimes planted by ponds, and some records mapped as native may refer to such introductions.

Eurosiberian Temperate element.

References: Atlas (358b), Hultén & Fries (1986), Jermy *et al.* (1982), Meusel *et al.* (1965), Preston & Croft (1997).

M. J. Y. FOLEY & M. S. PORTER

Carex pseudocyperus Cyperus Sedge

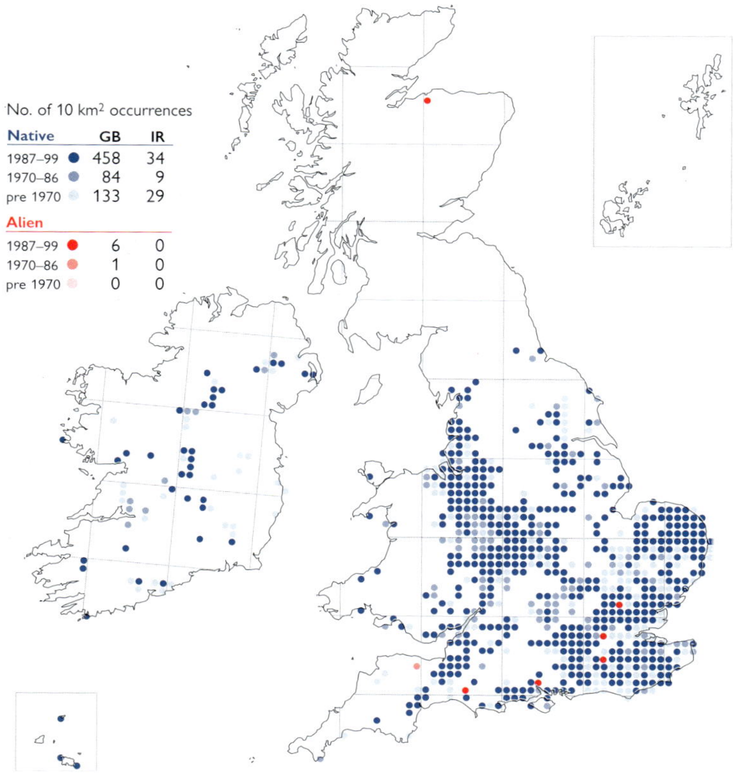

A pioneer species of wet mud and shallow water, found in a wide variety of habitats including the edges of lakes, ponds, reservoirs, rivers, ditches, canals and backwaters. It also occurs in reedswamps and tall-herb fens, and readily colonises sand-, gravel-, clay- and marl-pits. It is shade-tolerant, and is found in swampy *Alnus* woods and *Salix* carr. Reproduction is by seed. Lowland.

Native (change –0.27). *C. pseudocyperus* was lost from many squares in Ireland before 1930. It has declined across much of its British range, but particularly in S.E. and E. England, since the 1962 *Atlas*. It is sometimes planted at the edge of ponds.

Eurosiberian Temperate element; also in E. Asia and N. America.

References: Atlas (357a), Hultén & Fries (1986), Jermy *et al.* (1982), Meusel *et al.* (1965), Preston & Croft (1997).

M. J. Y. FOLEY & M. S. PORTER

Carex rostrata Bottle Sedge

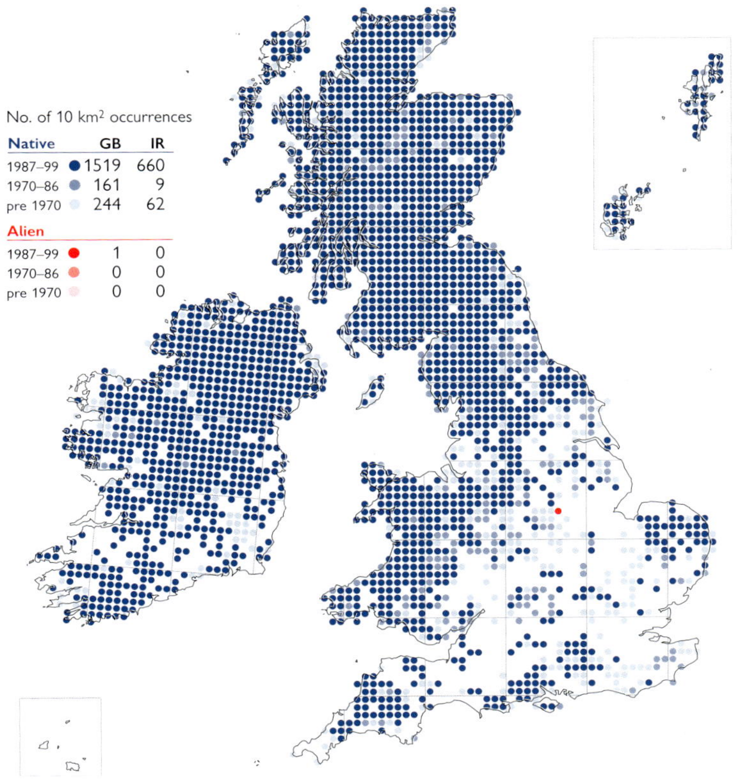

A rhizomatous perennial herb found in emergent stands on the edges of lakes and ponds, rivers and streams, in ditches, swamps, fens and bog pools, wet meadows, flush-bogs on hillsides, sea-cliff flushes, wet dune-slacks, and *Alder* and *Salix* carr. It usually grows in oligotrophic or mesotrophic, acidic waters, though it also occurs in nutrient-poor calcareous conditions. 0–1040 m (Creag Meagaidh, Westerness).

Native (change –0.19). The distribution of *C. rostrata* is stable except in S.E. England. Here, it had been lost from some sites before 1930 and further losses have occurred because of drainage and habitat reclamation.

Circumpolar Boreo-temperate element.

References: Atlas (357b), Hultén & Fries (1986), Jermy *et al.* (1982), Meusel *et al.* (1965), Preston & Croft (1997).

M. J. Y. FOLEY & M. S. PORTER

Carex rostrata × *C. vesicaria* (*C.* × *involuta*)

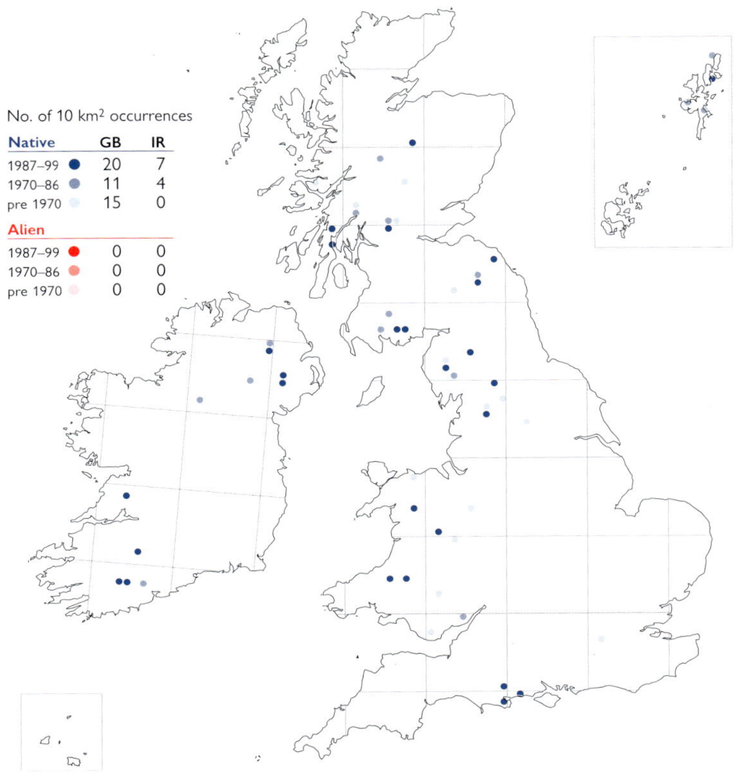

A plant of wet, marshy areas, especially by streams and lakes, often with both parents. It is, however, present in Shetland in the apparent absence of *C. vesicaria.* It is highly sterile, but may spread to form large stands. Generally lowland, but upper altitudinal limit unknown.

Native. This hybrid is probably under-recorded; it is very variable and many forms closely resemble one or other of the parent species. There is no evidence for any change in its distribution.

Recorded from the Nordic countries, and from scattered localities further south.

References: Jermy *et al.* (1982), Stace (1975).

M. J. Y. FOLEY & M. S. PORTER

Carex vesicaria Bladder-sedge

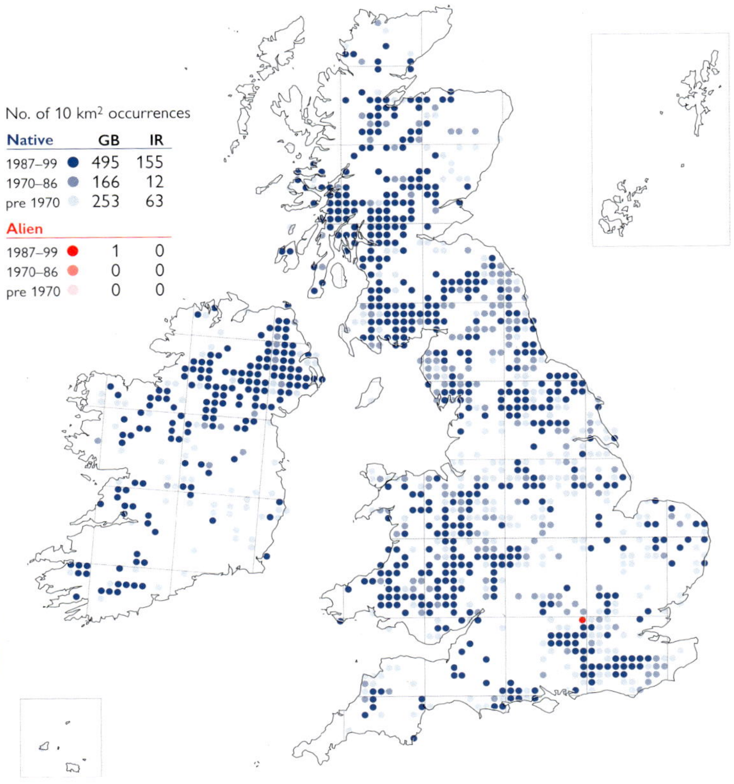

A perennial herb of wet habitats, mainly mesotrophic and at least slightly basic, occurring where the water table lies close to or above the soil surface. It is found by lakes, rivers, streams, ponds and canals, in marshes and swamps, ditches, wet meadows and depressions in pasture, and in wet woodland. It also colonises wet hollows in disused sand-, gravel- and clay-pits. 0–455 m (Llyn Gorast, Cards.).

Native (change –0.52). *C. vesicaria* has been lost from many sites in England and S.E. Ireland since the 1962 *Atlas.* Causes of this decline in Britain include drainage, falling water tables, ditch cleaning and eutrophication.

Circumpolar Boreo-temperate element.

References: Atlas (357c), Hultén & Fries (1986), Jermy *et al.* (1982), Preston & Croft (1997).

M. J. Y. FOLEY & M. S. PORTER

Carex saxatilis Russet Sedge

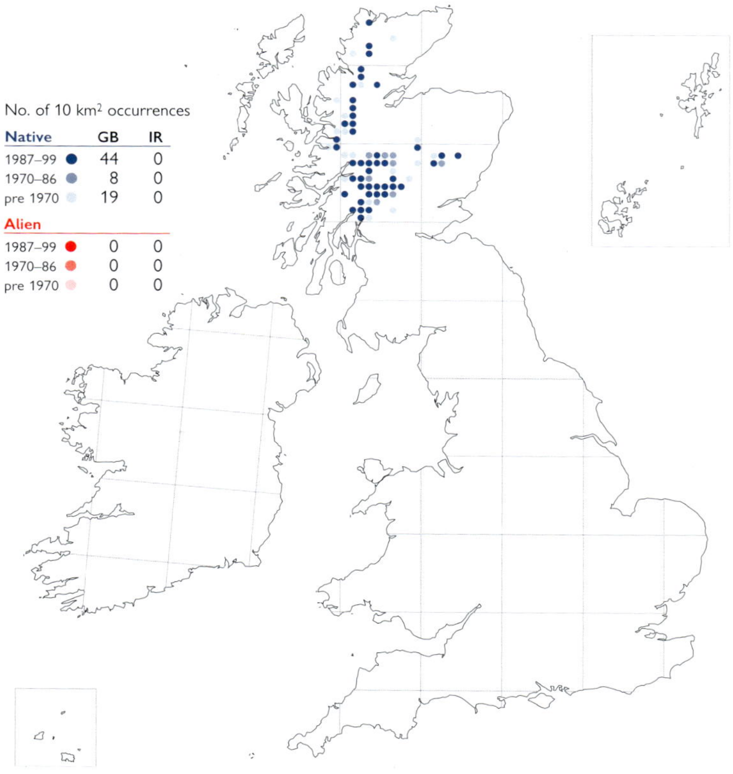

This rhizomatous perennial herb is usually, but not exclusively, found on base-rich substrates, in areas where there is little water movement. It occurs on flat mountain tops and gentle slopes, and in damp flushes and hollows where snow lies late. From 460 m in Glen Clunie (S. Aberdeen) to 1125 m on Ben Lawers (Mid Perth).

Native (change –0.35). The distribution of *C. saxatilis* is unlikely to have changed markedly, and it is possibly still present in many 10-km squares for which there are only pre-1987 records. It is under little threat, and is now better recorded than it was when mapped by Stewart *et al.* (1994).

Circumpolar Arctic-montane element; absent from mountains of C. Europe.

References: Atlas (358a), Hultén & Fries (1986), Jermy *et al.* (1982).

M. J. Y. FOLEY & M. S. PORTER

Carex pendula Pendulous Sedge

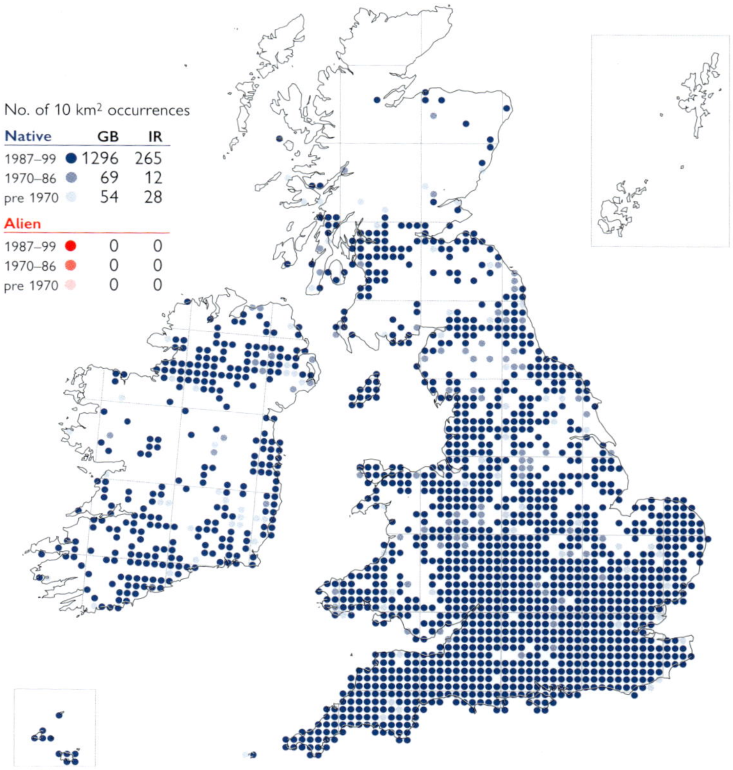

A perennial herb usually growing on damp base-rich, heavy (often clay) soils, in shaded habitats. It is found in deciduous woodland, by ditches, ponds, streams, in hedgerows and on track-sides. Reproduction by seed can be prolific. Lowland, but exceptionally at 410 m at Alston (Cumberland).

Native (change +1.30). *C. pendula* has shown a very pronounced northerly expansion since the 1962 *Atlas*. It is now widely grown in gardens as an ornamental plant, and often escapes, so that the limits of its native range are totally obscure; all records are mapped here as if they are native.

European Southern-temperate element.

References: Atlas (358d), Hultén & Fries (1986), Jermy *et al.* (1982), Meusel *et al.* (1965).

M. J. Y. FOLEY & M. S. PORTER

Carex sylvatica Wood-sedge

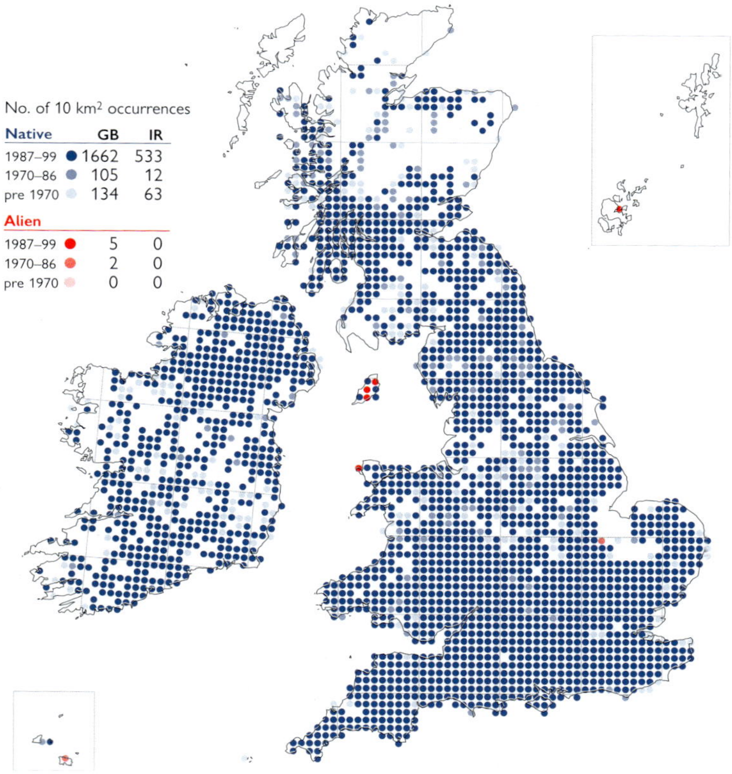

This perennial herb is found in a wide range of woodland habitats, generally preferring those where there is some base-enrichment and where the soil is moist and clayey. In many woods it is particularly frequent along the sides of paths or rutted tracks. It is occasionally found in open scrub and damp grassland, but likely to be a woodland relic in such places. Generally lowland, but reaching 565 m on Ben Bulben (Co. Sligo), and reportedly at 640 m in the Scottish Highlands.

Native (change +0.05). There has been no significant change in the distribution of *C. sylvatica* since the 1962 *Atlas*, and it is now better recorded.

Eurasian Temperate element.

References: Atlas (356b), Hultén & Fries (1986), Jermy *et al.* (1982), Meusel *et al.* (1965).

M. S. PORTER & M. J. Y. FOLEY

Carex capillaris Hair Sedge

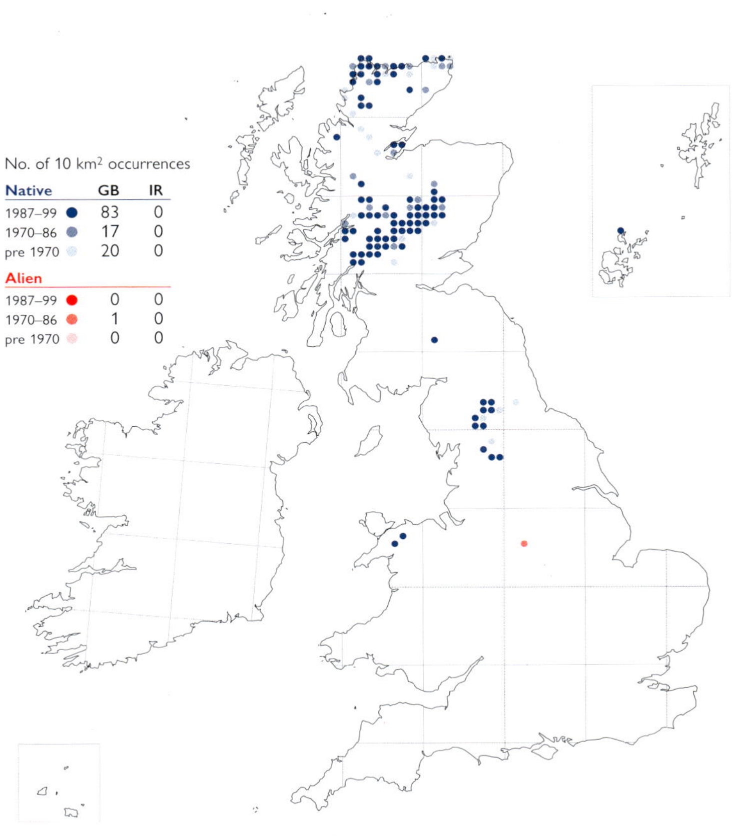

A perennial herb of base-rich upland grasslands, particularly those flushed by calcareous springs, and moist limestone or mica-schist crags, slopes and ledges. Though it can tolerate some shading, it is generally found in open situations and short vegetation, often in species-rich communities. It is usually found at moderate or high altitudes, reaching 1035 m on Ben Lawers (Mid Perth), but it descends to sea level in N. Scotland.

Native (change –0.35). The overall distribution of *C. capillaris* is unchanged since the 1962 *Atlas*.

Circumpolar Boreo-arctic Montane element.

References: Atlas (356c), Hultén & Fries (1986), Jermy *et al.* (1982), Stewart *et al.* (1994).

M. S. PORTER & M. J. Y. FOLEY

Carex strigosa Thin-spiked Wood-sedge

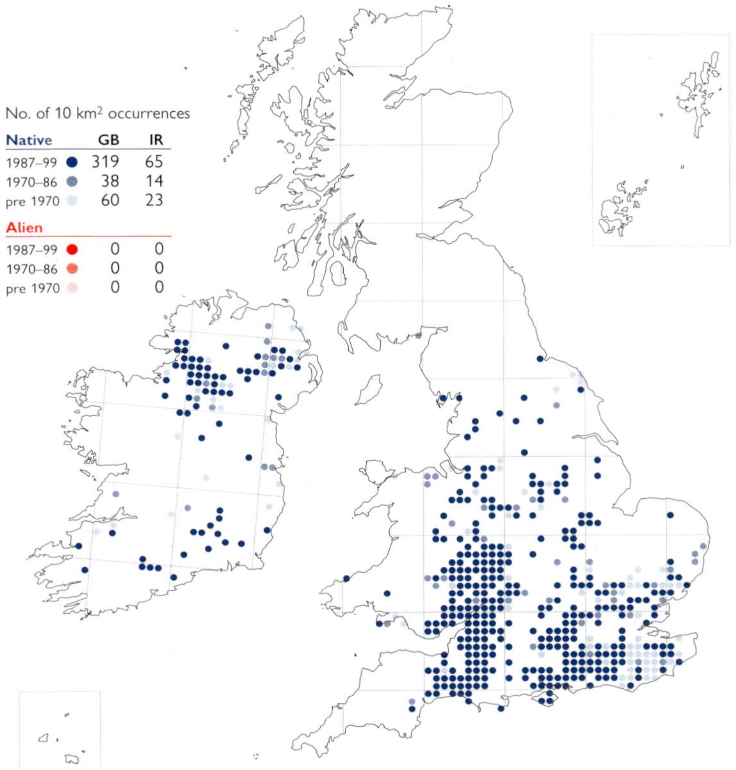

A perennial herb of moist, base-rich, sometimes clayey, soils in deciduous or mixed woodlands, often found near streams or seepages. It occurs most frequently in clearings and along tracks, but is sometimes found in considerable shade. Lowland.

Native (change +0.60). *C. strigosa* is easily overlooked, and it is better recorded now than in the 1962 *Atlas*. Many sites were lost before 1930, particularly in Ireland. Since 1962 its distribution has been stable other than in Sussex and Kent, where there has been a noteworthy decline, the reasons for which are unclear.

Suboceanic Temperate element.

References: Atlas (359a), Hultén & Fries (1986), Jermy *et al.* (1982).

M. J. Y. FOLEY & M. S. PORTER

Carex flacca Glaucous Sedge

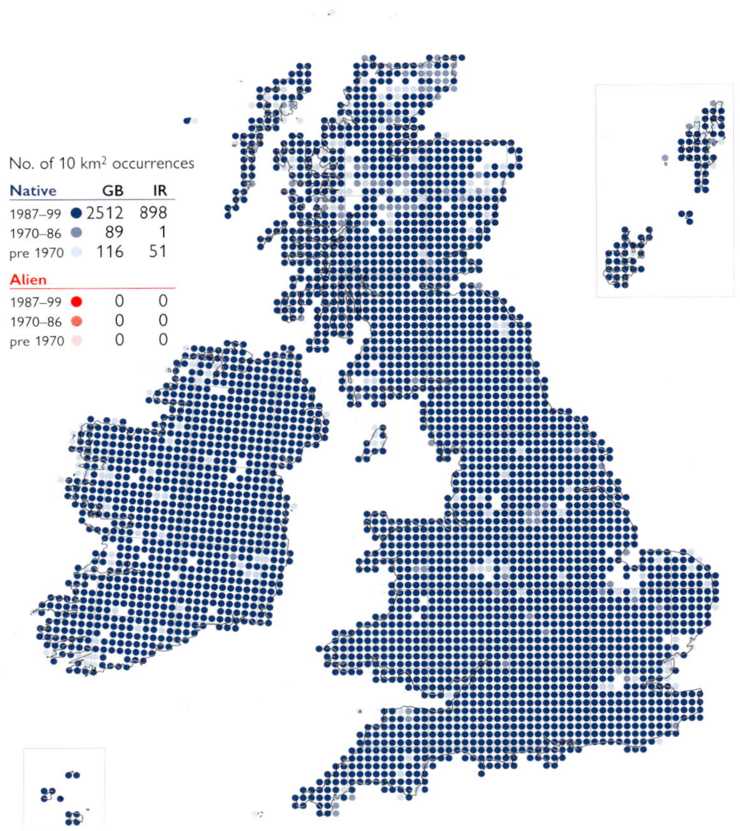

A rhizomatous perennial herb of unshaded neutral and calcareous grasslands over a wide range of substrates including chalk, limestone, sand and clay, and a frequent pioneer of disturbed bare areas. It is tolerant of both damp and dry conditions, and also occurs in wet meadows, on spray-drenched sea-cliffs, on the uppermost parts of saltmarshes, in base-rich mountain flushes and on rock ledges. Generally lowland, but reaching 790 m on Mt Brandon (S. Kerry).

Native (change +0.53). The distribution of *C. flacca* has not changed since the 1962 *Atlas*.

European Southern-temperate element; widely naturalised outside its native range.

References: Atlas (361a), Grime *et al.* (1988), Hultén & Fries (1986), Jermy *et al.* (1982), Taylor (1956).

M. J. Y. FOLEY & M. S. PORTER

Carex panicea Carnation Sedge

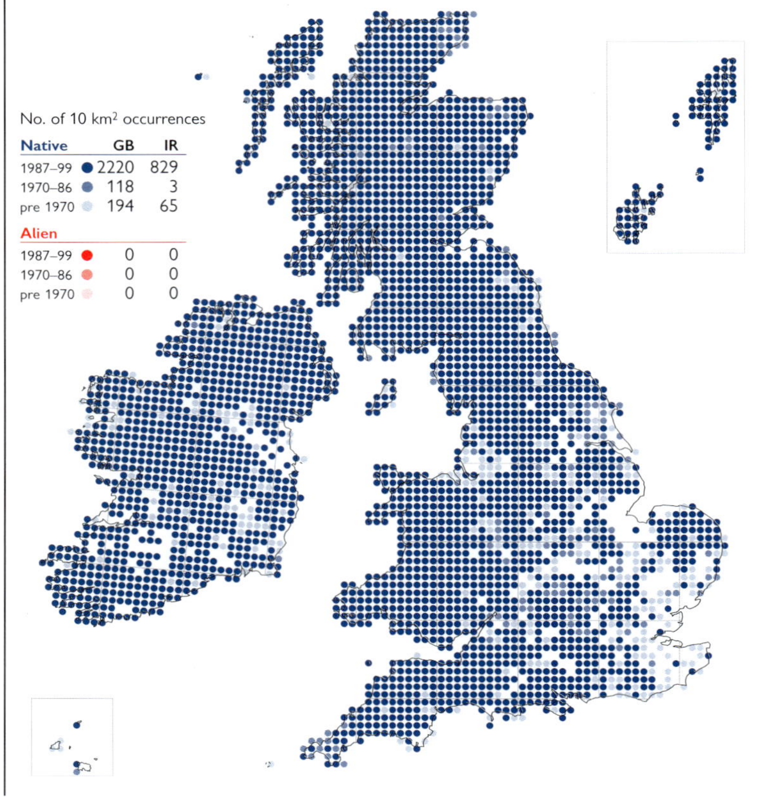

A perennial herb found in a wide range of damp or wet habitats on neutral to moderately acidic soils, including marshy grassland, hay- and water-meadows, mires, sea-cliff grassland, heaths and flushes in montane grassland; also, in N. & W. Britain and Ireland, in the uppermost parts of saltmarshes. 0–1125 m (Cairngorms).

Native (change –0.31). *C. panicea* was mapped as 'all records' in the 1962 *Atlas*. It has declined, particularly in S. and E. England, and analysis of the database reveals most losses have occurred since 1950. Its distribution in Ireland is more stable.

European Boreo-temperate element; also in C. Asia.

References: Atlas (359d), Grime *et al.* (1988), Hultén & Fries (1986), Jermy *et al.* (1982), Meusel *et al.* (1965).

M. J. Y. FOLEY & M. S. PORTER

Carex vaginata Sheathed Sedge

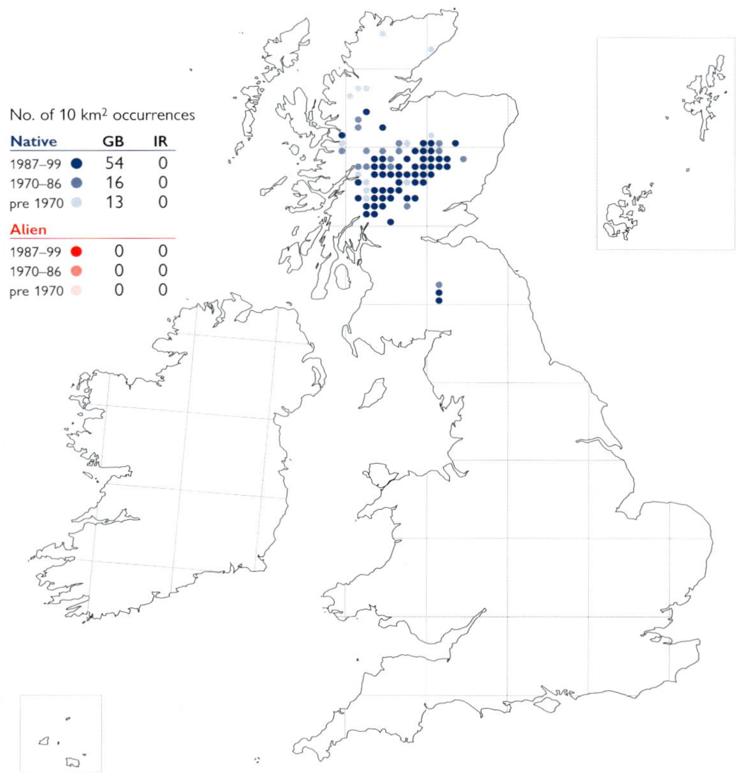

A rhizomatous perennial herb of flushed, moderately basic to slightly acidic mountain grassland; also found in flush bogs and on rock ledges. It can be locally abundant and form large patches by rhizomatous growth, but is a shy flowerer and for that reason may often be overlooked. From 370 m at Creag an Eirionnaich near Blair Atholl (Mid Perth) to 1150 m on Cairn Toul (S. Aberdeen), but usually found above 700 m.

Native (change +0.05). The overall distribution of *C. vaginata* is stable.

Circumpolar Boreo-arctic Montane element.

References: Atlas (360a), Hultén & Fries (1986), Jermy *et al.* (1982), Stewart *et al.* (1994).

M. J. Y. FOLEY & M. S. PORTER

Carex depauperata Starved Wood-sedge

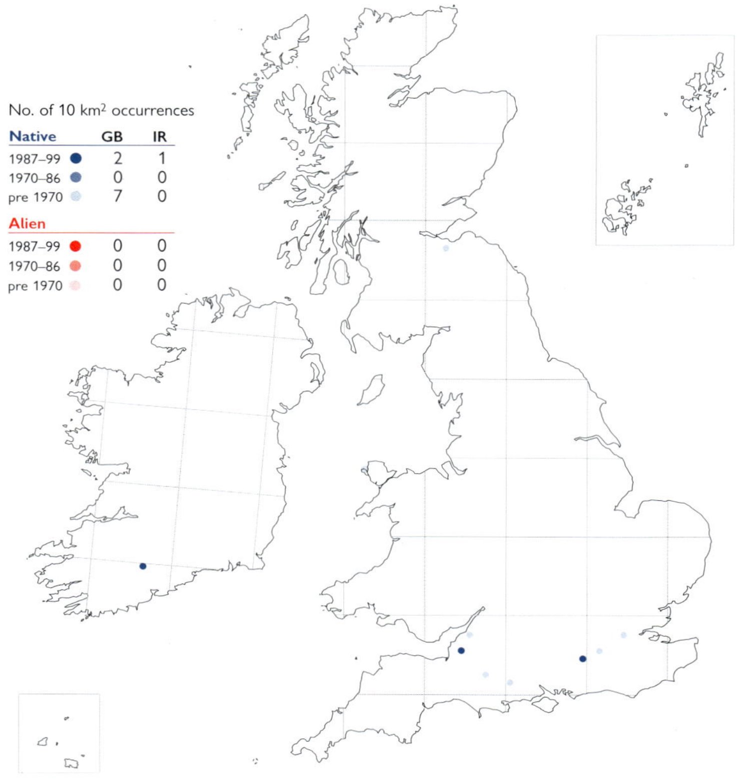

A perennial herb of dry, base-rich, free-draining soils. At its two extant sites in England it grows in semi-shade on laneside banks at the edge of deciduous woodland; in Ireland it grows in a partially shaded, wooded limestone area. It cannot tolerate deep shade. Lowland.

Native. Woodland removal or lack of woodland management probably account for the decline of this species in Britain. It responds well to coppicing and periodic disturbance. One English site has been augmented with re-introductions. The species has been recently added to the Scottish flora as a result of the discovery of old herbarium specimens (Rich *et al.*, 2000).

Submediterranean-Subatlantic element; also in C. Asia.

References: Atlas (356d), Curtis & McGough (1988), Jermy *et al.* (1982), O'Mahony (1976), Wigginton (1999).

M. S. PORTER & M. J. Y. FOLEY

Carex laevigata Smooth-stalked Sedge

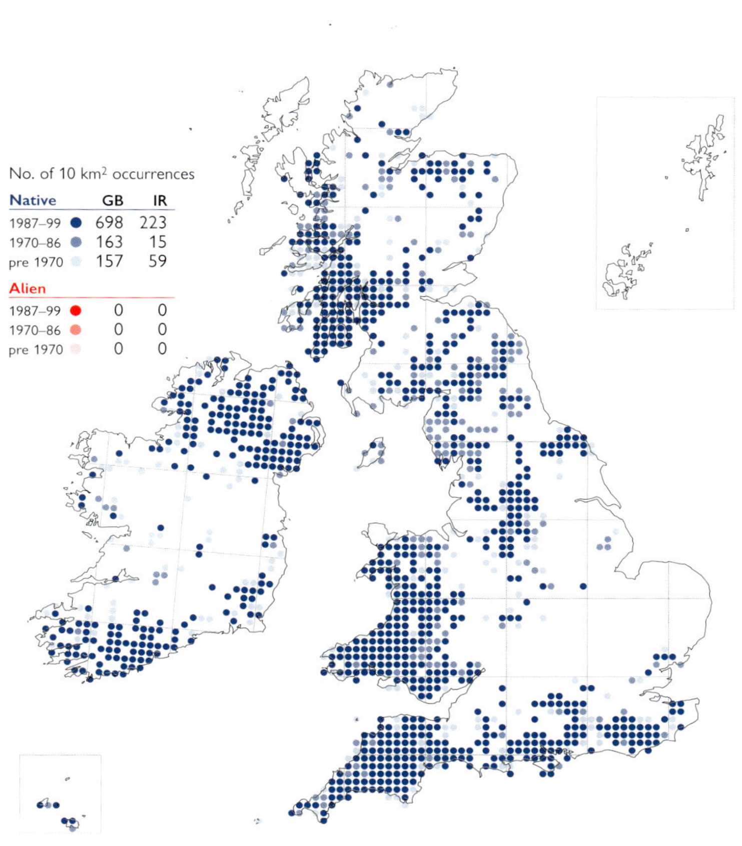

This perennial herb characteristically grows in moist woodlands on heavy clay soils, often where there is some flushing with base-rich water. Although most frequent in shaded sites, it is sometimes found in more open situations, such as on the edges of reedbeds, in open woodland on hillsides, or occasionally in open grassy flushes and damp meadows. It is a plant of low or moderate altitudes, but ascends to about 410 m in the Slaheny Valley (Co. Kerry).

Native (change –0.01). The distribution of this species has undergone little change since the 1962 *Atlas*.

Oceanic Temperate element.

References: Atlas (353d), Jermy *et al.* (1982).

M. S. PORTER & M. J. Y. FOLEY

Carex binervis Green-ribbed Sedge

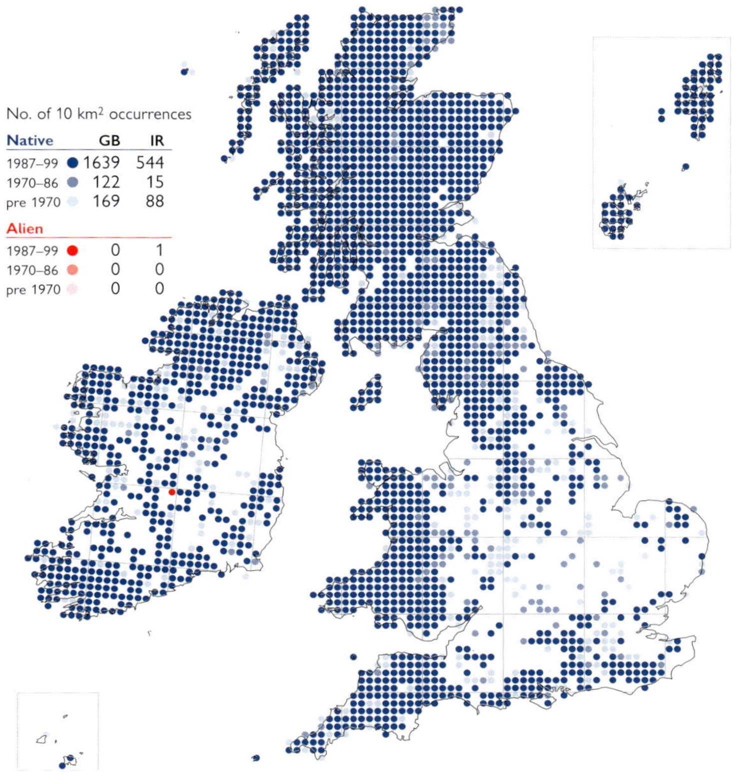

A perennial herb of both wet and dry acidic soils found in a wide range of habitats from lowland heaths and *Nardus* grassland to heather moors, rocky hillsides and mountain summits. Its preferred sites are generally open but it also occurs in shady places, including deciduous and mixed woodlands. It is also occasionally found on coastal cliffs. 0–930 m (Glyder Fach, Caerns.) and reportedly to 975 m in the Scottish Highlands.

Native (change –0.17). This species is common in suitable habitats, and its distribution appears to be stable. It was mapped as 'all records' in the 1962 *Atlas*.

Oceanic Temperate element.

References: Atlas (354d), Dupont (1962), Hultén & Fries (1986), Jermy *et al.* (1982).

M. S. PORTER & M. J. Y. FOLEY

Carex distans Distant Sedge

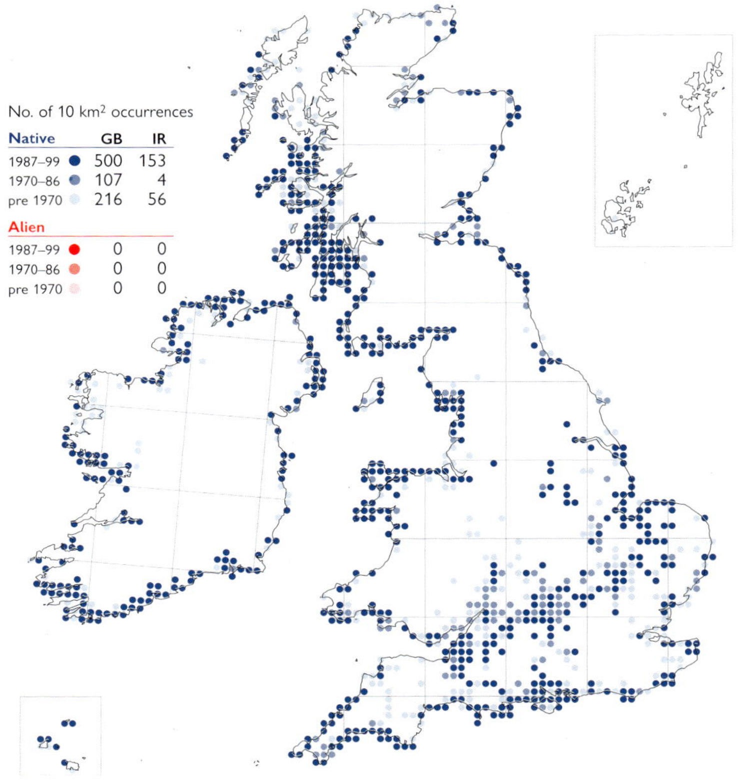

A perennial herb found on sea-cliffs, rocky shores, in coastal grasslands and the uppermost parts of saltmarshes; usually growing in bare rock-crevices or on free-draining sandy or gravelly substrates, but also in flushes and seepage zones on cliffs, and where streams debouch onto the shore. Inland, it occurs in wet meadows, marshes and fens on mineral-rich soils. Lowland.

Native (change –0.47). There has been little change in the distribution of *C. distans* since the 1962 *Atlas*, although it has declined inland due to the drainage and improvement of its wetland habitats.

European Southern-temperate element.

References: Atlas (354a), Hultén & Fries (1986), Jermy *et al.* (1982), Meusel *et al.* (1965).

M. J. Y. FOLEY & M. S. PORTER

Carex punctata Dotted Sedge

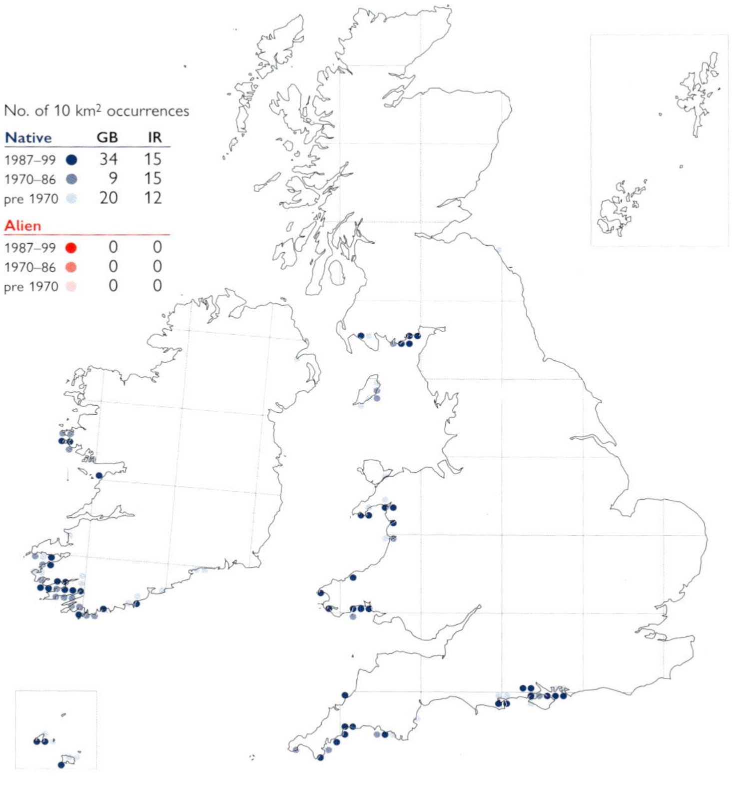

A perennial herb of sheltered rock ledges and clefts on sea-cliffs, invariably in seepage zones where freshwater trickles down the cliff-face. It is also found on wet sandy patches in saltmarshes, and amongst rocks or on sand where streams debouch onto the shore. It occurs in similar habitats to *C. distans*, and may sometimes be mistaken for it. Lowland.

Native (change +0.15). *C. punctata* tends to be rather mobile and is often present in small populations. However, the overall distribution appears to be largely stable, and recent recording has led to the discovery of many new sites within its known range.

Suboceanic Southern-temperate element.

References: Atlas (354b), Hultén & Fries (1986), Jermy *et al.* (1982), Stewart *et al.* (1994).

M. S. PORTER & M. J. Y. FOLEY

Carex extensa Long-bracted Sedge

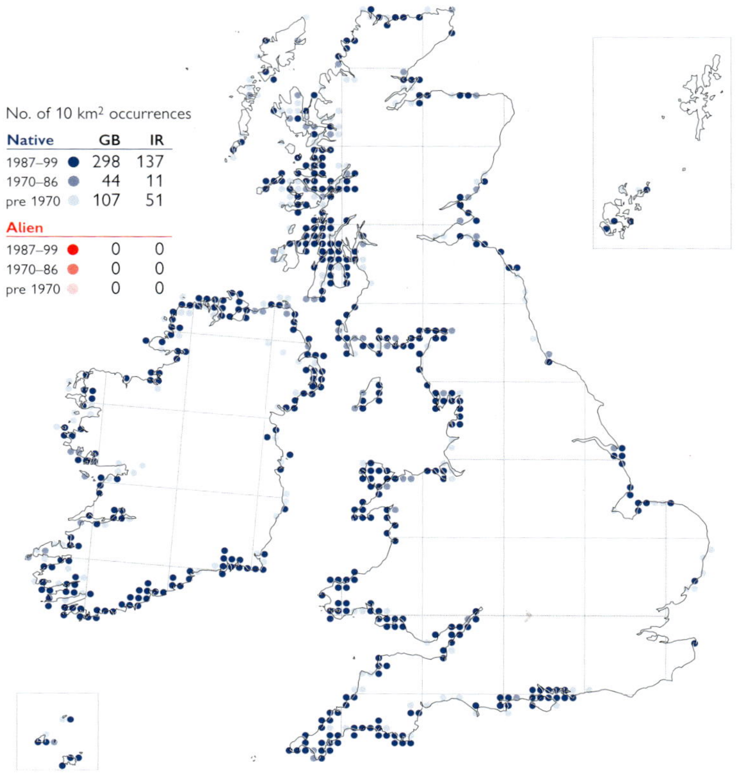

A perennial herb, mainly confined to areas within reach of sea water or spray. It is found on muddy or sandy estuarine flats, at the uppermost levels of saltmarshes and the edges of brackish ditches, and on moist coastal rocks and low cliffs. It often grows with *C. distans*. Lowland.

Native (change –0.23). There has been little change in the distribution of *C. extensa* since the 1962 *Atlas*.

European Southern-temperate element.

References: Atlas (356a), Hultén & Fries (1986), Jermy *et al.* (1982), Meusel *et al.* (1965).

M. J. Y. FOLEY & M. S. PORTER

Carex hostiana Tawny Sedge

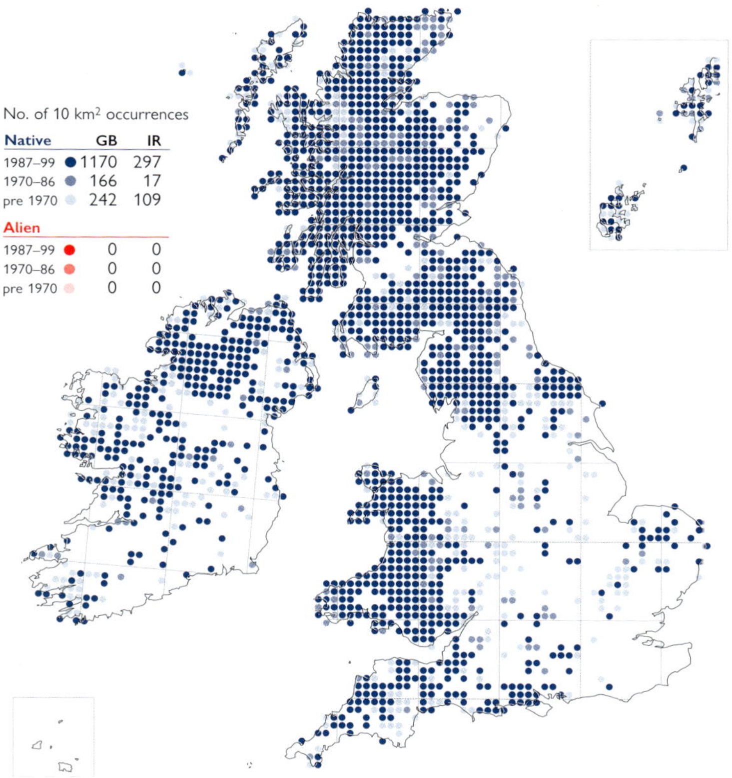

A perennial herb of damp, base-rich grassland and flushes. It occurs in fens, flushed valley bogs and mires, wet meadows and marshes. In lowland areas of Ireland it tolerates more acidic sites (Webb & Scannell, 1983). 0–630 m (Melmerby High Scar, Cumberland) and 760 m in W. Ross.

Native (change –0.05). *C. hostiana* was mapped as 'all records' in the 1962 *Atlas*. It has declined, particularly in S. England and S. Ireland, and analysis of the database reveals most losses have occurred since 1950. In upland areas, the distribution is more stable.

European Temperate element; also in N. America.

References: Atlas (354c), Hultén & Fries (1986), Jermy *et al.* (1982), Meusel *et al.* (1965).

M. S. PORTER & M. J. Y. FOLEY

Carex hostiana × *C. viridula* (*C.* × *fulva*)

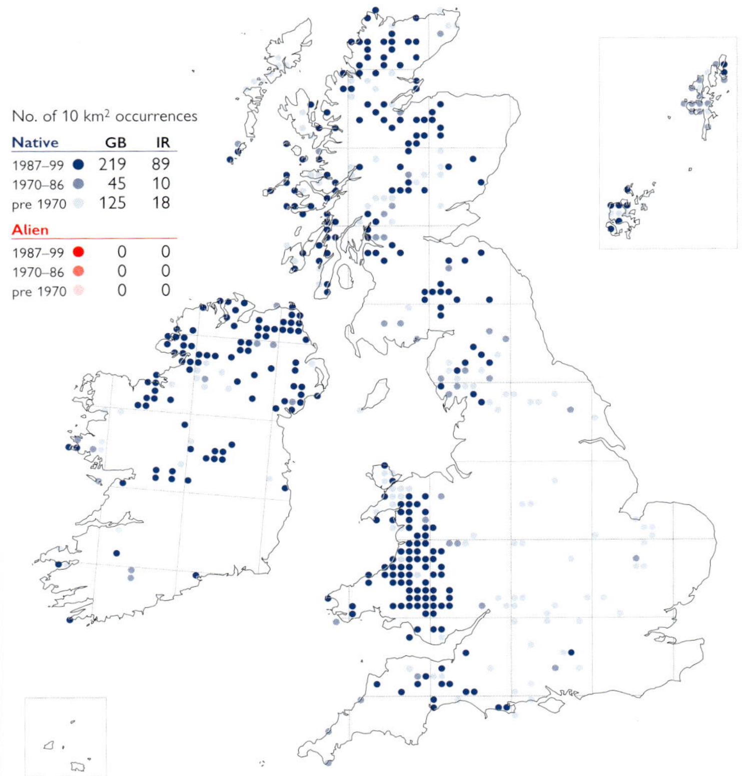

C. hostiana forms sterile hybrids with all three subspecies of *C. viridula*. These hybrids may be found in fens, wet meadows, heathland and moorland, and on the margins of lakes, especially where there is flushing with base-rich water. The hybrid with *C. viridula* subsp. *viridula*, the rarest of the three, is also found in dune-slacks. 0–440 m (Ulpha, Cumberland) and certainly higher in Scotland.

Native. *C.* × *fulva* is the most frequent of all sedge hybrids in the British Isles and is undoubtedly still under-recorded. Plants with *C. viridula* subsp. *oedocarpa* as one parent are most numerous, and sometimes occur in the absence of *C. hostiana*.

Widespread in temperate Europe.

References: Jermy *et al.* (1982), Stace (1975).

M. S. PORTER & M. J. Y. FOLEY

Carex flava Large Yellow-sedge

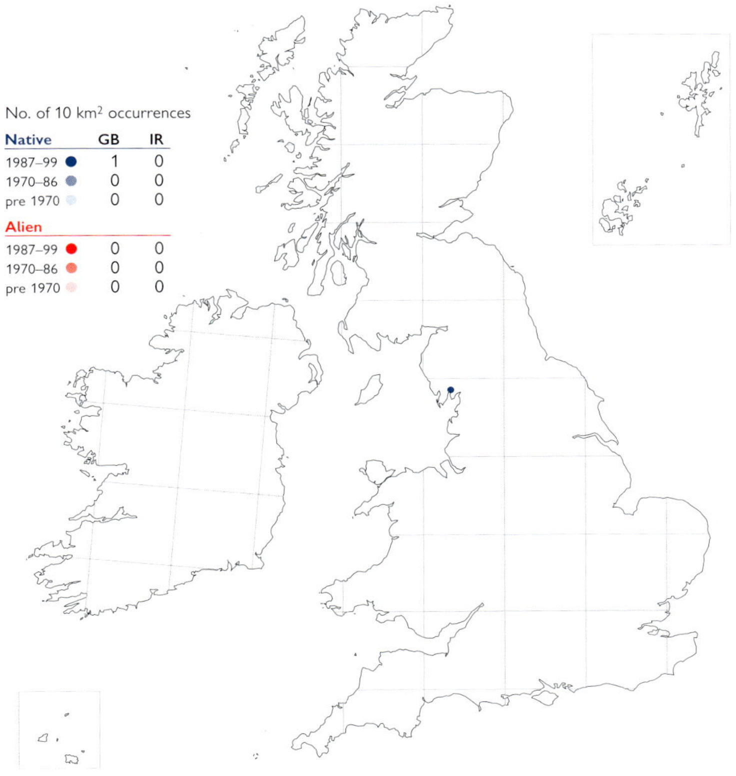

This perennial is found in the transition zone between woodland and raised mire, growing on peaty soil flushed by calcareous water from adjacent limestone outcrops. An open canopy or light shade is preferred. Lowland.

Native. *C. flava* has been confirmed from only one site, Roudsea Wood (Westmorland). Its favoured conditions of dappled shade are maintained by judicious thinning of the tree canopy. Hybrids between *C. flava* and *C. viridula* have been recorded from Greywell Moors (N. Hants.), Malham Tarn (Mid-W. Yorks.) and R. Corrib near Menlough (N.E. Galway), suggesting that *C. flava* might formerly have been more widespread.

European Boreo-temperate element; also in C. Asia and N. America.

References: Atlas (355a), Hultén & Fries (1986), Jermy *et al.* (1982), Wigginton (1999).

M. S. PORTER & M. J. Y. FOLEY

Carex viridula subsp. *brachyrrhyncha*

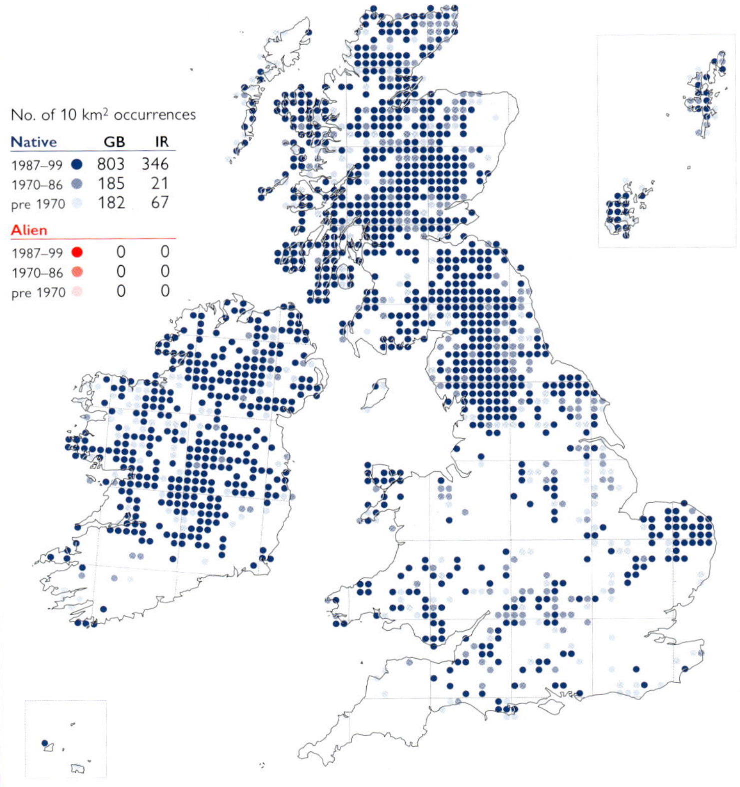

A perennial herb of fens and calcareous mires, especially in areas of winter flooding. It is also frequent base-rich flushes and wet ledges on sea-cliffs, hills and mountainsides, usually in short, open vegetation. 0–775 m (Coire nan Laogh, Glen Banchory, Easterness).

Native (change for species –0.01). The map shows many more records of this subspecies than in the 1962 *Atlas* but there have been local losses, particularly in S. England, presumably arising from drainage and agricultural improvements. Populations in many lowland areas are small.

European Temperate element; also in eastern N. America.

References: Atlas (355b), Hultén & Fries (1986), Jermy *et al.* (1982), Meusel *et al.* (1965).

M. S. PORTER & M. J. Y. FOLEY

Carex viridula subsp. *oedocarpa*

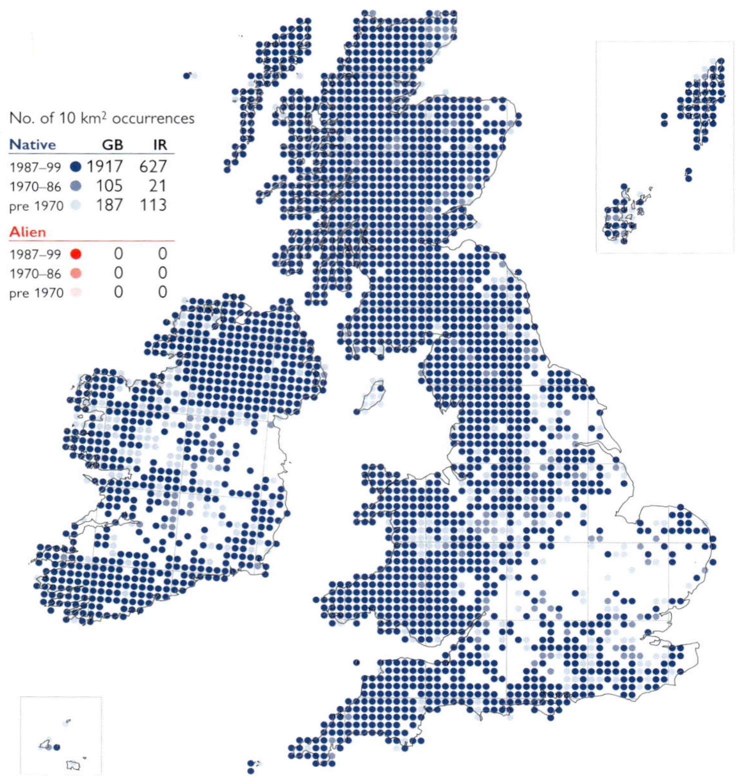

This perennial grows on wet ground on soils ranging from acidic to neutral or occasionally mildly alkaline. It is found in a wide variety of open habitats, including sandy shores, tracks, streamsides, wet fields and heathland, the margins of ponds and lakes, wet rocky hillsides, flushes and seepage areas. 0–930 m (Beinn Heasgarnich, Mid Perth).

Native (change for species –0.01). This, the most widespread of the *C. viridula* subspecies, was mapped as 'all records' in the 1962 *Atlas*. It is now much better recorded but it has declined at the edges of its range in S. England, and in the English and Irish Midlands, almost certainly due to drainage and habitat loss.

Suboceanic Boreo-temperate element; also in eastern N. America.

References: Atlas (355c), Hultén & Fries (1986), Jermy *et al.* (1982).

M. S. PORTER & M. J. Y. FOLEY

Carex viridula subsp. *viridula*

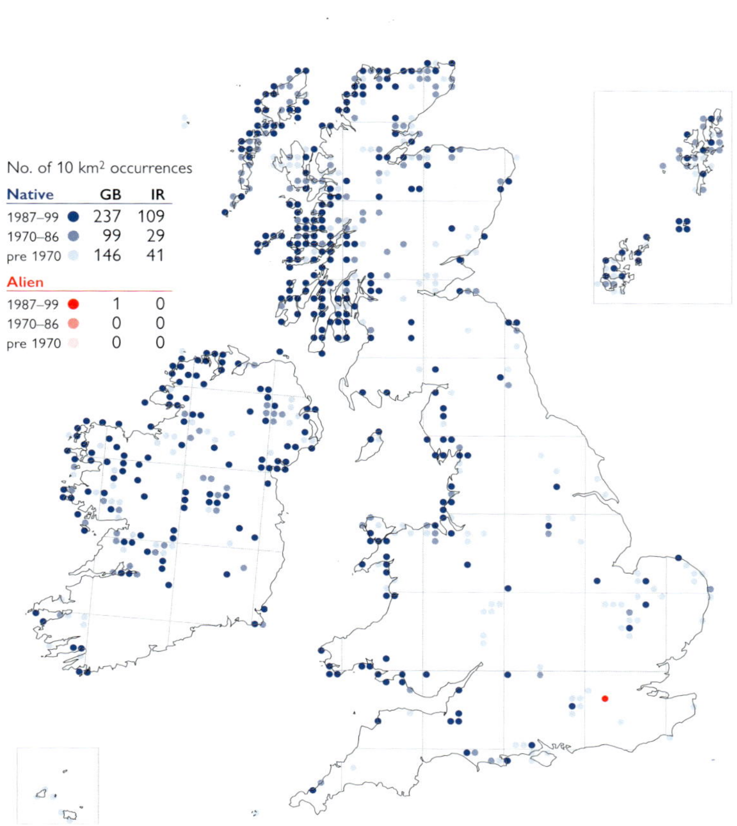

This perennial herb occurs in open, damp or wet habitats, including dune-slacks, the upper edge of saltmarshes and on the stony margins of lakes or pools, and in open fens and marshes. It usually occurs on flushed acidic soils, but also locally on base-rich substrates. Lowland, but possibly to 400 m in Perthshire.

Native (change for species –0.01). A diminutive plant which is easily overlooked, *C. viridula* subsp. *viridula* is known to be more frequent than when mapped by Jermy *et al.* (1982) and is probably still under-recorded in W. Scotland and Ireland.

European Boreo-temperate element; also in E. Asia and widespread in N. America.

References: Atlas (355d), Hultén & Fries (1986).

M. S. PORTER & M. J. Y. FOLEY

Carex pallescens Pale sedge

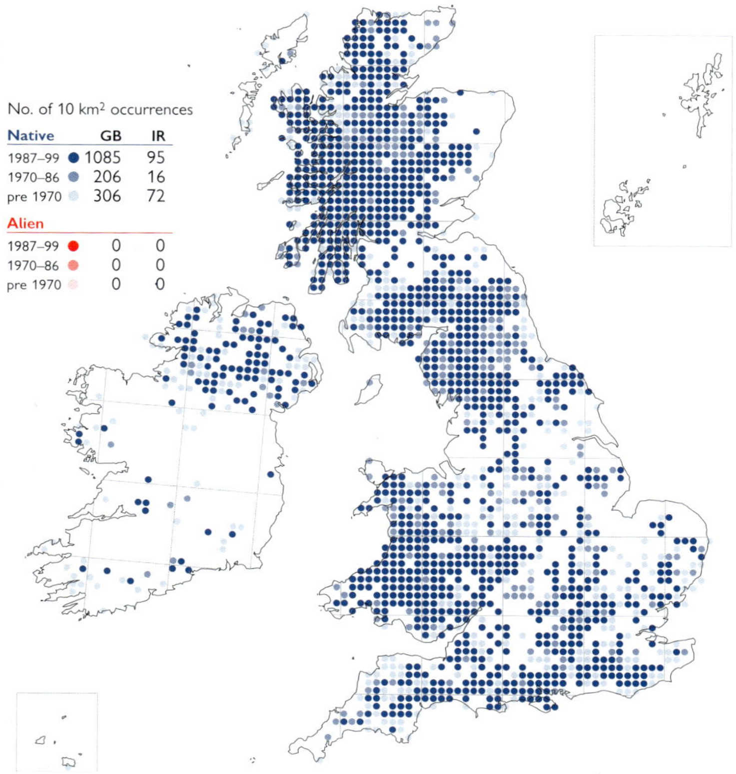

A perennial herb of damp grassland, woodland rides and clearings and stream banks; usually on moist, mildly acidic to neutral soils. In the Scottish mountains and some of the hills of northern England it is found on acidic, wet grassy slopes. Lowland to 790 m in Atholl (E. Perth).

Native (change –0.51). *C. pallescens* was mapped as 'all records' in the 1962 *Atlas*. There has clearly been a marked decline in many areas; analysis of the database reveals that most of the losses have occurred since 1950.

Eurosiberian Boreo-temperate element; also in N. America.

References: Atlas (359b), Hultén & Fries (1986), Jermy *et al.* (1982), Meusel *et al.* (1965).

M. J. Y. FOLEY & M. S. PORTER

Carex digitata Fingered Sedge

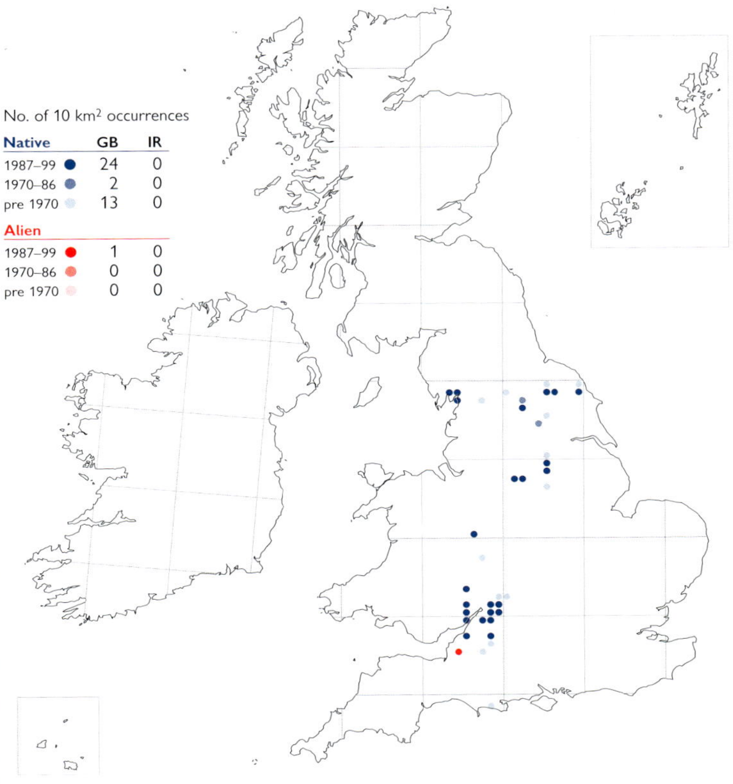

No. of 10 km² occurrences

Native		GB	IR
1987–99	●	24	0
1970–86	●	2	0
pre 1970	●	13	0
Alien			
1987–99	●	1	0
1970–86	●	0	0
pre 1970	●	0	0

A perennial herb, mainly occurring in open deciduous woodland over limestone, but also found on adjacent scree, rock outcrops, and sheltered limestone pavement. It seeds freely in lightly shaded habitats but ceases to flower and dies out if the canopy becomes too closed or the site overgrown. New plants can appear from dormant seed following disturbance. Lowland.

Native (change +0.04). The distribution of *C. digitata* seems stable, and it is now much better recorded than in the 1962 *Atlas*.

European Boreo-temperate element, with a continental distribution in W. Europe; also in E. Asia.

References: Atlas (363a), David (1978a), Hultén & Fries (1986), Jermy *et al.* (1982), Meusel *et al.* (1965), Stewart *et al.* (1994).

M. J. Y. FOLEY & M. S. PORTER

Carex ornithopoda Bird's-foot Sedge

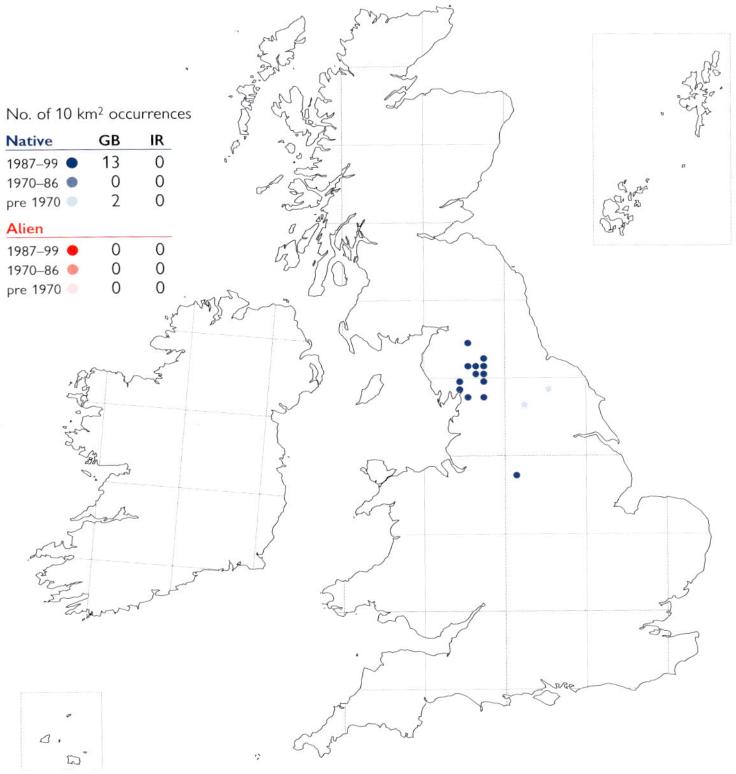

No. of 10 km² occurrences

Native		GB	IR
1987–99	●	13	0
1970–86	●	0	0
pre 1970	●	2	0
Alien			
1987–99	●	0	0
1970–86	●	0	0
pre 1970	●	0	0

A perennial herb of skeletal, well-drained soils on S.-facing slopes overlying Carboniferous limestone. It grows in open grassland, on rocky outcrops, screes, crags and limestone pavements, and occasionally also in partial shade in open limestone woodland. Lowland to 600 m on Long Fell, Warcop (Westmorland).

Native (change +0.28). *C. ornithopoda* is still very localised, but its known range has been extended since the 1962 *Atlas* by discoveries in Cumberland and Yorkshire. Populations may be increasing in Westmorland, but plants are often vulnerable to intensive grazing by sheep.

European Boreal-montane element.

References: Atlas (362d), Corner & Roberts (1989), David (1980), Hultén & Fries (1986), Jermy *et al.* (1982), Meusel *et al.* (1965), Porter & Roberts (1997), Wigginton (1999).

M. S. PORTER & M. J. Y. FOLEY

Carex humilis Dwarf Sedge

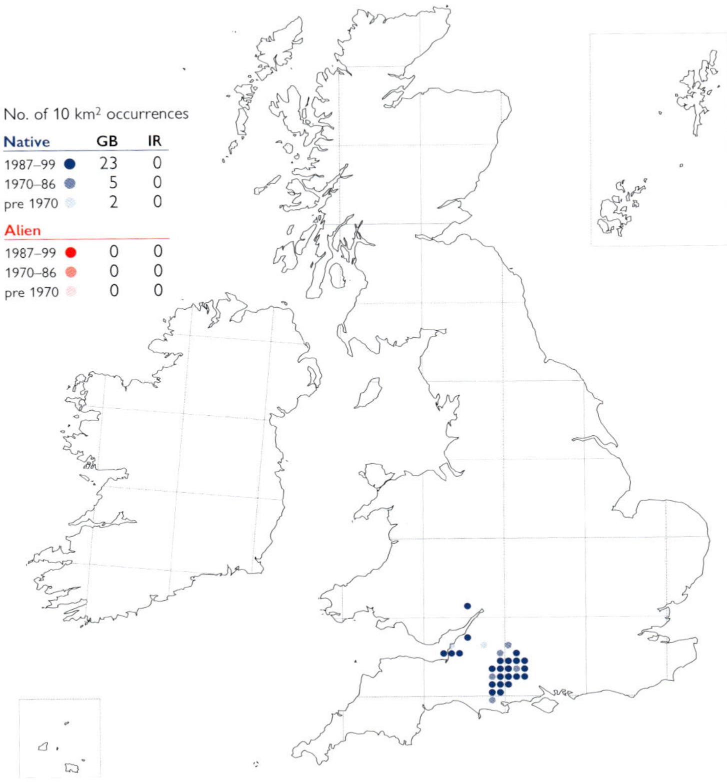

No. of 10 km² occurrences

Native		GB	IR
1987–99	●	23	0
1970–86	●	5	0
pre 1970	●	2	0
Alien			
1987–99	●	0	0
1970–86	●	0	0
pre 1970	●	0	0

A perennial herb of closely-grazed, calcareous grassland, especially on steep slopes on chalk downland. Over limestone it also occurs locally in grazed pastures as well as on field margins, track-sides and rock outcrops. It fruits freely, but regeneration from seed and spread into new areas has only rarely been reported. Lowland.

Native (change –0.01). The overall distribution of *C. humilis* is stable, but detailed studies in Dorset have revealed that over 10% of its recorded sites have been ploughed up (Pearman, 1997). Many extant sites now have some form of statutory protection, but the smaller ones are often ungrazed.

Eurasian Temperate element, with a continental distribution in W. Europe.

References: Atlas (362d), Jermy *et al.* (1982), Meusel *et al.* (1965), Stewart *et al.* (1994).

M. J. Y. FOLEY & M. S. PORTER

Carex caryophyllea Spring-sedge

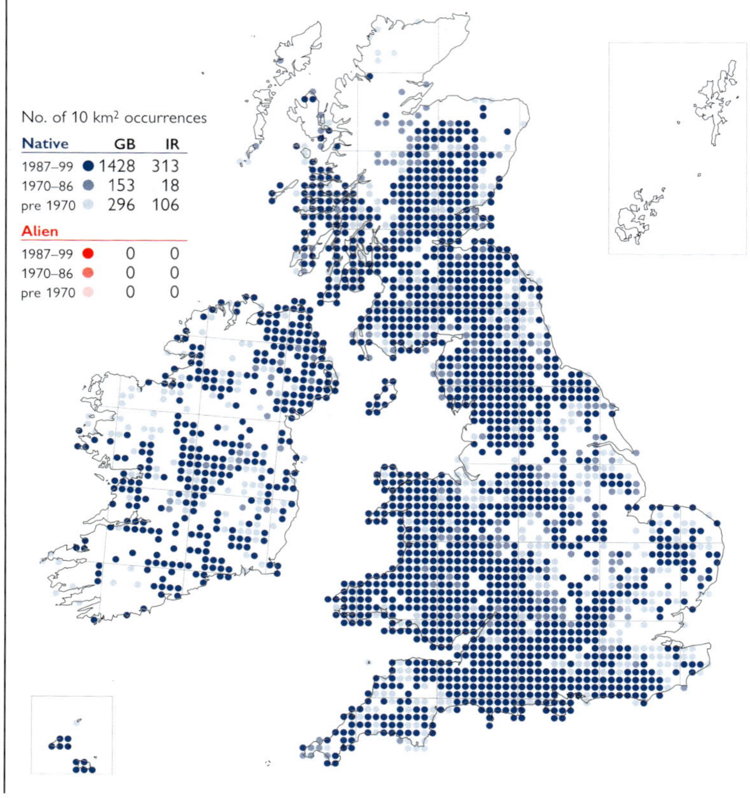

A shortly rhizomatous perennial typically found in meadows and pastures on dry, calcareous or base-rich soils; also occurring in grasslands and heaths on mildly acidic soils, and in well-drained pastures, ledges and flushes in the uplands. 0–765 m (Knock Fell, Westmorland).

Native (change –0.20). *C. caryophyllea* was mapped as 'all records' in the 1962 *Atlas*. It has declined, especially in Devon, the English Midlands and S.E. England, and analysis of the database reveals that most of the losses have occurred since 1950.

Eurosiberian Temperate element.

References: Atlas (362b), Grime *et al.* (1988), Hultén & Fries (1986), Jermy *et al.* (1982), Meusel *et al.* (1965).

M. J. Y. FOLEY & M. S. PORTER

Carex filiformis Downy-fruited Sedge

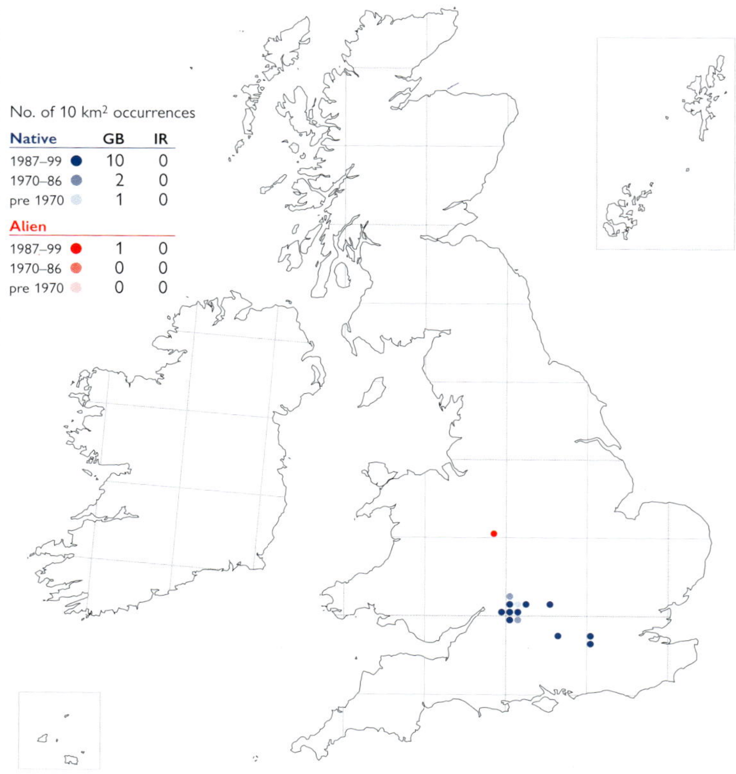

This rhizomatous perennial herb occurs on calcium-rich soils in a wide range of habitats. It is most characteristic of damp meadows, but is also found in dry grassland, grassy woodland rides and along roadsides. Lowland.

Native (change +0.23). *C. filiformis* is vulnerable to agricultural improvement, and some sites have been lost since the 1962 *Atlas*, though the situation now seems to have stabilised and all of its 13 extant sites have some degree of protection.

Eurosiberian Temperate element, with a continental distribution in W. Europe.

References: Atlas (359c), Hultén & Fries (1986), Jermy *et al.* (1982), Wigginton (1999).

M. S. PORTER & M. J. Y. FOLEY

Carex ericetorum Rare Spring-sedge

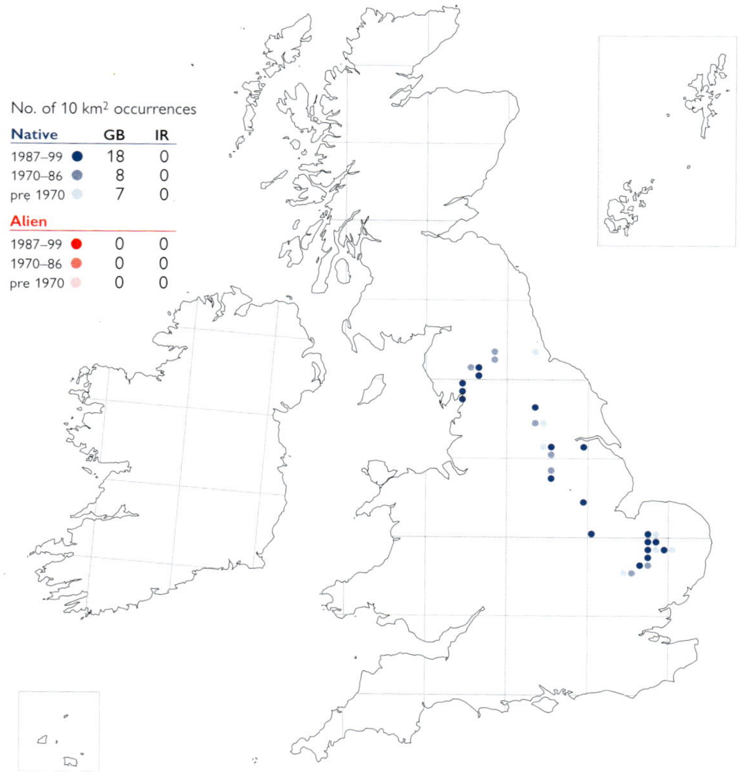

A perennial herb restricted to dry, grazed grasslands; it occurs on infertile calcareous soils overlying limestone, chalk or chalky boulder-clay. It is a poor competitor and soon disappears if under-grazing allows the sward to become too rank. It reproduces vegetatively and by seed, albeit sparingly. Lowland, reaching 400 m near Shap (Westmorland).

Native (change –0.46). *C. ericetorum* may be under-recorded, being very similar to *C. caryophyllea*. However, it has clearly declined, losses being due to ploughing, lack of grazing and nutrient enrichment from adjoining arable land.

Eurosiberian Boreal-montane element, with a continental distribution in W. Europe.

References: Atlas (362a), David (1981), Hultén & Fries (1986), Jermy *et al.* (1982), Meusel *et al.* (1965), Stewart *et al.* (1994).

M. J. Y. FOLEY & M. S. PORTER

Carex montana Soft-leaved Sedge

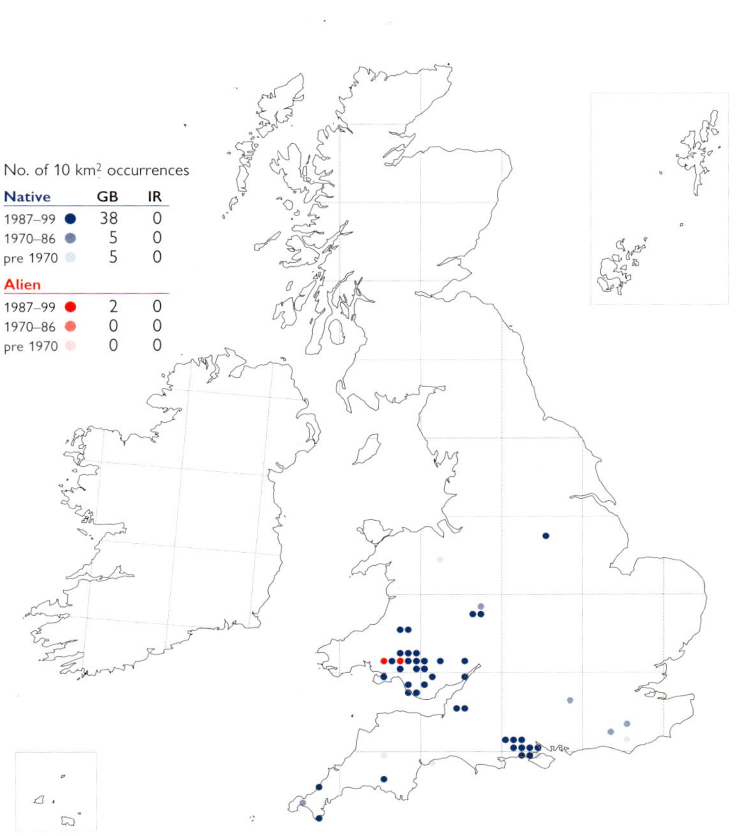

This perennial herb was thought to be confined to rough, open grassland on limestone. However, recent studies have shown that it grows at these sites only where non-calcareous drift overlays the calcareous bedrock, and it can in fact thrive in neutral to acidic grassland, on heathland and in woodland rides, often in partial shade. Generally lowland, but reaching 560 m at Carreg yr Ogof (Carms.).

Native (change +0.68). The recognition of the wider ecological amplitude of *C. montana* has led to its discovery in many additional 10-km squares since the 1962 *Atlas*.

European Temperate element, with a continental distribution in W. Europe; also in C. and E. Asia.

References: Atlas (362c), Hultén & Fries (1986), Jermy *et al.* (1982), Kay & John (1994), Meusel *et al.* (1965), Stewart *et al.* (1994).

M. J. Y. FOLEY & M. S. PORTER

Carex pilulifera Pill Sedge

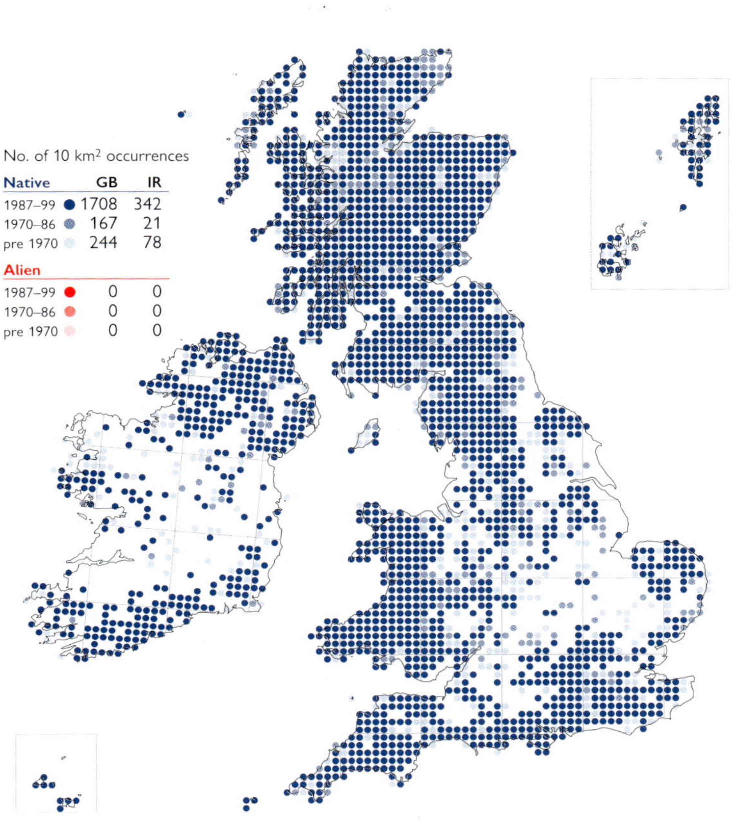

A perennial herb of dry sites on base-poor, usually acidic soils; its habitats include sandy heaths, dune grassland, open woodlands, upland grassland and moorland. 0–1140 m (Beinn a' Bhuird, S. Aberdeen).

Native (change –0.04). *C. pilulifera* was mapped as 'all records' in the 1962 *Atlas*. Its overall distribution appears to be little changed.

European Temperate element.

References: Atlas (361d), Grime *et al.* (1988), Hultén & Fries (1986), Jermy *et al.* (1982), Meusel *et al.* (1965).

M. S. PORTER & M. J. Y. FOLEY

Carex atrofusca Scorched Alpine-sedge

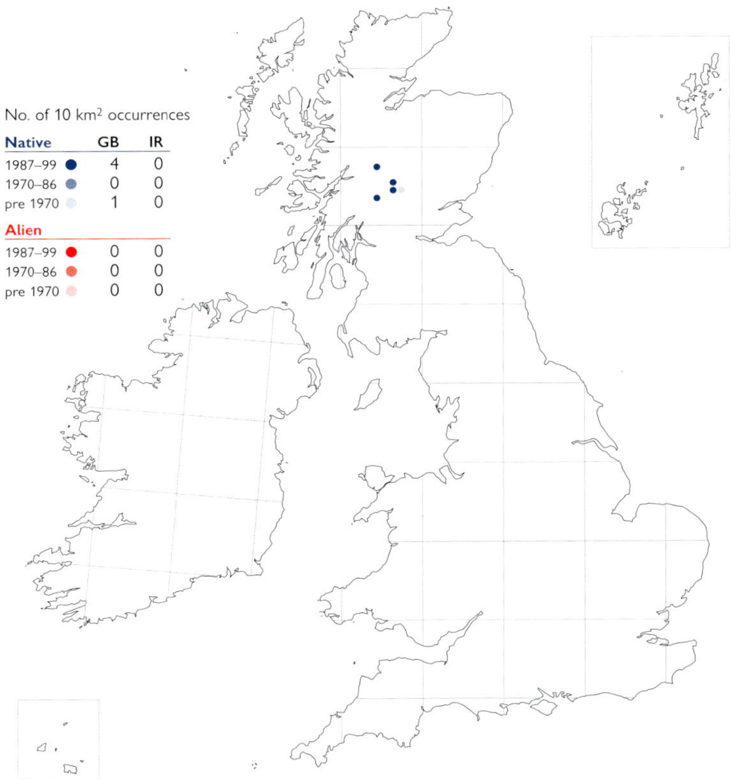

This perennial herb grows in stony, calcareous, usually micaceous flushes (especially at the point of emergence of springs) and bogs, and rarely on wet mountain ledges and crags. It often grows on or amongst grassy tussocks. Formerly recorded below 600 m, it is currently known from five localities between 680 m (Ben Lawers, Mid Perth) and 1000 m (Beinn Heasgarnich, Mid Perth).

Native (change –0.11). There has been little change in the distribution of this species since the 1962 *Atlas*. A record from Beinn an Dothaidh (Main Argyll) in 1976 has not been confirmed and is now regarded as dubious.

Circumpolar Arctic-montane element.

References: Atlas (363d), Hultén & Fries (1986), Jermy *et al.* (1982), Meusel *et al.* (1965), Wigginton (1999).

M. S. PORTER & M. J. Y. FOLEY

Carex limosa Bog-sedge

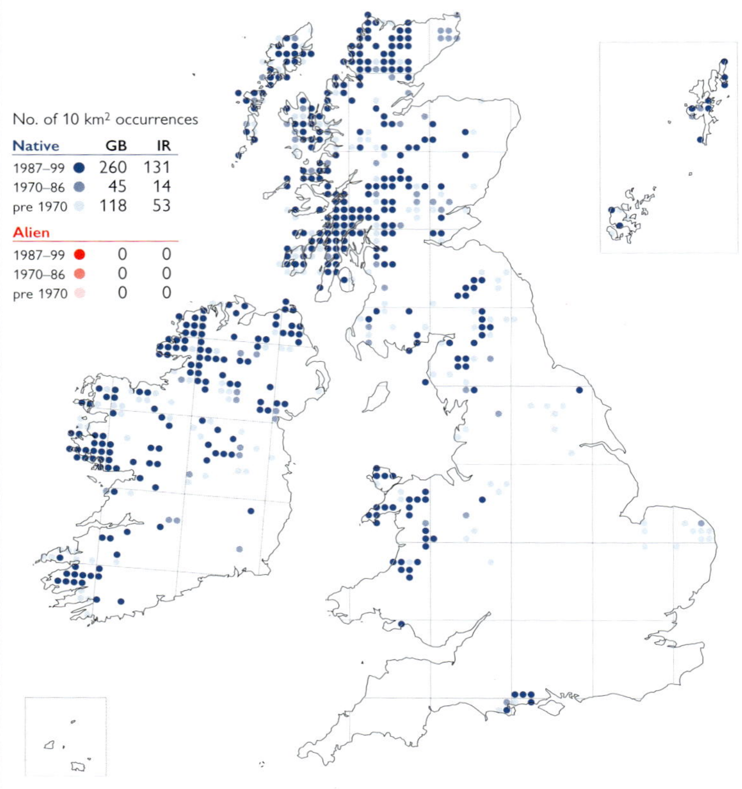

A perennial herb of *Sphagnum* mires and the wet, peaty margins of pools, often growing in standing water. Most of its sites are acidic and oligotrophic but, unlike *C. magellanica*, it tends to occur in areas subject to some mineral enrichment. Generally lowland, but reaching 830 m on Meall nan Tarmachan (Mid Perth).

Native (change +0.14). *C. limosa* is now much better recorded than it was in the 1962 *Atlas*, and the distribution is probably stable, at least in Wales, Scotland and W. & N. Ireland. In England and C. Ireland it was lost from many lowland sites before 1930, and this decline has continued due to drainage, afforestation and, in Ireland, peat extraction.

Circumpolar Boreal-montane element.

References: Atlas (360b), Hultén & Fries (1986), Jermy *et al.* (1982), Preston & Croft (1997).

M. S. PORTER & M. J. Y. FOLEY

Carex rariflora Mountain Bog-sedge

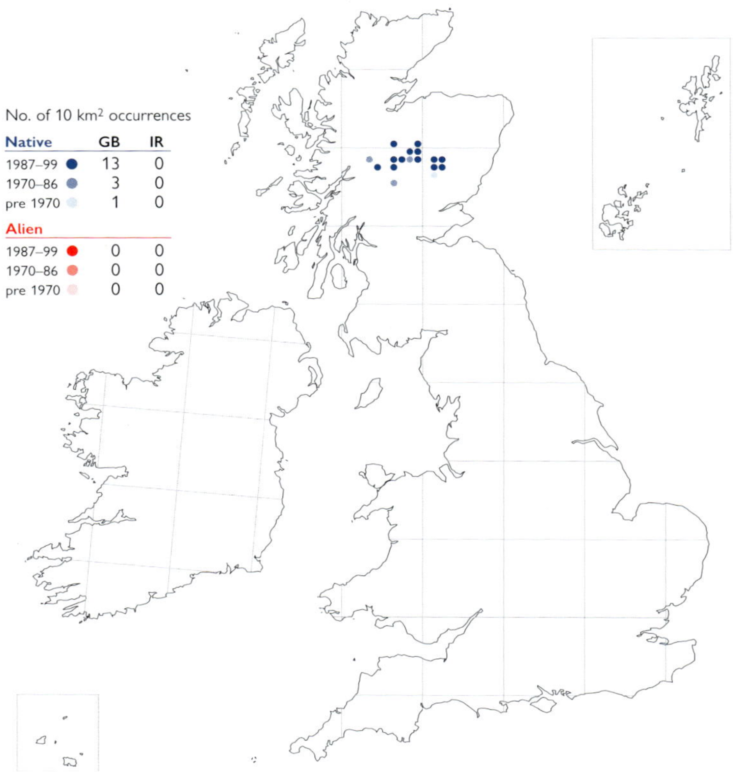

A perennial herb of wet, base-poor peaty substrates, mainly occurring in flush-bogs on gentle slopes, often in areas of late snow-lie or by streams flowing from them. It also occurs by pools, and on terraces by the side of incised burns. From 790 m in Drumochter Forest (Westerness) to 1125 m at Lochan Bhuidhe and Ben Macdui (both Banffs.).

Native (change +0.28). Many new sites for *C. rariflora* have been discovered since the 1962 *Atlas*, some of them very extensive, and more than thirty populations are now known.

Circumpolar Arctic-montane element; absent from mountains of C. Europe.

References: Atlas (360d), Hultén & Fries (1986), Jermy *et al.* (1982), Wigginton (1999).

M. S. PORTER & M. J. Y. FOLEY

Carex magellanica Tall Bog-sedge

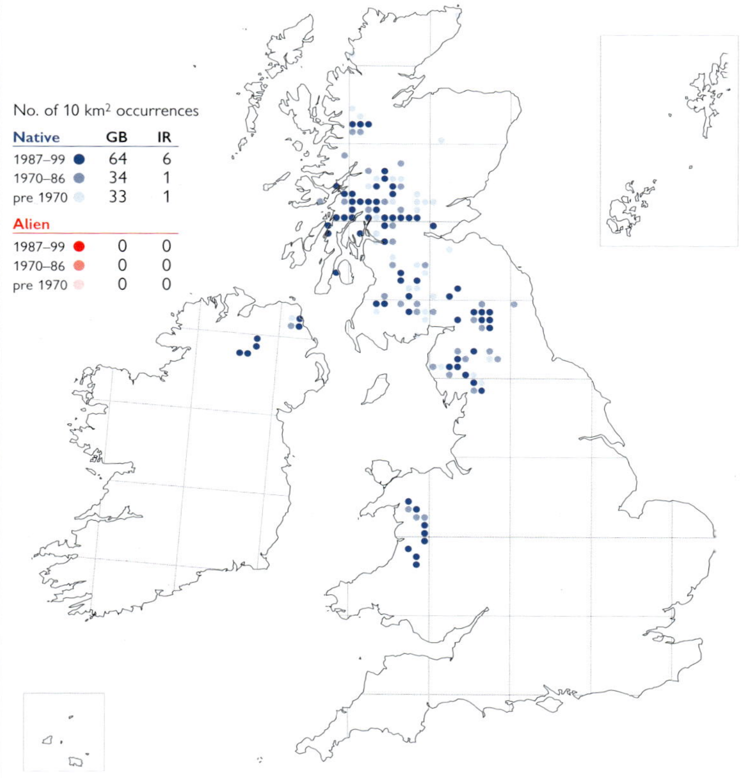

A perennial of wet ground, pools and hummocks in *Sphagnum* bogs, or at the edges of gently sloping mires where there is slight lateral water movement; such sites often occur on watersheds. It generally occurs in open ground, but sometimes persists in carr. From 30 m (Shian, Main Argyll), but generally upland, reaching 685 m (Ben Lui, Mid Perth).

Native (change −0.02). *C. magellanica* is thinly scattered in suitable habitats. Some colonies have been lost as a result of drainage and afforestation but many new sites have been found, particularly in Wales, where it was collected *c.* 1835 but not refound until 1963, in Main Argyll and in N. Ireland.

Circumpolar Boreal-montane element.

References: Atlas (360c), Curtis & McGough (1988), Hultén & Fries (1986), Jermy *et al.* (1982), Stewart *et al.* (1994).

M. S. PORTER & M. J. Y. FOLEY

Carex atrata Black Alpine-sedge

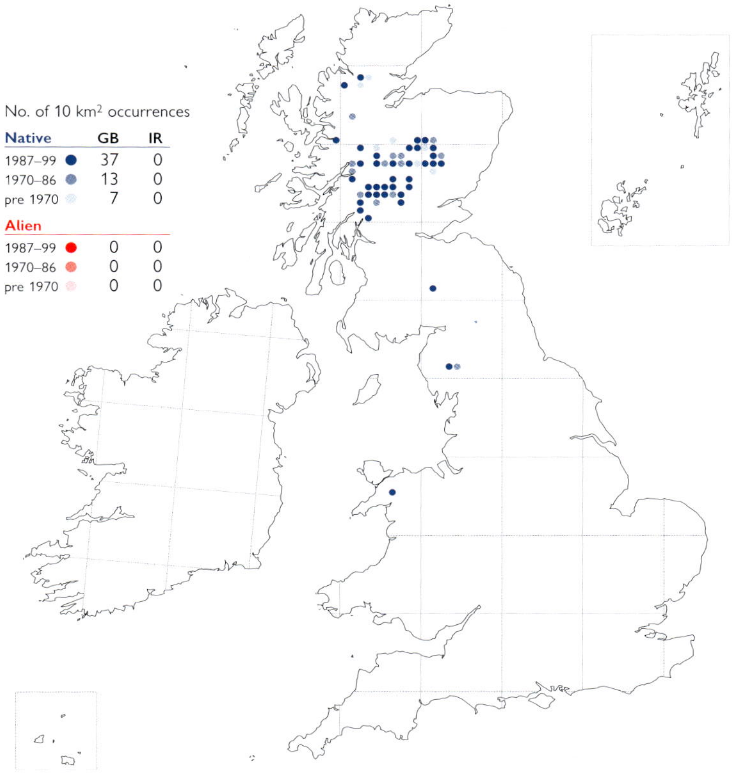

A perennial herb found on ungrazed faces and ledges of wet or dry calcareous cliffs. It grows in short vegetation or amongst tall herbs, or in dwarf *Salix* scrub. Populations are probably maintained by vegetative growth. From 550 m in Coire Ghamhnain (Main Argyll) and usually above 700 m, reaching 1095 m on Ben Lawers (Mid Perth).

Native (change –0.02). The distribution of *C. atrata* is stable.

Circumpolar Boreo-arctic Montane element.

References: Atlas (363c), Hultén & Fries (1986), Jermy *et al.* (1982), Meusel *et al.* (1965), Stewart *et al.* (1994).

M. J. Y. FOLEY & M. S. PORTER

Carex buxbaumii Club Sedge

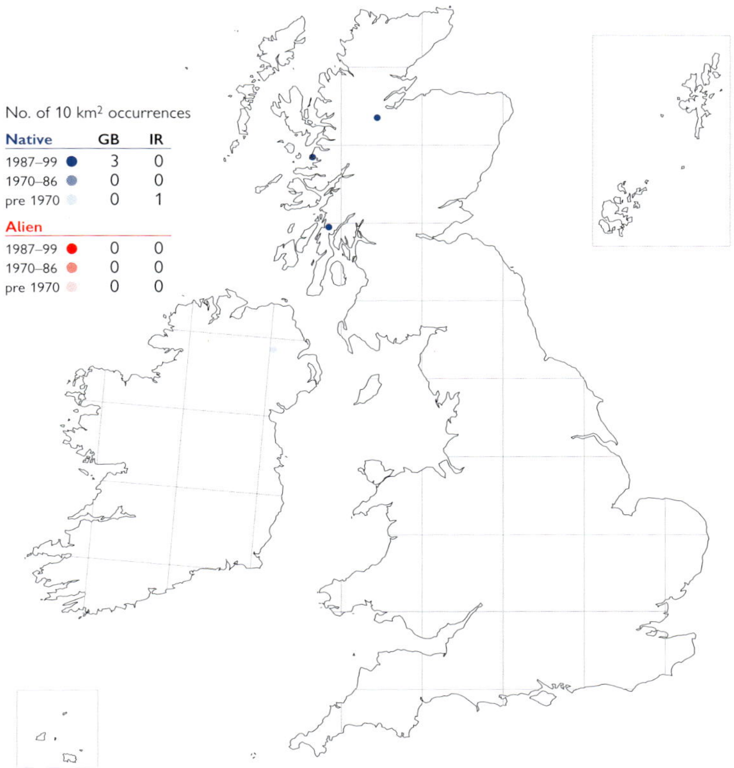

A perennial herb found in mesotrophic fens on or near the margins of lakes, often bordering outflow streams, and always in areas which are periodically inundated. It is often subject to light grazing by cattle. Lowland.

Native. *C. buxbaumii* is known from four sites in Scotland, two of which have been discovered since the 1962 *Atlas*. All colonies appear to be thriving. In Lough Neagh, it was last recorded on an island in 1886, shortly after the 'almost impenetrable thicket of scrubby wood' in which it grew was cleared and the area grazed as pasture.

Eurosiberian Boreal-montane element, with a continental distribution in W. Europe; also in E. Asia and N. America.

References: Atlas (363b), Harron (1986), Hultén & Fries (1986), Jermy *et al.* (1982), Meusel *et al.* (1965), Wigginton (1999).

M. S. PORTER & M. J. Y. FOLEY

Carex norvegica Close-headed Alpine-sedge

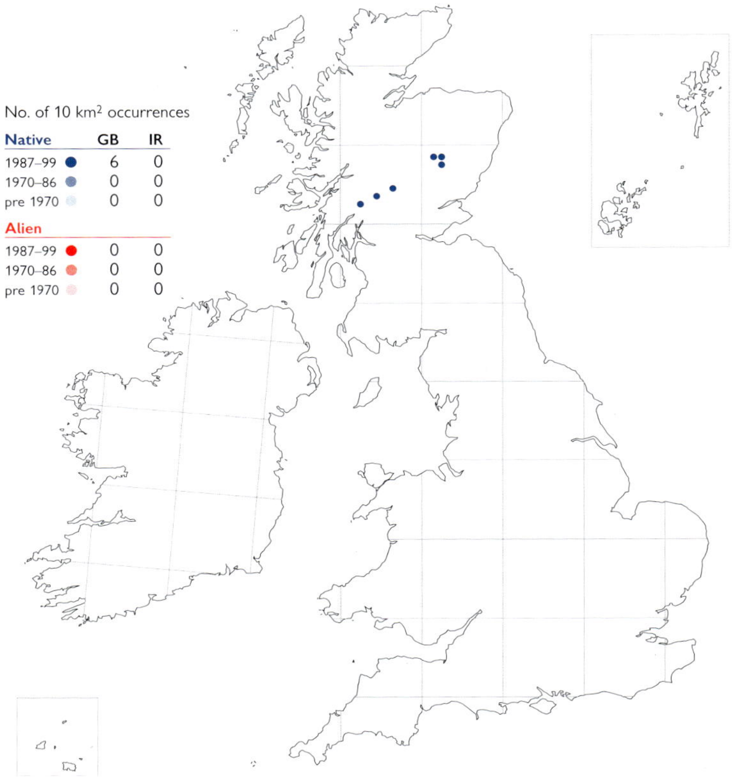

C. norvegica is a perennial herb of wet, stony slopes, ledges and turf over basic rock and with base-rich run-off. All the sites have a mainly N.-facing aspect, and occur in places where snow lies late. Populations are usually quite small and of very limited extent. From 700 m at Corrie Fee (Angus) to 975 m on Beinn Heasgarnich (Mid Perth).

Native. The distribution of *C. norvegica* is stable, but it has recently been discovered in a new locality in W. Perth.

Circumpolar Arctic-montane element.

References: Atlas (363b), Hultén & Fries (1986), Jermy *et al.* (1982), Meusel *et al.* (1965), Wigginton (1999).

M. J. Y. FOLEY & M. S. PORTER

Carex recta Estuarine Sedge

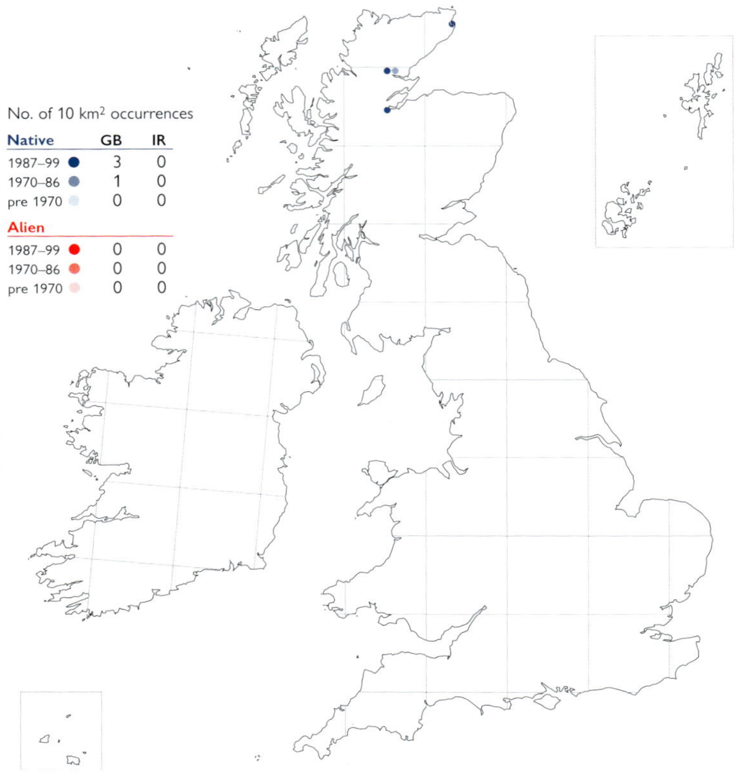

A rhizomatous perennial found in marshes along the lower reaches and estuaries of the Wick River (Caithness), the River Beauly and the Kyle of Sutherland (both E. Ross), growing in places where silt is periodically deposited or where the water-table fluctuates. It sets few viable seeds and reproduction is mostly vegetative. Lowland.

Native. The distribution of *C. recta* is unchanged since the 1962 *Atlas*, but the population on the Beauly River has declined considerably since the 1980s. Faulkner (1972) considered that this taxon probably arose through hybridisation of *C. aquatilis* and the Scandinavian species *C. paleacea*.

Oceanic Boreal-montane element; also in N. America.

References: Atlas (364a), Hultén & Fries (1986), Jermy *et al.* (1982), Preston & Croft (1997), Wigginton (1999).

M. S. PORTER & M. J. Y. FOLEY

Carex aquatilis Water Sedge

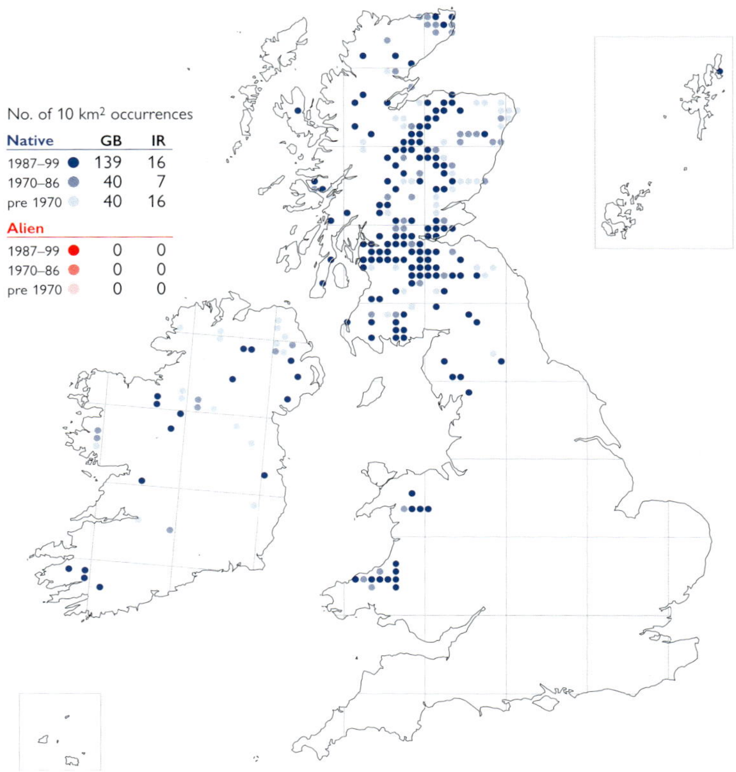

A morphologically variable, rhizomatous perennial. In the lowlands, robust plants grow on river banks and the margins of lakes, mires and reed-swamps. In its upland sites, it is a shorter plant and often grows on deep, wet, gently sloping peat. 0–975 m (Glas Maol, Angus).

Native (change +0.76). Because of better recognition and recording, *C. aquatilis* is now known from very many more 10-km squares than it was in the 1962 *Atlas*, although it may remain overlooked in some areas.

Circumpolar Boreo-arctic Montane element; absent from mountains of C. Europe.

References: Atlas (364c), Hultén & Fries (1986), Jermy *et al.* (1982), Preston & Croft (1997), Stewart *et al.* (1994).

M. S. PORTER & M. J. Y. FOLEY

Carex acuta Slender Tufted-sedge

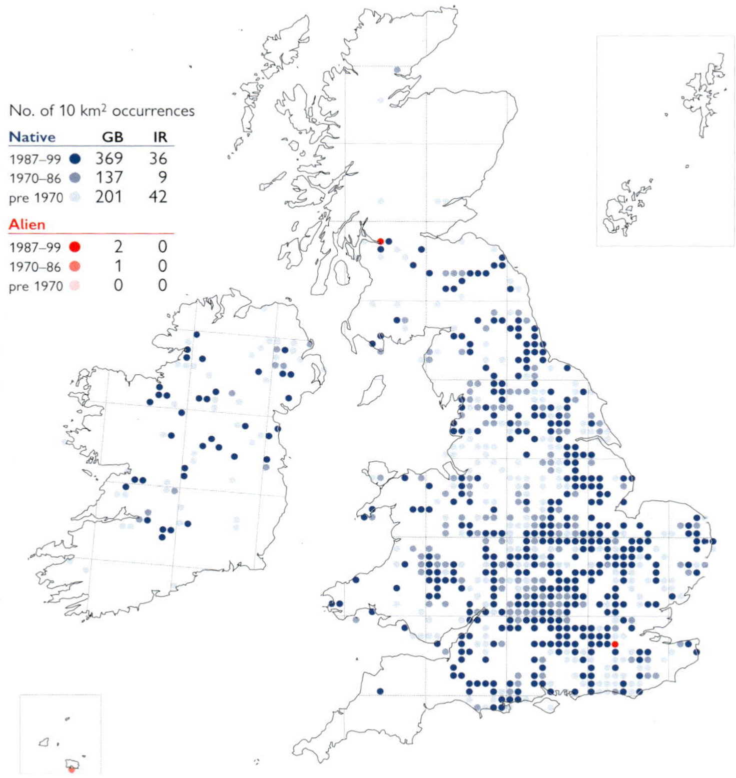

A rhizomatous perennial of shallow water or wet ground at the edges of rivers, streams, canals, lakes and ponds, in swamps, ditches and unimproved flood meadows and in marshland. It usually grows in calcareous and mesotrophic or eutrophic conditions, in areas subject to frequent flooding. It is shade-tolerant and sometimes grows under riverside trees or in wet woodland. Lowland, but formerly at 335 m at Greystoke (Cumberland).

Native (change –0.46). *C. acuta*, though now better recorded than for the 1962 *Atlas*, has declined in many areas, principally because of drainage and canalisation of rivers and streams.

Eurosiberian Boreo-temperate element.

References: Atlas (364b), Hultén & Fries (1986), Jermy *et al.* (1982), Meusel *et al.* (1965), Preston & Croft (1997).

M. J. Y. FOLEY & M. S. PORTER

Carex trinervis Three-nerved Sedge

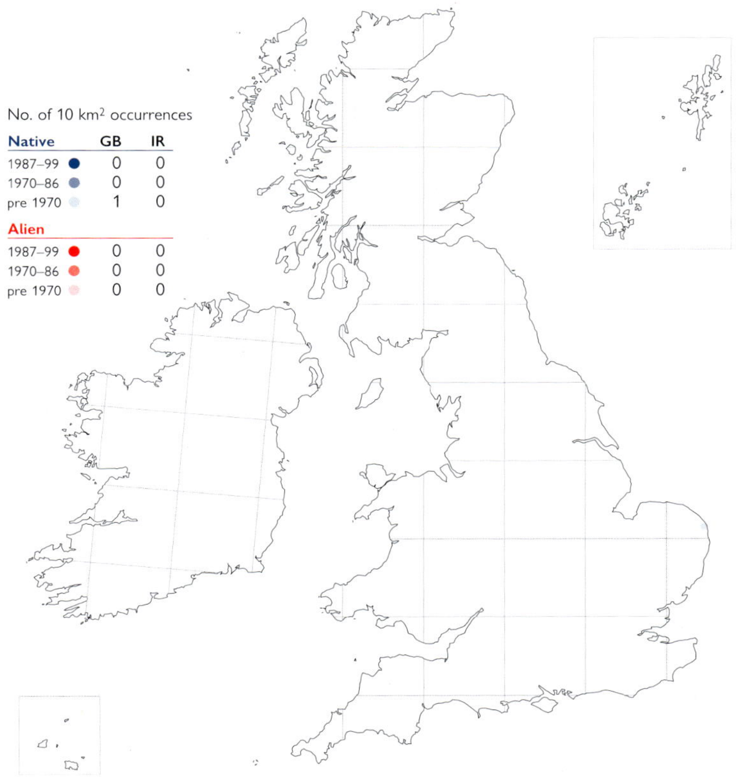

This species was known in our area only from Ormesby (E. Norfolk). Precise details of the habitat are unknown, but it must have been different from the damp dune-slacks and heaths in which it grows on the coast of W. Europe. Lowland.

Native. The Ormesby specimens were collected by H.G. Glasspoole in 1869; the species has never been re-found there. Some of the specimens are morphologically close to *C. nigra*, with which *C. trinervis* is known to hybridise freely. Their identification as *C. trinervis* is not absolutely certain.

Oceanic Temperate element.

References: Dupont (1962), Hultén & Fries (1986), Jermy *et al.* (1982).

M. J. Y. FOLEY & M. S. PORTER

Carex nigra Common Sedge

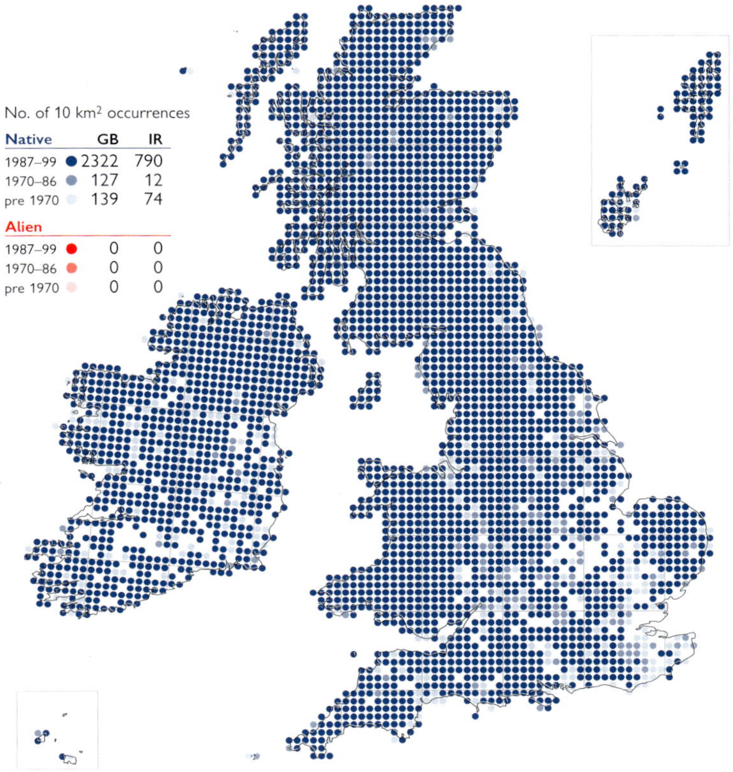

C. nigra occurs in a wide range of wet habitats, avoiding only those which are extremely basic or acidic. It is found in fens, fen-meadows, bogs, streamsides and flushes, wet grassland and dune-slacks. This sedge is morphologically very variable: tussock forms may occur in stagnant, acidic sites, and rhizomatous tufted variants are known from calcareous mires. 0–1005 m (Beinn Heasgarnich, Mid Perth).

Native (change –0.01). Mapped as 'all records' in the 1962 *Atlas*, the distribution of *C. nigra* seems stable other than in S.E. England, where it is has declined because of drainage.

Eurosiberian Boreo-temperate element; also in N. America.

References: Atlas (364d), Grime *et al.* (1988), Hultén & Fries (1986), Jermy *et al.* (1982), Meusel *et al.* (1965).

M. J. Y. FOLEY & M. S. PORTER

Carex elata Tufted-sedge

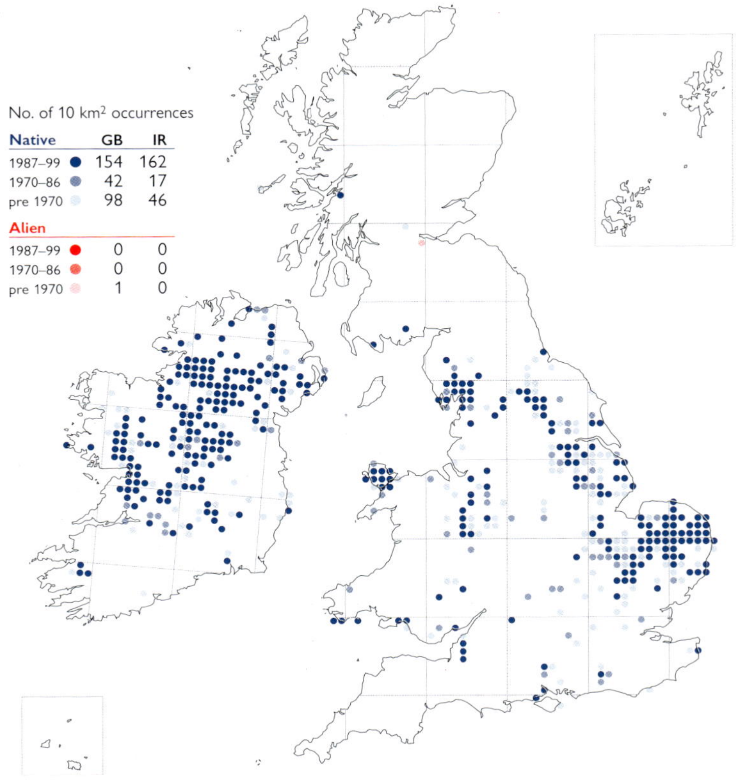

A tussock-forming perennial herb of oligotrophic or mesotrophic (sometimes eutrophic) marshy habitats, often calcareous, including fens, the margins of lakes, ponds, rivers and canals, ditches prone to seasonal flooding, and wet *Alnus* or *Salix* woodland. Lowland.

Native (change –0.32). *C. elata* is now much better recorded than for the 1962 *Atlas*, especially in Ireland. There have been losses, particularly in East Anglia, from drainage and ditch widening.

Eurasian Temperate element.

References: Atlas (364a), Hultén & Fries (1986), Jermy *et al.* (1982), Meusel *et al.* (1965), Preston & Croft (1997), Stewart *et al.* (1994).

M. J. Y. FOLEY & M. S. PORTER

Carex bigelowii Stiff Sedge

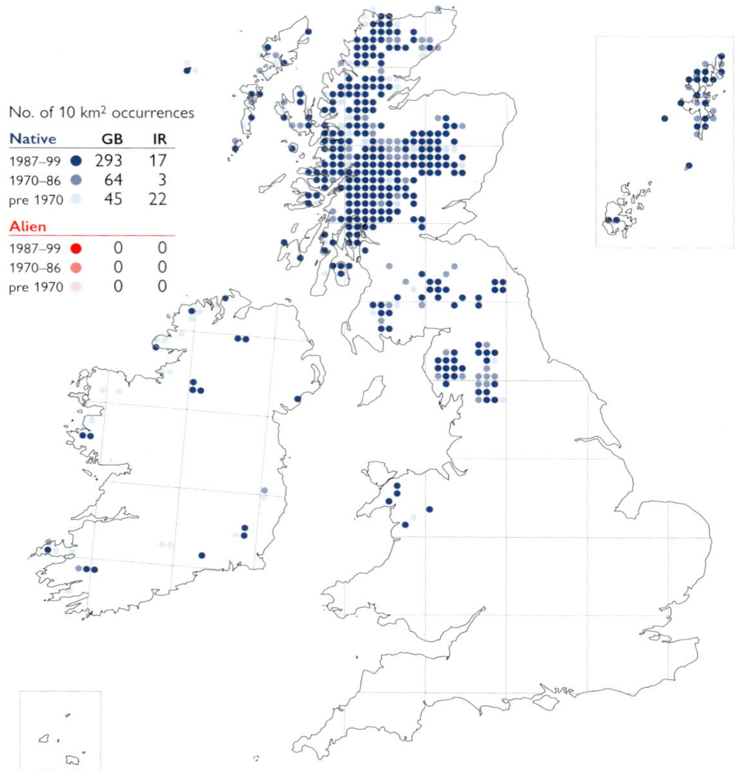

No. of 10 km² occurrences

Native	GB	IR
1987–99 ●	293	17
1970–86 ●	64	3
pre 1970 ●	45	22
Alien		
1987–99 ●	0	0
1970–86 ●	0	0
pre 1970 ●	0	0

A perennial herb of well-drained montane grassland and sedge-heath, of open stony ground, in corries where snow lies late, and gullies subject to flushing. From 15 m at Tressa Ness (Shetland), but generally an upland species, reaching 1305 m on Ben Macdui (S. Aberdeen).

Native (change –0.20). Apart from a few pre-1930 losses, particularly in Ireland, the distribution of *C. bigelowii* is stable.

Circumpolar Arctic-montane element.

References: Atlas (365a), Hultén & Fries (1986), Jermy *et al.* (1982).

M. J. Y. FOLEY & M. S. PORTER

Carex microglochin Bristle Sedge

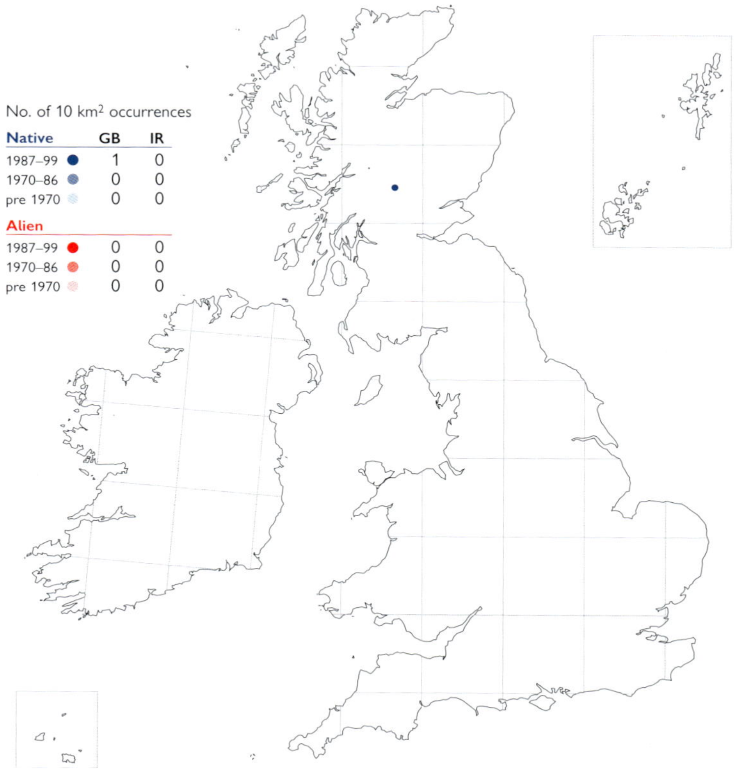

No. of 10 km² occurrences

Native	GB	IR
1987–99 ●	1	0
1970–86 ●	0	0
pre 1970 ●	0	0
Alien		
1987–99 ●	0	0
1970–86 ●	0	0
pre 1970 ●	0	0

A shortly-rhizomatous perennial herb of base-rich flushes on gently sloping or 'stepped' ground on micaceous silt or gravel. Also, occasionally, on steep burn-sides downstream of large colonies. Reproduction appears to be mainly vegetative. Between 610 m and 975 m on Ben Lawers (Mid Perth).

Native. Regular monitoring since the 1980s has shown little change in the distribution of the many colonies of *C. microglochin* within the only 10-km square from which it is recorded.

European Arctic-montane element; also in C. Asia and N. & S. America.

References: Atlas (370b), Hultén & Fries (1986), Jermy *et al.* (1982), Meusel *et al.* (1965), Wigginton (1999).

M. S. PORTER & M. J. Y. FOLEY

Carex pauciflora Few-flowered Sedge

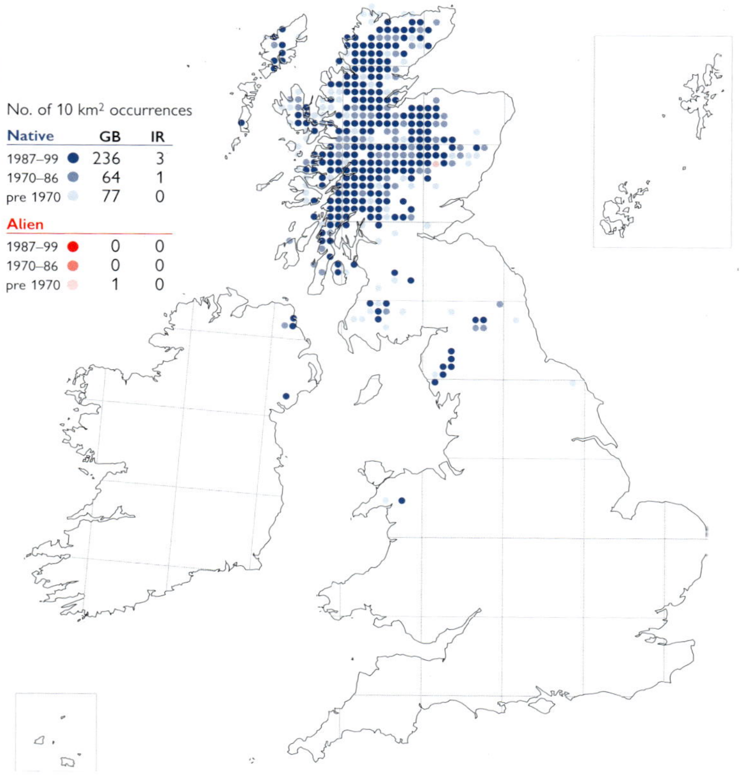

No. of 10 km² occurrences

Native	GB	IR
1987–99 ●	236	3
1970–86 ●	64	1
pre 1970 ●	77	0
Alien		
1987–99 ●	0	0
1970–86 ●	0	0
pre 1970 ●	1	0

A perennial herb of wet, acidic, oligotrophic raised and blanket bogs, often growing on and around hummocks and usually in association with *Sphagnum* species. Lowland to 650 m (Beinn Heasgarnich, Mid Perth) and 820 m elsewhere in Scotland.

Native (change –0.59). The range of *C. pauciflora* has changed very little since the 1962 *Atlas*. It is a very inconspicuous species and is easily overlooked, so it is probably present in many 10-km squares for which only pre-1987 records are available.

Circumpolar Boreal-montane element.

References: Atlas (370c), Curtis & McGough (1988), Hultén & Fries (1986), Jermy *et al.* (1982).

M. J. Y. FOLEY & M. S. PORTER

Carex rupestris Rock Sedge

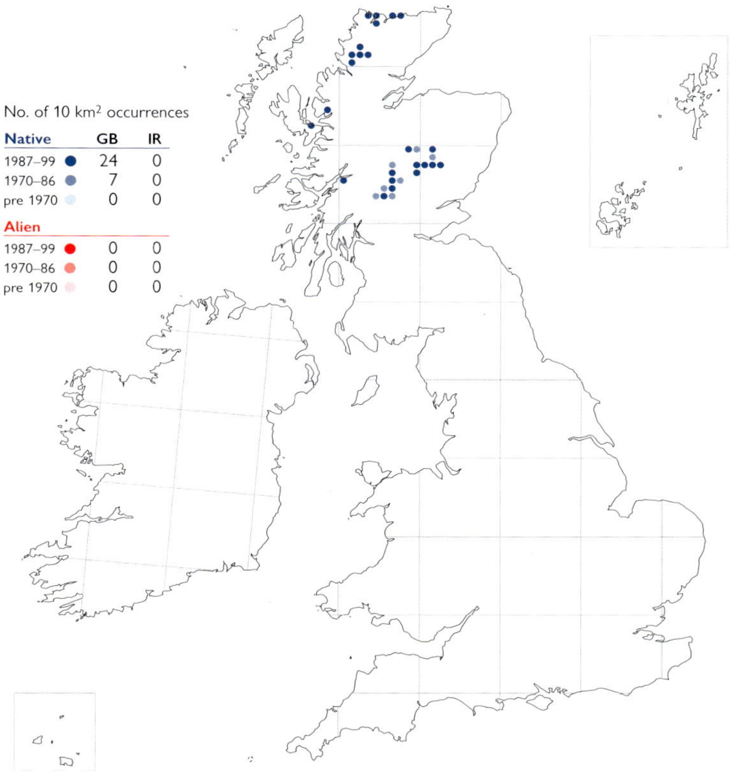

A perennial herb of basic substrates on cliff ledges and crevices, and on broken rocky or grassy slopes, always over base-rich rocks. It is often a shy flowerer, and it sometimes grows with *C. pulicaris* with which it can be easily confused. Found near sea level in N.W. Scotland, but usually from 600 m, to 935 m on Ben Lawers (Mid Perth).

Native (change +0.27). *C. rupestris* is now better recorded than it was for the 1962 *Atlas*, particularly in N.W. Scotland.

Circumpolar Arctic-montane element.

References: Atlas (370a), Hultén & Fries (1986), Jermy *et al.* (1982), Meusel *et al.* (1965), Stewart *et al.* (1994).

M. J. Y. FOLEY & M. S. PORTER

Carex pulicaris Flea Sedge

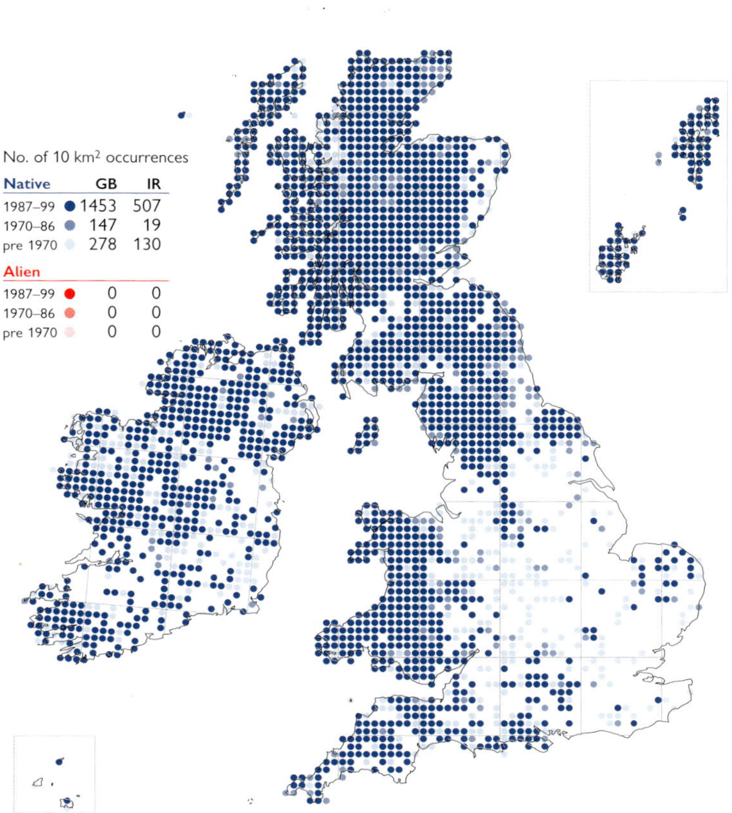

A perennial herb of damp or wet neutral or calcareous soils, and of more acidic soils where these are flushed by mineral-enriched groundwater. Habitats include short-sedge mires, damp meadows and pastures, fen-meadows, wet heaths, flushes and springs, montane rock ledges and *Dryas* heath. It also occurs, albeit rarely, on N.-facing slopes in chalk and limestone districts. 0–915 m (Macgillycuddy's Reeks, S. Kerry, and W. Ross).

Native (change −0.51). The distribution of *C. pulicaris* in N. and W. Britain and Ireland is largely unchanged. Elsewhere, losses from the lowlands, already apparent in the 1962 *Atlas*, have accelerated through drainage and the destruction of damp meadows and pastures.

Suboceanic Temperate element.

References: Atlas (370d), Hultén & Fries (1986), Jermy *et al.* (1982).

M. J. Y. FOLEY & M. S. PORTER

Sasa palmata Broad-leaved Bamboo

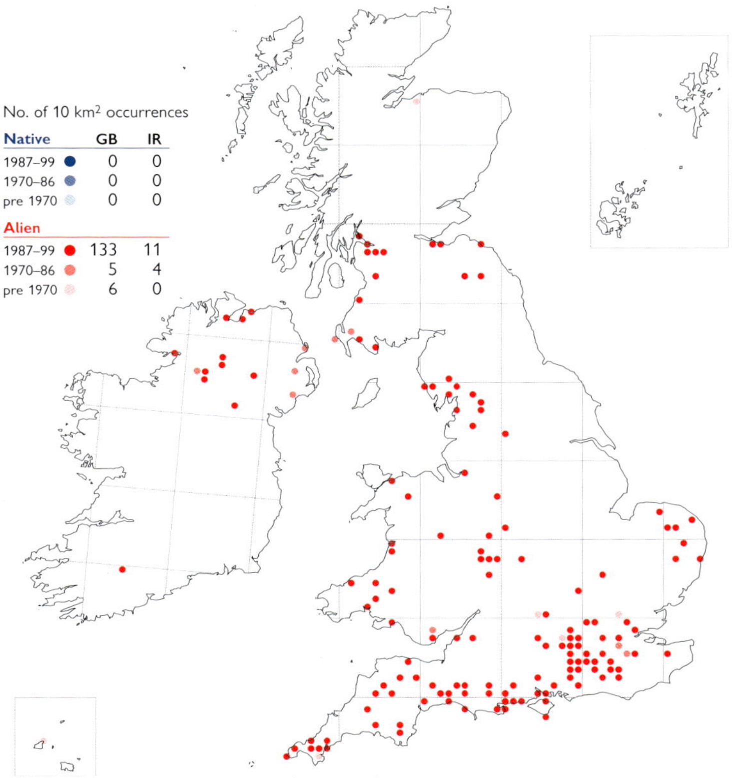

This bamboo is becoming widely naturalised in abandoned gardens, damp woodlands and along shaded, overgrown stream banks. Lowland.

Neophyte. *S. palmata* was introduced to Britain in about 1889 and is commonly grown in gardens and the grounds of country houses. It was recorded from the wild by 1964 and is certainly increasing, mostly due to deliberate planting.

Native of Japan and Sakhalin.

Reference: Ryves *et al.* (1996).

S. J. LEACH

Pseudosasa japonica Arrow Bamboo

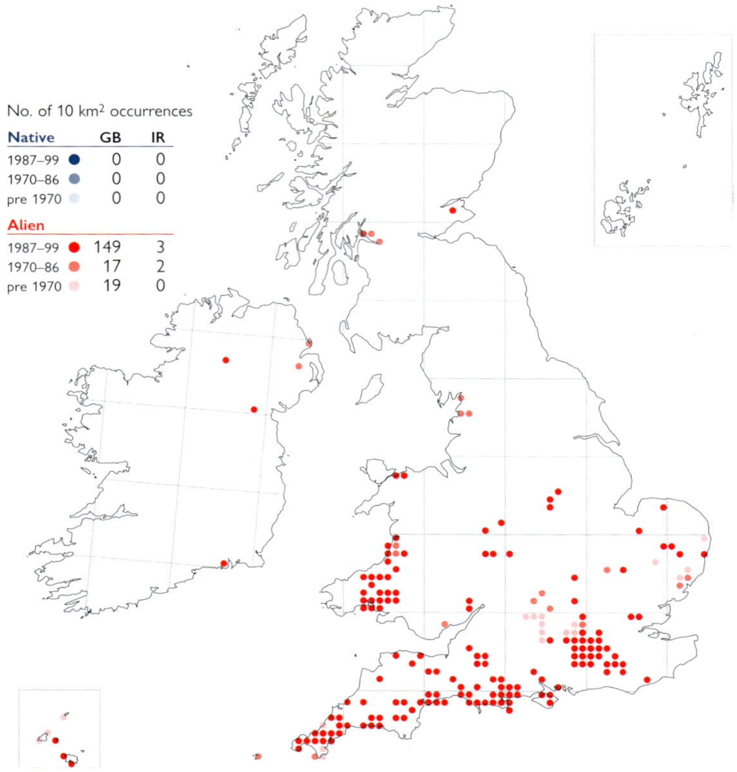

The most commonly grown bamboo in Britain, frequently persisting in old gardens and along wooded streamsides in the grounds of country houses, and now becoming well-established in the wild in a variety of usually moist, shaded habitats. Lowland.

Neophyte. *P. japonica* has been grown in British gardens since 1850, and was first recorded in the wild in 1955. It is certainly increasing, mostly due to deliberate planting.

Native of Japan and Korea.

References: French *et al.* (1999), Ryves *et al.* (1996).

S. J. LEACH

Leersia oryzoides Cut-grass

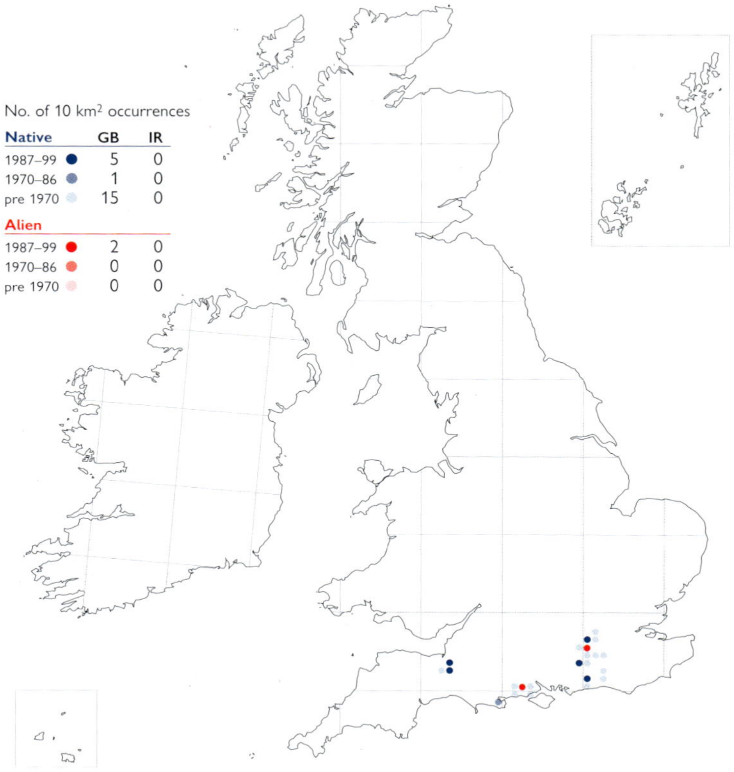

A rhizomatous perennial herb of nutrient-rich mud around the cattle-trampled margins of lakes and ponds, in ditches, on canal banks and riversides; also formerly in wet meadows. Lowland.

Native (change –0.40). *L. oryzoides* has decreased since the 1962 *Atlas*. Several populations have been lost in the last twenty years, including those in Somerset which were last seen in the early 1990s, and it is now apparently restricted to just three native sites in S.E. England, though it has been re-introduced into two others. The drainage and infilling of ponds and ditches, and possibly the over-zealous maintenance of canal banks, are thought to have contributed to its decline.

European Temperate element; also in C. and E. Asia and N. America.

References: Atlas (371b), Hultén & Fries (1986), Meusel *et al.* (1965), Wigginton (1999)

S. J. LEACH

Nardus stricta Mat-grass

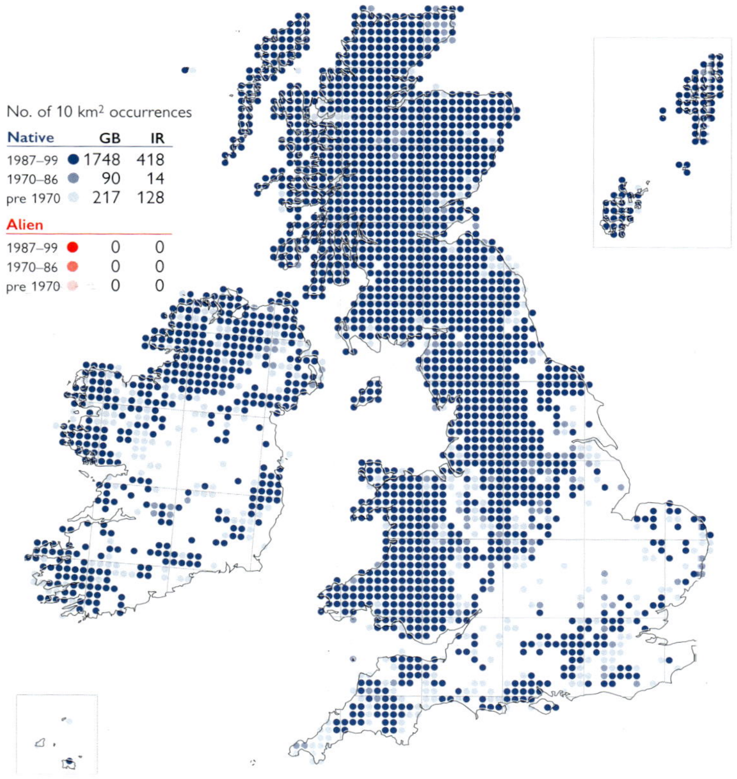

A densely tufted, shortly rhizomatous perennial found on winter-wet, base-poor, infertile and peaty soils, occurring in great quantity on upland hill-slopes, moorland and mountains, including snow-bed communities, but also found on lowland mires, heaths and acidic grasslands, and even sometimes in the upper reaches of saltmarsh turf. 0–1250 m (Ben Macdui, S. Aberdeen).

Native (change –0.68). Although the range of *N. stricta* has not changed appreciably since the 1962 *Atlas*, it continues to decline locally in lowland Britain because of habitat destruction.

European Boreo-temperate element; also in C. Asia and widely naturalised outside its native range.

References: Atlas (405a), Chadwick (1960), Grime *et al.* (1988), Hultén & Fries (1986), Meusel *et al.* (1965).

F. H. PERRING

Milium effusum Wood Millet

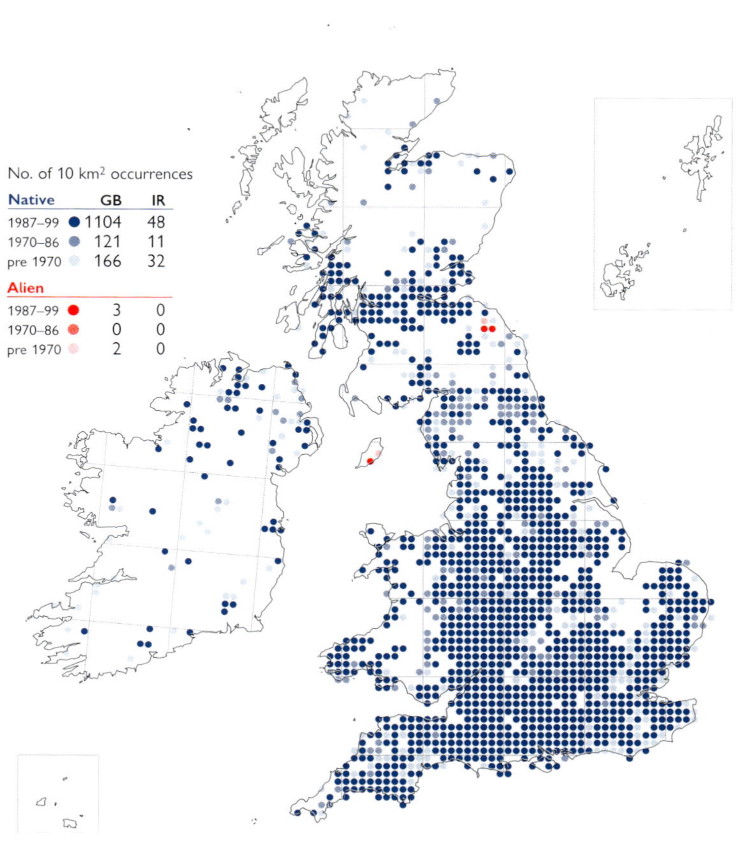

A perennial herb of damp, deciduous woods and shaded banks, where it grows on winter-wet, calcareous to mildly acidic clay and loam soils, and also over rocks in W. Scotland. It is regarded as an indicator of ancient woodland in some parts of E. England. However, it is able to colonise open sites which are disturbed by felling or fire, and there is evidence of spread to more recent woodland in some upland areas. 0–380 m (W. of Dockray, Cumberland).

Native (change +0.31). This species was mapped as 'all records' in the 1962 *Atlas*. It is now better recorded, and there is no evidence of a genuine change in its overall distribution.

Circumpolar Boreo-temperate element, with a disjunct distribution.

References: Atlas (403a), Grime *et al.* (1988), Hultén & Fries (1986), Meusel *et al.* (1965).

F. H. PERRING

Milium vernale Early Millet

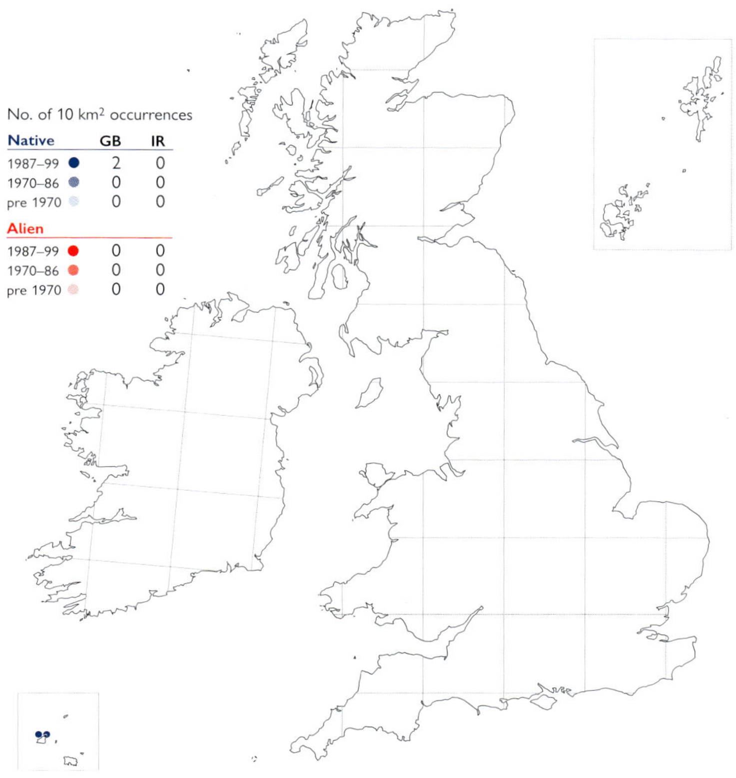

An inconspicuous, early-flowering, annual grass growing in small patches in the short, nearly closed turf of fixed dunes. Lowland.

Native. In our area, this species is known only from two localities on the north coast of Guernsey, where it was first recorded in 1899 but then remained unrecorded for fifty years. The Guernsey plant is a genetically distinct prostrate population which was described as subsp. *sarniense* by McClintock (1986).

Mediterranean-Atlantic element; it reaches its northern limit on the island of Terschelling, Netherlands.

References: Atlas (403a), McClintock (1975), Tutin (1950).

F. H. PERRING

Festuca pratensis Meadow Fescue

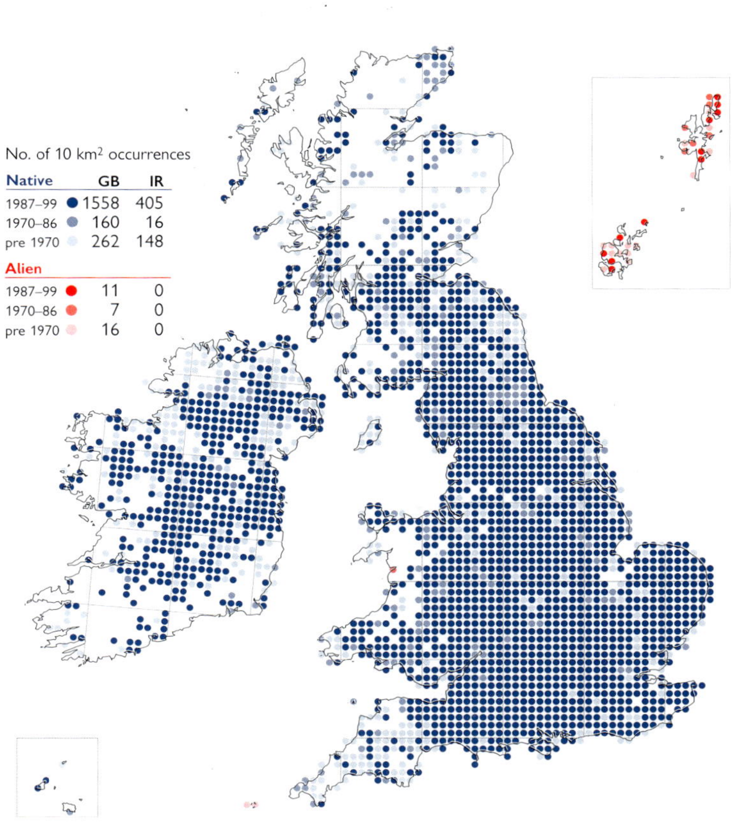

A short-lived perennial found in a wide range of neutral grasslands, usually on fertile soils, including pastures, hay- and water-meadows. It is often sown for fodder and has become naturalised on roadsides, railway banks and waste ground. 0–575 m (Hartside, Cumberland) and exceptionally at 845 m on Great Dun Fell (Westmorland).

Native (change –0.16). The native distribution of *F. pratensis* has been obscured by sowing, and it may only occur in N. & W. Britain and W. Ireland as a relic of cultivation. It may be decreasing due to the loss of wet meadows and a decline in its popularity in grass mixtures.

Eurosiberian Boreo-temperate element, but widely naturalised so that distribution is now Circumpolar Boreo-temperate.

References: Atlas (373b), Grime *et al.* (1988), Hultén & Fries (1986), Meusel *et al.* (1965).

F. H. PERRING

Festuca arundinacea Tall Fescue

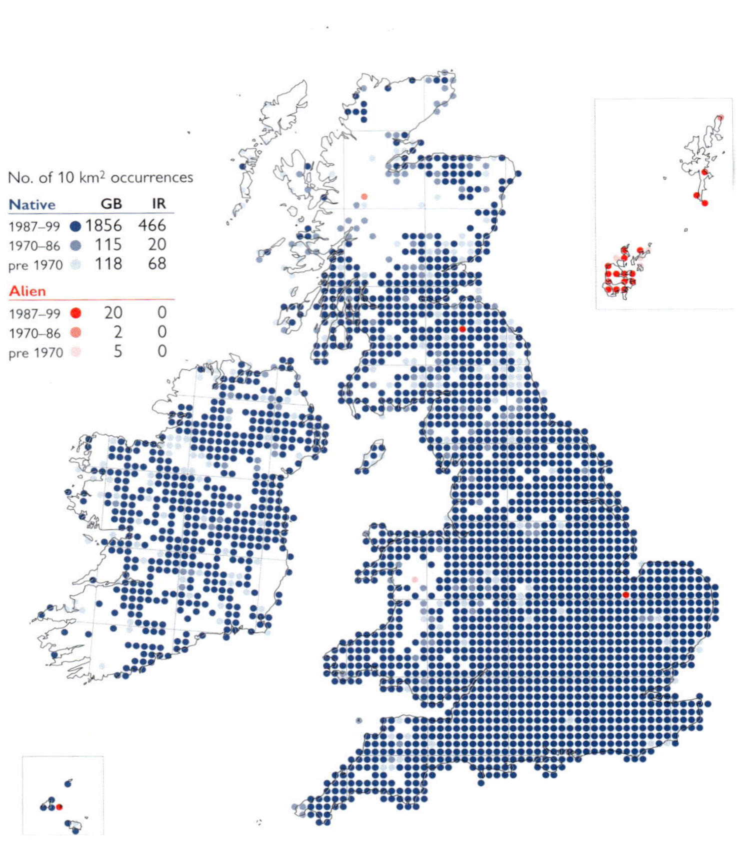

A robust perennial of scrub and woodland margins, hedgerows, pastures and meadows, river gravel, roadsides, railway banks and waste ground, on neutral or basic soils. It is also found along the banks of tidal rivers in places liable to inundation by brackish or sea water, and on slumping sea-cliffs. 0–430 m (N. of Alston, Cumberland).

Native (change +1.71). *F. arundinacea* was formerly used in seed mixtures and is only a relic of cultivation in some areas. However, it appears to have become much more frequent since the 1962 *Atlas*, although it may previously have been confused with *F. pratensis*. It can persist in swards that are infrequently mown.

Eurosiberian Southern-temperate element; widely naturalised outside its native range.

References: Atlas (373c), Gibson (2001), Hultén & Fries (1986), Meusel *et al.* (1965).

F. H. PERRING

Festuca gigantea Giant Fescue

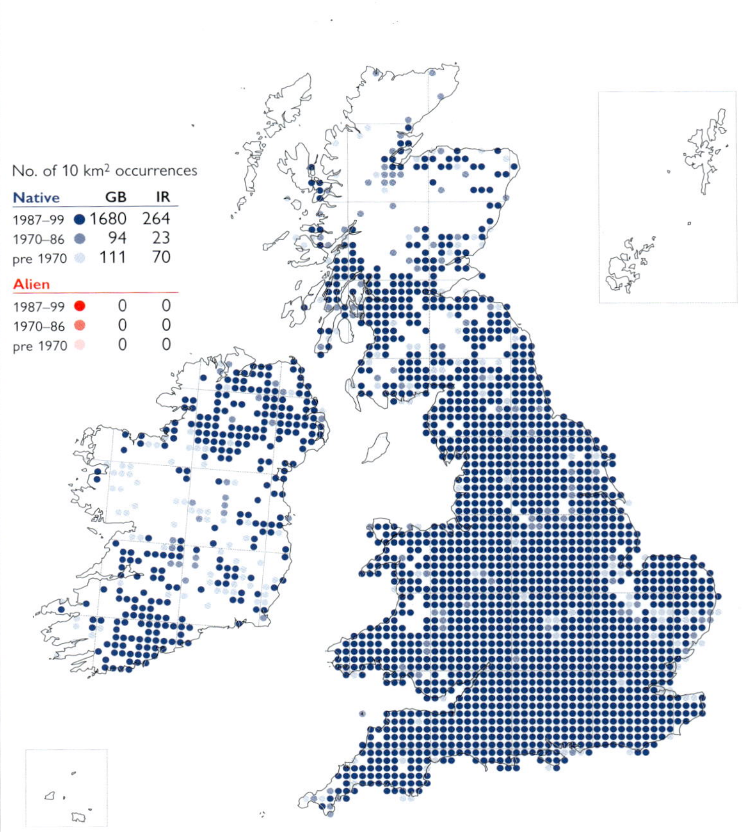

A tufted perennial herb of moist woodland on neutral to base-rich soils, often associated with *Brachypodium sylvaticum* and *Bromopsis ramosa*. It is particularly frequent beside streams, along rides, in disturbed areas in clearings and on woodland margins, and as a colonist of secondary woodland. It is dispersed by its fruits, which adhere to fur and clothing. 0–370 m (near Nenthead, Cumberland).

Native (change +0.46). The distribution of *F. gigantea* is stable or slightly increasing.

European Temperate element; also in C. Asia.

References: Atlas (373d), Grime *et al.* (1988), Hultén & Fries (1986), Meusel *et al.* (1965), Peterken (1981).

F. H. PERRING

Festuca altissima Wood Fescue

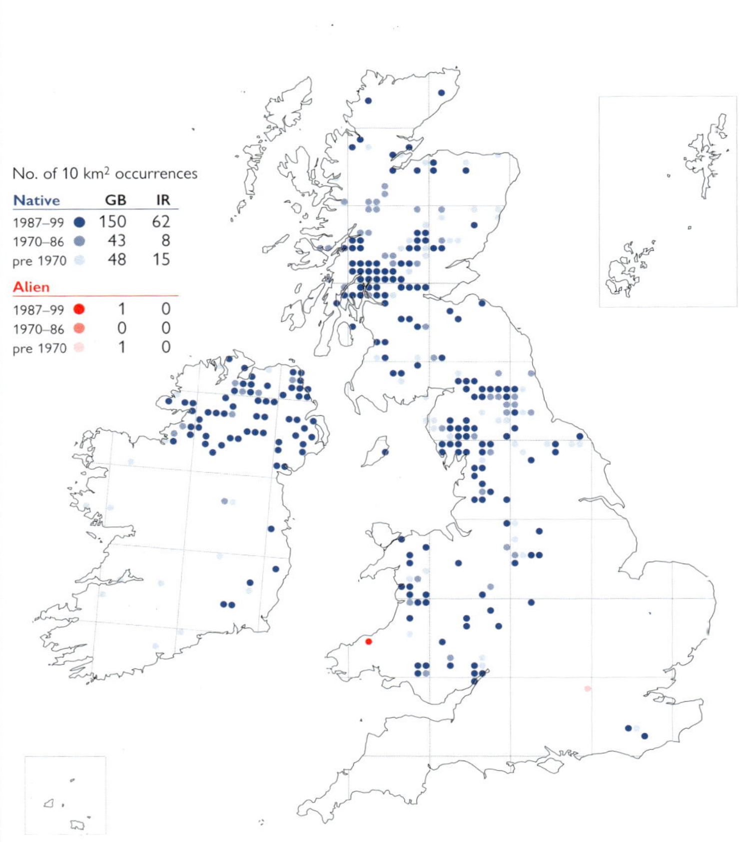

A long-lived perennial herb of moist, wooded valleys, on rocky slopes, deciduous wood margins and streamsides, especially on seepage lines or by waterfalls. It grows on soils of a moderate base status, often with *Luzula sylvatica*. 0–330 m (Haweswater, Westmorland).

Native (change +0.83). This species grows in habitats which are sometimes virtually inaccessible, and happens to be frequent in some areas of N. England and Scotland which were not well-recorded for the 1962 *Atlas*. More thorough recording has revealed that it is much more frequent than previously thought.

European Temperate element.

References: Atlas (374a), Hultén & Fries (1986), Meusel *et al.* (1965), Stewart *et al.* (1994).

F. H. PERRING

Festuca heterophylla Various-leaved Fescue

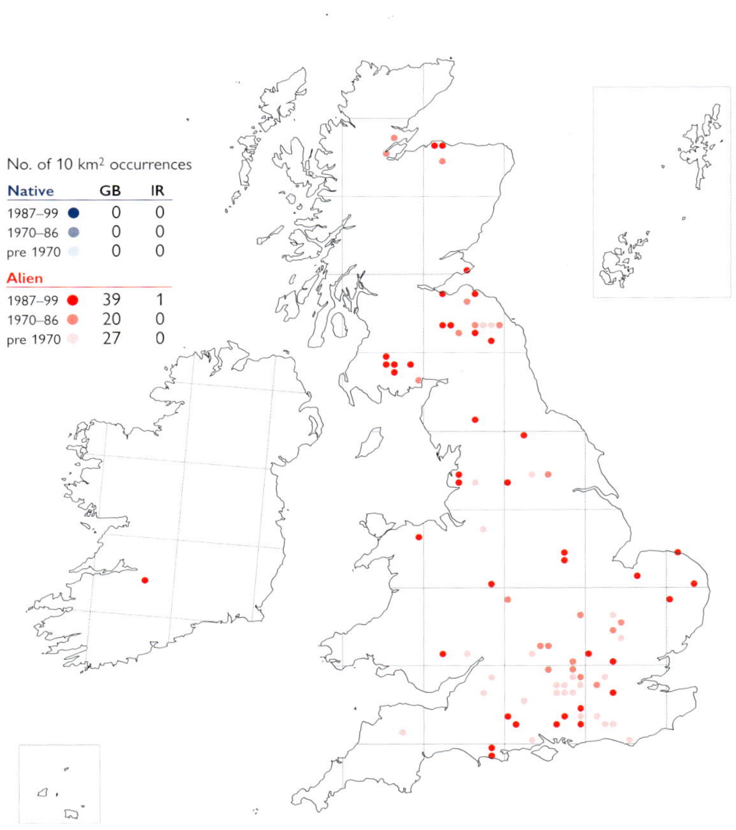

A rather tall, densely tufted perennial herb, widely naturalised in woods, spinneys and wood-borders on light soils. Lowland.

Neophyte (change +0.20). This species became available in Britain in 1812 and was originally planted for ornament or ground cover and possibly for fodder. It was a constituent of Victorian grass-seed mixtures and was frequently sown in woodland rides, beneath tree-belts and in parks and gardens. It was first recorded in the wild in 1874. More recently, it has occurred as a contaminant of grass-seed mixtures. There has been little change in the distribution since the 1962 *Atlas*, although there are new records from Scotland.

A European Temperate species.

References: Atlas (374b), Brewis *et al.* (1996), Hultén & Fries (1986), Ryves *et al.* (1996), Meusel *et al.* (1965).

S. J. LEACH

Festuca rubra sens. lat. Red Fescues

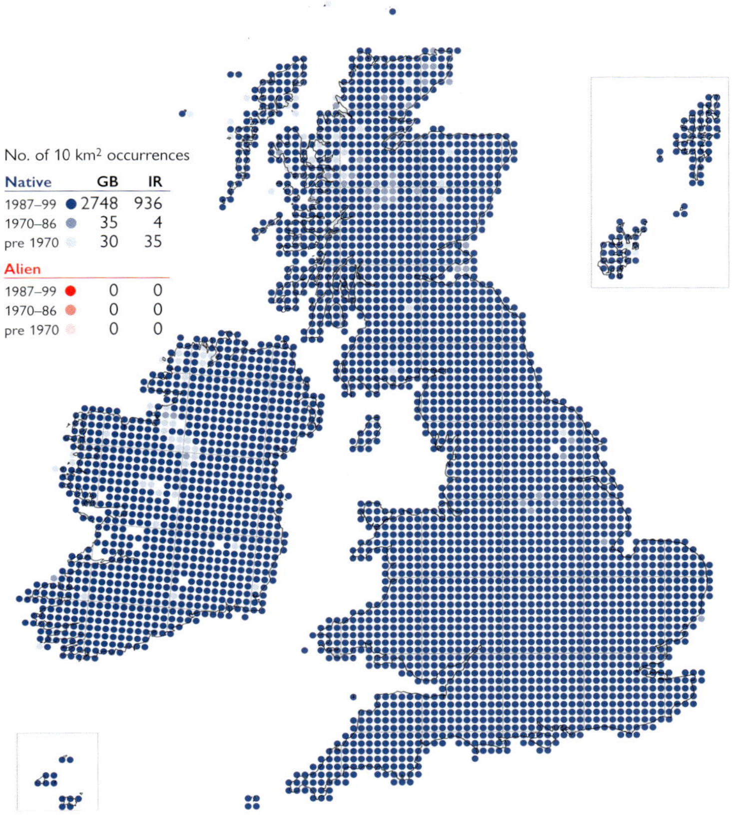

These extremely variable tufted or rhizomatous perennials are found in all kinds of grassy habitats, including lowland meadows and pastures, saltmarshes, sea-cliffs, sand dunes, hill grasslands, mountain slopes and rock ledges. 0–1080 m (Snowdon, Caerns.).

Native (change +2.96). There has been little change in the range of this ubiquitous taxon since the 1962 *Atlas*. *F. rubra sens. lat.* includes *F. arenaria* and *F. rubra*, the latter divided into seven subspecies. Of these, *F. rubra* subsp. *rubra* occurs throughout the range of the species. The other taxa are mapped separately.

Circumpolar Wide-boreal element; widely naturalised outside its native range.

References: Atlas (374c), Al-Bermani (1991), Grime *et al.* (1988), Hultén & Fries (1986), Stace *et al.* (1992).

S. J. LEACH

Festuca arenaria Rush-leaved Fescue

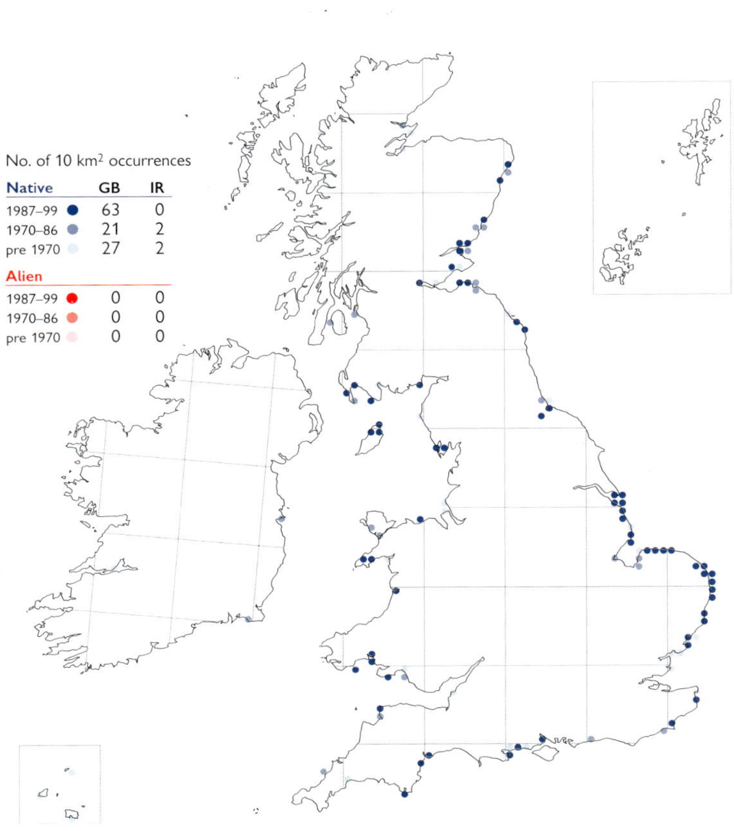

An extensively rhizomatous perennial found on sand dunes and open sandy shingle; also, more rarely, on cliff-tops, ledges and rough ground near the sea. On sand dunes it typically occurs on semi-mobile foredunes dominated by *Ammophila arenaria* or *Leymus arenarius*. Lowland.

Native. The map in the 1962 *Atlas* underestimated the distribution of this species which now includes both *F. juncifolia* (the taxon mapped in the 1962 *Atlas*) and *F. rubra* subsp. *arenaria*.

Oceanic Temperate element.

References: Atlas (374d), Auquier (1971b), Freijsen & Heeres (1972), Stewart *et al.* (1994).

S. J. LEACH

Festuca rubra subsp. *juncea*

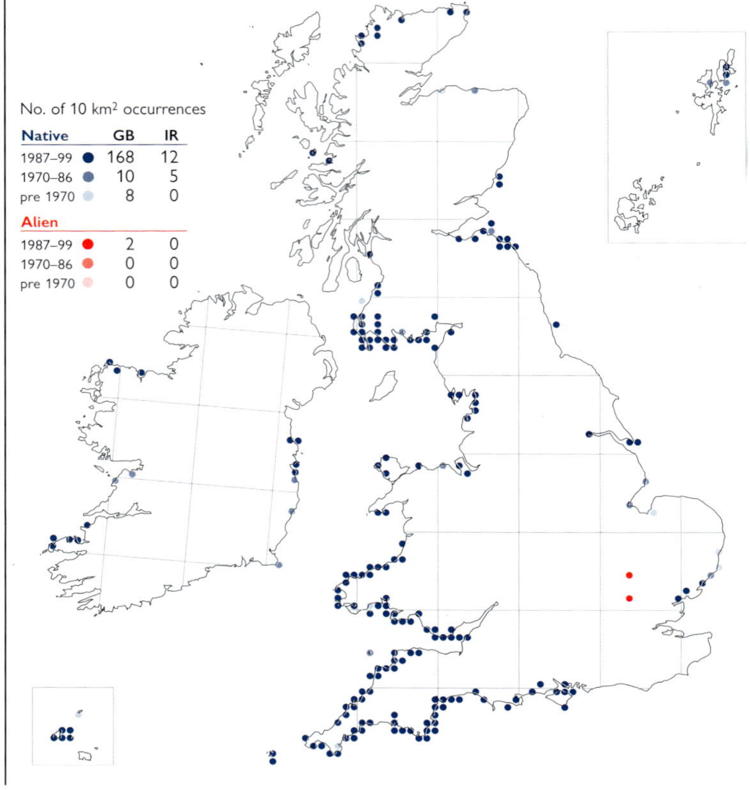

No. of 10 km² occurrences

Native	GB	IR
1987–99	168	12
1970–86	10	5
pre 1970	8	0
Alien		
1987–99	2	0
1970–86	0	0
pre 1970	0	0

A densely tufted, shortly rhizomatous perennial of sea-cliffs and coastal rock outcrops, occasionally found inland in rocky places. Lowland.

Native. Subsp. *juncea*, as presently delimited, incorporates coastal populations previously assigned to *F. rubra* subsp. *pruinosa*. Neither taxon was mapped in the 1962 *Atlas*, and there are insufficient data available to allow an assessment of changes in distribution. In common with other *F. rubra* subspecies, subsp. *juncea* is probably under-recorded. Non-pruinose populations, in particular, can be easily overlooked.

Widespread in Europe; wider distribution uncertain.

References: Auquier (1971a), Sell & Murrell (1996).

S. J. LEACH

Festuca rubra subsp. *litoralis*

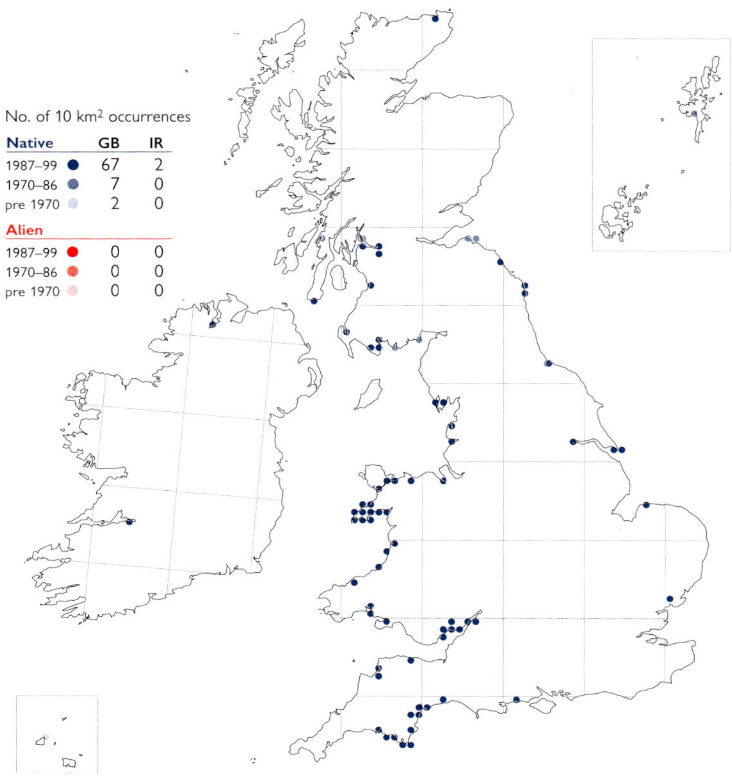

No. of 10 km² occurrences

Native	GB	IR
1987–99	67	2
1970–86	7	0
pre 1970	2	0
Alien		
1987–99	0	0
1970–86	0	0
pre 1970	0	0

A shortly rhizomatous, mat-forming perennial of saltmarshes, often dominating large areas in the middle and upper marsh; also found in brackish grazing marshes and other sandy or muddy saline areas. Lowland.

Native. This taxon was not mapped in the 1962 *Atlas* and there is no information available on any changes in its distribution.

European Boreo-temperate element; recorded from the Atlantic coast of Europe and the Baltic region.

S. J. LEACH

Festuca rubra subsp. *commutata*

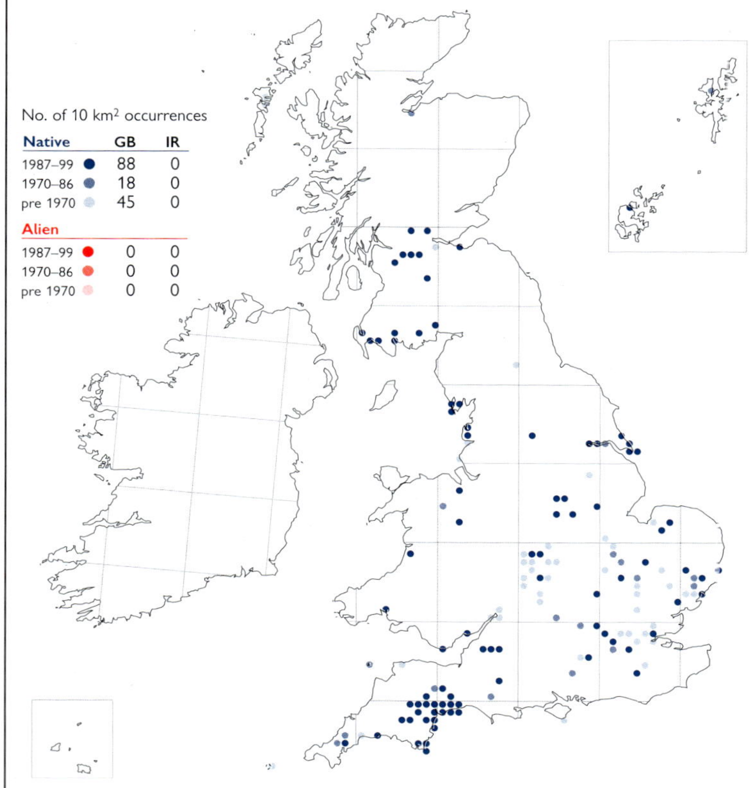

No. of 10 km² occurrences

Native	GB	IR
1987–99	88	0
1970–86	18	0
pre 1970	45	0
Alien		
1987–99	0	0
1970–86	0	0
pre 1970	0	0

A densely tufted perennial occurring in all kinds of grassy places, especially on well-drained soils. Under the name of 'Chewing's Fescue', this taxon is an important constituent of grass-seed mixtures. Many records from roadsides, amenity grasslands and garden lawns are undoubtedly introductions. Mainly lowland.

Native. There is no information available on trends in the distribution of this taxon. It is certainly under-recorded in most areas. It is not feasible to distinguish between native and introduced populations with any accuracy, and all records are mapped as if they are native.

European Temperate element.

S. J. LEACH

Festuca rubra subsp. arctica

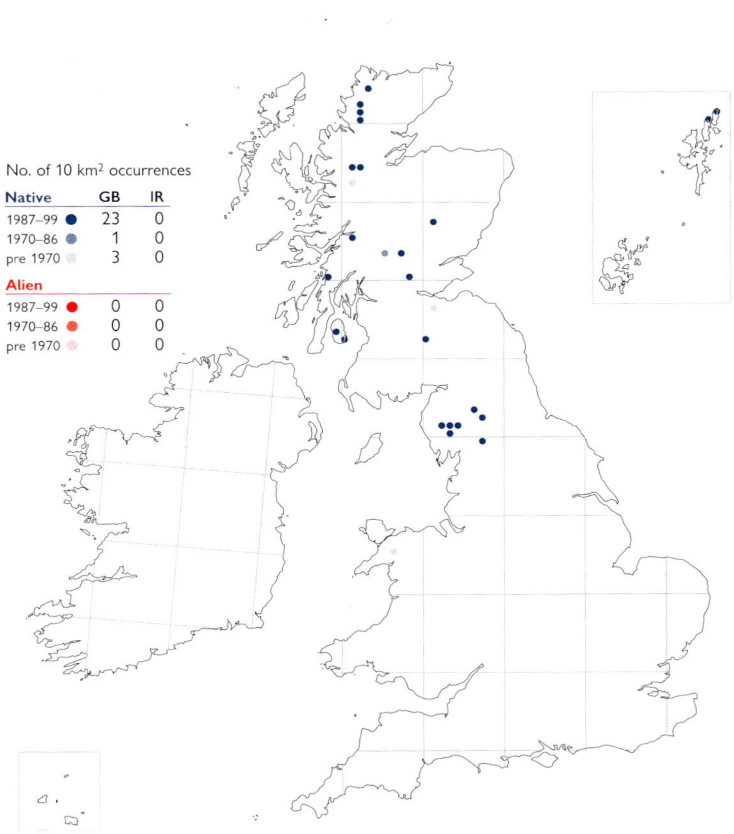

A rhizomatous perennial which grows on wet mountain slopes and gullies, rock ledges and flushes, and also on serpentine at low altitudes. 0–915 m (Meall nan Tarmachan, Mid Perth).

Native. *F. rubra* subsp. *arctica* (formerly named *F. richardsonii*) is under-recorded. For example, the first reports of it in England were only recently published (Halliday, 1995). It is not possible to ascertain trends in distribution on the basis of the scant existing data.

Circumpolar Arctic-montane element.

S. J. LEACH

Festuca rubra subsp. scotica

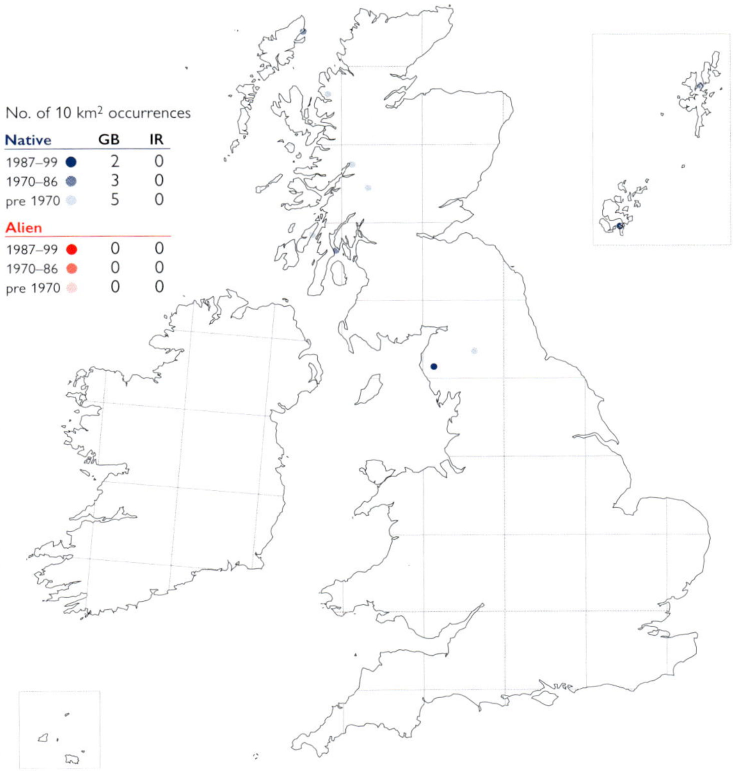

A rhizomatous perennial of grassy, usually wet, habitats, around rock outcrops and on cliff ledges. From sea level to at least 825 m in the Scottish Highlands.

Native. *F. rubra* subsp. *scotica* was first recognised by C. E. Hubbard from an unnamed location in the Cairngorms but not named formally until 1991 (Al-Bermani & Stace, 1991). The map is almost certain to underestimate the distribution of this little known and easily overlooked taxon.

Suboceanic Boreal-montane element; recorded from Scandinavia and Iceland as well as Britain, and perhaps more widespread.

Reference: Halliday (1995).

S. J. LEACH

Festuca rubra subsp. megastachys

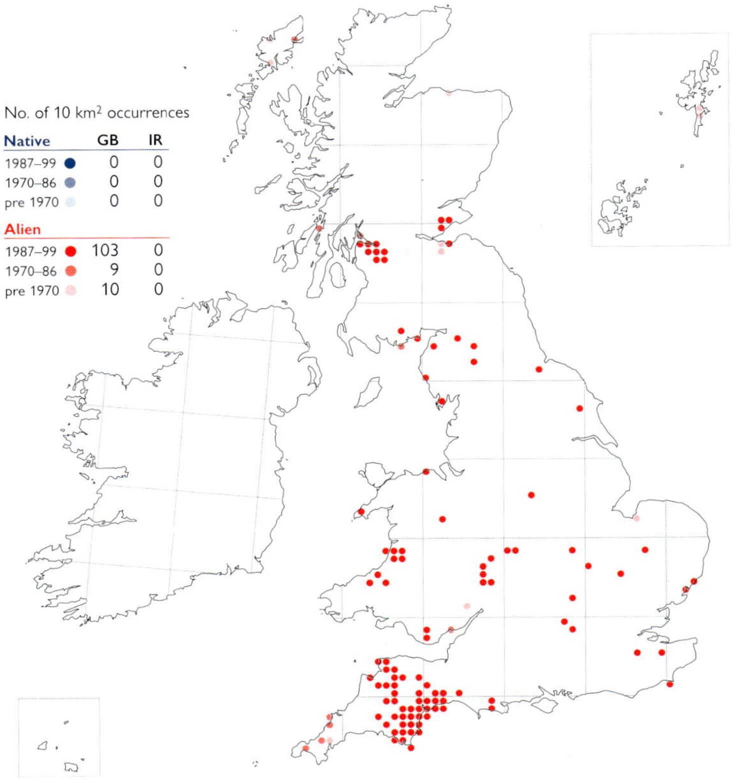

A rhizomatous, patch-forming perennial herb of waysides and other grassy places, including re-seeded grasslands and garden lawns. Lowland.

Neophyte. This taxon was not mapped in the 1962 *Atlas*. It was first recorded in the wild in Britain in 1966 (Shetland) and is under-recorded, having been ignored by most recorders.

Wider native distribution uncertain.

S. J. LEACH

Festuca ovina agg. Sheep's-fescues

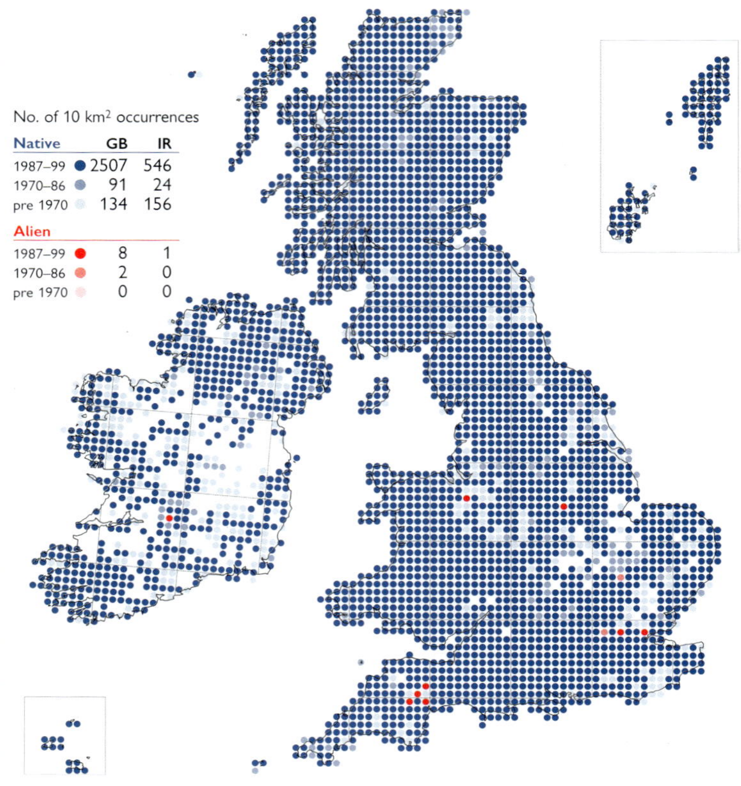

These perennial herbs grow in a variety of well-drained acidic and basic, usually infertile, habitats, including open woodland, grasslands, sand dunes, upland heaths and moors, rock ledges and sea-cliffs. Some segregates are common in grass-seed mixtures. 0–1305 m (Ben Macdui, S. Aberdeen).

Native (change –0.15). The distribution of *F. ovina* agg. is stable. All eight species forming this aggregate are mapped separately. Some of these are poorly known and many botanists continue to record only *F. ovina* agg.

This aggregate includes *F. vivipara* and a complex of other taxa which have a Eurasian Boreo-temperate distribution but have been widely introduced elsewhere.

References: Atlas (375a), Grime *et al.* (1988), Hultén & Fries (1986), Meusel *et al.* (1965), Stace *et al.* (1992), Wilkinson & Stace (1991).

S. J. LEACH

Festuca ovina Sheep's-fescue

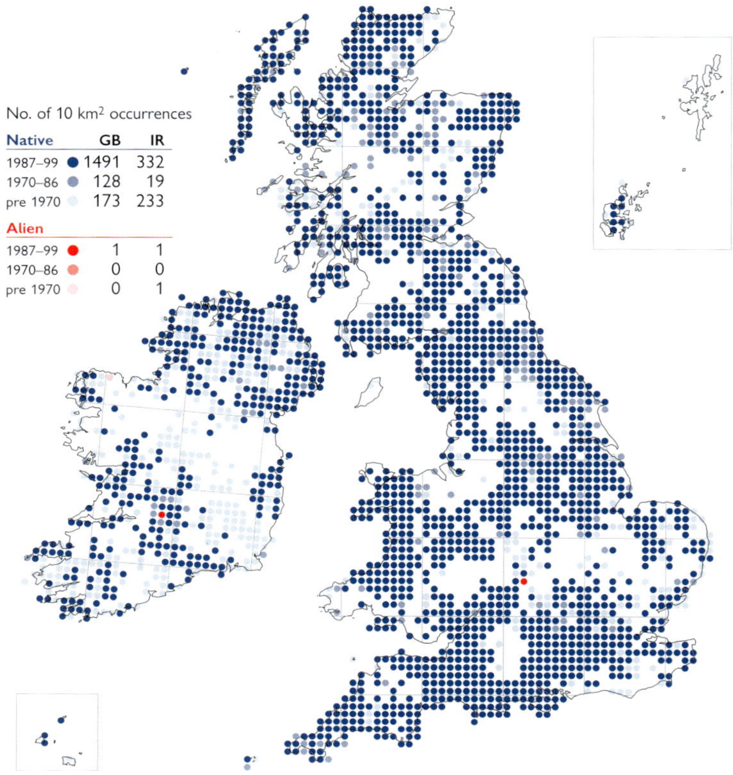

This morphologically variable, densely tufted perennial herb occurs in a wide range of unproductive, usually well-drained grassy habitats, including lowland calcareous grasslands, upland heaths and moors, mountain slopes and rock ledges, and sea-cliffs. 0–1305 m (Ben Macdui, S. Aberdeen).

Native. The distribution of *F. ovina* is stable, but it is very unevenly recorded. The map shows records of the species and its three component subspecies. This grass is of enormous agricultural and socio-economic importance, being one of the predominant constituents of hill pastures across great swathes of upland Britain and Ireland.

Eurasian Boreo-temperate element; widely naturalised outside its native range.

References: Grime *et al.* (1988), Hultén & Fries (1986), Meusel *et al.* (1965), Wilkinson & Stace (1991).

S. J. LEACH

Festuca vivipara Viviparous Sheep's-fescue

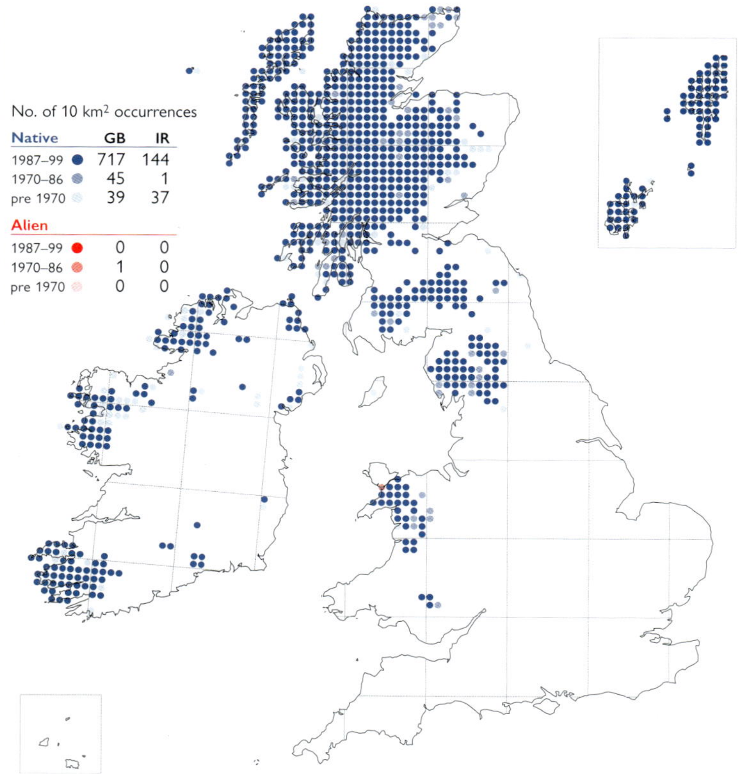

A tufted perennial herb of upland heathy pastures, open *Betula* and *Quercus* woodland, rock ledges and crevices, and a wide range of mountain slope and plateau communities including areas of late snow-lie; also found in the drier parts of bogs and on stream banks. It grows on both basic and acidic substrates. From sea level in W. Scotland and Ireland, to 1215 m on Ben Macdui (S. Aberdeen).

Native (change +0.13). The range of *F. vivipara* has remained stable since the 1962 *Atlas*, but it is now better recorded in England and Wales.

Circumpolar Boreo-arctic Montane element.

References: Atlas (375b), Hultén & Fries (1986), Wilkinson & Stace (1991).

F. H. PERRING

Festuca filiformis Fine-leaved Sheep's-fescue

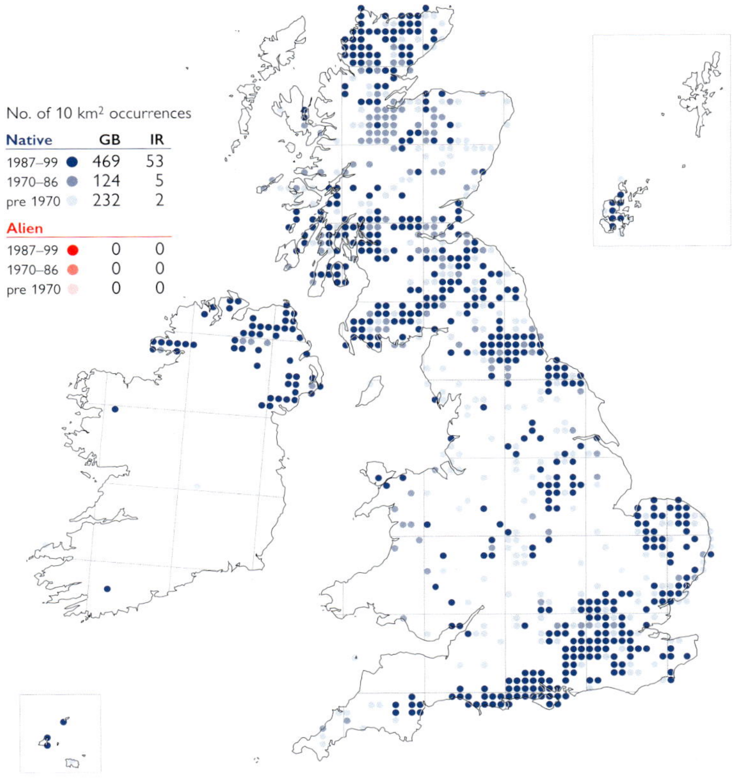

This densely tufted perennial grass occurs on heaths and moors, in open woodland and parkland and other grassy places, usually on acidic, sandy, well-drained soils. 0–1035 m (Meall Ghaordie, Mid Perth).

Native. The map in Perring & Sell (1968) was based on material seen by C. E. Hubbard or other reliable recorders. Since then *F. filiformis* has become more widely known, and this explains its apparent increase. There has, however, been much confusion between it and other taxa within with *F. ovina* species-complex (especially short-awned variants of *F. ovina*), and it is probably still over-recorded in some areas and overlooked in others.

Suboceanic Temperate element.

References: Atlas Supp. (148b), Meusel *et al.* (1965), Wilkinson & Stace (1991).

S. J. LEACH

Festuca armoricana Breton Fescue

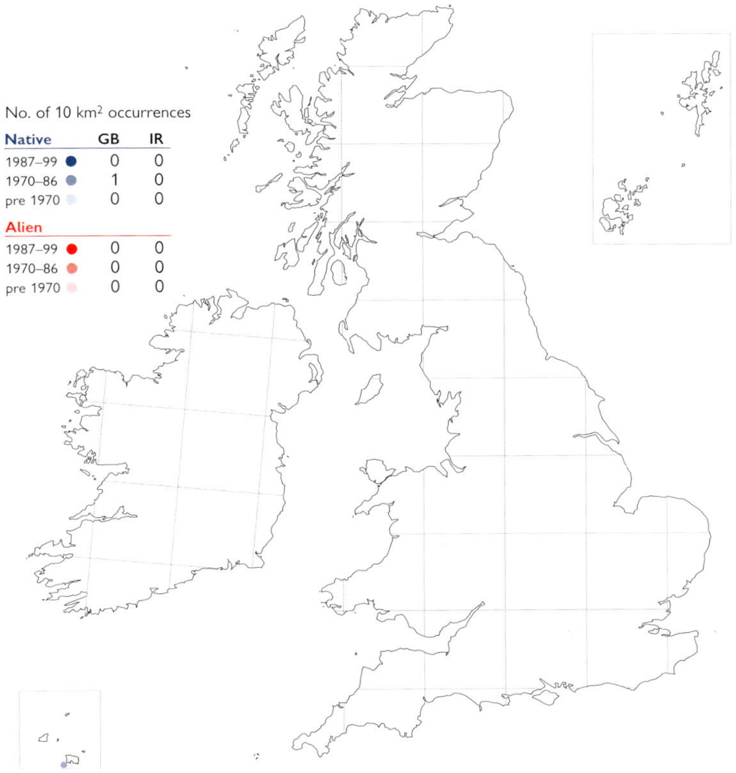

A densely tufted perennial herb apparently restricted to fixed sand dunes at St Ouen's and St Brelade's Bays, Jersey. Lowland.

Native. This species, a member of the *F. ovina* complex, was not mapped in the 1962 *Atlas*. Few British botanists are familiar with it and it may have been overlooked elsewhere in the Channel Islands, and possibly also in S. England.

Oceanic Temperate element.

Reference: Wilkinson & Stace (1991).

S. J. LEACH

Festuca huonii Huon's Fescue

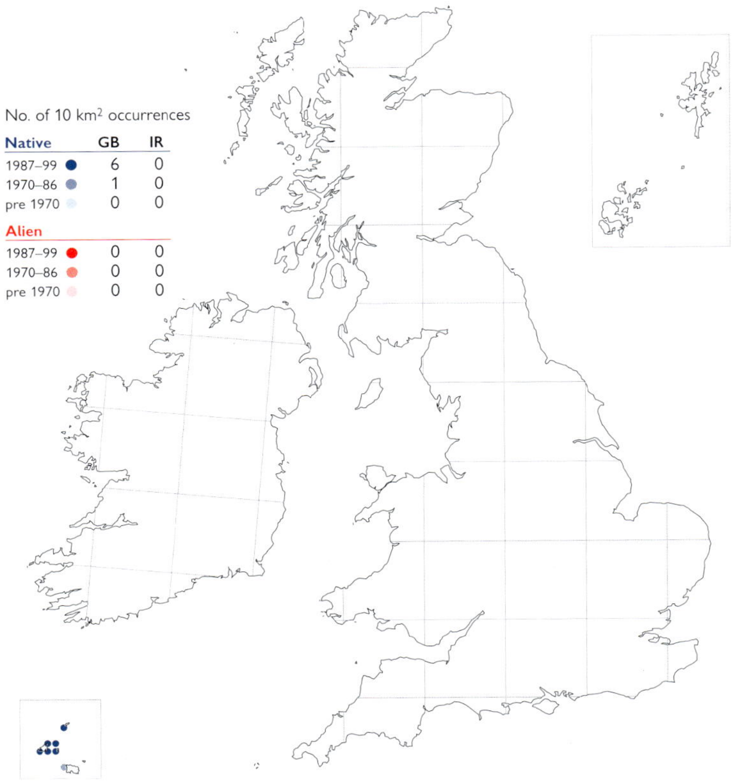

A densely tufted perennial herb apparently restricted to grassy cliff-tops and ledges on acid rocks in the Channel Islands. Lowland.

Native. This member of the *F. ovina* complex was not recognised at the time of the 1962 *Atlas*. However, it is frequent within its restricted range and there are no indications of any recent change in distribution. It was reported from near Prawle Point, S. Devon, in 1992 (Takagi-Arigho, 1995). However, Stace (1997) suggests that the record 'needs confirming', and it may have been based on depauperate specimens of *F. longifolia*. It has not been mapped.

Oceanic Temperate element; confined to Brittany and the Channel Islands.

References: Auquier (1973), Wilkinson & Stace (1991).

S. J. LEACH

Festuca lemanii Confused Fescue

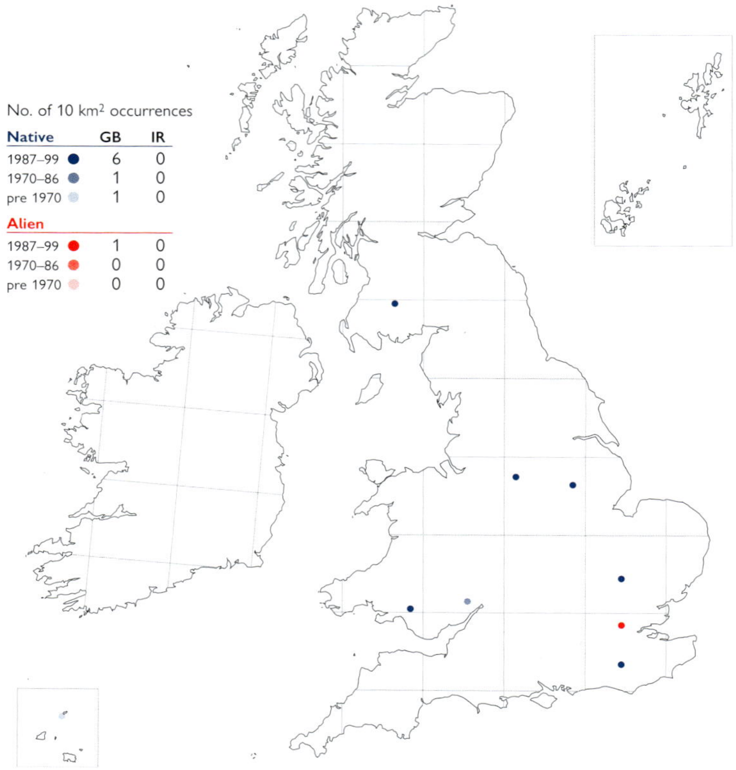

A tufted perennial herb of grassy places on well-drained, acidic or calcareous soils, often occurring with *F. ovina*. It is characteristic of limestone cliff rock crevices in Derbyshire. Lowland.

Native or alien. This species was not separately mapped in the 1962 *Atlas*. At that time, *F. lemanii* and *F. brevipila* were collectively known by the misapplied name '*F. longifolia*', and there continues to be much confusion between *F. lemanii* and other taxa within the *F. ovina* species-complex, including *F. ovina* itself. It is little known to British botanists, can easily be overlooked and is probably much under-recorded.

Oceanic Temperate element.

References: Atlas (375c, see above), Sell & Murrell (1996), Wilkinson & Stace (1989, 1991).

S. J. LEACH

Festuca longifolia Blue Fescue

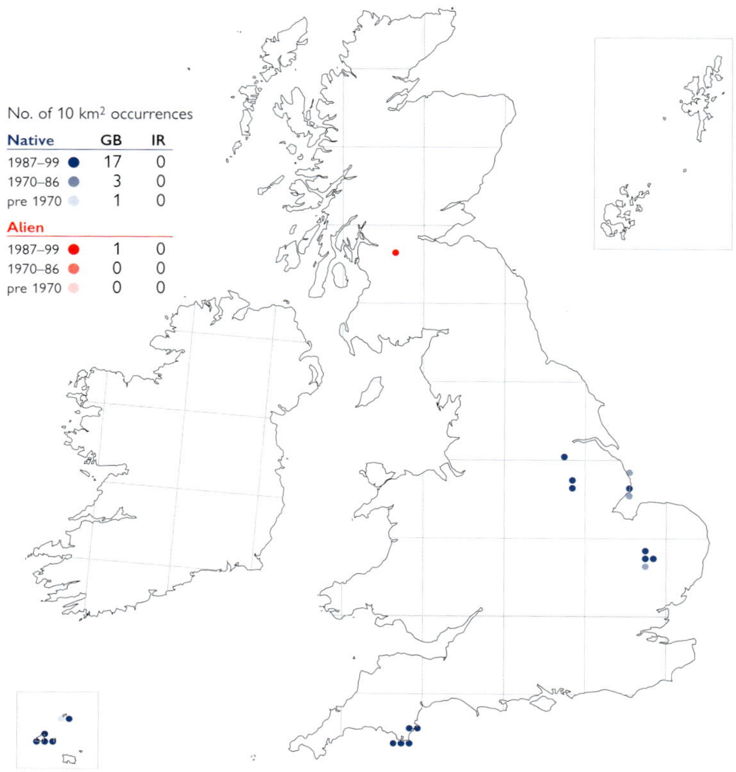

A densely tufted perennial herb occurring on dry, rabbit-grazed heaths, sandy roadside banks and, in S. Devon and the Channel Islands, on maritime cliff-tops and ledges. Lowland.

Native. Past confusion between *F. longifolia* and other glaucous taxa of *Festuca*, and a consequent lack of reliable historical data, means that changes in distribution remain unclear. Several colonies in Breckland have been lost since the early 1980s. Most surviving populations there, and in Nottinghamshire and Lincolnshire, are small and at risk of being ousted by rank grass or shaded out by trees. Coastal populations are less threatened.

Oceanic Temperate element.

References: Meusel *et al.* (1965), Trist (1973), Wigginton (1999), Wilkinson & Stace (1989).

S. J. LEACH

Festuca brevipila Hard Fescue

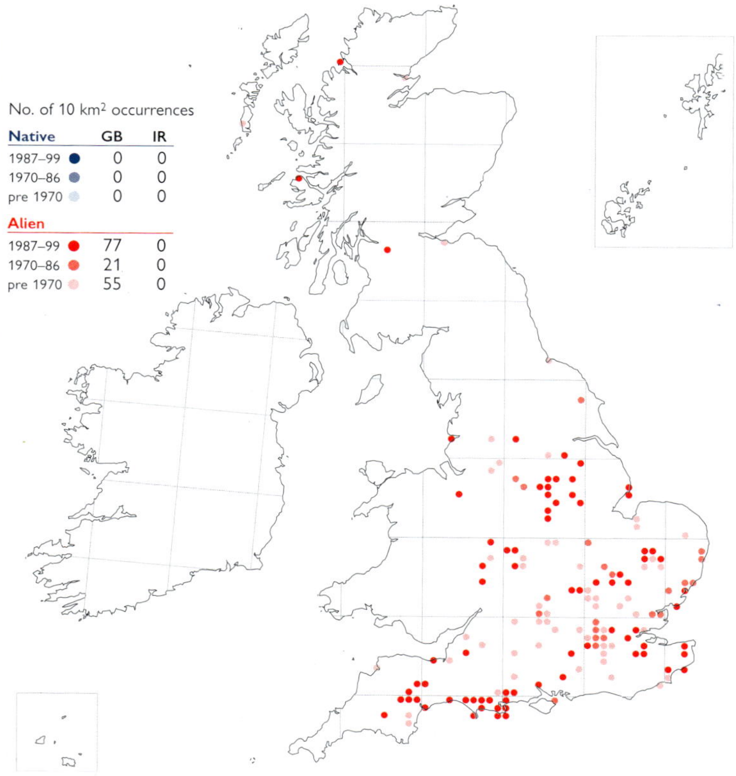

A tufted perennial herb, introduced in turf-grass and seed mixtures (as *F. duriuscula*) and frequently naturalised on roadsides, railway banks, commons, golf courses and other amenity grasslands, especially on well-drained, acidic soils. Generally lowland, but reaching 365 m at Taddington (Derbys.).

Neophyte. *F. brevipila* and *F. lemanii* were collectively mapped under the misapplied name '*F. longifolia*' in the 1962 *Atlas*, and there continues to be much confusion between *F. brevipila* and other taxa within the *F. ovina* species-complex. It has been known in the wild since about 1830 (Middlesex), and is probably much under-recorded.

Native of C. Europe.

References: Atlas (375c, see above), Ryves *et al.* (1996), Wilkinson & Stace (1989, 1991).

S. J. LEACH

Festuca pratensis × *Lolium perenne* (× *Festulolium loliaceum*) Hybrid Fescue

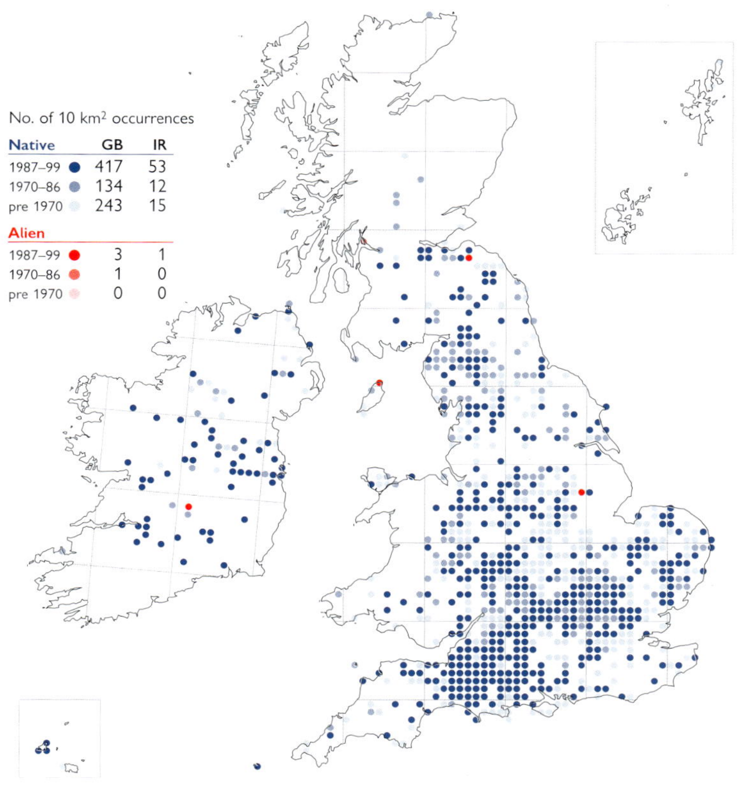

This rather variable hybrid occurs in pastures and meadows, in marshy grassland beside rivers and ponds, and on waysides. It usually grows on damp, fertile, rather heavy soils. Generally lowland, but reaching 470 m at Tyne Head (Cumberland).

Native. This is the most widespread of the *Festuca* × *Lolium* hybrids, and is certainly more frequent than the map in Perring & Sell (1968) suggests. It is probably still under-recorded, but it may sometimes have been recorded erroneously for other × *Festulolium* hybrids or even for variants of *L. perenne*. It is unclear whether the absence of recent records in many squares reflects a lack of recording or a genuine decline.

Widespread in temperate Europe.

References: Atlas Supp. (148c), Stace (1975).

S. J. LEACH

Lolium perenne Perennial Rye-grass

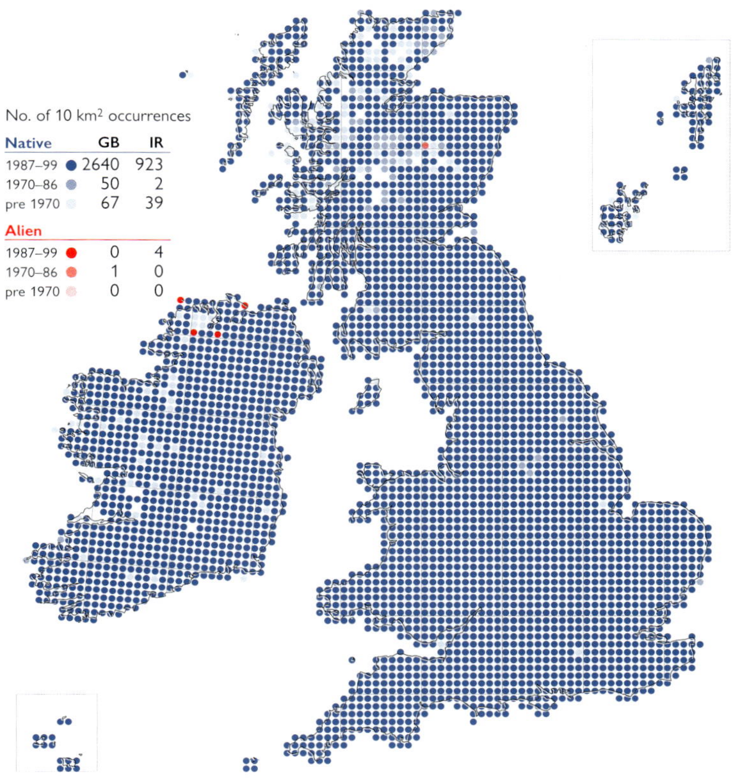

L. perenne is predominantly a species of improved lowland pasture, leys, and hay meadows, but is found widely in other habitats, including downland, rush-pasture, inundated grasslands, amenity grassland and road verges; also in open ruderal habitats. It favours fertile, heavy, neutral soils, but is also found on those which are mildly acidic or basic. 0–570 m (Alston Moor, Cumberland).

Native (change –0.29). *L. perenne* has been cultivated since at least the 17th century, and cultivars developed by plant breeders are currently amongst the most commonly sown agricultural grasses. Its distribution is stable.

European Southern-temperate element; widely naturalised outside its native range.

References: Atlas (375d), Beddows (1967), Grime *et al.* (1988), Hultén & Fries (1986), Meusel *et al.* (1965).

F. H. PERRING

Lolium multiflorum Italian Rye-grass

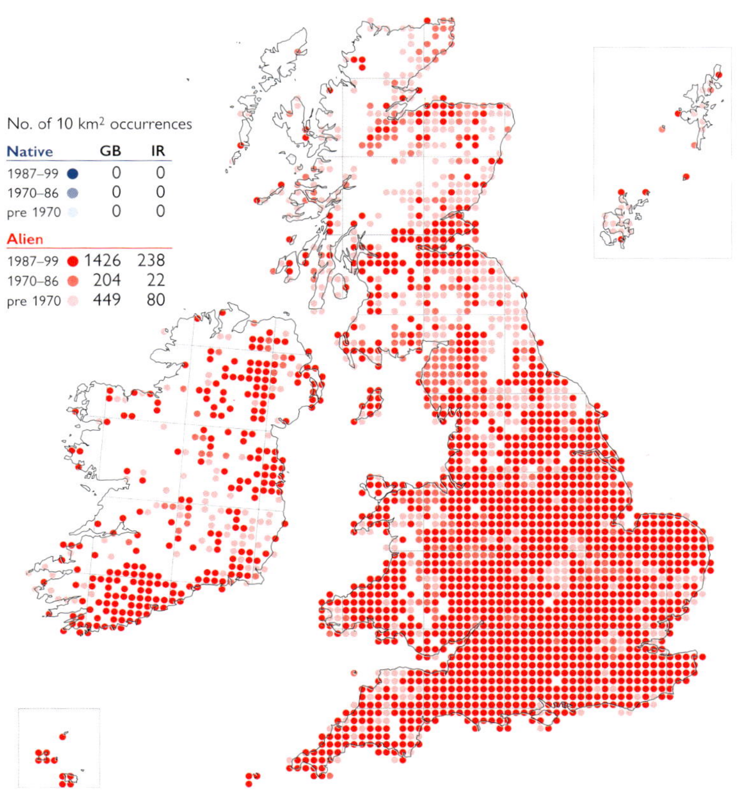

An annual or short-lived perennial found in leys, on field margins, in gateways, along farm tracks, on roadsides and rough ground. It often persists for a few years, but rarely becomes naturalised. Generally lowland, but reaching 410 m above Garrigill (Cumberland).

Neophyte (change –1.06). *L. multiflorum* began to be used in quantity in agriculture after about 1830, and until recently it was often grown in temporary leys for hay or silage. It was known from the wild by 1840. Its range is similar to that in the 1962 *Atlas*, but it is more frequent in Ireland than previously thought.

Probably native of the Mediterranean region, but widely used for fodder and now widespread in Europe and other temperate regions.

References: Atlas (376a), Beddows (1973), Hultén & Fries (1986), Ryves *et al.* (1996), Rich & Jermy (1998).

F. H. PERRING

Lolium multiflorum × L. perenne (*L. × boucheanum*)

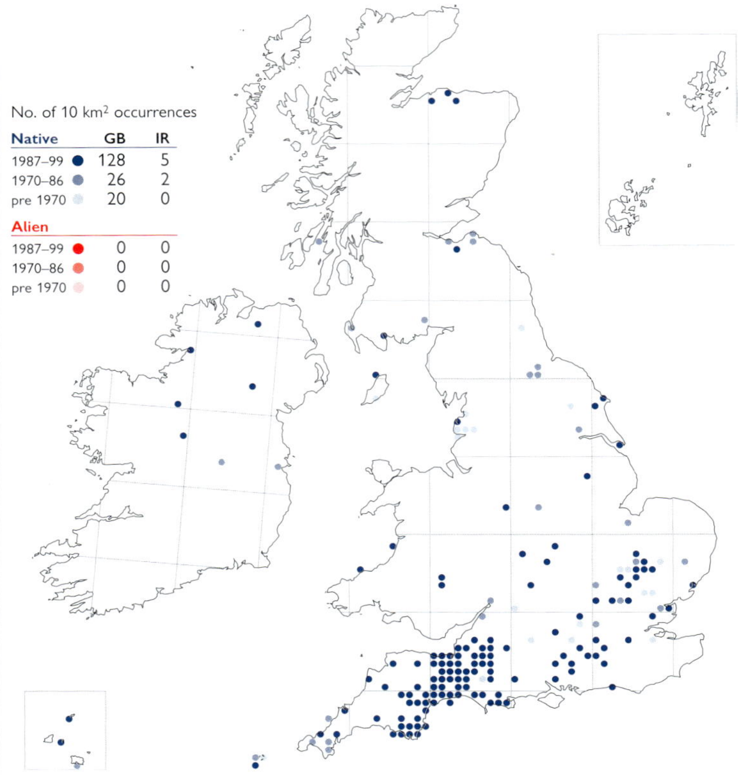

An annual or short-lived perennial herb of temporary leys, field margins, roadsides and waste ground. This hybrid is fertile, and can set seed. Populations are usually casual, but can persist for a few years. Probably lowland.

A spontaneous hybrid between native and alien parents. *L. × boucheanum* occasionally arises spontaneously, but occurs much more frequently when planted in fields and grassland. The parents are completely inter-fertile, and many cultivars of the hybrid have been developed. It was, up to about 1980, probably the most commonly sown *Lolium* but is dramatically under-recorded; it is often mistaken for its parents, but it is known to be frequent in some areas where botanists are familiar with it.

Widespread in temperate Europe.

References: Ryves *et al.* (1996), Stace (1975).

D. A. PEARMAN

Lolium temulentum Darnel

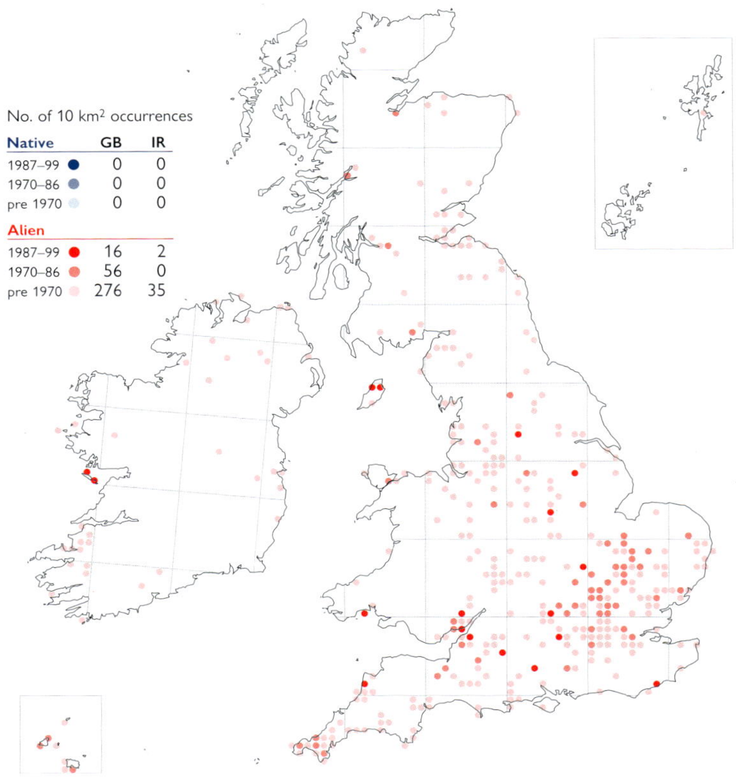

An annual, formerly often a persistent weed of arable land. It is now a rare casual of waste places, originating from grain, bird-seed and wool shoddy. Lowland.

Archaeophyte (change –4.05). *L. temulentum*, first recorded in Britain by 1548, was formerly a serious weed of arable land. It had almost disappeared from this habitat before the Second World War. It continues to be recorded as a casual, but less frequently now than in the 1950s and 1960s. It is sometimes toxic to humans and livestock (Cooper & Johnson, 1998).

Perhaps native of the Mediterranean region and S.W. Asia, but native range obscured by its spread with cultivation; once widespread in Europe and naturalised in Japan, N. & S. America, Australasia and elsewhere.

References: Atlas (376b), Curtis & McGough (1998), Hultén & Fries (1986).

D. A. PEARMAN

Vulpia fasciculata Dune Fescue

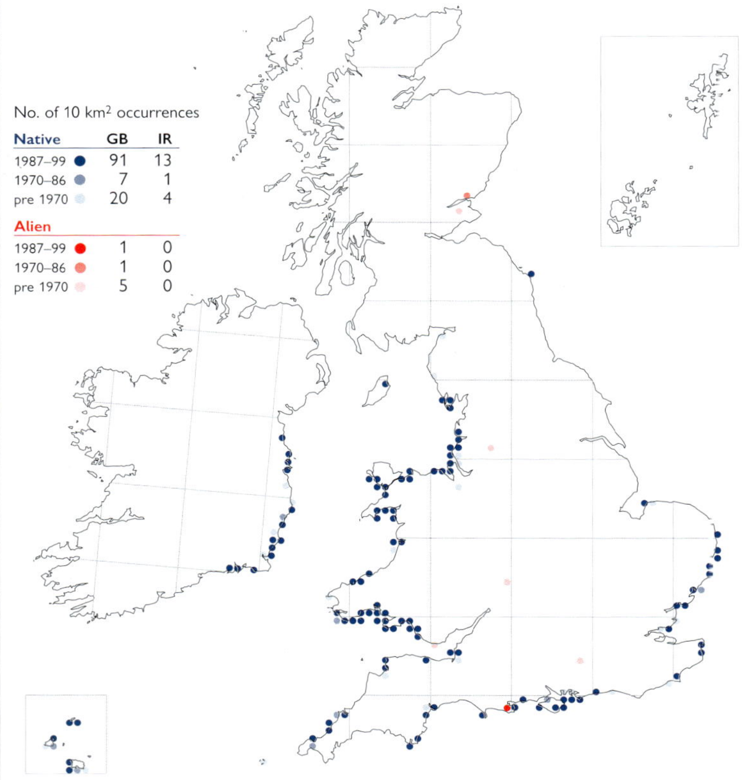

An annual of sand dunes, particularly open, disturbed parts of fixed dunes, and sandy shingle, frequently associated with other winter-annuals. Lowland.

Native (change +0.37). This species appears to have increased since the 1962 *Atlas*, especially in East Anglia and N. Wales. It is not clear, however, whether it was missed by earlier recorders. Its abundance on some sites increased following reduction of rabbit populations due to myxomatosis in the mid-1950s.

Mediterranean-Atlantic element.

References: Atlas (376c), Stewart *et al.* (1994), Watkinson (1978).

S. J. LEACH

Vulpia bromoides Squirreltail Fescue

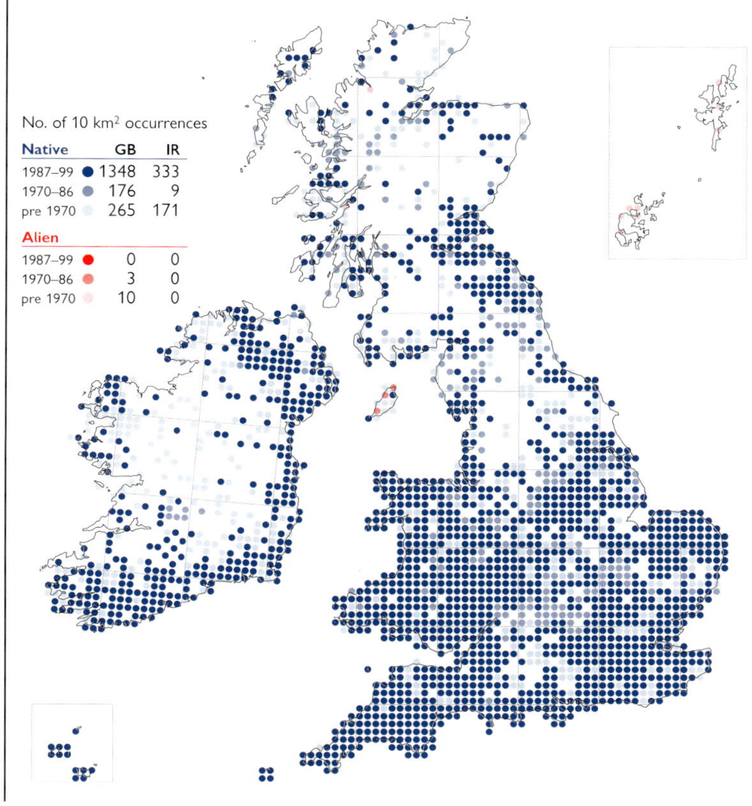

An annual of open grasslands, heaths, cliff-tops and sand dunes. It also grows in artificial habitats such as quarries, wall-tops, by railways and on waste ground in built-up areas. It was a frequent introduction from wool shoddy. *V. bromoides* favours well-drained soils, often growing abundantly on drought-prone S.-facing banks and slopes, but appears to be indifferent to soil pH. 0–490 m (Fanna Hill, Roxburghs.).

Native (change +0.18). There has been little change in the range of *V. bromoides* since the 1962 *Atlas*, though it now appears to be more frequent in many areas.

Submediterranean-Subatlantic element; widely naturalised outside its native range.

References: Atlas (376d), Hultén & Fries (1986), Meusel *et al.* (1965).

S. J. LEACH

Vulpia myuros Rat's-tail Fescue

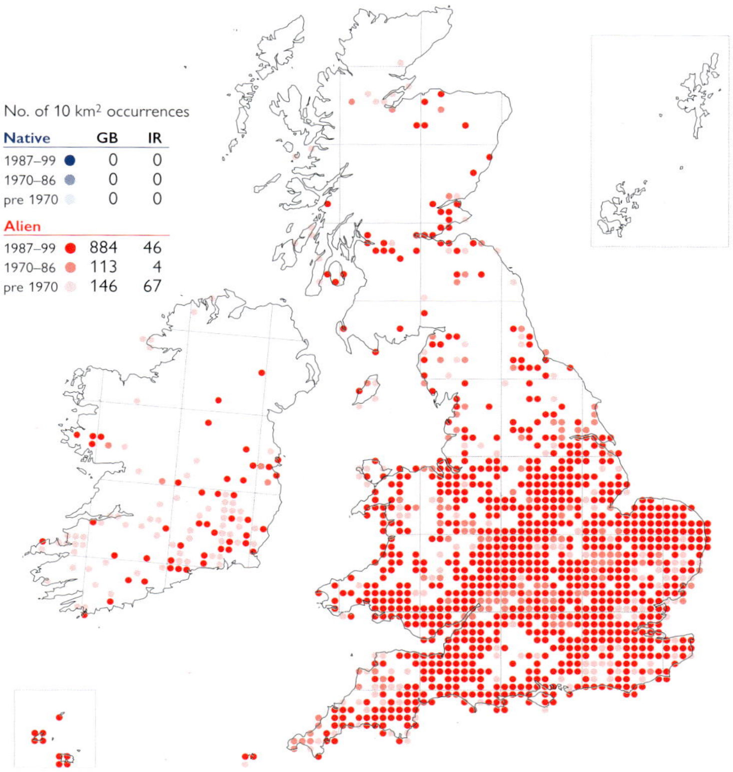

An annual growing by railways, on walls and waysides, in pavement cracks and on waste ground in built-up areas. Occasionally found as a weed of cultivation and as an introduction from wool shoddy, grain and grass-seed mixtures. Lowland.

Archaeophyte (change +1.55). Even allowing for the possibility that this species was under-recorded in the past, it is clear that *V. myuros* has become increasingly frequent across much of its range since the 1962 *Atlas*. It is likely that the plant has colonised many areas via the rail network.

As an archaeophyte *V. myuros* has a Eurosiberian Southern-temperate distribution; it is widely naturalised outside this range.

References: Atlas (377a, b), Meusel *et al.* (1965).

S. J. LEACH

Vulpia ciliata Bearded Fescue

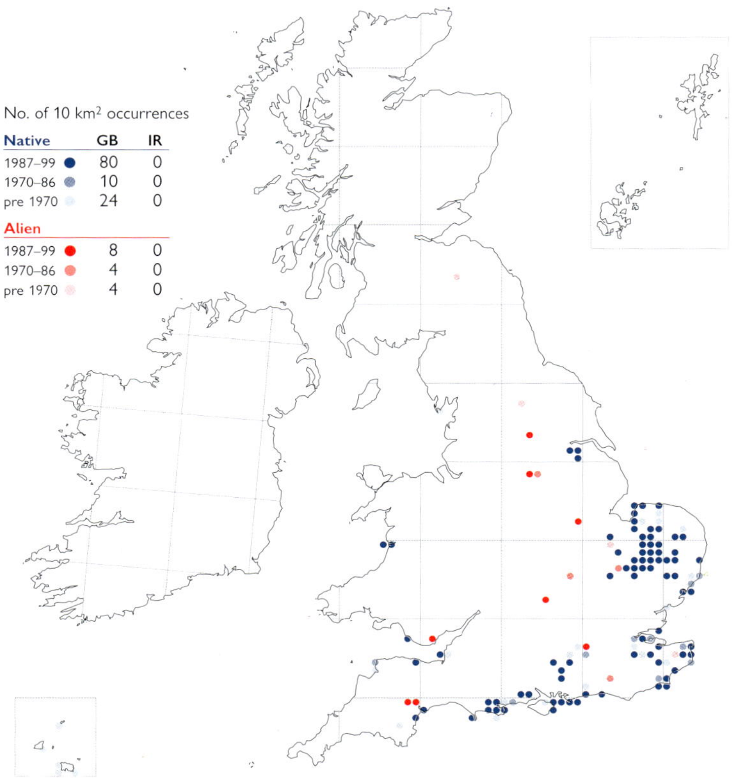

This annual occurs in disturbed sandy places. The native subsp. *ambigua* is found on tracks and paths through coastal dunes, and inland on sandy heaths, along roadsides and in patches of open grassland. The introduced subsp. *ciliata* is a rare casual from grain and wool shoddy. Lowland.

Native (change +0.78). *V. ciliata* was probably under-recorded in the 1962 *Atlas*. However, several new roadside populations of subsp. *ambigua* have been found since 1980, suggesting that it may be spreading in some areas (e.g. New Forest, S. Hants.). It was discovered in N. Wales in 1991.

Mediterranean-Atlantic element; subsp. *ambigua* is restricted to Britain, Belgium & N. France.

References: Atlas (377c), Carey *et al.* (1995), Stewart *et al.* (1994), Watkinson *et al.* (1998).

S. J. LEACH

Vulpia unilateralis Mat-grass Fescue

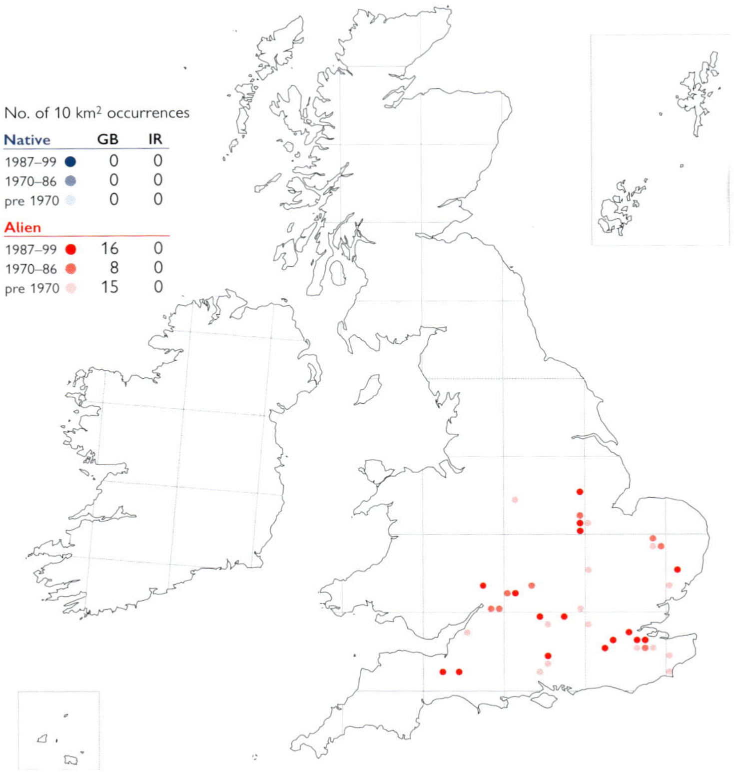

An annual of bare stony ground, dry banks and grassy tracks on chalk and limestone; also on railway ballast, walls and rubbish tips. Lowland.

Neophyte (change –0.56). *V. unilateralis* was first recorded in the British Isles in 1903. This inconspicuous species has been discovered in some new 10-km squares since the 1962 *Atlas*, though whether this means the species is increasing is uncertain. It has been much overlooked in the past, and is probably still under-recorded.

A Submediterranean-Subatlantic species; also recorded in C. Asia.

References: Atlas (379c), Stace (1961), Stewart *et al.* (1994).

S. J. LEACH

Cynosurus cristatus Crested Dog's-tail

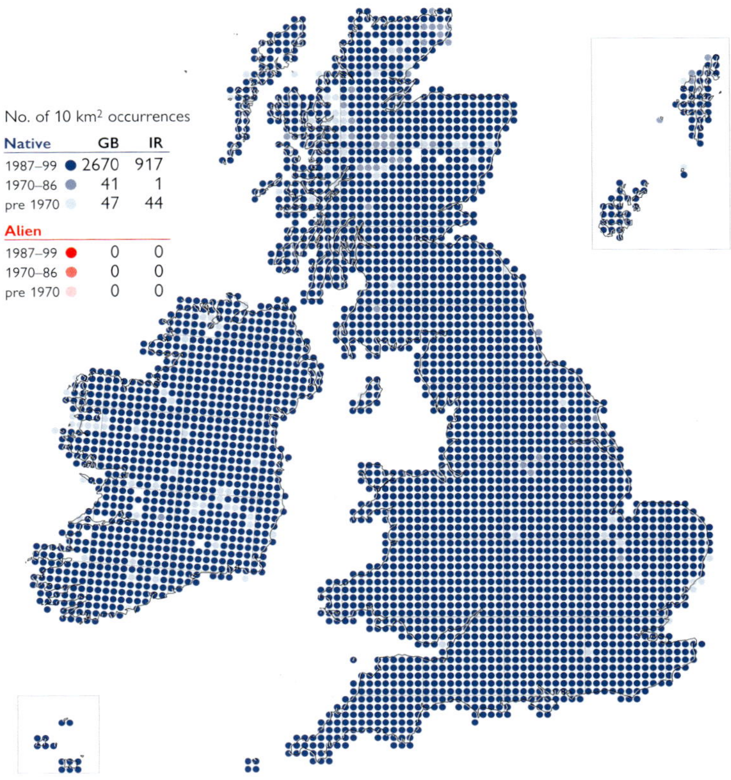

A short-lived perennial herb growing in a wide variety of grasslands, particularly short and heavily grazed swards. It grows in a range of neutral to base-rich, fairly well-drained or damp soils, avoiding only the extremes of base-status, waterlogging, drought and disturbance. Generally lowland, but to 660 m on Burnhope Seat (Cumberland) and exceptionally at 845 m on Great Dun Fell (Westmorland).

Native (change +0.02). *C. cristatus* was a frequent constituent of seed mixtures until the 1940s, and is still used in amenity sowings and possibly in upland leys on poor soils. There has been no change in its distribution since the 1962 *Atlas*.

European Temperate element; widely naturalised outside its native range.

References: Atlas (382d), Grime *et al.* (1988), Hultén & Fries (1986), Lodge (1959), Meusel *et al.* (1965).

D. A. PEARMAN

Cynosurus echinatus Rough Dog's-tail

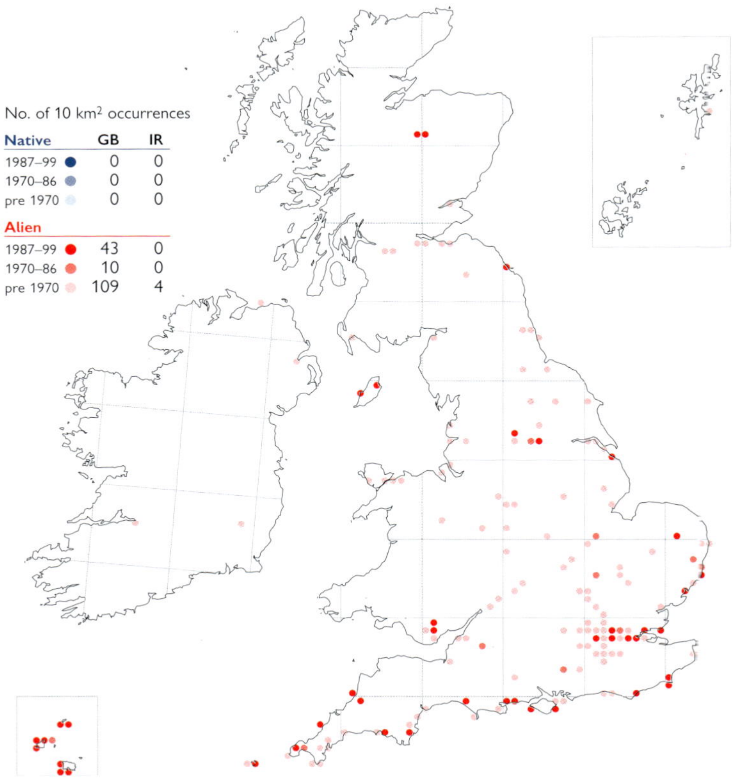

An annual grass naturalised on open sandy soils in the Channel Islands and the Isles of Scilly where it can sometimes be a pest in bulb fields, and in a few localities in S. England. Elsewhere, it is found as a grain and wool-shoddy casual in waste areas, tips and occasionally on arable land. Lowland.

Neophyte (change –1.47). *C. echinatus* has been recorded in the wild since 1778 but it has become less frequent as a casual since the 1962 *Atlas*. It is sometimes grown as an ornamental grass for drying.

Native of S. Europe, the Mediterranean region and S.W. Asia; widely naturalised further north in Europe and in other continents.

References: Atlas (383a), Meusel *et al.* (1965).

D. A. PEARMAN

Puccinellia maritima Common Saltmarsh-grass

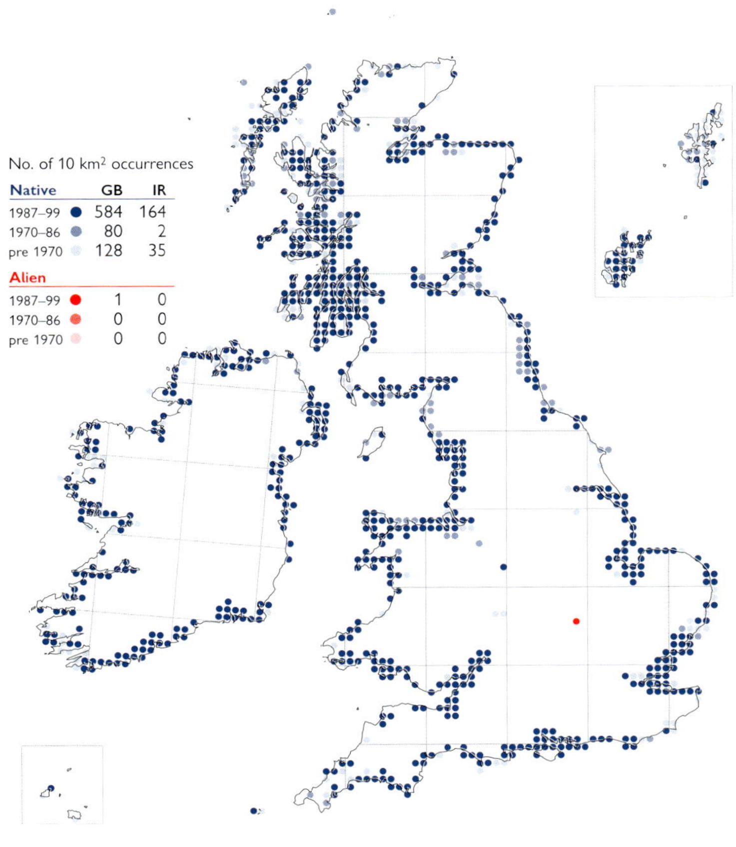

A stoloniferous perennial herb of saltmarshes, often dominant over large areas in the lower and middle marsh, and in pans and depressions in the upper marsh; also locally on bare saline soils above the tidal limit, on sea walls and beside grazing marsh ditches. Rarely, it occurs in saline areas inland, and as a colonist by salt-treated roads. Lowland.

Native (change –0.27). The coastal distribution of *P. maritima* has remained largely unchanged since the 1962 *Atlas*. Since the early 1970s it has been recorded occasionally alongside salt-treated roads in England.

Oceanic Boreo-temperate element; also in N. America.

References: Atlas (377d), Gray & Scott (1977), Hultén & Fries (1986), Matthews & Davison (1976).

S. J. LEACH

Puccinellia distans Reflexed Saltmarsh-grass

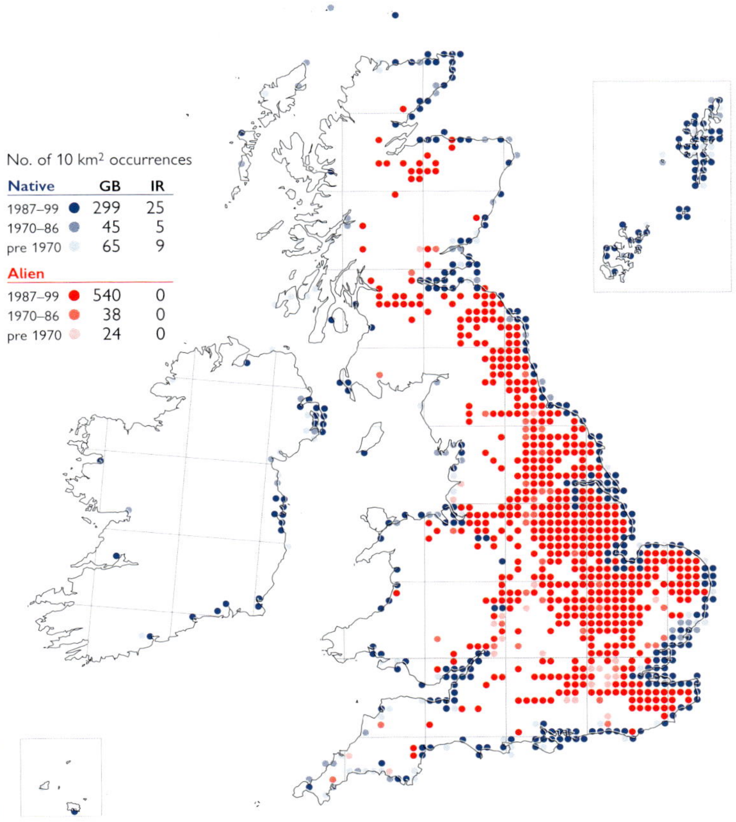

A perennial herb growing on barish muddy ground near the sea, along the upper edges of saltmarshes, on sea walls and amongst coastal rocks; also in saline areas inland, and as a colonist by salt-treated roads. It favours compacted, poorly-drained, heavy soils. Lowland, with a roadside record at 520 m at Holme Moss (Cheshire).

Native (change +3.02). As in the 1962 *Atlas*, this map includes records for subsp. *distans* and subsp. *borealis*; the latter is also mapped separately. Whilst its coastal distribution is stable (apart from being much better recorded in N. Scotland), it has spread rapidly along roadsides since 1970.

Eurosiberian Boreo-temperate element; widely naturalised outside its native range.

References: Atlas (378a), Coombe (1994), Hultén & Fries (1986), Scott (1985), Scott & Davison (1982).

S. J. LEACH

Puccinellia distans subsp. *borealis*

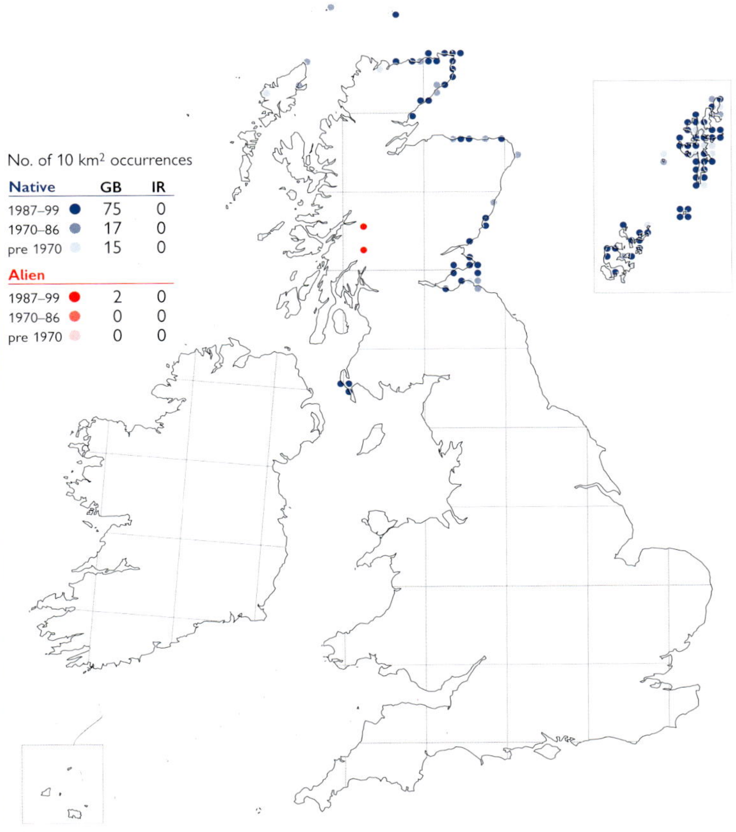

A tufted perennial herb of rocky shores, growing on low cliffs and amongst rocks and boulders on pebble beaches, on the stonework of harbour walls and slipways, and only rarely in saltmarshes. Lowland.

Native. This taxon, at one time regarded as a species in its own right (*P. capillaris*), has only recently had its taxonomy and distribution clearly defined (Trist & Butler, 1995). Because there are few historical data, changes in its distribution cannot be assessed.

European Boreo-arctic Montane element; restricted to the coasts of N. Europe.

S. J. LEACH

Puccinellia fasciculata Borrer's Saltmarsh-grass

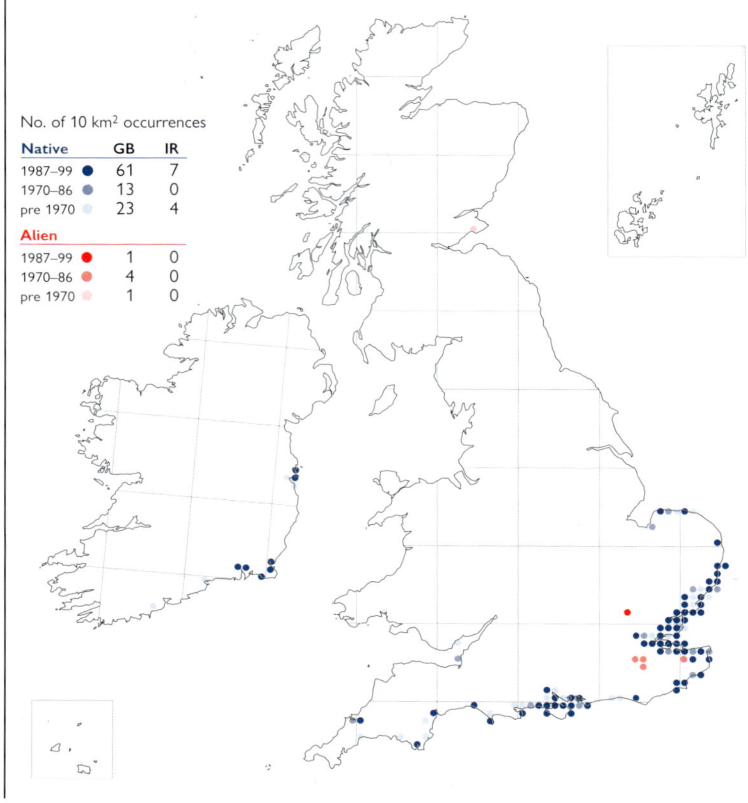

A tufted perennial herb of bare places by the sea, in grazing marshes around cattle-poached pools and depressions, on earthen sea walls, vehicle tracks and the mud dredged from ditches. It also occurs rarely beside salt-treated roads inland. Lowland.

Native (change –0.51). This species has probably decreased in some areas as a result of the infilling of pools and ditches, the upgrading of sea walls and the conversion of coastal grazing marshes to arable. It now includes *P. pseudodistans* (as *P. fasciculata* var. *pseudodistans*), which was mapped separately in the 1962 *Atlas*.

Suboceanic Southern-temperate element; also in N. America.

References: Atlas (378b, c), Curtis & McGough (1988), Kitchener (1983), Stewart *et al.* (1994).

S. J. LEACH

Puccinellia rupestris Stiff Saltmarsh-grass

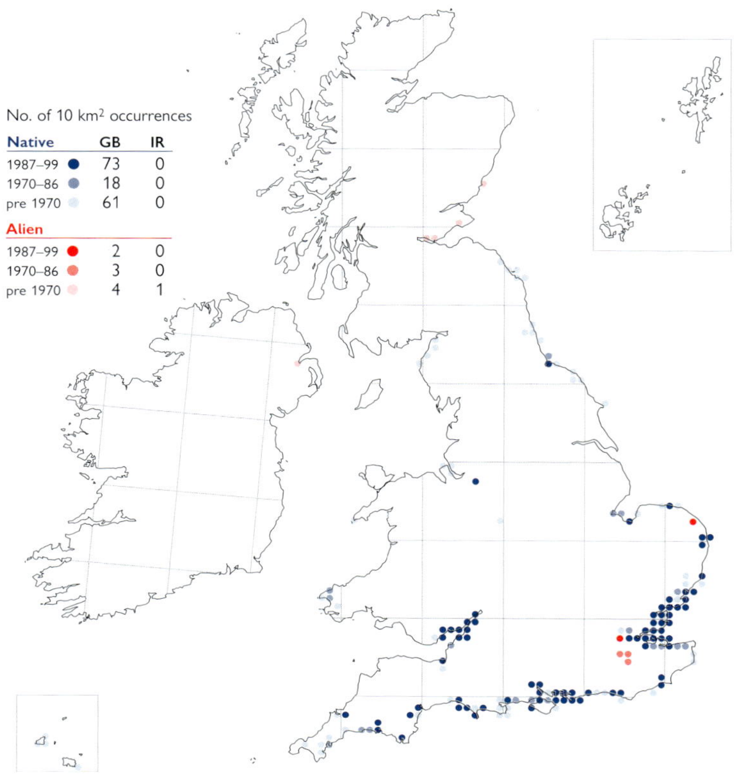

An annual or biennial herb growing on bare saline soils above the tidal limit, behind sea walls, on tracks and in grazing marshes around cattle-trodden pools and depressions, and sometimes on firm muddy shingle and in rock crevices. *P. rupestris* occurs rarely inland by saline springs and salt-treated roads. Lowland.

Native (change –0.40). *P. rupestris* was declining in N. England before 1930, and is now practically extinct north of The Wash. It has also declined in S. England, probably due to the infilling of pools and ditches, upgrading of sea walls and the conversion of coastal grazing marshes to arable. It was recorded before 1930 in S.E. Scotland as a ballast alien.

Oceanic Southern-temperate element.

References: Atlas (378d), Kitchener (1983), Stewart *et al.* (1994).

S. J. LEACH

Briza media Quaking-grass

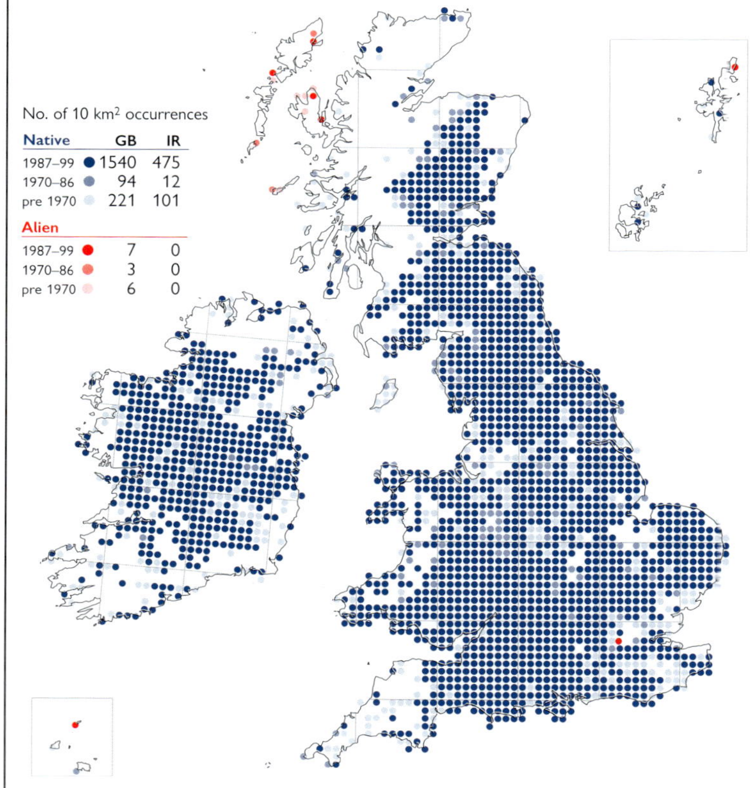

A shortly rhizomatous perennial grass, most frequently found in unimproved, species-rich, well-grazed grassland on infertile, calcareous soils and favouring well-drained slopes. However, it also occurs in old meadows and pastures on neutral and sometimes acidic soils, in the drier parts of fens, and occasionally in soligenous mires. 0–720 m (Knock Fell, Westmorland).

Native (change –0.75). There has been little change in the range of *B. media* since the 1962 *Atlas*. However, the map suggests some decline at the 10-km scale and local floras suggest that the ploughing and improvement of grasslands has led to a decline in frequency in many areas.

European Temperate element.

References: Atlas (383b), Grime *et al.* (1988), Hultén & Fries (1986), Meusel *et al.* (1965).

D. A. PEARMAN

Briza minor Lesser Quaking-grass

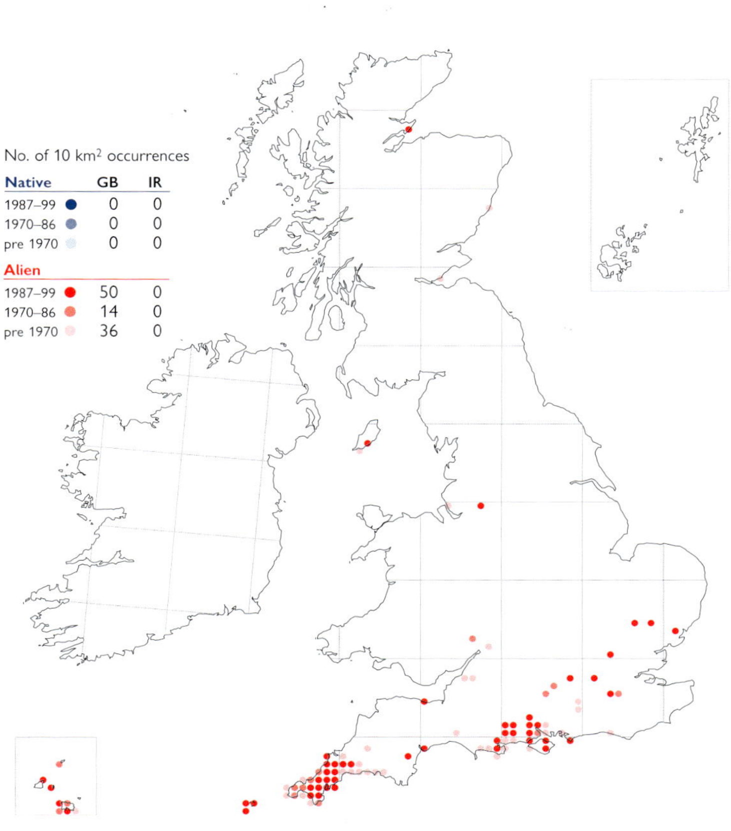

An annual mainly found on light, base-poor soils in arable habitats, including bulb-fields. It is also found, as a casual, by roadsides, on walls and, rarely, on waste land or rubbish tips. Lowland.

Archaeophyte (change +0.28). *B. minor* was first recorded from the wild in Britain in 1696 (Jersey). It is susceptible to herbicides and is unable to survive in highly fertilised crops, but persists in favourable, open habitats under non-intensive farming regimes.

A Mediterranean-Atlantic species. It was introduced with cultivation in the northern part of this range and is widely naturalised in warm-temperate regions throughout the world.

References: Atlas (383c), Ryves *et al.* (1996), Stewart *et al.* (1994).

D. A. PEARMAN

Briza maxima Greater Quaking-grass

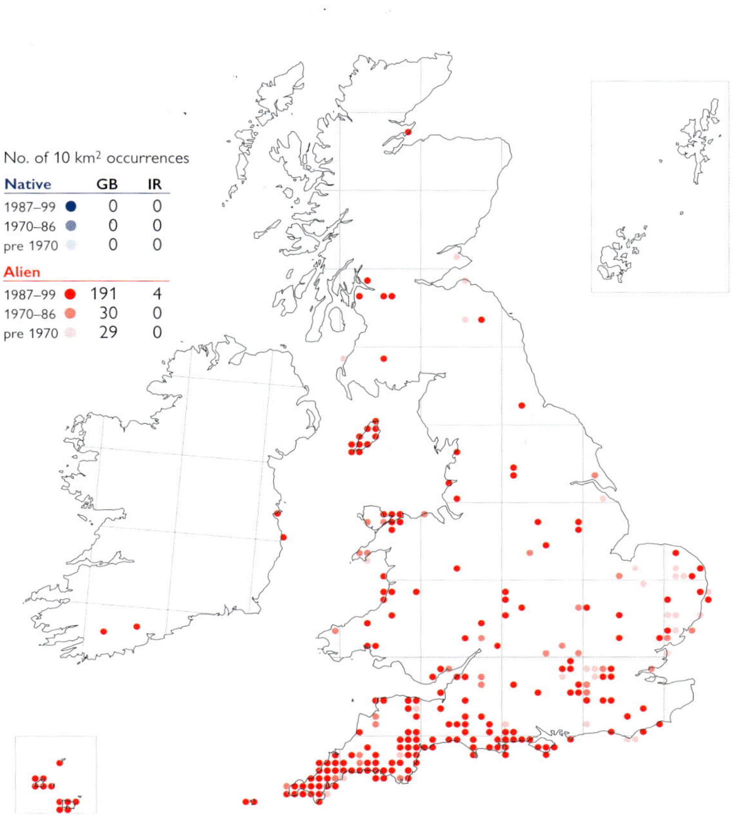

An annual, naturalised or occurring casually on dry, bare banks and field margins, in cultivated ground including gardens, bulb-fields, on sand dunes, sea-cliffs, rubbish tips, waste ground and wall-tops, and in pavement cracks. Lowland.

Neophyte. *B. maxima* was introduced into Britain by 1633 and recorded in the wild by 1860 (Jersey). It appears to be increasing, especially in S.W. England and the Channel Islands. It is frequently grown for ornamental purposes, and is increasing as a garden escape. It is also introduced with wool shoddy and esparto.

Native of the Mediterranean region; widely naturalised in warm-temperate regions throughout the world.

References: French *et al.* (1999), Ryves *et al.* (1996).

S. J. LEACH & D. A. PEARMAN

Poa infirma Early Meadow-grass

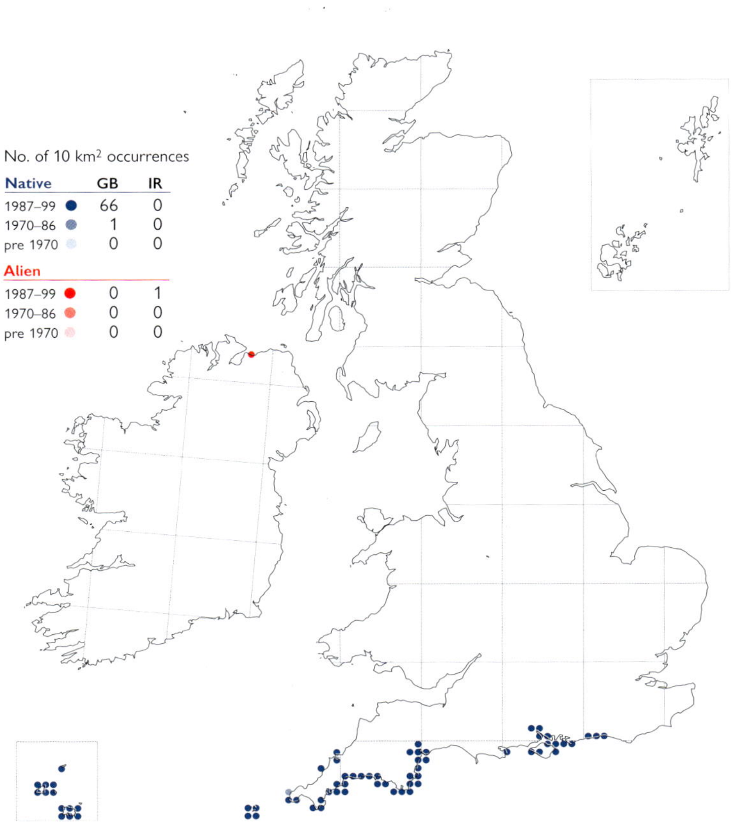

An annual growing near the sea in open, trampled grassland, on cliff-top paths, track-sides, picnic sites, lawns and car parks, and in stabilised dunes and other sandy places. Lowland.

Native (change +1.33). At the time of the 1962 *Atlas*, *P. infirma* was thought to be restricted to W. Cornwall, the Isles of Scilly and the Channel Islands. However, there has been a very considerable extension of range eastwards in the last 10–15 years (Takagi-Arigho, 1994). This can probably be attributed to better recording, as botanists are now more familiar with this species, which is only apparent in early spring and closely resembles *P. annua*. In Ireland, it was found in a garden in 1987 (Co. Londonderry) and in the wild in W. Cork in 2000.

Mediterranean-Atlantic element.

References: Atlas (380a), Wigginton (1999).

S. J. LEACH

Poa annua Annual Meadow-grass

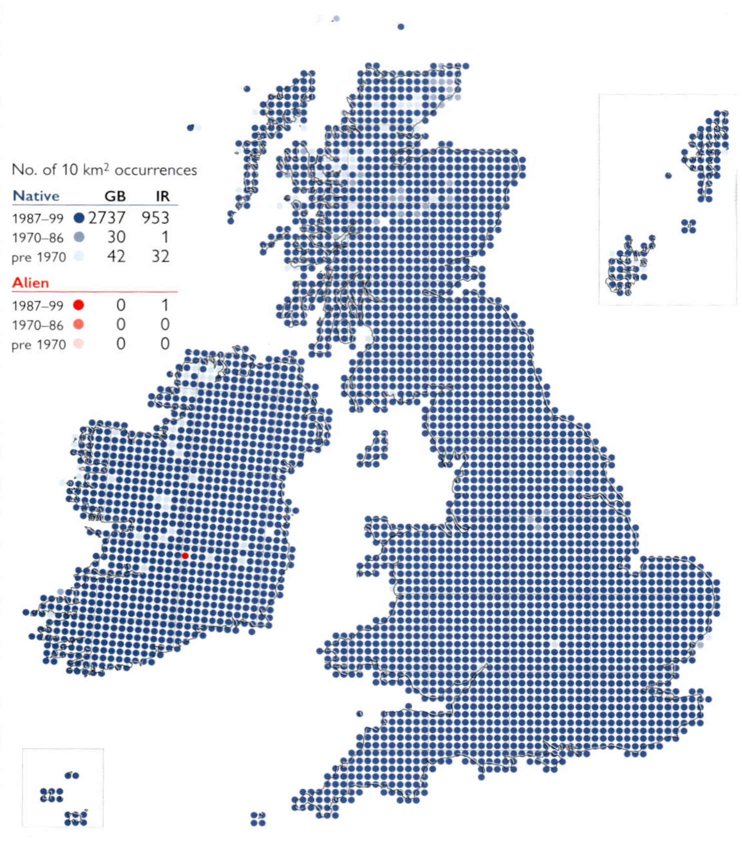

An annual growing in a wide range of disturbed and man-made habitats, including over-grazed and trampled grasslands, lawns, arable fields, waste ground, paths, waysides and wall-tops. Perennial variants occur in montane and coastal grassland. *P. annua* is also a common wool and bird-seed alien, and a familiar garden weed throughout the British Isles. 0–1210 m (Ben Lawers, Mid Perth).

Native (change +0.83). *P. annua* is ubiquitous, and its range has not changed since the 1962 *Atlas*.

Eurosiberian Wide-temperate element, but widely naturalised so that distribution is now Circumpolar Wide-temperate.

References: Atlas (379d), Hultén & Fries (1986), Hutchinson & Seymour (1982), Grime *et al.* (1988), Warwick (1979).

S. J. LEACH

Poa trivialis Rough Meadow-grass

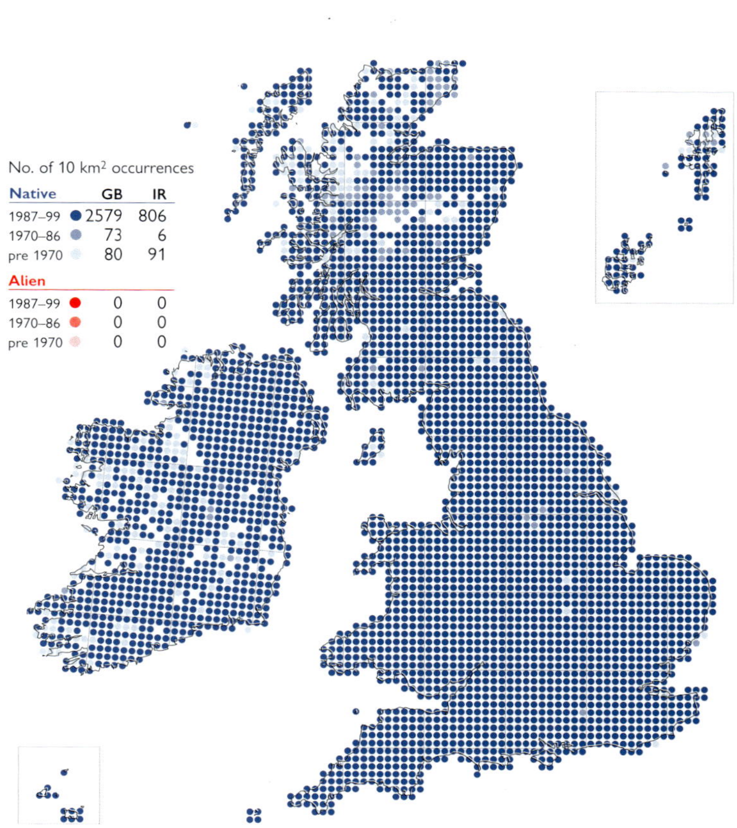

A stoloniferous perennial herb of open woodland, meadows, pastures, walls, waste ground, waysides and cultivated land; it also grows in marshes and beside ponds, ditches and streams. It was formerly included in commercial grass-seed mixtures, and is still used in amenity and wild-flower grasslands. It is a common wool alien. Generally lowland, but reaching 1065 m on Carn Eige, Glen Affric (Easterness).

Native (change +1.10). There has been no change in the range of *P. trivialis* since the 1962 *Atlas*. It is very widespread, being absent only from montane areas.

Eurosiberian Wide-temperate element, but widely naturalised so that distribution is now Circumpolar Wide-temperate.

References: Atlas (381c), Grime *et al.* (1988), Hultén & Fries (1986).

S. J. LEACH

Poa pratensis sens. lat.

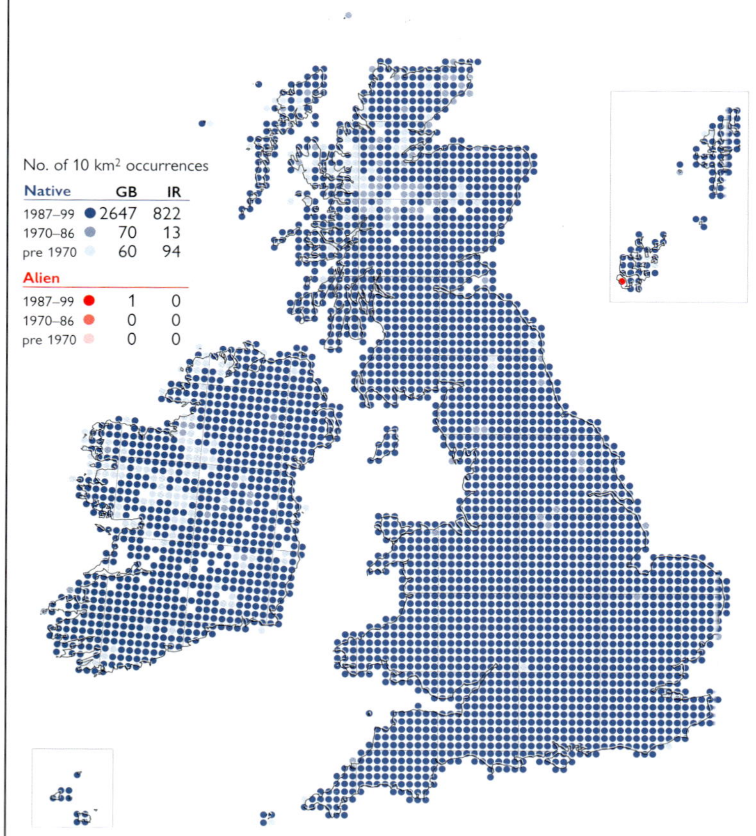

These rhizomatous perennial herbs are found in a wide range of habitats, from wet to dry, infertile to fertile, somewhat acidic to strongly calcareous. They grow in grasslands, on sand dunes, river banks, mountain slopes, roadsides, waste ground and wall-tops. 0–1065 m (Ben Lawers, Mid Perth).

Native (change +0.60). There is no evidence of any change in the range of *P. pratensis sens. lat.* since the 1962 *Atlas*. The aggregate includes three species (*P. pratensis sens. str.*, *P. humilis* and *P. angustifolia*), all of which are mapped separately.

The three species in our area have a Circumpolar Wide-temperate distribution; the fourth European species, *P. alpigena*, extends the range to the high Arctic.

References: Atlas (381b), Barling (1959, 1962, 1967), Grime *et al.* (1988), Hultén & Fries (1986).

S. J. LEACH

Poa humilis Spreading Meadow-grass

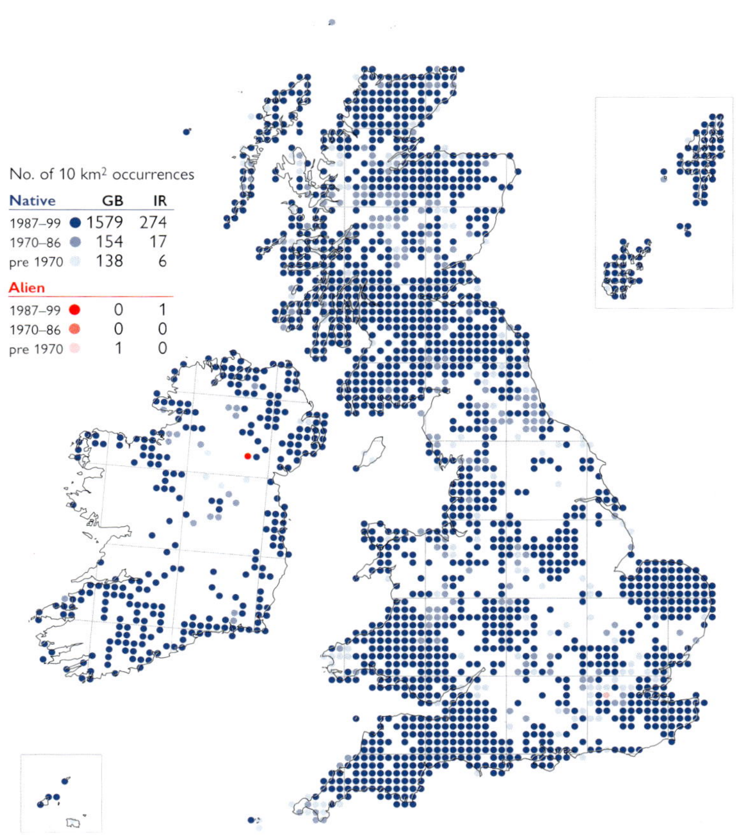

A rhizomatous perennial herb found in a wide array of grassland habitats, including neutral meadows, calcareous grassland, sand dunes, roadsides, wall-tops, river banks and mountain slopes. Morphologically and cytologically this is a highly variable species; some of its variants may be apomictic, and its ecology is poorly understood. 0–670 m (Knock Fell, Westmorland) and probably higher elsewhere.

Native. In the 1962 *Atlas*, *P. humilis* was not distinguished from other taxa within the *P. pratensis* group, but it was mapped by Perring & Sell (1968). It is still ignored by some recorders. The paucity of reliable historical data makes distributional trends difficult to assess.

World distribution uncertain.

References: Atlas Supp. (150b), Barling (1962), Trist (1989).

S. J. LEACH

Poa pratensis sens. str. Smooth Meadow-grass

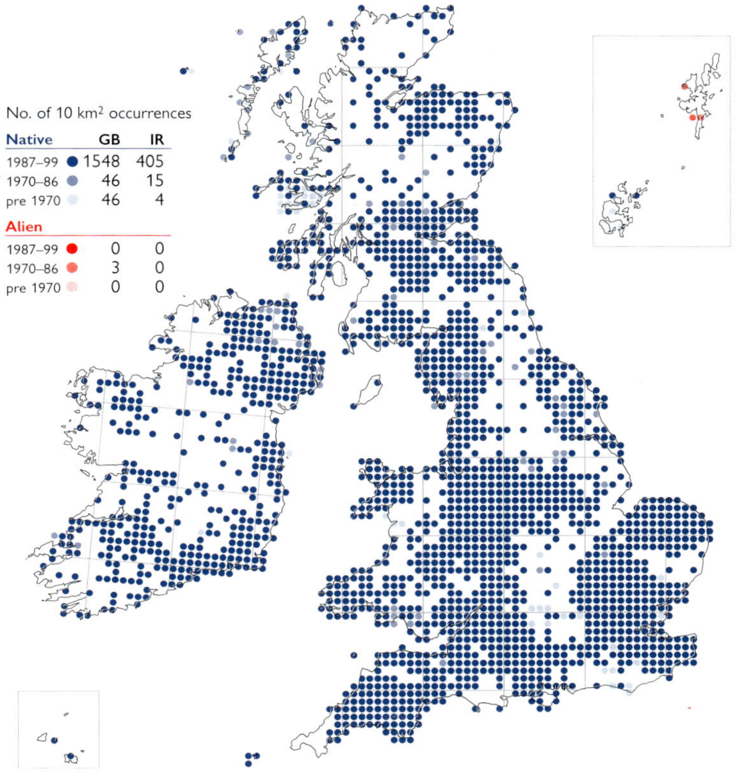

A rhizomatous perennial herb of meadows, pastures, waysides and waste places; formerly an important constituent of commercial seed mixtures, and still used in the sowing of amenity and wild-flower grasslands. It is a versatile grass, preferring well-drained, neutral soils of moderate to high fertility, and tolerant of grazing and trampling. Generally lowland, but upper altitudinal limit unknown.

Native. There are no reliable historical data to allow an assessment of whether the distribution of this taxon is changing. However, the *P. pratensis* aggregate appears to have a stable distribution in many lowland areas where this is the most common segregate. It is unevenly recorded.

Circumpolar Wide-temperate element; widely naturalised outside its native range.

References: Barling (1967), Grime *et al.* (1988).

S. J. LEACH

Poa angustifolia Narrow-leaved Meadow-grass

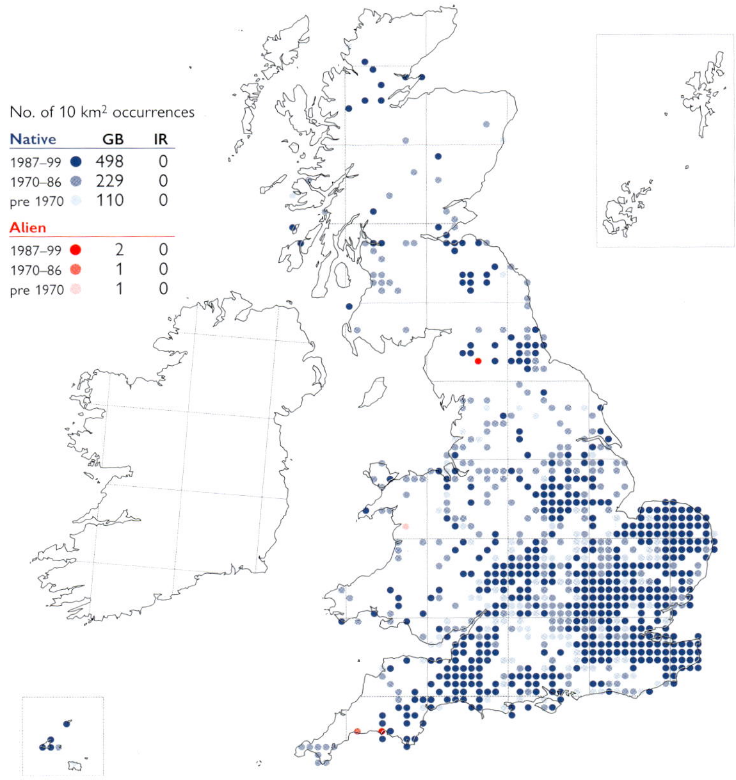

A rhizomatous perennial herb of dry grassland, wall-tops, rough ground and railway banks, typically on relatively infertile chalky, sandy or gravelly soils. In parts of S. England it often occurs in partial shade under *Fagus* trees. Lowland.

Native. Historically, *P. angustifolia* has been confused with other members of the *Poa pratensis* group. Its distribution was mapped by Perring & Sell (1968) and investigated by Sargent *et al.* (1986). It is probably decreasing in semi-natural habitats, but trends are difficult to assess given the lack of historical data. The map, apart from the under-recording in Sussex and Wiltshire, is probably a reasonable reflection of its distribution.

Circumpolar Southern-temperate element.

References: Atlas Supp. (150a), Barling (1959, 1967), Hultén & Fries (1986).

S. J. LEACH

Poa chaixii Broad-leaved Meadow-grass

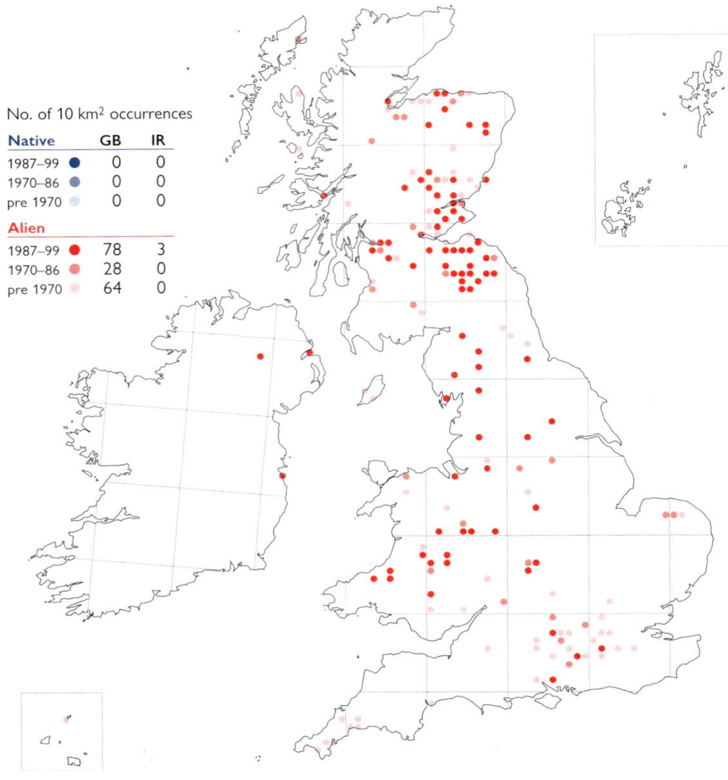

A stout, densely tufted perennial herb naturalised in copses and open woodland. It is occasionally found growing with *Festuca heterophylla*, another ornamental species admired by Victorian gardeners. Generally lowland, but reaching 395 m (The Quiraing, N. Ebudes).

Neophyte (change –0.05). *P. chaixii* has been grown in gardens for ornament or ground cover since 1802, and was much planted in the 19th century in the grounds of large estates, especially in Scotland. It was recorded in the wild by 1852. There has been little change in distribution since the 1962 *Atlas*, but it is possibly under-recorded.

Native of montane woods in C. & S. Europe; widely naturalised further north.

References: Atlas (382a), Hultén & Fries (1986), Meusel *et al.* (1965), Ryves *et al.* (1996).

S. J. LEACH

Poa flexuosa Wavy Meadow-grass

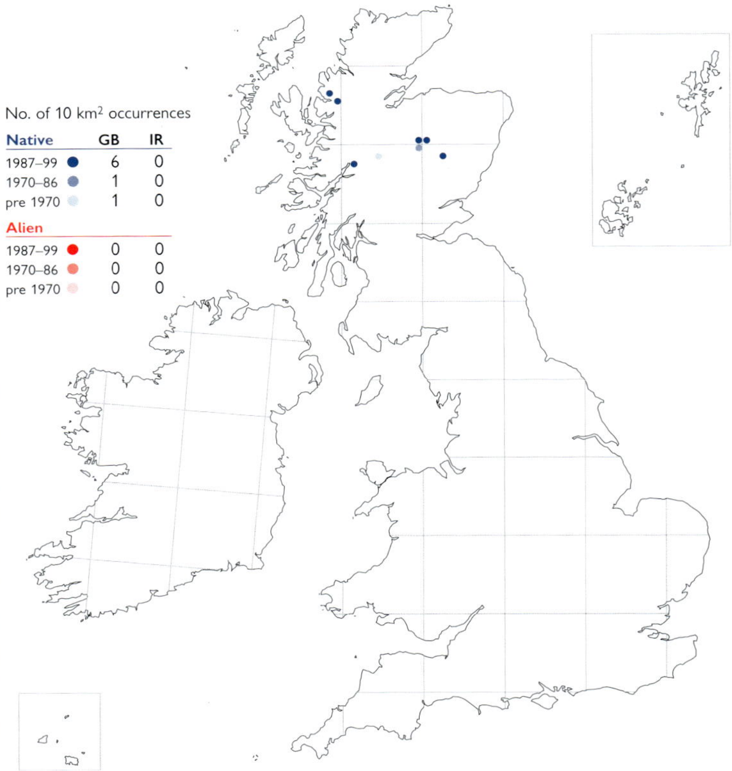

A tufted perennial herb of acidic rock ledges, screes and stony mountain plateaux. From 760 m to 1100 m (Ben Nevis, Westerness).

Native. *P. flexuosa* is now known from more 10-km squares than were mapped in the 1962 *Atlas*. Since 1970 it has been recorded from about a dozen sites, most of these supporting only very small populations. *P.* × *jemtlandica* (*P. flexuosa* × *P. alpina*) occurs with *P. flexuosa* at some sites, and may perhaps have given rise to some mis-identifications.

European Arctic-montane element; also in N. America.

References: Atlas (380b), Hultén & Fries (1986), Meusel *et al.* (1965), Wigginton (1999).

S. J. LEACH

Poa compressa Flattened Meadow-grass

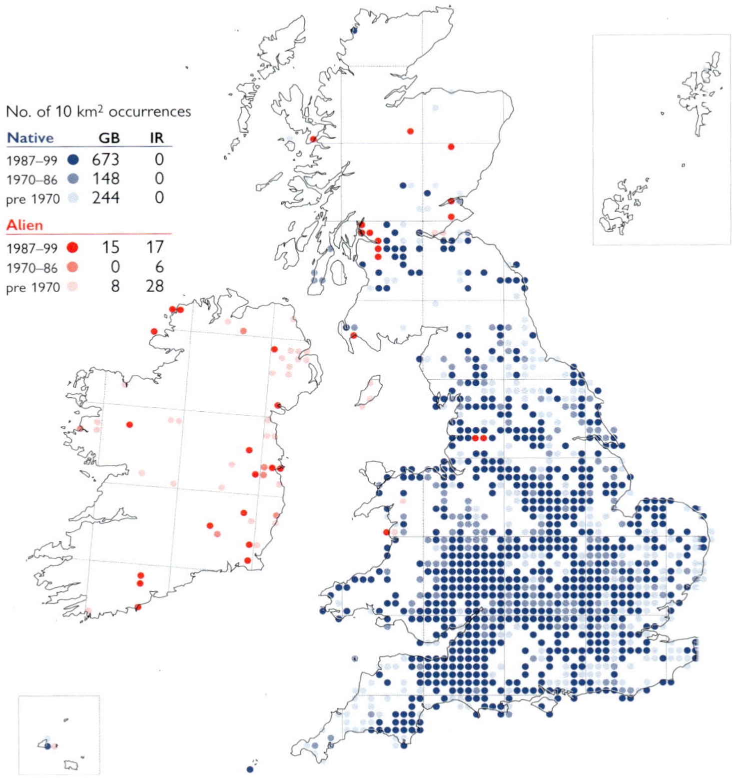

A rhizomatous perennial herb of rough or stony ground, cinders, dry grassy banks, waysides and walls. Some populations on rubbish tips and waste ground are probably introductions from wool shoddy and other sources. Generally lowland, but reaching 365 m at Mallerstang (Westmorland).

Native (change +0.21). There is no reliable information on trends in the distribution of *P. compressa*, and it is not possible to make a clear distinction between native and introduced populations, particularly in Scotland. It is regarded as alien in Ireland (Scannell & Synnott, 1987). The species is sometimes confused with other *Poa* species, particularly *P. humilis*, and is probably under-recorded.

European Temperate element; widely naturalised outside its native range.

References: Atlas (381a), Hultén & Fries (1986).

S. J. LEACH

Poa palustris Swamp Meadow-grass

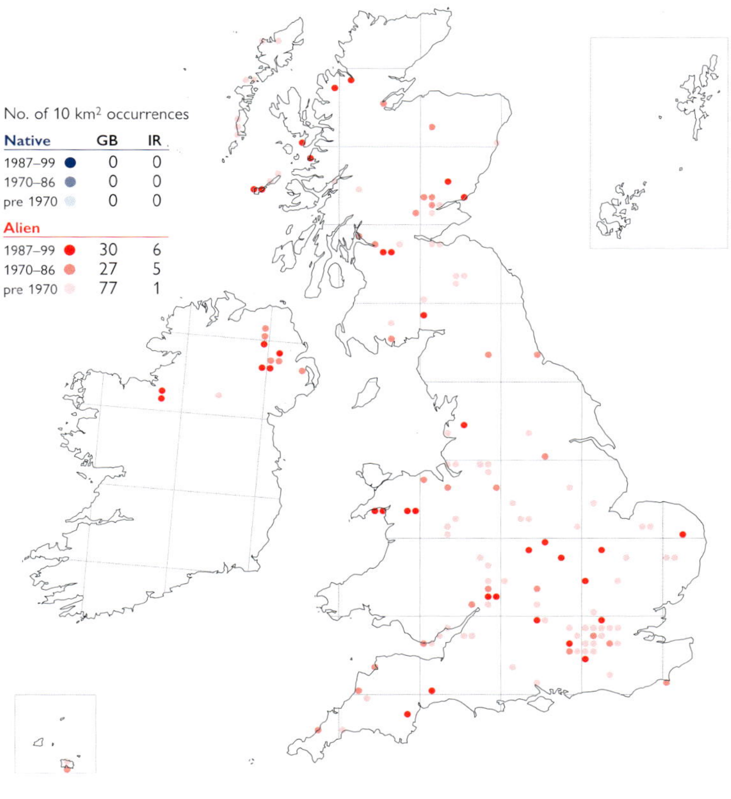

A short-lived tufted perennial growing in marshes, fens, ditches, *Salix* carr and at the edges of rivers, canals, lakes and ponds; occasionally around docks, by railways and on waste ground. Generally lowland, but upper altitudinal limit unknown.

Neophyte (change –1.55). *P. palustris* is thought to have been introduced in the 19th century as a fodder grass, and was recorded from the wild in 1879. The map suggests a substantial decline in the last forty years, though at many old localities its occurrence was probably only transient. It is easily overlooked.

A Circumpolar Boreo-temperate species, but absent as a native from much of W. Europe.

References: Atlas (381d), Curtis & McGough (1988), Hultén & Fries (1986), Meusel *et al.* (1965), Stewart *et al.* (1994).

S. J. LEACH

Poa glauca Glaucous Meadow-grass

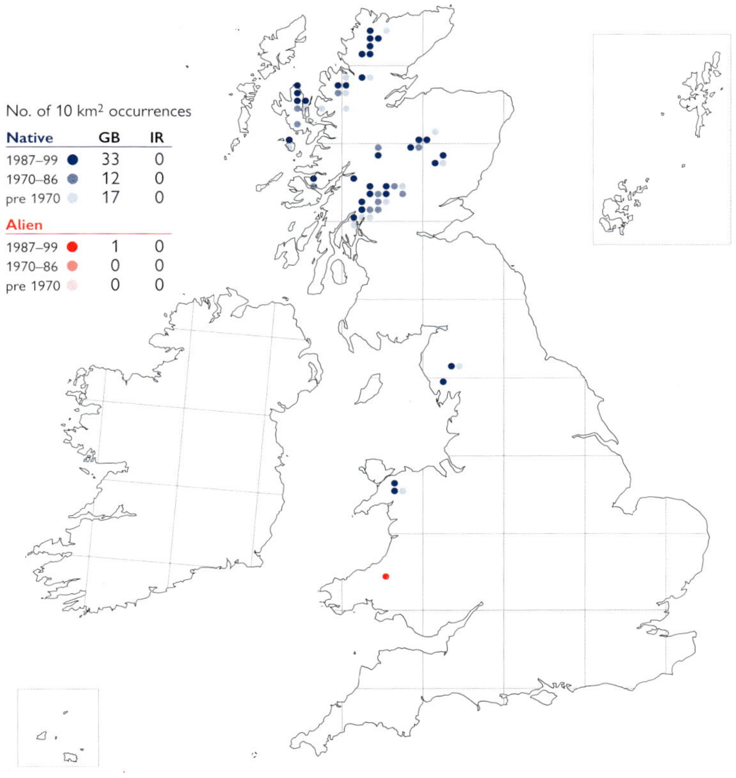

A tufted perennial herb of damp mountain rock faces, open ledges, screes and rocky slopes on calcareous substrates, often with *Poa alpina*. It is also known as a casual by a disused railway in Cardiganshire. From 305 m on Trotternish, Skye (N. Ebudes) to 1110 m on Lochnagar (S. Aberdeen).

Native (change –0.48). *P. glauca* appears to have declined since the 1962 *Atlas*, though it may be present in many sites in remote mountain areas for which there are only pre-1987 records. It is thought to be sensitive to grazing, which might explain its decline in some areas.

Circumpolar Boreo-arctic Montane element.

References: Atlas (380d), Hultén & Fries (1986), Meusel *et al.* (1965), Stewart *et al.* (1994).

S. J. LEACH

Poa nemoralis Wood Meadow-grass

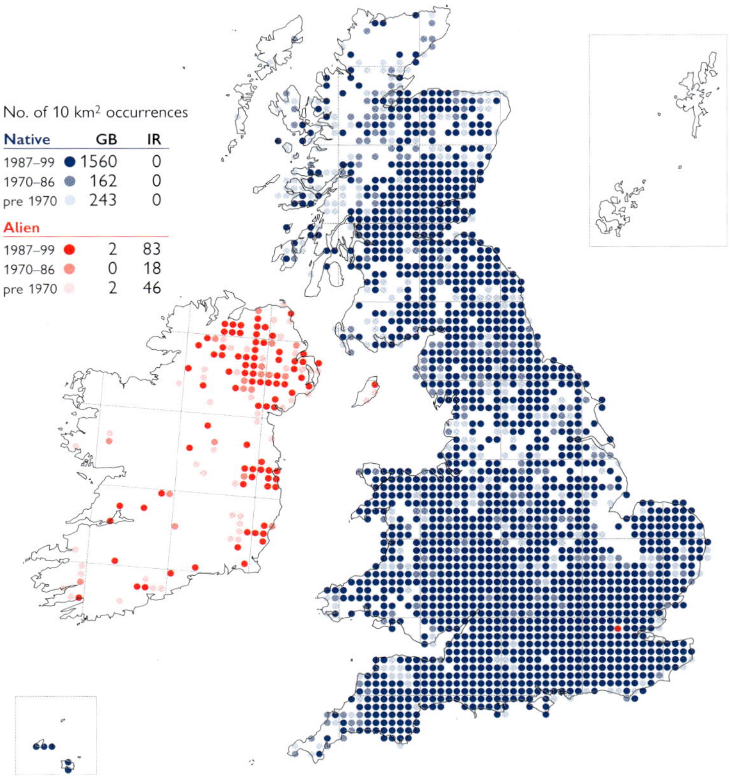

A tufted perennial herb of woodland rides and glades, hedgerows and other shaded places; also locally on walls, and in the mountains on dry rock ledges. It was occasionally sown in woodlands and parks for its ornamental value, while in some areas it may have been introduced with wool shoddy, grass-seed or soil. Generally lowland, but reaching 915 m (Sgurr na Lappaich, Glen Farrar, Easterness).

Native (change +0.27). The native distribution of *P. nemoralis* is impossible to delimit with any degree of certainty. It was mapped as 'all records' for the 1962 *Atlas*, and may be spreading in parts of N.W. Britain and Ireland, where it was probably originally introduced.

Circumpolar Boreo-temperate element.

References: Atlas (380c), Hultén & Fries (1986), Meusel *et al.* (1965), Scannell & Synnott (1987).

S. J. LEACH

Poa bulbosa Bulbous Meadow-grass

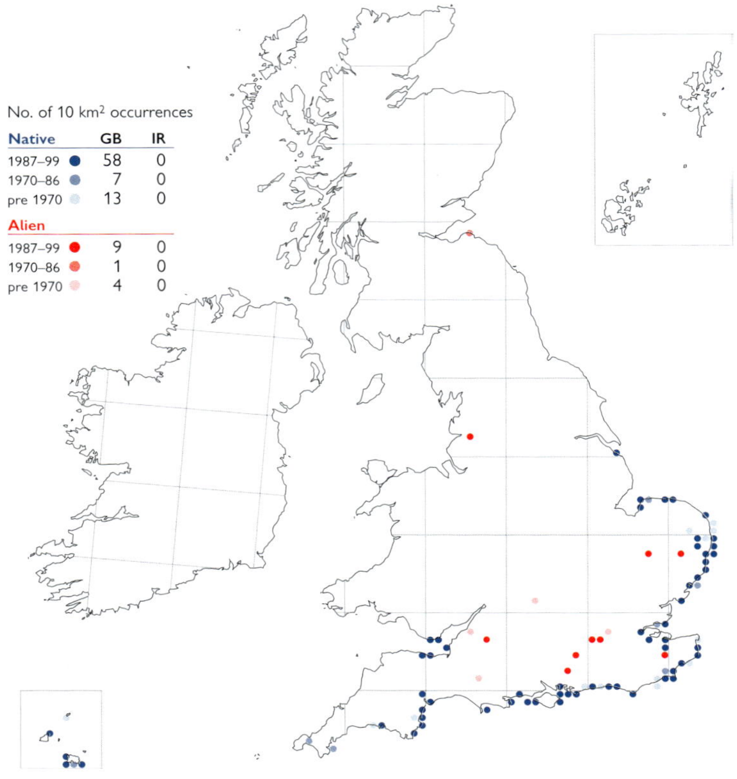

A tufted bulbous-based perennial herb of open grassland and barish sandy or rocky places near the sea; mainly on sand dunes and stabilised shingle, but also on bare chalk and limestone. Some populations are wholly or partially proliferous. Lowland.

Native (change +0.63). The distribution of *P. bulbosa* is probably stable; there are few losses and most of the newly recorded populations are probably long-established. The origins of the inland colonies are uncertain: they could be previously overlooked native populations, recent arrivals as a result of a natural extension of range, or introductions with sand and ballast.

Eurosiberian Southern-temperate element; widely naturalised outside its native range.

References: Atlas (380b), Hultén & Fries (1986), Meusel *et al.* (1965), Stewart *et al.* (1994).

S. J. LEACH

Poa alpina Alpine Meadow-grass

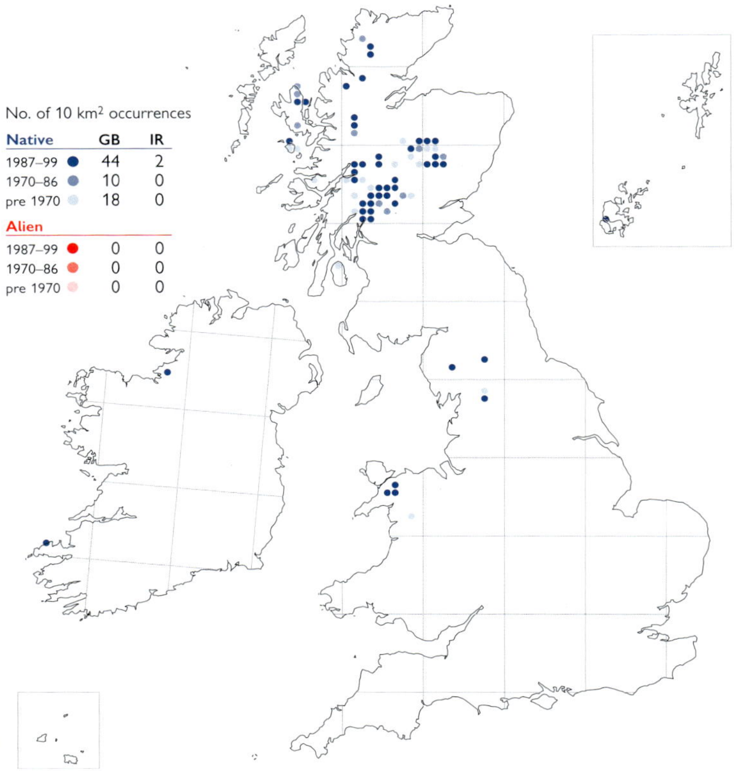

A perennial herb of damp mountain rock faces, open ledges and rocky slopes on calcareous substrates, often with *P. glauca*. Most populations are wholly or partially proliferous. From 580 m at High Cup Nick (Westmorland) to 1190 m on Aonach Beag (Westerness).

Native (change –0.31). *P. alpina* has gone from some sites, with several outlying populations having suffered from injudicious collecting. However, some of its remotest stations are seldom visited by botanists, and it is probably present in many squares for which only pre-1987 records are available.

Circumpolar Arctic-montane element, with a disjunct distribution.

References: Atlas (380a), Curtis & McGough (1988), Hultén & Fries (1986), Meusel *et al.* (1965), Stewart *et al.* (1994).

S. J. LEACH

Dactylis glomerata Cock's-foot

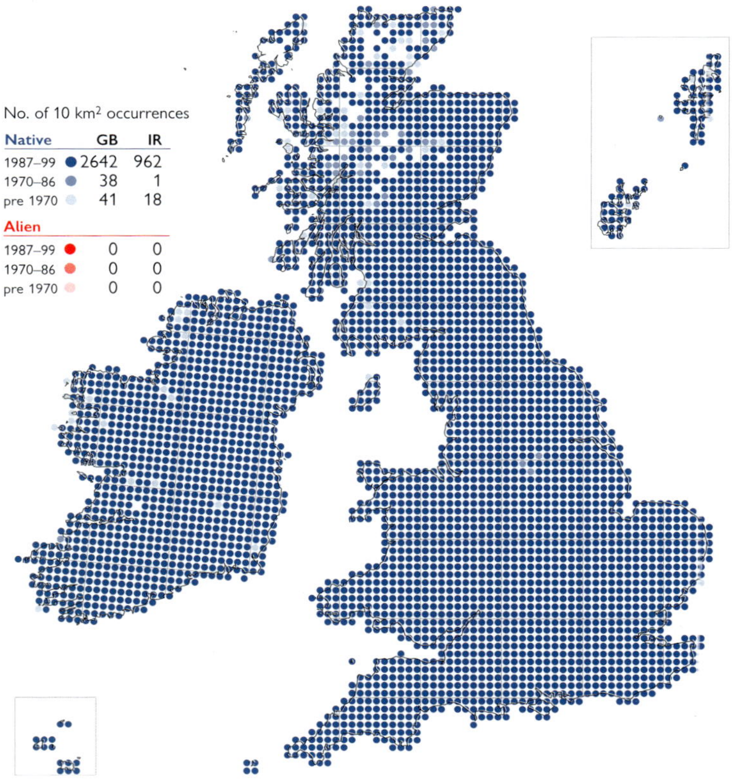

A tufted perennial herb of woods, meadows and pastures, downland and hill-slopes, maritime cliff grasslands, fixed dunes, field margins, roadsides and waste ground on a wide range of fertile, neutral and basic soils. 0–685 (Breadalbanes, Mid Perth) and exceptionally at 845 m on Great Dun Fell (Westmorland).

Native (change –0.06). There has been no change in the distribution of *D. glomerata* since the 1962 *Atlas*. It is tolerant of grazing and is a common constituent of grass-seed mixtures for leys, with much seed coming from Denmark. It is often a relic of cultivation in the extreme north and west.

Eurosiberian Southern-temperate element, but widely naturalised so that distribution is now Circumpolar Southern-temperate.

References: Atlas (382c), Beddows (1959), Grime *et al.* (1988), Hultén & Fries (1986).

F. H. PERRING

Catabrosa aquatica Whorl-grass

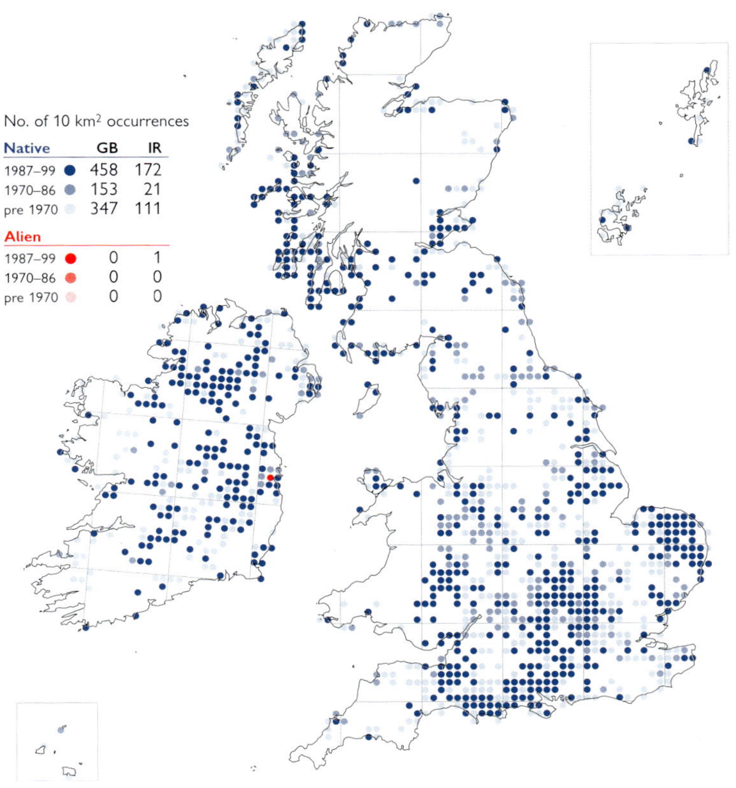

A stoloniferous herb of muddy pond margins, cattle-poached ditches, canals and sluggish streams; also, as var. *uniflora*, on wet open sand by the sea. Almost entirely lowland, but recorded in flushes at 710 m on Little Fell (Westmorland).

Native (change –0.69). *C. aquatica* has much declined since the 1962 *Atlas*, due to the drainage and infilling of ponds, and the canalisation of lowland watercourses. However, var. *uniflora* is widespread in W. and N.W. Scotland, perhaps also in W. Ireland, and is undoubtedly more common than indicated by the map in Perring & Sell (1968).

European Boreo-temperate element; also in C. Asia and N. America.

References: Atlas (382b), Atlas Supp. (151a), Hultén & Fries (1986), Preston & Croft (1997).

S. J. LEACH

Catapodium rigidum Fern-grass

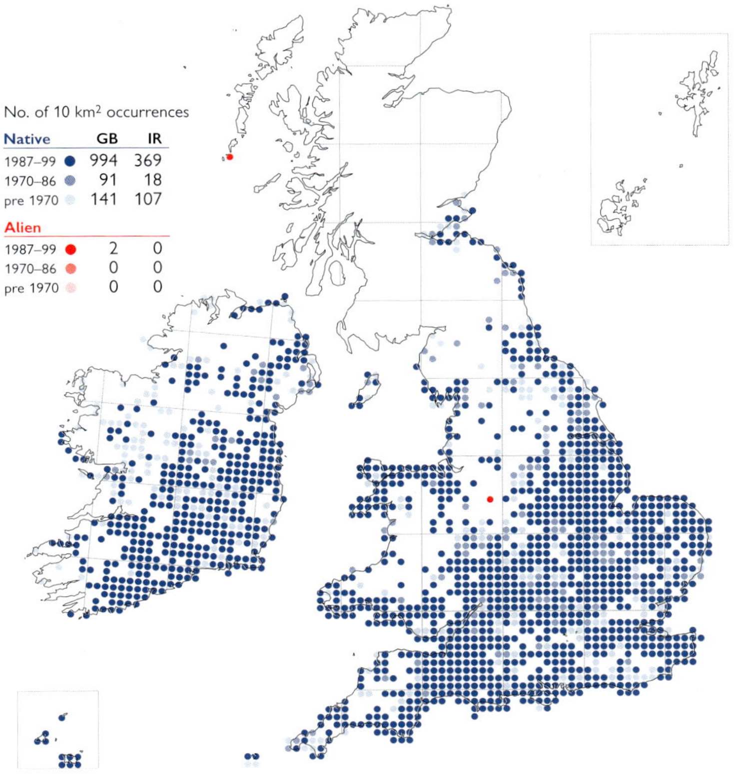

An annual of dry, barish places on sandy banks, stabilised shingle and around rock outcrops, usually preferring calcareous substrates; also in artificial habitats such as quarries, walls, pavements and railway ballast. 0–355 m (Llangollen, Denbs.).

Native (change +0.35). There has been little change in the distribution of *C. rigidum* since the 1962 *Atlas*. Subsp. *majus*, a coastal plant of S.W. Britain, Ireland and the Channel Islands, was mapped by Perring & Sell (1968), but it is treated as a variety by Stace (1997) and is not mapped separately here.

Submediterranean-Subatlantic element.

References: Atlas (379a), Atlas Supp. (149a), Clark (1974), Grime *et al.* (1988), Meusel *et al.* (1965).

S. J. LEACH

Catapodium marinum Sea Fern-grass

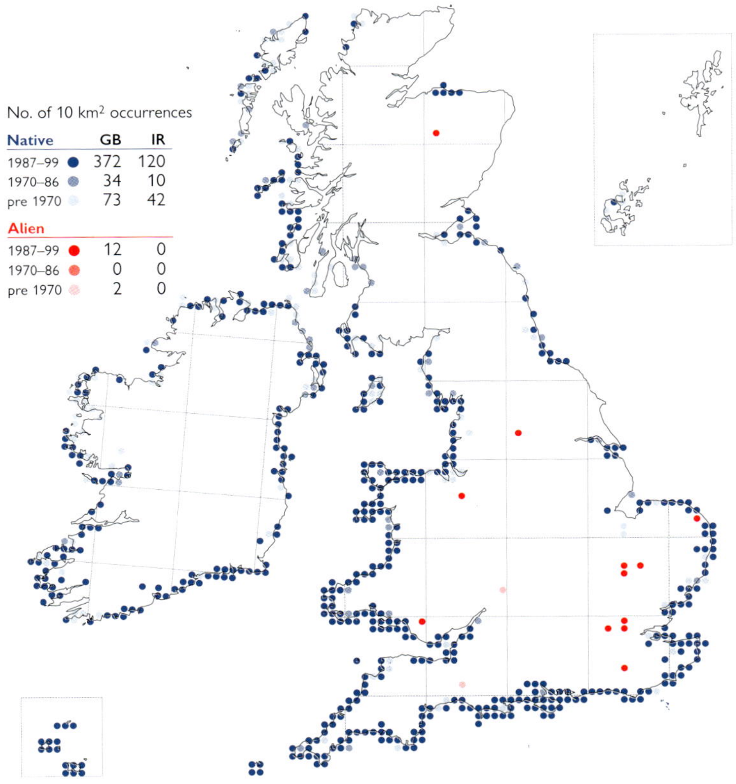

An annual of dry bare places by the sea, rock crevices, grassy banks, cliff-tops, sand dunes and stabilised shingle; also in artificial habitats such as walls and pavements, and increasingly inland by salt-treated roads. Lowland.

Native (change +0.52). There has been little change in the distribution of *C. marinum* since the 1962 *Atlas*, apart from its recent colonisation of roadsides near the coast, mainly in S. England and along the M4 motorway in S. Wales. It was first recorded in that habitat in the mid-1980s, though only since the mid-1990s has it begun to spread appreciably.

Mediterranean-Atlantic element.

Reference: Atlas (379b).

S. J. LEACH

Sesleria caerulea Blue Moor-grass

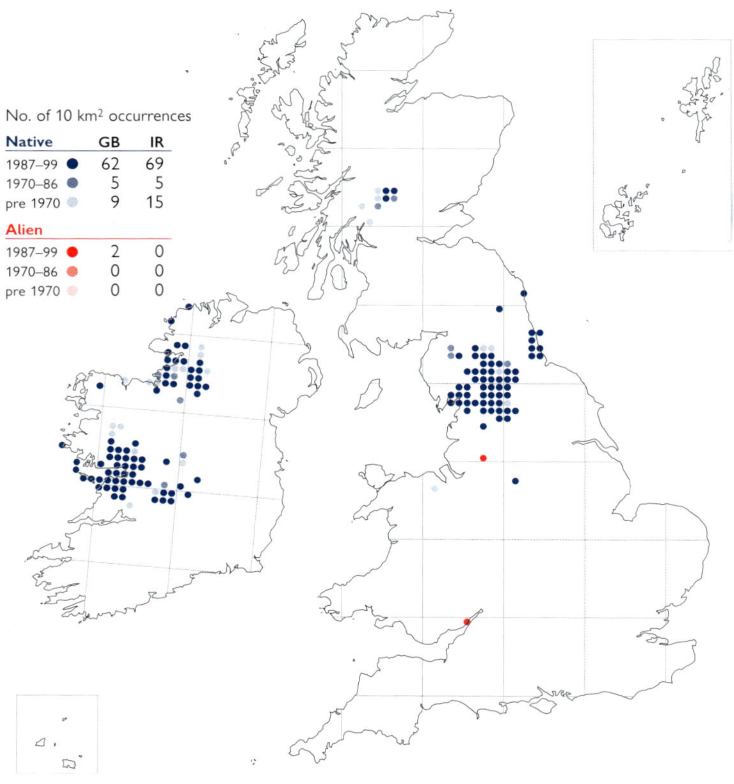

A tufted, rhizomatous perennial of well-drained, mainly open habitats on limestone, including grassland and heath, screes and cliffs, and the grikes and clint-hollows of limestone pavement. It extends locally into open woodland in Ireland and N. England, and is found on sandy loams over micaceous schists in Perthshire. 0–1005 m (Ben Lawers, Mid Perth).

Native (change –0.09). Because *S. caerulea* is not very palatable to sheep, it becomes dominant in heavily grazed areas, forming a species-poor turf. It is still abundant in areas of Carboniferous limestone in N. England and the Burren and it was discovered in the Derbyshire Dales in 1989.

European Boreo-temperate element.

References: Atlas (384b), Dixon (1982), Halliday (1997), Hultén & Fries (1986), Meusel *et al.* (1965), Stewart *et al.* (1994).

F. H. PERRING

Parapholis strigosa Hard-grass

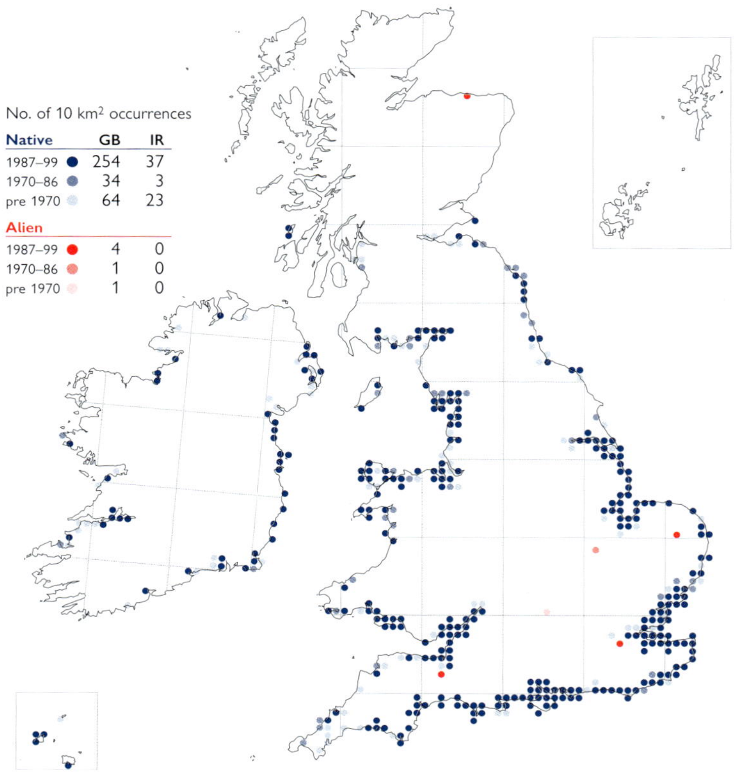

An annual of damp barish places by the sea; especially characteristic of the upper parts of grazed *Festuca rubra-Puccinellia maritima* saltmarshes, but also on mud banks, shingle ridges, saltmarsh-sand dune transitions and sea walls. In W. Britain and Ireland it occurs along rocky coasts in beach-head saltmarshes. Rarely, it grows inland by salt-treated roads. Lowland.

Native (change +0.14). There is no evidence of any appreciable change in the distribution of *P. strigosa* since the 1962 *Atlas*. Apparent gains in some areas, especially in Ireland, may be due to it having been overlooked by earlier recorders. The species is inconspicuous and is probably still somewhat under-recorded.

Suboceanic Southern-temperate element.

References: Atlas (404c), Hultén & Fries (1986).

S. J. LEACH

Parapholis incurva Curved Hard-grass

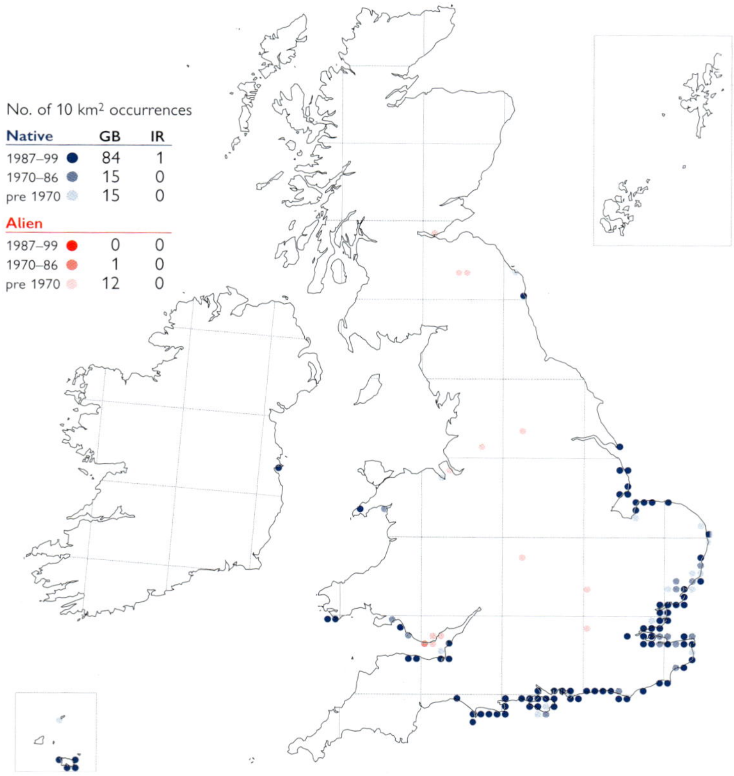

An annual of bare places by the sea, including gravelly mud banks, shingle ridges, rock ledges and cliff-tops, and the uppermost parts of saltmarshes; also in artificial habitats such as sea walls and wooden mooring stays. There are rare occurrences around docks and inland as a wool and ballast alien. Lowland.

Native (change +0.09). *P. incurva* is an extremely inconspicuous species which is much better recorded than it was for the 1962 *Atlas*. It was discovered in Ireland in 1979. It has been lost from some sites due to coastal reclamation and the upgrading of sea defences.

Mediterranean-Atlantic element.

References: Atlas (404d), Akeroyd (1984), Stewart *et al.* (1994).

S. J. LEACH

Glyceria maxima Reed Sweet-grass

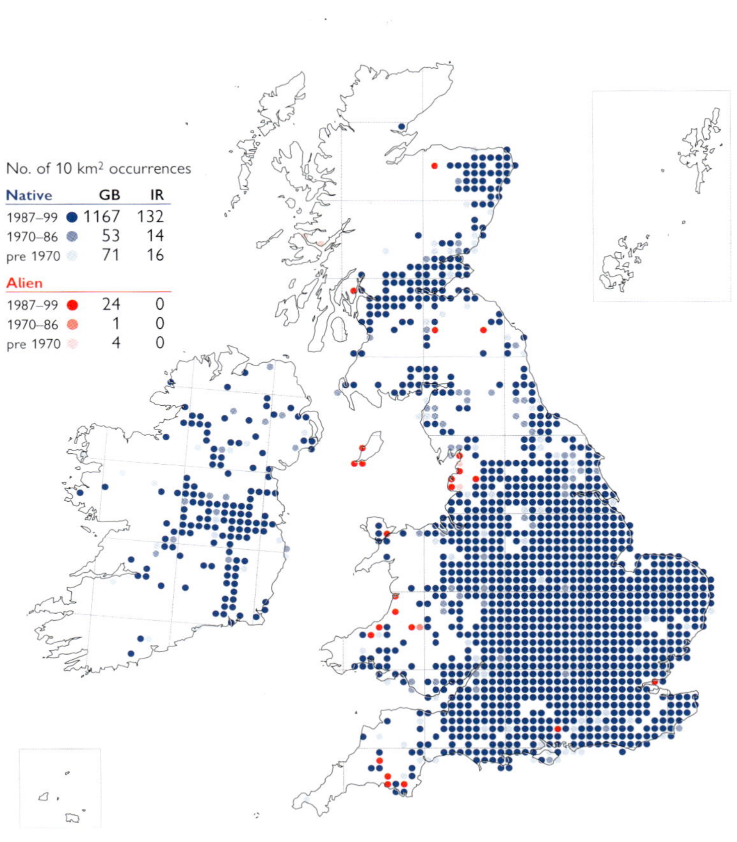

A rhizomatous perennial herb growing in ditches, canals, lakes and ponds, either rooted on the bank or in the water, and often forming floating rafts; also in seasonally-flooded grasslands. It was formerly cultivated as a fodder crop, and is much planted in ponds. Generally lowland, but reaching 600 m at Sprinkling Tarn (Cumberland).

Native (change +0.65). *G. maxima* has increased since the 1962 *Atlas*, particularly in the north, either by deliberate introductions or by natural spread from sites where it was probably originally planted. Many of these populations are shown as native on the map.

Circumpolar Temperate element; widely naturalised outside its native range.

References: Atlas (373a), Grime *et al.* (1988), Hultén & Fries (1986), Lambert (1947), Meusel *et al.* (1965), Preston & Croft (1997).

S. J. LEACH

Glyceria fluitans Floating Sweet-grass

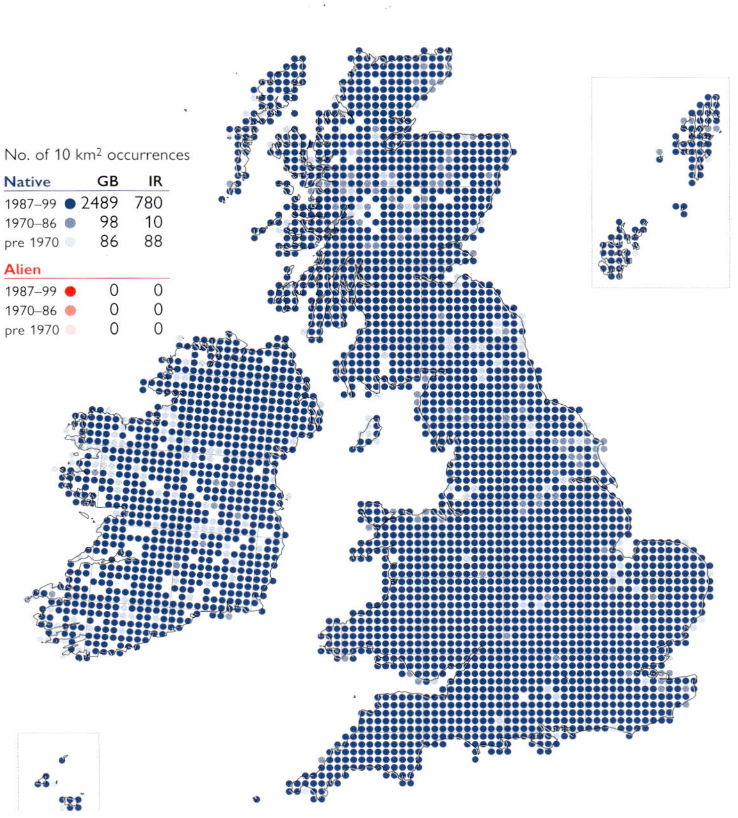

A perennial herb of marshes, swamps and muddy pond margins, and forming floating rafts in shallow water by ditches, rivers, ponds and lakes; tolerant of high levels of disturbance and nutrient-enrichment. 0–720 m (Knock Fell, Westmorland).

Native (change +0.89). This species was rather unevenly recorded in the 1962 *Atlas*, but there is no evidence of any appreciable change in range over the last forty years.

European Temperate element; widely naturalised outside its native range.

References: Atlas (372b), Grime *et al.* (1988), Hultén & Fries (1986), Meusel *et al.* (1965), Preston & Croft (1997).

S. J. LEACH

Glyceria fluitans × *G. notata* (*G.* × *pedicellata*) Hybrid Sweet-grass

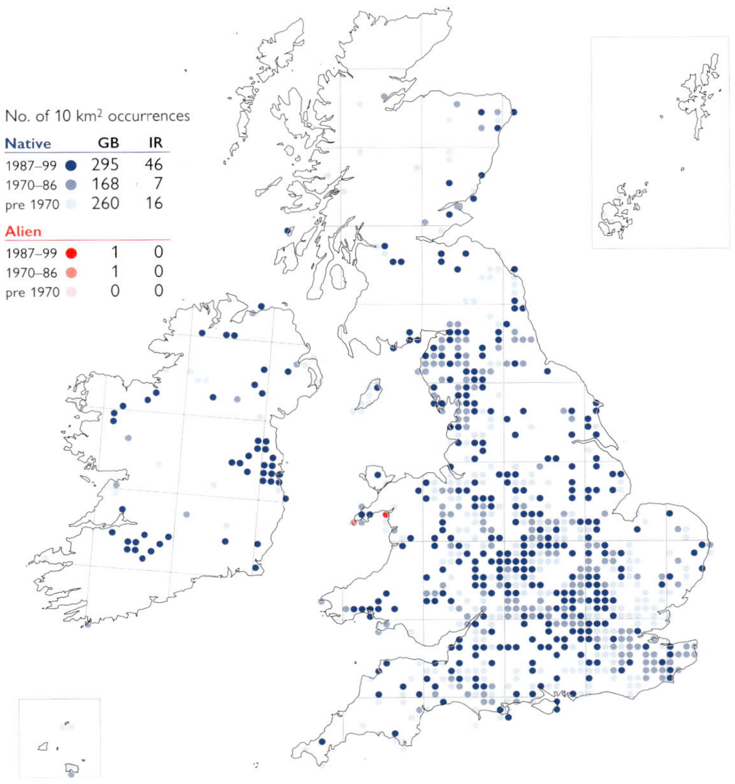

A stoloniferous perennial herb of ponds, ditches, streams and swampy depressions in pastures, growing with one or both parents, or with neither. It is highly sterile, but spreads vegetatively and detached ramets may be carried by water to new sites. 0–550 m (Moor House, Westmorland).

Native. The map in Perring & Sell (1968) under-estimated the distribution of this taxon. The apparent extension of range in some areas, particularly in Ireland, is likely to be attributable to better recording. It is probably still underrecorded, and there is insufficient evidence to indicate whether its distribution is changing.

Widespread in temperate Europe.

References: Atlas Supp. (148a), Borrill (1956), Preston & Croft (1997), Stace (1975).

S. J. LEACH

Glyceria declinata Small Sweet-grass

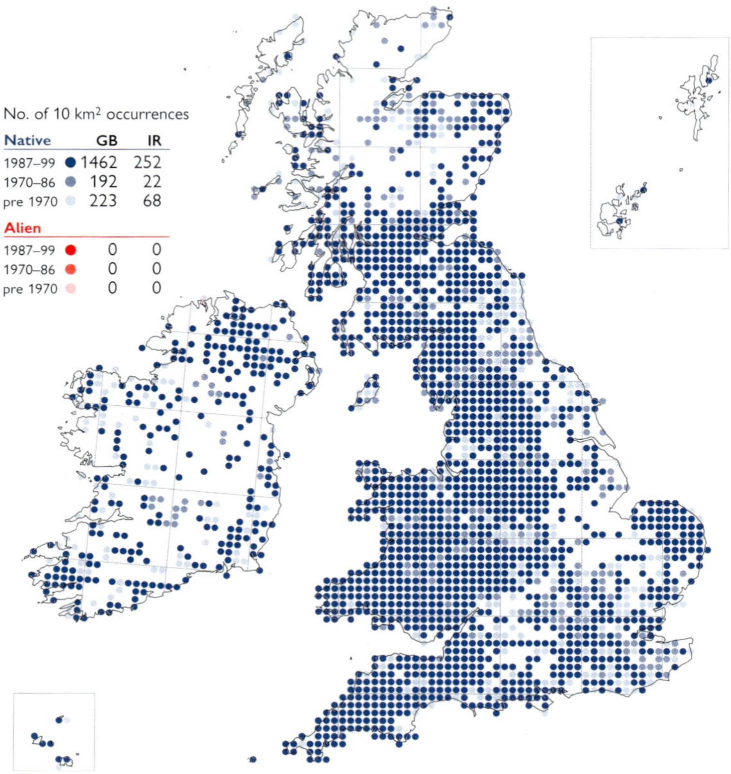

No. of 10 km² occurrences

Native	GB	IR
1987–99 ●	1462	252
1970–86 ●	192	22
pre 1970 ●	223	68
Alien		
1987–99 ●	0	0
1970–86 ●	0	0
pre 1970 ●	0	0

A perennial herb of muddy pond margins, cattle-trampled ditches and marshy fields; also in shallow water by ponds, rivers and canals. 0–500 m (Llyn Crugnant, Cards.).

Native (change +1.79). This species is easily overlooked, but its small stature and glaucous leaves give it a distinctive appearance. It was very under-recorded in the 1962 *Atlas*, and this is probably still true in some areas, especially in Ireland. It has evidently declined in some parts of E. England.

Suboceanic Temperate element.

References: Atlas (372d), Hultén & Fries (1986), Preston & Croft (1997).

S. J. LEACH

Glyceria notata Plicate Sweet-grass

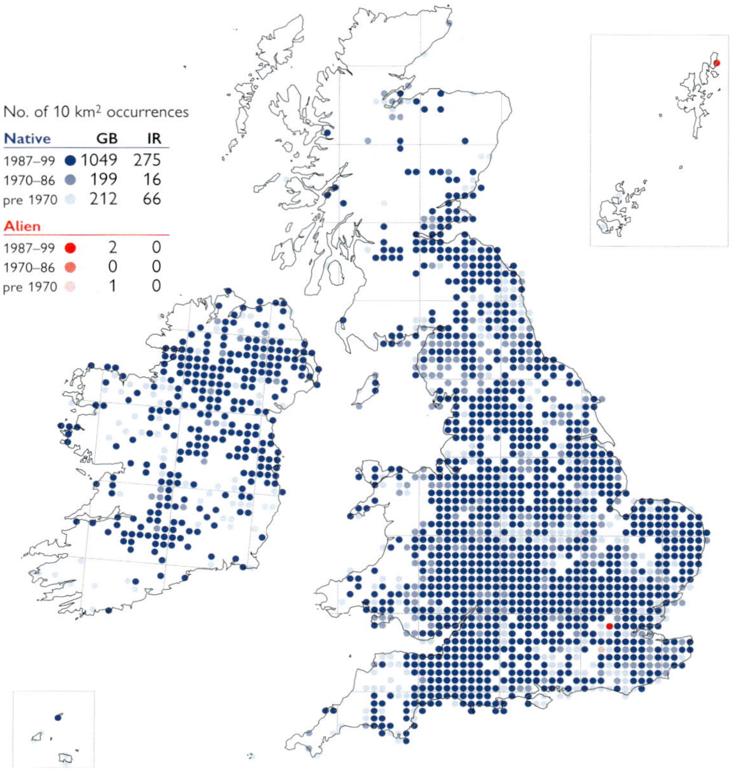

No. of 10 km² occurrences

Native	GB	IR
1987–99 ●	1049	275
1970–86 ●	199	16
pre 1970 ●	212	66
Alien		
1987–99 ●	2	0
1970–86 ●	0	0
pre 1970 ●	1	0

A stoloniferous perennial herb of ditches, streams and muddy pond margins, occurring on more calcareous substrates than other British *Glyceria* species. It reproduces by seed and by detached stolons. 0–380 m (Malham Tarn, Mid-W. Yorks.).

Native (change +0.31). *G. notata* was mapped as 'all records' in the 1962 *Atlas*. Although it was probably under-recorded in the past, there are sufficient older records to suggest that it has declined in parts of S. England because of drainage.

European Temperate element; also in C. Asia.

References: Atlas (372c), Hultén & Fries (1986), Preston & Croft (1997).

S. J. LEACH

Melica nutans Mountain Melick

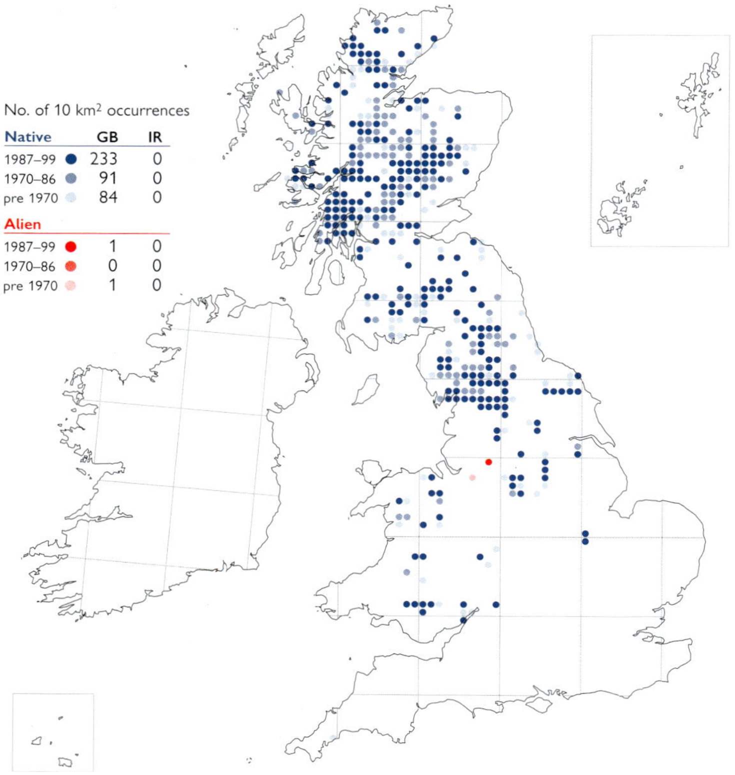

No. of 10 km² occurrences

Native	GB	IR
1987–99 ●	233	0
1970–86 ●	91	0
pre 1970 ●	84	0
Alien		
1987–99 ●	1	0
1970–86 ●	0	0
pre 1970 ●	1	0

A rhizomatous, perennial grass of basic soil over limestone and other base-rich rocks, occurring in shady places in deciduous woodland, on woodland margins, in the grikes of limestone pavement and on rock ledges. Lowland to 820 m in Glen Isla (Angus).

Native (change –0.17). The 1962 *Atlas* showed some decline in *M. nutans* before 1930, especially in S. Scotland, but other than local losses the distribution is unchanged. It is now much better recorded.

Eurasian Boreo-temperate element, with a continental distribution in W. Europe.

References: Atlas (384a), Hultén & Fries (1986), Meusel *et al.* (1965).

F. H. PERRING

Melica uniflora Wood Melick

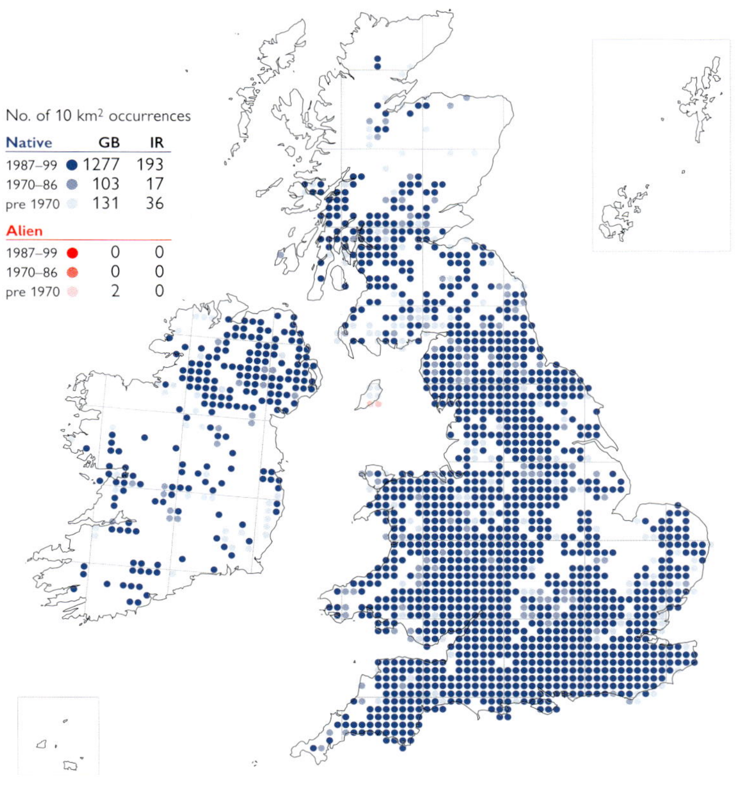

A rhizomatous perennial grass of woodland rides and margins, of shady hedge banks and rock ledges, mainly on free-draining, base-rich soils. It often grows in localised patches, suggesting that regeneration is mainly by rhizomatous spread. 0–395 m (Ysbyty Ifan, Denbs.), and up to 485 m in the Scottish Highlands.

Native (change –0.04). There has been little change in the distribution of *M. uniflora* since the 1962 *Atlas*, although the increase in records from N. Ireland is a remarkable example of improved recording. It has been lost from the Isle of Man, where it may have been originally introduced.

European Temperate element.

References: Atlas (383d), Grime *et al.* (1988), Hultén & Fries (1986), Meusel *et al.* (1965).

F. H. PERRING

Helictotrichon pubescens Downy Oat-grass

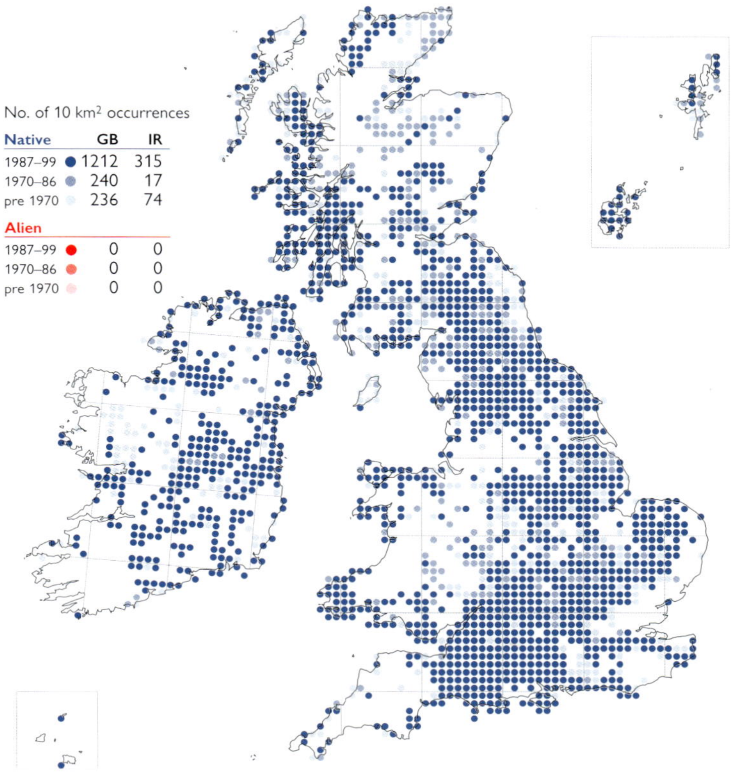

A perennial grass growing in a wide range of moist or dry, neutral and calcareous grasslands. It is found in meadows and pastures, on roadsides and railway banks, in open woodland and rides, coastal cliffs, fixed dunes, mildly acidic grassland over gravel, and some less acidic heaths. It tolerates modest mowing, grazing and manuring, but not artificial fertilisers or competition from vigorous fodder-grasses. 0–550 m (Moor House, Cumberland).

Native (change +0.35). Although *H. pubescens* is now known to be much more frequent than when mapped in the 1962 *Atlas*, there have clearly been some losses throughout its range.

European Temperate element; also in C. Asia and widely naturalised outside its native range.

References: Atlas (393d), Dixon (1991), Grime *et al.* (1988), Hultén & Fries (1986), Meusel *et al.* (1965).

F. H. PERRING

Helictotrichon pratense Meadow Oat-grass

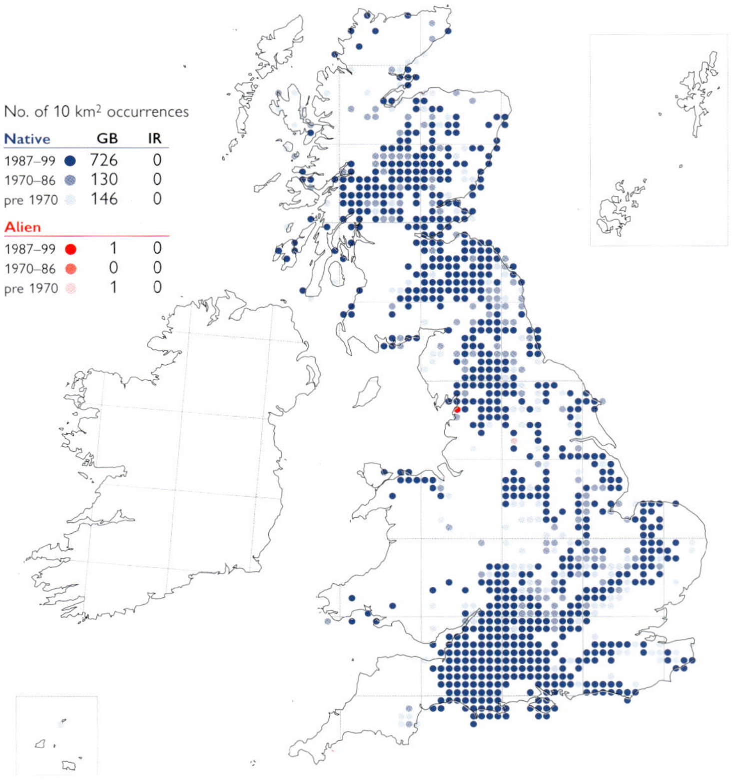

A perennial herb of calcareous rendzina and brown earth soils, usually over chalk or limestone, but also over glacial deposits and basic igneous rocks. It is characteristic of well-grazed chalk and limestone downland, but is also found on screes, cliffs and limestone pavements, and occasionally in open *Fraxinus* woods, on sand dunes and on montane ledge communities with *Dryas octopetala*. 0–835 m (Breadalbanes, Mid Perth).

Native (change +0.31). The distribution of *H. pratense* is stable, though the apparent losses in S.E. England may be significant indicators of a decline in the lowlands. Montane populations with larger spikelets may be taxonomically distinct.

European Temperate element.

References: Atlas (393c), Dixon (1991), Grime *et al.* (1988), Hultén & Fries (1986), Meusel *et al.* (1965).

F. H. PERRING

Arrhenatherum elatius False Oat-grass

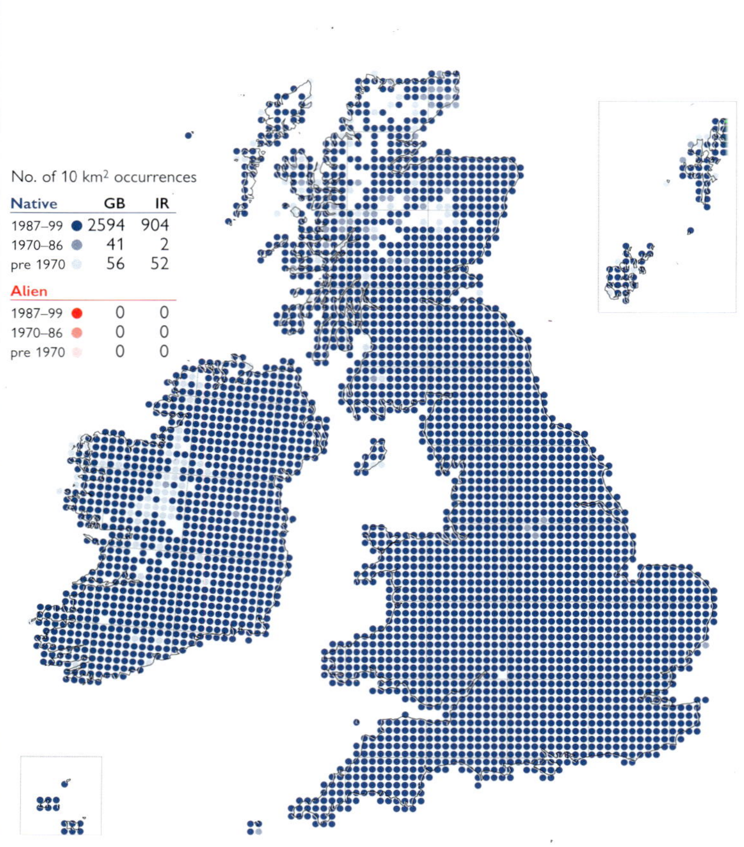

A tufted perennial grass found in a very wide range of neutral to base-rich habitats. It is frequent in grasslands, but is especially common on road verges, hedge banks, and river banks. It is an important stabiliser of limestone scree, and a coloniser of bare, muddy, calcareous cliffs and maritime shingle. Var. *bulbosum* is widely distributed on roadside banks and as an arable weed on light soils. 0–550 m (E. Perth).

Native (change +0.37). The distribution of *A. elatius* is similar to that mapped in the 1962 *Atlas*. Var. *bulbosum* is still frequent in arable fields, and is encouraged by current direct drilling practices.

European Temperate element; widely naturalised outside its native range.

References: Atlas (394a), Grime *et al.* (1988), Hultén & Fries (1986), Meusel *et al.* (1965), Pfitzenmeyer (1962).

D. A. PEARMAN

Avena strigosa Bristle Oat

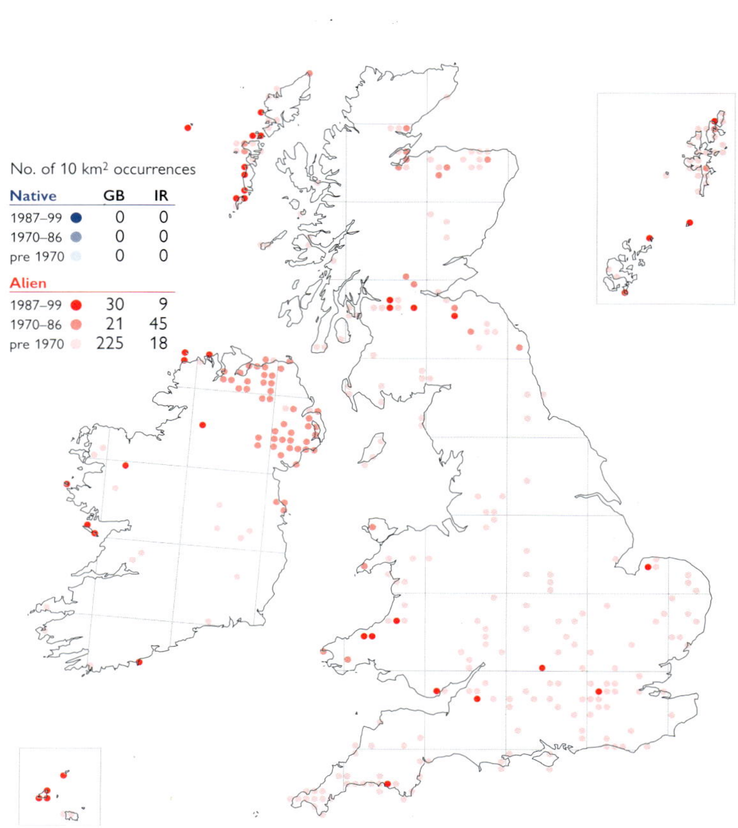

An annual grass once extensively cultivated, and also occurring as a field weed in N. & W. Britain, and in Ireland. It is an infrequent casual elsewhere. Lowland.

Casual (change −3.01). *A. strigosa* was for many centuries the main cereal cultivated in those areas of the north and west where conditions were unfavourable for *A. sativa*; it was first recorded in the wild in Britain in 1790. Since 1945 cultivation of *A. strigosa* has almost entirely ceased, new strains of *A. sativa* having replaced it. This is reflected in the declining frequency of the species as a casual.

Origin uncertain, but probably arose in cultivation; found as a crop or weed in N., W. & C. Europe.

References: Atlas (393b), Chater (1993), Hultén & Fries (1986).

D. A. PEARMAN

Avena fatua Wild-oat

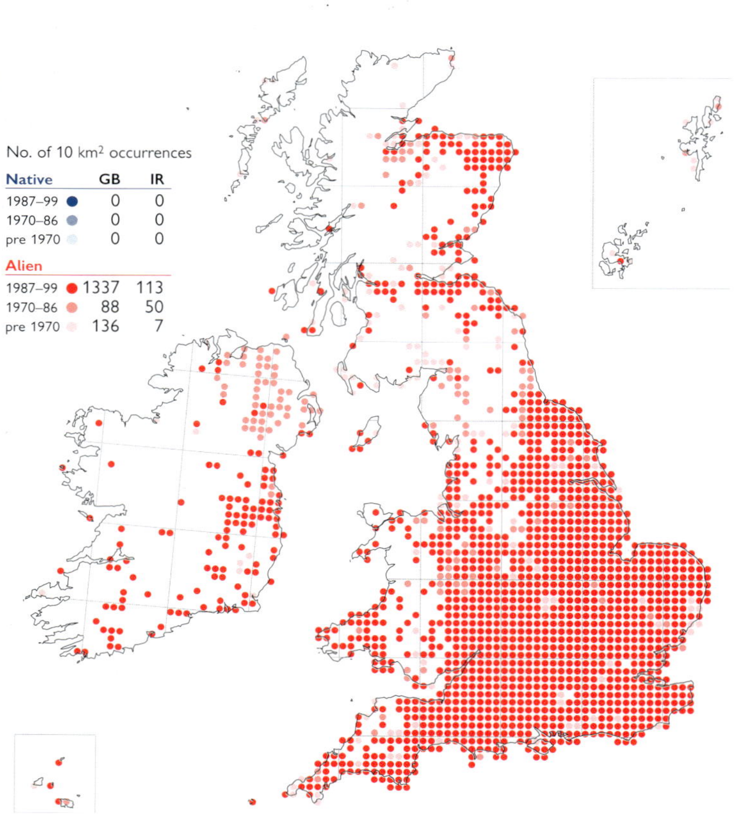

An annual grass which is a common weed on arable land, especially in cereals, seeding before the cultivated crops. It is also found on roadsides and waste ground. Generally lowland, but reaching 300 m at Ffair-Rhos (Cards.).

Archaeophyte (change +1.17). *A. fatua* has expanded northwards and westwards in Britain since the 1962 *Atlas*. Within its former range it has dramatically increased in frequency since 1945, and has proved resistant to some herbicides, although specific controls are now available. However, weeding by hand ('rogueing') is still a common practice.

Perhaps native to the E. Mediterranean region and the Near East, but has spread with cultivation and now has a virtually worldwide distribution outside the tropics.

References: Atlas (392d), Hultén & Fries (1986), Sharma & Vanden Born (1978).

D. A. PEARMAN

Avena sterilis Winter Wild-oat

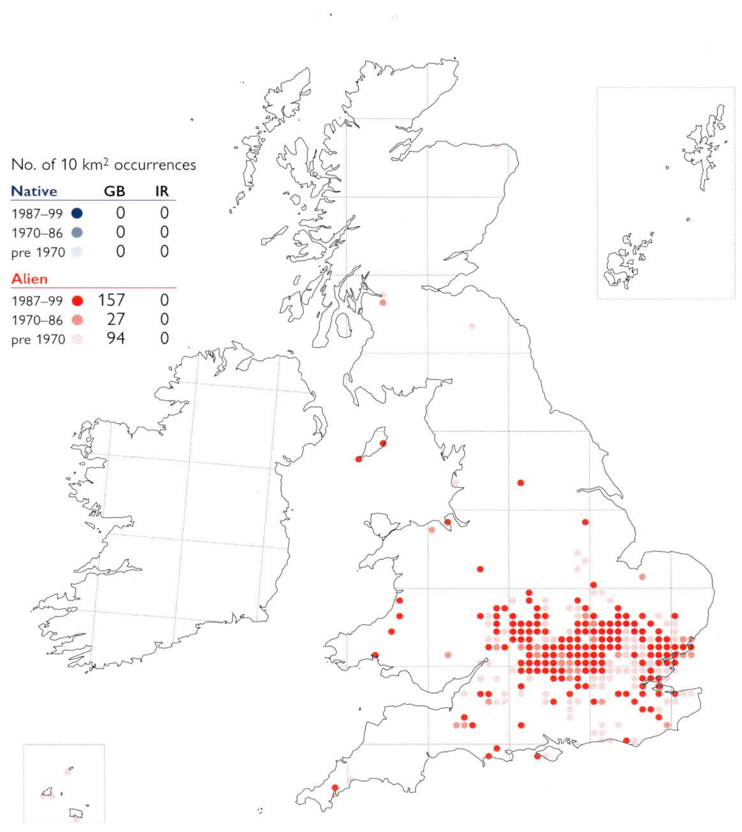

An annual weed of winter cereal crops, predominately found on heavier clay soils. It also occurs as a wool or grain alien on waste ground. Lowland.

Neophyte (change +0.14). *A. sterilis*, which was cultivated in Britain by 1640, was first recorded in the wild in 1910 (Port Meadow, Oxon) and has apparently spread from Oxfordshire to neighbouring counties. It is now less frequent than twenty years ago. Almost all of the records of *A. sterilis* are for subsp. *ludoviciana;* subsp. *sterilis* is a rare casual, occasionally cultivated for ornament.

Native of the Mediterranean region, S.W. & C. Asia; widely naturalised north of its native range in Europe.

References: Atlas (393a), Thurston (1954).

D. A. PEARMAN

Avena sativa Oat

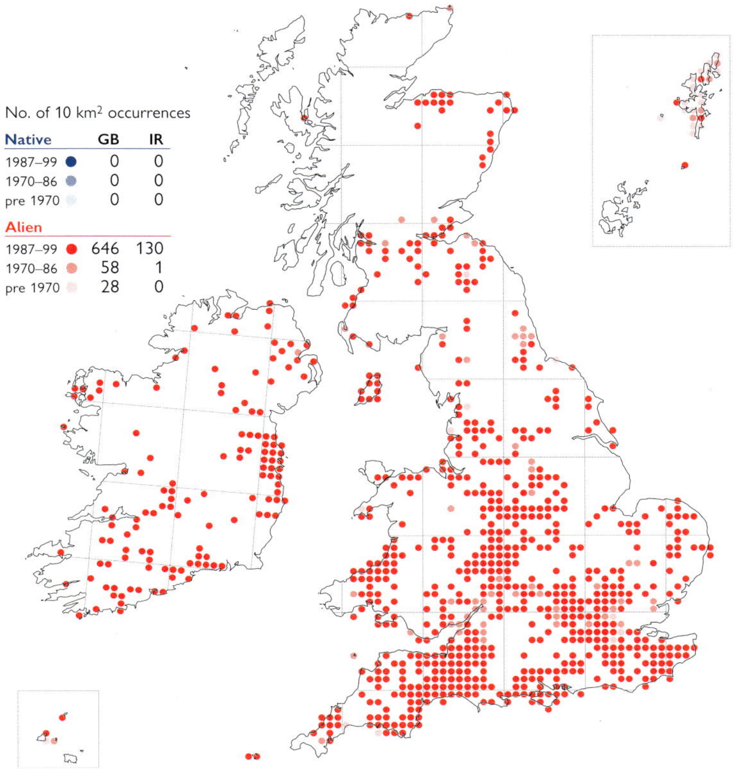

This annual grass is a frequent relic of arable crops, and an occasional casual on field edges, roadsides, tips and waste ground. It does not persist and is never naturalised. Lowland.

Casual. *A. sativa* was not cultivated until relatively recently compared to other cereals, perhaps evolving in Europe in the 2nd millennium BC and not reaching Britain until the Iron Age. It is now much less frequently grown as a crop than it was before the advent of mechanical transport, when there was more demand for horse food. However, botanists have only recently begun to record it as a casual; the first record in the wild was published in 1908.

A. sativa probably originated in cultivation, by selection from *A. fatua.*

Reference: Zohary & Hopf (2000).

D. A. PEARMAN

Gaudinia fragilis French Oat-grass

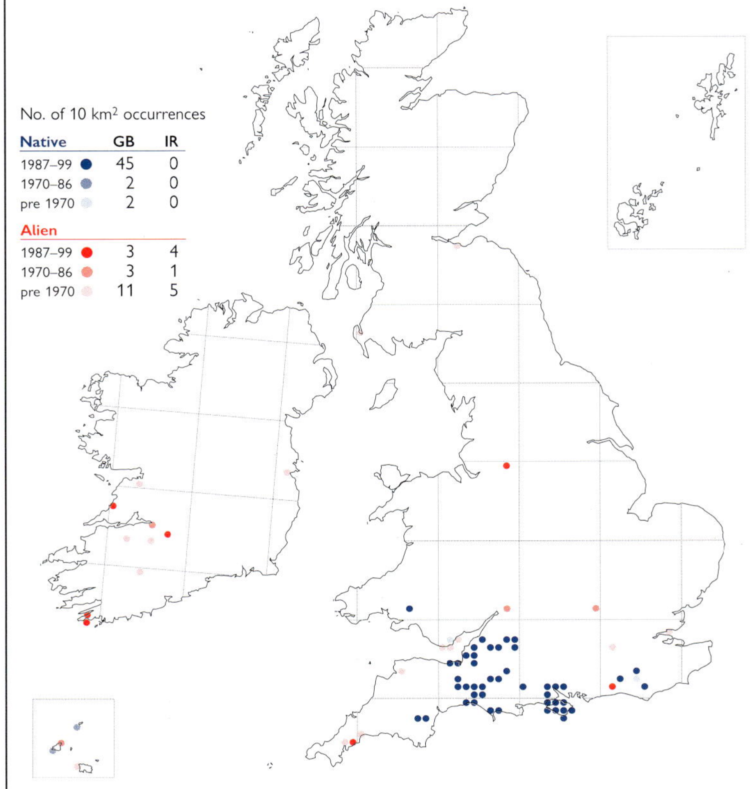

An annual or short-lived perennial herb of meadows, pastures and waysides on calcareous clay soils; also an occasional casual around docks and on tips. Lowland.

Native or alien. *G. fragilis* was cultivated in Britain by 1770 and was first recorded in the wild in 1903. It appears to have increased since 1980, though many new records are of well-established populations that must have been previously overlooked. It has a strikingly similar distribution to *Oenanthe pimpinelloides* and this, together with its preference for old meadows, has led some to consider that it might be native, at least in its core areas. Alternatively, it may have been introduced in the late 19th and early 20th centuries with grass seed imported from S. Europe.

Submediterranean-Subatlantic element.

References: McClintock (1972), Marren (1999).

S. J. LEACH

Trisetum flavescens Yellow Oat-grass

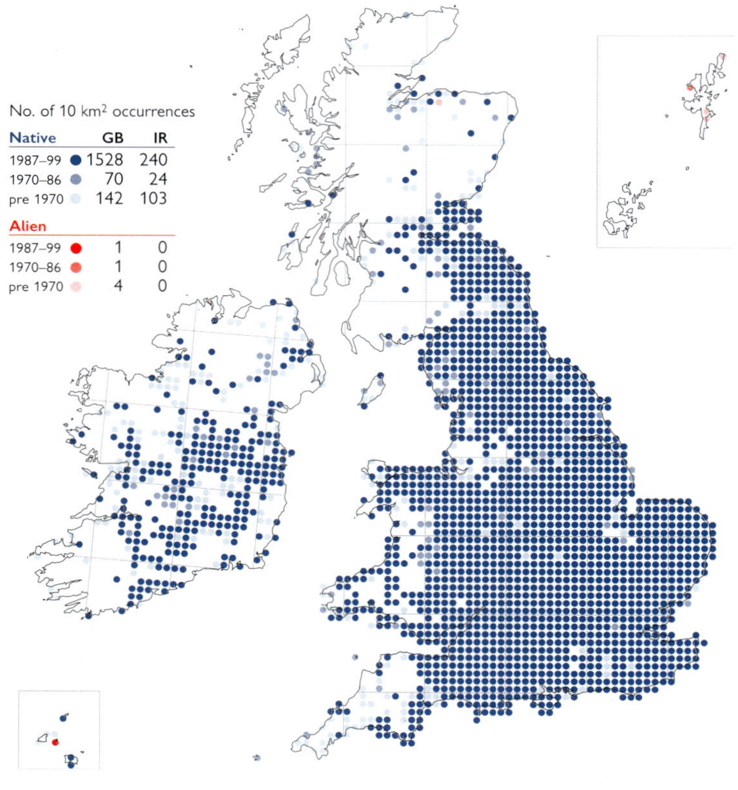

A perennial of well-drained neutral and calcareous grassland, found in lowland pasture and hay meadows, on downland, banks and roadsides, and occasionally rocks. It is most abundant in old, ungrazed hay meadows. It is highly palatable to stock and susceptible to damage by heavy trampling. 0–550 m (Moor House, Westmorland).

Native (change –0.13). The distribution of *T. flavescens* is stable. It is alien in Shetland and its status is doubtful in other areas of N. and W. Scotland, and in N. Wales and S.W. England. The alien subsp. *purpurascens* appears to be widely sown in seed mixtures on verges and in grassland.

European Temperate element; widely naturalised outside its native range.

References: Atlas (392c), Dixon (1995), Grime *et al.* (1988), Hultén & Fries (1986), Meusel *et al.* (1965).

F. H. PERRING

Koeleria vallesiana Somerset Hair-grass

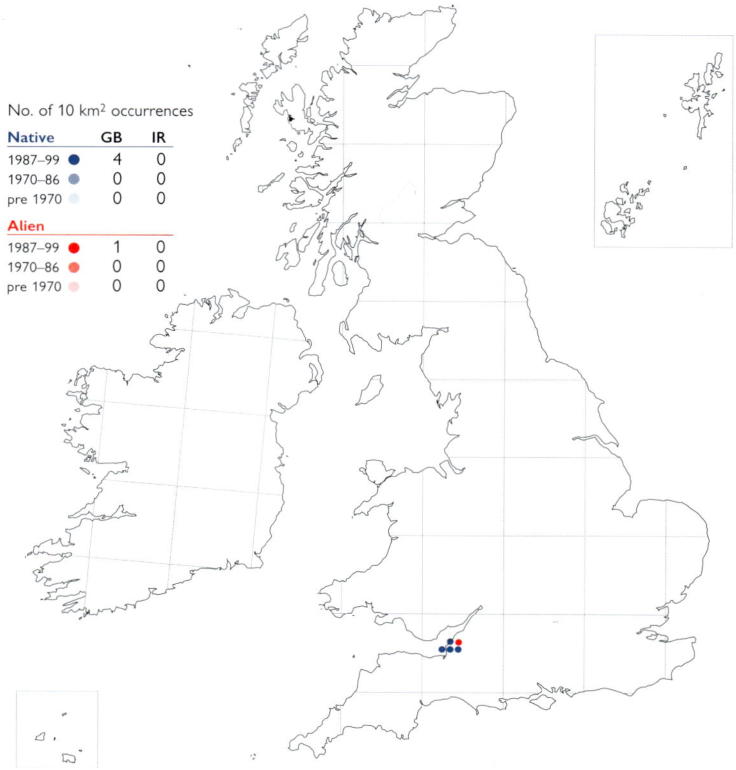

A tufted perennial grass of open, sheep-grazed turf around rock outcrops on S.-facing Carboniferous limestone slopes. Its sites are characterised by high levels of insolation and summer drought. Lowland.

Native. In the British Isles *K. vallesiana* has always been restricted to the Mendip hills (N. Somerset). It was first collected in this area in 1726, but not recognised until it was rediscovered by Druce in 1904. It was introduced to Goblin Combe (N. Somerset). Its distribution is stable.

Suboceanic Southern-temperate element.

References: Atlas (392b), Green *et al.* (1997), Sell & Murrell (1997), Wigginton (1999).

S. J. LEACH

Koeleria macrantha Crested Hair-grass

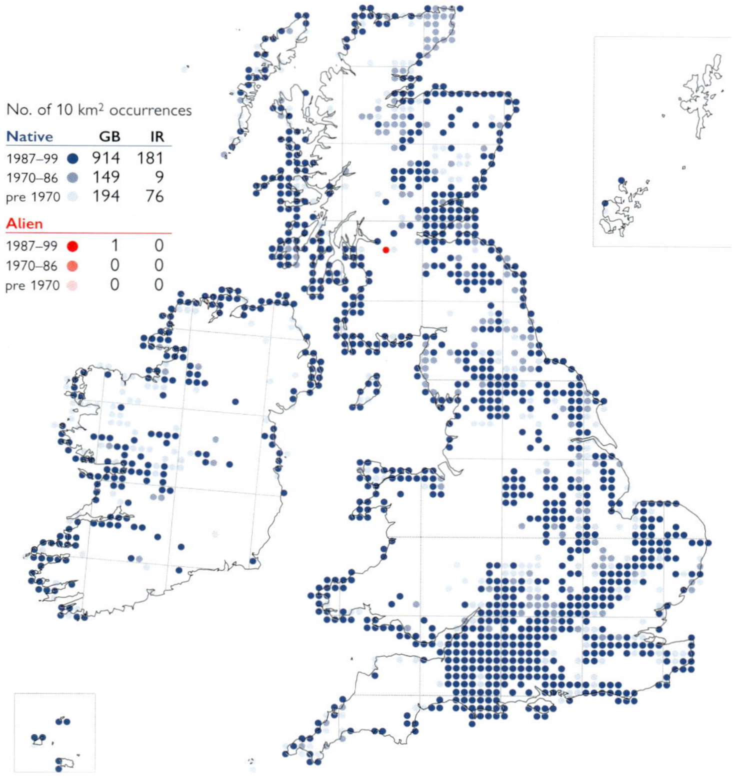

A perennial herb of grasslands on infertile, mainly calcareous substrates; also around rock outcrops, screes, quarry heaps and old lead workings. In many areas it is most frequent in dry, sandy, base-rich grassland on cliff-tops and dunes. Generally lowland, but reaching at least 680 m on Mt Brandon (S. Kerry).

Native (change –0.29). The map suggests little appreciable change in the distribution of *K. macrantha* since the 1962 *Atlas*, where it was mapped as 'all records', but *K. macrantha* is known to have declined in many squares due to habitat destruction and intensification of grassland management.

Circumpolar Temperate element.

References: Atlas (392a), Dixon (2000), Grime *et al.* (1988), Hultén & Fries (1986), Looman (1978), Meusel *et al.* (1965).

S. J. LEACH

Deschampsia cespitosa Tufted Hair-grass

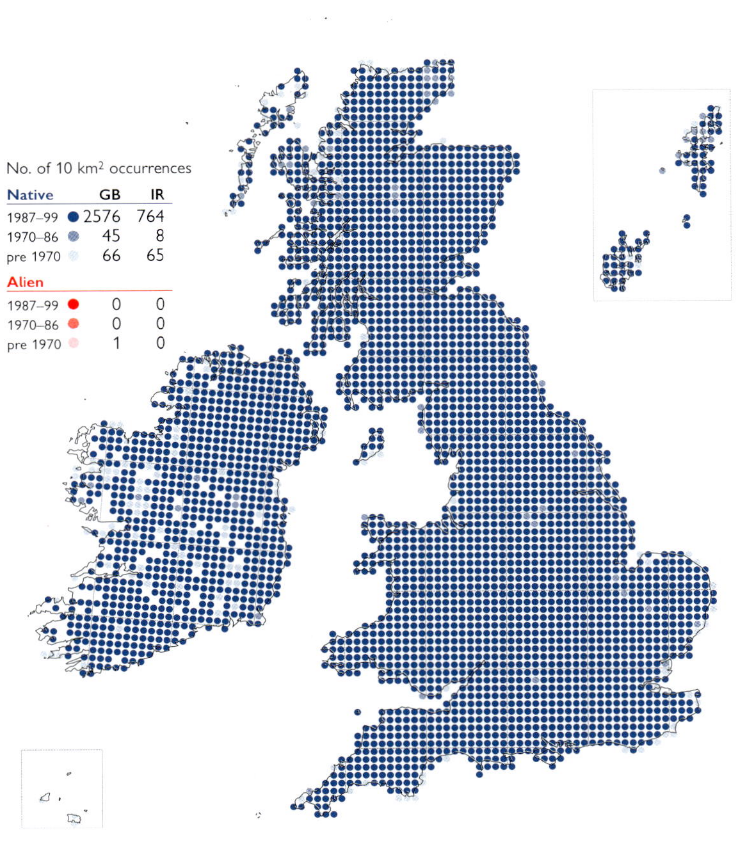

A tufted perennial growing on poorly-drained, mildly acidic, neutral or basic soils in woodland, rough and marshy grasslands, fen-meadows, grass-heath, and a wide range of montane habitats. It can rapidly colonise bare ground and tolerates some disturbance. It is morphologically and cytologically variable and either proliferates or sets copious seed. 0–1235 m (Ben Macdui, S. Aberdeen).

Native (change –0.09). The distribution of *D. cespitosa* is stable. It often persists in over-grazed pasture because its unpalatable leaves have a high silica content. Subsp. *cespitosa* is widespread; the other two subspecies are mapped separately.

Circumpolar Wide-boreal element; widely naturalised outside its native range.

References: Atlas (394d), Davy (1980), Grime *et al.* (1988), Hultén & Fries (1986), Meusel *et al.* (1965).

D. A. PEARMAN

Deschampsia cespitosa subsp. *parviflora*

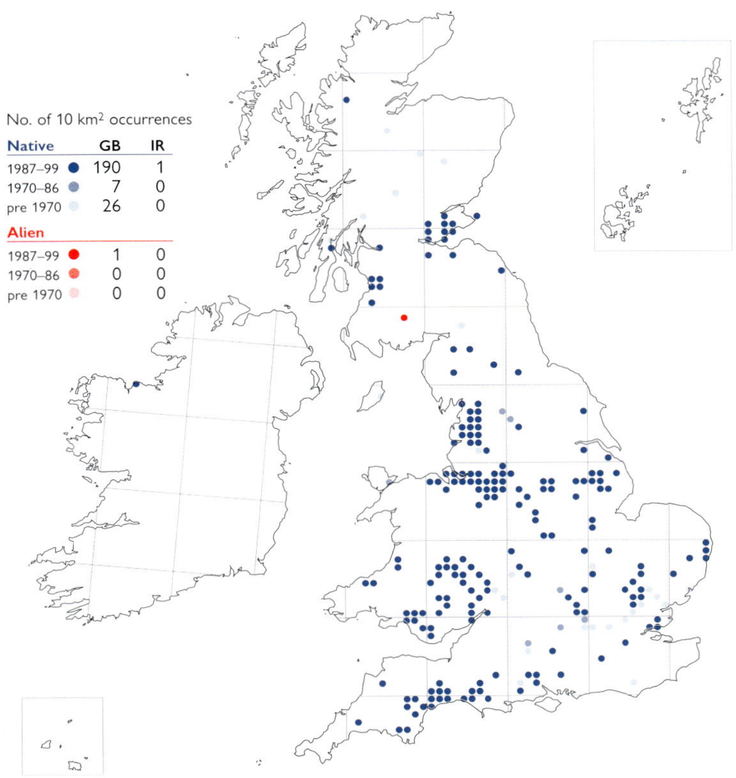

D. cespitosa subsp. *parviflora* is found in damp woodland and woodland rides on poorly-drained, heavy soils. The persistent seed bank allows it to survive the cycle of dense shade of mature coppice. Lowland.

Native. The presence of this taxon in our area was only recently brought to the attention of British and Irish botanists (Rich & Rich, 1988), although it was mentioned briefly in earlier works (e.g. Hubbard, 1954). It is therefore very poorly recorded. It was first recorded in Ireland in 1999, in Co. Sligo.

Also recorded from C. Europe.

Reference: Chiapella (2000).

D. A. PEARMAN

Deschampsia cespitosa subsp. *alpina*

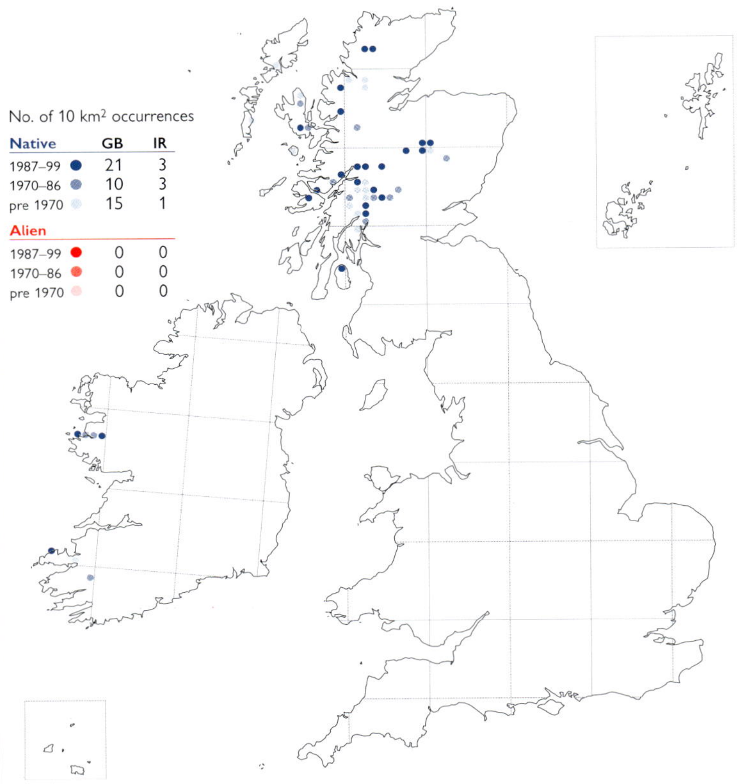

A tufted perennial of very open montane habitats, growing on rock ledges, slumped soil and gravelly flushes, often in areas of late snow-lie. It reproduces vegetatively by proliferous spikelets. From 800 m to 1235 m (Ben Macdui, S. Aberdeen).

Native. Subsp. *alpina* is poorly known and much under-recorded. It is sometimes confused with proliferous forms of subsp. *cespitosa* (for example in N. Wales) and records mapped are those reported as subsp. *alpina*. There are numerous unmapped Scottish records that may represent either subsp. *alpina* or proliferous subsp. *cespitosa*.

European Arctic-montane element; also recorded from Greenland, E. Canada and E. Asia but absent from the mountains of C. Europe.

References: Atlas (395a), Hultén & Fries (1986), Meusel *et al.* (1965), Stewart *et al.* (1994).

D. A. PEARMAN

Deschampsia setacea Bog Hair-grass

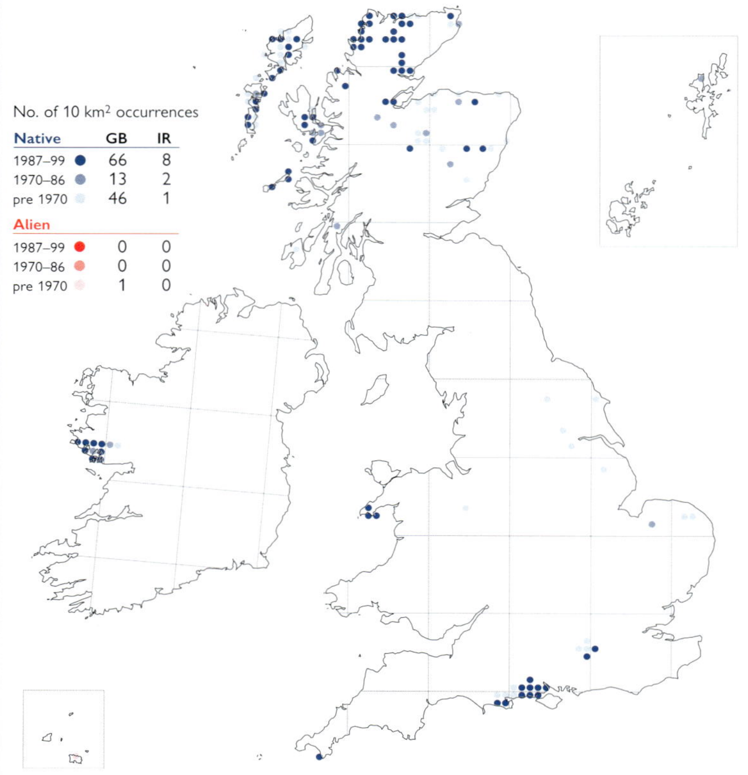

A densely tufted perennial of peaty or stony margins of lochs, shallow pools and seasonally inundated depressions on heaths, and on acid bogs. It appears to favour bare areas that are flooded in winter but dry in summer, and possibly where there is some lateral water movement. 0–320 m (Loch Morlich, Easterness).

Native (change –0.04). *D. setacea* has been lost since the 1930s from many lowland sites in E. England and Scotland through habitat destruction or undergrazing of heathland. In the Outer Hebrides, it may be extant in many squares for which there are only pre-1987 records. It is now much better recorded in Ireland than in the 1962 *Atlas*.

Oceanic Temperate element.

References: Atlas (395c), Curtis & McGough (1988), Dupont (1962), Hultén & Fries (1986), Meusel *et al.* (1965), Stewart *et al.* (1994).

F. H. PERRING

Deschampsia flexuosa Wavy Hair-grass

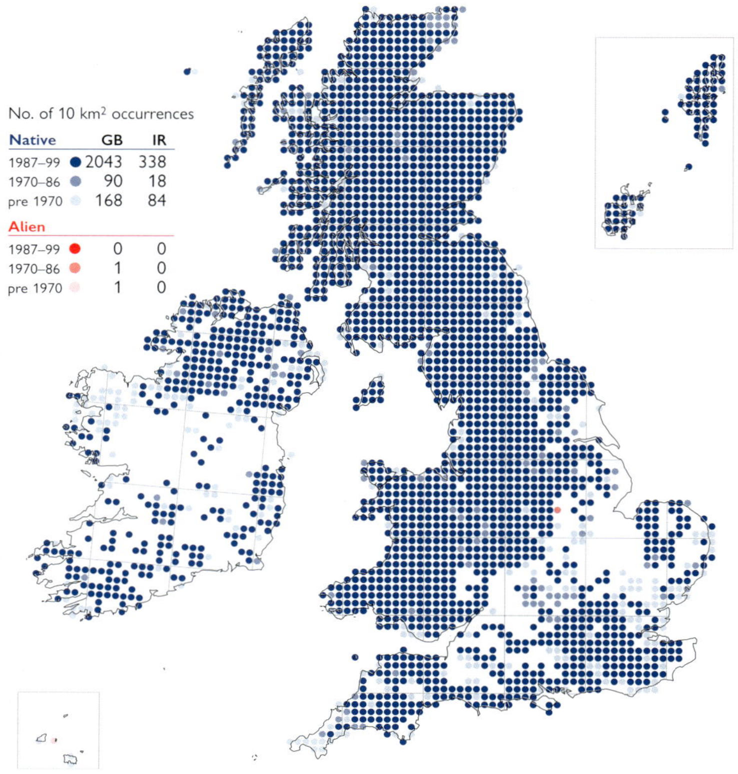

A loosely to densely tufted, clump or carpet-forming perennial of acid heaths, moorland, hill-pasture and open woodland, usually of *Betula* or *Quercus*. It grows on a wide range of freely-draining base-poor substrates, including leached soils over basic rocks. It can survive in sheep-grazed woodland. 0–1220 m (Ben Macdui, Banffs.).

Native (change –0.22). Though showing no general change from the 1962 *Atlas*, this species is declining locally in lowland areas due to the destruction of heathland and acidic grassland.

European Boreo-temperate element; also in E. Asia and N. America.

References: Atlas (395b), Grime *et al.* (1988), Hultén & Fries (1986), Meusel *et al.* (1965), Scurfield (1954).

F. H. PERRING

Holcus lanatus Yorkshire-fog

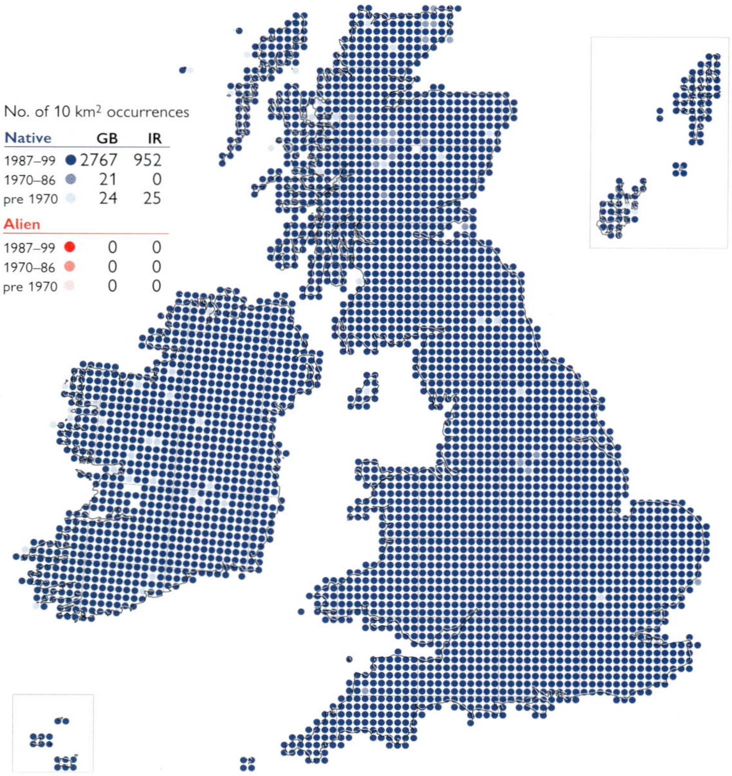

A short-lived, tufted perennial found in a wide range of grasslands, including hay meadows, pastures, chalk and limestone grassland, and also occurring in hedge banks, open woodland and moorland. It grows in dry to winter-wet, acidic to calcareous soils, and is most vigorous in moist but not waterlogged habitats. It tolerates mowing and grazing, but not heavy trampling. 0–650 m (Cross Fell, Cumberland), and exceptionally at 845 m on Great Dun Fell (Westmorland).

Native (change +1.34). There has been no appreciable change in the range of *H. lanatus* since the 1962 *Atlas*.

European Southern-temperate element; widely naturalised outside its native range.

References: Atlas (394b), Beddows (1961), Grime *et al.* (1988), Hultén & Fries (1986), Meusel *et al.* (1965), Thompson & Turkington (1988).

F. H. PERRING

Holcus mollis Creeping Soft-grass

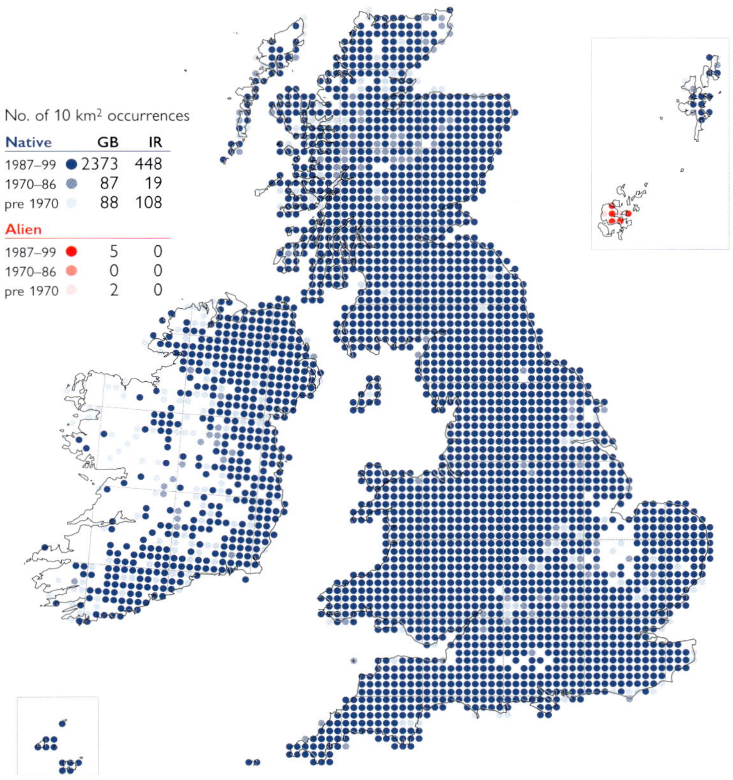

A creeping, rhizomatous perennial of well-drained acidic or neutral soils, found in *Quercus* and *Betula* woods, open conifer plantations, hedge banks, on heathland, and under *Pteridium*. It is capable of spreading into damper grassland and is locally a troublesome weed of arable land. *H. mollis* tolerates occasional mowing or light grazing but is a poor competitor with more vigorous grasses. 0–580 m (Greygarth Fell, W. Lancashire).

Native (change +0.80). There has been no appreciable change in the range of *H. mollis* since the 1962 *Atlas*.

European Temperate element.

References: Atlas (394c), Grime *et al.* (1988), Hultén & Fries (1986), Meusel *et al.* (1965), Ovington & Scurfield (1956).

F. H. PERRING

Corynephorus canescens Grey Hair-grass

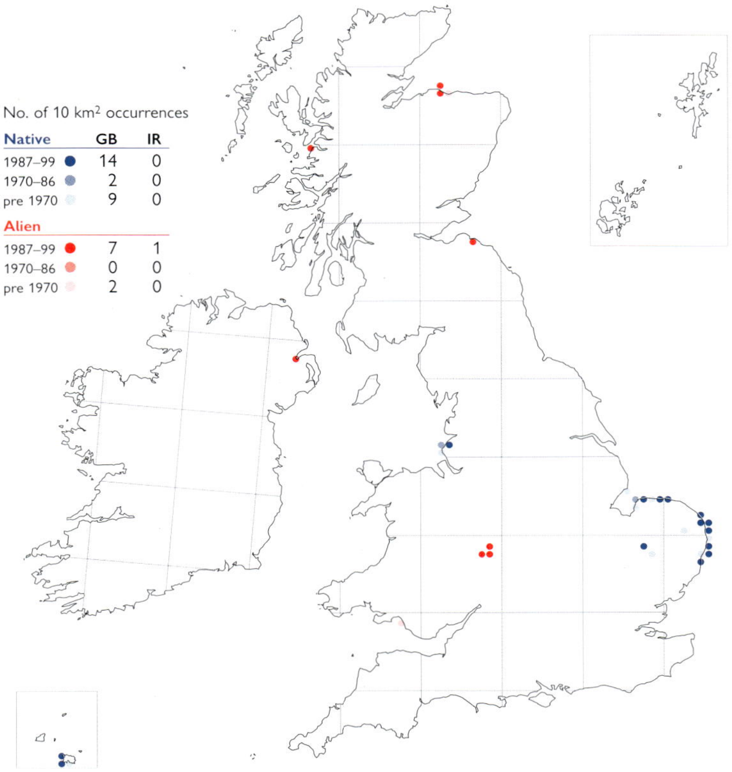

A perennial herb growing in open areas on consolidated dunes, on sandy shingle and on open sand. It also occurs on dunes and sandy heathland on acidic soils inland. It requires mobile sand for its survival; mature tufts are reinvigorated by partial burial. Lowland.

Native (change +0.01). The distribution of *C. canescens* is now probably stable, following losses before 1930. Its status is uncertain in some areas. In Lancashire, Savidge *et al.* (1963) regard it as native but Sell & Murrell (1996) treat it as probably introduced. In Scotland, Trist (1998) describes it as alien, but Ryves *et al.* (1996) suggest that its recent arrival may be an expansion of its natural range.

European Southern-temperate element.

References: Atlas (396b), Hultén & Fries (1986), Marshall (1967), Meusel *et al.* (1965), Stewart *et al.* (1994), Wigginton (1999).

S. J. LEACH

Aira caryophyllea Silver Hair-grass

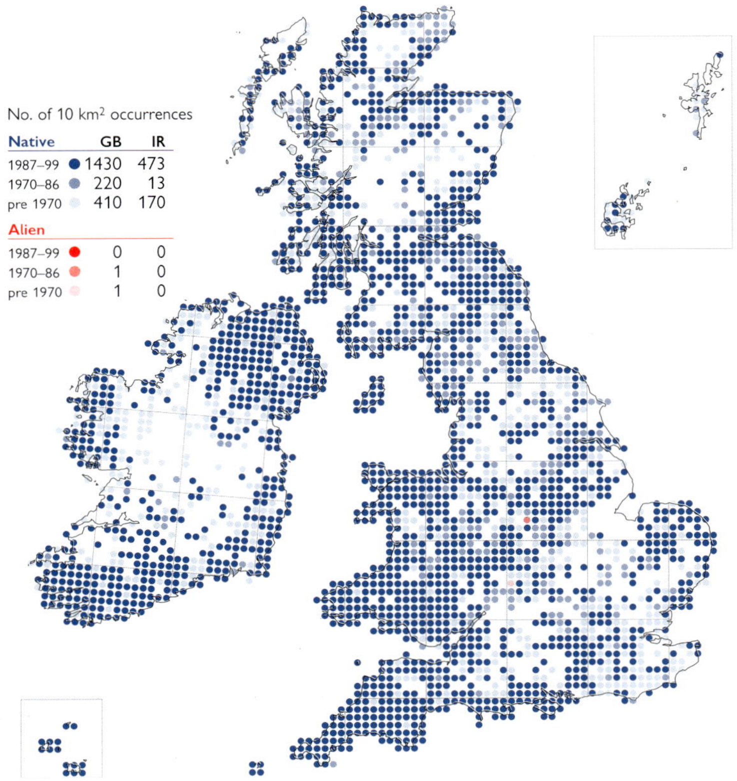

An annual of well-drained sandy and rocky places, cliff-tops, heaths, summer-parched grasslands, anthills and stabilised sand dunes; also on stone walls and railway ballast. Rarely, it is recorded as a wool casual. 0–560 m (Mourne Moutains, Co. Down).

Native (change –0.52). *A. caryophyllea* was mapped as 'all records' in the 1962 *Atlas*. It appears to have declined throughout its range, especially in S.E. England. Analysis of the database indicates that most of the losses have occurred since 1950.

European Southern-temperate element; widely naturalised outside its native range.

References: Atlas (396a), Hultén & Fries (1986), Meusel *et al.* (1965).

S. J. LEACH

Aira praecox Early Hair-grass

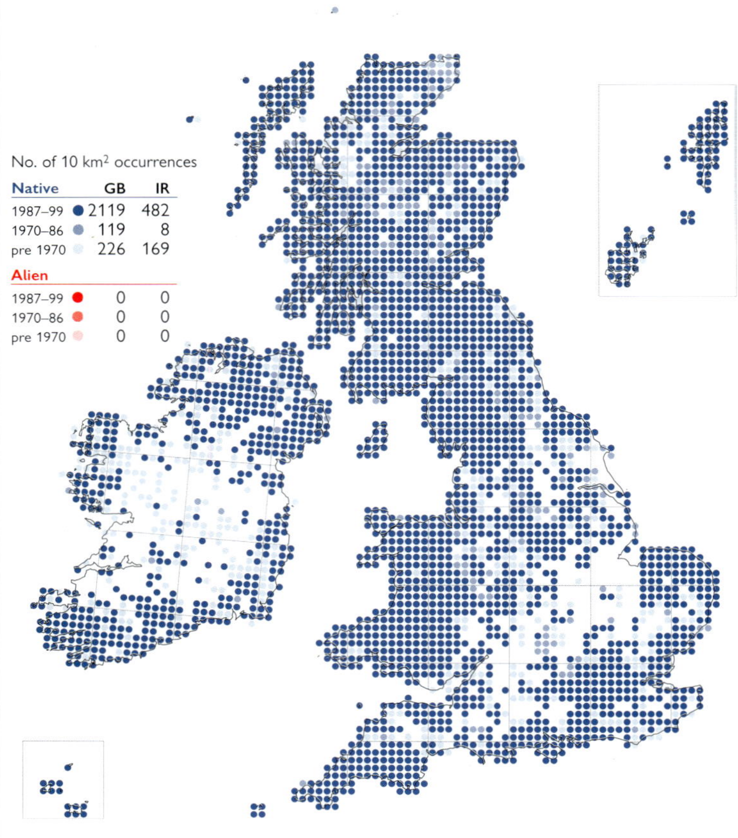

An annual of sandy, gravelly and rocky places, commonly on thin acidic soils around rock outcrops, on walls, cliff-tops, heaths and sand dunes. It is rarely recorded as a wool casual. Generally lowland, but reaching 685 m on Mangerton Mountain (S. Kerry).

Native (change –0.19). *A. praecox* was mapped as 'all records' in the 1962 *Atlas*. It seems to have declined in C. Ireland and England, but not to the extent of *A. caryophyllea*.

Suboceanic Southern-temperate element; widely naturalised outside its native range.

References: Atlas (395d), Grime *et al.* (1988), Hultén & Fries (1986), Meusel *et al.* (1965), Rich & Woodruff (1990).

S. J. LEACH

Hierochloe odorata Holy-grass

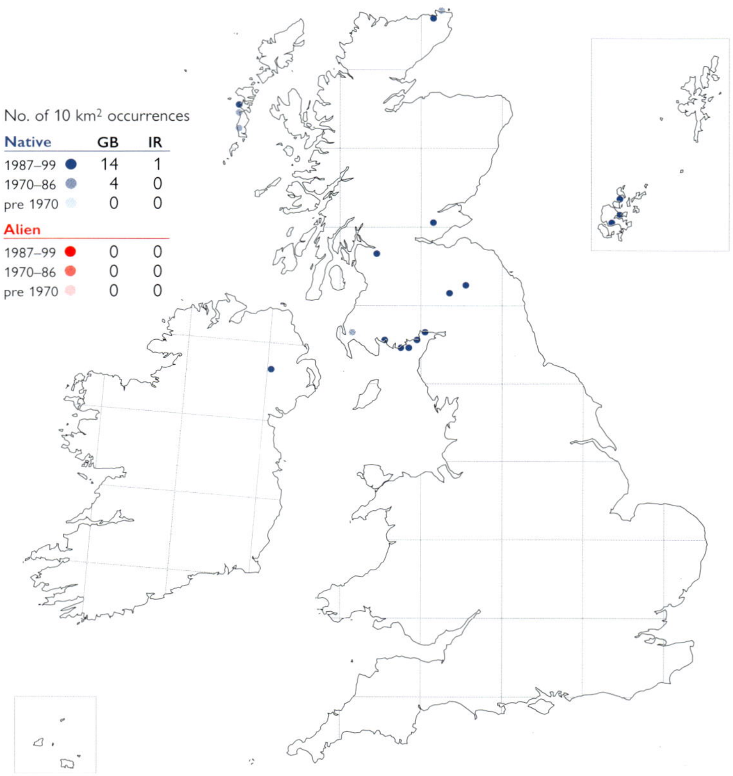

A rhizomatous perennial herb occurring in a range of wetland habitats, including lakeside reed-beds, sedge swamps, *Salix* carr, river banks and wet meadows; also, in S.W. Scotland, at the base of coastal cliffs where streams emerge and along the upper edge of fringing saltmarshes. 0–300 m (Clearburn Loch, Selkirks.).

Native (change +0.39). There has been little change in the distribution of *H. odorata* since the 1962 *Atlas*, though it is now known to be frequent in Orkney, where it was first recorded in 1980, and where all sites are near Norse church sites. It is abundant at its only site in Ireland, on Lough Neagh.

Circumpolar Boreal-montane element, with a continental distribution in W. Europe.

References: Atlas (403b), Curtis & McGough (1988), Hultén & Fries (1986), Meusel *et al.* (1965), Wigginton (1999).

S. J. LEACH

Anthoxanthum odoratum Sweet Vernal-grass

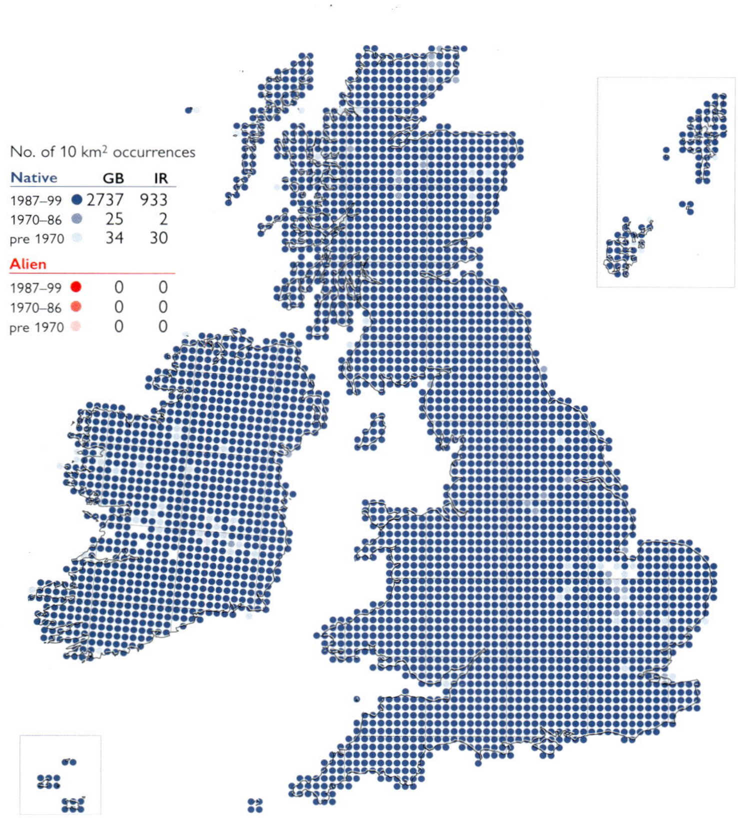

A short-lived perennial herb which occurs in a wide variety of grassland habitats, including old pastures and meadows, hill grassland, heaths, the drier parts of mires and on sand dunes. It is most frequent on acidic soils, and avoids drought-prone or water-logged sites. Reproduction is by seed. 0–1030 m (Cairngorms).

Native (change +0.90). *A. odoratum* was, until the 1920s, a component of grass-seed mixtures, but is no longer sown and has been lost from many pastures that have been improved. Nonetheless it is still ubiquitous at the 10-km square scale.

Eurosiberian Wide-temperate element, but widely naturalised so that distribution is now Circumpolar Wide-temperate.

References: Atlas (403c), Grime *et al.* (1988), Hultén & Fries (1986), Meusel *et al.* (1965), Wu & Jain (1980).

D. A. PEARMAN

Anthoxanthum aristatum Annual Vernal-grass

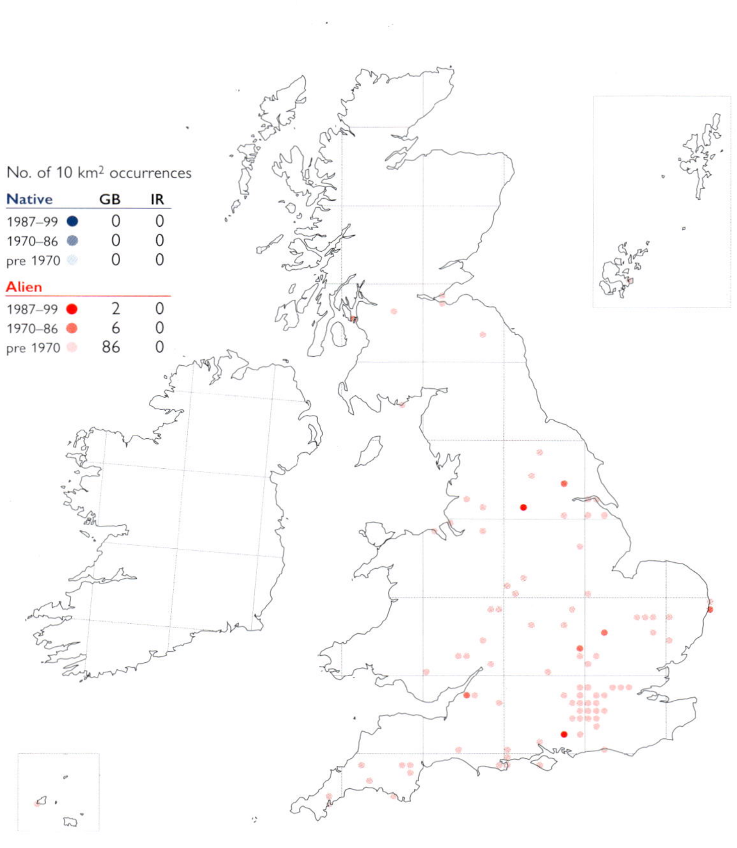

This annual was formerly a persistent weed of cereal crops on sandy or gravelly soils in S.E. England and East Anglia, but is now found only as a rare casual of cultivated ground and in waste places, and around paper mills and dock quaysides. Lowland.

Neophyte (change −2.65). *A. aristatum* is thought to have been introduced from France in the latter half of the 19th century as an impurity in fodder-plant seed. It was first recorded in the wild in 1872. Its decline was noticeable in the 1962 *Atlas*, with many losses before 1930. This decline is probably due to improved grain-cleaning techniques, which have reduced its incidence as an arable weed.

Native of S. Europe; widespread as a naturalised or casual introduction further north.

References: Atlas (403d), Meusel *et al.* (1965), Ryves *et al.* (1996).

S. J. LEACH

Phalaris arundinacea Reed Canary-grass

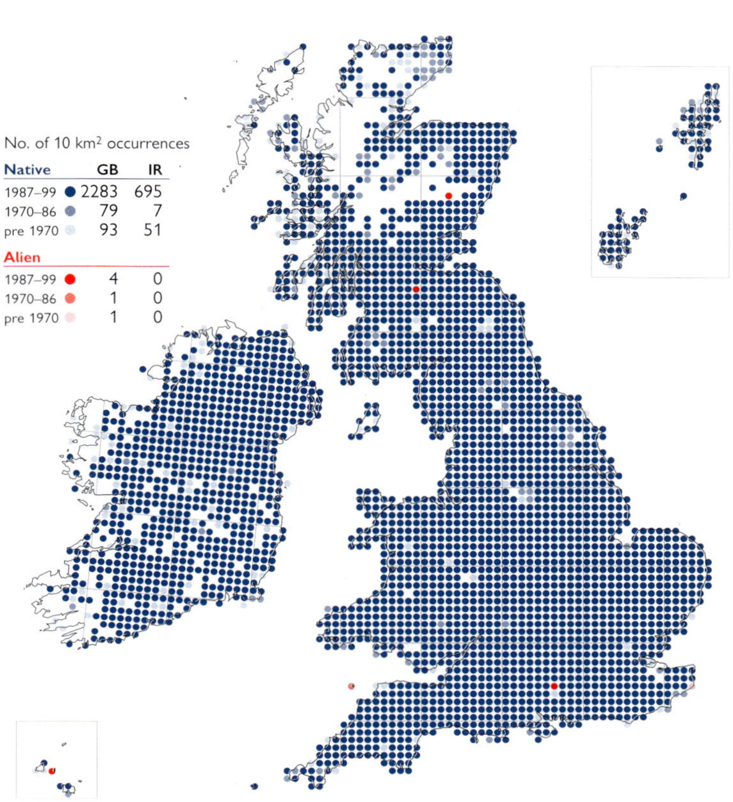

A rhizomatous perennial herb of ditches, river banks, *Alnus-Salix* carr and the margins of canals, reservoirs, lakes and ponds, growing especially well where the water-table fluctuates widely. It prefers slightly drier sites than *Glyceria maxima* and *Phragmites australis*, and can occur on roadsides. 0–475 m (Llyn y Figyn, Cards.) and exceptionally at 845 m on Great Dun Fell (Westmorland).

Native (change +0.23). There is no evidence of any change in the distribution of this species since the 1962 *Atlas*. Its ability to tolerate summer-dry conditions has enabled it to survive in the modern landscape.

Circumpolar Boreo-temperate element; widely naturalised outside its native range.

References: Atlas (404a), Grime *et al.* (1988), Hultén & Fries (1986), Preston & Croft (1997).

S. J. LEACH

Phalaris aquatica Bulbous Canary-grass

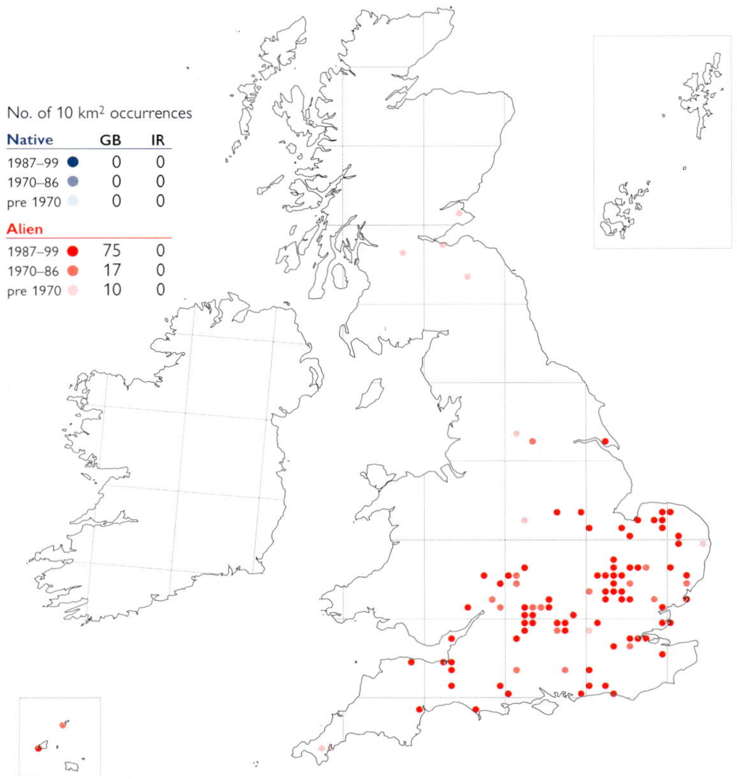

A shortly rhizomatous, perennial herb naturalised in woodland glades, field-borders, roadsides and waste places. It also occurs as a casual on rubbish tips and waste ground as a wool, esparto and bird-seed alien, and as a contaminant of grass-seed. Lowland.

Neophyte. *P. aquatica* was first cultivated in Britain in 1778. It is sown as cover and food for pheasants, more rarely for grazing or silage, sometimes as robust Australian strains. It was recorded in the wild by 1912 and seems to be increasing, especially in East Anglia and S.E. England. It is probably under-recorded.

Native of the Mediterranean region and S.W. Asia; widely introduced by cultivation elsewhere.

References: Anderson (1961), Ryves *et al.* (1996).

S. J. LEACH

Phalaris canariensis Canary-grass

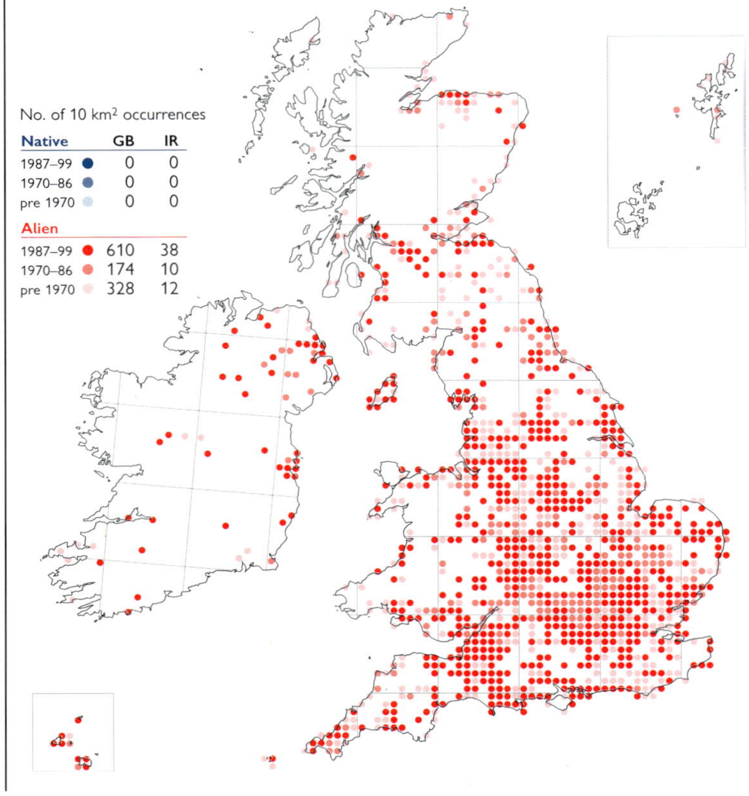

An annual of waste ground, rubbish tips, walls, roadsides and pavement cracks, especially in built-up areas. It is a casual, rarely persisting, from bird-seed, grain and, possibly, wool shoddy. Generally lowland, but reaching 430 m at Nenthead (Cumberland).

Neophyte (change −0.32). *P. canariensis* was recorded in the wild as early as 1632. Comparison with the 1962 *Atlas* suggests that losses have been balanced by new records.

Perhaps native to N.W. Africa and the Canary Islands; widely naturalised in the Mediterranean region and in other continents.

References: Atlas (404b), Anderson (1961).

S. J. LEACH

Phalaris minor Lesser Canary-grass

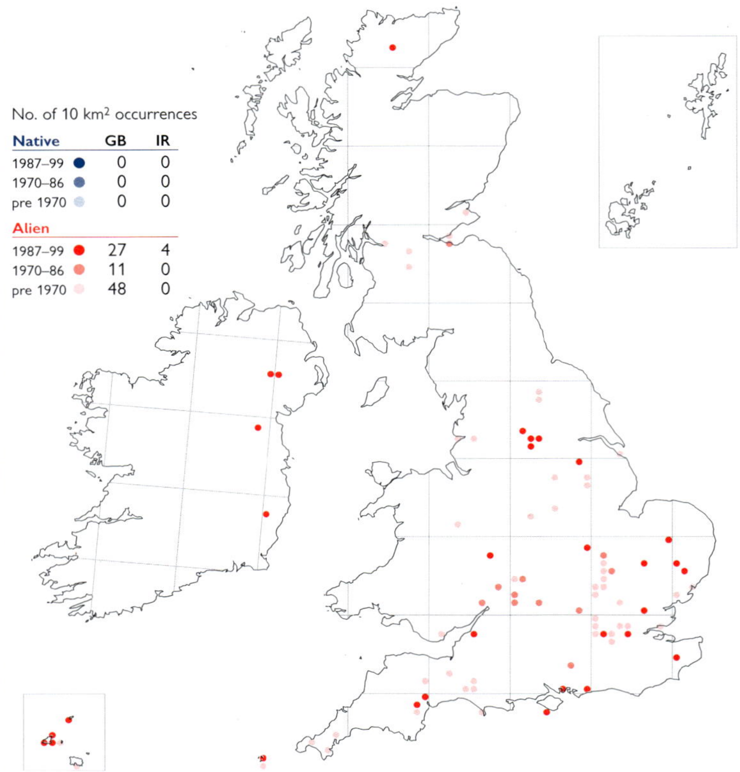

This annual is usually found as a casual on rubbish dumps and waste ground, but is sometimes established amongst arable crops, including bulb-fields in the Isles of Scilly and carrot fields in East Anglia. It mainly originates from grain, bird-seed, wool and esparto. Lowland.

Neophyte. *P. minor* was noted by Ryves *et al.* (1996) as possibly native in the Channel Islands, where it was known in the wild as early as 1791, but most *Phalaris* species were spreading by then and it is probably alien there. It is almost certainly under-recorded.

A Mediterranean-Atlantic species; very widely naturalised outside its native range.

References: Anderson (1961), McClintock (1975).

S. J. LEACH

Phalaris paradoxa Awned Canary-grass

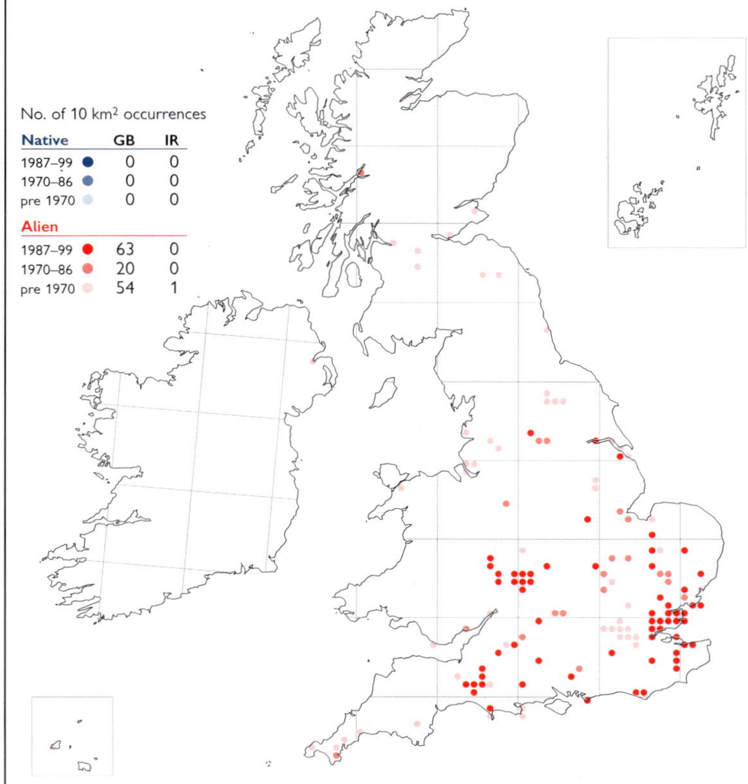

An annual occurring as a casual from bird-seed, grain, wool, esparto and other sources. It is included as a constituent of game-bird seed mixtures. *P. paradoxa* is found on tips and waste ground, and as a weed in arable fields and newly sown grass leys. Lowland.

Neophyte. *P. paradoxa* was apparently cultivated in 1687. It was recorded as a casual by 1859 (Surrey) and appears to be an increasing species, becoming well-established as an arable weed in S. Britain. It has probably been under-recorded.

Native of the Mediterranean region and S.W. Asia; widely naturalised or present as a casual elsewhere.

References: Anderson (1961), Ryves *et al.* (1996).

S. J. LEACH

Agrostis capillaris Common Bent

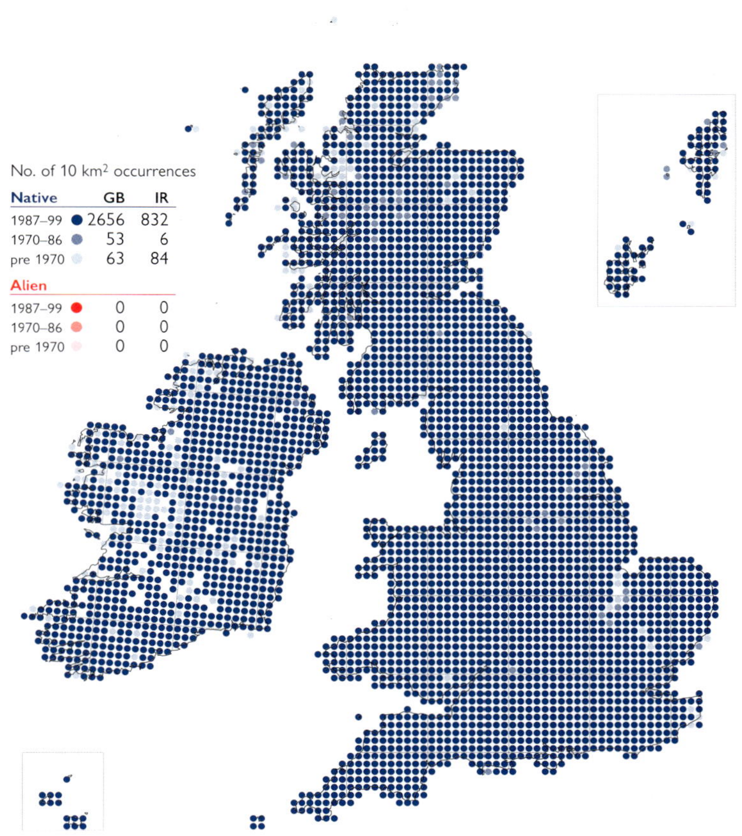

A rhizomatous, occasionally stoloniferous, perennial herb occurring mainly on dry or damp, neutral to moderately acidic soils; its habitats include lowland pastures, hay meadows, upland hill-pasture, heaths, open woodland and scrub, sand dunes and a range of ruderal habitats including areas contaminated by heavy metals. It is extensively used as a lawn grass, on its own or with other species. 0–1210 m (Ben Lawers, Mid Perth).

Native (change +1.28). There has been no change in the range of *A. capillaris* since the 1962 *Atlas*.

Eurosiberian Boreo-temperate element; widely naturalised outside its native range.

References: Atlas (398a), Grime *et al.* (1988), Hultén & Fries (1986).

D. A. PEARMAN

Agrostis gigantea Black Bent

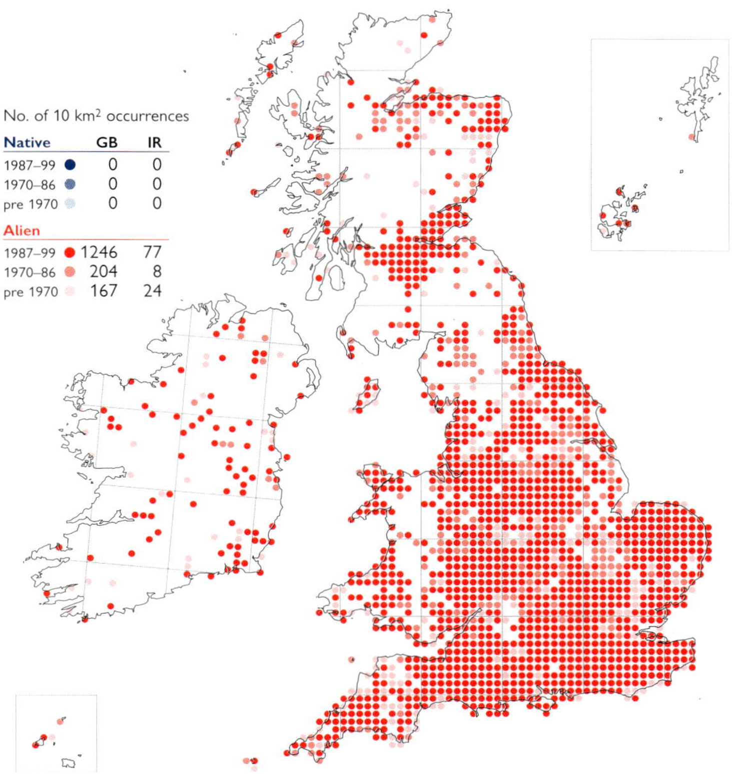

A sprawling, perennial herb, behaving as a rampant weed in cornfields and neglected arable land, particularly on lighter soils, where it spreads by seed and by rhizomes. In wetter habitats, where it is much rarer, it can persist in taller, closed vegetation by rhizomatous growth. Lowland.

Archaeophyte (change +1.39). *A. gigantea* may have been overlooked in the past, or confused with *A. stolonifera*. Despite being vulnerable to herbicides, it has shown a marked increase since the 1962 *Atlas*. It is, however, almost certainly still under-recorded.

As an archaeophyte *A. gigantea* has a Eurasian Southern-temperate distribution, but it is naturalised in N. America so its distribution is now Circumpolar Southern-temperate.

References: Atlas (398b), Hultén & Fries (1986).

D. A. PEARMAN

Agrostis castellana Highland Bent

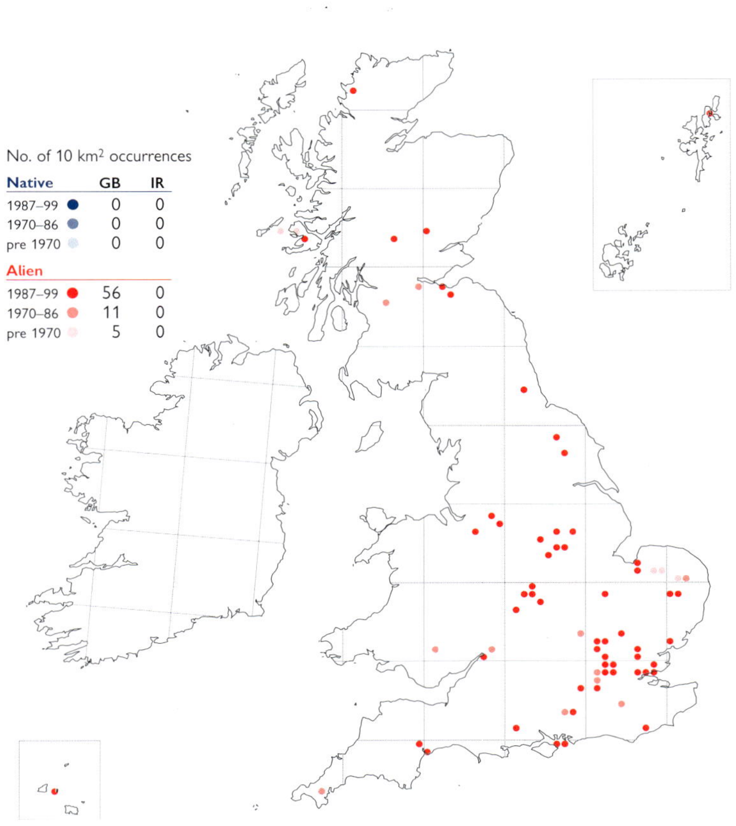

A perennial herb of roadsides, amenity grassland, temporary leys and cultivated land. It is often sown in grass-seed mixtures, and is also introduced with wool shoddy and in wild-flower mixtures. Lowland.

Neophyte. This species was first recorded in Britain, as a casual, in 1924. It is not known when *A. castellana* was first imported as an intentional constituent of grass-seed mixtures. Its distribution is uncertain and some authorities regard it as under-recorded (e.g. Stace, 1997).

Native of the Mediterranean region.

Reference: Ryves *et al.* (1996).

D. A. PEARMAN

Agrostis stolonifera Creeping Bent

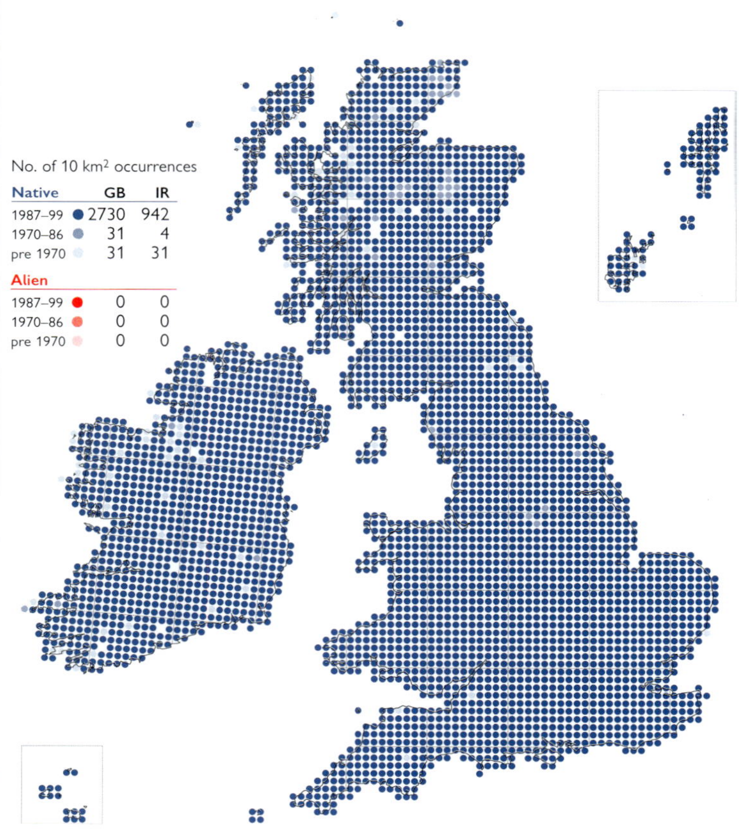

A stoloniferous perennial herb, found in many habitats on heavy and light, moist and dry soils. It occurs in permanent grassland (including inundation and brackish communities), in upper saltmarsh and dune-slacks, on sand dunes and sandy flats, on cliffs and in mires, springs, flushes and ditches. It also grows on spoil heaps and a wide range of open and disturbed habitats, and is a weed in arable sites. 0–945 m (Ben Lawers, Mid Perth).

Native (change +3.66). There has been no change in the range of *A. stolonifera* since the 1962 *Atlas*. It is very variable in form and distinct ecotypes have evolved in response to newly available habitats.

Circumpolar Wide-temperate element; widely naturalised outside its native range.

References: Atlas (398c), Grime *et al.* (1988), Hultén & Fries (1986).

D. A. PEARMAN

Agrostis curtisii Bristle Bent

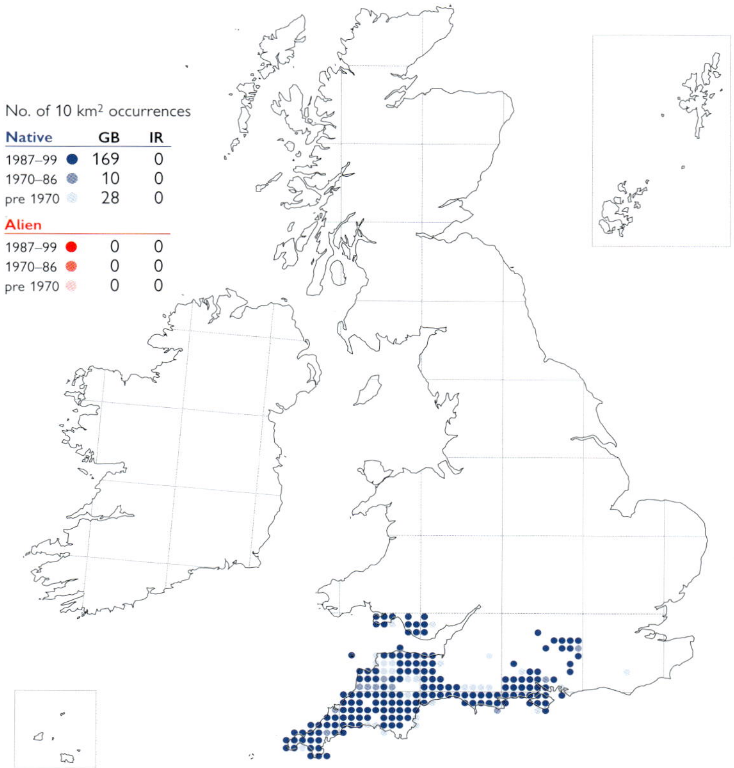

A tufted perennial herb of the drier parts of impoverished, sandy and peaty heaths, especially those with impeded drainage, occasionally extending to more consistently waterlogged ground. In closed communities, it occurs as scattered plants, but it can seed prolifically into burnt or disturbed ground and rapidly form an almost continuous cover. It also occurs in open acidic woodland over gravel and sand. Generally lowland, but reaching 610 m on Dartmoor (S. Devon).

Native (change –0.26). *A. curtisii* has been lost from some 10-km squares on the edges of its range through habitat destruction. However, these losses are balanced to some extent by its ability to colonise open ground rapidly.

Oceanic Southern-temperate element.

References: Atlas (397c), Dupont (1962), Ivimey-Cook (1959).

D. A. PEARMAN

Agrostis canina Velvet Bent

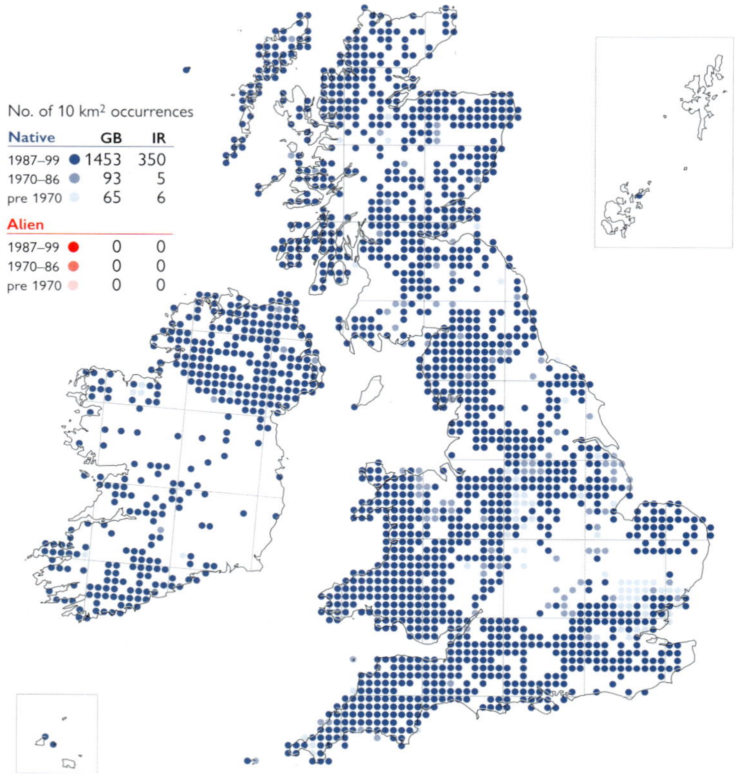

A stoloniferous, perennial grass of infertile, acidic, peaty soils in mires, wet heath, fens and fen-meadows, springs and soakways, swamps and water margins. It may exploit vegetation gaps or act as a dominant. It spreads by stolon fragments. 0–1035 m (Macgillycuddy's Reeks, S. Kerry).

Native. The map of *A. canina* in the 1962 *Atlas* included both this species and *A. vinealis*. Although these have long been recognised at varietal or subspecific rank, they have only recently been treated as species in British floras (Clapham *et al.*, 1987). *A. canina sens. str.* is under-recorded in areas for which only records of the aggregate are available.

Circumpolar Boreo-temperate element, with a disjunct distribution.

Reference: Grime *et al.* (1988).

D. A. PEARMAN

Agrostis vinealis Brown Bent

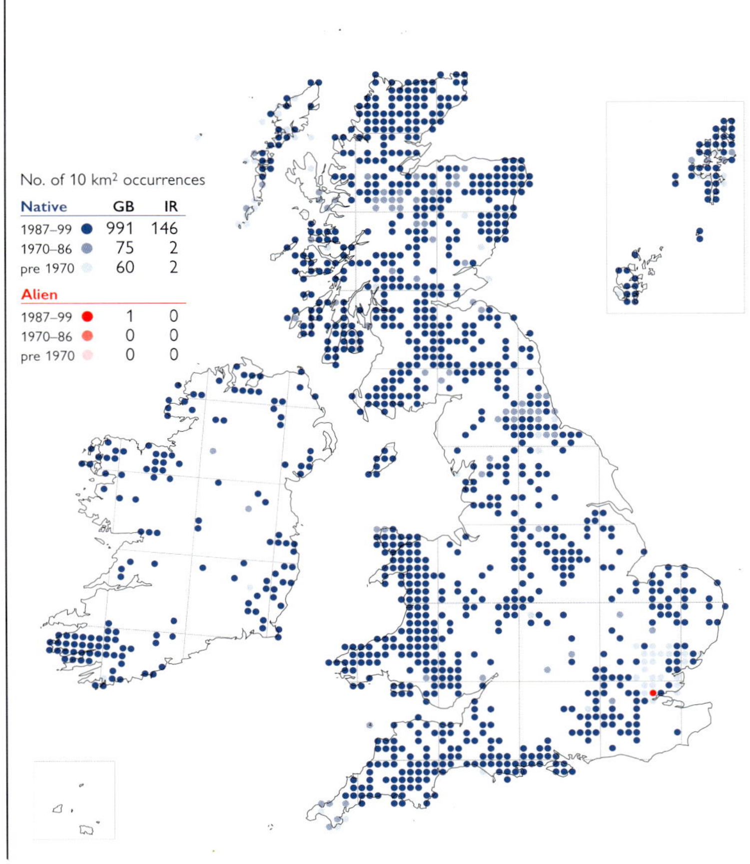

A shortly rhizomatous perennial herb, mainly of dry or free-draining, acidic, sandy or peaty soils on heaths, in acidic grass-heath, in open woodland (especially *Betula*, *Pinus* and *Quercus*), in woodland clearings and on rides. On some lowland heaths *A. vinealis* may grow in damp situations, but unlike *A. canina* it avoids waterlogged soils. It is a drought-resistant lawn grass. 0–845 m (Little Dun Fell, Westmorland), and probably higher in Scotland.

Native. Until recently this species was treated as a variety or subspecies of *A. canina*; it has not been recorded systematically and the existing map is incomplete.

European Temperate element; also in N. America.

Reference: Grime *et al.* (1988).

D. A. PEARMAN

Calamagrostis epigejos Wood Small-reed

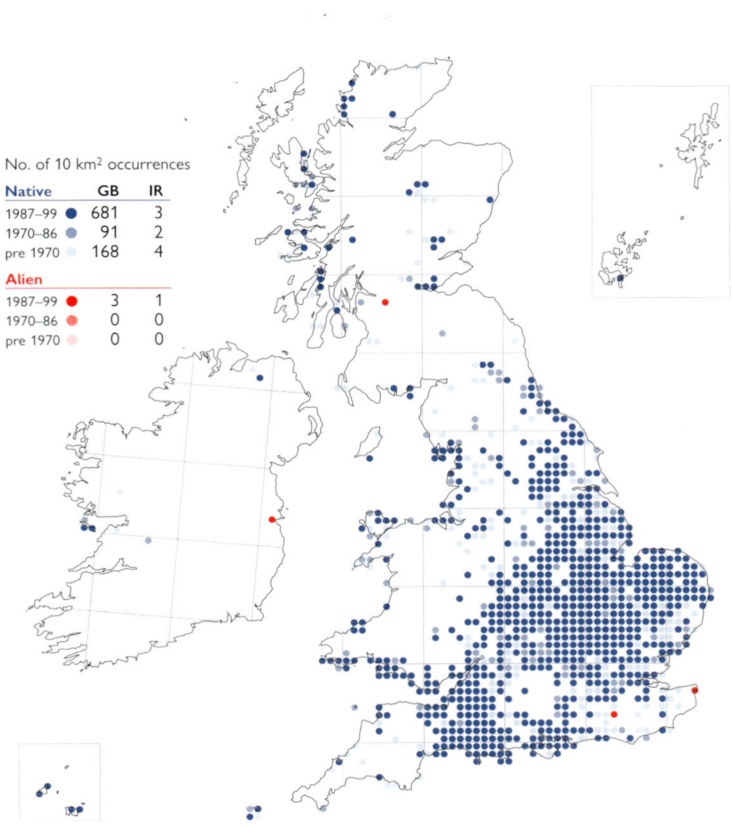

A tufted rhizomatous perennial, occurring in damp woods, ditches, fens, ungrazed or lightly grazed grasslands, and on sheltered sea-cliffs and sand dunes; also as a colonist of artificial habitats such as old quarries, roadsides and railway banks. It usually grows on light sands or heavy clays. 0–370 m (Great Asby, Westmorland).

Native (change +0.47). There has been some change since the 1962 *Atlas*, with *C. epigejos* either having been better recorded, or increasing its range. Counties in which local floras recorded it rarely in the past now have more records and it has spread locally because of a relaxation of grazing.

Eurasian Boreo-temperate element.

References: Atlas (396d), Curtis & McGough (1988), Hultén & Fries (1986), Meusel *et al.* (1965).

S. J. LEACH

Calamagrostis canescens Purple Small-reed

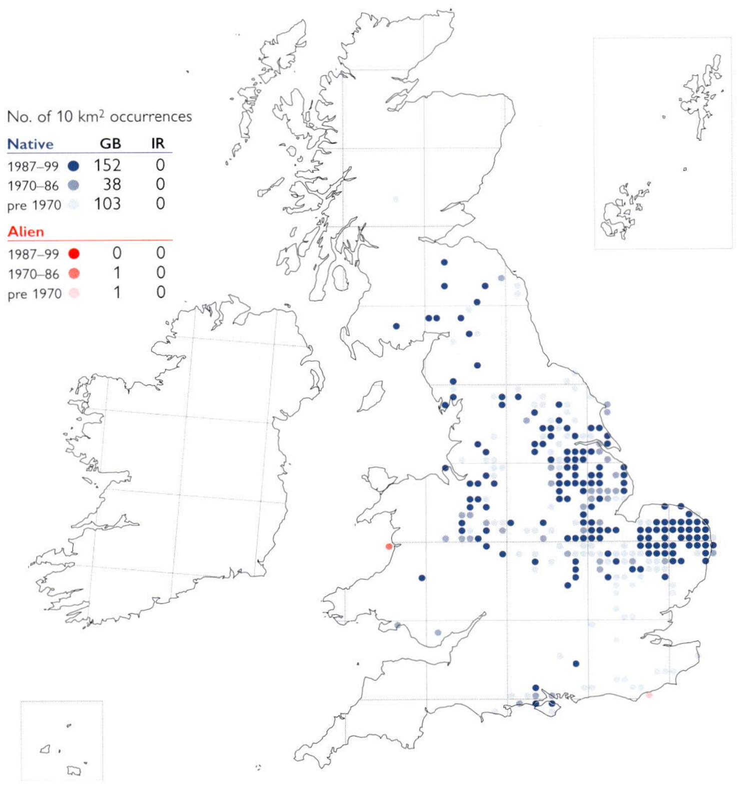

A perennial herb of lakeside marshes, fen-meadows, tall-herb fens and *Alnus* or *Salix* carr, often in extremely species-rich vegetation and sometimes forming extensive stands. Generally lowland, but reaching 335 m on Malham Moor (Mid-W. Yorks.).

Native (change –0.33). The retreat of *C. canescens* from its peripheral localities was apparent in the 1962 *Atlas*, and has continued in some areas. Even within its core range, succession and falling water tables may be reducing the extent of wetland habitats available for this species. On the other hand, all records in the Scottish Borders have been made since 1962.

Eurosiberian Boreo-temperate element, with a continental distribution in W. Europe.

References: Atlas (397a), Hultén & Fries (1986), Meusel *et al.* (1965).

S. J. LEACH

Calamagrostis purpurea Scandinavian Small-reed

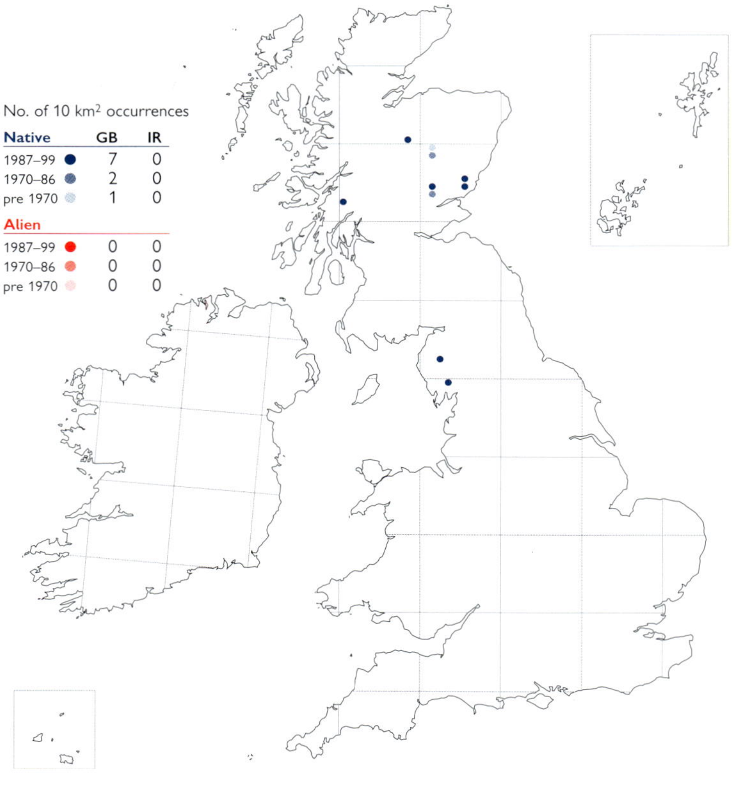

No. of 10 km² occurrences

Native		GB	IR
1987–99	●	7	0
1970–86	●	2	0
pre 1970	●	1	0
Alien			
1987–99	●	0	0
1970–86	●	0	0
pre 1970	●	0	0

An apomictic rhizomatous perennial herb of wet *Salix* carr, especially in areas that are flooded in winter. It also occurs in marshes, ditches and old peat diggings. Generally lowland, but upper altitudinal limit unknown.

Native. The presence of this taxon in Britain was not recognised until 1980, although it had been collected before that date (such as from near Braemar, S. Aberdeen, in 1941). It may be under-recorded.

Eurosiberian Boreal-montane element.

References: Hultén & Fries (1986), Stewart (1989), Wigginton (1999).

S. J. LEACH

Calamagrostis stricta Narrow Small-reed

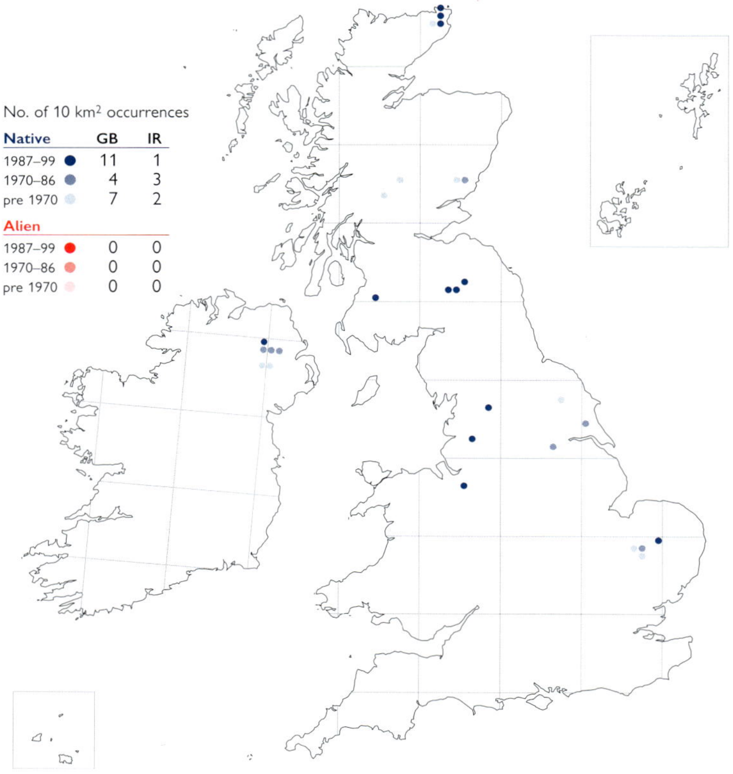

No. of 10 km² occurrences

Native		GB	IR
1987–99	●	11	1
1970–86	●	4	3
pre 1970	●	7	2
Alien			
1987–99	●	0	0
1970–86	●	0	0
pre 1970	●	0	0

A tufted rhizomatous perennial herb of near-neutral mires and lake margins. 0–340 m (Kingside Loch, Selkirks.).

Native (change –0.74). It is hard to assess trends in the distribution of *C. stricta* as it has been confused in the past with *C. scotica, C. purpurea* and hybrids of *C. canescens* (including *C. canescens* × *C. stricta*). It has certainly been lost from some sites through drainage, but is easily overlooked and may still be present in several squares for which only pre-1987 records are available.

Circumpolar Boreo-arctic Montane element.

References: Atlas (397b), Crackles (1995, 1997), Curtis & McGough (1988), Hultén & Fries (1986), Meusel *et al.* (1965), Stewart *et al.* (1994), Wigginton (1999).

S. J. LEACH

Calamagrostis scotica Scottish Small-reed

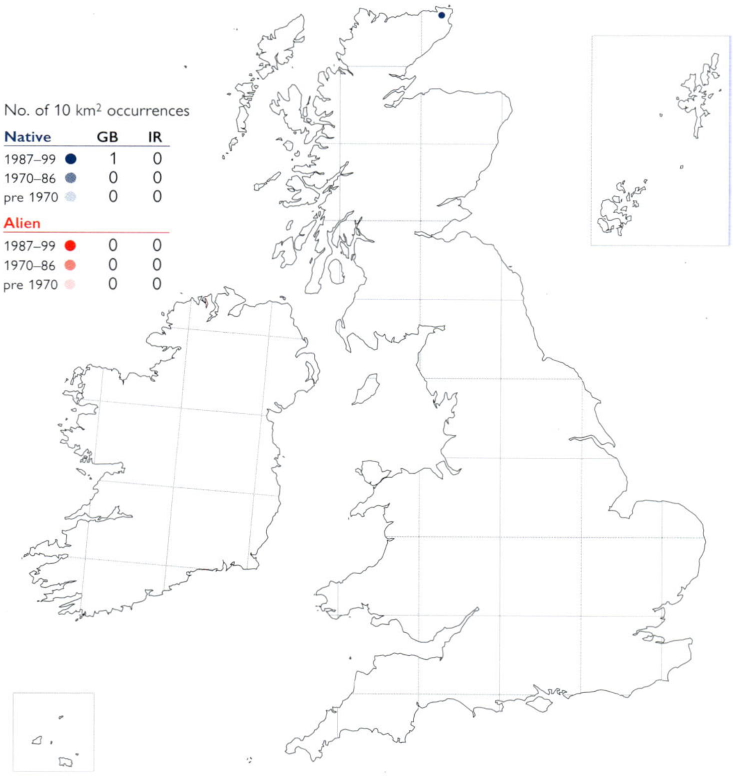

No. of 10 km² occurrences

Native		GB	IR
1987–99	●	1	0
1970–86	●	0	0
pre 1970	●	0	0
Alien			
1987–99	●	0	0
1970–86	●	0	0
pre 1970	●	0	0

A tufted rhizomatous perennial herb, restricted to a single site, where it occurs in areas of *Juncus* dominated pasture, along drainage channels and in *Salix* carr and tall-herb fen. Lowland.

Native. This taxon was not mapped separately in the 1962 *Atlas*, being grouped with *C. stricta* as *C. neglecta* agg. Reports of *C. scotica* occurring elsewhere in N. Scotland and in Roxburghshire have never been substantiated and are probably errors for *C. stricta*.

Endemic.

Reference: Wigginton (1999).

S. J. LEACH

Ammophila arenaria Marram

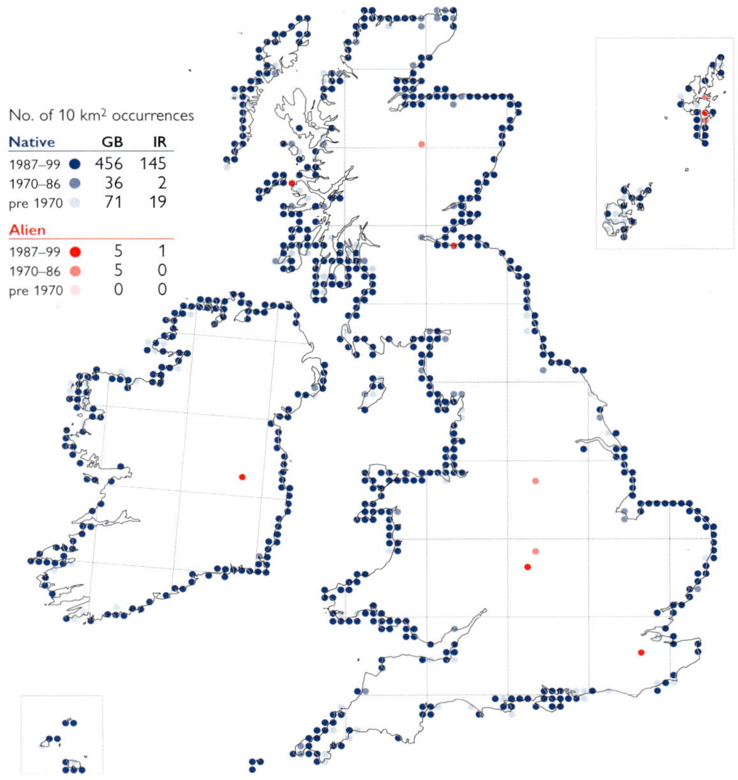

A rhizomatous perennial herb of coastal sand dunes. *A. arenaria* is an important species in the stabilisation of mobile dunes and blow-outs, and is widely planted as a sand binder. Inland, it is a rare casual, though several attempts have been made in recent years to establish it on inland golf courses. Lowland.

Native (change –0.26). There is no evidence of appreciable change in the distribution of *A. arenaria* since the 1962 *Atlas*. Introductions inland and the widespread planting of *A. arenaria* on eroding dunes have had little impact on its overall distribution.

European Southern-temperate element; widely naturalised outside its native range.

References: Atlas (396c), Huiskes (1979), Hultén & Fries (1986).

S. J. LEACH

Gastridium ventricosum Nit-grass

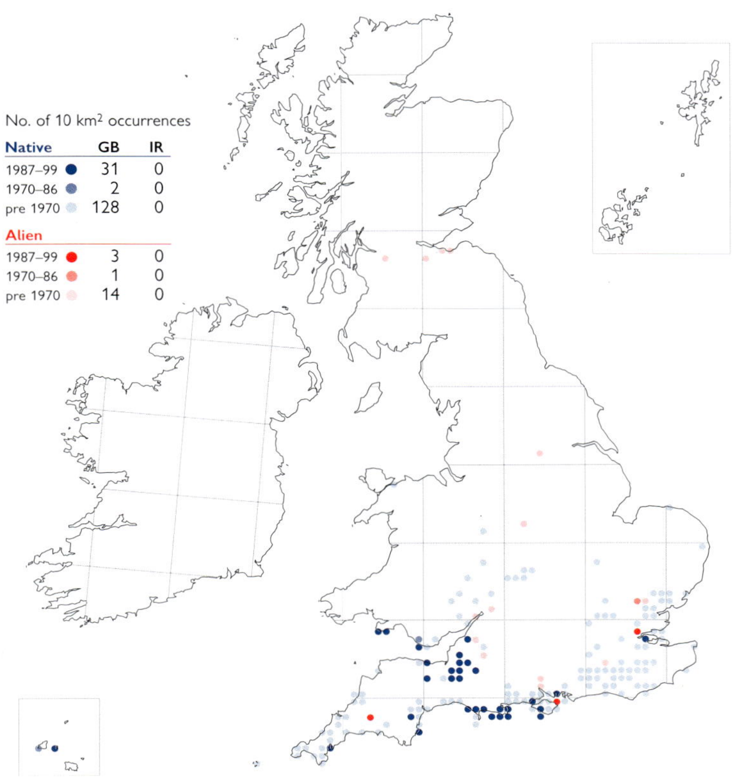

An annual of well-drained grasslands on calcareous substrates, especially on S.-facing slopes where the turf is broken by patches of crumbling soil. It used to be a frequent weed of cornfields, and is an occasional wool and grain casual. Lowland.

Native or alien (change –0.48). *G. ventricosum* was first recorded in 1690 and is possibly native in grassland in S.W. Britain and in the Channel Islands, where it has been found to be much more widespread than shown in the 1962 *Atlas*. Formerly it was regarded as an arable weed, but it had largely vanished from such habitats by 1930. As habitat details are not available for many old records, they are mapped as if they are native.

Mediterranean-Atlantic element.

References: Atlas (400a), Ryves *et al.* (1996), Trist (1986), Wigginton (1999).

S. J. LEACH

Lagurus ovatus Hare's-tail

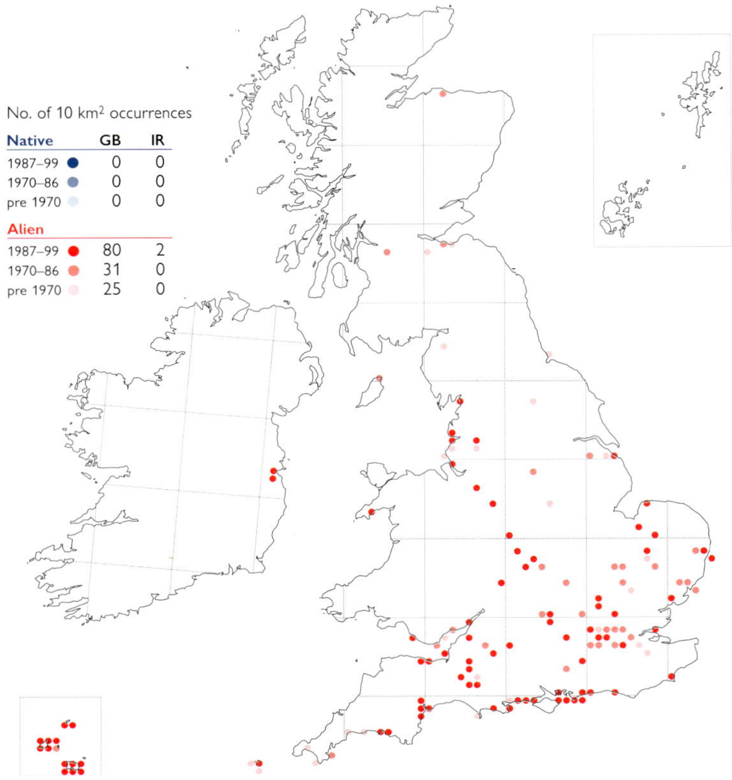

This annual grass is naturalised on sand dunes in the Channel Islands and S. England. Elsewhere it occurs as a casual garden escape or introduction from wool shoddy or grain, on walls, pavements, roadsides, rubbish tips and in car parks. Lowland.

Neophyte (change +1.35). *L. ovatus* was introduced to cultivation in Britain by 1640 and is widely grown in gardens. It was first recorded in the wild in 1791 in Guernsey, where it has long been naturalised. Deliberate attempts to establish it in Jersey were eventually successful in the 1860s. It appears to be increasing as an established alien on sand dunes in S. England.

A Mediterranean-Atlantic species; widely naturalised in N. & S. America, S. Africa, Australia and elsewhere.

References: Atlas (400b), McClintock (1975), Ryves *et al.* (1996).

S. J. LEACH

Apera spica-venti Loose Silky-bent

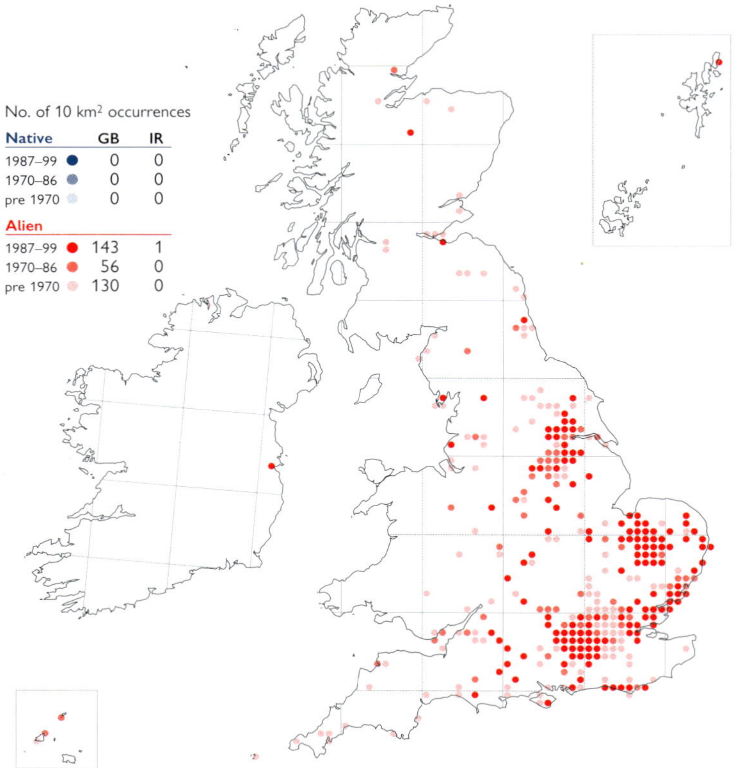

An annual of open habitats, mostly in arable fields, where it can be a troublesome weed, but also in a wide variety of waste places, including sandy tracks and roadsides. Lowland, but casual at 690 m on Ingleborough (Mid-W. Yorks.).

Archaeophyte (change –0.21). *A. spica-venti* has been treated as a native by some authors, but its artificial habitat and the transient nature of its populations suggests otherwise. It was grown in the London area in the early 19th century as an ornamental grass for drying.

As an archaeophyte *A. spica-venti* has a Eurosiberian Boreo-temperate distribution; it is widely naturalised outside this range.

References: Atlas (399a), Hultén & Fries (1986), Meusel *et al.* (1965), Stewart *et al.* (1994), Warwick *et al.* (1985).

D. A. PEARMAN

Apera interrupta Dense Silky-bent

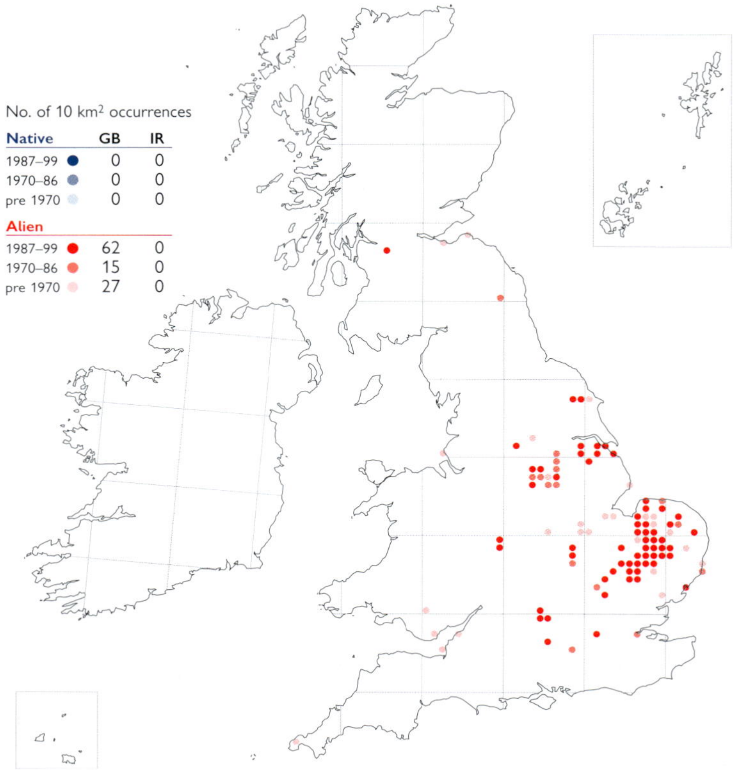

An annual predominantly found in arable fields, where it can be a weed, but also recorded from road verges, trackways and quarries. It has a more permanent niche in some tightly grazed grassy or grass-heath habitats. It occurs in waste ground as a casual from wool shoddy and imported aggregates, and as a seed impurity. Lowland.

Neophyte (change +0.80). The distribution of *A. interrupta*, which was first recorded in the wild in 1848, seems to be stable or even increasing, at least at the 10-km square scale. The seed is long-lived and disturbance is the key to its survival.

A Eurosiberian Southern-temperate species.

References: Atlas (399b), Hultén & Fries (1986), Meusel *et al.* (1965), Stewart *et al.* (1994).

D. A. PEARMAN

Mibora minima Early Sand-grass

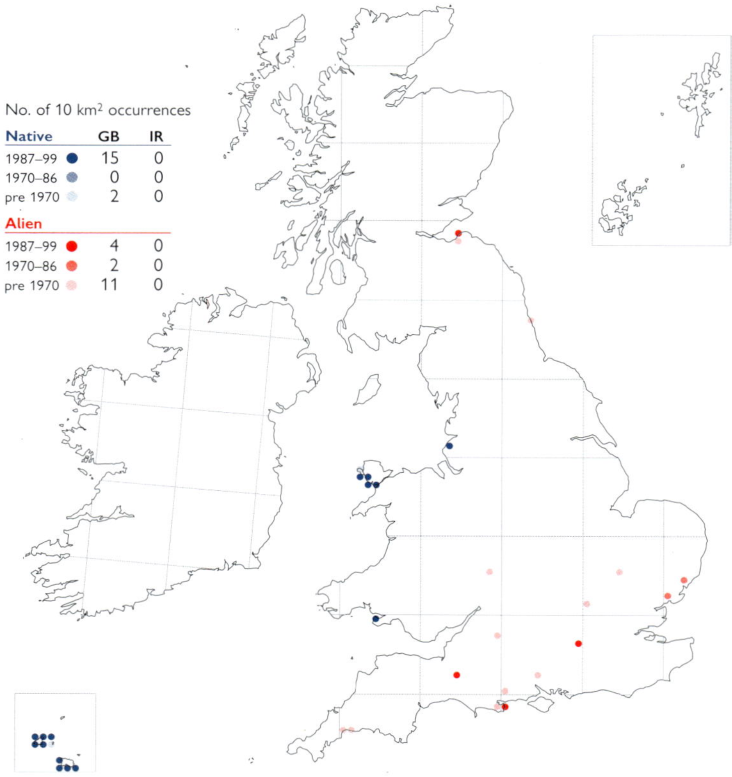

A diminutive winter-annual of coastal sand dunes in Wales, on open nutrient-poor substrates which are free-draining but damp in winter; also, in the Channel Islands, on barish gravelly cliff-slopes. It is recorded at other dune sites in England and Scotland, and inland as a casual plant in nurseries and gardens. Lowland.

Native (change –0.01). Several new populations of *M. minima* have been discovered on dunes since 1980, and it is almost impossible to say whether they are native or alien. A Lancashire population, found in 1996, seems most likely to be the result of spread from the known native sites in Wales.

Suboceanic Southern-temperate element.

References: Atlas (399d), Meusel *et al.* (1965), Wigginton (1999).

S. J. LEACH

Polypogon monspeliensis Annual Beard-grass

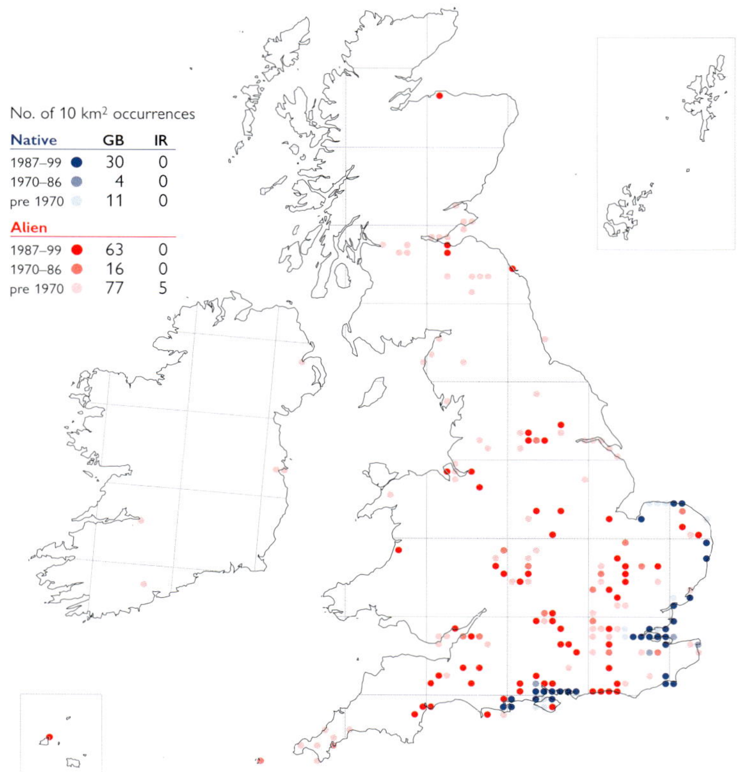

An annual of barish places by the sea, in damp, cattle-trodden grazing marshes, at the edges of dried-up brackish pools and ditches, and in the uppermost parts of saltmarshes. Also around docks and inland as a casual from wool, bird-seed and other sources. Lowland.

Native (change +0.60). The native distribution of *P. monspeliensis* is broadly similar to that mapped in the 1962 *Atlas*, and it occurs with similar frequency as a casual. However, several local floras (e.g. Brewis *et al.*, 1996) report that pond and ditch infilling, and the drainage and conversion of coastal marshes to arable, are causing declines in native populations. It may be becoming established inland.

Mediterranean-Atlantic element; widely naturalised outside its native range.

References: Atlas (399c), Stewart *et al.* (1994).

S. J. LEACH

Polypogon viridis Water Bent

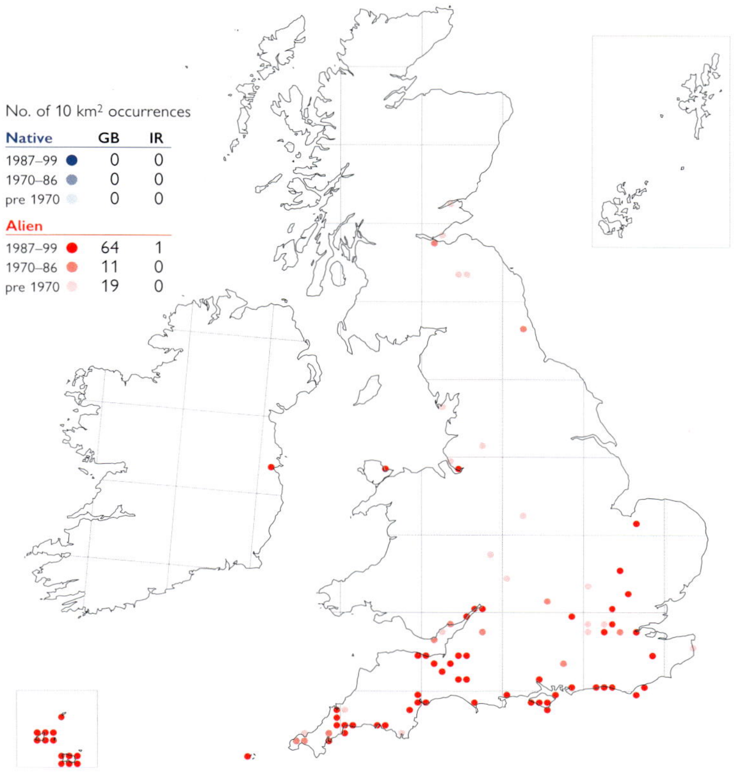

An annual or perennial herb which is well-naturalised in the Channel Islands on roadsides and by pools. In England it grows on tips and damp waste ground, and is spreading as a weed of nurseries, gardens and pavement cracks. Lowland.

Neophyte (change +1.28). *P. viridis* was introduced into cultivation in 1800 and was first noted in the wild in Cardiff in 1876. It was recorded in Guernsey in 1897, and in Jersey in 1906. It has spread in Jersey since the 1960s, and also appears to be increasing in England, at least locally; for example, it was first recorded in Somerset in 1989 and is now known from nine 10-km squares.

Native of S. Europe, S.W. & C. Asia & N. Africa; widely naturalised elsewhere.

References: Atlas (398d), Le Sueur (1984), McClintock (1975, 1987a).

S. J. LEACH

Alopecurus pratensis Meadow Foxtail

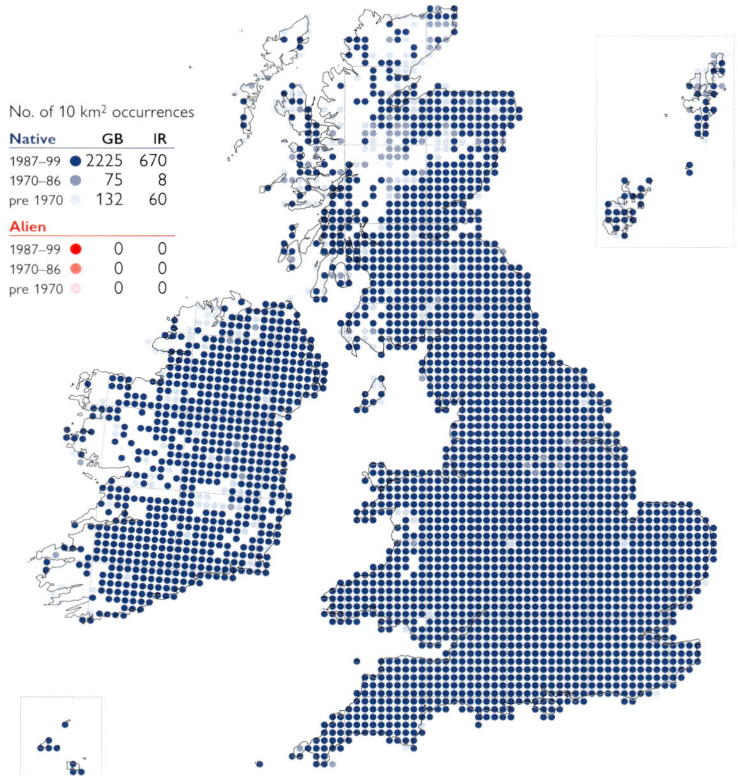

A tufted perennial herb of a wide range of grasslands, particularly those with moist, fertile soils. It also occurs on roadsides and woodland margins. It avoids waterlogged habitats, and is absent from light and dry soils, including arable fields. Generally lowland, but reaching 610 m on Ettrick Pen (Dumfriess.), and exceptionally at 845 m on Great Dun Fell (Westmorland).

Native (change +0.09). *A. pratensis* was frequently sown in seed mixtures up to 1950. Many local strains still occur, particularly in pastures that have not been re-sown recently. Its distribution is stable.

Eurosiberian Boreo-temperate element, but widely naturalised so that distribution is now Circumpolar Boreo-temperate.

References: Atlas (401d), Grime *et al.* (1988), Hultén & Fries (1986), Meusel *et al.* (1965).

D. A. PEARMAN

Alopecurus geniculatus Marsh Foxtail

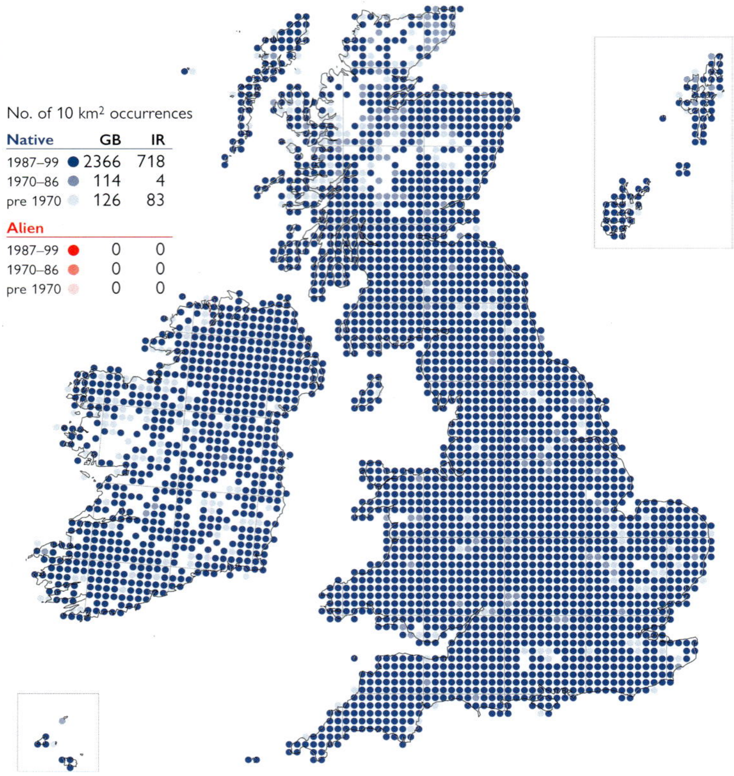

A perennial herb, rooting at the nodes, which is frequent in fertile sites which are flooded in winter such as muddy ditch-sides, wet arable fields, pond margins and grazing marshes. It readily colonises bare mud and disturbed areas, where it can become very lush, but avoids both the most acidic and the most alkaline sites. Generally lowland, but reaching 595 m on Highfield (S. Northumb.), and exceptionally at 845 m on Great Dun Fell (Westmorland).

Native (change +0.83). The range of *A. geniculatus* is probably stable, and has not significantly changed since the 1962 *Atlas*.

European Boreo-temperate element; widely naturalised outside its native range.

References: Atlas (402a), Grime *et al.* (1988), Hultén & Fries (1986).

D. A. PEARMAN

Alopecurus geniculatus × *A. pratensis* (*A.* × *brachystylus*)

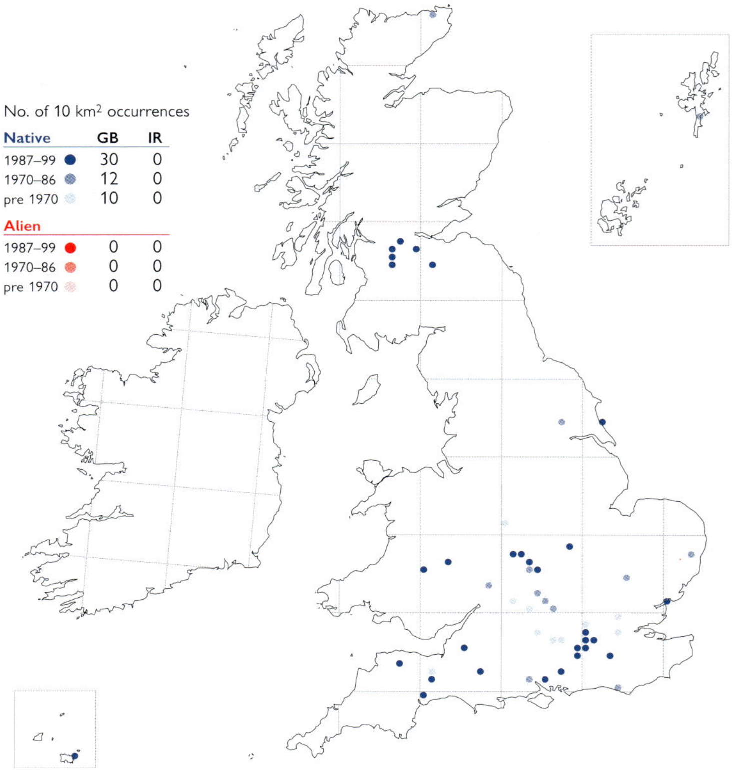

A perennial herb of wet meadows, pond margins and streamsides, usually where the parents grow together. It is almost invariably sterile. Its formation is facilitated by the fact that flowers of the parents are always markedly protogynous, and may therefore be cross-pollinated in the period before self-pollination is possible. Generally lowland, but reaching 425 m at Mole's Chamber (N. Devon).

Native. This hybrid is almost certainly under-recorded.

Widespread in boreal and temperate Europe.

Reference: Stace (1975).

D. A. PEARMAN

Alopecurus bulbosus Bulbous Foxtail

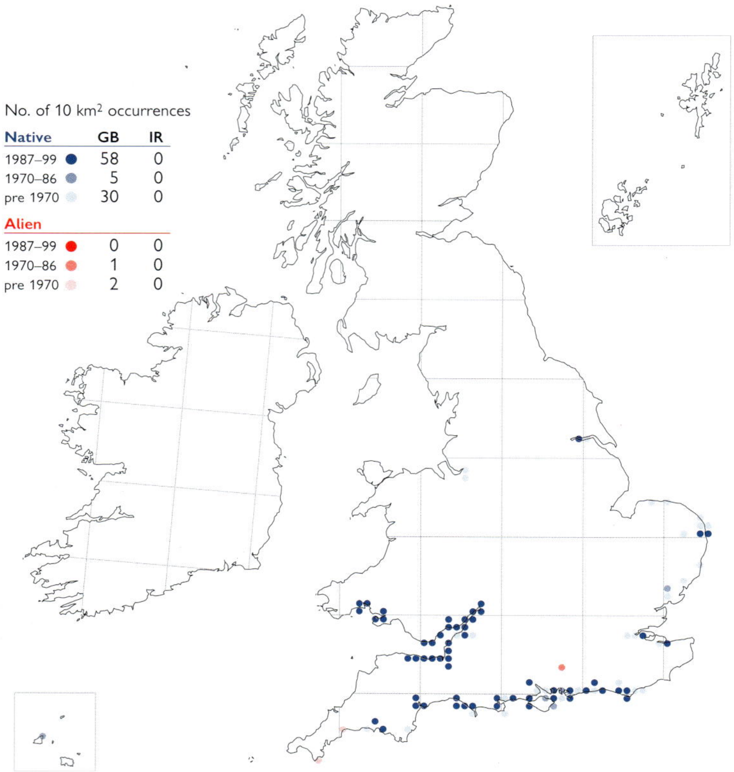

A perennial herb of periodically flooded brackish grassland in unimproved coastal grazing marshes, at the edges of ditches and in trampled ground at the base of sea walls; also locally in the uppermost parts of saltmarshes. Lowland.

Native (change +0.30). The 1962 *Atlas* suggested a marked decline of this species, but it was much under-recorded until the 1980s, and is now known to be still present (and sometimes abundant) at many sites from which it was thought to have been lost. Nevertheless, many populations have declined due to drainage and improvement of its habitat. Where salinity is reduced by sea-defence schemes it readily hybridises with *A. geniculatus* and eventually disappears, leaving the hybrid behind.

Suboceanic Southern-temperate element.

References: Atlas (402c), Stewart *et al.* (1994).

D. A. PEARMAN

Alopecurus aequalis Orange Foxtail

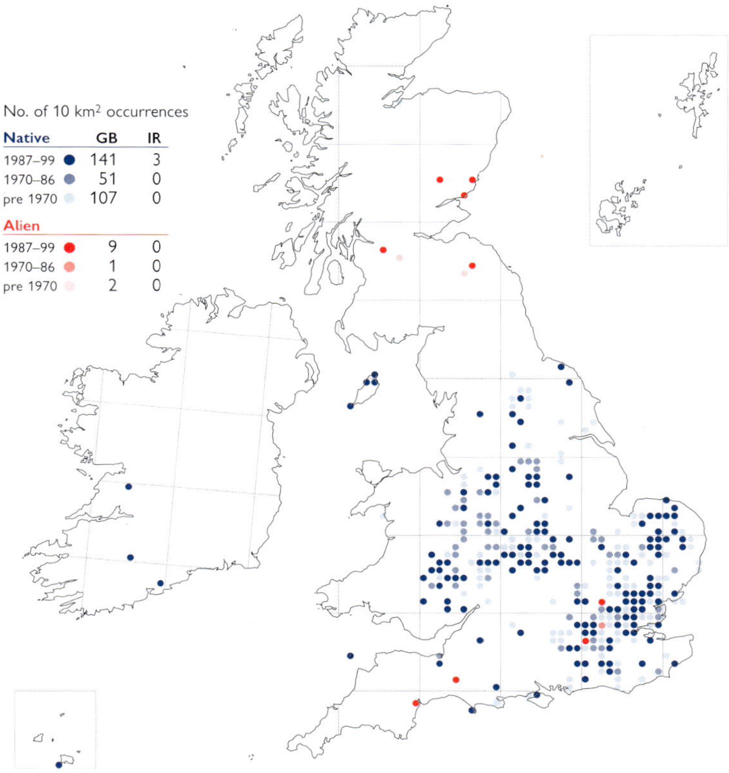

No. of 10 km² occurrences

Native	GB	IR
1987–99	141	3
1970–86	51	0
pre 1970	107	0
Alien		
1987–99	9	0
1970–86	1	0
pre 1970	2	0

An annual, most frequent on drying mud but found in a wide variety of habitats associated with freshwater, including the margins of ponds, ditches, reservoirs, turloughs and flooded gravel-pits. It has also recently been found as a weed in aquatic garden centres. Lowland.

Native (change –0.33). *A. aequalis* has been lost from many suitable sites, but seems readily to colonise new open habitats. Population sizes are very variable, and it may not appear when water levels remain high, possibly leading to under-recording. It may sometimes be overlooked as, or mistaken for, *A. geniculatus*. It was not discovered in Ireland until 1992.

Circumpolar Boreo-temperate element.

References: Atlas (402b), Goodwillie (1999b), Hultén & Fries (1986), Stewart *et al.* (1994).

D. A. PEARMAN

Alopecurus borealis Alpine Foxtail

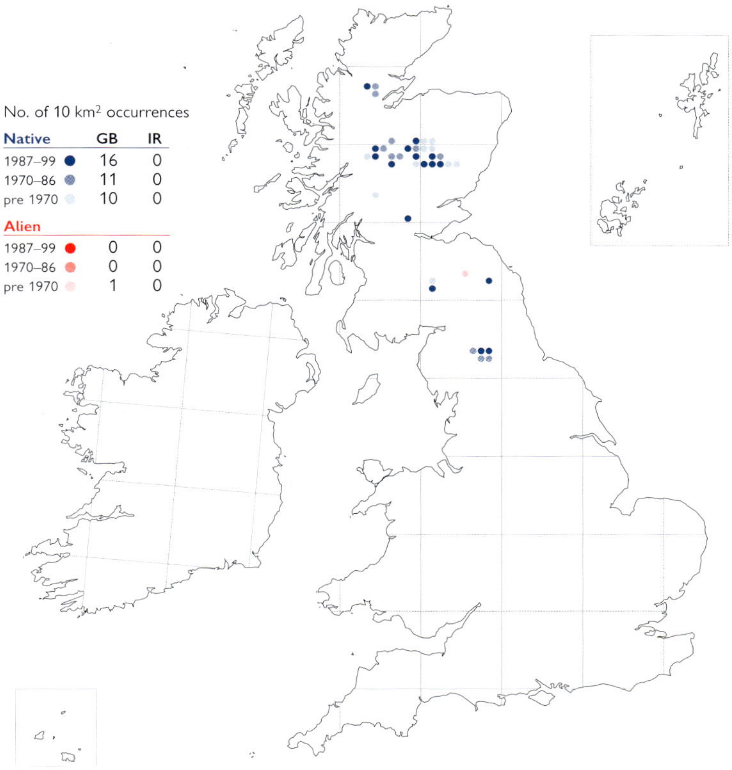

No. of 10 km² occurrences

Native	GB	IR
1987–99	16	0
1970–86	11	0
pre 1970	10	0
Alien		
1987–99	0	0
1970–86	0	0
pre 1970	1	0

A shortly rhizomatous perennial herb, found in oligotrophic springs and flushes, or on their borders, often associated with late snow-beds. It occurs on a wide range of acidic or slightly basic rocks. From 450 m on Widdybank Fell (Durham) to 1220 m on Braeriach (S. Aberdeen).

Native (change –0.24). This species has not only been recorded at new localities within its existing range since the 1962 *Atlas*, but has also been found in new areas of C. Scotland. It is an inconspicuous grass, especially as it is shy-flowering and any inflorescences produced are often grazed off.

Circumpolar Arctic-montane element; in Europe restricted to Britain, Svalbard, arctic Russia and the Urals.

References: Atlas (402d), Hultén & Fries (1986), Stewart *et al.* (1994).

D. A. PEARMAN

Alopecurus myosuroides Black-grass

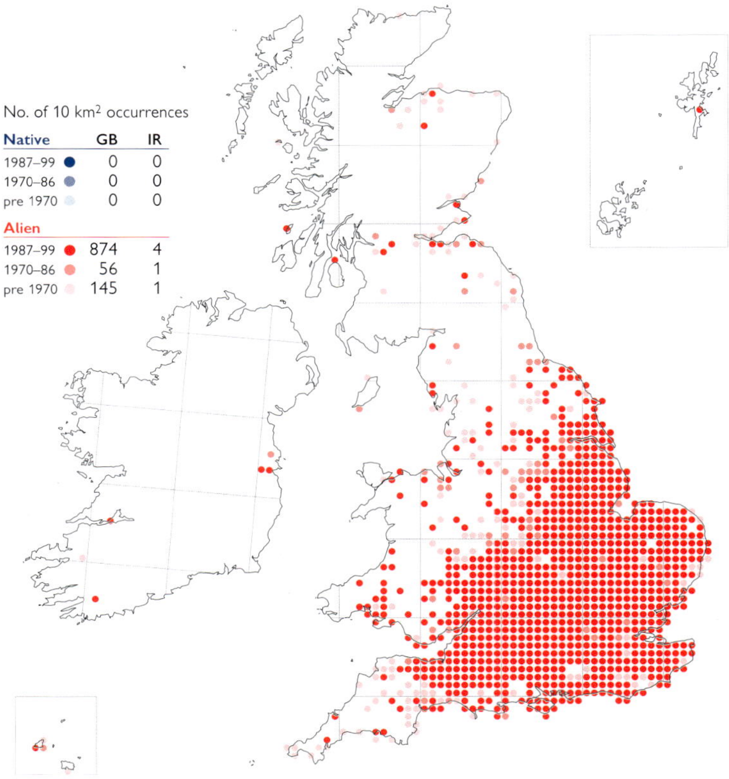

No. of 10 km² occurrences

Native	GB	IR
1987–99	0	0
1970–86	0	0
pre 1970	0	0
Alien		
1987–99	874	4
1970–86	56	1
pre 1970	145	1

An annual of rank and neglected grassland and arable land, rapidly increasing by seed to become a pest, particularly of cereal crops. It grows on both light and heavy soils. Lowland.

Archaeophyte (change +0.42). *A. myosuroides* has maintained, or slightly increased, its core distribution since the 1962 *Atlas*. It is still a frequent and troublesome weed of arable land. Some strains have evolved specific resistance to several commonly used herbicides, especially in E. Britain, and the species has benefited from the increased planting of winter cereal crops and the decline of stubble burning.

As an archaeophyte *A. myosuroides* has a European Southern-temperate distribution; it is widely naturalised outside this range.

References: Atlas (401c), Hultén & Fries (1986), Meusel *et al.* (1965), Naylor (1972).

D. A. PEARMAN

Phleum pratense sens. lat.

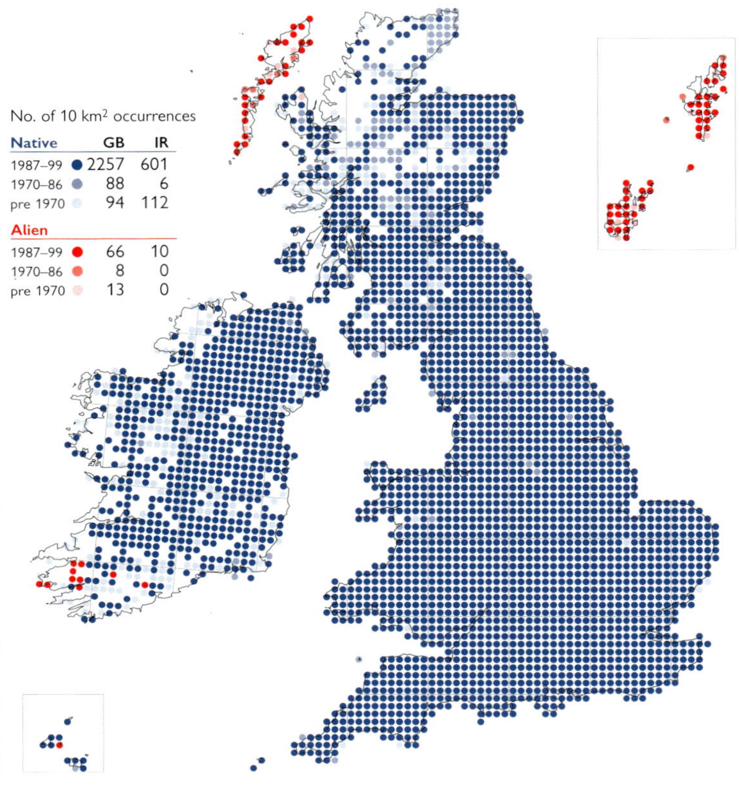

These grasses grow in a wide range of grassy habitats, such as meadows and pastures, rough grassland, field margins, verges and waste places. They are common in sown grasslands and frequent casuals from wool shoddy, bird-seed and other sources. 0–445 m (Clun Forest, Salop), and exceptionally at 845 m on Great Dun Fell (Westmorland).

Native (change –0.33). This aggregate comprises *P. pratense sens. str.* and *P. bertolonii.* There has been little change in its distribution since the 1962 *Atlas*. Much seed sown as leys comes from N. America. It has not been possible to distinguish native and introduced records, except in S.W. Ireland and the Northern Isles.

The distributions of the two component species are given in the following accounts.

References: Atlas (400c), Grime *et al.* (1988), Hultén & Fries (1986), Ryves *et al.* (1996).

S. J. LEACH

Phleum pratense Timothy

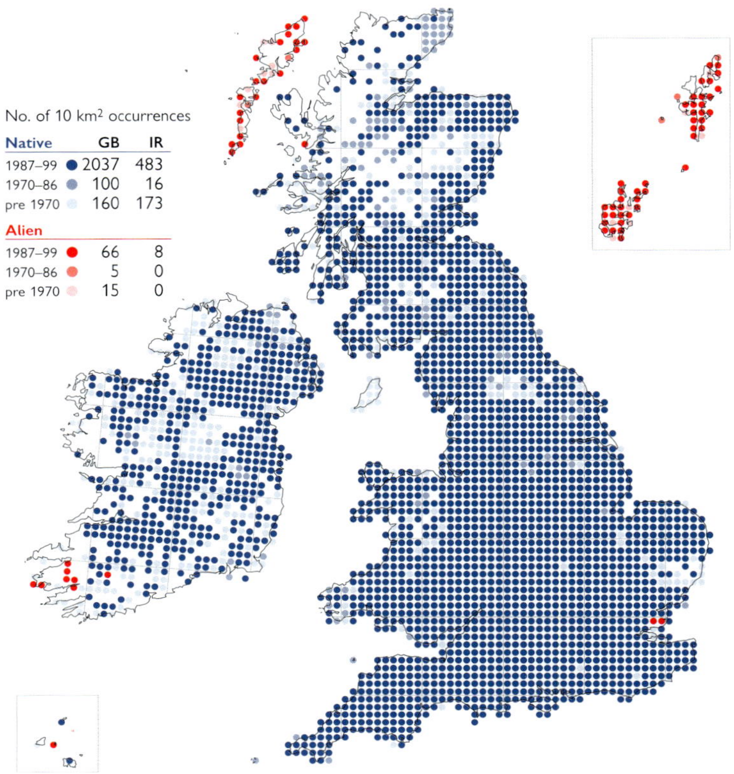

A perennial herb growing in a wide range of grassy habitats, including meadows, pastures, rough grassland, field margins and waysides; also sown in grasslands and as a casual from wool shoddy, bird-seed and other sources. *P. pratense sens. str.* tends to occur on rather heavier, damper soils than *P. bertolonii.* 0–445 m (Clun Forest, Salop).

Native. *P. pratense sens. str.* was not mapped in the 1962 *Atlas*, being included with *P. bertolonii* within the map of *P. pratense sens. lat.* These two taxa are still much confused. Native and alien records have seldom been distinguished.

Eurosiberian Temperate element, but widely naturalised so that distribution is now Circumpolar Temperate.

References: Grime *et al.* (1988), Ryves *et al.* (1996).

S. J. LEACH

Phleum bertolonii Smaller Cat's-tail

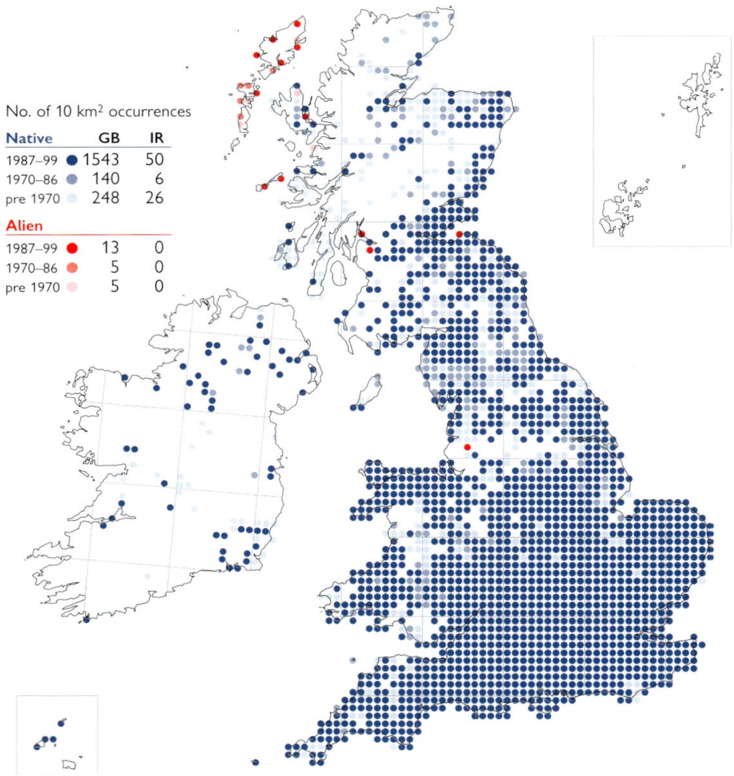

A perennial herb of old meadows and pastures, downs, roadside-banks and waste places; often with *P. pratense*, but showing a preference for slightly thinner swards and drier, less fertile soils. It occasionally occurs as a wool-alien, on shoddy fields and rubbish tips. Lowland.

Native. This taxon was not separately mapped in the 1962 *Atlas*. There is much confusion between *P. pratense* and *P. bertolonii*, and distributional trends cannot be assessed using existing data; robust agricultural strains of *P. bertolonii* are distinguishable from small *P. pratense* only by their differing chromosome number (Halliday, 1997). It is particularly under-recorded in Ireland.

European Southern-temperate element; widely naturalised outside its native range.

References: Grime *et al.* (1988), Hultén & Fries (1986), Ryves *et al.* (1996).

S. J. LEACH

Phleum alpinum Alpine Cat's-tail

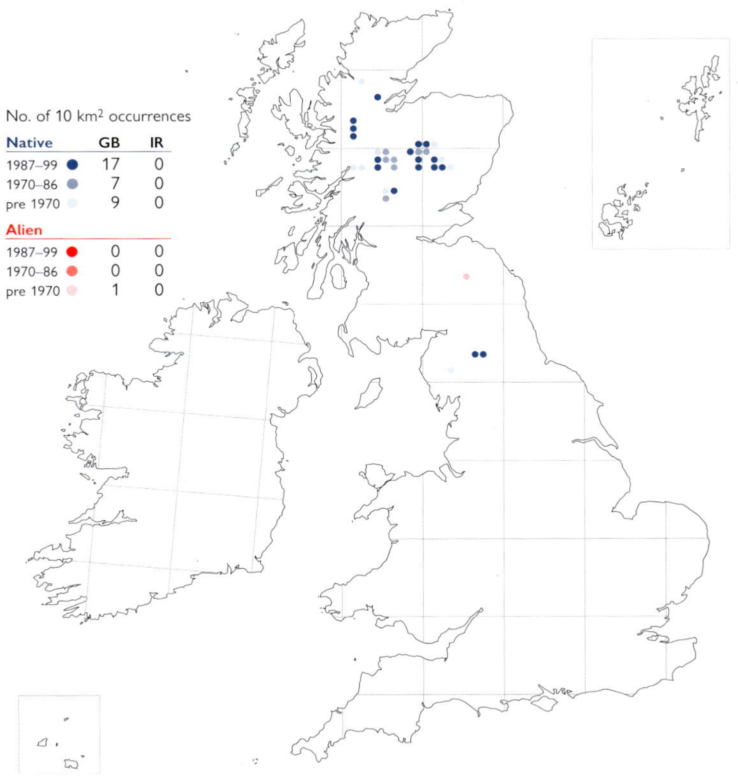

A loosely tufted perennial herb usually found over damp, calcareous or base-enriched substrates in the mountains. It occurs in a variety of habitats, including moist or dry cliff-faces, corrie rock ledges and wet grassy slopes. From 610 m on Braeriach (S. Aberdeen) to 1220 m on Cairntoul (S. Aberdeen).

Native (change –0.30). *P. alpinum* normally occurs in lightly grazed or ungrazed habitats, and heavy grazing may be partly responsible for its present restriction in some areas. However, there is no evidence of widespread decline, and it probably still persists undetected in some squares for which there are only pre-1987 records.

Circumpolar Boreo-arctic Montane element.

References: Atlas (400d), Hultén & Fries (1986), Meusel *et al.* (1965), Ryves *et al.* (1996), Stewart *et al.* (1994).

S. J. LEACH

Phleum phleoides Purple-stem Cat's-tail

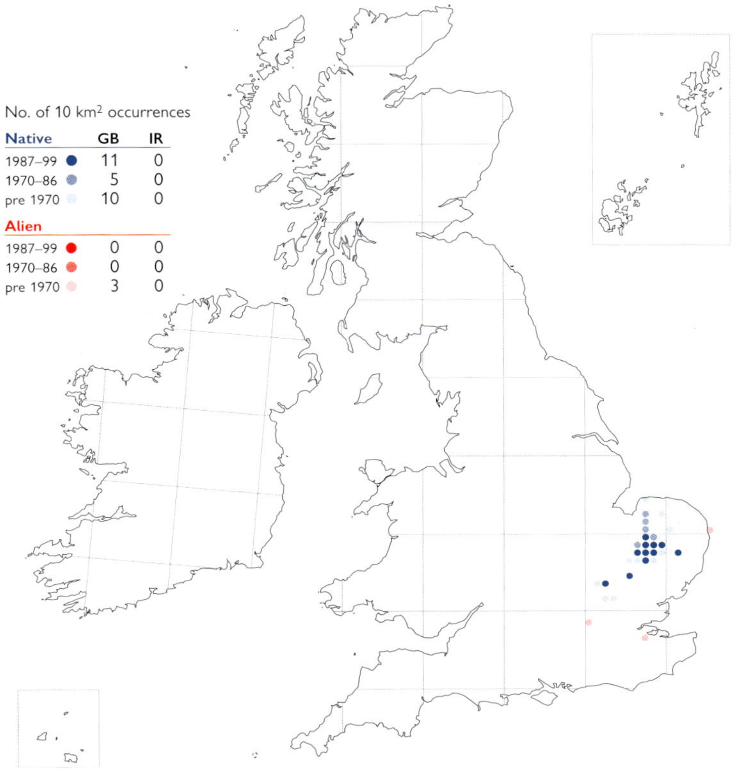

A perennial herb of open habitats on free-draining sandy or chalky soils, especially in Breckland where it occurs on grazed grass-heaths, road verges and track-side banks, and in the vicinity of pits, rabbit warrens and other disturbed places. Lowland.

Native (change –0.10). Most losses of *P. phleoides* occurred before 1930. Since then, its known range has been extended eastwards; its easternmost site, at Stuston Common (E. Suffolk), was first discovered in 1991. It is thriving in open habitats in Breckland.

Eurosiberian Temperate element, with a continental distribution in W. Europe.

References: Atlas (401a), Hultén & Fries (1986), Meusel *et al.* (1965), Wigginton (1999).

S. J. LEACH

Phleum arenarium Sand Cat's-tail

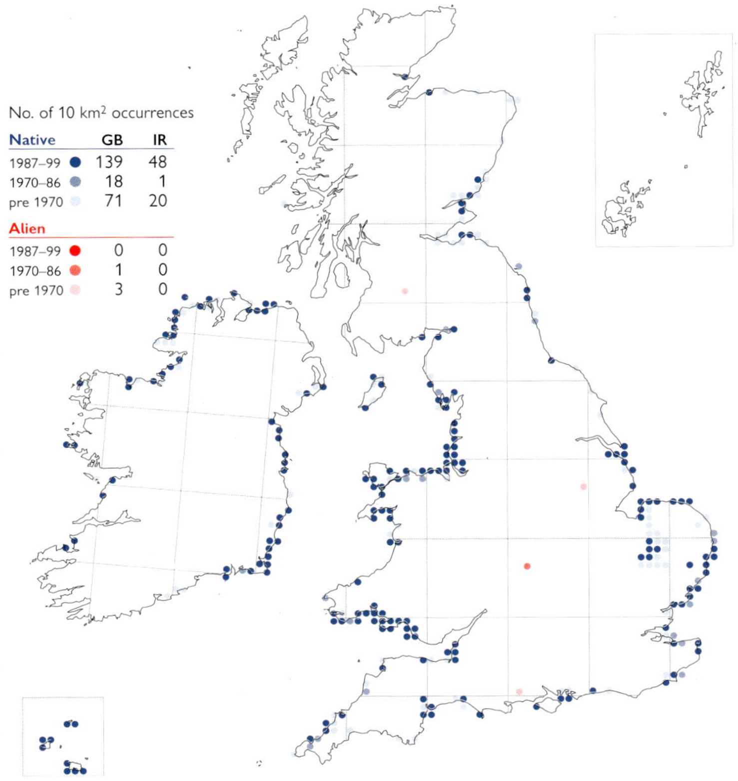

An annual of coastal sand dunes and sandy shingle, usually on mobile or semi-fixed *Ammophila* dunes and frequently associated with winter-annuals such as *Aira praecox*, *Myosotis ramosissima* and *Vulpia fasciculata*; also inland in Breckland, on open grass-heaths, wind-blown sand banks and other disturbed, open, sandy areas. It is a rare casual inland, sometimes arriving with imported sea-sand. Lowland.

Native (change –0.56). There has been little change in the distribution of *P. arenarium* since the 1962 *Atlas*, other than a continuing decline in the Breckland which may be caused by the decrease of mobile, sandy habitats.

European Southern-temperate element.

References: Atlas (401b), Ernst & Malloch (1994), Hultén & Fries (1986), Meusel *et al.* (1965).

S. J. LEACH

Bromus arvensis Field Brome

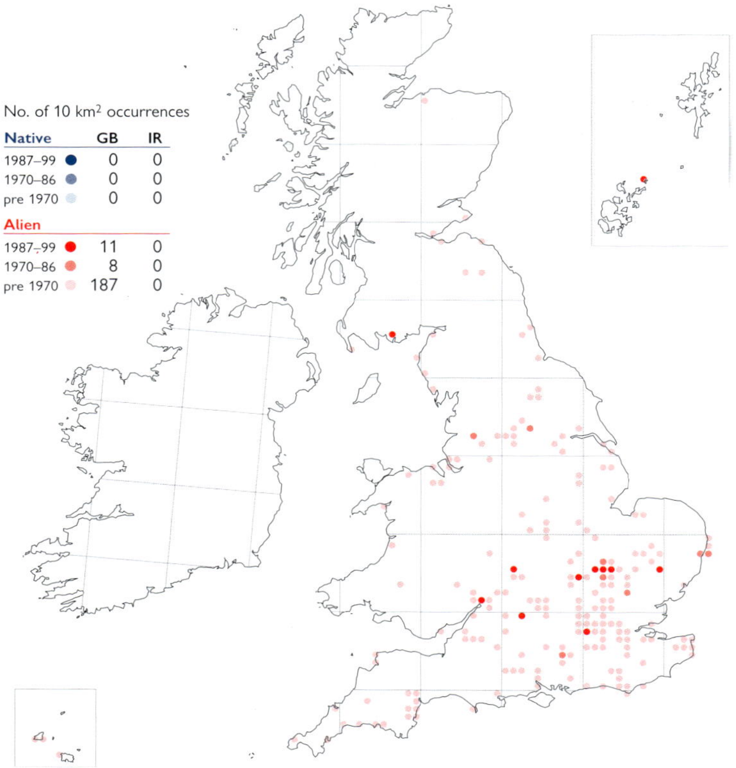

No. of 10 km² occurrences

		GB	IR
Native			
1987–99	●	0	0
1970–86	●	0	0
pre 1970	●	0	0
Alien			
1987–99	●	11	0
1970–86	●	8	0
pre 1970	●	187	0

This annual wool, grain and agricultural seed alien is found as casual or shortly persistent populations in arable and cultivated fields and on tracksides, docks and waste ground, mostly on nutrient-poor, light or sandy soils. Lowland.

Neophyte (change –3.15). *B. arvensis* may originally have been sown as a hay-crop, on its own or with other fodder plants. It was first recorded in 1763 and was formerly frequent, but underwent a dramatic decline during the 20th century, probably due to improved seed cleaning methods, increased use of herbicide and the decline in wool shoddy. Most pre-1970 squares refer to records made before 1930.

Native range uncertain; *B. arvensis* has spread with cultivation to attain a European Temperate distribution.

References: Atlas (388a), Hultén & Fries (1986), Ryves *et al.* (1996).

S. J. LEACH

Bromus commutatus Meadow Brome

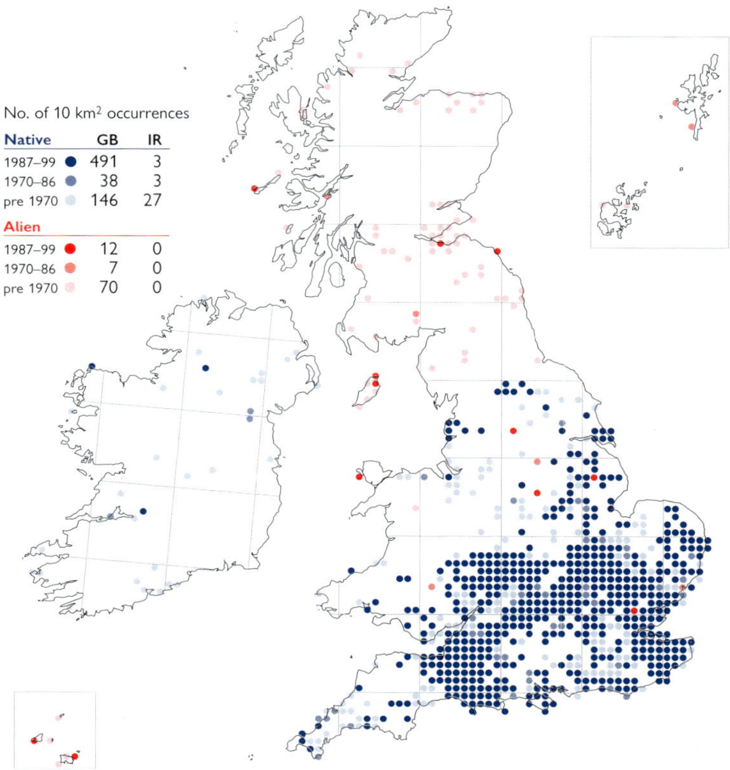

No. of 10 km² occurrences

		GB	IR
Native			
1987–99	●	491	3
1970–86	●	38	3
pre 1970	●	146	27
Alien			
1987–99	●	12	0
1970–86	●	7	0
pre 1970	●	70	0

An annual of unimproved damp meadows, also found on waysides, road verges and the borders of fields and tracks. Many recent records are of casual occurrences arising from grass-seed impurities. Lowland.

Native (change +1.07). *B. commutatus* is much better recorded than it was for the 1962 *Atlas*, and it is now possible to see a coherent distribution. It has declined since 1960 because of agricultural improvements to its meadow habitats. It is very closely allied to *B. racemosus* and easily confused with it; indeed, Stace (1997) suggests it might be better treated as a subspecies of that species.

European Temperate element; widely naturalised outside its native range.

References: Atlas (387c), Hultén & Fries (1986).

D. A. PEARMAN

Bromus racemosus Smooth Brome

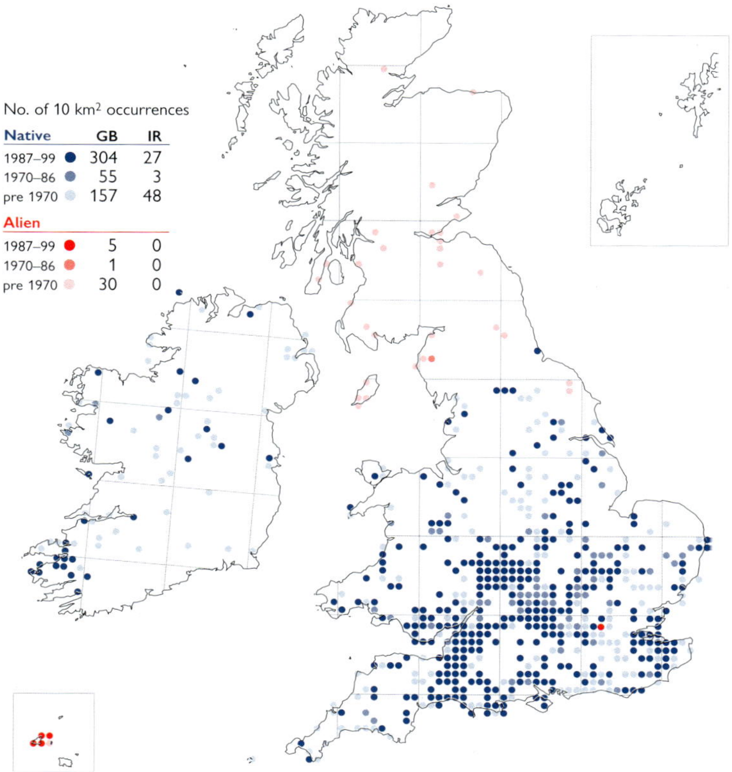

No. of 10 km² occurrences

		GB	IR
Native			
1987–99	●	304	27
1970–86	●	55	3
pre 1970	●	157	48
Alien			
1987–99	●	5	0
1970–86	●	1	0
pre 1970	●	30	0

An annual of unimproved hay- and water-meadows, usually on damp, periodically flooded alluvial soils. It is most frequent on the drier margins of fields, sometimes growing on the dredgings from the ditches bordering them. It is also found as a grass-seed casual in arable margins and on verges. Generally lowland, with an old record of 365 m in the Scottish Highlands.

Native (change +0.74). *B. racemosus* is now much better recorded than it was in the 1962 *Atlas*. It has declined everywhere since 1930 through draining and improvement of its habitats, though it may occasionally have been mistaken for the closely related and perhaps conspecific *B. commutatus*.

European Temperate element; widely naturalised outside its native range.

References: Atlas (387b), Curtis & McGough (1988), Hultén & Fries (1986).

D. A. PEARMAN

Bromus hordeaceus Soft-brome

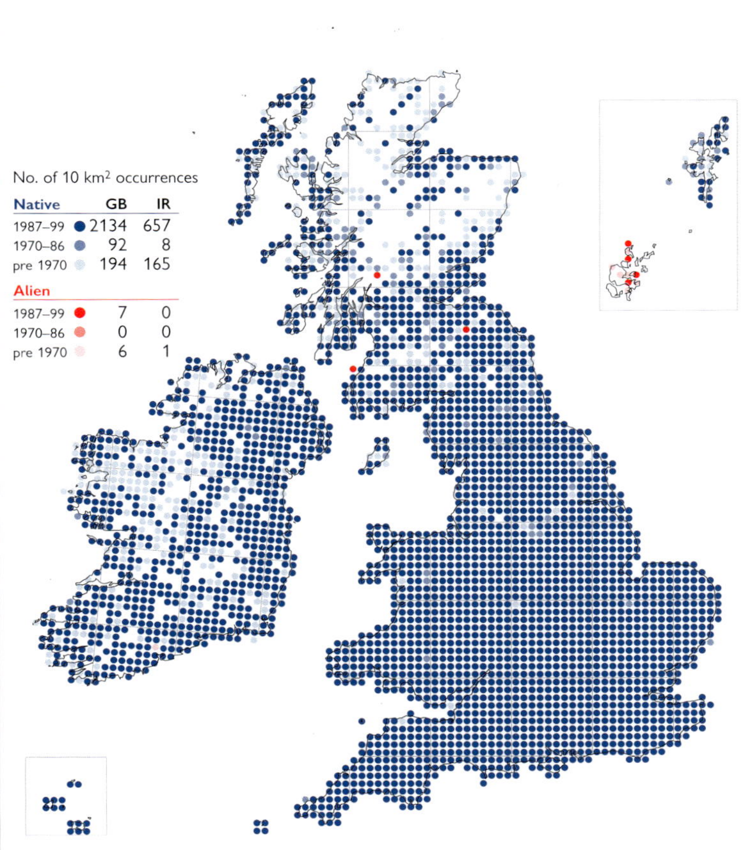

A winter-annual of moderately fertile neutral soils, particularly favouring disturbed or open habitats. It occurs on coastal cliffs, in pastures and hay meadows; also introduced as an impurity in grass-seed to arable fields, tracksides and waste ground. It does not form a persistent seed bank, and is intolerant of heavy grazing or frequent mowing. It tends to avoid wet or very acidic sites. 0–550 m (Kilhope, Co. Durham).

Native (change –0.37). The distribution of this variable species is unchanged. Subsp. *hordeaceus* occurs throughout the range of the species, subsp. *divaricatus* is a rare casual, and the other two subspecies are mapped separately.

European Southern-temperate element; widely naturalised outside its native range.

References: Atlas (386d), Grime *et al.* (1988), Hultén & Fries (1986), Smith (1968).

D. A. PEARMAN

Bromus hordeaceus subsp. *ferronii*

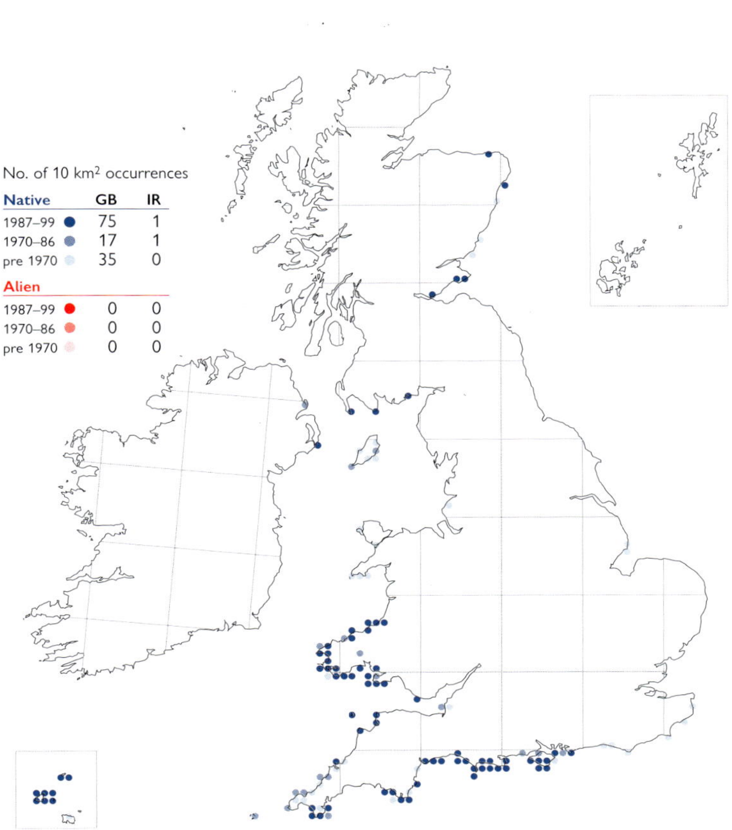

A winter-annual of bare areas on exposed coastal cliff-slopes, shingle beaches and banks, and occasionally sand dunes. It flowers and sets seed in spring, before its habitats become severely droughted. Lowland.

Native. The distribution of subsp. *ferronii* is stable. It might occasionally be over-recorded for dwarf forms of *B. hordeaceus* subsp. *hordeaceus*, as in Cornwall where some of the pre-1970 records are thought to refer to the latter subspecies.

Oceanic Temperate element; restricted to the Atlantic coast of N.W. Europe.

References: Atlas Supp. (151b), Smith (1968).

D. A. PEARMAN

Bromus hordeaceus subsp. *thominei*

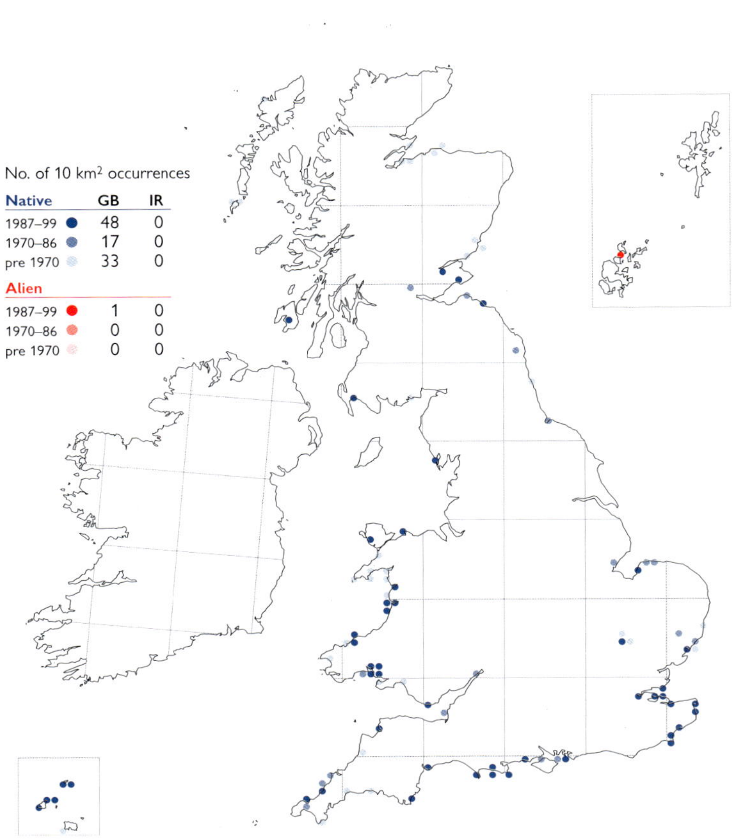

A winter-annual of coastal fixed dunes, grassy cliff-tops and slopes, and other sandy places by the sea. Inland, it is recorded from sandy soils and ruderal habitats. Lowland.

Native. This taxon was redefined by Smith (1968); the map in Perring & Sell (1968) represents the earlier, broader circumscription. Only confirmed inland records have been accepted; the remainder have been treated as probable errors for *B.* × *pseudothominei*.

Suboceanic Temperate element.

D. A. PEARMAN

Bromus hordeaceus × *B. lepidus* (*B.* × *pseudothominei*) Lesser Soft-brome

Native	GB	IR
1987–99 ●	0	0
1970–86 ●	0	0
pre 1970 ●	0	0
Alien		
1987–99 ●	340	2
1970–86 ●	112	2
pre 1970 ●	93	7

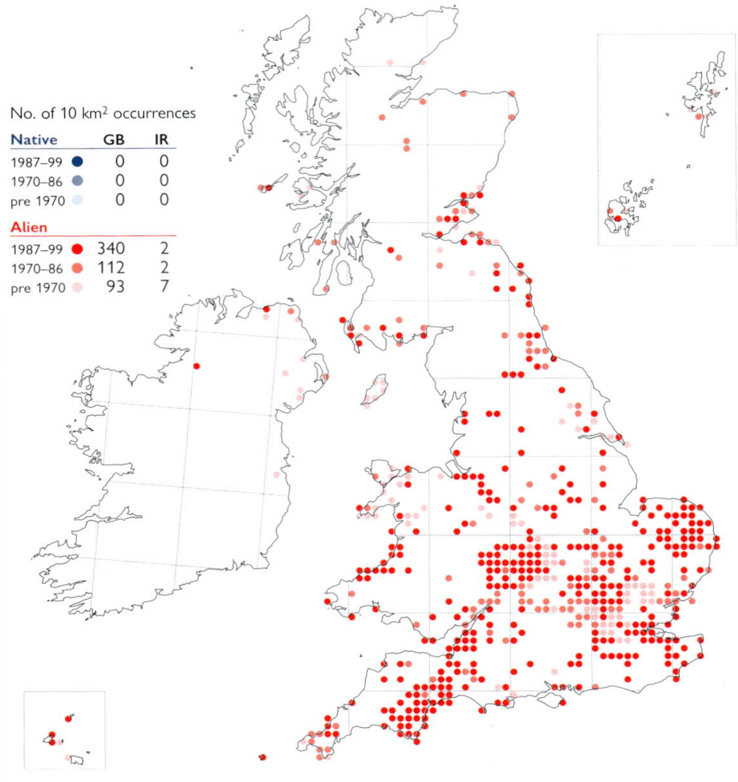

An annual which is most frequent in sown grassland, but is also recorded from a wide variety of ruderal habitats. It is fully fertile, and often occurs in the absence of both its putative parents. Lowland.

Neophyte. *B.* × *pseudothominei* was described by Smith (1968) as a hybrid of *B. hordeaceus* subsp. *hordeaceus* and *B. lepidus*. It was previously confused with *B. hordeaceus* subsp. *thominei* and was mapped with that taxon by Perring & Sell (1968). It is almost certainly under-recorded.

Widespread in Europe, but with a poorly documented distribution.

Reference: Stace (1975).

D. A. PEARMAN

Bromus lepidus Slender Soft-brome

Native	GB	IR
1987–99 ●	0	0
1970–86 ●	0	0
pre 1970 ●	0	0
Alien		
1987–99 ●	158	6
1970–86 ●	116	9
pre 1970 ●	502	60

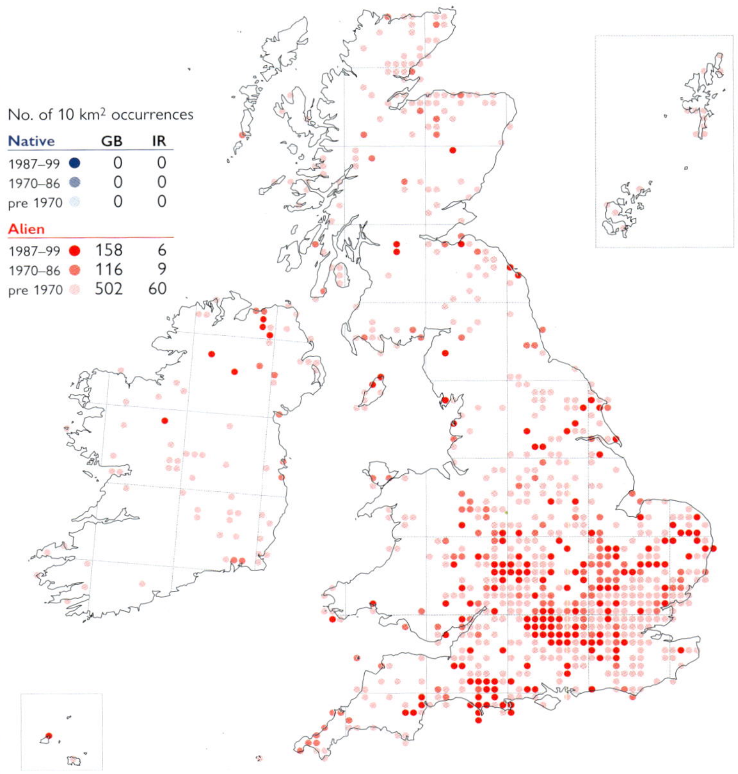

An annual of improved grasslands, sown with *Lolium multiflorum* and *L. perenne*, and a frequent contaminant of other grass-seed mixtures. It also occurs in arable fields and waste places. Lowland.

Neophyte (change –3.91). *B. lepidus* was first recorded from the wild in 1836, not long after the first major imports of seeds for grass leys, and it seems to be an introduced species. The reason for the paucity of modern records is uncertain; it may be less common, or recent recorders may have failed to distinguish it from *B. hordeaceus*, or it may have been over-recorded in the past.

Native range unknown; the species is recorded as an apparent introduction in N.W. & N.C. Europe and S. Scandinavia. It may have evolved in its man-made habitat.

References: Atlas (387a), Hultén & Fries (1986), Smith (1968).

D. A. PEARMAN

Bromus interruptus Interrupted Brome

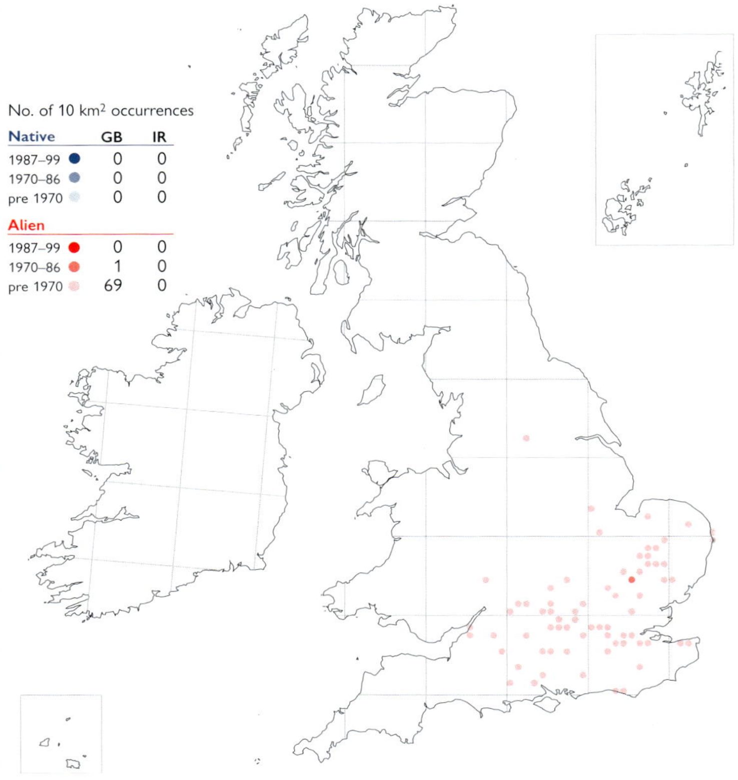

An annual of light soils in cultivated fields, especially in crops of *Onobrychis*, *Lolium* or *Trifolium*, and on waysides. Lowland.

Neophyte (change –1.73). *B. interruptus* was first collected in 1849 at Odsey (Cambridgeshire or Hertfordshire) and was last seen in 1972, beside a farm track at Pampisford, Cambridgeshire. Though extinct in the wild, it is retained in cultivation in botanic gardens.

B. interruptus is not known outside Britain, although it did appear as a casual in Holland, possibly imported with English fodder. Its history, habitat and scattered distribution in Britain suggest that it may have been introduced as a seed contaminant from an unknown native range rather than having evolved here as a neo-endemic.

References: Atlas (387d), Donald (1980), Marren (1999).

D. A. PEARMAN

Bromus secalinus Rye Brome

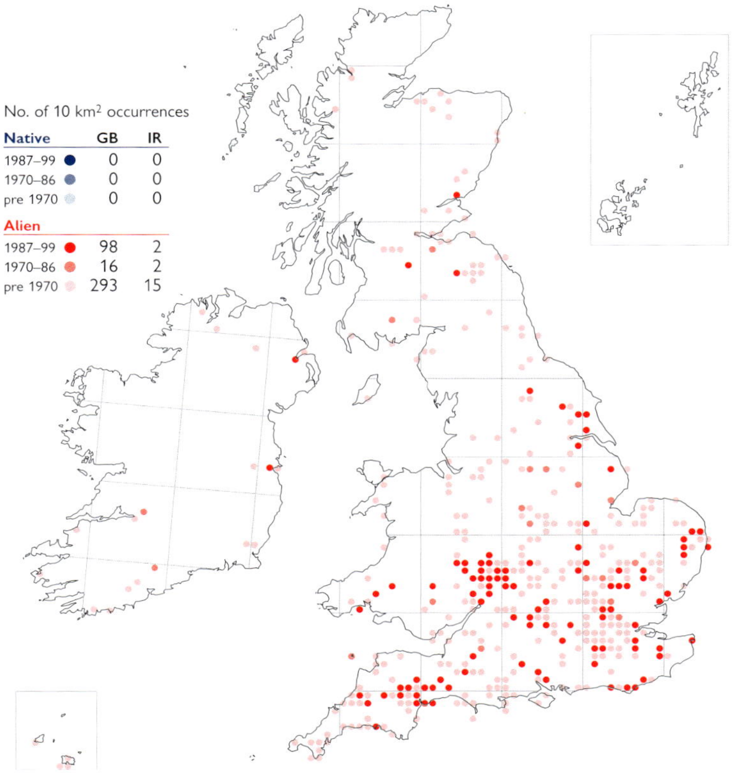

No. of 10 km² occurrences

Native	GB	IR
1987–99	0	0
1970–86	0	0
pre 1970	0	0
Alien		
1987–99	98	2
1970–86	16	2
pre 1970	293	15

An annual or biennial of cereal fields, which is also found as a casual on waste ground, and occasionally in improved leys. Lowland.

Archaeophyte (change −1.15). *B. secalinus* has probably been present in Britain since prehistoric times, initially as an arable weed or perhaps as an alternative source of grain when the main crop failed. It was frequent in the 19th and early 20th centuries, but has since undergone a dramatic decline. However, it appears to be making a comeback in a few areas, such as Norfolk and Worcestershire. *B. secalinus* is sometimes confused with *B. pseudosecalinus*.

B. secalinus evolved in cultivation or spread in prehistory as a weed, perhaps a mimic of rye, from an unknown native range.

References: Atlas (388b), Beckett *et al.* (1999), Hultén & Fries (1986).

S. J. LEACH & D. A. PEARMAN

Bromopsis ramosa Hairy-brome

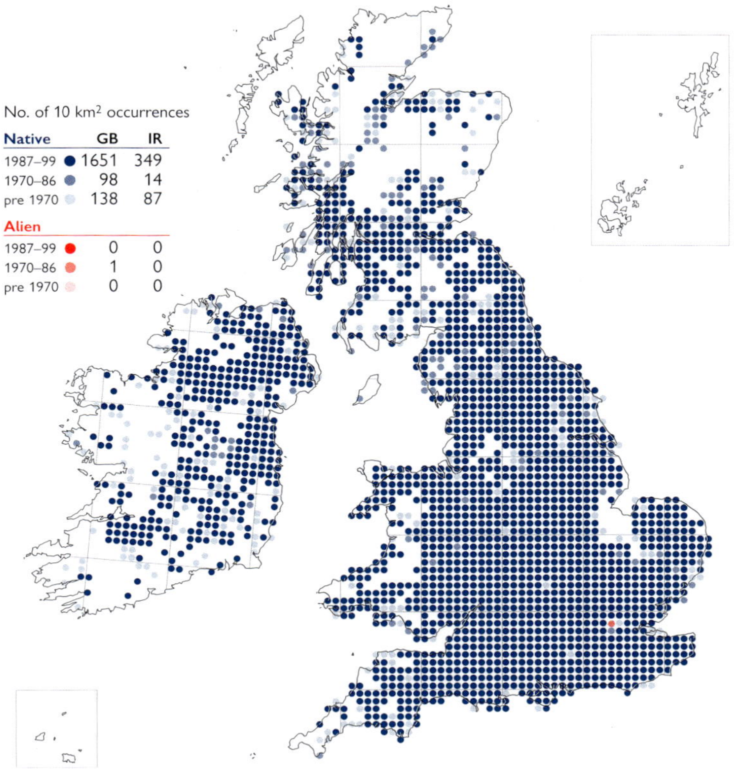

No. of 10 km² occurrences

Native	GB	IR
1987–99	1651	349
1970–86	98	14
pre 1970	138	87
Alien		
1987–99	0	0
1970–86	1	0
pre 1970	0	0

A tufted perennial herb of shaded habitats on moist, moderately base-rich soils, including woodlands and hedgerows; it occasionally persists on sites of former woodland. Some bare soil is necessary for successful establishment from seed. Almost entirely lowland, but reaching 420 m at Garrigill (Cumberland).

Native (change −0.18). The distribution of *B. ramosa* is unchanged since the 1962 *Atlas*.

European Temperate element.

References: Atlas (384d), Grime *et al.* (1988), Hultén & Fries (1986), Meusel *et al.* (1965).

D. A. PEARMAN

Bromopsis benekenii Lesser Hairy-brome

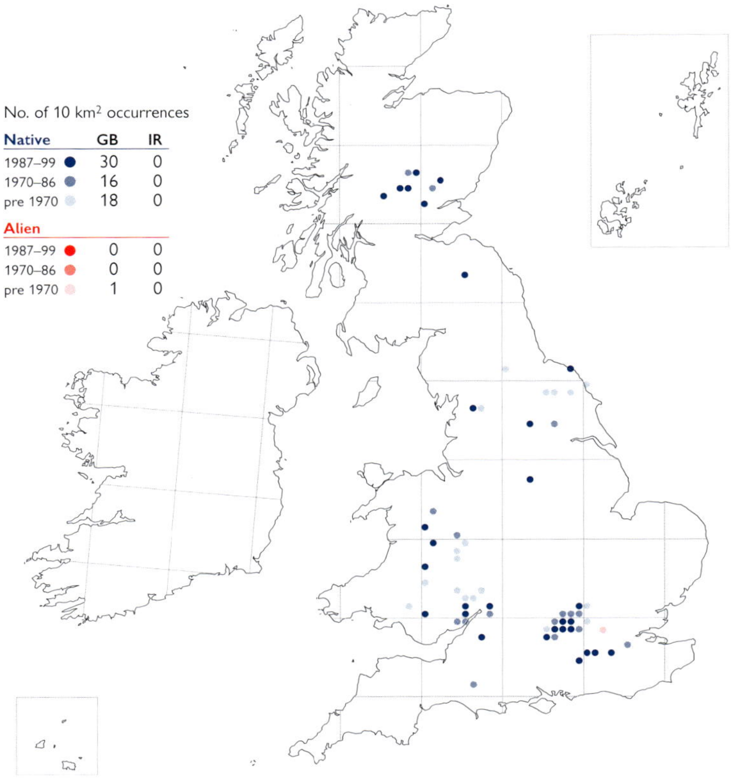

No. of 10 km² occurrences

Native	GB	IR
1987–99	30	0
1970–86	16	0
pre 1970	18	0
Alien		
1987–99	0	0
1970–86	0	0
pre 1970	1	0

A tufted perennial herb of lightly shaded places in woodland, especially beech woods; also found in scrub and hedgerows. It grows mainly in humus-rich but shallow calcareous soils, often on a slight slope. Lowland.

Native (change +0.25). *B. benekenii* is now much better recorded than it was for the 1962 *Atlas*, although it is almost certainly still under-recorded. It may be a subspecies of the very similar *B. ramosa*, with which it often grows.

European Temperate element, with a continental distribution in W. Europe; also in C. Asia.

References: Atlas (385a), Hultén & Fries (1986), Meusel *et al.* (1965), Stewart *et al.* (1994).

D. A. PEARMAN

Bromopsis erecta Upright Brome

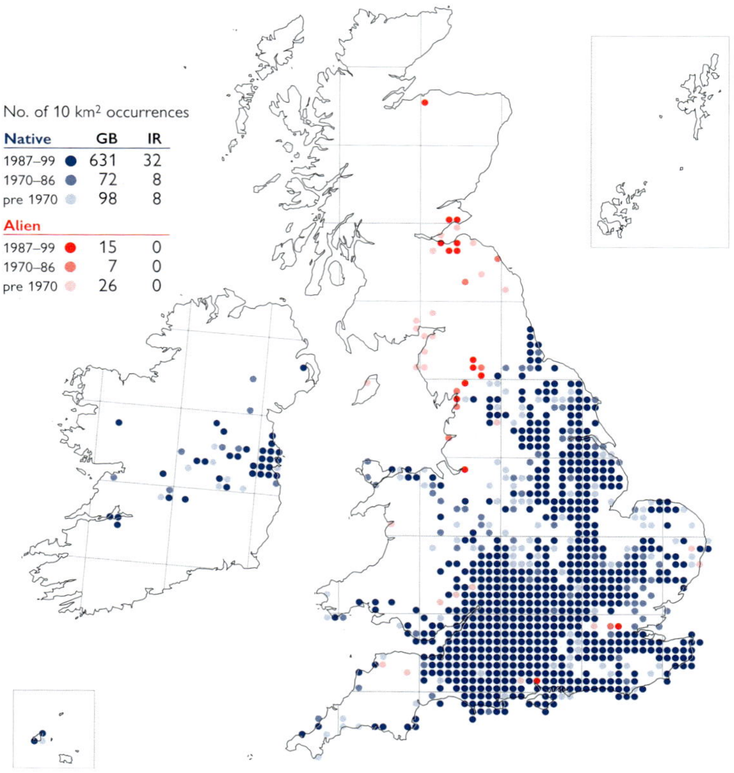

A tufted winter-green perennial herb of dry, relatively infertile calcareous soils, growing in ungrazed or undergrazed chalk and limestone grasslands, where it often forms dense stands with *Brachypodium pinnatum*. It also occurs on calcareous sand dunes, roadside banks, quarry spoil and occasionally waste ground, but avoids wet or arable sites. It spreads by seed. Lowland.

Native (change –0.01). *B. erecta* has increased in grasslands where grazing pressures have relaxed, especially after the reduction of the rabbit population through myxomatosis in the 1950s.

European Temperate element; widely naturalised outside its native range.

References: Atlas (384c), Grime *et al.* (1988), Hultén & Fries (1986), Meusel *et al.* (1965).

D. A. PEARMAN

Bromopsis inermis Hungarian Brome

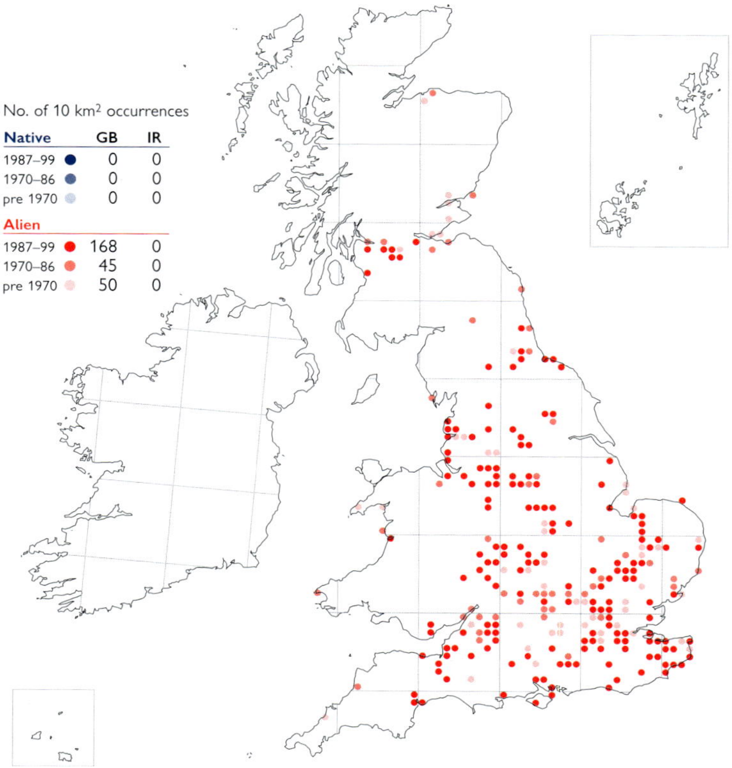

A rhizomatous perennial herb of rough grassy places, road verges and field margins, which is resistant to drought and persists on sandy, well-drained soils. Elsewhere it occurs as a casual. Lowland.

Neophyte (change +1.71). *B. inermis*, which was first cultivated in Britain in 1794, was formerly sown as a fodder grass, but it is now introduced only as a seed contaminant. It has been known in the wild since 1890, and is spreading quickly in some areas on verges and waste ground.

B. inermis subsp. *inermis* is a Eurosiberian Temperate species; subsp. *pumpelliana* is native to N. America.

References: Atlas (385b), Hultén & Fries (1986), Meusel *et al.* (1965), Ryves *et al.* (1996).

D. A. PEARMAN

Anisantha diandra Great Brome

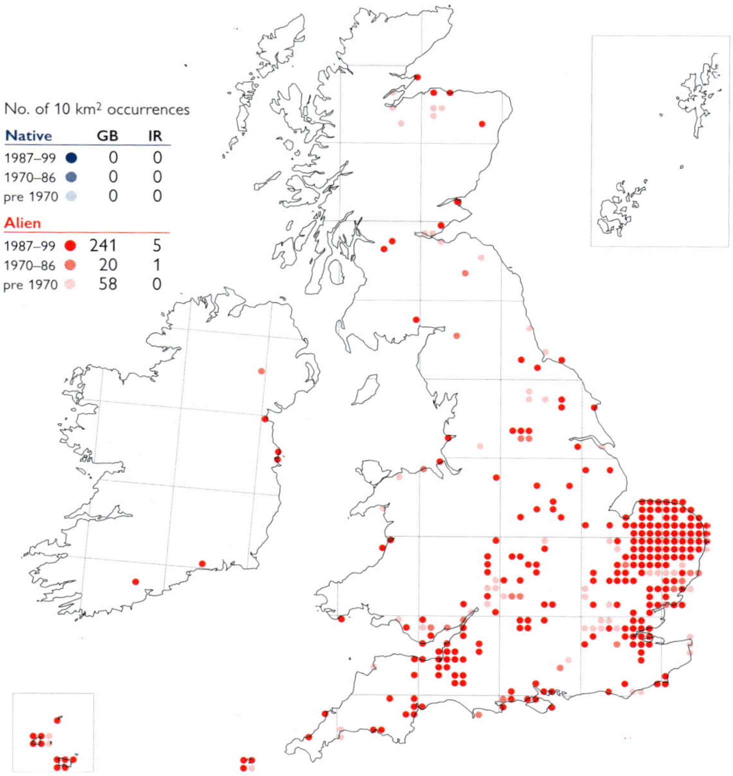

An annual of arable fields, waste ground and roadsides, and in open grassland and heathland on sandy soils. It is sometimes well-established on dunes. *A. diandra* is a grain, birdseed and wool alien. Lowland.

Neophyte (change +1.50). *A. diandra* was introduced to Britain, apparently from Morocco, in 1804 and was first recorded from the wild in 1835 (Fife). In the 1962 *Atlas* it was treated as possibly native in the Channel Islands, though native status there is now considered most unlikely. It has increased in S. England since 1962, and has considerably consolidated its distribution in East Anglia.

Native of the Mediterranean region and S.W. Asia; naturalised in W. Europe, N. & S. America, S. Africa and elsewhere.

References: Atlas (386a), Ryves *et al.* (1996).

S. J. LEACH

Anisantha rigida Ripgut Brome

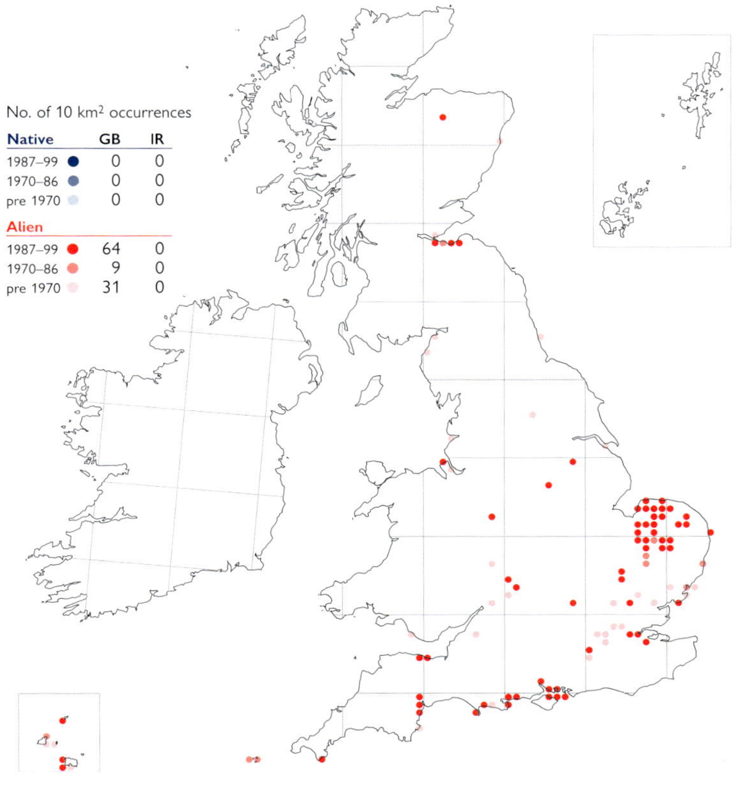

An annual of waysides, open grassland and disturbed or cultivated ground on light soils, usually not persisting but well-established on sand dunes and other sandy places near the sea in S. England and the Channel Islands. Lowland.

Neophyte (change +1.13). *A. rigida* is a wool, grain and agricultural seed alien, and was first recorded in the wild in 1834. It has previously been treated as native in the Channel Islands, although it is more likely to have been introduced there. It has increased since the 1962 *Atlas*, although it may have been confused in some areas with *A. diandra*.

Native of the Mediterranean region; closely related to and perhaps only subspecifically distinct from *A. diandra*.

References: Atlas (386b), Ryves *et al.* (1996).

S. J. LEACH

Anisantha sterilis Barren Brome

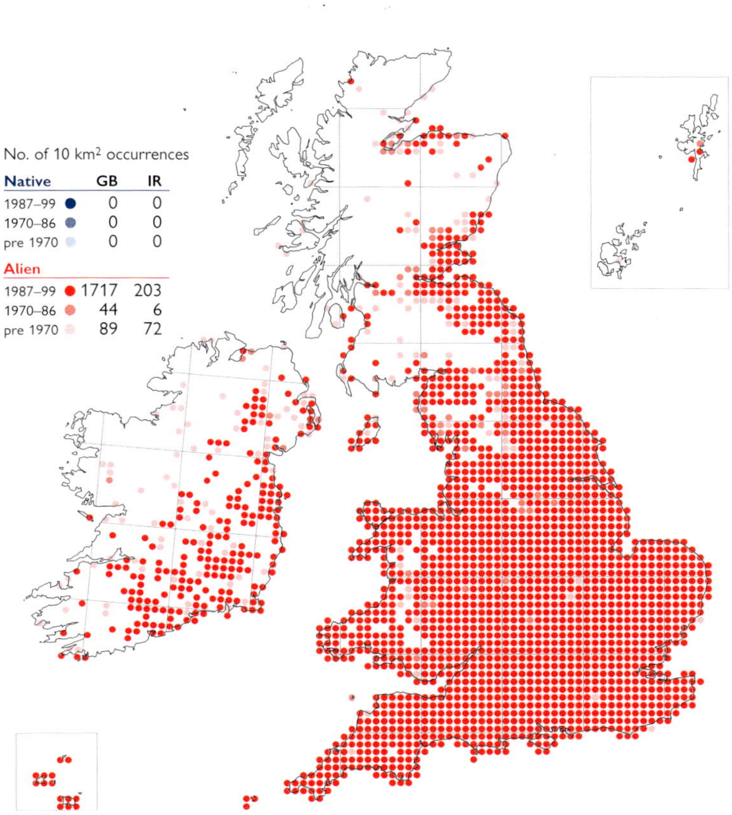

An annual of roadsides, railway banks, open grassland, gardens and waste ground. It can be a serious weed in fields where winter cereals are grown repeatedly using minimum cultivation techniques, especially in S. Britain (Grime *et al.*, 1988). *A. sterilis* has frequently been introduced with wool shoddy. Generally lowland, but reaching 365 m in Derbyshire.

Archaeophyte (change +0.05). The distribution of *A. sterilis* is stable.

As an archaeophyte *A. sterilis* has a European Southern-temperate distribution; it is widely naturalised outside this range.

References: Atlas (385c), Gray (1981), Hultén & Fries (1986).

S. J. LEACH

Anisantha tectorum Drooping Brome

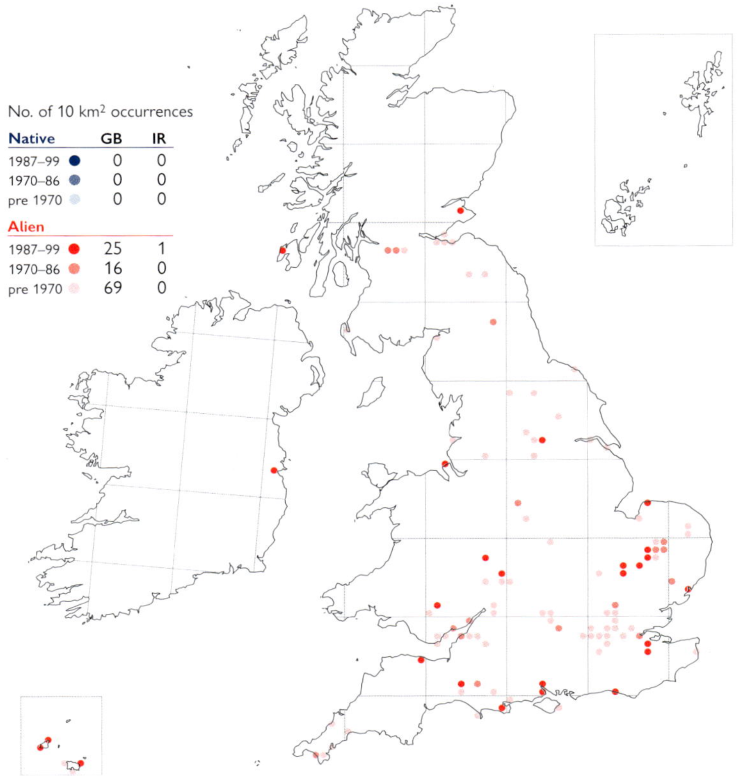

An annual of waste ground, roadsides and grassy places on sandy soils. It is a casual from grain, wool shoddy and grass-seed, but can be naturalised on sandy banks, field margins and grass heaths in Breckland. Many recent records are from newly sown road verges. Lowland.

Neophyte (change –0.17). *A. tectorum* was first grown in Britain in 1776 and has been known in the wild since 1863. It appears to have declined in Breckland, where it was formerly abundant between Brandon and Thetford but has not been seen there since the early 1980s (Beckett *et al.*, 1999).

A Eurosiberian Southern-temperate species, extending as a native to N. France and S. Scandinavia; naturalised in N. America, Australasia and elsewhere.

References: Atlas (386c), Hultén & Fries (1986), Meusel *et al.* (1965), Ryves *et al.* (1996), Upadhyaya *et al.* (1986).

S. J. LEACH

Anisantha madritensis Compact Brome

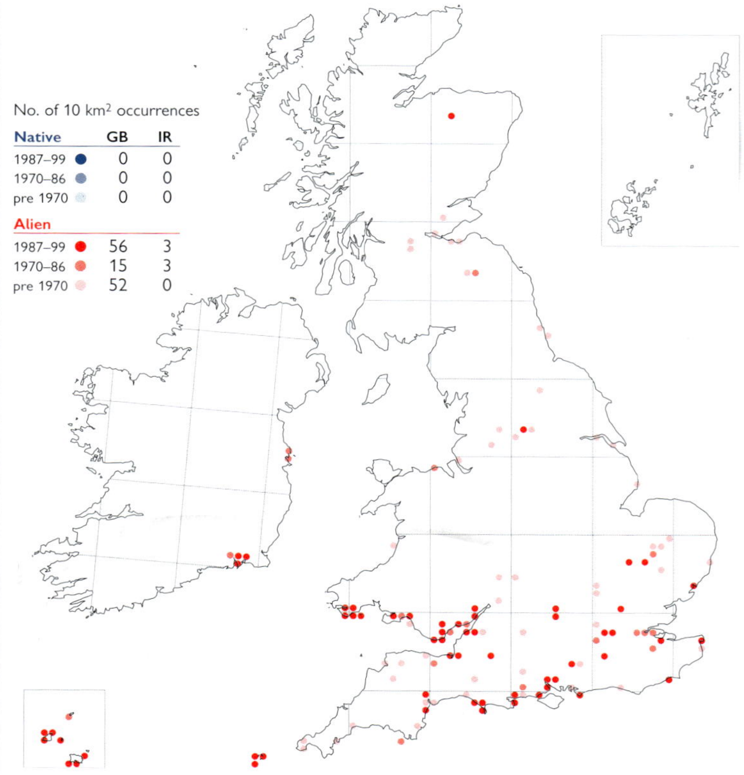

An annual of waste land, cultivated ground, roadside banks, walls and ruins; also around docks and on tips, on sand dunes and in other sandy or rocky places near the sea. *A. madritensis* is long-established (and considered by some to be possibly native) at a few sites in S.W. Britain and the Channel Islands. Otherwise, it is widespread as a casual, especially from wool shoddy. Lowland.

Neophyte (change +0.58). *A. madritensis*, known from the wild in Britain since 1716, has changed little in distribution since the 1962 *Atlas*.

Native of the Mediterranean region & S.W. Asia; naturalised in N. & S. America, Australia and elsewhere.

References: Atlas (385d), Wigginton (1999).

S. J. LEACH

Ceratochloa carinata California Brome

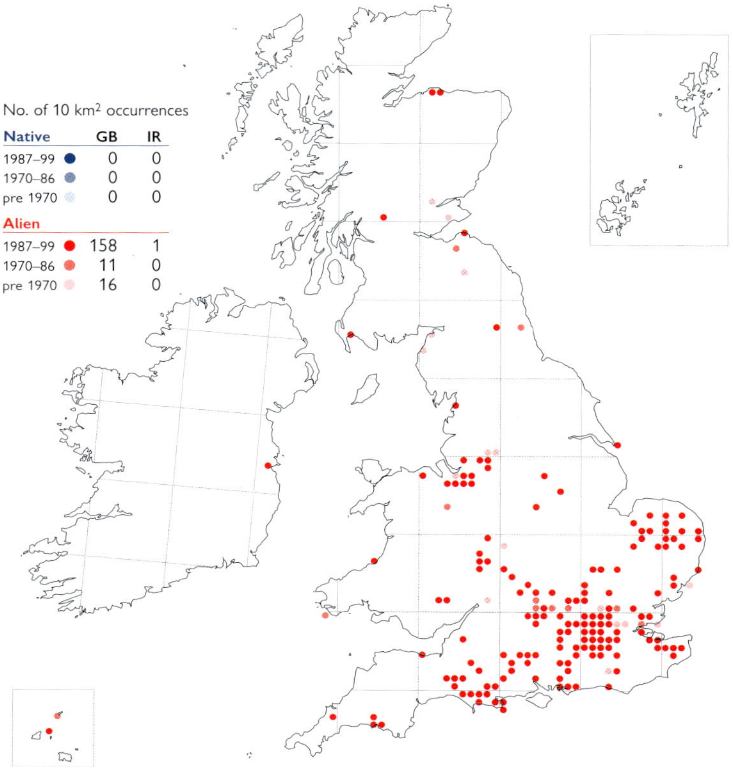

A vigorous perennial herb, which is often short-lived or annual in Britain, found on road verges, towpaths, beside paths, on field-borders, waste ground and by rivers. It is cleistogamous, and seed can be produced almost all year round. Lowland.

Neophyte (change +2.09). *C. carinata* was apparently originally introduced as a fodder grass but is also a contaminant of agricultural seed. It was first recorded as an escape from the Royal Botanic Gardens, Kew, in about 1919, but it did not begin to spread until 1945. There were still only a very few records in the 1962 *Atlas*. Since then it has spread rapidly, particularly on lighter soils. The new cultivar 'Deborah' was developed in *c.* 1975.

Native of western N. America.

References: Atlas (388c), Clement (1981a), Ryves *et al.* (1996).

D. A. PEARMAN

Ceratochloa cathartica Rescue Brome

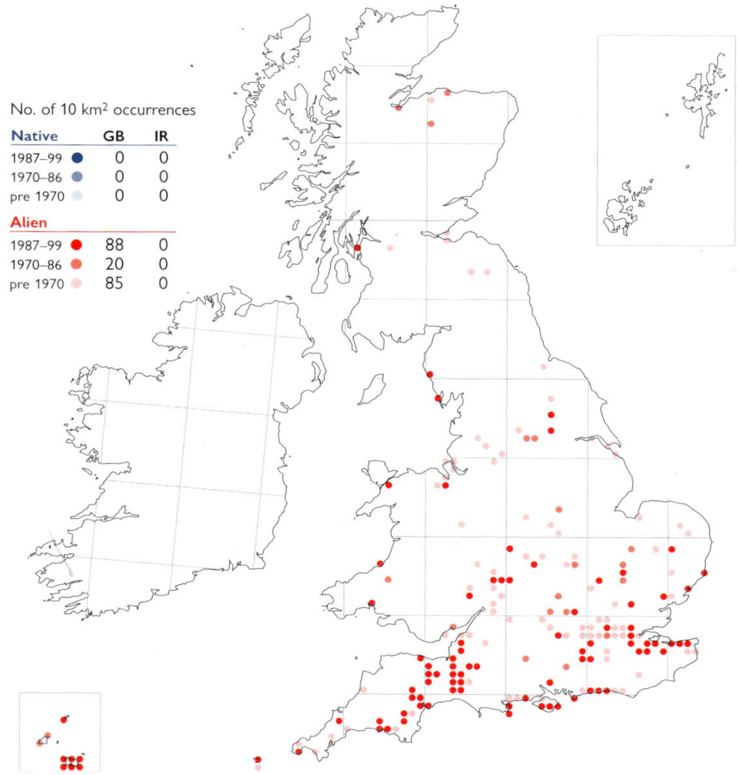

A usually short-lived perennial, found on roadsides, waste ground and the borders of fields, and in arable crops. It normally occurs as a casual, but can become naturalised, especially in the Channel Islands and the Isles of Scilly. Lowland.

Neophyte (change +0.63). *C. cathartica* was introduced as a fodder grass in 1788, and also occurs as a grain- and wool-alien. It was first recorded in the wild in 1870, and may be increasing slowly.

Native of N. & S. America.

References: Atlas (388d), Ryves *et al.* (1996).

D. A. PEARMAN

Brachypodium pinnatum Tor-grass

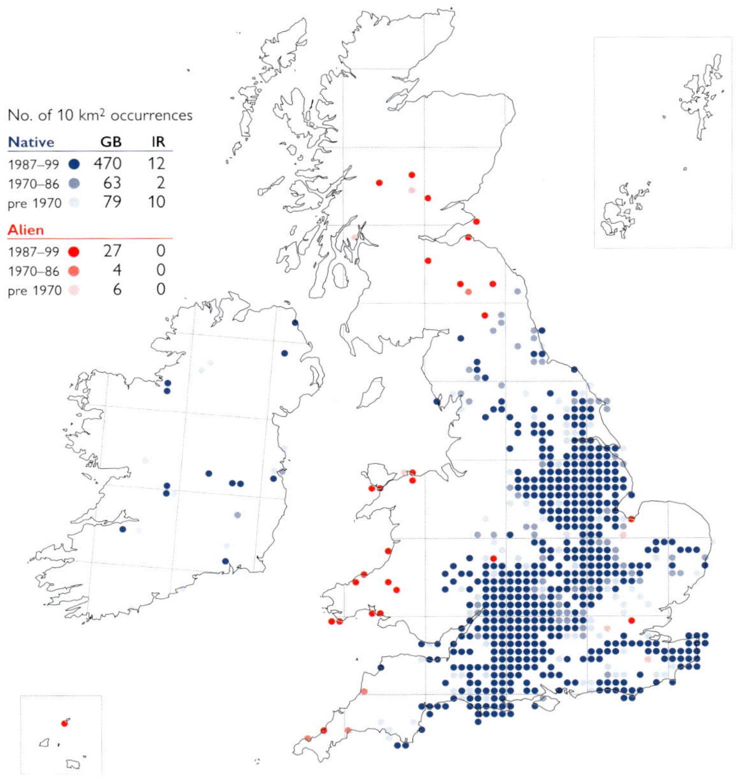

A perennial herb of dry, relatively infertile calcareous soils. In chalk and limestone grassland it is often dominant over large areas. It also occurs in scrub, quarries, and on railway banks and roadsides, where it may be accidentally introduced. It spreads by rhizomes and has a poor seed-set. Generally lowland, but reaching 305 m at Chelmorton (Derbys.).

Native (change +0.15). *B. pinnatum* has spread in and around its core areas since 1950, especially in East Anglia, due to a relaxation of grazing by stock and rabbits. Once established, it is extremely difficult to eliminate; attempts to control it by burning and herbicides have largely failed. It is also increasing as an alien.

Eurosiberian Temperate element.

References: Atlas (389b), Grime *et al.* (1988), Hultén & Fries (1986), Meusel *et al.* (1965).

D. A. PEARMAN

Brachypodium sylvaticum False Brome

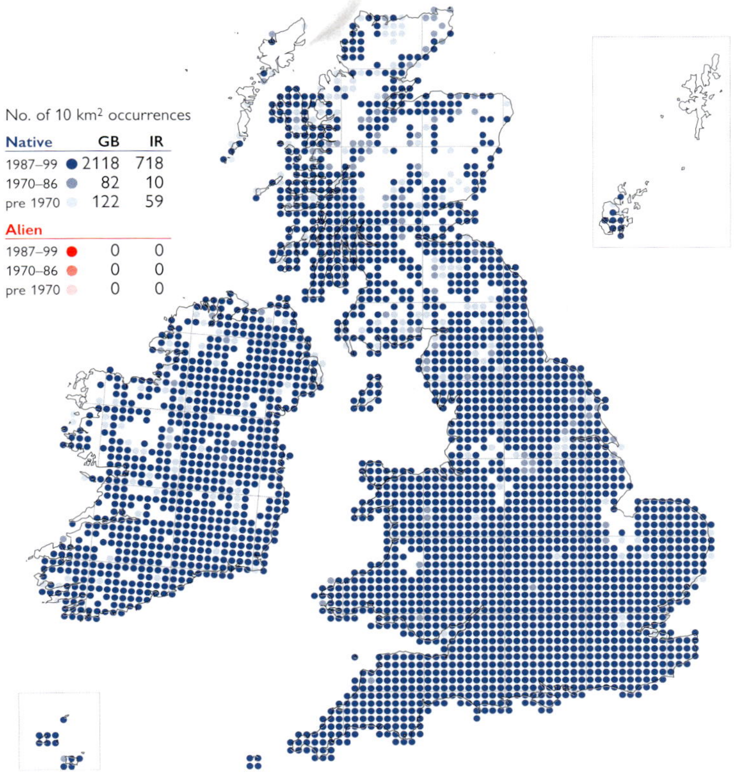

A tufted perennial herb of well-drained neutral to calcareous soils. In the lowlands, it is predominantly a plant of woodland and other shady habitats, including hedgerows, railway banks and roadsides, and it has colonised chalk and limestone downland following scrub invasion. It can persist in areas of former woodland or scrub. Above *c.* 200 m, it can occur in more open situations including limestone grassland and pavements, cliffs and screes. 0–465 m (Highfolds Scar, Malham, Mid-W. Yorks.).

Native (change –0.17). The distribution of *B. sylvaticum* is stable.

European Temperate element; also in C. and E. Asia and widely naturalised outside its native range.

References: Atlas (389a), Grime *et al.* (1988), Hultén & Fries (1986), Meusel *et al.* (1965).

D. A. PEARMAN

Elymus caninus Bearded Couch

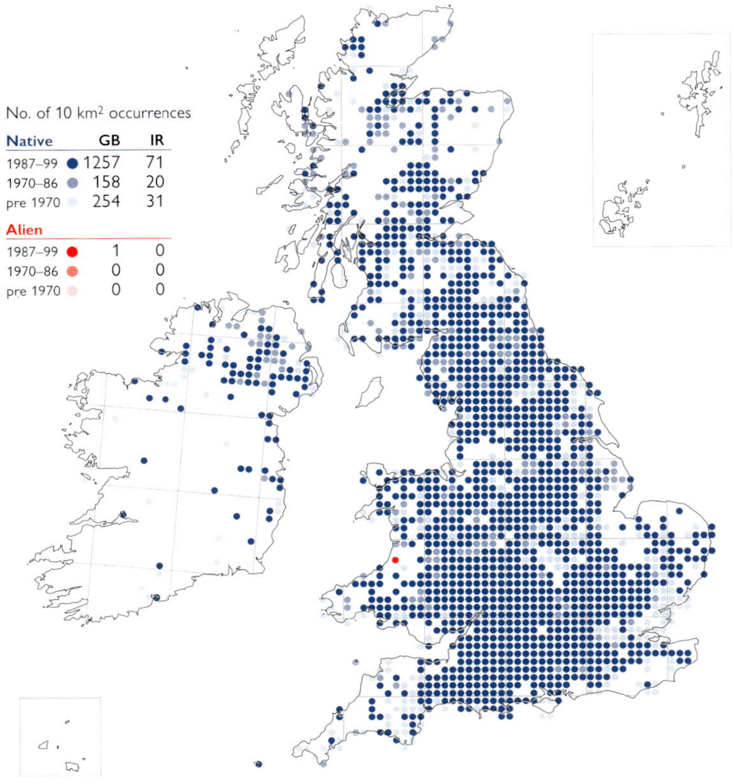

A loosely tufted perennial herb, intolerant of grazing, found in partially shaded sites in woodland, on river banks and roadside margins on free-draining, mainly base-rich, soils. It is also found in mountain areas in gullies, on cliffs and rock-ledges. 0–810 m (Creag na Caillich, Mid Perth).

Native (change +0.27). The overall distribution of *E. caninus* is stable. A variant with unawned lemmas, formerly described as *Agropyron donianum*, occurs on limestone rocks in C. and N. Scotland.

Eurosiberian Boreo-temperate element.

References: Atlas (389c, d), Grime *et al.* (1988), Hultén & Fries (1986), Meusel *et al.* (1965).

F. H. PERRING

Elytrigia repens Common Couch

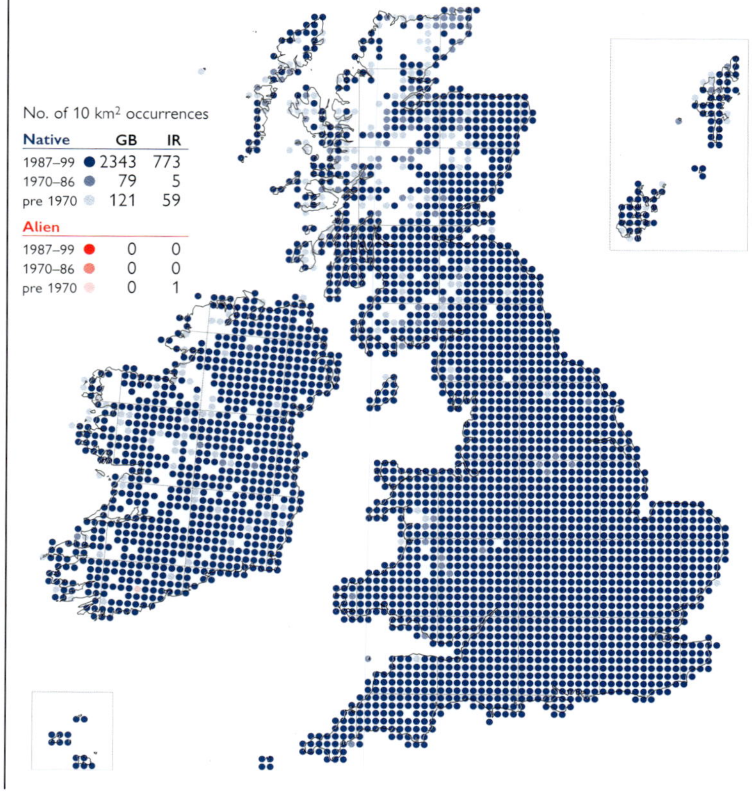

A rhizomatous herb, found in a wide range of fertile, disturbed habitats including waste ground, roadsides, railway banks, arable land and rough grassland; also in coastal areas on sand dunes, shingle, sea walls and the margins of saltmarshes. It is a notorious weed of gardens and agricultural land. 0–430 m (Nenthead, Cumbria) and exceptionally at 845 m on Great Dun Fell (Westmorland).

Native (change –0.01). The distribution of *E. repens* is stable. Two subspecies occur in our area; subsp. *repens* occurs throughout the range of the species, subsp. *arenosa* is mapped separately.

Eurosiberian Wide-temperate element, but widely naturalised so that the distribution is now Circumpolar Wide-temperate.

References: Atlas (390a), Grime *et al.* (1988), Hultén & Fries (1986), Palmer & Sagar (1963), Werner & Rioux (1977).

S. J. LEACH

Elytrigia repens subsp. *arenosa*

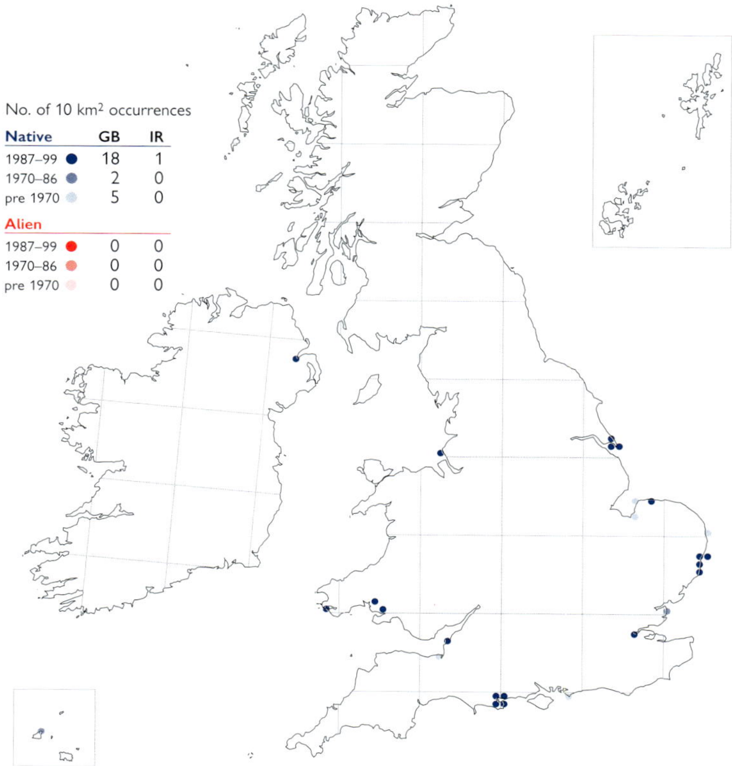

A perennial rhizomatous herb of coastal sand dunes and shingle. It also grows on firm sand at the rear of sandy beaches, adjoining sea walls and at the margins of saltmarshes, often in places subject to occasional tidal overflow. Lowland.

Native. This taxon was not mapped in the 1962 *Atlas* and trends in its distribution are difficult to assess. Its presence in Britain has only recently been confirmed and it is probably under-recorded.

Suboceanic Temperate element.

References: Trist (1995), Hultén & Fries (1986).

S. J. LEACH

Elytrigia atherica Sea Couch

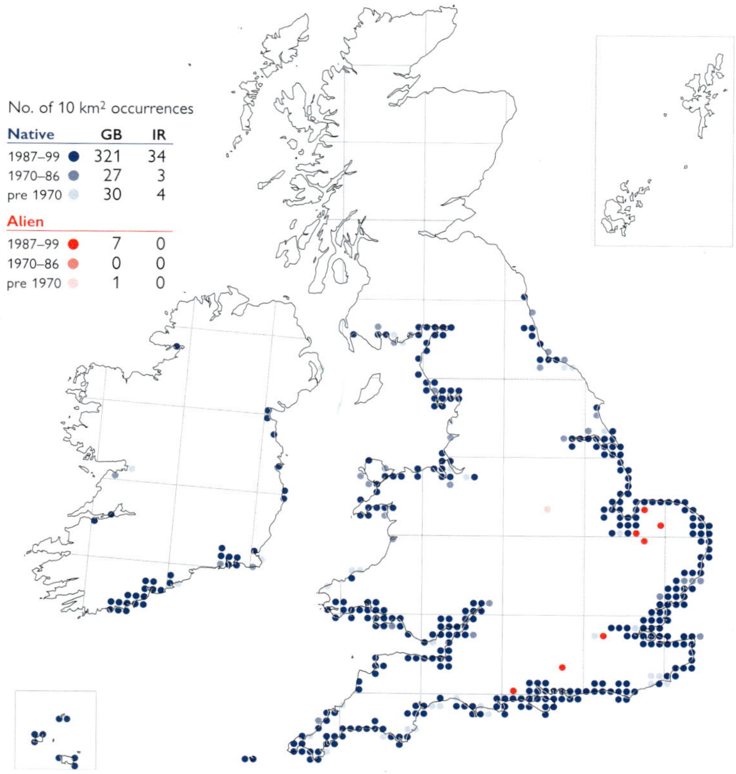

A rhizomatous perennial herb, growing on the margins of brackish creeks, in saltmarshes, saltmarsh-sand dune transitions, and on shingle banks and sea walls. In ungrazed situations it can form dense stands covering large areas to the almost total exclusion of other species. Lowland.

Native (change +0.32). There has been little change in distribution since the 1962 *Atlas*. Judging by the experience of Halliday (1997), it is likely that the sterile hybrid with *E. repens* (*E.* × *oliveri*), not mapped here, has sometimes been mis-recorded as *E. atherica*, thus obscuring the northern limit of this species.

European Southern-temperate element.

Reference: Atlas (390b).

S. J. LEACH

Elytrigia atherica × *E. juncea* (*E.* × *obtusiuscula*)

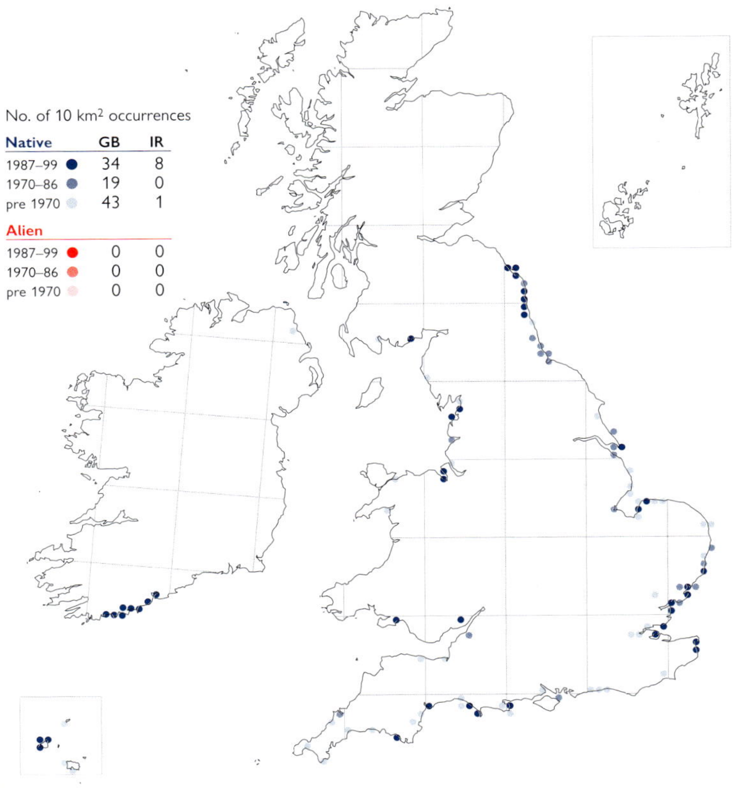

The hybrid *E.* × *obtusiuscula* occurs with the parents on sandy shores, shingle and the transition zone between saltmarshes and sand dunes. Lowland.

Native. There is no reliable information on trends in the distribution of this taxon. It is under-recorded, as many fieldworkers overlook it. There may also be some confusion with *E.* × *laxa* (*E. repens* × *E. juncea*).

Widespread on the Mediterranean and Atlantic coasts of Europe.

Reference: Stace (1975).

S. J. LEACH

Elytrigia juncea Sand Couch

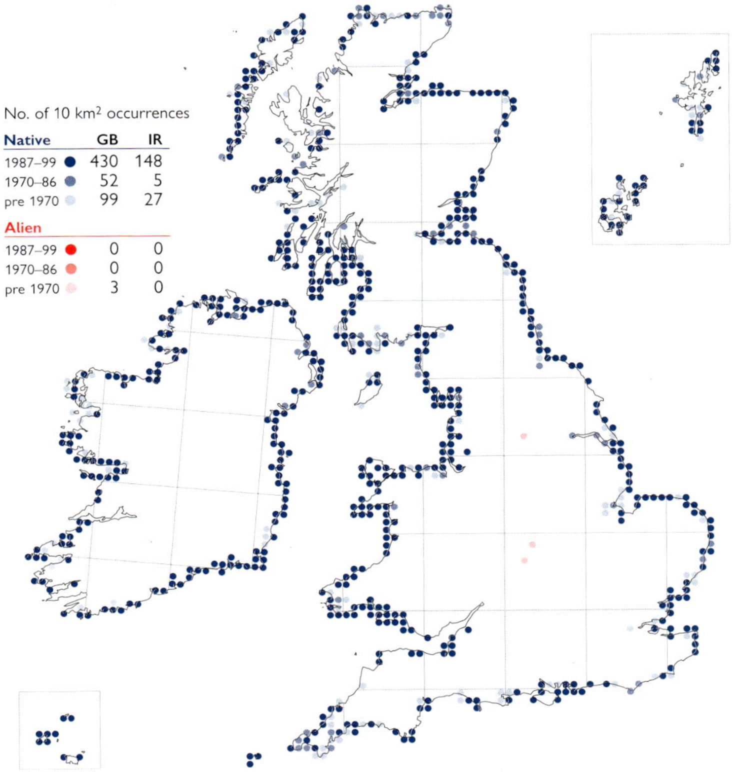

A rhizomatous perennial herb growing on or just above the strandline in loose sand, sometimes also on shingle. It is well known as a sand stabiliser on dune systems, forming low hummocky fore-dunes on the seaward side of the main *Ammophila arenaria* dunes. Lowland.

Native (change –0.28). There has been little change in the distribution of *E. juncea* since the 1962 *Atlas*.

European Southern-temperate element.

References: Atlas (390c), Hultén & Fries (1986), Meusel *et al.* (1965).

S. J. LEACH

Elytrigia juncea × *E. repens* (*E.* × *laxa*)

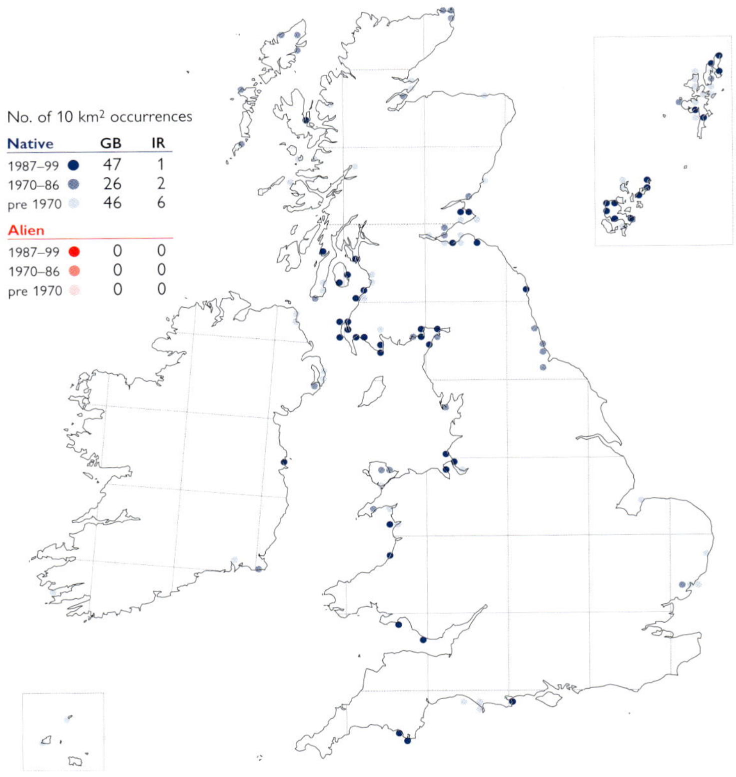

The hybrid *E.* × *laxa* occurs with the parents on sandy shores and shingle. Lowland.

Native. This taxon was not mapped in the 1962 *Atlas* and there is no reliable information on trends in its distribution. It is certainly under-recorded, as many fieldworkers are likely to overlook it. There may also be some confusion with *E.* × *obtusiuscula* (*E. atherica* × *E. juncea*).

Widespread on the Mediterranean and Atlantic coasts of Europe.

Reference: Stace (1975).

S. J. LEACH

Leymus arenarius Lyme-grass

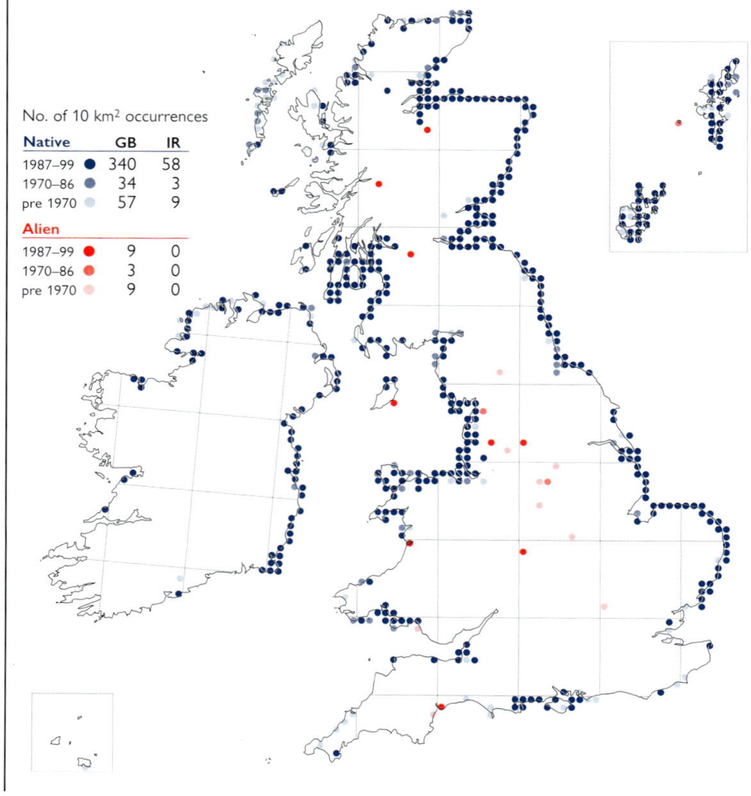

A rhizomatous perennial herb growing on coastal sand dunes, sometimes also on fine shingle; it is well known as an important species in the stabilisation of mobile dunes and widely planted as a sand binder. It is a rare casual or naturalised garden escape inland. Lowland.

Native (change +0.27). There is no evidence of appreciable change in the distribution of *L. arenarius* since the 1962 *Atlas*, though there are some losses in S. England. The widespread planting of *L. arenarius* on eroding dunes appears to have had little impact on its overall distribution.

European Boreo-arctic Montane element; also in E. Asia and N. America.

References: Atlas (390d), Bond (1952), Hultén & Fries (1986), Meusel *et al.* (1965).

S. J. LEACH

Hordelymus europaeus Wood Barley

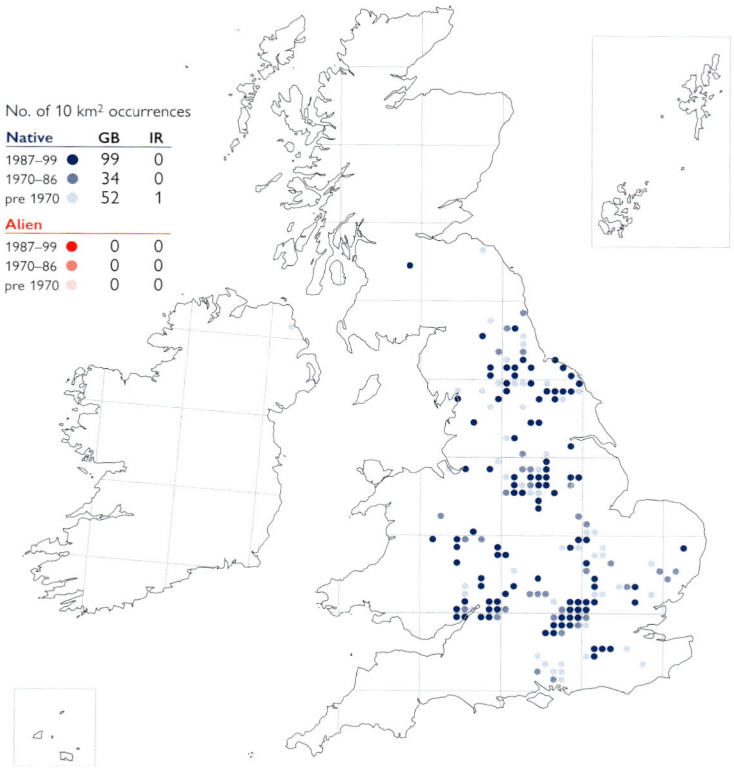

A perennial herb of woods and copses on calcareous soils, especially in sheltered beech woodlands and along medieval boundary banks and old hedgerows. Generally lowland, but reaching 440 m at Brough (Westmorland).

Native (change +0.12). Although now better recorded since the 1962 *Atlas*, *H. europaeus* has declined across much of its range mainly due to the removal of hedge banks and the coniferisation of deciduous woodland. It has not been seen in Ireland since 1949.

European Temperate element.

References: Atlas (391d), Curtis & McGough (1988), Hackney (1992), Hultén & Fries (1986), Meusel *et al.* (1965), Stewart *et al.* (1994).

S. J. LEACH

Hordeum distichon sens. lat. Barley

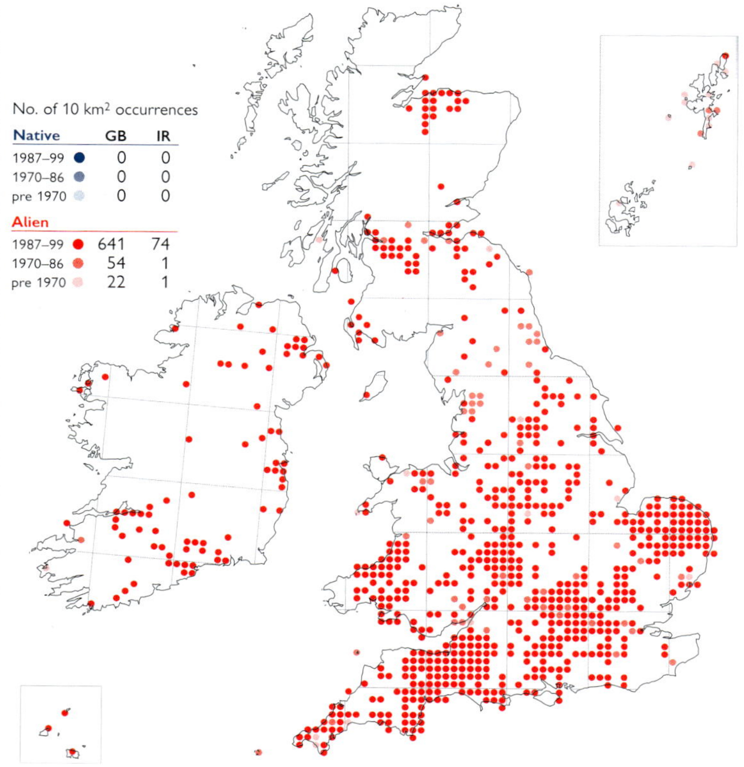

The common cultivated barley, occurring frequently but rarely persisting in waste places, fields, waysides and around farm buildings. It is mainly a relic of cultivation, but also occurs on roadsides from spilt grain and as a casual from wool shoddy and bird-seed. Lowland.

Casual. Barley has been cultivated in Britain for thousands of years, but was not recorded from the wild until 1908. This map includes records of the commonly cultivated *H. distichon sens. str.* and the rarer *H. vulgare*, which is also mapped separately. The species have sometimes been ignored by recorders, and are certainly under-recorded in some areas.

Unknown as a wild plant; widely cultivated in temperate regions throughout the world.

Reference: Ryves *et al.* (1996).

S. J. LEACH

Hordeum vulgare Six-rowed Barley

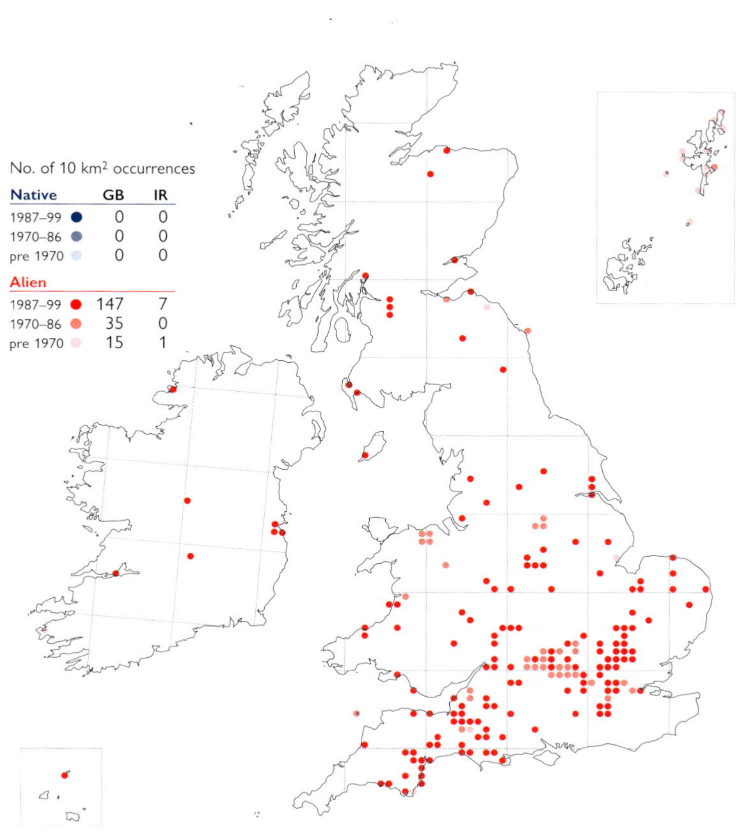

H. vulgare is sometimes found on arable land as a relic of cultivation, and on roadsides and waste ground as a bird-seed and grain casual. Lowland.

Casual. *H. vulgare* was known from the wild in Britain by at least 1905. It is now seldom grown as a crop in our area, and has thus probably decreased as a casual, though its decline in arable habitats may have been offset to some extent by recent introductions as a bird-seed casual. It is almost certainly ignored by many recorders; cereal relics were rarely recorded before 1970, but must have been just as frequent then as now.

Originated in cultivation; grown in temperate climates worldwide.

References: Ryves *et al.* (1996), Zohary & Hopf (2000).

S. J. LEACH

Hordeum murinum Wall Barley

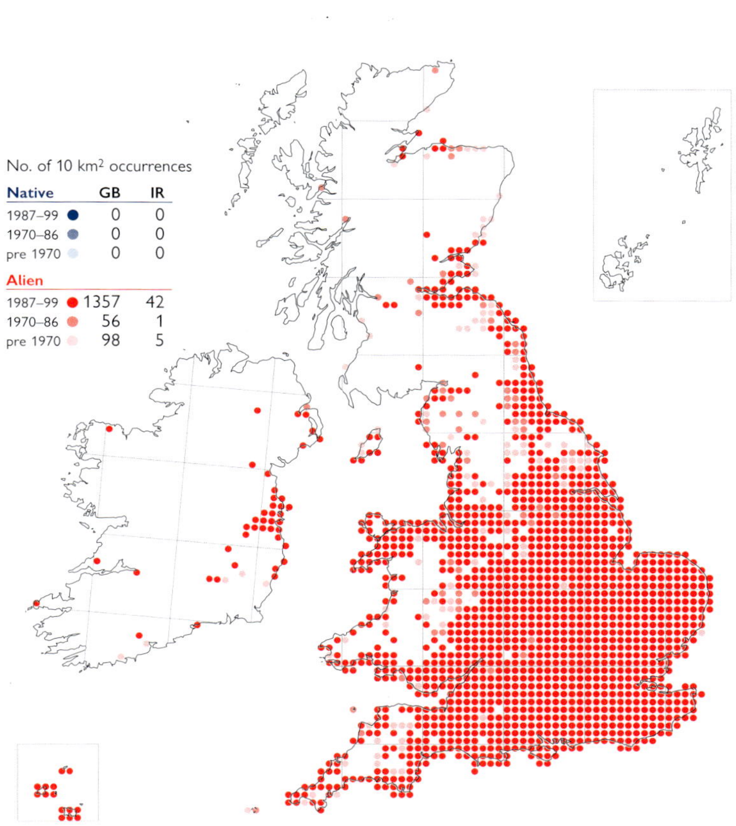

An annual growing in all kinds of fertile, disturbed ground, on roadsides, pavements, walls, railway banks and rough grassland. Generally lowland, but reaching 450 m on the Kirkstone Pass (Westmorland).

Archaeophyte (change –0.04). The distribution of this species is stable in Britain, but shows some increase in Ireland. Three subspecies occur in our area: subsp. *murinum* is common throughout the range, while subsp. *glaucum* and subsp. *leporinum* are infrequent casuals from wool shoddy and esparto.

As an archaeophyte *H. murinum* has a Eurosiberian Southern-temperate distribution, but it is widely naturalised so that its distribution is now Circumpolar Southern-temperate.

References: Atlas (391b), Davison (1970, 1971), Hultén & Fries (1986), Ryves *et al.* (1996).

S. J. LEACH

Hordeum jubatum Foxtail Barley

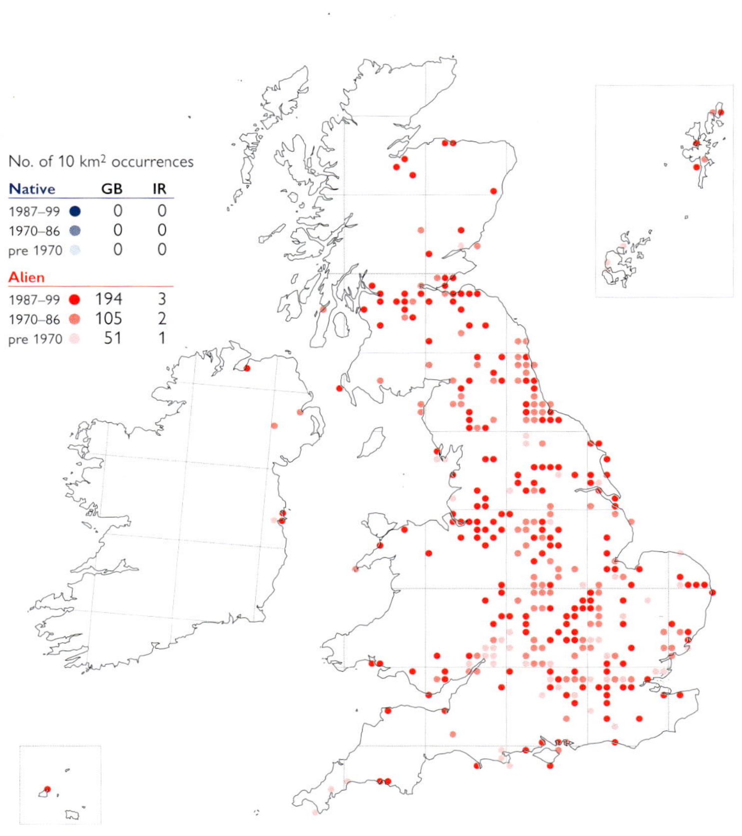

A short-lived perennial herb of roadsides, re-sown grassland and waste places, occurring as a casual from wool shoddy and bird-seed, and as a contaminant of commercial grass-seed mixtures. It is often found with maritime species beside salt-treated roads. Generally lowland, but reaching 410 m (Devil's Beef Tub, Dumfriess.).

Neophyte. *H. jubatum* was introduced as a fodder grass in 1782 and has been recorded from the wild since 1890. It was not mapped in the 1962 *Atlas*, but it has increased since 1970, particularly along roadsides.

Native of N. America and E. Asia; widely naturalised in N. & W. Europe.

References: Best *et al.* (1978), Coombe (1994), Hultén & Fries (1986), Ryves *et al.* (1996).

S. J. LEACH

Hordeum secalinum Meadow Barley

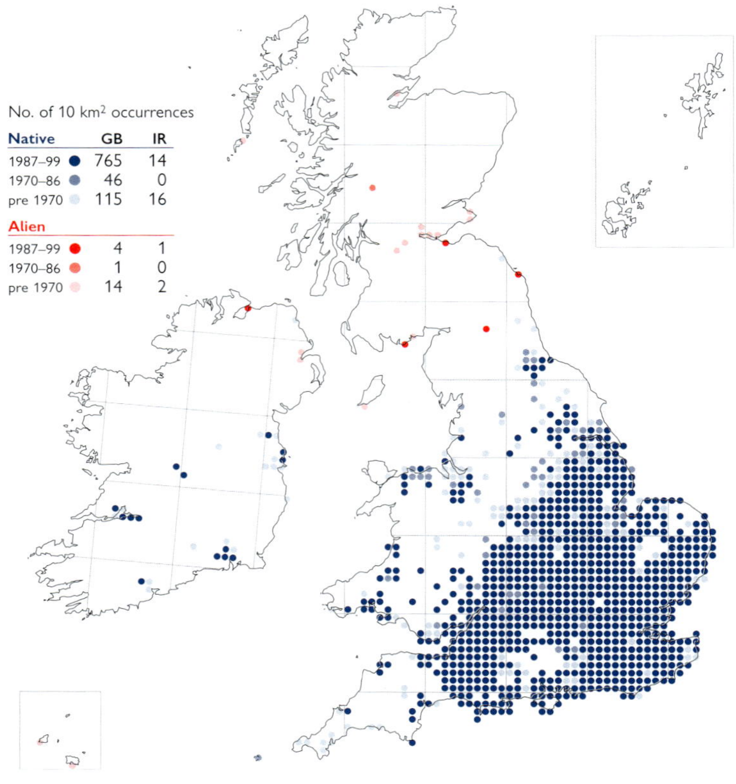

A perennial herb of meadows, pastures and roadsides, often in river valley floodplains and showing a strong preference for sticky clay soils. In coastal areas it is frequently abundant in grazing marsh grasslands and on earthen sea walls. Lowland.

Native (change –0.19). The map suggests little apparent change in the distribution of *H. secalinum* since the 1962 *Atlas*, apart from losses in a few outlying squares in N. and W. Britain. However, in some counties it is reported to be decreasing due to drainage, re-seeding and the conversion of grassland to arable. It can, however, withstand modest improvement. Its overall distribution is stable in Ireland.

European Temperate element.

References: Atlas (391a), Curtis & McGough (1988), Hultén & Fries (1986), Meusel *et al.* (1965).

S. J. LEACH

Hordeum marinum Sea Barley

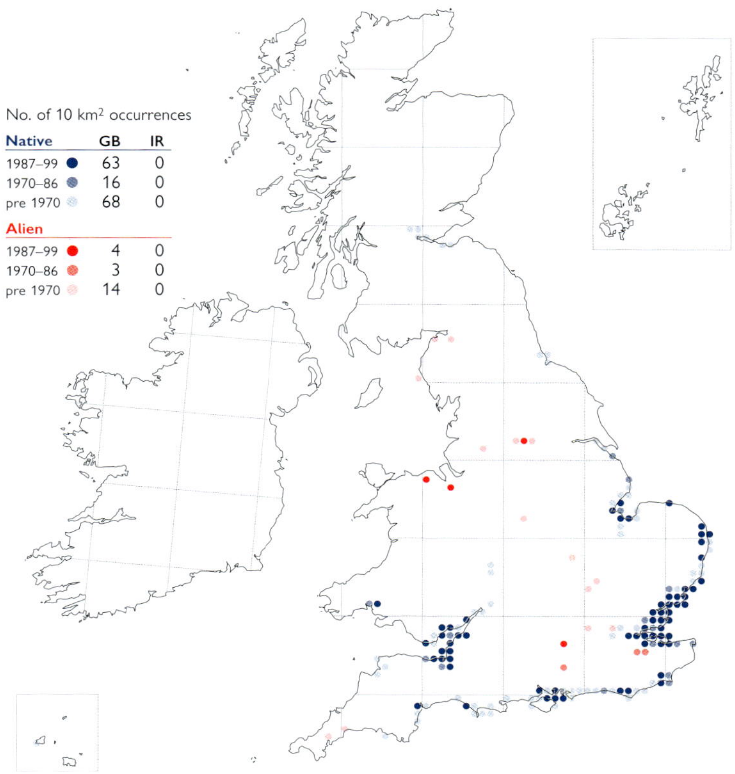

An annual of barish places by the sea, on the trampled margins of dried-up pools and ditches in grazing marshes, on tracks and sea walls, and in the uppermost parts of salt-marshes; also, very locally, beside salt-treated roads inland. Lowland.

Native (change –0.85). *H. marinum* has decreased in Britain, particularly along the S. coast and the E. coast from the Wash northwards. This is thought to be due to the rebuilding of sea defences, infilling of pools and ditches, the wholesale conversion of coastal grazing marshes to arable land, and, locally, the cessation of grazing.

Mediterranean-Atlantic element; also in C. Asia and widely naturalised outside its native range.

References: Atlas (391c), Hultén & Fries (1986), Stewart *et al.* (1994).

S. J. LEACH

Secale cereale Rye

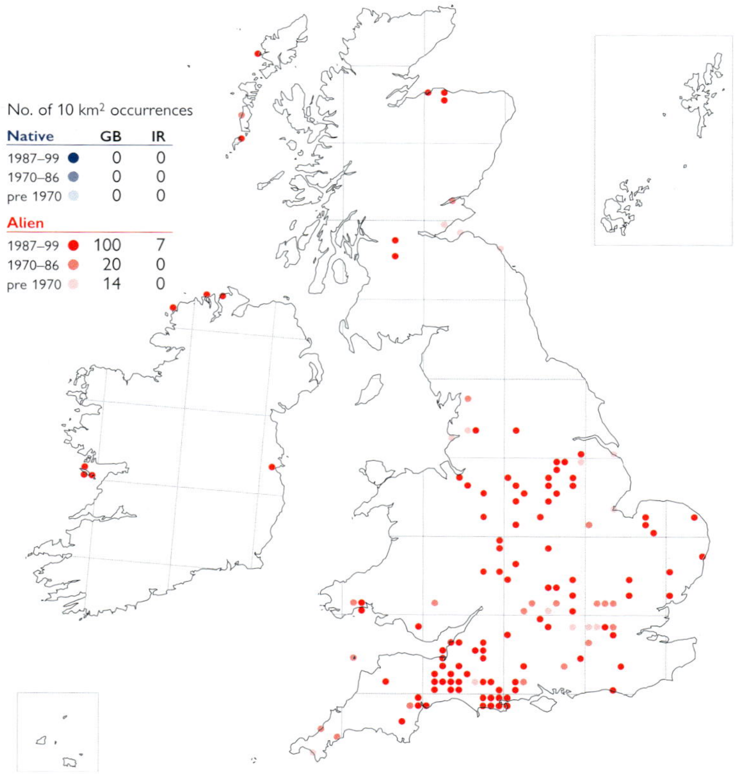

An annual, sometimes occurring in wheat and barley fields as a grain alien or relic of cultivation. It is also found, usually as a casual from grass-seed or bird-seed, on roadsides, manure heaps, rubbish tips, waste ground and even the cracks in pavements. Lowland.

Casual. *S. cereale* was formerly widely cultivated, but is now seldom grown in the British Isles. It was recorded in the wild by 1865, but current distributional trends are unclear.

Only known in cultivation; widely grown in temperate regions in N. & S. hemispheres.

References: Ryves *et al.* (1996), Vaughan & Geissler (1997), Zohary & Hopf (2000).

S. J. LEACH

Triticum aestivum Bread Wheat

Native	**GB**	**IR**
1987–99 ●	0	0
1970–86 ●	0	0
pre 1970	0	0
Alien		
1987–99 ●	705	59
1970–86 ●	32	0
pre 1970	8	0

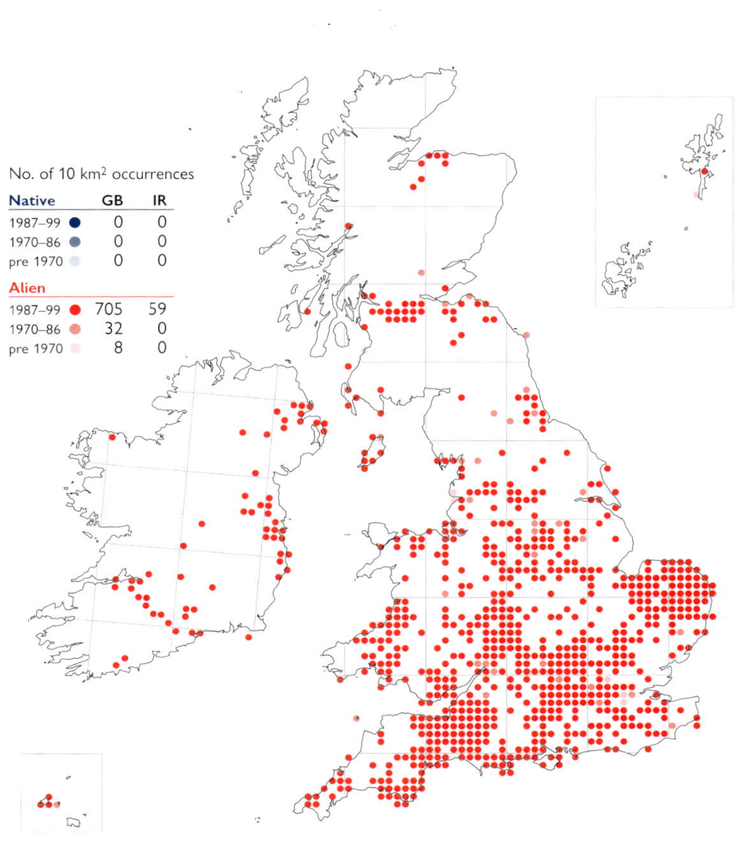

A self-pollinating annual which occurs as a casual in cereal fields and in all manner of waste places, both as a residue of crops and as a bird-seed alien. Lowland.

Casual. *T. aestivum* is one of the commonest crops of lowland, arable land. It usually occurs as single plants, never persists, and is often ignored by recorders. It was not noted in the wild until 1927; cereal relics were rarely recorded before 1970, but must have been just as frequent then.

Not known as a wild plant; originated in cultivation and now grown in suitable climates throughout the world.

References: Lupton (1985), Ryves *et al.* (1996), Vaughan & Geissler (1997), Zohary & Hopf (2000).

D. A. PEARMAN

Danthonia decumbens Heath-grass

Native	**GB**	**IR**
1987–99 ●	1908	627
1970–86 ●	190	18
pre 1970	282	169
Alien		
1987–99 ●	0	0
1970–86 ●	0	0
pre 1970	0	0

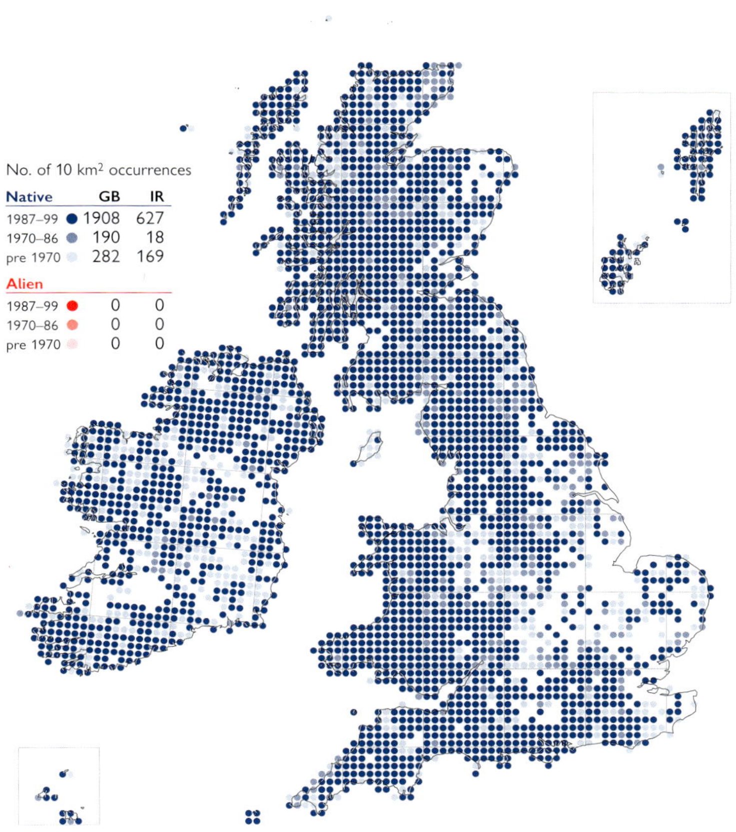

A densely tufted perennial herb of pastures, heathy grassland and moorland, favouring mildly acidic soils. It is also found in calcareous swards, including chalk and limestone grassland, but is then rooted into more acidic, superficial or leached horizons. It is frequent too in damp montane grassland. It spreads by seed and seems to have a persistent seed bank. 0–595 m (Tal-y-fan, Caerns.) and possibly at 1040 m on Macgillycuddy's Reeks (S. Kerry).

Native (change –0.40). *D. decumbens*, mapped as 'all records' in the 1962 *Atlas*, has declined in many areas of England and Ireland since 1950, presumably with the loss of permanent pastures.

European Temperate element.

References: Atlas (372a), Grime *et al.* (1988), Hultén & Fries (1986), Meusel *et al.* (1965).

D. A. PEARMAN

Cortaderia selloana Pampas-grass

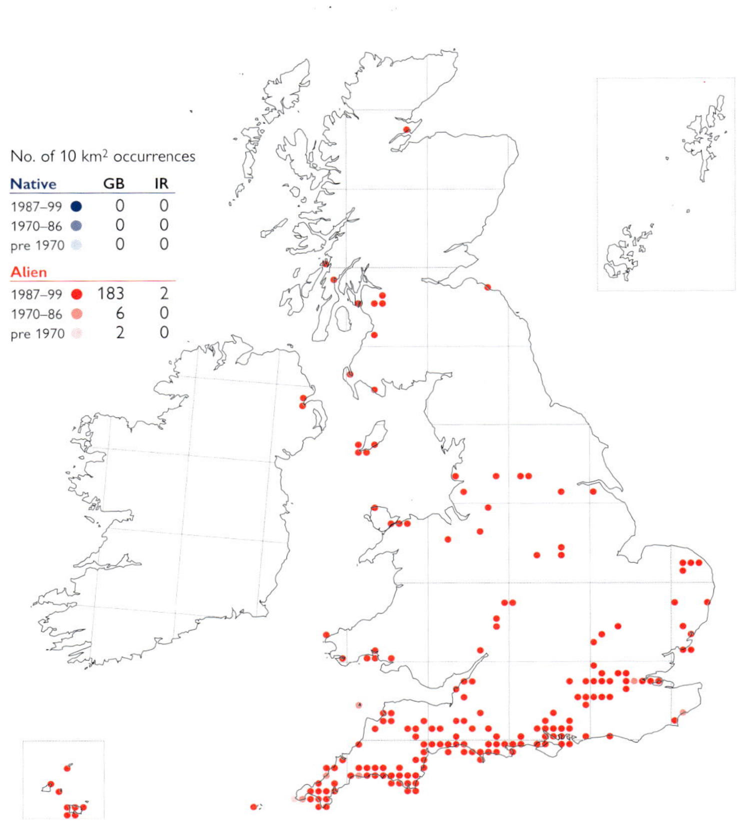

A large, tussock-forming, dioecious perennial grass which has become naturalised on roadsides, railway banks and rubbish dumps, and in rough grassland on sheltered sea-cliffs and sand dunes. Lowland.

Neophyte. *C. selloana* has been cultivated in Britain since 1848 and is commonly grown in municipal parks and gardens. It was first recorded in the wild in 1925. Most wild populations originate from garden throw-outs or deliberate planting. It is also reported to regenerate from seed, but such records require confirmation as the similar gynodioecious *C. richardii* seeds freely. It is increasing, particularly in S.W. England.

Native of S. America.

References: Grounds (1989), Ryves *et al.* (1996).

S. J. LEACH

Molinia caerulea Purple Moor-grass

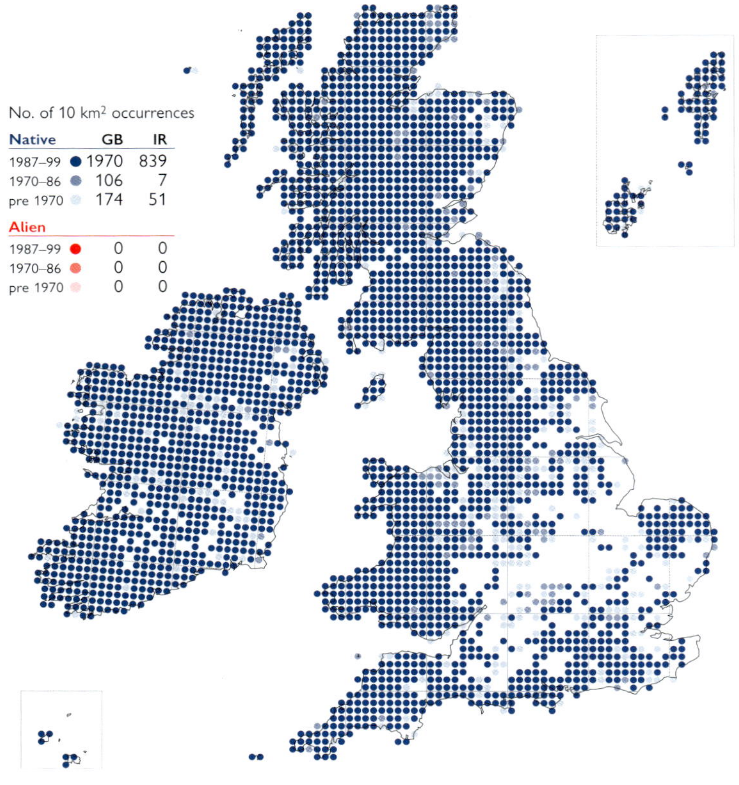

This deciduous perennial herb is found in a wide range of habitats, especially open heaths, moors, bogs and fens, but also in open birchwoods, mountain grassland and cliffs and stony lake margins. It is found on mildly basic to strongly acidic peats and mineral soils which are permanently or seasonally wet. 0–870 m (Meikle Kilrannoch, Angus).

Native (change –0.34). *M. caerulea*, mapped as 'all records' in the 1962 *Atlas*, has declined in some lowland areas through habitat destruction. Conversely, it has increased in frequency where grazing has ceased. Subsp. *caerulea* is found throughout the range of the species, subsp. *arundinacea* is widespread but scattered in fens and by rivers and canals.

Eurosiberian Boreo-temperate element.

References: Atlas (371d), Grime *et al.* (1988), Hultén & Fries (1986), Meusel *et al.* (1965), Taylor *et al.* (2001), Trist & Sell (1988).

F. H. PERRING

Phragmites australis Common Reed

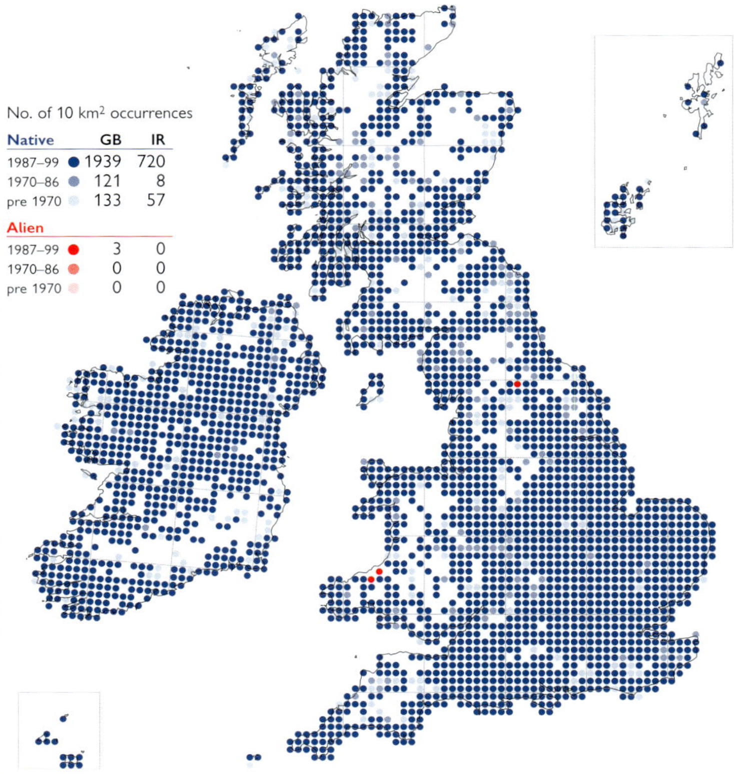

A rhizomatous and stoloniferous herb of swamps and fens, forming large stands in shallow water in ditches, rivers, lakes and ponds; also in brackish swamps and lagoons, and in freshwater seepages on sheltered sea-cliffs. It is frequently planted beside artificial water bodies. Generally lowland, but reaching 470 m on Brown Clee Hill (Salop).

Native (change +0.43). The distribution is broadly stable, though local losses have been noted, such as the die-back of reed-beds in Broadland in the last fifty years. However, it has increased in some areas due to eutrophication, and some reed-beds have been created for breeding birds.

Circumpolar Wide-temperate element.

References: Atlas (371c), Grime *et al.* (1988), Haslam (1972), Haslam & McDougall (1972), Hultén & Fries (1986), Preston & Croft (1997).

S. J. LEACH

Cynodon dactylon Bermuda-grass

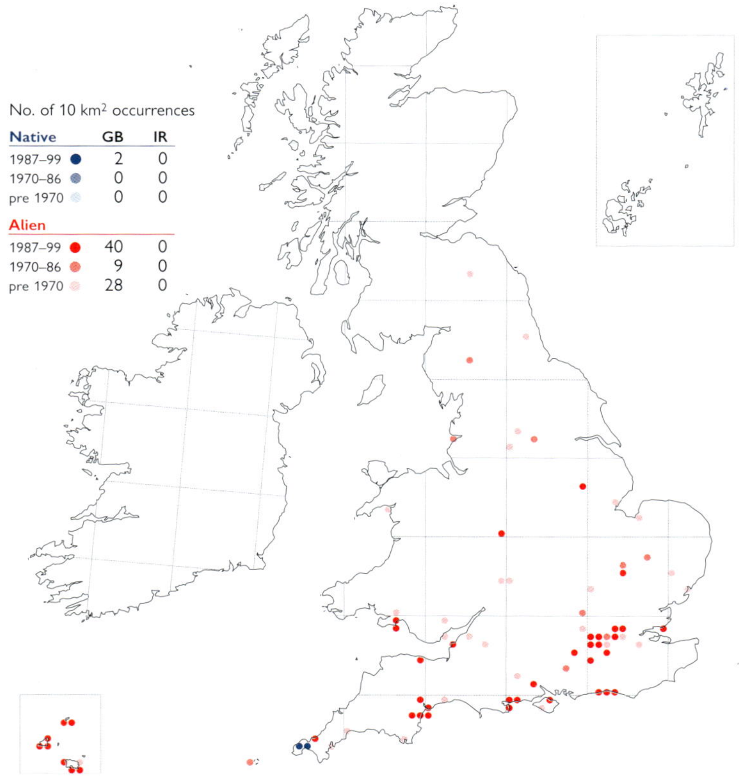

A mat-forming perennial herb, found on eroding fore-dunes in dry sandy places and short grassland by the sea, including lawns and sea-front promenades, and inland on lawns and roadside verges. It is also found around docks, on rubbish tips and as a casual from wool shoddy and other sources. Lowland.

Native or alien (change –0.10). *C. dactylon* was first recorded in the wild in 1688. Although sometimes considered to be an alien in our area, it may be native on fore-dunes in W. Cornwall and possibly in the Channel Islands (Wigginton, 1999). There has been little change in its distribution over the last thirty to forty years.

Eurasian Southern-temperate element.

Reference: Atlas (406a).

S. J. LEACH

Spartina maritima Small Cord-grass

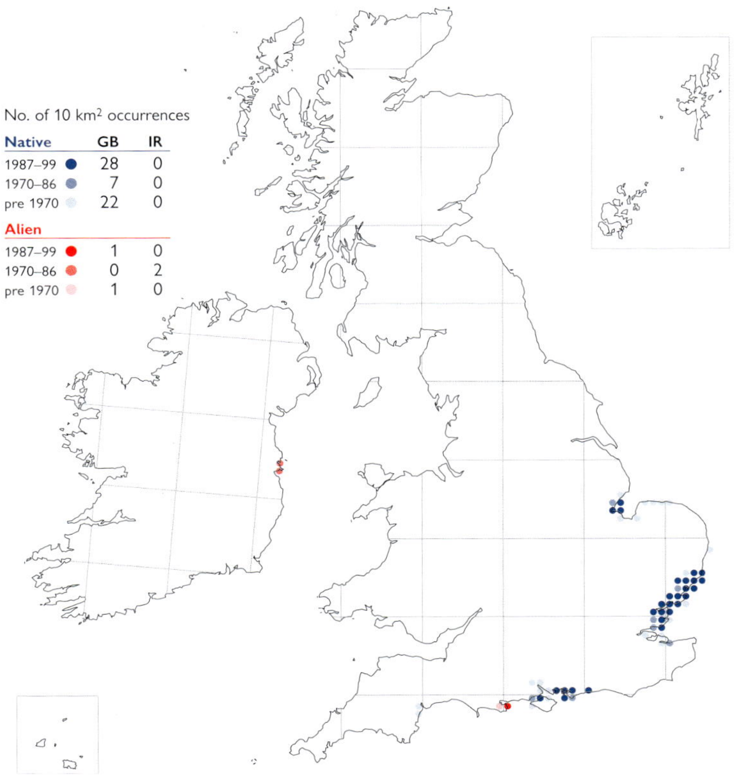

A perennial herb of tidal mud-flats, generally at higher elevations than *S. anglica*, such as in saltmarsh creeks and pans, and on bare ground behind sea walls. It rarely sets seed in Britain, relying instead on vegetative spread by means of extensively creeping rhizomes. Lowland.

Native (change –0.55). The range of this species has contracted during the past century, due partly to the spread of *S. anglica*. However, habitat loss – whether through successional changes, coastal erosion or land-claim – may have been important factors in the demise of many populations.

Suboceanic Southern-temperate element.

References: Atlas (405b), Adam (1990), Marchant & Goodman (1969), Stewart *et al.* (1994).

S. J. LEACH

Spartina alterniflora × *S. maritima* (*S.* × *townsendii*) Townsend's Cord-grass

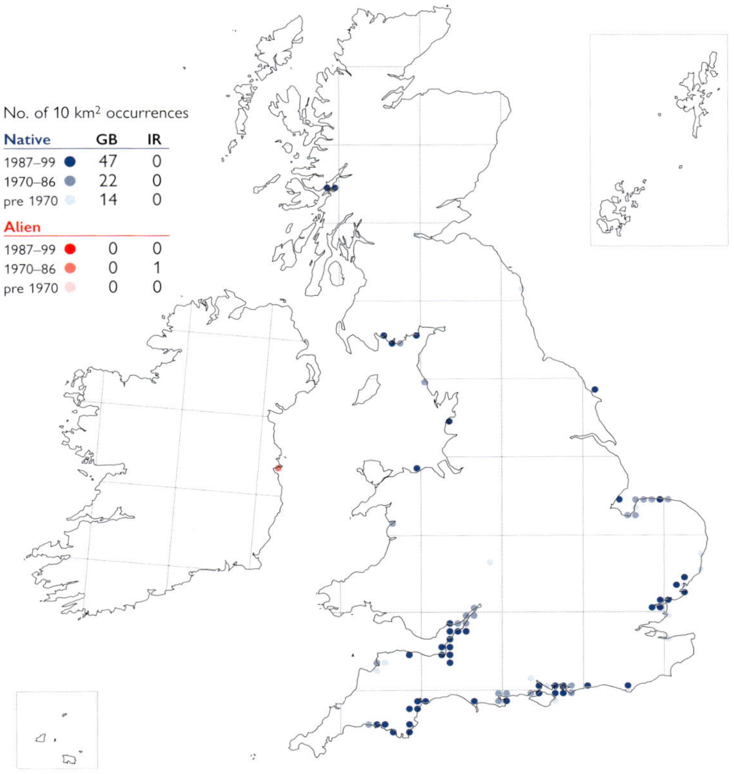

This rhizomatous, stand-forming perennial herb of tidal mud-flats and saltmarshes is planted as a mud-binder. It is sterile and reproduces vegetatively. Lowland.

A spontaneous hybrid between native and alien parents. The map in the 1962 *Atlas* included both this taxon and *S. anglica*, though *S.* × *townsendii* was separately mapped by Perring & Sell (1968). Attempts have been made to limit its spread, while in S. England it appears to have declined due to 'die-back' in many sites. It is widely recorded in Ireland but all material examined, apart from plants from Bull Island (Co. Dublin), is *S. anglica* (Curtis & McGorry, *in litt.*).

This hybrid originated in Southampton Water (S. Hants.) in about 1870.

References: Atlas Supp. (152c), Adam (1990), Goodman *et al.* (1959, 1969), Stace (1975).

S. J. LEACH

Spartina anglica Common Cord-grass

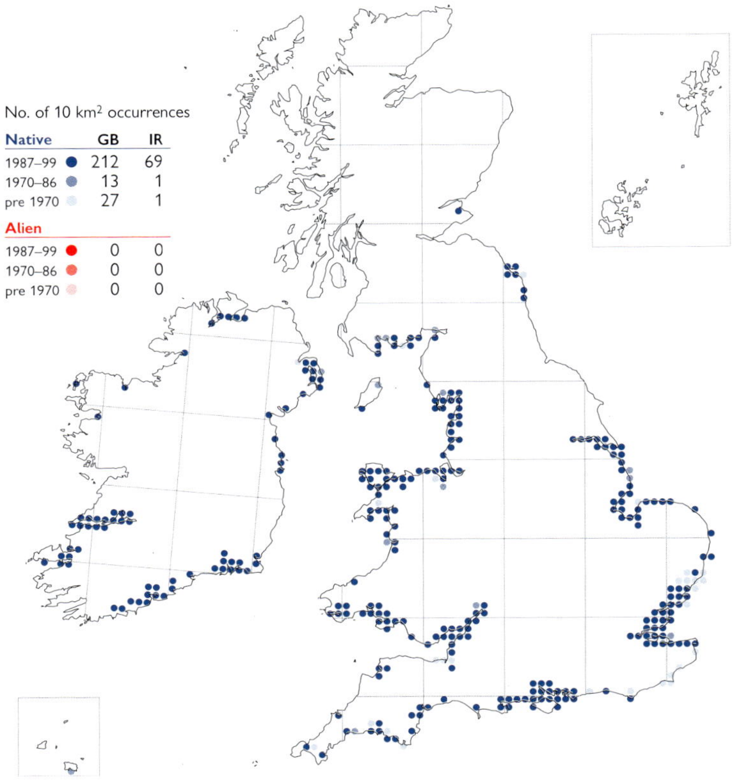

A rhizomatous perennial herb of tidal mud-flats and saltmarshes, much planted as a mud-binder and forming extensive stands in many estuaries. It originated in Southampton Water (S. Hants.) in about 1890 as an amphidiploid derivative of *S.* × *townsendii*. Lowland.

Native (change +0.11). The taxon was included within *S.* × *townsendii* in the 1962 *Atlas*, by which time it had become widespread both through deliberate planting and natural colonisation. It is not possible to separate native and planted populations and all are mapped as if they are native.

Endemic as a native to Britain, but widely planted in W. Europe.

References: Adam (1990), Ferris *et al.* (1997), Goodman *et al.* (1959, 1969), Goodman & Williams (1961), Gray & Benham (1990), Raybould *et al.* (1991).

S. J. LEACH

Panicum capillare Witch-grass

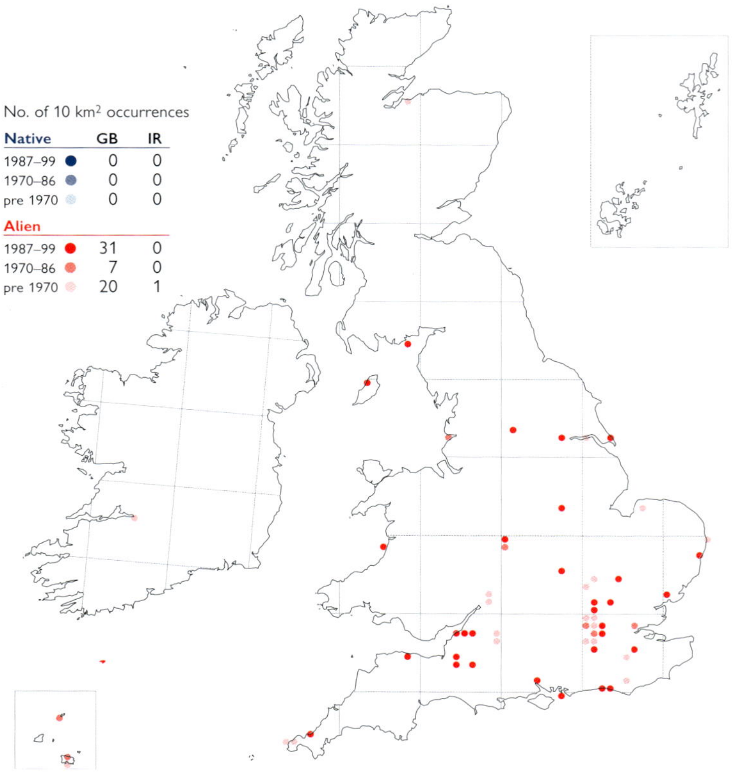

An annual of waste places, docks and rubbish tips, arising from bird- and oil-seed, wool shoddy and as a contaminant of agricultural seed; also occurring as a garden escape. Lowland.

Casual. *P. capillare* was cultivated in Britain by 1758 and was first recorded in the wild in 1867. It is increasingly grown in gardens for its beautiful panicles and, although still a rare casual, may be increasing in the wild.

Native of N. America; widely naturalised in C. & S. Europe and elsewhere.

Reference: Ryves *et al.* (1996).

D. A. PEARMAN

Panicum miliaceum Common Millet

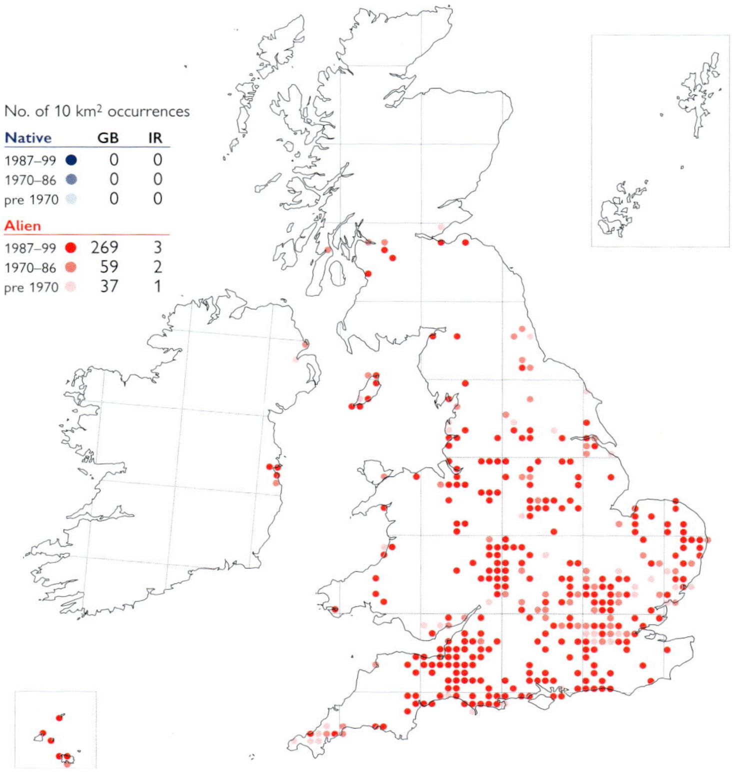

This large, tufted annual is a frequent bird- and oil-seed casual on rubbish tips, waste ground and in woodland around pheasant feeding areas; also occurring as a grain contaminant in arable crops, especially maize. Lowland.

Casual. *P. miliaceum* was introduced to cultivation by 1596 and was first recorded in the wild in Britain in 1872. It appears to be increasing in many areas, but probably remains under-recorded.

Originally domesticated in C. & E. Asia but now grown in warm-temperate and tropical regions throughout the world.

References: Green *et al.* (1997), Ryves *et al.* (1996), Vaughan & Geissler (1997), Zohary & Hopf (2000).

S. J. LEACH

Echinochloa crus-galli Cockspur

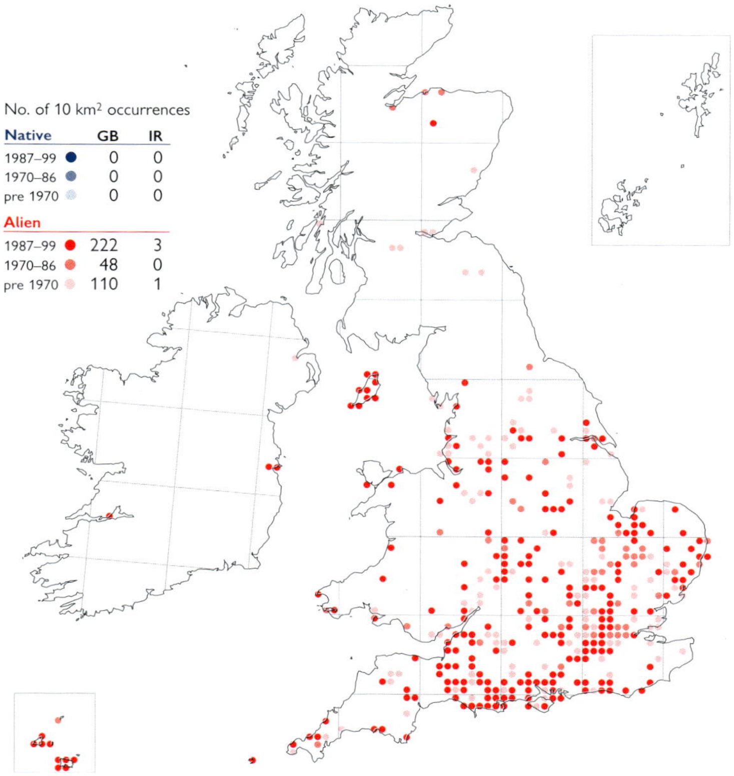

A rather stout annual occurring as a casual of rubbish tips, waste places and cultivated ground, mainly from bird-seed but also from wool shoddy, soya-bean and other sources; sometimes sown as food for game (Beckett *et al.*, 1999). Lowland.

Neophyte (change +0.75). *E. crus-galli* has been known in the British Isles since at least 1690, but increased after the Second World War when it was introduced with N. American seed (Rodwell, 2000). The distribution is now probably stable, and the species is reported to be persistent at some sites in S. England.

Native of warm-temperate and tropical regions of Europe, Asia & N. America; widely introduced as a fodder grass elsewhere.

References: Atlas (406b), Maun & Barrett (1986), Ryves *et al.* (1996).

S. J. LEACH

Echinochloa esculenta Japanese Millet

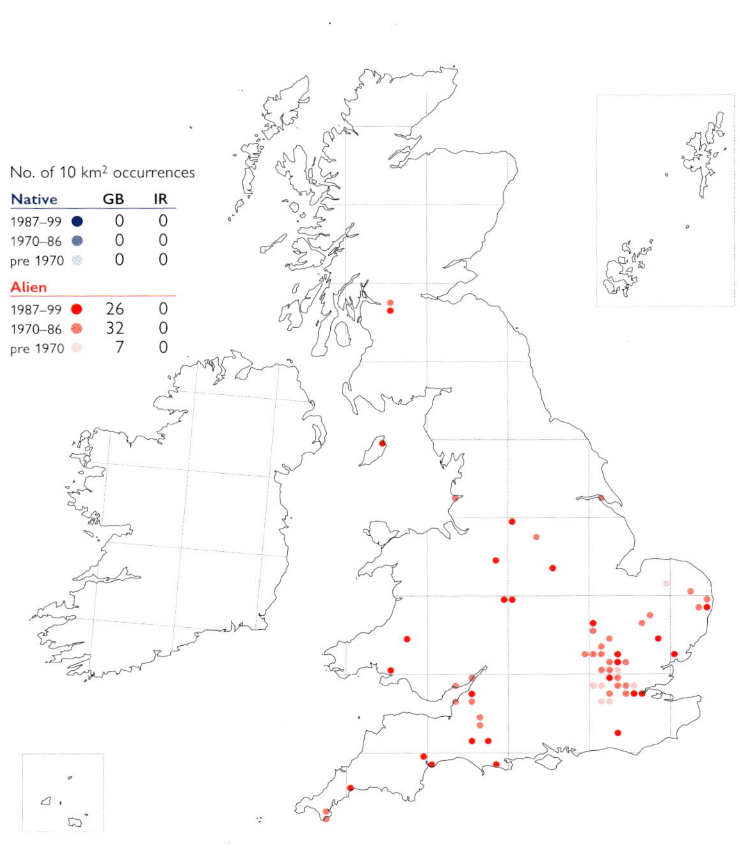

An annual found as a casual on rubbish tips and waste ground, arising as a constituent of bird-seed mixtures. It is also sown as a food-source for game. Lowland.

Casual. It is difficult to establish the date of the first record of this species in the wild due to taxonomic confusion, but it was certainly known by 1971. It may be increasing, although some older records may represent misidentifications of *E. frumentacea*.

Originated in cultivation in Japan as a derivative of *E. crus-galli*.

Reference: Ryves *et al.* (1996).

D. A. PEARMAN

Setaria pumila Yellow Bristle-grass

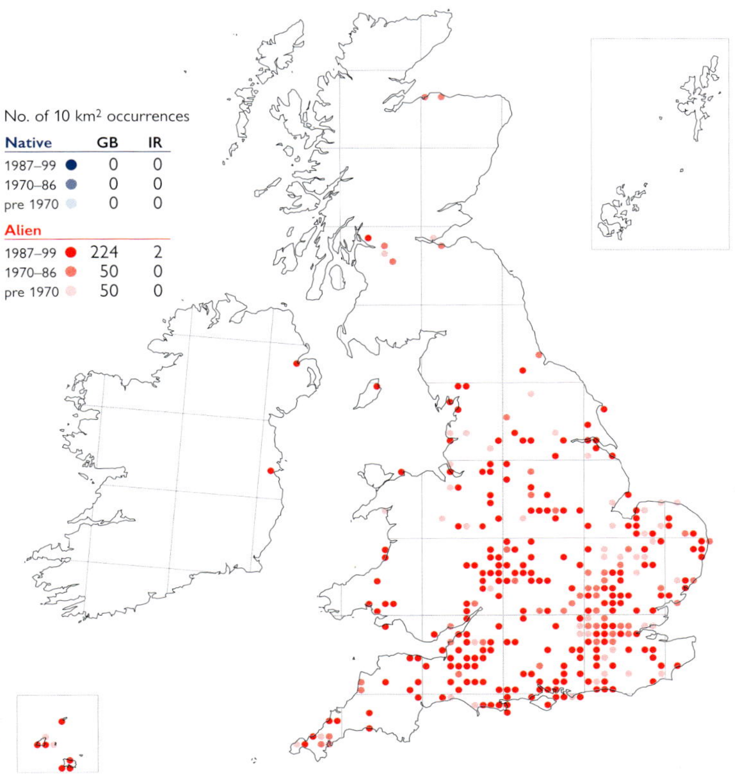

An annual occurring as a casual on rubbish tips, cultivated and waste ground, mainly from wool and bird-seed but also from grain, agricultural seed and oil-seed. Lowland.

Neophyte. *S. pumila* was first cultivated in Britain in 1819 and was known from the wild by 1867. It is probably under-recorded, but seems to be increasing.

Native range uncertain but probably centred on the Mediterranean region and S.W. Asia; the species is now naturalised in warm-temperate and subtropical areas throughout the N. hemisphere and elsewhere.

References: Hultén & Fries (1986), Ryves *et al.* (1996).

S. J. LEACH

Setaria verticillata Rough Bristle-grass

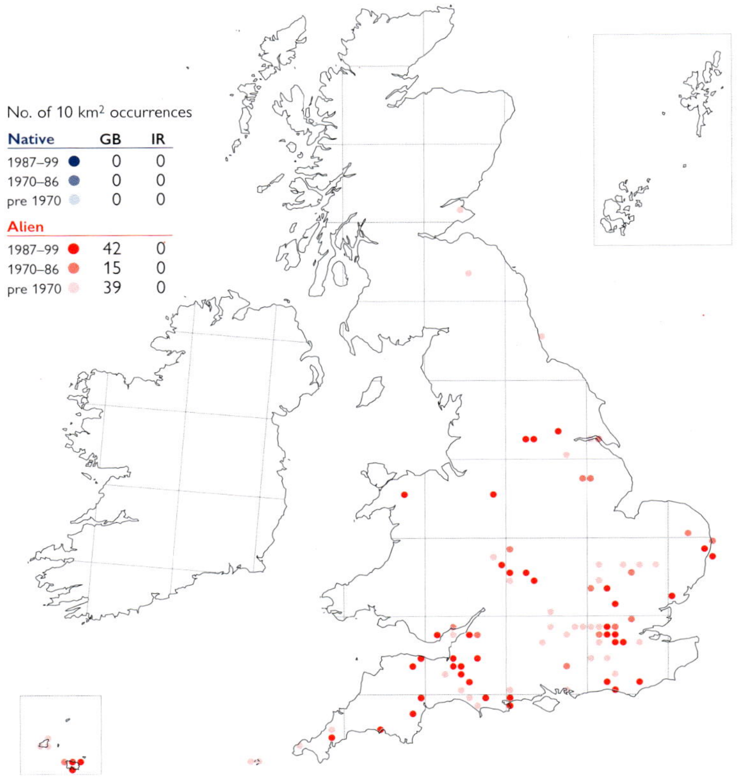

An annual of rubbish tips, dock quaysides, verges and waste ground; also, rarely, as a weed of arable land or garden centres. *S. verticillata* is a bird-seed, oil-seed, wool, cotton and esparto alien, usually occurring as a casual but sometimes persisting for a few years in S. England. Lowland.

Casual. *S. verticillata* was recorded from the wild in 1666. It is unclear whether its distribution is changing.

Native of Eurasia, but precise range obscured by its spread as a weed; it now grows in warm-temperate and subtropical areas of Eurasia, C. & S. America, Africa, Australia and elsewhere.

References: Ryves *et al.* (1996), Steel *et al.* (1983).

S. J. LEACH

Setaria viridis Green Bristle-grass

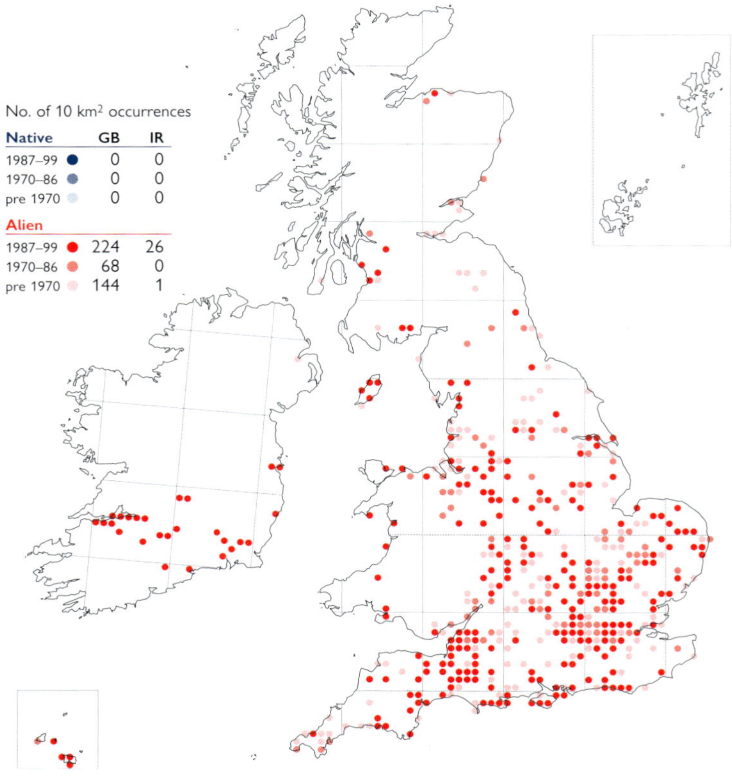

An annual occurring as a casual of cultivated and waste ground, road verges and rubbish tips, mainly introduced in bird-seed but also from wool shoddy, oil-seed, esparto and grain. Lowland.

Neophyte (change +0.80). *S. viridis* was first recorded in the wild in Britain in 1666. It is now much better recorded than for the 1962 *Atlas*. Any apparent increases are probably due to increased recording effort in urban areas, and a widening interest in alien taxa, although it is reported as increasing in Somerset (Green *et al.*, 1997).

A native of Eurasia, which now occurs in temperate and subtropical regions throughout the N. hemisphere; this spread has obscured its putative native range.

References: Atlas (406c), Douglas *et al.* (1985), Hultén & Fries (1986), Ryves *et al.* (1996).

S. J. LEACH

Setaria italica Foxtail Bristle-grass

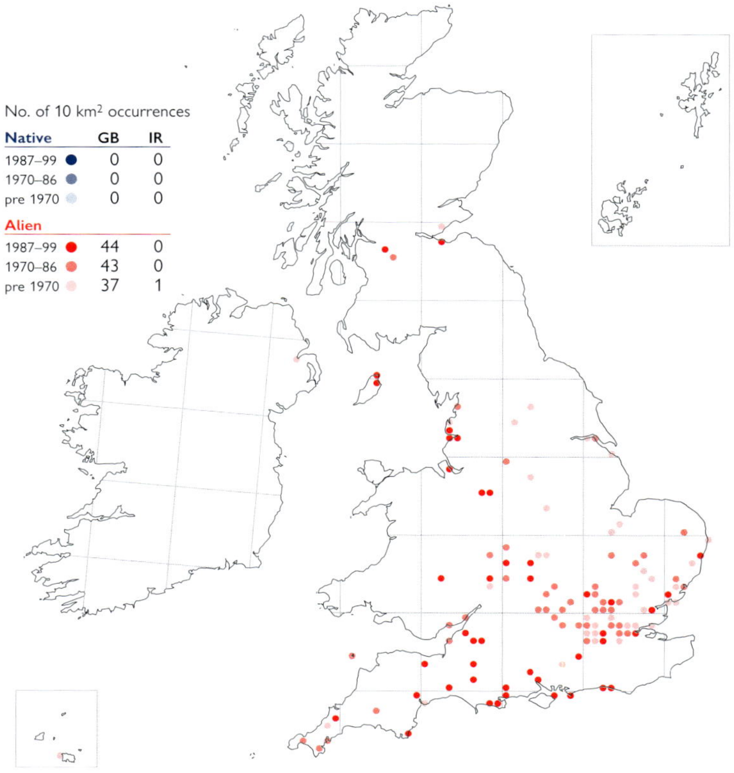

An annual of rubbish tips, docks and waste ground, mainly introduced in bird-seed and as a contaminant of grain. Lowland.

Casual. *S. italica* was being cultivated in Britain by 1739 and was known from the wild by at least 1905.

Originated in cultivation, perhaps by selection from *S. viridis*; now cultivated and often naturalised in warm-temperate and subtropical regions throughout the world.

References: Ryves *et al.* (1996), Zohary & Hopf (2000).

S. J. LEACH

Digitaria sanguinalis Hairy Finger-grass

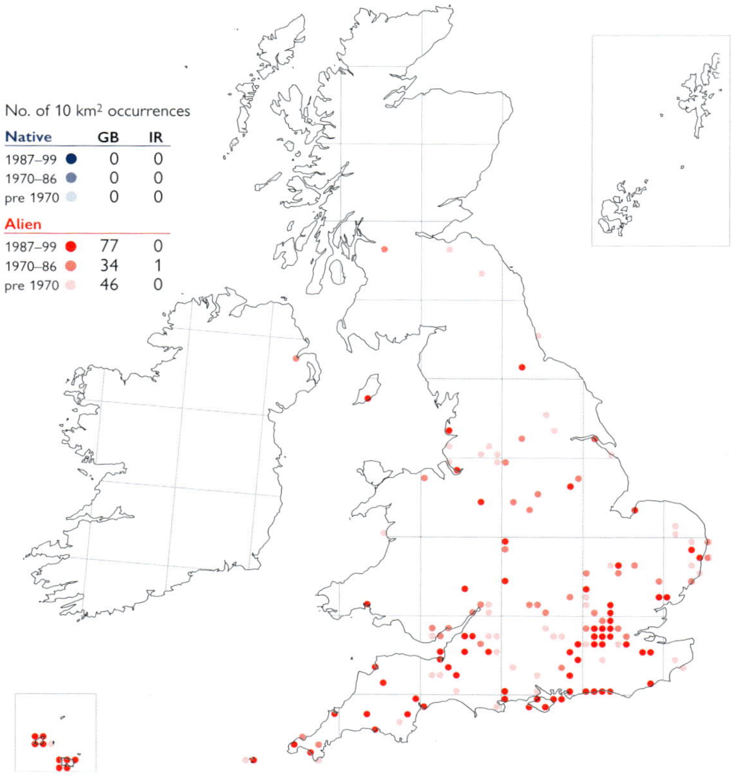

An annual of bulb-fields and waste places on light soils in the Channel Islands and the Isles of Scilly. On the mainland it is spreading as a weed of garden centres, flower borders and pavement cracks; also, as a casual, on rubbish tips. Lowland.

Neophyte. *D. sanguinalis* is a wool, agricultural seed, grain and bird-seed alien, and was first recorded in the wild in 1690. It was not mapped in the 1962 *Atlas*, but is increasing, particularly in S. England.

Native of S. Europe, the Mediterranean region and S.W. Asia; now an almost cosmopolitan weed in warm-temperate and subtropical areas.

References: Hultén & Fries (1986), Ryves *et al.* (1996).

S. J. LEACH

Sorghum halepense Johnson-grass

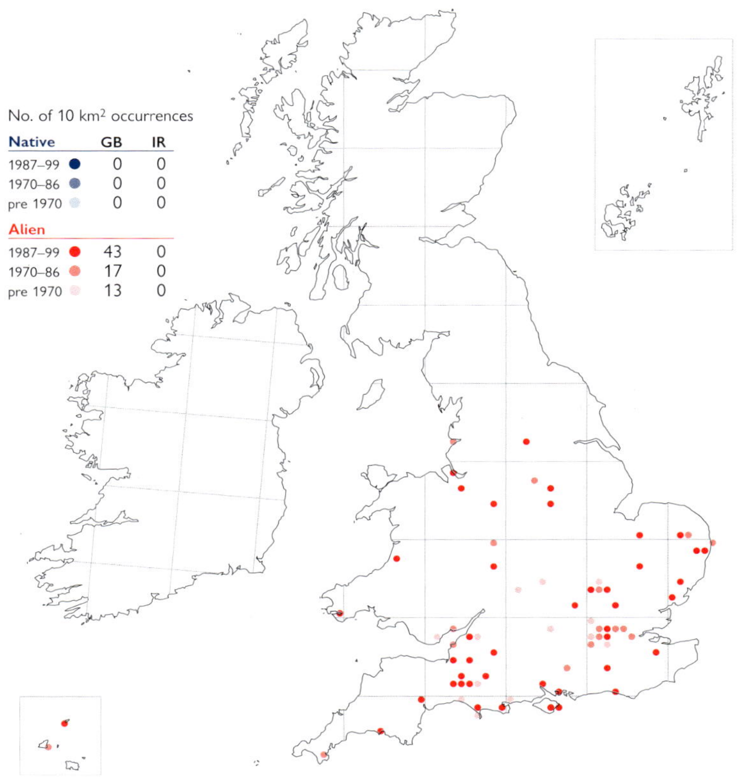

This tall rhizomatous perennial herb is a casual of rubbish tips and waste ground, introduced from bird-seed, wool, grain, soya-bean waste and other sources. Lowland.

Neophyte. *S. halepense* was cultivated in Britain by 1691, and was recorded from the wild by 1924. It is probably increasing, though some of the apparent increase could be due to the recent upsurge of interest in alien species.

Perhaps native to the Mediterranean region, but spread worldwide in cultivation or as a weed in warm-temperate and tropical areas and as a casual in temperate zones.

References: Ryves *et al.* (1996), Vaughan & Geissler (1997).

S. J. LEACH

Zea mays Maize

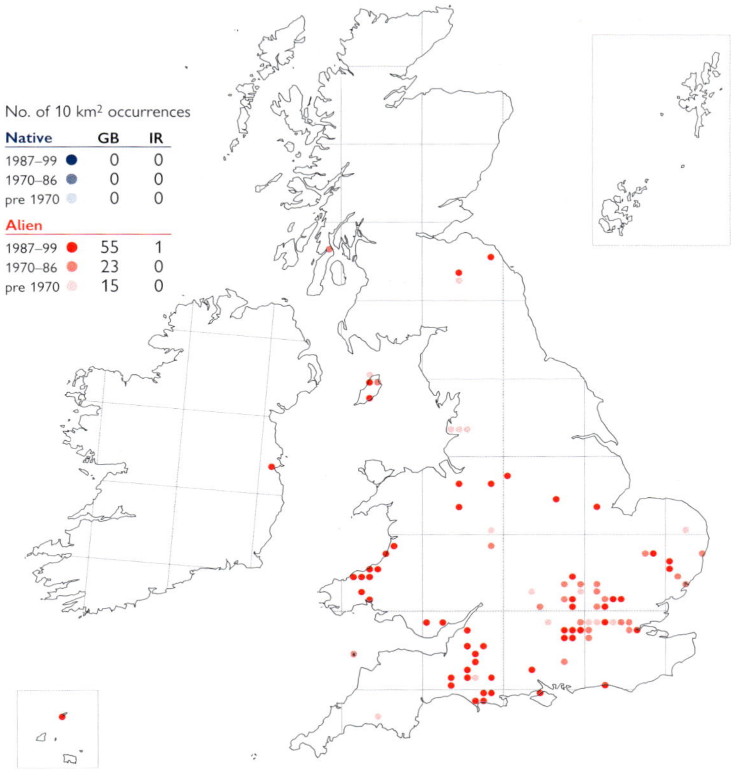

A relic or escape from cultivation found on field edges, rubbish tips and waste ground, and also as a casual of bird-seed and kitchen waste in these habitats. The male and female flowers are borne on separate inflorescences; pollination is by wind and the seeds are not dispersed but are retained on the cob. Lowland.

Casual. *Z. mays* was being grown in Britain by 1562, but only extensively cultivated in S. Britain since the 1970s, mainly for livestock feed and, on a smaller scale, for human consumption. It was recorded from the wild by 1876, and is certainly increasing.

Not known in the wild, but long cultivated in the Americas and now grown in some temperate and most tropical and subtropical regions.

References: Robinson & Treharne (1985), Ryves *et al.* (1996), Smart & Simmonds (1995), Vaughan & Geissler (1997).

S. J. LEACH

Sparganium erectum Branched Bur-reed

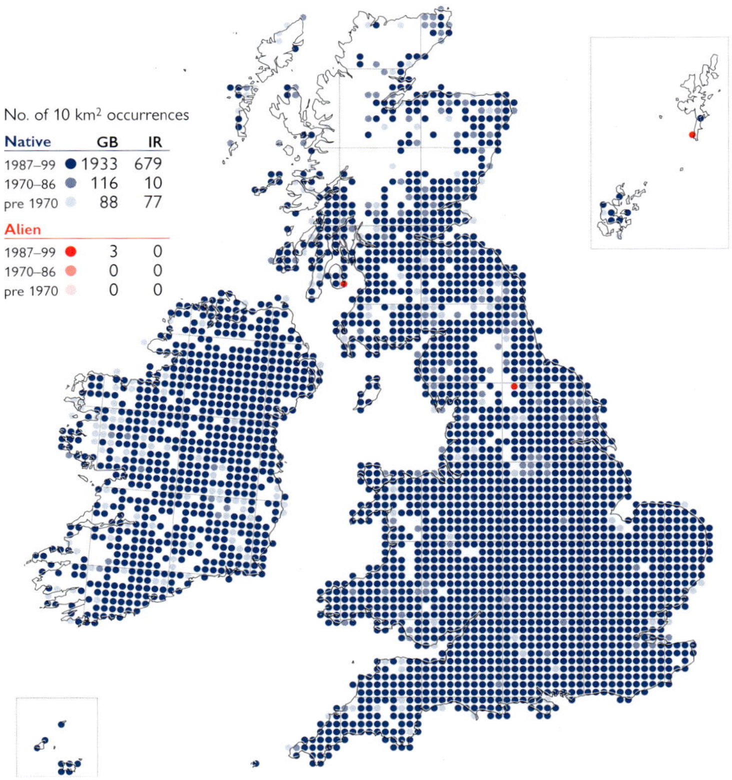

A rhizomatous perennial emergent which grows in shallow water in lakes, rivers, streams, canals and ditches. Although it usually occurs in a narrow band at the water's edge, it is sometimes found as larger stands in swamps. It grows in mesotrophic or eutrophic habitats, and is very tolerant of eutrophication. Cattle will eat it readily, and it is often absent or rare on grazed lake shores. 0–425 m (Nant Groes, Cards.).

Native (change +0.48). The distribution of this species is essentially similar to that mapped in the 1962 *Atlas*, with additional records probably attributable to more effective recording.

Circumpolar Temperate element, but absent from eastern N. America.

References: Atlas (345a), Atlas Supp. (144–145), Grime *et al.* (1988), Cook (1962), Hultén & Fries (1986), Preston & Croft (1997).

C. D. PRESTON

Sparganium emersum Unbranched Bur-reed

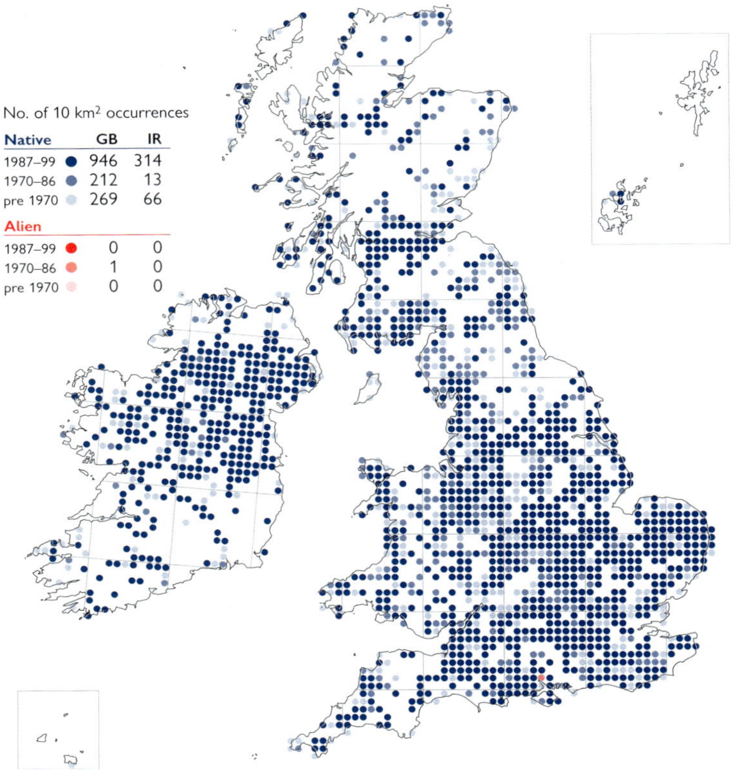

A perennial herb of still or slowly flowing, mesotrophic or eutrophic waters in lakes, ponds, rivers, streams, canals and ditches. Like *S. erectum*, it is a rhizomatous perennial but it usually grows in deeper water. It is tolerant of disturbance and may be frequent even in heavily managed rivers. 0–500 m (Crook Burn, Cumberland).

Native (change +0.61). The map shows that *S. emersum* is more frequent than was apparent in the 1962 *Atlas*, when the taxonomy of the floating-leaved *Sparganium* species was probably less well-understood by many recorders.

Circumpolar Boreo-temperate element.

References: Atlas (345b), Hultén & Fries (1986), Preston & Croft (1997).

C. D. PRESTON

Sparganium angustifolium Floating Bur-reed

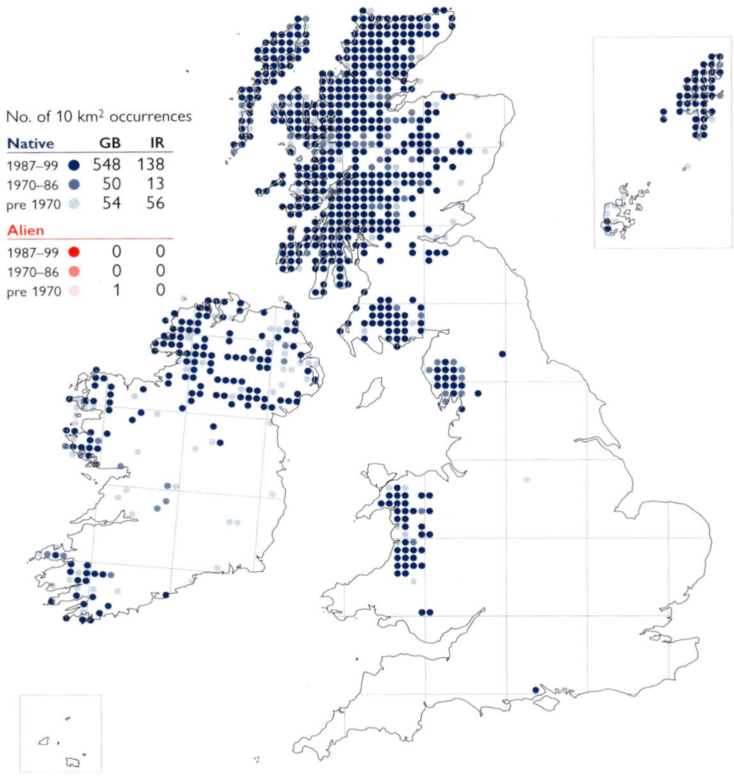

A perennial herb of clear, oligotrophic water, only rarely extending into mesotrophic conditions. It is most frequent in upland lakes but also grows in pools, rivers, streams, canals and ditches. Many sites are exposed to strong winds, but it prefers water 0.3–1.5 m deep, away from the most exposed shallows. 0–1005 m (Beinn Heasgarnich, Mid Perth).

Native (change +1.66). Because of taxonomic uncertainty, and its presence in areas which were poorly recorded, this species was seriously under-recorded in the 1962 *Atlas*. Some populations in N.E. Ireland and along the eastern edge of the Scottish range have been lost, possibly owing to eutrophication.

European Boreal-montane element; also in E. Asia and N. America.

References: Atlas (345c), Hultén & Fries (1986), Meusel *et al.* (1965), Preston & Croft (1997).

C. D. PRESTON

Sparganium natans Least Bur-reed

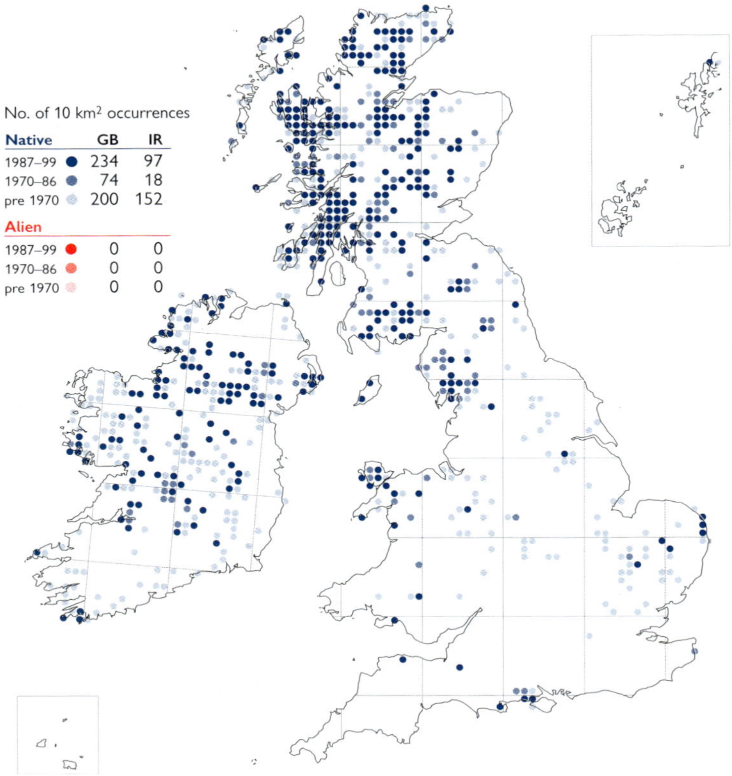

S. natans grows in shallow, sheltered waters at the edges of lakes, or in ponds, slowly flowing streams and drainage ditches. It is found in mesotrophic, highly calcareous to acidic waters. Its rhizomes are short and it usually reproduces by seed. 0–650 m (Lochan Achlarich, Mid Perth).

Native (change –0.13). The contraction of the range of this species, with the loss of most populations in the lowlands, was already apparent by 1930, and this decline has continued. These losses are attributable to the drainage of wetlands in historic times, and perhaps to eutrophication of surviving sites. The distribution elsewhere appears to be more stable.

Circumpolar Boreo-temperate element.

References: Atlas (345d), Hultén & Fries (1986), Meusel *et al.* (1965), Preston & Croft (1997).

C. D. PRESTON

Typha latifolia Bulrush

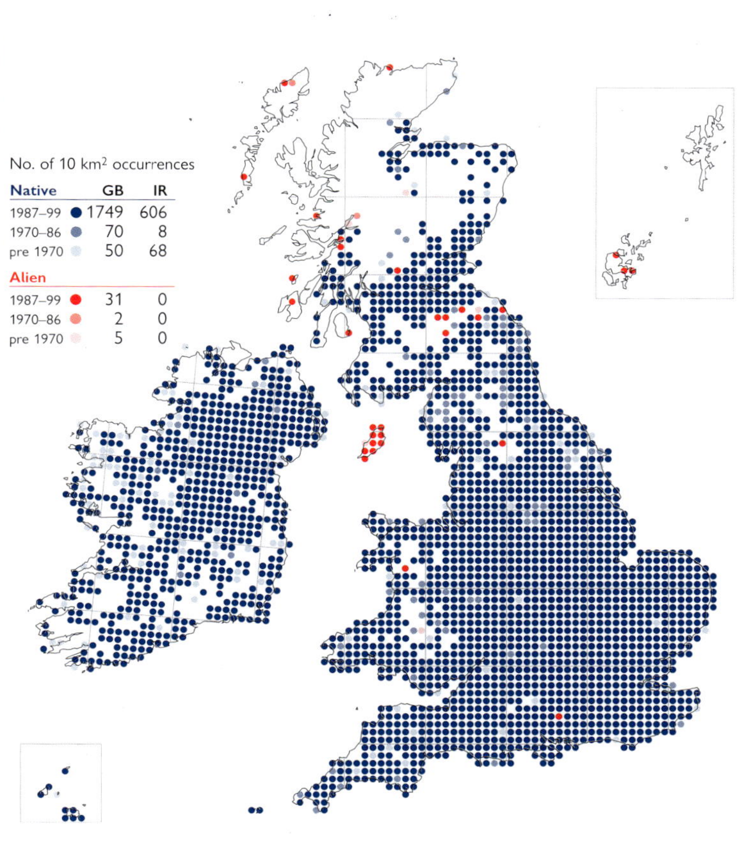

A rhizomatous perennial growing as an emergent in shallow water or on exposed mud at the edge of lakes, ponds, canals and ditches and (less frequently) by streams and rivers. It favours nutrient-rich sites. It spreads by wind-dispersed fruits, often colonising newly excavated ponds and ditches and subsequently spreading by vegetative growth. 0–500 m (Brown Clee Hill, Salop).

Native (change +1.01). There is some evidence that *T. latifolia* increased in frequency in the 20th century in many areas, for reasons that are not entirely clear. It is now much more frequently recorded in Wales, N. England and Scotland than it was in the 1962 *Atlas*.

Circumpolar Southern-temperate element.

References: Atlas (346a), Grace & Harrison (1986), Grime *et al.* (1988), Hultén & Fries (1986), Preston & Croft (1997).

C. D. PRESTON

Typha angustifolia Lesser Bulrush

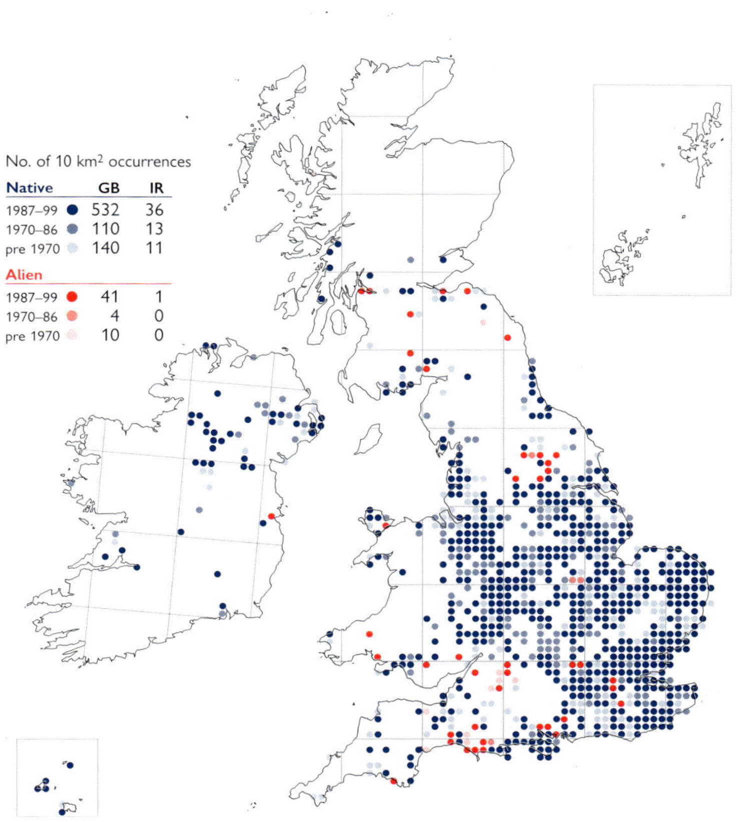

This tall, rhizomatous perennial grows as an emergent in mesotrophic or eutrophic water in lakes, ponds, ditches and flooded mineral workings. It tends to grow in deeper water than *T. latifolia*, and tolerates mesotrophic as well as eutrophic conditions. It also grows as floating rafts at some sites. Lowland.

Native (change +0.35). There is little evidence for any change in the distribution of this species since the 1962 *Atlas*. It is often cultivated as an ornamental aquatic, and some populations may have arisen from garden escapes.

Eurosiberian Temperate element; also in N. America.

References: Atlas (346b), Grace & Harrison (1986), Hultén & Fries (1986), Preston & Croft (1997).

C. D. PRESTON

Typha angustifolia × *T. latifolia* (*T.* × *glauca*)

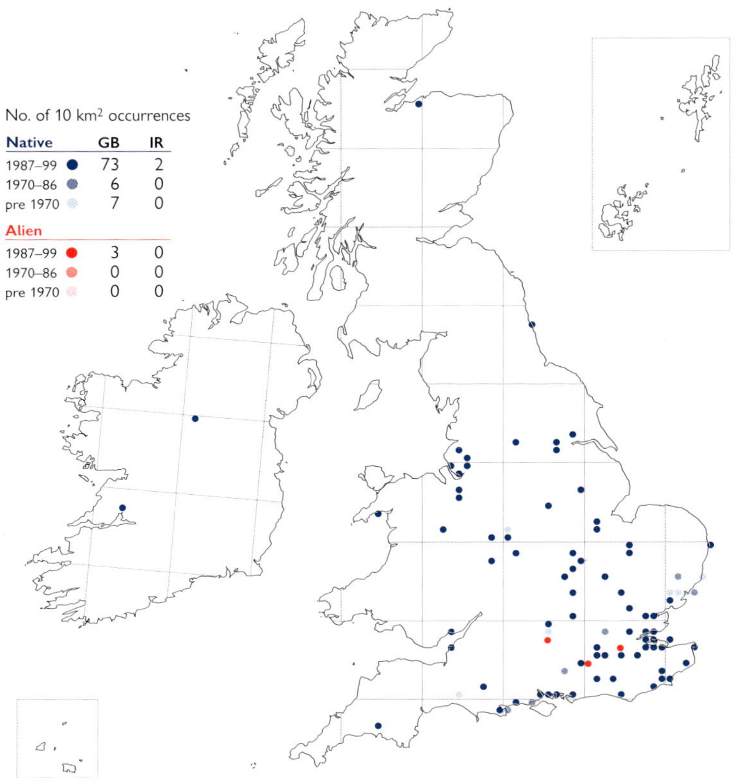

This vigorous rhizomatous perennial may form extensive clones in shallow water at the edge of eutrophic standing waters, sometimes in the absence of at least one parent. Research in N. America suggests that it is particularly successful in disturbed habitats with fluctuating water levels. F_1 hybrids are highly sterile. Lowland.

Native. Although this hybrid was clearly described by Lousley (1947), Tutin (1947) and Stace (1975) it was not mentioned by Clapham *et al.* (1952, 1962) and overlooked by most British botanists until Leslie (1984) drew attention to it. Records have increased gradually since then, but it must still be under-recorded. Lowland.

Widespread in Europe and N. America.

Reference: FNAEC (2000).

C. D. PRESTON

Tofieldia pusilla Scottish Asphodel

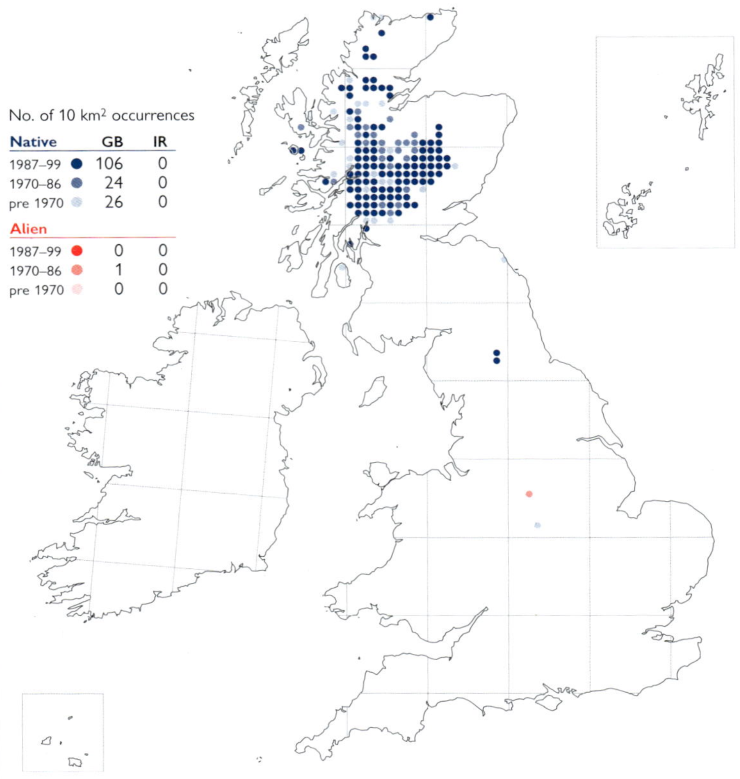

A rhizomatous, perennial herb growing by streams and in calcareous flushes, requiring constant moisture but not waterlogged conditions. Mainly upland, but from near sea level at Durness (W. Sutherland) to 975 m on Ben Lawers (Mid Perth).

Native (change –0.32). The distribution of *T. pusilla* appears to be stable. Since the 1962 *Atlas* its known range in Scotland has increased slightly, and it may still be present in squares for which there are only pre-1987 records. It was discovered new to science in 1671 in Berwickshire, and recorded from Charnwood Forest (Leics.) in 1828.

Circumpolar Arctic-montane element.

References: Atlas (311d), Hultén & Fries (1986), Meusel *et al.* (1965), Stewart *et al.* (1994).

I. TAYLOR

Narthecium ossifragum Bog Asphodel

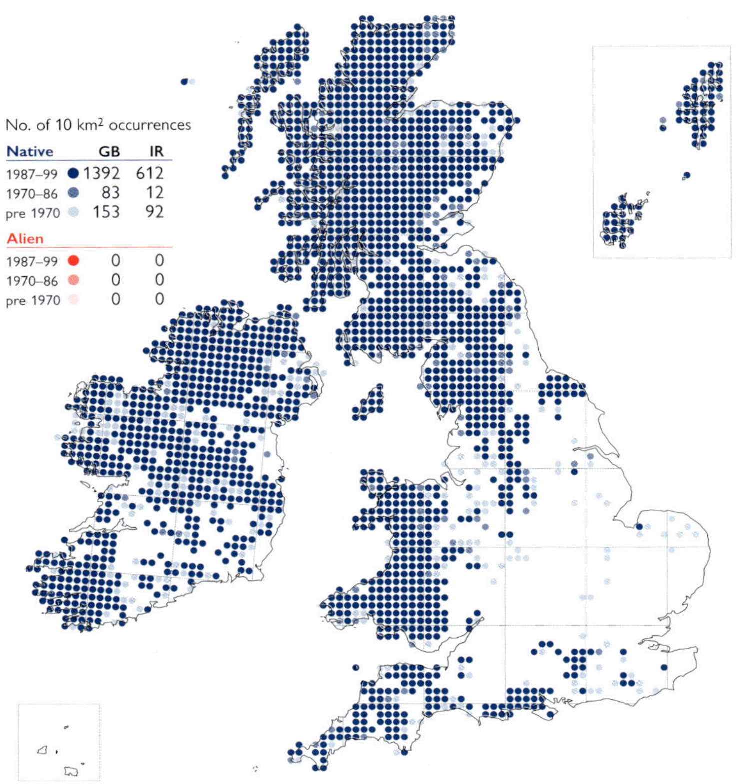

A rhizomatous perennial herb of wet, moderately basic to strongly acidic mineral soils and peats, in a wide range of raised, valley and blanket mire communities, and in wet heaths and flushes, especially where there is some water movement. It is intolerant of shade. Although it is slightly toxic, plants in the uplands may be heavily grazed. 0–1005 m (Beinn Heasgarnich, Mid Perth) but reportedly to 1130 m elsewhere in Scotland.

Native (change –0.32). Many sites for *N. ossifragum* were lost from lowland England before the 1962 *Atlas*. There have since been further local losses, notably from Surrey and Sussex, but the distribution is stable elsewhere.

Oceanic Boreo-temperate element.

References: Atlas (312a), Dupont (1962), Hultén & Fries (1986), Meusel *et al.* (1965), Summerfield (1974).

I. TAYLOR

Simethis planifolia Kerry Lily

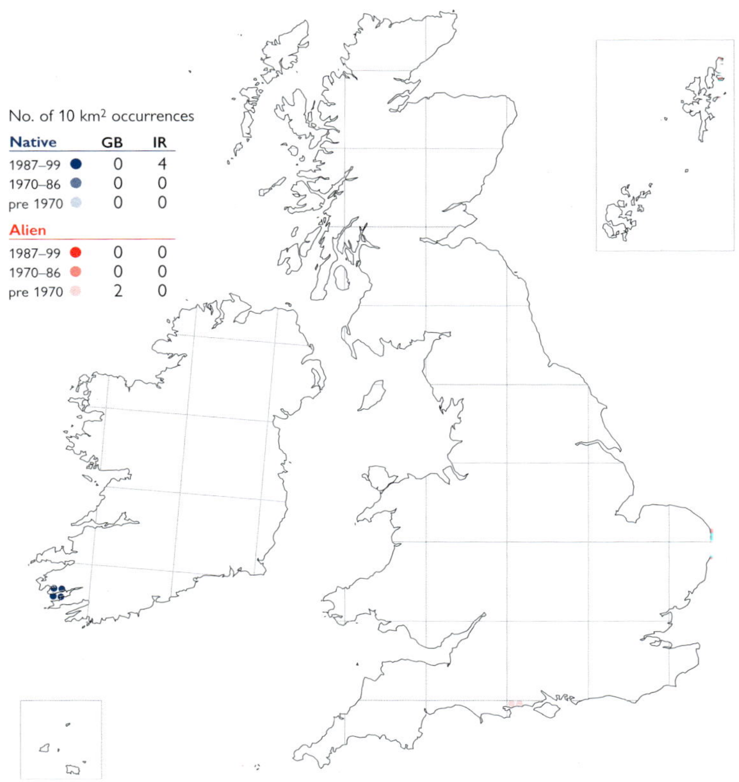

A rhizomatous, perennial herb found only in dry, rocky, maritime heath near Derrynane on the coast of S. Kerry and the Beara peninsula, W. Cork. It was formerly naturalised near the coast in S. England. Lowland.

Native. There has been some limited damage to the Irish population of *S. planifolia* from housing development, but its distribution appears to be relatively stable overall. Uncontrolled fires have affected some populations, but these may actually favour the species. The English plants may have been imported with *Pinus pinaster*. They were known in Dorset from 1847 until *c.* 1914, and in S. Hampshire in 1915.

Suboceanic Southern-temperate element.

References: Atlas (312b), Curtis & McGough (1988), Dupont (1962).

I. TAYLOR

Hemerocallis fulva Orange Day-lily

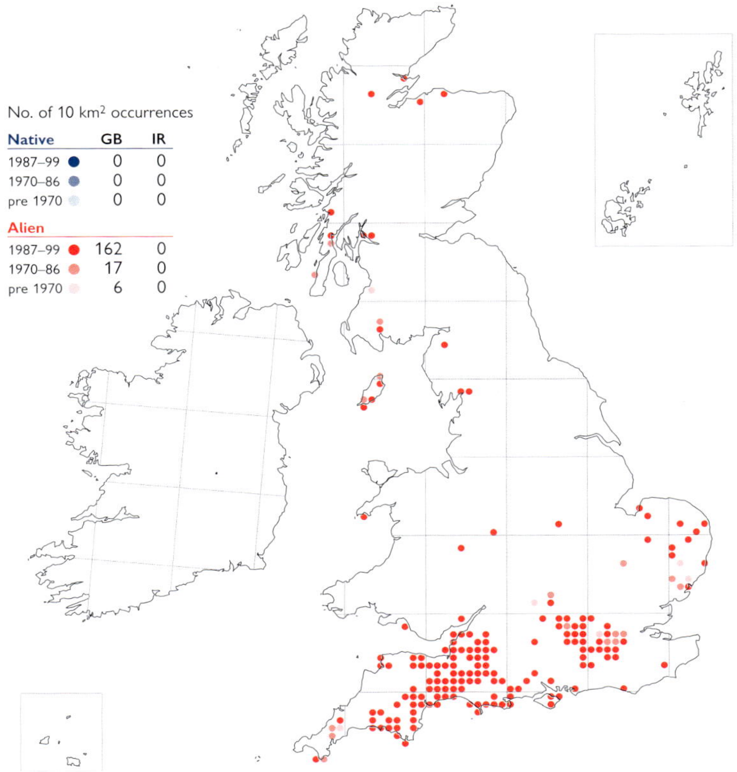

A rhizomatous perennial herb of open situations in dunes, on rough and waste ground and on roadsides. Lowland.

Neophyte. *H. fulva* was introduced into cultivation in Britain before 1596. It is now very commonly cultivated in gardens, and has become naturalised when discarded in suitable habitats. It was known to occur in the wild by at least 1905, and is probably increasing in range and frequency through repeated introductions.

A cultivated plant of garden origin.

I. TAYLOR

Kniphofia uvaria Red-hot-poker

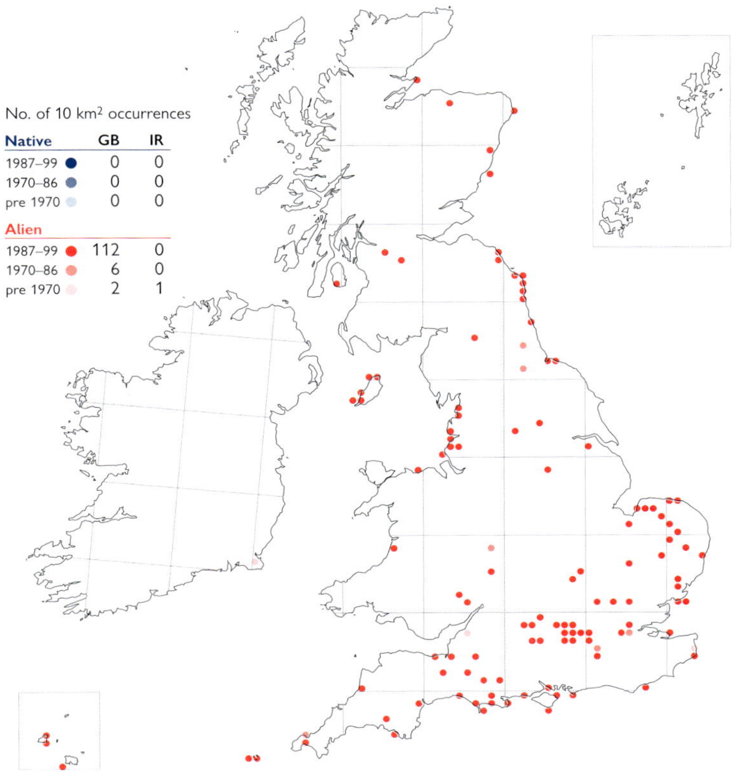

A rhizomatous perennial herb of sand dunes, river banks, roadsides, quarries and waste ground. It frequently occurs as persistent isolated clumps arising from discarded or deliberately planted garden stock. Seed is occasionally set and large populations can form in this way, especially on well-drained soils in southern and coastal localities. Lowland.

Neophyte. *K. uvaria* has been cultivated in Britain since 1705, but was not recorded from the wild until 1950. There is little evidence to suggest that it is increasing, although it may be becoming more frequent inland in S. England. A range of hybrids is cultivated and other taxa have almost certainly been recorded as *K. uvaria*.

Native of S. Africa.

I. TAYLOR

Colchicum autumnale Meadow Saffron

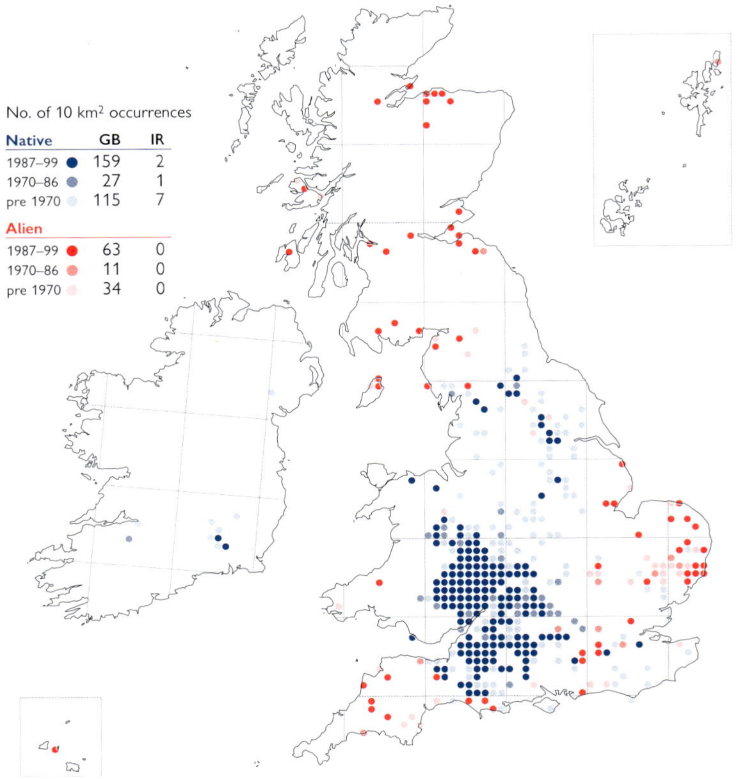

A cormous perennial herb of damp grassy places, including damp meadows and river banks, but most frequently encountered in clearings and rides within woodland, as it is toxic to livestock and often destroyed when found in grazed situations. Lowland.

Native (change –0.14). *C. autumnale* had been lost from most of its outlying sites before 1930 because of loss of habitat. Further losses have occurred since then, especially from meadows, but it remains frequent within its core area. This showy flower is popular in gardens, and readily becomes naturalised when planted or discarded into suitable habitats.

European Temperate element.

References: Atlas (317a), Butcher (1954), Curtis & McGough (1988), Mabey (1996), Meusel *et al.* (1965).

I. TAYLOR

Lloydia serotina Snowdon Lily

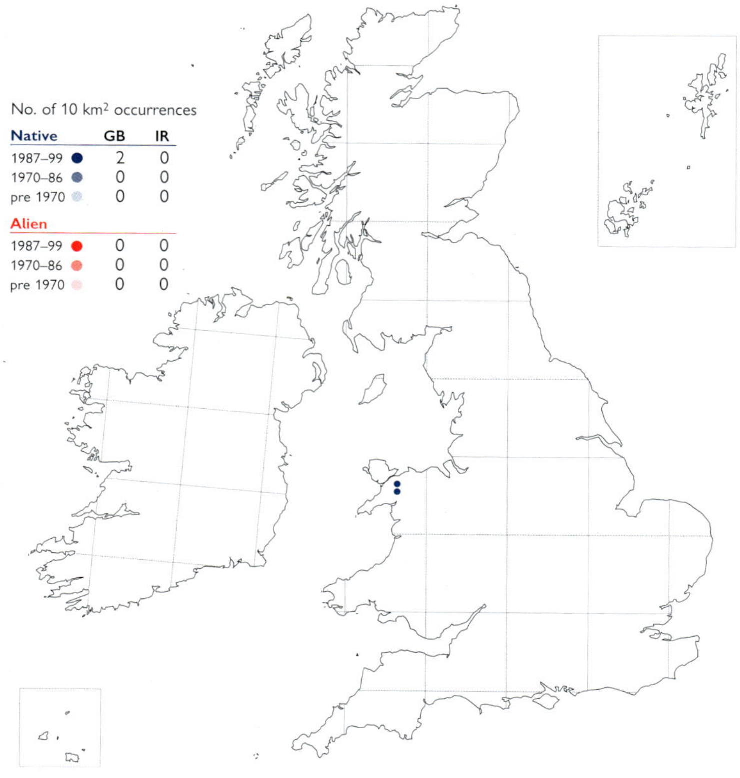

A bulbous perennial herb confined to damp ledges and crevices of mildly acidic rocks, usually on N.- to N.E.-facing cliffs, frequently in shaded sites and often sheltered by overhangs. Seeds are rarely produced and most reproduction is by rhizome-like structures. 550–760 m (Snowdonia, Caerns.).

Native. *L. serotina* appears to be a glacial relic in the British Isles. Populations have long been restricted to relatively inaccessible ground, perhaps because of over-grazing by sheep, and have been further reduced by collection since its discovery in the late 17th century. They now seem stable.

Circumpolar Arctic-montane element, but absent from the European Arctic and from eastern N. America.

References: Atlas (315a), Jones & Gliddon (1999), Meusel *et al.* (1965), Woodhead (1951a), Wigginton (1999).

I. TAYLOR

Gagea lutea Yellow Star-of-Bethlehem

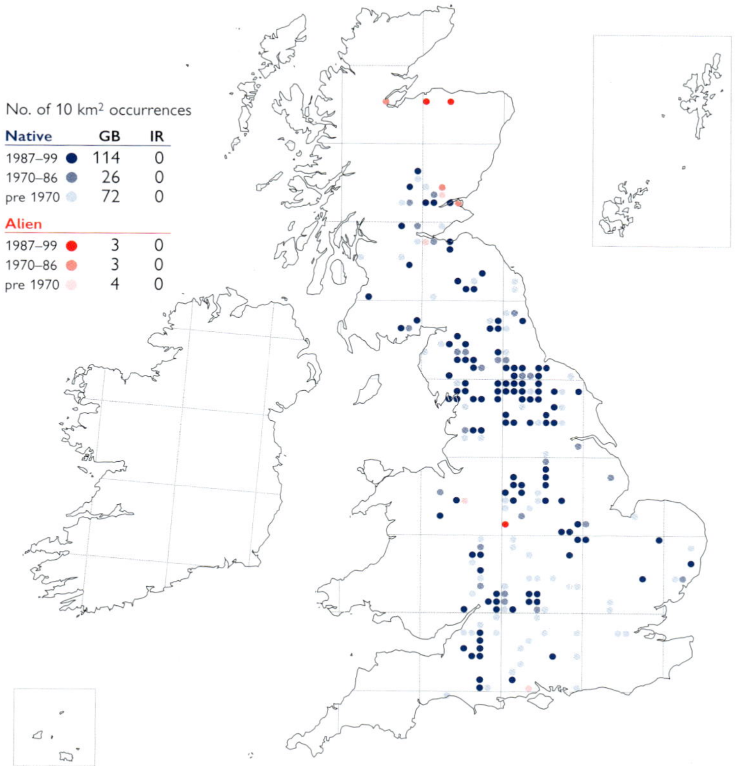

A bulbous perennial herb of moist, base-rich, shady habitats including woods, hedgerows, limestone pavements, pastures, riverbanks and stream banks; sometimes washed down on alluvium in riverine woodland subject to seasonal flooding. Mainly lowland, but to 340 m near Ribblehead (Mid-W. Yorks.).

Native (change +0.16). *G. lutea* is often shy-flowering, especially in its shadier localities, and many populations are small and widely scattered. These factors undoubtedly led to it being under-recorded in the 1962 *Atlas*. It is, therefore, difficult to assess trends, but the overall distribution appears to be relatively stable.

European Temperate element, with a continental distribution in W. Europe; also in C. and E. Asia.

References: Atlas (315a), Hultén & Fries (1986), Meusel *et al.* (1965), Stewart *et al.* (1994).

I. TAYLOR

Gagea bohemica Early Star-of-Bethlehem

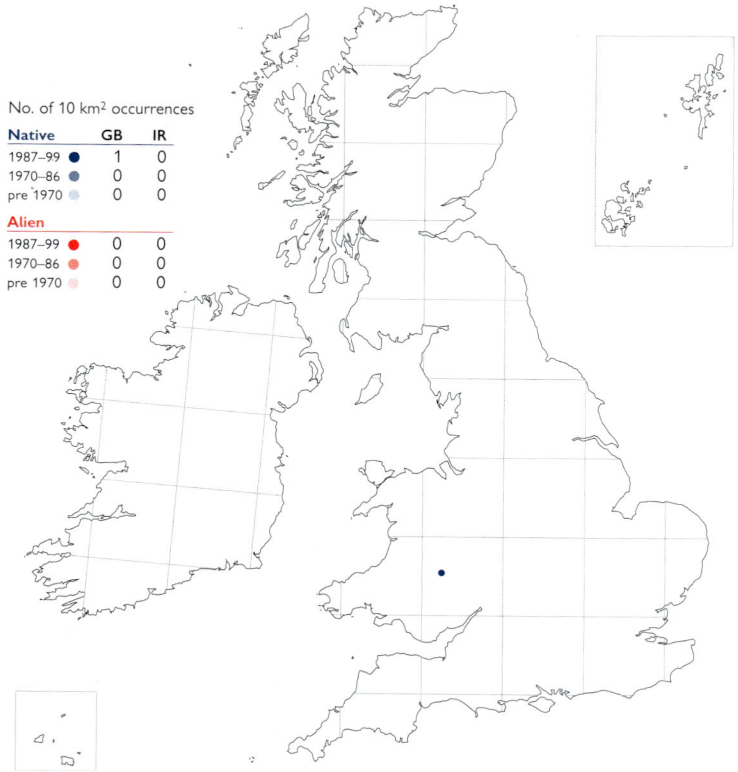

A bulbous perennial herb growing in shallow pockets of soil in crevices, on ledges and in small, grazed patches of turf on S.- and E.-facing slopes overlying dolerite. Between 210 and 330 m at Stanner Rocks (Rads.).

Native. *G. bohemica* rarely flowers in Britain, reproducing almost exclusively by bulbils. This, coupled with its very early flowering season and a long summer dormancy, led to the species remaining undetected until 1965 (and unrecognised as *G. bohemica* until 1975). However, the population is large and apparently stable.

European Southern-temperate element, with a continental distribution in W. Europe.

References: Kay & John (1995), Rix & Woods (1981), Slater (1990), Wigginton (1999).

I. TAYLOR

Tulipa sylvestris Wild Tulip

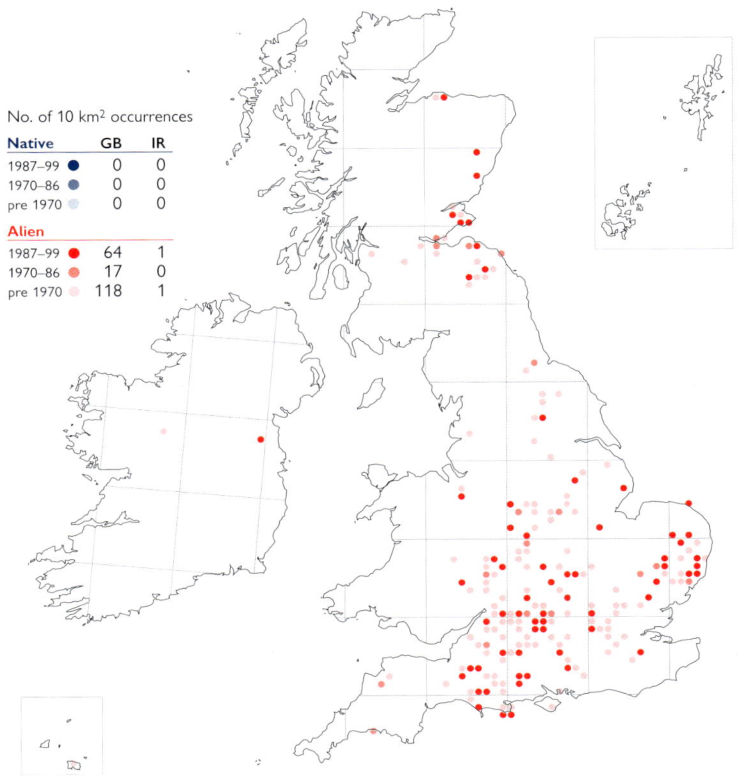

A bulbous perennial herb of open woodlands, orchards, hedgerows, riversides, chalk pits, grassy banks and waste ground. Populations can arise from discarded bulbs, deliberate planting in the wild or as relics of cultivation. Plants often flower poorly and set little seed, but populations can be very long-lived. Lowland.

Neophyte (change −0.69). *T. sylvestris* was in cultivation in Britain by 1596. It was formerly much cultivated, and was recorded from the wild by 1790. It appears to have been widely naturalised by the late 18th and 19th centuries but it had already declined dramatically by the time of the 1962 *Atlas*.

Native range uncertain; present in Europe, N. Africa and S.W. Asia.

References: Atlas (314d), Mabey (1996).

I. TAYLOR

Tulipa gesneriana Garden Tulip

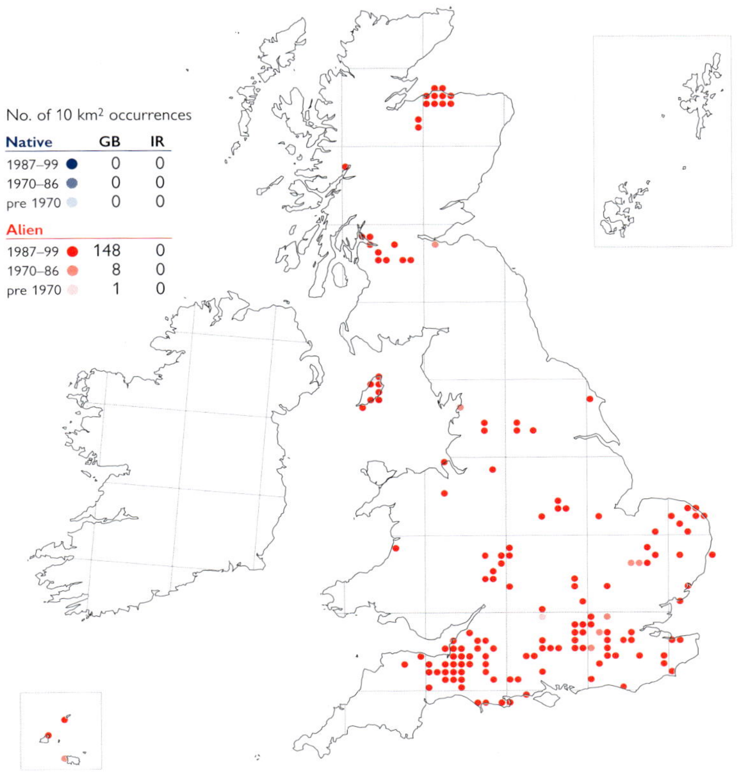

A bulbous perennial herb of rough ground, roadsides, quarries, churchyards and amenity grasslands. It often occurs as short-lived populations arising from discarded bulbs, but can persist when planted or thrown out in favourable habitats, or when occurring as a relic of cultivation. Lowland.

Neophyte. *T. gesneriana* was cultivated in Britain by 1577 and is very common in gardens; many forms and cultivars, developed and selected in Europe since the Renaissance, are available. It was not, however, recorded from the wild until 1955. It may be increasing, but might be over-enthusiastically recorded in some areas where it is scarcely present in the wild.

T. gesneriana is of uncertain, and probably garden, origin.

References: Grey-Wilson & Mathew (1981), Sell & Murrell (1996).

I. TAYLOR

Fritillaria meleagris Fritillary

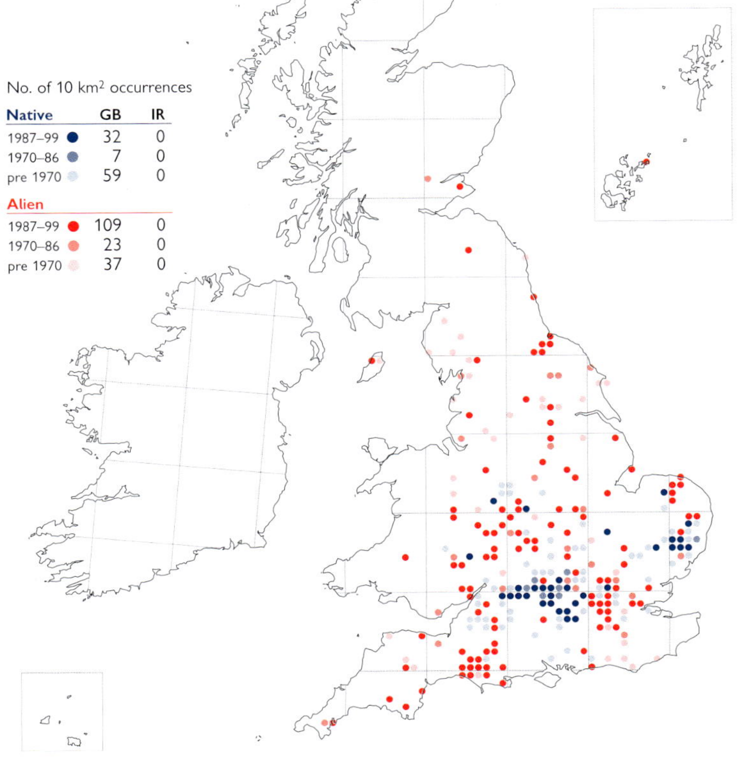

A bulbous perennial herb of damp, sometimes winter-flooded, neutral grasslands, usually those managed for hay with aftermath grazing. It is frequently planted in other grassland habitats and sometimes becomes naturalised. Lowland.

Native or alien (change +0.86). This species was cultivated in Britain by 1578 but only found in the wild in 1736; it has never been clear whether populations in traditionally managed floodplain meadows in C. and S.E. England are native. Many such populations have been lost through habitat destruction. Much of the decline, however, had occurred by the time of the 1962 *Atlas*, and losses since then have been limited by protection.

European Temperate element.

References: Atlas (314c), Gillam (1993), Harvey (1996), Hultén & Fries (1986), Oswald (1992), Mabey (1996), Stewart *et al.* (1994).

I. TAYLOR

Lilium martagon Martagon Lily

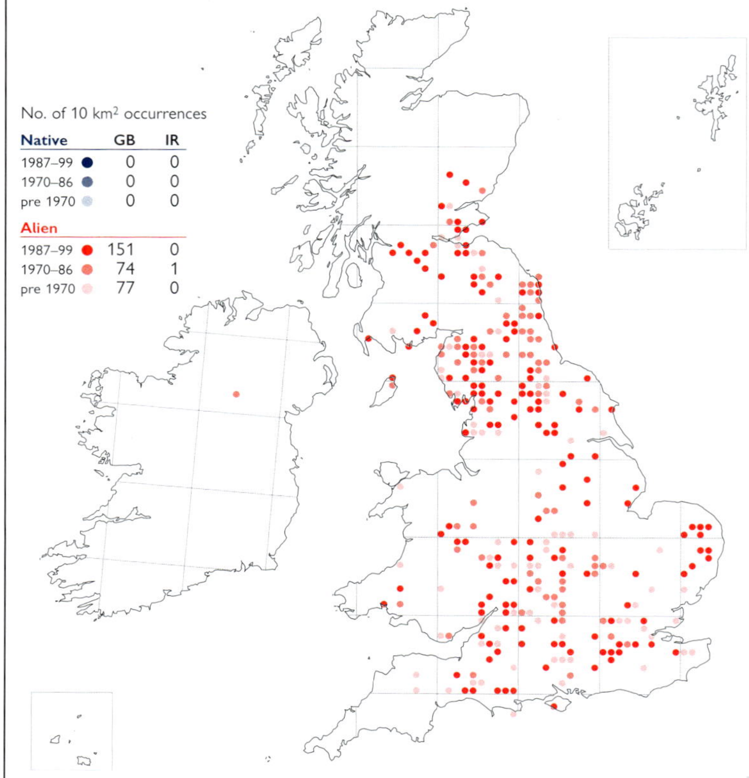

A bulbous perennial herb, usually occurring in small clumps near woodland edges or in coppiced woodland. It may be abundant in newly coppiced areas. It is also known from less natural sites, including old orchards and derelict gardens. The seeds are wind-dispersed but the bulbs may be carried by jays. Lowland.

Neophyte (change +0.83). *L. martagon* was introduced into British gardens by 1596. It was first recorded in the wild in 1782, and not until 1883 in the Wye Valley, where it was once considered to be native in ancient woodland. There is little evidence of significant distributional change since the 1962 *Atlas*.

A Eurosiberian Temperate species; naturalised in N. Europe north of its native range.

References: Atlas (314b), Harvey (1996), Lousley (1976a), Meusel *et al.* (1965).

I. TAYLOR

Lilium pyrenaicum Pyrenean Lily

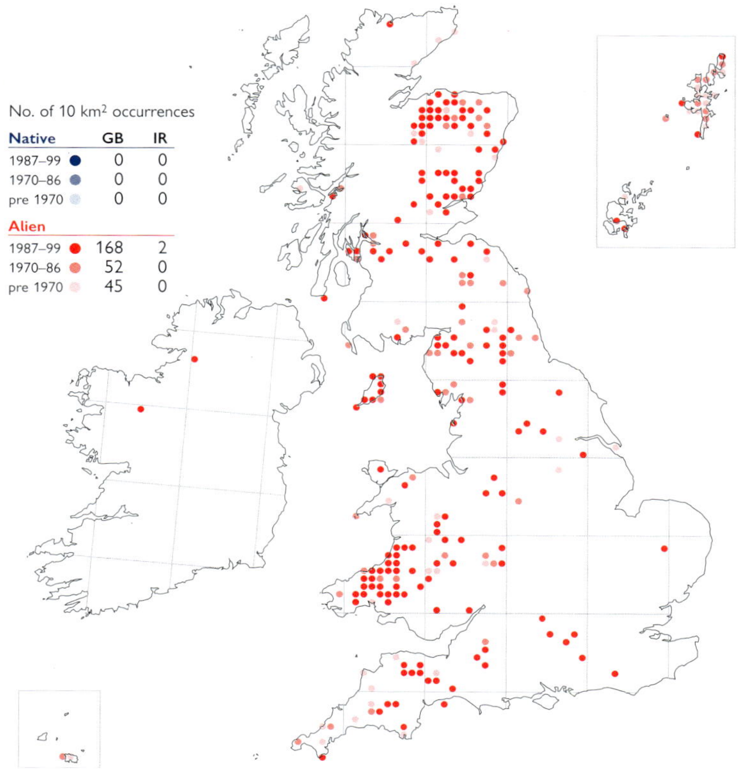

A bulbous perennial herb of woodlands, wood-borders, hedgerows and roadsides, often occurring as isolated clumps arising from discarded or deliberately planted garden stock. Relatively large, long-established and naturally regenerating populations are known from S.W. England, S. Wales and E. Scotland. 0–365 m (near Llanfair Clydogau, Cards.).

Neophyte. *L. pyrenaicum* has been cultivated in Britain since before 1596 and was recorded from the wild by 1853. There is little evidence of significant distributional change in recent years.

Native of the Pyrenees.

References: Dupont (1962), Meusel *et al.* (1965).

I. TAYLOR

Convallaria majalis Lily-of-the-valley

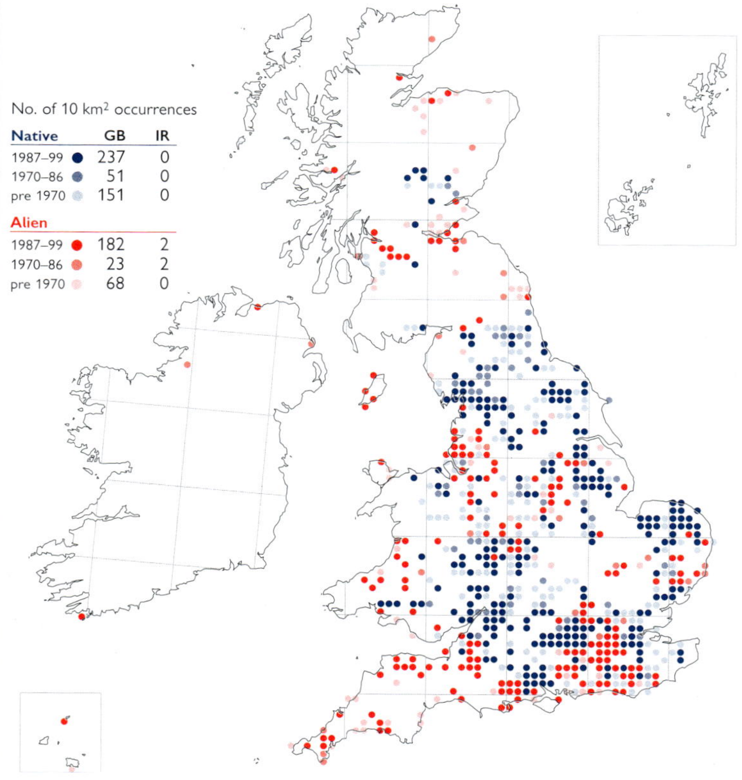

A rhizomatous perennial herb of freely-draining, nutrient-poor soils. It is most frequent in *Fraxinus* woods on limestone in N. and W. Britain, but also grows in the grikes of limestone pavement, in acidic woods in S.E. England, and in a fen in Cumbria. It is a common garden escape. 0–470 m (Carreg-goch, Brecs.).

Native (change +0.25). The native distribution of *C. majalis* is probably stable, but is now very largely obscured by naturalised plants. It is often difficult to separate native and alien plants and the map must be regarded as an approximation. Many of the losses had occurred by 1930.

European Boreo-temperate element; also in E. Asia and widely naturalised outside its native range.

References: Atlas (312c), Hultén & Fries (1986), Mabey (1996), Meusel *et al.* (1965), Peterken (1981).

I. TAYLOR

Polygonatum multiflorum Solomon's-seal

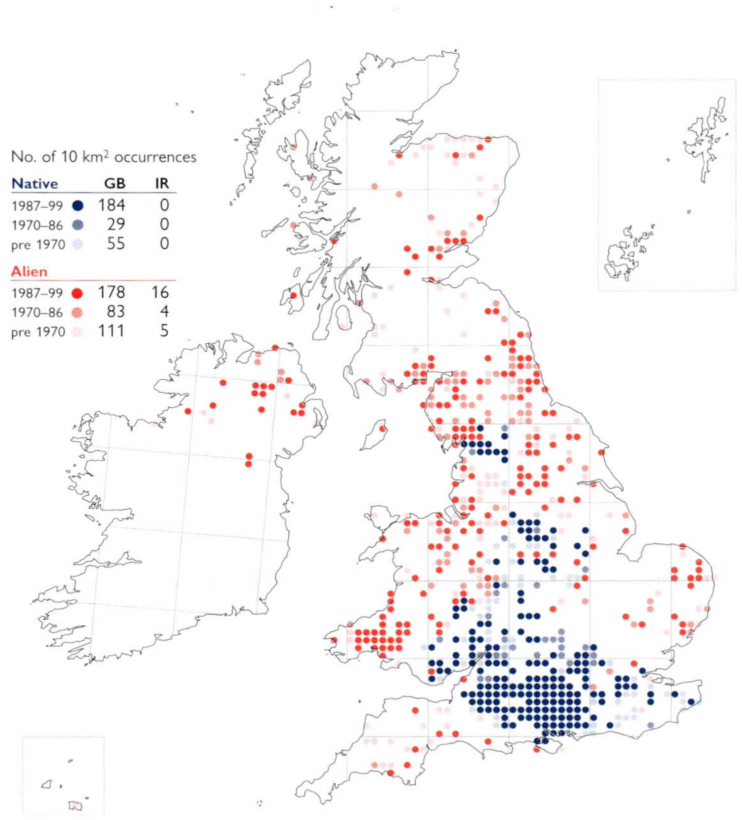

A rhizomatous perennial herb, native in *Fraxinus-Acer campestre* woods and thickets over chalk and limestone, and less frequently over non-acidic substrates. Lowland.

Native (change +0.27). *P. multiflorum* and *P.* × *hybridum* were not distinguished in the 1962 *Atlas*, so changes in their distribution are difficult to assess. *P. multiflorum* is often planted and its native range is difficult to define with certainty. Alien populations are certainly over-recorded for the hybrid *P.* × *hybridum*.

European Temperate element; also in C. and E. Asia.

References: Atlas (313b), Hultén & Fries (1986), Meusel *et al.* (1965).

I. TAYLOR

Polygonatum multiflorum × *P. odoratum* (*P.* × *hybridum*) Garden Solomon's-seal

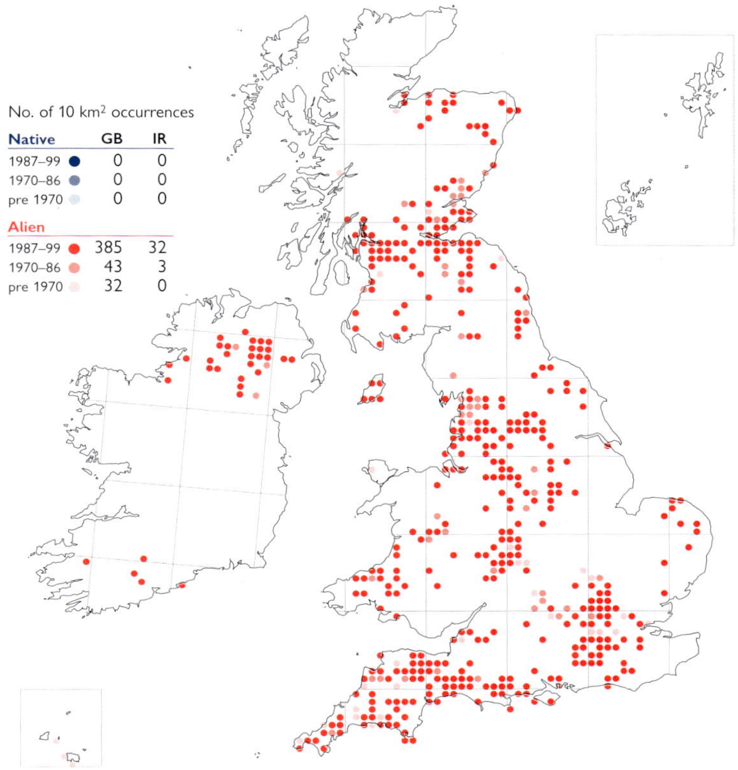

A rhizomatous perennial herb of woodland margins, churchyards, hedgerows, roadsides and rough ground. It is usually sterile. Lowland.

Neophyte. *P.* × *hybridum* is by far the most frequently grown *Polygonatum* in gardens, and is often outcast due to its vigour in cultivation. It has been recorded in the wild since at least 1867. Many records of naturalised plants previously identified as *P. multiflorum* are probably referable to this hybrid.

A hybrid of garden origin.

Reference: Stace (1975).

I. TAYLOR

Polygonatum odoratum Angular Solomon's-seal

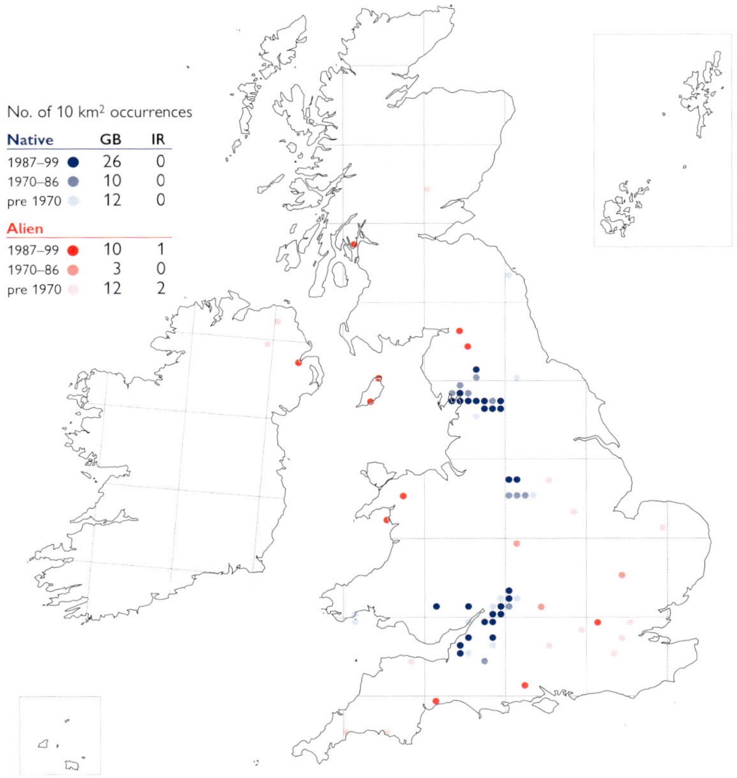

A rhizomatous, perennial herb of ancient *Fraxinus* woods, often growing in crevices and on outcrops of limestone. In N. England it is characteristic of grikes in limestone pavement. Generally lowland, but reaching 485 m on Craig y Cilau (Brecs.).

Native (change +0.34). Some populations of *P. odoratum* have been lost since the 1962 *Atlas* through the destruction of limestone pavement in the north of its range, but this threat is now much reduced and the distribution has probably stabilised. It is sometimes grown in gardens, and is occasionally naturalised when discarded or deliberately planted.

Eurasian Temperate element.

References: Atlas (313a), Hultén & Fries (1986), Meusel *et al.* (1965), Stewart *et al.* (1994).

I. TAYLOR

Polygonatum verticillatum
Whorled Solomon's-seal

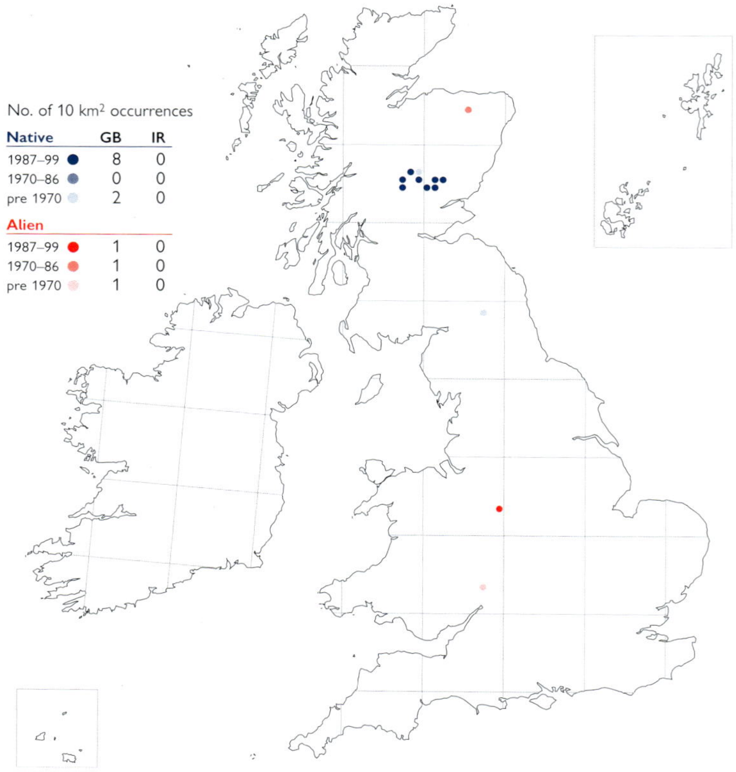

A rhizomatous, perennial herb usually found on moist, nutrient-rich, usually basic, soils in wooded gorges and on a wooded river bank. Plants reproduce vegetatively, by rhizomatous spread, but fruiting is generally poor, with recruitment from seed apparently very infrequent. Lowland.

Native. The distribution of *P. verticillatum* is stable, though some populations have been lost through erosion and habitat destruction. Flowering seems to be restricted by excessive shading at a number of localities and limited opportunities for cross-pollination further restrict seed production.

European Boreal-montane element.

References: Atlas (312d), Hultén & Fries (1986), Meusel *et al.* (1965), Wigginton (1999).

I. TAYLOR

Maianthemum bifolium May Lily

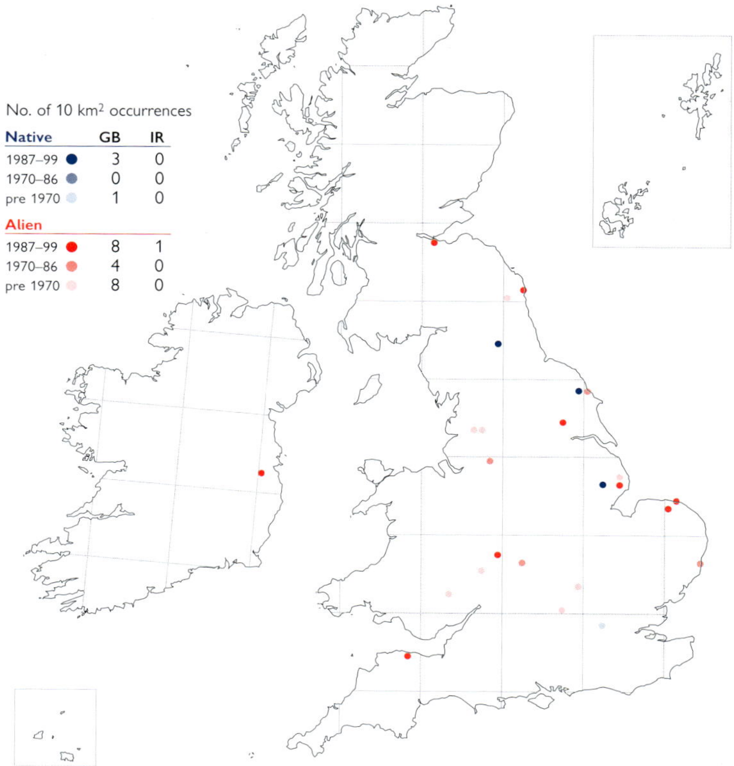

A rhizomatous, perennial herb of free-draining acidic soils in *Quercus-Betula* woodlands, but also persisting at sites replanted with conifers. Flowering and seed-set is poor. Lowland.

Native or alien (change +0.32). *M. bifolium* was known in both cultivation and the wild by 1597. Records from the 17th century suggest that it was formerly more abundant and widespread, but extant populations tend to be isolated and very small, although stable. It is native or long-established in N. & E. England, and is usually short-lived at those sites mapped as alien, other than in Norfolk where it has been known since 1955.

Eurasian Boreo-temperate element, with a continental distribution in W. Europe.

References: Atlas (313c), Hultén & Fries (1986), Meusel *et al.* (1965), Wigginton (1999).

I. TAYLOR

Paris quadrifolia Herb-Paris

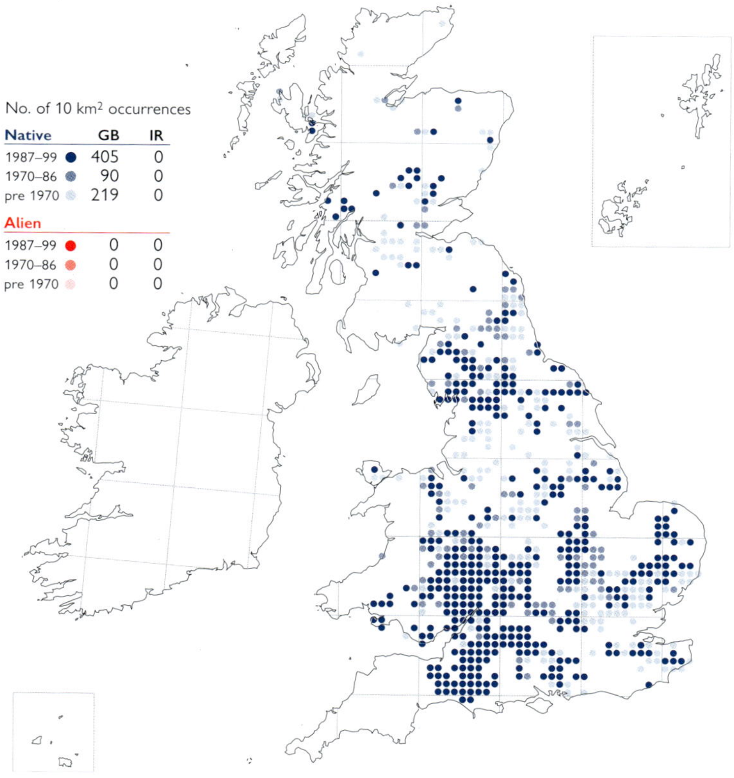

A rhizomatous, perennial herb of moist, calcareous, usually ancient, woodland, and occasionally found in grikes on open limestone pavement. It flowers and fruits most freely in the open stages of the coppice cycle, but persists in deep shade, and is well adapted to such conditions in managed woodland. Generally lowland, but reaching 360 m at Great Asby Scar (Westmorland) and Garrigill (Cumberland).

Native (change –0.68). Many sites for *P. quadrifolia* were lost in N. England and central Scotland before 1930, and since then there has been some decline in S.E. England through the destruction and coniferisation of woodland. It may spread into secondary woods which are adjacent to primary woodland.

Eurosiberian Boreo-temperate element.

References: Atlas (317b), Hultén & Fries (1986), Mabey (1996), Meusel *et al.* (1965).

I. TAYLOR

Ornithogalum pyrenaicum
Spiked Star-of-Bethlehem

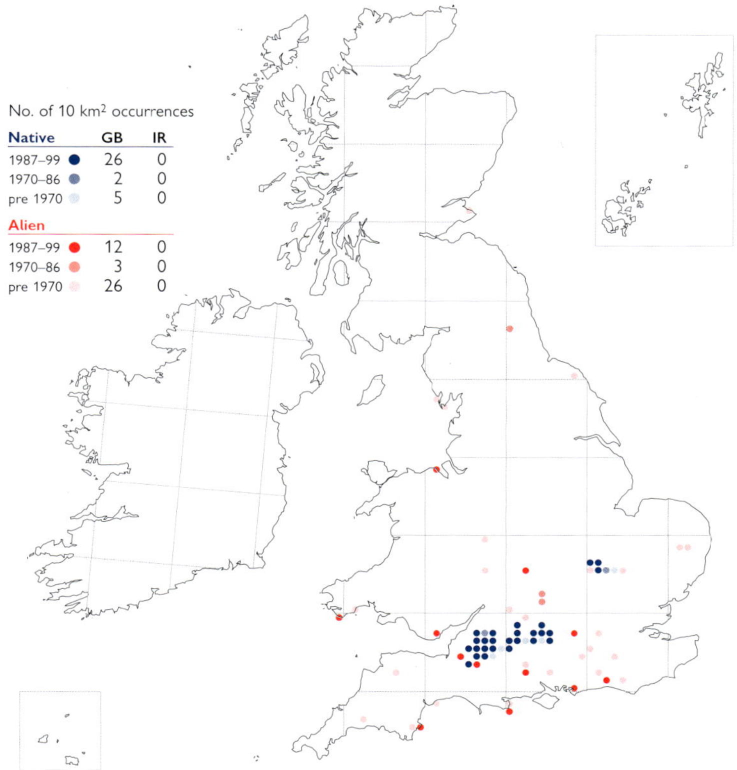

A bulbous perennial herb of *Fraxinus-Ulmus* woodland, hedgerows, road verges and rough grassy banks on calcareous soils, and a particular feature of green lanes on the borders of Somerset and Wiltshire. Lowland.

Native (change +0.14). There has been no significant change in the range of *O. pyrenaicum* since the 1962 *Atlas*, but some woodland populations appear to have succumbed to ecological changes following the loss of mature elms. It was formerly harvested, sometimes on a commercial scale, as a native equivalent to asparagus, and is occasionally grown in gardens, sometimes escaping and becoming naturalised.

Submediterranean-Subatlantic element.

References: Atlas (315d), Hill & Price (2000), Stewart *et al.* (1994).

I. TAYLOR

Ornithogalum angustifolium Star-of-Bethlehem

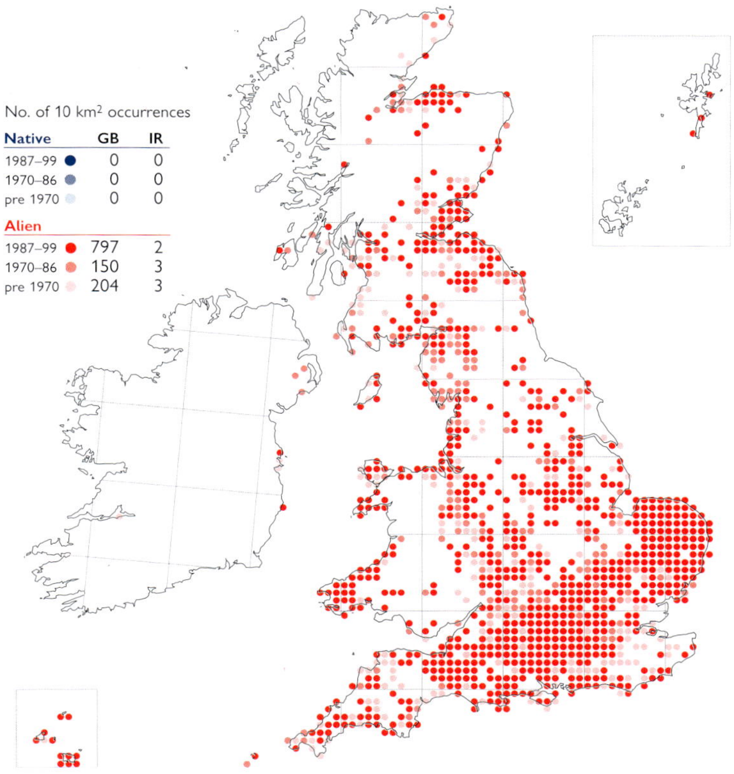

A bulbous perennial herb of rough pasture, dry, grassy banks and open woods. Lowland.

Neophyte (change +1.05). *O. angustifolium* is treated here as an introduction, although some authors (e.g. Trist, 1979; Petch & Swann, 1968) consider it to be native on sandy soils in Breckland. It was cultivated in Britain by 1548 and was first recorded in the wild in 1650, but not in Breckland until 1772. It has long been confused with the closely related *O. umbellatum*, which may be the more commonly cultivated species. The relative frequency of the two species in naturalised populations of *Ornithogalum* requires further clarification.

A European Southern-temperate species; widely naturalised outside its native range.

References: Atlas (315b), Beckett *et al.* (1999), Hultén & Fries (1986).

I. TAYLOR

Ornithogalum nutans
Drooping Star-of-Bethlehem

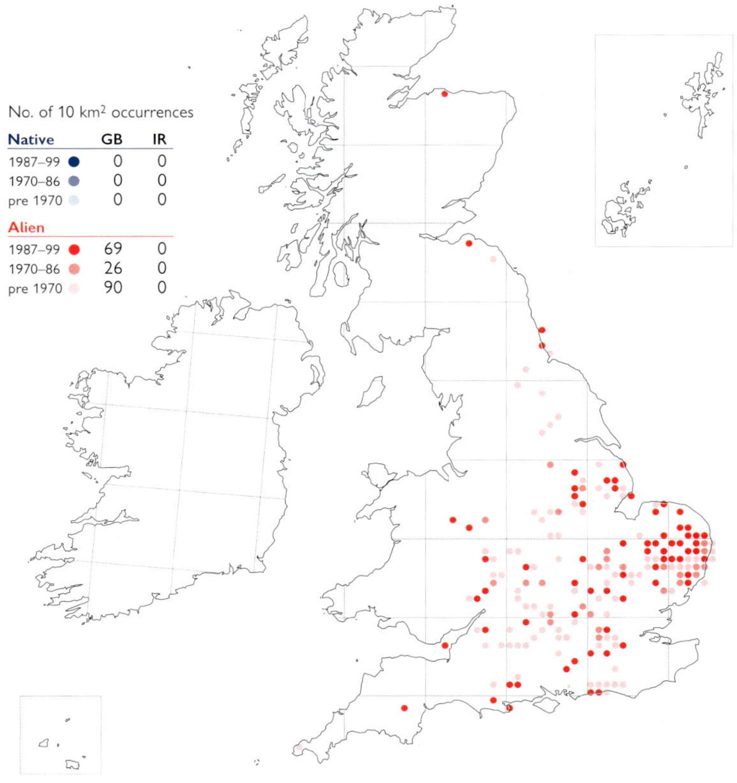

A bulbous perennial herb which occurs as a garden escape in scrub and on woodland edges, in churchyards, hedgerows, on roadsides and in rough grassy places. Lowland.

Neophyte (change –0.10). *O. nutans* was cultivated in Britain by 1648 and is frequently grown in gardens. It was recorded from the wild by 1821. Populations are usually small and short-lived though some are persistent, such as at Bodney churchyard (W. Norfolk) where it has been known since 1917.

Native of the Balkans (Bulgaria, Greece) and Turkey.

References: Atlas (315c), Beckett *et al.* (1999), Mabey (1996).

I. TAYLOR

Scilla siberica Siberian Squill

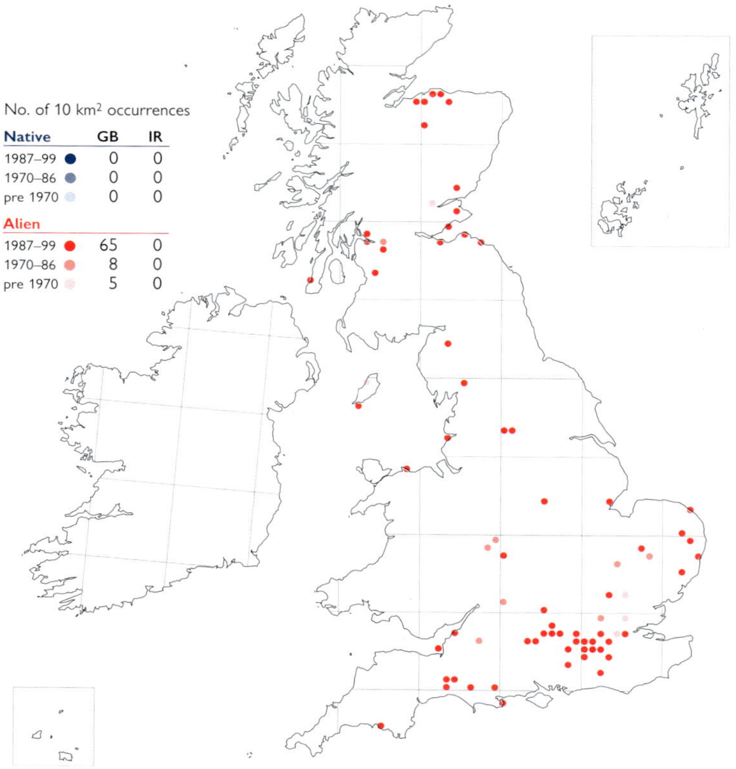

No. of 10 km² occurrences

Native	GB	IR
1987–99	0	0
1970–86	0	0
pre 1970	0	0
Alien		
1987–99	65	0
1970–86	8	0
pre 1970	5	0

A bulbous perennial herb of free-draining, often sandy soils in coastal grasslands, churchyards and open woodland, and on heaths, roadsides and waste ground. It often increases in abandoned gardens or when discarded or deliberately planted in suitable habitats. Lowland.

Neophyte. *S. siberica* was introduced in 1796 and is now commonly cultivated. However, it was not formally recorded from the wild until 1968.

Native of southern Russia, Turkey and the Caucasus.

Reference: Burton (1983).

I. TAYLOR

Scilla verna Spring Squill

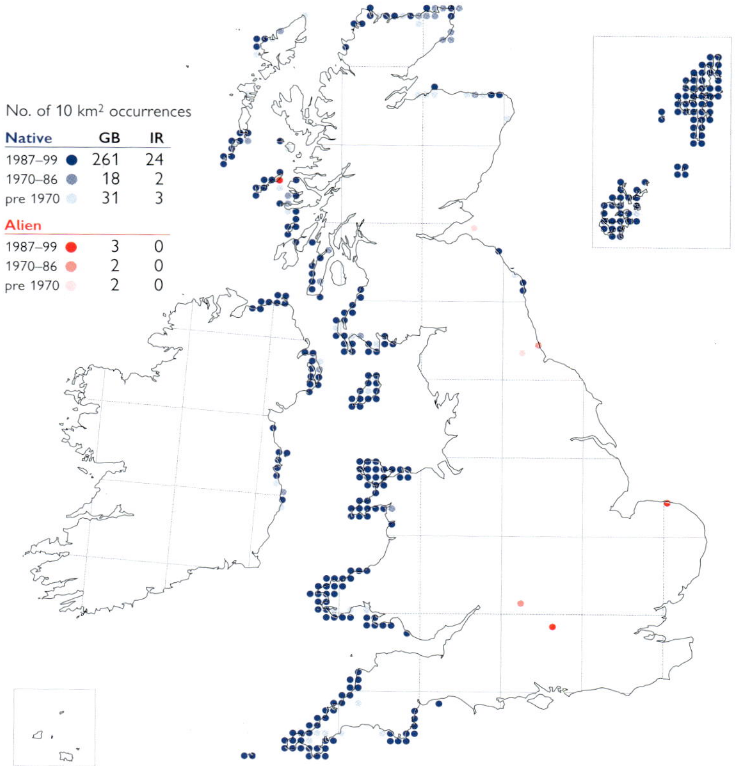

No. of 10 km² occurrences

Native	GB	IR
1987–99	261	24
1970–86	18	2
pre 1970	31	3
Alien		
1987–99	3	0
1970–86	2	0
pre 1970	2	0

A bulbous perennial herb of short turf and maritime heath on exposed cliff-tops and on rocky slopes near the sea, sometimes within the zone regularly affected by sea-water spray. In areas with a pronounced oceanic climate (e.g. Anglesey) it can occur on heathland well inland. Generally lowland, but reaching 415 m in Foula (Shetland).

Native (change +0.12). There is no evidence of any significant change in the distribution of *S. verna* since the 1962 *Atlas*. Within its highly restricted habitat *S. verna* is often present in large numbers, and its sites are usually too exposed for any major threat from successional or land-use change.

Oceanic Temperate element.

References: Atlas (316a), Dupont (1962), Hepburn (1952), Hultén & Fries (1986).

I. TAYLOR

Scilla autumnalis Autumn Squill

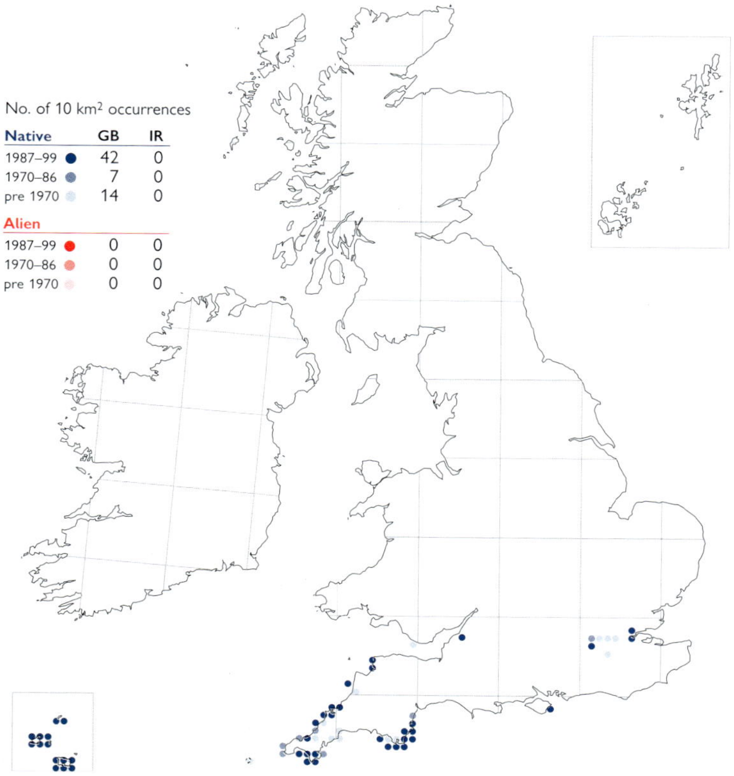

No. of 10 km² occurrences

Native	GB	IR
1987–99	42	0
1970–86	7	0
pre 1970	14	0
Alien		
1987–99	0	0
1970–86	0	0
pre 1970	0	0

A bulbous perennial herb of open, drought-prone grasslands and heathy vegetation in rocky or sandy places near the sea; also on terrace gravels in the lower Thames valley. Lowland.

Native (change −0.37). There were some losses of this species before 1930, but there is little evidence of significant change in distribution or abundance since the 1962 *Atlas*. *S. autumnalis* is tetraploid almost throughout its British range, the exception being on the S. coast of Cornwall and on Guernsey where a hexaploid race occurs.

Mediterranean-Atlantic element.

References: Atlas (316b), Stewart *et al.* (1994).

I. TAYLOR

Hyacinthoides non-scripta Bluebell

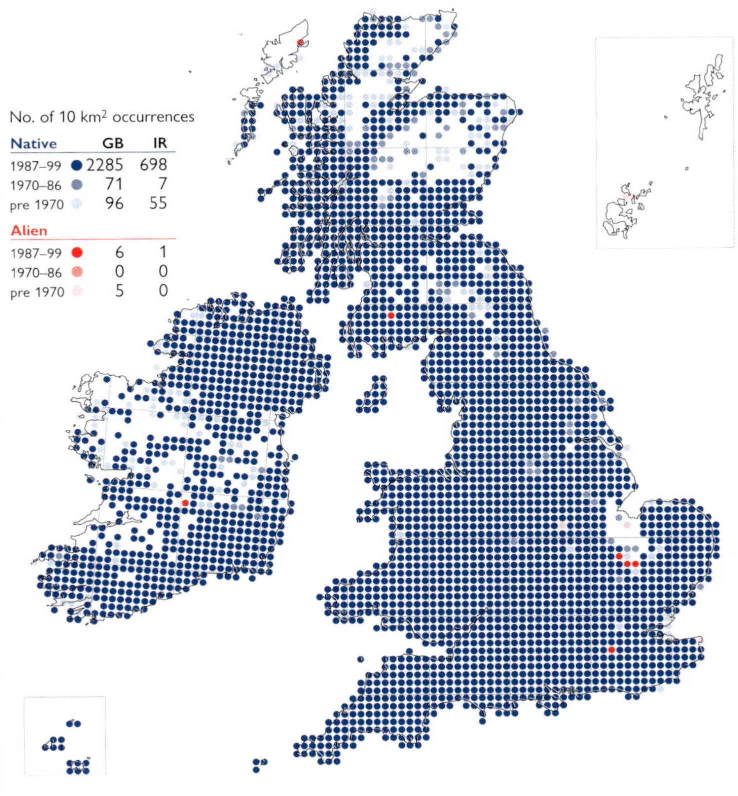

A bulbous perennial herb occurring, sometimes abundantly, in a wide variety of deciduous woodlands, in hedgerows, on shady banks and, especially in western and upland areas, in meadows, under *Pteridium* and on cliffs. It also occurs as a naturalised garden escape. It is sensitive to long-term grazing. Generally lowland, but reaching 685 m on Craig-yr-Ysfa (Caerns.).

Native (change –0.41). The overall distribution of *H. non-scripta* is stable and it remains abundant in suitable habitats throughout its range. A very few populations (e.g. in Norfolk) have been damaged recently by large-scale commercial collection.

Oceanic Temperate element.

References: Atlas (316c), Blackman & Rutter (1954), Dupont (1962), Grime *et al.* (1988), Meusel *et al.* (1965), Rackham (1980).

I. TAYLOR

Hyacinthoides hispanica Spanish Bluebell

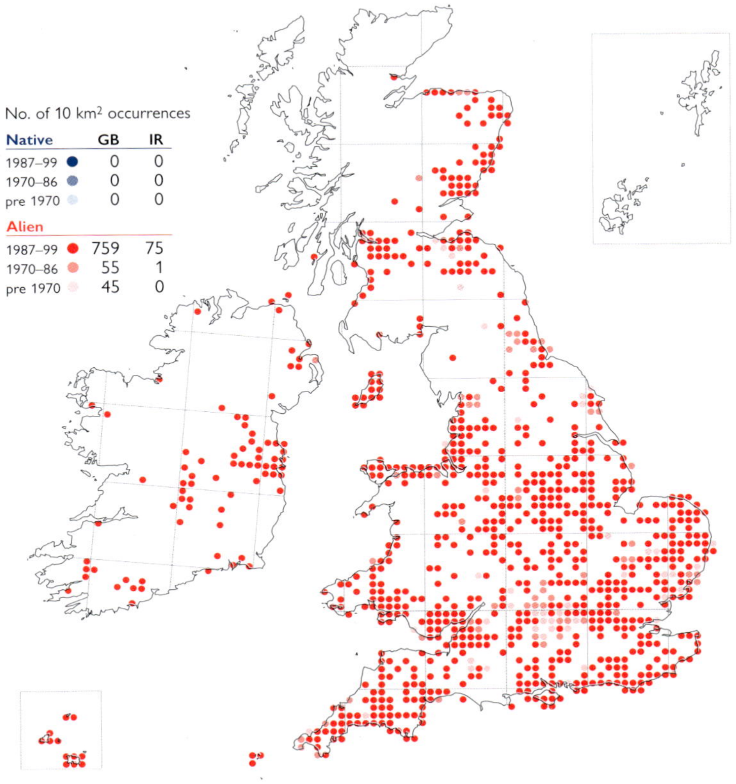

A bulbous perennial herb of woodland edges, hedgerows, churchyards and shady roadside-banks. Lowland.

Neophyte. *H. hispanica* was introduced into British gardens by 1683 and known from the wild by 1909. It is sometimes cultivated in gardens and occasionally becomes naturalised when planted or discarded in suitable habitats. It may be continuing to increase slowly, but it has long been confused with *H. hispanica* × *H. non-scripta* and probably remains somewhat over-recorded in error for the hybrid. It is apparently absent from some areas where recorders consider that all plants are referable to this hybrid.

Native of Portugal and W. Spain; naturalised elsewhere in S. & W. Europe.

I. TAYLOR

Hyacinthoides hispanica × *H. non-scripta*

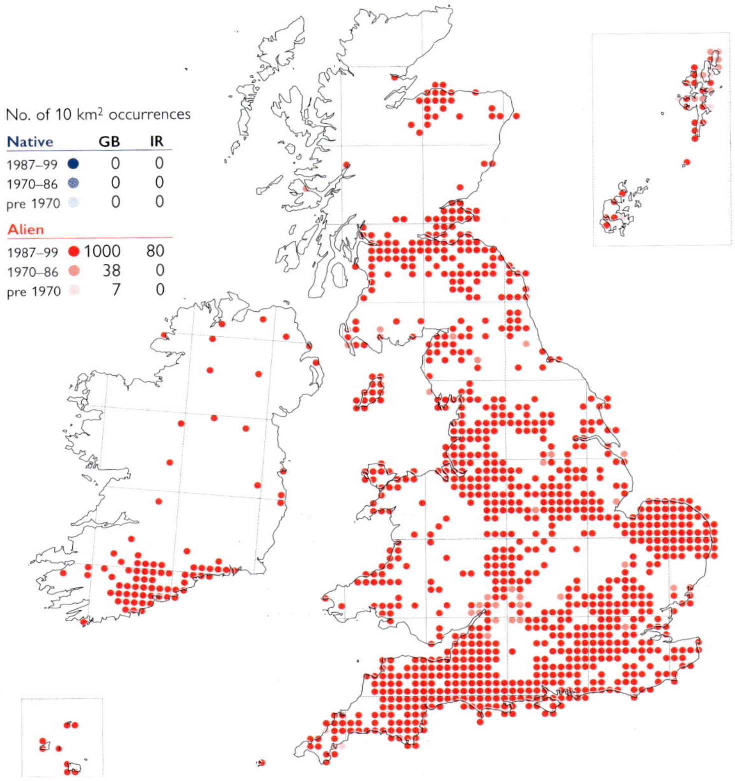

A bulbous perennial herb of woodlands, hedgerows, churchyards and shady roadsides, rough ground and waste places, probably most frequent in the entrances to amenity woods. Lowland.

Neophyte. This fertile hybrid is perhaps the commonest cultivated bluebell in gardens. Discarded bulbs readily become naturalised and introgression with *H. non-scripta* occurs. Although first recorded in the wild in 1963, its presence was only widely appreciated from about 1987 onwards. Its range and frequency are increasing, but it is still unevenly recorded.

This hybrid arises spontaneously where the native ranges of the parents meet, suggesting that the latter are only subspecifically distinct (Sell & Murrell, 1996); it is naturalised elsewhere in W. Europe.

References: Page (1987), Rich & Jermy (1998), Stace (1975).

I. TAYLOR

Hyacinthus orientalis Hyacinth

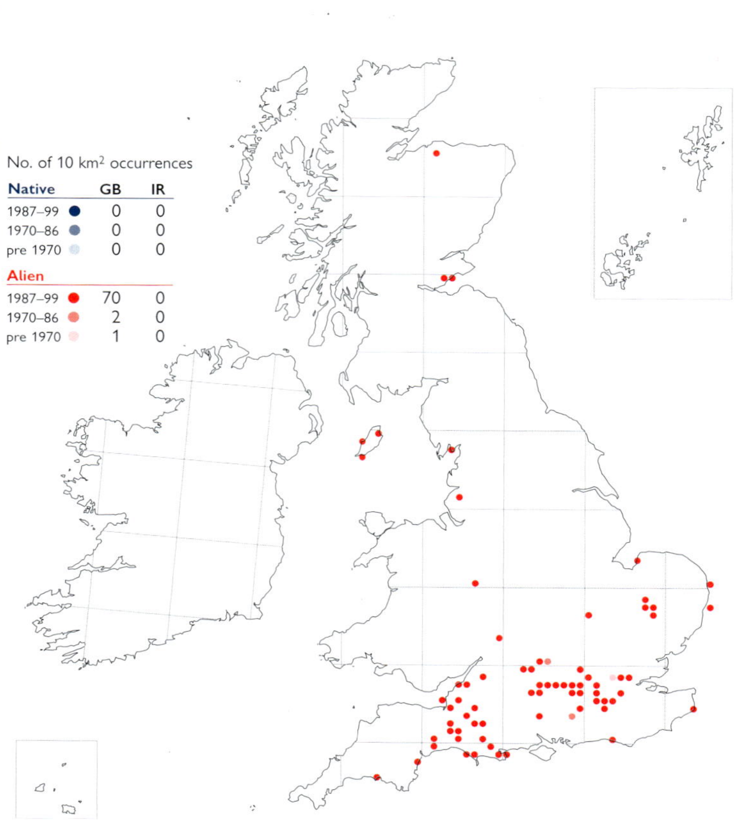

A bulbous perennial herb of open woodland, roadsides, hedge banks, waste ground and sand dunes, usually occurring as isolated clumps arising from discarded or deliberately planted garden stock. Populations can be persistent, especially in warm sites subject to summer drought. Lowland.

Neophyte. *H. orientalis* was being cultivated in Britain by 1596 and is very commonly grown as a winter-flowering house-plant. Bulbs are sometimes discarded or planted in the wild following flowering, and it was known from the wild by 1957. It is probably increasing.

Native of S.W. Asia, but modified in cultivation.

I. TAYLOR

Chionodoxa forbesii Glory-of-the-snow

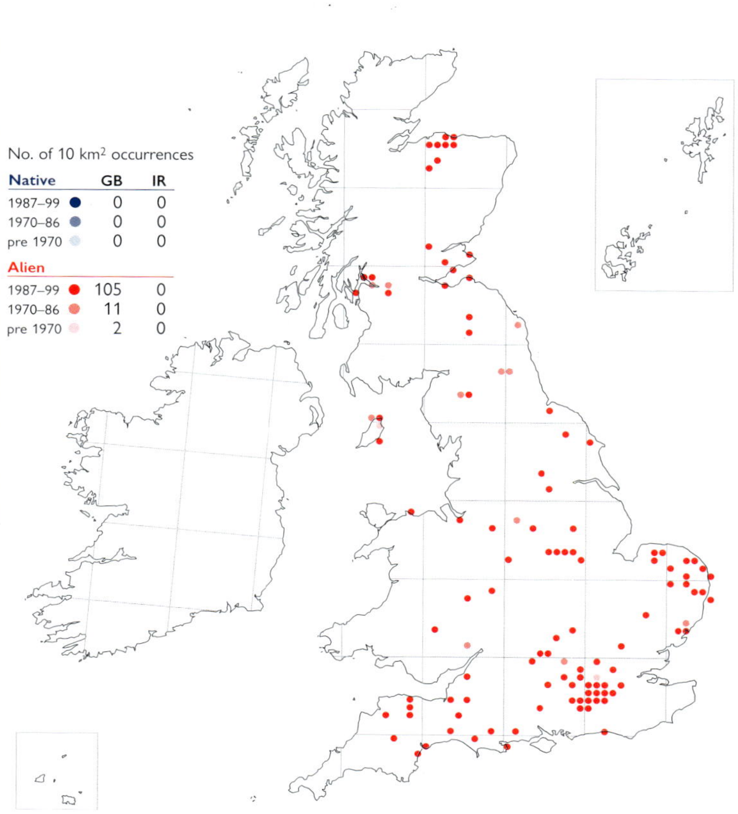

A bulbous perennial herb of free-draining soils on roadsides and waste ground, and in churchyards and other rough grassy places. It readily becomes naturalised by seed when discarded or deliberately planted in suitable habitats. Lowland.

Neophyte. *C. forbesii* has been cultivated in British gardens since 1877, but plants were not recorded from the wild until 1968. The considerable confusion with *C. luciliae* makes an assessment of distributional change difficult.

Native of the mountains of W. & S.W. Turkey.

I. TAYLOR

Muscari neglectum Grape-hyacinth

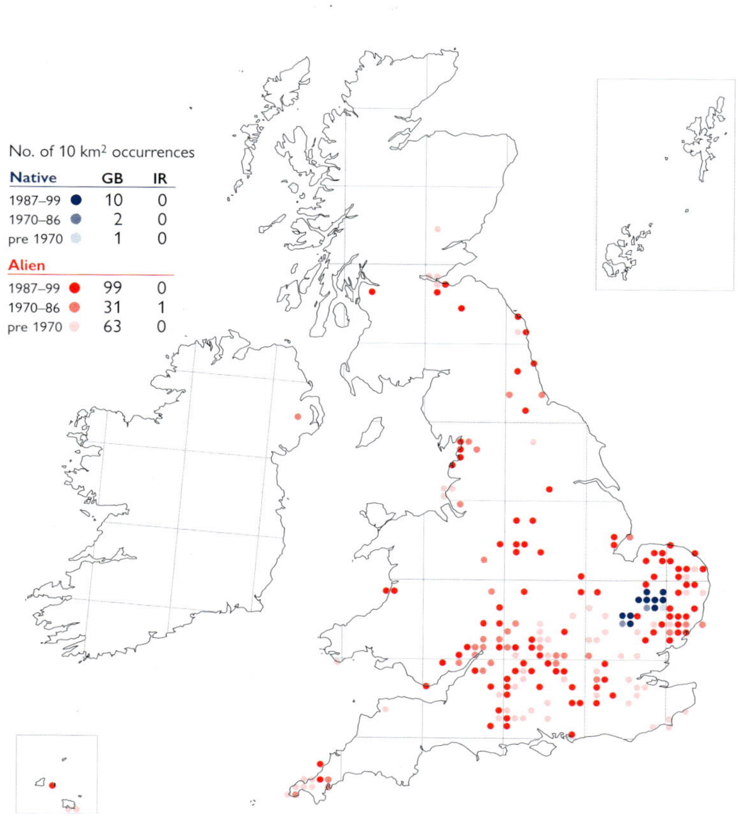

A bulbous perennial herb on free-draining soils, native or long-naturalised in grasslands, hedgerows, pine plantations and rough ground, and on roadsides on a wide range of nutrient-poor soils. It is also a short-lived garden escape or outcast near habitation, on roadsides, allotments and waste ground. Lowland.

Native or alien (change +1.55). *M. neglectum* was first recorded in Britain in 1776. It is sometimes considered native in E. England, where its range has contracted in recent years, mainly as a result of development. As an alien this species has been confused with, and over-recorded for, *M. armeniacum*, which is more common in gardens and as an escape.

Eurosiberian Southern-temperate element.

References: Atlas (316d), Wigginton (1999).

I. TAYLOR

Muscari armeniacum Garden Grape-hyacinth

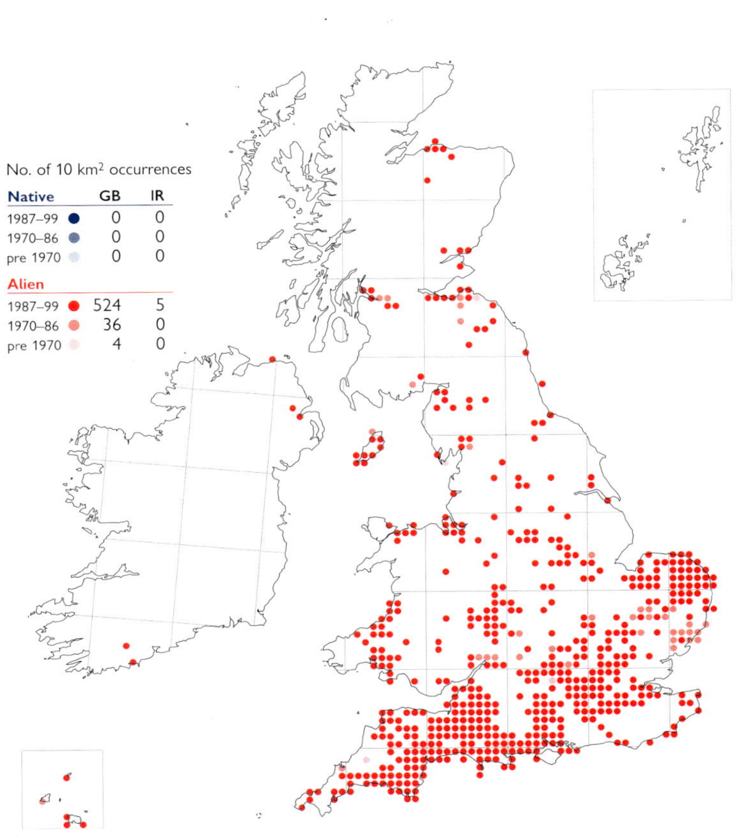

A bulbous perennial herb of free-draining soils in grasslands and hedgerows and on sand dunes, roadsides, walls and waste ground. It is often discarded from gardens and sometimes deliberately planted in the wild. Populations can rapidly spread vegetatively and by seed in suitable habitats. Lowland.

Neophyte. This species has been cultivated in Britain since 1878 and was recorded in the wild by 1892. It has long been confused with *M. neglectum* and, to a lesser extent, *M. botryoides*, and it has been under-recorded in some areas because of this. *M. armeniacum* is by far the commonest Grape-hyacinth in gardens and therefore the most frequently discarded. It is possibly increasing due to repeated introduction.

Native of the Balkans, Turkey and the Caucasus.

Reference: Rich & Jermy (1998).

I. TAYLOR

Muscari comosum Tassel Hyacinth

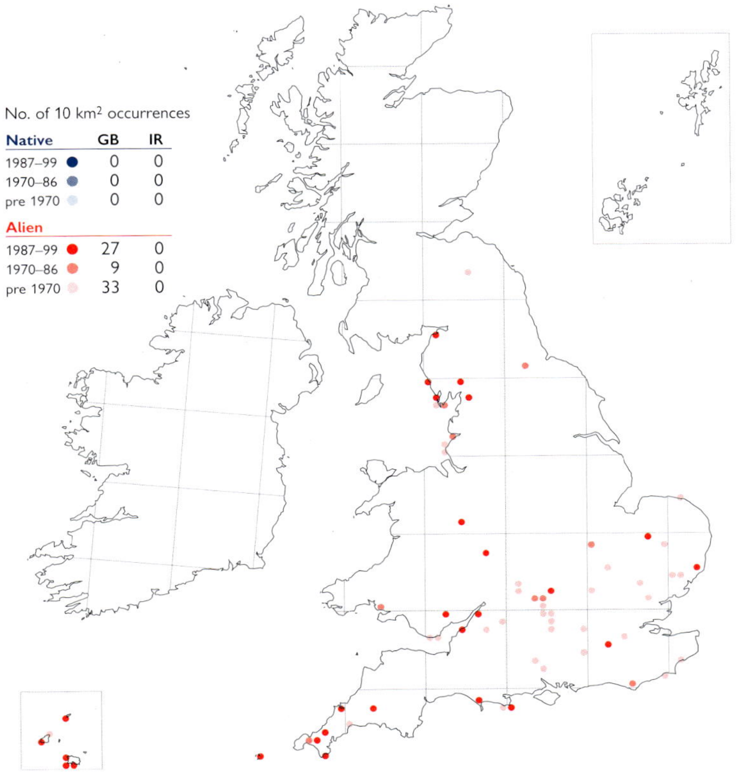

A bulbous perennial herb of sand dunes, sandy grasslands and cultivated and waste ground. Populations are usually short-lived when discarded into the wild, but those on dunes, especially in the warmer parts of Britain, can become well-established. Lowland.

Neophyte. *M. comosum* has been cultivated in Britain since 1596 and was known in the wild by 1888.

A European Southern-temperate species.

I. TAYLOR

Allium schoenoprasum Chives

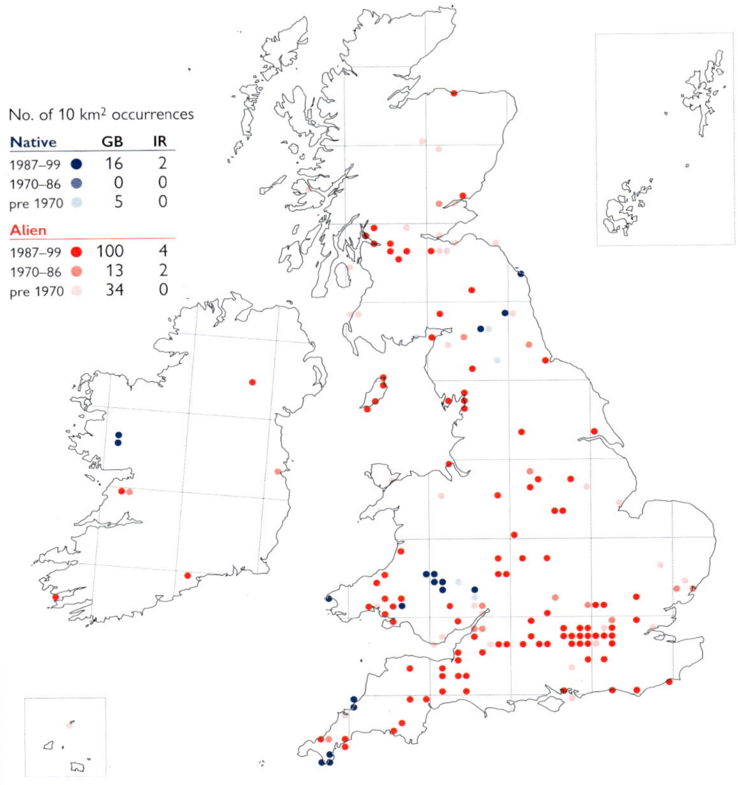

A bulbous perennial herb found as a native in a range of habitats, usually on thin soils over limestone, serpentine and basic igneous rocks; it sometimes grows in rank grass on deeper soils, and in crevices of riverside bedrock. As an alien it grows on roadsides and rubbish tips. Lowland.

Native (change +1.69). The overall native distribution of this species is unchanged since the 1962 *Atlas*, but alien occurrences have increased. It is regarded by Scannell & Synnott (1987) as probably introduced in Ireland, but Curtis & McGough (1988) suggest that it is native on limestone pavement at L. Mask (E. Mayo).

Assigned to the Circumpolar Boreo-arctic Montane element, but with an anomalous distribution; widely naturalised outside its native range.

References: Atlas (326c), Hultén & Fries (1986), Stewart *et al.* (1994).

I. TAYLOR

Allium roseum Rosy Garlic

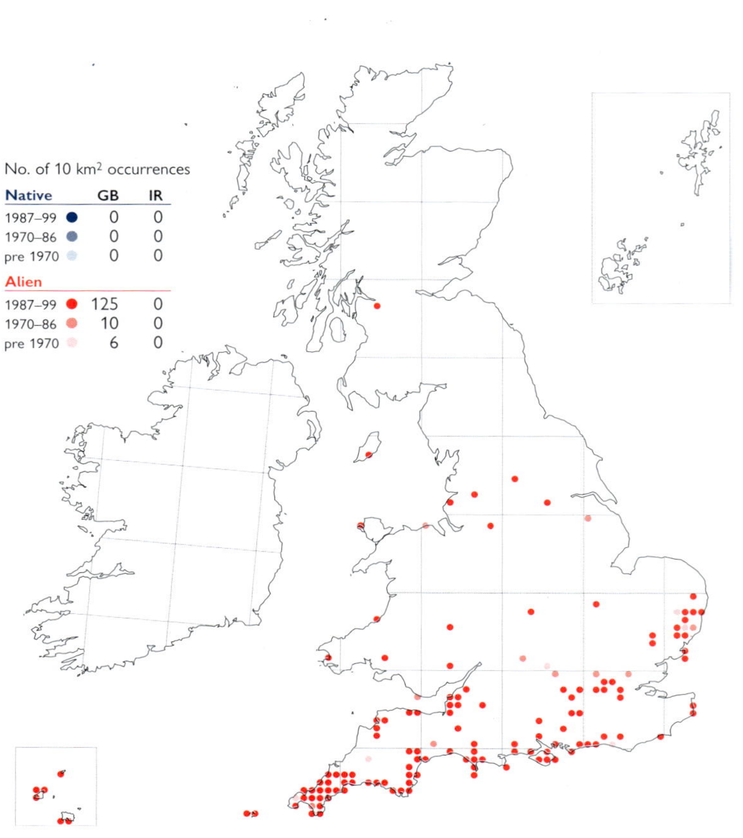

A bulbous perennial herb of rough and waste ground, open rocky slopes, hedge banks and roadsides. Lowland.

Neophyte. *A. roseum* was introduced into cultivation by 1752, and first recorded in the wild in 1837. Both var. *bulbiferum* and var. *roseum* are commonly cultivated in gardens; var. *bulbiferum*, in which some of the flowers in the umbel are replaced by bulbils, is particularly likely to be naturalised when planted or discarded in suitable habitats. *A. roseum* is spreading, especially in S.W. England.

A variable species, native of the Mediterranean region.

Reference: Oswald (1993).

I. TAYLOR

Allium triquetrum Three-cornered Garlic

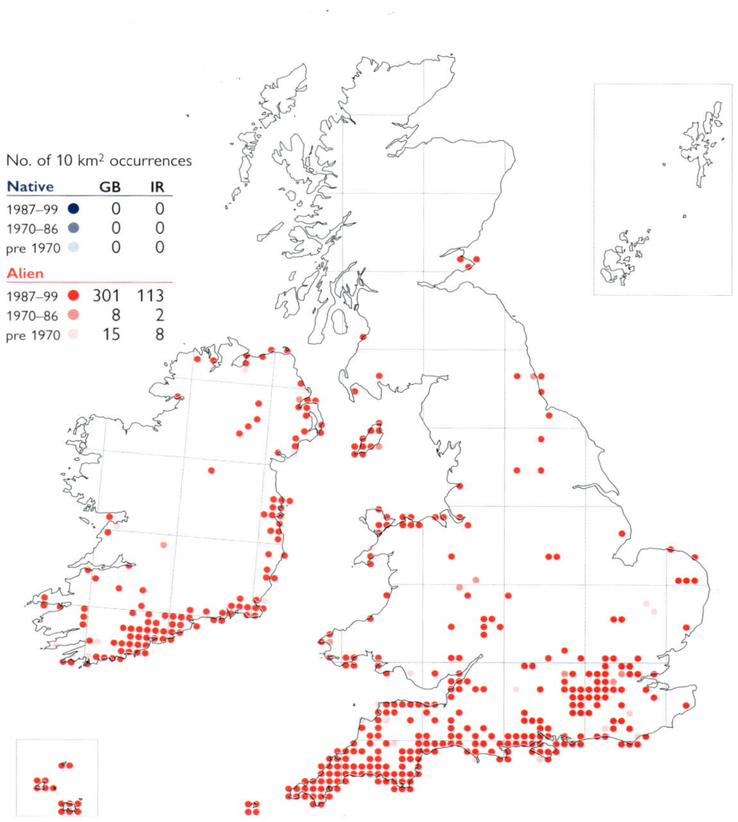

A bulbous perennial herb, spreading by ant-dispersed seed on roadsides, in hedge banks, on field margins and in rough and waste ground. Lowland.

Neophyte (change +2.46). *A. triquetrum* was introduced into cultivation by 1759 and noted as established here by 1849, initially in Guernsey. It is now thoroughly naturalised and increasingly abundant and widespread in milder areas with scattered, sometimes short-lived, populations elsewhere. It has considerably increased in numbers and range since it was mapped in the 1962 *Atlas*.

Native of the W. & C. Mediterranean region.

Reference: Atlas (326d).

I. TAYLOR

Allium paradoxum Few-flowered Garlic

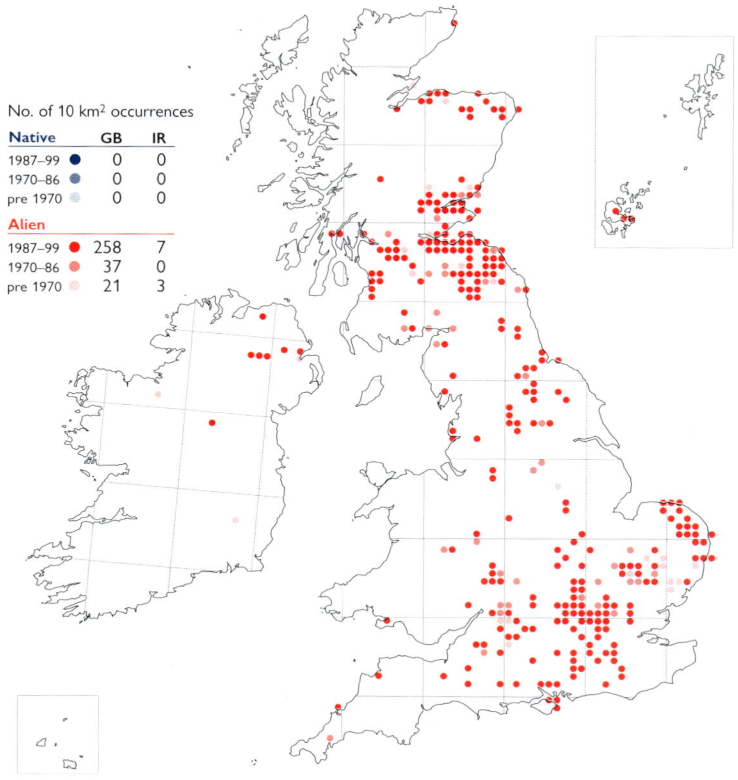

A bulbous perennial herb, spreading by means of bulbils in a wide variety of, usually ungrazed, situations such as river banks, roadsides, field margins, other rough and waste ground, and in woodland. Generally lowland, but reaching 375 m at Carter Bar (Roxburghs.).

Neophyte (change +1.83). *A. paradoxum* was introduced into cultivation in 1823 and was first recorded in the wild near Edinburgh in 1863. It can be very invasive in disturbed habitats, and is increasingly abundant throughout its range, especially in S. Scotland. Its predominantly eastern distribution contrasts with that of *A. triquetrum*.

Native of the Caucasus and Iran.

References: Atlas (327a), Oswald (1993).

I. TAYLOR

Allium ursinum Ramsons

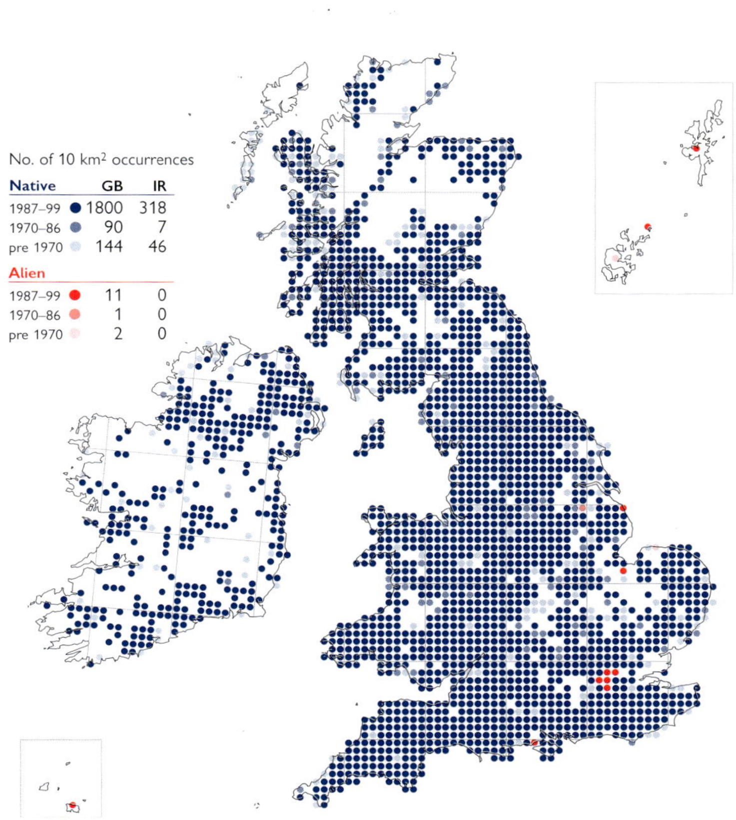

A bulbous perennial herb of moist woodlands, sometimes growing in more open situations such as riversides and hedge banks and occasionally in rock crevices, in scree and on coastal cliff ledges. Regeneration is primarily by seed. Generally lowland, but reaching *c.* 450 m at Great Clowder, Malham (Mid-W. Yorks.).

Native (change +0.24). There is little evidence of significant distributional change in this species since the 1962 *Atlas*, where it was mapped as 'all records'.

European Temperate element.

References: Atlas (327b), Grime *et al.* (1988), Hultén & Fries (1986), Mabey (1996), Meusel *et al.* (1965), Tutin (1957).

I. TAYLOR

Allium oleraceum Field Garlic

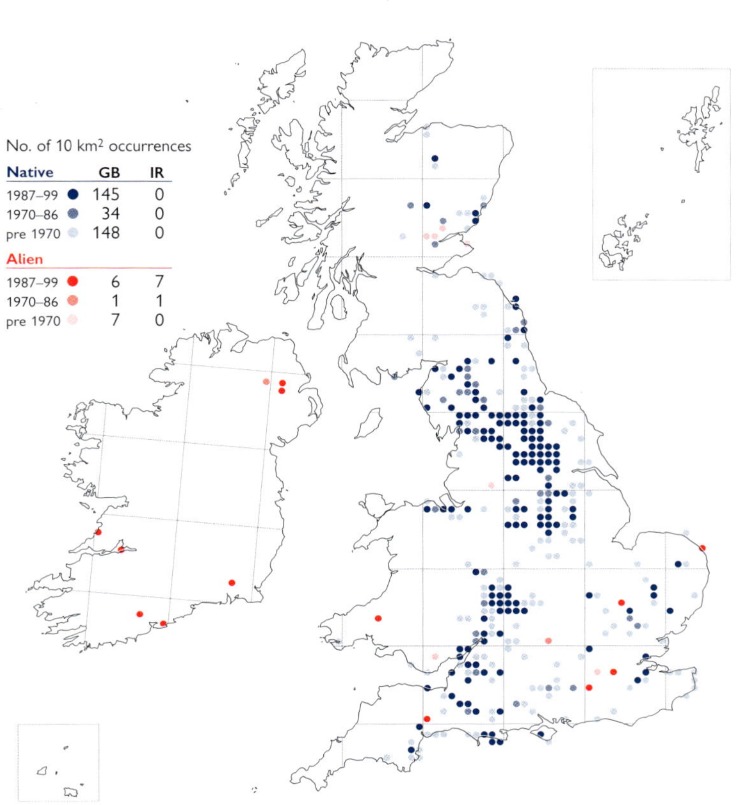

A bulbous perennial herb of dry, usually steeply sloping, calcareous grasslands, and on open sunny banks in river floodplains. 0–365 m (Dovedale, Derbys.).

Native (change –0.24). There has undoubtedly been some confusion with *A. scorodoprasum* in the past. There are now many more records of *A. oleraceum* than in the 1962 *Atlas*, although it is perhaps still under-recorded due to difficulties in locating plants in the field from early summer onwards. There are indications of some decline throughout its range.

European Temperate element; widely naturalised outside its native range.

References: Atlas (326a), Hultén & Fries (1986), Meusel *et al.* (1965), Stewart *et al.* (1994).

I. TAYLOR

Allium carinatum Keeled Garlic

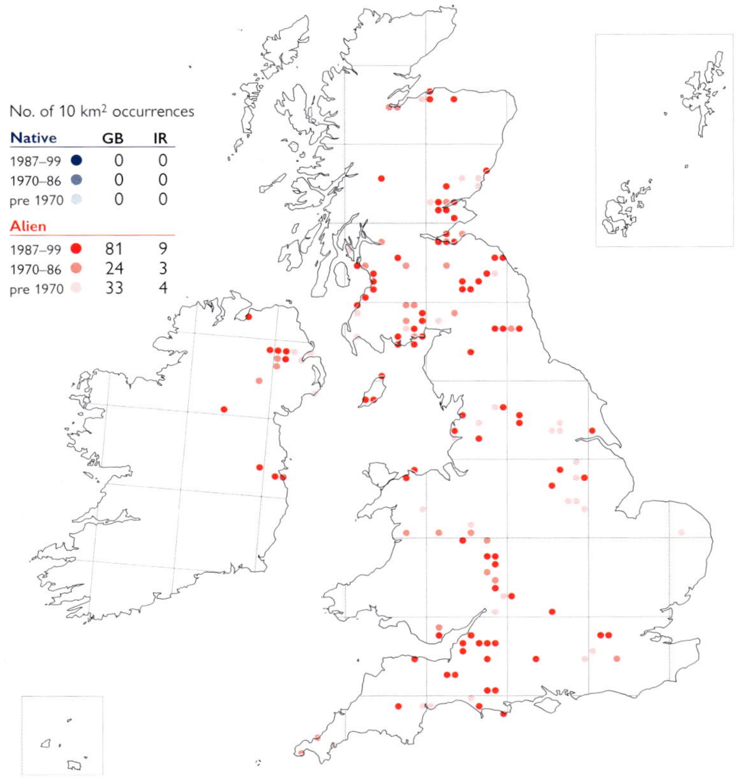

A bulbous perennial herb which becomes naturalised by seed (var. *pulchellum*) or, perhaps more frequently, by bulbils (var. *carinatum*) in churchyards, rough and waste ground and on roadsides. Lowland.

Neophyte (change +0.64). *A. carinatum* was cultivated in Britain by 1789 and is frequently grown in gardens. It has been naturalised in our area since at least 1806. There is little evidence of a marked increase in range since the 1962 *Atlas*.

Native of C. & S.E. Europe; naturalised north and west of its native range.

References: Atlas (326b), Hultén & Fries (1986), Meusel *et al.* (1965), Oswald (1993).

I. TAYLOR

Allium ampeloprasum Wild Leek

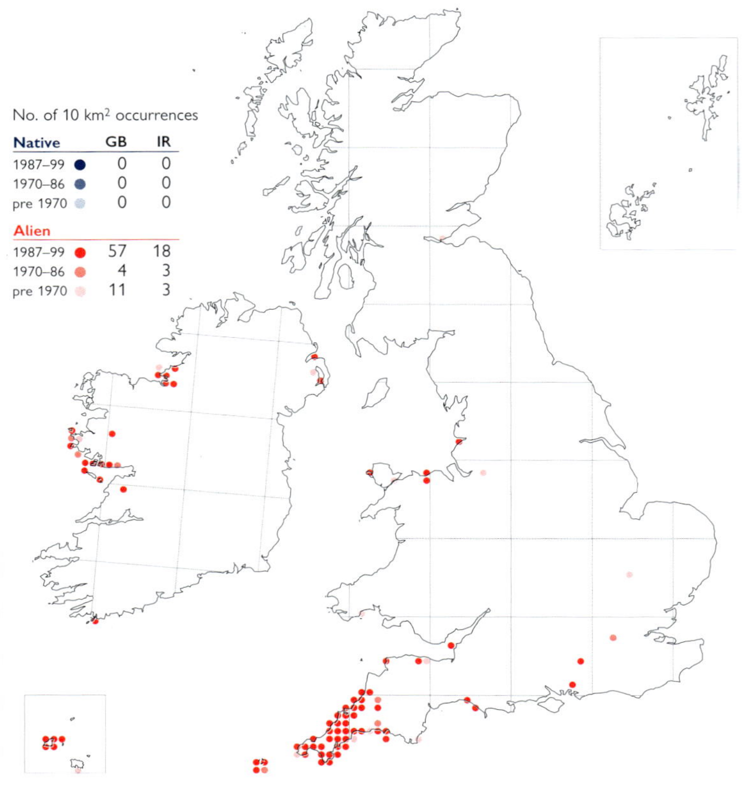

A robust, bulbous perennial herb of rank vegetation in sandy and rocky places near the sea, especially in old fields and hedge banks, on sheltered cliff-slopes, by paths and tracks and in drainage ditches and other disturbed places. Var. *ampeloprasum* reproduces mainly by seed, whereas the other varieties spread mainly by bulbils. Lowland.

Archaeophyte (change +0.77). Var. *babingtonii* has extended its range and become more frequent since 1930, var. *bulbiferum* is confined to the Channel Islands where it is relatively stable, and var. *ampeloprasum* occurs in small populations in S.W. England and Wales and may be declining. The species may sometimes be mis-recorded for *A. sativum*.

As an archaeophyte *A. ampeloprasum* has a Mediterranean-Atlantic distribution.

References: Atlas (325a, b), Mathew (1996), Wigginton (1999).

I. TAYLOR

Allium scorodoprasum Sand Leek

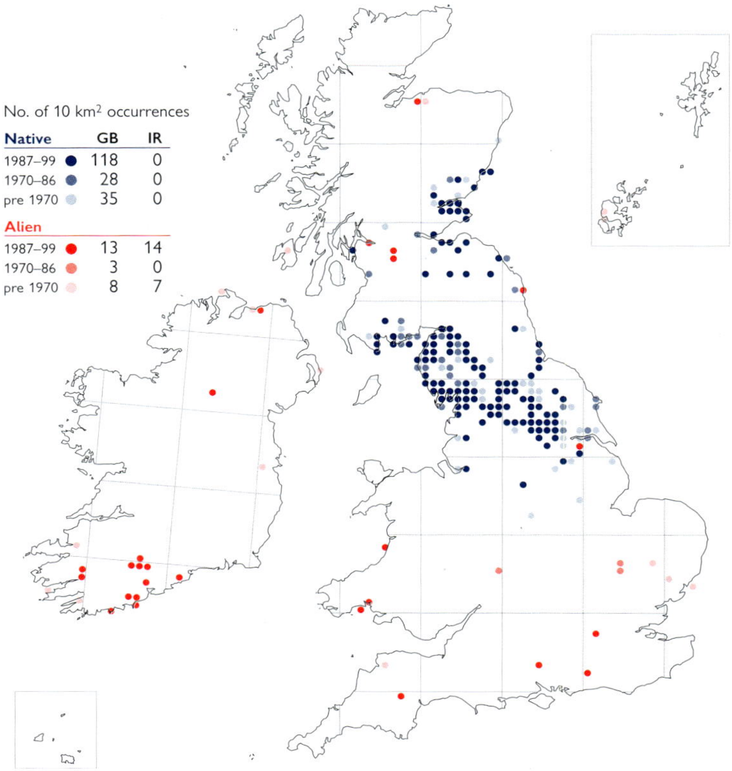

A bulbous, perennial herb spreading mainly by bulbils in rough grassland and waste ground, on road verges and track sides and by railways. It sometimes occurs in more natural habitats such as sandy river banks, open woodlands on well-drained soils and a variety of coastal situations. Lowland.

Native (change +0.30). *A. scorodoprasum* is much better recorded than it was for the 1962 *Atlas*, and its distribution seems to be stable.

European Temperate element, with a continental distribution in W. Europe; widely naturalised outside its native range. Our plant is sometimes considered to be a horticulturally derived variant of the S. European *A. scorodoprasum* subsp. *rotundum*.

References: Atlas (325c), Hultén & Fries (1986), Mathew (1996), Meusel *et al.* (1965), Stewart *et al.* (1994).

I. TAYLOR

Allium sphaerocephalon Round-headed Leek

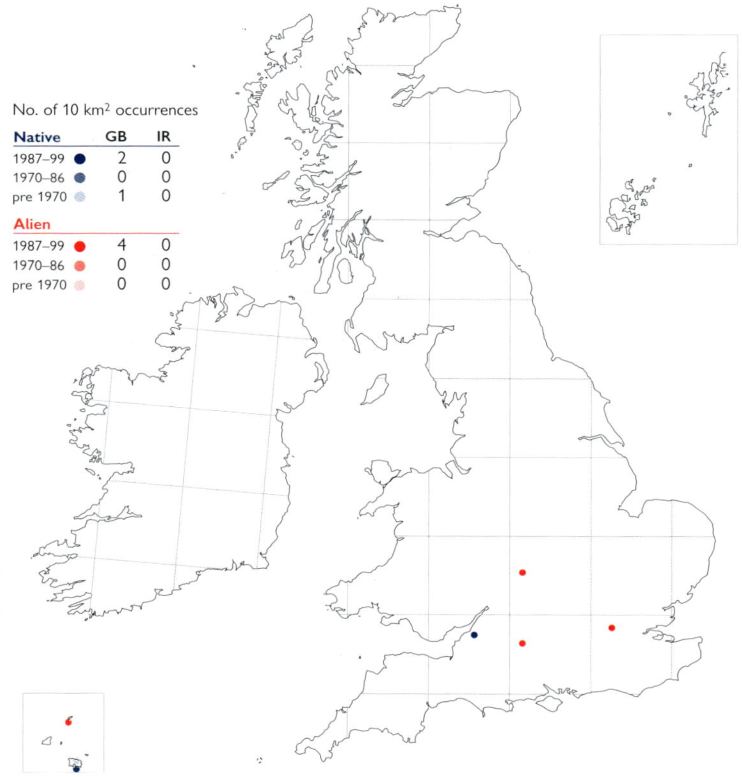

A bulbous perennial herb found as a possible native at only two localities: on dry, rocky S.- and W.-facing slopes in the Avon Gorge (W. Gloucs.), and on rough, sandy ground by the sea at St. Aubin's Bay, Jersey. Both populations are very small. Lowland.

Native or alien. *A. sphaerocephalon* was first cultivated in Britain in 1759, and was first recorded in the wild in Jersey in 1836 and in the Avon Gorge in 1847, where it is normally thought to be native but may be alien (Lovatt, 1982). Recreational pressure and safety works within the Avon Gorge may have contributed to a gradual decline in recent years. It appears to be stable in Jersey. As it is now widely grown in gardens, it can be expected to spread as an escape.

European Southern-temperate element.

References: Atlas (325c), Mathew (1996), Wigginton (1999).

I. TAYLOR

Allium vineale Wild Onion

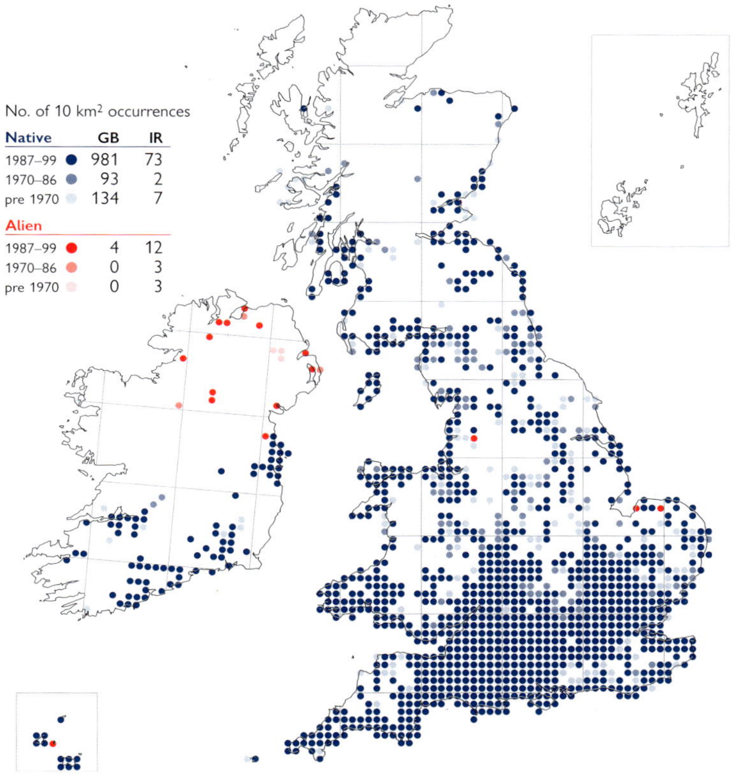

No. of 10 km² occurrences

Native	GB	IR
1987–99 ●	981	73
1970–86 ●	93	2
pre 1970 ●	134	7
Alien		
1987–99 ●	4	12
1970–86 ●	0	3
pre 1970 ●	0	3

A bulbous perennial herb of dry, neutral or calcareous soils, generally occurring in summer-dry grasslands, hedgerows, roadsides and cultivated ground, and formerly a serious weed of cereal crops in S.E. England. It is also found on coastal cliff ledges in W. Scotland. Generally lowland, but reaching 455 m in Wensleydale (N.W. Yorks.).

Native (change +0.90). There is little evidence of change in the range of this species since the 1962 *Atlas*. There are few available data to assess the relative abundance of the three varieties recognised in our area, although the exclusively floriferous var. *capsuliferum* appears to be rare.

European Temperate element; widely naturalised outside its native range.

References: Atlas (325d), Hultén & Fries (1986), Mathew (1996), Meusel *et al.* (1965), Richens (1947).

I. TAYLOR

Tristagma uniflorum Spring Starflower

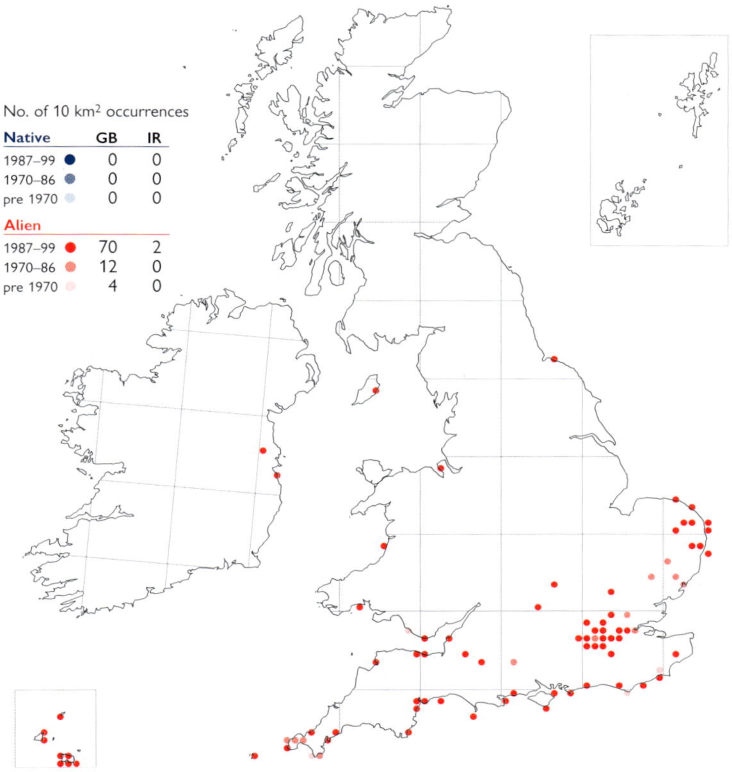

No. of 10 km² occurrences

Native	GB	IR
1987–99 ●	0	0
1970–86 ●	0	0
pre 1970 ●	0	0
Alien		
1987–99 ●	70	2
1970–86 ●	12	0
pre 1970 ●	4	0

A bulbous perennial herb found on cultivated ground, roadsides and waste ground, and in churchyards; also as a relic of cultivation. It readily becomes established in suitable habitats, especially on sandy soils in mild areas such as in the Isles of Scilly and the Channel Islands. Reproduction is mostly vegetative. Lowland.

Neophyte. *T. uniflorum* was first cultivated in 1832 and is frequently grown in gardens. It was recorded in the wild in 1921 (Jersey), and was known in the Isles of Scilly in 1952; it may be increasing in some areas. It is perhaps better known to gardeners as *Ipheion uniflorum*, and is still marketed under this name.

Native of S. America.

Reference: Lousley (1971).

I. TAYLOR

Leucojum aestivum Summer Snowflake

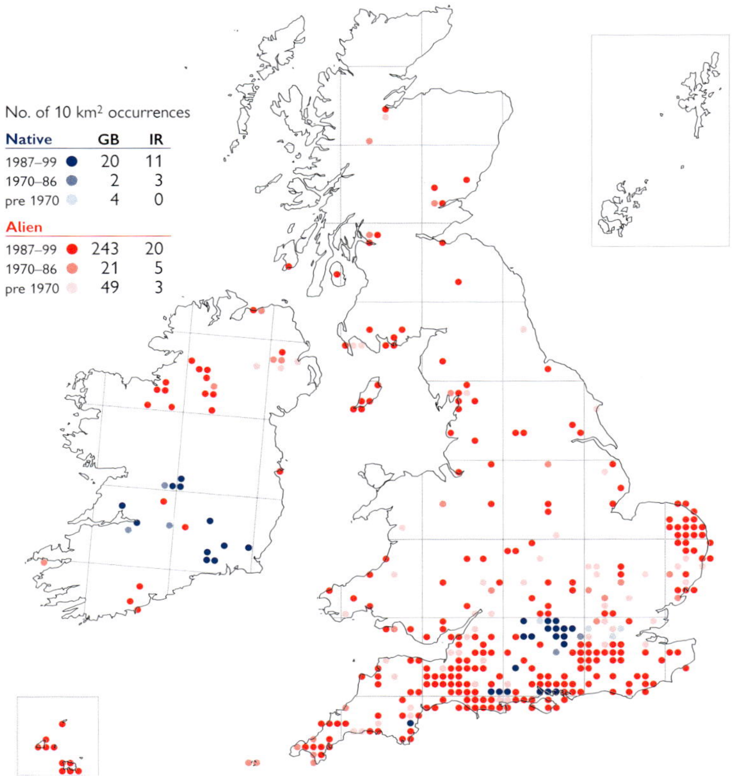

No. of 10 km² occurrences

Native	GB	IR
1987–99 ●	20	11
1970–86 ●	2	3
pre 1970 ●	4	0
Alien		
1987–99 ●	243	20
1970–86 ●	21	5
pre 1970 ●	49	3

A bulbous perennial herb, occurring as a native mostly in winter-flooded riverside *Alnus* or *Salix* carr, but occasionally found in other damp habitats such as meadows and woodland rides. It also occurs as a garden escape near habitation and on rubbish tips. Lowland.

Native (change +2.42). *L. aestivum* subsp. *aestivum* is native in some areas and subsp. *pulchellum* is an alien (which was in cultivation by 1596) but they have long been confused and the true subspecific identity of some records is in doubt. However, the presumed native distribution of subsp. *aestivum* in Britain and Ireland is stable. As aliens, both subspecies readily become naturalised in suitable situations and appear to be increasing.

European Southern-temperate element.

References: Atlas (327d), Farrell (1979), Gillam (1993), Wigginton (1999).

I. TAYLOR

Leucojum vernum Spring Snowflake

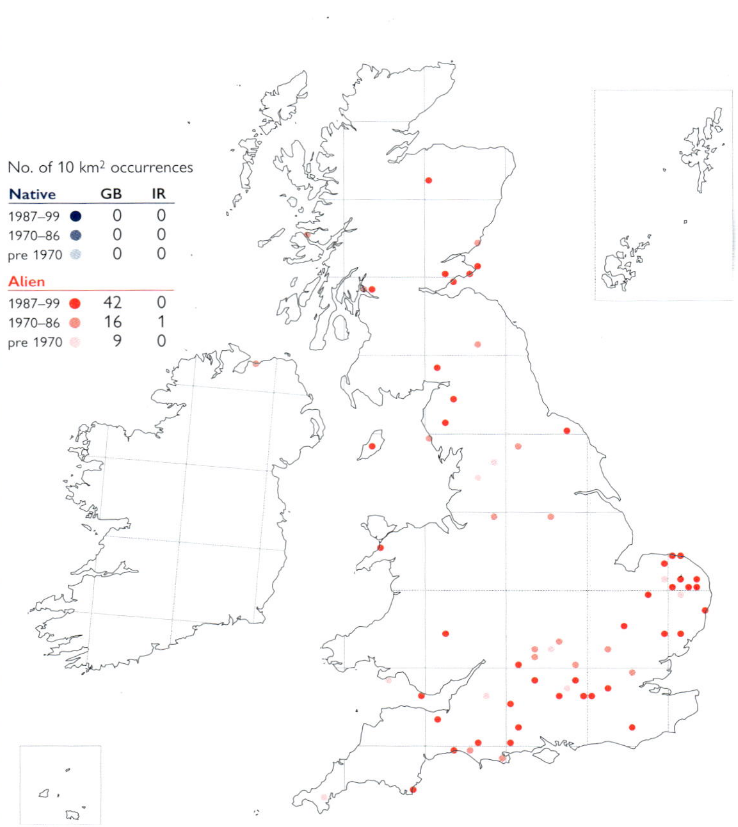

A bulbous perennial herb of damp soils, thoroughly naturalised in woodland and scrub, by watercourses and in grassy places. Lowland.

Neophyte (change +1.23). Although sometimes considered native at two localities in S.W. England, all British populations are almost certainly relics of cultivation or result from surplus garden plants being discarded into suitable habitats; it was being cultivated in Britain by 1596 and was first recorded in the wild in 1866. There has been little change in its distribution since the 1962 *Atlas*.

A European Temperate species, absent as a native from much of W. Europe.

References: Atlas (327c), Meusel *et al.* (1965).

I. TAYLOR

Galanthus nivalis Snowdrop

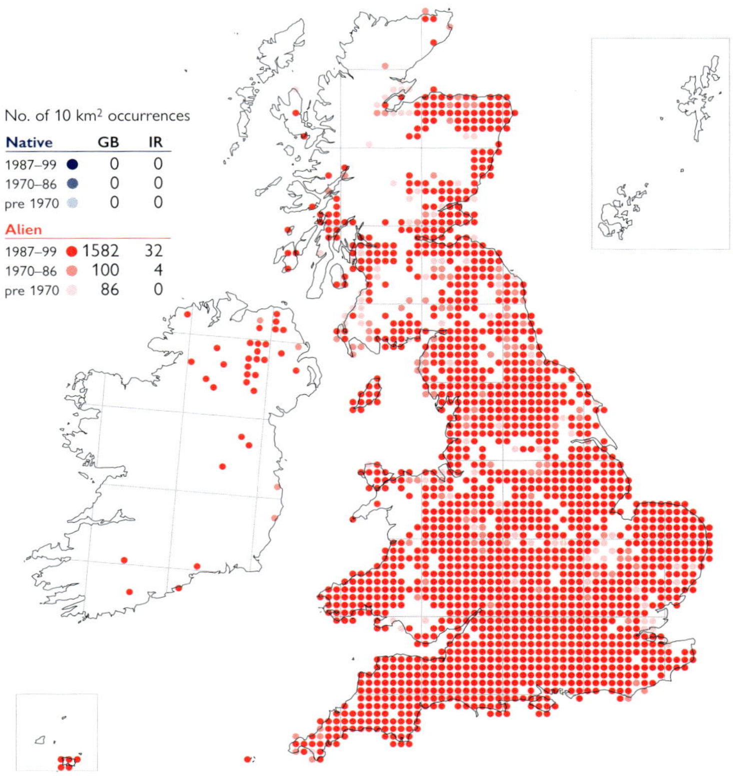

A bulbous perennial herb of moist woodlands and other shaded places. It is especially frequent in parks, large gardens and churchyards, but also occurs on road verges, by watercourses and in damp grassland. Spread is mainly by division of the bulbs, as seed production is poor. 0–370 m (Nenthall, Cumberland).

Neophyte (change +3.01). This species was known in cultivation in Britain in 1597 but was not recorded in the wild until 1778. Although it was formerly sometimes regarded as native, it is now considered to be alien. It was probably under-recorded in the 1962 *Atlas*, and there is little reason to suspect a real change in its distribution.

A European Southern-temperate species; widely naturalised outside its native range.

References: Atlas (328a), Davis (1999), Mabey (1996), Meusel *et al.* (1965).

I. TAYLOR

Galanthus plicatus Pleated Snowdrop

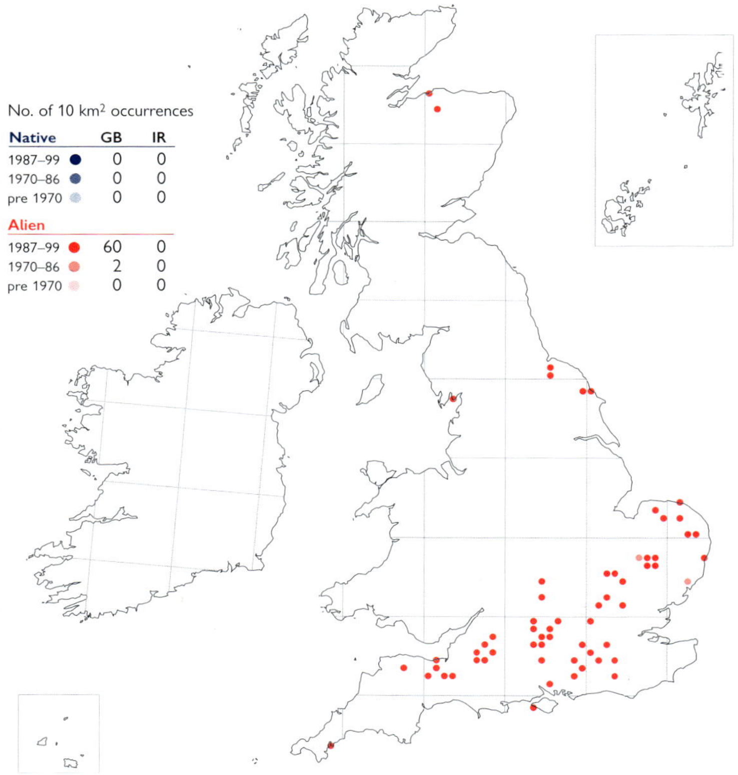

A bulbous perennial herb of deciduous woodland, hedgerows, roadsides, churchyards, cemeteries and parkland; it is also found as a relic of cultivation. Most established populations are associated with informal plantings in the wild. Unlike *G. nivalis*, *G. plicatus* often produces abundant seedlings and populations can be maintained in this way. Lowland.

Neophyte. *G. plicatus* has been grown in British gardens since 1818, but was not recorded from the wild until 1947 (E. Suffolk). It is less commonly cultivated than *G. nivalis* and is much less frequent in the wild. It may sometimes be overlooked as *G. nivalis*, but some records mapped as *G. plicatus* may be the hybrid between these species. *G. plicatus* may be increasing in some areas.

Native of S.E. Europe and N.W. Turkey.

Reference: Davis (1999).

I. TAYLOR

Galanthus elwesii Greater Snowdrop

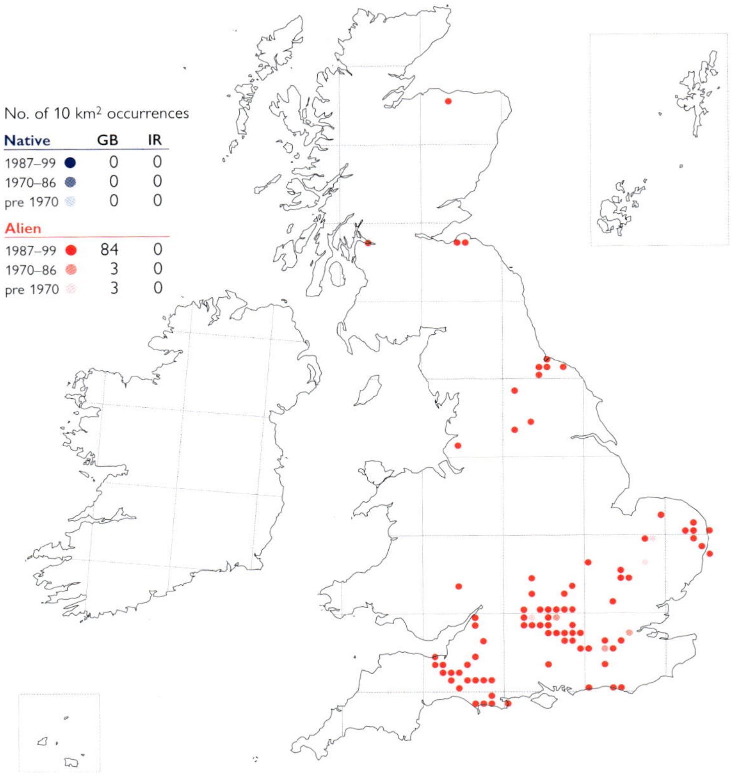

No. of 10 km² occurrences

Native	GB	IR
1987–99	0	0
1970–86	0	0
pre 1970	0	0
Alien		
1987–99	84	0
1970–86	3	0
pre 1970	3	0

A bulbous perennial herb of open woodland, roadsides, parkland, cemeteries and churchyards, generally in sunnier, warmer and somewhat drier situations than *G. nivalis* and *G. plicatus*. As with *G. plicatus*, most established populations are associated with informal plantings in the wild. Seed production is, however, less abundant than with *G. plicatus* although some populations are probably maintained and may be expanding by this means. Lowland.

Neophyte. *G. elwesii* has been grown in British gardens since 1875 and is now quite commonly cultivated, although far less so than *G. nivalis*. Consequently, it is much less frequent in the wild, from where it was first recorded in about 1957 (Surrey).

Native of the mountains of S.E. Europe and Turkey.

Reference: Davis (1999).

I. TAYLOR

Narcissus agg. Daffodils

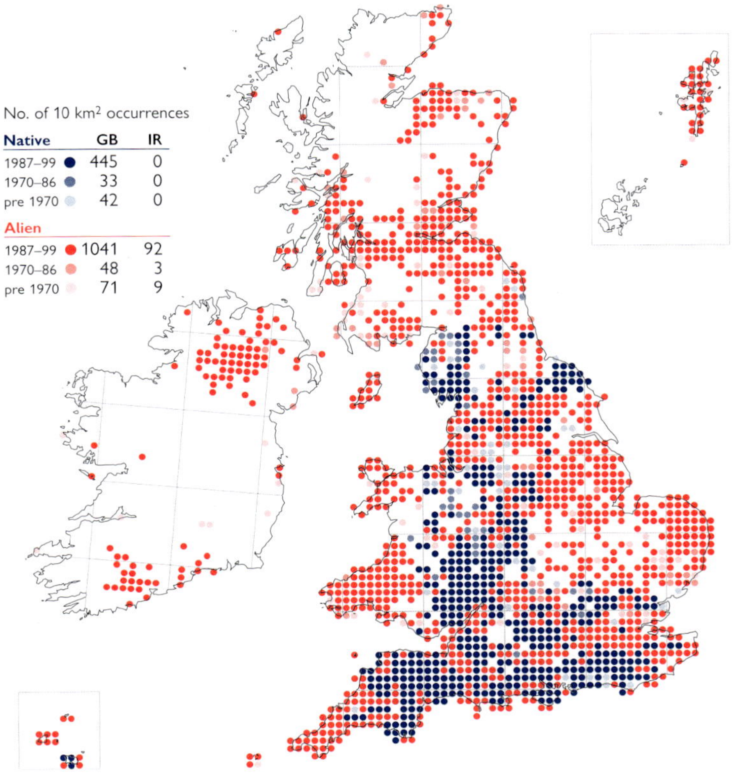

No. of 10 km² occurrences

Native	GB	IR
1987–99	445	0
1970–86	33	0
pre 1970	42	0
Alien		
1987–99	1041	92
1970–86	48	3
pre 1970	71	9

A genus of bulbous perennial herbs, found in woodlands, pastures, parks, churchyards and amenity grasslands, and on road verges, pathsides and waste ground; also as relics of cultivation. Plants frequently become naturalised when discarded or deliberately planted. 0–375 m (Plynlimon, Cards.).

Native. Many species, cultivars and hybrids are found in the wild, although only *N. pseudonarcissus* subsp. *pseudonarcissus* is native. Plants are often difficult to identify and inconsistencies in recording make interpretation of some maps difficult.

A genus centred on the Mediterranean region and most diverse in S.W. Europe. Its long history in cultivation with undocumented hybridisation render its taxonomy almost intractable.

References: Blanchard (1990), Sell & Murrell (1996).

I. TAYLOR

Narcissus poeticus Pheasant's-eye Daffodil

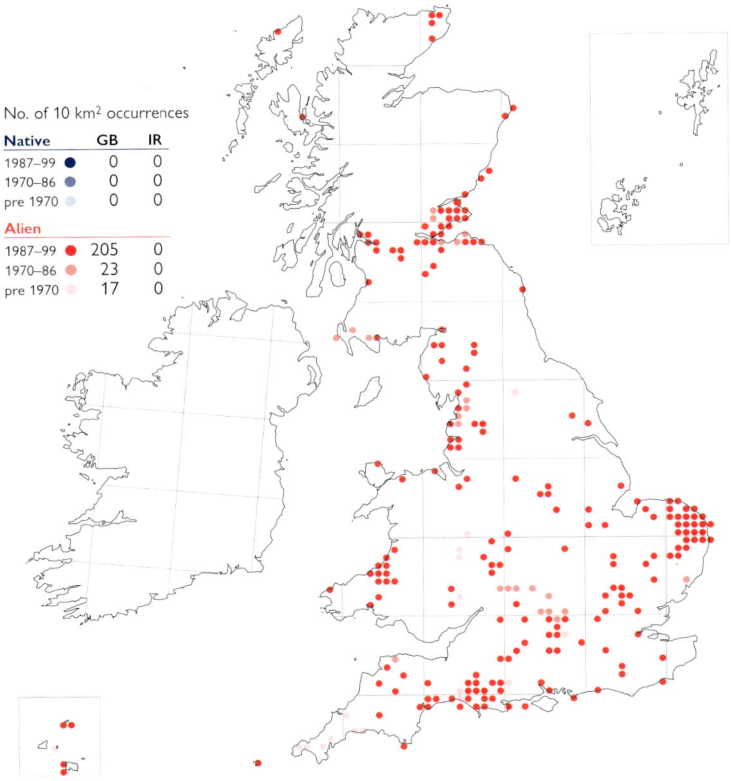

No. of 10 km² occurrences

Native	GB	IR
1987–99	0	0
1970–86	0	0
pre 1970	0	0
Alien		
1987–99	205	0
1970–86	23	0
pre 1970	17	0

A bulbous perennial herb naturalised in hedgerows, on roadsides, by tracks and paths and on rough and waste ground from discarded stock, and occasionally found in old fields and field margins as a relic of cultivation. Lowland.

Neophyte. *N. poeticus* was cultivated in Britain by 1538 and is common in gardens. It is frequently discarded and was known from the wild by 1795. The taxonomy of *N. poeticus* and its allies is complex, but most records are probably referable to subsp. *poeticus*. In the Channel Islands and the Isle of Man, however, subsp. *radiiflorus* is frequently found, and other variants and cultivars, especially the widely cultivated 'Actaea', are represented throughout its range. The species is unevenly recorded.

Native of the mountains of S. Europe.

References: Blanchard (1990), Sell & Murrell (1996).

I. TAYLOR

Narcissus poeticus × *N. pseudonarcissus* (*N.* × *incomparabilis*) Nonesuch Daffodil

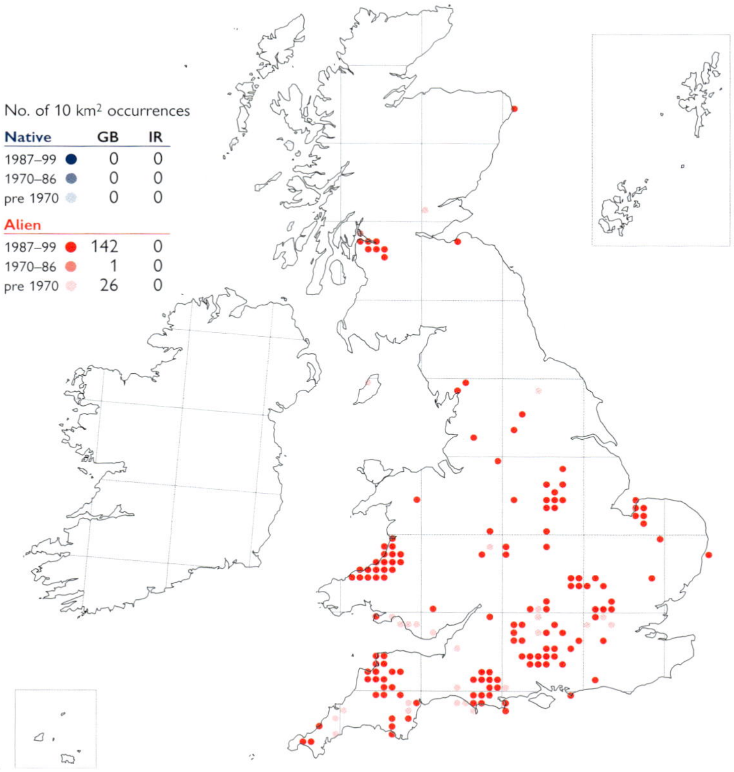

A bulbous perennial herb found naturalised in hedgerows, on roadsides, by tracks and paths and on waste ground, usually from discarded or deliberately planted stock; it is sometimes found on field margins as a relic of commercial cut-flower production. Lowland.

Neophyte. This taxon, which has been grown in British gardens since the 16th century, was first recorded in the wild in 1711. Further hybridisation with other introduced taxa and backcrossing with *N. pseudonarcissus* may occur in the wild. It is under-recorded, and sometimes mistaken for cultivars of *N. pseudonarcissus*.

This hybrid probably occurs naturally in France but the naturalised plants are of garden origin.

References: Blanchard (1990), Stace (1975).

I. TAYLOR

Narcissus poeticus × *N. tazetta* (*N.* × *medioluteus*) Primrose-peerless

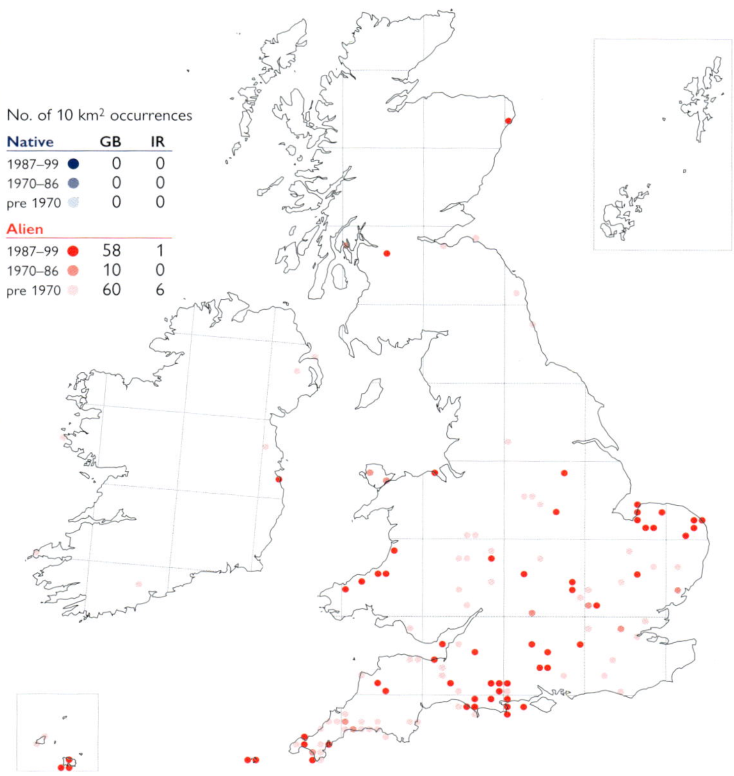

A bulbous perennial herb found in old bulb-fields as a relic of commercial cut-flower production, and also grown in gardens and readily becoming naturalised from discarded bulbs in hedgerows, on roadsides, by tracks and paths and on waste ground. Lowland.

Neophyte. This hybrid was recorded from the wild in 1737. This taxon includes a variety of mono- and bi-coloured, single- and double-flowered cultivars which are collectively known to gardeners as the 'Poetaz' narcissi. As with all alien *Narcissus* taxa, they are under-recorded.

This hybrid occurs naturally in France but the naturalised plants are of garden origin.

References: Blanchard (1990), Stace (1975).

I. TAYLOR

Narcissus pseudonarcissus subsp. *pseudonarcissus* Daffodil

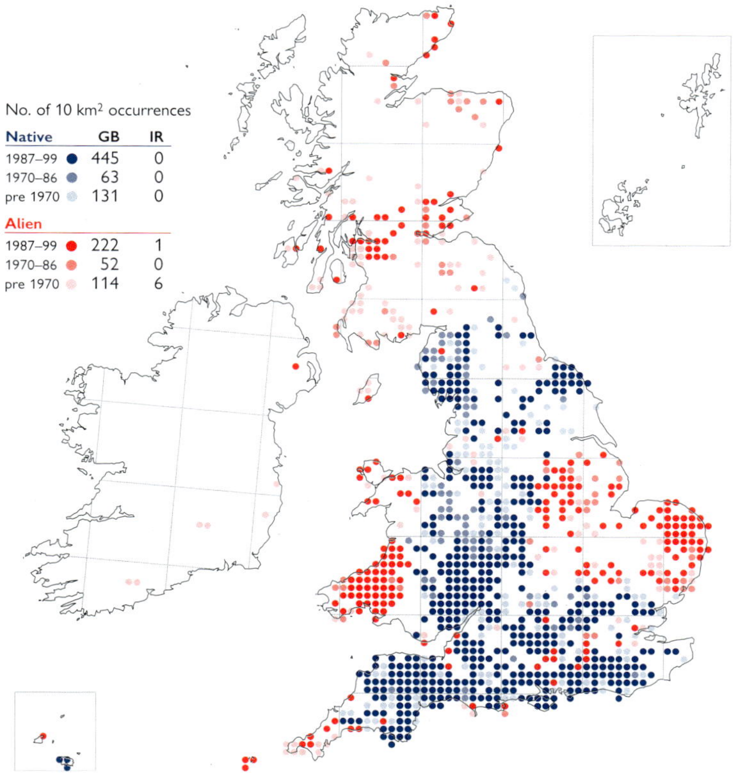

A bulbous perennial herb of *Fraxinus* and *Quercus* woods, *Pteridium* stands, scrubby banks and pastures. In woodland it responds favourably to coppicing, but plant vigour decreases after hot, dry summers. Lowland.

Native (change for species +0.87). There has been some decline in the distribution of subsp. *pseudonarcissus* recently, probably mainly due to changes in land use and management. The limits of its native range in our area are unclear; populations in semi-natural habitats in England and Wales are generally mapped as native unless evidence points to the contrary, but it is undoubtedly alien in Ireland and usually regarded as such in Scotland.

Suboceanic Temperate element; naturalised in Europe outside its native range.

References: Atlas (328b), Barkham (1992), Blanchard (1990), Caldwell & Wallace (1955), Mabey (1996), Sell & Murrell (1996).

I. TAYLOR

Narcissus pseudonarcissus subsp. *obvallaris*
Tenby Daffodil

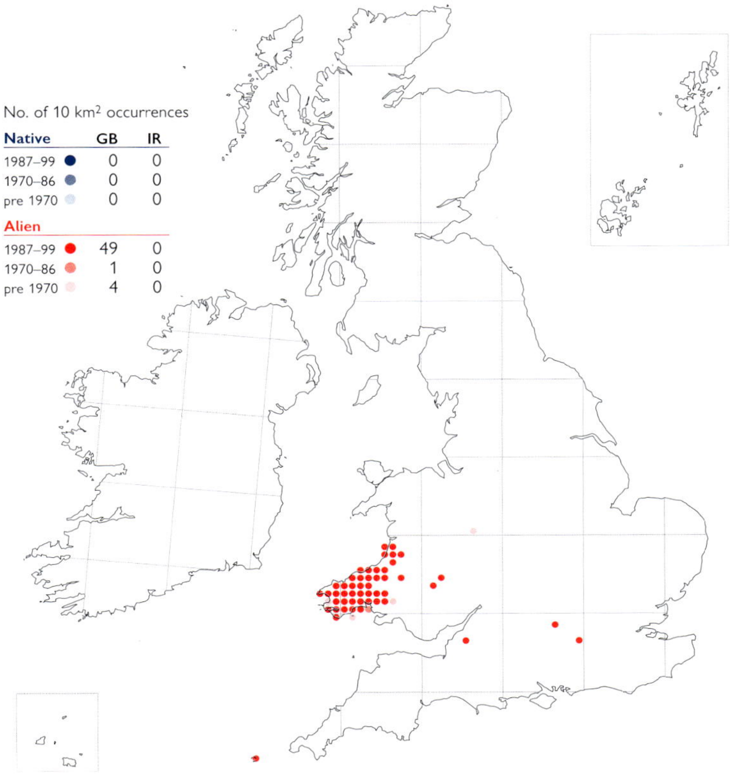

No. of 10 km² occurrences

Native		GB	IR
1987–99	●	0	0
1970–86	●	0	0
pre 1970	●	0	0
Alien			
1987–99	●	49	0
1970–86	●	1	0
pre 1970	●	4	0

A bulbous perennial herb of hedgerows, churchyards, roadsides and other grassy places, now usually closely associated with human habitation although formerly more widespread across farmland in S.W. Wales. Lowland.

Neophyte. An enigmatic taxon usually considered to have been derived from *N. pseudonarcissus* subsp. *major* in cultivation by Medieval times, although specimens closely resembling subsp. *obvallaris* occur in the wild in Spain. The Welsh population was much reduced by commercial bulb collectors in the early 19th century. It is better recorded now than in the 1962 *Atlas*.

A plant of uncertain origin.

References: Atlas (328c), Blanchard (1990), Mabey (1996), Sell & Murrell (1996).

I. TAYLOR

Narcissus pseudonarcissus subsp. *major*
Spanish Daffodil

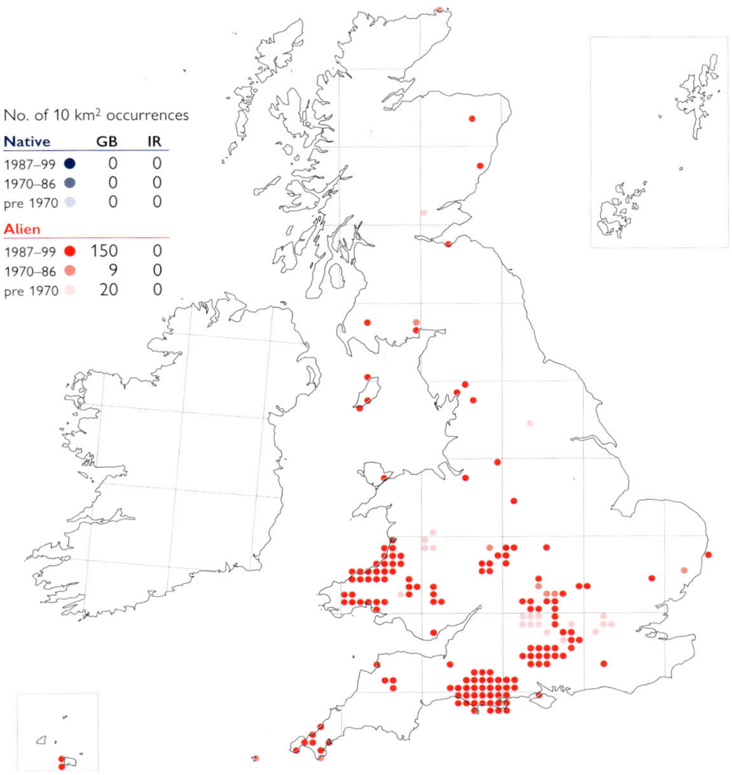

No. of 10 km² occurrences

Native		GB	IR
1987–99	●	0	0
1970–86	●	0	0
pre 1970	●	0	0
Alien			
1987–99	●	150	0
1970–86	●	9	0
pre 1970	●	20	0

A bulbous perennial herb readily naturalised from discarded or deliberately planted bulbs in woodlands, hedgerows, on roadsides, along tracks and paths and on waste ground. It is also found as a relic of commercial cut-flower production. Lowland.

Neophyte. This taxon was cultivated in British gardens by 1629, and was recorded in the wild in 1813. Hybridisation with other subspecies of *N. pseudonarcissus*, especially subsp. *pseudonarcissus*, complicates recording. Confusion also exists between this taxon and the larger single-coloured cultivars of *N.* × *incomparabilis*. It is certainly under-recorded.

Native of S. France and the Iberian peninsula.

Reference: Blanchard (1990).

I. TAYLOR

Asparagus officinalis subsp. *officinalis*
Garden Asparagus

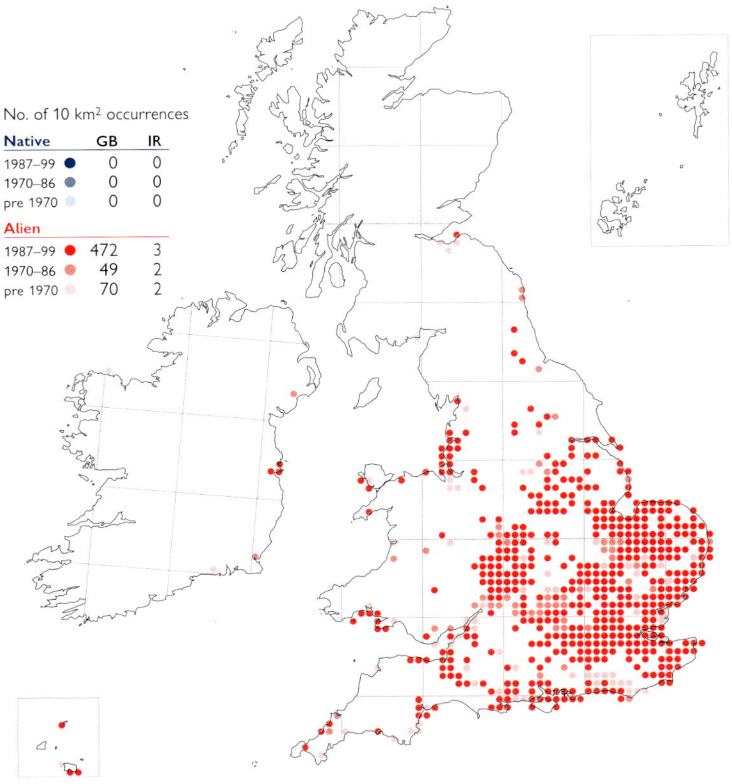

No. of 10 km² occurrences

Native		GB	IR
1987–99	●	0	0
1970–86	●	0	0
pre 1970	●	0	0
Alien			
1987–99	●	472	3
1970–86	●	49	2
pre 1970	●	70	2

A dioecious, rhizomatous perennial herb of freely-draining sandy soils. *A. officinalis* subsp. *officinalis* readily becomes naturalised, probably mainly by bird-dispersed seed, on grassy heaths and dunes. Lowland.

Archaeophyte (change for species +1.78). *A. officinalis* subsp. *officinalis* was cultivated in Britain in Roman times. It is now widely grown commercially and in gardens. Since the 1962 *Atlas*, records of subsp. *officinalis* have increased considerably, perhaps reflecting a genuine increase combined with better recording.

A Eurosiberian Temperate taxon; widely naturalised outside its native range.

References: Atlas (313d), Hultén & Fries (1986), Zohary & Hopf (2000).

I. TAYLOR

Asparagus officinalis subsp. *prostratus*
Wild Asparagus

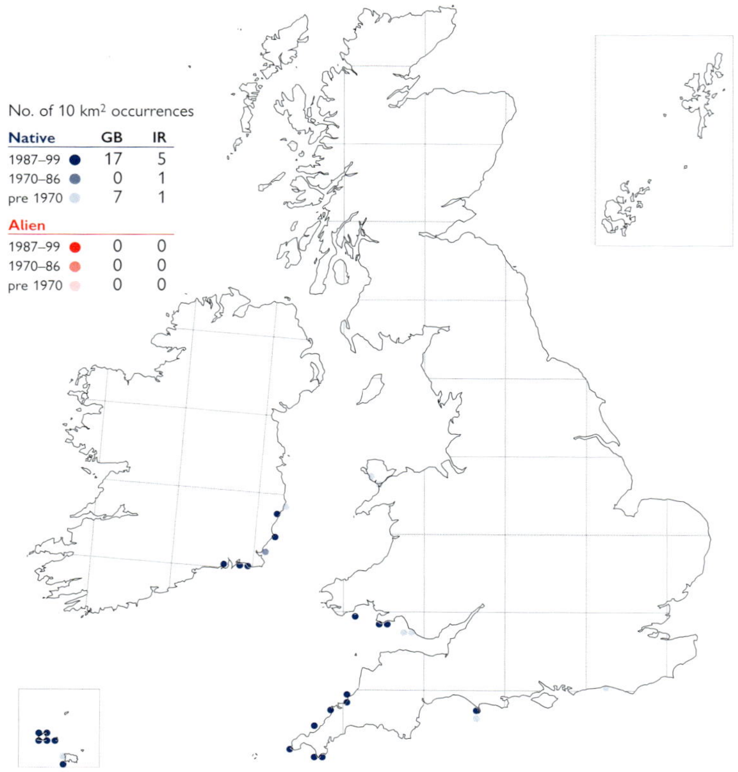

A dioecious, rhizomatous perennial herb of free-draining sea-cliffs and sand dunes. On sea-cliffs, the plants often grow through a dense mat of *Festuca rubra*, usually in very rocky soils. On dunes, it grows in open or closed turf, often by paths. Lowland.

Native. Although plants of this subspecies can be long-lived, recruitment appears to be poor and there is some evidence of decline, in both range and frequency. Remaining populations are scattered and many are very small and single-sexed. The prostrate habit of this subspecies remains constant in cultivation; it is cytologically distinct from subsp. *officinalis* and there is a strong case for treating it as a full species (Kay, 1997).

Oceanic Temperate element.

References: Atlas (313d), Curtis & McGough (1988), Kay & John (1995), Wigginton (1999).

I. TAYLOR

Ruscus aculeatus Butcher's-broom

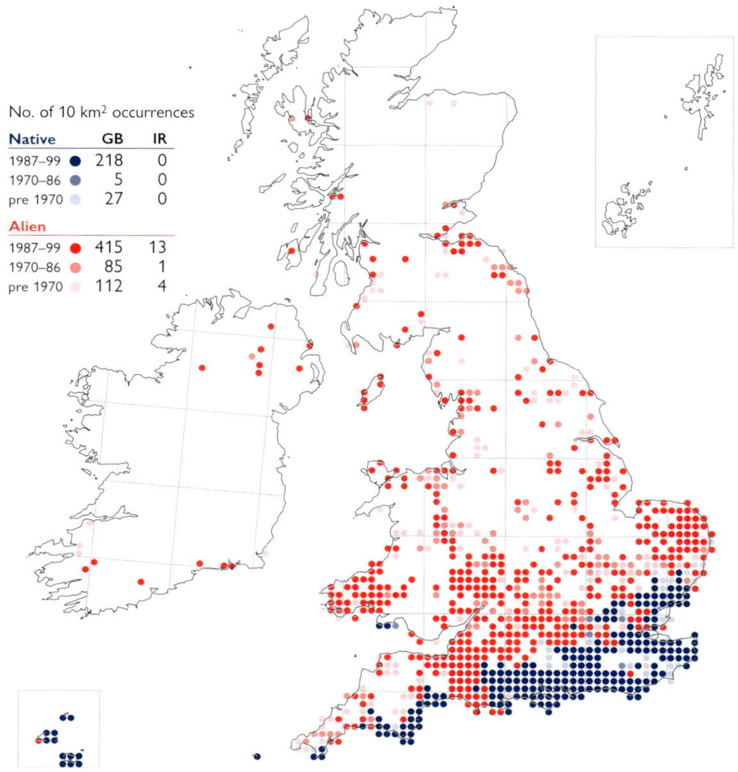

A dioecious, evergreen, rhizomatous shrub, found as a native in dry woods and hedgerows, and on cliffs and rocky ground near the sea. It is also naturalised in similar situations, and in churchyards and near habitation. It reproduces vegetatively by creeping rhizomes, and by seed, which may be bird-sown. Lowland.

Native (change +0.74). *R. aculeatus* is frequently grown in gardens, and may become established where discarded or deliberately planted, or where bird-sown. The overall distribution is stable since the 1962 *Atlas*, but the line between native and alien occurrences is becoming increasingly blurred and it is probably much planted within the mapped native range.

Submediterranean-Subatlantic element.

References: Atlas (314a), Mabey (1996), Meusel *et al.* (1965).

I. TAYLOR

Sisyrinchium bermudiana Blue-eyed-grass

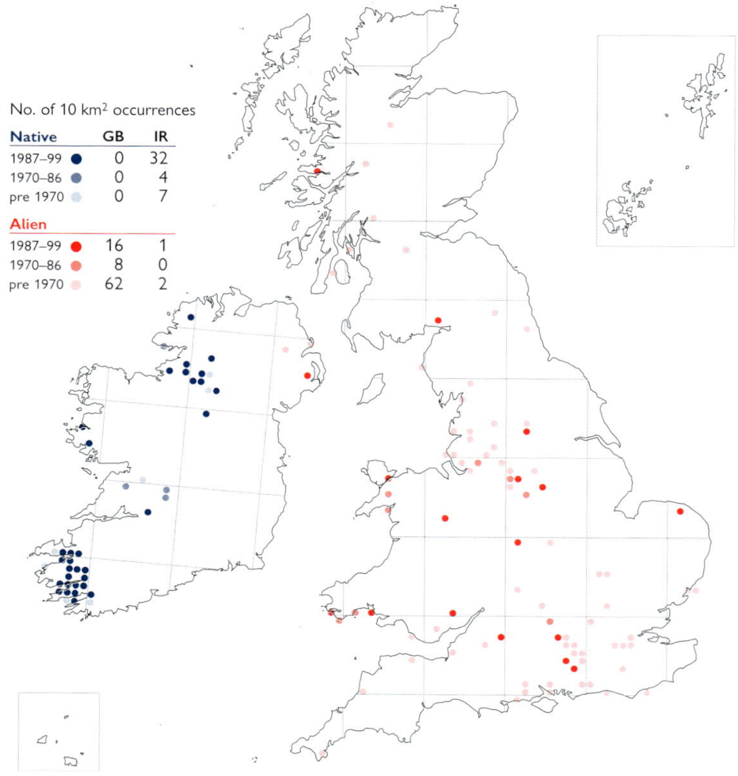

A perennial herb of wet meadows, ditches and lake-shores. Lowland.

Native (change −1.80). There is little evidence of significant change in the distribution of this species in Ireland. Naturalised plants have been reported in Britain, but *S. bermudiana* has been confused with *S. montanum* and records need to be checked (Stace, 1997), so any changes in its alien distribution cannot be easily assessed.

Oceanic Wide-temperate element; in Europe restricted to Ireland as a native but widespread in N. America.

References: Atlas (328d), Curtis & McGough (1988).

I. TAYLOR

Sisyrinchium montanum
American Blue-eyed-grass

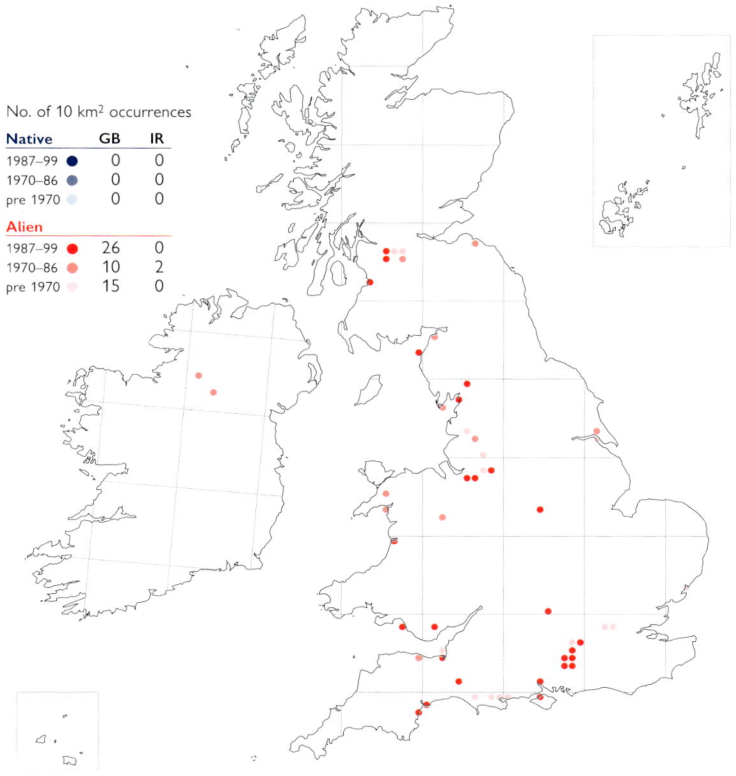

A perennial herb of open waste ground and rough grassy habitats, commonly grown in gardens and spreading freely by seed when plants are discarded into suitable habitats. Lowland.

Neophyte. *S. montanum* has been cultivated in Britain since 1693 and was known in the wild by 1871. Considerable confusion with *S. bermudiana* makes assessment of changes in their respective distributions in Britain difficult. The identity of some naturalised *Sisyrinchium* populations in Ireland also needs to be checked.

Native of N. America; widely naturalised in C. Europe.

I. TAYLOR

Sisyrinchium striatum Pale Yellow-eyed-grass

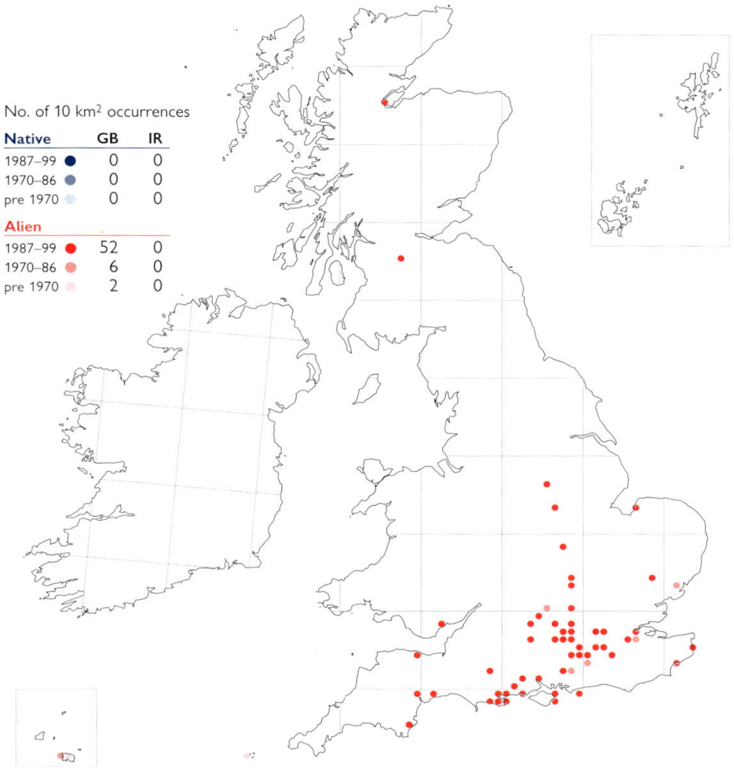

A fibrous-rooted evergreen perennial herb of open waste ground, frequently occurring as short-lived populations arising from discarded garden plants. Occasionally it is more persistent, especially in old quarries and on heathy waste ground where it self-sows prolifically and detached side-shoots can become established. Lowland.

Neophyte. *S. striatum* has been cultivated in Britain since 1788 and has gained popularity as a garden plant in recent years. It was recorded from the wild by 1928 and is probably increasing.

Native of S. America.

I. TAYLOR

Iris germanica Bearded Iris

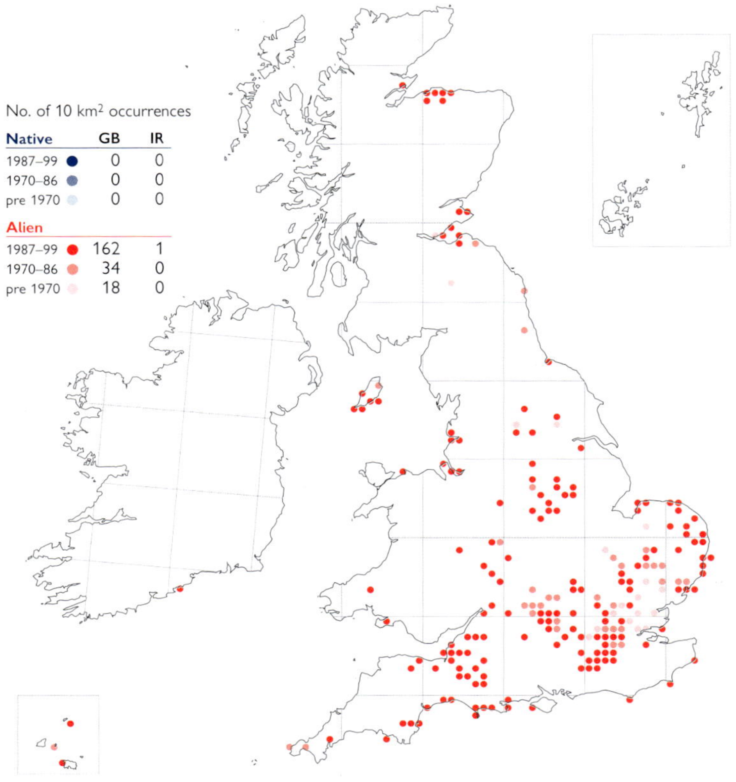

A rhizomatous perennial herb found in rough and waste ground, on roadsides and railway banks and in derelict gardens. It rarely produces seed in Britain, but spreads vegetatively. Lowland.

Neophyte. *I. germanica* has been grown in gardens since at least 995 (Harvey, 1981). It is frequently cultivated in gardens and becomes naturalised when planted or discarded in suitable habitats. It has been known in the wild since at least 1905.

This cultivated species is probably a hybrid of garden origin, but if so its parentage is very obscure.

Reference: Meusel *et al.* (1965).

I. TAYLOR

Iris sibirica Siberian Iris

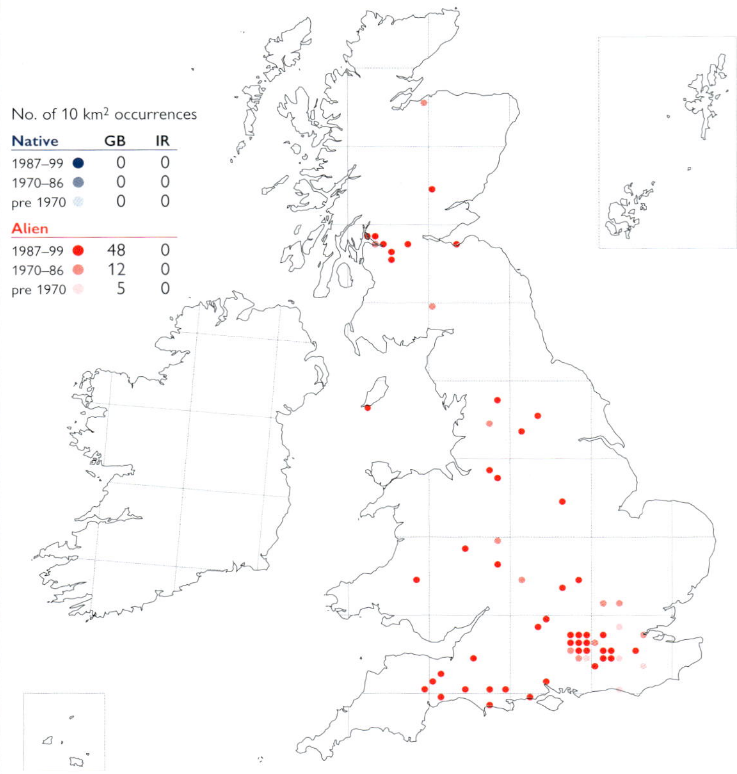

A rhizomatous perennial herb of moisture-retentive rough grassland, open damp woodland and waste ground. *I. sibirica* is often long-lived when discarded from gardens into suitable habitats, where it sometimes spreads by seed. Lowland.

Neophyte. *I. sibirica* was being cultivated in British gardens by 1596, and was recorded from the wild by 1928. Its popularity and ease of growth appear to be responsible for a steady increase in records since then.

Native of C. & E. Europe and W. Asia.

References: Hultén & Fries (1986), Meusel *et al.* (1965).

I. TAYLOR

Iris pseudacorus Yellow Iris

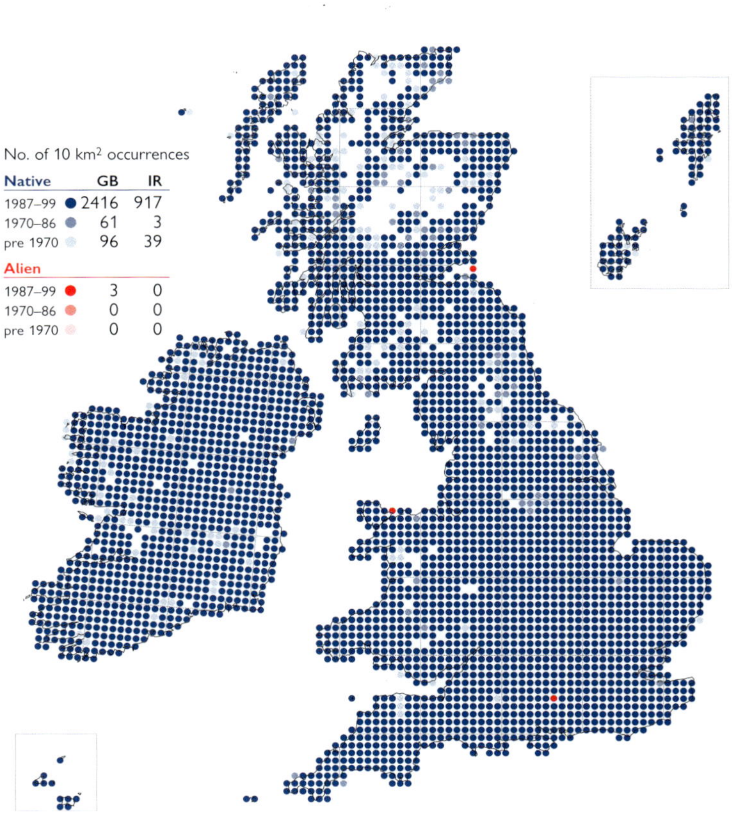

A rhizomatous perennial herb of wet meadows, wet woods, fens, the margins of lakes, ponds and watercourses, wet dune-slacks, and, in N. and W. Britain, also of coastal streams, shingle, upper saltmarshes and raised beaches. It reproduces by seed and by vegetative spread. Plants have occasionally been planted in the wild or have escaped from cultivation. Generally lowland, but reaching 480 m at Nenthead (Cumberland).

Native (change +0.16). The 10-km distribution of *I. pseudacorus* has been stable since the 1962 *Atlas*.

European Southern-temperate element; widely naturalised outside its native range.

References: Atlas (329c), Hultén & Fries (1986), Meusel *et al.* (1965), Preston & Croft (1997), Sutherland (1990).

I. TAYLOR

Iris foetidissima Stinking Iris

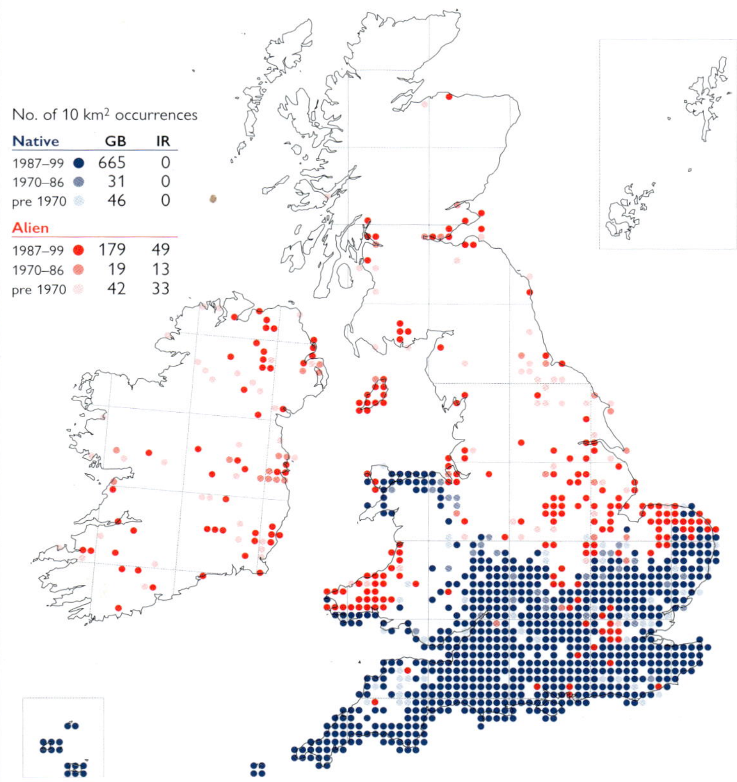

A perennial herb, highly tolerant of drought and shade, found in hedge banks and woods, and on sheltered scrubby sea-cliffs, mostly on calcareous substrates. Reproduction is by seed and by rhizomatous extension. Lowland.

Native (change +1.47). The native range of *I. foetidissima* is much obscured by garden escapes and introductions. It is treated here as alien throughout Ireland. Its overall distribution has not changed since the 1962 *Atlas*, although it is now much better recorded.

Suboceanic Southern-temperate element.

References: Atlas (329b), Mabey (1996), Mathew (1981).

I. TAYLOR

Romulea columnae Sand Crocus

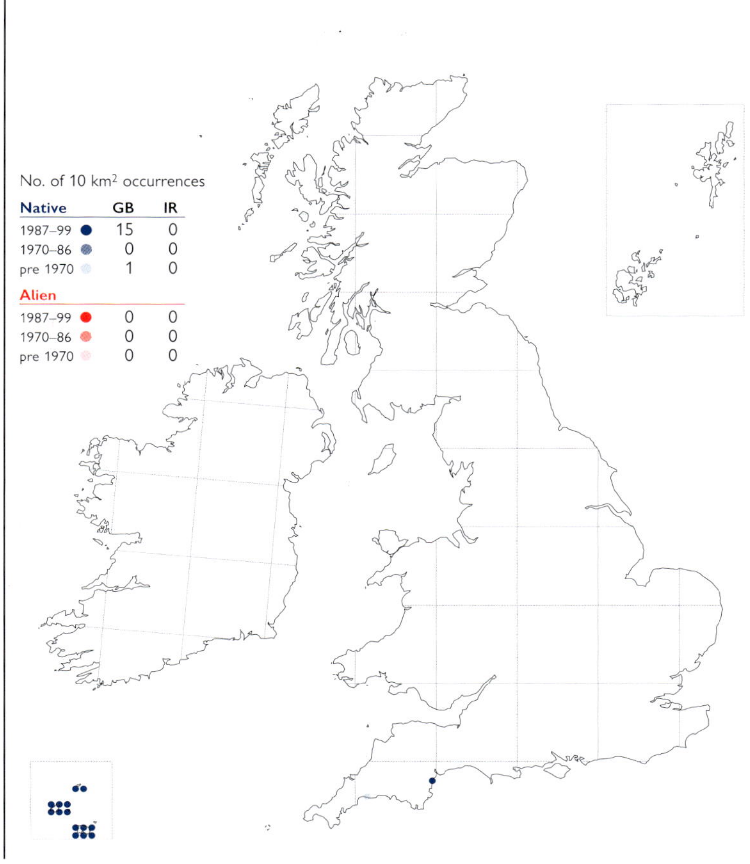

A cormous perennial herb of short, open turf on freely-draining sandy ground and cliff-slopes near the sea. Reproduction is mostly by seed, with division of the corm apparently much less significant. Lowland.

Native. There is no indication of change in the distribution of *R. columnae* since the 1962 *Atlas*, and its populations are stable in the Channel Islands (where it is common in suitable habitats) and at its sole British locality at Dawlish Warren (S. Devon). It was last reported from Cornwall in 1881.

Mediterranean-Atlantic element.

References: Atlas (330b), Wigginton (1999).

I. TAYLOR

Crocus vernus Spring Crocus

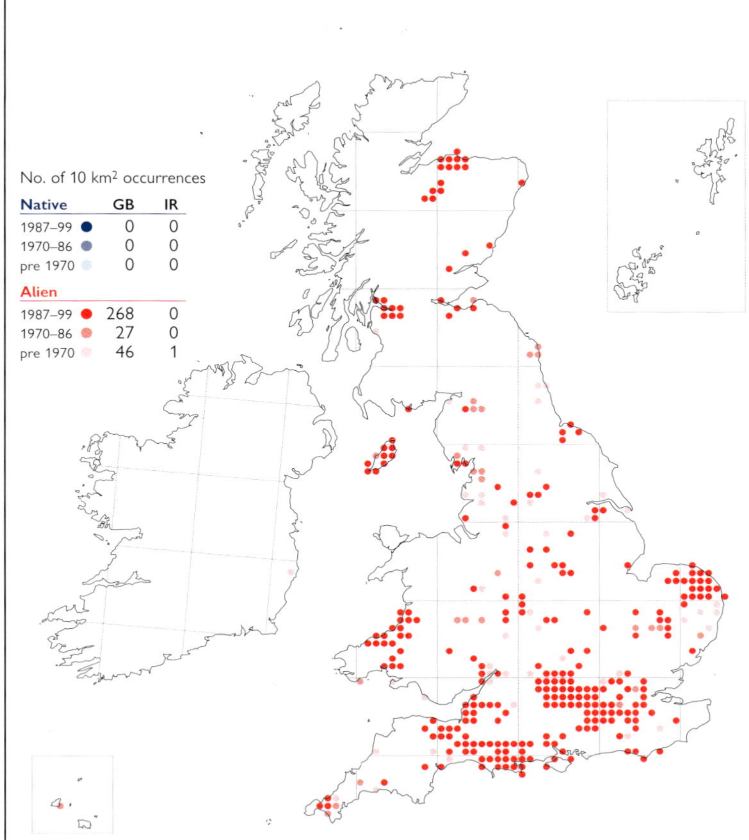

A cormous perennial herb, well-naturalised in a wide variety of grassy habitats, especially in churchyards and amenity grasslands, and on roadside verges. Lowland.

Neophyte (change +2.99). *C. vernus*, introduced into cultivation in Britain before 1600, is very common in parks and gardens and was first recorded in the wild in 1763. Some populations may be relics of cultivation as a substitute for saffron, and it is still found in great abundance at a few long-established sites such as at Inkpen (Berks.). It is unevenly recorded.

Native of upland and montane C. & S. Europe.

References: Atlas (330a), Mabey (1996), Mathew (1982).

I. TAYLOR

Crocus tommasinianus Early Crocus

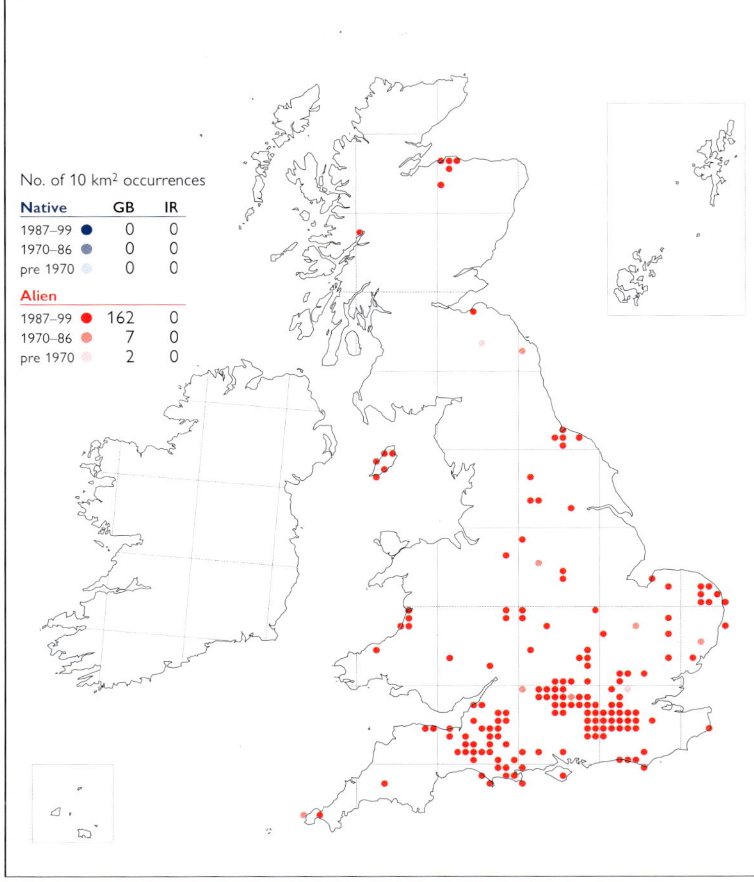

A cormous perennial herb of open deciduous woodland, churchyards, roadsides, parks and amenity grasslands. It can become invasive in gardens and is often discarded on to road- and track-sides. It frequently becomes naturalised, spreading both vegetatively and by seed. Lowland.

Neophyte. *C. tommasinianus* has been cultivated in Britain since 1847. It has been confused in the past with *C. vernus*, making assessment of changes in its distribution difficult. It was not recorded in the wild until 1963. However, its popularity as a garden plant and ease of cultivation probably mean that it is increasing.

Native of S. Jugoslavia, S. Hungary and N.W. Bulgaria.

Reference: Mathew (1982).

I. TAYLOR

Crocus angustifolius × *C. flavus* (*C.* × *stellaris*) Yellow Crocus

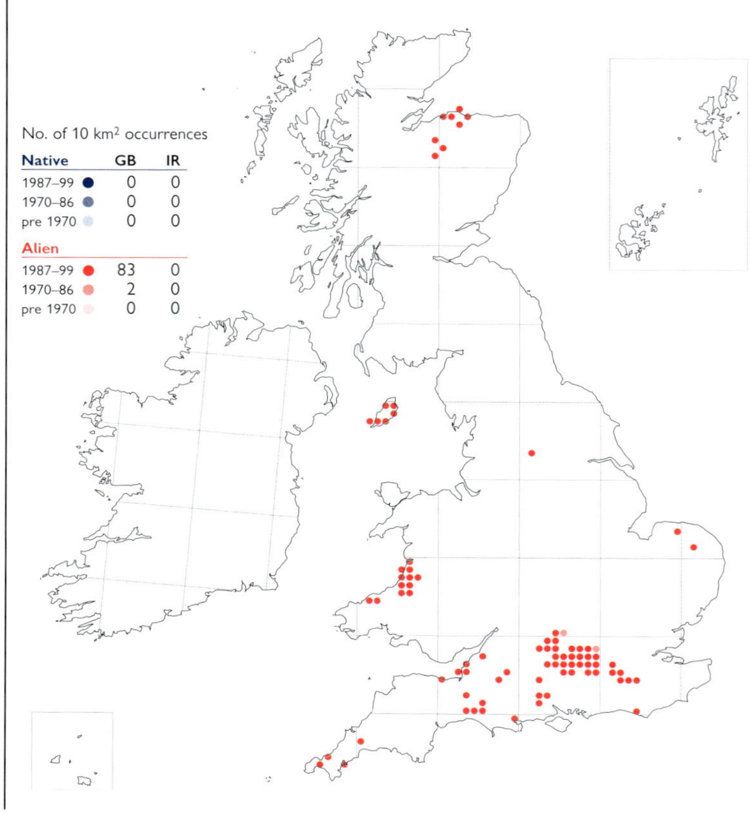

A cormous perennial herb of churchyards, roadsides, parks and other amenity grasslands. It is often long-lived when discarded into suitable habitats, where it spreads vegetatively. Lowland.

Neophyte. *C.* × *stellaris* is commonly grown in gardens as the sterile cultivar 'Golden Yellow' (also known as 'Dutch Yellow' or 'Yellow Giant'). This cultivar was raised over two hundred years ago and its parent species have been cultivated in Britain since the 16th century. *C.* × *stellaris* was first recorded in the wild in 1848. It appears to be steadily increasing, but it has been confused with other taxa, especially cultivars of its parent species and of *C. chrysanthus*, making changes in distribution difficult to assess.

A hybrid of garden origin.

Reference: Mathew (1982).

I. TAYLOR

Crocus nudiflorus Autumn Crocus

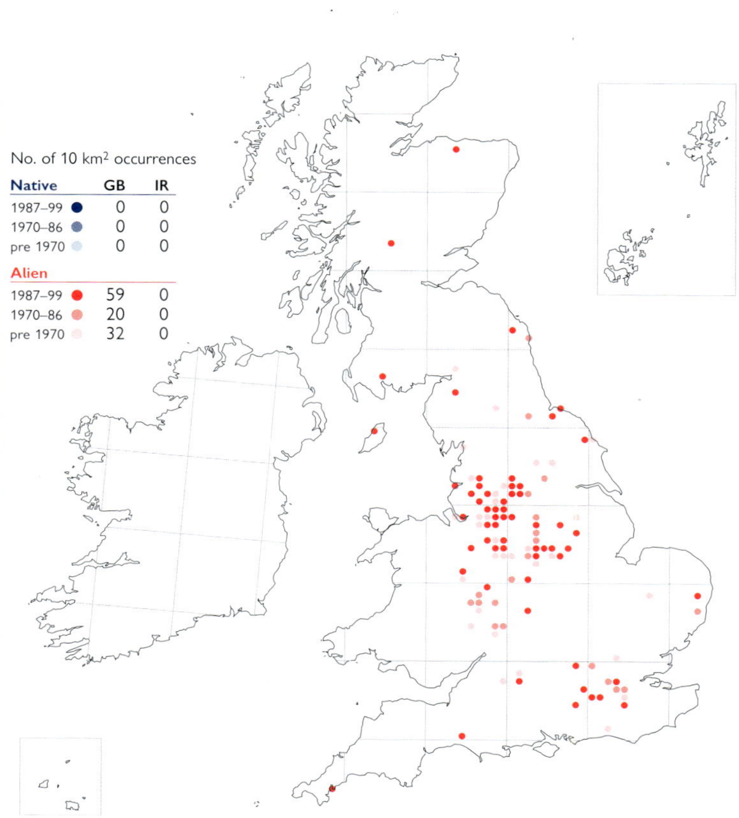

A cormous perennial herb found naturalised in meadows, pastures, amenity grasslands and on roadsides. It spreads vegetatively by means of rhizomes. Lowland.

Neophyte (change +0.60). This species, introduced into cultivation before 1600, is commonly grown in gardens and is still sometimes found as a relic of cultivation as a substitute for saffron. It was first recorded in the wild in 1738 and its distribution is probably stable.

Native of S.W. France and N. Spain.

References: Atlas (329d), Mabey (1996), Mathew (1982).

I. TAYLOR

Gladiolus illyricus Wild Gladiolus

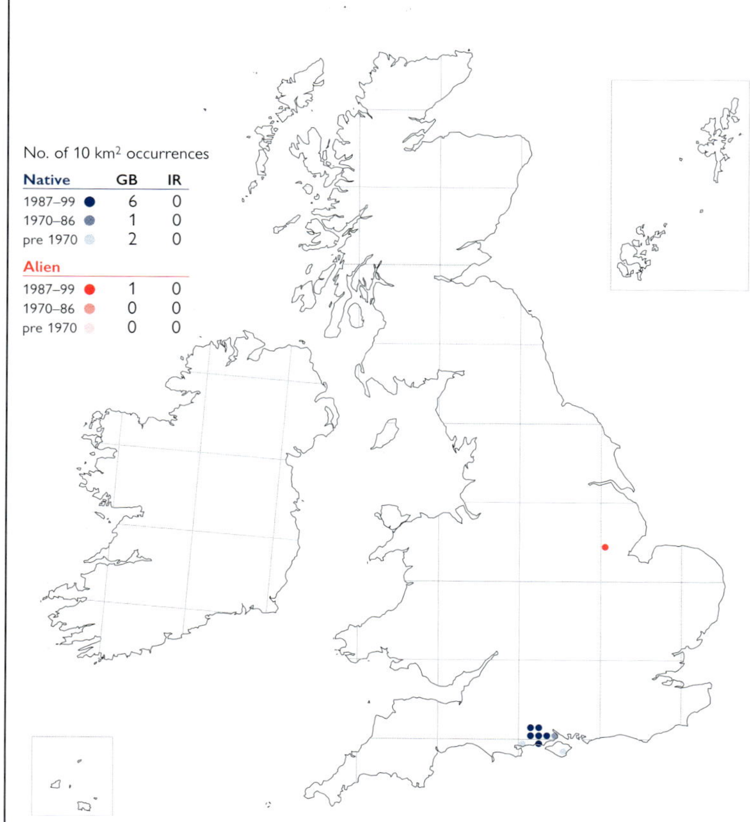

A cormous perennial of acidic, brown-earth soils, found on grass-heaths, usually in association with *Pteridium* which may afford the plant some protection from grazing. It reproduces primarily by offsets, as flowering and seed production appear to be limited. Lowland.

Native (change –0.10). First recorded in the wild in 1856, there has been little recent change in the overall distribution of this species, but populations have declined locally, probably as a result of recent changes in management such as *Pteridium* control and excessive or poorly managed burning.

Mediterranean-Atlantic element. Our plant may be an endemic subspecies.

References: Atlas (330d), Meusel *et al.* (1965), Wigginton (1999).

I. TAYLOR

Gladiolus communis Eastern Gladiolus

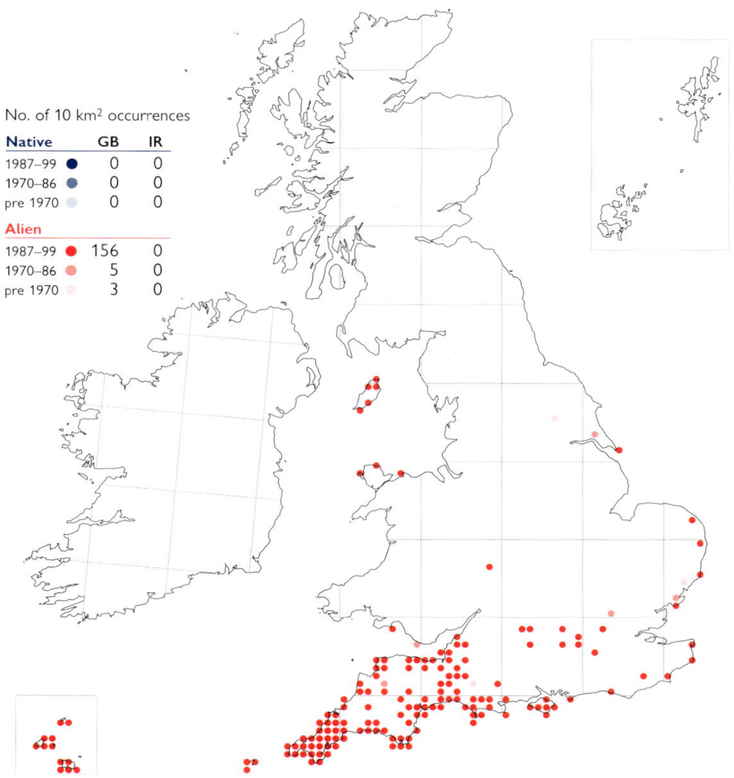

A cormous perennial of bulb fields, field margins, roadsides and rough ground. It is naturalised only in lowland areas experiencing very mild winters, as the corm is susceptible to winter frost. Lowland.

Neophyte. *G. communis*, introduced from S. Europe by 1596, is usually a persistent relic of cultivation. It was first recorded in the wild in 1862, but its distribution has probably changed little in recent years.

Native of the Mediterranean region.

Reference: Lousley (1971).

I. TAYLOR

Crocosmia paniculata Aunt-Eliza

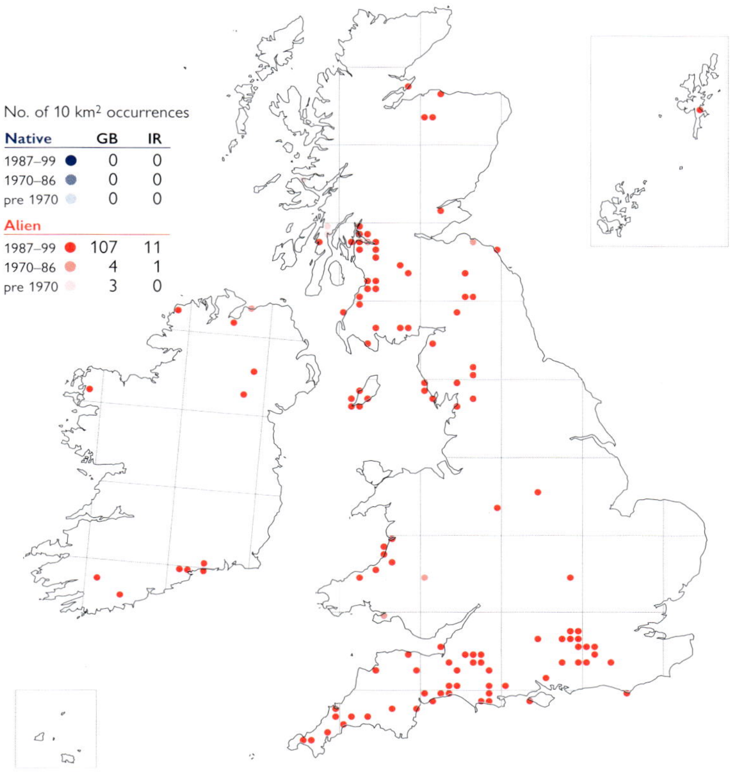

A cormous perennial herb of moist soils on roadsides, woodland margins, quarries and waste ground. *C. paniculata* is sometimes grown in gardens and may persist when discarded into suitable habitats, especially in milder areas. Generally lowland, but reaching 390 m N. of Carter Bar (Roxburghs.).

Neophyte. This species was cultivated in Britain by 1904 and first recorded in the wild in 1961. It appears to be steadily increasing its range. However, an assessment of distributional trends is complicated by confusion with cultivars which are often of hybrid origin, usually involving *C. masoniorum*. The name Aunt-Eliza is a corruption of its former genus *Antholyza*.

Native of S. Africa.

I. TAYLOR

Crocosmia aurea × *C. pottsii* (*C.* × *crocosmiiflora*) Montbretia

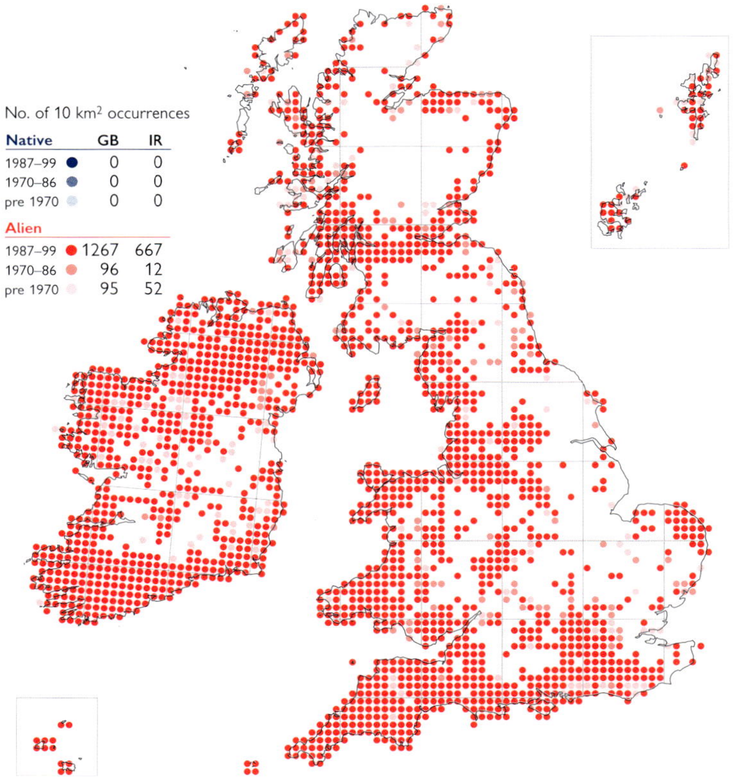

A cormous perennial herb spreading vegetatively by means of rhizomes to form dense clumps in woods and hedge banks, by roadsides and on waste ground. Although some viable seed is produced, most wild populations have probably arisen from discarded garden plants and subsequent vegetative spread. Generally lowland, but reaching 340 m N. of Carter Bar (Roxburghs.).

Neophyte (change +3.11). This hybrid, which is extremely common in gardens, was raised in France in 1880 and reached Britain the same year. It was first recorded from the wild in 1911; since the 1962 *Atlas* it has dramatically extended its range eastwards and considerably consolidated its distribution in W. Britain and Ireland.

A hybrid of garden origin between two S. African parents.

References: Atlas (330c), Nelson (1993).

I. TAYLOR

Phormium tenax New Zealand Flax

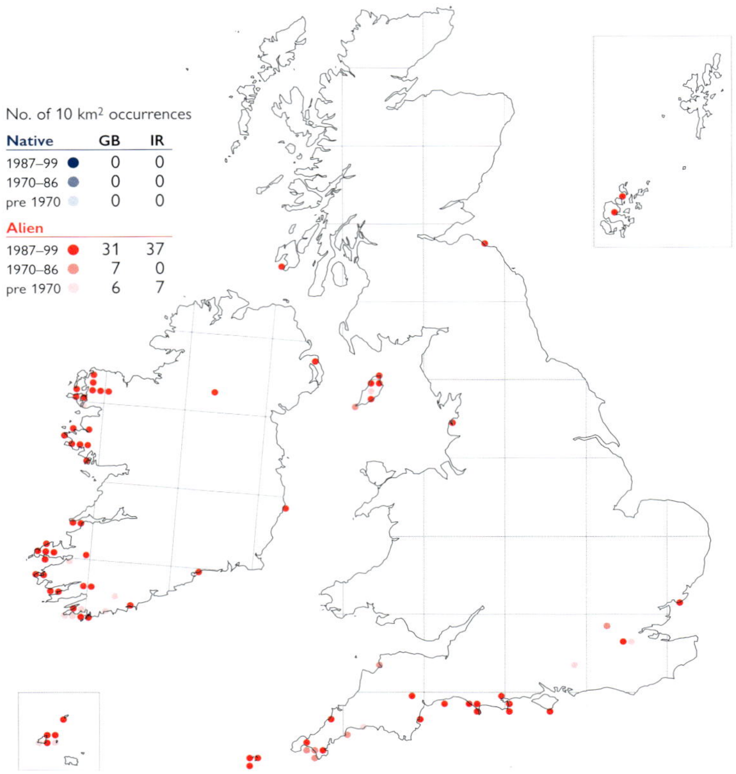

A large, long-lived, evergreen perennial herb of coastal cliffs, rocks and sand dunes; rarely inland on waste ground. It is often planted as a windbreak in maritime areas. Plants reproduce by seed and can become naturalised in suitable areas. Lowland.

Neophyte. *P. tenax* was introduced into cultivation in 1789 and is now very popular in gardens. It was formerly grown in the Isles of Scilly and Isle of Man as a fibre crop. It was recorded in the wild in 1898 (Isles of Scilly), and is probably increasing through deliberate plantings and the discarding of garden material.

Native of New Zealand and Norfolk Island.

T. D. DINES

Tamus communis Black Bryony

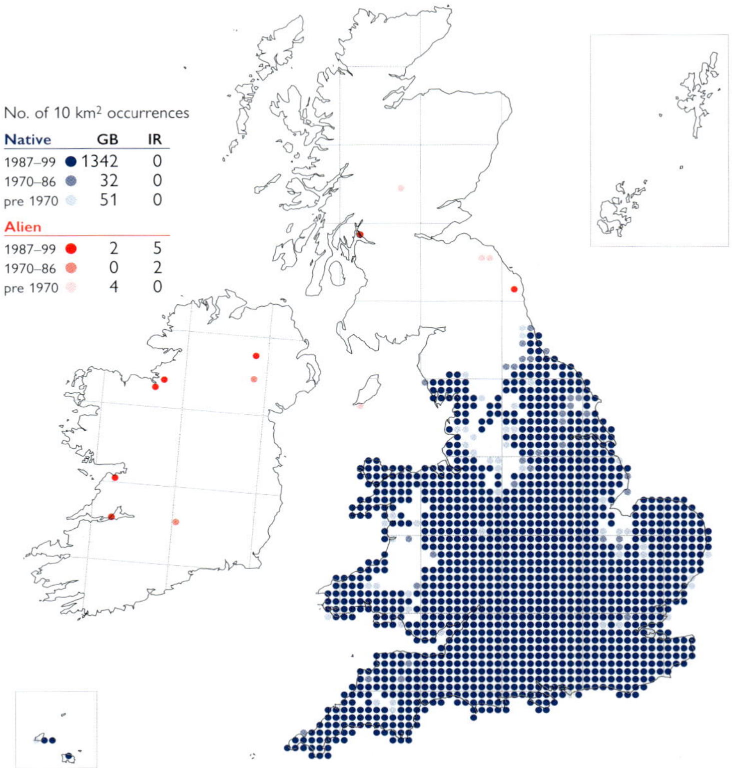

A dioecious, tuberous liane, found mostly on neutral to calcareous, well-drained soils, particularly those overlying chalk and limestone, but also on clay. It can be luxuriant in hedgerows, woodland edges and along paths and in waste land, but is often found in a depauperate, non-flowering state in woodland. It possesses a very large tuber and therefore avoids shallow or waterlogged soil. It is bird-sown, but is not a good colonist. Lowland.

Native (change –0.41). There is no change in the overall distribution of *T. communis* since the 1962 *Atlas*.

Submediterranean-Subatlantic element.

References: Atlas (331a), Burkill (1944), Grime *et al.* (1988), Meusel *et al.* (1965).

A. J. RICHARDS

Cypripedium calceolus Lady's-slipper

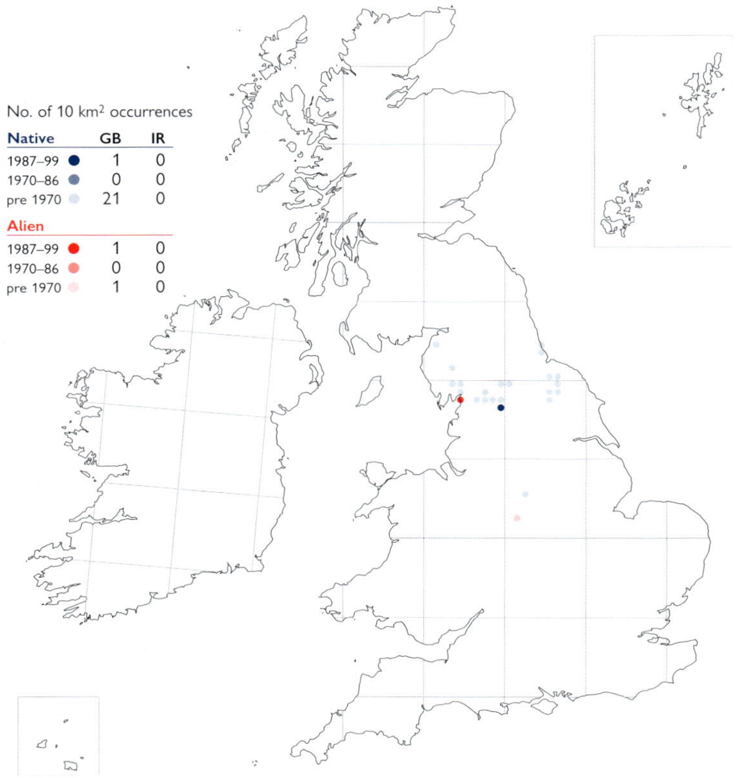

A rhizomatous perennial herb found on well-drained calcareous soils derived from limestone, in herb-rich grassland or, formerly, in open woodland. Plants are long-lived but seed-set may be poor. Generally lowland, but upper altitudinal limit unknown.

Native. This species suffered many losses due to collecting, mostly during the 19th century. The single native colony is heavily protected, although damage by rodents has been known. A recovery programme, including introduction of the species to new and former sites, is underway; details of these sites are not available.

Circumpolar Boreo-temperate element, with a continental distribution in W. Europe.

References: Atlas (331b), Hultén & Fries (1986), Kull (1999), Meusel *et al.* (1965), Ramsey & Stewart (1998), Wigginton (1999).

P. D. CAREY & T. D. DINES

Cephalanthera damasonium White Helleborine

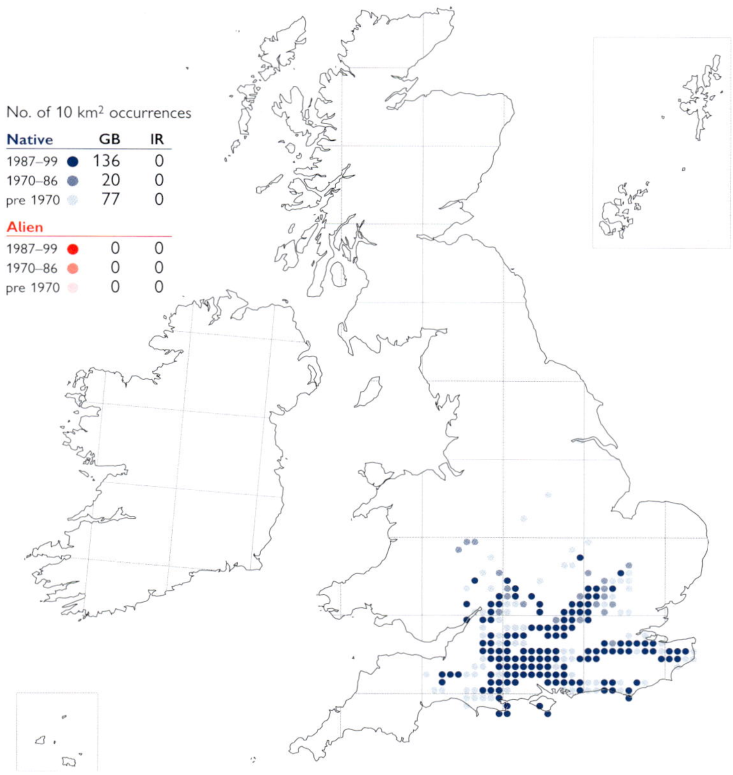

A shade-loving rhizomatous perennial herb usually found in woods with little ground cover, especially those of *Fagus*, but also extending into chalk scrub. It is restricted to well-drained soils on chalk and oolitic limestone. It is usually self-pollinated and the flowers often fail to open entirely. Lowland.

Native (change –0.94). Clearance of woodland, particularly since 1930, has resulted in the loss of many sites for *C. damasonium*. It is still plentiful, however, in suitable habitats, and its ability to colonise young *Fagus* plantations means that new populations are occasionally found.

European Temperate element.

References: Atlas (331c), Hultén & Fries (1986), Jenkinson (1991).

P. D. CAREY & T. D. DINES

Cephalanthera longifolia Narrow-leaved Helleborine

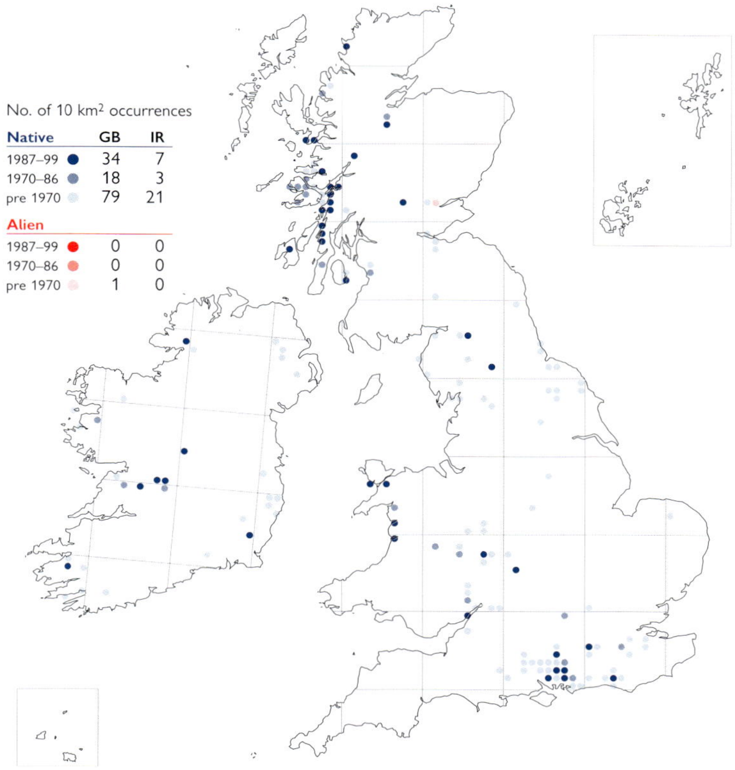

A rhizomatous perennial herb found in a variety of woodland types on calcareous soils, usually on chalk and hard limestone but also on calcareous schist in Scotland. It prefers permanent patches of light and is most frequent on steep, rocky slopes with an open tree canopy, but is also found along woodland edges and rides, and in scrub. Lowland.

Native (change –0.77). This species declined markedly in the 19th and 20th centuries, especially before 1970. Although collecting may have contributed to some losses (Graham, 1988), cessation of woodland management and coniferisation, both leading to denser canopies, are much more significant.

European Temperate element; also in C. Asia.

References: Atlas (331d), Allan & Woods (1993), Curtis & McGough (1988), Hultén & Fries (1986), Stewart *et al.* (1994).

P. D. CAREY & T. D. DINES

Cephalanthera rubra Red Helleborine

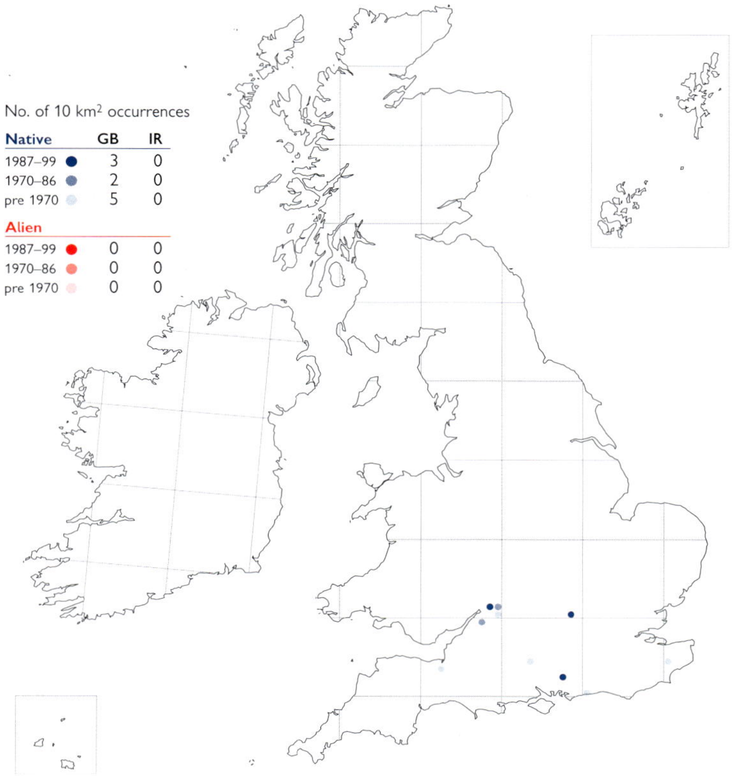

A long-lived rhizomatous perennial herb of well-drained sloping sites in deciduous woods, particularly those of *Fagus*, on calcareous soils. It is a poor competitor and does not form large colonies. Flowers are rarely produced and seed-set is very low, possibly due to the rarity of suitable pollinators. Lowland.

Native. This elusive species has always been rare and is now found at only three sites, with a total of perhaps only thirty plants. Current conservation management includes removal of shrubs and trees. Native populations may soon be augmented with plants grown *ex situ*.

European Temperate element.

References: Atlas (332a), Hultén & Fries (1986), Meusel *et al.* (1965), Wigginton (1999).

P. D. CAREY & T. D. DINES

Epipactis palustris Marsh Helleborine

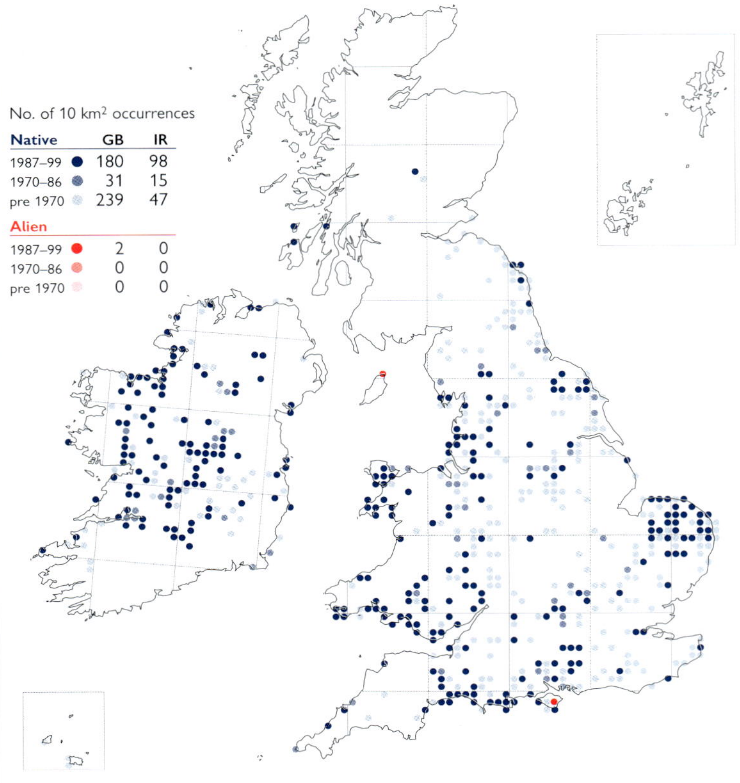

A rhizomatous perennial herb of neutral to calcareous fens, marshes, damp pastures, meadows and dune-slacks. It prefers flushed or seasonally inundated areas where competition from other vegetation is reduced. Other habitats include slumped terraces of wet, calcareous sea-cliffs and disused gravel-pits. Lowland.

Native (change –0.39). Drainage of fens and marshes has caused a considerable decline in this species, with most losses occurring before 1930. It is vulnerable to changes in water-level and to nutrient enrichment. The increase in Irish records since the 1962 *Atlas* is a result of improved recording.

Eurosiberian Temperate element.

References: Atlas (332b), Allan & Woods (1993), Curtis & McGough (1988), Hultén & Fries (1986), Jenkinson (1991), Meusel *et al.* (1965), Sanford (1991).

P. D. CAREY & T. D. DINES

Epipactis atrorubens Dark-red Helleborine

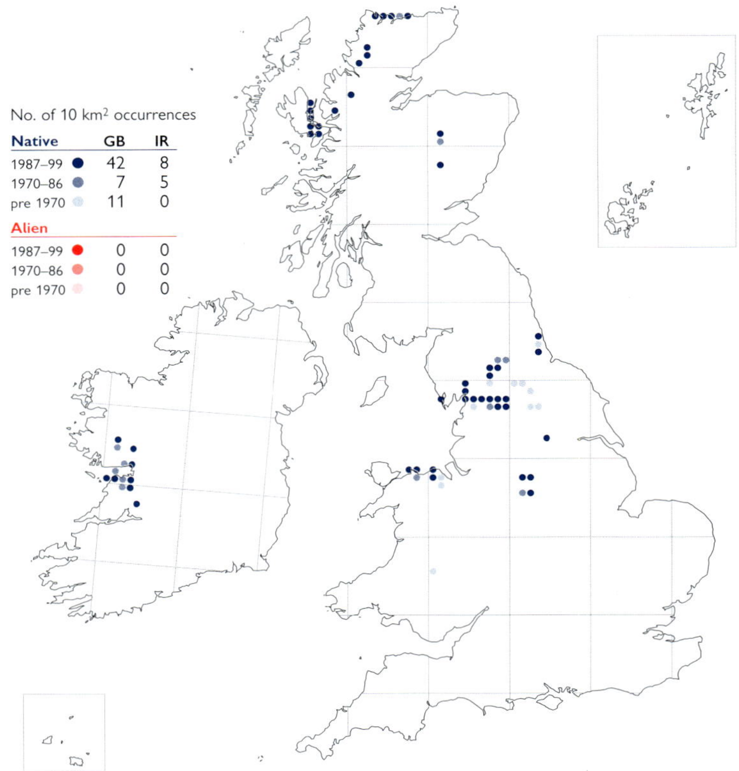

A perennial herb found mostly on bare rock or well-drained skeletal soils overlying limestone. Habitats include exposed scree slopes, open grassy banks, shaded grikes of limestone pavements and ledges on cliff and quarry faces. Reproduction is by seed but most populations are small and often include many non-flowering plants. 0–610 m (Gleann Beag, E. Perth).

Native (change +0.16). The overall distribution of *E. atrorubens* is stable. It is now recorded from more sites than ever before, but some populations have been lost as a result of quarrying activities and others are at risk from overgrazing by deer and rabbits.

Eurosiberian Boreo-temperate element.

References: Atlas (333c), Allan & Woods (1993), Hultén & Fries (1986), Meusel *et al.* (1965), Stewart *et al.* (1994).

P. D. CAREY & T. D. DINES

Epipactis purpurata Violet Helleborine

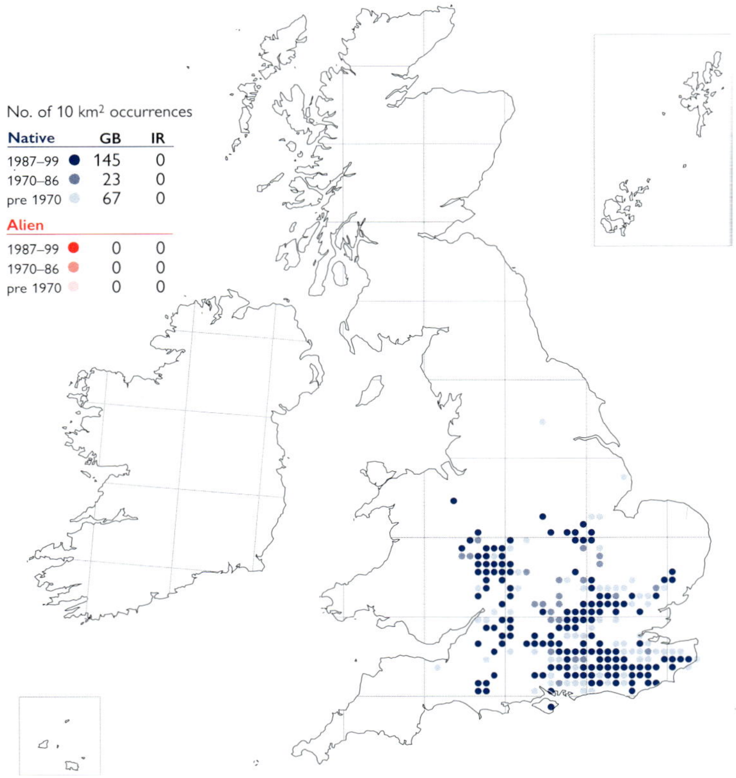

A perennial herb found in a range of woodland types. It is most frequent in densely shaded *Fagus* woods, particularly those on 'clay-with-flints' deposits, but is also found on calcareous, and occasionally acidic, sands and clays in mixed woodland and coppices of *Corylus* and *Carpinus*. Lowland.

Native (change –0.08). Woodland clearance caused a decline in *E. purpurata* before 1930, and this may have continued (to a lesser degree) since the 1962 *Atlas*. It is easily overlooked, however, and is now much better recorded. It requires a particular level of shade; it has increased in Hertfordshire due to a cessation of coppicing.

European Temperate element.

References: Atlas (332d), Bateman (1981), Hultén & Fries (1986), Jenkinson (1991), Sanford (1991).

P. D. CAREY & T. D. DINES

Epipactis helleborine Broad-leaved Helleborine

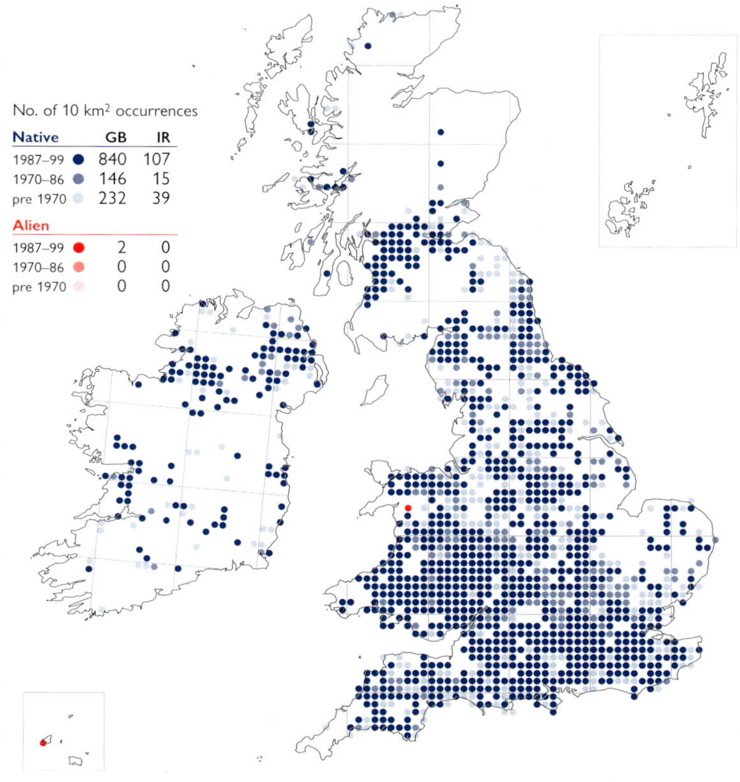

A rhizomatous perennial herb of calcareous to slightly acidic soils. Habitats include coniferous and deciduous woodland, hedgerows, shady banks, streamsides, roadsides, *Alnus* carr, dune-slacks, limestone pavement and screes. It may invade secondary woodland and also occurs in urban habitats, particularly abandoned gardens; it is said to be more common in Glasgow than anywhere else in Britain (Allan & Woods, 1993). 0–350 m (Ystradfellte, Brecs.).

Native (change +0.08). This species was mapped as 'all records' in the 1962 *Atlas*. The overall distribution is stable.

Eurasian Temperate element; widely naturalised outside its native range.

References: Atlas (332c), Hultén & Fries (1986), Jenkinson (1991), Light & MacConaill (1991), Sanford (1991).

P. D. CAREY & T. D. DINES

Epipactis youngiana Young's Helleborine

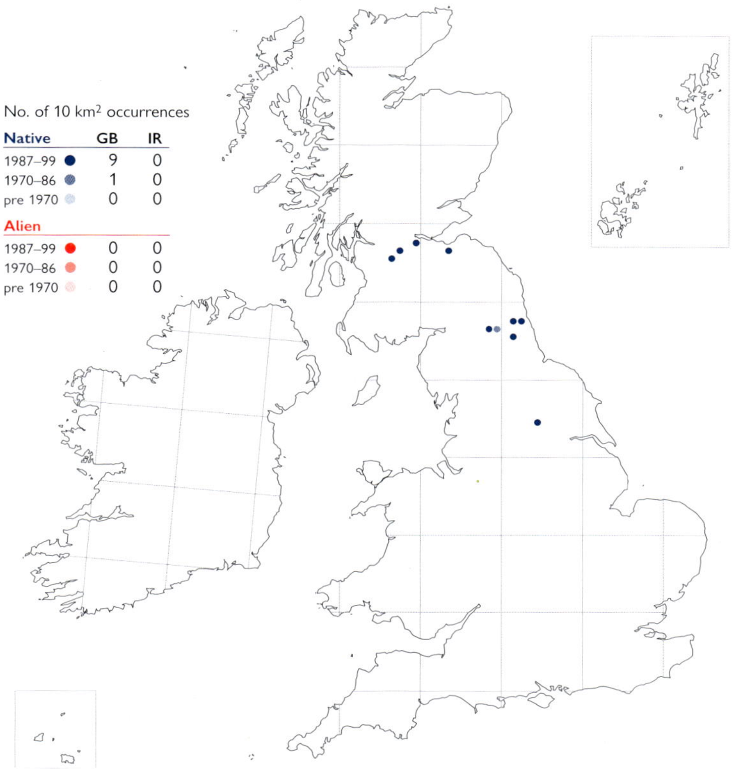

A rhizomatous perennial herb originally found growing in an oak wood on clay soil, and on mildly acidic soils polluted with zinc and lead metals. It has since been found on the steep, lightly wooded slopes of coal-waste bings in Scotland. It is self-pollinated. Lowland.

Native. This species was first identified in the 1970s, but not described until 1982. It is thought to result from hybridisation between *E. helleborine* and either *E. leptochila* (in England and Wales) or *E. phyllanthes* (in Scotland), with the new species emerging as a colonist of man-made habitats. Two colonies have been lost to woodland clearance. It is probably under-recorded.

Endemic.

References: Allan & Woods (1993), Richards & Porter (1982), Wigginton (1999).

P. D. CAREY & T. D. DINES

Epipactis leptochila Narrow-lipped Helleborine

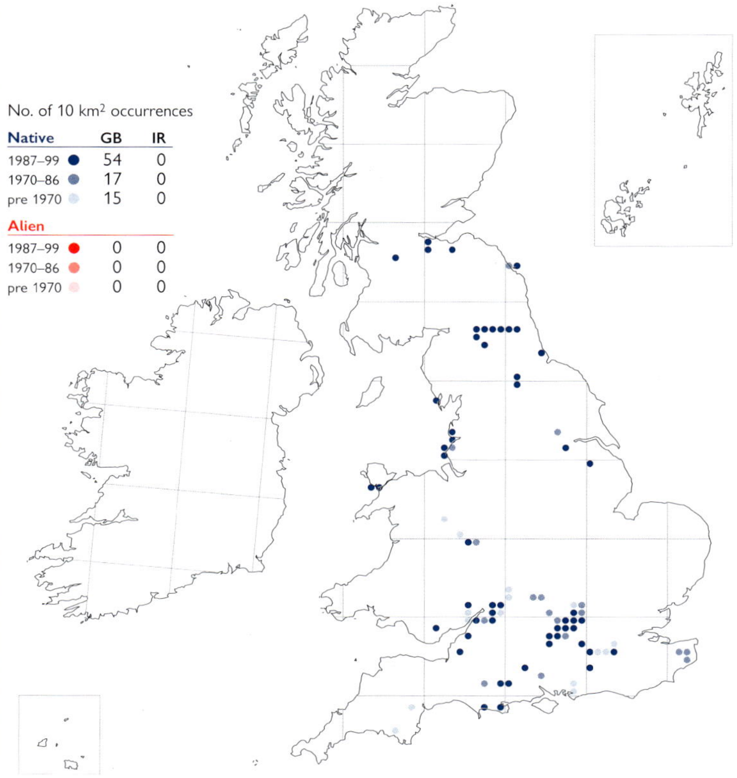

A rhizomatous perennial herb found in three distinct habitats: the deep shade of *Fagus* woods on calcareous substrates; under *Betula* on well-drained stony soils and spoil, often polluted with lead and zinc; and on the edges of dune-slacks where it grows amongst *Salix repens* and can spread into neighbouring conifer plantations. Lowland.

Native (change +0.26). This species is much better recorded now than it was in the 1962 *Atlas*, particularly in N. England and S. Scotland. Coastal dune populations are sensitive to both under- and over-grazing by rabbits, while inland sites have been lost to woodland clearance and coniferisation.

European Temperate element.

References: Atlas (333a), Allan & Woods (1993), Hultén & Fries (1986), Jenkinson (1991), Stewart *et al.* (1994).

P. D. CAREY & T. D. DINES

Epipactis phyllanthes
Green-flowered Helleborine

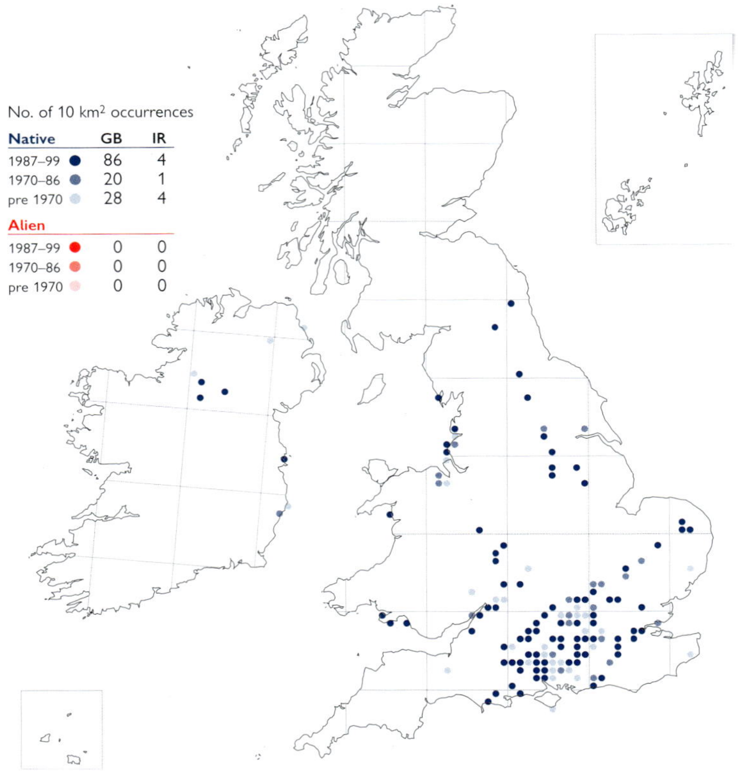

A rhizomatous perennial herb typically found in sparsely vegetated, shaded places on dry, acidic, humus-poor substrates. Habitats include *Fagus* woods on flinty clays or sandstones, *Pinus* and *Betula* scrub on the Bagshot sands, *Corylus* coppice on sandy alluvium, and on sand dunes. Lowland.

Native (change +0.19). Populations of *E. phyllanthes* tend to be sporadic, often not persisting for longer than about thirty years. It has declined in S. England, mostly through woodland clearance, but new sites have been found further north, some of which are increasing in size.

European Temperate element.

References: Atlas (333b), Curtis & McGough (1988), Hultén & Fries (1986), Jenkinson (1991), Sanford (1991), Stewart *et al.* (1994).

P. D. CAREY & T. D. DINES

Epipogium aphyllum Ghost Orchid

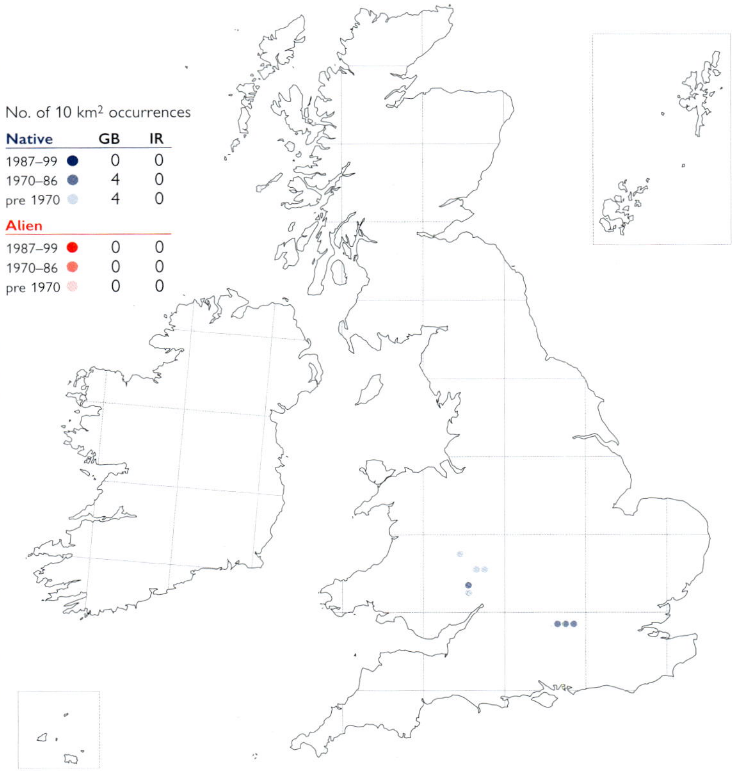

A saprophytic herb usually growing in deep leaf-litter in *Fagus* woods on chalk, with little or no associated ground flora. It is also occasionally recorded from *Quercus* woodland. The underground rhizomes have considerable longevity but the stems are short-lived and may not be produced annually. Although its flowers are pollinated by bees, seed is rarely produced. Lowland.

Native. *E. aphyllum* has appeared only sporadically since 1970 and, although there have been rumours of sightings, it has not been reliably recorded in Britain since 1986. The site at which it was last seen is now a commercial forest planted with conifers.

Eurasian Boreal-montane element.

References: Atlas (333d), Hultén & Fries (1986), Meusel *et al.* (1965), Wigginton (1999).

P. D. CAREY & T. D. DINES

Neottia nidus-avis Bird's-nest Orchid

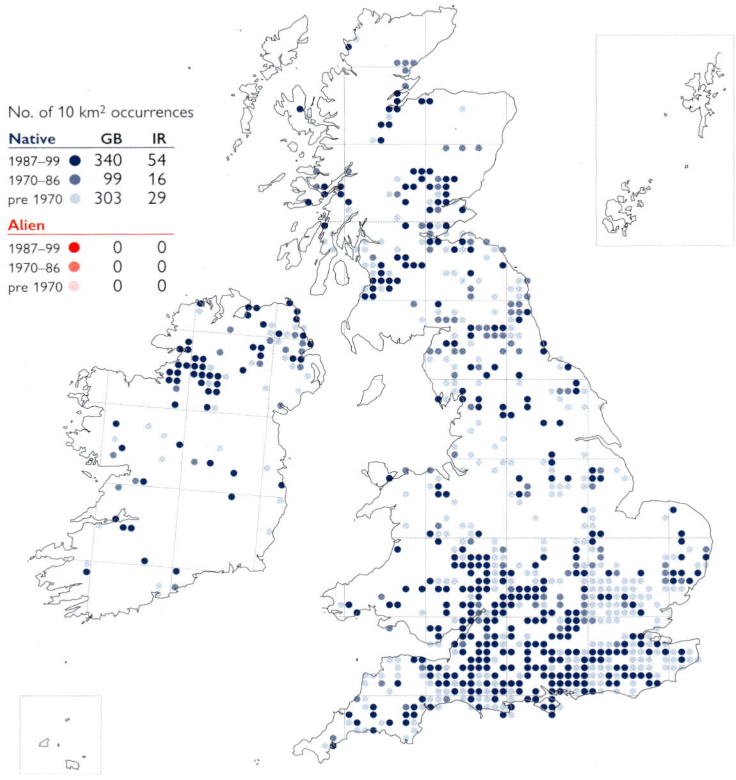

This saprophytic herb is most frequent in the deep humus of densely shaded *Fagus* woods on chalky soils. Less commonly it occurs in mixed deciduous woodland and mature *Corylus* coppices, on soils derived from limestones and base-rich clays and sands. Lowland.

Native (change –0.91). *N. nidus-avis* suffered a considerable decline throughout the 20th century, but particularly between 1930 and 1970, and especially in S.E. England. It is very vulnerable to habitat disruption, and most losses are probably due to changes in woodland management and coniferisation. It has not declined to the same extent in Ireland.

Eurosiberian Temperate element; also in E. Asia.

References: Atlas (335a), Allan & Woods (1993), Curtis & McGough (1988), Hultén & Fries (1986), Jenkinson (1991), Meusel *et al.* (1965), Sanford (1991).

P. D. CAREY & T. D. DINES

Listera ovata Common Twayblade

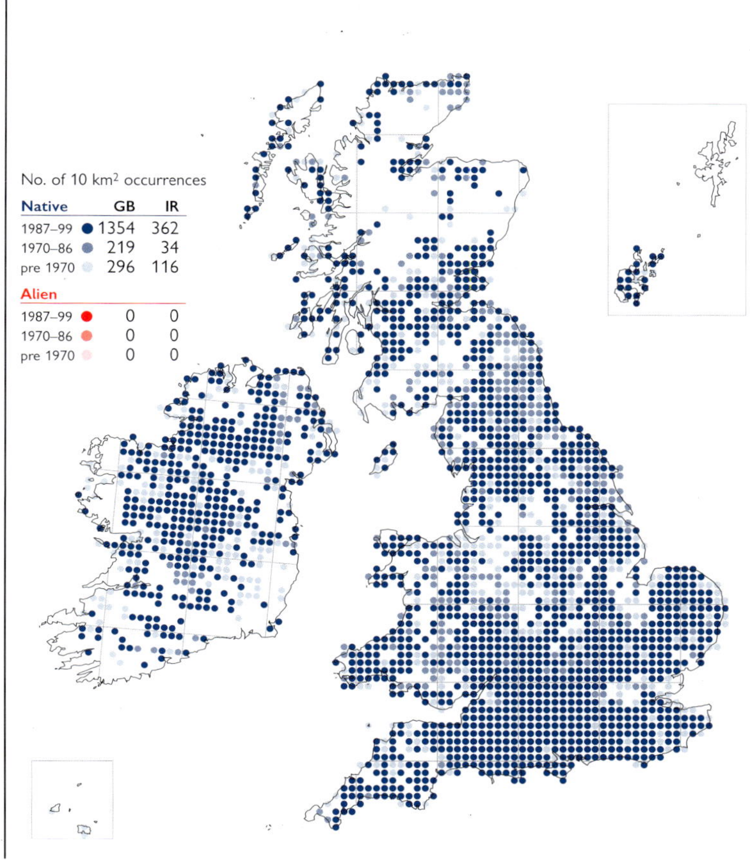

A rhizomatous perennial herb found on a wide range of calcareous to mildly acidic soils. Its habitats include grassland, woodland, hedgerows, scrub, sand dunes, dune-slacks, limestone pavement and heathland; in Anglesey and Ireland it also grows in fens. It frequently occurs, sometimes as large colonies, on railway banks and in disused quarries and sand-pits. 0–670 m (Ben Lawers, Mid Perth).

Native (change –0.54). There has been little change in the distribution of this species since the 1962 *Atlas*, although some lowland sites have been lost to agricultural activities, including ploughing of grassland and hedgerow removal.

Eurosiberian Boreo-temperate element.

References: Atlas (334c), Allan & Woods (1993), Hultén & Fries (1986), Jenkinson (1991), Sanford (1991).

P. D. CAREY & T. D. DINES

Listera cordata Lesser Twayblade

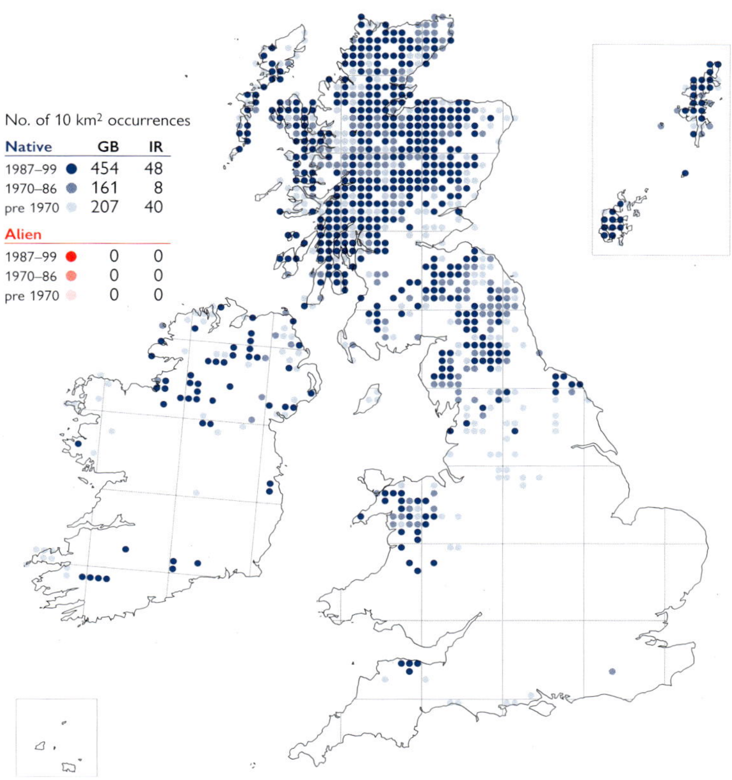

A perennial herb found in moorland and on peat bogs, often growing in *Sphagnum* or in the moss layer beneath *Calluna* and *Vaccinium*, always in wet, acidic conditions. It also grows in moss in damp, heavily shaded wood-carr, and occasionally under *Pinus*. Reproduction is by seed and root-buds. 0–1065 m (Stob Coire an Easain, Westerness).

Native (change –0.32). Although lost from many lowland sites before 1930, presumably due to habitat destruction, this species is much better recorded than it was in the 1962 *Atlas*. It is an inconspicuous plant which may still be under-recorded.

Circumpolar Boreal-montane element, with a disjunct distribution.

References: Atlas (334d), Allan & Woods (1993), Hultén & Fries (1986), Meusel *et al.* (1965).

P. D. CAREY & T. D. DINES

Spiranthes spiralis Autumn Lady's-tresses

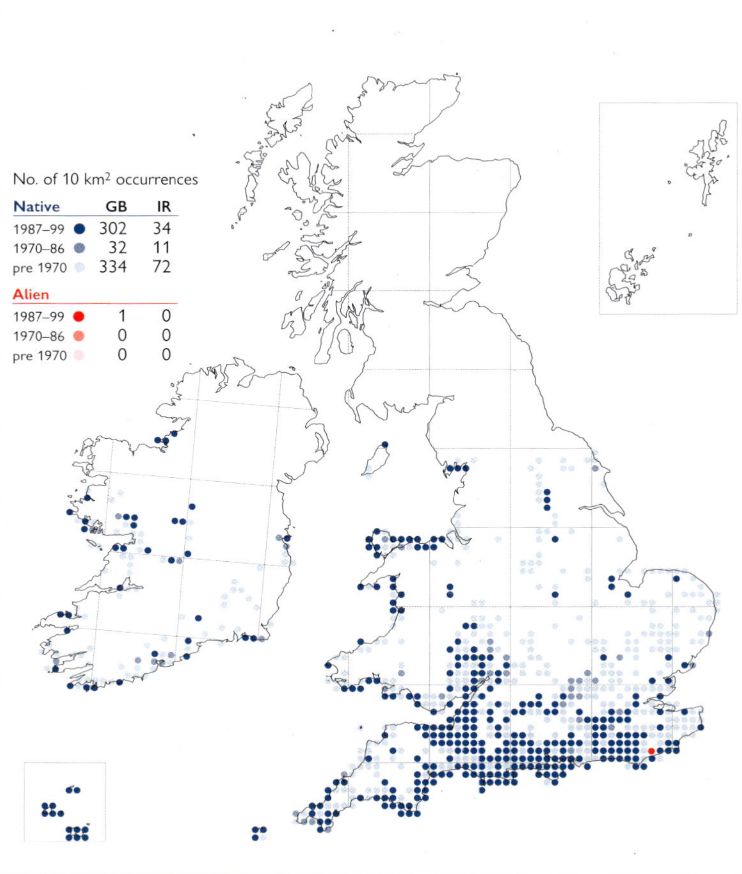

A rhizomatous herb of unimproved, well-grazed grassland on dry calcareous soils, especially on chalk and limestone, and on cliff-tops and sand dunes; also on lawns and, rarely, on less acidic heathland. It can persist for many years without flowering, appearing in hundreds when grazing or mowing ceases. Lowland.

Native (change –0.95). This species declined considerably, particularly before 1930, when many pastures were re-sown or converted to arable. The 1962 *Atlas* included the results of a survey conducted in the hot summer of 1955, when it flowered profusely on uncut lawns (Perring, 1956). Losses due to agricultural intensification and undergrazing still occur.

European Southern-temperate element.

References: Atlas (334a), Hultén & Fries (1986), Jenkinson (1991), Sanford (1991), Wells (1967).

P. D. CAREY & T. D. DINES

Spiranthes aestivalis Summer Lady's-tresses

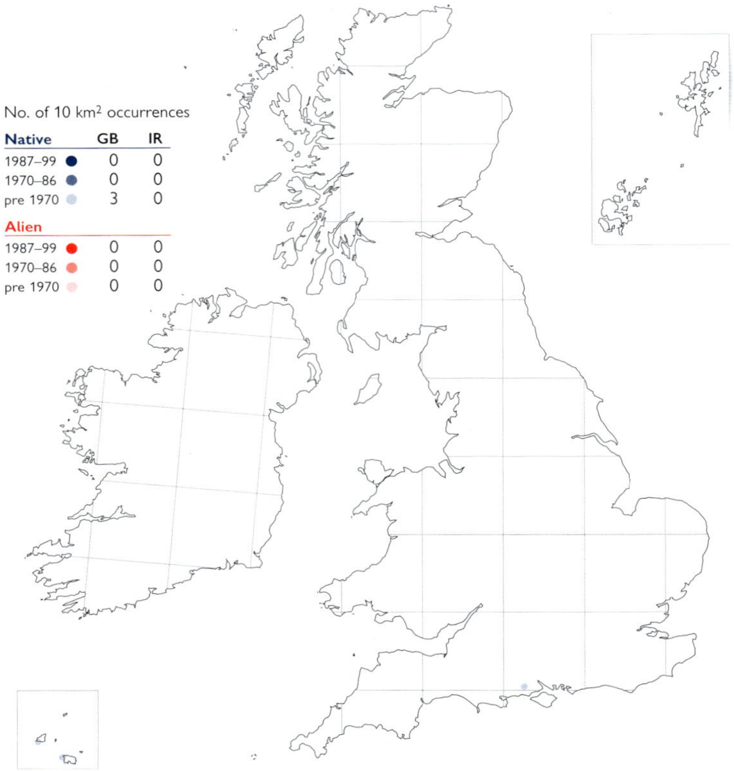

A rhizomatous perennial herb formerly occurring amongst *Sphagnum* in valley bogs in the New Forest (S. Hants.), in a *Sphagnum* bog on Guernsey, and on wet, sandy ground beside St. Ouen's Pond on Jersey. Lowland.

Native. This species is extinct in our area. The last record was in 1959 from the New Forest (S. Hants.), where it suffered from a combination of drainage and over-collecting. It was lost from the Channel Islands for the same reasons, becoming extinct in Guernsey in 1914 and last recorded in Jersey in *c.* 1925.

European Temperate element.

References: Atlas (334b), Brewis *et al.* (1996), Le Sueur (1984), McClintock (1975).

P. D. CAREY & T. D. DINES

Spiranthes romanzoffiana Irish Lady's-tresses

Native	**GB**	**IR**
1987–99	17	17
1970–86	3	11
pre 1970	1	16
Alien		
1987–99	0	0
1970–86	0	0
pre 1970	0	0

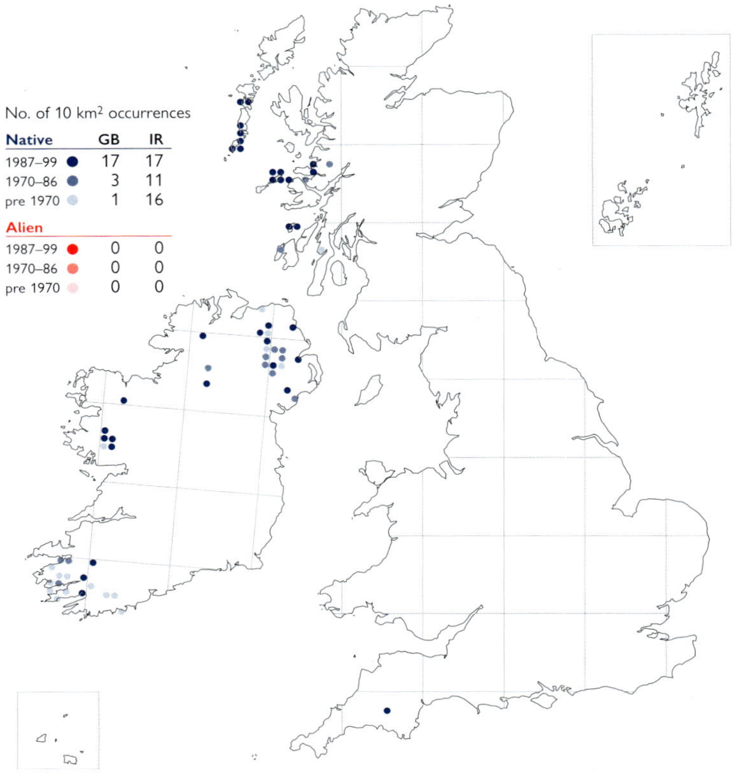

A rhizomatous herb of acidic, nutrient poor, periodically flooded or flushed vegetation, often growing on peaty soils by rivers, streams and lake margins. It frequently occurs amongst *Molinia caerulea* in pastures grazed by cattle or ponies. Reproduction is mostly vegetative in our area. Lowland.

Native (change +0.45). This species is now known from many more sites in Britain and N. Ireland, thanks to a deliberate recording effort, but may still be under-recorded. In Britain some sites have been lost to drainage and reclamation.

Oceanic Boreal-montane element; in Europe restricted to Britain and Ireland but widespread in N. America.

References: Atlas (334b), Allan & Woods (1993), Curtis & McGough (1988), Gulliver (1996), Hultén & Fries (1986), Pearman & Preston (2000), Stewart *et al.* (1994).

P. D. CAREY & T. D. DINES

Goodyera repens Creeping Lady's-tresses

Native	**GB**	**IR**
1987–99	104	0
1970–86	27	0
pre 1970	55	0
Alien		
1987–99	0	0
1970–86	0	0
pre 1970	2	0

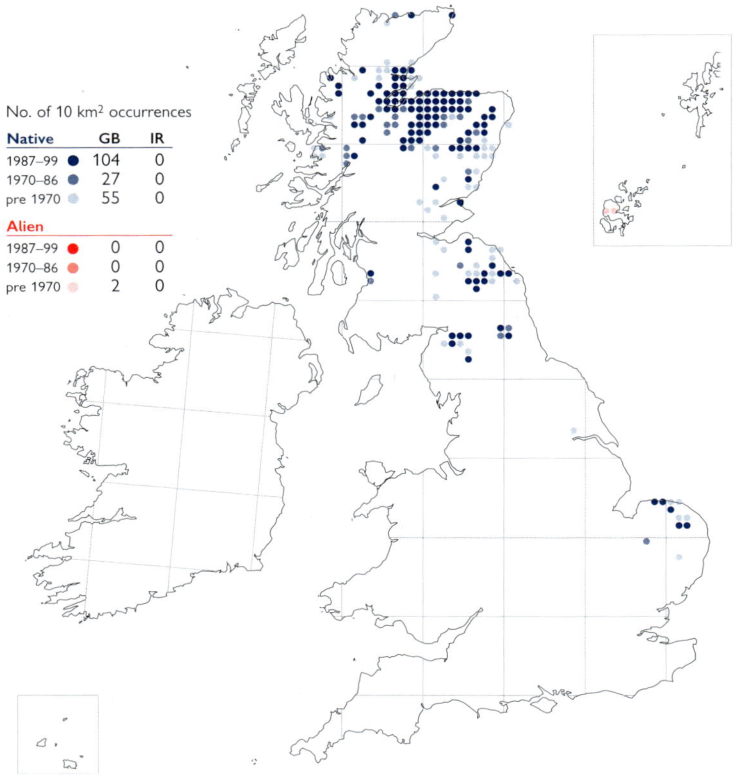

A creeping, evergreen perennial herb of semi-natural and planted coniferous woodland, usually of *Pinus sylvestris*, where it grows in slight to moderate shade in moist layers of moss and pine-needles. It is also found under *Pinus* on old sand dunes. 0–335 m (Morinsh, Banffs.).

Native (change –0.34). Felling and replanting of *P. sylvestris* has caused losses of *G. repens* from many sites, as has increased shading and scrub encroachment. Sites in East Anglia and Cumbria may result from natural colonisation by wind-blown seed, but plants could have been introduced with seedling pines from Scotland.

Circumpolar Boreal-montane element.

References: Atlas (335b), Allan & Woods (1993), Halliday (1997), Hultén & Fries (1986), Sanford (1991), Stewart *et al.* (1994).

P. D. CAREY & T. D. DINES

Liparis loeselii Fen Orchid

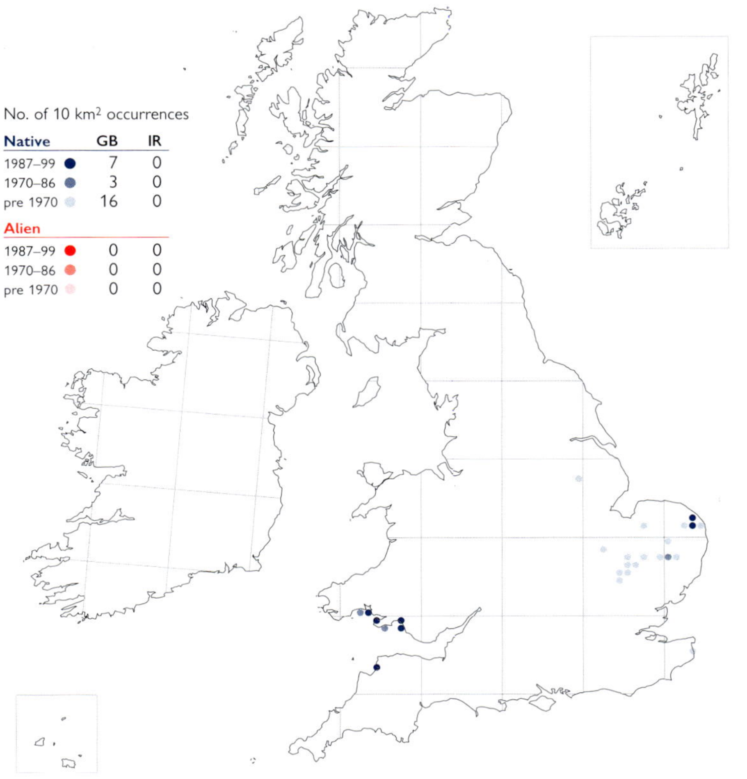

In East Anglia this pseudobulbous perennial herb is restricted to species-rich fens on infertile soils, and to old peat cuttings. Elsewhere, it grows in young dune-slacks. Lowland.

Native (change –0.38). This species declined greatly, especially before 1930, due to habitat destruction and, in East Anglia, scrub encroachment and the cessation of peat cutting. At its dune-slack sites coastal management and under-grazing stabilise dunes and reduce the number of young slacks available for colonisation.

Eurosiberian Temperate element, with a continental distribution in W. Europe; also in E. Asia and N. America.

References: Atlas (335d), Hultén & Fries (1986), Jones (1998), Kay & John (1995), Meusel *et al.* (1965), Sanford (1991), Wheeler *et al.* (1998), Wigginton (1999).

P. D. CAREY & T. D. DINES

Hammarbya paludosa Bog Orchid

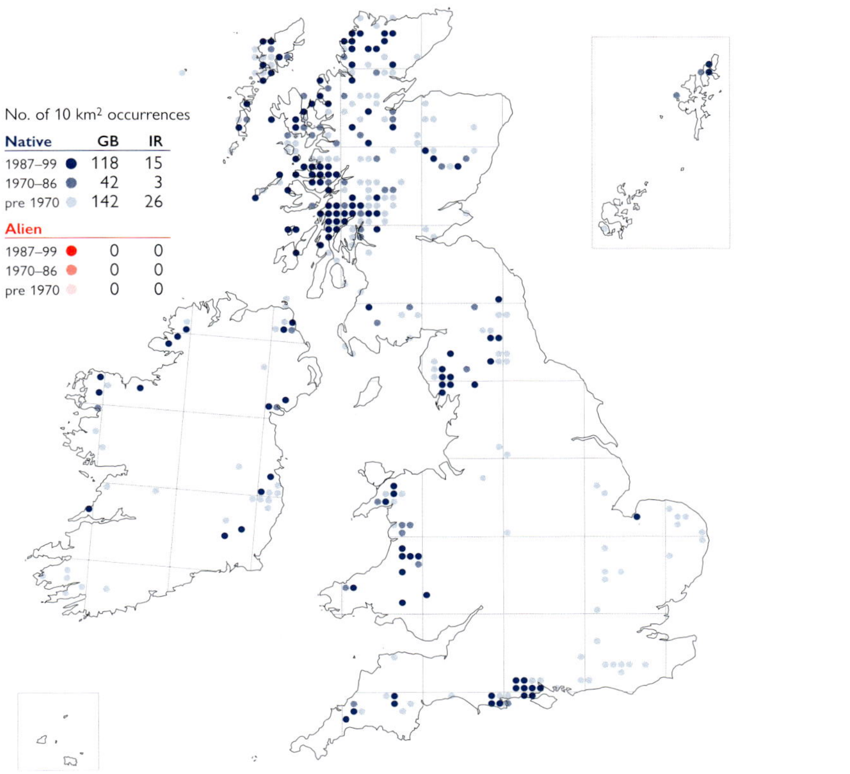

A pseudobulbous herb of boggy areas where the water is usually acidic but subject to some lateral movement. Typically it grows amongst saturated *Sphagnum*, but also on peaty mud and among grasses on the edges of runnels and flushes. 0–500 m (Llyn Anafon, Caerns.).

Native (change –0.32). Drainage of bogs, particularly in the lowlands, has caused a dramatic decline of this species, especially before 1930. Overgrazing may have caused losses in the uplands. It is inconspicuous and under-recorded in some areas; intensive searches in others have shown it to be more frequent than previously thought.

Circumpolar Boreal-montane element, with a disjunct distribution.

References: Atlas (335c), Allan & Wood (1993), Curtis & McGough (1988), Hultén & Fries (1986), Meusel *et al.* (1965), Stewart *et al.* (1994).

P. D. CAREY & T. D. DINES

Corallorhiza trifida Coralroot Orchid

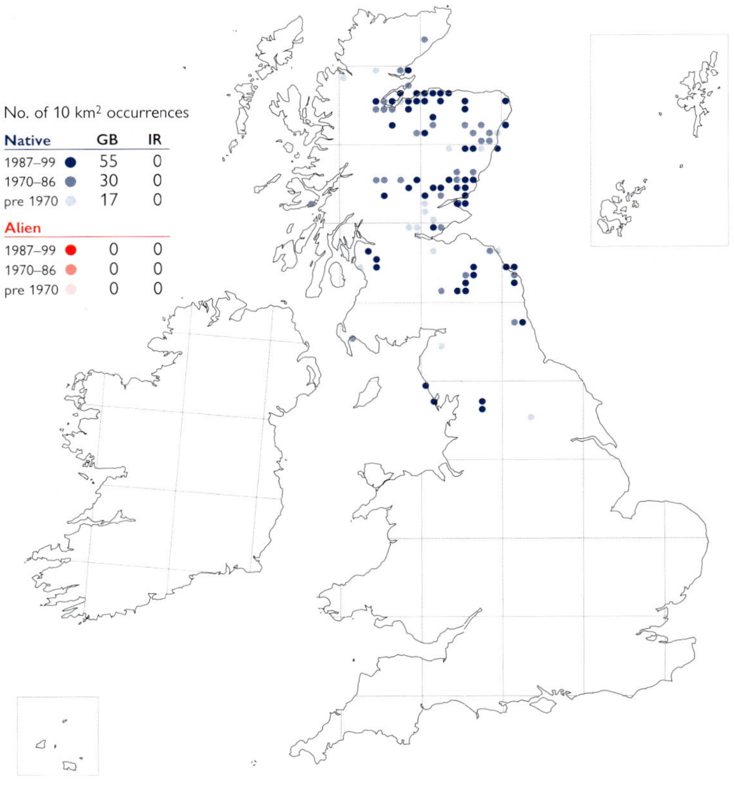

A saprophytic herb usually found in shaded, damp, *Alnus* and *Salix* carr on raised mires and lake margins, but which also occurs in dune-slacks with *Salix repens*. More rarely, it grows in tall-herb fen, in *Betula* and *Pinus* woods (amongst *Sphagnum*) and on moorland. It may colonise secondary habitats, including plantations and quarries. 0–365 m (Braemar, S. Aberdeen).

Native (change +0.61). *C. trifida* is easily overlooked and new sites are still being found; it is much better recorded now than in the 1962 *Atlas*.

Circumpolar Boreal-montane element, with a continental distribution in W. Europe.

References: Atlas (336a), Allan & Woods (1993), Hultén & Fries (1986), Meusel *et al.* (1965), Stewart *et al.* (1994).

P. D. CAREY & T. D. DINES

Herminium monorchis Musk Orchid

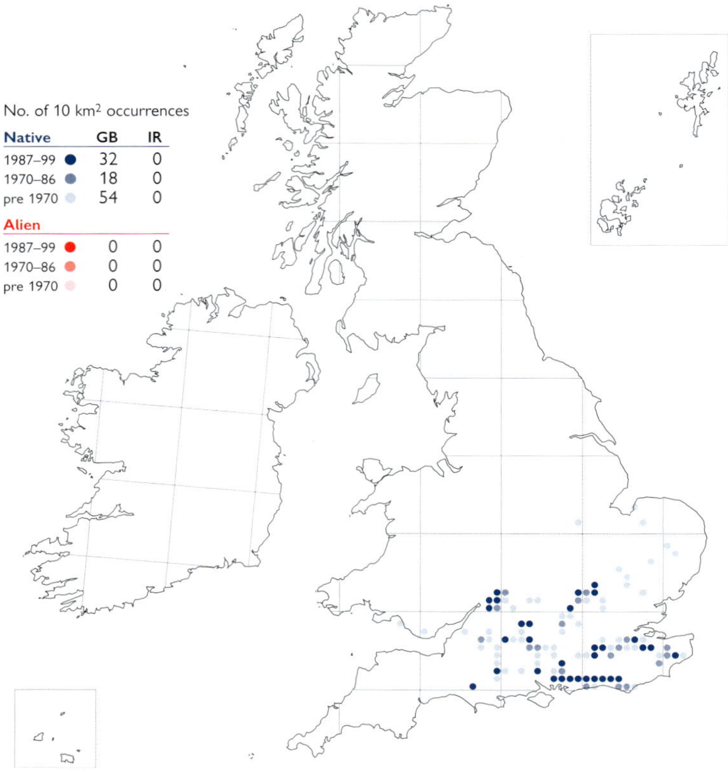

No. of 10 km² occurrences

Native	GB	IR
1987–99 ●	32	0
1970–86 ●	18	0
pre 1970 ●	54	0
Alien		
1987–99 ●	0	0
1970–86 ●	0	0
pre 1970 ●	0	0

A tuberous perennial herb of short turf on calcareous soils overlying chalk or oolitic limestone, particularly on the small terracettes of steep slopes. It also grows on quarry floors and on old lime kiln spoil heaps. Lowland.

Native (change –0.93). By 1930 this species had been lost from many sites, especially in East Anglia, through ploughing and increased grazing. Losses have continued, mostly due to a lack of grazing and consequent scrub encroachment, but have been partly offset by its ability to colonise new sites such as quarries.

Eurasian Temperate element, with a continental distribution in W. Europe.

References: Atlas (336b), Hultén & Fries (1986), Meusel *et al.* (1965), Sanford (1991), Stewart *et al.* (1994), Wells *et al.* (1998).

P. D. CAREY & T. D. DINES

Platanthera chlorantha Greater Butterfly-orchid

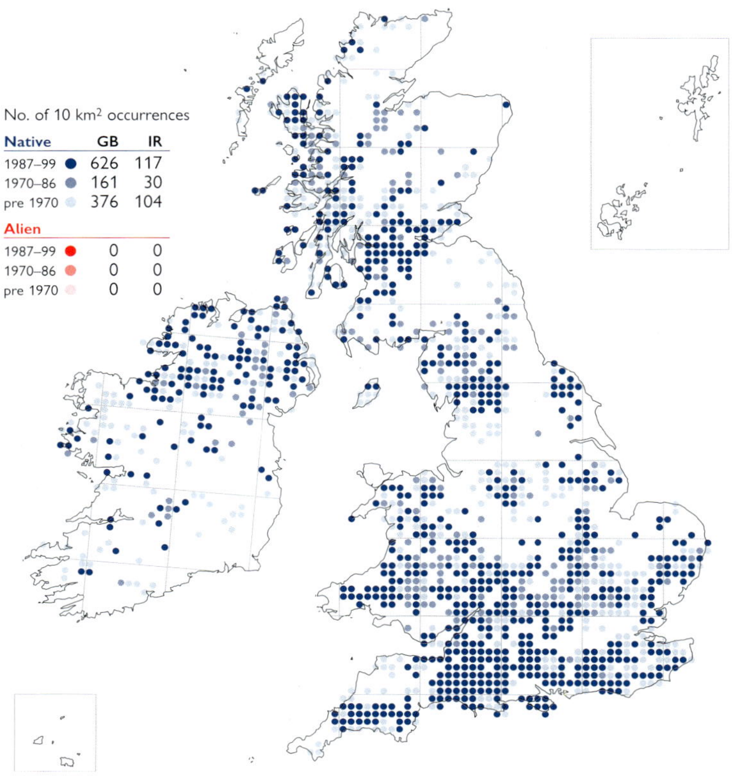

No. of 10 km² occurrences

Native	GB	IR
1987–99 ●	626	117
1970–86 ●	161	30
pre 1970 ●	376	104
Alien		
1987–99 ●	0	0
1970–86 ●	0	0
pre 1970 ●	0	0

This perennial herb is found in a wide variety of habitats, usually on well-drained calcareous soils. Typical habitats include downland, rough pasture, hay meadows, scrub, woodland and young plantations. It sometimes occurs on sand dunes and railway banks. Rarely, it grows on slightly acidic soils in moorland and wet, heathy pasture. 0–460 m (Harwood Dale, S. Northumb.).

Native (change –0.88). *P. chlorantha* was lost from many sites during the 20th century. Reasons include the felling, disturbance and coniferisation of woodland, and the agricultural improvement of pasture and scrub. It may be lost from woodland if the canopy becomes too dense.

European Temperate element.

References: Atlas (337b), Allan & Woods (1993), Hultén & Fries (1986), Jenkinson (1991), Sanford (1991).

P. D. CAREY & T. D. DINES

Platanthera bifolia Lesser Butterfly-orchid

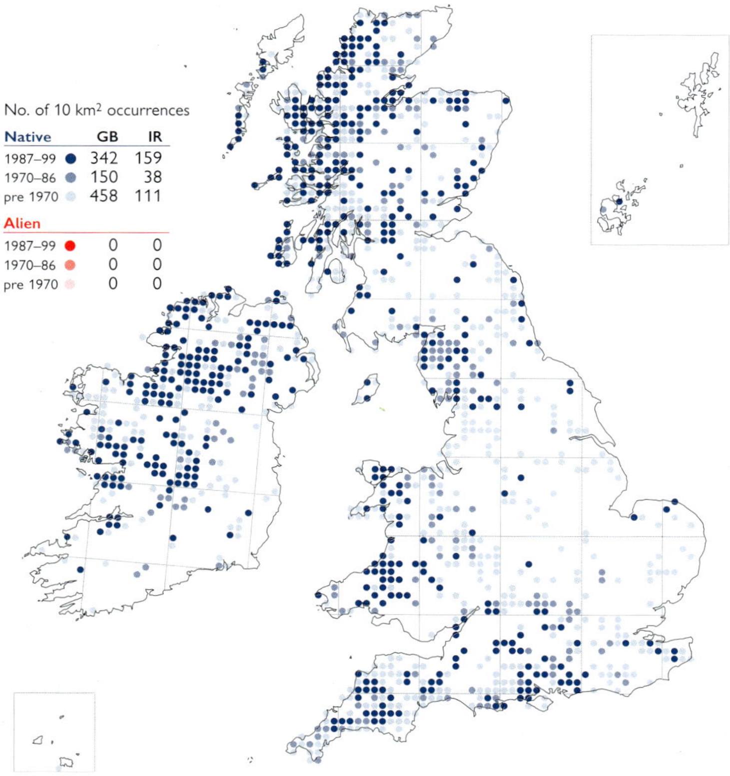

No. of 10 km² occurrences

Native	GB	IR
1987–99 ●	342	159
1970–86 ●	150	38
pre 1970 ●	458	111
Alien		
1987–99 ●	0	0
1970–86 ●	0	0
pre 1970 ●	0	0

A perennial herb of heathy pastures, grassland, open scrub, woodland edges and rides, and on moorland, often amongst *Pteridium*; it is found on a wide variety of acidic and calcareous soils overlying sands, gravels and clays. It is tolerant of considerable soil moisture, also being found in acidic bogs and calcareous fens. 0–365 m (Glenfeshie, Easterness).

Native (change –1.67). This species has suffered a considerable decline, with many losses in England occurring before 1930. In the lowlands it has been lost through drainage, woodland disturbance and agricultural intensification, while upland populations have also been lost to increased grazing.

Eurasian Boreo-temperate element.

References: Atlas (337c), Allan & Woods (1993), Hultén & Fries (1986), Jenkinson (1991), Sanford (1991).

P. D. CAREY & T. D. DINES

Anacamptis pyramidalis Pyramidal Orchid

Native	GB	IR
1987–99	681	241
1970–86	54	22
pre 1970	114	87

Alien		
1987–99	3	0
1970–86	0	0
pre 1970	0	0

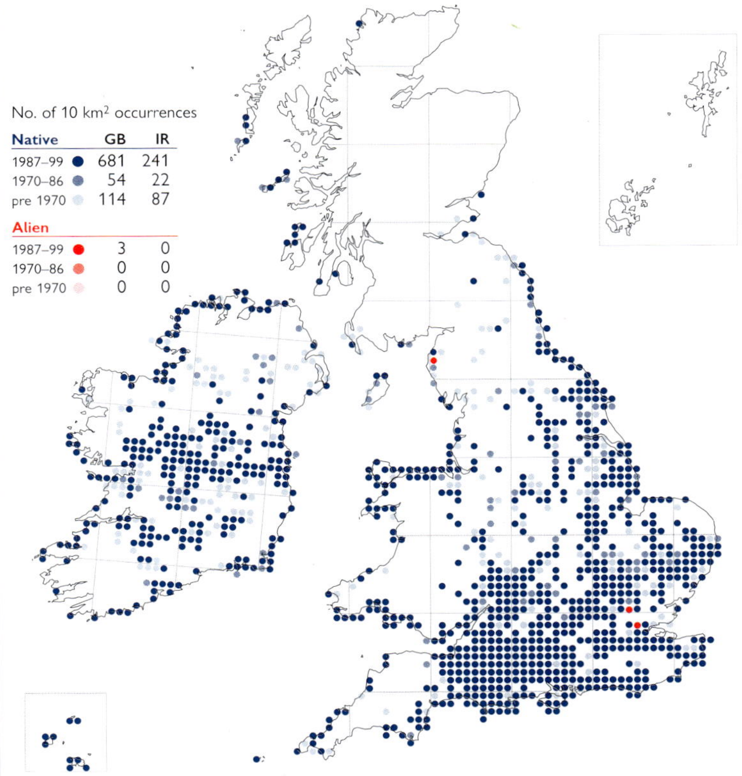

A tuberous perennial herb of well-drained calcareous soils. It is found in shortly grazed downland, dune-slacks and on cliff-tops, and also in the longer grass of semi-stable dunes, scrub, roadside verges and churchyards. It also grows in the grikes of limestone pavement, and can colonise the disturbed ground of abandoned quarries, industrial waste land and railway banks. 0–350 m (Brough, Westmorland).

Native (change +0.55). The overall distribution of this species is stable, although it has been lost from some sites, particularly in Ireland, largely due to agricultural improvement.

European Southern-temperate element.

References: Atlas (342d), Allan & Woods (1993), Hultén & Fries (1986), Jenkinson (1991), Sanford (1991).

P. D. CAREY & T. D. DINES

Pseudorchis albida Small-white Orchid

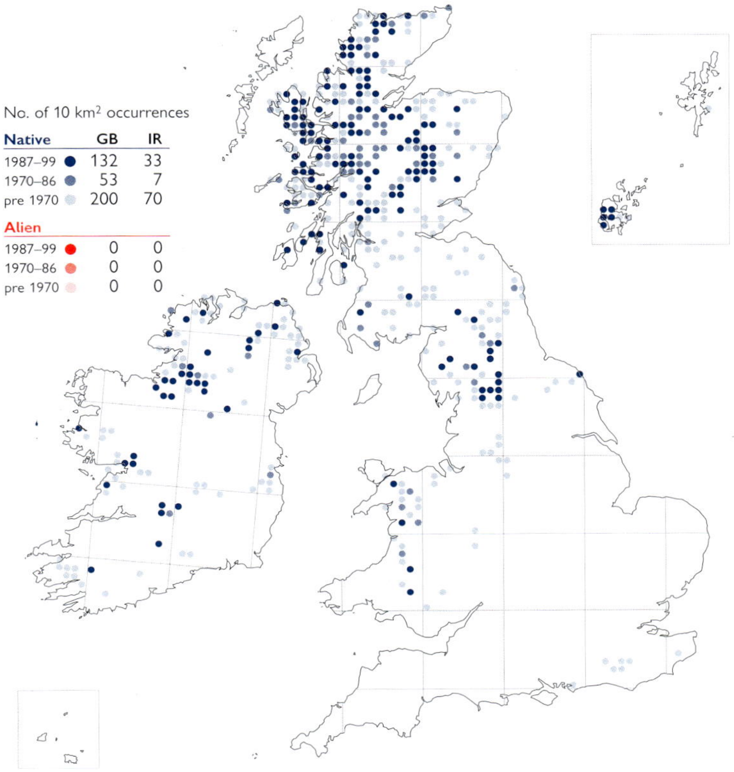

A tuberous perennial herb of well-drained hill pastures, streamsides, mountain grasslands and cliff ledges on a range of dry, acidic or calcareous soils; also on recently burnt moorland, but not persisting when *Calluna* regrows. Rarely, it occurs in acidic *Quercus* woodland. 0–550 m (Ben Chaisteil, Main Argyll).

Native (change –0.88). Many lowland sites for *P. albida* were lost before 1930 due to habitat destruction, agricultural improvement and overgrazing. There have been continued losses since then, although it is now much better recorded. However, it is inconspicuous and may be under-recorded in some areas, especially in the uplands.

European Boreal-montane element; also in N. America.

References: Atlas (337a), Allan & Woods (1993), Curtis & McGough (1988), Hultén & Fries (1986), Meusel *et al.* (1965).

P. D. CAREY & T. D. DINES

Gymnadenia conopsea Fragrant Orchid

Native	GB	IR
1987–99	810	253
1970–86	178	23
pre 1970	353	89

Alien		
1987–99	1	0
1970–86	0	0
pre 1970	0	0

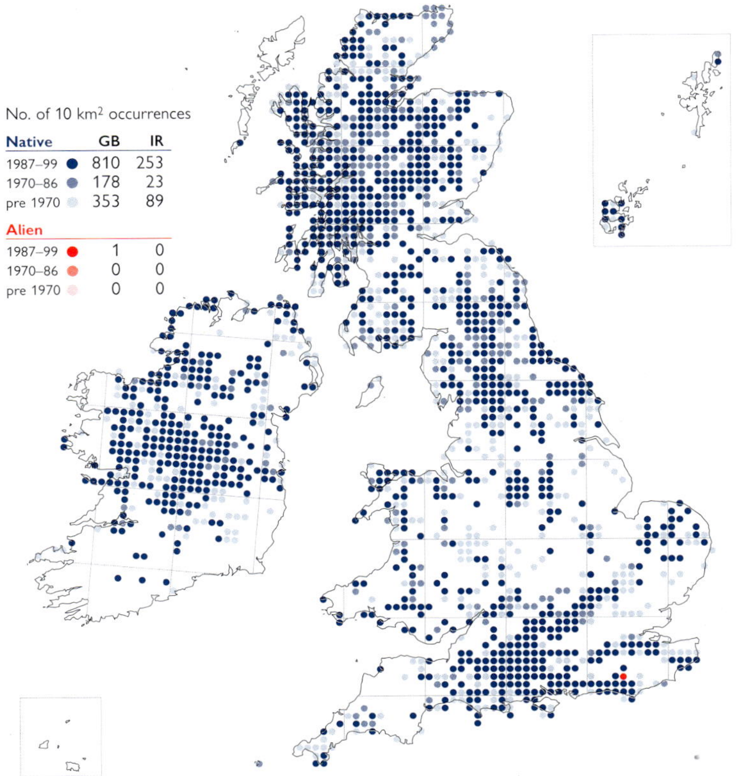

A tuberous perennial herb of both dry and moist chalk and limestone grasslands, limestone pavement, less acidic heaths, and wetter habitats such as base-rich fens. It also grows in artificial habitats, including quarries and railway banks. 0–610 m (Ben Lawers, Mid Perth).

Native (change –0.76). This species was mapped as 'all records' in the 1962 *Atlas*. It has declined throughout its range, probably due to habitat destruction, ploughing of pastures and drainage. Three subspecies are currently recognised, although each may deserve specific rank (Bateman *et al.*, 1997); they are mapped separately.

Eurasian Boreo-temperate element.

References: Atlas (336d), Allan & Woods (1993), Hultén & Fries (1986), Jenkinson (1991), Meusel *et al.* (1965), Sanford (1991).

P. D. CAREY & T. D. DINES

Gymnadenia conopsea subsp. *conopsea*

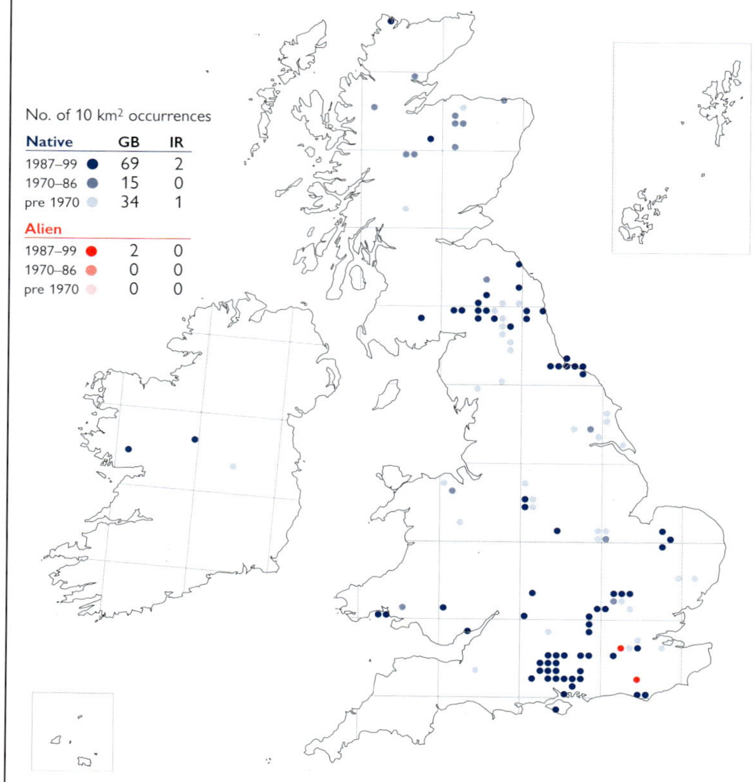

A tuberous perennial herb of species rich, dry chalk and limestone grasslands where the vegetation is not too dense, and occasionally in base-rich fens. It also grows in artificial habitats, such as abandoned quarries, lime waste and railway banks. 0–365 m (near Whitely Shield, S. Northumb.).

Native. Many recorders have not distinguished the subspecies of *G. conopsea* and this, the type subspecies, has been under-recorded in some areas and over-recorded in others. Even so, it is likely to have declined widely as a result of agricultural improvement, especially the conversion of pasture to arable in the 19th and 20th centuries.

Widespread in Europe, and extending eastwards throughout the Eurasian range of the species.

References: Allan & Woods (1993), Sanford (1991).

P. D. CAREY & T. D. DINES

Gymnadenia conopsea subsp. *densiflora*

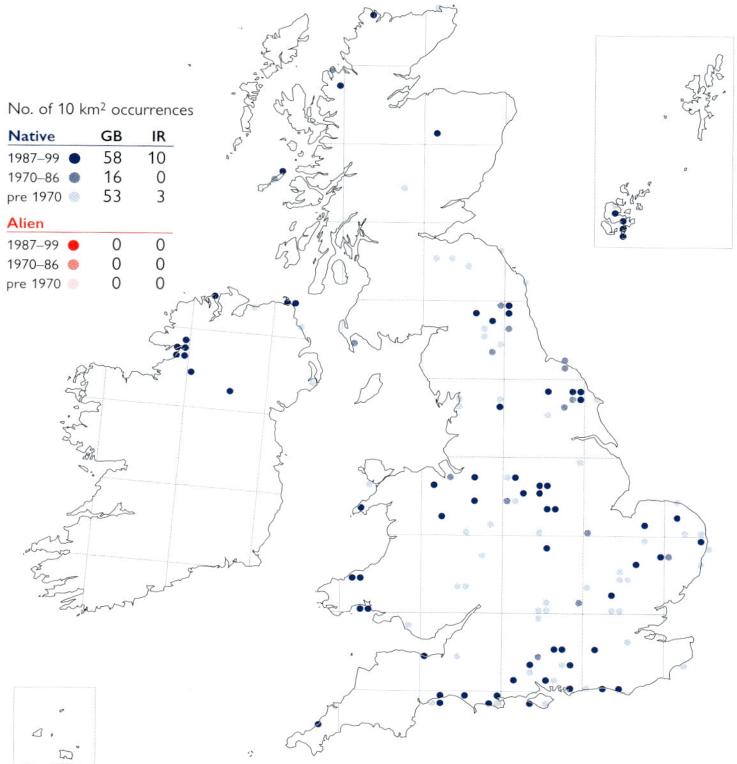

A tuberous perennial herb characteristically found in base-rich wet meadows, fens and ditches. Occasionally it is recorded from N.-facing chalk and limestone grassland. It sometimes grows with *G. conopsea* subsp. *borealis* and is the most common subspecies in Ireland. 0–310 m (Tomintoul, Banffs.).

Native. Although this taxon was mentioned as var. *densiflora* by Clapham *et al.* (1952) and mapped at subspecific rank by Perring & Sell (1968), it has not been recorded consistently. It is very sensitive to drainage and eutrophication, and has declined, particularly in fens. Ploughing of grassland has also caused the loss of some populations.

Widespread in Europe.

References: Atlas Supp. (141a), Allan & Woods (1993), Sanford (1991).

P. D. CAREY & T. D. DINES

Gymnadenia conopsea subsp. *borealis*

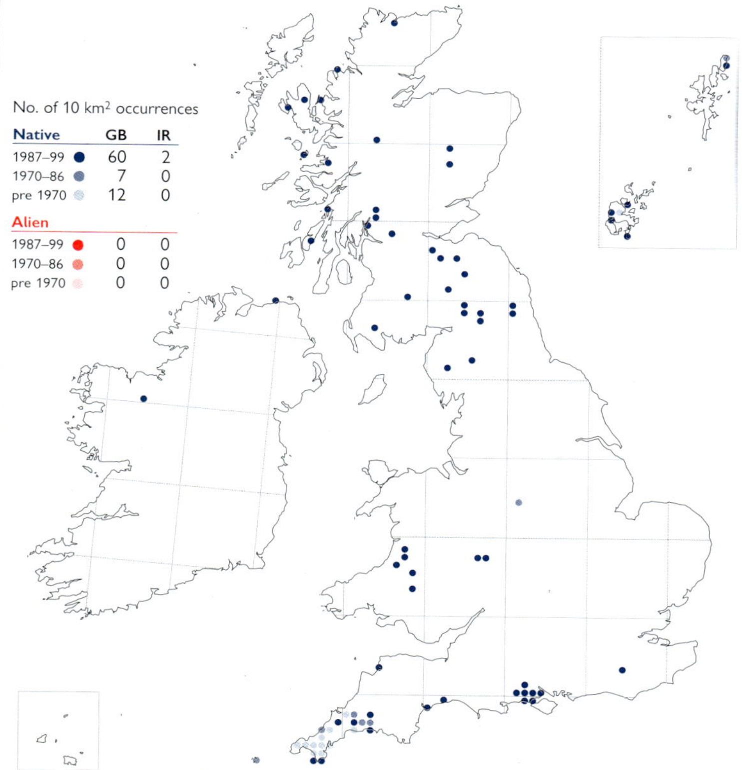

A tuberous perennial herb growing in base-rich to mildly acidic grasslands, heaths, moors, and in flushes on soils derived from substrates including sands, limestones and clays. It is also frequent on roadside verges and abandoned quarries. It sometimes grows with subsp. *densiflora*. 0–610 m (Ben Lawers, Mid Perth).

Native. This taxon was brought to the attention of British and Irish botanists by Rich & Rich (1988) and Stace (1991), and is certainly under-recorded; it is reported as being the most frequent *G. conopsea* subspecies in Scotland (Allan & Woods, 1993). It may have declined through habitat destruction, but lack of records makes any change impossible to assess.

A little-known taxon with a poorly documented distribution.

P. D. CAREY & T. D. DINES

Dactylorhiza fuchsii × *Gymnadenia conopsea* (X *Dactylodenia st-quintinii*)

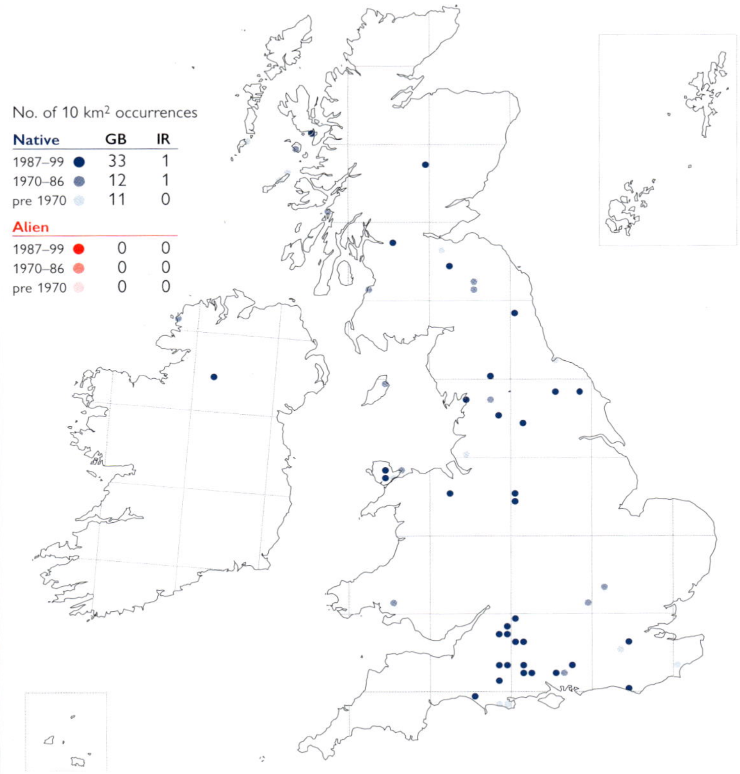

A tuberous perennial herb found occasionally when the parents grow together. Since all three subspecies of *G. conopsea* can be involved, it grows in a wide variety of habitats including chalk and limestone grasslands, less acidic heaths, and wetter habitats such as base-rich fens and flushes. Populations are usually small, and often consist of a single plant. 0–410 m (Ben Lawers, Mid Perth).

Native. Although this is the most common × *Dactylodenia* hybrid, it is still probably under-recorded.

This hybrid is also known on the mainland of W. Europe.

References: Bateman & Farrington (1989), Stace (1975).

T. D. DINES

Coeloglossum viride Frog Orchid

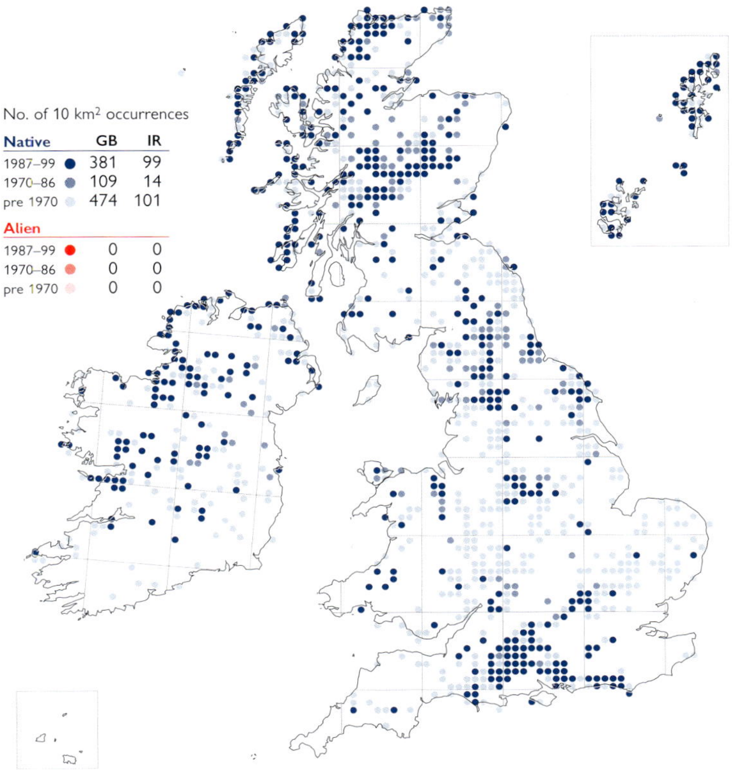

A tuberous perennial herb restricted in S. Britain to dry, well-grazed, base-rich grassland such as chalk downland and dunes, and in chalk pits. Elsewhere it grows in a wider range of calcareous grasslands, flushes, limestone pavement, scree, rocky ledges, roadsides and quarries. 0–915 m (Glen Doll, Angus).

Native (change −1.34). *C. viride* has declined considerably, particularly in C. England and East Anglia. Many losses occurred before 1930, but have continued since then, and are largely due to the ploughing and improvement of pastures. Molecular evidence suggests that this species should be included within *Dactylorhiza* (Bateman *et al.*, 1997).

Circumpolar Boreal-montane element.

References: Atlas (336c), Allan & Woods (1993), Hultén & Fries (1986), Meusel *et al.* (1965), Willems & Melser (1998).

P. D. CAREY & T. D. DINES

Dactylorhiza fuchsii Common Spotted-orchid

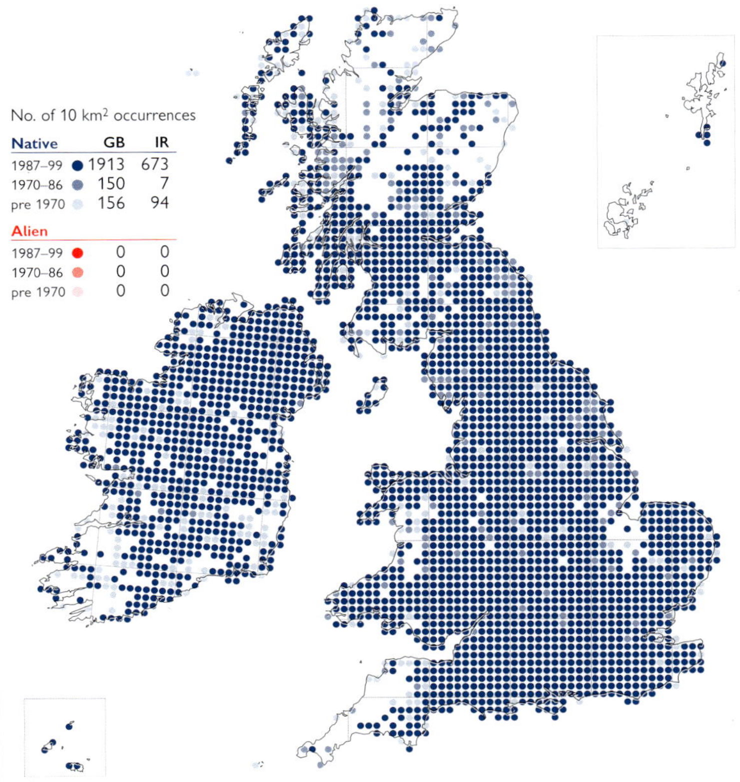

This tuberous perennial herb grows on neutral or base-rich soils in a wide range of habitats, including deciduous woodland, scrub, roadsides, chalk grassland, meadows, marshes, dune-slacks, fens and mildly acidic heaths. It can become abundant in artificial habitats such as waste ground, abandoned gravel-pits, quarries and railway banks. 0–530 m (Garrigill, Cumberland).

Native (change +0.33). There has been little change in the distribution of this species since the 1962 *Atlas*, and losses have been balanced by its ability to colonise newly available, often man-made, habitats.

Eurosiberian Temperate element.

References: Atlas (340d), Atlas Supp. (141b, c), Allan & Woods (1993), Bateman & Denholm (1989), Jenkinson (1991), Leeson *et al.* (1991), Sanford (1991), Sell & Murrell (1996).

P. D. CAREY & T. D. DINES

Dactylorhiza fuchsii × *D. incarnata* (*D.* × *kerneriorum*)

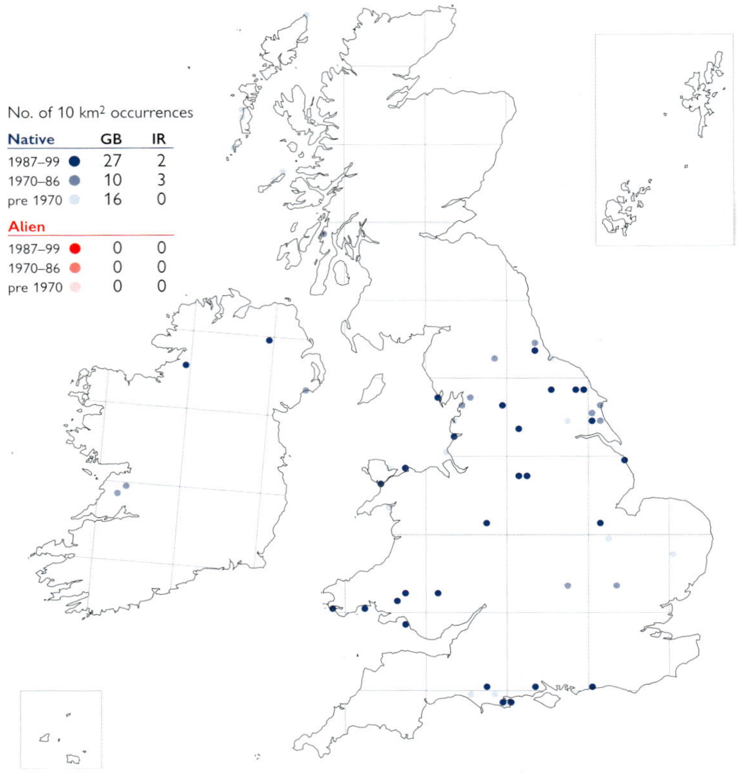

A tuberous perennial herb usually found with the parents in habitats such as calcareous marshes, fens, dune-slacks and wet meadows. It may persist in the absence of *D. incarnata* if drainage has caused the loss of this species from the site. Although both parents are diploid, this hybrid is usually sterile and populations usually consist of only a few plants. Lowland.

Native. Given the frequency of the parents in suitable habitat, this is a surprisingly uncommon hybrid, although it is probably under-recorded. Populations tend to be naturally transient, disappearing even if the habitat remains unaltered, and this probably accounts for most of the losses.

Widespread in temperate Europe.

References: Sanford (1991), Stace (1975).

T. D. DINES

Dactylorhiza fuchsii × *D. maculata* (*D.* × *transiens*)

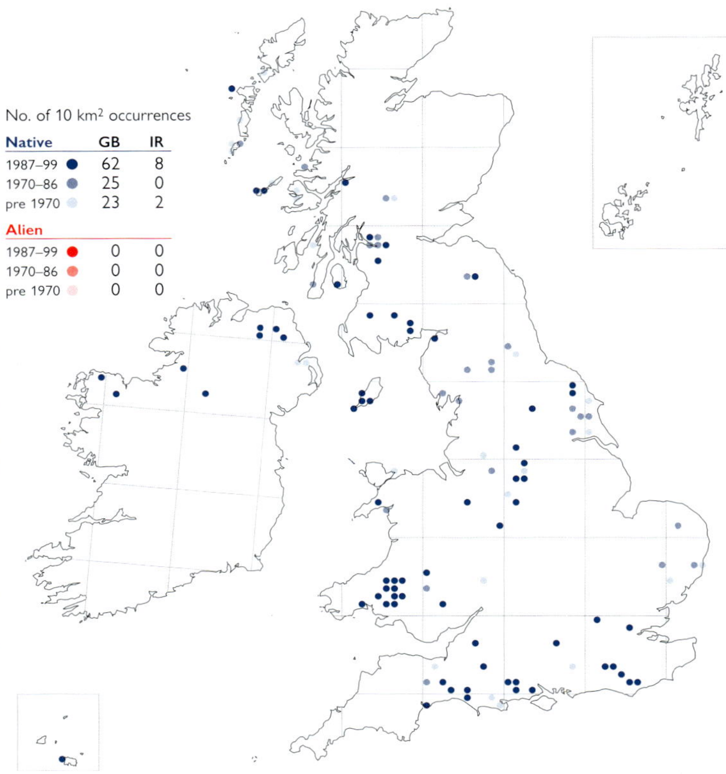

A tuberous perennial herb occasionally found where base-rich and base-poor soils coexist, allowing the parents to grow together, such as where sandstones are interbedded with limestones or where limestones are overlain by acidic peaty deposits. Habitats include grassland, heathland, marshes, wet meadows and woodland margins. It is triploid, sterile and usually only single or very few plants are found. Lowland.

Native. An uncommon hybrid which may sometimes be recorded in error for variants of either parent. Some losses are due to habitat destruction, but most are probably due to the transient nature of the hybrid.

This hybrid is also recorded from France.

References: Sanford (1991), Stace (1975).

T. D. DINES

Dactylorhiza fuchsii × *D. praetermissa* (*D.* × *grandis*)

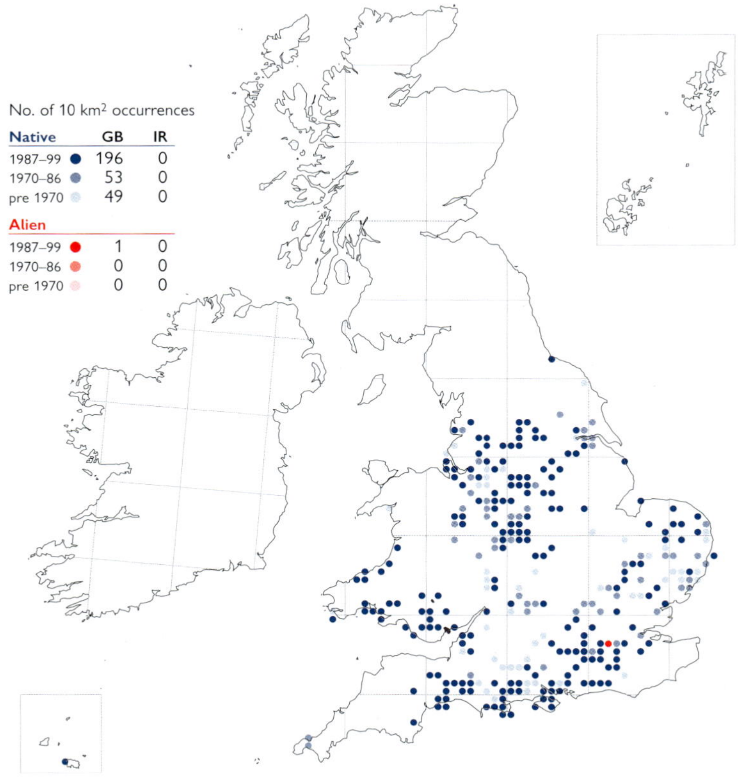

A tuberous perennial herb commonly found where the parents grow together, typically in habitats such as marshes, fens and wet meadows, but also on dry chalk downland. It is especially frequent in man-made habitats such as quarries and industrial waste land, and may persist in the absence of *D. praetermissa* if drainage has caused the loss of this species from the site. This hybrid is triploid, partially fertile and frequently back-crosses with the parents, creating long-lived hybrid swarms. Generally lowland, but reaching 380 m at Buxton (Derbys.).

Native. Although this is the most frequent *Dactylorhiza* hybrid, it is still probably under-recorded. Its overall distribution is likely to be stable.

This hybrid is also recorded from France.

References: Sanford (1991), Stace (1975).

P. D. CAREY & T. D. DINES

Dactylorhiza fuchsii × *D. purpurella* (*D.* × *venusta*)

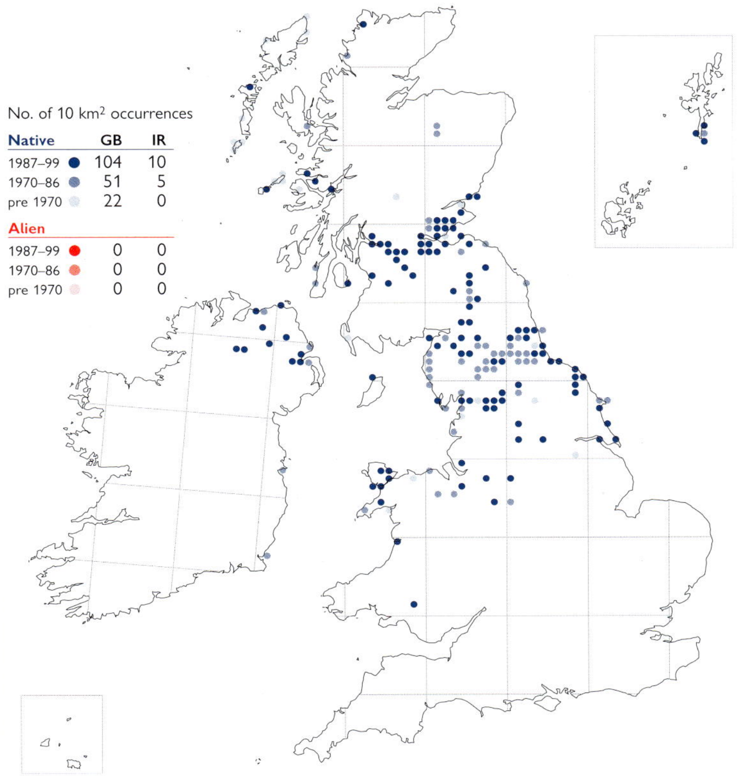

This tuberous perennial herb is frequent wherever the parents grow together, such as in marshes, fens and wet meadows. It is especially frequent in man-made habitats such as quarries and industrial waste land. It may persist in the absence of *D. purpurella* if drainage has caused the loss of this species from the site. This partially fertile triploid hybrid can back-cross with its parents, and is often present in long-lived hybrid swarms. 0–360 m (Nenthead, Cumberland).

Native. A frequent *Dactylorhiza* hybrid, but probably still under-recorded. Its overall distribution is likely to be stable.

Described from Britain; wider distribution uncertain.

Reference: Stace (1975).

P. D. CAREY & T. D. DINES

Dactylorhiza maculata Heath Spotted-orchid

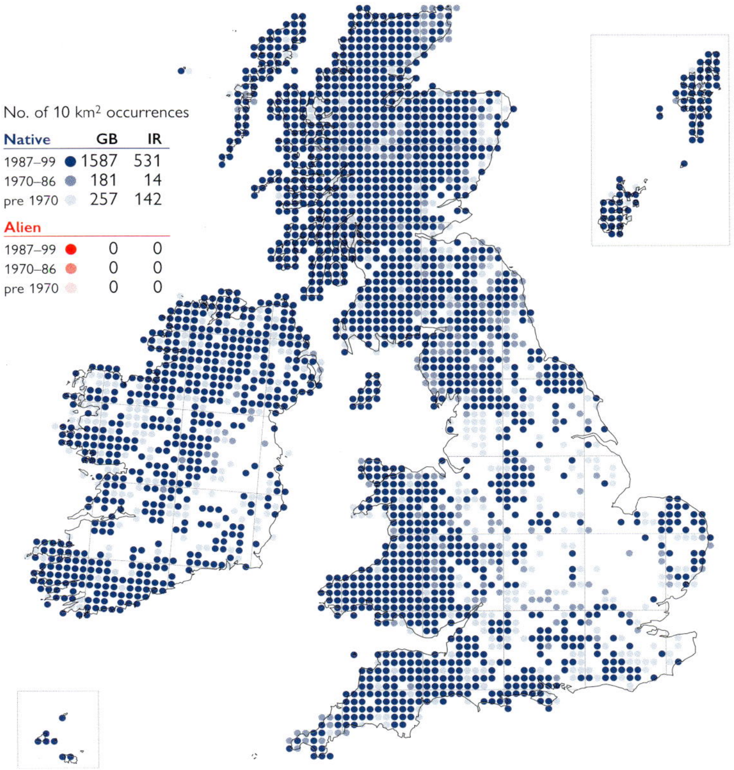

This tuberous perennial herb is found on a range of well-drained or wet acidic soils in a wide variety of habitats, including grasslands, moors, heaths, flushes and bogs. It also occurs in pockets of peat on limestone and, more rarely, in open woodland. 0–915 m (Ben Lawers, Mid Perth).

Native (change –0.42). This species has declined, particularly in lowland heaths and bogs, through habitat destruction and drainage. Cessation of grazing on heaths, causing reversion to woodland, also accounts for some losses. It can still be very abundant, however, in suitable habitats.

Eurosiberian Boreo-temperate element.

References: Atlas (341a), Allan & Woods (1993), Bateman & Denholm (1989), Hultén & Fries (1986), Jenkinson (1991), Meusel *et al.* (1965), Sanford (1991), Sell & Murrell (1996).

P. D. CAREY & T. D. DINES

Dactylorhiza maculata × *D. praetermissa* (*D.* × *hallii*)

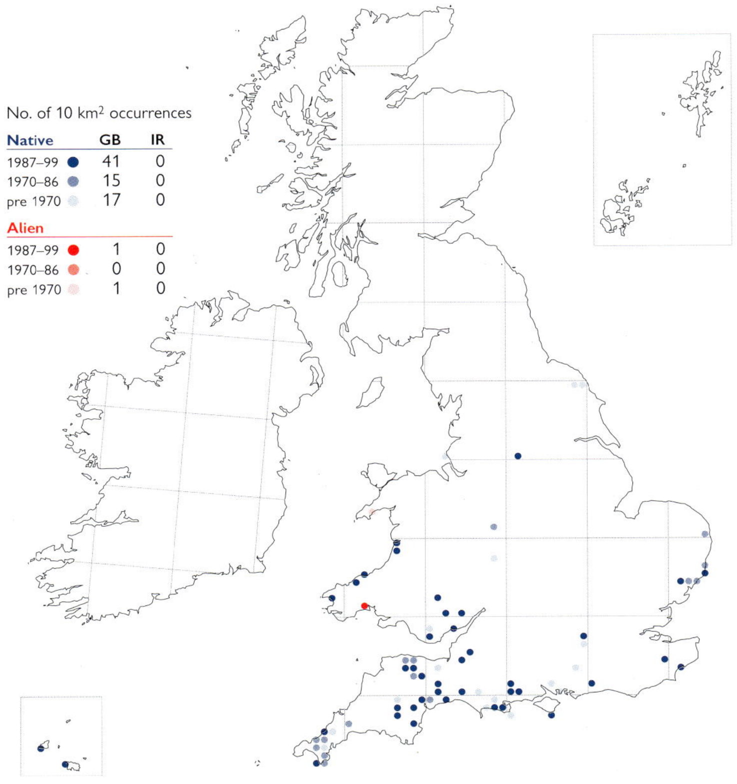

A tuberous perennial herb of calcareous to mildly acidic soils in marshes, damp grassland and disused quarries where the parents grow together. Since the parents differ markedly in their tolerance to calcium, they do not meet frequently, and the hybrid, which is tetraploid, is therefore uncommon. It is, however, relatively fertile and hybrid swarms can be formed through back-crossing with the parents. Lowland.

Native. Although uncommon, this hybrid is likely to be under-recorded. Its overall distribution is probably stable.

This hybrid is also recorded from the Netherlands.

References: Sanford (1991), Stace (1975).

T. D. DINES

Dactylorhiza maculata × *D. purpurella* (*D.* × *formosa*)

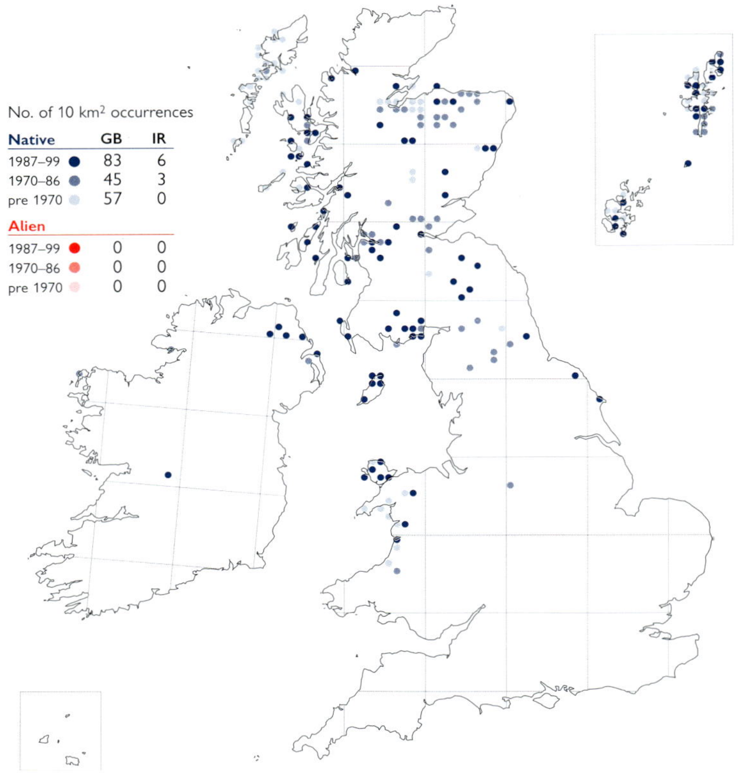

A tuberous perennial herb of calcareous to mildly acidic soils in marshes, damp grassland and wet heaths. Usually, but not always, it occurs in the presence of the parents. It is tetraploid, fully fertile and frequently back-crosses with the parents, creating long-lived hybrid swarms. 0–410 m (Ben Lawers, Mid Perth).

Native. A frequent hybrid, but almost certainly still under-recorded. Its overall distribution is likely to be stable.

This hybrid is also recorded from Norway.

Reference: Stace (1975).

P. D. CAREY & T. D. DINES

Dactylorhiza incarnata Early Marsh-orchid

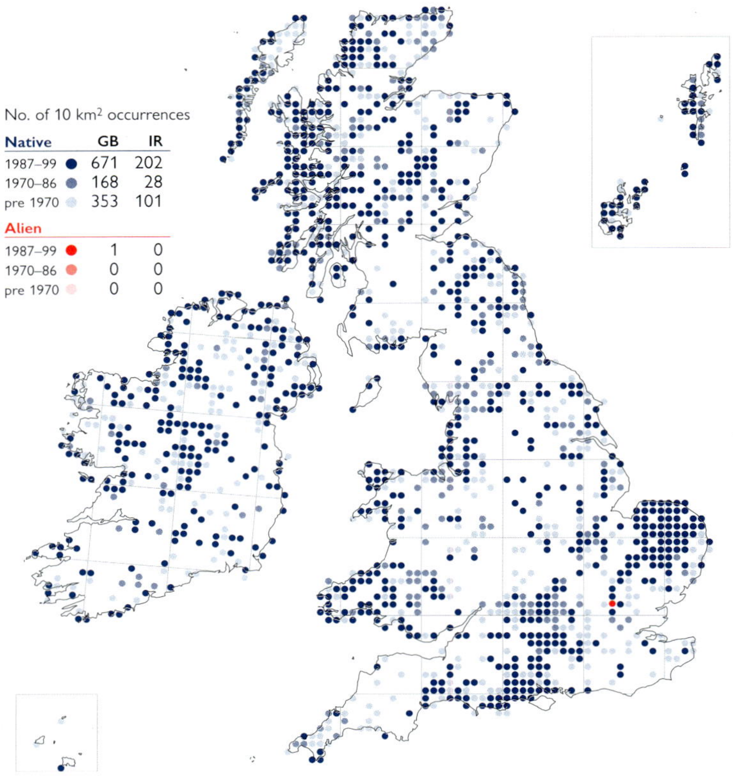

A tuberous perennial herb growing on damp or wet calcareous soils, in meadows, marshes, ditches, fens, flushes and dune-slacks, and also on more acidic soils in bogs and damp heaths. 0–610 m (Atholl, E. Perth, and Caenlochan, Angus).

Native (change –0.33). *D. incarnata* was mapped as 'all records' in the 1962 *Atlas*. It has declined due to drainage and agricultural improvement, and is often the first *Dactylorhiza* to be lost when a habitat begins to dry out. Six subspecies are recognised; all are mapped separately except subsp. *gemmana*, which is of uncertain status and has been neglected by recorders.

Eurosiberian Boreo-temperate element.

References: Atlas (341b), Allan & Woods (1993), Bateman & Denholm (1985), Hultén & Fries (1986), Jenkinson (1991), Sanford (1991), Sell & Murrell (1996).

P. D. CAREY & T. D. DINES

Dactylorhiza incarnata subsp. *incarnata*

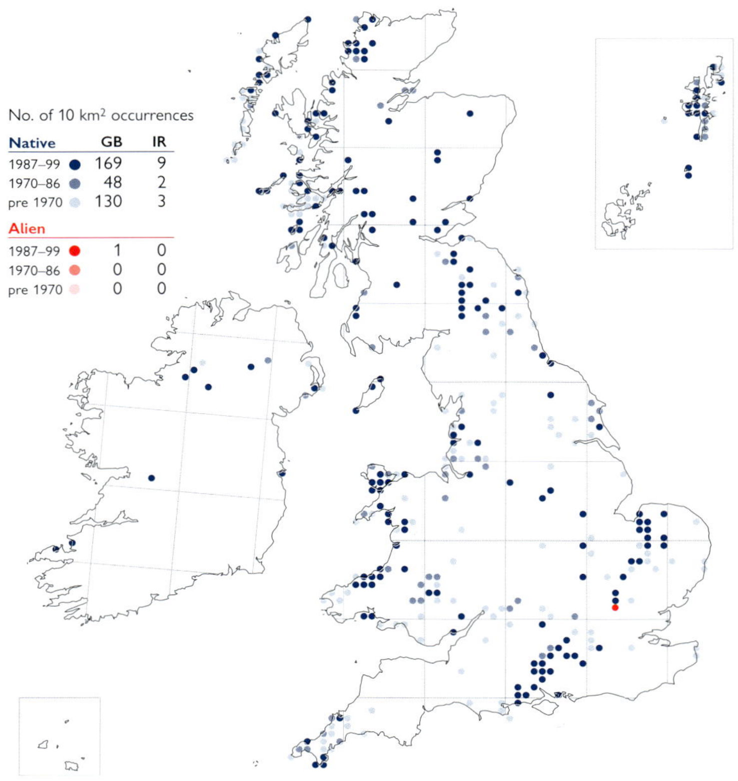

A tuberous perennial herb of calcareous fens, ditches, marshes, wet meadows and upland flushes on base-rich or calcareous soils. It can be a robust plant, especially in the south, and may be abundant in suitable habitats. 0–440 m (Ben Lawers, Mid Perth) and probably higher elsewhere in Scotland.

Native. This is the most widespread and variable subspecies and, being the type subspecies, has been very inconsistently recorded. There are, however, many more records now than when mapped by Perring & Sell (1968). It has suffered losses due to drainage and agricultural improvement of fens and marshes.

Eurosiberian Boreo-temperate element; subsp. *incarnata* occurs throughout much of the range of the species.

References: Atlas Supp. (143a), Allan & Woods (1993), Bateman & Denholm (1985), Jenkinson (1991).

P. D. CAREY & T. D. DINES

Dactylorhiza incarnata subsp. *coccinea*

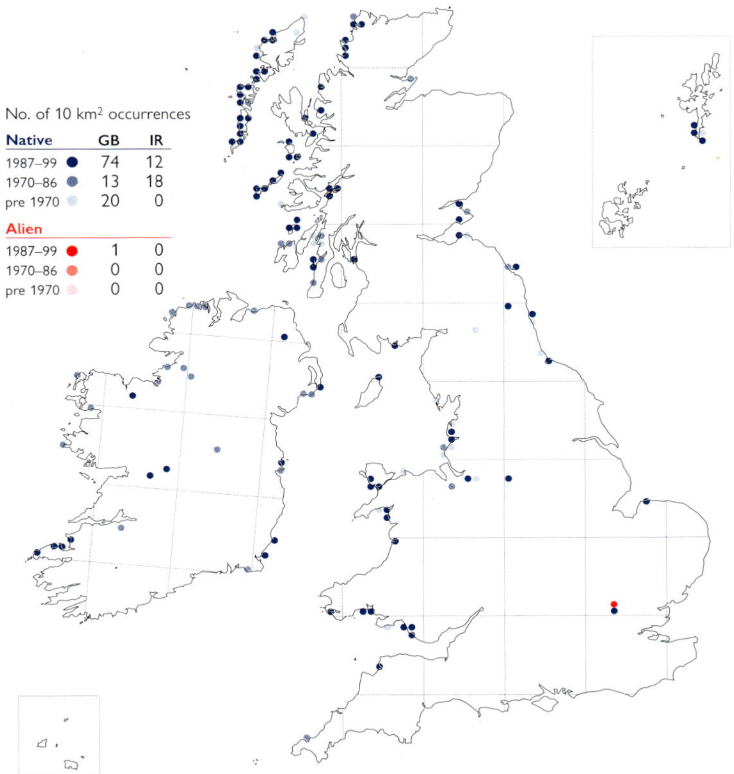

This subspecies is found in machair grassland and damp dune-slacks, where it can be extremely abundant. It also occasionally grows on the wet terraces of slumped sea-cliffs, and inland in calcareous fens, flushes and on highly saline fly-ash waste from power stations. Lowland.

Native. Subsp. *coccinea* was first mapped by Perring & Sell (1968). It is now much better recorded, particularly in W. Scotland. Its overall range is stable, but it has declined in some areas due to coastal developments and scrub encroachment on sand dunes.

Endemic.

References: Atlas Supp. (143c), Allan & Woods (1993), Bateman & Denholm (1985).

P. D. CAREY & T. D. DINES

Dactylorhiza incarnata subsp. *pulchella*

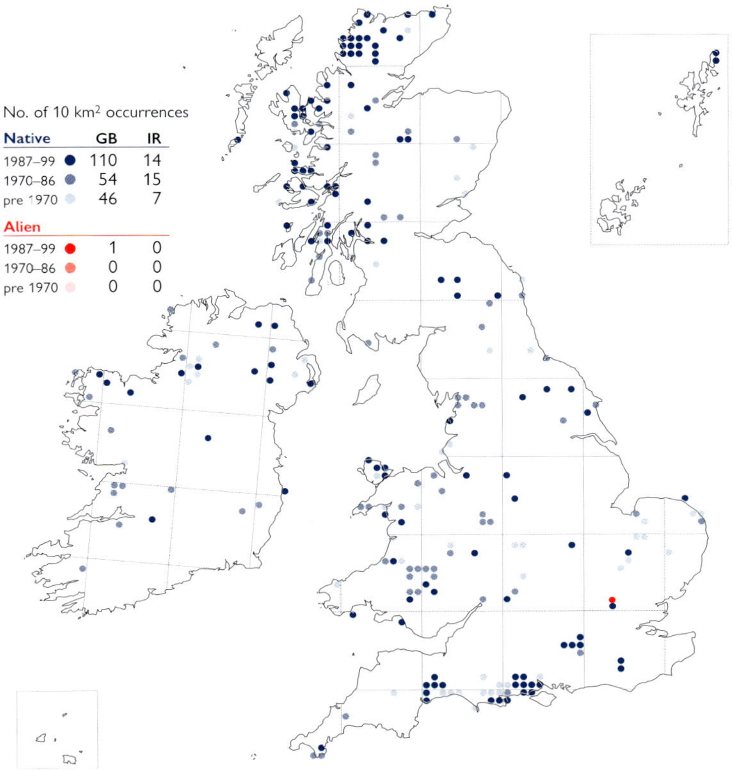

A tuberous perennial herb of acidic valley bogs, marshes and damp heathland, often growing with *Sphagnum*. It also occurs, however, in marshes on more neutral substrates and, occasionally, in fens. 0–395 m (Teesdale, Co. Durham).

Native. This subspecies was first mapped by Perring & Sell (1968), but is now much better recorded in some areas, especially in W. Scotland. It appears to have declined in parts of England and Wales, but this may simply reflect recent under-recording rather than genuine losses.

Subsp. *pulchella* has been reported from mainland Europe, but its distribution is uncertain.

References: Atlas Supp. (143b), Allan & Woods (1993), Bateman & Denholm (1985), Sanford (1991).

T. D. DINES

Dactylorhiza incarnata subsp. *cruenta*

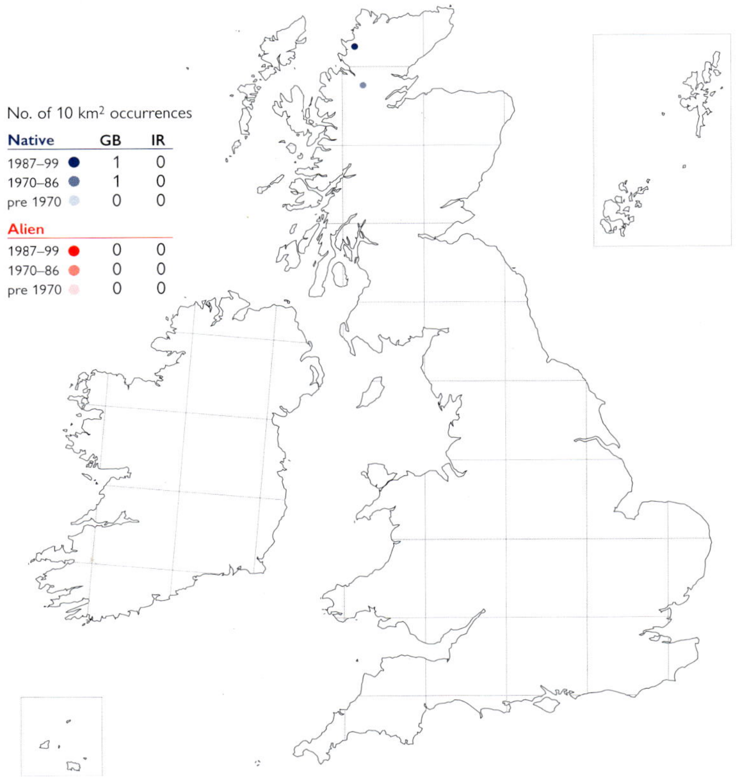

A tuberous perennial herb which grows in W. Ross on gently sloping, lightly grazed, neutral to slightly alkaline flushed grassland and, in W. Sutherland, in a *Carex lasiocarpa* mire. 140 m (W. Sutherland) and 300–450 m (W. Ross).

Native. This subspecies was first mapped by Perring & Sell (1968), on the basis of records from Ireland which have now been identified as spotted-leaved variants of subsp. *pulchella*. The true subsp. *cruenta* was first recorded in Scotland in 1984, and the unremarkable nature of the sites suggest that it is possibly being overlooked elsewhere.

Eurosiberian Boreal-montane element.

References: Atlas Supp. (143d), Allan & Woods (1993), Bateman & Denholm (1985), Hultén & Fries (1986), Wigginton (1999).

P. D. CAREY & T. D. DINES

Dactylorhiza incarnata subsp. *ochroleuca*

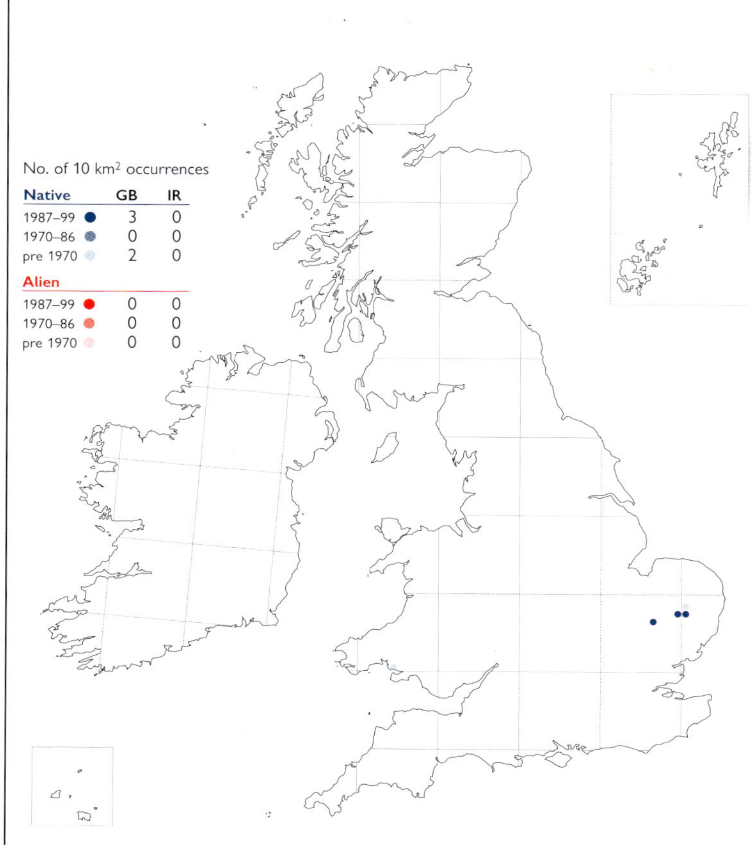

A tuberous perennial herb restricted to moist, periodically inundated calcareous fens, preferring areas of low competition that are slowly but only partially drying out. Lowland.

Native. This subspecies, first mapped by Perring & Sell (1968), has declined over the last fifty years and is now reduced to just two populations. Even these have decreased dramatically in size in the last ten years and are now close to extinction. The plant is vulnerable to changes in water level, to deer grazing, and to scrub encroachment. It is often confused with albino variants of other *D. incarnata* subspecies.

European Temperate element.

References: Atlas Supp. (143d), Bateman & Denholm (1983b, 1985), Sanford (1991), Wigginton (1999).

P. D. CAREY & T. D. DINES

Dactylorhiza praetermissa Southern Marsh-orchid

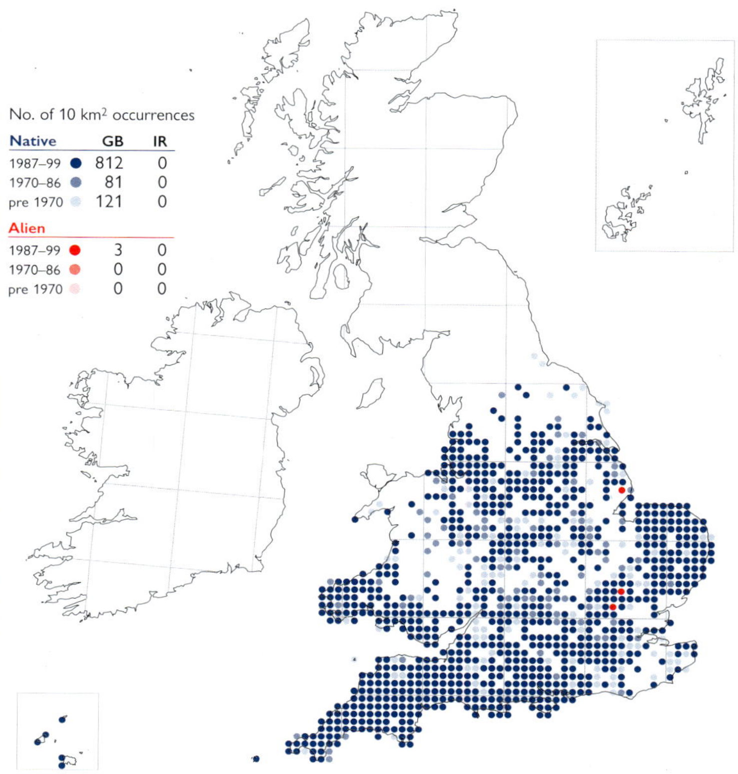

A tuberous perennial herb of calcareous marshes, fens, damp meadows, roadsides and dune-slacks, and also less acidic bogs and wet heathland. It avoids areas that are inundated for long periods, and is rarely found in drier sites such as chalk downland. It also colonises artificial habitats such as quarries and industrial waste ground. Lowland.

Native (change +0.84). This species is much better recorded now than in the 1962 *Atlas*, especially in the English Midlands. Although there have been local losses to drainage and habitat destruction, mostly since 1930, it is still frequent in suitable habitats in many areas and may have increased in man-made habitats.

Oceanic Temperate element.

References: Atlas (341c), Bateman & Denholm (1983a), Jenkinson (1991), Sanford (1991), Sell & Murrell (1996).

P. D. CAREY & T. D. DINES

Dactylorhiza purpurella Northern Marsh-orchid

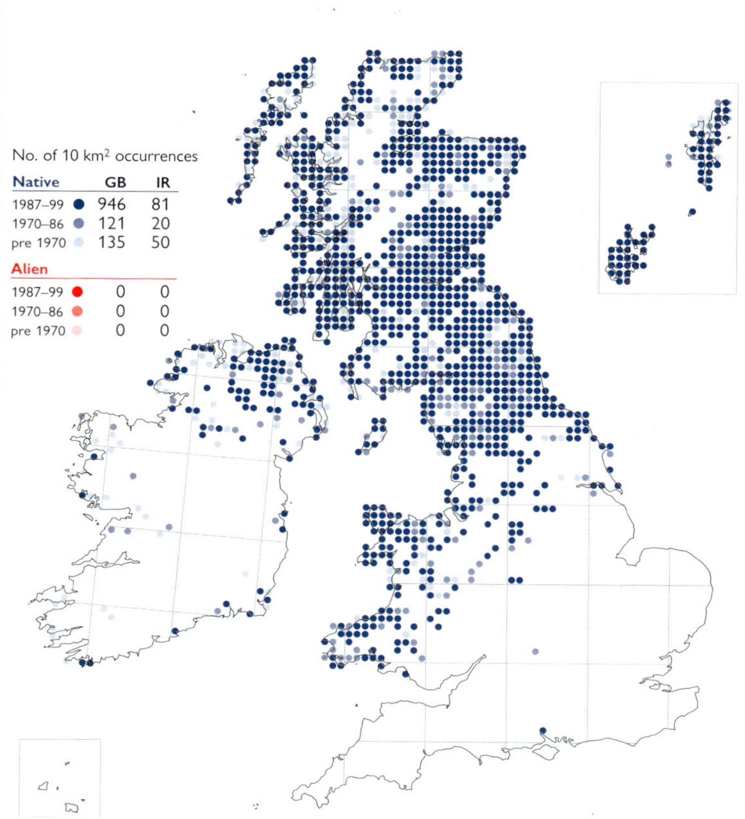

A tuberous perennial herb of neutral to base-rich soils, occurring in dune-slacks, fens, marshes, wet meadows, flushes, ditches and on roadside verges. It is also frequent in old quarries and urban waste land, where it can colonise drier sites such as rubble, and it occasionally grows in neglected gardens. 0–610 m (Creag Dhuba, Loch Ericht, Westerness).

Native (change +0.47). The overall distribution of *D. purpurella* is stable in Britain, although there have been local losses due to habitat destruction and drainage. In Ireland, however, it appears to have declined, especially in the south. It is sometimes confused with *D. majalis*.

Oceanic Boreal-montane element.

References: Atlas (341d), Allan & Woods (1993), Bateman & Denholm (1983a), Hultén & Fries (1986), Sell & Murrell (1996).

P. D. CAREY & T. D. DINES

Dactylorhiza majalis Western Marsh-orchid

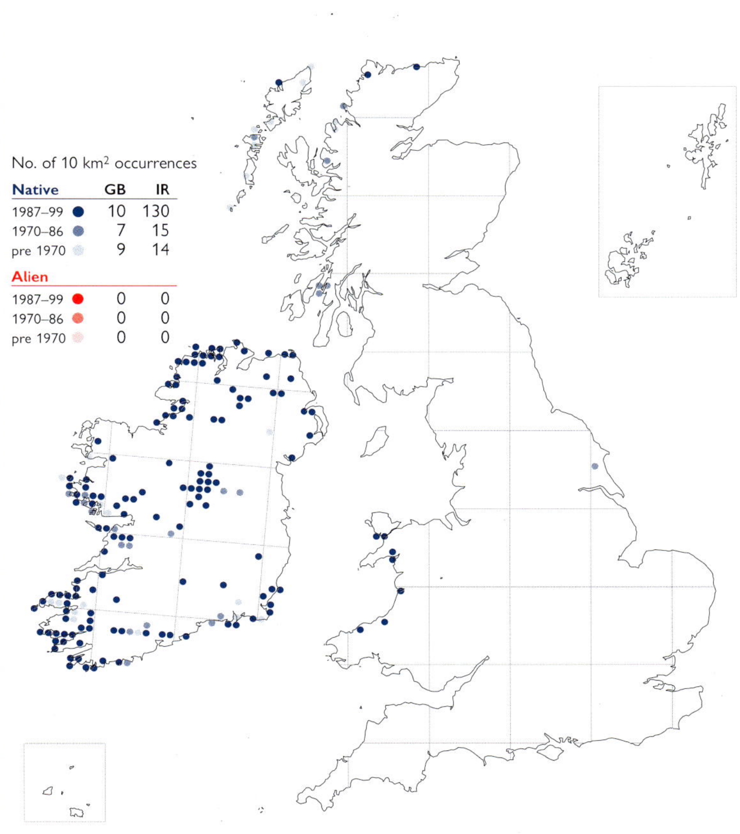

A tuberous perennial herb of damp, calcareous soils, growing in fens, marshes, wet meadows and dune-slacks. Lowland.

Native (change –0.41). This species has been much better recorded since the 1962 *Atlas*, particularly in Ireland. Its overall distribution is stable, but its taxonomy is not. Stace (1997) assigned plants in Ireland and North Uist to subsp. *occidentalis* and all others to subsp. *cambrensis*; recent work indicates that this last is better treated as a subspecies of *D. purpurella*, that subsp. *occidentalis* should be raised to specific rank, and that the North Uist plants are probably referable to *D. traunsteineri* (R. M. Bateman, *in litt.*).

European Temperate element.

References: Atlas (342a), Allan & Woods (1993), Bateman & Denholm (1983a), Crackles (1986), Hultén & Fries (1986), Sell & Murrell (1996).

P. D. CAREY & T. D. DINES

Dactylorhiza traunsteineri Narrow-leaved Marsh-orchid

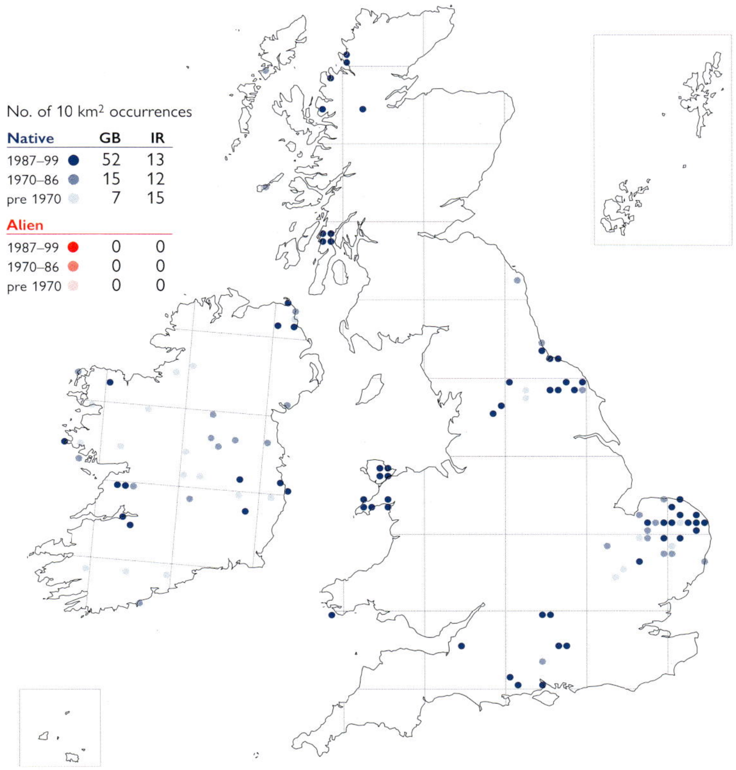

A tuberous perennial herb of damp, base-rich habitats such as marshes, water-meadows, flushes and fens. It sometimes prefers more open flushed or very wet areas with reduced competition. Lowland.

Native (change +0.78). This species is much better recorded than in the 1962 *Atlas*, especially in Scotland, and the overall distribution appears to be stable. There have been losses, however, mainly due to disturbance and lowering of water-tables. It may be under-recorded in some areas, especially Ireland.

Eurosiberian Boreo-temperate element.

References: Atlas (342b), Allan & Woods (1993), Bateman & Denholm (1983a), Curtis & McGough (1988), Hultén & Fries (1986), Jenkinson (1991), Sanford (1991), Sell & Murrell (1996), Stewart *et al.* (1994).

P. D. CAREY & T. D. DINES

Dactylorhiza lapponica Lapland Marsh-orchid

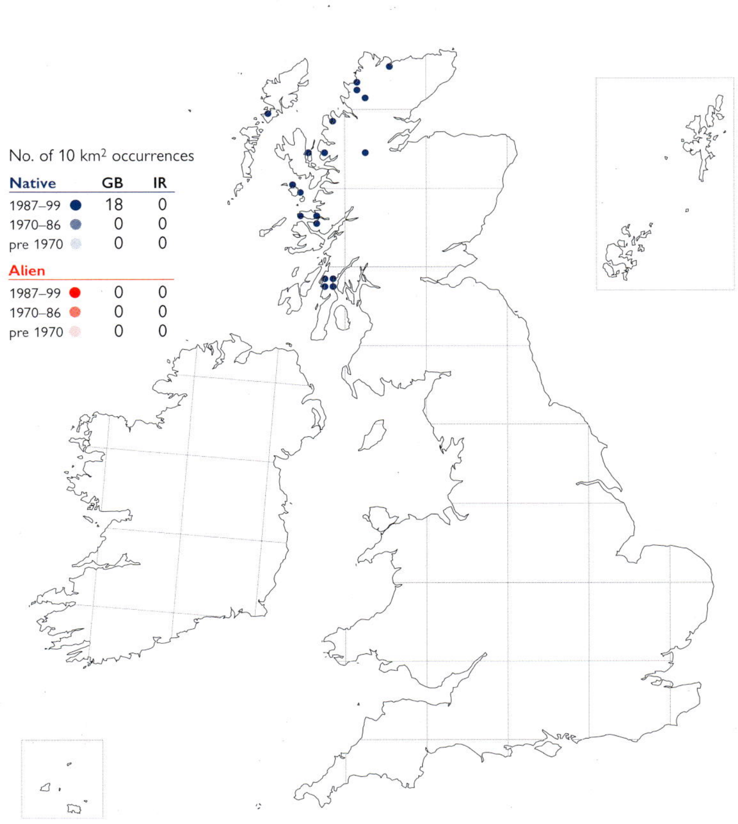

A tuberous perennial herb of base-rich hill flushes associated with superficially acidic and peaty soils; sometimes also spreading into nearby areas of more acidic wet heath. Most sites are moderately or heavily grazed. 0–310 m (Knapdale, Westerness).

Native. This taxon, first discovered in Britain in 1967, is almost certainly still under-recorded. Populations tend to be small but are apparently stable, despite annual fluctuations in numbers. It may have been over-looked as *D. traunsteineri* in N. Ireland; *D. lapponica* may best be treated as a subspecies or variety of *D. traunsteineri* (Bateman, 2001, Hedrén, 1996, Sell & Murrell, 1996).

European Boreal-montane element.

References: Allan & Woods (1993), Bateman & Denholm (1983a), Cowie & Sydes (1995), Wigginton (1999).

P. D. CAREY & T. D. DINES

Neotinea maculata Dense-flowered Orchid

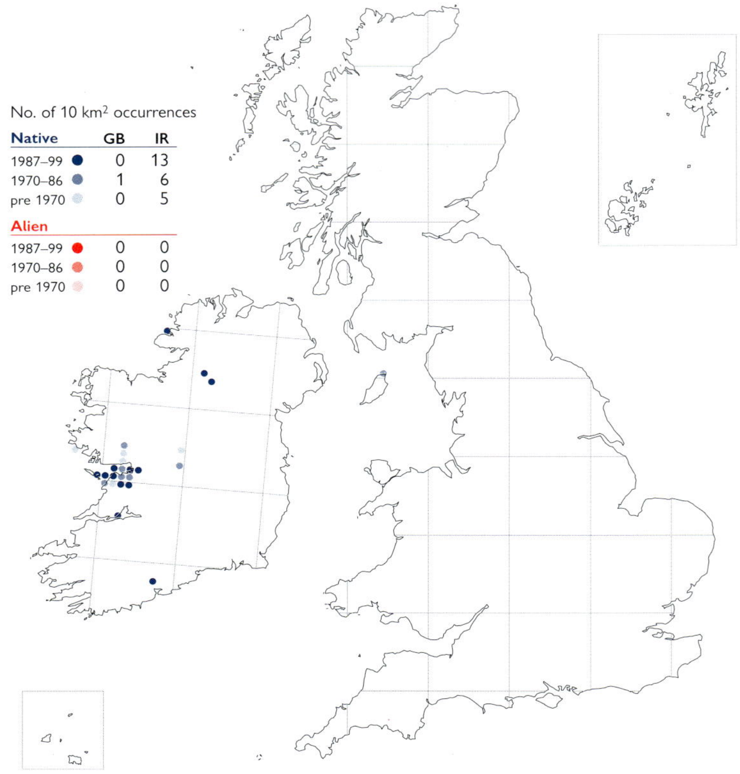

This tuberous perennial herb grows in a wide range of habitats on base-rich rocky or gravelly substrates. It can be found in the crevices of limestone pavement, in old pastures, hill grasslands, dunes and on road verges. Occasionally, it occurs on peat overlying more acidic rocks, and in *Corylus-Fraxinus* woodland. Lowland.

Native. Although this species has been lost from many sites in Ireland, it has recently been found at several new sites and its overall distribution is stable. It was found in the Isle of Man in 1967, but has not been recorded from there since 1986.

Mediterranean-Atlantic element.

References: Atlas (337d), Allen (1984), Webb & Scannell (1983).

P. D. CAREY & T. D. DINES

Orchis laxiflora Loose-flowered Orchid

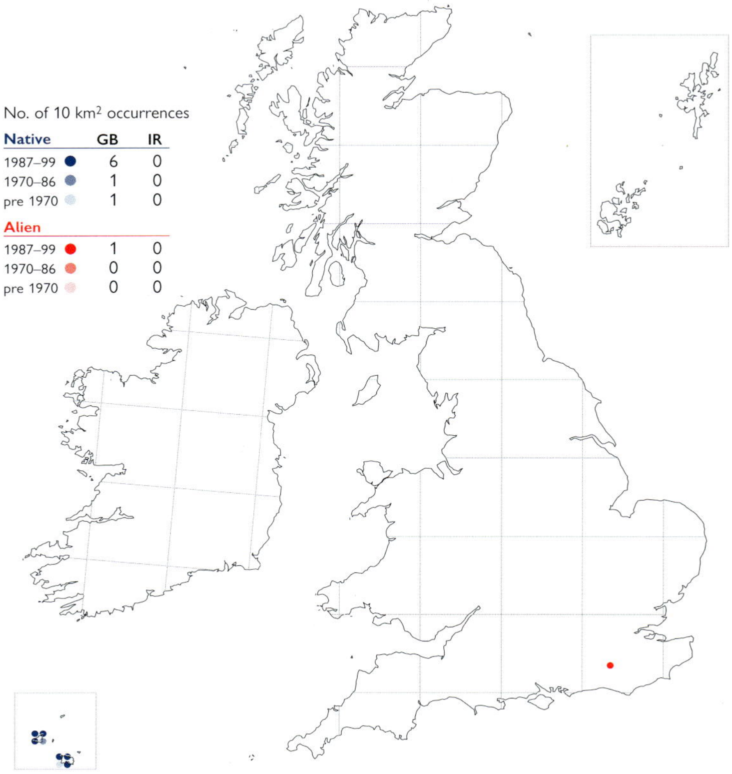

A tuberous perennial herb of base-rich or calcareous wet meadows and marshy fields. Lowland.

Native. *O. laxiflora* has decreased since the early 20th century due to the intensification of agriculture and development, but sites in both Jersey and Guernsey are now protected and their populations have been stable since 1970. Recent molecular studies indicate that this species may be more appropriately treated under *Anacamptis* (Bateman *et al.*, 1997).

Eurosiberian Southern-temperate element, with a continental distribution in W. Europe.

References: Atlas (339d), Hultén & Fries (1986), Le Sueur (1984), McClintock (1975), Meusel *et al.* (1965).

P. D. CAREY & T. D. DINES

Orchis mascula Early-purple Orchid

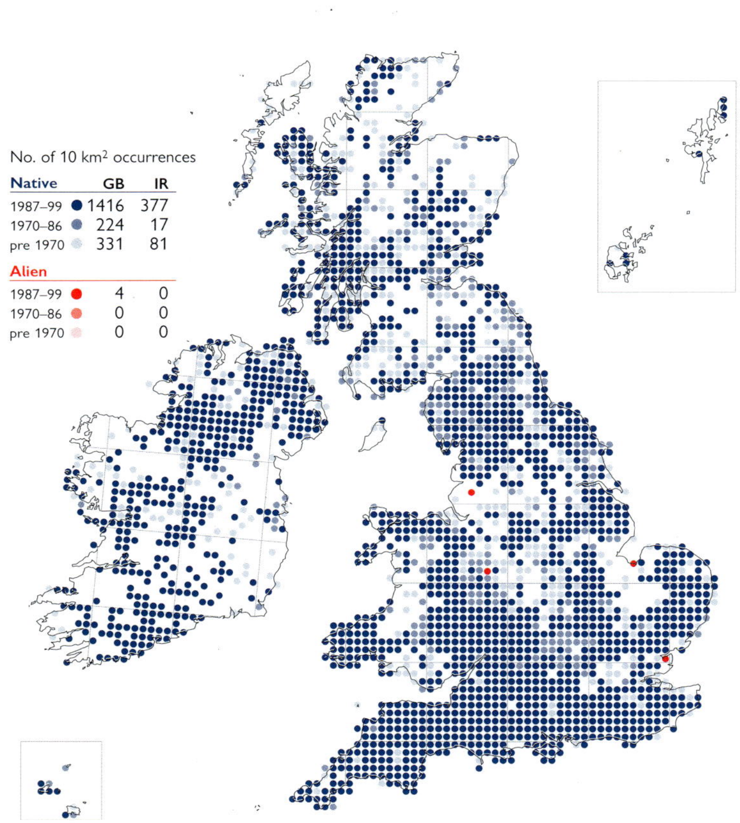

This tuberous perennial herb grows on a variety of neutral and calcareous soils, and is most frequent in woodland, coppices and calcareous grassland. However, it also occurs in hedgerows, scrub, on roadsides and railway banks and on limestone pavement and moist cliff ledges. 0–880 m (Caenlochan, Angus).

Native (change –0.72). This species, mapped as 'all records' in the 1962 *Atlas*, has declined in some areas, particularly C. England and parts of Scotland. Most losses are due to woodland felling and coniferisation, intensification of grassland management and ploughing. However, it is much better recorded now, especially in Ireland.

European Temperate element.

References: Atlas (340c), Allan & Woods (1993), Hultén & Fries (1986), Jenkinson (1991), Meusel *et al.* (1965), Sanford (1991).

P. D. CAREY & T. D. DINES

Orchis morio Green-winged Orchid

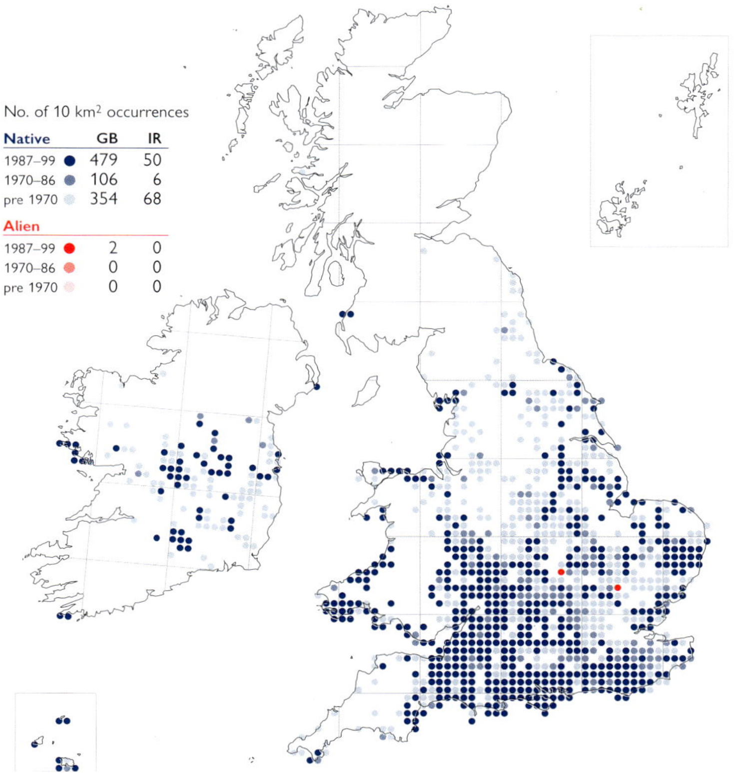

No. of 10 km² occurrences

Native		GB	IR
1987–99	●	479	50
1970–86	●	106	6
pre 1970	●	354	68
Alien			
1987–99	●	2	0
1970–86	●	0	0
pre 1970	●	0	0

A tuberous perennial herb of damp to dry, base-rich to mildly acidic soils. It is most frequent in hay meadows and pastures, but also grows on sand dunes, heaths and road-sides, and in quarries, gravel-pits, churchyards and lawns. 0–305 m (Co. Dublin).

Native (change –0.98). The steady decline of this species due to the ploughing and improvement of grasslands has taken place throughout the 19th and 20th centuries. It is often present in only small numbers in 10-km squares where it was once more frequent. Molecular studies indicate that it may be more appropriately placed in *Anacamptis* (Bateman *et al.*, 1997).

European Temperate element.

References: Atlas (340b), Curtis & McGough (1988), Hultén & Fries (1986), Meusel *et al.* (1965), Sanford (1991), Stewart *et al.* (1994), Wells *et al.* (1998).

P. D. CAREY & T. D. DINES

Orchis ustulata Burnt Orchid

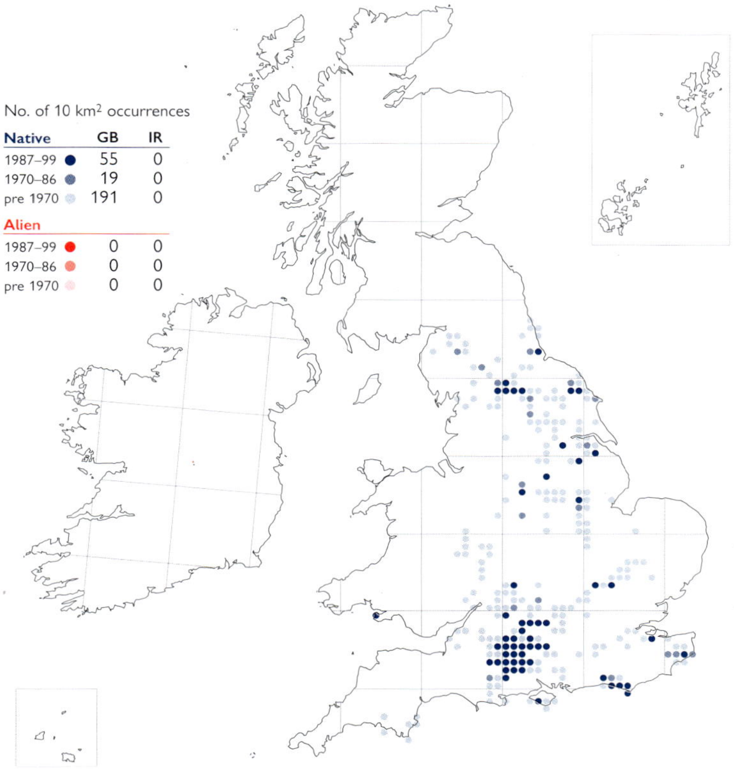

No. of 10 km² occurrences

Native		GB	IR
1987–99	●	55	0
1970–86	●	19	0
pre 1970	●	191	0
Alien			
1987–99	●	0	0
1970–86	●	0	0
pre 1970	●	0	0

This tuberous perennial herb requires warm, dry conditions and is often found in tightly grazed chalk and limestone grassland on S.-facing slopes. It also occurs on sandy and gravelly soils in river meadows and on sand dunes. Lowland.

Native (change –1.77). The spectacular decline of this species has been well-documented. Losses have been largely due to changes in agricultural practices, such as ploughing and the cessation of grazing, and through habitat destruction by building and quarrying. Recent molecular studies indicate that it may be more appropriately treated under *Neotinea* (Bateman *et al.*, 1997).

European Temperate element.

References: Atlas (340a), Hultén & Fries (1986), Jenkinson (1991), Meusel *et al.* (1965), Sanford (1991), Stewart *et al.* (1994).

P. D. CAREY & T. D. DINES

Orchis purpurea Lady Orchid

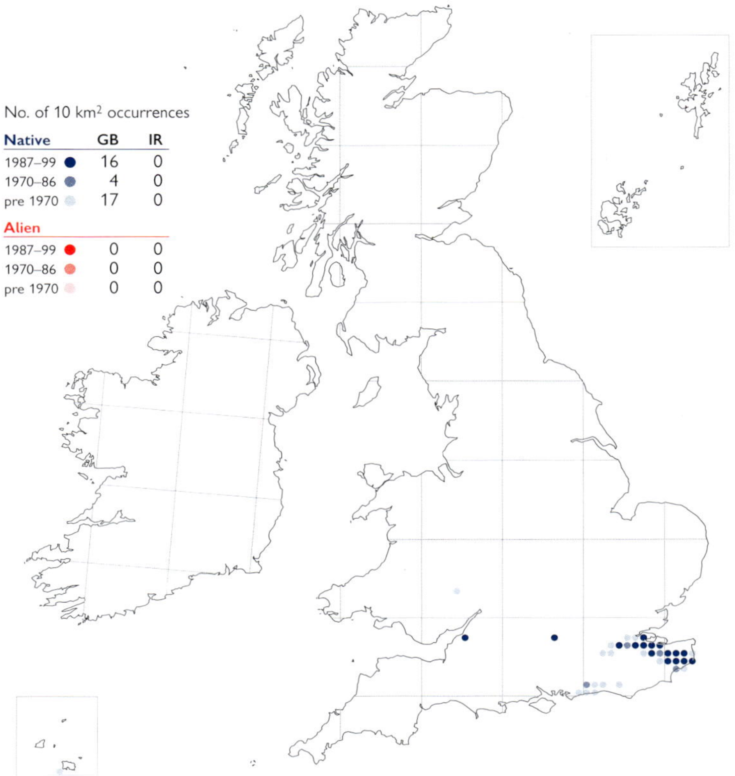

No. of 10 km² occurrences

Native		GB	IR
1987–99	●	16	0
1970–86	●	4	0
pre 1970	●	17	0
Alien			
1987–99	●	0	0
1970–86	●	0	0
pre 1970	●	0	0

This long-lived tuberous perennial herb is found on thin calcareous soils, typically over chalk but also on clay, ragstone and Carboniferous limestone. It grows in open *Corylus*, *Fagus* or *Fraxinus* woodland and scrub and, more rarely, in open grassland. Lowland.

Native (change –0.56). This species suffered losses in its core Kent area, largely before 1930, but now appears to be stable there. It was last recorded from Sussex in 1976. Elsewhere, it was recorded once in Herefordshire in 1967, it appears to be flourishing at its Oxfordshire site, where it was found in 1961, and it was discovered in W. Gloucestershire in 1991.

European Temperate element.

References: Atlas (339b), Hultén & Fries (1986), Meusel *et al.* (1965), Rose (1948), Stewart *et al.* (1994).

P. D. CAREY & T. D. DINES

Orchis militaris Military Orchid

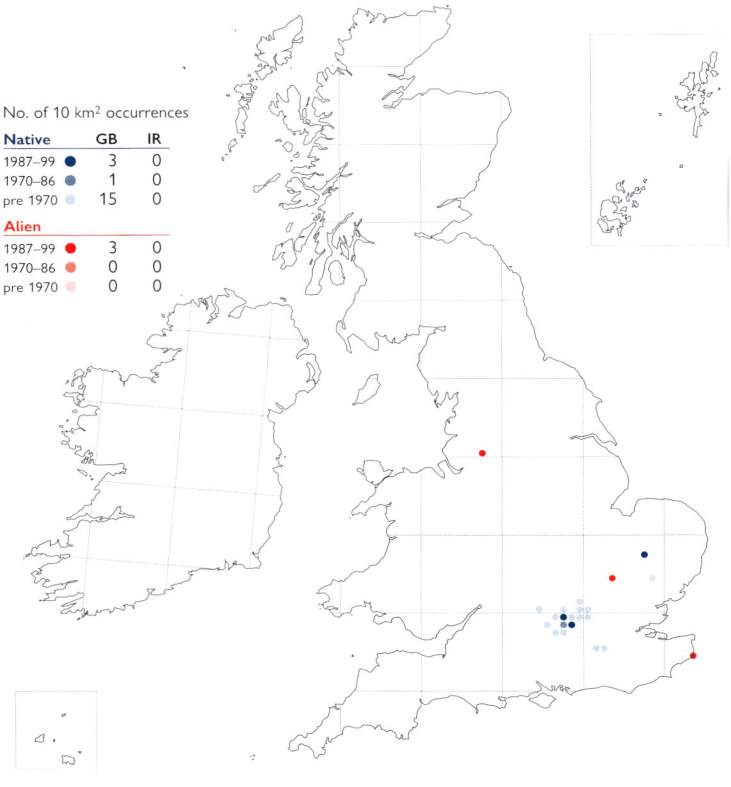

No. of 10 km² occurrences

Native	GB	IR
1987–99 ●	3	0
1970–86 ●	1	0
pre 1970 ●	15	0
Alien		
1987–99 ●	3	0
1970–86 ●	0	0
pre 1970 ●	0	0

A tuberous perennial herb found on chalk in grassland, scrub, woodland glades and a chalk pit. It was planted on industrial waste ground in S. Lancashire, and in new sites in Cambridgeshire and Kent. Lowland.

Native. Habitat destruction and collecting caused a decline in this species until it was considered extinct in the 1920s. It was re-found in Buckinghamshire in 1947 and Suffolk in 1954, where populations are stable, and it appears sporadically in two Oxfordshire sites.

Eurosiberian Temperate element, with a continental distribution in W. Europe.

References: Atlas (339c), Farrell (1985), Hultén & Fries (1986), Hutchings *et al.* (1998), Meusel *et al.* (1965), Sanford (1991), Sell & Murrell (1996), Waite & Farrell (1998), Wigginton (1999).

P. D. CAREY & T. D. DINES

Orchis simia Monkey Orchid

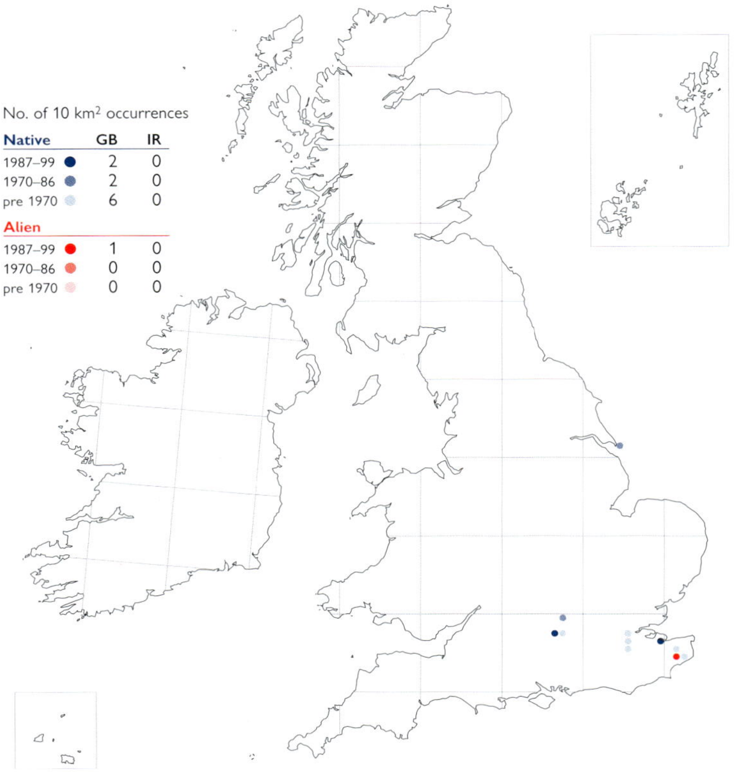

No. of 10 km² occurrences

Native	GB	IR
1987–99 ●	2	0
1970–86 ●	2	0
pre 1970 ●	6	0
Alien		
1987–99 ●	1	0
1970–86 ●	0	0
pre 1970 ●	0	0

This perennial, tuberous herb is found on S.-facing banks in grazed chalk grassland. It is tolerant of some degree of shade from scrub, and can also grow on woodland edges. Lowland.

Native. First recorded in 1777 in Kent, this species declined in the 19th century through the ploughing of downland, and was thought to have become extinct in Britain until the extant native Kent population was found in 1955. Seed from this site was used to established a second population in Kent. In 1974, a few plants appeared in S.E. Yorkshire, but these only persisted until 1983. The extant Oxfordshire and Kent populations are increasing, assisted by management of grazing.

European Southern-temperate element.

References: Atlas (339d), Meusel *et al.* (1965), Wigginton (1999).

P. D. CAREY & T. D. DINES

Aceras anthropophorum Man Orchid

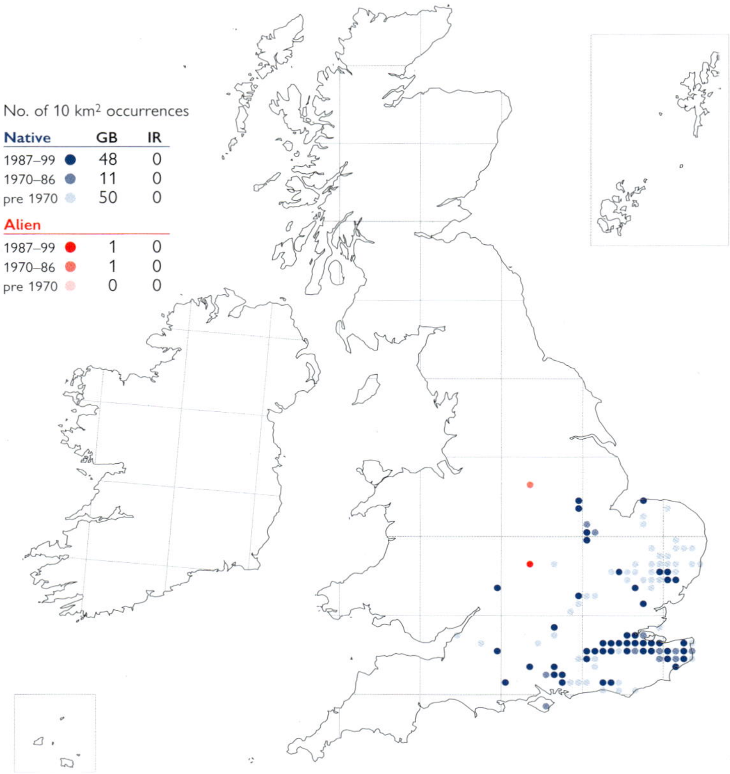

No. of 10 km² occurrences

Native	GB	IR
1987–99 ●	48	0
1970–86 ●	11	0
pre 1970 ●	50	0
Alien		
1987–99 ●	1	0
1970–86 ●	1	0
pre 1970 ●	0	0

This tuberous perennial herb is found in old chalk-pits and limestone quarries, calcareous grassland and on road verges. It tolerates considerable shade and is often found at the edge of scrub with grasses such as *Brachypodium pinnatum*. Continuous heavy grazing is detrimental, eventually causing its demise. Lowland.

Native (change –0.76). By 1930 most East Anglian populations of this species had been destroyed by ploughing. Similar fates have since eliminated more sites, and others have been lost to scrub encroachment, spray drift and inappropriate roadside cutting regimes. Morphological and molecular (Bateman *et al.*, 1997) evidence suggests that *Aceras* belongs in the genus *Orchis*.

Mediterranean-Atlantic element.

References: Atlas (342c), Meusel *et al.* (1965), Sanford (1991), Stewart *et al.* (1994).

P. D. CAREY & T. D. DINES

Himantoglossum hircinum Lizard Orchid

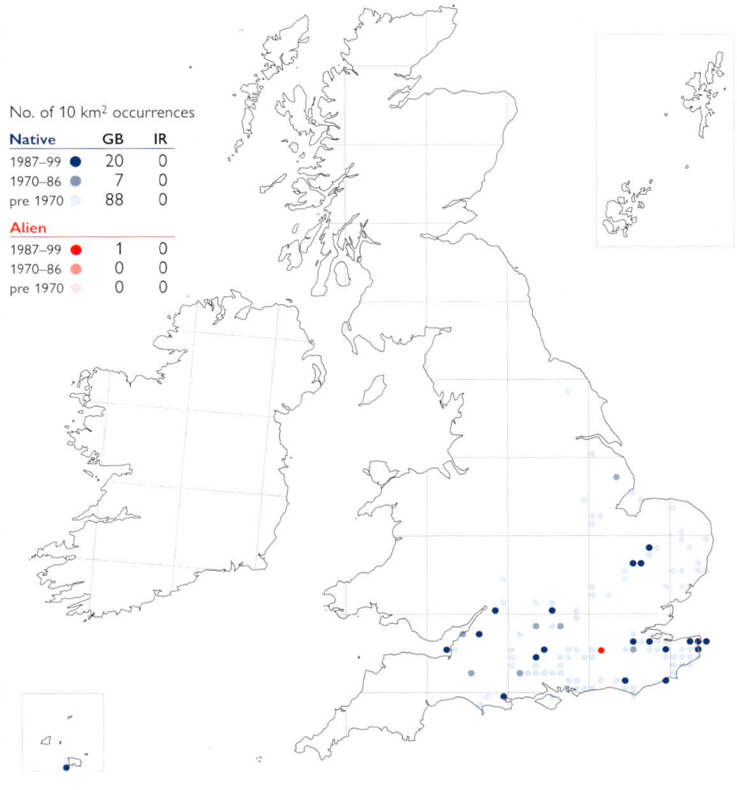

A tuberous, winter-green perennial herb growing on chalk and, rarely, limestone in open grassland, on roadsides and in quarries, and occasionally on calcareous sand dunes and heathland. Lowland.

Native (change –2.40). This species was restricted to Kent until the early 1900s, when it underwent a remarkable expansion, reaching as far north as Yorkshire. After 1934, it declined rapidly but maintained this range. Many occurrences are of monocarpic plants; with 9–11 populations recorded each year the map over-emphasises losses as historical records accumulate. Only two populations are definitely self-sustaining, although two more may be so.

Submediterranean-Subatlantic element.

References: Atlas (339a), Carey (1998, 1999), Meusel *et al.* (1965), Sanford (1991), Wigginton (1999).

P. D. CAREY & T. D. DINES

Serapias parviflora Small-flowered Tongue-orchid

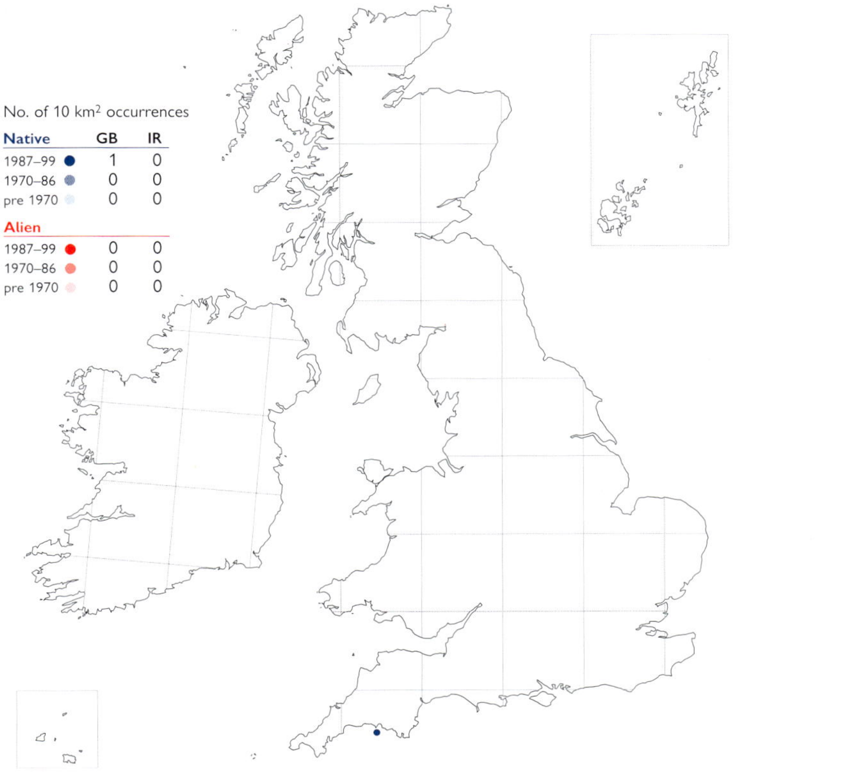

A small, tuberous perennial herb found growing in rabbit grazed grassland in *Ulex europaeus* and *Rubus fruticosus* agg. scrub on S.-facing coastal cliffs. Lowland.

Native or alien. Several *Serapias* species have been recorded for Britain, and this is perhaps the most convincingly native. It was first discovered in 1989, and the single known population flowered intermittently until 1998. It may have arisen from wind blown seed and is regarded as native by Madge (1994) and Rich (1997c), but Stace (1997) and Lang (in Rich & Jermy, 1998) are more cautious and regard it as alien. The original population has been augmented by plants raised from seed collected at the site.

Mediterranean-Atlantic element.

References: French *et al.* (1999), Murphy (1994).

T. D. DINES

Ophrys insectifera Fly Orchid

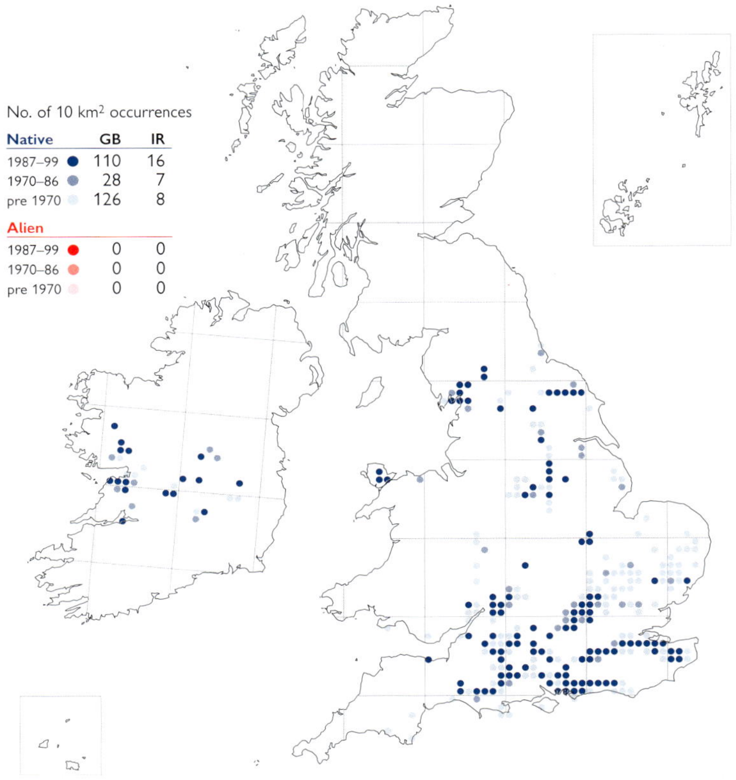

A shade-tolerant tuberous herb usually found on chalk and limestone soils in open deciduous woodland and scrub, but also recorded from grassland, chalk-pits, limestone pavement, disused railways, spoil heaps and, rarely, unstable coastal cliffs. In Ireland and Anglesey it is found only in open calcareous flushes and fens. 0–390 m (Helbeck Wood, Westmorland).

Native (change –1.34). This species declined dramatically before 1930, especially in East Anglia. Since then the losses have continued, but at a reduced rate. Most losses are due to scrub encroachment, the closing of woodland canopies, woodland clearance and drainage of fens.

European Temperate element.

References: Atlas (338d), Hultén & Fries (1986), Meusel *et al.* (1965), Sanford (1991).

P. D. CAREY & T. D. DINES

Ophrys sphegodes Early Spider-orchid

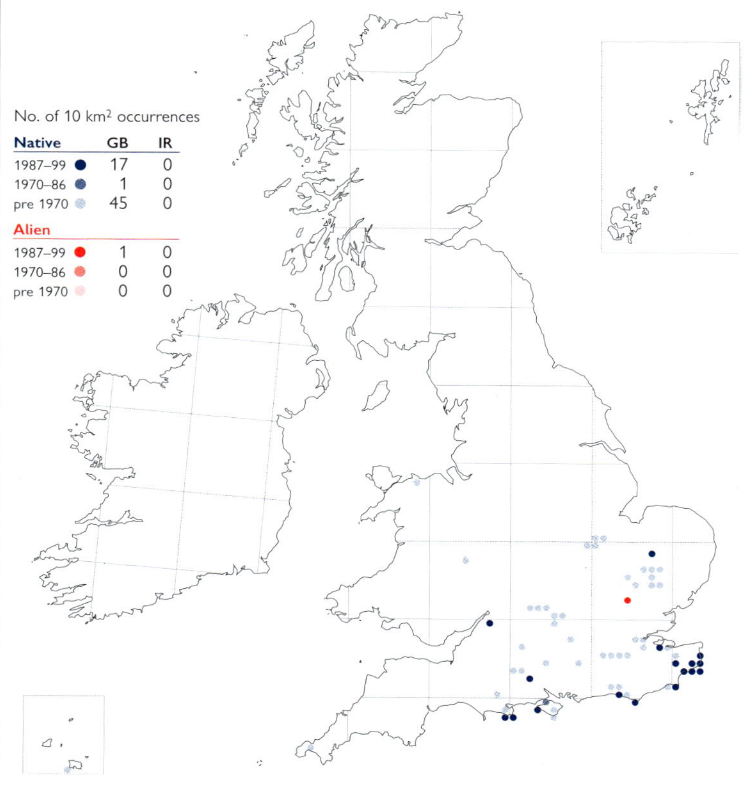

A winter-green, short-lived perennial tuberous herb of ancient, species-rich, heavily grazed grassland on chalk and Purbeck limestone. It can, however, tolerate taller grassland, and has also been found colonising disturbed ground in limestone quarries, old spoil heaps and by tracks. Lowland.

Native (change –0.11). *O. sphegodes* declined severely before 1930, becoming extinct in twelve vice-counties. Most losses were due to ploughing of grassland and changes in grazing regimes, to which it is particularly vulnerable. Its distribution is now stable, and new sites are occasionally found.

Submediterranean-Subatlantic element.

References: Atlas (338c), Jenkinson (1991), Meusel *et al.* (1965), Sanford (1991), Sanger & Waite (1998), Wigginton (1999).

P. D. CAREY & T. D. DINES

Ophrys apifera Bee Orchid

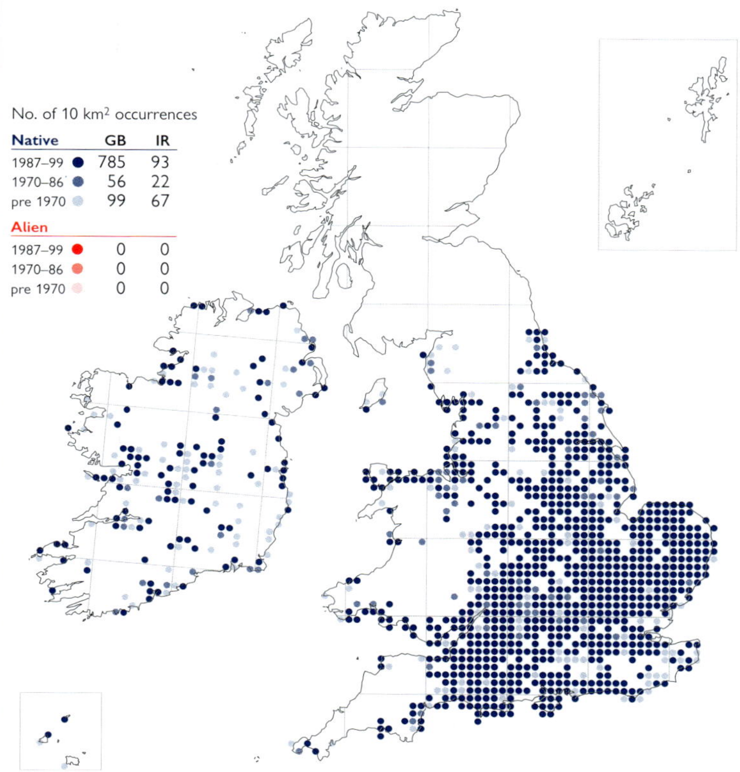

A tuberous perennial herb of calcareous, well-drained soils. Habitats include grasslands, scrub, railway banks, roadsides, lawns, sand dunes and limestone pavement; also disturbed sites such as quarries, gravel-pits and industrial waste ground. 0–335 m (near Parsley Hay, Derbys.).

Native (change +0.83). The overall distribution of this species in Britain is stable, but it has declined in Ireland, where many losses were before 1930. Declines are due to habitat destruction, especially ploughing of grassland and the in-filling of quarries. It is, however, much better recorded now, and readily colonises newly available sites.

Submediterranean-Subatlantic element.

References: Atlas (338a), Curtis & McGough (1988), Jenkinson (1991), Meusel *et al.* (1965), Sanford (1991), Wells & Cox (1989, 1991).

P. D. CAREY & T. D. DINES

Ophrys fuciflora Late Spider-orchid

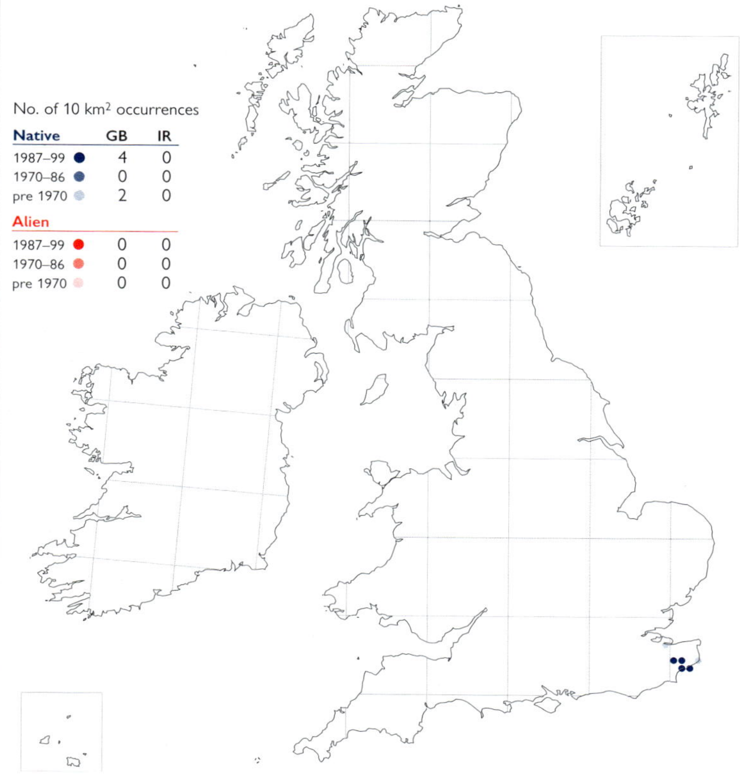

This long-lived, tuberous perennial herb grows on well-drained calcareous, chalky soil in species-rich, closely grazed grassland; it cannot compete in longer grassland. Some colonies are on previously disturbed areas, but rarely spread to adjacent sites. Lowland.

Native. Historically, sites for *O. fuciflora* have been lost due to grassland improvement, ploughing and the cessation of grazing. It was lost from two 10-km squares before 1950. Remaining populations, of which only five now produce plants regularly, are all protected and carefully managed.

Submediterranean-Subatlantic element.

References: Atlas (338b), Meusel *et al.* (1965), Wigginton (1999).

P. D. CAREY & T. D. DINES

CHAPTER 9
List of Contributors

The following botanists contributed to the *New Atlas* project by acting as Vice-county Recorders, by contributing a significant number of recent field records or by helping in other ways.

Mrs P. P. Abbott, Miss F. Abraham, T. A. Abrehart, P. Acton, Ms M. Adam, B. A. Adams, Dr J. Adams, Dr K. J. Adams, Rev. R. A. Addington, M. Adkins, Mrs P. M. Aglen, Dr J. R. Akeroyd, A. S. Alcorn, E. Alker, Ms D. R. Allan, E. Allan, Mrs M. M. Allan, Dr D. E. Allen, Dr D. J. Allen, Ms M. Allen, Miss M. D. B. Allen, Dr J. M. Allinson, M. Allison, A. G. Amphlett, G. Anderson, Mrs S. Andrews, J. Angus, Dr F. Anstey, Dr T. R. ap Rheinallt, Ms V. J. Appleby, G. Appleyard, A. Arbon, J. K. Archer, M. A. Arnold, M. Ashby, Dr P. A. Ashton, Mrs A. Ashwell, J. Atkins, A. Atkinson, Mrs M. G. C. Atkinson, M. H. Atkinson, Dr B. Attock, Dr F. M. Aungier, Mrs A. Austen, S. Aylward, Mrs D. Ayres.

H. E. E. Babb, Miss J. A. Bailey, Dr J. P. Bailey, Mrs P. Baker, Mrs P. M. Baker, A. Balbi, N. Baldock, K. Balkow, G. H. Ballantyne, A. Bamford, N. Bamford, R. Bamford, P. R. Banham, Mrs J. Banks, A. R. Barker, Mrs K. Barnacle, Mrs P. M. Barnes, A. L. Barnett, K. Barnett, Mrs D. Barrass, Ms H. Barri, Mrs M. Barron, Miss G. M. Barter, Dr T. P. Bartlett, Miss A. M. Bassett, N. Batchelor, M. Bates, G. H. Battershall, Mrs I. Battershall, B. D. Batty, Mrs P. Batty, Mrs G. Beckett, K. A. Beckett, J. A. R. Beckford, J. Bedford, S. Beesley, Dr H. Belcher, Mrs C. E. Bell, Mrs M. Bell, Mrs S. Bell, Dr G. Bellamy, R. M. Belringer, Mrs C. Belshaw, Mrs B. Benfield, Mrs J. Benfield, I. J. Bennallick, Mrs C. M. Bennett, H. F. Bennett, M. C. Bennett, P. M. Benoit, N. Benson, Miss M. Benstead, Mrs S. Berrett, Mrs J. Berry, N. H. Bertrand, Mrs A. L. Best, B. J. Best, D. Bevan, J. Bevan, J. M. S. Bevan, P. Billinghurst, Mrs A. Binding, J. Bingham, Mrs J. Birch, T. Birch, R. G. N. Bird, Dr E. L. Birse, Mrs B. Bishop, G. Bishop, J. G. Bishop, S. H. Bishop, W. Bishop, J. M. Blackburn, Mrs A. B. Blackley, Miss E. Blackman, T. H. Blackstock, R. A. Blades, Dr K. Bland, J. Blane, Mrs I. Blatchley, Dr A. J. Bleasdale, D. Bloomfield, K. Bloor, P. Boddington, Mrs A. E. Bolton, B. Bond, A. D. Boniface, B. Bonnard, Dr J. Bonnard, I. R. Bonner, C. R. Boon, A. Booth, C. Booth, Mrs C. Booth, F. Booth, B. N. Boothby, S. D. S. Bosanquet, Mrs A. M. Boucher, J. P. Bowdrey, Dr H. J. M. Bowen, K. Bowey, R. P. Bowman, J. C. Bowra, J. Box, D. C. Boyce, W. R. Brackenridge, Ms S. Bradley, H. Bradshaw, Dr M. E. Bradshaw, Miss C. Brady, N. Bragg, P. J. Brain, M. E. Braithwaite, Mrs P. F. Braithwaite, Mrs D. M. Bramley, J. Branscombe, A. J. P. Branson, C. Breen, A. Bremner, Lady Anne Brewis, Mrs M. A. Brewster, A. Brickstock, Mrs M. Briggs, Miss R. A. Brind, G. Bristowe, M. Broadbridge, Dr J. Brodie, Ms C. Brook, B. S. Brookes, Mrs D. L. Brookman, Mrs E. R. Brooks, Ms M. Broomfield, J. Brough, D. Broughton, C. Brown, Miss E. Brown, Miss L. Brown, I. Brown, Mrs J. Brown, M. Brown, Mrs M. Brown, N. H. Brown, Mrs P. Brown, Mrs R. Brown, E. S. Bruce, N. A. Bruce, Dr R. K. Brummitt, C. J. Bruxner, K. M. Bryant, Mrs J. Buchanan, P. T. Buchanan, Mrs S. Buckingham, Mrs J. Buckley-Earl, A. L. Bull, Miss E. R. Bullard, Mrs J. Bulleid, Dr R. J. Bullock, Dr S. J. Bungard, Mrs M. M. Burgoine, Mrs S. Burn, Miss M. J. Burnhill, Mrs M. Burnip, Ms A. Burns, Mrs J. E. Burnwell, B. Burrow, Mrs M. Burrow, Mrs L. B. Burt, P. Burton, R. M. Burton, Mrs S. Burton, A. A. Butcher, J. K. Butler, Mrs S. I. Butler, Mrs M. Button, B. K. Byrne.

Dr C. J. Cadbury, Miss H. M. Caddick, Dr J. Caffrey, Mrs M. Caine, T. E. Cairns, J. J. Callister, Mrs A. S. Campbell, J. M. Campbell, D. C. G. Cann, M. Cant, D. E. Carder, A. Carey, Rev. W. Carmichael, J. V. Carrington, F. Carroll, T. Carruthers, Ms A. B. Carter, Miss E. Carter, Mrs D. Casey, Mrs B. Cassels, K. Cassels, Ms G. E. Castle, M. Catt, Miss C. L. Cauldwell, Dr K. Cavanagh, Mrs P. Cavanagh, Dr B. V. Cave, Ms P. Cawthorne, Mrs B. K. Chadwick, N. L. Chadwick, Mrs C. Chaffey, S. P. Chambers, R. J. Chancellor, Mrs S. Chandler, G. Charter, Ms E. Charter, Mrs R. Charter, A. O. Chater, C. Chatters, Dr R. G. Chaytor, Dr C. Cheffings, J. Chester, Dr E. Chicken, Mrs J. Child, W. Chisholm, C. Christie, Mrs J. Christie, A. R. Church, Mrs J. W. Clark, Mrs J. H. Clarke, M. C. Clarke, J. Clayfield, R. Cleaver, E. J. Clement, Mrs B. Clough, P. Clough, R. Clough, P. R. Cobb, Ms C. M. Cockerill, Mrs D. I. Cockerill, E. Cockerill, Mrs E. M. Coe, R. L. Cole, A. Coles, Mrs J. E. Collar, Miss H. Colls, A. Colston, Miss J. Colthup, Mrs M. Combe, J. R. Comley, J.C. Comont, Miss E. R. T. Conacher, Dr J. Conaghan, Mrs P. Condry, W. M. Condry, Miss A. P. Conolly, Mrs A. Cook, N. Cook, Mrs P. Cook, P. J. Cook, R. J. Cooke, P. Cool, Dr D. E. Coombe, Mrs D. Cooper, Mrs M. Cooper, R. J. Cooper, Dr T. A. Cope, Miss V. Copp, A. Copping, Mrs P. Copson, R. Copson, P. Corbett, C. J. Cornell, Dr R. W. M. Corner, W. R. Corner, Mrs C. Costie, Dr D. C. F. Cotton, M. Coulson, D. Counsell, Dr N. Cowie, P. Coxhead, Miss V. Cozens, Dr F. E. Crackles, Mrs C. Craster, Mrs M. Craster, G. Crawford, Dr I. Crawford, Prof. R. R. Crawford, Prof. M. J. Crawley, I. M. Crewe, J. D. Crewe, M. D. Crewe, Mrs J. M. Croft, Mrs G. Crompton, C. S. Crook, T. S. Crosby, Ms J. Crosher, A. M. Cross, Dr J. R. Cross, S. Cross, J. Crossley, Miss G. A. Crouch, Dr H. J. Crouch, M. Crowley, N. Crowther, Mrs J. Cull, S. Cuming, D. Curran, C. W. Curtis, J. P. Curtis, Dr M. S. Curtis, Dr T. G. F. Curtis, B. Cuttell.

J. Daffern, V. Dale, P. Dalley, Mrs A. P. Daly, E. T. Daniels, P. Darby, Dr T. Dargie, P. Daunt, L. Davey, Dr C. T. David, D. E. David, R. W. David, Miss M. M. Davidson, Mrs M. M. Davidson, Mrs B. Davies, D. Davies, J. N. Davies, Dr K. L. Davies, Dr R. H. Davies, T. Davies, Dr B. N. K. Davis, J. Davis, D. Davison, S. P. Davison, Dr G. Davy, Mrs R. A. Dawes, H. J. Dawson, Mrs J. Dawson, Miss N. Dawson, T. Dawson, G. V. Day, J. J. Day, P. Day, R. L. d'Ayala, A. Dean, Miss E. R. Dean, T. Dean, M. Dearden, Dr I. G. Denholm, Dr M. P. Denne, Ms F. Dennis, Miss M. Devereau, Miss V. Devereau, Mrs F. D. Devery, K. J. S. Devonald, J. Dews, J. Dicker, Prof. J. H. Dickson, Dr T. D. Dines, T. J. Dingle, Miss J. E. Dinwiddie, A. Dixon, C. Dixon, J. R. Doe, W. R. Dolling, Miss F. Donald, T. Doncaster, D. M. Donovan, Mrs P. M. Donovan, Mrs C. M. Dony, Dr J. G. Dony, Dr D. A. Doogue, Ms C. S. Douglas, Mrs D. M. Downing, M. J. D'Oyley, Ms J. Drage, D. R. Drewett, Ms M. Dromey, Dr R. C. Dryden, Miss L. C. Dudley, Mrs D. S. Dudley-Smith, A. A. Dudman, Miss K. Duff, Miss A. J. Dunn, T. W. J. D. Dupree, J. Durkin, Dr R. B. Dwyer.

R. A. Eades, Mrs J. Eagle, Mrs P. Eaglesfield, D. P. Earl, J. J. Earley, Miss S. L. Eastwood, G. M. S. Easy, J. R. Edelsten, Dr S. M. Eden, Prof. J. A. Edgington, M. Edmondson, Mrs S. Edmondson, Prof. M. Edmunds, B. J. Edwards, Ms M. Edwards, Mrs S. Edwards, T. Edwards, T. Eggeling, Dr H. A. Ellis, Mrs P. Ellis, R. G. Ellis, R. W. Ellis, P. J. Ellison, R. E. Elloway, Mrs J. E. Emerson, D. W. Emley, Mrs S. M. Emley, Mrs S. E. Erskine, Mrs A. E. Evans, Dr B. Evans, Ms F. Evans, I. M. Evans, Ms J. Evans, Mrs P. A. Evans, Dr P. G. H. Evans, S. A. Evans, S. B. Evans, T. G. Evans, Mrs U. T. Evans, W. Evans, Ms F. Everingham, Ms E. Everiss.

Mrs A. Fairchild, K. Fairclough, Mrs K. M. Fallowfield, Miss L. Farrell, F. J. L. Farrow, G. Farwell, Mrs A. Faulkner, C. Faulkner, Dr J. S. Faulkner, G. Fenton, Dr J. Fenton, Mrs F. Fenwick, R. E. C. Ferreira, Dr C. Ferris, P. R. Ferris, Mrs K. Ferrousat, G. D. Field, Dr R. Finch, S. Fisher, S. R. Fisher, R. S. R. Fitter, Lady Rosemary FitzGerald, Miss R. Fitzgerald, Ms J. Fleming, Miss M. Flexen, A. Flynn, P. Foley, Mrs G. Fookes, Dr R. S. Forbes, Dr A. W. Ford, Mrs H. Formaggia, G. H. Forster, C. M. Forster Brown, Mrs J. Foster, B. R. W. Fowler, A. P. Fowles, Dr I. S. Francis, Dr J. C. Frankland, Miss A. Franks, Miss

A. R. Franks, A. J. L. Fraser, J. Fraser, R. Fraser, Mrs A. Freeman, D. Freeman, Dr C. N. French, J. Frew, Mrs A. P. Fritchley, Miss M. B. Fuller, P. J. Furze, R. Fussell, Dr K. Futter.

Mrs J. E. Gaffney, Dr P. A. Gainey, J. Gallagher, Ms J. Galley, M. J. Galliott, Dr A. Galton, P. H. Gamble, Ms L. R. Gander, G. G. Garbett, Mrs P. Gardam, A. Gardner, Rev. H. J. Gardner, Mrs J. Gardner, Mrs R. Garforth, W. Garforth, D. Garner, P. G. Garner, D. Garnet, Dr L. S. Garrad, J. Garside, L. W. Gaskell, P. S. Gateley, S. Gauld, P. H. Gay, Mrs G. M. Gent, Dr A. M. George, Dr D. M. George, D. Getty, P. Ghua, N. Gibbons, Mrs R. Gibbons, Dr C. Gibson, J. A. R. Gibson, Mrs L. Gibson, Miss C. J. Giddens, Dr O. Gilbert, T. Gilfoyle, Dr M. A. V. Gill, Miss B. Gillam, Dr M. E. Gillham, R. Gillibrand, Miss K. J. Gilmour, B. Goater, Mrs M. Goddard, A. Godfrey, M. Godfrey, Mrs M. Godfrey, Sqn.Ldr. M. F. Godfrey, B. Goggin, Dr J. A. Good, R. F. Goodchild, Miss R. Goode, Miss P. M. Goodhind, R. N. Goodwillie, P. R. Gordon, Miss V. Gordon, R. W. Gould, A. N. Graham, Rev. G. G. Graham, J. Graham, Mrs P. S. Graham, D. Grant, D. R. Grant, P. J. Grant, J. B. Grasse, Mrs S. Grasse, Dr E. Gray, M. Gray, Mrs V. Gray, Ms R. Grayer, D. E. Green, E. E. Green, I. P. Green, Miss J. Green, Mrs J. Green, Mrs J. A. Green, Mrs P. Green, P. R. Green, R. Green, Ms F. Greenshields, C. S. Greenway, E. F. Greenwood, Mrs M. Greenwood, P. Greenwood, B. J. Gregory, Mrs J. Gregory, Mrs M. Gregory, Ms D. E. Grey, Mrs H. E. M. Gribble, Ms A. Grierson, K. Griffin, Dr D. Griffith, Dr J. D. Griffith, R. S. L. I. Griffith, Mrs H. Grill, Dr R. H. Gritten, G. Grogan, Dr H. Grogan, R. E. Groom, S. Grover, D. Guest, J. P. Guest, Miss L. Guinan, Mrs A. Guiver, Mrs M. C. Gulliver, Dr R. L. Gulliver, J. G. Gunn, Mrs N. J. Guppy, P. G. Gutteridge.

P. Hackney, E. C. M. Haes, Dr V. Halcrow, Miss V. M. Halcrow, A. D. Hale, C. R. Hall, E. G. Hall, Dr J. Hall, Mrs J. Hall, Ms K. Hall, L. Hall, P. C. Hall, Dr G. Halliday, S. F. Halton, Dr D. J. Hambler, Dr A. C. Hamilton, Mrs J. Hammond, M. Hampton, S. Hancock, Dr A. Hannah, G. Hannah, Mrs J. Hannah, C. G. Hanson, D. A. Harding, M. Harding, R. A. Hardinge, M. Hardstaff, Mrs D. Hardy, T. Hardy, Dr A. D. R. Hare, P. A. Harmes, J. Harper, Ms J. Harral, Dr T. Harrington, Dr T. J. Harrington, Rev. B. G. Harris, G. Harris, J. Harris, Ms K. Harris, R. Harris, Miss Z. Harris, Mrs C. Harrison, D. Harrison, K. K. Harrison, W. J. Harron, D. Hart, Mrs D. Hart, Mrs J. Hartley, R. F. Hartley, S. Hartley, Mrs J. Hartwright, Mrs C. Harvey, Mrs H. G. Harvey, P. Harvey, Miss B. Harvie, T. R. Harwood, Mrs B. Haslam, D. Haslam, Miss S. M. Hastings, C. Hatch, P. Hatch, J. R. Hawell, C. J. Hawes, D. Hawker, G. Hawker, M. Hawking, Mrs C. D. Hawkins, R. D. Hawkins, J. E. Hawksford, C. C. Haworth, Ms S. Hay, W. Hay, C. Hayes, P. Hayter, Mrs D. Haythornthwaite, P. Haywood, Mrs J. Hazelhurst, Miss V. Headland, Dr A. D. Headley, Ms C. Heardman, J. J. Heath, Mrs K. Heath, Mrs P. Heathcote, P. C. Heathcote, B. Hedley, Dr S. Hedley, S. Heery, Mrs C. Helyar, Mrs R. C. Hemming, Prof D. M. Henderson, Mrs P. Henderson, Mrs M. Henriksen, Mrs E. Hesselgreaves, Miss M. C. Hewitt, R. Hewitt, M. J. Heywood, Mrs H. Heywood-Waddingham, R. J. Higgens, C. Higginbottom, Mrs J. A. Higgins, G. Higginson, Mrs M. Hignett, D. W. Hildred, A. G. Hill, Dr D. A. W. Hill, M. J. Hill, R. Hill, H. Hillier, Dr C. R. Hipkin, Mrs H. Hipkin, G. Hirons, Mrs J. Hirsh, A. G. Hoare, S. Hobbs, Mrs S. M. Hocking, M. Hodd, R. Hodd, Mrs T. Hodd, T. Hodd, J. B. S. Hodge, Mrs D. Hodgson, Dr R. M. H. Hodgson, A. J. Hodson, Miss P. Hodson, W. G. Hoff, Mrs B. G. Hogarth, Miss C. A. Hogarth, D. J. Hogarth, Mrs A. I. Hold, D. J. Holding, D. A. Holland, Mrs L. Holland, Mrs S. C. Holland, Dr P. M. Holligan, J. R. W. Hollins, D. A. Holland, Mrs L. K. Holland, Mrs L. Holloway, Miss M. E. Holmes, Dr D. T. Holyoak, L. Hooper, Miss J. Hooson, Mrs S. W. Hooton, J. R. Hope, I. J. Hopkins, Ms J. Hopkinson, Mrs V. Hopkinson, B. B. Hopton, Miss H. Horder, A. J. Horne, Miss A. Horsfall, F. Horsman, G. Hounsome, Mrs F. Houseman, Ms L. Houston, S. Howard, Dr M. J. Howe, M. Howells, J. Howitt, D. Howlett, Mrs J. Hudd, Mrs H. Hughes, Miss J. Hughes, M. Hughes, N. Hughes, Dr M. G. B. Hughes, Dr D. R. Humphreys, J. Humphreys, B. M. Humphries, Mrs J. M. Humphris, C. Hurford, C. Hutchinson, Dr G. Hutchinson, R. Hutchinson, Miss A. M. Hutchison, Mrs E. M. Hyde, Miss P. A. Hyde, R. Hyde, G. Hylands.

A. J. Iles, J. Iliff, Mrs M. Iliff, Mrs J. Ing, Dr M. J. P. Ingrouille, M. M. L. Innes, R. Irvine, Mrs V. D. Irvine, P. Irving, J. J. Ison, C. C. Ivin.

Mrs S. Jack, Dr T. A. Jacks, Mrs B. Jackson, Miss E. H. Jackson, Miss H. Jackson, Ms J. Jackson, J. K. Jackson, B. E. Jackson, Mrs E. Jackson, Mrs C. M. James, R. James, T. James, D. C. Jardine, Dr M. H. P. Jebb, N. Jee, M. B. Jeeves, Dr R. Jefferson, Miss B. Jenkins, Prof. B. Jennett, P. Jepson, A. C. Jermy, Dr R. F. John, Mrs P. Johns, Mrs C. Johnson, I. G. Johnson, Dr L. C. G. Johnson, N. P. Johnson, R. Johnson, W. L. Johnson, Mrs V. Johnston, Mrs A. Jones, A. W. Jones, D. Jones, Dr D. G. Jones, G. Jones, Ms J. Jones, Mrs J. R. Jones, Lady M. Jones, M. Jones, O. Jones, Dr P. A. Jones, Dr P. S. Jones, R. Jones, R. A. Jones, V. Jones, J. Jones, Mrs I. Joyce.

S. Karley, G. M. Kay, Dr Q. O. N. Kay, E. Kearns, Mrs O. Kearns, Mrs G. Keech, M. Keene, Frau M. Kees, Mrs M. J. Keirnen, Dr D. L. Kelly, Dr M. L. Kelly, Miss J. Kendrew, F. E. Kenington, A. G. Kenneth, I. Kennett, D. H. Kent, Mrs J. Kepl, Ms L. A. Kergon, Ms S. Kerry, Ms L. J. Kerslake, Dr R. Key, J. G. Keylock, Mrs S. Kightley, H. J. Killick, Dr J. J. King, Mrs L. King, Miss R. King, P. Kingsbury, Dr S. Kingsbury, P. K. Kinnear, P. Kirby, Dr P. Kirby, Dr J. Kirschner, Mrs C. Kitchen, M. A. R. Kitchen, G. D. Kitchener, Dr A. G. Knapp, J. Kneeshaw, Mrs L. Knight, L. Knight, P. Knight, T. D. Knight, R. Knowles, P. M. Knox, Dr P. Kohn.

Dr A. J. Lack, J. Lamb, Dr J. G. D. Lamb, Mrs A. Lambert, Miss D. S. Lambert, S. Lambert, P. W. Lambley, F. Lammiman, N. Lamming, D. A. Lang, D. C. Lang, Dr R. Langston, Mrs J. Langton, R. V. H. Lansdown, Dr D. R. Larner, Mrs B. Last, J. B. Latham, Miss S. M. Laver, A. Law, M. Lawley, Ms S. D. Lawley, Mrs J. M. Lawman, I. Lawrence, I. C. Lawrence, P. G. Lawson, Mrs J. Le Huquet, Mrs F. Le Sueur, D. Lea, Ms C. Leach, S. J. Leach, D. C. Leadbetter, A. Leak, Dr R. M. Leaney, R. Learmouth, R. Leavett, D. Ledsham, Dr A. R. Leech, Miss R. E. Lees, Mrs E. Legg, R. Leishman, A. Leitch, D. J. Leonard, Mrs Y. J. Leonard, Dr A. C. Leslie, Mrs J. F. Leslie, D. Lester, G. Levy, A. S. Lewis, D. E. Lewis, N. Lewis, R. Lewis, Mrs V. Lewis, G. Lilley, C. Lindley, Dr E. L. S. Lindsay, Ms H. Lindsay, G. Lines, Mrs O. Linford, F. Little, I. D. Little, Mrs M. E. P. Little, L. A. Livermore, Mrs P. D. Livermore, Ms M. Lloyd, J. M. Lock, L. Lock, Dr N. D. Lockhart, Mrs A. L. Locksley, Mrs A. R. Locksley, A. J. Lockton, D. Lockwood, Mrs P. Lockwood, Dr D. G. Long, Dr D. R. S. Long, M. Long, Mrs M. L. Long, Mrs S. Longrigg, Mrs M. Loukes, J. Lovell, Ms T. Lovering, Mrs P. Lovick, C. Lowe, Mrs H. Lowell, Dr J. Lowell, A. B. Loy, A. Lucas, Mrs M. J. Lucas, Mrs R. Lucas, Ms F. E. Lucy, Mrs H. Luft, P. Luke, B. Lumsden, Dr F. Lupton, P. S. Lusby, N. Lusmore, Miss R. W. Lynam, R. Lynch, J. P. Lyons, J. Lyth.

R. Mabey, I. A. Macdonald, Dr M. Macdonald, M. A. Macdonald, R. Mackechnie, Dr P. Mackie, Mrs L. MacLellan, M. MacLeod, Mrs A. C. Macpherson, Dr P. Macpherson, S. C. Madge, L. Magee, Mrs C. Mahaddie, J. Mallabar, D. Mallon, Dr J. R. Manning, Mrs B. Marcan, D. Mardon, L. J. Margetts, R. J. Marklow, Mrs A. Markwick, Miss S. K. Marner, Mrs M. Marples, P. J. Marriott, Mrs M. H. Marsden, Mrs M. V. Marsden, A. Marsh, A. D. Marshall, A. G. Marshall, Mrs H. Marshall, Mrs J. Marshall, Ms M. Marshall, T. Marshall, P. Marshman, J. Martin, J. P. Martin, Mrs K. Y. Martin, Miss L. M. E. Martin, Mrs M. E. R. Martin, Dr S. R. Martin, J. P. Martin, G. S. Marvin, R. Maskew, J. Mason, Mrs J. Mason, P. Mason, H. W. Matcham, Mrs B. Mathias, Dr D. Mathias, C. Matthews, Dr G. C. Matthews, R. Matthews, Dr P. D. L. Maurice, D. Mawer, Mrs D. Maxwell, J. F. Maxwell, S. A. Maxwell, Mrs D. Maxwell, Dr K. Maybury, R. Maycock, Mrs J. M. Maynard, Dr H. A. McAllister, R. T. McAndrew, Mrs R. McCance, Mrs W. N. McCarthy, D. McClintock, D. J. McCosh, D. E. McCutcheon, Dr A. W. McDonald, Mrs E. J. McDonnell, Dr H. N. McGough, Ms C. McHardy, J. W. McIntosh, D. R. McKean, N. D. McKee, Dr J. McLauchlin, Ms E. McLaughlin, A. McLay, G. McMichael, Ms C. McMullin, A. McNeill, D. McNeill, I. McNeill, Ms N. McNicholas, G. McQuade, Mrs C. McQuitty, Miss E. S. McTeague, A. M. McVeigh, Mrs R. Mead, W. Meagher, Dr B. Meatyard, G. R. Medcalf, T. F. Medd, E. Meek, W. R. Meek, R. D. Meikle, J. W. Meiklejohn, Dr S. L. Melville, Dr G. H. J. Meredith, Miss H. M. Meredith, O. J. Merne, D. Messenger, Dr J. Mewton, Dr E. C. Mhic Daeid, R. Middleton, Dr C. J. Miles, R. Mileto, Dr R. R. Mill, N. Millar, D. J. P. Miller, G. B. Miller, Mrs J. Miller, N. Miller, J. Milligan, Mrs P. M. Millman, Mrs D. J. Millward, B. Milne, D. Mitchell, Dr D. M. Mitchell, Dr D. N. Mitchell, J. Mitchell, W. Mitchell, Mrs M. Mobsby, P. Mobsby, Miss C. P. Mockridge, Dr B. Moffat, Ms K. Monahan, R. Monahan, J. M. Montgomery, Dr J. R. Moon, Dr E. Mooney, Ms A. Moorby, Prof. J. J. Moore, Ms L. Moore, Mrs S. R. Moore, Dr E. Moorkens, G. H. Morgan, I. K. Morgan, Miss V. Morgan, Miss M. Morris, C. M. Morrison, Miss B. Mortimer, Dr A. Morton, Miss A. Morton, Mrs P. Moses, Mrs C. Mosley, G. Moss, Mrs S. Moss, G. S. Motley, J. O. Mountford, N. J. Moyes, J. M. Mullin, A. R. G. Mundell, F. K. Murgatroyd,

Miss R. J. Murphy, C. E. Murray, Mrs C. W. Murray, Mrs J. Murray, Mrs M. Murray, Mrs M. E. Murray, Miss J. Muscott, A. W. B. Mutch.

Mrs M. Nagle, Mrs M. Nalder, Mrs D. G. Nash, Dr D. W. Nash, Miss S. Nash, Miss C. Neil, Dr E. C. Nelson, Dr S. E. Nelson, Mrs W. Nelson, J. A. Newbould, Mrs J. P. Newbould, Prof. P. J. Newbould, J. Newman, A. Newton, A. L. Newton, Mrs J. Newton, Ms J. Newton, Dr P. Newton, C. Nicholas, Mrs R. A. Nicholson, Ms R. Nickerson, D. J. Nicolle, Miss V. Nixon, Mrs B. A. Noaks, R. Noble, Mrs E. Norman, Mrs J. Norris, Mrs H. J. Northridge, R. H. Northridge, J. A. Norton, Miss M. P. Norton, Mrs L. Nottage, E. Nugent.

M. V. O'Brien, Dr M. O'Connor, R. O'Connor, Dr C. Ó Críodáin, Mrs E. O'Donnell, S. O'Donnell, T. O'Donoghue, Dr G. O Donovan, Dr M. A. Ogilvie, Mrs A. Ohlenschlager, M. O'Leary, Dr J. E. Oliver, Mrs M. Oliver, T. O'Mahony, J. L. O'Malley, Dr H. O'Reilly, G. Osmond, Mrs J. Ostler, Dr A. O'Sullivan, P. H. Oswald, Mrs J. Ottery, J. Ounsted, Mrs M. Overton, Mrs B. J. Ozanne.

K. W. Page, Mrs M. Page, D. Paice, J. R. Palmer, R. C. Palmer, Dr A. F. Pankhurst, Dr R. J. Pankhurst, T. Pankhurst, Mrs J. Pape, A. Parker, Mrs J. Parker, Mrs P. Parker, R. Parker, R. M. Parker, S. Parker, S. J. Parker, Dr J. A. N. Parnell, P. J. Parr, Dr S. Parr, Miss A. F. Parry, Mrs R. E. Parslow, Dr J. W. Partridge, Mrs S. Pashby, Mrs K. Patrick, B. R. Pawson, K. Payne, R. Payne, R. M. Payne, S. R. Payne, Mrs O. Peacock, Mrs E. M. Pearce, Mrs A. V. Pearman, D. A. Pearman, G. Peck, Mrs H. Peddle, L. Peddle, Mrs J. Pedlow, W. M. Peet, Mrs A. M. Pell, Mrs V. Pennell, R. A. Pennington, D. W. Perkins, Dr F. H. Perring, Dr L. Perrins, A. R. Perry, Dr C. L. Perry, Dr M. E. R. Perry, Mrs S. Perry, A. Phillips, B. W. Phillips, J. Phillips, Mrs P. Phillips, Ms P. M. Phillips, S. Phillips, E. G. Philp, Dr A. Pickering, Dr C. D. Pigott, N. Pike, R. S. Pile, Miss E. A. Pilling, R. Piper, S. Plant, Ms A. Plumb, R. Plummer, Miss A. P. Pockson, J. Poingdestre, Mrs J. A. Poll, Mrs J. Pollard, T. D. Pollard, Ms K. Pollock, Dr R. Pollock, M. D. Pool, Dr D. Poore, Dr C. R. Pope, P. J. Pope, P. S. Pope, R. D. Porley, Mrs C. Port, P. J. Port, Mrs C. Porter, Dr D. Porter, M. Porter, M. S. Porter, P. Porter, Miss A. De Potier, D. Poulter, Dr M. Powell, M. Powell, Dr A. H. Powling, Mrs S. Poyser, Rev. E. A. Pratt, Mrs M. Pratt, Dr T. Prescott, J. Presland, Dr C. D. Preston, Mrs M. Preston, W. V. Prestwood, D. T. Price, S. Price, W. J. H. Price, Rev. Dr A. L. Primavesi, Dr M. C. F. Proctor, B. Proctor, Miss H. M. Proctor, R. D. Pryce, Dr J. Puckett, Mrs B. Pullen, G. A. Pyman, K. Pyne, T. Pyner, M. Quirke.

Miss R. D. Rabey, I. Rabjohns, Mrs R. Race, Dr O. Rackham, Dr E. Radford, S. G. Rae, K. Raistrick, I. L. Ralphs, G. Randall, R. D. Randall, R. E. Randall, Dr J. C. A. Rathmell, J. N. Rayner, Mrs P. M. Reason, B. Redman, D. Redmore, E. J. Redshaw, D. K. Reed, Miss A. Rees, J. Rees, Mrs J. Rees, Dr J. S. Rees, Dr L. J. Rees, A. W. Reid, Dr E. Reid, Mrs M. A. Reid, G. Reilly, J. Reilly, Mrs M. Reilly, P. Reilly, Mrs J. Reynolds, Dr J. D. Reynolds, P. Reynolds, Mrs S. Reynolds, Mrs R. Rhodes, B. Ribbands, B. W. Ribbons, Dr T. C. G. Rich, Dr A. J. Richards, Ms G. M. Richards, Ms M. Richards, J. B. Richardson, Mrs R. Richens, D. I. Richmond, Mrs R. M. Richmond, N. Riddiford, Dr R. Riddington, Mrs G. Ridgway, Dr D. H. Riley, Mrs V. Riley, I. Rippey, Mrs S. A. Rippin, Mrs I. Ritchie, K. Rivett, Mrs M. Rivett, J. I. Robbins, Mrs B. M. Roberts, Mrs E. Roberts, F. J. Roberts, H. A. Roberts, J. R. Roberts, Mrs L. A. Roberts, R. H. Roberts, R. R. Roberts, D. Robertson, Mrs J. Robertson, K. Robertson, Mrs M. Robertson, Mrs W. Robertson, Mrs J. Robinson, Miss K. Robinson, K. J. Robinson, Mrs L. Robinson, A. W. Robson, Dr C. M. Roden, Capt. R. G. B. Roe, E. V. Rogers, P. H. Rollinson, C. C. Romer, Mrs C. Roper, Mrs J. Roper, Dr F. Rose, Dr H. Rose, I. C. Rose, J. Roskell, G. P. Rothero, Miss F. Rout, J. Rowe, M. W. Rowe, R. Rowlands, N. M. Rumens, Dr F. J. Rumsey, J. Ruppersbery, J. S. Russell, N. Russell, T. Russell, Miss A. Rutherford, T. Ryall, C. Ryan, Mrs P. I. Ryan.

Mrs M. Saag, G. Salisbury, B. Samson, N. A. Sanderson, M. N. Sanford, Miss J. Saunders, R. E. Saville, M. Sayce, Mrs A. J. Sayle, Ms M. J. P. Scannell, Mrs S. C. Schofield, Mrs F. E. Schumann, D. J. Scott, Mrs G. E. Scott, J. F. Scott, M. Scott, Mrs P. Scott, Ms R. Scott, Mrs V. M. Scott, W. Scott, Ms S. Scott-White, C. E. K. Scouller, R. W. H. Scroggs, Mrs H. Sealy-Lewis, Miss E. J. Searle, J. Secker, Ms A. Seddon, S. Segal, P. J. Selby, P. D. Sell, Mrs R. J. Sells, W. Semple, M. J. Senior, Mrs H. Sergeant, G. Sharkey, Mrs M. L. Shattock, A. I. Shaw, B. T. Shaw, D. Shaw, Dr M. C. Sheahan, Miss A. Shearcroft, D. H. D. Shearer, Dr M. Sheehy Skeffington, R. Sheppard, M. Sherlock, Mrs N. E. Sherlock, R. Sherlock, N. Shilton, R. Shoubridge, Dr A. J. Showler, Dr A. J. Silverside, Mrs S. Simkin, Mrs O. Simmonds, Mrs F. Simper, I. Simper, B. Simpson, G. Simpson, R. Singleton, Miss C. Sisam, Miss A. Skrimshire, A. A. Slack, Mrs L. E. Slack, G. R. Sloman, C. Small, I. S. Small, Miss G. J. B. Smart, P. Smiddy, Ms A. Smith, A. Smith, A. W. Smith, Miss B. Smith, Mrs C. J. Smith, Miss C. N. Smith, D. Smith, D. P. J. Smith, G. G. Smith, G. P. Smith, Mrs J. E. Smith, L. Smith, L. G. Smith, M. Smith, Mrs M. Smith, Dr M. E. B. Smith, P. A. Smith, Mrs P. E. Smith, Dr P. H. Smith, R. Smith, Dr R. A. H. Smith, R. E. N. Smith, S. Smith, T. Smith, Mrs D. Snaddon, Mrs P. Sneyd, M. J. Southam, J. Southey, J. F. Southey, L. M. Spalton, A. Spiers, Dr B. M. Spooner, K. L. Spurgin, Prof. C. A. Stace, Miss H. E. Stace, Mrs J. Stafford, Mrs S. Stafford, Mrs E. Stancer, J. M. Stanley, P. D. Stanley, Mrs M. M. W. Stapleton, Dr P. G. Stapleton, M. O. Stead, P. Stebbings, D. Steeden, J. Steeden, Miss L. Steel, G. Steele, Mrs E. Stenger, M. Stephens, M. L. Stephens, Mrs E. M. Stephenson, K. Stephenson, R. C. Stern, G. Steven, Dr D. P. Stevens, Dr R. A. Stevens, C. R. Stevenson, Ms A. Stewart, B. Stewart, Mrs E. W. Stewart, F. J. Stewart, Dr N. F. Stewart, Mrs O. M. Stewart, Mrs R. F. Stewart, Ms S. E. Stille, A. McG. Stirling, J. Stobart, Mrs J. Stobart, Mrs J. H. Stobbs, Miss M. Stockes, R. M. Stokes, J. M. Stone, Mrs S. J. Stone, M. Storey, Dr M. W. Storey, Dr I. Strachan, L. Street, Dr R. N. Stringer, Dr D. M. Stroud, F. Strudwick, Mrs M. Strudwick, Miss B. M. Sturdy, Mrs E. Sturt, Dr N. J. H. Sturt, Mrs R. Suddaby, Mrs A. Summers, Dr B. Sumner, M. Sutcliffe, Mrs S. J. Sutcliffe, M. D. Sutton, Dr E. Swale, Dr M. Swan, Prof G. A. Swan, Mrs E. Sweeney, P. Sweeney, Mrs P. Swettenham, R. J. Swindells, Dr J. B. Sykes, Mrs M. Sykes, Mrs N. Sykes, T. Sykes, D. M. Synnott.

J. Taggart, R. Takagi-Arigho, Mrs T. Tarpey, R. Tateson, Mrs J. Tavender, R. W. Tavender, A. Taylor, Mrs H. C. Taylor, J. Taylor, Ms J. Taylor, N. Taylor, N. W. Taylor, T. Taylor, W. G. Teagle, Mrs M. O. Tebble, Ms T. Teearu, Mrs M. I. Tempest, D. J. Tennant, I. Tew, J. A. Thickitt, I. R. Thirlwell, Ms C. Thomas, D. Thomas, D. M. Thomas, I. F. Thomas, Mrs J. P. H. Thomas, Dr R. E. Thomas, S. Thomas, B. H. Thompson, Mrs H. Thompson, J. A. Thompson, W. A. Thompson, B. Thomson, I. M. Thomson, P. Thomson, Mrs S. E. Thomson, Mrs K. Thorley, Dr A. K. Thorne, Mrs G. A. C. Thornton, Dr B. P. Thurlow, Miss J. Thurston, M. Tickner, M. Tickner, Ms S. Timms, D. J. Tinston, P. Tipping, Lady N. Tirard, C. Titcombe, R. Tofts, Miss M. P. G. Tolfree, Mrs E. Tolson, J. Tolson, Mrs P. E. Tompsett, A. Toomey, G. Toone, Comd. J. M. W. Topp, Miss T. Treacy, B. A. Tregale, K. Trewren, P. J. O. Trist, M. J. Trotman, S. Trotter, M. Troy, Prof. I. C. Trueman, Mrs K. Tucker, W. H. Tucker, Mrs M. Tuffs, Mrs M. Tulloh, Dr N. Turnbull, Dr C. Turner, Dr J. Turner, Mrs J. M. Turner, C. Tyas, Mrs J. Tyler, M. W. Tyler, Dr S. J. Tyler, P. J. Tymkow.

Mrs M. Uí Chonchubhair, A. Underhill, D. J. Upton, Dr M. B. Usher.

A. H. Vaughan, Dr R. M. Veall, J. D. R. Vernon, Miss B. J. Villiers, D. L. Vinall, Dr G. Visser.

Miss T. M. Wabeke, J. Waddell, N. Waddingham, W. Wagstaff, Mrs M. Wainwright, Dr S. J. Wainwright, A. J. Wake, J. R. Wakely, J. C. Wakerley, Dr A. Walker, Mrs A. Walker, A. G. Walker, C. Walker, P. E. G. Walker, Mrs J. Wall, Mrs A. Wallace, E. Wallace, I. Wallace, R. M. Walls, Dr W. G. Walsh, J. Walshe, Dr S. M. Walters, J. Wann, J. D. Wann, Dr S. Ward, R. H. Wardell, G. Waterhouse, R. A. Waterman, T. Waterman, Mrs B. Watkinson, C. Watson, C. P. Watson, Dr G. K Watson, K. Watson, K. J. Watson, Mrs S. Watson, Dr A. M. Watt, Dr D. Watts, Prof. W. A. Watts, Dr M. Way, Mrs C. Weaver, G. Weaver, Prof. D. A. Webb, F. H. Webb, Mrs J. Webb, B. Webster, R. B. Weddle, Dr D. A. Weir, Dr D. Welch, D. A. Wells, Mrs S. E. Wells, T. C. E. Wells, Miss C. Welsh, Mrs J. Welsh, S. Westerberg, C. Westhall, Mrs I. Weston, Mrs W. Weston, B. Westwood, Ms K. S. Westwood, Dr R. S. Weyl, Mrs J. Whatmough, G. S. Wheeldon, Mrs G. W. Wheeldon, Ms F. Wheeler, Mrs A. White, Mrs A. E. White, Mrs P. H. White, R. Whitehead, S. Whitworth, Ms S. J. Whild, M. Wilcox, C. E. Wild, J. Wilde, M. F. Wildish, Mrs V. Wilkin, Ms A. S. Wilkinson, Miss L. K. Wilkinson, A. Williams, Mrs D. Williams, G. Williams, J. C. Williams, L. R. Williams, P. Williams, S. Williams, Mrs V. Williams, Mrs V. A. Williams, Miss B. Williamson, Mrs H. Williamson, J. Williamson, Ms D. Willment, G. T. D. Wilmore, Dr A. Willmot, C. Wilson, Mrs D. Wilson, Mrs M. Wilson, Dr P. J. Wilson, R. Wilson, Miss T. Wilson, Mrs E. Wiltshire, J. Winham, P. Wisniewski, R. Wistow, A. E. Wohlgemuth, J. Wohlgemuth, Ms A. S. Wolfe, S. A. Wolfe-Murphy, L. J. Wolstenholme, A. Wood, D. C. Wood, D. J. Wood, Mrs E. G. Wood, Mrs H. Wood, J. A. Wood, Dr S. R. J. Woodell, Mrs J. Woodgate, B. E. Woodhams, Mrs F. A. Woodhead, Miss B. J. Woodliff,

J. P. Woodman, Dr P. M. Woodruffe, A. Woods, Ms J. A. Woods, M. Woods, R. G. Woods, Ms E. Woodward, Mrs J. M. Woodward, S. F. Woodward, D. H. Woolrich, Mrs E. M. Woolrich, G. Worrall, Mrs J. Wort, P. C. H. Wortham, Miss R. Wright, B. S. Wurzell, A. Wylie, Dr G. Wynne, Mrs J. Wynne-Jones, Mrs D. Wyse Jackson, Mrs L. M. Wyse Jackson, Dr M. B. Wyse Jackson, Dr P. S. Wyse Jackson.

M. J. Yates, Mrs P. Yates, Miss S. Yates, C. S. V. Yeates, S. J. Youell, Mrs A. Young, Mrs D. M. Young, G. Young, Miss M. E. Young, R. Youngman, Mrs M. P. Yule.

Dr J. A. Zoer, Dr B. Zonfrillo.

We would also like to acknowledge the following organisations for their assistance in making records available, allowing access to property, nature reserves or herbaria or for helping in other ways.

Armagh Field Naturalists' Society
Arnside and District Natural History Society
Ashmolean Natural History Society of Oxfordshire
Avon Flora Project
Bedfordshire Flora Group
Belfast Naturalists' Field Club
Berkshire, Buckinghamshire & Oxfordshire Wildlife Trust
Beverley Naturalists
Bord Na Mona, Boora, Co. Offaly
Botanical Society of Scotland (Botany of the Lothians Survey)
Botany Section, La Société Guernesiaise
Botany Section, La Société Jersiaise
Bradford Botany Group, West Yorkshire
Bristol Naturalists' Society
Bristol Region Environmental Records Centre
Buckinghamshire Environmental Records Centre
Cambridgeshire Flora Group
Cambridgeshire Wildlife Trust
Cardiff Naturalists
Centre for Ecology and Hydrology (Bangor)
Central Area Recording System for the Environment
Central Environmental Surveys
Central Fisheries Board, Dublin
Cheshire Flora Group
Cheshire Wildlife Trust
Cleveland Naturalists' Field Club
Colchester Natural History Society
Cornwall Wildlife Trust
Coventry and District Natural History Society
Countryside Council for Wales
Darlington and Teesdale Naturalists'Field Club
Dartmoor National Park Authority
Department of Agriculture and Regional Development, Northern Ireland
Department of Biological Sciences, Polytechnic South West, Plymouth
Department of Plant Sciences,
University of Cambridge
Derby Museum & Art Gallery (Derbyshire BRC)
Derby Natural History Society
Derbyshire Wildlife Trust
Devon Wildlife Trust
Devonshire Association (Botany Section)
Doncaster Museum Service
Dorset Environmental Records Centre
Dublin Naturalists' Field Club
Dúchas, the Heritage Service
Durham Wildlife Trust
East Norfolk Flora Group
East Yorkshire Botany Group
Ecological Advisory Service, Bradford Met. Council
Edinburgh Natural History Society
Elan Valley Estate, Radnorshire
English Nature
Environment and Heritage Service
Essex Field Club
Exmoor Natural History Society
Fife Nature Biological Records Centre
Fife Ranger Service
Forestry Commission
Frieth Natural History Society
Glamorgan Wildlife Trust
Glasgow Natural History Society
Gloucestershire Flora Committee
Gloucestershire Environmental Records Centre
Gloucestershire Naturalists Society
Habitat Survey Team, Environment and Heritage Service (Northern Ireland)
Hamilton Natural History Society
Hampshire Biological Record at Hampshire County Council
Hampshire Wildlife Trust (Flora Group)
Helensburgh and District Natural History Society
Herefordshire Botanical Society
Hertfordshire Biological Records Centre
Hertfordshire Natural History Society
Hertfordshire Outdoors
High Batts Nature Reserve
Hull Natural History Society
Huntingdonshire Fauna and Flora Society
Inverness Botany Group
Islay Natural History Trust
Isle of Wight Natural History & Archaeological Society
Isles of Scilly Enviromental Trust
Joint Nature Conservation Committee
Kettering & District Natural History Society
Kingswood Abbey Women's Institute
Lancashire County Coucil
Lancashire Wildlife Trust
Leverhulme Trust
Lincolnshire Naturalists Union
Lincolnshire Wildlife Trust
Llanelli Naturalists
London Ecology Unit
London Natural History Society
Lothian Wildlife Information Centre (Scottish Wildlife Trust)
Lyme Natural History Recording Group
Mar Lodge Ranger Service
Mid Antrim Field Club
Mid-Somerset Natural History Society
Milton Keynes Natural History Society
Aiuthorypack Society
Montgomeryshire Wildlife Trust
Mournes Advisory Council
National Botanic Gardens, Glasnevin
National Museums and Galleries of Wales
National Trust
Natural History Museum, London
Natural History Society of Northumbria
Naturebase at Dundee City Museum & Art Gallery
Norfolk and Norwich Naturalists' Society
Norfolk Wildlife Trust
North Hertfordshire Museums Service
North Wales Wildlife Trust
Northants Flora Group
Northants Wildlife Trust
Northern Naturalists' Union
Northumberland Wildlife Trust
Nottingham Biological & Geological Records Centre
Nottinghamshire Wildlife Trust
Orkney Biodiversity Records Centre

Orkney Field Club
Pembrokeshire Coast National Park Authority
Perthshire Society of Natural Science, Botanical Section
Plantlife
Pond Life Project
Potteries Museum & Art Gallery, Stoke-on-Trent, Staffordshire
Radnorshire Wildlife Trust
Royal Botanic Garden, Edinburgh
Royal Botanic Gardens, Kew
Royal Society for the Protection of Birds
Scottish Borders Biological Record Centre
Scottish Natural Heritage
Scottish Wildlife Trust
Shared Earth Trust, Ceredigion
Shropshire Botanical Society
Sir Tatton Sykes, Sledmere Estate
Snowdon National Park Authority
Somerset Rare Plant Group
Sorby Natural History Society, Sheffield
South Holderness Countryside Society
South Somerset Natural History Society
Southampton Natural History Society
Staffordshire Wildlife Trust
Suffolk Biological Records Centre
Suffolk Naturalists' Society
Suffolk Wildlife Trust
Surrey Flora Committee
Sussex Botanical Recording Society
Swarthmore Botany Club
Tees Valley Wildlife Trust
Trinity College, Dublin
Tyne and Wear Museums
U3A Botany Group, Cambridge
Ulster Museum
University of Birmingham
University of Derby
Warton Botany Group
Warwickshire Flora 2000 Recorder Team
Warwickshire Habitat Biodiversity Audit
West Norfolk Flora Group
Wild By Design
Wildlife Trust West Wales
Wiltshire Archaeological & Natural History Society
Wiltshire Botanical Society
Wiltshire Flora Mapping Project
Wiltshire Wildlife Trust
Worcestershire Flora Project
Worcestershire Wildlife Trust
Yeovil and District Natural History Society
Yoredale Natural History Society
Yorkshire Naturalists' Union

CHAPTER 10

Glossary

The following glossary includes words that may be unfamiliar, or familiar words used in this Atlas in an unusually restricted sense. The words cited relate to the main themes of the introductory material or species accounts. We have made no attempt to provide a comprehensive glossary to the ecological or other terms used in the species accounts, as these are included in standard reference works.

10-km square: a 10 × 10 km grid square of the British and Irish national grids or the UTM grid (Channel Islands), used as the basic unit for distribution mapping in this Atlas (see p. 47). 10-km squares are sometimes simply referred to as 'squares' in the text.

1962 Atlas: *Atlas of the British Flora* (Perring & Walters, 1962).

Aggregate (abbreviated to **agg.**): a group of closely related *taxa* which are treated together for recording purposes because the component *segregates* cannot always be identified.

Alien: a plant which was introduced to our area by man, or arrived naturally from an area in which it was present as an alien (see p. 10; cf *native*). A synonym of *introduction*. Alien plants can be divided into *archaeophytes* and *neophytes*.

Archaeophyte: a plant which was introduced to our area by man (or arrived naturally from an area in which it was present as an introduction) and became naturalised before AD 1500 (see p. 10; cf *neophyte*).

Arctic-montane: a *major biome category* which includes plants occurring north of the tree-line or (on mountains) above the tree-line, or both (see pp. 27, 29).

Atlas: the abbreviation used in the References paragraph of the species accounts to refer to the *Atlas of the British Flora* (Perring & Walters, 1962).

Atlas Supp.: the abbreviation used in the References paragraph of the species accounts to refer to the *Critical Supplement to the Atlas of the British Flora* (Perring & Sell, 1968).

BRC: Biological Records Centre (see Table 1.1, p. 2).

BSBI: Botanical Society of the British Isles (see Table 1.1, p. 2).

Biome: an area defined by its potential vegetation type, which is ultimately controlled by climate. North of the tropics, the major biomes form latitudinal belts (*zonobiomes*). However, these patterns are complicated by the presence of mountains, as increasing altitude has a similar effect to increasing latitude, leading to the presence (for example) of vegetation characteristic of the Boreal biome at high altitudes in the Temperate biome.

Boreal-montane: a *major biome category* which includes plants occurring in the coniferous forest zone, either in the Boreal *zonobiome* or on mountains further south, or both (see pp. 27, 29).

Boreo-arctic Montane: a *major biome category* which includes plants occurring in both the Arctic-montane and the Boreal-montane biomes (see pp. 27, 29).

Boreo-temperate: a *major biome category* which includes plants occurring in both the Boreal-montane and the Temperate biomes (see pp. 27, 29).

CCW: Countryside Council for Wales (see Table 1.1, p. 2).

CEH: Centre for Ecology and Hydrology (see Table 1.1, p. 2).

Casual: used to describe an alien plant (or a population of such plants) which persists in an area for only a brief period, at most approximately five years (see p. 11).

Change index: an index of relative changes in the British range size (as measured by the number of occupied *10-km squares*) of species mapped in the 1962 Atlas and in this Atlas (see Chapter 7). In the species accounts, the change index is cited as '(change . . .)'.

Circumpolar: an *eastern limit category* which includes plants which are found in Europe, Asia and North America (see pp. 27, 29).

Critical: used to describe a plant which is difficult to identify because the distinctions between it and related *taxa* are slight, or are obscured by phenotypic variation, hybridisation or other factors. Certain identification of critical taxa requires experience and expertise, and is sometimes impossible.

Critical Supplement: *Critical Supplement to the Atlas of the British Flora* (Perring & Sell, 1968).

Cultivar: a variant of a cultivated plant. Cultivars may be given special names, which are cited within inverted commas, e.g. *Brachyglottis* 'Sunshine' (p. 658).

DEFRA: Department for Environment, Food and Rural Affairs (see Table 1.1, p. 2).

EHS: Environment and Heritage Service (see Table 1.1, p. 2).

EN: English Nature (see Table 1.1, p. 2).

Eastern limit category: in the classification devised by Preston & Hill (1997), this is one component of the *floristic element*, the other being the *major biome category*. Plants are classified into eastern limit categories on the basis of their longitudinal distribution (see p. 27).

Established: used to describe a population of alien plants which has persisted in an area for at least five years, spreading vegetatively or reproducing by seed (see p. 11). A synonym of *naturalised*.

Eurasian: an *eastern limit category* which includes plants which reach their eastern limit east of 120°E (see pp. 27, 29).

European: an *eastern limit category* which includes plants with a mainly European distribution; they may extend east to the Caucasus, Pontic Asia and the Middle East but do not occur east of 60°E (see pp. 27, 29).

Eurosiberian: an *eastern limit category* which includes plants which reach their eastern limit between 60°E and 120°E (see pp. 27, 29).

Flora: used in three senses: (1) as a collective name for the plants occurring in a specific geographical area; (2) for a book designed to enable the user to identify the plants in a particular area; (3) for a book which lists the plants in a particular area and details their habitats and localities (as in 'county Flora' or 'local Flora'). Some authors distinguish the last two usages from the first by an initial capital F.

Floristic element: a group of *taxa* with similar geographical distributions at the present day in a defined area. The floristic elements cited in this Atlas are based on those defined by Preston & Hill (1997), which classify species by the *major biomes* in which they occur and their *eastern limits* (see p. 27).

Frequency: the proportion of *10-km squares* occupied by a species within its *range*.

Hyperoceanic: describes a distribution which is markedly western within the *Oceanic* zone.

Infraspecific taxon: any *taxon* below the rank of species (e.g. subspecies, variety).

Introduction: a plant which was introduced to our area by man, or arrived naturally from an area in which it was present as an introduction (see p. 10; cf *native*). A synonym of *alien*. Introductions can be divided into *archaeophytes* and *neophytes*.

JNCC: Joint Nature Conservation Committee (see Table 1.1, p. 2).

Lowland: used in the species accounts to describe a plant occurring at altitudes between sea-level and 300 m (cf *montane*, *upland*).

Major biome category: in the classification devised by Preston & Hill (1997), this is one component of the *floristic element*, the other being the *eastern limit category*. The major biome category describes the biome(s) in which the plant occurs (see p. 27).

Mediterranean-Atlantic: a *floristic element* which includes plants which are confined to the Southern biome in eastern Europe but extend into the Temperate biome in western Europe, and thus tend to occur along the Mediterranean and Atlantic seaboards (see p. 27).

Mediterranean-montane: a *floristic element* which includes plants which in Europe have a montane distribution in the Southern biome but occur at low altitudes in Temperate regions (see p. 27).

Montane: used in the descriptions of the altitudinal range of species to indicate altitudes above 600 m (cf *lowland*, *upland*).

NERC: Natural Environment Research Council (see Table 1.1, p. 2).

Native: a plant which arrived in our area without the intentional or unintentional intervention of man (see p. 10; cf *alien*, *introduced*).

Naturalised: used to describe a population of alien plants which has persisted in an area for at least five years and is spreading vegetatively or reproducing by seed. A synonym of *established* (see p. 11).

Neophyte: a plant which was introduced to our area by man (or arrived naturally from an area in which it was present as an introduction) and became naturalised after AD 1500 (see p. 10; cf *archaeophyte*).

Oceanic: an *eastern limit category* which includes plants occurring in the Atlantic zone of Europe; they do not (or only just) extend eastwards to Germany and have an western distribution in Scandinavia (see pp. 27, 29).

RSPB: Royal Society for the Protection of Birds (see Table 1.1, p. 2).

Range: the total geographical spread of a species.

Rare: in Britain, used both in the general sense and (usually as 'nationally rare') in a more restricted sense to describe plant species recorded recently in 15 or fewer 10-km squares (cf *scarce*).

SNH: Scottish Natural Heritage (see Table 1.1, p. 2).

Scarce: in Britain, used both in the general sense and (usually as 'nationally scarce') in a more restricted sense to describe plant species recorded recently in 16-100 10-km squares (cf *rare*).

Segregate: an individual member of an *aggregate.*

Sensu lato (abbreviated to ***sens. lat.***): applied to a *taxon* which is interpreted 'in the broad sense', i.e. as including two or more closely related taxa (cf *sensu stricto*). Thus *Pyrus communis sens. lat.*, mapped on p. 357, includes *P. communis sens. str.* and the closely related *P. pyraster*.

Sensu stricto (abbreviated to ***sens. str.***): applied to a *taxon* which is interpreted 'in the narrow sense', i.e. as excluding closely related taxa (cf *sensu lato*).

Southern: a *major biome category* which includes plants occurring in the warm-temperate zone south of the broad-leaved deciduous forest zone (see pp. 27, 29). In Europe the Southern zone is the Mediterranean region.

Southern-temperate: a *major biome category* which includes plants occurring in both the Temperate and the Southern zones (see pp. 27, 29).

Spontaneous hybrid: a hybrid which arises in our area by natural cross-pollination, rather than a hybrid which is introduced or one which arises by deliberate pollination by man.

Submediterranean-subatlantic: a *floristic element* which includes plants which are widespread in the Southern biome in eastern Europe and the Southern and Temperate biome in western Europe. Members of this floristic element have more extensive distributions than the species in the Mediterranean-Atlantic element (see p. 27).

Suboceanic: an *eastern limit category* which includes plants with a predominantly western distribution in Europe; their eastern limits are reached in the western Mediterranean area, western Central Europe and Sweden (see pp. 27, 29).

Taxon (plural **taxa**): any rank of the taxonomic hierarchy (e.g. family, species, subspecies), although generally used in this Atlas (as by many other authors) as a collective name for species and infraspecific taxa.

Temperate: a *major biome category* which includes plants occurring in the cool-temperate, broad-leaved deciduous forest zone (see pp. 27, 29).

Tetrad: one of the twenty-five 2 × 2 km grid squares in a *10-km square*. Tetrads are often used as mapping units for county or local Floras (see Chapter 2).

Upland: used in descriptions of the altitudinal range of species to indicate altitudes between 300 m and 600 m (cf *lowland, montane*).

Vice-county: an area used for biological recording purposes. The vice-counties are based on former administrative counties, but their boundaries are never changed. For a map and list of the vice-counties, see Chapter 8.

Vice-county Recorder: an honorary official of the Botanical Society of the British Isles charged with compiling plant records from a *vice-county.*

WFS: Wild Flower Society (see Table 1.1, p. 2).

Wide-boreal: A *major biome category* including plants which occur in the Arctic-montane, Boreal-montane and Temperate zones (see pp. 27, 29).

Wide-temperate: A *major biome category* including plants which occur in the Boreal-montane, Temperate and Southern zones (see pp. 27, 29).

Wild-flower mixture: a mixture of seeds, often containing (or supposedly containing) only seeds of native provenance, which is sold for sowing in gardens, on roadside verges, in country parks, on field edges or elsewhere.

Wool-shoddy alien: a plant brought to our area as a seed attached to wool of foreign origin and found in the wild when waste wool ('shoddy') is spread on fields as a soil conditioner (a common practice in some areas in the 1940s and 1950s). See Lousley (1961) for further details.

Zonobiome: see *biome.*

CHAPTER 11
Bibliography

References cited in the CD-ROM as well as those cited in the text are included in this bibliography.

Aarssen, L. W. 1981. The biology of Canadian weeds. 50. *Hypochoeris radicata* L. *Canadian Journal of Plant Science*, 61: 365–381.

Aarssen, L. W., Hall, I. V. & Jensen, J. I. N. 1986. The biology of Canadian weeds. 76. *Vicia angustifolia* L., *V. cracca* L., *V. sativa* L., *V. tetrasperma* (L.) Schreb. and *V. villosa* Roth. *Canadian Journal of Plant Science*, 66: 711–737.

Abbott, R. J., James, J. K., Irwin, J. A. & Comes, H. P. 2000. Hybrid origin of the Oxford Ragwort, *Senecio squalidus* L. *Watsonia*, 23: 123–138.

Abraham, F. & Rose, F. 2000. Large-leaved limes on the South Downs. *British Wildlife*, 12: 86–90.

Adam, P. 1990. *Saltmarsh ecology.* Cambridge: Cambridge University Press.

Adams, A. W. 1955. Biological Flora of the British Isles. No. 52. *Succisa pratensis* Moench. *Journal of Ecology*, 43: 709–718.

Agnew, A. D. Q. 1961. The ecology of *Juncus effusus* in North Wales. *Journal of Ecology*, 49: 83–102.

Agnew, A. D. Q. 1968. The interspecific relationships of *Juncus effusus* and *J. conglomeratus* in Britain. *Watsonia*, 6: 377–388.

Aiken, S. G., Newroth, P. R. & Wile, I. 1979. The biology of Canadian weeds. 34. *Myriophyllum spicatum* L. *Canadian Journal of Plant Science*, 59: 201–215.

Aiton, J. 1789. *Hortus Kewensis.* 3 vols. London: G. Nicol.

Akeroyd, J. R. 1984. *Parapholis incurva* (L.) C. E. Hubbard – a grass overlooked in Ireland. *Irish Naturalists' Journal*, 21: 228–230.

Akeroyd, J. R. 1991. *Anthyllis vulneraria* L. subsp. *polyphylla* (DC.) Nyman, an alien kidney-vetch in Britain. *Watsonia*, 18: 401–403.

Akeroyd, J. R. 1993. The distribution and status of *Rumex pulcher* L. in Ireland. *Irish Naturalists' Journal*, 24: 284–285.

Akeroyd, J. [R.] (ed.). 1996a. *The wild plants of Sherkin, Cape Clear and adjacent islands of West Cork.* Sherkin Island: Sherkin Island Marine Station.

Akeroyd, J. R. 1996b. Coastal ecotypic variants of two vetches, *Vicia sepium* L. and *V. sylvatica* L. (Fabaceae), in Britain and Ireland. *Watsonia*, 21: 71–78.

Akeroyd, J. R. & Beckett, G. 1995. *Petrorhagia prolifera* (L.) P. W. Ball & Heywood (Caryophyllaceae), an overlooked native species in eastern England. *Watsonia*, 20: 405–407.

Akeroyd, J. R. & Briggs, D. 1983a. Genecological studies of *Rumex crispus* L. I. Garden experiments using transplanted material. *New Phytologist*, 94: 309–323.

Akeroyd, J. R. & Briggs, D. 1983b. Genecological studies of *Rumex crispus* L. II. Variation in plants grown from wild-collected seed. *New Phytologist*, 94: 325–343.

Akeroyd, J. R. & Clarke, K. 1993. *Dianthus armeria* L. new to Ireland and other rare plants in West Cork. *Watsonia*, 19: 185–187.

Akeroyd, J. R. & Doogue, D. 1988. *Plantago major* L. subsp. *intermedia* (DC.) Arcangeli (Plantaginaceae) in Ireland. *Irish Naturalists' Journal*, 22: 441–443.

Akeroyd, J. R. & Preston, C. D. 1984. *Halimione portulacoides* (L.) Aellen on coastal rocks and cliffs. *Watsonia*, 15: 95–103.

Akeroyd, J. R., Warwick, S. L. & Briggs, D. 1978. Variations in four populations of *Senecio viscosus* L. as revealed by a cultivation experiment. *New Phytologist*, 81: 391–400.

Aksoy, A., Dixon, J. M. & Hale, W. H. G. 1998. Biological Flora of the British Isles. No. 199. *Capsella bursa-pastoris* (L.) Medikus (*Thlaspi bursa-pastoris* L., *Bursa bursa-pastoris* (L.) Shull, *Bursa pastoris* (L.) Weber). *Journal of Ecology*, 86: 171–186.

Al-Bermani, A.-K. K. A. 1991. *Taxonomic, cytogenetic and breeding relationships of* Festuca rubra sensu lato. Ph.D. thesis, University of Leicester.

Al-Bermani, A.-K. K. A. & Stace, C. A. 1991. A new subspecies of *Festuca rubra* L. *Watsonia*, 18: 315–316.

Allan, B. & Woods, P. 1993. *Wild orchids of Scotland.* Edinburgh: HMSO.

Allan, H. H. 1961. *Flora of New Zealand. Vol. 1, Indigenous Tracheophyta.* Wellington: P. D. Hasselberg.

Allen, D. E. 1957. *Leontodon autumnalis* var. *salinus* (Aspegren) Lange. *Proceedings of the Botanical Society of the British Isles*, 2: 240–241.

Allen, D. E. 1984. *Flora of the Isle of Man.* Douglas: Manx Museum and National Trust.

Allen, D. E. 1986. *The botanists.* Winchester: St Paul's Bibliographies.

Alliende, M. C. & Harper, J. L. 1989. Demographic studies of a dioecious tree, 1. Colonization, sex and age structure of a population of *Salix cinerea. Journal of Ecology*, 77, 1029–1047.

Alston, A. H. G. 1949. *Equisetum ramosissimum* as a British plant. *Watsonia*, 1: 149–153.

Amor, R. L. & Richardson, R. G. 1980. The biology of Australian weeds. 2. *Rubus fruticosus* agg. *Journal of the Australian Institute of Agricultural Science*, 46: 87–97.

Anderson, D. E. 1961. Taxonomy and distribution of the genus *Phalaris. Iowa State Journal of Science*, 36: 1–96.

Anderson, R. C. & Beare, M. H. 1983. Breeding system and pollination ecology of *Trientalis borealis* (Primulaceae). *American Journal of Botany*, 70: 408–415.

Anderson, R. C., Dhillon, S. S. & Kelley, T. M. 1996. Aspects of the ecology of an invasive plant, garlic mustard (*Alliaria petiolata*), in Central Illinois. *Restoration Ecology*, 4(2): 181–191.

Antrobus, S. & Lack, A. J. 1993. Genetics of colonising and established populations of *Primula veris. Heredity*, 71: 252–258.

Arnold, H. R. & Preston C. D. [1997]. *Fieldwork for Atlas 2000. 3. List of taxa covered by the project.* [London: Botanical Society of the British Isles.]

Ash, G. M. 1947. *Epilobium adenocaulon* Hausskn. × *obscurum* Schreb. *Report of the Botanical Society and Exchange Club of the British Isles*, 13: 160.

Asher, J., Warren, M., Fox, R., Harding, P., Jeffcoate, G. & Jeffcoate, S. 2001. *The Millenium atlas of butterflies.* Oxford: Oxford University Press.

Atkinson, M. D. 1992. Biological Flora of the British Isles. No. 175. *Betula pendula* Roth and *B. pubescens* Ehrh. *Journal of Ecology*, 80: 837–870.

Atkinson, M. D. 1996. The distribution and naturalisation of *Lathraea clandestina* L. (Orobanchaceae) in the British Isles. *Watsonia*, 21: 119–128.

Atlas see Perring & Walters (1962).

Atlas Supp. see Perring & Sell (1968).

Auquier, P. 1971a. *Festuca rubra* L. subsp. *pruinosa* (Hack.) Piper: morphologie, écologie, taxonomie. *Lejeunia*, 56: 1–16.

Auquier, P. 1971b. Le problème de *Festuca rubra* L. subsp *arenaria* (Osb.) Richt. et de ses relations avec *F. juncifolia* St-Amans. *Lejeunia*, 57: 1–24.

Auquier, P. 1973. Une fétuque nouvelle de Bretagne: *Festuca huonii. Candollea*, 28: 15–19.

Babington, C. C. 1834. *Flora Bathoniensis.* London: Longman & Co.

Bailey, J. P. & Connolly, A. P. 2000. Prize-winners to pariahs – A history of Japanese Knotweed *s.l.* (Polygonaceae) in the British Isles. *Watsonia*, 23: 93–110.

Bailey, J. P., Child, L. E. & Conolly, A. P. 1996. A survey of the distribution of *Fallopia* × *bohemica* (Chrtek & Chrtková) J. Bailey (Polygonaceae) in the British Isles. *Watsonia*, 21: 187–198.

Bain, J. F. 1991. The biology of Canadian weeds. 96. *Senecio jacobaea* L. *Canadian Journal of Plant Science*, 71: 127–140.

Baker, H. G. 1947. Biological Flora of the British Isles. No. 19. *Melandrium* [genus] (Roehling emend.) Fries (pp. 271–274); *Melandrium album* (Mill.) Garcke (pp. 274–282); *Melandrium dioicum* (L. emend.) Coss. & Germ. (pp. 283–292). *Journal of Ecology*, 35: 271–292.

Baker, H. G. 1955. *Geranium purpureum* Vill. and *G. robertianum* L. in the British Flora: I. *Geranium purpureum. Watsonia*, 3: 160–167.

Baker, K., Richards, A. J. & Tremayne, M. 1994. Fitness constraints on flower number, seed number and seed size in the dimorphic species *Primula farinosa* L. and *Armeria maritima* (Miller) Willd. *New Phytologist*, 128: 563–570.

Ball, P. W. & Tutin, T. G. 1959. Notes on annual species of *Salicornia* in Britain. *Watsonia*, 4: 193–205.

Ballantyne, G. H. 1985. *The flowering plants of Kinross*, edn 2. Kirkaldy: Scottish Wildlife Trust.

Ballantyne, G. H. 1992. *Orobanche alba* Steph. ex Willd. in Fife (v.c. 85). *Watsonia*, 19: 39–41.

Balme, O. E. 1954. Biological Flora of the British Isles. No. 40. *Viola lutea* Huds. *Journal of Ecology*, 42: 234–240.

Bangerter, E. B. 1964. *Veronica peregrina* L. in the British Isles. *Proceedings of the Botanical Society of the British Isles*, 5: 303–313.

Bangerter, E. B. 1966. Further notes on *Veronica peregrina* L. *Proceedings of the Botanical Society of the British Isles*, 6: 215–220.

Bangerter, E. B. & Kent, D. H. 1957. *Veronica filiformis* Sm. in the British Isles. *Proceedings of the Botanical Society of the British Isles*, 2: 197–217.

Bangerter, E. B. & Kent, D. H. 1962. Further notes on *Veronica filiformis. Proceedings of the Botanical Society of the British Isles*, 4: 384–397.

Bangerter, E. B. & Kent, D. H. 1965. Additional notes on *Veronica filiformis. Proceedings of the Botanical Society of the British Isles*, 6: 113–118.

Bannister, P. 1965. Biological Flora of the British Isles. No. 100. *Erica cinerea* L. *Journal of Ecology*. 53: 527–542.

Bannister, P. 1966. Biological Flora of the British Isles. No. 102. *Erica tetralix* L. *Journal of Ecology*, 54: 795–813.

Barkham, J. P. 1992. Population dynamics of the wild daffodil (*Narcissus pseudonarcissus*). IV. Clumps and gaps. *Journal of Ecology*, 80: 797–808.

Barling, D. M. 1955. *Tragopogon pratensis* in the central Cotswolds. *Watsonia*, 3: 210–212.

Barling, D. M. 1959. Biological studies in *Poa angustifolia. Watsonia*, 4: 147–168.

Barling, D. M. 1962. Studies in the biology of *Poa subcaerulea* Sm. *Watsonia*, 5: 163–173.

Barling, D. M. 1967. *Poa pratensis* L., *P. angustifolia* L. and *P. subcaerulea* Sm. *Proceedings of the Botanical Society of the British Isles*, 6: 363–364.

Barr, C. J., Bunce, R. G. H., Clarke, R. T., Fuller, R. M., Furse, M. T., Gillespie, M. K., Groom, G. B., Hallam, C. J., Hornung, M., Howard, D. C. & Ness, M. J. 1993. *Countryside Survey 1990 Main Report.* London: Department of the Environment.

Barry, R. & Wade, P. M. 1986. Biological Flora of the British Isles. No. 162. *Callitriche truncata* Guss. *Journal of Ecology*, 74: 289–294.

Barty-King, H. 1977. *A tradition of English wine.* Oxford: Oxford Illustrated Press.

Bassett, I. J. & Crompton, C. W. 1975. The biology of Canadian weeds. 11. *Ambrosia artemisiifolia* L. and *A. psilostachya* DC. *Canadian Journal of Plant Science*, 55: 463–476.

Bassett, I. J. & Crompton, C. W. 1978. The biology of Canadian weeds. 32. *Chenopodium album* L. *Canadian Journal of Plant Science*, 58: 1061–1072.

Bassett, I. J. & Munro, D. B. 1985. The biology of Canadian weeds. 67. *Solanum ptycanthum* Dun., *S. nigrum* L. and *S. sarrachoides* Sendt. *Canadian Journal of Plant Science*, 65: 401–414.

Bassett, I. J. & Munro, D. B. 1987. The biology of Canadian weeds. 81. *Atriplex patula* L., *A. prostrata* Boucher ex DC, and *A. rosea* L. *Canadian Journal of Plant Science*, 67: 1069–1082.

Bassett, I. J., Crompton, C. W. & Woodland, D. W. 1977. The biology of Canadian weeds. 21. *Urtica dioica* L. *Canadian Journal of Plant Science*, 57: 491–498.

Bateman, R. [M.] 1981. The Hertfordshire Orchidaceae. *Transactions of the Hertfordshire Natural History Society*, 28(4): 56–79.

Bateman, R. M. 2001. Evolution and classification of European orchids: insights from molecular and morphological characters. *Journal Europäische Orchideen*, 33: 501–568.

Bateman, R. M. & Denholm, I. 1983a. A reappraisal of the British and Irish dactylorchids, 1. The tetraploid marsh-orchids. *Watsonia*, 14: 347–376.

Bateman, R. M. & Denholm, I. 1983b. *Dactylorhiza incarnata* (L.) Soó. subsp. *ochroleuca* (Boll) P. F. Hunt & Summerhayes. *Watsonia*, 14: 410–411.

Bateman, R. M. & Denholm, I. 1985. A reappraisal of the British and Irish dactylorchids, 2. The diploid marsh-orchids. *Watsonia*, 15: 321–355.

Bateman, R. M. & Denholm, I. 1989. A reappraisal of the British and Irish dactylorchids, 3. The spotted-orchids. *Watsonia*, 17: 319–349.

Bateman, R. M. & Farrington, O. S. 1989. Patterns of variation in bigeneric orchid hybrids: British × *Dactylogymnadenia. American Journal of Botany* 76 (supp.): 241.

Bateman, R. M., Pridgeon, A. M. & Chase, M. W. 1997. Phylogenetics of subtribe Orchidinae (Orchidoideae, Orchidaceae) based on nuclear ITS sequences. 2. Infrageneric relationships and reclassification to achieve monophyly of *Orchis sensu stricto. Lindleyana*, 12: 113–141.

Bean, W. J. 1970. *Trees and shrubs hardy in the British Isles*, edn 8, I. *A-C.* London: John Murray.

Bean, W. J. 1973. *Trees and shrubs hardy in the British Isles*, edn 8, II. *D-M.* London: John Murray.

Bean, W. J. 1976. *Trees and shrubs hardy in the British Isles*, edn 8, III. *N-Rh.* London: John Murray.

Bean, W. J. 1980. *Trees and shrubs hardy in the British Isles*, edn 8, IV. *Ri-Z.* London: John Murray.

Beattie, A. J. 1972. The pollination ecology of *Viola.* 2, Pollen loads of insect-visitors. *Watsonia*, 9: 13–25.

Beckett, G., Bull, A. & Stevenson, R. 1999. *A flora of Norfolk.* Privately published.

Beddows, A. R. 1959. Biological Flora of the British Isles. No. 68. *Dactylis glomerata* L. *Journal of Ecology*, 47: 223–239.

Beddows, A. R. 1961. Biological Flora of the British Isles. No. 77. *Holcus lanatus* L. *Journal of Ecology*, 49: 421–430.

Beddows, A. R. 1967. Biological Flora of the British Isles. No. 107. *Lolium perenne* L. *Journal of Ecology*, 55: 567–587.

Beddows, A. R. 1973. Biological Flora of the British Isles. No. 131. *Lolium multiflorum* Lam. *Journal of Ecology*, 61: 587–600.

Beerling, D. J. 1998. Biological Flora of the British Isles. No. 202. *Salix herbacea* L. *Journal of Ecology*, 86, 872–895.

Beerling, D. J. & Perrins, J. M. 1993. Biological Flora of the British Isles. No. 177. *Impatiens glandulifera* Royle. *Journal of Ecology*, 81: 367–382.

Beerling, D. J. Bailey, J. P. & Conolly, A. P. 1994. Biological Flora of the British Isles. No. 183. *Fallopia japonica* (Houtt.) Ronse Decraene. *Journal of Ecology*, 82: 959–979.

Beesley, S. & Wilde, J. 1997. *Urban flora of Belfast.* Belfast: Institute of Irish Studies.

Bell, J. N. B. & Tallis, J. H. 1973. Biological Flora of the British Isles. No. 130. *Empetrum nigrum* L. *Journal of Ecology*, 61: 289–305.

Benoit, P. & Richards, M. 1963. *A contribution to the flora of Merioneth*, edn 2. Haverfordwest: West Wales Naturalists' Trust.

Best, K. F. 1977. The biology of Canadian weeds. 22. *Descurainia sophia* (L.) Webb. *Canadian Journal of Plant Science*, 57: 499–507.

Best, K. F. & McIntyre, G. I. 1975. The biology of Canadian weeds. 9. *Thlaspi arvense* L. *Canadian Journal of Plant Science*, 55: 279–292.

Best, K. F., Banting, J. D. & Bowess, G. G. 1978. The biology of Canadian weeds. 31. *Hordeum jubatum* L. *Canadian Journal of Plant Science*, 58: 699–708.

Best, K. F., Bowes, G. G., Thomas, A. G. & Maw, M. G. 1980. The biology of Canadian weeds. 39. *Euphorbia esula* L. *Canadian Journal of Plant Science*, 60: 651–663.

Bevis, J., Kettell, R. E. & Shepard, B. 1978. *Flora of the Isle of Wight.* Newport: Isle of Wight Natural History and Archaeological Society.

Bezzant, L. 1992. *Lychnis flos-cuculi* – dwarf form. *Rock garden*, 22: 423.

Birkinshaw, C. R. & Sanford, M. N. 1996. *Pulmonaria obscura* Dumort. (Boraginaceae) in Suffolk. *Watsonia*, 21: 169–178.

Birse, E. M. 1997. Creeping spearwort, *Ranunculus reptans* L., at the Loch of Strathbeg. *BSBI News*, 74: 17–19.

Bishop, G. F. & Davy, A. J. 1994. Biological Flora of the British Isles. No. 180. *Hieracium pilosella* L. *Journal of Ecology*, 82: 195–210.

Bittrich, V. & Kadereit, J. 1988. Cytogenetical and geographical aspects of sterility in *Lysimachia nummularia. Nordic Journal of Botany*, 8: 325–328.

Blackman, G. E. & Rutter, A. J. 1954. Biological Flora of the British Isles. No. 45. *Endymion non-scriptus* (L.) Garcke. *Journal of Ecology*, 42: 629–638.

Blackstock, T. H. 1981. The distribution of *Juncus filiformis* L. in Britain. *Watsonia*, 13: 209–214.

Blackstock, T. H. & Roberts, R. H. 1986. Observations on the morphology and fertility of *Juncus × surrejanus* Druce ex Stace & Lambinon in north-western Wales. *Watsonia*, 16: 55–63.

Blackstock, T. H. & Roberts, R. H. 1998. *Trifolium occidentale* D. E. Coombe (Fabaceae) in Anglesey (v.c. 52). *Watsonia*, 22: 182–184.

Blackstock, T. H., Howe, E. A. & Rimes, C. A. 1991. Whorled Caraway, *Carum verticillatum* (L.) Koch in Lleyn. *Welsh Bulletin of the Botanical Society of the British Isles*, 51: 8–10.

Blanchard, J. W. 1990. Narcissus – *A guide to wild daffodils.* Woking: Alpine Garden Society.

Blomgren, E. 1992. Portlakmålla *Halimione portulacoides*, funnen i Bohuslån. [*Halimione portulacoides* found on the western coast of Sweden]. *Svensk Botanisk Tidskrift*, 86: 61–62.

Böcher, T. W. 1940. Studies on the plant-geography of the North-Atlantic heath-formation I. The heaths of the Faroes. *Kongelige Danske Videnskabernes Selskabs, Biologiske Meddelelser*, 15(3): 1–64.

Bolòs, O. de & Vigo, J. 1984. *Flora dels Països Catalans*, I. *Introducció. Licopodiàcies-Capparàcies.* Barcelona: Editorial Barcino.

Bolòs, O. de & Vigo, J. 1990. *Flora dels Països Catalans*, II. *Crucíferes-Amarantàcies.* Barcelona: Editorial Barcino.

Bolòs, O. de & Vigo, J. 1995. *Flora dels Països Catalans*, III. *Pirolàcies-Compostes.* Barcelona: Editorial Barcino.

Bond, T. E. T. 1952. Biological Flora of the British Isles. No. 35. *Elymus arenarius* L. *Journal of Ecology*, 40: 217–227.

Bonnard, B. 1988. *Flora of Alderney.* Alderney: privately published.

Boon, C. R. 1999. British and Irish floristic elements applied to the Bedfordshire flora. *Bedfordshire Naturalist*, 52: 78–91.

Boorman, L. A. 1967. Biological Flora of the British Isles. No. 106. *Limonium vulgare* Mill., *Limonium humile* Mill. *Journal of Ecology*, 55: 221–229, 229–232.

Boorman, L. A. & Fuller, R. M. 1984. The comparative ecology of two sand dune biennials: *Lactuca virosa* L. & *Cynoglossum officinale* L. *New Phytologist*, 96: 609–629.

Booth, E. M. 1979. *The flora of County Carlow.* Dublin: Royal Dublin Society.

Borrill, M. 1956. A biosystematic study of some *Glyceria* species in Britain. 1. Taxonomy. *Watsonia*, 3: 291–298.

Bowen, H. J. M. 1968. *The flora of Berkshire.* Oxford: privately published.

Bowen, H. J. M. 1986. Changes in the Berkshire flora. *Watsonia*, 16: 111–112.

Bowen, H. J. M. 2000. *The flora of Dorset.* Newbury: Pisces Publications.

Bowman, R. P. & Sell, P. D. 1993. *Pilosella × floribunda* (Wimmer & Grab.) Arvet-Touvet (Asteraceae) in the British Isles. *Watsonia*, 19: 187–188.

Bowra, J. C. 1992a. Prickly Lettuce (*Lactuca serriola*) – a population explosion in Warwickshire. *BSBI News*, 60: 12–16.

Bowra, J. C. 1992b. Hybridization of *Oenothera* L. subgenus *Oenothera* in Britain. *BSBI News*, 61: 19–33.

Bowra, J. C. 1997. Hybridization of *Oenothera* subgenus *Oenothera* in Britain II. *BSBI News*, 76: 64–71.

Bowra, J. C. 1999. *Oenothera* (evening-primroses) – the way forward. *BSBI News*, 81: 24–26.

Boyce, P. 1993. *The genus* Arum. London: HMSO.

Boyd, W. E., Laing, A. I., Steven, G. & Dickson, J. H. 1988. The history and present management of two rare endemic trees on the island of Arran, Scotland. *Environmental Conservation*, 15: 65–66.

Bradshaw, M. E. 1962. The distribution and status of five species of the *Alchemilla vulgaris* L. aggregate in Upper Teesdale. *Journal of Ecology*, 50: 681–706.

Braithwaite, M. [E.] 1991. The Scottish cabbage patch. New Zealand Bittercress, *Cardamine uniflora*. *BSBI News*, 58: 38–39.

Braithwaite, M. E. & Long, D. G. 1990. *The botanist in Berwickshire.* Berwickshire Naturalists Club.

Brenan, J. P. M. 1948. *Senecio squalidus* L. × *vulgaris* L. *Report of the Botanical Society and Exchange Club of the British Isles*, 13: 364.

Brenan, J. P. M. 1950. *Artemisia verlotorum* Lamotte and its occurrence in Britain. *Watsonia*, 1: 209–223.

Brenan, J. P. M. 1961. *Amaranthus* in Britain. *Watsonia*, 6: 261–280.

Brenan, J. P. M. 1965. *Amaranthus hybridus × retroflexus = A. × adulterinus* Thell. *Proceedings of the Botanical Society of the British Isles*, 6: 122–123.

Brewis, A., Bowman, R. P. & Rose, F. 1996. *The flora of Hampshire.* Colchester: Harley Books.

Briggs, D., Block, M. & Jennings, S. 1989. The possibility of determining the age of colonies of clonally propagating herbaceous species from historic records: the case of *Aster novi-belgii* L. (first recorded as *A. salignus* Willd.) at Wicken Fen Nature Reserve, Cambridgeshire, England. *New Phytologist*, 112: 577–584.

Briggs, J. 1999. *Kissing Goodbye to Mistletoe?* London: Plantlife and the Botanical Society of the British Isles.

Briggs, M. 1990. *Sussex plant atlas: Selected supplement.* Brighton: Booth Museum of Natural History, Borough of Brighton.

Briggs, M. 1997a. *Phacelia tanacetifolia. BSBI News*, 74: 48.

Briggs, M. 1997b. Non-native *Mentha pulegium* (Pennyroyal). *BSBI News*, 74: 50.

Brightmore, D. 1968. Biological Flora of the British Isles. No. 112. *Lobelia urens* L. *Journal of Ecology*, 56: 613–620.

Brightmore, D. 1979. Biological Flora of the British Isles. No. 146. *Frankenia laevis* L. *Journal of Ecology*, 67: 1097–1107.

Brightmore, D. & White, P. H. F. 1963. Biological Flora of the British Isles. No. 94. *Lathyrus japonicus* Willd. *Journal of Ecology*, 51: 795–801.

Brock, T. C. M., Mielo, H. & Oostermeijer, G. 1989. On the lifecycle and germination of *Hottonia palustris* L. in a wetland forest. *Aquatic Botany*, 35: 153–166.

Brockie, W. B. 1970. Artificial hybridisation in *Epilobium* involving New Zealand, European, and North American species. *New Zealand Journal of Botany*, 8: 94–97.

Broderick, D. H. 1990. The biology of Canadian weeds. 93. *Epilobium angustifolium* L. (Onagraceae). *Canadian Journal of Plant Science*, 70: 247–259.

Brown, I. R., Kennedy, D. & Williams, D. A. 1982. The occurrence of natural hybrids between *Betula pendula* Roth and *B. pubescens* Ehrh. *Watsonia*, 14: 133–145.

Brown, J. M. B. 1953. *Studies on British beechwoods.* Forestry Commission Bulletin No. 20. London: HMSO.

Brullo, S. 1980. Taxonomic and nomenclatural notes on the genus *Limonium* in Sicily. *Botaniska Notiser*, 133: 281–293.

Brummitt, R. K. 1996. Two subspecies of *Calystegia silvatica* (Kit.) Griseb. Convolvulaceae in the mediterranean region. *Lagascalia*, 18: 338–340.

Brummitt, R. K. & Chater, A. O. 2000. *Calystegia* (Convolvulaceae) hybrids in West Wales. *Watsonia*, 23: 161–165.

Brummitt, R. K. & Heywood, V. H. 1960. Pink-flowered Calystegiae of the *Calystegia sepium* complex in the British Isles. *Proceedings of the Botanical Society of the British Isles*, 3: 384–388.

Brysting, A. K., Gabrielsen, T. M., Sørlibråten, O., Ytrehorn, O. & Brochmann, C. 1996. The Purple Saxifrage, *Saxifraga oppositifolia*, in Svalbard: two taxa or one? *Polar Research*, 15: 93–105.

Bullard, E. R. 1995. *Wildflowers in Orkney, a new checklist.* Kirkwall: privately published.

Bullard, E. R., Shearer, H. D. H., Day, J. D. & Crawford, R. M. M. 1987. Survival and flowering of *Primula scotica* Hook. *Journal of Ecology*, 75: 589–602.

Bungard, S. J. & Leach, S. [J.] 1991. *Atriplex littoralis* by the way. *BSBI News*, 59: 11–12.

Bunting, E. S. 1988. Exploited plants. Oilseed Rape. *Biologist*, 35: 95–100.

Burbidge, F. W. & Colgan, N. 1902. A new *Senecio* hybrid. *Irish Naturalists' Journal*, 11: 311–317.

Burdon, J. J. 1983. Biological Flora of the British Isles. No. 154. *Trifolium repens* L. *Journal of Ecology*, 71: 307–330.

Burkill, I. H. 1944. Biological Flora of the British Isles. No. 12. *Tamus communis* L. *Journal of Ecology*, 32: 121–129.

Burnett, J. H. (ed.). 1964. *The vegetation of Scotland.* Edinburgh & London: Oliver & Boyd.

Burnett, J. [H.] 1997. Notes on *Veronica anagallis-aquatica* agg. *BSBI News*, 75: 15–17.

Burton, R. M. 1978. *Bidens connata. BSBI News*, 18: 15–16.

Burton, R. M. 1979. Two species of *Senecio* L. from E. Kent. *Watsonia*, 12: 392.

Burton, R. M. 1983. *Flora of the London area.* London: London Natural History Society.

Burton, R. M. 1998. Botanical records for 1997. Introduction. *The London Naturalist*, 77: 225–228.

Burtt, B. L. 1950. *Lapsana intermedia* in Britain. *Watsonia*, 1: 234–237.

Butcher, R. W. 1947. Biological Flora of the British Isles. No. 18. *Atropa belladonna* L. *Journal of Ecology*, 34: 345–353.

Butcher, R. W. 1954. Biological Flora of the British Isles. No. 42. *Colchicum autumnale* L. *Journal of Ecology*, 42: 249–257.

Butterfield, L. 1999. Boscregan – last refuge of the Purple Viper's-bugloss? *British Wildlife*, 10: 166–171.

Byatt, J. I. 1975. Hybridization between *Crataegus monogyna* Jacq. and *C. laevigata* (Poiret) DC. in south-eastern England. *Watsonia*, 10: 253–264.

Byfield, A. & Pearman, D. 1996. *Dorset's disappearing heathland flora.* London: Plantlife and Sandy: Royal Society for the Protection of Birds.

Cadbury, D. A., Hawkes, J. G. & Readett, R. C. 1971. *A computer-mapped flora: a study of the county of Warwickshire.* London: Academic Press.

Caldwell, J. & Wallace, T. J. 1955. Biological Flora of the British Isles. No. 48. *Narcissus pseudonarcissus* L. *Journal of Ecology*, 43: 331–341.

Callaghan, D. A. 1998. Biological Flora of the British Isles. No. 203. *Lythrum hyssopifolia* L. *Journal of Ecology*, 86: 1065–1072.

Callaghan, T. V., Svensson, B. M. & Headley, A. D. 1986. The modular growth of *Lycopodium annotinum*. *Fern Gazette*, 13: 65–76.

Cannell, M. G. R. 1984. Exploited plants. Sitka spruce. *Biologist*, 31: 255–261.

Cannon, J. F. M. 1964. Infraspecific variation in *Lathyrus nissolia* L. *Watsonia*, 6: 28–35.

Carey, P. D. 1998. Modelling the spread of *Himantoglossum hircinum* (L.) Spreng. at a site in the south of England. *Botanical Journal of the Linnean Society*, 126: 159–172.

Carey, P. D. 1999. Changes in the distribution and abundance of *Himantoglossum hircinum* (L.) Sprengel (Orchidaceae) over the last 100 years. *Watsonia*, 22: 353–364.

Carey, P. D., Watkinson, A. R. & Gerard, F. F. O. 1995. The determinants of the distribution and abundance of the winter annual grass *Vulpia ciliata* ssp. *ambigua*. *Journal of Ecology*, 83: 177–187.

Carlisle, A. & Brown, A. H. F. 1968. Biological Flora of the British Isles. No. 109. *Pinus sylvestris* L. *Journal of Ecology*, 56: 269–307.

Carter, R. N. & Prince, S. D. 1985. The geographical distribution of Prickly Lettuce (*Lactuca serriola*). I. A general survey of its habitats and performance in Britain. *Journal of Ecology*, 73: 27–38.

Catling, P.M. & Dobson, I. 1985. The biology of Canadian weeds. 69. *Potamogeton crispus* L. *Canadian Journal of Plant Science*, 65: 655–668.

Cavers, P. B. & Harper, J. L. 1964. Biological Flora of the British Isles. No. 98. *Rumex obtusifolius* L., *Rumex crispus* L. *Journal of Ecology*, 52: 737–754, 754–766.

Cavers, P. B. & Harper, J. L. 1967. The comparative biology of closely related species living in the same area. IX. *Rumex*: the nature of adaptation to a sea-shore habitat. *Journal of Ecology*, 55: 73–82.

Cavers, P. B., Bassett, I. J. & Crompton, C. W. 1980. The biology of Canadian weeds. 47. *Plantago lanceolata* L. *Canadian Journal of Plant Science*, 60: 1269–1282.

Cavers, P. B., Heagy, M. I. & Kokron, R. F. 1979. The biology of Canadian weeds. 35. *Alliaria petiolata* (M. Bieb) Cavara and Grande. *Canadian Journal of Plant Science*, 59: 217–229.

Chadwick, M. J. 1960. Biological Flora of the British Isles. No. 73. *Nardus stricta* L. *Journal of Ecology*, 48: 255–267.

Chalk, D. 1986. Hedge Veronica (*Hebe × franciscana* (Eastwood) Souster). *BSBI News*, 43: 17–18.

Challice, J. & Kovanda, M. 1978. Chemotaxonomic survey of the genus *Sorbus* in Europe. *Naturwissenschaften*, 65: 111–112.

Chamberlain, D. E., Fuller, R. J., Bunce, R. G. H., Duckworth, J. C. & Shrubb, M. 2000. Changes in the abundance of farmland birds in relation to the timing of agricultural intensification in England and Wales. *Journal of Applied Ecology*, 37: 771–788.

Chapman, V. J. 1947a. Biological Flora of the British Isles. No. 20. *Suaeda maritima* (L.) Dum. *Journal of Ecology*, 35: 293–302.

Chapman, V. J. 1947b. Biological Flora of the British Isles. No. 21. *Suaeda fruticosa* Forsk. *Journal of Ecology*, 35: 303–310.

Chapman, V. J. 1950. Biological Flora of the British Isles. No. 29. *Halimione portulacoides* (L.) Aell. *Journal of Ecology*, 38: 214–222.

Chater, A. O. 1992. *Laburnum anagyroides* and *L. alpinum* as hedge plants in Cardiganshire, v.c. 46. *BSBI Welsh Bulletin*, 52: 4–5.

Chater, A. O. 1993. *Avena strigosa*, Bristle Oat, and other cereals as crops and casuals in Cardiganshire, v.c. 46. *Botanical Society of the British Isles Welsh Bulletin*, 55: 7–14.

Chater, A. O. 1996. *Ulmus laevis* naturalised in Cards, VC 46. *BSBI News*, 75: 63.

Chater, A. O. [1997]. *Fieldwork for Atlas 2000. 2. Collecting and pressing specimens.* [London: Botanical Society of the British Isles.]

Chater, A. O. & Rich, T. C. G. 1995. *Rorippa islandica* (Oeder ex Murray) Borbás (Brassicaceae) in Wales. *Watsonia*, 20: 229–238.

Chiapella, J. 2000. The *Deschampsia cespitosa* complex in central and northern Europe: a morphological analysis. *Botanical Journal of the Linnean Society*, 134: 495–512.

Chittenden, F. J. 1956. *The Royal Horticultural Society Dictionary of gardening*, edn 2 by P. M. Synge. 4 vols. Oxford: Clarendon Press.

Chmelar, J. 1984. Weeping willows. *International Dendrology Society Yearbook*, 1983: 107–110.

Church, T. & Smith, T. 2000. *Arran's flora. Revised edition.* Arran Natural History Society.

Clabby, G. & Osborne, B. A. 1999. Biological Flora of the British Isles. No. 204. *Mycelis muralis* (L.) Dumort. (*Lactuca muralis* L.). *Journal of Ecology*, 87: 156–172.

Clapham, A. R. 1951. A proposal for mapping the distribution of British vascular plants. *In*: J. E. Lousley (ed.), *The study of the distribution of British plants*, pp. 110–117. Oxford: Botanical Society of the British Isles.

Clapham, A. R. (ed.). 1969. *Flora of Derbyshire.* Derby: County Borough of Derby Museums and Art Gallery.

Clapham, A. R., Pearsall, W. H. & Richards, P. W. 1942. Biological Flora of the British Isles. No. 8. *Aster tripolium* L. *Journal of Ecology*, 30: 385–395.

Clapham, A. R., Tutin, T. G. & Moore, D. M. 1987. *Flora of the British Isles*, edn 3. Cambridge: Cambridge University Press.

Clapham, A. R., Tutin, T. G. & Warburg, E. F. 1952. *Flora of the British Isles.* Cambridge: Cambridge University Press.

Clapham, A. R., Tutin, T. G. & Warburg, E. F. 1962. *Flora of the British Isles* edn 2. Cambridge: Cambridge University Press.

Clapham, A. R., Tutin, T. G. & Warburg, E. F. 1981. *Excursion flora of the British Isles*, edn 3. Cambridge: Cambridge University Press.

Clark, S. C. 1974. Biological Flora of the British Isles. No. 136. *Catapodium rigidum* (L.) C. E. Hubbard. *Journal of Ecology*, 62: 937–958.

Clarke, P. M. & Clarke, J. 1991. *The Flowering Plants of Colonsay and Oransay.* Colonsay: privately printed.

Clarke, W. A. 1900. *First records of British flowering plants*, edn 2. London: West, Newman & Co.

Clarke, W. G. 1925. *In Breckland Wilds.* London: Robert Scott.

Clement, E. J. (comp.). 1978a. Adventive News 11. *BSBI News*, 19: 12–17.

Clement, E. J. (comp.). 1978b. Adventive News 12. *BSBI News*, 20: 9–14.

Clement, E. J. 1979. *Sisymbrium volgense* Bieb. ex E. Fourn. in Britain. *Watsonia*, 12: 311–314.

Clement, E. J. 1980. *Potentilla rivalis* Nutt. ex Torrey & Grey new to Britain. *Watsonia*, 13: 49.

Clement, E. J. 1981a. Sweet Bromegrass in Britain. *BSBI News*, 28: 12–14.

Clement, E. J. 1981b. Cockleburs in Britain. *BSBI News*, 29: 13.

Clement, E. J. 1983. Berry Catchfly, *Cucubalus baccifer* L., in Britain. *BSBI News*, 34 34.

Clement, E. J. 1985a. Hedge Veronica (*Hebe* × *franciscana*) and allies in Britain. *BSBI News*, 41: 18.

Clement, E. J. 1985b. Selfheals (*Prunella* spp.) in Britain. *BSBI News*, 41: 20.

Clement, E. J. 1993a. *Eragrostis curvula* established in S. Hants (v.c. 11). *BSBI News*, 63: 28–30.

Clement, E. J. 1993b. Buttonweed (*Cotula coronopifolia* L.), new to S. Hants (v.c. 11). *BSBI News*, 64: 43–46.

Clement, E. J. 1997. Can *Euphorbia robbiae* be revived? *BSBI News*, 76: 58–60.

Clement, E. J. 1999. Misconceptions about *Amsinckia lycopsoides* Lehm. *BSBI News*, 80: 44–45.

Clement, E. J. 2000. *Ludwigia* × *kentiana* E. J. Clement: a new hybrid aquatic. *Watsonia*, 23: 167–172.

Clement, E. J. & Foster, M. C. 1994. *Alien plants of the British Isles.* London: Botanical Society of the British Isles.

Clifford, H. T. 1959. On putative hybrids between *Juncus inflexus* L. and *Juncus effusus* L. *Kew Bulletin*, 13: 392–395.

Coats, A. M. 1963. *Garden shrubs and their histories.* London: Vista Books.

Cody, W. J. & Crompton, C. W. 1975. The biology of Canadian weeds. 15. *Pteridium aquilinum* (L.) Kuhn. *Canadian Journal of Plant Science*, 55: 1059–1072.

Cody, W. J. & Wagner, V. 1980. The biology of Canadian weeds. 49. *Equisetum arvense* L. *Canadian Journal of Plant Science*, 61: 123–133.

Coker, P. D. 1962. Biological Flora of the British Isles. No. 86. *Corrigiola litoralis* L. *Journal of Ecology*, 50: 833–840.

Coker, P. D. 1966. Biological Flora of the British Isles. No. 104. *Sibbaldia procumbens* L. *Journal of Ecology*, 54: 823–831.

Coker, P.D. & Coker, A. M. 1973. Biological Flora of the British Isles. No. 133. *Phyllodoce caerulea* (L.) Bab. *Journal of Ecology*, 61: 901–913.

Coleman, M., Hollingsworth, M. L. & Hollingsworth, P. M. 2000. Application of RAPDs to the critical taxonomy of the English endemic elm *Ulmus plotii* Druce. *Botanical Journal of the Linnean Society*, 133: 241–262.

Colgan, N. 1904. *Flora of the County Dublin.* Dublin: Hodges, Figgis & Co.

Compton, R. H. 1913. Further notes on *Epilobium* hybrids. *Journal of Botany*, 51: 79–85.

Compton, S. G. & Key, R. S. 2000. Biological Flora of the British Isles. No. 211. *Coincya wrightii* (O.E. Schulz) Stace. *Journal of Ecology*, 88: 535–548.

Conolly, A. P. 1977. The distribution and history in the British Isles of some alien species of *Polygonum* and *Reynoutria*. *Watsonia*, 11: 291–311.

Conolly, A. P. 1991. *Polygonum lichiangense* W. Smith: rejected as a naturalized British species. *Watsonia*, 18: 351–358.

Conway, V. M. 1942. Biological Flora of the British Isles. No. 6. *Cladium mariscus* (L.) R. Br. *Journal of Ecology*, 30: 211–216.

Cook, C. D. K. 1962. Biological Flora of the British Isles. No. 82. *Sparganium erectum* L. *Journal of Ecology*, 50: 247–255.

Cook, C. D. K. 1966. A monographic study of *Ranunculus* subgenus *Batrachium* (DC.) A. Gray. *Mitteilungen der Botanischen Staatssammlung München*, 6: 47–237.

Cook, C. D. K. & Urmi-König, K. 1983. A revision of the genus *Stratiotes* (Hydrocharitaceae). *Aquatic Botany*, 16: 213–249.

Cook, P. J. 1997. 'Summer Cypress' (*Bassia scoparia*) on Yorkshire roadsides. *BSBI News*, 74: 48–49.

Cook, P. J. 1998. *Bassia scoparia* in v.cc. 28, 53 and 54. A possible connection with east coast ports. *BSBI News*, 78: 63.

Cooke, A. S. 1994. Colonisation by Muntjac deer *Muntiacus reevesi* and their impact on vegetation. *In*: M. E. Massey & R. C. Welch (eds), *Monks Wood National Nature Reserve: the Experience of 40 Years 1953–93*, pp. 45–61. Peterborough: English Nature.

Coombe, D. E. 1956a. Biological Flora of the British Isles. No. 60. *Impatiens parviflora* DC. *Journal of Ecology*, 44: 701–713.

Coombe, D. E. 1956b. Notes on some British plants seen in Austria. *Veröffentlichungen des Geobotanisches Institutes Rübel in Zürich*, 35: 128–137.

Coombe, D. E. 1961. *Trifolium occidentale*, a new species related to *T. repens* L. *Watsonia*, 5: 68–87.

Coombe, D. E. 1973. The prostrate junipers at Gew Graze. *The Lizard*, 5(1): 7–12.

Coombe, D. E. 1994. 'Maritime' plants of roads in Cambridgeshire (v.c. 29). *Nature in Cambridgeshire*, 36: 37–60.

Cooper, M. R. & Johnson, A. W. 1988. *Poisonous plants in Britain and their effects on animals and man.* London: Her Majesty's Stationery Office.

Cope, T. A. & Stace, C. A. 1978. The *Juncus bufonius* L. aggregate in western Europe. *Watsonia*, 12: 113–128.

Cope, T. A. & Stace, C. A. 1983. Variation in the *Juncus bufonius* L. aggregate in western Europe. *Watsonia*, 14: 263–272.

Cope, T. A. & Stace, C. A. 1985. Cytology and hybridization in the *Juncus bufonius* L. aggregate in western Europe. *Watsonia*, 15: 309–320.

Corner, R. W. M. 1969. Plant Notes: *Carex appropinquata* Schumach. – in Scotland. *Proceedings of the Botanical Society of the British Isles*, 7: 562.

Corner, R. W. M. 1989. Observations on inland populations of *Viola canina* L. in south-eastern Scotland and north-western England. *Watsonia*, 17: 351–352.

Corner, R. W. M. 1997. A Pennine 'saltmarsh' flora. *BSBI News*, 77: 40–41.

Corner, R. W. M. 1999. Observations on a low-altitude site for *Draba norvegica* and *Poa glauca* in West Sutherland v.c. 108. *Botanical Journal of Scotland*, 51: 127–129.

Corner, R. W. M. & Roberts, F. J. 1989. *Carex ornithopoda* Willd. in Cumberland. *Watsonia*, 17: 437–438.

Cotton, D. C. F. 1999. Least duckweed *Lemna minuta* Kunth, in Ireland. *Irish Naturalists' Journal*, 26: 199–200.

Cottrell, J. E., Forrest G. I. & White, I. M. S. 1997. The use of RAPD analysis to study diversity in British black poplar (*Populus nigra* L. subsp. *betulifolia* (Pursh) W. Wettst. (Salicaceae)). *Watsonia*, 21: 305–312.

Cowie, N. R. & Sydes, C. 1995. *Status, distribution, ecology and management of Lapland marsh Orchid* Dactylorhiza lapponica. Scottish Natural Heritage Review no. 42. Edinburgh: Scottish Natural Heritage.

Cox, J. 1997. Hampshire purslane found in Dorset. *Recording Dorset*, 7: 20–21.

Crackles, F. E. 1986. *Dactylorhiza majalis* (Reichb.) P. F. Hunt & Summerhayes subsp. *cambrensis* (R. H. Roberts) R. H. Roberts in S. E. Yorkshire. *Watsonia*, 16: 78–80.

Crackles, [F.] E. 1990. *Flora of the East Riding of Yorkshire.* Hull: Hull University Press.

Crackles, F. E. 1995. A graphical analysis of the characters of *Calamagrostis stricta* (Timm) Koeler, *C. canescens* (Wigg.) Roth and their hybrid populations in S. E. Yorks, v.c. 61, northern England. *Watsonia*, 20: 397–404.

Crackles, F. E. 1997. Variation in some populations of *Calamagrostis stricta* (Timm) Koeler in the British Isles and the putative past hybridization with *C. canescens* (Wigg.) Roth. *Watsonia*, 21: 341–354.

Crawley, M. J., Harvey, P. H. & Purvis, A. 1996. Comparative ecology of the native and alien floras of the British Isles. *Philosophical Transactions of the Royal Society of London*, 351B: 1251–1259.

Crisp, P. C. 1972. *Cytotaxonomic studies in the section* Annui *of* Senecio. Ph.D. thesis, University of London.

Croft, J. [M.] 2000. Fen Violet *Viola persicifolia* at Wicken Fen: a reinforcement population. *Nature in Cambridgeshire*, 42: 27–33.

Croft, J. M. & Preston, C. D. 1999. Recording studies and interpretative projects in the British Isles, 1983–97. *Acta Botanica Fennica*, 162: 35–41.

Crompton, C. W., Hall, I. V., Jensen, K. I. N. & Hildebrand, P. D. 1988. The biology of Canadian weeds. 83. *Hypericum perforatum* L. *Canadian Journal of Plant Science*, 68: 149–162.

Crompton, G. & Whitehouse, H. L. K. 1983. *Annotated checklist of the flora of Cambridgeshire.* Cambridge: privately published.

Crook, C. S. 1996. *BSBI data transfer standards for computerised botanical records.* [London: Botanical Society of the British Isles.]

Cross, J. R. 1975. Biological Flora of the British Isles. No. 137. *Rhododendron ponticum* L. *Journal of Ecology*, 63: 345–364.

Crowder, A. A., Pearson, M. C., Grubb, P. J. & Langlois, P. H. 1990. Biological Flora of the British Isles. No. 167. *Drosera* [genus] L. (p. 233); *Drosera rotundifolia* L. (pp. 233–252); *Drosera anglica* Huds. (pp. 252–257); *Drosera intermedia* Drev. & Hayne (pp. 257–267). *Journal of Ecology*, 78: 233–267.

Cullen, J. 1986. *Anthyllis* in the British Isles. *Notes from the Royal Botanic Garden Edinburgh*, 43: 277–281.

Cunningham, M. H. & Kenneth, A. G. 1979. *The flora of Kintyre.* Wakefield: EP Publishing.

Curtis, T. G. F. 1981. A further station for *Mercurialis perennis* L. in the Burren with comments on its status there. *Irish Naturalists' Journal*, 20: 184–185.

Curtis, T. G. F. & FitzGerald, R. A. 1994. The re-discovery of *Carex divisa* Hudson, Divided Sedge, in Ireland. *Irish Naturalists' Journal*, 24: 496–498.

Curtis, T. G. F. & McGough, H. N. 1988. *The Irish Red Data Book. 1. Vascular Plants.* Dublin: Stationery Office.

Dalby, K. 1989. An experts approach to the *Salicornia* problem. *BSBI News*, 53: 9–10.

Dalby, D. H. & Rich, T. C. G. 1995. The history, taxonomy, distribution and ecology of Mountain scurvy-grass (*Cochlearia micacea* Marshall). Back from the Brink Project Report no. 42. (Contractor: Plantlife). Edinburgh: Scottish Natural Heritage.

Dale, H. M. 1974. The biology of Canadian weeds. 5. *Daucus carota. Canadian Journal of Plant Science*, 54: 673–685.

Dandy, J. E. 1958. *List of British vascular plants.* London: British Museum (Natural History) & Botanical Society of the British Isles.

Dandy, J. E. 1969. *Watsonian vice-counties of Great Britain.* London: Ray Society.

Daniels, R. E., McDonnell, E. J. & Raybould, A. F. 1998. The current status of *Rumex rupestris* Le Gall (Polygonaceae) in England and Wales and threats to its survival and genetic diversity. *Watsonia*, 22: 33–39.

Davey, A. J. 1961. Biological Flora of the British Isles. No. 80. *Epilobium nerterioides* A. Cunn. *Journal of Ecology*, 49: 753–759.

David, R. W. 1978a. The distribution of *Carex digitata* L. in Britain. *Watsonia*, 12: 47–49.

David, R W. 1978b. The distribution of *Carex elongata* L. in the British Isles. *Watsonia*, 12: 158–160.

David, R. W. 1980. The distribution of *Carex ornithopoda* Willd. in Britain. *Watsonia*, 13: 53–54.

David, R. W. 1981. The distribution of *Carex ericetorum* Poll. in Britain. *Watsonia*, 13: 225–226.

David, R. W. & Kelcey, J. G. 1985. Biological Flora of the British Isles. No. 159. *Carex muricata* L. aggregate (pp. 1021–1022), *Carex spicata* Huds. (pp. 1022–1028), *Carex muricata* L. [sensu stricto] (pp. 1028–1032), *Carex divulsa* Stokes (pp. 1032–1039). *Journal of Ecology*, 73: 1021–1039.

Davie, J. H. & Akeroyd, J. R. 1983. *Pachyphragma macrophyllum* (Hoffm.) Busch (Cruciferae), a Caucasian species naturalized in Co. Avon, England. *Botanical Journal of the Linnean Society*, 87: 77–82.

Davis, A. 1992. Exploited plants. White clover. *Biologist*, 39: 129–133.

Davis, A. P. 1999. The genus *Galanthus*. Portland, Oregon: Timber Press.

Davis, T. A. W. 1970. *Plants of Pembrokeshire.* Haverfordwest: West Wales Naturalists' Trust.

Davison, A. W. 1970. The ecology of *Hordeum murinum* L. I. Analysis of the distribution in Britain. *Journal of Ecology*, 58: 453–466.

Davison, A. W. 1971. The ecology of *Hordeum murinum* L. II. The ruderal habitat. *Journal of Ecology*, 59: 493–506.

Davy, A. J. 1980. Biological Flora of the British Isles. No. 149. *Deschampsia caespitosa* (L.) Beauv. *Journal of Ecology*, 68: 1075–1096.

Davy, A. J. & Bishop, G. F. 1991. Biological Flora of the British Isles. No. 172. *Triglochin maritima* L. *Journal of Ecology*, 79: 531–555.

Dawson, H. J. & Ingrouille, M. J. 1995. A biometric survey of *Limonium vulgare* Miller and *L. humile* Miller in the British Isles. *Watsonia*, 20: 239–254.

Dearnley, T. C. & Duckett, J. G. 1999. Juniper in the Lake District National Park. A review of condition and regeneration. *Watsonia*, 22: 261–267.

Desrochers, A. M., Bain, J. F. & Warwick, S. I. 1988. The biology of Canadian weeds. 89. *Carduus nutans* L. and *Carduus acanthoides* L. *Canadian Journal of Plant Science*, 68: 1053–1068.

Dickson, J. H., Hunter, R. & Walters, S.M. 1993. *Alchemilla acutiloba* Opiz new to Scotland. *Botanical Journal of Scotland*, 46: 499–502.

Dickson, J. H., Macpherson, P. & Watson, K. 2000. *The changing flora of Glasgow.* Edinburgh: Edinburgh University Press.

Dietrich, W. 1991. The status of *Oenothera cambrica* Rostański and *O. novae-scotiae* Gates (Onagraceae). *Watsonia*, 18: 407–408.

Dietrich, W., Wagner, W. L. & Raven, P. H. 1997. Systematics of *Oenothera* section *Oenothera* subsection *Oenothera* (Onagraceae). *Systematic Botany Monographs*, 50: 1–234.

Dines, T. D. 1996a. Atlas 2000: the official launch. *BSBI News*, 72: 3–5.

Dines, T. D. [1996b]. *Atlas 2000. Instruction Booklet.* [London: Botanical Society of the British Isles.]

Dines, T. D. [1997]. *Fieldwork for Atlas 2000. 4. A beginners guide to recording.* [London: Botanical Society of the British Isles.]

Dines, T. D. [1998]. *Fieldwork for Atlas 2000. 5. Criteria for mapping and provisional list of taxa to be mapped.* [London: Botanical Society of the British Isles.]

Dines, T. D. [1999]. *Strategy for the final year.* [London: Botanical Society of the British Isles.]

Dines, T. D. & Preston, C. D. 2000. *Eleocharis parvula* discovered in Scotland. *Watsonia*, 23: 341–342.

Dixon, J. M. 1982. Biological Flora of the British Isles. No. 151. *Sesleria albicans* Kit. ex Schultes. *Journal of Ecology*, 70: 667–684.

Dixon, J. M. 1991. Biological Flora of the British Isles. No. 173. *Avenula pratensis* (L.) Dumort. (pp. 829–846), *Avenula pubescens* (Hudson) Dumort. (pp. 846–865). *Journal of Ecology*, 79: 829–865.

Dixon, J. M. 1995. Biological Flora of the British Isles. No. 187. *Trisetum flavescens* (L.) Beauv. *Journal of Ecology*, 83: 895–909.

Dixon, J. M. 2000. Biological Flora of the British Isles. No. 212. *Koeleria macrantha* (Ledeb.) Schultes. *Journal of Ecology*, 88: 709–726.

Dodds, J. G. 1953. Biological Flora of the British Isles. No. 38. *Plantago coronopus* L. *Journal of Ecology*, 41: 467–478.

Donald, D. 1980. *Bromus interruptus* (Hack.) Druce: dodo or phoenix? *Nature in Cambridgeshire*, 23: 48–50, 51.

Donaldson, F., Donaldson, F. & McMillan, N. F. 1976. The present status of tear-thumb (*Polygonum sagittatum* L.) in Ireland, and notes on some other Kerry plants. *Irish Naturalists' Journal*, 18: 331–332.

Donaldson, F., Donaldson, F. & McMillan, N. F. 1978. *Polygonum sagittatum* L. its status in Kerry. *Irish Naturalists' Journal*, 19: 168.

Dony, J. G. 1967. *Flora of Hertfordshire.* Hitchin: Hitchin Museum.

Dony, J. G. 1976. *Bedfordshire plant atlas.* Luton: Borough of Luton Museum and Art Gallery.

Dony, J. G. 1977. Change in the flora of Bedfordshire, England, from 1798 to 1976. *Biological Conservation*, 11: 307–320.

Doogue, D., Nash, D., Parnell, J., Reynolds, S. & Wyse Jakson, P. (comps & eds). 1998. *Flora of County Dublin.* Dublin: Dublin Naturalists' Field Club.

Doohan, D. J. & Monaco, T. J. 1992. The biology of Canadian weeds. 99. *Viola arvensis* Murr. *Canadian Journal of Plant Science*, 72: 187–201.

Douglas, B. J., Thomas, A. G., Morrison, I. N. & Maw, M. G. 1985. The biology of Canadian weeds. 70. *Setaria viridis* (L.) Beauv. *Canadian Journal of Plant Science*, 65: 669–690.

Druce, G. C. 1903. *Campanula persicifolia* L. in Britain. *Journal of Botany*, 41: 289–290.

Druce, G. C. 1908. *List of British plants.* Oxford: Clarendon Press.

Druce, G. C. 1927. *Flora of Oxfordshire*, edn 2. Oxford: Clarendon Press.

Druce, G. C. 1928. *British plant list*, edn 2. Arbroath: T. Buncle & Co.

Druce, G. C. 1930. *The flora of Northamptonshire.* Arbroath: T. Buncle & Co.

Druce, G. C. 1932. *The comital flora of the British Isles.* Arbroath: T. Buncle & Co.

Dudman, A. A. & Richards, A. J. 1997. *Dandelions of Great Britain and Ireland.* Botanical Society of the British Isles Handbook no. 9. London: Botanical Society of the British Isles.

Duncan, U. K. 1980. *Flora of East Ross-shire.* Edinburgh: Botanical Society of Edinburgh.

Dunn, A. J. 1997. Biological Flora of the British Isles. No. 196. *Stachys germanica* L. *Journal of Ecology*, 85: 531–539.

Dunn, S. T. 1905. *Alien Flora of Britain.* London: West, Newman & Co.

Dupont, P. 1962. *La flore atlantique européenne.* Documents pour les cartes des productions végétales serie Europe-Atlantique, vol. 1. Toulouse: Faculté des Sciences.

Dyer, A. F., Parks, J. C. & Lindsay, S. 2000. Historical review of the uncertain taxonomic status of *Cystopteris dickieana* R. Sim (Dickie's Bladder Fern). *Edinburgh Journal of Botany*, 57: 71–81.

Easy, G. 1979. *Malva alcea* L. and *M. moschata* L. in Cambridgeshire. *Watsonia*, 23: 24–25.

Ebinger, J. E. 1962. *Luzula* × *borreri* in England. *Watsonia*, 5: 251–254.

Edees, E. S. 1972. *Flora of Staffordshire.* Newton Abbot: David & Charles.

Edees, E. S. & Newton, A. 1988. *Brambles of the British Isles.* London: Ray Society.

Edgington, M. J. 1999. *Erica ciliaris* L. (Ericaceae) discovered in the Blackdown Hills, on the Somerset–Devon border (v.c. 3). *Watsonia*, 22: 426–428.

Elkington, T. T. 1963. Biological Flora of the British Isles. No. 91. *Gentiana verna* L. *Journal of Ecology*, 51: 755–767.

Elkington, T. T. 1964. Biological Flora of the British Isles. No. 96. *Myosotis alpestris* F. W. Schmidt. *Journal of Ecology*, 52: 709–722.

Elkington, T. T. 1971. Biological Flora of the British Isles. No. 124. *Dryas octopetala* L. *Journal of Ecology*, 59: 887–905.

Elkington, T. T. & Woodell, S. R. J. 1963. Biological Flora of the British Isles. No. 92. *Potentilla fruticosa* L. *Journal of Ecology*, 51: 769–781.

Ellenberg, H. 1988. *Vegetation ecology of central Europe*, translation of edn 4. Cambridge: Cambridge University Press.

Ellis, R. G. 1983. *Flowering plants of Wales*. Cardiff: National Museum of Wales.

Ernst, W. H. O. & Malloch, A. J. C. 1994. Biological Flora of the British Isles. No. 181. *Phleum arenarium* L. *Journal of Ecology*, 82: 403–413.

Everett, S. (comp.). 1999. Alder death cause revealed. *British Wildlife*, 10: 365.

FNAEC see Flora of North America Editorial Committee.

Farmer, A. M. 1989. Biological Flora of the British Isles. No. 165. *Lobelia dortmanna* L. *Journal of Ecology*, 77: 1161–1173.

Farrell, L. 1979. The distribution of *Leucojum aestivum* L. in the British Isles. *Watsonia*, 12: 325–332.

Farrell, L. 1985. Biological Flora of the British Isles. No. 160. *Orchis militaris* L. (*O. galatea* Poir, *O. rivini* Gouan, *O. tephrosanthes* Willd. & Sw.). *Journal of Ecology*, 73: 1041–1053.

Faulkner, J. S. 1972. Chromosome studies on *Carex* section *Acutae* in north-west Europe. *Botanical Journal of the Linnean Society*, 65: 271–301.

Fearn, G. M. 1973. Biological Flora of the British Isles. No. 134. *Hippocrepis comosa* L. *Journal of Ecology*. 61: 915–926.

Fearn, G. M. 1987. Exploited plants. Sainfoin. *Biologist*, 34: 93–97.

Ferguson, I. K. & Ferguson, L. F. 1974. *Polygonum maritimum* L. new to Ireland. *Irish Naturalists' Journal*, 18: 95.

Ferris, C., King, R. A. & Gray, A. J. 1997. Molecular evidence for the maternal parentage in the hybrid origin of *Spartina anglica* C. E. Hubbard. *Molecular Ecology*, 6: 185–187.

Fiala, J. L. 1988. *Lilacs*. London: Christopher Helm.

Field, M. H. 1994. The status of *Bupleurum falcatum* L. (Apiaceae) in the British flora. *Watsonia*, 20: 115–117.

Firbank, L. G. 1988. Biological Flora of the British Isles. No. 165. *Agrostemma githago* L. *Journal of Ecology*, 76: 1232–1246.

Fischer, M. 1975. The *Veronica hederifolia* group: taxonomy, ecology and phylogeny. *In*: S. M. Walters (ed.), *European floristic and taxonomic studies*, pp. 48–60. BSBI Conference Report no. 15. Faringdon: E. W. Classey.

Fischer, M. A. 1987. On the origin of *Veronica persica* (Scrophulariaceae) – a contribution to the history of a neophytic weed. *Plant Systematics and Evolution*, 155: 105–132.

Fitter, A. H. & Peat, H. J. 1994. The Ecological Flora Database. *Journal of Ecology*, 82: 415–425.

FitzGerald, R. 1990. *Rare Plant Survey of South-West England. Vol. 3. Cornwall*. CSD Report, no. 1060. Peterborough: Nature Conservancy Council.

FitzGerald, R. 1998. Althaea hirsuta – *Hairy Mallow – status of British records between 1792 and 1997*. Back from the Brink Project Report No. 98. London: Plantlife.

FitzGerald, R. & Jermy, C. 1987. *Equisetum ramosissimum* in Somerset. *Pteridologist*, 1: 178–181.

FitzGerald, R. & Stewart, N. F. 2000. *Three-lobed Water Crowfoot*, Ranunculus tripartitus – *report for 1999*. Plantlife Report No. 157. London: Plantlife.

Fletcher, R. & Stace, C. A. 2000. A new section and species of *Festuca* (Poaceae) naturalized in England. *Watsonia*, 23: 173–177.

Flora of North America Editorial Committee. 1993a. *Flora of North America north of Mexico. Vol. 1. Introduction*. New York: Oxford University Press.

Flora of North America Editorial Committee. 1993b. *Flora of North America north of Mexico. Vol. 2. Pteridophytes and Gymnosperms*. New York: Oxford University Press.

Flora of North America Editorial Committee. 1997. *Flora of North America north of Mexico. Vol. 3. Magnoliophyta: Magnoliidae and Hamamelidae*. New York: Oxford University Press.

Flora of North America Editorial Committee. 2000. *Flora of North America north of Mexico. Vol. 22. Magnoliophyta: Alismatidae, Arecidae, Commelinidae (in part), and Zingiberidae*. New York: Oxford University Press.

Fogg, G. E. 1950. Biological Flora of the British Isles. No. 31. *Sinapis arvensis* L. *Journal of Ecology*, 38: 415–429.

Foley, M. J. Y. 1993. *Orobanche reticulata* Wallr. populations in Yorkshire (north-east England). *Watsonia*, 19: 247–257.

Foley, M. J. Y. 2000. A morphological comparison between some British *Orobanche* species (Orobanchaceae) and their closely-related non-British counterparts from continental Europe: *Orobanche reticulata* Wallr. *s.l. Watsonia*, 23: 257–267.

Foley, M. J. Y. & Porter, M. S. 2000. *Carex muricata* L. subsp. *muricata* (Cyperaceae) – a review of its present status in Britain. *Watsonia*, 23: 279–286.

Fone, A. L. 1989. Competition in mixtures of the annual *Hypochoeris glabra* and perennial *H. radicata*. *Journal of Ecology*, 77: 484–494.

Forbes, R. 2000. Assessing the status of *Stratiotes aloides* L. (Water-soldier) in Co. Fermanagh, Northern Ireland (v.c. H33). *Watsonia*, 23: 179–196.

Forestry Authority Scotland. 1998. *Caledonian pinewood inventory*. Edinburgh: Forestry Commission (published on disk).

Foss, P. J. & Doyle, G. J. 1988. Why has *Erica erigena* (the Irish heather) such a markedly disjunct European distribution? *Plants Today*, 1: 161–168.

Foss, P. J. & O'Connell, C. A. 1985. Notes on the ecology of *Sarracenia purpurea* L. on Irish peatlands. *Irish Naturalists' Journal*, 21: 440–443.

Foust , C. M. 1992. *Rhubarb: the wondrous drug*. Princeton, New Jersey: Princeton University Press.

Fowler, N., Zasada, J. & Harper, J. L. 1983. Genetic components of morphological variation in *Salix repens*. *New Phytologist*, 95: 121–131.

Fraser, J. & Hemsley, A. (eds). 1917. *Johnson's Gardeners' Dictionary and Cultural Instructor*. London: G. Routledge.

Fraser, L., Turkington, R. & Chanway, C. P. 1993. The biology of Canadian weeds. 102. *Gaultheria shallon* Pursh. *Canadian Journal of Plant Science*, 73: 1233–1247.

Freijsen, A. H. J. & Heeres, E. 1972. Welke soort rood zwenkgras (*Festuca rubra* s.l.) komt voor in de jonge kustduinen van Voorne en die van overig Nederland? *Gorteria*, 6: 57–60.

French, C. N., Murphy, R. J. & Atkinson, M. G. C. 1999. *Flora of Cornwall*. Camborne: Wheal Seton Press.

Frost, L. C., Houston, L., Lovatt, C. M. & Beckett, A. 1991. *Allium sphaerocephalon* L. and introduced *A. carinatum* L., *A. roseum* L. and *Nectaroscordum siculum* (Ucria) Lindley on St Vincent's Rocks, Avon Gorge, Bristol. *Watsonia*, 18: 381–385.

Fryer, J. & Hylmö, B. 1994. The native British *Cotoneaster* – Great Orme Berry – renamed. *Watsonia*, 20: 61–63.

Fryer, J. & Hylmö, B. 1997. Five new species of *Cotoneaster* Medik. (Rosaceae) naturalized in Britain. *Watsonia*, 21: 335–349.

Gardiner, A. S. 1974. A history of the taxonomy and distribution of the native oak species. *In*: M. G. Morris & F. H. Perring (eds), *The British Oak*, pp. 13–26. BSBI Conference Report no. 14. Faringdon: Classey.

Geltman, D. V. 1992. *Urtica galeopsifolia* Wierzb. ex Opiz (Urticaceae) in Wicken Fen (E. England). *Watsonia*, 19: 127–129.

Gent, G., Wilson, R. *et al.* 1995. *The Flora of Northamptonshire and the Soke of Peterborough*. Kettering and District Natural History Society and the Northamptonshire Flora Group.

George, A. S. 1984. *Flora of Australia. Vol. 4, Phytolaccaceae to Chenopodiaceae*. Canberra: Australian Government Printing Service.

Gerarde, J. 1633. *The herball or generall Historie of Plantes*, revised by T. Johnson. London.

Gibbons, D. W., Reid, J. B. & Chapman, R. A. 1993. *The new atlas of breeding birds in Britain and Ireland: 1988–91*. London: T. & A. D. Poyser.

Gibbons, E. J. 1975. *The flora of Lincolnshire*. Lincoln: Lincolnshire Naturalists' Union.

Gibbons, E. J. & Lousley, J. E. 1958. An inland *Armeria* overlooked in Britain. Part 1. *Watsonia*, 4: 125–135.

Gibbons, E. J. & Weston, I. 1985. *Supplement to the flora of Lincolnshire*. Lincolnshire Naturalists' Union.

Gibson, C. 2000. Notes on Essex specialities. 3: Annual Sea-purslane *Atriplex pedunculata* L. *Essex Naturalist* n.s., 17: 129–132.

Gibson, D. J. 2001. Biological Flora of the British Isles. No. 217. *Festuca arundinacea* Schreber (*F. elatior* L. ssp. *arundinacea* (Schreber) Hackel). *Journal of Ecology*, 89: 304–324.

Gilbert, J. L. 1965. *Flora of Huntingdonshire: Wildflowers*. Peterborough: Peterborough Museum Society.

Gilbert, O. L. 1970. Biological Flora of the British Isles. No. 118. *Dryopteris villarii* (Bellardi) Woynar. *Journal of Ecology*, 58: 301–313.

Gilbert, O. L. 1990. Wild figs by the River Don, Sheffield. *Watsonia*, 18: 84–85.

Gilbert, O. L. 1995. Biological Flora of the British Isles. No. 184. *Symphoricarpos albus* (L.) S. F. Blake. *Journal of Ecology*, 83: 159–166.

Gillam, B. (ed.). 1993. *The Wiltshire Flora*. Newbury: Pisces Publications.

Gimingham, C. H. 1960. Biological Flora of the British Isles. No. 74. *Calluna vulgaris* (L.) Hull. *Journal of Ecology*, 48: 455–483.

Godwin, H. 1943. Biological Flora of the British Isles. No. 11. Rhamnaceae (pp. 66–68), *Rhamnus cathartica* L. (pp. 69–76), *Frangula alnus* Miller (*Rhamnus Frangula* L.) (the genus *Rhamnus* sensu lato, L.) (pp. 77–92). *Journal of Ecology*, 31: 66–92.

Godwin, H. 1975. *The history of the British Flora*, edn 2. Cambridge: Cambridge University Press.

Goodman, P. J. 1969. Biological Flora of the British Isles. No. 116. *Spartina* [genus] Schreb. *Journal of Ecology*, 57: 285–287.

Goodman, P. J., Braybrooks, E. M. & Lambert, J. M. 1959. Investigations into 'die-back' in *Spartina townsendii* agg. I. The present status of *Spartina townsendii* in Britain. *Journal of Ecology*, 47: 651–677.

Goodman, P. J., Braybrooks, E. M., Marchant, C. J. & Lambert, J. M. 1969. Biological Flora of the British Isles. No. 116. 4. *Spartina* × *townsendii* H. & J. Groves sensu lato, *Spartina alterniflora* × *S. maritima*. *Journal of Ecology*, 57: 298–313.

Goodman, P. J. & Williams, W. T. 1961. Investigations into 'die-back' in *Spartina townsendii* agg. III. Physiological correlates of 'die-back'. *Journal of Ecology*, 49: 391–398.

Goodwillie, R. 1999a. *Campanula trachelium* L. in Clare (H9), new to the Burren. *Irish Naturalists' Journal*, 26: 286.

Goodwillie, R. 1999b. *Alopecurus aequalis* Sobol. new to Clare (H9) and S. E. Galway (H15). *Irish Naturalists' Journal*, 26: 286–287.

Gorer, R. 1970. *The development of garden flowers*. London: Eyre & Spottiswoode.

Gornall, R. J. 1988. The coastal ecodeme of *Parnassia palustris* L. *Watsonia*, 17: 139–143.

Grace, J. B. & Harrison, J. S. 1986. The biology of Canadian weeds. 73. *Typha latifolia* L., *Typha angustifolia* & *Typha* × *glauca* Godr. *Canadian Journal of Plant Science*, 66: 361–379.

Graham, G. G. 1988. *The flora & vegetation of County Durham*. Durham Flora Committee & Durham County Conservation Trust.

Graham, G. G. & Primavesi, A. L. 1993. *Roses of Great Britain and Ireland*. Botanical Society of the British Isles Handbook no. 7. London: Botanical Society of the British Isles.

Graham, R. A. 1950. Mint notes, II. *Mentha gracilis* Sole, and its relationship to *Mentha cardiaca* Baker. *Watsonia*, 1: 276–278.

Graham, R. A. 1951. Mint notes, IV. *Mentha piperita* L. and the British peppermints. *Watsonia*, 2: 30–35.

Graham, R. A. 1958. Mint notes, VIII. A new mint from Scotland. *Watsonia*, 4: 119–121.

Grant, M. & Miller D. 1998. *Lavatera olbia* × *L. thuringiaca* in gardens. *BSBI News*, 79: 67.

Grassly, N. C., Harris, S. A. & Cronk, Q. C. B. 1996. British *Apium repens* (Jacq.) Lag. (Apiaceae) status assessed using random amplified polymorphic DNA (RAPD). *Watsonia*, 21: 103–111.

Gray, A. J. & Benham, P. E. M. 1990. Spartina anglica – *a research review*. ITE Research Publication no. 2. London: HMSO.

Gray, A. J. & Scott, R. 1977. Biological Flora of the British Isles. No. 140. *Puccinellia maritima* (Huds.) Parl. *Journal of Ecology*, 65: 699–716.

Gray, G. R. 1981. *Aspects of the ecology of Barren Brome (*Bromus sterilis *L.)*. D.Phil. thesis, University of Oxford.

Green, I. P. Higgins, R. J., Kitchen, C. & Kitchen, M. A. R. 2000. *The flora of the Bristol Region*. Newbury: Pisces Press.

Green, P. R. 1989. *Haloragis micrantha* – Creeping Raspwort. *BSBI News*, 51: 48.

Green, P. R, Green I. P. & Crouch G. A. 1997. *The Atlas Flora of Somerset*. Wayford & Yeovil: privately published.

Green, P. S. 1954. *Stellaria nemorum* L. subspecies *glochidisperma* Murbeck in Britain. *Watsonia*, 3: 122–126.

Green, P. S. 1973. *Hebe* × *franciscana* (Eastwood) Souter, not *H.* × *lewisii* – naturalised in Britain. *Watsonia*, 9: 371–372.

Greig-Smith, P. 1948. Biological Flora of the British Isles. No. 23. *Urtica* [genus] L. (pp. 339–343), *Urtica dioica* L. (pp. 343–351), *Urtica urens* L. (pp. 351–355). *Journal of Ecology*, 36: 339–355.

Greig-Smith, J. & Sagar, G. R. 1981. Biological causes of local rarity in *Carlina vulgaris*. *In*: H. Synge (ed.), *The Biological Aspects of Rare Plant Conservation*, pp. 389–400. Chichester: John Wiley & Sons.

Grenfell, A. L. (comp.). 1982. Adventive News 24. *BSBI News*, 33: 10–13.

Grenfell, A. L. 1983. Adventive ferns – 1. *BSBI News*, 35: 12.

Grenfell, A. L. 1984. Cucurbitaceae in Britain. *BSBI News*, 38: 13–18.

Grey-Wilson, C. 1997. *Cyclamen: a guide for gardeners, horticulturalists and botanists*. London: B. T. Batsford.

Grey-Wilson, C. & Mathew, B. 1981. *Bulbs*. London: Collins.

Grieve, M. 1974. *A Modern Herbal*, new edition by C. F. Leyel. London: Jonathan Cape.

Griffiths, M. E. & Proctor, M. C. F. 1956. *Helianthemum canum* (L.) Baumg. *In*: M. C. F. Proctor, Biological Flora of the British Isles. No. 58. *Helianthemum* Mill. *Journal of Ecology*, 44: 677–682.

Grigson, G. 1955. *The Englishman's Flora*. London: Phoenix House.

Grime, J. P., Hodgson, J. G. & Hunt, R. 1988. *Comparative Plant Ecology*. London: Unwin Hyman.

Groot, W. J. de, Thomas, P. A. & Wein, R. W. 1997. Biological Flora of the British Isles. No. 194. *Betula nana* L. and *B. glandulosa* Michx. *Journal of Ecology*, 85: 241–264.

Grose, D. 1957. *The Flora of Wiltshire*. Devizes: Natural History Section, Wiltshire Archaeological and Natural History Society.

Gross, K. L. & Werner, P. A. 1978. The biology of Canadian weeds. 28. *Verbascum thapsus* L. and *V. blattaria* L. *Canadian Journal of Plant Science*, 58: 401–413.

Gross, R. S., Werner, P. A. & Hawthorn, W. R. 1980. The biology of Canadian weeds. 38. *Arctium minus* (Hill) Bernh. and *A. lappa* L. *Canadian Journal of Plant Science*, 60: 621–634.

Grounds, R. 1989. *Ornamental grasses*. London: C. Helm.

Gulliver, R. L. 1996. The status of *Spiranthes romanzoffiana* Cham. (Orchidaceae), Irish Lady's Tresses, on Colonsay (v.c. 102) in 1995; with special reference to associated plant communities. *Watsonia*, 21: 202–204.

Gulliver, R. L. 1997. Pyramidal Bugle (*Ajuga pyramidalis*) on Colonsay (VC 102). *Glasgow Naturalist*, 23(2): 55.

Gulliver, R. L. 1998. Population sizes of *Gentianella uliginosa* (Willd.) Boerner, Dune Gentian, on Colonsay (v.c. 102), in 1996. *Watsonia*, 22: 111–113.

Gynn, E. G. & Richards, A. J. 1985. Biological Flora of the British Isles. No. 161. *Acaena novae-zelandiae* T. Kirk. *Journal of Ecology*, 73: 1055–1063.

Hackney, P. (ed.). 1992. *Stewart & Corry's Flora of the North-east of Ireland*, edn 3. Belfast: Queen's University.

Hadley, G. (ed.). 1985. *A map Flora of mainland Inverness-shire*. Edinburgh: Botanical Society of Edinburgh & Botanical Society of the British Isles.

Haines-Young, R., Barr, C. J., Black, H. I. J., Briggs, D. J., Bunce, R. G. H., Clarke, R. T., Cooper, A., Dawson, F. H., Firbank, L. G., Fuller, R. M., Furse, M. T., Gillespie, M. K., Hill, R., Nornung, M., Howard, D. C., McCann, T., Morecroft, M. D., Petit, S., Sier, A. R. J., Smart, S. M., Smith, G. M., Stott, A., Stuart, R. C. & Watkins, J. W. 2000. *Accounting for nature: assessing habitats in the UK countryside*. London: Department of the Environment, Transport and the Regions.

Hall, P. C. 1980. *Sussex plant atlas*. Brighton: Booth Museum of Natural History, Borough of Brighton.

Hall, I. V., Steiner, E., Threadgill, P. & Jones, R. W. 1988. The biology of Canadian weeds. 84. *Oenothera biennis* L. *Canadian Journal of Plant Science*, 68: 163–173.

Halliday, G. 1995. Two subspecies of *Festuca rubra* L. new to England. *Watsonia*, 20: 412.

Halliday, G. 1997. *A Flora of Cumbria*. Lancaster: Centre for North-West Regional Studies, University of Lancaster.

Handley, R. J. & Davy, A. J. 2000. Discovery of male plants of *Najas marina* L. (Hydrocharitaceae) in Britain. *Watsonia*, 23: 331–334.

Hanson, C. G. & Mason, J. L. 1985. Bird seed aliens in Britain. *Watsonia*, 15: 237–252.

Harding, P. T. & Sheail, J. 1992. The Biological Records Centre: a pioneer in data gathering and retrieval. *In*: P. T. Harding (ed.), *Biological recording of changes in British wildlife*, pp. 5–19. London: HMSO.

Harley, R. M. 1956. *Rubus arcticus* L. in Britain. *Watsonia*, 3: 237–238.

Harmes, P. A. & Spiers, A. 1993. *Polygonum maritimum* L. in East Sussex (v.c. 14). *Watsonia*, 19: 271–273.

Harper, J. L. 1957. Biological Flora of the British Isles. No. 61. *Ranunculus acris* L. (pp. 289–314), *Ranunculus repens* L. (pp. 314–325), *Ranunculus bulbosus* L. (pp. 325–342). *Journal of Ecology*, 45: 289–342.

Harper, J. L. & Wood, W. A. 1957. Biological Flora of the British Isles. No. 62. *Senecio jacobaea* L. *Journal of Ecology*, 45: 617–637.

Harper, M. 2000. *At war with aliens*. London: Plantlife.

Harris, E. M. H. 1997. *Ulmus laevis* – European White Elm. *BSBI News*, 76: 57–58.

Harris, G. R. & Lovell, P. H. 1980. Localised spread of *Veronica filiformis*, *V. agrestis* and *V. persica*. *Journal of Applied Ecology*, 17: 815–826.

Harrison, S. G. 1968. A New Zealand Willow-herb in Wales. *Nature in Wales*, 11: 74–78.

Harrold, P. 1978. A glabrous variety of *Sagina subulata* (Sw.) C. Presl. in Britain. *Transactions of the Botanical Society of Edinburgh*, 43: 1–5.

Harron, J. 1986. *Flora of Lough Neagh*. Belfast: Irish Naturalists' Journal Committee & Coleraine: University of Ulster.

Harron, J. 1992. The present distribution of the dark-leaved willow *Salix myrsinifolia* Salisb., in north-east Ireland. *Irish Naturalists' Journal*, 24: 8–11.

Hartog, C. den. 1970. *The sea-grasses of the world*. Amsterdam: North-Holland Publishing Company.

Harvey, H. J. & Meredith, T. C. 1981. The biology and conservation of Milk-Parsley, *Peucedanum palustre* at Wicken Fen. *Nature in Cambridgeshire*, 24: 38–42.

Harvey, J. [H.] 1981. *Mediaeval gardens*. London: B. T. Batsford.

Harvey, J. H. 1996. Fritillary and martagon – wild or garden? *Garden History*, 24: 30–38.

Harvey, M. J. 1966. An experiment with *Epilobium angustifolium*. *Proceedings of the Botanical Society of the British Isles*, 6: 229–231.

Haskell, G. 1960. The Raspberry wild in Britain. *Watsonia*, 4: 238–255.

Haslam, S. M. 1972. Biological Flora of the British Isles. No. 128. *Phragmites communis* Trin. *Journal of Ecology*, 60: 585–610.

Haslam, S. M. & McDougall, D. S. A. 1972. *The Reed ('Norfolk Reed')*, edn 2. Norfolk Reed Growers' Association.

Hawkes, J. G. 1990. *The potato: evolution, biodiversity and genetic resources*, edn 3. London: Belhaven Press.

Hawthorn, W. R. 1974. The biology of Canadian weeds. 4. *Plantago major* and *P. rugelli*. *Canadian Journal of Plant Science*, 54: 383–396.

Hayward, C. M. 1995. *The spread of* Epilobium brunnescens, *a non-indigenous species, in Cwm Idwal NNR, its effects on the local indigenous flora and the possible effect of climate change on its spread*. M.Sc. thesis, University of Wales, Bangor.

Headley, A. D. & Callaghan. T. V. 1990. Modular growth of *Huperzia selago* (Lycopodiaceae: Pteridophyta). *Fern Gazette*, 13: 361–372.

Hedberg, K. O. 1992. Taxonomic differentiation in *Saxifraga hirculus* L. (Saxifragaceae), a circumpolar Arctic-Boreal species of Central Asiatic origin. *Botanical Journal of the Linnean Society*, 109: 377–393.

Hedrén, M. 1996. Genetic differentiation, polyploidization and hybridization in northern European *Dactylorhiza* (Orchidaceae): evidence from allozyme markers. *Plant Systematics and Evolution*, 210: 31–55.

Heinze, B. 1997. A PCR marker for a *Populus deltoides* allele and its use in studying introgression with native European *Populus nigra*. *Belgian Journal of Botany*, 129: 123–130.

Henderson, D. M. 1991. *Annotated checklist of the flora of West Ross*. Inverewe: privately published.

Hepburn, I. 1952. *Flowers of the coast*. London: Collins.

Hepper, F. N. 1956. Biological Flora of the British Isles. No. 59. *Silene nutans* L. *Journal of Ecology*, 44: 693–700.

Heslop-Harrison, Y. 1955a. Biological Flora of the British Isles. No. 49. *Nuphar* [genus] Sm. (pp. 342–343), *Nuphar lutea* (L.) Sm. (pp. 344–355), *Nuphar pumila* (Timm) DC. (pp. 355–360), *Nuphar* × *intermedia* Ledeb. (pp. 360–364). *Journal of Ecology*, 43: 342–364.

Heslop-Harrison, Y. 1955b. Biological Flora of the British Isles. No. 53. *Nymphaea* L. em. Sm. [genus] (pp. 719–721), *Nymphaea alba* L. (pp. 722–734). *Journal of Ecology*, 43: 719–734.

Hewett, D. G. 1964. Biological Flora of the British Isles. No. 97. *Menyanthes trifoliata* L. *Journal of Ecology*, 52: 723–735.

Hibberd, D. 1994. *Hardy Geraniums*. London: Cassell & Royal Horticultural Society.

Hignett, M. & Lacey, W. S. 1977. *Plants of Montgomeryshire. The field records of Janet Macnair.* Welshpool: Montgomeryshire Field Society & North Wales Naturalists' Trust.

Hill, D. J. & Price, B. 2000. Biological Flora of the British Isles. No. 210. *Ornithogalum pyrenaicum* L. *Journal of Ecology*, 88: 354–365.

Hill, M. O., Mountford, J. O., Roy, D. B. & Bunce, R. G. H. 1999. *Ellenberg's indicator values for British plants.* Huntingdon: Institute of Terrestrial Ecology.

Hill, M. O., Preston, C. D. & Smith, A. J. E. (eds). 1991–94. *Atlas of the bryophytes of Britain and Ireland.* 3 vols. Colchester: Harley Books.

Hobson, D. D. 1991. The status of *Populus nigra* L. in the Republic of Ireland. *Watsonia*, 18: 303–305.

Hobson, D. D. 1993. *Populus nigra* L. in Ireland – an indigenous species? *Irish Naturalists' Journal*, 24: 244–247.

Hocking, P. J. 1982. Salt and mineral nutrient levels in fruits of two strand species, *Cakile maritima* and *Arctotheca populifera*, with special reference to the effect of salt on the germination of *Cakile*. *Annals of Botany*, n.s. 50: 335–343.

Holland, S. C. (ed.). 1986. *Supplement to the Flora of Gloucestershire.* Bristol: Grenfell Publications.

Hollick, K. M. & Patrick, S. [1980]. *Supplement to Flora of Derbyshire 1969. Additional records received 1974–9.* Derby: Derby City Council, Museums and Art Gallery.

Hollings, E. & Stace, C. A. 1978. Morphological variation in the *Vicia sativa* L. aggregate. *Watsonia*, 12: 1–14.

Hollingsworth, M. L. & Bailey, J. P. 2000. Hybridisation and clonal diversity in some introduced *Fallopia* species (Polygonaceae). *Watsonia*, 23: 111–121.

Hollingsworth, P. M. & Swan, G. A. 1999. Genetic differentiation and hybridisation among subspecies of Deergrass (*Trichophorum cespitosum* (L.) Hartman) in Northumberland. *Watsonia*, 22: 235–242.

Hollingsworth, P. M., Tebbitt, M., Watson, K. J. & Gornall, R. J. 1998. Conservation genetics of an arctic species, *Saxifraga rivularis* L., in Britain. *Botanical Journal of the Linnean Society*, 128: 1–14.

Holyoak, D. T. 1999. *Gentianella uliginosa* (Willd.) Börner (Gentianaceae) rediscovered in north Devon. *Watsonia*, 22: 428–429.

Hopkins, J. J. 1996. Scrub ecology and conservation. *British Wildlife*, 8: 28–36.

Horrill, A. D. 1972. Biological Flora of the British Isles. No. 125. *Melampyrum cristatum* L. *Journal of Ecology*, 60: 235–244.

Howard, H. W. & Lyon, A. G. 1952. Biological Flora of the British Isles. No. 36. *Nasturtium officinale* R. Br. (pp. 228–238), *Nasturtium microphyllum* Boenningh. ex Rchb. (pp. 239–245). *Journal of Ecology*, 40: 228–245.

Howarth, S. E. & Williams, J. T. 1968. Biological Flora of the British Isles. No. 110. *Chrysanthemum leucanthemum* L. *Journal of Ecology*, 56: 585–595.

Howarth, S. E. & Williams, J. T. 1972. Biological Flora of the British Isles. No. 127. *Chrysanthemum segetum* L. *Journal of Ecology*, 60: 573–584.

Howitt, R. C. L. & Howitt, B. 1963. *A Flora of Nottinghamshire.* Nottingham: privately published.

Howitt, R. C. L. & Howitt, B. M. 1990. Willows. *In*: *A Guide to Some Difficult Plants*, pp. 28–40. London: Wild Flower Society.

Hubbard, C. E. 1954. *Grasses.* Harmondsworth: Penguin Books.

Huiskes, A. H. L. 1979. Biological Flora of the British Isles. No. 144. *Ammophila arenaria* (L.) Link. *Journal of Ecology*, 67: 363–382.

Hull, P. & Smart, G. J. B. 1984. Variation in two *Sorbus* species endemic to the Isle of Arran, Scotland. *Annals of Botany*, 53: 641–648.

Hultén, E. 1968. *Flora of Alaska and neighboring territories.* Stanford: Stanford University Press.

Hultén, E. 1971. The circumpolar plants. II. Dicotyledons. *Kungliga Svenska Vetenskapsakademiens Handlingar, Fjärde Serien*, 13: 1–463.

Hultén, E. & Fries, M. 1986. *Atlas of north European vascular plants north of the Tropic of Cancer.* 3 vols. Königstein: Koeltz Scientific Books.

Hultgård, U.-M. 1987. *Parnassia palustris* L. in Scandinavia. *Symbolae Botanicae Upsaliensis*, 28(1): 1–128.

Hume, L., Martinez, J. & Best, K. 1983. The biology of Canadian weeds. 60. *Polygonum convolvulus* L. *Canadian Journal of Plant Science*, 63: 959–971.

Hurry, J. B. 1930. *The woad plant and its dye.* Oxford: Oxford University Press.

Hurst, C. P. 1901. The range of *Diotis candidissima* Desf., in England and Wales, and in Ireland. *Memoirs and Proceedings of the Manchester Literary and Philosophical Society*, 46(1): 1–8.

Hutchings, M. J. & Price, E. A. C. 1999. Biological Flora of the British Isles. No. 205. *Glechoma hederacea* L. (*Nepeta glechoma* Benth., *N. hederacea* (L.) Trev.). *Journal of Ecology*, 87: 347–364.

Hutchings, M. J. , Mendoza, A. & Havers, W. 1998. Demographic properties of an outlier population of *Orchis militaris* L. (Orchidaceae). *Botanical Journal of the Linnean Society*, 126: 95–107.

Hutchinson, C. S. & Seymour, G. B. 1982. Biological Flora of the British Isles. No. 153. *Poa annua* L. *Journal of Ecology*, 70: 887–901.

Hutchinson, J., Colosi, J. & Lewin, R. A. 1984. The biology of Canadian weeds. 63. *Sonchus asper* (L.) Hill and *S. oleraceus* L. *Canadian Journal of Plant Science*, 64: 731–744.

Hutchinson, T. C. 1968. Biological Flora of the British Isles. No. 115. *Teucrium scorodonia* L. *Journal of Ecology*, 56: 901–911.

Ingram, M. 1958. The ecology of the Cairngorms, IV. The *Juncus* zone: *Juncus trifidus* communities. *Journal of Ecology*, 46: 707–737.

Ingram, R. & Noltie, H. J. 1981. *The flora of Angus.* Dundee: Dundee Museum and Art Galleries.

Ingram, R. & Noltie, H. J. 1995. Biological Flora of the British Isles. No. 186. *Senecio cambrensis* Rosser. *Journal of Ecology*, 83: 537–546.

Ingrouille, M. J. 1981. A newly discovered *Limonium* in East Sussex. *Watsonia*, 13: 181–184.

Ingrouille, M. J. 1985. The *Limonium auriculae-ursifolium* (Pourret) Druce group (Plumbaginaceae) in the Channel Isles. *Watsonia*, 15: 221–229.

Ingrouille, M.[J.] 1989. A non-expert's approach to the *Salicornia* problem. *BSBI News*, 53: 11–12.

Ingrouille, M. J. & Pearson, J. 1987. The pattern of morphological variation in the *Salicornia europaea* L. aggregate (Chenopodiaceae). *Watsonia*, 16: 269–281.

Ingrouille, M. J. & Smirnoff, N. 1986. *Thlaspi caerulescens* J. & C. Presl (*T. alpestre* L.) in Britain. *New Phytologist*, 102: 219–233.

Ingrouille, M. J. & Stace, C. A. 1986. The *Limonium binervosum* aggregate (Plumbaginaceae) in the British Isles. *Botanical Journal of the Linnean Society*, 92: 177–217.

Ivimey-Cook, R. B. 1959. Biological Flora of the British Isles. No. 71. *Agrostis setacea* Curt. *Journal of Ecology*, 47: 697–706.

Ivimey-Cook, R. B. 1963. Biological Flora of the British Isles. No. 88. *Hypericum linarifolium* Vahl. *Journal of Ecology*, 51: 727–732.

Ivimey-Cook, R. B. 1969. Investigations into the phenetic relationships between species of *Ononis* L. *Watsonia*, 7: 1–23.

Ivimey-Cook, R. B. 1984. *Atlas of the Devon Flora.* Exeter: Devonshire Association.

Jackson, A. 1995. The Plymouth pear – the recovery programme for one of Britain's rarest trees. *British Wildlife*, 6: 273–278.

Jackson, M. T. 1986. Exploited plants. The potato. *Biologist*, 33: 161–167.

Jacquemart, A.-L. 1996. Biological Flora of the British Isles. No. 193. *Vaccinium uliginosum* L. *Journal of Ecology*, 84: 771–785.

Jacquemart, A.-L. 1997. Biological Flora of the British Isles. No. 195. *Vaccinium oxycoccos* L. and *Vaccinium microcarpum* (Turcz. ex Rupr.) Schmalh. *Journal of Ecology*, 85: 381–396.

Jacquemart, A.-L. 1998. Biological Flora of the British Isles. No. 200. *Andromeda polifolia* L. *Journal of Ecology*, 86: 527–541.

Jalas, J. & Suominen, J. (eds). 1972–99. *Atlas Florae Europaeae, Volume 1* (1972), *2* (1973), *3* (1976), *4* (1979), *5* (1980), *6* (1983), *7* (1986), *8* (1989), *9* (1991), *10* (1994), *11* (1996), *12* (1999). Helsinki: Committe for Mapping the Flora of Europe and Societas Biologica Fennica Vanamo.

James, N. D. G. 1982. *The Forester's Companion.* Oxford: Blackwell.

James, R., Mitchell, S. C., Kett, J. & Leaton, R. 1981. The natural history of *Quercus ilex* L. in Norfolk. *Watsonia*, 13: 271–286.

James, T. J. 1997. The changing flora of Hertfordshire. *Transactions of the Hertfordshire Natural History Society*, 33: 62–84.

James, T. J. & Goldsmith, F. B. 1993. The decline and extinction of Common Cotton-grass (*Eriophorum angustifolium* Honckeny) in Hertfordshire. *Transactions of the Hertfordshire Natural History Society*, 31: 353–363.

Jenkinson, M. N. 1991. *Wild orchids of Dorset.* Gillingham: Orchid Sundries.

Jerling, L. 1988. Population dynamics of *Glaux maritima* (L.) along a distributional cline. *Vegetatio*, 74: 161–170.

Jerling, L. & Elmgren, G. 1996. Phenotypic variation in clonal growth of *Glaux maritima* along an environmental cline. *Acta Botanica Neerlandica*, 45: 367–380.

Jermy, A. C. 1989. The history of *Diphasiastrum issleri* (Lycopodiaceae) in Britain and a review of its taxonomic status. *Fern Gazette*, 13: 257–265.

Jermy, A. C. & Camus, J. C. 1991. *The illustrated field guide to ferns and allied plants of the British Isles.* London: Natural History Museum.

Jermy, A. C. & Crabble, J. A. (eds). 1978. *The island of Mull: a survey of its flora and environment.* London: British Museum (Natural History).

Jermy, A. C., Arnold, H. R., Farrell, L. & Perring, F. H. 1978. *Atlas of ferns of the British Isles.* London: Botanical Society of the British Isles & British Pteridological Society.

Jermy, A. C., Chater, A. O. & David, R. W. 1982. *Sedges of the British Isles.* Botanical Society of the British Isles Handbook no. 1, edn 2. London: Botanical Society of the British Isles.

Jermyn, S. T. 1974. *Flora of Essex.* Colchester: Essex Naturalists' Trust

Jobling, J. 1990. *Poplars for wood production and amenity.* Forestry Commission Bulletin no. 92. London: HMSO.

John, R. F. 1992. *Genetic variation, reproductive biology and conservation in isolated populations of rare plant species.* Ph.D. thesis, University of Wales, Swansea.

Jones, A. 1991. Welsh Mudwort? *BSBI Welsh Bulletin*, 52: 6–8.

Jones, B. & Gliddon, C. 1999. Reproductive biology and genetic structure in *Lloydia serotina*. *Plant Ecology*, 141: 151–161.

Jones, B. M. G. 1959. Distribution of *Bunias orientalis* in Britain. *Proceedings of the Botanical Society of the British Isles*, 3: 330.

Jones, D. A. & Turkington, R. 1986. Biological Flora of the British Isles. No. 163. *Lotus corniculatus* L. *Journal of Ecology*, 74: 1185–1212.

Jones, E. M. 1975. Taxonomic studies of the genus *Atriplex* (Chenopodiaceae). *Watsonia*, 10: 233–251.

Jones, E. W. 1945. Biological Flora of the British Isles. No. 13. *Acer* L. [genus] (pp. 215–219), *Acer pseudo-platanus* L. (pp. 220–237), *Acer platanoides* L. (p. 238), *Acer campestre* L. (pp. 239–252). *Journal of Ecology*, 32: 215–252.

Jones, E. W. 1959. Biological Flora of the British Isles. No. 67. *Quercus* L. [genus], *Quercus robur* L. and *Q. petraea* (Matt.) Liebl. (pp. 169–216), *Quercus borealis* Mich. var. *maxima* Sarg. (p. 216), *Quercus cerris* L. (pp. 216–217), *Quercus ilex* L. (pp. 218–222). *Journal of Ecology*, 47: 169–222.

Jones, M. 1984. The plant remains. *In*: B. Cunliffe, *Danebury: an Iron Age hillfort in Hampshire. Vol. 2. The excavations, 1969–78: the finds*, pp. 483–495. Council for British Archaeology Research Report no. 52.

Jones, M. 1987. *Orobanche hederae* Duby in the British Isles. *In*: H.C. Weber & W. Forstreuter, (eds), *Parasitic Flowering Plants*, pp. 457–471. Proceedings of the 4th International Symposium on Parasitic Flowering Plants. Marburg.

Jones, P. S. 1998. Aspects of the population biology of *Liparis loeselii* (L.) Rich. var. *ovata* Ridd. ex Godfery (Orchidaceae) in the dune slacks of South Wales, UK. *Botanical Journal of the Linnean Society*, 126: 123–139.

Jones, V. & Richards, P. W. 1954. Biological Flora of the British Isles. No. 46. *Juncus acutus* L. *Journal of Ecology*, 42: 639–650.

Jones, V. & Richards, P. W. 1956. Biological Flora of the British Isles. No. 57. *Saxifraga oppositifolia* L. *Journal of Ecology*, 44: 300–316.

Jones, V. & Richards, P. W. 1962. Biological Flora of the British Isles. No. 83. *Silene acaulis* (L.) Jacq. *Journal of Ecology*, 50: 475–487.

Jong, T. G. de, Klinkhamer, P. G. L. & Boorman, L. A. 1990. Biological Flora of the British Isles. No. 170. *Cynoglossum officinale* L. *Journal of Ecology*, 78: 1123–1144.

Jonsell, B. 1968. Studies in the north-west European species of *Rorippa* s. str. *Symbolae Botanicae Upsaliensis*, 19(2): 1–222.

Jonsell, B. (ed.). 2000. *Flora Nordica 1. Lycopodiaceae – Polygonaceae.* Stockholm: Bergius Foundation, Royal Swedish Academy of Sciences.

Jonsell, B., Nordal, I. & Roberts, F. J. 2000. *Viola rupestris* and its hybrids in Britain. *Watsonia*, 23: 269–278.

Jowsey, W. H. (comp.). 1978. *Botanical atlas of the Harrogate district.* Harrogate: Harrogate and District Naturalists' Society.

Kadereit, J. W. 1983. *Senecio vernalis* Waldst. & Kit. in Britain. *BSBI News*, 35: 8.

Kadereit, J. W. 1986a. *Papaver somniferum* L. (Papaveraceae): a triploid hybrid? *Botanische Jahrbücher für Systematik, Pflanzengeschichte und Pflanzengeographie*, 106: 221–244.

Kadereit, J. W. 1986b. A revision of *Papaver* section *Argemonidium*. *Notes from the Royal Botanic Garden Edinburgh*, 44: 25–43.

Kadereit, J. W. 1987. Experimental evidence on the affinities of *Papaver somniferum* (Papaveraceae). *Plant Systematics and Evolution*, 156: 189–195.

Kadereit, J. W. 1989. A revision of *Papaver* L. section *Rhoeadium* Spach. *Notes from the Royal Botanic Garden Edinburgh*, 45: 225–286.

Kay, G. M. 1998. A tale of two *Bidens*. *BSBI News*, 78: 64–65.

Kay, Q. O. N. 1969. The origin and distribution of diploid and tetraploid *Tripleurospermum inodorum* (L.) Schultz Bip. *Watsonia*, 7: 130–141.

Kay, Q. O. N. 1971a. Biological Flora of the British Isles. No. 122. *Anthemis cotula* L. *Journal of Ecology*, 59: 623–636.

Kay, Q. O. N. 1971b. Biological Flora of the British Isles. No. 123. *Anthemis arvensis* L. *Journal of Ecology*, 59: 637–648.

Kay, Q. O. N. 1972. Variation in sea mayweed (*Tripleurospermum inodorum* (L.) Koch) in the British Isles. *Watsonia*, 9: 81–107.

Kay, Q. O. N. 1985. Hermaphrodites and subhermaphrodites in a reputedly dioecious plant, *Cirsium arvense* (L.) Scop. *New Phytologist*, 100: 457–472.

Kay, Q. O. N. 1994. Biological Flora of the British Isles. No. 182. *Tripleurospermum inodorum* (L.) Schultz Bip. *Journal of Ecology*, 82: 681–697.

Kay, Q. O. N. 1997. *Review of the taxonomy, biology, geographical distribution and European conservation status of* Asparagus prostratus *Dumort. (*A. officinalis *subsp.* prostratus*), sea asparagus.* South Area Report. Cardiff: Countryside Council for Wales.

Kay, Q. O. N. 1998. Genetic variation, origins, history and conservation of *Sorbus domestica*, *S. leyana* and *S. minima*. *In*: A. Jackson & A. Flanagan (eds), *The Conservation Status of* Sorbus *in the UK*, pp. 5–14. London: Royal Botanic Gardens, Kew.

Kay, Q. O. N. & Harrison, J. 1970. Biological Flora of the British Isles. No. 119. *Draba aizoides* L. *Journal of Ecology*, 58: 877–888.

Kay, Q. O. N. & John, R. 1994. *Population genetics and demographic ecology of some scarce and declining vascular plants of Welsh lowland grassland and related habitats.* Science Report No. 93. Bangor: Countryside Council for Wales.

Kay, Q. O. N. & John, R. 1995. *The conservation of scarce and declining plant species in lowland Wales: population genetics, demographic ecology and recommendations for future conservation in 32 species of lowland grassland and related habitats.* Science Report No. 110. Bangor: Countryside Council for Wales.

Kay, Q. O. N., John, R. F. & Jones, R. A. 1999. Biology, genetic variation and conservation of *Luronium natans* (L.) Raf. in Britain and Ireland. *Watsonia*, 22: 301–315.

Kelly, D. L. 1990. *Cornus sericea* L. in Ireland: an incipient weed of wetlands. *Watsonia*, 18: 33–36.

Kelly, J. (ed.). 1995. *The Hillier gardener's guide to trees and shrubs.* Newton Abbot: David & Charles.

Kennedy, D. & Brown, I. R. 1983. The morphology of the hybrid *Betula pendula* Roth × *B. pubescens* Ehrh. *Watsonia*, 14: 329–336.

Kent D. H. 1956. *Senecio squalidus* in the British Isles – 1, early records (to 1877). *Proceedings of the Botanical Society of the British Isles*, 2: 115–118.

Kent D. H. 1957. *Senecio squalidus* L. in the British Isles – 3, East Anglia. *Transactions of the Norfolk and Norwich Naturalists' Society*, 18: 30–31.

Kent D. H. 1960. *Senecio squalidus* L. in the British Isles – 2, the spread from Oxford (1879–1939). *Proceedings of the Botanical Society of the British Isles*, 3: 375–379.

Kent, D. H. 1963. *Pteris cretica* L. *Proceedings of the Botanical Society of the British Isles*, 5: 121.

Kent D. H. 1964a. *Senecio squalidus* L. in the British Isles – 4, Southern England (1940 →). 5, The Midlands (1940 →). 6, Northern England (1940 →). *Proceedings of the Botanical Society of the British Isles*, 5: 210–219.

Kent D. H. 1964b. *Senecio squalidus* L. in the British Isles – 9, Ireland. *Irish Naturalists' Journal*, 14: 203–205.

Kent, D. H. 1965. The taxonomy of *Asplenium trichomanes* in Europe. *British Fern Gazette*, 9: 147–160.

Kent, D. H. 1968. *Echinops. Proceedings of the Botanical Society of the British Isles*, 7: 243–244.

Kent, D. H. 1975. *The historical flora of Middlesex.* London: Ray Society.

Kent, D. H. 1992. *List of vascular plants of the British Isles.* London: Botanical Society of the British Isles.

Kent, D. H. 1996. *List of vascular plants of the British Isles by D. H. Kent (1992). Supplement 1 (December 1996).* London: Botanical Society of the British Isles.

Kent, D. H. 2000. *List of vascular plants of the British Isles by D. H. Kent (1992). Supplement 2 (7th August 2000).* London: Botanical Society of the British Isles.

Kenward, H. K. & Hall, A. R. 1995. Biological evidence from Anglo-Scandinavian deposits at 16–22 Coppergate. *The Archaeology of York*, 14: 435–797 + i–xxiii.

Kenworthy, J. B. (ed.). 1976. *John Anthony's Flora of Sutherland.* Edinburgh: Botanical Society of Edinburgh.

Kerney, M. 1999. *Atlas of the land and freshwater molluscs of Britain and Ireland.* Colchester: Harley Books.

Killick, H. J. 1975. The decline of *Vicia sativa* L. *sensu stricto* in Britain. *Watsonia*, 10: 288–289.

Killick, J., Perry, R. & Woodell, S. 1998. *The Flora of Oxfordshire.* Newbury: Nature Conservation Bureau.

Kirchner, F. & Bullock, J. M. 1999. Taxonomic separation of *Ulex minor* Roth. and *U. gallii* Planch.: morphometrics and chromosome counts. *Watsonia*, 22: 365–376.

Kirschner, J. & Rich, T. C. G. 1996. *Luzula multiflora* subsp. *hibernica*, a new tetraploid taxon of *Luzula* sect. *Luzula* (Juncaceae) from Ireland. *Watsonia*, 21: 89–97.

Kitchener, G. 1983. Maritime plants on inland roadsides of West Kent. *Transactions of the Kent Field Club*, 9(2): 87–94.

Kitchener, G. D. & McKean, D. R. 1998. Hybrids of *Epilobium brunnescens* (Cockayne) Raven & Englehorn (Onagraceae) and their occurrence in the British Isles. *Watsonia*, 22: 49–60.

Klinkhamer, P. G. L. & Jong, T. G. de. 1993. Biological Flora of the British Isles. No. 176. *Cirsium vulgare* (Savi) Ten. *Journal of Ecology*, 81: 177–191.

Kull, T. 1999. Biological Flora of the British Isles. No. 208. *Cypripedium calceolus* L. *Journal of Ecology*, 87: 913–924.

Kytövuori, I. 1969. *Epilobium davuricum* Fisch. (Onagraceae) in Eastern Fennoscandinavia compared with *E. palustre* L. A morphological, ecological and distributional study. *Annales Botanici Fennici*, 6: 35–58.

Kytövuori, I. 1972. The Alpinae group of the genus *Epilobium* in northernmost Fennoscandinavia. A morphological, taxonomical and ecological study. *Annales Botanici Fennici*, 9: 163–203.

Lambert, J. M. 1947. Biological Flora of the British Isles. No. 17. *Glyceria maxima* (Hartm.) Holmb. *Journal of Ecology*, 34: 310–344.

Lansdown, R. V. 1999. A terrestrial form of *Callitriche truncata* Guss. subsp. *occidentalis* (Rouy) Braun-Blanquet (Callitrichaceae). *Watsonia*, 22: 283–286.

Lansdown, R. V. & Bruinsma J. 1999. *Callitriche palustris* L. new for Britain and Ireland. *BSBI News*, 82: 18–19.

Lavin, J. C. & Wilmore, G. T. D. 1994. *The West Yorkshire Plant Atlas.* Bradford: City of Bradford Metropolitan Council.

Le Huquet, J. 1993. A Flora of Herm and its off-islets. *Report and Transactions, La Société Guernesiaise*, 23: 493–559.

Le Sueur, F. 1981. *Crassula* aff. *radicans* (Haw.) Dietr. in Jersey. *BSBI News*, 29: 26–27.

Le Sueur, F. 1984. *Flora of Jersey.* Société Jersiaise.

Leach, S. J. 1984. Notes on the distribution of *Artemisia maritima* L. in eastern Scotland. *Watsonia*, 15: 36–38.

Leach, S. J. 1986. The rediscovery of *Carex maritima* Gunn. on the fairways at St. Andrews Links, Fife. *Watsonia*, 16: 80–81.

Leach, S. J. 1988. Rediscovery of *Halimione pedunculata* (L.) Aellen in Britain. *Watsonia*, 17: 170–171.

Leach, S. J. 1989. *Halimione portulacoides* (L.) Aellen in Co Down. *Irish Naturalists' Journal*, 23: 74–75.

Leach, S. J. 1990. *Cochlearia danica* on inland roadsides. *BSBI News*, 55: 20–21.

Leach, S. J. 1997. *Phacelia tanacetifolia* as a 'green manure'. *BSBI News*, 75: 40–41.

Leach, S. J. 1999. *Atriplex littoralis* on inland roadsides. *BSBI News*, 81: 36–37.

Leather, S. R. 1996. Biological Flora of the British Isles. No. 189. *Prunus padus* L. *Journal of Ecology*, 84: 125–132.

Lee, J. A. 1977. The vegetation of British inland salt marshes. *Journal of Ecology*, 65: 673–698.

Leeson, E., Haynes, C. & Wells, T. C. E. 1991. Studies of the phenology and dry matter allocation of *Dactylorhiza fuchsii*. *In*: T. C. E. Wells & J. H. Willems (eds), *Population Ecology of Terrestrial Orchids*, pp. 125–138. The Hague: SPB Academic Publishing.

Lehmann, E. & Schwemmle, J. 1927. Genetische Untersuchungen in der Gattung *Epilobium*. *Bibliotheca Botanica*, 95: 1–156.

Lemna, W. K. & Messersmith, C. G. 1990. The biology of Canadian weeds. 94. *Sonchus arvensis* L. *Canadian Journal of Plant Science*, 70: 509–532.

Leslie, A. C. 1976. *Dipsacus strigosus* Willd. in Cambridgeshire, v.c. 29. *Watsonia*, 11: 67–74.

Leslie, A. C. 1980a. A new alien *Pulmonaria*. *BSBI News*, 25: 16–17.

Leslie, A. C. 1980b. *Solanum nigrum* L. subsp. *schultesii* (Opiz) Wessely in Britain. *BSBI News*, 24: 21–22.

Leslie, A. C. 1980c. Further records of *Dipsacus strigosus* Willd. in Cambridgeshire. *Watsonia*, 13: 126–128.

Leslie, A. C. 1981a. A note on naturalized *Doronicum* in Britain. *BSBI News*, 27: 22–23.

Leslie, A. C. 1981b. *Polystichum munitum* Presl – a new British alien. *BSBI News*, 28: 24.

Leslie, A. C. 1982. A new alien *Symphytum*. *BSBI News*, 30: 16–17.

Leslie, A. C. 1984. Some new *Typha* records. *Watsonia*, 15: 168.

Leslie, A. C. 1987. *Flora of Surrey, supplement and checklist*. Guildford: privately published.

Lewin, R. A. 1948. Biological Flora of the British Isles. No. 22. *Sonchus* L. [genus] (pp. 203), *Sonchus oleraceus* L. emend Gouan (pp. 204–216), *Sonchus asper* (L.) Hill (pp. 216–223). *Journal of Ecology*, 36: 203–223.

Lewis, G. P. 1980. *Cytisus nigricans* L. *BSBI News*, 26: 20–21.

Light, M. H. S. & MacConaill, M. 1991. Patterns of appearance in *Epipactis helleborine* (L.) Crantz. *In*: T. C. E. Wells & J. H. Willems (eds), *Population Ecology of Terrestrial Orchids*, pp. 77–87. The Hague: SPB Academic Publishing.

Liljefors, A. 1955. Cytological studies in *Sorbus*. *Acta Horti Bergiani*, 17(4): 47–113.

Livermore, L. A. & Livermore, P. D. 1987. *The flowering plants and ferns of North Lancashire*. Preston: privately published.

Lodge, R. W. 1959. Biological Flora of the British Isles. No. 70. *Cynosurus cristatus* L. *Journal of Ecology*, 47: 511–518.

Looman, J. 1978. Biological Flora of the Canadian Prairie Provinces. V. *Koeleria gracilis* Pers. *Canadian Journal of Plant Science*, 58: 459–466.

Loudon, Mrs (ed.). 1855. *Loudon's encyclopaedia of plants*, edn 2. London: Longman, Brown, Green, and Longmans.

Lousley, J. E. 1944. Notes on British rumices: II. *Report of the Botanical Society and Exchange Club of the British Isles*, 12: 547–585.

Lousley, J. E. 1946. A new hybrid *Senecio* from the London area. *Report of the Botanical Society and Exchange Club of the British Isles*, 12: 869–874.

Lousley, J. E. 1947. *Typha angustifolia* L. × *latifolia* L. *Report of the Botanical Society and Exchange Club of the British Isles*, 13: 174.

Lousley, J. E. 1948a. *Calystegia sylvestris* (Willd.) R. & S. *Report of the Botanical Society and Exchange Club of the British Isles*, 13: 265–268.

Lousley, J. E. 1948b. *Ficus carica* L. in Britain. *Report of the Botanical Society and Exchange Club of the British Isles*, 13: 330–333.

Lousley, J. E. 1953. The recent influx of aliens into the British flora. *In*: Lousley, J. E. (ed.), *The changing flora of Britain*, pp. 140–159. Oxford: Botanical Society of the British Isles.

Lousley, J. E. 1958. *Plantago intermedia* in Britain? *Proceedings of the Botanical Society of the British Isles*, 3: 33–36.

Lousley, J. E. (comp.). 1961. A census list of wool aliens found in Britain, 1946–60. *Proceedings of the Botanical Society of the British Isles*, 4: 221–247.

Lousley, J. E. 1961 bis. The status of *Cucubalus baccifer* in England. *Proceedings of the Botanical Society of the British Isles*, 4: 262–268.

Lousley, J. E. 1968. A glabrous perennial *Sonchus* in Britain. *Proceedings of the Botanical Society of the British Isles*, 7: 151–157.

Lousley, J. E. 1971. *Flora of the Isles of Scilly*. Newton Abbot: David & Charles.

Lousley, J. E. 1976a. *Flora of Surrey*. London and Newton Abbot: David & Charles.

Lousley, J. E. 1976b. Three species of Polygonaceae established in Britain. *Watsonia*, 11: 144–146.

Lousley, J. E. & Kent, D. H. 1981. *Docks and knotweeds of the British Isles*. Botanical Society of the British Isles Handbook no. 3. London: Botanical Society of the British Isles.

Lovatt, C. M. 1982. *The history, ecology and status of the rare plants and the vegetation of the Avon Gorge, Bristol*. Ph.D. thesis, University of Bristol.

Lovett-Doust, J., Lovett-Doust, L. & Groth, A. T. 1990. The biology of Canadian weeds. 95. *Ranunculus repens* L. *Canadian Journal of Plant Science*, 70: 1123–1141.

Lovett-Doust, L., Mackinnon, A. & Lovett-Doust, J. 1985. Biology of Canadian weeds. 71. *Oxalis stricta* L., *O. corniculata* L., *O. dillenii* Jacq. ssp. *dillenii* and *O. dillenii* Jacq. ssp. *filipes* (Small) Eiten. *Canadian Journal of Plant Science*, 65: 691–709.

Lovis, J. D. 1955. The problem of *Asplenium trichomanes*. *In*: J.E. Lousley (ed.), *Species studies in the British Flora*, pp. 99–103. London: Botanical Society of the British Isles.

Lucas, M. J. & Middleton, J. 1985. *Flowers and ferns around Huddersfield*. Huddersfield: Kirklees Metropolitan Council.

Lupton, F. G. H. 1985. Exploited plants. Wheat. *Biologist*, 32: 97–105.

Lusby, P. 1998. On the extinct plants of Scotland. *In*: R. A. Lambert (ed.), *Species history in Scotland*, pp. 45–62. Edinburgh: Scottish Cultural Press.

Lusby, P. & Wright, J. 1996. *Scottish Wild Plants: their history, ecology and conservation*. Edinburgh: The Stationery Office.

Mabey, R. 1996. *Flora Britannica*. London: Sinclair-Stevenson.

McAllister, H. A. 1999. *Lysimachia punctata* L. and *L. verticillaris* Sprengel (Primulaceae) naturalised in the British Isles. *Watsonia*, 22: 279–281.

McAllister, H. A. & Rutherford, A. 1990. *Hedera helix* L. and *H. hibernica* (Kirchner) Bean (Araliaceae) in the British Isles. *Watsonia*, 18: 7–15.

McCallum Webster, M. 1978. *Flora of Moray, Nairn & East Inverness*. Aberdeen: Aberdeen University Press.

McClintock, D. 1972. Short Notes. *Gaudinia fragilis* (L.) Beauv. *Watsonia*, 9: 143–146.

McClintock, D. 1975. *The wildflowers of Guernsey*. London: Collins.

McClintock, D. 1986. The Guernsey Millet: a new subspecies. *Report and Transactions, La Société Guernesiaise*, 21: 698–700.

McClintock, D. 1987a. *Supplement to The Wild Flowers of Guernsey (Collins, 1975)*. St Peter Port: La Société Guernesiaise.

McClintock, D. 1987b. The dwarf spurrey of Jersey and Guernsey. *Report and Transactions, La Société Guernesiaise*, 22: 110–111.

McCollin, D., Moore, L. & Sparks, T. 2000. The flora of a cultural landscape: environmental determinants of change revealed using archival sources. *Biological Conservation*, 92: 249–263.

McCosh, D. J. 1988. Local floras – a progress report. *Watsonia*, 17: 81–89.

MacDonald, M. A. & Cavers, P. B. 1991. The biology of Canadian weeds 97. *Barbarea vulgaris* R. Br. *Canadian Journal of Plant Science*, 71: 149–166.

McDonnell, E. J. 1995. *The status of Toadflax-leaved St. John's-wort (*Hypericum linariifolium* Vahl) in Britain in 1994*. Back from the Brink Project Report no. 40. London: Plantlife.

McHaffie, H. [S.] 1999. *Athyrium distentifolium* var. *flexile*: an endemic variety. *Botanical Journal of Scotland*, 51: 227–236.

McKean, D. R. (comp. & ed.). 1989. *A check-list of the flowering plants and ferns of Midlothian*, edn 2. Edinburgh: Botanical Society of Edinburgh.

McKee, F. 1999. The vascular plant database for Northern Ireland (NI) and NI Atlas 2000 project. *BSBI News*, 80: 10–12.

McKee, J. & Richards, A. J. 1998. The effect of temperature on reproduction in five *Primula* species. *Annals of Botany*, 82: 359–374.

McNaughton, I. H. & Harper, J. L. 1964. Biological Flora of the British Isles. No. 99. *Papaver rhoeas* L. (pp. 767–779), *Papaver dubium* L. (pp. 780–783), *Papaver lecoqii* Lamotte (pp. 783–786), *Papaver argemone* L. (pp. 786–789), *Papaver hybridum* L. (pp. 789–793). *Journal of Ecology*, 52: 767–793.

McNeill, J. 1977. The biology of Canadian weeds. 25. *Silene alba* (Miller) E. H. L. Krause. *Canadian Journal of Plant Science*, 57: 1103–1114.

McNeill, J. 1980. The biology of Canadian weeds. 46. *Silene noctiflora* L. *Canadian Journal of Plant Science*, 60: 1243–1253.

Macpherson, P. 1997. Plant status nomenclature and Atlas 2000. *BSBI News*, 7–8.

Macpherson, P. & Clement, E. J. 1986. *Montia parviflora* – new to Scotland. *BSBI News*, 44: 17–19.

Macpherson, P., Dickson, J. H., Ellis, R. G., Kent, D. H. & Stace, C. A. 1996. Plant status nomenclature. *BSBI News*, 72: 13–16.

McVean, D. N. 1953. Biological Flora of the British Isles. No. 37. *Alnus glutinosa* (L.) Gaertn. *Journal of Ecology*, 41: 447–466.

McVean, D. N. 1955a. Ecology of *Alnus glutinosa* (L.) Gaertn. I. Fruit formation. *Journal of Ecology*, 43: 46–60.

McVean, D. N. 1955b. Ecology of *Alnus glutinosa* (L.) Gaertn. II. Seed distribution and germination. *Journal of Ecology*, 43: 61–71.

McVean, D. N. 1961. Post-glacial history of juniper in Scotland. *Proceedings of the Linnean Society of London*, 172: 53–55.

McVean, D. N. & Ratcliffe, D. A. 1962. *Plant communities of the Scottish Highlands*. Monographs of the Nature Conservancy no. 1. London: Her Majesty's Stationery Office.

Madge, S. 1994. The status of *Serapias parviflora* Parl in Britain. *Botanical Cornwall*, 6: 51–52.

Mal, T. K., Lovett-Doust, J., Lovett-Doust, L. & Mulligan, G. A. 1992. The biology of Canadian weeds. 100. *Lythrum salicaria*. *Canadian Journal of Plant Science*, 72: 1305–1330.

Malik, N. & Vanden Born, W. H. 1988. The biology of Canadian weeds. 86. *Galium aparine* L. and *Galium spurium* L. *Canadian Journal of Plant Science*, 68: 481–499.

Malloch, A. J. C. & Okusanya, O. T. 1979. An experimental investigation into the ecology of some maritime cliff species I. Field observations. *Journal of Ecology*, 67: 283–292.

Manchester, S. J. & Bullock, J. M. 2000. The impacts of non-native species on UK biodiversity and the effectiveness of control. *Journal of Applied Ecology*, 37: 845–864.

Marchant, C. J. & Goodman, P. J. 1969. Biological Flora of the British Isles. No. 116. *Spartina maritima* (Curtis) Fernald (pp. 287–291), 2. *Spartina alterniflora* Loisel. (pp. 291–295), 3. *Spartina glabra* Muhl. (pp. 295–297). *Journal of Ecology*, 57: 287–297.

Mardon, D. K. 1990. Conservation of montane willow scrub in Scotland. *Transactions of the Botanical Society of Edinburgh*, 45: 427–436.

Margetts, L. J. & David, R. W. 1981. *A review of the Cornish flora*. Redruth: Institute of Cornish Studies.

Margetts, L. J. & Spurgin, K. L. 1991. *The Cornish Flora Supplement 1981–90*. St. Ives: Trendrine Press.

Marren, P. R. 1971. The Lundy Cabbage. *Annual Report of the Lundy Field Society*, 22: 27–31.

Marren, P. R. 1972. Addenda to The Lundy Cabbage. *Annual Report of the Lundy Field Society*, 23: 51–52.

Marren, P. [R.] 1981. A possible origin of *Carum verticillatum* (L.) Koch in north-eastern Scotland. *Watsonia*, 13: 323.

Marren, P. R. 1983. The history of Dickie's fern in Kincardineshire. *Transactions of the Botanical Society of Edinburgh*, 44: 157–164.

Marren, P. R. 1988. The past and present distribution of *Stachys germanica* L. in Britain. *Watsonia*, 17: 59–68.

Marren, P. [R.] 1999. *Britain's rare flowers*. London: T. & A.D. Poyser.

Marren, P. R., Payne, A. G. & Randall, R. E. 1986. The past and present status of *Cicerbita alpina* (L.) Wallr. in Britain. *Watsonia*, 16: 131–142.

Marsden, M. H. 1995. A check list of flowering plants and ferns wild on Sark and its off islets. *Report and Transactions, La Société Guernesiaise*, 23: 754–783.

Marsden-Jones, E. M. & Turrill, W. B. 1954. *British Knapweeds*. London: Ray Society.

Marsden-Jones, E. M. & Turrill, W. B. 1957. *The Bladder Campions*. London: Ray Society.

Marshall, J. K. 1967. Biological Flora of the British Isles. No. 105. *Corynephorus canescens* (L.) Beauv. *Journal of Ecology*, 55: 207–220.

Martin, J. 1993. *Cotula coronopifolia* established in Yorkshire. *BSBI News*, 64: 42–43.

Martin, M. H. & Frost, L. C. 1980. Autecological studies of *Trifolium molinerii* at the Lizard Peninsula, Cornwall. *New Phytologist*, 86: 329–344.

Martinsson, K. 1991a. *Callitriche* in Sweden: case studies of reproductive biology and intraspecific variation in a semi-aquatic plant genus. *Acta Universitatis Upsaliensis*, 327.

Martinsson, K. 1991b. Geographical variation in fruit morphology in Swedish *Callitriche hermaphroditica* (Callitrichaceae). *Nordic Journal of Botany*, 11: 497–512.

Mathew, B. 1981. *The Iris*. London: Batsford.

Mathew, B. 1982. *The Crocus*. London: Batsford.

Mathew, B. 1996. *A review of* Allium *Section* Allium. Kew: Royal Botanic Gardens.

Matthews, P. & Davison, A. W. 1976. Maritime species on roadside verges. *Watsonia*, 11: 146–147.

Maun, M. A. & Barrett, S. C. H. 1986. The biology of Canadian weeds. 77. *Echinochloa crus-galli* (L.) Beauv. *Canadian Journal of Plant Science*, 66: 739–759.

Maw, M. G., Thomas, A. G. & Stahevitch, A. 1985. The biology of Canadian weeds. 66. *Artemisia absinthum* L. *Canadian Journal of Plant Science*, 65: 389–400.

Maxted, N. & Trueman, I. C. 1983. An investigation into the occurrence of *Salix × meyerana* Rostk. ex Willd. in Shropshire. *Watsonia*, 14: 337–346.

May, R. F. 1967. *A list of the flowering plants and ferns of Carmarthenshire*. Haverfordwest: West Wales Naturalists' Trust.

Meade, M. 1989. Year-by-year observations of *Selinum carvifolia*, *Parnassia palustris* and other species on Sawston Hall Moor. *Nature in Cambridgeshire*, 31: 43–45.

Meikle, R. D. 1984. *Willows and poplars of Great Britain and Ireland*. Botanical Society of the British Isles Handbook no. 4. London: Botanical Society of the British Isles.

Meikle, R. D. 1992. British willows; some hybrids and some problems. *Proceedings of the Royal Society of Edinburgh*, 98B: 13–20.

Melville, R. 1940. Contributions to the study of British elms, III. The Plot Elm, *Ulmus plotii* Druce. *Journal of Botany*, 78: 181–192.

Mercer, R. J. 1981. *Grimes Graves, Norfolk excavations 1971–2: volume 1*. Department of the Environment Archaeological Reports no. 11. London: Her Majesty's Stationery Office.

Meredith, H. M. 1994. *Scrophularia scorodonia* L. (balm-leaved figwort) – an enigma. *Botanical Cornwall*, 6: 21–36.

Meredith, T. C. & Grubb, P. J. 1993. Biological Flora of the British Isles. No. 179. *Peucedanum palustre* (L.) Moench. *Journal of Ecology*, 81: 813–826.

Messenger, G. 1971. *Flora of Rutland*. Leicester: Leicester Museums.

Meusel, H. & Jäger, E. J. 1992. *Vergleichende Chorologie der zentraleuropäischen Flora. Volume 3*. 2 vols. Jena: Gustav Fischer.

Meusel, H., Jäger, E., Rauschert, S. & Weinert, E. 1978. *Vergleichende Chorologie der zentraleuropäischen Flora. Volume 2*. 2 vols. Jena: Gustav Fischer.

Meusel, H., Jäger, E. & Weinert, E. 1965. *Vergleichende Chorologie der zentraleuropäischen Flora. Volume 1*. 2 vols. Jena: Gustav Fischer.

Millar, J. M. 1993. *Flowers of Iona*. Iona: New Iona Press.

Millward, D. 1988. *A Flora of Wensleydale*. Yoredale Natural History Society.

Milne-Redhead, E. 1990. The BSBI Black Poplar survey, 1973–88. *Watsonia*, 18: 1–5.

Mitchell, A. F. 1972. *Conifers in the British Isles. A descriptive handbook*. London: Her Majesty's Stationery Office.

Mitchell, A. [F.] 1996. *Alan Mitchell's Trees of Britain*. London: Harper Collins.

Mitchell, J. 1981. *Elatine hydropiper* at Kilmannan Reservoir. *Glasgow Naturalist*, 20: 185–186.

Mitchell, N. D. & Richards, A. J. 1979. Biological Flora of the British Isles. No. 145. *Brassica oleracea* L. ssp. *oleracea*. *Journal of Ecology*, 67: 1087–1096.

Miyanishi, K., Eriksson, O. & Wein, R. W. 1991. The biology of Canadian weeds. 98. *Potentilla anserina* L. *Canadian Journal of Plant Science*, 71: 791–801.

Moore, D. M. 1958. Biological Flora of the British Isles. No. 66. *Viola lactea* Sm. *Journal of Ecology*, 46: 527–535.

Moore, J. J. 1966. *Minuartia recurva* (All.) Schinz and Thell. new to the British Isles. *Irish Naturalists' Journal*, 15: 130–132.

Moore, R. J. 1975. The biology of Canadian weeds. 13. *Cirsium arvense* (L.) Scop. *Canadian Journal of Plant Science*, 55: 1033–1048.

Morris, M. G. & Perring, F. H. 1974. *The British Oak*. BSBI Conference Report no. 14. Faringdon: Classey.

Morton, J. K. 1973. A cytological study of the British Labiatae (excluding *Mentha*). *Watsonia*, 9: 239–246.

Mountford, J. O. 1994. Floristic change in English grazing marshes: the impact of 150 years of drainage and land-use change. *Watsonia*, 20: 3–24

Mulligan, G. A. & Bailey, L. G. 1975. The biology of Canadian weeds. 8. *Sinapis arvensis* L. *Canadian Journal of Plant Science*, 55: 171–183.

Mulligan, G. A. & Findlay, J. N. 1974. The biology of Canadian weeds. 3. *Cardaria draba*, *C. chalepensis* and *C. pubescens*. *Canadian Journal of Plant Science*, 54: 149–160.

Mulligan, G. A. & Munro, D. B. 1981. The biology of Canadian weeds. 48. *Cicuta maculata* L., *C. douglasii* (DC.) Coult. & Rose and *C. virosa* L. *Canadian Journal of Plant Science*, 61: 93–105.

Murphy, J. P. 1981. *Senecio × albescens* Burbidge & Colgan at Killiney, Co. Dublin: a seventy-eight years old population. *Watsonia*, 13: 303–311.

Murphy, R. J. 1994. Progress report. *Botanical Cornwall*, 6: 1–7.

Murray, C. W. 1980. *The botanist in Skye*, edn 2. London: Botanical Society of the British Isles.

Muscott, J. (comp. & ed.). 1989. *A checklist of the flowering plants and ferns of West Lothian*. Edinburgh: Botanical Society of Edinburgh.

Myerscough, P. J. 1980. Biological Flora of the British Isles. No. 148. *Epilobium angustifolium* L. *Journal of Ecology*, 68: 1047–1074.

Myerscough, P. J. & Whitehead, F. H. 1966. Comparative biology of *Tussilago fafara* L., *Chamaenerion angustifolium* (L.) Scop., *Epilobium montanum* L., and *Epilobium adenocaulon* Hausskn. I. General biology and germination. *New Phytologist*, 65: 192–210.

Myerscough, P. J. & Whitehead, F. H. 1967. Comparative biology of *Tussilago farfara* L., *Chamaenerion angustifolium* (L.) Scop., *Epilobium montanum* L., and *Epilobium adenocaulon* Hausskn. II. Growth and ecology. *New Phytologist*, 66: 785–823.

Naylor, R. E. L. 1972. Biological Flora of the British Isles. No. 129. *Alopecurus myosuroides* Huds. *Journal of Ecology*, 60: 611–622.

Nelson, E. C. 1977. The discovery in 1810 and subsequent history of *Phyllodoce caerulea* (L.) Bab. in Scotland. *The Western Naturalist*, 6: 45–72.

Nelson, E. C. 1993. Who was the author of *Montbretia crocosmiflora*? *Watsonia*, 19: 265–267.

Nelson, E. C. (comp. & ed.). 2000a. *An annotated topographical checklist of the flowering plants, conifers, ferns and fern allies of the Burren Region*. Wisbech: privately published.

Nelson, E. C. 2000b. A history, mainly nomenclatural, of St Dabeoc's Heath. *Watsonia*, 23: 47–58.

Nethercott, P. J. M. 1998. Conservation status of *Sorbus* in the Avon Gorge. *In*: A. Jackson & M. Flanagan (eds), *The conservation status of* Sorbus *in the UK*, pp. 40–43. London: Royal Botanic Gardens, Kew.

New, J. K. 1961. Biological Flora of the British Isles. No. 76. *Spergula arvensis* L. *Journal of Ecology*, 49: 205–215.

Newcombe, M. 1991. Sea Heath and other plants on a degraded cliff face. *Transactions of the Kent Field Club*, 11(2): 93

Newlands, C. & Smith, H. 1998. Management and conservation status of sites with *Orobanche reticulata* Wallr. populations. *The Naturalist*, 123: 70–75.

Newton, A. 1971. *Flora of Cheshire*. Chester: Cheshire Community Council.

Newton, A. [1991]. *Supplement to Flora of Cheshire*. Leamington Spa: privately published.

Nicolle, D. 1991. *Scilla bithynica*. *BSBI News*, 58: 40–41.

Nijs, J. C. M. den. 1984. Biosystematic studies of the *Rumex acetosella* complex (Polygonaceae). VIII. A taxonomic revision. *Feddes Repertorium*, 95: 43–66.

Nixon, C. J. & Worrell, R. 1999. *The potential for natural regeneration of conifers in Britain*. Forestry Commission Bulletin no. 120. Edinburgh: Forestry Commission.

Noble, J. C. 1982. Biological Flora of the British Isles. No. 152. *Carex arenaria* L. *Journal of Ecology*, 70: 867–886.

Norman, E. 1987. An alien *Silene* in Scotland. *BSBI News*, 45: 26.

O'Donovan, J. T. & Sharma, M. P. 1987. The biology of Canadian weeds. 78. *Galeopsis tetrahit* L. *Canadian Journal of Plant Science*, 67: 787–796.

O'Leary, M. 1989. The habitat of *Selinum carvifolia* in Cambridgeshire. *Nature in Cambridgeshire*, 31: 36–43.

O'Mahony, T. 1976. *Carex depauperata* Curt. in NE Cork (H5) a sedge new to Ireland. *Irish Naturalists' Journal*, 18: 296–298.

O'Mahony, T. 1985. The history of *Geranium purpureum* Vill. in the Irish flora. *Irish Naturalists' Journal*, 21: 517–521.

Ockendon, D. J. 1968. Biological Flora of the British Isles. No. 113. *Linum perenne* L. ssp. *anglicum* (Miller) Ockendon. *Journal of Ecology*, 56: 871–882.

Ockendon, D. J. & Walters, S. M. 1970. Studies in *Potentilla anserina* L. *Watsonia*, 8: 135–144.

Ockendon, D. J., Walters, S. M. & Whiffen, T. P. 1969. Variation within *Centaurea nigra* L. *Proceedings of the Botanical Society of the British Isles*, 7: 549–552.

Ogilvie, M. 1995. *The wild flowers of Islay*. Islay: Lochindaal Press.

Okusanya, O. T. 1979a. An experimental investigation into the ecology of some maritime cliff species II. Germination studies. *Journal of Ecology*, 67: 293–304.

Okusanya, O. T. 1979b. An experimental investigation into the ecology of some maritime cliff species III. Effects of water on growth. *Journal of Ecology*, 67: 579–590.

Okusanya, O. T. 1979c. An experimental investigation into the ecology of some maritime cliff species IV. Cold sensitivity and competition studies. *Journal of Ecology*, 67: 591–600.

Oleskevich, C., Shamoun, S. F. & Punja, Z. K. 1996. The biology of Canadian weeds. 105. *Rubus strigosus* Michx., *Rubus parviflorus* Nutt., and *Rubus spectabilis* Pursh. *Canadian Journal of Plant Science*, 76: 187–201.

Oliver, J. 1998. North American asters in Wiltshire. *Wiltshire Archaeological and Natural History Magazine*, 91: 128–138.

Oliver, J. 2000. *Quercus × rosacea* in Savernake Forest. *BSBI News*, 84: 31–34.

Oswald, P. [H.] 1992. The Fritillary in Britain – a historical perspective. *British Wildlife*, 3: 200–210.

Oswald, P. [H.] 1993. Native and naturalised garlics in the Cambridge University Botanic Garden. *Nature in Cambridgeshire*, 35: 67–75.

Oswald, P. H. 2000. Historical records of *Lactuca serriola* L. and *L. virosa* L. in Britain, with special reference to Cambridgeshire (v.c. 29). *Watsonia*, 23: 149–159.

Ovington, J. D. & Scurfield, G. 1956. Biological Flora of the British Isles. No. 54. *Holcus mollis* L. *Journal of Ecology*, 44: 272–280.

Packham, J. R. 1978. Biological Flora of the British Isles. No. 141. *Oxalis acetosella* L. *Journal of Ecology*, 66: 669–693.

Packham, J. R. 1983. Biological Flora of the British Isles. No. 155. *Lamiastrum galeobdolon* (L.) Ehrend. & Polatschek. *Journal of Ecology*, 71: 975–997.

Page, C. N. 1982. The history and spread of bracken in Britain. *Proceedings of the Royal Society of Edinburgh*, 81B: 3–10.

Page, C. N. 1997. *The ferns of Britain and Ireland*, edn 2. Cambridge: Cambridge University Press.

Page, C. N. & Barker, M. A. 1985. Ecology and geography of hybridisation in British and Irish horsetails. *Proceedings of the Royal Society of Edinburgh*, 86B: 265–272.

Page, K. [W.] 1987. Hybrid bluebells. *BSBI News*, 46: 9.

Page, K. W. 1988. *Prunus pensylvanicus* L. fil. – a new alien record. *BSBI News*, 50: 34.

Palin, M. A. 1988. Biological Flora of the British Isles. No. 164. *Ligusticum scoticum* L. *Journal of Ecology*, 76: 889–902.

Palmer, J. H. & Sagar, G. R. 1963. Biological Flora of the British Isles. No. 93. *Agropyron repens* (L.) Beauv. *Journal of Ecology*, 51: 783–794.

Palmer, J. R. 1990. *Solanum chenopodioides* Lamarck, established in S.E. London. *BSBI News*, 54: 35–36.

Palmer, J. R. 1991. Mexican orange, *Choisya ternata*, naturalised in a west Kent woodland. *BSBI News*, 58: 41.

Palmer, J. R. 1994. *Cerastium brachypetalum* – status in W. Kent. *BSBI News*, 65: 21–22.

Palmer, M. A. & Bratton, J. H. 1995. *A sample survey of the flora of Britain and Ireland*. U.K. Conservation no. 8. Peterborough: Joint Nature Conservation Committee.

Pan, J.-T. 1988. A conspectus of the genus *Bergenia* Moench. *Acta Phytotaxonomica Sinica*, 26: 120–129 [in Chinese with an English summary].

Pankhurst, R. J. & Mullin, J. M. 1991. *Flora of the Outer Hebrides*. London: Natural History Museum Publications.

Parker, D. M. 1979. *Saxifraga rosacea* and *S. hypnoides* in the British Isles. *BSBI News*, 21: 22–23.

Parker, D. M. 1981. The re-introduction of *Saxifraga cespitosa* to North Wales. *In*: H. Synge (ed.), *The Biological Aspects of Rare Plant Conservation*, Appendix 1, pp. 506–508. Chichester: J. Wiley & Sons.

Parker, D. [M.] 1996. Tufted Saxifrage. *British Wildlife*, 7: 201.

Parks, J. C., Dyer, A. F. & Lindsay, S. 2000. Allozyme, spore and frond variation in some Scottish populations of the ferns *Cystopteris dickieana* and *Cystopteris fragilis*. *Edinburgh Journal of Botany*, 57: 83–105.

Parnell, J. [A. N.] & Needham, M. 1998. Morphometric variation in Irish *Sorbus* L. (Rosaceae). *Watsonia*, 22: 153–161.

Parnell, J. A. N. 1985. Biological Flora of the British Isles. No. 157. *Jasione montana* L. *Journal of Ecology*, 73: 341–358.

Paton, A. 1967. True service trees of Worcestershire. *Proceedings of the Botanical Society of the British Isles*, 7: 9–13.

Patrick, S. & Hollick, K. M. [1974]. *Supplement to Flora of Derbyshire 1969. Additional records received 1969–1974*. Derby: Derby Borough Council, Museums and Art Gallery.

Paul, A. M. 1987. The status of *Ophioglossum azoricum* (Ophioglossaceae : Pteridophyta) in the British Isles. *Fern Gazette*, 13: 173–187.

Pearman, D. A. 1994. *Sedges and their allies in Dorset*. Dorchester: Dorset Environmental Records Centre.

Pearman, D. A. 1996. Atlas 2000. *BSBI News*, 72: 3.

Pearman, D. [A.] 1997. *Carex humilis* Leysser in Dorset (v.c. 9). *Watsonia*, 21: 368–374.

Pearman, D. A. & Preston, C. D. 1996. Atlas 2000 – a new atlas of flowering plants and ferns. *British Wildlife*, 7: 305–308.

Pearman, D. A. & Preston, C. D. 2000. *A Flora of Tiree, Gunna and Coll*. Dorchester: privately published.

Pearson, M. C. & Rogers, J. A. 1962. Biological Flora of the British Isles. No. 85. *Hippophaë rhamnoides* L. *Journal of Ecology*, 50: 501–513.

Perring, F. H. 1956. *Spiranthes spiralis* (L.) Chevall. in Britain, 1955. *Proceedings of the Botanical Society of the British Isles*, 2: 6–9.

Perring, F. H. 1960. Report on the survey of *Arctium* L. agg. in Britain, 1959. *Proceedings of the Botanical Society of the British Isles*, 4: 33–37.

Perring, F. H. 1970. The last seventy years. *In*: F.[H.] Perring (ed.), *The flora of a changing Britain*, pp. 128–135. Hampton: E. W. Classey.

Perring, F. H. 1973. Mistletoe. *In*: P. S. Green (ed.), *Plants wild and cultivated*, pp. 139–145. Hampton: E. W. Classey.

Perring, F. H. 1974. Changes in our native vascular plant flora. *In*: D. L. Hawksworth (ed.), *The changing flora and fauna of Britain*, pp. 7–25. London: Academic Press.

Perring, F. H. 1992. BSBI distribution maps scheme – the first forty years. *In*: P.T. Harding (ed.), *Biological recording of changes in British wildlife*, pp. 1–4. London: HMSO.

Perring, F. H. 1994. *Symphytum* – Comfrey. *In*: A. R. Perry & R. G. Ellis (eds), *The common ground of wild and cultivated plants*, pp. 65–70. BSBI Conference Report no. 22. Cardiff: National Museum of Wales.

Perring, F. H. & Farrell, L. 1977. *British Red Data Books: 1. Vascular Plants*. Lincoln: Society for Nature Conservation.

Perring, F. H. & Farrell, L. 1983. *British Red Data Books: 1. Vascular Plants*, edn 2. Lincoln: Royal Society for Nature Conservation.

Perring, F. H. & Sell, P. D. (eds). 1968. *Critical supplement to the Atlas of the British Flora*. London: Thomas Nelson & Sons.

Perring, F. H. & Walters, S. M. (eds). 1962. *Atlas of the British Flora*. London: Thomas Nelson & Sons.

Perring, F. H. & Walters, S. M. (eds). 1976. *Atlas of the British Flora*, 2nd ed. Wakefield: EP Publishing.

Perring, F. H. & Walters, S. M. (eds). 1982. *Atlas of the British Flora*, 3rd ed. Wakefield: EP Publishing.

Perring, F. H., Sell, P. D., Walters, S. M. & Whitehouse, H. L. K. 1964. *A Flora of Cambridgeshire*. Cambridge: Cambridge University Press.

Petch, C. P. & Swann, E. L. 1968. *Flora of Norfolk*. Norwich: Jarrold & Sons.

Peterken, G. F. 1974. A method for assessing woodland flora for conservation using indicator species. *Biological Conservation*, 6: 239–245.

Peterken, G. F. 1975. Holly survey. *Watsonia*, 10: 297–299.

Peterken, G. F. 1981. *Woodland conservation and management*. London: Chapman & Hall.

Peterken, G. F. & Lloyd, P. S. 1967. Biological Flora of the British Isles. No. 108. *Ilex aquifolium* L. *Journal of Ecology*, 55: 841–858.

Peterson, S.-R. 1969. Biology of the mouse-ear chickweed, *Cerastium vulgatum*. *The Michigan Botanist*, 8: 151–157.

Pfitzenmeyer, C. D. C. 1962. Biological Flora of the British Isles. No. 81. *Arrhenatherum elatius* (L.) J. & C. Presl *Journal of Ecology*, 50: 235–245.

Phillips, E. N. M. 1977. *Carduus pycnocephalus* L. on Plymouth Hoe, S. Devon. *Watsonia*, 11: 384–385.

Phillips, M. E. 1954. Biological Flora of the British Isles. No. 43. *Eriophorum angustifolium* Roth. *Journal of Ecology*, 42: 612–622.

Phillips, R. & Rix, M. 1993. *Vegetables*. London: Pan Books.

Philp, E. G. 1982. *Atlas of the Kent flora*. Maidstone: Kent Field Club.

Pigott, C. D. 1955. Biological Flora of the British Isles. No. 50. *Thymus* L. [genus] (pp. 365–368), *Thymus drucei* Ronniger emend Jalas (pp. 369–379), *Thymus serpyllum* Linn. emend Mill subsp. *serpyllum* (pp. 379–382), *Thymus pulegioides* Linn. (pp. 383–387). *Journal of Ecology*, 43: 365–387.

Pigott, C. D. 1958. Biological Flora of the British Isles. No. 65. *Polemonium caeruleum* L. *Journal of Ecology*, 46: 507–525.

Pigott, C. D. 1968. Biological Flora of the British Isles. No. 111. *Cirsium acaulon* (L.) Scop. *Journal of Ecology*, 56: 597–612.

Pigott, C. D. 1969. The status of *Tilia cordata* and *T. platyphyllos* on the Derbyshire limestone. *Journal of Ecology*, 57: 491–504.

Pigott, C. D. 1981. The status, ecology and conservation of *Tilia platyphyllos* in Britain. *In*: H. Synge (ed.), *The Biological Aspects of Rare Plant Conservation*, pp. 305–317. Chichester: John Wiley & Sons.

Pigott, C. D. 1991. Biological Flora of the British Isles. No. 174. *Tilia cordata* Miller. *Journal of Ecology*, 79: 1147–1207.

Pigott, [C.] D. 1992. The clones of comon lime (*Tilia* × *vulgaris* Hayne) planted in England during the seventeenth and eighteenth centuries. *New Phytologist*, 121: 487–493.

Piqueras, J. & Klimes, L. 1998. Demography and modelling of clonal fragments in the pseudoannual plant *Trientalis europaea*. *Plant Ecology*, 136: 213–227.

Porley, R. D. 1999. Separation of *Carex vulpina* L. and *C. otrubae* Podp. (Cyperaceae) using transverse leaf sections. *Watsonia*, 22: 431–432.

Porter, M. S. & Roberts, F. J. 1997. Bird's-foot Sedge (*Carex ornithopoda* Willd.) in Cumbria. *The Carlisle Naturalist*, 5: 18–23.

Praeger, R. L. 1934. *The Botanist in Ireland*. Dublin: Hodges, Figgis, & Co.

Preston, C. D. 1988. The spread of *Epilobium ciliatum* Raf. in the British Isles. *Watsonia*, 17: 279–288.

Preston, C. D. 1990. An index and bibliography to distribution maps published between 1962 and 1989. *In*: F. H. Perring & S. M. Walters (eds), *Atlas of the British Flora*, edn 3 (reprint), pp. 426–434. London: Botanical Society of the British Isles.

Preston, C. D. 1993. The distribution of the Oxlip *Primula elatior* (L.) Hill in Cambridgeshire. *Nature in Cambridgeshire*, 35: 29–60.

Preston, C. D. 1995. *Pondweeds of Great Britain and Ireland*. Botanical Society of the British Isles Handbook no. 8. London: Botanical Society of the British Isles.

Preston, C. D. [1996]. *Fieldwork for Atlas 2000. 1. Notes on identification works and some difficult and under-recorded taxa*. [London: Botanical Society of the British Isles.]

Preston, C. D. 1997. The genus *Rosa* in Cambridgeshire (v.c. 29): an interim account. *Nature in Cambridgeshire*, 39: 40–53.

Preston, C. D. 2000. Engulfed by suburbia or destroyed by the plough: the ecology of extinction in Middlesex and Cambridgeshire. *Watsonia*, 23: 59–81.

Preston, C. D. & Croft, J. M. 1997. *Aquatic plants in Britain and Ireland*. Colchester: Harley Books.

Preston, C. D. & Hill, M. O. 1997. The geographical relationships of British and Irish vascular plants. *Botanical Journal of the Linnean Society*, 124: 1–120.

Preston, C. D. & Pearman, D. A. 1998a. Aquatic plants at high altitudes in the Breadalbane Mountains (V.C. 88), Scotland. *Watsonia*, 22: 187–190.

Preston, C. D. & Pearman, D. A. 1998b. J. E. Dandy's & G. Taylor's unpublished study of *Potamogeton* × *sudermanicus* Hagstr. in Britain, with an account of the current distribution of the hybrid. *Watsonia*, 22: 163–172.

Preston, C. D. & Pearman, D. A. 2000. *Centarium littorale* as a saltmarsh species in Scotland. *BSBI Scottish Newsletter*, 22: 14–16.

Preston, C. D. & Sell, P. D. 1989. The Aizoaceae naturalised in the British Isles. *Watsonia*, 17: 217–245.

Preston, C. D. & Whitehouse, H. L. K. 1986. The habitat of *Lythrum hyssopifolia* L. in Cambridgeshire, its only surviving English locality. *Biological Conservation*, 35: 41–62.

Prime, C. T. 1954. Biological Flora of the British Isles. No. 41. *Arum neglectum* (Towns.) Ridley. *Journal of Ecology*, 42: 241–248.

Prime, C. T. 1960. *Lords & Ladies*. London: Collins.

Primavesi, A. L. & Evans, P. A. (eds). 1988. *Flora of Leicestershire*. Leicester: Leicestershire Museums, Art Galleries and Records Service.

Prince, S. D. & Hare, A. D. R. 1981. *Lactuca saligna* and *Pulicaria vulgaris* in Britain. *In*: H. Synge (ed.), *The biological aspects of rare plant conservation*, pp. 379–388. Chichester: John Wiley & Sons.

Pring, M. E. 1961. Biological Flora of the British Isles. No. 78. *Arabis stricta* Huds. *Journal of Ecology*, 49: 431–437.

Pritchard, N. M. 1960. *Gentianella* in Britain. II. *Gentianella septentrionalis* (Druce) E. F. Warb. *Watsonia*, 4: 218–237.

Proctor, M. C. F. 1956. Biological Flora of the British Isles. No. 58. *Helianthemum* Mill. [genus] (pp. 675–677), *Helianthemum chamaecistus* Mill. (pp. 683–688), *Helianthemum apenninum* (L.) Mill. (pp. 688–692). *Journal of Ecology*, 44: 675–692.

Proctor, M. C. F. 1960. Biological Flora of the British Isles. No. 72. *Tuberaria guttata* (L.) Fourreau. *Journal of Ecology*, 48: 243–253.

Proctor, M. C. F. 1965. The distinguishing characters and geographical distributions of *Ulex minor* and *Ulex gallii*. *Watsonia*, 6: 177–187.

Proctor, M. C. F. & Groenhof, A. C. 1992. Peroxidase isoenzyme and morphological variation in *Sorbus* L. in South Wales and adjacent areas, with particular reference to *S. porrigentiformis* E. F. Warb. *Watsonia*, 19: 21–37.

Proctor, M. C. F., Yeo, P. F. & Lack, A. J. 1996. *The natural history of pollination*. London: HarperCollins.

Pryce, R. D. 1999. *Carmarthenshire rare plant register*. Llanelli: privately published.

Pugsley, H. W. 1924. A new *Statice* in Britain. *Journal of Botany*, 62: 129–134.

Pugsley, H. W. 1936. The British Robertsonian Saxifrages. *Journal of the Linnean Society of London (Botany)*, 50: 267–289.

Pyne, K. 1997. *Mespilus germanica* in southern Britain. *BSBI News*, 75: 49–50.

Rackham, O. 1975. *Hayley Wood*. Cambridge: Cambridgeshire and Isle of Ely Naturalists' Trust.

Rackham, O. 1980. *Ancient woodland: its history, vegetation and uses in England*. London: Edward Arnold.

Rackham, O. 1990. *Trees and woodland in the British Landscape*, edn 2. London: J. M. Dent & Sons.

Rackham, O. 1999. The woods 30 years on: where have the Primroses gone? *Nature in Cambridgeshire*, 41: 73–87.

Ramsey, M. M. & Stewart, J. 1998. Re-establishment of the lady's slipper orchid (*Cypripedium calceolus* L.) in Britain. *Botanical Journal of the Linnean Society*, 126: 173–181.

Randall, R. E. 1977. The past and present status and distribution of Sea Pea, *Lathyrus japonicus* Willd., in the British Isles. *Watsonia*, 11: 247–251.

Randall, R. E. 1988. A field survey of *Mertensia maritima* (L.) Gray, Oyster plant, in Britain during 1986 and 1987. Contract Surveys no. 20. Peterborough: Nature Conservancy Council.

Randall, R. E. & Thornton, G. 1996. Biological Flora of the British Isles. No. 191. *Peucedanum officinale* L. *Journal of Ecology*, 84: 475–485.

Raspé, O., Findlay, C. & Jacquemart, A.-L. 2000. Biological Flora of the British Isles. No. 214. *Sorbus aucuparia* L. *Journal of Ecology*, 88: 910–930.

Ratcliffe, D. 1959. Biological Flora of the British Isles. No. 69. *Hornungia petraea* (L.) Rchb. *Journal of Ecology*, 47: 241–247.

Ratcliffe, D. 1960. Biological Flora of the British Isles. No. 75. *Draba muralis* L. *Journal of Ecology*, 48: 737–744.

Ratcliffe, D. A. 1959. The habitat of *Koenigia islandica* L. in Scotland. *Transactions and Proceedings of the Botanical Society of Edinburgh*, 37: 272–275.

Raven, P. H. 1963. *Circaea* in the British Isles. *Watsonia*, 5: 262–272.

Raven, P. H. & Raven, T. E. 1976. The genus *Epilobium* (Onagraceae) in Australasia: a systematic and evolutionary study. *New Zealand Department of Scientific and Industrial Research Bulletin*, 216.

Raybould, A.F., Gray, A. J., Lawrence, M. J. & Marshall, D. F. 1991. The evolution of *Spartina anglica* C. E. Hubbard (Gramineae) – origin and genetic variability. *Biological Journal of the Linnean Society*, 43: 111–126.

Rechinger, K. H. 1961. Notes on *Rumex acetosa* L. in the British Isles. (Beitrag zur Kenntnis von Rumex, no. XV). *Watsonia*, 5: 64–66.

Reid, J. A. 1975. The distinction between *Oxalis corniculata* L. and *O. exilis* A. Cunn. *Watsonia*, 10: 290–291.

Reilly, P. A. 1993. The flowering plants and ferns of The Phoenix Park, Dublin. *Glasra*, 2: 5–72.

Reynolds, S. 1994. *Chenopodium capitatum* in Ireland. *BSBI News*, 66: 36.

Reynolds, S. 1999. *Lactuca serriola* L. in Dublin (H21). *Irish Naturalists' Journal*, 26: 285–286.

Rich, T. C. G. 1987. The genus *Barbarea* R. Br. (Cruciferae) in Britain and Ireland. *Watsonia*, 16: 389–396.

Rich, T. C. G. 1988a. Cabbage patch – IV. *Cardaria chalepensis* (L.) Handel-Mazzetti in the British Isles. *BSBI News*, 48: 12–14.

Rich, T. [C. G.] 1988b. A little cabbage patch V. Food for thought or the rape of mustard and cress. *BSBI News*, 49: 12–13.

Rich, T. C. G. 1991. *Crucifers of Great Britain and Ireland*. Botanical Society of the British Isles Handbook no. 6. London: Botanical Society of the British Isles.

Rich, T. C. G. 1994a. *Luzula pallidula* Kirschner in Ireland. *Irish Botanical News*, 4: 26–28.

Rich, T. C. G. 1994b. *Pedicularis sylvatica* L. subsp. *hibernica* D. A. Webb (Scrophulariaceae) new to Wales. *Watsonia*, 20: 70–71.

Rich, T. C. G. 1994c. Ragweeds (*Ambrosia* L.) in Britain. *Grana*, 33: 38–43.

Rich, T. C. G. 1996. Is *Gentianella uliginosa* (Willd.) Boerner (Gentianaceae) present in England? *Watsonia*, 21: 208–209.

Rich, T. C. G. 1997a. Early gentian (*Gentianella anglica* (Pugsley) E.F.Warb.) present in Wales. *Watsonia*, 21: 289–290.

Rich, T. C. G. 1997b. *The management of semi-natural lowland grassland for selected rare and scarce vascular plants: a review*. English Nature Research Report no. 216. Peterborough: English Nature and the Wildlife Trusts.

Rich, T. C. G. 1997c. Wildlife Reports: Flowering Plants, England. *British Wildlife*, 9: 124–125.

Rich, T. C. G. 1999a. Conservation of Britain's biodiversity: *Filago lutescens* Jordan (Asteraceae), Red-tipped cudweed. *Watsonia*, 22: 251–260.

Rich, T. C. G. 1999b. The potential for seed dispersal by sea water in *Coincya wrightii* (O. E. Schulz) Stace and *C. monensis* (L.) W. Greuter & Burdet subsp. *monensis*. *Watsonia*, 22: 422–423.

Rich, T. C. G. 1999c. Conservation of Britain's biodiversity IV: *Filago pyramidata* (Asteraceae), Broad-leaved cudweed. *Edinburgh Journal of Botany*, 56: 61–73.

Rich, T. C. G. 1999d. Conservation of Britain's biodiversity: *Cyperus fuscus* L. (Cyperaceae), Brown Galingale. *Watsonia*, 22: 397–403.

Rich, T. [C. G.] & Baeker, M. 1998. The distribution of *Sorbus lancastriensis* E. F. Warburg. *In*: A. Jackson & M. Flanagan (eds), *The Conservation Status of* Sorbus *in the UK*, pp. 44–47. London: Royal Botanic Gardens, Kew.

Rich, T. C. G. & Brown, N. 2000. *Suaeda vera* Forssk. ex J. F. Gmel. (Chenopodiaceae), shrubby sea-blite, present in Anglesey (v.c. 52), Wales. *Watsonia*, 23: 343–344.

Rich, T. C. G. & Jermy, A. C. 1998. *Plant crib 1998*. London: Botanical Society of the British Isles.

Rich, T. C. G. & Lewis, J. 1999. Use of herbarium material for mapping the distribution of *Erophila* (Brassicaceae) taxa *sensu* Filfilan & Elkington in Britain and Ireland. *Watsonia*, 22: 377–385.

Rich, T. C. G. & Rich, M. D. B. 1988. *Plant crib*. London: Botanical Society of the British Isles.

Rich, T. C. G. & Woodruff, E. R. 1990. *The BSBI Monitoring Scheme, 1987–1988*. 2 vols. Nature Conservancy Council Report no. 1265. London: Botanical Society of the British Isles.

Rich, T. C. G. & Woodruff, E. R. 1996. Changes in the vascular plant floras of England and Scotland between 1930–1960 and 1987–8: the BSBI monitoring scheme. *Biological Conservation*, 75: 217–229.

Rich, T. [C. G.], Donovan, P., Harmes, P., Knapp, A., McFarlane, M., Marrable, C., Muggeridge, N., Nicholson, R., Reader, M. & P., Rich, E. & White, P. 1996. *Flora of Ashdown Forest*. Sussex Botanical Recording Society.

Rich, T. C. G., FitzGerald, R. & Sydes, C. 1998. Distribution and ecology of Small Cow-wheat (*Melampyrum sylvaticum* L.; Scrophulariaceae) in the British Isles. *Botanical Journal of Scotland*, 50: 29–46.

Rich, T. C. G., Gibson, C. & Marsden, M. 1999. Re-establishment of the extinct native plant *Filago gallica* L. (Asteraceae), narrow-leaved cudweed, in Britain. *Biological Conservation*, 91: 1–8.

Rich, T. C. G., Holyoak, D. T., Margetts, L. J. & Murphy, R. J. 1997. Hybridisation between *Gentianella amarella* (L.) Boerner and *G. anglica* (Pugsley) E. F. Warb. (Gentianaceae). *Watsonia*, 21: 313–325.

Rich, T. C. G., Jones, R. A. & Jebb, M. 2000. Three new British sites for *Carex depauperata* With. (Cyperaceae) represented in the Irish National Herbarium, Glasnevin. *Watsonia*, 23: 340–341.

Rich, T. C. G., Kay, G. M. & Sydes, C. 1999. Distribution and ecology of Pyramidal Bugle (*Ajuga pyramidalis* L., Lamiaceae) in the British Isles. *Botanical Journal of Scotland*, 51: 181–193.

Rich, T. C. G., Lambrick, C. R., Kitchen, C. & Kitchen, M. A. R. 1998. Conserving Britain's biodiversity. 1. *Thlaspi perfoliatum* L. (Brassicaceae), Cotswold Pennycress. *Biology in Conservation*, 7: 915–926.

Rich, T. C. G., Lambrick, C. R. & McNab, C. 1999a. Conservation of Britain's biodiversity: *Cyperus fuscus* L. (Cyperaceae), Brown Galingale. *Watsonia*, 22: 397–403.

Rich, T. C. G., Lambrick, C. R. & McNab, C. 1999b. Conservation of Britain's biodiversity: *Salvia pratensis* L. (Lamiaceae), Meadow Clary. *Watsonia*, 22: 405–411.

Rich, T. C. G., Richardson, S. J. & Rose, F. 1995. Tunbridge Filmy-fern, *Hymenophyllum tunbrigense* (Hymenophyllaceae: Pteridophyta), in South-East England in 1994/1995. *Fern Gazette*, 15: 51–63.

Richards, A. J. 1973. An upland race of *Potentilla erecta* (L.) Räusch. in the British Isles. *Watsonia*, 9: 301–317.

Richards, [A]. J. 1989. *Primulas of the British Isles*. Princes Risborough: Shire Publications.

Richards, A. J. 1993. *Primula*. London: B. T. Batsford.

Richards, A. J. & Porter, A. F. 1982. On the identity of a Northumberland *Epipactis*. *Watsonia*, 14: 121–128.

Richards, P. W. 1943a. Biological Flora of the British Isles. No. 9. *Juncus macer* S. F. Gray. *Journal of Ecology*, 31: 51–59.

Richards, P. W. 1943b. Biological Flora of the British Isles. No. 10. *Juncus filiformis* L. *Journal of Ecology*, 31: 60–65.

Richards, P. W. & Clapham, A. R. 1941a. Biological Flora of the British Isles. No. 1. *Juncus* [genus] L. *Journal of Ecology*, 29: 362–368.

Richards, P. W. & Clapham, A. R. 1941b. Biological Flora of the British Isles. No. 2. *Juncus inflexus* L. *Journal of Ecology*, 29: 369–374.

Richards, P. W. & Clapham, A. R. 1941c. Biological Flora of the British Isles. No. 3. *Juncus effusus* L. *Journal of Ecology*, 29: 375–380.

Richards, P. W. & Clapham, A. R. 1941d. Biological Flora of the British Isles. No. 4. *Juncus conglomeratus* L. *Journal of Ecology*, 29: 381–384.

Richards, P. W. & Clapham, A. R. 1941e. Biological Flora of the British Isles. No. 5. *Juncus subnodulosus* Schrank. *Journal of Ecology*, 29: 385–391.

Richards, P. W. & Evans, G. B. 1972. Biological Flora of the British Isles. No. 126. *Hymenophyllum tunbrigense* (L.) Sm. (pp. 245–258), *Hymenophyllum wilsonii* Hooker (258–268). *Journal of Ecology*, 60: 245–268.

Richens, R. H. 1947. Biological Flora of the British Isles. No. 16. *Allium vineale* L. *Journal of Ecology*, 34: 209–226.

Richens, R. H. 1961. Studies on *Ulmus*, IV. The village elms of Huntingdonshire and a new method for exploring taxonomic discontinuity. *Forestry*, 34: 47–64.

Richens, R. H. 1983. *Elm*. Cambridge: Cambridge University Press.

Richens, R. H. 1987. The history of the elms in Wales. *Nature in Wales*, 5: 3–11.

Rickard, M. H. 1989. Two spleenworts new to Britain – *Asplenium trichomanes* subsp. *pachyrachis* and *Asplenium trichomanes* nothosubsp. *staufferi*. *Pteridologist*, 1: 244–248.

Ritchie, J. C. 1954. Biological Flora of the British Isles. No. 44. *Primula scotica* Hook. *Journal of Ecology*, 42: 623–628.

Ritchie, J. C. 1955. Biological Flora of the British Isles. No. 51. *Vaccinium vitis-idaea* L. *Journal of Ecology*, 43: 701–708.

Ritchie, J. C. 1956. Biological Flora of the British Isles. No. 56. *Vaccinium myrtillus* L. *Journal of Ecology*, 44: 291–299.

Rix, E. M. & Woods, R. G. 1981. *Gagea bohemica* (Zauschner) J. A. & J. H. Schultes in the British Isles, and a general review of the *G. bohemica* species complex. *Watsonia*, 13: 265–270.

Roach, F. A. 1985. *Cultivated fruits of Britain: their origin and history*. Oxford: Basil Blackwell.

Roberts, F. J. 1977. *Viola rupestris* Schmidt and *Juncus alpinus* Vill. in Mid-W. Yorkshire. *Watsonia*, 11: 385–386.

Roberts, H. A. & Boddrell, J. E. 1983a. Seed survival and periodicity of seedling emergence in ten species of annual weeds. *Annals of Applied Biology*, 102: 523–532.

Roberts, H. A. & Boddrell, J. E. 1983b. Seed survival and periodicity of seedling emergence in eight species of *Cruciferae*. *Annals of Applied Biology*, 103: 301–309.

Roberts, H. A. & Boddrell, J. E. 1984. Seed survival and periodicity of seedling emergence in four weedy species of *Papaver*. *Weed Research*, 24: 195–200.

Roberts, H. A. & Feast, P. M. 1973. Emergence and longevity of seeds of annual weeds in cultivated and undisturbed soil. *Journal of Applied Ecology*, 10: 133–143.

Roberts, R. H. 1964. *Mimulus* hybrids in Britain. *Watsonia*, 6: 70–75.

Roberts, R. H. 1968. The hybrids of *Mimulus cupreus*. *Watsonia*, 6: 371–376.

Roberts, R. H. 1982. *Flowering plants and ferns of Anglesey*. Cardiff: National Museum of Wales.

Robinson, J. B. D. & Treharne, K. J. 1985. Exploited plants. Maize. *Biologist*, 32: 199–207.

Robson, N. K. B. 1958a. *Hypericum maculatum* in Britain and Europe. *Proceedings of the Botanical Society of the British Isles*, 3: 99–100.

Robson, N. K. B. 1958b. *Hypericum undulatum* Schousb. ex Willd. in Ireland? *Irish Naturalists' Journal*, 12: 269.

Robson, N. K. B. 1985. Studies in the genus *Hypericum* L. (Guttiferae) 3. Sections 1. *Campylosporus* to 6a. *Umbraculoides*. *Bulletin of the British Museum (Natural History), Botany*, 12: 163–325.

Robson, N. K. B. 1996. Studies in the genus *Hypericum* L. (Guttiferae) 6. Sections 20. *Myriandra* to 28. *Elodes*. *Bulletin of the British Museum (Natural History), Botany*, 26: 75–217.

Rodwell, J. S. (ed.). 1991a. *British plant communities*, 1. *Woodlands and scrub*. Cambridge: Cambridge University Press.

Rodwell, J. S. (ed.). 1991b. *British plant communities*, 2. *Mires and heaths*. Cambridge: Cambridge University Press.

Rodwell, J. S. (ed.). 1992. *British plant communities*, 3. *Grasslands and montane communities*. Cambridge: Cambridge University Press.

Rodwell, J. S. (ed.). 1995. *British plant communities*, 4. *Aquatic communities, swamps and tall-herb fens*. Cambridge: Cambridge University Press.

Rodwell, J. S. (ed.). 2000. *British plant communities*, 5. *Maritime communities and vegetation of open habitats*. Cambridge: Cambridge University Press.

Roe, R. G. B. 1978. *Veronica crista-galli* Stev. in the British Isles. *Watsonia*, 12: 129–132.

Roe, R. G. B. 1981. *The flora of Somerset*. Taunton: Somerset Archaeological and Natural History Society.

Roper, P. 1993. The distribution of the Wild Service Tree, *Sorbus torminalis* (L.) Crantz, in the British Isles. *Watsonia*, 19: 209–229.

Rose, F. 1948. Biological Flora of the British Isles. No. 25. *Orchis purpurea* Huds. *Journal of Ecology*, 36: 366–377.

Rose, F. 1988. Plants to look for in the British Isles some of which might be expected to occur as natives. *BSBI News*, 49: 11–12.

Rose, F. 1989. Key to annual *Salicornia* species of South England and North France. *BSBI News*, 53: 12–16.

Rose, R. J., Bannister, P. & Chapman, S. B. 1996. Biological Flora of the British Isles. No. 192. *Erica ciliaris* L. *Journal of Ecology*, 84: 617–628.

Rostański, K. 1982. The species of *Oenothera* L. in Britain. *Watsonia*, 14: 1–34.

Rothero, G. & Thompson, B. 1994. *An annotated checklist of the flowering plants and ferns of Main Argyll*. Oban: Argyll Flora Project.

Rougemont, G. M. de 1989. *A field guide to the crops of Britain and Europe*. London: Collins.

Rowell, T. A. 1984. Further discoveries of the Fen Violet (*Viola persicifolia* Schreber) at Wicken Fen, Cambridgeshire. *Watsonia*, 15: 122–123.

Rowell, T. A., Walters, S. M. & Harvey, H. J. 1982. The rediscovery of the Fen Violet, *Viola persicifolia* Schreber, at Wicken Fen, Cambridgeshire. *Watsonia*, 14: 183–184.

Rumsey, F. J. 1997. *Asplenium viride* Hudson (Aspleniaceae) in Greater London. *Watsonia*, 21: 376–378.

Rumsey, F. [J.] & Headley, A. 1998. Are British *Orobanche* species in decline? *The Naturalist*, 123: 76–85.

Rumsey, F. J. & Jury, S. L. 1991. An account of *Orobanche* L. in Britain and Ireland. *Watsonia*, 18: 257–295.

Rumsey, F. J., Jermy, A. C. & Sheffield, E. 1998. The independent gametophytic stage of *Trichomanes speciosum* Willd. (Hymenophyllaceae), the Killarney Fern and its distribution in the British Isles. *Watsonia*, 22: 1–19.

Rumsey, F. J., Sheffield, E. & Farrar, D. R. 1990. British filmy-fern gametophytes. *Pteridologist*, 2: 40–42.

Rushton, B. S. 1978a. *Quercus robur* L. and *Quercus petraea* (Matt.) Liebl.: a multivariate approach to the hybrid problem, 1. Data acquisition, analysis and interpretation. *Watsonia*, 12: 81–101.

Rushton, B. S. 1978b. *Quercus robur* L. and *Quercus petraea* (Matt.) Liebl.: a multivariate approach to the hybrid problem, 2. The geographical distribution of population types. *Watsonia*, 12: 209–224.

Rutherford, A. & Stirling, A. McG. 1987. Variegated archangels. *BSBI News*, 46: 9–11.

Ryves, T. B., Clement, E. J. & Foster, M. C. 1996. *Alien grasses of the British Isles*. London: Botanical Society of the British Isles.

Saber, M. A., Clements, D. R., Hall, M. R., Doohan, D. J. & Crompton, C. W. 1995. The biology of Canadian weeds. 105. *Linaria vulgaris* Mill. *Canadian Journal of Plant Science*, 75: 525–537.

Sagar, G. R. & Harper, J. L. 1964. Biological Flora of the British Isles. No. 95. *Plantago major* L. (pp. 189–205), *Plantago media* L. (pp. 205–210), *Plantago lanceolata* L. (pp. 211–218). *Journal of Ecology*, 52: 189–221.

Salisbury, E. J. 1952. *Downs and dunes*. London: G. Bell & Sons.

Salisbury, E. [J.] 1953. A changing flora as shown in the study of weeds of arable land and waste places. *In*: J. E. Lousley (ed.), *The changing flora of Britain*, pp. 130–139. Oxford: Botanical Society of the British Isles.

Salisbury E. [J.] 1961. *Weeds & Aliens*. London: Collins.

Salisbury, E. J. 1963. Fertile seed production and self-incompatability of *Hypericum calycinum* in England. *Watsonia*, 5: 368–376.

Salisbury E. J. 1964. *Weeds & Aliens*, edn 2. London: Collins.

Salisbury, E. J. 1969a. A note on fertile seed production by *Hypericum calycinum*. *Watsonia*, 7: 24.

Salisbury, E. J. 1969b. The reproductive biology and occasional seasonal dimorphism of *Anagallis minima* and *Lythrum hyssopifolia*. *Watsonia*, 7: 25–39.

Salisbury, E. J. 1974. The reproduction of *Juncus tenuis* and its dispersal. *Transactions of the Botanical Society of Edinburgh*, 42: 187–190.

Salmon, C. E. 1926. A new *Myosotis* from Britain. *Journal of Botany*, 64: 289–295.

Sandwith, N. Y. 1951. *Ranunculus marginatus*. *Watsonia*, 2: 104.

Sanford, M. 1991. *The orchids of Suffolk*. Ipswich: Suffolk Naturalists' Society.

Sanger, N.P. & Waite, S. 1998. The phenology of *Ophrys sphegodes* (the early spider orchid): what annual censuses can miss. *Botanical Journal of the Linnean Society*, 126: 75–81.

Sargent, C., Mountford, O. & Greene, D. 1986. The distribution of *Poa angustifolia* L. in Britain. *Watsonia*, 16: 31–36.

Savidge, J. P., Heywood, V. H. & Gordon, V. (eds). 1963. *Travis's Flora of South Lancashire*. Liverpool: Liverpool Botanical Society.

Scannell, M. J. P. 1973. *Juncus planifolius* R. Br. in Ireland. *Irish Naturalists' Journal*, 17: 308–309.

Scannell, M. J. P. 1975. The known distribution of *Juncus planifolius* R. Br. in Ireland. *Watsonia*, 10: 418–419.

Scannell, M. J. P. & Jebb, M. H. P. 2000. Flora of Connemara and the Burren – records from 1984. *Glasra*, 4: 7–45.

Scannell, M. J. P. & Synnott, D. M. 1987. *Census catalogue of the flora of Ireland*, edn 2. Dublin: Stationery Office.

Scannell, M. J. P. & Webb, D. A. 1976. The identity of the Renvyle *Hydrilla*. *Irish Naturalists' Journal*, 18: 327–331.

Schotsman, H. D. 1967. *Les Callitriches: espèces de France et taxa nouveaux d'Europe*. Paris: Éditions Paul Lechevalier.

Schroeder, F.-G. 1970. Exotic *Amelanchier* species naturalised in Europe, and their occurrence in Great Britain. *Watsonia*, 8: 155–162.

Scott, G. A. M. 1963a. Biological Flora of the British Isles. No. 89. *Mertensia maritima* (L.) S. F. Gray. *Journal of Ecology*, 51: 733–742.

Scott, G. A. M. 1963b. Biological Flora of the British Isles. No. 90. *Glaucium flavum* Crantz. *Journal of Ecology*, 51: 743–754.

Scott, G. A. M. & Randall, R. E. 1976. Biological Flora of the British Isles. No. 139. *Crambe maritima* L. *Journal of Ecology*, 64: 1077–1091.

Scott, M., Scott, S. & Sydes, C. 1999. A Scottish perspective on the conservation of pillwort. *British Wildlife*, 10: 297–302.

Scott, N. E. 1985. The updated distribution of maritime species on British roadsides. *Watsonia*, 15: 381–386.

Scott, N. E. & Davison, A. W. 1982. De-icing salt and the invasion of road verges by maritime plants. *Watsonia*, 14: 41–52.

Scott, W. 1968. *Pilosella flagellaris* (Willd.) Sell & C. West subsp. *bicapitata* Sell & C. West – in Zetland. *Proceedings of the Botanical Society of the British Isles*, 7: 192–193.

Scott, W. & Palmer, R. 1987. *The flowering plants and ferns of the Shetland Islands*. Lerwick: Shetland Times.

Scully, R. W. 1916. *Flora of County Kerry*. Dublin: Hodges, Figgis & Co.

Scurfield, G. 1954. Biological Flora of the British Isles. No. 39. *Deschampsia flexuosa* (L.) Trin. *Journal of Ecology*, 42: 225–233.

Scurfield, G. 1962. Biological Flora of the British Isles. No. 84. *Cardaria draba* (L.) Desv. (*Lepidium draba* L.). *Journal of Ecology*, 50: 489–499.

Sealy, J. R. & Webb, D. A. 1950. Biological Flora of the British Isles. No. 30. *Arbutus* L., *Arbutus unedo* L. *Journal of Ecology*, 38: 223–236.

Seavey, S. R. & Raven, P. H. 1977. Chromosomal evolution in *Epilobium* sect. *Epilobium* (Onagraceae). *Plant Systematics and Evolution*, 127: 107–119.

Sell, P. D. (ed.). 1967. Taxonomic and nomenclatural notes on the British flora. *Watsonia*, 6: 292–318.

Sell, P. D. 1981. *Lapsana intermedia* Bieb. or *Lapsana communis* L. subsp. *intermedia* (Bieb.) Hayek? *Watsonia*, 13: 299–302.

Sell, P. D. 1986. The genus *Cicerbita* Wallr. in the British Isles. *Watsonia*, 16: 121–129.

Sell, P. D. 1989. The *Sorbus latifolia* (Lam.) Pers. aggregate in the British Isles. *Watsonia*, 17: 385–399.

Sell, P. [D.] 1990. Hazels in Cambridgeshire. *Nature in Cambridgeshire*, 23: 50–53.

Sell, P. D. 1991. The cherries and plums of Cambridgeshire. *Nature in Cambridgeshire*, 33: 29–39.

Sell, P. D. 1992. More plums. *Nature in Cambridgeshire*, 34: 59–60.

Sell, P. D. 1994. *Ranunculus ficaria* L. sensu lato. *Watsonia*, 20: 41–50.

Sell, P. [D.] & Murrell, G. 1996. *Flora of Great Britain and Ireland*, 5. *Butomaceae-Orchidaceae*. Cambridge: Cambridge University Press.

Seymour, W. A. (ed.). 1980. *A history of the Ordnance Survey*. Folkestone: Wm Dawson & Sons.

Shamsi, S. R. A. & Whitehead, F. H. 1974. Comparative eco-physiology of *Epilobium hirsutum* L. and *Lythrum salicaria* L. I. General biology, distribution and germination. *Journal of Ecology*, 62: 279–290.

Sharma, M. P. & Vanden Born, W. H. 1978. The biology of Canadian weeds. 27. *Avena fatua* L. *Canadian Journal of Plant Science*, 58: 141–157.

Shaw, J. M. H. 1995. *Iochroma australis* – a new British alien. *BSBI News*, 68: 41–42.

Shaw, M. (ed.). 1988. *A Flora of the Sheffield Area*. Sheffield: Sorby Natural History Society.

Sheppard, A. W. 1991. Biological Flora of the British Isles. No. 171. *Heracleum sphondylium* L. *Journal of Ecology*, 79: 235–258.

Shirreffs, D. A. 1985. Biological Flora of the British Isles. No. 158. *Anemone nemorosa* L. *Journal of Ecology*, 73: 1005–1020.

Shivas, M. G. 1960. Contribution to the cytology and taxonomy of species of *Polypodium* in Europe and America, II. Taxonomy. *Journal of the Linnean Society (Botany)*, 58: 27–38.

Showler, A. J. & Rich, T. C. G. 1993. *Cardamine bulbifera* (L.) Crantz (Cruciferae) in the British Isles. *Watsonia*, 19: 231–245.

Silverside, A. J. 1990a. The nomenclature of some hybrids of the *Spiraea salicifolia* group naturalized in Britain. *Watsonia*, 18: 147–151.

Silverside, A. J. 1990b. A new hybrid binomial in *Mimulus*. *Watsonia*, 18: 210–212.

Silverside, A. J. 1990c. Dandelions and their allies. *In*: *A guide to some difficult plants*, pp. 41–67. Wild Flower Society.

Silverside, A. J. 1991. The identity of *Euphrasia officinalis* L. and its nomenclatural implications. *Watsonia*, 18: 343–350.

Silverside, A. J. 1994. *Mimulus*: 180 years of confusion. *In*: A. R. Perry & G. Ellis (eds), *The Common Ground of Wild & Cultivated Plants*, pp. 59–64. Cardiff: National Museum of Wales. [See also Perry, R. 1997. *Mimulus*: 180 years of confusion – a correction. *BSBI News*, 77: 56.]

Silverside, A. J. 1998. *Euphrasia*. *In*: T. C. G. Rich & A. C. Jermy (eds), *Plant Crib*, pp. 269–272. London: Botanical Society of the British Isles.

Silverside, A. J. & Jackson, E. H. (eds). 1988. *A check-list of the flowering plants and ferns of East Lothian*. Edinburgh: Botanical Society of Edinburgh.

Simmonds, N. W. 1945. Biological Flora of the British Isles. No. 14. *Polygonum* L. em. Gaertn. (pp. 117–120), *Polygonum persicaria* L. (pp. 121–131), *Polygonum lapathifolium* L. (*P. tomentosum* Schr. of many continental authors) (pp. 121–131), *Polygonum petecticale* (Stokes) Druce (*P. maculatum* Trimen & Dyer, *P. nodosum* Persoon, etc) (pp. 140–143). *Journal of Ecology*, 33: 117–143.

Simmonds, N. W. 1946. Biological Flora of the British Isles. No. 15. *Gentiana pneumonanthe* L. *Journal of Ecology*, 33: 295–307.

Simpson, B. B., Neff, J. L. & Seigler, D. S. 1983. Floral biology and floral rewards of *Lysimachia* (Primulaceae). *American Midland Naturalist*, 110: 249–256.

Simpson, D. A. 1986. Taxonomy of *Elodea* Michx in the British Isles. *Watsonia*, 15: 1–14.

Simpson, F. W. 1982. *Simpson's Flora of Suffolk*. East Bergholt: Suffolk Naturalists' Society.

Simpson, N. D. 1960. *A bibliographical index of the British flora*. Bournemouth: privately published.

Sinker, C. A., Packham, J. R., Trueman, I. C., Oswald, P. H., Perring, F. H. & Prestwood, W. V. 1985. *Ecological Flora of the Shropshire Region*. Shrewsbury: Shropshire Trust for Nature Conservation.

Skene, K. R., Sprent, J. I., Raven, J. A. & Herdman, L. 2000. Biological Flora of the British Isles. No. 215. *Myrica gale* L. *Journal of Ecology*, 88: 1079–1094.

Slater, F. M. 1990. Biological Flora of the British Isles. No. 168. *Gagea bohemica* (Zauschner) J. A. & J. H. Schultes (*G. saxatilis* Koch). *Journal of Ecology*, 78: 535–546.

Sledge, W. A. 1975. Further comments on the supposed occurrence of *Euphrasia salisburgensis* in Yorkshire. *The Naturalist*, no. 934: 87–89.

Small, E. & Cronquist, A. 1976. A practical and natural taxonomy for *Cannabis*. *Taxon*, 25: 405–435.

Smart, J. & Simmonds, N. W. (eds). 1995. *Evolution of crop plants*, edn 2. London: Longman.

Smit, P. G. 1973. A revisin of *Caltha* (Ranunculaceae). *Blumea*, 21: 119–150.

Smith, A. J. E. 1963. Variation in *Melampyrum pratense* L. *Watsonia*, 5: 336–367.

Smith, P. H. 1984. The distribution, status and conservation of *Juncus balticus* Willd. in England. *Watsonia*, 15: 15–26.

Smith, P. M. 1968. The *Bromus mollis* aggregate in Britain. *Watsonia*, 6: 327–344.

Smith, P. M. 1986. Native or introduced? Problems in the taxonomy and plant geography of some widely introduced annual brome-grasses. *Proceedings of the Royal Society of Edinburgh*, 89B: 273–281.

Smith, R. A. H., Stewart, N. F., Taylor, N. W. & Thomas, R. E. 1992. *Check-list of the plants of Perthshire*. Perth: Perthshire Society of Natural Science.

Smith, R. T. & Taylor, J. A. (eds). 1995. *Bracken: an environmental issue*. International Bracken Group Special Publication no. 2. University of Leeds Printing Service.

Smith, U. K. 1979. Biological Flora of the British Isles. No. 147. *Senecio integrifolius* (L.) Clairv. (*Senecio campestris* (Retz.) DC.). *Journal of Ecology*, 67: 1109–1124.

Snogerup, B. 1982. *Odontites litoralis* Fries subsp. *litoralis* in the British Isles. *Watsonia*, 14: 35–39.

Snogerup, S. 1993. A revision of *Juncus* subgen. *Juncus* (Juncaceae). *Willdenowia*, 23: 23–73.

Sobey, D. G. 1981. Biological Flora of the British Isles. No. 150. *Stellaria media* (L.) Vill. *Journal of Ecology*, 69: 311–335.

Soltis, D. E. 1984. Autopolyploidy in *Tolmiea menziesii* (Saxifragaceae). *American Journal of Botany*, 71: 1171–1174.

Sowter, F. A. 1949. Biological Flora of the British Isles. No. 27. *Arum* L., *Arum maculatum* L. *Journal of Ecology*, 37: 207–219.

Sparling, J. H. 1968. Biological Flora of the British Isles. No. 114. *Schoenus nigricans* L. (*Chaetospora nigricans* Kunth). *Journal of Ecology*, 56: 883–899.

Spencer, J. 2000. A bleak future for *Gaultheria shallon* in the New Forest. *BSBI News*, 84: 47–48.

Spicer, K. W. & Catling, P. M. 1988. The biology of Canadian weeds No. 88. *Elodea canadensis* Michx. *Canadian Journal of Plant Science*, 68: 1035–1051.

Spinage, C. 2000. The Wild Pear tree – one of Britain's rarest trees or a garden escape? *British Wildlife*, 11: 313–318.

Spooner, B. M. 1982. *Salvia reflexa* Hornem. in Britain. *BSBI News*, 31: 17–18.

Srutek, M. & Teckelmann, M. 1998. Review of biology and ecology of *Urtica dioica*. *Preslia (Praha)*, 70: 1–19.

Stace, C. A. 1961. *Nardurus maritimus* (L.) Murb. in Britain. *Proceedings of the Botanical Society of the British Isles*, 4: 248–261.

Stace, C. A. 1972. The history and occurrence in Britain of hybrids in *Juncus* subgenus *Genuini*. *Watsonia*, 9: 1–11.

Stace, C. A. (ed.). 1975. *Hybridization and the flora of the British Isles*. London: Academic Press.

Stace, C. A. 1991. *New Flora of the British Isles*. Cambridge: Cambridge University Press.

Stace, C. A. 1997. *New Flora of the British Isles*, edn 2. Cambridge: Cambridge University Press.

Stace, C. A., Al-Bermani, A.-K. K. A. & Wilkinson, M. J. 1992. The distinction between the *Festuca ovina* L. and *Festuca rubra* L. aggregates in the British Isles. *Watsonia*, 19: 107–112.

Stead, J. 1980. Plant portraits, 1. *Salix reticulata*, the net-veined willow. *Quarterly Bulletin of the Alpine Garden Society*, 48: 59–60.

Stearn, L. F. 1975. *Supplement to the Flora of Wiltshire*. Devizes: Wiltshire Archaeological and Natural History Society.

Steel, M. G., Cavers, P. B. & Lee, S. M. 1983. The biology of Canadian weeds. No. 59. *Setaria glauca* (L.) Beauv. and *S. verticillata* (L.) Beauv. *Canadian Journal of Plant Science*, 63: 711–725.

Stevens, D. P. 1990. The distribution of gender in *Petasites hybridus* (Butterbur). *BSBI News*, 55: 10–11.

Stewart, A., Pearman, D. A. & Preston, C. D. (comps & eds). 1994. *Scarce plants in Britain*. Peterborough: Joint Nature Conservation Committee.

Stewart, N. F. 1988. A provisional list of vascular plants growing in Falkirk district. *Forth Naturalist and Historian*, 10: 53–79.

Stewart, O. [M.] 1987. *Juncus subulatus* at Grangemouth, vc 86. *Botanical Society of the British Isles Scottish Newsletter*, 9: 12.

Stewart, O. M. 1989. *Calamagrostis purpurea*. *Botanical Society of the British Isles Scottish Newsletter*, 11: 6.

Stewart, O. [M.] 1990. Flowering plants of Kirkcudbrightshire. *Transactions of the Dumfriesshire and Galloway Natural History and Antiquarian Society*, 3rd series, 65: 1–68.

Strid, A. (ed.). 1986. *Mountain Flora of Greece. Vol. 1*. Cambridge: Cambridge University Press.

Sturt, N. 1995. Wild Clary (*Salvia verbenaca*) in churchyards. *BSBI News*, 68: 28–29.

Styles, B. T. 1962. The taxonomy of *Polygonum aviculare* and its allies in Britain. *Watsonia*, 5: 177–214.

Styles, B. T. 1976. *Stachys annua*. *BSBI News*, 12: 12.

Summerfield, R. J. 1974. Biological Flora of the British Isles. No. 135. *Narthecium ossifragum* (L.) Huds. *Journal of Ecology*, 62: 325–339.

Sutherland, W. J. 1990. Biological Flora of the British Isles. No. 169. *Iris pseudacorus* L. *Journal of Ecology*, 78: 833–848.

Svensson, R. & Wigren, M. 1986. Sminkrotens historia och biologi i Sverige. (History and biology of *Lithospermum arvense* in Sweden). *Svensk Botanisk Tidskrift*, 80: 107–131.

Swan, G. A. 1993. *Flora of Northumberland*. Newcastle upon Tyne: Natural History Society of Northumbria.

Swan, G. A. 1999. Identification, distribution and a new nothosubspecies of *Trichophorum cespitosum* (L.) Hartman (Cyperaceae) in the British Isles and N. W. Europe. *Watsonia*, 22: 209–233.

Swann, E. L. 1975. *Supplement to the Flora of Norfolk*. Norwich: privately published.

Swanton, C. J., Cavers, P. B., Clements, D. R. & Moore, M. J. 1992. The biology of Canadian weeds. 101. *Helianthus tuberosus* L. *Canadian Journal of Plant Science*, 72: 1367–1382.

Sykes, N. 1993. *Wild plants and their habitats in the North York Moors*. Helmsley: North York Moors National Park.

Synnott, D. M. 1983. Notes on *Salix phylicifolia* L. and related Irish willows. *Glasra*, 7: 1–10.

Synnott, D. [M.] 1986. An outline of the flora of Mayo. *Glasra*, 9: 13 -117.

Tabbush, P. & Beaton, A. 1998. Hybrid poplars: present status and potential in Britain. *Forestry*, 71, 355–364.

Takagi-Arigho, R. 1994. *Poa infirma* – flourishing? . . . or fleeing? *BSBI News*, 65: 14–18.

Takagi-Arigho, R. 1995. Plant records. *Festuca huonii*. *Watsonia*, 20: 299.

Tansley, A. G. 1939. *The British Islands and their vegetation*. Cambridge: Cambridge University Press.

Tarpey, T. & Heath, J. 1990. *Wild flowers of North East Essex*. Colchester: Colchester Natural History Society.

Taschereau, P. M. 1977. *Atriplex praecox* Hülphers: a species new to the British Isles. *Watsonia*, 11: 195–198.

Taschereau, P. M. 1985a. Taxonomy of *Atriplex* species indigenous to the British Isles. *Watsonia*, 15: 183–209.

Taschereau, P. M. 1985b. Field studies, cultivation experiments and the taxonomy of *Atriplex longipes* Dejer in the British Isles. *Watsonia*, 15: 211–219.

Taschereau, P. M. 1989. Taxonomy, morphology and distribution of *Atriplex* hybrids in the British Isles. *Watsonia*, 17: 247–264.

Taylor, F. J. 1956. Biological Flora of the British Isles. No. 55. *Carex flacca* Schreb. *Journal of Ecology*, 44: 281–290.

Taylor, K. 1971. Biological Flora of the British Isles. No. 121. *Rubus chamaemorus* L. *Journal of Ecology*, 59: 293–306.

Taylor, K. 1997a. Biological Flora of the British Isles. No. 197. *Geum urbanum* L. *Journal of Ecology*, 85: 705–720.

Taylor, K. 1997b. Biological Flora of the British Isles. No. 198. *Geum rivale* L. *Journal of Ecology*, 85: 721–731.

Taylor, K. 1999a. Biological Flora of the British Isles. No. 207. *Galium aparine* L. *Journal of Ecology*, 87: 713–730.

Taylor, K. 1999b. Biological Flora of the British Isles. No. 209. *Cornus suecica* L. (*Chamaepericlymenum suecicum* (L.) Ascherson & Graebner). *Journal of Ecology*, 87: 1068–1077.

Taylor, K. & Markham, B. 1978. Biological Flora of the British Isles. No. 142. *Ranunculus ficaria* L. *Journal of Ecology*, 66: 1011–1031.

Taylor, K., Rowland, A. P. & Jones, H. E. 2001. Biological Flora of the British Isles. No. 216. *Molinia caerulea* (L.) Moench. *Journal of Ecology*, 89: 126–144.

Taylor, P. 1989. *The genus* Utricularia*: a taxonomic monograph*. Kew Bulletin Additional Series XIV. London: Her Majesty's Stationery Office.

Telfer, M. G., Preston, C. D. & Rothery, P. (in press). A general method for the calculation of relative change in range size from biological atlas data. *Biological Conservation*, 107.

Tennant, D. J. 1996. *Cystopteris dickieana* R. Sim in the central and eastern Scottish Highlands. *Watsonia*, 21: 135–139.

Thakur, V. 1965. Biosystematics of some species of *Epilobium*. Ph.D. thesis, University of Durham.

Thirsk, J. 1997. *Alternative agriculture*. Oxford: Oxford University Press.

Thomas, G. S. 1990. *Perennial garden plants*, edn 3. London: J. M. Dent.

Thompson, J. D. & Turkington, R. 1988. The biology of Canadian weeds. 82. *Holcus lanatus* L. *Canadian Journal of Plant Science*, 68: 131–147.

Thompson, K. 1994. Predicting the fate of temperate species in response to human disturbance and global change. *In*: T. J. B. & C. E. B. Boyle (eds), *Biodiversity, temperate ecosystems, and global change*, pp. 61–76. Berlin: Springer-Verlag.

Thor, G. 1979. *Utricularia* i Sverige, speciellt de förbisedda arterna *U. australis* och *U. ochroleuca*. [*Utricularia* in Sweden, especially the overlooked species *U. australis* and *U. ochroleuca*.] *Svensk Botanisk Tidskrift*, 73: 381–395.

Thor, G. 1987. Sumpbläddra, *Utricularia stygia*, en ny svensk art. [*Utricularia stygia* Thor, a new *Utricularia* species in Sweden.] *Svensk Botanisk Tidskrift*, 81: 273–280.

Thor, G. 1988. The genus *Utricularia* in the Nordic countries, with special emphasis on *U. stygia* and *U. ochroleuca*. *Nordic Journal of Botany*, 8: 213–225.

Thurston, J. M. 1954. Survey of Wild Oats (*Avena fatua* and *A. ludoviciana*) in England and Wales in 1951. *Annals of Applied Biology*, 41: 619–636.

Tiley, G. E. D., Dodd, F. S. & Wade, P. M. 1996. Biological Flora of the British Isles. No. 190. *Heracleum mantegazzianum* Sommier & Levier. *Journal of Ecology*, 84: 297–319.

Timson, J. 1963. The taxonomy of *Polygonum lapathifolium* L., *P. nodosum* Pers. and *P. tomentosum* Schrank. *Watsonia*, 5: 386–395.

Timson, J. 1966. Biological Flora of the British Isles. No. 103. *Polygonum hydropiper* L. *Journal of Ecology*, 54: 815–821.

Tofts, R. 1999. Biological Flora of the British Isles. No. 206. *Cirsium eriophorum* (L.) Scop. *Journal of Ecology*, 87: 529–542.

Townsend, C. C. 1962. Some notes on *Galeopsis ladanum* L. and *G. angustifolia* Ehrh. ex Hoffm. *Watsonia*, 5: 143–149.

Tremayne, M. & Richards, A. J. 1997. The effects of breeding system and seed weight on plant fitness in *Primula scotica* Hooker. *In*: T. E. Tew, T. J. Crawford, J. W. Spencer, D. P. Stevens, M. B. Usher & J. Warren (eds), *The role of genetics in conserving small populations*, pp. 133–142. Peterborough: JNCC.

Trewick, S. & Wade, P. M. 1986. The distribution and dispersal of two alien species of *Impatiens*, waterway weeds in the British Isles. *Proceedings EWRS / AAB 7th Symposium on Aquatic Weeds 1986*: 351–356.

Trieste, L., Greef, B. de, Bondt, R. de, Vanden Bossche, D., D'Haeseleer, M., Slycken, J. van & Coart, E. 1997. Use of RAPD markers to estimate hybridization in *Salix alba* and *Salix fragilis*. *Belgian Journal of Botany*, 129: 140–148.

Trist, P. J. O. 1971. *A survey of the agriculture of Suffolk*. London: Royal Agricultural Society of England.

Trist, P. J. O. 1973. *Festuca glauca* Lam. and its var. *caesia* (Sm.) K. Richt. *Watsonia*, 9: 257–262.

Trist, P. J. O. (ed.). 1979. *An ecological Flora of Breckland*. East Ardsley: EP Publishing.

Trist, P. J. O. 1986. The distribution, ecology, history and status of *Gastridium ventricosum* (Gouan) Schinz & Thell. in the British Isles. *Watsonia*, 16: 43–54.

Trist, P. J. O. 1989. Spreading meadow-grass *Poa subcaerulea* Sm. *Nature in Cambridgeshire*, 31: 57–60.

Trist, P. J. O. 1995. *Elytrigia repens* (L.) Desv. ex Nevski subsp. *arenosa* (Spenner) A. Löve in northwestern Europe. *Watsonia*, 20: 385–390.

Trist, P. J. O. 1998. The distribution and status of *Corynephorus canescens* (L.) P. Beauv. (Poaceae) in Britain and the Channel Islands with particular reference to its conservation. *Watsonia*, 22: 41–47.

Trist, P. J. O. & Butler, J. K. 1995. *Puccinellia distans* (Jacq.) Parl. subsp. *borealis* (O. Holmb.) W. E. Hughes (Poaceae) in mainland Scotland and the Outer Isles. *Watsonia*, 20: 391–396.

Trist, P. J. O. & Sell, P. D. 1988. Two subspecies of *Molinia caerulea* (L.) Moench in the British Isles. *Watsonia*, 17: 153–157.

Trueman, I., Morton, A. & Wainwright, M. 1995. *The flora of Montgomeryshire*. Welshpool: Montgomeryshire Field Society and the Montgomeryshire Wildlife Trust.

Turkington, R. & Aarssen, L. W. 1983. Biological Flora of the British Isles. No. 156. *Hypochoeris radicata* L. *Journal of Ecology*, 71: 999–1022.

Turkington, R. & Burdon, J. J. 1983. The biology of Canadian weeds. 57. *Trifolium repens* L. *Canadian Journal of Plant Science*, 63: 243–266.

Turkington, R. & Cavers, P. B. 1979. The biology of Canadian weeds. 33. *Medicago lupulina* L. *Canadian Journal of Plant Science*, 59: 99–110.

Turkington, R. & Franko, G. D. 1980. The biology of Canadian weeds. 41. *Lotus corniculatus* L. *Canadian Journal of Plant Science*, 60: 965–979.

Turkington, R. A., Cavers, P. B. & Rempel, E. 1978. The biology of Canadian weeds. 29. *Melilotus alba* Desv. and *M. officinalis* (L.) Lam. *Canadian Journal of Plant Science*, 58: 523–537.

Turkington, R., Kenkel, N. C. & Franko, G. D. 1980. The biology of Canadian weeds. 42. *Stellaria media* (L.) Vill. *Canadian Journal of Plant Science*, 60: 981–992.

Turrill, W. B. 1962. *Hypericum elatum*. *Curtis's Botanical Magazine* n.s., 173: t. 376.

Tutin, T. G. 1936. New species of *Zostera* from Britain. *Journal of Botany*, 74: 227–230.

Tutin, T. G. 1942. Biological Flora of the British Isles. No. 7. *Zostera* [genus] L. (pp. 217), *Zostera marina* L. (pp. 217–224), *Zostera hornemanniana* Tutin (pp. 224–226). *Journal of Ecology*, 30: 217–226.

Tutin, T. G. 1947. *Typha angustifolia* L. × *latifolia* L. *Report of the Botanical Society and Exchange Club of the British Isles*, 13: 173–174.

Tutin, T. G. 1950. *Milium scabrum* Merlet. *Watsonia*, 1: 345–348.

Tutin, T. G. 1957. Biological Flora of the British Isles. No. 63. *Allium ursinum* L. *Journal of Ecology*, 45: 1003–1010.

Tutin, T. G. 1980. *Umbellifers of the British Isles*. Botanical Society of the British Isles Handbook no. 2. London: Botanical Society of the British Isles.

Tutin, T. G. & Chater, A. O. 1974. *Asperula occidentalis* Rouy in the British Isles. *Watsonia*, 10: 170–171.

Tutin, T. G., Burges, N. A., Chater, A. O., Edmondson, J. R., Heywood, V. H., Moore, D. M., Valentine, D. H., Walters, S. M. & Webb, D. A. (eds). 1993. *Flora Europaea*, 1, edn 2. *Psilotaceae to Platanaceae*. Cambridge: Cambridge University Press.

Tutin, T. G., Heywood, V. H., Burges, N. A., Moore, D. M., Valentine, D. H., Walters, S. M. & Webb, D. A. (eds). 1968. *Flora Europaea*, 2. *Rosaceae to Umbelliferae*. Cambridge: Cambridge University Press.

Tutin, T. G., Heywood, V. H., Burges, N. A., Moore, D. M., Valentine, D. H., Walters, S. M. & Webb, D. A. (eds). 1972. *Flora Europaea*, 3. *Diapensiaceae to Myoporaceae*. Cambridge: Cambridge University Press.

Tutin, T. G., Heywood, V. H., Burges, N. A., Moore, D. M., Valentine, D. H., Walters, S. M. & Webb, D. A. (eds). 1976. *Flora Europaea*, 4. *Plantaginaceae to Compositae (and Rubiaceae)*. Cambridge: Cambridge University Press.

Tutin, T. G., Heywood, V. H., Burges, N. A., Moore, D. M., Valentine, D. H., Walters, S. M. & Webb, D. A. (eds). 1980. *Flora Europaea*, 5. *Alismataceae to Orchidaceae (Monocotyledones)*. Cambridge: Cambridge University Press.

Tutin, T. G., Heywood, V. H., Burges, N. A., Valentine, D. H., Walters, S. M. & Webb, D. A. (eds). 1964. *Flora Europaea*, 1. *Lycopodiaceae to Platanaceae*. Cambridge: Cambridge University Press.

Uotila, P. 1978. Variation, distribution and taxonomy of *Chenopodium suecicum* and *C. album* in N. Europe. *Acta Botanica Fennica*, 108: 1–35.

Upadhyaya, M. K., Tilsner, H. R. & Pitt, M. D. 1988. The biology of Canadian weeds. 87. *Cynoglossum officinale* L. *Canadian Journal of Plant Science*, 68: 763–774.

Upadhyaya, M. K., Turkington, R. & McIlvride, D. 1986. The biology of Canadian weeds. 75. *Bromus tectorum* L. *Canadian Journal of Plant Science*, 66: 689–709.

Valentine, D. H. 1947. The distribution of sexes in Butterbur. *North Western Naturalist*, 22: 111–114.

Valentine, D. H. 1980. Ecotypic and polymorphic variation in *Centaurea scabiosa* L. *Watsonia*, 13: 103–109.

Valverde, T. & Silvertown, J. 1997. A metapopulation model for *Primula vulgaris*, a temperate forest understorey herb. *Journal of Ecology*, 85: 193–210.

Vaughan, J. G. & Geissler, C. A. 1997. *The New Oxford Book of Food Plants*. Oxford: Oxford University Press.

Verdcourt, B. 1948. Biological Flora of the British Isles. No. 24. *Cuscuta* [genus] L. (pp. 356–358), *Cuscuta europaea* L. (pp. 358–365). *Journal of Ecology*, 36: 356–365.

Vogel, J. C., Rumsey, F. J., Russell, S. J., Cox, C. J., Holmes, J. S., Bujnoch, W., Stark, C., Barrett, J. A. & Gibby, M. 1999. Genetic structure, reproductive biology and ecology of isolated populations of *Asplenium csikii* (Aspleniaceae, Pteridophyta). *Heredity*, 83: 604–612.

Vogt, R. 1987. Die Gattung *Cochlearia* L. (Cruciferae) auf der Iberischen Halbinsel. *Mitteilungen der Botanischen Staatssammlung München*, 23: 393–421.

Wade, A. E. 1970. *The flora of Monmouthshire*. Cardiff: National Museum of Wales.

Wade, A. E., Kay, Q. O. N., Ellis, R. G. & National Museum of Wales. 1994. *Flora of Glamorgan*. London: HMSO.

Waite, S. & Farrell, L. 1998. Population biology of the rare military orchid (*Orchis militaris* L.) at an established site in Suffolk, England. *Botanical Journal of the Linnean Society*, 126: 109–121.

Waldren, S. & Scally, L. 1993. Ecological factors controlling the distribution of *Saxifraga spathularis* and *S. hirsuta* in Ireland. *In*: M. J. Costello & K. S. Kelly (eds), *Biogeography of Ireland: past, present, and future*, pp. 45–55. Occasional publication of the Irish Biogeographical Society no. 2. Dublin: Irish Biogeographical Society.

Walker, K. J. 2000. The distribution, ecology and conservation of *Arenaria norvegica* subsp. *anglica* Halliday (Caryophyllaceae). *Watsonia*, 23: 197–208.

Walters, S. M. 1946. Observations on varieties of *Viola odorata* L. *Report of the Botanical Society and Exchange Club of the British Isles*, 12: 834–839.

Walters, S. M. 1949. Biological Flora of the British Isles. No. 26. *Eleocharis* [genus] R.Br. (pp. 192–194), *Eleocharis palustris* (L.) R. Br. em. R.& S. (pp. 194–202), *Eleocharis uniglumis* (Link) Schult. (pp. 203–206). *Journal of Ecology*, 37: 192–206.

Walters, S. M. 1953. *Montia fontana* L. *Watsonia*, 3: 1–6.

Walters, S. M. 1954. The distribution maps scheme. *Proceedings of the Botanical Society of the British Isles*, 1: 121–130.

Walters, S. M. 1963. *Eleocharis austriaca* Hayek, a species new to the British Isles. *Watsonia*, 5: 329–335.

Walters, S. M. 1979. *Epilobium lanceolatum* Seb. & Mauri – a plant to look for in your garden. *Watsonia*, 12: 399.

Walters, [S.] M. 1993. *Wild & garden plants*. London: HarperCollins.

Ward, L. K. 1973. The conservation of juniper I. Present status of juniper in southern England. *Journal of Applied Ecology*, 10: 165–188.

Ward, L. K. 1981. The demography, fauna and conservation of *Juniperus communis* in Britain. *In*: H. Synge (ed.), *The Biological Aspects of Rare Plant Conservation*, pp. 319–329. Chichester: John Wiley & Sons.

Wardle, D. A. 1987. The ecology of ragwort (*Senecio jacobaea* L.) – a review. *New Zealand Journal of Ecology*, 10: 67–76.

Wardle, P. 1961. Biological Flora of the British Isles. No. 79. *Fraxinus excelsior* L. *Journal of Ecology*, 49: 739–751.

Warwick, S. I. 1979. The biology of Canadian weeds. 37. *Poa annua* L. *Canadian Journal of Plant Science*, 59: 1053–1066.

Warwick, S. I. & Black, L. 1982. The biology of Canadian weeds. 52. *Achillea millefolium* L. s.l. *Canadian Journal of Plant Science*, 62: 163–182.

Warwick, S. I. & Black, L. D. 1988. The biology of Canadian weeds. 90. *Abutilon theophrasti*. *Canadian Journal of Plant Science*, 68: 1069–1085.

Warwick, S. I. & Sweet, R. D. 1983. The biology of Canadian weeds. 58. *Galinsoga parviflora* and *G. quadriradiata* (=*G. ciliata*). *Canadian Journal of Plant Science*, 63: 695–709.

Warwick, S. I., Black, L. D. & Zilkey, B. F. 1985. Biology of Canadian weeds. 72. *Apera spica-venti*. *Canadian Journal of Plant Science*, 65: 711–721.

Watkinson, A. R. 1978. Biological Flora of the British Isles. No. 143. *Vulpia fasciculata* (Forskål) Samp. *Journal of Ecology*, 66: 1033–1049.

Watkinson, A. R., Newsham, K. K. & Forrester, L. 1998. Biological Flora of the British Isles. No. 201. *Vulpia ciliata* Dumort. ssp. *ambigua* (Le Gall) Stace & Auquier. *Journal of Ecology*, 86: 690–705.

Watson, M. F. 1997. *Oxalis*. *In*: J. Cullen, *et al.*, (eds), *European Garden Flora, Vol. 5, Dicotyledons (Part III)*, pp. 18–26. Cambridge: Cambridge University Press.

Watt, A. S. 1931. Preliminary observations on Scottish beechwoods. Introduction and Part I. *Journal of Ecology*, 19: 137–157.

Watt, A. S. 1934. The vegetation of the Chiltern Hills, with special reference to the beechwoods and their seral relationships. *Journal of Ecology*, 22: 445–507.

Weaver, S. E. & Lechowicz, M. J. 1983. The biology of Canadian weeds. 56. *Xanthium strumarium* L. *Canadian Journal of Plant Science*, 63: 211–225.

Weaver, S. E. & McWilliams, E. L. 1980. The biology of Canadian weeds. 44. *Amaranthus retroflexus* L., *A. powellii* S. Wats. and *A. hybridus* L. *Canadian Journal of Plant Science*, 60: 1215–1234.

Weaver, S. E. & Riley, W. R. 1982. The biology of Canadian weeds. 53. *Convolvulus arvensis* L. *Canadian Journal of Plant Science*, 62: 461–472.

Weaver, S. E. & Warwick, S. I. 1984. The biology of Canadian weeds. 64. *Datura stramonium* L. *Canadian Journal of Plant Science*, 64: 979–991.

Webb, C. J. 1990. New Zealand species of *Hydrocotyle* (Apiaceae) naturalised in Britain and Ireland. *Watsonia*, 18: 93–95.

Webb, D. A. 1950. Biological Flora of the British Isles. No. 28. *Saxifraga* [genus] L. (Section *Dactyloides* Tausch) (pp. 185–194), *Saxifraga cespitosa* L. (incl. *S. groenlandica* L.) (pp. 194–197), *Saxifraga hartii* Webb (pp. 197–199), *Saxifraga rosacea* Moench. (pp. 199–206), *Saxifraga hypnoides* L. (emend Webb) (pp. 206–213). *Journal of Ecology*, 38: 185–213.

Webb, D. A. 1955a. The distribution maps scheme: a provisional extension to Ireland of the British national grid. *Proceedings of the Botanical Society of the British Isles*, 1: 316–318.

Webb, D. A. 1955b. Biological Flora of the British Isles. No. 47. *Erica mackaiana* Bab. *Journal of Ecology*, 43: 319–330.

Webb, D. A. 1956. A new subspecies of *Pedicularis sylvatica* L. *Watsonia*, 3: 239–241.

Webb, D. A. 1957. *Hypericum canadense* L., a new American plant in Western Ireland. *Irish Naturalists' Journal*, 12: 113–116.

Webb, D. A. 1958. *Hypericum canadense* L. in Western Ireland. *Watsonia*, 4: 140–144.

Webb, D. A. 1980. The biological vice-counties of Ireland. *Proceedings of the Royal Irish Academy*, 80B: 179–196.

Webb, D. A. 1984. *Polygonum mite* Schrank in Ireland. *Irish Naturalists' Journal*, 21: 283–286.

Webb, D. A. 1985. What are the criteria for presuming native status? *Watsonia*, 15: 231–236.

Webb, D. A. & Gornall, R. J. 1989. *Saxifrages of Europe*. London: Christopher Helm.

Webb, D. A. & Halliday, G. 1973. The distribution, habitat and status of *Hypericum canadense* L. in Ireland. *Watsonia*, 9: 333–344.

Webb, D. A. & Scannell, M. J. P. 1983. *Flora of Connemara and the Burren*. Dublin: Royal Dublin Society & Cambridge University Press.

Webb, D. A., Parnell, J. & Doogue, D. 1996. *An Irish flora*, edn 7. Dundalk: Dundalgan Press (W. Tempest).

Webster, S. D. 1988. *Ranunculus penicillatus* (Dumort.) Bab. in Great Britain and Ireland. *Watsonia*, 17: 1–22.

Wegmüller, S. 1971. A cytotaxonomic study of *Lamiastrum galeobdolon* (L.) Ehrend. & Polatschek in Britain. *Watsonia*, 8: 277–288.

Wein, R. W. 1973. Biological Flora of the British Isles. No. 132. *Eriophorum vaginatum* L. *Journal of Ecology*, 61: 601–615.

Welch, D. 1966. Biological Flora of the British Isles. No. 101. *Juncus squarrosus* L. *Journal of Ecology*, 54: 535–548.

Welch, D. 1967. Notes on *Myosotis scorpioides* agg. *Watsonia*, 6: 276–279.

Welch, D. 1993. *Flora of North Aberdeenshire*. Banchory: privately published.

Welch, D. 1995. An early Scottish record of *Rubus arcticus* L. (Rosaceae). *Watsonia*, 20: 418.

Welch, D. & Welch, M. J. 1998. Colonisation by *Cochlearia officinalis* L. (Brassicaceae) and other halophytes on the Aberdeen–Montrose main road in north-east Scotland. *Watsonia*, 22: 190–193.

Welch, D. & Innes, M. 1999. Southward recolonisation by *Mertensia maritima* (L.) Gray on the coast of north-eastern Scotland. *Watsonia*, 22: 424–426.

Welch, D., Scott. D. & Doyle, S. 2000. Studies on the paradox of seedling rarity in *Vaccinium myrtillus* L. in NE Scotland. *Botanical Journal of Scotland*, 52: 17–30.

Wells, T. C. E. 1967. Changes in a population of *Spiranthes spiralis* (L.) Chevall. at Knocking Hoe National Nature Reserve, Bedfordshire, 1962–65. *Journal of Ecology*, 55: 83–99.

Wells, T. C. E. 1976. Biological Flora of the British Isles. No. 138. *Hypochoeris maculata* L. *Journal of Ecology*, 64: 757–774.

Wells, T. C. E. & Barling, D. M. 1971. Biological Flora of the British Isles. No. 120. *Pulsatilla vulgaris* Mill. *Journal of Ecology*, 59: 275–292.

Wells, T. C. E. & Cox, R. 1989. Predicting the probability of the bee orchid (*Ophrys apifera*) flowering or remaining vegetative from the size and number of leaves. *In*: H. W. Pritchard (ed.), *Modern methods in orchid conservation*, pp. 127–139. Cambridge: Cambridge University Press.

Wells, T. C. E. & Cox, R. 1991. Demographic and biological studies on *Ophrys apifera*: some results from a 10 year study. *In*: T. C. E. Wells & J. H. Willems (eds), *Population ecology of terrestrial orchids*, pp. 47–61. The Hague: SPB Academic Publishing.

Wells, T. C. E., Rothery, P., Cox, R. & Bamford, S. 1998. Flowering dynamics of *Orchis morio* L. and *Herminium monorchis* (L.) R. Br. at two sites in eastern England. *Botanical Journal of the Linnean Society*, 126: 39–48.

Wentworth, J. E. & Gornall, R. J. 1996. Cytogenetic evidence for autopolyploidy in *Parnassia palustris*. *New Phytologist*, 134: 641–648.

Werner, P. A. 1975. The biology of Canadian weeds. 13. *Cirsium arvense* (L.) Scop. *Canadian Journal of Plant Science*, 55: 1033–1048.

Werner, P. A. & Rioux, R. 1977. The biology of Canadian weeds. 24. *Agropyron repens* (L.) Beauv. *Canadian Journal of Plant Science*, 57: 905–919.

Werner, P. A. & Soule, J. D. 1976. The biology of Canadian weeds. 18. *Potentilla recta* L., *P. norvegica* L. and *P. argentea* L. *Canadian Journal of Plant Science*, 56: 591–603.

Werner, P. A., Bradbury, I. K. & Gross, R. S. 1980. The biology of Canadian weeds. 45. *Solidago canadensis* L. *Canadian Journal of Plant Science*, 60: 1393–1409.

Westerhoff, D. & Clark, M. J. 1992. *The New Forest heathlands, grasslands and mires. A management review and strategy*. Lyndhurst: English Nature.

Wheeler, B. R. & Hutchings, M. J. 1999. The history and distribution of *Phyteuma spicatum* L. (Campanulaceae) in Britain. *Watsonia*, 22: 387–395.

Wheeler, B. D., Lambley, P. W. & Geeson, J. 1998. *Liparis loeselii* (L.) Rich. in eastern England: constraints on distribution and population development. *Botanical Journal of the Linnean Society*, 126: 141–158.

White, J. (ed.). 1982. *Studies on Irish Vegetation*. Dublin: Royal Dublin Society.

White, J. E. J. 1992. Ornamental uses of willow in Britain. *Proceedings of the Royal Society of Edinburgh*, 98B: 183–192.

White, J. W. 1912. *The Flora of Bristol*. Bristol: John Wright & Sons.

Whitehead, L. E. 1976. *Plants of Herefordshire – a handlist*. Hereford: Herefordshire Botanical Society.

Whittington, G. & Edwards, K. T. 2000. *Illecebrum verticillatum* in the Outer Hebrides. *Botanical Journal of Scotland*, 52: 101–104.

Wigginton, M. J. (comp. & ed.). 1999. *British Red Data Books. 1. Vascular plants*, edn 3. Peterborough: Joint Nature Conservation Committee.

Wigginton, M. J. & Graham, G. G. 1981. *Guide to the identification of some of the more difficult vascular plant species*. Nature Conservancy Council England Field Unit Occasional Paper no. 1. Banbury: Nature Conservancy Council.

Wigston, D. L. 1979. *Nothofagus* Blume in Britain. *Watsonia*, 12: 344–345.

Wilcock, C. C. & Jones, B. M. G. 1974. The identification and origin of *Stachys* × *ambigua* Sm. *Watsonia*, 10: 139–147.

Wilkinson, M. J. & Stace, C. A. 1989. The taxonomic relationships and typification of *Festuca brevipila* Tracey and *F. lemanii* Bastard (Poaceae). *Watsonia*, 17: 289–299.

Wilkinson, M. J. & Stace, C. A. 1991. A new taxonomic treatment of the *Festuca ovina* L. aggregate (Poaceae) in the British Isles. *Botanical Journal of the Linnean Society*, 106: 347–397.

Willems, J. H. & Melser, C. 1998. Population dynamics and life-history of *Coeloglossum viride* (L.) Hartm.: an endangered orchid species in The Netherlands. *Botanical Journal of the Linnean Society*, 126: 83–93.

Williams, J. T. 1963. Biological Flora of the British Isles. No. 87. *Chenopodium album* L. *Journal of Ecology*, 51: 711–725.

Williams, J. T. 1969. Biological Flora of the British Isles. No. 117. *Chenopodium rubrum* L. *Journal of Ecology*, 57: 831–841.

Williams, L. R. 2000. Annual variations in the size of a population of *Cardamine impatiens* L. *Watsonia*, 23: 209–212.

Williamson, M. & Gaston, K. J. 1999. A simple transformation for sets of range sizes. *Ecography*, 22: 674–680.

Williamson, R. 1978. *The Great Yew Forest: the natural history of Kingley Vale*. London: Macmillan.

Willis, A. J. & Davies, E. W. 1960. *Juncus subulatus* Forsk. in the British Isles. *Watsonia*, 4: 211–217.

Wilson, A. 1956. *The altitudinal range of British plants*, edn 2. Arbroath: T. Buncle & Co.

Wilson, G. B., Houston, L., Whittington, W. J. & Humphries, R. N. 2000. Biological Flora of the British Isles. No. 213. *Veronica spicata* L. ssp. *spicata* and ssp. *hybrida* (L.) Gaudin. *Journal of Ecology*, 88: 890–909.

Wilson, G. B., Whittington, W. J. & Humphries, R. N. 1995. Biological Flora of the British Isles. No. 185. *Potentilla rupestris* L. *Journal of Ecology*, 83: 335–343.

Wilson, G. B., Wright, J., Lusby, P., Whittington, W. J. & Humphries, R. N. 1995. Biological Flora of the British Isles. No. 188. *Lychnis viscaria* L. *Journal of Ecology*, 83: 1039–1051.

Wilson, P. J. 1990. *The ecology and conservation of rare arable weed species and communities.* Ph.D. thesis, University of Southampton.

Wilson, P. J. 1991. Britain's arable weeds. *British Wildlife*, 3: 149–161.

Wilson, P. J. 1997. *The status, ecology and conservation of Martin's Ramping Fumitory* (Fumaria reuteri *Boiss.*). Eastleigh: Hampshire & Isle of Wight Wildlife Trust.

Wilson, P. J. 1999. *The status and distribution of* Dianthus armeria *L. in Britain. Report for 1998.* Back from the Brink Report No. 117. London: Plantlife.

Wilson, P. J. & Aebischer, N. J. 1995. The distribution of dicotyledonous arable weeds in relation to distance from the field edge. *Journal of Applied Ecology*, 32: 295–310.

Wilson, Z. A., Dawson, J., Russell, J. & Mulligan, B. J. 1991. Exploited plants. *Arabidopsis thaliana. Biologist*, 38: 163–169.

Winfield, M. & Parker, J. 2000. A molecular analysis of *Gentianella* in Britain. Report no. 155. Cambridge: Cambridge University Botanic Garden.

Wolff, K. & Morgan-Richards, M. 1999. The use of RAPD data in the analysis of population genetic structure: case studies in *Alkanna* (Boraginaceae) and *Plantago* (Plantaginaceae). *In*: P. M. Hollingsworth, R. M. Bateman & R. J. Gornall (eds), *Molecular Systematics and Plant Evolution*, pp. 51–73. London: Taylor & Francis.

Woodell, S. R. J. 1958. Biological Flora of the British Isles. No. 64. *Daboecia cantabrica* K. Koch. *Journal of Ecology*, 46: 205–216.

Woodell, S. R. J. 1965. Natural hybridization between the Cowslip (*Primula veris* L.) and the Primrose (*P. vulgaris* Huds.) in Britain. *Watsonia*, 6: 190–202.

Woodell, S. R. J. & Dale, A. 1993. Biological Flora of the British Isles. No. 178. *Armeria maritima* (Mill.) Willd. *Journal of Ecology*, 81: 573–588.

Woodhead, F. 1994. *Flora of the Christchurch Area.* Bournemouth: privately published.

Woodhead, N. 1951a. Biological Flora of the British Isles. No. 32. *Lloydia serotina* (L.) Rchb. *Journal of Ecology*, 39: 198–203.

Woodhead, N. 1951b. Biological Flora of the British Isles. No. 33. *Lobelia* [genus] L. (pp. 456–457), and *Lobelia dortmanna* L. (pp. 458–464). *Journal of Ecology*, 39: 456–464.

Woodhead, N. 1951c. Biological Flora of the British Isles. No. 34. *Subularia aquatica* L. *Journal of Ecology*, 39: 465–469.

Woods, R. G. 1993. *Flora of Radnorshire.* Cardiff: National Museum of Wales & Bentham-Moxon Trust.

Woods, R. G. 1998. The conservation of some *Sorbus* species in Wales. *In*: A. Jackson & M. Flanagan (eds), *The conservation status of* Sorbus *in the UK*, pp. 32–39. London: Royal Botanic Gardens, Kew.

Woodward, F. I. 1997. Life at the edge: a 14 year study of a *Verbena officinalis* population's interactions with climate. *Journal of Ecology*, 85: 899–906.

Woodward, F. I. 1975. The climatic control of the altitudinal distribution of *Sedum rosea* (L.) Scop. and *S. telephium* L. II. The analysis of plant growth in controlled environments. *New Phytologist*, 74: 335–348.

Woodward, F. I. & Pigott, C. D. 1975. The climatic control of the altitudinal distribution of *Sedum rosea* (L.) Scop. and *S. telephium* L. I. Field observations. *New Phytologist*, 74: 323–334.

Worrell, R. & Malcolm, D. C. 1998. Anomolies in the distribution of Silver Birch (*Betula pendula* Roth) populations in Scotland. *Botanical Journal of Scotland*, 50: 1–10.

Wright, J. A. & Lusby, P. S. 1999. The past and present status of *Moneses uniflora* (L.) Gray (Pyrolaceae) in Scotland. *Watsonia*, 22: 343–352.

Wu, L. & Jain, S. 1980. Self-fertility and seed set in natural populations of *Anthoxanthum odoratum* L. *Botanical Gazette*, 141: 300–304.

Wurzell, B. 1988. *Conyza sumatrensis* (Retz.) E. Walker established in England. *Watsonia*, 17: 145–148.

Wurzell, B. 1990. Truth is stranger than fitches. *BSBI News*, 54: 24–25.

Wurzell. B. 1991. Aliens and adventives records. *BSBI News*, 58: 36–37.

Wurzell, B. 1992a. Spring flowering crocuses. *BSBI News*, 60: 36–38.

Wurzell, B. 1992b. Foreign *Crataegus* in Britain: a thorny problem. *BSBI News*, 61: 42–45.

Wurzell, B. 1994a. A history of *Conyza* in London. *BSBI News*, 65: 34–39.

Wurzell, B. 1994b. Year of the Bullwort. *BSBI News*, 66: 32.

Wynne, G. 1993. *Flora of Flintshire.* Denbigh: Gee & Son.

Wyse Jackson, P. S. 1991. A note on *Cochlearia scotica* Druce (Cruciferae). *Botanical Journal of the Linnean Society*, 106: 118–119.

Wyse Jackson, P. [S.] & Sheehy Skeffington, M. 1984. *The flora of Inner Dublin.* Dublin: Royal Dublin Society.

Yeo, P. F. 1966. A revision of the genus *Bergenia* Moench (Saxifragaceae). *Kew Bulletin*, 20: 113–148.

Yeo, P. F. 1968a. *Euphrasia stricta* Lehm. in Guernsey. *Proceedings of the Botanical Society of the British Isles*, 7: 383–385.

Yeo, P. F. 1968b. A contribution to the taxonomy of the genus *Ruscus. Notes from the Royal Botanic Garden Edinburgh*, 28: 237–264.

Yeo, P. F. 1973. The species of *Acaena* with spherical heads cultivated and naturalized in the British Isles. *In*: P. S. Green (ed.), *Plants wild and cultivated*, Appendix III, pp. 193–221. Hampton: E. W. Classey.

Yeo, P. F. 1975. The Yorkshire records of *Euphrasia salisburgensis. The Naturalist*, no. 934: 83–87.

Yeo, P. F. 1978. A taxonomic revision of *Euphrasia* in Europe. *Botanical Journal of the Linnean Society*, 77: 223–334.

Yeo, P. F. 1985. *Hardy Geraniums.* London: Croom Helm.

Young, D. P. 1958. *Oxalis* in the British Isles. *Watsonia*, 4: 51–69.

Ziburski, A., Kadereit, J. W. & Leins, P. 1986. Quantitative aspects of hybridization in mixed populations of *Rumex obtusifolius* L. and *R. crispus* L. (Polygonaceae). *Flora*, 178: 233–242.

Zohary, D. & Hopf, M. 2000. *Domestication of plants in the Old World*, edn 3. Oxford: Oxford University Press.

Index to the species accounts

R. G. Ellis

Accepted Latin names of all taxa that are mapped in the *New Atlas* or on the CD-ROM are in **bold** type; Latin names of all other taxa mentioned in the captions to the maps, whether accepted names or synonyms, are in *italics*. Vernacular names are in roman, and when they are the same as a Latin genus name they are always placed after the Latin name. All these entries are followed by the appropriate page number or by 'CD'.

Latin synonyms that do not occur in the captions are in *italics* and are of two types:

- synonyms used in the *Atlas of the British Flora* or the *Critical Supplement to the Atlas of the British Flora*, which are linked to the accepted name by the equivalent sign '≡'
- all other synonyms, which are linked to the accepted name by the word 'see'

These synonyms are included to enable the reader to link names which have changed to the accepted Latin name used in this Atlas. It should not be assumed that they are exact taxonomic synonyms in all instances, as they may (for example) be the names of taxa which were formerly treated as species but are now regarded as varieties of mapped species.

ABIES
alba 81
grandis 82
nordmanniana CD
procera 82
Abraham-Isaac-Jacob 501
ABUTILON
theophrasti 221
ACACIA
melanoxylon CD
ACAENA
anserinifolia CD
× **A. inermis** CD
inermis CD
novae-zelandiae 338
ovalifolia CD
pusilla, see **A. anserinifolia** CD
Acaena, Spineless CD
Two-spined CD
ACANTHACEAE 580 CD
ACANTHUS
mollis 580
spinosus CD
ACER
campestre 439
cappadocicum 438
negundo 440
platanoides 438
pseudoplatanus 439
saccharinum 439
ACERACEAE 438
ACERAS
anthropophorum 854
ACHILLEA
distans CD
ligustica CD
millefolium 647
ptarmica 647
Acinos
arvensis ≡ **Clinopodium acinos** 523
Aconite, Winter 96
ACONITUM
anglicum ≡ **A. napellus** 97
× **cammarum** (**A. napellus** × **A. variegatum**) 97
'Bicolor' 97
napellus *sens. lat.* 97
subsp. *napellus* 97
× **A. variegatum** (**A.** × **cammarum**) 97
napellus sens. str. 97
ACORUS
calamus 683
gramineus CD
ACROPTILON
repens CD
ACTAEA
spicata 98
Adder's-tongue 60
Least 61
Small 60
ADIANTACEAE 62
ADIANTUM
capillus-veneris 62
ADONIS
annua 111
ADOXA
moschatellina 602
ADOXACEAE 602
AEGOPODIUM
podagraria 460
AEONIUM
cuneatum CD
Aeonium CD
AESCULUS
carnea 438
hippocastanum 437
indica CD
pavia 438
AETHEORHIZA
bulbosa CD
AETHUSA
cynapium 464
subsp. **agrestis** CD 464
subsp. *cynapium* 464
AGAPANTHUS
praecox CD
AGAVACEAE 834 CD
AGAVE
americana CD
AGERATUM
houstonianum CD
AGRIMONIA
eupatoria 336
var. *sepium* 336
odorata ≡ **A. procera** 336
procera 336
Agrimony 336
Bastard CD
Fragrant 336
Agropyron
caninum ≡ **Elymus caninus** 793
donianum 793
junceiforme ≡ **Elytrigia juncea** 795
pungens ≡ **Elytrigia atherica** 794
repens ≡ **Elytrigia repens** 794
AGROSTEMMA
githago 175
AGROSTIS
avenacea CD
canina 776
subsp. *montana*, see **A. vinealis** 777
AGROSTIS (*cont.*)
capillaris 775
castellana 775
curtisii 776
gigantea 775
hyemalis CD
lachnantha CD
scabra CD
semiverticillata ≡ **Polypogon viridis** 781
setacea ≡ **A. curtisii** 776
stolonifera 776
tenuis ≡ **A. capillaris** 775
vinealis 777
AILANTHUS
altissima 440
AIRA
caryophyllea 771
praecox 772
AIZOACEAE 136 CD
Ajowan CD
AJUGA
chamaepitys 519
pyramidalis 519
reptans 519
Ake-ake CD
Akiraho CD
ALCEA
rosea 221
ALCHEMILLA
acutiloba 340
alpina 338
conjuncta 338
filicaulis subsp. **filicaulis** 341
filicaulis subsp. **vestita** 341
glabra 342
glaucescens 339
glomerulans 342
gracilis, see **A. micans** 340
micans 340
minima 341
mollis 343
monticola 339
subcrenata 339
tytthantha CD
vestita, see **A. filicaulis** subsp. **vestita** 341
wichurae 342
xanthochlora 340
Alder 134
Grey 134
Italian 135
Red CD
Alexanders 458
Perfoliate CD
ALISMA
gramineum 668
lanceolatum 668
plantago-aquatica 668
ALISMATACEAE 667 CD
Alison, Golden 265
Hoary 265
Small 265
Sweet 266
Alkanet 499
False CD
Green 500
Yellow CD
ALLIARIA
petiolata 251
ALLIUM
ampeloprasum 822
var. *ampeloprasum* 822
var. *babingtonii* 822
var. *bulbiferum* 822
babingtonii ≡ *A. ampeloprasum* var. *babingtonii* 822
carinatum 821
var. *carinatum* 821
var. *pulchellum* 821
cepa CD
moly CD
neapolitanum CD
nigrum CD
oleraceum 821
paradoxum 820
pendulinum CD
porrum CD
roseum 820
var. *bulbiferum* 820
var. *roseum* 820
sativum CD
schoenoprasum 819
scorodoprasum 822
subsp. *rotundum* 822
sphaerocephalon 822
subhirsutum CD
triquetrum 820
unifolium CD
ursinum 821
vineale 823
var. *capsuliferum* 823
Allseed 436
Four-leaved 171
Almond CD
ALNUS
cordata 135
glutinosa 134
incana 134
rubra CD

ALOPECURUS
aequalis 783
alpinus ≡ **A. borealis** 783
borealis 783
× **brachystylus** (**A. geniculatus** × **A. pratensis**) 782
bulbosus 782
× *A. geniculatus* 782
geniculatus 782
× **A. pratensis** (**A.** × **brachystylus**) 782
myosuroides 783
pratensis 781
Alpine-sedge, Black 735
Close-headed 735
Scorched 733
ALSTROEMERIA
aurea CD
Altar-lily CD
ALTHAEA
hirsuta 220
officinalis 220
rosea, see **Alcea rosea** 221
ALYSSUM
alyssoides 265
saxatile 265
Amaranth, Common 149
Dioecious CD
Green 149
Indehiscent CD
Mucronate CD
Powell's CD
Purple CD
AMARANTHACEAE 149 CD
AMARANTHUS
albus 149
blitoides CD
blitum CD
bouchonii CD
capensis CD
caudatus CD
cruentus CD
deflexus CD
graecizans CD
hybridus 149
× **A. retroflexus** (**A.** × **ozanonii**) CD
hybridus sens. lat. CD
hypochondriacus CD
lividus, see **A. blitum** CD
× **ozanonii** (**A. hybridus** × **A. retroflexus**) CD
palmeri CD
powellii CD
quitensis CD
retroflexus 149
standleyanus CD
thunbergii CD
Amaryllidaceae, see LILIACEAE 808
AMARYLLIS
belladonna CD
AMBROSIA
artemisiifolia 661
coronopifolia, see **A. psilostachya** CD
psilostachya CD
trifida 662
AMELANCHIER
lamarckii 365
American-spikenard CD
AMMI
majus 471
visnaga CD
AMMOPHILA
arenaria 779
subsp. *breviligulata*, see **A. breviligulata** CD
breviligulata CD
Amomyrtus
luma, see **Luma apiculata** *CD*
AMSINCKIA
calycina, see **A. micrantha** 502
intermedia, see **A. micrantha** 502
lycopsoides 502
menziesii, see **A. micrantha** 502
micrantha 502
ANACAMPTIS
pyramidalis 843
ANACARDIACEAE 440
ANAGALLIS
arvensis 303
subsp. *arvensis* 303
subsp. *arvensis* f. *azurea* 303
subsp. *caerulea*, see **A. arvensis** subsp. **foemina** 303
subsp. **foemina** 303
ANAGALLIS (*cont.*)
foemina, see **A. arvensis** subsp. **foemina** 303
minima 303
tenella 302
ANAPHALIS
margaritacea 633
ANCHUSA
arvensis 500
azurea 500
ochroleuca CD
officinalis 499
Anchusa, Garden 500
ANDROMEDA
polifolia 289
ANEMONE
apennina 99
blanda 99
hupehensis × **A. vitifolia** (**A.** × **hybrida**) 100
× **hybrida** (**A. hupehensis** × **A. vitifolia**) 100
nemorosa 98
ranunculoides 99
Anemone, Balkan 99
Blue 99
Japanese 100
Wood 98
Yellow 99
ANETHUM
graveolens 465
Angel's-tears CD
Angel's-trumpets CD
Angelica-tree, Chinese CD
Japanese CD
ANGELICA
archangelica 474
pachycarpa CD
sylvestris 473
Angelica, Garden 474
Portugese CD
Wild 473
ANISANTHA
diandra 790
madritensis 792
rigida 791
rubens CD
sterilis 791
tectorum 791
ANOGRAMMA
leptophylla 62
ANTENNARIA
dioica 633
ANTHEMIS
arvensis 648
cotula 648
punctata CD
subsp. *cupaniana* CD
subsp. *punctata* CD
tinctoria 649
Antholyza 833
ANTHOXANTHUM
aristatum 773
odoratum 772
puelii ≡ **A. aristatum** 773
ANTHRISCUS
caucalis 457
cerefolium 457
sylvestris 456
ANTHYLLIS
vulneraria 375
subsp. **carpatica** 376
subsp. *carpatica* var. *pseudovulneraria* 376
subsp. *corbierei* 375
subsp. **lapponica** 376
subsp. **polyphylla** CD 375
subsp. *vulgare* var. *pseudovulneraria* ≡ *A. vulneraria* subsp. *carpatica* var. *pseudovulneraria* 376
subsp. *vulneraria* 375
ANTIRRHINUM
majus 546
APERA
interrupta 780
spica-venti 780
APHANES
arvensis 343
arvensis agg. 343
australis 344
inexspectata, see **A. australis** 344
microcarpa ≡ **A. australis** 344
APIACEAE 454 CD
APIUM
graveolens 468
inundatum 469
× **A. nodiflorum** (**A.** × **moorei**) 470
× *A. repens* 469
× **moorei** (**A. inundatum** × **A. nodiflorum**) 470
nodiflorum 469
repens 469
APOCYANACEAE 484 CD
APONOGETON
distachyos 671
APONOGETONACEAE 671
Apple 357
Crab 357
Apple-mint 527
False 528
Apple-of-Peru 485
Apples 357
APTENIA
cordifolia CD
AQUIFOLIACEAE 425
AQUILEGIA
pyrenaica CD
vulgaris 111
ARABIDOPSIS
thaliana 252
ARABIS
alpina 263
arenosa CD
brownii ≡ **A. hirsuta** 263
caucasica 263
collina CD
glabra 262
hirsuta 263
petraea 262
scabra 264
stricta ≡ **A. scabra** 264
turrita CD
Arabis, Garden 263
ARACEAE 683 CD
ARALIA
chinensis CD
elata CD
racemosa CD
ARALIACEAE 453 CD
ARAUCARIA
araucana 92
ARAUCARIACEAE 92
ARBUTUS
unedo 290
Archangel, Yellow 511
ARCTIUM
lappa 609
minus *sens. lat.* 609
minus 609
subsp. *nemorosum*, see **A. minus** *sens. lat.* 609
subsp. *pubens*, see **A. minus** *sens. lat.* 609
nemorosum 609
pubens, see **A. minus** *sens. lat.* 609
ARCTOSTAPHYLOS
alpinus 290
uva-ursi 290
ARCTOTHECA
calendula CD
Arctous
alpinus ≡ **Arctostaphylos alpinus** 290
AREMONIA
agrimonioides CD
ARENARIA
balearica 154
ciliata 154
leptoclados ≡ **A. serpyllifolia** subsp. **leptoclados** 153
montana CD
norvegica subsp. **anglica** 154
norvegica subsp. **norvegica** 153
serpyllifolia 152
subsp. **leptoclados** 153
subsp. *lloydii* 153
subsp. *macrocarpa*, see **A. serpyllifolia** subsp. **serpyllifolia** 153
subsp. **serpyllifolia** 153
ARGEMONE
mexicana CD
Argentine-pear CD
ARISARUM
proboscideum CD
ARISTEA
ecklonii CD
ARISTOLOCHIA
clematitis 93
rotunda CD
ARISTOLOCHIACEAE 93 CD
ARMERIA
alliacea, see **A. arenaria** 209
arenaria 209
maritima 208
subsp. **elongata** 209
subsp. *maritima* 208
ARMORACIA
rusticana 259
ARNOSERIS
minima 618
ARONIA
arbutifolia CD
melanocarpa CD
ARRHENATHERUM
elatius 766
var. *bulbosum* 766
Arrowgrass, Marsh 672
Sea 672
Arrowhead 667
Canadian CD
Narrow-leaved CD
ARTEMISIA
abrotanum CD
absinthium 645
annua CD
biennis CD
campestris 646
dracunculus CD
maritima ≡ **Seriphidium maritimum** 644
norvegica 646
stelleriana CD
verlotiorum 645
vulgaris 645
Arthrocnemon
perenne, see **Sarcocornia perennis** 144
Artichoke, Globe CD
Jerusalem 664
ARUM
italicum 684
subsp. **italicum** 685
subsp. **neglectum** 685
maculatum 684
Arum, Bog CD
Dragon CD
ARUNCUS
dioicus 323
Asarabacca 93
ASARINA
procumbens CD
ASARUM
europaeum 93
Ash 538
ASPARAGUS
officinalis
subsp. **officinalis** 827
subsp. **prostratus** 828
Asparagus, Garden 827
Wild 828
Aspen 234
ASPERUGO
procumbens 502
ASPERULA
arvensis 590
cynanchica 590
subsp. *cynanchica* 590
subsp. *occidentalis* 590
occidentalis, see *A. cynanchica* subsp. *occidentalis* 590
taurina CD
Asphodel, Bog 808
Scottish 808
White CD
ASPHODELUS
albus CD
Aspidiaceae, see DRYOPTERIDACEAE 75
ASPLENIACEAE 67
ASPLENIUM
adiantum-nigrum 68
subsp. *onopteris* ≡ **A. onopteris** 68
× *A. onopteris* (*A.* × *ticinense*) 68
billotii, see **A. obovatum** 68
ceterach, see **Ceterach officinarum** 71
csikii 70
cuneifolium 68
marinum 69
obovatum 68
onopteris 68
ruta-muraria 71

ASPLENIUM (*cont.*)
scolopendrium, see **Phyllitis scolopendrium** 67
septentrionale 71
× *ticinense* (*A. adiantum-nigrum* × *A. onopteris*) 68
trichomanes 69
subsp. **pachyrachis** 70
subsp. **quadrivalens** 70
subsp. **trichomanes** 69
trichomanes-ramosum, see **A. viride** 70
viride 70
ASTER agg. 638
alien N. American taxa 638
laevis CD 638
× **A. novi-belgii** (**A.** × **versicolor**) 639
lanceolatus 640
× **A. novi-belgii** (**A.** × **salignus**) 640
linosyris 641
novae-angliae 639
novi-belgii 639
× **salignus** (**A. lanceolatus** × **A. novi-belgii**) 640
schreberi CD 638
tripolium 640
var. *discoideus* 640
× **versicolor** (**A. laevis** × **A. novi-belgii**) 639
Aster, China CD
Goldilocks 641
Mexican CD
Sea 640
ASTERACEAE 607 CD
ASTILBE
× **arendsii** (**A. ? chinensis** × **A. japonica**) CD
? chinensis × **A. japonica** (**A.** × **arendsii**) CD
japonica CD
rivularis CD
ASTRAGALUS
alpinus 374
cicer CD
danicus 373
glycyphyllos 374
odoratus CD
ASTRANTIA
major 455
Astrantia 455
Athyriaceae, see WOODSIACEAE 72
ATHYRIUM
alpestre s.l. ≡ **A. distentifolium** 72 & **A. flexile** 73
distentifolium 72
var. *flexile* ≡ **A. flexile** 73
filix-femina 72
flexile 73
ATRIPLEX
glabriuscula 141
× *A. longipes* (*A.* × *taschereaui*) 141
halimus CD
hastata ≡ **A. prostrata** 141
hortensis 140
laciniata 143
littoralis 142
longipes 141
subsp. *praecox* 142
patula 142
pedunculata 143
portulacoides 143
praecox 142
prostrata 141
suberecta CD
× *taschereaui* (*A. glabriuscula* × *A. longipes*) 141
ATROPA
belladonna 486
Aubretia 264
AUBRIETA
deltoidea 264
AUCUBA
japonica 423
Aunt-Eliza 833
Auricula CD
AVENA
barbata CD
fatua 766
ludoviciana ≡ *A. sterilis* subsp. *ludoviciana* 767
sativa 767
sterilis 767
subsp. *ludoviciana* 767
subsp. *sterilis* 767
strigosa 766
Avens, Hybrid 335
Large-leaved CD
Mountain 336
Avens (*cont.*)
Water 335
Wood 335
Avenula
pratensis, see **Helictotrichon pratense** 765
pubescens, see **Helictotrichon pubescens** 765
Awlwort 278
Azalea, Trailing 288
Yellow 287
AZOLLA
filiculoides 81
AZOLLACEAE 81

Baby's-breath CD
BACCHARIS
halimiifolia CD
BALDELLIA
ranunculoides 667
BALLOTA
nigra 510
subsp. *foetida*, see **B. nigra** 510
subsp. *meridionalis*, see **B. nigra** 510
Balm 521
Bastard 516
Balm-of-Gilead 235
Balsam, Indian 453
Orange 452
Small 452
Touch-me-not 452
BALSAMINACEAE 452
Balsamita
major, see **Tanacetum balsamita** CD
Balsam-poplar, Eastern CD
Hybrid CD
Western 236
Bamboo, Arrow 740
Broad-leaved 739
Dwarf CD
Hairy CD
Maximowicz's CD
Narihira CD
Simon's CD
Square-stemmed CD
Veitch's CD
Baneberry 98
BARBAREA
intermedia 256
stricta 255
verna 256
vulgaris 255
Barberry 113
Box-leaved CD
Chinese CD
Clustered CD
Darwin's 113
Gagnepain's CD
Great CD
Hedge 114
Mrs Wilson's CD
Thunberg's 113
Barley 796
Antarctic CD
Argentine CD
Foxtail 797
Little CD
Meadow 798
Mediterranean CD
Sea 798
Six-rowed 797
Wall 797
Wood 796
BARTSIA
alpina 572
Bartsia, Alpine 572
French CD
Red 571
Yellow 572
Basil, Wild 523
BASSIA
scoparia 140
Bastard-toadflax 424
Bay 92
Bayberry CD
Beadplant CD
Beak-sedge, Brown 712
White 711
Bean, Broad 385
French CD
Runner CD
Bear's-breech 580
Bear's-breech, Spiny CD
Bearberry 290
Arctic 290
Beard, Old Man's 100
Beard-grass, Annual 781
BECKMANNIA
syzigachne CD
Bedstraw, Fen 591
Heath 594
Hedge 593
Lady's 592
Limestone 594
Northern 591
Slender 593
Tree CD
Wall 595
Beech 129
Beet, Caucasian CD
Fodder 144
Foliage CD
Leaf CD
Root 144
Sea 144
Spinach CD
Sugar 144
Beetle-grass, Brown CD
Beetroot 144
Beggarticks 665
Bellflower, Adria 585
Chimney CD
Clustered 585
Cornish CD
Creeping 586
Giant 586
Italian CD
Ivy-leaved 587
Milky CD
Nettle-leaved 586
Peach-leaved 584
Rampion 584
Spreading 583
Trailing 585
BELLIS
perennis 643
Bent, African CD
Black 775
Bristle 776
Brown 777
Common 775
Creeping 776
Highland 775
Rough CD
Small CD
Velvet 776
Water 781
BERBERIDACEAE 113 CD
BERBERIS
aggregata CD
buxifolia CD
darwinii 113
× **B. empetrifolia** (**B.** × **stenophylla**) 114
gagnepainii CD
glaucocarpa CD
julianae CD
× **stenophylla** (**B. darwinii** × **B. empetrifolia**) 114
thunbergii 113
vulgaris 113
wilsoniae CD
BERGENIA
crassifolia 313
Bermuda-buttercup CD
Bermuda-grass 800
African CD
BERTEROA
incana 265
BERULA
erecta 461
BETA
trigyna CD
vulgaris subsp. **cicla** CD
var. *cicla* CD
var. *flavescens* CD
vulgaris subsp. **maritima** 144
vulgaris subsp. **vulgaris** 144
Betonica
officinalis ≡ **Stachys officinalis** 507
Betony 507
BETULA
× **aurata** (**B. pendula** × **B. pubescens**) 133
nana 134
pendula 133
× **B. pubescens** (**B.** × **aurata**) 133
pubescens 133
BETULACEAE 133 CD
BIDENS
bipinnata CD
cernua 665
connata CD
frondosa 665
pilosa CD
tripartita 665
Bilberry 294
Bog 294
Bindweed, Field 490
Hairy 492
Hedge 491
Large 492
Sea 490
Birch, Downy 133
Dwarf 134
Silver 133
Bird's-foot 378
Orange 378
Bird's-foot-trefoil, Common 377
Greater 377
Hairy 377
Narrow-leaved 376
Slender 378
Bird's-nest, Yellow 297
Bird-in-a-bush 119
Birthwort 93
Bistort, Alpine 184
Amphibious 184
Common 183
Red 183
Bitter-cress, Hairy 262
Large 260
Narrow-leaved 261
New Zealand CD
Wavy 261
Bittersweet 488
Bitter-vetch 385
Wood 379
Black-bindweed 190
Black-eyed-Susan CD
Black-grass 783
Black-jack CD
Black-poplar 234
Hybrid 235
BLACKSTONIA
perfoliata 481
Blackthorn 353
Blackwood, Australian CD
Bladder-fern, Brittle 74
Dickie's 74
Mountain 74
Bladdernut CD
Bladder-sedge 723
Bladderseed 466
Bladder-senna 373
Orange CD
Bladderwort 582
Greater 582
Lesser 583
Blanketflower CD
BLECHNACEAE 81 CD
BLECHNUM
cordatum CD
spicant 81
Bleeding-heart 119
Blinks 151
Small-leaved CD
Blood-drop-emlets 545
Blown-grass CD
Bluebell 817
Italian CD
Spanish 817
Blueberry CD
Blue-eyed-grass 828
American 829
Blue-eyed-Mary 506
Blue-gum, Southern CD
Blue-sow-thistle, Alpine 625
Common 625
Hairless CD
Pontic CD
BLYSMUS
compressus 709
rufus 709
Bogbean 493
Bog-laurel CD
Bog-myrtle 129
Bog-rosemary 289
Bog-rush, Black 711
Brown 711

Bog-sedge 734
Mountain 734
Tall 734
BOLBOSCHOENUS
maritimus 706
Borage 501
Slender CD
BORAGINACEAE 494 CD
BORAGO
officinalis 501
pygmaea CD
Boston-ivy CD 434
BOTRYCHIUM
lunaria 61
Box 426
Carpet CD
BRACHIARIA
platyphylla CD
BRACHYGLOTTIS
? **compacta** × **B. laxifolia** (**B. 'Sunshine'**) 658
monroi CD
repanda CD
'Sunshine' (? **B. compacta** × **B. laxifolia**) 658
BRACHYPODIUM
distachyon CD
pinnatum 793
sylvaticum 793
Bracken 66
Bramble, Arctic 326
Chinese 325
Stone 325
White-stemmed CD
Brambles 327
BRASSICA
carinata CD
elongata CD
juncea 280
napus 279
subsp. *napobrassica*, see *B. napus* subsp. *rapifera* 279
subsp. *oleifera* 279
subsp. *rapifera* 279
nigra 280
oleracea 279
var. *oleracea* 279
rapa 280
subsp. *campestris* 280
subsp. *oleifera* 280
subsp. *rapa* 280
subsp. *sylvestris*, see *B. rapa* subsp. *campestris* 280
tournefortii CD
BRASSICACEAE 249 CD
Bridal-spray CD
Bridewort 322
Billard's 322
Confused 323
Intermediate 322
Pale CD
Brideworts 321
Bristle-grass, Adherent CD
Foxtail 804
Green 804
Knotroot CD
Nodding CD
Rough 803
Yellow 803
BRIZA
maxima 755
media 754
minor 755
Broadleaf, New Zealand 424
Brome, Barren 791
California 792
Compact 792
Drooping 791
False 793
Field 786
Foxtail CD
Great 790
Hungarian 790
Interrupted 788
Large-headed CD
Meadow 786
Patagonian CD
Rescue 792
Ripgut 791
Rye 789
Smith's CD
Smooth 786
Southern CD
Brome (*cont.*)
Stiff CD
Thunberg's CD
Upright 790
Western CD
BROMELIACEAE CD
BROMOPSIS
benekenii 789
erecta 790
inermis 790
subsp. **pumpelliana** CD
pumpelliana, see **B. inermis** subsp. **pumpelliana** CD
ramosa 789
BROMUS
arvensis 786
benekenii ≡ **Bromopsis benekenii** 789
carinatus ≡ **Ceratochloa carinata** 792
commutatus 786
diandrus ≡ **Anisantha diandra** 790
erectus ≡ **Bromopsis erecta** 790
ferronii ≡ **B. hordeaceus** subsp. **ferronii** 787
hordeaceus 787
subsp. **divaricatus** CD 787
subsp. **ferronii** 787
subsp. *hordeaceus* 787
subsp. **thominei** 787
× **B. lepidus** (**B.** × **pseudothominei**) 788
inermis ≡ **Bromopsis inermis** 790
interruptus 788
japonicus CD
lanceolatus CD
lepidus 788
madritensis ≡ **Anisantha madritensis** 792
marginatus, see **Ceratochloa marginata** CD
molliniformis, see **B. hordeaceus** subsp. **divaricatus** CD
mollis ≡ **B. hordeaceus** 787
pseudosecalinus CD
× **pseudothominei** (**B. hordeaceus** × **B. lepidus**) 788
pumpellianus, see **Bromopsis inermis** subsp. **pumpelliana** CD
racemosus 786
subsp. *commutatus*, see **B. commutatus** 786
ramosus ≡ **Bromopsis ramosa** 789
rigidus ≡ **Anisantha rigida** 791
rubens, see **Anisantha rubens** *CD*
secalinus 789
stamineus, see **Ceratochloa staminea** CD
sterilis ≡ **Anisantha sterilis** 791
tectorum ≡ **Anisantha tectorum** 791
thominei ≡ **B. hordeaceus** subsp. **thominei** 787
unioloides ≡ **Ceratochloa cathartica** 792
willdenowii, see **Ceratochloa cathartica** 792
Brooklime 554
Brookweed 304
Broom 404
Black CD
Hairy-fruited 404
Montpellier CD
Mount Etna CD
Prostrate 404
Spanish 405
White CD
Broomrape, Bean CD
Bedstraw 578
Common 580
Greater 577
Ivy 579
Knapweed 578
Oxtongue 579
Thistle 579
Thyme 578
Yarrow 577
BRUNNERA
macrophylla 499
BRYONIA
cretica subsp. *dioica*, see **B. dioica** 233
dioica 233
Bryony, Black 834
White 233
Buck's-beard 323
Buckler-fern, Broad 80
Crested 79
Hay-scented 78
Narrow 79
Northern 80
Rigid 79
Scaly 78
Buckthorn 433
Alder 433
Mediterranean CD
Buckwheat 186
Tall CD
BUDDLEJA
alternifolia CD
davidii 537
× **B. globosa** (**B.** × **weyeriana**) CD
globosa 537
× **weyeriana** (**B. davidii** × **B. globosa**) CD
BUDDLEJACEAE 537 CD
Buffalo-bur 489
Red CD
Bugle 519
Pyramidal 519
Bugle-lily CD
Bugloss 500
Buglossoides
arvensis, see **Lithospermum arvense** 495
purpureocaerulea, see **Lithospermum purpureocaeruleum** 494
Bugseed CD
Bullace 353
Bullrush 807
Lesser 807
Bullwort 471
BUNIAS
orientalis 252
BUNIUM
bulbocastanum 459
BUPLEURUM
baldense 467
falcatum 466
fruticosum CD
rotundifolium 467
subovatum 468
tenuissimum 467
Burdock, Greater 609
Lesser 609
Bur-grass, African CD
Australian CD
European CD
Bur-marigold, London CD
Nodding 665
Trifid 665
Burnet-saxifrage 460
Greater 459
Burnet, Fodder 337
Great 337
Salad 337
White CD
Bur-reed, Branched 805
Floating 806
Least 806
Unbranched 806
Butcher's-broom 828
Spineless CD
BUTOMACEAE 666
BUTOMUS
umbellatus 666
Butterbur 659
Giant 660
White 660
Buttercup, Aconite-leaved CD
Bulbous 101
Celery-leaved 103
Corn 102
Creeping 101
Goldilocks 103
Hairy 102
Jersey 103
Meadow 101
Rough-fruited CD
Small-flowered 102
St Martin's CD
Butterfly-bush 537
Alternate-leaved CD
Weyer's CD
Butterfly-orchid, Greater 842
Lesser 842
Butterwort, Alpine 581
Common 581
Large-flowered 581
Pale 580
Button-grass CD
Buttonweed 651
Annual CD
BUXACEAE 426 CD
BUXUS
sempervirens 426
Cabbage 279
Bastard 284
Isle of Man 282
Lundy 283
Pale CD
Steppe CD
Wallflower 282
Cabbage-palm CD
CABOMBA
caroliniana CD
CABOMBACEAE CD
CAKILE
maritima 283
CALAMAGROSTIS
canescens 777
× *C. stricta* 778
epigejos 777
neglecta agg. 778
purpurea 778
stricta 778
Calamint, Common 522
Lesser 522
Wood 522
Calamintha
ascendens ≡ **Clinopodium ascendens** 522
nepeta ≡ **Clinopodium calamintha** 522
sylvatica ≡ **Clinopodium menthifolium** 522
subsp. *ascendens* ≡ **Clinopodium ascendens** 522
CALCEOLARIA
chelidonioides CD
CALENDULA
arvensis CD
officinalis 661
CALLA
palustris CD
CALLISTEPHUS
chinensis CD
CALLITRICHACEAE 531
CALLITRICHE
brutia 534
hamulata *sens. lat.* 534
hermaphroditica 532
intermedia ≡ **C. hamulata** *sens. lat.* 534
obtusangula 533
palustris 533
platycarpa 533
stagnalis *sens. lat.* 532
stagnalis sens. str. 532
truncata 532
CALLUNA
vulgaris 291
CALOTIS
cuneifolia CD
CALTHA
palustris 95
var. *radicans* 95
CALYSTEGIA
× **lucana** (**C. sepium** × **C. silvatica**) 491
pulchra 492
sepium 491
subsp. *pulchra*, see **C. pulchra** 492
subsp. *sepium* 491
subsp. **roseata** 491
subsp. *silvatica*, see **C. silvatica** 492
subsp. *spectabilis* 491
× **C. silvatica** (**C.** × **lucana**) 491
silvatica 492
subsp. *disjuncta* 492
subsp. *silvatica* 492
soldanella 490
CAMELINA
microcarpa CD
sativa 271
CAMPANULA
alliariifolia CD
fragilis CD
glomerata 585
lactiflora CD
latifolia 586
medium 584
patula 583
persicifolia 584
portenschlagiana 585
poscharskyana 585
pyramidalis CD
rapunculoides 586
rapunculus 584
rhomboidalis CD
rotundifolia 587
trachelium 586
CAMPANULACEAE 583 CD

Campion, Alpine CD
Bladder 176
Moss 177
Red 178
Rose 174
Sea 176
White 178
Canary-grass 774
Awned 774
Bulbous 773
Confused CD
Lesser 774
Reed 773
Candytuft, Garden 274
Perennial 273
Wild 274
CANNABACEAE 126
CANNABIS
sativa 126
Canterbury-bells 584
Cape-gooseberry CD
Cape-lily, Powell's CD
Cape-pondweed 671
CAPRIFOLIACEAE 596 CD
CAPSELLA
bursa-pastoris 271
× *C. rubella* CD
rubella CD
CAPSICUM
annuum CD
Caraway 472
Whorled 472
CARDAMINE
amara 260
bulbifera 260
corymbosa CD
flexuosa 261
heptaphylla CD
hirsuta 262
impatiens 261
pratensis 261
raphanifolia 260
trifolia CD
Cardaminopsis
petraea ≡ **Arabis petraea** 262
Cardaria
draba ≡ **Lepidium draba** 277
subsp. *chalepense*, see **Lepidium draba** subsp. **chalepense** CD
CARDUUS
acanthoides ≡ **C. crispus** 610
crispus 610
× **C. nutans** (**C.** × **stangii**) 610
× *dubius*, see **C.** × **stangii** 610
nutans 611
pycnocephalus CD
× **stangii** (**C. crispus** × **C. nutans**) 610
tenuiflorus 610
CAREX
acuta 736
acutiformis 722
appropinquata 713
aquatilis 736
× *C. paleacea* 736
arenaria 717
atrata 735
atrofusca 733
bigelowii 738
binervis 727
× *boenninghauseniana*, see **C.** × **boenninghausiana** 713
× **boenninghausiana** (**C. paniculata** × **C. remota**) 713
buchananii CD
buxbaumii 735
capillaris 725
caryophyllea 732
chordorrhiza 717
curta 721
davalliana 720
demissa ≡ **C. viridula** subsp. **oedocarpa** 730
depauperata 726
diandra 714
digitata 731
dioica 719
distans 727
disticha 717
divisa 718
divulsa subsp. **divulsa** 716
divulsa subsp. **leersii** 716
echinata 719
elata 737
CAREX (*cont.*)
elongata 720
ericetorum 732
extensa 728
filiformis 732
flacca 725
flava 729
subsp. *brachyrrhyncha*, see **C. viridula** subsp. **brachyrrhyncha** 729
subsp. *oedocarpa*, see **C. viridula** subsp. **oedocarpa** 730
subsp. *pulchella*, see **C. viridula** subsp. **viridula** 730
subsp. *scotica*, see **C. viridula** subsp. **brachyrrhyncha** 729
subsp. *serotina*, see **C. viridula** subsp. **viridula** 730
× *C. viridula* 729
× **fulva** (**C. hostiana** × **C. viridula**) 729
hirta 721
hostiana 728
× **C. viridula** (**C.** × **fulva**) 729
× *C. viridula* subsp. *brachyrrhyncha* 729
× *C. viridula* subsp. *oedocarpa* 729
× *C. viridula* subsp. *viridula* 729
humilis 731
× **involuta** (**C. rostrata** × **C. vesicaria**) 723
lachenalii 720
laevigata 727
lasiocarpa 721
leersii ≡ **C. divulsa** subsp. **leersii** 716
lepidocarpa ≡ **C. viridula** subsp. **brachyrrhyncha** 729
subsp. *scotica*, see **C. viridula** subsp. **brachyrrhyncha** 729
limosa 734
magellanica 734
maritima 718
microglochin 738
montana 733
muricata subsp. **lamprocarpa** 716
muricata subsp. *leersii*, see **C. divulsa** subsp. **leersii** 716
muricata subsp. **muricata** 715
nigra 737
× *C. trinervis* 737
norvegica 735
ornithopoda 731
otrubae 714
× **C. remota** (**C.** × **pseudoaxillaris**) 715
ovalis 719
pallescens 730
panicea 726
paniculata 713
× **C. remota** (**C.** × **boenninghausiana**) 713
pauciflora 738
paupercula ≡ **C. magellanica** 734
pendula 724
pilulifera 733
polyphylla 716
pseudocyperus 722
× **pseudoaxillaris** (**C. otrubae** × **C. remota**) 715
pulicaris 739
punctata 728
rariflora 734
recta 736
remota 718
riparia 722
rostrata 723
× **C. vesicaria** (**C.** × **involuta**) 723
rupestris 739
saxatilis 724
scandinavica, see **C. viridula** subsp. **viridula** 730
serotina ≡ **C. viridula** subsp. **viridula** 730
spicata 715
strigosa 725
sylvatica 724
tomentosa, see **C. filiformis** 732
trinervis 737
vaginata 726
vesicaria 723
viridula subsp. **brachyrrhyncha** 729
viridula subsp. **oedocarpa** 730
viridula subsp. **viridula** 730
vulpina 714
vulpinoidea CD
CARLINA
vulgaris 608
CARPINUS
betulus 135
CARPOBROTUS
acinaciformis CD
edulis 136 CD
glaucescens CD
Carrot 478 CD
Australian CD
Moon 461
Sea 478
Wild 478
CARTHAMUS
lanatus CD
tinctorius 617
CARUM
carvi 472
verticillatum 472
CARYOPHYLLACEAE 152 CD
CASTANEA
sativa 130
Castor-oil-plant CD
Cat's-ear 619
Smooth 619
Spotted 619
Cat's-tail, Alpine 785
Purple-stem 785
Sand 785
Smaller 784
CATABROSA
aquatica 761
subsp. *minor* ≡ *C. aquatica* var. *uniflora* 761
var. *uniflora* 761
CATANANCHE
caerulea CD
CATAPODIUM
marinum 761
rigidum 761
subsp. *majus* 761
var. *majus*, see *C. rigida* subsp. *majus* 761
Catchfly, Alpine 175
Berry CD
Italian CD
Night-flowering 177
Nodding CD
Nottingham 175
Sand 179
Small-flowered 179
Spanish 176
Sticky 174
Sweet-William 177
Caterpillar-plant CD
Cat-mint 520
Garden 520
Caucasian-stonecrop 310
Lesser CD
Cedar, Atlas 85
Cedar-of-Lebanon 85
CEDRUS
atlantica 85
deodara 85
libani 85
Celandine, Greater 118
Lesser 106
CELASTRACEAE 425 CD
CELASTRUS
orbiculatus CD
Celery, Wild 468
CENCHRUS
echinatus CD
CENTAUREA
aspera CD
calcitrapa 616
cyanus 616
debeauxii subsp. *nemoralis*, see *C. nigra* subsp. *nemoralis* 617
diluta 617
macrocephala CD
melitensis CD
montana 615
nemoralis ≡ *C. nigra* subsp. *nemoralis* 617
nigra 617
subsp. *nemoralis* 617
subsp. *nigra* 617
scabiosa 615
var. *succisifolia* 615
solstitialis 616
CENTAURIUM
capitatum ≡ **C. erythraea** 479
erythraea 479
littorale 480
pulchellum 480
scilloides 479
tenuiflorum 480
Centaury, Common 479
Guernsey 479
Lesser 480
Perennial 479
Seaside 480
Slender 480
Yellow 478
CENTRANTHUS
calcitrapae CD
ruber 605
Centunculus
minimus, see **Anagallis minima** 303
Centuryplant CD
CEPHALANTHERA
damasonium 835
longifolia 835
rubra 835
CEPHALARIA
gigantea 606
CERASTIUM
alpinum 162
arcticum 162
subsp. *edmondstonii*, see **C. nigrescens** 162
arvense 161
atrovirens ≡ **C. diffusum** 164
brachypetalum 164
cerastoides 161
diffusum 164
fontanum 163
subsp. *glabrescens*, see *C. fontanum* subsp. *holosteoides* 163
subsp. *holosteoides* 163
subsp. **scoticum** 163
subsp. *vulgare* 163
glomeratum 163
holosteoides ≡ **C. fontanum** 163
nigrescens 162
pumilum 164
semidecandrum 165
tomentosum 161
CERATOCAPNOS
claviculata 120
CERATOCHLOA
brevis CD
carinata 792
'Deborah' 792
cathartica 792
marginata CD
staminea CD
CERATOPHYLLACEAE 94
CERATOPHYLLUM
demersum 94
submersum 95
CETERACH
officinarum 71
CHAENOMELES
japonica CD 356
× *C. speciosa* 356
× *C. speciosa* (*C.* × *superba*) CD
speciosa 356
× *superba* (*C. japonica* × *C. speciosa*) CD
CHAENORHINUM
minus 547
origanifolium CD
CHAEROPHYLLUM
aureum CD
hirsutum CD
temulentum ≡ **C. temulum** 456
temulum 456
Chaffweed 303
CHAMAECYPARIS
lawsoniana 89
nootkatensis CD
× **Cupressus macrocarpa** (× **Cupressocyparis leylandii**) 89
pisifera 90
'Squarrosa' 90
CHAMAEMELUM
nobile 648
Chamaenerion
angustifolium ≡ **Chamerion angustifolium** 418
Chamaepericlymenum
suecicum, see **Cornus suecica** 423
CHAMERION
angustifolium 418
Chamomile 648
Corn 648
Sicilian CD
Stinking 648
Yellow 649
Chard, Swiss CD

Charlock 281
CHASMANTHE
bicolor CD
Chasmanthe CD
Checkerberry CD
Cheiranthus
cheiri ≡ **Erysimum cheiri** 253
CHELIDONIUM
majus 118
CHENOPODIACEAE 136 CD
CHENOPODIUM
album agg. 140
ambrosioides CD
berlandieri CD
× **bontei** (**C. carinatum** × **C. cristatum**) CD
bonus-henricus 136
botryodes ≡ **C. chenopodioides** 137
bushianum CD
capitatum CD
carinatum CD
× **C. cristatum** (**C.** × **bontei**) CD
chenopodioides 137
cristatum CD
desiccatum CD
ficifolium 139
giganteum CD
glaucum 136
hircinum CD
hybridum 138
multifidum CD
murale 139
nitrariaceum CD
opulifolium 139
polyspermum 137
pratericola CD
probstii CD
pumilio CD
rubrum 137
strictum CD
suecicum CD
urbicum 138
vulvaria 138
section *Orthosporum* CD
Cherlearia
sedoides ≡ **Minuartia sedoides** 157
Cherry, Bird 355
Dwarf 354
Fuji CD
Japanese CD
Morello 354
Pin CD
Rum 355
St Lucie CD
Wild 354
Chervil, Bur 457
Garden 457
Golden CD
Hairy CD
Rough 456
Chestnut, Sweet 130
Chickweed, Common 158
Greater 159
Jagged 160
Lesser 158
Upright 165
Water 165
Chickweed-wintergreen 302
Chicory 618
Chilean-iris CD
Lesser CD
CHIMONOBAMBUSA
quadrangularis CD
CHIONODOXA
forbesii 818
luciliae CD 818
sardensis CD
Chives 819
CHLORIS
divaricata CD
truncata CD
virgata CD
CHOISYA
ternata CD
Chokeberry, Black CD
Red CD
CHRYSANTHEMUM
coronarium CD
leucanthemum ≡ **Leucanthemum vulgare** 649
parthenium ≡ **Tanacetum parthenium** 644
segetum 649
serotinum, see **Leucanthemella serotina** CD
vulgare ≡ **Tanacetum vulgare** 644
CHRYSOCOMA
coma-aurea CD
tenuifolia CD
CHRYSOSPLENIUM
alternifolium 321
oppositifolium 320
Cicely, Sweet 458
CICENDIA
filiformis 478
CICER
arietinum CD
reticulatum CD
CICERBITA
alpina 625
bourgaei CD
macrophylla 625
subsp. *macrophylla* 625
subsp. *uralensis* 625
plumieri CD
CICHORIUM
intybus 618
CICUTA
virosa 471
Cineraria CD
Cinquefoil, Alpine 330
Brook CD
Creeping 332
Grey CD
Hoary 329
Marsh 328
Rock 328
Russian 329
Shrubby 327
Spring 330
Sulphur 329
Ternate-leaved 330
Cinquefoils, Hybrid 332
CIRCAEA
alpina 421
× **C. lutetiana** (**C.** × **intermedia**) 421
alpina agg. 421
× **intermedia** (**C. alpina** × **C. lutetiana**) 421
lutetiana 421
CIRSIUM
acaule 613
arvense 614
dissectum 612
× **C. palustre** (**C.** × **forsteri**) 612
eriophorum 611
erisithales CD
× **forsteri** (**C. dissectum** × **C. palustre**) 612
helenioides, see **C. heterophyllum** 613
heterophyllum 613
oleraceum CD
palustre 613
tuberosum 612
vulgare 611
CISTACEAE 223
CITRULLUS
lanatus CD
CLADIUM
mariscus 712
CLARKIA
amoena CD
unguiculata CD
Clarkia CD
Clary CD
Annual 531
Meadow 530
Sticky CD
Whorled 531
Wild 530
CLAYTONIA
perfoliata 150
sibirica 150
Cleavers 594
Corn 595
False 595
CLEMATIS
flammula CD
montana CD
orientalis CD
tangutica CD
tibetana CD
vitalba 100
viticella CD
Clematis, Himalayan CD
Orange-peel CD
Purple CD
CLINOPODIUM
acinos 523
ascendens 522
CLINOPODIUM (*cont.*)
calamintha 522
menthifolium 522
vulgare 523
Cloudberry 325
Clover, Alsike 395
Bird's-foot 394
Bur CD
Clustered 395
Crimson 399
Egyptian CD
Hare's-foot 401
Hedgehog CD
Hungarian CD
Knotted 400
Long-headed 399
Narrow CD
Nodding CD
Red 398
Reversed 396
Rose CD
Rough 400
Sea 401
Starry CD
Strawberry 396
Subterranean 401
Suffocated 395
Sulphur 399
Twin-headed 400
Upright 396
Western 394
White 394
Woolly CD
Zigzag 398
Clubmoss, Alpine 54
Fir 53
Interrupted 54
Issler's 54
Kraus's 55
Lesser 55
Marsh 53
Stag's-horn 53
Club-rush, Bristle 708
Common 707
Floating 709
Grey 707
Round-headed 706
Sea 706
Sharp 708
Slender 708
Triangular 707
Wood 706
CLUSIACEAE 210 CD
COCHLEARIA
alpina ≡ *C. pyrenaica* subsp. *alpina* 269
anglica 269
atlantica, see **C. officinalis** agg. 269
danica 270
islandica, see **C. officinalis** agg. 269
megalosperma CD
micacea 270
officinalis agg. 269
subsp. **scotica** 270
pyrenaica 269
subsp. *alpina* 269
subsp. *pyrenaica* 269
scotica ≡ **C. officinalis** subsp. **scotica** 270
Cock's-eggs CD
Cock's-foot 760
Slender CD
Cocklebur, Argentine CD
Rough 662
Spiny 662
Cockspur 802
Cockspurthorn CD
Broad-leaved 371
Hairy CD
Large-flowered CD
Pear-fruited CD
Round-fruited CD
COELOGLOSSUM
viride 845
COINCYA
monensis subsp. **cheiranthos** 282
subsp. **monensis** 282
subsp. *recurvata*, see **C. monensis** subsp. **cheiranthos** 282
wrightii 283
COLCHICUM
autumnale 809
Colt's-foot 659
Purple 661
Columbine 111
Pyrenean CD
COLUTEA
arborescens 373
× **C. orientalis** (**C.** × **media**) CD
× **media** (**C. arborescens** × **C. orientalis**) CD
Comfrey, Bulbous CD
Caucasian CD
Common 497
Creeping 498
Crimean CD
Hidcote 498
Rough CD
Russian 497
Tuberous 498
White 499
COMMELINACEAE CD
Compositae, see ASTERACEAE 607
Coneflower CD
CONIUM
maculatum 466
CONOPODIUM
majus 459
CONRINGIA
orientalis 278
CONSOLIDA
ajacis 98
ambigua, see **C. ajacis** 98
orientalis 98
regalis 98
CONVALLARIA
majalis 812
CONVOLVULACEAE 490 CD
CONVOLVULUS
arvensis 490
CONYZA
bonariensis CD
canadensis 642
sumatrensis 643
COPROSMA
repens CD
Copse-bindweed 191
Coralbells CD
Coralberry CD
Hybrid 599
CORALLORHIZA
trifida 841
Coral-necklace 171
Coralroot 260
Pinnate CD
Cord-grass, Common 801
Prairie CD
Small 801
Smooth CD
Townsend's 801
CORDYLINE
australis CD
COREOPSIS
grandiflora CD
Coriander 458
CORIANDRUM
sativum 458
CORISPERMUM
hyssopifolium CD
leptopterum CD
CORNACEAE 422
Corncockle 175
Cornel, Dwarf 423
Cornelian-cherry 423
Cornflower 616
Perennial 615
Corn-lily, Blue CD
Red CD
Tubular CD
Cornsalad, Broad-fruited 603
Common 602
Hairy-fruited 603
Keeled-fruited 602
Narrow-fruited 603
CORNUS
alba 422
mas 423
sanguinea 422
sericea 422
stolonifera, see **C. sericea** 422
suecica 423
CORONILLA
scorpioides CD
valentina CD
varia ≡ **Securigera varia** 379

CORONOPUS
didymus 277
squamatus 277
CORREA
backhousiana CD
CORRIGIOLA
litoralis 170
CORTADERIA
richardii CD 799
selloana 799
CORYDALIS
bulbosa, see **C. cava** CD
cava CD
cheilanthifolia CD
claviculata ≡ **Ceratocapnos claviculata** 120
lutea ≡ **Pseudofumaria lutea** 119
solida 119
Corydalis, Climbing 120
Fern-leaved CD
Pale CD
Yellow 119
Corylaceae, see BETULACEAE 133
CORYLUS
avellana 135
× **C. maxima** CD
colurna CD
maxima CD
CORYNEPHORUS
canescens 771
COSMOS
bipinnatus CD
Costmary CD
COTONEASTER
adpressus CD
var. *praecox* CD
affinis CD
amoenus CD
apiculatus CD
ascendens CD
astrophoros CD
'Donard Gem' CD
atropurpureus CD
bacillaris CD
boisianus CD
bullatus 369
cambricus 368
cashmiriensis CD
cochleatus CD
congestus CD
conspicuus CD
? × **C. dammeri** (**C.** × **suecicus**) CD
cooperi CD
dammeri CD
× **C. salicifolius** (**C. 'Hybridus Pendulus'**) CD
dielsianus 370
divaricatus 368
ellipticus CD
fangianus CD
franchetii 369
frigidus 366
× **C. salicifolius** (**C.** × **watereri**) 366
'Gloire de Versailles' CD
henryanus CD
'Highlight' CD
hissaricus CD
hjelmqvistii CD
horizontalis 367
hsingshangensis CD
hummelii CD
'Hybridus Pendulus' (**C. dammeri** × **C. salicifolius**) CD
hylmoei CD
ignotus CD
insculptus CD
integerrimus 368
integrifolius 367
lacteus 367
laetevirens CD
linearifolius CD
lucidus CD
mairei CD
marginatus CD
microphyllus agg. 367
microphyllus *sens. str.* CD
monopyrenus CD
moupinensis CD
mucronatus CD
nanshan CD
nitens CD
nitidus CD
COTONEASTER (*cont.*)
obscurus CD
obtusus CD
pannosus CD
prostratus CD
pseudoambiguus CD
rehderi 369
rotundifolius CD
salicifolius 366
sherriffii CD
simonsii 368
splendens CD
sternianus 370
× **suecicus** (**C. ? conspicuus** × **C. dammeri**) CD
'Coral Beauty' CD
'Skogholm' CD
tengyuehensis CD
tomentellus CD
transens CD
villosulus CD
vilmorinianus CD
wardii CD
× **watereri** (**C. frigidus** × **C. salicifolius**) 366
zabelii CD
Cotoneaster, Ampfield CD
Apiculate CD
Ascending CD
Bearberry CD
Beautiful CD
Black-grape CD
Bois's CD
Bullate 369
Cherryred CD
Circular-leaved CD
Congested CD
Cooper's CD
Creeping CD
Dartford CD
Diel's 370
Distichous CD
Dwarf CD
Engraved CD
Fang's CD
Few-flowered CD
Franchet's 369
Fringed CD
Godalming CD
Henry's CD
Himalayan 368
Hjelmqvist's CD
Hollyberry 369
Hsing-Shan CD
Hummel's CD
Hylmö's CD
Kangting CD
Kashmir CD
Late 367
Lindley's CD
Lleyn CD
Maire's CD
Moupin CD
Mucronate CD
Obscure CD
One-stoned CD
Open-fruited CD
Procumbent CD
Purpleberry CD
Purple-flowered CD
Round-leaved CD
Sherriff's CD
Shiny CD
Short-felted CD
Showy CD
Silverleaf CD
Small-leaved CD
Spreading 368
Starry CD
Stern's 370
Swedish CD
Tengyueh CD
Thyme-leaved CD
Tibetan CD
Tree 366
Vilmorin's CD
Wall 367
Ward's CD
Waterer's 366
Weeping CD
Wild 368
Willow-leaved 366
Yunnan CD
Cotoneasters, Small-leaved 367
Cottongrass, Broad-leaved 701
Common 701
Hare's-tail 702
Slender 701
Cottonweed 647
COTULA
australis CD
coronopifolia 651
dioica CD
squalida CD
Couch, Australian CD
Bearded 793
Common 794
Sand 795
Sea 794
Cowbane 471
Cowberry 294
Cowherb 180
Cowslip 298
Japanese CD
Sikkim CD
Tibetan CD
Cow-wheat, Common 562
Crested 561
Field 562
Small 562
Crab, Purple CD
Crack-willow 236
Hybrid 238
Weeping CD
CRAMBE
cordifolia CD
maritima 284
Cranberry 293
American CD
Small 293
Crane's-bill, Alderney CD
Armenian CD
Bloody 446
Caucasian CD
Cut-leaved 446
Dove's-foot 448
Druce's 444
Dusky 449
French 443
Glandular CD
Hedgerow 447
Himalayan CD
Knotted 444
Long-stalked 446
Meadow 445
Munich CD
Pencilled 444
Purple 447
Rock 448
Round-leaved 445
Shining 448
Small-flowered 447
Wood 445
CRASSULA
aquatica 308
decumbens CD
helmsii 308
pubescens CD
subsp. *radicans* CD
tillaea 307
CRASSULACEAE 307 CD
CRATAEGUS
coccinioides CD
crus-galli CD 371
heterophylla CD
laciniata, see **C. orientalis** CD
laevigata 372
× **C. monogyna** (**C.** × **media**) 372
× **Mespilus germanica** (× **Crataemespilus grandiflora**) CD
× *macrocarpa*, see **C.** × **media** 372
× **media** (**C. laevigata** × **C. monogyna**) 372
monogyna 371
orientalis CD
oxyacanthoides ≡ **C. laevigata** 372
pedicellata CD
persimilis 371
submollis CD
succulenta CD
× **CRATAEMESPILUS**
grandiflora (**Crataegus laevigata** × **Mespilus germanica**) CD
Creeping-Jenny 300
CREPIS
biennis 627
capillaris 627
CREPIS (*cont.*)
foetida 628
subsp. *commutata* 628
subsp. *foetida* 628
subsp. *rhoeadifolia* 628
mollis 627
nicaeensis CD
paludosa 626
praemorsa 629
setosa 628
tectorum CD
vesicaria 628
Cress, Garden 274
Hoary CD
Rosy CD
Shepherd's 272
Thale 252
Tower CD
Trefoil CD
Violet CD
Crinitaria
linosyris ≡ **Aster linosyris** 641
CRINUM
bulbispermum × **C. moorei** (**C.** × **powellii**) CD
× **powellii** (**C. bulbispermum** × **C. moorei**) CD
CRITHMUM
maritimum 461
CROCOSMIA
aurea × **C. pottsii** (**C.** × **crocosmiiflora**) 833
× **crocosmiiflora** (**C. aurea** × **C. pottsii**) 833
masoniorum CD 833
paniculata 833
× *C. masoniorum* 833
pottsii CD
CROCUS
ancyrensis CD
angustifolius × **C. flavus** (**C.** × **stellaris**) 832
biflorus CD
subsp. *adamii* CD
subsp. *biflorus* CD
× **C. chrysanthus** CD
chrysanthus CD 832
kotschyanus CD
longiflorus CD
nudiflorus 832
pulchellus CD
purpureus ≡ **C. vernus** 831
serotinus CD
sieberi CD
speciosus CD
× **stellaris** (**C. angustifolius** × **C. flavus**) 832
'Dutch Yellow' 832
'Golden Yellow' 832
'Yellow Giant' 832
tommasinianus 831
vernus 831
Crocus, Ankara CD
Autumn 832
Bieberstein's CD
Early 831
Golden CD
Hairy CD
Italian CD
Kotschy's CD
Late CD
Sand 831
Sieber's CD
Silvery CD
Spring 831
Yellow 832
Crosswort 596
Caucasian CD
Crowberry 286
Crowfoot, Ivy-leaved 107
Round-leaved 107
Three-lobed 107
CRUCIATA
chersonensis ≡ **C. laevipes** 596
laevipes 596
Cruciferae, see BRASSICACEAE 249
CRYPTOGRAMMA
crispa 62
Cryptogrammaceae, see ADIANTACEAE 62
CRYPTOMERIA
japonica 88
Cuckooflower 261
Greater 260
CUCUBALUS
baccifer CD
Cucumber CD
Squirting CD

CUCUMIS
melo CD
sativus CD
CUCURBITA
maxima CD
pepo 233
CUCURBITACEAE 233 CD
Cudweed, American CD
Broad-leaved 632
Cape CD
Common 631
Dwarf 634
Heath 634
Highland 633
Jersey 635
Marsh 634
Narrow-leaved 632
Red-tipped 631
Small 632
Cumin CD
CUMINUM
cyminum CD
Cup-grass, Perennial CD
Cupidone, Blue CD
CUPRESSACEAE 89 CD
× **CUPRESSOCYPARIS**
leylandii (**Chamaecyparis nootkatensis** × **Cupressus macrocarpa**) 89
'Leighton Green' 89
CUPRESSUS
macrocarpa 89
Currant, Black 306
Buffalo CD
Downy 306
Flowering 306
Mountain 307
Red 305
Curtonus
paniculatus, see **Crocosmia paniculata** 833
CUSCUTA
campestris CD
epithymum 493
europaea 492
CUSCUTACEAE 492 CD
Cut-grass 740
CYCLAMEN
coum CD
hederifolium 300
repandum CD
Cyclamen 300
CYDONIA
oblonga CD
CYMBALARIA
hepaticifolia CD
muralis 547
subsp. **visianii** CD
pallida 548
CYNARA
cardunculus CD
CYNODON
dactylon 800
incompletus CD
CYNOGLOSSUM
germanicum 507
officinale 506
CYNOGLOTTIS
barrelieri CD
CYNOSURUS
cristatus 752
echinatus 752
CYPERACEAE 701 **CD**
CYPERUS
eragrostis 710
fuscus 710
longus 710
Cyphel 157
Cypress, Lawson's 89
Leyland 89
Monterey 89
Nootka CD
Sawara 90
CYPRIPEDIUM
calceolus 834
CYRTOMIUM
falcatum CD
CYSTOPTERIS
dickieana 74
fragilis 74
montana 74
CYTISUS
multiflorus CD
nigricans CD
CYTISUS (*cont.*)
scoparius 404
subsp. **maritimus** 404
subsp. *scoparius* 404
striatus 404

DABOECIA
cantabrica 288
DACTYLIS
glomerata 760
subsp. *aschersoniana*, see **D. polygama** CD
polygama CD
DACTYLOCTENIUM
radulans CD
× **DACTYLODENIA**
st-quintinii (**Dactylorhiza fuchsii** × **Gymnadenia conopsea**) 845
× *Dactylogymnadenia*
cookei, see × **Dactylodenia st-quintinii** 845
Dactylorchis
fuchsii ≡ **Dactylorhiza fuchsii** 845
incarnata ≡ **Dactylorhiza incarnata** 848
maculata ≡ **Dactylorhiza maculata** 847
majalis ≡ **Dactylorhiza majalis** 851
praetermissa ≡ **Dactylorhiza praetermissa** 850
purpurella ≡ **Dactylorhiza purpurella** 850
traunsteineri ≡ **Dactylorhiza traunsteineri** 851
DACTYLORHIZA
comosa, see **D. majalis** 851
subsp. *cambrensis*, see *D. majalis* subsp. *cambrensis* 851
subsp. *occidentalis*, see *D. majalis* subsp. *occidentalis* 851
× **formosa** (**D. maculata** × **D. purpurella**) 848
fuchsii 845
× **D. incarnata** (**D.** × **kerneriorum**) 846
× **D. maculata** (**D.** × **transiens**) 846
× **D. praetermissa** (**D.** × **grandis**) 846
× **D. purpurella** (**D.** × **venusta**) 847
× **Gymnadenia conopsea** (× **Dactylodenia st-quintinii**) 845
× **grandis** (**D. fuchsii** × **D. praetermissa**) 846
× **hallii** (**D. maculata** × **D. praetermissa**) 847
incarnata 848
subsp. **coccinea** 849
subsp. **cruenta** 849
subsp. *gemmana* 848
subsp. **incarnata** 848
subsp. **ochroleuca** 850
subsp. **pulchella** 849
× **kerneriorum** (**D. fuchsii** × **D. incarnata**) 846
lapponica 851
maculata 847
subsp. *fuchsii*, see **D. fuchsii** 845
× **D. praetermissa** (**D.** × **hallii**) 847
× **D. purpurella** (**D.** × **formosa**) 848
majalis 851
subsp. *cambrensis* 851
subsp. *majalis* var. *brevifolia*, see **D. purpurella** 850
subsp. *occidentalis* 851
subsp. *praetermissa*, see **D. praetermissa** 850
subsp. *purpurella*, see **D. purpurella** 850
subsp. *traunsteinerioides*, see **D. traunsteineri** 851
praetermissa 850
purpurella 850
× **transiens** (**D. fuchsii** × **D. maculata**) 846
traunsteineri 851
subsp. *lapponica*, see **D. lapponica** 851
× **venusta** (**D. fuchsii** × **D. purpurella**) 847
Daffodil 826
Bunch-flowered CD
Cyclamen-flowered CD
Hoop-petticoat CD
Lesser CD
Nonesuch 826
Paper-white CD
Pheasant's-eye 825
Sea CD
Spanish 827
Tenby 827
Winter CD
Daffodils 825
DAHLIA
pinnata CD
Dahlia CD
Daisy 643
Bur CD
Crown CD
Oxeye 649
Seaside 641
Shasta 650
Daisy-bush CD
Mangrove-leaved CD
DAMASONIUM
alisma 669
Dame's-violet 253
Dandelions 626
DANTHONIA
decumbens 799
DAPHNE
laureola 411
mezereum 411
DARMERA
peltata 313
Darnel 750
DATURA
ferox CD
stramonium 489
DAUCUS
carota 478
subsp. *carota* 478
subsp. **gummifer** 478
subsp. **sativus** CD 478
glochidiatus CD
Day-lily, Orange 809
Yellow CD
Dead-nettle, Cut-leaved 513
Henbit 514
Northern 513
Red 513
Spotted 512
White 512
Deergrass 702
Cotton 702
DELAIREA
odorata CD
DENNSTEADTIACEAE 66
Deodar 85
DESCHAMPSIA
alpina ≡ **D. cespitosa** subsp. **alpina** 769
cespitosa 769
subsp. **alpina** 769
subsp. *cespitosa* 769
subsp. **parviflora** 769
var. *parviflora*, see **D. cespitosa** subsp. **parviflora** 769
flexuosa 770
setacea 770
DESCURAINIA
sophia 251
Desmazeria
marina, see **Catapodium marinum** 761
rigida, see **Catapodium rigidum** 761
DEUTZIA
scabra CD
Deutzia CD
Dewberry 327
Dewplant, Deltoid-leaved CD
Pale CD
Purple CD
Rosy CD
Shrubby CD
Sickle-leaved CD
DIANTHUS
armeria 182
barbatus 182
caesius CD
caryophyllus CD
× **D. gratianopolitanus** CD
× **D. plumarius** CD
deltoides 181
gallicus CD
gratianopolitanus 181
plumarius 181
DIAPENSIA
lapponica 297
Diapensia 297
DIAPENSIACEAE 297
DICENTRA
exima 119
× *D. formosa* 119
formosa 119
DICHONDRA
micrantha CD
DICKSONIA
antarctica CD
DICKSONIACEAE CD
DIGITALIS
lutea CD
purpurea 551
DIGITARIA
ciliaris CD
ischaemum CD
sanguinalis 804
Dill 465
DIOSCOREACEAE 834
DIPHASIASTRUM
alpinum 54
complanatum 54
subsp. *issleri*, see **D. complanatum** 54
issleri, see **D. complanatum** 54
DIPLOTAXIS
erucoides CD
muralis 279
tenuifolia 278
DIPSACACEAE 605 CD
DIPSACUS
fullonum 605
laciniatus CD
pilosus 606
sativus 605
strigosus CD
DISPHYMA
crassifolium CD
Dittander 276
DITTRICHIA
graveolens CD
viscosa CD
Dock, Aegean CD
Argentine CD
Broad-leaved 198
Clustered 196
Curled 194
Fiddle 198
Golden 198
Greek 194
Hooked CD
Marsh 198
Northern 193
Obovate-leaved CD
Patience CD
Russian CD
Scottish 193
Shore 197
Water 194
Willow-leaved CD
Wood 196
Dodder 493
Greater 492
Yellow CD
Dog's-tail, Crested 752
Golden CD
Rough 752
Dog's-tooth-violet CD
Dog-rose 347
Glaucous 349
Hairy 349
Round-leaved 350
Dog-violet, Common 226
Early 226
Heath 227
Pale 228
Dogwood 422
Red-osier 422
White 422
DORONICUM
columnae CD
× **D. pardalianches** × **D. plantagineum** (**D.** × **excelsum**) CD
× **excelsum** (**D. columnae** × **D. pardalianches** × **D. plantagineum**) CD 658, 659
pardalianches 658
× **D. plantagineum** (**D.** × **willdenowii**) CD
plantagineum 659
× **willdenowii** (**D. pardalianches** × **D. plantagineum**) CD 658, 659
DOWNINGIA
elegans CD
Downy-rose, Harsh 350
Sherard's 351
Soft 351

DRABA
aizoides 266
incana 267
muralis 267
norvegica 266
Dracunculus
vulgaris CD
Dragon's-teeth CD
Dropseed, African CD
Dropwort 324
Drosanthemum
floribundum CD
Drosera
anglica 222
× **D. rotundifolia** (**D.** × **obovata**) 222
intermedia 223
longifolia, see *D. anglica* 222
× **obovata** (**D. anglica** × **D. rotundifolia**) 222
rotundifolia 222
DROSERACEAE 222
DRYAS
octopetala 336
DRYOPTERIDACEAE 75 CD
DRYOPTERIS
abbreviata ≡ **D. oreades** 77
aemula 78
affinis 77
× *D. expansa* 78
× **D. filix-mas** (**D.** × **complexa**) 78
assimilis, see **D. expansa** 80
borreri ≡ **D. affinis** 77
carthusiana 79
× **D. dilatata** (**D.** × **deweveri**) 80
cristata 79
× **complexa** (**D. affinis** × **D. filix-mas**) 78
× **deweveri** (**D. carthusiana** × **D. dilatata**) 80
dilatata 80
expansa 80
filix-mas 77
lanceolatocristata ≡ **D. carthusiana** 79
oreades 77
pseudomas, see **D. affinis** 77
remota 78
submontana 79
× *tavelii*, see **D.** × **complexa** 78
villarii ≡ **D. submontana** 79
subsp. *montana*, see **D. submontana** 79
subsp. *submontana*, see **D. submontana** 79
DUCHESNEA
indica 334
Duck-potato CD
Duckweed, Common 686
Fat 686
Greater 685
Ivy-leaved 686
Least 687
Rootless 687
Dysentery-herb, Short-fruited CD

ECBALLIUM
elaterium CD
ECHINOCHLOA
colona CD
crus-galli 802
esculenta 803
frumentacea CD *803*
utilis, see **E. esculenta** 803
ECHINOPS
bannaticus 608
'Taplow Blue' 608
exaltatus 608
ritro 608
sphaerocephalus 607
ECHIUM
lycopsis ≡ **E. plantagineum** 496
pininana CD
plantagineum 496
rosulatum CD
vulgare 495
Eelgrass 682
Dwarf 683
Narrow-leaved 683
EGERIA
densa CD
EHRHARTA
stipoides CD
ELAEAGNACEAE 408 CD
ELAEAGNUS
× *ebbingei* (*E. macrophylla* × *E. pungens*) CD
macrophylla CD
× *E. pungens* (*E.* × *ebbingei*) CD
pungens CD
umbellata CD
ELATINACEAE 210
ELATINE
hexandra 210
hydropiper 210
Elder 597
American CD
Dwarf 597
Red-berried 596
Elecampane 635
ELEOCHARIS
acicularis 705
austriaca 704
multicaulis 704
palustris 703
subsp. *microcarpa* ≡ **E. palustris** subsp. **palustris** 703
subsp. **palustris** 703
subsp. *vulgaris* 703
parvula 705
quinqueflora 705
uniglumis 704
ELEOGITON
fluitans 709
Elephant-ears 313
ELEUSINE
indica CD
subsp. **africana** CD
subsp. **indica** CD
multiflora CD
tristachya CD
Elm, Dutch 125
English 125
Huntingdon 124
Plot's 126
Wych 124
ELODEA
callitrichoides CD
canadensis 670
ernstiae, see **E. callitrichoides** CD
nuttallii 670
ELYMUS
arenarius ≡ **Leymus arenarius** 796
athericus, see **Elytrigia atherica** 794
caninus 793
subsp. *donianus*, see *Agropyron donianum* 793
farctus, see **Elytrigia juncea** 795
junceus, see **Elytrigia juncea** 795
× *laxus*, see **Elytrigia** × **laxa** 795
× *obtusiusculus*, see **Elytrigia** × **obtusiuscula** 795
pycnanthus, see **Elytrigia atherica** 794
repens, see **Elytrigia repens** 794
scabrus CD
ELYTRIGIA
atherica 794
× **E. juncea** (**E.** × **obtusiuscula**) 795
× *E. repens* (*E.* × *oliveri*) 794
campestris subsp. *maritima*, see **E. repens** subsp. **arenosa** 794
juncea 795
× **E. repens** (**E.** × **laxa**) 795
× **laxa** (**E. juncea** × **E. repens**) 795
× **obtusiuscula** (**E. atherica** × **E. juncea**) 795
× *oliveri* (*E. atherica* × *E. repens*) 794
repens 794
subsp. **arenosa** 794
subsp. *repens* 794
EMPETRACEAE 286
EMPETRUM
hermaphroditum ≡ **E. nigrum** subsp. **hermaphroditum** 287
nigrum 286
subsp. **hermaphroditum** 287
subsp. *nigrum* 286
Enchanter's-nightshade 421
Alpine 421
Upland 421
Endymnion
non-scriptus ≡ **Hyacinthoides non-scripta** 817
EPILOBIUM
adenocaulon, see **E. ciliatum** 415
adnatum ≡ **E. tetragonum** 414
× **aggregatum** (**E. montanum** × **E. obscurum**) 413
EPILOBIUM (*cont.*)
alsinifolium 417
anagallidifolium 417
angustifolium, see **Chamerion angustifolium** 418
brunnescens 417
ciliatum 415
× **E. hirsutum** (**E.** × **novae-civitatis**) 415
× **E. montanum** 415
× **E. obscurum** 416
× **E. parviflorum** 416
hirsutum 412
× *interjectum*, see **E. ciliatum** × **E. montanum** 415
komarovianum CD
lamyi ≡ **E. tetragonum** 414
lanceolatum 413
× **limosum** (**E. montanum** × **E. parviflorum**) 413
montanum 412
× **E. obscurum** (**E.** × **aggregatum**) 413
× **E. parviflorum** (**E.** × **limosum**) 413
nerterioides ≡ **E. brunnescens** 417
× **novae-civitatis** (**E. ciliatum** × **E. hirsutum**) 415
obscurum 414
palustre 416
parviflorum 412
pedunculare CD
roseum 414
tetragonum 414
× *vicinum*, see **E. ciliatum** × **E. obscurum** 416
EPIPACTIS
atrorubens 836
dunensis, see **E. leptochila** 837
helleborine 837
× *E. leptochila* 837
× *E. phyllanthes* 837
leptochila 837
muelleri, see **E. leptochila** 837
var. *leptochila*, see **E. leptochila** 837
palustris 836
phyllanthes 838
purpurata 836
youngiana 837
EPIPOGIUM
aphyllum 838
EQUISETACEAE 56
EQUISETUM
arvense 58
× **E. fluviatile** (**E.** × **litorale**) 58
fluviatile 58
hyemale 56
× **E. variegatum** (**E.** × **trachyodon**) 57
× **litorale** (**E. arvense** × **E. fluviatile**) 58
palustre 59
pratense 59
ramosissimum 57
sylvaticum 59
telmateia 60
subsp. *braunii* 60
× **trachyodon** (**E. hyemale** × **E. variegatum**) 57
variegatum 57
ERAGROSTIS
cilianensis CD
curvula CD
minor CD
parviflora CD
pilosa CD
tef CD
ERANTHIS
hyemalis 96
EREPSIA
heteropetala CD
ERICA
arborea CD
carnea × **E. erigena** (**E.** × **darleyensis**) CD
ciliaris 291
cinerea 292
× **darleyensis** (**E. carnea** × **E. erigena**) CD
erigena 292
lusitanica CD
mackaiana 291
× *E. tetralix* (*E.* × *stuartii*) 291
mediterranea ≡ **E. erigena** 292
praegeri, see *E.* × *stuartii* 291
× *stuartii* (*E. mackaiana* × *E. tetralix*) 291
terminalis CD
tetralix 292
vagans 293
ERICACEAE 287 CD
ERIGERON
acer 642
annuus CD
borealis 641
canadensis, see **Conyza canadensis** 642
glaucus 641
karvinskianus 642
mucronatus ≡ **E. karvinskianus** 642
neglectus 641
philadelphicus CD
ERINUS
alpinus 551
ERIOCAULACEAE 687
ERIOCAULON
aquaticum 687
septangulare ≡ **E. aquaticum** 687
ERIOCHLOA
pseudoacrotricha CD
ERIOPHORUM
angustifolium 701
gracile 701
latifolium 701
vaginatum 702
ERODIUM
botrys CD
brachycarpum CD
chium CD
cicutarium agg. 450
cicutarium 450
subsp. *bipinnatum*, see **E. lebelii** 451
subsp. *dunense* 451
crinitum CD
cygnorum CD
subsp. **cygnorum** CD
subsp. **glandulosum** CD
glutinosum ≡ **E. lebelii** 451
lebelii 451
malachoides CD
manescavii CD
maritimum 450
moschatum 450
EROPHILA
glabrescens 268
majuscula 268
verna 268
subsp. *spathulata* 267
verna agg. 267
ERUCA
vesicaria 281
ERUCASTRUM
gallicum 282
ERYNGIUM
amethystinum CD
campestre 456
giganteum CD
'Silver Ghost' CD
maritimum 455
planum CD
Eryngo, Blue CD
Field 456
Italian CD
Tall CD
ERYSIMUM
cheiranthoides 253
cheiri 253
decumbens × **E. perofskianum** (**E.** × **marshallii**) CD
× **marshallii** (**E. decumbens** × **E. perofskianum**) CD
dens-canis CD
ESCALLONIA
macrantha 305
rubra see **E. macrantha** 305
var. *macrantha*, see **E. macrantha** 305
Escallonia 305
Escalloniaceae, see GROSSULARIACEAE 305
ESCHSCHOLZIA
californica 118
EUCALYPTUS
globulus CD
gunnii CD
pulchella CD
urnigera CD
viminalis CD
EUONYMUS
europaeus 425
japonicus 425
latifolius CD
EUPATORIUM
cannabinum 666

EUPHORBIA
amygdaloides 432
subsp. *amygdaloides* 432
subsp. **robbiae** 432
characias 432
subsp. **characias** CD 432
subsp. **wulfenii** CD 432
coralliodes CD
cyparissias 431
× **E. esula** (**E.** × **pseudoesula**) CD
× **E. waldsteinii** (**E.** × **gayeri**) CD
dulcis 428
esula 431
× **E. waldsteinii** (**E.** × **pseudovirgata**) 431
exigua 429
× **gayeri** (**E. cyparissias** × **E. waldsteinii**) CD
helioscopia 429
hyberna 427
lathyris 429
maculata CD
oblongata CD
paralias 430
peplis 427
peplus 430
var. *minima* 430
platyphyllos 428
portlandica 430
× **pseudoesula** (**E. cyparissias** × **E. esula**) CD
× **pseudovirgata** (**E. esula** × **E. waldsteinii**) 431
serrulata 428
stricta ≡ **E. serrulata** 428
uralensis, see **E.** × **pseudovirgata** 431
waldsteinii CD
EUPHORBIACEAE 426 CD
EUPHRASIA
anglica 564
arctica sens. lat. 565
subsp. **arctica** 565
subsp. **borealis** 565
× **E. confusa** 565
× *E. nemorosa* 565
borealis 565
brevipila 565
cambrica 568
campbelliae 570
× *E. confusa* 570
× *E. micrantha* 570
× *E. nemorosa* 570
confusa 567
× **E. nemorosa** 567
× **E. scottica** 567
curta auct., see **E. ostenfeldii** 569
eurycarpa ≡ **E. ostenfeldii** 569
foulaensis 568
× *E. ostenfeldii* 569
frigida 568
heslop-harrisonii 571
marshallii 569
× *E. nemorosa* 569, 570
micrantha 570
× *E. scottica* 570
nemorosa 566
officinalis agg. 563
officinalis sens str. 563
subsp. *monticola* 563
subsp. *officinalis* 563
ostenfeldii 569
var. *rupestris* 569
pseudokerneri 566
f. *elongata* 566
rhumica ≡ **E. micrantha** 570
rivularis 564
rostkoviana subsp. **montana** 563
rostkoviana subsp. **rostkoviana** 563
rotundifolia 569
salisburgensis 571
var. *hibernica* 571
scottica 570
stricta CD
tetraquetra 566
vigursii 564
Evening-primrose, Common 419
Fragrant 420
Intermediate 419
Large-flowered 419
Small-flowered 420
Evening-primroses 418
Everlasting, Mountain 633
Pearly 633
Everlastingflower, New Zealand CD
Everlasting-pea, Broad-leaved 387
Narrow-leaved 387
Norfolk CD
Two-flowered 387
EXACULUM
pusillum 479
Eyebrights 563

FABACEAE 372 CD
FAGACEAE 129 CD
FAGOPYRUM
dibotrys CD
esculentum 186
FAGUS
sylvatica 129
FALCARIA
vulgaris 472
FALLOPIA
aubertii, see **F. baldschuanica** 190
baldschuanica 190
× **bohemica** (**F. japonica** × **F. sachalinensis**) 189
convolvulus 190
dumetorum 191
japonica 189
× **F. sachalinensis** (**F.** × **bohemica**) 189
sachalinensis 190
False-acacia 372
False-buck's-beard CD
Red CD
Tall CD
FARGESIA
spathacea CD
FASCICULARIA
bicolor CD
Fat-hen 140
FATSIA
japonica CD
Fatsia CD
Fennel 464
False CD
Giant CD
Hog's 474
Fen-sedge, Great 712
Fenugreek CD
Blue CD
Sickle-fruited CD
Fern, Beech 67
Jersey 62
Kangaroo CD
Killarney 64
Lemon-scented 67
Limestone 73
Maidenhair 62
Marsh 66
Oak 73
Ostrich 72
Parsley 62
Ribbon CD
Royal 61
Sensitive CD
Water 81
Fern-grass 761
Sea 761
FERULA
communis CD
Fescue, Bearded 751
Blue 748
Breton 747
Chewing's 744
Confused 748
Dune 750
Giant 742
Hard 748
Huon's 747
Hybrid 749
Mat-grass 752
Meadow 741
Rat's-tail 751
Red 745
Rush-leaved 743
Spiky CD
Squirreltail 751
Tall 742
Various-leaved 743
Wood 742
Fescues, Red 743
FESTUCA
altissima 742
arenaria 743
subsp. *oraria*, see *F. juncifolia* 743
FESTUCA (*cont.*)
armoricana 747
arundinacea 742
brevipila 748
diffusa, see **F. rubra** subsp. **megastachys** 745
duriuscula 748
filiformis 747
gautieri CD
gigantea 742
guestfalica, see **F. ovina** agg. 746
heterophylla 743
huonii 747
juncifolia 743
lemanii 748
longifolia 748
nigrescens, see **F. rubra** subsp. **commutata** 744
ovina 746
ovina agg. 746
pratensis 741
× **Lolium perenne** (× **Festulolium loliaceum**) 749
richardsonii 745
rubra *sens. lat.* 743
subsp. **arctica** 745
subsp. *arenaria* 743
subsp. **commutata** 744
subsp. **juncea** 744
subsp. **litoralis** 744
subsp. **megastachys** 745
subsp. *pruinosa* 744
subsp. *rubra* 743
subsp. **scotica** 745
tenuifolia ≡ **F. filiformis** 747
trachyphylla, see **F. brevipila** 748
vivipara 746
× **FESTULOLIUM**
loliaceum (**Festuca pratensis** × **Lolium perenne**) 749
Feverfew 644
FICUS
carica 127
Fiddleneck, Common 502
Field-rose 344
Short-styled 347
Field-speedwell, Common 558
Crested CD
Green 557
Grey 558
Fig 127
Figwort, Balm-leaved 543
Cape CD
Common 542
Green 543
Water 542
Yellow 543
Filaginella
uliginosa, see **Gnaphalium uliginosum** 634
FILAGO
apiculata ≡ **F. lutescens** 631
gallica 632
germanica ≡ **F. vulgaris** 631
lutescens 631
minima 632
pyramidata 632
spathulata ≡ **F. pyramidata** 632
vulgaris 631
Filbert CD
FILIPENDULA
kamtschatica CD
ulmaria 324
vulgaris 324
Filmy-Fern, Tunbridge 63
Wilson's 63
Finger-grass, Hairy 804
Smooth CD
Tropical CD
Water CD
Fir, Caucasian CD
Douglas 82
Giant 82
Noble 82
Firethorn 370
Asian CD
Flat-sedge 709
Saltmarsh 709
Flax 435
Fairy 435
New Zealand 834
New Zealand, Lesser CD
Pale 434
Perennial 435
Fleabane, Alpine 641
Argentine CD
Blue 642
Canadian 642
Common 636
Guernsey 643
Hairy CD
Irish 635
Mexican 642
Small 637
Stinking CD
Tall CD
Woody CD
Fleawort, Field 657
Marsh 658
Flixweed 251
Flossflower CD
Flowering-rush 666
Fluellen, Round-leaved 548
Sharp-leaved 548
FOENICULUM
vulgare 464
subsp. *piperitum* 464
subsp. *vulgare* 464
Fool's-water-cress 469
Forget-me-not, Alpine 504
Bur 506
Changing 505
Creeping 503
Early 505
Field 505
Great 499
Jersey 504
Pale 503
Tufted 503
Water 502
White CD
Wood 504
FORSYTHIA
× **intermedia** (**F. suspensa** × **F. viridissima**) 538
suspensa × **F. viridissima** (**F.** × **intermedia**) 538
Forsythia 538
Fountain-bamboo, Chinese CD
Indian CD
Fox-and-cubs 630
Irish CD
Yellow CD
Foxglove 551
Fairy 551
Straw CD
Foxglove-tree CD
Fox-sedge, American CD
False 714
True 714
Foxtail, Alpine 783
Bulbous 782
Marsh 782
Meadow 781
Orange 783
FRAGARIA
× **ananassa** 334
moschata 334
muricata, see **F. moschata** 334
vesca 333
FRANGULA
alnus 433
FRANKENIA
laevis 232
FRANKENIACEAE 232
FRAXINUS
excelsior 538
FREESIA
× **hybrida** CD
Freesia CD
Fringecups 320
FRITILLARIA
meleagris 811
Fritillary 811
Frogbit 669
FUCHSIA
'Corallina' CD
magellanica 420
'Riccartonii' 420
Fuchsia 420
Large-flowered CD
FUMARIA
bastardii 121
capreolata 120
subsp. *babingtonii* 120
subsp. *capreolata* 120

FUMARIA (*cont.*)
densiflora 122
martinii ≡ **F. reuteri** 121
micrantha ≡ **F. densiflora** 122
muralis 121
occidentalis 120
officinalis 122
subsp. *officinalis* 122
subsp. *wirtgenii* 122
parviflora 123
purpurea 122
reuteri 121
vaillantii 123
FUMARIACEAE 119 CD
Fumitory, Common 122
Dense-flowered 122
Few-flowered 123
Fine-leaved 123

GAGEA
bohemica 810
lutea 810
GAILLARDIA
aristata × G. pulchella (**G. × grandiflora**) CD
× **grandiflora** (**G. aristata × G. pulchella**) CD
GALANTHUS
caucasicus CD
elwesii 825
× **G. plicatus** CD
ikariae CD
nivalis 824
× *G. plicatus* 824
plicatus 824
subsp. **byzantinus** CD
subsp. **plicatus** CD
reginae-olgae CD
GALEGA
× *hartlandii* (*G. officinalis* × *G. patula*) 373
officinalis 373
× *G. patula* (*G.* × *hartlandii*) 373
Galeobdolon
luteum ≡ Lamiastrum galeobdolon 511
GALEOPSIS
angustifolia 514
bifida 516
segetum 514
speciosa 515
tetrahit 515
tetrahit *sens. lat.* 515
Galingale 710
Brown 710
Pale 710
GALINSOGA
ciliata ≡ **G. quadriradiata** 664
parviflora 664
quadriradiata 664
GALIUM
album, see *G. mollugo* subsp. *erectum* 593
aparine 594
boreale 591
constrictum 592
cruciata, see **Cruciata laevipes** 596
debile ≡ **G. constrictum** 592
elongatum, see *G. palustre* subsp. *elongatum* 592
fleurotii 593
mollugo 593
subsp. *erectum* 593
subsp. *mollugo* 593
× **G. verum** (**G. × pomeranicum**) 593
odoratum 591
palustre 592
subsp. *elongatum* 592
subsp. *palustre* 592
parisiense 595
× **pomeranicum** (**G. mollugo × G. verum**) 593
pumilum 593
saxatile 594
spurium 595
sterneri 594
tricornutum 595
uliginosum 591
verum 592
var. *maritimum* 592
Gallant-soldier 664
Garlic CD
American CD
Few-flowered 820
Field 821

Garlic (*cont.*)
Hairy CD
Honey CD
Italian CD
Keeled 821
Neapolitan CD
Rosy 820
Three-cornered 820
Yellow CD
GASTRIDIUM
phleoides CD
ventricosum 779
GAUDINIA
fragilis 767
GAULTHERIA
mucronata 289
× **G. shallon** (**G. × wisleyensis**) CD
procumbens CD
shallon 289
× **wisleyensis** (**G. mucronata × G. shallon**) CD
GAZANIA
rigens CD
GENISTA
aetnensis CD
anglica 406
hispanica 406
monspessulana CD
pilosa 406
tinctoria 405
subsp. **littoralis** 405
subsp. *tinctoria* 405
Gentian, Alpine 484
Autumn 482
Chiltern 482
Dune 483
Early 483
Field 481
Fringed 481
Marsh 483
Spring 484
Trumpet CD
Willow CD
GENTIANA
asclepiadea CD
clusii CD
nivalis 484
pneumonanthe 483
verna 484
GENTIANACEAE 478 CD
GENTIANELLA
amarella 482
subsp. *amarella* 482
subsp. *druceana* 482
subsp. *hibernica* 482
subsp. **septentrionalis** 482
× *G. anglica* 482
× *G. germanica* (*G.* × *pamplinii*) 482
× *G. uliginosa* 482, 483
anglica 483
subsp. *cornubiensis* 483
× *G. amarella* (*G.* × *davidiana*) 483
campestris 481
ciliata 481
× *davidiana* (*G. anglica* × *G. amarella*) 483
germanica 482
× *pamplinii* (*G. amarella* × *G. germanica*) 482
uliginosa 483
GERANIACEAE 443 CD
GERANIUM
columbinum 446
dissectum 446
endressii 443
'Wargrave Pink' 443
× **G. versicolor** (**G. × oxonianum**) 444
himalayense CD
× *G. pratense* 'Johnson's Blue' CD
ibericum CD
× **G. platypetalum** (**G. × magnificum**) 447 CD
lucidum 448
macrorrhizum 448
'Album' 448
maderense CD
× **magnificum** (**G. ibericum × G. platypetalum**) 447
molle 448
× **monacense** (**G. phaeum × G. reflexum**) CD
nodosum 444

GERANIUM (*cont.*)
× **oxonianum** (**G. endressii × G. versicolor**) 444
'A.T. Johnson' 444
'Claridge Druce' 444
'Winscombe' 444
phaeum 449
'Album' 449
var. *lividum* 449
var. *phaeum* 449
× **G. reflexum** (**G. × monacense**) CD
platypetalum CD
pratense 445
psilostemon CD
purpureum 449
subsp. *forsteri* 449
subsp. *purpureum* 449
pusillum 447
pyrenaicum 447
robertianum 449
subsp. *celticum* 449
subsp. *maritimum* 449
rotundifolium 445
rubescens CD
sanguineum 446
submolle CD
sylvaticum 445
versicolor 444
Geranium, Peppermint-scented CD
Germander, Cut-leaved 518
Wall 518
Water 518
German-ivy CD
GESNERIACEAE CD
GEUM
× **intermedium** (**G. rivale × G. urbanum**) 335
macrophyllum CD
rivale 335
× **G. urbanum** (**G. × intermedium**) 335
urbanum 335
Giant-rhubarb 410
Brazilian CD
Gipsywort 525
GLADIOLUS
communis 833
illyricus 832
subsp. *brittanicus* 832
Gladiolus, Eastern 833
Wild 832
Glasswort, Common 146
Glaucous 146
Long-spiked 147
One-flowered 145
Perennial 144
Purple 145
Shiny 146
Yellow 147
Glassworts 145
GLAUCIUM
flavum 118
GLAUX
maritima 304
GLECHOMA
hederacea 520
Globeflower 95
Globe-thistle 608
Blue 608
Glandular 607
Glory-of-the-snow 818
Boissier's CD
Lesser CD
GLYCERIA
declinata 764
fluitans 763
× **G. notata** (**G. × pedicellata**) 763
maxima 763
notata 764
× **pedicellata** (**G. fluitans × G. notata**) 763
plicata ≡ **G. notata** 764
GLYCINE
max CD
soja CD
GNAPHALIUM
luteoalbum 635
norvegicum 633
purpureum CD
supinum 634
sylvaticum 634
uliginosum 634
undulatum CD

Goat's-beard 622
Goat's-rue 373
Godetia CD
Goldenrod 637
Canadian 638
Early 638
Grass-leaved CD
Rough-stemmed CD
Golden-samphire 636
Golden-saxifrage, Alternate-leaved 321
Opposite-leaved 320
Goldilocks, Fine-leaved CD
Shrub CD
Gold-of-pleasure 271
Lesser CD
Good-King-Henry 136
GOODYERA
repens 840
Gooseberry 307
Goosefoot, Clammy CD
Crested CD
Fig-leaved 139
Foetid CD
Grey 139
Keeled CD
Many-seeded 137
Maple-leaved 138
Nettle-leaved 139
Nitre CD
Oak-leaved 136
Pitseed CD
Probst's CD
Red 137
Saltmarsh 137
Scented CD
Slimleaf CD
Soyabean CD
Stinking 138
Striped CD
Swedish CD
Upright 138
Gorse 407
Dwarf 408
Spanish 406
Western 407
Gramineae, see POACEAE 739
Grape-hyacinth 818
Compact CD
Garden 819
Grape-vine 433
Grass-of-Parnassus 321
Grass-poly 410
False CD
Greengage 353
Greenweed, Dyer's 405
Hairy 406
GRINDELIA
stricta CD
GRISELINIA
littoralis 424
GROENLANDIA
densa 680
Gromwell, Common 495
Field 495
Purple 494
GROSSULARIACEAE 305 CD
Ground-elder 460
Ground-ivy 520
Ground-pine 519
Groundsel 656
Eastern CD
Heath 656
Sticky 657
Tree CD
Welsh 656
Guelder-rose 597
GUIZOTIA
abyssinica 663
Gum, Cider CD
Ribbon CD
Urn-fruited CD
Gumplant, Coastal CD
GUNNERA
manicata CD 410
tinctoria 410
GUNNERACEAE 410 CD
Guttiferae, see CLUSIACEAE 210
GYMNADENIA
conopsea 843
subsp. **borealis** 844
subsp. **conopsea** 844
subsp. **densiflora** 844
var. *densiflora* 844

GYMNOCARPIUM
dryopteris 73
robertianum 73
Gymnogrammaceae, see ADIANTACEAE 62
GYPSOPHILA
paniculata CD

HAINARDIA
cylindrica CD
Hair-grass, Bog 770
Crested 768
Early 772
Grey 771
Mediterranean CD
Silver 771
Somerset 768
Tufted 769
Wavy 770
Hairy-brome 789
Lesser 789
Halimione
pedunculata ≡ **Atriplex pedunculata** 143
portulacoides ≡ **Atriplex portulacoides** 143
HALORAGACEAE 408 CD
HALORAGIS
micrantha CD
HAMMARBYA
paludosa 841
Hampshire-purslane 418
Hard-fern 81
Chilean CD
Hard-grass 762
Curved 762
One-glumed CD
Hardhack CD
Hare's-ear, Shrubby CD
Sickle-leaved 466
Slender 467
Small 467
Hare's-tail 779
Harebell 587
Broad-leaved CD
Harlequinflower, Plain CD
Hart's-tongue 67
Hartwort 476
Hawk's-beard, Beaked 628
Bristly 628
French CD
Leafless 629
Marsh 626
Narrow-leaved CD
Northern 627
Rough 627
Smooth 627
Stinking 628
Tuberous CD
Hawkbit, Autumn 620
Lesser 620
Rough 620
Scaly CD
Hawkweeds 631
Hawthorn 371
Midland 372
Oriental CD
Various-leaved CD
Hazel 135
Turkish CD
Heath, Blue 288
Cornish 293
Corsican CD
Cross-leaved 292
Darley Dale CD
Dorset 291
Irish 292
Mackay's 291
Portuguese CD
Prickly 289
St Dabeoc's 288
Tree CD
Heather 291
Bell 292
Heath-grass 799
HEBE
barkeri CD
brachysiphon CD
dieffenbachii CD
elliptica × **H. salicifolia** (**H.** × **lewisii**) CD
elliptica × **H. speciosa** (**H.** × **franciscana**) 561
× **franciscana** (**H. elliptica** × **H. speciosa**) 561
'Blue Gem' 561
HEBE (*cont.*)
× **lewisii** (**H. elliptica** × **H. salicifolia**) CD
salicifolia 560
Hebe, Barker's CD
Dieffenbach's CD
Hooker's CD
Lewis's CD
HEDERA
algeriensis CD
canariensis, see **H. algeriensis** CD
colchica 453
helix 453
'Hibernica' 454
subsp. **helix** 454
subsp. **hibernica** 454
Hedge-parsley, Knotted 477
Spreading 477
Upright 477
HEDYPNOIS
cretica CD
HELENIUM
autumnale CD
HELIANTHEMUM
apenninum 224
× *H. nummularium* (*H.* × *sulphureum*) 224
canum ≡ *H. oelandicum* subsp. *incanum* 224
subsp. *levigatum*, see *H. oelandicum* subsp. *levigatum* 224
subsp. *piloselloides*, see *H. oelandicum* subsp. *piloselloides* 224
chamaecistus ≡ **H. nummularium** 223
nummularium 223
oelandicum 224
subsp. *incanum* 224
subsp. *levigatum* 224
subsp. *piloselloides* 224
× *sulphureum* (*H. apenninum* × *H. nummularium*) 224
HELIANTHUS
annuus 663
× **H. decapetalus** (**H.** × **multiflorus**) CD
× **laetiflorus** (**H. pauciflorus** × **H. tuberosus**) 663
× **multiflorus** (**H. annuus** × **H. decapetalus**) CD
pauciflorus 663
× **H. tuberosus** (**H.** × **laetiflorus**) 663
petiolaris CD
rigidus 663
tuberosus 664
HELICHRYSUM
bellidioides CD
HELICTOTRICHON
neesii CD
pratense 765
pubescens 765
Heliotrope, Winter 660
Hell, Tree of 440
Hellebore, Corsican CD
Green 96
Stinking 96
Helleborine, Broad-leaved 837
Dark-red 836
Green-flowered 838
Marsh 836
Narrow-leaved 835
Narrow-lipped 837
Red 835
Violet 836
White 835
Young's 837
HELLEBORUS
argutifolius CD
foetidus 96
orientalis CD
viridis 96
Helxine
soleirolii, see **Soleirolia soleirolii** 128
HEMEROCALLIS
fulva 809
lilioasphodelus CD
Hemlock 466
Hemlock-spruce, Western 83
Hemp 126
Hemp-agrimony 666
Hemp-nettle, Bifid 516
Common 515
Downy 514
Large-flowered 515
Red 514
Hemp-nettles, Common 515
Henbane 486
HEPATICA
nobilis CD
HERACLEUM
mantegazzianum 476
sphondylium 476
Herb-Paris 814
Herb-Robert 449
Giant CD
Greater CD
HERMINIUM
monorchis 842
HERMODACTYLUS
tuberosus CD
HERNIARIA
ciliolata 171
subsp. *ciliolata* 171
subsp. *subciliata* 171
glabra 170
hirsuta CD
HESPERIS
matronalis 253
HEUCHERA
sanguinea CD
HIBISCUS
trionum 221
HIERACIUM agg. 631
aurantiacum, see **Pilosella aurantiaca** 630
brunneocroceum, see *Pilosella aurantiaca* subsp. *carpathicola* 630
caespitosum, see **Pilosella caespitosa** *CD*
flagellare subsp. *bicapitatum*, see **Pilosella flagellaris** subsp. **bicapitata** 630
flagellare subsp. *flagellare*, see **Pilosella flagellaris** subsp. **flagellaris** 630
peleterianum, see **Pilosella peleteriana** 629
pilosella, see **Pilosella officinarum** 629
pilosella sens. lat. 631
praealtum, see **Pilosella praealta** CD
HIEROCHLOE
odorata 772
HIMANTOGLOSSUM
hircinum 855
HIPPOCASTANACEAE 437 CD
HIPPOCREPIS
comosa 379
emerus CD
HIPPOPHAE
rhamnoides 408
HIPPURIDACEAE 531
HIPPURIS
vulgaris 531
HIRSCHFELDIA
incana 283
Hoary Cress 277
Hogweed 476
Giant 476
HOHERIA
populnea CD
HOLCUS
lanatus 770
mollis 771
Hollow-root CD
Holly 425
Highclere 426
New Zealand 643
Holly-fern 76
House CD
Hollyhock 221
Australian CD
HOLODISCUS
discolor CD
Holoschoenus
vulgaris, see **Scirpoides holoschoenus** 706
HOLOSTEUM
umbellatum 160
Holy-grass 772
HOMOGYNE
alpina 661
HONCKENYA
peploides 155
Honesty 264
Honewort 468
Honeybells CD
Honeysuckle 601
Box-leaved 600
Californian CD
Fly 600
Garden CD
Honeysuckle (*cont.*)
Henry's CD
Himalayan 599
Japanese 601
Perfoliate 601
Tartarian CD
Wilson's 600
Hop 126
HORDELYMUS
europaeus 796
HORDEUM
distichon *sens. lat.* 796
distichon sens. str. 796
euclaston CD
geniculatum CD
hystrix, see **H. geniculatum** CD
jubatum 797
marinum 798
murinum 797
subsp. **glaucum** CD 797
subsp. **leporinum** CD 797
subsp. *murinum* 797
pubiflorum CD
pusillum CD
secalinum 798
vulgare 797
Horehound, Black 510
White 516
Hornbeam 135
Horned-poppy, Yellow 118
HORNUNGIA
petraea 272
Hornwort, Rigid 94
Soft 95
Horse-chestnut 437
Indian CD
Red 438
Horse-nettle CD
Horse-radish 259
Horsetail, Branched 57
Field 58
Great 60
Mackay's 57
Marsh 59
Rough 56
Shady 59
Shore 58
Variegated 57
Water 58
Wood 59
Hottentot-fig 136
HOTTONIA
palustris 299
Hound's-tongue 506
Green 507
House-leek 309
Cobweb CD
Huckleberry, Garden CD
HUMULUS
lupulus 126
HUPERZIA
selago 53
Hutchinsia 272
Hutera
cheiranthos, see **Coincya monensis** subsp. **cheiranthos** 282
monensis, see **Coincya monensis** subsp. **monensis** 282
wrightii, see **Coincya wrightii** 283
Hyacinth 818
Tassel 819
HYACINTHOIDES
hispanica 817
× **H. non-scripta** 817
italica CD
× *massartiana*, see **H. hispanica** × **H. non-scripta** 817
non-scripta 817
× *variabilis*, see **H. hispanica** × **H. non-scripta** 817
orientalis 818
HYDRANGEA
macrophylla CD
Hydrangea CD
HYDRANGEACEAE 304 CD
HYDRILLA
verticillata 670
HYDROCHARIS
morsus-ranae 669
HYDROCHARITACEAE 669 CD
Hydrocotylaceae, see APIACEAE 454

HYDROCOTYLE
moschata CD
novae-zeelandiae CD
ranunculoides CD
vulgaris 454
HYDROPHYLLACEAE 494
HYMENOPHYLLACEAE 63
HYMENOPHYLLUM
tunbrigense 63
wilsonii 63
HYOSCYAMUS
niger 486
Hypericaceae, see CLUSIACEAE 210
HYPERICUM
androsaemum 211
× **H. hircinum** (**H.** × **inodorum**) 211
calycinum 210
canadense 216
× **desetangsii** (**H. maculatum** × **H. perforatum**) 213
elatum ≡ **H.** × **inodorum** 211
elodes 216
forrestii CD
'Hidcote' CD
hircinum 211
hirsutum 215
humifusum 214
× *H. linariifolium* 214
× **inodorum** (**H. androsaemum** × **H. hircinum**) 211
linariifolium 214
maculatum 212
subsp. **maculatum** 212
subsp. *maculatum* × *H. perforatum* 213
subsp. **obtusiusculum** 213
subsp. *obtusiusculum* × *H. perforatum* 213
× **H. perforatum** (**H.** × **desetangsii**) 213
montanum 215
nummularium CD
perforatum 212
pseudohenryi CD
pulchrum 215
var. *procumbens* 215
tetrapterum 214
undulatum 213
xylosteifolium CD
HYPOCHAERIS
glabra 619
maculata 619
radicata 619
Hypolepidaceae *pro parte*, see DENNSTEADTIACEAE 66
Hyssop CD
HYSSOPUS
officinalis CD

IBERIS
amara 274
sempervirens 273
umbellata 274
Iceland-purslane 186
Ice-plant, Heart-leaf CD
ILEX
× **altaclerensis** (**I. aquifolium** × **I. perado**) 426
aquifolium 425
× **I. perado** (**I.** × **altaclerensis**) 426
Illecebraceae, see CARYOPHYLLACEAE 152
ILLECEBRUM
verticillatum 171
IMPATIENS
capensis 452
glandulifera 453
noli-tangere 452
parviflora 452
Indian-rhubarb 313
INULA
conyzae 636
crithmoides 636
helenium 635
oculus-christi CD
salicina 635
IOCHROMA
australe CD
Ipheion
uniflorum 823
IPOMOEA
hederacea CD
lacunosa CD
purpurea CD
IRIDACEAE 828 CD
IRIS
ensata CD
filifolia × **I. tingitana** (**I.** × **hollandica**) CD
foetidissima 830
germanica 829
× **hollandica** (**I. filifolia** × **I. tingitana**) CD
latifolia CD
orientalis CD
pseudacorus 830
× **robusta** (**I. versicolor** × **I. virginica**) CD
sibirica 830
spuria CD
unguicularis CD
versicolor CD
× **I. virginica** (**I.** × **robusta**) CD
xiphioides, see **I. latifolia** CD
xiphium CD
Iris, Algerian CD
Bearded 829
Blue CD
Dutch CD
English CD
Japanese CD
Purple CD
Siberian 830
Snake's-head CD
Spanish CD
Stinking 830
Turkish CD
Windermere CD
Yellow 830
ISATIS
tinctoria 252
subsp. *tinctoria* 252
tinctoria sens. lat. 252
ISOETACEAE 55
ISOETES
echinospora 56
× *I. lacustris* 56
histrix 56
lacustris 55
ISOLEPIS
cernua 708
setacea 708
IVA
xanthiifolia CD
Ivy 453
Algerian CD
Atlantic 454
Common 454
Persian 453
IXIA
campanulata CD
paniculata CD

Jacob's-ladder 494
Japanese-lantern 486
JASIONE
montana 588
Jasmine, Red CD
Summer CD
Winter CD
JASMINUM
beesianum CD
nudiflorum CD
officinale CD
Johnson-grass 805
JONOPSIDIUM
acaule CD
Jonquil, Campernelle CD
JUGLANDACEAE 128 CD
JUGLANS
regia 128
JUNCACEAE 688 CD
JUNCAGINACEAE 672
JUNCUS
acutiflorus 692
× **J. articulatus** (**J.** × **surrejanus**) 692
acutus 695
alpinoarticulatus 691
ambiguus 690
articulatus 692
balticus 695
biglumis 693
bufonius *sens. lat.* 689
bufonius *sens. str.* 690
bulbosus 693
capitatus 691
castaneus 694
Juncus (*cont.*)
compressus 688
conglomeratus 697
× **diffusus** (**J. effusus** × **J. inflexus**) 697
dudleyi, see **J. tenuis** 688
effusus 696
× **J. inflexus** (**J.** × **diffusus**) 697
filiformis 696
foliosus 690
gerardii 689
inflexus 696
kochii 693
maritimus 694
mutabilis ≡ **J. pygmaeus** 693
pallidus CD
planifolius CD
pygmaeus 693
ranarius, see **J. ambiguus** 690
squarrosus 688
subnodulosus 691
subulatus 695
subuliflorus, see **J. conglomeratus** 697
× **surrejanus** (**J. acutiflorus** × **J. articulatus**) 692
tenuis 688
trifidus 689
triglumis 694
Juneberry 365
Juniper, Common 90
JUNIPERUS
communis 90
subsp. *alpina*, see **J. communis** subsp. **nana** 91
subsp. **communis** 91
subsp. **hemisphaerica** 91
subsp. **nana** 91

KALMIA
angustifolia CD
latifolia CD
polifolia CD
Kangaroo-apple CD
Karo CD
Kelch-grass CD
KERRIA
japonica 324
Kerria 324
Ketmia, Bladder 221
Kew Weed 664
KICKXIA
elatine 548
spuria 548
Kidneyweed CD
Knapweed, Common 617
Giant CD
Greater 615
Russian CD
Knautia
arvensis 606
Knawl, Annual 169
KNIPHOFIA
× **praecox** CD
uvaria 809
Knotgrass 188
Cornfield 189
Equal-leaved 188
Northern 188
Ray's 187
Sea 187
Knotgrasses 187
Knotweed, Alpine CD
Chinese CD
Giant 190
Himalayan 183
Japanese 189
Lesser 182
Soft CD
KOBRESIA
simpliciuscula 712
Kochia
scoparia, see **Bassia scoparia** 140
KOELERIA
cristata ≡ **K. macrantha** 768
glauca, see **K. macrantha** 768
macrantha 768
vallesiana 768
KOELREUTERIA
paniculata CD
KOENIGIA
islandica 186
Kohlrauschia
nanteuilii, see **Petrorhagia nanteuilii** 180
prolifera ≡ **Petrorhagia nanteuilii** & **P. prolifera** 180
Kohuhu CD
Koromiko 560

Labiatae, see LAMIACEAE 507
Labrador-tea CD
LABURNUM
alpinum 403
× **L. anagyroides** (**L.** × **watereri**) CD
anagyroides 403
× *L. alpinum* (*L.* × *watereri*) 403
× **watereri** (**L. alpinum** × **L. anagyroides**) CD 403
Laburnum 403
Scottish 403
LACTUCA
saligna 625
sativa CD
var. *capitata* CD
var. *crispa* CD
serriola 624
f. *integrifolia* 624
tatarica CD
subsp. *pulchella* CD
virosa 624
Lady's-mantle, Alpine 338
Silver 338
Lady's-slipper 834
Lady's-tresses, Autumn 839
Creeping 840
Irish 840
Summer 840
Lady-fern 72
Alpine 72
Newman's 73
LAGAROSIPHON
major 671
LAGURUS
ovatus 779
LAMARCKIA
aurea CD
Lamb's-ear 508
LAMIACEAE 507 CD
LAMIASTRUM
galeobdolon subsp. **argentatum** 512
galeobdolon subsp. **galeobdolon** 511
galeobdolon subsp. **montanum** 511
LAMIUM
album 512
amplexicaule 514
confertum 513
hybridum 513
maculatum 512
molucellifolium ≡ **L. confertum** 513
purpureum 513
LAMPRANTHUS
falciformis CD
roseus CD
LAPPULA
squarrosa 506
LAPSANA
communis 618
subsp. *communis* 618
subsp. **intermedia** CD 618
Larch, European 84
Hybrid 84
Japanese 84
LARIX
decidua 84
× **L. kaempferi** (**L.** × **marschlinsii**) 84
× *eurolepis*, see **L.** × **marschlinsii** 84
× *henryana*, see **L.** × **marschlinsii** 84
kaempferi 84
leptolepis, see **L. kaempferi** 84
× **marschlinsii** (**L. decidua** × **L. kaempferi**) 84
Larkspur 98
LATHRAEA
clandestina 577
squamaria 576
LATHYRUS
annuus CD
aphaca 388
cicera CD
grandiflorus 387
heterophyllus CD
hirsutus 388
japonicus 385

LATHYRUS (*cont.*)
latifolius 387
linifolius 385
montanus ≡ **L. linifolius** 385
niger CD
nissolia 388
odoratus CD
palustris 386
var. *pilosus* 386
pratensis 386
sativus CD
sylvestris 387
tuberosus 386
LAURACEAE 92
Laurel, Cherry 356
Portugal 355
LAURUS
nobilis 92
Laurustinus 598
LAVANDULA
angustifolia 529
× **L. latifolia** (**L.** × **intermedia**) 529
× **intermedia** (**L. angustifolia** × **L. latifolia**) 529
LAVATERA
arborea 219
cretica 219
olbia, see **L. thuringiaca** 220
× *L. thuringiaca* 220
plebeia CD
thuringiaca *sens. lat.* 220
trimestris CD
Lavender, Garden 529
Lavender-cotton 646
LEDUM
palustre CD
Leek CD
Broad-leaved CD
Round-headed 822
Sand 822
Wild 822
LEERSIA
oryzoides 740
LEGOUSIA
hybrida 587
speculum-veneris CD
Leguminosae, see FABACEAE 372
LEMNA
gibba 686
minor 686
minuscula, see **L. minuta** 687
minuta 687
polyrhiza ≡ **Spirodela polyrhiza** 685
trisulca 686
LEMNACEAE 685
LENS
culinaris CD
orientalis CD
Lenten-rose CD
LENTIBULARIACEAE 580
Lentil CD
LEONTODON
autumnalis 620
hispidus 620
var. *glabratus* 620
saxatilis 620
taraxacoides ≡ **L. saxatilis** 620
LEONURUS
cardiaca 511
cardiaca sens. lat. 511
Leopardplant CD
Przewalski's CD
Leopard's-bane 658
Eastern CD
Harpur-Crewe's CD
Plantain-leaved 659
Willdenow's CD
LEPIDIUM
africanum CD
bonariense CD
campestre 275
densiflorum, see **L. virginicum** 275
divaricatum CD
draba 277
subsp. **chalepense** CD
subsp. *draba* CD
graminifolium CD
heterophyllum 275
hyssopifolium CD
latifolium 276
neglectum, see **L. virginicum** 275
perfoliatum 276
LEPIDIUM (*cont.*)
ramosissimum, see **L. virginicum** 275
ruderale 276
sativum 274
virginicum 275
Leptinella CD
Hairless CD
LEPTOCHLOA
fusca CD
LEPTOSPERMUM
lanigerum CD
scoparium CD
Lettuce, Blue CD
Cabbage CD
Cos CD
Garden CD
Great 624
Least 625
Prickly 624
Wall 626
LEUCANTHEMELLA
serotina CD
LEUCANTHEMUM
lacustre × **L. maximum** (**L.** × **superbum**) 650
maximum, see **L.** × **superbum** 650
× **superbum** (**L. lacustre** × **L. maximum**) 650
vulgare 649
LEUCOJUM
aestivum 823
subsp. *aestivum* 823
subsp. *pulchellum* 823
vernum 824
Leucorchis
albida ≡ **Pseudorchis albida** 843
LEVISTICUM
officinale 474
LEYCESTERIA
formosa 599
LEYMUS
arenarius 796
LIBERTIA
elegans CD
formosa CD
LIGULARIA
dentata CD
przewalskii CD
LIGUSTICUM
scoticum 473
LIGUSTRUM
ovalifolium 539
vulgare 539
Lilac 538
LILIACEAE 808 CD
LILIUM
martagon 812
pyrenaicum 812
Lily, African CD
Jersey CD
Kerry 808
Martagon 812
May 814
Peruvian CD
Pyrenean 812
Snowdon 810
Lily-of-the-valley 812
False CD
Lime 217
Large-leaved 216
Small-leaved 217
LIMNANTHACEAE 451
LIMNANTHES
douglasii 451
LIMONIUM
auriculae-ursifolium 200
bellidifolium 200
binervosum agg. 201
binervosum subsp. **anglicum** 202
binervosum subsp. **binervosum** 201
binervosum subsp. **cantianum** 201
binervosum subsp. **mutatum** 202
binervosum subsp. **sarniense** 203
binervosum subsp. **saxonicum** 202
britannicum subsp. **britannicum** 204
britannicum subsp. **celticum** 205
britannicum subsp. **coombense** 205
britannicum subsp. **transcanalis** 205
dodartiforme 207
humile 199
hyblaeum CD
latifolium CD
loganicum 206
LIMONIUM (*cont.*)
× *neumanii* (*L. humile* × *L. vulgare*) 199
normannicum 200
paradoxum 203
parvum 206
procerum subsp. **cambrense** 204
procerum subsp. **devoniense** 204
procerum subsp. **procerum** 203
var. *medium* 206
recurvum subsp. **humile** 208
recurvum subsp. **portlandicum** 207
recurvum subsp. **pseudotranswallianum** 208
recurvum subsp. **recurvum** 207
transwallianum 206
vulgare 199
× *L. humile* (*L.* × *neumanii*) 199
LIMOSELLA
aquatica 546
australis 546
subulata ≡ **L. australis** 546
LINACEAE 434
LINARIA
arenaria CD
dalmatica CD
genistifolia subsp. *dalmatica*, see **L. dalmatica** CD
maroccana 551
pelisseriana 550
purpurea 549
repens 549
× **L. vulgaris** (**L.** × **sepium**) 550
× **sepium** (**L. repens** × **L. vulgaris**) 550
supina 550
vulgaris 549
LINNAEA
borealis 599
LINUM
anglicum ≡ *L. perenne* subsp. *anglicum* 435
bienne 434
catharticum 435
perenne 435
subsp. *anglicum* 435
usitatissimum 435
LIPARIS
loeselii 841
Liquorice, Wild 374
LISTERA
cordata 839
ovata 839
LITHOSPERMUM
arvense 495
officinale 495
purpureocaeruleum 494
Little-Robin 449
LITTORELLA
uniflora 537
Liverleaf CD
LLOYDIA
serotina 810
LOBELIA
dortmanna 589
erinus 589
urens 589
Lobelia, Californian CD
Garden 589
Heath 589
Lawn CD
Water 589
Lobeliaceae, see CAMPANULACEAE 583
LOBULARIA
maritima 266
Loganberry CD
Logfia
gallica, see **Filago gallica** 632
minima, see **Filago minima** 632
LOISELEURIA
procumbens 288
LOLIUM
× **boucheanum** (**L. multiflorum** × **L. perenne**) 750
× *hybridum*, see **L.** × **boucheanum** 750
multiflorum 749
× **L. perenne** (**L.** × **boucheanum**) 750
perenne 749
subsp. *multiflorum*, see **L. multiflorum** 749
subsp. *perenne*, see **L. perenne** 749
remotum CD
rigidum CD
temulentum 750
Lombardy-poplars 235
Londonpride 315
False 316
Lesser CD
Scarce CD
London-rocket 249
False 250
Longleaf 472
LONICERA
caprifolium 601
× **L. etrusca** (**L.** × **italica**) CD 601
henryi CD
involucrata CD
× **italica** (**L. caprifolium** × **L. etrusca**) CD 601
japonica 601
ledebourii, see **L. involucrata** CD
nitida 600
periclymenum 601
pileata 600
tatarica CD
xylosteum 600
Loosestrife, Dotted 301
Fringed 301
Lake CD
Tufted 302
Yellow 301
Lophochloa
cristata, see **Rostraria cristata** *CD*
Loranthaceae, see VISCACEAE 424
Lords-and-Ladies 684
Italian 684
LOTUS
angustissimus 378
corniculatus 377
glaber 376
hispidus ≡ **L. subbiflorus** 377
pedunculatus 377
subbiflorus 377
tenuis ≡ **L. glaber** 376
uliginosus ≡ **L. pedunculatus** 377
Lousewort 576
Marsh 575
Lovage 474
Scots 473
Love-grass, African CD
Jersey CD
Small CD
Weeping CD
Love-in-a-mist 97
Love-lies-bleeding CD
Lucerne 392
Sand 392
LUDWIGIA
× *kentiana* 418
palustris 418
LUMA
apiculata CD
LUNARIA
annua 264
subsp. *annua* 264
subsp. *pachyrhiza* 264
Lungwort 496
Mawson's CD
Narrow-leaved 497
Red CD
Suffolk 496
Lupin, False CD
Garden 402
Narrow-leaved CD
Nootka 403
Russell 402
Tree 402
White CD
LUPINUS
albus CD
angustifolius CD
arboreus 402
× **L. polyphyllus** (**L.** × **regalis**) 402
hartwegii 402
mutabilis 402
nootkatensis 403
polyphyllus 402
× **regalis** (**L. arboreus** × **L. polyphyllus**) 402
LURONIUM
natans 667
LUZULA
arcuata 700
× **borreri** (**L. forsteri** × **L. pilosa**) 698
campestris 699
forsteri 697
× **L. pilosa** (**L.** × **borreri**) 698

LUZULA (*cont.*)
luzuloides 699
multiflora 699
subsp. *congesta* 699
subsp. *multiflora* 699
pallescens ≡ **L. pallidula** 700
pallidula 700
pilosa 698
spicata 700
sylvatica 698
LYCHNIS
alpina 175
chalcedonica CD
coronaria 174
flos-cuculi 174
var. *congesta* 174
viscaria 174
LYCIUM agg. 485
barbarum 485
chinense 485
LYCOPERSICON
esculentum 487
LYCOPODIACEAE 53
LYCOPODIELLA
inundata 53
LYCOPODIUM
alpinum ≡ **Diphasiastrum alpinum** 54
annotinum 54
clavatum 53
inundatum ≡ **Lycopodiella inundata** 53
selago ≡ **Huperzia selago** 53
Lycopsis
arvensis ≡ **Anchusa arvensis** 500
LYCOPUS
europaeus 525
Lyme-grass 796
LYSICHITON
americanus 684
camtschatcensis CD
LYSIMACHIA
ciliata 301
nemorum 300
nummularia 300
punctata 301
terrestris CD
thyrsiflora 302
verticillaris 301
vulgaris 301
LYTHRACEAE 410 CD
LYTHRUM
hyssopifolium 410
junceum CD
portula 411
salicaria 410

MACLEAYA
cordata CD
× **M. microcarpa** (**M.** × **kewensis**) CD
× **kewensis** (**M. cordata** × **M. microcarpa**) CD
Madder, Field 590
Wild 596
Madwort 502
MAHONIA
aquifolium 114
× **M. repens** (**M.** × **decumbens**) CD
× **decumbens** (**M. aquifolium** × **M. repens**) CD
repens CD
MAIANTHEMUM
bifolium 814
kamtschaticum CD
Maize 805
MALCOLMIA
maritima 254
Male-fern 77
Mountain 77
Scaly 77
Mallow, Chinese CD
Common 218
Dwarf 219
French CD
Greek CD
Least 218
New Zealand CD
Prairie CD
Prickly CD
Royal CD
Small 218
Maltese-Cross CD
MALUS
atrosanguinea × **M. niedzwetzkvana** (**M.** × **purpurea**) CD
domestica 357
sylvestris *sens lat.* 357
sylvestris sens. str. 357
subsp. *mitis*, see *M. domestica* 357
MALVA
alcea CD
× *M. moschata* CD
moschata 217
neglecta 219
nicaeensis CD
parviflora 218
pusilla 218
sylvestris 218
verticillata CD
MALVACEAE 217 CD
Mangel-wurzel 144
Maple, Ashleaf 440
Cappadocian 438
Field 439
Norway 438
Silver 439
Mare's-tail 531
Marigold, African CD
Corn 649
Dwarf CD
Field CD
French 666
Pot 661
Southern CD
Marjoram, Wild 523
Marram 779
American CD
Marrow 233
MARRUBIUM
vulgare 516
Marsh-bedstraw, Common 592
Slender 592
Marsh-elder CD
Marsh-mallow 220
Rough 220
Marsh-marigold 95
Marsh-orchid, Early 848
Lapland 851
Narrow-leaved 851
Northern 850
Southern 850
Western 851
Marshwort, Creeping 469
Lesser 469
MARSILEACEAE 63
Masterwort 475
Mat-grass 740
MATRICARIA
discoidea 650
maritima, see **Tripleurospermum maritimum** 651
matricarioides ≡ **M. discoidea** 650
perforata, see **Tripleurospermum inodorum** 651
recutita 650
MATTEUCCIA
struthiopteris 72
MATTHIOLA
incana 254
longipetala 255
odoratissima 255
sinuata 254
Mayweed, Scented 650
Scentless 651
Sea 651
Meadow-foam 451
Meadow-grass, Alpine 760
Annual 756
Broad-leaved 758
Bulbous 760
Early 755
Flattened 758
Glaucous 759
Narrow-leaved 757
Rough 756
Smooth 757
Spreading 757
Swamp 759
Wavy 758
Wood 759
Meadow-rue, Alpine 112
Chinese CD
Common 112
French CD
Lesser 112
Meadowsweet 324
Giant CD
MECONOPSIS
cambrica 117
MEDICAGO
arabica 393
falcata ≡ **M. sativa** subsp. **falcata** 392
laciniata CD
lupulina 391
minima 393
polymorpha 393
praecox CD
sativa subsp. **falcata** 392
sativa subsp. **sativa** 392
× subsp. *falcata* 392
sativa subsp. **varia** 392
truncatula CD
× *varia*, see **M. sativa** subsp. **varia** 392
Medick, Black 391
Bur 393
Early CD
Sickle 392
Spotted 393
Strong-spined CD
Tattered CD
Toothed 393
Medlar 371
MELAMPYRUM
arvense 562
cristatum 561
pratense 562
sylvaticum 562
MELICA
nutans 764
uniflora 765
Melick, Mountain 764
Wood 765
Melilot, Furrowed CD
Ribbed 391
Small 391
Tall 390
White 390
MELILOTUS
albus 390
altissimus 390
indicus 391
officinalis 391
sulcatus CD
MELISSA
officinalis 521
melissophyllum 516
Melon CD
Water CD
MENTHA
aquatica 526
× **M. arvensis** (**M.** × **verticillata**) 526
× **M. arvensis** × **M. spicata** (**M.** × **smithiana**) 526
× *M. longifolia* (*M.* × *dumetorum*) 527
× **M. spicata** (**M.** × **piperita**) 527
arvensis 525
× **M. spicata** (**M.** × **gracilis**) 525
× *dalmatica* 525
× *dumetorum* (*M. aquatica* × *M. longifolia*) 527
× *gentilis* ≡ **M.** × **gracilis** 525
× **gracilis** (**M. arvensis** × **M. spicata**) 525
longifolia 527
× **M. spicata** (**M.** × **villosonervata**) 528
× **M. suaveolens** (**M.** × **rotundifolia**) 528
× *niliaca* 527
× **piperita** (**M. aquatica** × **M. spicata**) 527
pulegium 529
requienii 529
× **rotundifolia** (**M. longifolia** × **M. suaveolens**) 528
rotundifolia 528
scotica 527
× **smithiana** (**M. aquatica** × **M. arvensis** × **M. spicata**) 526
spicata 527
× **M. suaveolens** (**M.** × **villosa**) 527
suaveolens 528
× **verticillata** (**M. aquatica** × **M. arvensis**) 526
× **villosa** (**M. spicata** × **M. suaveolens**) 527
× **villosonervata** (**M. longifolia** × **M. spicata**) 528
MENYANTHACEAE 493
MENYANTHES
trifoliata 493
MERCURIALIS
annua 426
perennis 426
Mercury, Annual 427
Dog's 426
MERTENSIA
maritima 501
MESPILUS
germanica 371
MEUM
athamanticum 465
Mexican-stonecrop, Greater CD
Lesser CD
Mexican-tea CD
Mezereon 411
MIBORA
minima 780
Michaelmas-daisies 638
Michaelmas-daisy, Common 640
Confused 639
Glaucous CD
Hairy 639
Late 639
Narrow-leaved 640
Nettle-leaved CD
Mignonette, Corn CD
Garden CD
White 286
Wild 286
MILIUM
effusum 741
scabrum ≡ **M. vernale** 741
vernale 741
subsp. *sarniense* 741
Milk-parsley 475
Cambridge 473
Milk-vetch, Alpine 374
Chick-pea CD
Lesser CD
Purple 373
Milkwort, Chalk 437
Common 436
Dwarf 437
Heath 436
Millet, Autumn CD
Common 802
Early 741
Great CD
Japanese 803
Shama CD
Transvaal CD
White CD
Wood 741
MIMOSACEAE CD
MIMULUS agg. 544
× **burnetii** (**M. cupreus** × **M. guttatus**) 545
cupreus 545
× **M. guttatus** (**M.** × **burnetii**) 545
× **M. guttatus** × **M. luteus** CD
× **M. luteus** (**M.** × **maculosus**) CD
guttatus 544
× **M. luteus** (**M.** × **robertsii**) 545
luteus 545
× **maculosus** (**M. cupreus** × **M. luteus**) CD
moschatus 544
× **robertsii** (**M. guttatus** × **M. luteus**) 545
Mind-your-own-business 128
Mint, Bushy 525
Corn 525
Corsican 529
Round-leaved 528
Sharp-toothed 528
Spear 527
Tall 526
Water 526
Whorled 526
Mintweed CD
MINUARTIA
hybrida 157
recurva 155
rubella 156
sedoides 157
stricta 156
verna 156
MISOPATES
calycinum CD
orontium 547
Mistletoe 424
Mock-orange 304
Hairy 305

MOEHRINGIA
trinervia 155
MOENCHIA
erecta 165
caerulea 800
subsp. *altissima*, see *M. caerulea* subsp. *arundinacea* 800
subsp. *arundinacea* 800
subsp. *caerulea* 800
litoralis, see *M. caerulea* subsp. *arundinacea* 800
MONESES
uniflora 296
Moneywort, Cornish 561
Monk's-hood 97
Hybrid 97
Monk's-rhubarb 192
Monkeyflower 544
Coppery 545
Hybrid 545
Scottish CD
Monkeyflowers 544
Monkey-puzzle 92
MONOTROPA
hypophegea, see **M. hypopitys** 297
hypopitys 297
MONOTROPACEAE 297
MONSONIA
brevirostrata CD
Montbretia 833
Giant CD
Potts' CD
MONTIA
fontana 151
subsp. **amporitana** 152
subsp. **chondrosperma** 152
subsp. **fontana** 151
subsp. *minor*, see **M. fontana** subsp. **chondrosperma** 152
subsp. **variabilis** 151
parvifolia CD
perfoliata, see **Claytonia perfoliata** 150
sibirica, see **Claytonia sibirica** 150
Moonwort 61
Moor-grass, Blue 762
Purple 800
MORACEAE 127 CD
Morning-glory, Common CD
Ivy-leaved CD
White CD
MORUS
nigra CD
Moschatel 602
Motherwort 511
Mountain-laurel CD
Mountain-pine, Dwarf CD
Mouse-ear, Alpine 162
Arctic 162
Common 163
Dwarf 164
Field 161
Grey 164
Little 165
Sea 164
Shetland 162
Starwort 161
Sticky 163
Mouse-ear-hawkweed 629
Shaggy 629
Shetland 630
Tall CD
Mousetail 111
Mousetailplant CD
Mudwort 546
Welsh 546
MUEHLENBECKIA
complexa CD
Mugwort 645
Annual CD
Chinese 645
Hoary CD
Norwegian 646
Slender CD
Mulberry, Black CD
Mullein, Broussa CD
Caucasian CD
Dark 541
Dense-flowered 540
Great 541
Hoary 541
Hungarian CD
Moth 539
Nettle-leaved CD
Mullein (*cont.*)
Orange 540
Purple CD
Twiggy 540
White 542
Mung-bean CD
MUSCARI
armeniacum 819
atlanticum ≡ **M. neglectum** 818
botryoides CD 819
comosum 819
neglectum 818
Musk 544
Musk-mallow 217
Greater CD
Mustard, Ball 271
Black 280
Chinese 280
Garlic 251
Hare's-ear 278
Hedge 251
Hoary 283
Russian CD
Tower 262
White 281
Mustard-and-Cress 274
MYCELIS
muralis 626
MYOSOTIS
alpestris 504
arvensis 505
var. *sylvestris* 505
brevifolia 503
caespitosa ≡ **M. laxa** 503
discolor 505
laxa 503
ramosissima 505
scorpioides 502
secunda 503
sicula 504
stolonifera 503
sylvatica 504
MYOSOTON
aquaticum 165
MYOSURUS
minimus 111
MYRICA
gale 129
pensylvanica CD
MYRICACEAE 129 CD
MYRIOPHYLLUM
alterniflorum 409
aquaticum 409
spicatum 409
verticillatum 408
MYRRHIS
odorata 458
MYRTACEAE CD
Myrtle, Chilean CD

Naiad, Slender 681
Holly-leaved 682
NAJADACEAE 681
NAJAS
flexilis 681
marina 682
NARCISSUS agg. 825
bulbocodium CD
cyclamineus CD
× **N. pseudonarcissus** CD
'February Gold' CD
hispanicus, see **N. pseudonarcissus** subsp. **major** 827
× **incomparabilis** (**N. poeticus** × **N. pseudonarcissus**) 826
× **N. pseudonarcissus** 826
jonquilla × **N. pseudonarcissus** (**N.** × **odorus**) CD
× **medioluteus** (**N. poeticus** × **N. tazetta**) 826
minor CD
obvallaris ≡ **N. pseudonarcissus** subsp. **obvallaris** 827
× **odorus** (**N. jonquilla** × **N. pseudonarcissus**) CD
papyraceus CD
'Paper White' CD
'Poetaz' 826
poeticus 825
'Actaea' 825
subsp. *poeticus* 825
subsp. **radiiflorus** CD 825
NARCISSUS poeticus (*cont.*)
× **N. pseudonarcissus** (**N.** × **incomparabilis**) 826
× **N. tazetta** (**N.** × **medioluteus**) 826
pseudonarcissus subsp. **major** 827
× subsp. *pseudonarcissus* 827
pseudonarcissus subsp. **obvallaris** 827
pseudonarcissus subsp. **pseudonarcissus** 826
tazetta CD
triandrus CD
Nardurus
maritimus ≡ **Vulpia unilateralis** 752
NARDUS
stricta 740
NARTHECIUM
ossifragum 808
Nasturtium
microphyllum, see **Rorippa microphylla** 257
officinale, see **Rorippa nasturtium-aquaticum** 257
× *sterile*, see **Rorippa** × **sterilis** 257
Nasturtium 451
Flame CD
Navelwort 308
NECTAROSCORDUM
siculum CD
Needle-grass, American CD
Mediterranean CD
NEOTINEA
maculata 852
NEOTTIA
nidus-avis 838
NEPETA
cataria 520
× **faassenii** (**N. nepetella** × **N. racemosa**) 520
nepetella × **N. racemosa** (**N.** × **faassenii**) 520
racemosa 520
NERTERA
granadensis CD
NESLIA
apiculata 271
paniculata 271
paniculata sens. lat. 271
Nettle, Common 127
Small 127
NICANDRA
physalodes 485
NICOTIANA
alata 489 CD
× **N. forgetiana** (**N.** × **sanderae**) CD 489
forgetiana CD
rustica CD
× **sanderae** (**N. alata** × **N. forgetiana**) CD 489
tabacum CD
NIGELLA
damascena 97
Niger 663
Nightshade, Black 487
Deadly 486
Green 487
Leafy-fruited 488
Red CD
Small CD
Tall CD
Ninebark CD
Nipplewort 618
Nit-grass 779
Eastern CD
NOTHOFAGUS
alpina, see **N. nervosa** 130
nervosa 130
× **N. obliqua** CD
obliqua 129
procera, see **N. nervosa** 130
NOTHOSCORDUM
borbonicum CD
gracile, see **N. borbonicum** CD
NUPHAR
advena CD
lutea 94
× *N. pumila* (*N.* × *spenneriana*) 94
pumila 94
× *spenneriana* (*N. lutea* × *N. pumila*) 94
NYMPHAEA
alba 93
NYMPHAEACEAE 93 CD
NYMPHOIDES
peltata 493

Oak, Algerian CD
Evergreen 131
Lucombe 131
Pedunculate 132
Red 132
Sessile 131
Turkey 130
Oat 767
Bristle 766
Slender CD
Oat-grass, Downy 765
False 766
French 767
Meadow 765
Yellow 768
Oceanspray CD
OCHAGAVIA
carnea CD
ODONTITES
jaubertianus CD
vernus 571
subsp. **litoralis** 572
subsp. *pumilus* 572
subsp. *serotinus* 571
subsp. *vernus* 571
OEMLERIA
cerasiformis CD
OENANTHE
aquatica 464
crocata 463
fistulosa 462
fluviatilis 464
lachenalii 463
pimpinelloides 462
silaifolia 462
OENOTHERA agg. 418
biennis 419
× *O. cambrica* 420
× **O. glazioviana** (**O.** × **fallax**) 419
cambrica 420
× *O. glazioviana* 419, 420
erythrosepala ≡ **O. glazioviana** 419
× **fallax** (**O. biennis** × **O. glazioviana**) 419
glazioviana 419
parviflora, see **O. cambrica** 420
stricta 420
OLEACEAE 538 CD
OLEARIA
avicenniifolia CD
× **O. moschata** (**O.** × **haastii**) CD
× **haastii** (**O. avicenniifolia** × **O. moschata**) CD
macrodonta 643
paniculata CD
traversii CD
Oleaster, Broad-leaved CD
Spreading CD
Omalotheca
norvegica, see **Gnaphalium norvegicum** 633
supina, see **Gnaphalium supinum** 634
sylvatica, see **Gnaphalium sylvaticum** 634
OMPHALODES
verna 506
ONAGRACEAE 412 CD
Onion CD
Wild 823
Oniongrass CD
ONOBRYCHIS
viciifolia 375
ONOCLEA
sensibilis CD
ONONIS
alopecuroides CD
baetica CD
mitissima CD
natrix CD
reclinata 389
repens 390
subsp. *maritima* 390
spinosa 389
var. *horrida* 389
ONOPORDUM
acanthium 614
nervosum CD
OPHIOGLOSSACEAE 60

OPHIOGLOSSUM
azoricum 60
lusitanicum 61
vulgatum 60
subsp. *ambiguum* ≡ **O. azoricum** 60
OPHRYS
apifera 856
fuciflora 856
insectifera 855
sphegodes 856
Orache, Australian CD
Babington's 141
Common 142
Early 142
Frosted 143
Garden 140
Grass-leaved 142
Long-stalked 141
Shrubby CD
Spear-leaved 141
Orange, Mexican CD
Orange-ball-tree 537
Orchid, Bee 856
Bird's-nest 838
Bog 841
Burnt 853
Coralroot 841
Dense-flowered 852
Early-purple 852
Fen 841
Fly 855
Fragrant 843
Frog 845
Ghost 838
Green-winged 853
Lady 853
Lizard 855
Loose-flowered 852
Man 854
Military 854
Monkey 854
Musk 842
Pyramidal 843
Small-white 843
ORCHIDACEAE 834
ORCHIS
laxiflora 852
mascula 852
militaris 854
morio 853
purpurea 853
simia 854
ustulata 853
Oregon-grape 114
Newmarket CD
OREOPTERIS
limbosperma 67
ORIGANUM
vulgare 523
ORNITHOGALUM
angustifolium 815
nutans 815
pyrenaicum 815
umbellatum ≡ **O. angustifolium** 815
ORNITHOPUS
compressus CD
perpusillus 378
pinnatus 378
sativus CD
OROBANCHACEAE 576 CD
OROBANCHE
alba 578
artemisiae-campestris 579
caryophyllacea 578
crenata CD
elatior 578
hederae 579
loricata, see **O. artemisiae-campestris** 579
maritima, see *O. minor* var. *maritima* 580
minor 580
var. *maritima* 580
var. *minor* 580
picridis ≡ **O. artemisiae-campestris** 579
purpurea 577
rapum-genistae 577
reticulata 579
subsp. *pallidiflora*, see *O. reticulata* subsp. *procera* 579
subsp. *procera* 579
subsp. *reticulata* 579
Orpine 310
ORTHILIA
secunda 296
ORYZOPSIS
miliacea CD
OSCULARIA
deltoides CD
Osier 240
Broad-leaved 242
Eared 245
Fine 243
Shrubby 246
Silky-leaved 244
OSMUNDA
regalis 61
OSMUNDACEAE 61
Osoberry CD
OTANTHUS
maritimus 647
OXALIDACEAE 441 CD
OXALIS
acetosella 442
articulata 442
corniculata 441
corymbosa, see **O. debilis** 442
debilis 442
decaphylla CD
dillenii CD
europaea ≡ **O. stricta** 441
exilis 441
incarnata 443
latifolia 443
megalorrhiza CD
pes-caprae CD
rosea CD
stricta 441
tetraphylla CD
valdiviensis CD
Oxeye, Autumn CD
Yellow 637
Oxlip 298
Oxtongue, Bristly 621
Hawkweed 621
OXYRIA
digyna 199
OXYTROPIS
campestris 375
halleri 374
Oxytropis, Purple 374
Yellow 375
Oysterplant 501
PACHYSANDRA
terminalis CD
PAEONIA
lactiflora CD
mascula CD 209
officinalis 209 CD
PAEONIACEAE 209 CD
Pampas-grass 799
Early CD
PANCRATIUM
maritimum CD
PANICUM
capillare 802
dichotomiflorum CD
miliaceum 802
schinzii CD
Pansy, Dwarf 232
Field 231
Garden 231
Horned 229
Mountain 230
Wild 230
PAPAVER
argemone 117
atlanticum 115
bracteatum 114
dubium 116
subsp. **dubium** 116
subsp. **lecoqii** 116
hybridum 117
lecoqii ≡ **P. dubium** subsp. **lecoqii** 116
orientale 114
pseudoorientale 114
rhoeas 115
somniferum 115
subsp. **setigerum** CD 115
subsp. *somniferum* 115
PAPAVERACEAE 114 CD
Papilionaceae, see FABACEAE 372
PARAPHOLIS
incurva 762
strigosa 762
PARENTUCELLIA
viscosa 572
PARIETARIA
diffusa ≡ **P. judaica** 128
judaica 128
officinalis CD
PARIS
quadrifolia 814
PARNASSIA
palustris 321
Parnassiaceae, see SAXIFRAGACEAE 313
Parrot's-feather 409
Parsley, Corn 470
Cow 456
Fool's 464
Garden 470
Stone 471
Parsley-piert 343
Slender 344
Parsley-pierts 343
Parsnip, Wild 475
PARTHENOCISSUS
inserta 434
quinquefolia 434
tricuspidata CD 434
PASPALUM
distichum CD
Pasqueflower 100
PASTINACA
sativa 475
PAULOWNIA
tomentosa CD
Pea, Black CD
Chick CD
Fodder CD
Garden 389
Indian CD
Marsh 386
Scurfy CD
Sea 385
Sweet CD
Tuberous 386
Peach CD
Pear, Cultivated 357
Plymouth 356
Wild 357
Pearlwort, Alpine 167
Annual 167
Heath 166
Knotted 166
Procumbent 167
Sea 168
Snow 166
Pears 357
PEDICULARIS
palustris 575
sylvatica 576
subsp. **hibernica** 576
subsp. *sylvatica* 576
PELARGONIUM
tomentosum CD
Pellitory-of-the-wall 128
Eastern CD
Peltiphyllum
peltatum, see **Darmera peltata** 313
Penny-cress, Alpine 273
Caucasian CD
Field 272
Garlic CD
Perfoliate 273
Pennyroyal 529
Pennywort, Floating CD
Hairy CD
Marsh 454
New Zealand CD
PENTAGLOTTIS
sempervirens 500
Peony CD
Garden 209
Peplis
portula ≡ **Lythrum portula** 411
Pepper, Sweet CD
Peppermint 527
Peppermint-gum, White CD
Pepper-saxifrage 465
Pepperwort, African CD
Argentine CD
Field 275
Least 275
Narrow-leaved 276
Perfoliate 276
Smith's 275
Tall CD
PERICALLIS
hybrida CD
Periwinkle, Greater 485
Intermediate CD
Lesser 484
Pernettya
mucronata 289
PERSICARIA
alpina CD
amphibia 184
amplexicaulis 183
bistorta 183
campanulata 182
hydropiper 185
lapathifolia 185
laxiflora, see **P. mitis** 185
maculosa 184
minor 186
mitis 185
mollis CD
nepalensis CD
pensylvanica CD
sagittata CD
vivipara 184
wallichii 183
weyrichii CD
Persicaria, Nepal CD
Pale 185
PETASITES
albus 660
fragrans 660
hybridus 659
japonicus 660
PETRORHAGIA
nanteuilii 180
prolifera 180
saxifraga CD
PETROSELINUM
crispum 470
segetum 470
PETUNIA
axillaris × **P. integrifolia** (**P.** × **hybrida**) 490
× **hybrida** (**P. axillaris** × **P. integrifolia**) 490
Petunia 490
PEUCEDANUM
officinale 474
ostruthium 475
palustre 475
PHACELIA
tanacetifolia 494
Phacelia 494
PHALARIS
aquatica 773
arundinacea 773
brachystachys CD
canariensis 774
minor 774
paradoxa 774
Phanerophlebia
falcatum, see **Cyrtomium falcatum** CD
PHASEOLUS
coccineus CD
vulgaris CD
Pheasant's-eye 111
PHEGOPTERIS
connectilis 67
PHILADELPHUS
coronarius 304
? × **P. microphyllus** × **P. pubescens** (**P.** × **virginalis**) 305
× **virginalis** (? **P. coronarius** × **P. microphyllus** × **P. pubescens**) 305
PHLEUM
alpinum 785
arenarium 785
bertolonii 784
phleoides 785
pratense 784
pratense *sens. lat.* 784
subsp. *bertolonii*, see **P. bertolonii** 784
subsp. *pratense*, see **P. pratense** 784
subsp. *serotinum*, see **P. bertolonii** 784
PHLOMIS
fruticosa CD
russeliana CD
PHLOX
paniculata CD
Phlox CD

PHORMIUM
colensoi, see **P. cookianum** CD
cookianum CD
tenax 834
PHOTINIA
davidiana CD
PHRAGMITES
australis 800
communis ≡ **P. australis** 800
PHUOPSIS
stylosa CD
PHYGELIUS
capensis CD
PHYLLITIS
scolopendrium 67
PHYLLODOCE
caerulea 288
PHYMATOSORUS
diversifolius CD
PHYSALIS
alkekengi 486
ixocarpa CD
peruviana CD
philadelphica CD
PHYSOCARPUS
opulifolius CD
PHYSOSPERMUM
cornubiense 466
PHYTEUMA
orbiculare 588
scheuchzeri CD
spicatum 588
tenerum ≡ **P. orbiculare** 588
PHYTOLACCA
acinosa CD
polyandra CD
PHYTOLACCACEAE CD
PICEA
abies 83
engelmannii CD
glauca CD
omorika CD
sitchensis 83
Pick-a-back-plant 320
Pickerelweed CD
PICRIS
echioides 621
hieracioides 621
Pigmyweed 308
Jersey CD
New Zealand 308
Scilly CD
Pignut 459
Great 459
Pigweed, Cape CD
Guernsey CD
Indehiscent CD
Perennial CD
Prostrate CD
Short-tepalled CD
Thunberg's CD
White 149
Pillwort 63
PILOSELLA
aurantiaca 630
subsp. *aurantiaca* 630
subsp. *carpathicola* 630
caespitosa CD
× **P. lactucella** (**P.** × **floribunda**) CD
flagellaris subsp. **bicapitata** 630
flagellaris subsp. **flagellaris** 630
× **floribunda** (**P. caespitosa** × **P. lactucella**) CD
officinarum 629
peleteriana 629
praealta CD
subsp. *arvorum* CD
subsp. **praealta** CD
subsp. *spraguei* CD
subsp. **thaumasia** CD
PILULARIA
globulifera 63
Pimpernel, Blue 303
Bog 302
Scarlet 303
Yellow 300
PIMPINELLA
major 459
saxifraga 460
PINACEAE 81 CD
Pine, Austrian 86
Bhutan CD
Corsican 86
Pine (*cont.*)
Lodgepole 86
Macedonian CD
Maritime 87
Monterey 87
Scots 86
Weymouth 87
Pineappleweed 650
PINGUICULA
alpina 581
grandiflora 581
lusitanica 580
vulgaris 581
Pink, Clove CD
Jersey CD
Pink-sorrel 442
Annual CD
Four-leaved CD
Garden 443
Large-flowered 442
Pale 443
Ten-leaved CD
Pink 181
Cheddar 181
Childing 180
Deptford 182
Maiden 181
Proliferous 180
Pinkweed CD
PINUS
contorta 86
mugo CD
nigra 86
subsp. *laricio* 86
subsp. *nigra* 86
peuce CD
pinaster 87
ponderosa CD
radiata 87
strobus 87
sylvestris 86
wallichiana CD
Pipewort 687
Piptatherum
miliaceum, see **Oryzopsis miliacea** CD
Pirri-pirri-bur 338
Bronze CD
PISUM
sativum 389
Pitcherplant CD
PITTOSPORACEAE CD
PITTOSPORUM
crassifolium CD
tenuifolium CD
PLAGIOBOTHRYS
scouleri CD
Plane, London 123
Plant, Poached-egg 451
PLANTAGINACEAE 534 CD
PLANTAGO
afra CD 536
arenaria 536 CD
coronopus 534
lanceolata 536
major 535
subsp. **intermedia** 535
subsp. *major* 535
maritima 535
media 536
Plantain, Branched 536
Buck's-horn 534
Glandular CD
Greater 535
Hoary 536
Ribwort 536
Sea 535
PLATANACEAE 123
PLATANTHERA
bifolia 842
chlorantha 842
PLATANUS
× **hispanica** (**P. occidentalis** × **P. orientalis**) 123
hybrida, see **P.** × **hispanica** 123
occidentalis × **P. orientalis** (**P.** × **hispanica**) 123
orientalis 123
PLEIOBLASTUS
chino CD
pygmaeus CD
simonii CD
Ploughman's-spikenard 636
Plum 353
Cherry 353
Wild 353
PLUMBAGINACEAE 199 CD
Plume-poppy, Hybrid CD
POA
alpigena 756
alpina 760
× *P. flexuosa* (*P.* × *jemtlandica*) 758
angustifolia 757
annua 756
bulbosa 760
chaixii 758
compressa 758
flabellata CD
flexuosa 758
glauca 759
humilis 757
infirma 755
× *jemtlandica* (*P. alpina* × *P. flexuosa*) 758
nemoralis 759
palustris 759
pratensis *sens. lat.* 756
pratensis *sens. str.* 757
subcaerulea ≡ **P. humilis** 757
trivialis 756
POACEAE 739 CD
Poached-egg Plant 451
Pokeweed, Chinese CD
Indian CD
POLEMONIACEAE 494 CD
POLEMONIUM
caeruleum 494
Polyanthus 298
POLYCARPON
tetraphyllum 171
POLYGALA
amara ≡ **P. amarella** 437
amarella 437
calcarea 437
serpyllifolia 436
vulgaris 436
POLYGALACEAE 436
POLYGONACEAE 182 CD
POLYGONATUM
× **hybridum** (**P. multiflorum** × **P. odoratum**) 813
multiflorum 813
× **P. odoratum** (**P.** × **hybridum**) 813
odoratum 813
verticillatum 814
POLYGONUM
amphibium ≡ **Persicaria amphibia** 184
amplexicaule, see **Persicaria amplexicaulis** 183
arenarium CD
arenastrum 188
aviculare 188
aviculare agg. 187
baldschuanicum, see **Fallopia baldschuanica** 190
bistorta ≡ **Persicaria bistorta** 183
boreale 188
campanulatum, see **Persicaria campanulata** 182
convolvulus ≡ **Fallopia convolvulus** 190
cuspidatum ≡ **Fallopia japonica** 189
dumetorum ≡ **Fallopia dumetorum** 191
hydropiper ≡ **Persicaria hydropiper** 185
lapathifolium ≡ **Persicaria lapathifolia** 185
maritimum 187
minus ≡ **Persicaria minor** 186
mite ≡ **Persicaria mitis** 185
neglectum 187
nodosum ≡ **Persicaria lapathifolia** 185
oxyspermum 187
subsp. *oxyspermum* 187
subsp. *raii* 187
patulum CD
persicaria ≡ **Persicaria maculosa** 184
polystachyum ≡ **Persicaria wallichii** 183
raii ≡ **P. oxyspermum** 187
rurivagum 189
sachalinense ≡ **Fallopia sachalinensis** 190
viviparum ≡ **Persicaria vivipara** 184
POLYPODIACEAE 64 CD
Polypodies 64
POLYPODIUM
australe, see **P. cambricum** 66
cambricum 66
interjectum 65
× **P. vulgare** (**P.** × **mantoniae**) 65
POLYPODIUM (*cont.*)
× **mantoniae** (**P. interjectum** × **P. vulgare**) 65
vulgare 65
vulgare *sens. lat.* 64
subsp. *prionodes* ≡ **P. interjectum** 65
subsp. *serrulatum* ≡ **P. cambricum** 66
subsp. *vulgare* ≡ **P. vulgare** 65
Polypody 65
Intermediate 65
Southern 66
POLYPOGON
monspeliensis 781
viridis 781
POLYSTICHUM
aculeatum 76
× **P. setiferum** (**P.** × **bicknellii**) 76
× **bicknellii** (**P. aculeatum** × **P. setiferum**) 76
lonchitis 76
munitum CD
setiferum 75
Pond-sedge, Greater 722
Lesser 722
Pondweed, American 676
Blunt-leaved 678
Bog 673
Bright-leaved 675
Broad-leaved 672
Curled 679
Fen 673
Fennel 680
Flat-stalked 677
Grass-wrack 679
Hairlike 678
Horned 682
Lesser 677
Loddon 673
Long-leaved 675
Long-stalked 676
Opposite-leaved 680
Perfoliate 676
Red 675
Sharp-leaved 679
Shetland 677
Shining 674
Slender-leaved 680
Small 678
Various-leaved 674
Willow-leaved 674
PONTEDERIA
cordata CD
PONTEDERIACEAE CD
Poplar, Berlin CD
Generous CD
Grey 234
White 233
Poppy, Atlas 115
Californian 118
Common 115
Long-headed 116
Mexican CD
Opium 115
Oriental 114
Prickly 117
Rough 117
Welsh 117
POPULUS
alba 233
× **P. tremula** (**P.** × **canescens**) 234
balsamifera CD 235
× **P. deltoides** (**P.** × **jackii**) 235
× **P. trichocarpa** (**P. 'Balsam Spire'**) CD
'Balsam Spire' (**P. balsamifera** × **P. trichocarpa**) CD
× **berolinensis** (**P. laurifolia** × **P. nigra 'Italica'**) CD
× **canadensis** (**P. deltoides** × **P. nigra**) 235
× *candicans* 235
× **canescens** (**P. alba** × **P. tremula**) 234
deltoides × **P. nigra** (**P.** × **canadensis**) 235
deltoides × **P. trichocarpa** (**P.** × **generosa**) CD
× **generosa** (**P. deltoides** × **P. trichocarpa**) CD
'Beaupré' CD
gileadensis, see **P.** × **jackii** 235
× **jackii** (**P. balsamifera** × **P. deltoides**) 235
laurifolia × **P. nigra 'Italica'** (**P.** × **berolinensis**) CD
nigra (fastigiate cultivars) 235
'Gigantea' 235
'Italica' 235
'Plantierensis' 235
subsp. **betulifolia** 234

POPULUS (*cont.*)
nigra sens. lat. 234
tremula 234
trichocarpa 236
'Trichobel' 236
PORTULACA
oleracea 150
subsp. *nitida* 150
subsp. *oleracea* 150
subsp. *sativa* 150
PORTULACACEAE 150 CD
POTAMOGETON
acutifolius 679
alpinus 675
berchtoldii 678
coloratus 673
compressus 679
crispus 679
epihydrus 676
filiformis 680
friesii 677
gramineus 674
× **P. lucens** (**P.** × **zizii**) 675
× **P. perfoliatus** (**P.** × **nitens**) 675
lucens 674
× **P. perfoliatus** (**P.** × **salicifolius**) 674
natans 672
× **nitens** (**P. gramineus** × **P. perfoliatus**) 675
nodosus 673
obtusifolius 678
pectinatus 680
perfoliatus 676
polygonifolius 673
praelongus 676
pusillus 677
rutilus 677
× **salicifolius** (**P. lucens** × **P. perfoliatus**) 674
trichoides 678
× **zizii** (**P. gramineus** × **P. lucens**) 675
POTAMOGETONACEAE 672
Potato 488
Purple CD
Potatoes of Canada 664
POTENTILLA
anglica 331
× **P. erecta** (**P.** × **suberecta**) 332
× *P. reptans* (*P.* × *mixta* sens. str.) 332
× **P. reptans** & **P. erecta** × **P. reptans** (**P.** × **mixta**) 332
anserina 328
argentea 329
crantzii 330
erecta 331
subsp. *erecta* 331
subsp. **strictissima** 331
× *P. reptans* (*P.* × *italica*) 332
× **P. reptans** & **P. anglica** × **P. reptans** (**P.** × **mixta**) 332
fruticosa 327
inclinata CD
intermedia 329
× *italica* (*P. erecta* × *P. reptans*) 332
× **mixta** (**P. anglica** × **P. reptans** & **P. erecta** × **P. reptans**) 332
× *mixta* sens. str. (*P. anglica* × *P. reptans*) 332
neumanniana 330
norvegica 330
palustris 328
recta 329
reptans 332
rivalis CD
rupestris 328
sterilis 333
× **suberecta** (**P. anglica** × **P. erecta**) 332
tabernaemontani ≡ **P. neumanniana** 330
POTERIUM
polygamum ≡ **Sanguisorba minor** subsp. **muricata** 337
sanguisorba ≡ **Sanguisorba minor** subsp. **minor** 337
PRATIA
angulata CD
Pride-of-India CD
Primrose-peerless 826
Primrose 297
Bird's-eye 299
Scottish 299
PRIMULA
auricula CD
× *P. hirsuta* (*P.* × *pubescens*) CD
elatior 298
farinosa 299
florindae CD
japonica CD
× **polyantha** (**P. veris** × **P. vulgaris**) 298
× *pubescens* (*P. auricula* × *P. hirsuta*) CD
scotica 299
sikkimensis CD
× *tommasinii*, see **P.** × **polyantha** 298
veris 298
× **P. vulgaris** (**P.** × **polyantha**) 298
vulgaris 297
PRIMULACEAE 297 CD
Prince's-feather CD
Privet, Garden 539
Wild 539
PRUNELLA
laciniata 521
× *P. vulgaris* 521
vulgaris 521
PRUNUS
avium 354
cerasifera 353
cerasus 354
davidiana CD
domestica 353
subsp. *domestica* 353
subsp. *insititia* 353
subsp. *italica* 353
× **P. spinosa** (**P.** × **fruticans**) 354
dulcis CD
× **fruticans** (**P. domestica** × **P. spinosa**) 354
incisa CD
× *italica*, see *P. domestica* subsp. *italica* 353
laurocerasus 356
lusitanica 355
mahaleb CD
padus 355
pensylvanica CD
persica CD
serotina 355
serrulata CD
'Kanzan' CD
spinosa 353
PSEUDOFUMARIA
alba CD
lutea 119
Pseudognaphalium
luteoalbum, see **Gnaphalium luteoalbum** 635
undulatum, see **Gnaphalium undulatum** CD
PSEUDORCHIS
albida 843
PSEUDOSASA
japonica 740
PSEUDOTSUGA
menziesii 82
PSORALEA
americana CD
PTERIDACEAE CD
PTERIDIUM
aquilinum 66
PTERIS
cretica CD
PTEROCARYA
fraxinifolia CD
× *P. stenocarpa* (*P.* × *rehderiana*) CD
× *rehderiana* (*P. fraxinifolia* × *P. stenocarpa*) CD
PUCCINELLIA
capillaris 753
distans 753
subsp. **borealis** 753
subsp. *distans* 753
fasciculata 754
var. *pseudodistans* 754
maritima 753
pseudodistans 754
rupestris 754
PULICARIA
dysenterica 636
vulgaris 637
PULMONARIA
longifolia 497
'**Mawson's Blue**' CD
obscura 496
officinalis 496
rubra CD
PULSATILLA
vulgaris 100
Pumpkin CD
Purple-loosestrife 410
Purslane, Common 150
Pink 150
PYRACANTHA
coccinea 370
rogersiana CD
Pyrenean-violet CD
PYROLA
media 295
minor 295
rotundifolia subsp. **maritima** 296
rotundifolia subsp. **rotundifolia** 295
PYROLACEAE 295
PYRUS
communis 357
communis *sens. lat.* 357
cordata 356
pyraster 357

Quaking-grass 754
Greater 755
Lesser 755
Queensland-hemp CD
QUERCUS
borealis, see **Q. rubra** 132
canariensis CD
cerris 130
× **Q. suber** (**Q.** × **crenata**) 131
× **crenata** (**Q. cerris** × **Q. suber**) 131
× *hispanica*, see **Q.** × **crenata** 131
ilex 131
petraea 131
× **Q. robur** (**Q.** × **rosacea**) 132
× *pseudosuber*, see **Q.** × **crenata** 131
robur 132
× **rosacea** (**Q. petraea** × **Q. robur**) 132
rubra 132
Quillwort 55
Land 56
Spring 56
Quince CD
Chinese 356
Japanese CD

RADIOLA
linoides 436
Radish, Garden 285
Mediterranean CD
Sea 285
Wild 284
Ragged-Robin 174
Ragweed 661
Giant 662
Perennial CD
Ragwort, Broad-leaved 652
Chamois CD
Chinese CD
Common 653
Fen 653
Golden CD
Hedge CD
Hoary 654
Magellan 653
Marsh 654
Monro's CD
Narrow-leaved CD
Oxford 655
Purple CD
Shoddy CD
Shrub 658
Silver 652
Woad-leaved CD
Wood CD
RAMONDA
myconi CD
Ramping-fumitory, Common 121
Martin's 121
Purple 122
Tall 121
Western 120
White 120
Rampion, Oxford CD
Round-headed 588
Spiked 588
Ramsons 821
Rannoch-rush 671
RANUNCULACEAE 95 CD
RANUNCULUS
aconitifolius CD
flore pleno CD
acris 101
aquatilis 108
arvensis 102
auricomus 103
baudotii 108
bulbosus 101
circinatus 110
ficaria 106
subsp. *bulbifer* ≡ **R. ficaria** subsp. **bulbilifer** 106
subsp. **bulbilifer** 106
subsp. **chrysocephalus** CD
subsp. **ficaria** 106
subsp. **ficariiformis** CD
flammula 104
subsp. *flammula* 104
subsp. **minimus** 104
subsp. **scoticus** 105
× *R. reptans* (*R.* × *levenensis*) 105
fluitans 110
hederaceus 107
lenormandii ≡ **R. omiophyllus** 107
× *levenensis* (*R. flammula* × *R. reptans*) 105
lingua 104
marginatus CD
muricatus CD
× *novae-forestae*, see *R. omiophyllus* × *R. tripartitus* 107
omiophyllus 107
× *R. tripartitus* 107
ophioglossifolius 105
paludosus 103
parviflorus 102
peltatus 109
penicillatus 109
subsp. **penicillatus** 109
subsp. **pseudofluitans** 110
var. *calcareus*, see **R. penicillatus** subsp. **pseudofluitans** 110
var. *vertumnus*, see **R. penicillatus** subsp. **pseudofluitans** 110
repens 101
reptans 105
sardous 102
sceleratus 103
trichophyllus 108
tripartitus 107
Rape 279
Ethiopian CD
Long-stalked CD
Oil-seed 279
RAPHANUS
maritimus ≡ **R. raphanistrum** subsp. **maritimus** 285
raphanistrum ≡ **R. raphanistrum** subsp. **raphanistrum** 284
subsp. **landra** CD
subsp. **maritimus** 283
subsp. **raphanistrum** 284
sativus 285
RAPISTRUM
orientale ≡ **R. rugosum** 284
perenne CD
rugosum 284
Raspberry 326
Purple-flowered CD
Raspwort, Creeping CD
Rauli 130
Red-cedar, Japanese 88
Western 90
Red-hot-poker 809
Greater CD
Red-knotgrass CD
Lesser CD
Redshank 184
Redwood, Coastal 88
Reed, Common 800
REINECKEA
carnea CD
Reineckea CD
RESEDA
alba 286
lutea 286
luteola 285
odorata CD
phyteuma CD
RESEDACEAE 285 CD

Restharrow, Andalucian CD
Common 390
Mediterranean CD
Salzmann's CD
Small 389
Spiny 389
Yellow CD
Reynoutria
japonica, see **Fallopia japonica** 189
sachalinensis, see **Fallopia sachalinensis** 190
RHAMNACEAE 433 CD
RHAMNUS
alaternus CD
cathartica 433
RHEUM
× **hybridum** 191
palmatum CD
RHINANTHUS
angustifolius 573
groenlandicus 575
minor 573
subsp. **borealis** 575
subsp. *borealis* × subsp. *monticola* 575
subsp. *borealis* × subsp. *stenophyllus* 575
subsp. **calcareus** 574
subsp. **lintonii** 575
subsp. **minor** 573
subsp. **monticola** 574
subsp. **stenophyllus** 574
Rhodes-grass, Australian CD
Feathery CD
Rhodiola
rosea, see **Sedum rosea** 309
RHODODENDRON
luteum 287
ponticum 287
Rhododendron 287
Rhodostachys CD
Tresco CD
Rhubarb 191
Ornamental CD
RHUS
hirta, see **R. typhina** 440
typhina 440
Rhynchosinapis
cheiranthos ≡ **Coincya monensis** subsp. **cheiranthos** 282
monensis ≡ **Coincya monensis** subsp. **monensis** 282
wrightii ≡ **Coincya wrightii** 283
RHYNCHOSPORA
alba 711
fusca 712
RIBES
alpinum 307
nigrum 306
odoratum CD
rubrum 305
sanguineum 306
spicatum 306
sylvestre ≡ **R. rubrum** 305
uva-crispa 307
RICINUS
communis CD
RIDOLFIA
segetum CD
Robin's-plantain CD
ROBINIA
pseudoacacia 372
Roblé 129
Rock-cress, Alpine 263
Bristol 264
Hairy 263
Northern 262
Sand CD
Rocket, Eastern 250
French CD
Garden 281
Hairy 282
Perennial CD
Sea 283
Tall 250
Rock-rose, Common 223
Hoary 224
Spotted 223
White 224
RODGERSIA
podophylla CD
Rodgersia CD
ROMULEA
columnae 831
rosea CD
RORIPPA
amphibia 259
× **R. sylvestris** (**R.** × **anceps**) 259
× **anceps** (**R. amphibia** × **R. sylvestris**) 259
austriaca CD
islandica 258
islandica auct., see *R. palustris* 258
microphylla 257
× **R. nasturtium-aquaticum** (**R.** × **sterilis**) 257
nasturtium-aquaticum 257
nasturtium-aquaticum agg. 256
palustris 258
× **sterilis** (**R. microphylla** × **R. nasturtium-aquaticum**) 257
sylvestris 258
ROSA
afzeliana, see **R. caesia** subsp. **glauca** 349
agrestis 352
× **alba** (**R. arvensis** and/or **R. canina** × **R. gallica**) CD
× **andegavensis** (**R. canina** × **R. stylosa**) 348
arvensis 344
× **R. canina** (**R.** × **verticillacantha**) 345
× **R. rugosa** (**R.** × **paulii**) CD
arvensis and/or **R. canina** × **R. gallica** (**R.** × **alba**) CD
caesia subsp. **caesia** 349
caesia subsp. **glauca** 349
caesia subsp. *vosagiaca*, see **R. caesia** subsp. **glauca** 349
caesia × **R. canina** (**R.** × **dumalis**) 350
canina 347
× **R. obtusifolia** (**R.** × **dumetorum**) 347
× **R. rubiginosa** (**R.** × **nitidula**) 348
× **R. sherardii** (**R.** × **rothschildii**) 348
× **R. stylosa** (**R.** × **andegavensis**) 348
× **R. tomentosa** (**R.** × **scabriuscula**) 349
canina agg. 347
coriifolia, see **R. caesia** subsp. **caesia** 349
× *curvispina*, see **R.** × **scabriuscula** 349
× **dumalis** (**R. caesia** × **R. canina**) 350
dumalis auct. ≡ **R. caesia** 349
× **dumetorum** (**R. canina** × **R. obtusifolia**) 347
dumetorum auct., see **R. canina** 347
elliptica, see **R. agrestis** 352
ferruginea 346
gallica CD
glauca, see **R. ferruginea** 346
'Hollandica' 346
× **involuta** (**R. pimpinellifolia** × **R. sherardii**) 345
× *irregularis*, see **R.** × **verticillacantha** 345
× *latens*, see **R.** × **nitidula** 348
luciae CD
micrantha 352
mollis 351
× **R. pimpinellifolia** (**R.** × **sabinii**) 351
multiflora 344
× **nitidula** (**R. canina** × **R. rubiginosa**) 348
obtusifolia 350
× **paulii** (**R. arvensis** × **R. rugosa**) CD
pimpinellifolia 345
× **R. sherardii** (**R.** × **involuta**) 345
× **rothschildii** (**R. canina** × **R. sherardii**) 348
rubiginosa 352
rugosa 346
× **sabinii** (**R. mollis** × **R. pimpinellifolia**) 351
× **scabriuscula** (**R. canina** × **R. tomentosa**) 349
setigera CD
sherardii 351
stylosa 347
thibetanus. CD
tomentosa 350
× **verticillacantha** (**R. arvensis** × **R. canina**) 345
villosa ≡ **R. mollis** 351
wheldonii, see **R.** × **verticillacantha** 345
ROSACEAE 321 CD
Rose, Burnet 345
Dutch 346
Japanese 346
Many-flowered 344
Memorial CD
Prairie CD
Red, (of Lancaster) CD
Red-leaved 346
White, (of York) CD
Rosemary 530
Rose-of-heaven CD
Rose-of-Sharon 210
Roseroot 309
ROSMARINUS
officinalis 530
'Alba' 530
ROSTRARIA
cristata CD
Rowan 358
RUBIA
peregrina 596
RUBIACEAE 590 CD
RUBUS
arcticus 326
caesius 327
× *R. fruticosus* agg. 327
chamaemorus 325
cockburnianus CD
× **fraseri** (**R. odoratus** × **R. parviflorus**) CD
fruticosus agg. 327
series *Corylifolii* 327
idaeus 326
× *R. vitifolius* CD
loganobaccus CD
odoratus CD
× **R. parviflorus** (**R.** × **fraseri**) CD
parviflorus CD
phoenicolasius CD
saxatilis 325
spectabilis 326
tricolor 325
subgenus *Rubus* 327
RUDBECKIA
hirta CD
laciniata CD
RUMEX
acetosa 192
subsp. *acetosa* 192
subsp. **ambiguus** CD 192
subsp. *biformis* 192
subsp. **hibernicus** 192
acetosella 191
subsp. *acetosella* 191
subsp. *pyrenaicus* 191
alpinus ≡ **R. pseudoalpinus** 192
angiocarpus, see *R. acetosella* subsp. *pyrenaicus* 191
aquaticus 193
× *R. obtusifolius* 193
brownii CD
confertus CD
conglomeratus 196
crispus 194
subsp. *crispus* 194
subsp. **littoreus** 195
subsp. **uliginosus** 195
× **R. longifolius** (**R.** × **propinquus**) 195
× **R. obtusifolius** (**R.** × **pratensis**) 196
cristatus 194
dentatus CD
× **dufftii** (**R. obtusifolius** × **R. sanguineus**) 198
frutescens CD
hibernicus 192
× **hybridus** (**R. longifolius** × **R. obtusifolius**) 193
hydrolapathum 194
longifolius 193
× **R. obtusifolius** (**R.** × **hybridus**) 193
maritimus 198
obovatus CD
obtusifolius 198
× **R. sanguineus** (**R.** × **dufftii**) 198
palustris 198
patientia CD
× **pratensis** (**R. crispus** × **R. obtusifolius**) 196
× **propinquus** (**R. crispus** × **R. longifolius**) 195
pseudoalpinus 192
pulcher 198
rupestris 197
salicifolius CD
sanguineus 196
scutatus CD
tenuifolius ≡ **R. acetosella** 191
triangulivalvis, see **R. salicifolius** CD
RUPPIA
cirrhosa 681
maritima 681
spiralis ≡ **R. cirrhosa** 681
RUPPIACEAE 681
Rupturewort, Fringed 171
Hairy CD
Smooth 170
RUSCHIA
caroli CD
RUSCUS
aculeatus 828
hypoglossum CD
Rush, Alpine 691
Baltic 695
Blunt-flowered 691
Broad-leaved CD
Bulbous 693
Chestnut 694
Compact 697
Dwarf 691
Frog 690
Hard 696
Heath 688
Jointed 692
Leafy 690
Pigmy 693
Round-fruited 688
Saltmarsh 689
Sea 694
Sharp 695
Sharp-flowered 692
Slender 688
Somerset 695
Thread 696
Three-flowered 694
Three-leaved 689
Toad 690
Two-flowered 693
Rushes, Toad 689
Russian-vine 190
Rustyback 71
RUTACEAE CD
Rye 798
Rye-grass, Flaxfield CD
Italian 749
Mediterranean CD
Perennial 749
RYTIDOSPERMA
racemosum CD

Safflower 617
Downy CD
Saffron, Meadow 809
Sage, Jerusalem CD
Turkish CD
Wood 517
SAGINA
apetala 167
subsp. **apetala** 168
subsp. **erecta** 168
apetala auct. ≡ *S. apetala* subsp. *erecta* 168
ciliata ≡ **S. apetala** subsp. **apetala** 168
intermedia, see **S. nivalis** 166
maritima 168
var. *alpina* 168
nivalis 166
nodosa 166
procumbens 167
saginoides 167
subulata 166
SAGITTARIA
latifolia CD
rigida CD
sagittifolia 667
subulata CD
Sainfoin 375
SALICACEAE 233 CD
SALICORNIA agg. 145
dolichostachya 147
europaea 146
fragilis 147
lutescens, see **S. fragilis** 147
nitens 146
obscura 146
perennis ≡ **Sarcocornia perennis** 144
pusilla 145
ramosissima 145
SALIX
acutifolia CD
alba 237
var. *caerulea* 237
× **S. babylonica** (**S.** × **sepulcralis**) 237
× **S. fragilis** (**S.** × **rubens**) 238
× **S. pentandra** (**S.** × **ehrhartiana**) CD

SALIX (*cont.*)
× **ambigua** (**S. aurita** × **S. repens**) 245
arbuscula 248
arenaria, see *S. repens* var. *argentea* 247
atrocinerea, see **S. cinerea** subsp. **oleifolia** 243
aurita 244
× **S. caprea** (**S.** × **capreola**) 244
× **S. caprea** × **S. viminalis** (**S.** × **stipularis**) 245
× **S. cinerea** (**S.** × **multinervis**) 245
× **S. repens** (**S.** × **ambigua**) 245
× **S. viminalis** (**S.** × **fruticosa**) 246
babylonica 237
× **S. fragilis** (**S.** × **pendulina**) CD
× **calodendron** (**S. caprea** × **S. cinerea** × **S. viminalis**) 241
caprea 240
subsp. *caprea* 240
subsp. **sphacelata** 241
var. *sphacelata*, see **S. caprea** subsp. **sphacelata** 241
× **S. cinerea** (**S.** × **reichardtii**) 241
× **S. cinerea** × **S. viminalis** (**S.** × **calodendron**) 241
× **S. viminalis** (**S.** × **sericans**) 242
× **capreola** (**S. aurita** × **S. caprea**) 244
cinerea 242
subsp. **cinerea** 242
subsp. **oleifolia** 243
× **S. phylicifolia** (**S.** × **laurina**) 243
× **S. purpurea** × **S. viminalis** (**S.** × **forbyana**) 243
× **S. viminalis** (**S.** × **smithiana**) 244
daphnoides 239
× **ehrhartiana** (**S. alba** × **S. pentandra**) CD
elaeagnos 240
eriocephala CD
× **forbyana** (**S. cinerea** × **S. purpurea** × **S. viminalis**) 243
fragilis 236
× **S. pentandra** (**S.** × **meyeriana**) 237
× **fruticosa** (**S. aurita** × **S. viminalis**) 246
herbacea 249
lanata 248
lapponum 247
× **laurina** (**S. cinerea** × **S. phylicifolia**) 243
× **meyeriana** (**S. fragilis** × **S. pentandra**) 237
× **mollissima** (**S. triandra** × **S. viminalis**) 238
× **multinervis** (**S. aurita** × **S. cinerea**) 245
myrsinifolia 246
× **S. phylicifolia** (**S.** × **tetrapla**) 246
myrsinites 248
nigricans ≡ **S. myrsinifolia** 246
× **pendulina** (**S. babylonica** × **S. fragilis**) CD
pentandra 236
phylicifolia 247
purpurea 239
× **S. viminalis** (**S.** × **rubra**) 239
× **reichardtii** (**S. caprea** × **S. cinerea**) 241
repens 247
subsp. *argentea*, see *S. repens* var. *argentea* 247
subsp. *repens*, see *S. repens* var. *repens* 247
var. *argentea* 247
var. *fusca* 247
var. *repens* 247
reticulata 249
× **rubens** (**S. alba** × **S. fragilis**) 238
× **rubra** (**S. purpurea** × **S. viminalis**) 239
× **sepulcralis** (**S. alba** × **S. babylonica**) 237
× **sericans** (**S. caprea** × **S. viminalis**) 242
× **smithiana** (**S. cinerea** × **S. viminalis**) 244
× **stipularis** (**S. aurita** × **S. caprea** × **S. viminalis**) 245
× **tetrapla** (**S. myrsinifolia** × **S. phylicifolia**) 246
triandra 238
× **S. viminalis** (**S.** × **mollissima**) 238
udensis CD
'Sekka' CD
viminalis 240
Sally-my-handsome CD
Salmonberry 326
SALPICHROA
origanifolia CD
Salsify 622
Slender CD
SALSOLA
kali subsp. *iberica* ≡ **S. kali** subsp. **ruthenica** 148
kali subsp. **kali** 148
kali subsp. **ruthenica** 148
Saltmarsh-grass, Borrer's 754
Common 753
Reflexed 753
Stiff 754
Saltwort, Prickly 148
Spineless 148
SALVIA
glutinosa CD
horminoides ≡ **S. verbenaca** 530
pratensis 530
reflexa CD
sclarea CD
verbenaca 530
verticillata 531
viridis 531
SAMBUCUS
canadensis CD
ebulus 597
nigra 597
racemosa 596
SAMOLUS
valerandi 304
Samphire, Rock 461
Sandbur, Spiny CD
Sand-grass, Early 780
Sandwort, Arctic 153
English 154
Fine-leaved 157
Fringed 154
Mossy 154
Mountain 156
Recurved 155
Sea 155
Spring 156
Teesdale 156
Three-nerved 155
Thyme-leaved 152
SANGUISORBA
canadensis CD
minor subsp. **minor** 337
minor subsp. **muricata** 337
officinalis 337
Sanicle 455
SANICULA
europaea 455
SANTALACEAE 424
SANTOLINA
chamaecyparissus 646
SAPINDACEAE CD
SAPONARIA
ocymoides CD
officinalis 179
SARCOCORNIA
perennis 144
Sarothamnus
scoparius ≡ **Cytisus scoparius** 404
subsp. *maritimus* ≡ **Cytisus scoparius** subsp. **maritimus** 404
SARRACENIA
purpurea CD
SARRACENIACEAE CD
SASA
palmata 739
veitchii CD
SASAELLA
ramosa CD
SATUREJA
montana CD
SAUSSUREA
alpina 609
Savory, Winter CD
Saw-wort 615
Alpine 609
SAXIFRAGA
aizoides 317
cernua 317
cespitosa 319
cuneifolia CD
cymbalaria 314
var. *cymbalaria* 314
var. *huetiana* 314
× **geum** (**S. hirsuta** × **S. umbrosa**) CD
granulata 318
hartii ≡ **S. rosacea** subsp. **hartii** 319
hirculus 314
hirsuta 316
× **S. spathularis** (**S.** × **polita**) 316
× **S. umbrosa** (**S.** × **geum**) CD
SAXIFRAGA (*cont.*)
hypnoides 318
nivalis 314
oppositifolia 316
paniculata CD
× **polita** (**S. hirsuta** × **S. spathularis**) 316
rivularis 317
rosacea subsp. **hartii** 319
rosacea subsp. **rosacea** 318
rotundifolia CD
spathularis 315
× **S. umbrosa** (**S.** × **urbium**) 315
stellaris 315
stolonifera CD
'Tricolor' CD
tridactylites 319
umbrosa CD
× **urbium** (**S. spathularis** × **S. umbrosa**) 315
SAXIFRAGACEAE 313 CD
Saxifrage, Alpine 314
Celandine 314
Drooping 317
Highland 317
Irish 318
Kidney 316
Livelong CD
Marsh 314
Meadow 318
Mossy 318
Purple 316
Pyrenean CD
Round-leaved CD
Rue-leaved 320
Starry 315
Strawberry CD
Tufted 319
Yellow 317
SCABIOSA
atropurpurea CD
columbaria 607
Scabious, Devil's-bit 607
Field 606
Giant 606
Small 607
Sweet CD
SCANDIX
pecten-veneris 457
SCHEUCHZERIA
palustris 671
SCHEUCHZERIACEAE 671
SCHISMUS
barbatus CD
SCHKUHRIA
pinnata CD
SCHOENOPLECTUS
lacustris 707
subsp. *lacustris*, see **S. lacustris** 707
subsp. *tabernaemontani*, see **S. tabernaemontani** 707
pungens 708
tabernaemontani 707
× *S. triqueter* 707
triqueter 707
SCHOENUS
ferrugineus 711
nigricans 711
SCILLA
autumnalis 816
bifolia CD
bithynica CD
liliohyacinthus CD
messeniaca CD
peruviana CD
siberica 816
verna 816
SCIRPOIDES
holoschoenus 706
SCIRPUS
americanus ≡ **Schoenoplectus pungens** 708
cernuus ≡ **Isolepis cernua** 708
cespitosus ≡ **Trichophorum cespitosum** 702
subsp. *cespitosus*, see **Trichophorum cespitosum** subsp. **cespitosum** 703
subsp. *germanicus*, see *Trichophorum cespitosum* subsp. *germanicum* 702
fluitans ≡ **Eleogiton fluitans** 709
holoschoenus ≡ **Scirpoides holoschoenus** 706
hudsonianus ≡ **Trichophorum alpinum** 702
SCIRPUS (*cont.*)
lacustris ≡ **Schoenoplectus lacustris** 707
subsp. *lacustris*, see **Schoenoplectus lacustris** 707
subsp. *tabernaemontani* ≡ **Schoenoplectus tabernaemontani** 707
maritimus ≡ **Bolboschoenus maritimus** 706
setaceus ≡ **Isolepis setacea** 708
sylvaticus 706
tabernaemontani ≡ **Schoenoplectus tabernaemontani** 707
triquetrus ≡ **Schoenoplectus triqueter** 707
SCLERANTHUS
annuus 169
subsp. *annuus* 169
subsp. **polycarpos** 170
perennis subsp. **perennis** 169
perennis subsp. **prostratus** 169
polycarpos 170
SCOLYMUS
hispanicus CD
Scorpion-vetch, Annual CD
Shrubby CD
SCORPIURUS
muricatus CD
SCORZONERA
humilis 621
SCROPHULARIA
aquatica ≡ **S. auriculata** 542
auriculata 542
nodosa 542
scorodonia 543
umbrosa 543
vernalis 543
SCROPHULARIACEAE 539 CD
Scurvygrass, Common 269
Danish 270
English 269
Mountain 270
Pyrenean 269
Tall CD
SCUTELLARIA
altissima CD
galericulata 517
hastifolia CD
minor 517
Sea-blite, Annual 148
Shrubby 147
Sea-buckthorn 408
Sea-fig, Angular CD
Lesser CD
Sea-heath 232
Sea-holly 455
Sea-kale 284
Greater CD
Sea-lavender, Alderney 200
Broad-leaved 200
Common 199
Florist's CD
Lax-flowered 199
Matted 200
Rottingdean CD
Sea-lavenders, Rock 201
Sea-milkwort 304
Sea-purslane 143
Pedunculate 143
Sea-spurrey, Greater 172
Greek 173
Lesser 173
Rock 172
SECALE
cereale 798
× **TRITICUM** (× **TRITICOSECALE**) CD
SECURIGERA
varia 379
Sedge, Bird's-foot 731
Bottle 723
Bristle 738
Brown 717
Carnation 726
Club 735
Common 737
Curved 718
Cyperus 722
Davall's 720
Dioecious 719
Distant 727
Divided 718
Dotted 728
Downy-fruited 732
Dwarf 731
Elongated 720
Estuarine 736

Sedge (*cont.*)
False 712
Few-flowered 738
Fingered 731
Flea 739
Glaucous 725
Green-ribbed 727
Hair 725
Hairy 721
Hare's-foot 720
Long-bracted 728
Oval 719
Pale 730
Pendulous 724
Pill 733
Remote 718
Rock 739
Russet 724
Sand 717
Sheathed 726
Silver-spiked CD
Slender 721
Smooth-stalked 727
Soft-leaved 733
Spiked 715
Star 719
Stiff 738
String 717
Tawny 728
Three-nerved 737
Water 736
White 721
SEDUM
acre 311
album 312
anacampseros CD
anglicum 312
confusum CD
dasyphyllum 312
elegans, see **S. forsterianum** 311
forsterianum 311
hispanicum CD
kamtschaticum CD
lydium CD
nicaeense CD
praealtum CD
reflexum ≡ **S. rupestre** 310
rosea 309
rupestre 310
sexangulare 311
spathulifolium CD
'Cape Blanco' CD
'Purpureum' CD
spectabile 309
spurium 310
stoloniferum CD
telephium 310
villosum 313
SELAGINELLA
kraussiana 55
selaginoides 55
SELAGINELLACEAE 55
Selfheal 521
Cut-leaved 521
SELINUM
carvifolia 473
SEMIARUNDINARIA
fastuosa CD
SEMPERVIVUM
arachnoideum CD
tectorum 309
SENECIO
× **albescens** (**S. cineraria** × **S. jacobaea**) 652
aquaticus 654
× **S. jacobaea** (**S.** × **ostenfeldii**) 654
× **baxteri** (**S. squalidus** × **S. vulgaris**) 655
bicolor, see **S. cineraria** 652
cambrensis 656
cineraria 652
× **S. jacobaea** (**S.** × **albescens**) 652
congestus, see **Tephroseris palustris** 658
cruentus, see **Pericallis hybrida** CD
doria CD
doronicum CD
erucifolius 654
fluviatilis 652
glastifolius CD
grandiflorus CD
greyi 658
inaequidens CD
SENECIO (*cont.*)
integrifolius ≡ **Tephroseris integrifolia** 657
subsp. *maritimus*, see **Tephroseris integrifolia** subsp. **maritima** 657
jacobaea 653
× *londinensis*, see **S.** × **subnebrodensis** 655
mikanioides, see **Delairea odorata** CD
× **ostenfeldii** (**S. aquaticus** × **S. jacobaea**) 654
ovatus CD
paludosus 653
palustris ≡ **Tephroseris palustris** 658
pterophorus CD
smithii 653
squalidus 655
× **S. viscosus** (**S.** × **subnebrodensis**) 655
× **S. vulgaris** (**S.** × **baxteri**) 655
× **subnebrodensis** (**S. squalidus** × **S. viscosus**) 655
sylvaticus 656
tangutica, see **Sinacalia tangutica** CD
vernalis CD
viscosus 657
vulgaris 656
subsp. *denticulatus* ≡ *S. vulgaris* var. *denticulatus* 656
var. *denticulatus* 656
var. *hibernicus* 656
Senna, Scorpion CD
SEQUOIA
sempervirens 88
SEQUOIADENDRON
giganteum 88
SERAPIAS
parviflora 855
SERIPHIDIUM
maritimum 644
Serradella CD
Yellow CD
SERRATULA
tinctoria 615
Service-tree 357
Arran 358
Swedish CD
Wild 365
SESELI
libanotis 461
SESLERIA
albicans, see **S. caerulea** 762
caerulea 762
SETARIA
adhaerens CD
faberi CD
geniculata, see **S. parviflora** CD
glauca, see **S. pumila** 803
italica 804
lutescens, see **S. pumila** 803
parviflora CD
pumila 803
verticillata 803
viridis 804
Shaggy-soldier 664
Shallon 289
Sheep's-bit 588
Sheep's-fescue 746
Fine-leaved 747
Viviparous 746
Sheep's-fescues 746
Sheep-laurel CD
Shepherd's-needle 457
Shepherd's-purse 271
Pink CD
SHERARDIA
arvensis 590
Shield-fern, Hard 76
Soft 75
Shoreweed 537
SIBBALDIA
procumbens 333
Sibbaldia 333
SIBTHORPIA
europaea 561
SIDA
rhombifolia CD
spinosa CD
SIDALCEA
candida CD
malviflora CD
Sieglingia
decumbens ≡ **Danthonia decumbens** 799
SIGESBECKIA
jorullensis, see **S. serrata** CD
orientalis CD
serrata CD
Signal-grass, Broad-leaved CD
Sharp-flowered CD
SILAUM
silaus 465
SILENE
acaulis 177
alba ≡ **S. latifolia** 178
armeria 177
coeli-rosa CD
conica 179
dioica 178
× **S. latifolia** (**S.** × **hampeana**) 178
gallica 179
× **hampeana** (**S. dioica** × **S. latifolia**) 178
italica CD
latifolia 178
maritima ≡ **S. uniflora** 176
noctiflora 177
nutans 175
otites 176
pendula CD
pratensis, see **S. latifolia** 178
quadrifida CD
uniflora 176
vulgaris 176
subsp. *maritima*, see **S. uniflora** 176
subsp. *vulgaris*, see **S. vulgaris** 176
Silky-bent, Dense 780
Loose 780
Silver-fir, European 81
Silverweed 328
SILYBUM
marianum 614
SIMAROUBACEAE 440
SIMETHIS
planifolia 808
SINACALIA
tangutica CD
SINAPIS
alba 281
subsp. **dissecta** CD
arvensis 281
Sinarundinaria
nitida, see **Fargesia spathacea** CD
SISON
amomum 471
SISYMBRIUM
altissimum 250
erysimoides CD
irio 249
loeselii 250
officinale 251
orientale 250
strictissimum CD
volgense CD
SISYRINCHIUM
bermudiana 828
californicum CD
laxum CD
montanum 829
striatum 829
SIUM
latifolium 460
Skullcap 517
Lesser 517
Norfolk CD
Somerset CD
Skunk-cabbage, American 684
Asian CD
Slipperwort CD
Slough-grass, American CD
Small-reed, Narrow 778
Purple 777
Scandinavian 778
Wood 777
Smearwort CD
Smilo-grass CD
SMYRNIUM
olusatrum 458
perfoliatum CD
Snapdragon 546
Trailing CD
Sneezeweed CD
Sneezewort 647
Snowberry 598
Snowdrop 824
Caucasian CD
Greater 825
Green CD
Pleated 824
Queen Olga's CD
Snowflake, Spring 824
Summer 823
Snow-in-summer 161
Soapwort 179
Rock CD
Soft-brome 787
Lesser 788
Slender 788
Soft-grass, Creeping 771
Soft-rush 696
Great CD
SOLANACEAE 485 CD
SOLANUM
carolinense CD
chenopodioides CD
cornutum, see **S. rostratum** 489
dulcamara 488
var. *marinum* 488
laciniatum CD
nigrum 487
subsp. **schultesii** CD 487
physalifolium 487
rostratum 489
sarachoides 488
scabrum CD
sisymbriifolium CD
triflorum CD
tuberosum 488
vernei CD
villosum CD
subsp. **miniatum** CD
subsp. *puniceum*, see **S. villosum** subsp. **miniatum** CD
subsp. **villosum** CD
SOLEIROLIA
soleirolii 128
SOLIDAGO
altissima, see **S. canadensis** 638
canadensis 638
gigantea 638
graminifolia CD
rugosa CD
virgaurea 637
Solomon's-seal 813
Angular 813
Garden 813
Whorled 814
SONCHUS
arvensis 623
subsp. *uliginosus* 623
asper 624
oleraceus 623
palustris 623
SORBARIA
kirilowii CD
sorbifolia CD
tomentosa CD
Sorbaria CD
Chinese CD
Himalayan CD
SORBUS
anglica 360
aria 360
× **S. aucuparia** (**S.** × **thuringiaca**) 360
arranensis 358
aucuparia 358
bristoliensis 364
croceocarpa 364
decipiens CD
devoniensis 364
domestica 357
eminens 361
hibernica 362
hybrida CD
intermedia 359
lancastriensis 362
latifolia 365
leptophylla 361
leyana 359
minima 359
porrigentiformis 362
pseudofennica 358
rupicola 363
subcuneata 363
× **thuringiaca** (**S. aria** × **S. aucuparia**) 360

SORBUS (*cont.*)
torminalis 365
vexans 363
wilmottiana 361
SORGHUM
bicolor CD
halepense 805
Sorrel, Common 192
French CD
Mountain 199
Sheep's 191
Southernwood CD
Sowbread, Eastern CD
Spring CD
Sow-thistle, Marsh 623
Perennial 623
Prickly 624
Smooth 623
Soyabean CD
Spanish-dagger, Curved-leaved CD
Spanish-needles CD
SPARAXIS
grandiflora CD
SPARGANIACEAE 805
SPARGANIUM
angustifolium 806
emersum 806
erectum 805
minimum ≡ **S. natans** 806
natans 806
SPARTINA
alterniflora CD
× **S. maritima** (**S.** × **townsendii**) 801
anglica 801
maritima 801
pectinata CD
× **townsendii** (**S. alterniflora** × **S. maritima**) 801
SPARTIUM
junceum 405
Spatter-dock CD
Spearwort, Adder's-tongue 105
Creeping 105
Greater 104
Lesser 104
Speedwell, Alpine 552
American 557
Breckland 556
Corsican CD
Fingered 556
French CD
Garden 560
Germander 553
Heath 553
Ivy-leaved 559
Large CD
Marsh 554
Rock 553
Slender 558
Spiked 560
Spring 557
Thyme-leaved 552
Wall 556
Wood 554
SPERGULA
arvensis 172
var. *nana* 172
morisonii CD
SPERGULARIA
bocconei 173
marginata, see **S. media** 172
marina 173
media 172
rubra 173
rupicola 172
Spider-orchid, Early 856
Late 856
Spiderwort CD
Spignel 465
Spike-rush, Common 703
Dwarf 705
Few-flowered 705
Many-stalked 704
Needle 705
Northern 704
Slender 704
Spinach CD
New Zealand CD
Tree CD
SPINACIA
oleracea CD
Spindle 425
Evergreen 425
Large-leaved CD
SPIRAEA agg. 321
alba CD 321
× **S. douglasii** (**S.** × **billardii**) 322
× **S. salicifolia** (**S.** × **rosalba**) 322
× **arguta** (**S. multiflora** × **S. thunbergii**) CD
× **billardii** (**S. alba** × **S. douglasii**) 322
× **brachybotrys** (**S. canescens** × **S. douglasii**) CD
canescens CD 321
× **S. douglasii** (**S.** × **brachybotrys**) CD
cantoniensis × **S. trilobata** (**S.** × **vanhouttei**) CD
chamaedryfolia CD
douglasii 323
subsp. **douglasii** CD
subsp. **menziesii** CD
× **S. salicifolia** (**S.** × **pseudosalicifolia**) 323
japonica CD 321
'Fortunei' CD
media CD 321
multiflora × **S. thunbergii** (**S.** × **arguta**) CD
× **pseudosalicifolia** (**S. douglasii** × **S. salicifolia**) 323
× **rosalba** (**S. alba** × **S. salicifolia**) 322
salicifolia 322
tomentosa CD 321
× **vanhouttei** (**S. cantoniensis** × **S. trilobata**) CD
Spiraea, Elm-leaved CD
Himalayan CD
Japanese CD
Lange's CD
Russian CD
Van Houtte's CD
SPIRANTHES
aestivalis 840
romanzoffiana 840
spiralis 839
SPIRODELA
polyrhiza 685
Spleenwort, Black 68
Forked 71
Green 70
Irish 68
Lanceolate 68
Maidenhair 69
Sea 69
SPOROBOLUS
africanus CD
Spotted-laurel 423
Spotted-orchid, Common 845
Heath 847
Springbeauty 150
Spring-sedge 732
Rare 732
Spruce, Engelmann CD
Norway 83
Serbian CD
Sitka 83
White CD
Spurge, Balkan CD
Broad-leaved 428
Caper 429
Coral CD
Cypress 431
Dwarf 429
Figert's CD
Gáyer's CD
Irish 427
Leafy 431
Mediterranean 432
Petty 430
Portland 430
Purple 427
Sea 430
Spotted CD
Sun 429
Sweet 428
Twiggy 431
Upright 428
Waldstein's CD
Wood 432
Spurge-laurel 411
Spurrey, Corn 172
Pearlwort CD
Sand 173
Squill, Alpine CD
Autumn 816
Greek CD
Portugese CD
Pyrenean CD
Siberian 816
Spring 816
Turkish CD
Squinancywort 590
St John's-wort, Des Etangs' 213
Hairy 215
Imperforate 212
Irish 216
Marsh 216
Pale 215
Perforate 212
Round-leaved CD
Slender 215
Square-stalked 214
Toadflax-leaved 214
Trailing 214
Wavy 213
St Patrick's-cabbage 315
St Paul's-wort, Eastern CD
Western CD
STACHYS
alpina 508
× **ambigua** (**S. palustris** × **S. sylvatica**) 509
annua 510
arvensis 510
byzantina 508
germanica 508
lanata, see **S. byzantina** 508
officinalis 507
var. *nana* 507
palustris 509
× **S. sylvatica** (**S.** × **ambigua**) 509
recta CD
sylvatica 509
Staff-vine CD
STAPHYLEA
pinnata CD
STAPHYLEACEAE CD
Starflower, Spring 823
Starfruit 669
Star-of-Bethlehem 815
Drooping 815
Early 810
Spiked 815
Yellow 810
Star-thistle, Lesser 617
Maltese CD
Red 616
Rough CD
Yellow 616
Statice CD
Steeple-bush 323
STELLARIA
alsine ≡ **S. uliginosa** 160
graminea 160
holostea 159
media 158
neglecta 159
nemorum 157
subsp. *glochidisperma* ≡ **S. nemorum** subsp. **montana** 158
subsp. **montana** 158
subsp. *nemorum* 157
pallida 158
palustris 159
uliginosa 160
STERNBERGIA
lutea CD
Stink-grass CD
STIPA
capensis CD
neesiana CD
Stitchwort, Bog 160
Greater 159
Lesser 160
Marsh 159
Wood 157
Stock, Hoary 254
Night-scented 255
Sea 254
Virginia 254
Stonecrop, Biting 311
Butterfly 309
Colorado CD
English 312
Hairy 313
Kamchatka CD
Stonecrop (*cont.*)
Least CD
Love-restoring CD
Mossy 307
Pale CD
Reflexed 310
Rock 311
Spanish CD
Tasteless 311
Thick-leaved 312
White 312
Stork's-bill, Common 450
Eastern CD
Garden CD
Hairy-pitted CD
Mediterranean CD
Musk 450
Sea 450
Soft CD
Sticky 451
Three-lobed CD
Western CD
Stranvaesia CD
Strapwort 170
STRATIOTES
aloides 669
Strawberry, Barren 333
Garden 334
Hautbois 334
Wild 333
Yellow-flowered 334
Strawberry-blite CD
Strawberry-tree 290
SUAEDA
fruticosa ≡ **S. vera** 147
maritima 148
vera 147
SUBULARIA
aquatica 278
SUCCISA
pratensis 607
Succory, Lamb's 618
Sumach, Stag's-horn 440
Summer-cypress 140
Sundew, Great 222
Oblong-leaved 223
Round-leaved 222
Sunflower 663
Lesser CD
Perennial 663
Thin-leaved CD
Swede 279
Sweet-briar 352
Small-flowered 352
Small-leaved 352
Sweet-flag 683
Slender CD
Sweet-grass, Floating 763
Hybrid 763
Plicate 764
Reed 763
Small 764
Sweet-William 182
Swida
sanguinea, see **Cornus sanguinea** 422
sericea, see **Cornus sericea** 422
Swine-cress 277
Lesser 277
Sword-fern, Western CD
Sycamore 439
SYMPHORICARPOS
albus 598
var. *laevigatus* 598
× **chenaultii** (**S. microphyllus** × **S. orbiculatus**) 599
microphyllus × **S. orbiculatus** (**S.** × **chenaultii**) 599
orbiculatus CD 599
rivularis ≡ **S. albus** 598
SYMPHYTUM
asperum CD
× **S. officinale** (**S.** × **uplandicum**) 497
bulbosum CD
caucasicum CD
grandiflorum 498
× ? **S.** × **uplandicum** (**S. 'Hidcote Blue'**) 498
'Hidcote Blue' (**S. grandiflorum** × ? **S.** × **uplandicum**) 498
ibericum, see **S. grandiflorum** 498
officinale 497
orientale 499

SYMPHYTUM (*cont.*)
- **tauricum** CD
- **tuberosum** 498
- × **uplandicum** (**S. asperum** × **S. officinale**) 497

SYRINGA
- **vulgaris** 538

TAGETES
- **erecta** CD 666
- **minuta** CD
- **patula** 666

TAMARICACEAE 232 CD
Tamarisk 232
- African CD

TAMARIX
- **africana** CD
- *anglica* ≡ **T. gallica** 232
- **gallica** 232

TAMUS
- **communis** 834

TANACETUM
- **balsamita** CD
  - subsp. *balsamita* CD
  - subsp. *balsamitoides* CD
- **macrophyllum** CD
- **parthenium** 644
- **vulgare** 644

Tansy 644
- Rayed CD

Tapegrass CD
TARAXACUM agg. 626
Tare, Hairy 381
- Slender 381
- Smooth 382

Tarragon CD
Tasmanian-fuchsia CD
Tasselweed, Beaked 681
- Spiral 681

TAXACEAE 92
TAXODIACEAE 88
TAXUS
- **baccata** 92

Teaplants 485
Tear-thumb, American CD
Teasel, Cut-leaved CD
- Fuller's 605
- Small 606
- Wild 605
- Yellow-flowered CD

Tea-tree, Broom CD
- Woolly CD

TEESDALIA
- **nudicaulis** 272

Teff CD
TELEKIA
- **speciosa** 637

TELLIMA
- **grandiflora** 320

TEPHROSERIS
- *integrifolia* sens. lat. 657
  - subsp. **integrifolia** 657
  - subsp. **maritima** 657
- **palustris** 658

TETRAGONIA
- **tetragonioides** CD

Tetragoniaceae, see AIZOACEAE 136
TETRAGONOLOBUS
- **maritimus** CD

TEUCRIUM
- **botrys** 518
- **chamaedrys** 518
  - × *T. lucidum* 518
- **scordium** 518
- **scorodonia** 517

THALICTRUM
- **alpinum** 112
- **aquilegiifolium** CD
- **delavayi** CD
- **flavum** 112
- **minus** 112

Thelycrania
- *alba*, see **Cornus alba** 422
- *sanguinea* ≡ **Cornus sanguinea** 422
- *sericea*, see **Cornus sericea** 422

THELYPTERIDACEAE 66
THELYPTERIS
- *dryopteris* ≡ **Gymnocarpium dryopteris** 73
- *oreopteris* ≡ **Oreopteris limbosperma** 67
- **palustris** 66
- *phegopteris* ≡ **Phegopteris connectilis** 67

THELYPTERIS (*cont.*)
- *robertiana* ≡ **Gymnocarpium robertianum** 73
- *thelypteroides* subsp. *glabra*, see **T. palustris** 66

THERMOPSIS
- **montana** CD

THESIUM
- **humifusum** 424

Thimbleberry CD
Thistle, Cabbage CD
- Carline 608
- Cotton 614
- Creeping 614
- Dwarf 613
- Golden CD
- Marsh 613
- Meadow 612
- Melancholy 613
- Milk 614
- Musk 611
- Plymouth CD
- Reticulate CD
- Slender 610
- Spear 611
- Tuberous 612
- Welted 610
- Woolly 611
- Yellow CD

THLASPI
- **alliaceum** CD
- *alpestre* ≡ **T. caerulescens** 273
- **arvense** 272
- **caerulescens** 273
- **macrophyllum** CD
- **perfoliatum** 273

Thorn-apple 489
Thorow-wax 467
- False 468

Thrift 208
- Jersey 209

Throatwort CD
THUJA
- **plicata** 90

Thyme, Basil 523
- Breckland 524
- Garden CD
- Large 524
- Wild 524

THYMELAEACEAE 411
THYMUS
- *drucei* ≡ **T. polytrichus** 524
- **polytrichus** 524
- *praecox*, see **T. polytrichus** 524
  - subsp. *arcticus*, see **T. polytrichus** 524
- **pulegioides** 524
- **serpyllum** 524
- **vulgaris** CD

Tickseed, Large-flowered CD
TILIA
- **cordata** 217
  - × **T. platyphyllos** (**T.** × **europaea**) 217
- × **europaea** (**T. cordata** × **T. platyphyllos**) 217
- **platyphyllos** 216
- × *vulgaris*, see **T.** × **europaea** 217

TILIACEAE 216
Timothy 784
Toadflax, Annual 551
- Balkan CD
- Common 549
- Corsican CD
- Italian 548
- Ivy-leaved 547
- Jersey 550
- Malling CD
- Pale 549
- Prostrate 550
- Purple 549
- Sand CD
- Small 547

Tobacco CD
- Red CD
- Sweet 489
- Wild CD

TOFIELDIA
- **pusilla** 808

TOLMIEA
- **menziesii** 320

Tomatillo CD
Tomato 487
Tongue-orchid, Small-flowered 855
Toothpick-plant CD
Toothwort 576
- Purple 577

TORDYLIUM
- **maximum** 476

Tor-grass 793
TORILIS
- **arvensis** 477
- **japonica** 477
- **nodosa** 477

Tormentil 331
- Trailing 331

TRACHELIUM
- **caeruleum** CD

TRACHYSPERMUM
- **ammi** CD

TRACHYSTEMON
- **orientalis** 501

TRADESCANTIA
- **fluminensis** CD
- **virginiana** CD

TRAGOPOGON
- **hybridus** CD
- **porrifolius** 622
  - subsp. *porrifolius* 622
- **pratensis** 622
  - subsp. *minor* 622
  - subsp. **pratensis** 622

TRAGUS
- **australianus** CD
- **berteronianus** CD
- **racemosus** CD

Traveller's-joy 100
Treacle-mustard 253
Treasureflower CD
- Plain CD

Tree-fern, Australian CD
Tree-mallow 219
- Garden 220
- Smaller 219

Tree-of-heaven 440
Trefoil, Hop 397
- Large 397
- Lesser 397
- Slender 398

TRICHOMANES
- **speciosum** (gametophyte) 64 (sporophyte) 64

TRICHOPHORUM
- **alpinum** 702
- **cespitosum** 702
  - subsp. **cespitosum** 703
  - subsp. *cespitosum* × subsp. *germanicum* 703
  - subsp. *germanicum* 702

TRIENTALIS
- **europaea** 302

TRIFOLIUM
- **alexandrinum** CD
- **angustifolium** CD
- **arvense** 401
- **aureum** 397
- **bocconei** 400
- **campestre** 397
- **cernuum** CD
- *constantinopolitanum* CD
- **dubium** 397
- **echinatum** CD
- **fragiferum** 396
- **glomeratum** 395
- **hirtum** CD
- **hybridum** 395
  - subsp. *hybridum* 395
- **incarnatum** subsp. **incarnatum** 399
- **incarnatum** subsp. **molinerii** 399
- **lappaceum** CD
- **medium** 398
- **micranthum** 398
- **occidentale** 394
- **ochroleucon** 399
- **ornithopodioides** 394
- **pannonicum** CD
  - subsp. *elongatum* CD
  - subsp. *pannonicum* CD
- **pratense** 398
  - var. *sativum* 398
- **repens** 394
- **resupinatum** 396
- **scabrum** 400
- **squamosum** 401
- **stellatum** CD

TRIFOLIUM (*cont.*)
- **striatum** 400
- **strictum** 396
- **subterraneum** 401
- **suffocatum** 395
- **tomentosum** CD

TRIGLOCHIN
- **maritimum** 672
- **palustre** 672

TRIGONELLA
- **caerulea** CD
- **corniculata** CD
- **foenum-graecum** CD
- *procumbens* CD

TRINIA
- **glauca** 468

TRIPLEUROSPERMUM
- **inodorum** 651
- **maritimum** 651
  - subsp. *inodorum*, see **T. inodorum** 651

TRISETUM
- **flavescens** 768
  - subsp. *purpurascens* 768

TRISTAGMA
- **uniflorum** 823

Triticale CD
× **TRITICOSECALE** (**SECALE** × **TRITICUM**) CD
TRITICUM
- **aestivum** 799
- **turgidum** CD

Tritonia
- *crocosmiflora*, see **Crocosmia** × **crocosmiiflora** 833

TROLLIUS
- **europaeus** 95

TROPAEOLACEAE 451 CD
TROPAEOLUM
- **majus** 451
- **speciosum** CD

TSUGA
- **heterophylla** 83

TUBERARIA
- **guttata** 223

Tufted-sedge 737
- Slender 736

Tulip, Cretan CD
- Garden 811
- Wild 811

TULIPA
- **gesneriana** 811
- **saxatilis** CD
- **sylvestris** 811

Tunicflower CD
Turnip 280
Turritis
- *glabra* ≡ **Arabis glabra** 262

Tussac-grass CD
TUSSILAGO
- **farfara** 659

Tussock-sedge, Fibrous 713
- Greater 713
- Lesser 714

Tutsan 211
- Forrest's CD
- Irish CD
- Stinking 211
- Tall 211
- Turkish CD

Twayblade, Common 839
- Lesser 839

Twinflower 599
TYPHA
- **angustifolia** 807
  - × **T. latifolia** (**T.** × **glauca**) 807
- × **glauca** (**T. angustifolia** × **T. latifolia**) 807
- **latifolia** 807

TYPHACEAE 807

ULEX
- **europaeus** 407
  - × **U. gallii** 407
- **gallii** 407
- **minor** 408

ULMACEAE 124 CD
ULMUS
- *carpinifolia* ≡ **U. minor** 125
- × **elegantissima** (**U. glabra** × **U. plotii**) 124

ULMUS (*cont.*)
glabra 124
× **U. minor** (**U.** × **vegeta**) 124
? × **U. minor** × **U. plotii** (**U.** × **hollandica**) 125
× **U. plotii** (**U.** × **elegantissima**) 124
× **hollandica** (? **U. glabra** × **U. minor** × **U. plotii**) 125
var. *hollandica* ≡ **U.** × **hollandica** 125
var. *vegeta*, see **U.** × **vegeta** 124
'Dutch Elm' 125
laevis CD
minor 125
subsp. *minor* 124
plotii 126
procera 125
× **vegeta** (**U. glabra** × **U. minor**) 124
'Huntingdon Elm' 124
Umbelliferae, see APIACEAE 454
UMBILICUS
rupestris 308
UROCHLOA
panicoides CD
URTICA
dioica 127
galeopsifolia 127
urens 127
URTICACEAE 127 CD
UTRICULARIA
australis 582
intermedia *sens. lat.* 583
intermedia sens. str. 583
minor 583
neglecta ≡ **U. australis** 582
ochroleuca 583
stygia 583
vulgaris *sens. lat.* 582
vulgaris *sens. str.* 582

VACCARIA
hispanica 180
VACCINIUM
corymbosum CD
macrocarpon CD
microcarpum 293
myrtillus 294
oxycoccos 293
uliginosum 294
vitis-idaea 294
Valerian, Annual CD
Common 604
Marsh 604
Pyrenean 604
Red 605
VALERIANA
dioica 604
officinalis 604
subsp. *collina* 604
subsp. *sambucifolia* 604
pyrenaica 604
sambucifolia, see *V. officinalis* subsp. *sambucifolia* 604
VALERIANACEAE 602 CD
VALERIANELLA
carinata 602
dentata 603
eriocarpa 603
locusta 602
subsp. *dunensis* ≡ *V. locusta* var. *dunensis* 602
var. *dunensis* 602
rimosa 603
VALLISNERIA
spiralis CD
Velvetleaf 221
Venus's-looking-glass 587
Large CD
VERBASCUM
blattaria 539
bombyciferum CD
chaixii CD
densiflorum 540
lychnitis 542
nigrum 541
phlomoides 540
phoeniceum CD
pulverulentum 541
pyramidatum CD
speciosum CD
thapsus 541
virgatum 540
VERBENA
bonariensis CD
litoralis CD
officinalis 507
rigida CD
VERBENACEAE 507 CD
Vernal-grass, Annual 773
Sweet 772
VERONICA
acinifolia CD
agrestis 557
alpina 552
anagallis-aquatica 555
× **V. catenata** (**V.** × **lackschewitzii**) 555
arvensis 556
austriaca CD
beccabunga 554
catenata 555
ceratocarpa × *V. polita* 558
chamaedrys 553
crista-galli CD
filiformis 558
fruticans 553
hederifolia 559
subsp. **hederifolia** 559
subsp. **lucorum** 559
× **lackschewitzii** (**V. anagallis-aquatica** × **V. catenata**) 555
longifolia 560
× **V. spicata** CD
montana 554
officinalis 553
peregrina 557
persica 558
polita 558
praecox 556
repens CD
scutellata 554
serpyllifolia 552
subsp. **humifusa** 552
subsp. *serpyllifolia* 552
spicata 560
subsp. *hybrida* 560
subsp. *spicata* 560
sublobata, see. **V. hederifolia** subsp. **lucorum** 559
triphyllos 556
verna 557
Veronica, Hedge 561
Vervain 507
Argentinian CD
Slender CD
Vetch, Bithynian 384
Bush 382
Common 382
Crown 379
Fine-leaved 380
Fodder 381
Horseshoe 379
Hungarian CD
Kidney 375
Narbonne CD
Purple CD
Spring 384
Tufted 380
Wood 380
Vetchling, Grass 388
Hairy 388
Meadow 386
Yellow 388
VIBURNUM
lantana 598
× **V. rhytidophyllum** (**V.** × **rhytidophylloides**) CD
opulus 597
× **rhytidophylloides** (**V. lantana** × **V. rhytidophyllum**) CD
rhytidophyllum CD
tinus 598
Viburnum, Wrinkled CD
VICIA
angustifolia ≡ **V. sativa** subsp. **nigra** & subsp. **segetalis** 383
benghalensis CD
bithynica 384
cracca 380
faba 385
hirsuta 381
lathyroides 384
lutea 384
narbonensis CD
orobus 379
VICIA (*cont.*)
pannonica CD
parviflora 381
sativa 382
subsp. **nigra** 383
subsp. **sativa** 383
subsp. **segetalis** 383
sepium 382
sylvatica 380
var. *condensata* 380
tenuifolia 380
tenuissima ≡ **V. parviflora** 381
tetrasperma 382
villosa 381
VIGNA
radiata CD
VINCA
difformis CD
major 485
subsp. *hirsuta* 485
subsp. *major* 485
minor 484
VIOLA
arvensis 231
× **V. tricolor** (**V.** × **contempta**) 231
× **bavarica** (**V. reichenbachiana** × **V. riviniana**) 226
canina 227
subsp. *canina* 227
subsp. **montana** 227
× **V. riviniana** (**V.** × **intersita**) 227
× **contempta** (**V. arvensis** × **V. tricolor**) 231
cornuta 229
hirta 225
× **V. odorata** (**V.** × **scabra**) 225
× **intersita** (**V. canina** × **V. riviniana**) 227
kitaibeliana 232
lactea 228
× **V. riviniana** 228
× *lambertii*, see **V. lactea** × **V. riviniana** 228
lutea 230
odorata 224
palustris 229
subsp. **juressi** 229
subsp. *palustris* 229
persicifolia 228
reichenbachiana 226
× **V. riviniana** (**V.** × **bavarica**) 226
riviniana 226
× *V. rupestris* 225
rupestris 225
× **scabra** (**V. hirta** × **V. odorata**) 225
stagnina ≡ **V. persicifolia** 228
tricolor 230
subsp. **curtisii** 230
subsp. *tricolor* 230
× **wittrockiana** 231
VIOLACEAE 224
Violet, Fen 228
Hairy 225
Marsh 229
Sweet 224
Teesdale 225
Violet-willow, European 239
Siberian CD
Viper's-bugloss 495
Giant CD
Lax CD
Purple 496
Viper's-grass 621
Virgin's-bower CD
Virginia-creeper 434
False 434
VISCACEAE 424
VISCUM
album 424
VITACEAE 433 CD
VITIS
vinifera 433
VULPIA
ambigua ≡ **V. ciliata** 751
bromoides 751
ciliata 751
subsp. *ambigua* 751
subsp. *ciliata* 751
fasciculata 750
membranacea ≡ **V. fasciculata** 750
myuros 751
unilateralis 752
WAHLENBERGIA
hederacea 587
Wallaby-grass CD
Swamp CD
Wallflower 253
Siberian CD
Wall-rocket, Annual 279
Perennial 278
White CD
Wall-rue 71
Walnut 128
Wandering-jew CD
Warty-cabbage 252
Water-cress 257
Hybrid 257
Narrow-fruited 257
Water-cresses 256
Water-crowfoot, Brackish 108
Common 108
Fan-leaved 110
Pond 109
River 110
Stream 109
Thread-leaved 108
Water-dropwort, Corky-fruited 462
Fine-leaved 464
Hemlock 463
Narrow-leaved 462
Parsley 463
River 464
Tubular 462
Water-lily, Fringed 493
Least 94
White 93
Yellow 94
Water-milfoil, Alternate 409
Spiked 409
Whorled 408
Water-parsnip, Greater 460
Lesser 461
Water-pepper 185
Small 186
Tasteless 185
Water-plantain 668
Floating 667
Lesser 667
Narrow-leaved 668
Ribbon-leaved 668
Water-purslane 411
Water-shield, Carolina CD
Water-soldier 669
Water-speedwell, Blue 555
Pink 555
Water-starwort, Autumnal 532
Blunt-fruited 533
Common 532
Intermediate 534
Pedunculate 534
Short-leaved 532
Various-leaved 533
Water-violet 299
Waterweed, Canadian 670
Curly 671
Esthwaite 670
Large-flowered CD
Nuttall's 670
South American CD
Waterwort, Eight-stamened 210
Six-stamened 210
WATSONIA
borbonica CD
Wayfaring-tree 598
Weasel's-snout 547
Pale CD
Weeping-grass CD
WEIGELA
florida CD
Weigelia CD
Weld 285
Wellingtonia 88
Wheat, Bread 799
Rivet CD
Whin, Petty 406
White-elm, European CD
Whitlowgrass, Common 268
Glabrous 268
Hairy 268
Hoary 267
Rock 266
Wall 267
Yellow 266
Whitlowgrasses, Common 267

Whorl-grass 761
Wild-oat 766
  Winter 767
Willow, Almond 238
  Bay 236
  Creeping 247
  Cricket-bat 237
  Dark-leaved 246
  Downy 247
  Dwarf 249
  Eared 244
  Ehrhart's CD
  Goat 240
  Green-leaved 239
  Grey 242
  Heart-leaved CD
  Holme 241
  Laurel-leaved 243
  Mountain 248
  Net-leaved 249
  Olive 240
  Purple 239
  Sachalin CD
  Sharp-stipuled 238
  Shiny-leaved 237
  Tea-leaved 247
  Weeping 237
  White 237
  Whortle-leaved 248
  Woolly 248
Willowherb, Alpine 417
  American 415
  Broad-leaved 412
  Bronzy CD
  Chickweed 417
  Great 412
  Hoary 412
  Marsh 416
  New Zealand 417
  Pale 414
  Rockery CD
Willowherb (*cont.*)
  Rosebay 418
  Short-fruited 414
  Spear-leaved 413
  Square-stalked 414
Windmill-grass CD
Wineberry, Japanese CD
Wingnut, Caucasian CD
Winter-cress 255
  American 256
  Medium-flowered 256
  Small-flowered 255
Wintergreen, Common 295
  Intermediate 295
  One-flowered 296
  Round-leaved 295
  Serrated 296
Wireplant CD
Witch-grass 802
Woad 252
WOLFFIA
  arrhiza 687
Woodruff 591
  Blue 590
  Pink CD
Wood-rush, Curved 700
  Fen 700
  Field 699
  Great 698
  Hairy 698
  Heath 699
  Southern 697
  Spiked 700
  White 699
Wood-sedge 724
  Starved 726
  Thin-spiked 725
WOODSIA
  alpina 75
  ilvensis 75
Woodsia, Alpine 75
  Oblong 75
WOODSIACEAE 72 CD
Wood-sorrel 442
Wormwood 645
  Field 646
  Sea 644
Woundwort, Downy 508
  Field 510
  Hedge 509
  Hybrid 509
  Limestone 508
  Marsh 509

XANTHIUM
  ambrosioides CD
  echinatum, see **X. strumarium** 662
  spinosum 662
  strumarium 662
  strumarium group 662

Yard-grass CD
  American CD
  Fat-spiked CD
Yarrow 647
  Southern CD
  Tall CD
Yellow-cress, Austrian CD
  Creeping 258
  Great 259
  Hybrid 259
  Marsh 258
  Northern 258
Yellow-eyed-grass CD
  Pale 829
  Veined CD
Yellow-pine, Western CD
Yellow-rattle 573
  Greater 573
Yellow-sedge, Large 729
Yellow-sorrel, Chilean CD
  Fleshy CD
  Least 441
  Procumbent 441
  Sussex CD
  Upright 441
Yellow-vetch 384
Yellow-wort 481
Yellow-woundwort, Annual 510
  Perennial CD
Yew 92
Yorkshire-fog 770
YUCCA
  recurvifolia CD
YUSHANIA
  anceps CD

ZANNICHELLIA
  palustris 682
ZANNICHELLIACEAE 682
ZANTEDESCHIA
  aethiopica CD
ZEA
  mays 805
Zerna
  erecta, see **Bromopsis erecta** 790
  inermis, see **Bromopsis inermis** 790
  ramosa, see **Bromopsis ramosa** 789
    subsp. *benekenii*, see **Bromopsis benekenii** 789
ZOSTERA
  angustifolia 683
  hornemanniana 683
  marina 682
    var. *angustifolia*, see **Z. angustifolia** 683
  noltei 683
ZOSTERACEAE 682